Comprehensive Organic Functional Group Transformations

Comprehensive Organic Functional Group Transformations

Editors-in-Chief

Alan R. Katritzky, FRS

University of Florida, Gainesville, FL, USA

Otto Meth-Cohn

University of Sunderland, UK

Charles W. Rees, FRS

Imperial College of Science, Technology and Medicine, London, UK

Volume 5

SYNTHESIS: CARBON WITH TWO ATTACHED HETEROATOMS WITH AT LEAST ONE CARBON-TO-HETEROATOM MULTIPLE LINK

Volume Editor

Christopher J. Moody

Loughborough University of Technology, UK

PERGAMON

UK	Elsevier Science Ltd., The Boulevard, Langford Lane, Kidlington, Oxford OX5 1GB, UK
USA	Elsevier Science Inc., 660 White Plains Road, Tarrytown, New York 10591-5153, USA
JAPAN	Elsevier Science Japan, Tsunashima Building Annex, 3-20-12 Yushima, Bunkyo-ku, Tokyo 113, Japan

First edition 1995

Library of Congress Cataloging in Publication Data
Comprehensive organic functional group transformations / editors-in-chief, Alan R. Katritzky, Otto Meth-Cohn, Charles W. Rees. — 1st ed.
p. cm.
Includes indexes.
Contents: v. 1. Synthesis: carbon with no attached heteroatoms / edited by Stanley M. Roberts — v. 2. Synthesis: carbon with one heteroatom attached by a single bond / edited by Steven V. Ley — v. 3. Synthesis: carbon with one heteroatom attached by a multiple bond / edited by Gerald Pattenden — v. 4. Synthesis: carbon with two heteroatoms, each attached by a single bond / edited by Gordon W. Kirby — v. 5. Synthesis: carbon with two attached heteroatoms with at least one carbon-to-heteroatom multiple link / edited by Christopher J. Moody — v. 6. Synthesis: carbon with three or four attached heteroatoms / edited by Thomas L. Gilchrist — v. 7. Indexes.
1. Organic compounds—Synthesis. I. Katritzky, Alan R. II. Meth-Cohn, Otto. III. Rees, Charles W.
QD262.C534 1995
547'.2—dc20 95–31088

British Library Cataloguing in Publication Data
A catalogue record for this book is available from the British Library.

ISBN 0–08–040604–1 (set : alk. paper)
ISBN 0–08–042326–4 (Volume 5)

∞™ The paper used in this publication meets the minimum requirements of the American National Standard for Information Sciences—Permanence of Paper for Printed Library Materials, ANSI Z39.48–1984.

Printed and bound in Great Britain by Cambridge University Press, Cambridge, UK.

Contents

Preface

Some years ago the three of us met in a London club reviewing an ongoing publishing venture in Organic Synthesis. The conversation drifted to a consideration of volumes on the synthesis of key functional groups. No doubt the good wine helped since we actually broached the idea of a work on the synthesis of *all* functional groups. Would it be useful? Definitely. Would it be feasible? How would it be organized? Where do you start? We recognized that functionality was based on the coordination and heteroatom attachment of a carbon atom. But putting together a complete framework seemed particularly daunting. Two of us became very interested in the fascinating bouquet of the Muscat de Beaumes de Venise.

At our next dinner together Alan announced that he had solved the problems posed last time—problems that Charles and I hoped he had forgotten! He brought out a remarkable matrix analysis of *all* functional groups, analysed rigorously and logically. Even unknown functions were covered. Although we were all very impressed, the practicalities of the idea still seemed daunting. Those who know Alan's terrier instincts will appreciate that he would not give up such a challenge so easily. Our twice yearly club get-togethers, occasionally with friends from Pergamon, refined our thinking. Alan's cosmic vision was tempered by Charles's intuitive realism and fully supported by the publishers.

Another major problem remained: how to reduce our thinking into a practical handbook for authors—a dismaying task for three busy chemists. We settled on a seven-volume work and the indomitable ARK produced a rough breakdown to fit such a format. Putting flesh on these bones became feasible during a fortuitous three-month break between jobs by myself, and the largest handbook ever assembled by Pergamon (120 pages) was written and page allocations agreed—even for little or unknown functional groups. Sample chapters were commissioned and finally proved very encouraging, despite our first chosen topic uncovering virtually no known examples!

Contracts were defined and agreed, volume editors approached, and potential authors considered during a pleasant preconference stay in Grasmere. Following the sale of Pergamon to Elsevier Science Ltd there was a lull in the project but soon *Comprehensive Organic Functional Group Transformations* was back on track, and everyone adhered to a very businesslike timetable.

OTTO METH-COHN
Sunderland

CHARLES W. REES
London

ALAN R. KATRITZKY
Florida

Contributors to Volume 5

Dr K. Afarinkia
Department of Chemistry, Kings' College London, Strand, London WC2R 2LS, UK

Professor P. D. Bailey
Department of Chemistry, Heriot-Watt University, Riccarton, Edinburgh EH14 4AS, UK

Dr I. K. Boddy
AgrEvo UK Ltd, Chesterfield Park, Saffron Walden, Essex CB10 1XL, UK

Dr M. Butters
Process Research and Development, Pfizer Central Research, Sandwich, Kent CT13 9NJ, UK

Dr S. Challenger
Process Research and Development, Pfizer Central Research, Sandwich, Kent CT13 9NJ, UK

Dr B. P. Clark
Lilly Research Centre Ltd., Eli Lilly and Co., Erl Wood Manor, Sunninghill Road, Windlesham, Surrey GU20 6PH, UK

Dr I. D. Collier
Department of Chemistry, Heriot-Watt University, Riccarton, Edinburgh EH14 4AS, UK

Dr R. D. Connell
Pharmaceutical Division, Bayer Corporation, 400 Morgan Lane, West Haven, CT 06516-4175, USA

Professor D. P. Curran
Department of Chemistry, University of Pittsburgh, Pittsburgh, PA 15260, USA

Dr C. P. Dell
Lilly Research Centre Ltd., Eli Lilly and Co., Erl Wood Manor, Sunninghill Road, Windlesham, Surrey GU20 6PH, UK

Dr P. J. Dunn
Process Research and Development, Pfizer Central Research, Sandwich, Kent CT13 9NJ, UK

Dr M. J. Ford
AgrEvo UK Ltd, Cambridge, Essex CB2 5HU, UK

Dr P.T. Gallagher
Lilly Research Centre Ltd., Eli Lilly and Co., Erl Wood Manor, Sunninghill Road, Windlesham, Surrey GU20 6PH, UK

Dr J. Gilmore
Lilly Research Centre Ltd., Eli Lilly and Co., Erl Wood Manor, Sunninghill Road, Windlesham, Surrey GU20 6PH, UK

Dr A. Ishii
Department of Chemistry, Faculty of Science, Saitama University, Urawa, Saitama 338, Japan

Dr A. D. Jones
ZENECA Pharmaceuticals, Hurdsfield Industrial Estate, Macclesfield, Cheshire SK10 2NA, UK

Professor W. Kantlehner
Institut für Organische Chemie und Isotopenforschung, Universität Stuttgart, Pfaffenwaldring 55, D-7000 Stuttgart 80, Germany

Professor S. Kato
Department of Chemistry, Faculty of Engineering, Gifu University, Yanagido, Gifu 501-11, Japan

Dr W. W. Mergen
Carl Zeiss AG, Turnstrasse 27, D-73428 Aalen, Germany

Dr K. M. Morgan
Department of Chemistry, Heriot-Watt University, Riccarton, Edinburgh EH14 4AS, UK

Dr P. Molina
Departamento de Quimica Organica, Facultdad de Ciencias, Universidad de Murcia, Campus de Espinardo, 30071 Murcia, Spain

Dr T. Murai
Department of Chemistry, Faculty of Engineering, Gifu University, Yanagido, Gifu 501-11, Japan

Dr R. Muthyala
Dry Silver Technology Center, 3M Research Center 201-3N-04, St. Paul, Minnesota, MN 55144-1000, USA

Professor J. Mulzer
Fachbereich Chemie, Institut für Organische Chemie, Freie Universität Berlin, Takustrasse 3, D-14195 Berlin, Germany

Professor J. Nakayama
Department of Chemistry, Faculty of Science, Saitama University, Urawa, Saitama 338, Japan

Professor A. Ogawa
Department of Applied Chemistry, Faculty of Engineering, Osaka University, Suita, Osaka 565, Japan

Professor M. A. Ogliaruso
Department of Chemistry, Virginia Polytechnic Institute and State University, 221 Davidson Hall, Blacksburg, VA 24061-0132, USA

Dr W. M. Owton
Lilly Research Centre Ltd., Eli Lilly and Co., Erl Wood Manor, Sunninghill Road, Windlesham, Surrey GU20 6PH, UK

Dr H. Qi
Department of Chemistry, University of Pittsburgh, Pittsburgh, PA 15260, USA

Dr P. Quale
Department of Chemistry, University of Manchester, Manchester M13 9PL, UK

Professor M. Regitz
Fachbereich Chemie, Universität Kaiserslautern, Erwin-Schrödinger Strasse, D-6750 Kaiserslautern, Germany

Dr N. Sonoda
Department of Applied Chemistry, Faculty of Engineering, Osaka University, Suita, Osaka 565, Japan

Professor B. Stanovnik
Department of Chemistry and Chemical Technology, University of Ljubljana, Aškerčeva 5, P.O.B. 537, SLO-61001 Ljubljana, Slovenia

Dr A. Tarraga
Departamento de Quimica Organica, Facultdad de Ciencias, Universidad de Murcia, Campus de Espinardo, 30071 Murcia, Spain

Dr W.-G. Veek
Fachbereich Chemie, Universität Kaiserslautern, Erwin-Schrödinger Strasse, D-6750 Kaiserslautern, Germany

Dr M. V. Vinader
Glaxo-Wellcome Medicines Research Center, Stevenage, Hertfordshire SG1 2NY, UK

Dr P. Warner
ZENECA Pharmaceuticals, Mereside, Alderley Park, Macclesfield, Cheshire SK10 4TG, UK

Professor L. Weber
Anorganische Chemie, Fakultät für Chemie, Universität Bielefeld, 33615 Bielefeld Universitätstrasse, Postfach 100131, D-33501 Bielefeld, Germany

Dr A. C. Williams
Lilly Research Centre Ltd., Eli Lilly and Co., Erl Wood Manor, Sunninghill Road, Windlesham, Surrey GU20 6PH, UK

Dr J. M. J. Williams
Department of Chemistry, Loughborough University of Technology, Loughborough, Leicestershire LE11 3TU, UK

Dr J. F. Wolfe
Office of the Provost, Virginia Polytechnic Institute and State University, 201 Burrus Hall, Blacksburg, VA 24061-0132, USA

Contents of All Volumes

Introduction

OBJECTIVES, SCOPE, AND COVERAGE

Comprehensive Organic Functional Group Transformations (COFGT) aims to present the vast subject of organic synthesis in terms of the introduction and interconversion of functional groups. All organic structures can be considered as skeletal frameworks of carbon atoms to which functional groups are attached[a]; it is the latter which are mainly responsible for chemical reactivity and which are highlighted in COFGT. All known functional groups fit a logical and comprehensive pattern and this forms the basis for the detailed list of contents. The format of the present work was designed with the intention to cover systematically all the possible arrangements of atoms around a carbon, including those which are quite unfamiliar. The work also considers the possibility of as yet unknown functional groups which may be constructed in the future and prove to be important; thus COFGT also indicates what is not known and so points the way to new research areas.

The philosophy of the present work has been to rationalize this enormous subject within as logical and formal a framework as possible, in a scholarly and critical fashion. COFGT is designed to provide the first point of entry to the literature for synthetic organic chemists, together with an unrivalled source for anyone interested in less common, obscure, or unknown functional groups.

All functional groups are viewed as being carbon based (even if the group contains no carbon). Thus, a nitro compound is considered from the standpoint of the immediately attached carbon atom, whether di- (*sp*), tri- (sp^2), or tetracoordinated(sp^3). The work is organized on the basis of formation or rupture of bonds to a carbon atom and it is the nature of the carbon atom left after the transformation that determines the classification of the overall sequence. Several key criteria have been used to organize the work and to minimize overlap. These are, in order of priority:

1. the number of attached heteroatoms;
2. the coordination of the carbon atom involved in the functional group;
3. the nature of the immediately attached heteroatom(s); and
4. the Latest Placement Principle.

These four key principles have been used to determine the content of each volume, and to develop the detailed chapter breakdown within each volume.

Thus, according to the number of attached heteroatoms:

Volume 1 deals with synthetic reactions which result in the alteration of bonding at carbon atoms which are left with *no* attached heteroatoms.

Volume 2 deals with syntheses which result in carbon atoms attached to *one* heteroatom by a single bond.

Volume 3 deals with syntheses which result in carbon atoms attached to *one* heteroatom by a double or by a triple bond.

Volume 4 deals with syntheses which result in carbon atoms attached to *two* heteroatoms, each by a single bond.

Volume 5 deals with syntheses which result in carbon atoms attached to *two* heteroatoms by one single and one double bond, or by two double bonds, or by one single and one triple bond.

Volume 6 deals with syntheses which result in carbon atoms attached to *three* or *four* heteroatoms.

Volume 7 comprises the author and subject indexes.

Certain key principles apply to all the volumes because all functional groups are viewed as carbon based (e.g. a nitro group is either alkyl-, vinyl-, aryl-, or alkynyl-); these are:

(a) Volumes are subdivided according to the *coordination* of the carbon atom which is the product of the reaction, i.e., tetra- coming before tri- before di- before monocoordinated carbon functions.

---

[a] The major exception to this lies in heterocyclic compounds, where the cyclic heteroatoms are more logically considered as part of the framework. The subject of heterocycles has been treated elsewhere in the companion work *Comprehensive Heterocyclic Chemistry* published in 1984 with a second edition to be published in 1996.

In Volumes 1 and 6, reactions producing four-coordinated carbon are considered first, followed by three- and then two-coordinated carbon. The other volumes contain a more limited range of coordination types (Volumes 2 and 4 only four-coordinated, Volumes 3 and 5 only two- or three-coordinated). Each type of coordination is allocated a separate section in each volume.

(b) Attached *heteroatoms* are discussed in the following order of priority:

Halogens—F, Cl, Br, I
Chalcogens—O, S, Se, Te
Nitrogen—N
Other group 15 elements—P, As, Sb, Bi
Metalloids—B, Si, Ge
Main group metals—Sn, Pb, Al, Ga, In, Tl, Be, Mg, Ca, Sr, Ba, Li, Na, K, Rb, Cs
Transition metals—Cu, Ag, Au, Zn, Cd, Hg, Ti, Zr, Hf, Cr, Mo, W, Mn, Fe, Co, Ni, Pd, Pt, and others.

Higher coordination of heteroatoms is treated after lower. Thus, in sections dealing with iodo compounds, monocoordinate (e.g. iodides) are discussed before dicoordinate (e.g. iodoxyls) and tricoordinate functions.

(c) The *Latest Placement Principle* (or Last Position Principle) is used to avoid undue overlap in the work. Thus, the carbon attached to the heteroatom is discussed at the last possible position in the above prioritizing of heteroatoms. Examples of its application are noted later. On this basis, for example, when both C—C and C—H bonds are formed the reaction will appear in the latest chapter (i.e. Chapters 1.04–1.10 rather than in the earlier Chapters 1.01–1.03), and when both C—C and C═C bonds are formed, this will be found in the later Chapter 1.17. The Latest Placement Principle is particularly important in determining where to find electrocyclic reactions in Volume 1. If C—H, ═C—C, and C═C bonds are all formed in a reaction, then the latest appropriate chapter will deal with the reaction. Only if a change in heterofunction occurs is the reaction left to a later volume.

Exceptions to the above principles are rare. However the reactions of heteroarenes are mentioned along with those of arenes. If on reduction no change in the heterofunction occurs (e.g. in going from thiophene to tetrahydrothiophene or from pyridine to 2,3,4,5-tetrahydropyridine) the reaction is found in Volume 1. However, when the function changes (e.g. pyridine to piperidine), the conversion is considered in Volume 2. Conversion of methyl phenyl sulfone into methyl cyclohexyl sulfone appears in Volume 2, whereas the formation of cyclohexyl methyl ketone is treated in Volume 1, since the coordination of the carbon atom to the heteroatoms is changed in the first but not in the second hydrogenation.

Some further exceptions to the rigorous ordering of the work have been made for the purpose of easy reference. Thus, in Volume 1, a special chapter on ions, radicals, and carbenes is added: this chapter is limited to the treatment of species capable of more than a transitory existence. Throughout the work aspects of the Latest Placement Principle are occasionally ignored for reasons of clarity. Thus metal ligands that are incidental to the chemistry under discussion are not considered when prioritizing. Also, references to aromatic substituents, some of which involve a heteroatom (e.g. pyridyl, thienyl, etc.) but are incidental to the chemistry being described, are not viewed as changing the priority.

Within each section, we have endeavored to explain the influence of important secondary effects such as inclusion in a ring, degree of strain, degree of substitution, various types of activation, influence of stereochemistry, and so on, on the transformation under consideration. General synthetic methods are treated before specific methods.

Transient intermediates, as such, do not fall within the scope of this work. Although there is clearly no sharp division, we have attempted to restrict coverage of radicals, etc., to more stable, longer lived species. It is the aim of this work to consider all organic functional groups provided that the molecules which incorporate them, though they may be unstable, can have a finite lifetime and chemistry. The whole work deals with the generation and transformation of functional groups, *not* of molecules such as CO_2, COS, CS_2, ClCN, etc. Such simple carbon derivatives are not treated unless a further carbon is attached (e.g. RN═C═O).

VOLUME 1 SYNTHESIS: CARBON WITH NO ATTACHED HETEROATOMS

Volume 1 deals solely with the formation of nonheteroatom functional groups and as such is different in style to the remaining volumes.

In addition to the general principles, Volume 1 is further organized as follows:

1. By the type of bond formed (i.e. C—H before C—C).
2. By the type of reaction involved (i.e. substitution, then addition, then rearrangement). With C=C bond formation the order is addition, elimination, condensation, then electrocyclic and other methods. One rearrangement chapter only is devoted to each of the Parts I and II.
3. In Parts II and III the treatment of formation of ions, radicals, and carbenes is added at the end of the section dealing solely with those species with a significant rather than a transient lifetime.

In Volume 1, the heteroatom sequence is a secondary feature since only remote heteroatom functions are involved in the products: but the standard order pertains in reactants that contain heteroatoms (see, e.g. Chapters 1.01 and 1.02).

All the major structural influences that are treated throughout this work apply equally (or perhaps more importantly) in Volume 1. Thus the effects of conjugation, remote substituents, rings, stereochemistry, strain, kinetic or thermodynamic factors, solvation, primary, secondary and tertiary nature, etc., are mentioned whenever relevant.

VOLUME 2 SYNTHESIS: CARBON WITH ONE HETEROATOM ATTACHED BY A SINGLE BOND

Volume 2 is arranged in three parts: I, II and III, dealing respectively with sp^3, sp^2, and *sp* carbon linked to the heteroatom. In each chapter we have endeavored to explain important effects due to such features as the primary, secondary, tertiary nature, ring effects, strain activation, effect of beta, gamma, and more remote functionality, stereochemical effects, and so on. Methods that are common to a larger group are dealt with at their first appearance and suitably cross-referenced.

Volumes 2–6 all deal with the synthesis of functions involving at least one heteroatom. To avoid major overlap we have applied the Latest Placement Principle; that is, the chemistry is discussed at the last possible position based on the prioritization of the carbon attached to the heteroatom. Thus the compound CH_3ONH_2 is treated under "Alkyl Chalcogenides" in the subsection "Functions Based on the RON-Unit" (i.e. 2.02.6). However, CH_3ONHCH_3 appears under "Alkyl Nitrogen Compounds" (2.06.2.3) since the Latest Placement Principle prevails. Also, dialkyl ethers appear in Part I of Volume 2 (Functions Linked by a Single Bond to an sp^3 Carbon Atom), while alkyl aryl ethers appear in Part II of Volume 2 (Functions Linked by a Single Bond to an sp^2 Carbon Atom). Exceptions to the rule are:

(a) When a fully unsaturated heterocyclic substituent (e.g. thienyl, pyridyl, etc.) is used as an example of an aryl group, the ring heteroatom(s) is (are) not taken into account (e.g. 2-methoxypyridine should strictly appear in Volume 6, but is covered in Volume 2 along with 3- and 4-methoxypyridine).

(b) Carbon-based metal ligands that are incidental to the synthesis under discussion (e.g. carbonyls, cyclopentadienyls, etc.) are not taken into consideration.

VOLUME 3 SYNTHESIS: CARBON WITH ONE HETEROATOM ATTACHED BY A MULTIPLE BOND

Volume 3 follows the logical development indicated in Volume 2. Thus, according to the Last Placement Principle, the imines, RCH=N—R, appear in Volume 3 rather than in Volume 2 (where functions singly bonded to carbon are treated). Furthermore, acetophenone, $PhCOCH_3$, is treated under α,β-unsaturated ketones (3.05) rather than saturated ketones (3.04). Chloronitroacrylonitriles would appear under the section "α,β-Vinylic Nitriles with Nitrogen-based Substituents" (3.19.2.7), not under the related earlier section dealing with halo-substituents (3.19.2.3)

VOLUME 4 SYNTHESIS: CARBON WITH TWO HETEROATOMS, EACH ATTACHED BY A SINGLE BOND

Volume 4 is in three parts. Part I deals with tetracoordinated carbon bearing two heteroatoms, Part II with tricoordinated carbon bearing two heteroatoms, and Part III (a brief chapter) with stabilized radicals, ions, and the like bearing two heteroatoms. The material is arranged according to the Latest Placement Principle: thus, the synthesis of $CHBr_2CHI_2$ would appear in the section dealing with diiodo, not dibromo functions (i.e. in 4.01.5, not 4.01.4), and the synthesis of CF_3CH-BrCl is discussed in Volume 6 (carbons bearing three heteroatoms), rather than in Volume 4.

VOLUME 5 SYNTHESIS: CARBON WITH TWO ATTACHED HETEROATOMS WITH AT LEAST ONE CARBON-TO-HETEROATOM MULTIPLE BOND

Volume 5 is in three parts. Part I deals with functions with one doubly bonded and one singly bonded heteroatom, Part II with functions containing two doubly bonded heteroatoms and Part III with one triply bonded and one singly bonded heteroatom. Part I constitutes the bulk of Volume 5.

The arrangement of the chemistry in each part follows the same logical sequence. The multiply bonded heteroatom is focused on first and then the other heteroatom in a secondary classification, both following the priority rules already described. Each section excludes the coverage of the previous sections. Thus, all carbonyl derivatives will appear in Chapters 5.01–5.10 but not in Chapters 5.11, *et seq*.

According to the Latest Placement Principle structure RC(O)OC(S)R is discussed in the chapter dealing with carbons bearing a doubly bonded sulfur and singly bonded oxygen (5.12.3), *not* in that dealing with doubly and singly bonded oxygen (5.04.1). Another effect of the Latest Placement Principle is that the amides RCONMePh are discussed under *N*-arylalkanoamides (5.06.2.4), rather than *N*-alkylalkanoamides (5.06.2.2). Again, exceptions are made to the latest placement rules for: (a) hetaryl rings used as examples of aryl substituents which are not viewed as functional groups. Thus, 2-methylimidazole is not considered as an example of an amidine function and 2-methoxy-pyridine is not an example of a doubly bonded nitrogen, singly bonded oxygen function; (b) metal ligands that are incidental to the organic chemistry under discussion are not viewed as functions in priority considerations.

VOLUME 6 SYNTHESIS: CARBON WITH THREE OR FOUR ATTACHED HETEROATOMS

Volume 6 is in four parts. Part I deals with tetracoordinate carbons bearing three heteroatoms. Part II covers tetracoordinate compounds bearing four heteroatoms, i.e. substituted methanes, and Part III deals with tricoordinate systems bearing three heteroatoms, i.e. where one heteroatom is attached by a double bond. Part IV is brief and deals with stabilized radicals and ions. Not surprisingly, the coverage of Volume 6 is very large—and also shows that many gaps in the development of organic chemistry still exist.

The organization within the three sections not only follows the same broad logic developed in the previous volumes, but also has a structure unique to the multiheteroatom volume. According to the Latest Placement Principle $CF_3C(NR_2)_3$ appears in the section dealing with carbons bearing three nitrogens (6.05.1.1), not that dealing with carbons bearing three halogens (6.01.2), while $(CF_3CH_2O)_2CO$ appears in Part III, not in Part I.

In the chapter dealing with iminocarbonyl functions in Part III, the substituents on nitrogen are discussed in each appropriate subsection in the order outlined above. Thus, the RN= group would be first considered with R = H, then alkyl, alkenyl, aryl and hetaryl, alkynyl and then heteroatom substituents in the usual order.

In each relevant section, we have endeavored to explain the influence of important secondary effects on the synthesis such as structure (primary, secondary, etc.), ring effects, strain, activation, stereochemistry, remote substituent effects, etc.

The arrangement of the chemistry in each of Parts I–III follows a similar pattern. Thus, each section commences with functions containing at least one halogen. This section deals with all combinations of halogen with other heteroatoms in the described order. The next section deals with functions containing at least one chalcogen in combination with any other heteroatoms except halogens. Subsequent sections each exclude the previous title heteroatom functions.

VOLUME 7 INDEXES

Subject Indexes are included in each of Volumes 1–6 and Cumulative Subject and Author Indexes appear in Volume 7. Most entries in the Subject Index consist of two or three lines: the first line is the entry itself (e.g. Lactones) and the second line is descriptive of that entry (e.g. reduction); in many cases more detail is given (e.g. with 9-BBN).

REFERENCES

The references are handled by the system previously used successfully in *Comprehensive Heterocyclic Chemistry*. In this system reference numbers appear neither in the text, nor as footnotes, nor at the end of chapters. Instead, each time a reference is cited in the text there appears (in parentheses) a two-letter code assigned to the journal being cited, which is preceded by the year (tens and units only for twentieth-century references) and followed by the page number. For example: "It was shown ⟨80TL1327⟩ that . . .". In this phrase, "80" refers to 1980, "TL" to *Tetrahedron Letters*, and "1327" to the page number. For those journals which are published in parts, or which have more than one volume number per year, the appropriate part of the volume is indicated, e.g. as in ⟨73JCS(P2)1594⟩ or ⟨78JOM(162)611⟩, where the first example refers to *J. Chem. Soc., Perkin Trans 2*, 1973, page 1594, and the second to *J. Organomet. Chem.*, 1978, volume 162, page 611.

This reference system is adopted because it is far more useful to the reader than the conventional "superscript number" system. It enables readers to go directly to the literature reference cited, without first having to consult the bibliography at the end of each chapter.

References to the last century quote the year in full. Books have a prefix "B-" and if they are commonly quoted (e.g. *Organic Reactions*) they will have a code. Otherwise, as with uncommon journals, they are given a miscellaneous code (MI) and numbered arbitrarily *abb1*, *abb2*, etc., where *abb* refers to the volume and chapter number and *1*, *2*, etc., are assigned sequentially. Patents are assigned appropriate three-letter codes.

The references are given in full at the end of each volume. They include *Chemical Abstract* references when these are likely to help; in particular, they are given for all patents, and for less accessible sources such as journals in languages other than English, French, or German, company reports, obscure books, and theses.

5.01
Acid Halides

JONATHAN M. J. WILLIAMS
Loughborough University of Technology, Leicestershire, UK

5.01.1 SINGLY-BONDED HALOGEN FUNCTIONS—ACID HALIDES (RCOHal)

5.01.1.1 General Methods for Acid Halides

The chemistry and preparation of acid halides were reviewed in 1972 ⟨B-72MI 501-01⟩, and methods of preparation were discussed in 1977 ⟨B-77MI 501-01⟩. More recently, reviews containing information on the preparation of acid halides have also appeared ⟨85HOU(E5/1)587, 91COS(6)301⟩. The following sections consider the preparation of acid fluorides, chlorides, bromides, and iodides separately.

5.01.1.1.1 Acid halides from carboxylic acids

The preparation of acid halides may be achieved directly from the corresponding carboxylic acid with a variety of reagents. However, the nature of the reagents employed generally depends on the acid halide required. Although thionyl chloride and oxalyl chloride are the two most common reagents for the preparation of acid chlorides, the other thionyl halides and oxalyl halides have not generally been found to be convenient for the preparation of the other acid halides.

The use of 1-halo-*N*,*N*-2-trimethylpropenylamines stands out as a general method for the preparation of any acid halide direct from the carboxylic acid. 1-Chloro-*N*,*N*-2-trimethylpropenylamine is available by treatment of *N,N*-dimethylisobutyramide with phosgene, followed by triethylamine ⟨79OS(59)26⟩. The chloro compound is readily converted into the other α-haloenamines using potassium fluoride, dibromomethane, or potassium iodide, as illustrated in Scheme 1.

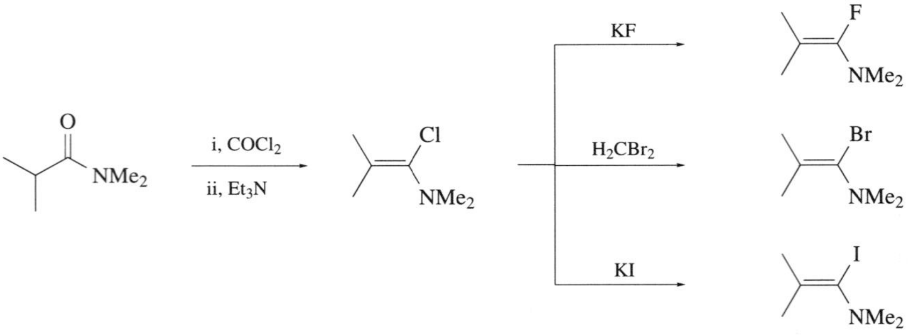

Scheme 1

Ghosez and co-workers have shown that 1-halo-*N,N*-2-trimethylpropenylamines can be used to convert carboxylic acids directly into acid halides at room temperature in dichloromethane ⟨79CC1180⟩. As illustrated in Equations (1)–(4), a number of acid halides can be conveniently prepared using this method.

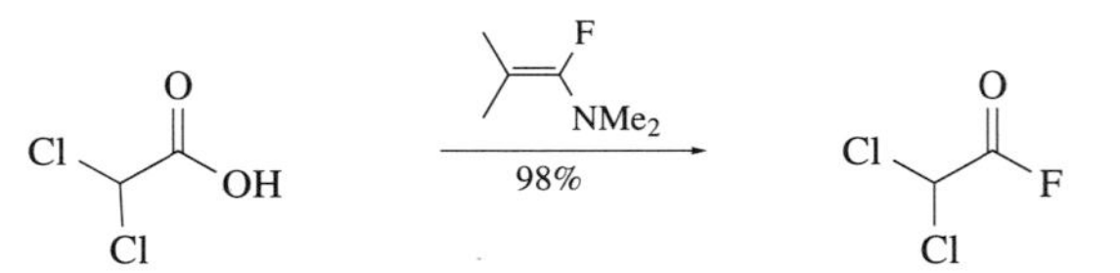

(1)

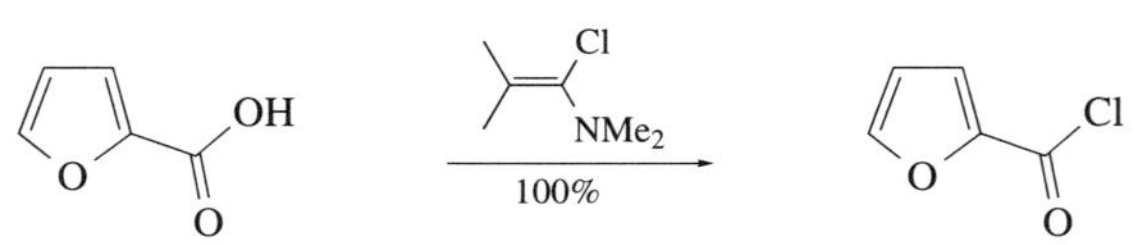

(2)

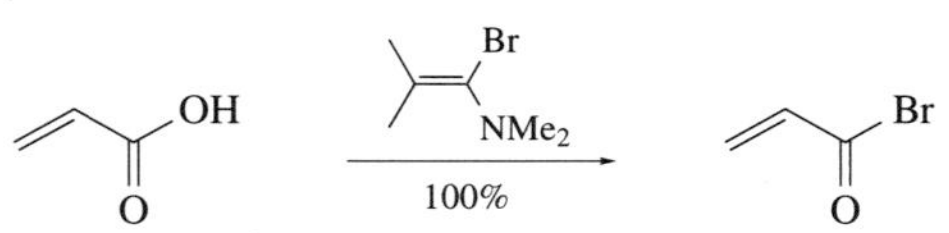

(3)

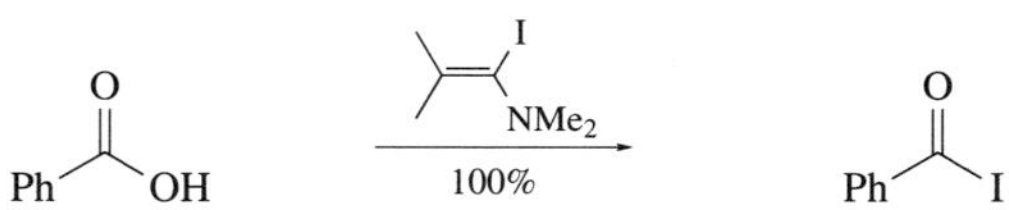

(4)

The reactions are believed to proceed as shown in Scheme 2 via the intermediate (**1**), which has been detected by low temperature NMR spectroscopy. In many cases, the acid halide was not isolated, and the product was derivatized *in situ*. The only by-product of the reaction is the *N,N*-dimethylisobutyramide, which does not normally interfere with subsequent reactions.

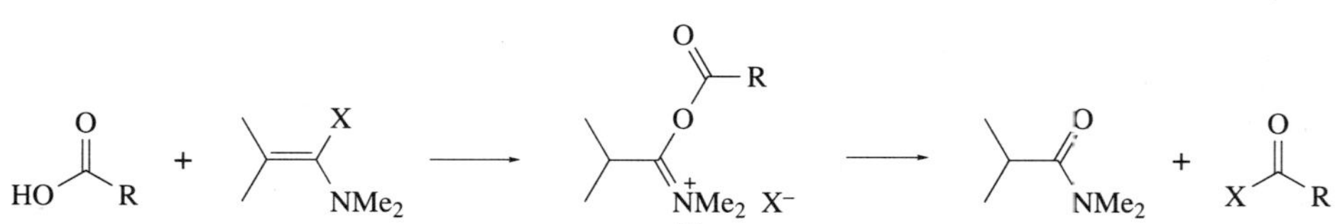

(**1**)

Scheme 2

When the separation of the product from *N,N*-dimethylisobutyramide by distillation is troublesome, the authors suggest the use of the alternative reagent (**2**), whose high molecular weight results in an easier separation by fractional distillation.

Thermally unstable acid halides were prepared at low temperature, and reacted *in situ*. These include HCOCl, HCOCOCl, HCOF, HCOBr, HCOI, and MeCOCOCl, all of which were isolated as the corresponding methyl esters.

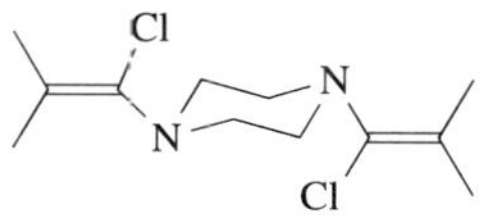

(2)

5.01.1.1.2 Acid halides from acid chlorides

The majority of well-documented preparations of acid halides are for acid chlorides, and for this reason, acid fluorides, acid bromides, and acid iodides are often more conveniently prepared from the acid chloride by an exchange process with a suitable source of halide (Scheme 3). Suitable reagents include potassium fluoride, tetrabutylammonium fluoride, trimethylsilyl bromide, sodium iodide, and diiodosilane. These reactions are equilibrium processes, and require some means to drive the reaction to completion, normally by the use of excess reagent or distillation of a lower boiling product or by-product. Specific reactions are discussed in further detail in the relevant section.

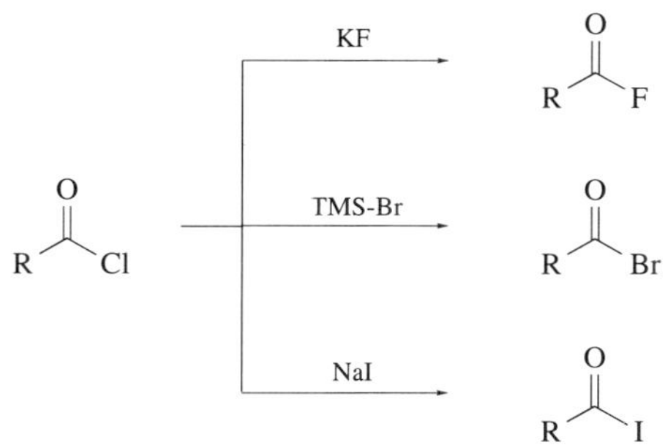

Scheme 3

5.01.1.1.3 Acid halides from carboxylic acid esters

Generally, carboxylic acid esters are only converted into acid chlorides under rather forcing conditions. However, for labile esters, milder conditions have been employed. Enol esters react with hydrogen chloride or hydrogen fluoride to afford the corresponding acid halide ⟨69JOC2486⟩. An earlier patent claims that this process is also viable for the formation of acid bromides and iodides ⟨49USP2475966⟩. Thus, treatment of isopropenyl octanoate with anhydrous hydrogen fluoride gas at 35 °C for 1 h affords octanoyl fluoride (Equation (5)). Similarly, stearoyl chloride was prepared in an analogous manner, although a higher temperature was employed (Equation (6)).

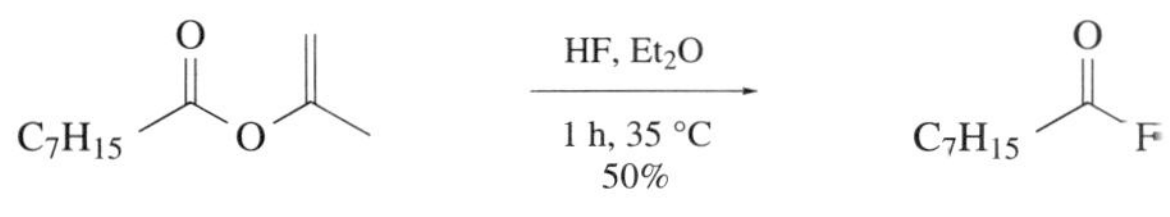

(5)

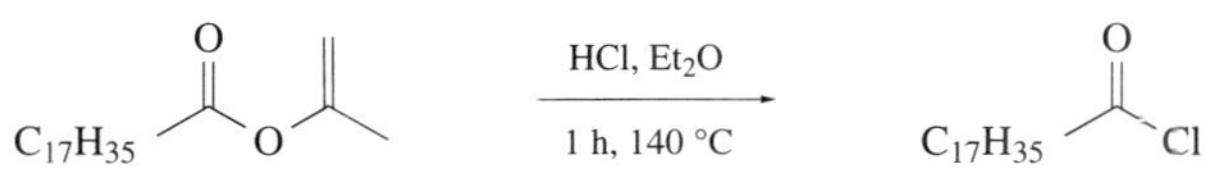

(6)

5.01.1.2 Acid Fluorides

Acid fluorides are generally more stable to hydrolysis than the other acid halides, and, indeed, many are isolated by an aqueous extractive workup procedure. Compounds containing functionalities which are not compatible with the other acid halides have been prepared, such as α-hydroxy acid fluorides.

5.01.1.2.1 *Acid fluorides from acid chlorides*

Acid fluorides are frequently made from acid chlorides. A number of reagents have been employed for this conversion, including KF ⟨66JOC2316⟩, KHF_2 ⟨56CB862⟩, KF/sulfinate ⟨58CB2553⟩, and other fluoride salts ⟨78BCJ1267⟩. These reactions are generalized by Equation (7). The majority of contemporary methods employ some method for enhancing the solubility of the fluoride, such as metal chelating agents, or phase transfer catalysis.

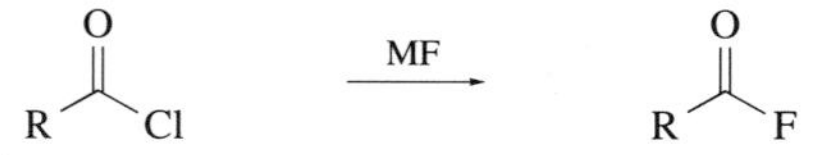

(7)

Treatment of acid chlorides with potassium fluoride and the phase transfer catalyst triethylbenzylammonium chloride results in the formation of acid fluorides, which are distilled out as the reaction progresses ⟨82SC513⟩. For example, pivaloyl fluoride and 3-chloropropionyl fluoride were prepared in fair yields (Table 1, entries 1 and 2). The alkyl chloride is not converted to the alkyl fluoride under these conditions (Table 1, entry 2). The quaternary ammonium fluoride salt is suggested to be the reactive phase transfer catalyst.

Table 1 Preparation of acid fluorides from acid chlorides.

Entry	*Substrate*	*Conditions*	*Product*	*Yield* (%)
1	Bu^tCOCl	a	Bu^tCOF	70
2	$ClCH_2CH_2COCl$	a	$ClCH_2CH_2COF$	58
3	PhCOCl	b	PhCOF	81
4	MeCOCl	c	MeCOF	100
5	Pr^nCOCl	c	Pr^nCOF	90
6	PhCOCl	c	PhCOF	90
7	MeCOCl	d	MeCOF	74
8	PhCOCl	d	PhCOF	71
9	*p*-$MeOC_6H_4CO_2H$	e	*p*-$MeOC_6H_4CO_2F$	83
10	EtCOCl	f	EtCOF	90
11	PhCOCl	f	PhCOF	92
12	Cl_3CCOCl	g	Cl_3CCOF	73
13	PhCOCl	g	PhCOF	70

[a] $KF/PhCH_2Et_3NCl$. [b] Bu_4NF. [c] KF/18-crown-6. [d] KF/CaF_2 sulfolane. [e] ZnF_2/C_5H_5N. [f] $C_5H_5NH(HF)nF$. [g] Morpholinosulfur trifluoride.

In a similar fashion, the phase transfer catalyst tetrabutylammonium fluoride has been used to prepare benzoyl fluoride in 81% isolated yield, after stirring for 1 h at 25°C (Table 1, entry 3) ⟨84JOC3216⟩.

Liotta and Harris found that the use of 18-crown-6 greatly enhanced the solubility of potassium fluoride, rendering it soluble in acetonitrile or benzene ⟨74JA2250⟩. Acetyl chloride was converted into acetyl fluoride quantitatively, and it was later shown by other workers that butanoyl fluoride and benzoyl fluoride could both be isolated in 90% yields from the corresponding acid chlorides (Table 1, entries 4–6) ⟨79JOC1016⟩.

It was also found that potassium fluoride, supported on calcium fluoride in sulfolane the solvent, allowed for the conversion of benzoyl chloride and acetyl chloride into the corresponding fluorides in fair yields (Table 1, entries 7 and 8) ⟨86CC791⟩.

Zinc fluoride was first reported to convert acid chlorides into acid fluorides in 1894 ⟨1894MI 501-01⟩. It was shown in 1978 that the use of zinc fluoride in pyridine is more efficient, especially for the formation of aroyl and arenesulfonyl fluorides ⟨78BCJ1267⟩. For example, *p*-methoxybenzoyl chloride was converted to *p*-methoxybenzoyl fluoride in good yield (Table 1, entry 9). It was suggested that pyridine accelerates the reaction by coordination to the zinc, and consequent weakening of the zinc–fluorine bond. However, this reagent was shown to be less useful for the preparation of aliphatic acid fluorides.

Pyridinium poly(hydrogen fluoride), a reagent developed by Olah and co-workers, is effective for the conversion of acid chlorides and acid anhydrides into acid fluorides ⟨79JOC3872⟩. Good yields were obtained, with the reaction generally taking only 15 min. For example, propionyl fluoride and benzoyl fluoride were prepared from the corresponding acid chlorides in good yields (Table 1, entries 10 and 11). The reagent is 30% pyridine and 70% hydrogen fluoride, by weight, which corresponds to about 9:1 equiv. hydrogen fluoride : pyridine. This reagent is prepared by the addition of condensed anhydrous hydrogen fluoride to pyridine at −78 °C.

Dialkylaminosulfur trifluorides have been used to convert acid chlorides to acid fluorides in reasonably good yields ⟨75S801⟩. These reagents are more selective than the previously reported use of sulfur tetrafluoride ⟨60JA543⟩. Morpholinosulfur trifluoride was found to be the most suitable reagent, and was used to convert trichloroacetyl chloride into trichloroacetyl fluoride in 73% yield, and benzoyl chloride to benzoyl fluoride in 70% yield (Table 1, entries 12 and 13).

5.01.1.2.2 Acid fluorides from carboxylic acids

Cyanuric fluoride, another reagent developed by Olah *et al.*, is a mild reagent which, in the presence of pyridine as a promoter, allows for the direct conversion of carboxylic acids into acid fluorides, but does not affect alkene, hydroxy, or aromatic functionalities ⟨73S487⟩. The mechanism proposed involves nucleophilic attack of the pyridinium carboxylate on cyanuric fluoride, and breakdown of the intermediate (**3**) to generate the acid fluoride (Scheme 4). This method was successfully applied to the conversion of 1-hydroxycycloheptane-1-carboxylic acid into the corresponding acid fluoride in 95% yield, and benzoic acid was converted into benzoyl fluoride in 97% yield (Table 2, entries 1 and 2). Cyanuric fluoride has also been used in the preparation of the new peptide coupling reagents, [(9-fluorenylmethyl)oxy]carbonyl (Fmoc) amino acid fluorides. Whereas for certain amino acids the corresponding acid chlorides undergo unwanted side reactions, including cyclization, the acid fluorides are readily isolable ⟨90JA9651⟩. This reflects the fact that acid fluorides are less reactive than acid chlorides to neutral oxygen nucleophiles. Interestingly, nitrogen nucleophiles and anionic nucleophiles are almost equally reactive to both acid chlorides and acid fluorides. Thus, while the protected aspartic acid (**4**) was cyclized to the anhydride (**5**) upon treatment with thionyl chloride, due to the reactivity of the intermediate acid chloride (**6**) cyanuric fluoride and pyridine were used to convert (**4**) directly into the acid fluoride (**7**), as illustrated in Scheme 5.

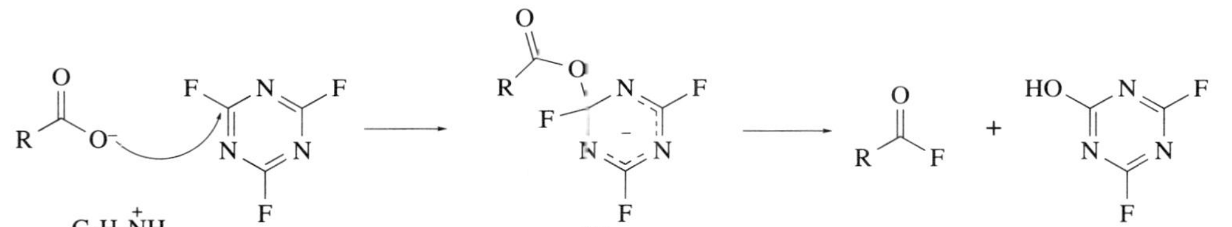

(**3**)

Scheme 4

Table 2 Preparation of acid fluorides from carboxylic acids.

Entry	*Substrate*	*Conditions*	*Product*	*Yield (%)*
1	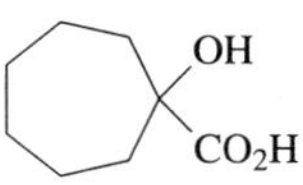	a	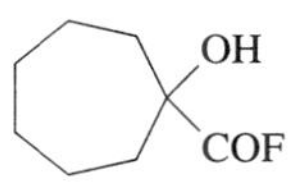	95
2	$PhCO_2H$	a	PhCOF	97
3	HCO_2H	b	HCOF	25
4	$MeCO_2H$	b	MeCOF	80
5	*c*-$C_6H_{11}CO_2H$	b	*c*-$C_6H_{11}COF$	90
6	$EtCO_2H$	c	EtCOF	61
7	$PhCO_2H$	c	PhCOF	60

[a] $(CNF)_3/C_5H_5N$. [b] SeF_4/C_5H_5N. [c] C_3F_6/Et_2NH.

Olah has used the reagent selenium tetrafluoride to convert carboxylic acids and anhydrides to acid halides ⟨74JA925⟩. Generally, the reactions were run with pyridine, which forms a complex with selenium tetrafluoride. Good yields were obtained for most of these conversions, as shown in Table 2 (entries 3–5). For acetic anhydride and propionic anhydride, 2 mol of product were formed for

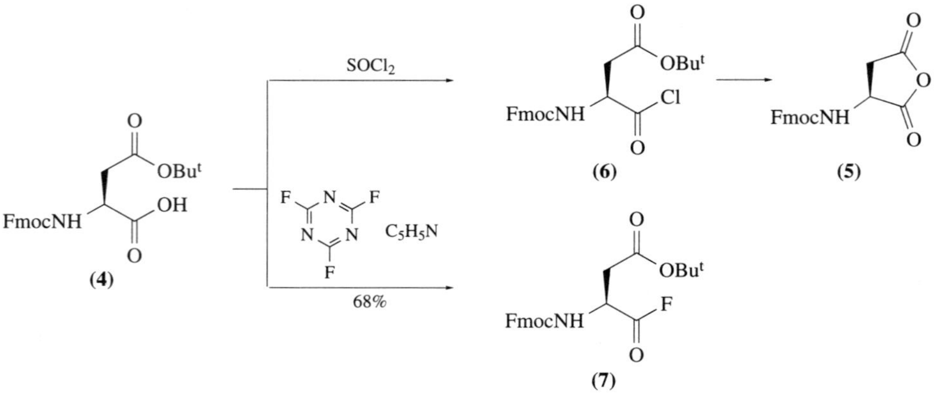

Scheme 5

each mole of selenium tetrafluoride employed, to afford the acid fluorides in 90% and 85% yields respectively.

A number of fluorinated alkenes have been used as a fluorine source in the conversion of carboxylic acids into acid fluorides. For example, the hexafluoropropene/diethylamine reagent was used to convert propionic and benzoic acids into the acid fluorides in reasonable yields ⟨79BCJ3377⟩, as shown in Table 1 (entries 6 and 7). Hexafluoropropene dimer in the presence of potassium carbonate was also used in the preparation of acid fluorides from carboxylic acids ⟨77TL2337, 77TL2893⟩. Polar solvents or complexing agents were required, and the formation of by-products proved to be a problem.

5.01.1.2.3 *Acid fluorides from alcohols and aldehydes*

The conversion of primary alcohols into acid fluorides has been achieved using caesium fluoroxysulfate (Equation (8)) ⟨89TL6095⟩. Similarly, aldehydes can be converted into acid fluorides using caesium fluoroxysulfate (Equation (9)). These reactions proceed in yields of 80–85%. However, there appear to be problems with the generality of these reactions. For example, phenylacetaldehyde affords a 2 : 1 mixture of benzyl fluoride and benzoyl fluoride, and 9-anthraldehyde afforded 9-fluoroanthracene as the major product. These workers have shown that radical inhibitors slowed down the reactions of the aldehydes, and therefore assume that the oxidation reaction proceeds via a radical mechanism.

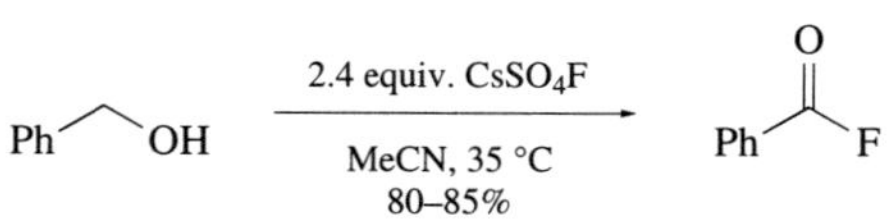

(8)

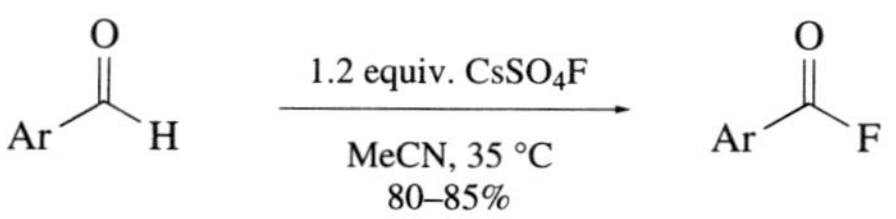

(9)

Uranium hexafluoride has been shown to oxidize a number of functional groups ⟨78JA5396⟩. Although carboxylic acids do not react with this reagent, a number of aldehydes were converted directly to acid fluorides in modest yields. Heptaldehyde was converted to heptanoyl fluoride in 47% yield, benzaldehyde to benzoyl fluoride in 40% yield, and 2-bromobenzaldehyde to 2-bromobenzoyl fluoride in 35% yield, as generalized by Equation (10).

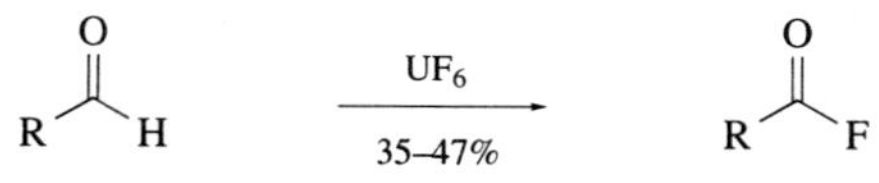

(10)

Primary alcohols were similarly oxidized to give a mixture of the aldehyde and acid fluoride products (Equation (11)).

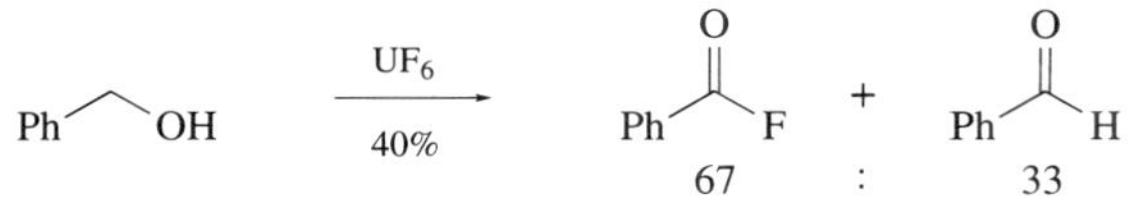

(11)

5.01.1.3 Acid Chlorides

The preparation and use of acid chlorides is more widespread than for any of the other acid halides. There are now numerous reagents and procedures for the synthesis of acid chlorides, frequently starting from the parent carboxylic acid, although other routes are also available. In many instances, preparations of acid chlorides have been designed such that these reagents may be prepared with sufficient purity that they are suitable for use in the next stage of a synthetic sequence without recourse to elaborate work-up procedures. Perhaps for this reason, the volatile reagents thionyl chloride and oxalyl chloride are often employed for the conversion of carboxylic acids, since they may be used in excess, and simply distilled out of the reaction mixture to afford the acid chloride product with good levels of purity. The only by-products are gaseous for both of these reagents, which represents an additional advantage.

5.01.1.3.1 Acid chlorides from carboxylic acids and their derivatives

(i) Using thionyl chloride

There are now numerous examples of the use of thionyl chloride as a reagent for the direct conversion of carboxylic acids to the corresponding acid chloride. Indeed, thionyl chloride is the most widely employed reagent for this conversion. The reaction occurs with the stoichiometry indicated in Equation (12). Typically, the acid is treated with a small excess of thionyl chloride, often in the presence of an inert solvent, and the reaction mixture is normally heated at reflux until the conversion is complete, generally about 2 or 3 h. The end point of the reaction is often readily determined for solid carboxylic acids with partial solubility in the chosen solvent. Since the acid chloride is generally soluble in the reaction medium, the reaction may be terminated once all of the solids have disappeared. Alternatively, particularly for larger scale reactions, once the evolution of gas (sulfur dioxide and hydrogen chloride) has stopped, this is often judged as the end point of the reaction. The acid chloride is obtained in crude form by the removal of excess thionyl chloride under reduced pressure, and for many applications is of sufficient purity to be used directly in the next step.

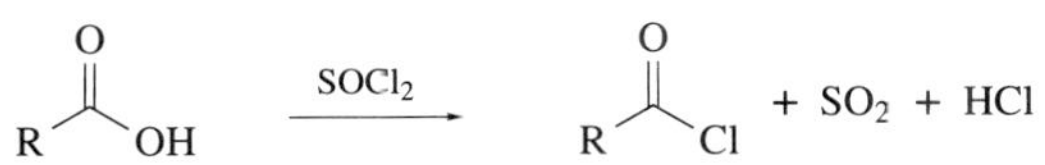

(12)

In order to remove traces of thionyl chloride from the crude acid chloride, the reaction mixture can be concentrated under reduced pressure, and then an inert solvent such as tetrachloromethane or benzene is added, and the solution reevaporated. This procedure may be repeated more than once. For example, Milewska and Poloński converted the protected malic acid derivative (**8**) to the acid chloride (**9**) by refluxing with thionyl chloride for 30 min, evaporation of excess reagent, treating with tetrachloromethane, and evaporating to dryness (Equation (13)). This procedure was repeated twice to remove traces of thionyl chloride ⟨88S475⟩. When possible, the acid chloride may of course be purified by distillation or crystallization. The addition of DMF as a catalyst is also popular, since the reaction may be conducted at room temperature, or for shorter reaction times. Typically, catalytic quantities of DMF are added, although, more conveniently, this may simply be one or two drops of DMF.

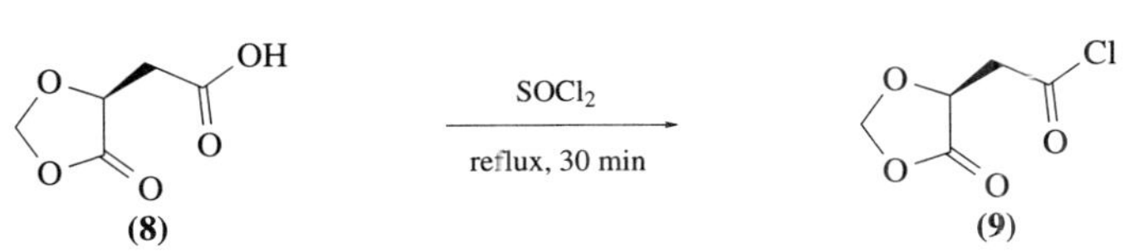

(13)

A large number of aromatic carboxylic acids have been converted into the acid chlorides by treatment with thionyl chloride at reflux. The excess reagent is simply removed under reduced pressure to afford the crude aromatic acid chlorides. Wallace *et al.* used this procedure to generate 23 aromatic acid chlorides (Equation (14)) ⟨90S1143⟩. Conversion of carboxylic acids into acid chlorides with thionyl chloride has been conducted in a number of solvents, including benzene ⟨88OSC(6)549⟩, and tetrachloromethane ⟨88OSC(6)190⟩.

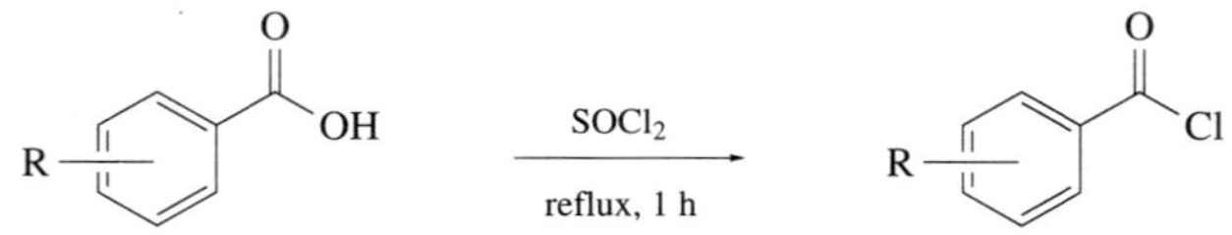

(14)

Similarly, Olah *et al.* have shown that the use of refluxing thionyl chloride is acceptable for the conversion of sterically hindered carboxylic acids to their acid chlorides, including formation of di(1-adamantyl)acetyl chloride (Equation (15)), 2-(1-adamantyl)-3,3-dimethylbutanoyl chloride, and 2-*t*-butyl-3,3-dimethylbutanoyl chloride ⟨89S566⟩. 2,6-Substituted arenecarboxylic acids are also converted into the acid chlorides by treatment with thionyl chloride. For example, Structure (**10**) was converted into the acid chloride (**11**) in 98% yield under these conditions ⟨88S525⟩ (Equation (16)). Similarly, mesitoic acid was readily converted into the acid chloride in 90–97% yield simply by refluxing in thionyl chloride until the evolution of sulfur dioxide and hydrogen chloride ceased ⟨55OSC(3)555⟩.

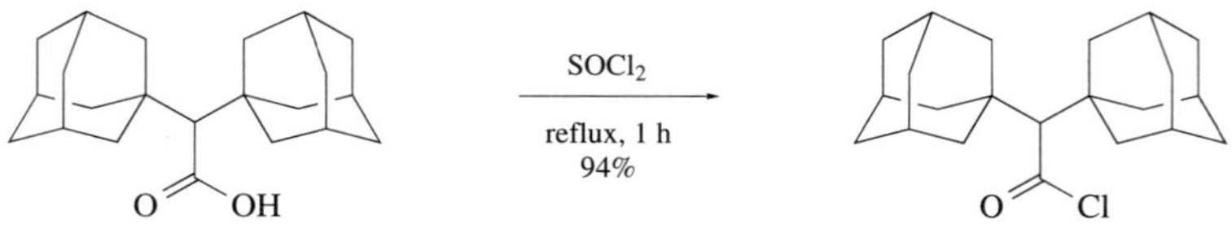

(15)

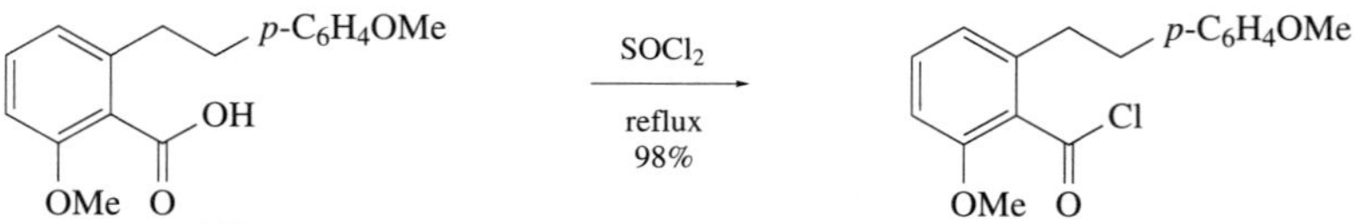

(16)

(**10**) (**11**)

Perhaps surprisingly, Hwang and Fowler found that conversion of carboxylic acids containing an alkene functionality was readily prepared by treatment with neat thionyl chloride at reflux ⟨85JOC2719⟩. This simple procedure was employed in the formation of the acid chloride (**12**) in good yield (Equation (17)). However, the reaction of a *cis*-α,β-unsaturated acid with thionyl chloride and pyridine at 75 °C for 4 h afforded the *trans*-α,β-unsaturated acid chloride as the only reported product in good yield (Equation (18)) ⟨89S598⟩.

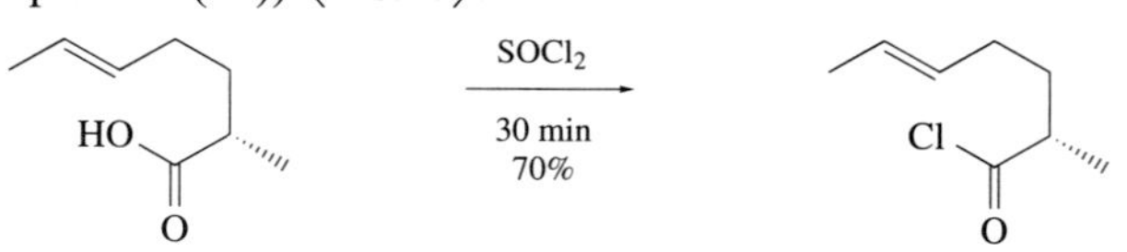

(17)

(**12**)

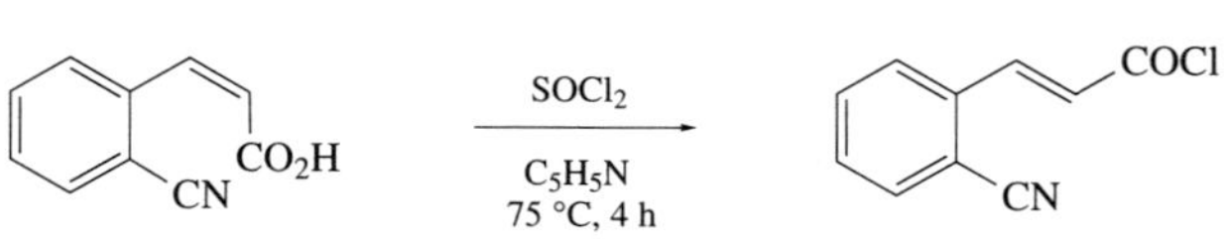

(18)

For sparingly soluble carboxylic acids, conversion to the acid chloride may be especially slow. Burdett showed that for biphenyl-4,4′-dicarboxylic acid the reaction with thionyl chloride in refluxing 1,2-dichloroethane gave a yield of less than 5% after 48 h under these conditions ⟨91S441⟩. However, addition of the phase transfer catalyst, benzyltriethylammonium chloride allowed the reaction to go to completion in 16–24 h, with >90% yield (Equation (19)). The formation of biphenyl-4,4′-dicarboxylic acid chloride can also be achieved using neat thionyl chloride, or with phosphorus pentachloride in refluxing xylene, although the workup was apparently easier for the phase transfer catalyzed reaction.

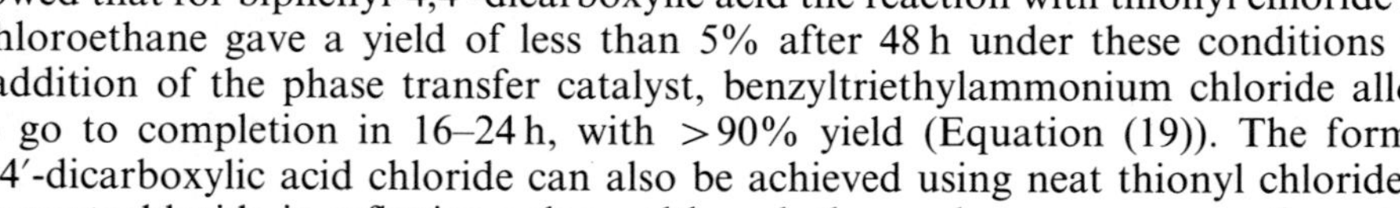

(19)

Matsuda *et al.* have noted that the conversion of carboxylic acids into acid chlorides with thionyl chloride in pyridine requires a reaction time of about 1 h, but when dichloromethane is employed as solvent and conversion is effected with pyridine and thionyl chloride, the reaction time is reduced to about 6 min ⟨85T3625⟩. Matsuda *et al.* also discovered that if the carboxylic acid is converted to the dicyclohexylammonium salt, then the reaction time is further reduced to within 1 min under similar conditions (Scheme 6). Generally, the dicyclohexylammonium salts were prepared by treating a solution of carboxylic acid in ethyl acetate with dicyclohexylamine, and isolating the precipitated salt. Conversion into the acid chloride with thionyl chloride and pyridine in dichloromethane was generally followed by *in situ* derivatization to amides. The use of dicyclohexylammonium salts of benzyloxycarbonyl amino acids was reported to give good results with regard to racemization in peptide coupling reactions.

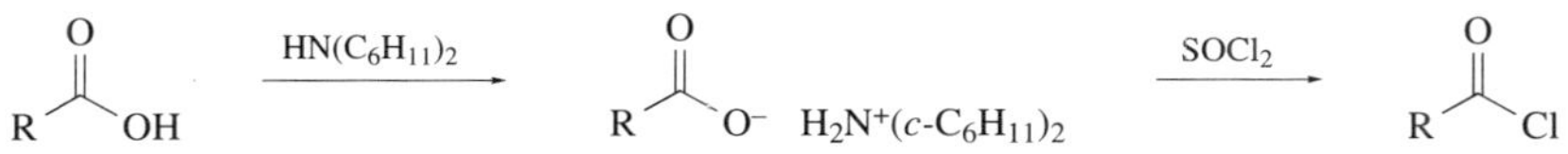

Scheme 6

For the conversion of sensitive carboxylic acids into acid chlorides, Wolfe *et al.* found that with careful control of stoichiometry, thionyl chloride could be used to form acid chlorides from penicillins ⟨68CJC2549⟩. Sufficient pyridine was required to ensure that the conditions were always basic. They employed at least 1 equivalent of pyridine for every equivalent of hydrogen chloride and sulfur dioxide formed, according to Scheme 7. Pyridine prevents destruction of the penicillin by acting as a buffer. Isolation of the penicillin acid chloride was achieved by addition of diethyl ether and precipitation. The preconversion of the carboxylic acid into the potassium salt allowed the use of only one equivalent of pyridine. Addition of a dichloromethane solution of thionyl chloride to this afforded the acid chloride. This procedure was used in the conversion of α-phenoxyethylpenicillin (**13**) into the acid chloride (**14**) (Equation (20)). In general, the technique of preforming the carboxylate salt is valuable for acid-sensitive substrates. The same authors noted that use of the stronger base triethylamine led to a side reaction taking place, affording a dehydrated rearranged product ⟨63JA643⟩.

$RCO_2H + SOCl_2$	⟶	$RCOCl + SO_2 + HCl$	Need 2 equiv. C_5H_5N
$RCO_2M + SOCl_2$	⟶	$RCOCl + SO_2 + MCl$	Need 1 equiv. C_5H_5N
$RCO_2H{\bullet}H_2O + 2SOCl_2$	⟶	$RCOCl + 2SO_2 + 3HCl$	Need 5 equiv. C_5H_5N

Scheme 7

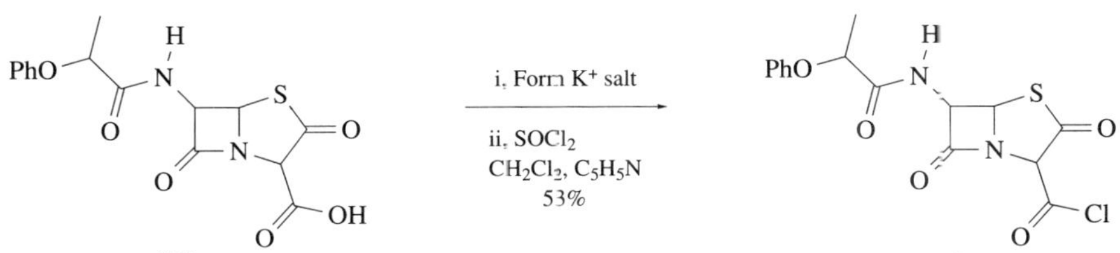

(**13**) (**14**) (20)

However, in some instances the use of triethylamine has also been successful. Brain *et al.* found that alkoxyquinolinecarboxylic acids were sensitive to hot thionyl chloride ⟨63JCS491⟩. The desired acid chloride products were obtained by treatment with thionyl chloride in cold chloroform or dichloromethane in the presence of triethylamine (Equation (21)).

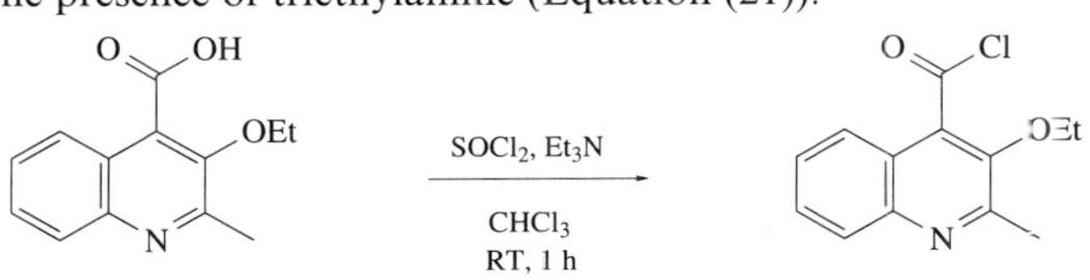

(21)

In 1959, Bosshard *et al.* discovered that sulfonic and carboxylic acids which do not react with thionyl chloride under normal conditions, such as *p*-nitrobenzoic acid, were converted to the acid chloride by using DMF as a catalyst ⟨59HCA1653⟩. The mechanism of catalysis involves the reaction of DMF and thionyl chloride (or other chlorinating reagent) to give the electrophilic chloroiminium species (**15**). Attack by the carboxylic acid forms the intermediate (**16**), which by reaction with chloride at the carbonyl affords the acid halide and regenerates DMF (Scheme 8).

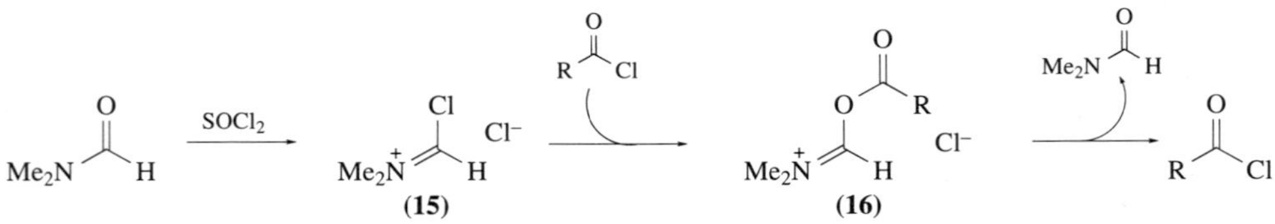

Scheme 8

Since the use of DMF allows for milder conditions and shorter reaction times to be employed, more sensitive substrates may be converted than in the absence of DMF, as illustrated in the conversion of isoxazolecarboxylic acid (**17**) into the acid chloride (**18**), which can also be made by treatment of the sodium salt with phosphorus pentachloride or thionyl chloride (Equation (22)) ⟨61JOC1514⟩.

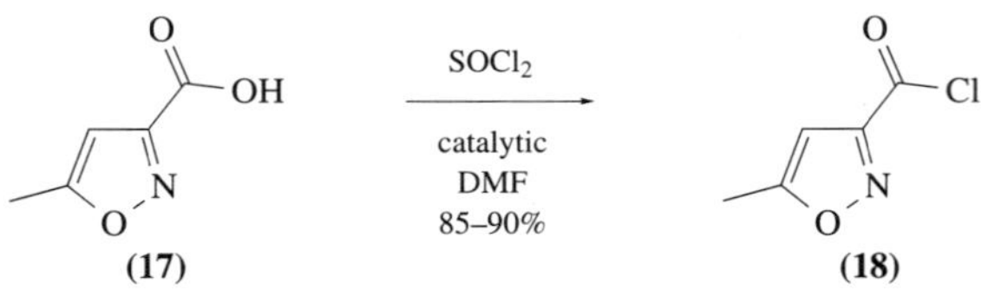

(22)

Normant *et al.* have employed hexamethylphosphoramide (HMPA) as a solvent for thionyl chloride reactions, and they assume the formation of the activated species (**19**) as the active intermediate ⟨69CR(C)1325⟩. The hydrogen chloride generated during the reaction is absorbed by the HMPA, and thereby affords mild conditions. Reaction occurs at −20 to −10 °C, for example, in the conversion of crotonic acid into crotonoyl chloride in 80% yield.

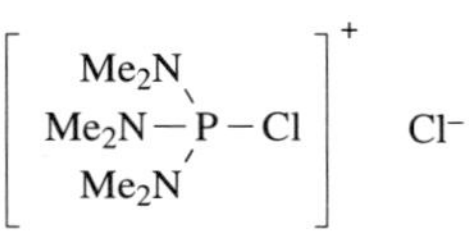

(19)

Olah and Wu have shown that the treatment of α-hydroxy acids with thionyl chlorides gives decarboxylation products ⟨91S1177⟩. For example, treatment of 2-hydroxy-2-phenylpropionic acid with thionyl chloride affords acetophenone (74%), and mandelic acid affords benzaldehyde. However, α-hydroxy acid fluorides have been prepared (see 5.01.1.2.2), and are reported to be stable.

Hartmann protected 1-hydroxy-1-cyclopropanecarboxylic acid as the acetate, and was able to form the acid chloride without rearrangement by using thionyl chloride and catalytic DMF (Equation (23)) ⟨89S272⟩. A slightly modified procedure was employed to avoid the formation of the anhydride, which in this case was not converted to the acid chloride by excess thionyl chloride. The acid was added dropwise to a solution of thionyl chloride and catalytic DMF in dichloromethane over 1 h, and subsequently stirred at 50 °C for a further 2 h. Attempts to purify by distillation led to decomposition of the product. A similar modification was employed in the preparation of menthoxyacetyl chloride from menthoxyacetic acid, although anhydride formation was not mentioned as a problem under the more standard conditions ⟨55OSC(3)547⟩. 1-Cyano-1-cyclopropanecarboxylic acid was treated with thionyl chloride in refluxing benzene for 2 h, and converted directly into the acid azide *in situ*. No special procedure was required in this instance ⟨91SL919⟩.

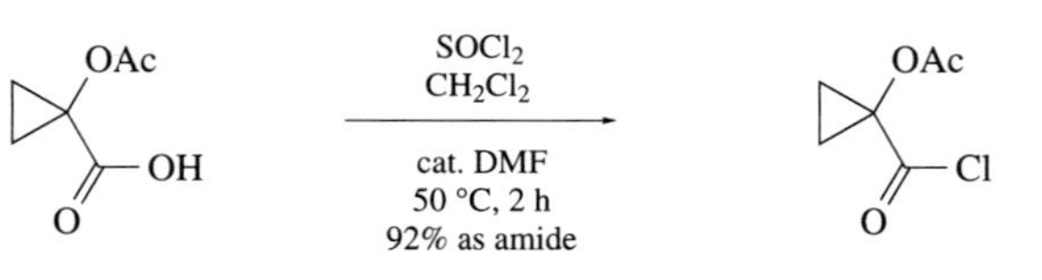

(23)

(ii) Using oxalyl chloride

In 1920, Adams and Ulich introduced oxalyl chloride as a reagent for the conversion of carboxylic acids to the corresponding acid chlorides ⟨20JA599⟩. The oxalyl chloride affords the acid chloride, with carbon monoxide, carbon dioxide, and hydrogen chloride as by-products (Equation (24)).

(24)

Although thionyl chloride is still generally employed as the first choice reagent for the preparation of acid chlorides, oxalyl chloride is normally a good alternative, and in some cases only oxalyl chloride is suitable for particularly sensitive substrates. Generally, reactions employing oxalyl chloride are performed at room temperature, although sometimes conditions such as refluxing benzene may be employed. Oxalyl chloride and thionyl chloride have similar boiling points, and so there is little difference in the convenience of the workup. Both reagents generate only gaseous by-products, including hydrogen chloride, and both reagents may be used in conjunction with catalytic amounts of DMF to accelerate the reaction. Oxalyl chloride is more expensive than thionyl chloride which detracts from its use in larger scale preparations.

In a typical procedure, the carboxylic acid and oxalyl chloride are stirred at room temperature in an inert solvent (normally benzene or dichloromethane). The reaction time appears to be very substrate-dependent, but reaction is generally complete after about 12 h. The addition of catalytic DMF accelerates the reaction, which is then generally complete in about 2 h. As the DMF is added to the reaction mixture, the evolution of gas is often observed. As with reactions employing thionyl chloride, the reaction may be judged to be complete when there is no further gas evolution, and for sparingly soluble carboxylic acids, upon dissolution of all the solids. The following examples illustrate the range of functional groups which are tolerant of the mild oxalyl chloride procedure for the preparation of acid chlorides.

Recently, a series of carboxylic acid substrates (**20**) containing pyrazole rings were smoothly converted into the acid chlorides (**21**) on exposure to oxalyl chloride in dichloromethane, and converted without isolation into *N*-methylhydroxamic acids (Equation (25)) ⟨91S18⟩. Benzene was employed as the solvent for the conversion of the unsaturated acid (**22**) into the acid chloride (**23**). The reaction was run at room temperature, with brief heating prior to isolation (Equation (26)) ⟨91S1173⟩. Other unsaturated carboxylic acids have been converted into acid chlorides by oxalyl chloride ⟨91OS(70)101⟩.

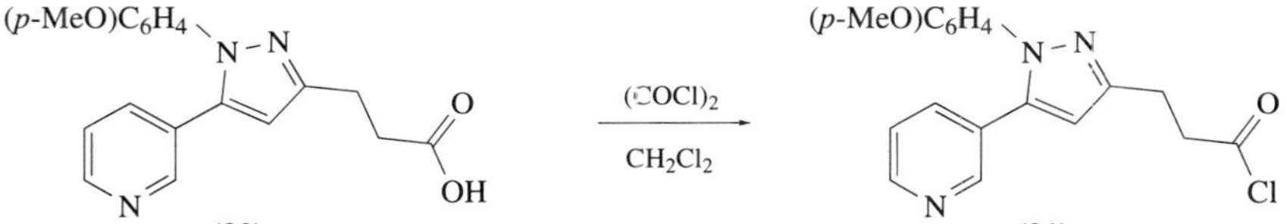

(25)

(20) **(21)**

converted to the *N*-methyl hydroxamic acid *in situ*

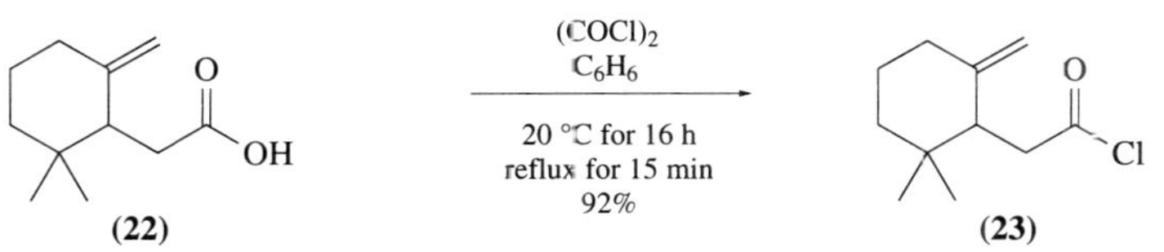

(26)

(22) **(23)**

The isoindolecarboxylic acid (**24**) was treated with oxalyl chloride in benzene for 4 h followed by concentration (Equation (27)). The product acid chloride (**25**) was used directly in the formation of the acid azide ⟨90MI 501-01⟩.

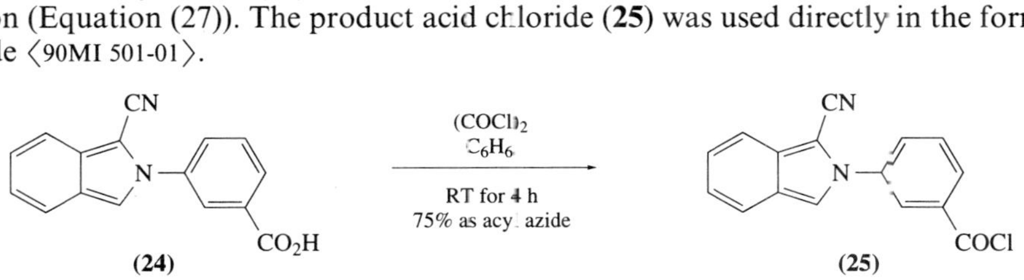

(27)

(24) **(25)**

A number of dicarboxylic acids have been converted into the diacid chlorides by treatment with oxalyl chloride in dichloromethane ⟨90JA9586⟩. The reaction was catalyzed by DMF, and judged to be complete once the acid had dissolved. The diacid chlorides (**26**) and (**27**) were both prepared, and derivatized *in situ* into amides (Equations (28) and (29)).

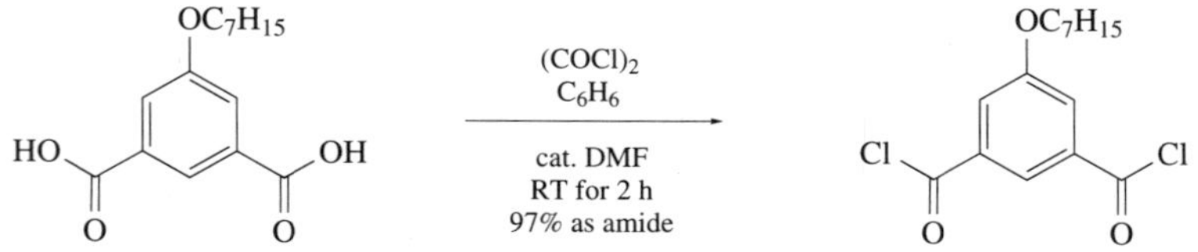

(26)

(28)

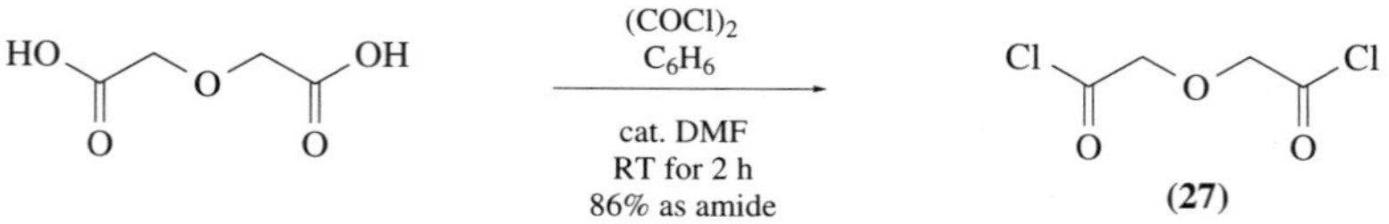

(29)

Carboxylic acids containing phosphonate groups ⟨89S506⟩ and silyl groups ⟨87OS(66)87⟩ have been readily converted into the corresponding acid chlorides with oxalyl chloride.

Several procedures have been reported for the preparation of 2-methoxy-2-trifluoromethyl-2-phenylacetic acid (MTPA-Cl) ⟨69JOC2543⟩. The reaction between MTPA and thionyl chloride in the presence of sodium chloride was found to give uncertain results. Although no racemization was found to occur in the process, variable amounts of methoxytrifluoromethylphenylchloromethane are formed by decarboxylative chlorination of MTPA or MTPA-Cl ⟨90TL75⟩. Since this reagent is often required on a small scale for the derivatization of chiral alcohols to the corresponding diastereomeric esters, Ward and Rhee have devised a useful method for the preparation of MTPA-Cl on a small scale ⟨91TL7165⟩. One of the problems with the normal method of adding oxalyl chloride and then one drop of DMF, is that when working on a small scale, it is difficult to add small amounts of DMF. The chloroiminium species (**15**) that is formed is more reactive than MTPA-Cl, and the alcohols are converted into formate esters (**28**) after workup (Scheme 9). Increasing the relative amount of DMF employed decreases the yield of the isolated MTPA ester. This problem is overcome by the formation of MTPA-Cl in hexane solution, in which the chloroiminium species (**15**) precipitates out. Filtration and concentration affords the crude MTPA-Cl, which was used directly by treatment with an alcohol in the presence of triethylamine and dmap. The method was found to be applicable to the preparation of other acid chlorides, although presumably the acid chloride in question must be sufficiently soluble in hexane for useful results.

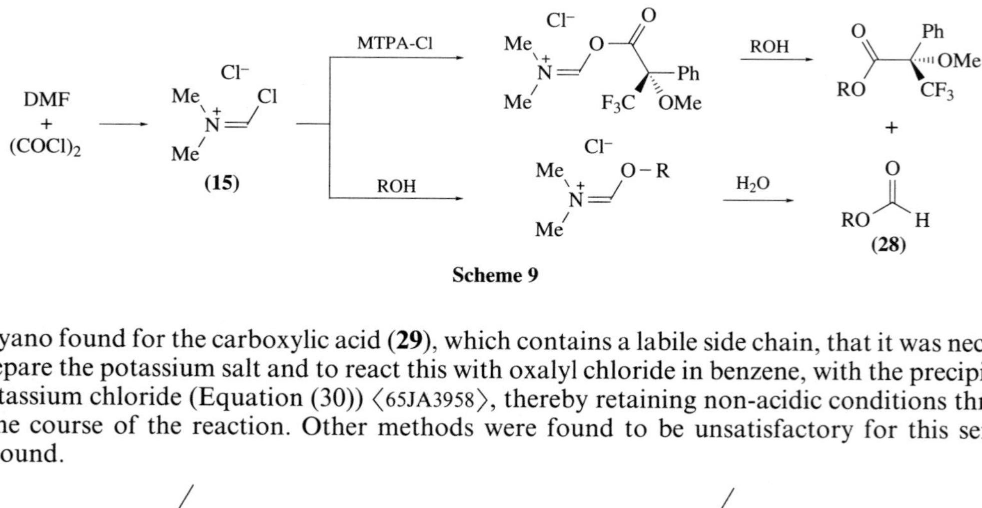

Scheme 9

Miyano found for the carboxylic acid (**29**), which contains a labile side chain, that it was necessary to prepare the potassium salt and to react this with oxalyl chloride in benzene, with the precipitation of potassium chloride (Equation (30)) ⟨65JA3958⟩, thereby retaining non-acidic conditions throughout the course of the reaction. Other methods were found to be unsatisfactory for this sensitive compound.

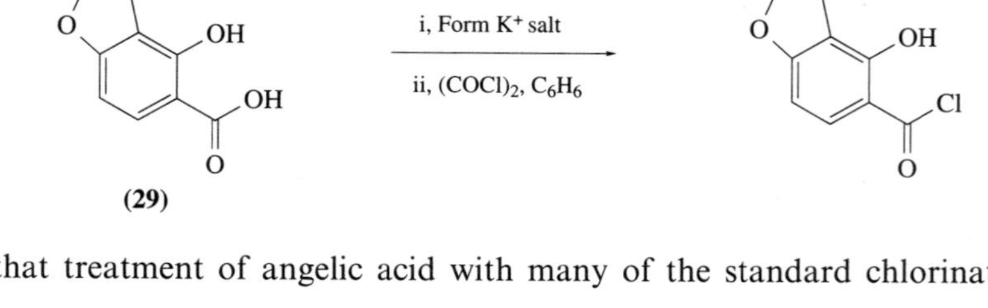

(30)

(**29**)

Beeby found that treatment of angelic acid with many of the standard chlorinating reagents, including phosphorus oxychloride, phosphorus trichloride, and thionyl chloride afforded tigloyl chloride (Equation (31)). Oxalyl chloride afforded a 2 : 1 ratio of angeloyl chloride to tigloyl chloride, and, finally, it was discovered that treatment of the potassium salt in diethyl ether with oxalyl chloride and catalytic DMF for 2 h at 0 °C afforded pure angeloyl chloride (Equation (32)) ⟨77TL3379⟩.

(31)

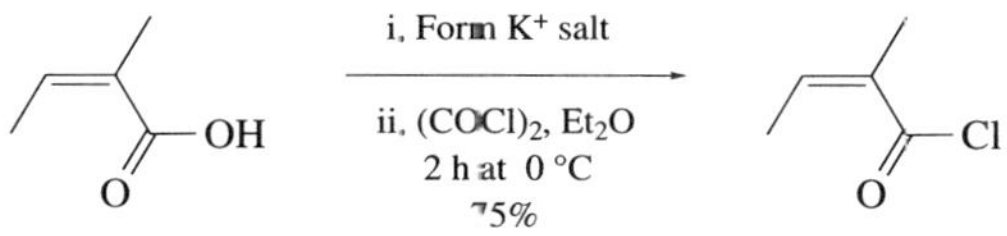

(32)

Conversion of the enantiomerically pure half ester (**30**) into the corresponding acid chloride (**31**) may be achieved without loss of stereochemical integrity by using oxalyl chloride (Equation (33)). However, rearrangement (via the anhydride) leading to racemization was found to occur with thionyl chloride, unless care was taken to use very pure thionyl chloride and a reaction temperature of 30 °C ⟨47JA2568⟩.

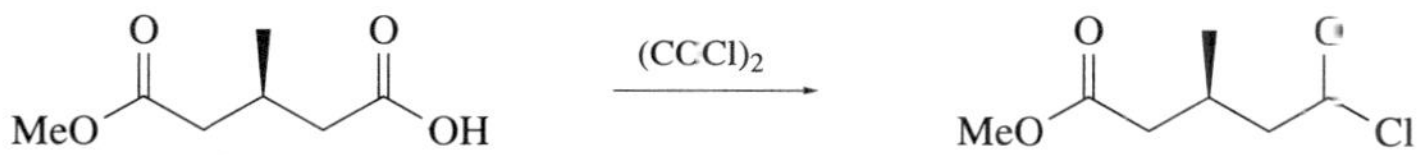

(30) **(31)** (33)

Carbon-14 labeled acid chlorides were prepared from the magnesium carboxylates, which had been formed from the reaction of Grignard reagents with labeled carbon dioxide (Scheme 10). In this instance the excess oxalyl chloride was destroyed by the addition of potassium stearate, since this circumvented the problem of separating volatile acid chlorides from excess oxalyl chloride. The labeled acid chlorides were isolated in 57–88% yields as their derived anilides ⟨85MI 501-01⟩.

$$\mathrm{RMgX} + {}^{14}\mathrm{CO_2} \longrightarrow \mathrm{R^{14}CO_2MgX} \xrightarrow{(\mathrm{COCl})_2} \mathrm{R^{14}COCl}$$

Scheme 10

(iii) Using phosphorus reagents

There are a number of phosphorus reagents which have been employed for the conversion of carboxylic acids to acid chlorides.

p-Nitrobenzoic acid has been converted in very high yield to the acid chloride by treatment with 10 equiv. phosphorus oxychloride (Equation (34)). However, the authors do not recommend the procedure over the more conventional methods of treatment with phosgene or thionyl chloride, because of problems associated with the removal of phosphorus-containing compounds ⟨79MI 501-01⟩.

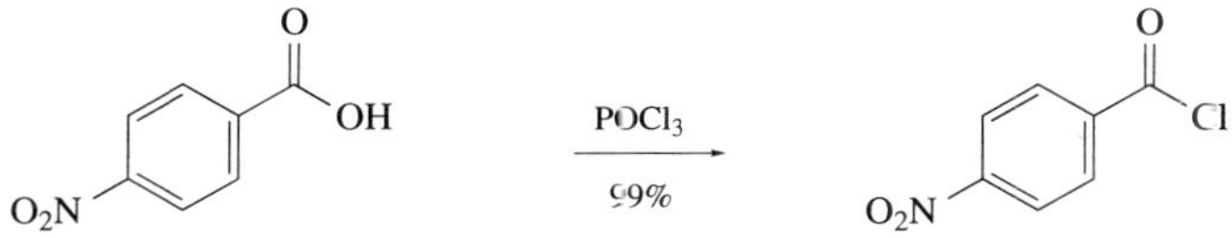

(34)

The combination of triphenylphosphine and tetrachloromethane was reported in 1966 to be useful for the generation of acid chlorides from the carboxylic acids (Equation (35)). Since no acidic or basic products are generated, neutral reactions can be performed with these reagents ⟨66JA3440⟩. One of the difficulties of using phosphorus-based reagents for any reaction is trouble with purification of the product and removal of triphenylphosphine oxide. To this end, a number of immobilized, polymer-bound phosphines have been employed. The use of polymer-supported triphenylphosphine and treatment with the carboxylic acid in tetrachloromethane converted the acids in the following yields to their acid chlorides: octanoic acid (63%), benzoic acid (90%), α-furoic acid (82%), β-naphthoic acid (84%), and cinnamic acid (77%) ⟨75CC622⟩.

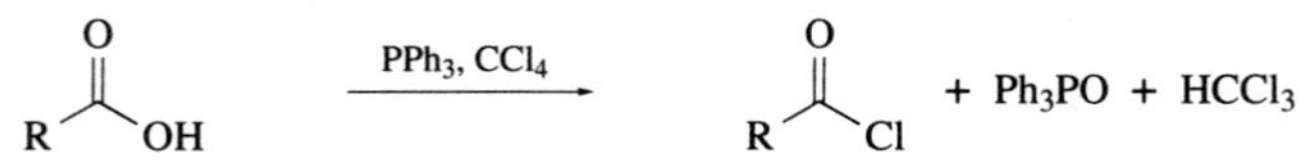

(35)

Similarly, the pentacoordinate dichlorophosphine polymer (**32**) was employed as beads; these were suspended in dry dichloromethane and the carboxylic acid added. Phenylacetic acid was treated in this manner, filtered to remove the polymer reagent, and dichloromethane removed *in vacuo* to give a quantitative yield of phenylacetyl chloride (Equation (36)) ⟨74JA6469⟩.

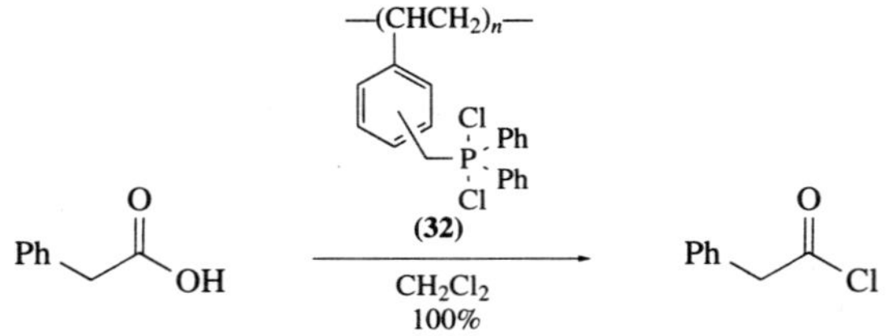

(36)

The direct conversion of carboxylic acid esters into the corresponding acid chlorides has been reported. Although high temperatures are required, good yields of the acid chloride have been obtained using phosphorus(V) reagents (Equation (37)) ⟨75JOC3026⟩.

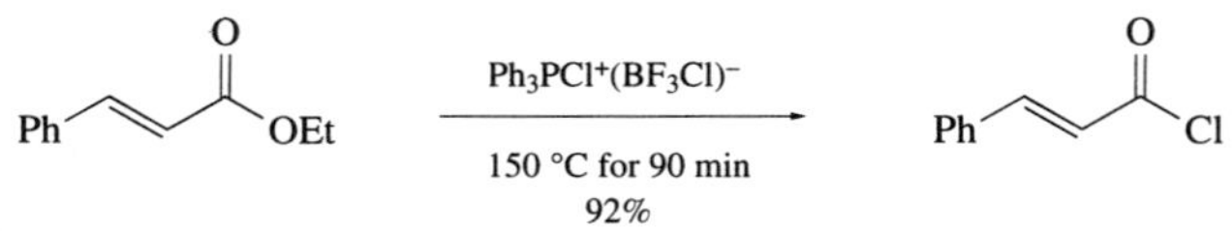

(37)

Phosphorus pentachloride in carbon tetrachloride has been employed to prepare acetylenedicarboxylic acid chloride (**33**) (Equation (38)). This reaction fails with thionyl chloride, since this reagent adds across the triple bond ⟨82CB804⟩.

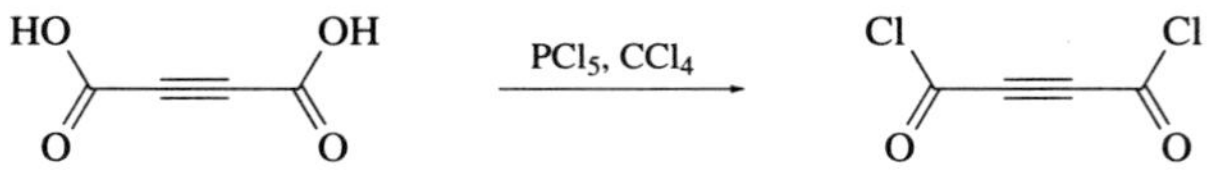

(**33**)

(38)

(iv) Using miscellaneous reagents

The preparation of α-keto acid chlorides has proved to be troublesome with most of the standard reagents, such as thionyl chloride, oxalyl chloride, and the phosphorus chlorides. However, the use of α,α-dichloromethyl ethers has been shown to be effective for the generation of these sensitive compounds ⟨83OS(61)1⟩. Neat pyruvic acid was treated with methyl (α,α-dichloromethyl) ether at 25–50 °C, to afford pyruvoyl chloride in fair yield (Equation (39)), although this did not prove to be general for all α-keto acids. The use of α,α-dichloromethyl ethers for the conversion of carboxylic acids into acid chlorides was first reported by Heslinga *et al.* in 1957 ⟨57RTC969⟩.

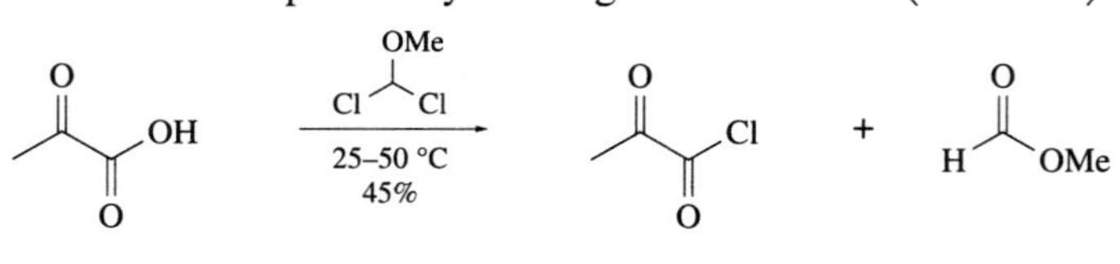

(39)

N,N-Dimethylchloromethyleneammonium chloride (**15**) has been prepared by the reaction of oxalyl chloride with DMF, and used to prepare acid chlorides from carboxylic acids ⟨87OS(66)121⟩. 6-Oxodecanoic acid was treated with (**15**) to give the acid chloride which was reduced *in situ* with lithium tri(*t*-butoxy)aluminium hydride to give the keto aldehyde in 85–93% yield (Equation (40)). Use of this reagent to produce acid chlorides which are reduced *in situ* with sodium borohydride into the corresponding alcohols has also been reported ⟨83CL835⟩.

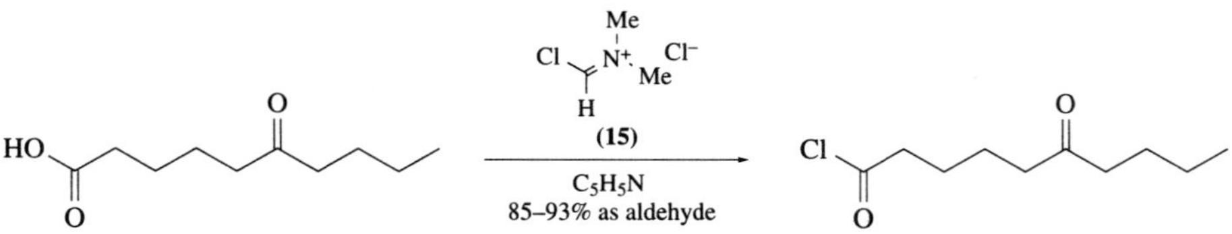

(40)

Schmidt *et al.* have converted α-amino acids protected with the Fmoc group to the corresponding amino acid chlorides by treatment with thionyl chloride in dichloromethane at 20–40 °C ⟨89S832⟩. In an alternative procedure, *N*-benzyloxycarbonyl amino acids were treated with 1-chloro-*N*,*N*-2-trimethyl-1-propen-1-amine at 0 °C in dichloromethane and stirred for 10 min, and the resulting amino acid chloride used directly for coupling to amino acid esters or hydroxy esters (Equation (41)) ⟨88S475⟩. There was no apparent racemization of *N*-benzyloxycarbonyl-(*S*)-phenylalanine after storage of up to 1 h, nor during the subsequent peptide forming step with the benzyl ester of (*R*)- or (*S*)-proline.

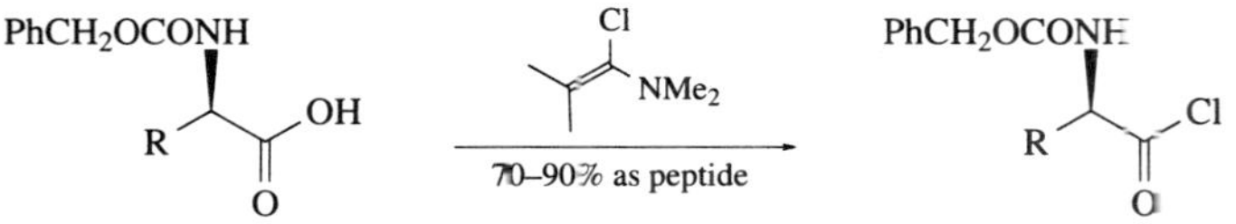

(41)

In 1938, Brown showed that volatile acid chlorides were readily prepared by exchange reaction of the carboxylic acid with benzoyl chloride. For example, acetyl chloride was prepared in 85% yield, monochloroacetyl chloride in 76% yield, crotonoyl chloride in 80% yield, isobutanoyl chloride in 88% yield, and caproyl chloride in 80% yield, as shown by Equation (42). Caproyl chloride (b.p. 150–153 °C) was the highest boiling acid chloride reported ⟨38JA1325⟩. Similarly, Kyrides had reported the preparation of butanoyl chloride in 90% yield by exchange of butanoic acid with phthaloyl chloride ⟨37JA206⟩. Carbon-13 labeled acetic acid was converted to the acid chloride by formation of the sodium salt followed by reaction with excess benzoyl chloride with no added solvent. The lower boiling acetyl chloride was distilled from the reaction mixture (Equation (44)) ⟨89RTC51⟩. This is a useful method for the preparation of low boiling acid chlorides.

(42)

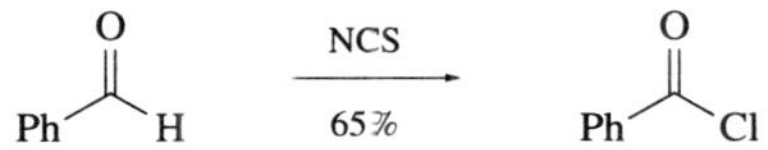

(43)

Cyanuric chloride has been employed for the conversion of carboxylic acids into acid chlorides ⟨78JOC3972⟩. The reagent behaves in a similar manner to cyanuric fluoride (see 5.01.1.2.2).

5.01.1.3.2 Acid chlorides from aldehydes

A number of methods have been reported for the direct oxidative conversion of aldehydes to the corresponding acid chloride. For example, it has been reported that aromatic acid chlorides may be prepared by treating an aromatic aldehyde with sulfuryl chloride (SO_2Cl_2) in the presence of a catalytic amount of phosphine, phosphine oxide, or phosphonium compound ⟨82MI 501-01⟩. Other reagents which have been employed for this process include *t*-butyl hypochlorite, *N*-chlorosuccinimide, and iodobenzene dichloride ⟨83T1187⟩. For example, treatment of benzaldehyde with *N*-chlorosuccinimide affords benzoyl chloride in 65% yield (Equation (43)) ⟨51BSB54⟩.

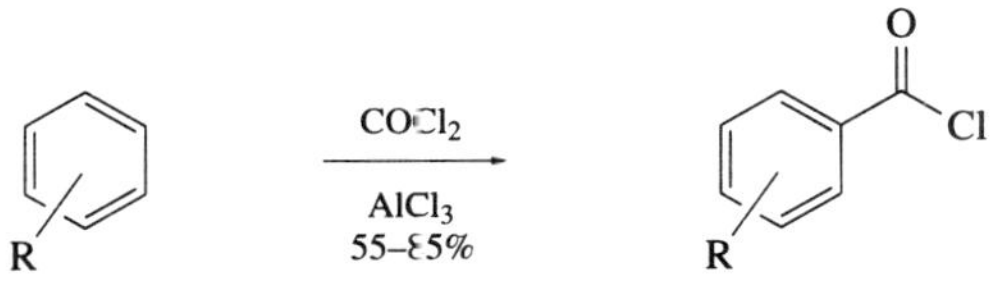

(44)

5.01.1.3.3 Acid chlorides by miscellaneous methods

The use of phosgene in Friedel–Crafts acylation reactions has been employed for the preparation of aromatic acid chlorides. However, the acid chloride product can react further to afford a symmetrical ketone. The use of excess phosgene helps to overcome this problem, and the generation of aromatic acid chlorides in 55–85% yields has been reported (Equation (44)) ⟨83OS(61)8⟩. The phosgene can be generated *in situ* by the action of aluminum trichloride on oxalyl chloride. The reaction of indole with oxalyl chloride in the absence of a Lewis acid affords the α-keto acid chloride (**34**) (Equation (45)) ⟨54JA6208⟩.

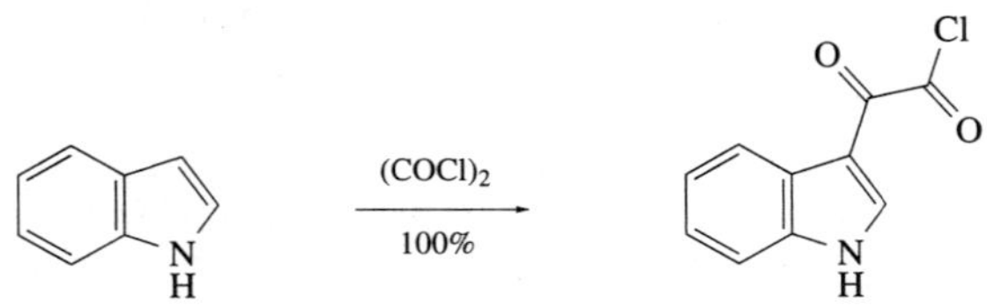

(34)

(45)

The direct introduction of the chlorocarbonyl group has been achieved, although this reaction is not frequently used. For example, Kharasch and Brown reported that cyclohexane was converted to cyclohexanecarboxylic acid chloride in fair yield by treatment with oxalyl chloride and the radical initiator dibenzoyl peroxide (Equation (46)) ⟨42JA329⟩. 1,1-Diphenylethylene was chlorocarbonylated by excess phosgene in 95% yield (Equation (47)).

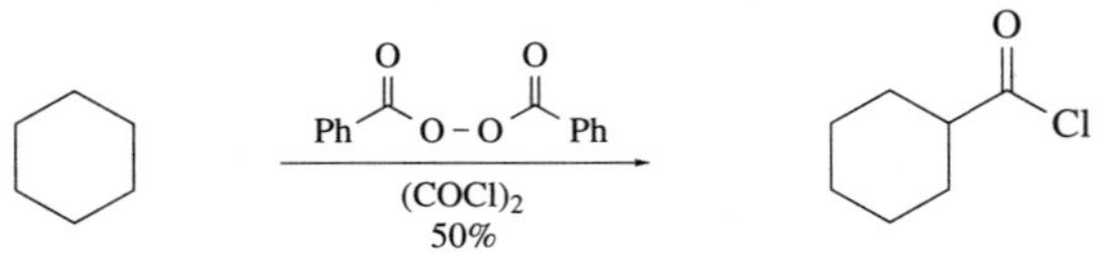

(46)

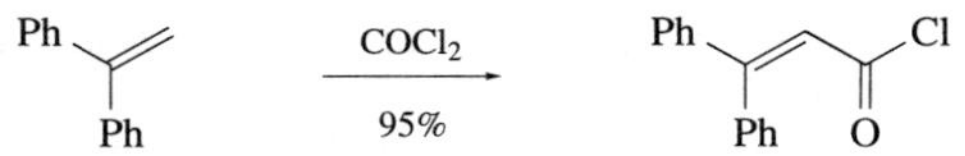

(47)

An alternative approach is the catalytic carbonylation of alkyl chlorides. The reaction works most efficiently for activated substrates such as vinyl chlorides or aryl chlorides, due to the relative ease of the oxidative addition process. For example, vinyl chloride was converted to acryolyl chloride over modified palladium catalysts (Equation (48)) ⟨B-76MI 501-01⟩. A number of examples of this process are known ⟨B-80MI 501-01⟩.

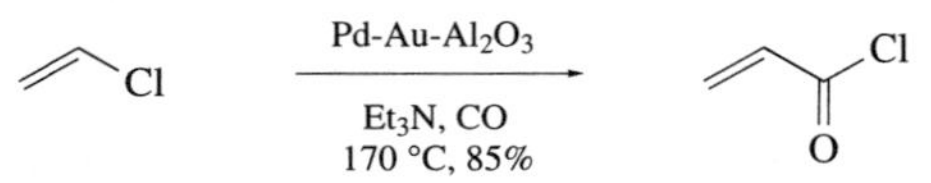

(48)

Several less common methods for the preparation are discussed in Patai ⟨B-72MI 501-01⟩, but are rather specific for certain substrates.

5.01.1.3.4 Summary of methods for acid chloride formation

Table 3 identifies some of the features of thionyl chloride, oxalyl chloride, phosphorus compounds, and benzoyl chloride as reagents for the formation of acid chlorides from carboxylic acids. Appropriate advantages and disadvantages are outlined.

Table 3 Comparison of reagents for the formation of acid chlorides.

	$SoCl_2$	*$(COCl)_2$*	*PPh_3/CCl_4*	*PhCOCl*
Boiling point (°C)	78–79	63–64	—	198–199
Molecular mass	118.97	126.93	—	140.57
Relative cost	1	15	11	3
By-products	SO_2, HCl	CO, CO_2, HCl	Ph_3PO, $HCCl_3$	$PhCO_2H$
Advantages	Low cost Simple workup DMF catalyzed Well documented	Simple workup DMF catalyzed Fairly mild Room temperature	Neutral conditions Fast reaction Also polymer bound	Good for low b.p. products
Disadvantages	Harsh conditions Not compatible with alkynes or acid-sensitive compounds	Pyridine to remove HCl if necessary Storage problems Expensive	Removal of Ph_3PO Expensive	Need excess reagent Not well documented Not good for high b.p. products

5.01.1.4 Acid Bromides

The chemistry of acid bromides is significantly less well documented than for acid chlorides. Consequently, the number of synthetic routes is also reduced, and often the starting point is the corresponding acid chloride, although other substrates have also proved to be suitable, as described in the following discussion.

5.01.1.4.1 Acid bromides from acid chlorides

Treatment of an acid chloride with trimethylsilyl bromide generates an equilibrium with the corresponding acid bromide and trimethylsilyl chloride. The trimethylsilyl chloride may be distilled out of the reaction mixture to displace the equilibrium, as illustrated in Equation (49). Distillation of the product may then be achieved directly ⟨81S216⟩. Using this methodology, benzoyl bromide was prepared from benzoyl chloride in 79% yield. Similarly, octanoyl bromide (93%) and oxalyl bromide (85%) were readily prepared. Hydrogen bromide has also been employed to convert an acid chloride to an acid bromide ⟨B-72MI 501-01⟩.

(49)

5.01.1.4.2 Acid bromides from esters

Treatment of esters with triphenylphosphine dibromide is reported to give the acid bromide and alkyl bromide ⟨73TL5121⟩. Reactions were either run neat or in benzonitrile or acetonitrile, at high temperatures. For example, butyl acetate was converted into acetyl bromide and butyl bromide in good yield (Equation (50)). This procedure was less successful when phenyl esters were employed as substrates, requiring higher temperatures, and affording lower yields.

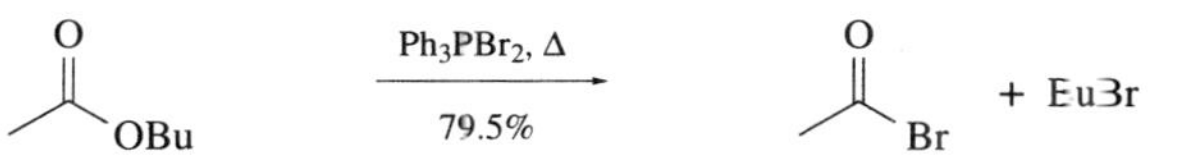

(50)

Similarly, trialkylsilyl esters are converted into the corresponding acid bromide on treatment with triphenylphosphine dibromide, although much milder reaction conditions may be employed, since products are obtained by stirring the silyl ester with triphenylphosphine dibromide at 15 °C for 15–20 min ⟨86JOC4941⟩. This reaction even works well for hindered silyl esters, including triisopropylsilyl, *t*-butyldiphenylsilyl, and *t*-butyldimethylsilyl esters. The reaction conditions are mild and neutral. The acid bromide could be isolated by distillation, although in many cases, the product was esterified *in situ* with a suitable alcohol, to give high yields of esters (ca. 90% over the two steps). For example, triisopropylsilyl phenylacetate and triisopropylsilyl *p*-nitrobenzoate were converted to the corresponding acid bromides in good yields, and isolated as the derived ethyl esters (Equations (51) and (52)).

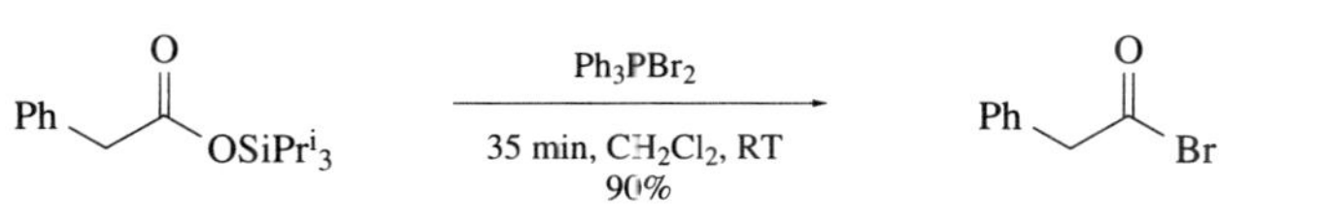

(51)

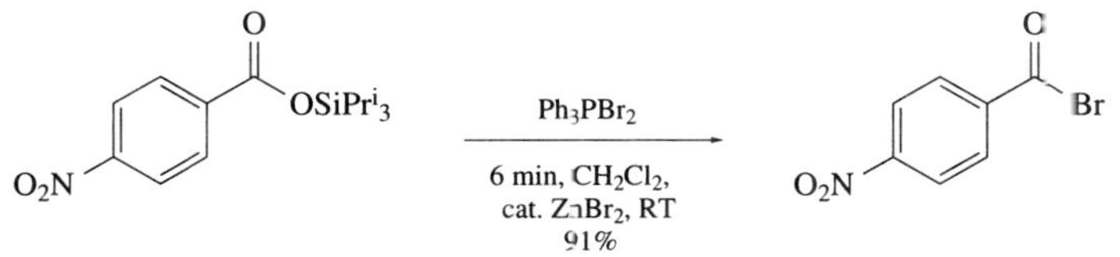

(52)

5.01.1.4.3 Acid bromides from carboxylic acids

Carboxylic acids have been directly converted into acid bromides by treatment with triphenylphosphine dibromide ⟨82S684⟩. This modification of a similar procedure reported earlier ⟨66LA(693)132⟩ allows for a good yield of the acid bromides after purification by distillation. The mechanism is believed to proceed via the phosphonium salt (**35**), as illustrated in Scheme 11 ⟨86SC1423⟩.

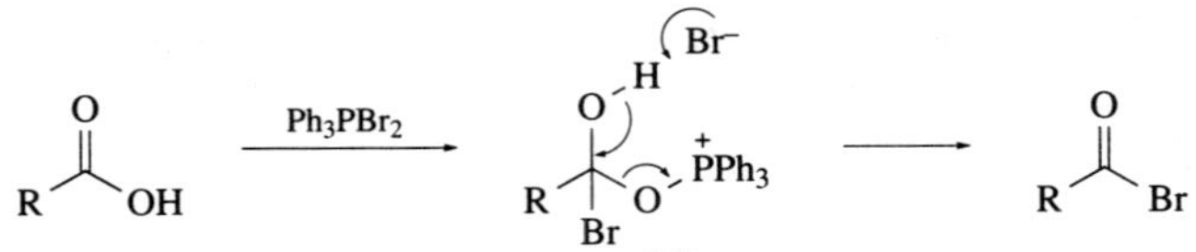

(**35**)

Scheme 11

For example, treatment of benzoic acid with triphenylphosphine dibromide in dichloromethane at 15°C for 15 min gave benzoyl bromide (Equation (53)). The procedure was also found to be effective for the generation of the diacid bromide derived from phenylmalonic acid (Equation (54)). The preparation of acid bromides of fatty acids using triphenylphosphine dibromide, and *in situ* conversion into methyl esters is simple and rapid. Bromine was added to a solution of triphenylphosphine in dichloromethane. Palmitic acid was added and stirred for 3 min. To this crude solution, methanol was added, and after 10 min, the methyl palmitate was isolated in 95% yield (Equation (55)) ⟨86SC1423⟩.

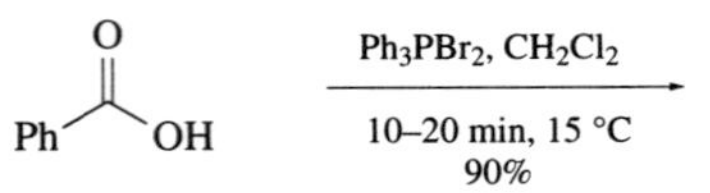

(53)

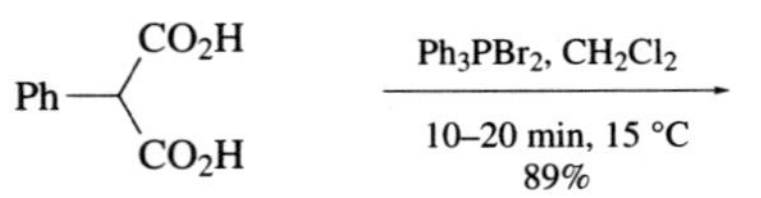

(54)

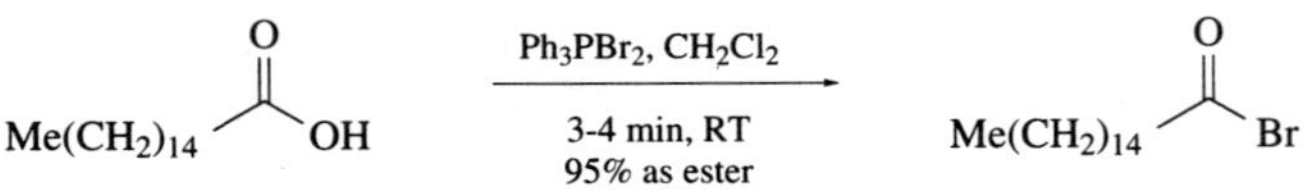

(55)

In some cases, prior conversion of the carboxylic acid into the silyl ester, followed by treatment of the crude solution with triphenylphosphine dibromide gave superior yields. This is presumably due to the formation of trimethylsilyl bromide as a by-product, rather than hydrogen bromide. For example, using this two-step procedure, cinnamic acid was converted into cinnamoyl bromide in 90% yield of purified product (Scheme 12). In several other cases, the initially formed acid bromide was converted, without isolation, into an ester in good overall yield from the acid (70–98% isolated yields).

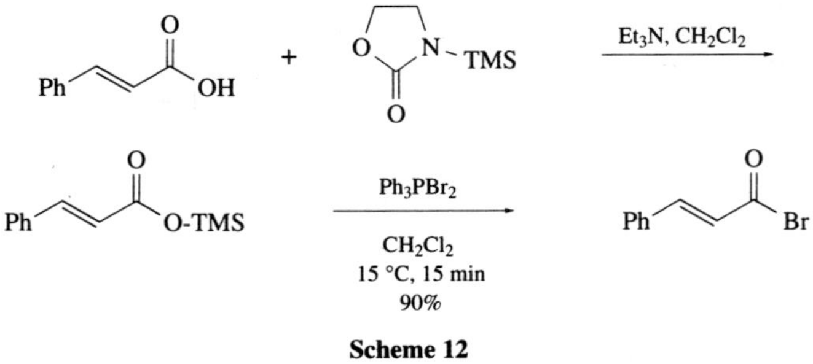

Scheme 12

5.01.1.4.4 Acid bromides from aldehydes

Treatment of some aldehydes with *N*-bromosuccinimide with simultaneous exposure to light directly affords the corresponding acid bromide ⟨79TL3809⟩. This procedure was found to be nonselective in the presence of an alkene functionality. Generally, the product was converted *in situ* into either the ester or the amide, to facilitate isolation. However, in one case, benzoic acid was converted into benzoyl bromide, isolated in 48% yield. Similar chemistry was reported by Marko and Mekhalfia ⟨90TL7237⟩. The aldehyde and 1 equiv. *N*-bromosuccinimide in tetrachloromethane gave a rapid reaction in the presence of a catalytic amount of 2,2′-azobisisobutyronitrile (AIBN). Filtration afforded the acid bromide, which could be purified by distillation (Equations (56) and (57)), but the authors obtained higher yields by *in situ* derivatization to amides.

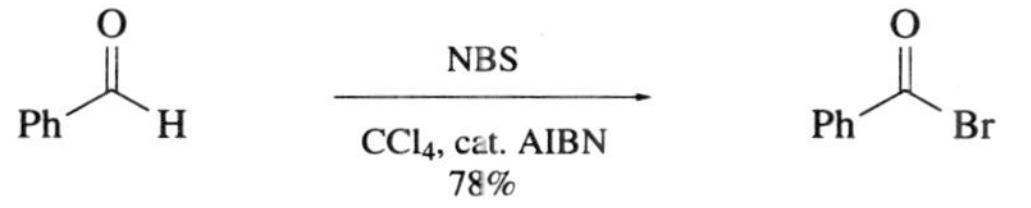

(56)

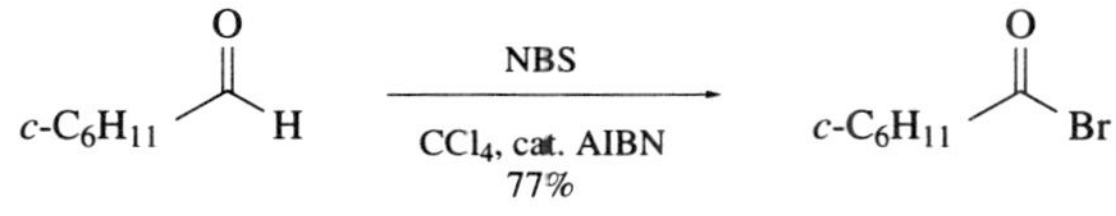

(57)

5.01.1.5 Acid Iodides

Acid iodides, as with acid bromides, have not been as widely employed as their acid chloride counterparts. There are certainly not many syntheses of acid iodides, and the majority of preparations are from the acid chlorides by exchange with some convenient source of iodide.

5.01.1.5.1 Acid iodides from acid chlorides

Various methods to effect the conversion of acid chlorides into acid iodides have been developed, the first example being in 1832, when Wöhler and Liebig reported the preparation of benzoyl iodide by treatment of benzoyl chloride with potassium iodide (Equation (58)) ⟨1832MI 501-01⟩. Staudinger and Authes reported a similar conversion using hydrogen iodide as the reagent (Equation (59)) ⟨13CB1423⟩.

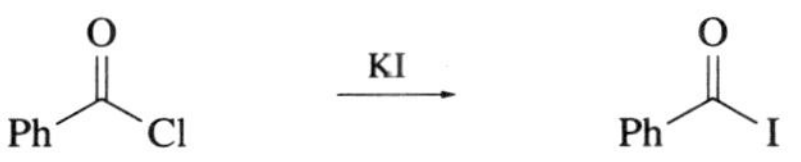

(58)

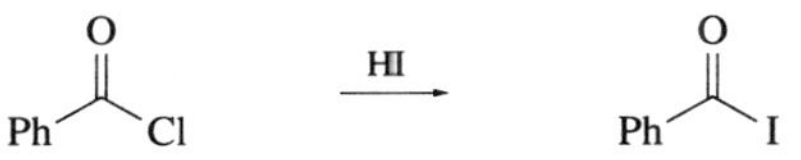

(59)

More recently, Hoffmann and co-workers have converted acid chlorides into acid iodides with sodium iodide in acetonitrile (Equation (60)), using a reactor–extractor apparatus; the acid iodide is continuously extracted from the acetonitrile with pentane, which enhances the isolated yield of the product ⟨81S715, 82S237⟩. This process affords good yields of the iodides, as shown in Table 4. Of particular note is the fact that exchange of the acid chloride occurs without disrupting alkyl chlorides or alkyl bromides (Table 4, entries 2 and 3). The same process was employed for the formation of dicarboxylic acid iodides (Table 4, entries 4 and 5).

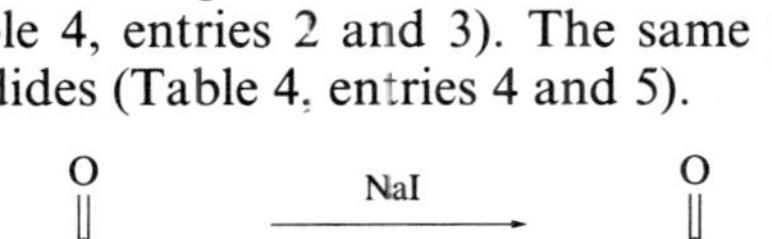

(60)

Similarly, trimethylsilyl iodide and diiodosilane have been employed to convert acid chlorides into the corresponding acid iodides (Table 4, entries 6 and 7) ⟨81S216⟩. During the exchange reaction, an equilibrium is established, which in this case is driven to completion by the continuous removal of volatile trimethylsilyl chloride.

5.01.1.5.2 Acid iodides from carboxylic acids and their derivatives

Until recently, the formation of acid iodides from the parent carboxylic acid had not been achieved in a direct reaction. However, acid anhydrides were used as acid iodide precursors in 1857 when treatment of acetic anhydride with phosphorus and iodine was reported to afford acetyl iodide (Equation (61)) ⟨1857LA(104)111⟩.

Table 4 Preparation of acid iodides from acid chlorides.

Entry	*Substrate*	*Method*	*Product*	*Yield (%)*
1	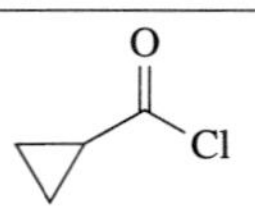	a	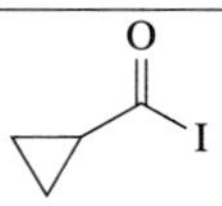	83
2	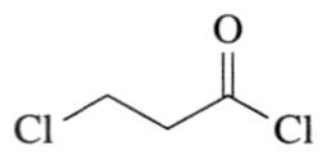	a	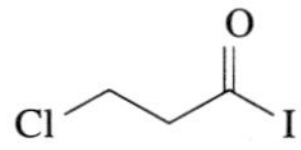	57
3	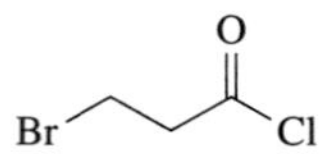	a	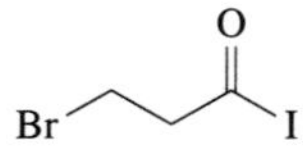	83
4	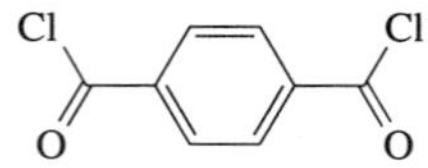	a	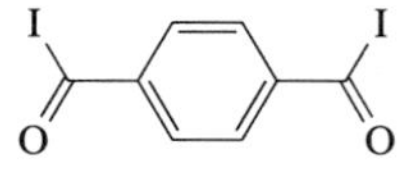	94
5	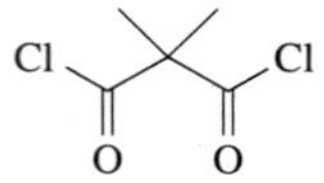	a	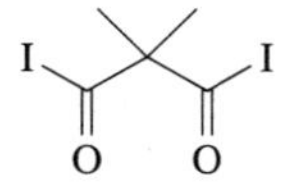	90
6	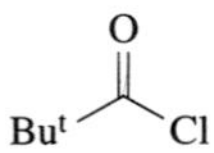	b	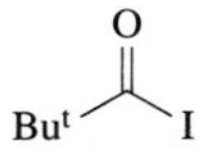	92
7	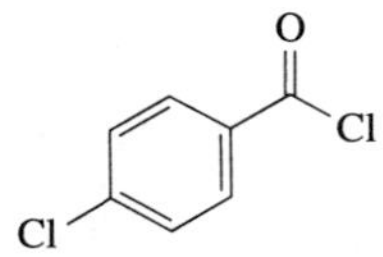	b	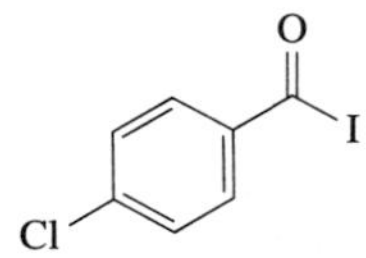	78

[a] NaI/pentane. [b] Trimethylsilyl iodide.

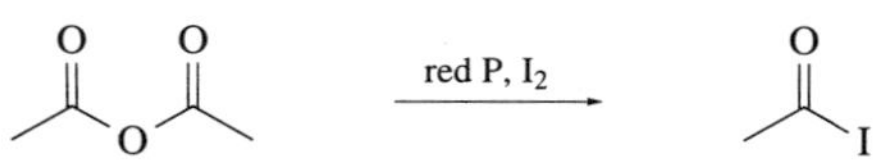

(61)

Ghosez has devised a route for the *in situ* derivatization of carboxylic acids with 1-iodo-*N*,*N*-2-trimethylpropenylamine, as discussed more fully in the section on acid halides (see 5.01.1.1).

Recently, Keinan and Sahai found that diiodosilane can be used to convert carboxylic acids directly into the corresponding acid iodides ⟨90JOC3922⟩. The initially formed silyl ester reacts fairly readily to form the acid iodide. The reaction is greatly accelerated by the addition of iodine. Although the reactions are essentially quantitative, purification by distillation reduces the yield to 92% for the conversion of dodecanoic acid to dodecanoyl iodide. This procedure was also successfully applied to acetic, butyric, and valeric acids (Equation (62)). An analogous procedure employing trimethylsilyl iodide and iodine afforded only very low yields, and it is believed that the intermediate silyl ester in this case is fairly stable to further reaction, as outlined in Scheme 13. This is contrary to the finding of Jung and Lyster who have reported the conversion of esters to acid iodides with trimethylsilyl iodide ⟨77JA968⟩. Ethyl hexadecanoate was converted into hexadecanoyl iodide after prolonged treatment with trimethylsilyl iodide at 75 °C. Presumably the more forcing conditions employed account for the conversion of the intermediate silyl ester into product in this case.

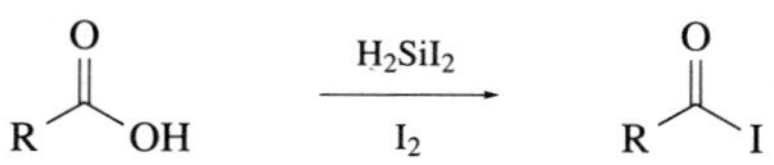

(62)

Keinan and Sahai ascribe the rate enhancement effect of added iodine to the formation of I^{3-}, which is a softer and more potent nucleophile than is I^-. Since no interaction between iodine and the silyl iodides was observed by ^1H NMR spectroscopy, the difference in reactivity between diiodosilane and trimethylsilyl iodide was attributed to a consequence of the greater oxophilicity of diiodosilane.

Esters are also converted into acid iodides with diiodosilane. For example, isopropyl acetate was converted into acetyl iodide in good yield (Equation (63)). For the reaction with esters, the use of iodine was essential, and the use of more than 1 equiv. iodine results in an even more effective

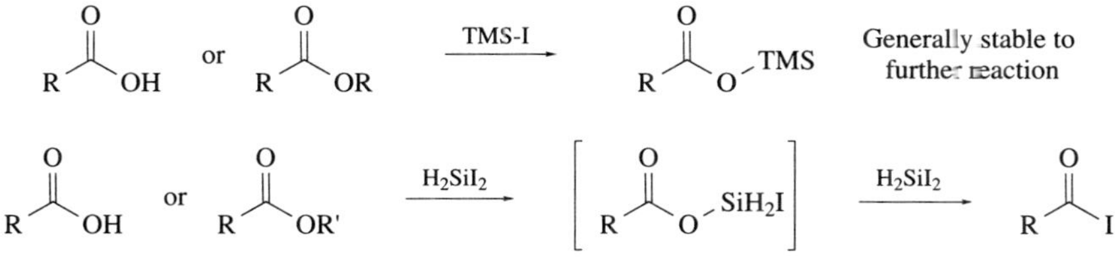

Scheme 13

conversion. Similarly, lactones may also be used as substrates, to yield the iodo-substituted acid iodide (Equation (64). Acid anhydrides were also found to be suitable substrates, and again the use of iodine as a coreagent accelerated the reaction (Equation (65)). Small amounts of silyl esters were also formed, depending on the exact conditions employed, although these would be expected to be converted into the acid iodides after longer reaction times.

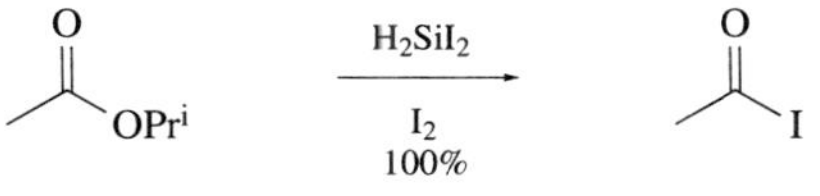

(63)

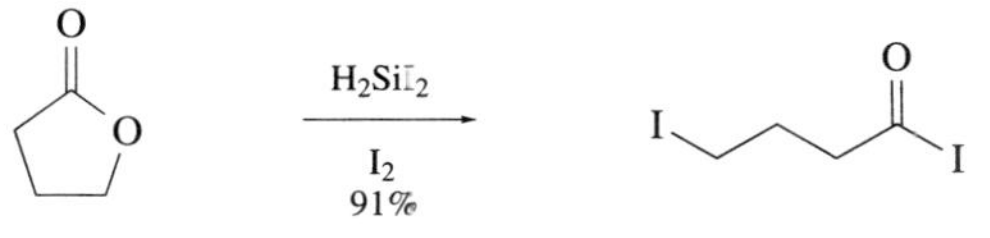

(64)

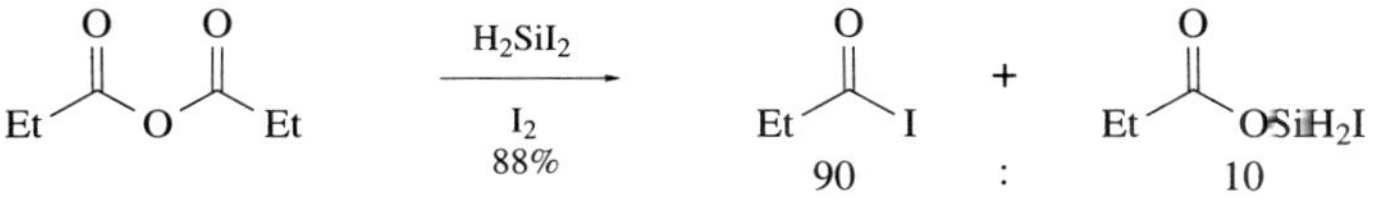

(65)

5.02
Carboxylic Acids

MICHAEL A. OGLIARUSO and JAMES F. WOLFE
Virginia Polytechnic Institute and State University, Blacksburg, VA, USA

5.02.1 GENERAL METHODS

Numerous review articles, monographs, and books describing the general synthetic methods used to prepare carboxylic acids have been published ⟨B-68MI 502-01, B-70MI 502-01, 76AR(B)199, B-77MI 502-01, 78GSM111, B-78MI 502-02, 79GSM67, B-79MI 502-01, 80GSM75, 81GSM87, 82GSM100, 82MI 502-02, 83GSM98, 85GSM96, 86GSM131, 87GSM130, 88GSM75, 89GSM89, 90GSM91, B-91MI 502-01, 92GSM79⟩. In addition, two specific reviews on the preparation of monobasic carboxylic acids have been published ⟨B-48MI 502-01, B-51MI 502-01⟩, as have individual reviews on, naphthenic acid (cyclohexanecarboxylic acid, ⟨42CRV97⟩), dehydroabietic acid and some derivatives ⟨58BSF17⟩, and the chemistry of formic acid and its simple derivatives ⟨69CRV673⟩.

At least nine general review articles on the synthesis of dibasic carboxylic acids have been published ⟨B-47MI 502-01, B-52MI 502-01, 59CRV89, 66MI 502-01, B-69MI 502-02, 72MI 502-01, B-75MI 502-01, 79MI 502-01, 79MI 502-02⟩ in addition to four more specific reviews on sebasic acid (1,8-octanedicarboxylic acid ⟨51C169⟩), tetrahydrophthalic acid ⟨57CRV641, 60CI(L)1096⟩, and the syntheses of alkylated alkanedioic acids ⟨59CRV89⟩.

The general syntheses of polyacids are represented by four review articles ⟨B-47MI 502-02, B-69MI 502-02, 72MI 502-02, 73RCR939⟩, while specific reviews on, the photochemistry of carboxylic acid derivatives ⟨78CRV97⟩, the preparation of ^{14}C-labeled carboxylic acids ⟨75MI 502-01⟩, the association of carboxylic acids ⟨53QR255⟩, and acids and their oxidation products ⟨B-57MI 502-01⟩ have also been published.

Other reviews more directly concerned with specific synthetic methods discussed in this chapter will be referenced where those methods are reported.

5.02.1.1 By Hydrolysis Reactions

The mechanism of the synthesis of carboxylic acids via hydrolysis ⟨54CI1154⟩ and solvolysis ⟨80AG169⟩ has been reviewed.

5.02.1.1.1 Hydrolysis of esters

Since ester formation is the result of an equilibrium in the reaction of an acid with an alcohol (Equation (1)), it is understandable that esters can be reconverted into their acid and alcohol precursors by treatment with water. For the majority of esters, such a hydrolysis procedure for the preparation of pure acids in good yields is not applicable, but a few esters do undergo extensive hydrolysis by simple treatment with cold water. Such esters include: methyl and ethyl oxalates; ethyl formate; and, esters of α-hydroxyacids such as glycollic, lactic, etc. ⟨1879LA1, 05JCS747, 09JCS1004⟩. Esters of the lower α-amino fatty acids ⟨01CB445⟩ and of β-alkylaminobutyric acids ⟨32M61, 32M65, 32M68⟩ have also been quantitatively hydrolyzed by refluxing with an excess of water for several hours. Partial hydrolysis of diesters using a catalytic amount of water in a miscible solvent and/or an emulsifer has also been reported (Equation (2)) ⟨83GEP3148422⟩.

$$ROH + RCO_2H \rightleftharpoons RCO_2R + H_2O \quad (1)$$

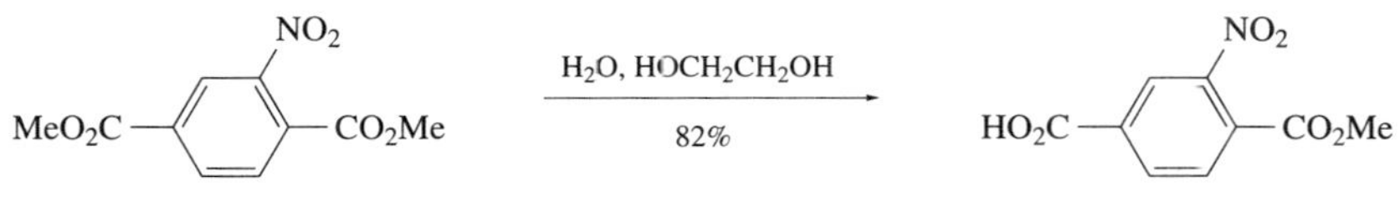

(2)

Although the hydrolysis of the majority of esters can best be accomplished using a variety of acidic or basic reagents, the structure of the molecule to be hydrolyzed as well as the products desired from the hydrolysis, are the two important criteria to consider when choosing whether to use an acidic or basic reagent. Thus, saponification by aqueous or alcoholic alkali metal hydroxides is used for the hydrolytic cleavage of unhindered esters which are not otherwise sensitive to hydroxide ion, while acid-catalyzed cleavage is the method of choice for esters which contain additional functionalities which are sensitive to aqueous alkali.

Acidic and basic reagents also differ in their mode of reaction since acidic hydrolysis (Equation (3)) is a catalytic process in which the products and reactants reach equilibrium concentrations unless the reaction is forced toward the desired acid, while basic hydrolysis (Equation (4)) is not a catalytic process, since a molar equivalent of hydroxide ion is consumed for each equivalent of ester converted to carboxylate salt.

$$R^1CO_2R^2 + H_2O \xrightleftharpoons{H^+} R^1CO_2H + R^2OH \qquad (3)$$

$$R^1CO_2R^2 + {}^-OH \longrightarrow R^1CO_2^- + R^2OH \qquad (4)$$

5.02.1.1.2 *Hydrolysis of nitriles*

Due to the ease with which a cyano group can be introduced into various molecules, such as by nucleophilic displacement, addition to carbonyl groups of aldehydes and ketones, hydrocyanation of α,β-unsaturated carbonyl compounds, and direct aromatic substitution, the nitrile function is an extremely useful precursor to carboxylic acids. The ultimate products of hydrolysis of a nitrile are a carboxylic acid and ammonia, but it is generally possible to regulate the conditions so that an amide can be isolated (Scheme 1).

$$RCN + H_2O \xrightarrow{H^+ \text{ or } HO^-} RCONH_2 \longrightarrow RCO_2H + NH_3$$

Scheme 1

Hydrolysis of nitriles can be accomplished either under acidic or basic conditions and in some cases by the use of water alone (Equation (5)) ⟨55OSC(3)221⟩.

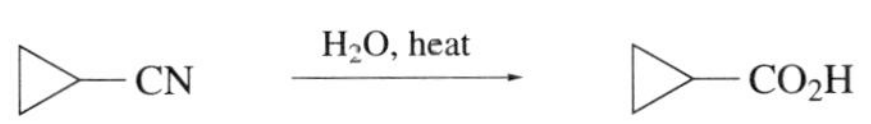

(5)

Alkaline hydrolysis of nitriles generally results in the formation of the salt of the carboxylic acid and ammonia, and is conveniently applied to the simple alkyl cyanides and also to nitriles such as α-cyanocarboxylic acids producing dicarboxylic acids which are unstable in hot mineral acid. In cases where the hydrolysis in boiling aqueous or alcoholic solutions is sluggish, higher-boiling solvents such as glycerol ⟨46JA1844, 55JA6682⟩, ethylene glycol ⟨50JA4091, 56JA450⟩, ethylene glycol monomethyl ether ⟨51JA436⟩, or diethylene glycol ⟨69CPB1564⟩ have been used to promote the hydrolysis.

Mineral acids which have been used to hydrolyze nitriles range from aqueous hydrochloric ⟨73JOC4044⟩ and hydrobromic acids ⟨86T2275⟩ to concentrated sulfuric ⟨55OSC(3)560⟩ and polyphosphoric acids ⟨54JA3039⟩, all of which can, under the appropriate conditions, first convert the nitrile into a primary amide, which may or may not be isolated and purified, before further hydrolysis with nitrous acid to the ultimate carboxylic acid product. The combined action of sulfuric acid and sodium nitrite on nitriles satisfactorily produces a carboxylic acid without the necessity of isolating the intermediate amide ⟨45JCS751⟩.

5.02.1.1.3 *Hydrolysis of amides*

Primary, secondary, and tertiary amides (Equation (6)), as well as lactams (Equation (7)) and imides (Equation (8)) undergo hydrolysis in the presence of alkali hydroxides or mineral acids in much the same manner as do esters and nitriles producing carboxylic acids.

$$RCON\langle \quad + \quad H_2O \xrightarrow{H^+ \text{ or } HO^-} RCO_2H \qquad (6)$$

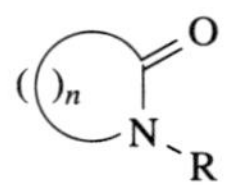

$$+ \quad H_2O \xrightarrow{H^+ \text{ or } HO^-} RNH(CH_2)_nCO_2H \qquad (7)$$

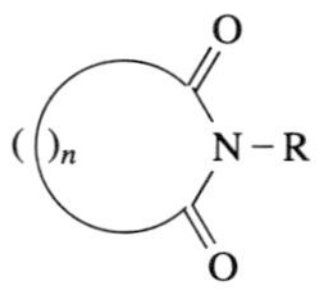

$$+ \quad H_2O \xrightarrow{H^+ \text{ or } HO^-} HO_2C(CH_2)_nCO_2H \qquad (8)$$

5.02.1.1.4 *Hydrolysis of acyl halides and anhydrides*

Acyl halides, mainly chlorides, can be converted into the corresponding carboxylic acid by treatment with water without the necessity of acidic or basic catalysts (Equation (9)). Chlorides of aliphatic acids hydrolyze upon contact with cold water producing a violent reaction, while the chlorides of the aromatic acids are more resistant toward hydrolysis. In general, aromatic acid chlorides hydrolyze slowly in cold water, but the most convenient method of converting them into carboxylic acids is by heating in aqueous alkali under reflux until hydrolysis is complete, followed by acidification. In general, acyl halides are prepared from the free acids in order to activate the acyl function toward nucleophilic acyl substitution and therefore are seldom employed in the hydrolytic preparation of acids.

$$RCOX \quad + \quad H_2O \xrightarrow[X\,=\,\text{halide}]{} RCO_2H \quad + \quad HX \qquad (9)$$

Compared to acid chlorides, anhydrides are relatively stable in water ⟨63JCS2918, 71JA5821⟩, so mineral acids ⟨60JCS3070⟩, alkali hydroxides ⟨73S416⟩, and tertiary amines ⟨53JCS1409⟩ have been used to accelerate their hydrolysis.

5.02.1.1.5 *Hydrolysis of di- and trihalides*

The most common dihalide function used as a precursor to carboxylic acids is the 1,1-dihalovinyl function, which upon treatment with concentrated sulfuric acid followed by quenching with water produces carboxylic acids, often in excellent yields (Equation (10)). A related, but considerably milder, process involves the reaction of carbonium ion precursors, such as secondary or tertiary alcohols, esters of tertiary alcohols or alkenes, with 1,1-dichloroethylene in the presence of concentrated sulfuric acid or a mixture of sulfuric acid and boron trifluoride. Hydrolysis of the reaction mixture affords the carboxylic acid with two or more carbon atoms than contained by the intermediate carbonium ion as shown in Scheme 2. A comprehensive review of this approach has been reported ⟨66AG(E)870⟩, while another report describes its intramolecular applications ⟨73TL1569⟩.

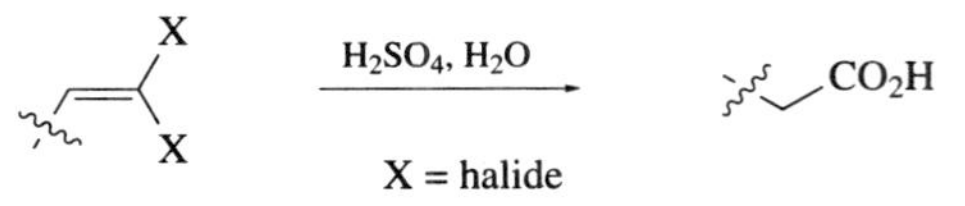

(10)

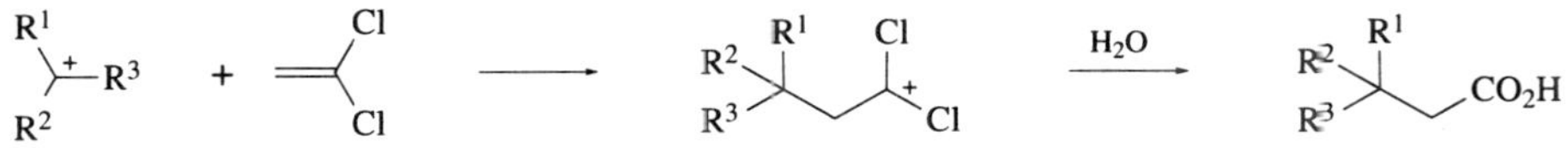

Scheme 2

Although the trihalomethyl group is a convenient precursor to the carboxyl function (Equation (11)), and the hydrolysis of benzotrichlorides is a convenient route to aromatic carboxylic acids, the trihalomethyl group is not as easily introduced into aliphatic molecules as it is into the aromatic series. Thus, most examples of this approach to the preparation of carboxylic acids involve aromatic molecules ⟨71S131⟩.

$$RCX_3 + H_2O \xrightarrow[X = \text{halide}]{H^+ \text{ or } HO^-} RCO_2H \qquad (11)$$

5.02.1.2 By Carbonation of Organometallic Reagents

A wide variety of organometallic reagents have been treated with both gaseous and solid carbon dioxide to produce high yields of carboxylic acids. The general reaction involved is shown in Scheme 3, and involves treatment of the organometallic reagent with some form of carbon dioxide followed by acidification. The mechanism of the reaction probably involves a rapid coordination of the metal atom with one of the oxygen atoms of carbon dioxide followed by a rate-determining nucleophilic attack of the R group on the carbon of the carbon dioxide (Scheme 4). A general review of the carbonation of organometallic reagents ⟨B-69MI 502-01⟩, as well as separate reviews on the use of organosodium reagents ⟨57CRV867⟩, organolithium reagents ⟨54OR258⟩, and organomagnesium reagents ⟨70S615⟩ for the production of carboxylic acids via carbonation have been published, and the ultimate conclusion which can be drawn from these reviews is, that even though the organolithium compounds are more reactive than the Grignard reagents, both of these reagents are the most frequently employed organometallic reagents used for carbonation.

$$RM + CO_2 \longrightarrow RCO_2M \xrightarrow{H_3O^+} RCO_2H$$

M = MgX, Li, Na, etc.

Scheme 3

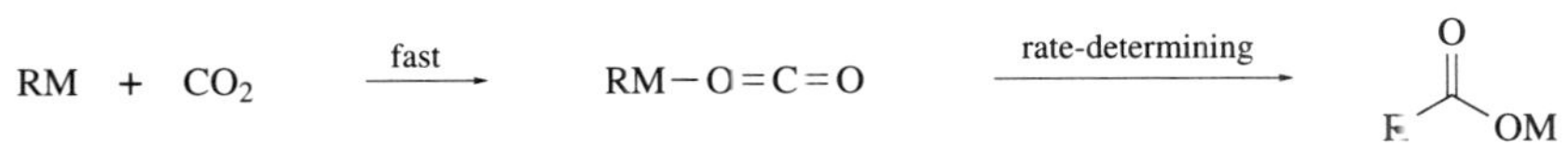

Scheme 4

Three other types of reactions which are similar to the organometallic-carbonation reaction, but which will not be extensively discussed in this chapter are, the carbonation of carbonium ions formed from mono-alkenes upon protonation during the Koch carboxylic acid synthesis ⟨63AG(E)719⟩, hydrocarboxylation reactions, which involve the introduction of the CO_2H group by treatment of unsaturated organic molecules with carbon monoxide in the presence of an appropriate (usually a transition metal), catalyst ⟨62CRV283, B-67MI 502-01, 68AG(E)329, 73ACR422, 73S509, 77MI 502-01, 78CZ1, 78MI 502-01, 81MI 502-01, 81YZ265, 85MI 502-01, 85S253⟩, and carbomethylation reactions, where metal enolates are treated with halo carboxylic acid or their salts (Equations (12) and (13)) ⟨70S628⟩.

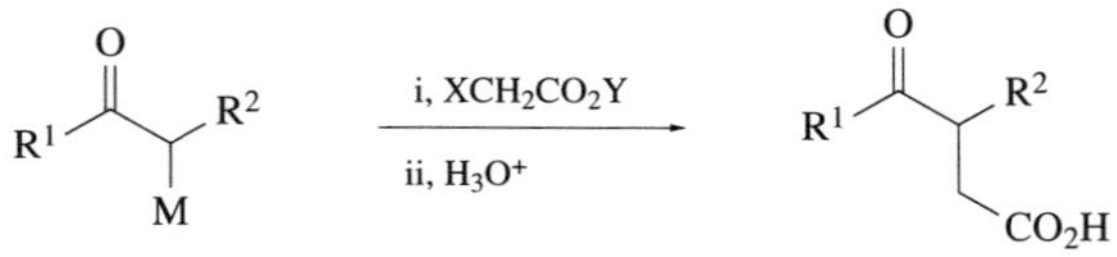

(12)

M = metal; X = halide; Y = salt cation

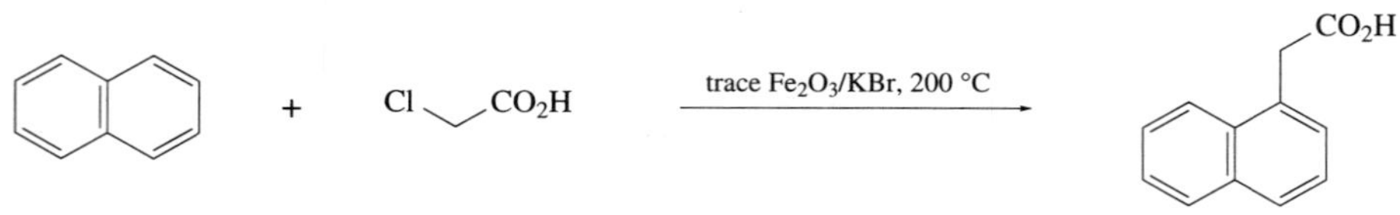

(13)

5.02.1.3 By Oxidation Reactions

The preparation of carboxylic acids can be accomplished by the oxidation of a number of functional groups using a wide variety of reagents. In the subsequent sections of this chapter the preparation of carboxylic acids is primarily characterized by the structure of the acid produced, the functionality of the starting material, and finally, by the oxidizing agent used.

The general mechanism of oxidation of organic compounds has been reviewed in two publications ⟨B-64MI 502-01, B-64MI 502-02⟩, while more specific discussions on the use of noble metal catalysts ⟨57AG600⟩, cerium(IV) ion ⟨73S347⟩, and, chromium-catalyzed oxidations ⟨92CRV113⟩ have also been published. Reviews of the kinetics and mechanism of permanganate oxidations ⟨58CRV403⟩ and the preparation, properties, and use of selenium dioxide as an oxidizing agent in preparative organic chemistry are also available ⟨41AG146, 45BSF506, 45CRV235⟩.

General reviews on the use of nickel peroxide ⟨75CRV491⟩, oxygen ⟨79MI 502-03, 81MI 502-03⟩, and ozone ⟨77MI 502-02, B-78MI 502-01, B-82MI 502-01⟩ in the preparation of monocarboxylic acids by oxidation of hydrocarbons have been published, but the largest number of books and review articles have been published on the use of periodate and periodic acids as an oxidizing agent in organic and bioorganic chemistry ⟨28BSF683, 28CR(186)382, 40AG321, 44OR(2)341, 54MI 502-01, 60AQ93, B-69MI 502-03, 74S229⟩.

Two reviews on the preparation of carboxylic acids by oxidations using lead tetraacetate ⟨40AG321, 58AG173⟩ have also been published.

5.02.1.4 By Miscellaneous Reactions

The miscellaneous reactions used to prepare carboxylic acids are specific to the structure of the carboxylic acids produced and can not be generally characterized.

5.02.2 ALKANOIC ACIDS

Specific reviews published on a variety of synthetic approaches to alkanoic acids include: the essential fatty acids ⟨B-47MI 502-04⟩; the industrial synthesis of fatty acids ⟨42MI 502-01⟩; the synthesis and chemistry of fatty acids and their derivatives ⟨B-48MI 502-02, B-55MI 502-01, B-57MI 502-02, B-60MI 502-01⟩; developments in the preparation of natural and synthetic straight-chain fatty acids ⟨53QR(7)175, 57CRV191⟩; some new aspects of fatty acid chemistry ⟨62CI(L)1304⟩; fatty acid oxidation ⟨B-63MI 502-01⟩; and, ozonolysis of fatty acids and their derivatives ⟨B-63MI 502-02⟩.

Synthesis of alkanoic acid via the Kolbe electrolysis has been reviewed ⟨52QR(6)380, B-60MI 502-02, 60RCP243, 67CRV623, 79MI 502-04⟩, as has the synthesis of functionalized carboxylic acids by coupling reactions ⟨49BSFD297, 79S633, 80AG(E)171, 82S521⟩, the synthesis of antiinflammatory α-arylalkanoic acids by a 1,2-aryl shift ⟨84AG(E)413⟩, and the synthesis of pyrethroid acids (cyclopropanecarboxylic acids, ⟨81AG(E)703⟩).

A review of the preparation of alkyl substituted di- and tricarboxylic acids ⟨70S99⟩, and of dicarboxylic acid derivatives from fatty acids ⟨79MI 502-05⟩ has also been published.

5.02.2.1 Unsubstituted Alkanoic Acids

5.02.2.1.1 By hydrolysis reactions

Acid hydrolysis of esters has been accomplished at room temperature (Equation (14)) ⟨67BSF2C11⟩ and at higher temperatures for shorter periods of time (Equation (15)) ⟨82GEP3116474⟩ and Equation (16) ⟨74S36⟩ with similar yields of acid products obtained. In a study of acid catalyzed hydrolysis of esters which failed to undergo alkaline hydrolysis or where alkaline hydrolysis would normally be considered to be the method of choice, it was found that a mixture of methanesulfonic acid in 90% formic acid as solvent gave excellent conversion of the esters into carboxylic acids under rather mild conditions (Equations (17) and (18)) ⟨64CI(L)193⟩.

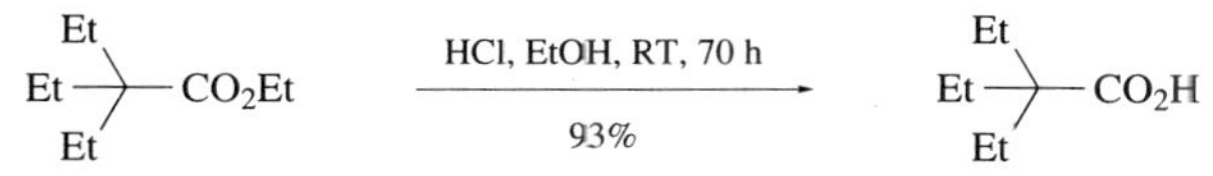

(14)

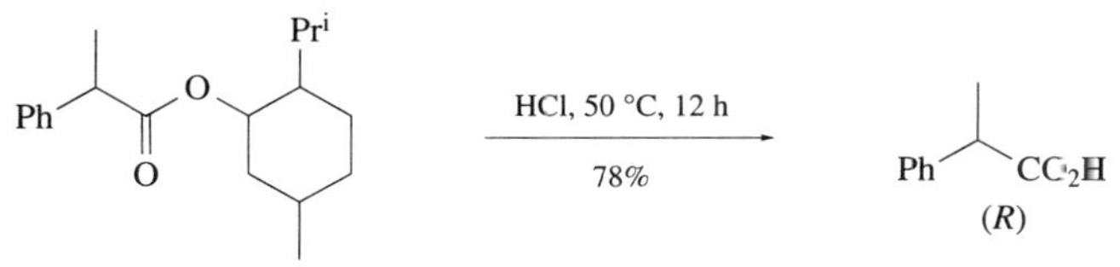

(15)

69:31 enantiomeric ratio

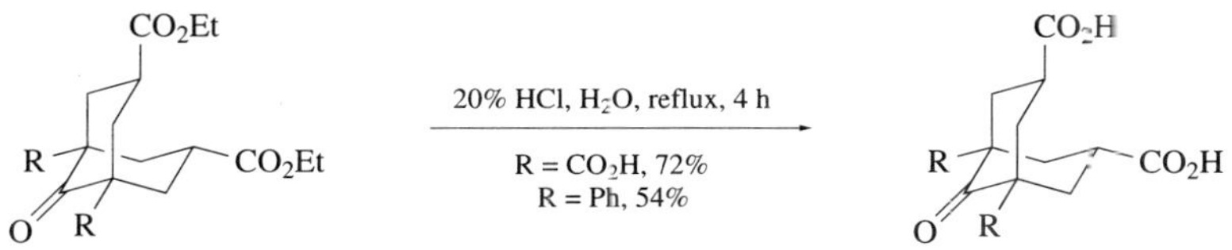

(16)

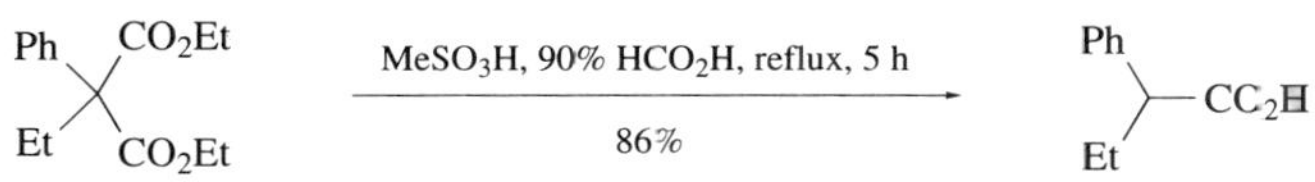

(17)

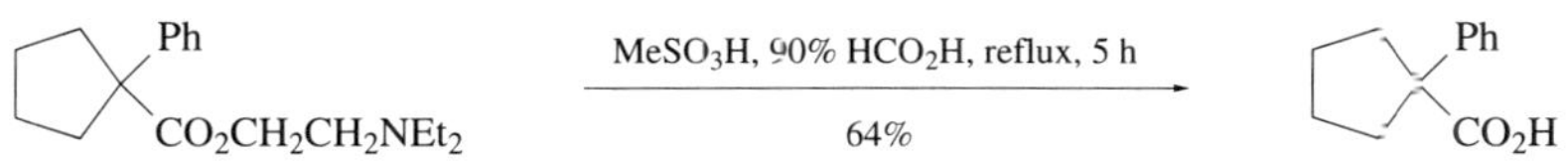

(18)

Although *t*-butyl esters are stable under neutral and basic conditions and are readily hydrolyzed under acidic conditions, two approaches to their hydrolysis which have proven very successful have been reported. The first involves refluxing the *t*-butyl alkanoic acid esters with concentrated hydrochloric acid in dioxane as a cosolvent (Equations (19) and (20)) ⟨59JA5817⟩, while the second approach involves cleavage of the *t*-butyl esters in a two-phase system at room temperature using aqueous mineral acids such as 10M H_2SO_4, 9M HBr or 12M HCl in the presence of a lipophilic quaternary onium salt such as hexadecyltributylphosphonium bromide (Equation (21)) ⟨82JOC154⟩.

$$RCO_2Bu^t \; + \; \text{conc. HCl} \xrightarrow{\text{dioxane, reflux, 2–5 h}} RCO_2H \qquad (19)$$

R = Pr^n, 94%; $Me_2CH(CH_2)_2$, 45%; $Me_2CH(CH_2)_3$, 92%; n-C_9H_{19}, 95%; $Ph(CH_2)_2$, 97%; $PhCH_2CH(Me)$, 93%; $PhCH_2CH(Et)$, 92%; $PhCH_2CH(Pr^i)$, 93%.

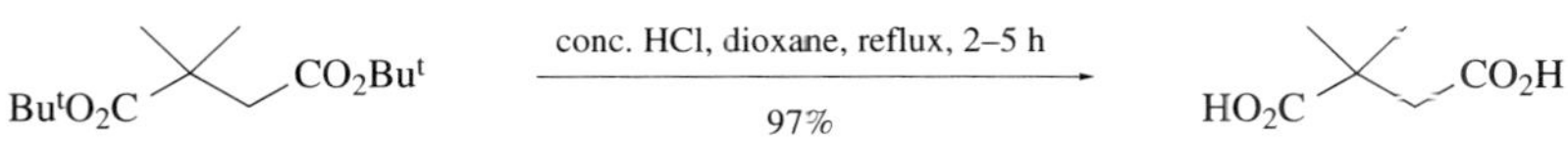

(20)

$$RCO_2Bu^t \xrightarrow[\text{87–94\%}]{\text{HX (aq.), cat., RT}} RCO_2H \qquad (21)$$

R = octyl, dodecyl; X = HSO_4, Br, Cl

Hydrolysis of ethyl and *n*-amyl esters of acetic acid has been accomplished at room temperature by treatment with acidic ion exchange resins such as Amberlite IR-100 (Equation (22)) ⟨52JCS1607⟩.

$$MeCO_2R \xrightarrow[R\,=\,Et,\ n\text{-}C_5H_{11}]{\text{Amberlite IR-100, } H_2O,\ 25\ °C} MeCO_2H \qquad (22)$$

Acid hydrolysis of *gem*-diesters is usually accompanied by decarboxylation and production of a monoacid attached to the original carbon (Equation (17)) ⟨64CI(L)193⟩, (Equation (23)) ⟨32OSC(1)290⟩, and (Equation (24)) ⟨32OSC(1)523⟩.

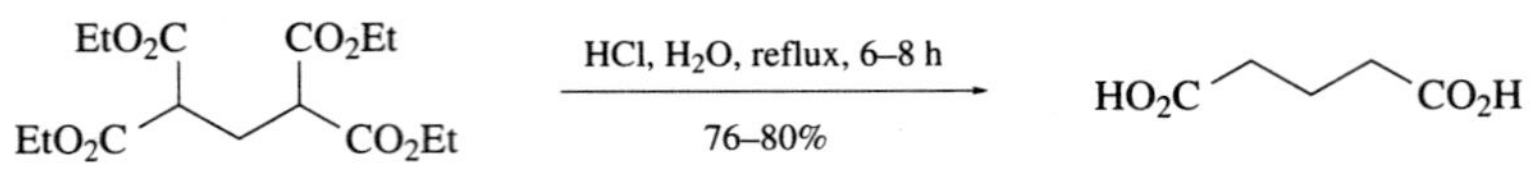

(23)

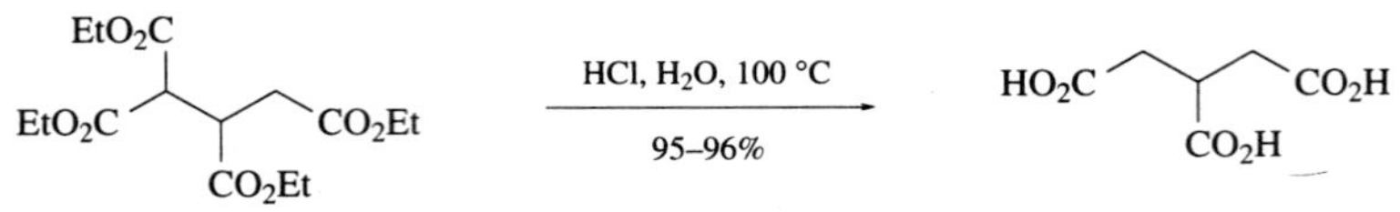

(24)

Due to its essential irreversibility, saponification by aqueous or alcoholic alkali metal hydroxides is the most widely used method for the hydrolytic cleavage of unhindered esters which are not otherwise sensitive to hydroxide ion. Some typical examples of such a process is the hydrolysis of *s*-butyl 3-methylheptanoate (Equation (25)) ⟨73OSC(5)762⟩, the ethyl and *t*-butyl esters of 2-arylpropanoic acid (Equation (26)) ⟨86SC499⟩, and trimyristin (Equation (27)) ⟨32OSC(1)379⟩.

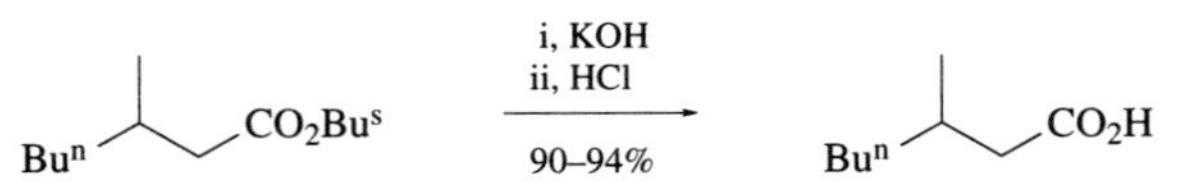

(25)

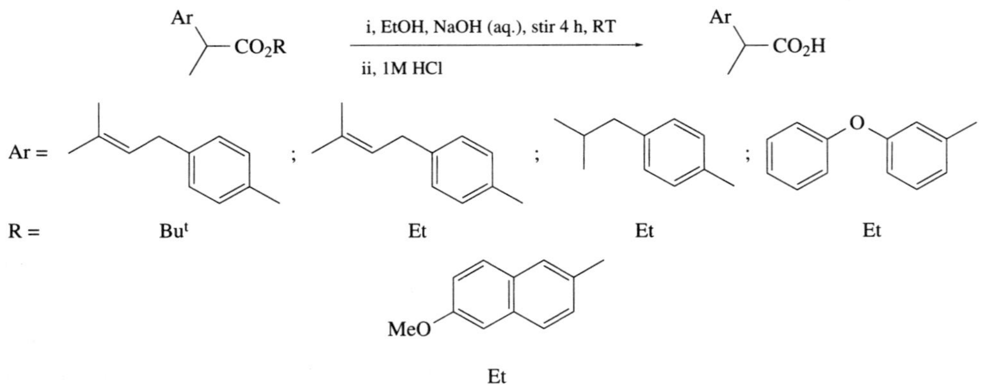

(26)

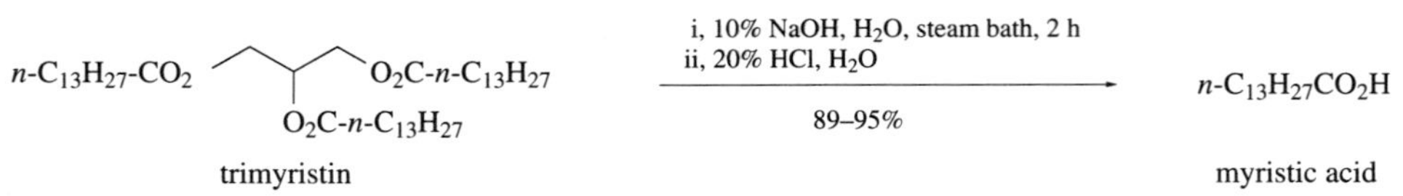

(27)

A variety of esters of cycloalkylcarboxylic acids of varying ring size have been saponified using alkali metal hydroxides without rupture of the ring system, examples include the cyanomethyl esters of three- and six-membered ring acids (Equation (28)) ⟨82TL2347⟩, the menthyl esters of cyclopropane 1,2-dicarboxylic acid (Equation (29)) ⟨88OS(67)76⟩, and dimethyl 3,3a,3b,4,6a,7a-hexahydro-3,4,7-metheno-7H-cyclopenta[*a*]pentalene-7,8-dicarboxylate (Equation (30)) ⟨89OS(68)198⟩.

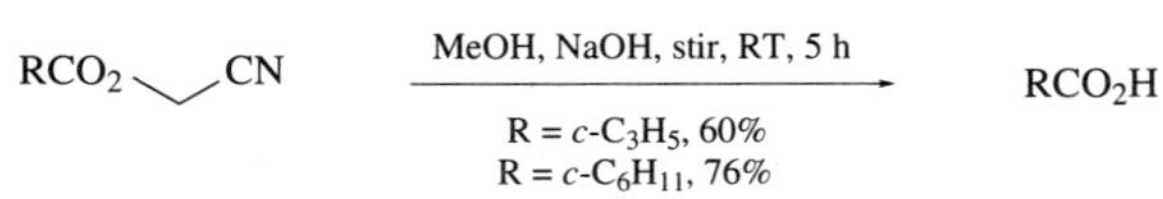

(28)

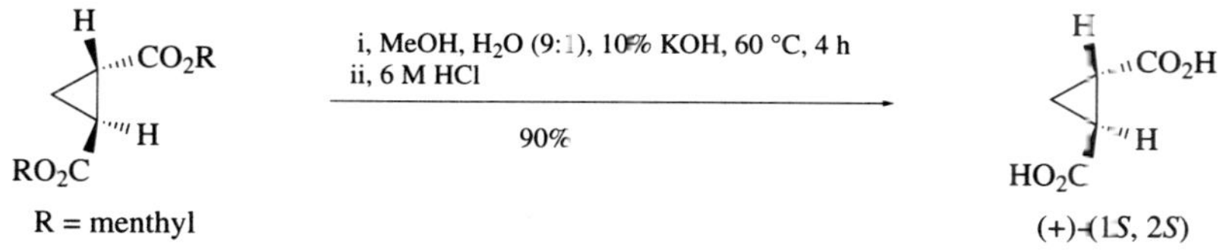

(29)

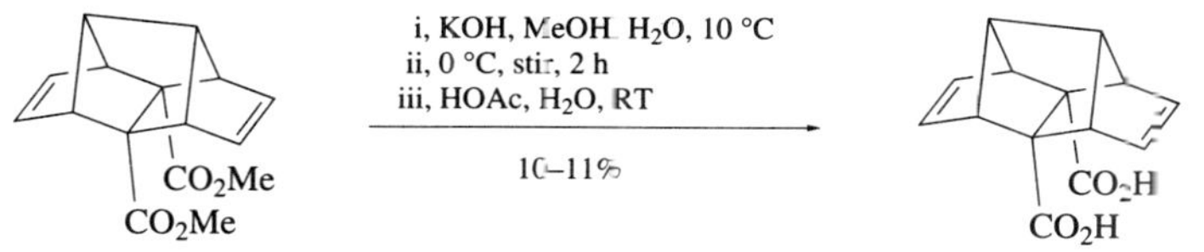

(30)

Two approaches which have been used to hydrolyze hindered esters of cycloalkylcarboxylic acids, also without ring opening, are the reaction with lithium in a liquid ammonia/tetrahydrofuran solvent mixture for saponification of the methyl ester of desoxypodocarpic acid (Equation (31)) ⟨58JA217⟩, and treatment with $KOBu^t$ in DMSO for saponification of methyl *O*-methylpodocarpate and methyl dehydroabietate (Equation (32)) ⟨64TL2969⟩. However, due to the presence of an α-keto function and the temperature used, basic hydrolysis of ethyl 2-oxocyclohexanecarboxylate results in ring opening to give pimelic acid (Equation (33)) ⟨43OSC(2)531⟩.

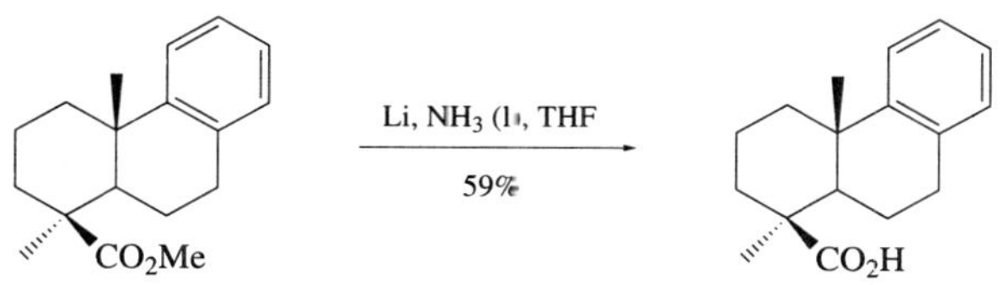

(31)

(32)

R^1	R^2	*Time* (h)	*Yield* (%)
MeO	H	2	97
H	Pr^i	1	94

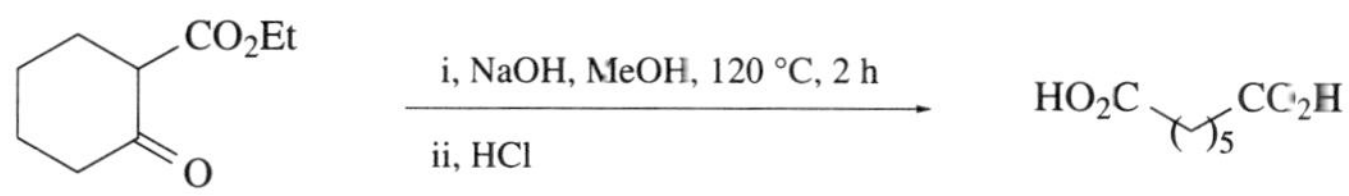

(33)

Bases other than aqueous and alcoholic alkali metal hydroxides which have been used to hydrolyze aliphatic carboxylate esters include trimethylamine (Equation (34)) ⟨33JA4079⟩, imidazole (Equation (35)) ⟨64JA837, 66BCJ852⟩, and amidines (Equation (36)) ⟨66BCJ852⟩.

$$MeCO_2Me \xrightarrow[\text{ii, HCl}]{\text{i, }Me_3N\text{, MeOH}} MeCO_2H \qquad (34)$$

$$MeCO_2R \xrightarrow[\text{ii, }H_3O^+]{\text{i, imidazole}} MeCO_2H \qquad (35)$$

R	*Ref.*	*R*	*Ref.*
Et	64JA837	*p*-$MeOC_6H_4$	64JA837
F_3CCH_2	64JA837	*p*-ClC_6H_4	64JA837
Ph	64JA837, 66BCJ852	*m*-$O_2NC_6H_4$	64JA837
		p-$O_2NC_6H_4$	64JA837
p-MeC_6H_4	64JA837, 66BCJ852	2,4-$(O_2N)_2C_6H_3$	64JA837

$$MeCO_2R \;+\; \text{amidines} \xrightarrow{85\%\ EtOH,\ H_2O,\ 30\ °C} MeCO_2H \qquad (36)$$

R	*Amidines used*
Et	Formamidine, acetamidine, guanidine
Ph	Formamidine, acetamidine, benzamidine
p-MeC_6H_4	benzamidine

A facile dealkylation of hindered aliphatic carboxylic esters has been accomplished using an excess of *n*-propyl lithium in ether under argon at 0°C (Equation (37)) ⟨79T2077⟩. This procedure produces the corresponding carboxylic acid very cleanly and in satisfactory yields.

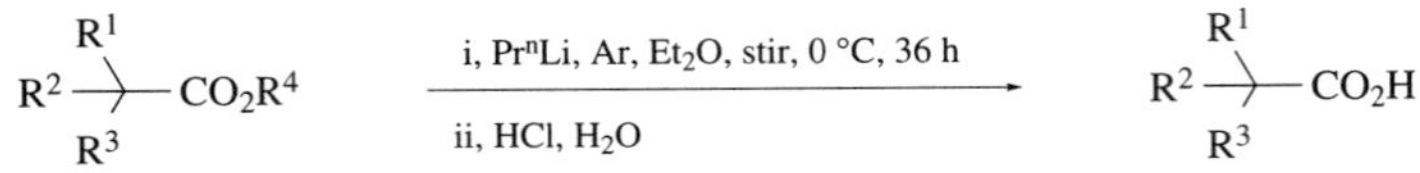

(37)

R^1	R^2	R^3	R^4	*Yield* (%)
Bu^t	Pr^i	Et	Et	61
Bu^t	Et	Et	Et	44
Bu^t	Pr^i	Me	Et	77
Bu^t	Pr^i	Me	Me	39
Bu^t	Pr^i	Me	Bu^t	82
Bu^t	Pr^i	Me	*n*-C_8H_{17}	72
Pr^i	Pr^i	Pr^i	Et	53
Pr^i	Pr^i	Et	Et	42
Pr^i	Pr^i	Me	Et	72

Asymmetric synthesis of aliphatic carboxylic acids has been accomplished by saponification of esters of sulfonamide-shielded alcohols (Scheme 5 and Table 1) ⟨85TL6051⟩. The asymmetry results from a face selective Michael addition to the corresponding enoate esters.

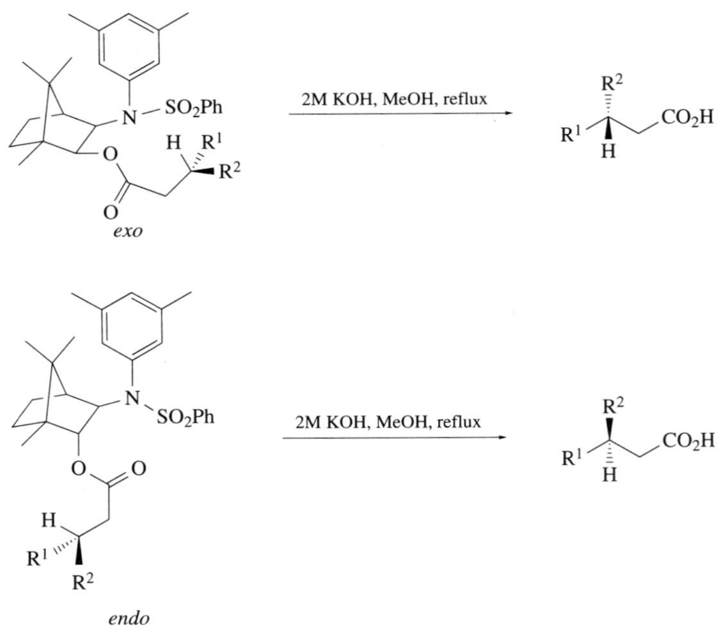

Scheme 5

Saponification of one ester function in a diester substrate has been reported to occur in diethyl malonate using potassium hydroxide in ethanol (Equation (38)) ⟨1884CB780⟩, in ethyl *t*-butyl 2-carbobenzyloxyaminopentane 1,5-dicarboxylate using sodium hydroxide in a dioxane–water mixture (Equation (39)) ⟨61HCA1991⟩, and in dimethyl 1,11-undecanoate using barium hydroxide in methanol (Equation (40)) ⟨63OSC(4)635⟩, all under rather mild conditions.

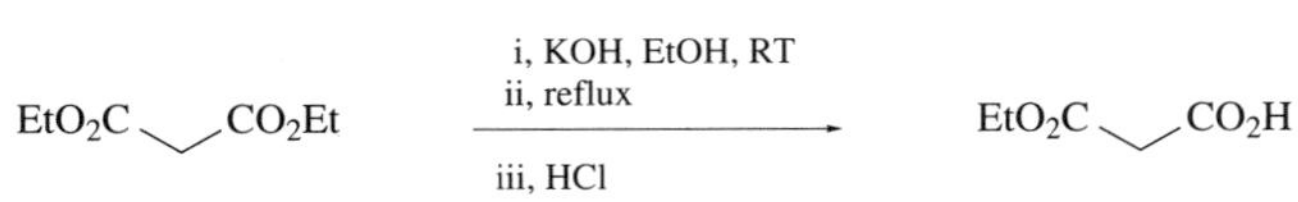

(38)

Table 1 Asymmetric synthesis of alkanoic acids.

R^1	R^2	*Configuration of acid obtained from* exo *adduct*	endo *adduct*
Me	Et		
Me	Pr^i		
Et	Me		
Et	Pr^i	S	R
Pr^i	Me		
Pr^i	Et	R	S
Me	$CH_2{=}CHCH_2$	S	R
Me	Ph	S	R
Me	*p*-MeC_6H_4		R

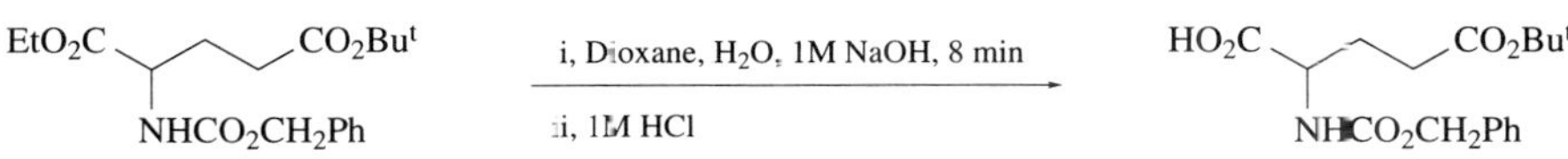

(39)

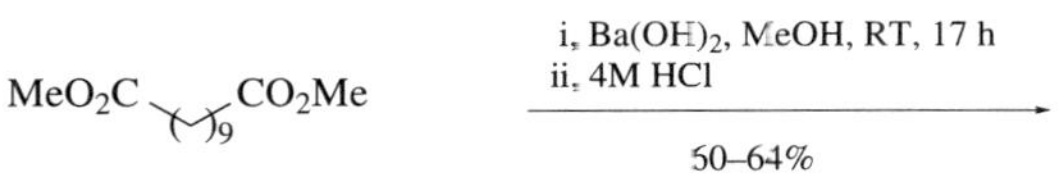

(40)

As was reported for the acid hydrolysis of *gem*-diesters, basic hydrolysis of such esters can also be accompanied by decarboxylation and production of a monoacid (Equation (41)) ⟨43OSC(2)416⟩, but the majority of such hydrolyses produced the diacids (Equation (42) and Table 2) ⟨43OSC(2)93, 43OSC(2)474, 55OSC(3)495, 55OSC(3)705⟩ even if the *gem*-diesters are attached to cyclobutane (Equation (43)) ⟨55OSC(3)213⟩.

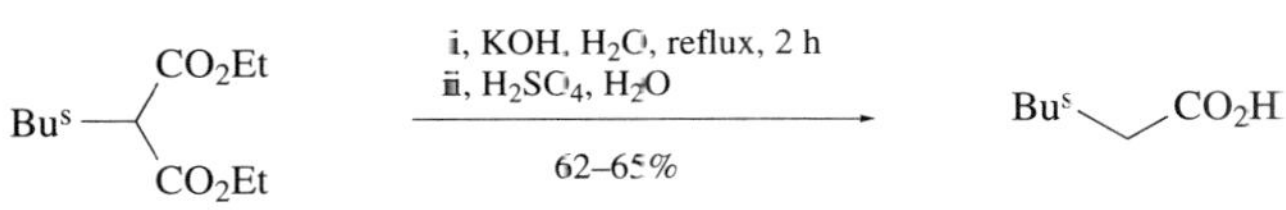

(41)

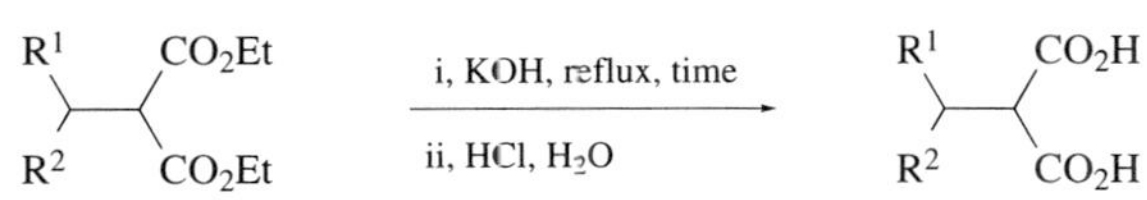

(42)

Table 2 Alkanoic *gem*-diacids by basic hydrolysis of *gem*-diesters.

R^1	R^2	*Time* (h)	*Ref.*
H	Bu^n	1	43OSC(2)93
H	Bu^i	1	43OSC(2)93
H	Bu^s	1	43OSC(2)93, 55OSC(3)495
H	*n*-C_6H_{13}	4–5	43OSC(2)474
H	Ph	3	55OSC(3)705
Me	Me	1	43OSC(2)93

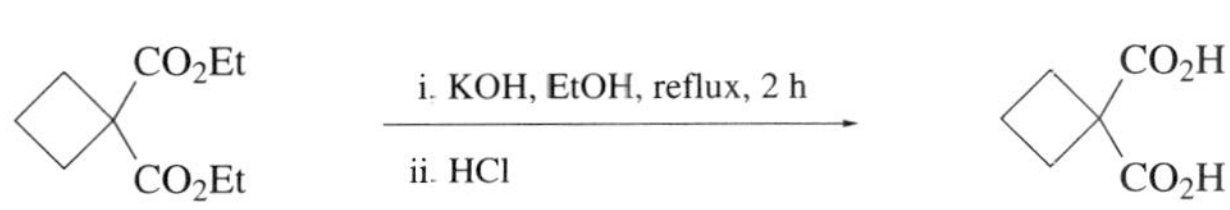

(43)

Nitriles, like esters, can be hydrolyzed to carboxylic acid using both acidic and basic reagents, and at least one nitrile, malonitrile, has been thoroughly reviewed ⟨69CRV(69)591⟩. When treated with 57% hydriodic acid at high temperature, 4-nitrophenylacetonitrile undergoes nitro group reduction as well as nitrile hydrolysis to produce 4-aminophenylacetic acid (Equation (44)) ⟨71ZC110⟩ in acceptable yields, but better yields of alkanoic acids are obtained using 100% phosphoric acid at similar temperatures (Equation (45)) ⟨27RTC600⟩.

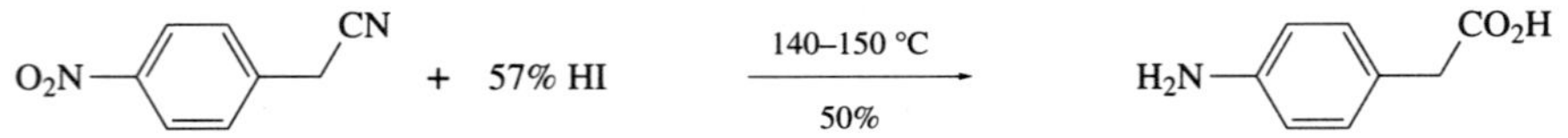
(44)

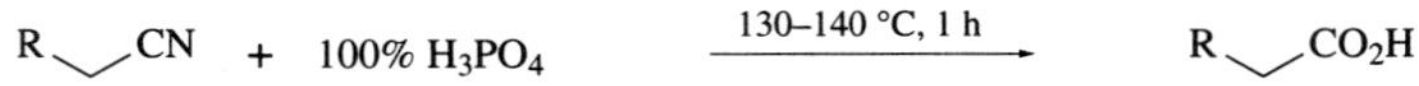
(45)

R	*Yield (%)*
H	80
Me	90

Refluxing hydrochloric acid has been used to produce phenylsuccinic acid from ethyl β-phenyl-β-cyanopropionate (Equation (46)) ⟨63OSC(4)804⟩, and glutaric acid from trimethylenecyanide (1,3-dicyanopropane, Equation (47)) ⟨32OSC(1)289⟩, while treatment of potassium 4-cyanobutanoate with hydrochloric acid at room temperature produces the intermediate glutaric acid monoamide which can be further hydrolyzed to glutaric acid by refluxing with fresh hydrochloric acid (Scheme 6) ⟨63OSC(4)496⟩.

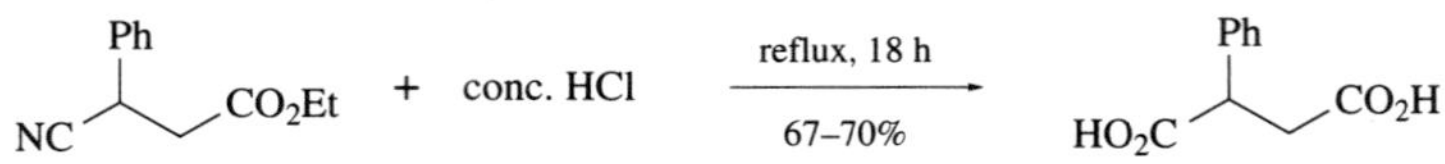
(46)

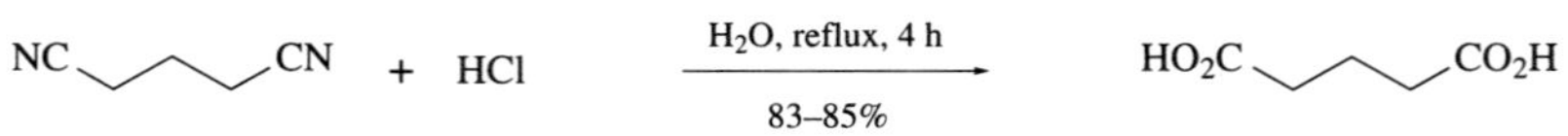
(47)

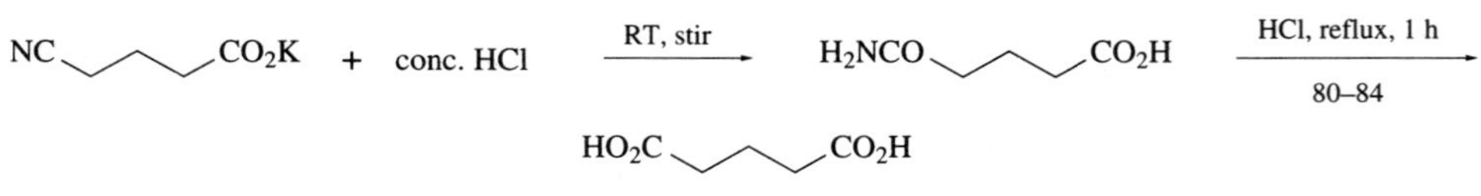

Scheme 6

Sulfuric acid hydrolysis of hindered nitriles can also lead to amide intermediates, which upon further treatment with sodium nitrite in fresh sulfuric acid produces good yields of the desired alkanoic acids (Scheme 7) ⟨56JA5416⟩, Scheme 8 ⟨48JA2824, 67BSF2011⟩. On the other hand, treatment of unhindered nitriles such as benzyl cyanide (Equation (48)) ⟨32OSC(1)436⟩, pentafluorobenzyl cyanide (Equation (48)) ⟨61JCS808⟩, and 2,4,6-trimethylbenzyl cyanide (Equation (48)) ⟨55OSC(3)557⟩ with sulfuric acid produces the corresponding carboxylic acid directly.

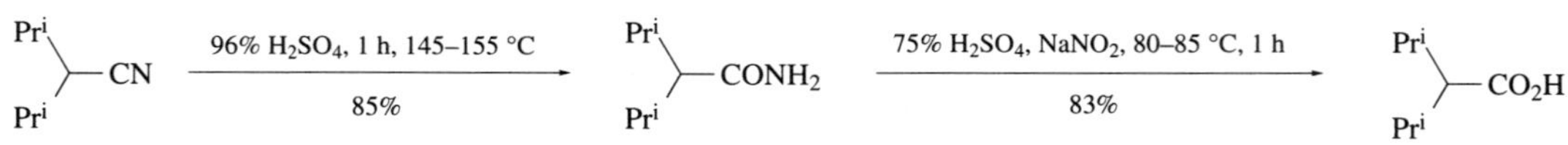

Scheme 7

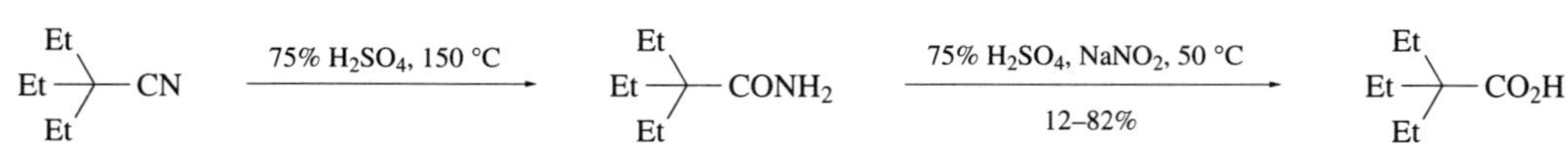

Scheme 8

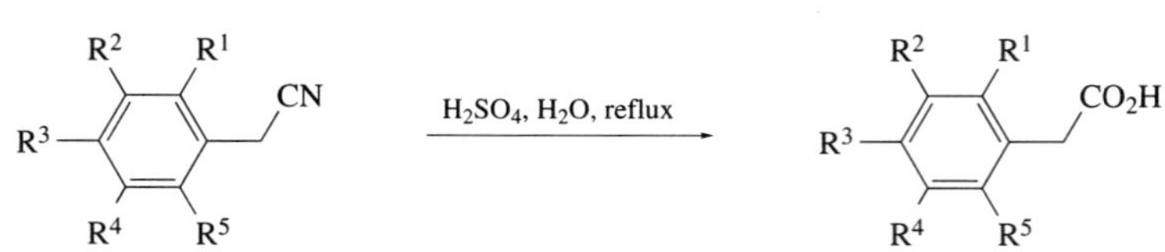
(48)

$R^1 = R^2 = R^3 = R^4 = R^5 = H$, reflux 3 h, 77.5% yield
$R^2 = R^4 = H$; $R^1 = R^3 = R^5 = Me$, reflux 6 h, 87% yield
$R^1 = R^2 = R^3 = R^4 = R^5 = F$, reflux 45 min with HOAc

2,4,6-Trimethylbenzyl cyanide has also been hydrolyzed to α-mesitylacetic acid using potassium hydroxide in ethylene glycol (Equation (49)) ⟨50JA4091⟩, a method also used to prepare 3-benzyl-3-methylvaleric acid from 3-benzyl-3-methylpentanenitrile (Equation (50)) ⟨63OSC(4)93⟩.

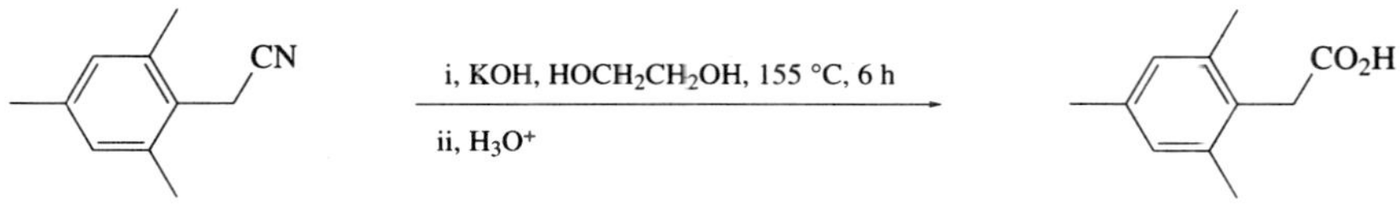

(49)

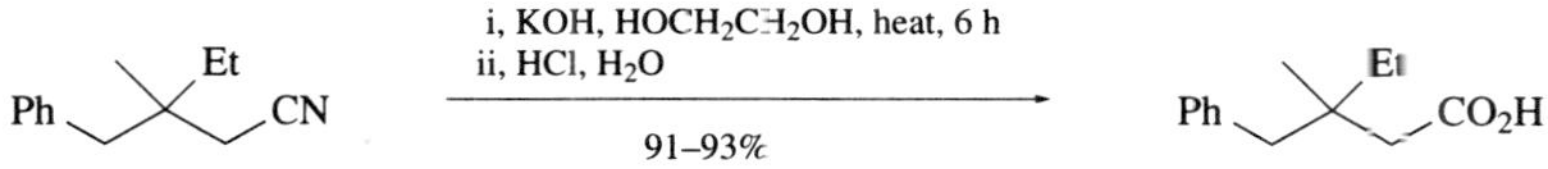

(50)

Aqueous potassium hydroxide has been used to produce *n*-tridecanoic acid from its precursor nitrile (Equation (51)) ⟨43OSC(2)292⟩, while alcoholic sodium hydroxide has been used to produce 2-aryl substituted propanoic acids from their nitrile precursors (Equation (52)) ⟨86SC499⟩.

$$n\text{-}C_{12}H_{25}CN \xrightarrow[\text{ii, HCl}]{\text{i, 90\% KOH, }H_2O\text{, reflux 30 h}} n\text{-}C_{12}H_{25}CO_2H \qquad (51)$$

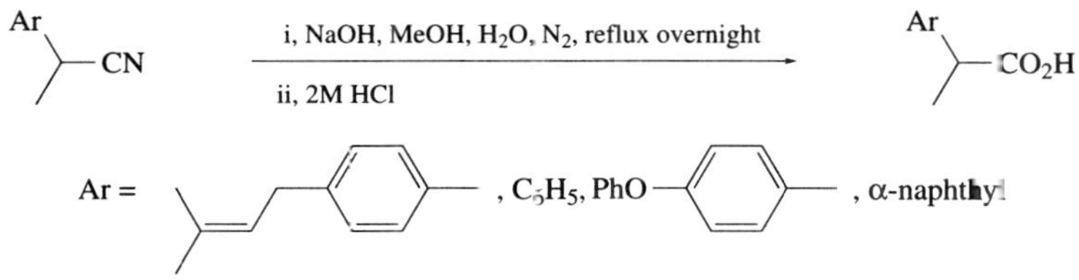

(52)

Sodium hydroxide has also been used to produce, *β*-(2,2′-dimethylbenzhydryl) glutaric acid from 2-(2,2′-dimethylbenzhydryl)-1,3-propanedinitrile by refluxing in ethylene glycol (Equation (53)) ⟨56JA450⟩, and the *gem*-dicarboxylic acid, malonic acid, from sodium α-cyanoethanoate, without accompanying decarboxylation, by refluxing in water (Equation (54)) ⟨43OSC(2)376⟩.

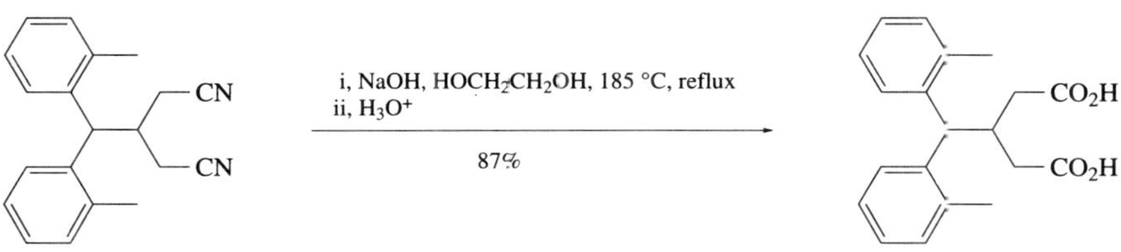

(53)

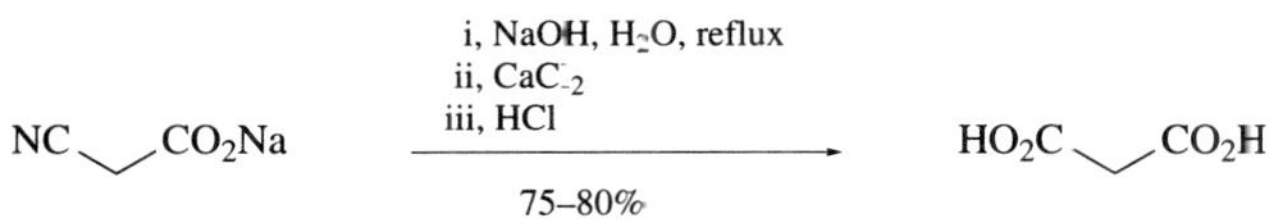

(54)

Acid hydrolysis of α-cyanocarboxylic acids with hydrochloric acid (Equation (55)) ⟨32OSC(1)451⟩, (Equation (56)) ⟨63OSC(4)790⟩, and of the *gem*-dinitrile, 1,1-dicyano-2-cyclohexylpentane, with sulfuric acid (Equation (57)) ⟨83S733⟩, produces *gem*-dicarboxylic acids initially which undergo decarboxylation to give the final products.

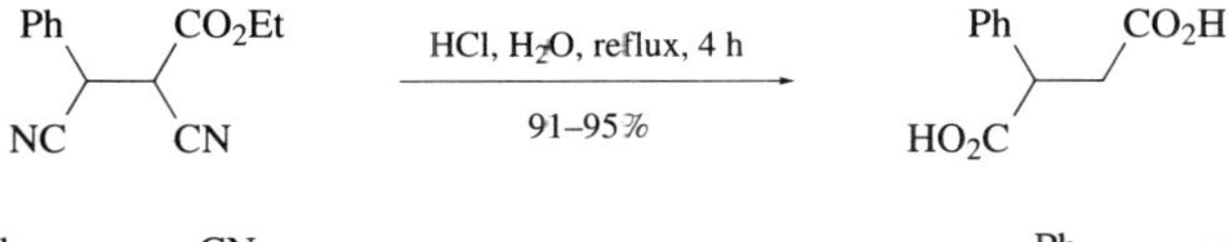

(55)

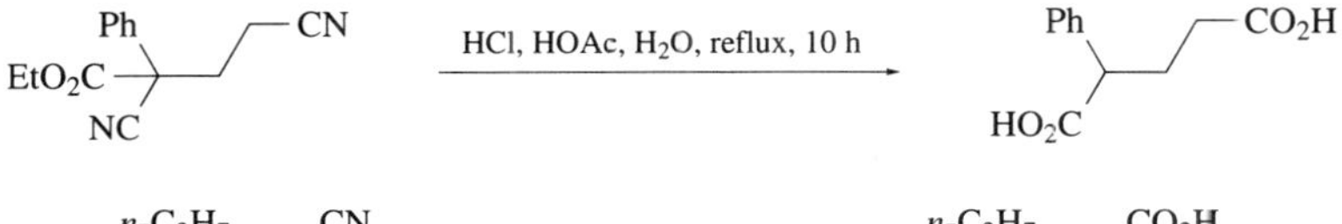

(56)

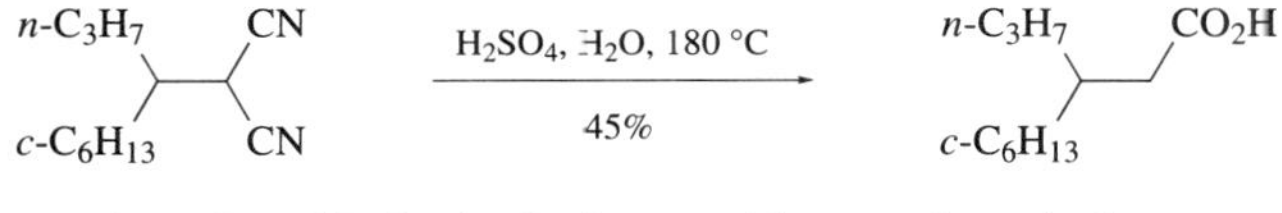

(57)

In addition to the examples of amide hydrolysis to acids mentioned above, several other examples are noteworthy and include: the reaction of primary amides with *n*-butyl nitrite in the presence of acetic acid and gaseous hydrogen chloride (Equation (58) and Table 3) ⟨48JA3091⟩; the reaction of secondary amides with sodium nitrite in acetic anhydride (Equation (59)) ⟨55JA6011⟩; and, the reaction of primary amides with the strongly acidic cation exchange agent Amberlyst 15 (Equation (60)) ⟨81JOC5351⟩. Although this last method utilizes a strongly acidic reagent to effect hydrolysis, it is considered a mild hydrolysis method since reaction of glycyl-DL-asparagine produces glycyl-DL-aspartic acid by selective hydrolysis of the unsubstituted carboxyamide group in the presence of a peptide bond which is left intact.

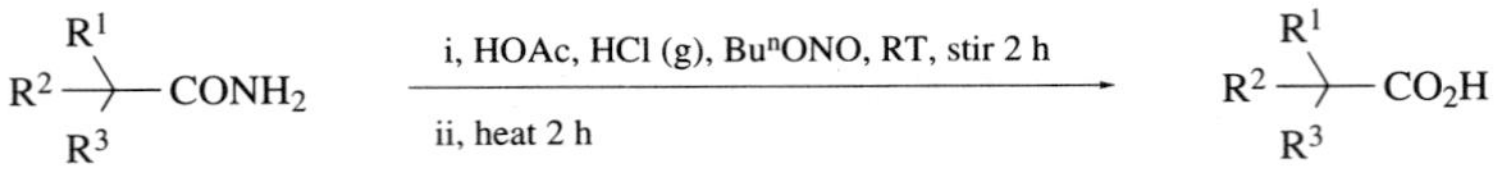

(58)

Table 3 Alkanoic acids from primary amides and *n*-butyl nitrite.

R^1	R^2	R^3	Yield (%)
Pr^n	Pr^n	Pr^n	82
Bu^n	Bu^n	Bu^n	80–90
Bu^n	$Pent^n$	$Pent^n$	50
$Pent^n$	$Pent^n$	$Pent^n$	94
Me	$Hept^n$	$Hept^n$	83
Bu^n	$Hept^n$	$Hept^n$	82
$c\text{-}C_5H_{10}N—CH_2$	Bu^n	Bu^n	60

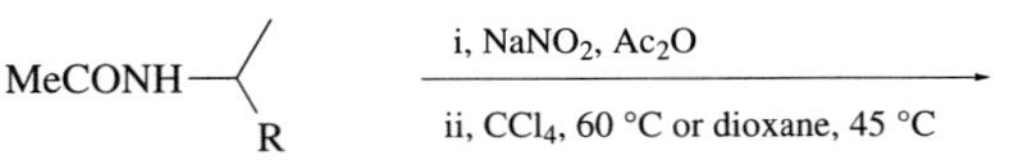

(59)

R	Yield (%)
Et	67
Ph	55

$$RCONH_2 \xrightarrow{\text{Amberlyst 15, } H_2O\text{, 100 °C, 4–72 h}} RCO_2H \quad (60)$$

R	Time (h)	Yield (%)
$PhCH_2$	4	84
2-pyrrole (L)	72	90
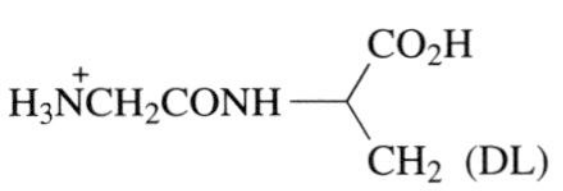	24	95

Treatment of primary or secondary amides with nitrosonium tetrafluoroborate has been reported to be effective for the preparation of simple and alkyl substituted alkanoic acids (Equation (61) and Table 4) ⟨65JOC2386⟩.

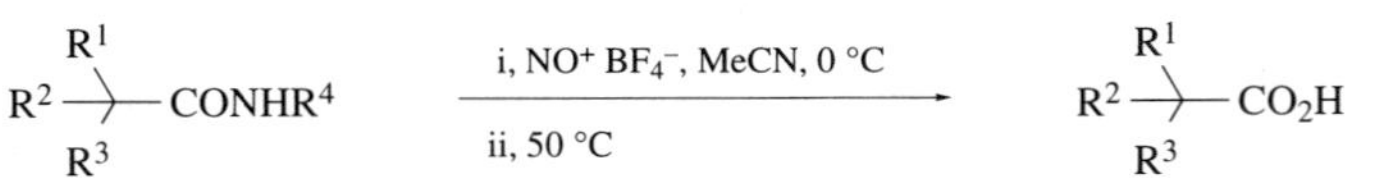

(61)

Table 4 Alkanoic acids from amides and nitrosonium tetrafluoroborate.

R^1	R^2	R^3	R^4	Yield (%)	R^1	R^2	R^3	R^4	Yield (%)
H	H	H	H	71	Ph	Ph	Ph	H	21
Me	Me	Me	H	79	Pr^n	Pr^n	Pr^n	H	61
H	Pr^i	Pr^i	H	69	Pr^i	Pr^i	Pr^i	H	50
H	H	Bu^i	H	80	H	H	Pr^i	Pr^i	73
H	Pr^i	Bu^t	Pr^i	63	H	H	Pr^n	H	90
H	Et	Bu^t	H	70					

Two interesting hydrolysis reactions of imides which produce carboxylic acids are the acid catalyzed reaction of α,α'-dicyano-β-ethyl-β-methylglutarimide producing 3-ethyl-3-methylglutaric acid (Equation (62)) ⟨63OSC(4)441⟩, and the base-catalyzed reaction of 2α-thiohomophthalimide which affords homophthalic acid (Equation (63)) ⟨73OSC(5)612⟩.

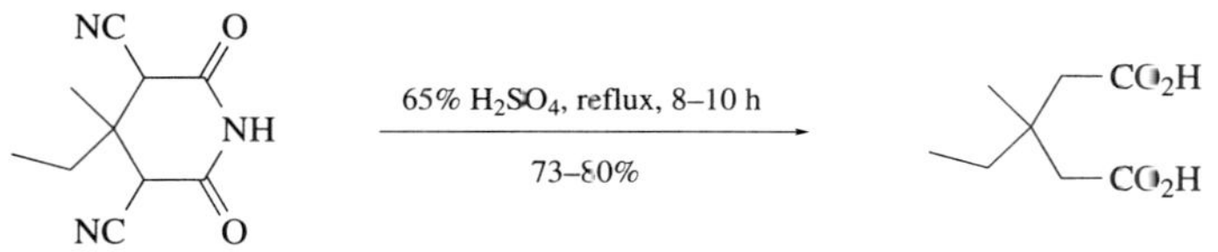

(62)

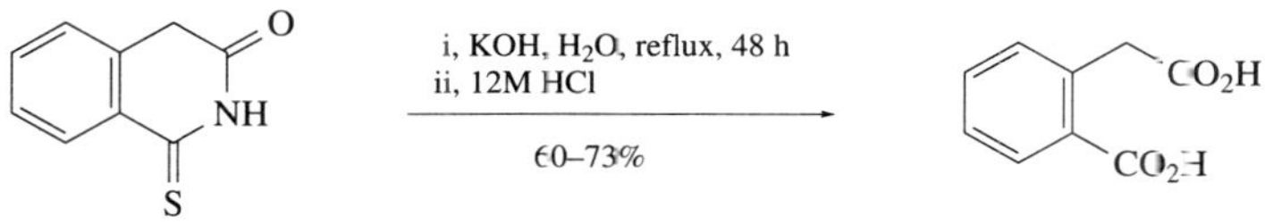

(63)

A study ⟨60JCS3070⟩ of the mechanism of the perchloric acid-catalyzed hydrolysis in aqueous dioxane of the symmetrical acetic anhydride and the unsymmetrical acetic benzoic anhydride using $H_2{}^{18}O$ to determine the position of bond fusion established that the reactions occur via an A2 mechanism (Scheme 9), while the unsymmetric acetic mesitoic anhydride reacts via an A1 mechanism (Scheme 9). A separate study ⟨53JCS1409⟩ of the hydrolysis of acetic anhydride in water at 0 °C with a series of heterocyclic tertiary amines (pyridines, 2-,3- or 4-picolines, 2,6-lutidine, quinoline and isoquinoline) as catalysts, showed that the presence of α-substituents in the amine catalysts decreased its catalytic activity. The mechanism proposed for the reaction involved transfer of the acetylium group ($MeCO^+$) from acetic anhydride to the catalyst in a rate-determining step, followed by rapid hydrolysis of the resulting cation.

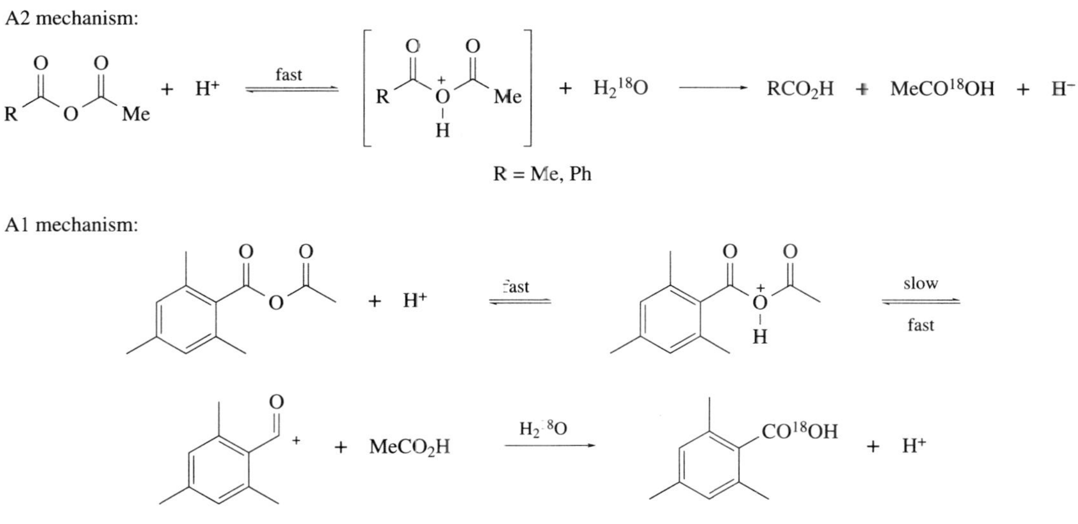

Scheme 9

An example of a base-catalyzed hydrolysis of an anhydride is the reaction of 6,7-dimethoxytetralin-2,3-dicarboxylic anhydride with aqueous sodium hydroxide (Equation (64)) ⟨73S416⟩, while a solvolytic cleavage of succinic anhydride by methanol is illustrated in Equation (65) which affords a mixed ester acid product ⟨55OSC(3)169⟩.

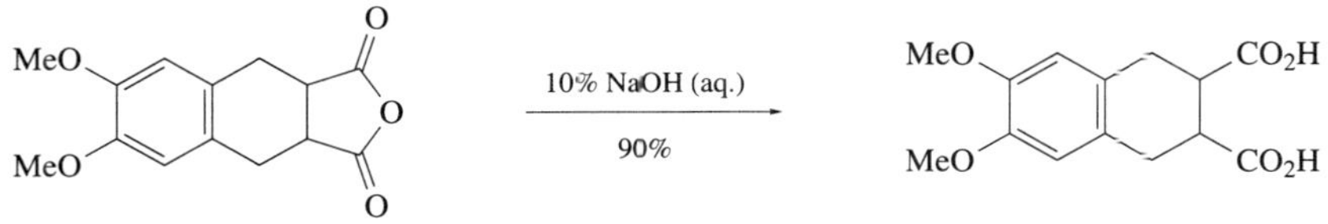

(64)

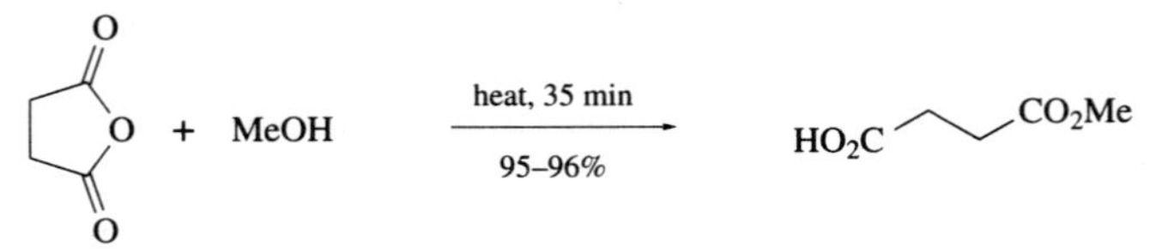

(65)

Trihalomethyl groups have been converted into carboxylic acids by use of the haloform reaction ⟨34CRV275⟩, and by basic hydrolysis, as in the preparation of di-(*p*-chlorophenyl) acetic acid (Equation (66)) ⟨55OSC(3)270⟩.

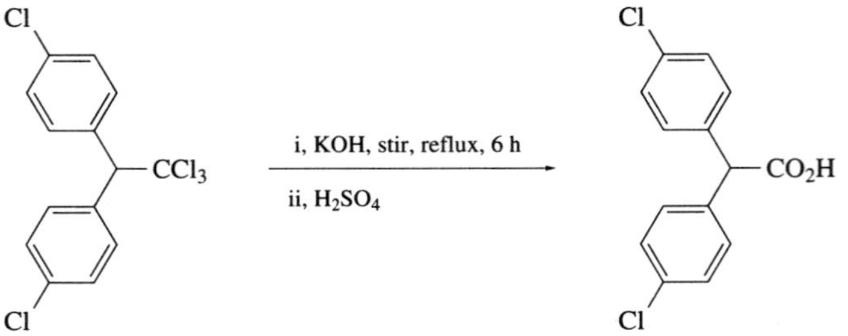

(66)

The synthesis of carboxylic acids by treatment of compounds containing a 1,1-dichlorovinyl group with acid has been reviewed ⟨57JGU2481, 66AG(E)870⟩, but other 1,1-dihalovinyl groups can also be used as precursors to carboxylic acids. For example, 1,1-dibromovinyl groups have been converted into the corresponding carboxylic acids upon treatment with palladium (1,2-bis(diphenylphosphino)ethane) (Pd(diphos)) in the presence of base (Equation (67) and Table 5) ⟨86JOC4354⟩, while 1,1-difluorovinyl groups have been converted into the corresponding carboxylic acids with reagents ranging from sulfuric acid alone to mercuric acetate in trifluoroacetic acid (Equation (68) and Table 6) ⟨80CL651⟩.

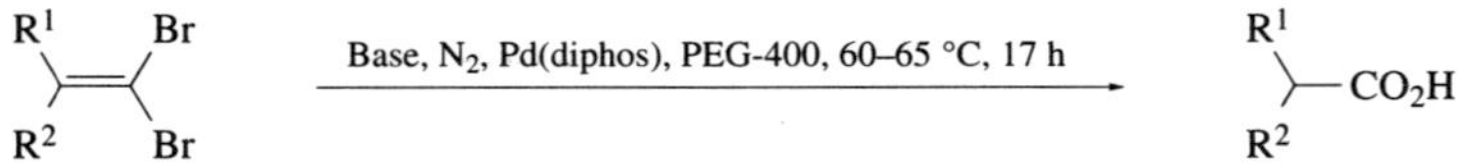

(67)

Table 5 Alkanoic acids by reaction of 1,1-dibromovinyl groups with palladium [1,2-bis(diphenylphosphino)ethane].

R[1]	*R*[2]	*Base*	*Yield* (%)
Ph	H	NaOH	95
o-Tol	H	NaOH	51
p-Tol	H	NaOH	77
p-ClC_6H_4	H	NaOH	35
p-ClC_6H_4	H	KOH	56
p-An	H	NaOH	85
PhCH=CH	H	NaOH	58
Ph	Me	NaOH	18
Ph	Me	KOH	23
$-(CH_2)_2-CH(Me)-(CH_2)_2-$		NaOH	42

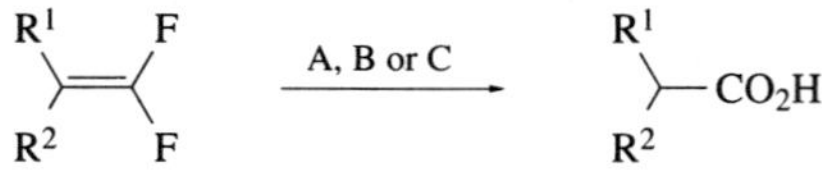

(68)

One direct method reported for the preparation of tertiary alkylacetic acids involves the condensation of 1,1-dichlorethene with a reagent (alkene, alcohol or alkyl halide), which will readily form a carbocation in concentrated sulfuric acid ⟨85S493⟩. The intermediate 1,1-dichloride products from this condensation are not isolated but are hydrolyzed immediately *in situ* to produce the tertiary alkylacetic acids (Equation (69) and Table 7).

Table 6 Alkanoic acids from 1,1-difluorovinyl groups using various reagents.

R[1]	*R*[2]	*Method*[a]	*Yield* (%)
H	Ph	A	93
H	Ph	B	80
H	*p*-$Pr^iC_6H_4$	A	70
H	*p*-$Pr^iC_6H_4$	B	81
Me	Ph	A	94
Me	*p*-$Bu^tC_6H_4$	A	80
H	*n*-C_6H_{13}	C	69
H	*n*-$C_{10}H_{21}$	C	88

[a] Method: A, conc. H_2SO_4, −10 °C, stir room temp. 3 h; B, conc. H_2SO_4 + alcohol; C, $Hg(OAc)_2$, CF_3CO_2H, then aqueous $NaHCO_3$ and H_2S.

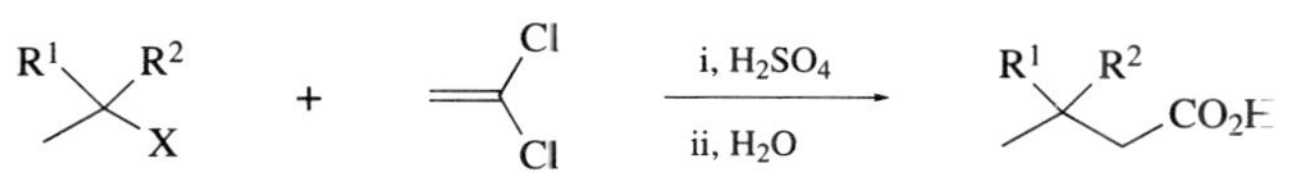

(69)

Table 7 Tertiary alkylacetic acids via condensation of 1,1-dichloroethane with alkenes, alcohols or alkyl halides.

R[1]	*R*[2]	*X*	*Total Rx time* (h)	*Temp.* (°C)	*Yield* (%)
Me	Me	OH	2.5[a]	2	90
Me	Me	Cl	4[a]	6	77
Me	Me	OMe	2.5[a]	5	55
Me	Et	OH	3[a]	6	53
Me	Et	OH	3[b]	4–11	72
Me	Et	OH	4.5[b]	2–11	75
Et	Et	OH	3.5[a]	4	13
Et	Et	OH	5[b]	4–17	30

[a] Using 99% sulfuric acid by weight. [b] Using 96–98% sulfuric acid by weight.

Several interesting approaches to the preparation of carboxylic acids by *O*-alkyl cleavage have been reported and should be mentioned. The first of these involves the use of lithium iodide in 2,4,6-collidine to deesterify 10-carboethoxy-1,1-dimethyl-*trans*-decalin (Equation (70)) ⟨63JOC2184⟩, or sodium iodide and methyl trichlorosilane in acetonitrile to deesterify methyl 2-methylproponate (Equation (71)) ⟨83JOC3667⟩.

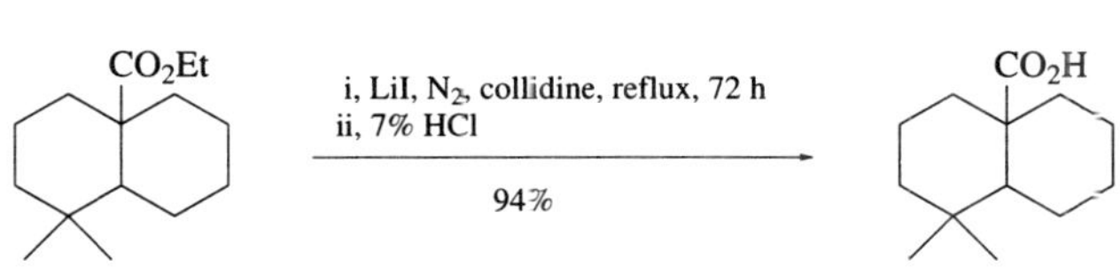

(70)

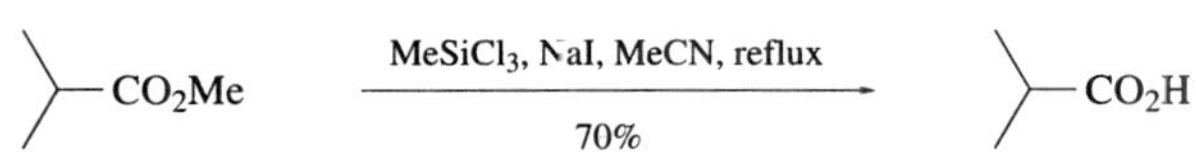

(71)

Two types of dithiolates in acetonitrile have been used to cleave 2-bromoethyl lauroate in excellent yields, sodium trithiocarbonate afforded an 86% yield (Scheme 10) ⟨74S715⟩, while sodium ethanedithiolate afforded an 92% yield of lauric acid (Scheme 10) ⟨75S510⟩. Another alkali thiolate which has been used to cleave esters is *n*-propylthiol in hexamethylphosphoramide (HMPA), which has also been found to produce excellent yields of the corresponding acids from methyl *O*-methylpodocarpate and methyl 3β-acetoxy-Δ^5-etienate under similar reaction conditions (Equation (72)) ⟨70TL4459⟩.

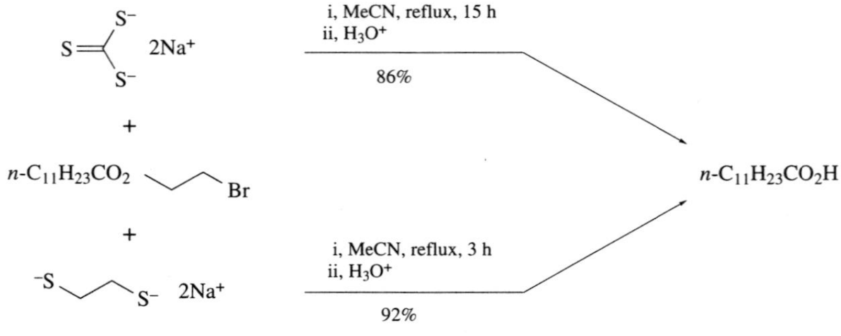

Scheme 10

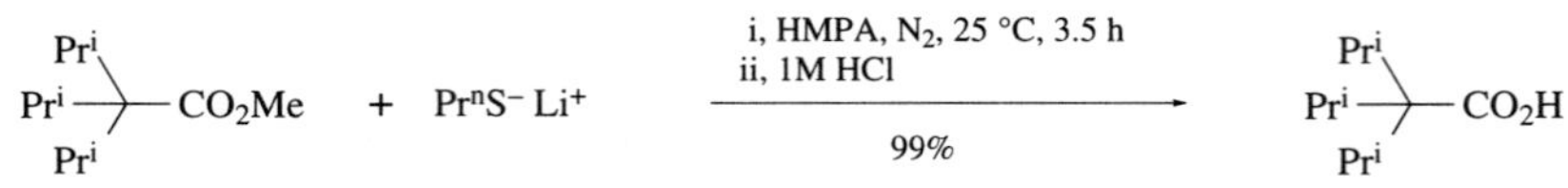 (72)

Other esters	*Reaction time* (h)	*% Yield acid*
Methyl *O*-methylpodocarpate	1.5	100
Methyl 3β-acetoxy-Δ^5-etienate	24	92

Two other methods have been reported to convert methyl *O*-methylpodocarpate, as well as several other hindered esters, into their corresponding acids. Both methods involve the use of Lewis acid, with the first utilizing boron trichloride (Equation (73)) ⟨71CC667⟩, while the second utilizes an aluminum halide in ethanethiol (Scheme 11 and Table 8) ⟨78TL5211⟩.

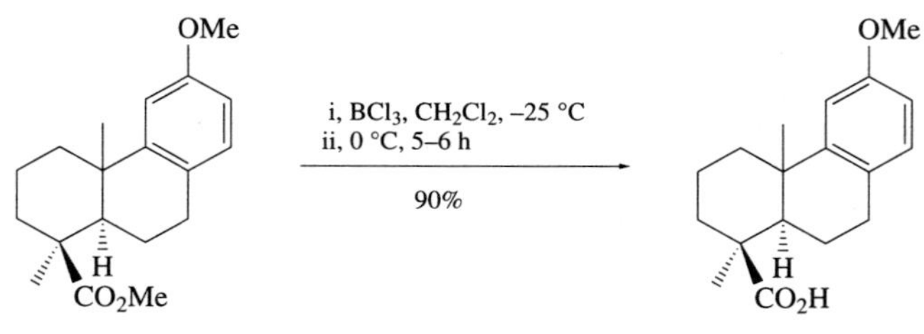 (73)

Other esters	*Product*	*Yield* (%)
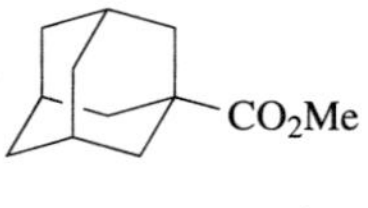	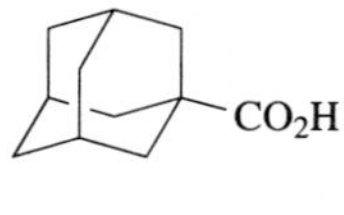	94
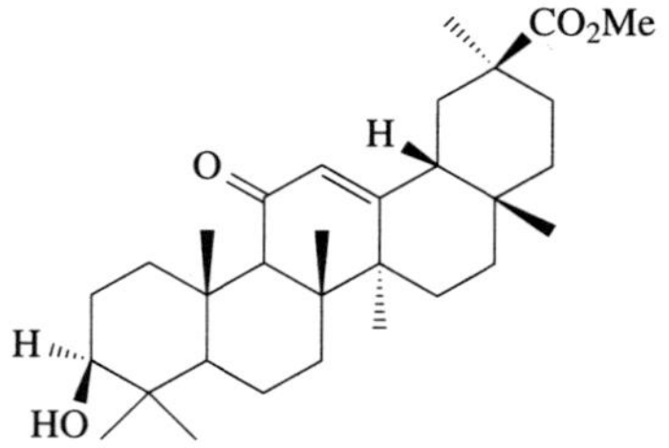	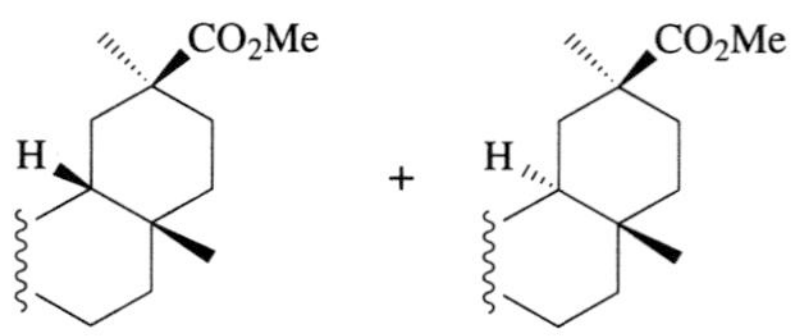 (epimeric at C-18)	40

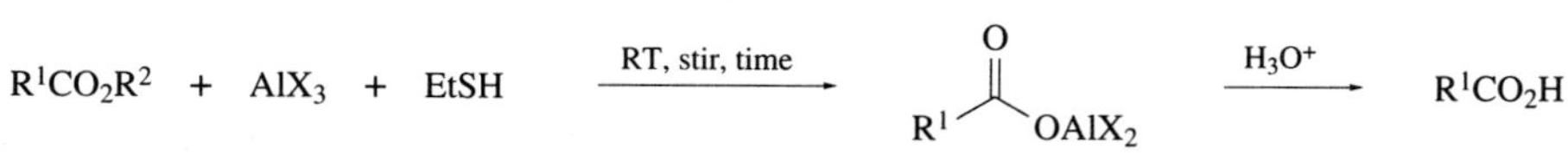

Scheme 11

An interesting preparation of cycloundecanecarboxylic acid via a ring contraction of cyclododecanone, which involves the intermediate formation of an enamine and its reaction with diphenyl phosphorazidate followed by basic hydrolysis in ethylene glycol, has been reported (Scheme 12)

Table 8 Conversion of hindered esters to their corresponding acids using aluminum halide in ethanethiol.

Substrate	*Lewis acid*	*Time* (h)	*Product*	*Yield* (%)
n-$C_{11}H_{23}CO_2CH_2Ph$	$AlCl_3$	1.5	n-$C_{11}H_{23}CO_2H$	98
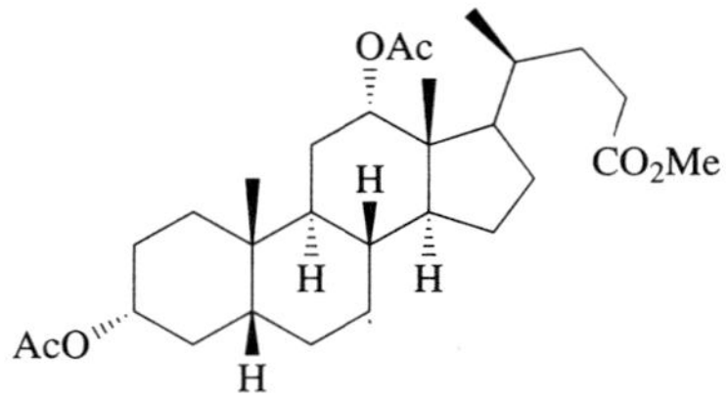	$AlBr_3$	9	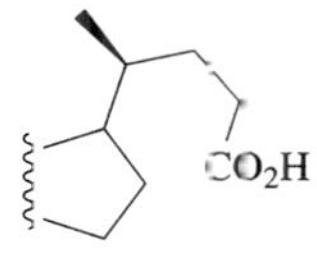	89
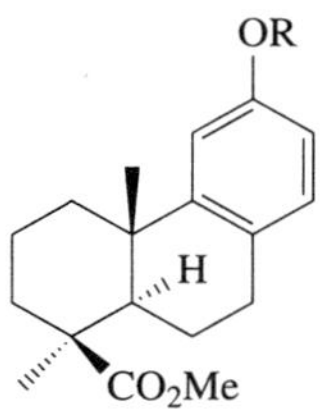			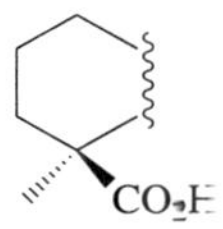	
R = H	$AlCl_3$	<20		98
R = H	$AlBr_3$	5.5		98
R = Me	$AlBr_3$	7.5		96
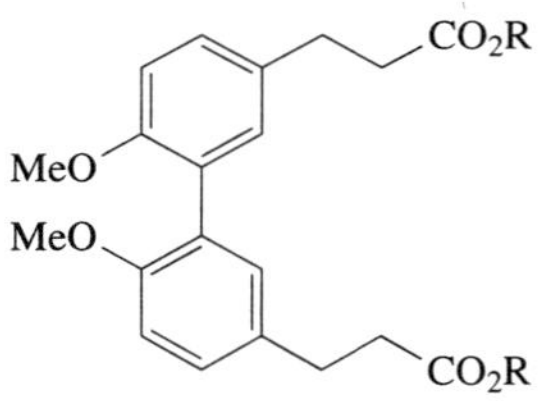			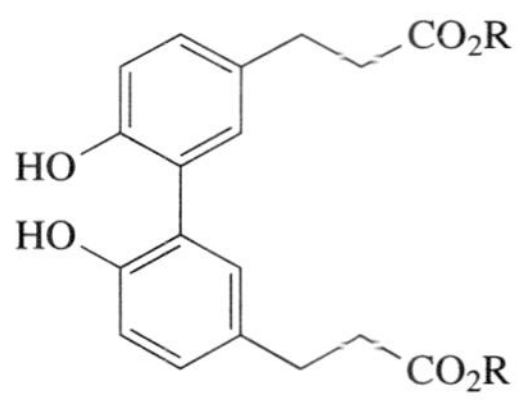 **(1)** 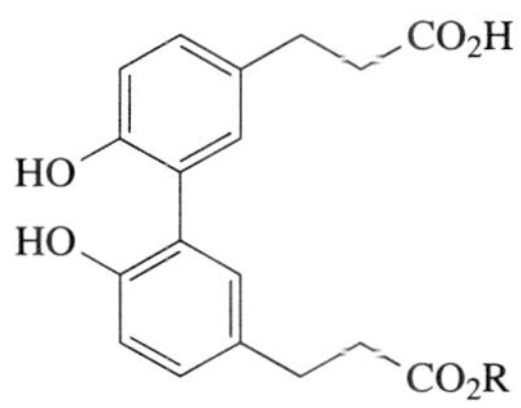 **(2)** 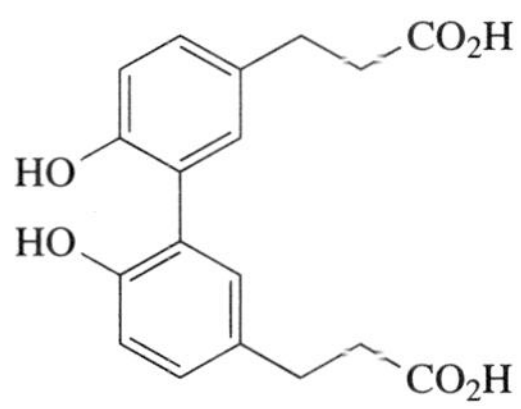 **(3)**	
R = Me	$AlBr_3$[a]	24	**(1)** + **(2)** + **(3)**	18 44 34
R = Et	$AlBr_3$	50	**(1)** + **(2)** + **(3)**	30 54 7

[a] Methylene chloride added to solvent.

⟨84OS(62)191⟩, while another phosphorus reagent, tetraethyl dimethylaminomethylenephosphonate has been used in conjunction with aldehydes to produce diethyl 1-dimethylaminoalkenyl phosphonates, which under acid hydrolysis produce substituted acetic acids (Scheme 13) ⟨68AG(E)391⟩.

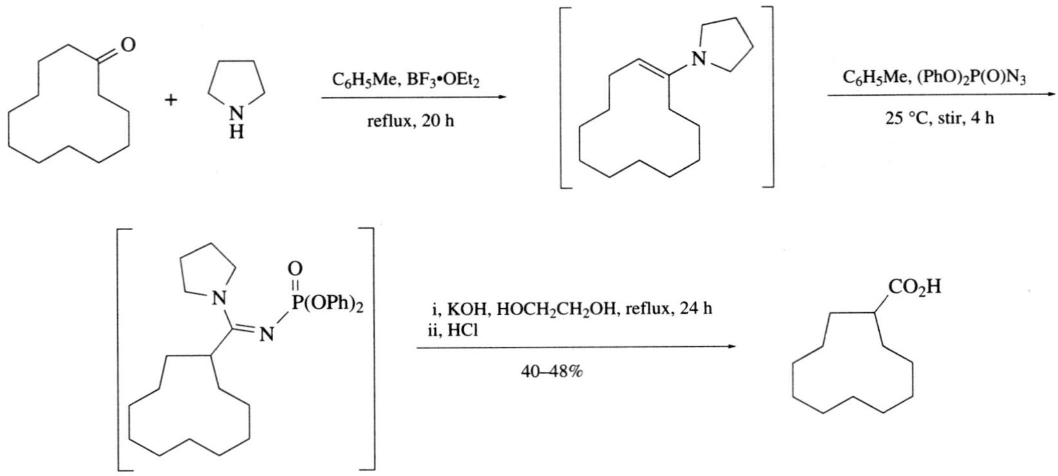

Scheme 12

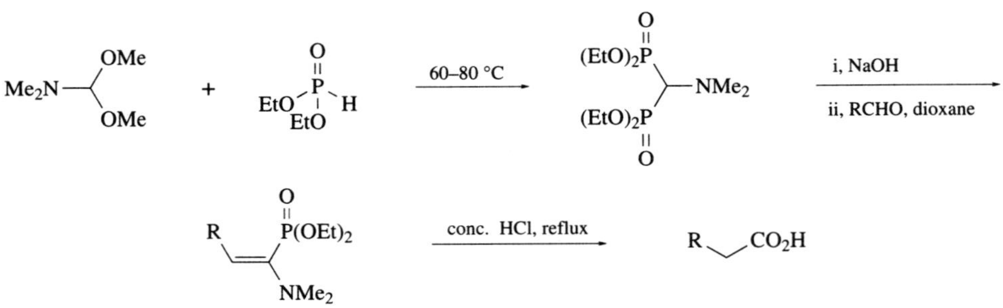

R	*Yield* (%)
Ph	75
p-$O_2NC_6H_5$	80
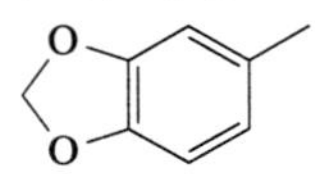	90

Scheme 13

5.02.2.1.2 By carbonation of organometallic reagents

While some examples of the general carbonation reaction of organometallic reagents used to prepare alkanoic acids are reported in Equation (74) and Table 9, one interesting result is the metalative dimerization which has been reported to occur when styrene or α-methylstyrene is treated with sodium in ethereal solvents (Scheme 14) ⟨61JOC307⟩. Carbonation of the intermediate dimeric dianion produces racemic meso-2,5-diphenyladipic acid and its 2,5-dimethyl analogue.

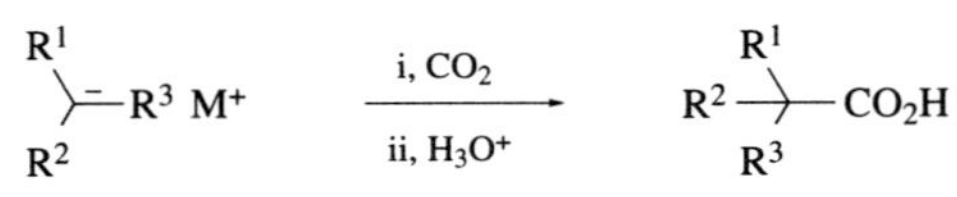

(74)

Table 9 Alkanoic acids from carbonation of organolithium, -sodium or Grignard reagents.

R¹	*R²*	*R³*	*M*	*Yield (%)*	*Ref.*
H	Ph	Ph	Li		72CC752
Ph	Ph	Ph	Li		72CC752
	c-C_6H_{11}		Li		77CC752
	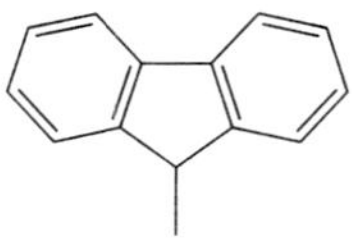		Li	75	43JA262
	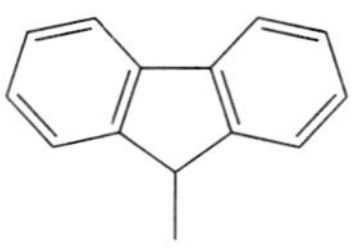		Na		43JA262
H	Me	Et	MgCl	76–86	32OSC(1)361
Me	Me	Me	MgCl	67	48JA2824, 67BSF2011
Me	Me	Me	MgCl	69–70	32OSC(1)524
Me	Me	Et	MgCl	40	48JA2824, 67BSF2011
Me	Et	Et	MgCl	17	48JA2824, 67BSF2011
Et	Et	Et	MgCl	5–7	48JA2824, 67BSF2011
Et	Et	Et	MgCl	75–80[a]	67BSF2011
	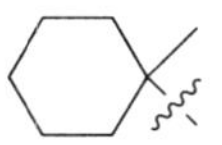		MgCl	15	48JA2824

[a] Reaction was run under pressure.

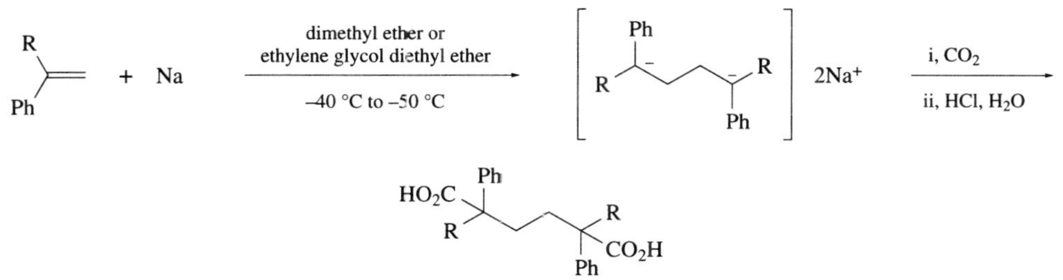

R	*Solvent*	*Yield (%)*
H	dimethyl ether	80–85
H	ethylene glycol diethyl ether	40
Me	ethylene glycol diethyl ether	50

Scheme 14

In the early 1970s it was found that dimethylarenes such as 1,3- and 1,8-dimethylnaphthalene can be dimetalated with *n*-amylsodium complexed with TMEDA ⟨65MI 502-01⟩ to produce disodio derivatives, which upon carbonation produce 1,3- or 1,8-naphthalene diacetic acid, respectively (Scheme 15) ⟨73JOC1491⟩.

One interesting approach to the carbonation of Grignard reagents is their reaction with iron pentacarbonyl to produce an intermediate alkoxy iron tetracarbonyl magnesium bromide complex which upon treatment with iodine in water produces the carboxylic acid (Scheme 16) ⟨81NKK1333⟩.

Carbonation of phosphoranes has been reported to be an effective method of carboxylic acid production which affords aliphatic as well as cyclic acids in yields ranging from 80–90% (Scheme 17) ⟨74TL1275⟩.

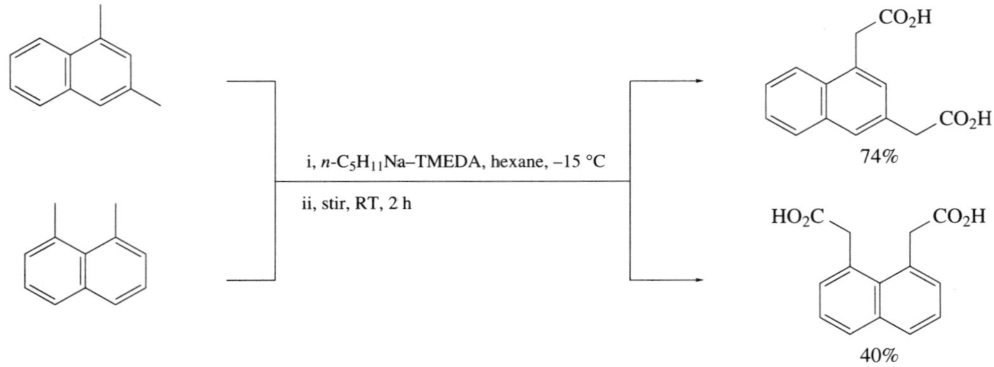

Scheme 15

$$RMgBr + Fe(CO)_5 \xrightarrow{THF} [ROFe(CO)_4]^- MgBr^+ \xrightarrow{I_2, H_2O} RCO_2H$$

R	*Yield* (%)
n-C_8H_{17}	65
n-C_6H_{13}	60
Bu^s	15
c-C_6H_{11}	30

Scheme 16

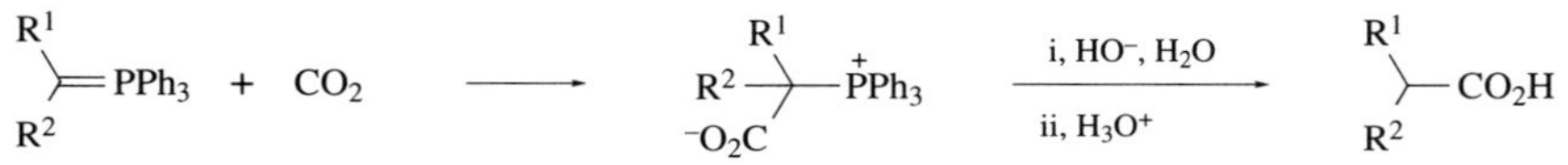

R[1]	*R*[2]	*Yield* (%)
$-(CH_2)_2-$		90
$-(CH_2)_3-$		89
$-(CH_2)_5-$		86
$-(CH_2)_6-$		89

R[1]	*R*[2]	*Yield* (%)
Me	Me	85
Me	Et	90
Pr^n	H	83
$PhCH_2$	H	81
Ph	H	80

Scheme 17

One report of the carboxylation of organoaluminum compounds to produce carboxylic acids involves the treatment of dialkylaluminum halides with gaseous carbon dioxide at various pressures followed by acid hydrolysis (Scheme 18) ⟨78IZV2094⟩. The various halides, carbon dioxide pressures, and other conditions used for this reaction are reported in Table 10 along with the yields obtained, and their results led the authors to report that the reactivity of the organoaluminum reagents towards carbonation follows the sequence: $R^1_3Al > R^1_2AlBr \sim R^1AlCl_2 > R^2_2\ AlCl > R^1_2AlCl$ (where $R^1 > R^2$).

$$R_2AlX + CO_2\,(g) \longrightarrow (RCO_2)RAlX \xrightarrow{H_2O,\ HCl} RCO_2H + RH + Al(OH)_3 + HX$$

Scheme 18

The use of electrocarbonylation of halides to prepare carboxylic acids has been studied in great detail ⟨85TL1509, 86GEP3522304, 86JCE1013⟩, and the following results have been reported: the anode used in all cases was made of magnesium; while the cathode was made from either platinum, gold, nickel, graphite or stainless steel, with stainless steel producing the best results. Aprotic solvents such as *N*-methyl 1-2-pyrrolidine (NMP), acetonitrile, acetone, DMF, THF, or a mixture of THF and HMPA have all been used, as have the following supporting electrolytes: tetra-*n*-butyl-

Table 10 Carbonation of organoaluminum compounds.

Organoaluminum substrate	*Temp.* (°C)	*Pressure* (atm)	*Time* (h)	*Solvent*	*Carboxylic acid*	*Selectivity*	*Yield* (%)
Et_2AlCl	140	1	4.5	*o*-Xylene	$EtCO_2H$	47	12.4
$(n\text{-}C_7H_{15})_2AlCl$	170	140	4	Dodecane	$n\text{-}C_7H_{15}CO_2H$	97	71
$(n\text{-}C_7H_{15})_2AlCl$	180	380	5	Decane–ether (1 : 2)	$n\text{-}C_7H_{15}CO_2H$	96	47.7
$(n\text{-}C_8H_{17})_2AlCl$	100	1	4	Decane	$n\text{-}C_8H_{17}CO_2H$	0	0
$(n\text{-}C_8H_{17})_2AlCl$	100	230	5	Decane–ether (1 : 2)	$n\text{-}C_8H_{17}CO_2H$	0	0
$(n\text{-}C_8H_{17})_2AlCl$	120	1	5	Decane	$n\text{-}C_8H_{17}CO_2H$	81	12.8
$(n\text{-}C_8H_{17})_2AlCl$	135	1	4.5	Decane	$n\text{-}C_8H_{17}CO_2H$	59	13.8
$(n\text{-}C_8H_{17})_2AlCl$	145	1	4.5	Decane	$n\text{-}C_8H_{17}CO_2H$	52.5	15.2
$(n\text{-}C_8H_{17})_2AlCl$	145	100	3	Decane	$n\text{-}C_8H_{17}CO_2H$	96	41.5
$(n\text{-}C_8H_{17})_2AlCl$	145	100	5	Decane	$n\text{-}C_8H_{17}CO_2H$	90	42
$(n\text{-}C_8H_{17})_2AlCl$	150	120	3.5	Decane–ether (1 : 2)	$n\text{-}C_8H_{17}CO_2H$		47
$(n\text{-}C_8H_{17})_2AlBr$	150	1	5	Dodecane	$n\text{-}C_8H_{17}CO_2H$	72.3	16
$(n\text{-}C_8H_{17})_2AlCl$	160	130	4	Dodecane	$n\text{-}C_8H_{17}CO_2H$	92	49
$(n\text{-}C_8H_{17})_2AlCl$	160–165	1	5	Decane	$n\text{-}C_8H_{17}CO_2H$	50	16
$(n\text{-}C_{10}H_{21})_2AlBr$	190	65	4	Dodecane	$n\text{-}C_{10}H_{21}CO_2H$		74
$(n\text{-}C_{10}H_{21})_2AlBr$	195	125	5	Dodecane	$n\text{-}C_{10}H_{21}CO_2H$	94.8	78.7
$(n\text{-}C_{10}H_{21})_2AlCl$	200	120	4	Dodecane–ether (1 : 1)	$n\text{-}C_{10}H_{21}CO_2H$	94	16

ammonium fluoride (tbaf), bromide, perchlorate, or tetrafluoroborate, tetraethylammonium chloride or lithium perchlorate. All halide electrocarboxylation reactions were performed using chlorides, bromides or iodides at temperatures ranging from −5°C to 25°C, using carbon dioxide at 1 atmosphere pressure. The general reactions involved in this approach to the preparation of carboxylic acids are shown in Scheme 19, while the results obtained are reported in Table 11.

Anode:	Mg	$\longrightarrow$	$Mg^{2+} + 2e^-$
Cathode:	$RX + CO_2 + 2e^-$	$\longrightarrow$	$RCO_2^- + X^-$
Overall:	$Mg + RX + CO_2$	$\longrightarrow$	$RCO_2^- + Mg^{2+} + X^-$

Scheme 19

Table 11 Electrocarboxylation of halides.

Halide	*Temp* (°C)	*Solvent*	*Acid product*	*Yield*[a] (%)
$PhCH_2Cl$	0	DMF	$PhCH_2CO_2H$	80–90
$PhCH_2Cl$	−5	THF	$PhCH_2CO_2H$	90
$PhCH_2Cl$	−5	Acetone	$PhCH_2CO_2H$	90
$PhCH_2Cl$	−5	MeCN	$PhCH_2CO_2H$	90
PhCHClMe	−5	DMF	$PhCHMeCO_2H$	80
$n\text{-}C_{10}H_{21}Br$	−5	DMF	$n\text{-}C_{10}H_{21}CO_2H$	75
$n\text{-}C_{18}H_{37}Br$	−5	THF–HMPA	$n\text{-}C_{18}H_{37}CO_2H$	~50[b]
$ClCH_2CO_2Et$	−5	THF	$HO_2C—CH_2—CO_2Et$	~90[b]
$m\text{-}Bu^tC_6H_4CHClMe$	−5	DMF	$m\text{-}Bu^tC_6H_4CHMeCO_2H$	85

[a] Yields of isolated acids versus the amount of consumed halide. [b] Acid not isolated, yield was determined by acidimetry or chromatography.

Because of the continuing interest in labeled compounds for reaction mechanism determinations both chemically and spectroscopically, three approaches to the preparation of these compounds are now described. The first method is straightforward and simply involves the reaction of $^{14}CO_2$ with Grignard reagents followed by acidification to form the labeled acids (Equation (75)) ⟨83MI 502-01⟩. The second method involves formation of an ester enolate, followed by reaction with $^{14}CO_2$ and treatment with acid (Scheme 20) ⟨86JCS(P1)521⟩. The final method is a hydrocarbonylation reaction and involves treatment of organoboranes with ^{13}C labelled carbon monoxide at 1 atmosphere pressure followed by oxidation using hydrogen peroxide (Equation (76)) ⟨82CC1273⟩.

$$RMgX + {}^{14}CO_2 \xrightarrow[\text{ii, } H_3O^+]{\text{i, stir, 15 h}} R^{14}CO_2H \qquad (75)$$

R	*Yield (%)*
Bu^n	32
Bu^t	76
$PhCH_2$	98

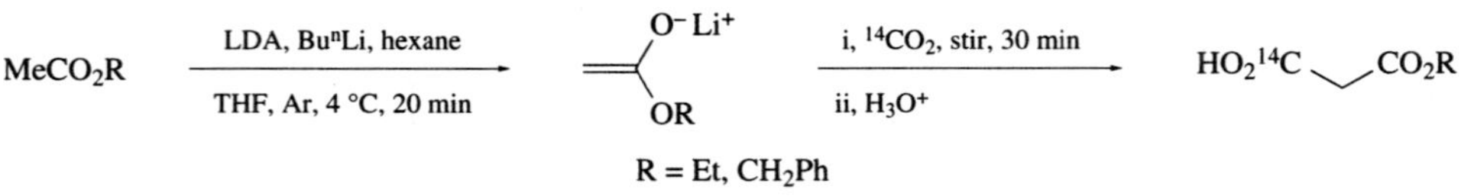

Scheme 20

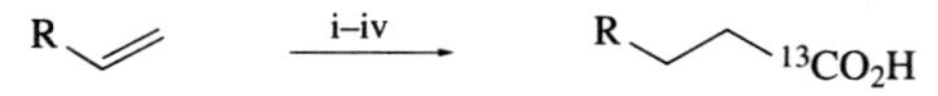

(76)

i, 9-BBN, THF; ii, ^{13}CO, $KBH(OPr^i)_2$; iii, H_2O_2, NaOAc; iv, $AgNO_3$, NaOH, 50 °C, 30 min

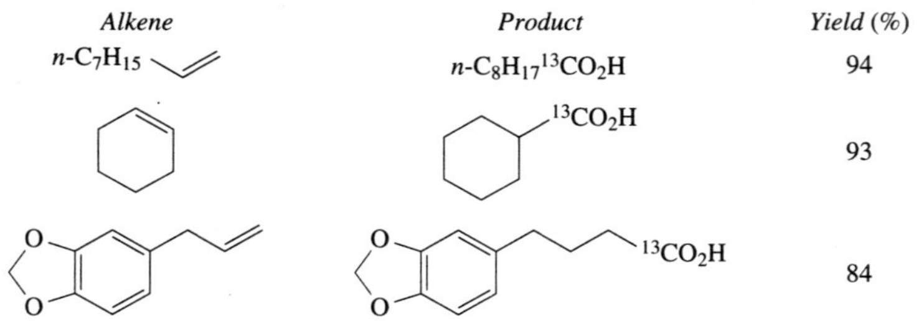

Although a thorough review is beyond the scope of this chapter, several reactions which should be mentioned as methods useful for the preparation of alkyl substituted alkanoic acids via anions are the so-called "name reactions". Some of the more useful of these reactions along with some leading references are: acetoacetic ester synthesis ⟨42OR(1)266⟩; Darzens condensation ⟨49OR(5)413, 55CRV283⟩; Dieckmann condensation ⟨42OR(1)266, 67OR(15)1⟩; malonic ester synthesis ⟨57OR(9)107, 69CRV591⟩; Michael reaction ⟨49OR(5)79, 58CRV895, 59OR(10)179, 62AG(E)626⟩; and the use of carboxylic acid dianions ⟨82S521⟩.

5.02.2.1.3 By oxidation reactions

At least three reviews have been written covering different aspects of the oxidation of alcohols to carboxylic acids which include the use of hydrogen peroxide ⟨B-62MI 502-01⟩, periodate oxidation of *vic*-diols ⟨44OR(2)341⟩, and phenyl derivatives of dihydric and polyhydric alcohols ⟨B-56MI 502-03⟩.

In the presence of sodium nitrite, nitric acid has been used to convert δ-hydroxyvaleraldehyde to glutaric acid (Equation (77)) ⟨63OSC(4)499⟩, while in the presence of ammonium vanadate it effectively oxidizes cyclohexanol to adipic acid (Equation (78)) ⟨32OSC(1)18⟩. Nitric acid alone has been used to convert a series of α,ω-diols into dicarboxylic acids (Equation (79)) ⟨70MI 502-01⟩.

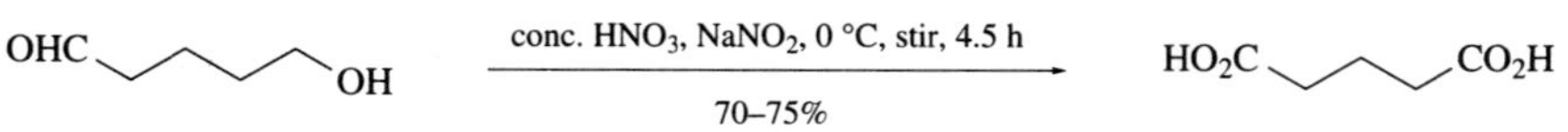

(77)

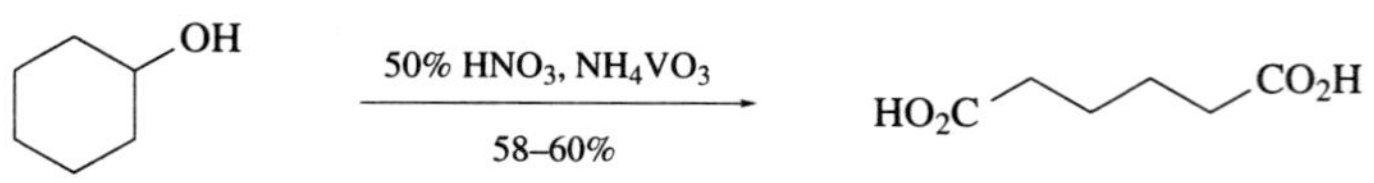

(78)

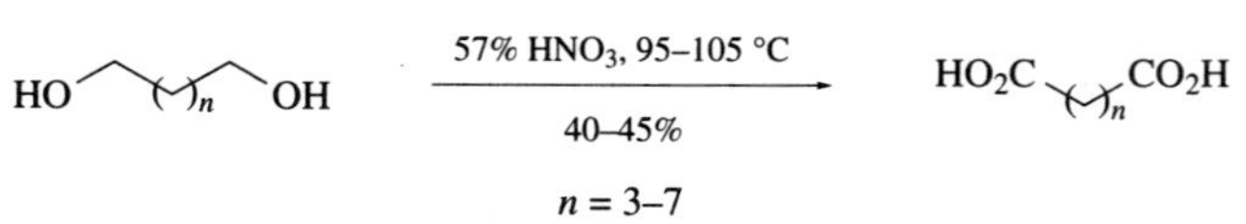

(79)

Chromium trioxide, in combination with several other reagents, has been used to oxidize alcohols to acids. In combination with acetic acid, chromium trioxide converts 3-phenyl-1-propanol-D into its corresponding carboxylic acid-D albeit in low yields, while better yields are obtained using potassium hydroxide at high temperatures (Equation (80)) ⟨48JA3484⟩. Together with *t*-butyl alcohol, chromium trioxide forms *t*-butyl chromate, which at low temperatures converts alcohols into alkanoic acids (Equation (81)) ⟨65BCJ893⟩. In the presence of sulfuric acid aqueous chromic acid produces the Jones' reagent, which can be used to oxidize 2-*t*-butyl-3,3-dimethylbutanol to its corresponding acid (Equation (82)) ⟨60JA2498⟩, and in the presence of potassium fluoride in acetone it can be used to convert silyl ethers of primary alcohols directly into their corresponding acids without prior cleavage of the silicon–oxygen bond (Equation (83)) ⟨85SC759⟩.

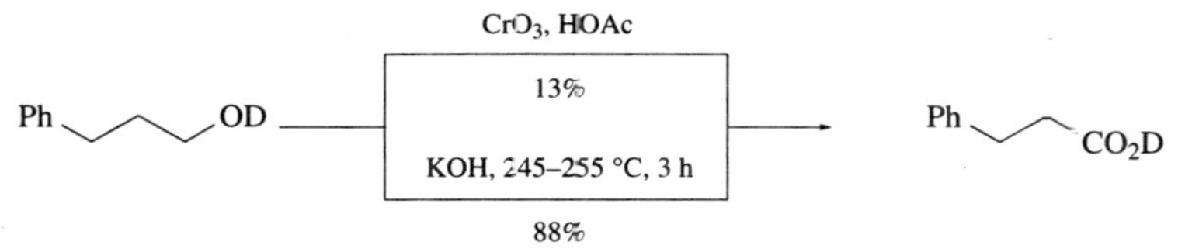

(80)

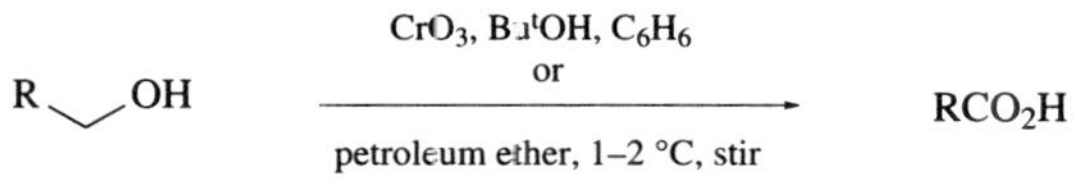

(81)

R = Me, Pr^n, Bu^n, n-C_5H_{11}, n-$C_{11}H_{23}$, n-$C_{15}H_{31}$, Pr^i, Bu^i, Bu^t, Me_2C=CH–CH_2CH_2C(Me)=CH–

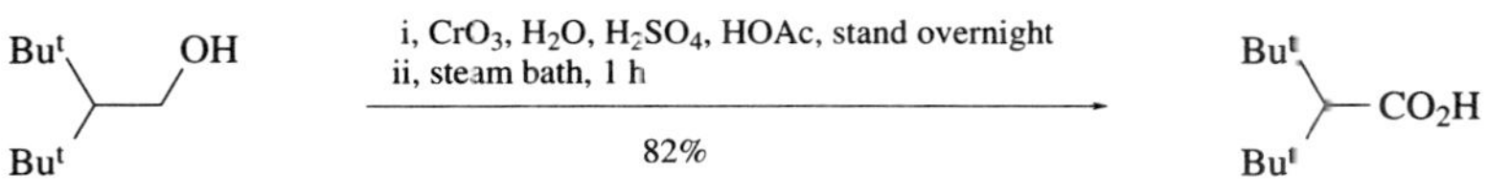

(82)

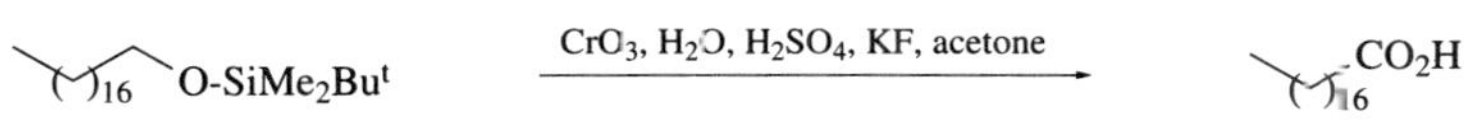

(83)

Although solid sodium permanganate monohydrate suspended in hexane effects heterogeneous oxidation of primary alcohols to carboxylic acids (Equation (84)) ⟨81TL1655⟩, a much more effective permanganate reagent is the so called "purple benzene", formed by the addition of potassium permanganate to benzene in the presence of a reagent which will assist solution of the permanganate by the benzene. Two reagents which have been used for this purpose have been tricaprylmethyl ammonium chloride ⟨74TL1511⟩ and 18-crown-6 ⟨72JA4024⟩. The resulting "purple benzene" reagent is a very convenient oxidant for various functional groups (alcohols, aldehydes, nitriles, double bonds and arenes) to carboxylic acids in good yields and under very mild conditions (Equation (85) and Table 12).

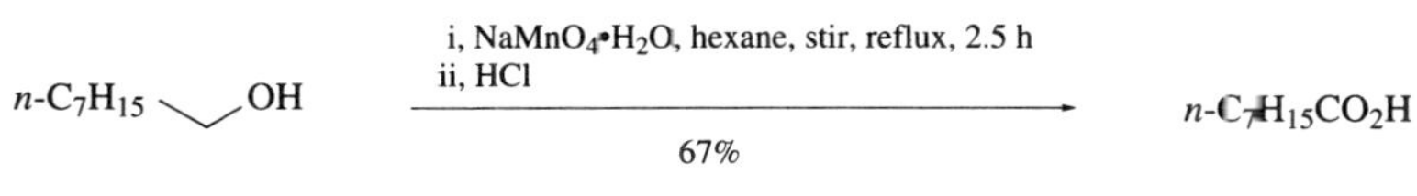

(84)

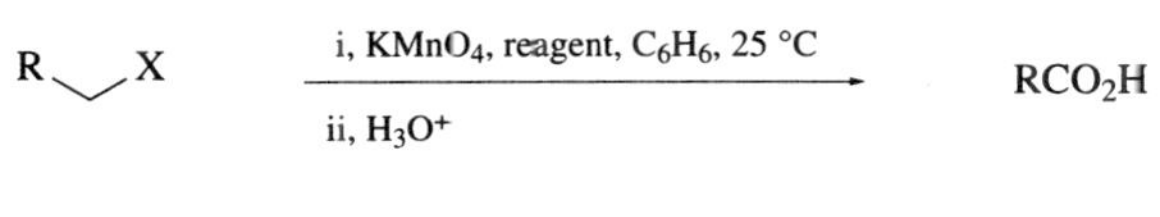

(85)

Table 12 Alkanoic acids from oxidation of various functional groups with "purple benzene".

R	*X*	*Reagent*	*Yield* (%)	*Ref.*
n-C_6H_{13}	OH	18-crown-6	70	72JA4024
n-C_7H_{15}	OH	(n-$C_9H_{19}CO)_3N^+$Me Cl^-	47	74TL1511
Ph	CN	18-crown-6	86	72JA4024
Ph	OH	(n-$C_9H_{19}CO)_3N^+$Me Cl^-	92	74TL1511
Ph	OH	18-crown-6	100	72JA4024
Ph	H	18-crown-6	78	72JA4024

The effective use of air in the presence of a platinum or platinum oxide catalyst as an oxidizing agent to effect conversion of alcohols to acids is illustrated for simple, long chain and dialcohols in Equation (86) and Table 13) ⟨60T67⟩, while similar results are obtained using nickel peroxide (Equation (87) and Table 14) ⟨62JOC1597, 64CPB403⟩. One interesting aspect of this latter reagent is whereas α-hydroxy ketones are cleaved to acid, double and triple bonds are not effected.

$$\text{alcohol or diol} \xrightarrow{\text{Pt cat., air, solvent, time, temp}} \text{mono- or diacid} \qquad (86)$$

Table 13 Oxidation of alcohols to carboxylic acids using catalyzed air.

Substrate	*Solvent*	*Catalyst*	*Time* (h)	*Temp* (°C)	*Product*	*Yield* (%)
EtOH	H_2O	Pt/C	0.5	20	$MeCO_2H$	100
Pr^nOH	H_2O	Pt/C	11	66	$EtCO_2H$	98
Bu^nOH	H_2O	Pt/C	17	80	Pr^nCO_2H	100
n-$C_6H_{13}OH$	H_2O	Pt/C	25	93	n-$C_5H_{11}CO_2H$	99
n-$C_{12}H_{25}OH$	n-heptane	PtO_2	2	60	n-$C_{11}H_{23}CO_2H$	96
$Ph(CH_2)_2OH$	H_2O	Pt/C	12	Reflux	$PhCH_2CO_2H$	90
$HO(CH_2)_2OH$	H_2O	Pt/C	6	Reflux	$HO_2C—CO_2H$	100
$HO(CH_2)_4OH$	H_2O	Pt/C	32	Reflux	$HO_2C—(CH_2)_2—CO_2H$	55
$PhCH_2OH$	H_2O	Pt/C	10	Reflux	$PhCO_2H$	97

$$\text{alcohol or diol} \xrightarrow[\text{ii, } H_3O^+]{\text{i, } NiO_2\text{, NaOH, stir, temp, time}} \text{carboxylic acid} \qquad (87)$$

Table 14 Oxidation of alcohols to carboxylic acids using nickel peroxide.

Substrate	*Temp* (°C)	*Time* (h)	*Product*	*Yield* (%)
EtOH	30	5	$MeCO_2H$	99
Pr^nOH	30	5	$EtCO_2H$	96
Bu^nOH	30	5	Pr^nCO_2H	97
Bu^iOH	30	5	Pr^iCO_2H	85
Bu^iOH	0	5	Pr^iCO_2H	40
n-$C_5H_{11}OH$	30	5	Bu^nCO_2H	97
$Me_2CH(CH_2)_2OH$	30	5	Bu^iCO_2H	78
n-$C_8H_{17}OH$	30	5	n-$C_7H_{15}CO_2H$	59
$HC{\equiv}CCH_2OH$	5	0.5	$HC{\equiv}CCO_2H$	50
$PhCH{=}CHCH_2OH$	50	6	$PhCH{=}CHCO_2H$	81
$Ph(CH_2)_3OH$	0	30	$Ph(CH_2)_2CO_2H$	71
$MeCH(OH)CH(OH)Me$	50	7	$MeCO_2H$	90
$MeCH(OH)COMe$	50	5	$MeCO_2H$	67
$Pr^nCH(OH)COPr^n$	50	5	Pr^nCO_2H	68
$MeCH(OH)CO_2H$	50	7	$MeCO_2H$	92
$MeCOCO_2H$	50	10	$MeCO_2H$	75

Several different methods used to oxidize aldehydes and ketones to carboxylic acids have been reviewed: the Canizzaro reaction ⟨44OR(2)94⟩, the Favorskii reaction ⟨50BSFD35, 60OR(11)261⟩, the use of manganese dioxide ⟨76S133⟩, the use of permanganate ⟨87S85⟩, the use of selenium dioxide ⟨49OR(5)331⟩, autooxidation of aldehydes ⟨48AG169⟩, and the production of long chain carboxylic acids starting with cyclohexane-1,3-diones ⟨55AG769⟩.

Although the haloform reaction has been mentioned previously, some examples of acid preparation from methyl ketones and secondary methyl carbinols via this method using sodium bromite ($NaOH + 3Br_2 \rightarrow NaBrO_2$) are reported in Equation (88) and Table 15 ⟨32OSC(1)526, 83CC392, 85S674⟩.

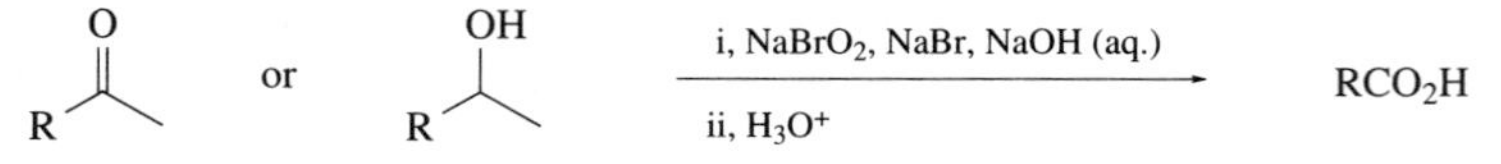

(88)

An interesting analogue of the haloform reaction is the reaction of ketones which contain both α- and α′-hydrogen atoms with a carbon tetrachloride solution of potassium hydroxide thus transforming the starting ketones into rearranged acids (Equations (89)–(91)) ⟨69JOC1454⟩.

Table 15 Alkanoic acids via the haloform reaction.

Ketone or Alcohol	*Product*	*Yield (%)*	*Ref.*
EtCOMe	$EtCO_2H$	[a]	83CC392
Pr^nCOMe	Pr^nCO_2H	[a]	83CC392
Bu^iCOMe	Bu^iCO_2H	95, 95	83CC392, 85S674
Bu^tCOMe	Bu^tCO_2H	71–74, 80	32OSC(1)526, 83CC392
n-$C_5H_{11}COMe$	n-$C_5H_{11}CO_2H$	93	83CC392
n-$C_6H_{13}CH(OH)Me$	n-$C_6H_{13}CO_2H$	40	83CC392

[a] Water-soluble product; yield not determined.

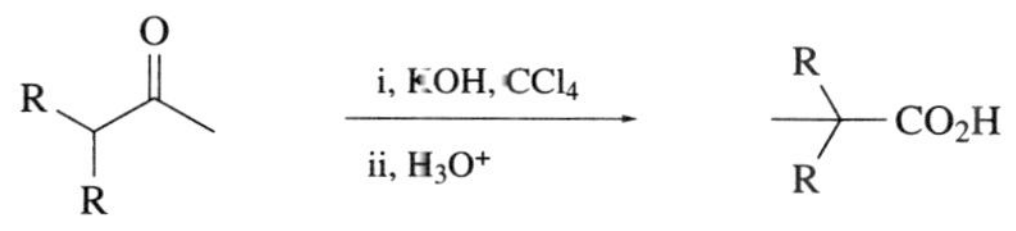

(89)

R	*Yield (%)*
Ph	70
Me	70

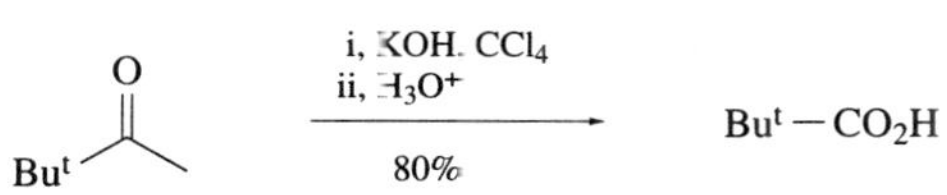

(90)

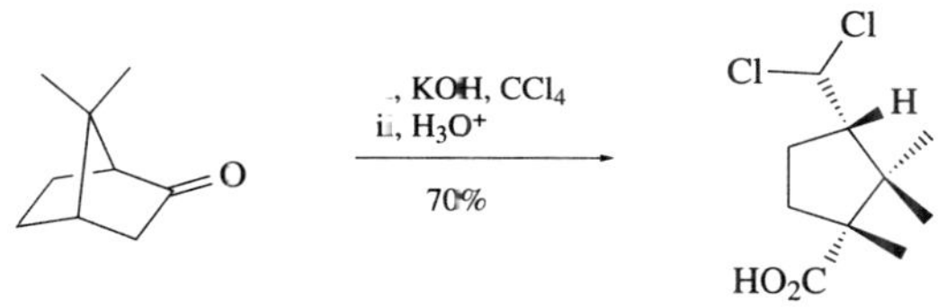

(91)

Solid sodium permanganate monohydrate suspended in hexane oxidizes aldehydes as well as alcohols to their corresponding carboxylic acids ⟨81TL1655⟩, as does potassium permanganate ⟨43OSC(2)315⟩.

Cerium(IV) ion in perchloric acid ⟨55TFS(51)1105⟩ or sulfuric acid ⟨50JCS3425, 55TFS1105⟩ oxidizes formaldehyde and acetaldehyde to formic acid, while methyl ethyl ketone, diethyl ketone, and methyl isopropyl ketone are all converted into acetic acid.

Periodate oxidation of α-hydroxy ketones (Scheme 21) ⟨36JCS1788⟩ and 1,3-diketones (Equation (92) and Table 16) ⟨56JA2489⟩ has been used to produce mono- and dicarboxylic acids. The study ⟨60JCS1593⟩ of the mechanism of periodate oxidation of acetoin (3-hydroxy-3-methylbutan-2-one) using ^{18}O as a tracer in ^{18}O-labeled periodate produced acetone and acetic acid, where the oxygen of the acetone came from the hydroxy group of the hydroxy ketone starting material and the additional oxygen atom of the acetic acid came from the periodate (Scheme 22) ⟨67QR3⟩. Periodate ions have also been reportedly used as a cooxidant with potassium permanganate (Lemieux–von Rudloff reagent (Scheme 23) ⟨76CB41⟩, ruthenium dioxide (Equation (93)) ⟨69JOC112, 69JOC116⟩, and ruthenium(III) chloride (Equation (94)) ⟨89OS(68)41⟩. In the latter reaction the ruthenium(III) chloride catalyzes the oxidation process for the cleavage of the β-chloro enol acetate intermediate which is formed by trapping the β-chloro enolate intermediate with acetic anhydride.

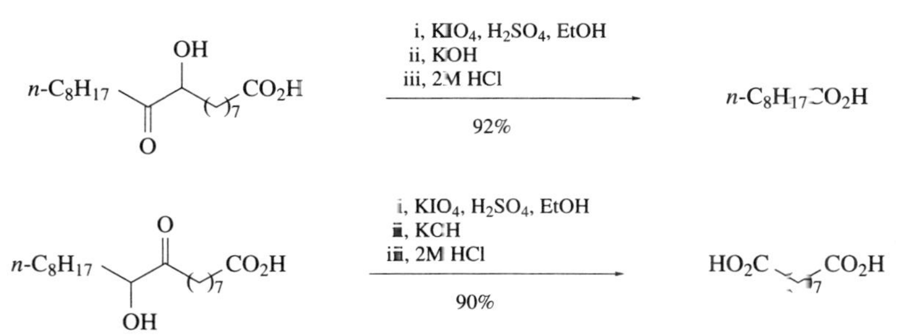

Scheme 21

$$\text{1,3-diketone} \xrightarrow[\text{iii, } H_3O^+]{\text{i, } IO_4^-,\ H_2SO_4;\ \text{ii, KOH}} \text{diacid} \qquad (92)$$

Table 16 Carboxylic acids via periodate oxidation of α-hydroxy- and 1,3-diketones.

1,3-Diketone	*Product*	*Yield (%)*
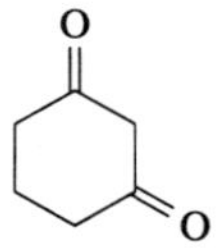		87
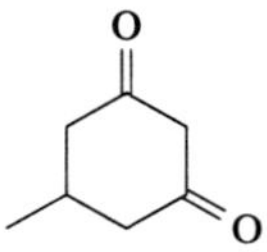	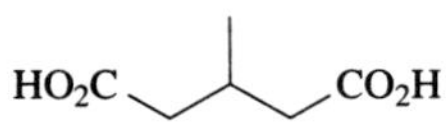	90
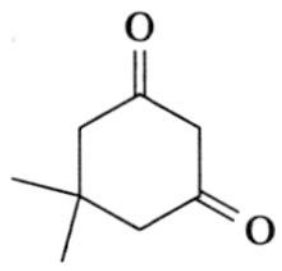		92
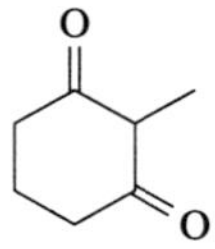	$MeCO_2H$ + 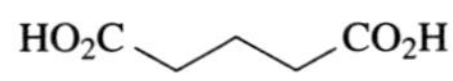	94 59
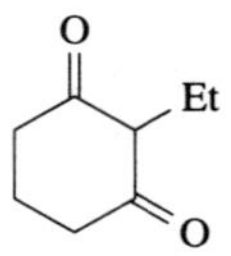	$EtCO_2H$ + 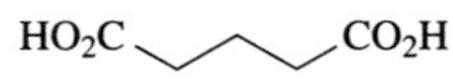	90 68
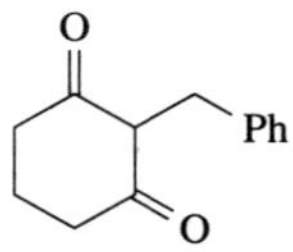	$PhCH_2CO_2H$ + 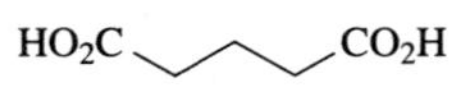	85 73
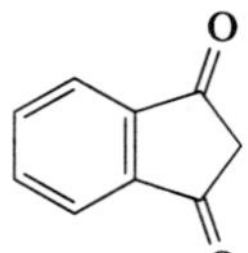	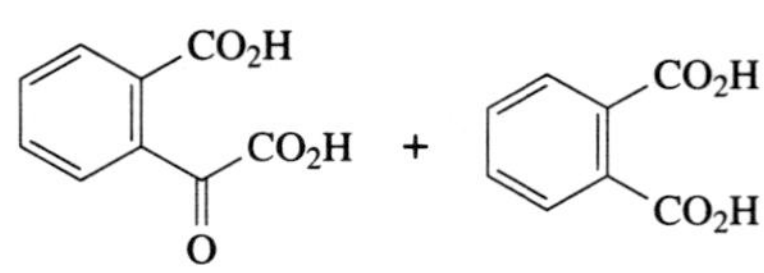	18 + 18

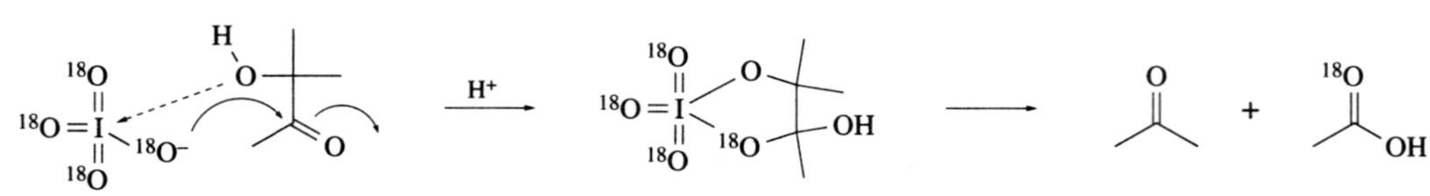

Scheme 22

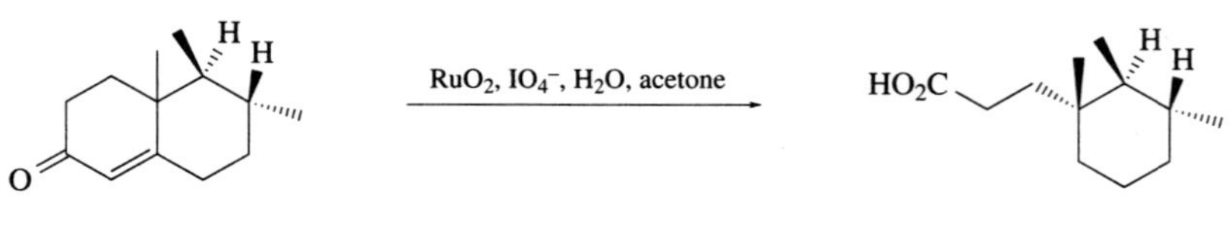

(93)

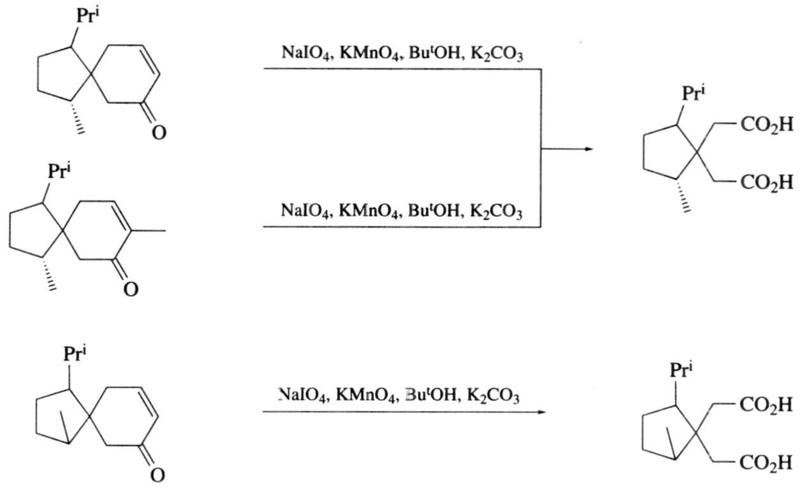

Scheme 23

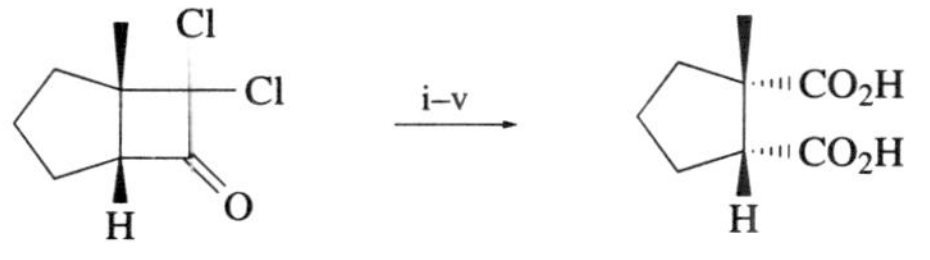

(94)

i, Bu^{n}Li, hexane, THF, N_2, stir 15 min; ii, $(MeCO)_2O$, RT, stir 1 h; iii, MeCN, CCl_4, H_2O, $NaIO_4$, $RuCl_3$, stir 5 h; iv, 10% NaOH (aq.); v, 2M HCl

Jones' reagent ($CrO_3 + H_2SO_4 + H_2O$) has been used to oxidize 2-adamantanecarbaldehyde to 2-adamantane carboxylic acid ⟨72SC79⟩, and to produce an 80% yield of callitrisic acid from callitrisaldehyde (Equation (95)) ⟨71CC10⟩. This latter reaction can also be accomplished using silver(I) oxide in methanol (Equation (95)) ⟨71CC10⟩.

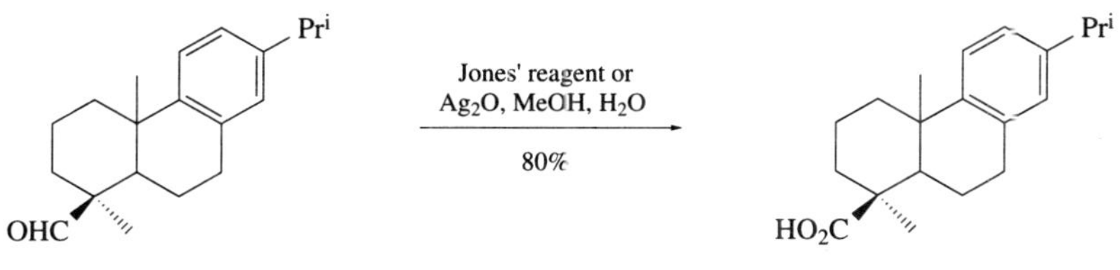

(95)

Silver oxide has also been used to oxidize decanal ^{13}C aldehyde to decanoic ^{13}C acid ⟨84JOC174⟩, and dodecanol and cyclohexa-3,4-diene-1-al to their corresponding acids (Equation (96) and Table 17) ⟨68JA5616⟩, while in this same report silver(II) oxide oxidizes a variety of aldehydes, unsaturated aldehydes and α,β-unsaturated cyanohydrins to their corresponding acids with no *cis*, *trans* isomerization of the alkenic linkage (Equation (96)).

$$\text{aldehyde or cyanohydrin} \xrightarrow{Ag_2O,\ NaCN,\ MeOH,\ 25\ ^\circ C,\ time} \text{carboxylic acid} \qquad (96)$$

Treatment of 3,4-dimethoxyphenylpyruvic acid with hydrogen peroxide causes elimination of carbon monoxide and the production of homoveratric acid (Equation (97)) ⟨43OSC(2)333⟩. In combination with 2 mol. % of selenium dioxide, hydrogen peroxide has also been effective in causing oxidative ring-contractions of cyclic ketones producing cyclic carboxylic acid (Scheme 24) ⟨57JOC1680⟩, while in combination with sodium hydroxide, 30% hydrogen peroxide converts α,β-unsaturated carbonyl compounds into mono- and dicarboxylic acids (Equation (98) and Table 18) ⟨70JOC1275⟩.

Table 17 Alkanoic acids via silver oxide oxidation of aldehydes.

Substrate	*Time* (h)	*Product*	*Yield* (%)
n-$C_{11}H_{23}CHO$	14[a]	n-$C_{11}H_{23}CO_2H$	90
PhCHO	6	$PhCO_2H$	
PhCH=CHCHO (*cis* or *trans*)	2	$PhCH{=}CHCO_2H$ (*cis* or *trans*)	90
PhCH=CHCH(OH)CN	3	$PhCH{=}CHCO_2H$	
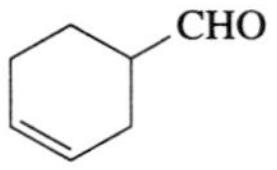	14[a]	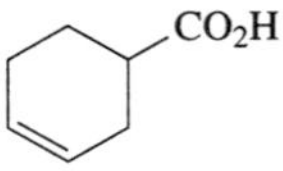	97

[a] Reaction conditions used were Ag_2O in $THF–H_2O$ (9:1)

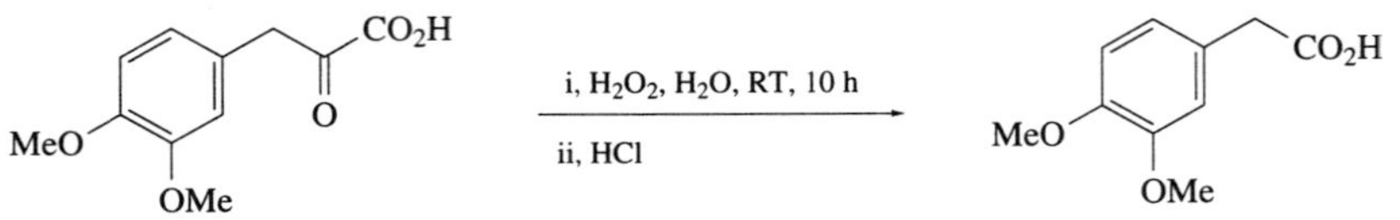

(97)

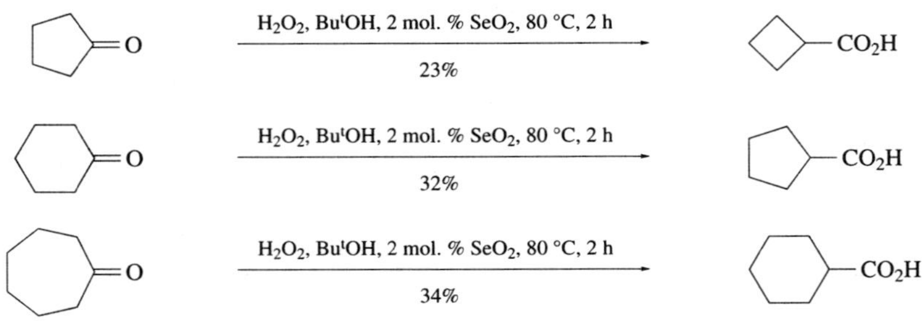

Scheme 24

$$\alpha,\beta\text{-unsaturated ketone} \xrightarrow[\text{ii, } H_3O^+]{\text{i, 30\% } H_2O_2\text{, NaOH, heat}} \text{carboxylic acid} \qquad (98)$$

Hexanoic acid can be prepared from hexanal in 79–96% yield by oxidation with *m*-chloroperbenzoic acid in THF ⟨67JA291⟩.

The oxidation of double or triple bonds to produce carboxylic acids has been accomplished using a variety of reagents, and at least one review of the oxidation of double bonds has been published ⟨49CRV385⟩. Among the reagents used to oxidize double bonds are two permanganate reagents, sodium permanganate in hexane, which has been used to produce dodecanoic acid from 1-tridecene in 13% yield ⟨81TL1655⟩, and potassium permanganate, which in base has been used to prepare azelaric acid from ricinolenic acid in 32–36% yield ⟨43OSC(2)53⟩, and which in acetic acid has been used to oxidatively cleave both double and triple bonds producing carboxylic acids (Equation (99) and Table 19) ⟨77JOC3749⟩.

$$\text{alkene or alkyne} \xrightarrow{KMnO_4\text{, HOAc, } H_2O} \text{carboxylic acid} \qquad (99)$$

"Purple benzene", previously discussed has been used to convert 1-octene into heptanoic acid in 81% yield ⟨74TL1511⟩ and cyclohexene to adipic acid in 100% yield ⟨72JA4024⟩.

Ruthenium tetroxide has been used to produce a monoester dicarboxylic acid from an unsaturated keto diester (Equation (100)) ⟨68TL2681⟩, while Sharpless and co-workers ⟨81JOC3936⟩ have reported an improved, general method for cleaving simple alkenes to carboxylic acids with a catalytic amount of ruthenium tetroxide in the presence of acetonitrile and a stoichiometric amount of sodium periodate. This method has been used to cleave oxidatively cyclic allylic alcohols and α,β-unsaturated

Table 18 Mono- and dicarboxylic acids from α,β-unsaturated carbonyl compounds using sodium hydroxide/hydrogen peroxide.

Substrate	*Product*	*Yield* (%)
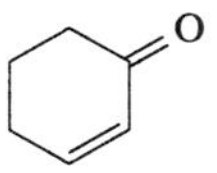	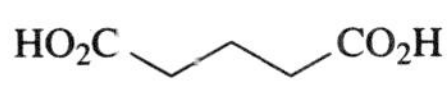	72
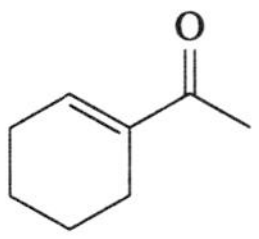	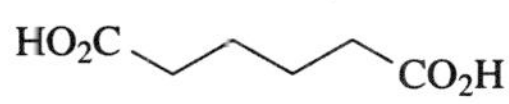	67
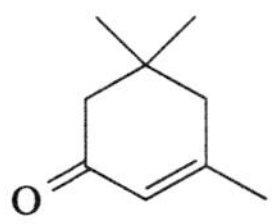		84
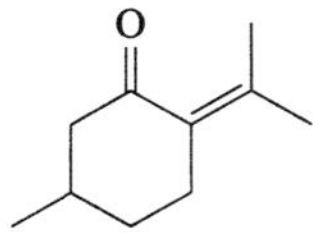	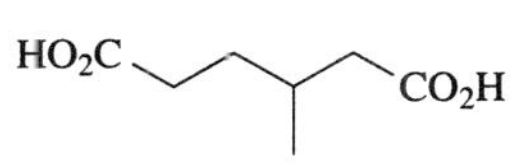	60
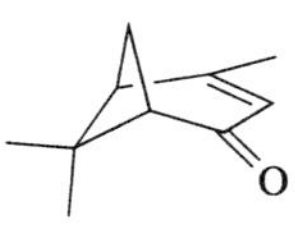	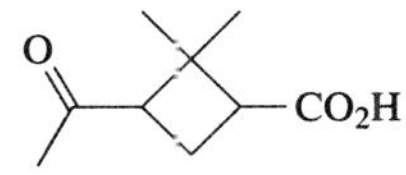 *cis* : *trans* = 1 : 1	85

Table 19 Alkanoic acids via oxidative cleavage of double or triple bonds using permanganate in acetic acid.

Substrate	*Conditions*	*Product*	*Yield* (%)
n-$C_6H_{13}CH{=}CH_2$	Stir 3 h	n-$C_6H_{13}CO_2H$	80
n-$C_8H_{17}CH{=}CH_2$	C_6H_6, PTC[a], 4 h	n-$C_8H_{17}CO_2H$	85
n-$C_9H_{19}CH{=}CH_2$	Stir overnight	n-$C_9H_{19}CO_2H$	84
n-$C_{10}H_{21}CH{=}CH_2$	Stir 4 h	n-$C_{10}H_{21}CO_2H$	90
n-$C_{12}H_{25}CH{=}CH_2$	Stir 3 h	n-$C_{12}H_{25}CO_2H$	83
n-$C_{14}H_{29}CH{=}CH_2$	Stir overnight	n-$C_{14}H_{29}CO_2H$	84
n-$C_{16}H_{33}CH{=}CH_2$	C_6H_6, PTC[a], 6 h	n-$C_{16}H_{33}CO_2H$	80
n-$C_{18}H_{37}CH{=}CH_2$	Aliquat 336, 3 h	n-$C_{18}H_{37}CO_2H$	86
n-$C_4H_9C{\equiv}CH$	Pentane, Aliquat 336, 3 h	n-$C_4H_9CO_2H$	66
n-$C_6H_{13}C{\equiv}CH$	Pentane, Aliquat 336, 5 h	n-$C_6H_{13}CO_2H$	70
n-$C_8H_{17}C{\equiv}CH$	Pentane, Aliquat 336, 3 h	n-$C_8H_{17}CO_2H$	70

[a] PTC = benzylhexadecyldimethylammonium chloride.

ketones producing carboxylic acids where one or more carbons have been excised from the starting material (Equation (101) and Table 20 ⟨87JOC689⟩.

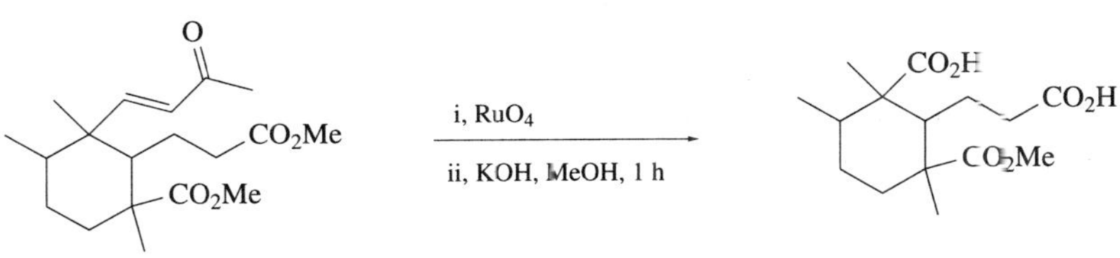

(100)

$$\text{alkenes, cyclic allylic alcohols or cyclic } \alpha,\beta\text{-unsaturated ketones} \xrightarrow{RuO_4,\ NaIO_4,\ MeCN,\ RT,\ stir} \text{carboxylic acids} \quad (101)$$

Table 20 Alkanoic acids via oxidative cleavage of cyclic allylic alcohols and α,β-unsaturated ketones using ruthenium tetroxide in acetonitrile.

Substrate	*Product*	*Yield (%)*
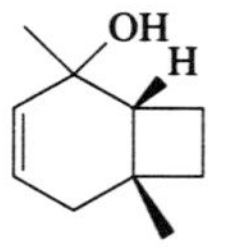	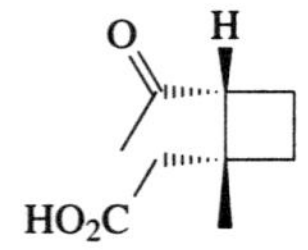	60
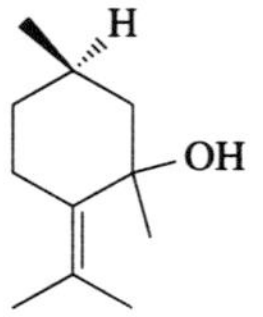	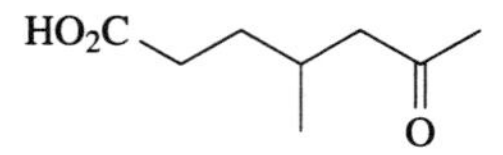	66
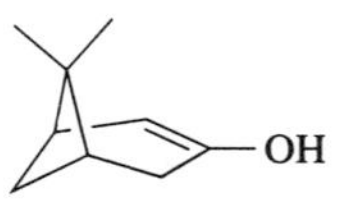	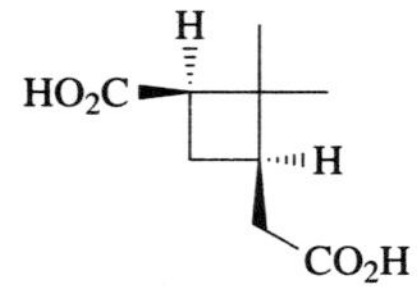	76
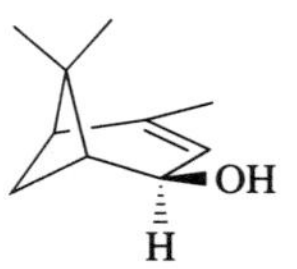	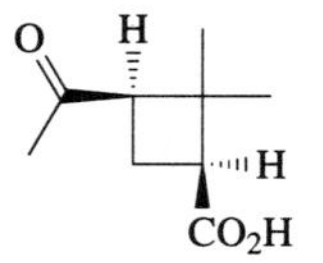	78
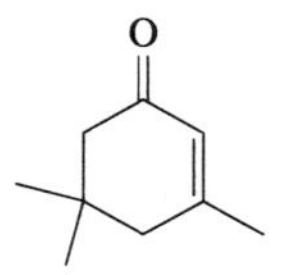	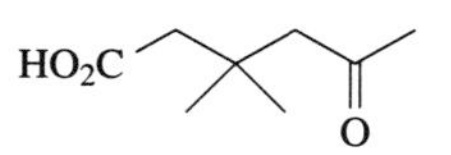	86
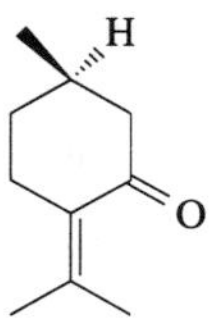	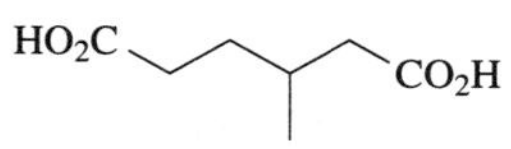	92

Oxygen in the presence of copper(I) chloride has been reported to effectively cleave the aromatic double bond in 1,2-dihydroxybenzene and produce *cis*, *cis*-monomethyl muconate in good yields (Equation (102)) ⟨87OS(66)180⟩.

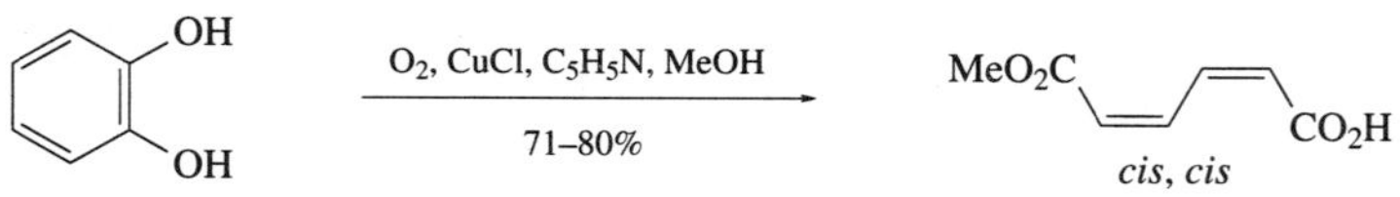

(102)

The conversion of aldehydes to aldoximes and their subsequent treatment with base in diethylene glycol (DEG) or 2 methoxyethanol (MOE) at 120–190 °C affords carboxylic acids in good to excellent yields (Scheme 25) ⟨62JOC629⟩.

5.02.2.2 Haloalkanoic Acids

The production and use of aliphatic carboxylic acids containing fluorine has been reviewed ⟨76CZ3⟩.

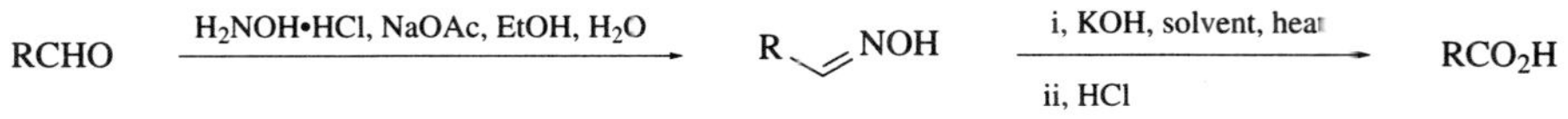

R	*Solvent*	*Time* (h)	*Temp* (°C)	*Yield* (%)
Me	DEG	3	190	63
Pr^i	DEG	3	170	88
Pr^i	MOE	8	120	89
Bu^t	DEG	3	170	38
$PhCH_2$	DEG	3	190	80
PhCH=CH (*anti*)	DEG	3	190	68

Scheme 25

5.02.2.2.1 By hydrolysis reactions

Acid hydrolysis of esters has been used to produce ω-fluoro substituted carboxylic acids. Thus, sulfuric acid hydrolyzes methyl 5-fluoropentanoate, 6-fluorohexanoate and 7-fluoroheptanoate to their corresponding ω-fluoro substituted carboxylic acids (Equation (103)) ⟨56JA2255⟩.

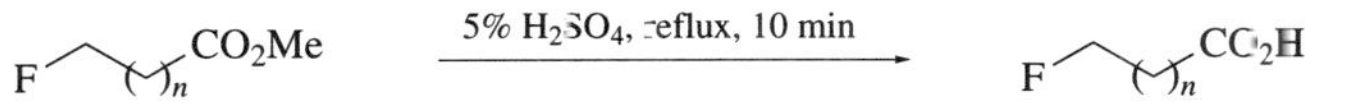

(103)

n	*Yield* (%)
3	77
4	92
5	

α-Chlorophenylacetic acid is obtained via acid hydrolysis of ethyl α-chlorophenylacetate (Equation (104)) ⟨63OSC(4)169⟩, using hydrochloric acid in acetic acid, and 2,3-dibromopropanoic acid is obtained by hydrobromic acid hydrolysis of methyl 2,3-dibromopropanoate (Equation (105)) ⟨40JA3495⟩.

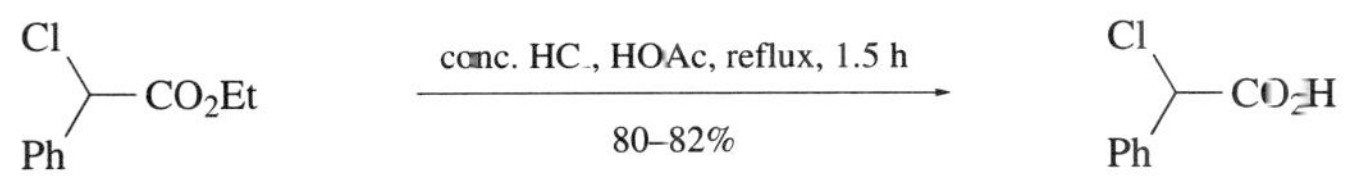

(104)

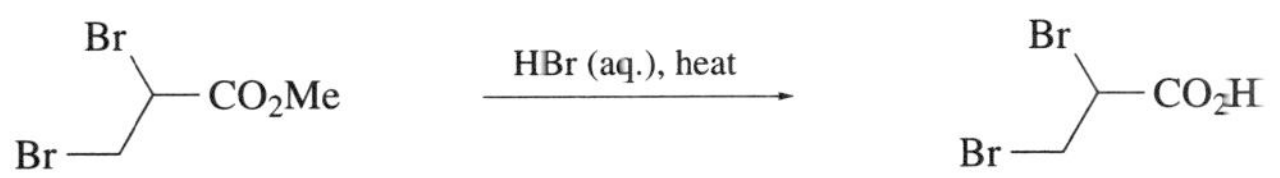

(105)

Two interesting base-catalyzed ester hydrolysis reactions which produce haloalkanoic acids are the ethanolic potassium hydroxide hydrolysis of diethyl 2-chloro-2-cyclopropyl-1,1-dicarboxylate which leads to the monoester product retaining the chloro substituent as well as the double bond (Equation (106)) ⟨87OS(66)173⟩, and the enzyme catalysis of racemic and enantiomerically pure (*S*)-ethyl 2-fluorohexanoate (Scheme 26) ⟨90OS(69)10⟩. This latter reaction is interesting because it demonstrates the kinetic resolution of racemic esters using lipase P-30 Amano, a relatively inexpensive reagent producing a rapid hydrolysis reaction which affords products with high enantiomeric excesses. Other biotransformations leading to carboxylic acids are discussed in Section 5.02.5.

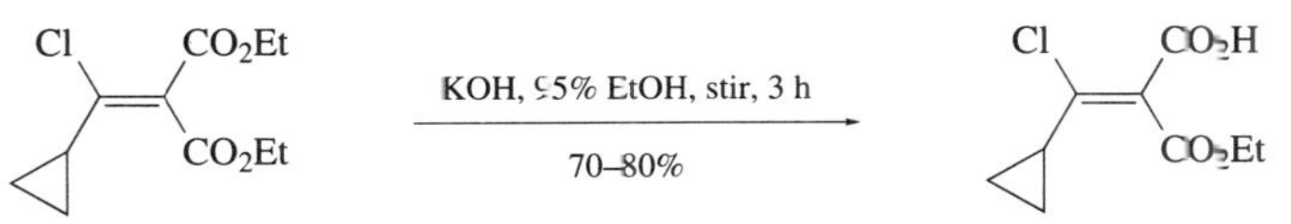

(106)

Hydrolysis of nitriles has been used to prepare C_5, C_6, and C_7 ω-fluoro substituted acids (Equation (107)) ⟨56JA2255⟩, α- (Equation (108)) ⟨62PCS117⟩, and β-chloro carboxylic acids (Equation (109)) ⟨56JA2255⟩, and α-chloro adipic acid (Equation (110)) ⟨62T69⟩. Preparation of the mixed halo acid

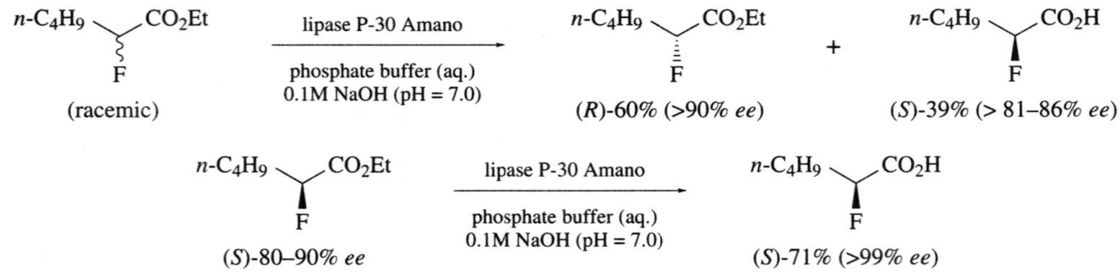

Scheme 26

3-chloro-2,2,3-trifluoropropionic acid is accomplished by *in situ* hydrolysis of the intermediate nitrile formed by condensation of chlorotrifluoroethylene with sodium cyanide (Scheme 27) ⟨73OSC(5)239⟩, while acid hydrolysis of ethylene cyanohydrin produces β-bromopropionic acid in good yields (Equation (111)) ⟨32OSC(1)131⟩. Acid hydrolysis of 2-substituted 1,1-dicyanoethylene oxide with hydrobromic acid produces good yields of substituted α-bromoacetic acids (Scheme 28) ⟨86T2275⟩.

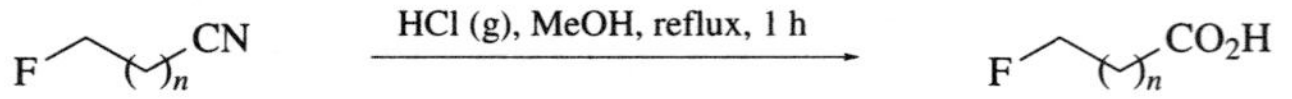 (107)

n	*Yield* (%)
3	83.5
4	68
5	

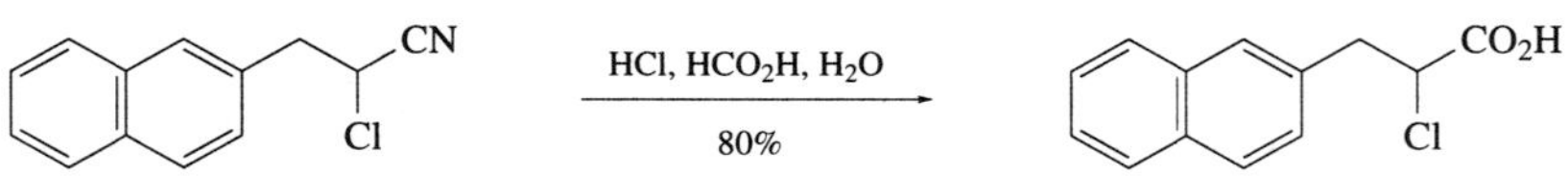 (108)

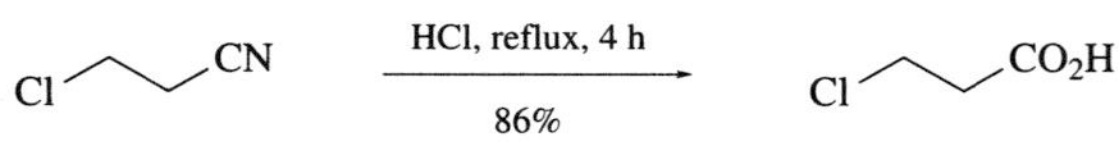 (109)

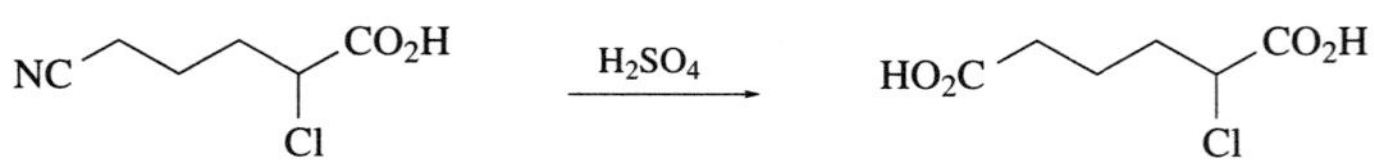 (110)

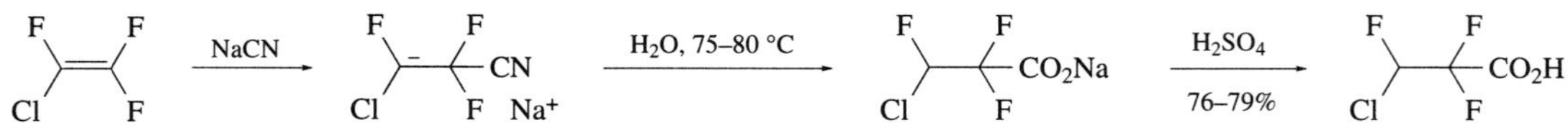

Scheme 27

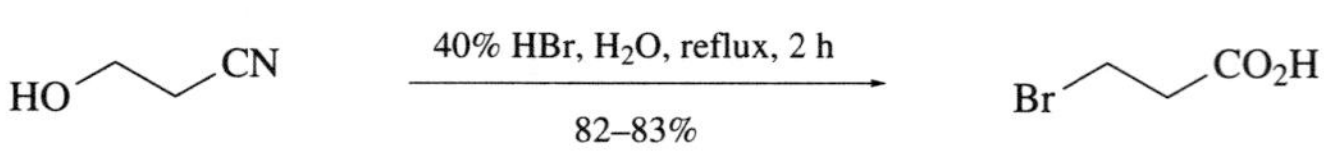 (111)

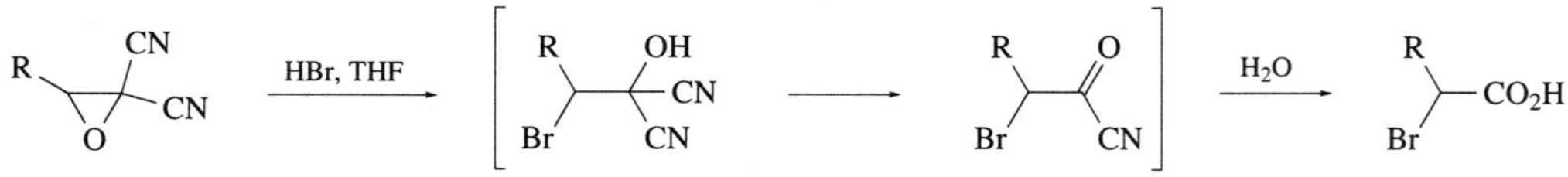

R	*Yield* (%)
Ph	70
p-ClC$_6$H$_4$	70

Scheme 28

Three examples of halo alkanoic acid preparation by hydrolysis of acid halides are worth reporting. In the first example various α,α,γ-trichloro acid chlorides are hydrolyzed at room temperature to afford excellent yields of the corresponding α,α,γ-trichloroalkanoic acids (Equation (112)) ⟨82CL1255⟩, while in the second example, bromination of ε-benzoylaminocaproic acid with bromine and phosphorus tribromide affords the intermediate α-bromo acid bromide which, upon hydrolysis with water, produces ε-benzoylamino-α-bromocaproic acid (Scheme 29) ⟨43CSC(2)74⟩. In the final example α-fluoro-α,β-unsaturated acids are formed upon hydrolysis of their precursor acid fluorides (Equation (113)) ⟨79MI 502-06⟩.

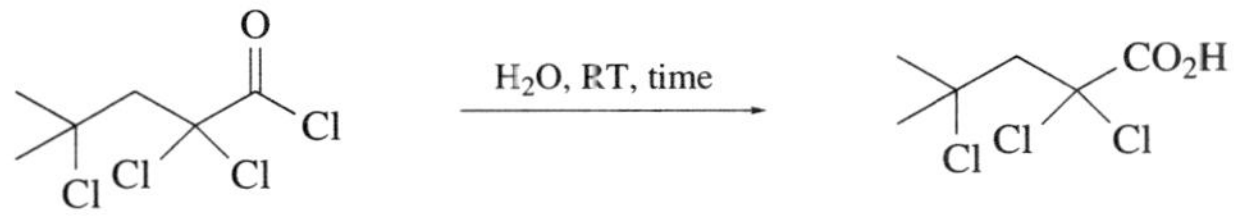

(112)

Products	*Time* (min)	*Yield* (%)
n-$C_6H_{13}CHClCH_2CCl_2CO_2H$	1	99
n-$C_8H_{17}CHClCH_2CCl_2CO_2H$	1	99
$PhCHClCH_2CCl_2CO_2H$	1	97
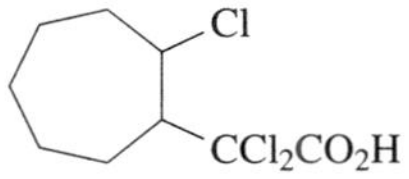	120	98

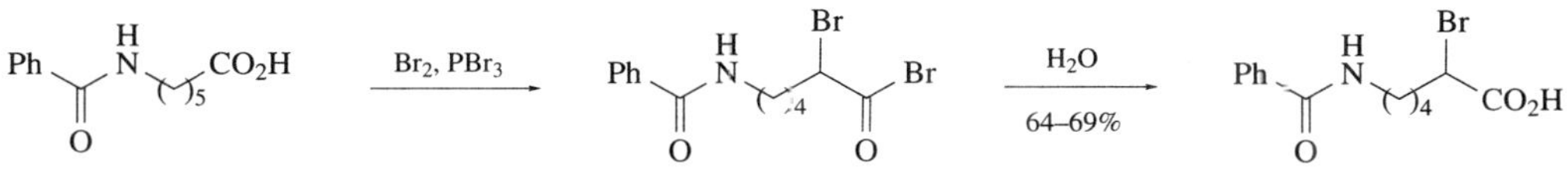

Scheme 29

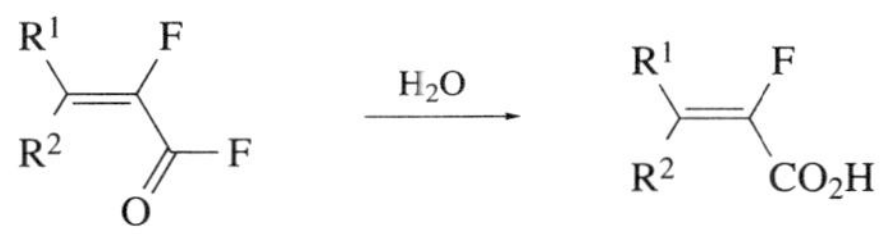

(113)

R^1 = Me, Ph, Ph; R^2 = Me, H, Me

Preparation of α-chloroglutaric acid by hydrolysis of 5,5-dichloro-4-pentenoic acid can be achieved if chlorine gas is bubbled through the reaction mixture (Equation (114)) ⟨63BCJ1264⟩. Whether hydrolysis of the vinyl dichloride occurs first followed by addition of chlorine, or addition of chlorine to the multiple bond precedes hydrolysis has not been established.

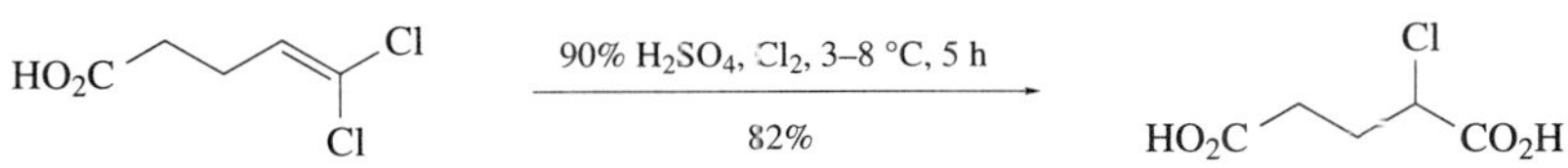

(114)

Some miscellaneous hydrolysis reactions which have been used to produce halo alkanoic acids include: the preparation of γ-bromobutyric acid by hydrobromic acid cleavage of γ-phenoxybutyric acid (Equation (115)) ⟨29JA260⟩; reaction of bromine with silver methyl azelate followed by acid hydrolysis of the resulting bromo ester to form 8-bromooctanoic acid (Scheme 30) ⟨56JA2255⟩; and, reaction of phosphorus pentachloride with a series of α,ω-dicyanoalkanes followed by acid hydrolysis of the α,α′-bis(trichlorophosphazo)-α,α,β,β,α′,α′,β′,β′-octachloroalkane intermediates to produce the corresponding α,α,β,β,α′,α′,β′,β′-octachloroalkanoic diacids (Scheme 31) ⟨73S545⟩.

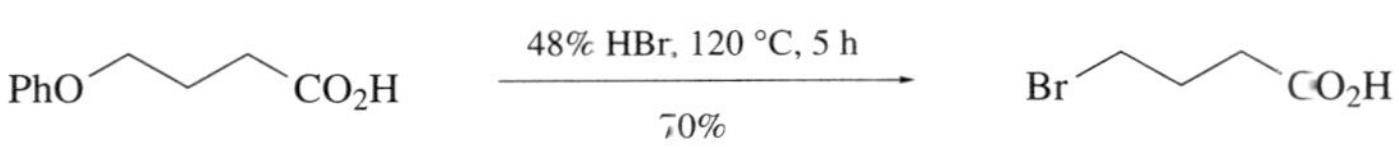

(115)

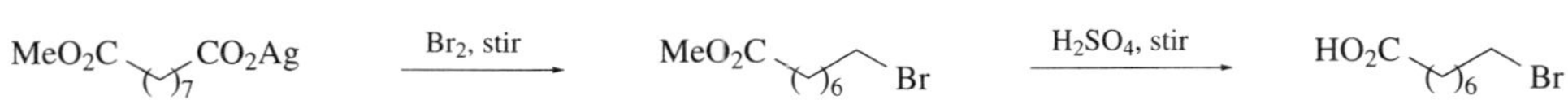

Scheme 30

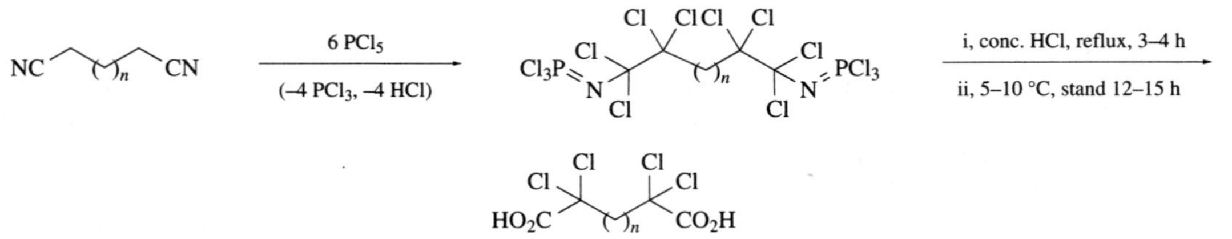

n	*Yield* (%)
1	25
2	56
3	82
5	68
6	89

Scheme 31

5.02.2.2.2 By carbonation of organometallic reagents

Treatment of perfluoroalkyl iodides with carbon dioxide in the presence of metal couples comprised of zinc and either copper, lead, cadmium or mercury produces the corresponding perfluorocarboxylic acids in fair yields (Equation (116) and Table 21) ⟨76CC885⟩. α,β-Unsaturated fluorine containing acids with either one or two fluorine atoms in the β-position can be prepared from the same lithium salt via the reaction sequence shown in Scheme 32 ⟨82S297⟩, while 5-aryl-2,3,4,5-tetrafluoro-2,4-pentadienoic acids can be prepared from a mixture of *trans*, *trans* : *trans*, *cis* : *cis* : *trans* 4-aryl-1-chloro-1,2,3,4-tetrafluoro-1,3-butadiene in a ratio of 1 : 1 : 1 by carbonation of the lithium salt formed using *n*-butyllithium (Equation (117)) ⟨85ZOR769⟩.

$$R_FI + Zn\text{–}Cu + CO_2\ (g,\ 40\ ml/min) \xrightarrow[\text{ii, HCl (aq.)}]{\text{i, DMSO}} R_FCO_2H \qquad (116)$$

Table 21 Perfluorocarboxylic acids from perfluoroalkyl iodides using carbon dioxide/metal couples.

Perfluoroalkyl iodide[a]	*Time* (h)	*Temp.* (°C)	*Yield acid*[b] (%)
C_4F_9I	3	25	40
C_4F_9I	1.5	30	42
$C_6F_{13}I$	3	25	45
$C_6F_{13}I$	1.5	30	63
$C_8F_{17}I$	3	25	47

[a] Molar ratio of metal couple to iodide = 1.5 : 1. [b] Based on the perfluoroalkyl iodide consumed.

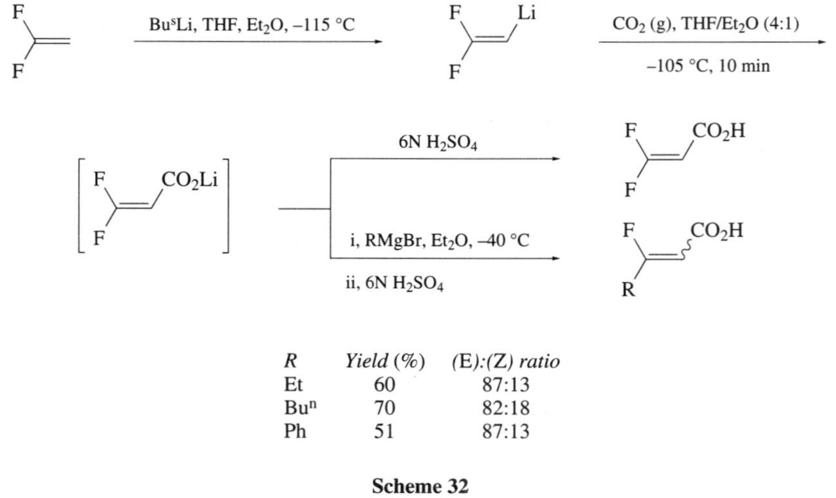

R	*Yield* (%)	(E):(Z) *ratio*
Et	60	87:13
Bu^n	70	82:18
Ph	51	87:13

Scheme 32

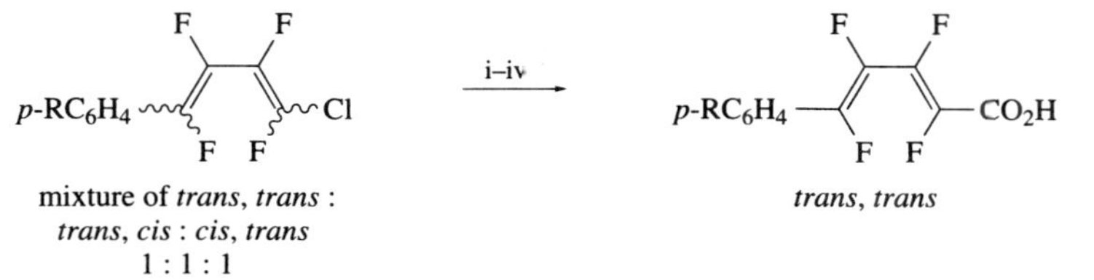

(117)

i, Bu^nLi, THF/hexane (1:1), –95 °C, stir, 30 min; ii, CO_2 (s), heat to 25 °C; iii, NaOH (aq.); iv, HCl (aq.)

R	*Yield* (%)
Me	25
MeO	23
CF_3	20
Me_2N	51

Carbonation of the lithium acetylide formed from 3-chloropropyne with *n*-butyllithium produces an 85% yield of chlorotetrolic acid (Scheme 33) ⟨66CR(C)761, 69BSF911⟩.

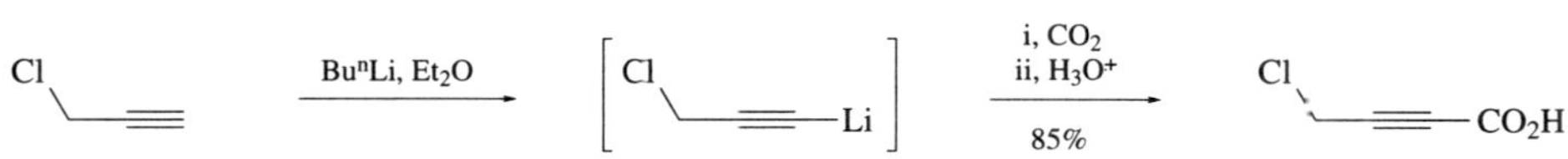

Scheme 33

A 72% yield of heptadecafluorooctanoic acid has been reportedly produced from heptadecafluoroactyl iodide by reaction with carbon dioxide catalyzed by zinc metal in combination with ultrasound ⟨84EUP115085⟩.

5.02.2.2.3 By oxidation reactions

Preparation of haloalkanoic acids by oxidation of alcohols has been accomplished using nitric acid, which oxidizes trimethylene chlorohydrin to β-chloropropionic acid in 79% yield ⟨32OSC(1)168⟩ and 6-bromo-1-hexanol to 6-bromohexanoic acid in 91% yield ⟨50JA5137⟩. Via a hydrolysis–oxidation reaction sequence in the presence of acidic acid, nitric acid converts 2-fluoroheptyl acetate, prepared by condensation of 1-bromo-2-fluoroheptane with sodium acetate, into 2-fluoroheptanoic acid (Scheme 34) ⟨73OSC(5)580⟩. Potassium permanganate oxidizes 1,1,1-trichloro-5-pentanol to 5,5,5-trichlorovaleric acid in 76–94% yield ⟨63BCJ1264⟩, and chromium trioxide or potassium dichromate which under a variety of conditions produce chloro, bromo and fluoro substituted acids from their alcohol precursors (Equation (118) and Table 22) ⟨56JA2255⟩.

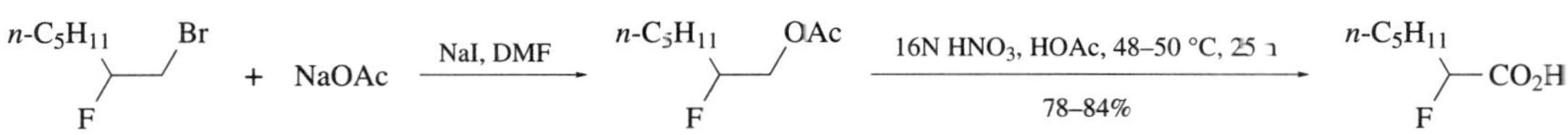

Scheme 34

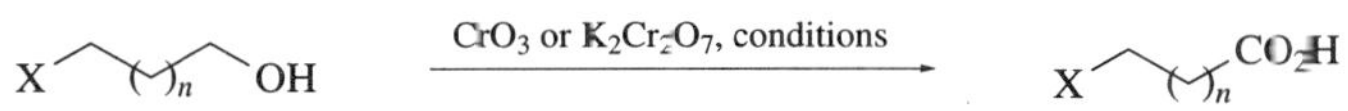

(118)

One interesting reaction which appears to bridge the oxidation of alcohols and the oxidation of aldehydes categories is the preparation of dichloroacetic acid from chloral hydrate using calcium carbonate in the presence of sodium cyanide (Scheme 35) ⟨43OSC(2)181⟩, while a true aldehyde oxidation is illustrated by the preparation of β-chloropropionic acid from β-chloropropionaldehyde in 60–65% yield using fuming nitric acid at 30–35 °C ⟨32OSC(1)166⟩.

Whereas unsaturated fluorine substituted carboxylic acids containing a terminal double bond can be prepared in 48–85% yields by reaction of polyfluorocycloalkenes with aqueous potassium hydroxide in the presence of *t*-butanol (Equation (119)) ⟨78DOK(239)330⟩, reaction of 1-chloro-2,3,3-trifluorocyclobutene with basic potassium permanganate produces the saturated 2,2-difluorosuccinic acid (Equation (120)) ⟨73OSC(5)393⟩.

Table 22 Halo-substituted carboxylic acids via oxidation of their alcohol precursors using chromium trioxide or potassium dichromate.

X	n	*Conditions*	*Yield* (%)
F	1	$K_2Cr_2O_7$, H_2SO_4, H_2O, 25 °C	74
F	2	$K_2Cr_2O_7$, H_2SO_4, H_2O, 25 °C	68
Cl	2	$K_2Cr_2O_7$, H_2SO_4, H_2O, 25 °C	56
Cl	3	i, CrO_3, HOAc, H_2O, ⟨5 °C ii, RT, 24 h	92
Cl	4	i, CrO_3, HOAc, H_2O, ⟨5 °C ii, RT, 24 h	72
F	6	i, CrO_3, HOAc, H_2O, ⟨5 °C, stir 2 h ii, RT, 24 h	78
F	7	i, CrO_3, HOAc, H_2O, ⟨5 °C, stir 2 h ii, RT, 24 h	87
F	8	i, CrO_3, HOAc, H_2O, ⟨5 °C, stir 2 h ii, RT, 24 h	93
Cl	8	i, CrO_3, HOAc, H_2O, ⟨5 °C, stir 2 h ii, RT, 24 h	51
Br	8	i, CrO_3, HOAc, H_2O, ⟨5 °C, stir 2 h ii, RT, 24 h	31
F	9	i, CrO_3, HOAc, H_2O, ⟨5 °C, stir 2 h ii, RT, 24 h	61
F	16	CrO_3, 90% HOAc, stand RT 24 h	97

Scheme 35

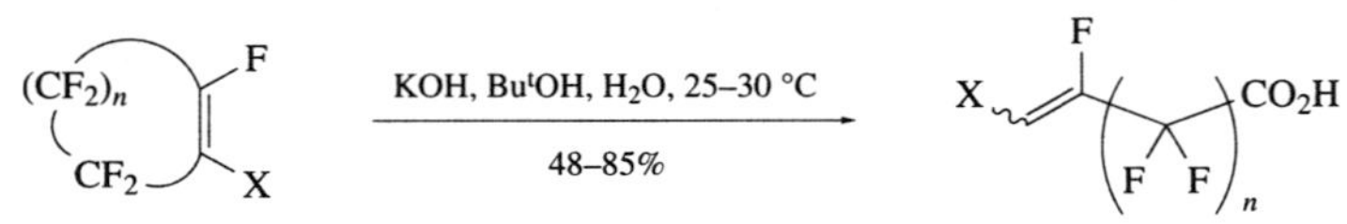

(119)

n	*X*
1–3	F
3	Cl

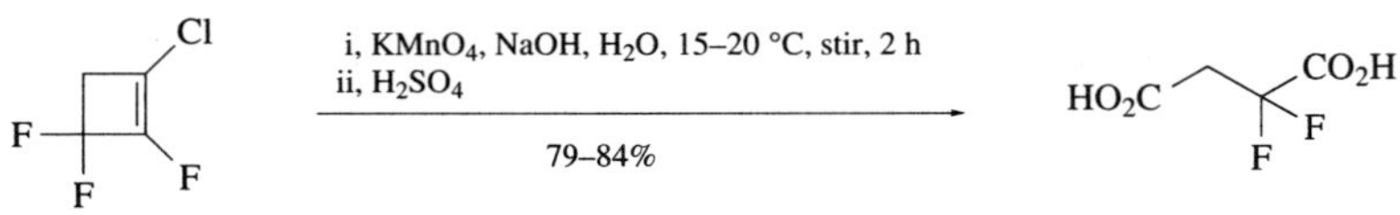

(120)

5.02.2.2.4 *By miscellaneous reactions*

One approach to the preparation of haloalkanoic acids is the halogenation of molecules already containing the carboxylic acid function and reactions utilizing this approach include: bromine addition to the double bond of fumaric acid producing α,*β*-dibromosuccinic acid in 72–84% yield ⟨43OSC(2)177⟩; α-bromination of *n*-caproic acid in the presence of phosphorus trichloride producing an 83–89% yield of α-bromocaproic acid ⟨32OSC(1)115⟩; the reaction of pyruvic acid with sulfuryl chloride producing a 96–98% yield of chloropyruvic acid ⟨73OSC(5)635⟩; and, the preparation of (*S*)-2-chloroalkanoic acids from their (*S*)-2-aminoalkanoic acid precursors via diazotization (Equation (121)) ⟨87OS(66)151⟩. α-Chlorocarboxylic acids have also been synthesized by chlorination of compounds containing a dichlorovinyl group if the reaction is run in acid medium (Equation (122) and Table 23) ⟨62T69⟩.

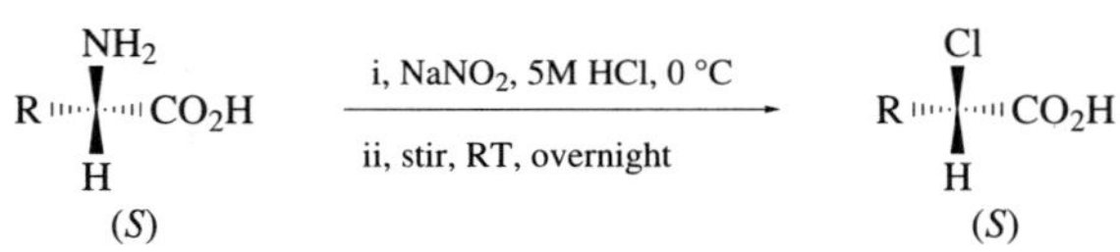

(121)

R	*Yield* (%)	ee (%)
Me	64	95.6
Pr^i	62	97.7
Bu^i	58	95.8
Bu^s	59	98.3

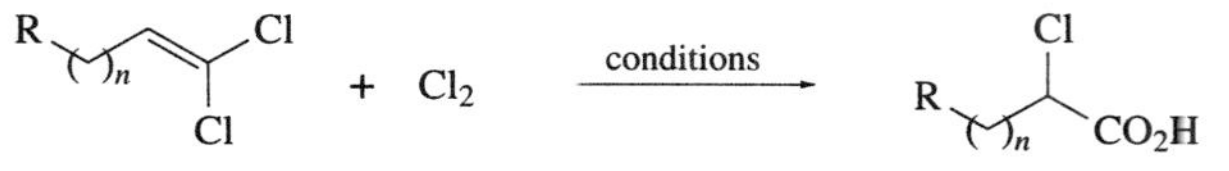

(122)

Table 23 Acid-catalyzed chlorination of a dichlorovinyl substituent producing the corresponding α-chlorocarboxylic acid.

R	n	*Conditions*	*Yield* (%)
Cl	1	HCO_2H, 30 °C	60
Cl	1	93% H_2SO_4, 15–20 °C	66
MeO	1	HCO_2H, 30 °C	60
MeO	1	HOAc, $Hg(OAc)_2$	25
Ph	1	HCO_2H, 30 °C	63
Ph	1	HOAc, $Hg(OAc)_2$	54
HO_2C	1	HCO_2H, 30 °C	73
MeO_2C	1	HOAc, $Hg(OAc)_2$	27
p-ClC_6H_4	1	93% H_2SO_4, 15–20 °C	83
p-$Cl_2C{=}CHCH_2C_6H_4$	1	HCO_2H, 30 °C	30[a]
Cl_3C	2	93% H_2SO_4, 15–20 °C	52
H	3	93% H_2SO_4, 15–20 °C	71
Cl	3	HCO_2H, 30 °C	69
Cl	3	HOAc, $Hg(OAc)_2$	62
Cl	3	HCl	0
Cl	3	70% $HClO_4$	36
Cl	3	93% H_2SO_4, 15–20 °C	78
HO_2C	3	93% H_2SO_4, 15–20 °C	77
HO_2C	3	HCO_2H, 30 °C	82
MeO_2C	3	HOAc, $Hg(OAc)_2$	55
NC	3	HOAc, $Hg(OAc)_2$	54[b]
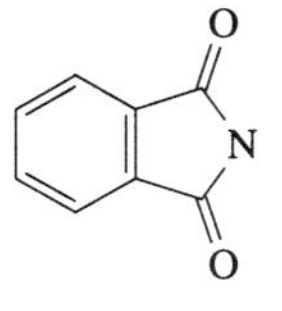	3	93% H_2SO_4, 15–20 °C	92
HO_2C	4	93% H_2SO_4, 15–20 °C	73
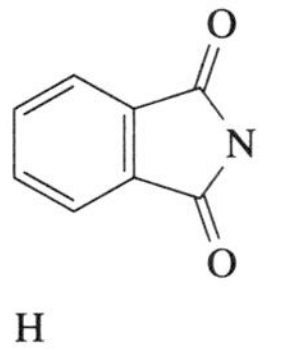	4	93% H_2SO_4, 15–20 °C	84
H	5	93% H_2SO_4, 15–20 °C	51
Cl	5	93% H_2SO_4, 15–20 °C	70
Cl	5	HCO_2H, 30 °C	85
HO_2C	5	93% H_2SO_4, 15–20 °C	69
Cl	7	HCO_2H, 30 °C	82

[a] Product is diacid *p*-$HO_2CCHClCH_2C_6H_4CH_2CHClCO_2H$.
[b] In a separate step, sulfuric acid hydrolysis of the nitrile produces the diacid $HO_2C(CH_2)_3CHClCO_2H$

Substitution of iodide for either chlorine or bromide in haloalkanoic acids has been accomplished by treatment with sodium iodide in acetone (Equation (123)) ⟨56JA2255⟩.

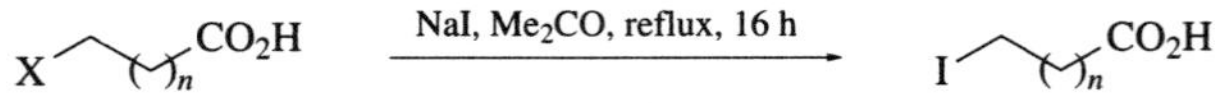

(123)

n	*X*	*Yield* (%)
2	Cl	61
3	Cl	67
4	Cl	71
8	Cl	67
9	Br	71

5.02.2.3 Oxygen Substituted Alkanoic Acids

A large number of books and reviews of oxygen-substituted alkanoic acids have been published which address the synthesis and chemistry of various classes of these types of compounds. These reviews include: a general report on hydroxy acids ⟨B-45MI 502-01⟩; the use of oxazolines in the synthesis of α-hydroxy acids ⟨85MI 502-02⟩; the synthesis and stereochemistry of β-hydroxy acids via aldol-type condensations ⟨B-80MI 502-01⟩; the stereoselective synthesis of β-hydroxy acids ⟨81MI 502-02⟩; hydroxy monocarboxylic acids and related compounds ⟨B-52MI 502-02⟩; carbonic acid and its derivatives ⟨B-52MI 502-03⟩; studies on phellonic acid (22-hydroxydocasanoic acid, ⟨63AC(P)315⟩); the preparation and properties of chromancarboxylic acids ⟨81CHE(36)315⟩; the chemistry and biology of malic acid and related substances ⟨B-60MI 502-03⟩; the preparation of α-alkoxy and α-hydroxy acids by reactions of aryl trichloromethyl carbinols with nucleophiles ⟨71S131⟩; dihydroxy and trihydroxy acids via oxidation of trihydric alcohols ⟨B-76MI 502-01⟩; a general report on tartaric acids ⟨42MI 502-02⟩; general review of pantothenic acid (2,4-dihydroxy-3,3-dimethyl-1-oxobutyl)-β-alanine ⟨41AG51, B-43MI 502-01, 45BSF17⟩; shikimic acid (3,4,5-trihydroxy-1-cyclohexene-1-carboxylic acid, ⟨65CRV435⟩; dihydric and polyhydric alcohols and their oxidation products with function groups in separate side chains ⟨B-56MI 502-02⟩; trihydric alcohols and their oxidation products ⟨B-52MI 502-04⟩; tetrahydric alcohols and their oxidation products ⟨B-52MI 502-05⟩; tetronic acids ⟨60QR292⟩; saccharinic acids ⟨57MI 502-01, 58MI 502-01⟩; D-glucuronic acid ⟨53MI 502-01⟩; pentahydric alcohols and their oxidation products ⟨B-52MI 502-06⟩; hexa- and polyhydric alcohols and their oxidation products ⟨B-52MI 502-07⟩; and, the nonulosaminic acids, neuraminic acids, and related compounds (sialic acids) ⟨56AG349, 58MI 502-02⟩.

Aldehyde acids have been reviewed in three publications, a general review ⟨B-45MI 502-04⟩, a specific review of methyl esters of tetrahydroxy aldehyde acids ⟨54MI 502-02⟩, and, a general review of aldehydo- and keto-acids ⟨B-52MI 502-08⟩, while a general review of ketonic acids ⟨B-45MI 502-05⟩, two specific reviews of α-keto acids ⟨47CRV585, 83CRV321⟩, and one specific review of β-keto acids ⟨82OPP249⟩ has also been published.

5.02.2.3.1 By hydrolysis reactions

Acid hydrolysis of esters has been used to prepare α-keto acids from unsaturated diesters (Equation (124)) ⟨B-76MI 502-02⟩, α-ketoglutaric acid from ethyl oxalylsuccinate in 75–77% yield ⟨55OSC(3)510⟩, and chelidonic acid from diethyl 2,4,6-trioxo-1,7-heptandioate in 76–79% yield (Equation (125)) ⟨43OSC(2)126⟩, while basic hydrolysis of esters has been used to prepare acetoacetic acid from ethyl acetoacetate ⟨63OSC(4)633⟩, 2,5-dimethylmandelic acid from ethyl 2,5-dimethylphenylhydroxymalonate in 63–70% yield ⟨55OSC(3)326⟩, 2,2-diphenyl-3-hydroxypropionic acid from its methyl ester ⟨43JA262⟩, 3-methyl-2-furoic acid from methyl 3-methyl-2-furoate in 90–93% yield ⟨63OSC(4)628⟩, and 1,3-dimethyl-5-oxobicyclo[2.2.2]octane-2-carboxylic acid in 67% yield from its methyl ester (Equation (126)) ⟨87OS(66)37⟩.

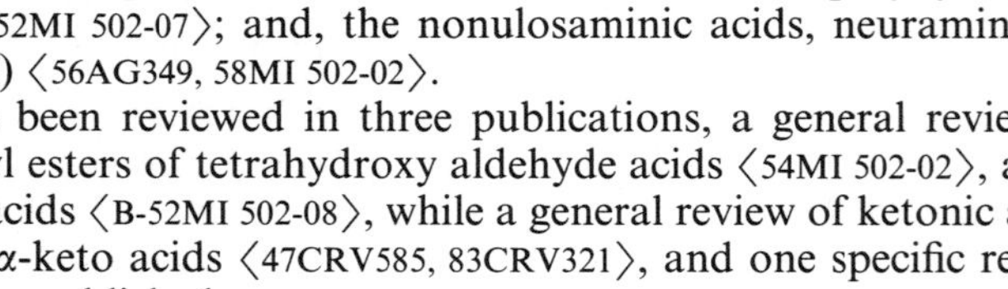

(124)

$R^1 = C_6–C_8$ (*n*-alkyl); R^2 = Et, Pr^n

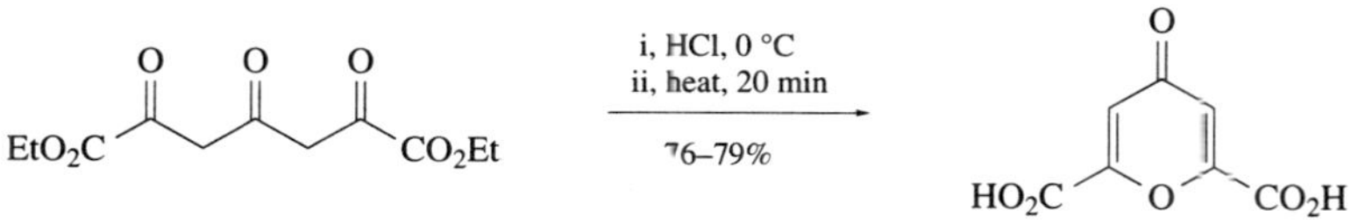

(125)

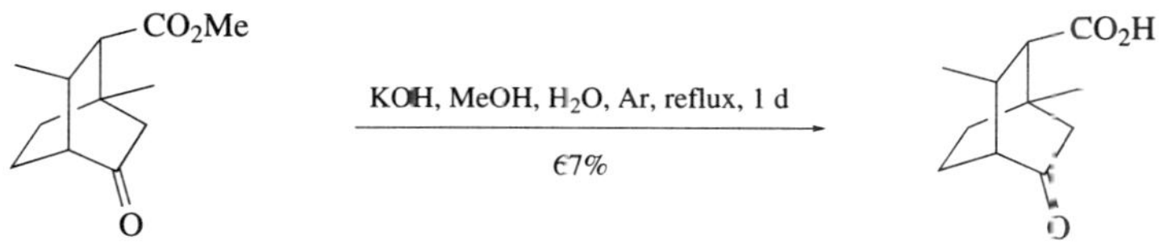

(126)

Asymmetric syntheses of oxygen containing alkanoic acids by base-catalyzed hydrolysis of esters are also represented in the literature by the synthesis of the α-hydroxy acids, α-hydroxydecanoic acid (Equation (127)) ⟨85HCA216⟩, and α-alkylmandelic acids (Equation (128)) ⟨82TL1259⟩, while the β-hydroxy acids are represented by *cis*-2-hydroxy-1-methylcyclohex-5-en-1-carboxylic acid (Equation (129)) ⟨78CC961⟩ and a series of optically pure β-hydroxy acids (Equation (130)) ⟨82BCJ2186⟩.

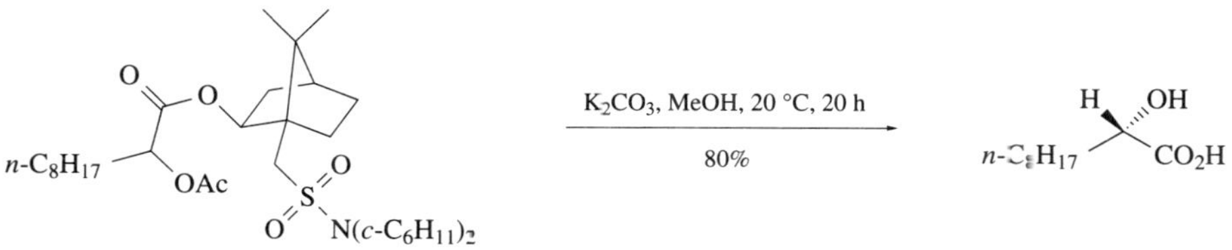

(127)

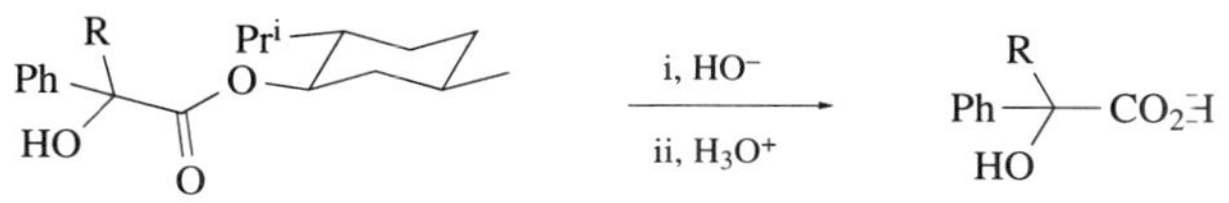

(128)

R	*Yield* (%)	ee (%)	*Absolute configuration*
Me	69–77	17–38	*R* + *S*
Et	72–76	0.2–49	*R*

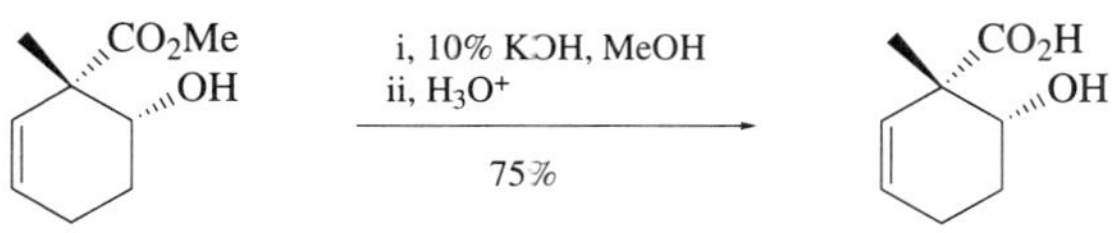

(129)

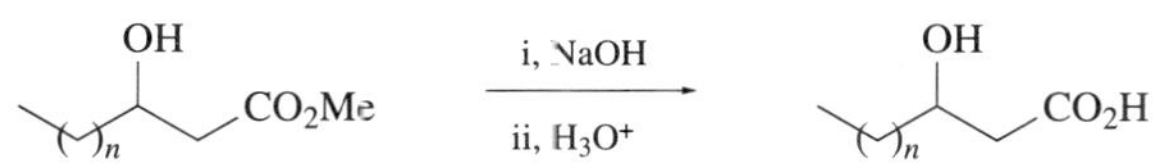

(130)

n	*Configuration*	*Yield* (%)
0	*R*	70
0	*S*	70
6	*R*	79
6	*S*	76
8	*R*	80
8	*S*	80
10	*R*	80
10	*S*	81
12	*R*	82
12	*S*	81

Base-catalyzed saponification of bicyclic γ- and δ-lactones produces *cis*-hydroxy acids (Scheme 36) ⟨84MI 502-01⟩, while similar treatment of 4,4-dimethyl-5-(isopropyl)dihydrofuran-2,3-dione produces 4-methyl-2-oxopentanoic acid in 90% yield ⟨85JAP6638⟩.

Acid hydrolysis of nitriles has been used to synthesize mandelic acid ⟨32OSC(1)336⟩, atrolactic acid (α-hydroxy-α-methylphenylacetic acid) ⟨63OSC(4)58⟩, pentafluoromandelic acid ⟨61JCS808⟩,

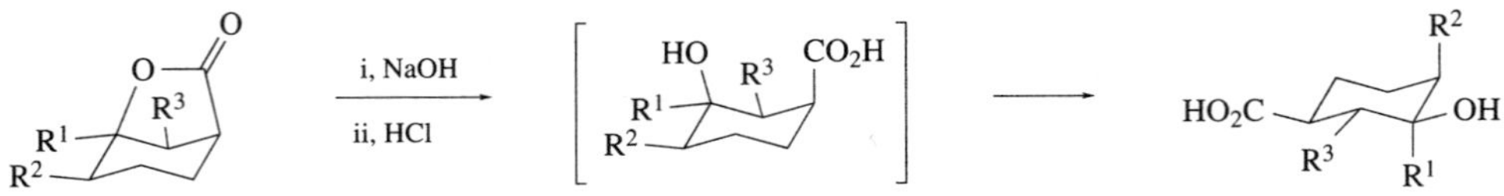

R^1	R^2	R^3	Yield (%)
H	H	H	98.4
Me	H	H	94.3
H	Me	H	94.9
H	H	Me	96.2

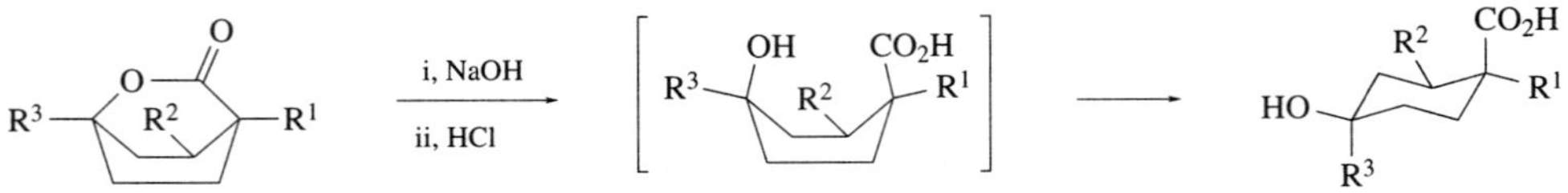

R^1	R^2	R^3	Yield (%)
Me	H	H	96.8
Me	Me	H	95.9
Me	H	Me	91

Scheme 36

γ-phenoxybutyric acid ⟨29JA260⟩, benzoylformic acid ⟨55OSC(3)114⟩, and a large variety of β-(substituted benzoyl) alkanoic acids (Equation (131) and Table 24) ⟨73JOC4044⟩.

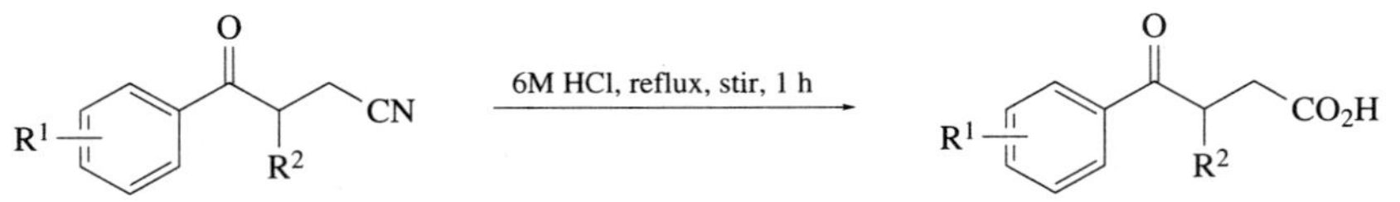

(131)

Table 24 β-(Substituted benzoyl)alkanoic acids via acid hydrolysis of corresponding nitriles.

R^1	R^2	Yield (%)	R^1	R^2	Yield (%)
p-NO_2	H	68	p-NH_2[a]	Me	73
p-Br	Me	86	p-F	Et	91
p-Cl	Me	82	p-NH_2[a]	Et	82
p-F	Me	87	p-NH_2[a]	Pr^n	69
m-NO_2	Me	87	p-F	Ph	63
H	Me	95	H	Ph	91

[a] R^1 group in starting material was the acetylated amine.

β-Hydroxypropionic acid can be prepared from ethylene cyanohydrin upon basic hydrolysis using aqueous sodium hydroxide ⟨32OSC(1)321⟩, while aqueous potassium hydroxide in the presence of hydrogen peroxide has been used to produce δ-oxoazelaic acid from 2,2-di(3-cyanopropyl)-1,3-dioxolan (Equation (132)) ⟨62JCS4722⟩. Treatment of the higher homologues of 2,2-disubstituted hydroxybutyronitriles with sodium hydroxide in ethylene glycol affords the corresponding hydroxy acids (Equation (133)) ⟨78JCR(S)198⟩, while for the hydrolytically resistant 1-cyano-1-(m-methoxyphenyl)-4-hydroxycyclohexane potassium hydroxide in diethylene glycol had to be used (Equation (134)) ⟨69CPB1564⟩.

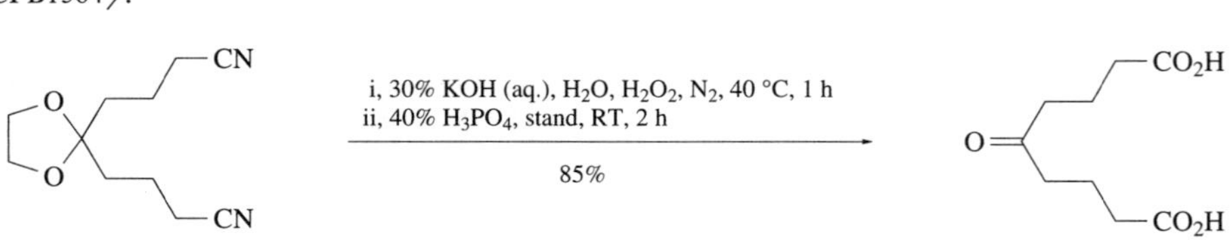

(132)

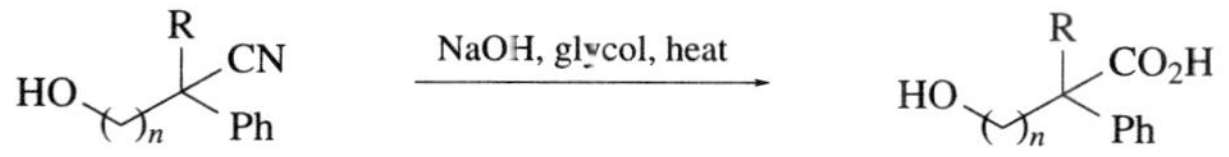

(133)

R	*n*	*Yield* (%)
Et	3	92[a]
Ph	3	90[a]
$PhCH_2$	3	85[a]
Ph	4	91
$PhCH_2$	4	90
Et	5	85
Ph	5	89
$PhCH_2$	5	90
Ph	6	88
$PhCH_2$	6	86

[a] Product cyclizes to corresponding lactone.

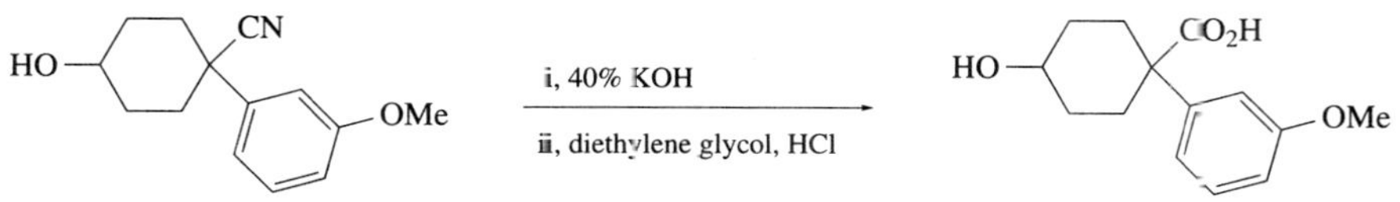

(134)

Sodium hydroxide hydrolysis of α- and β-aldoximes has shown that formation of the intermediate nitriles from the α-aldoximes is very slow compared to their β-analogues, and thus the yield of the acid final products is higher from the hydrolysis of the β-analogues (Scheme 37 and Table 25) ⟨36JA1304⟩.

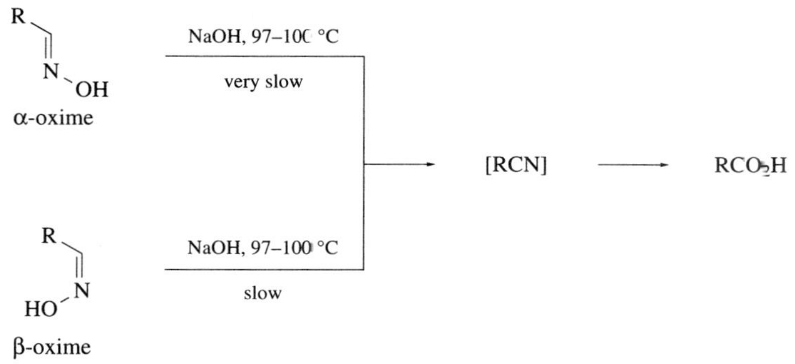

Scheme 37

Table 25 Carboxylic acids via sodium hydroxide hydrolysis of α,β-aldoximes.

R	*Oxime structure*	*Time* (%)	*Yield* (%)
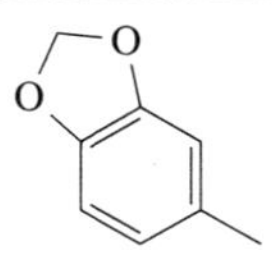	α	12	21
	β	12	58
Ph	α	4	10
	β	4	38
p-$MeOC_6H_4$	α	4	8
	β	4	39
m-$O_2NC_6H_4$	β	1.5	62
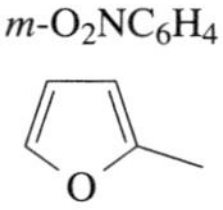	α	1.5	18
	β	1.5	49

Amides and related structures have also been hydrolyzed to acids using a variety of reagents. Sodium hydroxide hydrolyzes atrolactamide to atrolactic acid (α-hydroxy-2-methylbenzeneacetic acid) ⟨63OSC(4)58⟩, potassium hydroxide hydrolyzes mixtures of the chiral secondary amides (1′*R*, 3*R*)- and (1′*R*, 3*S*)-3-hydroxy-*N*-(2-hydroxy-1-phenylethyl)-3-substituted propionamides to 3-hydroxy-3-substituted propionic acids (Equation (135)) ⟨86CB2191⟩, while nitrous anhydride in base is required to convert the analogous chiral mixtures of (1′*S*, 3*R*)- and (1′*S*, 3*S*)-3-hydroxy-3-phenyl-*N*-(1-substituted)propionamides into (*R*)-3-hydroxy-3-phenylpropionic acids (Equation (136)) ⟨86CB2191⟩ and to convert acetylaldoamides to acetylaldonic acids without effecting cleavage of the ester functions (Equation (137)) ⟨38JA235⟩. Hydrochloric acid hydrolyzes α-acetamidocinnamic acid to phenylpyruvic acid in 88–94% yield ⟨43OSC(2)519⟩.

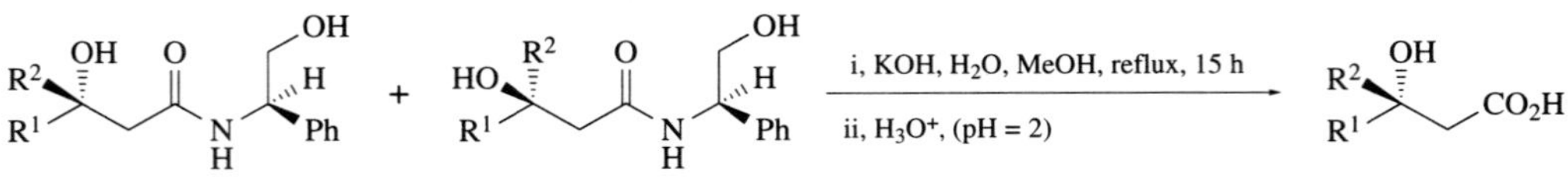

(135)

(1′*R*, 3*R*) (1′*R*, 3*S*)

R¹	*R²*
Ph	H
Bu^t	H
Pr^i	H
Pr^n	H
Ph	Me

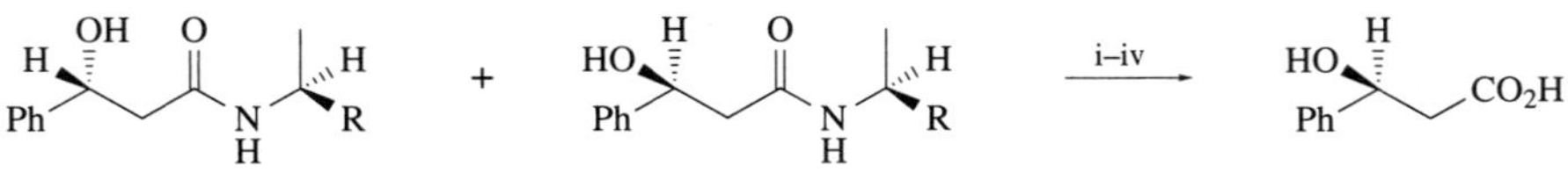

(136)

(1′*S*, 3*R*) (1′*S*, 3*S*)

i, N_2O_4, NaOAc, CCl_4, 0 °C; ii, RT, 4 d; iii, KOH, reflux, 1 h; iv, H_3O^+, (pH = 3)

R	*Yield* (%)
Ph	80
β-naphthyl	85

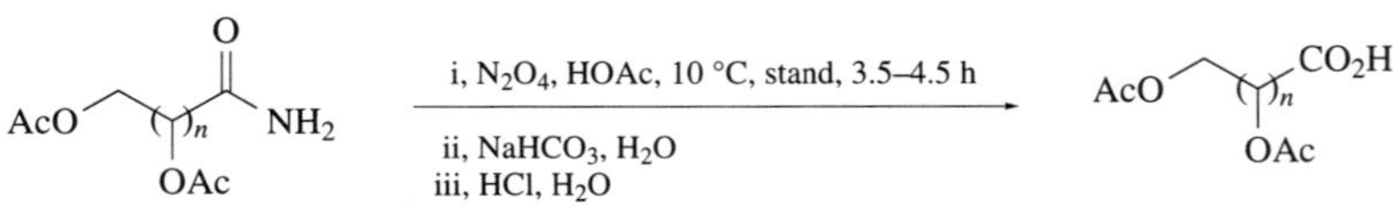

(137)

n	*Product*
3	tetraacetyl-*l*-arabonic acid
4	pentaacetyl-*d*-galactonic acid
4	pentaacetyl-*d*-gluconic acid

In the anhydride class of compounds, the mixed methyl anhydride of 3β-acetoxy-5-androstene-17β-carboxylic acid has been hydrolyzed to the free acid using pyridine ⟨73OSC(5)8⟩, while sulfuric acid has been used to convert (−)-(1*R*, 3*R*)-3-chlorocamphoric anhydride to (−)-(1*S*, 4*R*)-camphanic acid in 72% yield (Equation (138)) ⟨92OS(71)48⟩.

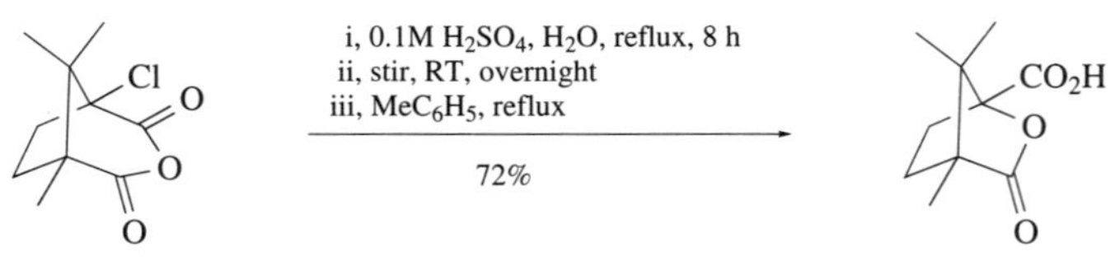

(138)

Preparation of oxygen-substituted alkanoic acid via hydrolysis of halides is represented in the dihalide series by the basic hydrolysis of dichloroacetophenone ⟨55OSC(3)538⟩ and *p*,α,α-tribromoacetophenone ⟨63OSC(4)110⟩ affording mandelic acid and *p*-bromomandelic acid in 85–90% and 69–83% yields, respectively. In the trihalide series, oxygen-substituted alkanoic acids are prepared by alcoholic potassium hydroxide hydrolysis of trichloromethylcarbinols to give α-alkoxyacetic acids (Equation (139)) ⟨48JA1153⟩ and (Equation (140)) ⟨50JA5012⟩, the structure of the alkoxy group formed being a function of the alcohol used as a solvent during the hydrolysis. A frequently encountered method for introduction of the trihalomethyl group into an aliphatic framework is the base-catalyzed condensation of aldehydes with chloro- or bromoform which affords the trihalomethylcarbinols, which are eventually hydrolyzed to acids. This approach has been used in a one-pot conversion of aliphatic aldehydes to α-methoxyaliphatic acids (Scheme 38) ⟨78JOC2702⟩ and α-hydroxyarylacetic acids (Equation (141)) ⟨68JOC2565⟩. By the use of triethylbenzylammonium chloride (TEBA) and phase-transfer catalysis conditions, this approach has also been extended to aliphatic (Equation (142)) ⟨79MI 502-07⟩ and aromatic (Equation (143)) ⟨82USP4347379⟩ ketones as well.

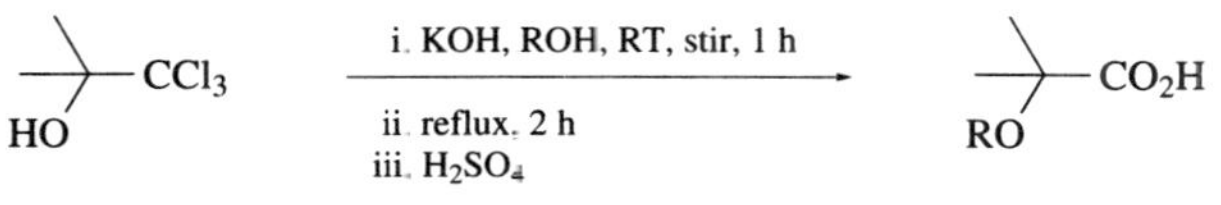

(139)

R	*Yield* (%)
Et	68
Bu^n	78
Bu^i	61
Bu^n–CH(Et)–CH_2	70
n-$C_{12}H_{25}$	24
H_2C=CH–CH_2	69
EtOCH_2CH_2	48
Bu^{n}OCH_2CH_2	50
Pr^i	44
t-C_5H_{11}	20

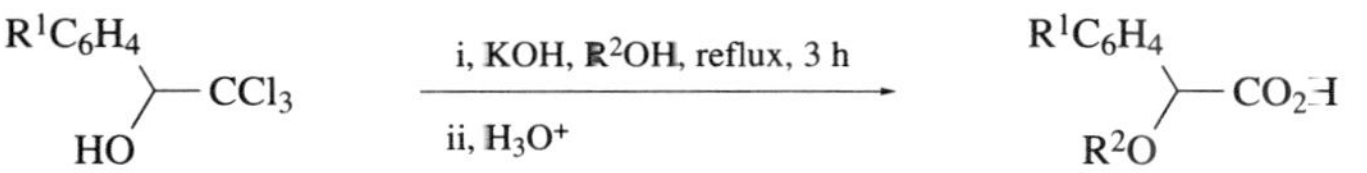

(140)

R^1	*R*^{2}
p-Me	Et
o-Cl	Et
m-MeO	Et
p-MeO	Et
H	Et
p-Me	Bu^n
o-Cl	Bu^n
o-MeO	Bu^n
m-MeO	Bu^n
H	Bu^n

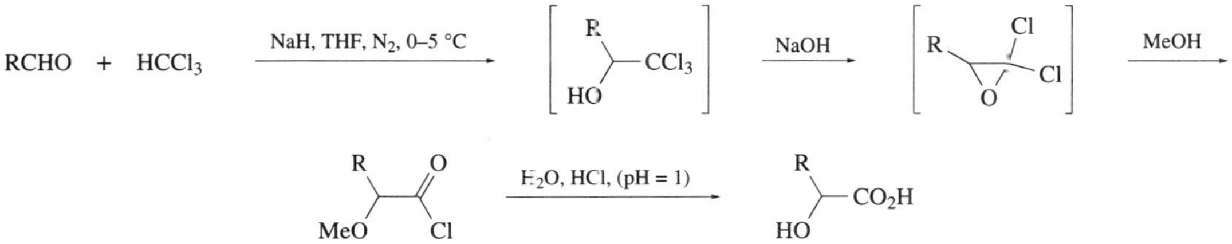

R	*Yield* (%)
n-C_5H_{11}	54
n-C_6H_{13}	53
(Pr^i)$_2$CEt	56
Pr^i	59
Pr^n	57
Bu^n	53
n-C_7H_{15}	54
n-C_8H_{17}	51
c-C_6H_{11}	63

Scheme 38

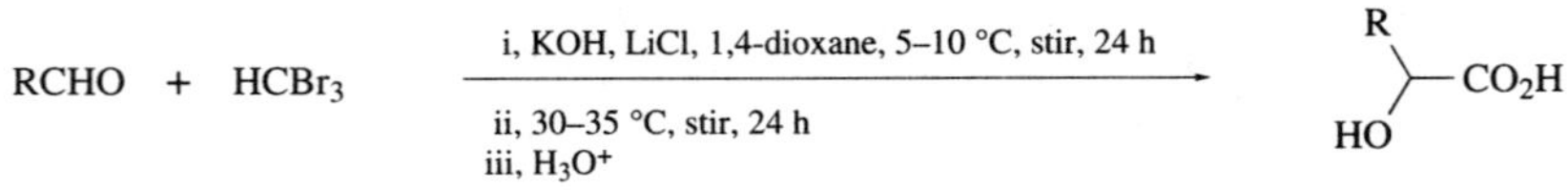

(141)

R	*Yield (%)*
Ph	90–92
o-ClC_6H_4	98–100
m-ClC_6H_4	99–100
p-ClC_6H_4	96–100
o-FC_6H_4	91–100
m-FC_6H_4	89–100
p-FC_6H_4	92–100
p-$MeOC_6H_4$	44–47
1-naphthyl	77–82
2-naphthyl	82–95

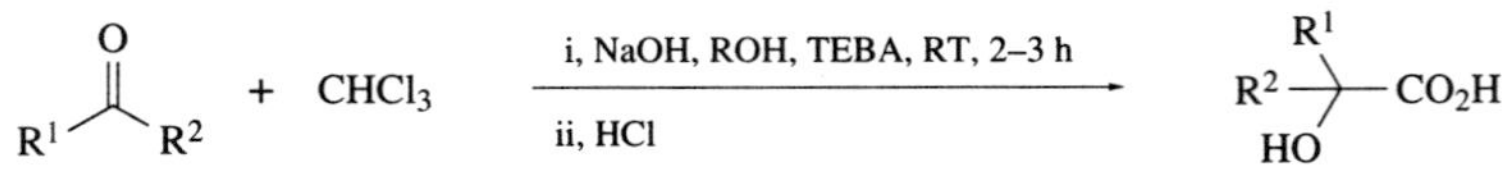

(142)

R^1	*R^2*
Me	Ph
Et	Ph
Ph	Ph
$PhCH_2$	$PhCH_2$

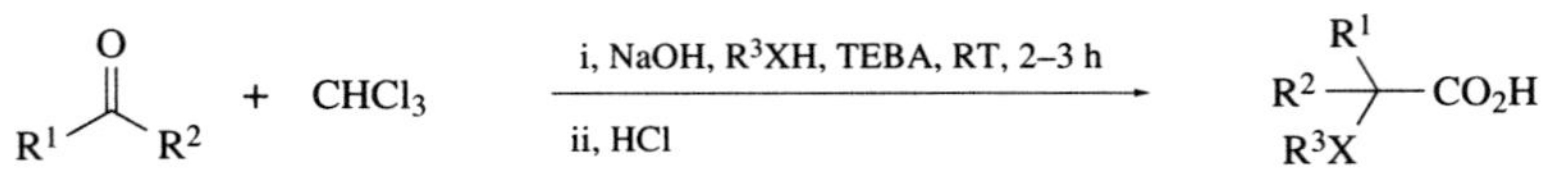

(143)

R^1	*R^2*	*R^3*	*X*
Me	Me	Me	O
Me	Ph	Me	O
Me	Bu^i	Et	O
Bu^i	OBu^i	Me	O
Me	Me	*c*-C_6H_{13}	S
Me	Et	Bu^i	O
−*c*-C_6H_{12}−		Bu^i	O
Et	*n*-C_5H_{11}	Bu^i	O
Me	Me	Et	O
Me	Me	*n*-$C_{12}H_{25}$	O
−*c*-C_6H_{12}−		Me	O
Me	Pr^n	Me	O
Me	Me	Ph	O
Me	Me	Ph	S

5.02.2.3.2 By carbonation of organometallic reagents

Numerous substrates, organometallic reagents, and reaction conditions have been used to produce oxygen-substituted alkanoic acids by carbonation. Examples of some typical carbonation reactions of organolithium reagents are illustrated in Scheme 39 ⟨73CC500⟩ and Scheme 40 ⟨74S443⟩. Carbonation of metallic aldimines, formed by addition of organometallic reagents to isocyanides, followed by hydrolysis of the intermediate α-imino acids produces α-keto acids (Scheme 41) ⟨74JOC600⟩. α-Keto acids have also been prepared by the iron–sulfur complex catalyzed reaction of carbon dioxide and phenylhydrazine with sulfur-containing acetates producing α-phenylhydrazones of the corresponding acids, which upon hydrolysis generates the corresponding α-keto acids (Scheme 42) ⟨83JAP5813530⟩. Aliphatic β-keto acids can be synthesized by carbonation of ketones possessing acidic hydrogens using the hindered base potassium 4-methyl-2,6-di-*t*-butylphenoxide (Scheme 43) ⟨77G499⟩, while aliphatic and alicyclic β-keto acids have been prepared in the same manner in good yields using the lithium salt (Equation (144)) ⟨73JOC4086⟩ of the hindered base.

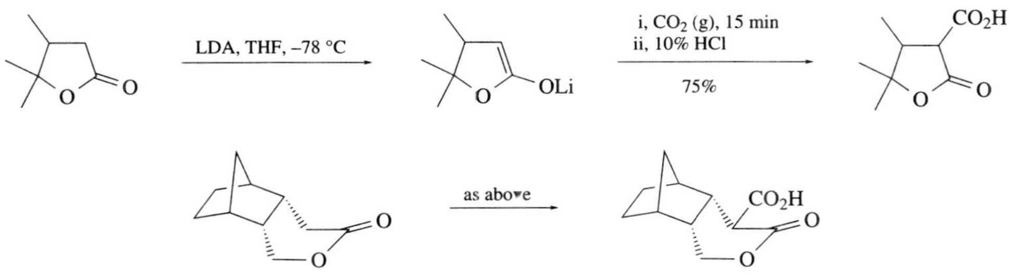

Scheme 39

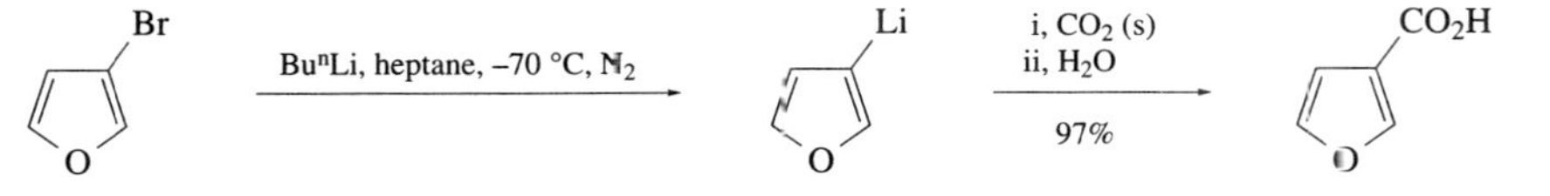

Scheme 40

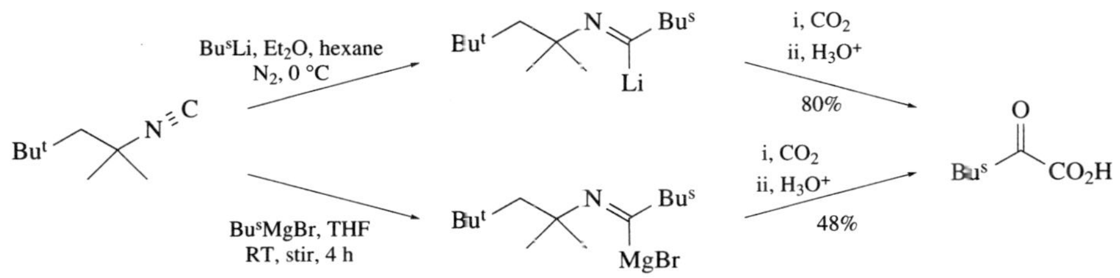

Scheme 41

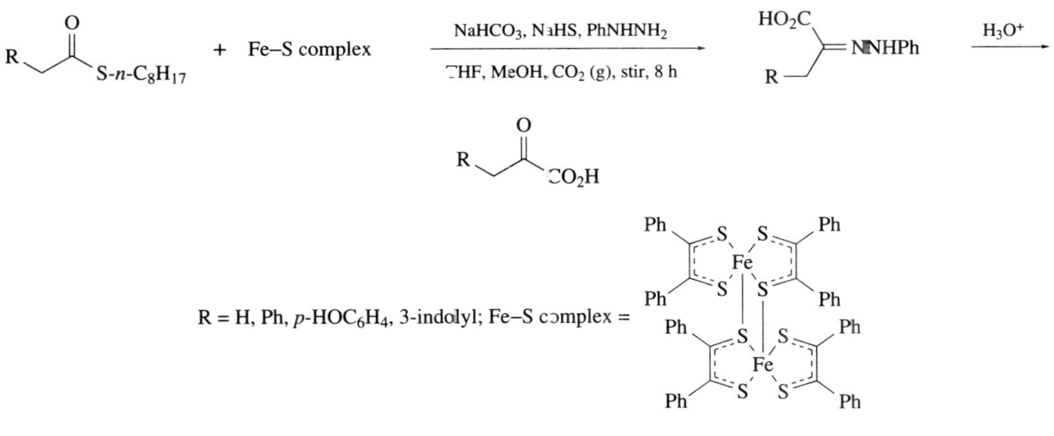

Scheme 42

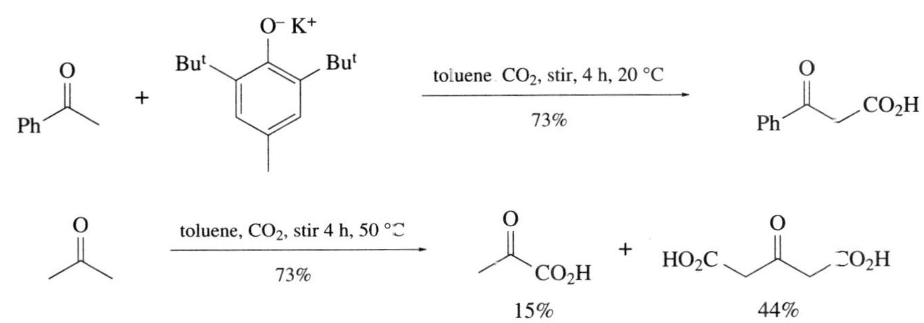

Scheme 43

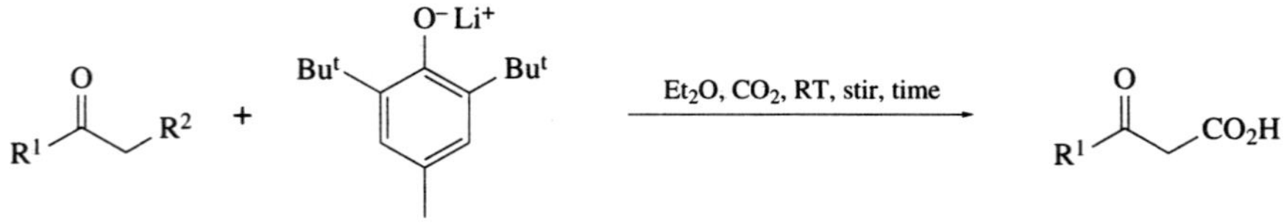

(144)

Ketone	*Time* (h)	*Product*	*Yield* (%)
PhCOMe	24	$PhCOCH_2CO_2H$	75
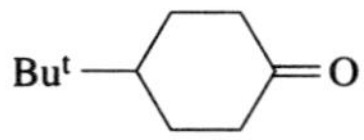	10	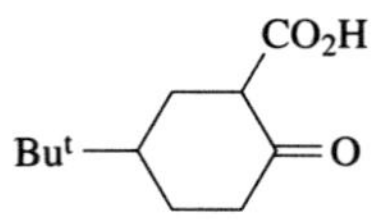	89
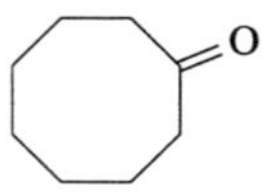	16	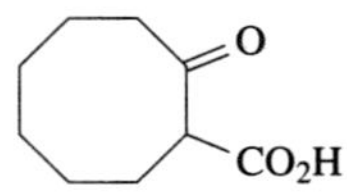	77

β-Keto acids have also been prepared from active methylene compounds using iron(III) ethoxide in DMF (Equation (145)) ⟨74CL1035⟩, Stile's reagent (methyl magnesium carbonate, ⟨63JA616⟩ with levulinic acid (Scheme 44) ⟨74JOC3144⟩) and γ-lactones (Equation (146)) ⟨74JOC1676⟩, and triethylamine and magnesium halide with ketones (Equation (147) and Table 26) ⟨85JOC4877⟩. Two publications have reported that heterocyclic magnesium halide carbon dioxide complexes can be used as carbon dioxide carriers when allowed to react with substrates possessing an active hydrogen atom such as ketones containing at least one α-proton. The general reaction which may be used to illustrate these transcarboxylation reactions is shown in Scheme 45 and involves the reaction of the heterocyclic magnesium halide complex with gaseous carbon dioxide to produce the carbon dioxide carrier, which upon reaction with an enolizable ketone affords the reduced complex and the corresponding β-keto acid. The heterocyclic magnesium halide or lithium carbon dioxide complexes studied (A–I) ⟨80CC327, 83CC1487⟩ are shown below and the results obtained using them are reported in Table 27. Finally, treatment of acetophenone with potassium carbonate in combination with benzyl tri-*n*-butylammonium chloride can be used to catalyze the addition of carbon dioxide and produce a 42% yield of benzoylacetic acid ⟨82NKK811⟩.

$$\text{Ketone} \xrightarrow[\text{ii, 80 °C, 2 h; iii, } H_2SO_4\text{, 0 °C}]{\text{i, } Fe(OEt)_3\text{, DMF, } CO_2 \text{ (5 atm.), RT, autoclave}} \beta\text{-keto acid} \qquad (145)$$

Ketone	*Product*	*Yield* (%)
PhCOMe	$PhCOCH_2CO_2H$	7
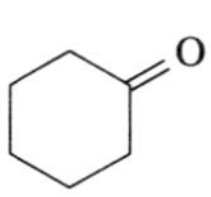	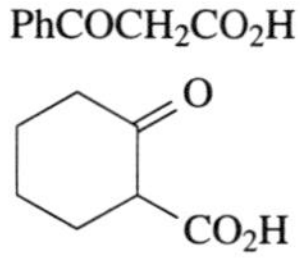	31–48

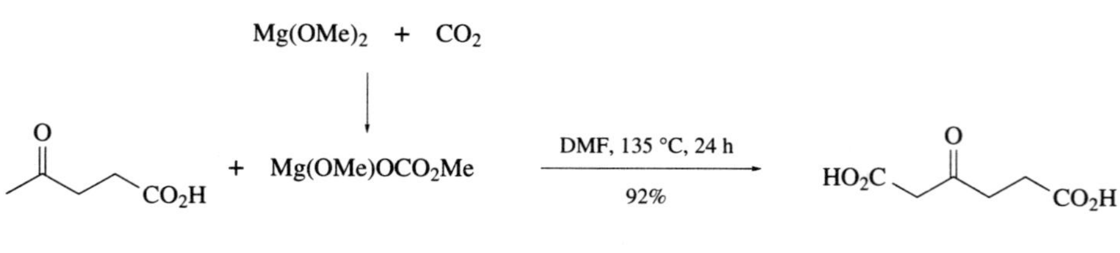

Scheme 44

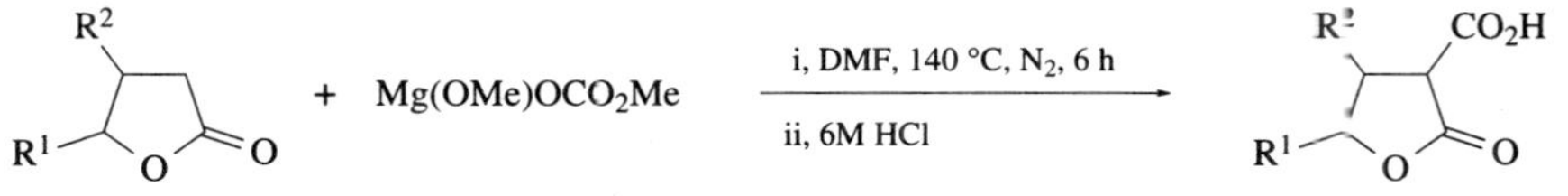

(146)

R^1	R^2
Bu^n	H
n-C_6H_{13}	H
Ph	H
n-$C_{13}H_{27}$	CO_2H (*trans*)

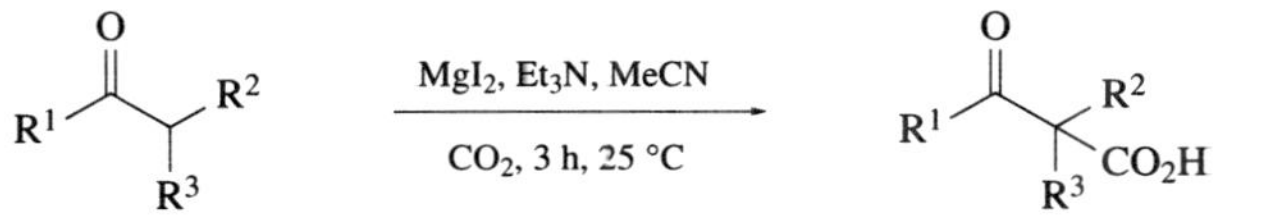

(147)

Table 26 β-Keto acids from ketones using magnesium halide/triethylamine.

R^1	*R^2*	*R^3*	*Yield* (%)	*R^1*	*R^2*	*R^3*	*Yield* (%)
Ph	H	H	75	Pr^i	Me	Me	65
Ph	H	Me	85		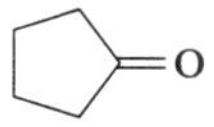		35
Ph	Me	Me	90		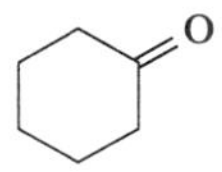		70
Et	H	Me	58		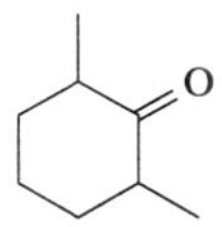		65
Pr^n	H	CH_2Me	75				

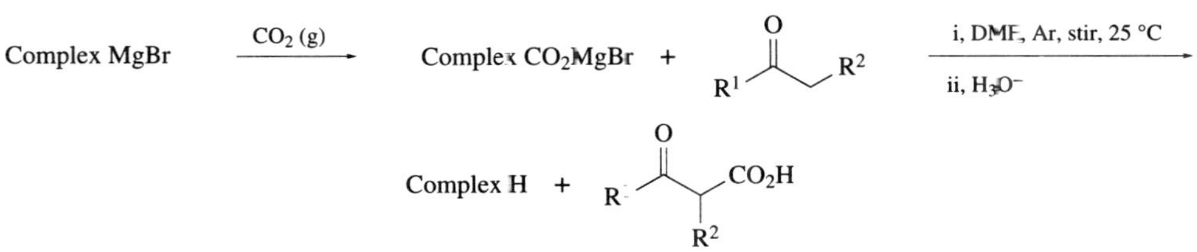

Scheme 45

β-Hydroxycarboxylic acids have been prepared by carbonation of β-hydroxyalkyl chlorides using n-butyllithium and α-naphthyllithium (Scheme 46) ⟨83JCS(P1)3019⟩, and by carbonation of trialkylvinylborate magnesium chlorides (Scheme 47 and Table 28) ⟨85CC1478⟩. Organoboranes have also been used to prepare ^{13}C labeled oxygen-substituted alkanoic acid via a similar carbonylation–oxidation sequence but using ^{13}C labeled carbon monoxide (Equation (148)) ⟨82CC1273⟩, while ^{14}C labeled analogues have been prepared by reaction of ^{14}C labeled carbon dioxide with Grignard reagents (Scheme 48 and Table 29) ⟨83MI 502-01⟩.

Table 27 *β*-Ketocarboxylic acids from ketones via transcarboxylation using heterocyclic magnesium halide complexes/carbon dioxide.

Ketone[a]	*Time* (h)	*β-Keto acid product*	*Yield using carrier*[c] (%) *A*	*B*	*C*	*D*	*E*	*F*	*G*	*H*	*I*	*Ref.*
$PhCOMe$	20	$PhCOCH_2CO_2H$				34	74	31	79	5	52	83CC1487
$PhCOMe$	40	$PhCOCH_2CO_2H$	NR	50	9							80CC327
$PhCOMe$	19[b]	$PhCOCH_2CO_2H$		58								80CC327
$PhCOCH_2Me$	20	$PhCOCHMeCO_2H$				2	49	2	80	3	12	83CC1487
$PhCH{=}CHCOMe$	19[b]	$PhCH{=}CHCOCH_2CO_2H$		76								80CC327
$PhCH{=}CHCOMe$	40	$PhCH{=}CHCOCH_2CO_2H$		51								80CC327
$PhCH{=}CHCOMe$	20	$PhCH{=}CHCOCH_2CO_2H$				25	78	11	78	5	71	83CC1487
$PhCH_2SCOMe$	65[b]	$PhCH_2SCOCH_2CO_2H$		54								80CC327
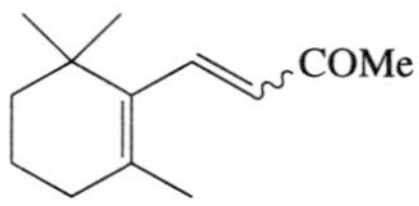	19[b]	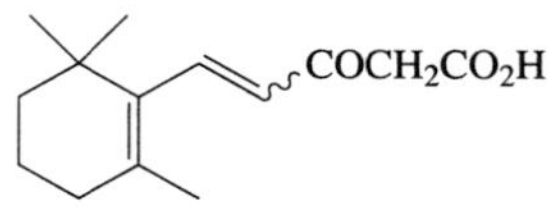	77									80CC327
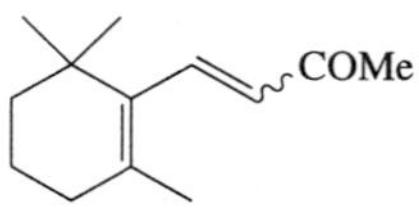	20	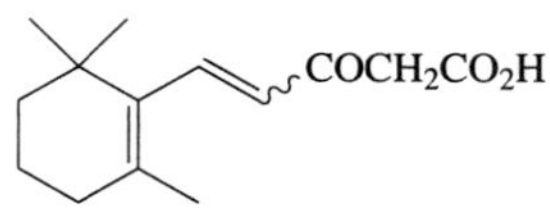				19	58	21	53	2	39	83CC1487

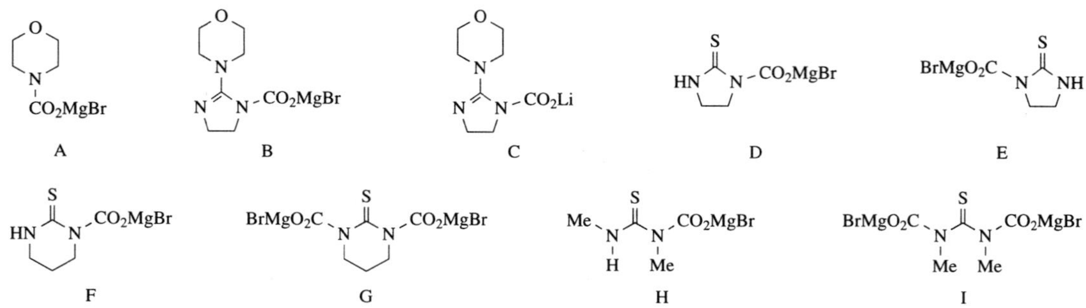

[a] Molar ratio of complex carrier to ketone = 4:1. [b] Reaction was run under a continuous stream of CO_2 at 3 kg cm^{-1}.
[c] See above for structure of complexes A–I used.

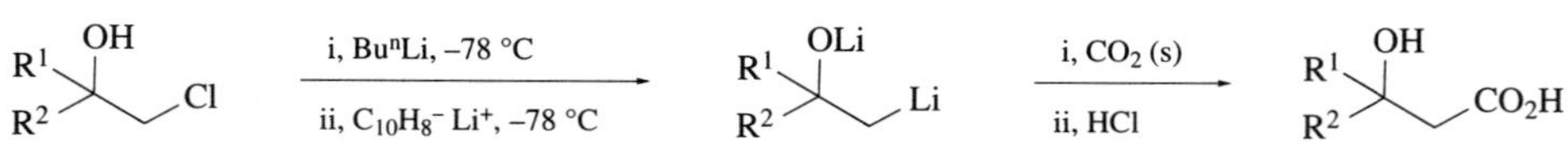

R[1]	*R[2]*	*Yield (%)*[a]
H	Ph	22
Me	$MeCH{=}CH_2$	46
$MeCH{=}CH_2$	$MeCH{=}CH_2$	64
$MeCH{=}CH_2$	Ph	92
Ph	Ph	60

[a] Isolated as methyl esters

Scheme 46

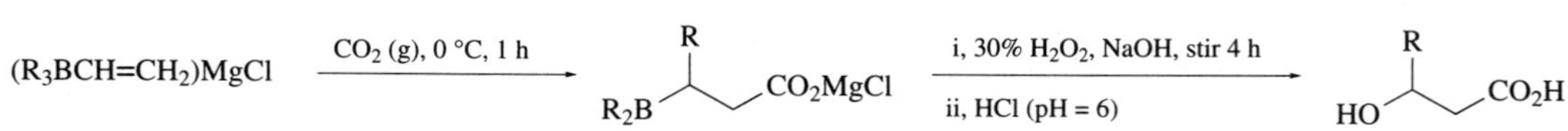

Scheme 47

Table 28 β-Hydrocarboxylic acids via carbonation of trialkylvinylborate magnesium chlorides.

R	*Product*	*Yield (%)*
n-C_6H_{13}	n-$C_6H_{13}CH(OH)CH_2CO_2H$	82
n-C_7H_{15}	n-$C_7H_{15}CH(OH)CH_2CO_2H$	81
n-C_8H_{17}	n-$C_8H_{17}CH(OH)CH_2CO_2H$	85
n-$C_{10}H_{21}$	n-$C_{10}H_{21}CH(OH)CH_2CO_2H$	78
n-$C_{12}H_{25}$	n-$C_{12}H_{25}CH(OH)CH_2CO_2H$	83
c-C_6H_{11}	c-$C_6H_{11}CH(OH)CH_2CO_2H$	74
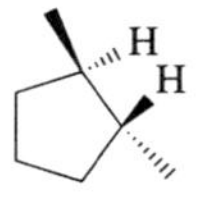	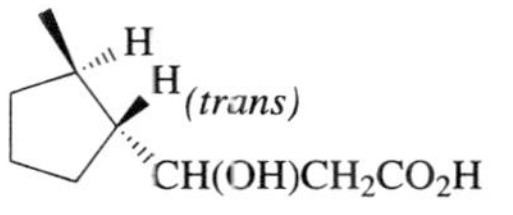	80

(148)

i, 9-BBN, THF; ii, ^{13}CO (1 atm), $KBH(OPr^i)_2$; iii, H_2O_2, NaOAc; iv, $AgNO_3$, NaOH, 50 °C, 30 min

R	*Yield (%)*
$HO(CH_2)_7$	94
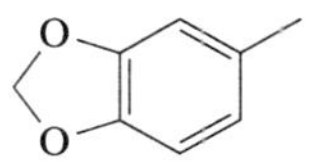	84

$$RMgX + {}^{14}CO_2 \xrightarrow{\text{stir, 15 h}} R^{14}CO_2Mg \xrightarrow{H_3O^+} R^{14}CO_2H$$

Scheme 48

Table 29 Oxygen substituted ^{14}C-labelled carboxylic acids via reaction of Grignard reagent with $^{14}CO_2$.

Halide	*Product*	*Yield (%)*
$MeOCH_2Cl$	$MeOCH_2{}^{14}CO_2H$	1
$MeO(CH_2)_2Cl$	N.R.	
$MeO(CH_2)_3Cl$	$MeO(CH_2)_3{}^{14}CO_2H$	33
$PhCH_2OCH_2Cl$	$PhCH_2OCH_2{}^{14}CO_2H$	<2
o-$MeOC_6H_4Br$	o-$MeOC_6H_4{}^{14}CO_2H$	7
p-$MeOC_6H_4Br$	p-$MeOC_6H_4{}^{14}CO_2H$	73

5.02.2.3.3 By oxidation reactions

Oxidation of oxygen-substituted alcohols produces the corresponding acids as exemplified by the conversion of 3-(3-hydroxypropyl)cyclopentanone to β-(3-oxocyclopentyl)propionic acid in 60% yield using ruthenium dioxide and potassium metaperiodate ⟨72JOC1947⟩, and by the conversion of the isomeric tetrahydropyranyloxy-2,2,4,4-tetramethylcyclobutan-1-carbinols into their corresponding isomeric hydroxy acids in 50% yield using chromium trioxide in acetone ⟨68CC206⟩. Many alcohol oxidation reactions which produce oxygen-substituted alkanoic acid are however accompanied by chain or ring cleavage as in the preparation of pyruvic acid from tartaric acid (Equation (149)) ⟨32OSC(1)475⟩, δ-acetyl n-valeric acid from 2-methylhexanol (Equation (150)) ⟨63OSC(4)19⟩, 2-carboxy-4,5-dimethoxy-phenylacetaldehyde from 2-hydroxy-5,6-dimethoxyindane-1-one (Equation (151)) ⟨50CB390⟩, and 2-carboxyhexanal from tetrahydroterrein, the reduction product of terrein (Scheme 49) ⟨35BJ(29)606⟩. Substrate reorganization has also been observed to occur in the benzilic acid rearrangement of benzoin to benzilic acid catalyzed by sodium periodate ⟨30JA3647⟩ or by sodium bromate ⟨32OSC(1)89⟩.

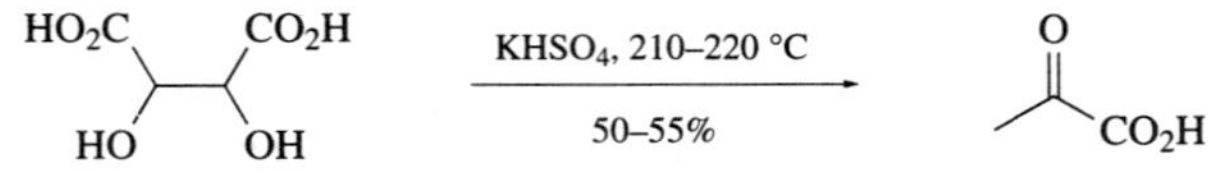

(149)

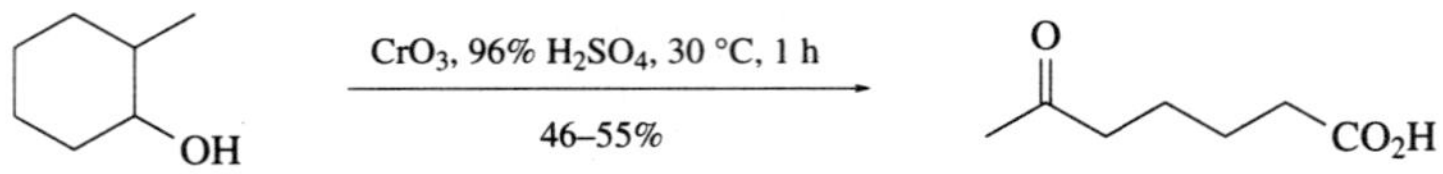

(150)

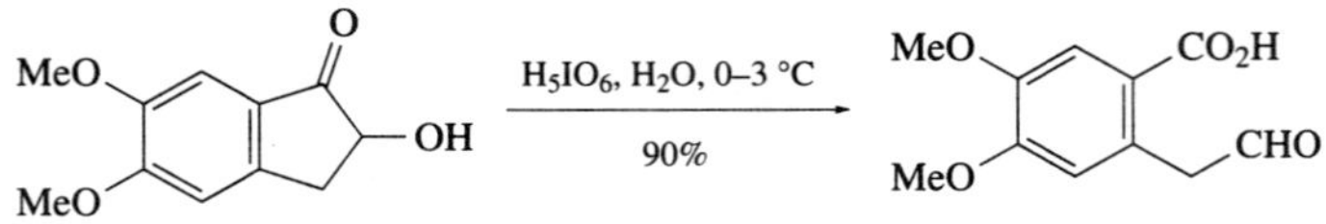

(151)

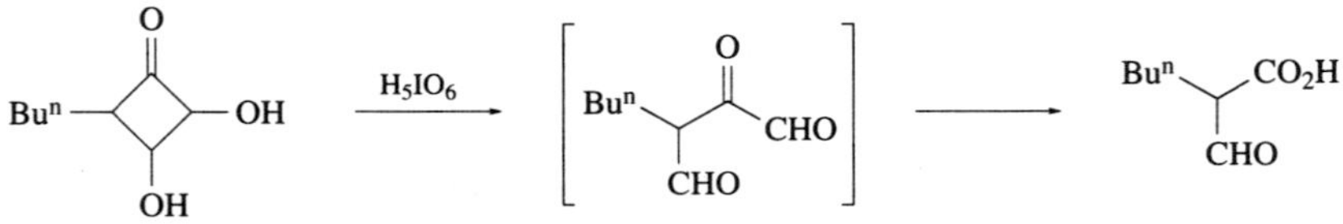

Scheme 49

A variety of oxidizing agents have been reported to convert aldehydes to oxygen-substituted alkanoic acids. Examples of these reagents include sodium hydroxide at room temperature which converts furfural to 2-furoic acid in 60–63% yield via a dismutative Cannizzaro reaction ⟨32OSC(1)276⟩, however an 86–90% yield of 2-furoic acid can be obtained by treatment of furfural with sodium hydroxide in the presence of a cuprous oxide-silver oxide catalyst ⟨63OSC(4)493⟩. Other reports include the oxidation of 5-methylfurfural to 5-methylfuroic acid using silver oxide in methanol ⟨58JCS1313⟩, oxidation of chromone-3-carboxaldehyde to chromone-3-carboxylic acid using Jones' reagent, a mixture of chromium trioxide and sulfuric acid ⟨73TL1995⟩, and oxidation of a series of aldehydes containing variously protected hydroxyl groups to their corresponding carboxylic acid using potassium permanganate in the presence of *t*-butyl alcohol and aqueous sodium dihydrogen phosphate without effecting cleavage of the protecting groups (Equation (152) and Table 30) ⟨86TL4537⟩.

$$\text{RCHO} \xrightarrow{\text{KMnO}_4,\ \text{Bu}^t\text{OH},\ 5\%\ \text{NaH}_2\text{PO}_4} \text{RCO}_2\text{H} \qquad (152)$$

An example of ketone oxidation to acids is the sodium hydroxide–sodium hypobromite oxidation of 3β-acetoxy-5-pregnen-20-one to 3β-hydroxyetienic acid (Equation (153)) ⟨73OSC(5)8⟩, an example of the haloform reaction on methyl ketones.

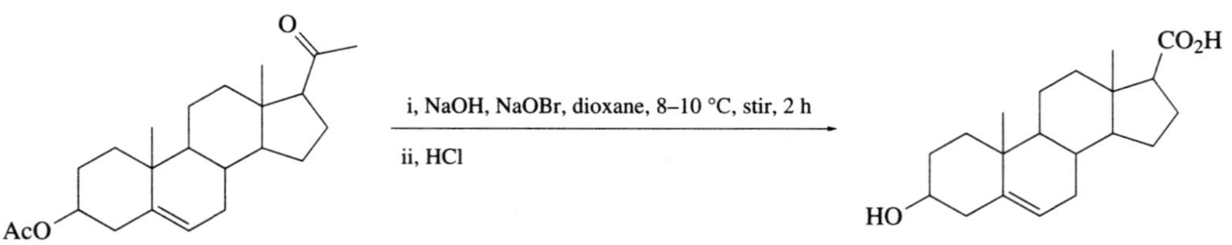

(153)

Potassium permanganate is the reagent of choice for the oxidative cleavage of double bonds and the formation of carboxylic acids, an example of which is the formation of 2-carboxyphenyl-2-oxoacetic acid (phthalonic acid) from naphthalene upon reaction with potassium permanganate in the presence of sodium hydroxide ⟨43OSC(2)523⟩. An indication of the improved oxidizing ability of potassium permanganate when it is used with 18-crown-6 is obtained when the results of oxidation of α-pinene with aqueous permanganate versus an 18-crown-6 solution of permanganate are compared, since the pisonic acid product is obtained in 40–60% yield using the former system and in 90% yield using the latter system (Scheme 50) ⟨72JA4024⟩.

Table 30 Oxidation of aldehydes containing protected hydroxyl groups using potassium permanganate/sodium dihydrogen phosphate in *t*-butyl alcohol.

Substrate	*Reaction time* (min)	*Yield* (%)
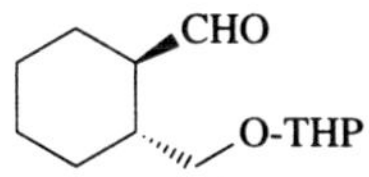	10	98
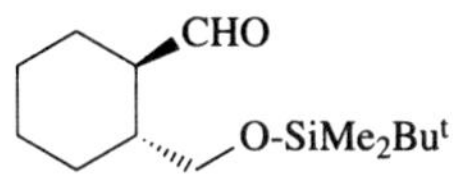	10	95
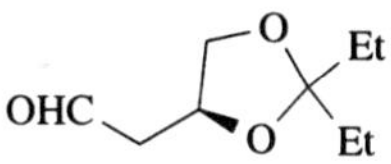	10	98
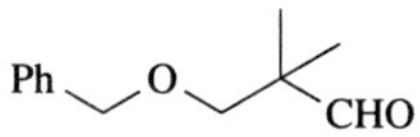	10	99
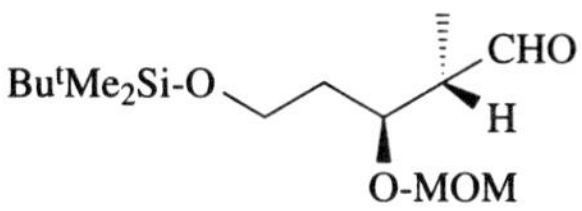	5	98
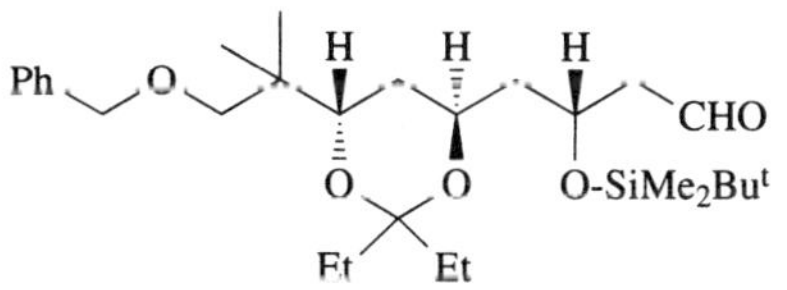	5	97
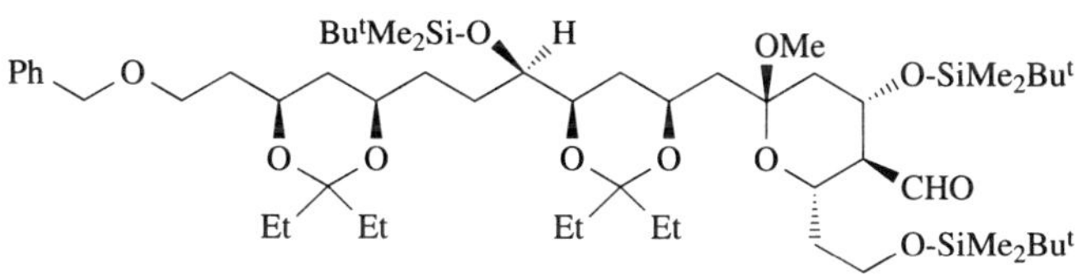	40	80–85

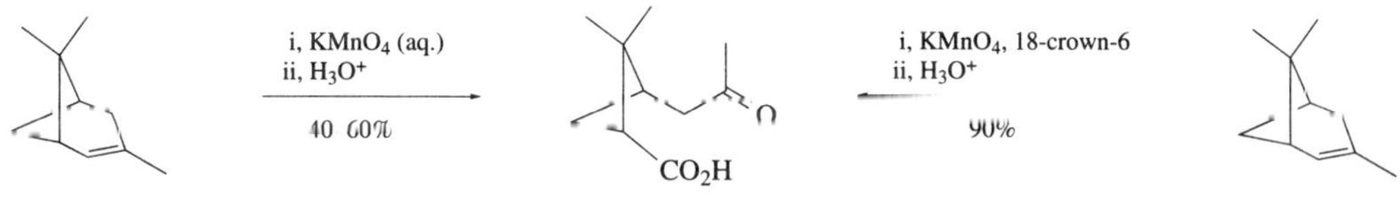

Scheme 50

Another reagent which has been used to oxidize oxygen-substituted double bonds to oxygen-substituted alkanoic acids is 30% hydrogen peroxide containing sodium hydroxide which converts the α,β-unsaturated cyclic ketone isophorone to 3,3-dimethyl-5-ketohexanoic acid in 84% yield ⟨70JOC1275⟩, while using the same oxidizing mixture in ethanol converts a series of 2-alkylidenecyclopentanones into a mixture of 5-oxoalkanoic acids and their corresponding ethyl esters (Equation (154)) ⟨69BSF2871⟩.

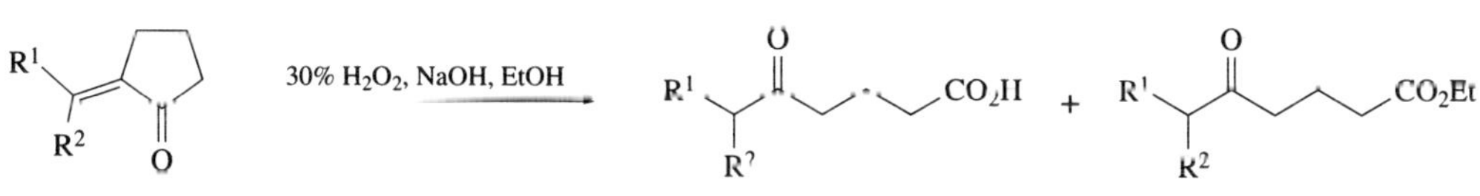

(154)

R^1	*R^2*
–$(CH_2)_4$–	
Me	Me
H	Pr^n
H	H
H	Ph
H	α-furanyl

Treatment of 5-methyl 1-(2′,6′-dimethylheptyl)- or 1-dodecyl-5-methyl-1-cyclohexene with ozone produces 4,8,12-trimethyl-6-ketoundecanoic acid and 4-methyl-6-ketosteric acid, respectively, in excellent yields (Equation (155)) ⟨43JA745⟩.

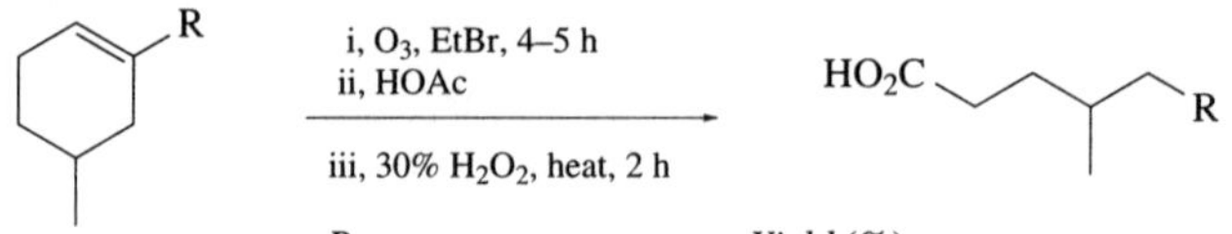

(155)

R	*Yield* (%)
n-$C_{12}H_{25}$	93
$CH_2CH(Me)(CH_2)_3Pr^i$	92

Many hydroxy, alkoxy, and keto substituted alkanoic acids have been prepared via malonic or acetoacetic acid syntheses and a representative selection of such anion preparations are presented in Table 31. An interesting anionic preparation of γ-hydroxy acids involves reaction of metalated acids with epoxides (Equations (156) and (157)) ⟨72JOC1907⟩, while long chain primary keto acids are prepared via alkylzinc chloride reaction with ω-carboethoxyacyl chlorides (Equation (158)) ⟨47JA2350⟩. Anions are also involved in the preparation of a series of α-aryl-β-hydroxy-propionic acids via an Ivanov reagent (Equation (159) and Table 32) ⟨52JA1730⟩.

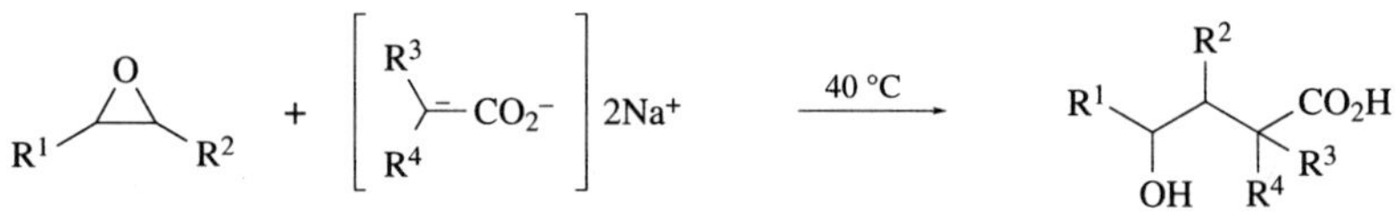

(156)

R^1	R^2	R^3	R^4
$-(CH_2)_4-$		Me	Me
Ph	H	Me	Me
Ph	H	H	Bu^t

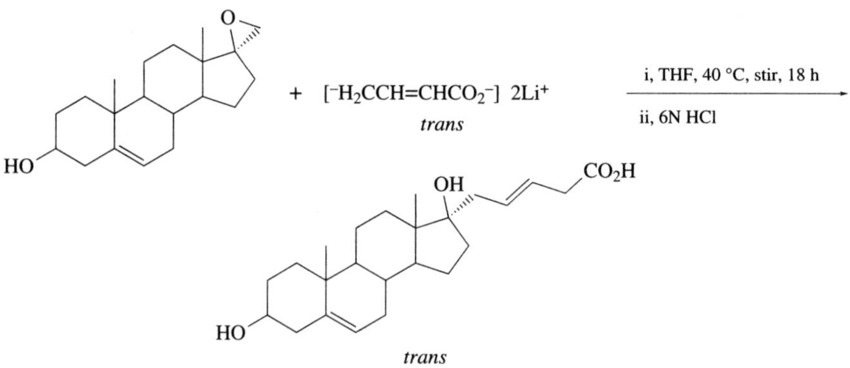

(157)

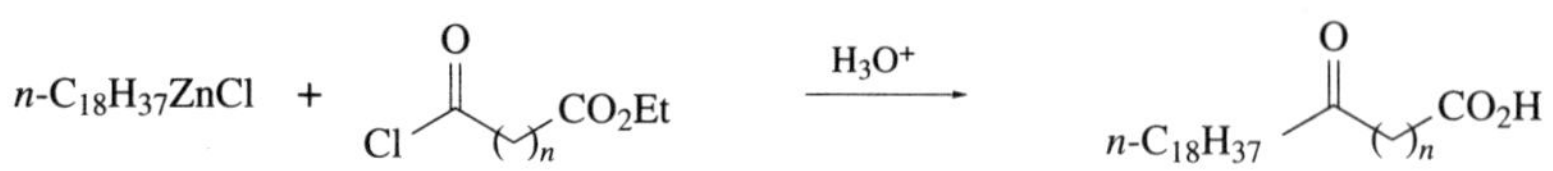

(158)

n	*Yield* (%)
8	77
9	92
10	79
11	89
12	76
13	83
14	77
15	80

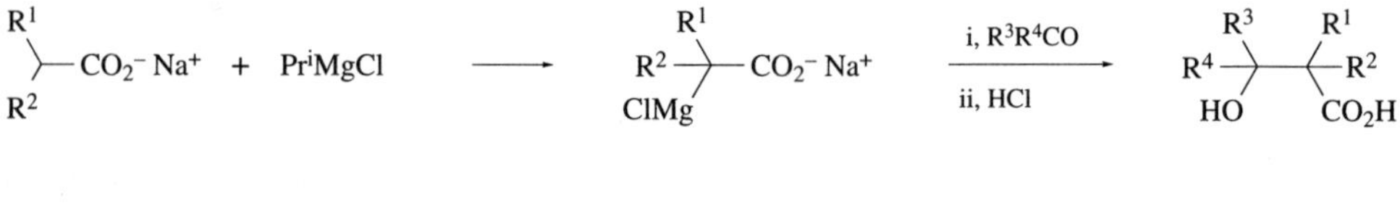

(159)

Table 31 Selected anion preparation of oxygen-substituted alkanoic acids.

Substrate	*Reactant*	*Base and conditions*	*Product*	*Yield (%)*	*Ref.*
$ClCH_2CO_2H$	NaOEt	i, NaOEt, EtOH ii, HCl	$EtOCH_2CO_2H$	74	43OSC(2)260
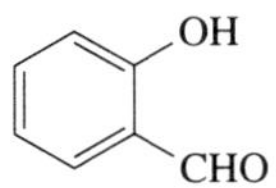	$ClCH_2CO_2H$	i, NaOH, H_2O, reflux, 3 h ii, HCl	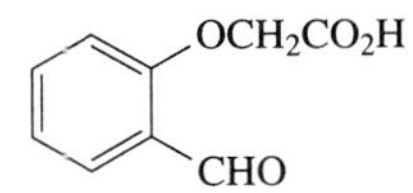	82–83	73OSC(5)251
$H_2C(CO_2Et)_2$	HCHO	$KHCO_3$, H_2O, 25–30 °C, 1 h	$(HOCH_2)_2C(CO_2Et)_2$	72–75	73OSC(5)381
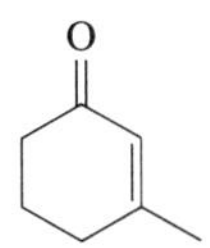	$H_2C(CO_2Et)_2$	i, NaOEt, EtOH, stir 9 d ii, KOH, stir overnight, reflux, 1 h iii, HCl	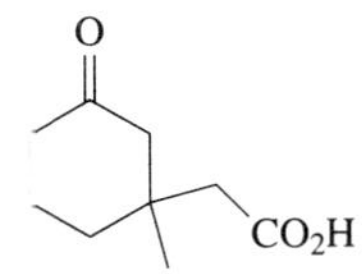	79	90OS(69)173
$PhCH_2O_2CCH_2CH(CO_2CH_2Ph)_2$	n-$C_{10}H_{21}COCl$	i, NaOEt ii, H_2, Pd, heat	n-$C_{10}H_{21}CO(CH_2)_2CO_2H$		50JCS325
$EtCH(CO_2Et)_2$	$(PhCO_2)_2$	i, NaH, C_6H_6, stir 2 h ii, C_6H_6, stir 30 min	$PhCO_2C(Et)(CO_2Et)_2$	75–78	73OSC(5)379
$MeCO_2Et$	Ph_2CO	i, $LiNH_2$, NH_3 ii, NH_4Cl	$Ph_2C(OH)CH_2CO_2Et$	75–84	73OSC(5)564
$MeCOCH_2CO_2Et$	$ClCO(CH_2)_3CO_2Et$	i, Na, ether, stir RT, overnight ii, NH_3 (g), ether	$EtO_2C(CH_2)_3COCH_2CO_2Et$	50–59	73OSC(5)384
$MeCOCH_2CO_2Et$	Bu^nBr	NaOEt, reflux	$MeCOCH(Bu^n)CO_2Et$	69–72	32OSC(1)248
$EtO_2C{-}CO_2Et$	Me_2CO	i, NaOEt, stir 40 °C, 1 h ii, H_2SO_4, H_2O	$MeCOCH_2COCO_2Et$	61–66	32OSC(1)238
$EtO_2C(CH_2)_2CO_2Et$	$EtO_2C{-}CO_2Et$	i, NaOEt, EtOH, RT, stand 12 h ii, HCl, reflux 4 h	$HO_2CCH_2COCO_2H$	73–83	73OSC(5)687

Table 32 α-Aryl-β-hydroxypropionic acid using Ivanov reagent.

R^1	R^2	R^3	R^4	Yield (%)
Ph	H	H	H	66
Ph	H	Me	H	50
Ph	H	Et	H	61
Ph	H	Pr^i	H	63
Ph	H	Et_2CH	H	74
Ph	H	Ph	H	80
Ph	H	Me	Me	77
Ph	H	Me	Et	73
Ph	H	Et	Et	82
Ph	H	Pr^i	Pr^i	45
Ph	H	$-(CH_2)_4-$		61
Ph	H	$-(CH_2)_5-$		84
Ph	H	Ph	Me	73
Ph	H	$-CH_2SCH_2CH_2-$		47
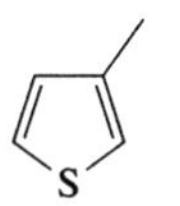	H	$-(CH_2)_5-$		74
Ph	H	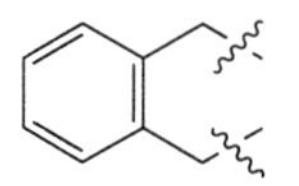		70
Ph	Ph	H	H	91
Ph	$PhCH_2$	H	H	22
Ph	PhO	H	H	26
Ph	$c\text{-}C_6H_{11}$	H	H	94
Ph	Et	$-(CH_2)_4-$		57
Ph	H	Ph	Ph	35
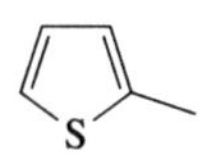	H	H	H	69
$p\text{-}C_6H_5C_6H_4$	H	H	H	78
$p\text{-}C_6H_5C_6H_4$	H	$-(CH_2)_5-$		83
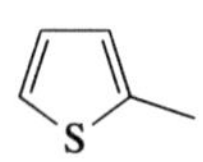		$-(CH_2)_4-$		87
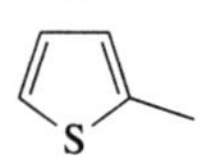		$-(CH_2)_5-$		77

5.02.2.3.4 By miscellaneous reactions

Several miscellaneous reactions which have been used to prepare oxygen-substituted alkanoic acids are worthy of discussion beginning with the hydrolysis of esters using a lithium salt nucleophile in a basic solvent ⟨60HCA113, 65JCS6655⟩. The most complete study of this approach to ester hydrolysis has been performed using the methyl ester shown in Equation (160) and Table 33, with the results indicating that lithium iodide in either 2,6-lutidine, 2,4,6-collidine or DMF produced the best yields ⟨65JCS6655⟩. Methyl ester cleavage to carboxylic acids has also been accomplished for methyl mesitoate, methyl *O*-methylpodocarpate, methyl triisopropyl acetate and methyl 3β-acetoxy-Δ^5-etienate using dbn in refluxing xylene ⟨72TL3987⟩.

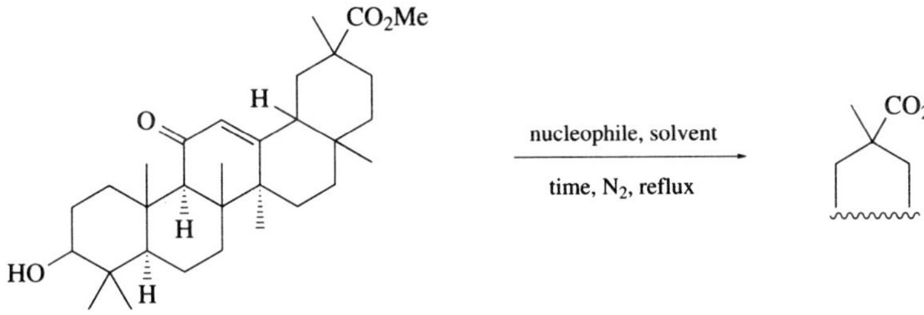

(160)

Table 33 Methyl ester hydrolysis of methyl glycyrrhetate using a lithium salt nucleophile in a basic solvent.

Nucleophile	*Solvent*	*Time* (h)	*Yield* (%)
Li I	C_5H_5N	8	30
Li I	2,6-Lutidine	8	100
Li I	2,4,6-Collidine	8	100
Li I	*N*-Methyl-2-pyrrolidone	3	86
Li I	DMF	1.5	100
Li Br	DMF	4	100
Li F	DMF	24	0
Li OH	DMF	16	100
Na I	DMF	16	40

Treatment of butenolides with zinc acetate causes elimination of the oxalyl moiety as zinc oxylate and formation of the corresponding 2-hydroxy-4-ketocarboxylic acids (Equation (161) and Table 34) ⟨83G421⟩. An interesting enantioselective preparation of α,α-disubstituted γ-keto esters, which can eventually be hydrolyzed to their corresponding acids, involves treatment of chiral bicyclic lactams with 10% sulfuric acid in *n*-butanol (Equation (162)) ⟨84JA1146⟩.

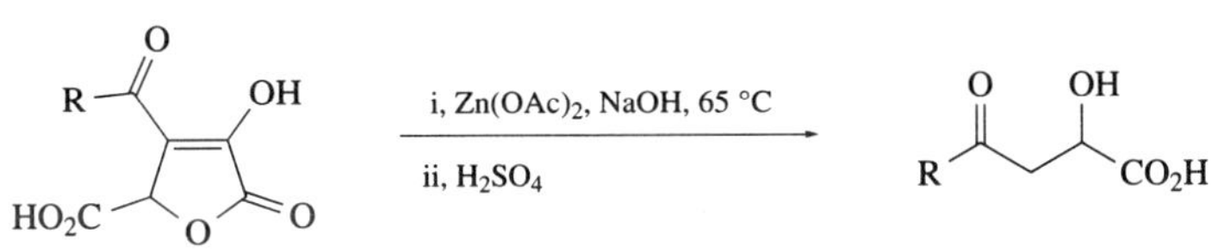

(161)

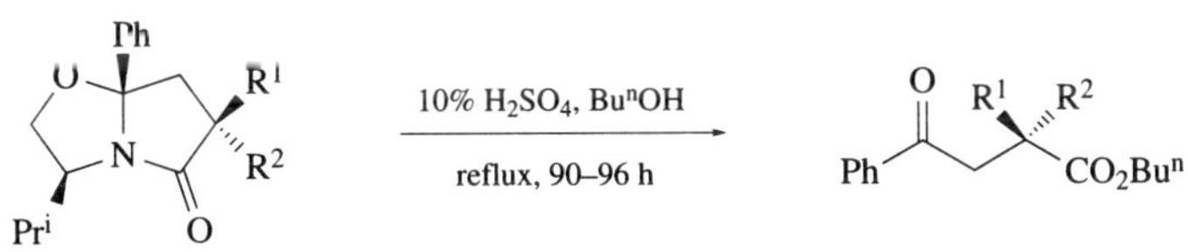

(162)

R^1	*R^2*	*Configuration*	*Yield* (%)
Me	$PhCH_2$	*S*	79
$PhCH_2$	Me	*R*	78
Me	Et	*S*	71
Et	Me	*R*	87
Me	*p*-$MeOC_6H_4CH_2$	*S*	88
p-$MeOC_6H_4CH_2$	Me	*R*	92

Novel preparation of α-keto acids include the basic hydrolysis of 5-ylidene-2,2-pentamethylene-1,3-dioxolan-4-ones (Equation (163)) ⟨84JCS(P1)1531⟩, and the similar hydrolysis of 4-alkyl-2,2-bis(trifluoromethyl)-2*H*-oxazol-5-ones (Equation (164)) ⟨79LA1547⟩.

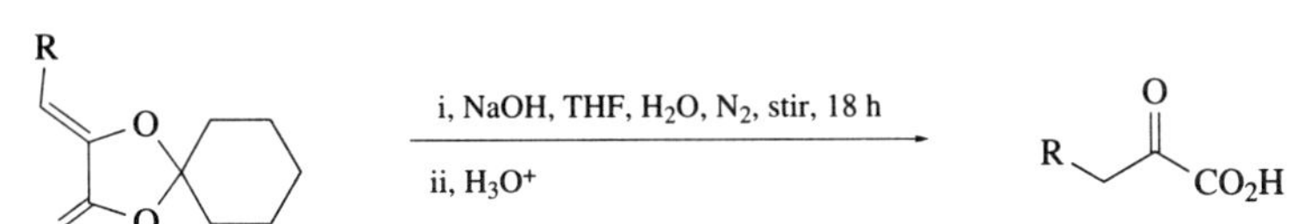

(163)

R	*Yield* (%)
Pr^n	76
Pr^i	73
Ph	89

Table 34 Conversion of butenolides to 2-hydroxy-4-ketocarboxylic acids using zinc acetate.

R	*Time* (h)	*Yield* (%)
Me	1.5	76
Pr^i	2	87
Bu^i	2	84
Ph	1	83
EtO	1.5	86
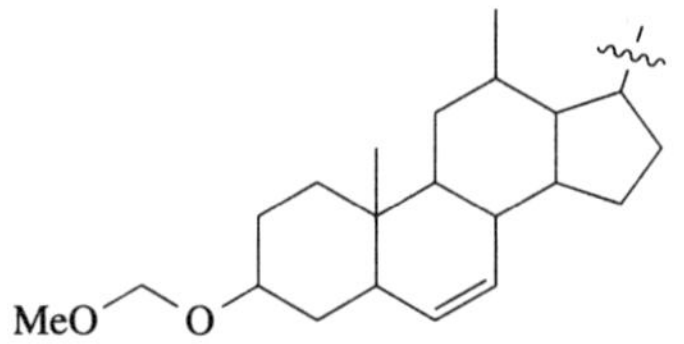	5	45

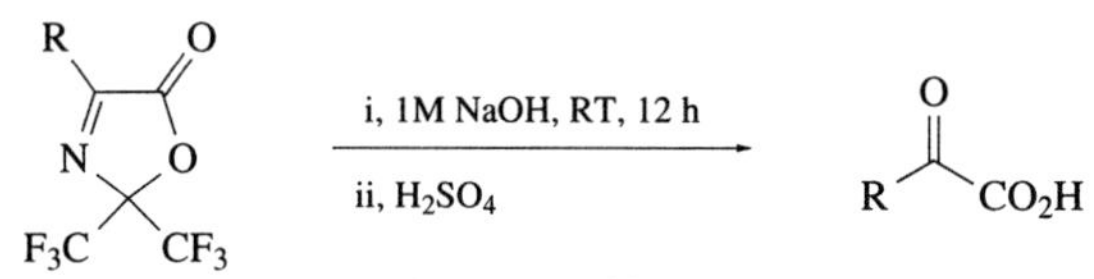

(164)

R	*Yield* (%)
Ph	91
$PhCH_2$	67

In the hydroxy acid series, α-substituted α-hydroxyphenylacetic acids have been prepared by the base-catalyzed hydrolysis of α-substituted α-hydroxyoxazolines (Equation (165) and Table 35) ⟨80JOC2785⟩, while (2*S*, 3*S*)-3-hydroxy-3-phenyl-2-methylpropanoic acid has been prepared by the lithium hydroxide–hydrogen peroxide catalyzed removal of the chiral auxiliary from (*S*)-3-(1-oxo-2-methyl-3-hydroxy-3-phenylpropyl)-4-(phenylmethyl)-2-oxazolidinone (Equation (166)) ⟨89OS(68)83⟩.

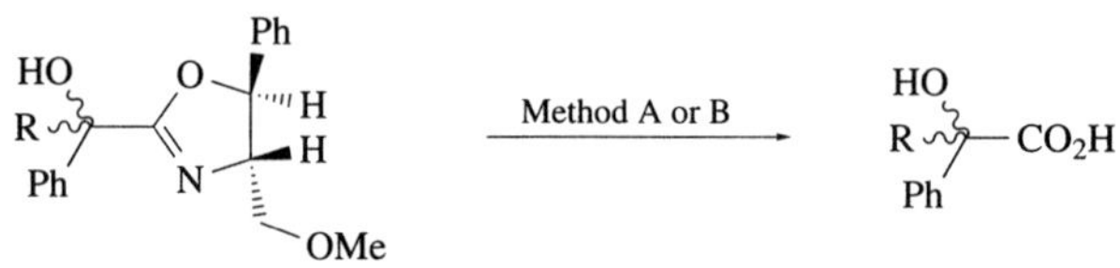

(165)

Method A = methyl iodide in base
i, MeI, Me_2SO, stir, RT, 18–30 h
ii, 2M KOH, reflux, 8 h
iii, 0 °C, 12M HCl

Method B = triethyloxonium fluoroborate
i, Et_3OBF_4, CH_2Cl_2, stir, RT, overnight
ii, 2M KOH, DMSO, reflux, 8 h

Table 35 α-Substituted α-hydroxyphenylacetic acids via base-catalyzed hydrolysis of α-substituted α-hydroxyoxazolines.

R	*Yield* (%)	*ee* %	*Configuration*
Me	70–76[a]	0–47.6[a]	R or S[a]
Et	65	33.0	S
Pr^n	62	39.0	S
Pr^i	57	41.0	S
Bu^i	55	50.0	S
p-MeC_6H_4	60	76.0	S
p-$MeOC_6H_4$	55	62.0	S
1-Naphthyl	62	65.0	S
2-Thienyl	73	87.0	S

[a] Different reaction conditions afford different values for pure R or S configurations.

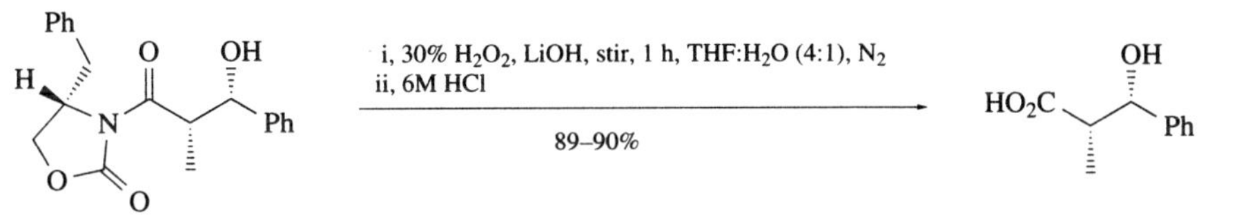

(166)

5.02.2.4 Sulfur-substituted Alkanoic Acids

Two general reviews on the synthesis of sulfur-substituted alkanoic acids have been published which cover the synthesis of monothio carboxylic acids ⟨B-78MI 502-03⟩ and the methods of synthesis of dithiocarboxylic acids and esters ⟨83S605⟩, while more specific reviews of lipoic acid (6,8-epidithio-octanoic acid, ⟨55MI 502-02, 57MI 502-04, 61MI 502-01, 65AG(E)846⟩ and the synthesis of L-2-oxothiazolidine-4-carboxylic acid using phenyl chloroformate ⟨58MI 502-03⟩ have also been published.

5.02.2.4.1 By hydrolysis reactions

Low temperature treatment of α-(*t*-butyl) D- and DL-α-phenoxymethylpenicilloate with gaseous hydrogen chloride in methylene chloride causes ester hydrolysis and formation of D- and DL-phenoxymethylpenicilloic acid hydrochloride (Equation (167)) ⟨59JA3089⟩. Trifluoroacetic acid in aromatic solvents has been found to be effective for ester hydrolysis of penicillanic esters without effecting the amide side chain or the bicyclic ring structure (Equation (168)) ⟨69JA5674⟩. Similar results have been obtained using a Lewis acid in methylene chloride (Equation (169)) ⟨78GEP2725519⟩ or sodium hydroxide in pyridine–water mixtures ⟨69JA5674⟩. In molecules which contain an acid substituent positioned adjacent to a methoxycarbonyl group ester hydrolysis can be accomplished in good to excellent yields using pyridinium chloride in pyridine (Equation (170)) ⟨81S487⟩.

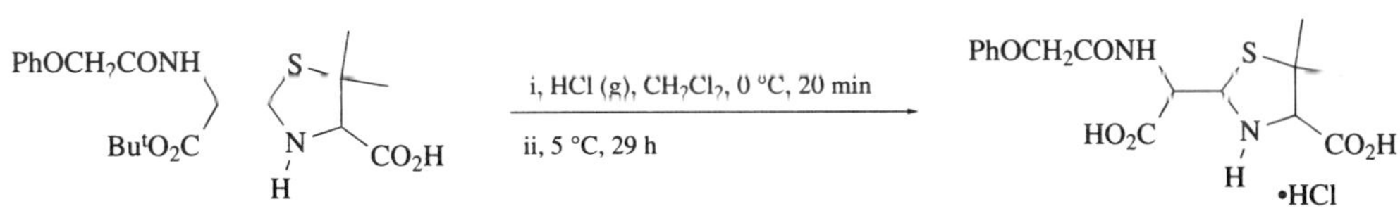

(167)

Configuration	*Yield (%)*
D	94
DL	96

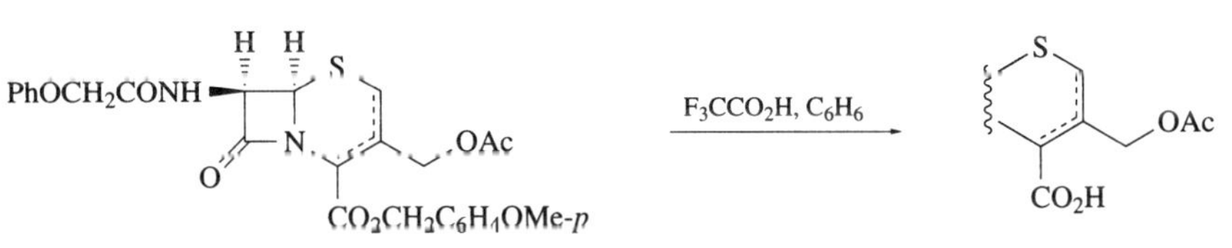

(168)

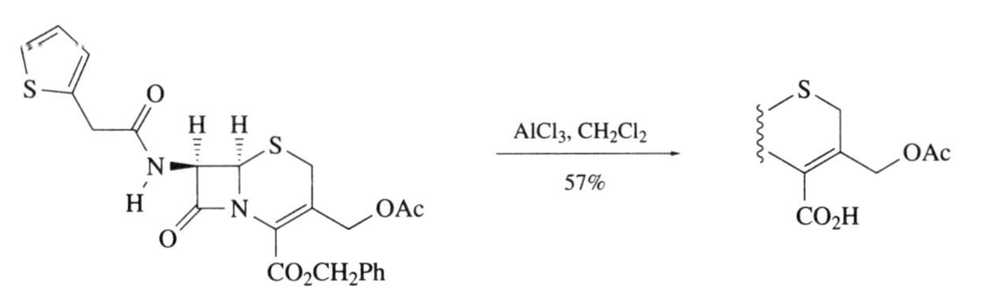

(169)

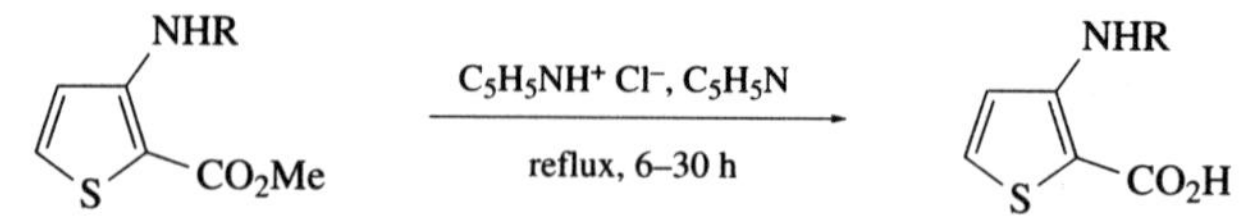

(170)

R	*Yield* (%)
COMe	51
$COCF_3$	67
$COBu^t$	80
COPh	83
SO_2Me	100
$SO_2C_6H_4Me$-*p*	78

In the amide class of compounds 2-carbamyl-4-hexadecylaminothiophene can be hydrolyzed to 4-hexadecylamino-2-thiophenecarboxylic acid by refluxing for 40 hours in a mixture of 6M hydrogen chloride in methyl cellosolve ⟨83USP4382143⟩.

5.02.2.4.2 By carbonation of organometallic reagents

Reports of carbonation of metallated substrates containing sulfur have included the carbonation of the lithium salt of *trans*-1,5-dimethyl-7-*p*-toluenesulfonyl-1,5-octadiene (Equation (171)) ⟨84S876⟩, and the mono- and dilithium salts of 2,2′;5′,2″-terthiophene (Equation (172)) ⟨83JOC4076⟩. An interesting carbonation reaction is observed to occur when 2-lithiothiophene is treated first with trialkylboron, forming a 2-trialkylboronthiophene lithium salt, followed by treatment with carbon dioxide gas. The final products from this sequence of reactions are 5-alkylthiophene-2-carboxylic acids which are formed via the mechanism shown in Scheme 51 ⟨86MI 502-01⟩. Another interesting carbonation reaction has been used to convert 6,11-dihydro[*b,e*]thiepin-11-on-3-yl acetic acid into its ^{14}C-acid analogue (Scheme 52) ⟨79MI 502-08⟩.

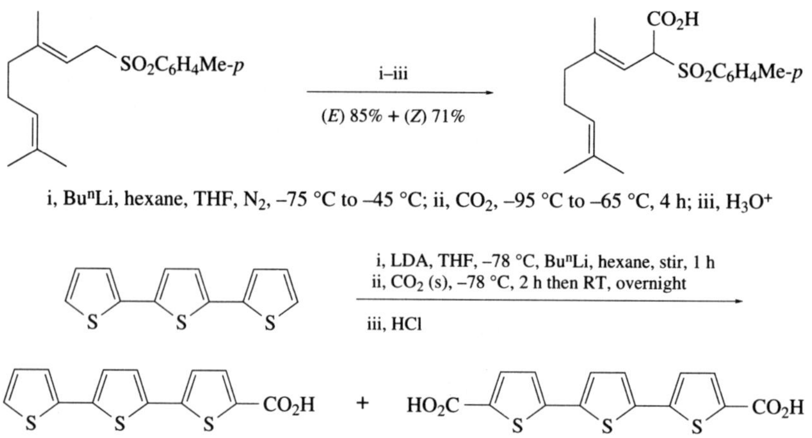

(171)

i, Bu^nLi, hexane, THF, N_2, –75 °C to –45 °C; ii, CO_2, –95 °C to –65 °C, 4 h; iii, H_3O^+

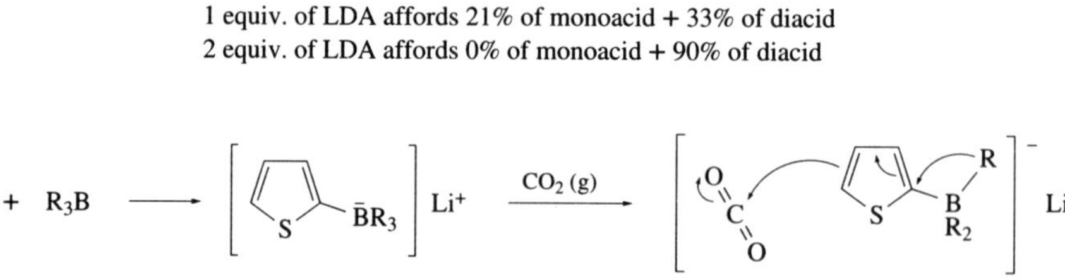

(172)

0.5 equiv. of LDA affords 69% of monoacid + 3% of diacid
1 equiv. of LDA affords 21% of monoacid + 33% of diacid
2 equiv. of LDA affords 0% of monoacid + 90% of diacid

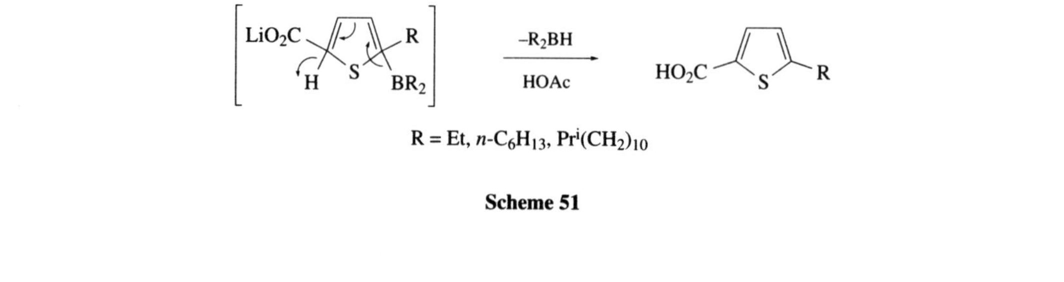

R = Et, *n*-C_6H_{13}, $Pr^i(CH_2)_{10}$

Scheme 51

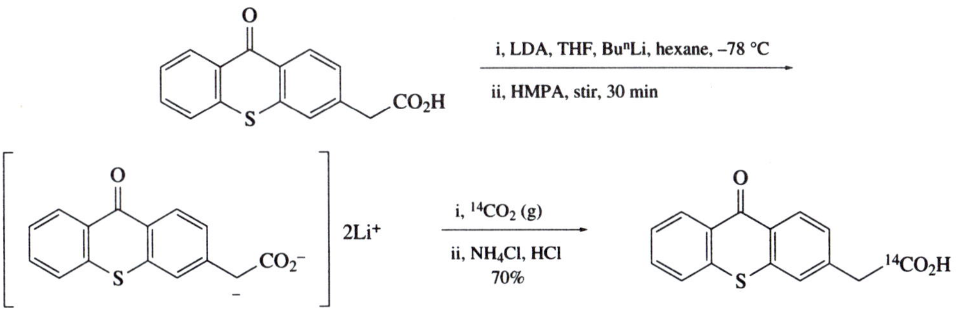

Scheme 52

5.02.2.4.3 By oxidation reactions

Haloform oxidation of 2-acetylthiophene using sodium bromite produces a 40% conversion to 2-thenoic acid ⟨83CC392⟩, while 3-thenoic acid has been prepared in 95–97% yield by oxidation of 3-thienaldehyde using silver oxide in sodium hydroxide ⟨63OSC(4)919⟩.

5.02.2.4.4 By miscellaneous reactions

An example of an anion preparation of a sulfur-substituted carboxylic acid is the reaction of potassium dithiobenzoate with sodium chloroacetate to form thiobenzoylthioglycolic acid (Equation (173)) ⟨73OSC(5)1046⟩.

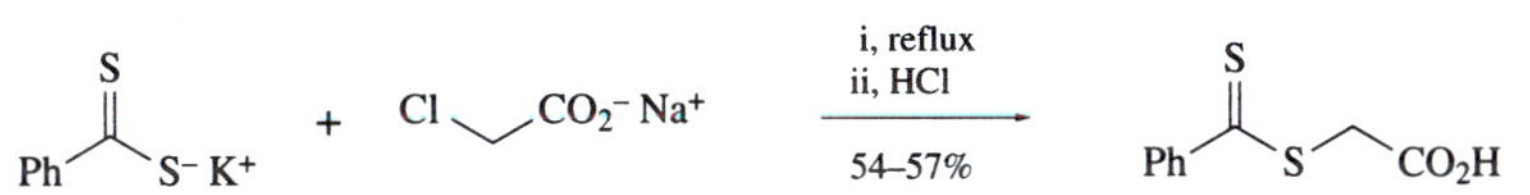

(173)

An interesting preparation of ω-alkyl- or arylthio acids involves the ring opening of lactones using aluminum halide and ethanethiol via the mechanism shown in Scheme 53 ⟨81JOC5163⟩. The structure of the acid product obtained depends upon the structure of the lactone substrate as illustrated in Equation (174) and Table 36.

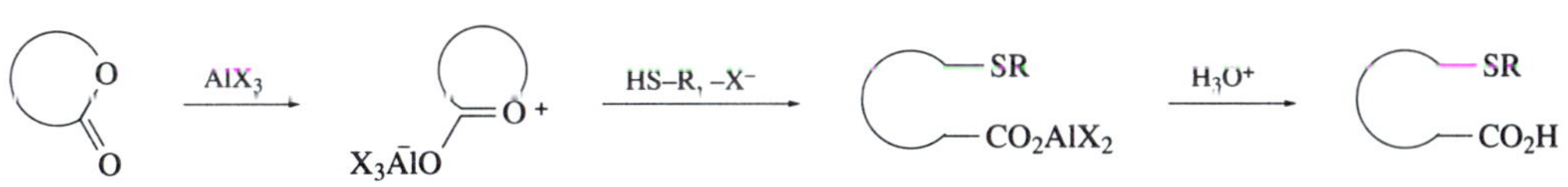

Scheme 53

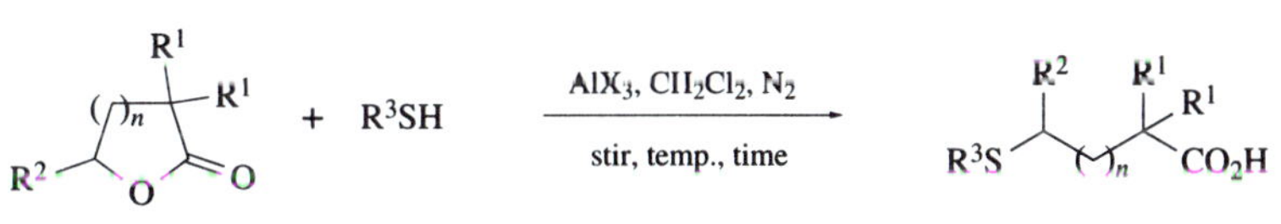

(174)

5.02.2.5 Amino Acids

By far more books and reviews have been written about this class of carboxylic acid than any other class of acids. General reviews have been published on: the chemistry of amino acids and proteins ⟨B-41MI 502-01, 44MI 502-01, 45MI 502-01, 46MI 502-01, 47MI 502-02, 48MI 502-03, 49MI 502-01, 49PCS2099, 50MI 502-01, 51MI 502-02, B-51MI 502-02, 52BSF821, 52MI 502-04, 53MI 502-02, 54MI 502-03, 55MI 502-03, B-55MI 502-02, 56MI 502-01, 57MI 502-05, 59MI 502-01, 60ANY(88)533, 60MI 502-02, B-61MI 502-01, B-61MI 502-02⟩; natural amino acids ⟨B-43MI 502-02⟩; the isolation and synthesis of the naturally occurring α-amino acids ⟨46CRV501⟩; new amino acids and keto acids in green plants, and the biosynthesis of amino acids ⟨55AG381⟩; new natural amino acids ⟨56AG313⟩; and, the biosynthesis of aromatic amino acids and its regulation ⟨68AG(E)350⟩. Other reviews have covered, activation of amino acids ⟨57MI 502-03⟩,

Table 36 ω-Alkyl- or arylthio acids via ring opening of lactones using aluminum halide/ethanethiol.

R¹	*R²*	*R³*	n	*X*	*Temp* (°C)	*Time* (h)	*Yield* (%)
H	H	Et	0	Cl	0	1.5	61
H	H	Ph	0	Cl	RT	2	81
H	H	Et	1	Br	RT	1.3	95
H	H	Pr^n	1	Br	RT	1	92
H	H	Pr^i	1	Br	RT	1.2	94
H	H	Bu^i	1	Br	RT	3	83
H	H	Ph	1	Br	RT	7	54
Ph	H	Et	1	Br	RT	21	84
Ph	H	Ph	1	Br	RT	50	30
H	Me	Et	1	Cl	RT	23	85
H	Me	Et	1	Br	RT	2	81
H	n-C_7H_{15}	Et	1	Cl	RT	37	91
H	n-C_7H_{15}	Et	1	Br	RT	13	91
	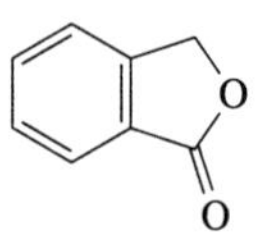	Et		Br	RT	23	10
	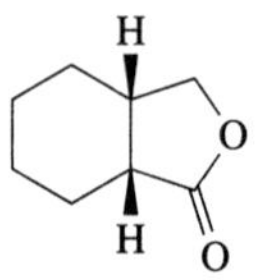 (*dl-cis*)	Et		Br	RT	11	49
H	H	Et	2	Br	RT	3.5	53
H	H	Ph	2	Br	reflux	0.5	32
	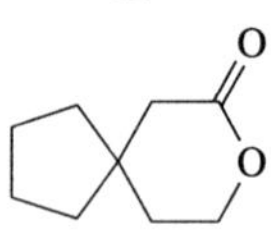	Ph		Br	RT	10 days	45
H	H	Et	3	Br	RT	13	45

the synthesis of optically active α-amino acids ⟨B-89MI 502-01⟩, stereochemistry of amino acids ⟨48MI 502-01⟩, the preparation and criteria of purity of the amino acids ⟨47MI 502-01⟩, resolution of racemic α-amino acids ⟨54MI 502-04⟩, and, displacement chromatography as a preparative method for amino acids ⟨50CI(L)383⟩.

Reviews of structural studies of amino acids and proteins include, x-ray studies ⟨48MI 502-02⟩, ultraviolet absorption spectra ⟨52MI 502-03⟩, infrared analysis ⟨52MI 502-02⟩, solid state infrared absorption spectra of optically active and racemic straight-chain α-amino acids ⟨57ANY(69)94⟩, mass spectrometry ⟨60C293⟩, and rotary dispersion ⟨61T176⟩.

More specialized review articles have been published on: the production and utilization of amino acids ⟨78AG(E)176⟩; synthesis and stereochemistry of β-hydroxy acids and β-amino acids and their derivatives via aldol-type condensations ⟨B-80MI 502-01⟩; hydroxy- and amino-monocarboxylic acids and related compounds ⟨B-52MI 502-02⟩; chemistry and biology of hydroxy-amino acids ⟨44AG135⟩; *N*-hydroxy-α-amino acids in organic chemistry ⟨86CRV697⟩; iodinated amino acids and iodo proteins ⟨55FOR349⟩; new synthesis of sulfur-containing amino acids ⟨48AG308⟩; synthetic study of α-amino acids and related compounds using isonitriles ⟨77NKK109⟩; and, chemical synthesis and resolution of glutamic acid ⟨63CRV625⟩.

The synthesis and chemistry of poly-amino acids and peptides has been covered in a large number of books and review articles ⟨51AG7, 54AG507, 57AG362, 57MI 502-02, 58MI 502-04, B-58MI 502-01, 60ANY(88)676, 61BSF2001, B-62MI 502-02, 62RCP3, 63AG(E)358, 63PAC335, 64MI 502-01⟩, while more specific reviews on terminal amino acids in peptides and proteins ⟨45MI 502-02⟩, poly-α-amino acids ⟨51MI 502-01⟩, poly-β-amides ⟨68AG(E)278⟩, synthesis of biologically active peptides ⟨62C385⟩, and the synthesis of peptides from anhydrides ⟨B-55MI 502-03⟩ have also been published. One review on the determination of the constitution of peptides and proteins has also appeared ⟨57AG189⟩.

Various aspects of the synthesis and chemistry of proteins have also been reviewed including: synthesis ⟨B-52MI 502-09⟩; discoveries in the evolution of proteins ⟨74AG(E)186⟩; amino acid sequences in proteins ⟨52MI 502-01, 59N(46)35⟩; structure of proteins ⟨49BSFD251, B-53MI 502-01, B-62MI 502-03, B-63MI 502-03⟩.

More specific reviews on the sulfur chemistry of proteins ⟨59MI 502-02⟩, and ribonucleic acid and protein synthesis ⟨61JCE217⟩ have been published, while reviews on the structural studies of proteins include: ultraviolet spectra of proteins and amino acids ⟨62MI 502-01⟩; optical rotatory dispersion of polypeptides and proteins ⟨61T134, 61T143⟩; and NMR spectra of proteins ⟨58ANY(70)870⟩.

5.02.2.5.1 By hydrolysis reactions

Most of the hydrolysis reactions used to prepare amino acids involve the hydrolysis of nitriles which are, in turn, prepared by addition of cyanide to aldehydes or ketones. Examples of this approach are the preparation of DL-alanine from acetaldehyde (Scheme 54) ⟨32OSC(1)21⟩, and α-aminoisobutyric acid ⟨43OSC(2)29⟩, diethylacetic acid ⟨55OSC(3)66⟩, and α-phenylpropionic acid ⟨55OSC(3)88⟩ from acetone, 3-pentanone and acetophenone, respectively (Scheme 55). Acid hydrolysis of γ-phthalimidobutyronitrile affords a 47–62% yield of γ-aminobutyric acid ⟨43OSC(2)25⟩, while basic hydrolysis of β-cyanoethylamine using barium hydroxide affords a 75–80% yield of β-alanine (Scheme 56) ⟨55OSC(3)34⟩. Barium hydroxide has also been used to produce glycine from the intermediate bisulfate salt of cyanomethylamine prepared from methyleneaminoacetonitrile ⟨32OSC(1)298⟩, but changing conditions converts the starting nitrile to glycine ethyl ester hydrochloride (Scheme 57) ⟨43OSC(2)310⟩.

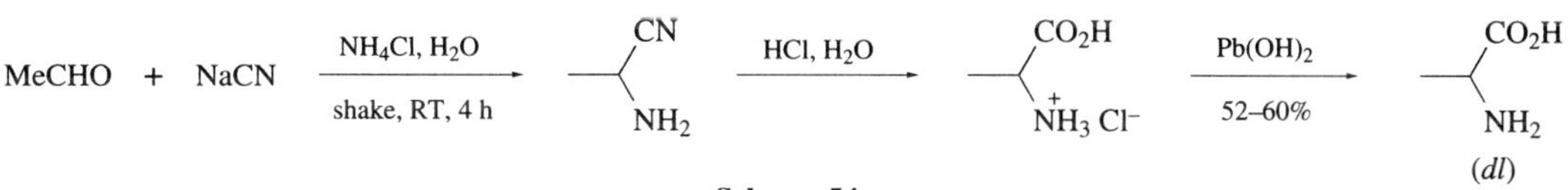

Scheme 54

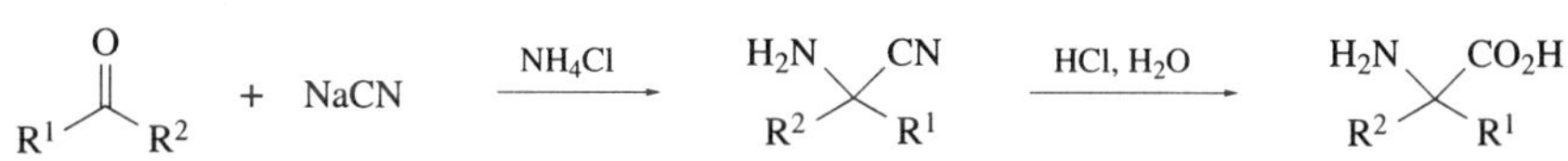

R^1	R^2	*Yield* (%)
Me	Me	30–33
Et	Et	39–43
Ph	Me	40

Scheme 55

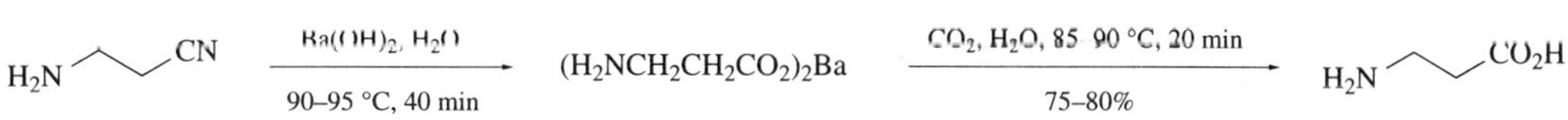

Scheme 56

Acid-catalyzed ring opening of the *N*-protected β-lactone *N*-(benzoyloxycarbonyl)-L-serine β-lactone using trifluoroacetic acid (Method A) or a mixture of trifluoroacetic acid and *p*-toluenesulfonic acid (Method B) produces stereochemically pure α-amino acids (Scheme 58, Table 37) ⟨88JA2237⟩, but if a similar substrate is hydrolyzed using pyrazole in acetonitrile, a 43% yield of *N*-(benzoyloxycarbonyl)-β-(pyrazol-1-yl)-2-alanine is obtained (Equation (175)) ⟨91OS(70)1⟩.

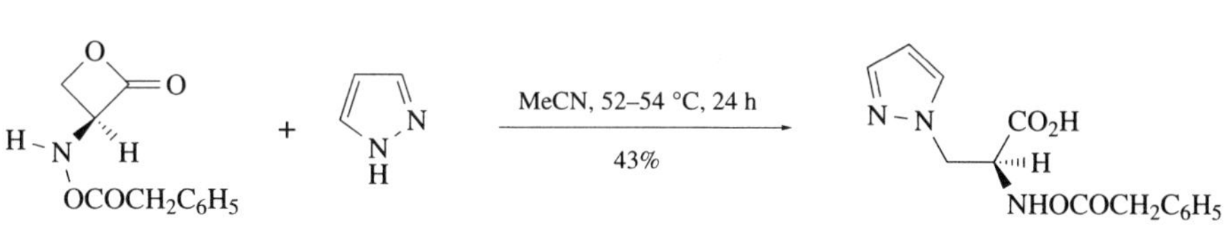

(175)

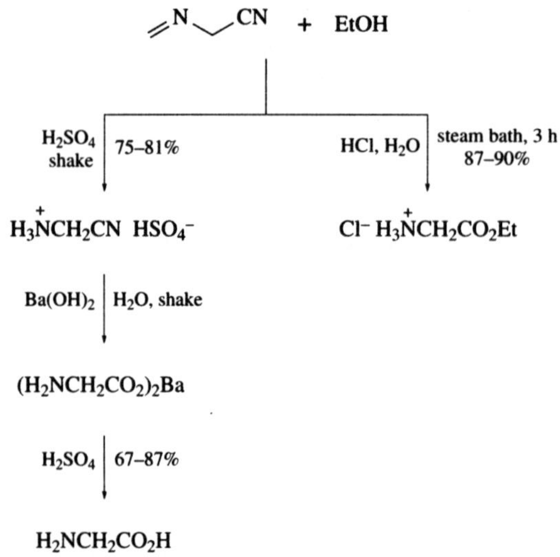

Scheme 57

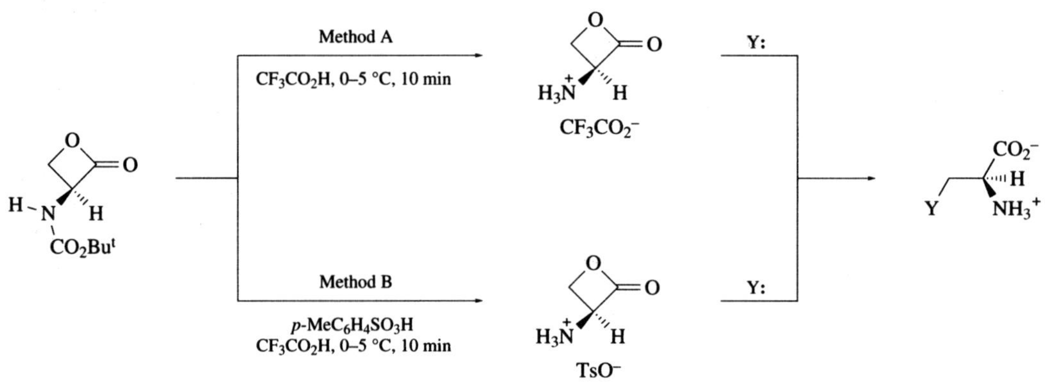

Scheme 58

Table 37 Stereochemically pure α-amino acids via acid-catalyzed ring opening of *N*-(benzoyloxycarbonyl)-L-serine β-lactone using various reagents.

Method	*Y (nucleophile)*	*Yield (%)*
A	$^-O_2CCH(\overset{+}{N}H_3)CH_2SH$	93
A	$Cl^-H_3\overset{+}{N}CH_2CH_2SH$	85
A	HS^- Li^+	88
A	$Na_2S_2O_3$	83
B	Me_2S	88
A	HCl	92
A	$Bu_4^nN^+$ CN^-	84
B	Pyrazole	77
B	NaN_3	96

Treatment of succinimide with potassium hydroxide and potassium hypobromite causes ring opening and formation of a 41–45% yield of β-alanine ⟨43OSC(2)19⟩.

ω-Aminoalkanoic acids of varying length have been prepared by the hydrolysis of lactams in acid (Equation (176)) ⟨43OSC(2)28, 63OSC(4)39⟩ or by hydrolysis in the presence of Li, Ca or Ba hydroxide and CO_2 (Equation (177)) ⟨81JAP39050⟩, the CO_2 being used to precipitate the metal in the form of its carbonate salt.

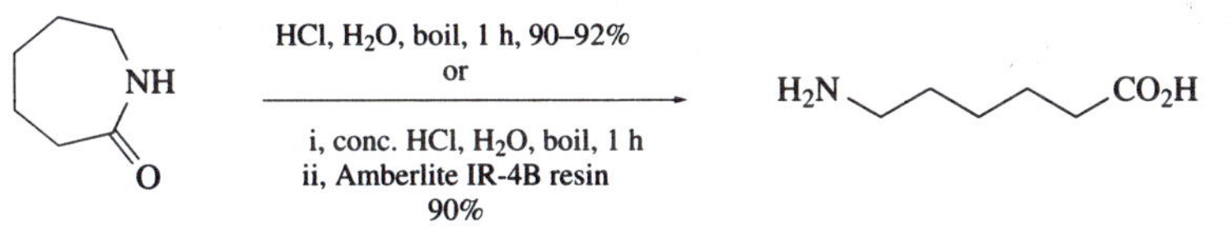

(176)

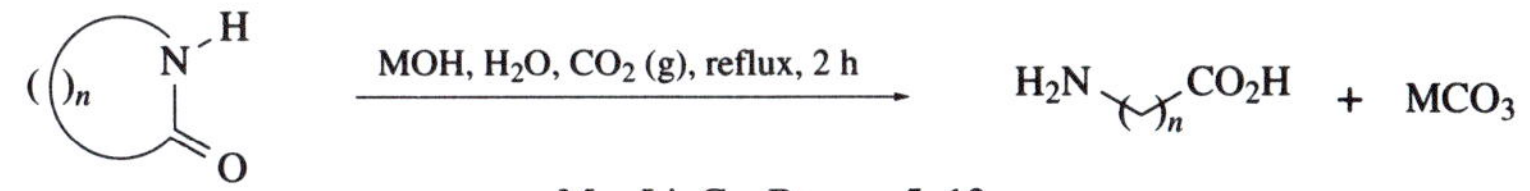

(177)

M = Li, Ca, Ba; n = 5–13

M	*n*	*Yield* (%)
Li	5	93
Ca	5	54
Ba	5	67

Hydrolysis of the azlactone of α-benzoylaminocinnamic acid produces good yields of DL-β-phenylalanine (Equation (178)) ⟨43OSC(2)489⟩, while hydrolysis of its 4-azaspiro analogue, (*Z*)- or (*E*)-7-oxo-5-phenyl-6-oxo-4-azaspiro[2,4]hept-4-enes, produces good yields of its cyclopropane α-amino acid analogues (Scheme 59) ⟨85S773⟩.

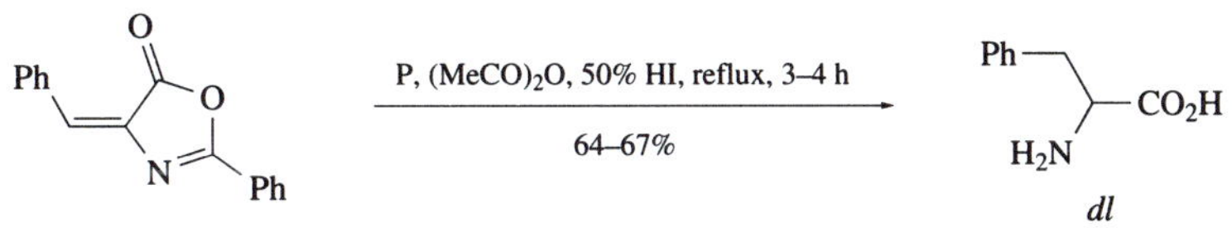

(178)

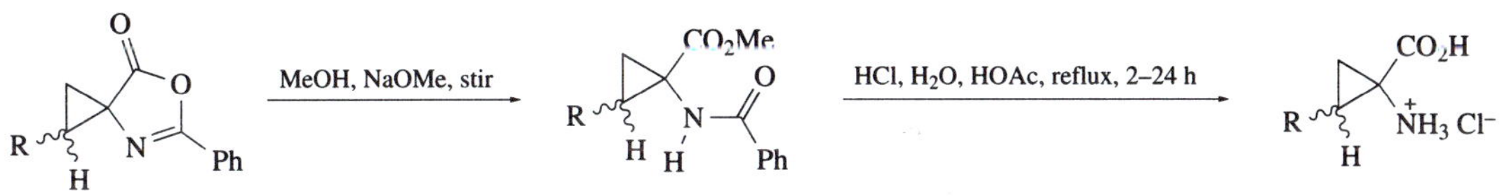

R	*Stereochemistry*	*Yield* (%)
Ph	(*E*)	85
Ph	(*Z*)	60
p-MeC$_6$H$_4$	(*E*)	80
p-MeC$_6$H$_4$	(*Z*)	55
p-MeOC$_6$H$_4$	(*Z*)	28
m-MeOC$_6$H$_4$	(*E*)	76
p-ClC$_6$H$_4$	(*E*)	82
p-ClC$_6$H$_4$	(*Z*)	50
o-ClC$_6$H$_4$	(*E*)	88
Me	(*E*)	90
Me	(*Z*)	84

Scheme 59

Both the *N*-carboalkyloxyamino ⟨57JA637⟩ and the *N*-benzylsulfonyl ⟨57JA645⟩ acid phenylhydrazides have been hydrolyzed using ferric chloride producing the corresponding *N*-substituted amino acids (Scheme 60 and Table 38), which can be further cleaved using sodium in liquid ammonia to produce the unprotected amino acid.

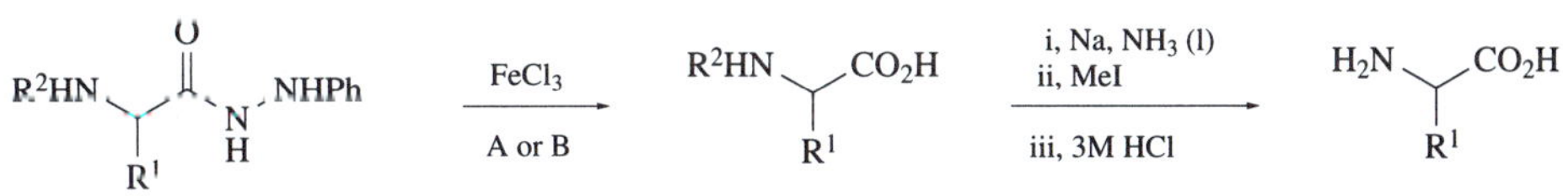

Conditions A: HCl, cellosolve, H_2O, 96 °C
Conditions B: i, Me_2CO, 35–40 °C; ii, reflux, 2 h; iii, 3M NaOH

Scheme 60

Excellent yields of α-hydrazino acids of ⩾98% optical purity and specific absolute configuration have been obtained upon saponification of the corresponding α-hydrazinoesters which upon hydro-

Table 38 *N*-Substituted amino acids from *N*-carboalkyloxyamino and *N*-benzylsulfonyl acid phenylhydrazines via ferric chloride hydrolysis.

R[1]	*R*[2]	*Conditions*	*Yield* %
H	$CH_2{=}CHCH_2CO_2$	A	
Me	$CH_2{=}CHCH_2CO_2$	A	29[a]
Bu^i	$CH_2{=}CHCH_2CO_2$	A	85[a]
Ph	$CH_2{=}CHCH_2CO_2$	A	37[a]
Bu^i	$PhCH_2SO_2$	B	87[a]
$MeSCH_2CH_2$	$PhCH_2SO_2$	B	83[a], 77[b]

[a] Yield of *N*-substituted amino acid. [b] Yield of unprotected amino acid.

genolysis using H_2/PtO_2 affords the corresponding α-amino acids in excellent yields (Scheme 61) ⟨86JA6394⟩.

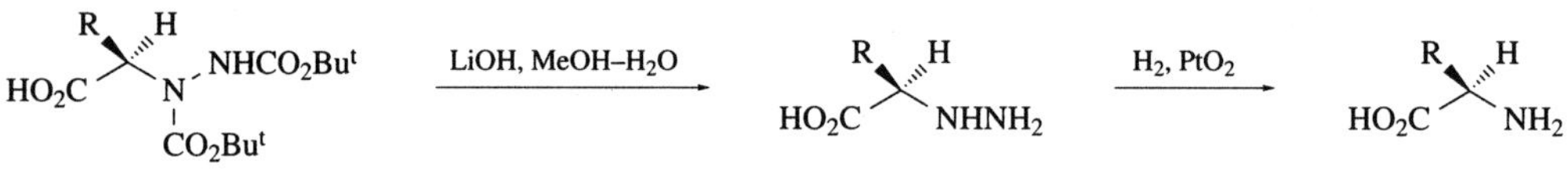

R	*Yield (hydrazino acid)*[a] (%)	*Yield (amino acid)*[a] (%)
Me	78	92
$PhCH_2$	81	89
Bu^i	81	91
Et	65	80
Bu^n	78	90

[a] All products had *R* absolute configuration and 98% *ee*

Scheme 61

ω-Aminoalkanoic acids have been prepared from ω-amino substituted vinyl 1,1-dichlorides upon hydrolysis using sulfuric acid (Equation (179)) ⟨57JGU(27)2481⟩.

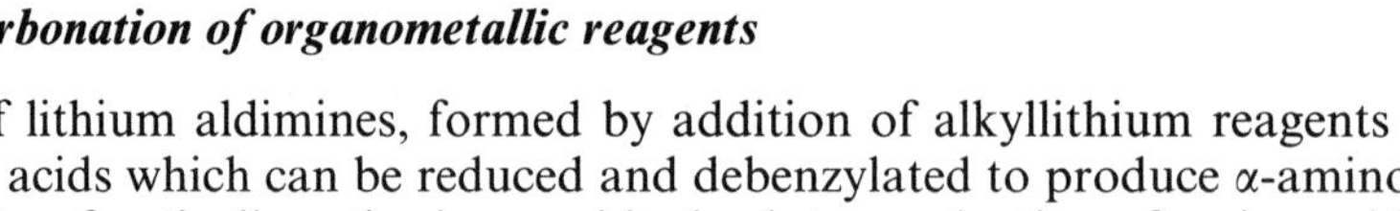

(179)

n	*Yield* (%)
6	86
8	84

5.02.2.5.2 *By carbonation of organometallic reagents*

Carbonation of lithium aldimines, formed by addition of alkyllithium reagents to isocyanides, produces α-imino acids which can be reduced and debenzylated to produce α-amino acids (Scheme 62) ⟨74JOC604⟩. Use of optically active isocyanides leads to production of amino acids of reasonably high optical purity. In a related reaction of *n*-butyllithium with benzylisocyanide which contains α-hydrogens, the initial α-lithio derivative formed can be carbonated to give the *N*-formyl α-amino acid which upon subsequent hydrolysis of the *N*-formyl group produces the free amino acid product α-phenylglycine (Scheme 63) ⟨72SC423⟩.

5.02.2.5.3 *By miscellaneous reactions*

The method most used for preparation of amino acids is the reaction of a haloacid with ammonia under a variety of conditions as illustrated in Equation (180) and Table 39 for some selected examples. In some cases this approach is combined with a deprotecting step for the production of diamino acids (Scheme 64) ⟨43OSC(2)374⟩.

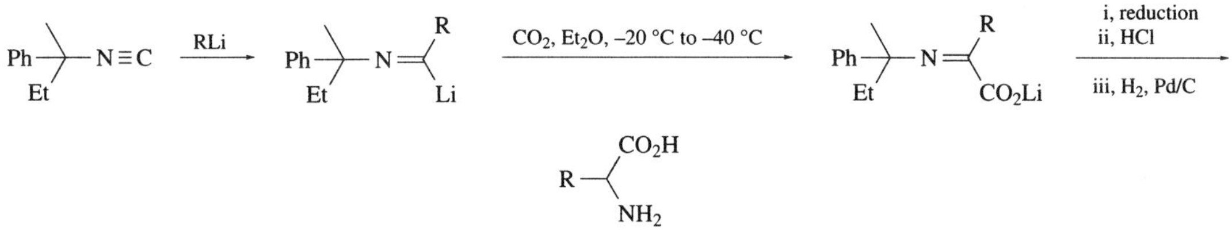

R	*Yield* (%)
Et	64
Pr^i	97
Bu^s	85

Scheme 62

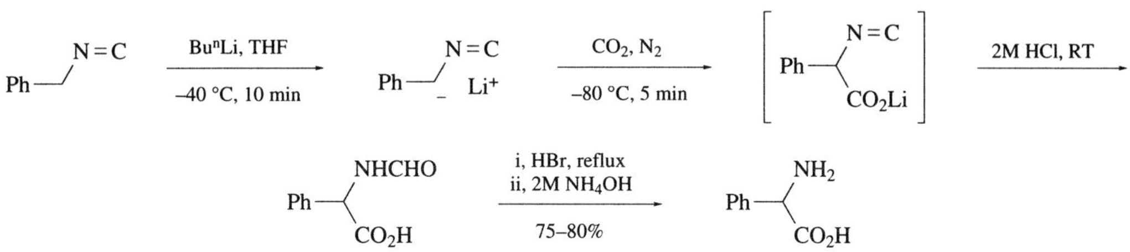

Scheme 63

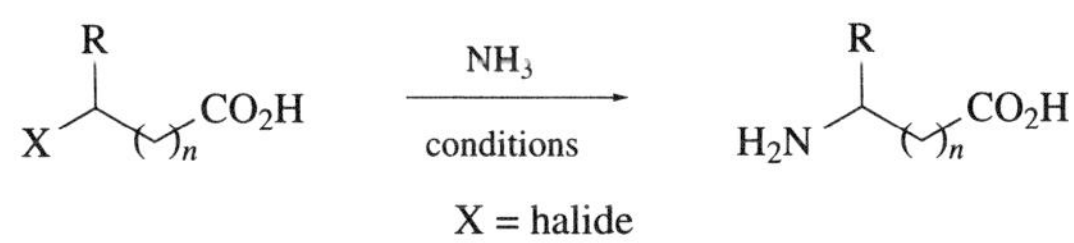

(180)

Table 39 Amino acids via reaction of haloacids with ammonia.

Halo acid	*Conditions*	*Product*	*Yield* (%)	*Ref.*
$ClCH_2CO_2H$	H_2O, RT, 48 h	$H_2NCH_2CO_2H$	64–65	32OSC(1)300 43OSC(2)382
$MeCHBrCO_2H$	H_2O, RT, 4 d	$MeCH(NH_2)CO_2H$	65–70	32OSC(1)23
$Bu^nCHBrCO_2H$	H_2O, RT, 20–30 h	$Bu^nCH(NH_2)CO_2H$	62–67	32OSC(1)48
$PhCHClCO_2H$	sealed tube	$PhCH(NH_2)CO_2H$	48–50	62PCS117
$PhCH_2CHBrCO_2H$	RT	$PhCH_2CH(NH_2)CO_2H$		55OSC(3)705
m-$F_3CC_6H_4CHClCO_2H$	sealed tube	*m*-$F_3CC_6H_4CH(NH_2)CO_2H$	42	62PCS117
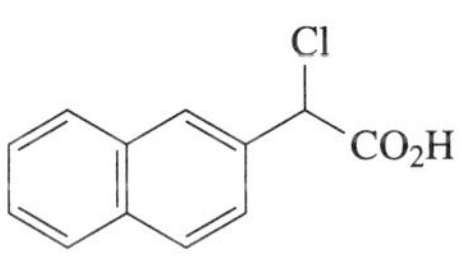	sealed tube	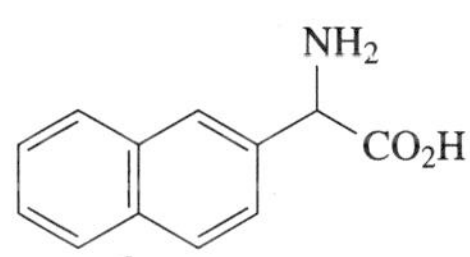	46–49	62PCS117
$HO_2CCH_2ClCO_2H$	autoclave, 70 °C, 10 h	$HO_2CCH_2CH(NH_2)CO_2H$	83	62T69
$HO_2C(CH_2)_2CHClCO_2H$	autoclave, 260–265 °C, 4 h	$HO_2C(CH_2)_2CH(NH_2)CO_2H$	87	63BCJ1264
$BrCH_2(CH_2)_3CHBrCO_2H$	H_2O, 55–60 °C, 4 d	$H_2NCH_2(CH_2)_3CH(NH_2)CO_2H$	62	50JA5137
$ClCH_2(CH_2)_5CO_2H$	autoclave, H_2O, 100 °C, 2 h	$H_2NCH_2(CH_2)_5CO_2H$	72	57JGU2481
$ClCH_2(CH_2)_7CO_2H$	autoclave, H_2O, 100 °C, 2 h	$H_2NCH_2(CH_2)_7CO_2H$		57JGU2481
$ClCH_2(CH_2)_9CO_2H$	autoclave, H_2O, 100 °C, 2 h	$H_2NCH_2(CH_2)_9CO_2H$		57JGU2481

The versatile malonic ester synthesis has been used to prepare pure D- or L-*N*-acyl- and *N*-benzoyloxycarbonylamino acids in a series of steps which include preparation of the amino substituted malonic half ester, decarboxylation, and resolution by enzymatic hydrolysis of the resulting

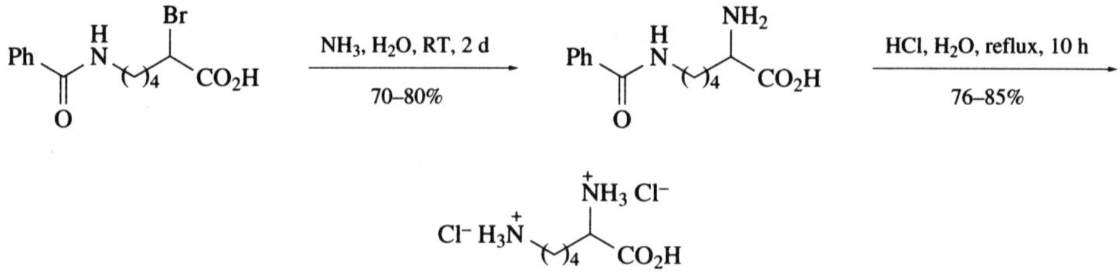

Scheme 64

ester (Scheme 65 and Table 40) ⟨73JOC457⟩, while the phthalimidomalonic acid synthesis, a combined malonic ester-Gabriel synthesis, has been used with ethyl chloracetate to prepare DL-aspartic acid (Scheme 66) ⟨63OSC(4)55⟩, or with β-chloro methylsulfide to prepare DL-methionine (Scheme 66) ⟨43OSC(2)384⟩. Another use of the malonic ester synthesis to prepare amino acids is in combination with a Lossen rearrangement (Scheme 67) ⟨56JOC1182⟩.

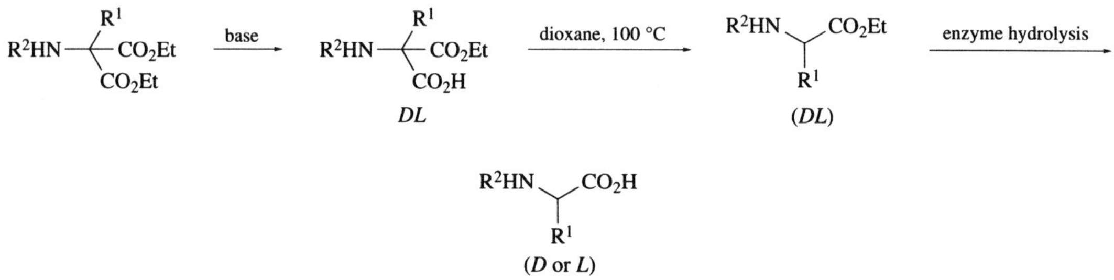

Scheme 65

Table 40 Malonic ester synthesis of pure D or L-*N*-acyl- and *N*-benzoyloxy carbonylamino acids.

R¹	*R²*	*Configuration*
$PhCH_2$	$PhCH_2OCO$	*L*
o-$MeC_6H_4CH_2$	MeCO	*L*
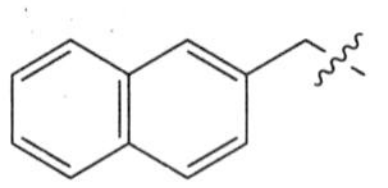	MeCO	*D*
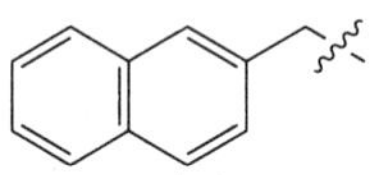	MeCO	*L*
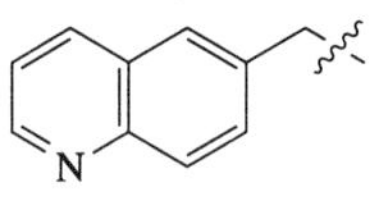	H	*L*

Reduction of a variety of functional groups has also been effective in producing amino acids, for example diethyl oximinomalonate affords diethyl aminomalonate ⟨73OSC(5)376⟩ or diethyl acetamidomalonate ⟨73OSC(5)373⟩ depending upon the reagents used (Scheme 68), while reduction of α-acetominocinnamic acid in acetic acid produces β-phenylalanine (Scheme 69) ⟨43OSC(2)491⟩. Azides may be reduced catalytically as in the preparation of glycine *t*-butyl ester in 78–82% yield from *t*-butyl azidoacetate ⟨73OSC(5)586⟩, or by treatment with triphenylphosphine followed by hydrolysis using 40% HBr (Scheme 70 and Table 41) ⟨55LA(591)117⟩.

5.02.2.6 Other Nitrogen-substituted Alkanoic Acids

One general review on nitrogen-substituted carboxylic acids ⟨79COC841⟩, one specific review on the synthesis of carpyrinic acid and related pyridines with long aliphatic chains ⟨72S464⟩, and

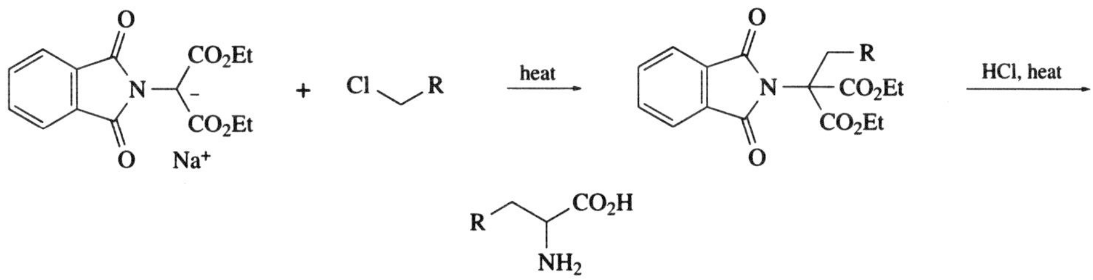

R	*Yield* (%)
CO_2H	42–4 3
$MeSCH_2$	84–85

Scheme 66

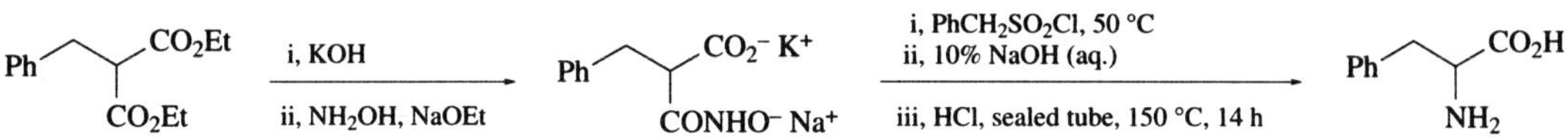

Scheme 67

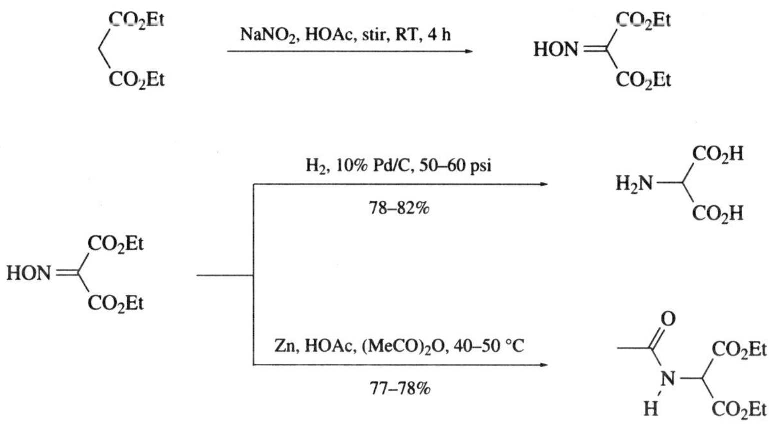

Scheme 68

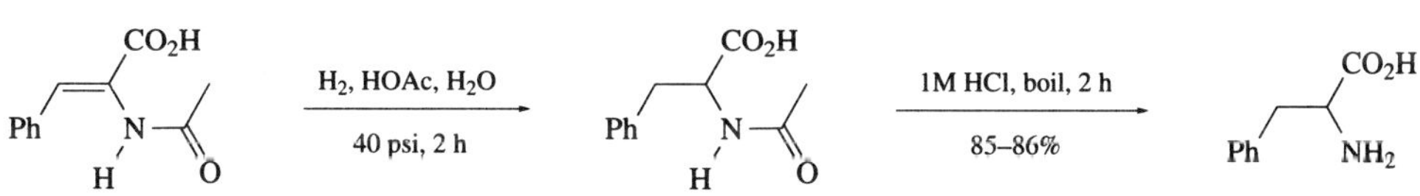

Scheme 69

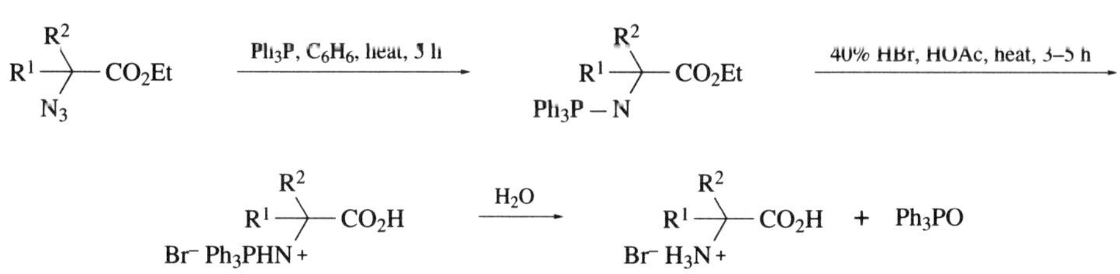

Scheme 70

Table 41 Catalytic reduction of azides to α-amino acids using triphenylphosphine/hydrolysis.

R[1]	*R*[2]	*Product*	*Yield* (%)
Me	Me	$MeC(\overset{+}{N}H_3)CO_2H\ Br^-$	88
CO_2Et	$PhCH_2$	$PhCH_2CH(\overset{+}{N}H_3)CO_2H\ Br^-$	95
CO_2Et	CH_2CO_2Et	$HO_2CCH_2CH(\overset{+}{N}H_3)CO_2H\ Br^-$	100
H	p-$O_2NC_6H_4$	p-$O_2NC_6H_4CH(NH_2)CO_2H$	
H	$PhCONH(CH_2)_4$	$PhCONH(CH_2)_4CH(NH_2)CO_2H$	

also

$$N_3CH_2CH_2CO_2Me \xrightarrow{\text{as above}} Br^-H_3\overset{+}{N}CH_2CH_2CO_2H$$

seven specific reviews on folic acid ⟨46ANY(48)255, 47AG(59)65, 47MI 502-03, B-47MI 502-03, 50JPP409, 60MI 502-01, 61AG449⟩ have been published.

5.02.2.6.1 By hydrolysis reactions

Acids used to hydrolyze nitrogen-substituted esters include, sulfuric acid used to prepare 4-hexadecylamino-1-methylpyrrole-2-carboxylic acid from its ethyl ester (Equation (181)) ⟨83USP4382143⟩, a mixture of methanesulfonic acid in 90% formic acid used to prepare 5-oxothiazolo[3,2-*a*]pyrimidine-6-carboxylic acid from its ethyl ester (Equation (182)) ⟨64CI(L)193⟩, and hydrochloric acid used to produce a variety of nitrogen-substituted acids including: *N*-alkyl-*N*-hydroxy-α-aminocarboxylic acids (Equation (183)) ⟨82CZ345⟩; *N*-(4,4-diphenylbut-3-enyl)-3-piperidinecarboxylic acid (Equation (184)) ⟨83EUP66456⟩; and α,α′-iminodicarboxylic acids (Equations (185)–(188)) ⟨86LA334⟩. An interesting hydrochloric acid hydrolysis has been reported using *t*-butyl[1-(*t*-butoxycarbonyl)-3-oxo-4-pentenyl]carbamate which produces 4-ketopipecolic acid hydrochloride via a series of steps including deprotection and finally ring closure (Equation (189)) ⟨92OS(71)200⟩.

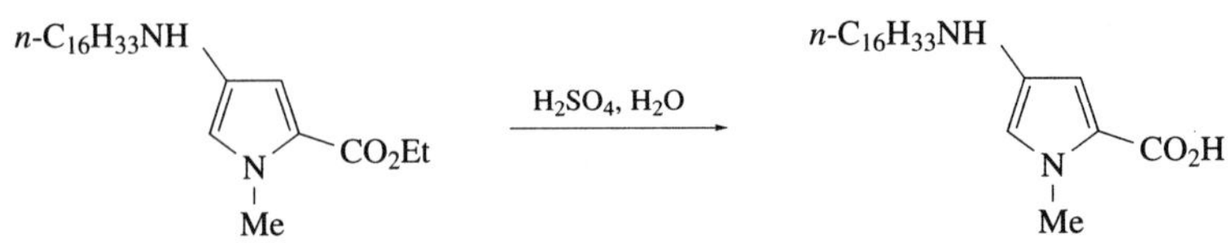

(181)

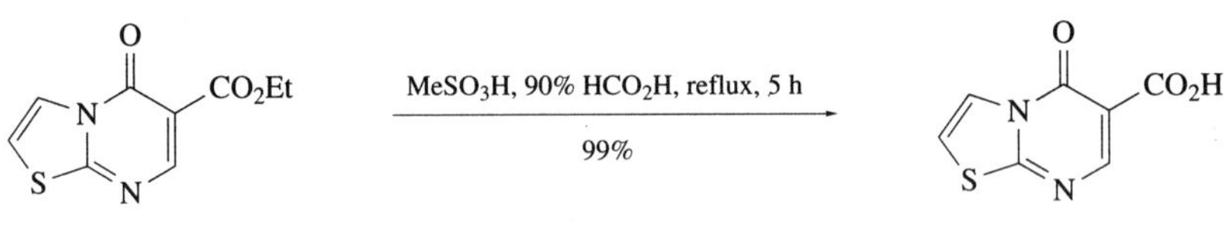

(182)

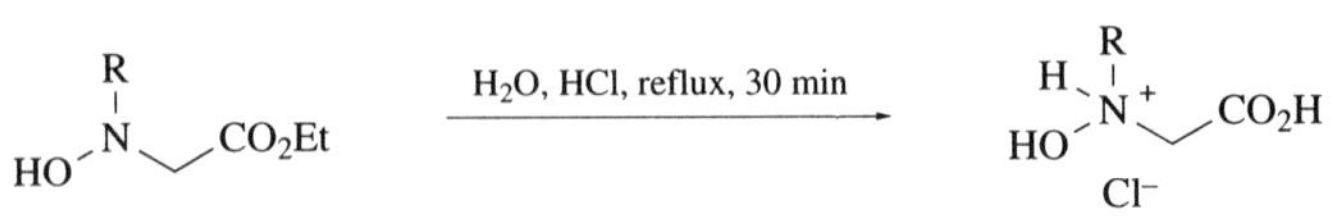

R	*Yield* (%)
Pr[i]	57
Bu[t]	35

(183)

(184)

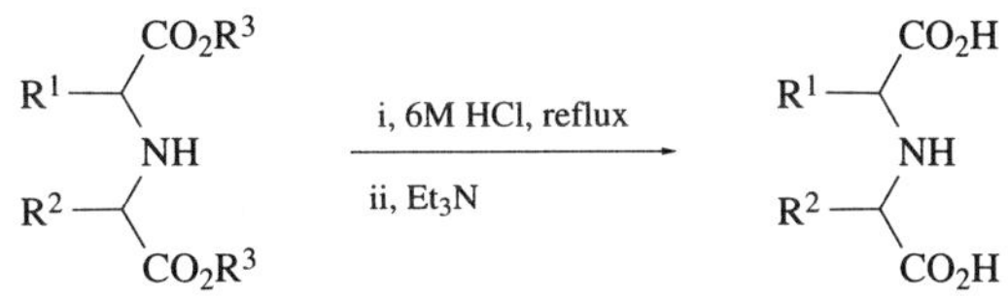

(185)

R^1	R^2	R^3
Me	Me	Et
Et	$PhCH_2$	Me
$PhCH_2$	$PhCH_2$	Me
Et	p-$HOC_6H_4CH_2$	Me
Et	p-$(Et_2OCCH(Me)O)C_6H_4$	Me

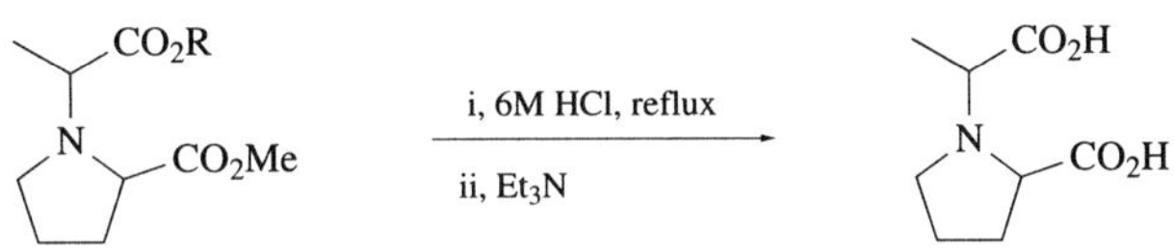

(186)

R = Et, Bu^i

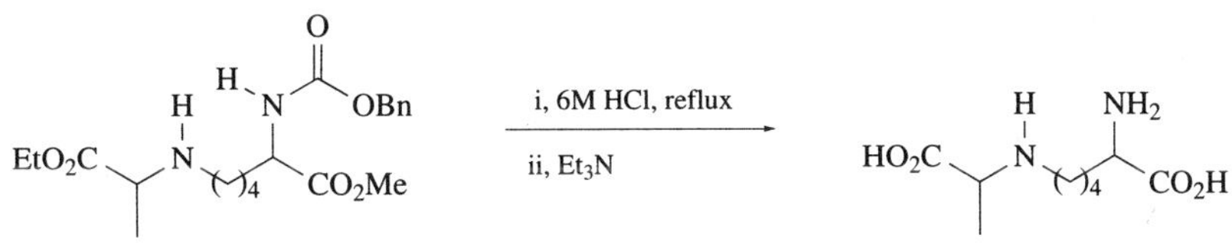

(187)

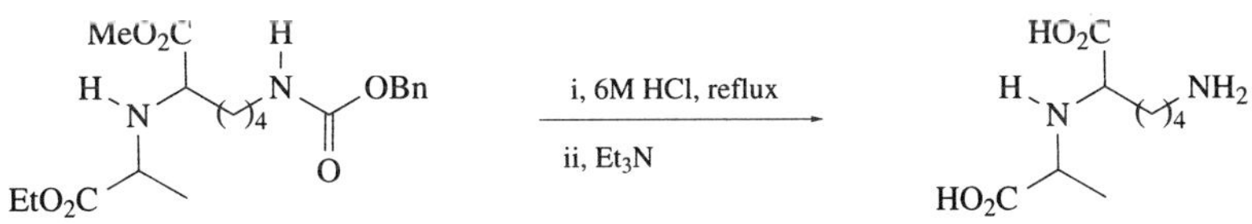

(188)

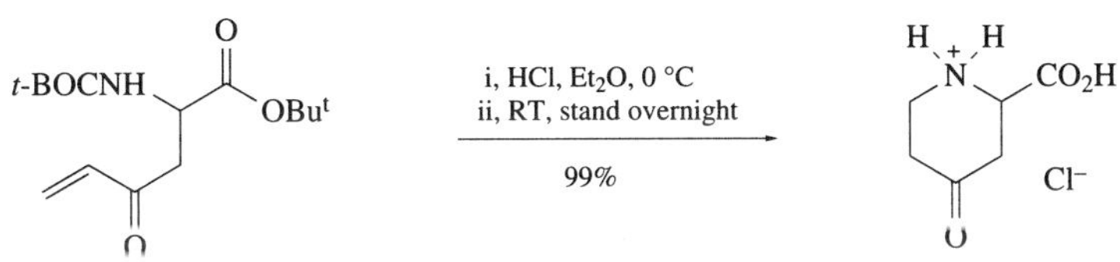

(189)

Examples of ester hydrolysis using a base include the preparation of α-substituted cyclic α-imino acids from their corresponding methyl ester hydrochlorides (Equation (190)) ⟨79CPB1931⟩, and DL-α-(isopropylideneaminooxy)propionic acid from its ethyl ester (Equation (191)) ⟨73OSC(5)1031⟩. Two ring closure reactions which occur during basic hydrolysis of nitrogen-substituted esters are the formation of dihydro-1,2,4,5-tetrazine-3,6-dicarboxylic acid from ethyl diazoacetate via a Diels–Alder dimerization (Scheme 71) ⟨91OS(70)79⟩, and the formation of 6-methyluracil from ethyl β-uraminocrotonate (Equation (192)) ⟨43OSC(2)422⟩.

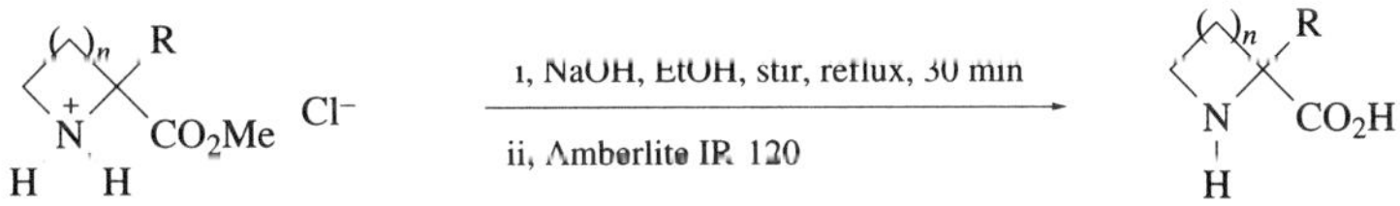

(190)

n	R	Yield (%)
2	Me	72
2	Ph	88
2	$PhCH_2$	79
3	Me	88
3	Ph	75
3	$PhCH_2$	61
4	Me	65
4	Ph	89
4	$PhCH_2$	61

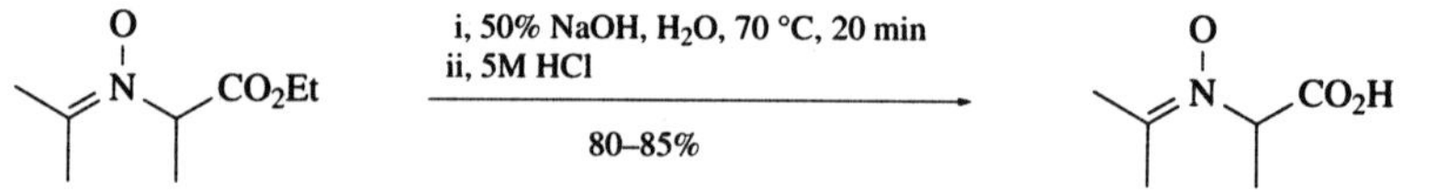

(191)

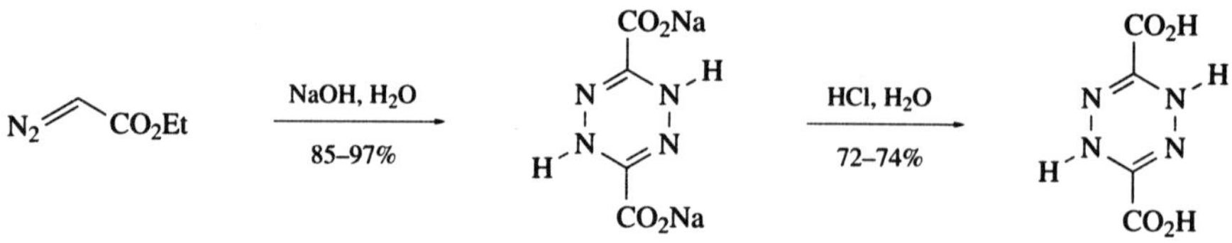

Scheme 71

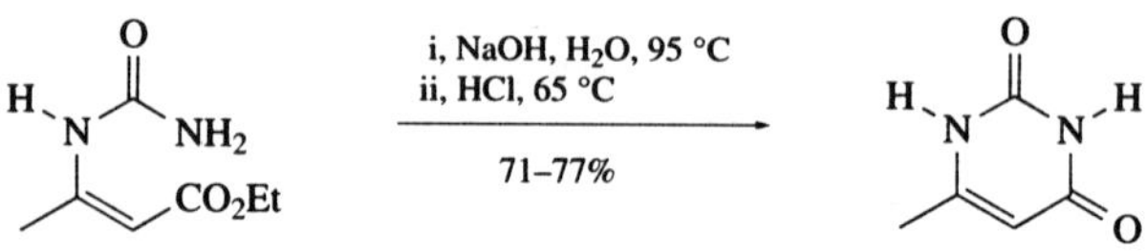

(192)

Sulfuric and hydrochloric acids have been the reagents of choice to effect hydrolysis of nitrogen-substituted nitriles. Using sulfuric acid *p*-nitrobenzyl cyanide has been converted to *p*-nitrophenylacetic acid in 92–95% yield ⟨32OSC(1)406⟩, and 4-nitro-1-methyl-5-cyanoimidazole has been converted into 4-nitro-1-methylimidazol-5-carboxylic acid in 92% yield ⟨45JCS751⟩. Hydrochloric acid has been used to convert *N*-benzoylaminoacrylic acid nitriles to 3-(*N*-benzoylamino)-3-substituted pyruvic acids (Equation (193)) ⟨79SC363⟩, 4-nitrazaalkanonitriles to 4-nitroazaalkanoic acids (Equation (194)) ⟨76AP(308)954⟩, and 1-ethyl-1,4-dihydro-4-oxocinnolin-3-nitriles to their corresponding acids (Equation (195)) ⟨78LA617⟩.

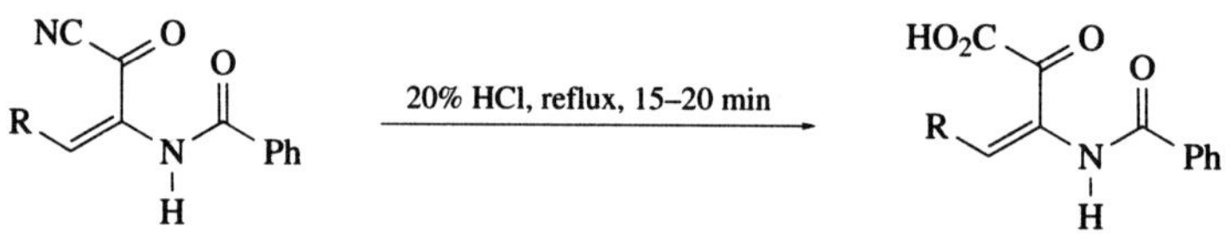

(193)

R	*Yield* (%)
Ph	62
p-$MeOC_6H_4$	48
PhCH=CH	39
o-HOC_6H_4	36
p-HOC_6H_4	36
p-MeC_6H_4	28
3,4-$(MeO)_2OC_6H_3$	64
o-$MeOC_6H_4$	54
piperidino	60
indolo	52
furfuralo	57

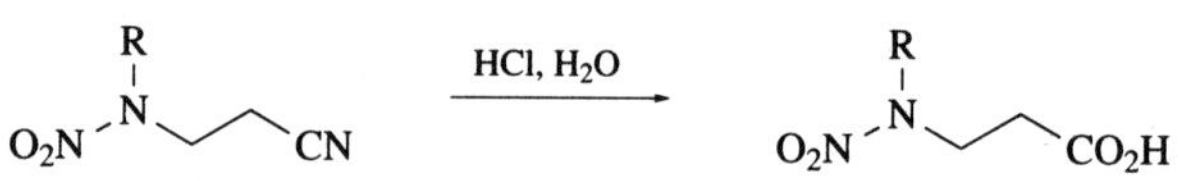

(194)

R	*Yield* (%)
Me	67
Et	81

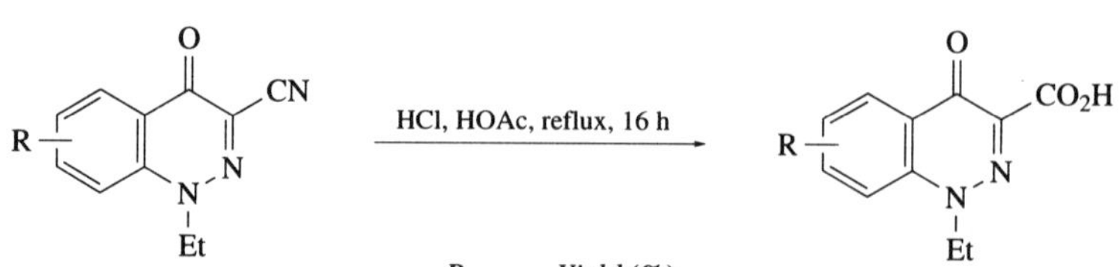

(195)

R	*Yield* (%)
H	92
6-Me	92
7-MeO	26
7-CN	51[a]

[a] Product is diacid

An example of amide hydrolysis producing a nitrogen-substituted carboxylic acid is the conversion of 1-benzoyl-1,2-dihydroquinaldamide to 1-benzoyl-1,2-dihydroquinaldinic acid in 61% yield using the acid ion-exchange resin Amberlite IR-120 (Equation (196)) ⟨57CI(L)736⟩.

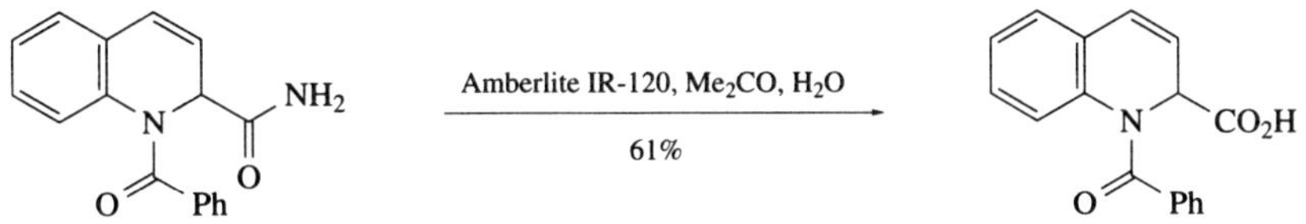

(196)

5.02.2.6.2 *By carbonation of organometallic reagents*

By conversion of 10-bromo-5-methyl-10,11-dihydro-5*H*-dibenz[*b*,*f*]azepine into a Grignard reagent, followed by treatment with solid carbon dioxide, a 50–60% yield of the corresponding carboxylic acid has been obtained ⟨78PHA(33)375⟩, but the production of α-nitrocarboxylic acids from primary nitroalkanes required the use of Stiles' reagent, methylmagnesium carbonate, formed by reaction of carbon dioxide with dimethoxymagnesium (Scheme 72) ⟨63JA616⟩. Another catalyst used to add carbon dioxide to primary nitroalkanes as well as to nitriles in phenoxide (Equation (197)) ⟨77G499⟩.

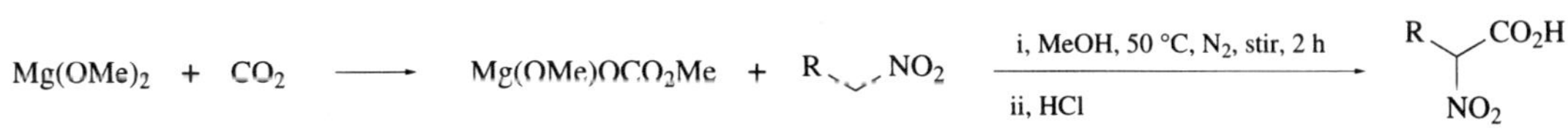

R	*Yield* (%)
H	63
Me	49
Bu^i	46

Scheme 72

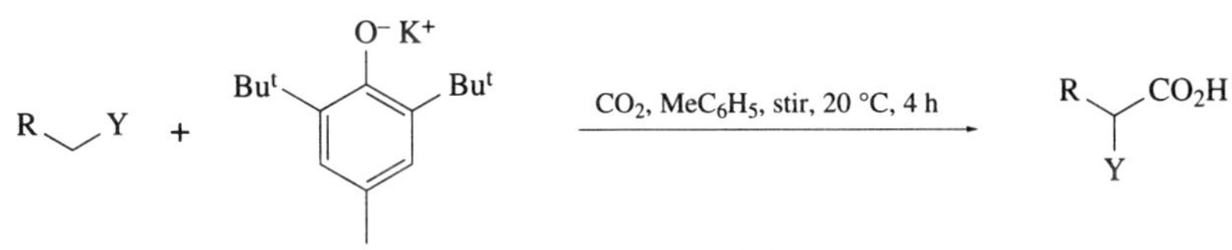

(197)

R	*Y*	*Yield* (%)
H	NO_2	26
Ph	CN	48

5.02.2.6.3 *By oxidation reactions*

An interesting preparation of nitrogen-substituted alkanoic acid via an oxidation reaction is the formation of nitrato carboxylic acids by treatment of cyclopentanone, cyclohexanone and norbornanone with ceric ammonium nitrate (can) in aqueous acetonitrile at 60 °C (Scheme 73) ⟨72CJC2047⟩.

5.02.2.6.4 *By miscellaneous reactions*

One preparation of cyanoalkanoic acids which has not realized its potential involves the treatment of the sodium salt of α halo acids with cyanide ion. Using sodium chloroacetate as a model, treatment with sodium cyanide affords sodium cyanoacetate ⟨32OSC(1)254⟩ which can be hydrolyzed directly to cyanoacetic acid or can be condensed with benzaldehyde and hydrolyzed to afford α-cyano-β-phenylacrylic acid (Scheme 74) ⟨32OSC(1)181⟩.

Condensation of indole with glycolic acid in the presence of potassium hydroxide at 250 °C affords an 87–93% yield of indol-3-acetic acid ⟨73OSC(5)654⟩, while condensation of maleic anhydride with aniline at room temperature produces an excellent yield of malenanilic acid (Equation (198)) ⟨73OSC(5)944⟩.

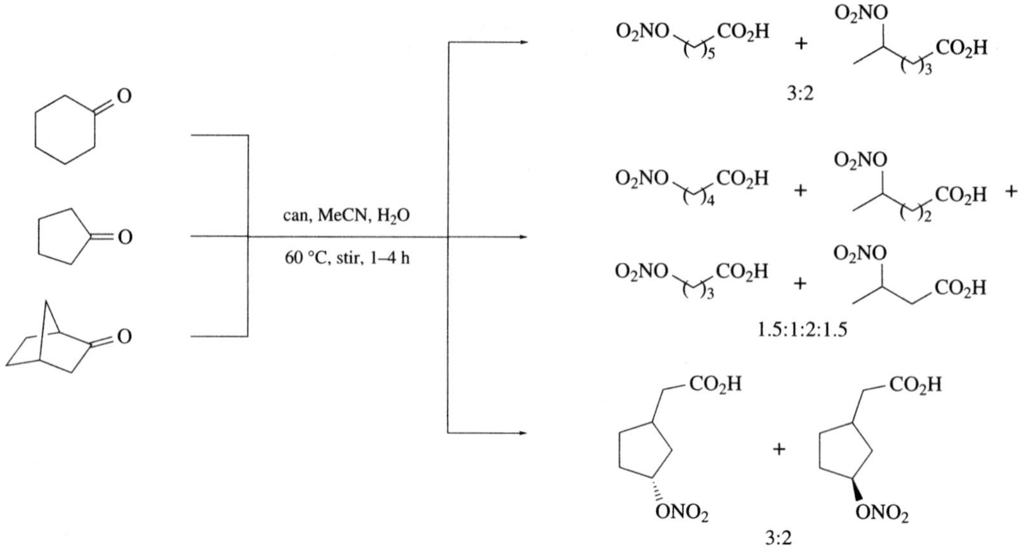

Scheme 73

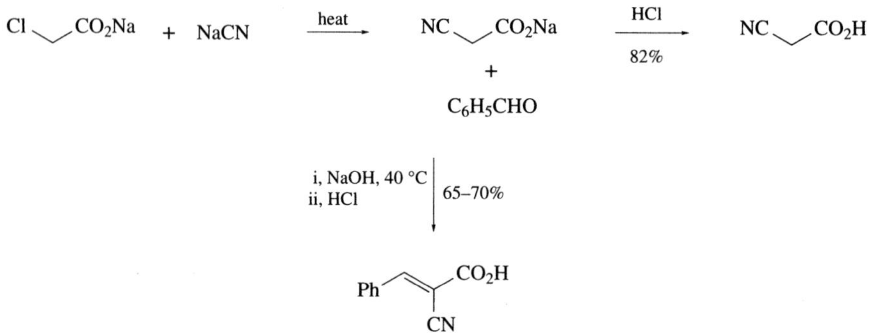

Scheme 74

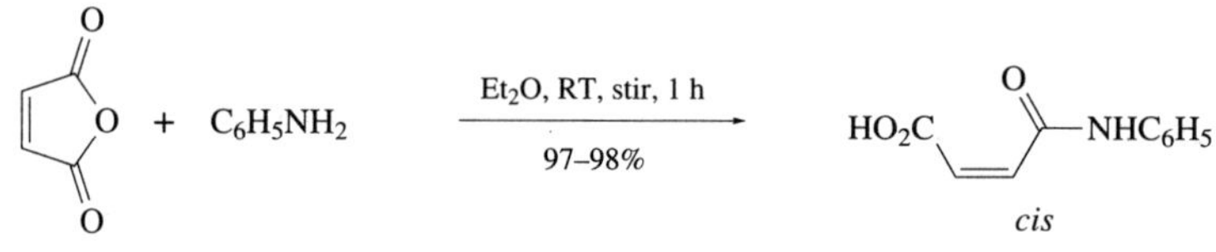 (198)

5.02.2.7 Other Heteroatom-substituted Alkanoic Acids

The synthesis of phosphorus-containing carboxylic acids has been reviewed ⟨B-83MI 502-01⟩.

5.02.2.7.1 By hydrolysis reactions

Saponification of methyl 2-methyl-2-(3-trimethylsilylpropyl)-3-butenoate affords the corresponding carboxylic acid retaining the trimethylsilyl function (Equation (199)) ⟨87OS(66)87⟩. Similar treatment of the methoxycarbonyl derivatives of cyclohexa-1,3-diene iron tricarbonyl substrates also produces the corresponding carboxylic acids retaining the iron tricarbonyl functions (Equation (200) and Table 42) ⟨82JCS(P1)1763⟩.

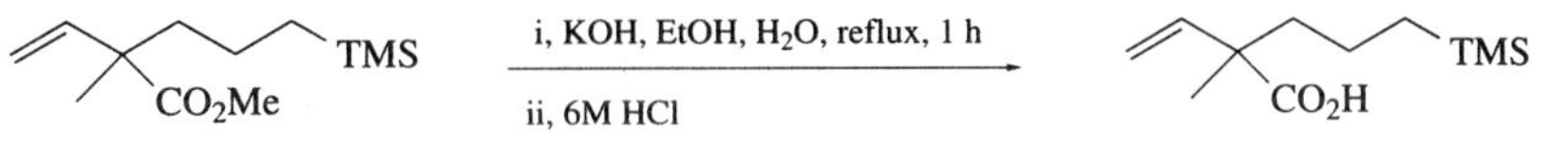

(199)

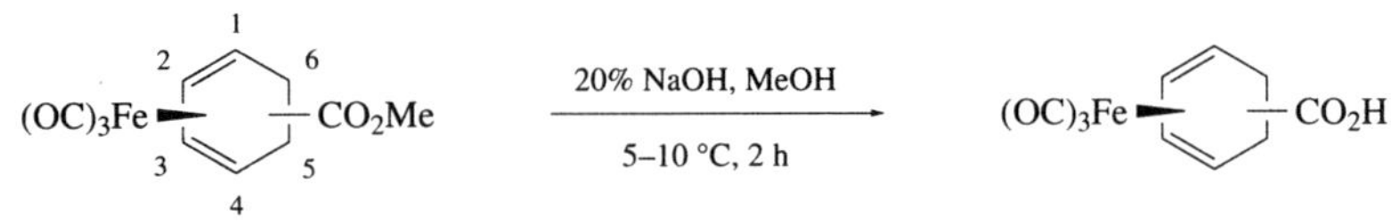

(200)

Table 42 Saponification of methoxycarbonyl derivatives of cyclohexa-1,3-diene iron tricarbonyl substrates.

Substituents, location, orientation substrate	*Product*	*Yield (%)*
2-CO_2Me	2-CO_2H	94
5-*trans*-CO_2Me	5-*trans*-CO_2H	85
5-*cis*-CO_2Me	5-*cis*-CO_2H + 5-*trans*-CO_2H	
1-CO_2Me, 6-*trans*-CO_2Me	1-CO_2Me, 6-*trans*-CO_2H	81
1,4-$(CO_2Me)_2$	1-CO_2Me, 4-CO_2H + 1,4-$(CO_2H)_2$	40, 42
1-CO_2Me, 6-*cis*-CO_2Me	1-CO_2Me, 6-*trans*-CO_2H[a]	74
5,6-*cis,cis*-$(CO_2Me)_2$	5,6-*cis,cis*-$(CO_2H)_2$	80
5-*cis*-CO_2Me, 6-*trans*-CO_2Me	5,6-*cis,cis*-$(CO_2H)_2$[a]	80
5-Me, 5-*cis*-CO_2Me	5-Me, 5-*cis*-CO_2H	40
5,5-$(CO_2Me)_2$	5-*cis*-CO_2Me, 5-*trans*-CO_2H[b]	84

[a] Epimerization occurs. [b] Preference for hydrolysis of α- over β-carbomethoxy group (*trans* over *cis*).

5.02.2.7.2 By carbonation of organometallic reagents

Lithium 4-methyl-2,6-di(*t*-butyl) phenoxide has been reported to catalyze the addition of carbon dioxide to sulfones which contain an acidic α-hydrogen (Equation (201)) ⟨73JOC4086⟩, while labeled carboxylic acids containing sulfur functions have been prepared by the standard addition of $^{14}CO_2$ to a Grignard reagent (Scheme 75) ⟨83MI 502-01⟩, or by a carbonylation–oxidation sequence of 9-borabicyclo[3.3.1]nonyl (9-BBN) organoborates with ^{13}CO (Equation (202)) ⟨82CC1273⟩.

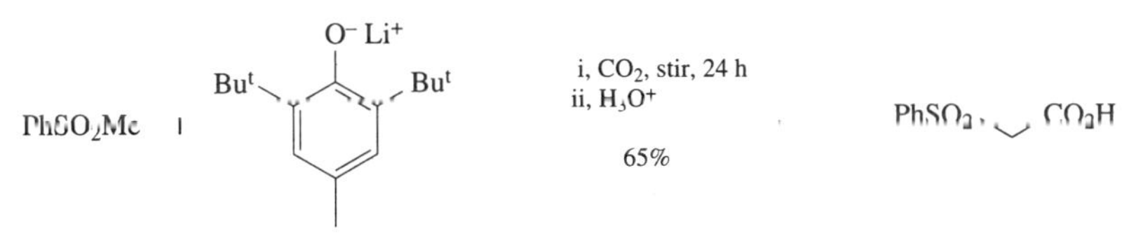

(201)

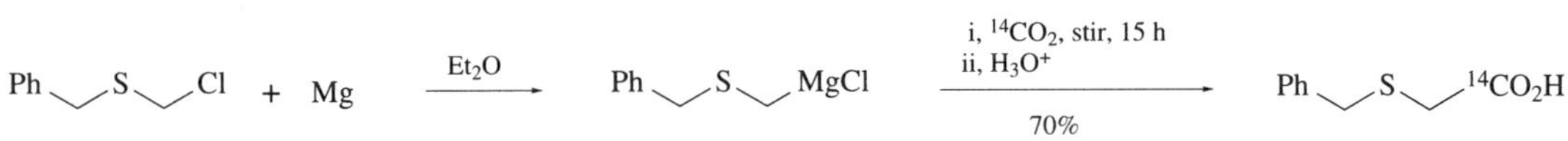

Scheme 75

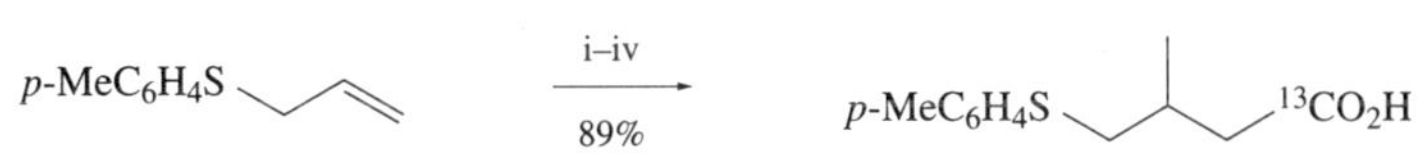

(202)

i, 9-BBN, THF; ii, ^{13}CO (1 atm), $KBH(OPr^i)_2$; iii, H_2O_2, NaOAc; iv, $AgNO_3$, NaOH, 50 °C, 30 min

Carbonation of lithium cyclopentadienyl iron alkyl- or aryldicarbonyl produces the corresponding carboxylic acids in 53–67% yield (Scheme 76) ⟨84JOM(267)309⟩.

5.02.3 α,β-UNSATURATED ACIDS

General reviews of unsaturated acids have discussed the structural relationships in the natural unsaturated higher fatty acids ⟨48PCS243⟩, the chemistry and biochemistry of unsaturated fatty acids ⟨76AG(E)61⟩, and methods of separation of long-chain unsaturated fatty acids ⟨63MI 502-01⟩.

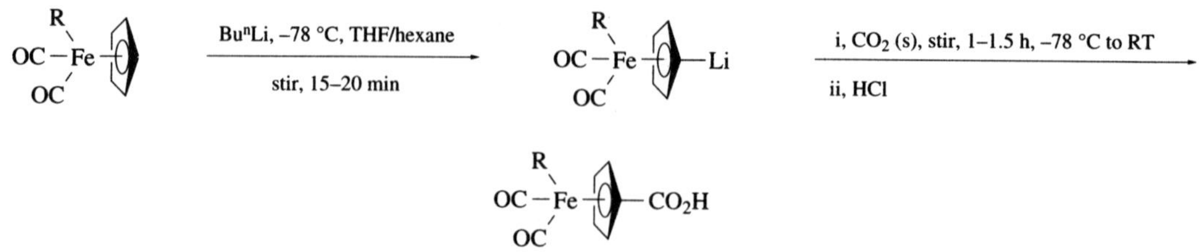

R	*Yield (%)*
$C_5H_4Mn(CO)_3$	53
Ph	63
CH_2Ph	67

Scheme 76

Reactions which produce unsaturated acids are usually active-hydrogen reactions in which carbanions are key intermediates and several of these name reactions have been reviewed. The general Perkin reaction ⟨42OR(1)210, 56CRV27, B-72MI 502-01⟩ which involves the condensation of an aryl aldehyde with acetic anhydride or an α-alkylacetic anhydride in the presence of a base such as the carboxylate salt corresponding to the anhydride or a tertiary amine is shown in Equation (203), while a specific reaction for the preparation of *m*-nitrocinnamic acid ⟨32OSC(1)398⟩ is shown in Equation (204).

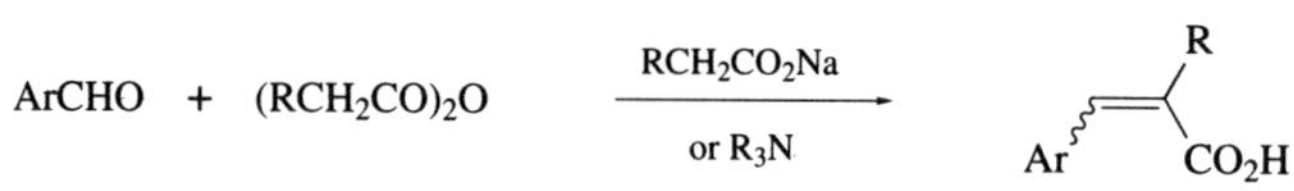

(203)

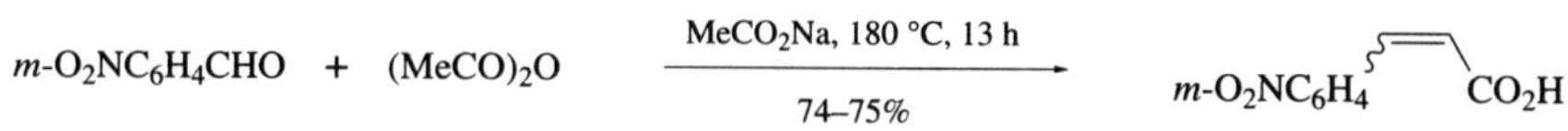

(204)

Reaction of malonic acid with an aldehyde using pyridine as the reaction solvent is referred to as the Doebner reaction (Equation (205)) ⟨49OR(5)229, 52JCS4521, B-72MI 502-01⟩ and permits the synthesis of both β-aryl- and β-alkyl-α,β-unsaturated acids.

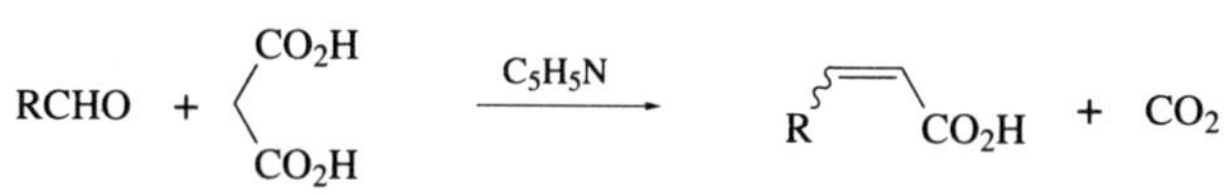

(205)

Base-catalyzed condensation of diethyl succinate with ketones or aldehydes produces the monoethyl esters of α-alkylidene- or α-aralkylidenesuccinic acid and is referred to as the Stobbe condensation ⟨49BSFD297, 51OR(6)1, 54OR(8)28, B-72MI 502-01⟩. The initially formed alkylidene or aralkylidene ethyl hydrogen succinates can be converted into β,γ-unsaturated acids (Scheme 77).

More specific reviews of ascorbic acid ⟨46MI 502-02, 55MI 502-01⟩ and tiglic and angelic acids ⟨55CRV659⟩ have also been published.

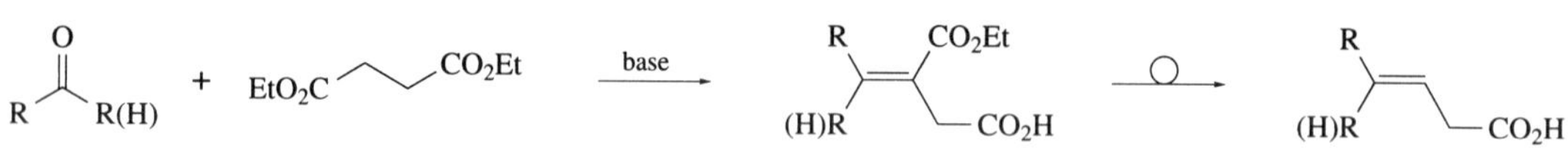

Scheme 77

5.02.3.1 By Hydrolysis Reactions

Examples of acid hydrolysis of esters which produce α,β-unsaturated acids include the trifluoroacetic acid catalyzed conversion of *t*-butyl 5,5-disubstituted-5-hydroxy-2-pentenoates into their corresponding acids (Equation (206)) ⟨76MI 502-01⟩. Acid also causes the dehydration of hydroxyacids, as in the conversion of citric acid into aconitic acid (Equation (207)) ⟨43OSC(2)12⟩. Examples of base hydrolysis of esters include the preparation of *trans*-2-methyl-2-dodecenoic acid from its methyl ester ⟨63OSC(4)608⟩, 4-ethyl-2-methyl-2-octenoic acid from its ethyl ester ⟨63OSC(4)444⟩,

2-methylene dodecanoic acid from its *t*-butyl ester (Equation (208)) ⟨63OSC(4)616⟩, and 2-hexenoic acid from its neopentyl ester ⟨52BJ163⟩.

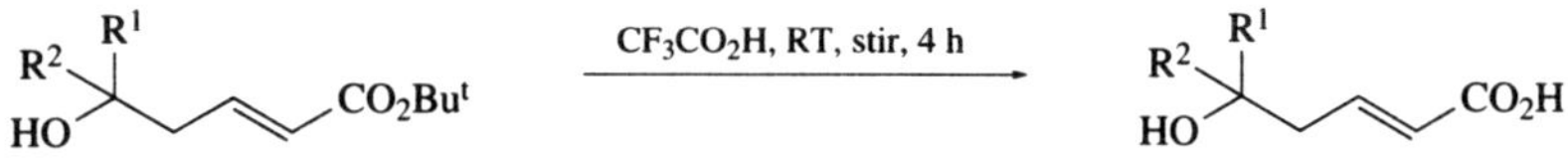

(206)

R^1	*R^2*
C_6H_5	C_6H_5
C_6H_5	Me
C_6H_5	*c*-C_6H_{11}

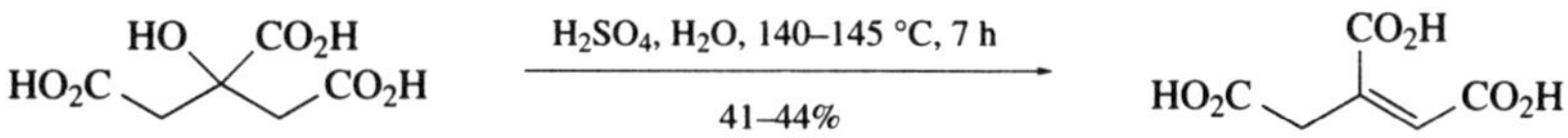

(207)

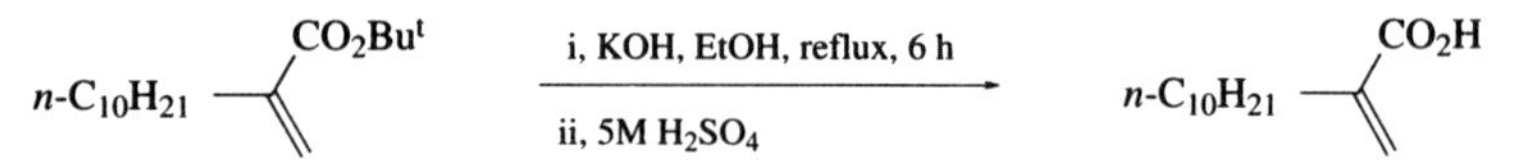

(208)

Base hydrolysis of the monoethyl esters of 2-substituted-2-chloro-1-ethene-1,1-dicarboxylates converts these substrates into ethyl 2-substituted-1-ethyne-1-carboxylates which can ultimately be converted into their corresponding acids (Equation (209)) ⟨87OS(66)173⟩.

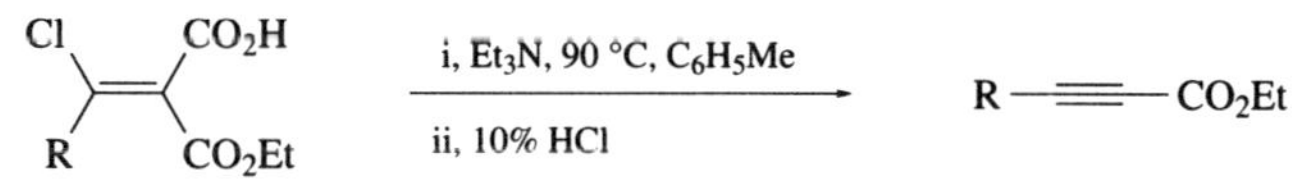

(209)

R	*Yield* (%)
Pr^i	67
Bu^t	70
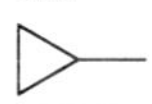	66–78
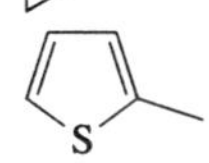	90
C_6H_5	80
p-$O_2NC_6H_4$	70

Hydrolysis of 3-methyl-4-nitro-5-(2-arylstyryl)isoxazoles causes isoxazole ring cleavage and formation of *trans*-cinnamic acids (Equation (210)) ⟨80JHC621⟩.

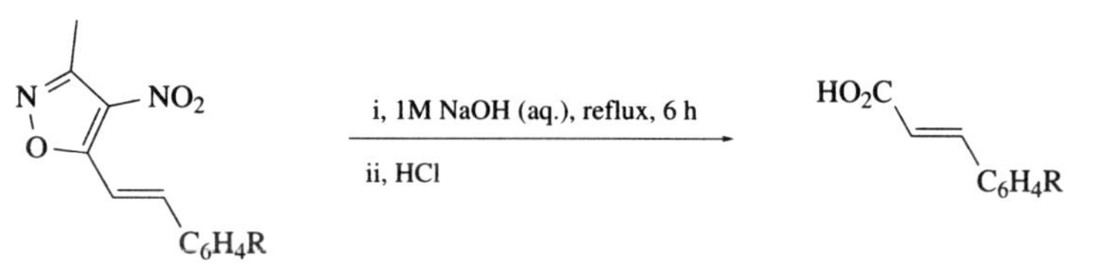

(210)

R	*Yield* (%)
H	65
p-Cl	67
o-Cl	90
m-Cl	90
2,4-Cl_2	81
p-Me	96
p-MeO	70

While refluxing an aqueous solution of itaconic anhydride without any catalyst produces a 24–39% yield of itaconic acid ⟨43OSC(2)368⟩, the product obtained from hydrolysis of citraconic anhydride depends upon the reagents employed. If the hydrolysis is performed in water alone with heat a 94% yield of citraconic acid is obtained (Scheme 78) ⟨43OSC(2)140⟩, whereas if an aqueous nitric acid solution is used a 43–52% yield of the *trans*-isomer, mesaconic acid, is obtained (Scheme 78) ⟨43OSC(2)382⟩.

When *gem*-difluoroalkenes containing a β-alkoxy and a γ-hydroxy group are treated with acid, α,β-unsaturated acids are obtained (Equation (211)) ⟨79MI 502-06⟩, while the trichloride, trichloromethyl

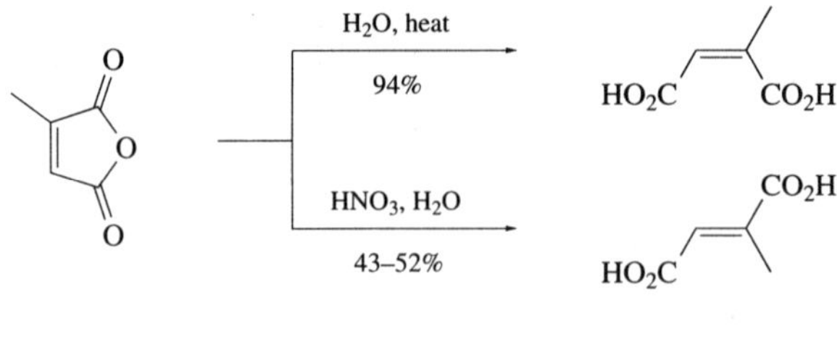

Scheme 78

pentafluorobenzene, is also converted into its corresponding acid by acid hydrolysis (Equation (212)) ⟨61JCS808⟩.

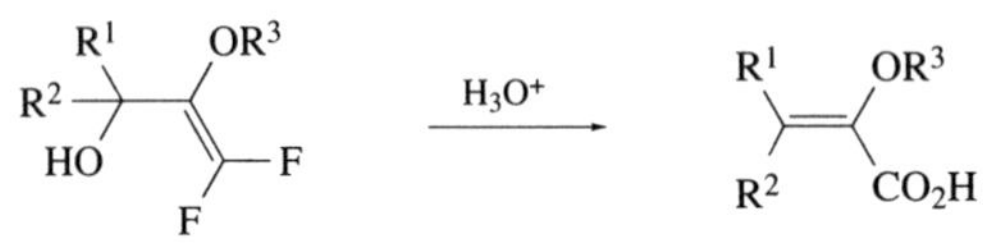

(211)

R^1	R^2	R^3
Me	Me	Ph
Ph	H	Ph

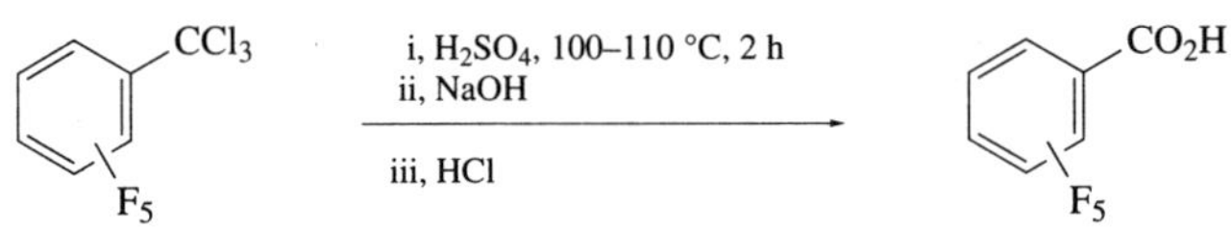

(212)

5.02.3.2 By Carbonation of Organometallic Reagents

At least one review of the synthesis of naturally occurring unsaturated fatty acids by sterically controlled carbonyl alkenation has been published ⟨64AG(E)250⟩.

Treatment of propyne with sodium amide produces the sodium acetylide which can be carbonated to produce tetrolic acid (Scheme 79) ⟨73OSC(5)1043⟩. Alkynes are also useful in the preparation of α,β-unsaturated alkenoic acids when they are added to the Grignard-derived alkylcopper complex, formed by addition of the dimethyl sulfide complex of copper(I) bromide to a Grignard reagent, followed by treatment with carbon dioxide (Scheme 80) ⟨85OS(64)1⟩.

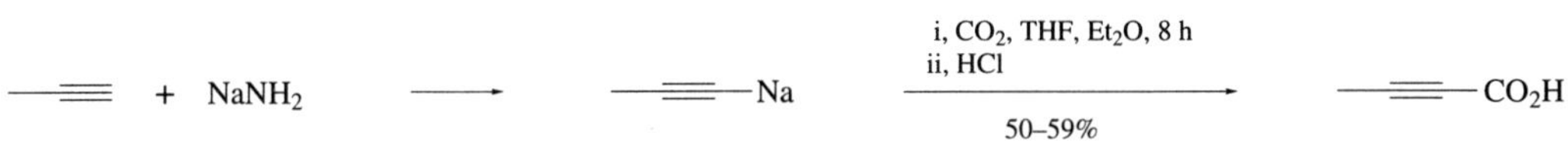

Scheme 79

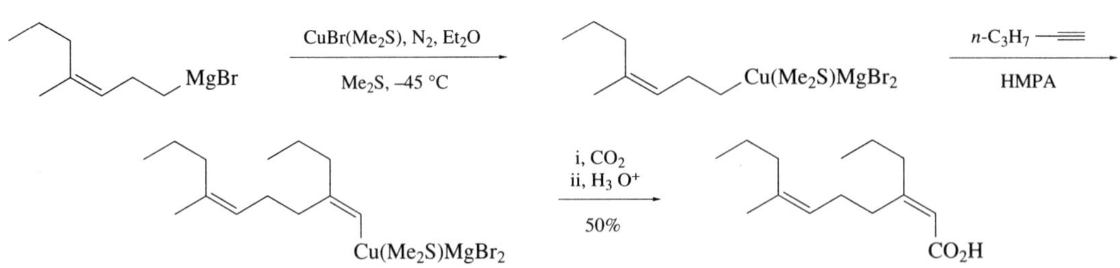

Scheme 80

Ketene bis(phenylthio) acetals, which are readily prepared by the reaction of carboxylic acids or esters with $Al(SPh)_3$, are reduced rapidly at low temperature by lithium naphthalenide to sulfur-stabilized vinyllithiums which react with carbon dioxide producing the corresponding carboxylic acids (Scheme 81) ⟨79JOC3601⟩.

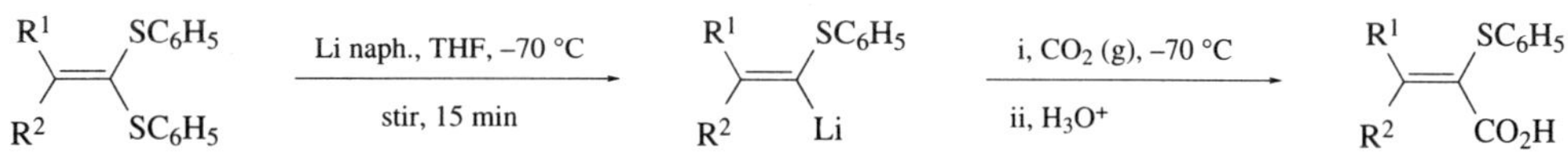

R¹	*R²*	*Yield (%)*
Me	Me	44
–(CH$_2$)$_5$–		74

Scheme 81

Reaction of alkynes with *n*-butyllithium produces 1,3-dilithioacetylides which afford allene 1,3-dicarboxylic acids upon treatment with a slurry of solid carbon dioxide in hexane (Scheme 82) ⟨79CC215, 80JCS(P1)2338⟩, however, by careful control of the amount of *n*-butyllithium reagent employed, alkynes can be converted into their monolithio salts which upon reaction with trialkylborates affords lithium trialkyalkynylborates. Upon subsequent treatment with carbon dioxide these lithium trialkylalkynylborates produce stereospecific *α,β*-unsaturated acids in good yield (Scheme 83) ⟨84TL1797⟩ via a mechanism similar to the one shown in Scheme 51.

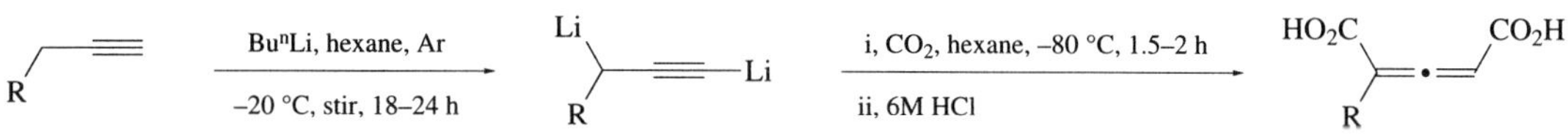

R	*Yield (%)*
Et	33
Pr^n	37
Bu^n	31.5
n-C$_5$H$_{11}$	13.5

Scheme 82

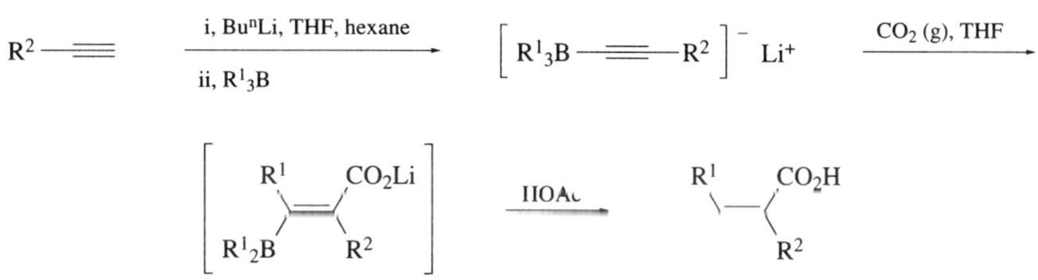

R¹	*R²*	*Yield (%)*
n-C$_6$H$_{13}$	Bu^n	82.3
n-C$_8$H$_{17}$	Bu^n	81.4
c-C$_5$H$_9$	Bu^n	83.7
c-C$_6$H$_{11}$	Bu^n	80.5
n-C$_6$H$_{13}$	C$_6$H$_5$	71.4
n-C$_8$H$_{17}$	C$_6$H$_5$	75.6

Scheme 83

The nickel(0) complex of 1,2-bis(dicyclohexylphosphino)ethane (DCPENi) has been reported to catalyze the addition of carbon dioxide to conjugated dienes to produce diene carboxylic acids ⟨84JOM(266)313⟩. The maleic anhydride used in the second step of Scheme 84 causes the intermediate nickel carboxylates to undergo reductive elimination.

5.02.3.3 By Oxidation Reactions

Oxidation of unsaturated alcohols to unsaturated acids has been accomplished using sodium ruthenate to produce cinnamic acid (Equation (213)) ⟨72CJC3741⟩, and with chromium trioxide to produce chlorotetrolic acid (Equation (214)) ⟨58CR(246)1877, 60AC(P)845⟩. In the oxidation using

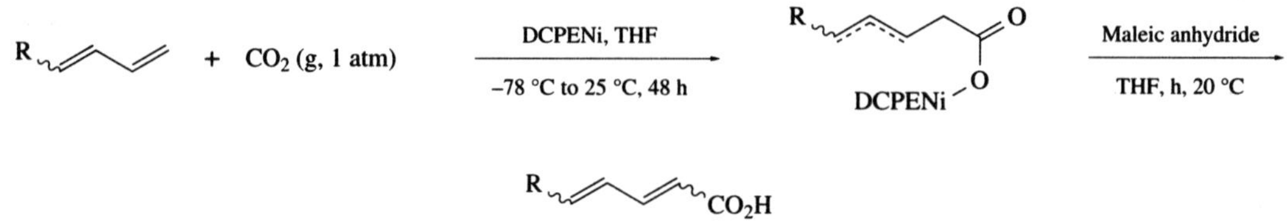

R	*Time* (h)	*Yield* (%)
H	48	25
Me	24	40

Scheme 84

sodium ruthenate it was found that low temperatures and short reaction times had no effect on the double bond, but if higher temperatures and longer reaction times were used, cleavage of the double bond occurred.

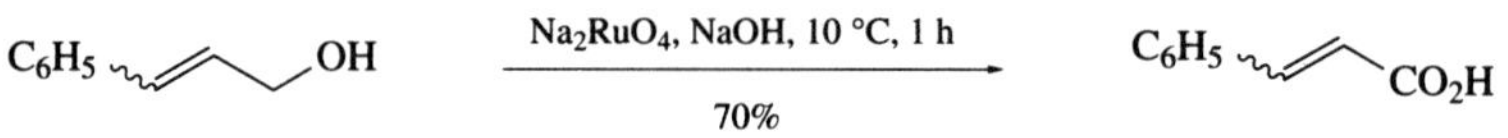

(213)

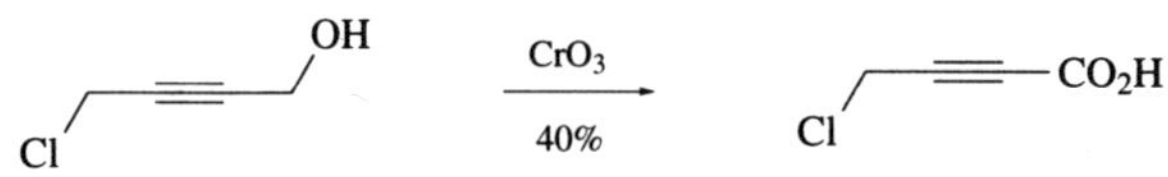

(214)

That the presence of double or triple bonds does not effect the course of aldehyde oxidation to acids using Jones' reagent was demonstrated by the oxidation of a series of *cis*-2-enediyn-1-als (Equation (215)) ⟨56CI(L)548, 58JCS1313⟩. Other reagents used to oxidize unsaturated aldehydes to unsaturated acids include: sodium chlorite in the presence of hydrogen peroxide, which oxidizes *trans*-2-hexenal to *trans*-2-hexenoic acid (Equation (216)) ⟨86JOC567⟩; and, alkaline silver(I) oxide, which oxidizes *trans*-dec-2-en-4,6-diynal to *trans*-dec-2-en-4,6-diynoic acid ⟨58JCS1313⟩, as well as the hindered keto aldehyde 2,4,4-trimethyl-3-cyclohexanal-1-one to its corresponding unsaturated acid (Equation (217)) ⟨81JOC5039⟩. Silver(I) oxide has also been used to oxidize 5-methylfurfuraldehyde to 5-methylfuroic acid ⟨58JCS1313⟩, while the stronger oxidizing agent sodium chlorate in the presence of vanadium pentoxide cleaves furfural and produces fumaric acid (Equation (218)) ⟨43OSC(2)302⟩.

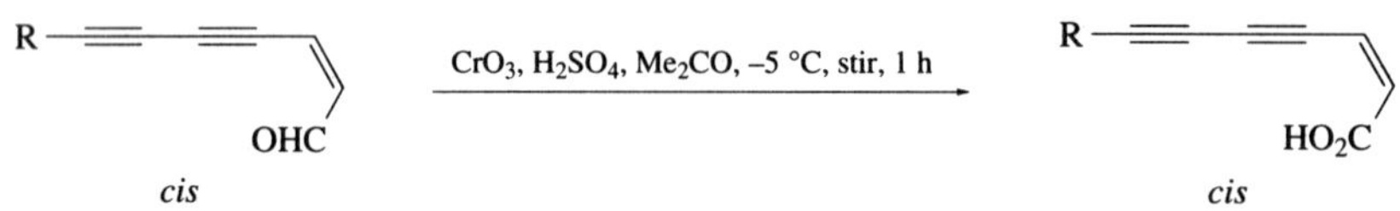

(215)

R = Pr^n, *cis*-MeCH=CH, MeC≡C

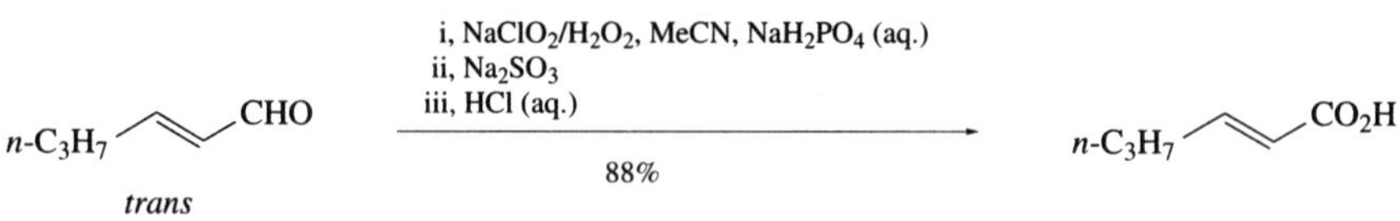

(216)

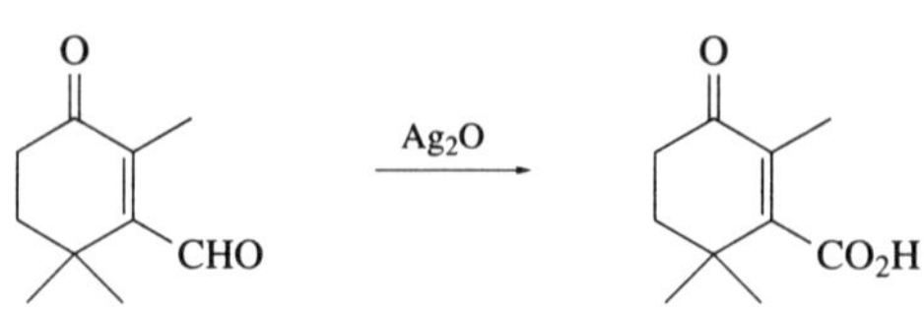

(217)

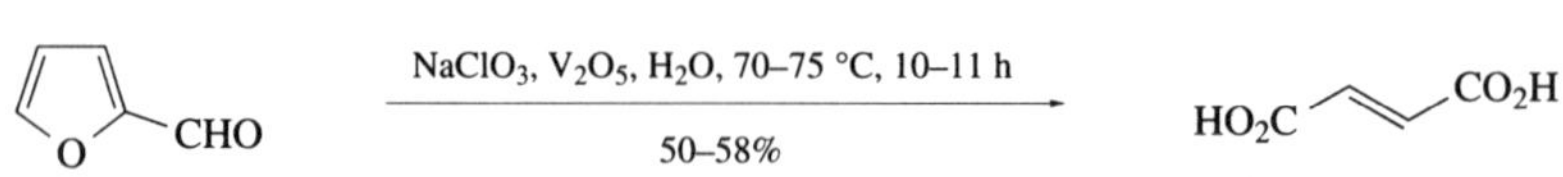

(218)

Oxidation of unsaturated ketones to produce unsaturated acids is usually accomplished using the haloform reaction on methyl ketones, as with the preparation of *trans*-dec-2-en-4,6-diynoic acid from *trans*-undec-3-en-5,7-diyn-2-one using sodium hypochlorite ⟨58JCS1313⟩, β,β-dimethylacrylic acid from 4-methyl-3-penten-2-one using potassium hypochlorite ⟨55OSC(3)302⟩ or sodium bromite ⟨83CC392⟩, and cinnamic acid from 4-phenyl-3-buten-2-one using sodium bromite ⟨83CC392⟩.

5.02.3.4 By Miscellaneous Reactions

One approach to the preparation of α,β-double or triple bond containing acids is the treatment of mono- or dihalide acids or esters with alcoholic potassium hydroxide. Using this approach a 37–43% yield of muconic acid has been obtained from diethyl α,δ-dibromoadipate (Equation (219)) ⟨55OSC(3)623⟩, a 73–88% yield of acetylenedicarboxylic acid from α,β-dibromosuccinic acid (Equation (220)) ⟨43OSC(2)10⟩, and a 77–81% yield of phenylpropiolic acid from ethyl α,β-dibromo-β-phenylpropionate (Equation (220)) ⟨43OSC(2)515⟩.

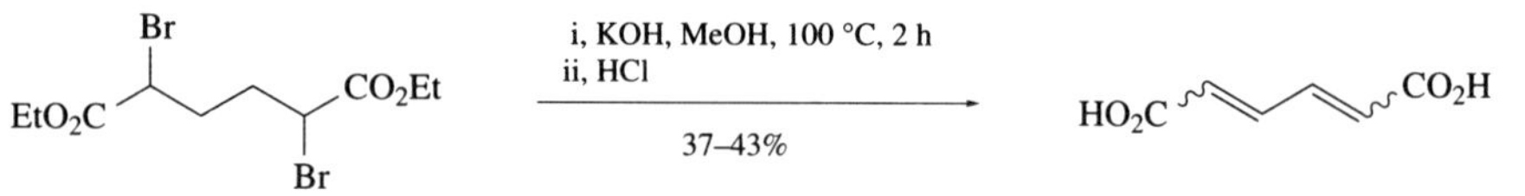

(219)

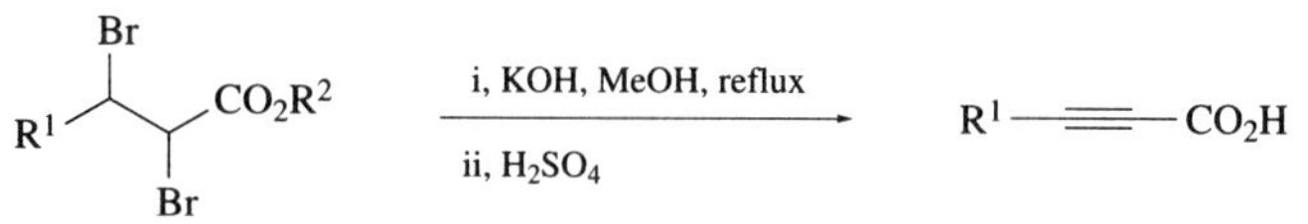

(220)

R¹	*R²*
CO_2H	H
C_6H_5	Et

Alkoxycarbonylation of lithium acetylides using methyl or ethyl chloroformate has been used to prepare ethyl tetrolate from propyne in 95–97% yield (Scheme 85) ⟨85OS(64)108⟩, and methyl 4-chloro-2-butynoate from propargyl chloride in 81–83% yield (Scheme 85) ⟨87OS(65)47⟩.

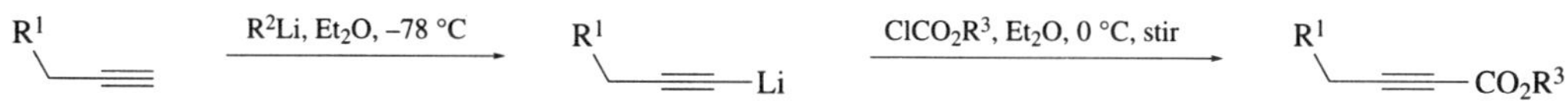

R¹	*R²*	*R³*	*Yield (%)*
H	Bu^n	Et	95–97
Cl	Me	Me	81–83

Scheme 85

During the 1970s and 1980s, Wittig type reagents allowed the conversion of a wide variety of carbonyl compounds into α,β-unsaturated acids. Some examples of the Wittig type reagents used and the acid produced include: triethyl phosphonoacetate used to convert paraformaldehyde into ethyl α-(hydroxymethyl)acrylate ⟨87OS(66)220⟩, benzophenone to β-phenylcinnamic acid ⟨73OSC(5)509⟩, and cyclohexanone to ethyl cyclohexylideneacetate ⟨73OSC(5)547⟩. Diethyl chlorophosphate in combination with lithium dimethylcuprate has been used to convert methyl 2-oxocyclohexanecarboxylate to methyl 2-methyl-1-cyclohexene-1-carboxylate (Scheme 86) ⟨84OS(62)14⟩, while carboxychloromethylphosphonate has been used as a carboxychloro alkenation reagent for the preparation of α-chloro-α,β-ethylenic carboxylic acids from a series of aldehydes and ketones (Scheme 87 and Table 43) ⟨78SC19⟩. Finally, ethyl (triphenylphosphoranylidene)acetate reacts with acid chlorides to produce α-allenic esters (Equation 221 and Table 44) ⟨84OS(62)202⟩.

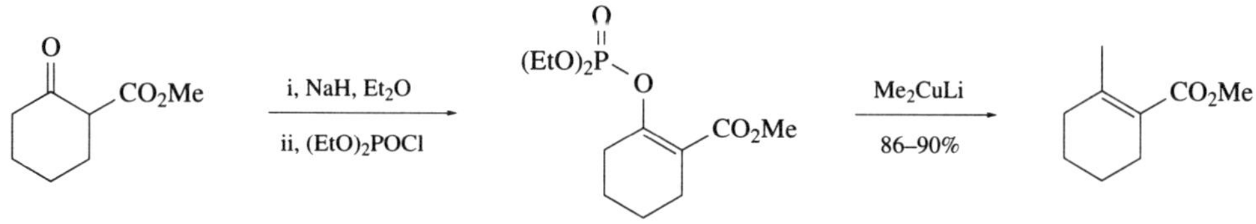

Scheme 86

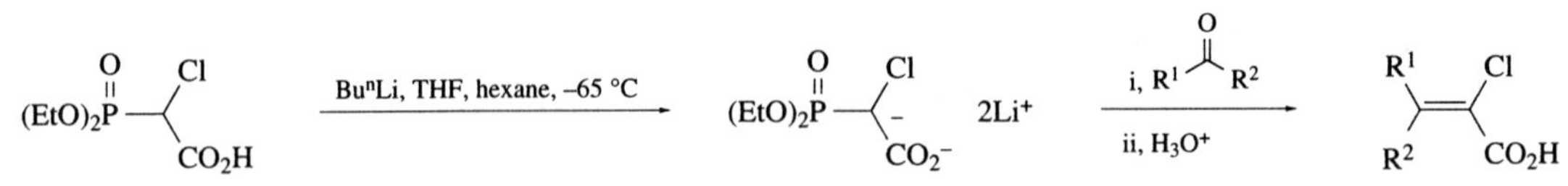

Scheme 87

Table 43 α-Chloro-α,β-ethylenic carboxylic acids from aldehydes and ketones.

R¹	*R²*	*Isomer*	*Yield (%)*
C_6H_5	H	(*Z*)	67
o-MeC_6H_4	H	(*Z*)	83
p-MeC_6H_4	H	(*Z*)	83
p-$Me_2NC_6H_4$	H	(*Z*)	86
p-ClC_6H_4	H	(*Z*)	86
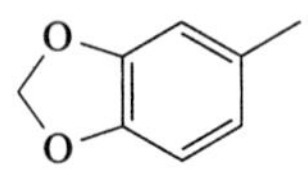	H	(*Z*)	88–91
3-Pyridyl	H	(*Z*)	41
2-thienyl	H	(*E*) + (*Z*)	85
$H_5C_6CH{=}CH$	H		96
C_6H_5	Me	(*E*) + (*Z*)	65
Bu^n	H	(*E*) + (*Z*)	88
Pr^i	H	(*E*) + (*Z*)	91
Bu^i	H	(*E*) + (*Z*)	85
Et	Me	(*E*) + (*Z*)	80
Pr^i	Me	(*E*) + (*Z*)	45
Et	Me	(*E*) + (*Z*)	80
$Me_2C{=}CH(CH_2)_2$	Me	(*E*) + (*Z*)	45–84

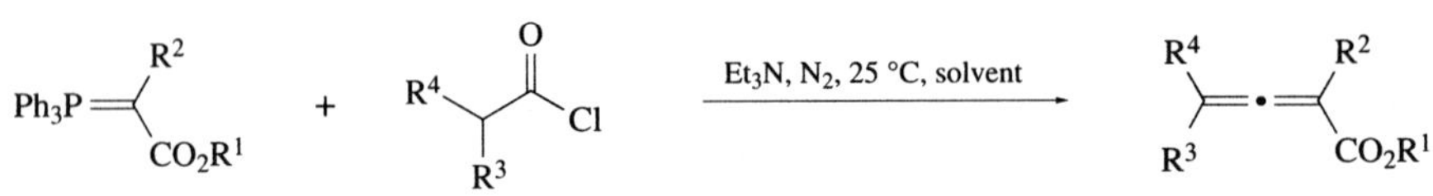

(221)

Table 44 α-Allenic esters from acid chlorides using ethyl (triphenylphosphoranylidene).

R¹	*R²*	*R³*	*R⁴*	*Solvent*	*Yield (%)*
Me	H	H	H	CH_2Cl_2	40
Et	H	Me	H	CH_2Cl_2	63
Et	H	Bu^t	H	CH_3CN	55
Me	H	C_6H_5	H	CH_3CN	23
Et	Me	H	H	CH_2Cl_2	59
Et	Me	Me	H	CH_2Cl_2	74
Et	Me	Me	Me	CH_2Cl_2	39
Me	Me	Bu^t	H	CH_3CN	66
Et	Me	C_6H_5	H	CH_2Cl_2	70

Several other preparations of α,β-unsaturated acids also involve intermediate anion formation for example, the Knoevenagel-type condensation of di-*t*-butyl malonate with paraformaldehyde in the presence of potassium and cupric acetate producing di(*t*-butyl)methylenemalonate (Equation (222)) ⟨85OS(64)63⟩, the sodium ethoxide catalyzed condensation of benzaldehyde and ethyl acetate producing ethyl cinnamate in 68–74% yield ⟨32OSC(1)252⟩, and the reaction of sodium glyoxylate with α-ketoesters (Equation (223) and Table 45) ⟨83G421⟩.

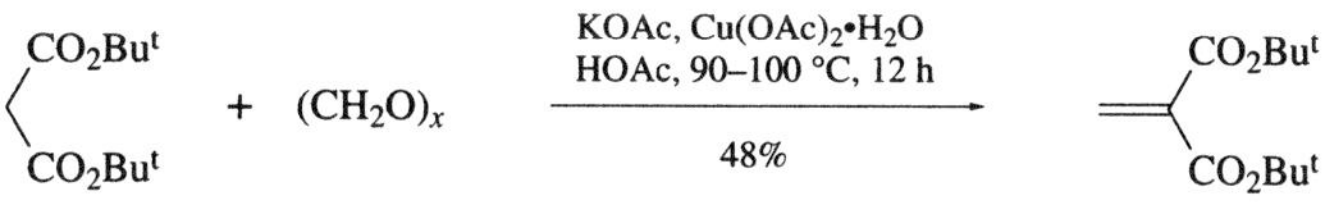

(222)

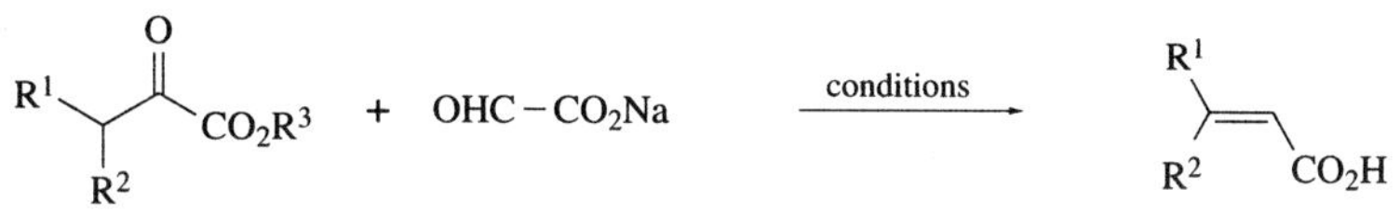

(223)

Table 45 α,β-Unsaturated acids via sodium glyoxylate reaction with α-ketoesters.

R[1]	*R*[2]	*R*[3]	*Conditions*	*Yield* (%)
EtO_2C	Me	Et	1N NaOH, H_2O, 0 °C, 10 min	77
MeO_2C	Bu^n	Me	1N NaOH, H_2O, 0 °C, 25 min	75
—$(CH_2)_3CO$—		Me	Na_2HPO_4, KH_2PO_4, 0–5 °C, stir 30 min	89
—$(CH_2)_4CO$—		Me	Na_2HPO_4, KH_2PO_4, 0–5 °C, stir 40 min	80

An interesting two step reaction, involving the intermediate (*E*)-(carboxyvinyl)trimethylammonium betaine, has been used to synthesize (*E*)-3-[(*E*)-3-phenyl-2-propenoxy]acrylic acid from ethyl propiolate and cinnamyl alcohol (Scheme 88) ⟨87OS(66)29⟩.

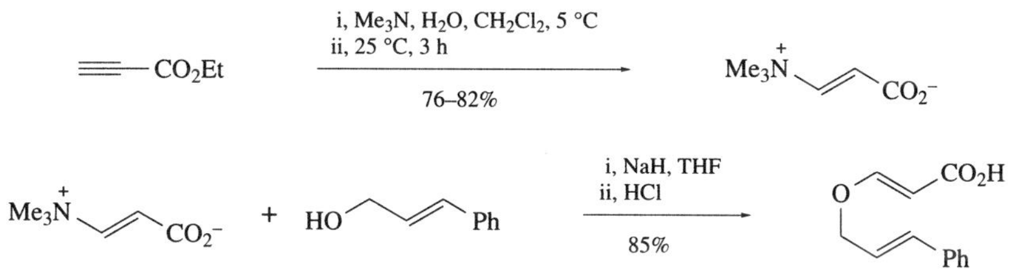

Scheme 88

5.02.4 HOMO- AND HETEROAROMATIC ACIDS

A variety of review articles have been written on the synthesis of different classes of aromatic carboxylic acids. At least one review on acid synthesis in Friedel–Crafts and related reactions ⟨B-64MI 502-03⟩, and one review on monocarboxylic acids of the benzene series ⟨B-54MI 502-01⟩ have been published. Several reviews have discussed the synthesis of phenolic/hydroxy carboxylic acids ⟨B-45MI 502-02, B-45MI 502-03, B-56MI 502-01, 57CRV583, B-61MI 502-03, B-76MI 502-03⟩, while at least four reviews of pyridine carboxylic acids have appeared ⟨62CHE179, 62CHE347, 79MI 502-09, 81MI 502-04⟩. Other specific reviews cover naphthoic acids ⟨B-49MI 502-01⟩, the synthesis of di- and tricarboxylic acids of aromatic ring systems by shifting of carboxylic groups ⟨58AG(70)1⟩, and the preparation of aromatic dibasic acids containing an azo bond via electrochemical synthesis ⟨80MI 502-01⟩.

5.02.4.1 By Hydrolysis Reactions

Partial hydrolysis of the diester dimethyl 2-nitrophthalate has been accomplished in the presence of a catalytic amount of water, a water miscible solvent and/or an emulsifier (Equation (224)) ⟨83GEP3148422⟩. One hundred percent sulfuric acid was required to hydrolyze the methyl esters of 2,4,6-trimethyl-, triethyl- and triisopropyl benzoates to produce a 90% yield of the corresponding trisubstituted benzoic acids ⟨41JA2431⟩, but cleavage of *t*-butyl benzoate occurred in a two phase system at room temperature with aqueous mineral acids such as 10M H_2SO_4, 9M HBr or 12M HCl in the presence of a lipophilic quaternary onium salt such as hexadecyltributylphosphonium bromide ⟨82JOC154⟩.

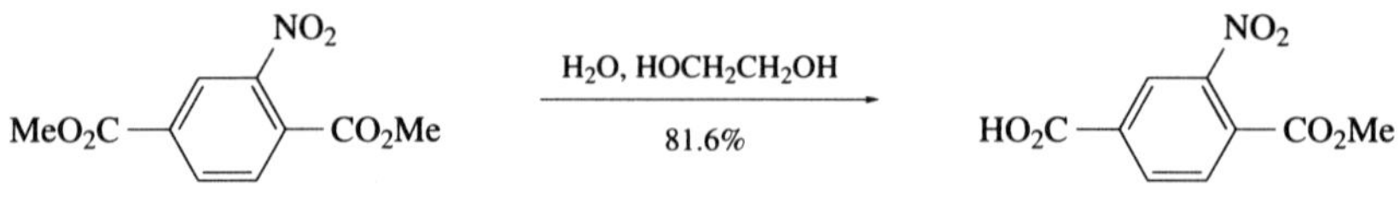

(224)

Base hydrolysis of aromatic esters has been accomplished using a variety of basic reagents. Sodium hydroxide successfully hydrolyzes methyl *m*-nitrobenzoate ⟨32OSC(1)391⟩, ethyl 7-chloro-4-hydroxy-3-quinolinecarboxylate ⟨55OSC(3)272⟩, and the methyl and ethyl esters of 1-benzyl-2,3-dimethyl-4-alkoxypyrrolo[2,3-*b*]pyridine-5-carboxylate (Equation (225)) ⟨85JHC1429⟩. Treatment of methyl 2-pyridone-5-carboxylate with 17% aqueous sodium hydroxide, followed by acidification, affords the tautomeric 6-hydroxynicotinic acid (Equation (226)) ⟨63OSC(4)532⟩.

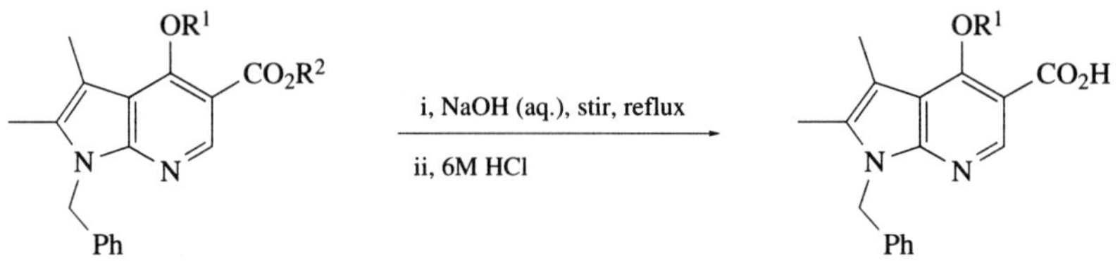

(225)

R^1	R^2	*Yield* (%)
Me	Me	54
Et	Me	54
Me	Et	100
Et	Et	100

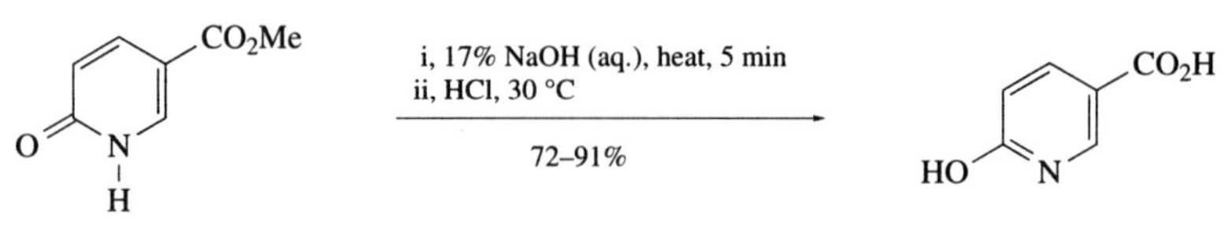

(226)

Sodium hydroxide in DMSO has been used to hydrolyze the methyl and ethyl esters of benzoic acid, as well as the sterically hindered mesitoic, 9-anthranoic, and *O*-methylpodocarpic acids ⟨63CI(L)775, 65JCS1290⟩.

Potassium hydroxide alone hydrolyzes ethyl *p*-fluorobenzoate ⟨43OSC(2)299⟩ and ethyl 3,5-dichloro-4-hydroxybenzoate ⟨55OSC(3)267⟩, but for hindered esters, such as the methyl, *t*-butyl or neopentyl esters of alkyl mesitoates, the complex formed between potassium hydroxide and 18-crown-6 in refluxing benzene or toluene causes saponification in excellent yields ⟨67JA7017⟩.

Other bases used to hydrolyze aromatic esters include, barium hydroxide in dioxane with ethyl 1,3-dihydroxy-2-naphthoate ⟨55OSC(3)637⟩, and trimethylamine in methanol with the methyl esters of *p*- and *o*-methyl, *p*- and *o*-nitro, *o*-chloro and unsubstituted benzoic acid as well as dimethyl phthalate ⟨33JA4079⟩. Several thiol related bases have also been found to be effective, including, sodium thioethoxide for hydrolysis of methyl *p*-anisate (Scheme 89) ⟨72AJC1731⟩, sodium thiophenoxide for hydrolysis of the benzyl and phenacyl esters of benzoic acid (Equation (217)) ⟨64JOC2006⟩, and a mixture of aluminum halide and ethanethiol for hydrolysis of the methyl and benzyl esters of benzoic acid (Equation (228)) ⟨78TL5211⟩.

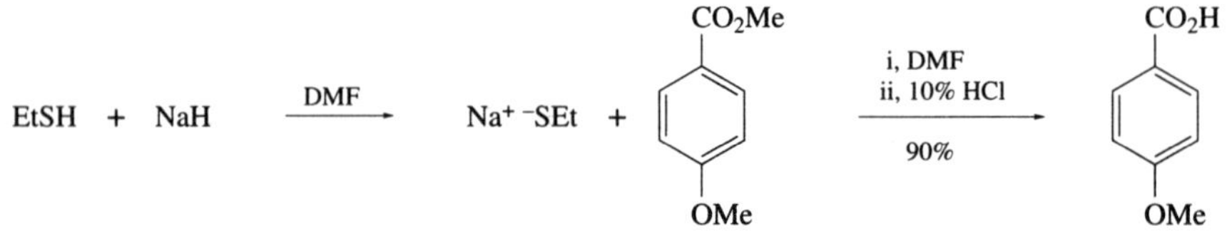

Scheme 89

$$PhCO_2R + PhS^- Na^+ \xrightarrow[\text{ii, } H_3O^+]{\text{i, DMF, 30 min, temp.}} PhCO_2H \quad (227)$$

R	*Temp.* (°C)	*Yield* (%)
$PhCH_2$	100	64
$PhCOCH_2$	25	84–87

$$PhCO_2R + AlX_3 \xrightarrow[\text{ii, } H_3O^+]{\text{i, EtSH, RT, stir}} PhCO_2H \quad (228)$$

R	*X*	*Yield* (%)
Me	Br	94
$PhCH_2$	Cl	91

Benzoate and phenylacetate esters have been oxidatively cleaved to the corresponding carboxylic acids by can in aqueous acetonitrile in 56 to 85% yield (Equation (229) and Table 46) ⟨86IJC433⟩.

$$R^1CO_2R^2 \xrightarrow{\text{can, MeCN, } H_2O\text{, reflux, 5–6 h, 80 °C}} R^1CO_2H \quad (229)$$

Table 46 Oxidative cleavage of benzoate and phenylacetic esters to carboxylic acids using can.

R[1]	*R*[2]	*Yield* (%)
Ph	Me	56
Ph	Et	59
Ph	Pr^n	61
p-ClC_6H_4	Et	58
m-ClC_6H_4	Et	58
p-$O_2NC_6H_4$	Et	62
p-BrC_6H_4	Et	65
Ph	$CH_2CH{=}CH_2$	78
p-ClC_6H_4	$CH_2CH{=}CH_2$	74
m-ClC_6H_4	$CH_2CH{=}CH_2$	70
p-$O_2NC_6H_4$	$CH_2CH{=}CH_2$	70
p-Tol	$CH_2CH{=}CH_2$	69
p-An	$CH_2CH{=}CH_2$	85
Ph	CH_2Ph	64
p-ClC_6H_4	CH_2Ph	65
p-Tol	CH_2Ph	70
$PhCH_2$	Et	68
$PhCH_2$	Pr^n	65

Quantitative hydrolysis of methyl benzoate has been achieved using a mixture of lithium iodide in DMF at 130 °C to 150 °C both with and without added nucleophiles to scavenge the methyl iodide formed (Equation (230)) ⟨72SC389⟩.

$$PhCO_2Me + LiI \xrightarrow[\text{100\%}]{\text{DMF, 24 h, conditions}} PhCO_2H \quad (230)$$

conditions: as above, 150 °C; as above plus NaOAc, 150 °C; as above plus NaCN, 130 °C

An excellent yield of 2,4,6-trimethylbenzoic acid has been obtained from its hindered methyl ester by reaction with the Lewis acid boron trichloride in methylene chloride ⟨71CC667⟩. Other *ortho*-substituted esters can be hydrolyzed to their corresponding acids using pyridinium chloride in pyridine at its boiling point so long as the *ortho*-substituent present is an acid substituent (Equation (231)) ⟨81S487⟩.

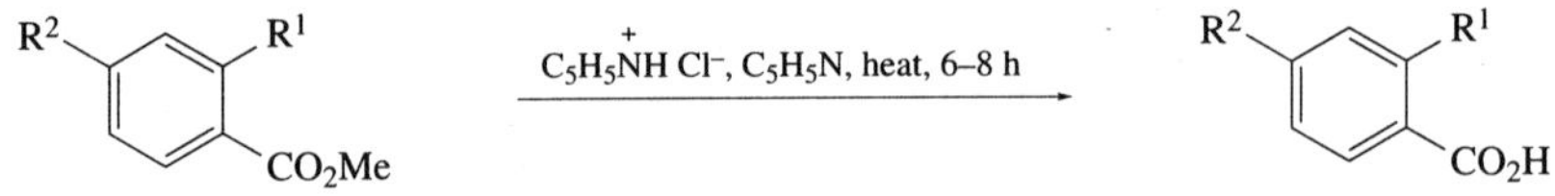

(231)

R^1	R^2	Yield (%)
OH	H	70
$NHSO_2C_6H_4Me$-*p*	H	89
CO_2Me	OH	80[a]
$NHSO_2C_6H_4Me$-*p*	CO_2Me	79[a]

[a] Reaction is run in C_5H_5N alone.

Aromatic nitriles have been hydrolyzed to their corresponding acids using both acids and bases. The two most commonly used acids have been sulfuric acid, which has been used to hydrolyze the nitriles shown in Equation (232) and Table 47, and phosphoric acid, which has been used to hydrolyze benzonitrile and α-cyanonaphthalene ⟨27RTC600⟩. The bases used to hydrolyze aromatic nitriles have been sodium hydroxide, which has been used to convert 3-chloro-4-cyanobiphenyl into 3-chlorobiphenyl-4-carboxylic acid in 81–90% yield ⟨66JCS(C)840⟩, and potassium hydroxide in glycerol which has been used to hydrolyze 6-methoxy-8-cyano- ⟨46JA1844⟩ and 6-cyano-4-methoxyquinoline ⟨55JA6682⟩ to their corresponding acids in 78 and 88% yields, respectively.

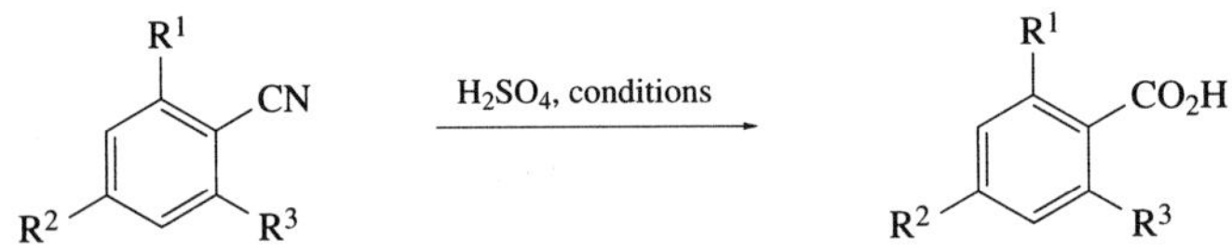

(232)

Table 47 Substituted benzoic acids from their corresponding nitriles via sulfuric acid hydrolysis.

R^1	R^2	R^3	Conditions	Yield (%)	Ref.
Me	H	H	i, 150 °C, 2 h ii, 190 °C, 1 h	80–89	43OSC(2)588
Cl	H	H	Heat	82	81MI 502-05
NO_2	NO_2	H	Heat	90	81MI 502-05
Me	H	Me	90–95 °C, 5 h	70	27RTC600

When 7,12-dicyanobenz[*k*]fluoroanthene was refluxed with potassium hydroxide in ethylene glycol monoethyl ether, instead of producing the dicarboxylic acid, the intermediate 7,12-dicarboxamide was formed which was converted into the diacid using either sulfuric acid or nitrous acid (Scheme 90) ⟨51JA436⟩. Other amides have also been hydrolyzed to carboxylic acids, and again both bases and acids have been used to effect the conversion. Treatment of 1-(*o*-amidophenyl)-1-(8-methyl-1-naphthyl)ethane with potassium hydroxide and ethylene glycol in a steel reaction vessel afforded 1-(*o*-carboxyphenyl)-1-(8-methyl-1-naphthyl)ethane in 94% yield (Scheme 91) ⟨50JOC617⟩. Using phosphoric acid at 145 °C to 160 °C converts the amides shown in Equation (233) into their corresponding acids ⟨27RTC600⟩, while the *N*-alkyl substituted amides shown in Equation (234) and Table 48 were converted to acetic or benzoic acid by nitrosation using nitrous acid, followed by thermal elimination of nitrogen from the intermediate *N*-alkyl-*N*-nitrosoamides ⟨55JA6011⟩. Nitrosonium fluoroborate in acetonitrile represents an attractive reagent for the hydrolysis, under anhydrous conditions, of hindered amides to their corresponding acids (Equation (235) and Table 49) ⟨65JOC2386⟩.

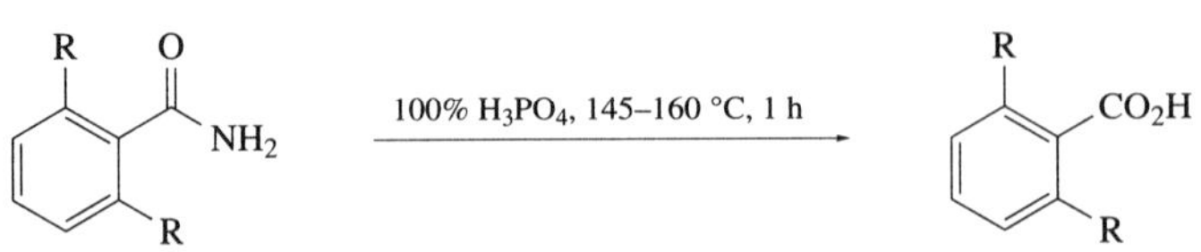

R = H, Me

(233)

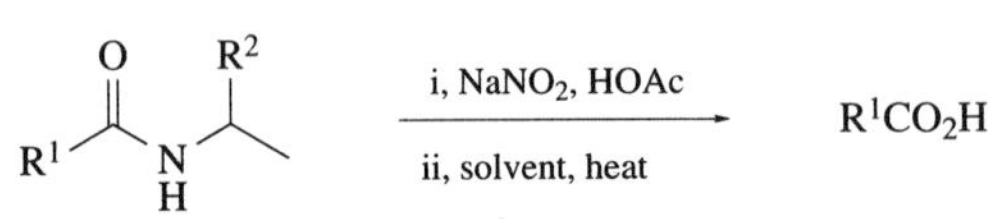

(234)

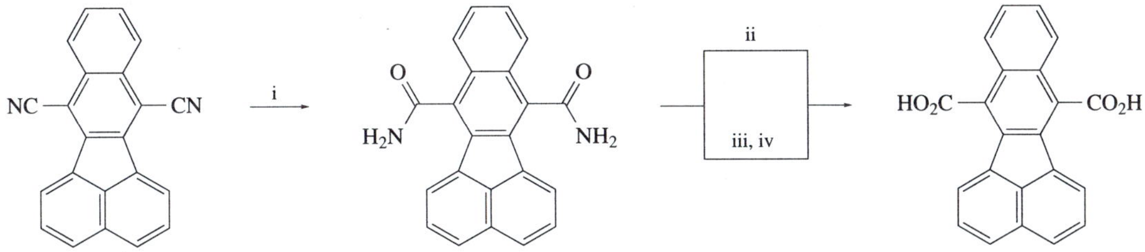

i, KOH, H_2O, ethylene glycol monomethyl ether, reflux, 48 h, copper flask, 83%; ii, H_2SO_4, HOAc, reflux, 48 h; iii, H_2SO_4, HOAc, $NaNO_2$, RT, 50 min; iv, heat, 30 min

Scheme 90

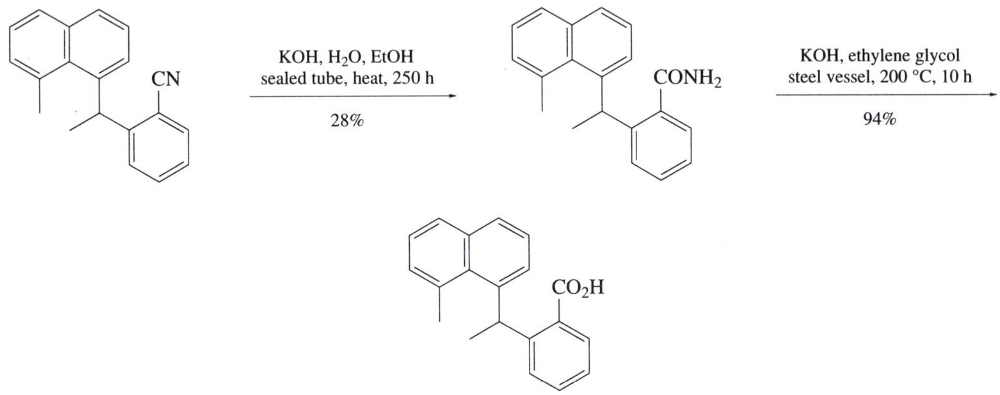

Scheme 91

Table 48 Conversion of *N*-alkyl-substituted amides to acetic or benzoic acid by nitrosation with nitrous acid and thermolysis.

R^1	R^2	*Solvent*	*Temp.* (°C)	*Yield* (%)
Me	Et	CCl_4	60	67
Me	Ph	Dioxane	45	55
Ph	Et	Pentane	25	64

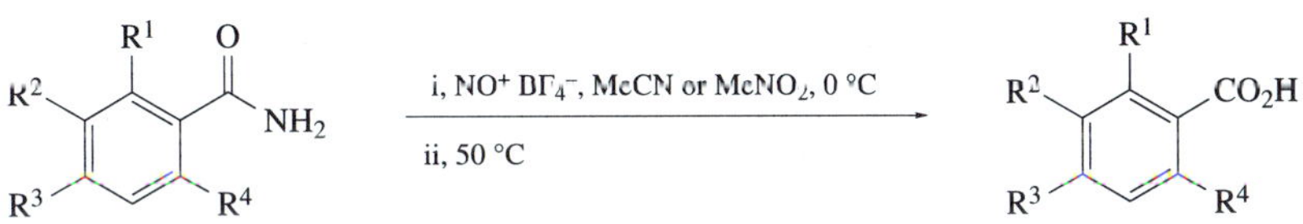

(235)

In the imide class of compounds *o*-sulfobenzoic imide has been hydrolyzed to the ammonium salt of *o*-sulfobenzoic acid in 91–94% yield using hydrochloric acid ⟨32OSC(1)14⟩, and 4-nitrophthalimide has been converted into 4-nitrophthalic acid in 96–99% yield using sodium hydroxide ⟨43OSC(2)457⟩.

Anhydrides have also been converted into aromatic acids by acid- and base-catalyzed reactions. For example, phthalic anhydride in a mixture of sulfuric and fuming nitric acids at 100–110 °C affords a mixture of 3- and 4-nitrophthalic acid ⟨32OSC(1)408⟩, while 1,8-naphthalic anhydride upon treatment with 85% aqueous potassium hydroxide at 60–70 °C affords 1,8-dipotassium naphthalate. This salt can be acidified to form the 1,8-diacid or heated to 400 °C to 430 °C and then acidified to afford the 2,6-naphthalene dicarboxylic acid (Scheme 92) ⟨73OSC(5)813⟩.

When reactive aromatic and heterocyclic molecules are treated with trifluoroacetic anhydride, α,α,α-trifluoroacetophenones are obtained, which upon treatment with aqueous sodium hydroxide in ethanol are hydrolyzed to the corresponding aromatic or heterocyclic carboxylic acids Scheme

Table 49 Conversion of hindered amides to their corresponding acids using trifluoroacetic anhydride.

R[1]	*R*[2]	*R*[3]	*R*[4]	*Yield* (%)
H	H	H	H	93
H	H	Me	H	95
Me	H	H	H	81
H	Bu^t	H	H	89
H	H	Bu^t	H	92
H	Et	H	H	87
H	H	Et	H	91
H	Pr^i	H	H	80
H	H	Pr^i	H	93
Me	H	Me	Me	78
Pr^i	H	Pr^i	Pr^i	70

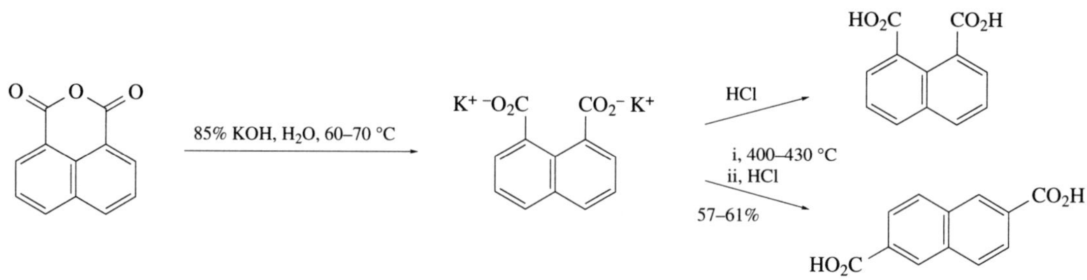

Scheme 92

93 and Table 50 ⟨77JFC(10)437⟩. Although not a true trihalide hydrolysis, since the resulting acids contain one carbon atom less than the acetophenone starting material, this trihalide hydrolysis–elimination does produce acids in good yields.

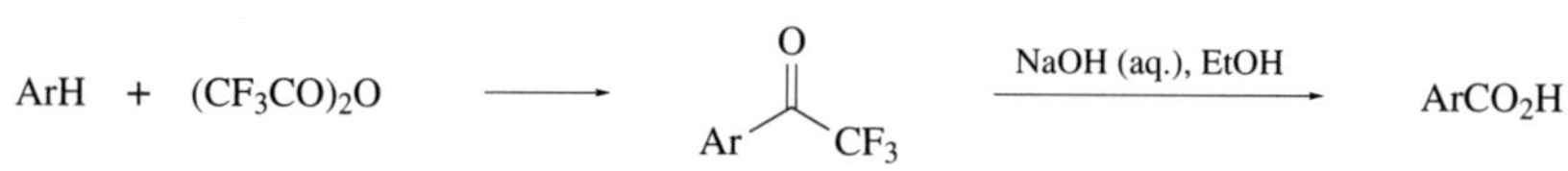

Scheme 93

5.02.4.2 By Carbonation of Organometallic Reagents

Normal carbonation reactions require the initial preparation of an organometallic reagent which is then treated with carbon dioxide. However, if the molecule to be carbonated contains an acidic enough hydrogen, a rather mild base can be used to produce the required initial anion and the use of organometallic reagents can be avoided. Such a reaction has been reported for resorcinol, which upon treatment with the mild base potassium bicarbonate followed by carbon dioxide affords a 57–60% yield of *β*-resorcyclic acid (2,4-dihydroxybenzoic acid) ⟨43OSC(2)557⟩.

Carbonation of aryllithium compounds prepared from halides have resulted in the preparation of benzoic acid ⟨72CC752⟩ and 1-carboxybenzo[*c*]phenanthrene ⟨61JOC1759⟩. Aryllithium reagents suitable for carbonation have also been prepared by direct ring metalation of benzene derivatives containing one or more alkoxyl functions. Lithiation is directed exclusively *ortho* to the ether function, and if followed by carbonation produces the corresponding *ortho*-carboxylic acid. Examples of this approach include the preparation of 5-methyl-2,3-methylenedioxybenzoic acid ⟨76S123⟩, homochroman-9-carboxylic acids (Scheme 94) ⟨74SC1⟩, and a series of mono- (Scheme 95 and Equations (236) and (237)) ⟨79JCS(P1)2079⟩ and dimethoxy (Scheme 96) ⟨73TL4839, 80JCS(P1)1658⟩ fluorobenzenes, in which case the lithiation is accompanied by replacement of fluorine in the same or in an adjacent position with the alkyl group of the lithium reagent. These results can be readily explained on the basis of a benzyne intermediate mechanism as illustrated in Schemes 95 and 96.

Table 50 Conversion of aromatics and heterocyclics to their corresponding acids using trifluoroacetic anhydride.

Ar	*Yield* (%)
p-$Me_2NC_6H_4$	75
p-$Et_2NC_6H_4$	72
2,4,6-$(MeO)_3C_6H_2$	0
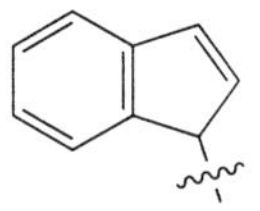	Not reported
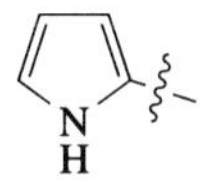	92
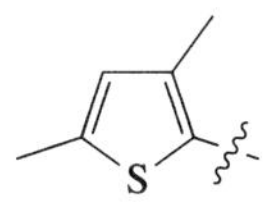	96
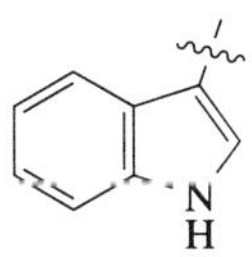	78
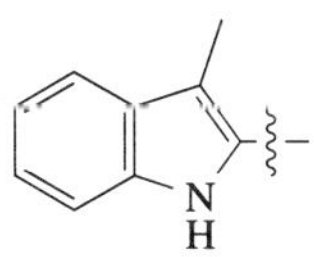	0

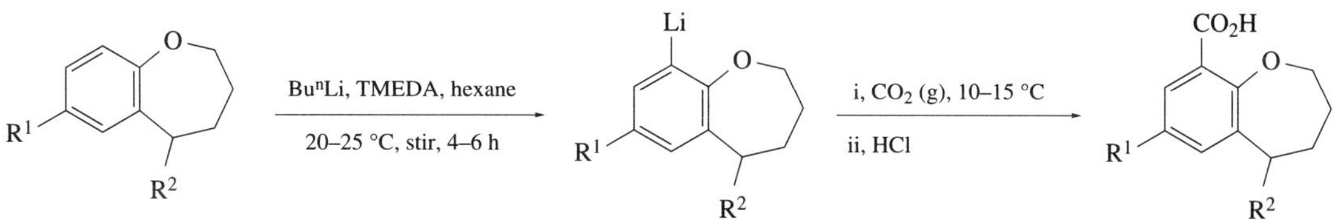

R¹	*R²*	*Yield* (%)
H	H	73
H	Me	68
MeO	Me	65
Ph	Me	48

Scheme 94

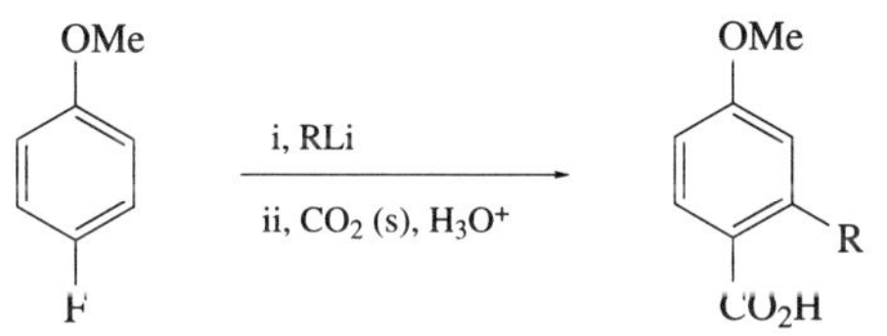 (236)

R	*Yield* (%)
n-$C_{12}H_{25}$	17
n-$C_{15}H_{31}$	43

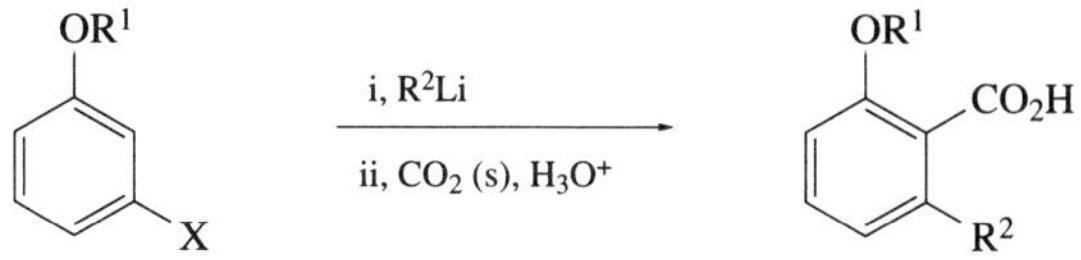 (237)

R¹	*X*	*R²*	*Yield* (%)
Me	Br	*n*-$C_{14}H_{29}$	
$PhCH_2$	F	Et	31

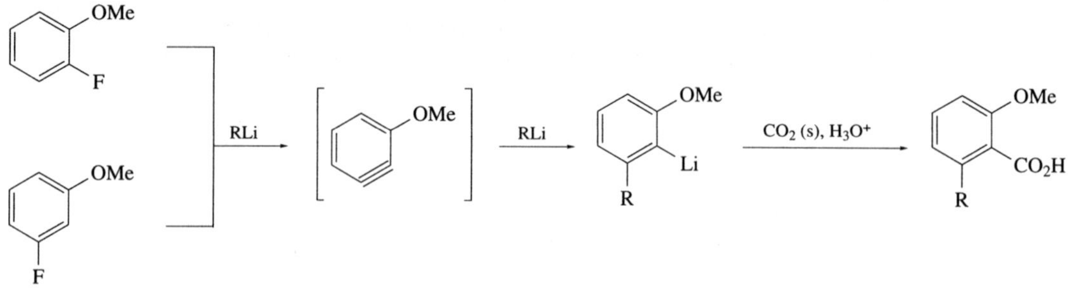

R	*Yield (%) from* ortho	*Yield (%) from* meta
Me	6	7
Et	29	54
Pr^n	40	84
Pr^i	0	0
Bu^n	16	25
n-$C_{12}H_{25}$	28	29
n-$C_{14}H_{29}$	39	37
n-$C_{15}H_{31}$	35	41

Scheme 95

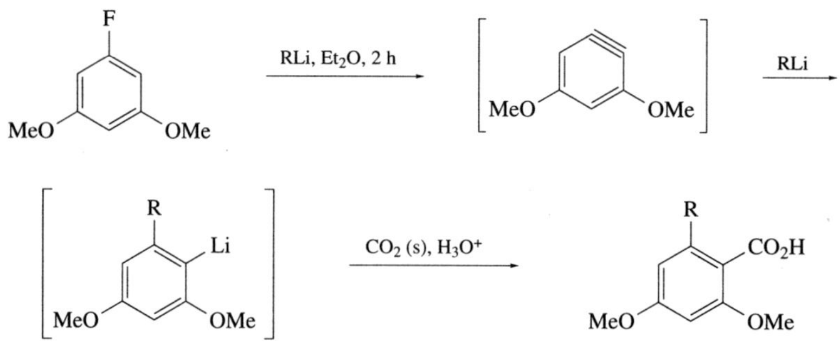

R	*Yield (%)*
Me	15
Et	45
Pr^n	
n-C_5H_{11}	37
n-C_7H_{15}	39
n-$C_{15}H_{31}$	34

Scheme 96

Carbonation of Grignard reagents has been used to prepare p-(4,4,4-trifluoro-n-butyl)-benzoic acid ⟨85ZOR2407⟩, mesitoic acid ⟨55OSC(3)553, 55OSC(3)555⟩, pentachlorobenzoic acid ⟨73OSC(5)890⟩, 4′-fluorobiphenyl-4-carboxylic acid ⟨66JA3318⟩, and α-naphthoic acid ⟨43OSC(2)425⟩. One extension of this general approach is the use of iron pentacarbonyl as a carbon monoxide source (Scheme 97) ⟨81NKK1333⟩.

$$p\text{-}RC_6H_4MgBr \;+\; Fe(CO)_5 \xrightarrow{\text{THF}} [\,p\text{-}RC_6H_4OFe(CO)_4]^- MgBr^+ \xrightarrow{I_2,\ H_2O} p\text{-}RC_6H_4CO_2H$$

R	*Yield (%)*
H	70
Me	78
MeO	59

Scheme 97

Other methods employed to produce aromatic carboxylic acids include the previously discussed use of Stiles' reagent, methylmagnesium carbonate, as a carboxylating agent (Equation (238)) ⟨69CC343⟩, and the use of a mixture of acetic acid–acetic anhydride–sodium acetate in the presence of palladium(II) chloride (Scheme 98) ⟨69TL1019⟩. The results obtained from the latter procedure apparently indicate that the carboxylation is an aromatic electrophilic substitution since with substituted benzenes containing electron-donating substituents *para*-isomers are obtained while with electron-withdrawing substituents *meta*-isomers result.

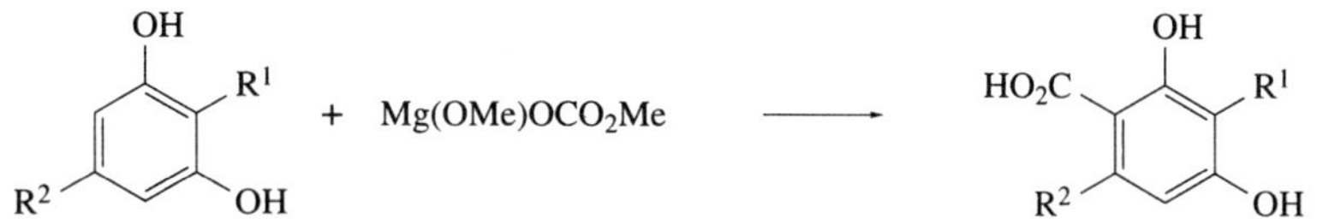

(238)

R^1	R^2	*Yield* (%)
H	H	45
H	OH	45
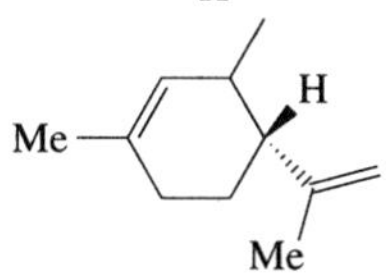	n-C_5H_{11}	

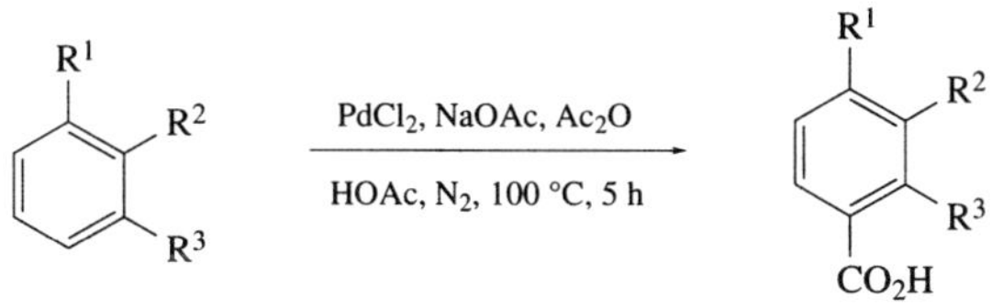

R^1	R^2	R^3
H	H	H
Me	H	H
MeO	H	H
MeO	H	MeO
MeO	MeO	H

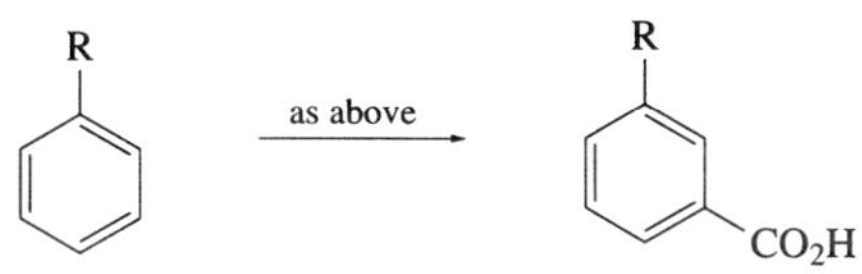

R = CO_2Me, NO_2

Scheme 98

5.02.4.3 By Oxidation Reactions

Aromatic substituted primary alcohols have been oxidized to aromatic acids using oxygen in the presence of potassium hydroxide and DME (Equation (239)) ⟨84IZV1567⟩, solid sodium permanganate monohydrate suspended in hexane (Equation (239)) ⟨81TL1655⟩, and nickel peroxide in the presence of sodium hydroxide (Equation (239)) ⟨62JOC1597⟩.

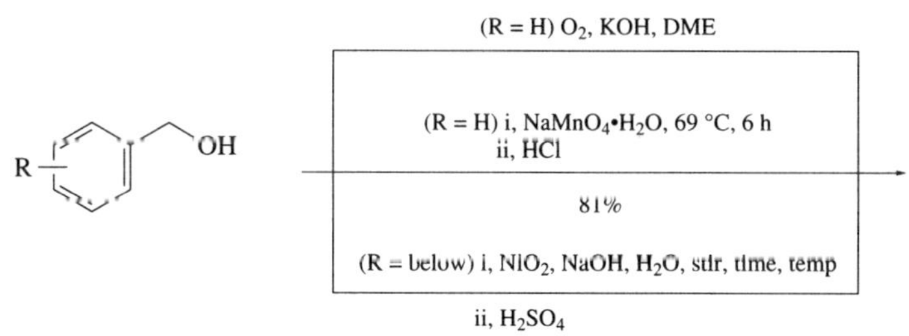

(239)

R	*Temp.* (°C)	*Time* (h)	*Yield* (%)
H	30	3	97
o-Me	30	3	97
m-Me	5	7	30
p-Me	5	7	81

Nickel peroxide has also been used to oxidatively cleave glycols, α-keto alcohols, and α-keto acids to aromatic acids (Scheme 99) ⟨64CPB403⟩, and a similar oxidative cleavage has been observed with sodium periodate in sulfuric acid with α-keto alcohols and α,β-diketones ⟨35JCS1467⟩.

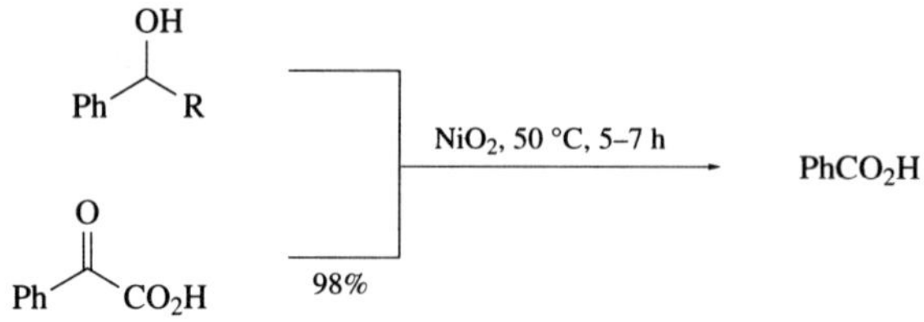

R	*Yield* (%)
CH(OH)Ph	97
COPh	99
CO_2H	90

Scheme 99

Aromatic aldehydes have also been oxidized to aromatic acids using a variety of reagents. For example: vanillin is oxidized to protocatechuic acid in 89–99% yield using potassium hydroxide at 240–245 °C ⟨55OSC(3)745⟩; potassium permanganate successfully converts *p*-(triphenylsilyl)benzyl alcohol into *p*-(triphenylsilyl)benzoic acid ⟨56JA1689⟩, piperonal to piperonylic acid ⟨43OSC(2)538⟩, *m*-(4-fluorophenyl)benzaldehyde to *m*-(4-fluorophenyl)benzoic acid ⟨66JA3318⟩, and 2,3,6-trichlorobenzaldehyde to 2,3,6-trichlorobenzoic acid ⟨51JCS1208⟩, which has also been accomplished using 30% nitric acid ⟨51JCS1208⟩. When used in conjunction with 18-crown-6 and benzene, potassium permanganate converts benzaldehyde to benzoic acid in almost quantitative yields ⟨72JA4024⟩.

Contrary to the results reported above using potassium hydroxide, when vanillin is oxidized using silver oxide in base vanillic acid, not protocatechuic acid, is formed in 83–95% yield ⟨63OSC(4)972⟩. Silver oxide has also been used to convert 9-anthracenecarboxaldehyde to 9-anthracenecarboxylic acid ⟨43JA262⟩, and *p*-tolualdehyde-α-D to *p*-toluic acid-α-D ⟨76JA1743⟩.

Haloform oxidation of acylaromatics to substituted benzoic acids can be accomplished for methyl β-naphthyl ketone using sodium hypochlorite alone ⟨43OSC(2)428⟩, or for 2,4,6-trimethylpropiophenone using sodium hypochlorite followed by sodium hydroxide in the presence of a phase-transfer catalyst (Scheme 100) ⟨82EUP46194⟩. The stable crystalline solid, sodium bromite has also been used to oxidize methyl ketones and secondary methyl carbinols to their corresponding aromatic acids (Equation (240) and Table 51) ⟨83CC392⟩.

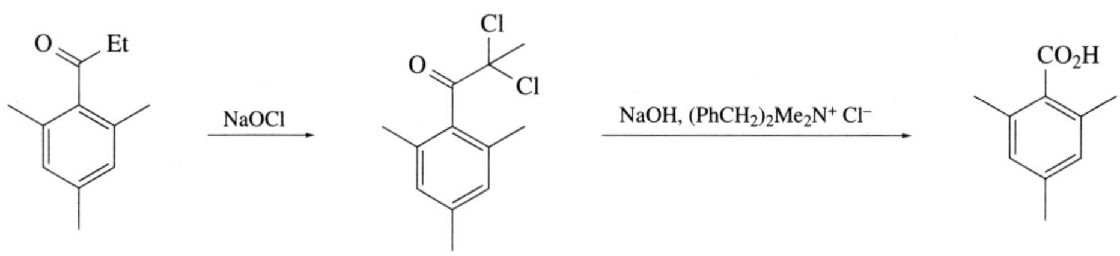

Scheme 100

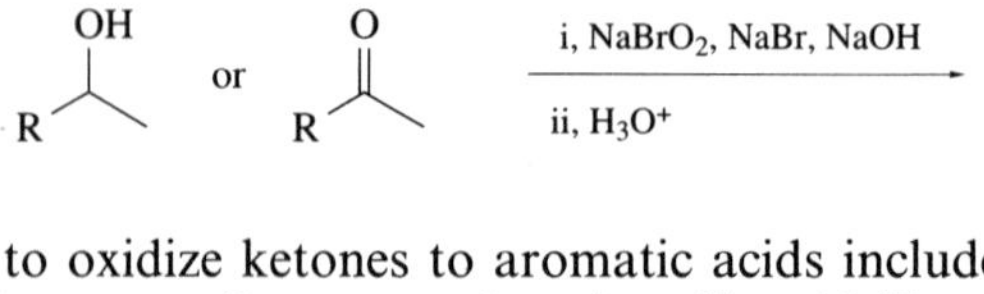

(240)

Other reagents used to oxidize ketones to aromatic acids include: sodium dichromate in acetic acid to oxidize 2-acetyfluorene to fluorenone-2-carboxylic acid (Equation (241)) ⟨55OSC(3)420⟩, 30% hydrogen peroxide to oxidize diethylaminomethyl phenyl ketone to benzoic acid ⟨83CC392⟩, nitric acid followed by potassium permanganate to convert *p*-methylacetophenone to terephthalic acid (Equation (242)) ⟨55OSC(3)791⟩, and potassium *t*-butoxide in anisole to oxidize benzophenone, *o*-, *m*- and *p*-dibenzoylbenzene all to benzoic acid ⟨71CB2637⟩.

Table 51 Aromatic acids via sodium bromite oxidation of methyl ketones and secondary methyl carbinols.

Ketone or alcohol	*Product*	*Yield* (%)
PhCOMe	$PhCO_2H$	94
p-MeC_6H_4COMe	*p*-$MeC_6H_4CO_2H$	96
p-ClC_6H_4COMe	*p*-$ClC_6H_4CO_2H$	90
p-$O_2NC_6H_4COMe$	*p*-$O_2NC_6H_4CO_2H$	87
2-NaphCOMe	2-$NaphCO_2H$	85
PhCH(OH)Me	$PhCO_2H$	92

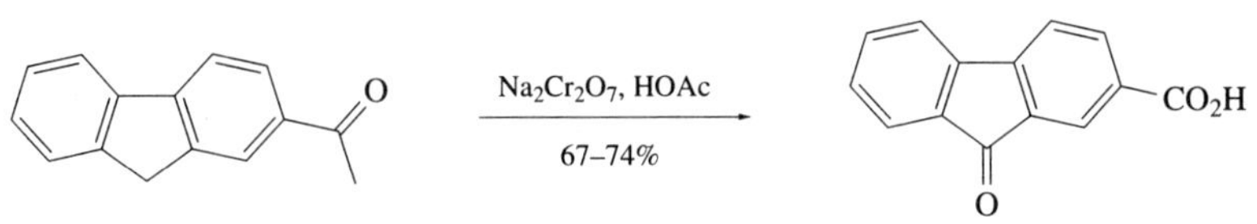

(241)

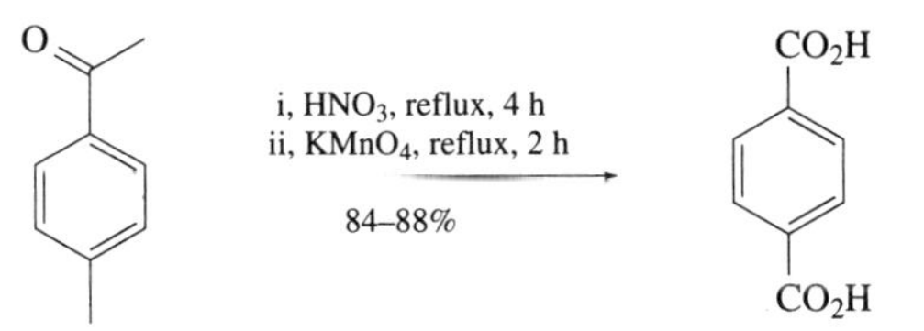

(242)

Reactions of double bonds with oxidizing agents which have led to aromatic acids include: quinoxaline with aqueous permanganate to give 75–77% of 2,3-pyrazinedicarboxylic acid 〈63OSC(4)824〉; styrene with "purple benzene" (potassium permanganate and tricaprylmethyl ammonium chloride in benzene) to give 95% of benzoic acid 〈74TL1511〉; and, phenanthrene and pyrene with ozone to give 81–88% of 2′-formyl-2-biphenylcarboxylic acid 〈73OSC(5)493〉 and 32–38% of 5-formyl-4-phenylanthroic acid 〈63OSC(4)484〉, respectively.

By the correct choice of reagents, mono- and polyalkylated benzenes have been oxidized to their corresponding carboxylic acids without effecting other ring substituents which may be present. Examples of these arenes oxidations include: conversion of 2-hydroxy-3-methylbenzoic acid to 2-hydroxyisophthalic acid in 46–61% yield using lead dioxide and potassium hydroxide 〈73OSC(5)617〉; conversion of 2,3,4-trichlorotoluene to 2,3,4-trichlorobenzoic acid 〈51JCS1208〉, *o*-xylene to *o*-toluic acid in 53–55% yield 〈55OSC(3)820〉, and *p*-cymene to *p*-toluic acid in 51% yield 〈55OSC(3)822〉 all using nitric acid.

Whereas chromium trioxide in sulfuric acid oxidizes triphenyl-(*p*-tolyl)silane to *p*-(triphenylsilyl)benzoic acid without effecting the *para* substituent (Equation (243)) 〈56JA1689〉, potassium dichromate in sulfuric acid oxidizes 2-nitroso-5-nitrotoluene to 2,5-dinitrobenzoic acid 〈55OSC(3)334〉 oxidizing both the alkyl and the nitroso substituents. A similar result occurs when 2,3-dimethylnaphthalene is treated with sodium dichromate in an autoclave at 250 °C, since both methyl groups are oxidized producing 2,3-naphthalenedicarboxylic acid in 87–93% yield 〈73OSC(5)810〉. Other arenes oxidized using sodium dichromate in sulfuric acid include *p*-nitrotoluene which produces 82–86% yield of *p*-nitrobenzoic acid 〈32OSC(1)392〉, and 2,4,6-trinitrotoluene which produces 57–59% yield of 2,4,6-trinitrobenzoic acid 〈32OSC(1)543〉. The series of 2-, 3-, 4-, 5- and 6-methylbenzo[*c*]-phenanthrenes have all been converted into their corresponding carboxybenzo[*c*]-phenanthrenes using aqueous sodium dichromate contained in a stainless steel autoclave which was rocked and heated at 250 °C for 65 hours which resulted in 64, 59, 71, 81, and 65% yields of acids, respectively 〈61JOC1759〉.

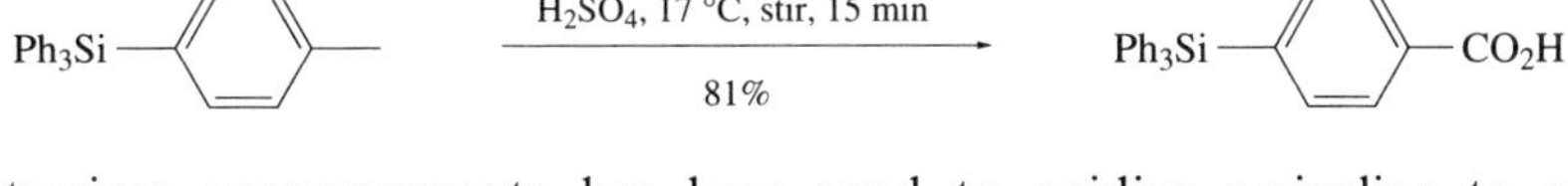

(243)

Aqueous potassium permangangate has been used to oxidize α-picoline to picolinic acid 〈55OSC(3)740〉, *o*-chlorotoluene to *o*-chlorobenzoic acid 〈43OSC(2)135〉, and *p*-tolylmercuric chloride to *p*-chloromercuribenzoic acid 〈32OSC(1)159〉, while a solution of potassium permanganate in

18-crown-6 has been used to convert toluene into benzoic acid in 78% yield and *p*-xylene to *p*-toluic acid in quantitative yields ⟨72JA4024⟩.

Fusaric acid and other 5-alkyl picolinic acids have been prepared by oxidation of 2-methyl-5-alkylpyridines with selenium dioxide in boiling pyridine (Equation (244)) ⟨68JIC1109⟩. The same reagent and conditions have also been used to oxidize bis(6-methyl-3-pyridyl)diacetylene to 1,4-bis(6-carboxypyridyl-3)butane in 36% yield ⟨68JIC1109⟩, and 2,4-dimethyl-6-phenylpyrimidine to 2-methyl-6-phenylpyrimidine-4-carboxylic acid ⟨79H(13)235⟩.

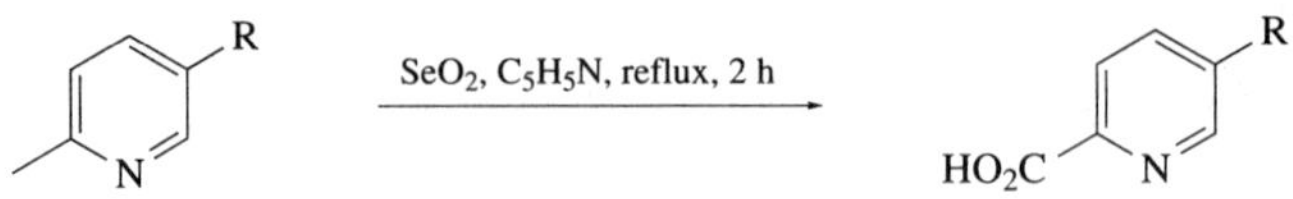

(244)

R	*Yield* (%)
Bu^n	40
n-C_6H_{13}	32

A somewhat mild oxidation procedure used to convert a series of substituted benzenes, naphthalenes and pyridines into their corresponding carboxylic acids involves the use of oxygen in the presence of sodium or potassium hydroxide or *t*-butoxide and 18-crown-6 in DME (Equation (245) and Table 52) ⟨84TL4989⟩.

$$\text{ArMe} \xrightarrow[\text{DME, temp., time}]{O_2\text{, base, 18-crown-6}} \text{ArCO}_2\text{H} \quad (245)$$

Table 52 Base-catalyzed oxidation of substituted benzenes, naphthalenes and pyridines in crown ether.

Ar	*Base*	*Temp* (°C)	*Time* (h)	*Yield* (%)
p-$O_2NC_6H_4$	KOH	20	3.5	94
o-$O_2NC_6H_4$	NaOH	20	5	50
2-Pyridyl	KOH	25	96	85
3-Pyridyl	Bu^tOK	60	12	25
4-Pyridyl	KOH	25	96	80
1-Naphthyl	KOH	60	96	25
2-Naphthyl	KOH	70	6	60
$C_6H_5Cr(CO)_3$	KOH	60	25	[a]

[a] Two products were obtained: 70% of benzoic acid and 17% of the chromium tricarbonyl complex of benzoic acid.

Haloalkyl benzene substituents such as bromomethyl, dibromomethyl, trichloroethyl, and dichlorovinyl have also been oxidized to the corresponding aromatic acid function. Thus, selenium dioxide at 160 °C converts 2-bromomethylheptafluoronaphthalene to heptafluoro-2-napthoic acid ⟨62JCS186⟩, and 3-dibromomethyl-4′-fluorobiphenyl-3-carboxylic acid ⟨66JA3318⟩, while alkaline potassium permanganate oxidizes 3-(*β*,*β*,*β*-trichloroethyl)-4-methylchlorobenzene to 4-chlorophthalic acid, and 3-(*β*,*β*-dichlorovinyl)-4-methylchlorobenzene to 5-chloro-2-methylbenzoic acid ⟨58JA1236⟩. Interestingly, if the latter two substrates are treated with chromium trioxide or sodium dichromate the methyl group is oxidized while the haloalkyl substituents are left uneffected. Using this approach 3-(*β*,*β*,*β*-trichloroethyl)-4-methylchlorobenzene is oxidized to 2-(*β*,*β*,*β*-trichloroethyl)-4-chlorobenzoic acid with chromium trioxide and 3-(*β*,*β*-dichlorovinyl)-4-methylchlorobenzene is oxidized to 2-(*β*,*β*,*β*-trichloroethyl)-4-chlorobenzoic acid with sodium dichromate ⟨58JA1236⟩.

5.02.4.4 By Miscellaneous Reactions

Two methods used to prepare aromatic acids not previously mentioned are the Friedel–Crafts acylation of mesitylene with oxalyl chloride followed by acid hydrolysis which produces mesitoic acid (Equation (246)) ⟨73OSC(5)706⟩, and the thermal decomposition of bis(3,4-dichlorobenzoyl) peroxide which produces 3,4-dichlorobenzoic acid (Equation (247)) ⟨73OSC(5)51⟩.

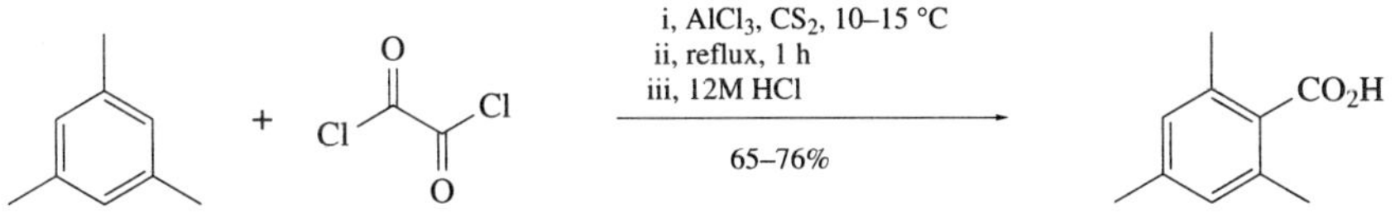

(246)

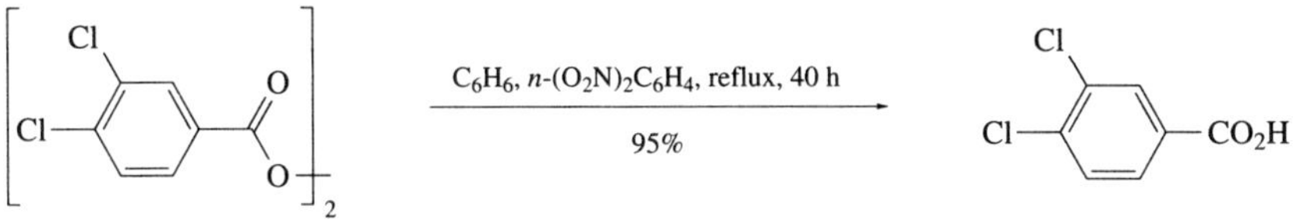

(247)

5.02.5 CARBOXYLIC ACIDS VIA BIOTRANSFORMATIONS

An indication of the increasing attention which has been given to the synthesis of carboxylic acids by hydrolytic and oxidative procedures utilizing either microbial agents directly or immobilized whole cell preparations containing hydrolytic or oxidative enzymes can be obtained by observing the increase in publications, especially in the patent literature, which has occurred during the 1980s and 1990s. Biotransformation procedures have been published for the preparation of carboxylic acids containing a wide variety of substituents. Thus, a review dealing with microbial oxidative metabolism of alkyl compounds for the synthesis of dibasic acids by yeasts and bacteria appeared in 1984 ⟨84MI 502-02⟩, while the use of immobilized biocatalysts for the preparation of pharmaceuticals has also been reviewed ⟨B-84MI 502-01⟩. The following examples represent a few applications of microbial hydrolysis and oxidation reactions to the synthesis of various carboxylic acids.

Mutant flavobacterium oxyclans cells in the presence of calcium carbonate have been reported to be an effective oxidizing agent for the conversion of disubstituted bis(hydroxymethyl)-1,3-dioxanes in 78–91% yields into disubstituted carboxyl hydroxymethyl-1,3-dioxanes (Equation (248)) ⟨73JOC1241⟩.

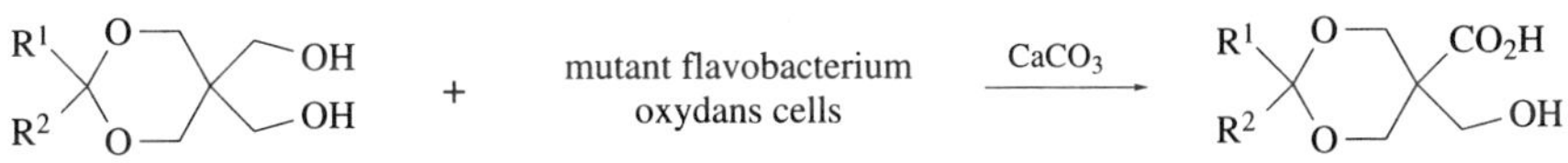

(248)

Optically pure L-α-hydroxy acids can be prepared in synthetically useful amounts from racemic 1,2-diols via oxidation using coimmobilized horse liver alcohol dehydrogenase (HLADH) and aldehyde dehydrogenase (AldDH) with NAD as cofactor (Scheme 101) ⟨85JOC1992⟩.

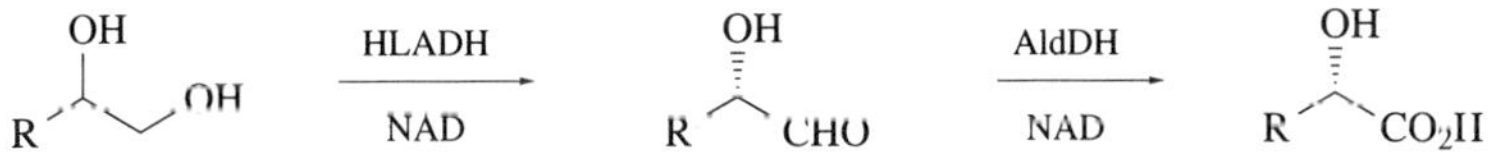

R = $HOCH_2$, FH_2C, ClH_2C, BrH_2C, Me, H_2NCH_2, $H_2C{=}CH$, Et, Pr^i

Scheme 101

n-Alkenes have been converted into *n*-alkene dicarboxylic acids on treatment with cultures or cells of *Candida* under aeration as reported for the preparation of *trans* 6 dodecenedioic acid from *trans*-6-dodecene (Equation (249)) ⟨83JAP165793⟩ and the conversion of *trans*-9-octadecenoic acid to *trans*-9-octandecenedioic acid (Equation (250)) ⟨83JAP165794⟩. Other Japanese patents describe the synthesis of ω-halocarboxylic acids by microorganism oxidation of alkyl or alkenyl halides (Equation 251) ⟨85JAP114195, 86JAP47675⟩.

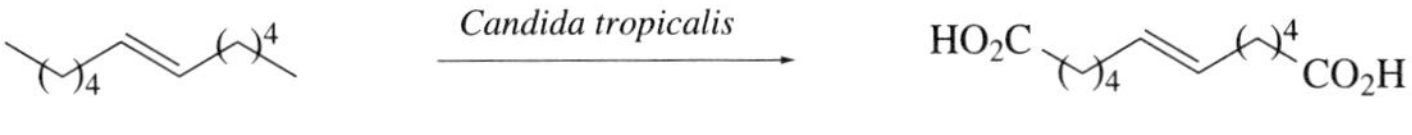

(249)

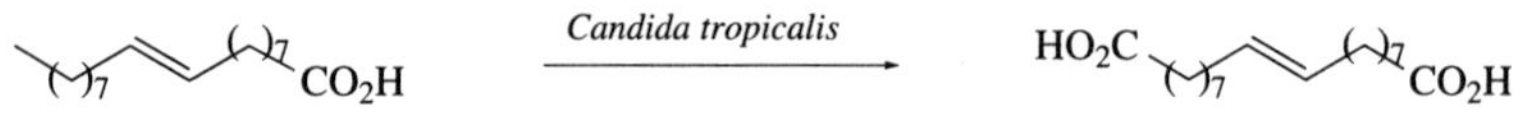

(250)

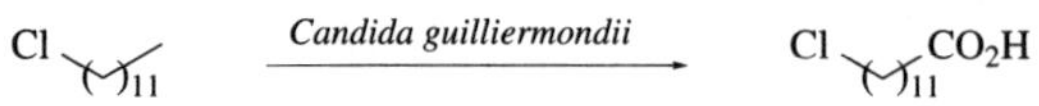

(251)

The important antiinflammatory agent (*R*)-(−)-2-(4′-isobutylphenyl)propanoic acid (ibuprofen) is obtained in 68% *ee*, along with two longer chain (*S*)-(+) acids, using microbial oxidation of (+)-1-isobutyl-4-(1′-methyloctylbenzene) by *Rhodococcus* s.p. BPM 1613 (Equation (252)) ⟨84MI 502-03⟩.

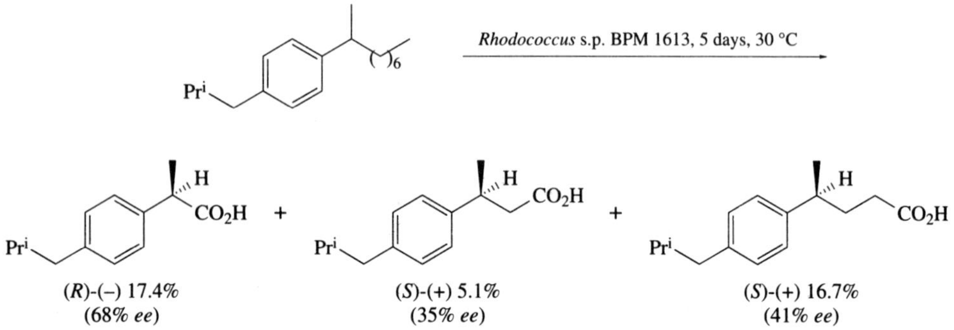

(252)

Results of a study of the production of β-hydroxycarboxylic acids from aliphatic carboxylic acids by microorganisms indicate that D-(−)- and L-(+)-β-hydroxyvaleric acids can be obtained from valeric acid by *Candida rugosa* IFO 0750 and IFO 1542, respectively, while β-hydroxyisovaleric acid was produced from isovaleric acid by *Endomyces reessii* CBS 179.60. The (−) and (+) enantiomers of α-hydroxymethylbutyric acid were produced from racemic α-methylbutyric acid by *Candida rugosa* IFO 0750 and *Trichosporon fermentans* CBS 2529, respectively; and β-hydroxyisocaproic acid was produced from isocaproic acid by *Endomyces reessii* CBS 179.60 ⟨81MI 502-06⟩.

Oxidation of a broad sampling of D-amino acids with immobilized whole cells of *Trigonopsis variabilis* produces α-keto acids (Equation (253)) ⟨81MI 502-07, 82MI 502-01⟩.

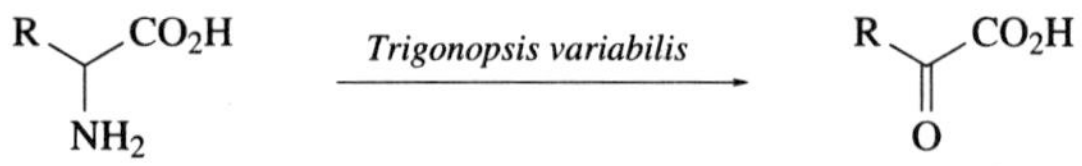

(253)

Hydrolysis of aliphatic and aromatic nitriles under varying but mild conditions has been accomplished using nitrilases available from a variety of industrial sources ⟨B-88MI 502-01, 90CC648, 90PAC1441, 90TL7223, 91TL341, 93T10099⟩. Using an immobilized nitrilase preparation (SP) available from Novo Industri A/S (Denmark) which consists of two enzymes, a hydratase, used to catalyze the conversion of a nitrile to the corresponding amide, and an amidase, used to catalyze the hydrolysis of the amide to the corresponding carboxylic acid, a variety of acyclic, cyclic and heterocyclic nitriles and dinitriles or nitriles containing base- or acid-sensitive functional groups have been converted to their corresponding carboxylic acids Scheme 102 and Table 53 ⟨90CC648, 90TL7223, 91TL341, 93T10099⟩.

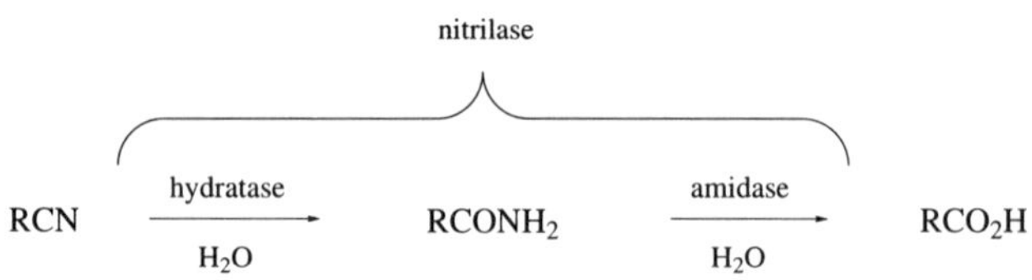

Scheme 102

It has also been reported ⟨84MI 502-04⟩ that stereospecific enzymatic hydrolysis has been used to produce carboxylic acids from esters, since lipase from *Candida cylindracea* hydrolyzes octyl *R*(+)-2-chloropropionate to *R*(+)-2-chloropropionic acid but does not hydrolyze octyl *S*(−)-2-chloropropionate. This same enzyme, however, exhibits no stereoselectivity in the hydrolysis of the methyl ester of the same acid, and thus both methyl *R*(+)- and *S*(−)-2-chloropropionate were hydrolyzed to *R*(+)- and *S*(−)-2-chloropropionic acid.

Table 53 Nitrile hydrolysis using immobilized nitrilase.

R	*Ref.*	*R*	*Ref.*
$HO(CH_2)_2$–	91TL341	$H_2C{=}CHCH_2$–	90TL7223
$Cl(CH_2)_3$–	90CC648		
$Bu^nCH(OH)CH_2$–	91TL341	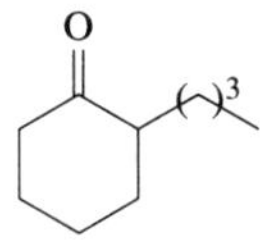	90TL7223
$Hex^nCH(OH)CH_2$–	91TL341	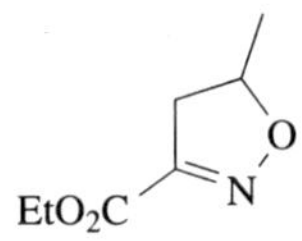	91TL341
c-C_3H_5–	90TL7223		
c-C_6H_{11}–	90TL7223	C_6H_5–	93T10099
$HOCH_2CH_2(OH)(CH_2)_3$–	91TL341	p-$H_2NC_6H_4$–	93T10099
$AcOCH_2CH_2(OAc)(CH_2)_3$–	91TL341	p-HOC_6H_4	93T10099
$BzOCH_2CH_2(OBz)(CH_2)_3$–	91TL341	p-ClC_6H_4	90CC648
$Pr^iC(OH)Me$	91TL341		93T10099
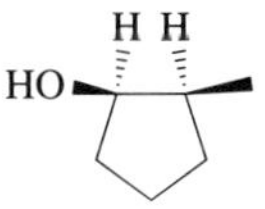	91TL341	m-MeC_6H_4	90CC648
		p-MeC_6H_4	90CC648
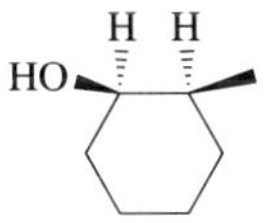	91TL341	m-$MeOC_6H_4$	90CC648
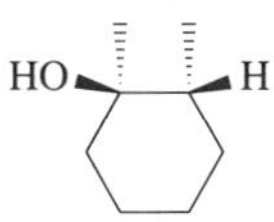	91TL341	p-$MeOC_6H_4$	90CC648 93T10099
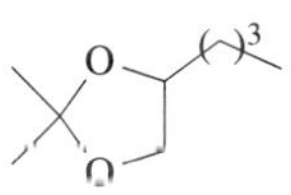	91TL341	p-$O_2NC_6H_4$	93T10099
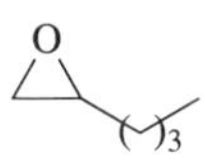	91TL341	m-$MeO_2CC_6H_4$–	90TL7223
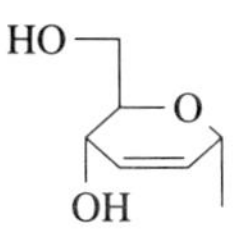	91TL341	$C_6H_5CH_2$–	90TL7223
		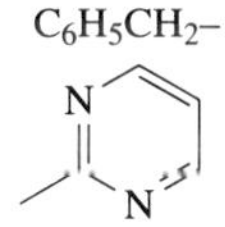	91TL341
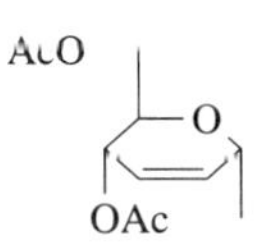	91TL341	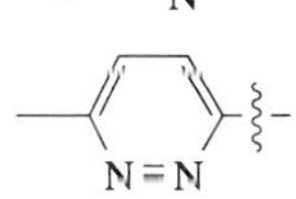	91TL341
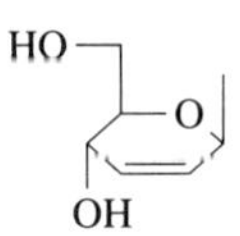	91TL341	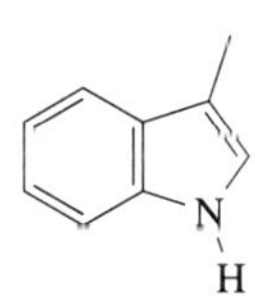	91TL341
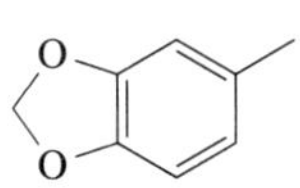	90TL7223	m-NCC_6H_4–	90TL7223[a]
		o-NCC_6H_4–	90TL7223[b]

(continued)

Table 53 Continued.

R	*Ref.*	*R*	*Ref.*
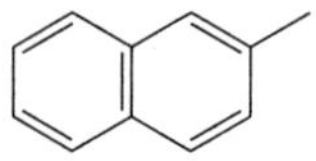	90TL7223	*p*-NCC_6H_4	90TL7223[a]
$NC(CH_2)_2$–	90TL7223[a]	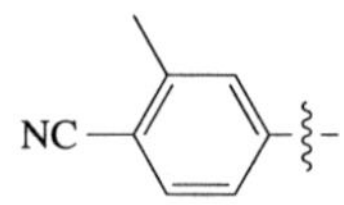	90TL7223[a]
$NC(CH_2)_3$–	90CC648[a] 90TL7223[c]	*m*-$NCCH_2C_6H_4CH_2$–	90TL7223[a]
NCCH=CH– (*trans*)	90TL7223[a]	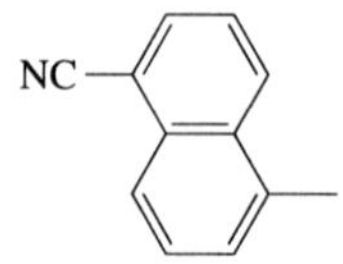	90TL7223[a]
NCCH=CH–CH=CH– (*cis, cis*)	90CC648[a]		

[a] Produces monocarboxylic acid product. [b] Produces dicarboxylic acid product.
[c] Produces a mixture of mono- and dicarboxylic acid products.

5.03
Carboxylic Esters and Lactones

JOHANN MULZER
Freie Universität Berlin, Germany

5.03.1 CARBOXYLIC ESTERS VIA ESTERIFICATION REACTIONS

The subject has been covered by a number of excellent reviews ⟨79COC(2)869, 80T2409, 85HOU(E5)659, 91COS(6)323, B-92MI 503-01⟩. Methods for the synthesis of esters and lactones are numerous; this chapter offers an overview of the most important and commonly used methods. Alcoholysis of acid chlorides and anhydrides is not treated here.

5.03.1.1 Direct Esterification Using Carboxylic Acid and Alcohol

This method is by far the most general and the cheapest. Carboxylic acid and alcohol are mixed in a suitable solvent (toluene, dichloromethane, etc.) and treated with a Brönsted or Lewis acid. An equilibrium is formed between the starting materials, ester and water. To shift this equilibrium towards the product side, the water can be removed with a molecular sieve or by azeotropic distillation. If an inexpensive acid or alcohol is used, this component can be applied in excess. Equations (1)–(9) show a selection of examples which demonstrate the broad applicability of this method. Acid-sensitive functional groups in the reactants like acetals are not tolerated ⟨55OSC(3)46, 57JA4759, 63JCS1927, 68CI(L)1568, 70TL4011, 71JOC2548, 71TL3453, 74JA8113, 75AP(308/12)954, 76MI 503-01, 81S413⟩. However, the esterification of lactic acid (Equation (4)) proceeds without racemization under the conditions indicated. Mechanistically this so-called 'Fischer–Speier esterification' proceeds via an $A_{AC}2$-type mechanism in which the O—acyl bond is newly created in the ester-forming step, which means that a carboxyl group is involved (Carboxyl Group Activation, CGA). The $A_{AC}1$-type mechanism (Scheme 1) is primarily observed for the esterification of aromatic carboxylic acids with bulky substituents around the CO_2H function. In general, the $A_{AC}2$-reaction (Scheme 2) is strongly disfavored by α- and β-branching in the acid (e.g., Structures (**1**), (**2**) and (**3**)). Moreover α-branched alcohols tend to eliminate under the acidic conditions. Therefore, in these cases alternative esterification methods are recommended (*vide infra*).

$$C_8H_{17}\text{–}CO_2H + Me\text{–}OH \xrightarrow[96\%]{H_2SO_4,\ MS\ 3Å} C_8H_{17}\text{–}CO_2Me \qquad (1)$$
68CI(L)1568

$$Pr\text{–}CO_2H + Ph\text{–}OH \xrightarrow[\text{azeotropic removal of } H_2O \text{ with xylene, } 86\%]{H_3BO_3,\ H_2SO_4} Pr\text{–}CO_2Ph \qquad (2)$$
71TL3453

$$Me\text{–}CO_2H + Pr^i\text{–}OH \xrightarrow[100\%]{\text{Amberlite IR 20, } CaSO_4} Me\text{–}CO_2Pr^i \qquad (3)$$
71JOC2548

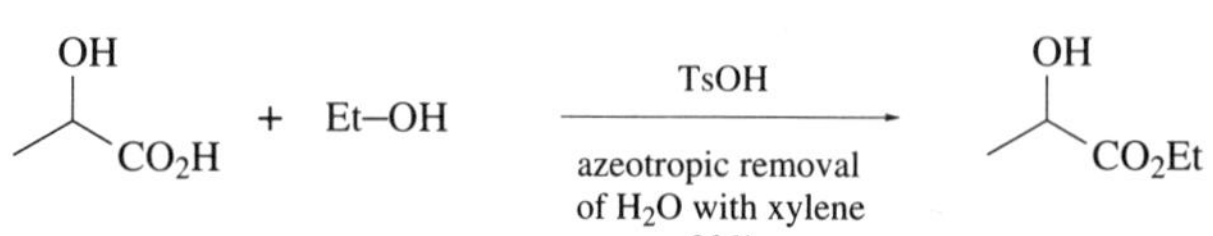

(4)
55OSC(4)46

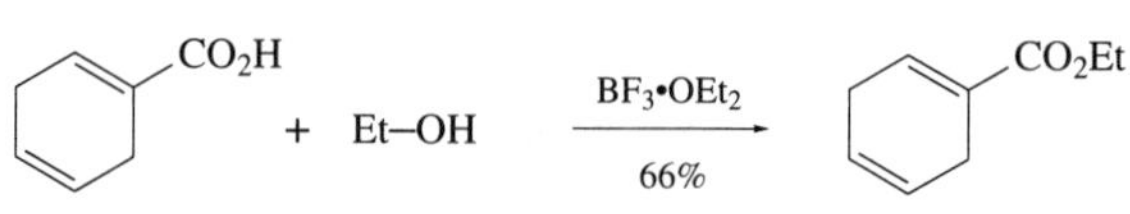

(5)
70TL4011

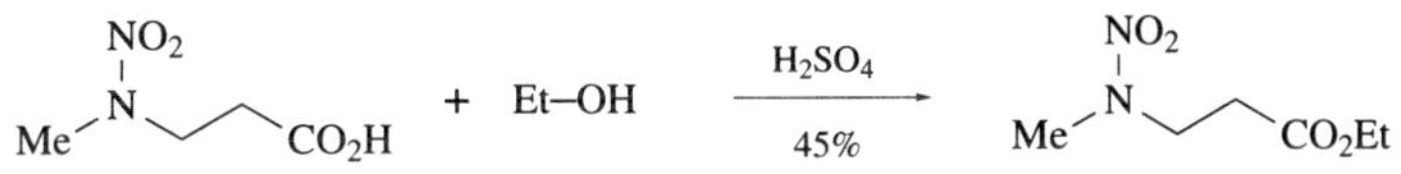

(6)

75AP954

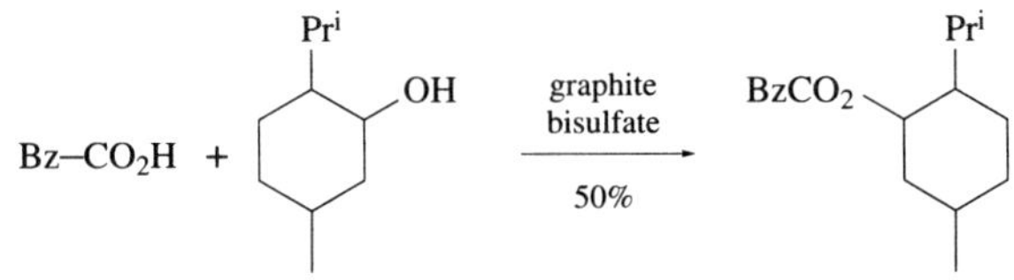

(7)

57JA4759
63JCS1927
74JA8113

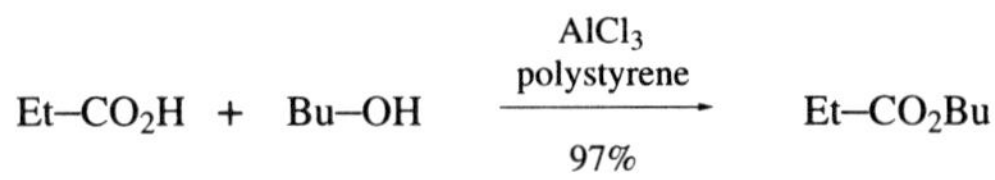

(8)

81S413

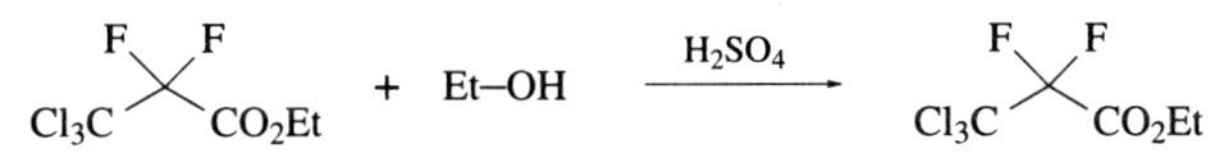

(9)

76MI 503-01

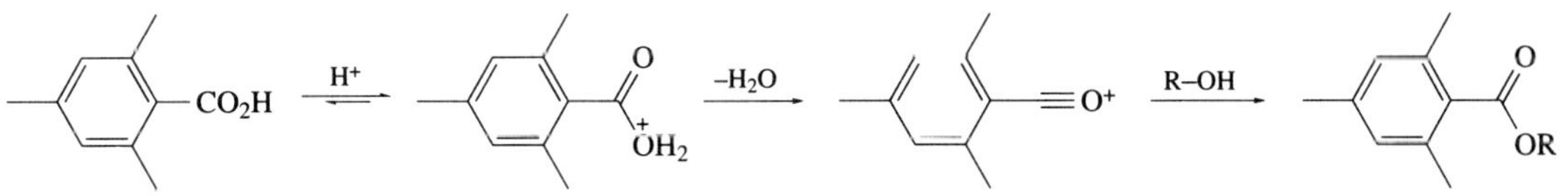

$A_{AC}1$-Esterification

Scheme 1

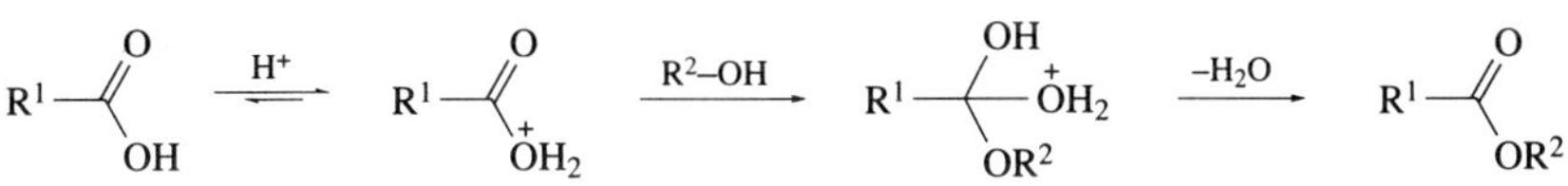

$A_{AC}2$-Esterification

Scheme 2

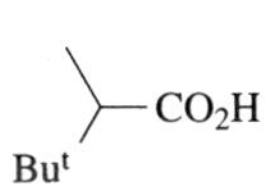

(1)

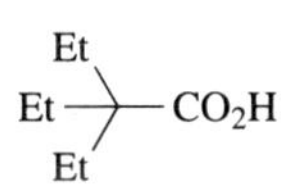

(2)

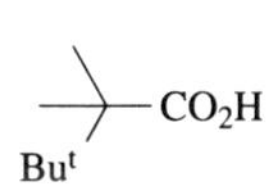

(3)

5.03.1.2 Esterification via Carboxylic Group Activated Intermediates (Mixed Anhydrides or Activated Esters)

This is the most important method when stoichiometric amounts of acid and alcohol are employed, and sensitive functionality is present in either component. The mechanism of the reaction involves the formation of a carboxyl group-activated intermediate (**4**) from the carboxylic acid and an activation agent X—Y (Scheme 3). The species (**4**) then acts as the acylating agent for the alcohol. In effect X—Y is transformed into H—Y and HOX during the reaction. Examples are given in Schemes 3 and 4 and Equations (10)–(18) which demonstrate the broad range of X—Y reagents that can be used. These may be trimethylsilyl chloride (Equation (10)) ⟨81BCJ1267⟩ or anhydrides (Equations (11) and (12)) ⟨65JCS4594, 65JOC927⟩ or tosyl chloride (Equations (13) and (14)) ⟨55JA6214, 75TL277⟩. Scheme 4 describes an interesting variation of the Mitsunobu process ⟨81S1, 92OR(42)335⟩ in which an oxyphosphonium species (**5**) is generated from the phosphine oxide and triflic anhydride. The oxyphosphonium salt then subsequently adds the acid and the alcohol to form the pentacovalent phosphorus intermediate (**6**) which decomposes to the ester and phosphine oxide, which may be recycled. Dicyclohexylcarbodiimide (dcc) is a very important reagent to form activated esters, especially under the catalysis of 4-*N*-pyrrolidinopyridine (ppy) or 4-dimethylaminopyridine (dmap) (Equations (16)–(19)) ⟨78TL4475, 85OS(63)183⟩. The yields of the esters are uniformly high; the only drawback of the method is the removal of dicyclohexylurea, which normally requires chromatography.

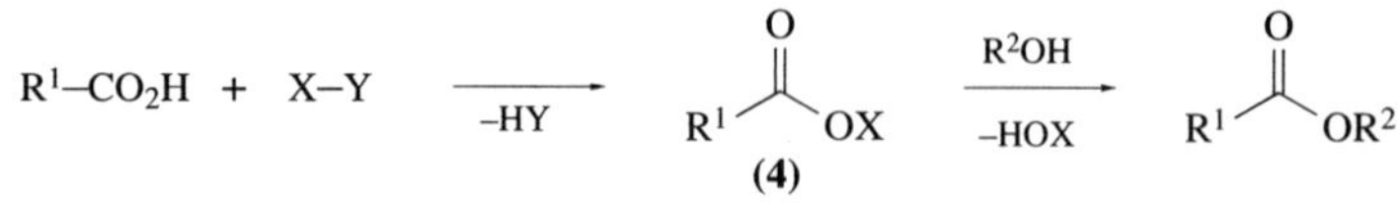

Scheme 3

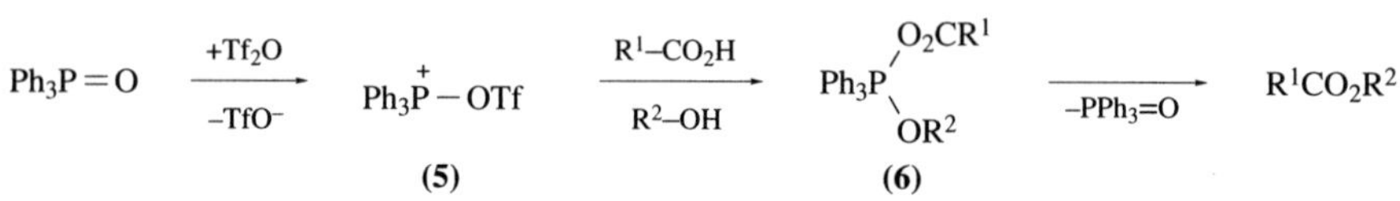

Scheme 4

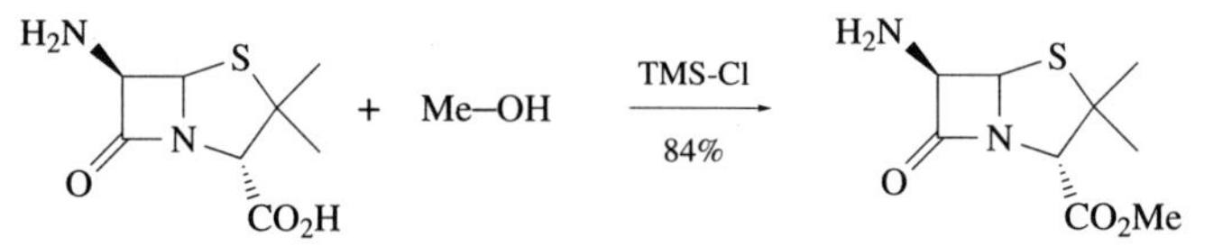
(10)

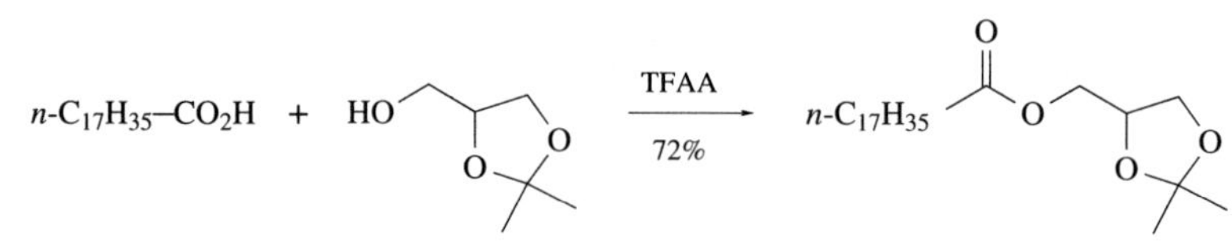
(11)

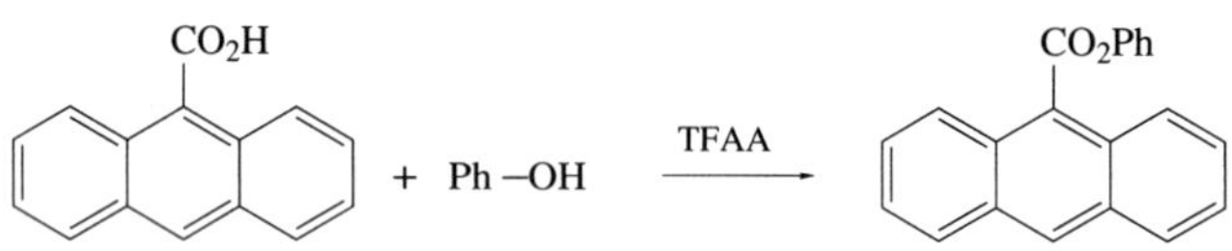
(12)

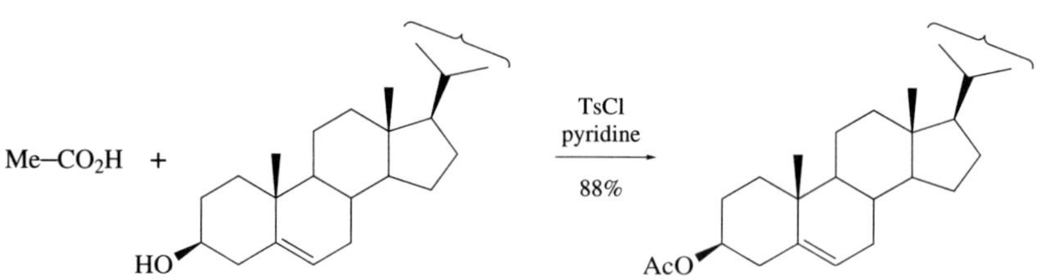
cholesterol
(13)

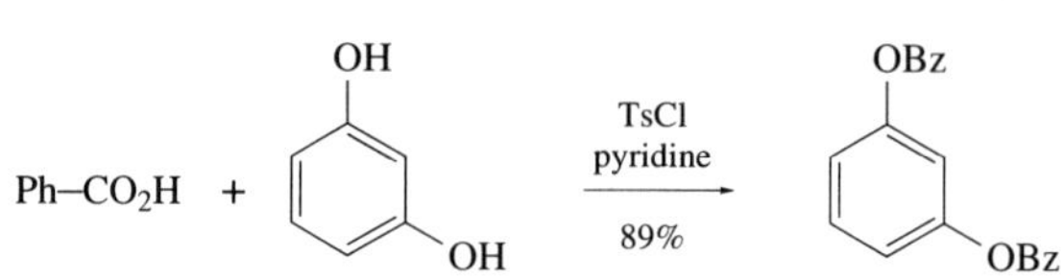
(14)

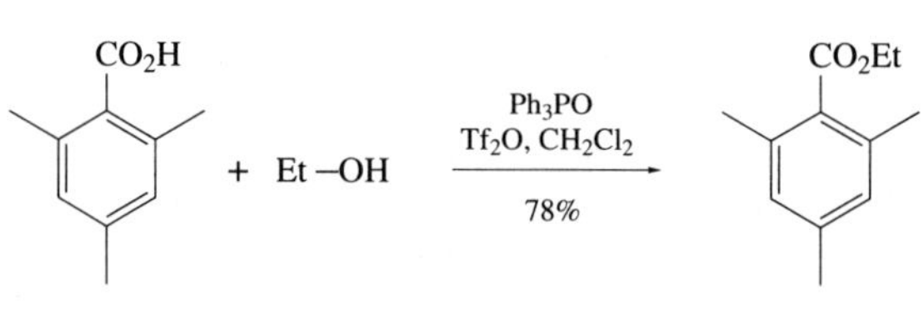
(15)

$$Me{-}CO_2H + Bu^t{-}OH \xrightarrow[90\%]{dcc,\ ppy} Me{-}CO_2Bu^t \quad (16)$$

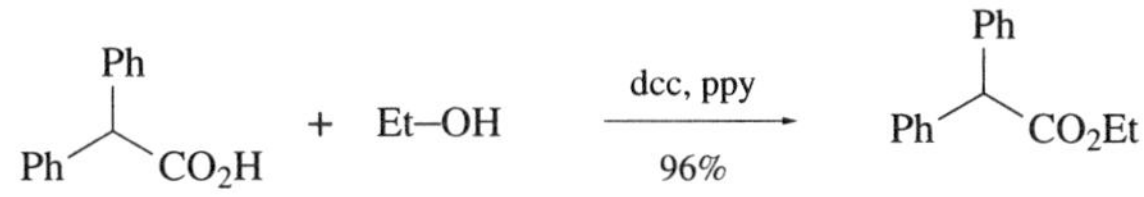

(17)

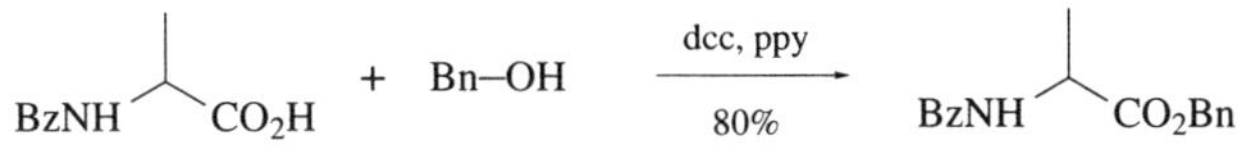

(18)

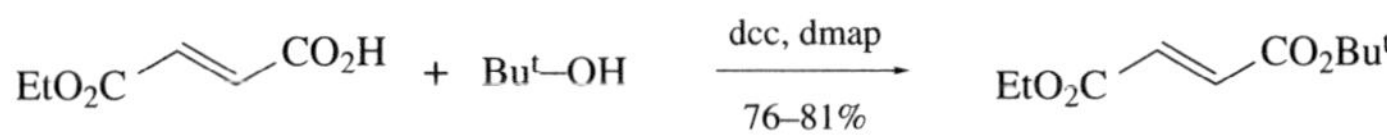

(19)

The use of mixed sulfonyl halides (Equations (20) and (21)) ⟨815790⟩ leads to volatile by-products but the method is not as common as using dcc. Carbonyldiimidazolide (CDI, Staab's reagent) (Equation (22)) ⟨82S833⟩ is particularly suitable for the synthesis of *t*-butyl esters; it gains in efficiency if bases like 1,5-diazabicyclo [5.4.0]undec-5-ene (dbu) are used to generate the carboxylate. In this way, the imidazolide (**7**) is formed (Scheme 5) which acts as the acylating agent proper. *t*-Butoxide as the base leads to α-deprotonation of the acid or the ester formed. The use of 2-bromopyridine (Equation (23)) ⟨77BCJ1863⟩ or *t*-butylisocyanide (Equation (24)) ⟨77JCR(5)119⟩ is less common and does not show any advantages over the methods mentioned previously.

$$Ph{-}CO_2H + Bn{-}OH \xrightarrow[80\%]{SO_2ClF,\ NEt_3} Ph{-}CO_2Bn \quad (20)$$

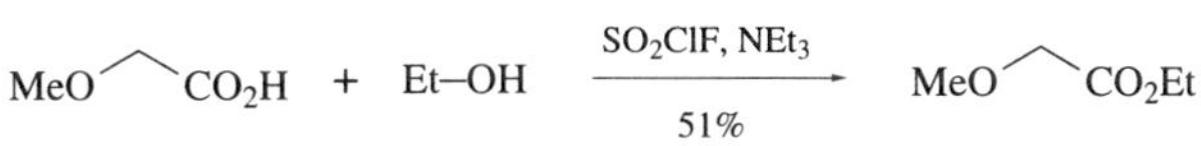

(21)

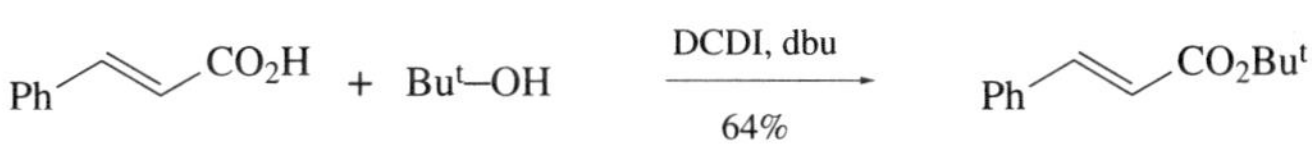

(22)

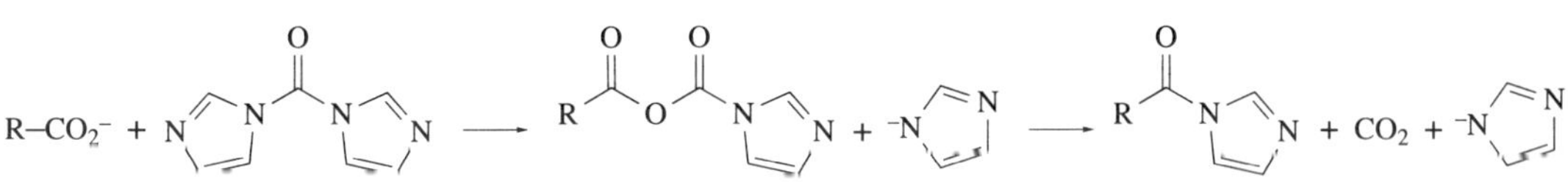

(**7**)

Scheme 5

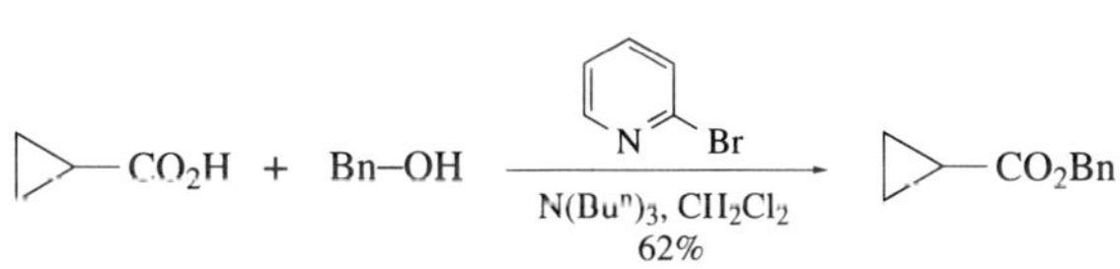

(23)

$$HO_2C{-}CO_2H + Bu^t{-}OH \xrightarrow[69\%]{Bu^tNC} HO_2C{-}CO_2Bu^t \quad (24)$$

Thiazolium salts (Equation (25)) are efficient reagents for the coupling of acids and alcohols, the mechanism of which is highly dependent on X ⟨76CL267⟩. With X = Cl or Br the carboxylate attacks the thiazolium ion to form a carbonyl group activated intermediate whereas in the case of X = F the alcohol substitutes the fluorine and a hydroxy group activated species is formed. In Equation (26) a particularly mild procedure is shown, which is compatible with functional groups like tetrazole in the acid ⟨75S456⟩. An O-acylated hydroxylamine species (**8**) is generated which then acylates the alcohol (Scheme 6). Yamaguchi's mixed anhydride procedure (Scheme 6 and Equation (27)) is the method of choice for many macrolactonizations (see 5.03.6.4) rather than for normal esterifications. Chlorosulfonyl isocyanate (Equation (28) and Scheme 7) adds the carboxylate to generate carbamate (**9**) which acylates the alcohol to the ester and easily removable by-products ⟨82S506⟩. 2-Bromomethylbenzoyl bromide can also be used as an activating agent (Equation (29)) ⟨87TL4711⟩. The mechanism of the reaction (Scheme 8) which involves the formation of an oxonium intermediate (**10**) which is attacked by the alcohol to form the ester and phthalide. Equations (30) and (31) show esterifications with mixed anhydrides of carboxylic acid and phosphonic acid chlorides ⟨80S547, 85JOC760⟩. For instance in Scheme 9 the reagent is generated from 1,3-oxazolidin-2-one and phosphorus pentachloride; the carboxylate substitutes the remaining chloride to form the active species (**11**). Equations (32) and (33) ⟨85CA(103)104568c, 93JA11434⟩ show two examples for the mild esterification procedures using *ortho*-esters or amide acetals whereas Equations (34) and (35) fit into the general scheme of mixed anhydride intermediates ⟨85CC473, 85JOC560⟩. Equations (36)–(39) sum up the procedures best suited for the esterification of acids with sterically hindered alcohols ⟨55JA6214, 57AG135, 65JOC927, 82S833, 83S135⟩.

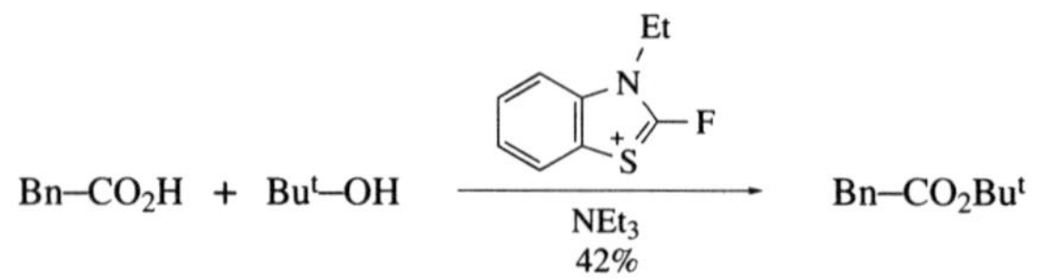

(25)

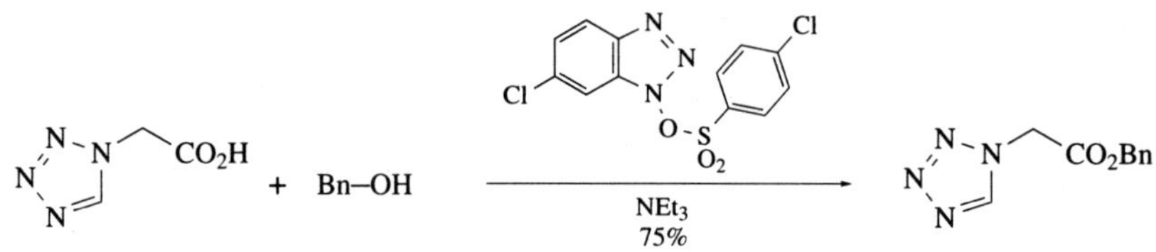

(26)

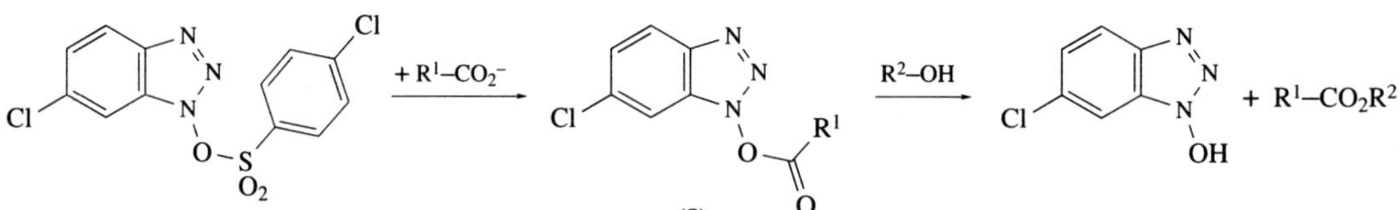

(8)

Scheme 6

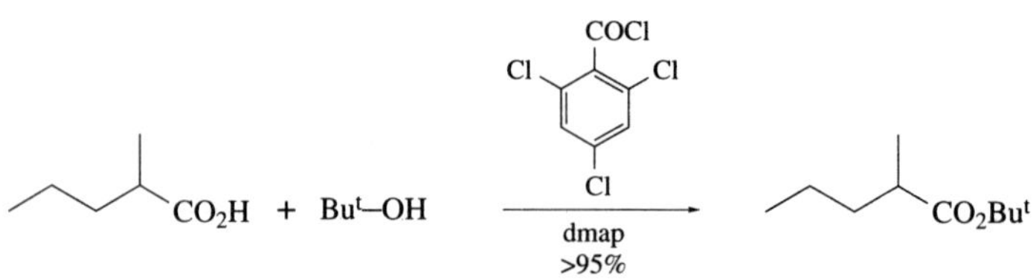

(27)
79BCJ1989

(28)

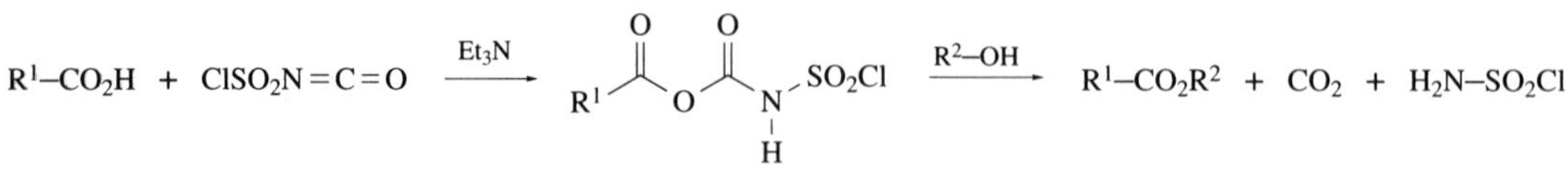

(9)

Scheme 7

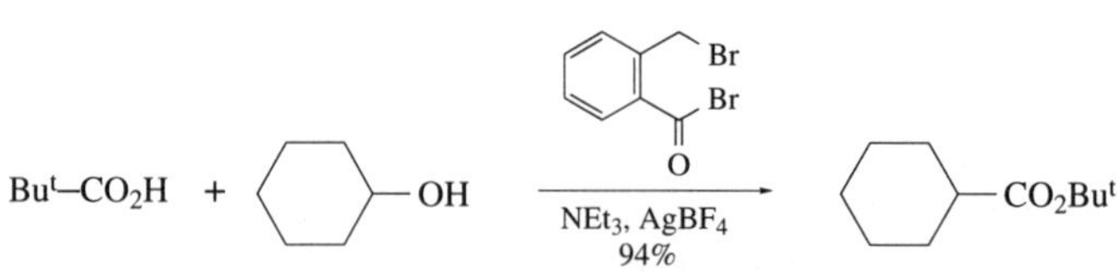

(29)

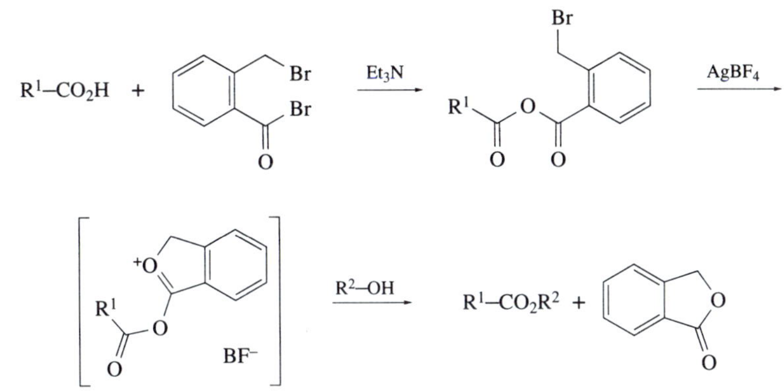

(10)

Scheme 8

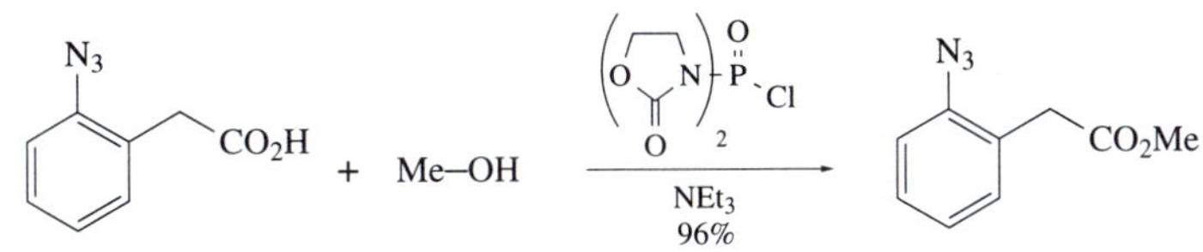

(30)

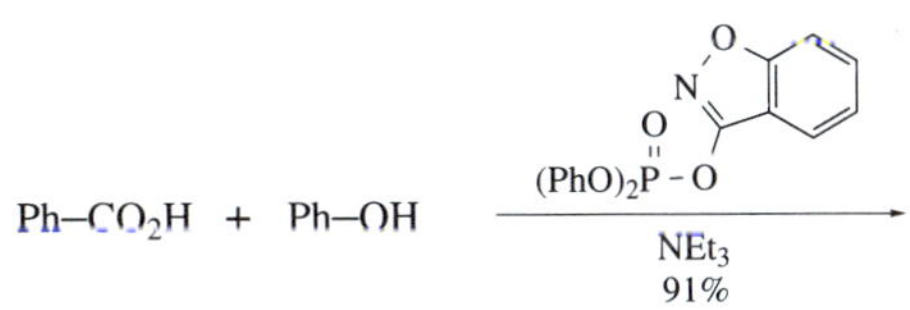

(31)

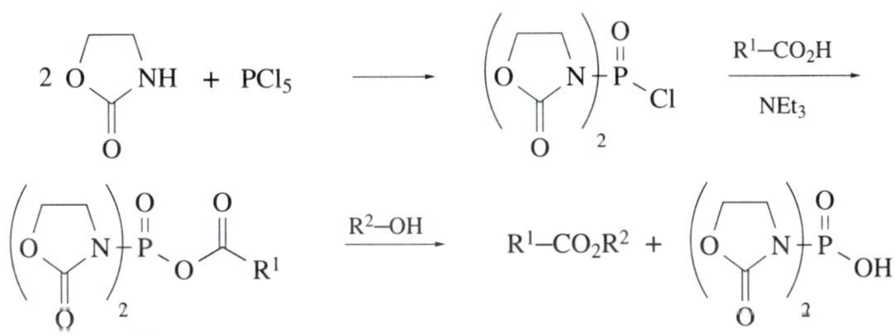

(11)

Scheme 9

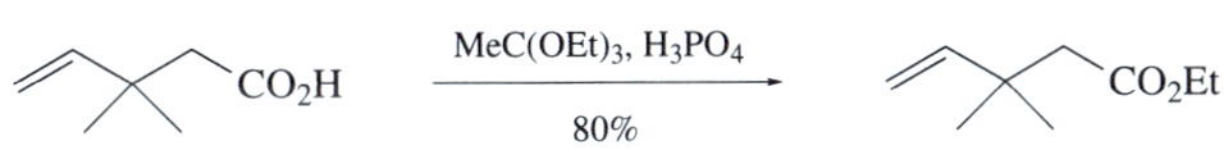

(32)

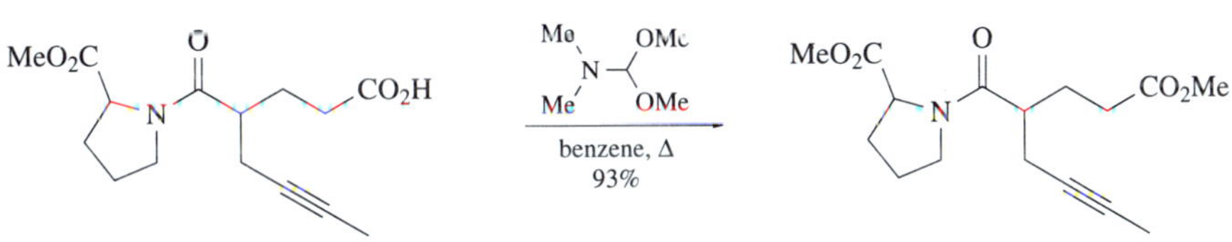

(33)

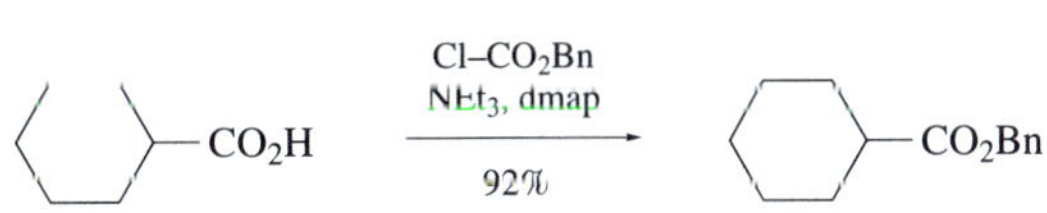

(34)

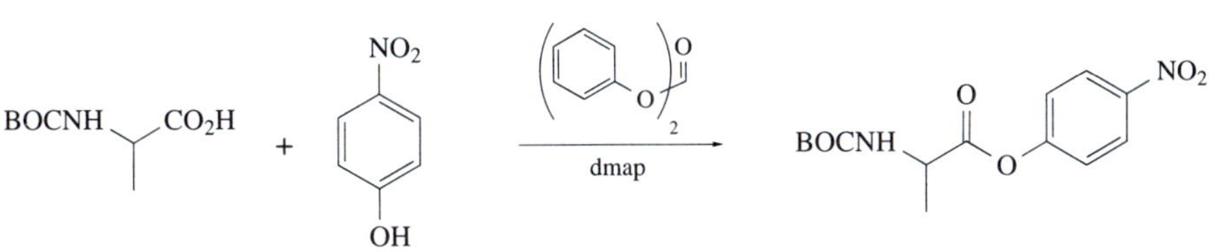

(35)

$$R^1\text{–}CO_2H + R^2\text{–}OH \xrightarrow{Tf_2O} R^1\text{–}CO_2R^2 \quad (36)$$

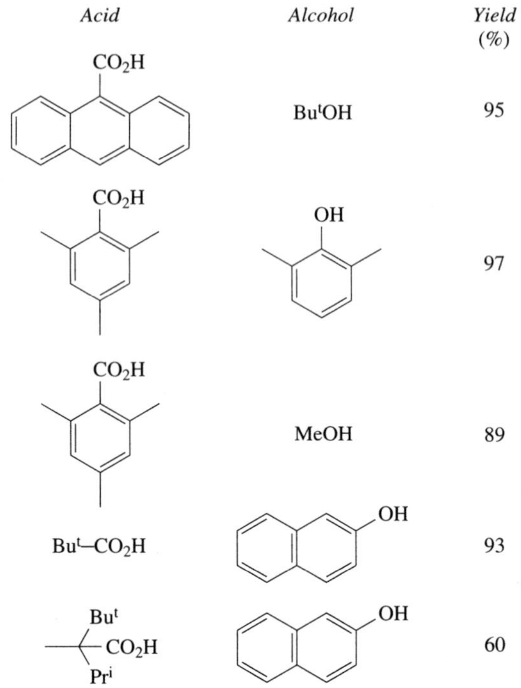

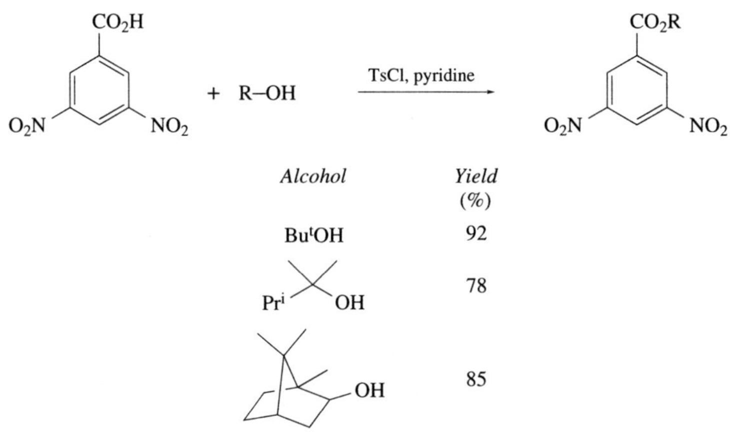

(37)

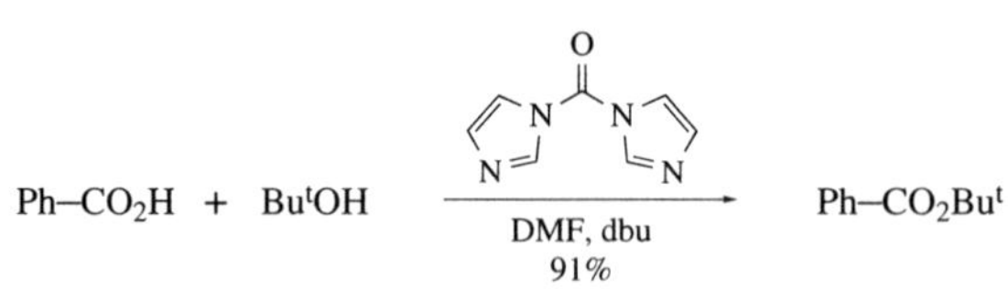

(38)

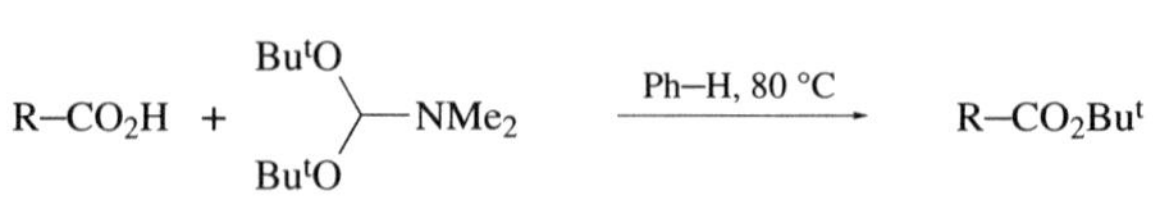

(39)

5.03.1.3 Esterification of Alcohols with Acid Halides and Anhydrides

In contrast to the A_{AC} esterifications this method makes use of preformed CGA species (**12**) and (**13**) to which the alcohol is added. The formation of acid chlorides ⟨B-72MI 503-01⟩ requires strong Lewis acids like thionyl or oxalyl chloride, which may react with acid sensitive functions in the molecule and also cause epimerisation of chiral α-centers in the substrate molecule. Alternatively, using anhydrides (**13**) for esterification means to lose 1 mol equiv. of the acid. Thus, both (**12**) and (**13**) should preferentially be used for inexpensive carboxylic acids with a low degree of functionalization. Typical and frequently encountered examples in multistep syntheses are benzoylation and acetylation of primary and secondary alcohols, whereas tertiary alcohols are generally too hindered to react, unless a strong base catalyst (e.g., calcium hydride 〈66M62〉) is added. The acetylations are normally performed in pyridine at 0–25 °C. Acid catalysis is less common. Acid chlorides can be used in pyridine or in dichloromethane with triethylamine as the base. If possible, the anhydride should be preferred, as the reaction mixture remains almost neutral throughout the reaction and the yields are higher, particularly if the acylation is catalyzed with *N*,*N*-dimethylaminopyridine (dmap, (**14**)) or, even better, 4-pyrrolidinopyridine (ppy, (**15**)) 〈78AG(E)569〉. Regular conditions are 0.05–0.2 mol equiv. of the catalyst in aprotic solvents (hexane, dichloromethane, THF, pyridine) at room temperature. Rate enhancements compared to the uncatalyzed reactions are several powers of ten, so that even hindered hydroxy functions (e.g., in (**16**) and (**17**)) are smoothly acylated. Some additional examples are given in Equations (40)–(43) and Scheme 10 〈94JA9009, 94JOC1160, 94JOC4749, 94JOC5532〉.

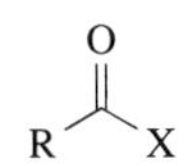

(12)

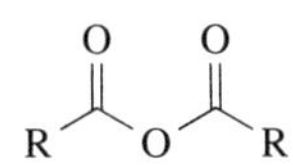

(13)

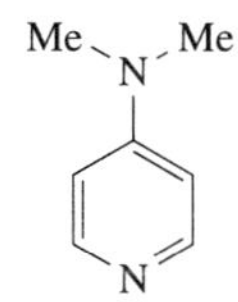

(14)

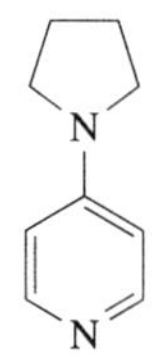

(15) ppy

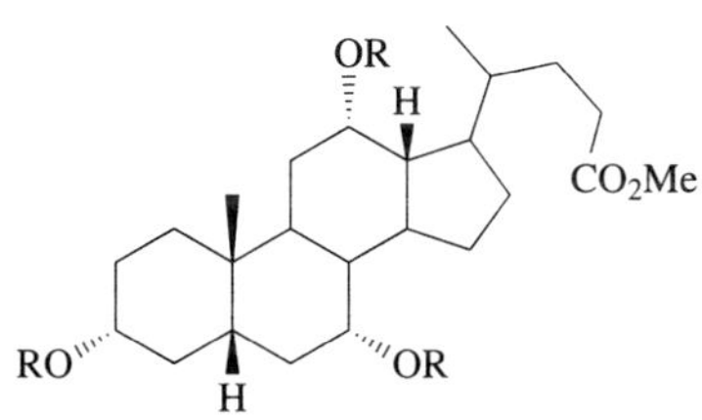

(16) **a**; R = H
b; R = COMe

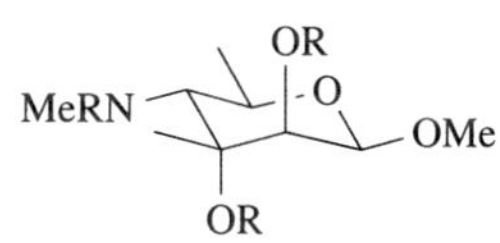

(17) **a**; R = H
b; R = COMe

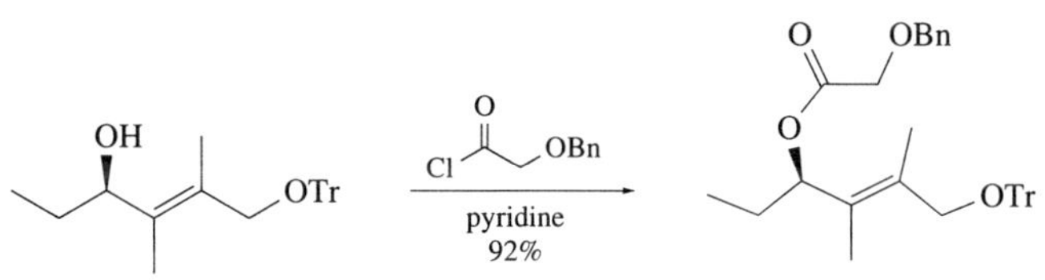

(40)

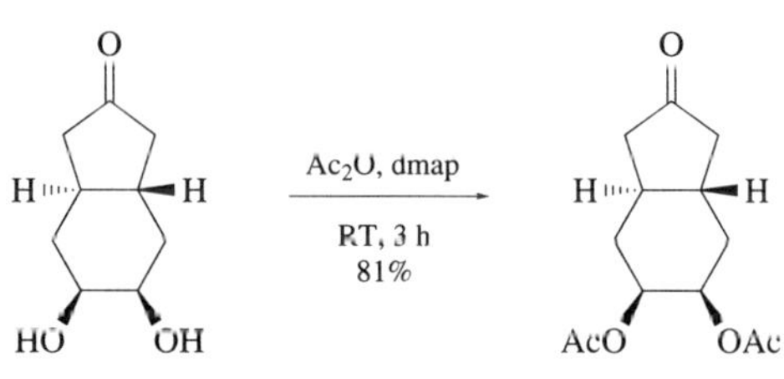

(41)

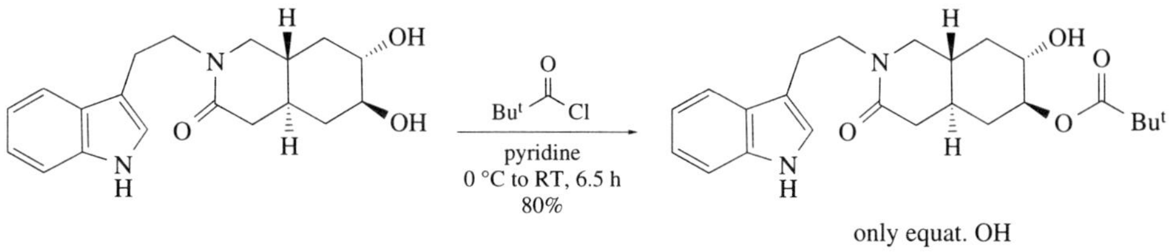

(42)

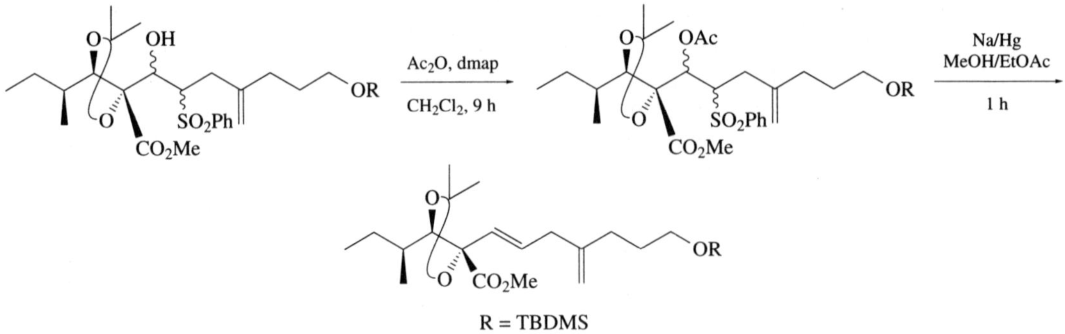

Scheme 10

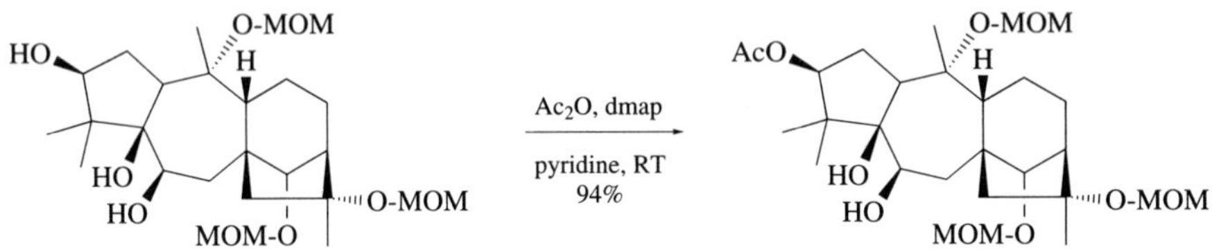

(43)

Mechanistically, the alcoholysis of acid chlorides may proceed via a one-step S_N2-like mechanism or more frequently, via the usual tetrahedral intermediate ⟨B-72MI 503-03⟩. The reaction is accelerated by electron-withdrawing substituents in the chloride and hardly affected by steric hindrance. On the other hand, benzoyl chlorides with electron-releasing substituents in the aromatic nucleus react via a S_N1-like mechanism, especially in protic solvents ⟨53JCS3352⟩. Anhydrides prefer the normal addition–elimination pathway, the rate-determining step being the formation of the tetrahedral intermediate. The drastic effect of the catalysts **(14)** and **(15)** may be attributed to the initial formation of the acyl pyridinium salt **(18)** in which X^- serves as the base for the deprotonation of the alcohol during its nucleophilic attack on the carbonyl group (Scheme 11). This explains why acid chlorides show much smaller rate enhancements with **(14)** and **(15)** than anhydrides, although the equilibrium concentration of **(18a)** is much higher than that of **(18b)**. However, chloride is a much weaker base than the carboxylate and cannot remove the proton from the alcohol with similar efficiency. Additionally, X^- must be close to the reaction site, which means that **(18)** is more efficient an acylating agent as a contact than as a solvent-separated ion pair. In fact, the rates are much higher in unpolar than in polar solvents.

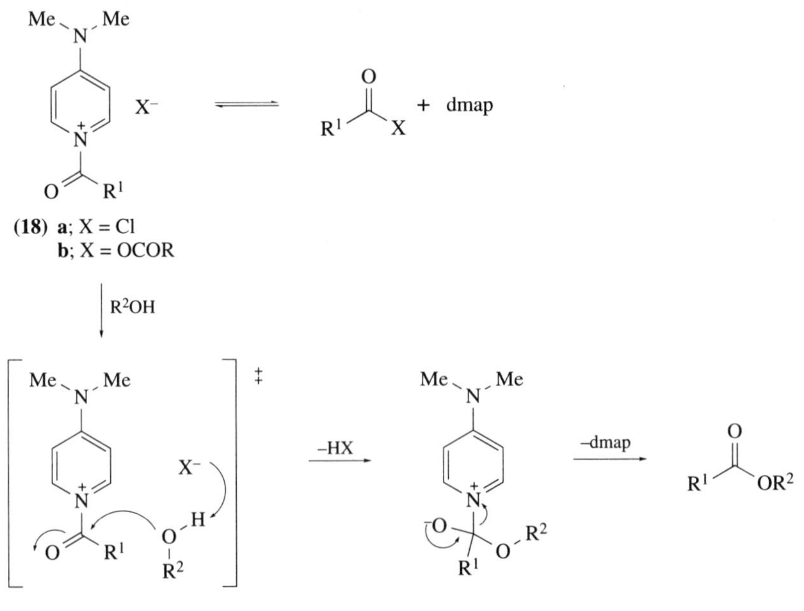

Scheme 11

5.03.1.4 Esterification via Hydroxyl Group Activation (HGA)

The mechanism of HGA esterification is shown in Scheme 12. The alcohol first reacts with a reagent X-Y under HY-elimination to form a hydroxyl group activated species (**19**), which is attacked by the carboxylate under S_N-substitution. In this mechanism, the esterification typically proceeds under formation of an O—alkyl bond under S_N2-inversion of configuration at R^2. This means that R^2 should be a primary or secondary, preferentially benzylic or allylic alkyl residue. In the latter case allylic transposition may also occur. Typical HGA esterifications are those with amidinium ion activation (Equation (44)) ⟨82CL1891⟩, or Appel-type reactions (Equation (47)) ⟨85JOC4991⟩. This mechanism partly proceeds under S_N1-conditions or subsequent deprotonation–reprotonation at the reaction center. Other examples are shown in Equations (45) and (46) ⟨83T1475⟩, (48) ⟨85JOC4991⟩, and (49) ⟨78TL4461⟩.

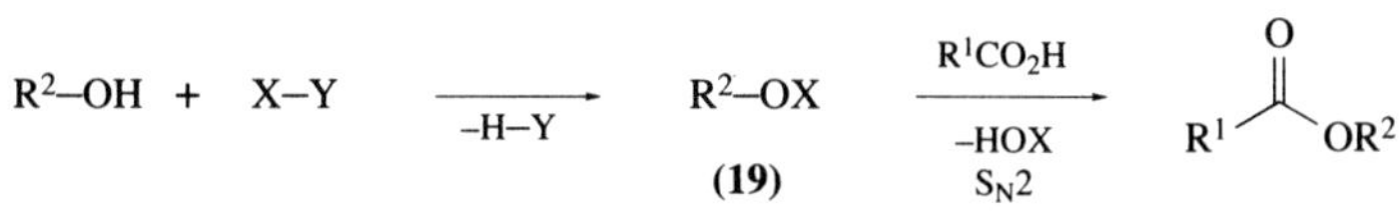

Scheme 12

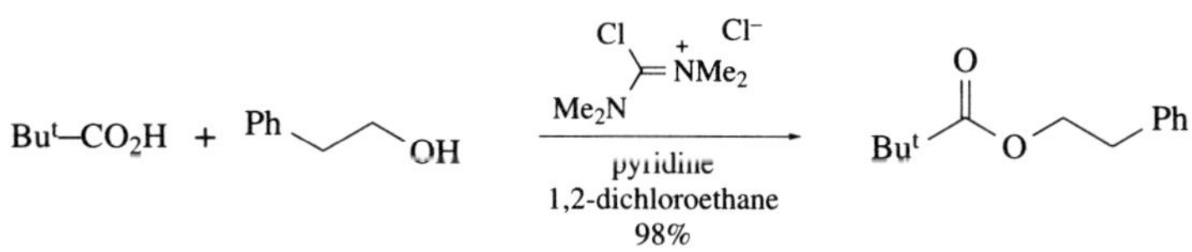
(44)

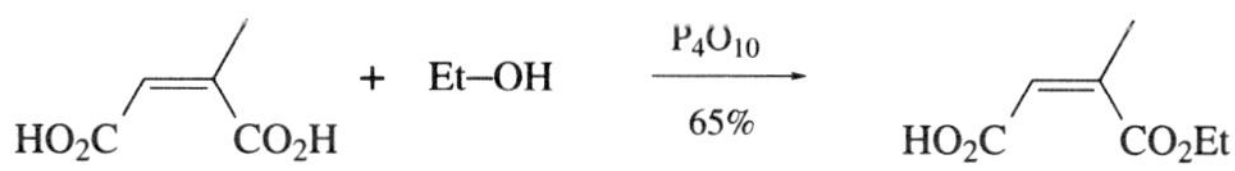
(45)

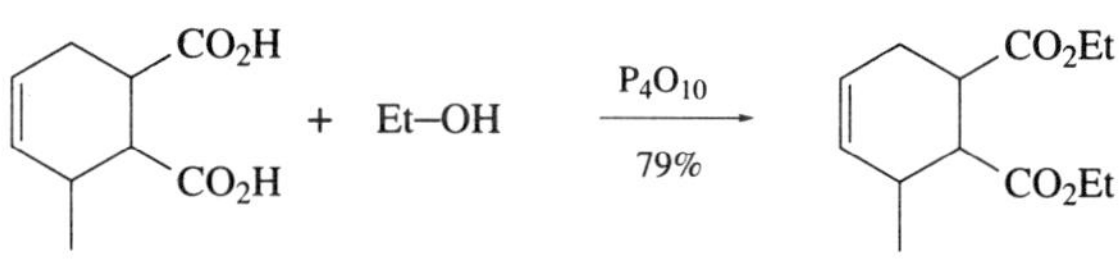
(46)

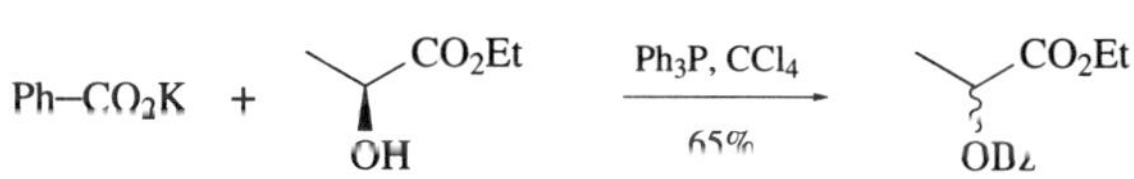
(47)

racemization

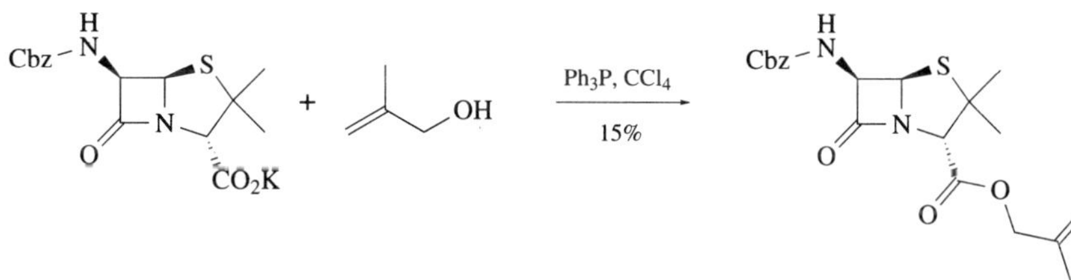
(48)

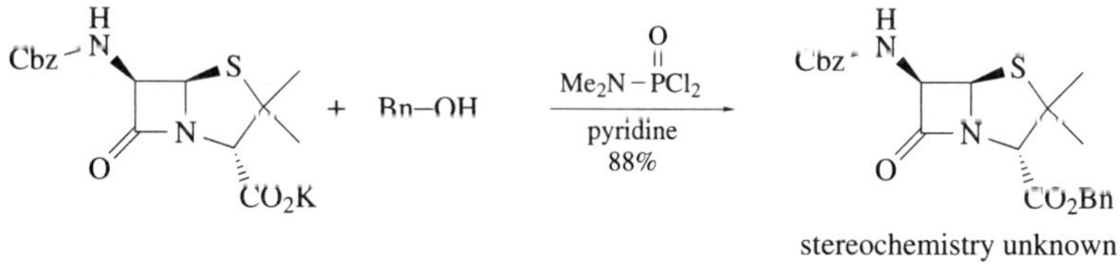
(49)

stereochemistry unknown

Mitsunobu-type reactions are very important ⟨81S1, 92OR(42)335⟩, for example Equations (50)–(52) show a clean S_N2-type inversion in most cases ⟨78TL4461, 81S1, 87JOC3468, 87JOC3784⟩. Allylic systems normally react without allylic rearrangement (Equation (52)). In Equations (53)–(54) esterification with amide acetals is shown; this also proceeds with clean S_N2-type inversion and under very mild conditions ⟨63AG(E)211, 63AG(E)212⟩.

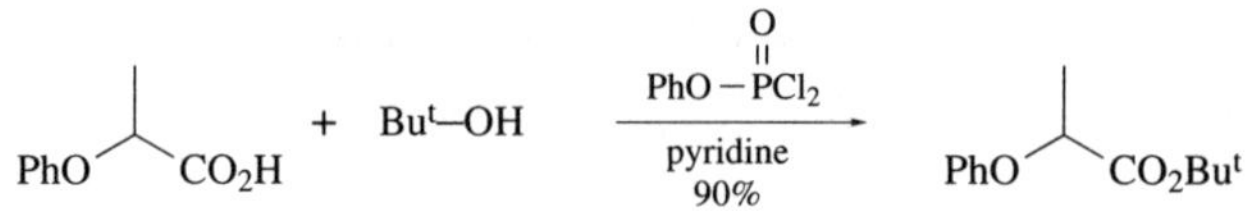
(50)

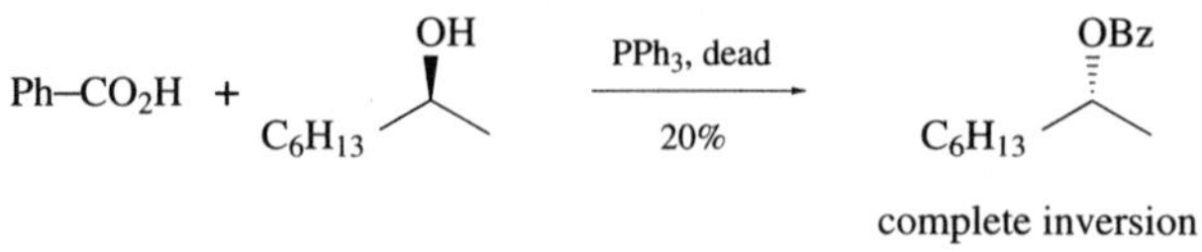
(51)

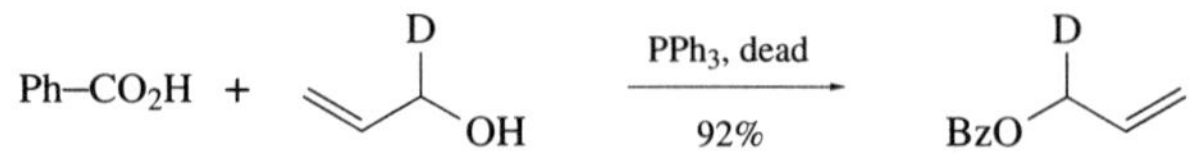
(52)

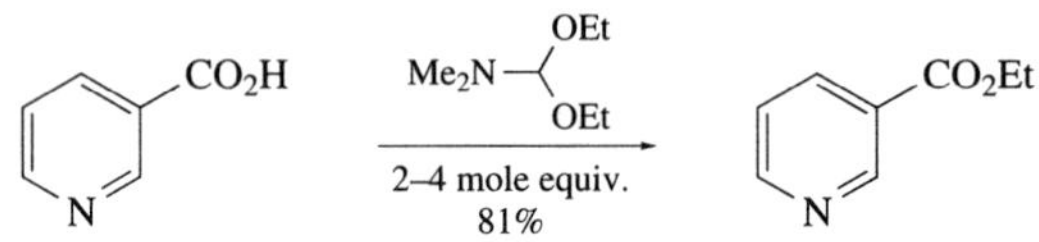
(53)

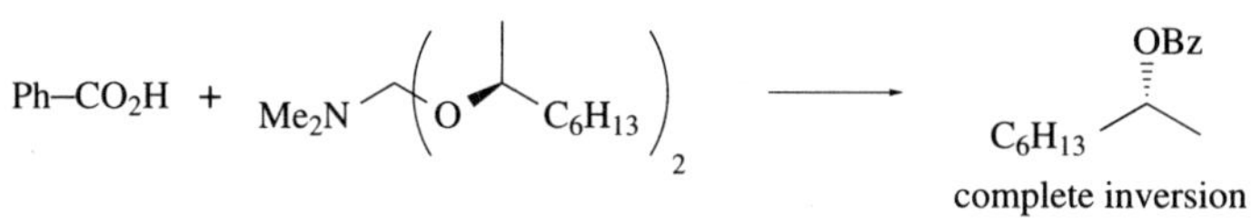
(54)

5.03.1.5 Alkylation of Carboxylate Anions

In effect these esterifications are very similar to the HGA esterification described in 5.03.1.4. The carboxylates are used in the form of sodium, potassium or caesium salts and are treated with alkyl bromides or iodides in DMF, hexamethylphosphoramide (HMPA), acetonitrile and/or water. The alkylating agent must show high S_N2 reactivity. If two leaving groups are present (Equation (63)) ⟨76CR(C)483⟩ the better one, which is bromine in this case, is substituted. Equation (67) ⟨86JOC546⟩ shows that under specified conditions, even more hindered alkyl components may be acceptable. Equation (68) provides an example of the synthesis of methyl esters with diazomethane or preferably, with the stable trimethylsilyl derivative ⟨81CPB1475⟩. For sensitive carboxylic acids this method is by far superior to other methylations, particularly if working on a small scale. Other examples are shown in Equations (55), (56) ⟨73OS(5)580⟩, (57) ⟨68JOC2143⟩, (58) ⟨74JOC1968⟩, (59) ⟨87JOC3777⟩, (60) ⟨83S649⟩, (61) ⟨78BCJ2401⟩, (62) ⟨82SC453⟩, (64) ⟨85S40⟩, (65) ⟨85TL4439⟩, and (66) ⟨81CI(C)164⟩.

$$R^1\text{–}CO_2^-Met^+ + R^2\text{–}X \longrightarrow R^1\text{–}CO_2R^2 + MetX \quad (55)$$

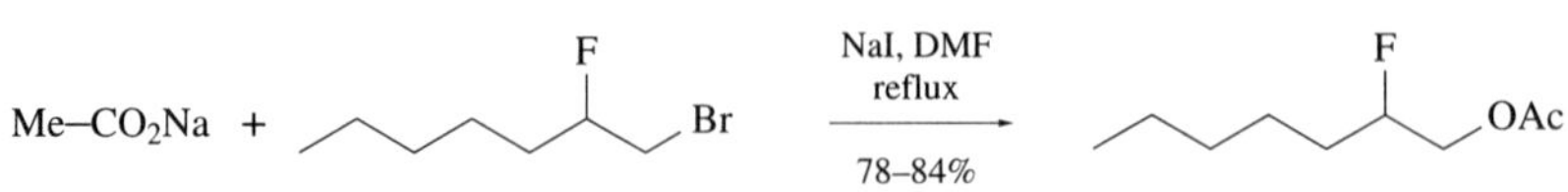
(56)

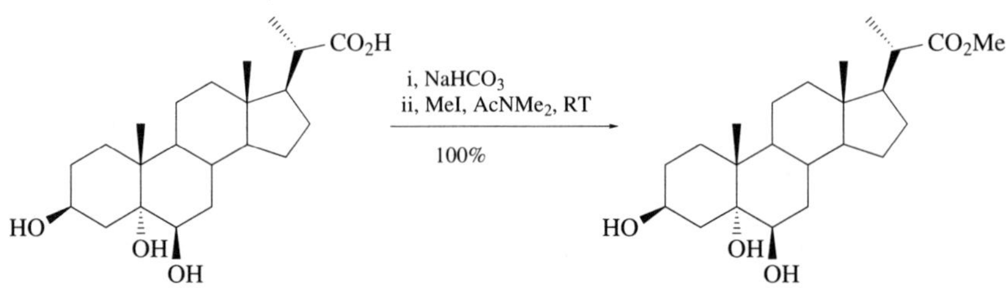
(57)

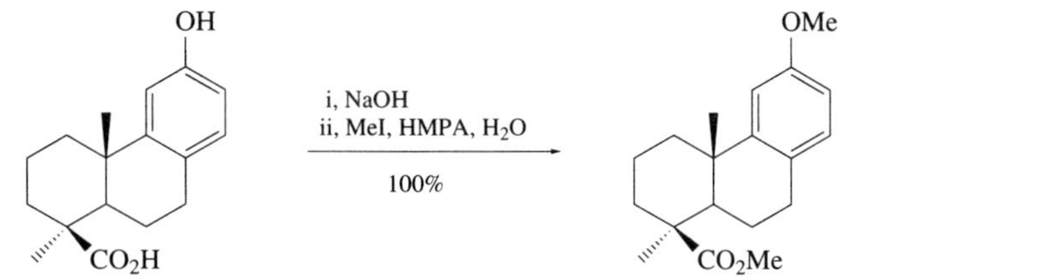

(58)

(59)

$$\text{Bn–CO}_2\text{H} + \text{Bn–Br} \xrightarrow[\text{iii, 1 } N \text{ KOH (aq.)}\ 60\%]{\text{i, HgO, HBF}_4\ \text{ii, Na}_2\text{CO}_3} \text{Bn–CO}_2\text{Bn} \qquad (60)$$

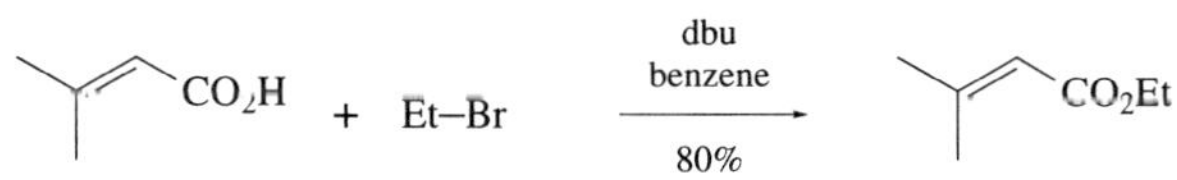

(61)

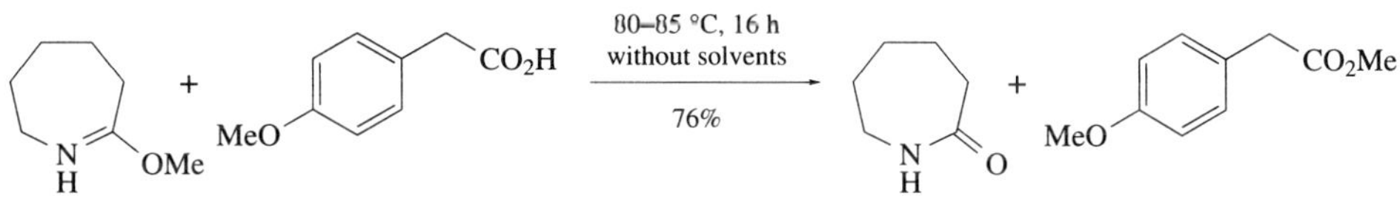

(62)

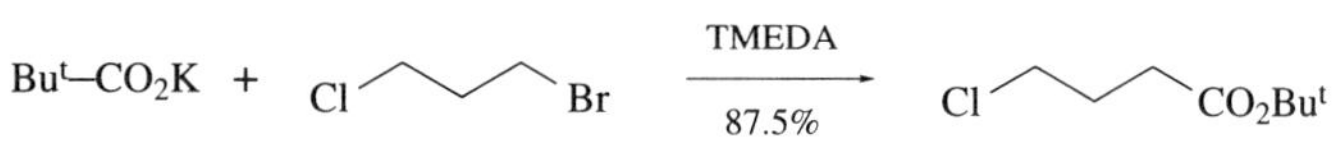

(63)

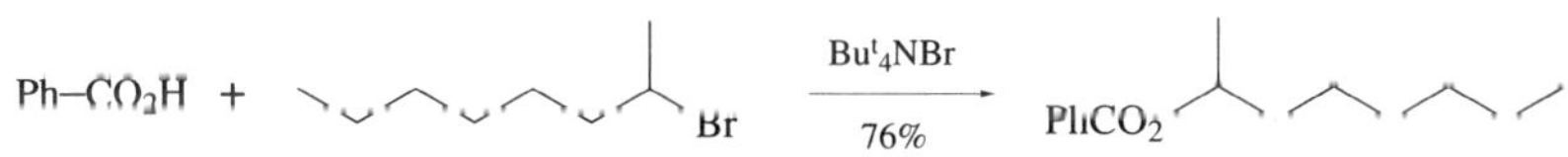

(64)

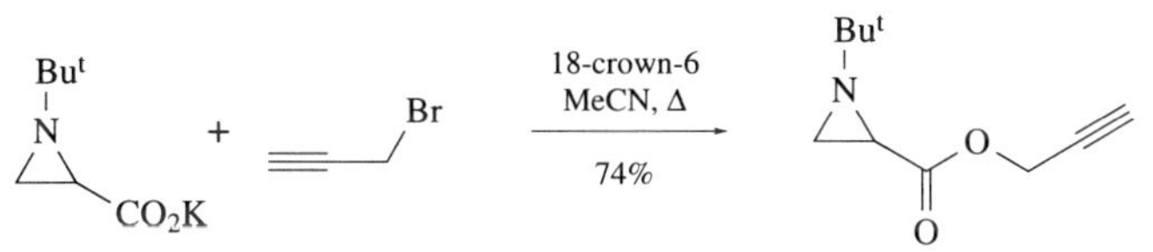

(65)

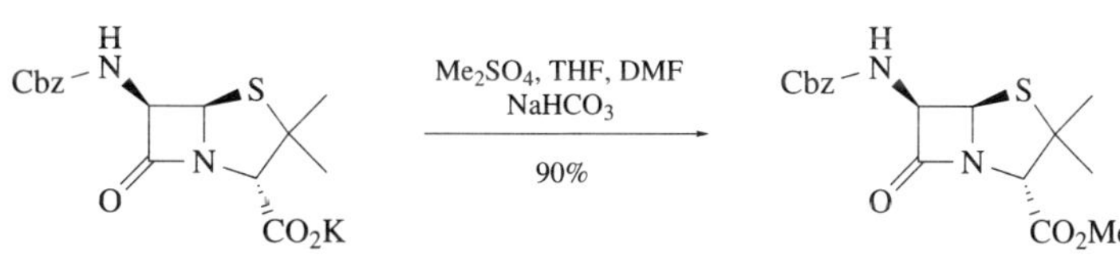

(66)

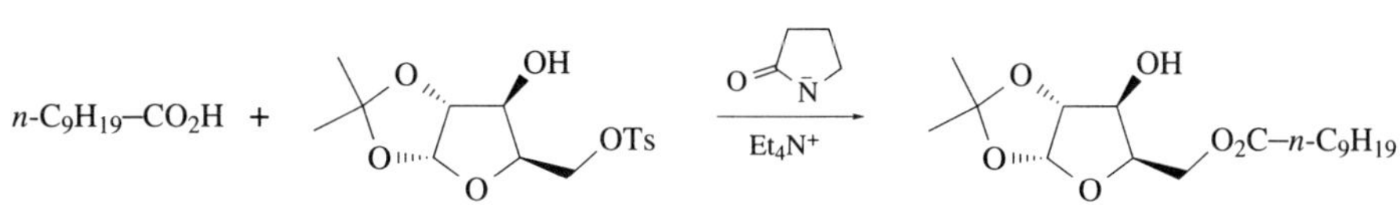

(67)

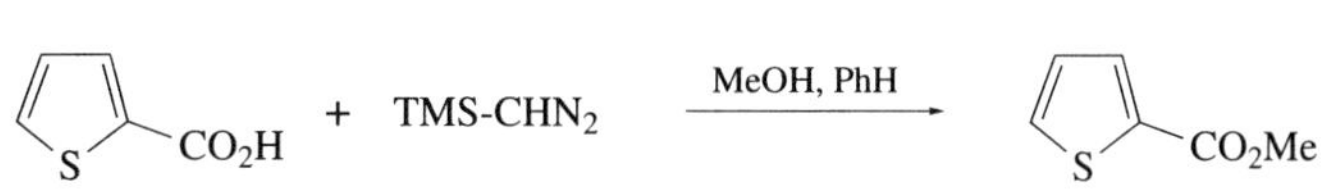

(68)

5.03.1.6 *Trans*-esterification

Trans-esterification is a good method if one of the components (ester or alcohol) is inexpensive, as equilibrium mixtures are formed and a large excess of the reagent has to be used to shift the equilibrium to the product side. A catalyst has to be added which may be any acid or base. Titanium *ortho*-esters are particularly efficient as shown in Equation (69) ⟨78GEP2822472⟩. Novel catalysts are the phosphonium ylides, which have been used in a large number of examples (Equation (70)) ⟨80TL2857⟩. The mechanism (Scheme 13) involves a cyclic intermediate (**20**), which serves to activate the ester carbonyl and to remove the proton from the alcohol ('double activation'). If the *trans*-esterification is performed with a vinyl ester, the reaction becomes irreversible, because the vinyl alcohol immediately undergoes keto–enol tautomerization to form acetaldehyde (Equation (71)) ⟨77JAP(K)77/102237⟩. The reaction is catalyzed by strong base. Acylated ketoximes (Equations (72) and (73)) are useful substrates for *trans*-esterification reactions, which may be performed either with the alcohols under $FeCl_3$- or $ZnCl_2$-catalysis (Equation (72)) or with the alkoxides without a catalyst (Equation (73)) ⟨86CL871⟩. The yields are uniformly good.

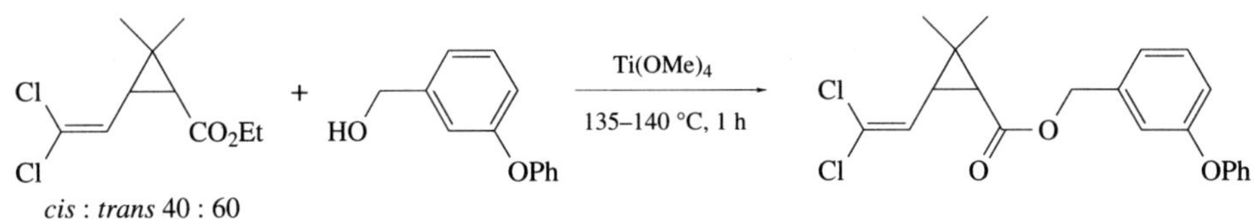

(69)

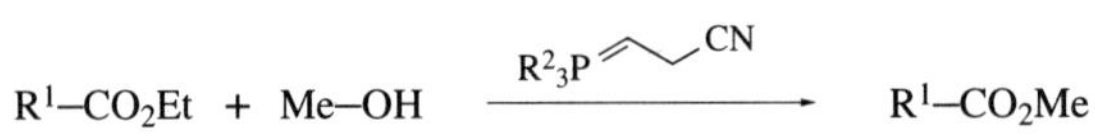

(70)

R¹	*R²*	*Time* (h)	*Temp.* (°C)	*Yield* (%)
Me	Ph	1	25	12
Me	Bu^n	0.5	30	94
p-NO_2Ph	Ph	24	25	71

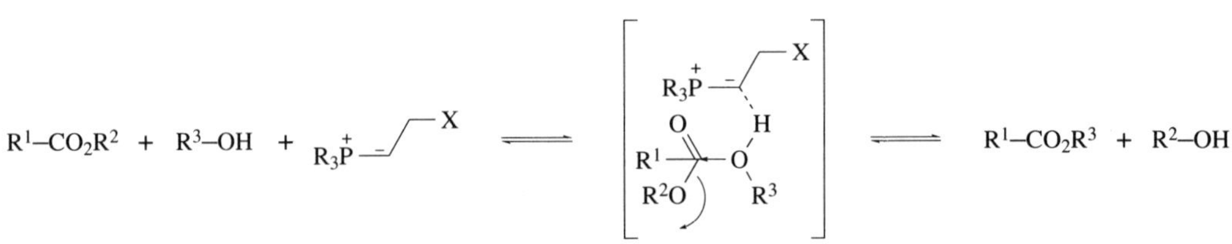

(**20**)

Scheme 13

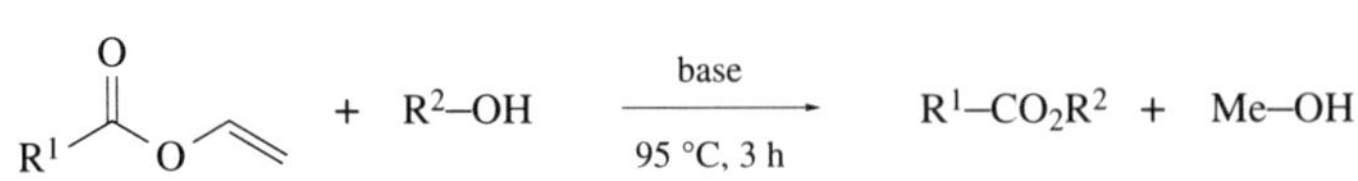

(71)

R¹	*R²*	*Base*	*Yield* (%)
Me	Ph	NaOPh, Na, KOH or KOPh	81–87
Pr^i	Ph	NaOPh	
Me	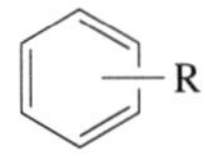		

R = *o*-Cl, *p*-Cl, *p*-Me, *p*-Bu^t
p-CO_2Me, *p*-NO_2

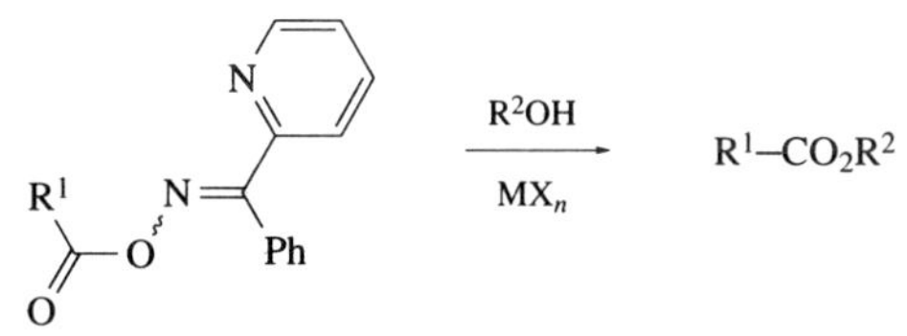

(72)

R^1	R^2	MX_n	Solvent	Time (h)	Yield (%)
n-$C_{15}H_{31}$	Bn		CH_2Cl_2	24	0
n-$C_{15}H_{31}$	Bn	$FeCl_3$	CH_2Cl_2	18	76
n-$C_{15}H_{31}$	Bn	$ZnCl_2$	CH_2Cl_2/MeCN	18	92
Ph	Bn	$ZnCl_2$	CH_2Cl_2/MeCN	18	96

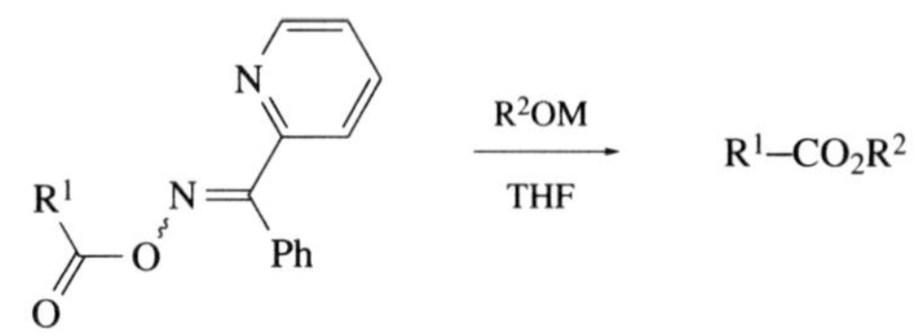

(73)

R^1	R^2OM	Time (min)	Yield (%)
Ph	BnOLi	<5	97
n-$C_{15}H_{31}$	BnONa	<5	96
Ph–CH=CH– (*trans*)	Bu^tOLi	<5	95
Bu^t	BnOLi	<5	97

5.03.1.7 Enzymatic *Trans*-esterifications

Enzymatic *trans*-esterification has gained enormous importance since the early 1980s, in particular for the optical resolution of carboxylic acids or (more frequently) alcohols. Equations (74)–(81) and Scheme 14 illustrate this point for a variety of substrates which demonstrate the broadness and versatility of this approach. In general, enzymatic *trans*-esterifications are performed in an organic solvent to avoid ester hydrolysis and, most advantageously, vinyl acetate or isopropenyl acetate are used as acyl donors to achieve irreversible *trans*-esterification (Equations (74) ⟨92JOC1605⟩, (78) ⟨90JA7434⟩, (79) ⟨90JCS(P1)2559⟩, (80), and (81) ⟨91LA47⟩). Another advantage of enzymatic acylations lies in their regioselectivity as demonstrated for the mono- and dipropionylation of castanospermine (Scheme 14) ⟨90JA2849⟩. In Equation (82) it is demonstrated that lipases can also differentiate between (*E*)- and (*Z*)-allylic alcohols ⟨90TL4875⟩. Geraniol is acylated preferentially over nerol. Scheme 15 shows how enantiotopic positions in 1,3-diols may be enantioselectively acylated via lipase-catalyzed direct acetylation or deacetylation ⟨83HCA2501, 84AG(E)142, 84TL2557⟩. Other examples include Equations (75), (76) ⟨84JA2687⟩, and (77) ⟨85JA7072⟩. In formulae (**21**)–(**29**) the enantiotopic selective hydrolysis of diesters and diacetates is demonstrated.

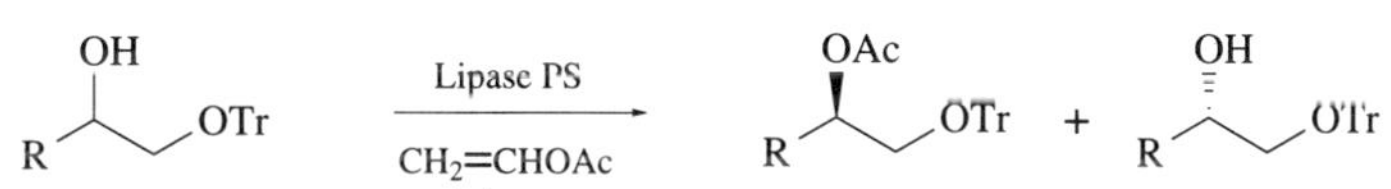

(74)

R = Me	*ee* > 98%	*ee* = 78%
R = Et	*ee* > 98%	*ee* > 98%
R = Pr	*ee* = 70%	*ee* = 56%

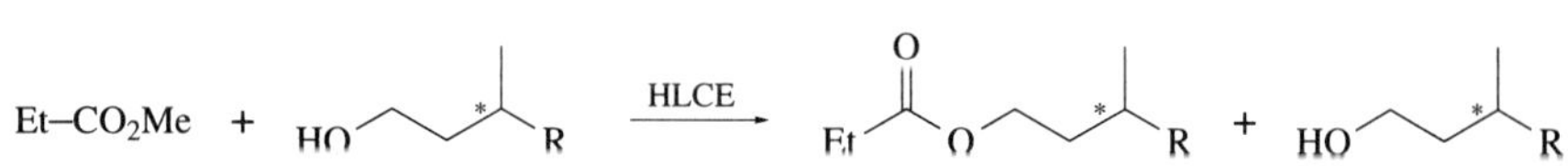

(75)

(*R*,*S*)-alcohol (*S*)-ester (*R*)-alcohol

R^1	(*S*)-Ester *ee* (%)	(*R*)-Alcohol *ee* (%)
OMe	98	
Et	94	94
$(SH_2)_3Pr^i$	95	97
$(CH_2)_3CH{=}CMe_2$	96	97

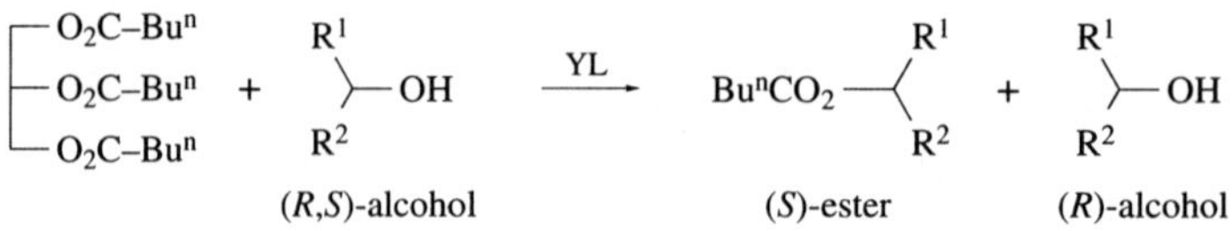

(76)

R^1	R^2	(S)-Ester ee (%)	(R)-Alcohol ee (%)
Me	Et	98	97
Me	C_6H_{13}	98	89
H	$CHCl{\bullet}CH_2Cl$	93	88

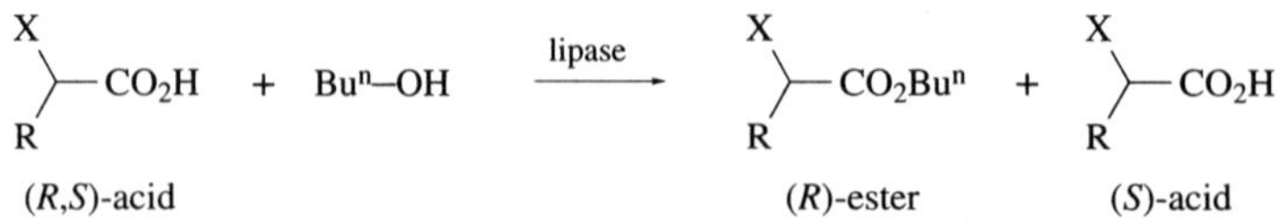

(77)

R	X	(R)-Ester ee (%)	(S)-Acid ee (%)
Me	Br	96	99.6
Me	Cl	95	95
Bu	Br	99	62
Ph	Cl	99	73

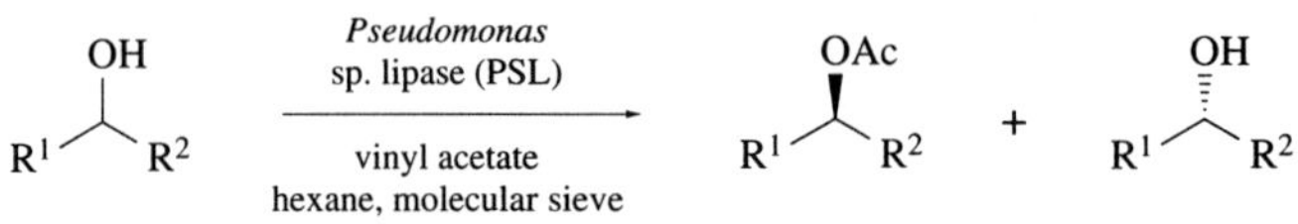

(78)

R^1	R^2	Acetate ee (%)	Alcohol ee (%)
Me	Ph–(C=CH_2)	>95	>95
vinyl	Ph	>95	46
vinyl	(*E*)-Ph–CH=CH	>95	>95
vinyl	Ph–C≡C	>95	>95
CH≡C	(*E*)-Ph–CH=CH	>95	68
Me	Bu^n–C≡C	>95	87
Me	Ph–C≡C	>95	>95
Me	TMS–C≡C	>95	>95

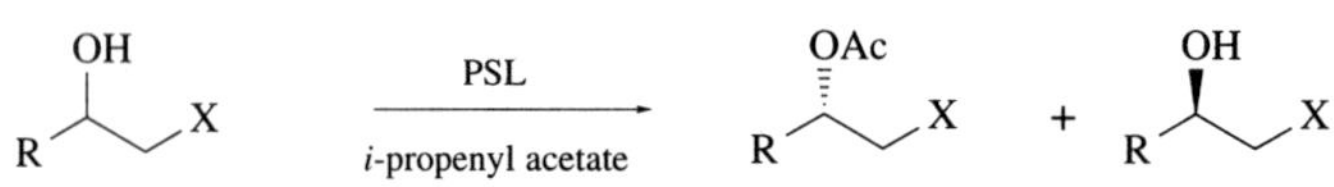

(79)

R	X	Solvent	Acetate ee (%)	Alcohol ee (%)
Ph	Cl	Pr^i_2O	92	97
2-naphthyl	Br	Pr^i_2O	95	80
p-Br-C_6H_4	Br	Pr^i_2O	95	94
TsO-CH_2	Cl	hexane	96	70
TsO-CH_2	Pr^i	hexane	86	66

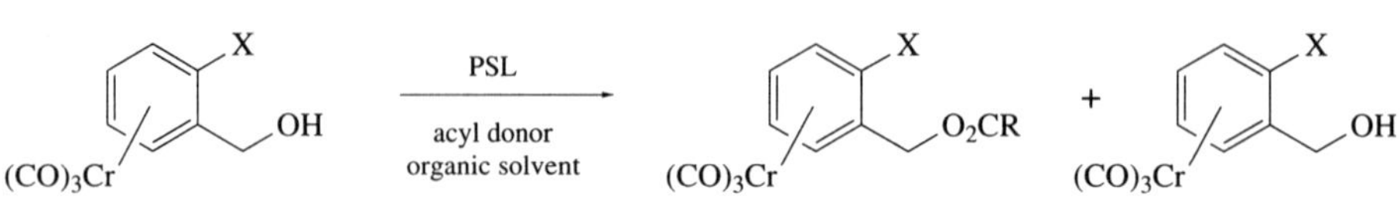

(80)

X	Acyl Donor	R	Solvent	Ester ee (%)	Alcohol ee (%)
Me	*i*-propenyl acetate	Me		98	100[a]
OMe	*i*-propenyl acetate	Me		97	95[a]
TMS	*i*-propenyl acetate	Me		84	85[a]
Me	vinyl benzoate	Ph	toluene	87	96[b]
Me	vinyl acetate	Me	toluene	88	61[b]

[a] 90TL3603. [b] 90TL3895.

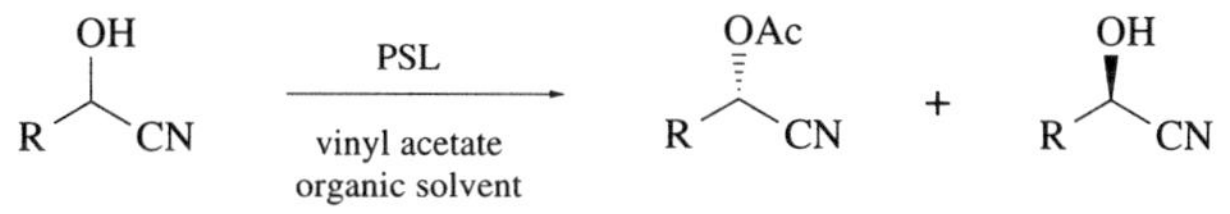

(81)

R	*Solvent*	*Acetate ee (%)*	*Cyanohydrin ee (%)*
Bn–CH_2		90	98
1-naphthyl–O–CH_2		91	96
BnO–CH_2		55	95
Pr^n	CH_2Cl_2	55	33
Ph		68	80
p-OH–C_6H_4		79	50

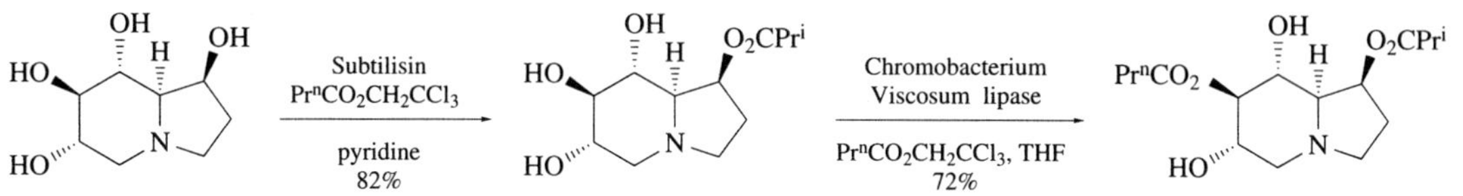

Scheme 14

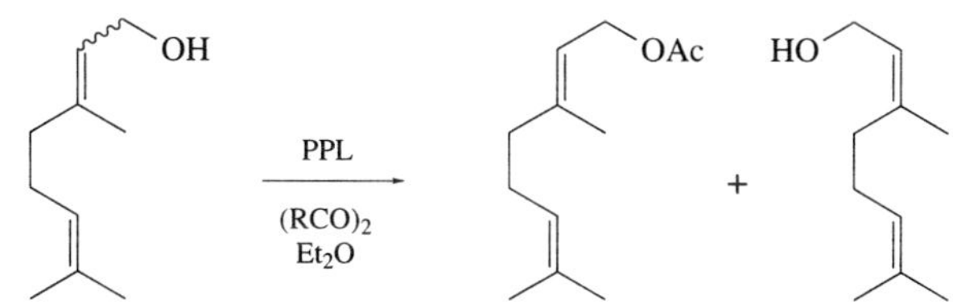

(82)

(*E*) Geraniol
(*Z*) Nerol

R	*Geranyl acetate (%)*	*Nerol (%)*
n-C_3H_7	85	15
n-C_5H_{11}	66	7
n-C_7H_{15}	72	7

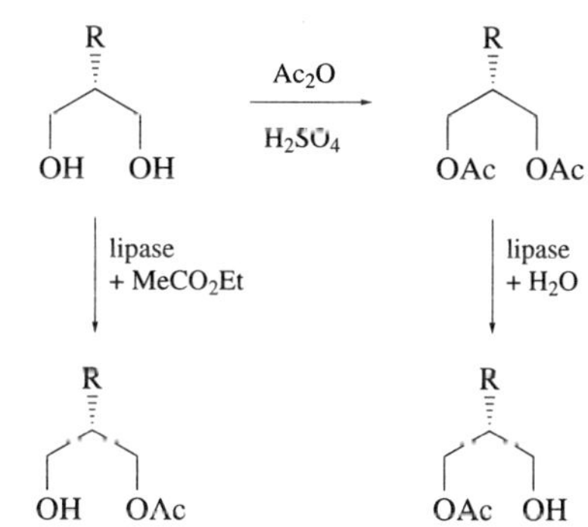

R	*ee (%)*	
$CH_2CH_2CH{=}CH_2$	90	>95
Ph	92	>95
Bn	13	61
C_6H_{11}	58	60

Scheme 15

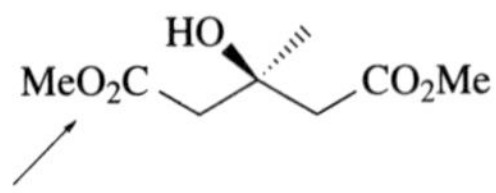

(21)[a] *ee* = 99%

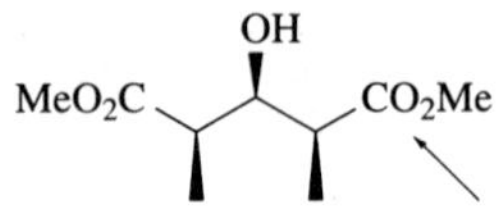

(22)[b] *ee* = 98%

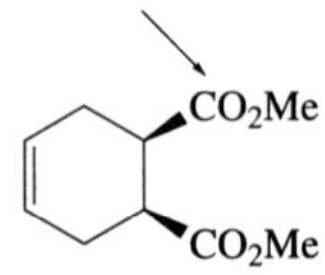

(23)[c] *ee* = 98%

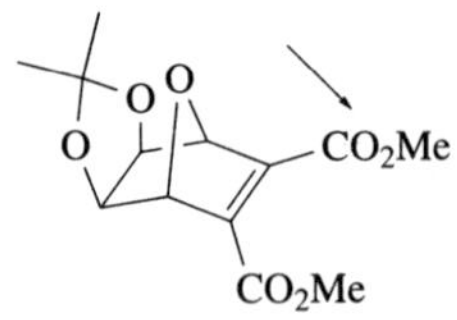

(24)[d] *ee* = 77%

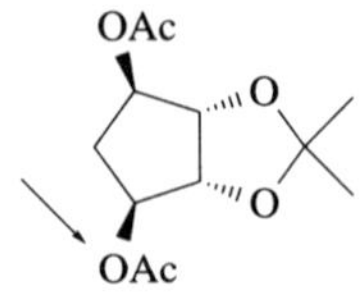

(25)[c] *ee* = 98%

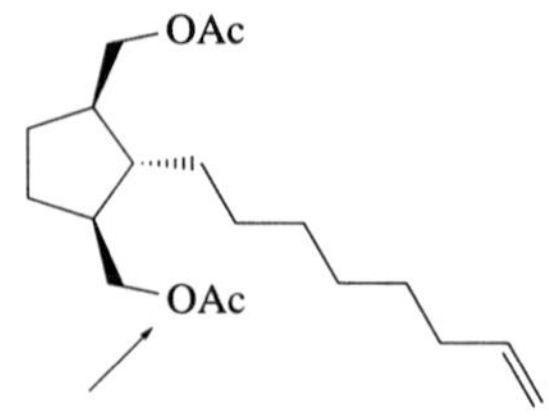

(26)[f] *ee* = 99%

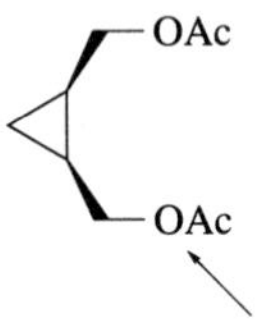

(27)[g] *ee* = 93%

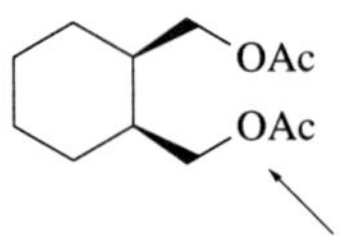

(28)[g] *ee* = 87%

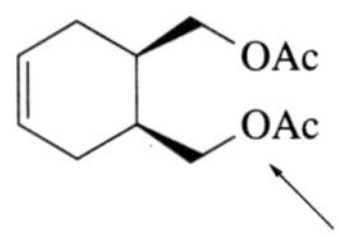

(29)[h] *ee* = 95%

[a] 75JA4144, 85T1857, 87ABC1345, 87JOC3477. [b] 83HCA2501 [c] 83HCA2501, 84TL2557, 84AG(E)142. [d] 84T145. [e] 86JA5655. [f] 87CPB1741. [g] 85CC1563. [h] 84JA7250

5.03.1.8 Monoacylation of Polyols

Monoacylation of polyols is one of the central problems in many natural product syntheses. In the simplest case the monoester is generated by S_N2 addition of the carboxylate anion to ethylene oxide (Equation (83)) 〈71MI 503-01〉. Selective monobenzoylation of diols can be achieved under Mitsunobu conditions (Equation (84)) 〈81S1〉. The table shows the ratio of mono- to dibenzoate for various substituents. 1,2-Diols are generally converted into the epoxide with retention of configuration (Equations (85) and (86)) 〈90MI 503-01〉 whereas in Equation (87) the monobenzoate is obtained, due to the *cis*-configuration of the hydroxyl group, which prohibits an S_N2-type ring closure. Selective monoacylation of the polyols **(30)** 〈91JA910〉 and **(31)** 〈88JA3670, 88JA3672〉 is easily achieved under the appropriate conditions, whereas this is only possible for the diols **(32)** 〈84TL5031〉, **(33)** 〈88JA3670〉, **(34)**, **(35)**, and **(36)** 〈85JOC1751〉 with varying selectivities. Generally, primary positions are preferred or in the case of secondary hydroxyls **(36)** and **(37)**, the steric influence of the environment is decisive. Diols **(37)**, **(38)** 〈85JOC1751〉, **(39)** 〈86T5961〉, and **(40)** 〈93TA457〉 are monoacylated at the unhindered hydroxyl position with almost complete selectivity. In addition, selective monoacylations, particularly with enantiodifferentiation, can be performed enzymatically (see 5.03.1.7).

2-OH-Alkyl ester

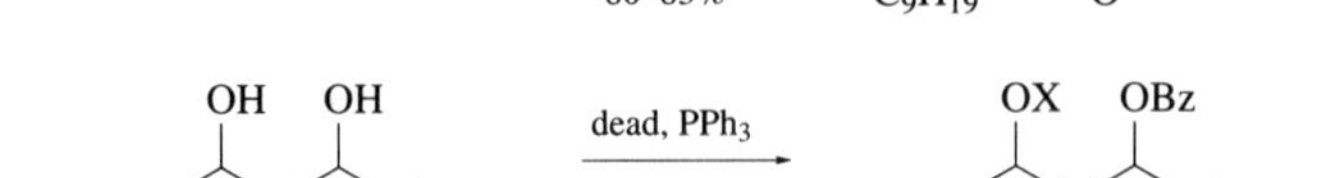

(83)

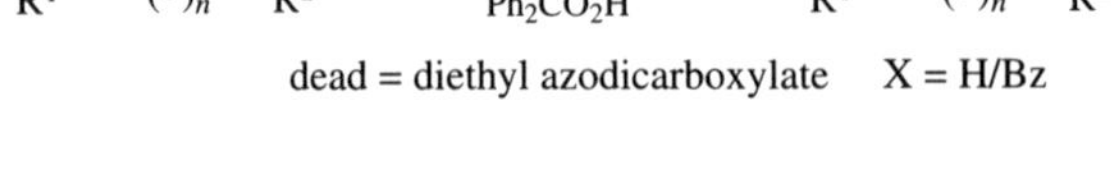

(84)

dead = diethyl azodicarboxylate X = H/Bz

n	*R*[1]	*R*[2]	*Yield* (%)
1	H	H	65/12
2	H	H	66/15
1	Me	H	70/7
1	Me	Me	57/20

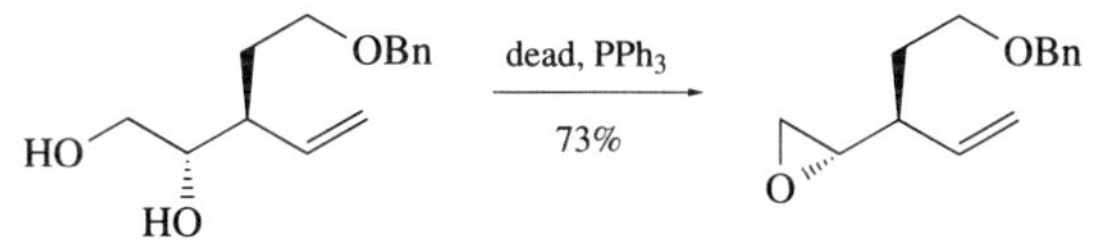

(85)

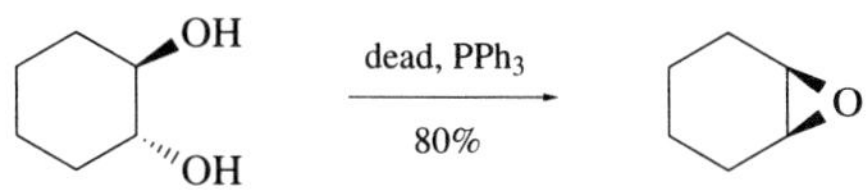

(86)

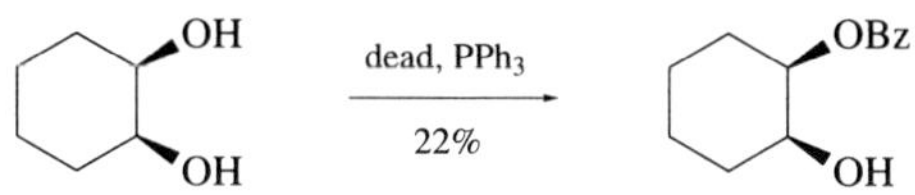

(87)

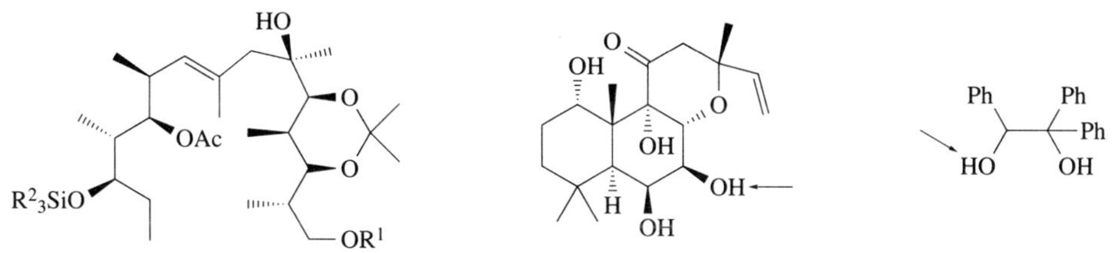

(30) R^1 = H, Ac **(31)** **(32)**

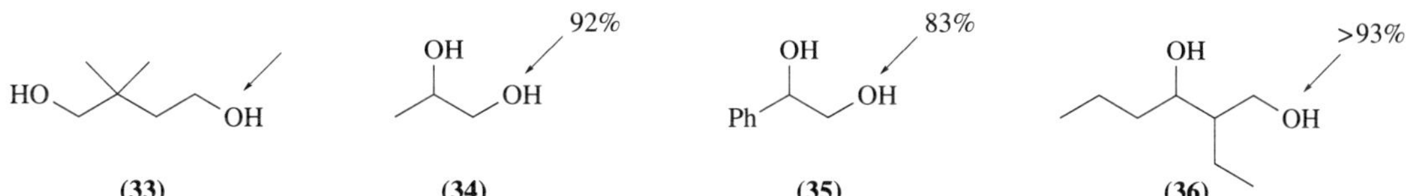

(33) **(34)** **(35)** **(36)**

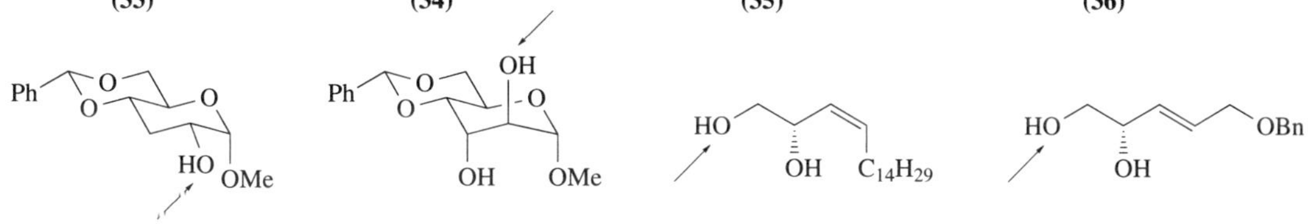

(37) **(38)** **(39)** **(40)**

5.03.2 ESTERS WITH POLYFUNCTIONAL ALCOHOL COMPONENTS

5.03.2.1 1- and 2-Haloalkyl Esters

1-Chloroalkyl esters (Equation (88)) ⟨76CR(C)483⟩ have been prepared from acyl chlorides and aldehydes, presumably via *O*-acylation of the aldehyde and subsequent addition of the chloride ion to the oxonium carbon. 2-Haloalkyl esters may be prepared in two different ways: either from cyclic ketene acetals with hydrogen chloride and subsequent Pinner cleavage (Scheme 16) ⟨48JA3781⟩ or via S_N2 type of an epoxide with a carboxylic acid under assistance of HX (Equation (89)) ⟨80MI 503-01⟩.

1-Cloroalkyl esters

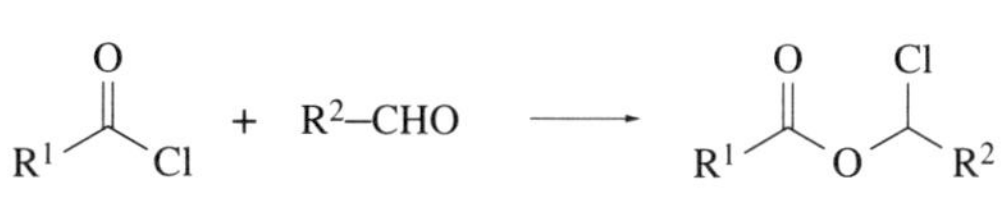

(88)

R^1 = Me, Ph, Ar
R^2 = Me, Pr^i, Bu^t, Ar

2-Haloalkyl esters

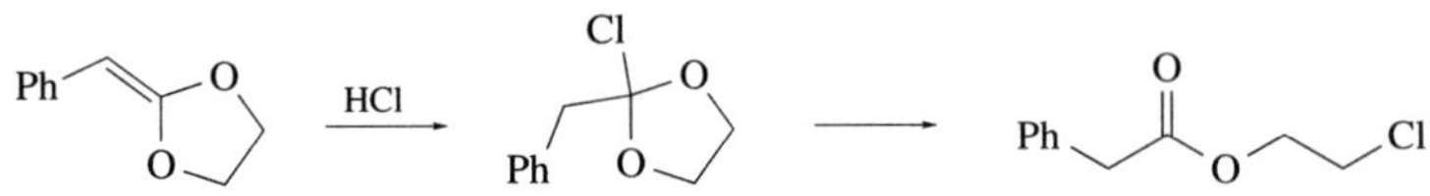

Scheme 16

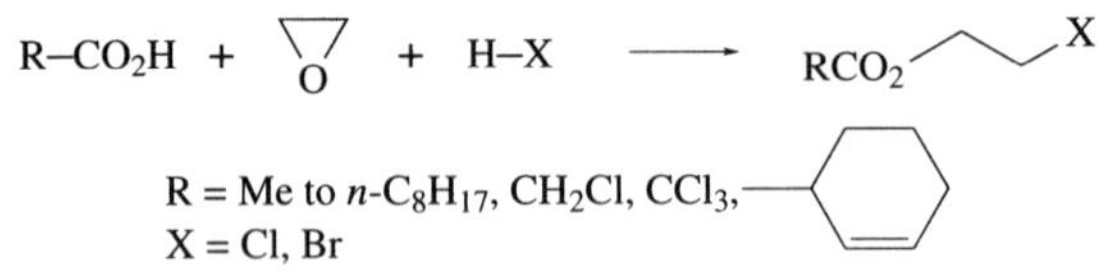

(89)

5.03.2.2 Nitrogen-substituted Alcohol Components

Aminoalcohols (Equation (90)) are first *N*-blocked by the addition of hydrogen chloride and then the free hydroxyl reacts with the acid chloride to form the ester ⟨47JA200⟩. Carbohydrate-derived amino alcohols (Equations (91) and (92)) ⟨70CB37⟩ may be converted into the ester via alkylation of the primary mesylate with a reactive carboxylate salt. Serine (Equation (93)) ⟨42JBC(144)203, 54JA2881, 56JA1367, 59JA1751⟩ is selectively *O*-acylated under acidic conditions analogous to Equation (90), and quite similarly the betaine acylation (Equation (94)) ⟨79OPP93⟩ can only occur at the free hydroxyl function. Equation (95) shows the S_N2-type alkylation of *p*-nitrobenzyl chloride to form the corresponding acetate ⟨55OSC(3)650⟩. The nitro group can subsequently be reduced to the amino function.

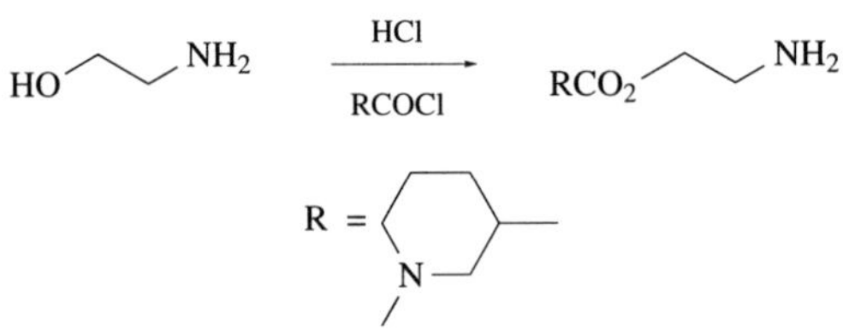

(90)

(91)

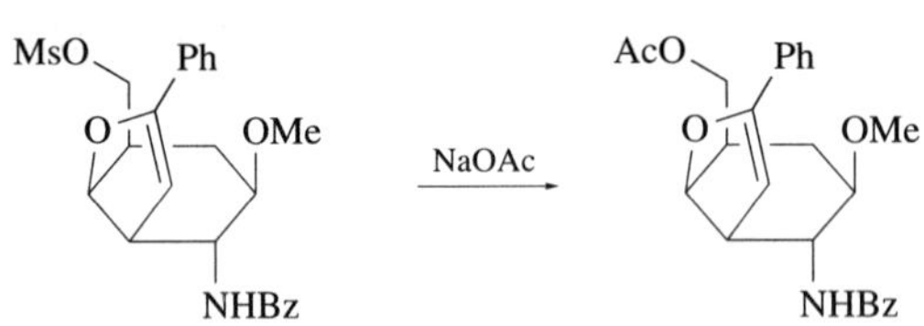

(92)

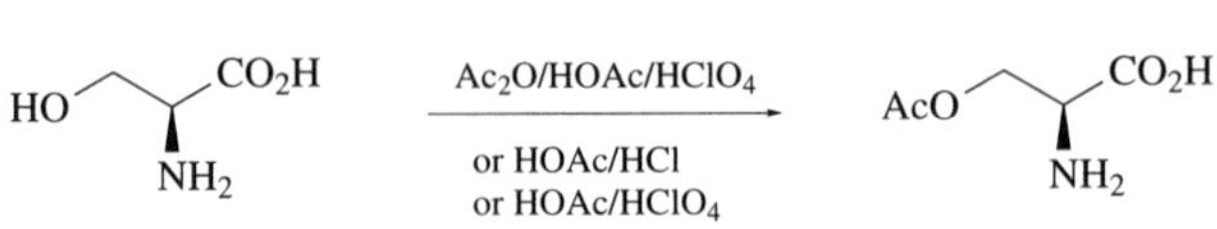

(93)

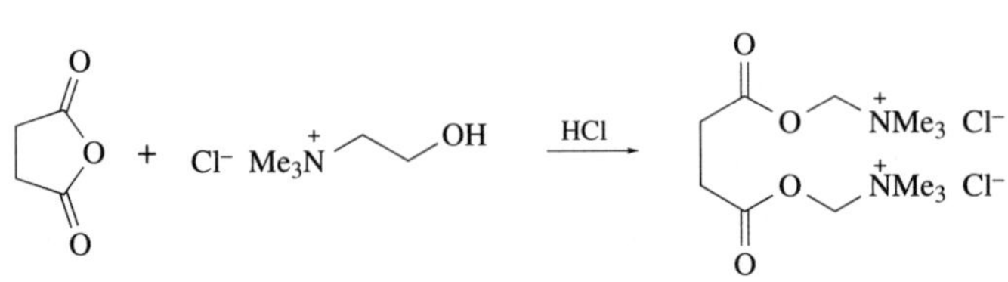

(94)

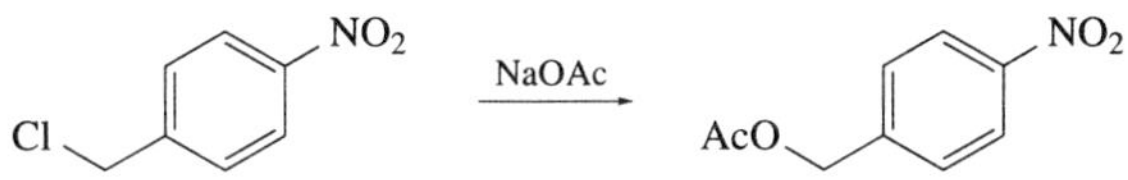

(95)

5.03.3 ESTERS VIA OXIDATION REACTIONS

5.03.3.1 Without Changing the Number of Carbons

Direct oxidation of alcohols to the corresponding carboxylic acids and *in situ* esterification of these acids with unreacted alcohol is possible with bromine and sodium bicarbonate in HMPA (Equation (96)) ⟨76S811⟩. A variation of this procedure involves NBS and TMS-chloride as an activator (Equation (97)) ⟨78JOC371⟩. Oxidation of aldehydes to carboxylic acids is a common transformation. However, the direct formation of esters is not so easily achieved, but using Caro's acid the desired alcohol may be obtained (Equation (98)) ⟨68JOC2525⟩. Esterification can also start from the aldehyde which is disproportionated according to the Claisen–Tishchenko protocol, catalyzed either with aluminum trialkoxide or with aluminum trihalides prepared *in situ* (Scheme 17) ⟨51HCA2172⟩.

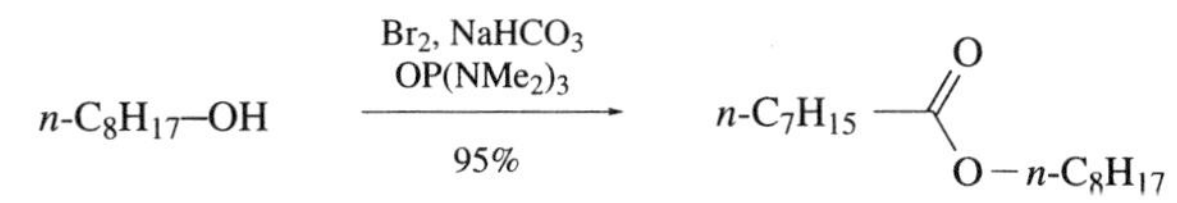

(96)

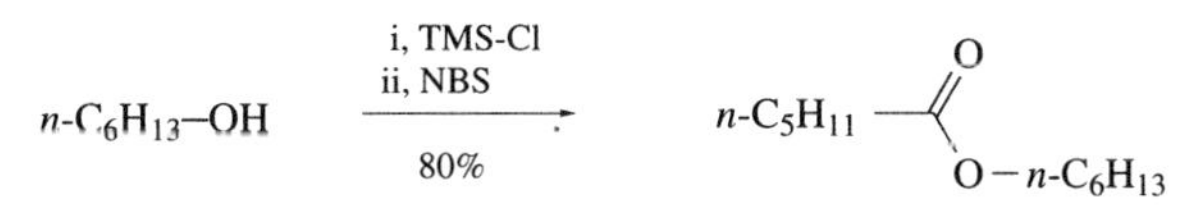

(97)

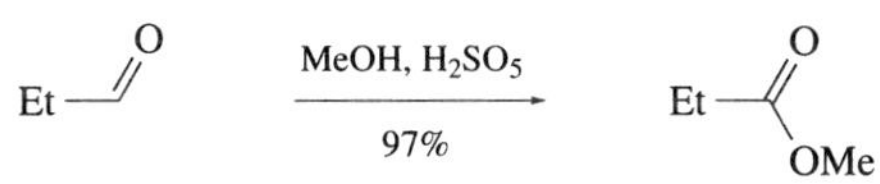

(98)

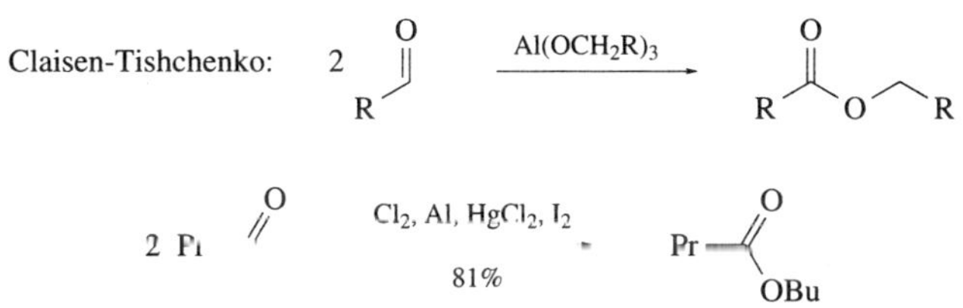

Scheme 17

5.03.3.2 Esters via Ozonolytic Cleavage of Double Bonds

This is very mild and general access to esters (Equation (99)) ⟨78AG(E)939⟩; however, if mixtures are to be avoided, the alkene must be symmetrical.

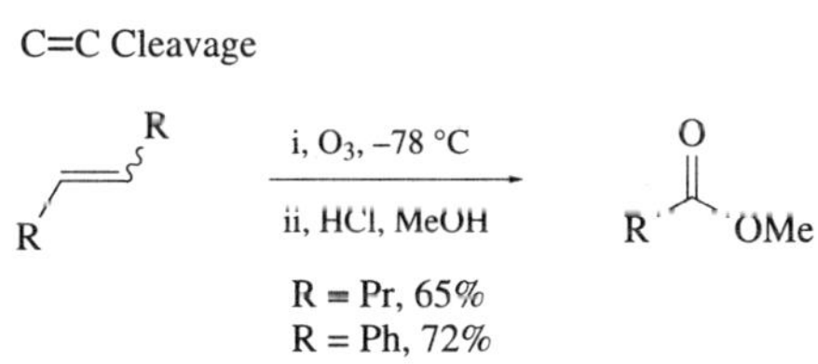

(99)

5.03.3.3 Esters via Baeyer–Villiger Oxidation

This reaction starts from symmetrical or unsymmetrical ketones, acyclic ketones furnish esters and cyclic ketones give lactones (Equation (100)) ⟨80HOU(6/1a)874, 81T2697⟩. For unsymmetrical ketones, the relative migratory aptitude is crucial for the formation of the product. The residue which migrates faster becomes the alcohol part of the ester. Thus, in Equation (101) ⟨64JOC2813⟩ isopropyl acetate and not methyl-2-methyl propionate is formed. Equations (102) ⟨73T1559⟩, (103) ⟨64JOC2813⟩, (104) ⟨60BSF1350⟩, and (105) ⟨73JCS(P1)2796⟩ demonstrate how one or two acetyl side-chains on ring systems, which mostly have served to effect the carbon—carbon-cyclization, may be subsequently degraded to acetate substituents with retention of the original configuration.

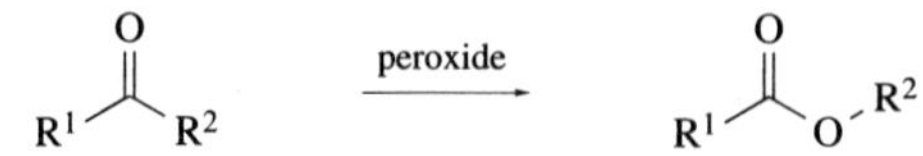

(100)

Migration aptitude: *t*-alkyl > cyclohexyl > *sec*-alkyl > aryl > *n*-alkyl > Me

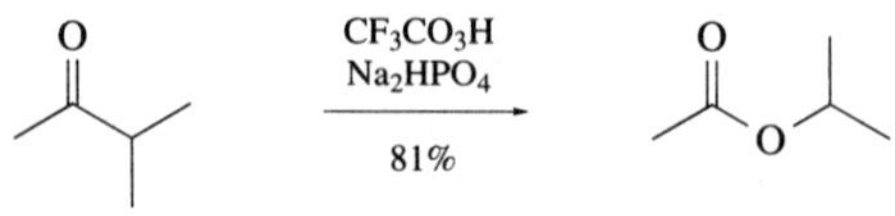

(101)

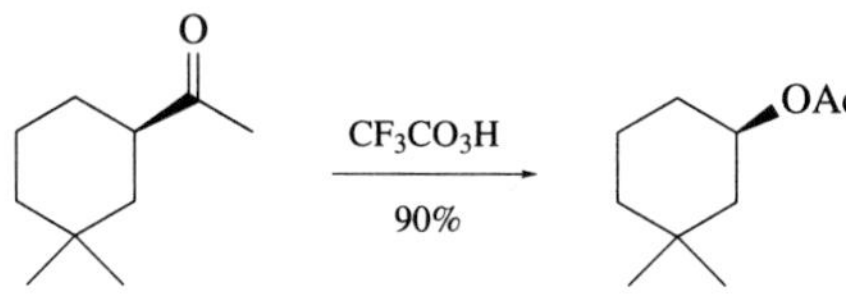

(102)

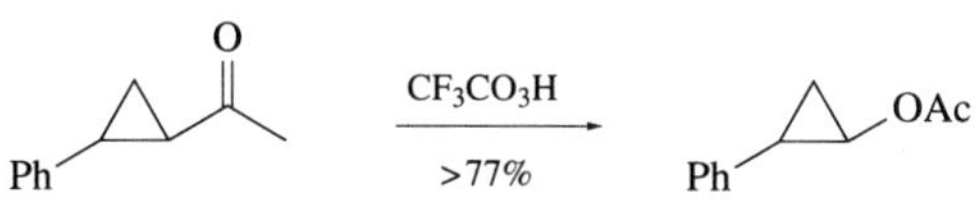

(103)

(104)

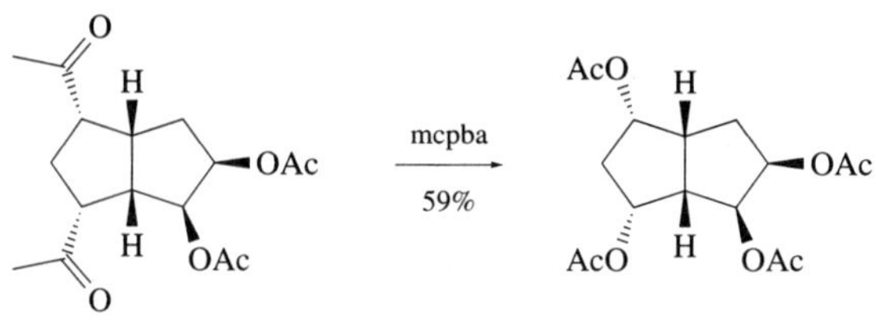

(105)

5.03.4 ESTERS VIA ONE CARBON CHAIN ELONGATION

5.03.4.1 Alkoxycarbonylation

Alkoxycarbonylation may be performed with α-alkenes, carbon monoxide, an alcohol, which is mostly used as a solvent, and a transition-metal complex as the catalyst. Equations (106) ⟨63TL1437, 68AG352, 73MI 503-02, 78MI 503-01, 79ICA345⟩, (107) ⟨84MIP84/1376A1⟩, (108) ⟨75JOC335, 75JOC3227, 77BCJ1639, 78AG(E)27⟩, and (109) ⟨94JA7455⟩ show typical reactions. For α-alkenes and palladium phosphine catalysts, problems of regiochemistry arise, which to a certain degree may be best solved by the ligands around the palladium (Equation (106)) or by choosing a platinum catalyst (Equation (107)), which provides the unbranched isomer selectively. Equation (108) shows how the organopalladium intermediate may be conveniently prepared via an oxymercuration of the alkene and subsequent metal–metal exchange. In Equation (110) the alkoxycarbonylation is combined with a cycloetherification reaction, to give a tetrahydrofuran with >95% stereoselectivity ⟨76JOC793⟩. Alternatively the alkoxycarbonylation may be started with a benzyl, aryl or vinyl halide which is converted into the organometallic intermediate by a platinum(II) catalyst (Equation (110)). This intermediate then reacts with carbon monoxide and the alcohol. The stereochemistry of the chain elongation of benzylic halides (Equation (111)) ⟨68TL633, 74JOC3318, 75BCJ2075, 75BCJ2091, 75TOC532, 79TL133, 80JA4193⟩ corresponds to an overall inversion of configuration (Equation (112)) ⟨76JA5832⟩. Equations (113) ⟨85JOM(285)293⟩, (114), (115) ⟨83EUP127276A2⟩, and (116) ⟨68TL633, 74JOC3318, 75BCJ2075, 75BCJ2091, 75JOC532, 79TL133, 80JA4193⟩ show some examples of the preparation of aromatic carboxylic esters from a variety of aromatic and heteroaromatic halides, and Equation (117) demonstrates the applicability of the method to acrylic esters where, however, the original alkene configuration is not totally maintained.

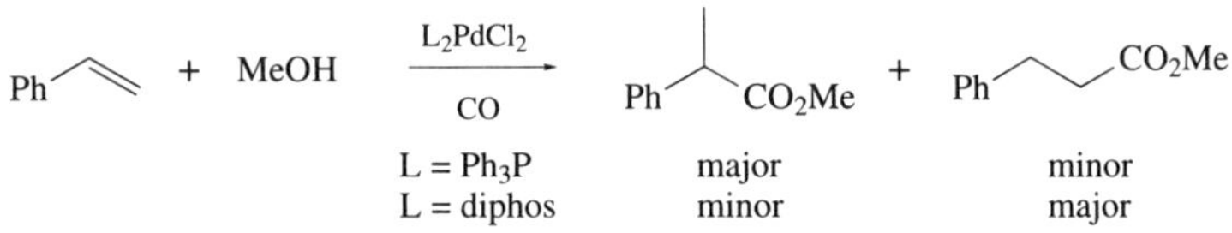

(106)

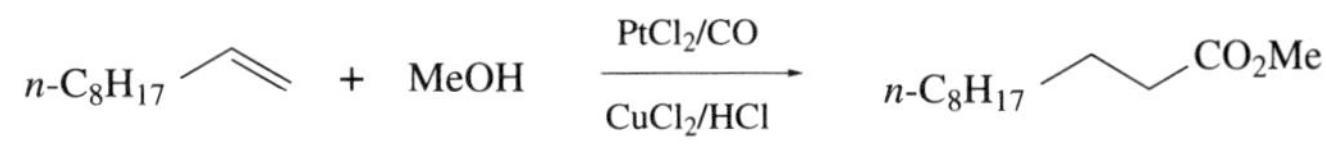

(107)

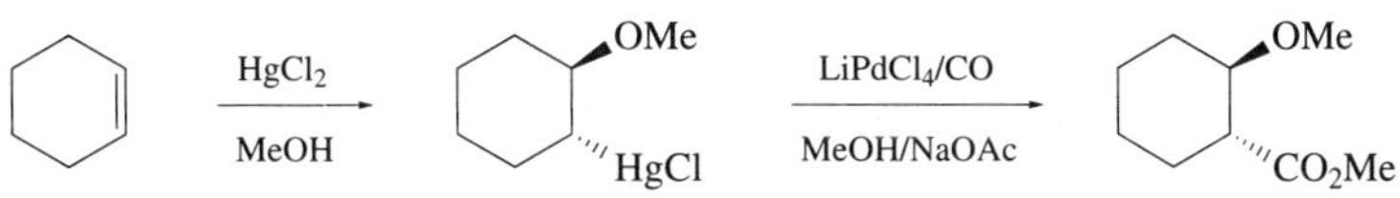

(108)

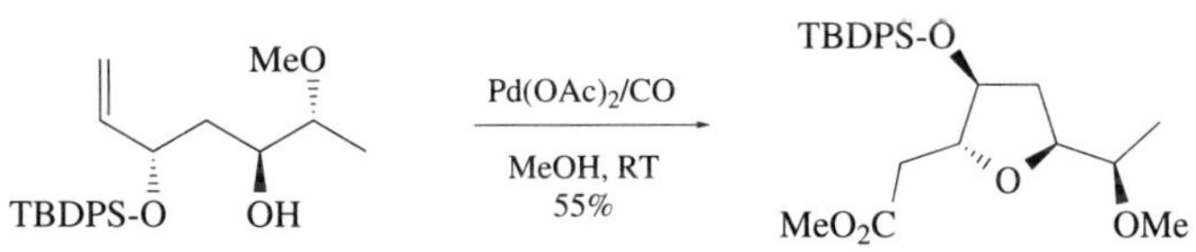

(109)

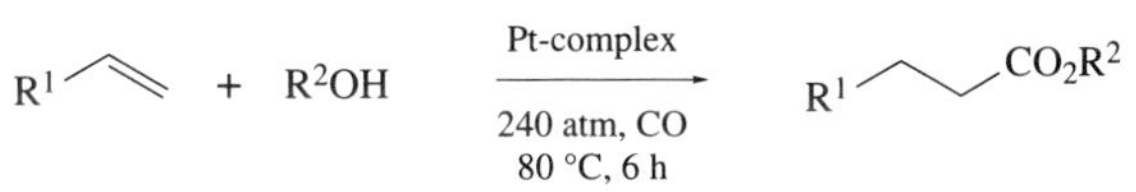

(110)

R^1 = Me, n-C_5H_{11}, n-$C_{12}H_{25}$, n-$C_{18}H_{33}$, EtMeCH, Pr^i–CH_2
R^2 = Me, Cl–CH_2–CH_2, Me_2CH, n-$C_{18}H_{37}$, Ph
Pt complex = $(Ph_3As)_2PtCl_2$•$SnCl_2$, $(Ph_3P)_2PtCl$•$SnCl_2$ etc.

$$\text{Bn–Cl} + \text{CO} + \text{ROH} \xrightarrow[\text{base}]{(Ph_3P)_2PdCl_2} \text{Bn–}CO_2R \qquad (111)$$

Stereochemistry:

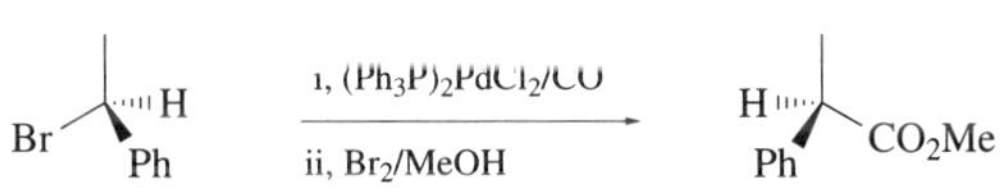

(112)

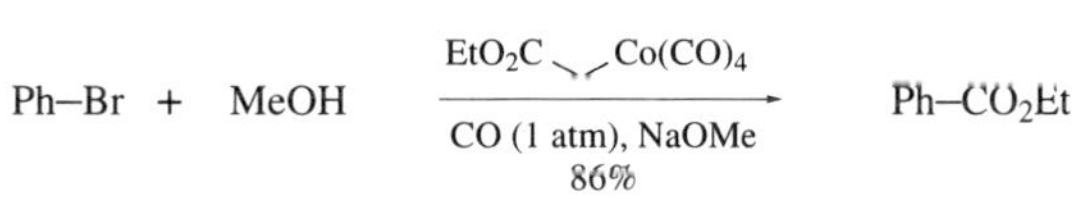

(113)

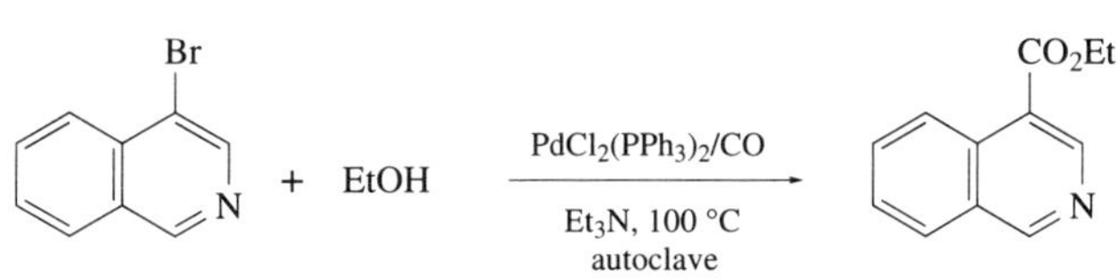

(114)

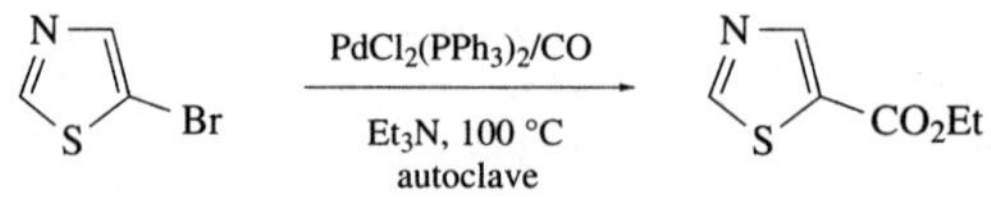

(115)

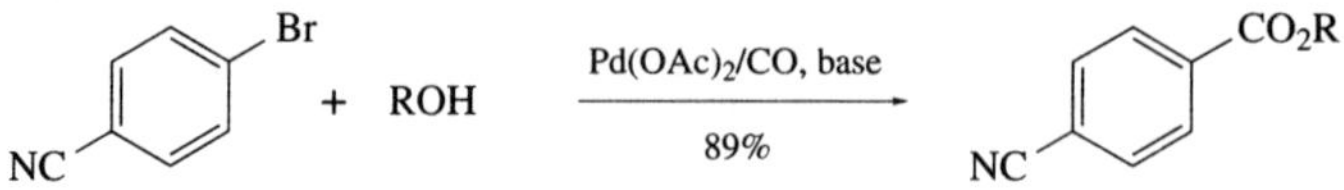

(116)

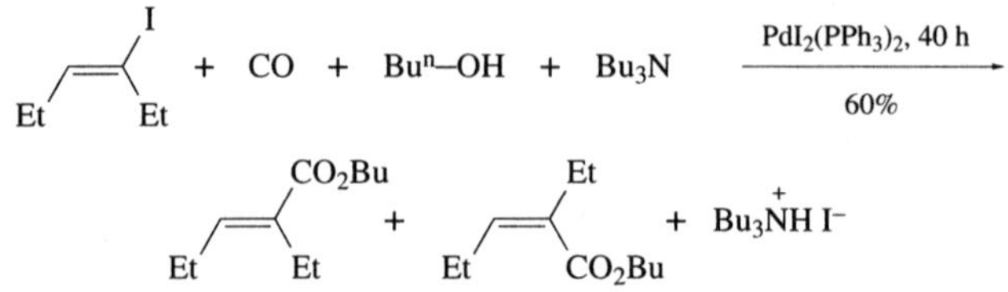

(117)

5.03.4.2 Arndt–Eistert Homologation

The general protocol of this reliable and well-established homologation is outlined in Scheme 18. A carboxylic acid is converted into the chloride and then treated with diazomethane to give the diazoketone, which is submitted to a Wolff rearrangement either under photolytic or thermal conditions using silver(I) catalysis. Table 1 shows various examples for aromatic or cycloaliphatic acids; a dicarboxylic acid (**41**) is bishomologated. All homologations proceed with retention of configuration.

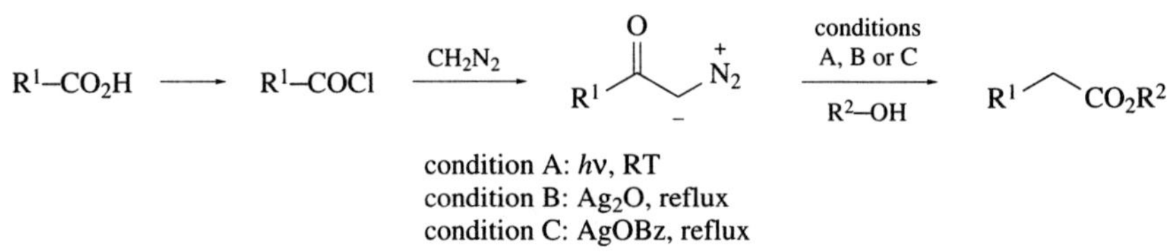

Scheme 18

Table 1 Arndt-Eistert reactions.

R^1-CO_2H	R^2	*Conditions*	*Yield (%)*	*Ref.*
α-naphthoic acid	Et	C	78	70OS77
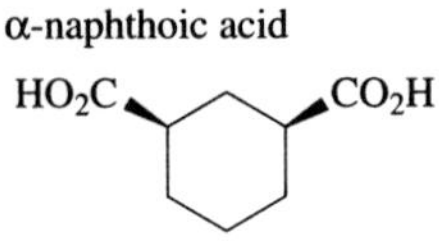 (**41**)	Me	C	80	63JOC3512
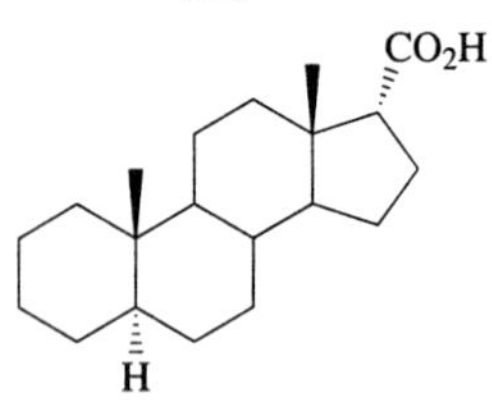	Me	B	85	49HCA647
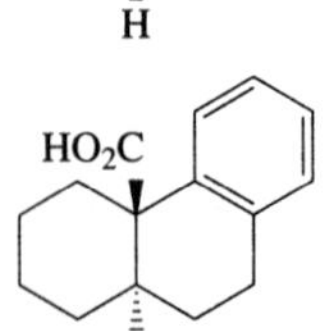	Me	A	86	81IJC(B)411

5.03.5 ESTERS OF PARTICULAR CARBOXYLIC ACIDS

5.03.5.1 Esters of Cycloalkanoic Acids

Cyclopropanoic esters have tremendous importance as insecticides (pyrethroids). Equation (118) ⟨24HCA390⟩, Scheme 19 ⟨78CB2206⟩, Scheme 20 ⟨85FRP2549470⟩, Equation (119) ⟨67BSF985⟩, and Schemes 21 ⟨83AG(E)63⟩, 22 ⟨59MI 503-01⟩, and 23 ⟨76GEP(O)2539048⟩, show the variety of methods that have been devised for the synthesis of such compounds. For instance, diazo compounds may be added to 1,3-dienes via a carbenoid (Equation (118)) or a 1,3-dipolar addition mechanism (Schemes 19 and 20). From the pyrazoline intermediate, nitrogen is eliminated to form the cyclopropane ring. Alternative methods involve the Michael-type addition of stabilized carbanionic reagents (Equation (119) and Scheme 21) to form an ester enolate which undergoes cyclization under elimination of sulfinate (Equation (119)) or triphenyl phosphine (Scheme 21). Scheme 21 constitutes an enantiocontrolled synthesis of chrysanthemic ester (**42**). A rather elaborate procedure is shown in Scheme 22. Claisen rearrangement generates keto esters (**43**), which are then converted into γ-halo esters (**44**) and cyclized with base. More directly cyclopropanoic esters are accessible via [2 + 2]-cycloaddition to form an α-chlorocyclobutanone (**45**) which undergoes ring contraction under Favorskii conditions (Scheme 23).

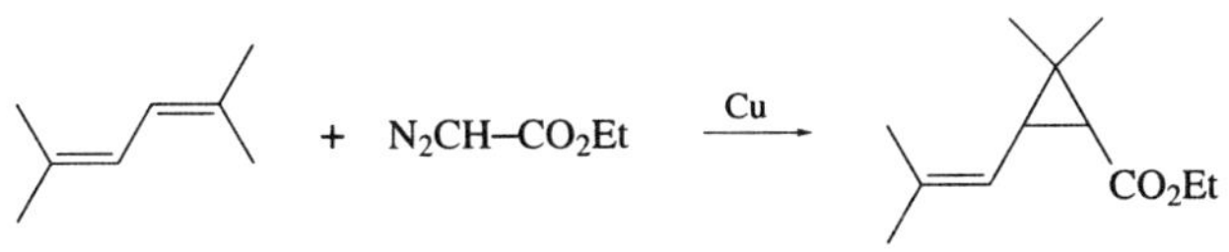

(118)

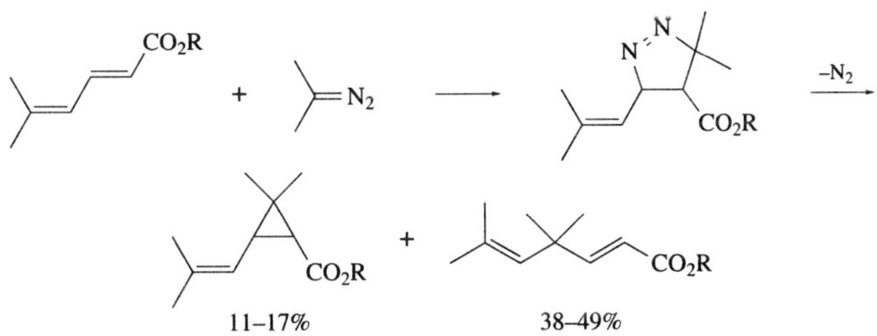

Scheme 19

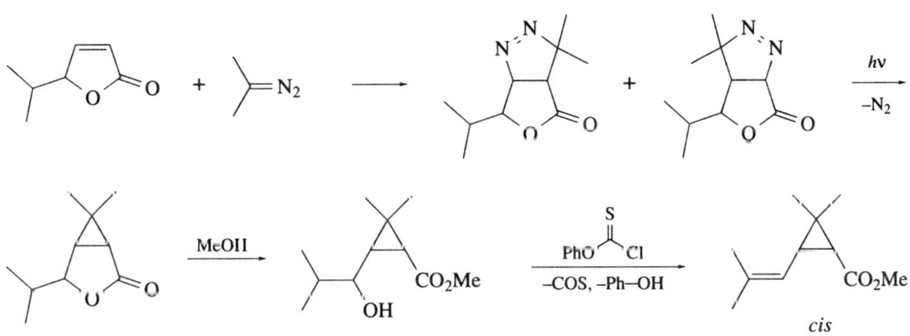

Scheme 20

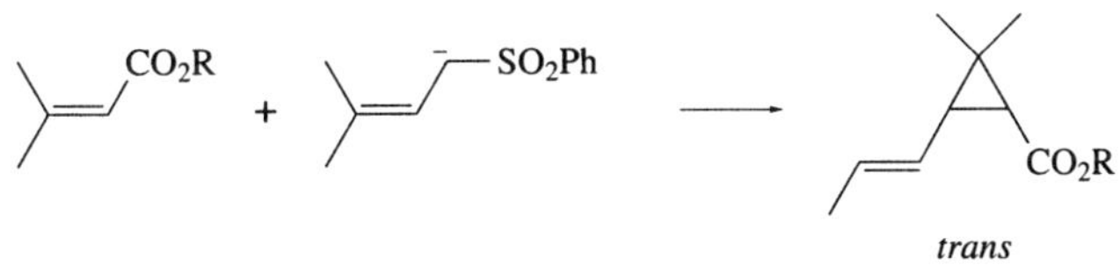

(119)

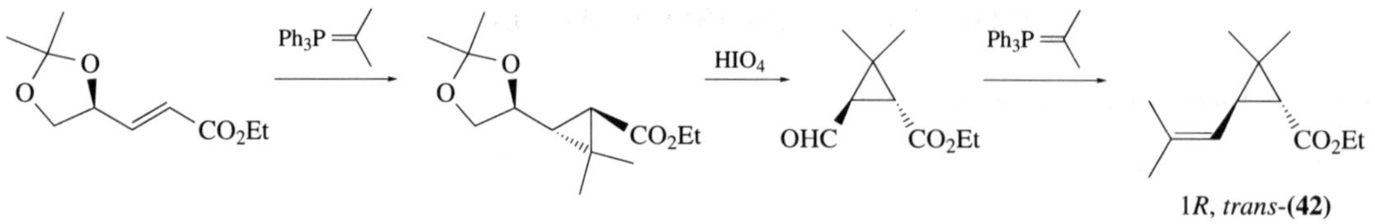

1*R*, *trans*-(**42**)

Scheme 21

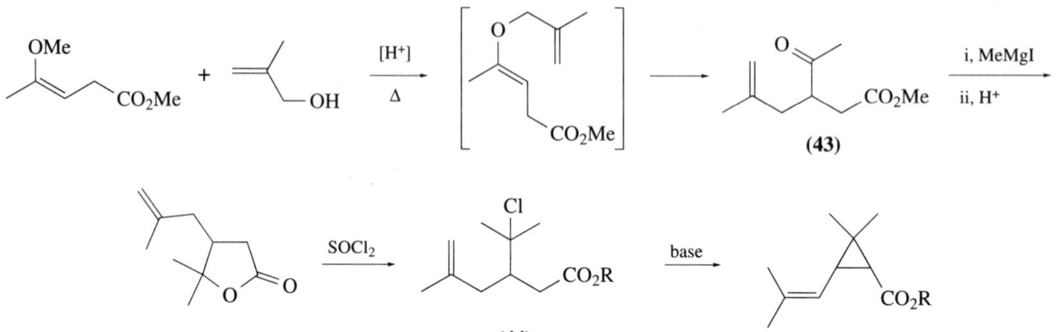

Scheme 22

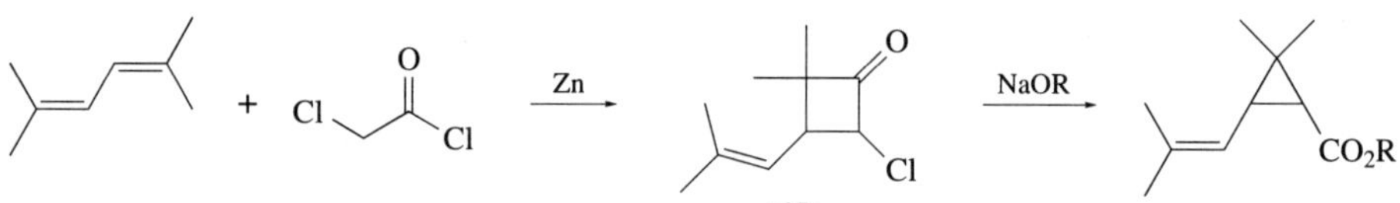

Scheme 23

There are many ways to generate cyclopentanoic esters. Some are shown in Equations (120)–(124). For example, reductive cyclization of 1,3-diiodides with acrylic ester furnishes (**46**), (Equation (120)) ⟨74JOC3273, 75S291⟩, whereas the standard ring contractions via Wolff or Favorskii rearrangements are more generally applicable (Equations (121) ⟨87CC83⟩ and (122)) ⟨59OS(39)37⟩. Dieckmann cyclization is highly suitable for annulation processes (Equation (123)) ⟨73JOC1941⟩. Palladium-mediated addition of a formal trimethylene methane synthon to acrylic esters furnishes cyclopentanoates with an *exo* methylene function suitable for further manipulation (Equation (124)) ⟨86JA284⟩. The method has been applied to a synthesis of brefeldin A. Another Favorskii six- to five-ring contraction is shown in Equation (125) ⟨84SC227⟩. Cyclohexanoates may be conveniently prepared via Dieckmann condensation (Equations (126) and (127)), which is particularly suitable for annulations, even in more sophisticated cases ⟨79JCS(P1)1255, 81JOC1538⟩. Of greater value is the Diels–Alder cycloaddition, because it is a connective method which proceeds under high diastereocontrol (suprafaciality with respect to both components and *endo*-selectivity as well as highly induced stereoselectivity) (Equations (128) and (129)) ⟨67JOC869, 85TL1631⟩.

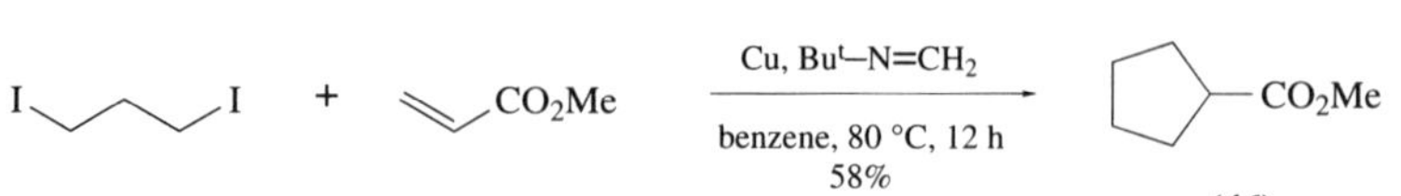

(**46**)

(120)

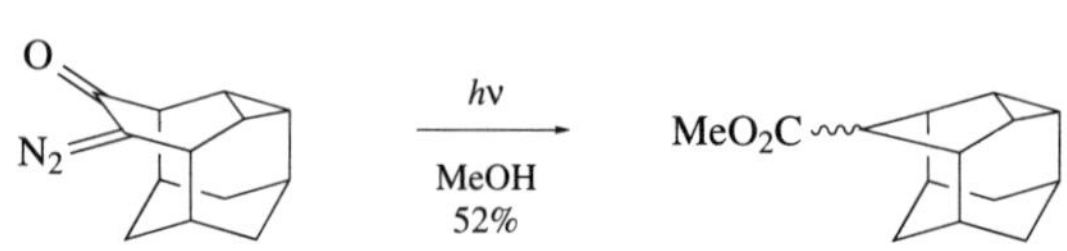

(121)

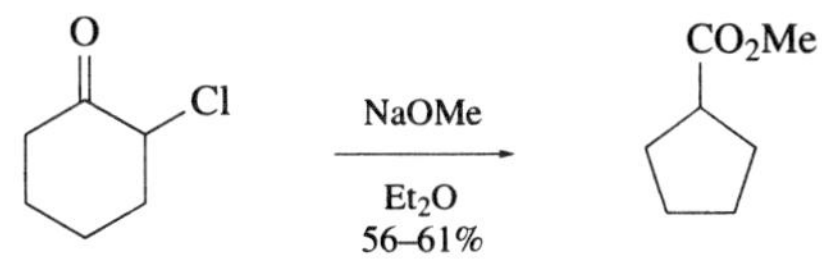

(122)

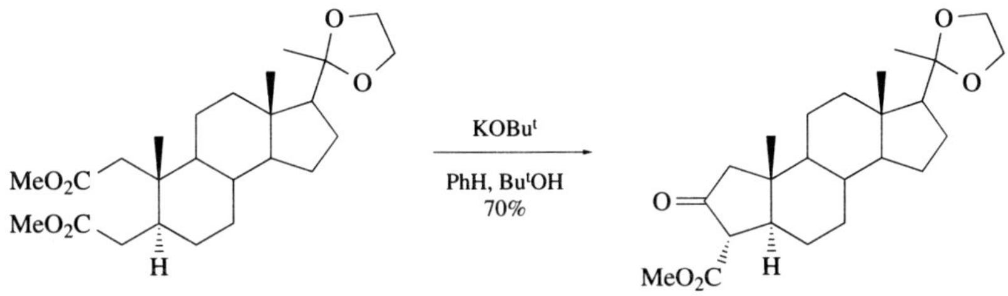

(123)

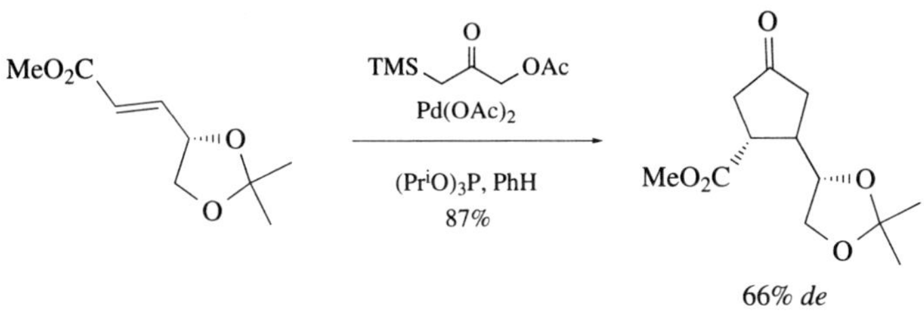

(124)

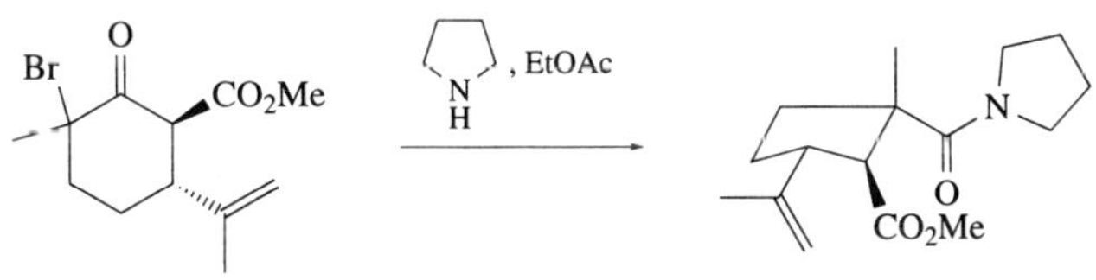

(125)

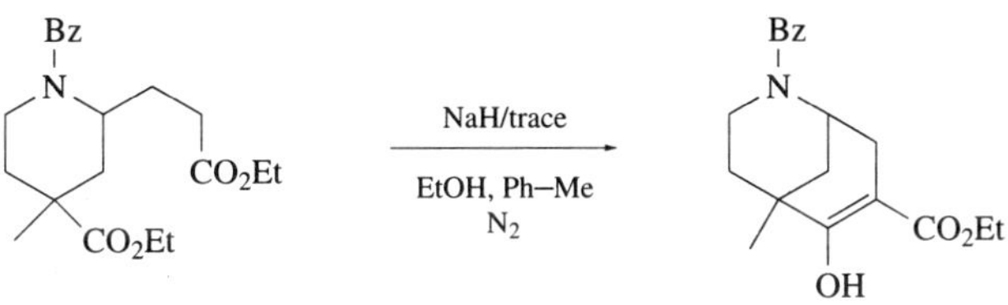

(126)

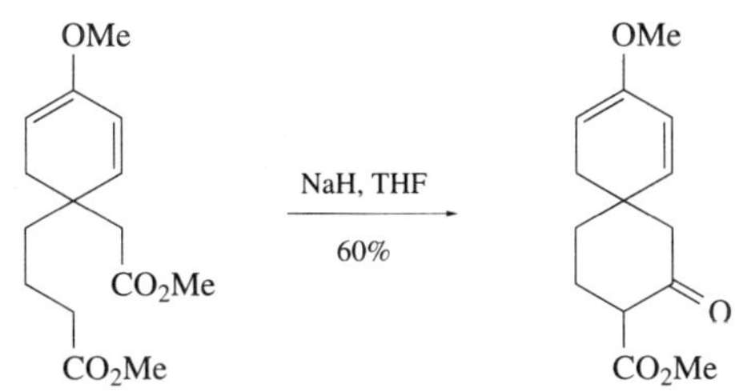

(127)

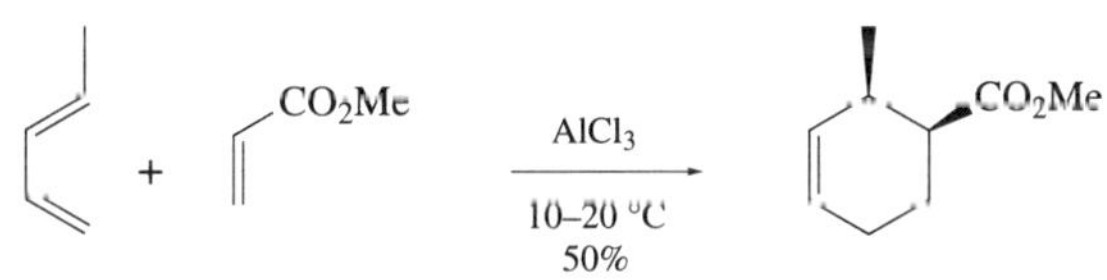

(128)

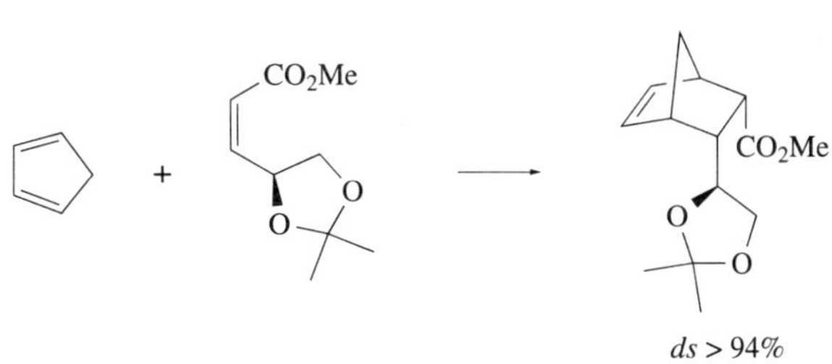

(129)

5.03.5.2 α- and β-C-Branched Alkanoic Esters

5.03.5.2.1 Enolate alkylation

The general procedure for enolate alkylation is shown in Scheme 24 ⟨82S521⟩. The ester is α-deprotonated with a strong hindered base such as lithium diisopropylamide (LDA), lithium cyclohexyl-isopropylamide (LICA) or potassium hexamethyldisilazide (KHMDS) to the enolate, as an (*E*/*Z*) mixture, which is then treated with an alkylating agent to form the α-branched ester. The yields are uniformly good if reactive alkylating components such as primary halides or tosylates are used. If diastereoselectivity is involved or desired in the alkylating step, the stereochemical induction may come from the alcohol component *O*-chiral ester enolate; (**47**) and (**48**) or from the carboxylic acid component *C*-chiral ester enolate; (**49**) and (**50**). The (*E*/*Z*)-geometry of the enolate intermediate only has an influence on the diastereoselectivity of the alkylation in the case of *O*-chiral species. There are only a few examples for *O*-chiral ester enolates. One is the camphor–diol-derived type shown in Scheme 25 where the sense of the asymmetric induction may be directed into both configurations A and B depending on the conditions of the enolate formation ⟨81AG(E)207⟩. Without HMPA the (*Z*)-enolate is formed, and with HMPA the (*E*)-enolate. As the alkylating component approaches the enolate from the outside direction, this leads to the formation of A and B, respectively.

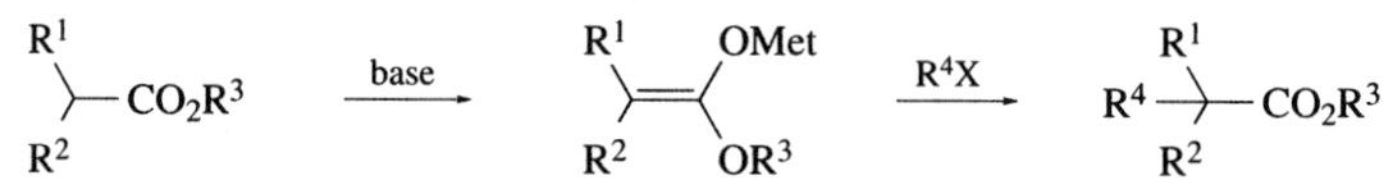

(*E*/*Z* mixture)

base: LDA, LICA, K-bis(TMS)amide
R^4X = MeI, BnBr, $MeO_2C–CH_2Br$, primary alkyl bromides, iodides, tosylates

R^1	*R^2*	*R^3*	*R^4–X*	*Yield (%)*
H	H	Bu^t	Bu^nI	85
H	Bu^n	Et	Pr^iI	50
Me	Me	Et	MeI	87
H	Et	Me	MOM-Cl	98
H	Et	Me	HC≡C–CH2Br	97

Scheme 24

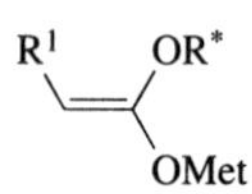

(*Z*)-(**47**)

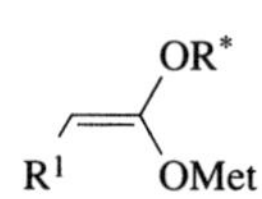

(*E*)-(**48**)

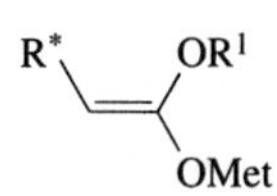

(*Z*)-(**49**)

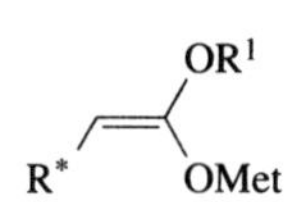

(*E*)-(**50**)

In Equation (130) a chiral malonate enolate is generated and submitted to a cycloalkylation forming a cyclopropane diester with high chiral induction ⟨80AG(E)1029⟩. A variation of the *O*-chiral ester enolate is the *N*-chiral amide, for example in the form of the very efficient Evans oxazolidinone (Scheme 26) ⟨81PAC1109⟩. The examples in the table show that normal alkylations proceed with reasonable inductions. The chiral auxiliary is removed with the desired Li-alkoxide to form the α-branched ester. Scheme 27 and Equations (131) ⟨79HCA2825⟩, (132) ⟨79HCA2829⟩, (133), (134) ⟨79HCA2825⟩, (135) ⟨80HCA1383⟩ and (136) ⟨81AG(E)971⟩ show examples of the *C*-chiral type. Particularly efficient with respect to diastereocontrol in the alkylation step is a β-hydroxyl function which is deprotonated first so that a chelated enolate dilithium species (**51**) is created (Scheme 27). This enolate is alkylated with high diastereoselectivity from the less-hindered side to form the *anti*-ester (**52**). Equations (131)–(135) show some examples, and in Equation (136) this principle is extended to an NH function. δ-Hydroxy-esters work similarly well (Equation (137)) ⟨86CL81⟩.

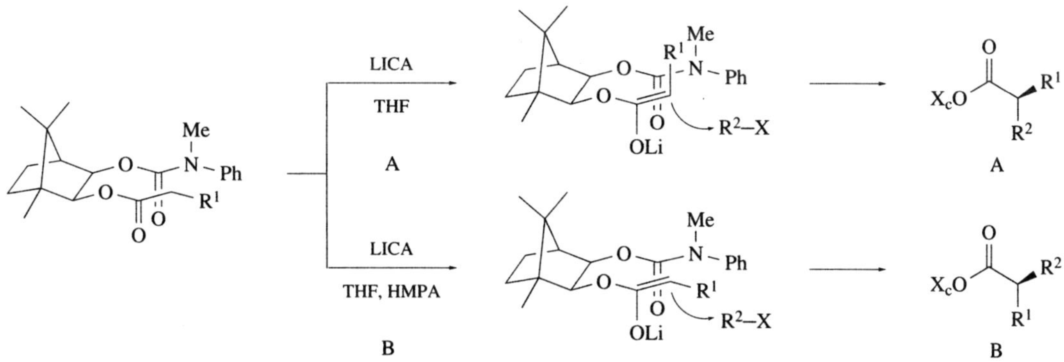

LICA = Lithium isopropylcyclohexylamide

R^1	*Solvent*	R^2X	*Yield* (%)	*Ratio A*:*B*
Bn	A	MeI	95	95 : 5
Me	A	BnBr	96	94 : 6
Me	B	BnBr	96	30 : 70
Bn	A	Bu^nI	93	90 : 10
Bn	B	Bu^nI	36 (90)	15 : 85
Me	A	Bu^iI	78 (90)	93 : 7
Me	B	Bu^iI	60 (90)	28 : 72

Scheme 25

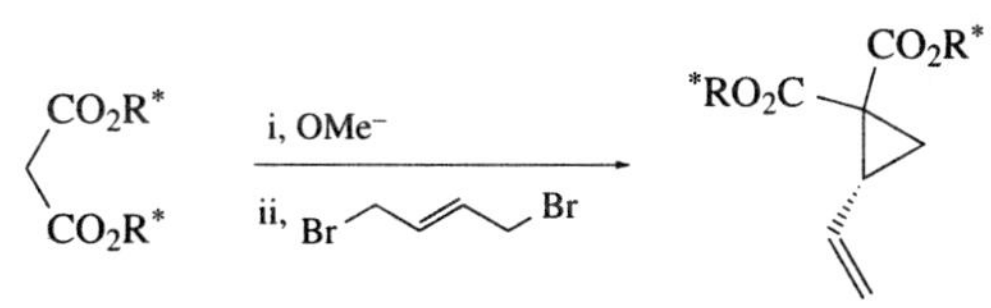

(130)

R* = 8-phenylmenthyl *ee* 76–80%

α-Chiral esters via Evans' amide

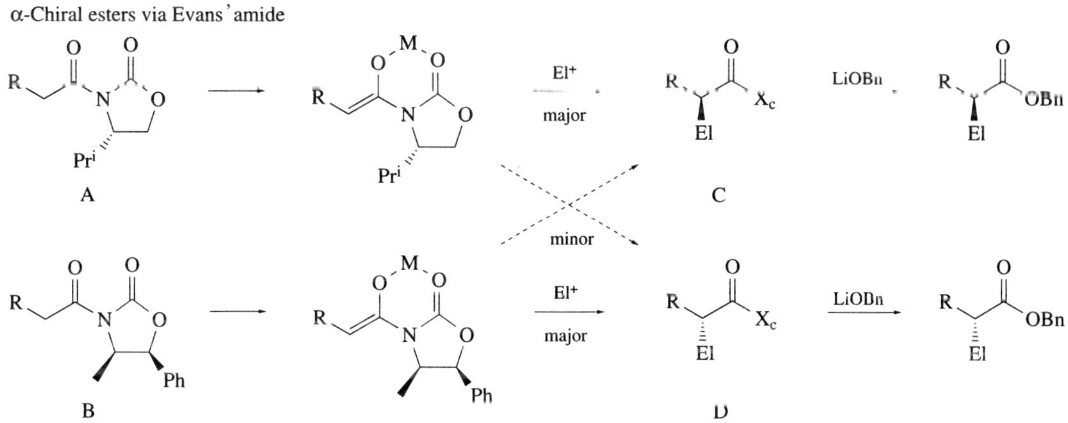

Imide	*El*	*Ratio C* : *D*[a]	*Yield* (%)
A (R = Me)	BnBr	99 : 1	92
B (R = Me)	BnBr	2 : 98	78
A (R = Et)	MeI	90 : 10 (91 : 9)	79
B (R = Et)	MeI	13 : 87 (7 : 93)	82
A (R = Me)	EtI	96 : 4	53
B (R = Me)	EtI	12 : 88 (4 : 96)	75
A (R = Me)	allyl-Br	98 : 2	75
B (R = Me)	allyl-Br	2 : 98	62

[a] Na-enolates are in parentheses

Scheme 26

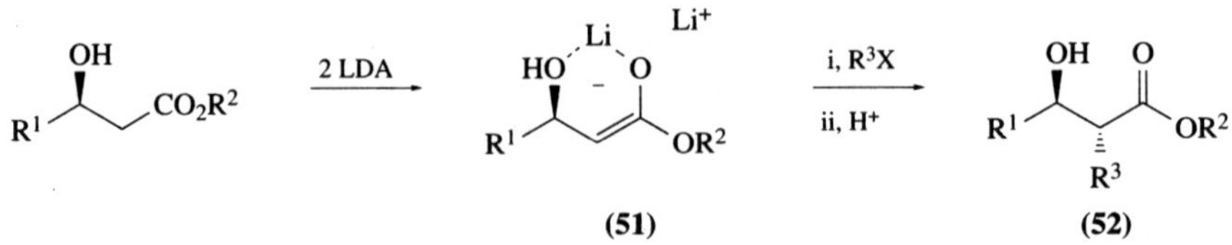

(51) **(52)**

Scheme 27

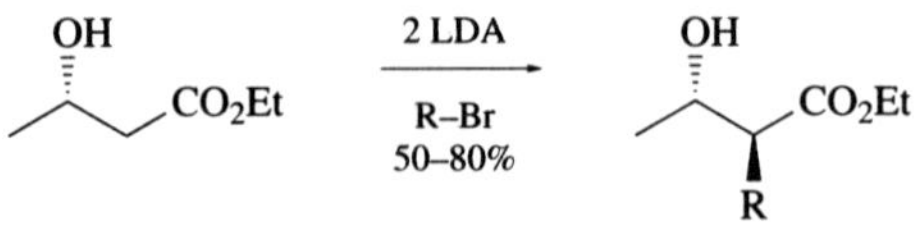

(131)

R–Br	*Ratio*
Allyl–Br	96 : 4
Bu^tBr	97 : 3

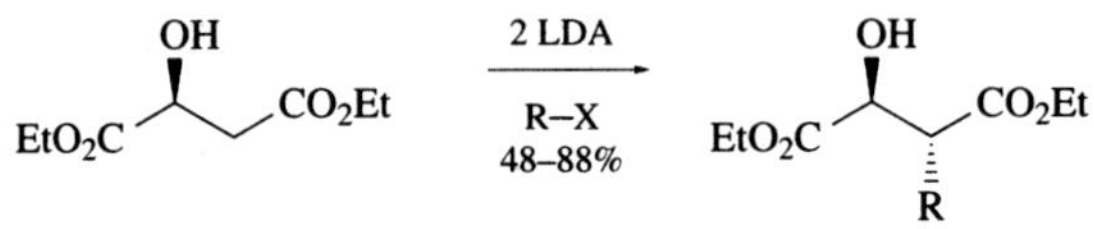

(132)

R–X	*Ratio*
MeI	>91 : 9
Allyl–Br	>95 : 5
BnBr	>91 : 9

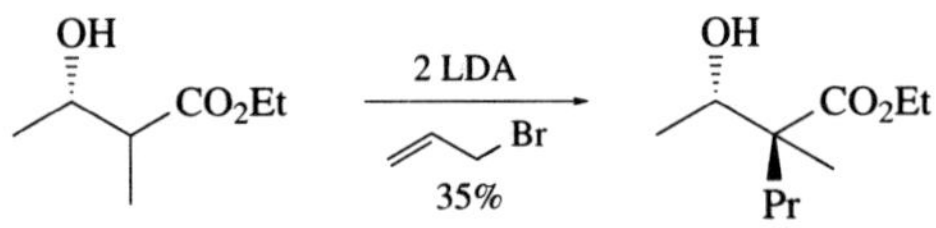

91 : 9 (133)

96 : 4 (134)

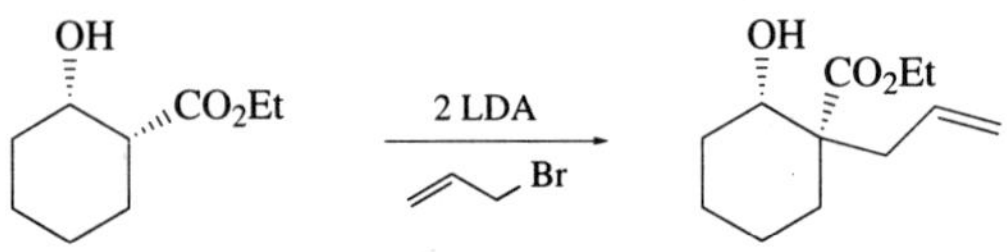

(135)

94.5 : 5.5

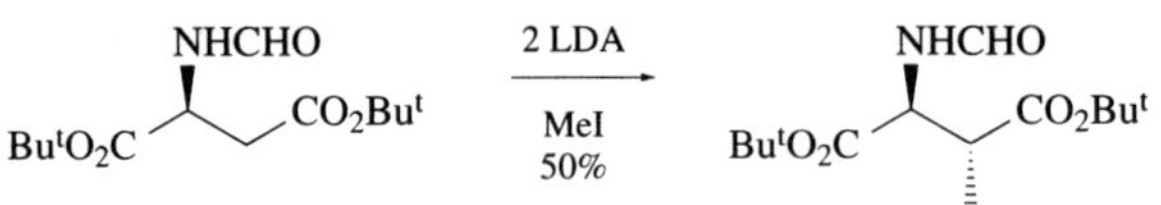

(136)

one isomer

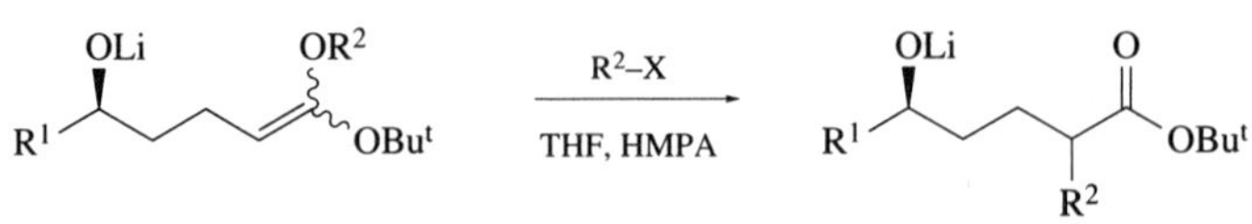

(137)

Scheme 28 and Equations (138)–(140) show *C*-chiral ester enolates which are incorporated into small rings; in Scheme 28 and Equations (138) and (139) this principle is combined with the presence of oxygen substituents which may be used for chelation of the lithium cation. Noteworthy in the examples shown in Scheme 28 ⟨83CB3413⟩ and Equation (138) ⟨82JA1735⟩ is the presence of a β-leaving group. However, due to stereoelectronic effects the oxygen-leaving group cannot be eliminated in Scheme 28 and due to ring strain effects this cannot happen in Equation (138). Enolate **(53)** is alkylated *syn* to the β-ethyl group, and in Equation (138) the alkylation occurs *anti* to the R^1-group. In Equation (139) the high stereoselectivity of the alkylation cannot be attributed to any chelating effect ⟨88JA4640⟩. Rather, the overall conformation of the ring system must be responsible. In Equation (140) it becomes clear that in conformationally anchored cyclohexane systems there is a preference for the equatorial alkylation of the enolate ⟨87JA2426⟩.

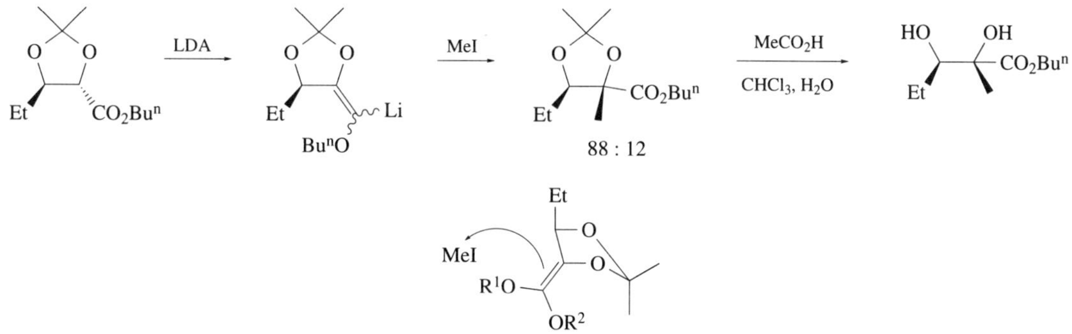

(53)

Scheme 28

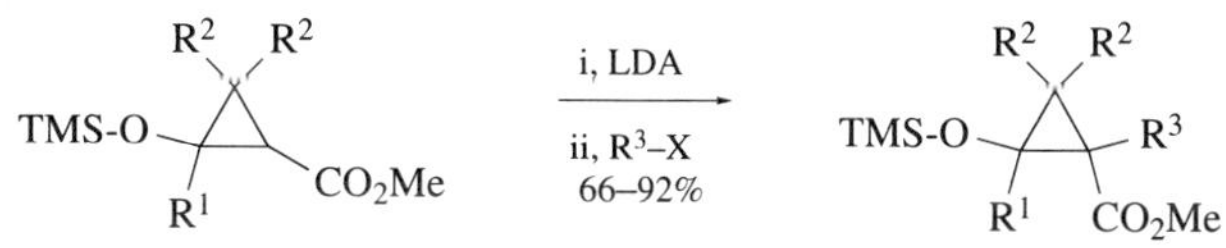

(138)

ds = 65 : 35 – 97 : 3

R^1 = Ph, Bu^t, CH_2=CH; R^2 = H, Me; R^3–X = MeI, Bu^nI, EtI, allylbromide, benzylbromide

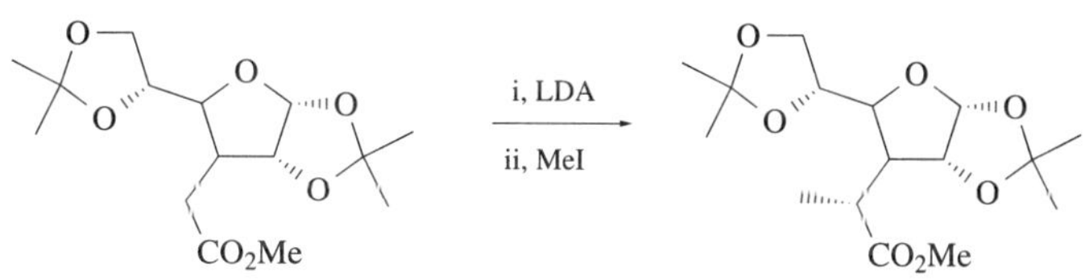

>99 : 1 (139)

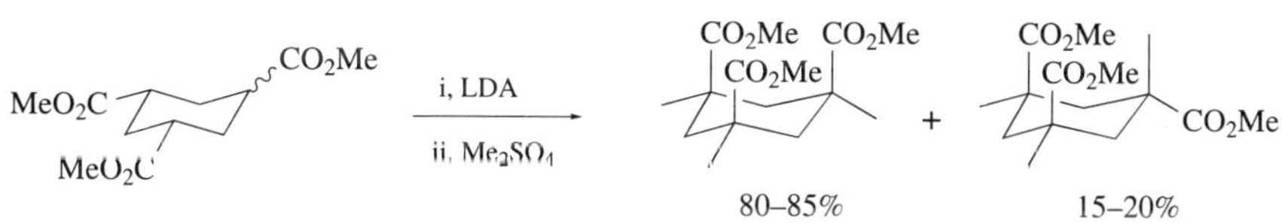

(140)

5.03.5.2.2 Conjugate addition to α,β-enoates

Organocopper derivatives add to acrylic esters with a wide variety of substituents (Equations (141) ⟨86ACS(B)466⟩, (142) ⟨87OM255⟩, (143) ⟨82T1509⟩, and (144) ⟨90T46⟩). The organometallic intermediate has enolate character and may thus be trapped with a carbon electrophile to form an α,β-dibranched ester in tandem fashion. The electrophile may be an alkyl halide or a second acrylate (Equation (145)) ⟨71HCA1939⟩. If the electrophile is an intramolecularly attached ester function, stereoselective cyclization to a cyclopentanone system occurs (Equation (146)) ⟨86JOC3376⟩.

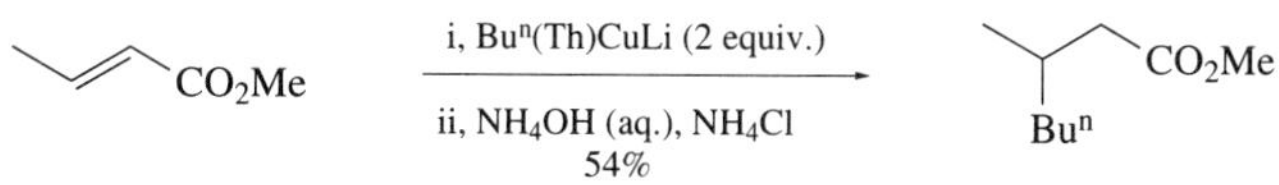

(141)

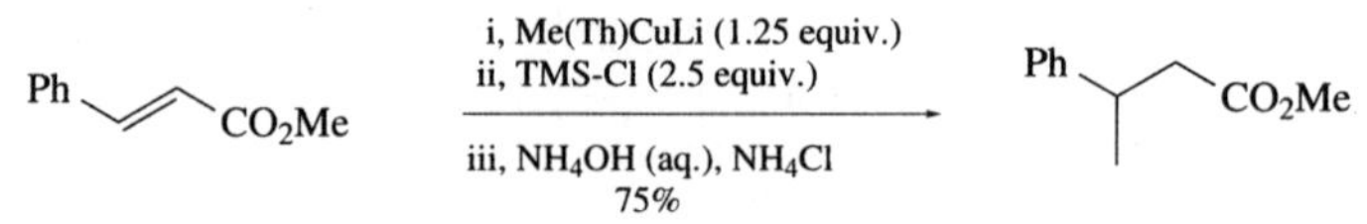

(142)

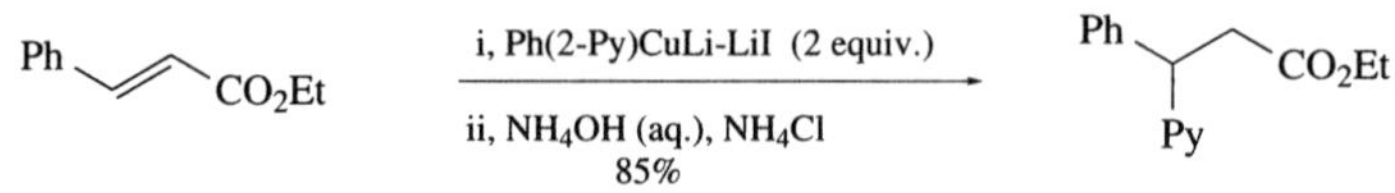

(143)

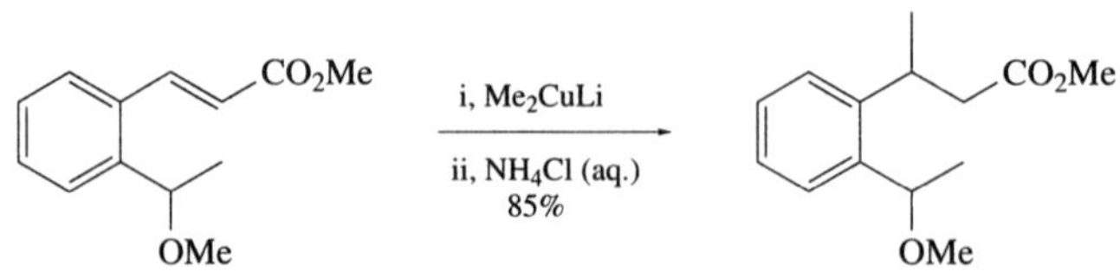

(144)

Enolate trapping reactions in the addition of organocopper reagents to alkenoic acid derivatives

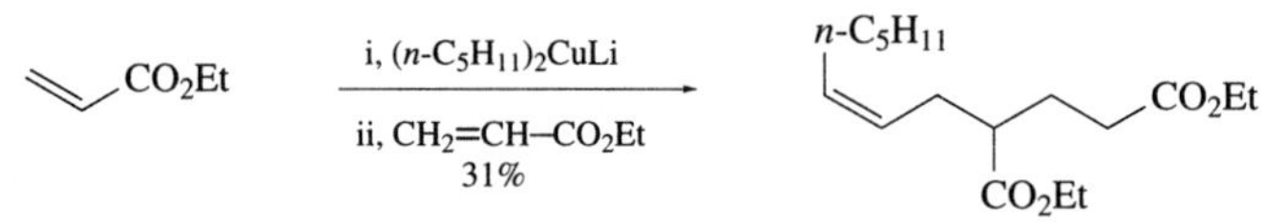

(145)

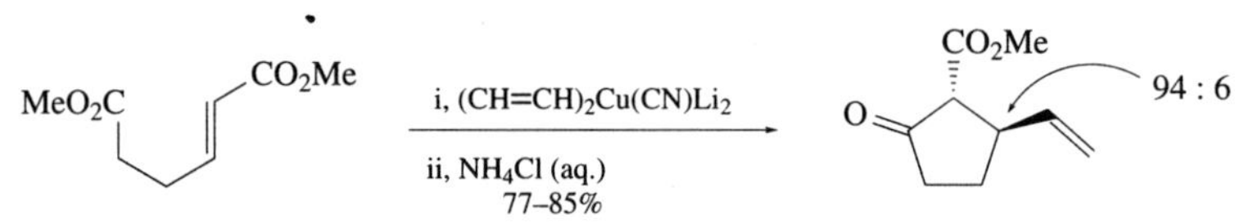

(146)

The stereochemistry of the cuprate addition may be controlled via a γ-substituent (preferably oxygen or nitrogen) or via a chiral alcohol or amine substituent in the ester (or amide). Equations (147)–(149) ⟨83TL2231, 86T2863, 87CC1523, 89CC31⟩, (150) ⟨83TL2231, 89JA2984⟩, (151), and (152) ⟨89AG(E)1706⟩ demonstrate the potential of an *O*-benzyl or *N,N*-dibenzyl ligand vicinal to the electrophilic site of the substrate. A different approach is shown for cinnamic esters in Equations (153) and (154) ⟨88TL4411⟩ where the stereocontrol is exerted by a chiral auxiliary in the *ortho*-position of the benzene ring. The diamine in Equation (154) works exceptionally well and can be easily removed by silica gel to form the aldehyde. Chirally substituted esters of the 8-phenylmenthyl type (**54a–e**) have been shown to add cuprates with high stereoselectivity (Scheme 29) ⟨81HCA2808⟩. In terms of the *ee* of the addition, the (*E*)-esters (**54a, b,** and **e**) are superior to the (*Z*)-isomers (**54c** and **d**). More conventional additions of enamines and enol esters to acrylates are shown in Equations (155)–(157) ⟨B-87MI 503-01, 87TL2087⟩. Equations (158) ⟨88JA1631⟩, (159) ⟨87OS(65)236⟩, (160) ⟨88TL5369⟩, (161) ⟨83JA7200⟩, and (162) ⟨83T2863⟩ show the conjugate addition of carbon-free radicals to α,β-enoates. These reactions proceed under mild conditions and show reasonable stereoselectivity, for example in the form of the α-C-glucoside derivative in Equation (159).

Stereoselective addition of organocopper reagents to γ-alkoxyalkenoates

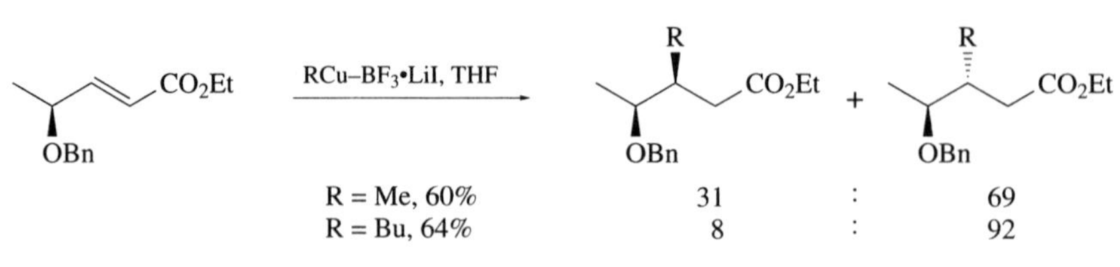

(147)

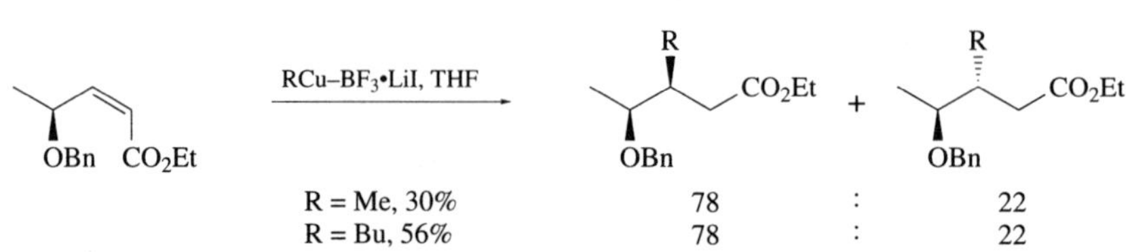

(148)

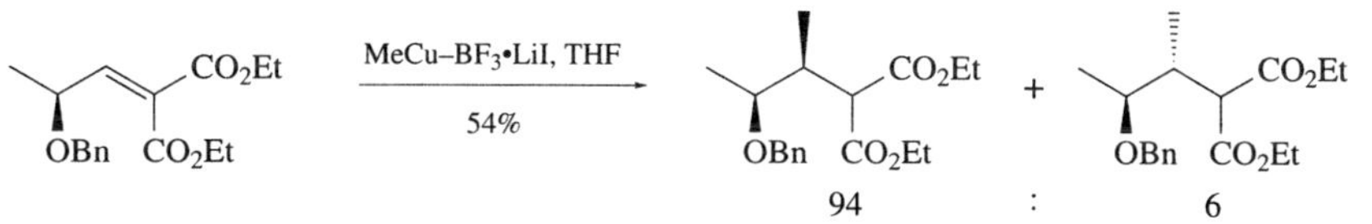
(149)

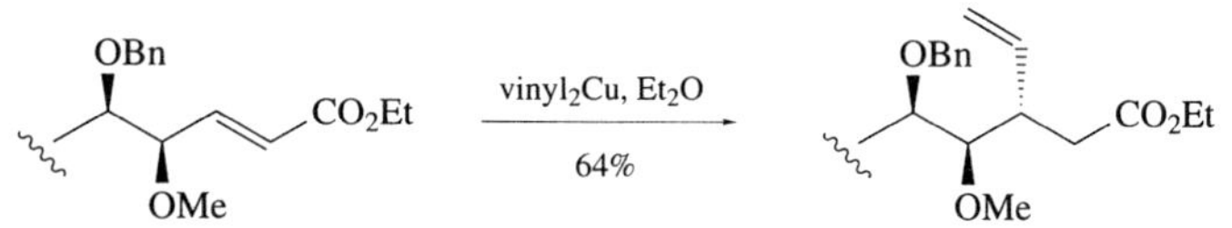
(150)

Stereoselective addition of organocuprates to protected γ-aminoalkenoates

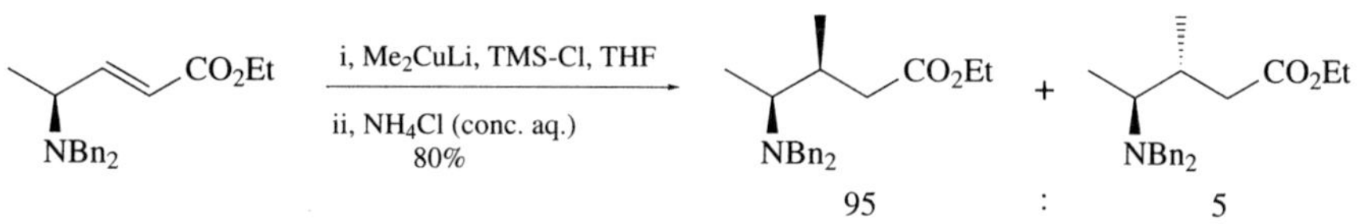
(151)

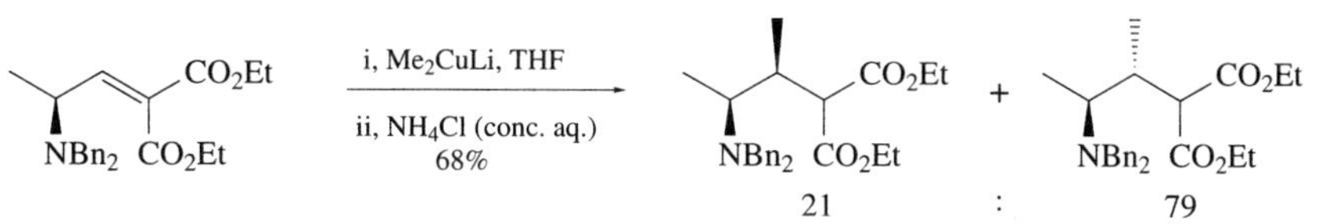
(152)

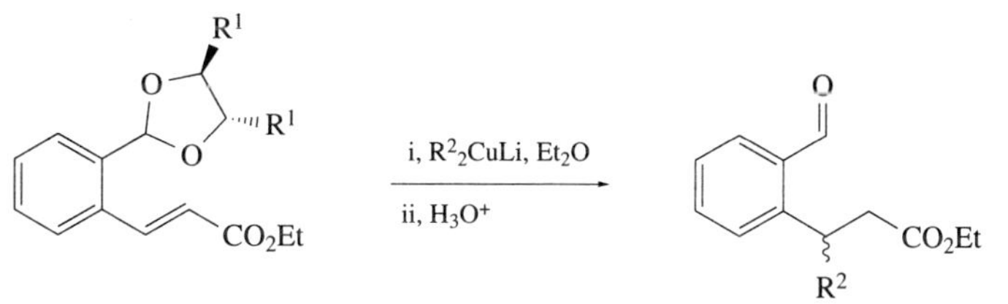
(153)

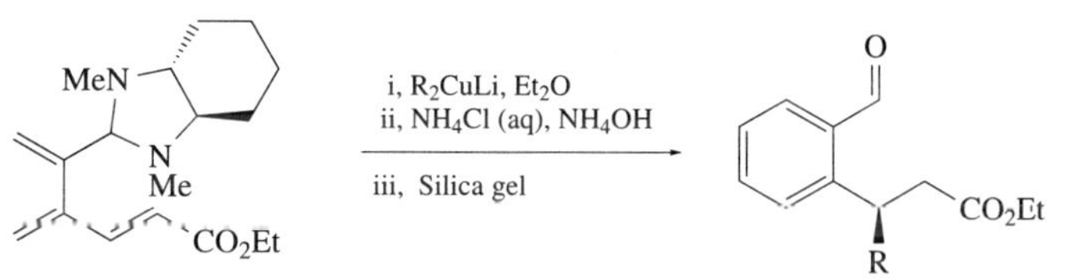
(154)

R	*Yield* (%)	*ee* (%)
Me	85	94
Bu	90	95
Ph	84	96
EtHC=CH	80	90

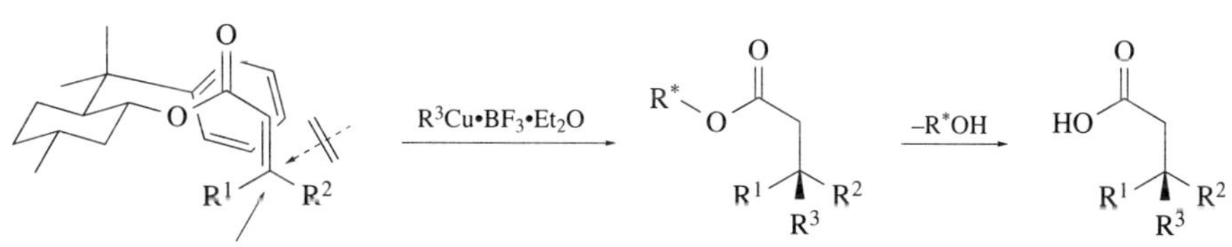

(54) **(55)** **(56)**

Entry	*R¹*	*R²*	*R³*	*Yield of* **(55)** (%)	*ee of* **(56)** (%)
a	H	Me	Ph	76	>99
b	H	Me	Bu^n	75	>99
c	Me	H	Ph	36	24
d	Me	H	Bu^n	76	70
e	H	Bu^n	Me	28	78

Scheme 29

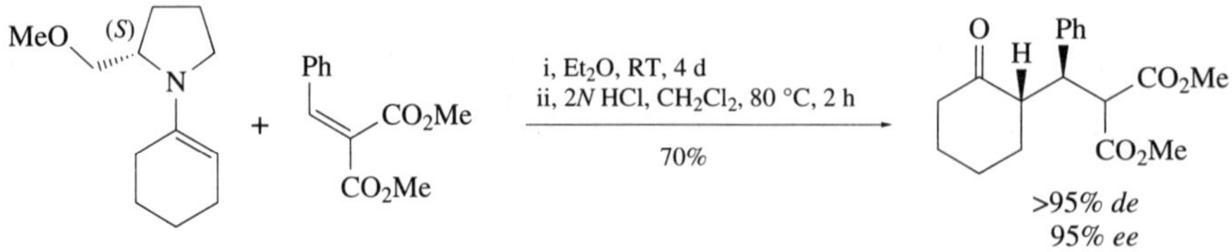

(155)

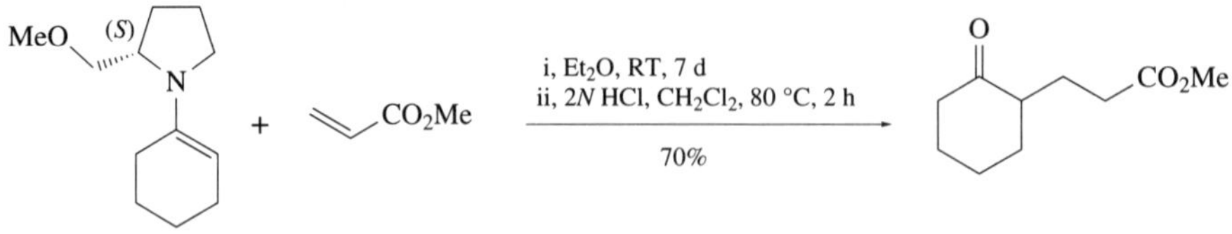

(156)

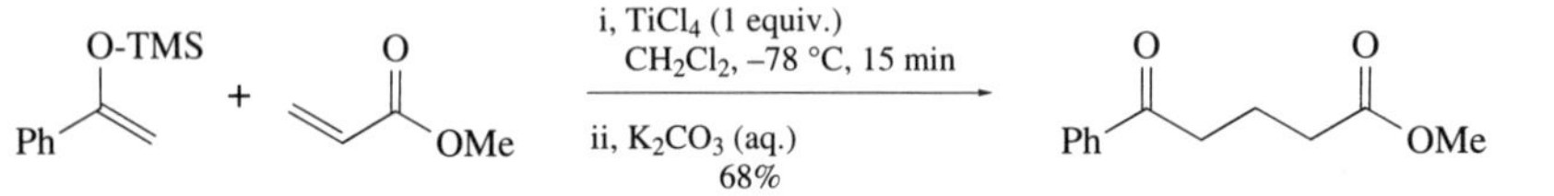

(157)

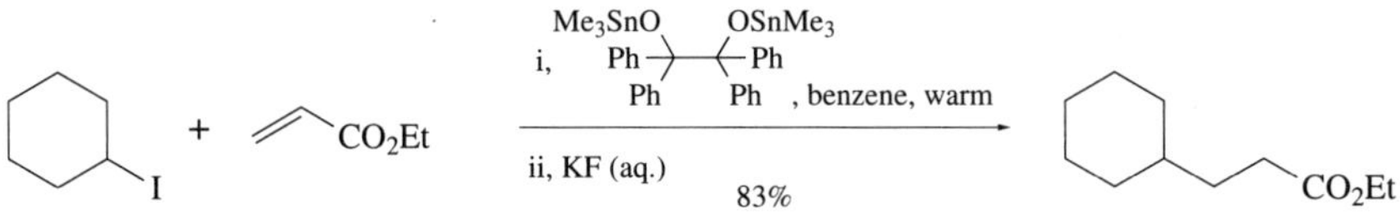

(158)

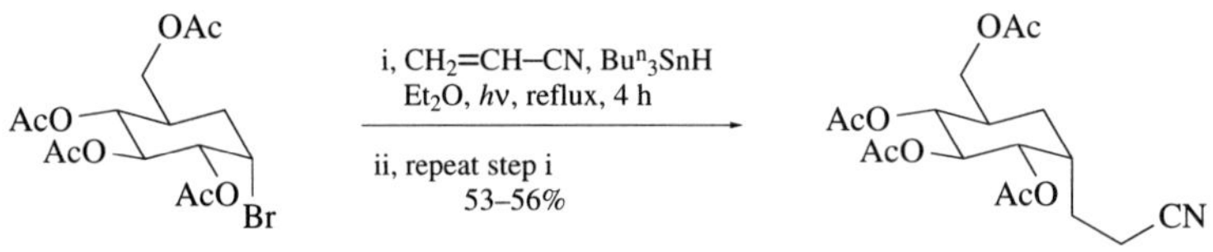

(159)

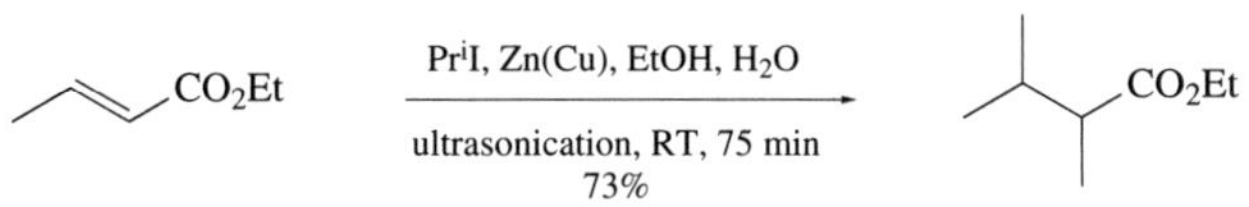

(160)

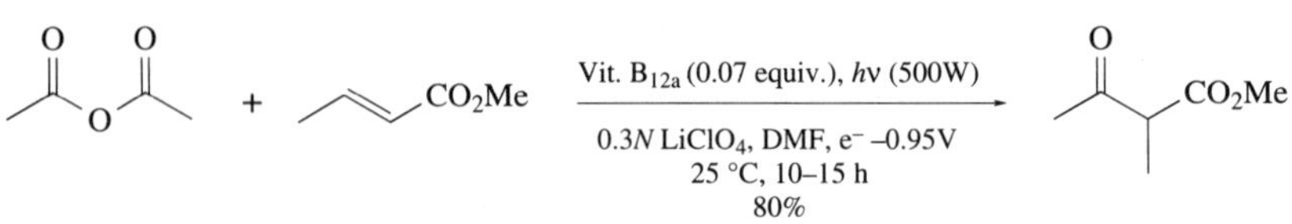

(161)

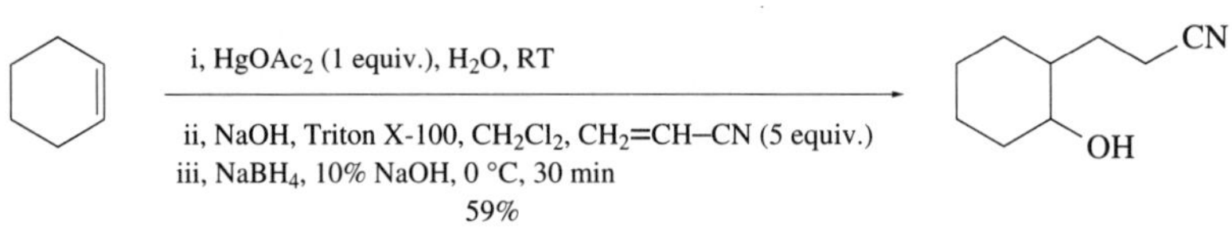

(162)

5.03.5.3 *α,β*-Unsaturated Esters

5.03.5.3.1 α,β-Unsaturated esters via redox reactions

Cyclic esters may be dehydrogenated via the enolate to form *α,β*-unsaturated esters (Saegusa reaction, Equation (163)) ⟨86T2971⟩. The direct oxidation of allylic alcohols to enoates may be effected by Corey's method as outlined in Scheme 30 ⟨68JA5616⟩. This procedure is particularly mild and has found application in the synthesis of acropia juvenile hormones. Alkynoic esters (**57**) are prepared by the addition of lithium propynoate to aldehydes followed by OH-protection. Ester (**57**) is transformed into (*Z*)-(**58**) by Lindlar hydrogenation and quantitative (*Z*)-(*E*)-isomerization to (*E*)-(**58**) can be achieved by reversible addition of thiolate (Scheme 31) ⟨84AG(E)143⟩.

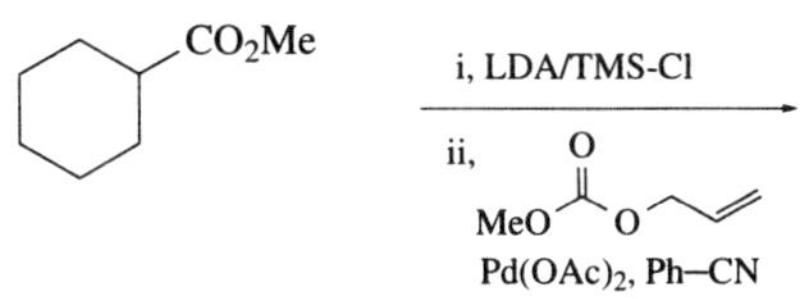

(163)

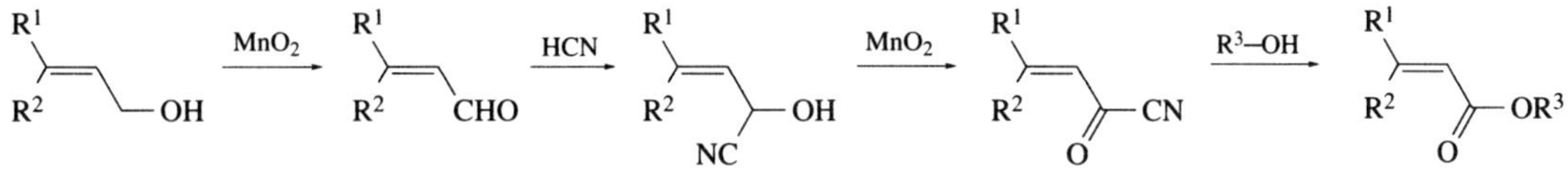

Scheme 30

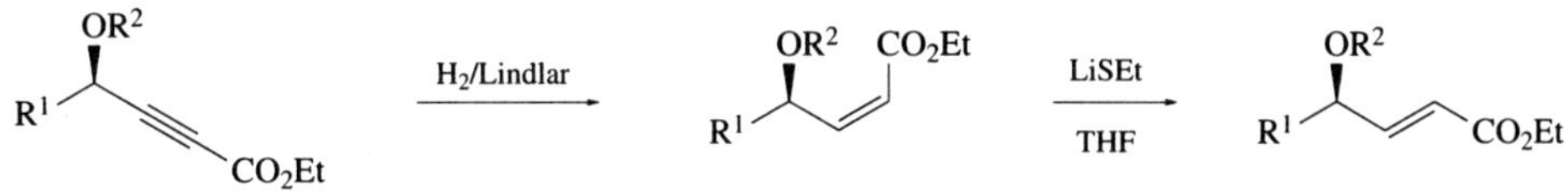

(57) a; R^1 = c-C_6H_{11}; R^2 = MEM (*Z*)-**(58)** (*E*)-**(58)**
b; R^1 = c-C_6H_{11}; R^2 = THP
c; R^1 = c-C_6H_{11}; R^2 = TBDMS
d; R^1 = c-C_6H_{11}; R^2 = Bz
e; R^1 = n-C_8H_{17}–CH=CH–$(CH_2)_7$(*E*); R^2 = TBDMS
f; R^1 = n-$C_{17}H_{35}$; R^2 = TBDMS

Scheme 31

5.03.5.3.2 α,β-Unsaturated esters via alkenation

Wittig or Horner alkenations are extremely useful and versatile for the synthesis of α,β-enoates. The geometry of the double bond can be controlled by an appropriate choice of the reagents and the solvent. The solvent effect is shown for the Wittig reaction in Equations (164) ⟨83AG(E)63⟩ and (165) ⟨79HCA2091⟩. The substrate also plays a major but not predictable role as demonstrated by Equations (166) and (167) ⟨88TL6823⟩. The Horner reaction shows a reliable (*E*)-selectivity in the formation of disubstituted alkenes (Equations (168) ⟨79JA259⟩ and (169) ⟨89TL4845⟩). (*Z*)-Alkenations can be achieved with Still's fluorinated phosphonate (Equation (170)), which is also extendable to trisubstituted α,β-enoates (Equation (171)). Equation (172) shows how the one and the same ketone may be alkenated (*E*)- or (*Z*)-selectively by an appropriate choice of the phosphonate ⟨84TL2183⟩.

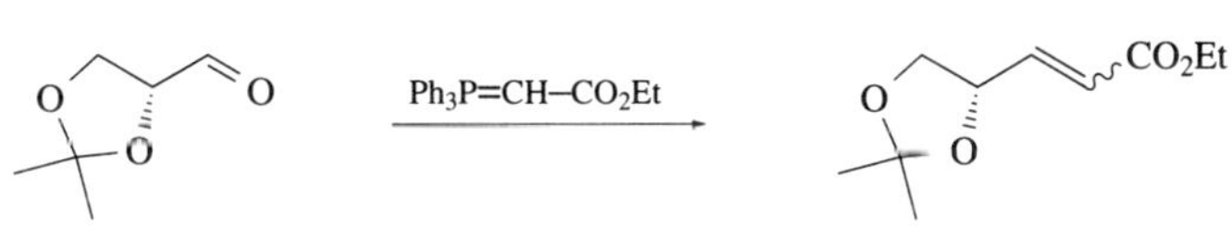

(164)

Solvent	(*E*) : (*Z*)
CH_2Cl_2	1 : 1
CH_2Cl_2/HOAc	>95 : 5
MeOH	8 : 92

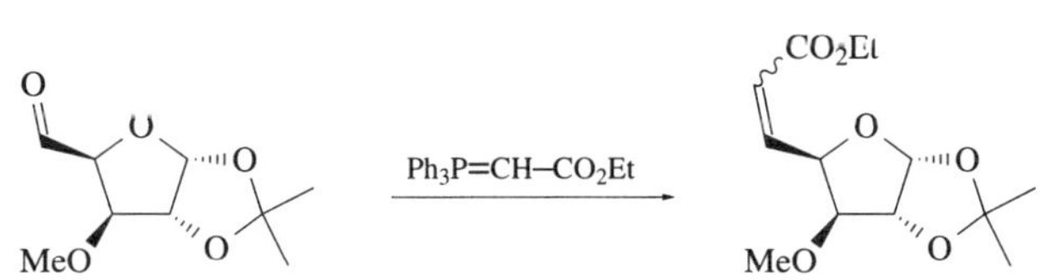

(165)

Solvent	(*E*) : (*Z*)
DMF	86 : 14
$CHCl_3$	40 : 60
MeOH	8 : 92

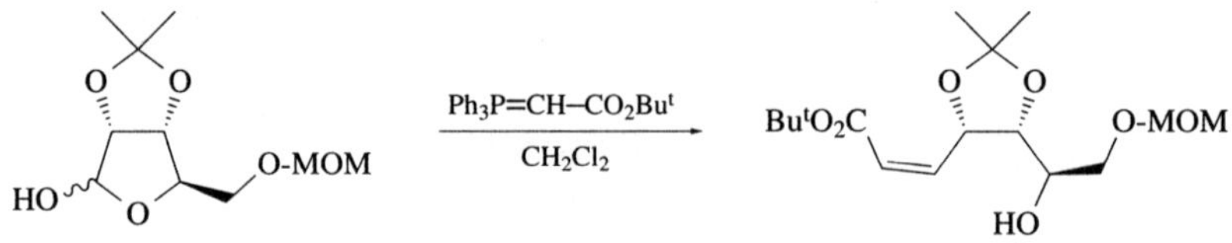

(166)

(*E*) : (*Z*) 1 : 10

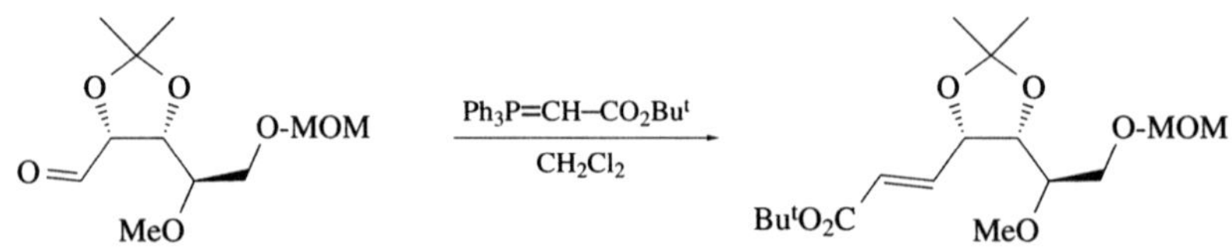

(167)

(*E*) : (*Z*) 19 : 1

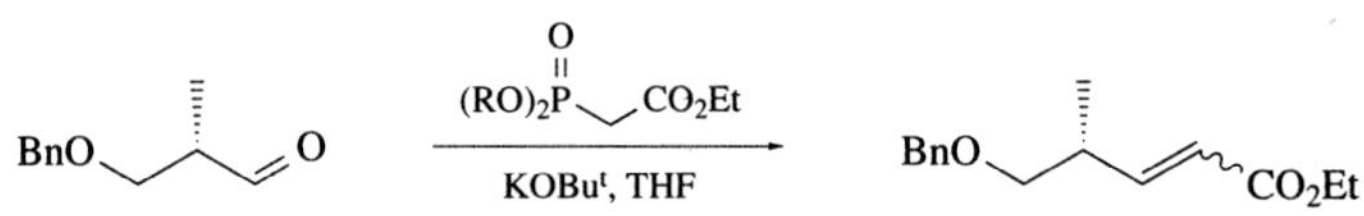

(168)

$R = Pr^i$ (*E*) : (*Z*) 95 : 5
R = Me (*E*) : (*Z*) 1 : 3

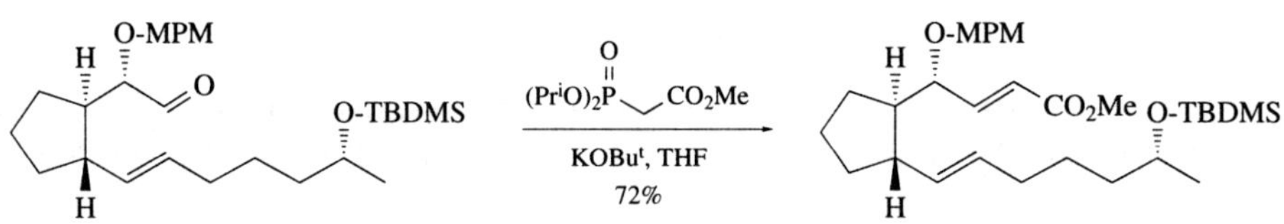

(169)

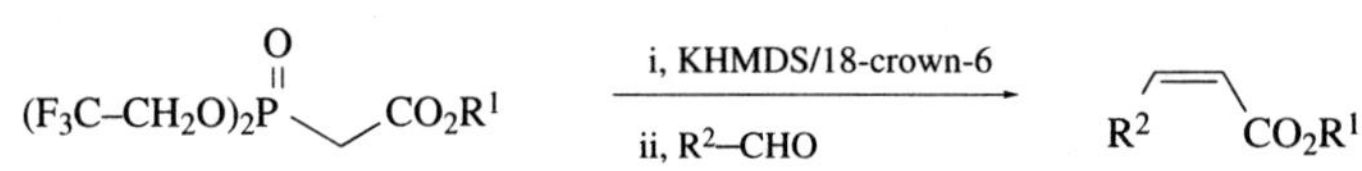

(170)

R^1 = Me, Et

R^2–CHO	*Yield (%)*	(*Z*) : (*E*)	*Ref.*
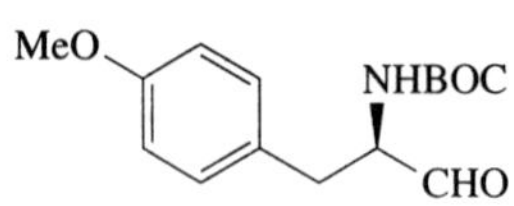	>80	all-(*Z*)	88TL4419 87CC311
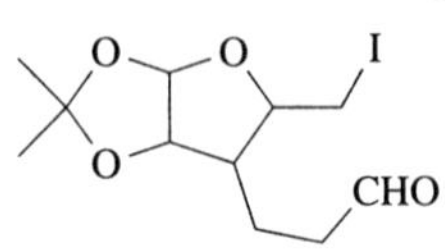	75	all-(*Z*)	89CC1160
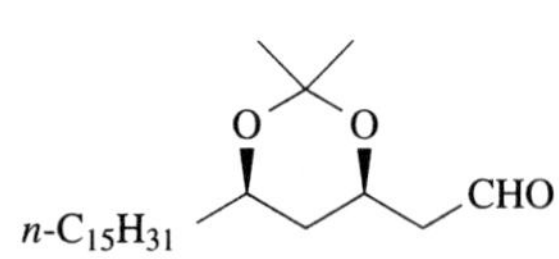	77	91 : 9	89TL6529
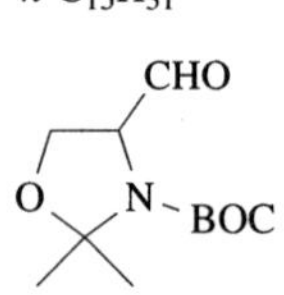	>87	>65 : 35	89TL3799

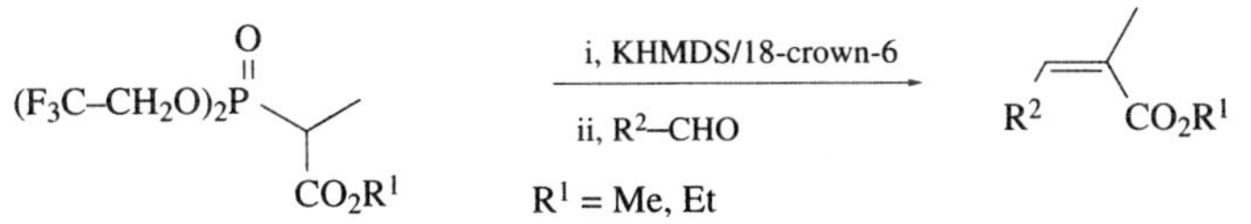

(171)

R^2–CHO	*Yield (%)*	*(Z) : (E)*	*Ref.*
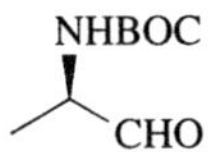	68	99 : 1	88H2077
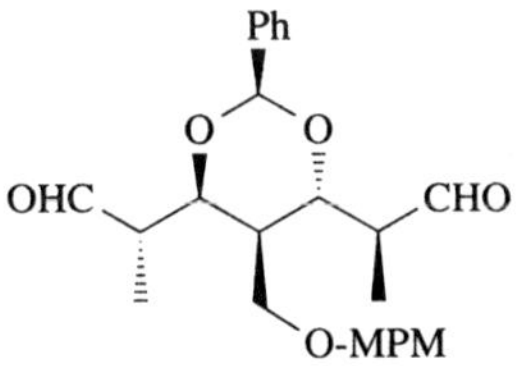	74	99 : 1	88TL4085

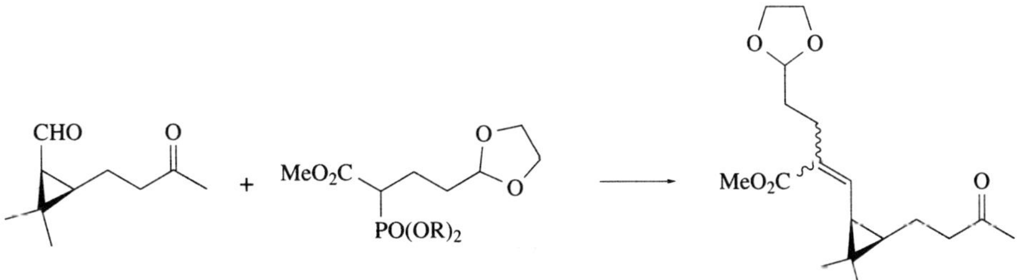

(172)

R = Me; LiHMDS, DME, 100%; (*E*) : (*Z*) = 86 : 14
R = CF_3CH_2; dbu, LiCl, MeCN, 100%; (*E*):(*Z*) = 25 : 75

5.03.5.3.3 α,β-Unsaturated esters via Heck coupling or alkoxy carbonylation

The Heck reaction is a reliable method for the synthesis of cinnamic esters with (*E*)-configuration (Equation (173)) ⟨77JOC3903, 77TL1037, 78JOC2952, 79JCS(P1)771, 79TL1133⟩. In Equation (174) the double bond migrates due to aromatization ⟨77JOC3903, 77TL1037, 78JOC2952, 79JCS(P1)771, 79TL1133⟩. Acetylenes can be hydroborated and then submitted to an alkoxycarbonylation to form (*E*)-enoates (Equation (175)) ⟨81CL879⟩.

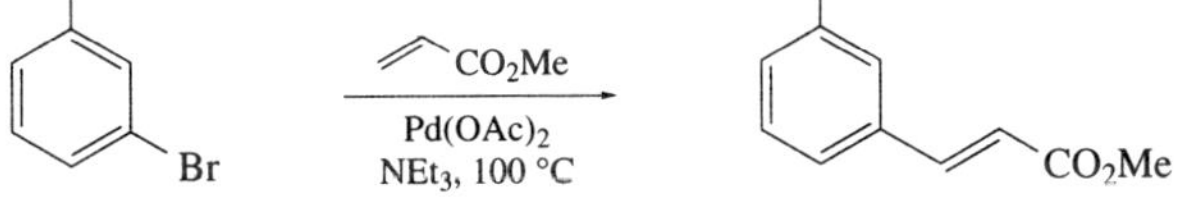

(173)

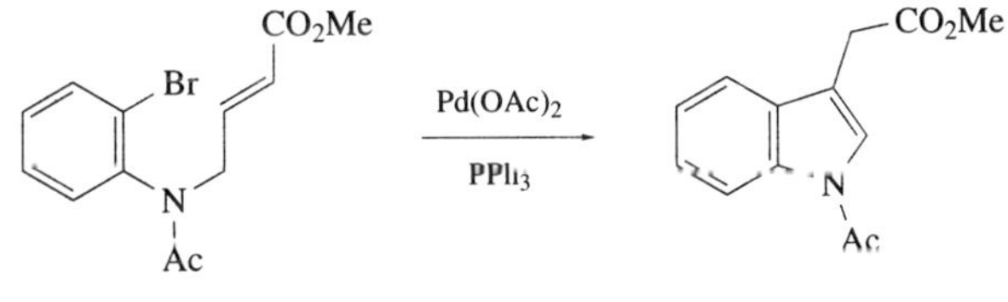

(174)

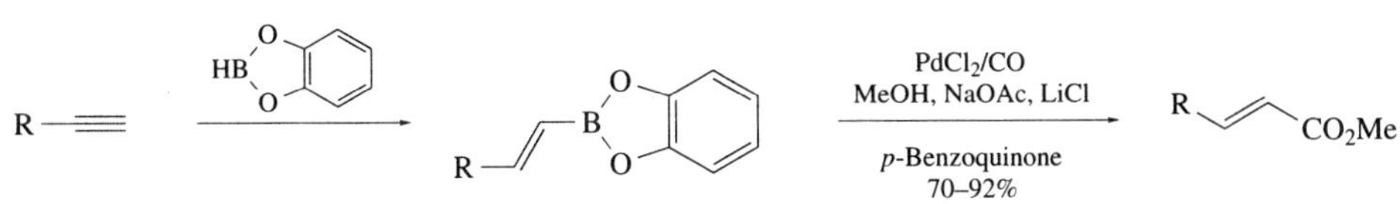

(175)

R = Bu^n, *n*-C_6H_{13}, Ph

5.03.5.4 β,γ-Unsaturated Esters

5.03.5.4.1 β,γ-Unsaturated esters via alkoxycarbonylation

Under normal carbonylation conditions, allylic acetates and dienes may be converted into β,γ-enoates as shown in Equations (176) ⟨64TL605⟩, (177) and Scheme 32 ⟨77BCJ553⟩.

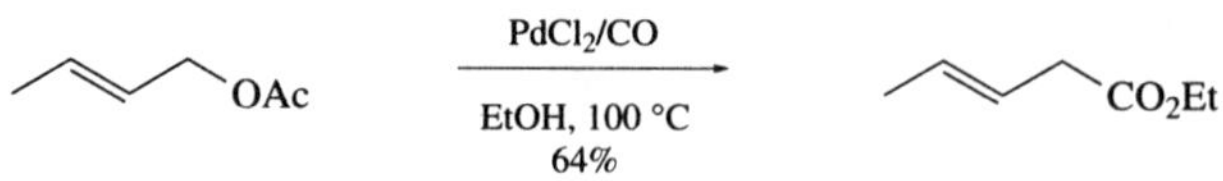

(176)

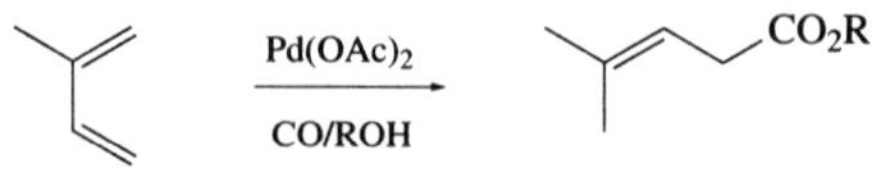

(177)

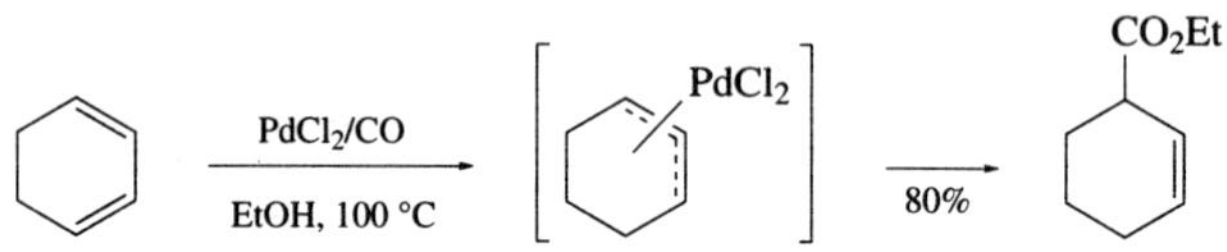

Scheme 32

5.03.5.4.2 β,γ-Unsaturated esters via S_N2' reaction

γ,δ-Dihydroxy-α,β-enoates in the form of the γ-mesylate undergo a stereoselective S_N2' reaction with cuprates to form the β,γ-enoates stereoselectively. The configuration of the α-center is controlled via the geometry of the double bond in starting material (Equations (178) and (179)) ⟨89JA4864⟩.

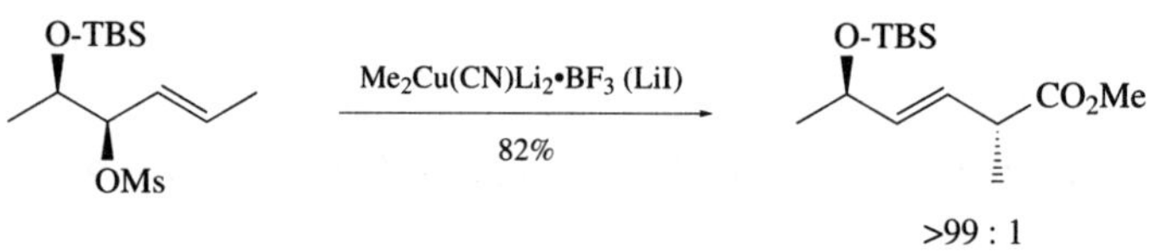

(178)

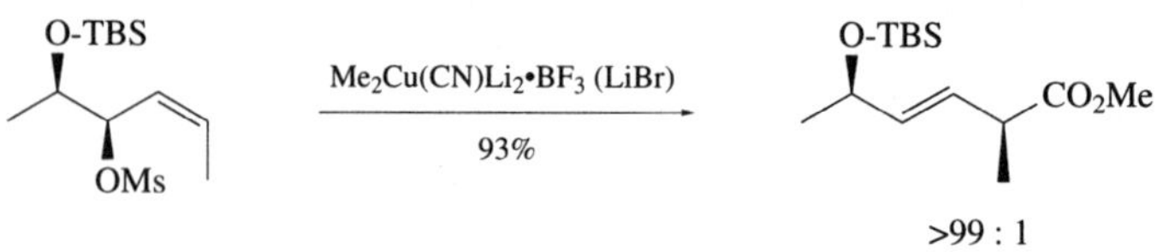

(179)

5.03.5.5 γ,δ-Unsaturated Esters

5.03.5.5.1 γ,δ-Unsaturated esters via Tsuji–Trost reaction

S_N2 Displacement reactions with malonate anions have been commonly used. Tsuji and Trost have developed a very mild variation of this reaction by adding malonate anions under Pd catalysis to allylic acetates. The reaction is versatile and compatible with many functional groups and proceeds with overall retention of configuration. However, frequently it is not regiocontrolled with respect to the allylic system (Scheme 33 and Equation (180)) ⟨77JA1649, 89AG(E)1173⟩.

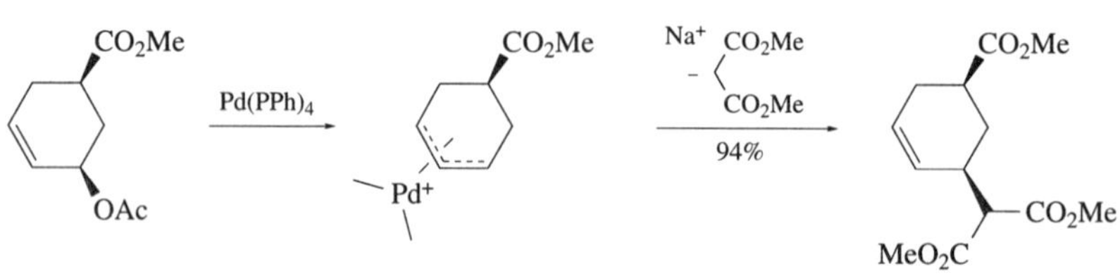

Scheme 33

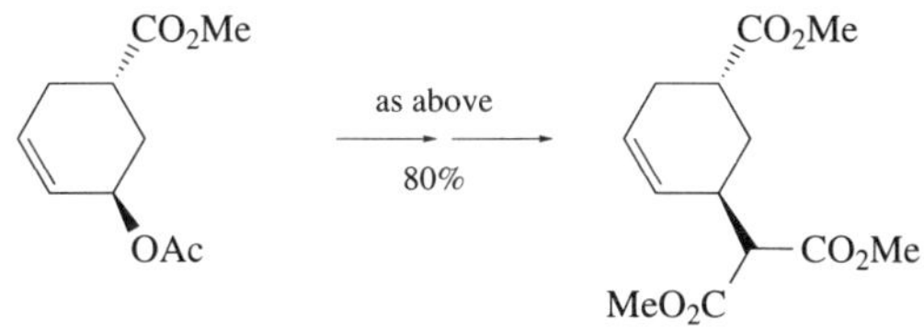

(180)

5.03.5.5.2 *γ,δ-Unsaturated esters via Claisen–Johnson-rearrangement*

This is the most efficient and stereochemically most reliable synthesis of γ,δ-unsaturated esters (see Equation (181) ⟨70JA741⟩ and Scheme 34 ⟨76JA1583⟩). The substrates are easily accessible; the *ortho*-ester and the allylic alcohol first form the mixed acetal which eliminates methanol to generate the enol ester which undergoes the rearrangement. The double bond in the γ,δ-unsaturated ester is created (*E*)-selectively, and the rearrangement proceeds suprafacially with respect to the allylic hydroxyl function 1,3-chirality transfer; Equation (182) ⟨88JOC1922⟩. The induced stereoselectivity is generally low (Scheme 35 ⟨88CC1531⟩ and Equation 183 ⟨88LA891⟩). Scheme 36 shows that an additional α,β-double bond may be inserted by a modification of the procedure ⟨90JOC3967⟩. Scheme 37 is a further example of this rearrangement ⟨80JOC2080⟩. Equation (184) demonstrates the applicability of the rearrangement to the propargylic system. The allenic esters are formed with varying diastereoselectivity ⟨88JOC4736⟩.

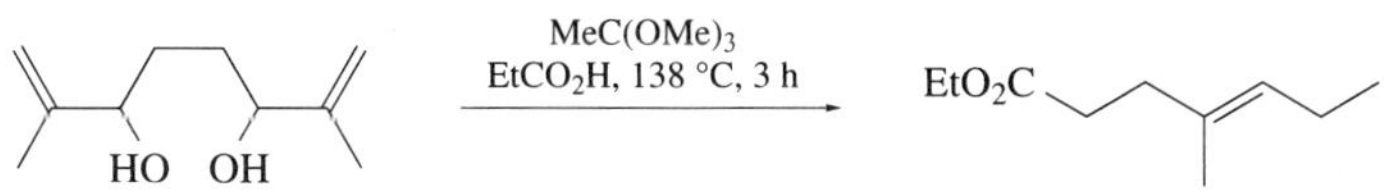

(181)

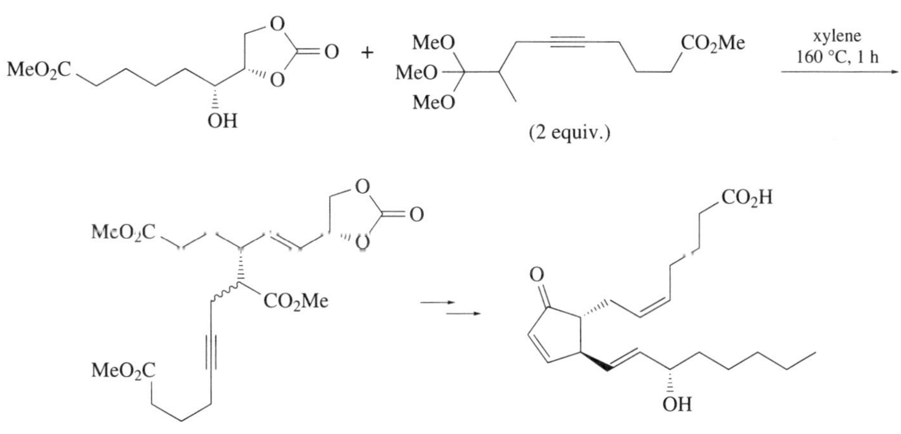

Scheme 34

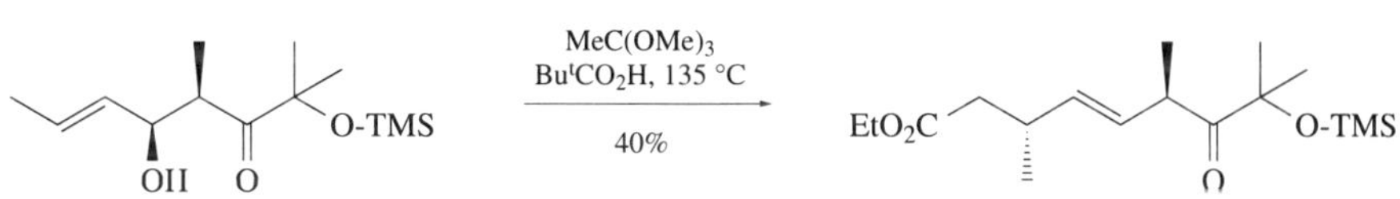

(182)

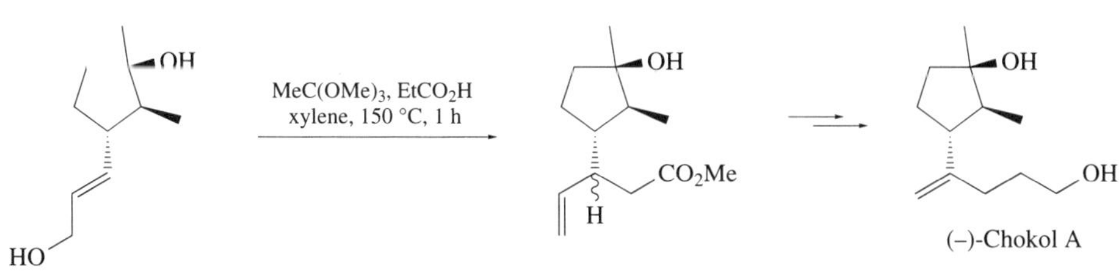

Scheme 35

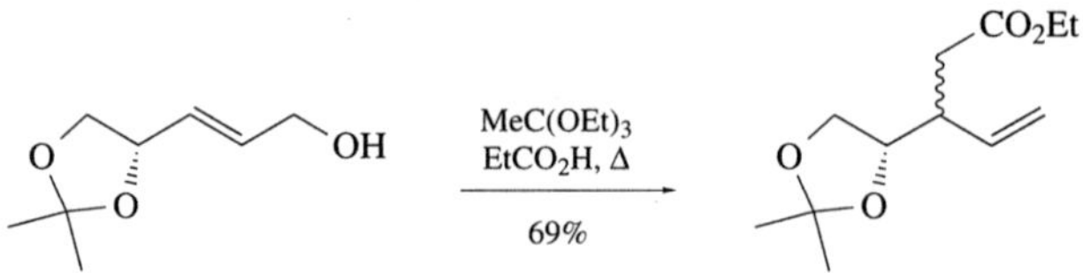

1 : 1 mixture

(183)

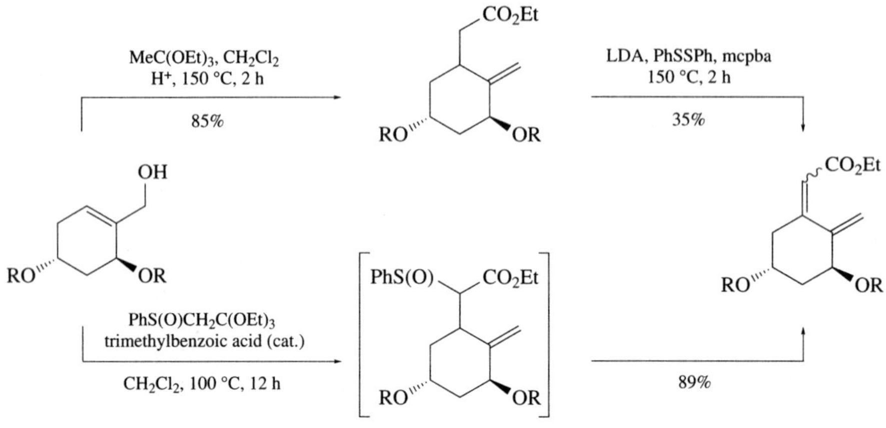

Scheme 36

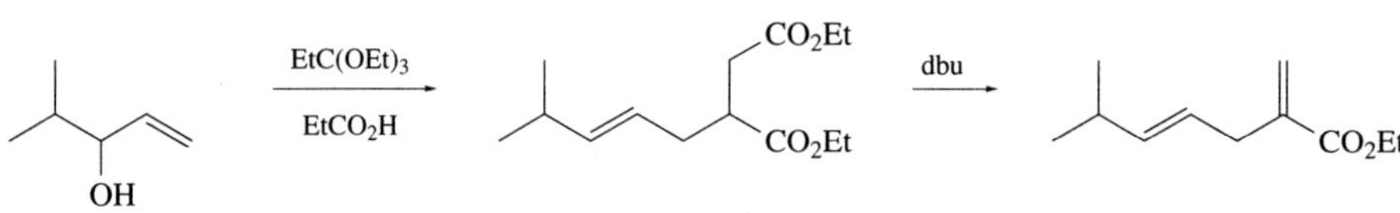

Scheme 37

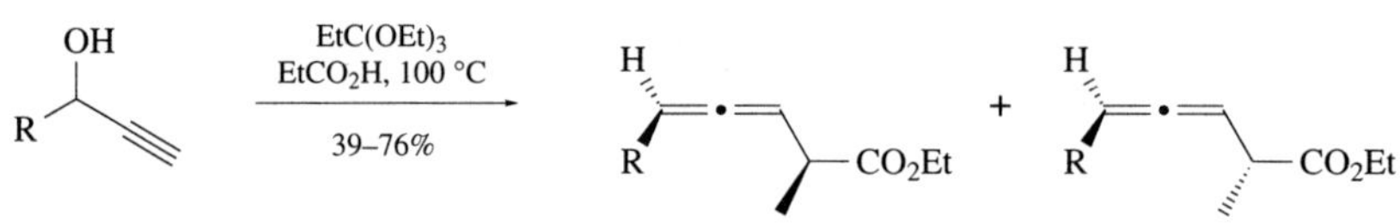

a b

(184)

R	*a* : *b*
Me	65 : 35
Pr^i	87 :13
Bu^t	89 : 11
cyclohexyl	82 : 18
Ph_3C	92 : 8
Ph	1 : 1
mesityl	1 : 1

5.03.5.6 α,β-Alkynoic Esters

Two one-carbon homologations are shown in Equation (185) ⟨80TL849⟩ and Scheme 38 ⟨89T363⟩. The first one is a conventional carbonylation reaction (Equation (185)) and the second one is the very useful Corey–Fuchs sequence which eventually converts an aldehyde into the alkynoate (Scheme 38).

α,β-Alkynoic esters via alkoxycarbonylation

$$R^1-\!\!\equiv \quad + \quad R^2OH \xrightarrow[\text{CuCl}_2\text{, NaOAc, RT; 59–74\%}]{\text{PdCl}_2\text{/CO (1 atm)}} R^1-\!\!\equiv\!\!-CO_2R^2 \qquad (185)$$

R^1 = Ph, n-C_5H_{11}, $PhOCH_2$
R^2 = Me, Pr^i

α,β-Alkynoic esters via Corey–Fuchs sequence

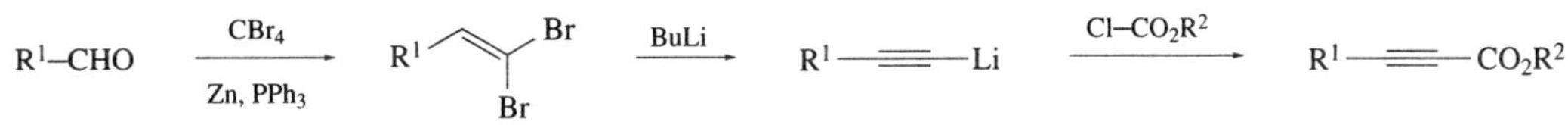

Scheme 38

5.03.6 LACTONES

In general lactones, except *β*-lactones, may be prepared from cycloalkanones via Baeyer–Villiger oxidation. Other methods are described in section 5.03.6.1.

5.03.6.1 *β*-Lactones

The smallest lactone rings which are synthetically useful are the *β*-lactones. There are extensive reviews on the application of *β*-lactones in synthesis ⟨93S441⟩. *β*-Lactones can be synthesized without changing the chain length by intramolecular S_N2 cyclization from *β*-bromocarboxylic acids (Equation (186)) ⟨50JA2998⟩ or by kinetically controlled iodolactonization of *β*,*γ*-unsaturated carboxylic acids (Equation (187)) ⟨72CC472⟩. The most general method is the dehydrative cyclization of *β*-hydroxycarboxylic acids which may be performed under retention (carboxyl group activation, CGA) or inversion (hydroxy group activation, HGA) of the *β*-OH function. For instance, the Mitsunobu reagent (Equation (188)) ⟨79AG(E)622⟩ can effect both mechanisms leading to the *cis-β*-lactone and the alkene (HGA) and *trans-β*-lactone (CGA), with *threo-β*-hydroxycarboxylic acid (**59**) as the substrate. Benzenesulfonyl chloride is the standard reagent for CGA cyclization (Scheme 39) ⟨72JA2000⟩ and has also been used extensively in more complex systems, for example, Equation (189) ⟨90TL7513⟩. Variations of this method are the cyclization of *β*-hydroxythiol esters with mercury(II) assistance (Equation (190)) ⟨76JA7874⟩. Danheiser *et al.* has combined the lactonization with the initial aldol-type addition in a one-pot synthesis of *β*-lactones (Scheme 40) ⟨91JOC1176⟩. Direct formation of *β*-lactones may be achieved via electrochemically-induced Reformatsky reactions (Equation (191)) ⟨93AG(E)1191⟩. An interesting dehydration of an optically active *β*-hydroxy acid under retention is shown in Scheme 41 ⟨69TL1047, 87HCA1320⟩. A similar aldol addition cyclization tandem process has been devised using acyl iron intermediates (Scheme 42) ⟨91SL781⟩.

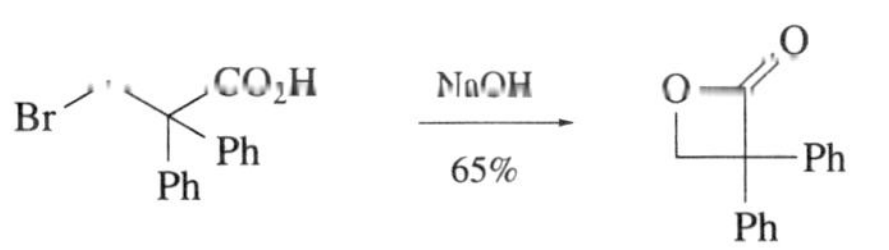

(186)

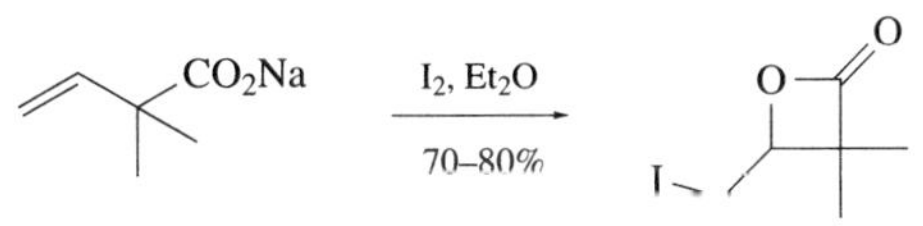

(187)

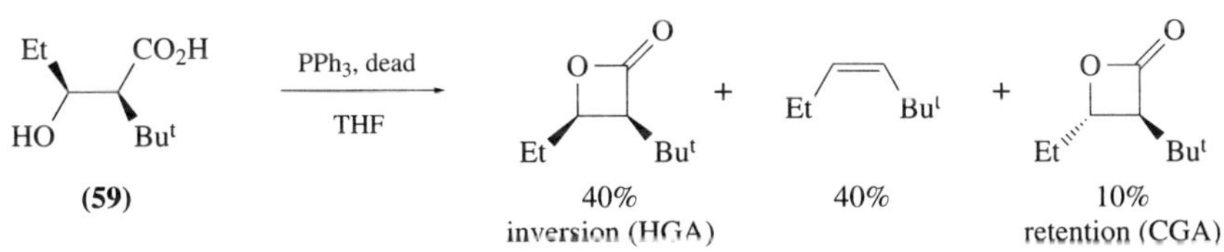

(188)

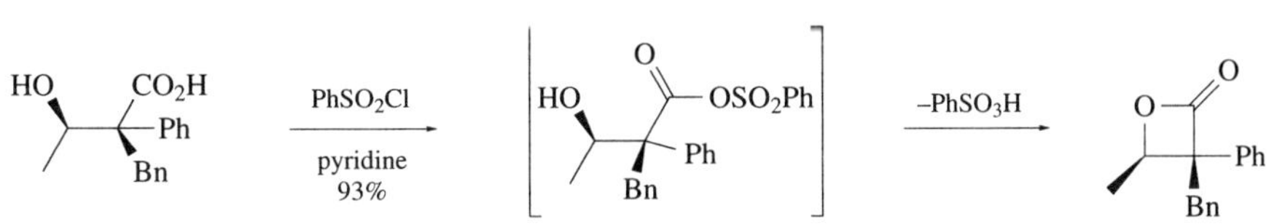

Scheme 39

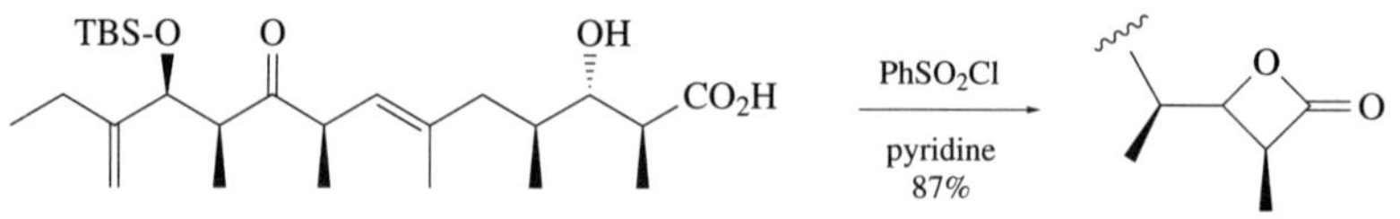

(189)

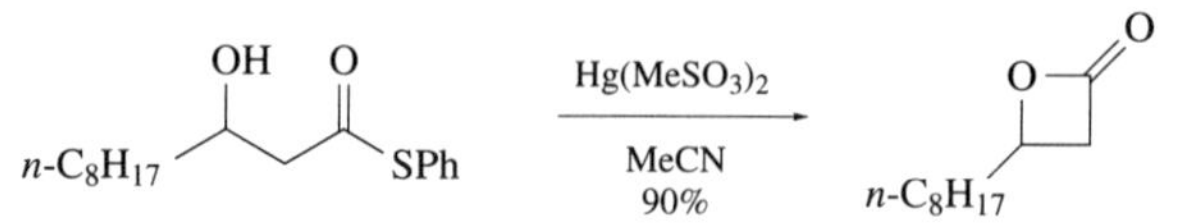

(190)

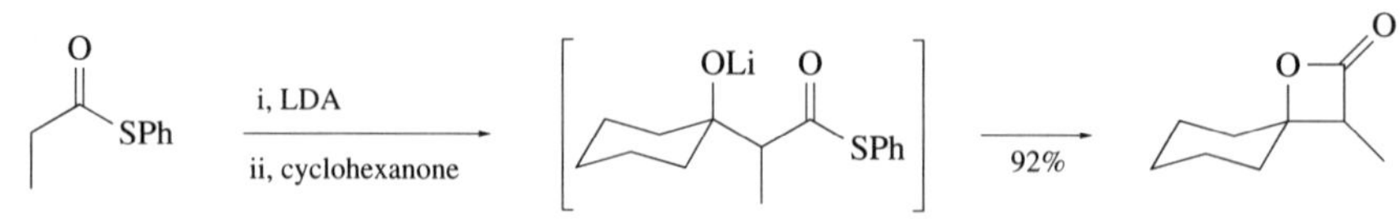

Scheme 40

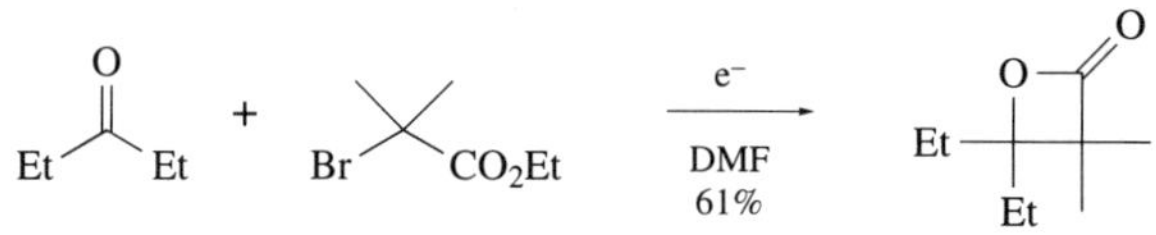

(191)

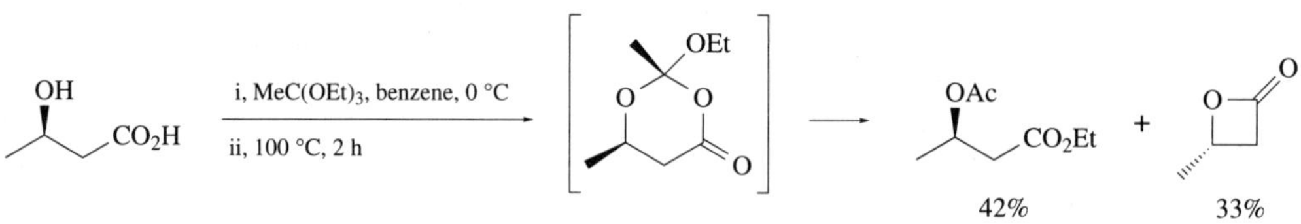

Scheme 41

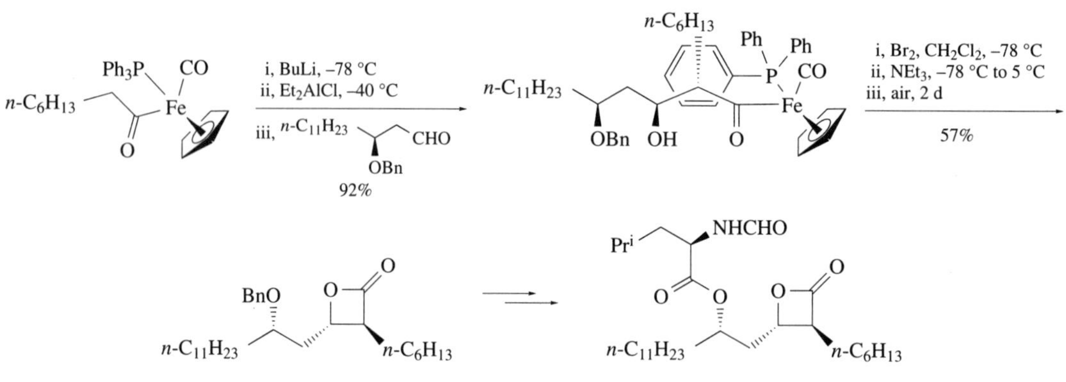

Scheme 42

Almost as general as the dehydrative cyclisation of β-hydroxycarboxylates are the Lewis acid induced [2 + 2]-cycloadditions of ketenes and carbonyl derivatives. For instance, as shown in Scheme 43 ⟨82TL5513⟩, the ketene is generated *in situ* by dehydrohalogenation of an acid chloride and trapped with a reactive ketone. The β-lactone rearranges under transketalization and ring expansion to a γ-lactone ring. The ketene chloral addition may be catalyzed enantioselectively (Equation (192)) ⟨82JA166⟩. TMS-ketene forms a silylated β-lactone suitable for further transformation (Equation (193)) ⟨78JGU1249⟩. An anionic access to ketene intermediates is shown in Scheme 44 ⟨75AG(E)765, 79LA219⟩. Isoxazoles may be 5-deprotonated, and undergo elimination of benzonitrile to form the alkynoxide which acts as a ketene anion equivalent towards aldehydes. The primary adduct **(60)** cyclizes to form the β-lactone enolate which is α-protonated or alkylated (Scheme 44). A ketene type intermediate may also be involved in the synthesis shown in Scheme 45 ⟨94JOC4972⟩.

A Fischer carbene complex is alkylated to form (**61**) which is deprotected to the aldehyde (**63**) and photocyclized to give the bicyclic β-lactone (**63**).

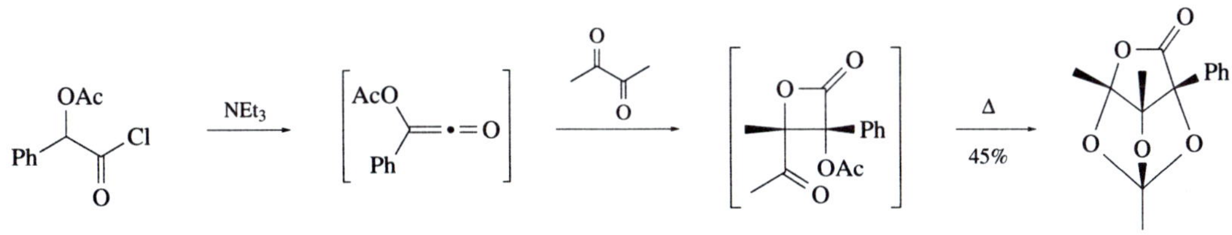

Scheme 43

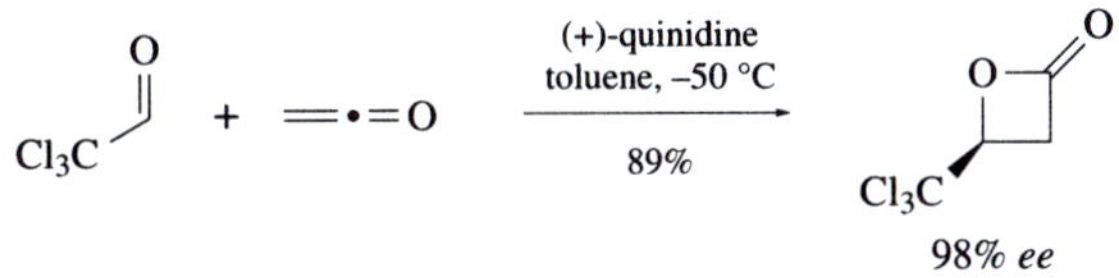

(192)

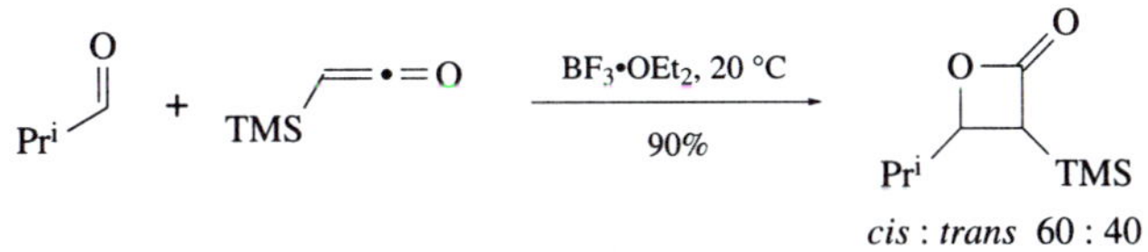

(193)

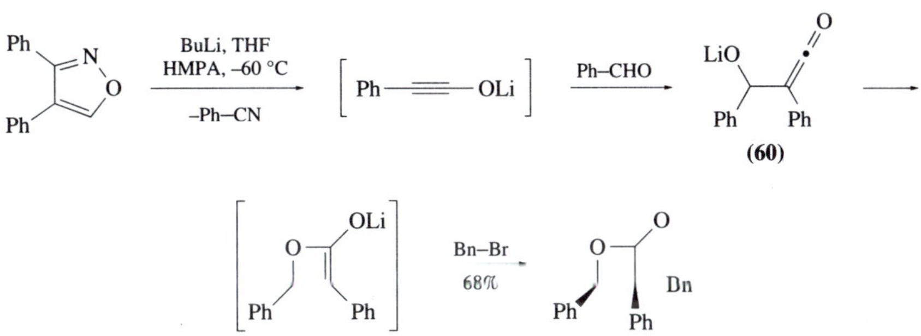

Scheme 44

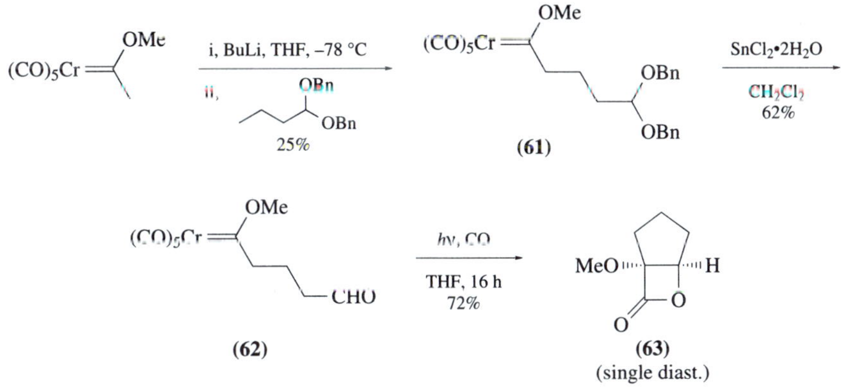

Scheme 45

β-Lactones may be also generated via carbonylation reactions. For instance, vinyl epoxide (**64**) undergoes CO insertion with $Fe_2(CO)_9$ to form the allylic acyl iron intermediate (**65**). Oxidative removal of the iron leads to C—C connection under formation of the *β*-lactone (**66**) with *trans*-configuration (Scheme 46) ⟨91T9929⟩. Simpler cases are shown in Scheme 47 which presumably proceed via CO insertion and formation of *β*-hydroxyacyl metal intermediates ⟨90CL1549⟩.

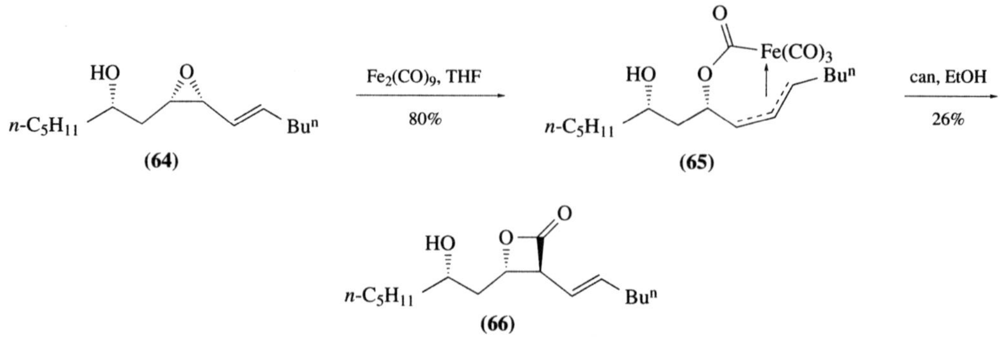

Scheme 46

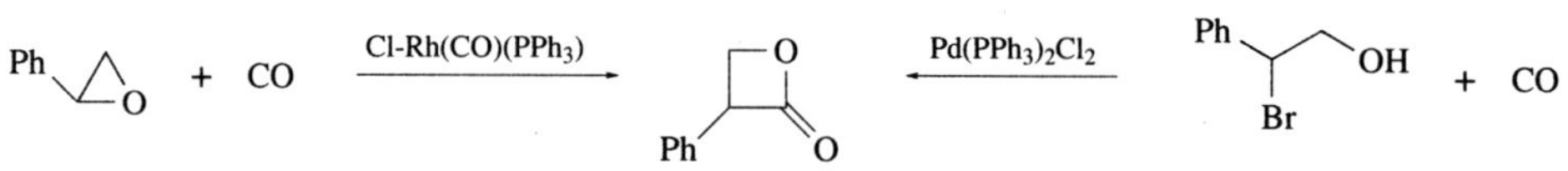

Scheme 47

5.03.6.2 *γ*-Lactones

5.03.6.2.1 γ-Lactones by direct cyclodehydration of γ-hydroxycarboxylates

Apart from particularly strained systems, *γ*-hydroxycarboxylates spontaneously cyclize to lactones, especially in the presence of the mineral acid (Equations (194) ⟨88JOC4098⟩, (195), (196), (197), and (198) ⟨61JA606⟩). If the cyclization causes ring strain then a dehydrating agent such as acetic anhydride is added (Equation (199)) ⟨58T1⟩. Sometimes an acetal ring cleavage may precede lactone formation thus demonstrating that the lactone is more stable than the acetal (Equations (200) ⟨88JA4640⟩ and (201)) ⟨80S403⟩. The hydroxycarboxylate may be generated *in situ* from carbonyl precursors (Equations (202), (203) ⟨71CC488⟩, Scheme 48 ⟨84TL2211⟩, and Equation (204) ⟨88AG(E)268⟩) and cyclizes spontaneously under these conditions. Normally these spontaneous lactonizations proceed with retention of configuration at the hydroxyl function. However, with certain dehydrating agents, HGA cyclization occurs under inversion of configuration (Equation (205) ⟨79JA1284⟩, Schemes 49 ⟨80JA3964⟩ and 50 ⟨72T1275⟩).

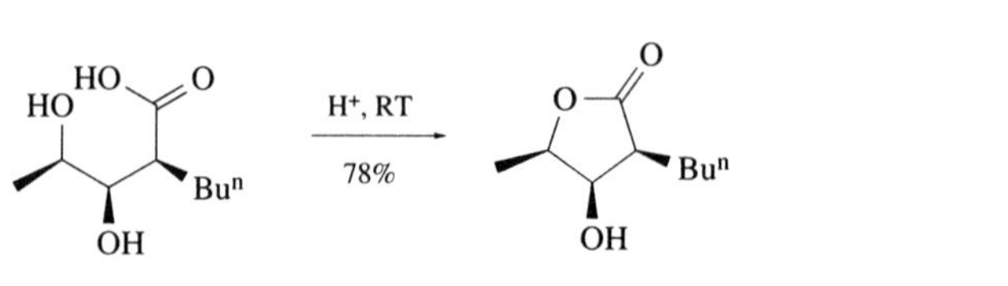

(194)

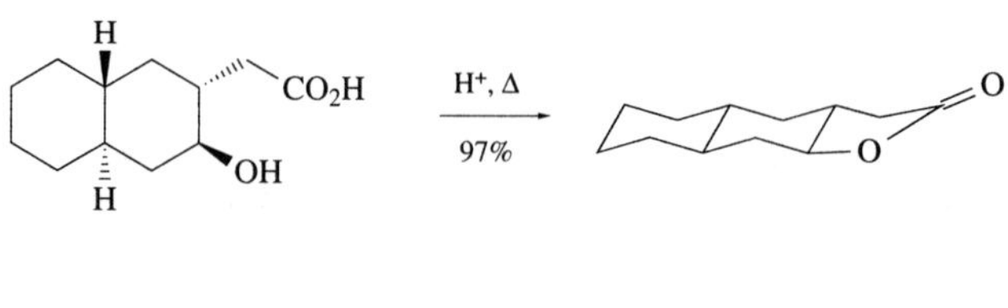

(195)

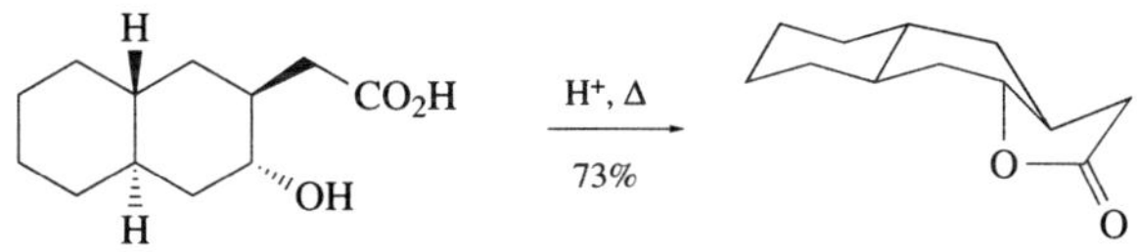

(196)

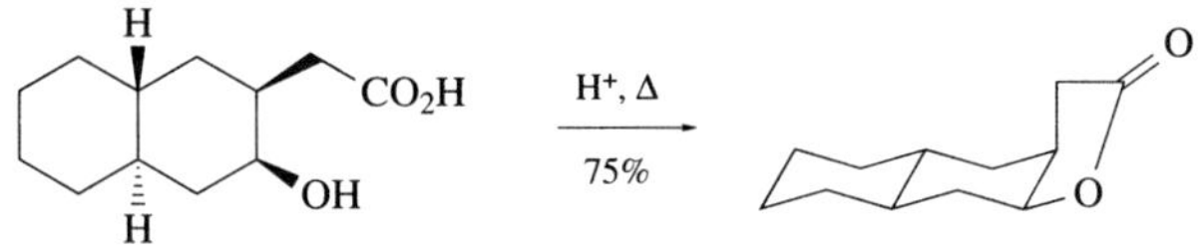

(197)

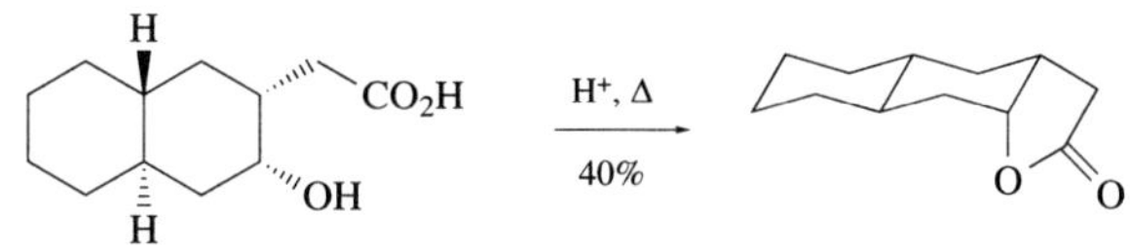

(198)

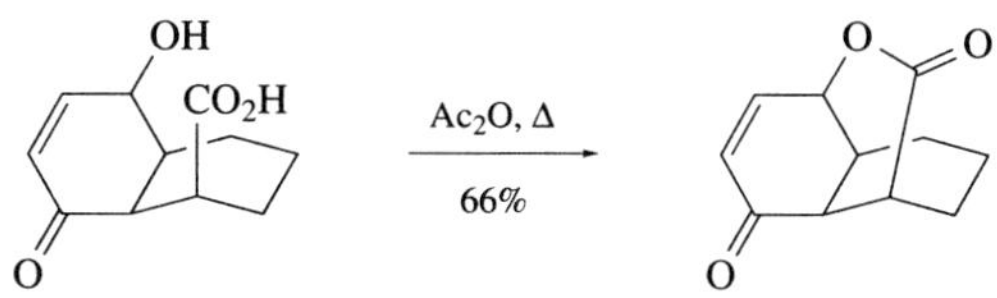

(199)

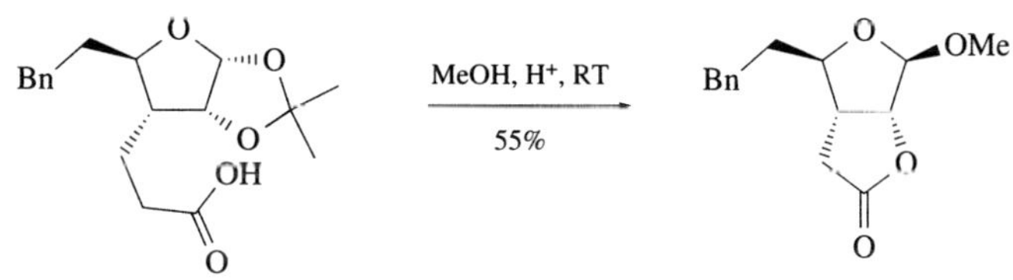

(200)

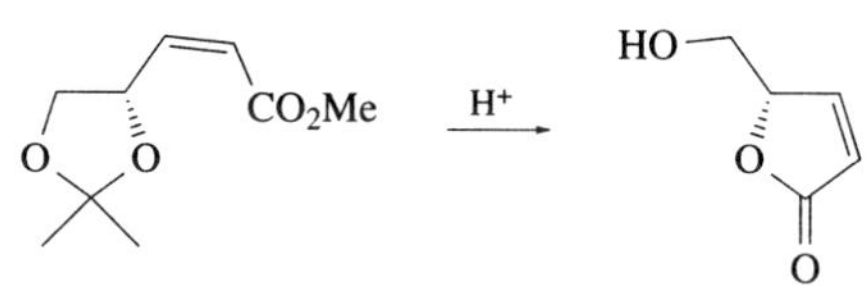

(201)

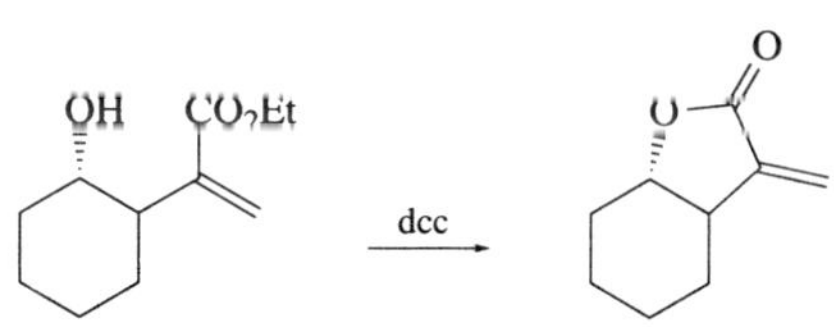

(202)

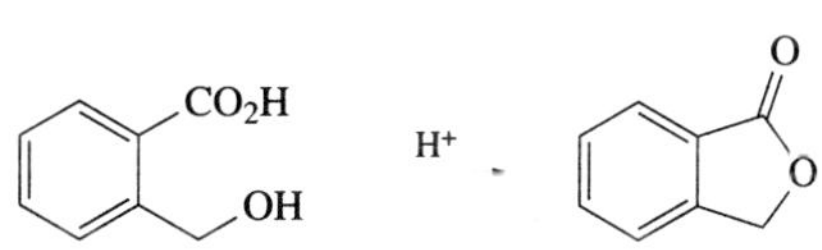

(203)

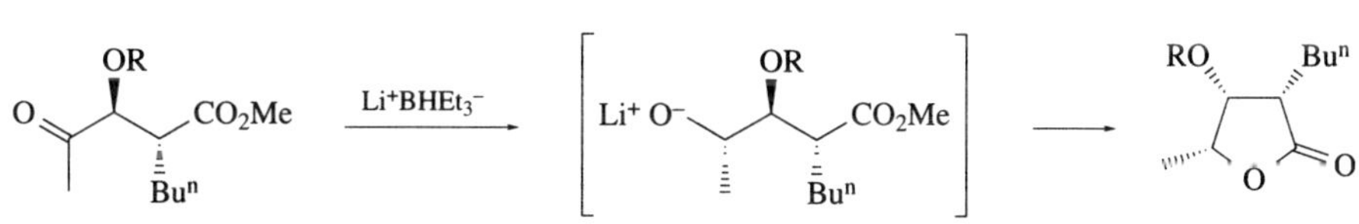

$R = -C(O)CH_2Pr^i$

Scheme 48

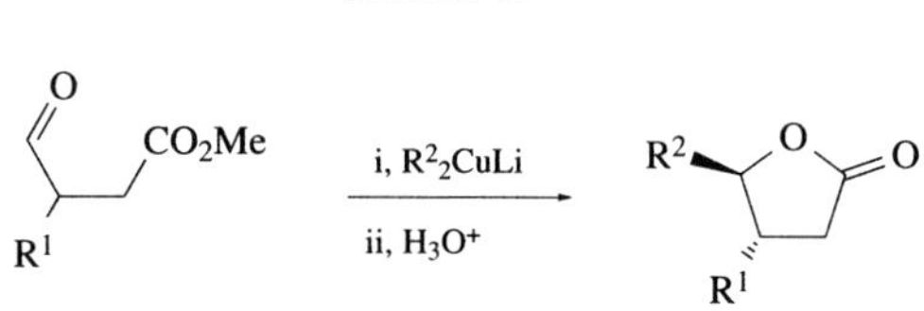

(204)

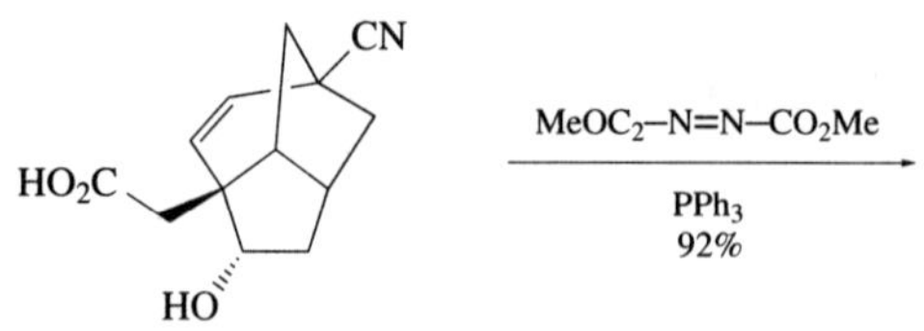

(205)

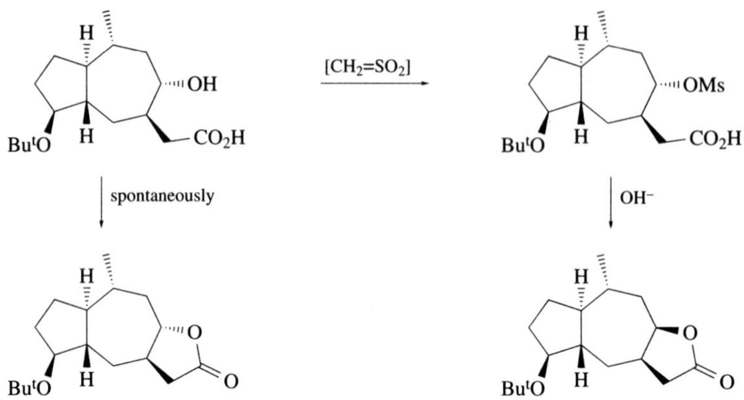

Scheme 49

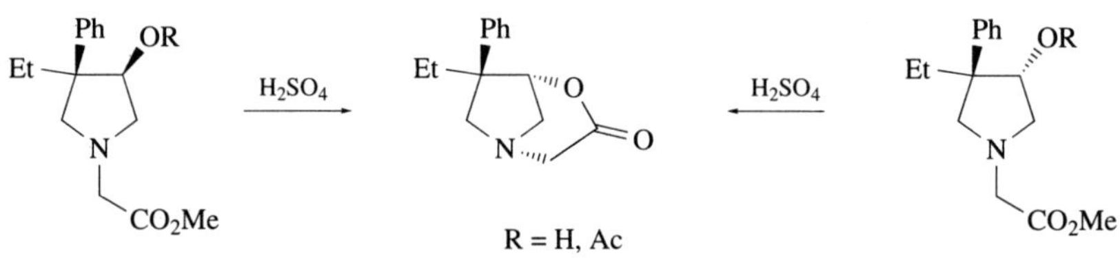

Scheme 50

5.03.6.2.2 γ-Lactones by enolate additions

In numerous cases the γ-hydrocarboxylate is generated by enolate addition, for example epoxides are opened by stabilized enolate anions to form γ-hydroxycarboxylates which cyclize on acidification, to give paraconic acid derivatives, such as nephromopsinic acid, for example (Scheme 51) ⟨50JA1357, 52JA4083, 58JA3079, 64JOC2810, 81H(16)381, 81H(16)951⟩. The reaction also works for sterically hindered epoxides (Equation (206)) ⟨72JOC1907⟩. An enantiocontrolled variation is shown (Scheme 52) utilising Meyers' oxazoline chemistry ⟨75JOC1186⟩. Aldol-type reactions are involved in the familiar Stobbe condensation (Scheme 53) ⟨51OR(6)1⟩ or in a more sophisticated way with β-lactone derived ester enolates (**67**) in the aldol addition step (Scheme 54) ⟨84T2211⟩. Depending on the reaction medium, monocyclic or bicyclic γ-lactones may be obtained. Aldol addition to the sulfone *ortho*-ester (**68**) furnishes butenolides after elimination of sulfinate (Scheme 55) ⟨87TL2135⟩. The spiro–Reformatsky addition is a powerful method to generate polycyclic systems like (**69**), an intermediate in a synthesis of lysergic acid (Equation (207)) ⟨70AG(E)457, 84JA1813⟩. The γ-hydroxycarboxylate may be generated by a Sakurai cyclization and forms the lactone under heating with acid (Scheme 56) ⟨87TL655⟩. Claisen rearrangement of *O*-vinyl-2,3-butene-1,4-diols furnishes the lactone (**70**) together with the regioisomer (**71**) in a one-pot operation (Scheme 57) ⟨69TL1591⟩. An elegant aldol-type addition retro aldol cleavage has been used in synthesis of paraconic acid derivatives (Scheme 58) ⟨88JOC2440⟩.

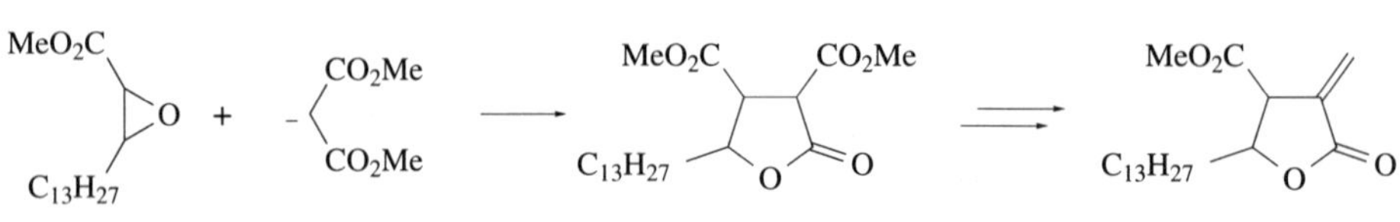

Scheme 51

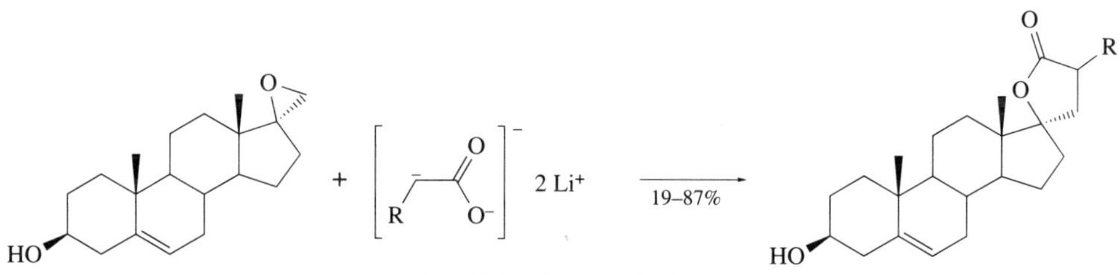
(206)

R = H, Me, Et, Bu^n, Ph, OMe

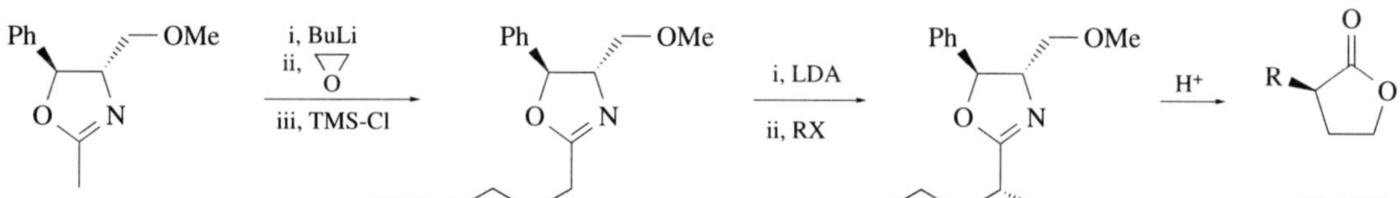
ee 60–73%

R = Me, Et, Pr^n, allyl, Bu^n

Scheme 52

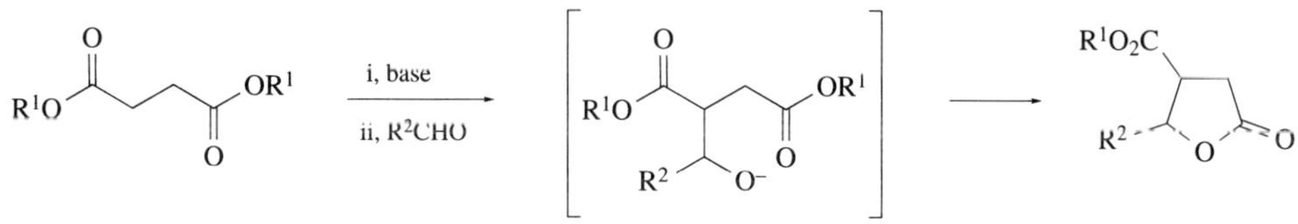

Scheme 53

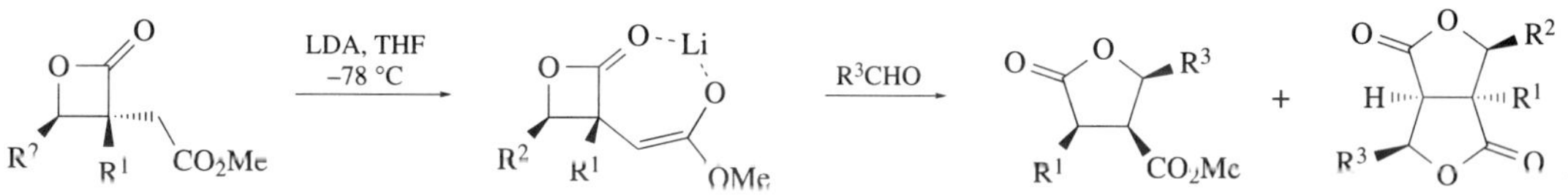
(67)

(in the presence of HMPA)

(in the presence of $ZnCl_2$)

Scheme 54

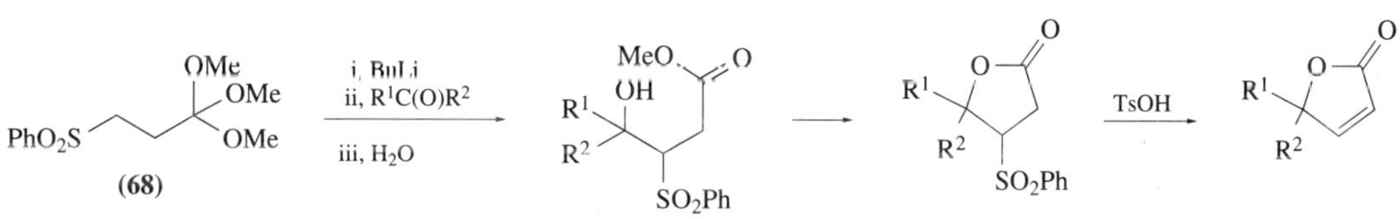
(68)

Scheme 55

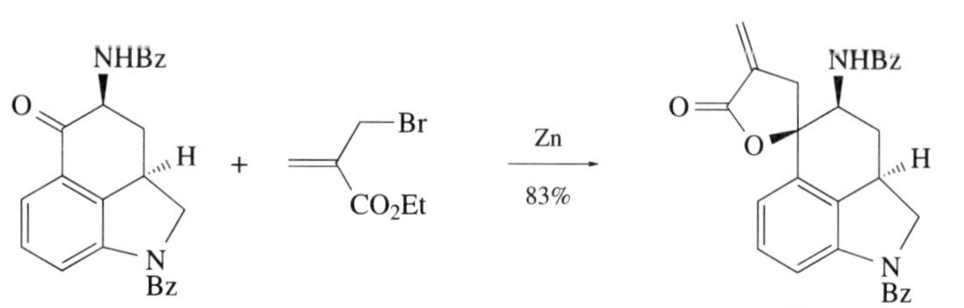
(207)

(69)

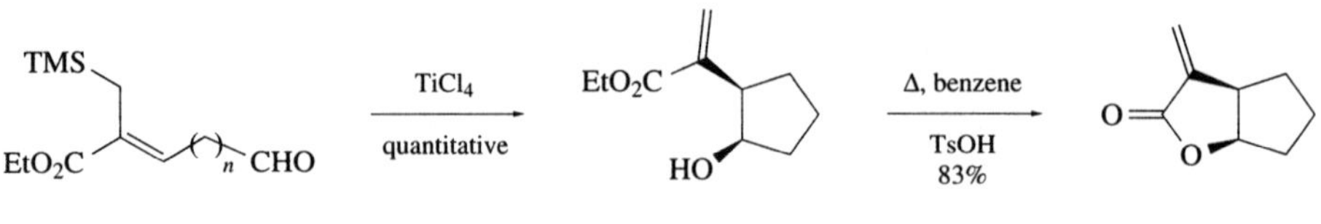

Scheme 56

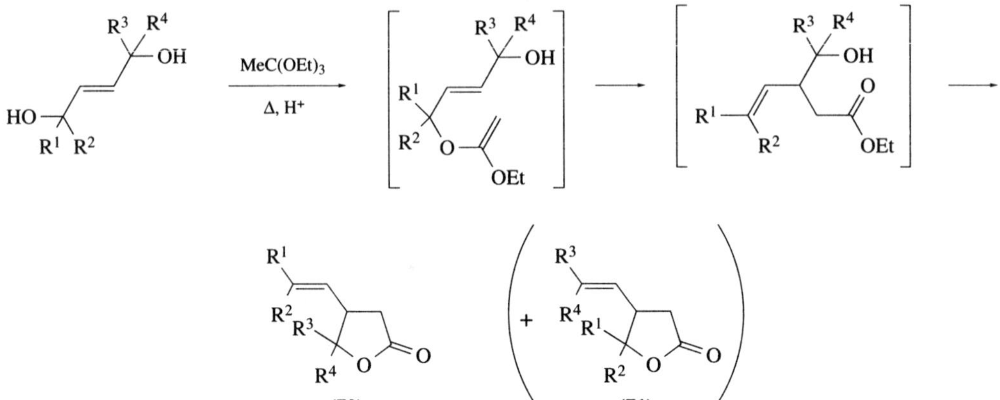

(70) (71)

Scheme 57

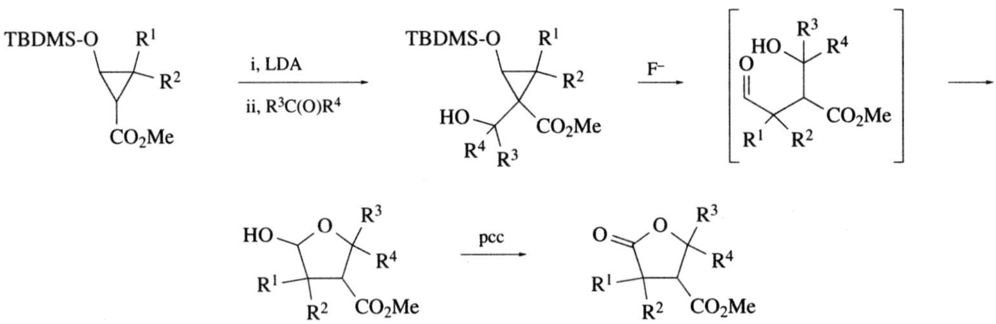

Scheme 58

5.03.6.2.3 γ-Lactones from β,γ- and γ-δ-ene-carboxylates

Direct lactonization of unsaturated carboxylic acids is a well-established method for the synthesis of γ-lactones (Equations (208) ⟨74JCS(P1)499⟩, (209) ⟨68JOC2819⟩, (210) ⟨73CC693⟩, and (211) ⟨50ACS1080, 57ACS582, 59JCS1313⟩). An additional functional group may be introduced via hydroxy-lactonisation (Schemes 59, 60 ⟨60BSF1412⟩, and 61 ⟨71JCS(C)2944⟩); the epoxy acid can also be generated from the diol and is then lactonized with acid (Equation (212) ⟨88JOC4098⟩). Halo-lactonizations of γ,δ-ene acids lead to the γ-lactone under kinetically controlled conditions, whereas thermodynamic control favors the δ-lactone (Scheme 62 ⟨69JA5675, 70JA397, 84MI 503-01, B-84MI 503-02⟩ and Equation (213) ⟨B-84MI 503-02⟩). In the case of β,γ-unsaturated acids it is thermodynamic control which generates the γ-lactone via the kinetically favored β-lactone (Scheme 63) ⟨74JCS(P1)1864⟩. An additional alternative is mercurylactonization (Equation (214)) ⟨59JCS227⟩.

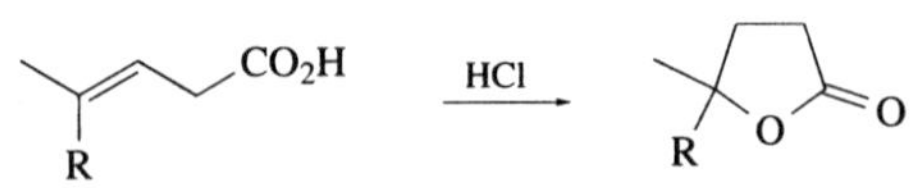

(208)

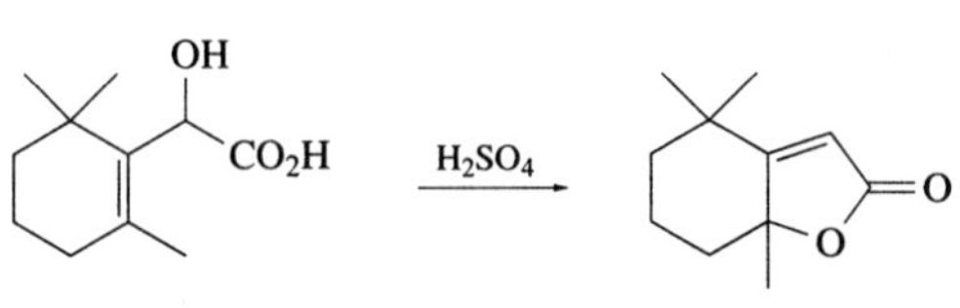

(209)

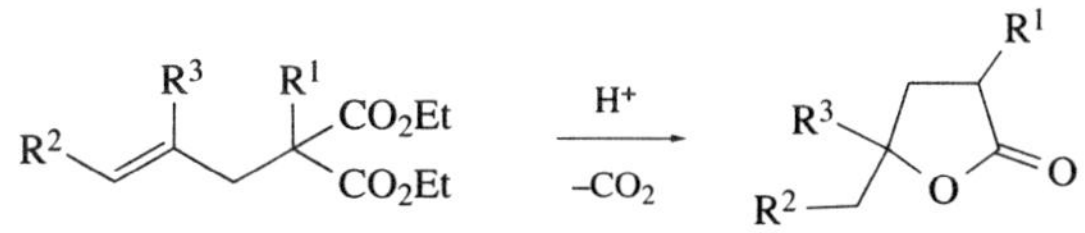

(210)

R^1 = H, C_4H_9
R^2 = H, C_4H_9, C_5H_{11}
R^3 = H, Me

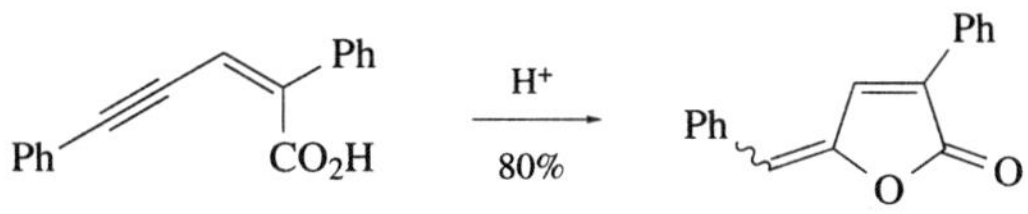

(211)

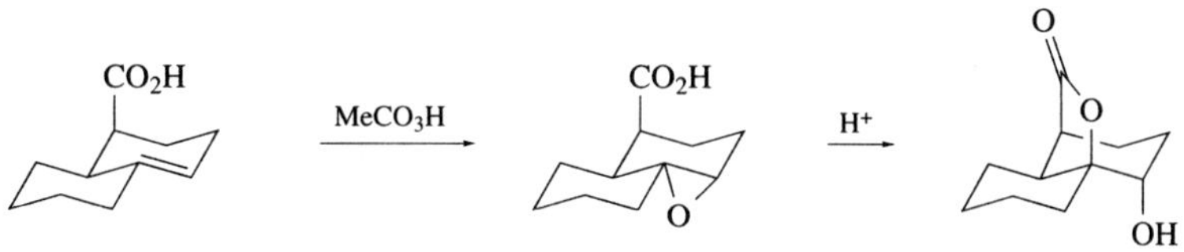

Scheme 59

Scheme 60

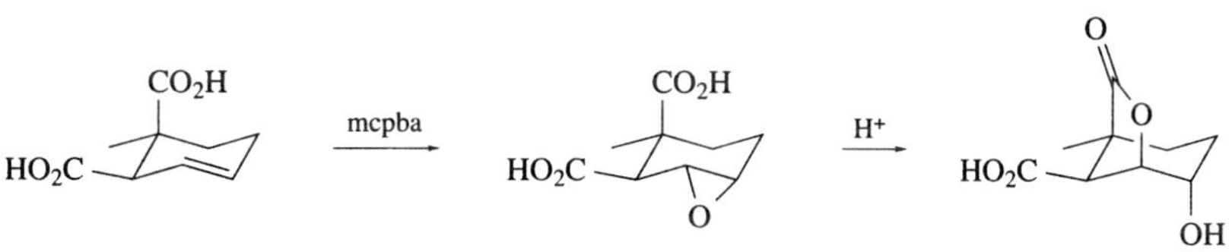

Scheme 61

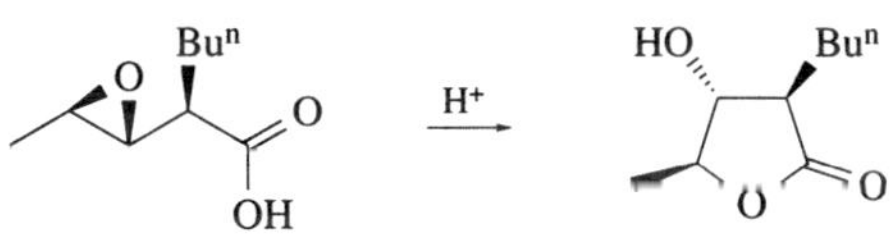

(212)

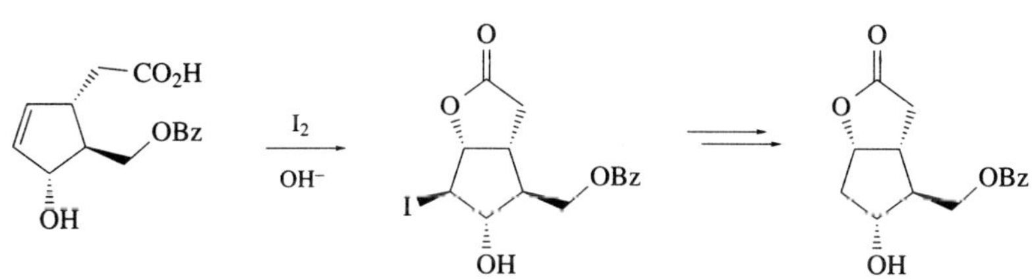

Scheme 62

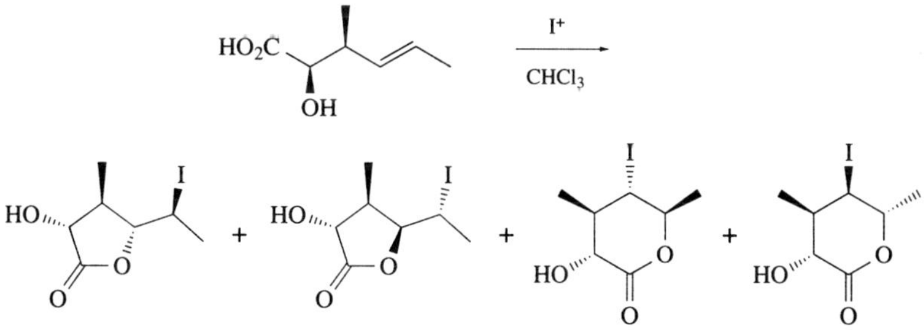

(213)

NIS	2	2	3	1
I_2	>20	1	0	0

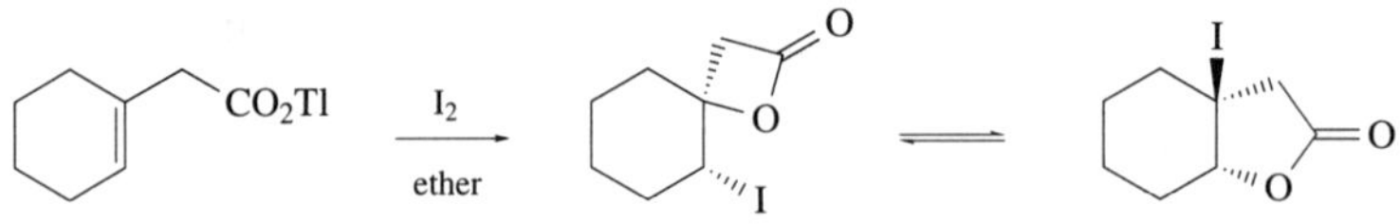

Scheme 63

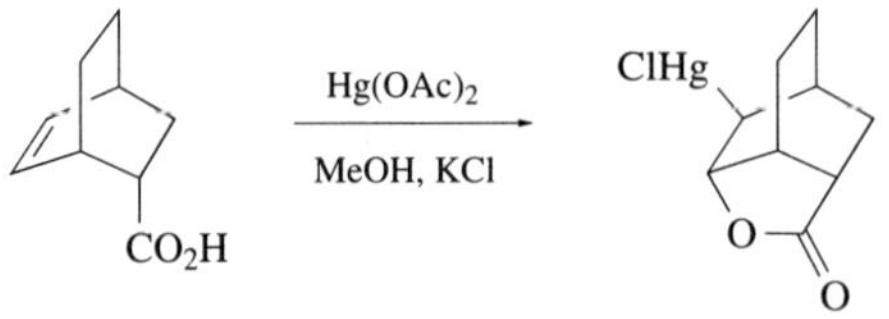

(214)

5.03.6.2.4 Ring enlargement of β-lactones

Under the influence of Lewis acids, β-lactones can undergo a stereocontrolled dyotropic ring enlargement to γ-lactones (Equation (215)) ⟨79AG(E)793⟩. The reaction is also suited for spiro- and fused annulation (Equations (216) and (217)) ⟨88JOC2371⟩.

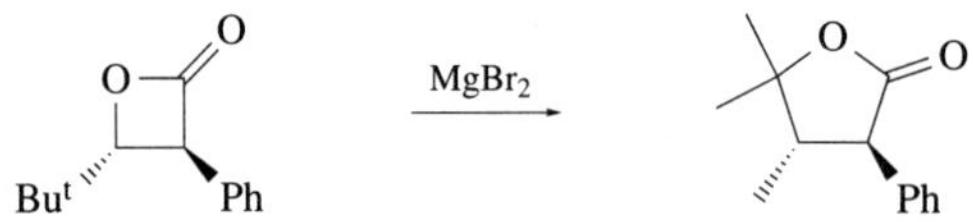

(215)

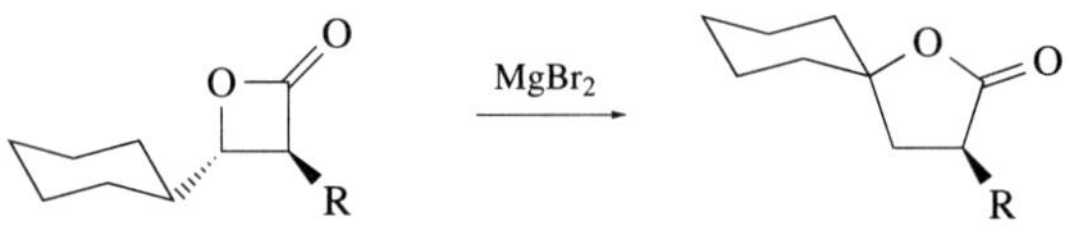

(216)

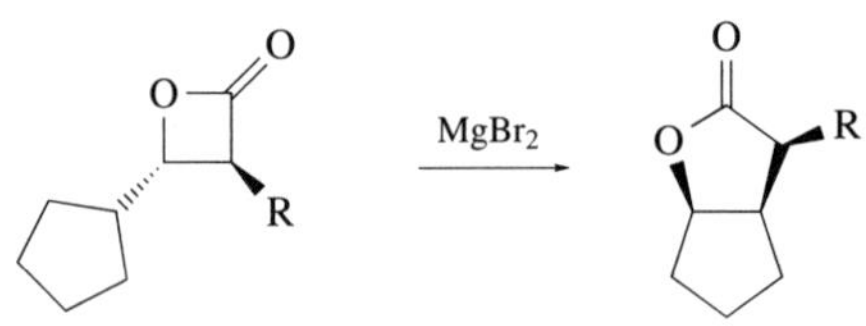

(217)

R = Ph, SPh, OPh

5.03.6.2.5 Carbonylation reactions

Allylic alcohols are transformed into γ-lactones by a palladium-catalyzed carbonylation reaction. If *ortho*-acetate is added, an ester group is incorporated to give γ-lactone (**72**) (Scheme 64) ⟨84JA1496, 85TL3207, 87TL325⟩. Carbonylation–lactonization sequences can also be performed with (*Z*)-vinyl iodides (Equation (218)) ⟨80JA4193⟩ or homopropargylic alcohols such as (**73**), which are converted into γ-lactones with an *exo* methylene group (Equation (219)) ⟨75TL51, 76CC907⟩.

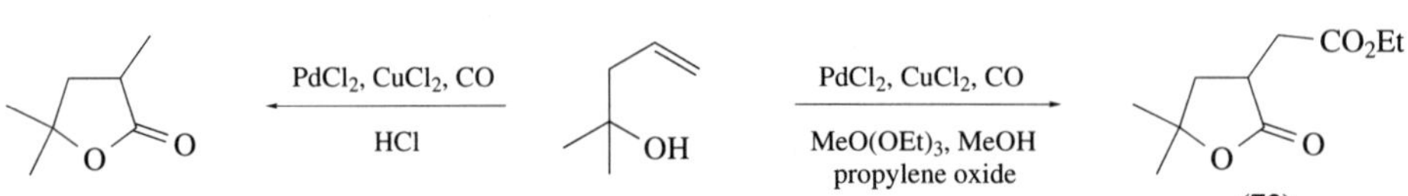

(72)

Scheme 64

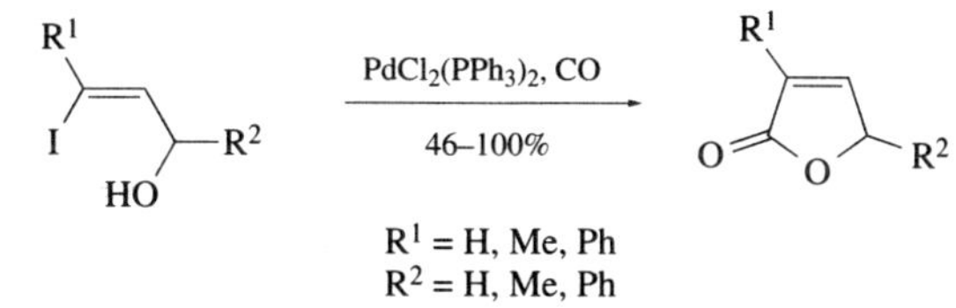

(218)

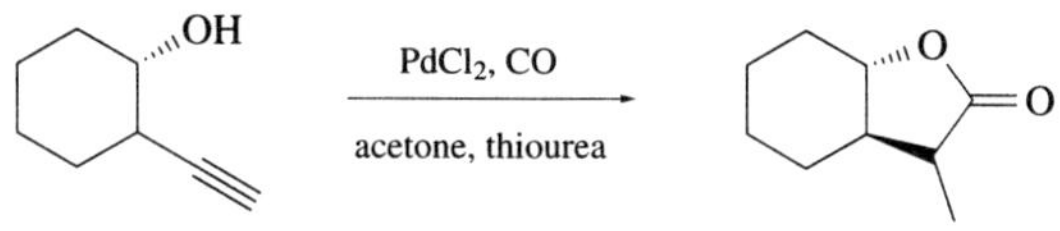

(73)

(219)

5.03.6.2.6 γ-Lactones via redox reactions

Furans can undergo a formal addition of water to give butenolides. This corresponds to an oxidation–reduction sequence. The choice of the reagents determines the regiochemical outcome of the reaction. Peracid/sodium borohydride generates the β-product, NBS, and zinc the α-product (Scheme 65) ⟨75JMC58⟩. Butane-1,4-diols undergo oxidative cyclization to γ-lactones with Fetizon's reagent (Scheme 66) ⟨75T171, 87JOC4319⟩. Particularly important in the γ-lactone series is the Baeyer–Villiger oxidation, which fits very well into a [2 + 2]-cyclization ring-enlargement sequence and constitutes a key transformation in Corey's prostaglandin synthesis (Scheme 67) ⟨70TL311, 74TL2439⟩. Also noteworthy in this sequence is the application of a halolactonization creating a second γ-lactone to establish the required *cis*-1,3-diol element. With camphorsulfonic acid, enantioselective lactonization is achieved (Scheme 68) ⟨85JA6404⟩.

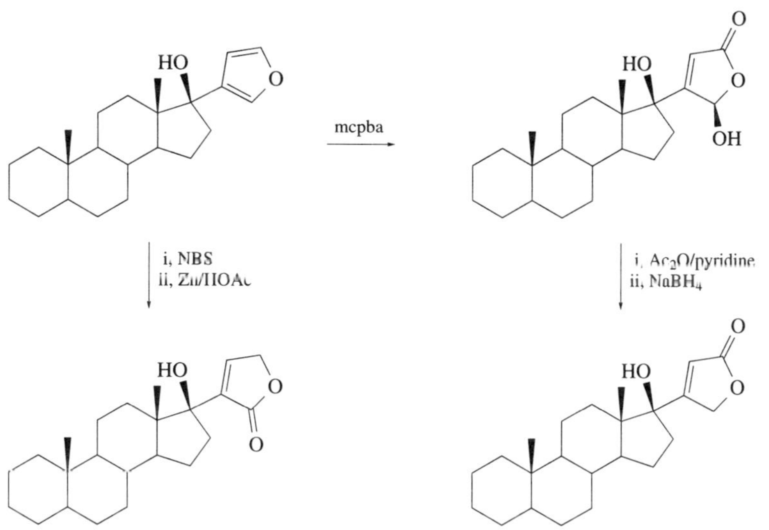

Scheme 65

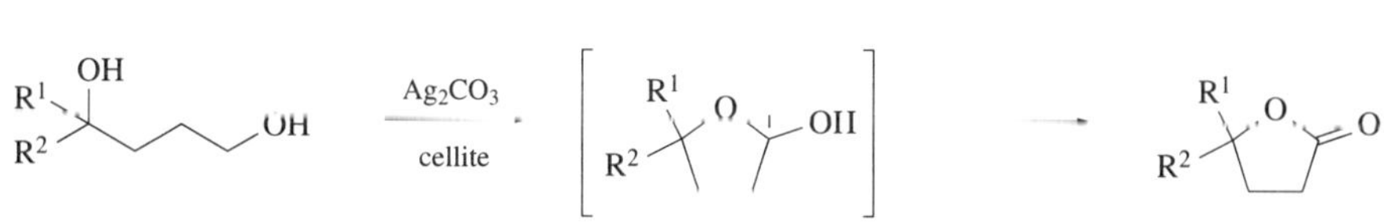

Scheme 66

5.03.6.3 δ-Lactones

The synthesis of δ-lactones is very similar to that of γ-lactones. Thus, direct cyclization of δ-hydroxy acids is quite common, though not as facile as in the γ-lactone case. Dehydration agents

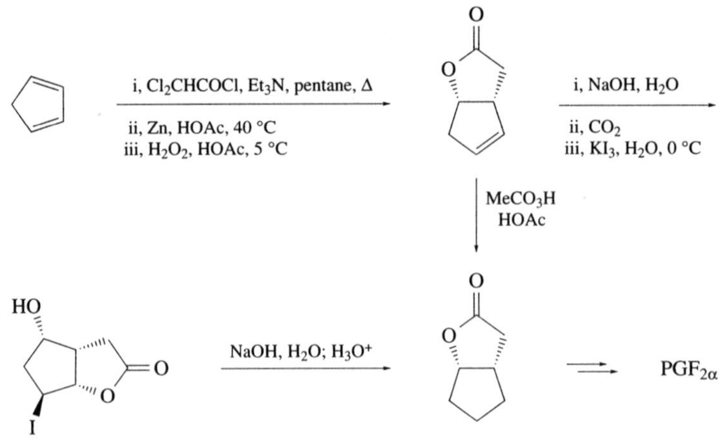

Scheme 67

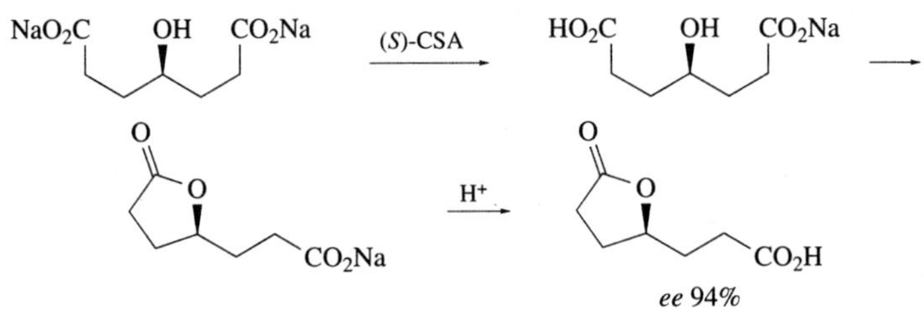

Scheme 68

(e.g., acetic anhydride) can be added to accelerate the lactonization. Meyers' method works also in the δ-lactone case (Scheme 69) ⟨B-84MI 503-01⟩ and halolactonizations and mercurylactonizations of δ,ε-unsaturated acids regioselectively furnish the δ-lactones (Equation (220)) ⟨B-84MI 503-02⟩. An interesting free-radical carbonylation δ-lactonization process has been described (remote functionalization, Equation (221)) ⟨94JA5473⟩. The ring opening of an δ,ε-epoxide which is generated via Payne rearrangement is shown in Scheme 70 ⟨88JOC4421⟩. Enantiotopic differentiation between two carboxyl groups is achieved via axially chiral 2,2′-binaphthyl amide (Scheme 71) ⟨87JA7188⟩.

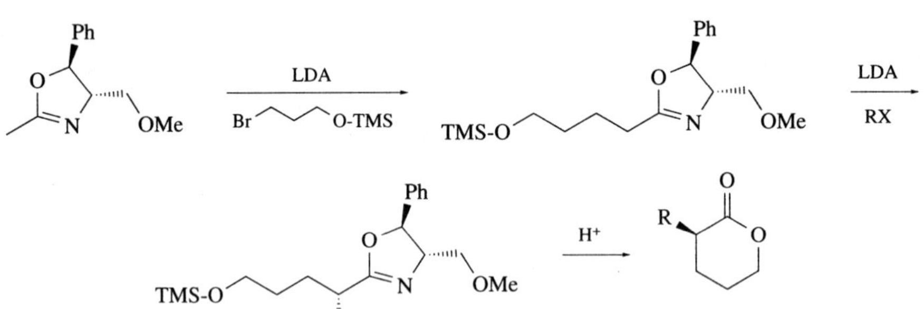

Scheme 69

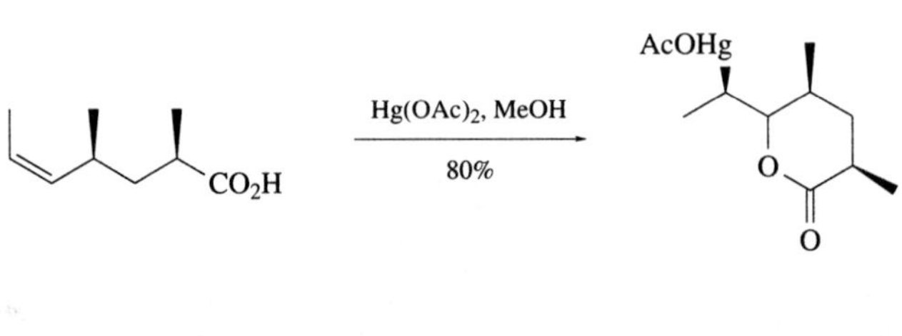

(220)

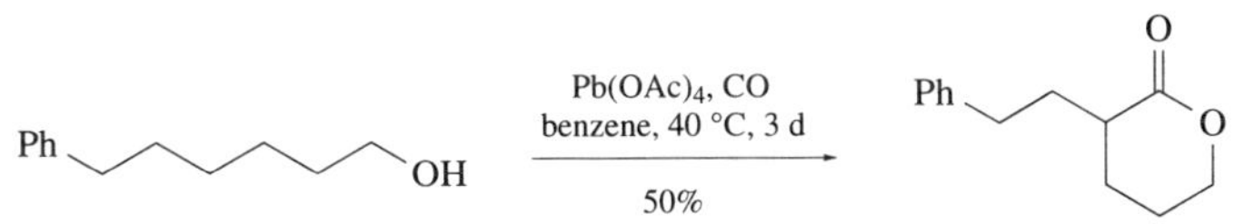

(221)

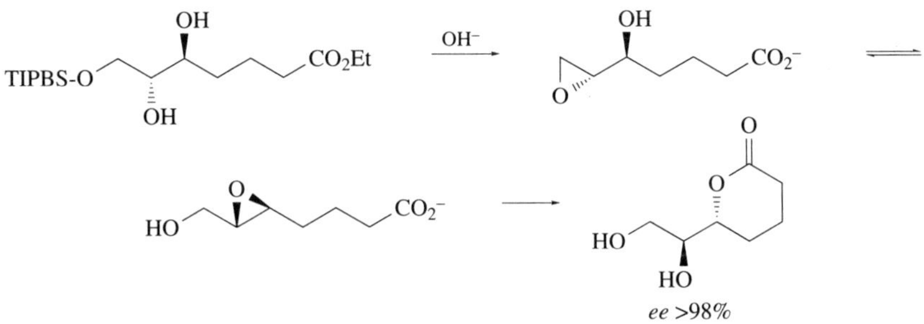

Scheme 70

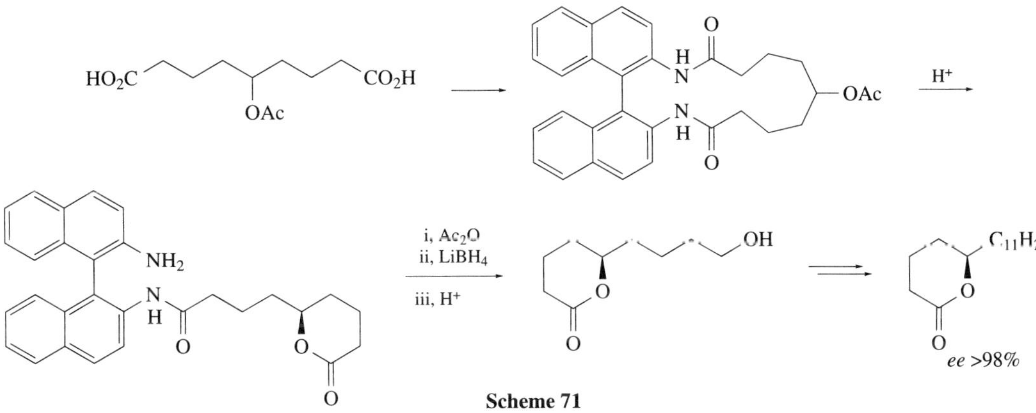

Scheme 71

5.03.6.4 Macrolactones

Macrolactones have an enormous importance in natural products synthesis. Consequently, a variety of lactonization methods have been developed since the early 1980s, all of which try to solve the basic problem of competition between macrolactonization and polymerization. High-dilution techniques have to be employed uniformly, but lactone ring sizes from 12 onward are accessible with respectable yields (generally $>50\%$).

Lactonizations may be subdivided into CGA and HGA procedures; some alternative methods are available, but they do not appear as generally applicable.

5.03.6.4.1 CGA methods

Using these methods is the most reliable route. A wide variety of CGA reagents have been introduced, the simplest being esters (Equation (222) ⟨82JA4015, 85CC1386⟩ and Scheme 72 ⟨84JA2954⟩) or protonation of the carboxyl group (Equation (223)) ⟨68T2443⟩. The CGA method of choice is Yamaguchi's procedure, using 2,4,6-trichlorobenzoyl chloride and 4-dimethylamino pyridine (dmap). This protocol works in many cases, for example, Equations (224) ⟨79BCJ1989, 79CL1019, 79CL1021, 81CL1415⟩, (225) ⟨87TL4569⟩, and (226) ⟨86AG(E)1028⟩. Phosphoryl chlorides are not so common (Equation (227)) ⟨82JA5521, 82JA5523⟩. For some time Corey's thiol ester method was very popular (Equations (228) ⟨84T2935⟩, (229) ⟨75JA2287⟩, (230) ⟨94JA5511⟩, and (231) ⟨78JA4618, 78JA4620, 79JA7131⟩) and has been modified by Masamune *et al.* (Equation (232)) ⟨81JA1568⟩ but apparently it was not found to be equivalent to Yamaguchi's method. Also recommended is Keck's procedure, using dicyclohexylcarbodiimide (dcc) as the dehydrating CGA reagent (Equations (233) ⟨82JA4686, 87JA1564, 87JA1565⟩ and (234) ⟨82JA4708⟩). The elegant ynamine method by Gais (Scheme 73)

⟨84AG(E)143, 84AG(E)145⟩ has not yet found general application, whereas Mukaiyama's pyridinium activation has proven its value in complex systems for example, Equations (235) ⟨84LA450, 84LA462, 84LA474⟩ and (236) ⟨78JA4858⟩. A special CGA activation is the photolytic generation of a ketene intermediate (**74**) which undergoes spontaneous cyclization to a 18-membered lactone (Scheme 74) ⟨87AG(E)362⟩. A photooxygenation of an oxazole ring to create an imide as the CGA species is shown in Scheme 75 ⟨81T4059⟩. Dibutyltin dihydroxide is a combined HGA and CGA reagent which closes lactone rings of different sizes efficiently (Scheme 76) ⟨83JA7130⟩.

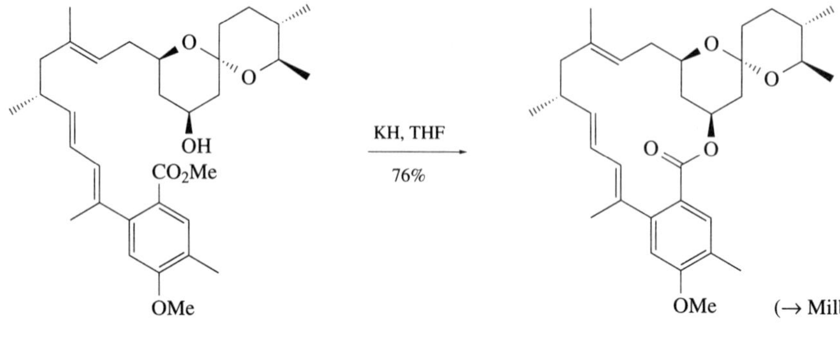

(222)

(→ Milbemycin β3)

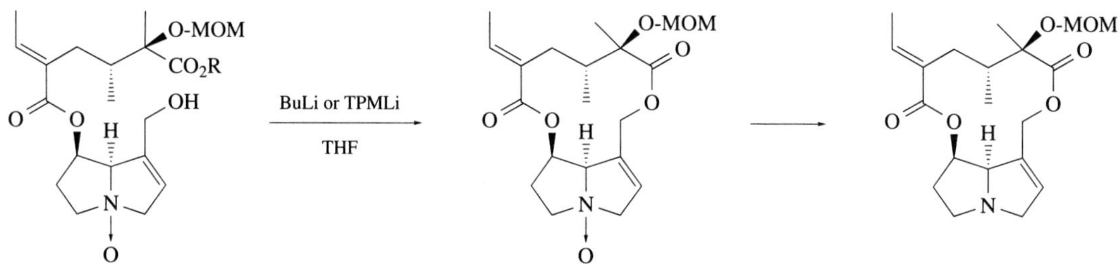

R = CH_2SMe, CH_2SO_2Me

Scheme 72

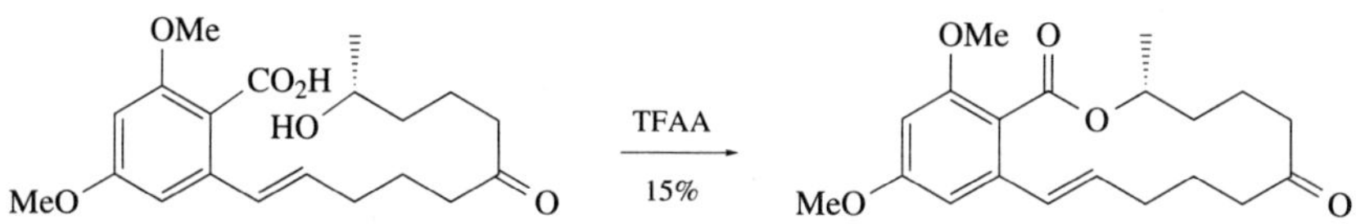

(223)

(→ Zearalenone)

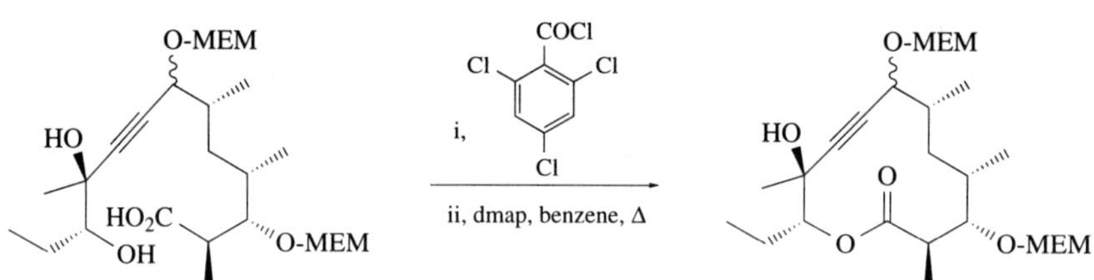

(224)

(→ Methynolide)

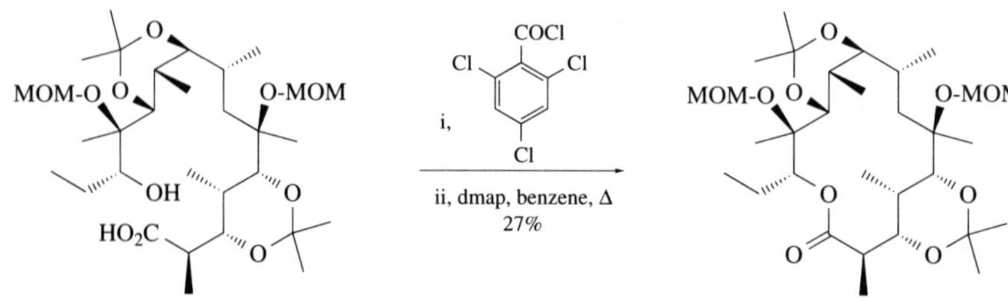

(225)

(→ (9S)-9-Dihydro-erythronolide A)

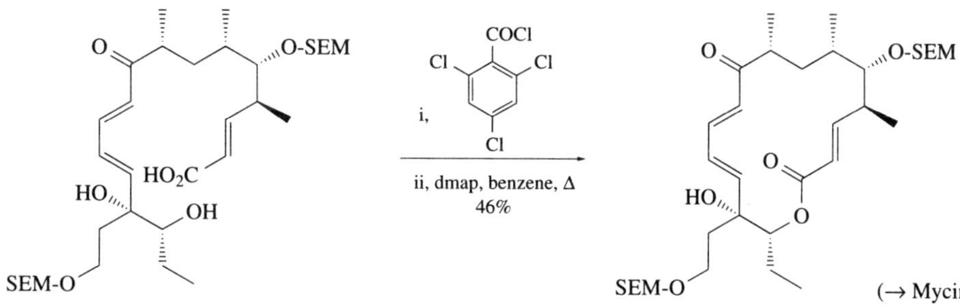

(226)

(→ Mycinolide V)

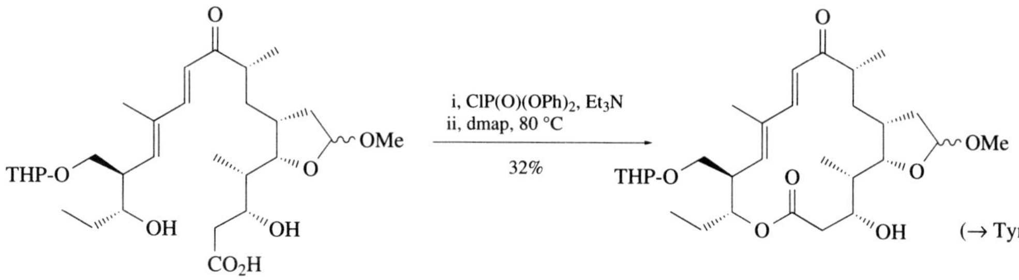

(227)

(→ Tynolide)

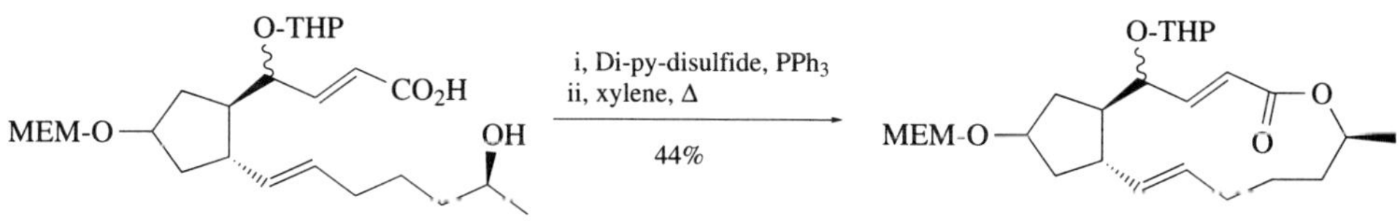

(228)

(→ Brefeldin A)

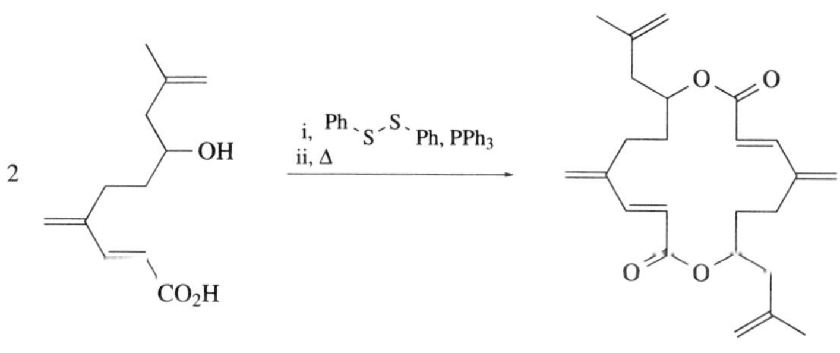

(229)

(→ Vermiculene)

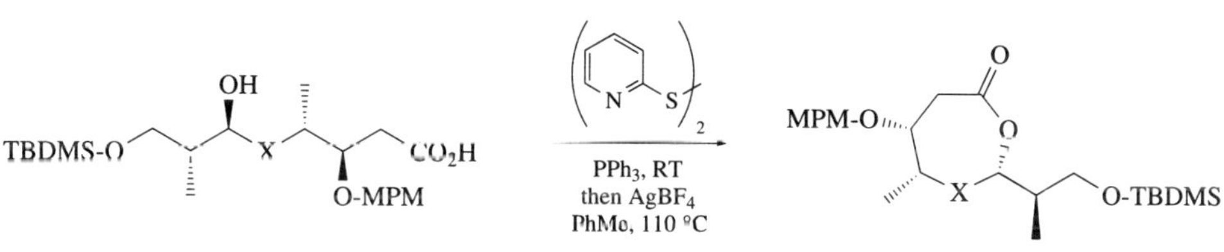

(230)

X = *cis*-CH=CH, CH_2–CH_2

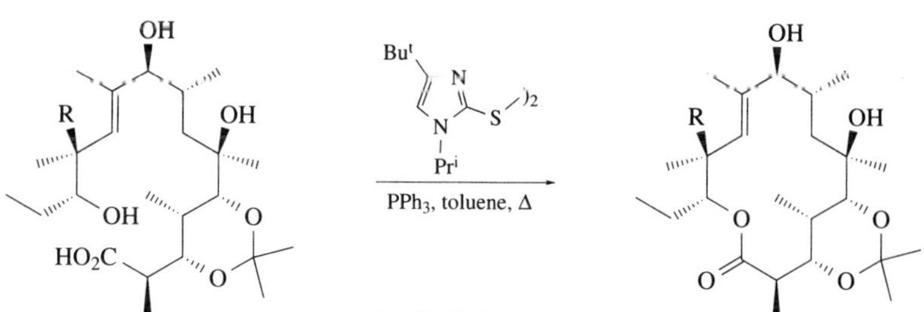

(231)

R = H, 50%
R = O-MTM, 30%

(→ Erythronolides A and B)

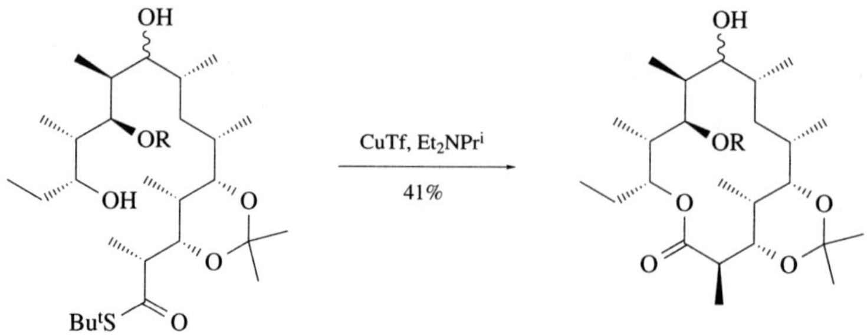

(232)

(→ 6-Deoxyerythronolide B)

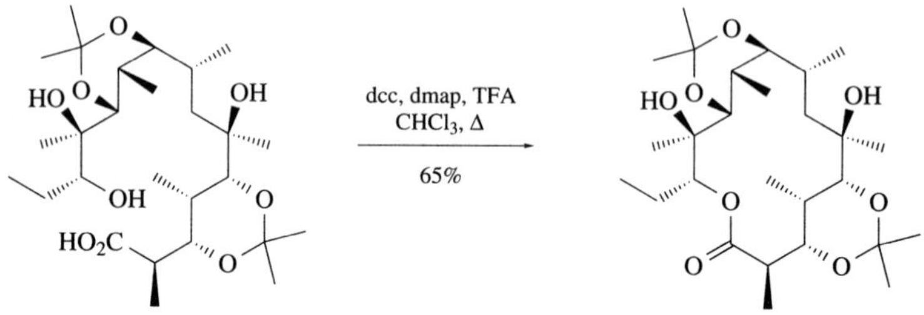

(233)

(→ Erythronolide A)

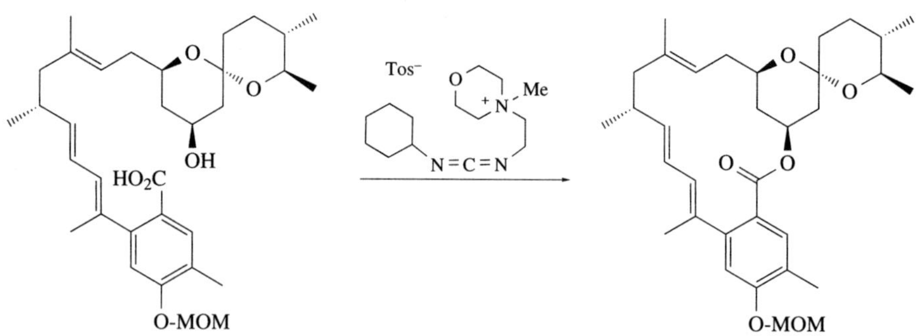

(234)

(→ Milbemycin β$_3$)

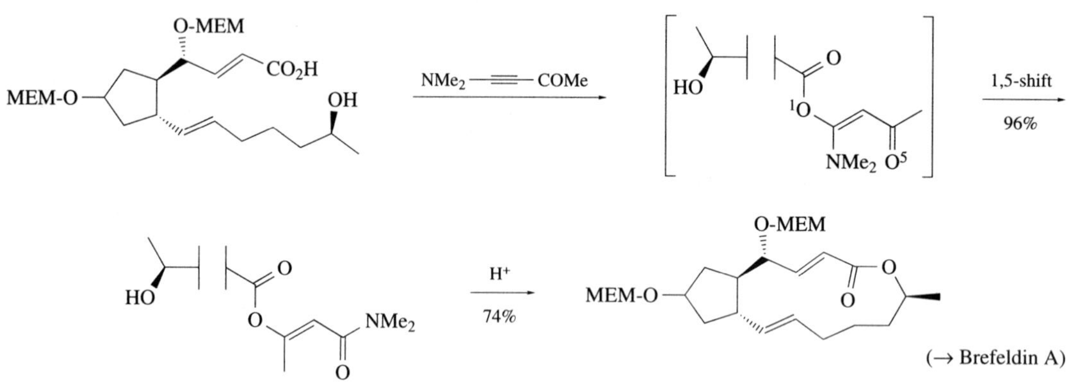

(→ Brefeldin A)

Scheme 73

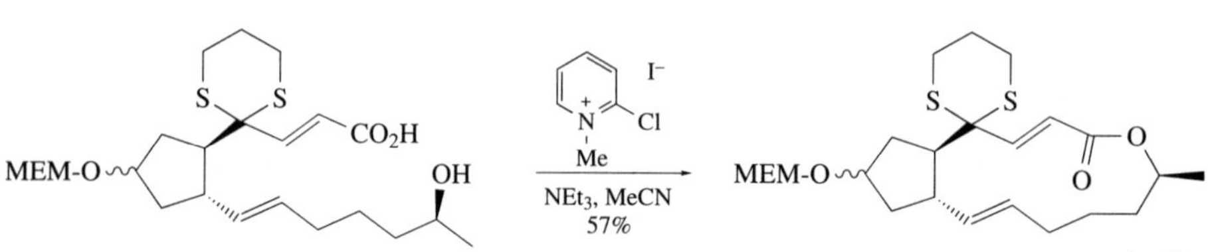

(235)

(→ Brefeldin A)

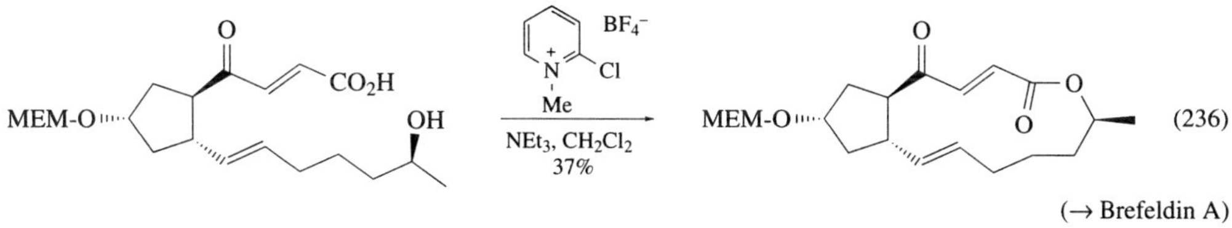

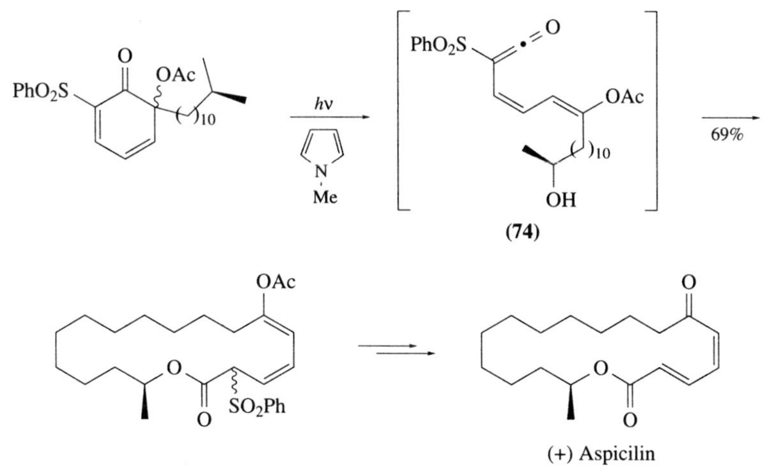

Scheme 74

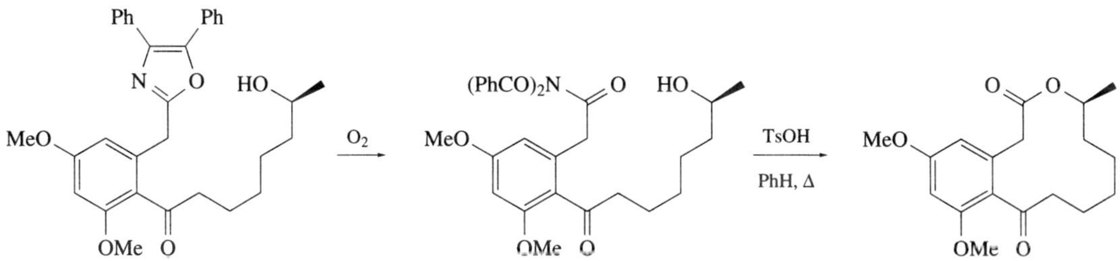

Scheme 75

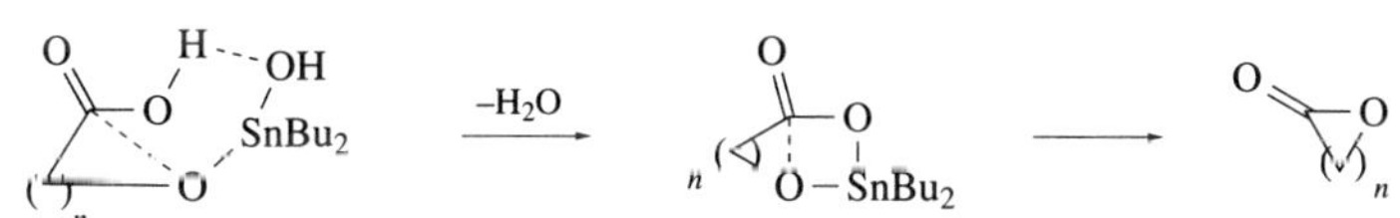

Scheme 76

5.03.6.4.2 HGA methods

This is the domain where the Mitsunobu reagent proves the most suitable macrolactonizing HGA reagent in many cases (e.g., Equations (237) ⟨77AG(E)264, 77JA646, 84TL2163⟩ and (238) ⟨86JOC4840⟩). Inversion of configuration is observed and the reagent is the best choice for diolide formation (Equation (237)). A more conventional procedure involves *O*-mesylation and S_N2-lactonization using caesium carbonate catalysis (Equations (239) ⟨81JA5183⟩ and (240) ⟨87JOC3937⟩). Table 2 contains some additional HGA and CGA lactonizations.

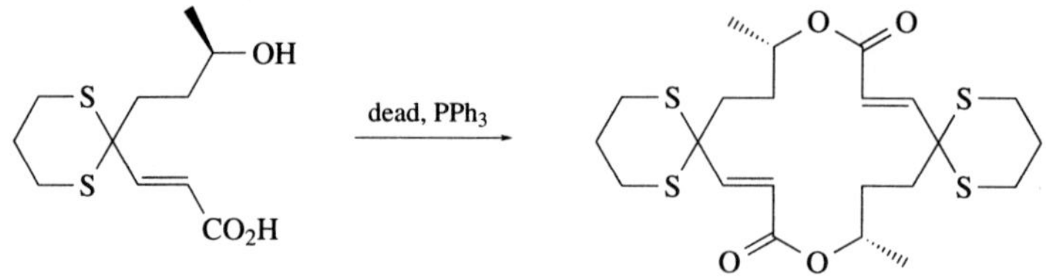

(237)

(→ Pyrenophorin)

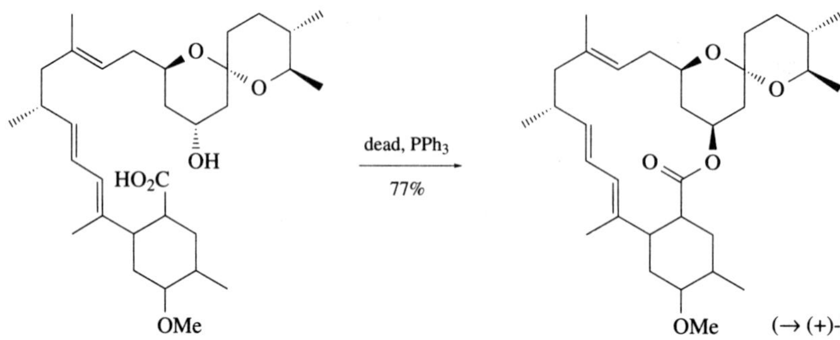

(238)

(→ (+)-Milbemycin β3)

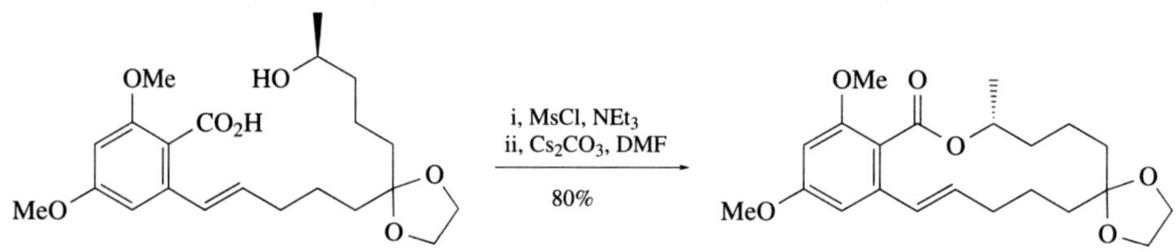

(239)

(→ Zearalenone)

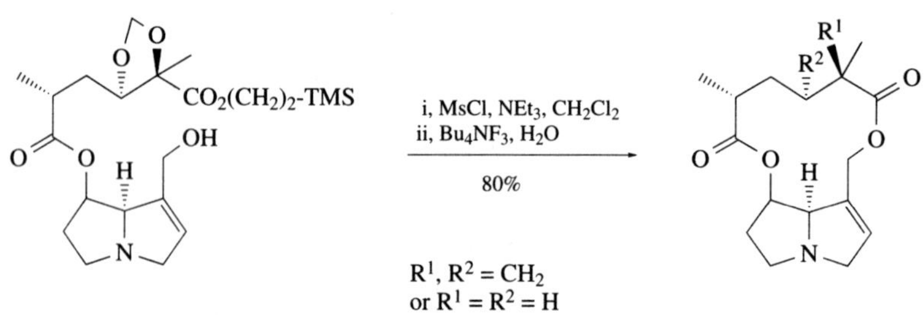

(240)

Table 2 Further CGA and HGA macrolactonizations.

Macrolide		*Type*	*Yield (%)*	*Ref.*
Verrucarin J	Bu^t–C(=O)–Cl, Et_3N	CGA	60	83JOC758
Narbanolide Vertaline	$(PhO)_2P(=O)$–Cl, Et_3N, dmap	CGA	54	85CPB532
Aplasmomycin (Diolide)	BOP–Cl, Et_3N	CGA	71	82JA6818
Gloesporone	dead, PPh_3	HGA	67	89LA1233
Latrurculine	dead, PPh_3	HGA		86JA2451
AB A 26771 B	Me_2N–CH(oneopenthyl)$_2$	HGA	39	80TL4611

5.03.6.4.3 *Special methods*

Triol (**75**) undergoes a glycol cleavage and oxidation to the carboxyl group which then lactonizes under the conditions (Equation (241)) ⟨82JOC3358⟩ and 9-membered lactones are best prepared not by direct lactonization, but by Ireland–Claisen rearrangement starting from precursor (**76**), which is converted into the ketene acetal by selenic acid elimination (Equation (242)) ⟨93JA5815⟩. Enzymatic macrolactonizations are rare in 1995, but may gain in importance (Equations (243) ⟨88TL5583⟩ and (244) ⟨89CL1775⟩).

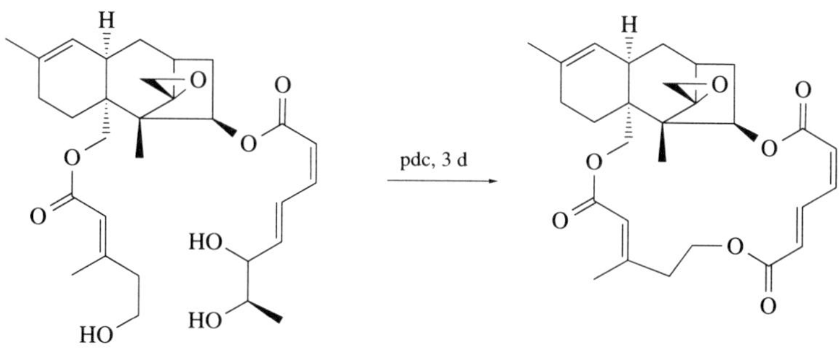

(**75**)

(241)

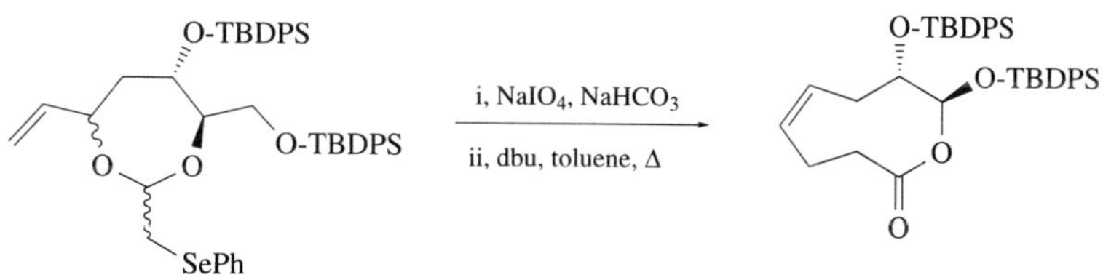

(**76**)

(242)

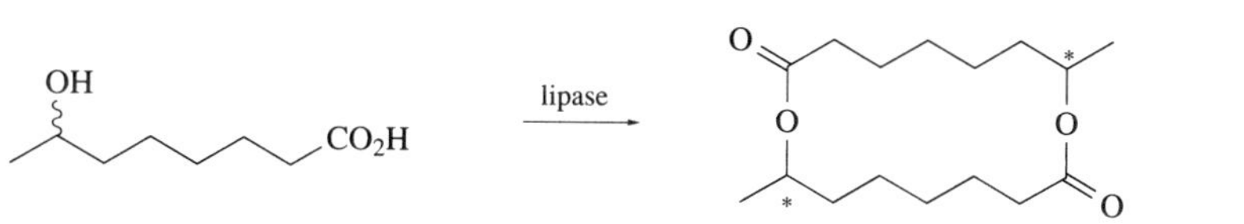

(243)

Enantioselectivity: $(R,R) : (S,S) = 96 : 4$
Diastereoselectivity: $(R, R + S, S) : (R, S) = 12 : 5$

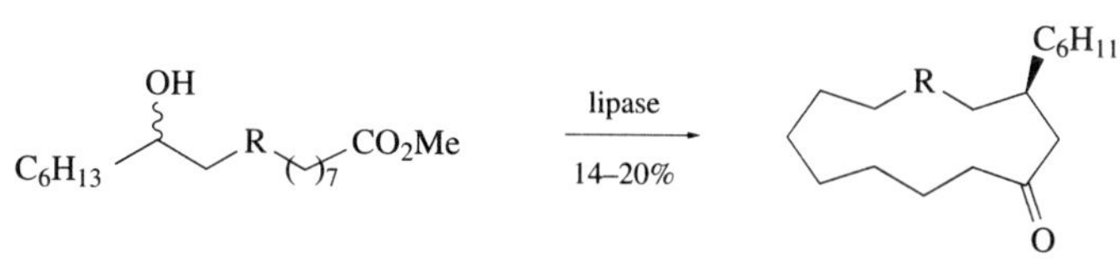

(244)

R = CH_2–CH_2, CH=CH (*E* and *Z*) *ee* >98%

5.04
Other Acyloxy Compounds

PETER T. GALLAGHER and JEREMY GILMORE
Lilly Research Centre Ltd., Windlesham, UK

5.04.1 ANHYDRIDES ($R^1COOCOR^2$)

5.04.1.1 General Methods of Synthesis

Reviews of the preparation of anhydrides are infrequent given the value of these reagents in synthesis. Of particular note are the reviews of Sandler and Karo and the section in *Methoden der Organischen Chemie* (*Houben-Weyl*) on the synthesis of anhydrides, as both contain selected preparative details ⟨B-86MI 504-01, 89HOU(5)633⟩. Four methods appear to be generally applicable to all types of anhydrides covered in this chapter and they are discussed in Section 5.04.1.1.1 to Section 5.04.1.1.4. These are by no means always the most efficient synthetic routes, and the headings by anhydride type should be consulted for more specific reagents.

5.04.1.1.1 From metal carboxylates and acyl chlorides

Methods for the synthesis of anhydrides have been documented since the 1850s when the reactions between acyl chlorides and anhydrous sodium acetate or potassium acetate were shown to produce anhydrides in 80–90% yield ⟨1853MI 504-01⟩. The olfactory properties of anhydrides were also appreciated very early on, benzoic acetic anhydride being described by Gerhardt as, 'et d'une agréable odeur de vin d'Espagne'. Apart from the early synthetic method development phase, the impetus for the synthesis of anhydrides has been the need to find appropriate reactive acylation reagents for transformations such as the Friedel–Crafts reaction, or in peptide coupling.

The sodium carboxylate–acyl chloride coupling procedure has proved to be applicable to the synthesis of both symmetrical and unsymmetrical anhydrides in excellent yields. Benzene has been used as a solvent ⟨85S681⟩, as has water when pyridine was used as a catalyst ⟨51JA4911, 64JCS755⟩. Examples of anhydrides which have been synthesised from sodium salts of acids and acyl chlorides are listed in Table 1.

Table 1 Anhydrides ($R^1COOCOR^2$) formed by reaction of sodium carboxylates (R^1CO_2Na) and acyl chlorides (R^2COCl).

R^1	R^2	*Reagents and conditions*	*Yield* (%)	*Ref.*
Ph	Me	RT	80–90	1853MI 504-01
(1)	**(2)**	C_6H_6, 3 h	77	51JA4911
Me	Ph	Chloroform, water	88	79CC179
Ph	Ph	Water, pyridine	96	64JCS755
$C_{15}H_{31}$	C_7H_{15}	90 °C, 2 h	85	60JOC1265

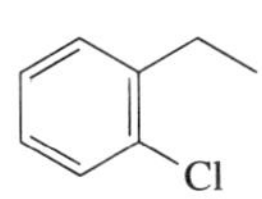

(1)

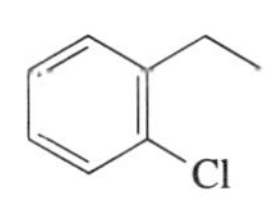

(2)

The reaction of acyl halides with sodium carboxylates has been modified by changing the metal to thallium when quantitative yields of the anhydride derived from thallium 2-methylpropanoate and 1,1-dimethylpropanoyl chloride were produced in ether at room temperature (Scheme 1) ⟨68JA2422⟩.

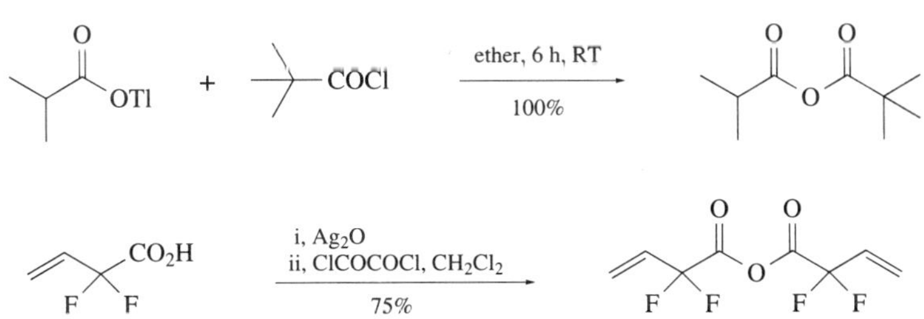

Scheme 1

Alternatively the silver salt of a carboxylic acid may be made from silver oxide, and then treatment with oxalyl chloride partially converts the silver salt into an acid chloride. Anhydride formation then results from coupling of the silver salt to the acid chloride (Scheme 1) ⟨86TL4437⟩.

Other modifications of the acyl chloride and sodium alkanoate method include the formation of the sodium salt *in situ* by the addition of an acid chloride to an aqueous solution of sodium hydrogencarbonate containing pyridine, or alternatively, as in the synthesis of C_{14}–C_{18} alkanoic acid anhydrides at −10 °C, using 20% aqueous sodium hydroxide and toluene in a two-phase reaction with tetra-*n*-butylammonium chloride, when yields of anhydride are 78–90% ⟨68AG(E)465, 83TL4195⟩. Similarly, anhydrides have been prepared from acid chlorides in benzene or acetonitrile in 52–99% yield using sodium hydrogen carbonate and tetra-*n*-butylammonium bromide in solid–liquid phase transfer catalytic reactions ⟨90JCR(S)84⟩. The synthesis of anhydrides from acid chlorides can only produce symmetrical anhydrides but the mechanism differs from the sodium alkanoate–acetyl chloride reaction in requiring initial hydrolysis of the acid chloride, before nucleophilic attack of the alkanoate on the acid chloride.

5.04.1.1.2 *From carboxylic acids and acyl chlorides in the presence of an organic base*

The preparative reactions of anhydrides which have been most widely reported are the reactions between acid chlorides and carboxylic acids in the presence of an organic base in an appropriate solvent, examples of which are shown in Table 2. The reaction of acid chlorides with carboxylic acids is of value as many acids are readily available and sensitive substrates are tolerated ⟨48JA2964⟩.

Anhydrous cobalt(II) chloride has been found to catalyse the formation of anhydrides in yields of 51–91% from acid chlorides and carboxylic acids in dichloromethane at 40 °C for 5–8 h ⟨92TL593⟩.

Table 2 Anhydrides ($R^1COOCOR^2$) formed by reaction of carboxylic acids (R^1CO_2H) and acyl chlorides (R^2COCl).

R^1CO_2H	*R^2COCl*	*Reagents and conditions*	*Yield (%)*	*Ref.*
n-$C_6H_{13}CO_2H$	n-$C_6H_{13}COCl$	Pyridine, C_6H_6, 60 °C	80	55OSC(3)28
n-$C_6H_{13}CO_2H$	n-$C_6H_{13}COCl$	Pyridine, ether, −10 °C	97	53JCS2117
(3)	**(4)**	Chloroform, pyridine	71	48JA2964
MeCOCl	MeCOCl	55 °C	50	1877LA191

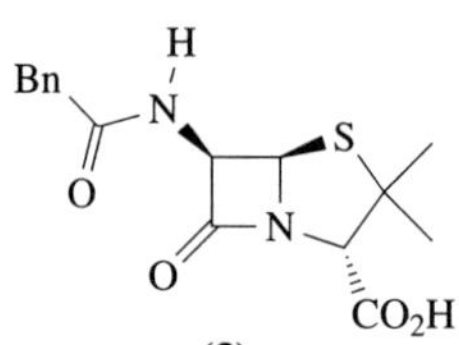

(3)

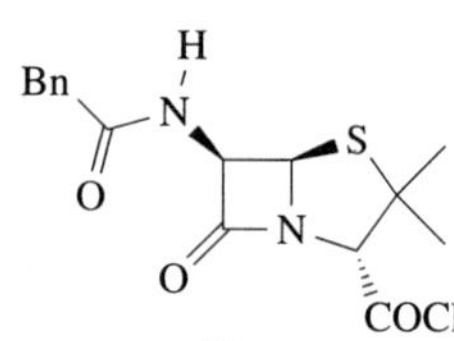

(4)

5.04.1.1.3 *Dehydration using acetic anhydride*

Acetic anhydride can be used as a dehydrating agent for the formation of anhydrides (Scheme 2). This may be achieved in the absence of catalysts or with acidic or basic catalysts ⟨09CB3484, 25CB1418, 57JA3508, 85LA1935⟩. Cyclic anhydrides are formed more rapidly than acyclic anhydrides ⟨66JIC615⟩. Conditions exist also for the synthesis of unsymmetrical acyclic anhydrides on heating in acetic anhydride, though, in the absence of catalyst, reaction time and reagent quantities are significant factors in determining the ratio of unsymmetrical to symmetrical anhydride formation ⟨87BCJ4325⟩. Autenrieth studied the preparation of anhydrides from carboxylic acids and acetic anhydride and found that improvements in yield of the unsymmetrical anhydrides resulted from increasing the ratio of acid to acetic anhydride from 1 : 1.3 to 1 : 3 and increasing the reaction time from 1 h to 8 h ⟨01CB168⟩.

5.04.1.1.4 *Polymer-assisted anhydride formation*

Fife *et al.* have demonstrated that commercially available poly-4-vinylpyridine (Poly-4-VP) catalyses the formation of both symmetrical and unsymmetrical anhydride formation in high yield, with short reaction times and at temperatures from 0 °C to 25 °C ⟨86JOC3744, 86TL4933, 86TL4937⟩. Symmetrical anhydrides may be synthesised by adding thionyl chloride to a solution of carboxylic acids in the presence of polymer (Scheme 3). Additionally, unsymmetrical anhydrides may be synthesised by mixing equimolar equivalents of acid and acid chloride in dichloromethane in the presence of the polyvinylpyridine catalyst. The mechanism of formation of anhydrides using the catalyst is depicted in Scheme 3.

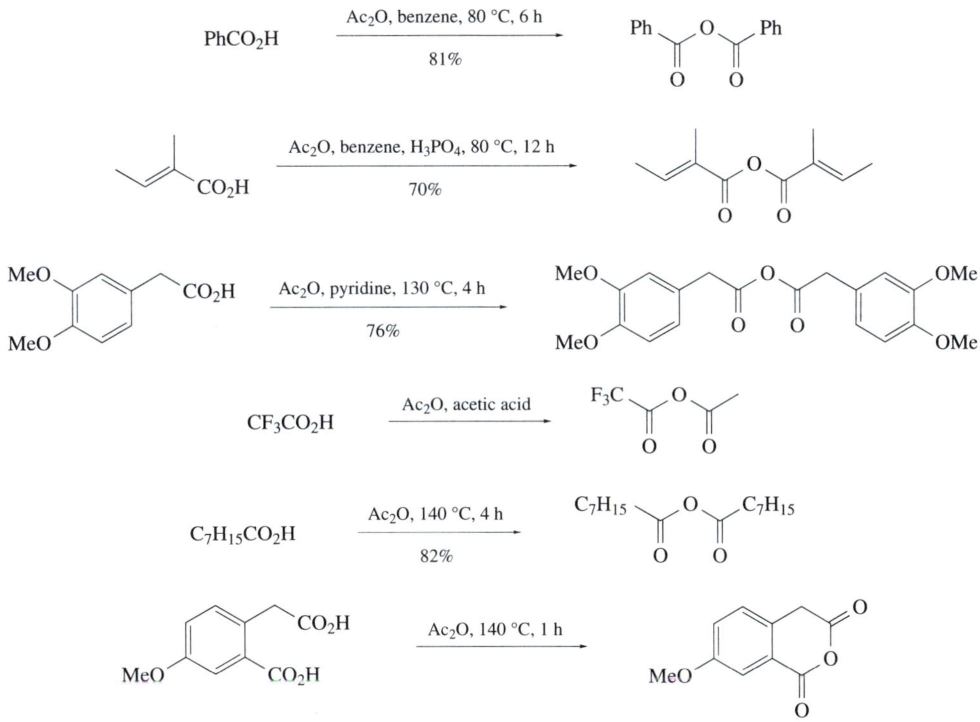

Scheme 2

5.04.1.2 Symmetrical Alkanoic Anhydrides

5.04.1.2.1 From metal carboxylates and acyl chlorides

Acid chlorides have been reacted with the sodium salts of carboxylic acids to produce symmetrical anhydrides (see Section 5.04.1.1.1).

5.04.1.2.2 From carboxylic acids or derivatives of carboxylic acids activated towards nucleophilic substitution

Carbodiimides as dehydrating agents have found widespread use in the formation of anhydrides (Scheme 4) ⟨85T5331, 87MIP8707604⟩.

The carbodiimide reagent most cited is dicyclohexylcarbodiimide (dcc) which has been used to produce symmetrical anhydrides as reactive intermediates in the synthesis of peptides ⟨63JA1997, 69ZN(B)314⟩. However, Yamashiro has isolated and characterised a number of peptide-related symmetrical anhydrides, representative examples of which are shown in Table 3 ⟨87MI 504-01⟩.

Benoiton and co-workers have introduced ammoniocarbodiimides (Scheme 5), which have simplified the isolation procedures leading to anhydrides since the ammoniourea by product is water soluble ⟨78S928⟩. Thus *N*-(dimethylaminopropyl)-*N*′-ethylcarbodiimide hydrochloride has been reacted with either *N*-benzyloxycarbonyl or *N*-*t*-butoxycarbonyl-protected amino acids in dichloromethane at 0°C to produce, after isolation, a range of characterised anhydrides in good yields (Table 4).

Methyl succinate has also been reacted with *N*-(dimethylaminopropyl)-*N*′-ethylcarbodiimide hydrochloride to produce the corresponding symmetrical anhydride ⟨85JOC1112⟩.

Ketene has been used as an activating and dehydrating agent in the formation of symmetrical anhydrides in yields greater than 80% (Equation (1)). The mechanism of this reaction involves initial mixed anhydride formation followed by nucleophilic displacement with a second equivalent of carboxylic acid, resulting in the formation of acetic acid ⟨55OSC(3)165⟩.

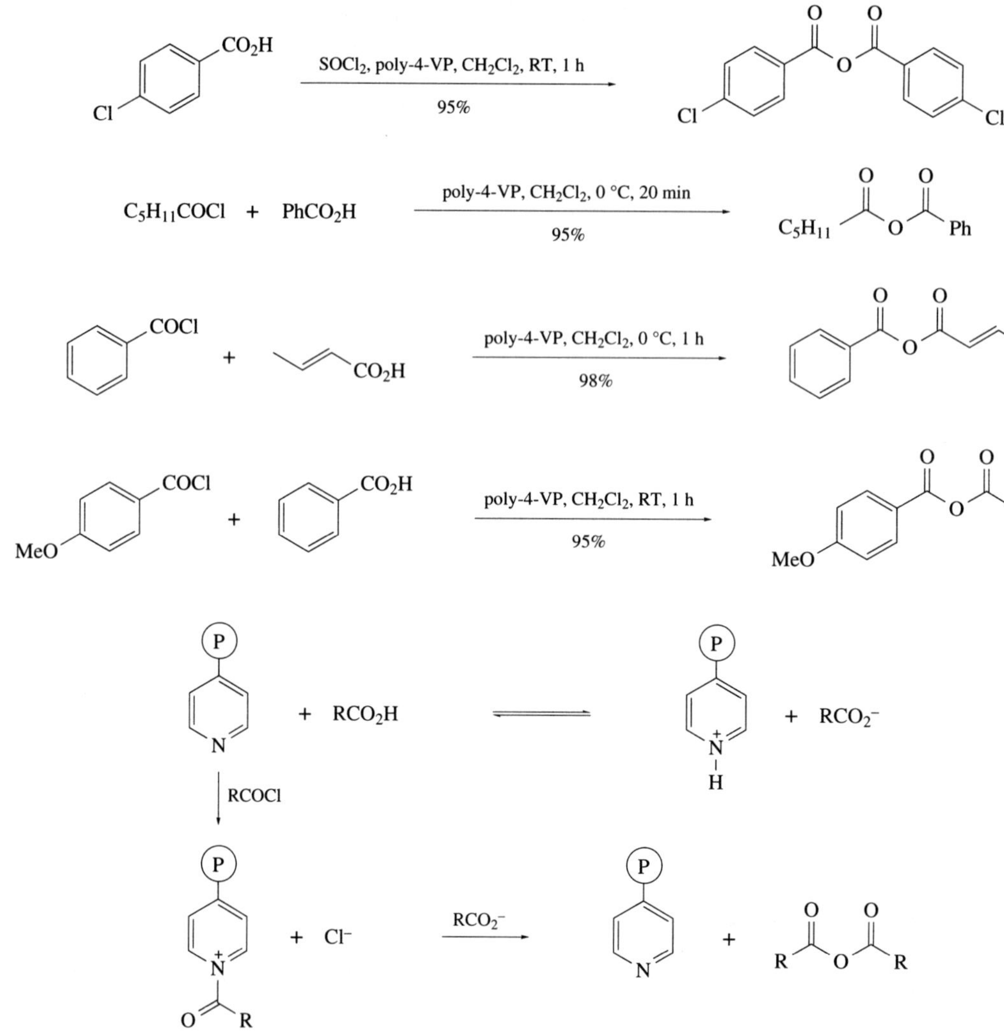

Scheme 3

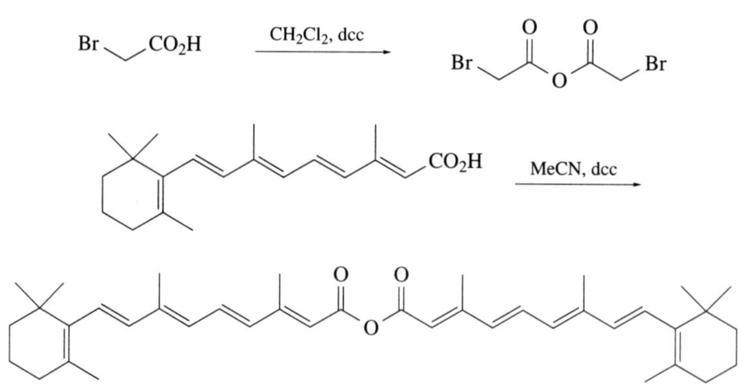

dcc = dicyclohexylcarbodiimide

Scheme 4

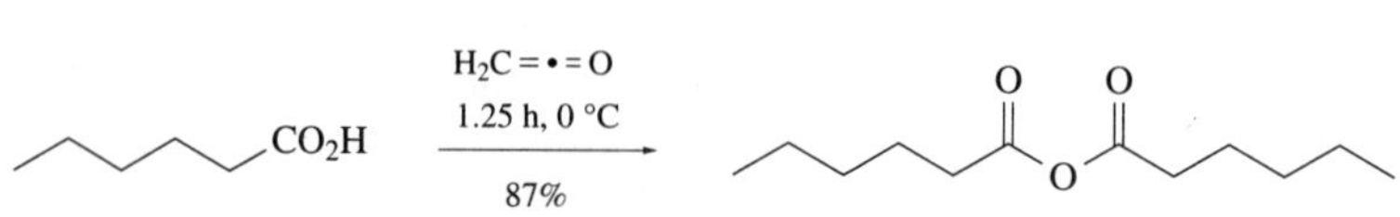 (1)

Table 3 Anhydrides formed by reaction of dcc with *t*-BOC-amino acids.

t-BOC-Amino acid[a]	*Reaction time and temp.*	*Yield of anhydride* (%)
(**5**)	1.5 h, −20 °C	99
(**6**)	1.5 h, −20 °C	100
(**7**)	1.5 h, −20 °C	96
(**8**)	1.5 h, 0 °C	97
(**9**)	1.5 h, −20 °C	100

[a] *t*-BOC = *t*-butoxycarbonyl.

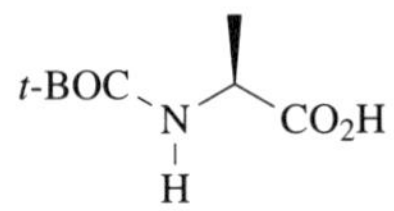

(**5**)

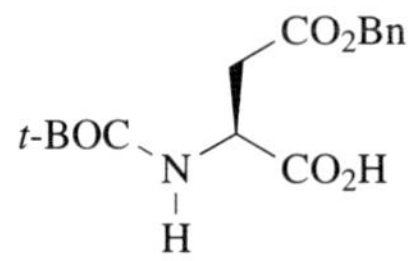

(**6**)

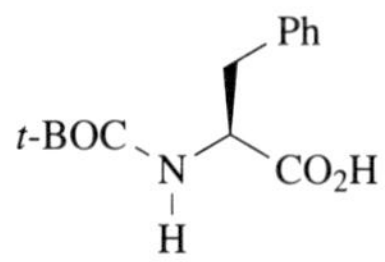

(**7**)

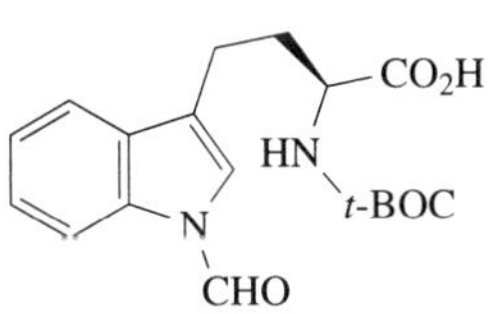

(**8**)

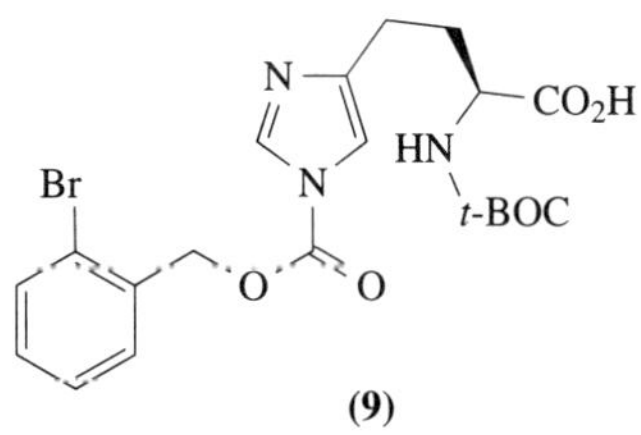

(**9**)

Table 4 Anhydrides formed by reaction of *N*-protected amino acids and *N*-(dimethylaminopropyl)-*N'*-ethylcarbodiimide hydrochloride.

(**10**) R^2	R^1 = *benzyl* *Yield* (%)	R^1 = *t-butyl* *Yield* (%)
H- (Gly)	80	60
Me- (L-Ala)	81	75
Me_2CH- (L-Val)	80	51
EtCHMe- (L-Ileu)	86	33
$PhCH_2$- (L-Phe)	77	76
[$CH_2CH_2CH_2$] (L-Pro)	90	80

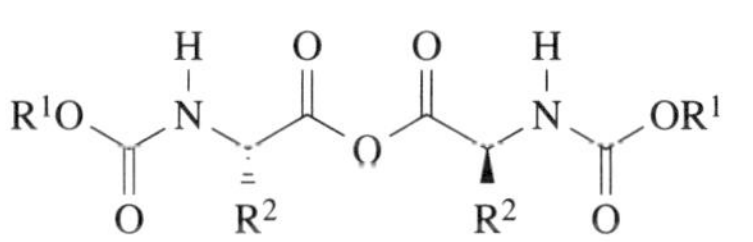

(**10**)

Acid chlorides may be heated with acetic anhydride (Equation (2)) to produce symmetrical anhydrides ⟨59JOC388⟩.

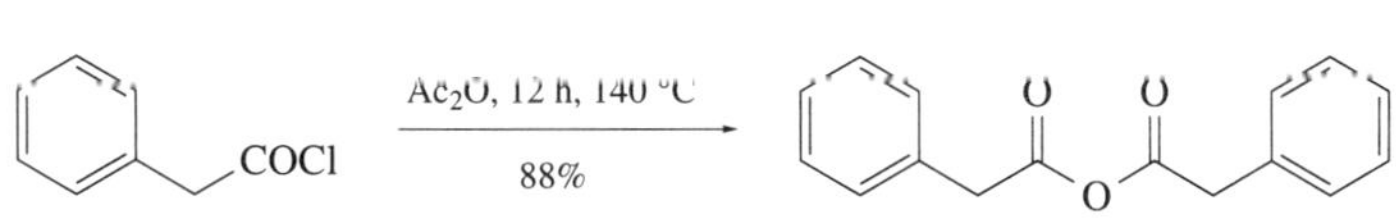

(2)

Acetic anhydride has also been frequently used as a dehydrating agent for anhydrides (see Section 5.04.1.1.3).

A widely applicable synthesis of symmetrical anhydrides from carboxylic acids is that due to Palomo-Coll and co-workers, who have used *N*,*N*-bis(2-oxo-3-oxazolidinyl)phosphorodiamidic

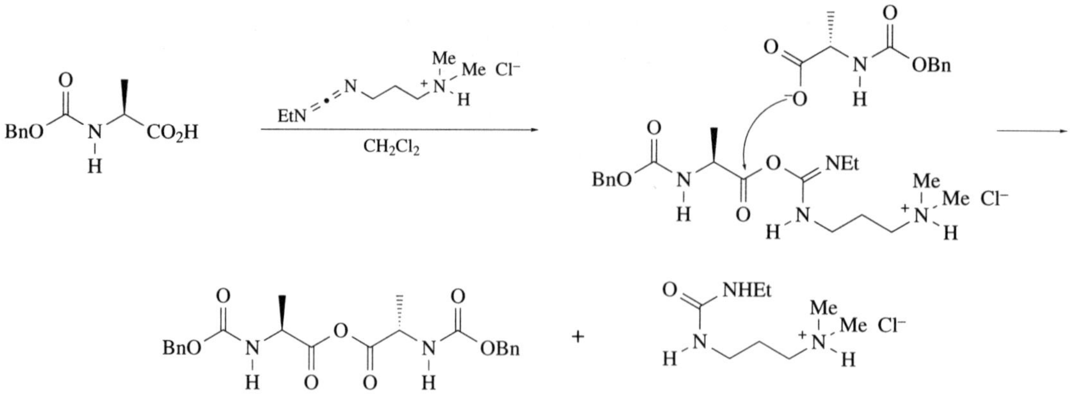

Scheme 5

chloride as an activating reagent ⟨81S616⟩. This method has several advantages, including short reaction times (typically 0.5 h) at room temperature with a reagent that is commercially available. The presence of electron-withdrawing or electron-donating groups in the acid does not affect the efficiency of the reaction and sterically hindered acids form anhydrides in high yields (Table 5).

Table 5 Anhydrides ($R^1COOCOR^1$) formed by reaction of carboxylic acids (R^1CO_2H), *N*,*N*-bis(2-oxo-3-oxazolidinyl)phosporodiamidic chloride and triethylamine in dichloromethane at RT.

R^1CO_2H	*Time* (h)	*Yield* (%)	*Anhydride*
Pr^nCO_2H	0.5	95	$Pr^nCOOCOPr^n$
$C_6H_{13}CO_2H$	0.5	97	$C_6H_{13}COOCOC_6H_{13}$
$C_{11}H_{23}CO_2H$	0.5	98	$C_{11}H_{23}COOCOC_{11}H_{23}$
$C_{17}H_{35}CO_2H$	0.5	97	$C_{17}H_{35}COOCOC_{17}H_{35}$
$BrCH_2CO_2H$	0.75	92	$BrCH_2COOCOCH_2Br$
Bu^tCO_2H	3	92	$Bu^tCOOCOBu^t$
(**11**)	0.5	98	(**12**)

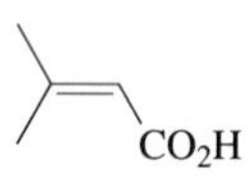

(**11**)

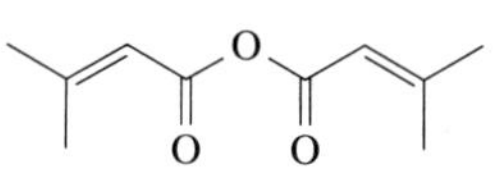

(**12**)

The mechanism of formation of anhydrides synthesised using the phosphorodiamidic chloride involves initial attack at phosphorus with carboxylate anion, then attack on the phosphoramidate ester with a second equivalent of carboxylate anion to produce the anhydride and a phosphoramidic acid (Scheme 6).

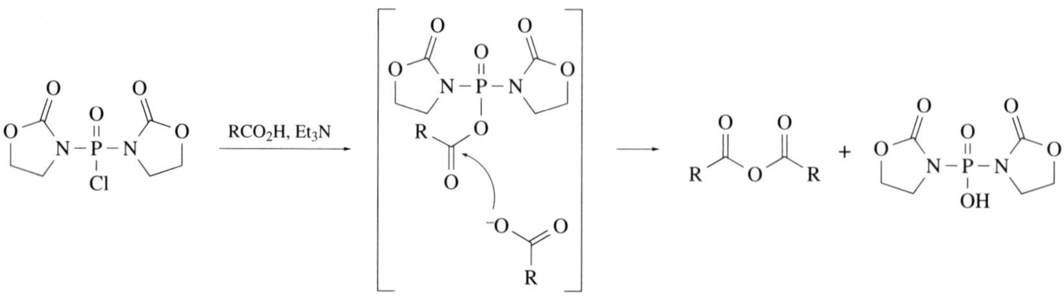

Scheme 6

Symmetrical anhydrides may also be synthesised from acids with either *N*-phenylphosphoramidochloridate or diphenylphosphorochloridate (Scheme 7) ⟨81S218⟩.

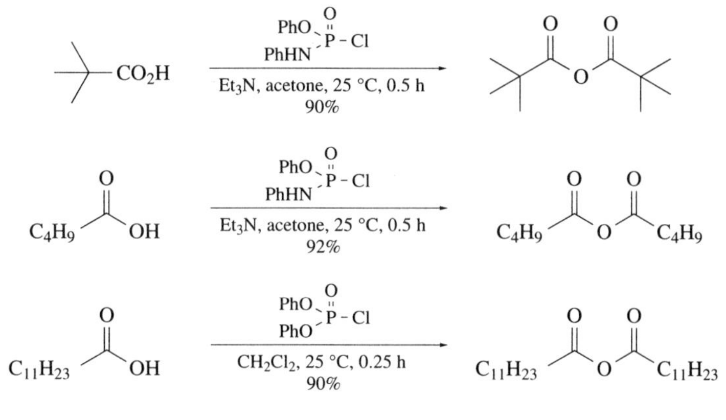

Scheme 7

Ethyl chloroformate and dialkyl carbonates have been used similarly to activate carboxylic acids to attack by a second equivalent of carboxylate anion ⟨63JOC1905, 63MI 504-01, 86ZOB613⟩. The use of dialkyl pyrocarbonates has been exemplified by Pozdnev, who has demonstrated that di-*t*-butylpyrocarbonate is an effective reagent for the formation of anhydrides of cinammic acid (93%) or furylacrylic acid (92%), as well as in the formation of anhydrides of *N*-protected amino acids ⟨92MI 504-01⟩.

Other activating groups include the imidazolide functionality generated from 1,1′-oxalyldiimidazole (Scheme 8) and, for example, hexadecanoic acid. Subsequent reaction with an equimolar quantity of hexadecanoic acid under methanesulfonic acid catalysis in acetonitrile has proved to be effective in the formation of hexadecanoic acid anhydride ⟨87CPB1262⟩.

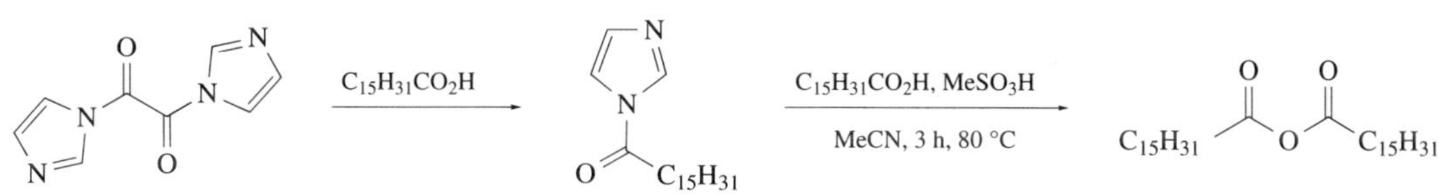

Scheme 8

Methanesulfonyl chloride (Equation (3)) has been used as an activating group for both alkyl and aryl carboxylic acids ⟨84JCR(S)100⟩. Reaction times are short (2–3 h), the isolation process is straightforward, and yields in the range 85–95% have been demonstrated for alkanoic acids (R = Et, Pr^i).

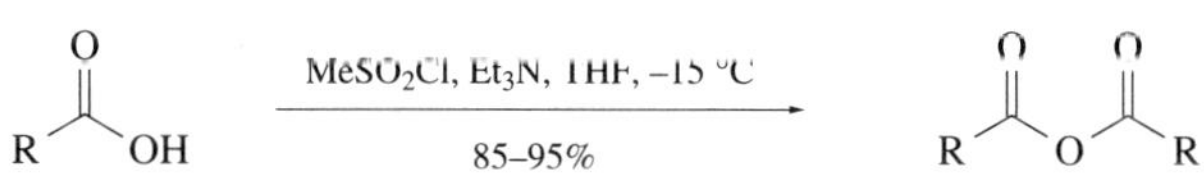

(3)

Similarly, reaction between carboxylic acids and chlorosulfonyl isocyanate has provided a variety of anhydrides ⟨82S506⟩. Anhydride formation is usually carried out in dichloromethane at 0°C in the presence of triethylamine, and yields of symmetrical anhydrides are generally in excess of 70%.

The use of trimethylsilyl esters and acid chlorides has been patented for the thermal production (140°C) of anhydrides in high yield (Table 6), trimethylsilylchloride being the by-product in this reaction ⟨83GEP3144791⟩.

Acyl trimethylsilylphosphates have been reacted with aqueous sodium hydrogen carbonate to produce a limited number of symmetrical anhydrides ⟨87IJC(B)407⟩.

Alkoxyalkynes have been used as dehydrating and activating agents for carboxylic acids in the formation of symmetrical anhydrides. Methoxyalkyne was originally used for this purpose and more recently Kita *et al.* have advocated the use of ethoxytrimethylsilylalkyne. Both reagents provide anhydrides in excellent yield (Scheme 9) ⟨54JCS1860, 84TL6027, 86JOC4150⟩.

Diethylaminopropyne has been used as an alternative to dicyclohexylcarbodimide in peptide

Table 6 Anhydrides formed by the reaction silyl esters of carboxylic acids (R^1CO_2TMS) and acyl chlorides (R^2COCl).

R^1	R^2	*Anhydride*	*Yield* (%)
Me	Me	$MeCO_2COMe$	85
Et	Et	$EtCO_2COEt$	96
Bu^t	Bu^t	$Bu^tCO_2COBu^t$	95
c-C_6H_{11}	c-C_6H_{11}	c-$C_6H_{11}CO_2COc$-C_6H_{11}	96

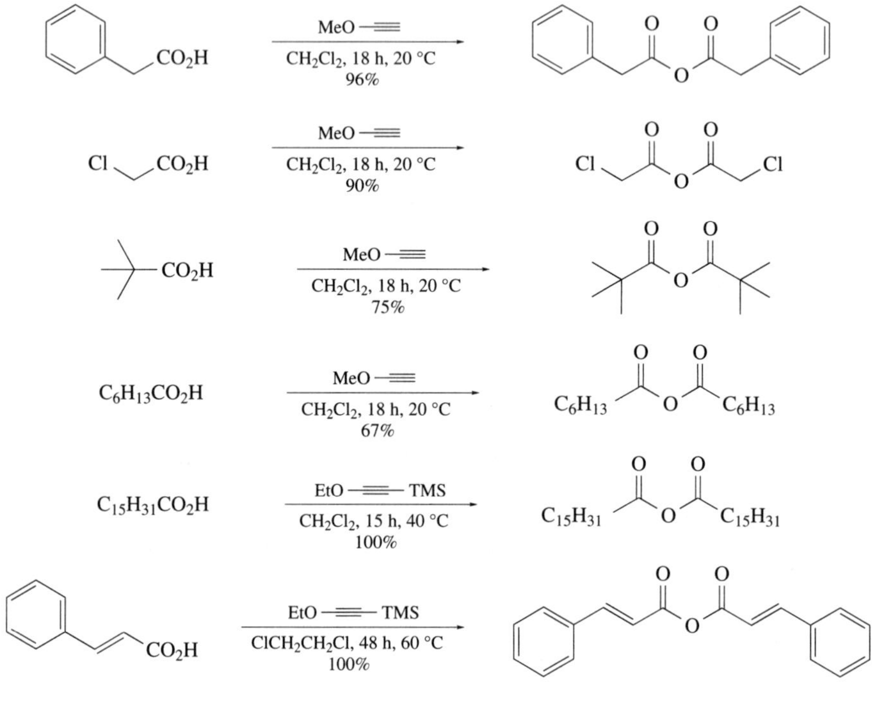

Scheme 9

couplings, where intermediate anhydride formation is required, prior to coupling to a peptidyl amine ⟨69ZN(B)314⟩.

The intermediates suggested (Scheme 10) in the formation of anhydrides from acids using alkoxyalkynes requires addition of the carboxylic acid to the alkyne to produce a ketene acetal, which is then attacked by a second equivalent of acid to form an *ortho*-ester. Elimination of either ethyl- or trimethylsilylacetate results in anhydride formation.

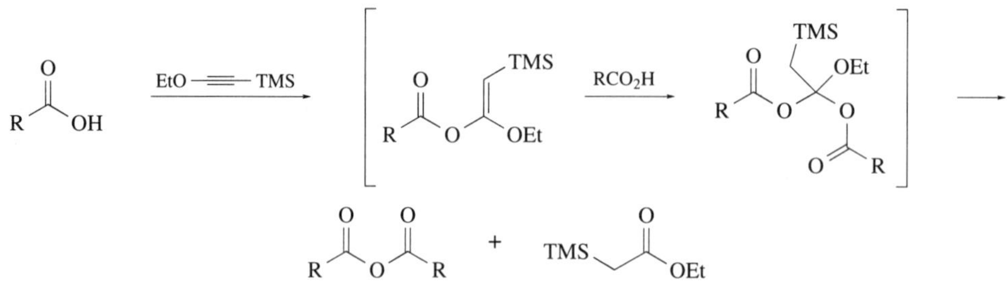

Scheme 10

An interesting synthesis of symmetrical C_3–C_5 anhydrides involves the reaction of dibenzoylethene with two equivs of a carboxylic acid in the presence of tributylphosphine ⟨64JOC1385⟩. The mechanism of the reaction (Scheme 11) is thought to include attack by tributylphosphine on dibenzoylethene to form an initial adduct which then deprotonates an equivalent of carboxylic acid to form a phosphonium carboxylate and dibenzoylethane. The phosphonium carboxylate is then attacked by a second equivalent of carboxylate anion, producing the anhydride and triphenylphosphine oxide.

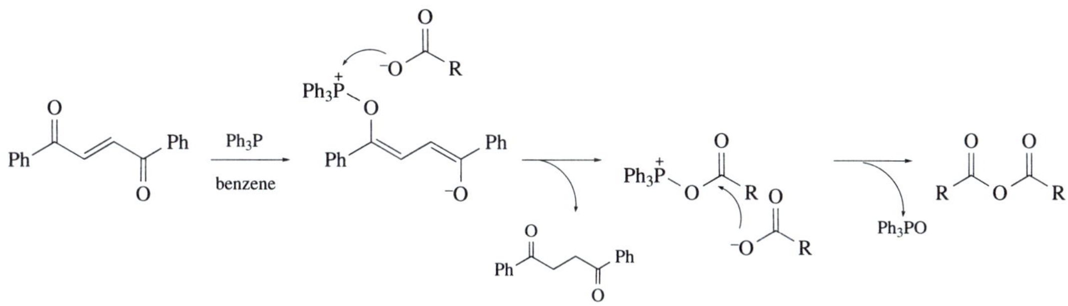

Scheme 11

Tetracyanoethylene and carboxylic acids with pyridine as activating base allow anhydride formation in moderate yields ⟨85TL1503⟩. However, this reaction is particularly good for the synthesis of adamantyl anhydrides; adamantane-1-carboxylic acid and adamantane-1-acetic acid anhydrides were both synthesised in 80% yield.

Cyanogen bromide is known to convert carboxylic acids to nitriles at high temperatures under sealed-tube conditions (Scheme 12). This reaction involves the intermediacy of a carboxylic cyanic anhydride. In anhydrous benzene the carboxylic cyanic anhydride may be trapped and, in the presence of pyridine, symmetrical anhydrides may be produced in yields of 78–95% ⟨73SC63⟩.

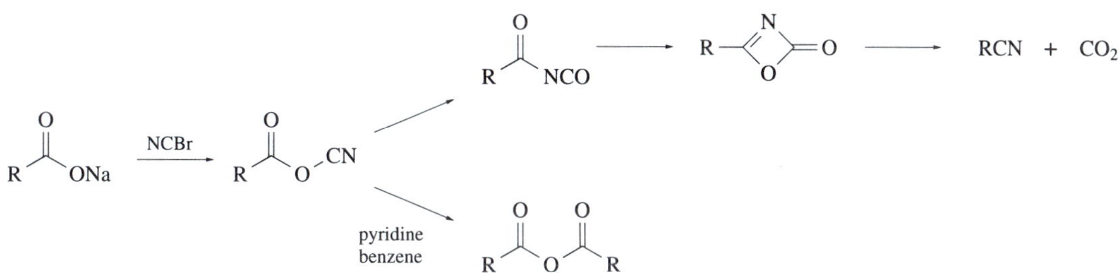

Scheme 12

Chloro(trisdimethylamino)phosphonium perchlorate (Scheme 13) was developed as a reagent for peptide coupling but forms anhydrides in good yield ⟨72TL4747⟩.

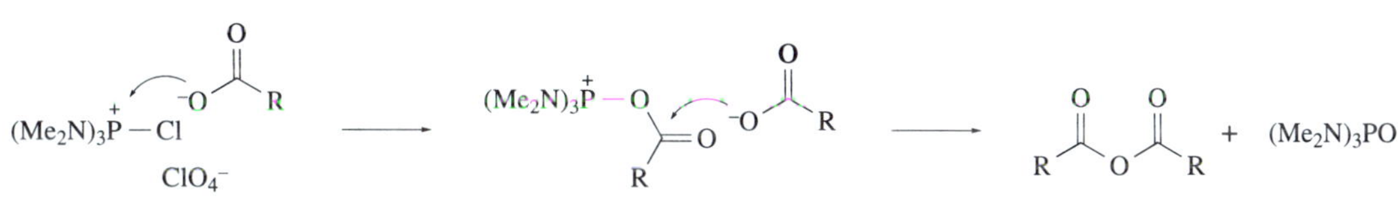

Scheme 13

p-Benzoquinonedibenzenesulfonamide reacts with triphenylphosphine to form a zwitterionic intermediate which then reacts with carboxylic acids to produce anhydrides (Scheme 14). The mechanism is probably similar to that found with chloro(trisdimethylamino)phosphonium perchlorate, though no base is required as the sulfonamide anion serves to deprotonate the carboxylic acid ⟨69TL4563⟩.

A patented process for the production of low-molecular-weight saturated haloanhydrides (Equation (4)) involves introducing, for example, chloroacetic acid in toluene to boiling acetic or propionic anhydride and immediately distilling from the reaction vessel either acetic or propionic acids with

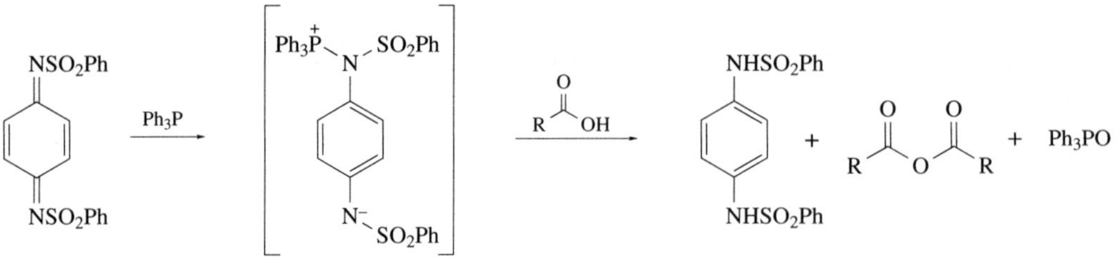

Scheme 14

toluene ⟨61GEP1100612⟩. Substituents may be bromine, chloromethyl, bromomethyl, trichloromethyl and iodomethyl and yields range from 78% to 97%.

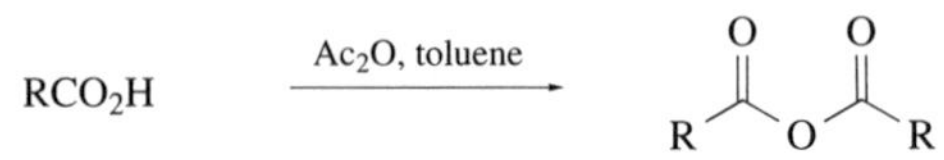

(4)

5.04.1.2.3 From alkene derivatives

Recent industrial syntheses of anhydrides have used the insertion of carbon monoxide into simple esters or alkenes over either rhodium or nickel–molybdenum catalysts. Thus, supported catalysts such as [1,2-bis(diphenylphosphino)butane]rhodium(I) chloride on silica pellets have been used to convert methyl acetate at 180°C and 15 bar in the presence of iodomethane and carbon monoxide to acetic anhydride with 98% selectivity ⟨86GEP3511050, 89GEP3811344⟩. Similarly, ethene has been carbonylated with a mixture of iodoethane, nickel iodide, molybdenum hexacarbonyl, triphenylphosphine and hydrogen at 170°C for 16 h also to prepare acetic anhydride ⟨82USP4335058⟩. Bicyclo[2.2.1]alkenes have been carbonylated (Scheme 15) with concomitant addition of phenylethene under palladium catalysis ⟨83TL813⟩.

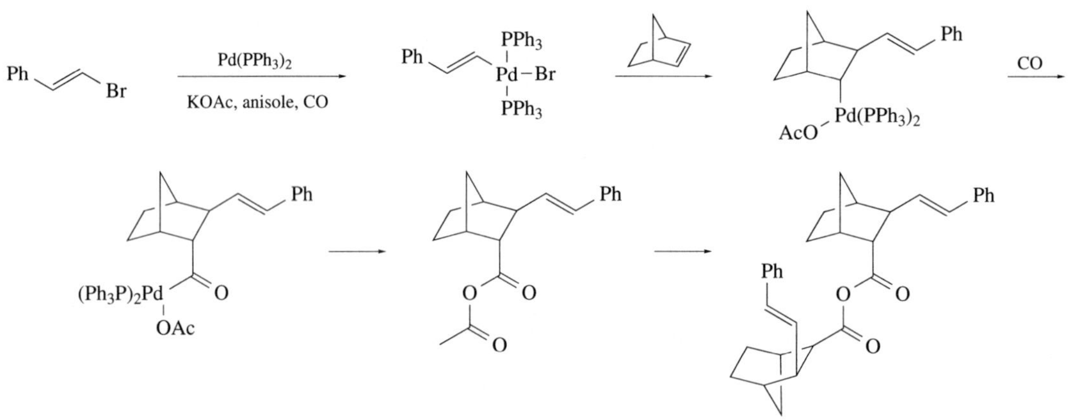

Scheme 15

5.04.1.3 Unsymmetrical Alkanoic Anhydrides

5.04.1.3.1 From metal carboxylates and acyl chlorides

One of the older methods used for the synthesis of unsymmetrical alkanoic anhydrides has been the reaction between metal carboxylate anions and acyl chlorides (see Section 5.04.1.1.1). This has been carried out in the absence of solvent at 90°C ⟨60JOC1265⟩ and is typified by Equation (5).

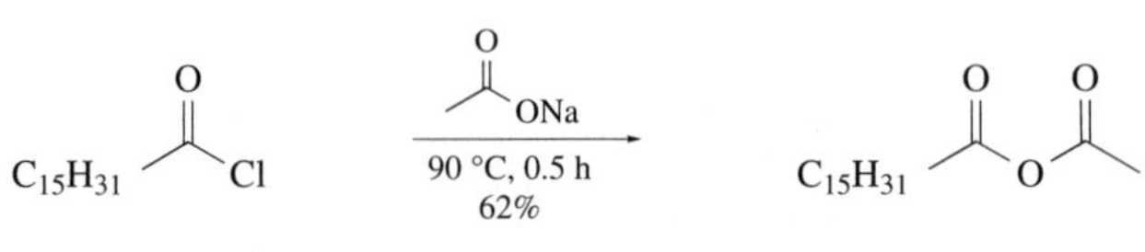

(5)

5.04.1.3.2 From carboxylic acids and acyl chlorides in the presence of an organic base

Palmitic acid may be coupled directly to C_{18} acid chloride at 100 °C under reduced pressure to produce the corresponding unsymmetrical anhydride in 87% yield ⟨58MI 504-01⟩.

As with symmetrical alkanoic anhydrides, the need to activate protected amino acids has driven the search for coupling procedures which selectively couple acyl fragments to amines to form peptides. Thus, *N*-benzyloxycarbonylglycine is activated by coupling to 4-methylpentanoyl chloride in toluene at 0 °C in the presence of triethylamine and then reacted, without isolation, with an amine ⟨51JA5553⟩. Similarly, 7-carboxymethylheptanoic acid has been coupled to chloroacetyl chloride in carbon tetrachloride (Scheme 16) ⟨86EUP268820⟩.

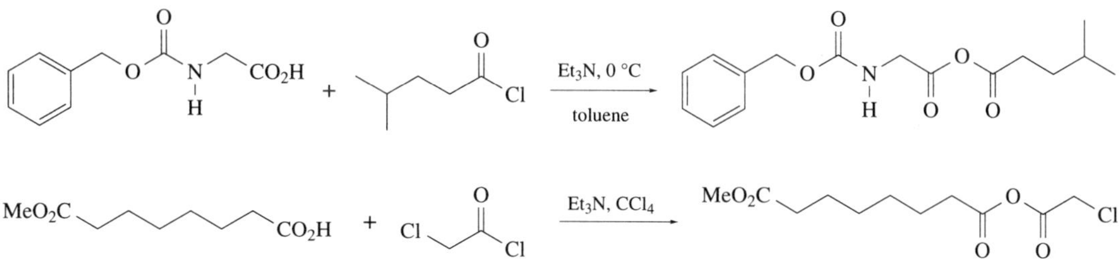

Scheme 16

Use of the reactivity of anhydrides has been made in the synthesis of the tetrazolyl cephalosporin cefazolin. Thus, tetrazolylacetic acid is first reacted with 2,2-dimethylpropanoyl chloride and triethylamine at −5 °C in dichloromethane to produce the mixed anhydride, which is then coupled to the aminocephalosporanic acid to produce cefazolin in an isolated yield of 90% (Scheme 17) ⟨85T5133⟩.

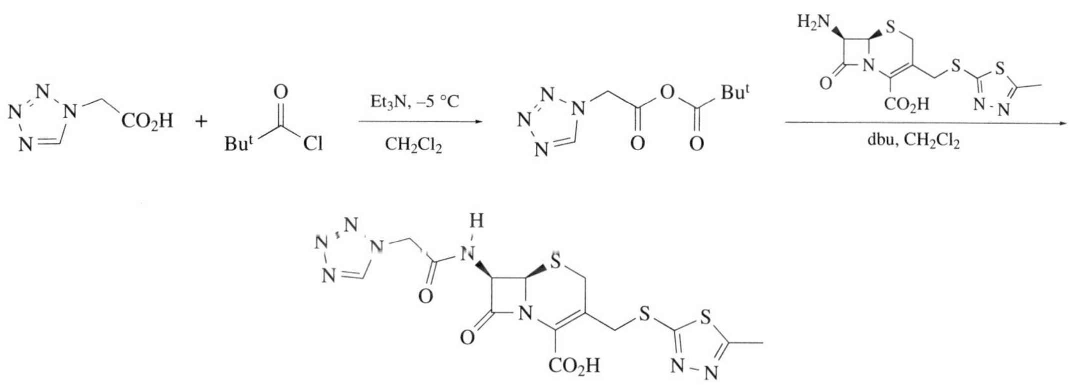

dbu = 1,5-diazabicyclo[5.4.0]undec-5-ene

Scheme 17

5.04.1.3.3 From anhydrides

The synthesis of certain unsymmetrical anhydrides can be effected by the heating of equimolar mixtures of anhydrides followed by distillation. Trifluoroacetic anhydride is particularly reactive under these conditions, forming acetyl trifluoroacetate in 86% yield ⟨54JCS2006⟩.

5.04.1.3.4 From carboxylic acids and ketenes

Ketene has been reacted at room temperature with chloroacetic acid to produce acetyl chloroacetate in 87% yield ⟨58ZOB1864⟩. Similarly, *N*-benzyloxycarbonylglycine has been reacted with diphenylketene to produce an activated anhydride for peptide coupling (Scheme 18) ⟨61CB1762⟩.

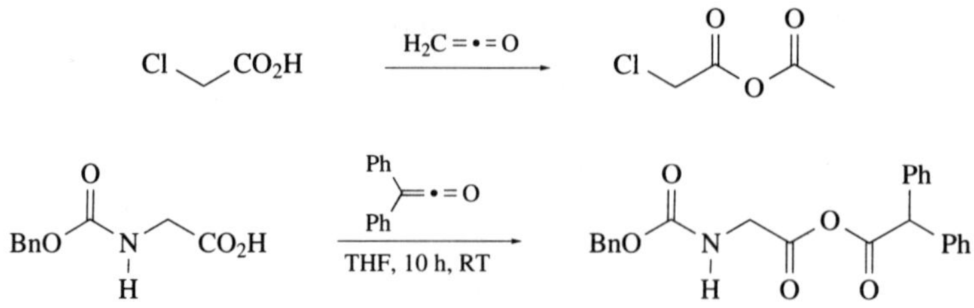

Scheme 18

5.04.1.3.5 From alkenes or imines

As with the synthesis of symmetrical anhydrides from alkenes (see Section 5.04.1.2.3) unsymmetrical anhydrides may be formed by the addition of carbon monoxide in the presence of acetic acid and acetic anhydride under copper and palladium catalysis to produce acetoxy anhydrides ⟨87JAP6289642, 88TL4435⟩. When R = Bn, Bu or nonyl, yields range from 84% to 95% (Equation (6)).

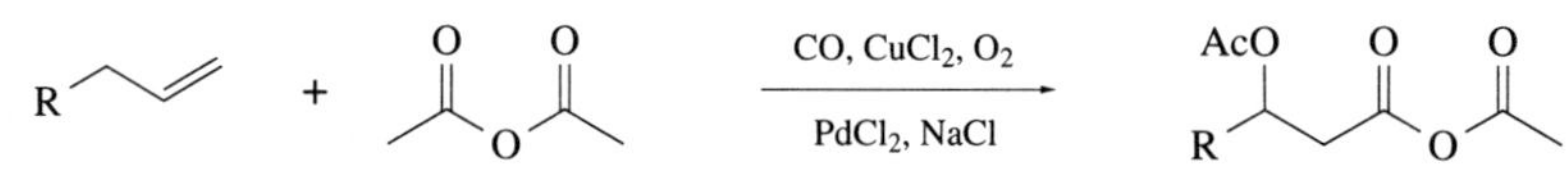

(6)

After extended reaction times, acetic anhydride adds to imines to produce aminopropanoic acetic anhydride derivatives (Equation (7)) ⟨50JA3874⟩.

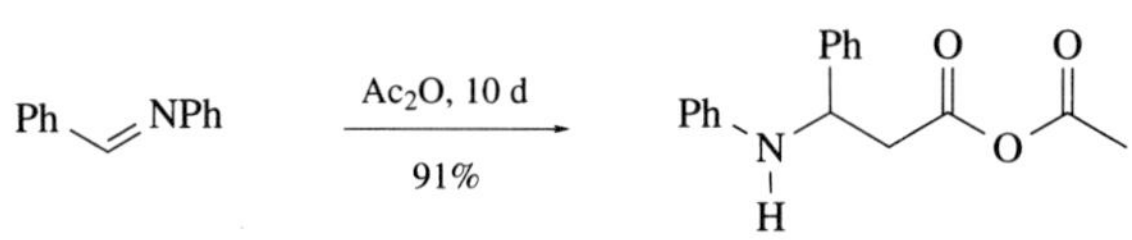

(7)

5.04.1.4 Symmetrical Aromatic and Heteroaromatic Anhydrides

After symmetrical alkanoic anhydrides, the largest group of anhydrides are symmetrical aromatic and heteroaromatic anhydrides. Many of the reactions of this section are similar to those in Section 5.04.1.2, the major differences arising when the enhanced stability of certain aromatic anhydrides allows their isolation under conditions which would result in hydrolysis of alkanoic anhydides (e.g., pyridine, water).

5.04.1.4.1 From metal carboxylates and acyl chlorides

This general method for the synthesis of anhydrides (see Section 5.04.1.1.1) has been exemplified most frequently and to greatest effect in the production of aromatic and heteroaromatic anhydrides. Thus, metal salts of arylcarboxylic acids can be reacted with arylcarboxylic acid chlorides in pyridine and water (Equation (8)) ⟨64JCS755⟩.

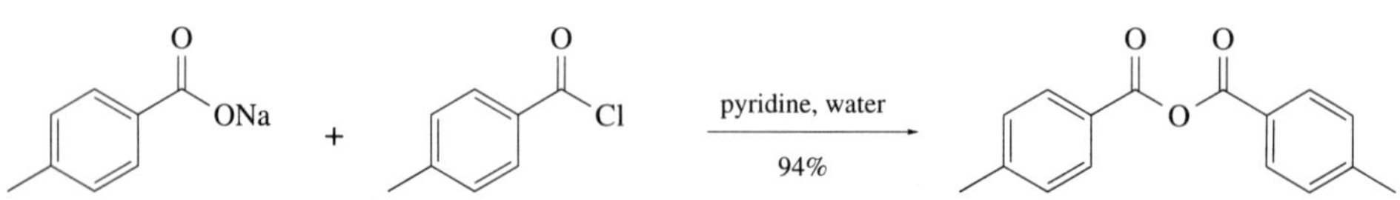

(8)

The sodium salts of arylcarboxylic acids may be partially converted *in situ* to acid chlorides with oxalyl chloride (Equation (9)); the remaining metal carboxylate then reacts with the newly generated acid chloride to provide anhydrides ⟨54JA5803⟩.

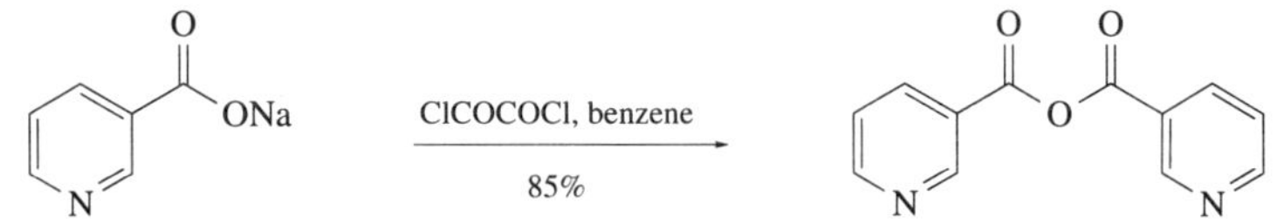

(9)

Similarly, acid chlorides may be subjected to partial hydrolysis with aqueous pyridine, and the carboxylate anion so produced then reacts rapidly to form anhydrides ⟨49JA2242⟩. This is successful for benzoyl, 2-chlorobenzoyl and 4-iodobenzoyl chlorides; yields range from 90% to 98%.

The synthesis of anhydrides from acid chlorides has been achieved under solid–liquid phase-transfer catalysis. Thus, sodium hydrogen carbonate, tetra-*n*-butylammonium bromide and acetonitrile are mixed to produce anhydrides in 82–99% yield in 3–5 h (Equation (10)) ⟨90JCR(S)84⟩.

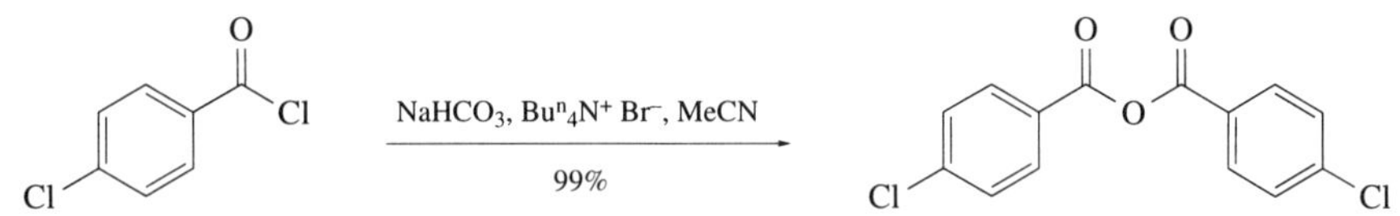

(10)

An exceptionally mild method for the formation of anhydrides is that developed by Kabalka and co-workers (Equation (11)). A catalytic quantity of cobalt(II) chloride when mixed with equimolar quantities of acid chloride and carboxylic acid at 40 °C produces anhydrides in 51–91% yields ⟨92TL593⟩.

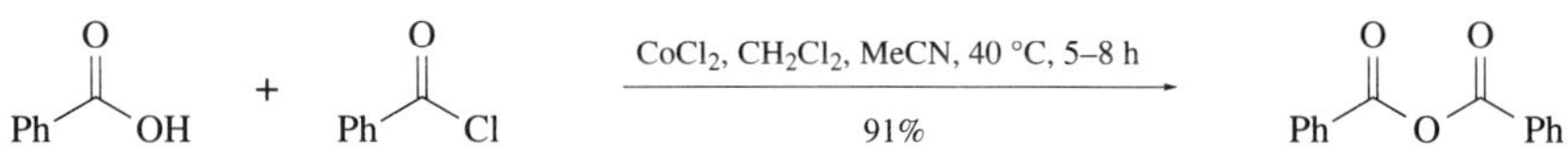

(11)

5.04.1.4.2 *From carboxylic acids or derivatives of carboxylic acids activated towards nucleophilic substitution*

The reaction of carboxylic acids with alkoxyalkynes (see Section 5.04.1.2.2) is also applicable to the synthesis of aromatic anhydrides (Equation (12)). Of particular note are the mild reaction conditions which allow the formation of anhydrides containing acid-sensitive functional groups ⟨86JOC4150⟩. The mechanism of the formation of anhydrides using alkynes is shown in Section 5.04.1.2.2.

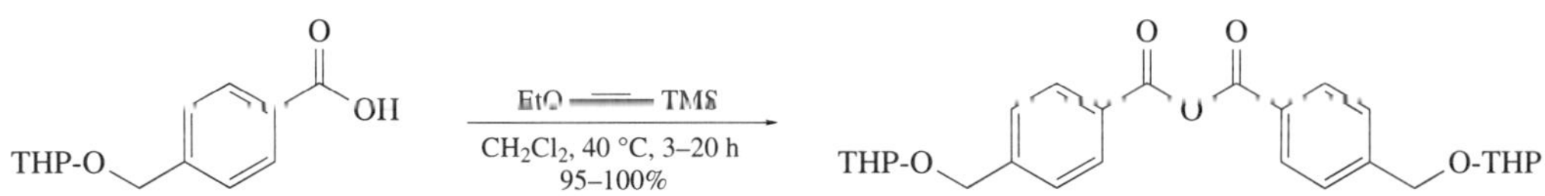

(12)

N,N-Bis(2-oxo-3-oxazolidinyl)phosphorodiamidic chloride has been demonstrated to be useful in the synthesis of alkanoic anhydrides (see Scheme 6). It is also a very effective reagent for the formation of aromatic anhydrides (Table 7), in yields ranging from 88% to 100% in 0.5–4 h ⟨81S616⟩.

Table 7 Aromatic and heteroaromatic anhydrides ($R^1COOCOR^1$) formed by reaction of carboxylic acids (R^1CO_2H) and *N,N*-bis(2-oxo-3-oxazolidinyl)phosphorodiamidic chloride.

R^1	*Reagents and conditions*	*Yield (%)*
4-$NO_2C_6H_4$	Et_3N, CH_2Cl_2, 0.5 h, RT	97
2,4,6-$Me_3C_6H_2$	Et_3N, MeCN, 0.5 h, RT	94
2,4-$Cl_2C_6H_3$	Et_3N, MeCN, 4 h, RT	91
3,4,5-$MeO_3C_6H_2$	*N*-Ethylmorpholine, acetone, 0.5 h, RT	95
3,5-$(NO_2)_2C_6H_3$	*N*-Ethylmorpholine, CH_2Cl_2, 0.5 h, RT	90
Pyrid-3-yl	Et_3N, THF, 0.5 h, RT	96
Pyrid-3-yl-1-oxide	*N*-Ethylmorpholine, CH_2Cl_2, 2.5 h, RT	88
2-Chloropyrid-3-yl	*N*-Ethylmorpholine, THF, 0.5 h, RT	99
3-(2-Chlorophenyl)-5-methylisoxazol-4-yl	N-Ethylmorpholine, CH_2Cl_2, 0.5 h, RT	98
Furan-2-yl	Et_3N, CH_2Cl_2, 0.5 h, RT	96

Carbomethoxysulfamoyl chloride, a stable crystalline reagent prepared by the addition of methanol to chlorosulfonyl isocyanate, has been used to dehydrate pyridinium salts of carboxylic acids in good yields (Table 8) ⟨82MI 504-01⟩.

Table 8 Aromatic and heteroaromatic anhydrides ($R^1COOCOR^1$) formed by reaction of pyridine salts of carboxylic acids (R^1CO_2H) with carbomethoxysulfamoyl chloride.

RCO_2H	*Reagents and conditions*	*Yield (%)*
Ph	Benzene, 0.1 h, RT	92
2-ClC_6H_4	THF, 0.5 h, RT	99
4-$NO_2C_6H_4$	CH_2Cl_2, 0.5 h, RT	83
4-MeC_6H_4	CH_2Cl_2, 0.5 h, RT	85
2,4,6-$Me_3C_6H_2$	CH_2Cl_2, 2 h, RT	45

Triphenylphosphine oxide has been reacted with triflic anhydride to form an intermediate phosphonium species ⟨89JOC1144⟩, which on reaction with acids give symmetrical anhydrides (Equation (13)).

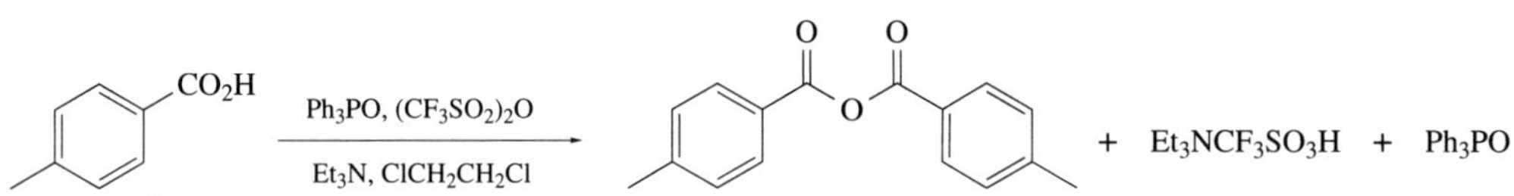

(13)

As for alkanoic anhydrides, tetracyanoethylene may be used for the synthesis of aromatic anhydrides in yields of 61–67% ⟨85TL1503⟩.

As for symmetrical alkanoic anhydrides, aromatic anhydrides may be prepared using chlorosulfonyl isocyanate as a dehydrating agent ⟨82S506⟩. Yields range from 79% to 88%.

Methanesulfonyl chloride has been used in the synthesis of aromatic anhydrides as for alkanoic anhydrides. Yields are excellent ⟨84JCR(S)100⟩, in the range 82% to 94%.

The pyrolytic formation of anhydrides from acid chlorides and trimethylsilyl esters provides high yields (84–99%) of easily purified anhydrides (Equation (14)) ⟨83GEP3144791⟩. The only by-product is the easily removed trimethylsilyl chloride. Heterocyclic anhydrides have also been synthesised by this method.

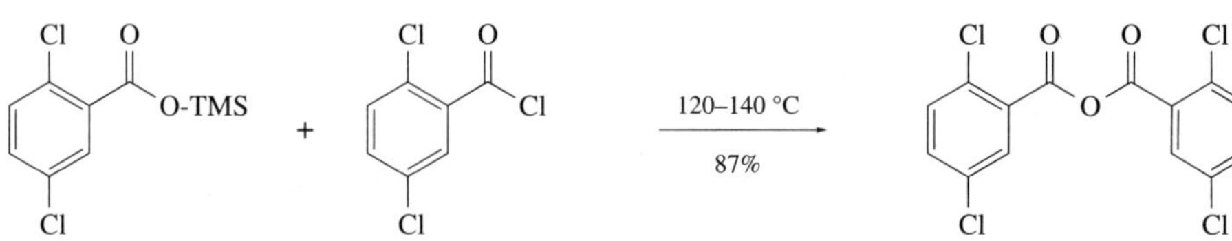

(14)

Staab *et al.* have prepared a number of reactive acylating agents including *N*-trifluoroacetylimidazole and *N*,*N*-carbonyldiimidazole, and demonstrated their use in the synthesis of cyclic anhydrides (79–92%) and benzoic anhydride (78%) ⟨62CB2073⟩. More recently, use has been made of 1,1′-oxalyldiimidazole, which reacts at 40 °C during 3–5 h with carboxylic acids under methanesulfonic acid catalysis (Equation (15)) ⟨87CPB1262⟩.

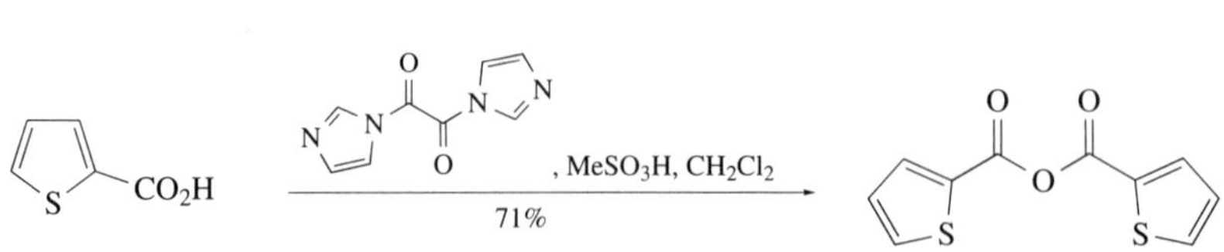

(15)

Trichlorotrifluoracetone readily hydrates in the presence of water, the resulting hydrate being easily esterified in the presence of pyridine (Equation (16)). The intermediate bisaroyloxy compounds may then collapse to form anhydrides in yields ranging from 80% to 97% ⟨86JOC3390⟩.

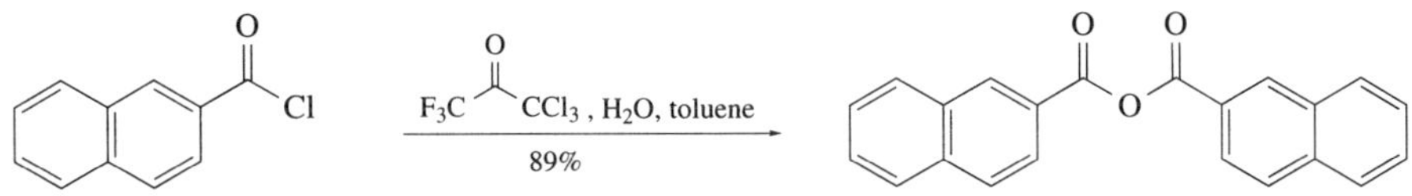

(16)

Phosphorus pentoxide on an inert mineral support has been used to dehydrate benzoic acid derivatives in yields ranging from 62% to 74% ⟨89SC3331⟩.

5.04.1.4.3 From arenes or arenediazonium tetrafluoroborates

Aromatic anhydrides may be synthesised by palladium-promoted carboxylation of arenes with carbon monoxide at 15 bar. Thus, benzoic anhydride was produced in 32% yield from benzene, whilst toluene and chlorobenzene resulted in regioisomeric mixtures of toluic acid anhydrides and chlorobenzoic acid anhydrides, respectively. Furan and thiophene were also converted to their respective three-substituted anhydrides (Scheme 19) ⟨82CC132⟩.

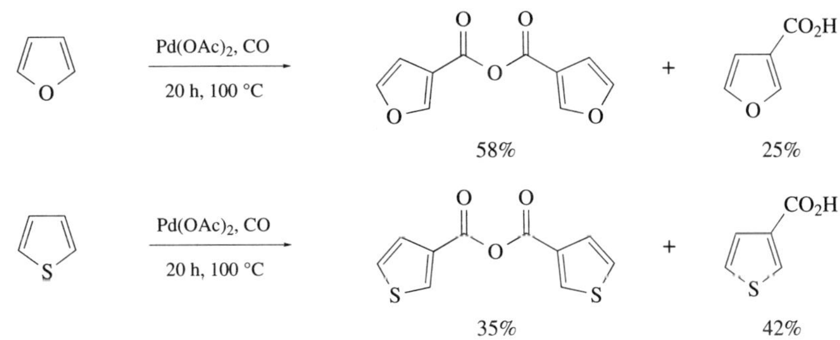

Scheme 19

Symmetrical aromatic anhydrides may be synthesised from aryl alkyl anhydrides by pyrolysis at 100–120 °C under reduced pressure (Scheme 20). Matsuda and co-workers have synthesised aryl alkyl anhydrides by palladium-catalysed insertion of carbon monoxide into diazonium tetrafluoroborates in the presence of sodium acetate. This produces the mixed anhydrides in 65% to 83% yield, which can then be pyrolysed to give the symmetrical aromatic anhydrides. Similarly, direct synthesis of symmetrical aromatic anhydrides can be achieved ⟨80TL2877⟩.

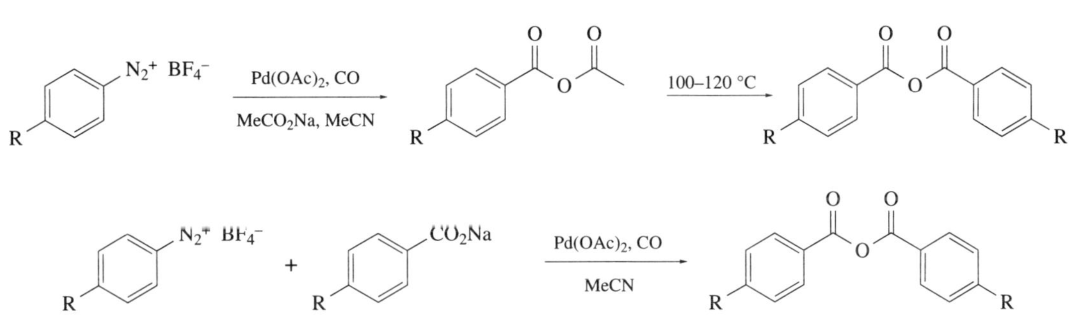

Scheme 20

5.04.1.4.4 Ozonolysis of oxazoles

Ozonolysis of diaryloxazoles has been demonstrated to produce anhydrides. This was confirmed when the unisolated anhydrides were subsequently converted into *N*-methylamides with methylamine ⟨88JCS(P1)529⟩.

5.04.1.5 Unsymmetrical Aromatic and Heteroaromatic Anhydrides

5.04.1.5.1 From carboxylic acids or metal carboxylates with acyl chlorides

Whilst investigating the Friedel–Crafts reaction, Zeavin *et al.* found that benzoyl chloride could be reacted with benzoic acid derivatives in the presence of pyridine to produce unsymmetrical aromatic anhydrides in good to excellent yields (Equation (17)). They observed that certain unsymmetrical anhydrides were unstable: thus, 4-bromobenzoyl benzoic acid anhydride melted at 82–83°C, solidified at 86°C and melted again at 165–182°C. On solidification, the resultant solid was separated and found to be a mixture of benzoic anhydride and 4-bromobenzoic anhydride ⟨32JA3738⟩.

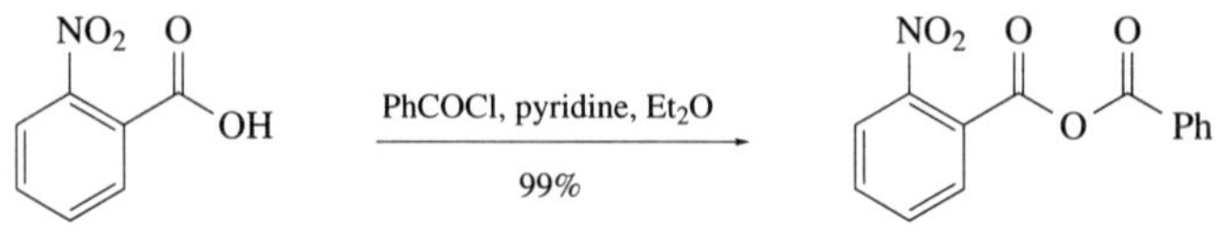

(17)

Anhydrous cobalt(II) chloride has also been used as an alternative catalyst to pyridine in the formation of unsymmetrical anhydrides ⟨92TL593⟩.

An efficient synthesis of unsymmetrical anhydrides involves the reaction of the metal carboxylate with the acyl chloride using pyridine as catalyst (Equation (18)). Other tertiary amine catalysts proved to be effective, including 3-aminopyridine and pyridine-*N*-oxide; however 2- or 4-aminopyridines, 2-fluoropyridine as well as pyridinecarboxylic acids were ineffective ⟨64JCS755⟩.

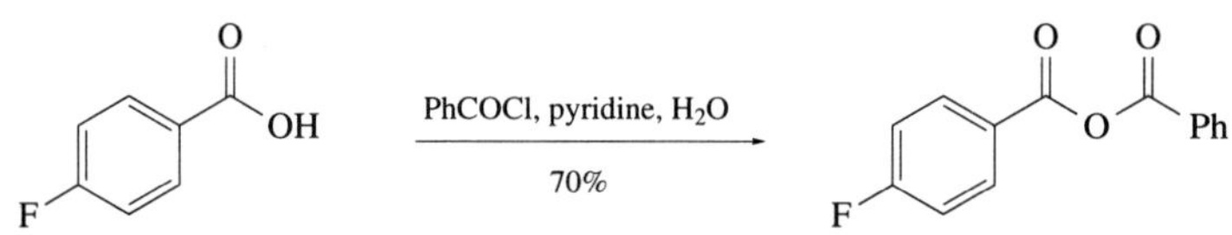

(18)

5.04.1.5.2 From arenediazonium tetrafluoroborates

As well as providing symmetrical anhydrides, the reaction of arenediazonium tetrafluoroborates with sodium carboxylates and carbon monoxide with palladium(II) acetate catalysis has been used to produce unsymmetrical anhydrides (Equation (19)) ⟨80TL2877, 81JOC4413⟩. Yields are moderate.

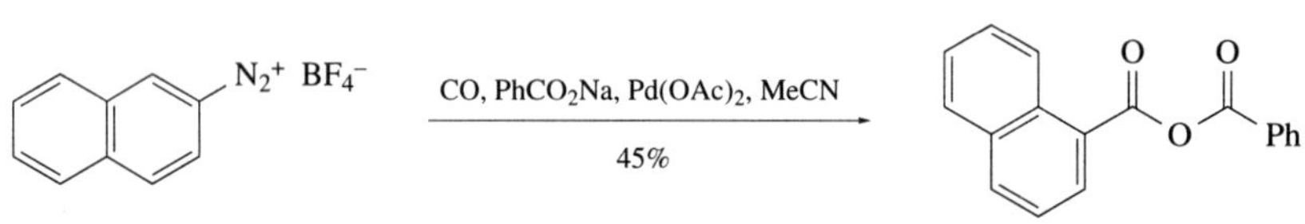

(19)

5.04.1.5.3 Polymer-assisted anhydride formation

Commercially available poly-4-vinylpyridine (Poly-4-VP) has also been used to produce unsymmetrical aromatic anhydrides (Equation (20)) in excellent yields (90–95%) ⟨86TL4933⟩.

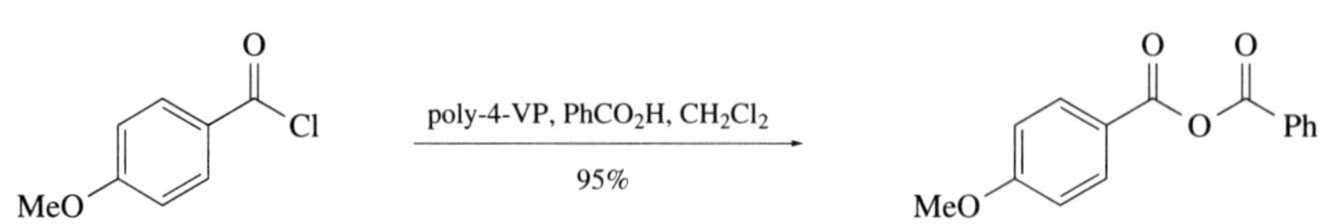

(20)

5.04.1.6 Other Anhydrides

The categories of anhydride synthesis considered in this section include the synthesis of alkyl aryl anhydrides, the synthesis of cyclic anhydrides and the modification of anhydrides to produce other anhydrides, for example, the Diels–Alder reaction of maleic anhydride.

5.04.1.6.1 Alkyl aryl anhydrides

The synthesis of alkyl aryl anhydrides follows the pattern of the symmetrical and unsymmetrical anhydrides in that the most used reactions are those between acid chlorides and carboxylic acids in the presence of a tertiary base, between metal carboxylates and acid chlorides, or the reaction of acetic anhydride (as a dehydrating agent) with carboxylic acids (Scheme 21) ⟨54JCS2006, 86JAN1257, 87TL4569, 88AQ120, 89H(28)565, 90TL709⟩.

Of particular note is the use of 2,4,6-trichlorobenzoyl chloride for forming activated anhydrides, especially prior to macrocyclic lactonisation. Examples include the intermediate anhydrides formed in the synthesis of patuolide A and (9S)-9-dihydroerythronolide ⟨86JAN1257, 87TL4569⟩.

Esters of thiocarboxylic acids are much more reactive than carboxylic esters and serve as biological acylating agents, for example, acetyl-CoA. Sakakibara and co-workers have used this principle to design an acylating reagent which will react in aqueous media (Scheme 22): thus, 2-benzoylthio-1-methylpyridinium chloride forms alkyl aryl anhydrides in two-phase water–chloroform conditions ⟨79CC179⟩ with yields ranging between 61% and 88%.

5.04.1.6.2 Cyclic anhydrides from dicarboxylic acids or derivatives of dicarboxylic acids

The most widely applied method of forming cyclic anhydrides, particularly when forming five- or six-membered rings, involves heating the appropriate diacid in acetic anhydride (Scheme 23) ⟨52JCS1363, 63OSC(4)242, 63OSC(4)790, 86JCS(P1)927, 87TL4441⟩.

Similarly, acetyl chloride ⟨43OSC(2)560, 57JCS4397, 58JA3711, 59CCC2748, 61ZOB665, 88TL1595, 92IJ(34)739⟩, dcc ⟨65CI(L)648, 86TL4609, 87CPB3453⟩, trifluoroacetic anhydride ⟨71CI(L)545, 90JOC4763⟩, alkoxyalkynes ⟨69JA3075, 84TL6027⟩, thionyl chloride ⟨54JCS3722, 64CI(L)1867, 90JOC1679⟩, *N*,*N*-bis(2-oxo-3-oxazolidinyl)phosphorodiamidic chloride ⟨81S616⟩ and montmorillonite and isoprenyl acetate under microwave irradiation ⟨93SC419⟩ have been used as dehydrating agents.

Formation of phthalic anhydrides may be achieved by directed *ortho*-metallation of *N*,*N*-diethyl benzamides and trapping of the *ortho*-lithiated species with carbon dioxide. Subsequent acidification and treatment of the acid amide with perchloric acid effects anhydride formation ⟨92T4863⟩.

A four-membered ring anhydride has been made by an intramolecular displacement of trifluoroacetate (Equation (21)) ⟨62JOC3146⟩.

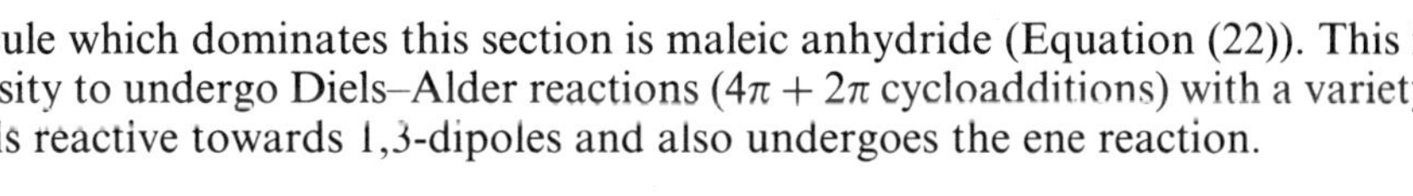

(21)

5.04.1.6.3 Cyclic anhydrides by modification of existing cyclic anhydrides

The molecule which dominates this section is maleic anhydride (Equation (22)). This is by virtue of its propensity to undergo Diels–Alder reactions ($4\pi + 2\pi$ cycloadditions) with a variety of dienes. Similarly, it is reactive towards 1,3-dipoles and also undergoes the ene reaction.

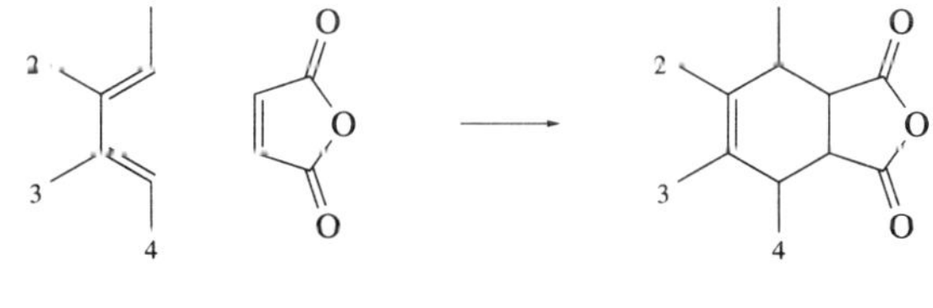

(22)

Categorised by atom substitution in the positions indicated, the following dienes have been reported to react with maleic anhydride: butadiene ⟨63OSC(4)890⟩; 1-*C* ⟨82JOC3450⟩; 1-*N* ⟨89TL3295⟩; 1-*O* ⟨50LA(568)1⟩; 2-*Si* ⟨86TL2913⟩; 1-*O*, 4-*N* ⟨82S958⟩; 1-*C*, 3-*O* ⟨87S850⟩; 1-*O*, 2-*O* ⟨89SL30⟩; 1-*O*, 2-*C* ⟨90S697⟩; 1-*C*, 2-*O* ⟨91T5481⟩; 2-*C*, 3-*C* ⟨49JCS1555⟩; 1-*O*, 3-*O* ⟨74JA7807, 82OS(61)147⟩; 1-*S*, 4-*C* ⟨81S753⟩; 1-*B*, 3-*C* ⟨87TL4169⟩; 2-*Si*, 3-*Si* ⟨86TL2761⟩. Cyclic dienes have also been reacted with maleic anhydride, the products usually being a mixture of *endo* and *exo* anhydrides (Scheme 24) ⟨55LA(597)95, 69TL2465, 81S753⟩.

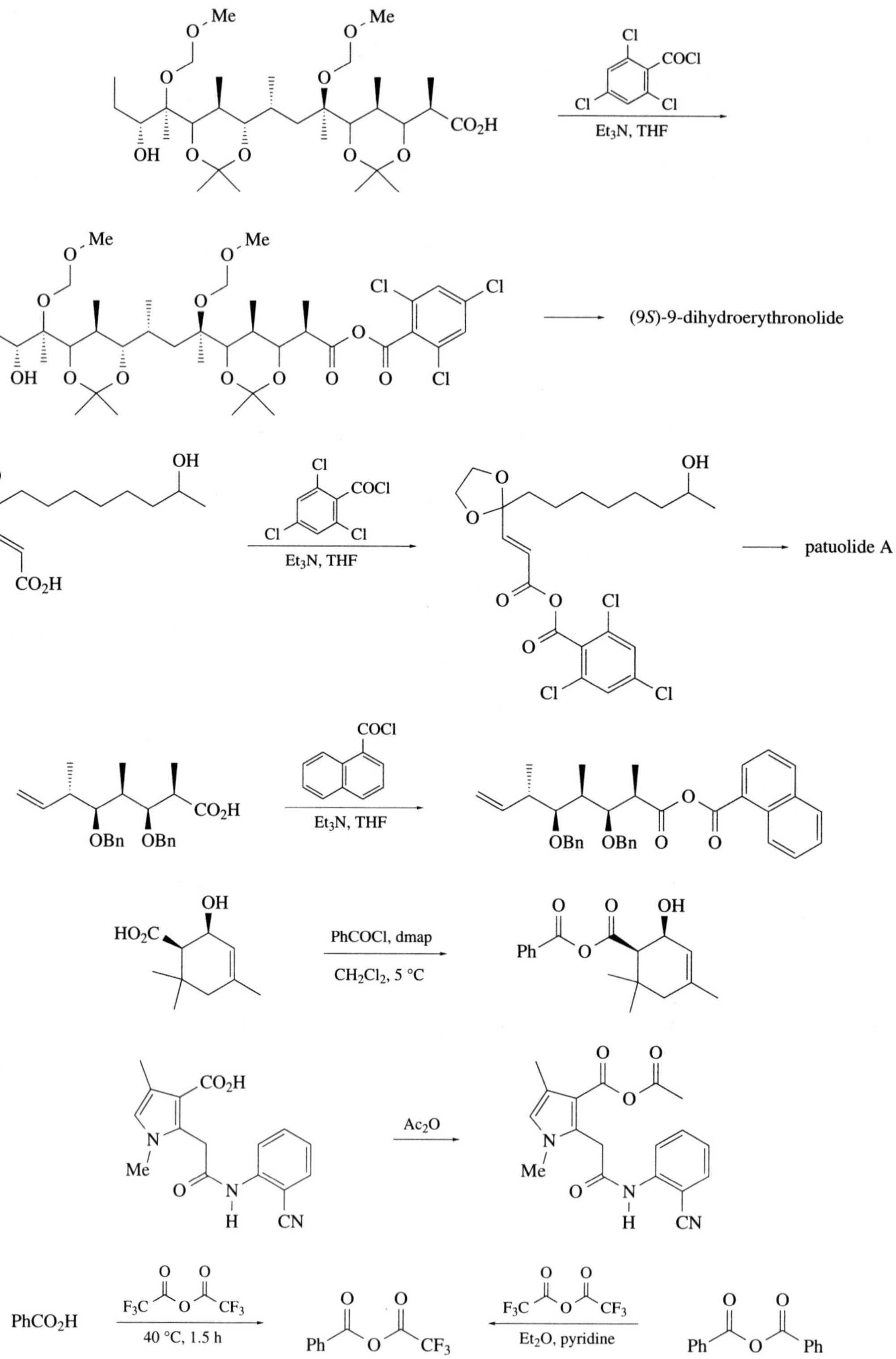

dmap = 4-dimethylaminopyridine

Scheme 21

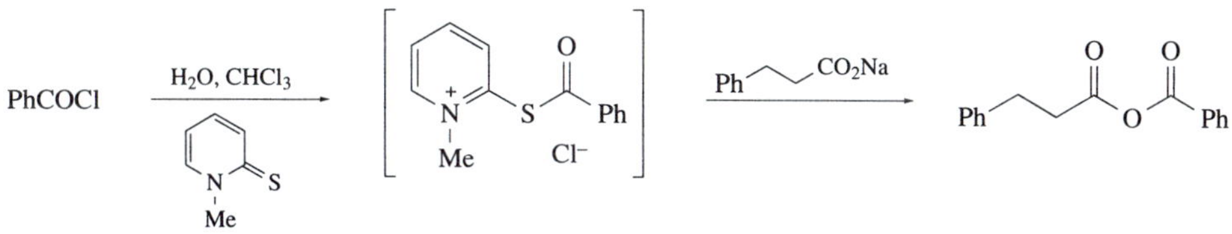

Scheme 22

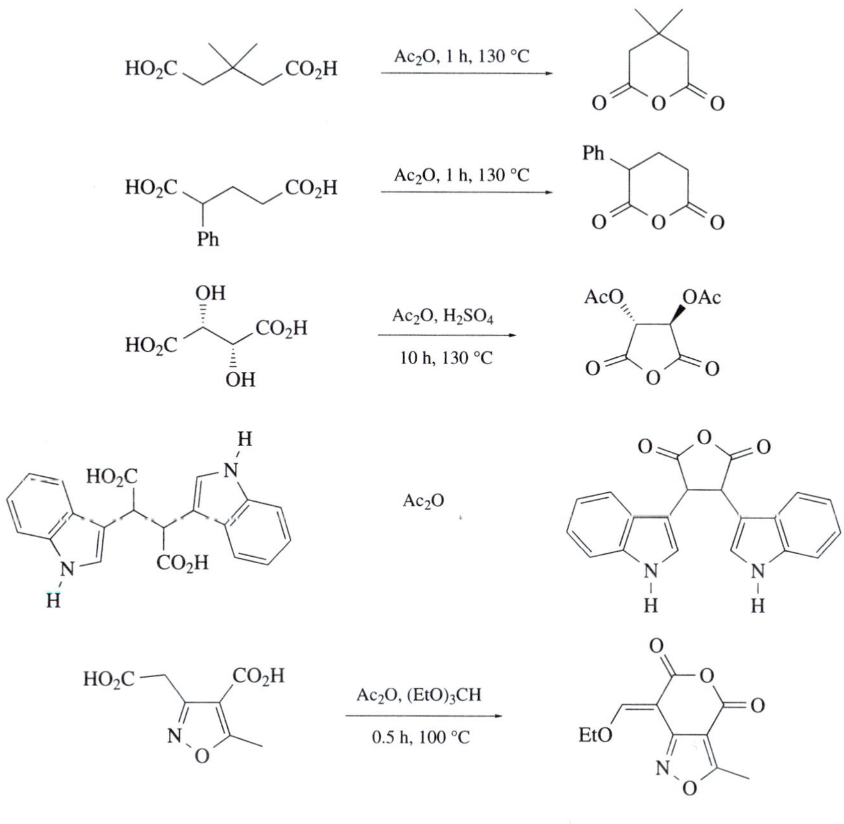

Scheme 23

Heterodienes have been used in the synthesis of heterocyclic anhydrides (Scheme 25) ⟨57HCA852, 59IZV484, 82IJC(B)589, 83T2869⟩.

1,3-Dipolar cycloadditions also feature in the reactions of maleic anhydride, though these are not as common as Diels–Alder reactions (Scheme 26) ⟨62T3, 75TL1739, 89T5979⟩.

As well as the Diels–Alder reaction, maleic anhydride also participates in the ene reaction (Scheme 27). Thus, alkenes where allylic protons are available for reaction react concertedly at temperatures between 120 °C and 220 °C to give substituted succinic anhydride derivatives ⟨57JA419, 63OSC(4)766, 64JA965⟩.

Substituted succinic anhydride derivatives may also be prepared by the thermal addition of maleic anhydride to aryl alkanes. The addition reactions are catalysed by *t*-butyl peroxide and are carried out at 150 °C. Yields range from 38% to 99% ⟨56JOC1473⟩.

More recently, electrophilic substitution of η^2-coordinated arenes has proved possible (Scheme 28). Thus, Harman and co-workers found that osmium complexes of electron-rich aromatic species (phenol, anisole) react with maleic anhydride in acetonitrile to produce arylated succinic anhydride derivatives ⟨91JA8972⟩.

5.04.2 ACYLHYPOHALITES ($RCOOX_1$; X = HALOGEN)

This section covers acylhypofluorites (fluorinating agents), acylhypochlorites, acylhypobromites (Hunsdiecker reaction intermediates) and acyloxy iodine derivatives (oxidising agents; e.g., the Dess–Martin reagent).

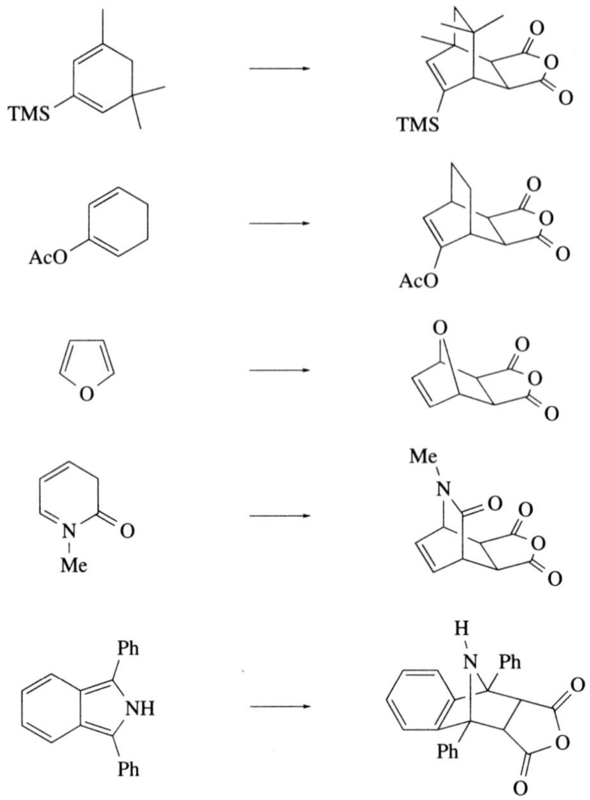

Scheme 24

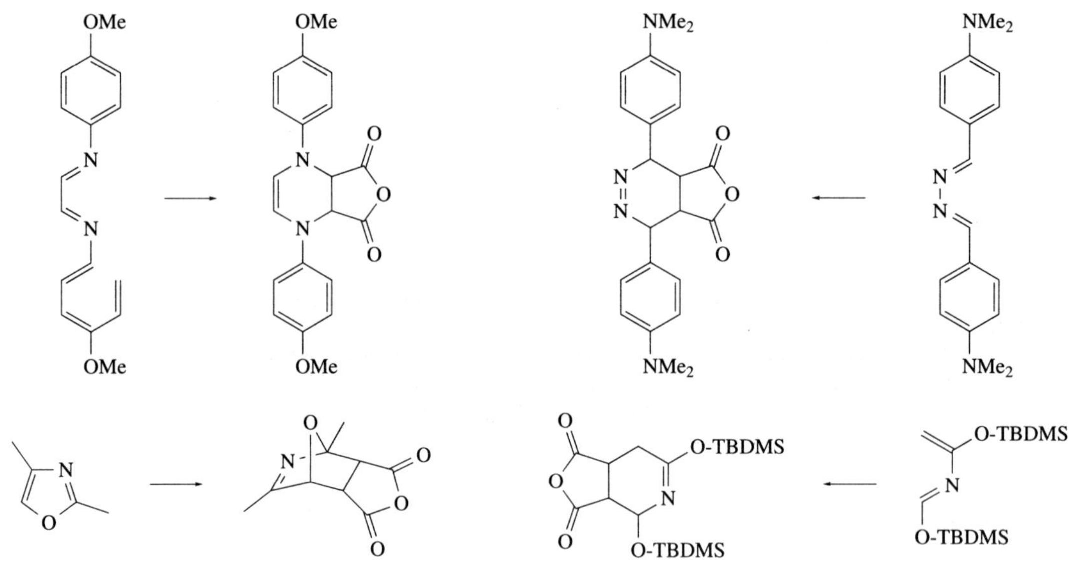

Scheme 25

5.04.2.1 Acylhypofluorites

Rozen and co-workers have reported the preparation of acetylhypofluorite by direct fluorination of sodium acetate ⟨80JFC(16)19, 81JOC4629, 84JOC806⟩ with elemental fluorine in difluorodichloromethane; trifluoroacetylhypofluorite has been prepared in a similar manner ⟨80JOC672⟩ and this synthetic strategy has been extended to a group of acylhypofluorites ⟨90JOC2621⟩. Lithium enolates

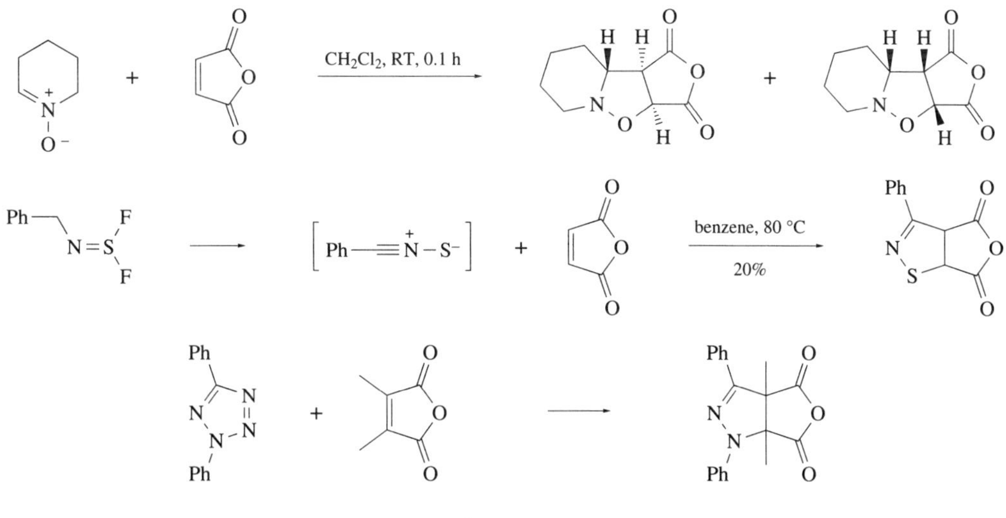

Scheme 26

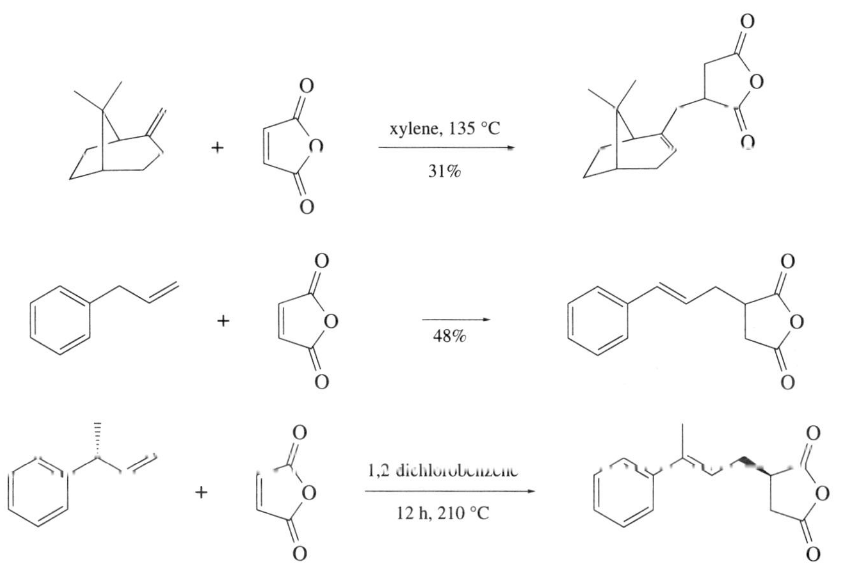

Scheme 27

of ketones have been fluorinated with acetylhypofluorite to produce fluorinated acylhypofluorites ⟨85S665⟩. Fluorine-18 has been used to produce radioactive acetylhypofluorite which in turn was used to prepare a fluorinated radiotracer ⟨88JMC362⟩.

Perfluorinated carboxylic acids react with fluorine to produce mixtures of perfluorinated acylhypofluorites and perfluorinated alkanes ⟨85JOC3698⟩.

5.04.2.2 Acylhypochlorites and Acylhypobromites

Acetylhypochlorite and acetylhypobromite have been synthesised by reacting silver acetate with elemental chlorine ⟨83JA3999⟩ and bromine ⟨64JOC2993, 70JOC2056⟩, respectively. Propionyl, isobutyryl and pivaloyl hypobromites are also reported to be accessible by bromination of their respective silver carboxylates ⟨83JA3999⟩.

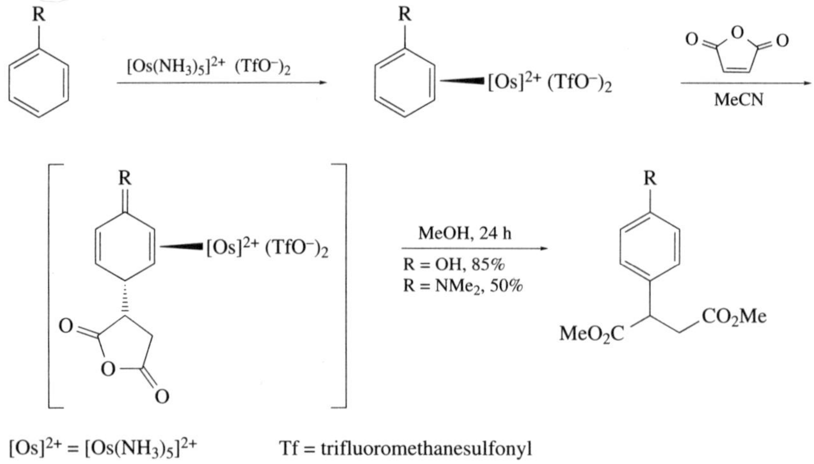

Scheme 28

5.04.2.3 Acyloxyiodides, Acyloxyiodoarenes and Acyloxyiodoalkanes

5.04.2.3.1 By oxidative acylation of iodine, iodoarenes and iodoalkanes

Bisacetoxyiodobenzene is prepared by oxidising iodobenzene with acetic anhydride and hydrogen peroxide ⟨53JCS107, 73OS(5)660⟩. Sodium perborate in acetic acid has been demonstrated to be an effective reagent for the oxidation of iodoarenes to acyloxyiodoarenes (Equation (23)) ⟨89T3299, 90JA5677⟩ in yields ranging from 71% to 100%. Iodoarenes may also be oxidised with fuming nitric acid in trifluoroacetic anhydride to bis(trifluoroacetoxyiodo)arenes ⟨77S574⟩.

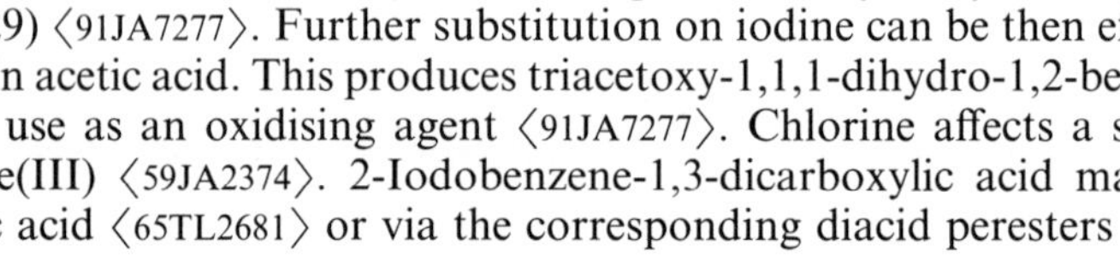

(23)

The substitution of a carboxylic acid *ortho* to iodine in an arene with potassium bromate in sulfuric acid allows facile intramolecular oxidative cyclisation to produce 1-hydroxy-1,2-benziodoxol-3(1H)-one 1-oxide (Scheme 29) ⟨91JA7277⟩. Further substitution on iodine can be then effected by heating with acetic anhydride in acetic acid. This produces triacetoxy-1,1,1-dihydro-1,2-benziodoxol-3(1H)-one which has found use as an oxidising agent ⟨91JA7277⟩. Chlorine affects a similar oxidation, though only to iodine(III) ⟨59JA2374⟩. 2-Iodobenzene-1,3-dicarboxylic acid may be oxidatively cyclised with peracetic acid ⟨65TL2681⟩ or via the corresponding diacid peresters in chlorobenzene ⟨74JA3319⟩.

Oxidation of alkyl iodides occurs, though this is rarer; thus, perfluorinated alkyl iodides may be

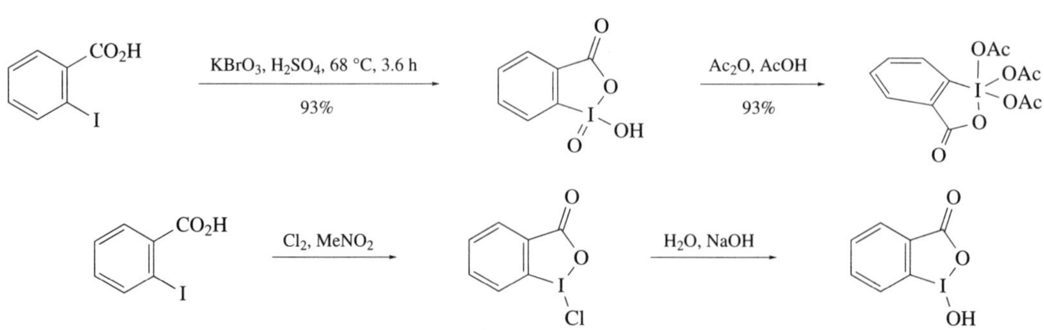

Scheme 29

oxidised with trifluoroperacetic acid ⟨86JFC(31)37⟩. Similarly, 2,2,2-trifluoroethyl iodide may be bistrifluoroacetoxylated (Equation (24)) ⟨87BCJ3307, 88ZOR593⟩. Iodine may be oxidised to iodine tristrifluoroacetate with fuming nitric acid and trifluoroacetic anhydride ⟨88JCS(P1)3103⟩.

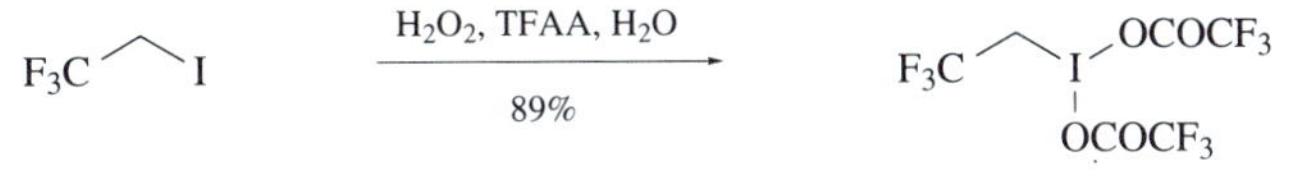

(24)

TFAA = trifluoroacetic anhydride

5.04.2.3.2 Acyloxylation of iodosoarenes and bis(halo)iodoarenes

Bis(acyloxy)iodoarenes may be synthesised by acyloxylation of iodosoarenes or bis(halo)-iodoarenes. Thus, iodosobenzene when treated with carboxylic acids in acetone produces symmetrical bis(acyloxy)iodobenzenes ⟨86CL967⟩, whilst trimethylsilyl triflate and trimethylsilyl acetate convert iodosobenzene to acetoxyiodobenzenetriflate in 78% yield ⟨90TL4821⟩. Similarly trifluoromethyliodine difluoride is converted to bis(trifluoroacetoxy)trifluoromethyl iodide with boron tris(trifluoroacetate) in acetonitrile ⟨89JFC(45)401⟩ and iodobenzene dichloride reacts with chloroacetic acid and mercury oxide in chloroform to produce bis(trichloroacetoxy)iodobenzene in 93% yield ⟨88TL2033⟩. Pentavalent iodine species, for example, tetrafluoroiodobenzene or iodylbenzene and close analogues substituted in the aromatic ring, may be converted to tetrakis(trifluoroacetoxy)-iodoarenes with trifluoroacetic acid ⟨83JA3999⟩.

Exchange reactions are also possible between acyloxyiodoarenes; thus, bis(trifluoroacetoxy)iodobenzene may be converted into a variety of bis(acyloxy)iodoarenes by treatment with sodium carboxylates in acetonitrile ⟨86JCS(P1)795⟩.

5.04.3 ACYL PEROXIDES (MONO AND DI)

5.04.3.1 Monoacyl Peroxides (Peracids) and Alkyl Acyl Peroxides (Peresters)

5.04.3.1.1 Monoacyl peroxides (peracids)

Monoacyl peroxides have been synthesised by direct oxidation of carboxylic acids, anhydrides and aldehydes ⟨57JA5982⟩, by hydrolysis of alkyl acyl peroxides or diacyl peroxides ⟨67T3327⟩ and by conversion of acid chlorides to acyl peroxides ⟨64JOC2099⟩.

(i) By oxidation of acids

Acetic acid and hydrogen peroxide at 50 °C produce acetyl peroxide ⟨51JA4958⟩; benzoic acid and hydrogen peroxide in methanesulfonic acid produce benzoyl peroxide ⟨73OSC(5)904⟩; and C_6–C_{16} carboxylic acids with hydrogen peroxide in sulfuric acid have been reacted to produce the respective acyl peroxides ⟨55JA4037⟩. Potassium persulfate under phase-transfer conditions has been demonstrated to be a versatile and high-yielding oxidant for the synthesis of acyl peroxides (Equation (25)) ⟨88SC2123⟩. The advantage of the phase-transfer approach to the synthesis of acyl peroxides is that the insolubility of oxidant in organic solvent is circumvented, thereby raising yields.

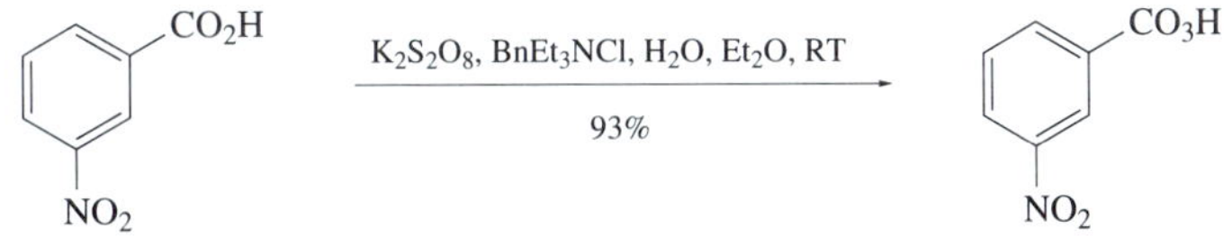

(25)

(ii) By oxidation of anhydrides

Oxidation of anhydrides may be achieved with oxygen or hydrogen peroxide and can provide monoacyl peroxides from alkanoic anhydrides and bis(monoacyl) peroxides from cyclic anhydrides (Scheme 30) ⟨56JA3819, 59JOC1354, 70JMC948, 73OSC(5)367⟩.

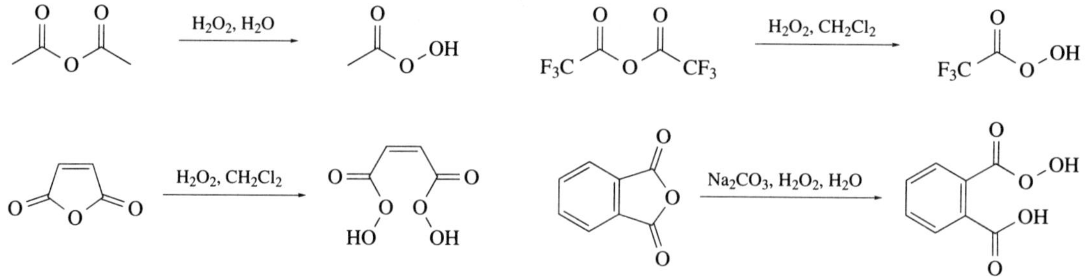

Scheme 30

(iii) By hydrolysis of alkyl monoacyl peroxides

Carboxylic acids may be converted into their respective 2-methoxyprop-2-yl acyl peroxides (see Section 5.04.3.1.2) with 2-methoxyprop-2-yl hydroperoxide and then hydrolysed to acyl peroxides ⟨92JOC1009⟩.

5.04.3.1.2 Alkyl acyl peroxides (peresters)

Alkyl acyl peroxides can be synthesised by dehydration of carboxylic acids and hydroperoxides, for which carbonyldiimidazole ⟨85BSB29⟩ is an effective reagent. Thus 4,6,6,6-tetrachlorohexanoic acid may be coupled in the presence of *N,N'*-carbonyldiimidazole to *t*-butylhydroperoxide in 90% yield. Alternatively, dcc in dichloromethane has been shown to dehydrate a variety of acids in the presence of 2-methoxyprop-2-yl hydroperoxide ⟨92JOC1009⟩. Acid chlorides also react with hydroperoxides under basic conditions to produce alkyl acyl peroxides (Scheme 31) ⟨85T4047, 87S517⟩.

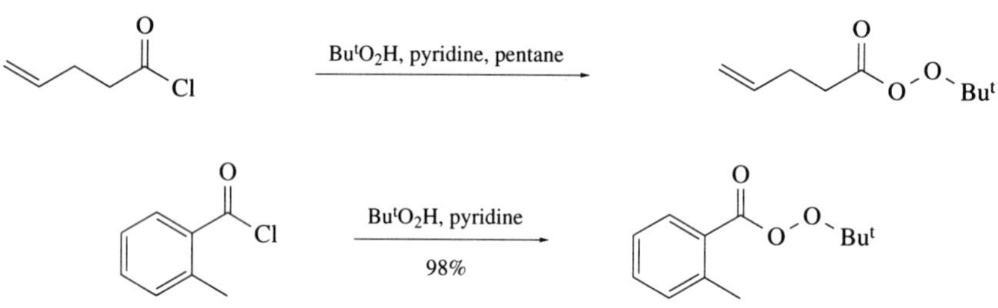

Scheme 31

5.04.3.2 Diacyl Peroxides

5.04.3.2.1 Coupling of monoacyl peroxides and acid chlorides

Coupling of monoacyl peroxides to acid chlorides has been achieved in the presence of sodium peroxide under aqueous conditions ⟨55OS(3)649, 88CPB178, 88USP4762945⟩. This method allows only the formation of symmetrical diacyl peroxides, as the acyl chloride is initially converted to the acyl peroxide and this then reacts with a second equivalent of acyl chloride (Scheme 32).

Urea has been used in combination with hydrogen peroxide; thus, dibenzoyl peroxide is synthesised in 62% yield from benzoyl chloride and hydrogen peroxide in ether ⟨55JA6370⟩. Similarly, pyridine and hydrogen peroxide have been used to form alkynoyl peroxides in ether (Equation (26)) ⟨70JPR(312)2⟩.

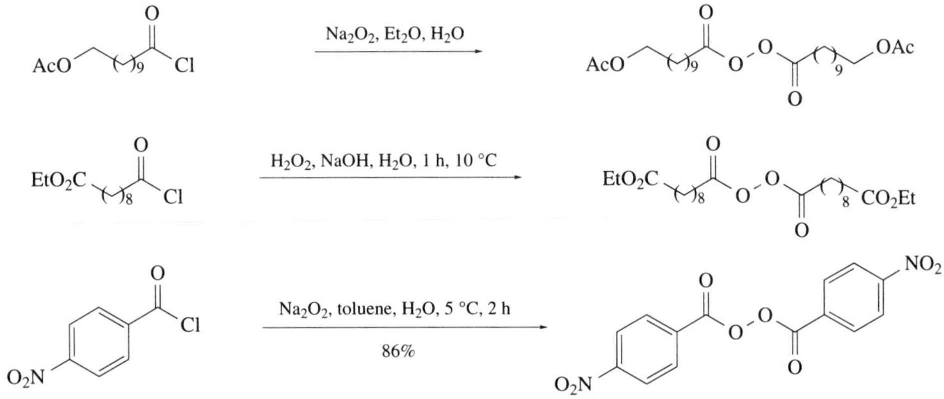

Scheme 32

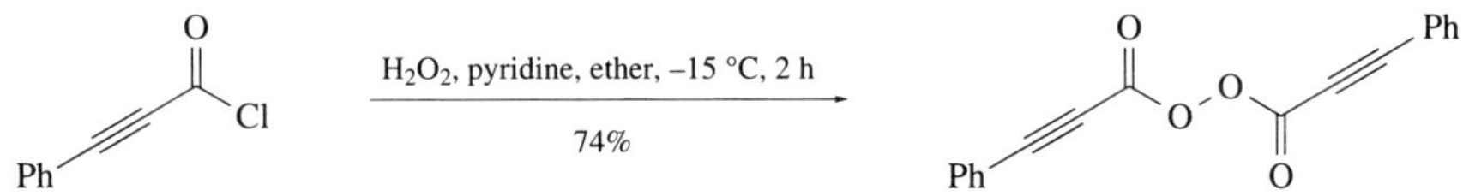

(26)

For the synthesis of unsymmetrical acyl peroxides, the sodium salts of monoacyl peroxides have been reacted with acyl chlorides and methyl carbonates (Scheme 33) ⟨50JA5426, 86TL5197⟩.

More usually a percarboxylic acid is reacted with an acyl chloride in the presence of an organic base (Scheme 34) ⟨85T4195, 85T4213, 87AG(E)1253, 87T5341⟩.

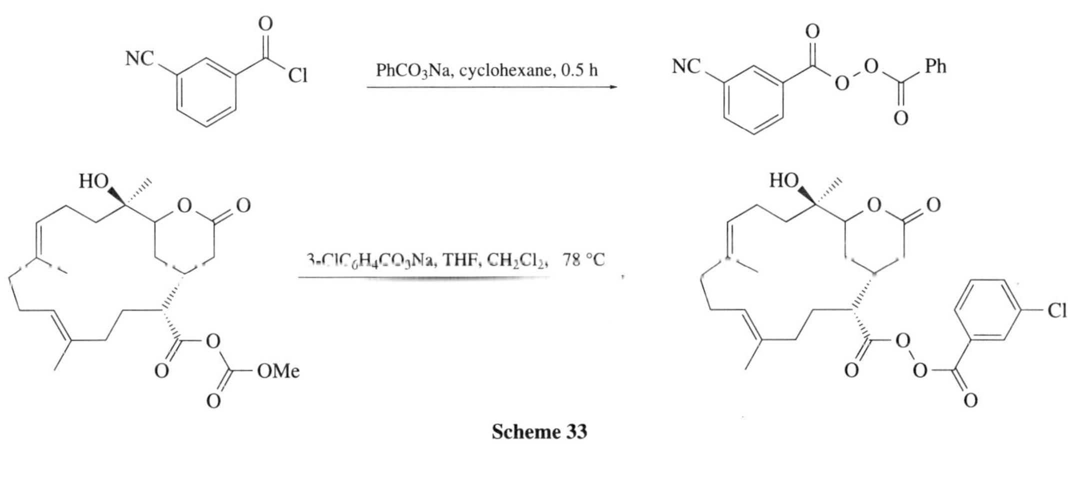

Scheme 33

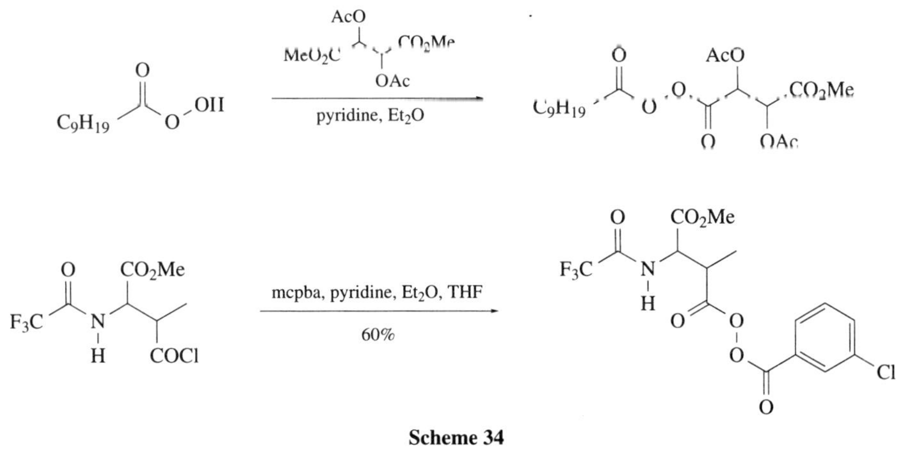

Scheme 34

5.04.3.2.2 *From peracids and acids with dcc*

In a reaction analogous to the formation of anhydrides from carboxylic acids, dicyclohexylcarbodiimide (dcc) dehydrates peracids in the presence of acids to form diacyl peroxides (Scheme 35) ⟨63JOC2168, 86T1285, 91CL61, 91JOC528⟩.

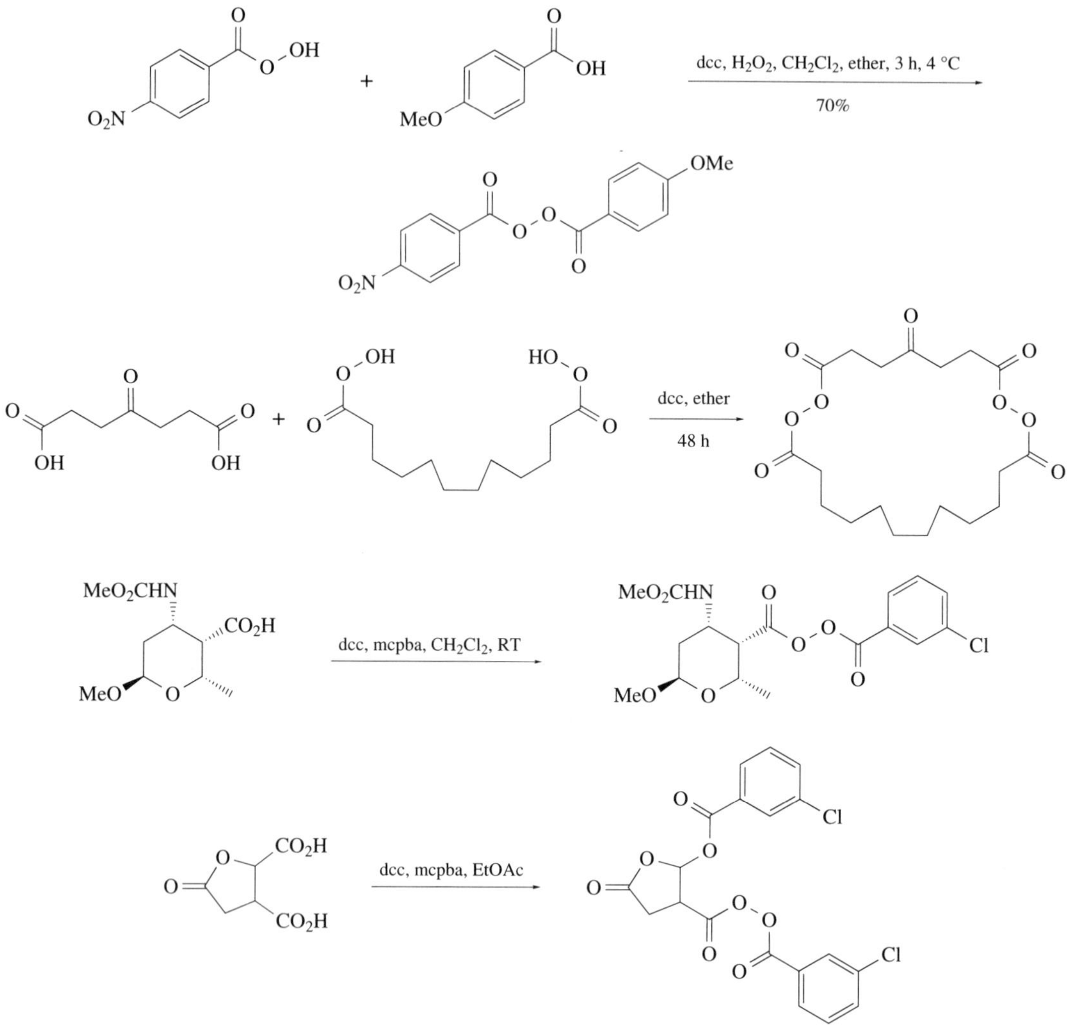

Scheme 35

5.04.3.2.3 *Other methods of synthesis*

Ozone, when bubbled through 4-methylbenzaldehyde in the presence of catalytic quantities of lithium chloride, causes an oxidative dimerisation to bis(-4-methylbenzoyl) peroxide in 50% yield ⟨83ZOR1764⟩. Similarly, irradiation in acetic anhydride of 4-chlorobenzaldehyde in the presence of air produces 4-chlorobenzoyl acetyl peroxide in 81% yield ⟨59DOK(128)1201⟩.

Anhydrides react in water or dilute acid conditions with peracids to produce mixed acyl peroxides in good yields ⟨64BSF542, 78CI(L)579⟩. Low-molecular-weight anhydrides such as acetic anhydride may be oxidised directly to peroxides. Thus, propionic anhydride may be oxidised to dipropionyl peroxide in 90% yield with basic hydrogen peroxide ⟨52LA(577)153, 59JOC107, 90JFC(46)423⟩.

5.04.4 *O*-ACYLSULFENIC, -SULFINIC AND -SULFONIC ACID DERIVATIVES

5.04.4.1 *O*-Acylsulfenic Acid Derivatives (R^1COOSR^2)

2,4-Dinitrobenzenesulfenyl acetate may be prepared by reacting 2,4-dinitrobenzenesulfenyl chlorides with sodium acetate in benzene ⟨56JA1207⟩.

5.04.4.2 *O*-Acylsulfinic Acid Derivatives ($R^1COOS(O)R^2$)

Methanesulfinyl acetate or trifluoroacetate may be synthesised by reacting silver acetate or silver trifluoroacetate respectively with methanesulfinyl chloride in carbon tetrachloride at $-20\,°C$ to $0\,°C$ ⟨81T3115⟩.

Reaction of DMSO with activating agents such as TFA or acetic anhydride produce acyloxysulfonium species which are capable of oxidising alcohols to ketones or aldehydes, intermediates that form the basis of the Swern oxidation ⟨81S165⟩.

Lead(IV) acetate oxidation of *N*-acyl-*N'*-arylsulfonylhydrazines results in the formation of acyl arylsulfonyldiazenes. The diazene is unstable and elimination of nitrogen occurs at room temperature with formation of *O*-acylsulfinic acid derivatives ⟨77AG(E)728⟩ which may be preferentially attacked at sulfur by amines to produce sulfinamides (Scheme 36).

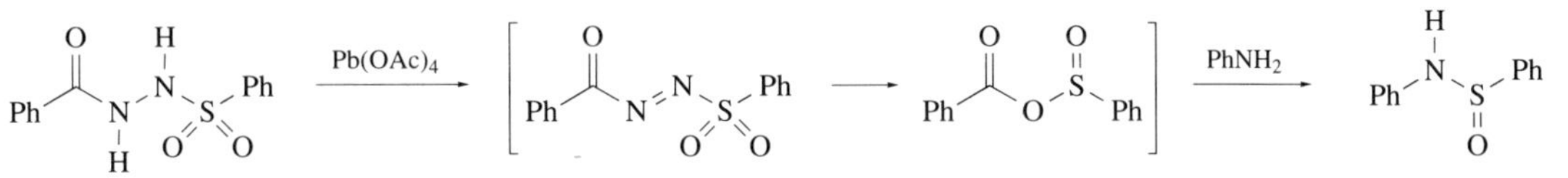

Scheme 36

5.04.4.3 *O*-Acylsulfonic Acid Derivatives

The most numerous isolated derivatives of acyloxysulfur are the acyloxysulfonates, which have been used as acylating agents ⟨83CC1113, 84JOC3875⟩. Thus, the silver salts of both methanesulfonic acid and trifluoromethanesulfonic acid react with a variety of carboxylic acid chlorides at room temperature ⟨83CB1183⟩ in the absence of solvent, or in liquid sulfur dioxide at $-30\,°C$, or in dichloromethane at $-50\,°C$. Alternatively, acyloxy trifluoromethanesulfonates may be synthesised by heating trifluoromethanesulfonic acid with acid chlorides in 1,2-dichloroethane (Scheme 37) ⟨83CB1183⟩.

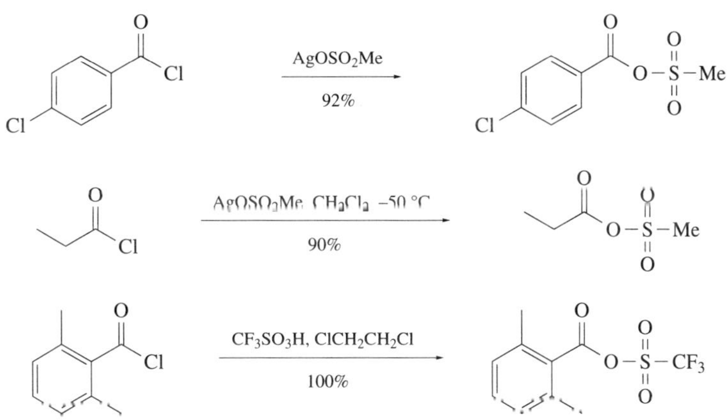

Scheme 37

Oxidising thioesters with ozone in dichloromethane results in formation of acyloxysulfonates (Scheme 38). The reaction goes via initial oxidation at sulfur followed by rearrangement to an acyloxysulfenate; further oxidation produces the acyloxysulfonate ⟨78TL2171⟩. Yields range from 50% to 86%.

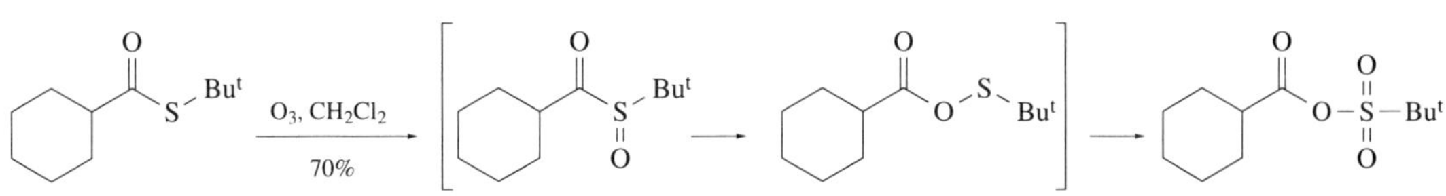

Scheme 38

5.04.5 ACYLOXYSELENIUM AND -TELLURIUM DERIVATIVES

5.04.5.1 Acyloxyselenium Derivatives

There are few reports in the literature of acyloxyselenium species. One of the earlier references is the acetoxylation of 4-methoxyanthraquinon-1-ylselenium bromide with silver acetate in acetic acid 〈52HCA1429〉 to provide the selenium(II) species (Scheme 39). More recently, oxidative approaches to acyloxyselenium derivatives have been attempted. Thus, both methylselenobenzene 〈73JOC3172〉 and 1-methylselenonaphthalene 〈92CL1437〉 have been acylated oxidatively with dibenzoyl peroxide in 92% and 100% yields, respectively.

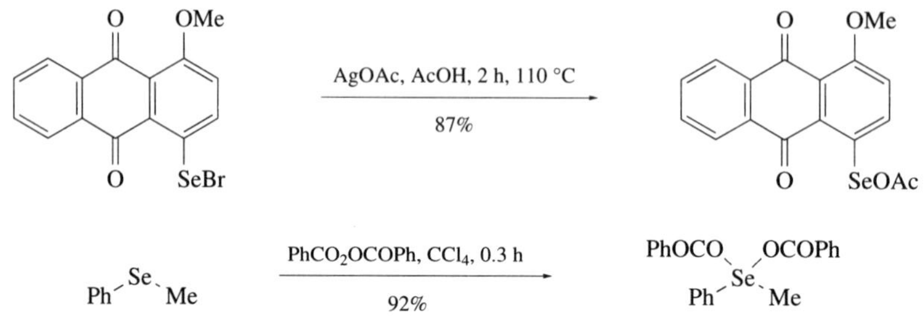

Scheme 39

5.04.5.2 Acyloxytellurium Derivatives

Diaryltellurium acetates may be prepared by acetoxylation of diaryl tellurides with one equiv. of lead(IV) acetate in benzene at room temperature (Scheme 40). Yields are >95% 〈72TL4779, 74JOM51〉. When three equivs. of lead(IV) acetate are reacted with diphenyl ditelluride, the only product is phenyltellurium triacetate, which may be readily hydrolysed to benzenetellurinyl acetate 〈72TL4779〉. Benzenetellurinyl acetate may also be prepared quantitatively by heating benzenetellurinic anhydride with acetic anhydride in chloroform under reflux for 0.5 h 〈87TL1281, 89JOC4391〉.

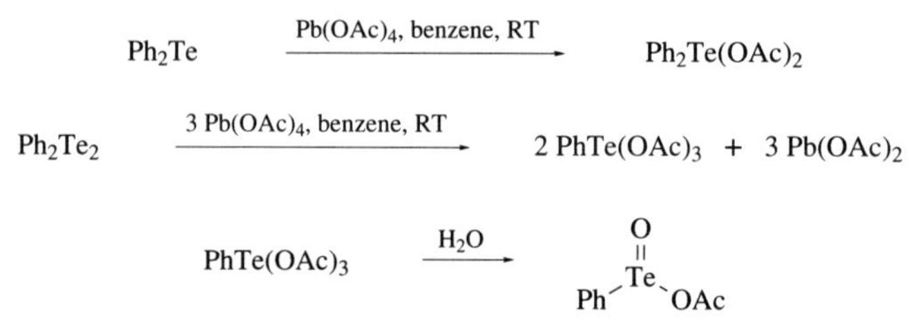

Scheme 40

Diaryltellurium acetates are synthesised by reacting diaryltellurium dichlorides with either silver acetate or sodium acetate (Equation (27)) 〈87IJC(A)260, 89IJC(A)1528〉; alternatively, diaryltelluroxides may be treated with carboxylic acid anhydrides 〈89IJC(A)1528〉.

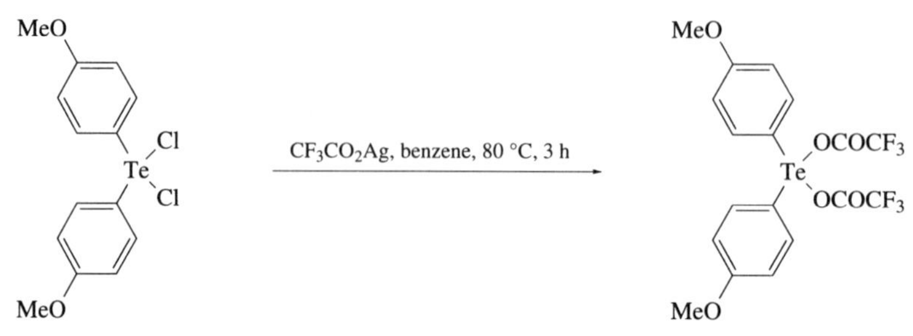

(27)

Dimethyltellurium diiodide may be converted into dimethyldimethoxytellurium with sodium methoxide in methanol. Removal of the methanol and replacement with benzene allows displacement of methoxide with anthranilic acid in an overall yield of 67% ⟨91MI 504-02⟩.

5.04.6 *O*-ACYLHYDROXYLAMINES (R_1COONR_2, $R_1COONRX$)

This section reviews principally three types of compounds: acylhydroxylamines, acylated hydroxamic acids and acylated amidoximes.

5.04.6.1 O-*Acylhydroxylamines*

5.04.6.1.1 Acylated hydroxylamines by direct acylation

Acylation of hydroxylamines derived from secondary amines is straightforward as only one product is usually possible. However, when hydroxylamines derived from primary amines are acylated, three possible products may arise (Equation (28)): the *O*-acylated hydroxylamine, the *N*-acylated hydroxylamine (hydroxamic acid, see Chapter 5.07.2) and the acylated hydroxamic acid (see Section 5.04.6.2).

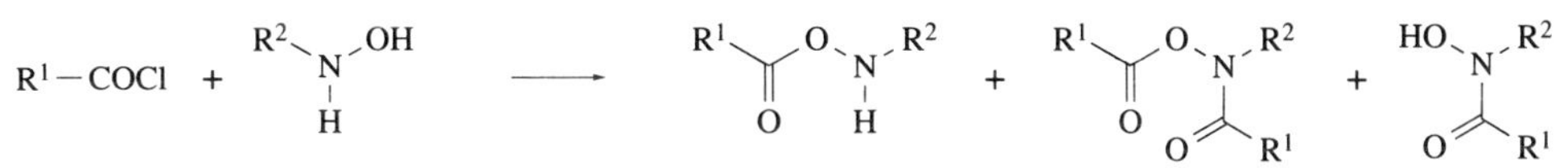

(28)

In this section the synthesis of *O*-acylated hydroxylamines is considered. Examples of mixed product distribution which can occur when either acid chlorides or anhydrides are reacted with hydoxylamines are shown in Scheme 41 ⟨86AP(319)1037, 87MI 504-02⟩.

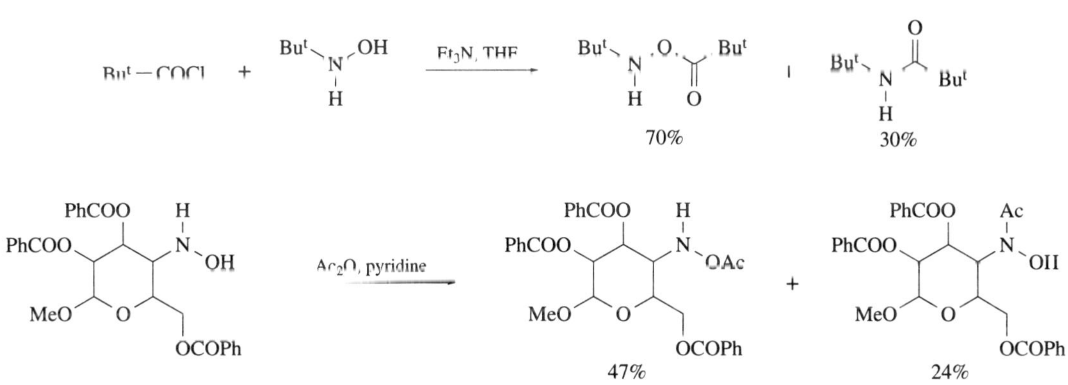

Scheme 41

O-Acylated hydroxylamines have been synthesised selectively; thus, using carbonyldiimidazole, *N*-methylhydroxylamine may be acylated with benzoic acid ⟨86CZ377⟩ or acetic acid (Equation (29)) ⟨86CB744⟩.

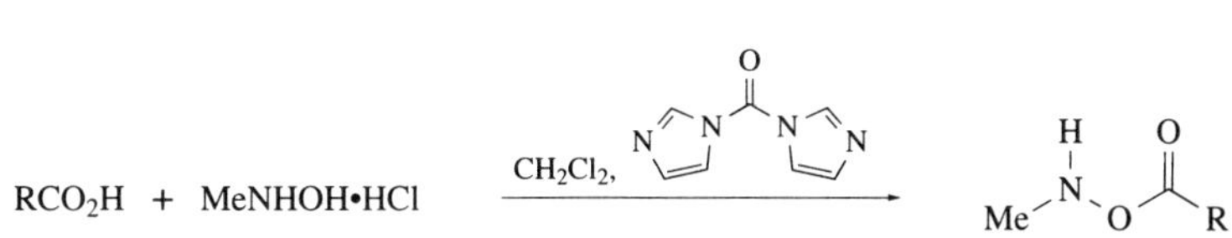

(29)

It has been reported that a hindered oximinohydroxylamine can be monoacylated on the hydroxylamine hydroxyl with acetic anhydride and potassium carbonate in ether ⟨85ZOR196⟩. Hydroxylamines derived from secondary amines are acylated under a variety of conditions (Scheme 42) ⟨73CR(C)1039, 85AP(318)977, 85CPB998, 85CZ389, 85HCA1730, 85CAR375⟩.

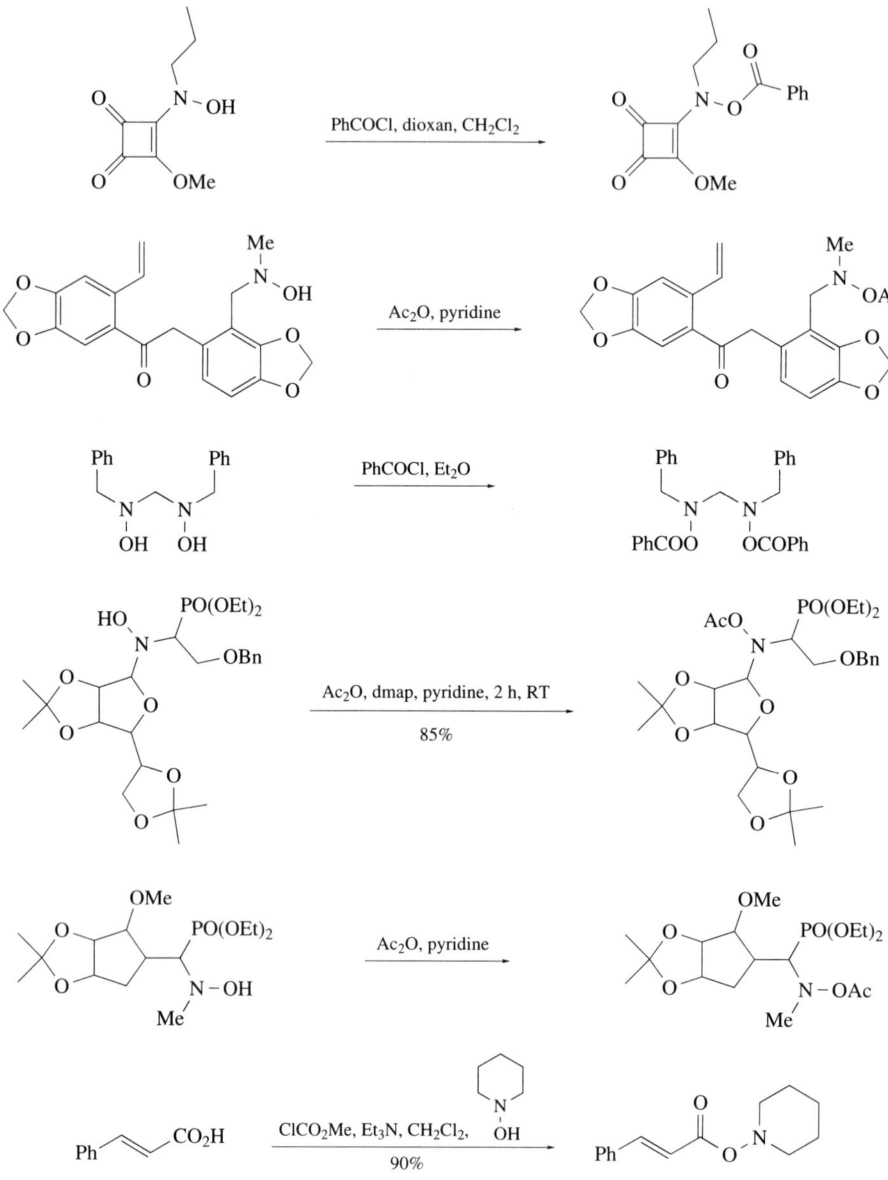

Scheme 42

Acyl cyanides have been demonstrated to be effective *O*-acylation reagents for arylhydroxylamines and have been used in the synthesis of 'ultimate carcinogens'. They are selective for oxygen and react with arylhydroxylamines at or around room temperature (Scheme 43) ⟨82TL1391, 89AG(E)337, 89AG(E)468⟩.

Driven by the need to find appropriate analogues of the biological acylating agent which produces the 'ultimate carcinogen', 2-acylthiazolium salts have been studied as selective agents for *O*-acylation (Equation (30)). The reaction proceeds by way of nucleophilic attack at the acyl carbonyl of the thiazolium salt by the hydroxylamine, followed by base-catalysed elimination of the thiazolium salt ⟨91CC1127⟩.

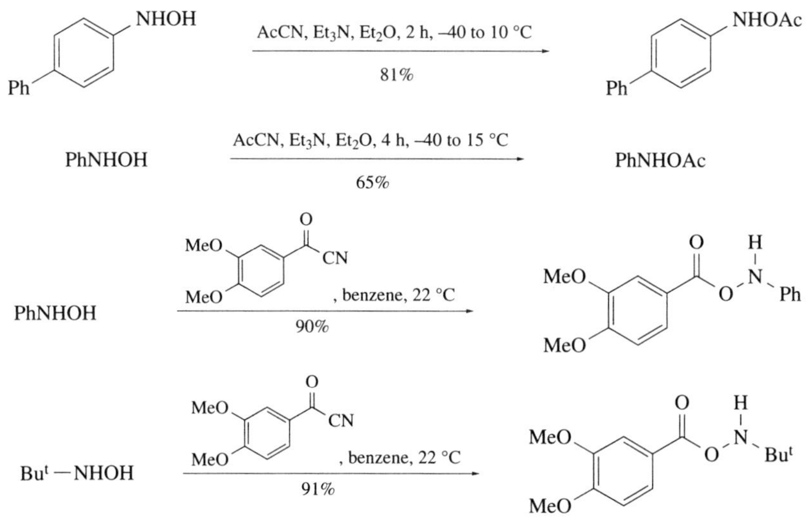

Scheme 43

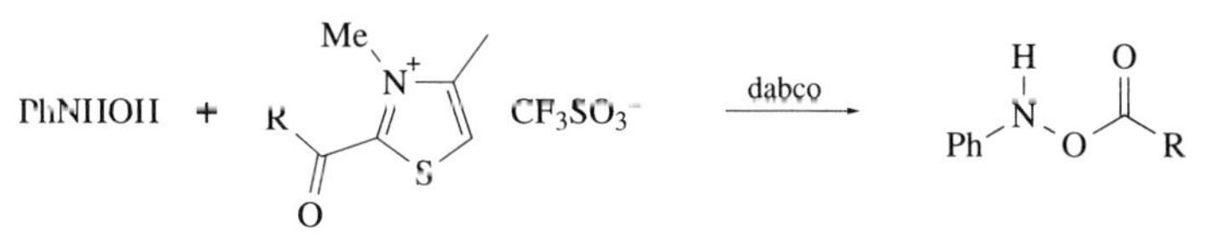 (30)

R = Ph, Me dabco = 1,4-diazabicyclo[2.2.2]octane

5.04.6.1.2 Oxidative acylations of amines

Dibenzoyl peroxide has been used to both oxidise and benzoylate both primary and secondary amines ⟨63AP(296)57, 63AP(296)420, 81JOC2148⟩. Poly-4-vinylpyridine has been shown to be an effective catalyst of this reaction (Equation (31)) ⟨83S537⟩.

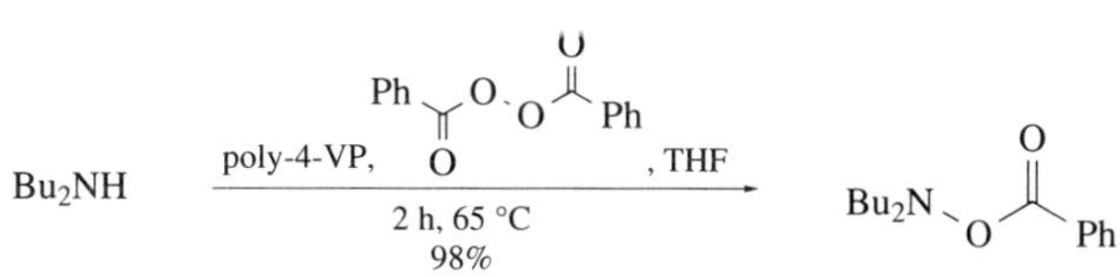 (31)

5.04.6.1.3 Reductive approaches to O-*acylhydroxylamines*

Phenyldimethylsilane has been used as a reducing agent for acyloximes in TFA (Scheme 44) ⟨88JOC5415⟩. A reductive acylation of tris(4-tosyloxymethyl)nitromethane has been reported to give an acylhydroxylamine (Scheme 44) ⟨85CAR375⟩.

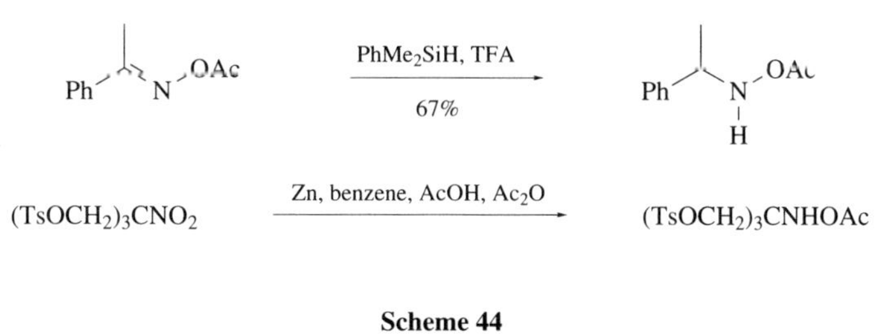

Scheme 44

5.04.6.1.4 *O-Acylhydroxylamines via Polonovskii rearrangements*

Piperidine-*N*-oxides undergo fragmentation via a Polonovskii rearrangement to give two substituted *N*-acetoxytryptamine derivatives (Equation (32)) ⟨81TL4405⟩. The uleine skeleton has been synthesised in an analogous fashion ⟨85LA2073⟩.

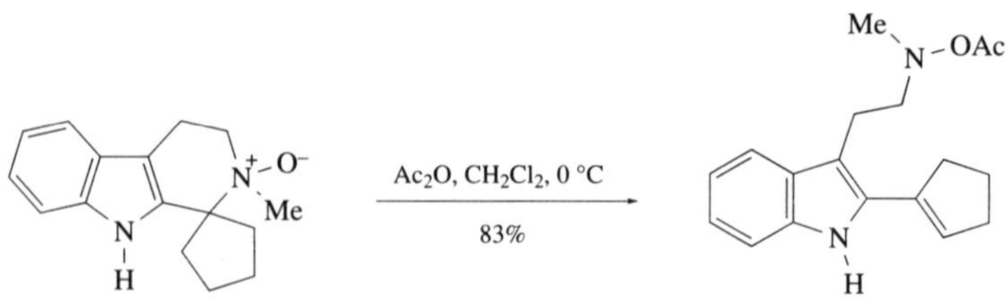

(32)

5.04.6.2 Acylated Hydroxamic Acids

5.04.6.2.1 *Bisacylation of hydroxylamines*

Acylated hydroxamic acids are relatively easy to prepare and a variety of conditions exist for their synthesis. Acetylation of hydroxylamines with an excess of acylating agent is a frequently adopted strategy, and acetic anhydride in pyridine ⟨85HCA1893, 86HCA236, 87JOC4933⟩, acetyl chloride with either triethylamine in dichloromethane ⟨88JMC1960⟩ or pyridine and dioxan ⟨53JA285⟩, benzoyl chloride in dioxan ⟨88IJC(B)1159⟩ or benzene containing pyridine ⟨89AJC2275⟩ have all been used successfully to acylate hydroxylamines.

5.04.6.2.2 *Acylation of hydroxamic acids*

The methods here are very similar to those of Section 5.04.6.2.1; acylations with acyl chlorides are the reactions most used (Scheme 45) ⟨46JCS25, 84S829, 86JOC167⟩, although ketene will also react with hydroxamic acids ⟨61JOC782⟩.

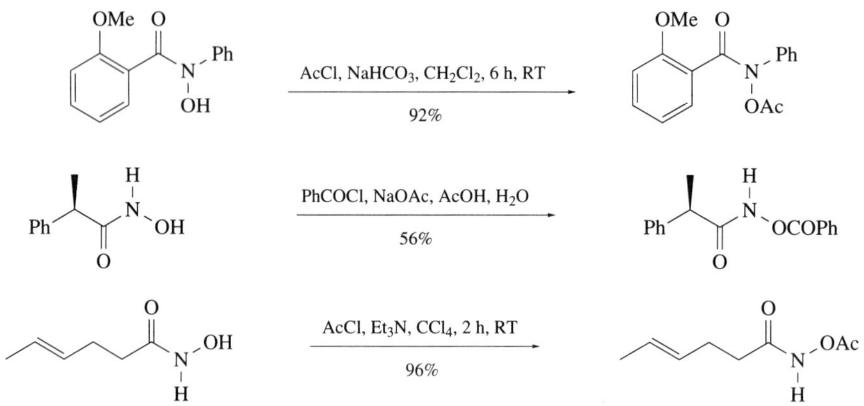

Scheme 45

5.04.6.2.3 *Other syntheses of acylated hydroxamic acids*

As for hydroxylamines, both oxidative and reductive approaches to the synthesis of acylated hydroxamic acids have been applied. Thus, aldehyde oximes have been reductively acylated with sodium cyanoborohydride and acetic anhydride in acetic acid ⟨88JMC499⟩, whilst benzoyl peroxide and acetyl chloride react with benzyl 6-aminohexanoate under phase-transfer conditions to produce benzyl *N*-acetyl-*O*-benzoyl-6-hydroxylaminohexanoate in 56% yield ⟨87AJC1919, 87TL1817⟩. Electrochemical reduction of azidoesters in acetonitrile in the presence of acetic anhydride produces acetylated hydroxamic acids (Scheme 46) ⟨86M(117)491⟩. Similarly 4-nitrostilbene has been reductively acylated (Scheme 46) ⟨86MI 504-01⟩.

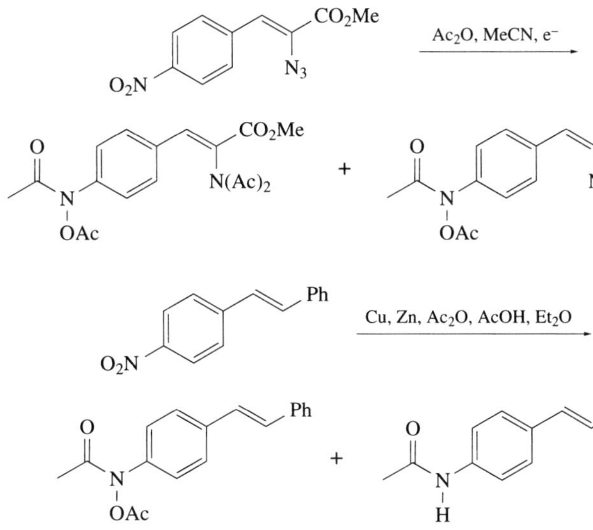

Scheme 46

5.04.7 *O*-ACYLOXIMES AND RELATED SYSTEMS

The requirement for the synthesis of acyl oximes ⟨68HOU(10/4)184⟩ derives from their utility as precursors in the Beckmann rearrangement ⟨60OR(11)1⟩ or as intermediates in the preparation of nitriles from aldehydes.

5.04.7.1 Acylation of Oximes with Acetic Anhydride

The most frequently encountered route to acyl oximes of ketones (ketoximes), is the direct acylation with acetic anhydride (Scheme 47) ⟨58JOC318, 84TL223, 86CPB4984, 86IZV2032, 87IJC(B)620⟩. Many oximes have been acylated in this fashion, usually in the presence of a base, typically pyridine. The reactions of acetic anhydride are facile, occurring around room temperature, and it is possible to selectively acylate ketoximes in the presence of phenols ⟨87IJC(B)620⟩. Mixtures of *syn*- and *anti*-oximes are also frequently encountered ⟨84TL223⟩.

The formation of acylated oximes, derived from aldehydes (aldoximes), with acetic anhydride is also straightforward (Scheme 48) ⟨86AP(319)325, 86IJC(B)373, 87JHC597, 90SC2901⟩. The majority of oximes are synthesised as precursors to nitriles.

5.04.7.2 Acylation of Oximes with Acyl Chlorides

As acid chlorides are generally easier to prepare and are more reactive than anhydrides, the method of choice when an oxime is to be acylated is that involving an acyl chloride and an organic base (Scheme 49) ⟨65JPS1373, 85CL701, 90S1129, 91JCS(P1)223⟩.

5.04.7.3 From Oximes and Carboxylic Acids with Dehydrating Agents

Dicyclohexylcarbodiimide (dcc) has been used as a dehydrating agent for the condensation of carboxylic acids with oximes (Scheme 50) ⟨68CR(C)1303, 86TL3239⟩. Similarly, trimethylsilyl-ethoxyacetylene, a dehydrating agent used in the formation of anhydrides, has been used to good effect in the formation of acylated oximes (Scheme 50) ⟨89S334⟩.

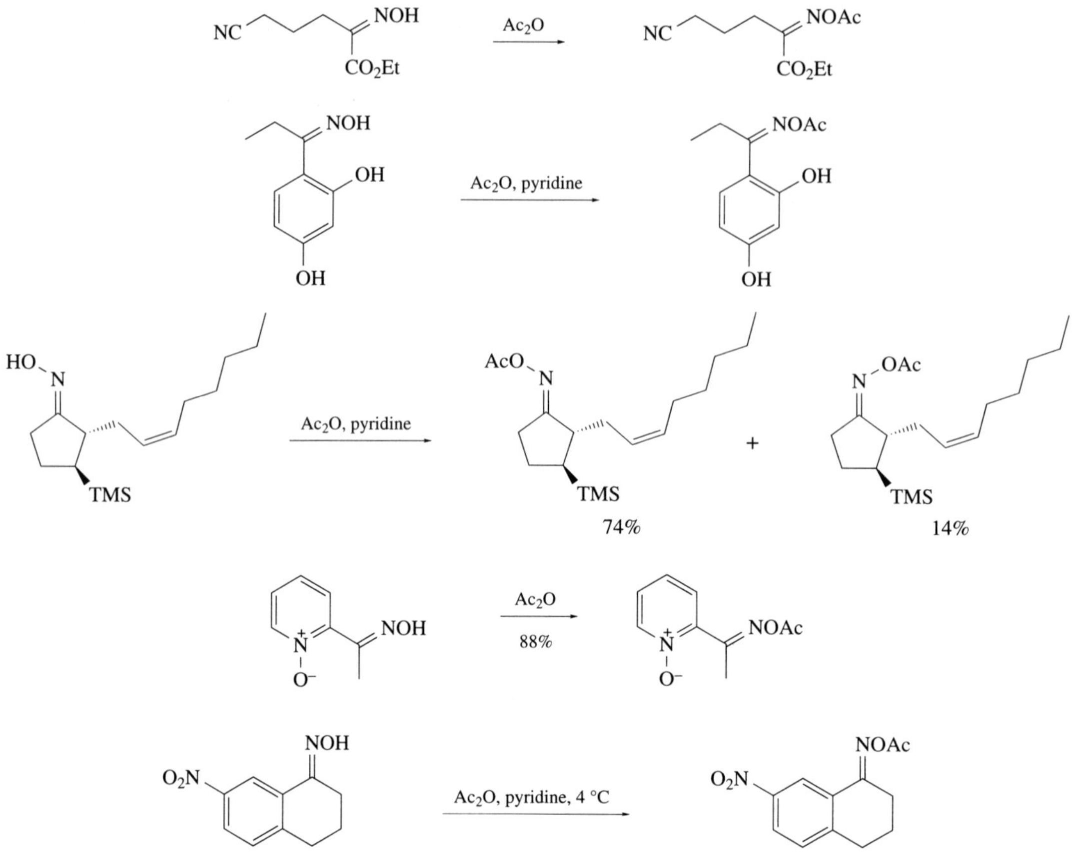

Scheme 47

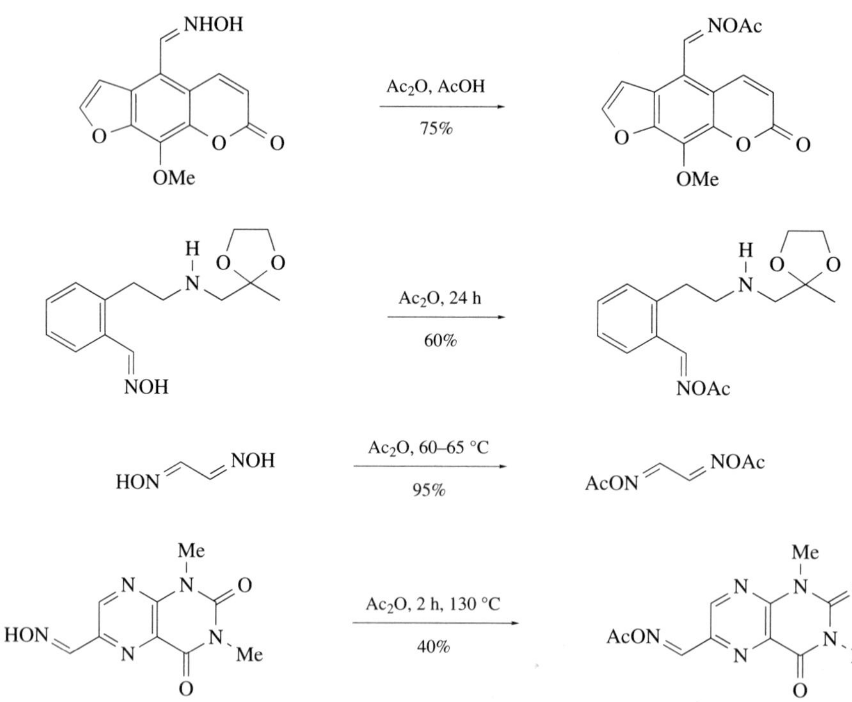

Scheme 48

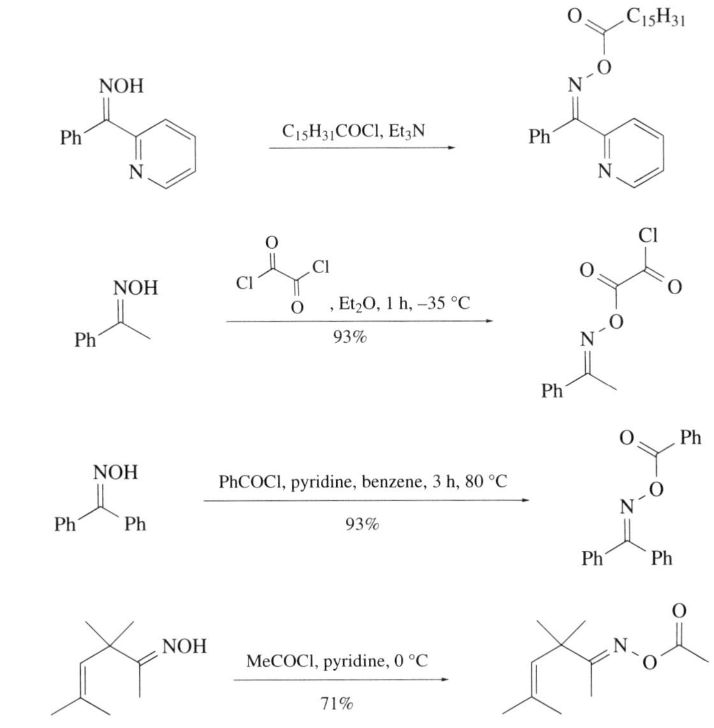

Scheme 49

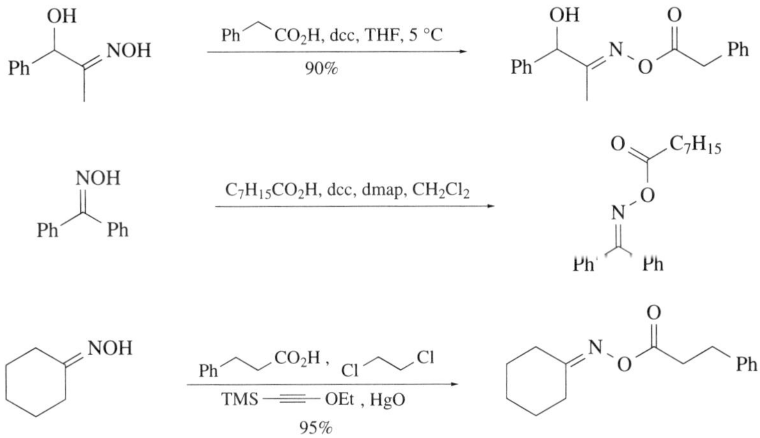

Scheme 50

5.04.7.4 Formation of Acylated Oximes with Diketene and Related Reactions

Benzaldehyde oxime was acylated with diketene in 56% yield in benzene in the absence of base. In the presence of triethylamine or 1,4-diazabicyclo[2.2.2]octane (dabco), the yields are raised to 86% and 90% respectively ⟨67JCED151⟩. A number of oximes have been acylated in this fashion; representative examples are shown in Table 9.

Oxime esters have been produced by the reaction of lipase Amano P with oximes in the presence of irreversible acyl transfer agents (vinyl carboxylates) in yields of 77–92% (Table 10). The reactions may be carried out in diisopropyl ether and reaction times are generally 4–12 h ⟨90SL690⟩.

4-Benzoyl-5-phenylfuran-2,3-dione may be reacted with oximes to produce, on thermolysis in xylene, an acylated oxime (Scheme 51). A ketene intermediate is proposed to account for the observed product ⟨87OPP52⟩.

Penicillin carboxylic acids may be activated with ethyl chloroformate in pyridine and acetone;

Table 9 Acylated oximes (R^1COCH_2COMe) formed by reaction of diketene and oximes.

R^1	*Solvent and catalyst*[a]	*Reaction time and temperature*	*Yield (%)*
(13)	Ether, dabco	1 h, 30 °C	95
(14)	Benzene, dabco	1 h, 30 °C	90
(15)	Ether, dabco	1 h, 30 °C	47
(16)	Ether, dabco	1 h, 30 °C	58
(17)	Ether, dabco	1 h, 30 °C	85

[a] dabco = 1,4-diazabicyclo[2.2.2]octane.

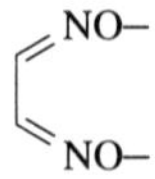

(13)

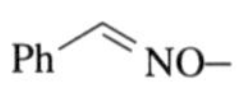

(14)

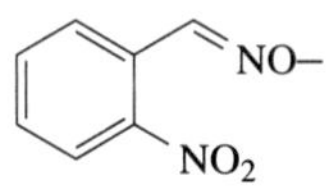

(15)

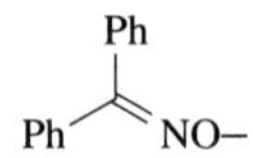

(16)

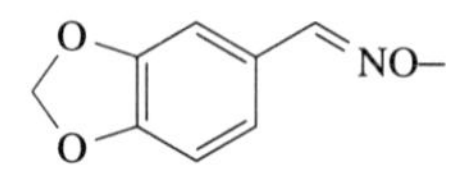

(17)

Table 10 Acylated oximes formed by reaction of vinyl carboxylates and oximes catalysed by lipase Amano P.

Acylated oxime	*Vinyl carboxylate*	*Reaction time and temperature*	*Yield (%)*
(18)	**(19)**	4 h, 25 °C	92
(20)	**(21)**	6 h, 25 °C	86
(22)	**(23)**	6 h, 25 °C	77
(24)	**(25)**	12 h, 25 °C	87
(26)	**(27)**	12 h, 25 °C	79

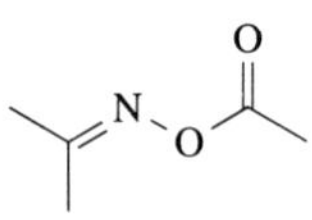

(18)

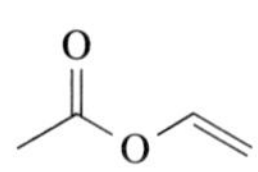

(19)

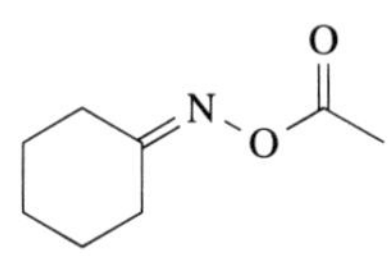

(20)

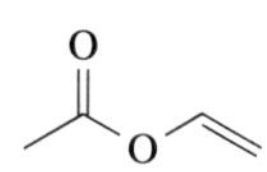

(21)

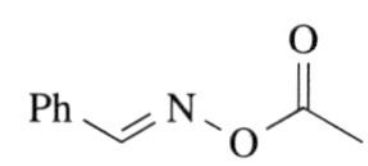

(22)

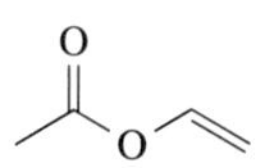

(23)

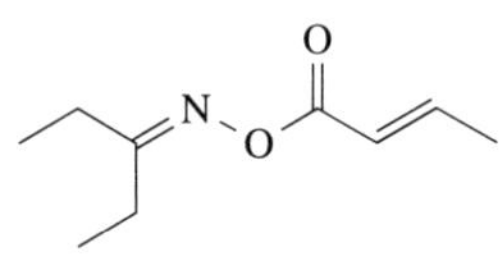

(24)

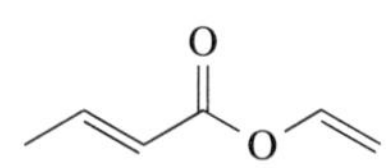

(25)

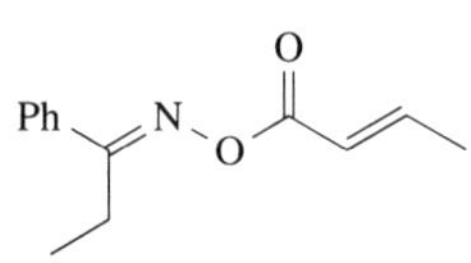

(26)

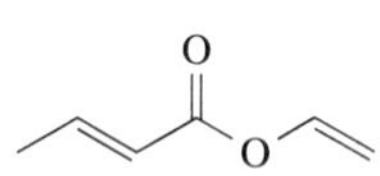

(27)

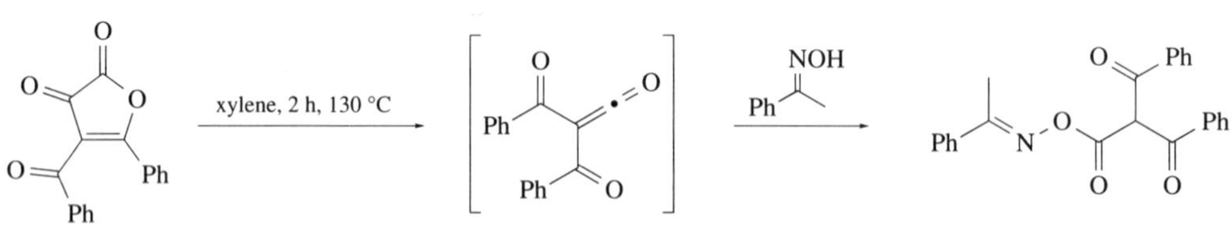

Scheme 51

this allows the reaction of benzaldehyde oxime to produce the acylated oxime (Equation (33)) ⟨71JCS(C)1917⟩.

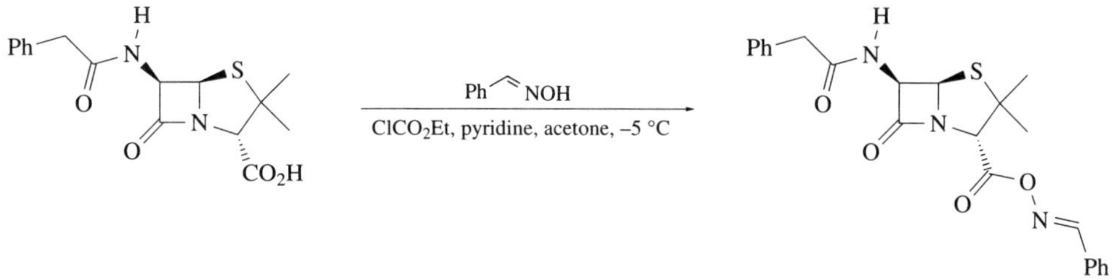

(33)

Benzophenone oxime has been acylated with 4-nitrobenzoic acid and triphenylphosphine in THF (Equation (34)) ⟨76JCS(P1)1708⟩.

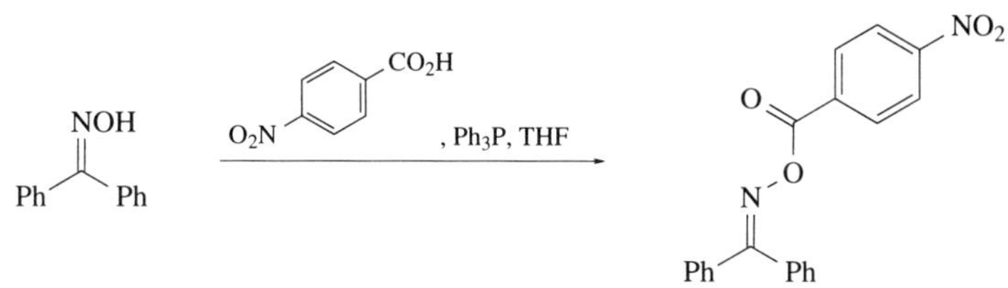

(34)

Bistrimethylsilylhydroxylamine and potassium hydride convert ketones to oximes and, without isolation of the oxime, an acid chloride may be introduced to acylate the oxime (Equation (35)) ⟨87S831⟩.

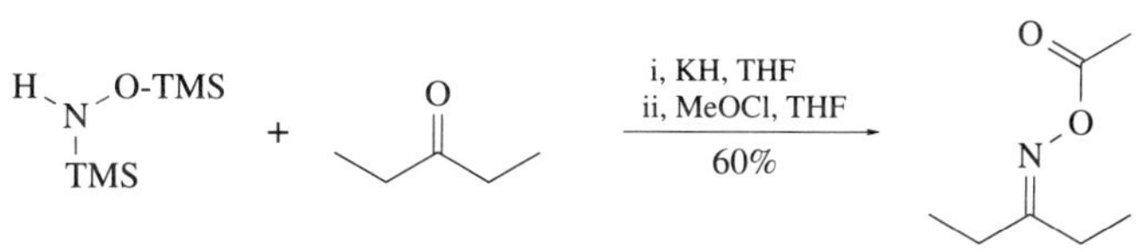

(35)

2-Aminobenzoylhydroxylamine has been used as a reagent for the one-step conversion of aldehydes to nitriles in the presence of boron trifluoride etherate. In the absence of boron trifluoride etherate, the intermediate *O*-(2-aminobenzoyl)-*N*-arylidene hydroxylamines may be isolated (Equation (36)) ⟨88SC2179⟩.

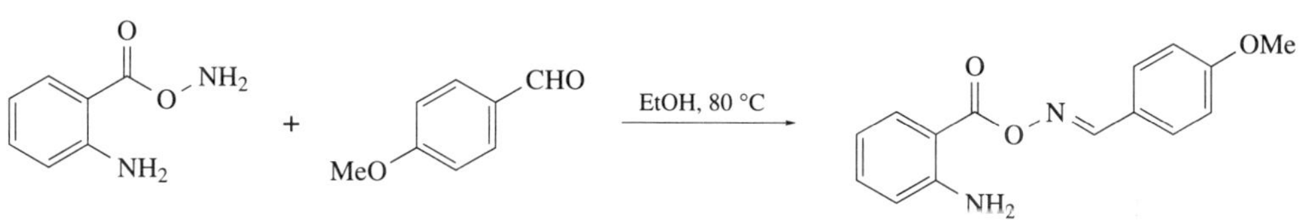

(36)

5.04.7.5 Hetero-substituted Acyl Oximes

Acylamidoximes constitute the largest group of compounds in this series. Their synthesis is frequently required for the synthesis of heterocycles. Amidoximes can be directly acylated on oxygen with acetic anhydride in either ether ⟨88CB723⟩, acetic acid ⟨87JHC325⟩, dichloromethane ⟨85CJC3089⟩ or dioxan (Equation (37)) ⟨62HCA441⟩.

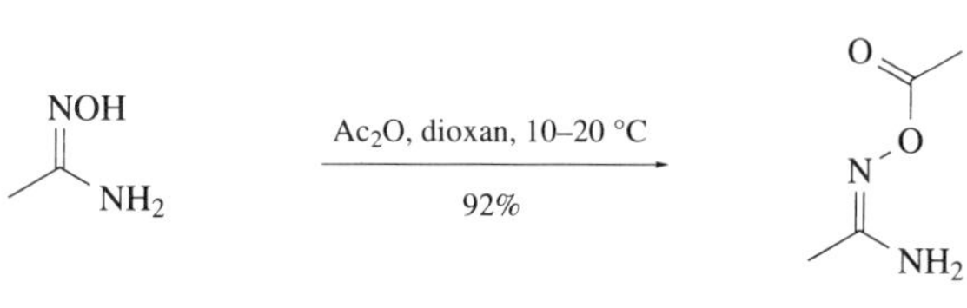

(37)

Acid chlorides acylate amidoximes in the presence of triethylamine in chloroform (Equation (38)) ⟨87H(26)163, 89CB1107⟩. Acetylation of phenylamidoxime with dcc and acetic acid gives a mixture of acetylated amidoxime (86%) and 3-phenyl-5-methyl-1,2,4-oxadiazole (7%) ⟨86JHC1657⟩.

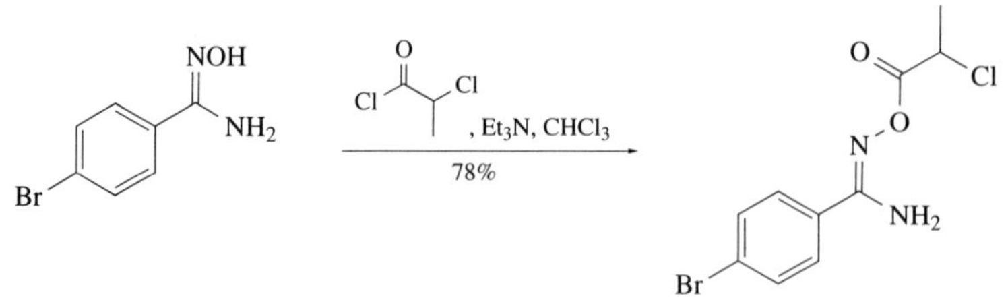

(38)

Benzyl glutamate *N*-carboxyanhydride may be used to acylate phenylamidoxime in quantitative yield (Equation (39)) ⟨66HCA1425⟩.

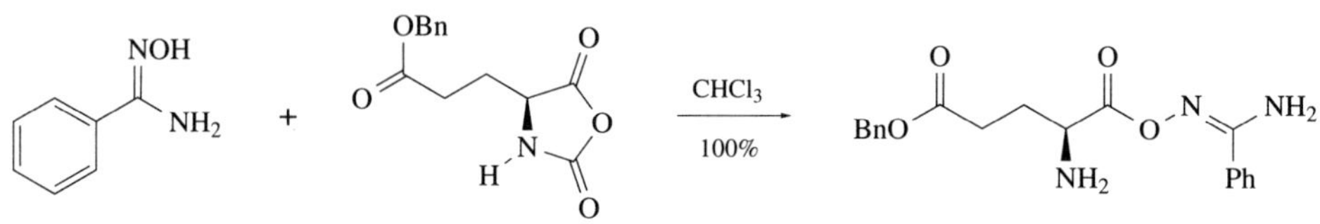

(39)

Azidooximes may be acetylated with acetic anhydride or aroylated with aroyl chlorides in aqueous potassium hydroxide solution (Equation (40)) ⟨86S843⟩. Yields are good.

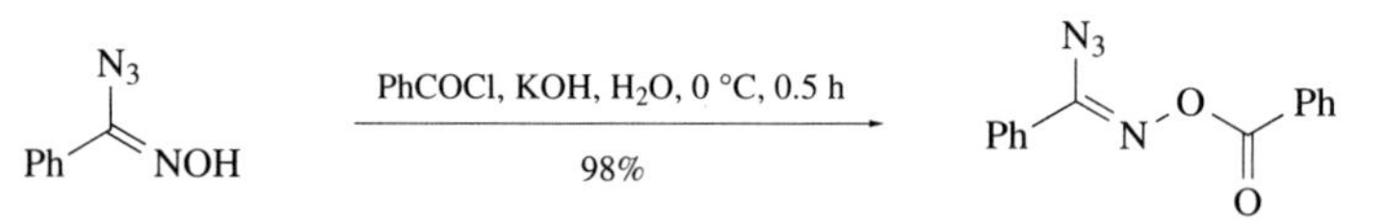

(40)

5.04.8 ACYLOXYPHOSPHORUS DERIVATIVES

5.04.8.1 Acyloxyphosphorus(III) Derivatives

The formation of acyloxyphosphorus(III) derivatives has been achieved by reaction of diphenylphosphinic chloride with the sodium salts of carboxylic acids in diethyl ether at room temperature (Scheme 52) ⟨81CB2272⟩, with carboxylic acids and triethylamine ⟨85JOM(294)C29⟩, and also with the silver or sodium salts of perfluorobenzoic acid derivatives ⟨86ZN(B)859⟩. Triphenylphosphine reacts with palladium(II) acetate to give phenylpalladium acetate and diphenylphosphinic acetate ⟨82JOM(235)243, 84TL5789⟩. Trifluoroacetic anhydride (TFAA) may also be reacted with diphenylphosphinic chloride in ether at $-40\,^{\circ}C$ to give diphenylphosphinic trifluoroacetate (Scheme 52). Diphenylphosphinic perfluorocarboxylates are thermally labile and rearrange irreversibly to perfluoroacyldiphenylphosphine oxides (Scheme 52) ⟨81CB2272⟩.

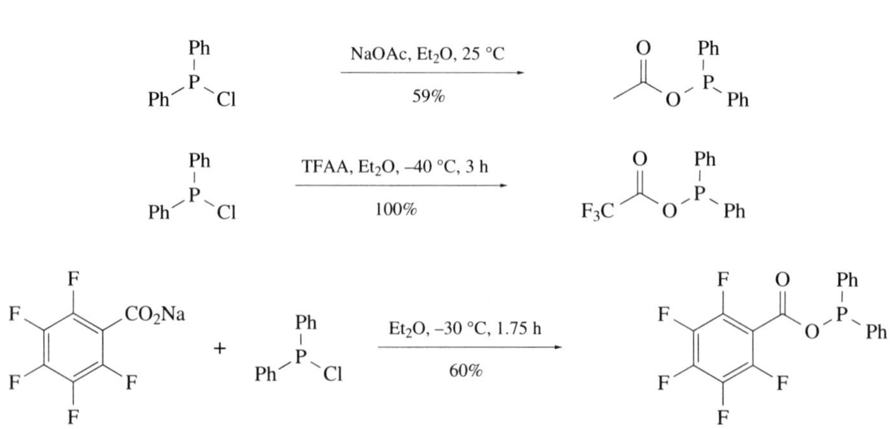

Scheme 52

Analogous nucleophilic substitution at phosphorus may be achieved with ethyldicyanophosphine and one of pivalic, isobutyric or methacrylic acid in the presence of catalytic quantities of sodium acetate, when one cyanide group may be displaced to produce acyloxyphosphites in 68–84% yields ⟨89IZV418⟩.

Acylation with acetic anhydride of bis(diethylamino)phenylphosphine similarly produces displacement of diethylamine to produce acetoxydiethylaminophenylphosphine in 42% yield ⟨86IZV2535⟩.

Bis(benzoyloxy)phenylphosphine and bis(benzoyloxy)methylphosphine have been prepared by reaction of dichlorophenylphosphine with metal carboxylates ⟨83CZ121⟩.

5.04.8.2 Acyloxyphosphorus(V) Derivatives

Diphenylphosphine acetate may be oxidised with oxygen in ether to give acetoxydiphenylphosphine oxide. Acyloxydiphenylphosphine oxides may also be synthesised by reacting phosphinic acids with perfluorocarboxylic acid anhydrides at −40 °C (Scheme 53) ⟨81CB2272⟩.

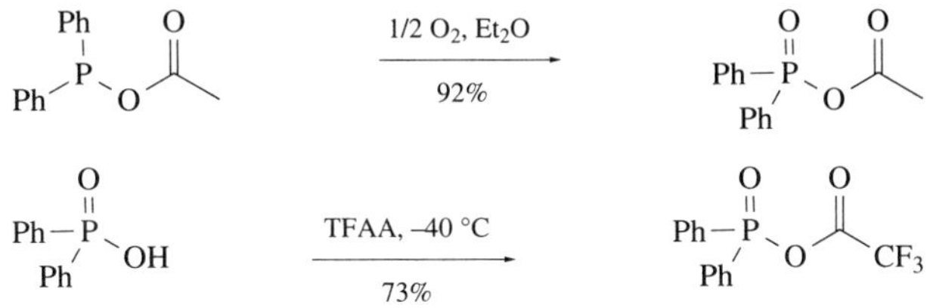

Scheme 53

Analogously to diphenylphosphinic chloride, dichlorophenylphosphine oxide may be reacted with 2 equivalents of silver benzoate or acetate in ether at room temperature to produce either bis(benzoyloxy)phenylphosphine oxide or bis(acetoxy)phenylphosphine oxide, respectively ⟨83CZ257, 83CZ363⟩. Oxidation of either bis(benzoyloxy)phenylphosphine or bis(benzoyloxy)methylphosphine in ether at room temperature with oxygen produces bis(benzoyloxy)methyl- and bis(benzoyloxy)-phenylphosphine oxides in 70–80% yields ⟨87CZ61⟩.

Dimethylphosphinothioyl chloride has been reacted with peptide carboxylic acids to form an activated dimethylphosphinothioic mixed anhydride which has then been coupled to form peptides ⟨88BCJ3653⟩. Similarly, amino sugars have been acylated selectively on nitrogen with dimethyphosphinothioates (Scheme 54) ⟨88BCJ4467⟩.

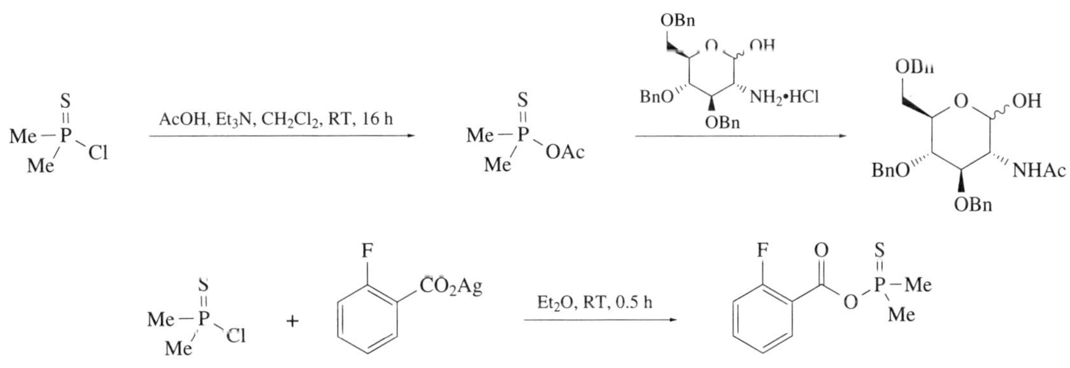

Scheme 54

The silver salts of fluorinated benzoic acid derivatives have been coupled with dimethylphosphinothioyl chloride in ether at 5 °C to provide benzoylated dimethylphosphinothioates (Scheme 54) ⟨85IFC(27)157⟩.

5.04.9 ACYLOXYARSENIC, -ANTIMONY AND -BISMUTH DERIVATIVES

5.04.9.1 Acyloxyarsenic Derivatives

Very few reactions exist to produce acyloxyarsenic derivatives. However, methylphenylchloroarsine may be substituted on arsenic with sodium acetate in methanol in 36% yield ⟨47JA925⟩;

similarly, dimethylchloroarsine has been reacted with silver trifluoroacetate to give methylphenyltrifluoroacetoxyarsine ⟨60CJC472⟩. More recently, dimethylaminodimethylarsine has been found to react smoothly with a range of carboxylic acids in yields of 80–95% and appears to be the method of choice for the formation of dimethylacyloxyarsines ⟨88CZ106⟩. Silanyloxyarsanes react with carboxylic acid chlorides or trifluoroacetic acid to give high yields of acyloxyarsanes (Scheme 55) ⟨90CZ71⟩.

$$Me_2AsOAc \xleftarrow{AcCl} Me_2AsOSiPh_3 \xrightarrow[75\%]{TFA} Me_2AsOCOCF_3$$

Scheme 55

5.04.9.2 Acyloxyantimony Derivatives

Of the reported derivatives of acyloxyantimony, the best represented group are the bis(acyloxy)triarylantimony(V) analogues. These acyloxyantimony derivatives have a long history ⟨1856LA(97)329⟩ and a wide variety of acyloxy substituents have been prepared by a general method which requires the heating of a carboxylic acid (for example, acetic acid) with a trialkyl- or triarylantimony oxide. When the carboxylic acid is a solid, the preferred alternative is to heat a solution of the carboxylic acid in chloroform in the presence of triphenylantimony(V) oxide with the removal of water ⟨65JOC1259, 89ZAAC(574)130, 90ZAAC(582)37, 92MI 504-02⟩.

Antimony(III) derivatives are rarer but have been reported. Thus diphenylantimony(III) chloride when heated in acetic acid gives a quantitative yield of diphenylantimony(III) acetate ⟨92JOM(426)49⟩. Triphenylantimony is converted to diphenylantimony(III) acetate with palladium(II) acetate ⟨77BCJ2021, 84TL5789⟩.

5.04.9.3 Acyloxybismuth Derivatives

The most used method for the synthesis of bismuth(V) derivatives is the reaction of bismuth carbonate with alkyl and aryl carboxylic acids ⟨51JA4470⟩. Alternatively, the triarylbismuth dichloride may be used in place of the carbonate ⟨51JA4470⟩ and in this fashion a large number of heterocyclic carboxylic acids have been reacted ⟨89ZAAC130, 90ZAAC37⟩.

Oxidative acylation with acetic anhydride and *t*-butyl hydroperoxide has been reported with triphenylbismuth and appears to be quite general in scope, though yields are only moderate (Equation (41)) ⟨88ZOB1934⟩.

$$Ph_3Bi + Bu^tO_2H \xrightarrow[65\%]{Ac_2O,\ Et_2O,\ RT,\ 90\ h} Ph_3Bi(OAc)_2 + Bu^tOH \qquad (41)$$

5.04.10 ACYLOXYSILANES

Acyloxysilanes have been synthesised principally as protecting groups for carboxylic acids. The three silyl esters most frequently encountered are trimethylsilyl (TMS), *t*-butyldimethylsilyl (TBDMS) and *t*-butyldiphenylsilyl. They have great utility as protecting groups because of their facile conversion to carboxylic acids and because of the variation in ease of hydrolysis, which is in the order TMS > TBDMS > *t*-butyldiphenylsilyl.

Silylation of carboxylic acids produces derivatives with enhanced volatility with respect to the carboxylic acid, and this property of silyl esters has been used for the mass-spectrometric characterisation of many natural products containing the carboxylic acid function.

The other reason for their synthesis is their role as intermediates in the synthesis of carboxylic acids in the Ireland–Claisen rearrangement.

5.04.10.1 From Carboxylic Acids and Trimethylsilyl Chloride

The conversion of carboxylic acids to TMS esters is usually accomplished by treating carboxylic acids with TMS-Cl and an organic base (Scheme 56), which is usually triethylamine ⟨86HCA44, 87CPB1903, 88JHC59, 88OM366, 89OM655⟩ or pyridine ⟨85JOM(294)139⟩ but may be sodium hydride ⟨86AG(E)994, 87T2045⟩ or lithium diisopropylamide (LDA) ⟨90CPB1836⟩, in an inert solvent. Yields are generally around 80%. Sodium salts of carboxylic acids may be used ⟨86TL4995⟩, as can carboxylic acids containing basic groups capable of catalysing silyl ester formation, when no additional base is required ⟨89LA387, 90ZOB1928⟩.

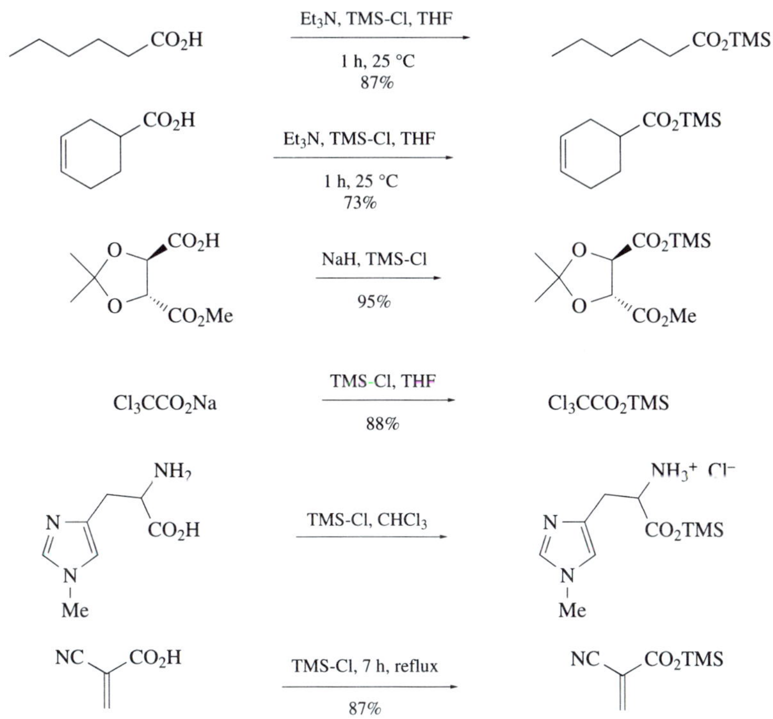

Scheme 56

5.04.10.2 From Carboxylic Acids and Other Silyl Chlorides

The use of TBDMS-Cl as a reagent for the protection of alcohols is well documented and from its introduction the synthetic utility of this reagent to protect carboxylic acids was recognised (Scheme 57) ⟨72JA6190, 85JCS(P1)2361, 87CJC2390, 87S801, 89TL5821⟩. Transformations which may be effected whilst the carboxylic acid group is protected as a TBDMS ester have been collated ⟨B-91MI 504-01⟩ together with methods of desilylation. The introduction of the TBDMS group is usually effected with TBDMS-Cl in the presence of a catalyst, usually imidazole, when the solvent of choice is DMF ⟨72JA6190⟩, though 1,5-diazabicyclo[5.4.0]undec-5-ene (dbu) has been demonstrated to be an effective alternative when reactions may be undertaken in benzene or dichloromethane at room temperature ⟨85TL475⟩.

t-Butyldiphenylsilyl chloride and dimethyl-(1,1,2-trimethylpropyl)silyl chloride are the other silyl chlorides which have been used as protecting groups for carboxylic acids. Pearson *et al.* have used *t*-butyldiphenylsilyl chloride to silylate an azetidinone carboxylic acid in 58% yield. Triethylamine was the catalyst and the reaction was carried out at 25°C in dichloromethane ⟨85JCS(P1)1927⟩. Similarly, a hydroxycyclopentanecarboxylic acid has been disilylated in the presence of silver nitrate and pyridine ⟨88TL4909⟩. Conditions which affect *t*-butyldimethylsilylation are also applicable; thus, *N*-tosylserine is disilylated with *t*-butyldiphenylsilyl chloride in DMF with imidazole as catalyst ⟨92SL41⟩.

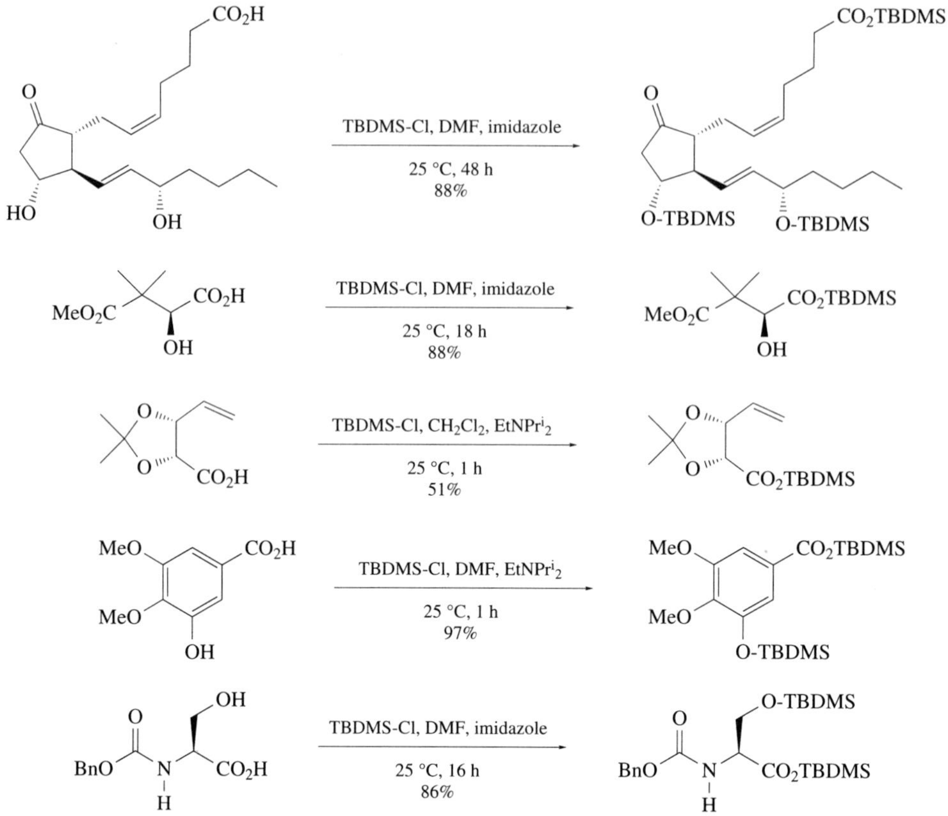

Scheme 57

Dimethyl-1,1,2-trimethylpropylsilyl chloride (theoxyldimethylsilyl chloride) has been recommended as an alternative to TBDMS-Cl as having the advantages of cheapness and ease of preparation. Dimethyl-1,1,2-trimethylpropylsilyl chloride has been shown to silylate phenylacetic acid in 61% with triethylamine in DMF ⟨85TL5515⟩.

Chloromethyldimethylchlorosilane reacts with sodium acetate to form either chloromethyldimethylacetoxysilane in refluxing THF or acetoxymethyldimethylacetoxysilane in DMF ⟨85SRI321⟩. Similarly, silyl dichlorides have also been shown to react with simple carboxylic acids (Equation (42)); thus, acetic acid in the presence of base reacts with dichloro(chloromethyl)[trimethylsilyl]methylsilane ⟨88JOM(334)139⟩.

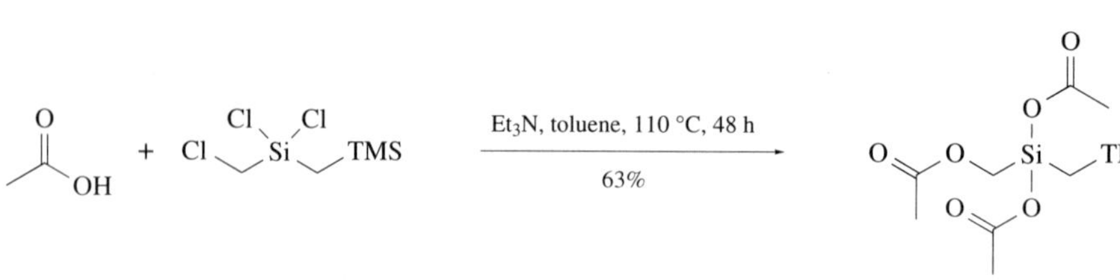

(42)

5.04.10.3 From Carboxylic Acids and Hexamethyldisilazane

Hexamethyldisilazane has been used to silylate carboxylic acids and a number of instances are reported in the literature. However, alone it is relatively unreactive and catalysis is required for high yields of TMS esters ⟨82JOC3966⟩. The catalyst of choice is saccharin (or its sodium salt) (Scheme 58), though bis(4-nitrophenyl)-*N*-(4-toluenesulphonyl) phosphoramidate and tetraphenyl iminodiphosphate were also found to be effective catalysts ⟨82JOC3966⟩.

Silylations with hexamethyldisilazane can also be catalysed with TMS-Cl ⟨87ZOB686, 88S990⟩ and may be complicated by the addition of the trimethylsilylamino group to electron-deficient alkenes ⟨89JOC4511⟩.

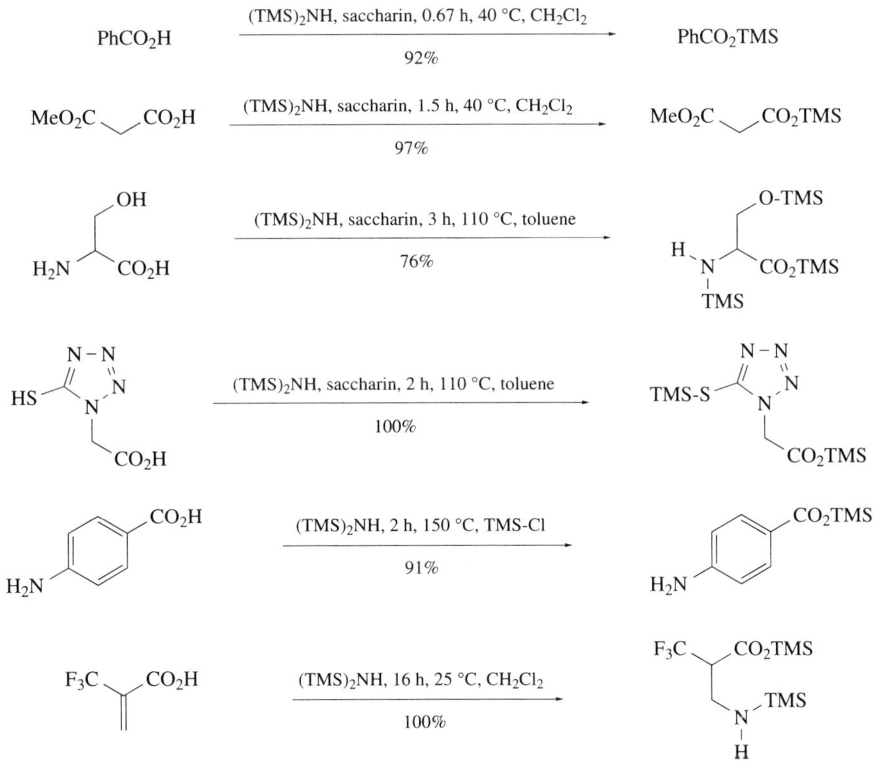

Scheme 58

5.04.10.4 From Carboxylic Acids and Organosilanes

Group 10 metals catalyse the cleavage of silicon–hydrogen bonds, allowing attack at silicon by nucleophiles containing NH or OH functionality. Thus, Yamamoto *et al.* have shown that carboxylic acid groups may be formed from *t*-butyldimethylsilane in the presence of catalytic 10% palladium-on-charcoal (Scheme 59). The reaction may be carried out in hexane, when the by-product is hydrogen. The isolation procedure is nonaqueous, thus avoiding any unwanted hydrolysis of the TBDMS ester, and the yields are high ⟨89BCJ2111⟩.

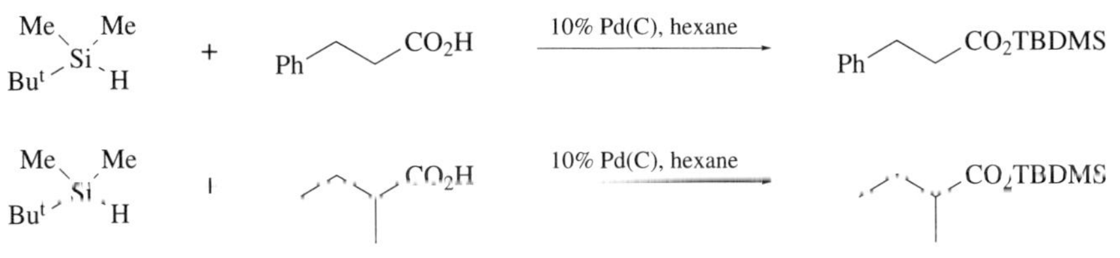

Scheme 59

Under palladium catalysis, methacrylic acid adds tributylsilane in yields between 83% and 85%. The reaction may be carried out in toluene or a mixture of DMF and toluene, the latter requiring triphenylphosphine ⟨90JAP280254⟩.

5.04.10.5 By Dealkylative Cleavage of Esters and Lactones

Trimethylsilyl iodide has been demonstrated to convert alkyl esters into acids via TMS esters (Scheme 60) ⟨76AG(E)774, 77JA968⟩ and lactones to iodoalkyl esters via TMS esters ⟨81SC763⟩. Trimethylsilyl bromide also reacts with lactones to produce the corresponding bromoalkyl TMS esters ⟨79AG(E)689⟩. Olah *et al.* found that a combination of TMS-Cl and sodium iodide in acetonitrile is effective in the dealkylative cleavage of esters and lactones via TMS esters, and have recommended the use of this combination of reagents as a cheaper alternative to trimethylsilyl iodide ⟨79JOC1247⟩.

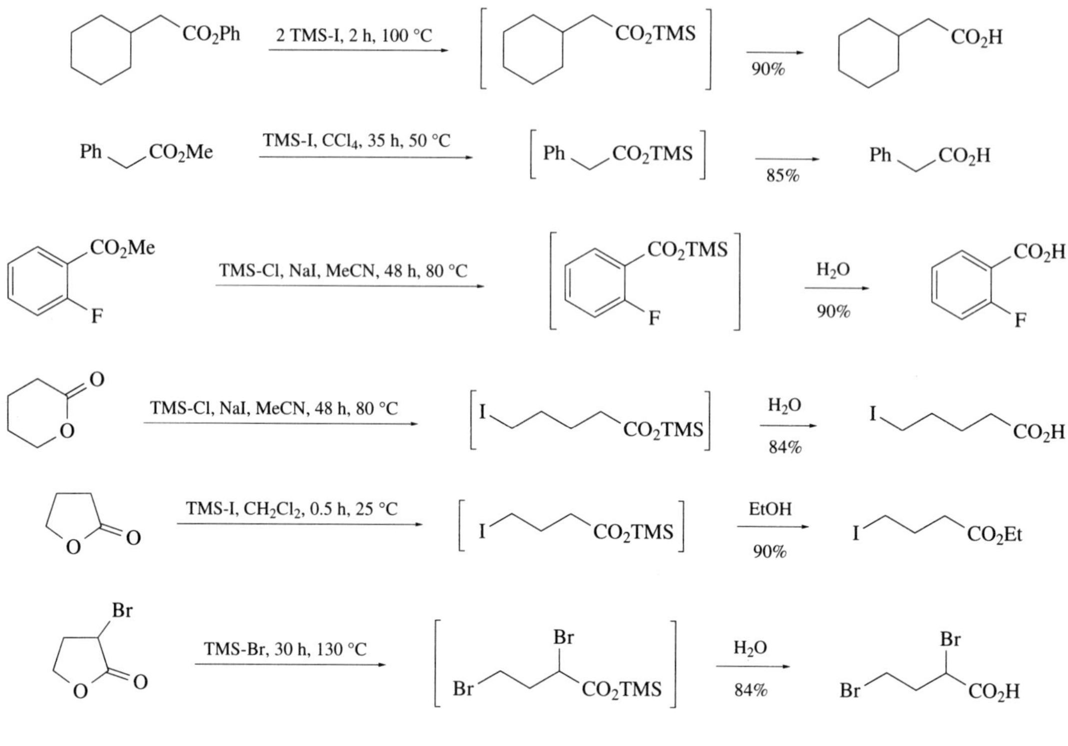

Scheme 60

5.04.10.6 Other Silylating Reagents

Alkylsilyl cyanides may be used to silylate carboxylic acids. Thus, *t*-butyldimethylsilyl cyanide silylates adamantane-1-carboxylic acid and 4-nitrobenzoic acid in 92% and 83% yields, respectively. Reaction times are short (5 min) and the reaction temperature is 80 °C ⟨86JOC3545⟩.

A comprehensive study of the complete mass spectra and general fragmentation pathways of TMS derivatives of protein amino acids has been made and this has facilitated the unequivocal identification of nonvolatile amino acids after appropriate derivatisation ⟨77JC355⟩. Kawashiro *et al.* have demonstrated the efficacy of *N*,*O*-bistrimethylsilyltrifluoroacetamide and advocated its use as a silylating agent for imino derivatives of alanine prior to mass spectrometry ⟨85BCJ1903⟩.

Ketene methyltrialkylsilyl acetals provide direct, mild and high-yielding silylating agents which require no catalysis, base or acid (Equation (43)) ⟨79TL4311⟩. These reagents also silylate primary, secondary and tertiary alcohols as well as phenols, thiols and acidic amides.

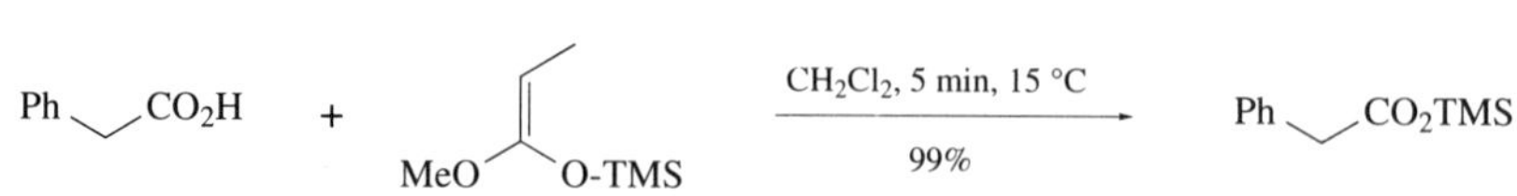

(43)

Trimethylsilyl iodide or trimethylsilyl triflate react with pyridine to produce trimethylsilylheteroarylium salts which effectively silylate arylcarboxylic or alkylcarboxylic acids in chloroform (Equation (44); $X^- = I^-$, $CF_3SO_3^-$) ⟨87S931⟩.

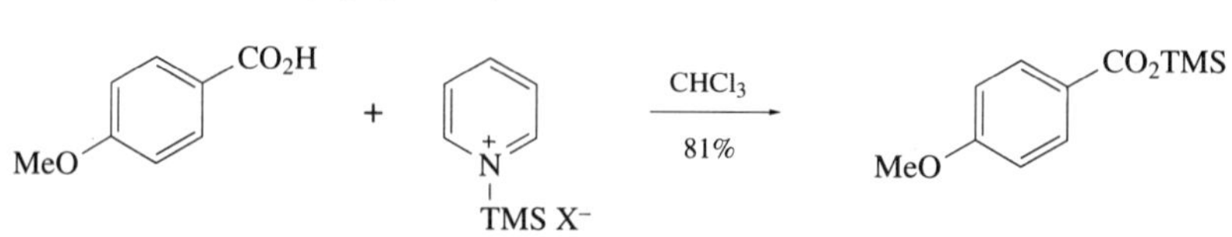

(44)

The problem of decarboxylation of malonic acid analogues has been overcome by reacting Meldrum's acid with a variety of silylated amines, alcohols and amides, resulting in the production of TMS ester derivatives of malonic acid with defined and selective functionality (Equation (45)) ⟨89TL3073⟩.

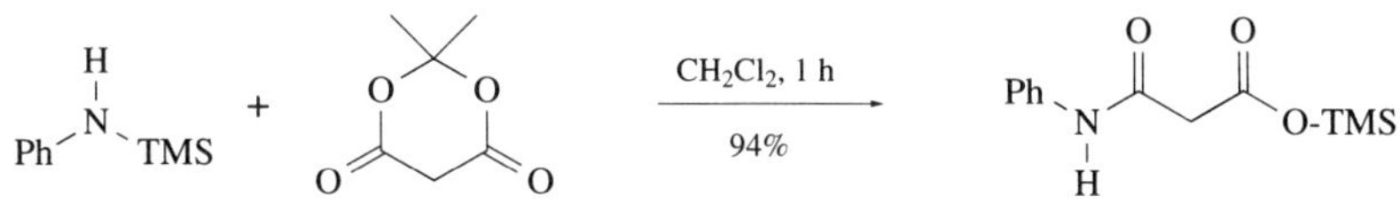

(45)

5.04.10.7 Photooxygenation of Siloxyalkenes

The photooxygenation of either 3-trimethylsiloxy- or 3-*t*-butyldimethylsiloxybenzofurans results in initial dioxetane formation, followed by rearrangement to α-silylperoxy ketones which then cleave to form the silylsalicylate esters (Scheme 61) ⟨90TL2259⟩. Similar oxidation and fragmentation has been shown to occur with 1,2-bistrimethylsiloxycyclohex-1-ene when a mixture of bis(silyl) adipates are produced ⟨90TL1245⟩.

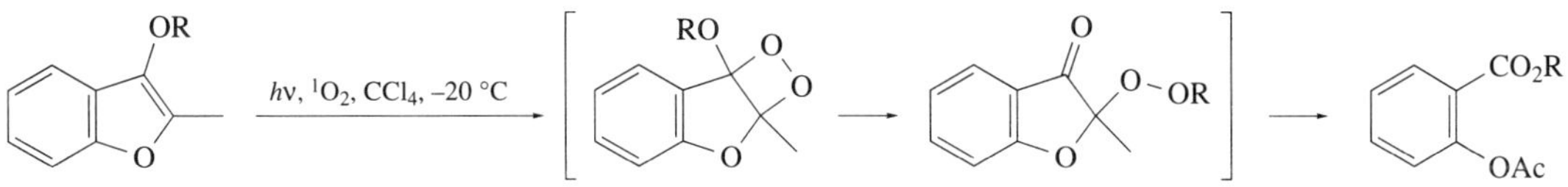

Scheme 61

5.04.10.8 The Ireland–Claisen Rearrangement

The Ireland–Claisen rearrangement of silylketene acetals has proved to be a versatile route to derivatives of pent-4-en-1-oic acid. Silylketene acetals are readily prepared by generating ester enolates with LDA and trapping the enolate with either TMS-Cl or TBDMS-Cl. The silylketene acetals are then thermally rearranged ([3,3]-sigmatropic), via a chair-like transition state, to TMS esters, which on mild hydrolysis provide pent-4-en-1-oic acids (Equation (46)) ⟨75TL3975, 76JA2868⟩. Examples demonstrating the scope of the reaction are shown in Table 11.

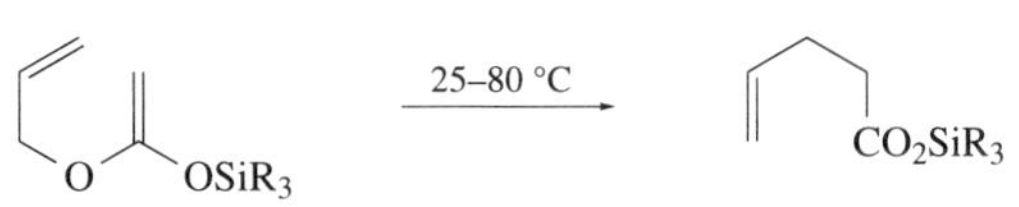

(46)

Table 11 Silyl alkenoates via the Ireland–Claisen rearrangement.

Silylketene acetal	*Silyl ester Yield* (%)	*Ref.*
(28)	**(29)** (68%)	91JOC353
(30)	**(31)** (75%)	89TL515
(32)	**(33)**	85H(23)549
(34)	**(35)**	85JOC707
(36)	**(37)** (77%)	84CC827

5.04.11 ACYLOXYBORANES

One of the earliest reports of the synthesis of acyloxyboranes is that of the reaction between triethylborane and acetic acid which produces acetoxydiethylborane ⟨37JPR251⟩.

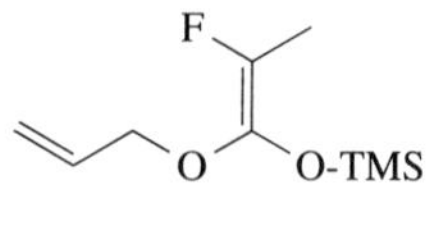

(28)

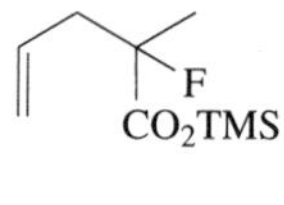

(29)

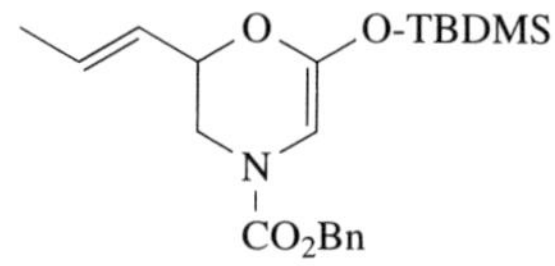

(30)

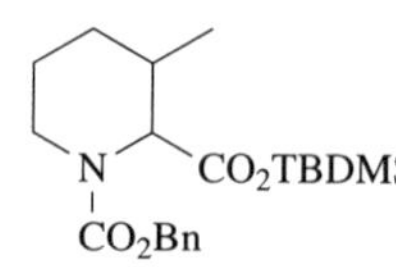

(31)

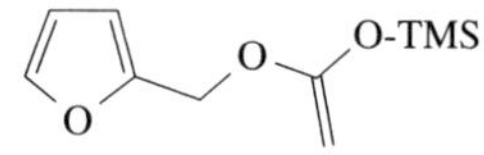

(32)

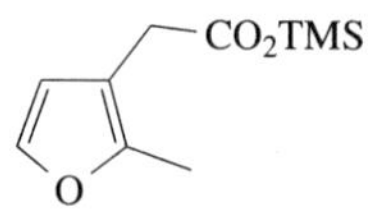

(33)

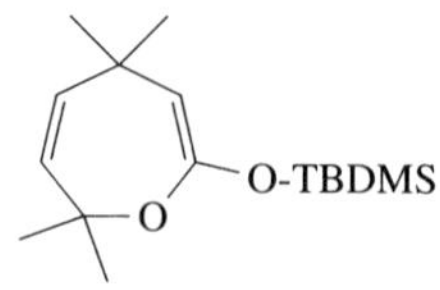

(34)

(35)

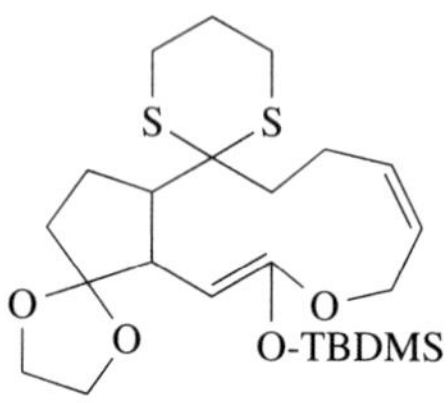

(36)

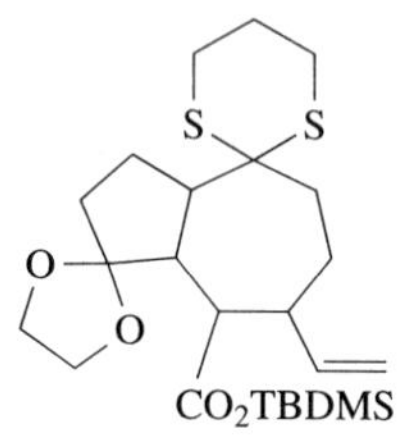

(37)

Diaryl- and dialkylboron chlorides have been reacted with carboxylic acids at room temperature to produce monoacyloxyboranes. Similar displacements of aryl- and alkylboron dichlorides provide the bisacyloxyboranes ⟨58JCS3648⟩. Displacements may be carried out either by the sodium salt, for example, sodium trifluoroacetate, or with alkanoic acids, when elimination of hydrogen chloride may be exothermic (Scheme 62).

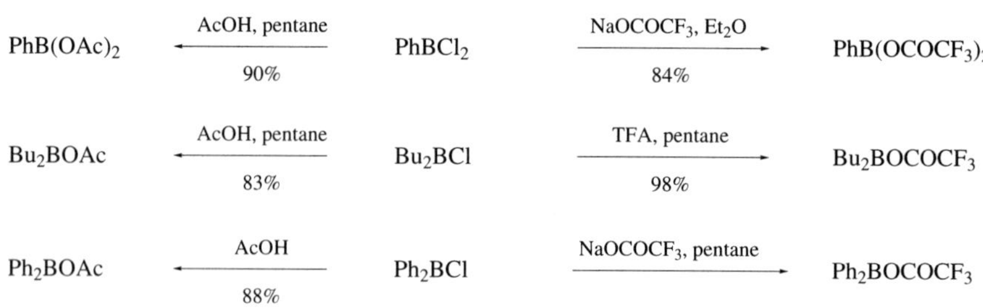

Scheme 62

Amino acids react with trialkylboranes, butyldiphenylborinates or sodium tetraphenylborate to produce acylated boron species which have strong intramolecular coordination of nitrogen to boron (Scheme 63) ⟨64JOC492, 70JOM(22)269, 74LA112, 83T2995⟩.

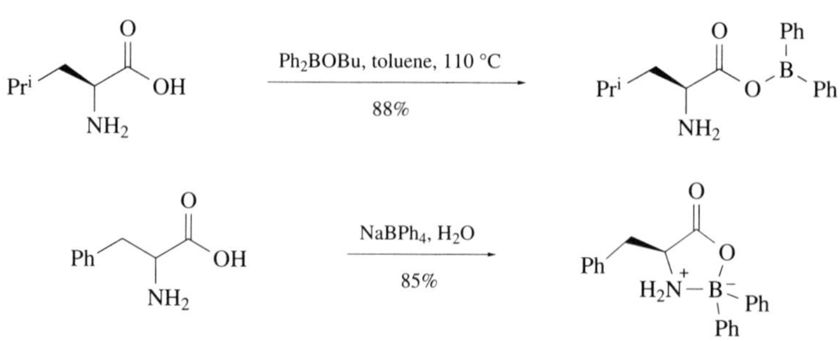

Scheme 63

The coordination of nitrogen to boron does not prevent nucleophilic reaction of the nitrogen of boroxazolidones; thus, imine formation results when aldehydes are condensed with boroxazolidones

⟨85T6063⟩. Similar coordination to trialkylboranes has been demonstrated with aziridine carboxylic acids ⟨92RCT211⟩.

β-Hydroxyacids react with phenylboronic acid to produce cyclic acyloxyboranes and this has been used to protect hydroxyacid functionality in total syntheses (Equation (47)) ⟨86TL5683⟩.

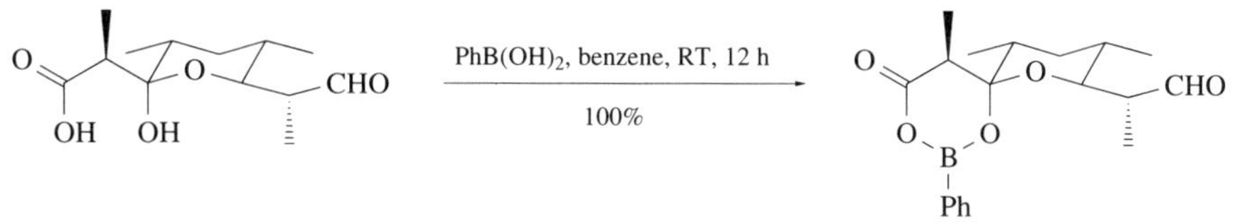

(47)

Acyloxyboranes are intermediates in the synthesis of alkenols from iodoalkynes. Thus, when borinane is added to 1-iodo-1-butyne, an alkenylborinane is produced which may then be ring-enlarged with sodium methoxide to produce a borocane. The seven-membered heterocyclic borocane is then ring-opened to produce an alkenylacetoxyborane (Scheme 64). This may then be oxidised to an alkenol ⟨84S920⟩.

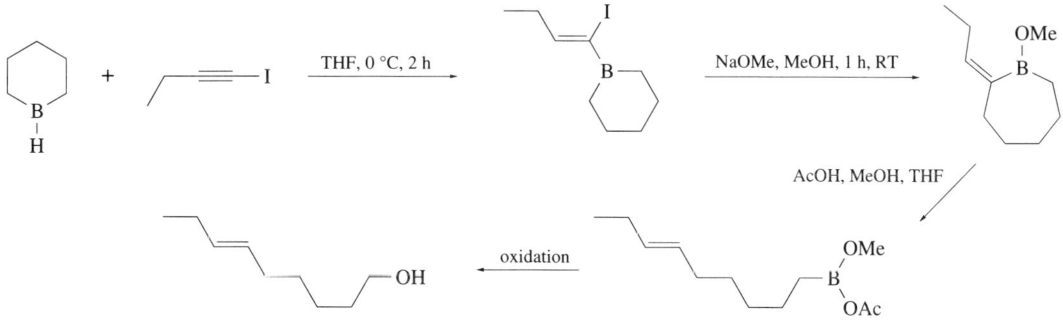

Scheme 64

5.04.12 ACYLOXY METALS

There are many inorganic salts of carboxylic acids and comprehensive coverage in this type of review is inappropriate. Acyloxy metals are well reviewed in Beilstein, where references to the preparation of the simple carboxylates of lithium, sodium, potassium, rubidium, copper, silver, beryllium, magnesium, calcium, strontium, barium, zinc, mercury, aluminum, scandium, the lanthanides, titanium, zirconium, thorium, tin, lead, vanadium, chromium, uranium, manganese, iron, cobalt and nickel and rhodium are to be found ⟨B-68MI 504-01⟩.

Of greater interest are the preparations of carboxylates encountered in synthetic organic procedures which are not included in the above list. Thus palladium(II) acetate may be synthesised by taking palladium sponge and heating in a mixture of acetic and nitric acid. Other carboxylates are also reported to be synthesised by this procedure, including palladium(II) benzoate and palladium(II) trifluoroacetate ⟨64CI(L)544⟩.

Silver benzoate is synthesised by taking aqueous solutions of silver nitrate and sodium benzoate and collecting the resultant precipitate ⟨50JA5163⟩.

Thallium(III) acetate may be synthesised in 61% by reacting thallium(III) oxide with acetic anhydride and acetic acid ⟨68JOC75⟩.

5.05
Acylsulfur, -selenium, or -tellurium Functions

AKIYA OGAWA and NOBORU SONODA
Osaka University, Japan

5.05.1 THIOCARBOXYLIC ACIDS AND THEIR SELENIUM ANALOGUES

5.05.1.1 Thiocarboxylic Acids

Thiocarboxylic acids are important parent compounds for the synthesis of a series of acylsulfur functions ⟨55HOU(9)745, 85HOU(E5)832⟩. Compared with carboxylic acids, the corresponding thiocarboxylic acids are more acidic: for example thioacetic acid ($pk_a = 3.33$) versus acetic acid ($pk_a = 4.76$); thiobenzoic acid ($pk_a = 2.48$) versus benzoic acid ($pk_a = 4.20$).

Although a number of synthetic methods for thiocarboxylic acids have been reported, this review deals only with high yield preparations. Some thiocarboxylic acids are commercially available.

The reaction of Grignard reagents with carbonyl sulfide leads to the formation of magnesium salts of thiocarboxylic acids, which, upon treatment with proton acids, provide the corresponding thiocarboxylic acids (Equation (1)) ⟨03CB1007, 75BSF1439⟩.

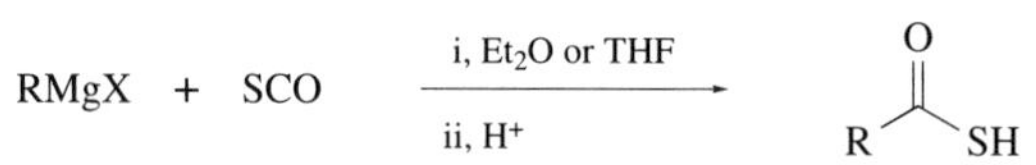

(1)

Thiocarboxylic acids can also be prepared by the reaction of hydrogen sulfide or its alkali metal salts with acid chlorides (Scheme 1) ⟨63OSC(4)924⟩, anhydrides (Equation (2)) ⟨63OSC(4)928⟩, or esters (Equation (3)) ⟨65BCJ320⟩.

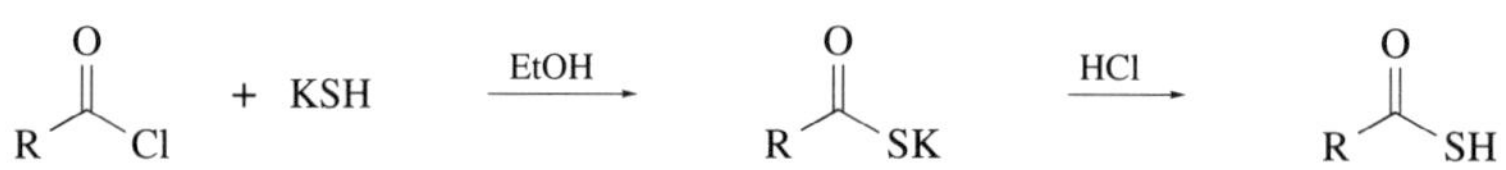

Scheme 1

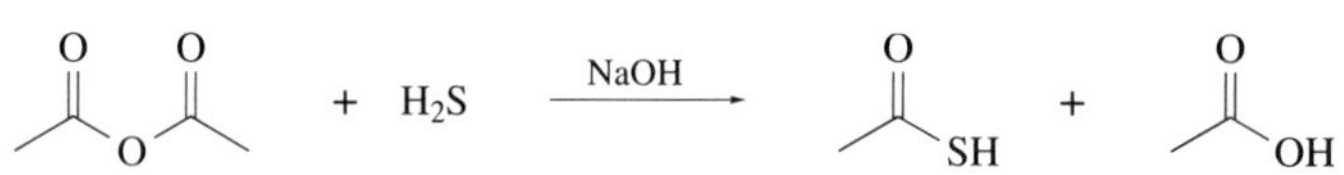

(2)

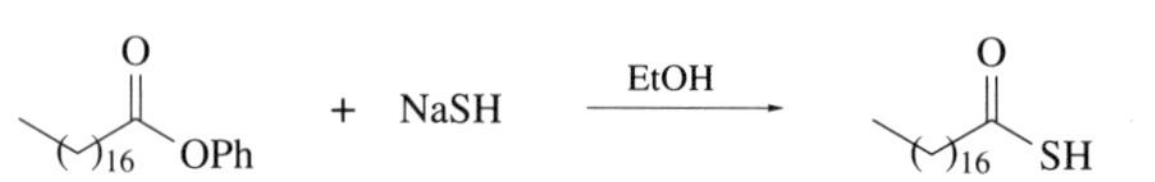

(3)

Anions generated from primary or secondary amides react with carbon disulfide in the presence of sodium hydride to give sodium thiocarboxylates with concomitant formation of NaSCN or R^1NCS (Scheme 2) ⟨73JA3440⟩.

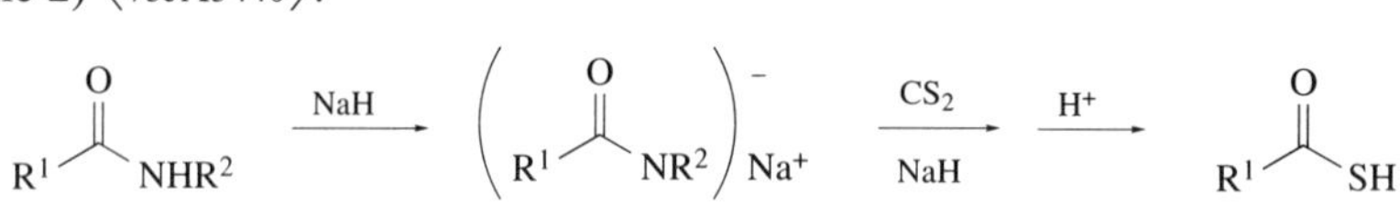

Scheme 2

Thiocarboxylic acids bearing α protons to the carbonyl undergo deprotonation by lithium diisopropylamide (LDA) to generate the corresponding dianions. The reaction with ketones affords β-hydroxy-substituted thiocarboxylic acids conveniently (Scheme 3) ⟨79TL703⟩.

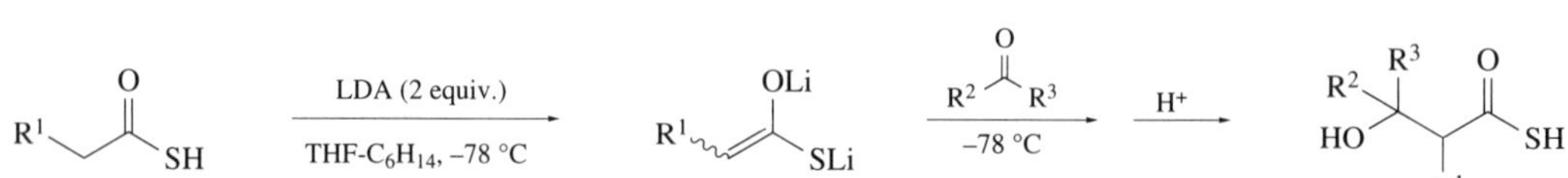

Scheme 3

5.05.1.2 Selenocarboxylic Acids

In contrast to thiocarboxylic acids, selenocarboxylic acids are very sensitive to air and are easily oxidized to give bis(acyl) diselenides ⟨72ACS1465⟩. Therefore, there are only very few examples, reported to date, of the isolation of selenocarboxylic acids.

Jensen *et al.* report the first synthesis of a selenocarboxylic acid by the reaction of benzoyl chloride with hydrogen selenide in pyridine under an inert atmosphere (Equation (4)) ⟨72ACS1465⟩.

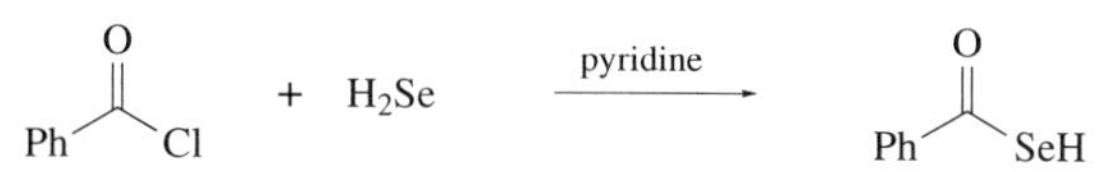

(4)

The reaction of bis(4-biphenylcarbonyl) diselenide with sodium methoxide, and subsequent hydrolysis with hydrochloric acid, leads to the successful isolation of 4-biphenylselenocarboxylic acid as red crystals (m.p. 39–40 °C) (Equation (5)) ⟨87NKK1430⟩. Similarly, 1-adamantylselenocarboxylic acid is isolated as a fairly stable solid (m.p. 28–29 °C).

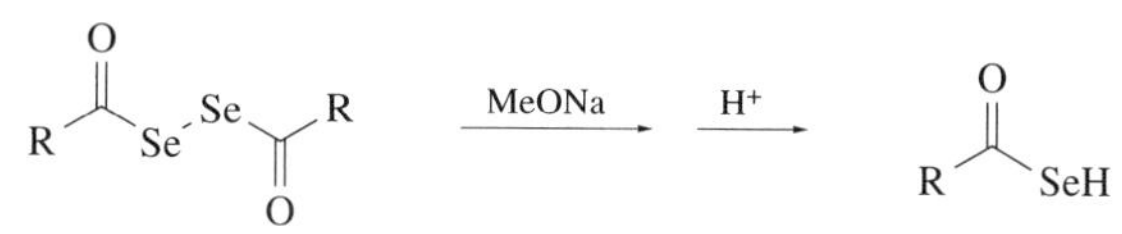

(5)

R = 4-biphenyl, 1-adamantyl

5.05.2 THIOCARBOXYLIC ESTERS AND THEIR SELENIUM AND TELLURIUM ANALOGUES

5.05.2.1 Thiocarboxylic Esters

In general, thiocarboxylic esters are synthesized mainly by the following five methods: (i) acylation of thiols or their metal salts with acyl halides; (ii) reaction of thiols or thiolate anions with carboxylic acids; (iii) reaction of thiols with esters; (iv) synthesis using silyl enol ethers or enolate anions of thiocarboxylic esters; (v) synthesis using carbon monoxide as a carbonyl source of thiocarboxylic esters. Several excellent reviews on the synthesis of thiocarboxylic esters have appeared ⟨85HOU(E5)832, 91COS(6)435⟩.

5.05.2.1.1 From acyl halides

Thiocarboxylic esters are synthesized conveniently by the reaction of thiols with acyl chlorides in the presence of tertiary amines such as triethylamine and pyridine ⟨89SC1533, 91S23, 91T8767⟩. As exemplified in Equations (6) and (7), this method can be employed for the preparation of multi-functionalized thiocarboxylic esters ⟨80TL2013, 80TL3085⟩.

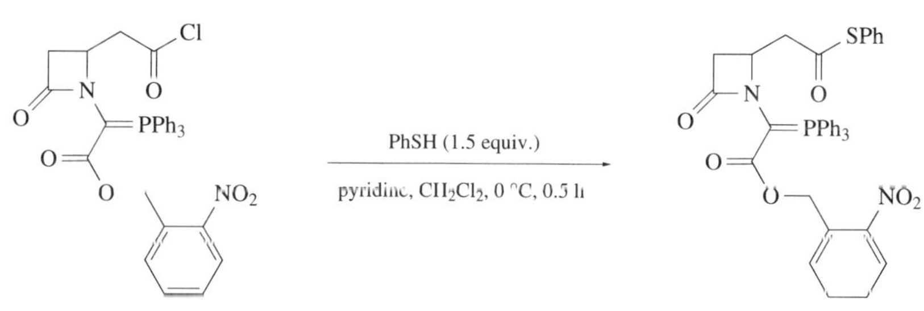

(6)

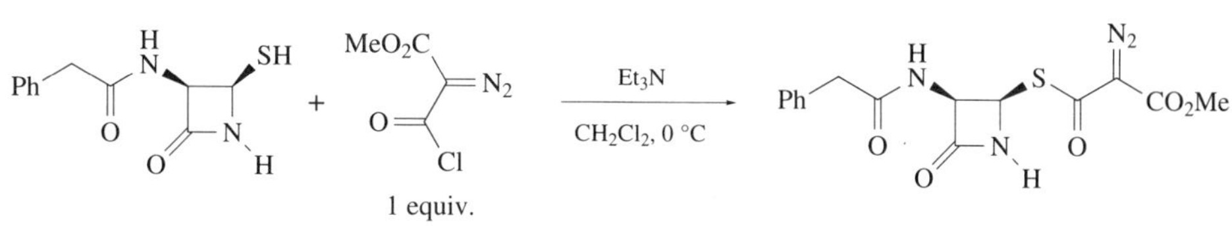

1 equiv.

(7)

Acyl fluorides, which can easily be derived from the corresponding carboxylic acids by using hexafluoropropylene oxide (Scheme 4), can also be used for acylation of thiols ⟨77CL483⟩.

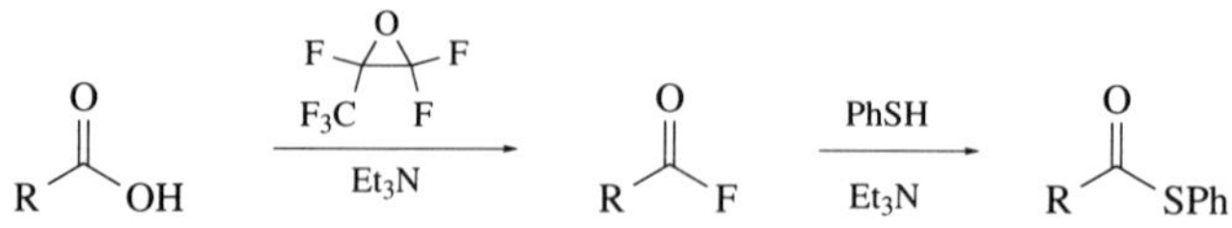

Scheme 4

However, thallium(I) thiolates ⟨75JA3515, 76TL4423⟩, trimethylsilyl sulfides ⟨82CL885⟩, and silver thiolates ⟨80CC1085⟩ are the preferred thiolating agents for acyl chlorides under mild conditions (Equation (8)).

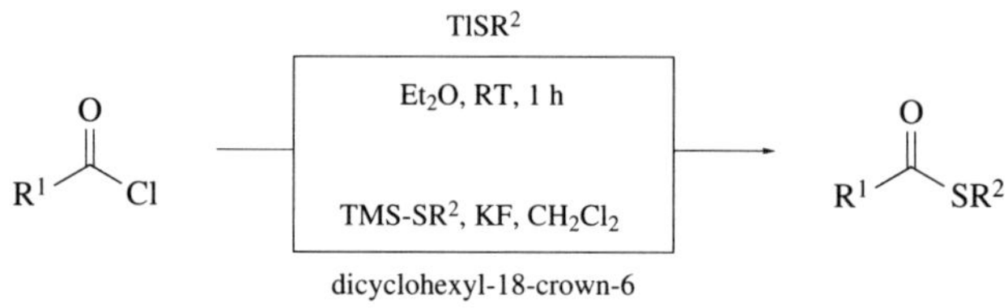

(8)

5.05.2.1.2 From carboxylic acids

Starting from carboxylic acids and thiols, thiocarboxylic esters can be synthesized by using dicyclohexylcarbodiimides (dcc) as exemplified in Equation (9) ⟨76SC453, 84TL2793, 89CC717⟩.

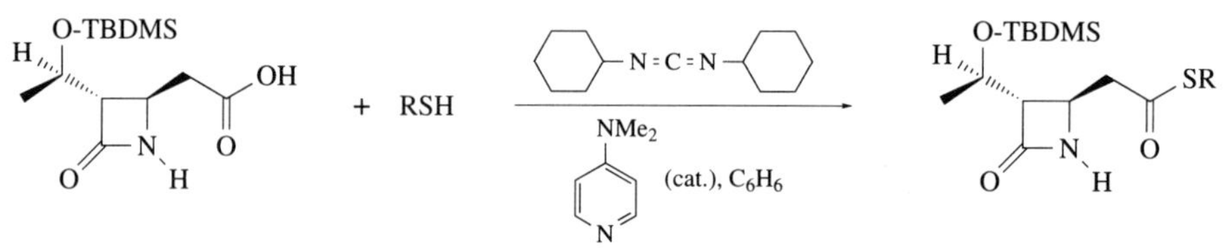

(9)

The method using disulfides and phosphines instead of thiols is much more convenient and is widely utilized for the synthesis of thiocarboxylic esters (Equation (10)) ⟨78JA4620, 78TL3523, 86TL4623, 91SC2097⟩.

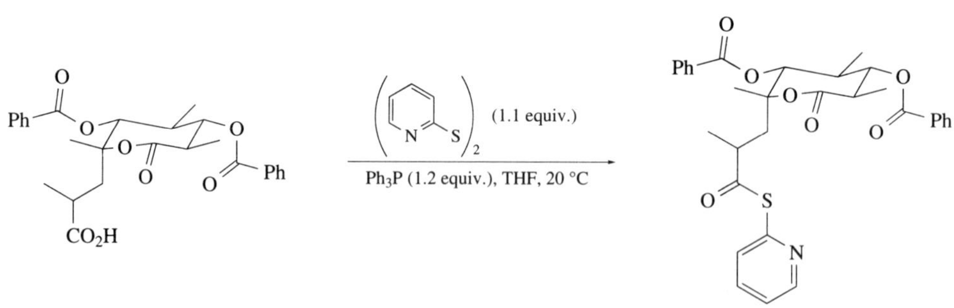

(10)

Tris(ethylthio)borane ($B(SEt)_3$) reacts directly with carboxylic acids to give thiocarboxylic esters in good yields ⟨91SC2097⟩. Similarly, the $SnCl_2/PhSH/Et_3N$ system is useful for the transformation of carboxylic acids into thiocarboxylic esters ⟨77JCS(P1)1672⟩.

Other methods involve the conversion of the —OH group of carboxylic acids into better leaving groups ⟨76CL741⟩. For example, thiocarboxylic esters are formed by the reaction of carboxylic acids with phosphinic chlorides in the presence of triethylamine, followed by the treatment with thiols

(Equation (11)) ⟨74JOC3302, 75CJC3693, 80CC429, 80TL5071, 88JOC4923, 91S920, 93JA5302⟩. Formation of thiocarboxylic esters via acylimidazoles ⟨92JA7287⟩, acyltriazoles ⟨77SC251, 85JOC273⟩, or acylcarbonates ⟨82JA5523⟩ (Schemes 5 and 6) has also been reported.

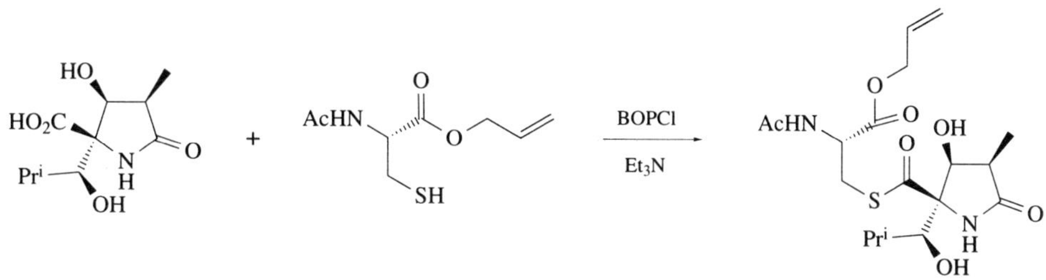

(11)

BOPCl = bis(2-oxo-3-oxazolidinyl)phosphinic chloride

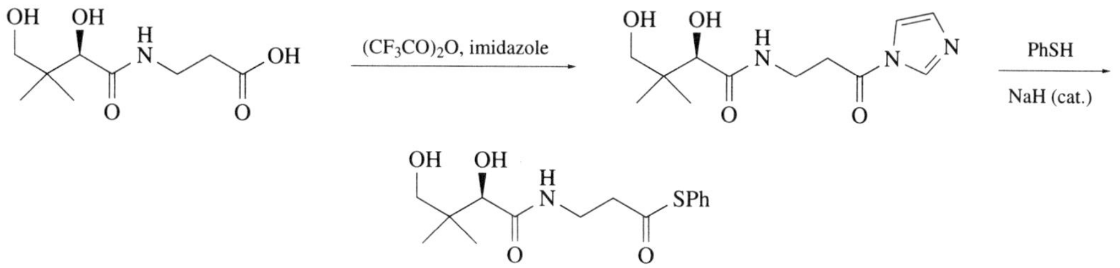

Scheme 5

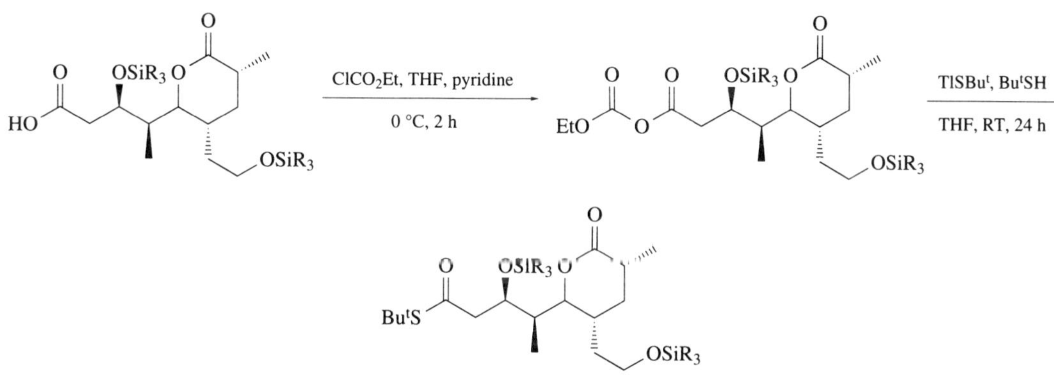

Scheme 6

5.05.2.1.3 From esters

Owing to the high oxophilicity of aluminum, aluminum sulfides like Me_2AlSR, $Al(SR)_3$ etc. react easily with a variety of esters to afford thiocarboxylic esters in good yields (Equation (12)). In the case of lactones, ω-hydroxy thiocarboxylic esters are obtained through ring-opening based on the cleavage of the carbonyl carbon–oxygen bonds of esters (Equation (13)) ⟨72JOM(42)33, 73JA5829, 77JOC3960, 78TL4319, 81JA861⟩.

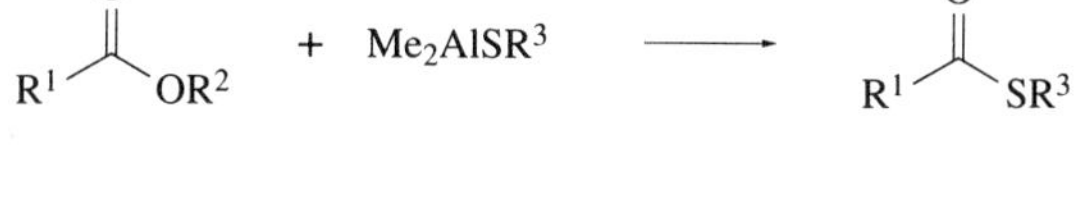

(12)

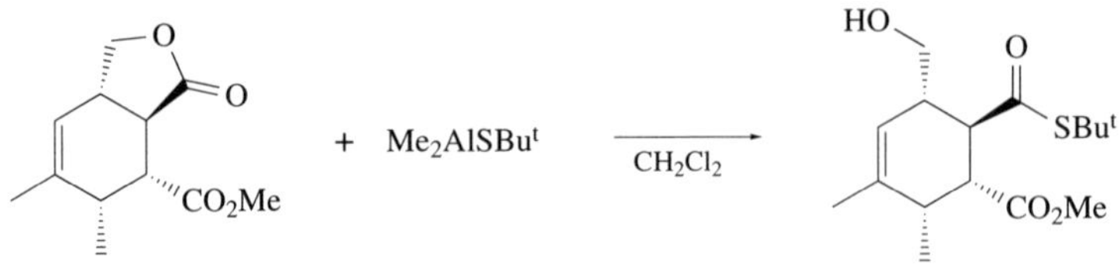
(13)

Instead of aluminum sulfides, the silyl sulfides/aluminum chloride systems are employed to synthesize thiocarboxylic esters (Equation (14)) ⟨74CL187⟩.

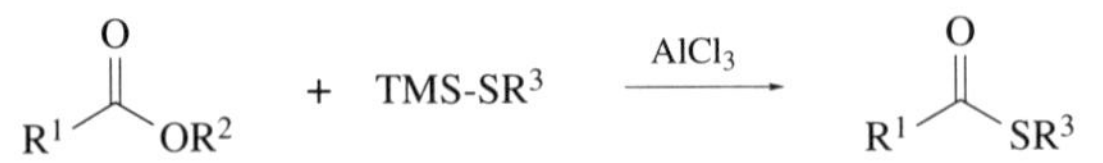
(14)

The reaction of activated esters with thiols or their alkali metal salts yields the desired thiocarboxylic esters under mild conditions (Scheme 7) ⟨78AG(E)267, 80CB2100⟩.

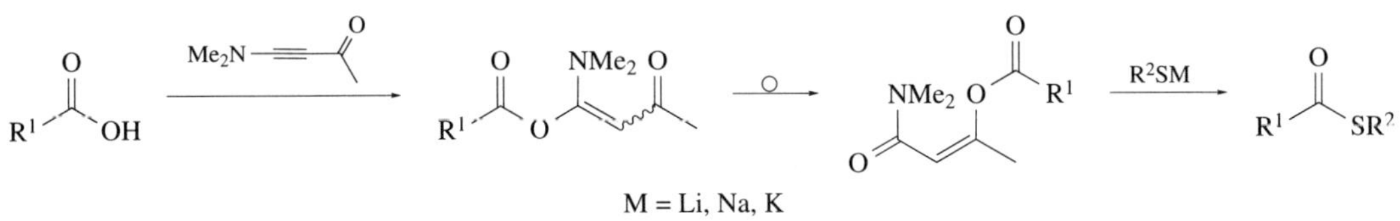

Scheme 7

5.05.2.1.4 *From thiocarboxylic esters*

Silyl enol ethers of thiocarboxylic esters react with a variety of electrophiles, for example aldehydes and enones, in the presence of Lewis acids to give the corresponding β-hydroxy thiocarboxylic esters ⟨86CL1817, 87CL743, 91TL6507⟩. Mukaiyama *et al.* report a series of asymmetric Aldol reactions of silyl enol ethers of thiocarboxylic esters with aldehydes by using chiral diamines as exemplified in Equation (15) ⟨89CL1001, 90CL1019, 91CL989, 91CL1901⟩.

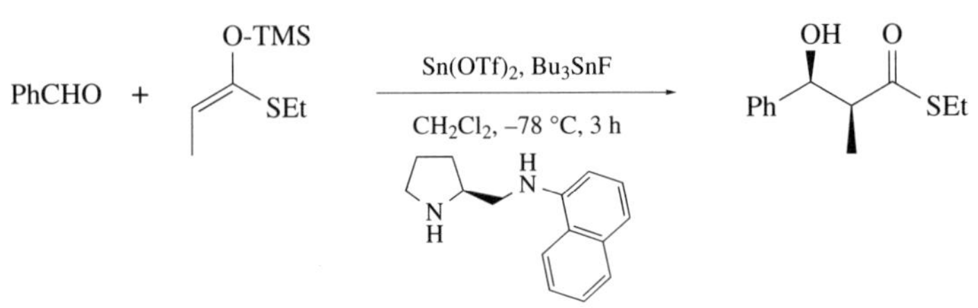
(15)

Tin(II) carboxylic thioesters, formed *in situ* by the addition of tin(II) thiolates to ketenes, react with imines in the presence of tin(II) triflate to yield β-aminocarboxylic thioesters in an *anti* selective manner (Scheme 8) ⟨86CL1013⟩.

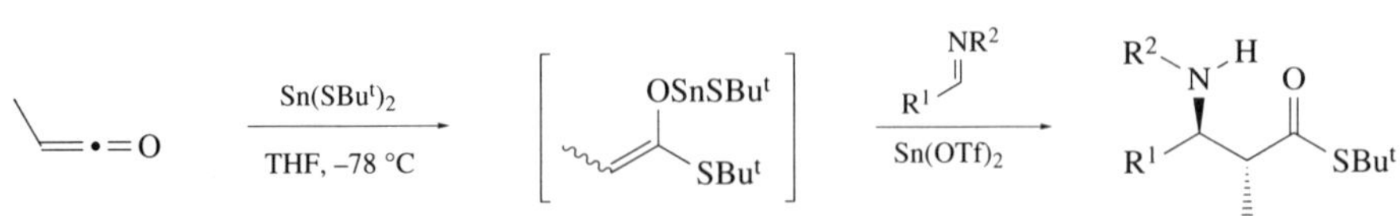

Scheme 8

Lewis acid-mediated condensation of 4-acetoxyazetidinones with silyl enol ethers is reported to give the corresponding thiocarboxylic esters (Equation (16)) ⟨85TL673, 88CJC1537⟩.

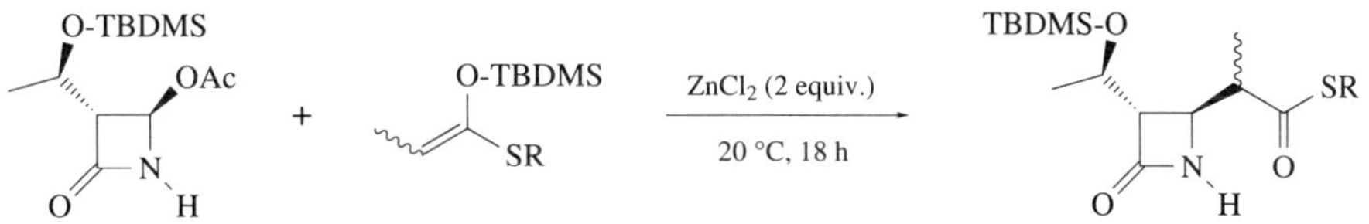

(16)

Proton abstraction from thiocarboxylic esters by nonnucleophilic bases (LDA, Bu^{t}OK etc.) leads to the generation of enolate anions of thiocarboxylic esters, which undergo alkylation with electrophiles, E^+, such as alkyl halides ⟨76TL2577⟩, aldehydes ⟨75TL3255⟩, imines ⟨86CL915, 87CL293⟩, and acyl halides ⟨87JOC4531⟩, affording the corresponding alkylated thiocarboxylic esters (Scheme 9).

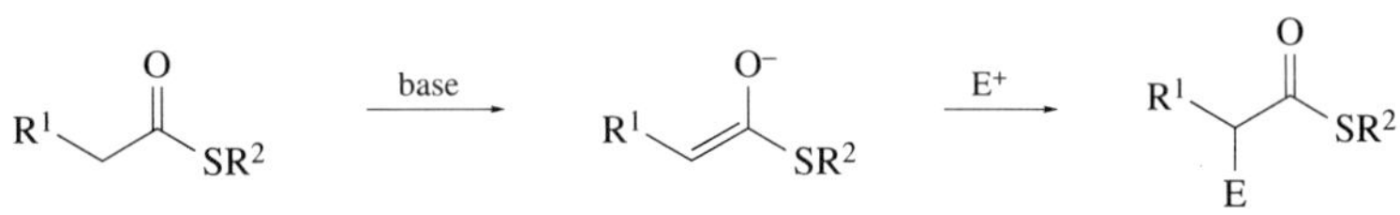

Scheme 9

The method can be applied to the asymmetric Aldol reaction as shown in Equation (17) ⟨86JA8279, 92JOC6339⟩.

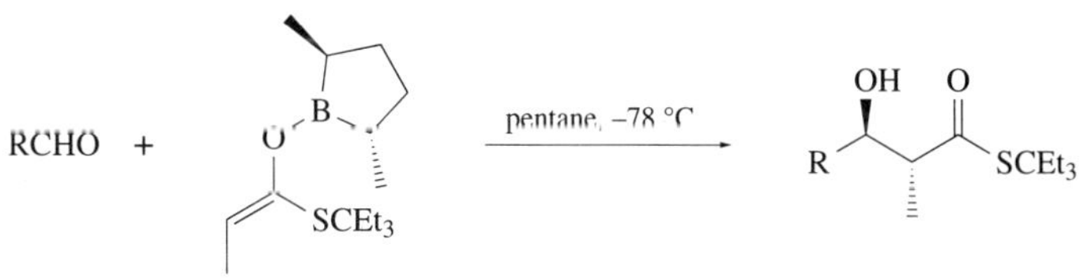

(17)

Claisen condensation of thiocarboxylic esters with isopropylmagnesium bromide as a base is also reported to proceed in good yields (Equation (18)) ⟨80JOC2766⟩.

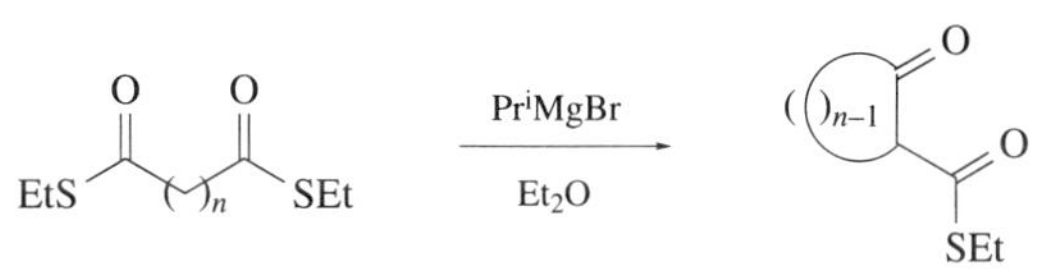

(18)

Lithiomethyl thioesters can be generated by treatment of aromatic thiocarboxylic esters with lithium tetramethylpiperidide (LITMP). The resulting dipole-stabilized carbanions can react with various electrophiles to give functionalized thiocarboxylic esters (Scheme 10) ⟨77TL1839⟩.

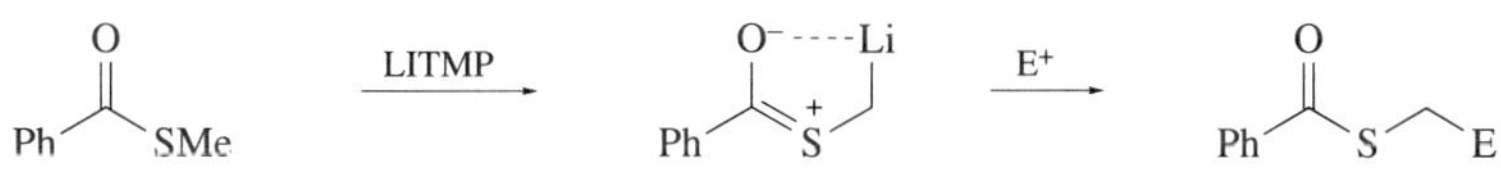

Scheme 10

5.05.2.1.5 *From carbon monoxide*

Diaryl or dibenzyl disulfides react with carbon monoxide in the presence of a catalytic amount of dicobalt octacarbonyl to give the corresponding thiocarboxylic esters (Equation (19)) ⟨85TL2609⟩. A possible mechanism includes the formation of $RCo(CO)_4$ by the elimination of SCO from $RSCo(CO)_4$, followed by quenching the acyl complex "$RC(O)Co(CO)_3$" with RSSR.

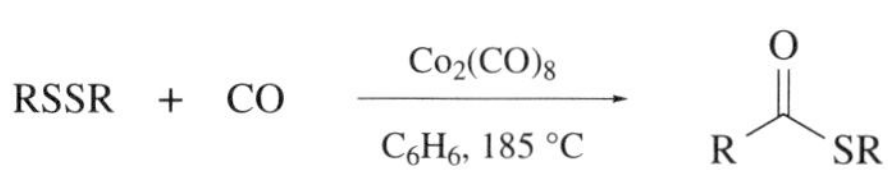

(19)

If thiols are used instead of disulfides, desulfurization takes place in preference to carbonylation. However, the addition of a diene to this $Co_2(CO)_8$-catalyzed reaction system results in the desired carbonylation to give thiocarboxylic esters (Equation (20)) ⟨86OM596⟩.

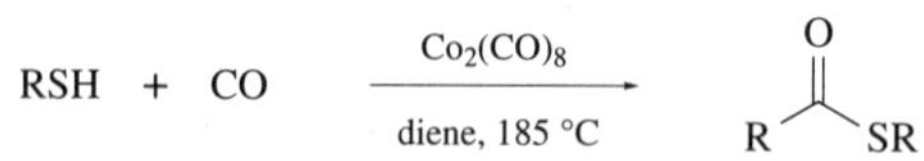

(20)

Diaryl disulfides react with terminal alkynes and carbon monoxide in the presence of tetrakis (triphenylphosphine)palladium to give β-arylthio-α,β-unsaturated thiocarboxylic esters with excellent regio- and stereoselectivity (Equation (21)) ⟨91JA9796⟩.

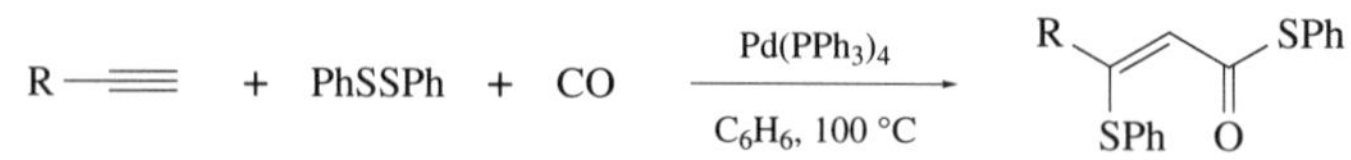

(21)

Alkyllithiums react with carbon monoxide at atmospheric pressure and low temperature to generate acyllithiums, which can be quenched with diorganyl disulfides giving thiocarboxylic esters (Equation (22)) ⟨84OM327⟩.

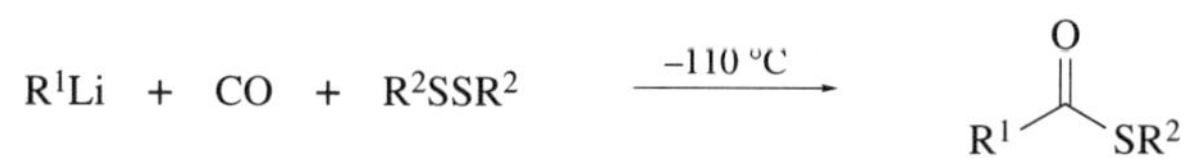

(22)

5.05.2.1.6 Miscellaneous

α-Nitro-substituted vinylic sulfides can also be employed as the precursors of thiocarboxylic esters as illustrated in Scheme 11 ⟨84CC670, 88CC282, 89JOC4723, 90T7429⟩.

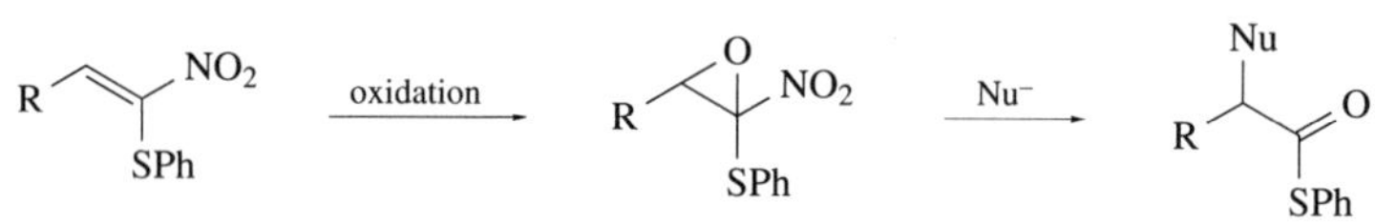

Scheme 11

5.05.2.2 Selenocarboxylic Esters

5.05.2.2.1 From acyl halides

A variety of methods for the preparation of selenocarboxylic esters have been developed. Among these, acylation of selenols and their metal salts is the most general method. The starting selenols can be prepared from elemental selenium and the corresponding Grignard reagents. For example, the reaction of the appropriate acyl chlorides with selenols in the presence of pyridine provides selenocarboxylic esters in good yields (Equation (23)) ⟨62BSB260⟩. Alkali metal salts of various selenols can also be employed for the synthesis of selenocarboxylic esters ⟨70CC1310, 85JOC811, 86T4577⟩. Similarly, the reaction of magnesium salts of arylselenols with aroyl halides gives aromatic selenoesters, which exhibit liquid crystal properties (Equation (24)) ⟨77AG(E)318⟩. The selenation of acyl halides with trimethylsilyl selenides is a convenient alternative to give selenocarboxylic esters (Equation (25)) ⟨77ZOR100⟩.

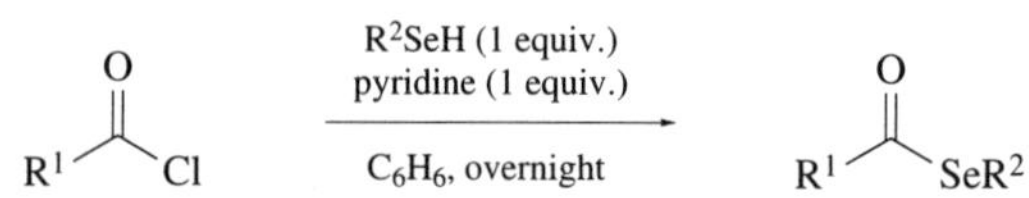

(23)

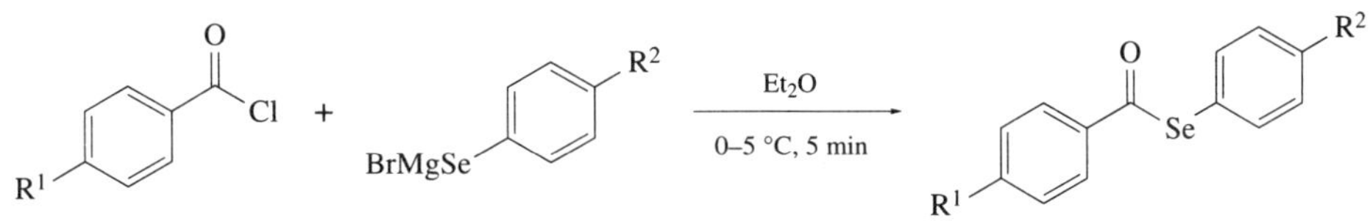

(24)

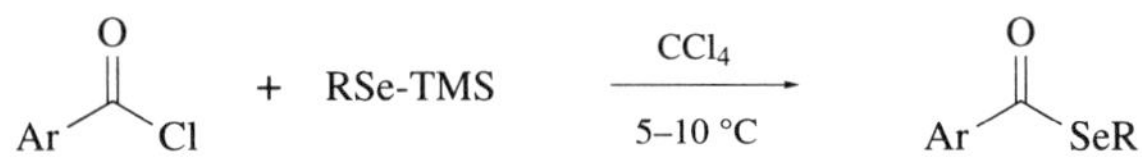

(25)

5.05.2.2.2 *From carboxylic acids*

Selenols can be acylated directly by carboxylic acids to give selenocarboxylic esters ⟨67JOC3929, 76JCS(P1)2452⟩. For example, heating of 4-selenobutyric acid at 160 °C resulted in ring closure to give γ-selenolbutyrolactone (Scheme 12) ⟨66JOC1202⟩.

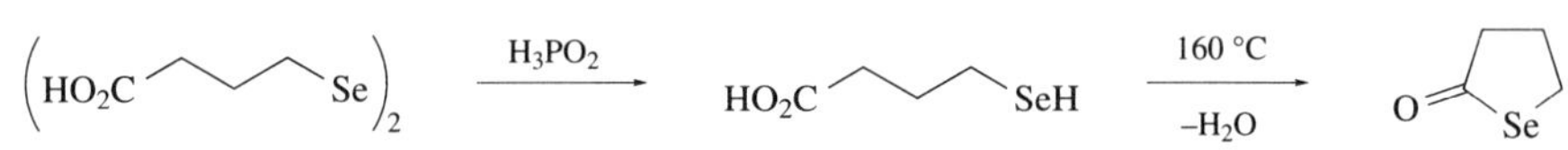

Scheme 12

Selenocyanates or *N*-phenylselenophthalimide (N-PSP) in the presence of trialkylphosphine are reported to act as excellent selenating reagents for various carboxylic acids ⟨78JOC1283, 81JOC1215, 85T4835⟩. Both processes may include the formation of selenophosphonium salts as the key intermediates. The N-PSP/Bu_3P method is superior in terms of ease in manipulation of reagent and high yield of selenocarboxylic esters (Scheme 13 and Equation (26)).

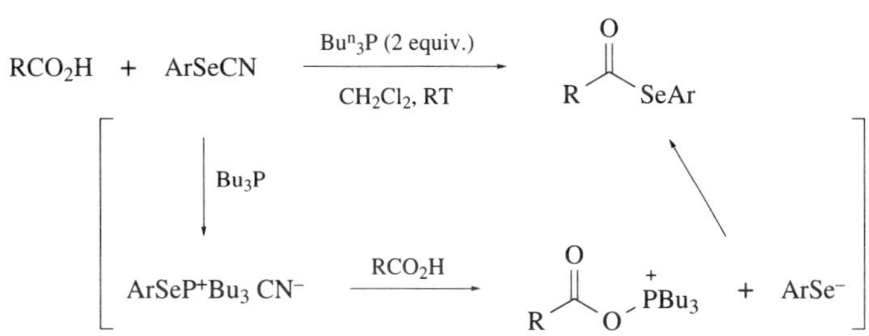

Scheme 13

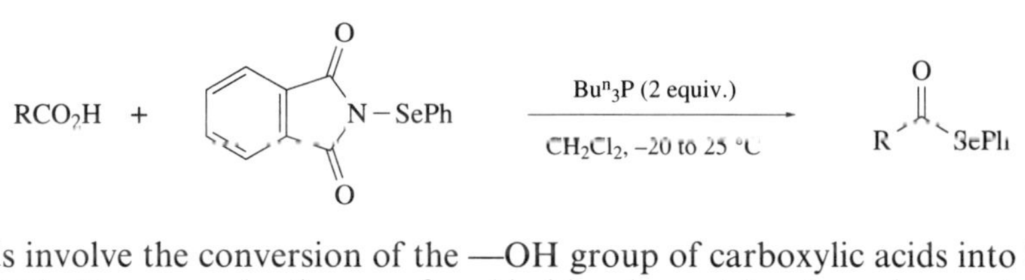

(26)

Other methods involve the conversion of the —OH group of carboxylic acids into better leaving groups ⟨78AG(E)267⟩. For example, the use of acyl imidazoles or triazoles as key acylating species is a mild and convenient method to give selenocarboxylic esters (Scheme 14) ⟨76TL4423, 77AG(E)244⟩. In particular, the reaction of acyl triazoles proceeds even in the absence of base. The reaction of carboxylic acids with phenyl dichlorophosphate, followed by the addition of benzeneselenol to the mixture, afforded selenocarboxylic esters (Scheme 15) ⟨85JA3285⟩.

5.05.2.2.3 *From esters*

Conversion of *O*-alkyl esters into the corresponding methylselenocarboxylic esters has been achieved by the reaction with dimethylaluminum methylselenolate, which is conveniently prepared by heating a toluene solution of trimethylaluminum with powdered selenium ⟨78JOC2735, 83TL4355, 85T4821⟩. This method can be applied to esters bearing various functional groups (Scheme 16).

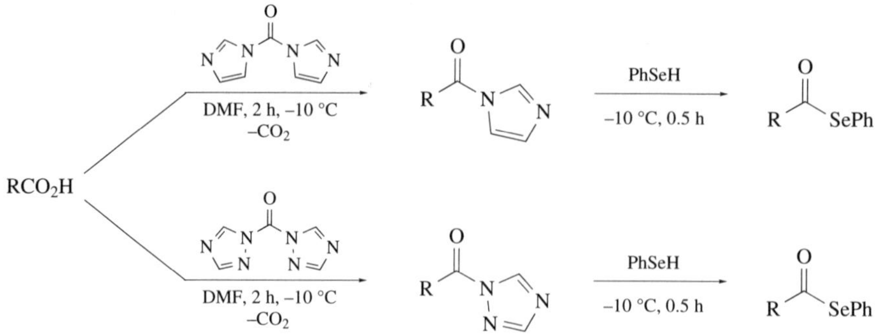

Scheme 14

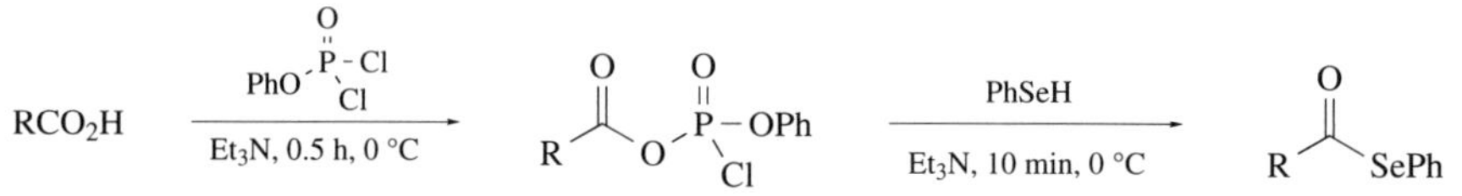

Scheme 15

Scheme 16

5.05.2.2.4 From selenocarboxylates

Owing to the difficulty of the preparation of selenocarboxylates, known examples of selenocarboxylic esters synthesized by alkylation of selenocarboxylates are rare ⟨72ACS1465⟩. Potassium selenocarboxylates, prepared by the reaction of bis(acyl) selenides with methanolic potassium hydroxide, can be used as the starting materials for the synthesis of selenocarboxylic esters (Scheme 17) ⟨76CL203, 78CL1007⟩. Amine salts of selenocarboxylic acids are the choice of the starting materials for the synthesis of selenocarboxylic esters ⟨77BCJ3007, 86S128⟩. A convenient method for preparation of amine salts of selenocarboxylic acids has been developed by the reaction of acid chlorides with Se, CO, and H_2O. Thus, alkylation of the amine salts of selenocarboxylic acids leads to a convenient method for the synthesis of selenocarboxylic esters (Scheme 18) ⟨91JOC3776⟩.

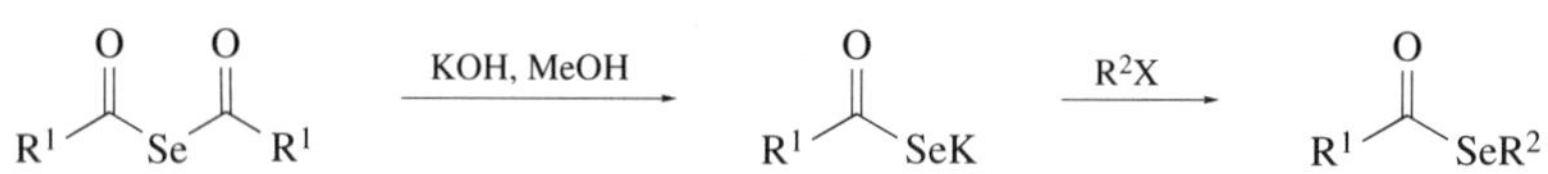

Scheme 17

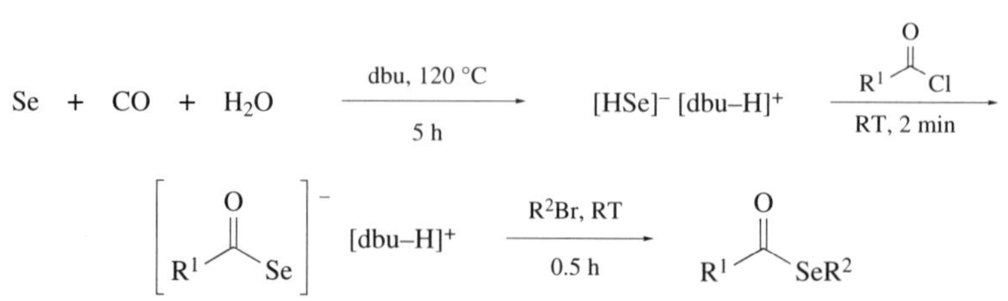

Scheme 18

5.05.2.2.5 *Miscellaneous*

Two examples of transition metal catalyzed synthesis of selenocarboxylic esters have been reported. One is the $Co_2(CO)_8$-catalyzed reaction of diaryl diselenides with carbon monoxide, which successfully affords aromatic selenocarboxylic esters in moderate to excellent yields (Equation (27)) ⟨87JOM(334)C45⟩. The other is the $Pd(PPh_3)_4$-catalyzed carbonylative addition of diaryl diselenides to terminal alkynes, which provides β-(arylseleno)-α,β-unsaturated selenocarboxylic esters in high yields (Equation (28)) ⟨91JA9796, 93MI 505-01⟩.

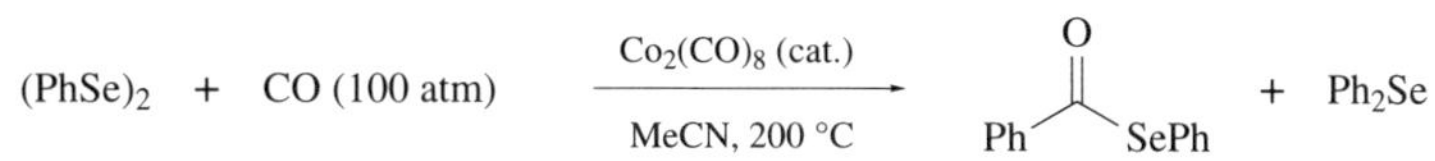

(27)

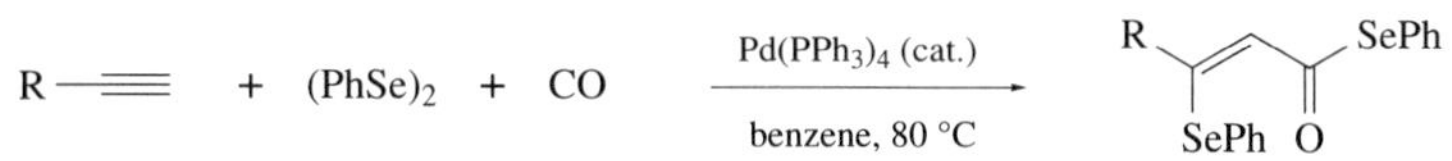

(28)

5.05.2.3 Tellurocarboxylic Esters

Tellurocarboxylic esters are synthesized mainly by the following two methods: (i) acylation of tellurolate anions with acyl halides; (ii) alkylation of tellurocarboxylate anions with alkyl halides (Equation (29)).

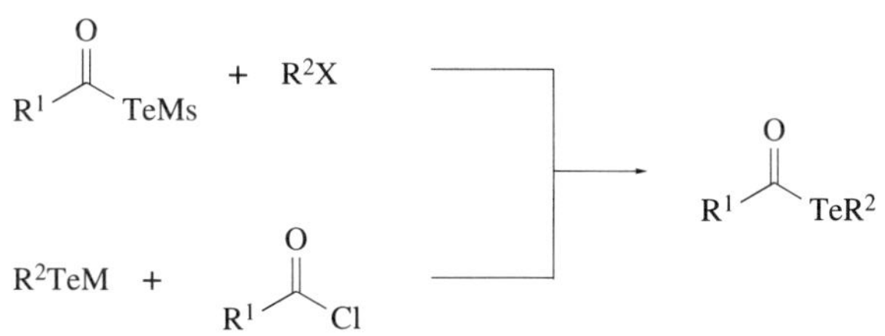

(29)

5.05.2.3.1 *From acyl halides*

The most general procedure for the synthesis of aromatic tellurocarboxylic esters ($R^1 = R^2 = Ar$ in Equation (29)) consists of the reduction of diaryl ditellurides with sodium borohydride, and the subsequent aroylation of the aryltellurolate anions thus formed with aroyl halides. Aromatic tellurocarboxylic esters are mostly orange–brown crystalline solids that can be purified easily by recrystallization from organic solvent and can be stored at room temperature under a nitrogen or argon atmosphere in the dark ⟨70BSB383, 79JOM(166)373, 80JOM(197)111, 81JOM(205)167, 92JA8313⟩. In some cases, aryltellurolate anions are prepared by the reaction of aryllithiums with metallic tellurium ⟨89JOM(377)281, 92SC1619⟩. For example, *Te*-ferrocenyl tellurobenzoate can be obtained by this method (Scheme 19) ⟨89JOM(377)281⟩. Acid anhydrides can also be employed for the acylation of tellurolate anions ⟨73SA(A)285, 80SA(A)769⟩.

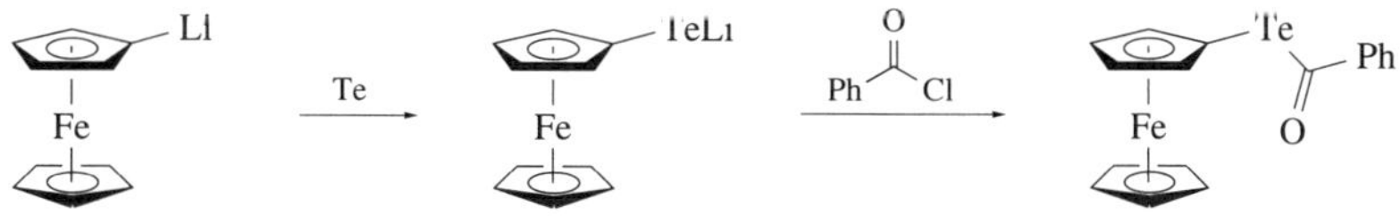

Scheme 19

A convenient and high yield preparation of tellurocarboxylic esters has been attained by the reaction of phenyl trimethylsilyl telluride with acid chlorides (Equation (30)) ⟨86CL977⟩.

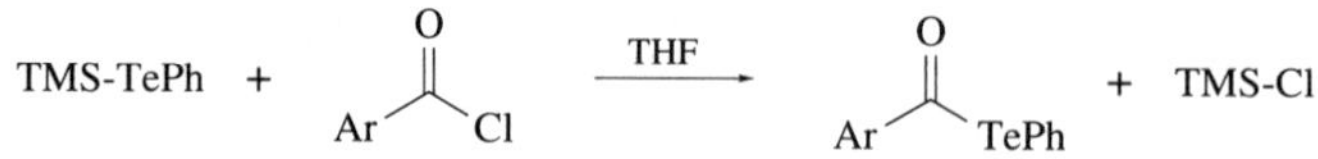

(30)

5.05.2.3.2 *From tellurocarboxylates*

The starting tellurocarboxylates are prepared from acid chlorides and tellurolate anions such as Na_2Te ⟨80ZN(B)217, 89TL1829⟩, NaTeH ⟨81JOC4194, 90JCR(S)56⟩, and Li_2Te ⟨92PS(67)27⟩. Alkylation of tellurocarboxylates thus formed by appropriate alkyl halides leads to the formation of the corresponding tellurocarboxylic esters in good yields. By this method a series of aliphatic tellurocarboxylic esters can be synthesized, which are very sensitive to molecular oxygen but which can be stored at −20 °C under an argon atmosphere in the dark for two weeks ⟨89TL1829, 90JCR(S)56, 92PS(67)27⟩.

Intramolecular alkylation by using phase-transfer catalysis gives rise to tellurophthalide as the first example of a γ-tellurolactone (Equation (31)) ⟨81JOC4194⟩.

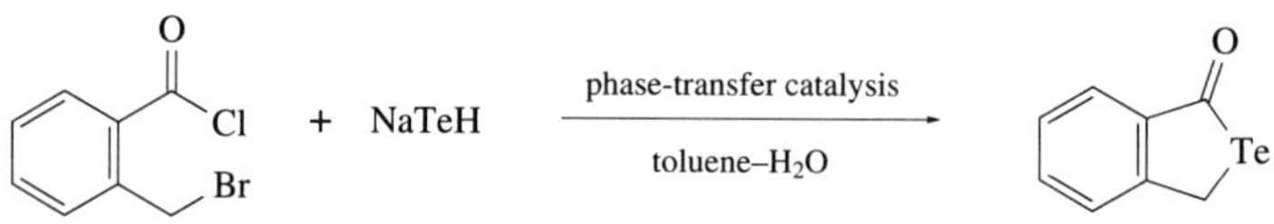

(31)

Bis(acyl) tellurides are the source of tellurocarboxylates, when treated with secondary amine (Scheme 20) ⟨87CL475⟩.

Alternatively, the $Co_2(CO)_8$-catalyzed carbonylation of diaryl ditellurides with carbon monoxide provides aromatic tellurocarboxylic esters in moderate yields ⟨87JOM(334)C43, 89JOM(361)63⟩.

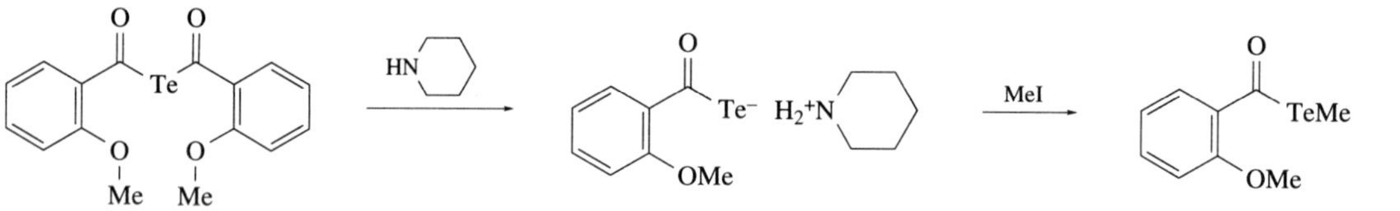

Scheme 20

5.05.3 BIS(ACYL) SULFIDES AND THEIR SELENIUM AND TELLURIUM ANALOGUES

5.05.3.1 Bis(acyl) Sulfides

Bis(acyl) sulfides have been synthesized by several methods. Among them, the most general method is the acylation of sodium and lithium sulfides (Na_2S and Li_2S) with acid chlorides (Equation (32)). In the case of sodium sulfide, the use of phase-transfer catalysis leads to high yield formation of bis(acyl) sulfides ⟨82SC897⟩. Lithium sulfide is synthesized by the selective reduction of elemental sulfur with lithium triethylborohydride ($LiBHEt_3$) ⟨78CC838, 79T2329⟩.

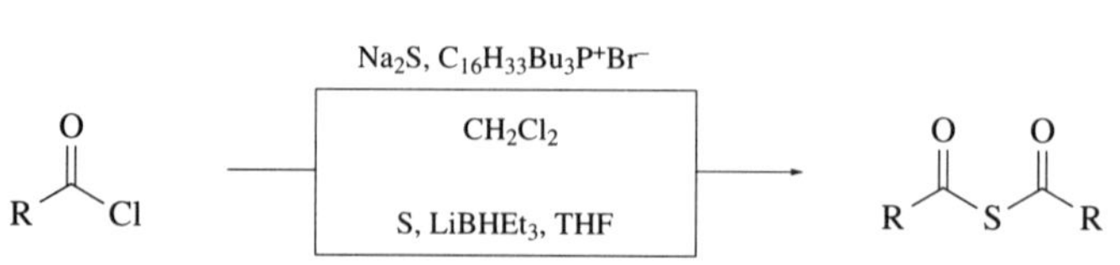

(32)

Instead of Na_2S or Li_2S, bis(trimethylsilyl) sulfide and bis(tributyltin) sulfide can be employed as an efficient sulfur transfer reagent (Equation (33)) ⟨82CL885, 87S1122⟩. In particular, the latter reagent enables high yield preparation of bis(acyl) sulfides.

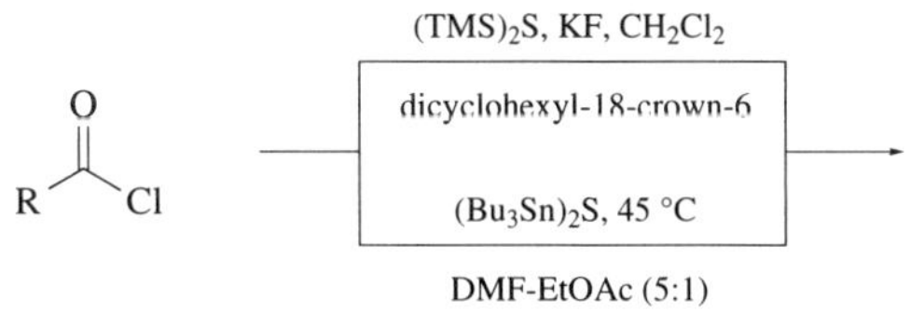

(33)

The alternative method for the synthesis of bis(acyl) sulfides is the acylation of thiocarboxylic acids or their amine salts (Equation (34)) ⟨50JA4270, 77JOM(129)189⟩.

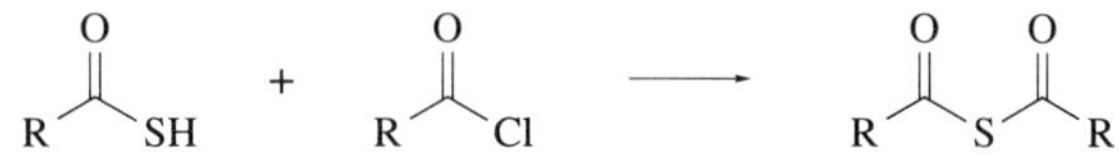

(34)

Other methods include the following: (i) the reaction of thiocarboxylic acids with dicyclohexylcarbodiimide (dcc) (Equation (35)) ⟨76JCS(P1)564⟩; (ii) the reaction of *O*-silyl thionesters with acid chlorides (Equation (36)) ⟨76JCS(P1)564⟩; (iii) the reaction of bis(acyl) disulfides with phosphines (Equation (37)) ⟨35CB163⟩.

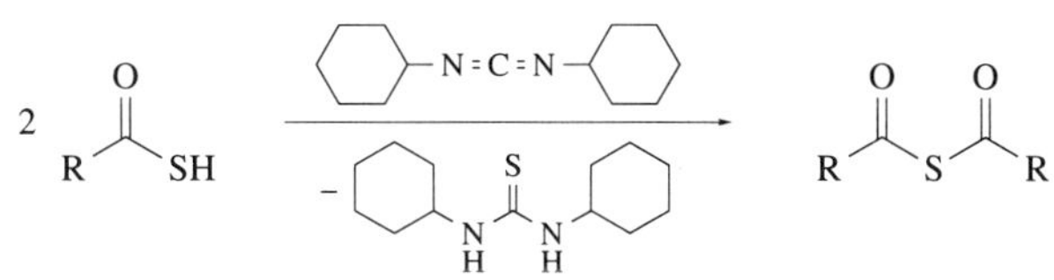

(35)

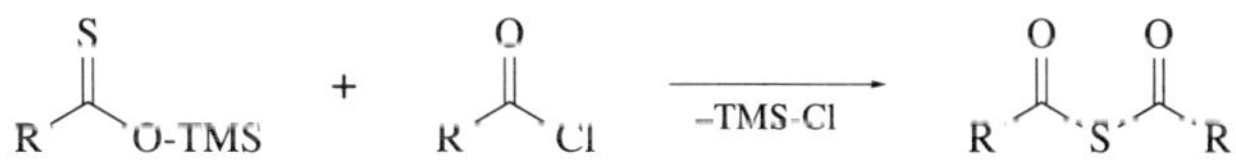

(36)

(37)

5.05.3.2 Bis(acyl) Selenides

Acylation of amine salts or alkali metal salts of hydrogen selenide using acid chlorides might be expected to be a general route to bis(acyl) selenides. However, the initial products in this reaction are not the desired bis(acyl) selenides but *O*-acyl selenonesters as shown in Schemes 21 and 22 ⟨91JOC3776⟩. In the case of aliphatic acid chlorides, the resulting *O*-acyl selenonesters easily isomerize to more stable bis(acyl) selenides ⟨72ACS1465, 89ZN(B)1050⟩. However, aromatic selenonesters are more stable than aliphatic ones, owing to conjugation between the aromatic ring and the selenocarbonyl group, and therefore do not isomerize to bis(acyl) selenides ⟨72ACS1465⟩.

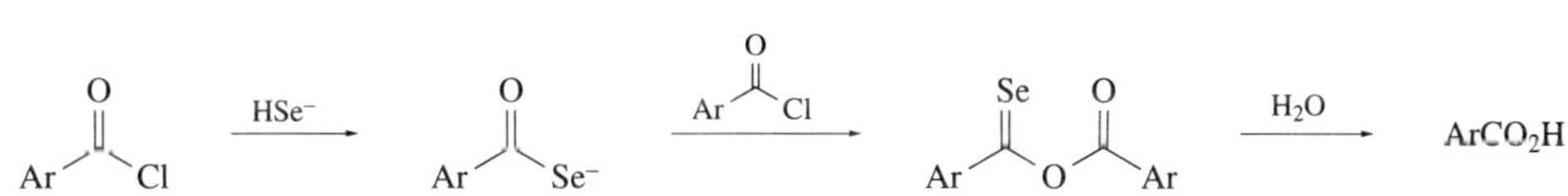

Scheme 21

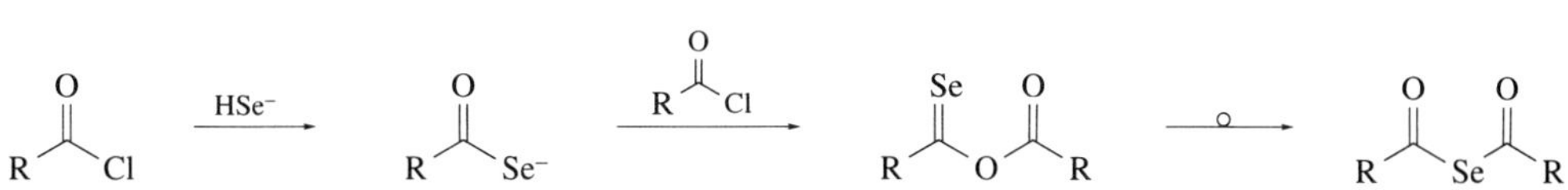

Scheme 22

Bis(acyl) selenides are also obtained by the oxidation of sodium selenocarboxylates and the subsequent treatment with triphenylphosphine (Scheme 23) ⟨77BCJ3007⟩.

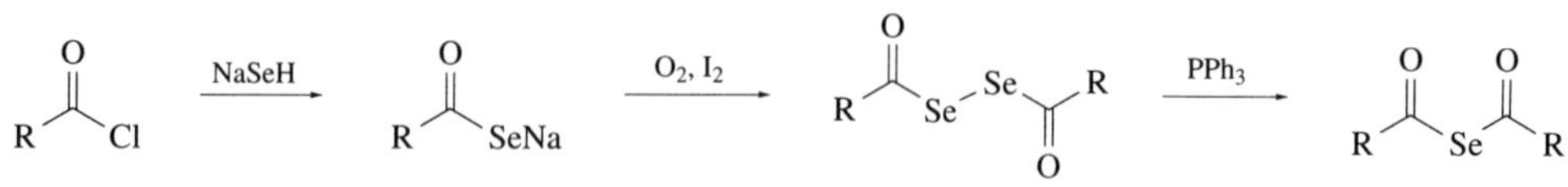

Scheme 23

A useful method for synthesis of bis(acyl) diselenides has been reported: the reaction of *O*-silyl selenonesters with acid chlorides provides bis(acyl) selenides in good yields (Equation (38)) ⟨94JCS(P1)1083⟩. This method can be applied to the preparation of unsymmetrical bis(acyl) selenides.

(38)

5.05.3.3 Bis(acyl) Tellurides

The synthesis of diacetyl telluride is accomplished by the reaction of acetyl chloride with less than stoichiometric amounts of bis(trimethylsilyl) telluride at −30 °C (Equation (39)) ⟨86ZC179⟩. Similarly, bis(pivaloyl) telluride is obtained from pivaloyl chloride.

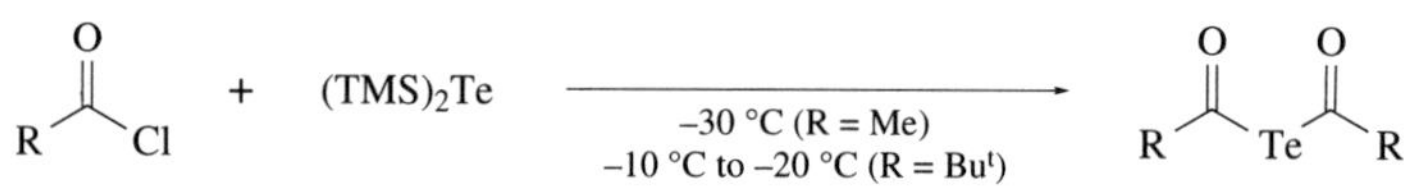

(39)

Bis(aroyl) tellurides can be prepared from aroyl chlorides and sodium telluride (Na_2Te) (Equation (40)) ⟨85AG(E)1041⟩. *O*-Phthaloyl chloride gives the corresponding cyclic bis(acyl) telluride, benzotellurophenedione ⟨78OPP289⟩. In general, bis(acyl) tellurides are extremely sensitive to air, moisture, and light, and easily decompose with precipitation of elemental tellurium.

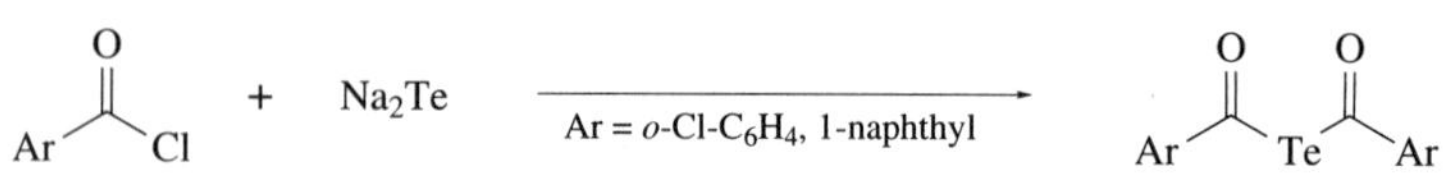

(40)

5.05.4 ACYLSULFENYL HALIDES

5.05.4.1 Acylsulfenyl Chlorides

The first example of isolation of acylsulfenyl chlorides was reported by Böhme *et al.*: the reaction of bis(acetyl) sulfide and bis(benzoyl) sulfide with chlorine successfully provides acetylsulfenyl chloride and benzoylsulfenyl chloride, respectively (Equation (41)) ⟨52LA(576)61⟩. Similar procedures can be employed for the synthesis of some aromatic sulfenyl chlorides ⟨69LA(730)121, 86T5969⟩ and methoxycarbonylsulfenyl chloride ⟨69LA(730)121⟩.

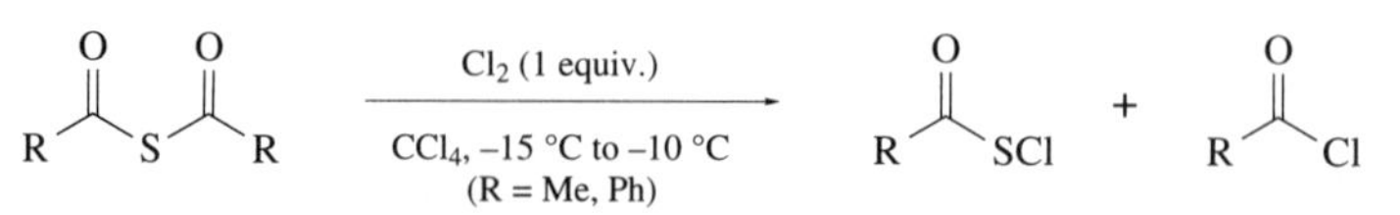

(41)

Instead of chlorine, sulfuryl chloride can also be employed as the chlorination reagent. Acetylsulfenyl chloride is prepared from bis(acetyl) sulfide or disulfide by the use of this reagent ⟨79JOC569, 85JOC4164⟩.

Chlorination of α-thiolacetanilides produces α-carbamoyl-α,α-dichlorosulfenyl chlorides, which undergo acid hydrolysis yielding carbamoylcarbonylsulfenyl chlorides (Scheme 24) ⟨72JOC1526⟩.

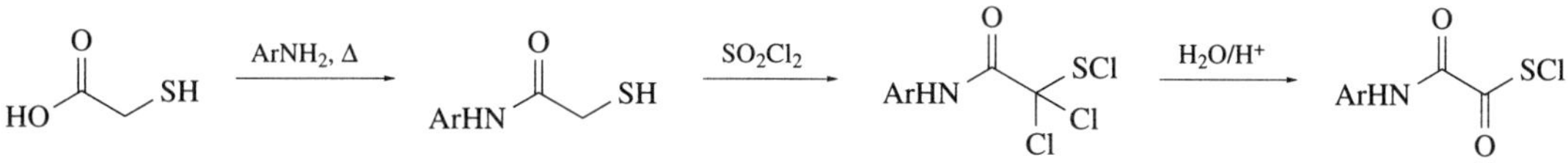

Scheme 24

Diphenyltin bis(thiocarboxylates) are useful precursors of acylsulfenyl chlorides (Equation (42)) ⟨83S814⟩. The reagent is conveniently prepared by the stoichiometric reaction of diphenyltin dichloride with the corresponding piperidinium or potassium thiocarboxylates.

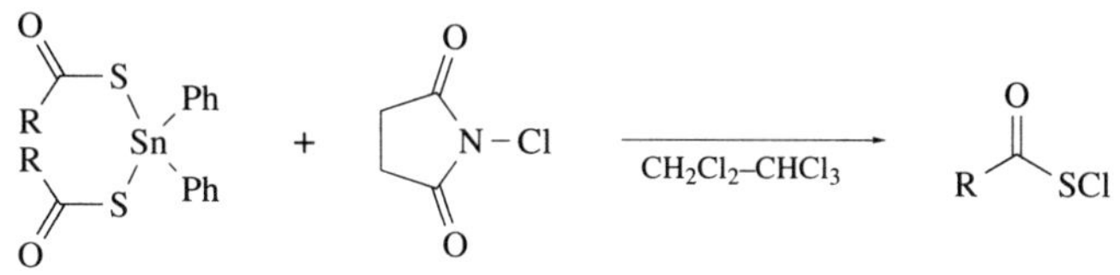 (42)

Acylsulfenyl chlorides are generally stable for 4–5 h at room temperature, but are gradually decomposed by moisture or in protic solvent.

5.05.4.2 Acylsulfenyl Bromides

Kato *et al.* report the first isolation of acylsulfenyl bromides from the reaction of phenylmercury thiocarboxylates with *N*-bromosuccinimide (NBS) (Equation (43)) ⟨82S1013⟩. Instead of phenylmercury thiocarboxylates, silver thiocarboxylates or diphenyltin, bis(thiocarboxylates) can also be utilized as a source of RC(O)S— ⟨83S552, 83S814⟩.

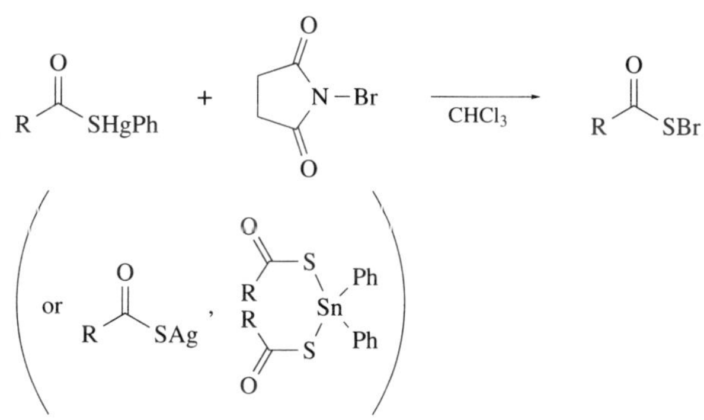 (43)

Thiocarboxylic acid *O*-silyl esters undergo novel thiophilic halogenation with NBS (Equation (44)) ⟨86T4593⟩. By this method, an aliphatic acylsulfenyl bromide was isolated for the first time.

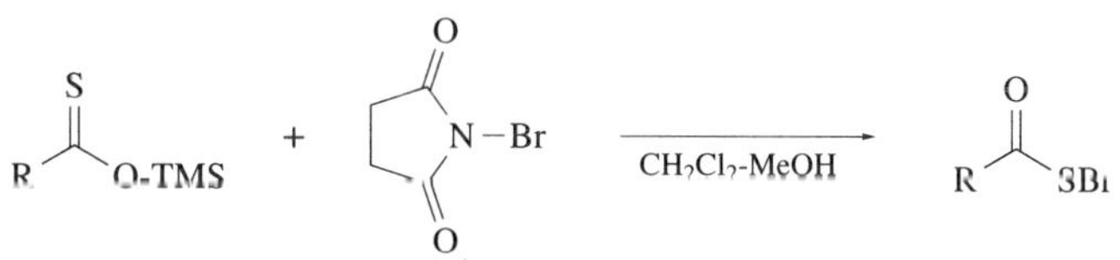 (44)

The acylsulfenyl bromides are stable for 2–3 h at room temperature, but, upon exposure to sunlight, they decompose to give bromine and the corresponding symmetrical disulfides.

5.05.4.3 Acylsulfenyl Iodides

Acylsulfenyl iodides are synthesized by similar procedures to those described for the synthesis of acylsulfenyl bromides. For example, iodination of silver or phenylmercury thiocarboxylates with iodine provides a high yield of the desired acylsulfenyl iodides ⟨82AG(E)150, 83S552⟩. Similarly, diphenyltin bis(thiocarboxylates) can be employed as the precursors of acylsulfenyl iodides ⟨83S814⟩.

Acylsulfenyl iodides are stable for several hours at room temperature, but decompose on exposure to air. Raman spectra of a series of acylsulfenyl halides are reported by Kato *et al.* ⟨86CL935⟩. In addition, some reports deal with the preparation of fluoro-substituted acylsulfenyl halides (X = Cl, Br, I) and their spectroscopic analysis ⟨69JOC4173, 75MI 505-01, 78JFC(11)327, 88ZAAC(566)151⟩.

5.05.5 ACYLSULFENIC ETHERS

Only very limited methods have been reported for the preparation of compounds containing the —C(O)—S—O— linkage. Among them, the alcoholysis of acylsulfenyl chlorides is the most general method for the synthesis of acylsulfenic ethers (Equation (45)) ⟨86T5969, 85TL4349, 85ZN(B)1563⟩. For example, a series of pyridylcarbonylsulfenic ethers is obtained by this method. Similarly, fluorocarbonylsulfenic trifluoroacetate can be synthesized from fluorocarbonylsulfenic acid chloride by treatment with silver trifluoroacetate (Equation (46)) ⟨73JFC(3)383⟩.

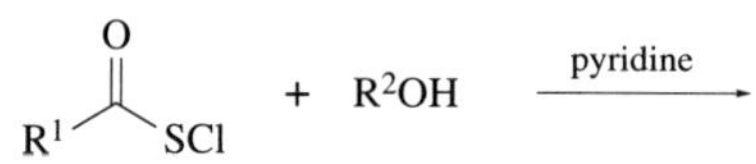

(45)

(46)

An acylsulfenic ether bearing a fluorenone structure has been reported to exhibit the properties of a ferroelectric liquid crystal. The acylsulfenic ether is obtained by the reaction of a fluorenone carboxylic acid derivative with thionyl chloride and a phenol derivative (Equation (47)) ⟨91BCJ2720⟩.

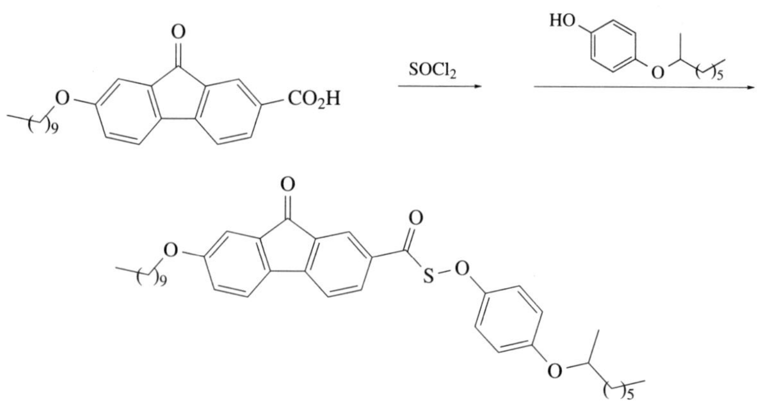

(47)

There are no examples reported of the synthesis of acylselenenic and acyltellurenic ethers.

5.05.6 ACYL-SUBSTITUTED DICHALCOGENIDES

As illustrated in Scheme 25, various types of acyl-substituted dichalcogenides have already been synthesized.

5.05.6.1 Monoacyl-substituted Dichalcogenides

5.05.6.1.1 Monoacyl-substituted disulfides

The synthesis of monoacyl-substituted disulfides can be divided into two modes of reaction: (i) reaction of acylsulfenyl chlorides with thiols; and (ii) reaction of thiocarboxylic acids with R^2SX (X = Cl, NR^3_2, SO_2R^3, $SC(O)OR^3$ etc.) (Equation (48)).

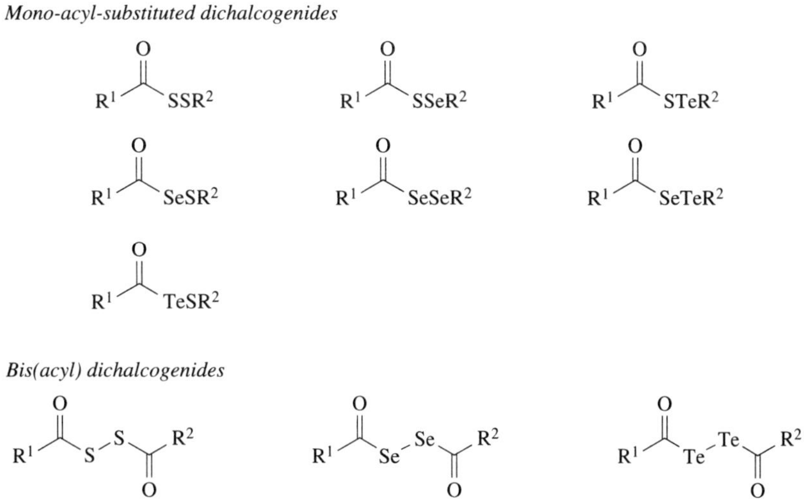

Scheme 25 Acyl-substituted dichalcogenides

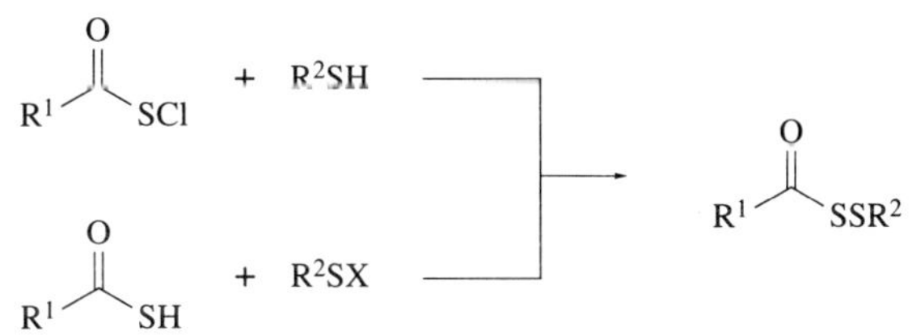

(48)

The former reaction is carried out by mixing thiols and a slight excess of acylsulfenyl chlorides at $-10\,^{\circ}C$ in anhydrous ether or CH_2Cl_2 ⟨52LA(576)61, 70BCJ1890, 71JOC2735, 71JOC3677, 92SUL199⟩. Instead of acylsulfenyl chlorides, acyl alkoxycarbonyl disulfides can be employed as shown in Equation (49) ⟨81JOC1374⟩.

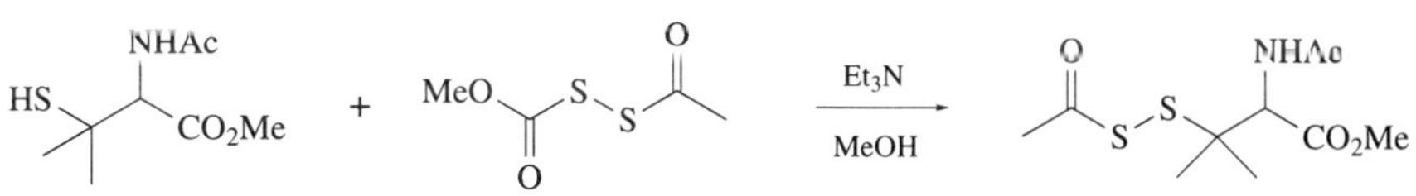

(49)

For example, sulfenyl chlorides and sulfenamides are conveniently employed for the sulfenylation of thiocarboxylic acids (Equation (50)) ⟨75BCJ2993, 75CL1231, 79AG(E)941⟩.

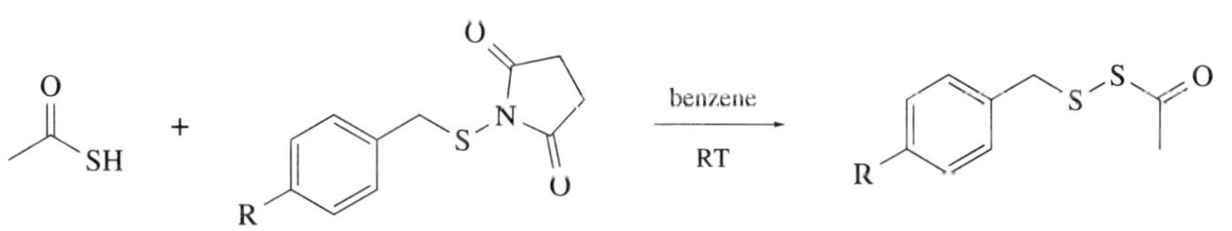

(50)

Thiolsulfonates ($R^1S—SO_2R^2$) ⟨67JOC3467⟩ and thiolsulfinates ($R^1S—S(O)R^2$) ⟨94S699⟩ are also appropriate sulfenylating reagents. Monoacyl-substituted disulfides can be prepared by the substitution reactions by using activated disulfides and thiocarboxylic acids (Equation (51) and Scheme 26) ⟨70JA7629, 76HCA1307, 91T6127⟩.

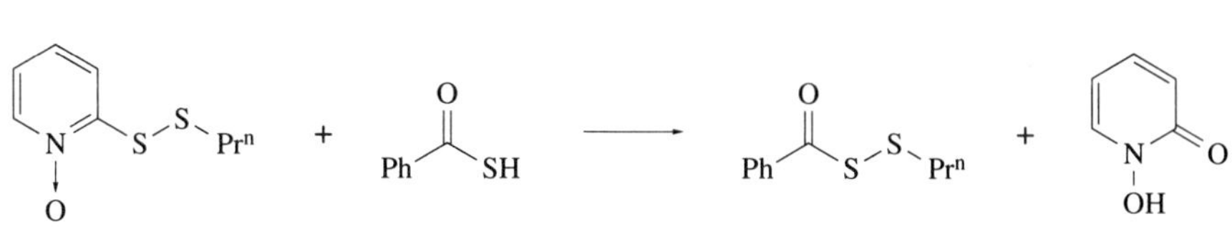

(51)

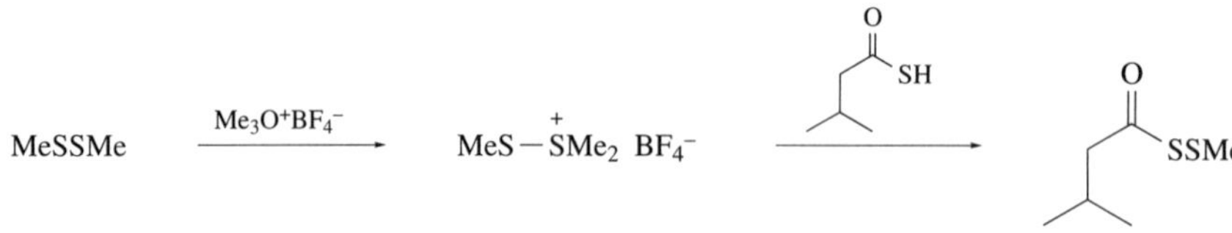

Scheme 26

Some other methods for the synthesis of monoacyl-substituted disulfides are reported. Addition of acetylthiosulfenyl chloride to carbon–carbon unsaturated compounds such as alkenes, conjugated dienes, and allenes gives rise to the desired monoacyl disulfides (Equation (52)) ⟨67JOC2925, 68JOC1533⟩. The addition involves episulfonium ions as key intermediates.

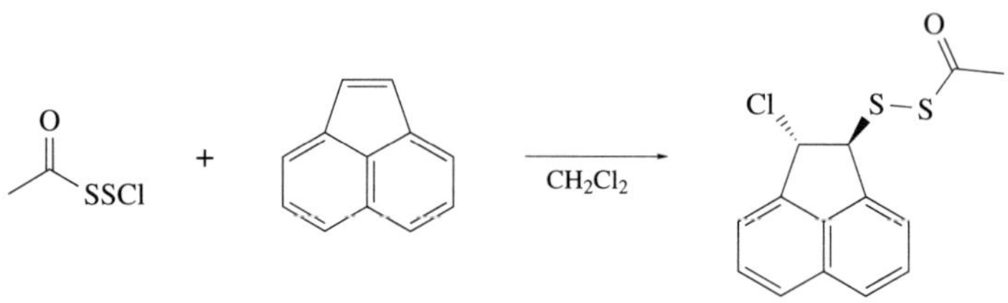

(52)

The reaction of acetylsulfenyl chlorides with activated aromatic compounds probably proceeds via tetravalent sulfur intermediates to give acetyl aryl disulfides (Equation (53)) ⟨71JOC3546⟩.

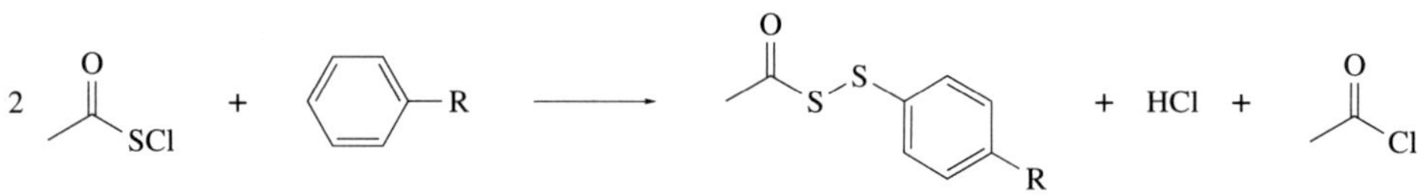

(53)

R = Me, MeO, HO

Oxidation of alkanedithioates with mcpba and the subsequent rearrangement of the resulting sulfines yields monoacyl disulfides as shown in Scheme 27 ⟨88CC390, 94S761⟩.

A perfluoro-substituted monoacyl disulfide is reported to be prepared from CF_3SSH and $CF_3C(O)X$ ⟨76JA6545⟩.

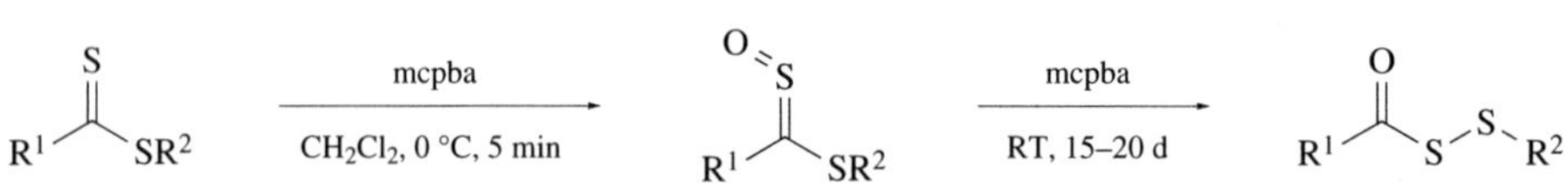

Scheme 27

5.05.6.1.2 Acylsulfenyl selenides

As well as monoacyl disulfides, acylsulfenyl selenides are prepared from acylsulfenyl halides or thiocarboxylic acid. The reaction of aroylsulfenyl bromides with diaryl diselenides provides aroylsulfenyl selenides in good yields (Equation (54)) ⟨85S519⟩.

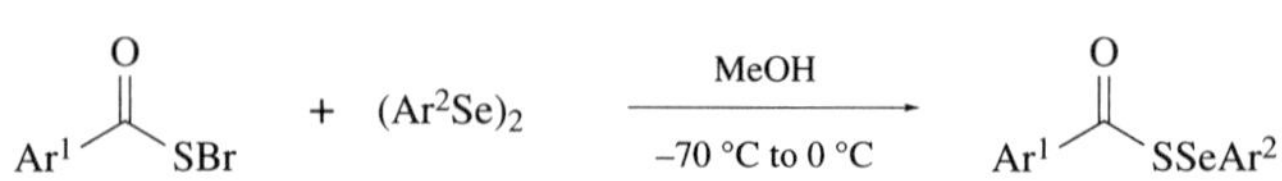

(54)

Selenenylation of thiocarboxylic acids with *N*-phenylselenophthalimide in CH_2Cl_2 is a useful alternative (Equation (55)) ⟨85S878, 87NKK1442⟩. Some metal salts of thiocarboxylic acid undergo selenenylation with selenenyl bromide to afford the desired acylsulfenyl selenides (Equation (56)) ⟨85S520⟩.

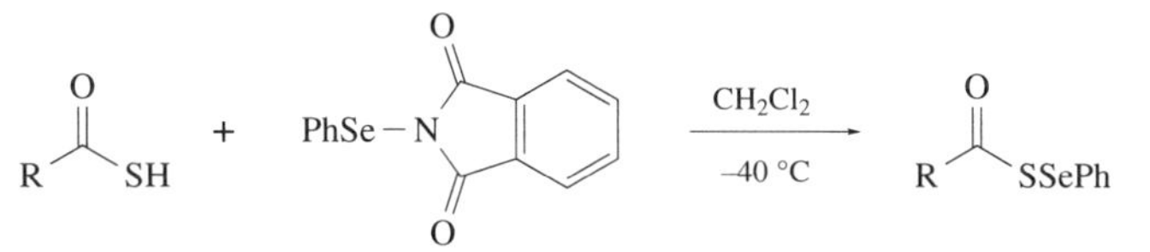

(55)

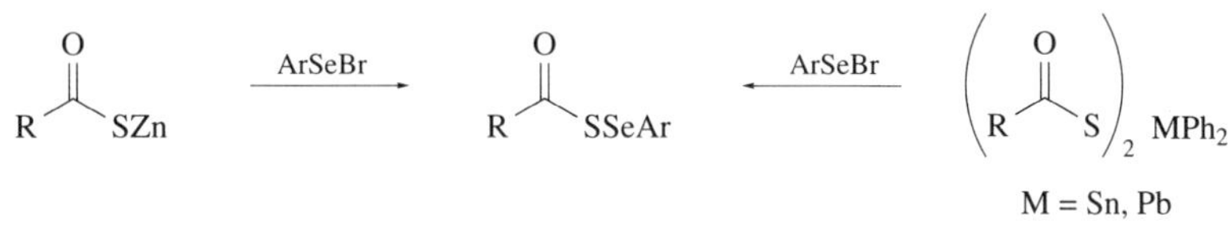

(56)

In the oxidation reaction of thionselenolesters (RC(=S)SePh) with mcpba, acylsulfenyl selenides are formed as a by-product probably via a similar mechanism as mentioned in Scheme 27 ⟨88BCJ3931⟩.

The synthetic utility of acylsulfenyl selenides has been demonstrated by Toru *et al.*: photoinitiated addition of acylsulfenyl selenides to alkenes provides a useful synthetic method for simultaneous introduction of both acylthio and seleno groups into organic molecules (Equation (57)) ⟨87NKK1442, 88BCJ2205, 89JCS(P1)1927⟩.

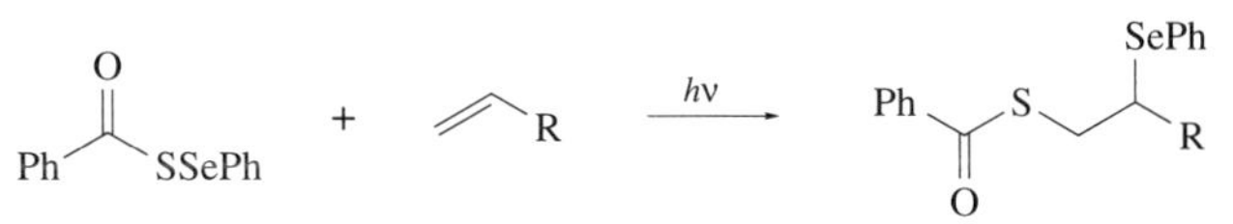

(57)

5.05.6.1.3 *Acylsulfenyl tellurides*

The synthesis of acylsulfenyl tellurides is performed mainly by the following two methods. One is the reaction of acylsulfenyl bromide with diphenyl ditelluride (Equation (58)) ⟨85S519⟩, and the other is the reaction of potassium or piperidinium thiocarboxylates with phenyltellurenyl bromides (Equation (59)) ⟨85S520⟩.

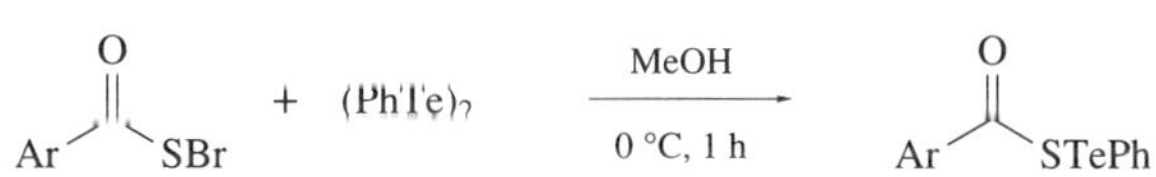

(58)

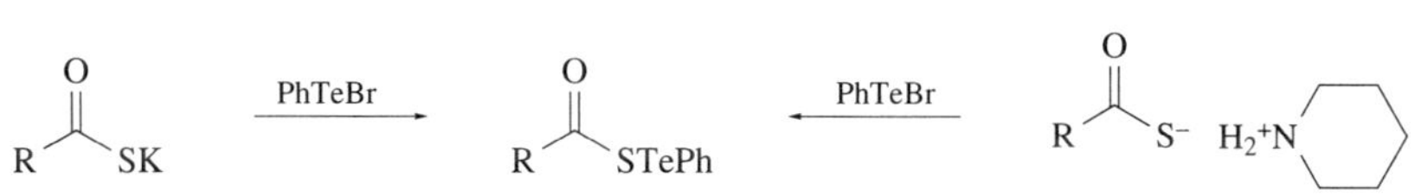

(59)

Although the yield is lower, a phenylsulfenyl telluride can be prepared by the reaction of thiobenzoic acid with benzenetellurenic acid (Equation (60)) ⟨87NKK1479⟩.

(60)

5.05.6.1.4 *Acylselenenyl sulfides*

The reaction of a triorganogermanium selenocarboxylate with an arenesulfenyl chloride is reported to provide the corresponding aroylselenenyl sulfide in moderate yield (Scheme 28) ⟨92ZN(B)558⟩.

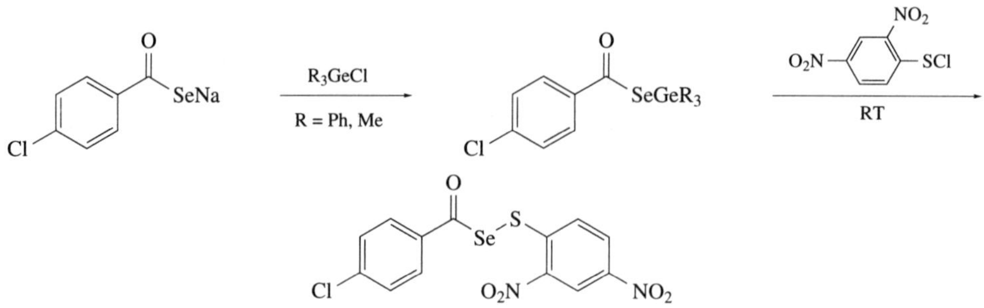

Scheme 28

5.05.6.1.5 Monoacyl-substituted diselenides

Monoacyl-substituted diselenides are synthesized by the selenenylation of metal salts of selenocarboxylates with areneselenenyl bromides. Potassium selenocarboxylates react with PhSeBr in ether at 0 °C to afford aroyl aryl diselenides in 49–61% yields (Equation (61)) ⟨87S371⟩. Use of organotin selenocarboxylates leads to increase in the yield of the desired monoacyl diselenides (Equation (62)) ⟨89S929⟩. Similarly, lead selenocarboxylates can be employed for the synthesis of monoacyl diselenides ⟨90JOM(386)313⟩.

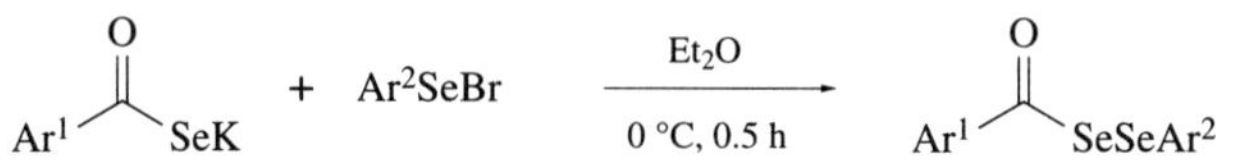

(61)

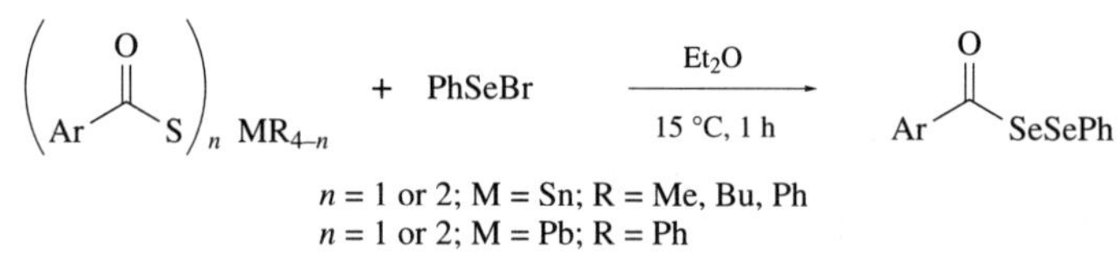

(62)

5.05.6.1.6 Acylselenenyl tellurides

Treatment of *O*-silyl selenocarboxylates with benzenetellurenyl iodide at room temperature yields aroylselenenyl phenyl tellurides in good yields (Equation (63)) ⟨94JCS(P1)1083⟩. *Se*-Organoarsanyl selenocarboxylates also react with PhTeI to give the corresponding acylselenenyl tellurides with the concomitant formation of arsanyl iodide (Ph_2AsI) (Equation (64)) ⟨94S282⟩. The acylselenenyl tellurides are relatively stable towards oxygen, and do not decompose upon exposure to air at room temperature for 2 h.

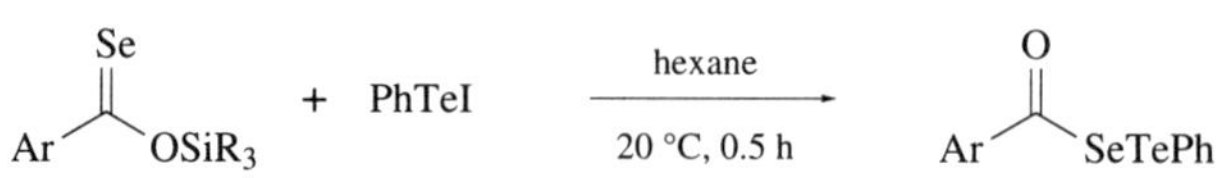

(63)

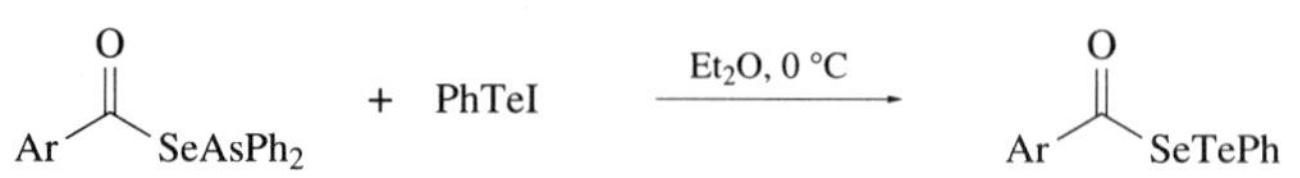

(64)

5.05.6.1.7 Acyltellurenyl sulfides

A piperidinium tellurocarboxylate, prepared by the reaction of the corresponding bis(acyl) telluride with piperidine, reacts with 2,4-dinitrobenzenesulfenyl chloride to provide 32% yield of acyltellurenyl sulfide as yellow crystals (Scheme 29) ⟨87CL475⟩.

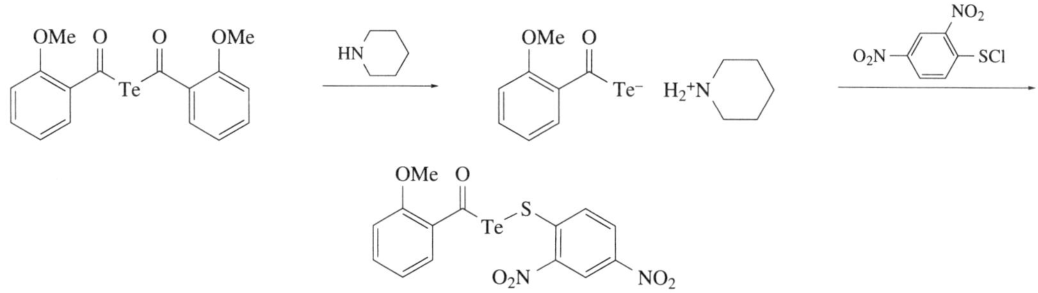

Scheme 29

5.05.6.2 Bis(acyl) Dichalcogenides

5.05.6.2.1 Bis(acyl) disulfides

Bis(acyl) disulfides are prepared by the following three methods: (i) oxidation of thiocarboxylic acids or their metal salts; (ii) acylation of disulfide anions (S_2^{2-}) with acyl halides; and (iii) reaction of metal thiocarboxylates with acylsulfenyl halides or acylsulfeneamides (Equation (65)).

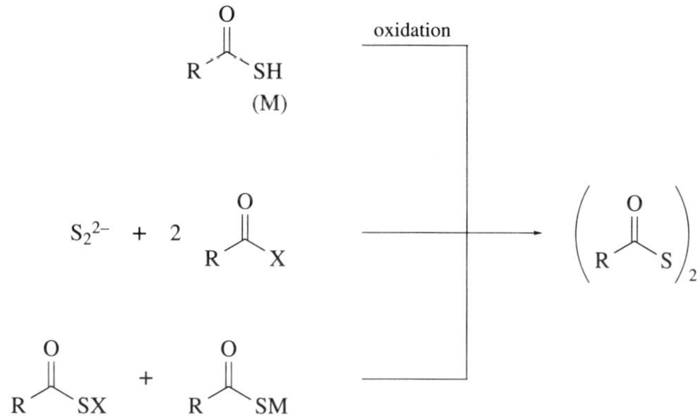

(65)

As to method (i), oxidation by iodine of potassium thiocarboxylates (prepared from acyl chlorides and potassium hydrogen sulfide) is generally utilized to synthesize bis(acyl) disulfides (Scheme 30) ⟨55OSC(3)116⟩. 2-Polyvinylpyridine/bromine complex is a useful oxidizing agent for thiocarboxylic acids, giving the corresponding bis(acyl) disulfides in excellent yields ⟨78S50⟩.

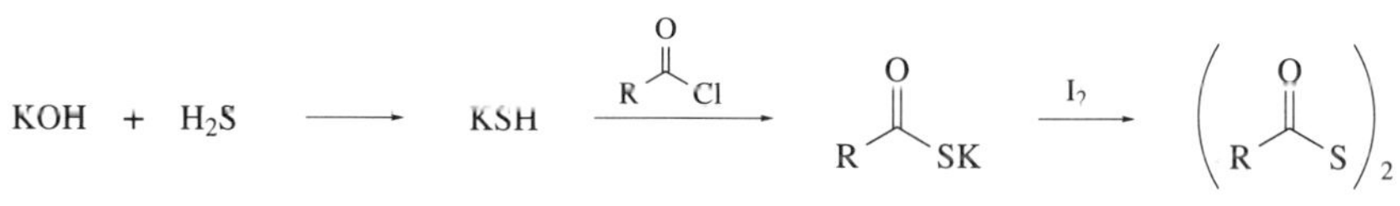

Scheme 30

Besides these, nickel peroxide ⟨80SC305⟩, thallium(III) acetate ⟨77BCJ220⟩, and an alkoxystannane/iron(III) chloride system ⟨90TL3591⟩ can be employed for the oxidation of thiocarboxylic acids. In the case of thallium(III) acetate, aliphatic thiocarboxylic acids are easily converted into bis(acyl) disulfides, whereas aromatic ones such as thiobenzoic acid react with $Tl(OAc)_3$ to give thallium(III) thiocarboxylates like $Tl(SC(O)Ph)_3$. The alkoxystannane/iron(III) chloride is a mild oxidizing system, and the reaction proceeds via thiostannanes as the intermediate (Equation (66)).

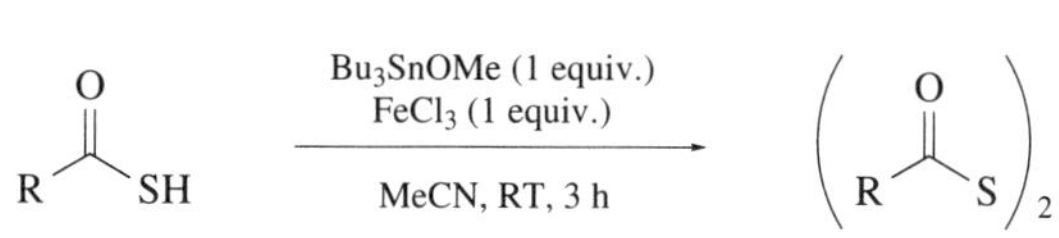

(66)

Bis(ferrocenecarbonyl) disulfide is obtained on treatment of the amine salts of thiocarboxylic acid with benzenesulfonyl chloride (Equation (67)) ⟨77JOM(129)189⟩.

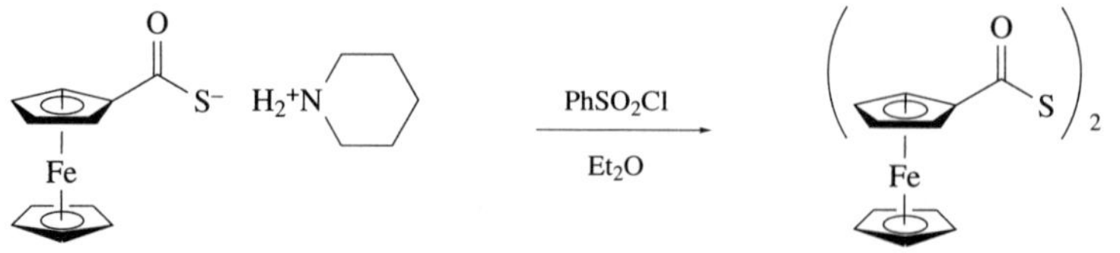

(67)

In method (ii), sodium disulfide, lithium disulfide, and hydrogen disulfide are employed for the acylation with acyl halides. For the acylation of sodium disulfide, the use of phase-transfer catalysts such as $(C_{16}H_{33})Bu_3P^+Br^-$ leads to the high yield formation of the desired bis(acyl) disulfides ⟨81S637, 90MI 505-01⟩. Gladysz *et al.* report the selective formation of Li_2S_2 by the reaction of elemental sulfur with $LiBHEt_3$. The acylation of Li_2S_2 with acid chlorides attains a convenient synthesis of bis(acyl) disulfides (Equation (68)) ⟨78CC838, 79T2329⟩.

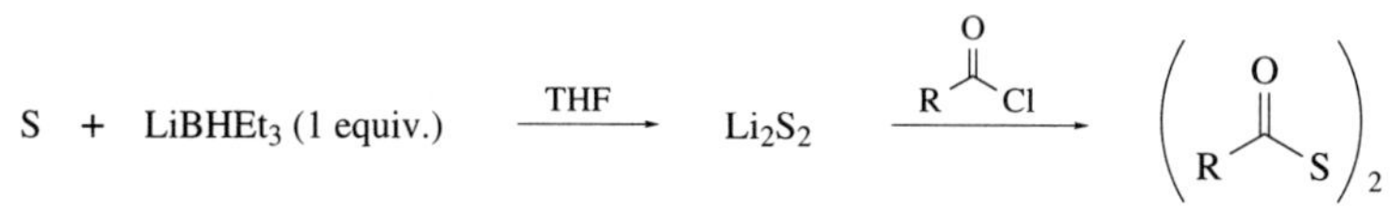

(68)

Solid-state reaction of acyl halides using benzyltriethylammonium tetrathiomolybdate also gives bis(acyl) disulfides in good yields ⟨94JCS(P1)767⟩.

Compared to the methods (i) and (ii), method (iii) is superior in terms of the synthesis of unsymmetrical bis(acyl) disulfides. For example, the reaction of *p*-methylbenzoylsulfenyl bromide with one equivalent of potassium thiobenzoate provides benzoyl *p*-methylbenzoyl disulfide in 88% yield (Equation (69)) ⟨82S1013⟩. *N*-(Aroylthio)succinimides (prepared from potassium thiocarboxylates and *N*-chlorosuccinimide) are useful reagents for the synthesis of unsymmetrical bis(acyl) disulfides (Scheme 31) ⟨80S721⟩.

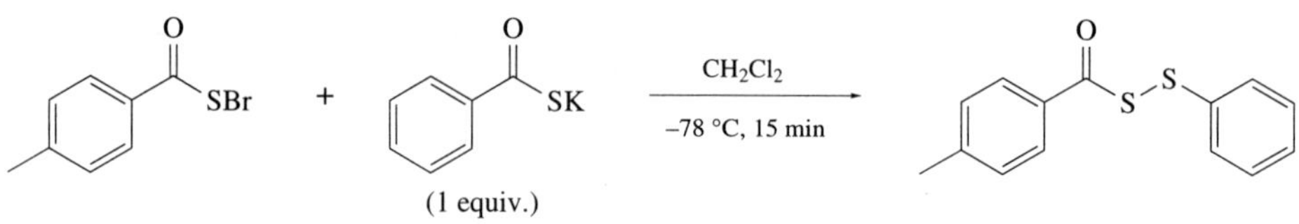

(69)

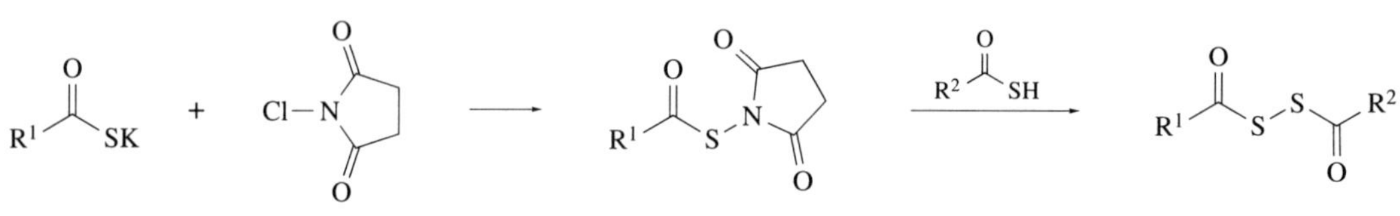

Scheme 31

The reaction of selenium or tellurium bis(thiocarboxylates) with halogens is also reported to yield bis(acyl) disulfides (Equation (70)) ⟨87NKK1457⟩.

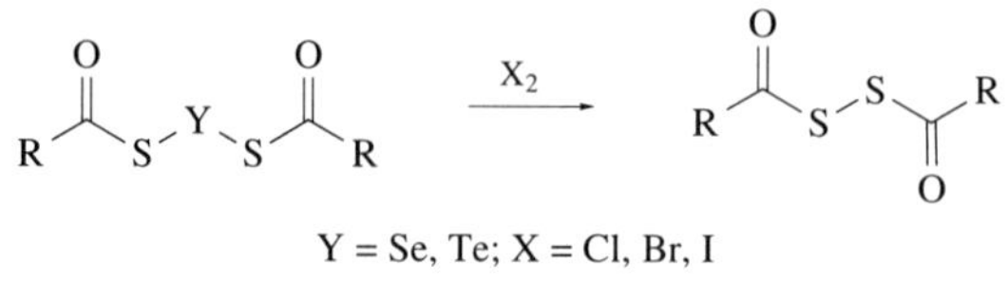

(70)

5.05.6.2.2 *Bis(acyl) diselenides*

Bis(acyl) diselenides can be synthesized by the oxidation of potassium or sodium selenocarboxylates with iodine, similar to the method for bis(acyl) disulfides (Equation (71)) ⟨76CL203, 77BCJ3007⟩. Amine salts of selenocarboxylic acids (generated by the reduction of metallic selenium with carbon monoxide and water in the presence of tertiary amine) also undergo air oxidation easily to give the corresponding bis(acyl) diselenides in good yields (Scheme 32) ⟨91JOC3776⟩.

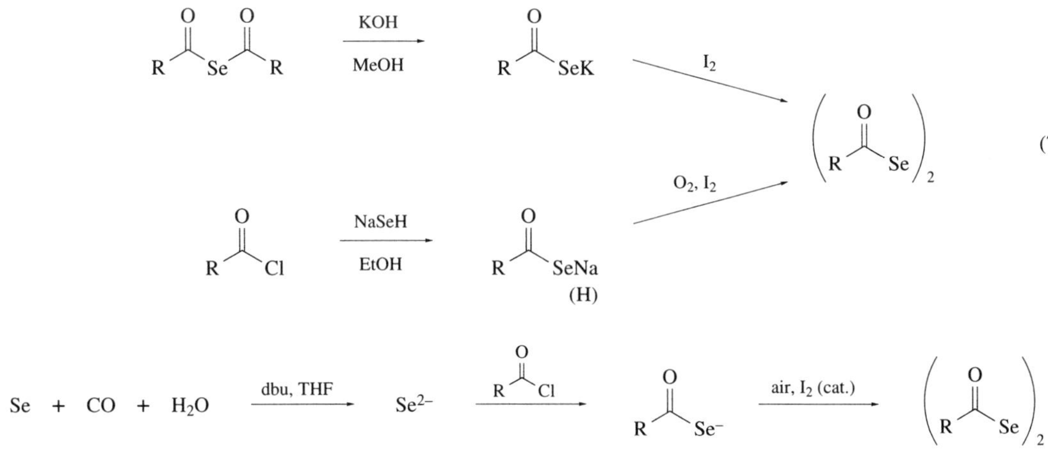

(71)

Scheme 32

Selective formation of Se_2^{2-} from elemental selenium, followed by acylation with acyl halides is an alternative method for preparing bis(acyl) diselenides. For example, selective reduction of selenium using sodium hydroxide, and the subsequent acylation, takes place successfully under phase-transfer conditions (Equation (72)) ⟨90JCR(S)230⟩.

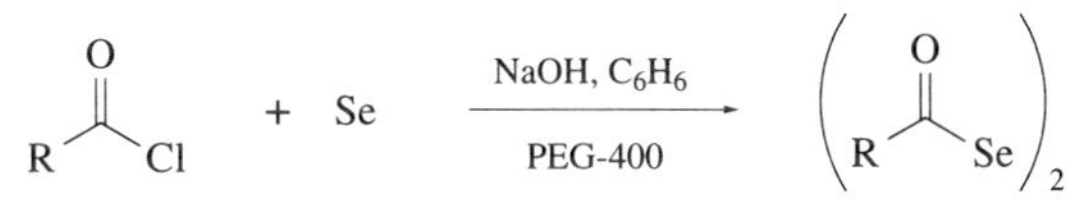

(72)

Samarium diiodide (SmI_2) can also be employed as a reducing agent for elemental selenium ⟨93SC1403⟩. The bis(acyl) diselenides are utilized as useful acylating reagents, because both acyl groups can be transferred into organic molecules effectively ⟨89CL1825⟩.

5.05.6.2.3 *Bis(acyl) ditellurides*

In contrast to bis(acyl) disulfides and diselenides, bis(acyl) ditellurides are extremely unstable, and therefore there are only two reports concerning the successful isolation of bis(acyl) ditellurides. Careful treatment of the mixture obtained from the reaction of sodium hydrogen telluride (NaTeH) with benzoyl chloride resulted in the isolation of bis(benzoyl) ditelluride in 9% yield ⟨87NKK1475⟩. Suzuki *et al.* report that bis(*O*-methoxybenzoyl) ditelluride can be isolated in 61% yield (Equation (73)) ⟨90JCR(S)56⟩. However, similar reaction using other aroyl chlorides results in a rapid decomposition of the desired products.

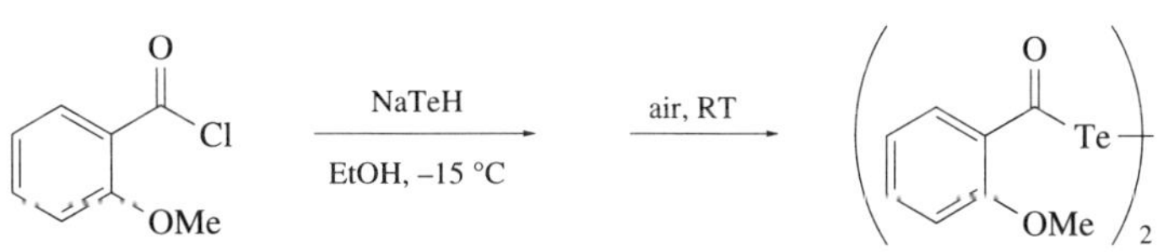

(73)

5.05.7 ACYLSULFENAMIDES

Amination of sodium thiocarboxylates with sodium hydroxylamine-*O*-sulfonate affords aroylsulfenamides ($ArC(O)SNH_2$), which, surprisingly, are stable to recrystallization and storage (Equation (74)) ⟨72JOC3820⟩. The resulting *N*-unsubstituted aroylsulfenamides can be converted into *N*-substituted derivatives upon treatment with isocyanates (Equation (75)). As mentioned in Section 5.05.6.2.1, *N*-(aroylthio)succinimides can be synthesized by the reaction of potassium thiocarboxylates with *N*-chlorosuccinimide, and are utilized as aroylsulfenylating reagents ⟨80S721, 81S370⟩.

(74)

(75)

Compared with aroylsulfenamides, acylsulfenamides are very unstable compounds. As indicated in Scheme 33, for example, the reaction of acetylsulfenyl chloride with aniline provides the corresponding acetylsulfenamide, which easily decomposes to give *N*-phenylacetamide and elemental sulfur ⟨52LA(576)61⟩.

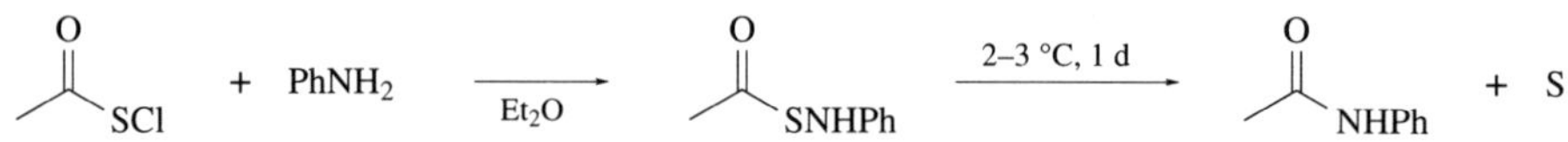

Scheme 33

5.05.8 ACYLTHIOSILANES AND THEIR SELENIUM AND TELLURIUM ANALOGUES

The first synthesis of acylthiosilanes has been reported by Gilman and Lichtenwalter: the preparation of lithium triphenylsilanethiolate from triphenylsilyllithium and elemental sulfur, followed by acylation of the resulting Ph_3SiSLi with benzoyl chloride provides (benzoylthio)triphenylsilane in good yield (Scheme 34) ⟨60JOC1064⟩.

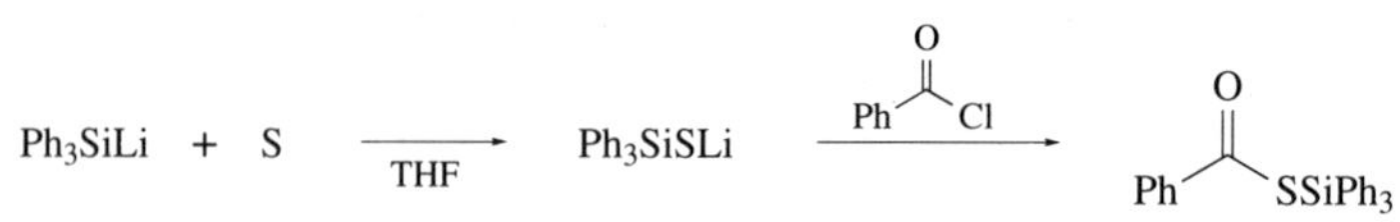

Scheme 34

It has been reported that the reaction of trifluoroacetylsulfenyl iodide with trimethylsilyl chloride affords (trifluoroacetylthio)trimethylsilane with deposition of molecular iodine (Equation (76)) ⟨88ZAAC(566)151⟩.

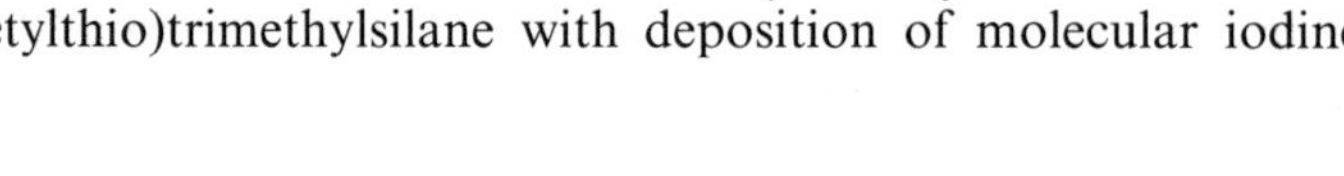

(76)

On the other hand, it might be expected that silylation of alkali metal salts of thiocarboxylic acids with chlorotriorganylsilanes would be the most general access to acylthiosilanes. However, the silylation takes place exclusively at the oxygen of thiocarboxylate anions to give thionoacyloxysilanes as illustrated in Equation (77) ⟨66JOC3439, 73BCJ244⟩.

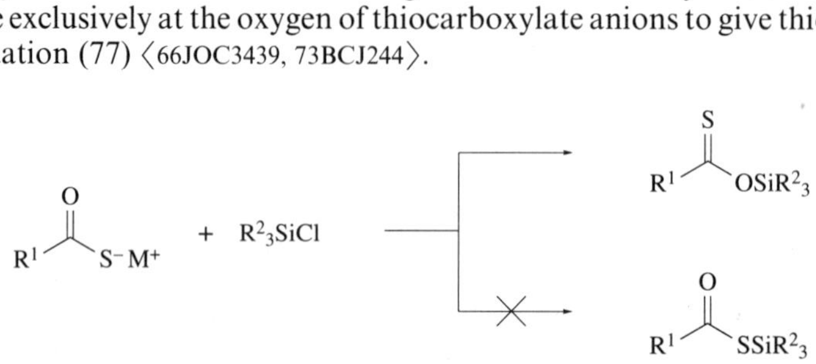

(77)

Similarly, the reaction of potassium selenobenzoate with trimethylsilyl chloride proceeds selectively via *O*-silylation to give only (selenobenzoyloxy)trimethylsilane in 68% yield (Equation (78)) ⟨72TL3751⟩.

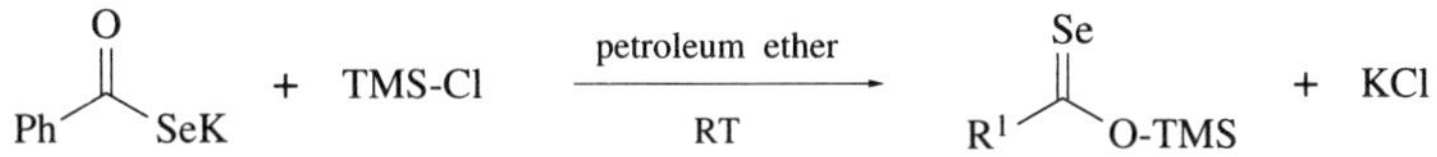

(78)

In contrast to alkali metal thiocarboxylates, the silylation of sodium tellurocarboxylates with *t*-butyldimethylsilyl chloride occurs initially at tellurium to form the desired *Te*-silyl ester, which then isomerizes slowly to the corresponding *O*-silyl ester via the migration of the silyl group (Scheme 35) ⟨90TL3587⟩.

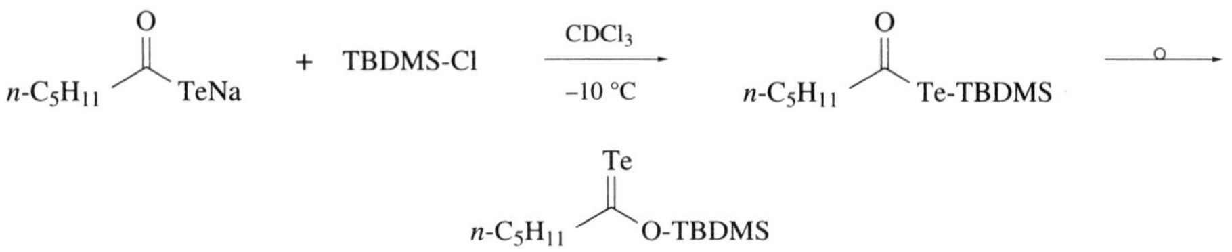

Scheme 35

A similar 1,3-silyl shift is observed in the reaction of pivaloyl chloride with equimolar amounts of bis(trimethylsilyl) telluride, which provides *Te*-silyl and *O*-silyl esters in 35% and 65% yields, respectively (Scheme 36) ⟨87CC820⟩. Furthermore, equimolar reaction of bis(trimethylsilyl) telluride and bis(pivaloyl) telluride leads to the formation of a 1 : 2 mixture of the *Te*-silyl ester and the *O*-silyl ester.

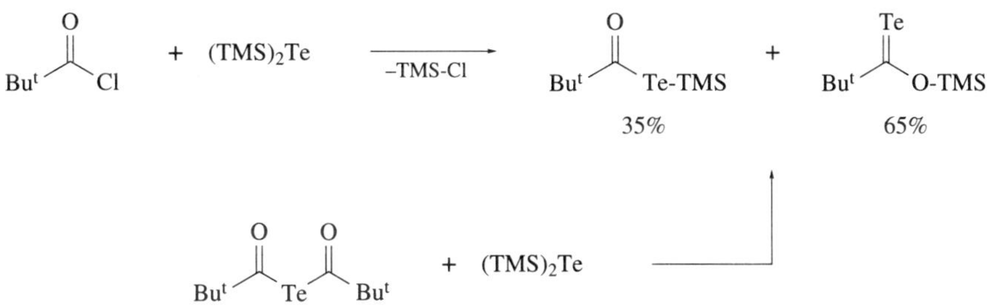

Scheme 36

5.06
Amides

PATRICK D. BAILEY, IAN D. COLLIER and KEITH M. MORGAN
Heriot-Watt University, Edinburgh, UK

5.06.1 GENERAL METHODS

5.06.1.1 Introduction

The amide unit is one of the most important and widely occurring functional groups. It is present in many natural products and is a key feature of many important man-made compounds. It can be an intermediate for the preparation of many other functional groups (e.g., amines, ketones and acids) and can be used for protecting amines. There are many synthetic methods for the preparation of the amide moiety, the chemistry often being driven by the stability of the amide bond due to conjugation of the nitrogen lone pair with the carbonyl group.

It would be inappropriate simply to discuss the breadth of substituted amides under the headings suggested by the nature of R^1, R^2 and R^3 (Structure (**1**)). Again and again, the same basic synthetic strategies would recur, and the principles behind the methodologies might be lost. Instead, this chapter is divided into two roughly equal parts; the first, Section 5.06.1 on General Methods, will summarise the main types of synthetic methodology that are employed in the preparation of amides, and some of the factors that might lead to the choice of one reagent over another. In the second part of this chapter (Sections 5.06.2–5.06.6), the different types of substituted amide will be considered systematically, so that the actual reagents successfully used for each type of amide bond formation

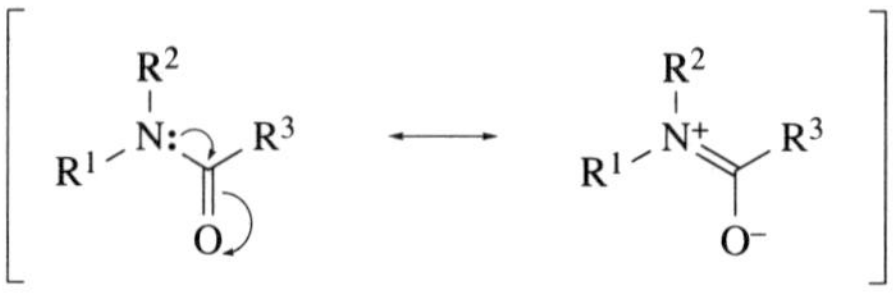

(**1**)

are surveyed; these sections contain the bulk of the specific references within this chapter. For more information on the mechanism, selectivity and practical aspects of the reagents, the reader should refer back to General Methods (Section 5.06.1).

Of course, there are tens of thousands of references in which amides are formed, and thousands of papers in which amide bond formation is the sole subject of the publication. The specific examples chosen in Sections 5.06.2–5.06.6 are therefore quite selective and guided to some extent by the material presented in *The Chemistry of Amides* ⟨B-70MI 506-01⟩, *Compendium of Organic Synthetic Methods, vols.* 1–5 ⟨B-84MI 506-01⟩ and *Comprehensive Organic Transformations* ⟨B-89MI 506-01⟩. Other reviews in this area are *The Organic Chemistry of Aliphatic Nitrogen Compounds* ⟨B-94MI 506-01⟩ and the review by North ⟨94MI 506-04⟩.

This first section has been subdivided into Acylation of Amines (Section 5.06.1.2), Hydrolysis of Nitriles (Section 5.06.1.3), Rearrangement Reactions (Section 5.06.1.4) and Other Methods (Section 5.06.1.5). However, before these synthetic strategies are discussed, the most important types of amides need to be considered briefly in order to be aware of the factors that will influence the preference for using one synthetic method over another.

5.06.1.1.1 N-*Unsubstituted amides*

Compounds of the type $RCONH_2$ or $ArCONH_2$ are common synthetic targets, for which derivatisation of the free acid or ester is often appropriate. Although most of the 'exotic' reagents would probably be successful, it is the cheap, simple, efficient procedures that dominate the literature, and the more unusual methods (e.g., rearrangements) are employed only when synthesis of the rest of the molecule influences the synthetic strategy for the simple amide. Thus, for the solid-phase synthesis of peptides possessing a C-terminal amide, the method of attachment to the polymeric support is usually chosen so that cleavage with ammonia liberates the peptide directly as the acid amide at the end of the synthesis.

5.06.1.1.2 Peptides

The commonest use of amide bond formation is in order to prepare peptides, and this has been reviewed extensively ⟨B-91MI 506-01, 93T11065⟩. As discussed under the following appropriate sections, the detailed substitution pattern can strongly influence the ease of formation of the peptide bond. However, there are a limited number of methods that have dominated the literature because of their efficiency and ease of use. In general, when carrying out a synthesis involving the formation of many peptide bonds, it is not possible to test a range of methods for each coupling reaction in order to optimise the overall yield. More often, a 'favourite' procedure is used for almost all of the couplings, with 'special' reagents being used only for the formation of 'difficult' amide bonds. For many peptides that are required in small quantities, solid-phase peptide synthesis is employed. As the impurities resulting from low efficiency couplings can accumulate rapidly, solid-phase synthesis requires routine couplings to be certainly >95% efficient, and usually >99% efficient. These methods usually employ a large excess of the (easily removed, but often expensive) acylating agent, and such methods are often inappropriate for nonpeptidic amides.

Finally, peptides are almost always prepared from the constituent amino acids, and racemisation of chiral centres is a major concern. In general, a protected amino acid or peptide is activated to give an acylating agent (Scheme 1, step 1), to which addition of the amino component leads to a new peptide bond (Scheme 1, step 2). The better the leaving group X, the faster is the coupling reaction. However, as the acylating agent becomes more reactive, so side-reactions can start to occur; the most problematic of these is probably racemisation, which usually occurs via the oxazolidine (Scheme 2). If R^1 is alkoxyl (urethanes), then the racemisation process is strongly suppressed. However, many of the procedures developed for peptide synthesis have been carefully designed to minimise this racemisation problem.

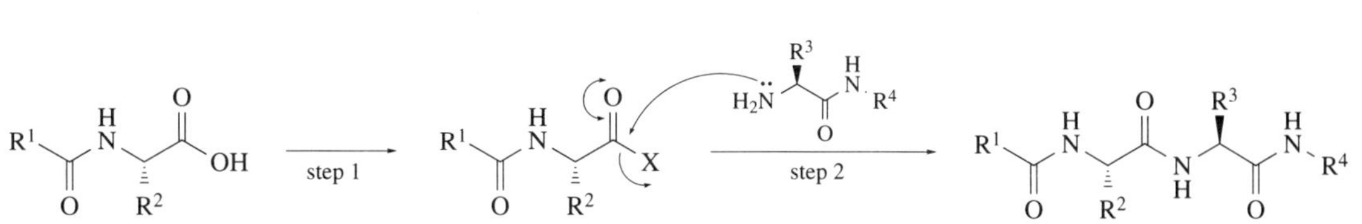

Scheme 1

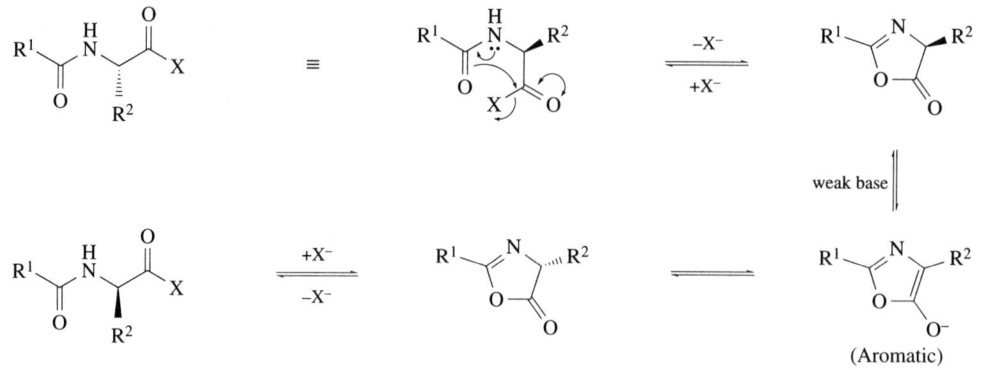

Scheme 2

5.06.1.1.3 Acyl protected amines

When the formation of an amide is required for protection of an amino group, then acylation is very effective because of the conjugative stabilisation of the nitrogen lone pair (see Structure (**1**)). In general, acylation of the free amine is again exploited, and simple, powerful acylating agents (e.g., Ac_2O, AcCl) are the norm. Silicon protection on nitrogen is too labile to be of practical use in most cases, but alkoxyacyl protection of nitrogen is commonly used; such urethane derivatives can be deprotected under milder conditions than simple amides, and the choice of alkoxy group R allows selective removal of nitrogen protection under conditions in which other functional groups (and other urethane derivatives) are unaffected (Equation (1)) ⟨B-91MI 506-02⟩.

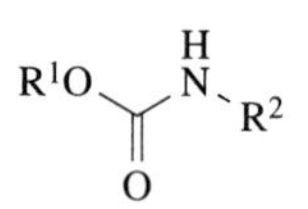 $\xrightarrow{\text{Reagent}}$ $R^1X + CO_2 + H_2NR^2$ (1)

R¹	*Reagent*
Bu^t	TFA
Fmoc	Piperidine, DMF
Bn	H_2, Pd-C

In the tabulated examples for R^1, each protecting group is stable to the conditions under which the other two are removed

Fmoc = fluorenylmethyloxycarbonyl

5.06.1.1.4 Cyclic amides

As with many other functional groups in this work, formation of an amide bond within a newly forming ring is a common occurrence. The ring itself can strongly influence both the ease of amide bond formation (e.g., with many cyclic peptides) and the actual methods that might be available. The appropriateness of different methods for ring formation will be discussed both in General Methods, and again under the appropriate substituted amide sections (Sections 5.06.2–5.06.6) for which the presence of a ring does not affect the category. Only for β-lactams, a class of compounds of immense biological importance, will completely different reactions for amide bond formation emerge.

5.06.1.2 Acylation of Amines

As indicated already, this is the most widely used method of forming amides, for which the basic reaction is shown in Scheme 3. The number of reagents that can be used to generate R—CO—X are quite bewildering, and the authors have tried to emphasise those procedures that are particularly successful and reliable.

In general, the rate of amide bond formation is governed by the stability of the tetrahedral intermediate (i.e., its formation is the rate-determining step), so it is the electron-withdrawing power of X (not the stability of X^-) that controls the overall rate of reaction. In practice, the stability of the leaving group usually parallels the stability of the intermediate (by reducing the δ^- charge on the tetrahedral carbon) (see Table 1).

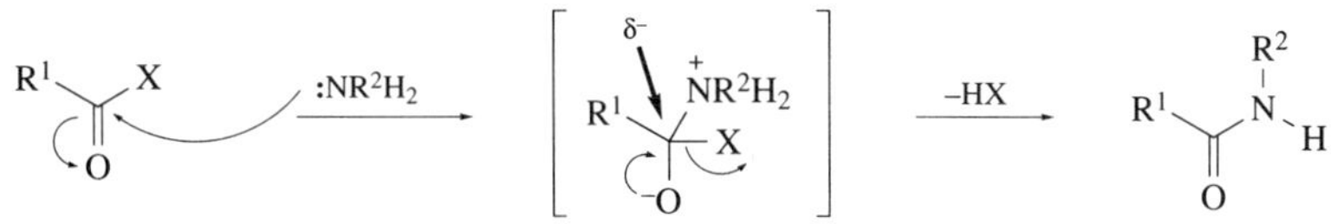

Scheme 3

Table 1 Stabilising effect of $-I/-M$ groups for acylation.

	–I	–M
Leaving group	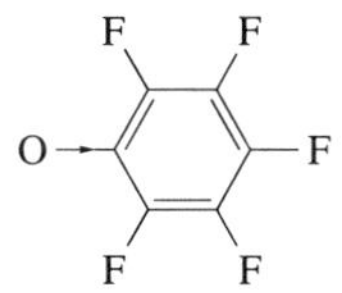	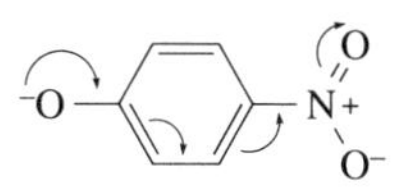
Stabilisation of negative charge developing on R	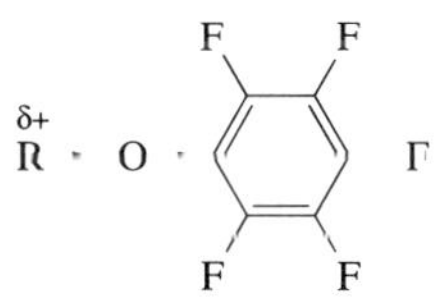	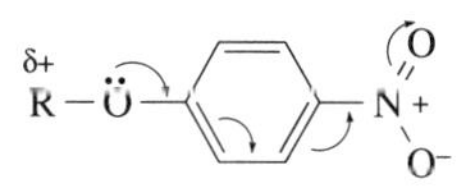

The acylation of an amine usually generates a mole equivalent of an acid (see Scheme 3). As amines are basic (whereas amides are not), the liberated acid can protonate all of the remaining amine when the reaction is only half complete; such acylations tend to grind to a halt at 50% conversion. It is therefore common to add a base which will neutralise the liberated acid (e.g., $NaHCO_3$), or act as a proton acceptor to allow free amine to be present in equilibrium (e.g., NEt_3) (Scheme 4).

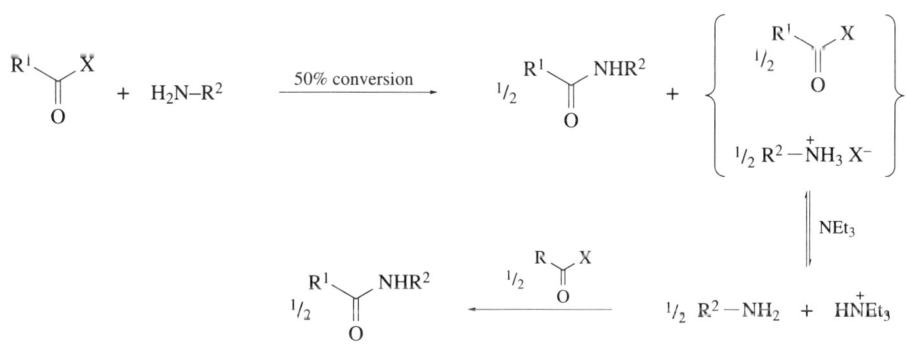

Scheme 4

In practical terms, acylation procedures can be divided into four categories.

(i) Direct reaction of an amine and an acid.

(ii) Reaction of a readily available acid derivative (e.g., simple methyl ester) with an amine.

(iii) Formation of an activated carbonyl compound (e.g., acid chloride, anhydride or active ester), followed by condensation with an amine.

(iv) One-pot reaction of a carboxylic acid with an amine, in which a coupling reagent activates the acid component *in situ*.

These practical distinctions are very important, despite the similarity in the structures of many of the acylating agents. For example, a simple amide might be prepared in good yield (~80%) from the methyl ester in several hours, but repetitive peptide couplings need to be much faster and more efficient than this; storable active esters or *in situ* coupling reagents are therefore widely used. An additional factor of great importance is the ease with which the by-products from the coupling procedure can be removed. In solid-phase peptide synthesis, immobilisation of the growing peptide on a polymeric support allows soluble by-products to be washed away; when amides are formed using standard solution-phase chemistry, the ease with which by-products can be removed sometimes dictates the procedure that is chosen.

While bearing these factors in mind, this chapter will consider acylation procedures under seven headings. (For an overview of acyl activation and for details on commonly used laboratory procedures specific to peptide synthesis see Jones ⟨B-91MI 506-01⟩, Bodansky and Bodansky ⟨B-94MI 506-03⟩.)

5.06.1.2.1 Directly from carboxylic acids

In certain special cases, it is possible to drive off water directly from an amine/carboxylic acid mixture. However, this procedure often fails since the acid can protonate the amine, thereby inactivating both partners (Equation (2)). If both components are stable to strong heating (extended reaction times at >150 °C), then direct reaction is possible. Conditions are slightly less forcing if a cyclic amide is being formed (Equation (3)) ⟨55HCA312, 78S752⟩. For primary amides of the type $RCONH_2$ or $ArCONH_2$, heating the carboxylic acid with formamide ($HCONH_2$), urea (H_2NCONH_2) or acetylurea ($CH_3CONHCONH_2$) does yield the desired product (see Table 2, Section 5.06.2.1.1), but the temperatures required are high, severely limiting the value of this approach.

(2)

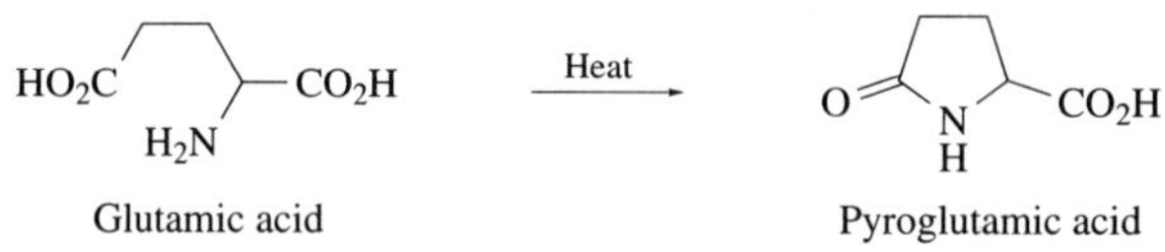

(3)

A much milder approach is the use of metalloamides that are able to coordinate to the oxygen of the carboxylic acid in order to promote nucleophilic attack. The aluminum amides seem to be particularly promising. They can be prepared from the amine using (cheap) trimethylaluminum (Equation (4)) ⟨70JOM(25)33⟩. Although most of the examples utilise fairly simple amines and acids, Equation (5) indicates the selectivity that is possible ⟨94MI 506-02⟩.

$$R^1R^2NH \quad + \quad AlMe_3 \longrightarrow R^1R^2NAlMe_2 \quad + \quad MeH\uparrow \qquad (4)$$

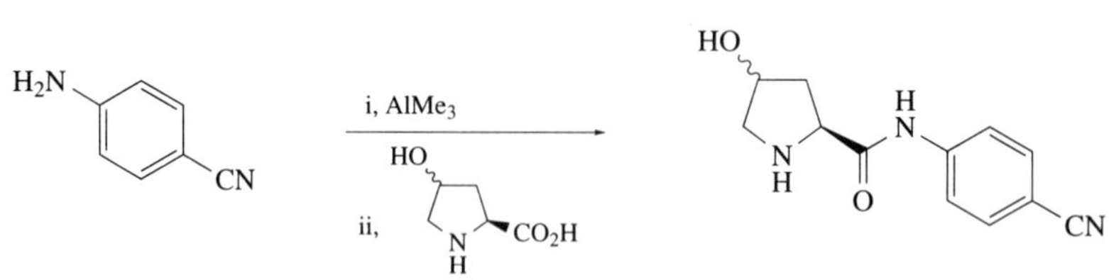

(5)

Peptides have also been prepared from an acid and amine by the use of hydrolytic enzymes being forced to work 'in reverse'. This catalytic approach is more or less restricted to peptides because of the availability of suitable enzymes. High concentrations of one component and careful control of pH and solvent are critical features, but because of the enantiospecificity of the enzymes, optically pure products can result from racemic substrates. (For a general overview of enzymes in organic synthesis, see Davies *et al.* ⟨B-89MI 506-02⟩ and Turner ⟨89MI 506-03⟩. More specific examples of peptide synthesis can be found in Jakubke ⟨B-89MI 506-04⟩ and Kullman ⟨B-89MI 506-05⟩.)

5.06.1.2.2 From simple esters

Simple methyl or ethyl esters are convenient synthetic intermediates but they only form amides reluctantly. Heating at around 100 °C for several hours is typical. Two examples that proceed smoothly are shown in Equations (6) ⟨77CCC560⟩ and (7) ⟨93JCS(P1)451⟩. In the first example, the oxazolidinone ring activates the carbonyl group, while in the second, six-membered ring formation facilitates the reaction.

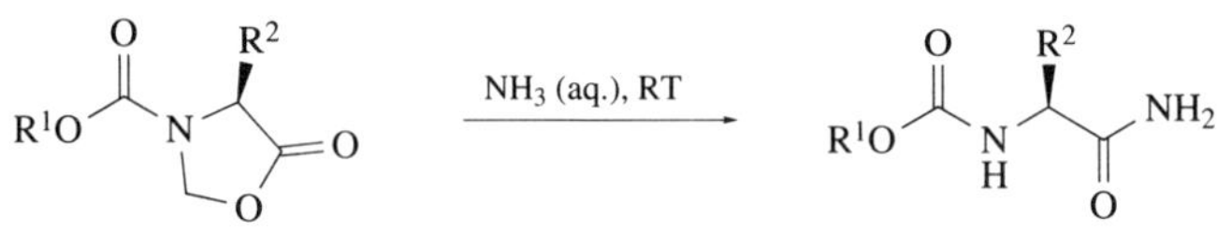

(6)

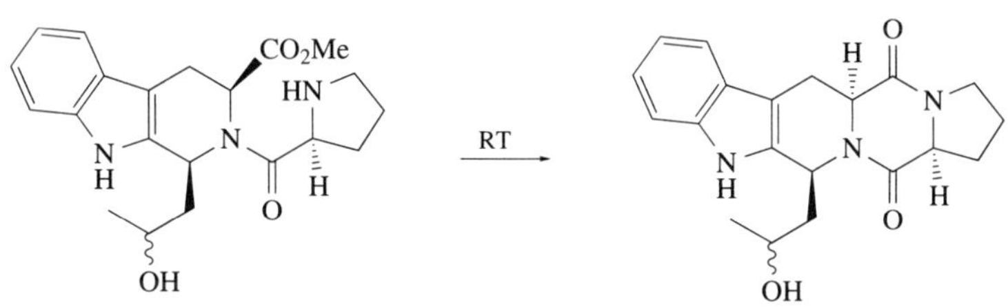

(7)

Other than simple heating of the amine and ester, the main trick is to make the nitrogen more nucleophilic. This has to be carefully balanced against making it more basic, which is often an unwanted property. Thus, examples of $LiNR_2$ acting as a useful nucleophile in the synthesis of amides of the type R^1CONR_2 are far outweighed by the use of lithium amides as powerful and efficient bases. However, a range of other metal amides have been used and the aluminum amides $R^1{}_2AlNR^2R^3$ seem particularly promising (see Equation (8) ⟨80OS(59)49⟩ and Table 5 (Section 5.06.2.2.2)).

(8)

The use of simple esters as substrates for enzymic couplings is rather more successful than using free acids, because the ester cannot be deprotonated to the unreactive carboxylate. This approach has been used to prepare both simple amides, using a lipase ⟨93T4007⟩, and peptides, using the proteolytic enzyme papain ⟨93TL1247⟩. However, in general, attempts to drive enzymes in reverse require substantial inhouse expertise.

5.06.1.2.3 From active esters

Rather than increasing the nucleophilicity of nitrogen, amine/ester couplings can be accelerated by making the acylating agent more electrophilic. This often requires only a slight increase in reactivity—a feature which can limit unwanted side reactions. Some commonly used active esters (shown in Figure 1) are usually prepared from the alcohol and acid using a diimide as a coupling agent, and generally form amides at room temperature in a few minutes or (at worst) a few hours (see Section 5.06.2.2.3).

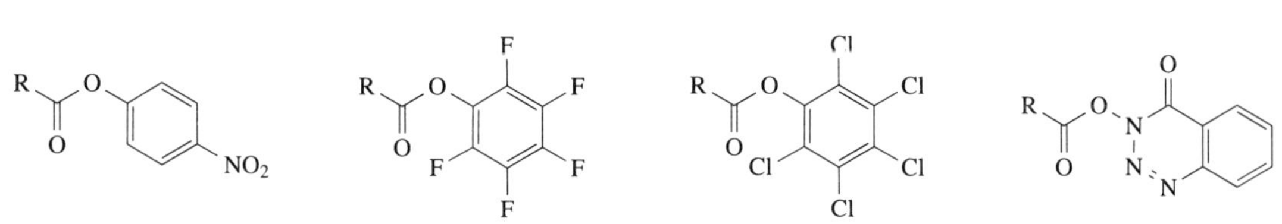

Figure 1 Commonly Used Active Esters

Active esters of *N*-hydroxysuccinimide and 1-hydroxybenzotriazole are also used widely, but they are normally prepared *in situ*. Active ester couplings are usually clean and convenient; they are a good, reliable way of preparing a wide range of amides. The PFP (pentafluorophenyl) esters are currently proving the most popular and in 1994 PFP thioesters were shown to activate very hindered acids ⟨94TL4865⟩.

5.06.1.2.4 From acid anhydrides

Being of similar reactivity to the active esters, acid anhydrides have proved popular for amide bond formation (see Table 6, Section 5.06.2.2.4). For cheap carboxylic acids, the symmetrical anhydride is often employed (e.g., acetic anhydride), even though half of the acylating agent is lost (Equation (9)). The efficiency of amide bond formation using symmetrical anhydrides led to their widespread use in peptide chemistry—but they have been largely superceded by active esters.

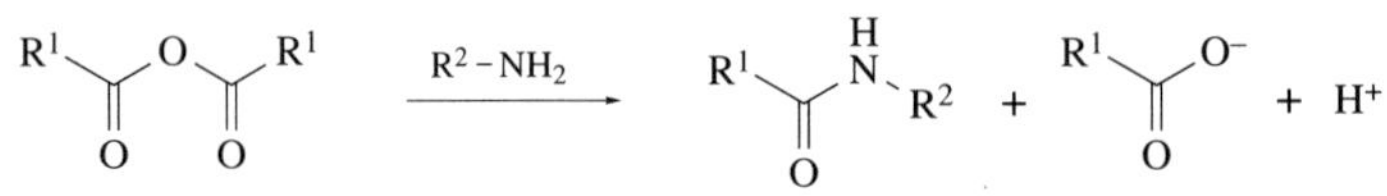

(9)

For expensive carboxylic acids, mixed anhydrides are more efficient. The two carbonyls must be strongly differentiated, so that only the desired amide is produced. This can be achieved using steric or electronic factors (Scheme 5), or by employing mixed phosphinic esters in which the P=O bond is very unreactive ⟨80AG(E)133, 85JCS(P1)461⟩. The mixed anhydride derived from ethyl chloroformate is particularly easy to prepare, and usually gives amides with high efficiency. With relatively stable anhydrides (e.g., from aryl acids), Schotten–Baumann conditions (shaking with aqueous sodium hydroxide), or the addition of catalyst such as pyridine ⟨63OSC(4)5⟩ or $CoCl_2$ ⟨87CC114⟩, can be used to accelerate the reaction.

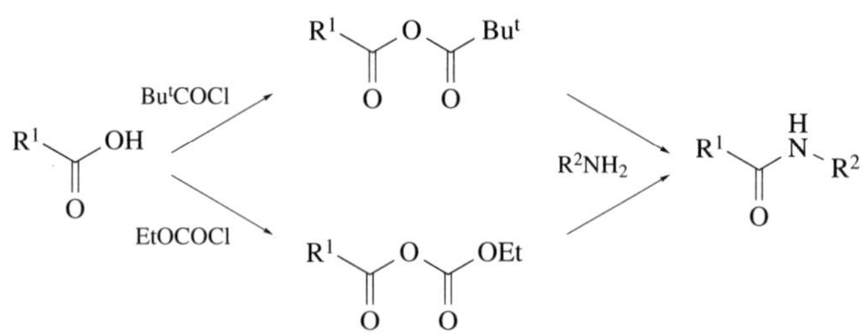

Scheme 5

5.06.1.2.5 From acyl halides

By far the most widely used acyl halides are the chlorides (see Table 7, Section 5.06.2.2.5). Although their high reactivity leads to a lack of selectivity (e.g., alcohols are often acylated too) and potential side reactions (e.g., racemisation of peptides), for simple amides with few other functional groups, the method is very reliable. The acid chlorides are readily prepared (again, dependent on other functional groups) using various reagents such as $SOCl_2$, PCl_3 and $POCl_3$. Less reactive acid chlorides (e.g., from aromatic acids) sometimes require forcing Schotten–Baumann conditions.

Acid bromides are occasionally used when they are particularly readily available. For example, α-bromo acid bromides are prepared from carboxylic acids by the Hell–Vollhardt–Zelinski reaction ⟨62CRV99⟩, and are highly efficient acylating agents. The α-bromine is then available for further transformations (Scheme 6).

Acyl fluorides are of limited practical use, despite reports in the early 1990s of their deployment in peptide synthesis ⟨92JOC6371⟩. A good example of their use, however, employs a trimethylsilylamide as the nucleophile (Equation (10)) ⟨87TL5099⟩. As the silyl group has a strong affinity for

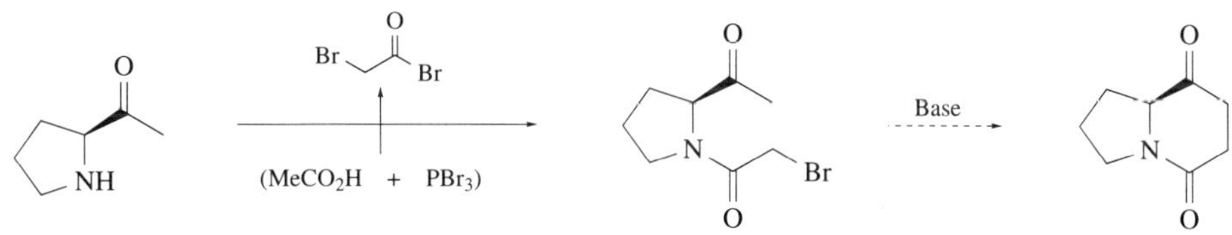

Pyrrolizidine

Scheme 6

fluorine, the reaction is autocatalytic, but the problems with preparing acyl fluorides are likely to limit the popularity of this approach.

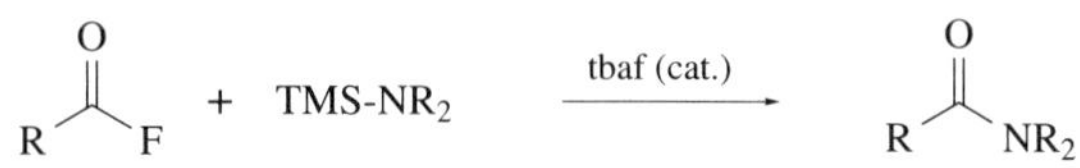

(10)

(tbaf = tetra-n-butylammonium fluoride)

5.06.1.2.6 *From acyl azides*

Azide is a moderately good leaving group and acyl azides have a special place in amide bond formation. It is probably the least active acylating agent that works adequately at room temperature—reaction times are typically several days at 0 °C (cf. methyl esters are usually too unreactive unless heated to ca. 100 °C). This low reactivity also minimises side-reactions that involve acylation, notably the racemisation of peptides possessing a chiral α-carbon. For such couplings, the acyl azide procedure is often the method of choice, particularly as the azides are easy to prepare, see Scheme 7, or by use of the reagent diphenylphosphoryl azide (DPPA) ⟨74T2151⟩.

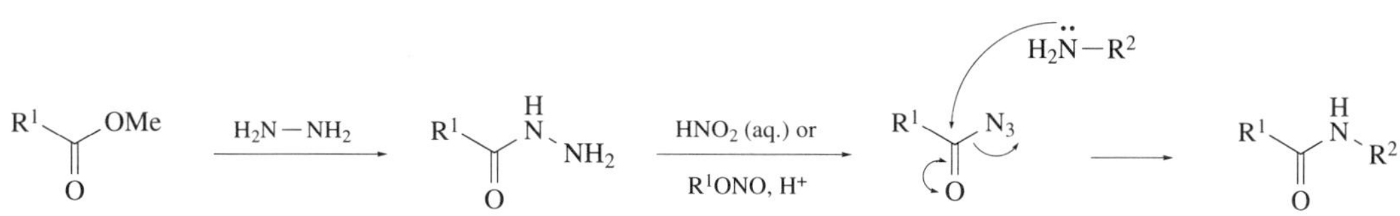

Scheme 7

In general, however, acyl azides are not used. This is because of the sluggishness of the reaction, and to the possibility of by-products resulting from the Curtius rearrangement. Unfortunately, accelerating acylation by heating (unwise for explosive reasons) also leads to an increase in rearranged by-products.

5.06.1.2.7 *Using* in situ *coupling reagents*

This is probably the most attractive method of amide formation. If the carboxylic acid component is very valuable, it is particularly advantageous if a separate activation step can be avoided. One-pot coupling procedures are rarely used for forming primary amides of the type $RCONH_2$ or $ArCONH_2$, presumably because the sources of ammonia do not usually lead to high yields. More importantly, the removal of by-products generated by *in situ* coupling reagents can be difficult. Many of the more recent reagents have been specifically designed to address this problem. Three classes of *in situ* coupling reagent have dominated amide bond formation in the 1980s and 1990s: diimides, phosphorus(V) reagents and uronium salts.

(i) Diimides

The mechanism by which diimides (shown in Table 8, Section 5.06.2.2.7(i)) induce coupling is shown in Scheme 8. Apparently, the amine can be present throughout the time when the carboxylic acid becomes activated, without itself attacking the diimide. For a primary amine at least, there is a possibility that the amine could become part of a mixed diimide (Scheme 9). However, such transposition of the amines seems to be rare, probably because protonation of the amine is substantially favoured over protonation of the diimide nitrogen, thereby slowing down the exchange reaction. This may be further affected by the addition of co-reagents such as 1-hydroxysuccinimide or 1-hydroxybenzotriazole (hobt), which keep the reaction slightly acidic. Although the additives maintain the required acidity, they are also believed to help moderate the reactivity of the *O*-acylurea intermediate that carries out the acylation reaction (Scheme 10). The *O*-acylurea is formed very rapidly (typically a few minutes), but its high reactivity can lead to side-reaction. However, *N*-hydroxy compounds are able to trap out the acylating agent as an active ester. When this procedure is used, the acid/imide/*N*-hydroxy additives are usually preactivated before addition of the amine. Oddly enough, although the active ester should form the amide bond less quickly, amides sometimes fail to form in the absence of the additive; it would thus appear that its influence on the acidity of the reaction mixture is a critical factor.

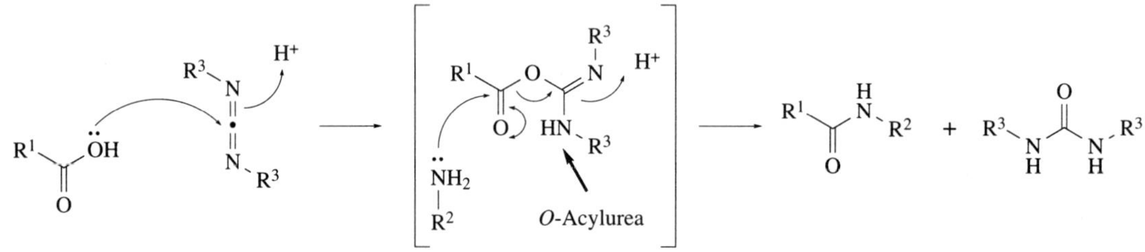

Scheme 8

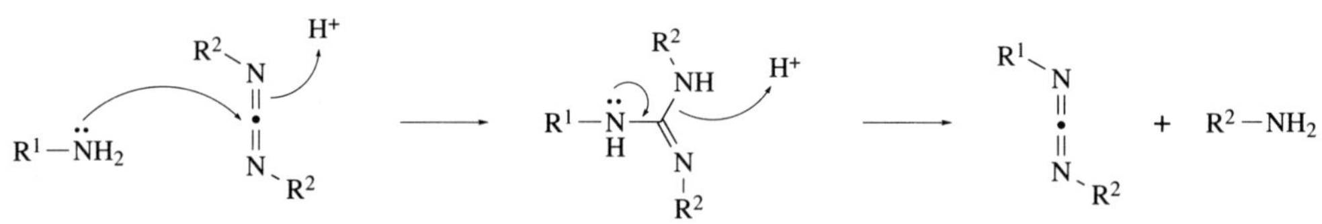

Scheme 9

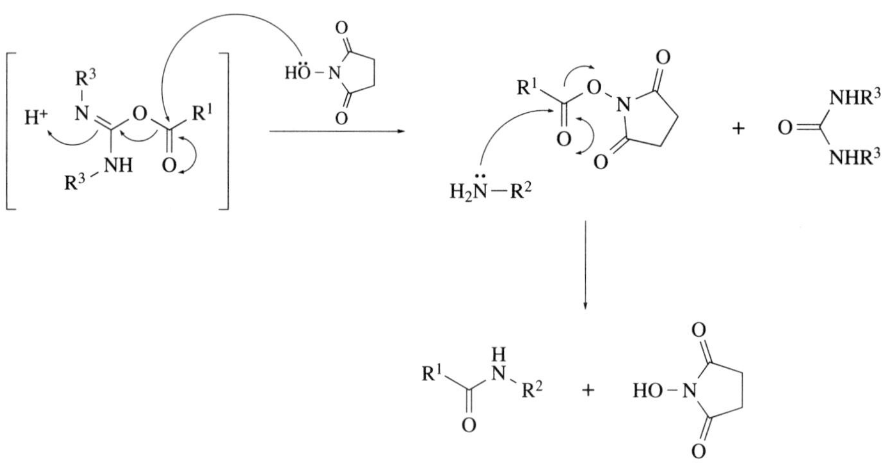

Scheme 10

Although the *N*-hydroxy additive is a catalyst, one mole equivalent is usually added (to complete the formation of the active ester during preactivation) and further additions are often made until the apparent pH is about 4–5. It is perhaps surprising that the *N*-hydroxy esters are such good acylating agents, but the —I effect of the nitrogen (which is always conjugated to further −M groups) clearly confers an almost ideal level of reactivity on the carbonyl group. With by-products resulting from both the diimide reagent (i.e., the urea) and the additive, purification of the amide product can be a problem. The leaving group ability (i.e., stability) of the $R_2N—O^-$ group means that the *N*-hydroxy additive can be easily removed by a base wash. However, the urea by-products can be more problematic, and have led to the development of water-soluble diimides, for which the urea by-product can be removed by an aqueous wash. These water-soluble diimides have the added feature that they work well in aqueous solution—useful when the acid/amine is only conveniently soluble in water.

(ii) Phosphorus(V) reagents

These one-pot reagents (shown in Table 9, Section 5.06.2.2.7(ii)) have started to figure prominently in the synthesis of cyclic peptides, but are less widely used for repetitive peptide synthesis. They are undoubtedly worth considering for any amide coupling that is proving problematical, particularly if it is a cyclisation reaction. Currently the most commonly used reagents are BOP ⟨75TL1219⟩ and BOP—Cl ⟨80S547⟩. The success of these phosphorus(V) reagents is a little odd, as they invariably generate acylating agents that can be formed by other methods, yet they often give excellent yields. Their basic mechanism is shown in Scheme 11, although transfer of X to form RCOX may be inter- or intramolecular.

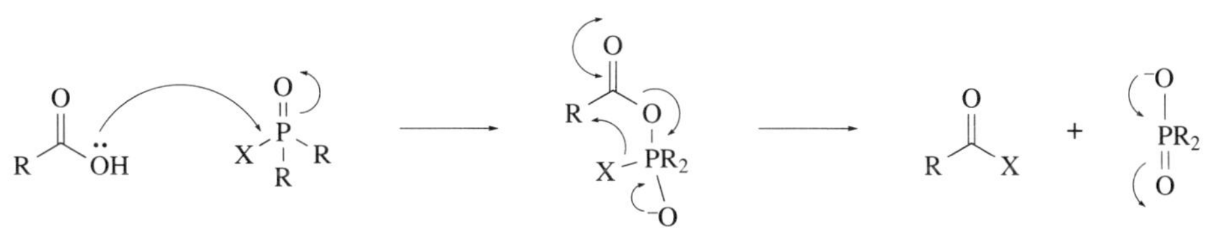

Scheme 11

(iii) Uronium salts

Uronium salts are gaining rapidly in popularity, and have the potential to lead the way in one-pot coupling reagents. (Table 10, Section 5.06.2.2.7(iii), Equation (11)). They effectively generate an active ester *in situ*, and one of their advantages over diimides is that no additive is required. Moreover, the by-products of the reaction (tetramethylurea and *N*-hydroxy compounds) are easily removed. They all operate by the mechanism shown in Scheme 12.

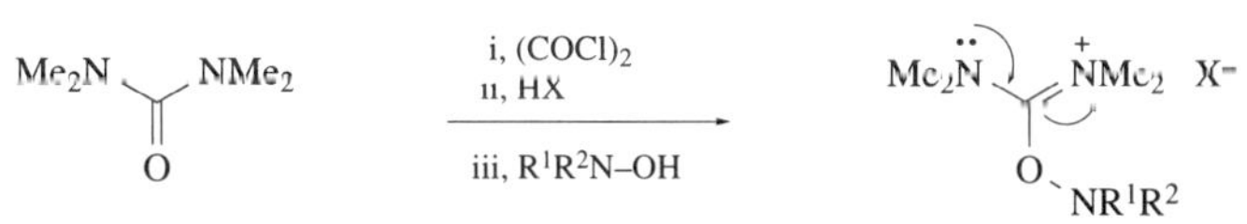

(11)

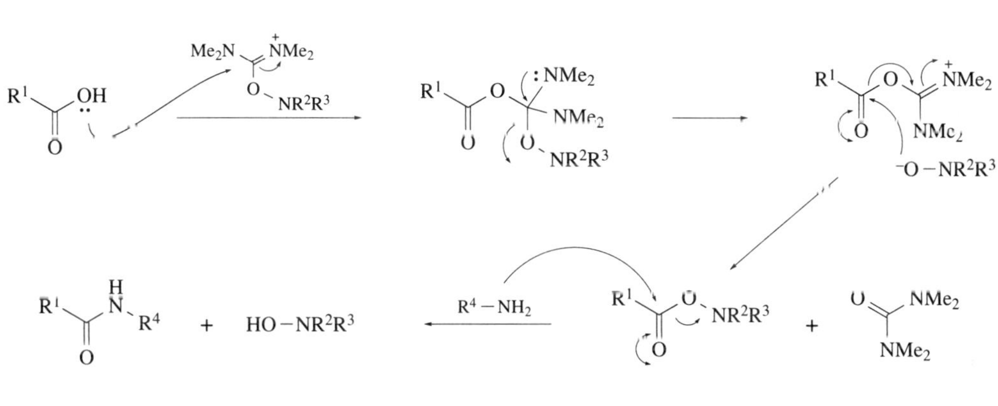

Scheme 12

(iv) Other in situ *coupling procedures*

Without exception, *in situ* coupling agents activate the carboxylic acid component without being irreversibly trapped by the nucleophilic amine. The three types of reagent discussed above are the most widely used—the obvious 'first choices' for one-pot coupling of a carboxylic acid and an amine. Other reagents such as those given in Tables 11 and 12 (Section 5.06.2.2.7(iv)) have been used on several occasions.

$PR^3_3/R^1X^1—X^2R^2$. A range of trialkylphosphines mixed with a compound containing a weak heteroatom–heteroatom bond (or equivalent, such as RS—CN) can be used. The mechanism is essentially an *in situ* generation of a phosphonium reagent similar to BOP (see 5.06.1.2(ii)) (Scheme 13). The mixtures that can allow this mechanism to operate are quite diverse (see Table 11), although it is perhaps an over-simplification to assert that this exact mechanism occurs in all cases.

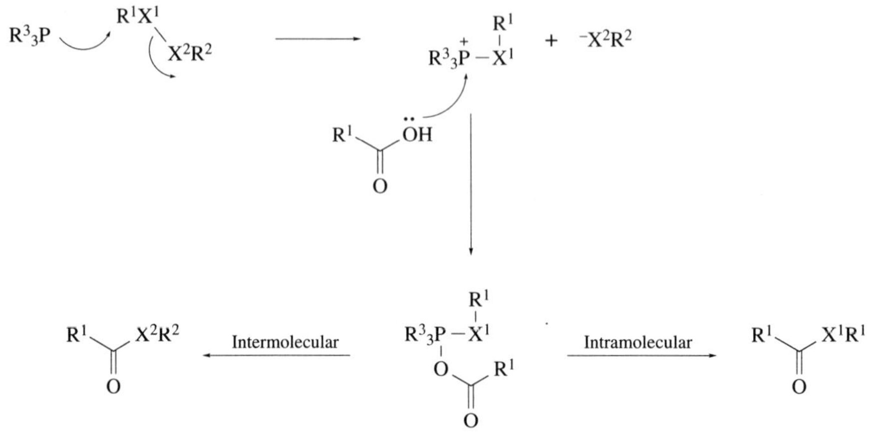

Scheme 13

RSO_2Cl. A range of sulfonyl chloride derivatives have also been used to effect amide bond formation, presumably via the sulfuric mixed anhydride in most cases (Table 23, Section 5.06.4.4.2). Tosylation of amines is a slow process, but must account for some loss of yield.

2-Halopyridinium salts. Several of these reagents have been developed (Table 19, Section 5.06.2.4.2), for which the general formula and presumed mode of action is indicated in Scheme 14.

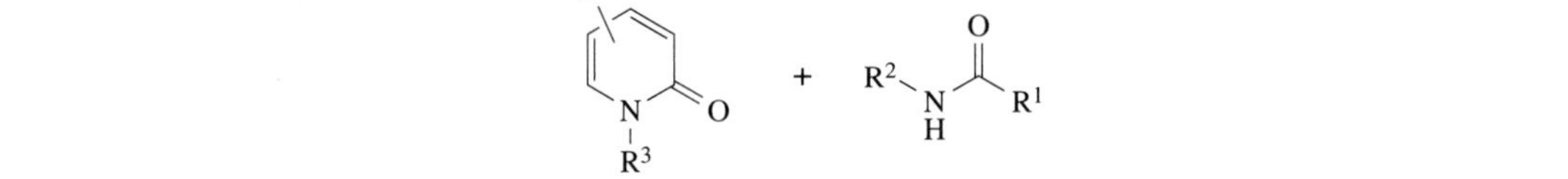

R^3 = Me, Et, Ph; R^4 = range of substituents; X = F, Cl, Br, I; Y = I, OTs, BF_4

Scheme 14

Boron reagents. Although not used widely, a diverse range of boron reagents have been developed (see Table 12). They presumably coordinate to the oxygen of the carboxylic acid, facilitating its departure.

5.06.1.3 Hydrolysis of Nitriles

Nitriles are at the same oxidation level as amides, and are often very accessible synthetic precursors to amides. The problem with hydrolysis of nitriles (given in Tables 4 and 20) is preventing it from proceeding all the way to the free carboxylic acid. That this reaction is hard to control is evident from the enormous number of reagents and variations that have been published, all claiming to have solved the problem of selective hydrolysis. Therefore the reader should be warned that it can be difficult to obtain high yields from this reaction. There are four types of procedure that have received almost all the attention: acid conditions, alkaline conditions, peroxide hydrolysis and enzyme hydrolysis. In general, *N*-unsubstituted amides are produced but *N*-alkylamides can be formed under certain conditions (see Table 13).

5.06.1.3.1 *Acidic hydrolysis*

The exact conditions for these reactions are essential, as the amidate ester intermediate can hydrolyse to amide or carboxylic acid (Scheme 15). Treatment with acid normally precedes hydrolysis with water, but carefully controlled water content can lead directly to the amide. Widely used acids include H_2SO_4, HCl, polyphosphoric acid (ppa) and BF_3/AcOH. With sulfuric acid/alcohol mixtures, the *N*-alkylated amide is apparently accessible directly (Equation (12)) ⟨73OSC(5)73⟩.

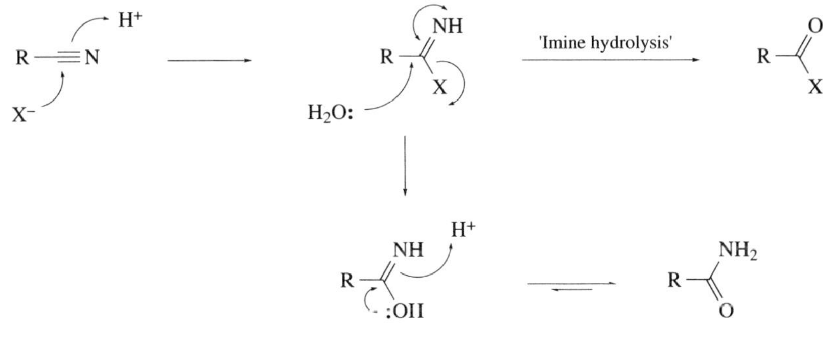

Scheme 15

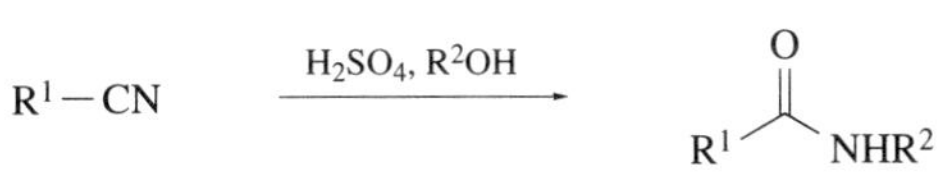

(12)

5.06.1.3.2 *Alkaline hydrolysis*

Nitriles are more electrophilic than amides, so in this manner selective hydrolysis should be possible. However, the use of a high concentration of KOH, in the presence of a high boiling alcohol [Bu^{t}OH or $(CH_2OH)_2$], seems almost essential. Since gentle heating is also required in most cases, the rest of the nitrile needs to be fairly robust. In the presence of an alkyl halide, the intermediate *N*-anion can be trapped to give *N*-alkylated amides (Scheme 16).

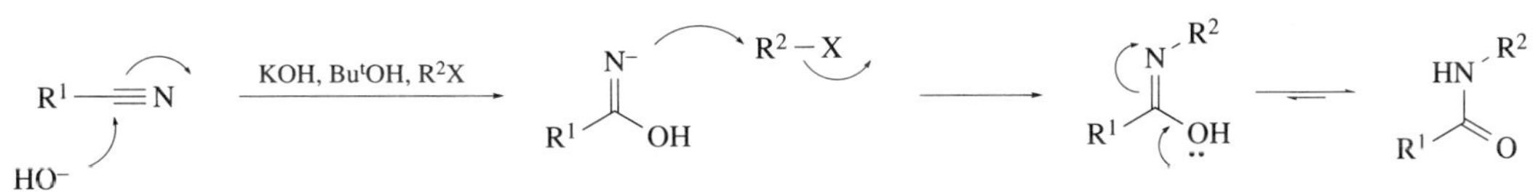

Scheme 16

5.06.1.3.3 *Peroxide hydrolysis*

Probably the most reliable of the hydrolytic procedures; typical conditions (and probable mechanism) for peroxide hydrolysis are given in Scheme 17.

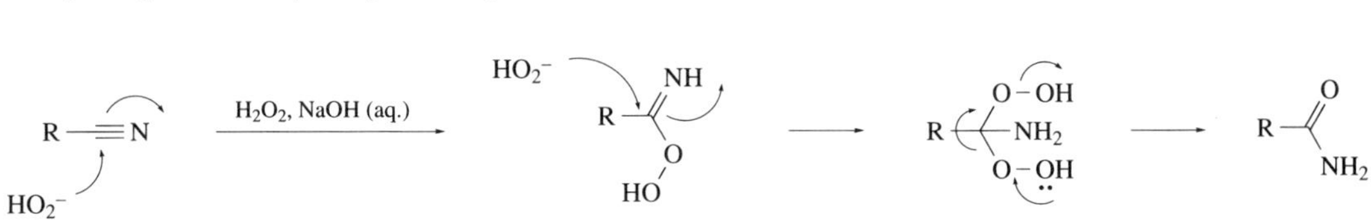

Scheme 17

5.06.1.3.4 *Enzymic hydrolysis*

Several nitrilases occur in nature, and these enzymes can give high yields of amides. Their selectivity means that the substitution range is fairly limited, but the enzymes are usually enantiospecific, so optically pure amides can be obtained from racemic nitriles. Whether the enzymic reaction is amenable to reasonable scale-up depends on the properties of the enzyme, particularly its tolerance to organic solvent. Thousands of tonnes per annum of acrylamide are produced from acrylonitrile in this way.

5.06.1.3.5 *Other methods*

A number of metal-based catalysts have been reported as effective in nitrile hydrolysis (see Table 13). One procedure that seems to be unique in its mechanism is $NOPF_6$ in the presence of a range of alkyl halides (1°, 2° or 3°), which yields the *N*-substituted amides R^1CONHR^2 ⟨79S274⟩. Provided the nitrile is otherwise inert to $[NO]^+$, this method would appear to be very valuable.

5.06.1.4 Rearrangement Reactions

There are several types of rearrangement reactions that provide important routes to amides. They are quite diverse in nature, and while some have been used extensively (e.g., Beckmann and Schmidt rearrangements), the scope and limitations of others has yet to be thoroughly explored.

5.06.1.4.1 *Beckmann rearrangement*

This is one of the most widely used rearrangement reactions ⟨88OR(35)1⟩. The key step is treatment of an oxime with a reagent that assists loss of the 'OH' group (usually heating with acid, or with TsCl/base). This triggers the migration of an alkyl or aryl group on to the nitrogen, and the amide is generated on workup.

As implied in Scheme 18, it is the substituent *anti* to the departing OH group that migrates on to the nitrogen, so the stereochemistry of the oxime controls the regiochemistry of the product. Importantly, the resonance stabilisation of the carbenium ion is not a prerequisite for success; oximes of small ring ketones usually undergo the Beckmann reaction cleanly, even though the linear $R^2—C{\equiv}N^+—R^1$ canonical is precluded. The importance of the Beckmann rearrangement lies in the wide availability of ketones; the two-step process of oxime formation and rearrangement rapidly leads to an amide in which an NH group has been inserted adjacent to the carbonyl group.

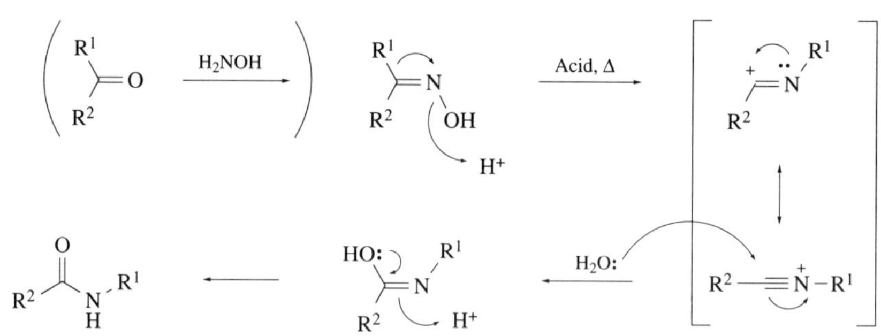

Scheme 18

5.06.1.4.2 *Schmidt rearrangement*

This reaction proceeds via a similar mechanism to the Beckmann rearrangement, but allows conversion of ketone to amide in a single step (see Scheme 19) ⟨71RCR835⟩. In general, the bulkier R group migrates, probably because it is more likely to be *anti* to the N_2 leaving group, and because the normal order of migration to the electron-deficient centres (Ar > 3°R > 2°R > 1°R) would favour this migration. The carbenium ion is an equilibrium mixture of stereoisomers. Nevertheless, mixtures of regioisomeric amides are usually formed (if $R^1 \neq R^2$), and these can be hard to separate—cf. separation of oxime stereoisomers, for the stereo/regio-specific Beckmann rearrangement. Despite the simplicity of the Schmidt procedure, hydrazoic acid is unpleasant to use, and azides can undergo unwanted (explosive) side-reactions.

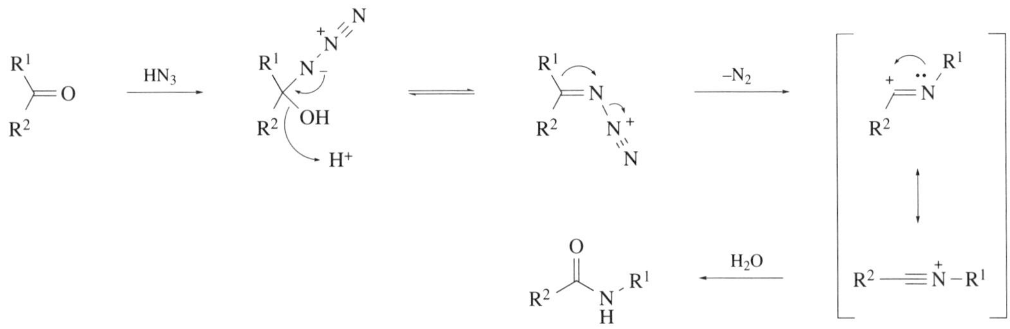

Scheme 19

5.06.1.4.3 *Claisen rearrangement of imidate esters*

The amides of allyl amines can be prepared by a very neat variation of the Claisen rearrangement, involving imidate esters. In practice, the procedure is normally used for the synthesis of allylamines via the allylamide, as only a limited range of acyl groups are really viable. The overall sequence of reactions is shown in Scheme 20; for specific examples see Schemes 36 ⟨76JA2352⟩ and 37 ⟨62JA4987, 63JA3288, 64OS(44)41⟩.

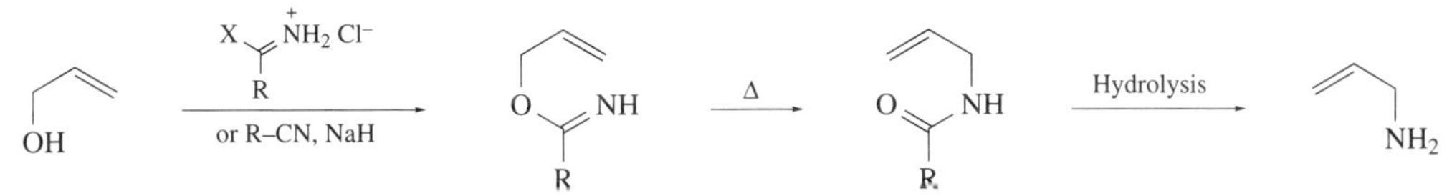

Scheme 20

The limited range of imidate esters that are readily accessible restricts the breadth of amides that are available directly. The rearrangement is accelerated by electron-withdrawing groups, with the most widely used R groups being Me < CCl_3 < CF_3. The attraction of this method is the high stereocontrol possible, because of the well-defined chair-like transition state for the Claisen rearrangement.

5.06.1.4.4 *Wolff rearrangement*

This silver(I) catalysed rearrangement of diazoketones generates amides, if carried out in the presence of an amine (Scheme 21) ⟨59CB528, 67RCR260⟩. The reaction is formally the acylation of the amine, but the carbenoid pathway can lead to possible side-reactions.

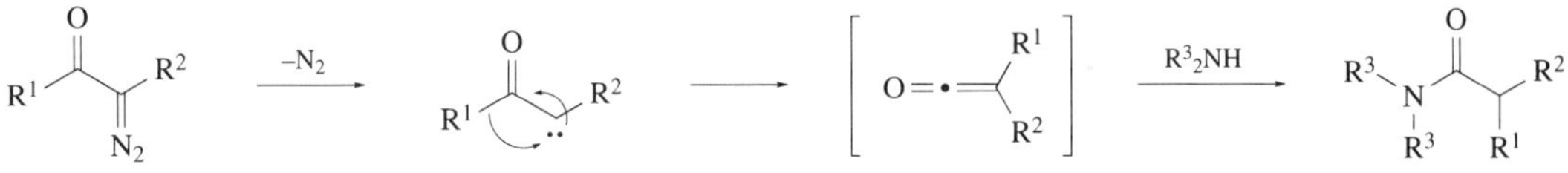

Scheme 21

5.06.1.4.5 Willgerodt rearrangement

This unusual oxidative rearrangement involves the reaction of a ketone with sulfur/ammonia. Here, an apparent migration of the carbonyl group occurs, and the ketone is oxidised to a primary amide (see Equation (13) below and Equations (19) and (20) ⟨45JA1783, 46JA525⟩ in Section 5.06.2.1.3.

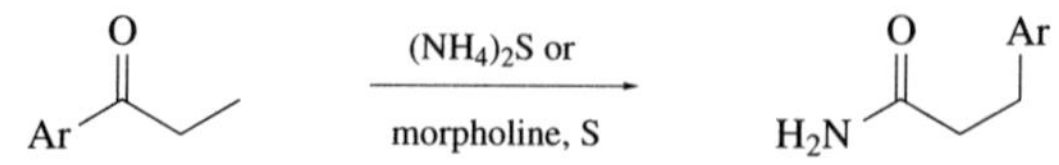

(13)

5.06.1.4.6 Rearrangement of oxaziridines

Oxaziridines rearrange to amides on heating or irradiation (Scheme 22 and Equation (30) ⟨57JA5739⟩). However, the reaction is rarely the method of choice, as side-reactions occur in the formation of the oxaziridines, and in the rearrangement process.

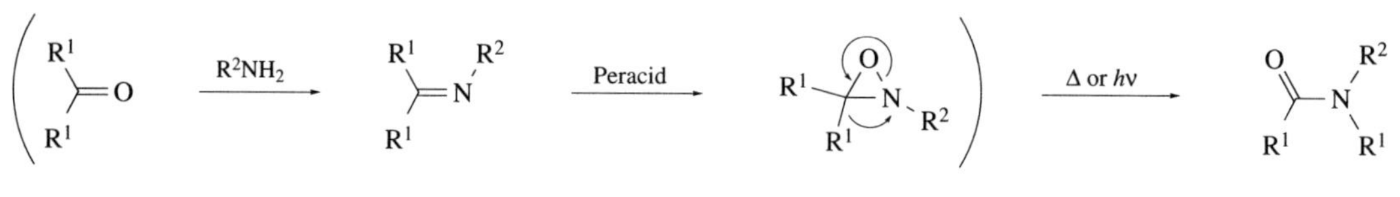

Scheme 22

5.06.1.4.7 Rearrangement of keto-amines

Some keto-amines can undergo a simple base-catalysed rearrangement process (e.g., Equation (14)). Prerequisites are that the amine can attack the carbonyl to give a ring of more than five atoms, and that the ketone has a *β*-electron-withdrawing group. Although the reaction may be quite general, the precursors are often hard to make; however, for ring expansion to give medium- to large-ring lactams, the method can be valuable (Equation (17); see also Equations (32) ⟨77JA182⟩ and (33) ⟨82HCA2299, 85HCA484⟩ and Scheme 30 ⟨92TL1455⟩).

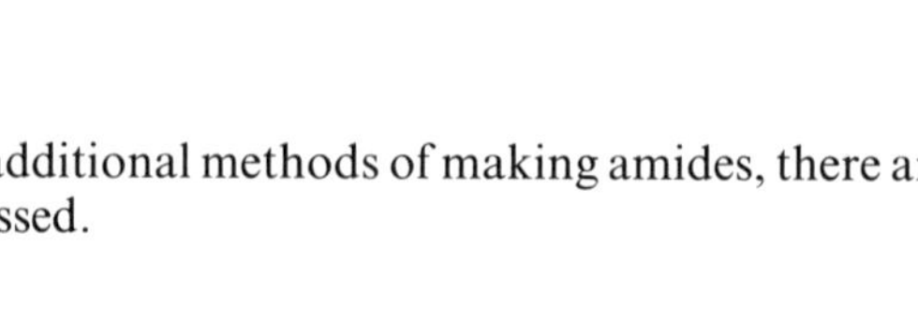

(14)

5.06.1.5 Other Methods

Although there are many additional methods of making amides, there are three general procedures that have not yet been discussed.

5.06.1.5.1 From alkyl halides

There are two tactics which have been employed:

(i) Convert the R^1—X into an organometallic reagent, and react this with a synthetic equivalent of $^+CONR_2$ (for example, isocyanates (see Equation (15) ⟨38JA540, 75TL981, 87TL3815⟩). The use of ultrasound ⟨86TL501⟩, or of titanium(IV) catalysts ⟨87TL3815⟩, have been found to aid the magnesium-mediated reactions.

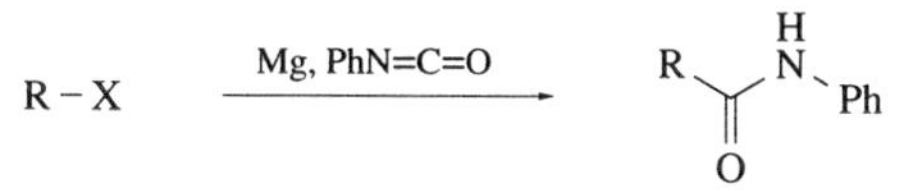

(15)

(ii) Use a synthetic equivalent of $^{-}CONR_2$ which can displace halide from R^1—X (for example, Equation (16) ⟨69JA1233, 71JOC2721, 74JOC3327⟩).

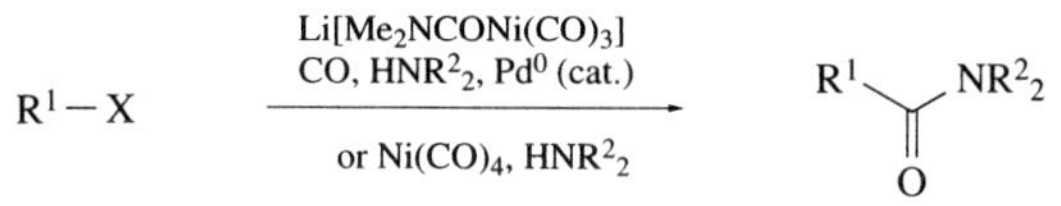

(16)

Although a wide variety of reagents have been employed (e.g., involving CO insertion from organometallic complexes), these two tactics are effectively used in almost all cases.

5.06.1.5.2 From amides to amides

The *N*-alkylation of amide anions is a useful way of gaining access to fully alkylated amides (see Tables 14 and 15, Section 5.06.2.2.10). However, despite the apparent simplicity of the transformation, the reaction can be fraught with problems. In particular, the amide anion can act as a base rather than as a nucleophile; alkylating agents that lack a β-proton (e.g., MeI, $PhCH_2Br$) are therefore more appropriate since elimination to the alkene cannot occur. Examples of bases are given in Equation (17) ⟨52JA1010, 73JOC1677, 79S527, 79T2169⟩.

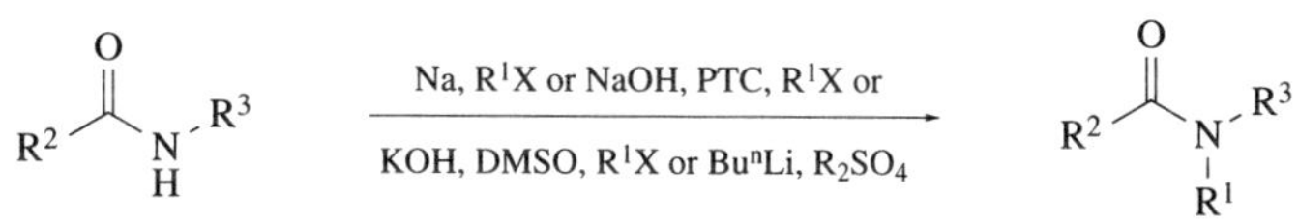

(17)

There are numerous examples of this process in intramolecular reactions of halo-amides, with 5- and 6-membered rings being formed particularly easily. In such cases, it may not be necessary to deprotonate the amide if the leaving group is sufficiently stable. Thus, Mitsunobu conditions give access to β-lactams, as shown in Equation (18) ⟨80JA7026, 81JOC1229⟩. Reductive alkylation is also sometimes possible, when amides are treated with reactive aldehydes, ketones or ketals; the success of the reaction depends upon the continuous reduction of the iminium intermediate (Scheme 23) ⟨62JOC2205, 76JOC725⟩.

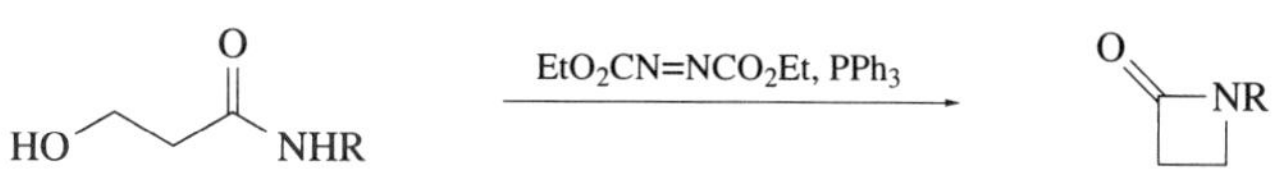

(18)

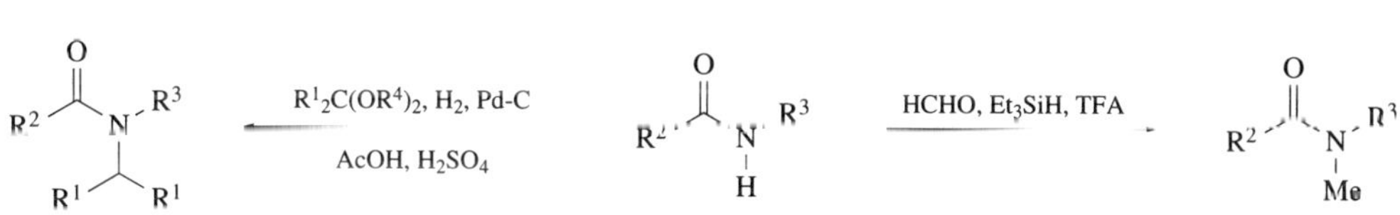

Scheme 23

5.06.1.5.3 β-Lactams

Most of the reasonable disconnections of the β-lactams have been explored for their synthesis, and these methods will generally be discussed under the appropriate heading for the type of amide being formed. It is perhaps worth noting, however, that of the many 'standard' procedures, a small number have dominated the literature (see Scheme 24, and Table 17 for reviews on the synthesis of β-lactams).

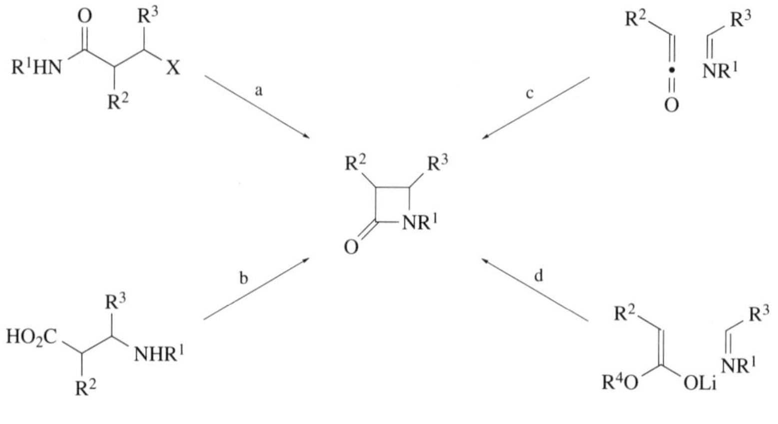

Scheme 24

Of the methods presented in Scheme 24, routes c and d have not yet been discussed, and they are, moreover, approaches only possible for the 4-membered β-lactams. Thus, ketenes are able to undergo a concerted [2 + 2]-cycloaddition, and ester enolates can condense, with imines to generate the β-lactam ring directly. These are both simple, efficient and versatile approaches that are very widely used. The range of ketenes, esters and imines that are available have ensured the success of these approaches (e.g., see Schemes 25 and 26, Equation (42) ⟨73JOC3437⟩, Schemes 32 ⟨80S933⟩, 33 ⟨87JA1798⟩, 34 ⟨92S439⟩ and 41 ⟨80JOC3413⟩ and Equation (70) ⟨86JA6054⟩).

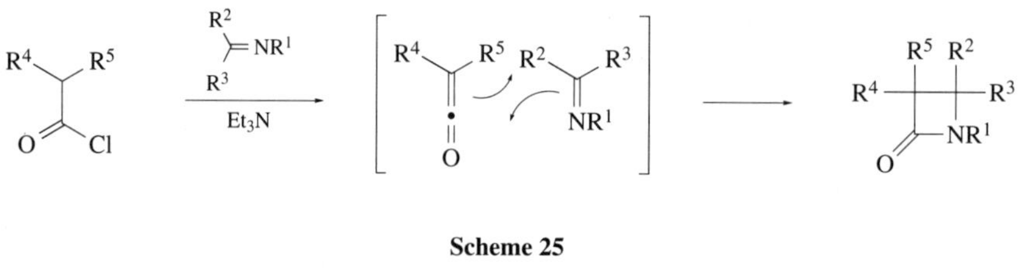

Scheme 25

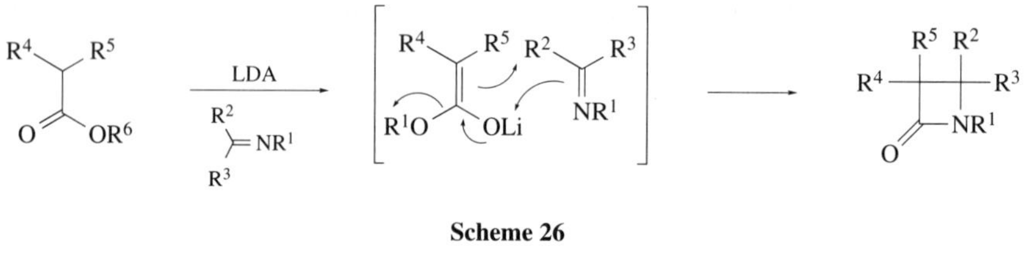

Scheme 26

In the following section, the different structural types of amides are discussed systematically. In all sections, a selection of the successful procedures are given, so that an appropriate method with some literature precedent can be chosen.

5.06.2 AMIDES OF ALKANOIC ACIDS

5.06.2.1 *N*-Unsubstituted Alkanoamides

5.06.2.1.1 From acids and acid derivatives

There are a number of examples in which primary amides have been prepared by the direct thermal reaction of a carboxylic acid with a source of ammonia. Examples are given in Table 2.

These methods have now been largely superceded by more efficient acylation techniques. A selection of conditions starting from simple esters, anhydrides and acyl halides are presented in

Table 2 Synthesis of primary amides directly from acids.

Acid	*Ammonia source*	*Ref.*
RCO_2H	NH_3, 190°C	31JA1879
RCO_2H	NH_2CONH_2, $HCONH_2$, $MeCONHCONH_2$	49JA2215
$C_6H_{13}CO_2H$	NH_2CONH_2	63OSC(4)513

Table 3. One procedure offers a one-pot conversion of amino acids into the corresponding amides ⟨83CB2037⟩, but it should be noted that the use of expensive active esters and specialist coupling procedures is generally unnecessary for the preparation of simple primary amides. Ammonia is normally the source of nitrogen, although in the final procedure a powerfully nucleophilic amide anion is employed to overcome steric crowding at the acyl centre.

Table 3 Synthesis of primary amides from carboxylic acid derivatives.

Acid derivative	*Reagents*	*Ref.*
RCO_2R^1	NH_3	41OSC(1)153, 55OSC(3)516, 87JOC4379
$(EtO)_2CHCH_2CO_2Et$	NH_3	73JMC1
RCO_2H	Ac_2O, cat. $CoCl_2$; then NH_3	87CC114
RCO_2H	$ClCO_2Et$; then NH_3	68JMC534
RCO_2H	CH_3SO_2Cl; then NH_3	82OPP396
RCO_2H	$P(NMe_2)_3$, $COCl_2$, NH_3	83CB2037
$RCOCl$	NH_3	49JA2215, 55OSC(3)490
$(Bu^t)_2CHCOCl$	NH_3, $NaNH_2$	60JA2498

5.06.2.1.2 *From the hydrolysis of nitriles*

As indicated in Section 5.06.1.3 there are a host of procedures for the hydrolysis of nitriles to amides. A selection of acidic, basic and neutral procedures is given in Table 4.

Table 4 Hydrolysis of nitriles to primary amides.

Reagents (acidic)	*Ref.*	*Reagents (basic/neutral)*	*Ref.*
H_2SO_4, Δ	57JA2530	H_2O_2, base	1885CB355
polyphosphoric acid, Δ	54JA3039		43OSC(2)44
	57JA725		55JA2519
H_2O, HCl	55OSC(3)66		71JOC3048
	55OSC(3)88	H_2O_2, NH_4OH	86JOC1065
	63OSC(4)58	H_2O_2, NaOH, PTC	80S243
	63OSC(4)496	KOH, Bu^tOH	76JOC3769
	63OSC(4)760	H_2O, catalyst Cu	82JOC4812
BF_3, HOAc	55JOC1448	$PdCl_2$, H_2O, Δ	74S574
	57JA725	MnO_2, DCM	66CC121
		KF, alumina	82SC177
		NaO_2, Me_2SO	79JOC4727

5.06.2.1.3 *From rearrangement reactions*

There are several examples of the preparation of *N*-unsubstituted amides by the Beckmann rearrangement of an aldoxime ⟨60OR(11)1⟩. Unfortunately, under normal Beckmann conditions these reactions are inefficient, producing nitriles and acids alongside the amide. Some improvements to earlier procedures using nickel catalysis ⟨46JCS599, 61JA1983⟩ or silica gel catalysis ⟨74T2899⟩ have been reported (Scheme 27) ⟨61JA1983, 74T2899⟩.

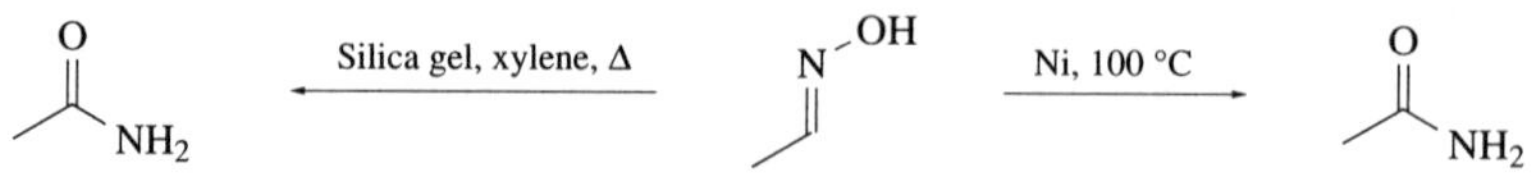

Scheme 27

A variation on the Arndt–Eistert homologation can also be used to prepare primary amides by the treatment of diazoketones with ammonia ⟨42OR(1)38⟩. One example is the synthesis of 4-methoxyphenylacetamide from 4-methoxybenzoyl chloride ⟨40JOC606⟩.

One useful rearrangement for the preparation of primary amides is the Willgerodt ⟨09JPR192⟩ reaction, which converts ketones into unsubstituted amides (Equations (19) ⟨45JA1783⟩ and (20) ⟨46JA525⟩. The scope and limitations of this intriguing reaction have been reviewed ⟨75S358⟩.

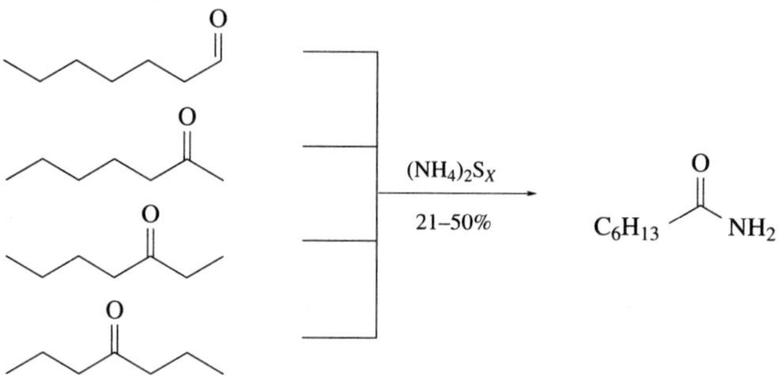

(19)

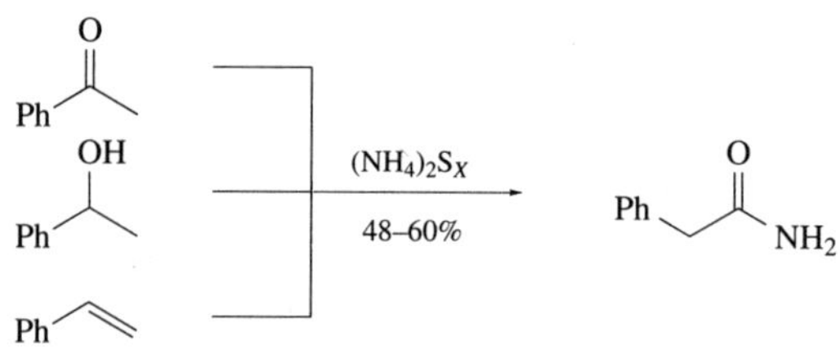

(20)

5.06.2.1.4 *From other methods*

One procedure to *N*-unsubstituted alkanoamides is a photochemical radical reaction which results in the addition of formamide across an unsaturated bond (Equation (21)) ⟨61AG621, 64JOC1855, 66JOC3829⟩. Unfortunately, the regioselectivity is generally poor ⟨65JOC3361⟩. Finally, the oxidation of aldehydes to the corresponding primary amide, using NBS, 2,2′-azobisisobutyronitrile (AIBN) and ammonia, has been reported ⟨90TL7237⟩.

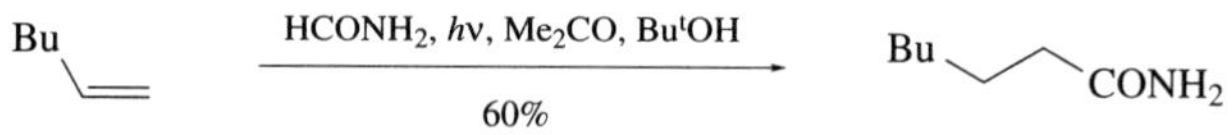

(21)

5.06.2.2 *N*-Alkylalkanoamides

5.06.2.2.1 *From acids*

There are a few reports of the synthesis of *N*-alkylalkanoamides directly from an acid and an amine under forcing conditions ⟨31JA1879, 37JA401, 49JA2215, 69JCS(C)874⟩. However, the milder conditions offered in Equations (22) and (23) are probably more synthetically useful ⟨61HCA1546, 80TL2443⟩. The first utilises an ion exchange resin as a catalyst ⟨61HCA1546⟩ and the second is a lactamisation of 6-amino alkanoic acid, catalysed by either alumina or silica gel ⟨80TL2443⟩.

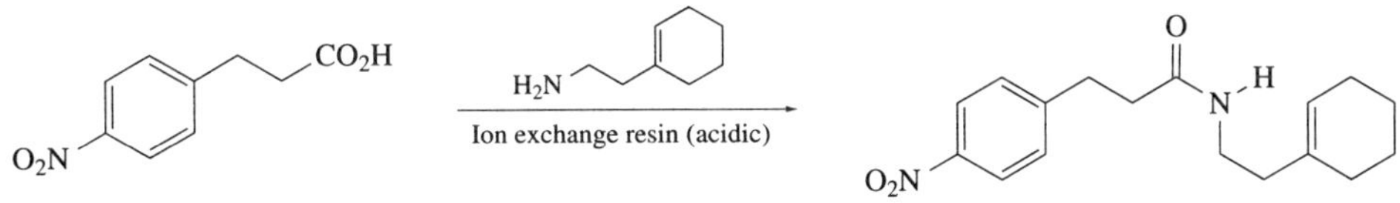

(22)

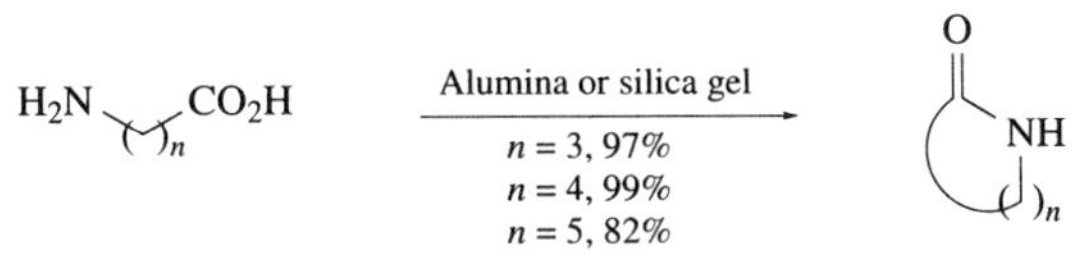

(23)

The preparation of other simple lactams has also been carried out by heating the appropriate amino acid with Bu_2SnO ⟨80JA7578⟩. This methodology has been applied to the synthesis of a complex macrolide ⟨83JA7130⟩.

5.06.2.2.2 *From esters*

A large number of methods are available for the preparation of amides from simple esters. The examples in Table 5 progress from simple heating to the use of more nucleophilic amide salts. Of the latter examples, the aluminum amides would appear to be the most useful and have been employed in total syntheses ⟨85JA1695, 85JA5732, 87JOC4665⟩. In contrast, the tin, titanium and magnesium reagents are generally restricted to the preparation of dimethylamides.

Table 5 Synthesis of alkylalkanoamides from esters

Reagents	*Ref.*	*Reagents*	*Ref.*
HNR_2, Δ	49JA2215	R_2AlNR_2	77TL4171, 80OS(59)49
HNR_2, $NaOCH_3$	63JOC2915	$XMgNR_2$	54JCS1188
HNR_2, cat NaCN	87JOC2033	R_3SnNR_2	69JCS(A)992, 69JCS(C)2565
$LiNR_2$	69TL1573, 70TL1791, 87JOC5745	$Ti(NR_2)_4$	69JCS(C)2565
		BBr_3, R_2NH	74TL3995
$NaNR_2$	71TL321		

5.06.2.2.3 *From active esters*

Section 5.06.1.2.3 discussed the use of active esters as a means to assist acylation. The development of active esters and diimide coupling methods is closely linked and both were driven by an urgent requirement for racemisation-free peptide bond formation. This simultaneous development has resulted in diimide chemistry being used to prepare many of the active esters. Early examples of active esters were derived from 4-nitrophenol, *N*-hydroxyphthalimide and *N*-hydroxysuccinimide ⟨64JA1839⟩. The phenolic ester type progressed through pentachlorophenols, to the now more familiar pentafluorophenols ⟨67JA183, 74TL1785⟩. *N*-Hydroxysuccinimide subsequently proved to be more useful as an additive in the 'one-pot' diimide coupling procedures (see Table 8).

Other active esters used in amide synthesis include thiazoline thione esters ⟨80TL841⟩ and thio esters. In one example (outlined in Equation (24)) an earlier macrolactonisation procedure ⟨75JA3515⟩ was adapted to prepare a range of *β*-lactams ⟨86CL915, 87CL293⟩. Related reports have used copper iodide ⟨87JOC4531⟩ or silver triflate ⟨87TL3019⟩ catalysis to activate the thio ester.

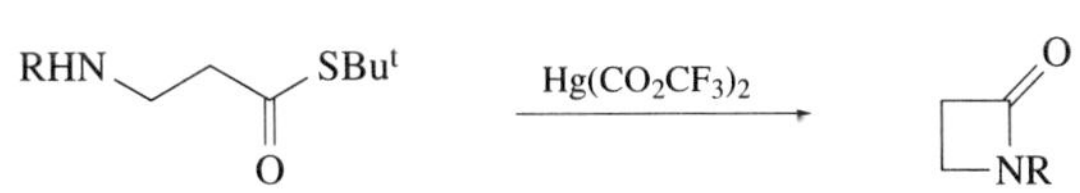

(24)

5.06.2.2.4 *From anhydrides*

Table 6 lists a range of *N*-acylation methods for alkyl amines. The early examples are for the preparation of simple acetamides using acetic anhydride or polymer-bound acetate. The later examples are more efficient *N*-acylation procedures using mixed anhydrides derived from the carboxylic acid RCO_2H and, for example, ethyl chloroformate ⟨71JCS(C)3540⟩. These methods are applicable to the preparation of many amides, but for peptide coupling reactions, they have been largely superceded by other techniques. However, the use of mixed anhydrides in peptide coupling has been reviewed ⟨62OR(12)157⟩.

Table 6 *N*-Acylation using anhydrides.

Reagents	*Ref.*	*Reagents*	*Ref.*
Ac_2O, Δ	43OSC(2)11	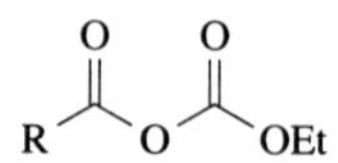 , Et_3N	71JCS(C)3540
Ac_2O, py	63OSC(4)5	RCO_2COBu^t, Et_3N	82JCS(P1)2909
polymer–OAc	82SC709	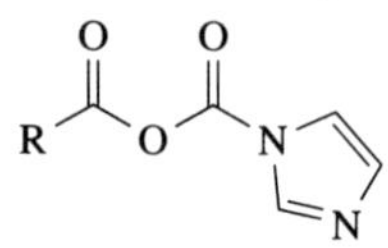	60T39
Ac_2O, $CoCl_2$ (cat.)	87CC114		62AG(E)351

5.06.2.2.5 *From acyl halides*

The reactions of aliphatic acyl halides including a large number of amide syntheses has been reviewed ⟨53CRV237⟩. Table 7 outlines a few selected procedures, for example, using excess amine to drive the reaction, or employing a modern acyl transfer catalyst.

Table 7 *N*-Acylation using acyl halides.

Reagents	*Ref.*
Δ, RCOCl	49JA2215, 63OSC(4)339, 86JA1039
NaOH, RCOCl	41OSC(1)99, 73OSC(5)387
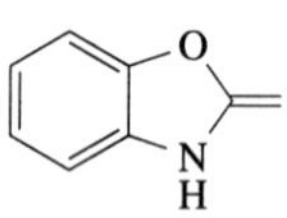 =S, Et_3N, RCOCl	81S991

5.06.2.2.6 *From azides*

Acyl azides have proved to be useful acylating agents, particularly in peptide coupling reactions ⟨74S549⟩. Azides are now most conveniently prepared using DPPA: the uses of this reagent have been described in full ⟨74T2151⟩. Equation (25) illustrates the synthetic potential of DPPA ⟨86CC344⟩.

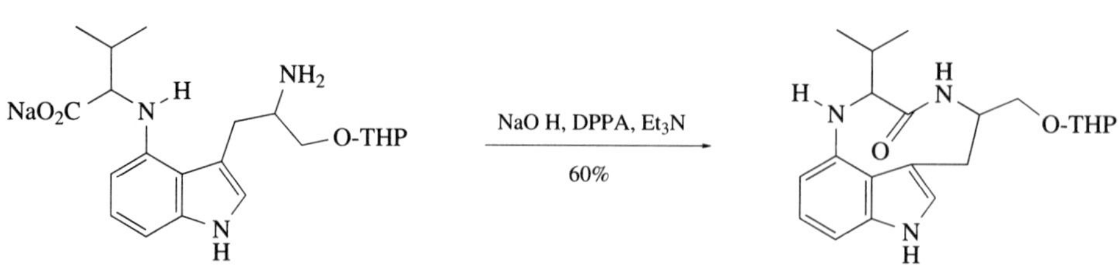

(25)

5.06.2.2.7 *Using* **in situ** *coupling reagents*

Finally, the largest section on acylating agents deals with *in situ* coupling reagents. These methods have been subdivided into four groups:

(i) carbodiimides,
(ii) phosphorus(V) reagents,
(iii) uronium salts,
(iv) other reagents.

The reagents and references for each group are tabulated separately.

(i) Carbodiimides

Dicyclohexylcarbodiimide was first reported as a peptide coupling agent in the mid-1950s and offered considerable improvement over the popular methods of the time. However, racemisation was still a problem until it was shown that the additive *N*-hydroxysuccinimide acted as a racemisation suppressant ⟨66CB110⟩. Later it was shown that hydroxybenzotriazole (hobt) was more efficient in this respect ⟨70CB788, 70CB2024, 71JA6318⟩ and dcc–hobt are now the reagents of choice in many peptide coupling reactions. A further improvement in racemisation suppression using $CuCl_2$ as a co-additive has been reported ⟨88CC419⟩, and in 1993 the additive 1-hydroxy-7-azabenzotriazole was reported as a superior alternative to hobt ⟨93JA4397⟩.

A study on a variety of other diimides was carried out by pioneering workers in this field ⟨56JOC439⟩ and resulted in the development of the 'water soluble' diimides ⟨61JOC2525⟩. These reagents offer the advantage of producing easily separated, water-soluble ureas as products (Table 8).

Table 8 Diimide coupling reagents.

Reagent	*Ref.*
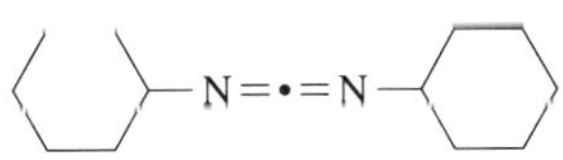	55CI(L)1087, 55JA1067 56JA1367, 59CB2813
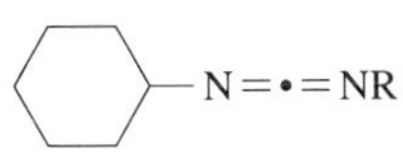	56JOC439
dcc[a], *N*-hydroxysuccinimide	66CB110, 67JA7151
dcc[a], hobt[b]	70CB788, 70CB2024, 71JA6318
dcc[a], hobt[b], $CuCl_2$	88CC419
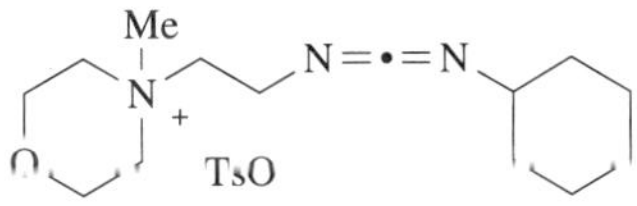	87TL3163
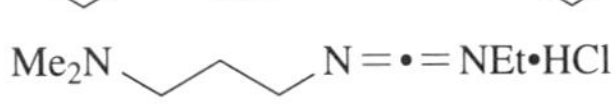	61JOC2525, 65JA2492
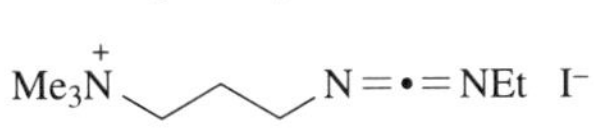	61JOC2525

[a] dcc = dicyclohexyl carbodiimide. [b] hobt = hydroxybenzotriazole.

(ii) Phosphorus(V) reagents

As outlined in Table 9, a wide range of phosphorus(V) reagents has been developed since the mid-1970s.

(iii) Uronium salts

The next generation of coupling reagents are the uronium salts, shown in Table 10. The first to be developed was HBTU ⟨78TL1269⟩ which was subsequently used in the synthesis of complex peptides ⟨84S572⟩. The rest of the series were prepared and subjected to a comparative study ⟨89TL1927⟩. The conclusions were that TBTU and HBTU were ideally suited to routine solid-phase peptide synthesis, TPTU/hobt was the best choice for segment condensations, TDBTU caused the least racemisation and TSTU and TNTU were useful for forming other active esters.

Table 9 Phosphorus(V) coupling reagents.

Reagents	*Ref.*	*Reagents*	*Ref.*
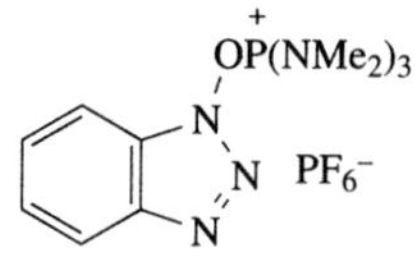 (BOP)	75TL1219	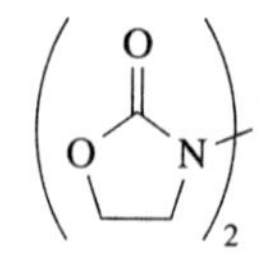 POCl, Et_3N	84S413 85JA4342
BOP–Cl	80S547		
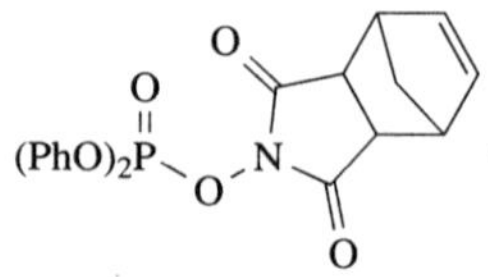, Et_3N	80CC1029	 POCl, *N*-methylmorpholine	84TL4825
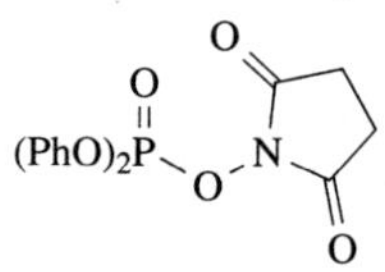, Et_3N	80TL1467	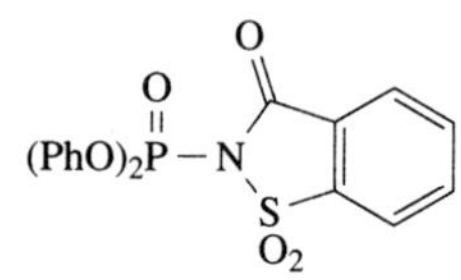, Et_3N	85CL123
KOH, Bu_4NX (X = SH or Br) (o-$NO_2C_6H_4O)_2POPh$	81CL285	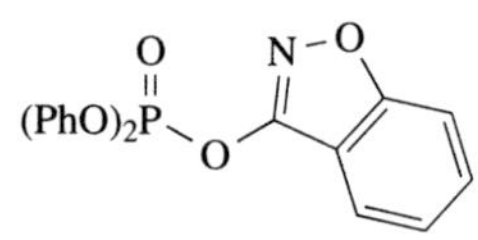, Et_3N	85JOC760
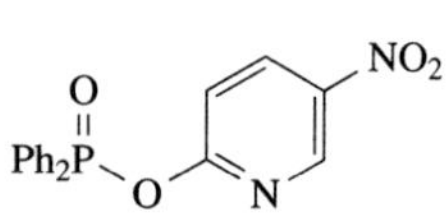, Et_3N	81CL1367	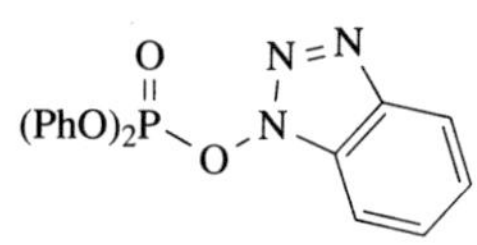, Et_3N	85TL1341
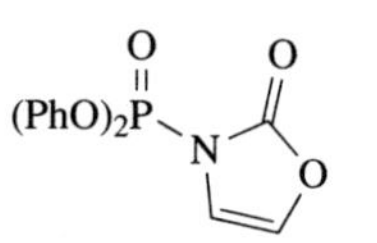	81TL1257	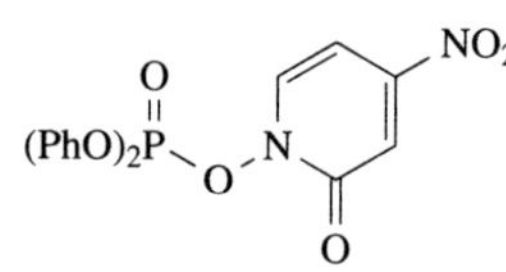, Et_3N	86CC719
(PhO)(PhNH)POCl, Et_3N	82S2882	$Ph_3P(OTf)_2$	75TL277 80JOC5052

(iv) Other coupling methods

Many other coupling procedures have been reported. Table 11 outlines a range of phosphorus(III) coupling systems while Table 12 outlines a range of alternative coupling procedures which are not so readily classified. The next two examples also involve the use of phosphorus(V) reagents, but lead exclusively to the formation of *N,N*-dimethyl amides (see Equations (26) ⟨66CI(L)1529⟩ and (27) ⟨79SC31⟩). Finally, the use of an acetylenic ester to form an active intermediate, and hence the amide, has been reported (Equation (28)) ⟨78HCA2428, 78HCA2437, 85OS(63)175⟩.

$$C_{11}H_{23}CO_2H \xrightarrow{\text{HMPA, }\Delta} C_{11}H_{23}CONMe_2 \qquad (26)$$

HMPA = hexamethyl phosphoramide

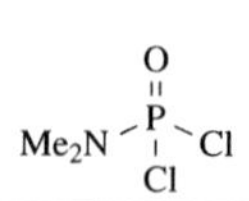

$$RCO_2H \longrightarrow RCONMe_2 \qquad (27)$$

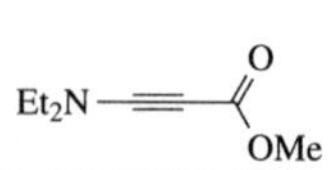

$$R^1CO_2H \xrightarrow[R^2_2NH]{} R^1CONR^2_2 \qquad (28)$$

Table 10 Uronium coupling reagents.

Reagents	*Ref.*	*Reagents*	*Ref.*
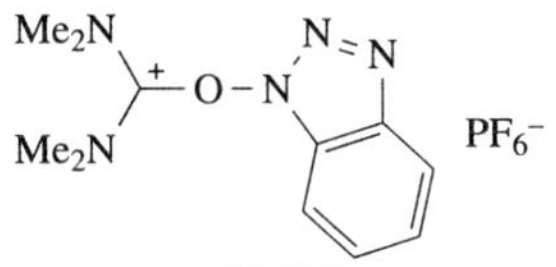 (HBTU)	78TL1269 89TL1927	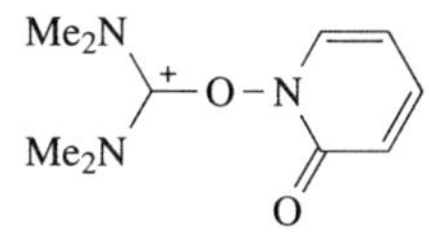 (TPTU)	89TL1927
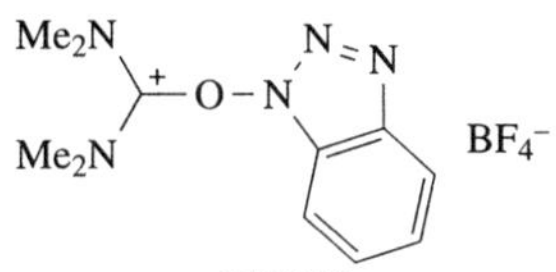 (TBTU)	89TL1927	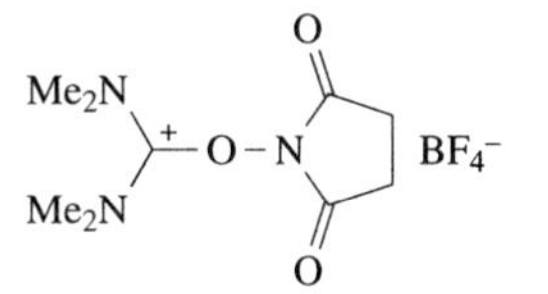 (TSTU)	89TL1927
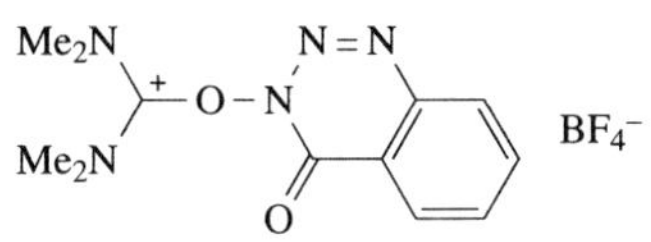 (TDBTU)	89TL1927	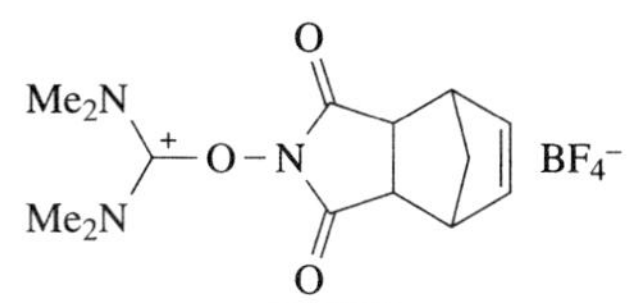 (TNTU)	89TL1927

Table 11 Phosphorus(III) coupling systems.

Reagents	*Ref.*	*Reagents*	*Ref.*
PPh_3, 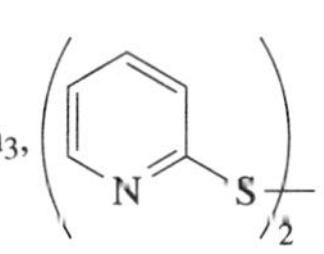	70TL1901	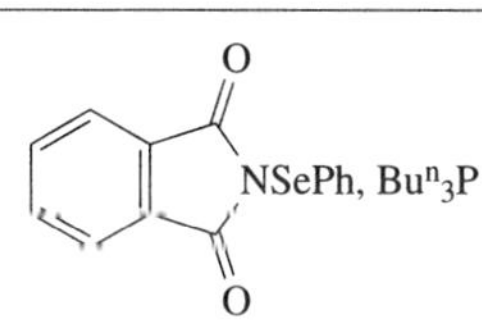	81JOC1215
Ph_3P, CCl_4 or $BrCCl_3$	71JOC1305	Ph_3P, Et_3N, 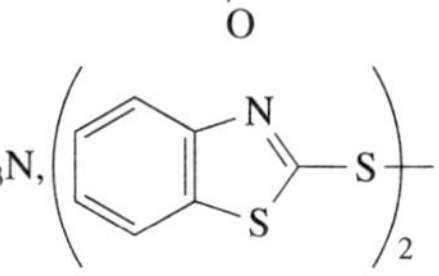	81S287
PR_3, CX_4 ($R = Bu$, Ph, NR_2; $X = Cl$, Br)	71TL3595		
o-$NO_2C_6H_4SCN$, Bu_3P	79JOC2945	P_2I_4, 2,6-lutidine, CCl_4	83CL449

5.06.2.2.8 From hydrolysis of nitriles

Section 5.06.2.1.2 showed that there are a large number of reagents and methods for the hydrolysis of nitriles to primary amides (see Table 4, Section 5.06.2.1.2). However, if the hydrolysis is carried out in the presence of a suitable electrophile, the hydrolytic intermediate can be trapped, thus producing *N*-alkylalkanoamides directly. The following examples in Table 13, which are all adaptations of the original Ritter reaction ⟨48JA4045⟩, illustrate the scope of this methodology.

Table 12 Other coupling reagents.

Reagents	*Ref.*	*Reagents*	*Ref.*
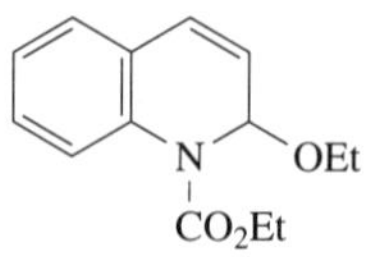	68JA1651	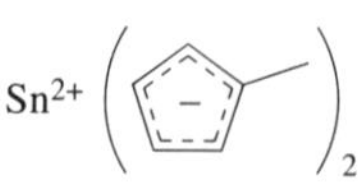	83CL683
$TiCl_4$	70CJC983	$BH_3{\bullet}NMe_3$	83S1013
$ClB(OMe)_2$	70T1539	TMS-CCOEt, HgO (cat.)	86JOC4150
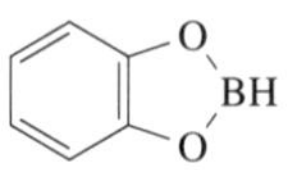	78JOC4393	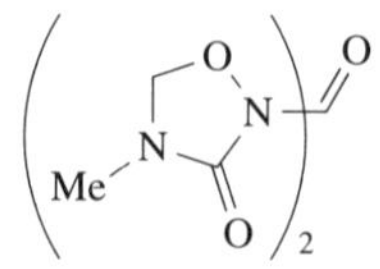	87TL5823
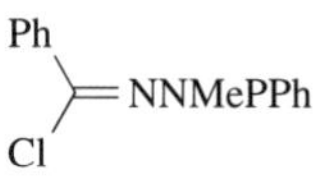	80JA4537		

Table 13 The formation of *N*-alkylalkanoamides from nitriles.

Electrophile	*Reagents*	*Ref.*
Adamantyl cation	(1) $Pb(OAc)_4$, TFA; (2) H_3O^+, MeCN or KCN, Br_2, H_2SO_4, MeCN	76S32, 77S632
R_3CH/R_3C^+	MeCN, BuOH, H_2SO_4, Br_2, MeCN, H_2O	64CB3234, 69OR(17)213
$C_{12}H_{25}CHCH_2$	NaCN, H_2SO_4, MeCN	48JA4045, 64MI 506-01, 69OR(17)213
$BuCHCH_2$	(1) $Hg(NO_3)_2$, MeCN; (2) $NaBH_4$, NaOH, H_2O	69JA5647
Cyclohexene	(1) PhSeCl, TFA, MeCN, H_2O; (2) H_2O_2	81CC546, 81JOC4727
ROH	H_2SO_4, RCN	73OSC(5)73
$PhCH_2C(CH_3)_2OH$	NaCN, H_2SO_4, HOAc	61JOC3002
t-Butylcyclohexanol	$SOCl_2$, MeCN	76JCS(P1)2205
1°, 2° or 3° RX	$NOPF_6$, MeCN	79S274
1° RX only	KOH, Bu^tOH, RCN	78S303
2-Iodobutane	NO_2BF_4, MeCN, H_2O	80JOC165
RCHO	Cl_3SiI, MeCN	91TL1825
R_2CO	R_3SiH, H_2SO_4, MeCN	74JOC2740
Bu^tCO_2K	Electrolyte, MeCN, H_2O	64ACS1567, 68BSF3657
R_3COMe	NO_2BF_4, MeCN, H_2O	80JOC165

5.06.2.2.9 *From rearrangement reactions*

As discussed in Section 5.06.1.4 there are a wide variety of synthetic procedures towards amides which utilise rearrangement reactions. The following sections deal with those rearrangements which have found synthetic applications in the synthesis of *N*-alkylalkanoamides.

(i) Beckmann rearrangement

The Beckmann rearrangement has been reviewed ⟨60OR(11)1, 88OR(35)1⟩, but several developments towards *N*-alkyl alkanoamides are outlined here. For instance, the one-pot procedure reported by Olah *et al.* (Equation (29)) ⟨79S537, 85OS(63)188⟩. More commonly the oxime is pre-formed and the rearrangement triggered by acid. Some modern alternatives involve the use of TsCl and pyridine ⟨84JA6414⟩ or a mixture of P_2O_5 and $MeSO_3H$ ⟨73JOC4071⟩ which acts like polyphosphoric acid (ppa) but is far easier to handle. An application of this methodology was reported by Jeffs *et al.* (Scheme 28) ⟨82JOC3876, 82JOC3881⟩. As a complementary reaction to the Beckmann rearrangement, it was found that when methoxylamine hydrochloride was employed in place of hydroxylamine a nitrone ring expansion could be accomplished, thus affording either regioisomer of the cyclic amide.

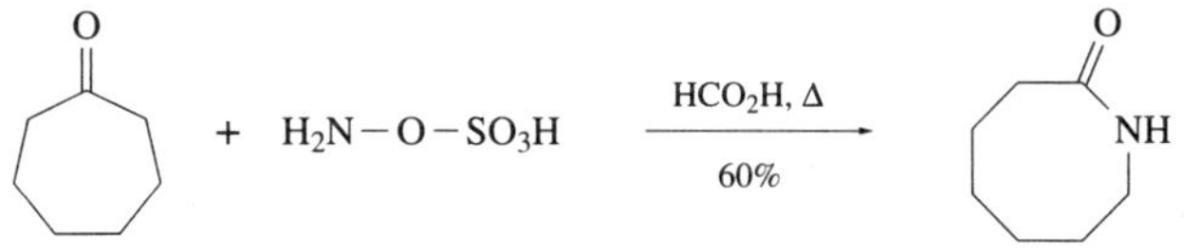

(29)

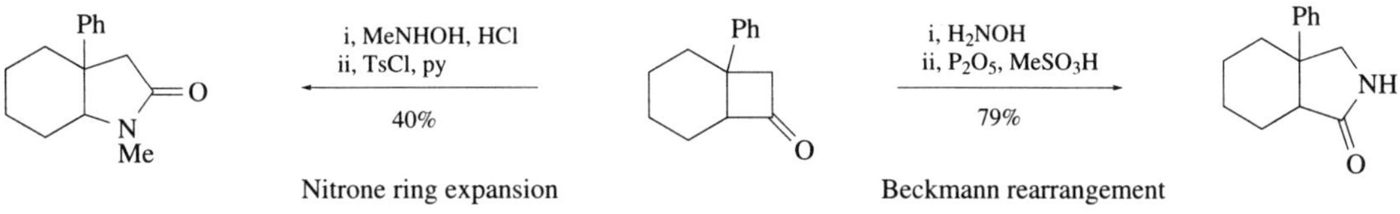

Scheme 28

(ii) Schmidt rearrangement

Treatment of a ketone with hydrazoic acid also leads to insertion of an adjacent NH group: this rearrangement has been reviewed ⟨46OR(3)307, 71RCR835⟩. In the early 1980s, regioselective lactam formation from bridged bicyclic ketones was reviewed ⟨81T1283⟩.

(iii) Oxaziridine–amide rearrangement

When heated, oxaziridines can undergo skeletal rearrangement to afford a range of *N*-alkyl alkanoamides, as exemplified in Scheme 29 ⟨67JCS(C)2128⟩ and Equation (30) ⟨57JA5739⟩; this reaction has been reviewed ⟨57JA5739⟩. The synthesis and reactions of 3-membered heterocycles containing two heteroatoms has also been reviewed ⟨79AHC(24)63⟩. A photochemical opening of an oxaziridine ring has been employed in a formal synthesis of aspartame (Equation (31)) ⟨87TL2595⟩.

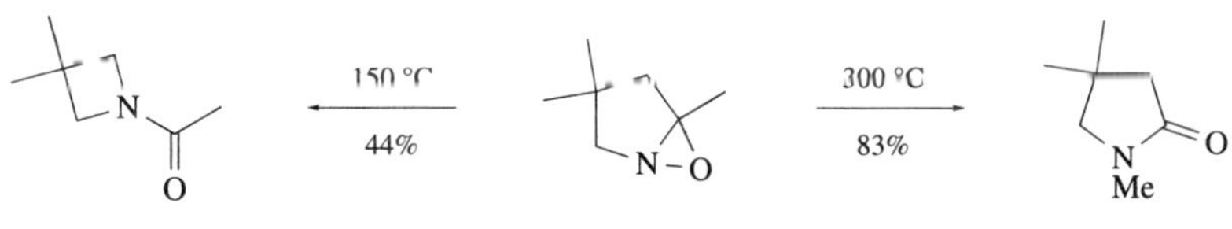

Scheme 29

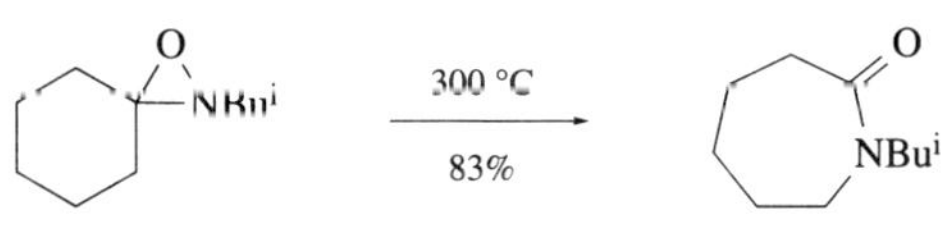

(30)

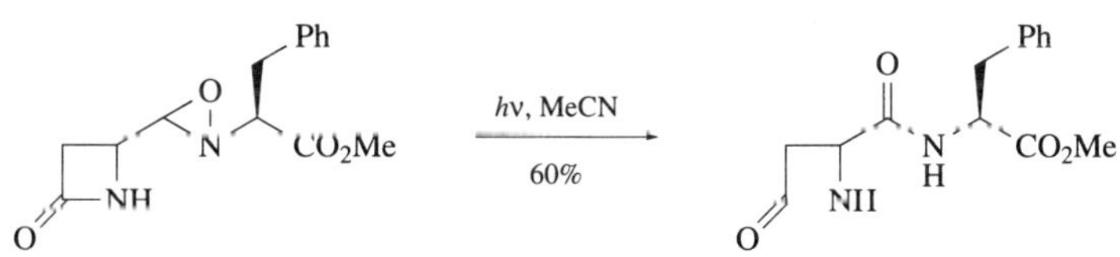

(31)

(iv) Miscellaneous rearrangements

A selection of miscellaneous rearrangements leading to *N*-alkyl alkanoamides are outlined in Scheme 30 ⟨92TL1455⟩ and Equations (32) ⟨77JA182⟩ and (33) ⟨82HCA2299, 35HCA434⟩. The examples include a macrocyclic ring expansion ⟨85HCA484⟩, and a one-pot homologation of a carbonyl compound to an amide ⟨77JA182⟩.

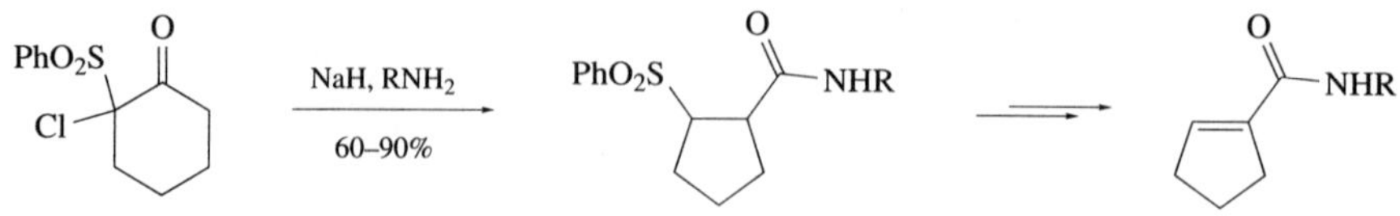

Scheme 30

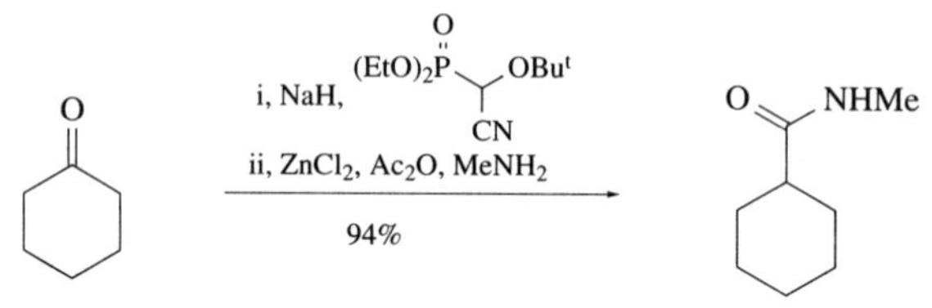

(32)

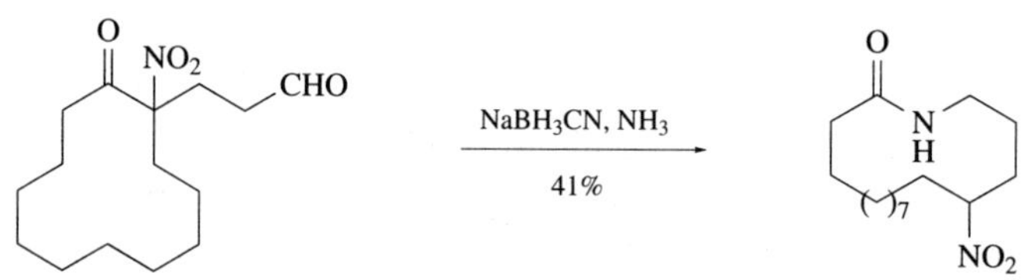

(33)

5.06.2.2.10 *Other methods*

(*i*) *Alkylation of amides*

In order to be reasonably selective, this section only includes *N*-alkylation reactions of the amide nitrogen (i.e., not *C*-alkylation adjacent to an amide carbonyl group). However, reactions involving *C*-alkylation that result in lactamisation have been included in the section on β-lactams. Tables 14, 15 and 16 list many of the various *N*-alkylation methods and are subdivided into groups of electrophiles: these are principally alkyl halides (Table 14) but also include alcohols, alkyl sulfates (Table 15) and alkenes (Equations (34) ⟨81CC670, 81CC1178⟩, (35) ⟨85JOC2220⟩, (36) ⟨87H(26)359⟩, (37) ⟨87CC1627⟩ and Table 16).

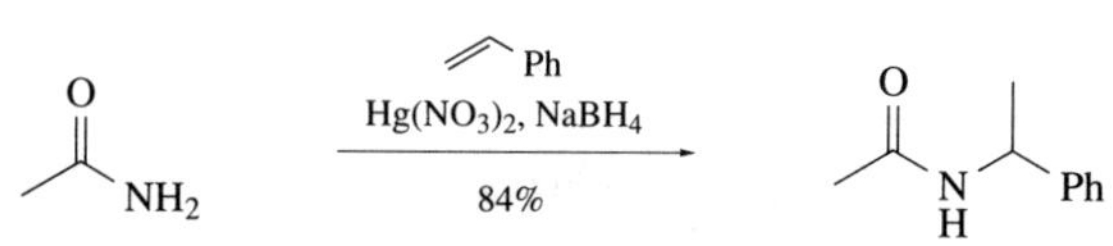

(34)

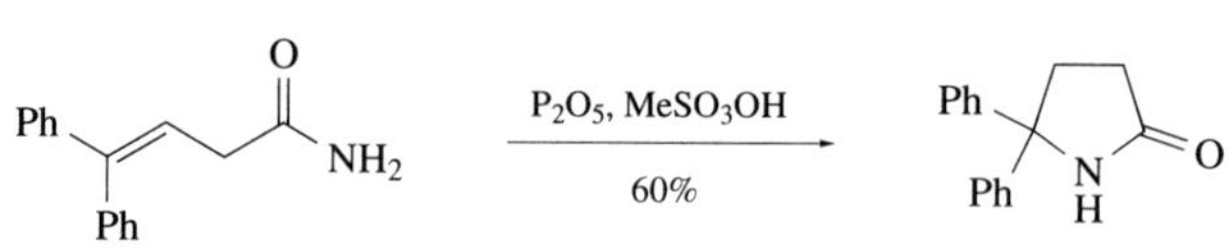

(35)

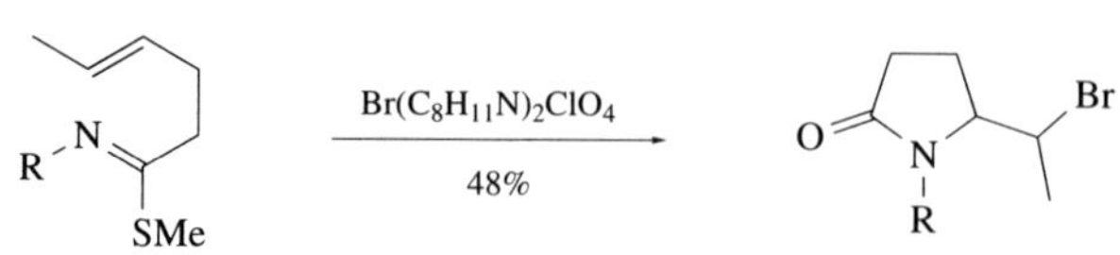

(36)

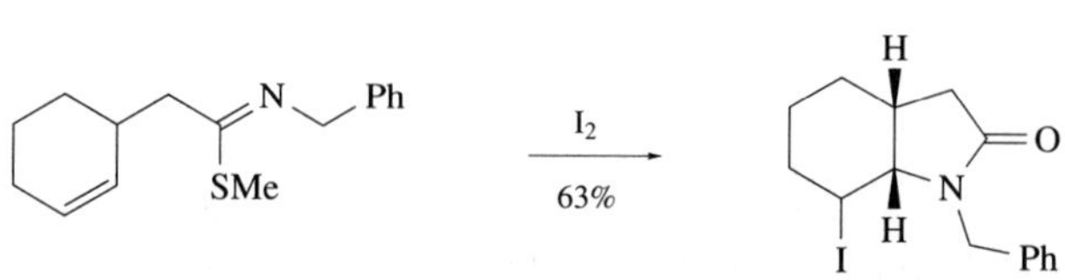

(37)

Table 14 Alkylation of amides with alkyl halides.

Reagents	*Ref.*
Na, RX (R = 1° alkyl, Br; X = Cl, Br, I)	1877CB327, 52JA1010
NaH, RX (R = 1°, 2° alkyl, allylic; X = Br, I)	49JOC1099, 52JA1010
NaH, MeI, DMF	61JA1492
NaH, MeI, DMSO	72CAR(23)251
NaOH, K_2CO_3, PhH, catalyst $(Bu^n{}_4N)HSO_4$, RX	79S527, 79S549
NaOH, $(Et_3NCH_2Ph)Cl$, PhH, RX	76S113
KOH, acetone, MeI	52JA1321
KOH, EtBr	1887CB3422
KOH, EtOH, RBr	52JA1010
KOH, DMSO, RX	71S266, 79T2169
$KOBu^t$, crown ether, RX	79SC757
KF, alumina, RX	81CL1143
NH_3, C_8H_{17}, $HCONH_2$	69JOC3204
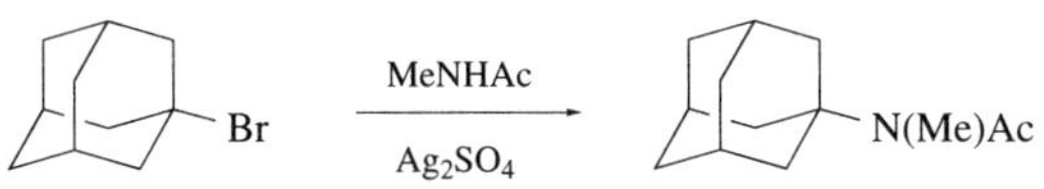	71JMC535

Table 15 Alkylation of amides with alcohols and alkyl sulfates.

Reagents	*Ref.*
Na, Me_2SO_4	35CB751
Bu^nLi, Me_2SO_4	73JOC1677
ROH, dead[a], PPh_3	81JOC1229, 80JA7026
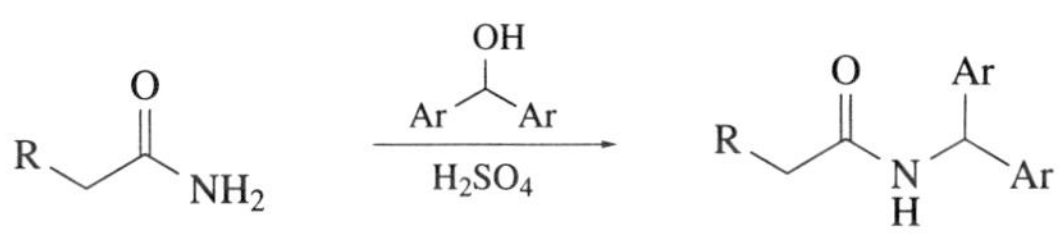	70CB2041
$R^1{}_2C(OR^2)_2$, H_2, catalyst Pd–C, HOAc, H_2SO_4	62JOC2205

[a] dead = diethyl azodicarboxylate

Table 16 Intramolecular alkylation of amides with alkenes.

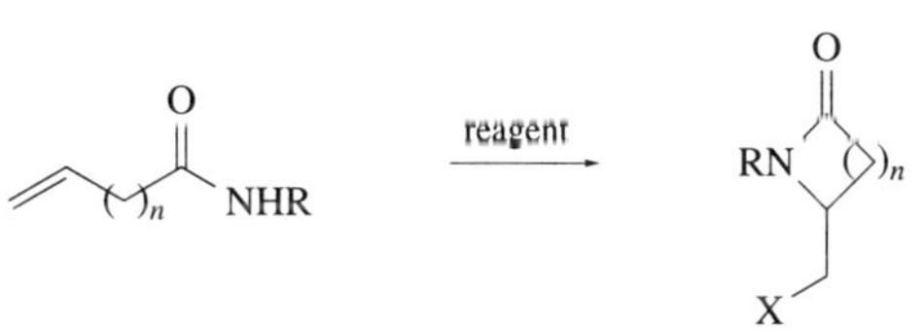

(a)

Reagent	*n*	*Ref.*
PhSeX (X = Cl, Br)	2,3	83JA7345, 86JCS(P1)1837, 86JOC1724, 87JOC2018
Br_2, K_2CO_3	1	82JA3233, 85TL5385, 87JOC4471, 87TL6257
PhSCl	1	83JA7345
TMS-OTf	2,3	85TL1803

(ii) Carbonylation reactions

In the above examples, amides were prepared by acylation of an amine or by *N*-alkylation of a simpler amide. The following reactions are somewhat different in that the amide is prepared via carbon–carbon bond formation. These are carbonylation reactions in which alkyl chains are homologated using either carbon monoxide directly (Equations (38) ⟨61AG621, 62CI(L)362, 70S99⟩, (39) ⟨82BCJ643⟩ and (40) ⟨86JOM(309)333⟩ or a synthetic equivalent (Equation (41) ⟨71JOC2721⟩ and Scheme 31 ⟨84JA2469⟩.

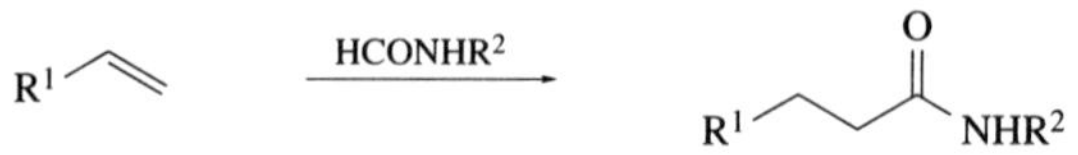

(38)

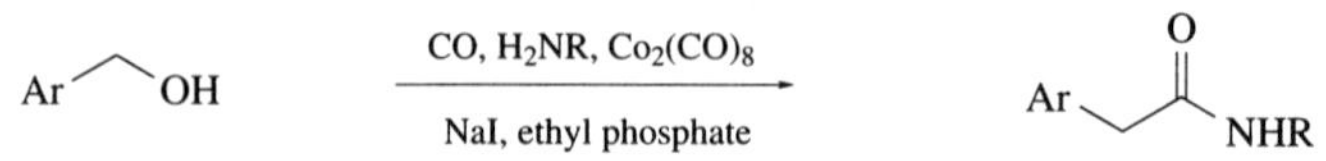

(39)

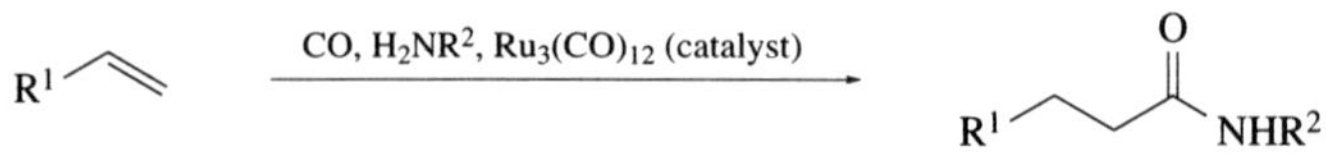

(40)

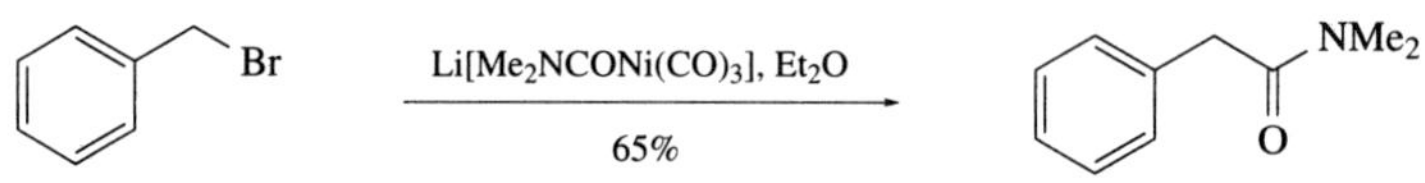

(41)

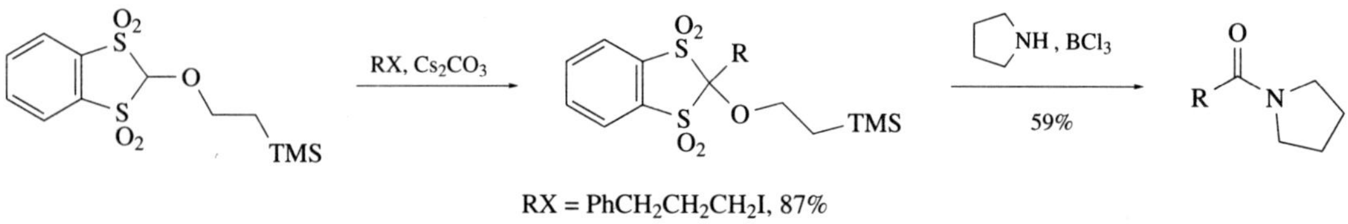

Scheme 31

5.06.2.2.11 β-Lactams

The synthesis of β-lactams (via [2 + 2]-cycloaddition) was first reported early this century ⟨17CB1035⟩, but interest in this class of compound was low until the therapeutic potential of penicillin was discovered in the 1940s. Since then several reviews have covered the burgeoning literature on the synthesis of β-lactams (otherwise known as azetidinones) (see Table 17).

Table 17 Selection of reviews on β-lactams.

Ref.	*Title*
57OR(9)388	Synthesis of β-lactams.
73S327	Synthesis of β-lactams.
76CSR181	Synthetic routes to β-lactams.
78T1731	β-Lactams: retrospect and prospect.
85AG(E)180	Developments in the field of β-lactam antibiotics.
87H(26)777	Advances in the chemistry of imines, in particular cycloaddition reactions.
88T5615	Applications of organometallic reagents in β-lactam chemistry.
89CRV1447	The ester enolate/imine condensation route to β-lactams.
89H(29)2225	Literature review of the ester enolate/imine condensation.
91T7503	Synthesis of 3-amino-2-azetidinones.

Examples of [2 + 2]-cycloaddition between an imine and a ketene (often generated *in situ*) are shown in Schemes 32 ⟨80S933⟩, 33 ⟨87JA1798⟩, 34 ⟨92S439⟩ and Equation (42) ⟨73JOC3437⟩. Other examples of this type of reaction have been reported where an *N*-aryl group is present and as such are included in Section 5.06.2.4.5.

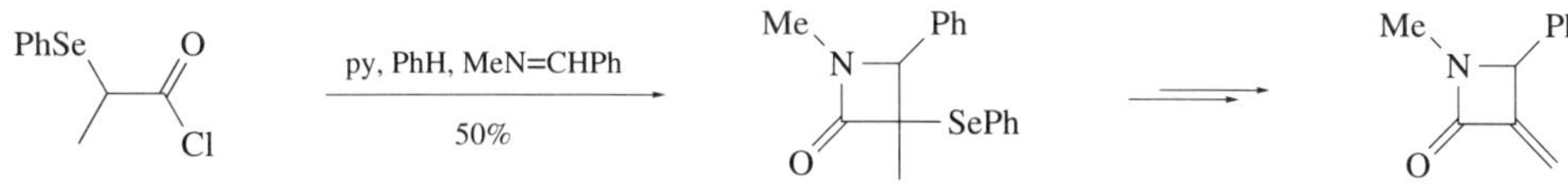

Scheme 32

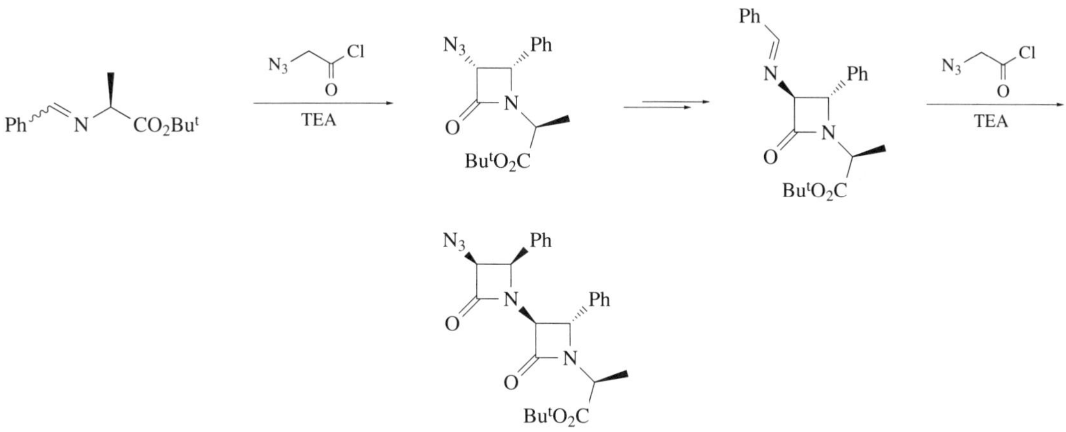

Scheme 33

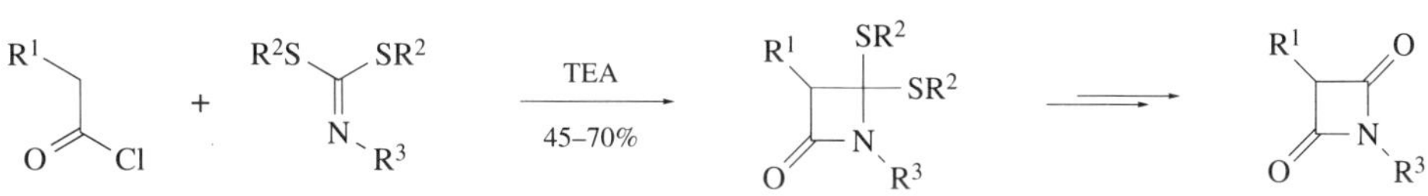

Scheme 34

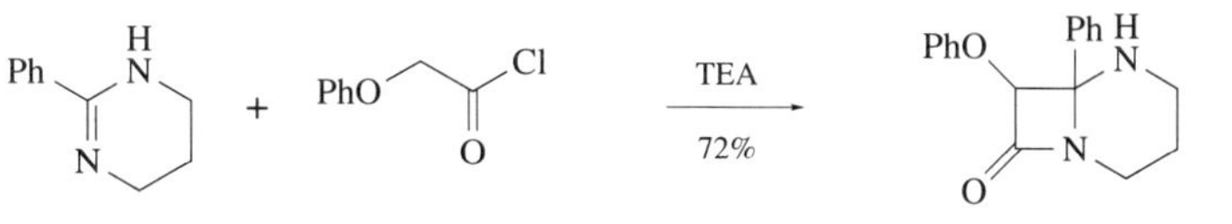

(42)

As reflected in the more recent reviews, the ester enolate/imine condensation is now commonly employed as a route to *N*-alkyl alkanoamide β-lactams. Several examples are outlined in Scheme 35 ⟨86TL1695⟩ and Equations (43) ⟨87TL3377⟩, (44) ⟨87TL5369⟩ and (45) ⟨91JOC1933⟩.

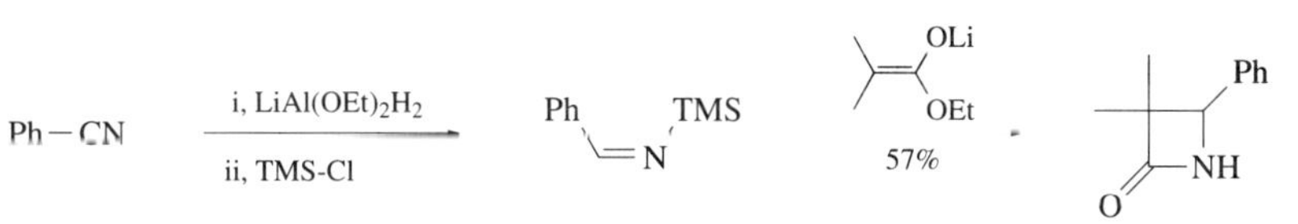

Scheme 35

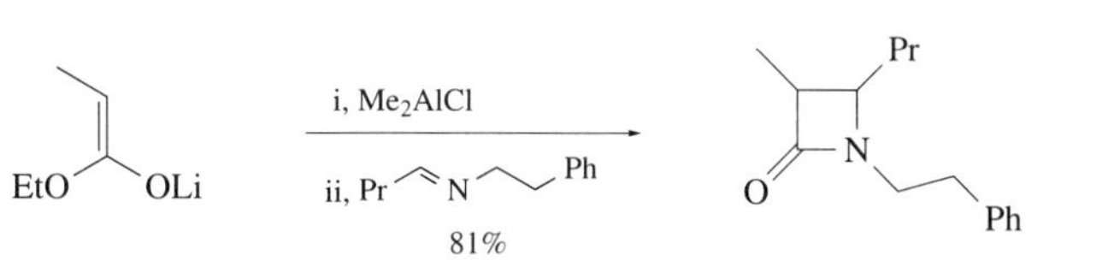

(43)

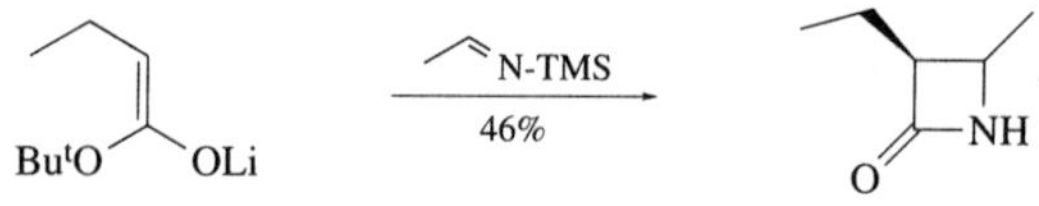

(44)

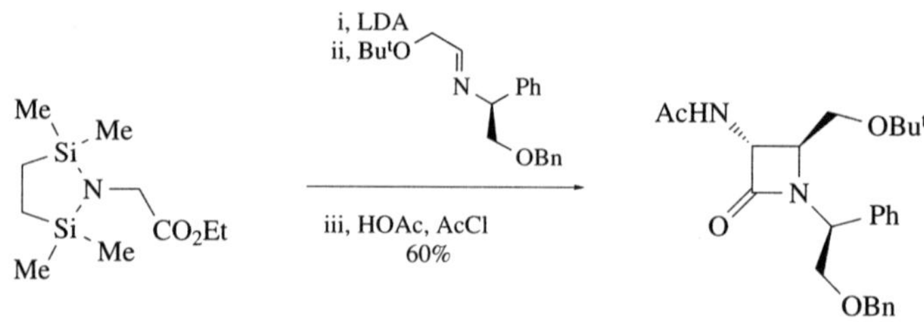

(45)

5.06.2.3 *N*-Alkenylalkanoamides–Enamides

The synthesis of enamides has been extensively reviewed ⟨78S489, 87S421⟩. This section outlines the key synthetic procedures adopted and describes some more recent developments.

5.06.2.3.1 *Acylation procedures*

In contrast to other sections of this chapter, most acylation routes to enamides start not from the α,β-unsaturated amine, but from the corresponding imine (Equations (46) ⟨60RTC401⟩, (47) ⟨81JHC691⟩ and (48) ⟨75HCA587⟩. Direct acylation of the enamine in the final example failed ⟨75HCA587⟩, and, presumably, many synthetic procedures towards enamides are complicated by both enamine–imine tautomerism and the potential nucleophilicity of the enamine α-carbon.

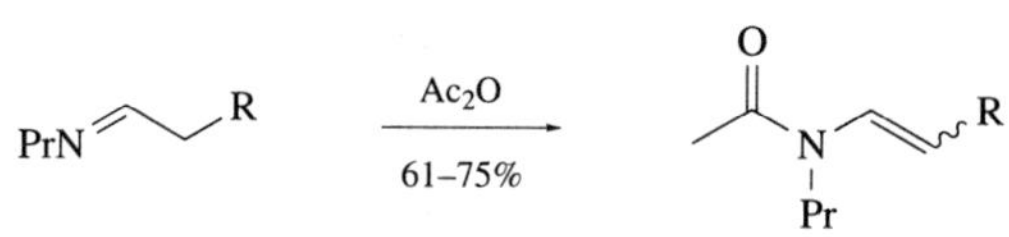

(46)

R = H, Me, Et

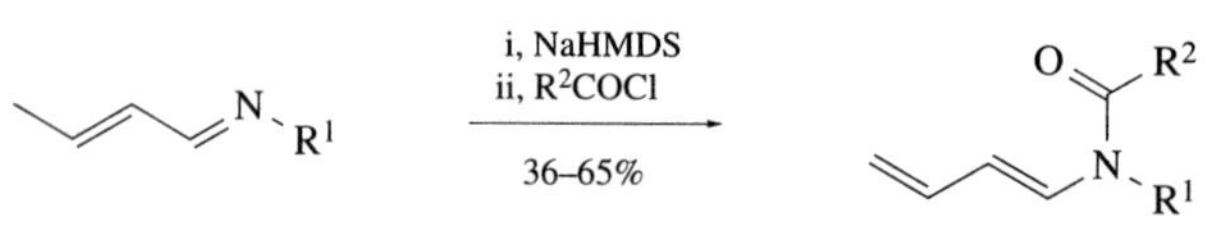

(47)

HMDS = hexamethyldisilazide

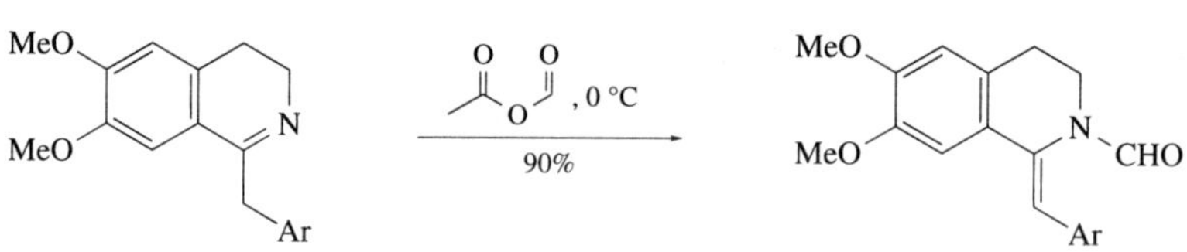

(48)

5.06.2.3.2 *From rearrangement reactions*

The Beckmann rearrangement has been reviewed ⟨60OR(11)1, 88OR(35)1⟩, but applications to the synthesis of enamides are relatively rare. Two examples are outlined in Equations (49) and (50) ⟨56JOC520, 85JCS(P1)831⟩. The Curtius rearrangement of α,β-unsaturated azides has also been employed in the synthesis of a range of enamides ⟨76TL3089, 88JCS(P1)2185⟩. Two other rearrangements are outlined in Schemes 36 ⟨76JA2352⟩ and 37 ⟨62JA4987, 63JA3288, 64OS(44)41⟩; the first is a thermal rearrangement of an acetimidic ester and the second is an unusual ring expansion that results from the reaction of phenoxide with chloramine.

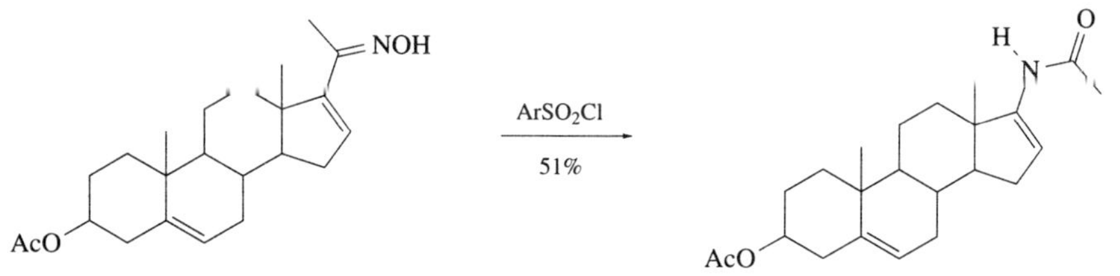

(49)

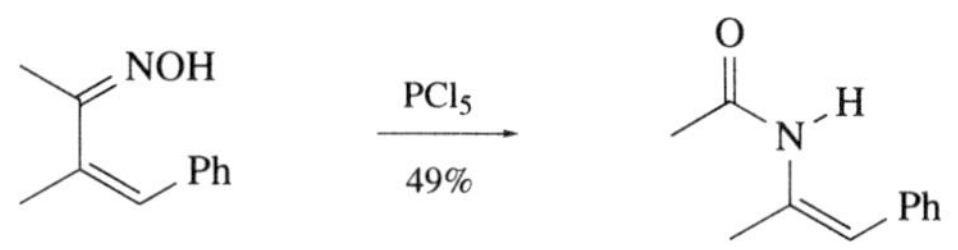

(50)

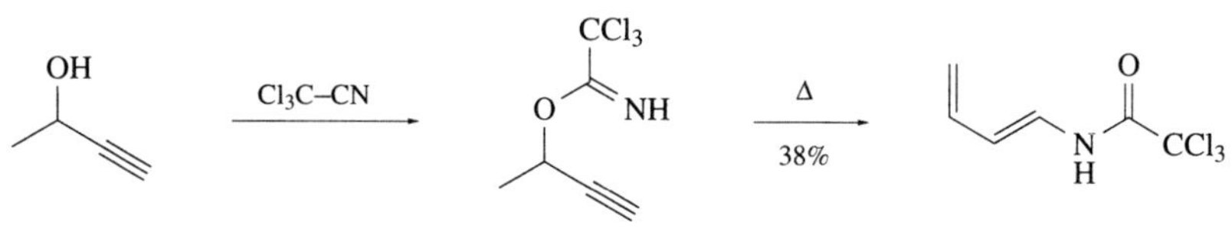

Scheme 36

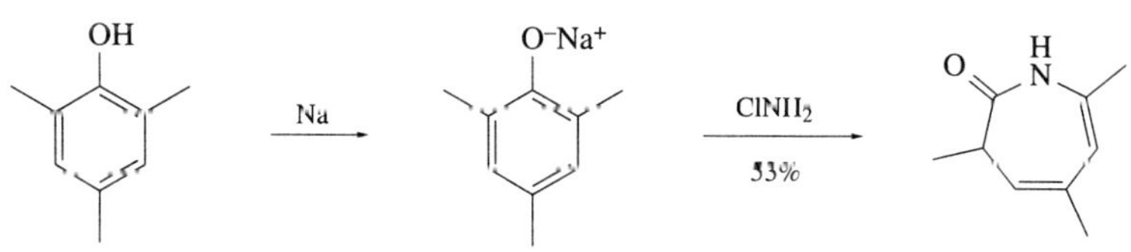

Scheme 37

5.06.2.3.3 *Other methods*

(i) Alkylation of amides

The alkylation of amides with a suitable electrophile, followed by elimination (often in one synthetic step) is another common route to *N*-alkenylalkanoamides (Scheme 38 ⟨57CB2291, 66CB2127⟩ and Equation (51) ⟨89TL5841⟩). A more direct method is the alkylation of an amide anion with an alkyne ⟨69T5745⟩ or a vinyl halide ⟨91CL1443⟩—this final method is stereospecific.

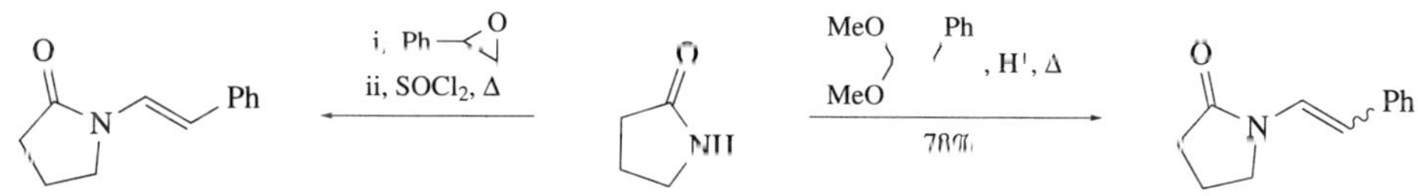

Scheme 38

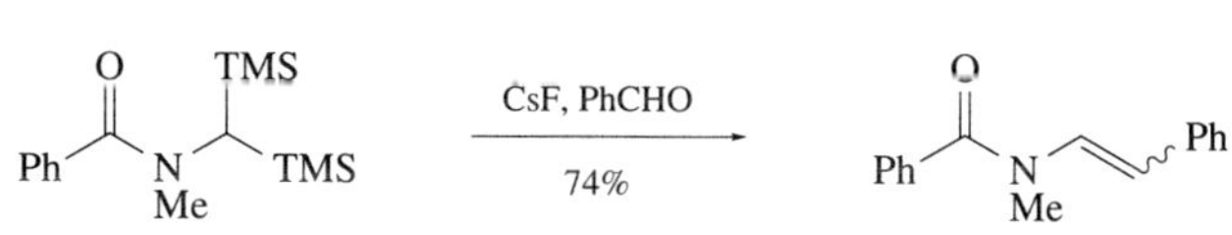

(51)

(ii) From oximes

The conversion of oximes to enamides has been reported by Barton and co-workers (see Equation (52) ⟨75JCS(P1)1237⟩). The mechanism proposed involved *N*- and *O*-acylation followed by homolytic

NO bond cleavage. Another route to enamides from oximes is base-catalysed fragmentation involving loss of acetonitrile (Equation (53) ⟨81TL4263⟩.

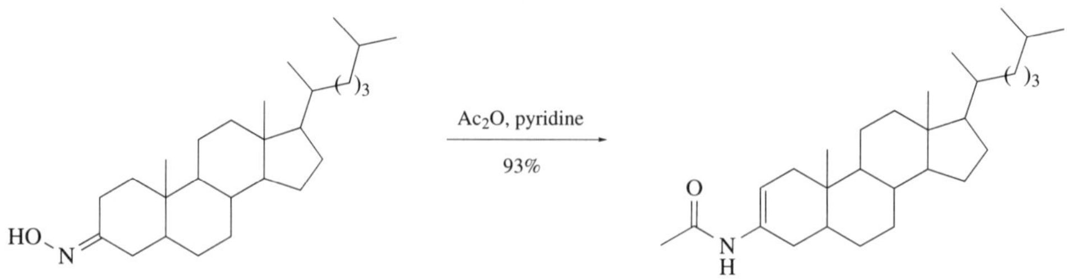

(52)

(53)

(iii) Miscellaneous procedures

As outlined in a review, a number of other procedures have been developed ⟨87S421⟩; for instance, the organometallic isomerisation of allylamides to enamides ⟨80JOC2139⟩ or Grignard addition to nitriles and acylation of the resultant iminomagnesium bromide ⟨65BSF1454⟩.

5.06.2.4 *N*-Aryl- and *N*-Heteroaryl–Alkanoamides

5.06.2.4.1 From acids and acid derivatives

Due to delocalisation of the nitrogen lone pair, aromatic amines are generally less basic than alkylamines. Many aromatic amines are also poor nucleophiles and more forcing conditions are generally required for amide formation. However, many of the techniques already discussed can be applied, and representative examples are outlined in Equations (54) ⟨41OSC(1)111⟩ and (55) ⟨55OSC(3)590⟩ and Table 18. A notable example is shown in Equation (56) ⟨71JOC3288⟩. Coupling with dicyclohexylcarbodiimide (dcc) and other 'standard' reagents had proved to be unsuccessful, but an efficient formylation of the 3-hydroxy-2-amino pyridine was eventually achieved by warming with phenyl formate.

(54)

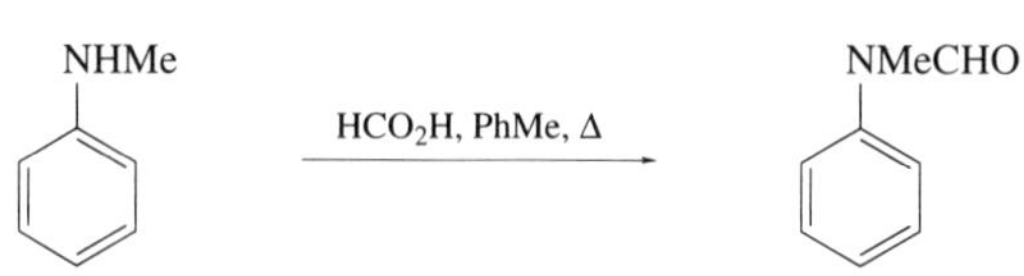

(55)

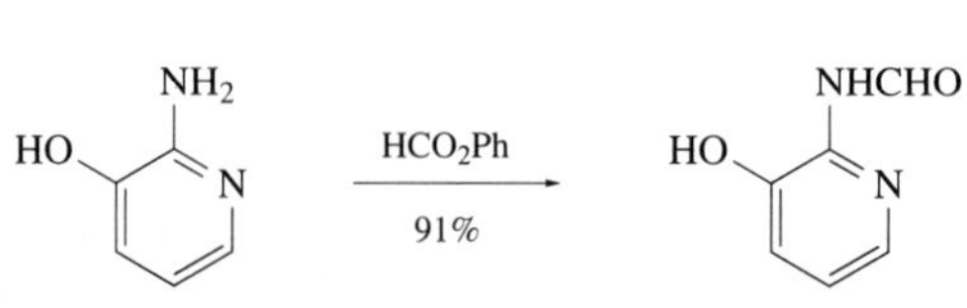

(56)

Table 18 Acylation of anilines.

Reagents	*Ref.*	*Reagents*	*Ref.*
$PhCOCH_2CO_2Et$, Δ	55OSC(3)108	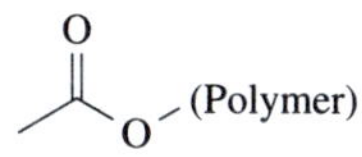	77SC57
$R^1CO_2R^2$, NaOMe	63JOC2915	AcCl	70JOC1219
EtOAc, NaH, DMSO	71TL321	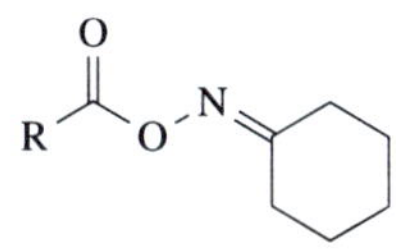	79S726
Ac_2O	55OSC(3)661	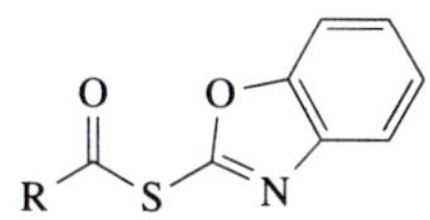	81S991

5.06.2.4.2 *Using* in situ *coupling agents*

As well as the simpler acylation methods, many of the *in situ* coupling reagents have also been utilised (see Table 19).

Table 19 Synthesis of *N*-arylalkanoamides via one-pot procedures.

Acid	*Amine*	*Reagent*	*Ref.*
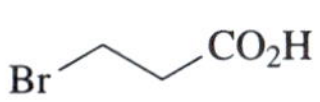	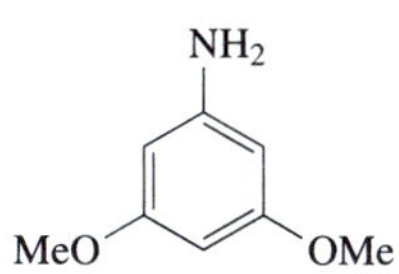	dcc	68JA4706
	Aniline	$SiCl_4$, pyridine	69JOC2766 (70%)
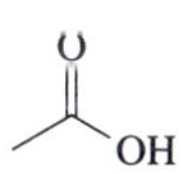	Aniline	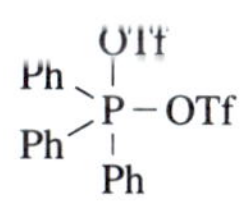	75TL277 (67%)
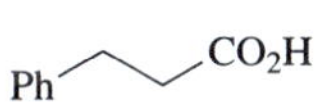	Aniline	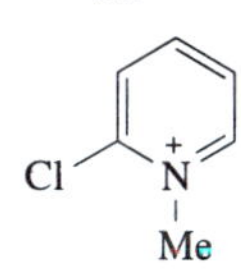	76CL57
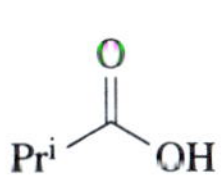	Aniline	 , TEA	78BCJ1866 (96%)
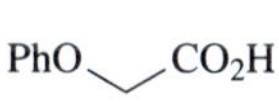	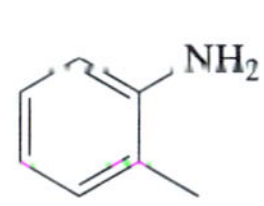	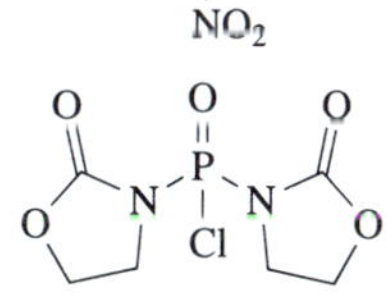	80S547 (92%)
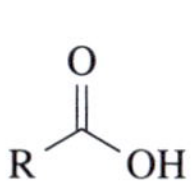	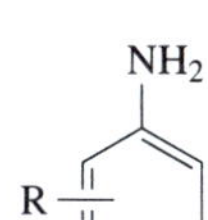	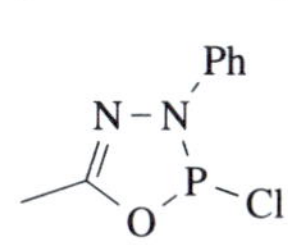	93BCJ927 (85–95%)

dcc = dicyclohexyl carbodiimide. Tf = trifluoromethane sulfonyl. TEA = triethylamine.

5.06.2.4.3 *From rearrangement reactions*

Aside from the usual acylation methods, compounds of this class can be prepared from various types of rearrangement reaction. For instance, anilides may be prepared by Beckmann rearrangement of phenyl ketones (Equation (57)); this and other examples can be found in an early review ⟨60OR(11)1⟩.

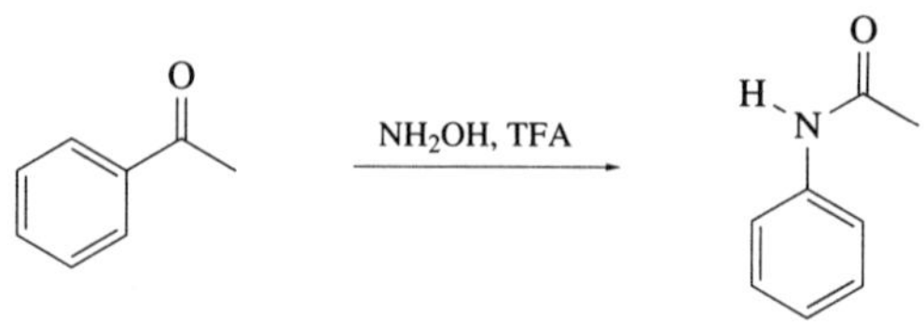

(57)

Similarly, anilines can be used to trap out the intermediate ketone formed in a Wolff rearrangement (Scheme 39) ⟨59CB528, 67RCR260⟩. The oxaziridine rearrangement has been used in an elegant preparation of a series of *N*-aryl lactams. 2,6-Dimethylaniline was reacted with a cyclic ketone to yield the corresponding imine, which was subsequently reacted with a peracid to generate the oxaziridine; irradiation then triggered rearrangement to the lactam (Scheme 40) ⟨58CB1057, 69TL2281⟩. Finally, the Smiles rearrangement is specific to the preparation of *N*-arylamides and involves conversion of a phenoxy amide into a hydroxy anilide via a 1,3 migration (Equation (58)) ⟨77S31, 77S33⟩.

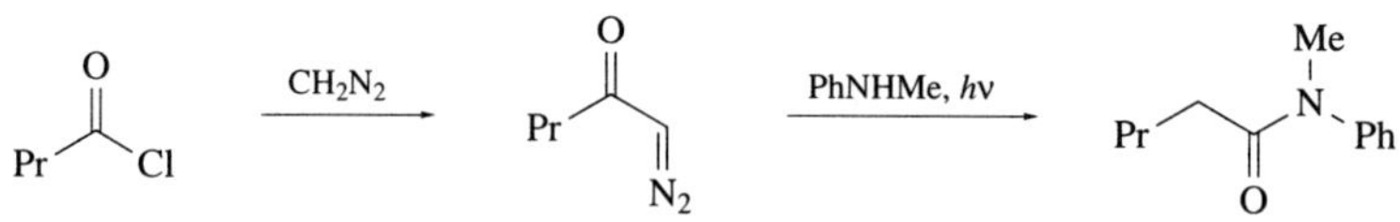

Scheme 39

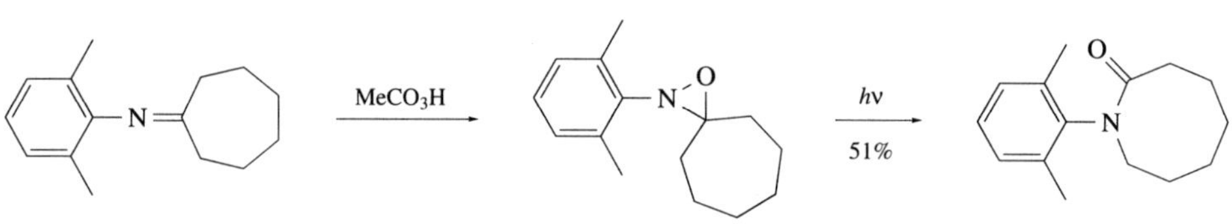

Scheme 40

(58)

5.06.2.4.4 *Other methods*

(i) N-*Alkylation of amides*

As would be expected, secondary *N*-arylamides can be alkylated to give the tertiary amides, and most of the methods cited in Section 5.06.2.2.10 should be applicable. The examples in Equations (59) and (60) are representative ⟨71S266, 81CPB1063⟩.

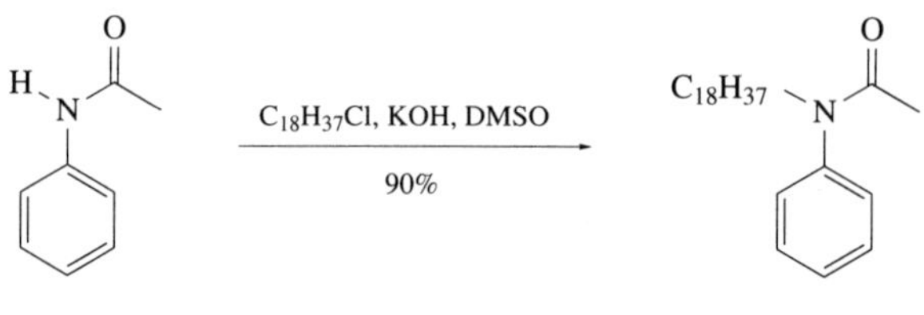

(59)

(60)

(ii) Carbonylation and oxidation reactions

As with their aliphatic counterparts, aromatic amines can participate in carbonylation reactions with alkenes (Equation (61)) ⟨52JA4496⟩ and in oxidation reactions (Equations (62) ⟨46JCS454⟩, (63) ⟨57JCS3032⟩ and (64) ⟨90TL7237⟩) to produce amides.

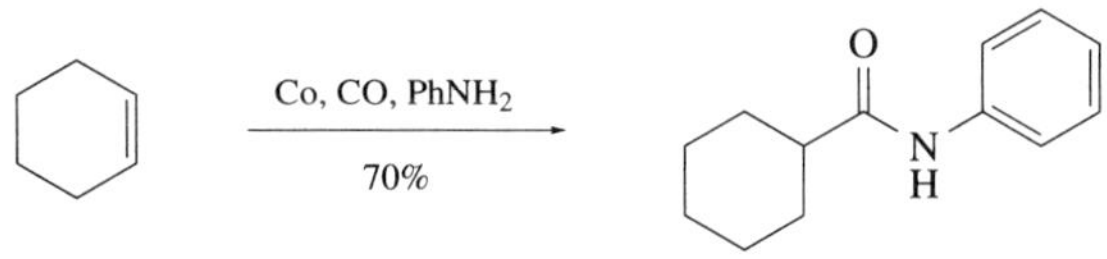

(61)

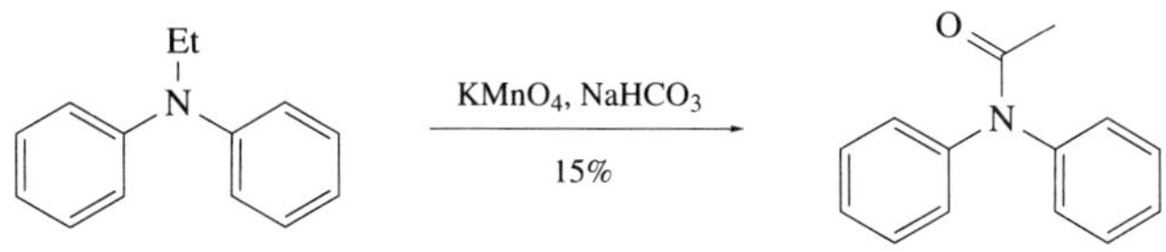

(62)

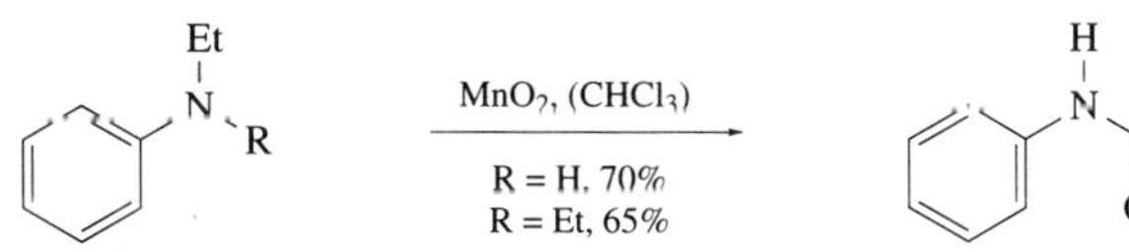

(63)

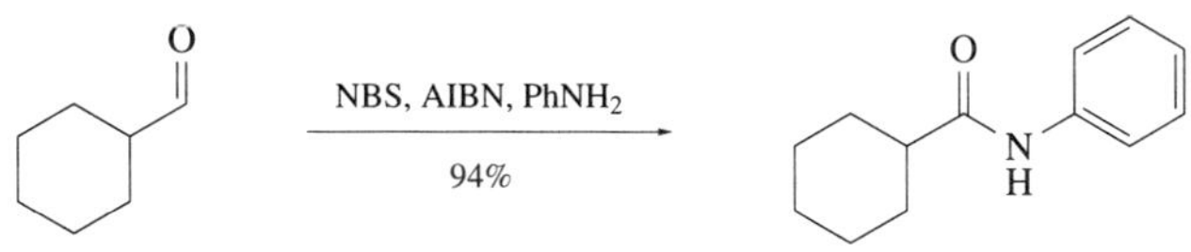

(64)

(iii) Amides from isocyanates

The addition of Grignard reagents to isocyanates are often high yielding, efficient reactions. As a consequence, phenyl isocyanate is an established assay for Grignard reagents ⟨47JA2007, 53JOC432⟩. Unfortunately, this method of amide synthesis is limited by the availability of isocyanates. An early example of this reaction and a more modern equivalent are shown in Equations (65) and (66) ⟨38JA540, 87TL3815⟩. Amides can also be prepared by the reaction of carboxylic acids with isocyanates; this method has been used to synthesise a series of amino acid anilides ⟨87CPB2698⟩.

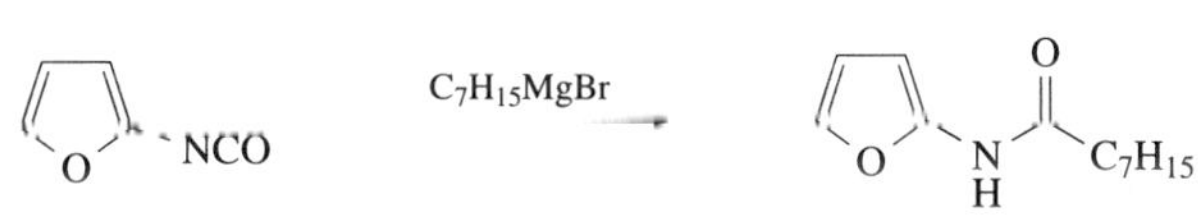

(65)

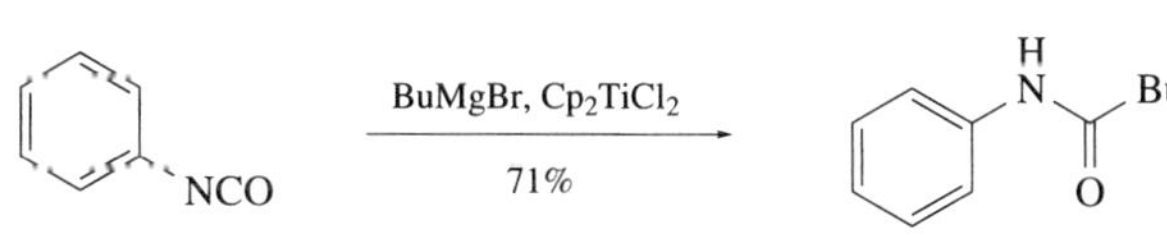

(66)

(iv) Miscellaneous reactions

The following examples are synthetic methods that are unique to the preparation of *N*-aryl or heteroarylamides. The first is the Semmler–Wolff aromatisation of cyclic oximes (Equations (67) and (68)) where the oxime is treated with a strong acylating agent and elimination of acetic acid yields the aromatic amide ⟨73JOC4073, 80S887⟩. Finally, Equation (69) outlines the amidation of an aromatic ring with a hydroxamic acid ⟨67JA7131, 81JOC4304⟩.

(67)

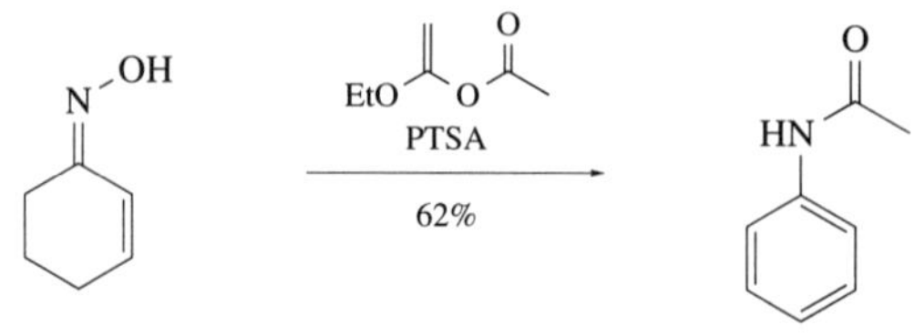

(68)

PTSA = *para*-toluene sulfonic acid

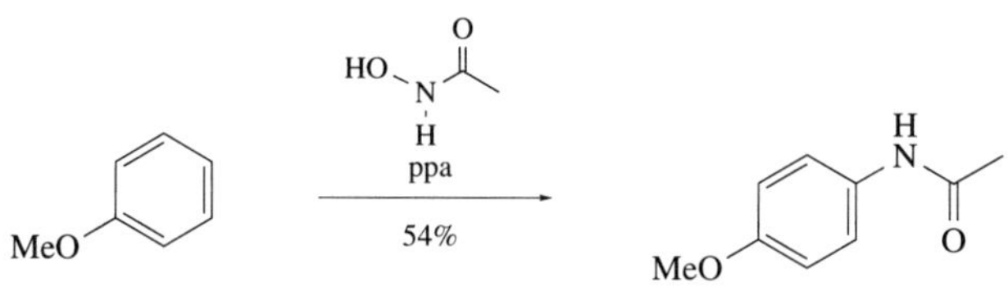

(69)

5.06.2.4.5 *β*-Lactams

The synthesis of *β*-lactams has been comprehensively reviewed (see Table 17, Section 5.06.2.2.11). There are two main synthetic strategies which are peculiar to *β*-lactams; these are [2 + 2]-cycloadditions and enolate/imine condensations. Both these reaction types often utilise an aryl imine and consequently produce *N*-arylalkanoamides as shown in Scheme 41 ⟨80JOC3413⟩ and Equations (70) ⟨86JA6054⟩, (71) ⟨74TL2633⟩ and (72) ⟨87TL1945⟩.

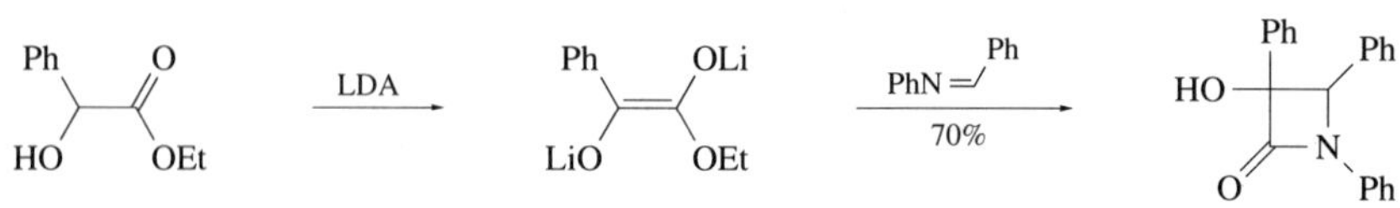

Scheme 41

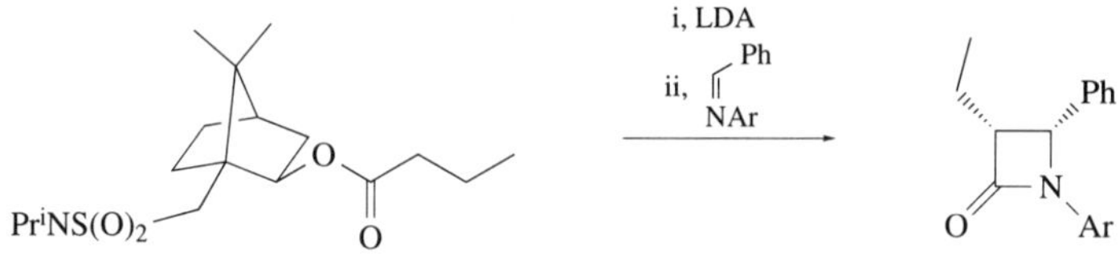

(70)

Ar = C_6H_4OMe, 88% yield, 92% *ee*

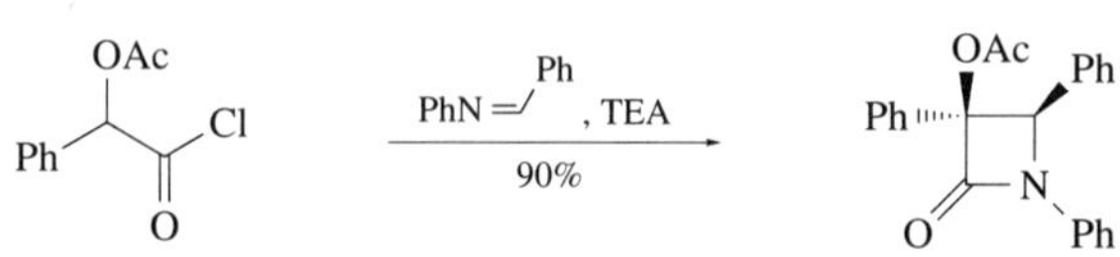

(71)

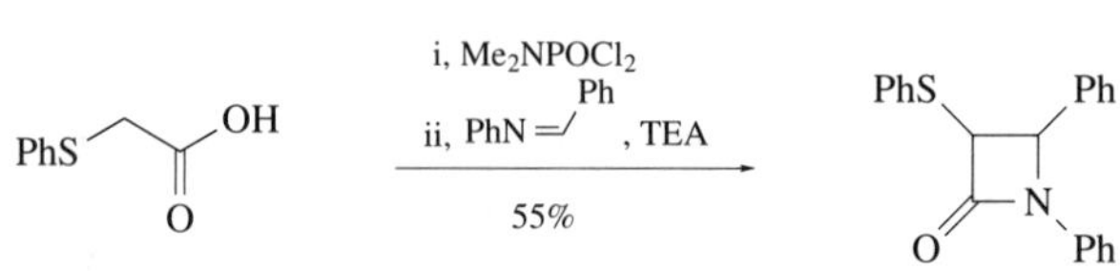

(72)

5.06.3 AMIDES OF ALKENOIC ACIDS

5.06.3.1 *N*-Unsubstituted Alkenoamides

There are relatively few reports which are specific to this class of compound, but their synthesis is often possible by the methods established in earlier sections: for example, the ammonolysis of an acyl halide (Equation (73)) ⟨62JCS2824⟩. In principle there is no reason why other acylation procedures cannot be employed, such as the reaction between ammonia and simple esters. However, the possibility of competitive [1,4]-addition reactions must be considered.

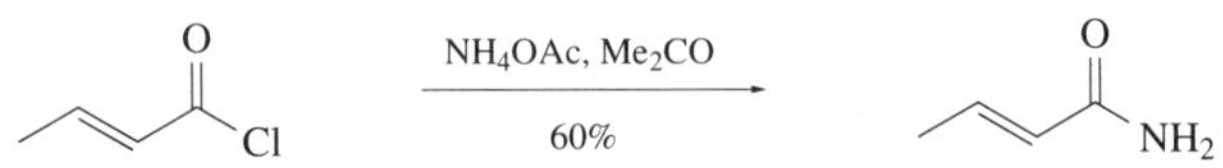

(73)

Various oxidations have been deployed in the syntheses of a range of *N*-unsubstituted alkenoamides (Scheme 42) ⟨66CC17, 71CC733⟩. Another possible route is the hydration of nitriles (see Table 4, Section 5.06.2.1.2). However, under some of these conditions epoxidation or hydration of the double bond can be a problem. The use of hydroquinone and copper(0) is an example of a hydration procedure which is highly regiospecific for the nitrile moiety ⟨82JOC4812⟩.

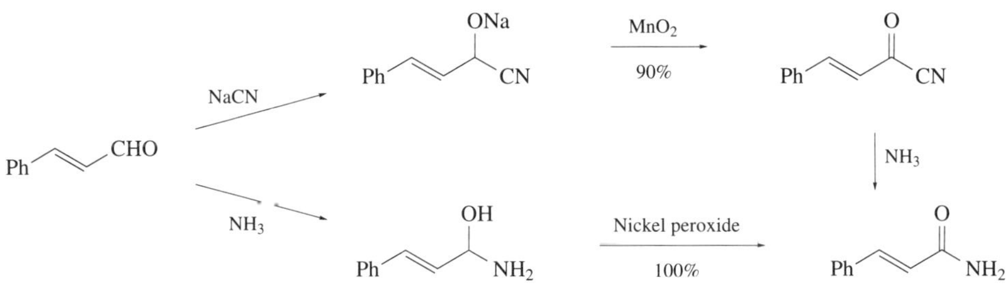

Scheme 42

5.06.3.2 *N*-Alkylalkenoamides

5.06.3.2.1 From acids and acid derivatives

This section only covers α,β-unsaturated alkenoic acid derivatives, since more remote unsaturation is unlikely to affect the reactivity of the acyl centre. The reaction of a suitably activated acid derivative and an amine would appear to offer a direct route to these compounds (Equation (74)) ⟨70JOC825⟩. However, α,β-unsaturation can reduce the electrophilicity of the carbonyl group, and thus drastically reduce the yield of established procedures such as dcc coupling (Scheme 43) or the mixed anhydride method (Scheme 44) ⟨80S385⟩. One way around this is the use of the phosphonate, shown in Scheme 45, which only gives the amide product and an easily removed phosphonic acid. The use of a more complex phosphonate ⟨84S413⟩ or 2-thiolthiazoline in conjunction with dcc ⟨80TL841⟩ have also been reported as viable routes to amides starting from α,β-unsaturated acids.

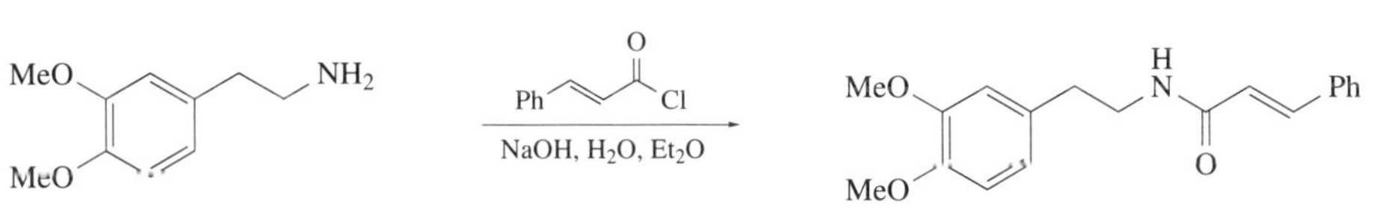

(74)

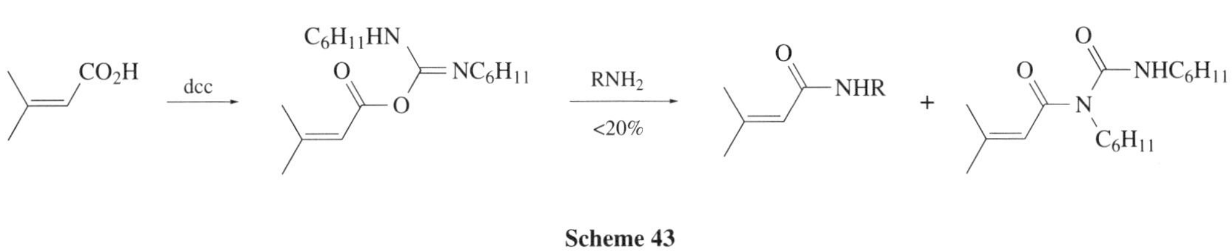

Scheme 43

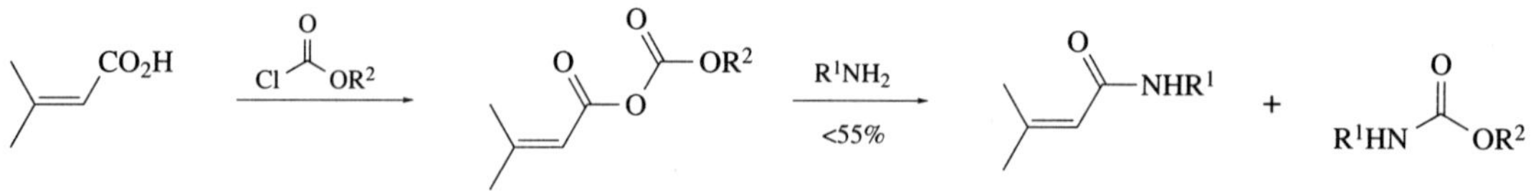

Scheme 44

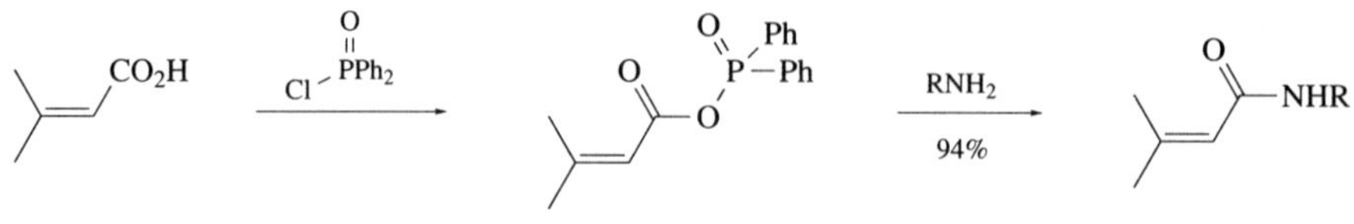

Scheme 45

5.06.3.2.2 From hydrolysis of nitriles

The hydrolysis of nitriles in the presence of an electrophile leads to *N*-substituted amides (see Table 13, Section 5.06.2.2.8), and this procedure can be effected with α,β-unsaturated nitriles such as acrylonitrile (Equation (75)) 〈73OSC(5)73〉.

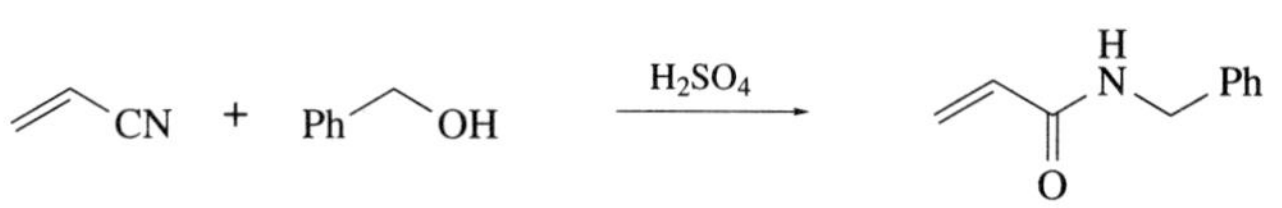

(75)

5.06.3.2.3 N-Alkylation of amides

Both intermolecular and intramolecular *N*-alkylation of alkenoamides have been reported (Equations (76) and (77)) 〈75CI(L)745, 76JOC725〉.

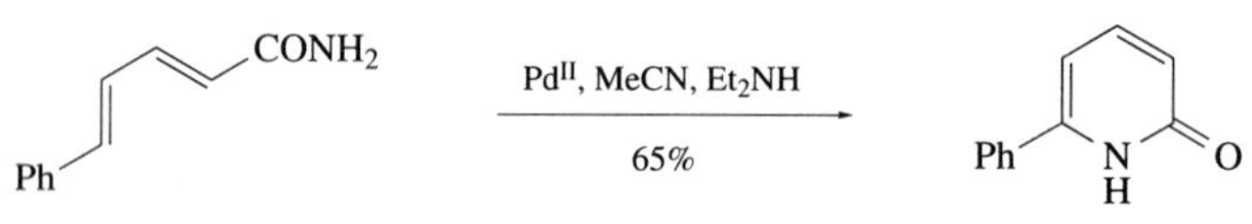

(76)

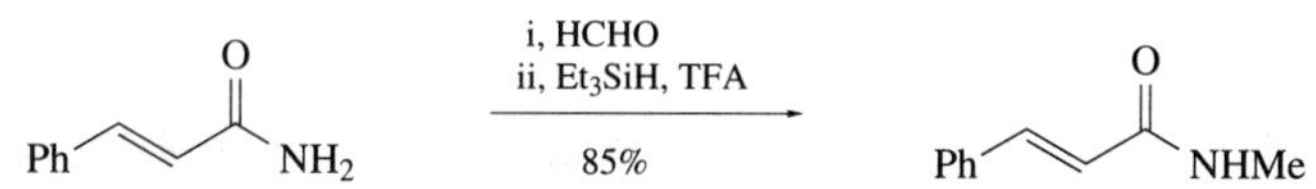

(77)

5.06.3.2.4 Using other methods

One group of pharmacologically important amides is the α-methylene lactams. These *N*-alkylalkenoamides have been approached in a number of ways, such as palladium-mediated cyclisation, organometallic addition to an imine followed by cyclisation of the intermediate amine or carbonylation (Equations (78) 〈83JOC4058〉, (79) 〈85SC1233, 87TL59〉 and (80) 〈86TL6339〉. In fact, a wide range of carbonylation reactions have been developed, as demonstrated by the examples outlined in Equations (81) 〈69JA1233〉, (82) 〈71JOC2721〉, (83) 〈74JOC3327〉 and (84) 〈85TL1109〉.

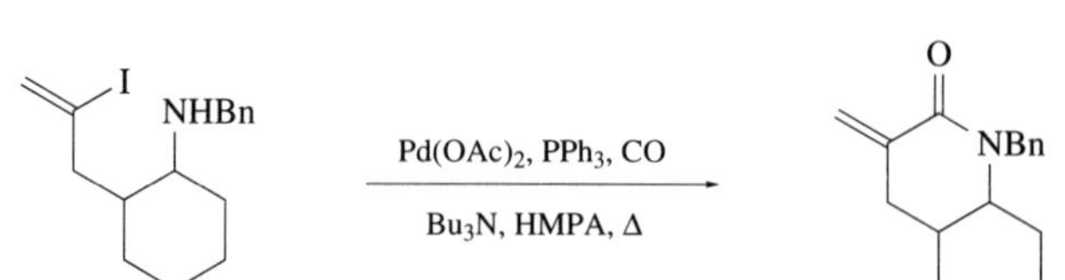

(78)

HMPA = hexamethylphosphoramide

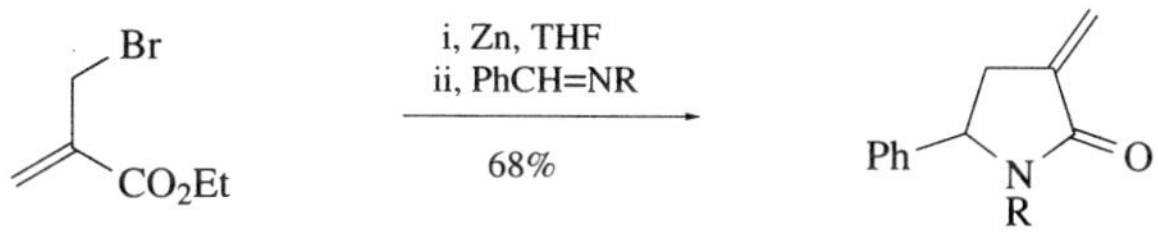

(79)

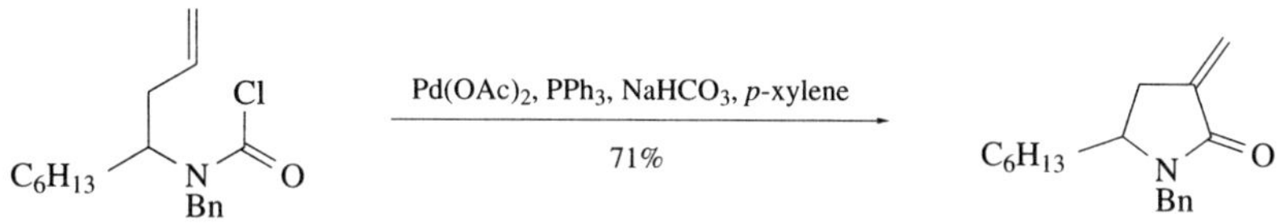

(80)

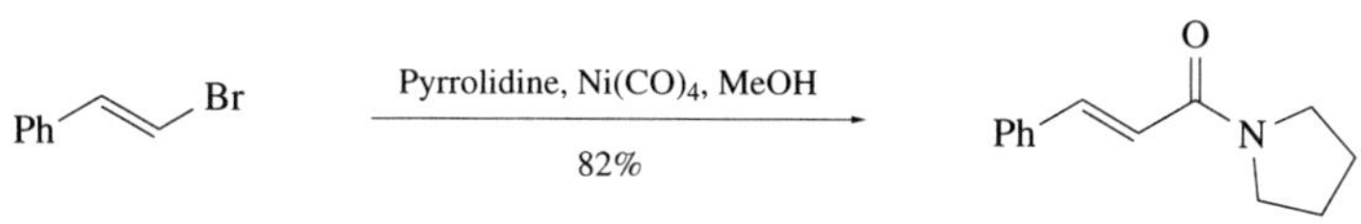

(81)

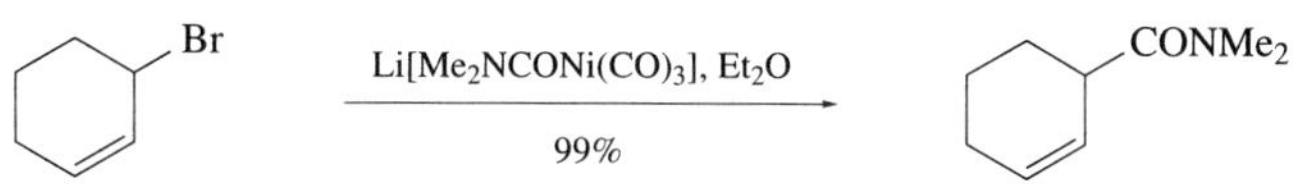

(82)

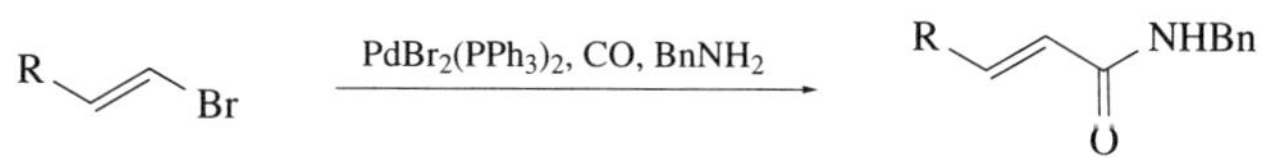

(83)

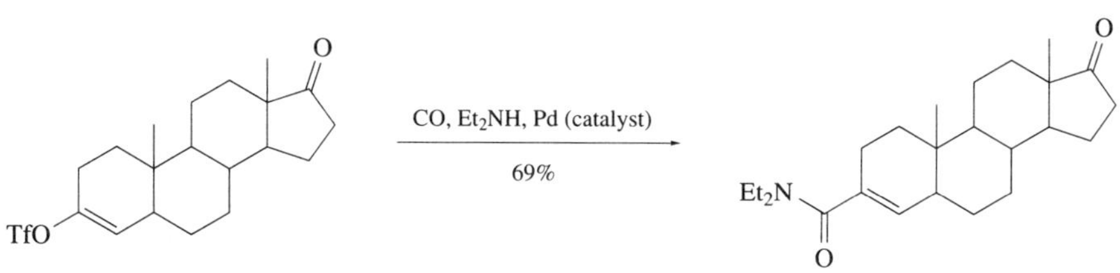

(84)

5.06.3.3 *N*-Alkenylalkenoamides

The synthesis of enamides has been reviewed ⟨78S489, 87S421⟩. However, there are very few examples which are specific to this class of amide.

5.06.3.4 *N*-Aryl– and *N*-Hetaryl–Alkenoamides

There are few specific reports on the preparation of this class of compounds, but one example arises from reaction between an alkyne and an isocyanate (Equation (85)) ⟨74BSF(2)1949⟩. One example of the acylation of an *N*-aryl amine has been shown in Equation (79) (where R is Ph) ⟨87TL59⟩. Some *N*-arylalkenoamides have also been prepared by more standard coupling procedures, for example using chlorosulfonyl isocyanate ⟨82S506⟩.

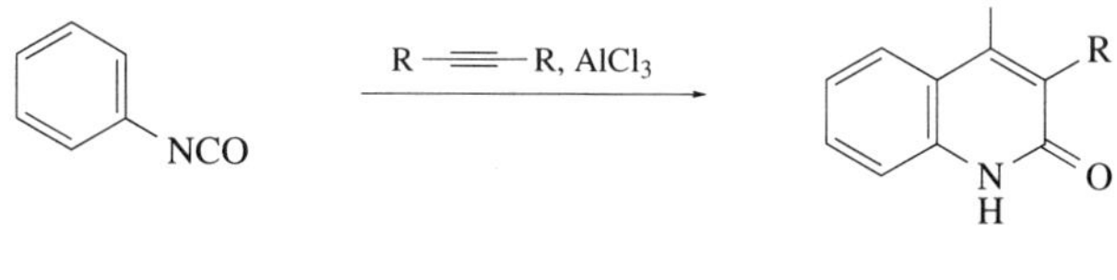

(85)

5.06.4 AMIDES OF AROMATIC AND HETEROAROMATIC ACIDS

5.06.4.1 *N*-Unsubstituted Arylamides

5.06.4.1.1 From acids and acid derivatives

Early approaches to primary arylamides were via the aminolysis of simple esters; for instance, with ammonium chloride in liquid ammonia ⟨38JA579⟩. One particularly difficult amide formation *en route* to some Rauwolfia alkaloids was effected by the more forcing conditions of excess $NaNH_2$ in liquid ammonia ⟨55JA469⟩.

5.06.4.1.2 From hydrolysis of nitriles

Many of the procedures outlined for the hydrolysis of alkyl nitriles (Table 4, Section 5.06.2.1.2) should be applicable to the hydrolysis of aromatic nitriles. Table 20 outlines several specific examples employed in the synthesis of arylamides.

Table 20 Hydrolysis of aromatic nitriles.

Nitrile	*Reagents*	*Ref.*
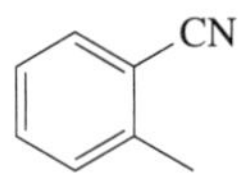	H_2O_2, NaOH, EtOH (aq.)	43OSC(2)586 50JOC800
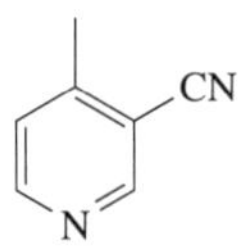	Ion exchange resin (basic), H_2O	48JA3945 60JOC560
PhCN	NaOH, Me_2SO	65JCS1290
PhCN	KOH, Bu^tOH	76JOC3769

5.06.4.1.3 From rearrangement reactions

As already discussed (Section 5.06.2.1.3) the standard Beckmann rearrangement of an aldoxime was not generally a viable route to *N*-unsubstituted amides ⟨60OR(11)1⟩. However, more recent developments include the use of nickel as a catalyst ⟨61JA1983⟩ or the use of silica gel as a catalyst, giving a range of aromatic primary amides from the corresponding aldoxime in 81–93% yield ⟨74T2899⟩.

5.06.4.1.4 Using other methods

One preparative method which is specific to aromatic amides is the Gatterman amide synthesis ⟨1888LA(249)244⟩, which involves carbamoylation of the aromatic ring and is mechanistically similar to Friedel–Crafts acylation procedures. Some developments are shown in Equations (86) ⟨60HCA1473⟩, (87) ⟨63TL77⟩, (88) ⟨68AG(E)172⟩ and (89) ⟨81S977⟩.

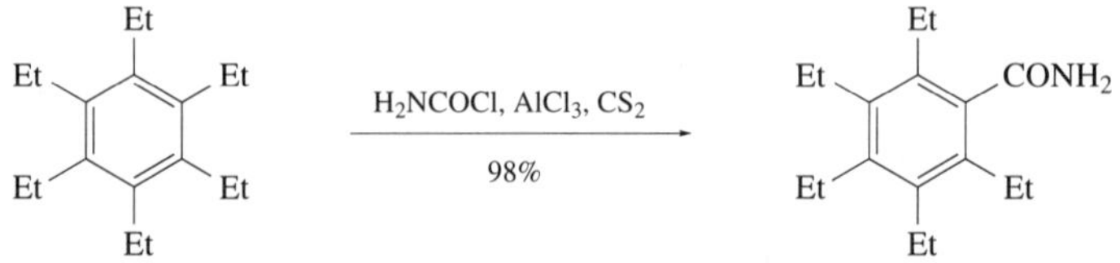

(86)

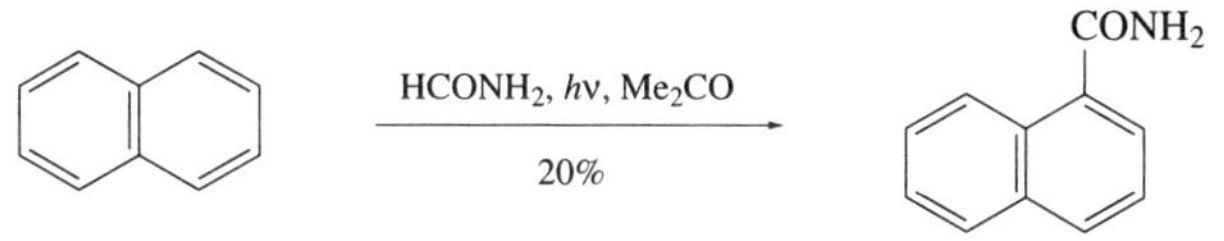

(87)

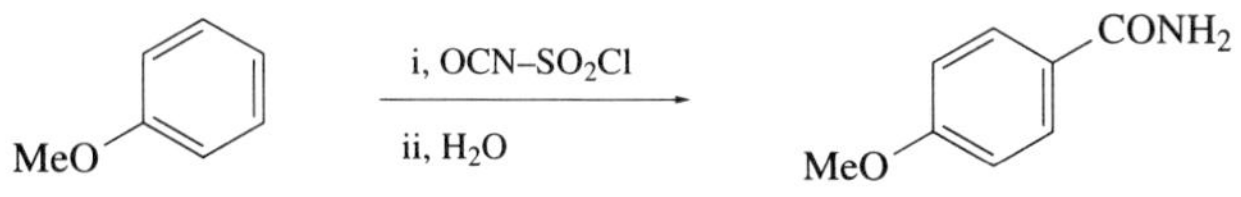

(88)

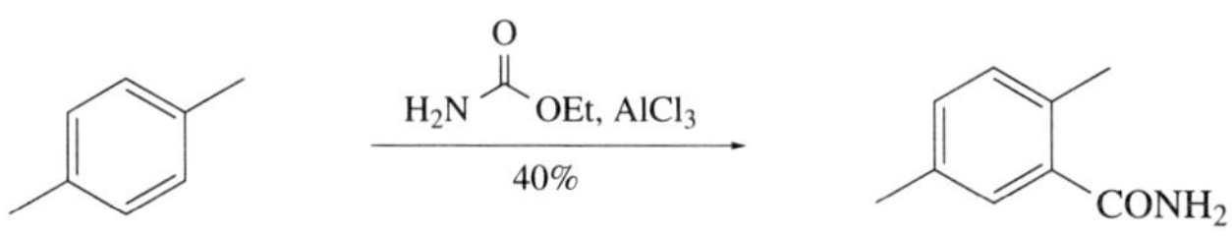

(89)

An alternative route towards primary amides of this type is the addition of aryl Grignard reagents to trimethylsilyl isocyanate (Equation (90)) ⟨75TL981⟩. Several oxidation procedures have been developed, as novel routes to a limited number of arylamides (Scheme 46 ⟨79CC970⟩ and Equations (91) ⟨66CC17⟩, (92) ⟨88CPB125⟩ and (93) ⟨90TL7237⟩.

$$\mathrm{PhMgBr} \xrightarrow[\text{49\%}]{\text{i, TMS-NCO; ii, } NH_4Cl,\ H_2O} \mathrm{PhCONH_2} \quad (90)$$

$$\mathrm{ArCHO} \xrightarrow[\text{58–89\%}]{\text{Ni peroxide, } NH_3,\ Et_2O} \mathrm{ArCONH_2} \quad (91)$$

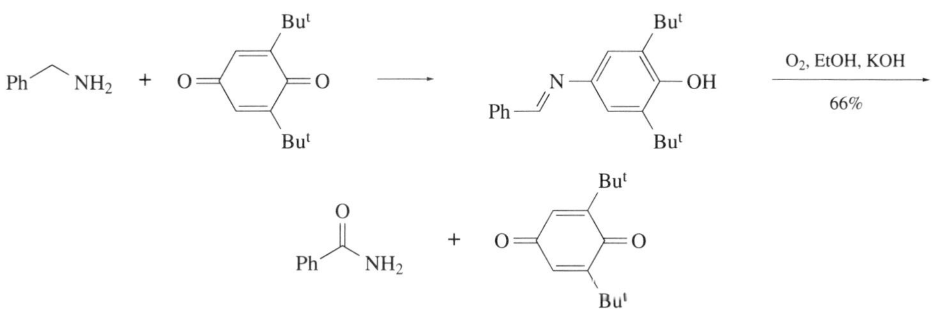

Scheme 46

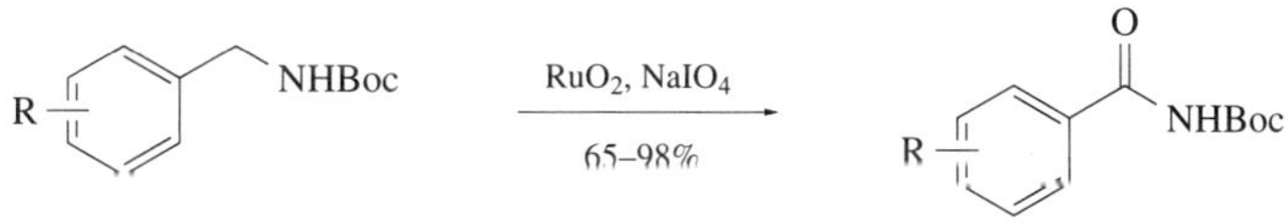

(92)

$$\mathrm{PhCHO} \xrightarrow[\text{78\%}]{\text{NBS, AIBN, } HNEt_2} \mathrm{PhCONEt_2} \quad (93)$$

5.06.4.2 *N*-Alkylarylamides

5.06.4.2.1 From acids and acid derivatives

Many of the procedures outlined in Section 5.06.2.2.1 for the acylation of amines are applicable here. Specific examples of mostly simple methods are shown in Equations (94) ⟨70TL1791⟩, (95) ⟨54JA1372⟩, (96) ⟨82BSF(2)167⟩ and (97) ⟨41OSC(1)99⟩ while Table 21 provides examples of coupling reagents employed to prepare this class of compound.

$$PhCO_2Et \xrightarrow[88\%]{Bu^nLi,\ Bn_2NH,\ THF} PhCONBn_2 \quad (94)$$

$$PhCOCl \xrightarrow[97\%]{DMF,\ 150\ ^\circ C} PhCONMe_2 \quad (95)$$

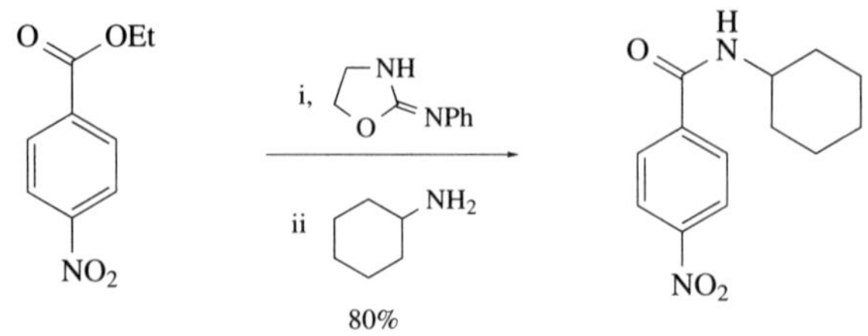

(96)

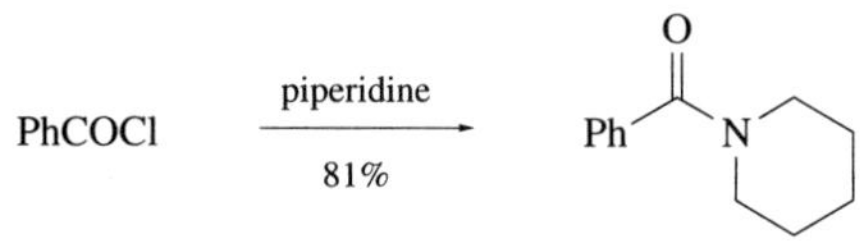

(97)

Table 21 Coupling reagents used to prepare *N*-alkylarylamides.

Reagents	*Ref.*	*Reagents*	*Ref.*
dcc	68JA4706	PPh_3, CCl_4	71JOC1305
dcc, 2-thiolthiazoline	80TL841	$NCPO(OEt)_2$, Et_3N	73TL1595
o-$NO_2C_6H_4SCN$, $Bu^n{}_3P$	79JOC2945	(PhO)(PhNH)POCl, Et_3N	82S288
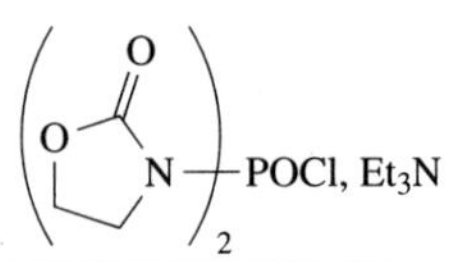	84S413	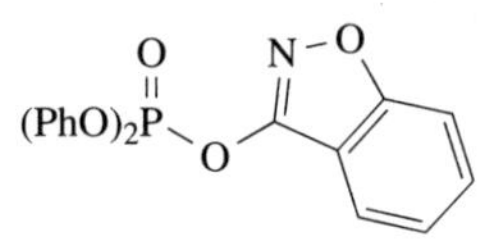	85JOC760

5.06.4.2.2 *From hydrolysis of nitriles*

Hydrolysis and alkylation starting from aromatic nitriles has been accomplished under relatively mild conditions, as shown in Equations (98) ⟨78S303⟩, (99) ⟨79S274⟩ and (100) ⟨81JOC4727⟩.

$$PhCN \xrightarrow[94\%]{i,\ KOH,\ Bu^tOH;\ ii,\ n\text{-}C_6H_{13}I} PhCONHC_6H_{13} \quad (98)$$

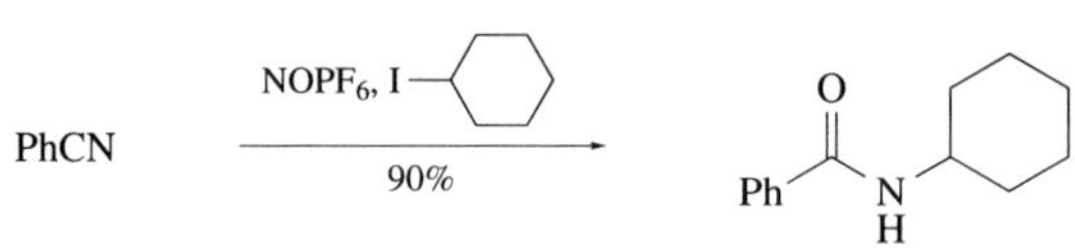

(99)

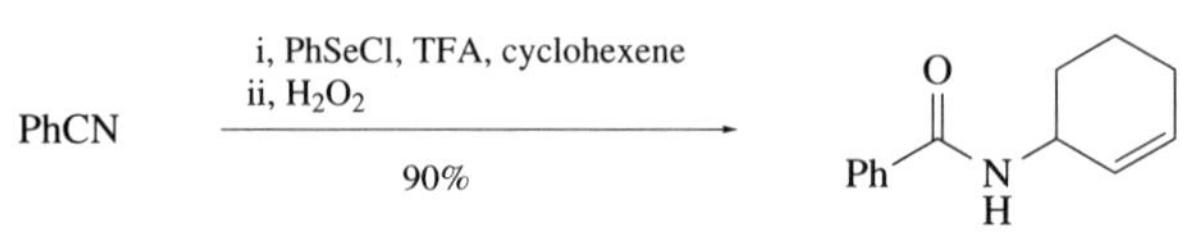

(100)

5.06.4.2.3 *Using other methods*

(i) N-*Alkylation of amides*

The transformation of one arylamide into another is an important route to many complex amides, as illustrated in Equations (101) ⟨54CB1294⟩, (102) ⟨75CC320⟩, (103) ⟨79T2169⟩ and (104) ⟨81S1005⟩.

One such procedure which gives access to dihydro isoquinolone derivatives is outlined in Equation (105) ⟨87JOC5378⟩. The ring closure generates a mole of base which deprotonates the isoquinolone, which can then be trapped with a range of electrophiles (in this case methyl iodide).

$$\mathrm{PhCONBu_2} \xrightarrow[67\%]{\mathrm{C_5H_5N{\cdot}HCl,\ 190\text{–}200\ ^\circ C}} \mathrm{PhCONHBu} \qquad (101)$$

$$\mathrm{PhCONH_2} \xrightarrow[80\%]{\text{i, ClCH}_2\text{SMe, MeSO}_3\text{H; ii, Raney Ni}} \mathrm{PhCONHMe} \qquad (102)$$

$$\mathrm{PhCONH_2} \xrightarrow[99\%]{\text{i, KOH, DMSO; ii, MeI}} \mathrm{PhCONMe_2} \qquad (103)$$

$$\mathrm{PhCONH_2} \xrightarrow[79\%]{\mathrm{Bu^nBr,\ NaOH,\ PTC}} \mathrm{PhCONBu_2} \qquad (104)$$

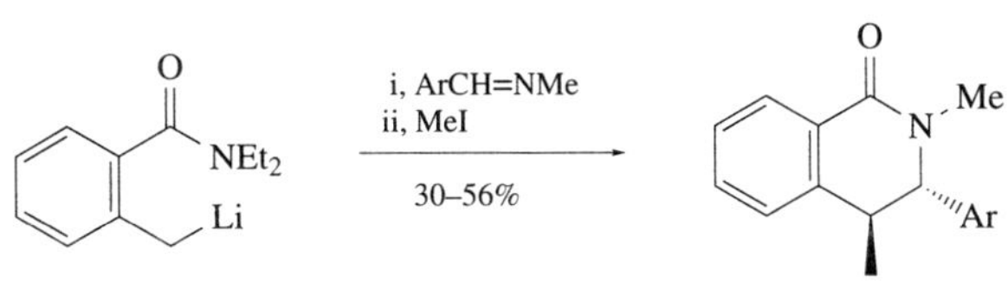

(105)

(ii) Carbonylation reactions

Several carbonylation procedures towards *N*-alkylarylamides have been developed as illustrated in Equations (106) ⟨74JOC3327⟩, (107) ⟨78JOC1684⟩, (108) ⟨86CC841⟩ and (109) ⟨86TL3931⟩. Other similar examples have also appeared ⟨82CC741, 82JOM(231)C12, 85JA4577⟩.

$$\mathrm{PhBr} \xrightarrow[79\%]{\mathrm{CO,\ PdBr_2(PPh_3)_2,\ BnNH_2}} \mathrm{PhCONHBn} \qquad (106)$$

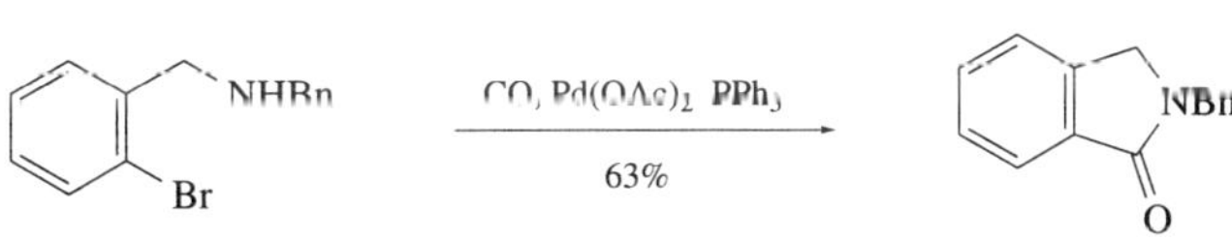

(107)

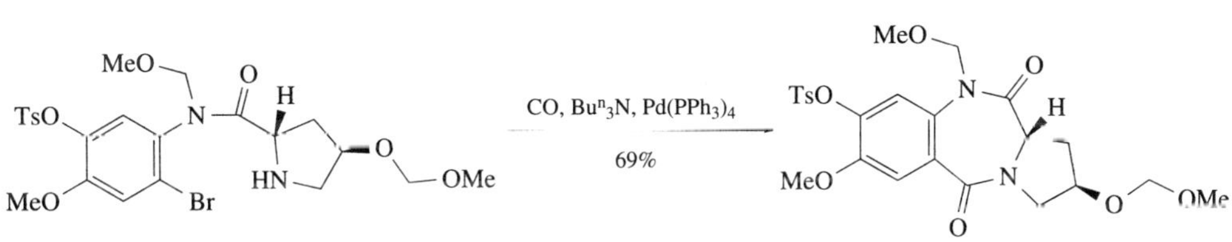

(108)

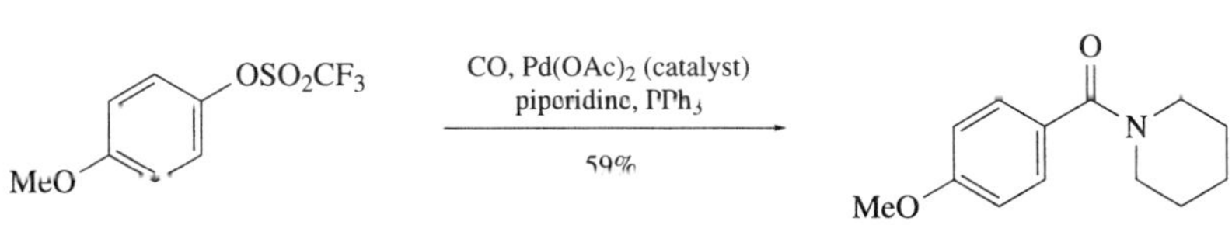

(109)

(iii) Miscellaneous reactions

Nucleophilic addition of an aryl Grignard reagent to an isocyanate is another route to *N*-alkylarylamides (Equations (110) ⟨68CB502, 86T885⟩ and (111) ⟨86TL501⟩. A similar procedure using carbamoyl chloride as the electrophile has also been reported ⟨87JA3402⟩. Finally, an example of the range of *N*-alkylarylamides produced from the corresponding aryl aldehyde, is outlined in Equation (112) ⟨82TL3255⟩.

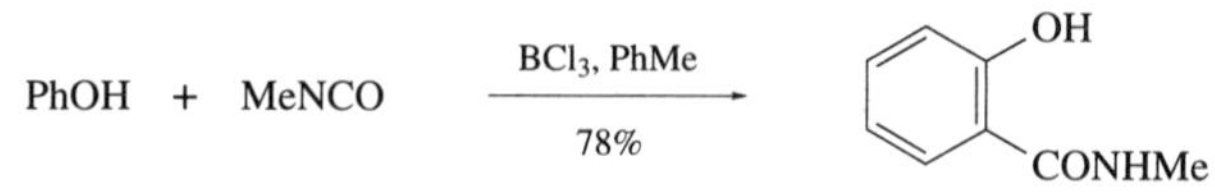

(110)

(111)

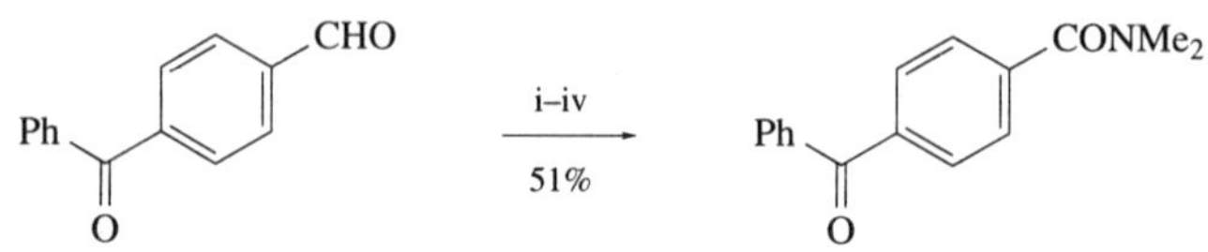

(112)

i, TMS-CN; ii, lithium diisopropylamide; iii, $Ph_2PO_2NMe_2$; iv, H_3O^+

5.06.4.3 *N*-Alkenylarylamides

The synthesis of enamides has been reviewed ⟨78S489, 87S421⟩. However, there are relatively few examples of this class of amide, and most of the procedures are variations on the *N*-alkylation theme (see Equations (113) ⟨81S729⟩, (114) ⟨82JOC3787⟩ and (115) ⟨91CL1443⟩).

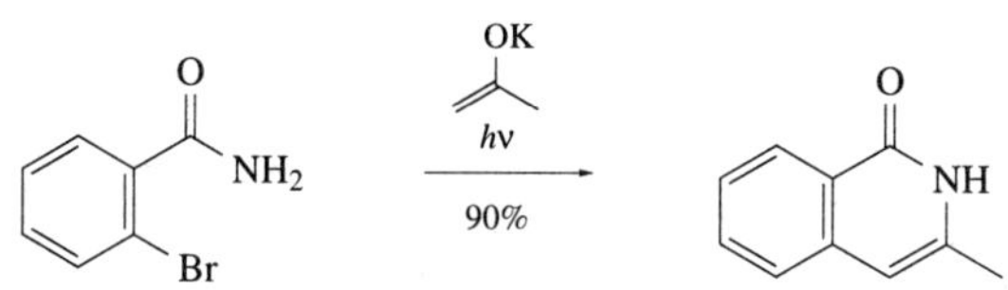

(113)

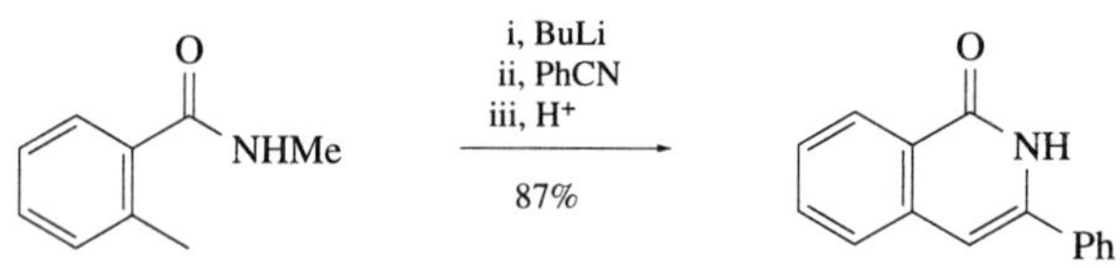

(114)

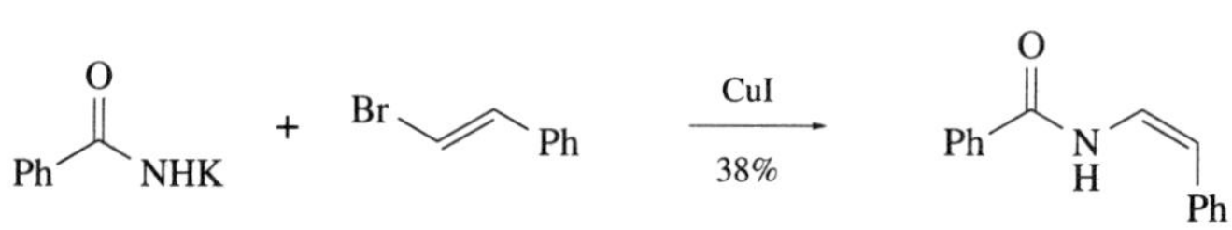

(115)

5.06.4.4 *N*-Arylarylamides

5.06.4.4.1 From acids and acid derivatives

These compounds should be accessible through most of the methods discussed previously; a range of simple procedures is outlined in Table 22 and Equations (116) ⟨55OSC(3)765⟩ and (117) ⟨55JA6214⟩.

Table 22 Acylation of $PhNH_2$.

Reagents	*Ref.*	*Reagents*	*Ref.*
$PhCO_2H$	41OSC(1)82	$PhCO_2Et$, LAH	62JOC1042
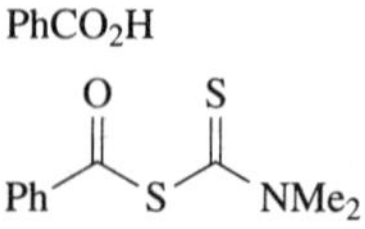	72TL4785	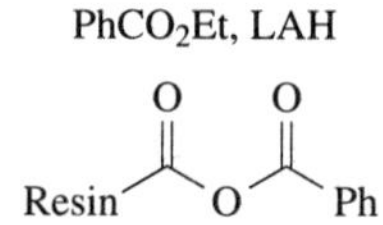	73TL1627

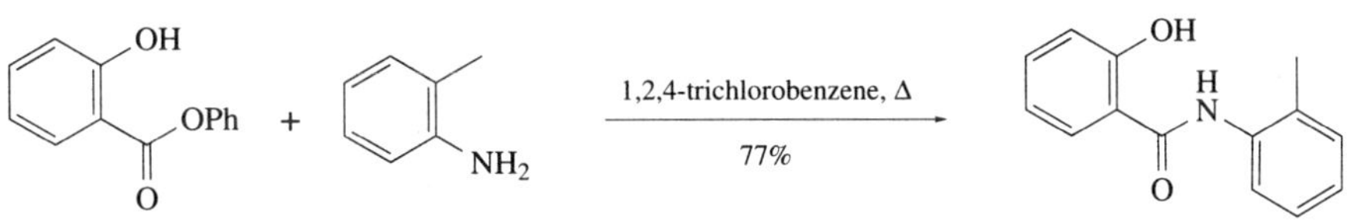

(116)

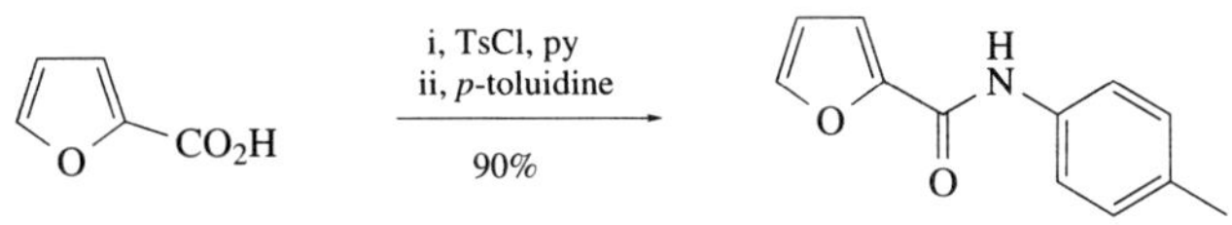

(117)

5.06.4.4.2 *Using* in situ *coupling reagents*

Table 23 lists coupling reagents used primarily for the coupling of aromatic acids with aromatic amines.

Table 23 Coupling reagents used to prepare ArCONHAr[1].

Reagents	*Ref.*	*Reagents*	*Ref.*
$(EtO)_2POCN$, Et_3N	73TL1595	$ClSO_2NCO$, Et_3N	82S506
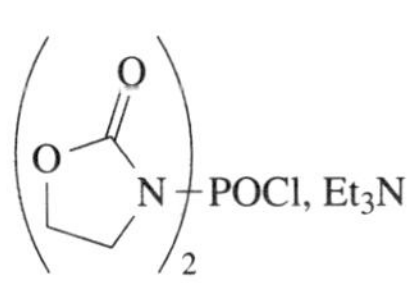	80S547	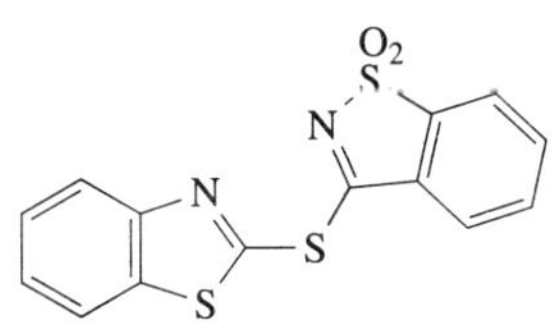	82S933
P_2I_4, 2,6-lutidine, CCl_4	83CL449	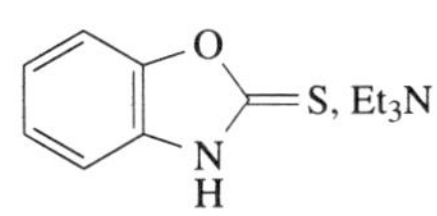	81S991
SO_2ClF, Et_3N	80S661	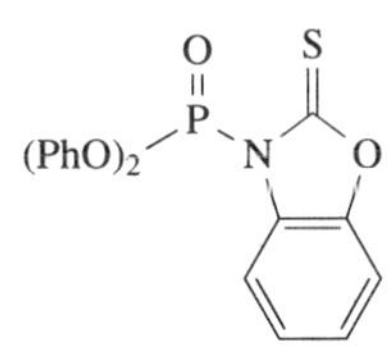	92BCJ1636
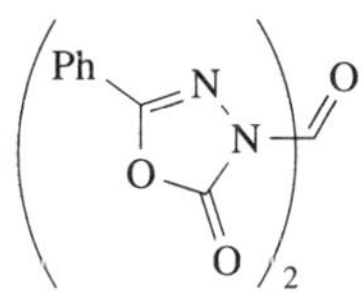	89BCJ539		

5.06.4.4.3 *Rearrangement reactions*

Equation (118) ⟨53JA5905⟩ and Schemes 47 ⟨65OR(14)1⟩ and 48 ⟨74JOC3924⟩ outline three examples of the diverse rearrangement chemistry towards *N*-arylarylamides. The second example is from a review of the Chapman rearrangement ⟨65OR(14)1⟩.

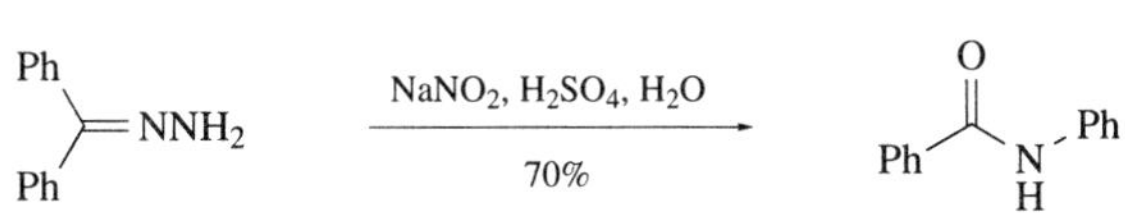

(118)

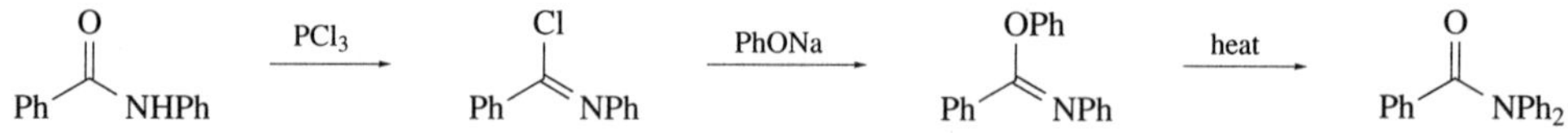

Scheme 47

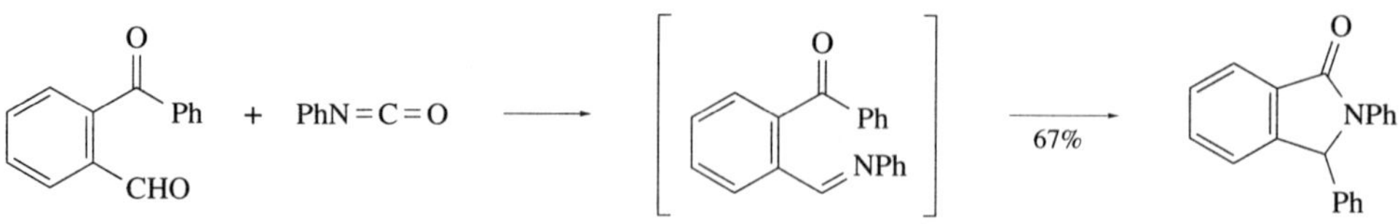

Scheme 48

5.06.4.4.4 *Miscellaneous reactions*

Finally, an early procedure for the Grignard addition to isocyanates ⟨38JA540⟩, an example of a Gatterman amide synthesis (see Section 5.06.4.1.4), and two carbonylation procedures which are all specific examples of *N*-arylarylamide synthesis are outlined in Scheme 49 ⟨38JA540⟩ and Equations (119) ⟨55CB301⟩, (120) ⟨74JOC3327⟩ and (121) ⟨75BCJ2091⟩.

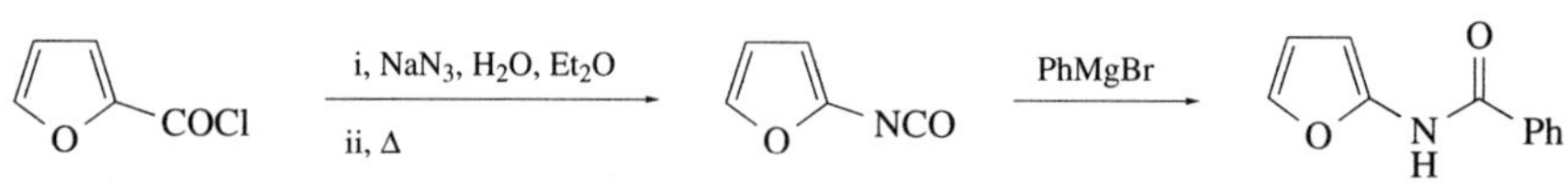

Scheme 49

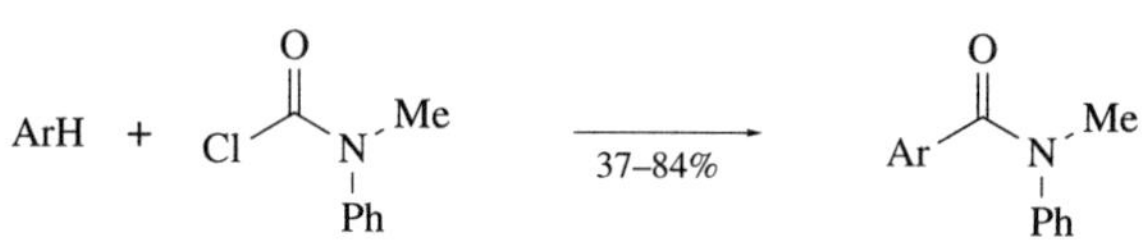
(119)

$$\text{PhBr} \xrightarrow[\text{94\%}]{\text{CO, PdBr}_2(\text{PPh}_3)_2\text{, PhNH}_2} \text{PhCONHPh} \qquad (120)$$

$$\text{PhI} \xrightarrow[\text{94–100\%}]{\text{CO, Pd (Black), PhNH}_2} \text{PhCONHPh} \qquad (121)$$

5.06.5 AMIDES OF ALKYNOIC ACIDS

There are few examples specific to the synthesis of this class of compound; as with alkenoic acids, there is the possibility of [1,4]-addition to alkynoic acids which may complicate some acylation procedures. One preparation of a range of alkynoic amides from the corresponding ester was reported to proceed in high yield in a methanolic solution of ammonia ⟨41JA1151⟩ and a similar procedure using 1 : 1 ethylene glycol: liquid ammonia has also been reported ⟨73PAC145⟩. Other simple examples have been prepared from an acyl halide and ammonia ⟨53CRV237⟩.

In 1993, the amide shown in Equation (122), which is an advanced intermediate in a natural product synthesis, was prepared using the BOP coupling reagent (see Table 9, Section 5.06.2.2.7) ⟨93TL2677⟩. The final examples given in Equations (123) ⟨75TL981⟩, (124) ⟨77S777⟩, (125) ⟨81LA1844⟩ and (126) ⟨88S621⟩ are specific to the preparation of alkynoic acid derivatives. These involve addition of acetylide anion equivalents to isocyanates, or a carbamoyl chloride.

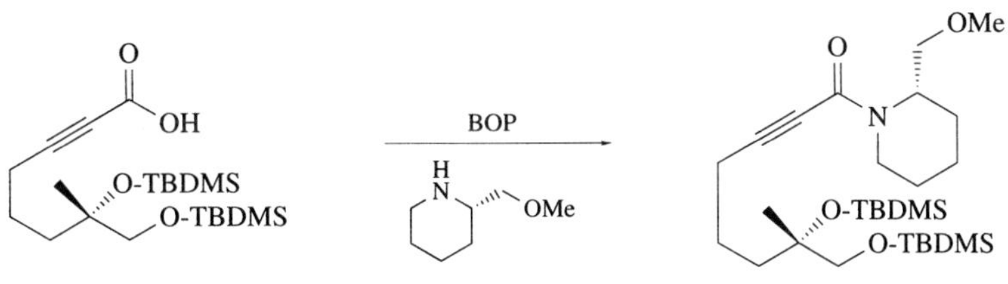

(122)

TBDMS = *t*-butyldimethylsilyl

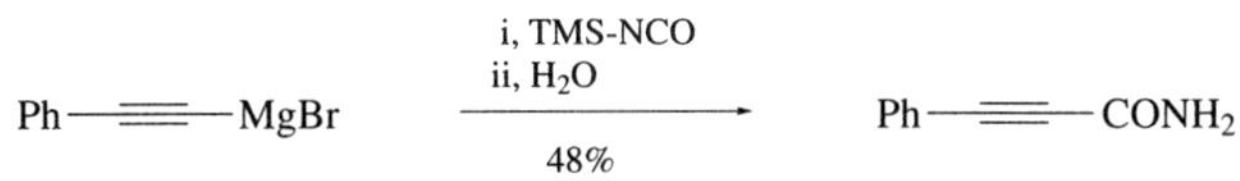

(123)

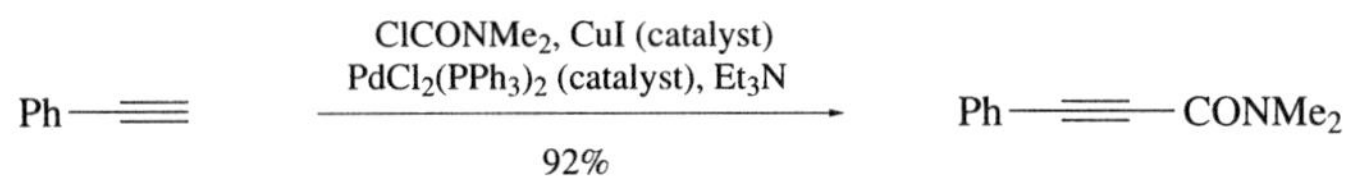

(124)

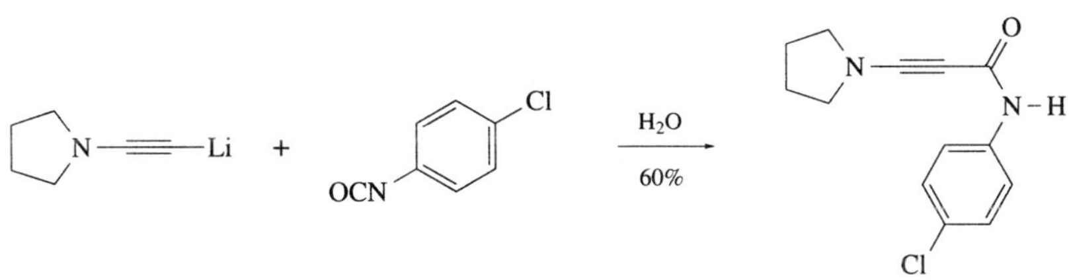

(125)

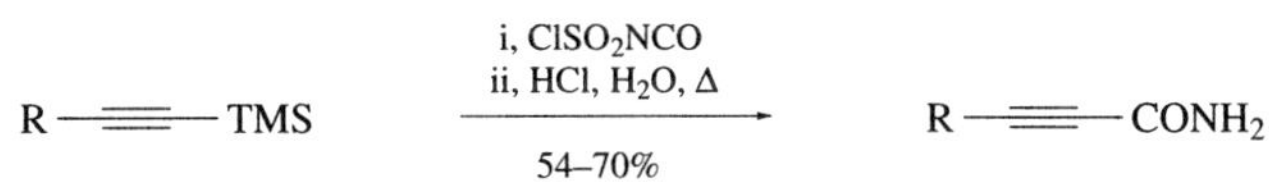

(126)

5.06.6 *N*-ACYL AMIDES

The principle methods for the preparation of imides closely parallel the early procedures used for amide preparation. These are the acylation of amines with dicarboxylic acids or cyclic anhydrides, and the cyclisation of amidic acids or esters; both these reactions often require fierce heating. Other methods include the heating of diamides, the hydrolysis of dinitriles and the reaction of isocyanates with anhydrides and amides. These methods have been comprehensively reviewed ⟨70CRV439, B-70MI 506-02, B-70MI 506-03⟩. A selected number of illustrative practical examples are given in Scheme 50 ⟨55OSC(3)151, 63OSC(4)106⟩ and Equations (127) ⟨41OSC(1)457⟩, (128) ⟨43OSC(2)562⟩, (129) ⟨63OSC(4)496⟩ and (130) ⟨73OSC(5)973⟩.

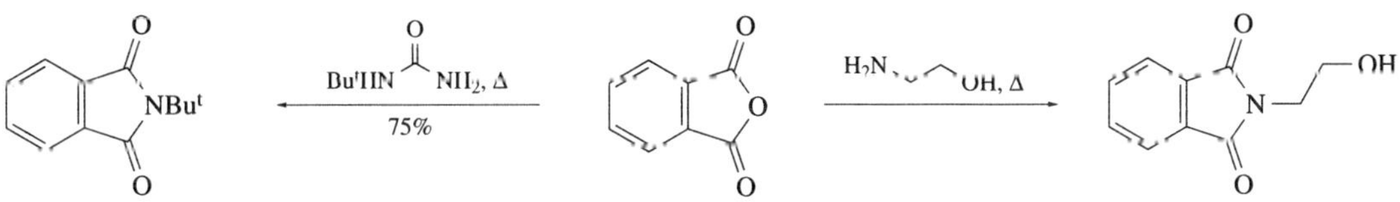

Scheme 50

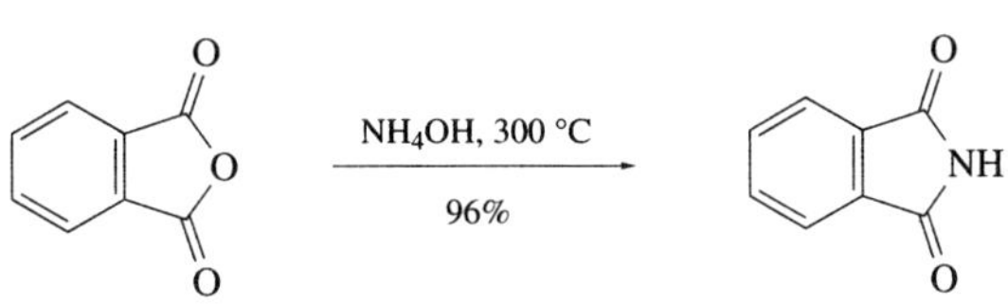

(127)

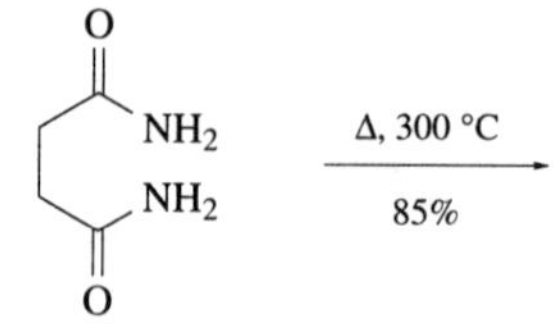

(128)

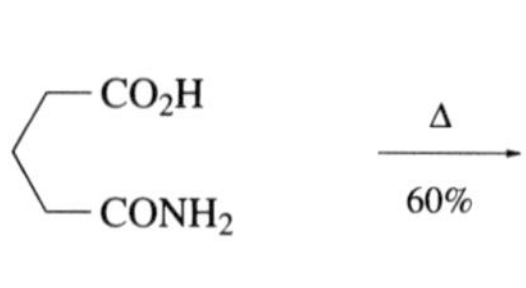

(129)

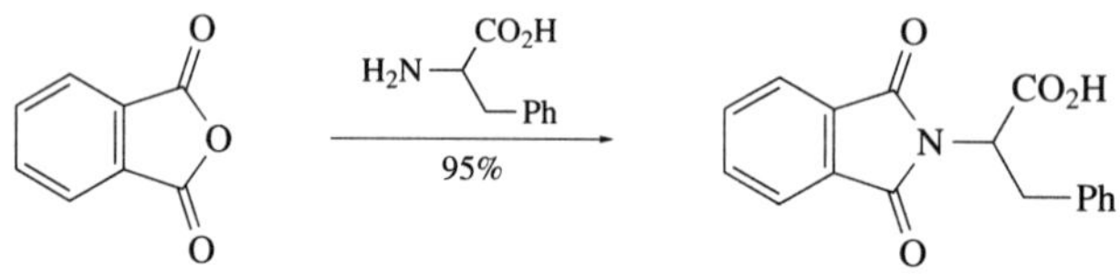

(130)

More recently, milder procedures which are more tolerant of other functional groups have been developed. These strategies include many familiar methods for acylation such as the use of mixed anhydrides (Equations (131)–(133)) ⟨73OSC(5)944, 87JMC498, 90JOC560⟩. Another established procedure for amide formation is the use of active esters. Scheme 51 and Equation (134) show two examples of imide syntheses via active esters ⟨86TL919, 90TL5201⟩. The latter example is notable in that the reagent hexamethyldisilazide (HMDS) is the source of ammonia and also activates the intermediate amidic acid as a silyl ester, hence facilitating cyclisation.

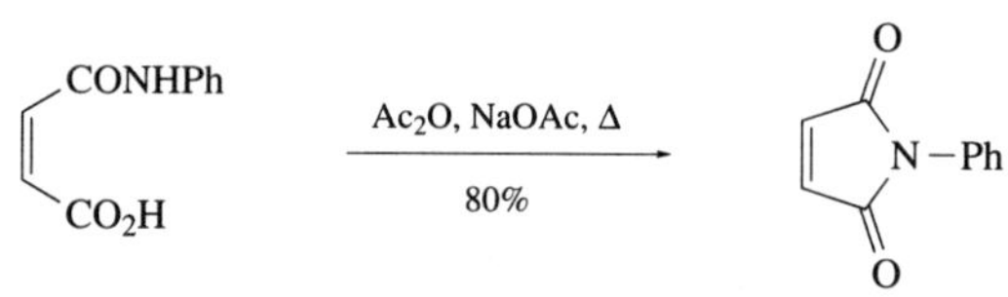

(131)

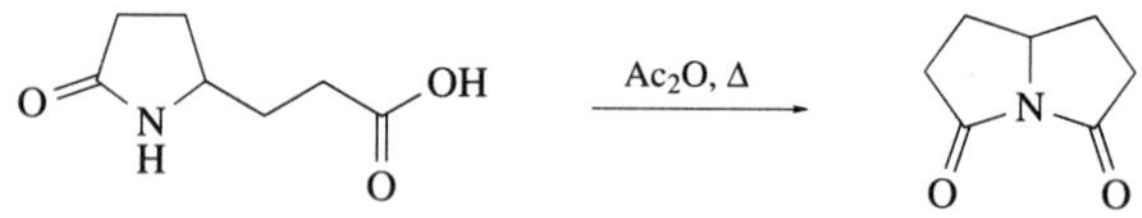

(132)

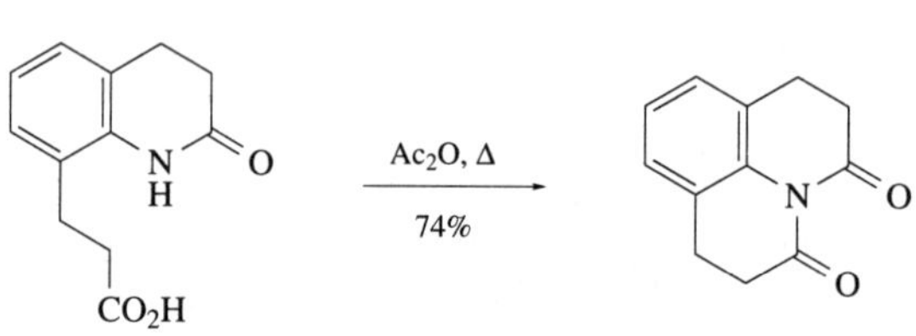

(133)

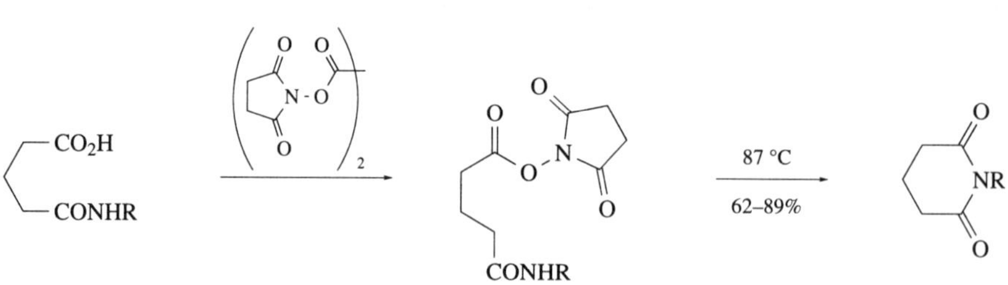

Scheme 51

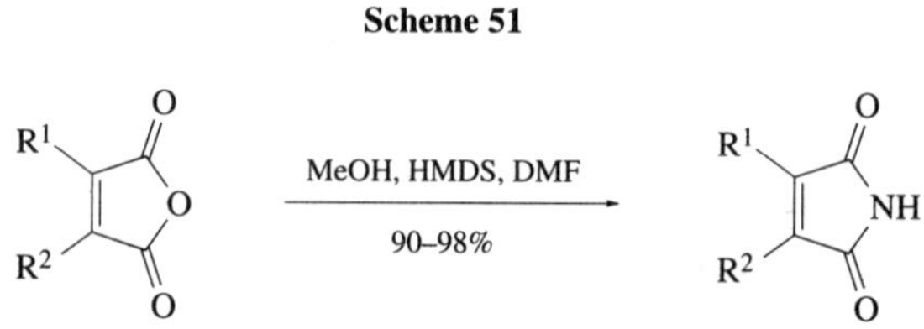

(134)

One final acylation method involves the reaction of an enolate with an isocyanate and has been used in the synthesis of macrocyclic imides (Equation (135)) ⟨89HCA1522, 92HCA62⟩. Another commonly used method for preparing imides, which is independent of acylation methodology, is the oxidation of amides (Equations (136)–(139)) ⟨58JA6682, 69JA3992, 75CC71, 85CPB5515⟩. Ruthenium tetroxide has been used for the preparation of acyclic imides in good yields ⟨87CPB364⟩. Although acyclic imides can also be prepared using acylation methods ⟨B-70MI 506-02⟩, many of these are susceptible to side-reactions such as over-acylation or decomposition to nitriles.

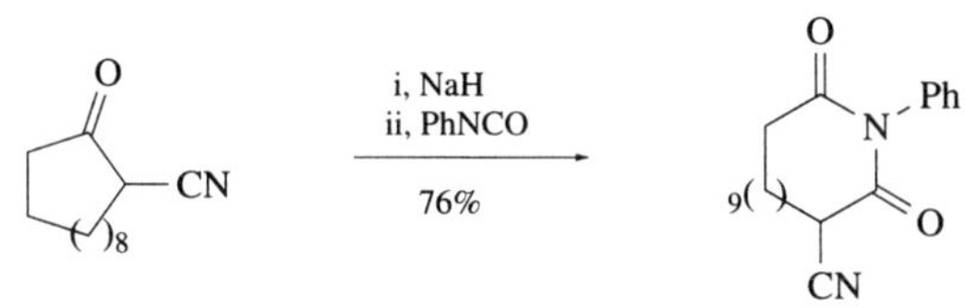

(135)

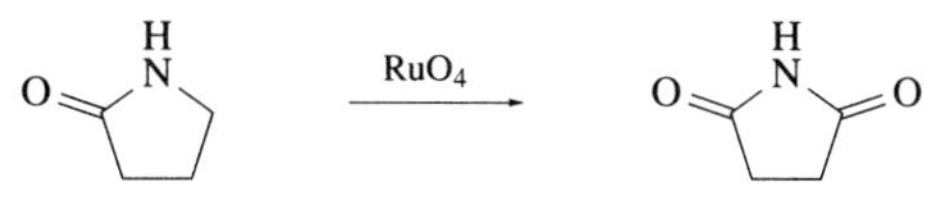

(136)

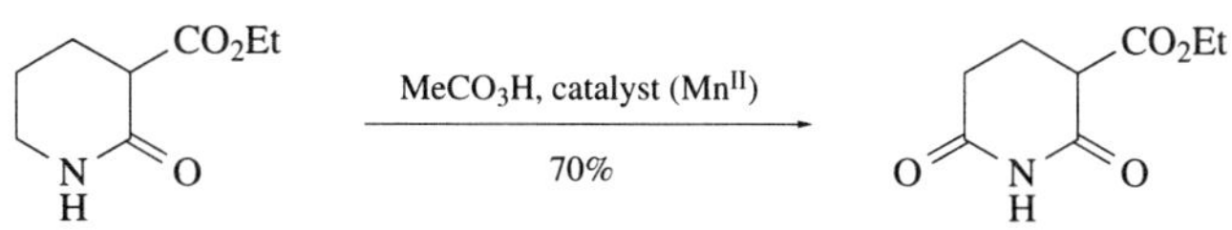

(137)

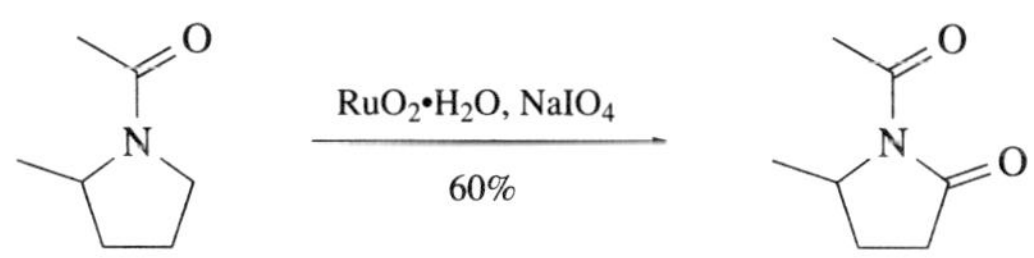

(138)

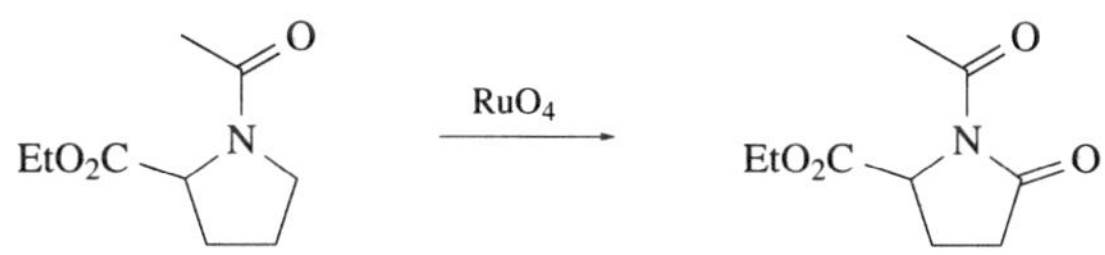

(139)

5.07
N-Heterosubstituted Amides

RICHARD D. CONNELL
Bayer Corporation, West Haven, CT, USA

5.07.1 AMIDES SUBSTITUTED BY FLUORINE, CHLORINE, BROMINE, OR IODINE

The synthesis of *N*-haloamides has been reviewed ⟨B-70MI 507-01, 85HOU(E5)1135⟩.

5.07.1.1 *N*-Fluoroamides

5.07.1.1.1 From amides

The first reported attempt to fluorinate amides such as acetamide or *N*-methylacetamide with fluorine gas resulted only in the formation of acetic acid, carbon dioxide, and fluoroamines ⟨66JCS(C)1514⟩. The first successful fluorination of formamides, amides, and lactams was described by Grakauskas in 1970 ⟨70JOC1545⟩.

Since then, a number of improvements on the fluorination of amides with both fluorine and fluorine transfer reagents have been reported. Barton was the first to investigate the use of alternate sources of fluorine for the preparation of *N*-fluoroamides ⟨74JCS(P1)732⟩. Although modest yields were obtained using perfluorofluoroxy reagents under the initially reported conditions, modification of these conditions by Hammer has lead to reproducibly good yields of *N*-fluoroamides and lactams ⟨84JA1543⟩. A summary of the various substrates, reagents, conditions and yields are found in Table 1.

5.07.1.1.2 From O-silyl enol ethers

The reaction of pyridine (**1**) with fluorine leads to formation of 1-fluoropyridone (**2**) and SiF (Equation (1)) ⟨83JOC761⟩.

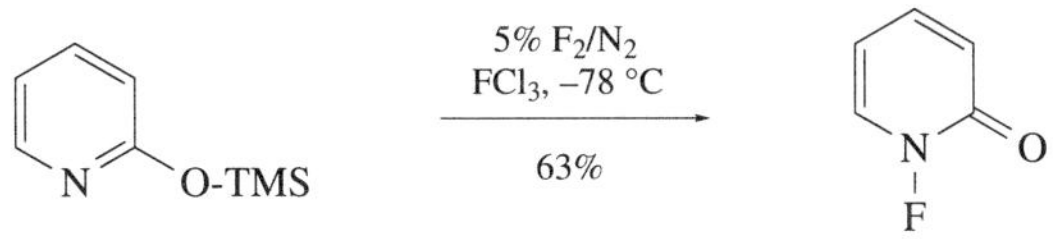

(1) **(2)** (1)

5.07.1.1.3 From perfluoromethanimine

Perfluoromethanimine (**3**) is readily attacked by fluoride, forming an intermediate perfluoromethanamine (**4**). Treatment of this intermediate with acyl fluorides provides the corresponding fluoroamides (Scheme 1) ⟨83JOC771⟩, see also ⟨74IC2705, 82POL129⟩.

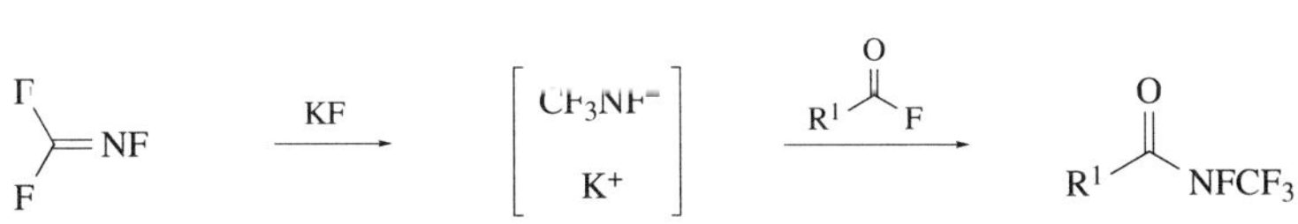

(3) **(4)**

R^1 = Me, 0%; R^1 = CF_3, 68%

Scheme 1

Table 1 Fluorination of amides and lactams.

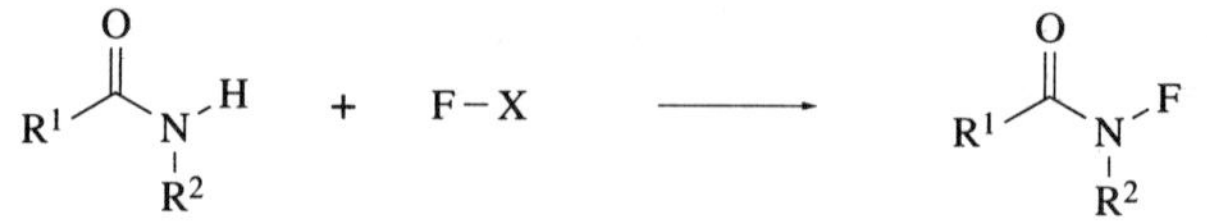

R[1]CONHR[2]	*Conditions*	*R[1]CONFR[2]*	*Ref.*
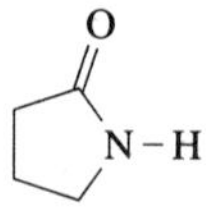	F_2, neat 0–5 °C (24%)	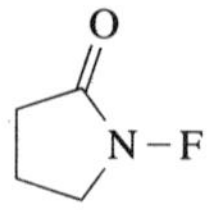	70JOC1545
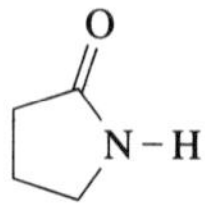	F_2, freon, 0 °C 15 min (76%)	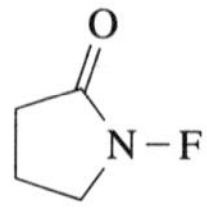	90JOC3374
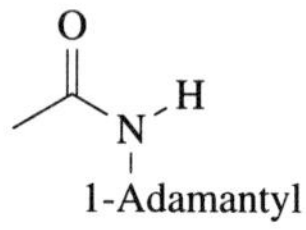	CF_3OF, $CHCl_3$ RT (4%)	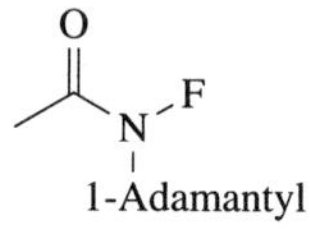	74JCS(P1)732
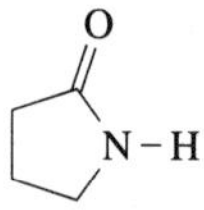	CF_3OF, $CHCl_3$ −78 °C (50–70%)	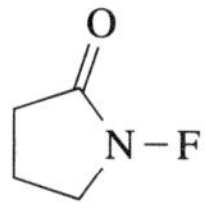	84JA1543
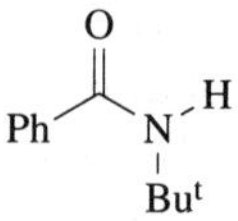	$(PhSO_2)_2NF$ KH, Et_2O, THF 30–40 °C (10%)	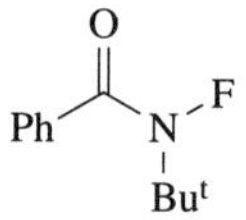	92T1595
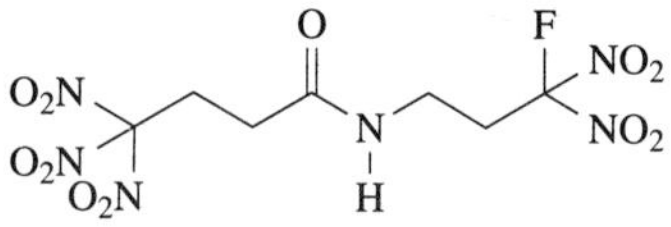	F_2, N_2, HOAc 14–16 °C (60%)	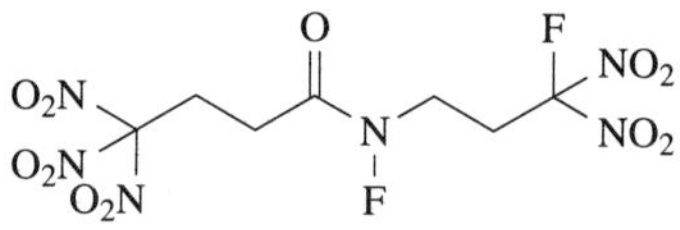	87IZV2857

5.07.1.1.4 From fluoroamines

Difluoroamine (HNF_2) reacts with tetramethyl-1,3-cyclobutanedione to provide *N*-fluoropyrrolidinedione ⟨69JOC1154⟩.

Cyclopentanone enol acetate was shown to react with tetrafluorohydrazine under acidic conditions, providing 3-difluoroamino-2-(*N*-fluoro)-azacyclohexanone ⟨69JA4778⟩.

5.07.1.2 *N*-Fluoroimides

N-Fluoroimides are more stable and easier to prepare than *N*-fluoroamides. The fluorination of imides has been achieved by treating the imide with caesium fluoroxysulfate (Equation (2)) ⟨91T7447⟩ or xenon difluoride ⟨80ZOR1330⟩.

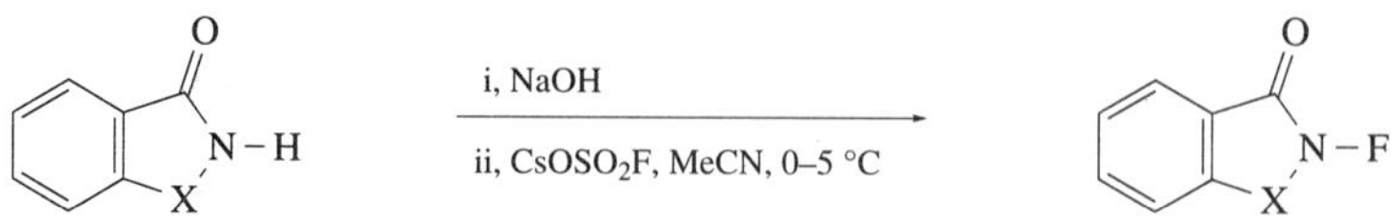

X = SO_2, 69%
X = CO, 48%

(2)

5.07.1.3 *N,N*-Difluoroamides

Freeman and co-workers have shown that α-diketones react under photolytic conditions with tetrafluorohydrazine (N_2F_4) to provide difluoroamides in good yields (Equation (3)) ⟨61JA3912⟩.

Acyl fluorides do not provide difluoroamides under these conditions. However, fluorinated acyl fluorides react with the HNF_2–KF adduct to provide difluoroamides ⟨71IC911⟩.

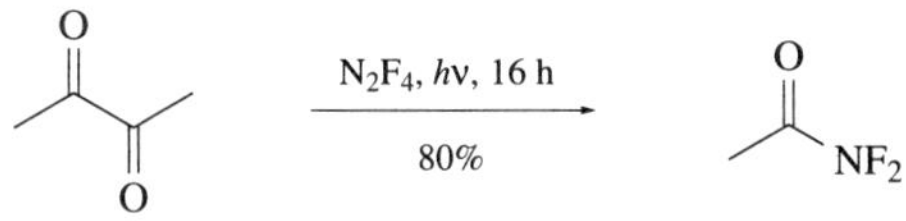

(3)

5.07.1.4 *N*-Chloroamides

N-Chloroamides are considerably more stable and easier to prepare than *N*-fluoroamides. However, certain classes of *N*-chloroamides are susceptible to rearrangement or decomposition. *N*-Chloroacetanilides ⟨84JOC5225⟩, for example, rearrange to chloroarylamides under certain conditions ⟨73MI 507-01⟩. Under basic conditions, *N*-chloroamides containing α-hydrogens have been reported to undergo Favorskii rearrangements ⟨61JA4469⟩, while *N*-chloroamides containing β-hydrogens are susceptible to the elimination of chloride under basic conditions ⟨76AG(E)294⟩. For *N*-acyl α-amino acids (**5**), this β-elimination provides a convenient route to α,β-unsaturated amino acids (**6**) (Scheme 2) ⟨77S457⟩.

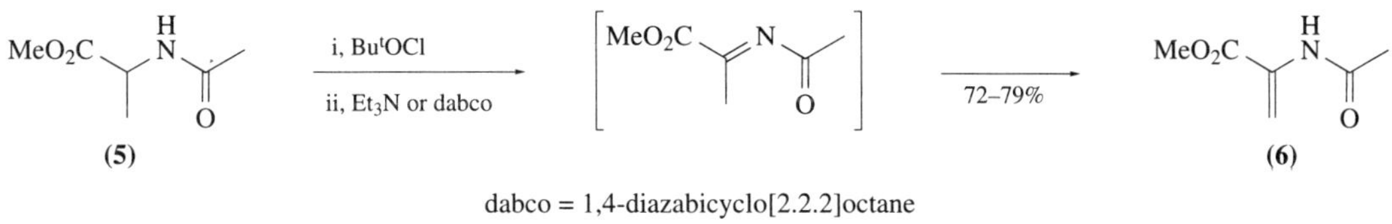

Scheme 2

5.07.1.4.1 From amides

Secondary and primary amides, lactams, and formamides are converted into the corresponding *N*-chloro derivatives under a variety of conditions using a variety of chlorinating reagents (Table 2). The use of chlorine gas allows for an economical synthesis that is amenable to large scale. Aside from chlorine gas, *t*-butyl hypochlorite (Bu^tOCl) is used most often for the chlorination of amides and lactams ⟨93SC2423⟩. The activity of Bu^tOCl has been reported to improve upon addition of borax ⟨78JOC3713, 82JOC4605, 94T11091⟩, sodium methoxide ⟨75CB2547⟩, or potassium *t*-butoxide ($KOBu^t$) ⟨76AG(E)294⟩. The use of NCS under basic conditions continues to emerge as a viable protocol for the chlorination of amides, especially for cases where the amide moiety contains electron-rich aromatic groups ⟨94H(38)159⟩.

5.07.1.4.2 From addition to imines

Cook has shown that ClF reacts with *N*-acyl imines (**7**) to provide the desired *N*-chloroimides (**8**) in quantitative yield (Scheme 3) ⟨91JOC5256⟩. The product is described as being thermally stable, but unstable in the presence of fluoride ion (and presumably other nucleophiles) owing to the facile displacement of the amine moiety.

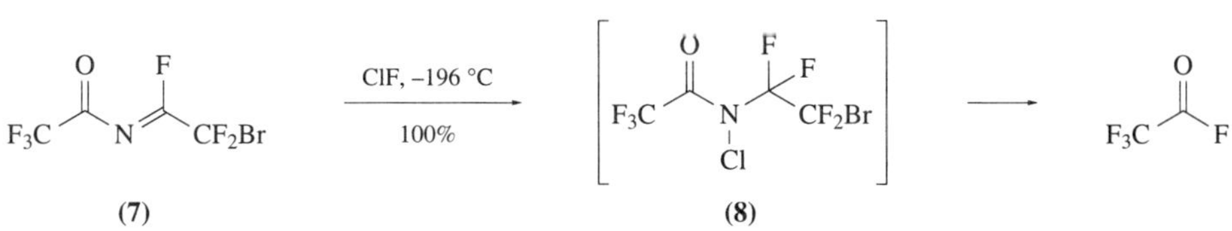

Scheme 3

Table 2 Chlorination of amides and lactams.

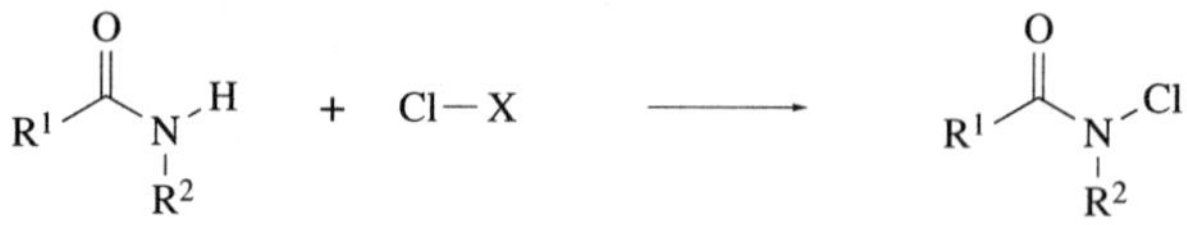

R^1CONHR^2	Conditions	$R^1CONClR^2$	Ref.
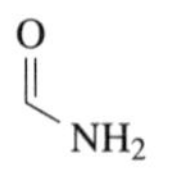	i, NaOCl, 0 °C ii, 2 mol l^{-1} H_2SO_4 CH_2Cl_2 (50%)	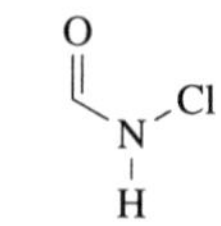	74JOC3136
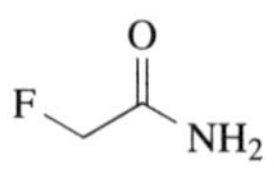	i, NaOCl, 0 °C ii, 2 mol l^{-1} H_2SO_4 CH_2Cl_2 (79%)	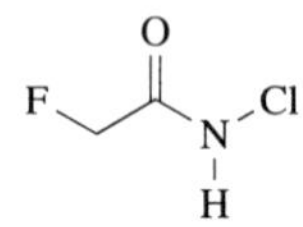	74JOC3136
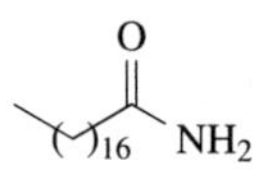	Cl_2, H_2O, 100 °C (90%)	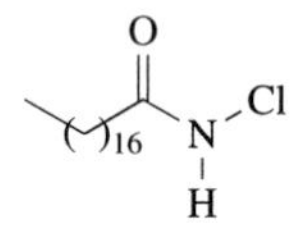	86OPP269
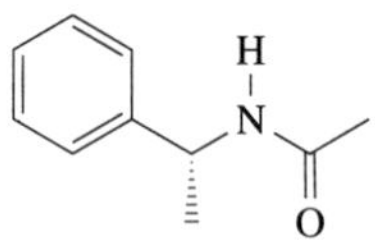	Bu^tOCl, MeOH RT, dark (100%)	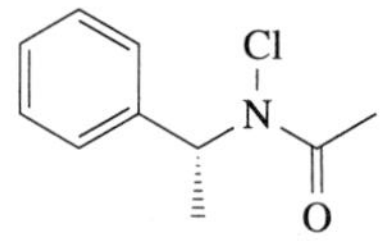	93SC2423
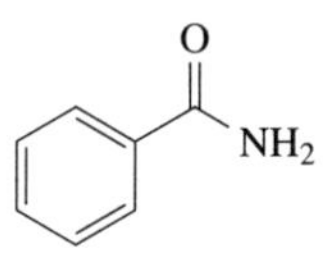	Bu^tOCl, MeOH 12–20 h (71%)	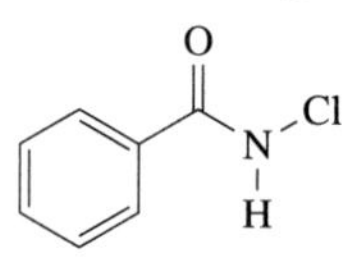	62JOC4532
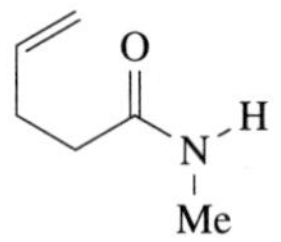	NaOCl, buffer $CHCl_3$, KOH (74%)	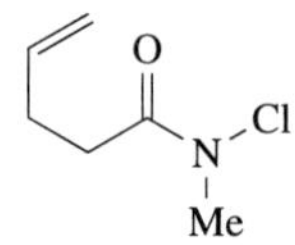	78JOC3750
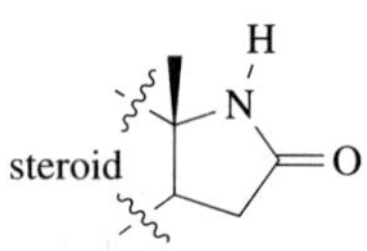	i, $KOBu^t$, THF ii, NCS, RT (76%)	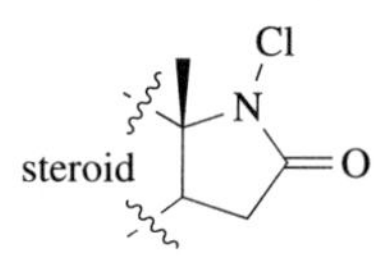	91H481
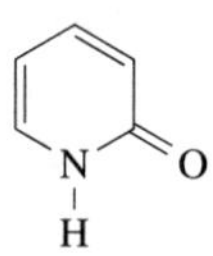	NaOCl, CH_2Cl_2, 30 s RT (59%)	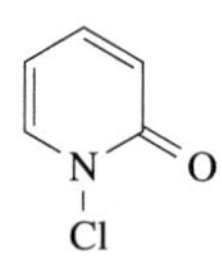	84JOC4784
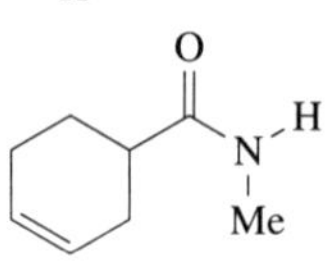	i, Bu^nLi, Et_2O, –78 °C ii, NCS (76%)	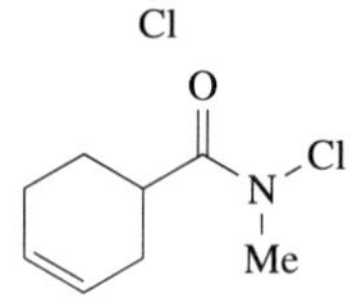	74JCS(P1)732

5.07.1.5 *N*-Chloroimides

The majority of methods used to prepare *N*-chloroimides have been directed toward the synthesis of *N*-chlorosuccinimide (NCS) and *N*-chlorophthalimide (Cl-Phth). These and other imide derivatives are almost always prepared by chlorination of the corresponding imide with chlorine ⟨22JCS2169, 79USP771219, 80USP781220⟩, *t*-butyl hypochlorite ⟨93SC2423⟩, trichloroisocyanuric acid ⟨75ZAAC(415)115, 91MI 507-05⟩, or chlorotricyanomethane (Equation (4)) ⟨77ZOB968⟩.

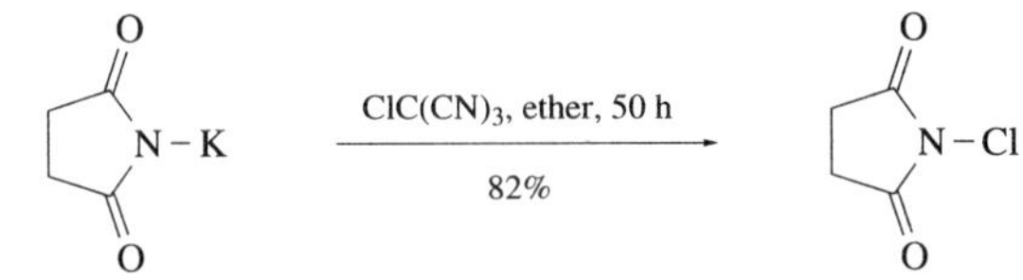

(4)

From a technology standpoint, the use of polymer-supported hypohalites ⟨87MI 507-03⟩ or chlorination by electrolysis using NaCl ⟨77MI 507-01, 80MI 507-02, 92MI 507-02⟩ may provide the potential for synthesis on a larger scale.

5.07.1.6 *N,N*-Dichloroamides

5.07.1.6.1 From amides

Treatment of amides with excess Bu^{t}OCl ⟨78ZOR1846⟩, sodium hypochlorite ⟨80CAR(79)205⟩, or calcium hypochlorite ⟨75JA1180⟩ provides the corresponding dichloroamides (Equation (5)). Care should be taken with these compounds since members of this family have been reported to detonate on standing ⟨73SA(A)613⟩.

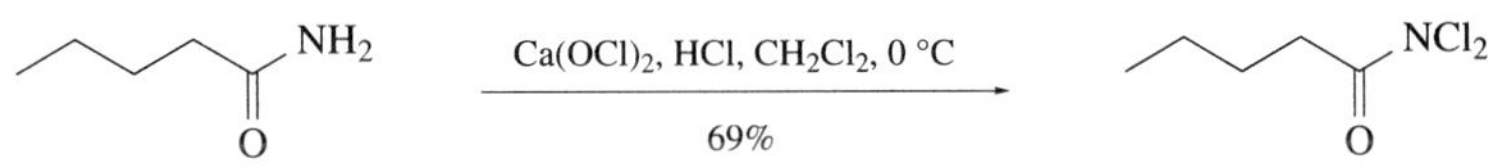

(5)

5.07.1.6.2 From perfluoroacyl iminosulfurdifluorides

Trifluoroacyl iminosulfurdifluorides (**9**) react with chlorine monofluoride to provide *N,N*-dichlorotrifluoroacetamide (**10**) (Equation (6)) ⟨71JCC788⟩.

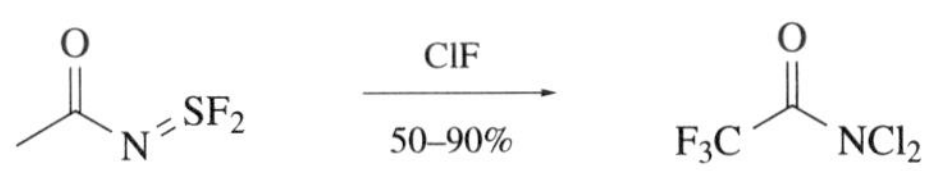

(6)

(9) **(10)**

5.07.1.7 *N*-Hetero-*N*-chloroamides

It has been shown that hydroxamates can be converted into the corresponding chloroamides using Bu^{t}OCl (Equation (7)). Attempts to prepare the corresponding *N*-bromo or *N*-iodo *N*-alkoxyamides were unsuccessful ⟨89JOC3394, 90T7247⟩.

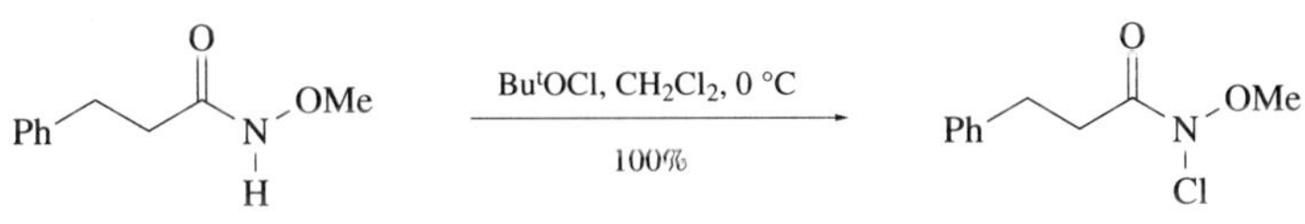

(7)

A similar reaction with other heteroamides and Bu^{t}OCl is not so straightforward. While the constrained azalactam (**11**) is converted in high yield to the corresponding chloro compound (**12**) (Equation (8)) ⟨85T575⟩, the acyclic acyl hydrazine (**13**) provides only the dimerized tetrazine (**14**) (Equation (9)). The reaction of silylamides with Bu^{t}OCl provides the isomeric *N*-chloroimidates rather than the desired *N*-chloro-*N*-silylamides ⟨87AJC701⟩.

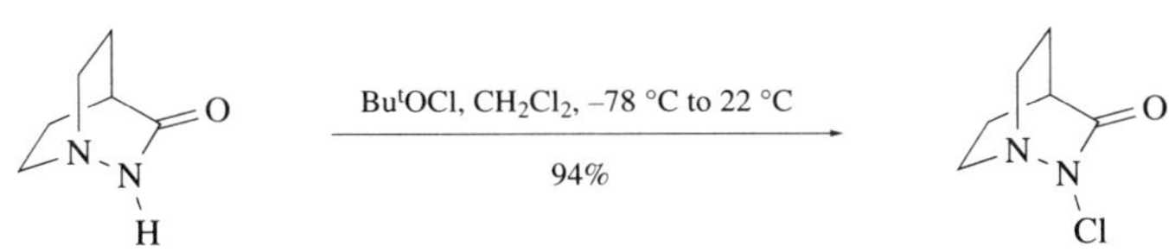

(8)

(11) **(12)**

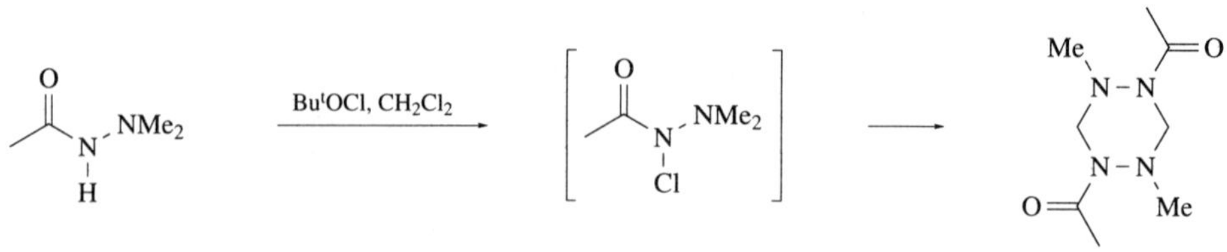

(13) (14) (9)

5.07.1.8 *N*-Bromoamides

The preparation of bromoamides is considerably more difficult than the preparation of chloroamides. The lower yields of isolated *N*-bromoamides are due to the (over)reactivity of most brominating agents and to the greater tendency of bromoamides to undergo Hofmann elimination.

N-Bromoamides are prepared almost exclusively from the corresponding amides. In a direct comparison of various brominating methods and conditions, acetyl hypobromite is reported to provide the highest yield of isolated bromoamide ⟨93SC2423⟩. However, the choice of brominating agent depends primarily on the substitution of the amide and on the ease of haloamide isolation (Table 3).

5.07.1.9 *N*-Bromoimides

5.07.1.9.1 From imides

Imides are converted to *N*-bromoimides upon treatment with sodium hypobromite ⟨61BSF2360⟩, sodium bromate ⟨93BCJ2426⟩, sodium bromite ⟨85BCJ769⟩, bromine chloride ⟨76S736⟩, or bromine under basic conditions ⟨50JCS747, 86JOC4959⟩. Imides can also be converted to *N*-bromoimides via electrolysis using aqueous sodium bromide solutions (Equation (10)). This process is especially practical considering that for every bromine atom that is discharged, one molecule of bromoimide is formed ⟨50JCS747⟩.

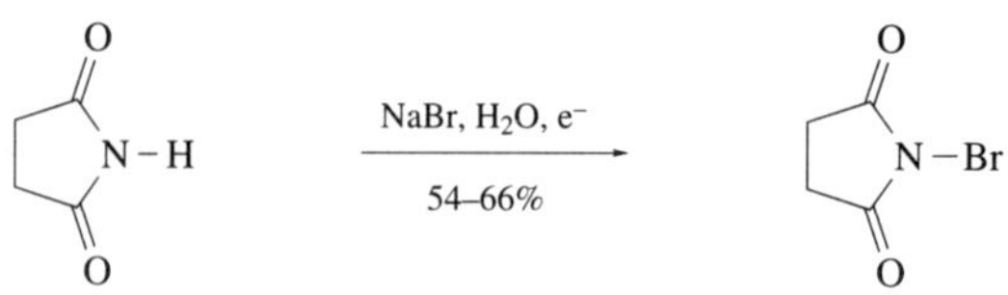

(10)

5.07.1.9.2 From N*-heteroimides*

Metalated imides react with bromine to provide the corresponding bromoimides and metal halides (Equation (11)) ⟨72JOM(35)303⟩. This reaction has worked successfully for converting antimony ⟨89POL581⟩, tin ⟨83JOM(255)295⟩, mercury ⟨03GI226⟩, and arsenic ⟨83IJC(A)120⟩ imides to *N*-bromoimides. In addition, phenyliodine(III) bis(phthalimide) reacts with bromine to provide *N*-bromophthalimide ⟨83S207⟩.

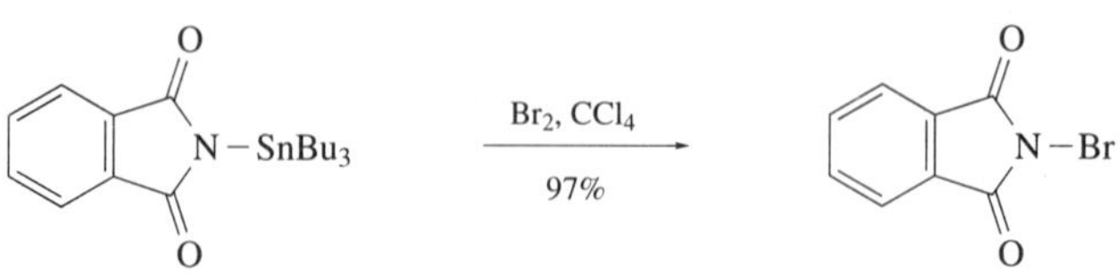

(11)

5.07.1.10 *N,N*-Dibromoamides

Gottardi has explored the synthesis and properties of *N,N*-dibromoamides prepared from the bromination of amides with dibromoisocyanuric acid (Equation (12)). The reaction is quite general

Table 3 Bromination of amides and lactams.

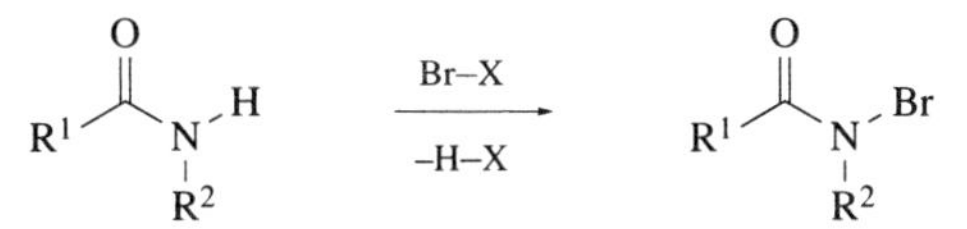

R¹CONHR²	*Conditions*	*R¹CONBrR²*	*Ref.*
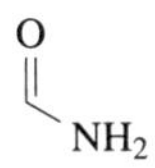	i, $NaBrO_3$, conc. H_2SO_4 ii, HBr, 30 min, RT (48%)	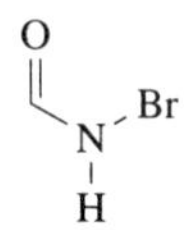	93BCJ2426
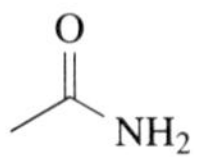	Br_2, 0 °C, KOH (41–50%)	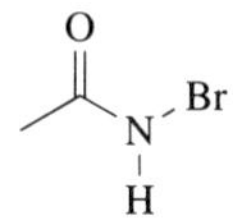	63OSC(4)104
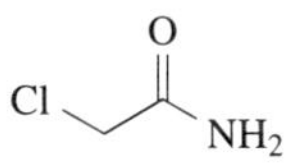	TFA, AgO Br_2 (61%)	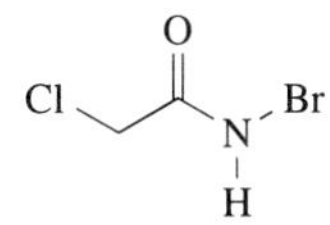	52JA2189
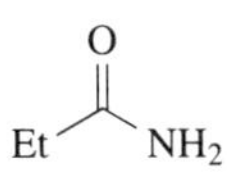	i, NaBr/NaOCl, 0 °C ii, amide, 2 h (87%)	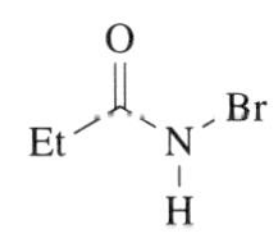	74JOC3136
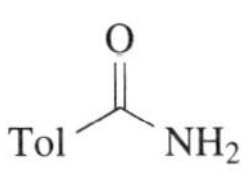	i, cold Br_2/NaOH ii, amide, 15 min (80–90%)	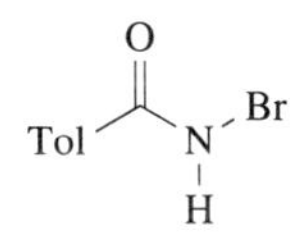	71BCJ1632
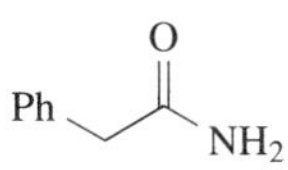	AcOH, $NaBrO_2$ (aq.) RT, 3–4 h (60%)	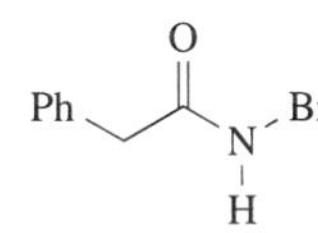	84CL2045
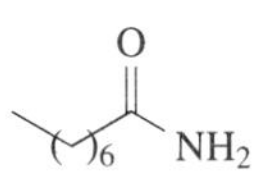	i, BTMA:Br_3,[a] NaOH, 0 °C ii, amide, 4 h (77%)	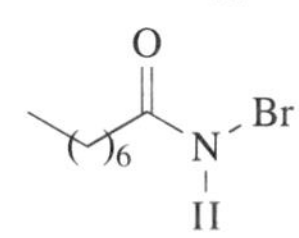	89JCS(P1)1702
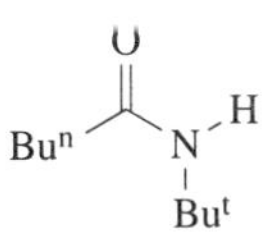	i, NaBr/Bu^{t}OH/NaOCl ii, amide, benzene/CCl_4 (87%)	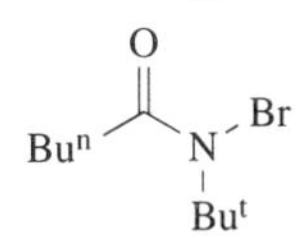	66JA3051
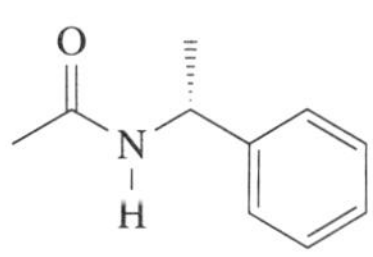	AcOBr/CCl_4, RT 1 h (>95%)	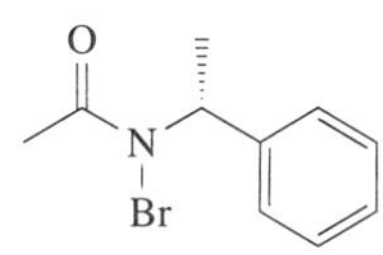	93SC2423
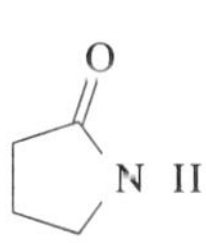	i, $NaBrO_3$, conc. H_2SO_4 ii, HBr, 30 min, RT (78%)	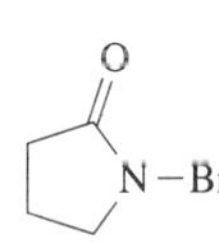	93BCJ2426
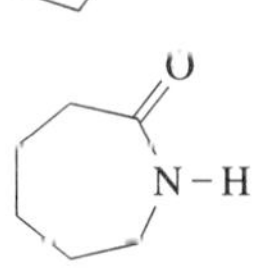	i, Br_2, 30–40 °C ii, K_2CO_3 (72%)	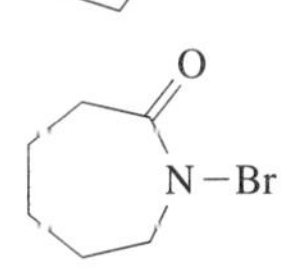	61CD480

[a] BTMA:Br_3, refers to benzyltrimethylammonium tribromide.

and is applicable to formamides, alkylamides, and benzamides ⟨73M421⟩. Other brominating agents such as bromine ⟨67JA5287⟩ and acetyl hypobromite ⟨70JOC2056⟩ have also been used to prepare dibromoamides from amides.

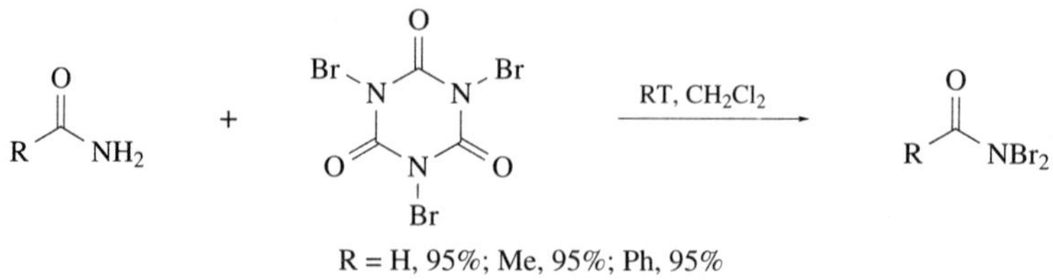

(12)

5.07.1.11 *N*-Iodo(I) Amides

N-Iodoamides are the least stable of the monohaloamides. As a result, there has not been the same volume of work in this area as with the *N*-bromo- and *N*-chloroamides.

5.07.1.11.1 From amides

N-Iodoamides are prepared by treating amides with iodine–HgO mixtures (Equation (13)) ⟨84AQ49⟩. They have also been prepared (but not isolated) by treating amides with sodium hydride followed by iodine ⟨91BCJ3732⟩.

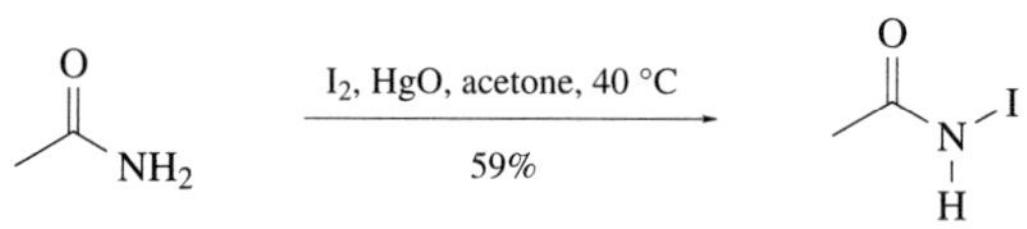

(13)

5.07.1.11.2 From N*-bromoamides*

N-Bromoamides can be converted to *N*-iodoamides upon treatment with iodine ⟨84AQ49⟩. The yield of desired *N*-iodoamides depends heavily on the reaction conditions (Equation (14)) ⟨75M1019⟩.

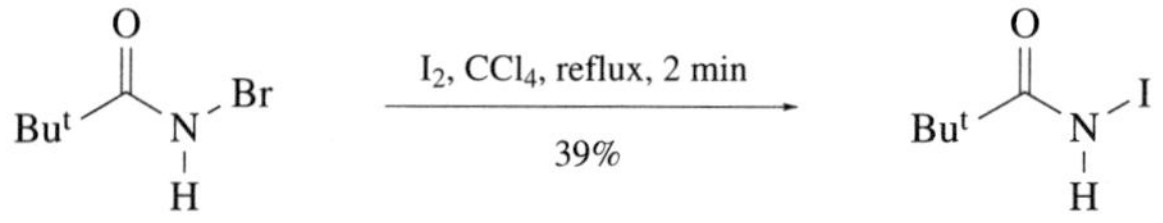

(14)

5.07.1.12 *N*-Iodo(I) Imides

5.07.1.12.1 From N*-metalimides*

Iodine has been reported to convert *N*-silver succinimide to the corresponding *N*-iodoimide (Equation (15)) ⟨73OSC(5)663⟩ see also ⟨53JA3493⟩. In related work, iodochloride (ICl) has been used to convert trialkylstannylsuccinimide and trialkylstannylphthalimide to the corresponding iodoimide ⟨83JOM(255)295⟩.

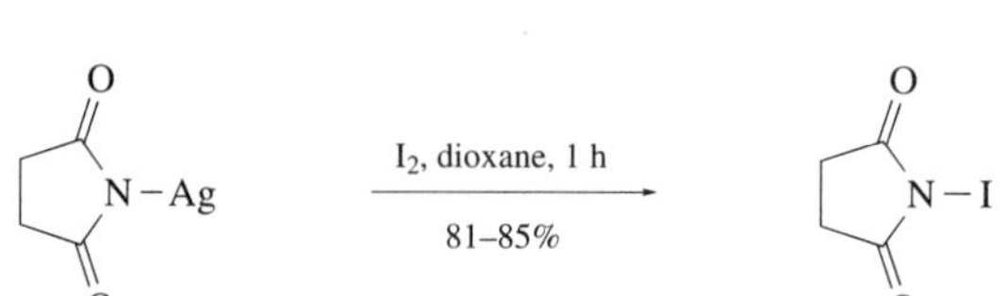

(15)

5.07.1.12.2 From N*-haloimides*

N-Iodosuccinimide (NIS) is prepared *in situ* from NCS using sodium iodide in acetone (Equation (16)) ⟨84TL233⟩. It can also be obtained in good yields by refluxing iodine and NBS in 1,2-dichloroethane for two days ⟨75M1019⟩, or by treating NBS with methyl iodide in dioxane ⟨75USP3914252⟩. Cleavage of phenyliodine(III) bis(phthalimide) with iodine also provides NIS ⟨83S207⟩.

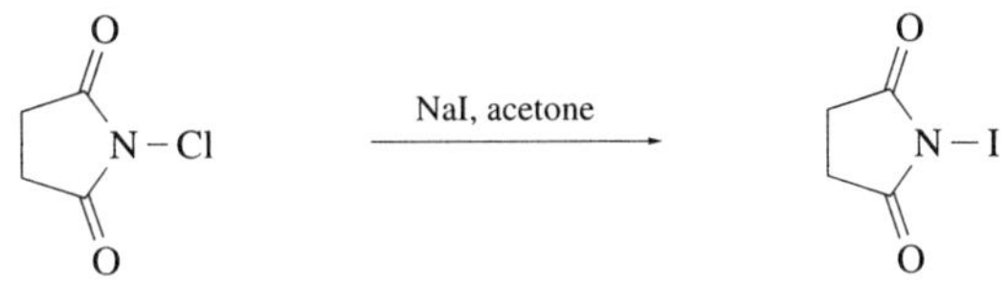

(16)

5.07.1.12.3 From imides

NIS has also been prepared from treatment of succinimide with mercuric oxide and iodine ⟨84AQ49⟩.

5.07.1.13 *N,N*-Diiodoamides

The synthesis of *N,N*-diiodoamides from *N,N*-dibromoamides using iodine was initially reported by Gottardi in 1973 ⟨73M421⟩, and later in greater detail ⟨74M611, 75M1019⟩. A variety of diiodoamides were prepared using this method (Equation (17)).

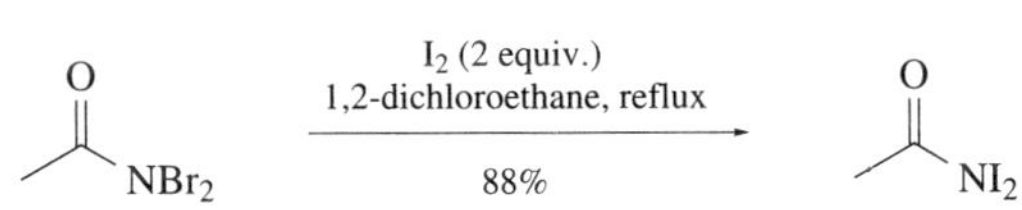

(17)

5.07.1.14 *N*-Iodo(III) Amides

5.07.1.14.1 From amides

A number of cyclic *N*-iodocarboxamides have been reported in the literature ⟨79JOC1447, 75JOC2129⟩. These compounds are usually prepared from oxidation of the corresponding iodobenzamide with chlorine gas (Equation (18)) ⟨65CC449⟩. With regard to acyclic derivatives, α-phenylacetamide reacts with [methoxy(tosyloxy)iodo]benzene to provide the corresponding *N*-iodo derivatives ⟨87JOC476⟩.

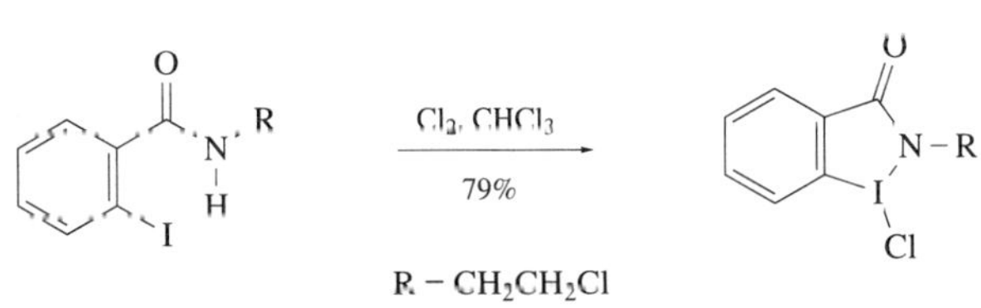

(18)

5.07.1.14.2 From N*-silylamides*

TMS-amides have been reported to react with trifluoromethyliodine(III) difluoride ⟨89JFC(45)401⟩ as well as with mixed iodonium sulfonates (**15**) (Scheme 4) ⟨90TL4821⟩ to provide the desired iodoamides (**16**).

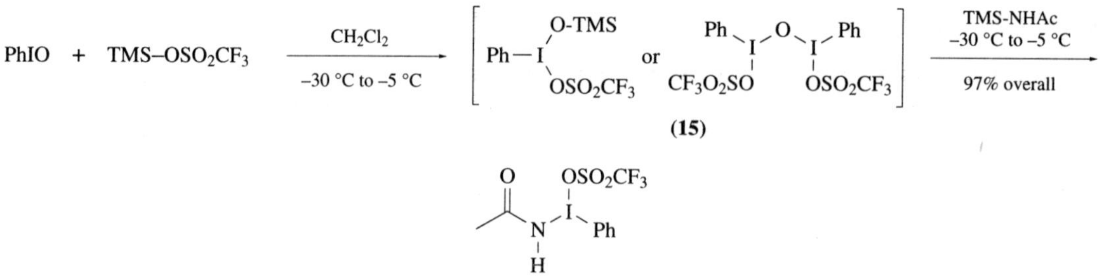

(15)

(16)

Scheme 4

5.07.1.15 *N*-Iodo(III) Imides

Phthalimide displaces both trifluoroacetates from phenyliodine(III) bis(trifluoroacetate) to provide the diphthalimide iodine derivative (**17**) (Equation (19)) ⟨83S207⟩.

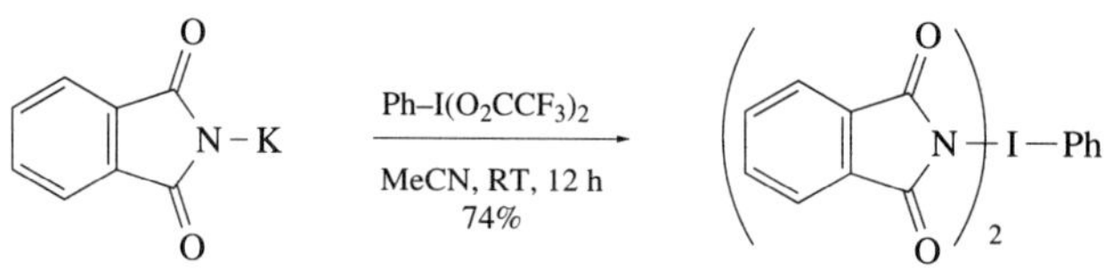

(19)

(17)

5.07.2 AMIDES SUBSTITUTED BY OXYGEN

The chemistry of oxygen-substituted amides is dominated by the hydroxy-substituted amide (hydroxamic acid) and the corresponding *O*-substituted derivatives (hydroxamate). The chemistry ⟨B-82MI 507-01, 85HOU(E5)1135, B-89MI 507-01⟩ and physical properties ⟨93BCJ421, 94JA916⟩ of hydroxamic acids have been reviewed by a number of groups. Miller has also prepared an excellent review covering the synthesis of hydroxamates ⟨86ACR49⟩.

5.07.2.1 Hydroxamic Acids

5.07.2.1.1 From acylation of hydroxylamine

Analogous to the formation of peptides from amines, the formation of hydroxamic acids from hydroxylamines usually requires activated acylating agents such as anhydrides ⟨92SC3067⟩, *N*-hydroxysuccinimide esters ⟨87JOC4471, 93JOC6083⟩, and acid chlorides ⟨84JOC106, 94TL5157⟩. Carboxylic acids can be activated *in situ* by other reagents, including *N,N'*-carbonyldiimidazole ⟨85SC1159⟩, and *N,N*-dimethylchloromethaniminium chloride ⟨85S929⟩. In contrast to amines, hydroxylamines react quite readily (under basic conditions) with a variety of esters to provide hydroxamic acids ⟨81JOC4188, 83T2571, 84JOC4541, 89JOC4088⟩. Amides ⟨62CRV155⟩ and lactams ⟨70S589⟩ can also be converted to hydroxamic acids, albeit not as easily.

In addition to traditional acylating agents, a number of heterocycles react with hydroxylamine to provide hydroxamic acids. Representative examples are found in Table 4.

The reaction between hydroxylamine and acylating agents is complicated by the competing side reaction of *O*-acylation as well as *O*- and *N*-bisacylation. The steric bulk both of the acylating agent as well as of the hydroxylamine plays an important role in determining the extent of *N*- vs. *O*-acylation. In cases of extremely hindered hydroxylamine derivatives, preferential *O*-acylation is observed (Equation (20)) ⟨86AP1037⟩.

Table 4 Hydroxamic acids from ring opening of heterocycles.

Starting Material	*Conditions*	*Hydroxamic acid*	*Ref.*
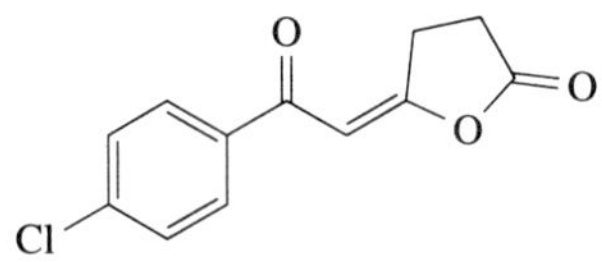	NH_2OH, Et_3N, 0 °C CH_2Cl_2 (69%)	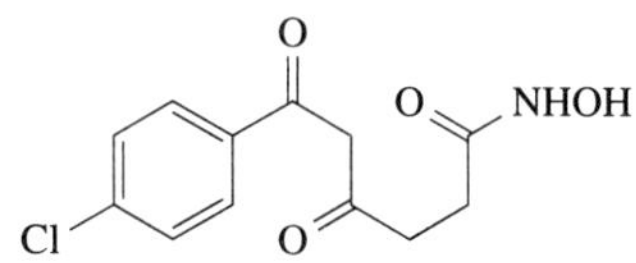	92JOC6662
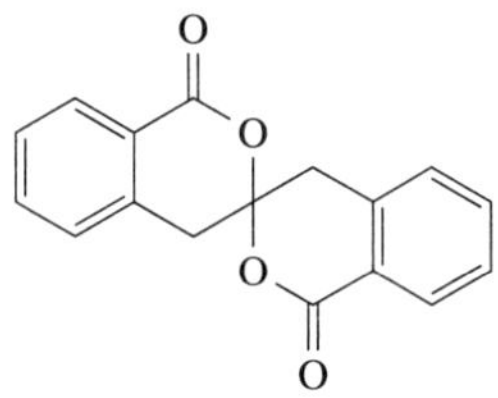	NH_2OH, MeOH (90%)	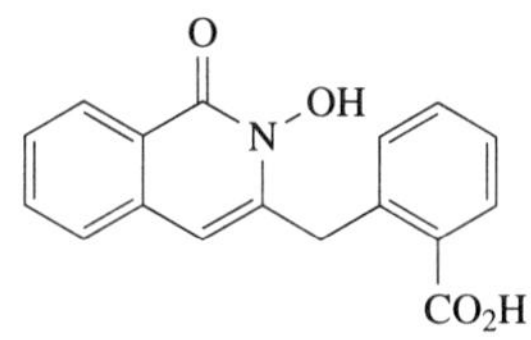	88AP41
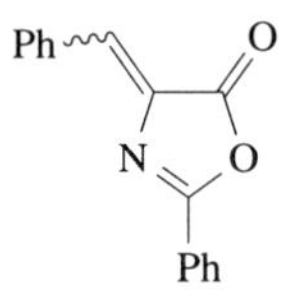	NH_2OH•HCl NaOMe, MeOH RT, 24 h (51%)	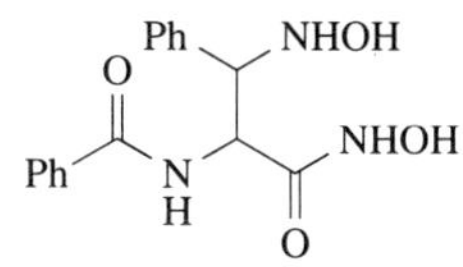	78SC497
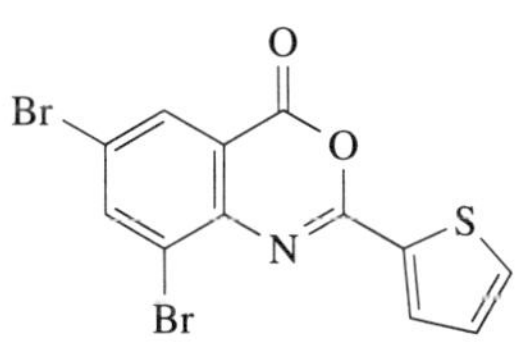	EtOH, NH_2OH (70%)	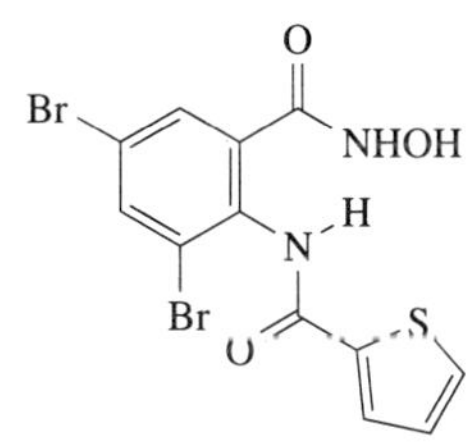	90ACH423
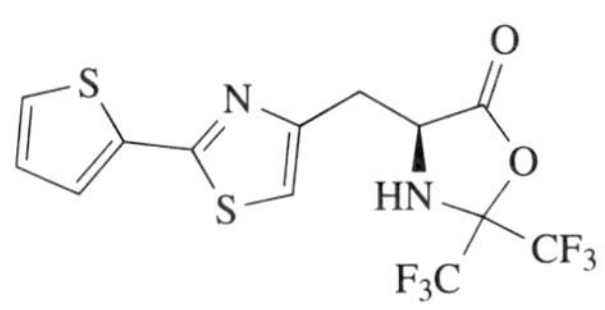	Et_2O, NH_2OH RT, 12 h (90%)	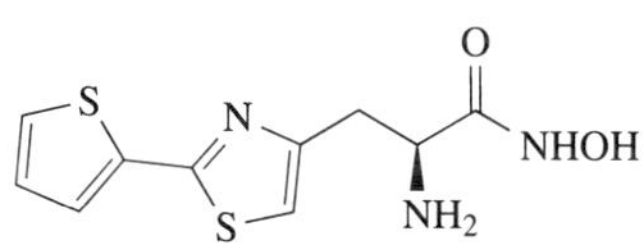	92S1145

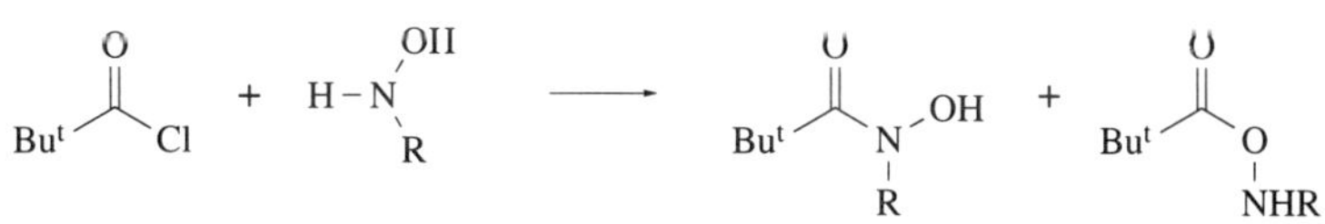

(20)

R = Bu^t	30%	70%
R = 1-adamantyl	10%	68%

The possibility of *O*-acylation should not overshadow the fact that the majority of acylating agents react with a wide variety of hydroxylamines to provide high yields of the corresponding hydroxamic acids.

To avoid the possibility of *O*-acylation, temporary protection of the hydroxyl group as a TMS-ether has been employed. Treatment of hydroxylamines with one equivalent of TMS-Cl followed by either an acid chloride ⟨88AP(321)311, 90AP(323)181, 90S1008⟩ or a mixed anhydride (Scheme 5) ⟨87S418⟩ provides, after aqueous workup, the corresponding hydroxamic acid. The use of *N*,*N*,*O*-tris(trimethylsilyl)amine has also been used to this end to provide unsubstituted hydroxamic acid upon hydrolytic workup ⟨83SC1053, 88ZOB1056, 88ZOB1299⟩.

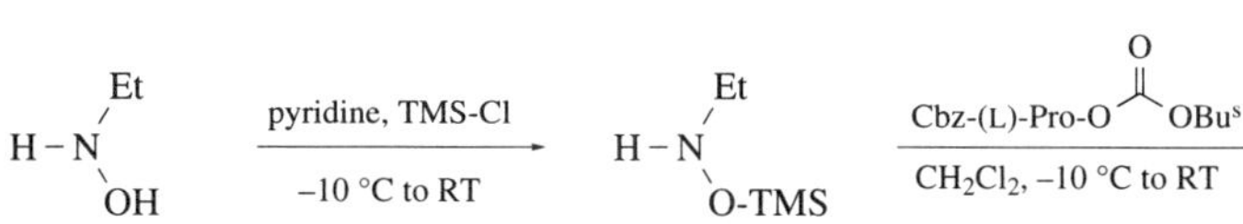

Scheme 5 (*continued*)

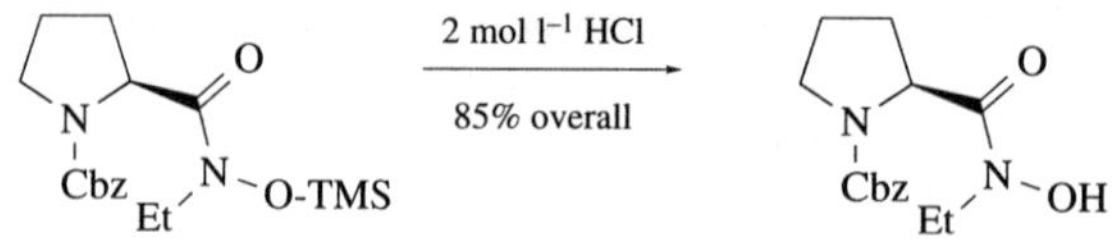

Cbz = Benzyl carbamate

Scheme 5

5.07.2.1.2 From reductive cyclization of nitro compounds

The reduction of nitro compounds can provide either amines or hydroxylamines, depending on the reaction conditions. When such nitro compounds contain an appropriately spaced acyl group, it is possible to convert these derivatives directly to cyclic hydroxamic acid derivatives (Equation (20a)) ⟨90SC1733⟩. Careful choice of reducing agent and reaction time are necessary, as hydroxamic acids themselves can be reduced to the corresponding lactams under the reaction conditions ⟨85S331⟩.

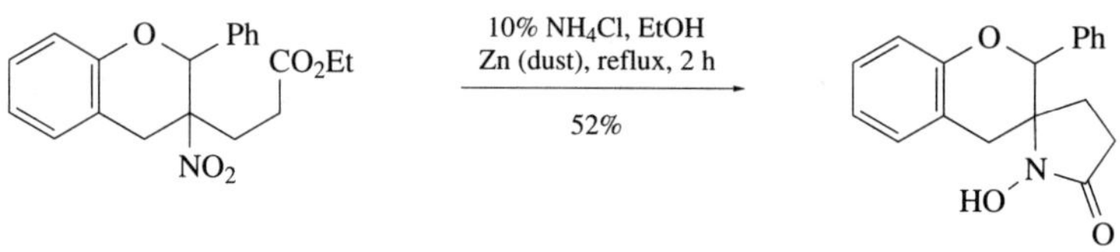

(20a)

5.07.2.1.3 From reduction of O*-acyl oximes*

The preparation of hydroxylamines from the borane-mediated reduction of oximes is a well-known transformation ⟨91CCC1317⟩. In 1976, Ganem reported an interesting variant of this procedure in which *O*-acyl oximes were reduced with borane in THF ⟨76TL1951⟩. Such a reduction provides intermediate *O*-acylated hydroxylamines which rearrange in an intermolecular fashion to the corresponding hydroxamic acids (Equation (21)).

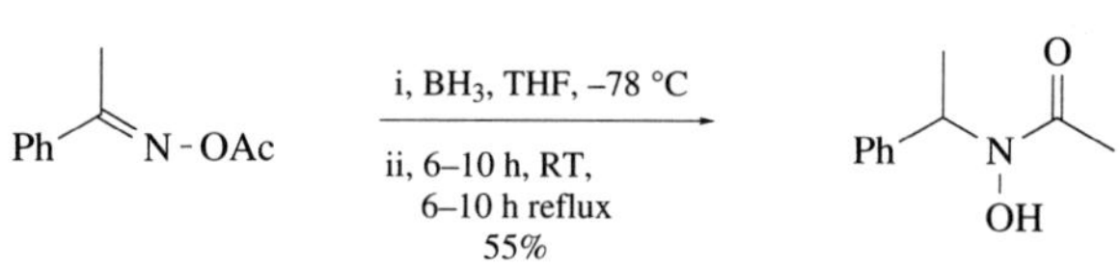

(21)

5.07.2.1.4 From ester enolates and oximes

The reaction of ester enolates with oximes provides *N*-hydroxy-*β*-lactams ⟨92BSB257⟩. The unprotected oximes are first treated with $AlCl_3$, followed by addition of the zinc enolate to provide, upon workup, the *N*-hydroxy-*β*-lactam (Equation (22)).

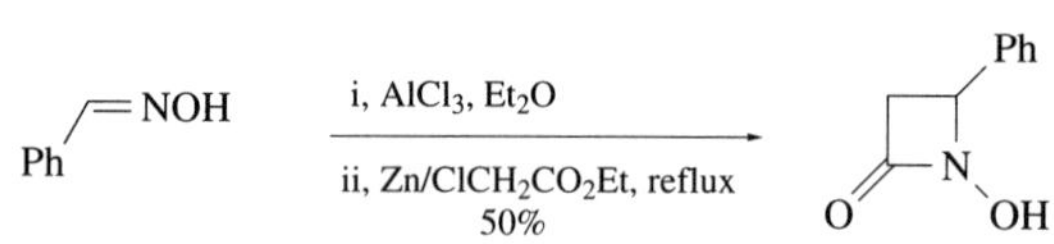

(22)

5.07.2.1.5 From oxidation of amines, hydroxylamines, oximes, and amides

The formation of hydroxamic acids from amines has usually required strong oxidants such as hydrogen peroxide/Na_2WO_4 ⟨87CC1471, 89BCJ2968, 90JOC1744, 91H(32)221, 92H(33)121⟩. In 1993, dimethyldioxirane (**18**) has been shown to effect this transformation ⟨93ACS1141⟩. This reagent is especially attractive to use since the reaction works for a variety of secondary amines, and excess oxidant is destroyed or removed upon evaporation of the solvent (Equation (23)).

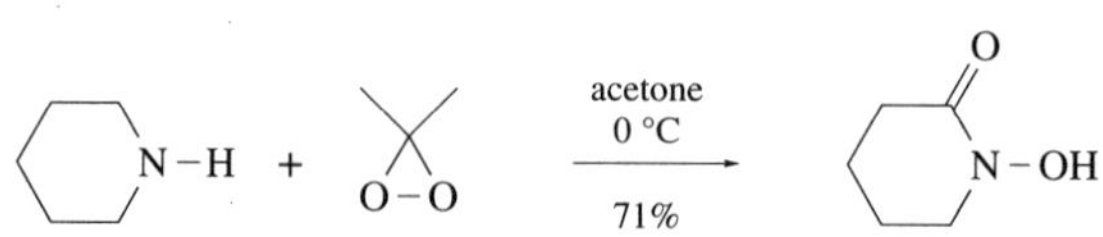

(18)

(23)

Hydroxylamines are also oxidized to hydroxamates upon treatment with oxidizing agents such as $Pb(OAc)_4$ ⟨88RTC381⟩, mcpba ⟨87JCS(P1)1635⟩, iron(III) chloride ⟨66JCS(C)1477⟩, or mercury(II)–EDTA complex ⟨88SC1331⟩. Hence, oxidation of *N*-hydroxyazetidine (**19**) with $Pb(OAc)_4$ provides the corresponding *N*-hydroxy-β-lactam (**20**) (Equation (24)) ⟨87H(26)437⟩.

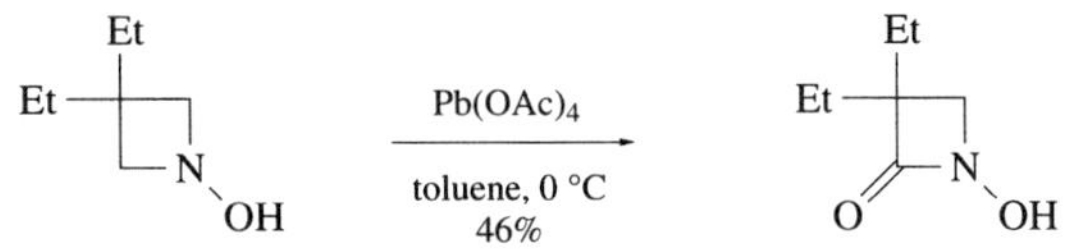

(19) **(20)** (24)

Hydroxamic acids have also been obtained from the oxidation of oximes with lead tetraacetate (Equation (25)) ⟨85JA5310⟩. Such products were not observed when sodium bicarbonate was present during the oxidation.

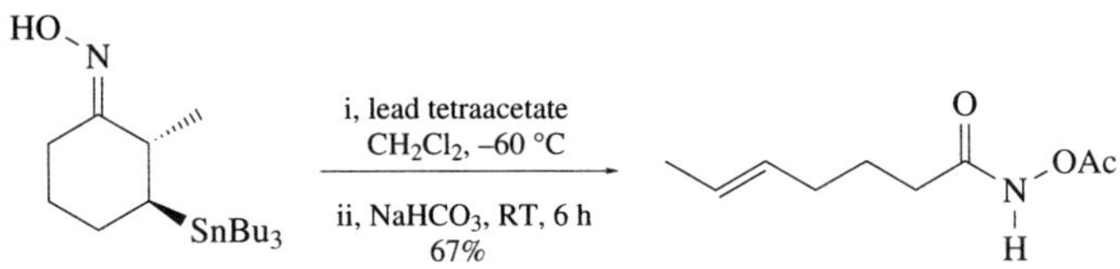

(25)

The selective oxidation of amides (and especially *N*-arylamides) to hydroxamic acids is often carried out by treating silylamide or silyllactams with MoOPh (Equation (26)) ⟨72CC1222⟩. The reaction works especially well for *N*-arylamides ⟨81IC1823, 89JOC5746, 89JOC5852, 91H(32)1687⟩.

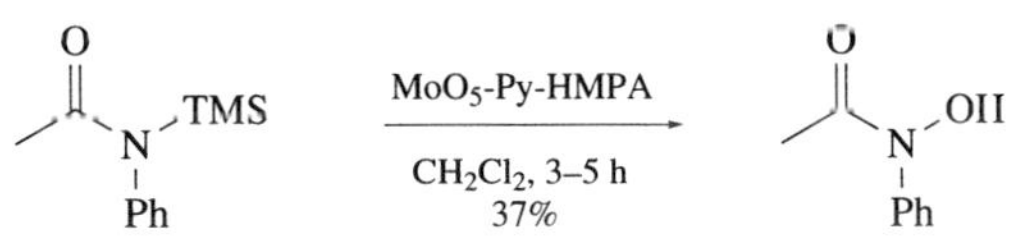

(26)

5.07.2.1.6 *From rearrangement of nitro compounds*

(i) From photolysis

The photolysis of nitro compounds under basic conditions leads to a rearrangement, resulting in the formation of hydroxamic acids. The reaction works equally well for acyclic as well as cyclic nitro derivatives ⟨82JOC5283⟩. In the case of unsymmetrical nitro compounds, the most-substituted β-carbon, or the β-carbon with an electron-withdrawing group, migrates preferentially (Equation (27)) ⟨90TL6547⟩. Yamada *et al.* have shown that the migration is stereospecific for chiral nitro compounds ⟨87JOC2327⟩.

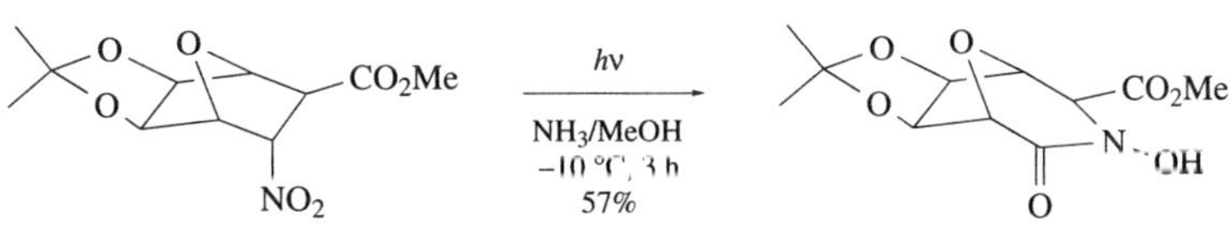

(27)

(ii) From acid hydrolysis

The reaction of nitrous acid with glyoxylic acids, such as camphor-3-glyoxylic acid (**21**), leads to the formation of the corresponding cyclic hydroxamate (**22**) (Equation (28)) ⟨67JOC1800⟩ see also ⟨B-88MI 507-02⟩.

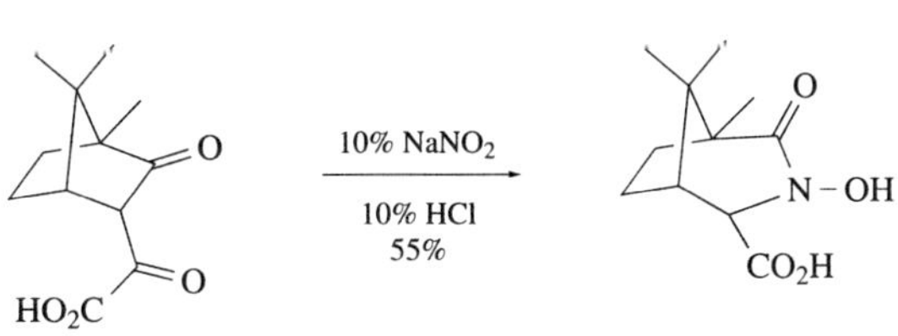

(21) **(22)** (28)

5.07.2.1.7 From ene reaction of acyl nitroso derivatives

Acyl nitroso compounds (**23**) are highly reactive intermediates generated from either the oxidation of hydroxamic acids (**24**) (Scheme 6, path a) ⟨89SL32⟩, the retro-Diels–Alder reaction of 9,10-dimethylanthracine adducts (**25**) (path b), or the thermal or rhodium-catalyzed decomposition of nitro diazo compounds (**26**) (path c) ⟨88TL5719, 89JOC3096⟩. Their use in synthesis has been the subject of reviews by both Kirby ⟨77CRV1⟩ and Weinreb ⟨82T3087⟩.

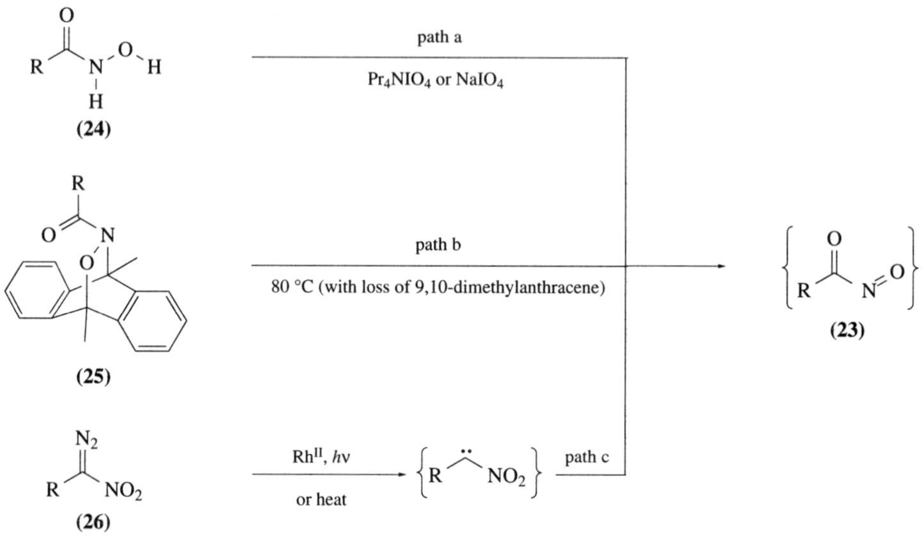

Scheme 6

The ene reaction of acyl nitroso derivatives was first observed by Kirby during studies on the Diels–Alder reactivity of these compounds ⟨77CRV1, 85JCS(P1)883⟩. Keck and co-workers elaborated on this work, improving upon the yields of the intermolecular version ⟨79TL4627⟩, and developing an intramolecular variation of the reaction (Equation (29)) ⟨79TL1185, 81JA3173⟩.

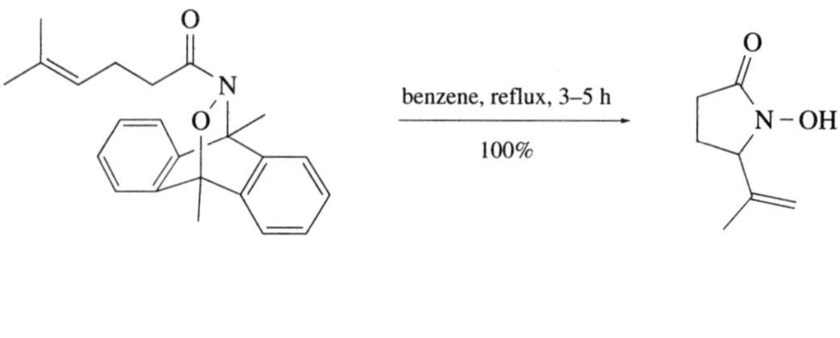

(29)

5.07.2.1.8 From rearrangement on 2-alkoxypyridine N*-oxides*

The 3,3-sigmatropic rearrangement of *O*-allylphenoxy derivatives is also applicable to *N*-oxide derivatives. Hence, 2-allyloxypyridine *N*-oxide (**27**) is smoothly transformed to 3-allyl-*N*-hydroxy-2-pyridone (**28**) upon refluxing in DMF (Equation (30)) ⟨90JCS(P1)1623⟩.

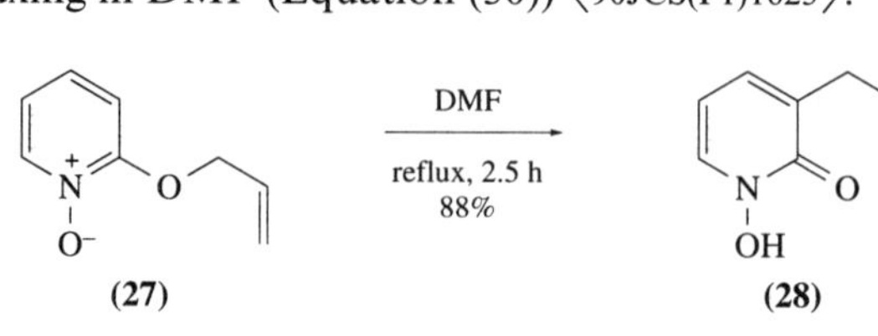

(30)

5.07.2.1.9 From deprotection of hydroxamates

As the complexity of the carboxylic acid moiety increases, the preparation of hydroxamic acids through deprotection of hydroxamates becomes increasingly necessary. This has led to a dramatic increase in the preparation of protected hydroxamic acid derivatives, as well as to a rise in novel or more selective methods for deprotection (Equation (31)) ⟨93T9137⟩. Table 5 lists a variety of examples in which hydroxamates are deprotected to the corresponding hydroxamic acid derivatives. In addition to these examples, the use of Pd/$BaSO_4$ is reported to be especially well suited for the deprotection of *O*-benzylhydroxamates ⟨95TL197⟩.

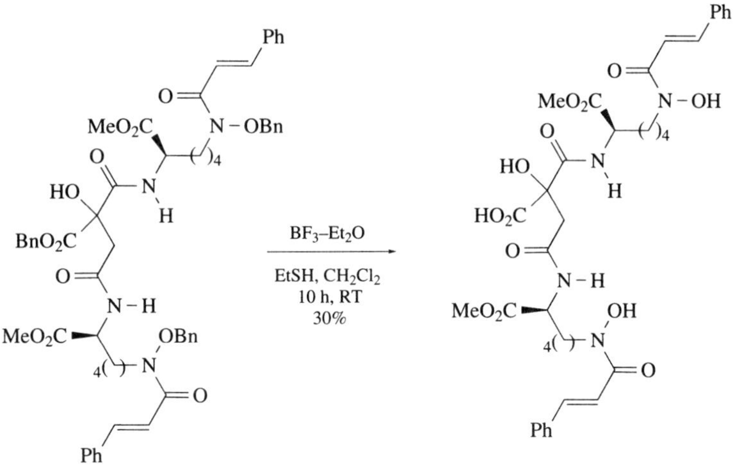

(31)

5.07.2.1.10 *From hydrolysis of thioimidate* N*-oxides*

S-Methyl *N*-methylthiobenzimidates (**29**) are hydrolyzed under basic conditions to benzoic acid and *N*-methylhydroxamic acid (**30**) (Equation (32)) ⟨86JOC5198⟩. When the reaction was stopped after only 20 min, the proportion of hydroxamic acid is increased to 83%. In a related method, the hydrolysis of 2-substituted pyridine *N*-oxides resulted in *N*-hydroxypyradones ⟨77JOC3301, 81JHC555, 87T2557⟩.

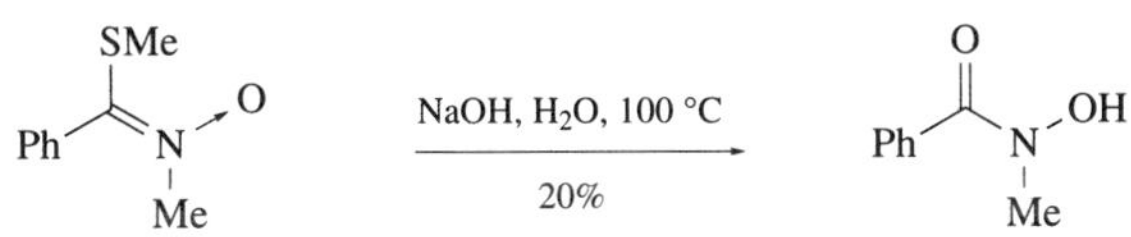

(29) **(30)** (32)

5.07.2.2 Hydroxamates

5.07.2.2.1 *From acylation of alkoxyamines and related derivatives*

Alkoxyamines react with many of the same acylating agents under primarily the same conditions as amines. It is therefore no surprise that many of the reagents used for peptide coupling are also useful for the synthesis of hydroxamates. Activation of the carboxylic acid by carbodiimides such as dcc ⟨85TL5385, 92SC871⟩, the water-soluble dcci ⟨93JA8898⟩, as well as carbonyl diimidazole ⟨88AP(321)235⟩, leads to high yields of hydroxamates. More recently, the use of triphenylphosphine and carbon tetrabromide has also been shown to provide high yields of *N*-methoxyamides from carboxylic acids ⟨90SC1105⟩.

Activation of carboxylic acids with chloroformates such as isopropenyl chloroformate ⟨93T9137⟩, isobutyl chloroformate ⟨89JOC3913⟩ or methyl chloroformate also provides high yields of hydroxamates (Equation (33)) ⟨89OS(67)69⟩.

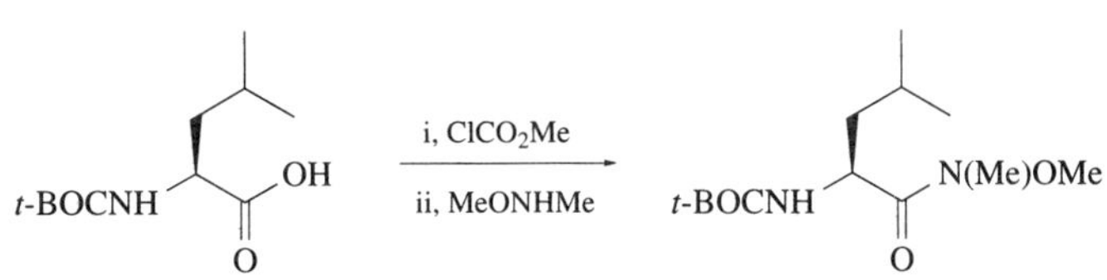

(31) **(32)** (33)

Table 5 Hydroxamic acids from deprotection of hydroxamates.

Starting Material	*Conditions*	*Product*	*Ref.*
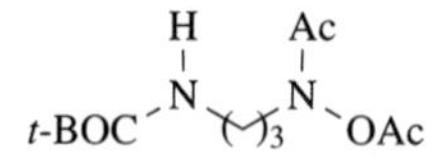	MeOH, NH_3, 5 h (100%)	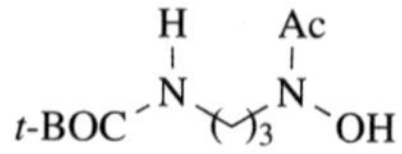	83JOC24
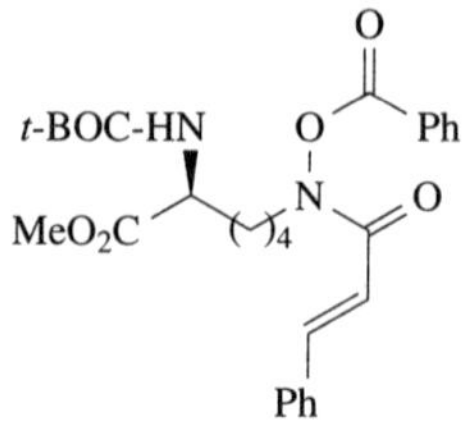	10% NH_3 in MeOH −23 °C, 3 h (93%)	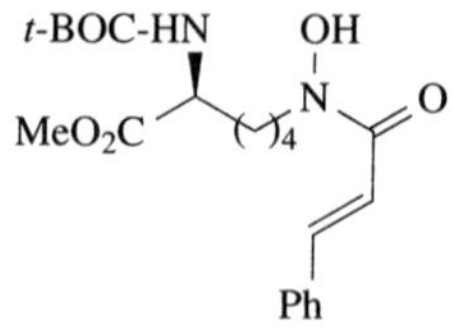	92JOC7140
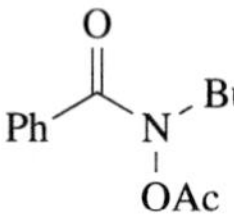	NH_2NH_2, EtOH (100%)	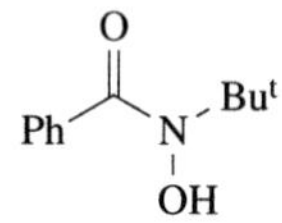	81S121
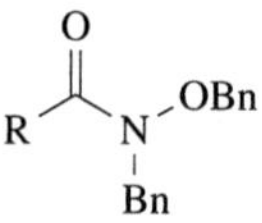	$Pd(OH)_2/C$, H_2 4 h, 26 °C (89%)	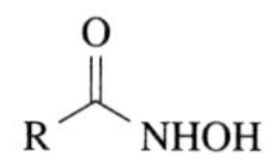	93JCS(P1)459
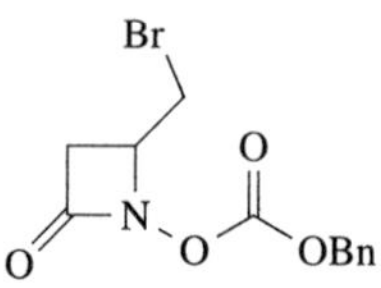	H_2, Pd/C, MeOH (>90%)	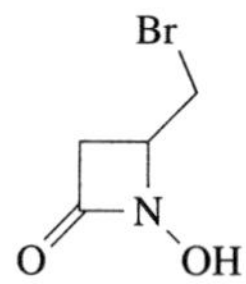	85TL5385
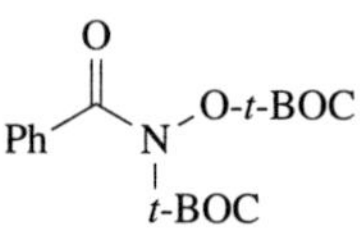	TFA, CH_2Cl_2, 0 °C 18 h (81%)	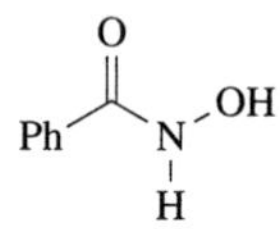	94TL6021
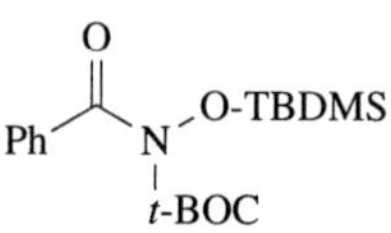	CeF, HOAc, MeCN (nearly quantitative)	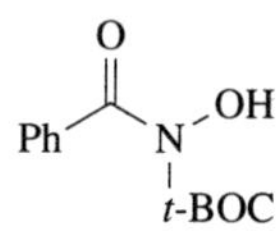	92TL5055
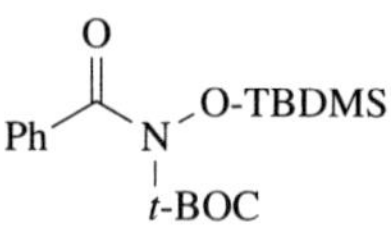	TFA, CH_2Cl_2 0 °C, 6 h (85%)	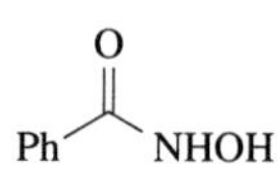	92TL5055
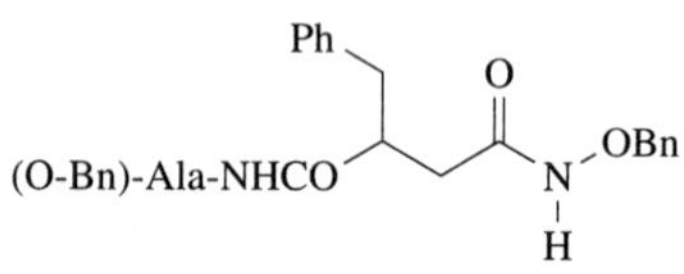	H_2, 10% Pd/C MeOH (91%)	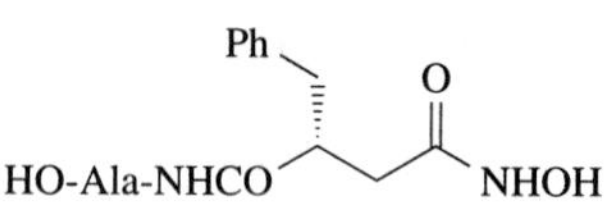	92SC871
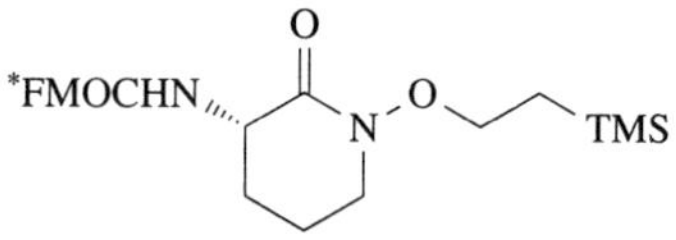	$BF_3 \cdot Et_2O$, MeCN RT, 0.5 h (99%)	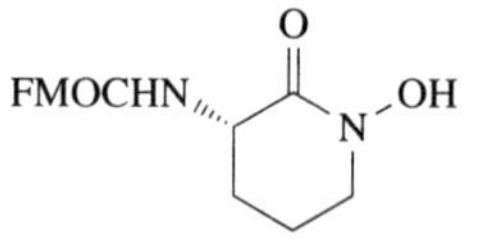	94JOC929
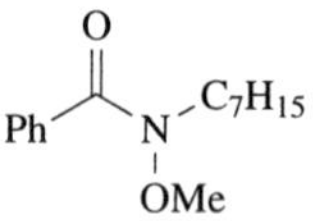	$AlCl_3$, EtSH, 0 °C RT, 43 h (89%)	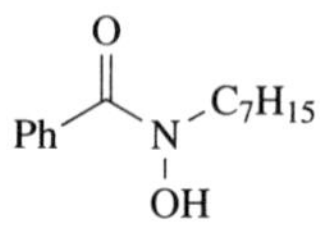	90SC887

*FMOC = 9-Fluorenylmethyl carbamate

The use of acid chlorides is another very reliable method for the preparation of hydroxamates from alkoxyamines. The effect of parameters such as solvent and base have been looked at in some detail ⟨89CCC2738⟩.

Esters and lactones react much less readily with alkoxyamines than do acid chlorides. Nevertheless, alkoxyamines do react with esters and lactones (albeit under refluxing alcohol conditions) to provide hydroxamate derivatives. To convert esters and lactones to hydroxamates under milder conditions, the use of aluminum-complexed alkoxyamines continues to be the method of choice. Although this procedure was originally developed for the preparation of amides ⟨82SC989, 94JOC1231⟩, this method is becoming increasingly popular for the synthesis of hydroxamates from esters and lactones ⟨90TL5297, 93TA1573, 94JOC4862, 94JOC5292⟩. In a related vein, tin-complexed alkoxyamines have also been used to prepare hydroxamates from esters (Equation (34)) ⟨93SL850⟩.

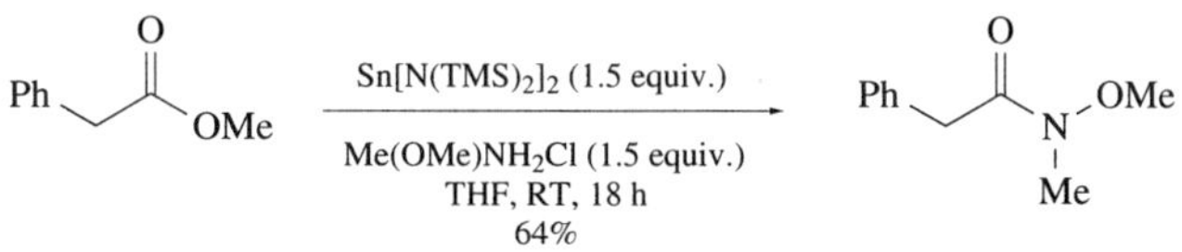

(34)

Certain amides and lactams also react with alkoxyamines to form hydroxamates. *N*-*t*-BOC-*β*-lactams (**33**) are normally resistant to ring opening by alkoxyamines and require addition of sodium ethanethiolate to effect ring opening (Equation (35)) ⟨90CC720⟩.

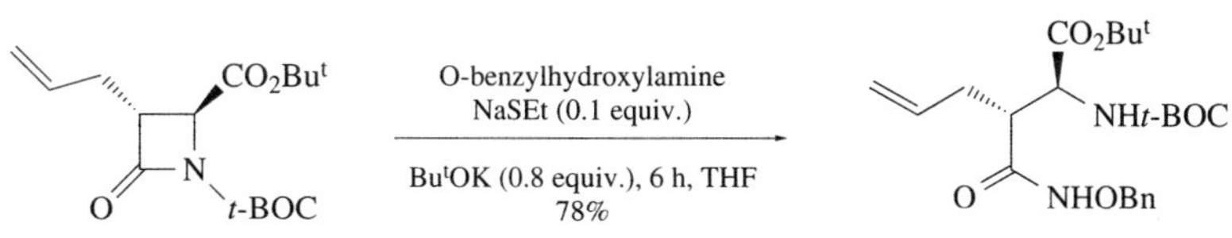

(35)

The activation of primary amides through exhaustive *t*-butoxycarbonylation has also been reported. These intermediates (**34**) react with *N*-methoxyamine to provide the corresponding hydroxamate (Scheme 7) ⟨91JOC5482⟩.

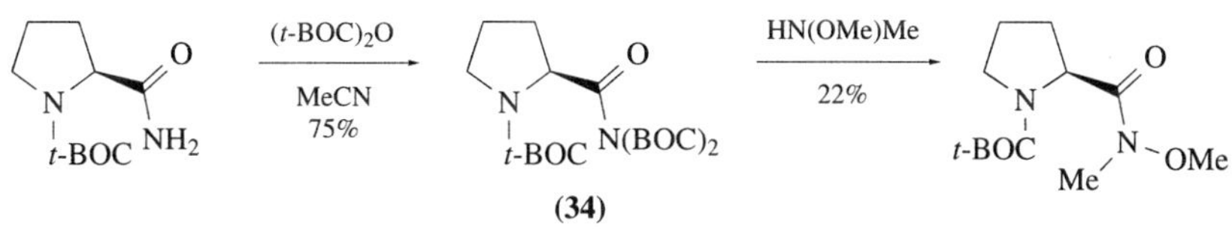

Scheme 7

A variety of nontraditional acylating agents react with alkoxyamines to produce hydroxamates. Some acylated chiral auxiliaries, such as the 1,3-oxazolidinine-2-thiones ⟨85CC1418⟩ and the thiazolidine-2-thiones ⟨85TL977, 85TL4855, 86JOC2391, 91TL2577⟩ react directly with alkoxyamines to provide hydroxamates. Others, like the Evans oxazolidinones ⟨86TL3119, 86TL4957, 86TL4961, 93TL7371, 94TL1247⟩, require co-treatment of the alkoxyamine with trimethylaluminum to provide optimal yields of hydroxamate. The iron acyls developed by Davies ⟨88TL6509, 93JCS(P1)459⟩ are first oxidatively cleaved with bromine, then treated with alkoxyamines to provide the desired hydroxamates. Several of these nontraditional acyl derivatives are listed in Table 6.

5.07.2.2.2 *From* O*-alkyloximes*

The alkylation of *O*-alkyloximes by silylketene acetals ⟨89CPB1179⟩ or ester enolates ⟨87JCS(P1)845⟩ provides alkoxylamines as intermediates. These *O*-alkylhydroxylamines cyclize under the reaction conditions to the corresponding *N*-alkoxy *β*-lactams (Equation (36)) ⟨84CL369⟩.

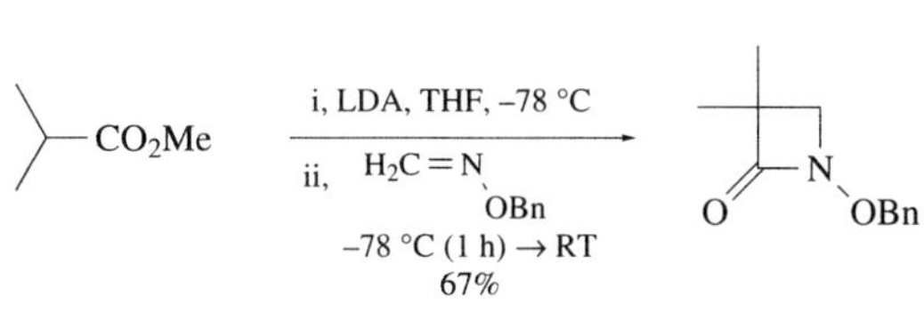

(36)

LDA = lithium diisopropylamide

Table 6 Reaction of alkoxyamines with assorted acyl derivatives.

Starting Material	*Conditions*	*Product*	*Ref.*
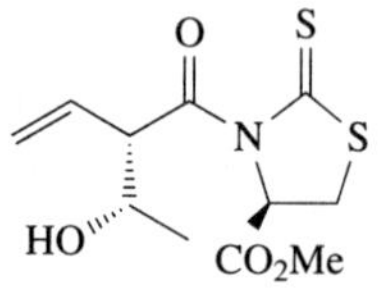	NH_2O-TBDPS dmap, MeCN (36%)	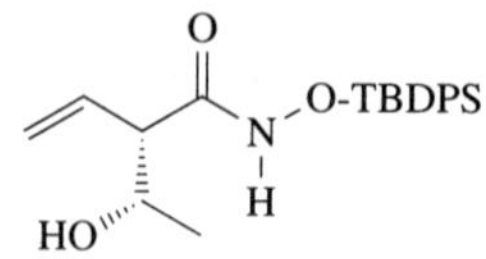	91TL2577
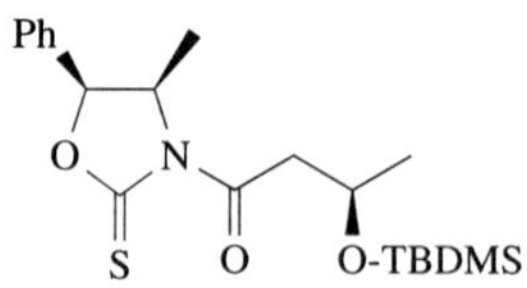	NH_2OBn, $CHCl_3$ (57%)	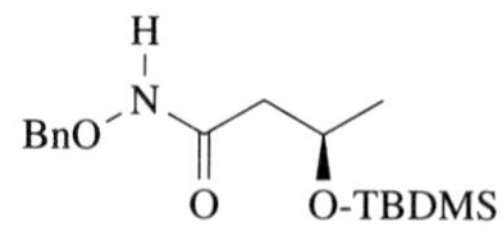	85CC1418
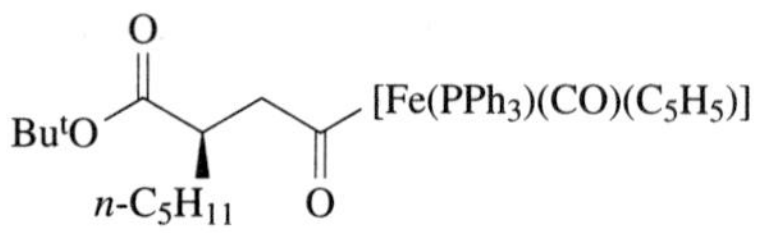	NBS, BnONHBn –40 °C to 25 °C CH_2Cl_2, 3 h (78%)	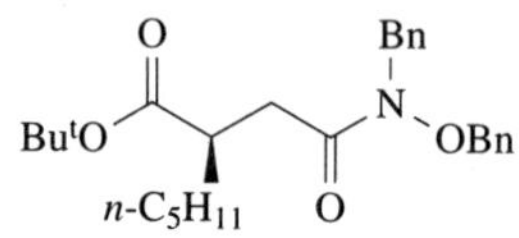	88TL6509
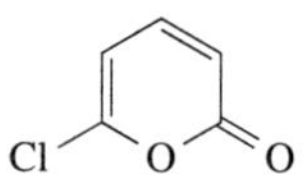	MeNHOMe (4 equiv.) benzene, 5 °C to RT (98%)	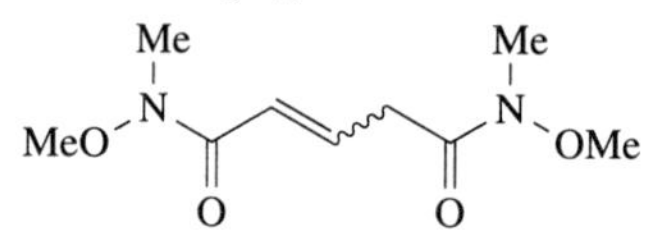	88JA6172
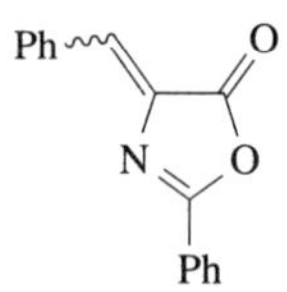	NH_2OBn, Et_2O $CHCl_3$, reflux, 1 h (78%)	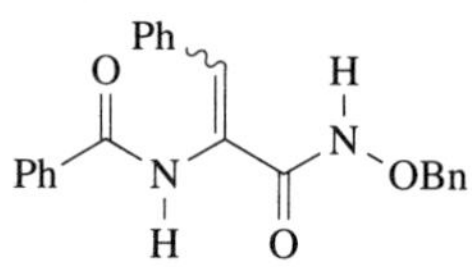	49JA1691
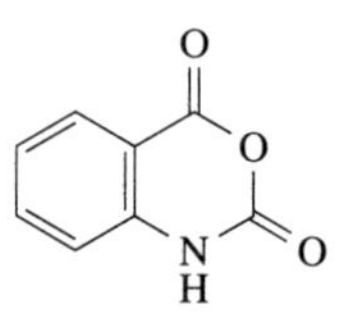	HN(OMe)Me, EtOH reflux, 1.5 h (75%)	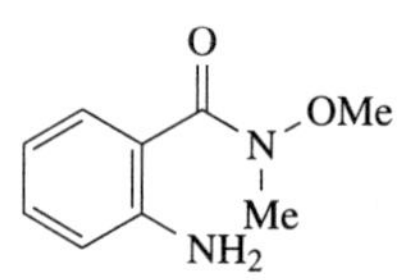	91JOC3750
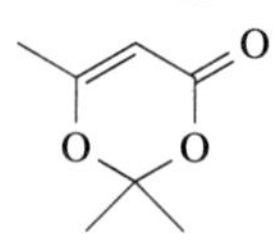	NH_2OBn xylene, reflux	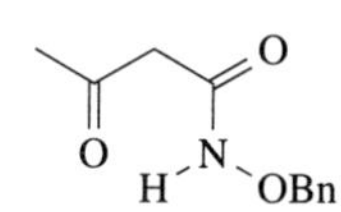	89CPB2952

TBDPS = *t*-butyldiphenylsilyl

5.07.2.2.3 From acyl nitroso derivatives

Acyl nitroso compounds (see Section 5.07.2.1.7) serve as transient dienophiles in Diels–Alder (DA) reactions. This area has been reviewed ⟨82T3087⟩, and a number of natural products have since been generated, in which an intermolecular ⟨85JOC1818, 86TL683, 87TL5937, 90TL2139, 91JCS(P1)1001, 92TL273⟩ or an intramolecular cyclization serves as the key synthetic step ⟨88JA2341, 89JOC4088, 90JA9601, 91CC1237, 91TL4325, 93JOC6083, 94JOC1358⟩. Recent advances in this area ⟨88TL6173, 89JOC4088, 89SL32, 89TL7061, 91JA5089, 91TA1209, 93JCS(P1)1397, 94JA562, 94JOC4602⟩ have focused upon improving the design of chiral acyl nitroso derivatives and on enhancing the stereoselectivity of intermolecular DA reactions (Equation (37)) ⟨92HCA109⟩.

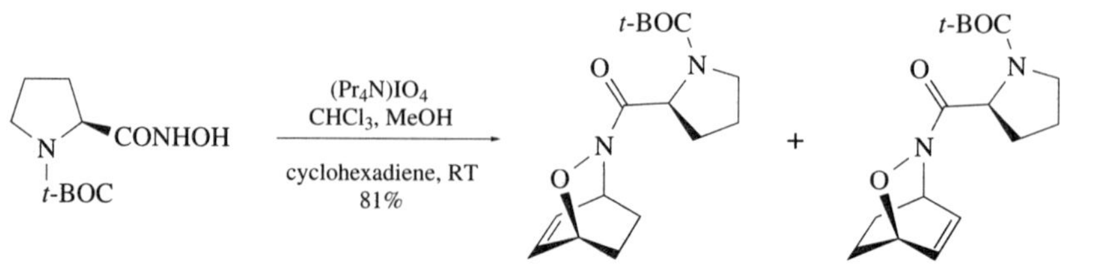

60 : 40 (37)

5.07.2.2.4 *From R—CO formation*

(i) From vinyl triflates

The conversion of vinyl triflates to α,β-unsaturated amides was first reported by Cacchi *et al.* ⟨85TL1109⟩. This methodology has since been utilized to prepare *N*-methyl-*O*-methylhydroxamic acids (Equation (38)) ⟨91JA5085⟩.

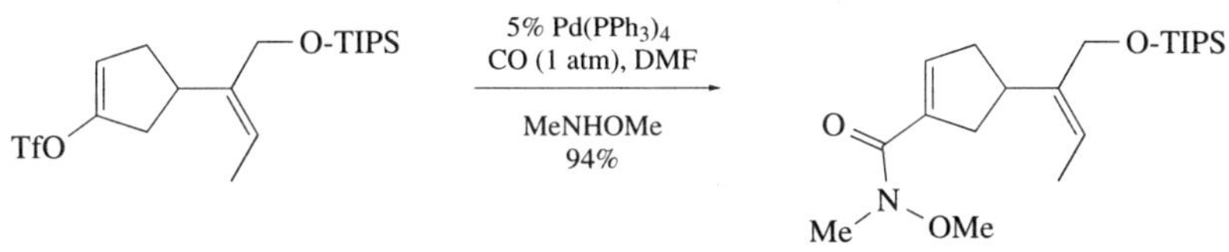

(38)

TIPS = Triisopropylsilyl

(ii) From alkylation of N,N′*-dimethoxyurea*

Hlasta and co-workers reported that *N,N′*-dimethoxyurea reacts with phenyllithium as well as with phenylmagnesium bromide to provide the corresponding *N*-methyl-*O*-methylhydroxamic acids ⟨89TL1773⟩. This methodology was later expanded by Wipple and Reich to include a variety of alkyllithium reagents (Equation (39)) ⟨91JOC2911⟩.

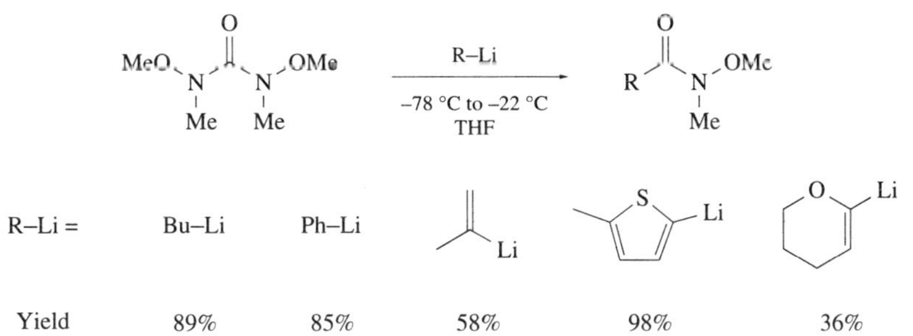

(39)

5.07.2.2.5 *From ring expansion of ketones*

Cyclohexanone (**35**) reacts with dimethoxyamine to provide the ring-enlarged cyclic hydroxamate (**36**) (Equation (40)) ⟨92TL3261⟩.

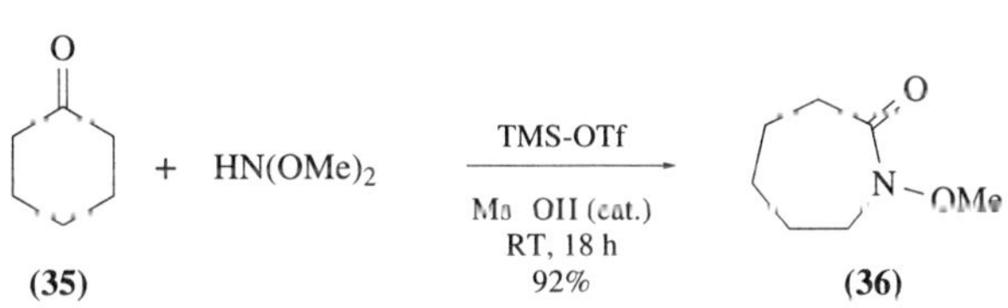

(40)

5.07.2.2.6 *From oxidation of nitrones*

Nitrones, synthesized by condensation of aldehydes with hydroxylamines, are oxidized with lead tetraacetate to the corresponding *N*-acyl amines (Equation (41)) ⟨73TL2711⟩.

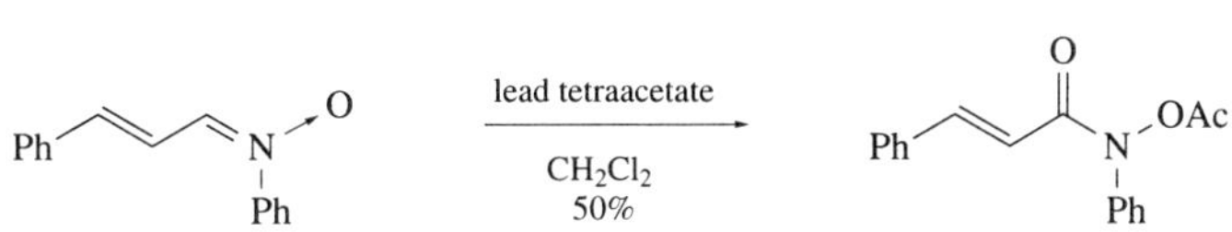

(41)

5.07.2.3 Hydroxy-substituted Imides

The chemistry of *N*-hydroxyimides has been reviewed by a number of groups ⟨74AG(E)419⟩.

5.07.2.3.1 From acylation of hydroxylamines or hydroxamic acids

Hydroxylamines are converted to *N*-hydroxyimides in the presence of excess acylating agent. The reaction works especially well for cyclic anhydrides, forming cyclic *N*-hydroxyimides (Equation (42)) ⟨91JMC387⟩.

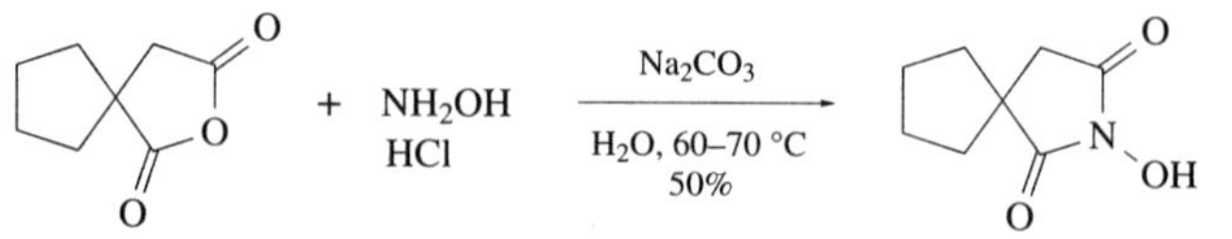

(42)

5.07.2.3.2 From α-nitro ketones

α-Nitroketones rearrange to *N*-hydroxyimides under acidic ⟨63JA827, 63JA2181⟩ as well as photolytic conditions. The photolytic rearrangement works well for a variety of monocyclic ketones such as α-nitro cyclohexanone (**37**) (Equation (43)) ⟨71CC1609⟩ but gives less predictable results for fused ring derivatives ⟨89JOC5945⟩.

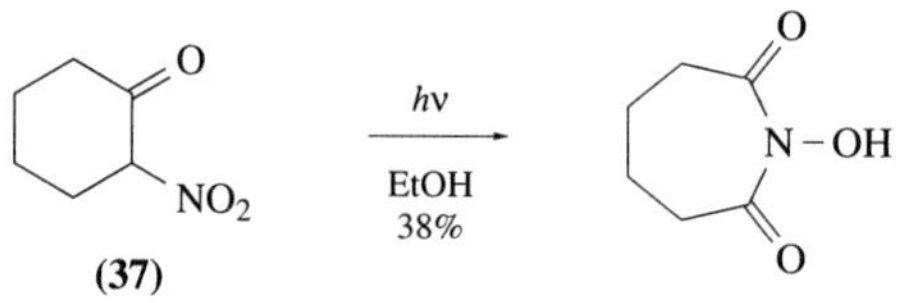

(**37**)

(43)

5.07.2.3.3 From deprotection of N*-alkoxyimides*

The cleavage of *O*-acyl-*N*-hydroxysuccinimides by amines provides the corresponding succinimides (Equation (44)). This reaction occurs under mild conditions and represents a fundamental transformation in the field of protein modification ⟨87MI 507-02⟩ as well as peptide synthesis (albeit for preparing amide bonds, not *N*-hydroxysuccinimide).

(**38**) (**39**)

(44)

The deprotection of several carboxy and carboalkoxy hydroxyimides has been reviewed ⟨80JCS(P1)2122⟩.

As with hydroxamates, *O*-deprotection of alkoxyimides provides the corresponding hydroxyimide. The deprotection of *O*-benzylsuccinimides is accomplished by palladium-catalyzed hydrogenation ⟨92CPB1177, 92HCA109⟩, while trityl-protected hydroxyimides can be deprotected by HCl in benzene ⟨71JOC3835⟩.

5.07.2.4 Alkoxy-substituted Imides

5.07.2.4.1 From acylation of alkoxyamines and hydroxamates

A review of the acylation of hydroxamates with acyl halides has been published ⟨89CCC2738⟩. In 1990, treatment of methoxyamine with excess acetic anhydride was shown to provide *N*-methoxydiacetamide, a useful acylating agent for amines and alcohols (Equation (45)) ⟨90TL243⟩.

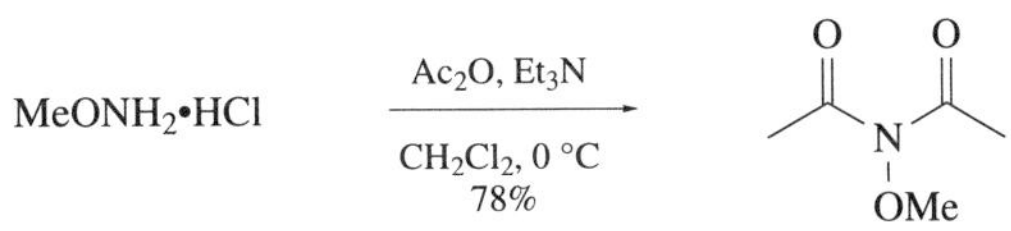

(45)

(40) **(41)**

Carboxylic acids react with the 2,2′-carbonylbis(3,5-dioxo-4-methyl-1,2,4-oxadiazolidine) **(42)** ⟨89JOC617⟩ to provide the corresponding acyl imide **(43)** (Equation (46)) ⟨88TL2661⟩.

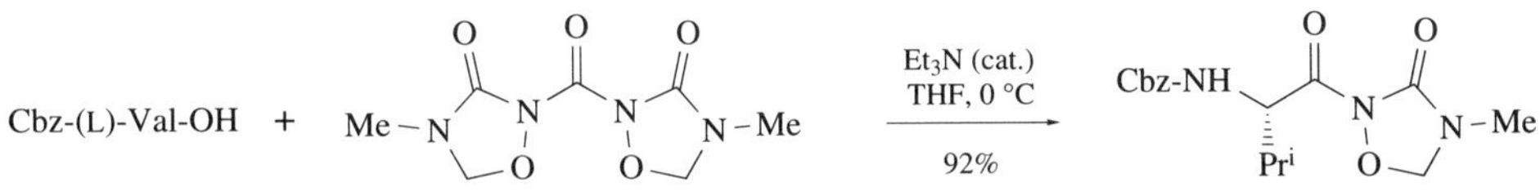

(46)

(42) **(43)**

More stable are the cyclic hydroxyimides formed from the reaction of alkoxyamines with cyclic anhydrides ⟨91JMC387, 92CPB1177⟩.

5.07.2.4.2 *From alkylation of hydroxyimides*

The hydroxy group of *N*-hydroxyimides can be alkylated by a variety of alkylating agents. While *N*-hydroxysuccinimide reacts easily with α-haloesters (RT, Et_3N, THF, 2–4 h), the sodium salts of *N*-hydroxyphthalimide and *N*-hydroxynaphthalimide are required in order to conduct the reaction at room temperature ⟨91JMC387⟩. Alcohols activated under Mitsunobu conditions (PPh_3, dead, $CHCl_3$) also serve as alkylating agents for the conversion of hydroxyimides to alkoxyimides ⟨94OPP111⟩. In addition, the OH group of *N*-hydroxyimides **(44)** is acidic enough to react with excess diazomethane, providing the corresponding *N*-methoxyimides **(45)** (Equation (47)) ⟨67JOC1800⟩.

(47)

(44) **(45)**

5.07.3 AMIDES SUBSTITUTED BY SULFUR, SELENIUM, OR TELLURIUM

There are substantial differences between the chemistry of oxygen and that of sulfur, with a more gradual variation through the sequence S, Se, Te, Po. The maximum coordination number is not limited to 4, nor is the valence limited to 2, as in the case of oxygen, since *d* orbitals can be utilized in bonding ⟨B-80 507-01⟩.

The organic chemistry of sulfur has been reviewed in great detail ⟨85HOU(E11)1⟩, as has the chemistry of compounds containing Se—N and Te—N bonds ⟨B-87MI 507-01⟩.

5.07.3.1 Amide Derivatives: Oxidation State +2, Dicoordinate

5.07.3.1.1 *From amides*

Amides and lactams react with a variety of sulfur transfer agents (RS—X) to provide, respectively, thioamides and thiolactams (Table 7). In most cases, the amide is deprotonated prior to the addition of the thio-transfer reagent in order to accelerate the reaction ⟨92JOC1948, 93TL6877⟩. It has been shown that the reactivity toward sulfination correlates nicely with the pK_a of the amide ⟨64JCS5679⟩.

Table 7 Reaction of amides and lactams with sulfur transfer agents.

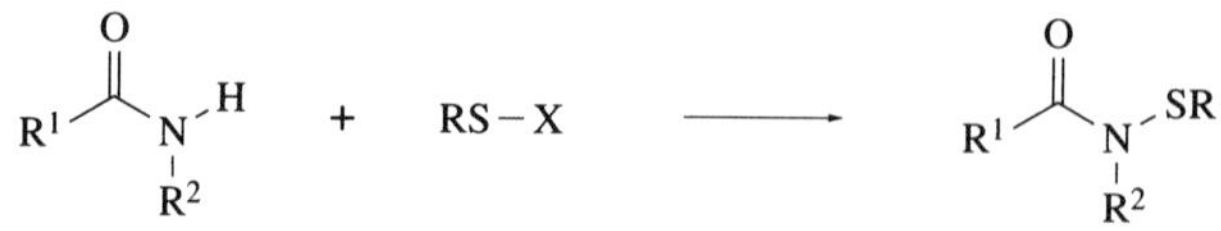

R¹CONHR²	*Conditions*	*Product*	*Ref.*
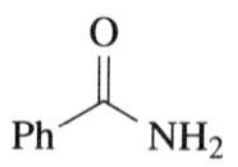	$ClSCClF_2$, pyridine, DMF −5 °C, 1 h to RT, 2 h (34%)	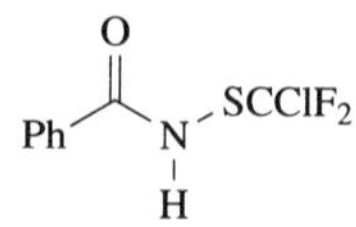	81ZOR533
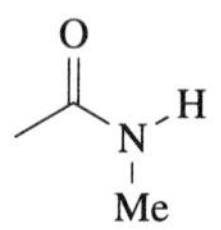	PhSCl, Et_3N, CCl_4 45 °C, 3 h (90%)	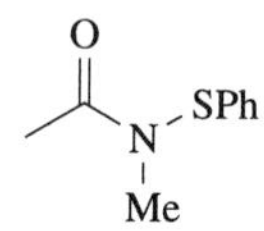	93TL6877
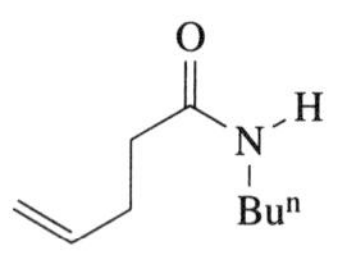	i, NaH, THF, amide ii, −78 °C, PhSCl (41–74%)	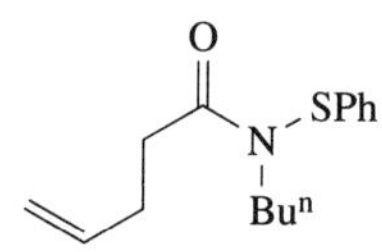	93TL6877
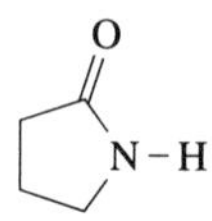	*c*-$C_6H_{11}SCl$, Et_3N, CCl_4 RT, 3 h (71%)	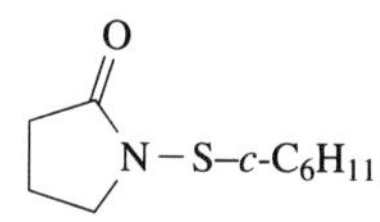	79S228
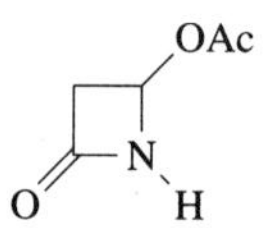	Phth-SCH_2CO_2Me Et_3N, benzene, RT 3 h (74%)	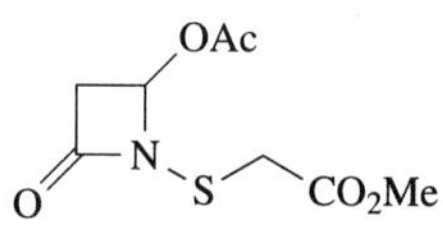	85TL3891
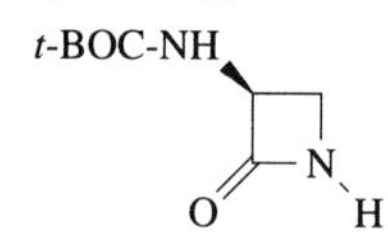	MeOC(O)SCl, DMF Et_3N, 0 °C (80%)	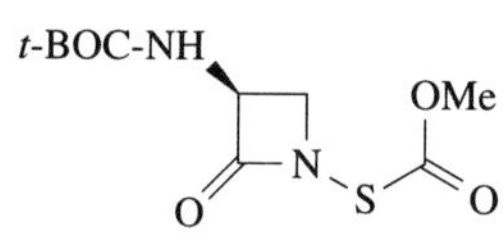	86JOC3133
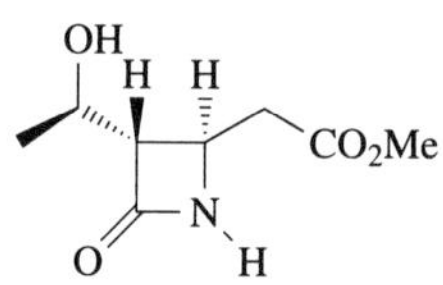	i, LDA, HMPA THF, −78 °C ii, $MeSSO_2Me$ (94%)	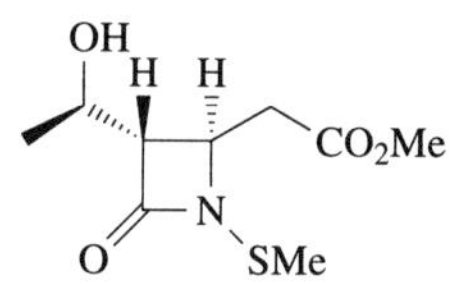	87H221

Cyclic thioamides can be prepared from the internal cyclization of amides containing an appropriately spaced disulfide ⟨85JCS(P1)153, 85JHC1353, 85JMC1661, 85JMC1772, 86JMC359, 88JHC23, 89JMC1024⟩ or thioalkyl group ⟨85JHC1635, 87H(25)569⟩. Such groups are initially activated via conversion to the corresponding halosulfide (**46**), which undergoes attack by the neighboring amide moiety (Scheme 8) ⟨83OPP315⟩. The analogous reaction takes place with selenium analogs ⟨87JA5549, 89BSB395, 91CZ135, 91SC85, 91T9053, 92TL3863⟩, but not with tellurium derivatives ⟨89JOC2964⟩.

Internal cyclizations involving the Pummerer rearrangement of sulfoxides have also been reported ⟨81CC510, 85JA5490, 92TL153⟩.

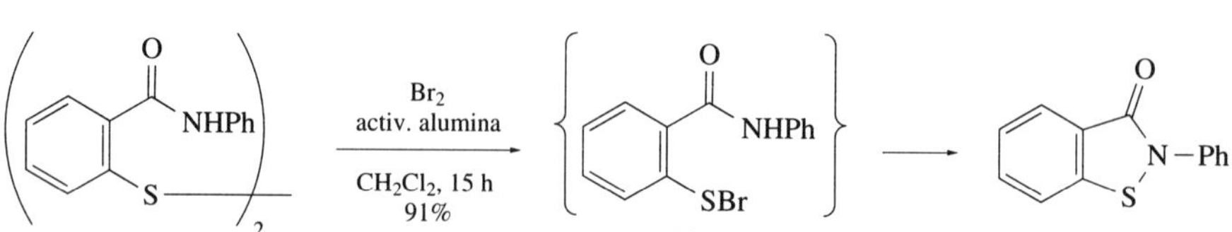

(**46**)

Scheme 8

5.07.3.1.2 *From* N-*haloamides*

N-Chlorolactams react with benzenethiol to provide the corresponding thiolactams. The formation of an orange-red color during the reaction indicates that the initial reaction is formation of phenylchlorosulfide and the lactam (**47**). The lactam then reacts with the chlorosulfide to form the thiolactam (**48**) (Scheme 9) ⟨87CC1233, 89JOC1904⟩. *N*-Iodoamides react with dialkyldithiocarbamates (presumably by a similar mechanism) to provide thioamides ⟨91BCJ3732⟩.

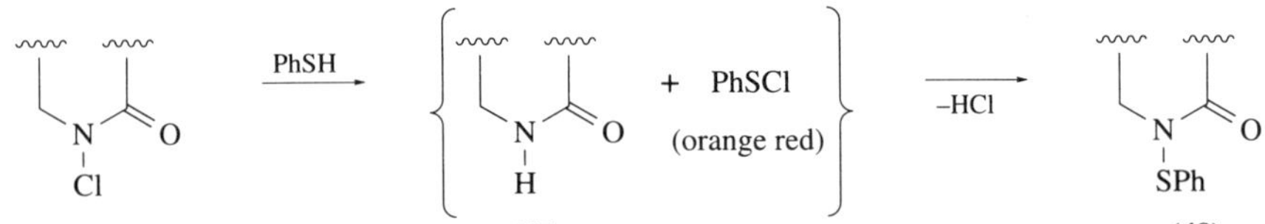

(**47**) (**48**)

Scheme 9

5.07.3.1.3 *From* N-*silylamides*

Primary ⟨86CJC308⟩ and secondary ⟨85S212⟩ TMS-amides react with phenylselenium chloride to provide the corresponding *N*-phenylselenoamides. The products could not be isolated, but rather were converted directly to other derivatives. *N*-(TMS) benzamide (**49**) reacts with an excess of sulfur dichloride to provide crystalline *N*-(amido)-*N*-sulfenyl chloride (**50**) (Equation (48)) ⟨73JA279⟩.

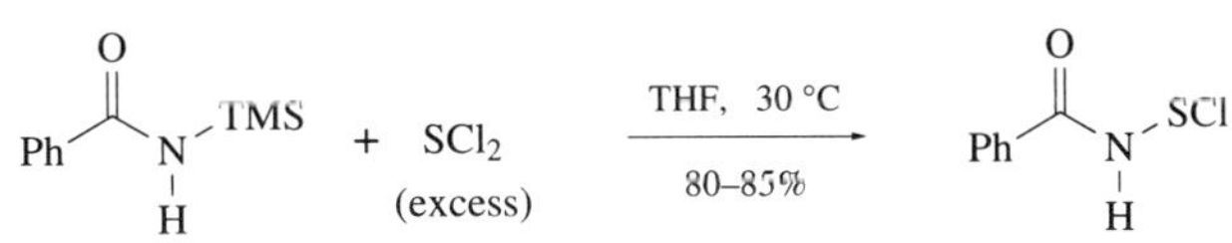

(**49**) (**50**) (48)

5.07.3.1.4 *From acylation of thioamines*

S-(*N*,*N*-Diethylthiocarbamoyl)sulfenamide (**51**) reacts with acid chlorides or acid anhydrides to provide the corresponding *N*-(thiocarbamoylthio)amides (**52**) (Equation (49)) ⟨86BCJ2017⟩. The internal cyclization of thioamines containing an appropriately spaced carboxy group also has been reported ⟨75JOC2029⟩.

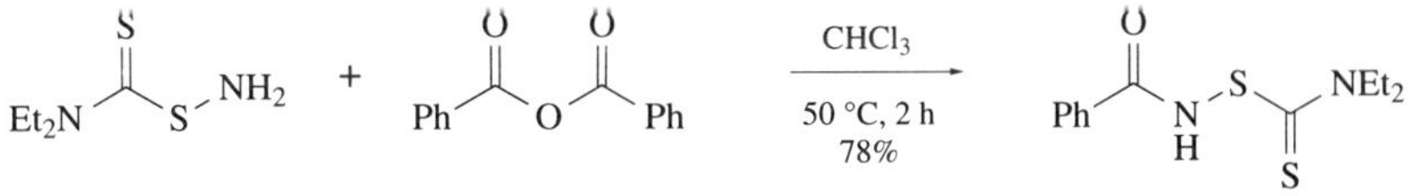

(**51**) (**52**) (49)

5.07.3.1.5 *From sulfenimines*

N-(Phenylsulfenyl)benzaldimine reacts with ester enolates to provide (upon loss of the thiophenol) low yields of *β*-lactams. The low yields and the loss of the thiophenol group indicate that enolate attack on sulfur occurs. The use of trityl-protected sulfenimine (**53**) under similar conditions provides trityl-protected, *N*-thio-*β*-lactam (**54**) (Equation (50)) ⟨86JOC1929⟩.

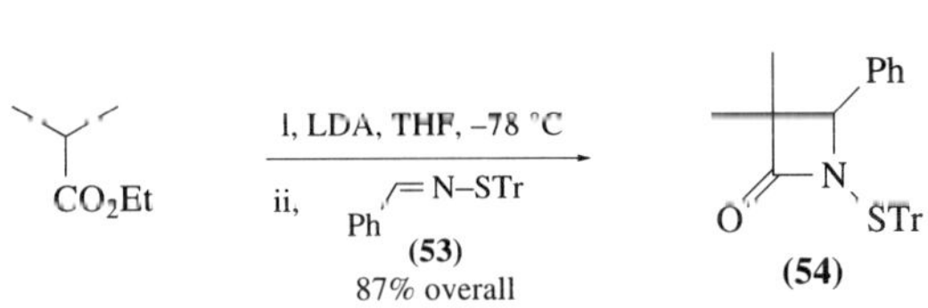

(**54**) (50)

5.07.3.1.6 *From sulfoxides*

The treatment of sulfoxides, appropriately spaced from amides, under Pummerer rearrangement conditions leads to *N*-thioamides (Equation (51)) ⟨85JA5490⟩.

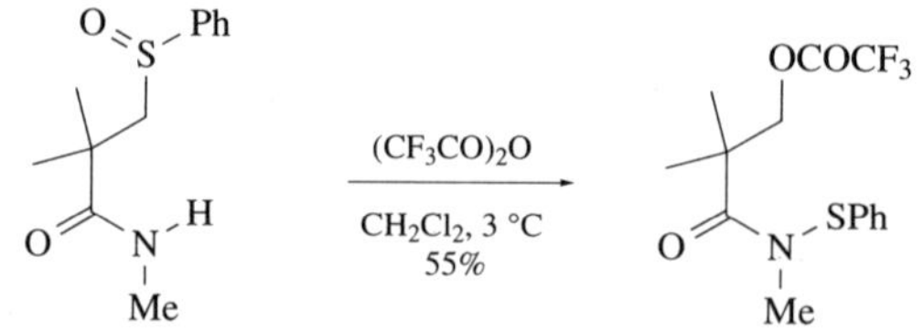

(51)

In a related vein, 2-(alkylsulfinyl)benzamides react with either trifluoroacetic acid ⟨92TL153⟩ or thionyl chloride ⟨81CC510⟩ to provide the corresponding benzisothiazol-3-ones.

5.07.3.1.7 *From 1,4-addition to thione* S-*imides*

Thione *S*-imides react with dry hydrochloric acid to provide the corresponding thioamides (Equation (52)) ⟨73JA279⟩.

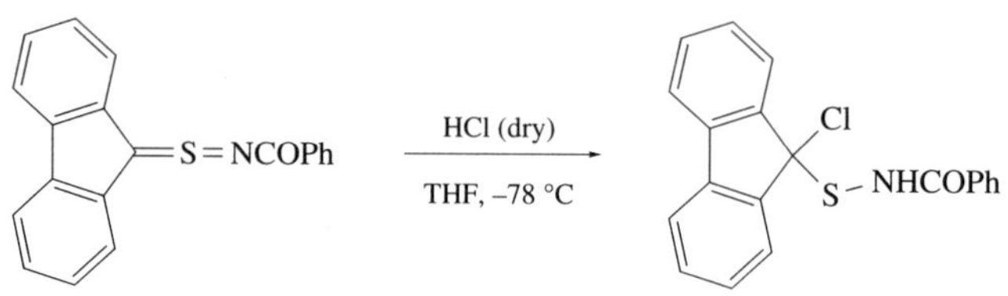

(52)

5.07.3.1.8 *From cycloaddition of thione* S-*imides*

Thione *S*-imides such as **(55)** react with *N*-isobutylpyrrolidine **(56)** to provide *N*-benzoyl-heterocycles **(57)** (Equation (53)) ⟨73JA279⟩.

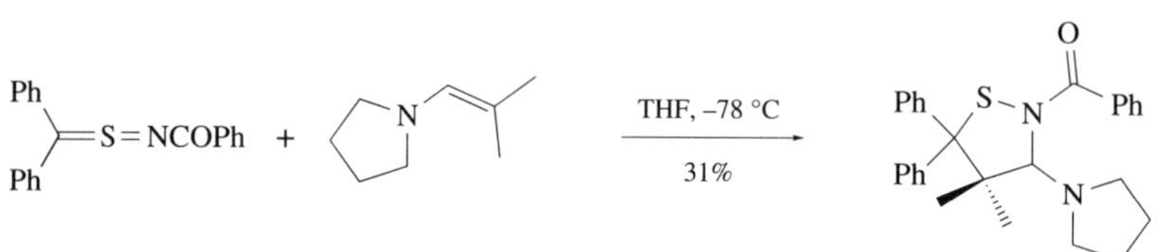

(55) (56) (57)

(53)

5.07.3.2 Imides Derivatives: Oxidation State +2, Dicoordinate

The history of thioimides formation was summarized in 1969 by Behforouz and Kerwood ⟨69JOC51⟩.

5.07.3.2.1 *From imides*

Imides are converted to *N*-thioimides and *N*-selenoimides by a number of transfer agents (Table 8). The resulting thioalkylimides and phenylseleniumimides are themselves being used more and more frequently as thioalkyl ⟨86TL3095, 90JOC2576⟩ and selenoaryl ⟨85T4835⟩ transfer reagents.

An improved preparation of *S,S′*-diimide derivatives from sulfur monochloride has been reported ⟨87JOC169⟩. Previous methods used to prepare this compound showed a strong solvent dependency (Scheme 10) ⟨66CJC2111⟩.

Care should be taken in the preparation of imides containing —CH_2E moiety on sulfur, where E is an electron-withdrawing group ⟨86JOC3133⟩. These compounds have been shown to eliminate phthalimide under relatively mild conditions. Similar reactivity has been found with the corresponding selenium phthalimides containing a CH_2CO_2R group, although higher temperatures were required to effect elimination ⟨86CC1152, 88JCS(P1)1913⟩.

5.07.3.2.2 *From* N-*haloimides*

N-Haloimides react with a wide assortment of sulfur derivatives to provide the corresponding thioimides. *N*-Chlorosuccinimide reacts with elemental sulfur to provide *N*-(chlorothio)succinimide

Table 8 Preparation of thioimides and selenoimides from imides.

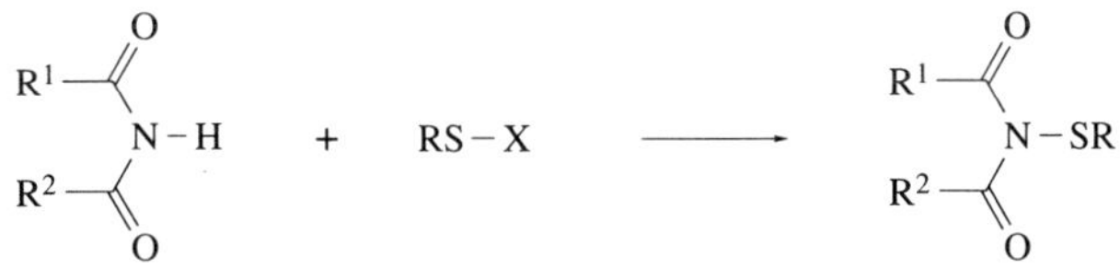

Imide	*RS–X*	*Conditions*	*Product*	*Ref.*
Phth-H	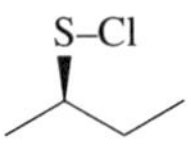	Et_3N (76%, 90% *ee*)	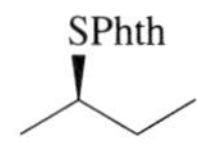	90TA141
Phth-K	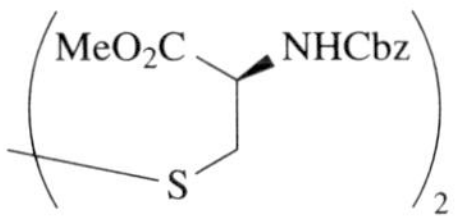	i, SO_2Cl_2, 0 °C 1,2-dichloroethane ii, phthalimide (61%)	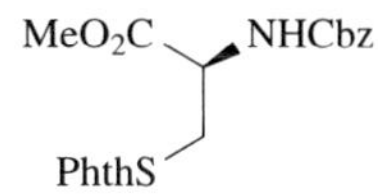	86JOC5332
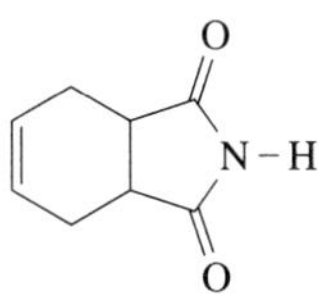	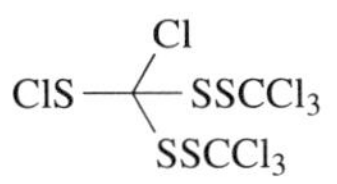	NaOH, H_2O CH_2Cl_2 (50%)	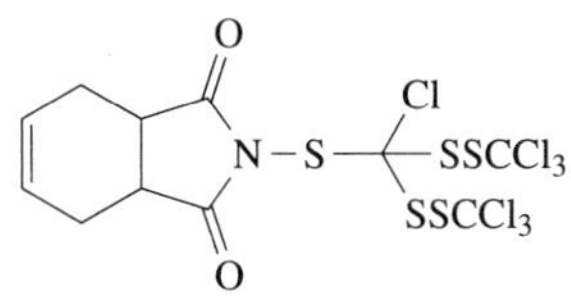	85T5145
Phth-H	$(c\text{-}C_6H_{11}S)_2$	NaBr, e^- 18 h (99%)	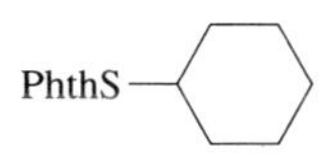	79JOC1554
Phth-H	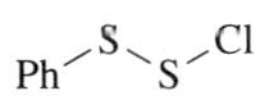	Et_3N, DMF 30 min (70%)	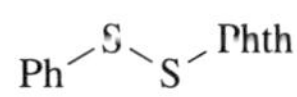	71JCS(C)3550
Phth-K	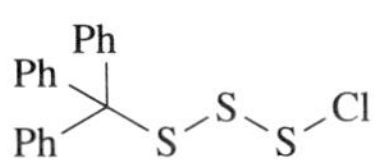	Et_2O, 4 h (47%)	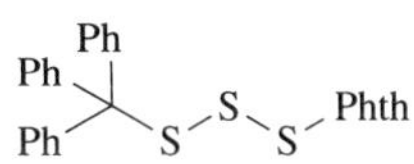	94JOC806
Phth-K	PhSeCl	hexane, 25 °C (90%)	Phth–SePh	85T4835

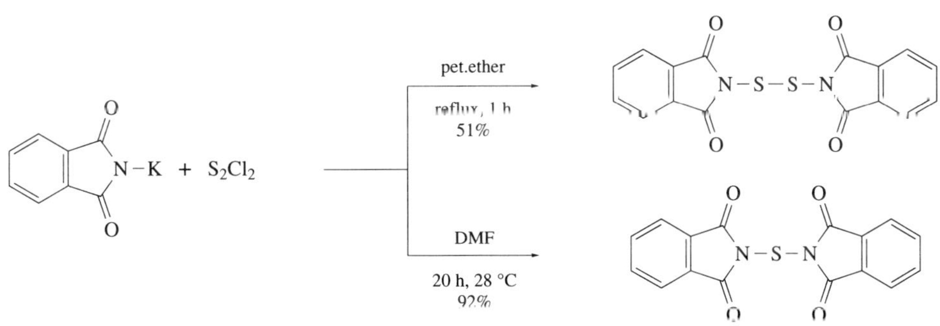

Scheme 10

⟨86ZOR100⟩. *N*-Bromo and *N*-chloro imides react with a variety of thio derivatives, including sodium dithiocarboxylates ⟨81S746⟩, thioacyl diphenylphosphino sulfides ⟨85CB1668⟩, dialkyl sulfides ⟨59CB2887⟩, disulfides ⟨67CB1248, 89AP(322)343⟩, and tin sulfides ⟨84S1037, 87S1122⟩ to provide thioimides. A similar reaction occurs when NCS is treated with allylphenyl selenide ⟨85T4835⟩ or diphenyl diselenide (**58**) (Equation (54)) ⟨90TL873⟩.

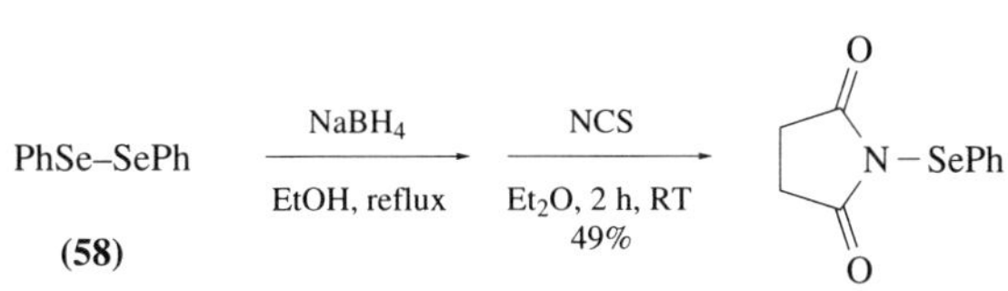

(**58**) (54)

5.07.3.2.3 *From* N-*silylimides*

Although it has been reported that TMS-phthalimide does not react with selenyl chloride ⟨86CJC308⟩, *N*-(TMS)succinimide (**59**) reacts with either sulfur dichloride or sulfur monochloride to provide high yields of *N,N′*-thiosuccinimide (**60**) or *N,N′*-dithiosuccinimide (**61**) (Scheme 11) ⟨78JA1222⟩.

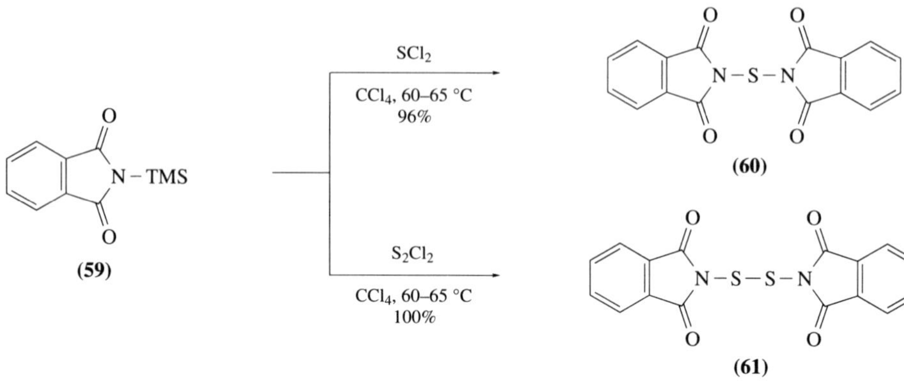

Scheme 11

5.07.3.2.4 *From reduction of disulfides*

Alkylphthalimido disulfides can be cleanly converted to thioalkyl phthalimides upon treatment with triphenylphosphine (Equation (55)) ⟨82JOC2785⟩. More nucleophilic phosphines such as tris(diethylamino)phosphine, $(Et_2N)_3P$, are less selective, providing mixtures of disulfide, phthalimide, as well as the desired thioalkyl phthalimide.

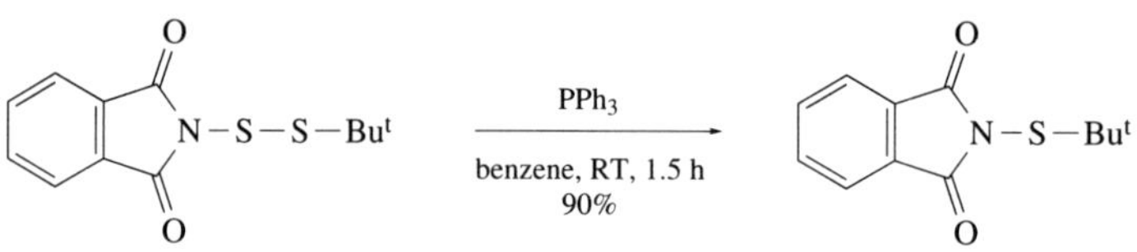

(55)

5.07.3.3 Dithioamides Derivatives Oxidation State +2, Dicoordinate

Heptasulfur imide (**62**) reacts with acetyl chloride in pyridine to provide the corresponding acyl imide (**63a**) ⟨52ZN(B)634⟩. The same product has been obtained from dcc coupling of the corresponding acid (Equation (56)) ⟨63JCS4918⟩.

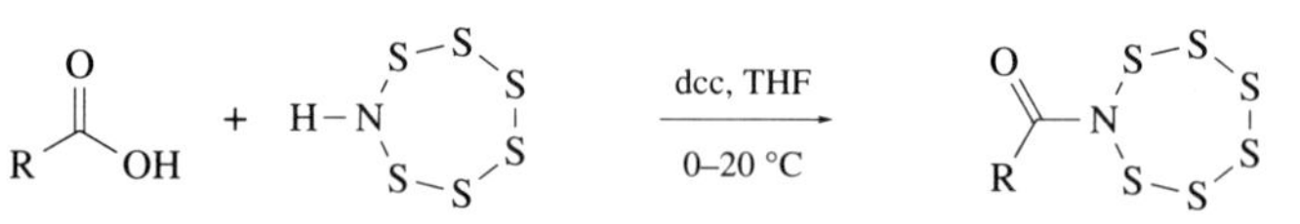

(56)

(**62**) (**63**) **a**; R = Me, 70%
b; R = Ph, 59%

dcc = dicyclohexylcarbodiimide

5.07.3.4 Amide Derivatives: Oxidation State +2, Tricoordinate

5.07.3.4.1 *From* N-*haloamides*

Treatment of dimethyl sulfide with the *N*-chloroamides ⟨77TL2235⟩ or *N*-bromoamides ⟨79JOC2863⟩ provides *N*-acyl sulfonium salts (**65**). Similar compounds can be prepared from the reaction between phenyliodonio tosylate of benzamide (**64**) and dimethylsulfide (Equation (57)) ⟨87TL5431⟩.

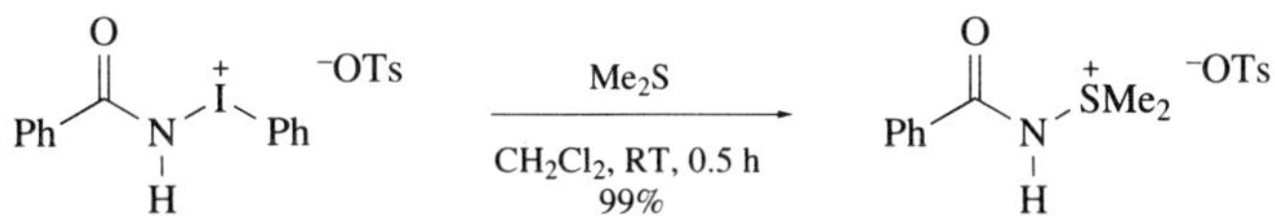

(57)

(64) **(65)**

5.07.3.4.2 *From sulfilimines*

Acid chlorides add to sulfilimines (**66**) to provide acyl aminosulfonium salts (Equation (58)) ⟨80LA1075⟩.

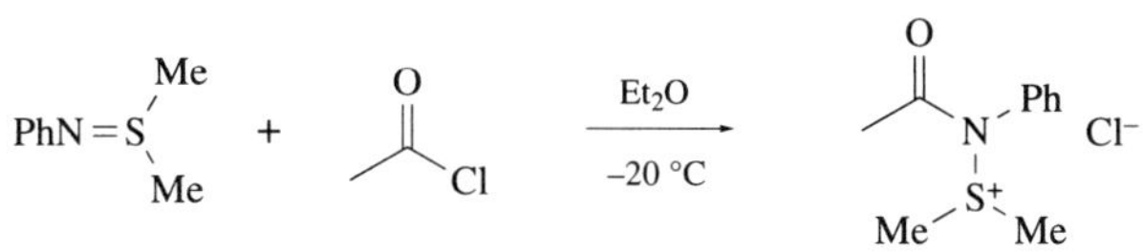

(58)

(66) **(65)**

5.07.3.4.3 *From nitrilium salts*

Nitrilium salts such as (**67**) react even at low temperatures with dimethyl sulfoxide to provide acyl sulfonium salts such as (**69**) ⟨80LA1064⟩. The reaction is believed to occur via rearrangement of an intermediate iminooxosulfonium salt (**68**) (Scheme 12).

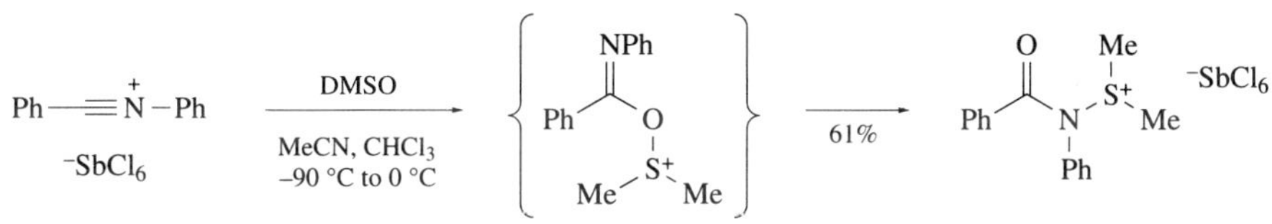

(67) **(68)** **(69)**

Scheme 12

5.07.3.5 Imide Derivatives: Oxidation State +2, Tricoordinate

N-Chlorosuccinimide reacts with dimethyl sulfide ⟨72JA7586⟩ as well as with *N*,*N*-dialkylsulfenamides ⟨76S308, 79S97⟩ to provide the corresponding *N*-succinimido sulfonium salts. Complexes such as (**70**) are stable at 0 °C, but decompose at higher temperatures. Complexes with counterions other than chloride (such as BF_4) are reported to be more stable (Scheme 13) ⟨76S308⟩.

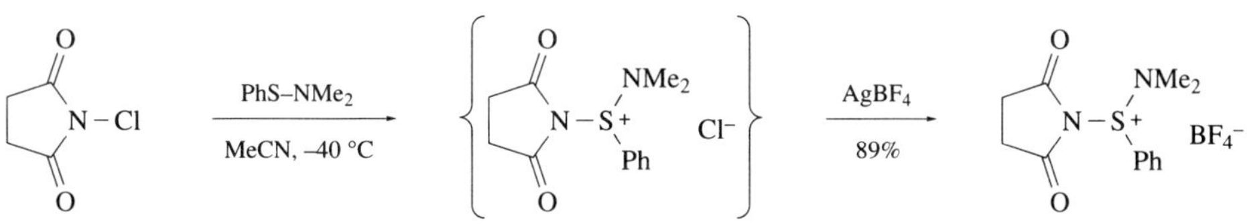

(70) **(71)**

Scheme 13

5.07.3.6 Amide Derivatives: Oxidation State +4, Tricoordinate

5.07.3.6.1 From thiodiimides

It has been reported that thiodiimides react with malonyl chloride to provide the ring-opened products ⟨69M959⟩. Fischer and Bolt have indicated that the product is, in fact, the cyclic oxathiazine (**72**), which reacts with aniline derivatives to provide thio-substituted diimides (**73**) (Scheme 14) ⟨85SUL11⟩.

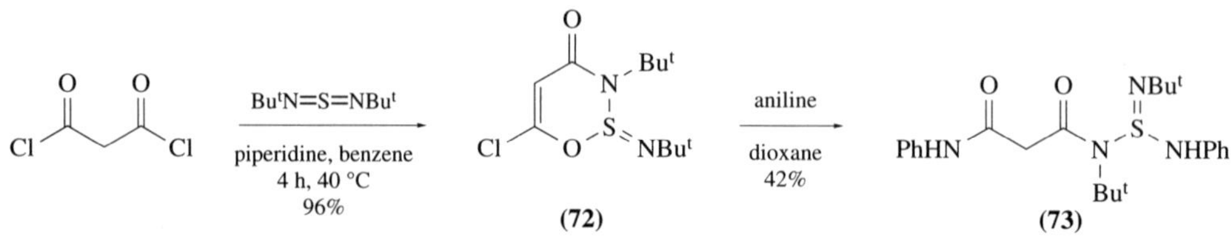

(72) **(73)**

Scheme 14

5.07.3.6.2 From oxidation of thioamides

Treatment of thioamides with the sodium salt of *N*-chlorosulfonamides provides a route to *N*-acyl-*N*′-aryl(alkyl)sulfonylsulfinamidines (**74**) (Equation (59)) ⟨81ZOR533⟩.

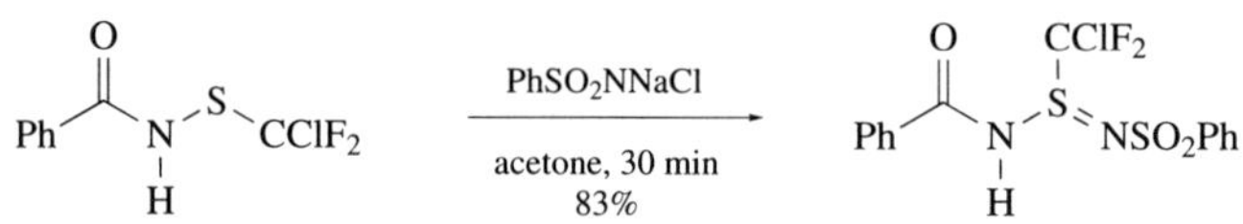

(74) (59)

The chemistry of sulfinic acids, esters, and amides has been reviewed by Tillett ⟨B-90MI 507-01⟩.

5.07.3.6.3 From amides

Treatment of sulfinic acid with thionyl chloride provides an intermediate sulfinyl chloride which reacts with amides to provide the desired sulfinamide (Equation (60)), ⟨91JCS(P1)3383, 93JCS(P1)1581⟩.

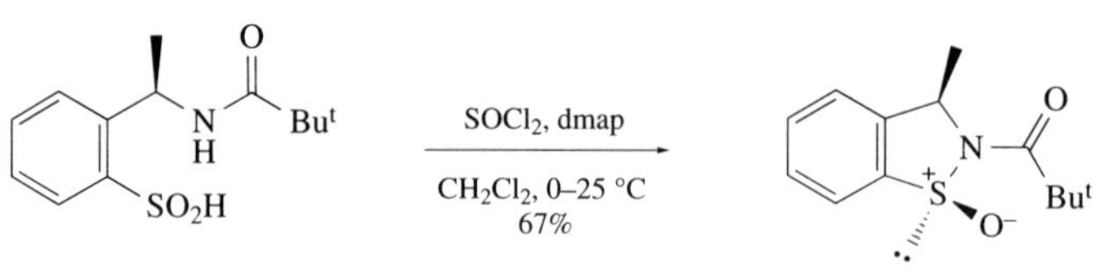

(60)

Other chalcogen derivatives prepared from the corresponding amides include the reaction of α- ⟨92T8615⟩ and β-aminoamides ⟨88CPB877⟩ with thionyl chloride to provide heterocyclic derivatives. Such derivatives can also be obtained by the reaction of diphenyldimethoxyselenium with trichloroacetamide ⟨89ZOB611⟩.

5.07.3.6.4 From N*-acyl-*N*-sulfinylamines*

N-Acyl-*N*-sulfinylamines are highly reactive dienophiles which undergo cycloaddition reactions with ketenes ⟨86CB1127⟩ and dienes ⟨86CB3405⟩ under mild conditions. α-Chloro-α-methoxy-*N*-sulfinylamine (**75**), prepared from iminoether and thionyl chloride, is converted quantitatively at room temperature to the corresponding *N*-acyl Diels–Alder adduct (**76**) upon treatment with diene (Scheme 15, ⟨88PS(35)1⟩).

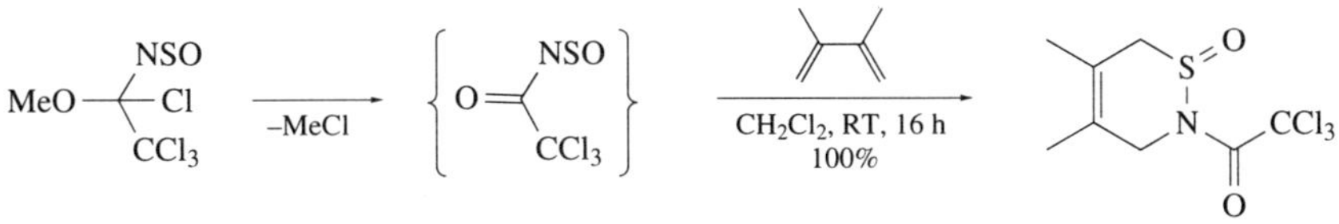

(75) **(76)**

Scheme 15

N-Phenylsulfonyl-*N'*-benzoylselenium diimide ($PhSO_2N{=}Se{=}NCOPh$) also undergoes regio-selective Diels–Alder reactions to provide the corresponding *N*-benzoyl heterocycles ⟨86ZOR1189⟩.

5.07.3.6.5 From oxidation of thioamides

Oxidation of cyclic sulfinylamides with mcpba ⟨81T2181⟩ or peroxide ⟨81CB80⟩ provide the corresponding sulfonamides (Equation (61)).

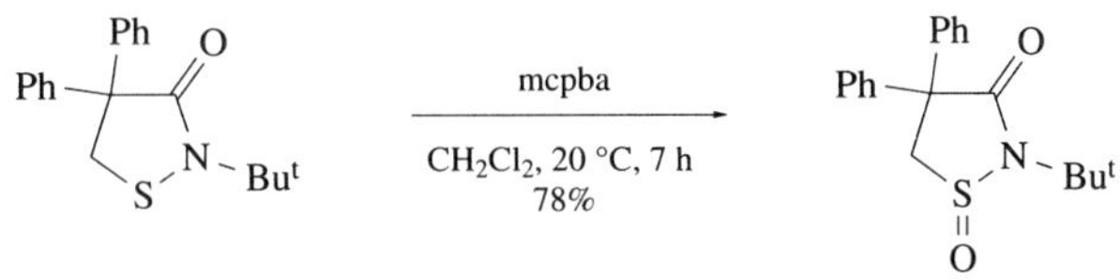

(61)

5.07.3.7 Amide Derivatives: Oxidation State +4, Tetracoordinate

5.07.3.7.1 From N*-haloamides*

The reaction of selenium with *N*-chloroamides provides the bis(acylamino)selenium dichlorides (**77**) (Equation (62)) ⟨78ZOR280⟩.

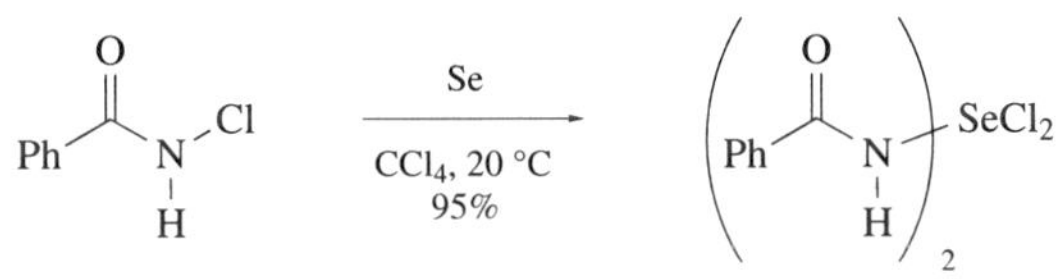

(62)

(77)

The reaction of *N*-chloroamides with arylselenium chloride provides the aryl(acylamino)selenium dichlorides ⟨74ZOR139⟩. Tellurium amides can be obtained through oxidative addition of diaryl-tellurium with bromoamides ⟨83IJC(A)503⟩.

5.07.3.7.2 From silylamides

Trimethylsilylamides react with arylselenium trichloride to provide selenoamide and TMS-Cl (Equation (63)) ⟨78ZOB280⟩.

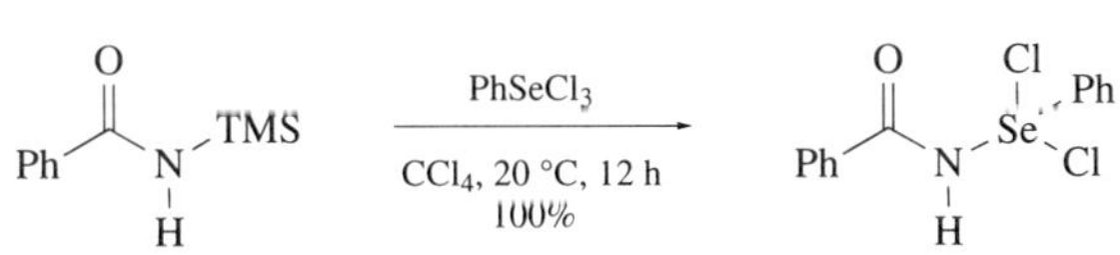

(63)

5.07.3.8 Imide Derivatives: Oxidation State +4, Tri- and Tetracoordinate

5.07.3.8.1 From haloimides

Arylselenium chloride inserts into the N—Cl bond of NCS to provide the imidoselenide Ar(Suc)$SeCl_2$ ⟨81ZOR529⟩. Dibenzoyltellurium adds to NBS ⟨86IJC(A)57⟩, and diaryltellurium oxidatively adds to bromo- and chloroimides, to provide the corresponding diarylhaloimides (Equation (64)) ⟨79ICAL99⟩ see also ⟨83IJC(A)503⟩.

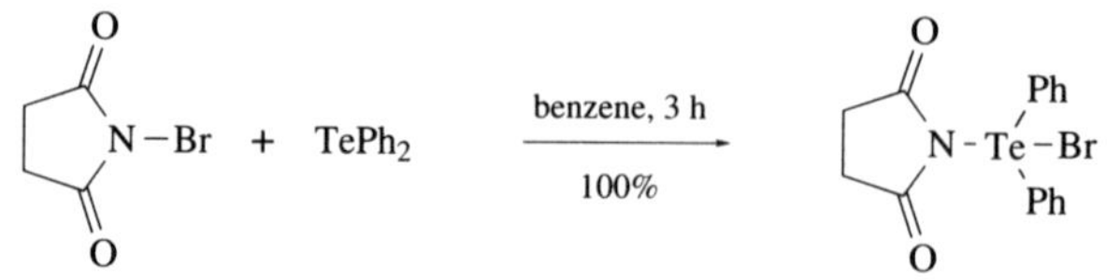

(64)

5.07.3.8.2 From oxidation of thioimides

Imidosulfonamides can also be prepared from the oxidation of thioimides with mcpba (Equation (65)) ⟨72TL5313, 73JOC4328⟩. The reaction is quite general, providing good yields of both alkyl and aryl sulfinylimides.

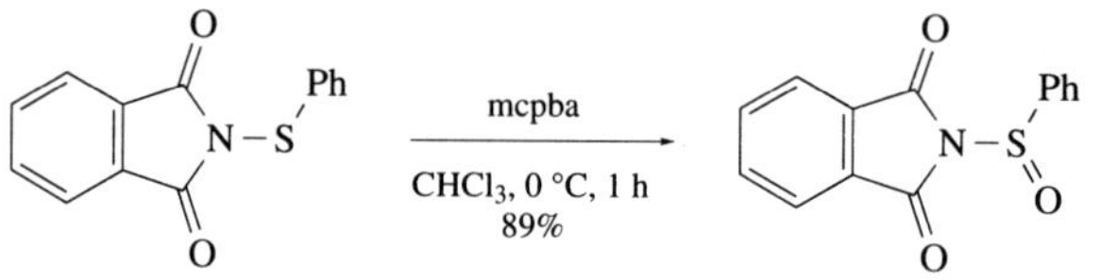

(65)

5.07.3.9 Amide Derivatives: Oxidation State +6, Tetracoordinate

A general review of the ester and amide derivatives of sulfonic acids has been published ⟨B-91MI 507-02⟩.

5.07.3.9.1 From acylation of sulfonamides

Sulfamates themselves are very poor nucleophiles that do not undergo acylation with standard coupling reagents. However, tetrabutylammonium sulfamate reacts with carboxylic acids activated by dcc (Equation (66)) ⟨82JOC176⟩.

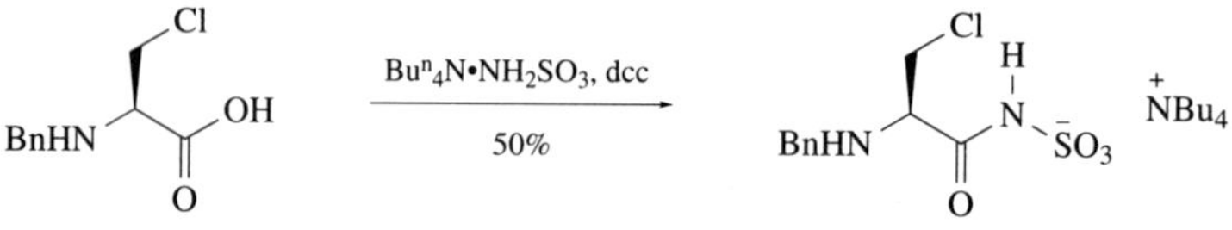

(66)

The acylation of sulfonamides provides a convenient route to *N*-acyl sulfonamides ⟨68M1289⟩. Sulfonamides react with anhydrides ⟨92JOC7001⟩ and acyl halides ⟨86S1021, 91BMC733, 92JOC5328, 93TL6821⟩, as well as with carboxylic acids which have been activated by dcc ⟨88T613, 91TL283⟩, dcci ⟨90JMC2621⟩, chloroformates ⟨89JCS(P1)1815⟩, or carbonyldiimidazole ⟨86TL3907, 88TL1653⟩. Ester acylation works well for activated ester derivatives as well as for cases of intramolecular acylation ⟨86JHC1253, 89TL3217⟩. With more hindered sulfonamides, precomplexation of the sulfonamide with trimethylaluminum can be used in order to effect coupling (Equation (67)) ⟨89TL6009, see also 87HCA1666, 93HCA957, 94HCA554⟩.

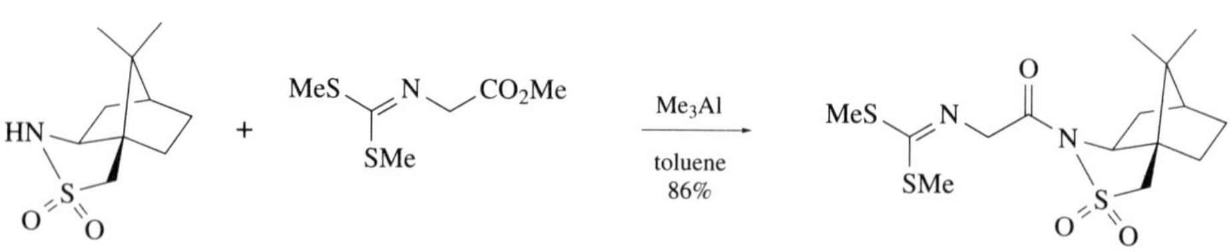

(67)

The coupling of sulfonamides with α,β-unsaturated acyl derivatives can lead to by-products resulting from 1,4-addition. The use of copper(I) chloride or copper powder has been reported to avoid this by-product ⟨89LA739⟩. Moreover, the choice of base used in the reaction has been shown to have a dramatic effect on the yield of 1,4-addition (Equation (68)) ⟨94SL197⟩.

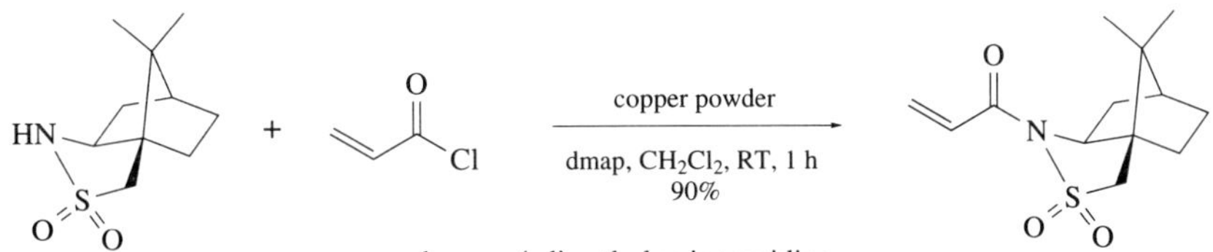

(68)

dmap = 4-dimethylaminopyridine

Ketenes react readily with sulfonamides to provide *N*-acyl amides ⟨80CB912⟩. An intramolecular variation of this reaction involves the photolysis of sulfonamides containing pendant α-diazoketones (**78**) (Equation (69)) ⟨88H(27)867⟩.

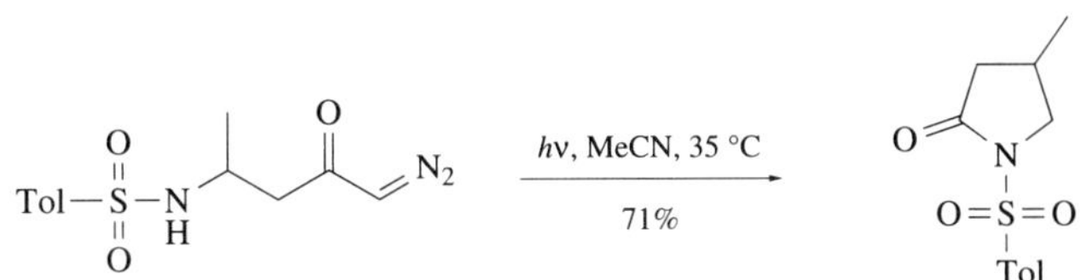

(**78**)

(69)

The heating of sultams with the 2,2,6-trimethyl-4H-1,3-dioxin-4-one (**79**) provides the acylated sultam (**80**) (Equation (70)) ⟨93TL5627⟩.

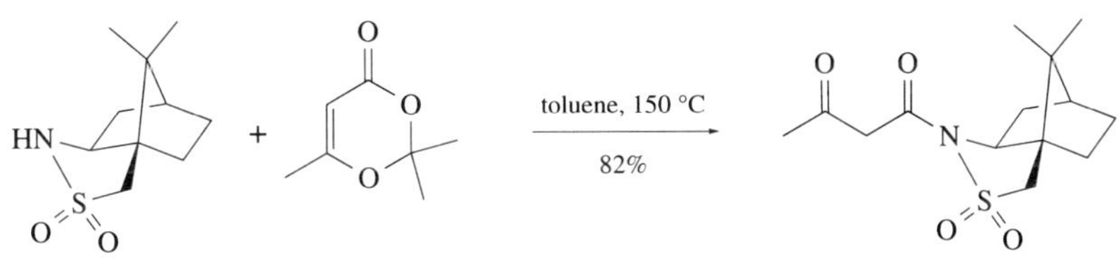

(**79**) (**80**)

(70)

Acyl silanes also have been found to react with *N*-methylsulfonamide under electrochemical oxidation to provide acyl sulfonamide (Equation (71)) ⟨92JOC4877⟩.

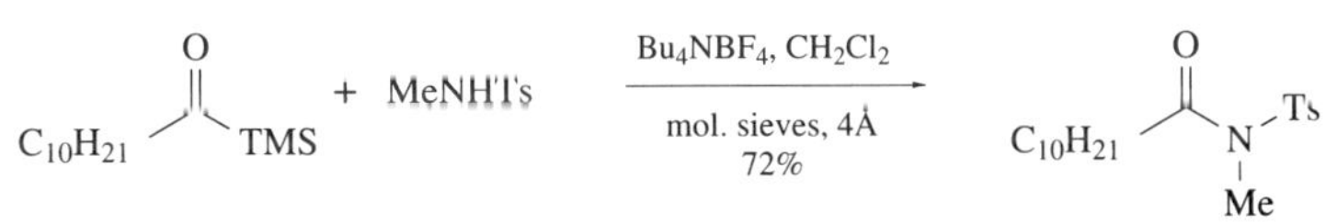

(71)

5.07.3.9.2 *From acylation of* N-*TMS sulfonamides*

The coupling of TMS-protected sulfonamides has been reported to eliminate 1,4-addition of amides during coupling with α,β-unsaturated acid chlorides (Equation (72)) ⟨92S582⟩.

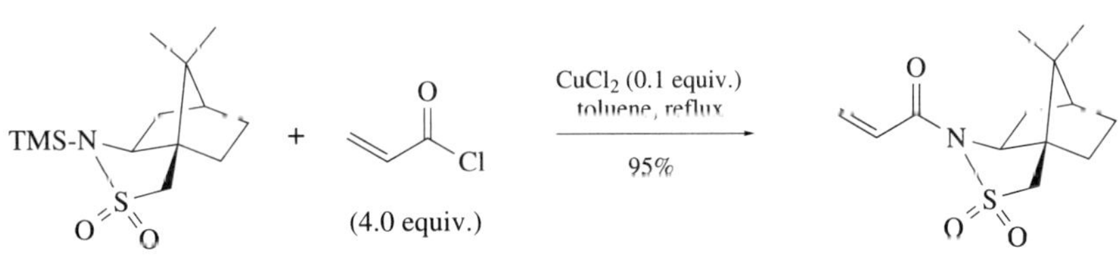

(72)

5.07.3.9.3 *From acylation of* N-*substituted sulfonamides*

A number of *N*-substituted sulfonamides can be converted to *N*-acyl sulfonamides via acylative cleavage. Examples of *N*-substituted sulfonamides which undergo such transformation include

N-sulfonylcyanamides ⟨49JCS3029⟩, *N*-nitrososulfonamides ⟨85BCJ3395⟩, and sulfonylureas (Equation (73)) ⟨85JOC4993, 86AP682⟩.

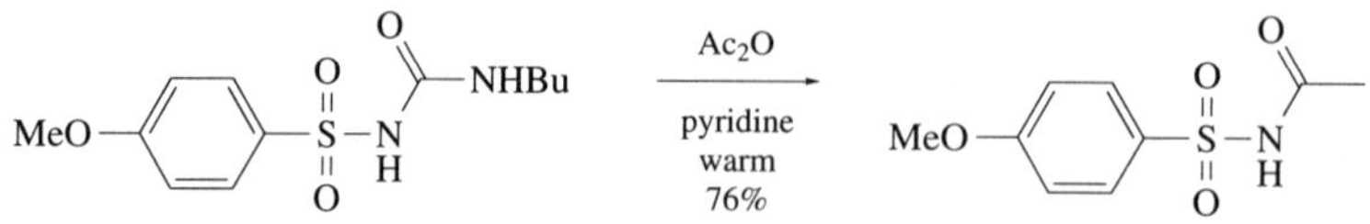

(73)

5.07.3.9.4 From sulfonyl isocyanates

The use of sulfonyl isocyanates for the preparation of *N*-sulfonylamides and lactams has been the subject of a number of reviews ⟨76CRV389, 81SR215, 86S437, 87H(26)1051⟩.

Isocyanates undergo cycloaddition reactions with alkenes ⟨85T2441, 91CAR283, 92SC2729⟩, allenes ⟨89SC3113⟩, thioketenes ⟨80TL4247, 88CB689⟩, benzocyclopropene ⟨89CL2181, 91BCJ106⟩, and vinylcyclopropanes ⟨76TL451⟩, providing a diverse array of *N*-sulfonyllactams.

Sulfonyl isocyanates react with enamines ⟨86FES539⟩, alkynes ⟨91CL1479⟩, ketones ⟨74LA561⟩, enol ethers ⟨85JOC169⟩, and vinylsilanes (Equation (74)) ⟨90JA462⟩, to provide a variety of acyclic *N*-acyl sulfonamides.

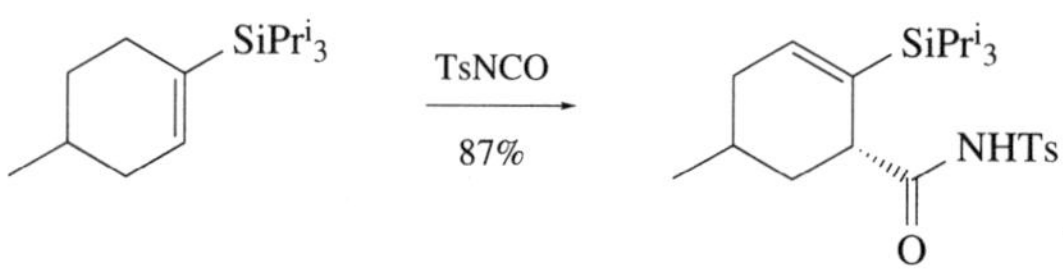

(74)

The reaction of arylstannanes (**81**) with isocyanates has been reported ⟨93JOC7022⟩. This methodology extends the regioselectivity of the Friedel–Crafts reaction between aryl derivatives and sulfonyl isocyanates (Equation (75)).

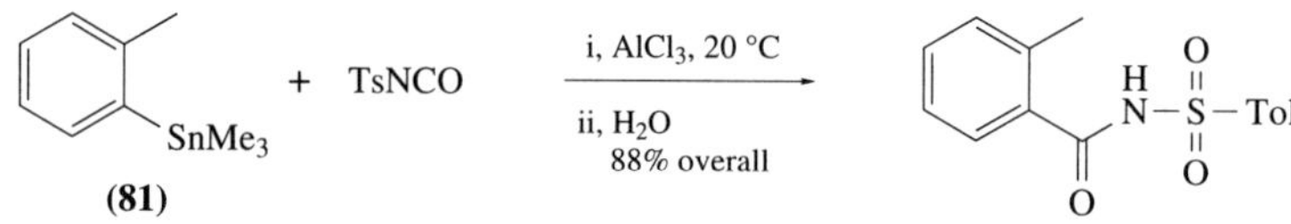

(75)

(81)

5.07.3.9.5 From acylation of **N-*tosylimines***

N-Tosyltriphenylbismuthimines (**82**) react with benzoyl chloride to provide *N*-tosylbenzenesulfonamide (**83**) (Equation (76)) ⟨91CL105⟩. The overall yield is low, and a substantial amount of polymeric bismuth-containing material is also obtained.

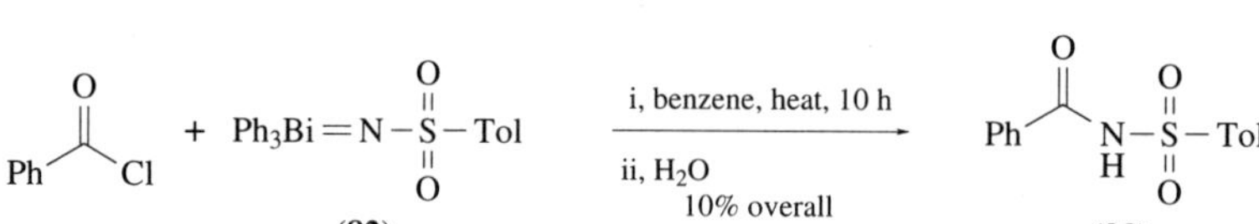

(76)

(82) **(83)**

5.07.3.9.6 From addition to **N-*tosylimines***

N-Tosylimines (**84**) react with cyclopropane acetals (**85**) to provide *N*-tosyllactams (Equation (77)) ⟨90CL905⟩.

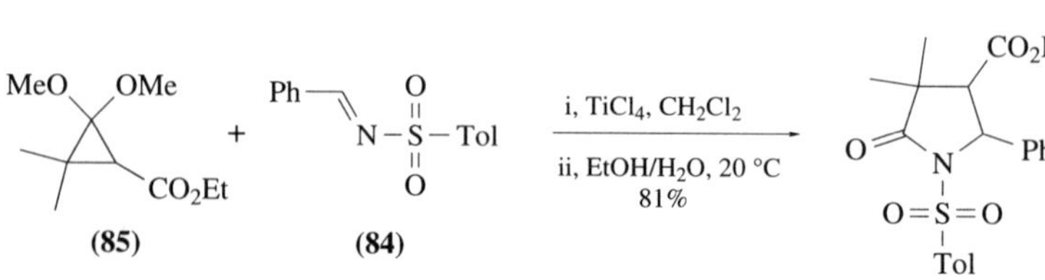

(77)

(85) **(84)**

5.07.3.9.7 From ring opening of N*-sulfonylaziridines*

The sodium salt of diethyl malonate reacts with *N*-sulfonylaziridines (**86**) to provide the *N*-sulfonyllactams (**87**) (Equation (78)) ⟨86CB2678⟩. The preference for ring opening is similar to that of styrene oxide in that malonate attack occurs regiospecifically at the benzyl position.

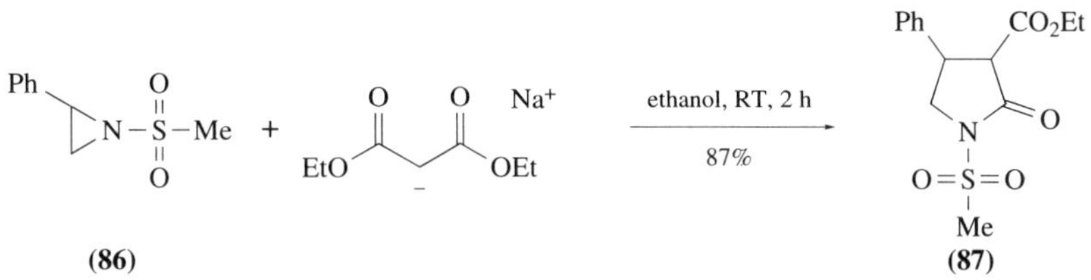

(78)

5.07.3.9.8 From ketones and N*-chlorosulfonamides*

The reaction of *N*-chlorosulfonamides with acetophenone initially leads to formation of the 2,2,2-trichloroacetophenone, which reacts further with the remaining sulfonamide to provide *N*-benzoylsulfonamide (Equation (79)) ⟨69ZOR279⟩. Based on this mechanism, it is not surprising that trichloromethylketones have been reported to serve as acylating agents for methylsulfonamides ⟨90SC2083⟩.

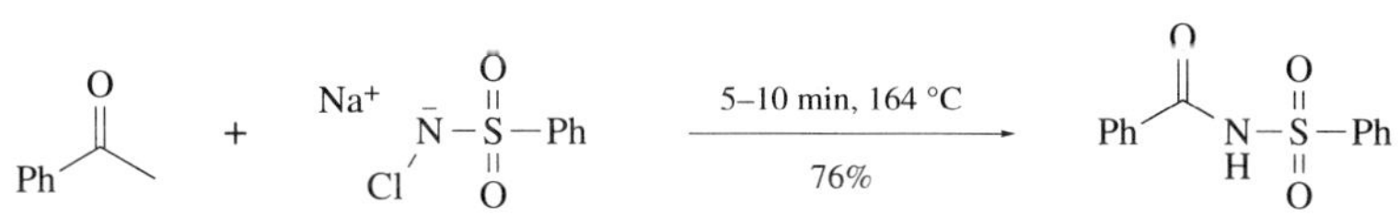

(79)

5.07.3.9.9 From aldehydes and N*-chlorosulfonamides*

Aldehydes react with a mixture of sodium benzenesulfonamide (**88**) and *N,N* dichlorobenzenesulfonamide (**89**) to provide the corresponding *N*-acyl benzenesulfonamide (Equation (80)) ⟨74ZOR667⟩.

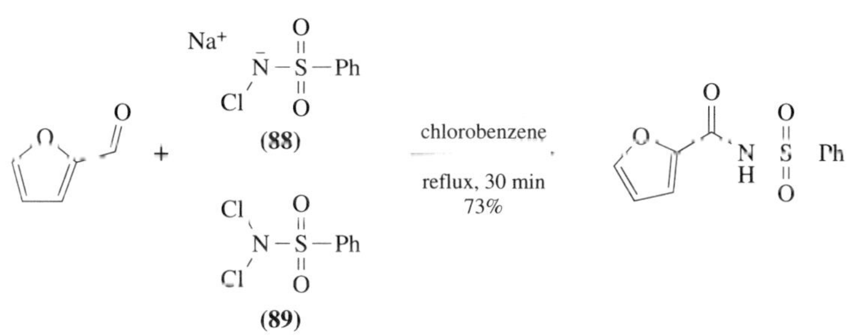

(80)

5.07.3.9.10 From sulfonylation of amides

Amides can be converted to acyl sulfonamides upon treatment with sulfonyl chlorides ⟨91JOC3210⟩. In many cases, it is necessary to deprotonate the amide prior to addition of the sulfonyl chloride ⟨84JA1133, 91TL283⟩.

The desire to sulfonate β-lactams led to the exploitation of pyridine-SO_3 ⟨88JOC5507, 89TL3219⟩,

DMF-SO_3 ⟨82JOC5160, 87TL4319⟩ and 2-picoline-SO_3 ⟨84H(21)191, 85JOC3462, 88JOC5507⟩ and other reagents ⟨83H(20)1531⟩ for the introduction of sulfite to amides and lactams (Equation (81)) ⟨82JOC179⟩.

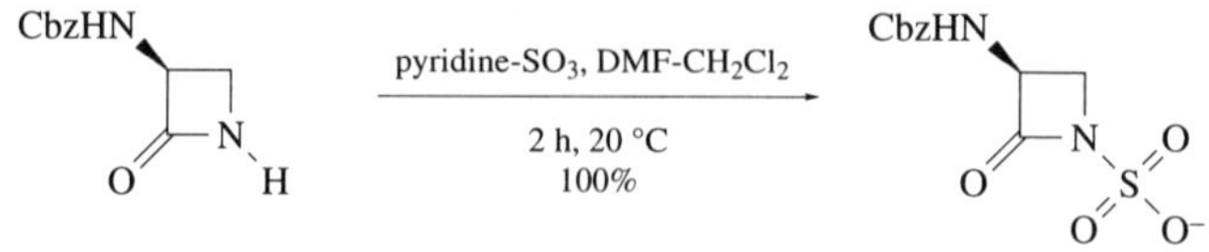

(81)

5.07.3.9.11 *From sulfonylation of lactim ethers*

Lactim ethers react with sulfonyl chlorides to provide the *N*-sulfonyllactams (Equation (82)) ⟨92OPP147⟩. The reaction is believed to proceed via initial formation of an *N*-acyl iminium salt followed by chloride attack at the methyl group.

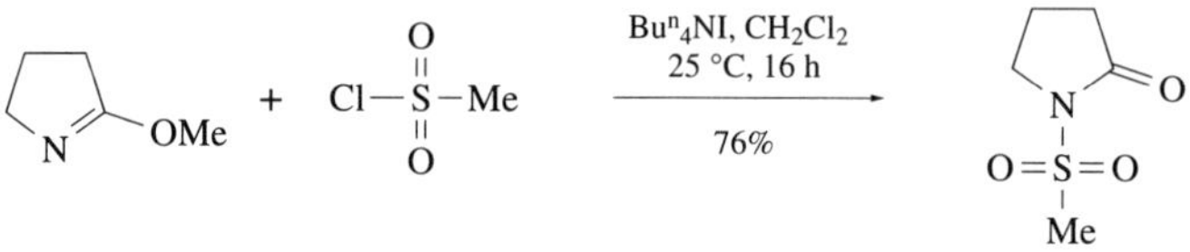

(82)

5.07.3.9.12 *From CO insertion*

The preparation of acyl sulfonamides through CO insertion continues to be a growing area of interest. The initial work in this area involved the nickel-mediated expansion of aziridines to β-lactams ⟨88CC710, 90JOC2943⟩. Methods have now been developed which allow for palladium- ⟨91TL7683⟩ or rhodium- ⟨91TL7431⟩ catalyzed conversion of (*N*-tosyl)propargylamines to *N*-tosyl-β-lactams. Palladium-mediated cyclocoupling of cyclic *N*-tosylcarbamates (**90**) to alkenes also provides *N*-tosyllactams in good yields (Equation (83)) ⟨90JA9646⟩.

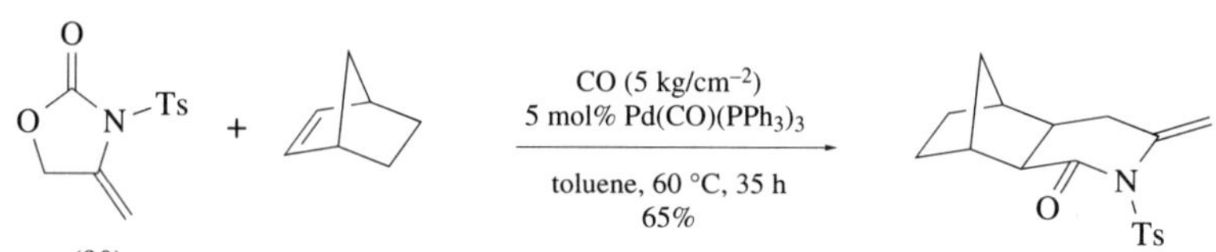

(90)

(83)

5.07.3.9.13 *From arylsulfonylazides and cycloenolethers*

Dipolar cycloaddition of arylsulfonylazides with sterically-demanding silyl enol ethers leads to sulfonamide formation with concomitant ring contraction (Equation (84)) ⟨73HCA1826, 82JOC5042⟩.

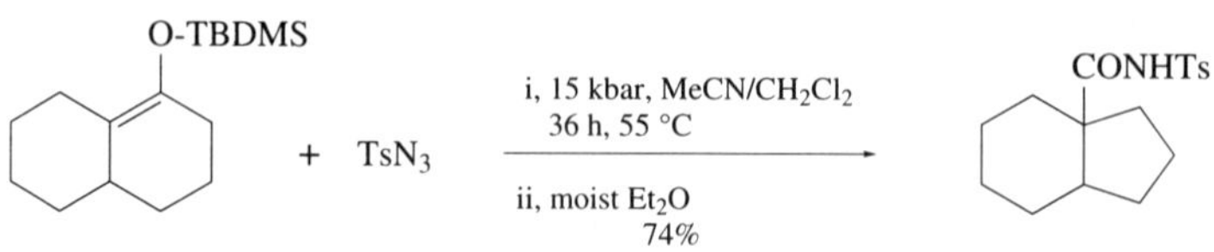

(84)

5.07.3.9.14 *From oxidative rearrangement of* N*-tosylamino acids*

N-Tosyl phenylglycine esters such as (**91**) can be oxidatively decarboxylated to produce *N*-benzoylsulfonamides (Equation (85)) ⟨81S610⟩. Application of higher oxygen pressure shortens reaction time and minimizes side reactions.

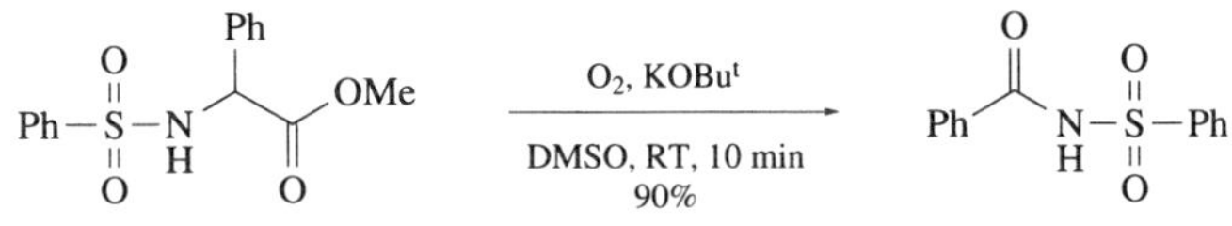

(91)

(85)

5.07.3.9.15 From oxidation of sulfinamides

O-Sulfinyl hydroxamic acids (**92**), prepared from hydroxamic acid and sulfinyl chloride, rearrange to methylsulfonamides under mild conditions (Equation (86)) ⟨86JCS(P2)1211⟩.

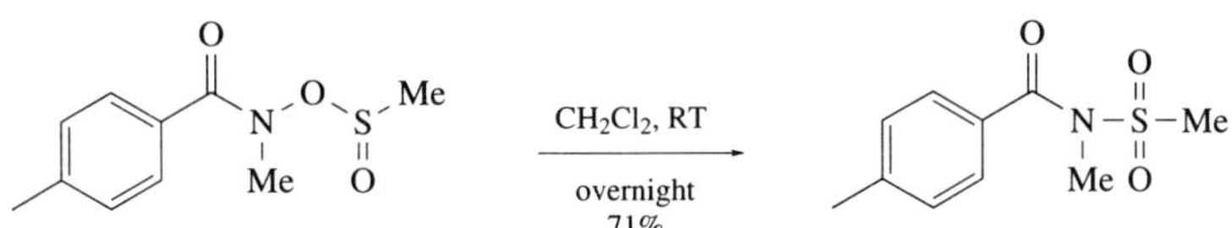

(92)

(86)

5.07.3.9.16 From oxidation of sulfinamides

Sultams can be obtained via oxidation of aminosulfides with peroxide ⟨78LA635⟩, hydroxide ⟨86FES676⟩, or permanganate ⟨84FES788⟩. The corresponding sulfinamides are also converted to sultams upon treatment with $NaIO_4$–$RuCl_3$ ⟨93JCS(P1)1581⟩, Equation (87) ⟨90TL4117⟩.

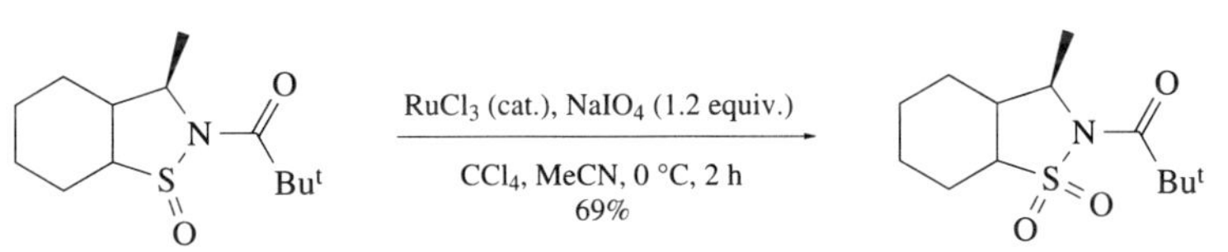

(87)

5.07.3.9.17 From α oxidation of sulfonamides

Zinc permanganate catalyzes the oxidation of *N*-sulfonamides to *N*-sulfonyl lactams (Equation (88)) ⟨83JA7755⟩. The related α-hydroxy-*N*-sulfonylamines can be oxidized by CrO_3 ⟨67JCS(C)58⟩, or MnO_2 ⟨86CPB4545⟩ to the desired sulfonamide product.

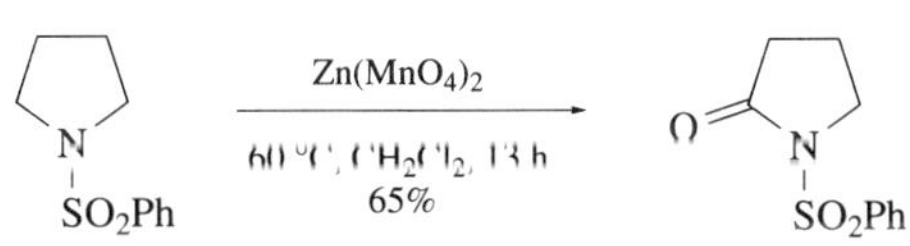

(88)

5.07.3.9.18 From ring opening of N*-tosyloxaziranes*

N-Tosyloxaziranes (**93**) rearrange photolytically to *N*-acyl sulfonamides (Equation (89)) ⟨77TL1721⟩. Minor products formed from triplet rather than singlet activation include benzaldehyde and 4-methoxysulfonamide.

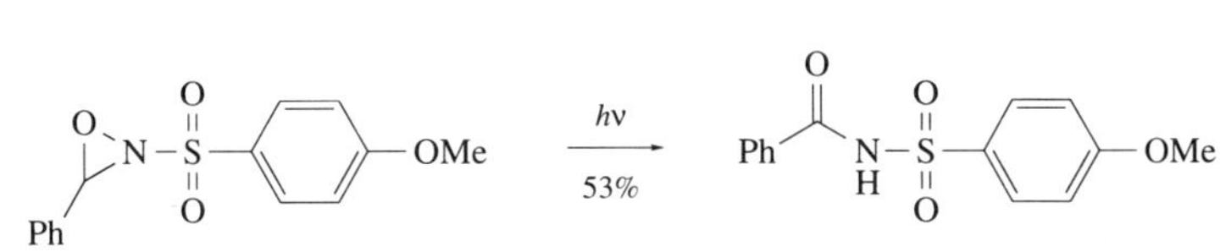

(93)

(89)

5.07.3.9.19 From hydrolysis of heterocycles

Acid hydrolysis of 1,4,3,5-oxathiadiazines ⟨73TL2783⟩, base hydrolysis of *N*-tosyloxathiazolinone ⟨70TL127⟩, and base hydrolysis of 6-methyl-5,6-dihydro-1,4,3-oxathiazine 4,4-dioxides (**94**) result in formation of highly functionalized acyclic sulfonamides (Equation (90)) ⟨85G265⟩.

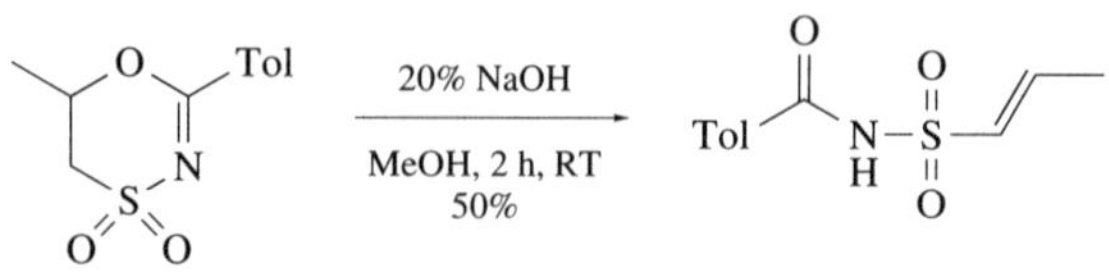

(**94**)

(90)

5.07.3.9.20 From hydrolysis of 2-iminoazetidine

2-*N*-(*p*-Toluenesulfonyl)iminoazetidines (**95**) undergo a ring-opening reaction under alkaline conditions to provide *β*-amino-*N*-acyl sulfonamides (**96**) (Equation (91)) ⟨94JOC5189⟩. It is interesting to note that no reaction was observed when the same molecule was refluxed in 10 equiv. of HCl for 4 days.

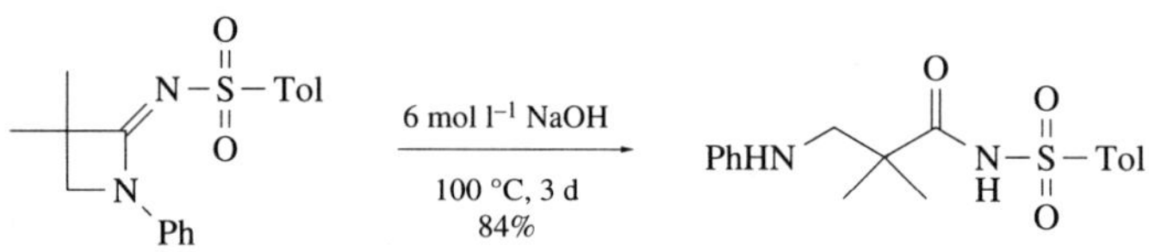

(**95**) (**96**)

(91)

5.07.3.9.21 From hydrolysis of N,N-*bis(sulfonyl)amides*

The acylation of silver dimesylamine (**97**) with acyl chlorides provides the corresponding *N*,*N*-bis(sulfonyl)amides (**98**). In contrast to the simple *N*-acyl derivative which suffers cleavage in water at room temperature, (**98**) undergoes sulfate cleavage providing *N*-acyl sulfonamides (**99**) (Scheme 16) ⟨88ZN(B)1495⟩.

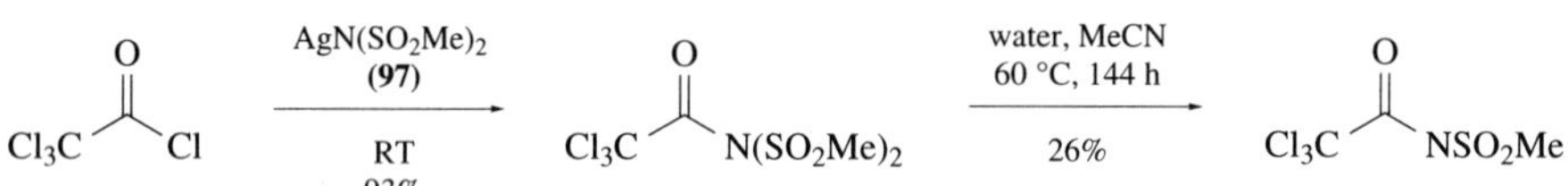

(**98**) (**99**)

Scheme 16

5.07.3.10 Imide Derivatives: Oxidation State +6, Tetracoordinate

5.07.3.10.1 From imides

Imides react with sulfene ($CH_2{=}SO_2$) ⟨78CJC1183⟩ and sulfonyl chlorides to provide *N*-sulfonylimides (Equation (92)) ⟨70JCED351, 92ACH357⟩. The yields of *N*-alkylsulfonyl phthalimides were consistently higher than those of the corresponding *N*-aryl derivatives.

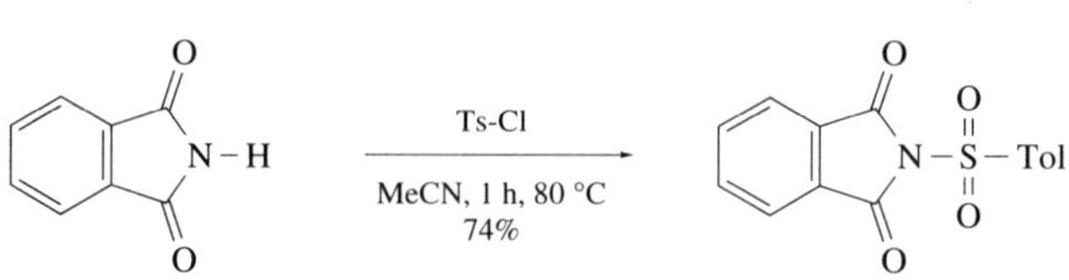

(92)

5.07.3.11 Amide Derivatives: Oxidation State +6, Hexacoordinate

Condensation of pentafluorotellurium isocyanate (**100**) onto freshly distilled acetic acid provides pentafluorotellurium acetamide (**101**) (Equation (93)) ⟨84IC2217⟩.

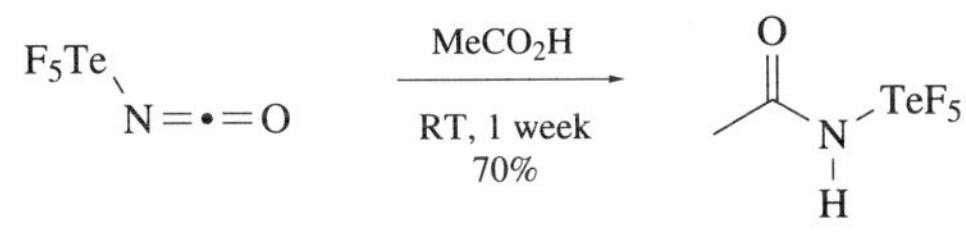

(100) (101)

(93)

5.07.3.12 Acyl Imide Derivatives (RCON=Chalc.): Oxidation State +4, Tricoordinate

5.07.3.12.1 From amides

Trifluoroacetamide reacts with diaryltellurium oxide to provide the tellurium-substituted *N*-acyl imine (Equation (94)) ⟨81ZOR401, see also 75TL649, 81PS(10)153⟩.

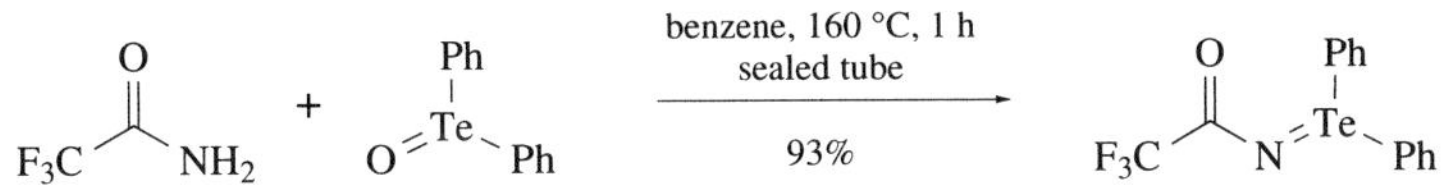

(94)

5.07.3.12.2 From N-*silylamides*

One equivalent of *N*,*O*-bis(trimethylsilyl)acetamide (bsa) reacts with tetrahalotellurium to provide the *N*-acyl tellurium imine (Scheme 17) ⟨77ZOR2055⟩. Prolonged heating and the use of two equivalents of bsa lead to formation of the diimide.

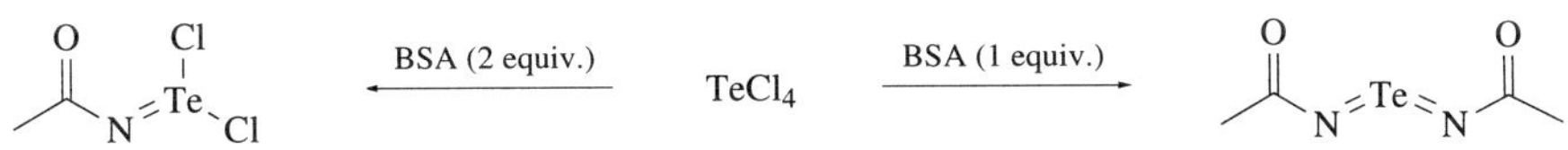

Scheme 17

5.07.3.12.3 From N-*haloamides*

The reaction between *N*,*N*-dichloroamides and *S*,*S*′-dialkyldisulfides leads to the formation of the desired *N*-acyl sulfinimidoyl chloride (Equation (95)) ⟨78ZOR1846⟩. A related reaction of sulfur with *N*-chloroamides also provides *N*-acyl dichlorosulfimides ⟨B-71MI 507-02⟩.

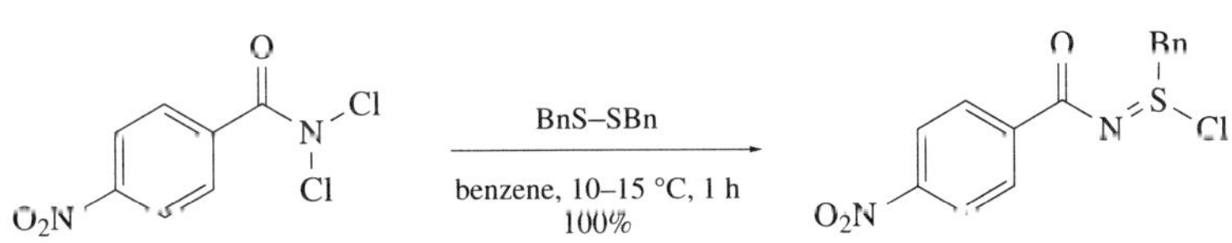

(95)

5.07.3.12.4 From ketenes

The reaction of *t*-butylcyanoketene, generated *in situ* from 2,5-diazo-3,6-di-*t*-butyl-1,4-benzoquinone (**102**), with *S*,*S*-diphenylsulfilimine (**103**) affords the acylated product (**104**) (Equation (96))

⟨85JOC2224⟩. The results are consistent with the fact that the diphenylsulfilimine is relatively acidic and can be acylated and alkylated with various electrophiles.

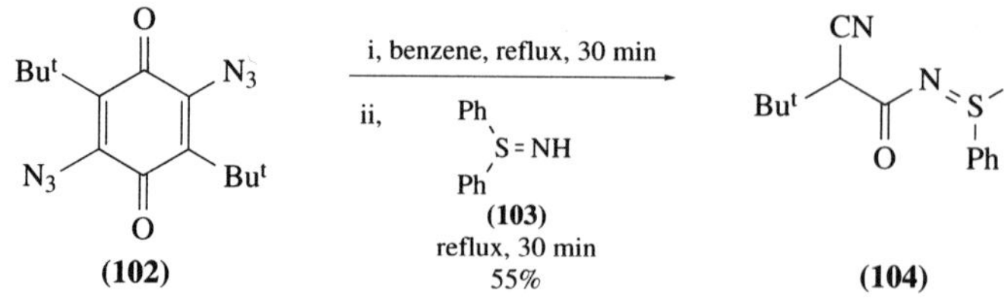

(96)

5.07.3.12.5 From acyl aminosulfenium salts

Primary acyl aminosulfenium salts (**105**) can be oxidized to acyl sulfilimines upon treatment with base (Equation (97)) ⟨71TL1761⟩. Treatment of the acyl sulfilimines with anhydrous HCl regenerates the starting sulfonium salt (with chloride as the counterion).

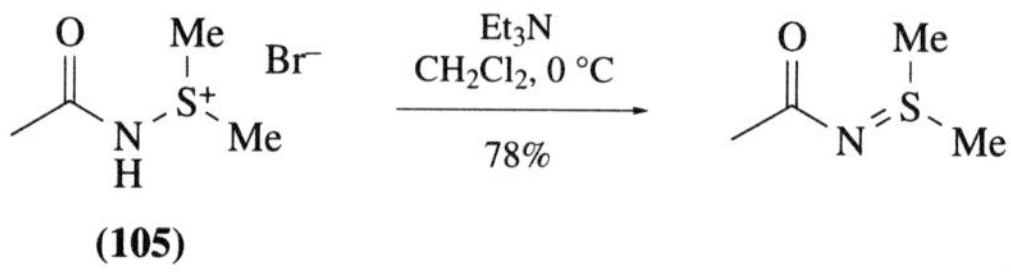

(97)

5.07.4 AMIDES SUBSTITUTED BY NITROGEN

The chemistry of *N*-amino amides has been reviewed in excellent detail ⟨67HOU(10/2)123, 85HOU(E5)1135⟩.

5.07.4.1 Acyl Hydrazine Derivatives

5.07.4.1.1 From acylation of hydrazine and its derivatives

The acylation of hydrazine itself is often accomplished by treating lactones ⟨86CCC2167, 92ZOR568⟩ or esters ⟨90SC1811⟩ with hydrazine hydrate in ethanol (Equation (98)). The yield of acyl hydrazine obtained by this method has been reported to be inversely proportional to the steric bulk of the acyl group ⟨58JA1895⟩.

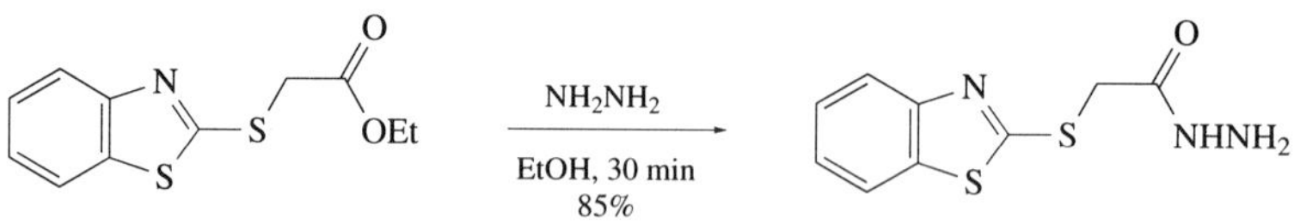

(98)

In cases where substituted hydrazines are used, acylation with ester derivatives is more difficult. To overcome this lack of reactivity, trimethylaluminum-complexed hydrazine derivatives have been used (Equation (99)) ⟨88TL739⟩.

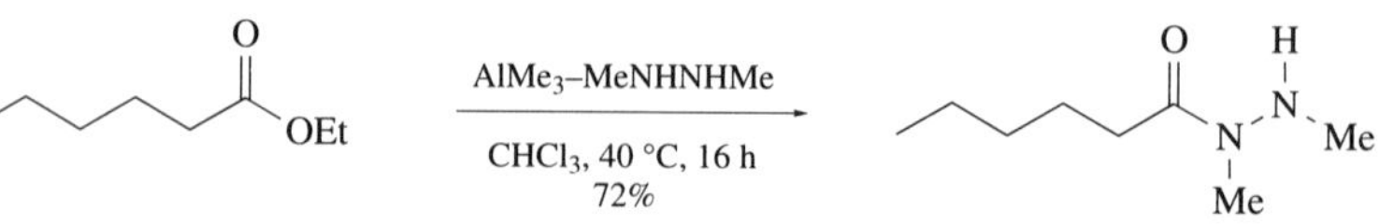

(99)

Acyl hydrazines can also be prepared from the usual acylating agents such as acid chlorides ⟨78JHC439, 93H(36)1263⟩ and anhydrides ⟨68IJ587⟩. In addition to the traditional acylating agents, hydrazine also reacts with a variety of nontraditional acylating agents to provide acyl hydrazines (Table 9).

The success in coupling of substituted hydrazines depends largely on the acylating agent and the substituents on hydrazine. When the hydrazine is *N*,*N*-disubstituted, acylation with acid chlorides ⟨88H(27)1881, 92SC1⟩ or activated carboxylic acid derivatives ⟨86JCS(P1)1655, 92S482⟩ provides only *N*,*N*-disubstituted-*N*′-acyl (Equation (100)) ⟨91JCS(P1)2435⟩.

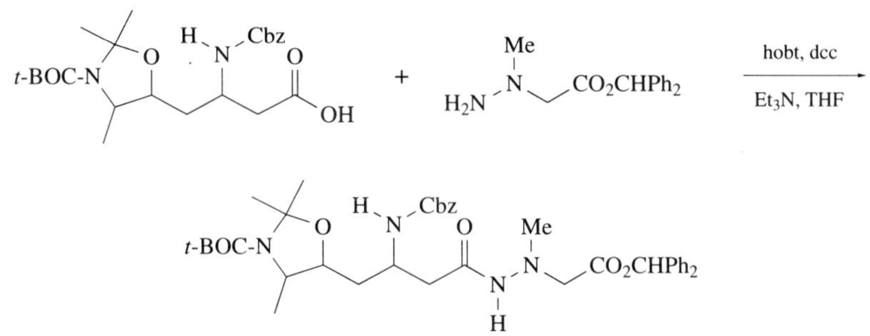

(100)

hobt = hydroxybenzotriazol

In the case of monosubstitution or *N*,*N*′-disubstitution, the regiochemistry of acylation is dependent on both electronic as well as steric factors. In the case of methyl hydrazine, acylation with acyl chlorides occurs primarily at the *N*-methyl nitrogen, providing the terminally unsubstituted acyl hydrazine (Equation (101)) ⟨91T10053⟩. With other unsymmetrically substituted hydrazines, mixtures of regioisomers and/or polyacylation are observed when a large difference in steric bulk or intrinsic reactivity does not exist between the two nitrogens.

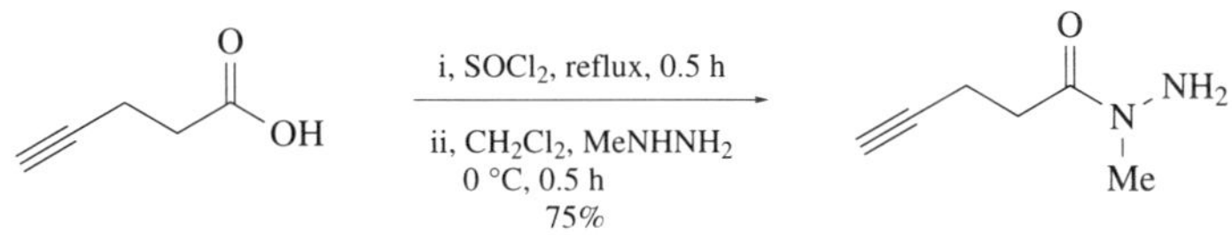

(101)

To effect acylation with unsymmetrical hydrazine derivatives, a protection/deprotection strategy is often employed. The protecting group should be chosen based on ease of introduction as well as removal (Scheme 18) ⟨91TL2766⟩. The *t*-BOC group ⟨85TL3171, 93JA8898⟩ is often the protecting group of choice, although acyl ⟨91TL133⟩, methylcarbamate ⟨90S1095⟩, and trichloroethylcarbamate ⟨92TL5717⟩ have also been used.

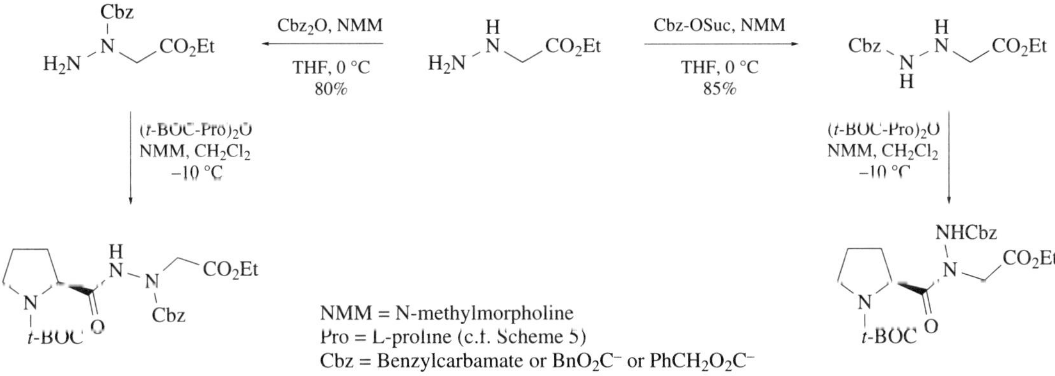

Scheme 18

The reaction of hydrazine and its derivatives with α,β-unsaturated esters is complicated by the ability of hydrazine to undergo Michael addition, leading to pyrazolidone formation ⟨90JOC6037⟩. Methods to avoid 1,4-addition include the use of more activated acylating agents, such as cyanomethyl esters ⟨68BCJ2521⟩, acid chlorides (Equation (102)) ⟨93TL457⟩, and acyl azides ⟨73IJC735⟩.

Table 9 Acyl hydrazines from diverse acylating agents.

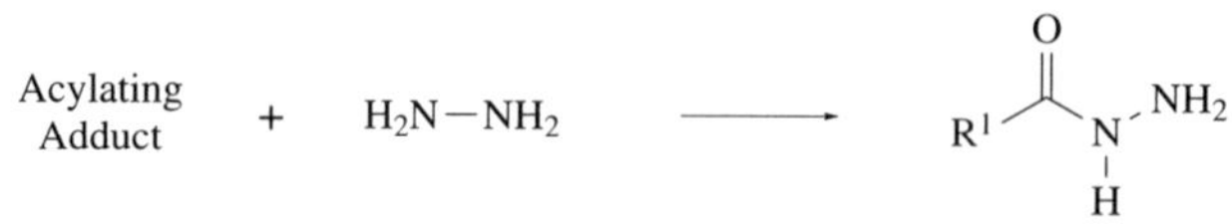

Acylating agent	*Conditions*	$R^1CONHNH_2$	*Ref.*
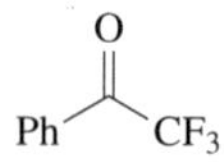	hydrazine	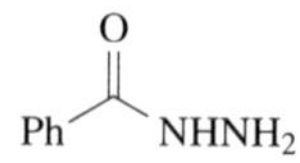	87SC1741
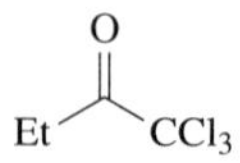	$PhNHNH_2$, 25 °C hexane (98%)	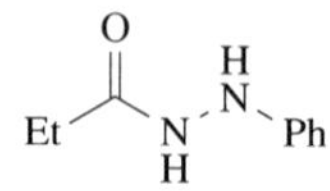	89SC1181
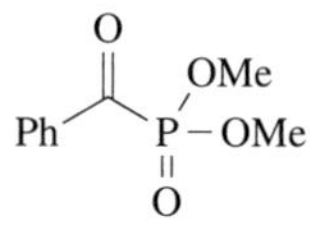	EtOH, NH_2NH_2 0 °C, 30 min (80%)	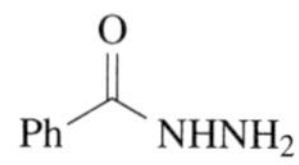	91PS239 see also 92PS163
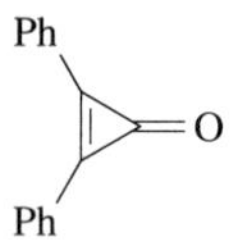	$PhNHNH_2$, NaOH EtOH (50%)	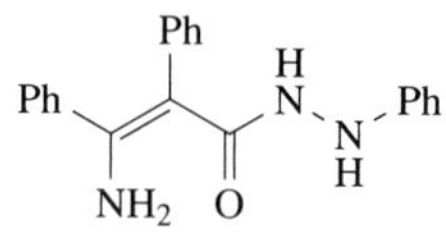	87H79
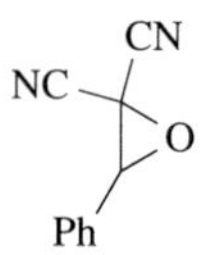	HBr, $NH_2NHCOPh$ MeCN, 6–24 h (70%)	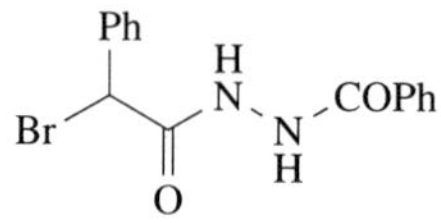	87S306
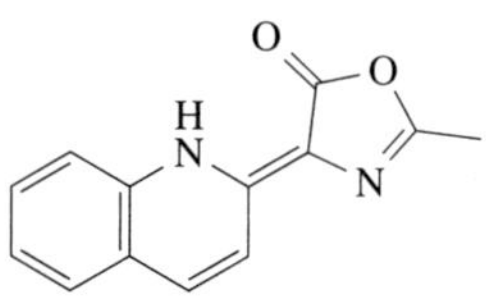	MeOH, NH_2NH_2 reflux, 1 h (55%)	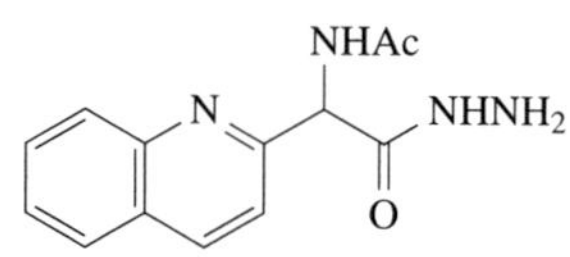	92H657
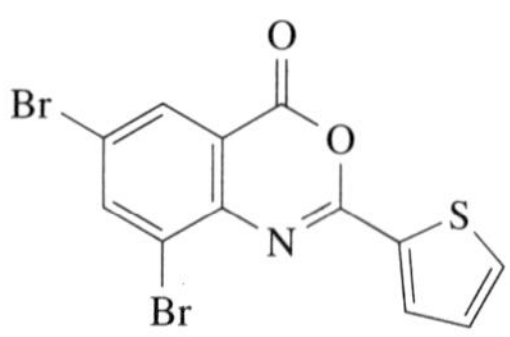	EtOH, NH_2NH_2 relux, 1 h (80%)	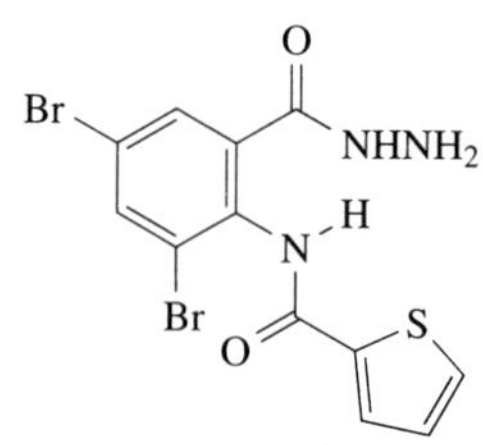	90ACH423
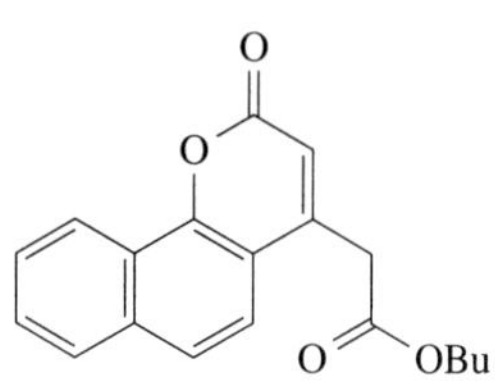	EtOH, NH_2NH_2 (70%)	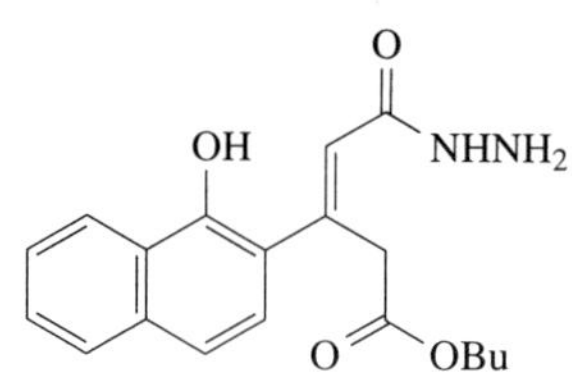	91ACH35
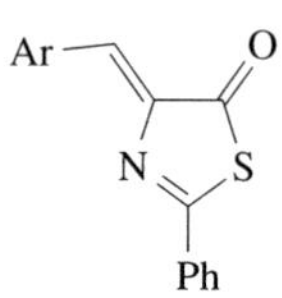	pyridine NH_2NH_2 (80%)	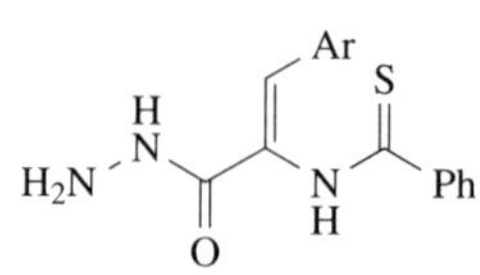	91PS159
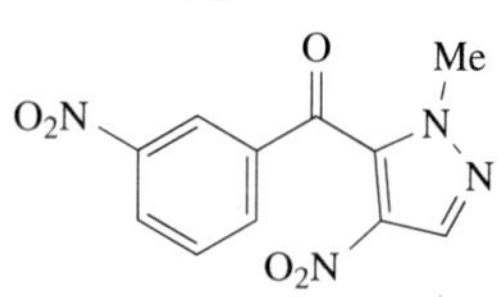	EtOH, NH_2NH_2 35 °C (61%)	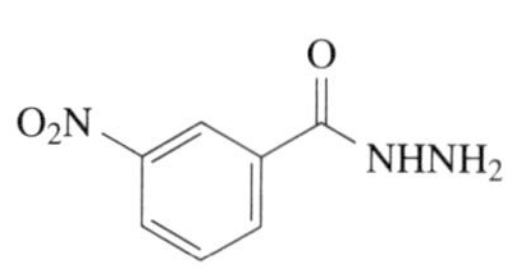	90KGS1430
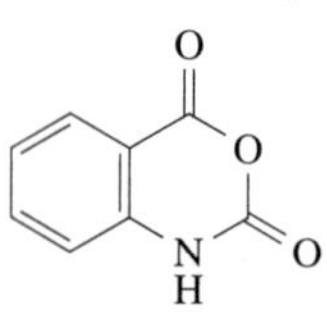	EtOH, NH_2NH_2 13 °C (59%)	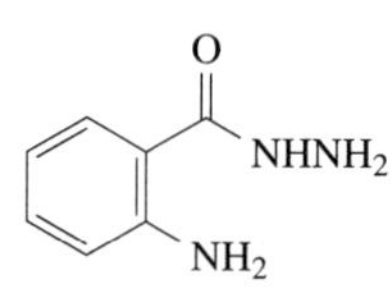	93H315

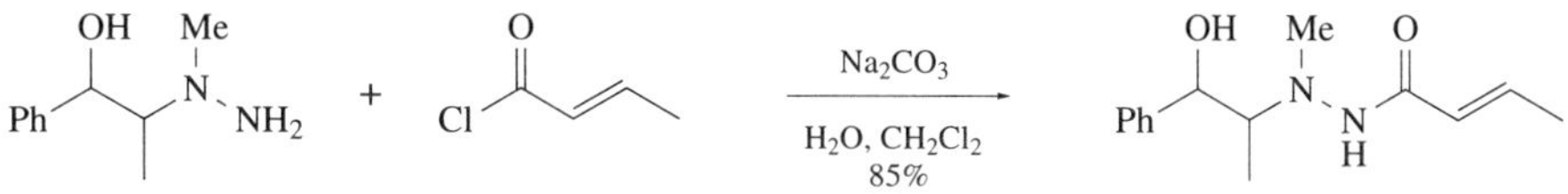

(102)

A relatively new approach to this problem is the use of enzymes to effect such transformations (Equation (103)) ⟨91S350⟩ see also ⟨92IJC(B)906⟩.

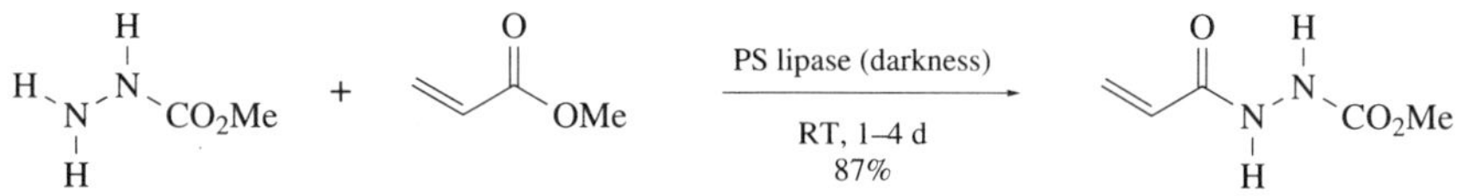

(103)

5.07.4.1.2 *From acylation of* N-*silylhydrazines*

In an example of solid-state synthesis, *N*-silylhydrazines prepared from the reaction of hydrazine with bsa react nearly quantitatively with acyl fluorides to provide acyl hydrazines (Equation (104)) ⟨94MIP03494⟩. It was noted that all other coupling conditions provided, at best, less than 20% acylation.

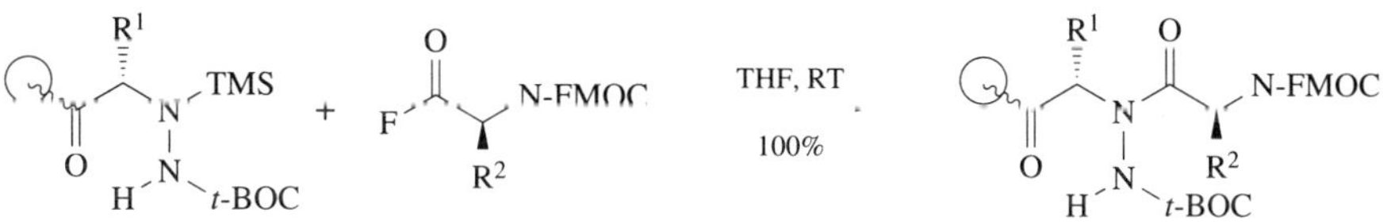

(104)

5.07.4.1.3 *From* N-*heteroamides*

The conversion of hydroxamic acids to acyl hydrazines has been reported to proceed by activation of the *N*-hydroxy group with either Tos-Cl or 2-chloro-1-methylpyridinium iodide, followed by displacement at the amide nitrogen by external amines ⟨87S183⟩. This transformation has been called into question by Hartmann, who insists that ureas and not acyl hydrazines are formed ⟨88S807⟩.

Unlike hydroxamic acid derivatives activated at oxygen, *N*-acyl oxaziridines (**106**) do react with amines such as piperidine to provide *N*-substituted acyl hydrazines (Equation (105)) ⟨65TL1857, 67JPR86⟩.

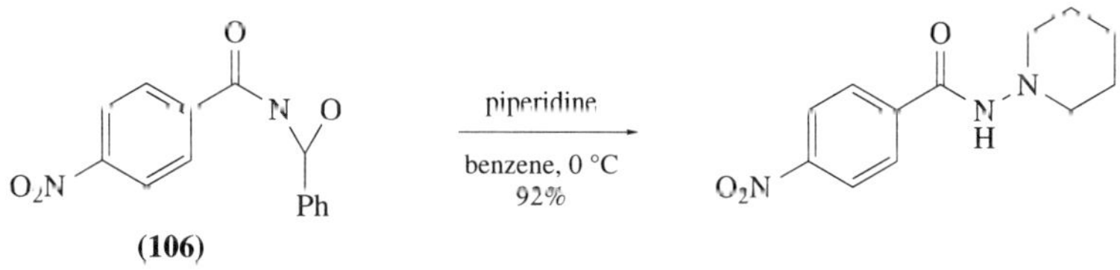

(105)

5.07.4.1.4 *From reduction of acyl nitrosoamines*

N-Nitrosoamides can be reduced by either zinc suspended in acetic acid, or by the combination of titanium tetrachloride/magnesium (Equation (106)) ⟨87TL341⟩. These conditions are reported to be very mild, and avoid overreduction to the parent amide. A review of the earlier work in this area has been reported by Müller and Bayer ⟨80HOU(4)736⟩.

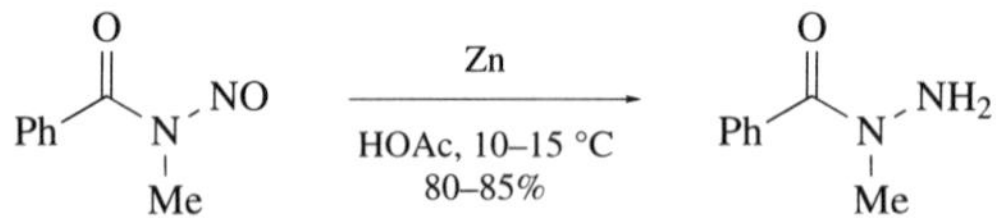

(106)

The reduction of nitrosoamines using acetic acid and/or acetic anhydride as a solvent provides a related route to acyl hydrazines (Equation (107)) ⟨75TL2701, see also 87TL341, 89CC1640⟩. This method avoids the isolation of primary hydrazines containing hydrazine-labile groups (such as esters).

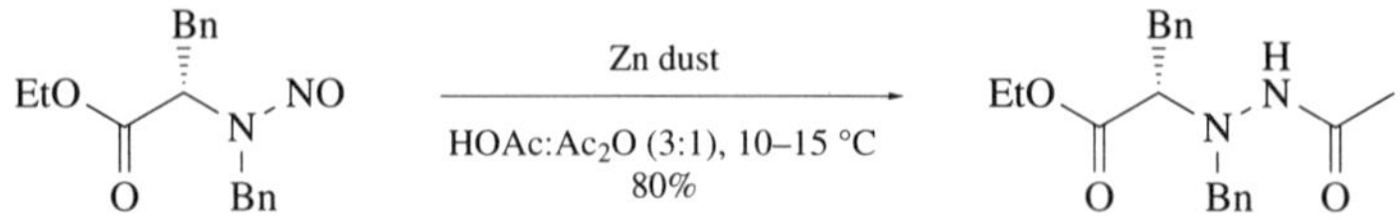

(107)

5.07.4.1.5 From amination of amides

N-Amino amides can also be obtained by treating amides with aminating agents, such as *O*-(2,4-dinitrophenyl) hydroxylamine ⟨77S487⟩, Mes-SO_3NH_2 ⟨73TL4091, 75CC725⟩, and chloroamine (Equation (108)) ⟨90IC309⟩ see also ⟨65JOC1311⟩.

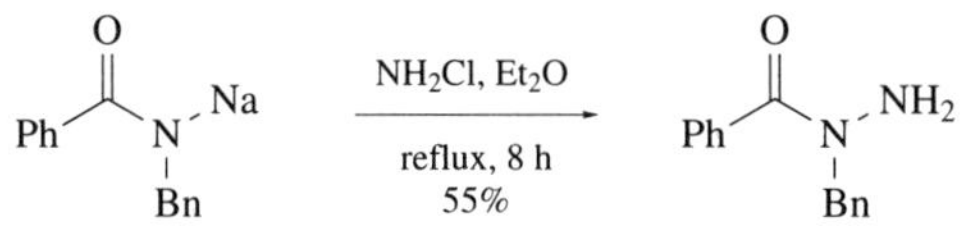

(108)

5.07.4.1.6 From solvolysis of heterocycles

A number of heterocycles provide acyl hydrazides upon solvolysis. Representative examples of such transformations are found in Table 10.

5.07.4.1.7 From reduction of acyl hydrazones and acyl diazines

Acyl hydrazones can be converted to acyl hydrazines by $LiAlH_4$ (Equation (109)) ⟨81ZOR2153⟩, $NaBH_4$ ⟨75LA1264, 91CZ215⟩, or hydrogenation using platinum ⟨53JA1933, 65ZC302, 71CJC1598⟩ or palladium as a catalyst ⟨85TL3167⟩. Enantioselective reductions using chiral rhodium complexes have also been reported ⟨92JA6266⟩.

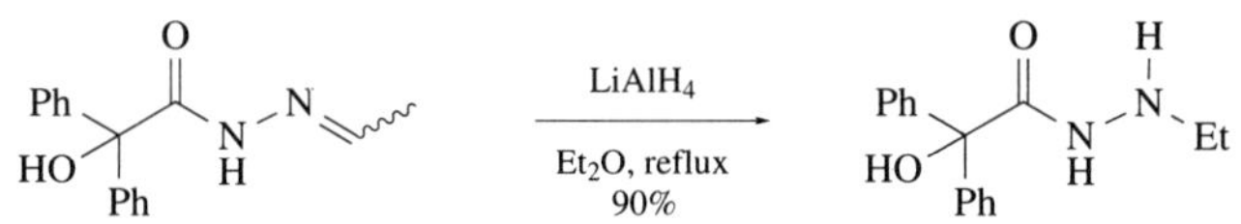

(109)

The reaction of acyl diazines with dabco leads to multiple products, with acyl hydrazine being the major compound (Equation (110)) ⟨87JHC773⟩. This transformation only takes place with electron-withdrawing groups on the acyl portion.

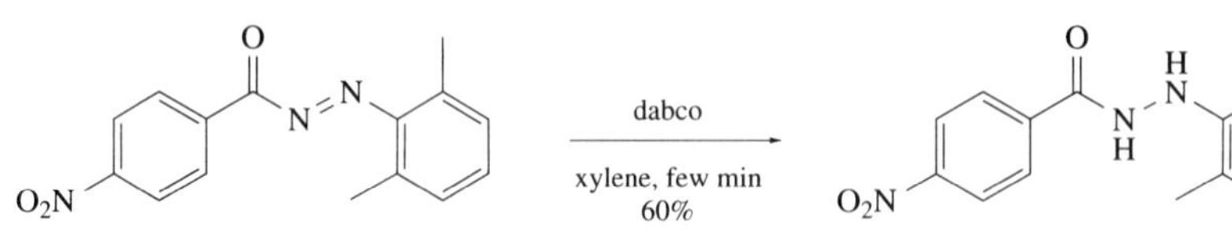

(110)

dabco = 1,4-diazabicyclo[2.2.2]octane

5.07.4.1.8 From alkylation of acyl diazines

Acyl diazines react with enamines ⟨79TL4761, 83JHC305, 85G663⟩, *β*-keto esters ⟨87T4249⟩, and Grignard reagents ⟨53JA1933⟩ to provide *β*-substituted acyl hydrazines. It is of interest to note that

Table 10 Solvolysis of heterocycles with substituted hydrazines.

Heterocycle	*Conditions*	*Product*	*Ref.*
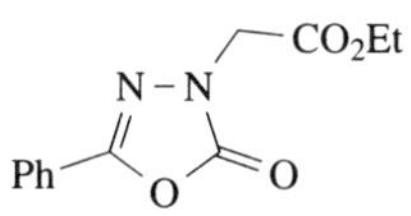	EtOH, NH_3 (aq.) 25 °C, 24 h (85%)	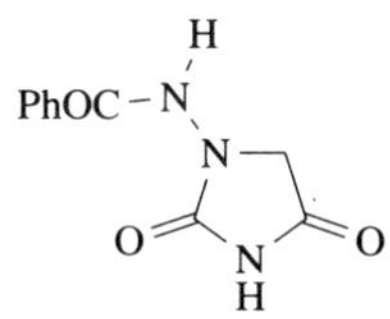	91JHC1511
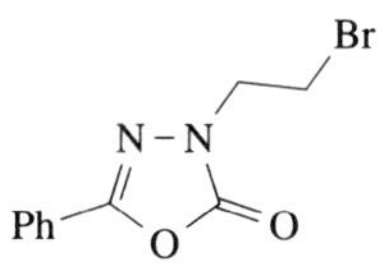	EtOH, $BnNH_2$ MeCN, RT (63%)	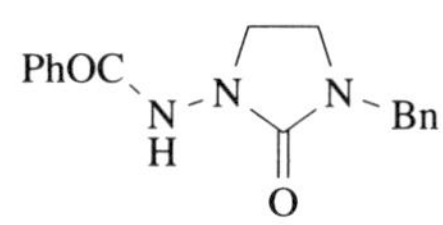	92JHC1081
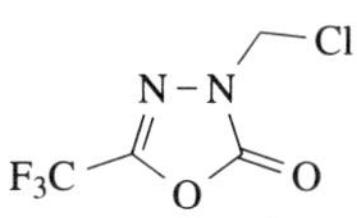	pyrrolidine, RT MeCN (51%)	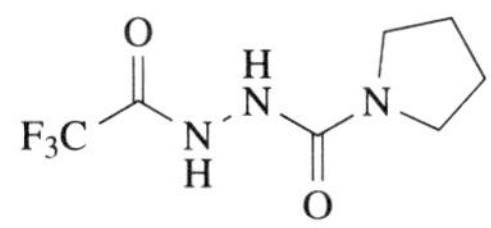	86HCA333
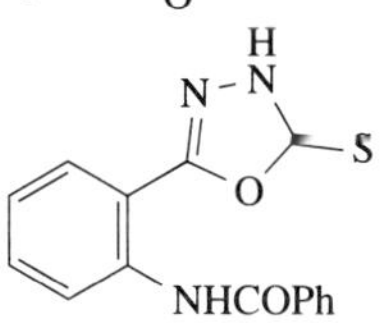	EtOH, Bu^nNH_2 (72%)	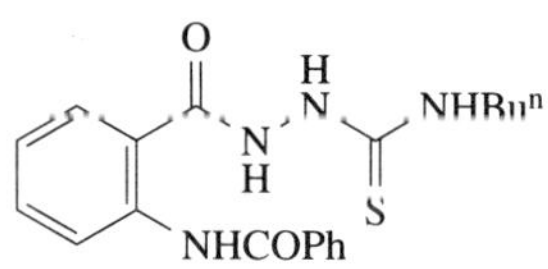	91PS373
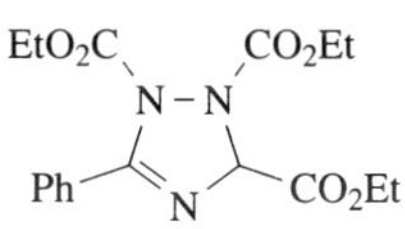	5% HCl dioxane	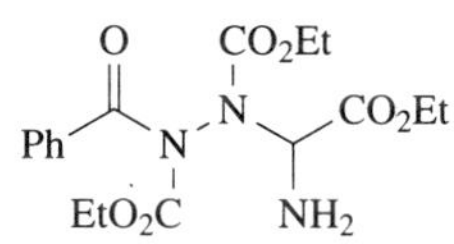	89T3535
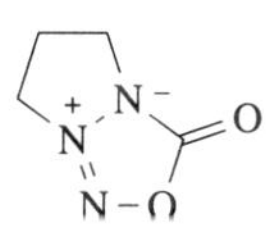	TFAA, Et_2O (50%)	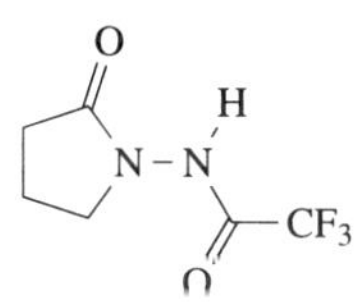	85TL5739

with 1-carboxy-2-carboalkoxydiazines (**107**), addition occurs β to the acyl group (Equation (111)), ⟨91G401⟩.

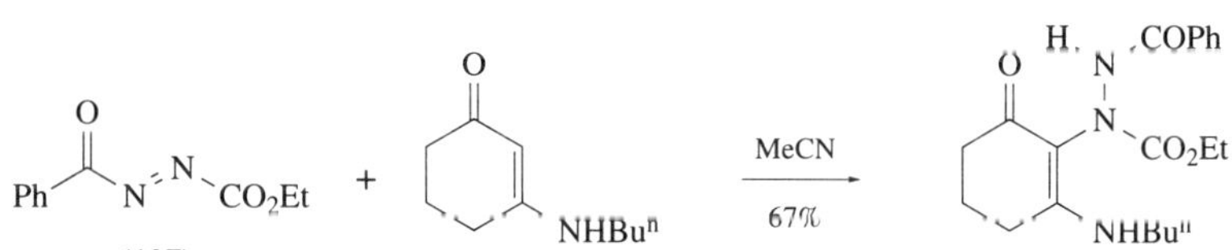

(**107**) (111)

5.07.4.1.9 *From 2 + 2 reactions of hydrazones*

Hydrazones also undergo 2 + 2 reactions with ketene derivatives, leading to formation of *N*-amino-β-lactams (**108**) (Equation (112)) ⟨90JOC2196⟩ see also ⟨91IJC(B)717, 91JHC593, 91T9599⟩. Although this reaction was initially reported to work only in the case of *N*,*N*-dialkylhydrazones, *N*-acyl hydrazones have been reported to undergo such a transformation ⟨93JHC61⟩.

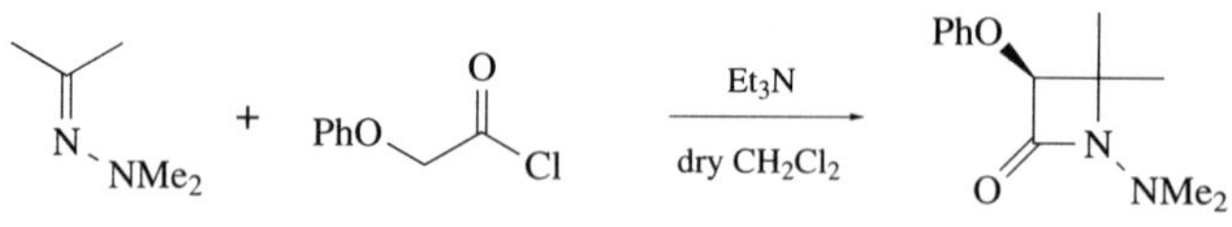

(108) (112)

5.07.4.1.10 *From CO insertion*

Treating the lithium salt of 1,1,2-tri-*n*-propylhydrazine (**109**) under a CO atmosphere results in formation of an acyl anion. This anion reacts with a variety of aldehydes or ketones to provide α-hydroxy acyl hydrazines (**110**) (Equation (113)) ⟨80AG(E)726⟩.

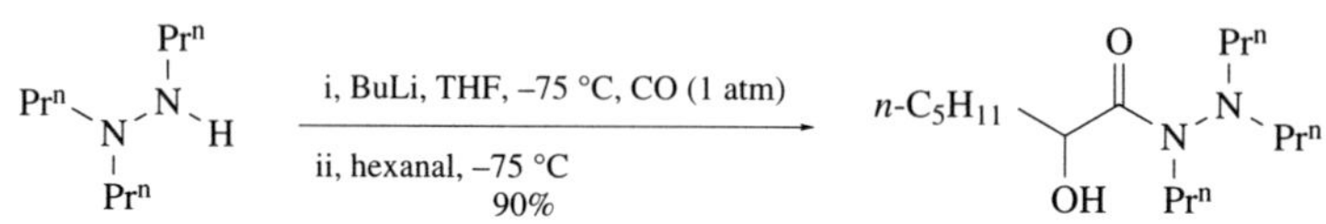

(109) (110) (113)

Azobenzenes (**111**) react with a stoichiometric amount of $Co(CO)_8$ and excess methyl iodide to provide 1,2-diaryl-2-acetyl hydrazines (**112**) (Equation (114)) ⟨90OM1245⟩.

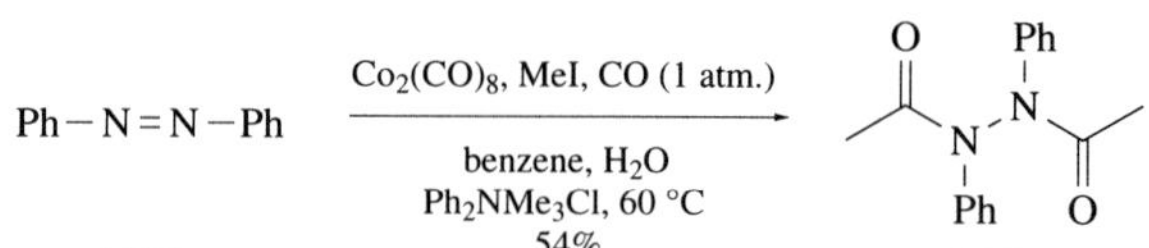

(111) (112) (114)

Related methodologies involving CO insertion include the reaction between azobenzene and chromium carbene ⟨89JA9194⟩, as well as the intermolecular cyclization of palladium azoaryls to form 3-oxoindazolines ⟨67JOM(10)511, 76S750⟩.

5.07.4.1.11 *From oxidation of hydrazones*

Hydrazones are converted to acyl hydrazines via oxidation with lead tetraacetate ⟨69JCS(C)2587, 82S824⟩, peracetic acid ⟨62JOC413⟩, singlet oxygen ⟨79TL2253⟩, and via phenyliodo(III) bisacetate (Equation (115)) ⟨93JCR(S)244⟩.

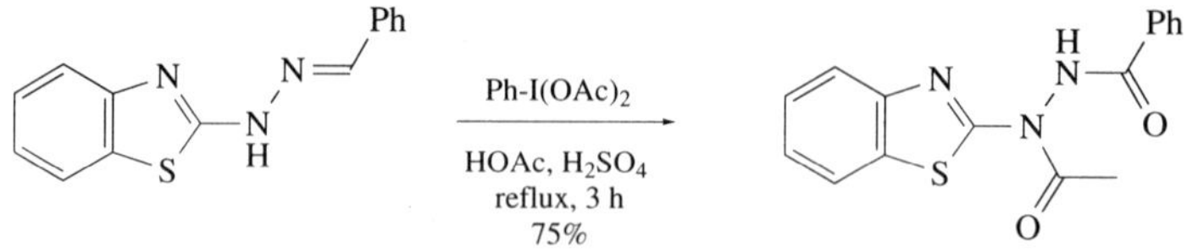

(115)

Bromine has been shown to convert hydrazones to bromoimidates. The intermediate bromoimidate (**113**), as well as chloroimidates and aminoimidates prepared via other methods ⟨75ZOR173, 75ZOR348, 76ZOB1557⟩, are hydrolyzed to the corresponding acyl hydrazine under aqueous conditions (Equation (116)) ⟨65JCS2999, see also 65TL841, 68JOC753⟩.

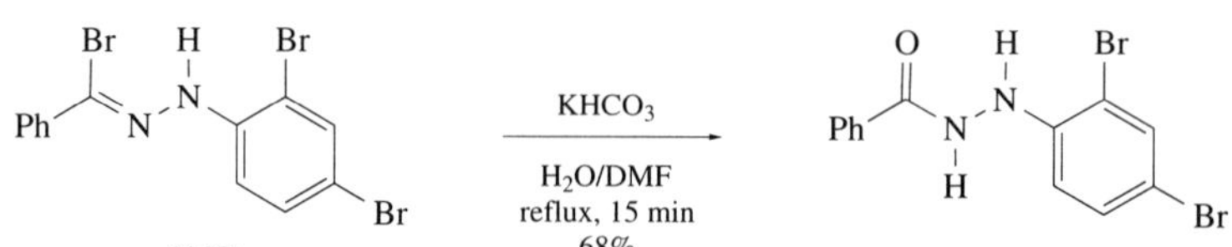

(113) (116)

5.07.4.1.12 From azo-alcohols

α-Azohydroperoxides (**114**) are easily prepared from the oxidation of phenylhydrazones. These intermediates can be reduced with dimethylsulfide to produce an azo-alcohol intermediate (**115**) which spontaneously rearranges to an acyl hydrazine (Scheme 19) ⟨82CL1607⟩.

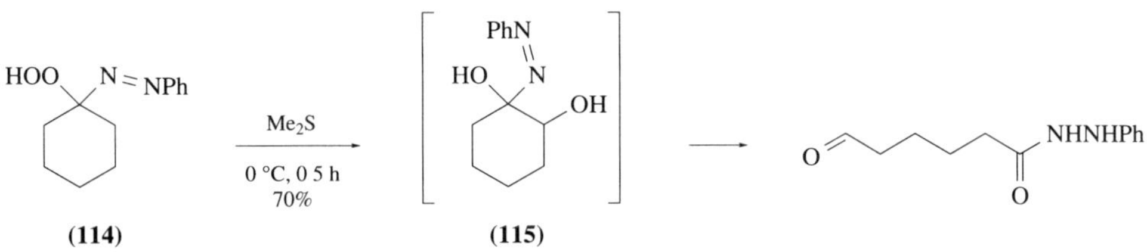

Scheme 19

The reaction of aryldiazo compounds with aldehydes or ketones leads to the formation of azo-alcohols. These intermediates tautomerize to the corresponding acyl hydrazine (Scheme 20) ⟨85JPR(327)893⟩.

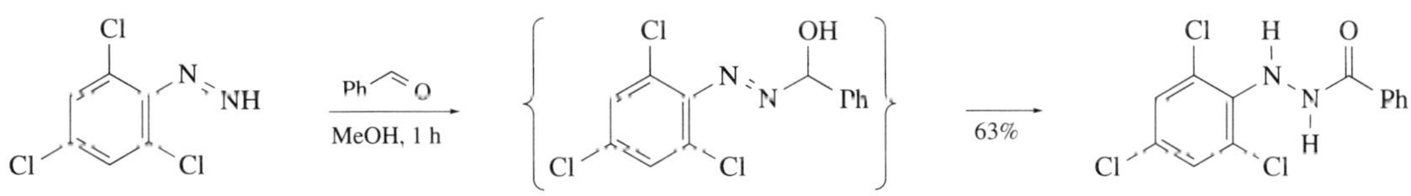

Scheme 20

The azo-alcohol derived from cyclohexanone can be used as a source of diazene for reaction with other aldehydes. Heating 1-hydroxy-1-azocyclohexane (**116**) in the presence of benzaldehyde provides the corresponding 1-benzoyl-2-phenylhydrazine (**117**) (Equation (117)) ⟨74ZC265⟩.

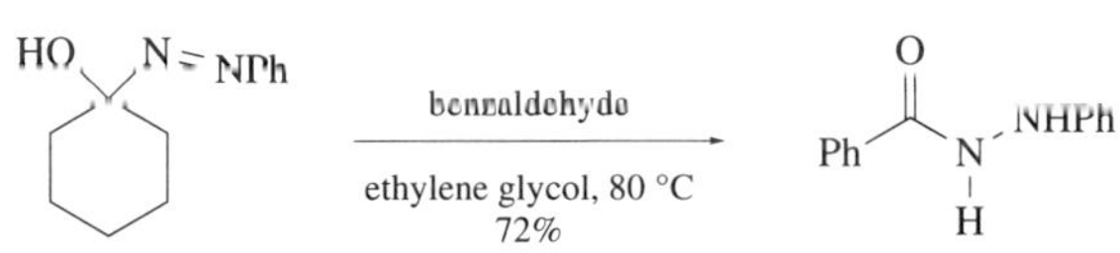

(116) **(117)** (117)

In the case of bridged cyclohexanol systems, only the isomer containing the hydroxy group in an *endo* orientation isomerizes in the desired acyl hydrazine ⟨87H(26)2081, 90TL7633⟩.

5.07.4.1.13 From oxidative cleavage of 3-pyrazolin-5-one

4-Amino-3-pyrazolin-5-one (**118**) is converted in near-quantitative yield to the corresponding 1,2-diacylhydrazides upon treatment with singlet oxygen in a buffered solution (Equation (118)) ⟨88AP(321)25⟩. The replacement of the 4-amino group with an alkyl group requires stronger oxidative conditions ($NaIO_4$) to effect ring opening ⟨88AP(321)551⟩.

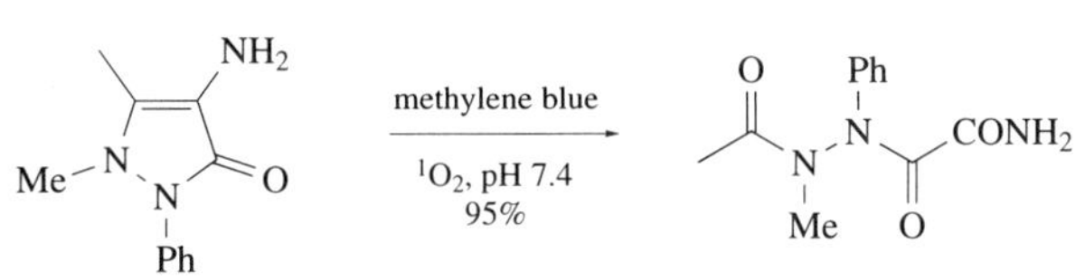

(118) (118)

5.07.4.1.14 From phenyldiazomethane

Phenyldiazomethane reacts with bis(trifluoromethyl)sulfine (**119**) at −25°C to provide the intermediate adduct (**120**) which is converted to the acylhydrazine (**121**) upon oxidation with mcpba (Scheme 21) ⟨88CB75⟩.

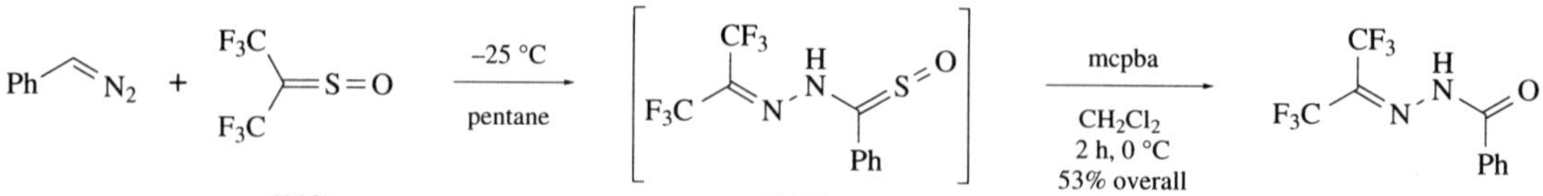

(119) **(120)** **(121)**

Scheme 21

5.07.4.1.15 From rearrangement of phenoxyhydrazones

The Chapman rearrangement involves the conversion of hydrazonyl aryl ethers (**122**) to 1-aroyl-2-*N*,*N*′-diarylhydrazines (**123**) (Equation (119)) ⟨73CJC4115⟩ see also ⟨72T5903, 75BCJ365, 88ZC24⟩. Electron-donating groups increase the rate of the reaction ⟨78BCJ512⟩. When the same material is rearranged under oxidative conditions, the 1,2-diaryl derivative is obtained ⟨73JCS(P2)1422⟩.

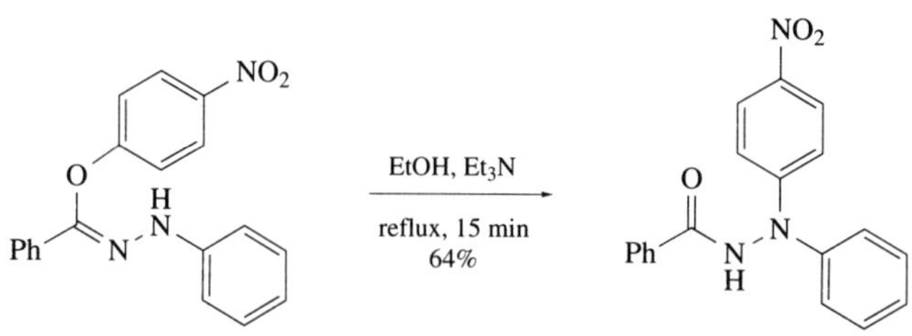

(122) **(123)** (119)

A related but different form of this reaction involves the thermolysis of tetrazoles in the presence of phenol. Extrusion of nitrogen provides an intermediate phenoxy imidate (**124**) which rearranges to the desired acyl hydrazine (Scheme 22) ⟨88ZC24⟩.

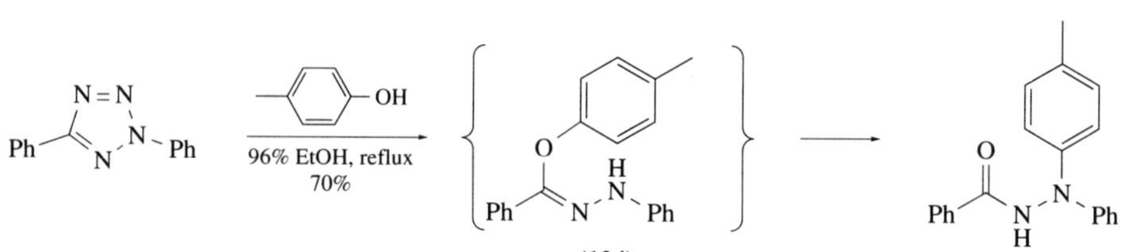

(124)

Scheme 22

5.07.4.1.16 From hydrolysis of thioacyl hydrazines

Thioacyl hydrazines (**125**) can be converted to corresponding acyl hydrazines in high yields upon treatment with trimethyloxonium tetrafluoroborate (Equation (120)) ⟨71CC1113⟩.

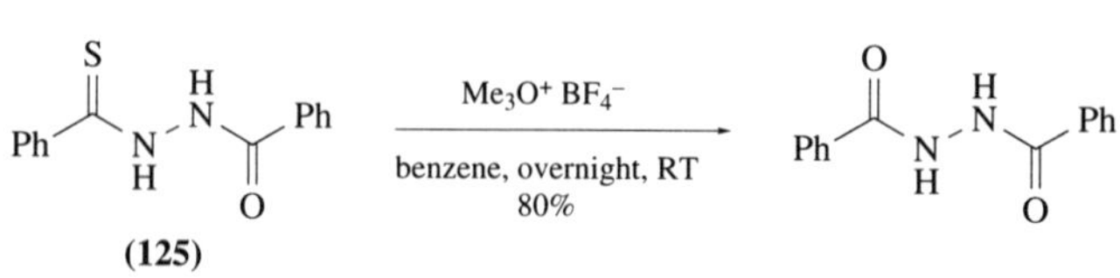

(125) (120)

5.07.4.1.17 From oxidation of hydrazine aminals

The hydrazine aminal derived from cyclopropanone and mesitylenesulfonylhydrazine (**126**) provides an *N*-substituted β-lactam (**127**) upon oxidation with MnO_2 (Equation (121)) ⟨73TL4091⟩.

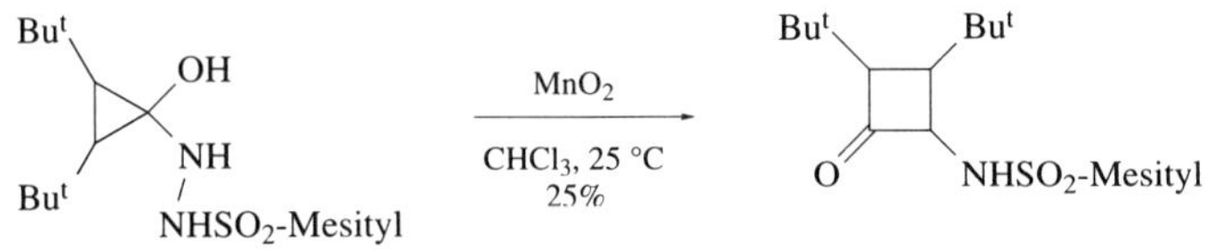

(**126**) (**127**) (121)

5.07.4.1.18 From ring opening of diaziridinone

Treatment of *N*,*N′*-di-*t*-butyldiaziridinone with the lithium salt of acetonitrile ⟨90TL5327⟩, or with the sodium salt of methyl α-cyanoacetate ⟨92JOC7359⟩, provides *N*,*N′*-di-*t*-butyl acyl hydrazines (Scheme 23).

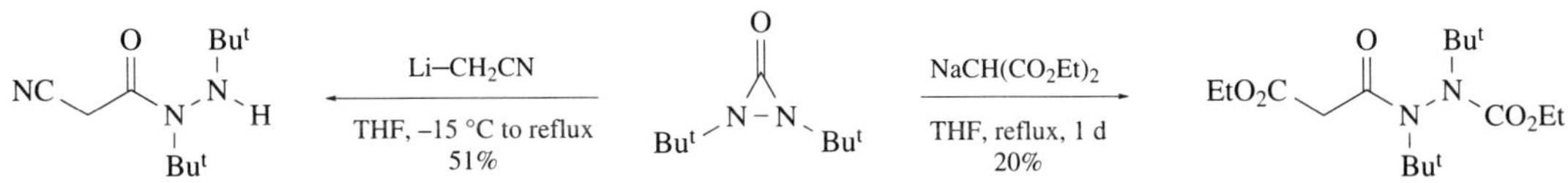

(**129a**) (**128**) (**129b**)

Scheme 23

5.07.4.2 1,1-Diacyl Hydrazines

5.07.4.2.1 From acylation of amines

The substitution on hydrazine is a critical factor in the formation of 1,1-diacyl hydrazine derivatives from hydrazines. The reaction with cyclic anhydrides can be complicated by 1,2-acylation in the case of hydrazine itself. When the hydrazine is unsymmetrical, treatment with acid anhydrides provides good overall yield of the desired *N*-amino imide (Equation (122)) ⟨70OPP275, 92SC1199⟩.

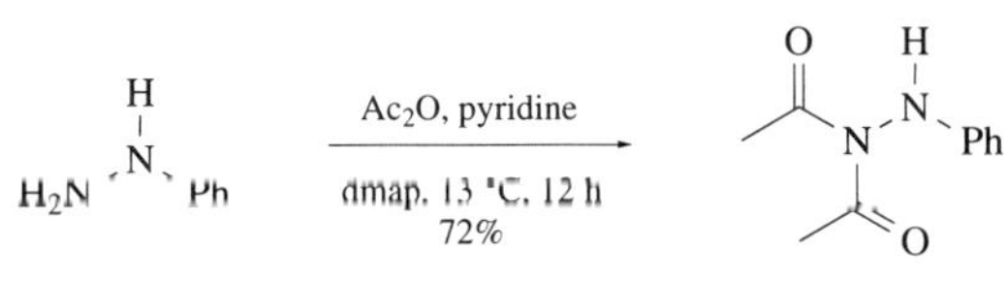

(122)

dmap = 4-dimethylaminopyridine

Other anhydride analogues used to prepare imides from hydrazine derivatives include the bis(cyano)enolacetate (**130**) (Equation (123)) ⟨91TL3449⟩.

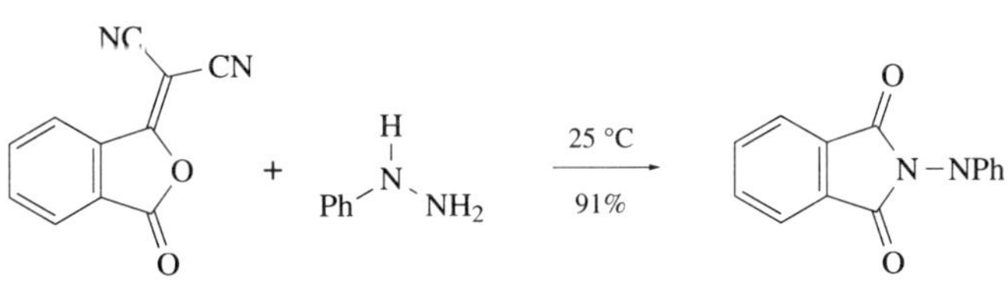

(**130**) (123)

As with monoacylhydrazines, enzymes are now being used for the preparation of *N*-aminoimides (Equation (124)) ⟨93S287⟩.

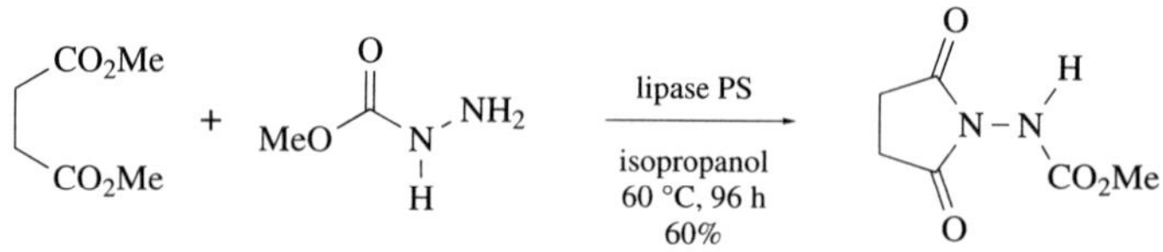

(124)

5.07.4.2.2 From amination of imides

O-Diphenylphosphinylhydroxylamine reacts with potassium phthalimide to provide *N*-aminophthalimide (Equation (125)) ⟨82S592⟩. Similar transformations with the sodium salt of phthalimide were less successful ⟨81JCS(P1)3284⟩.

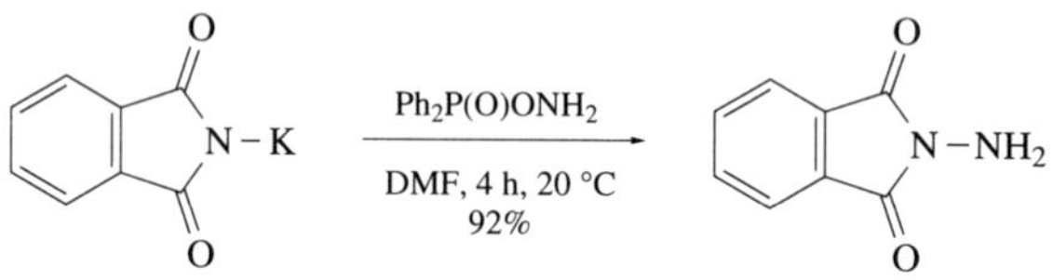

(125)

5.07.4.3 *N*-Nitroso Amides

The chemistry of *N*-nitrosoamides has been reviewed by a number of groups ⟨65HOU(10)67, B-70MI 507-01, B-86MI 507-01⟩.

5.07.4.3.1 From amides

The nitrosation of amides represents the most versatile method for the preparation of *N*-nitroso amides. Dinitrogen tetroxide (N_2O_4) is the most generally applicable reagent for this conversion ⟨86JCS(P1)2207, 89TL863, 90JCS(P1)3103, 91JA6318, 91JOC3744⟩. Other agents which have been reported to affect this transformation include N_2O_3 ⟨73OSC(5)650, 85TL215⟩, NOCl ⟨90CPB2352⟩, $NaNO_2$ ⟨89CB389⟩, $HOSO_2NO$ ⟨67CJC1539⟩, and $NOBF_4$ ⟨79S100⟩. For the preparation of hindered *N*-nitrosoamides, NO_2BF_4 has been reported to be especially useful (Equation (126)) ⟨89JOC3209⟩.

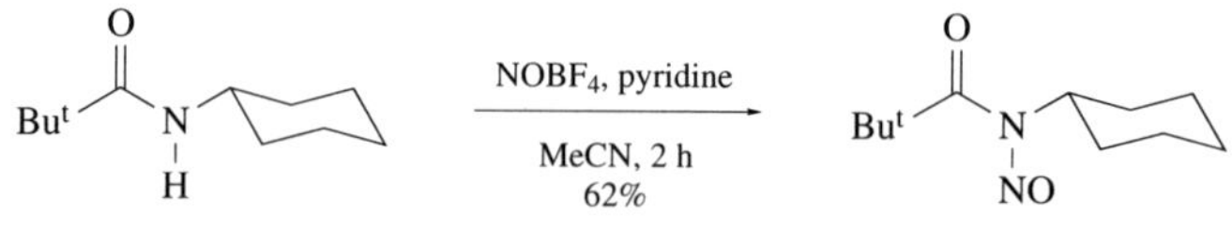

(126)

5.07.4.3.2 From nitrosoamines

Treatment of nitrosotrifluoromethane (**131**) with hydroxylamine followed by acid chlorides provides the corresponding *N*-nitrosoamides (**132**) (Scheme 24) ⟨83JFC(22)91⟩.

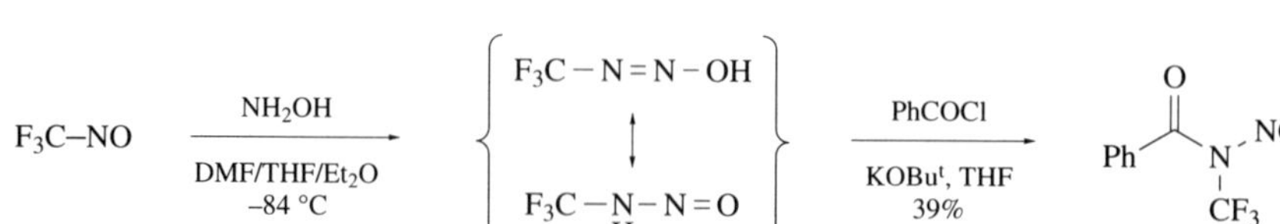

(131) **(132)**

Scheme 24

5.07.4.3.3 *From imidoyl chlorides*

The treatment of imidoyl chlorides with silver nitrite provides an alternate route to *N*-nitrosoamides (Equation (127)) ⟨85JCR(S)132⟩.

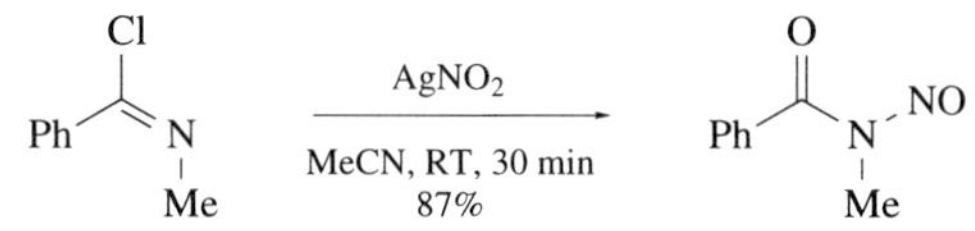

(133)

(127)

5.07.4.3.4 *From thioamides*

N-Propylthiobenzoylamide (**134**) reacts with N_2O_4 at 0 °C to provide the corresponding *N*-nitroso amides (Equation (128)) ⟨86S970⟩.

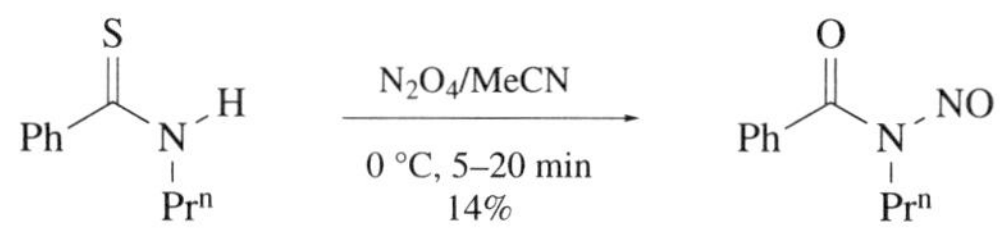

(134)

(128)

5.07.4.4 *N*-Nitroamides

The chemistry of *N*-nitroamides has been reviewed ⟨B-70MI 507-01⟩.

5.07.4.4.1 *From amides*

Amides are converted into N-nitroamides upon treatment with nitric acid/SO_3 ⟨61JA1191⟩, nitric acid/acetic anhydride ⟨77ZOR1144, 89TL863⟩, copper(II) nitrate/acetic anhydride ⟨61JA1191, 63JOC2294⟩, NO_2Cl ⟨74ZOR2301, 77ZN(B)98⟩, N_2O_5 ⟨76S241⟩, or NO_2BF_4 ⟨84JOC3322⟩. Other sources of electrophilic nitrate useful for *N*-nitration of amides include nitric acid/TFAA ⟨67CJC1539, 71JOC350⟩, NH_4NO_3/TFAA ⟨88S743, 89JCR(S)260⟩, morpholinium nitrate/TFAA ⟨91JOC7038⟩, morpholinium nitrate/heptafluorobutyric anhydride ⟨91JOC7038⟩, and $NO_3Bu^n{}_4N/Tf_2O$ (Equation (129)) ⟨93TL6669⟩.

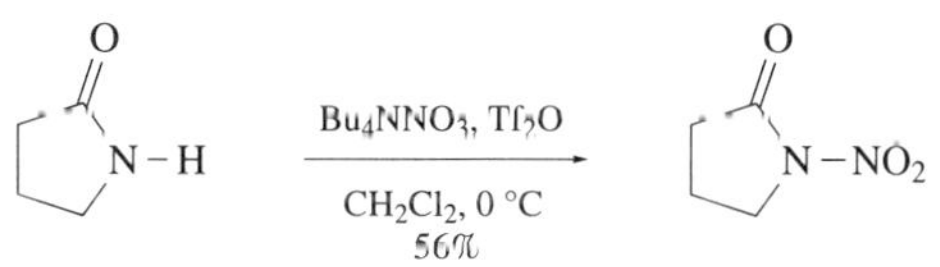

(129)

5.07.4.4.2 *From nitroamines*

The sodium salt of *N*-nitromethylamine (**135**) reacts with a variety of acid chlorides to provide the corresponding *N*-methyl-*N*-nitroamides (Equation (130)) ⟨76S241⟩. Similar reactions with more hindered nitroamines lead to very low yields of the desired product ⟨61JA1191, 66JOC3038⟩.

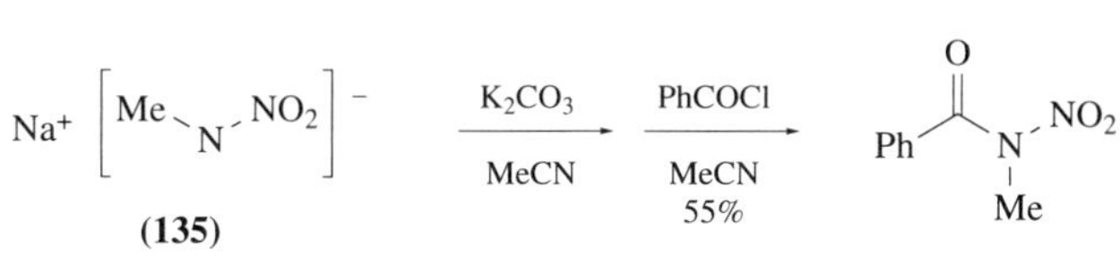

(135)

(130)

5.07.4.4.3 From imidoyl chlorides

Imidoyl chlorides react with silver nitrate to provide *N*-nitro amides ⟨92JCS(P2)281⟩. This reaction is especially useful for the nitration of electron-rich benzoyl derivatives which, under the usual conditions, leads preferentially to ring nitration (Equation (131)) ⟨85JCR(S)132⟩.

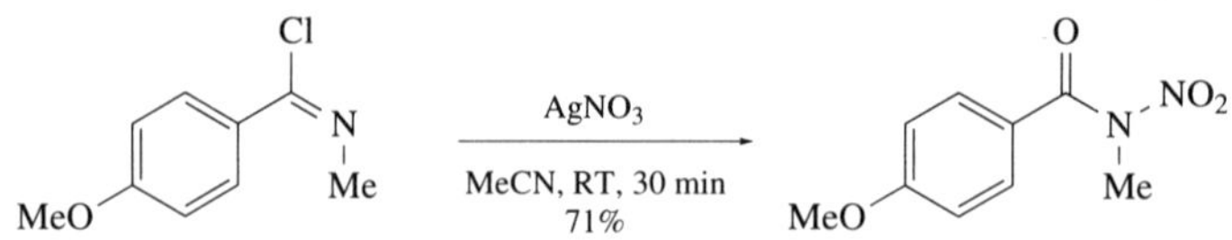

(131)

5.07.4.4.4 From silylamides

The 1,2-diacyl bis(TMS)hydrazine (**136**) reacts with N_2O_5 to provide the *N*-nitroso-1,2-diacyl hydrazine (**137**) (Equation (132)) ⟨91IZV1108, 91MI 507-06⟩.

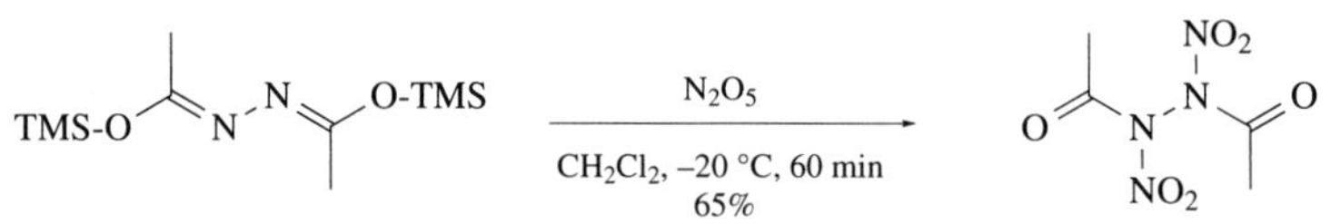

(136) **(137)** (132)

5.07.4.5 *N*-Nitroimides

Many of the reagents used to prepare *N*-nitroamides can be used to prepare *N*-nitroimides. Such reagents include $NO_3Bu^n{}_4N/Tf_2O$ ⟨93TL6669⟩, NH_4NO_3/TFAA ⟨88S743⟩, or N_2O_5 (Equation (133)) ⟨62JPR223⟩.

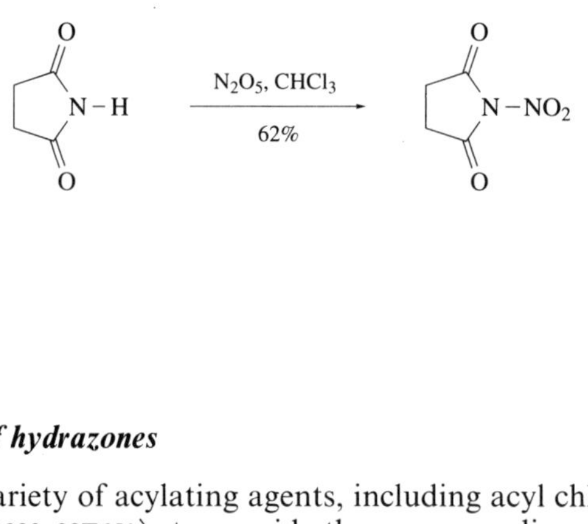

(133)

5.07.4.6 Acyl Hydrazones

5.07.4.6.1 From acylation of hydrazones

Hydrazones react with a variety of acylating agents, including acyl chlorides ⟨87JCS(P1)2511⟩ and anhydrides ⟨80BCJ3225, 92H(33)893, 92T651⟩, to provide the corresponding acyl hydrazones (Equation (134)). Such compounds can also be obtained by heating an equimolar mixture of α-diazoketones and ketohydrazones (**138**) (Equation (135)) ⟨84BCJ1667⟩.

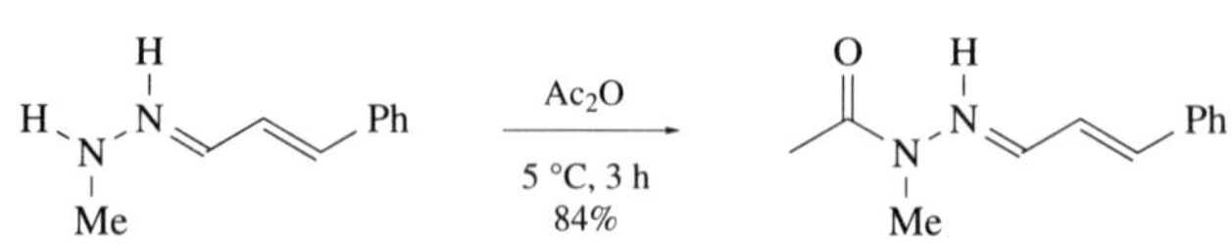

(134)

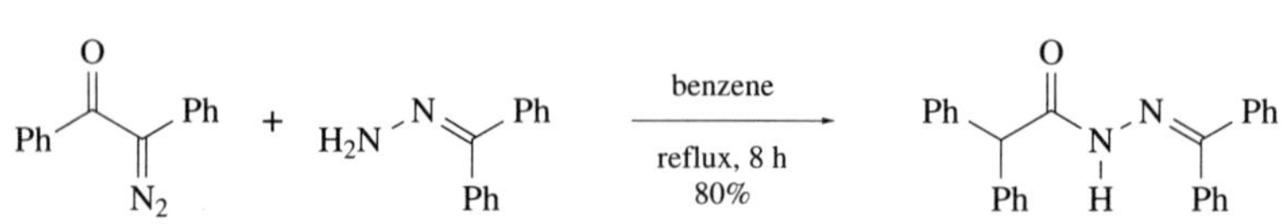

(135)

5.07.4.6.2 From acyl hydrazines and carbonyl derivatives

The reaction of 1-acyl hydrazines with aldehydes or ketones provides the corresponding acyl hydrazones (Equation (136)) ⟨91T10053, see also 92CC1186⟩. Acetals ⟨87TL285⟩, trichloromethylarenes ⟨91T447⟩, and imines derived from aniline ⟨86BCJ1575, 92T531⟩ have also been used to transfer aldehyde equivalents to acyl hydrazines.

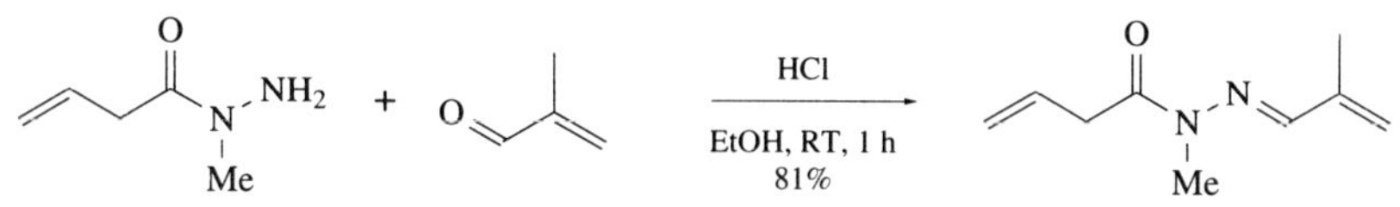

(136)

5.07.4.6.3 From N*-acyl-*N′*-triflylhydrazine*

N-Acyl-*N′*-triflylhydrazine (**139**) reacts with alkyl halides, tosylates, and nosylates to provide, upon loss of Tf-H, the acyl hydrazones (Scheme 25) ⟨75JOC3450⟩.

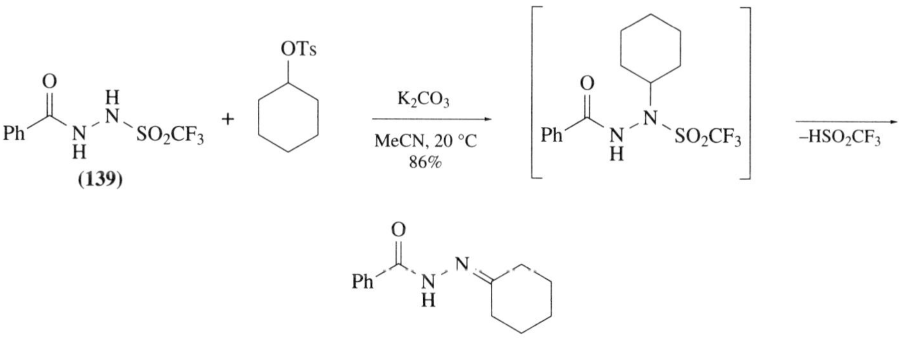

Scheme 25

5.07.4.6.4 From semicarbazones

Semicarbazones can be converted to acyl hydrazones by heating a mixture of acetic anhydride and acetic acid ⟨47JA875⟩. When only acetic anhydride is used, the 1,1-diacyl derivative is obtained (Equation (137)).

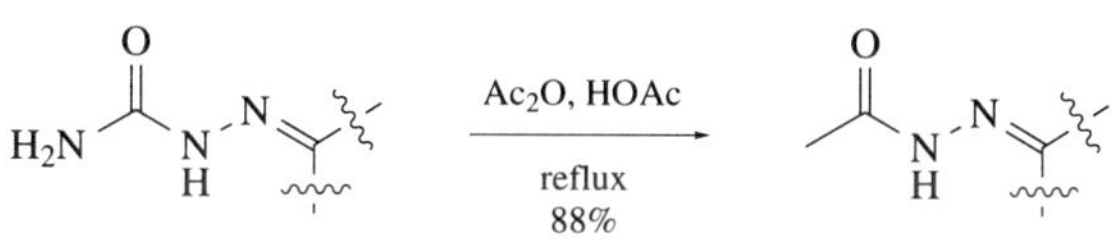

(137)

5.07.4.6.5 From tetrazoles

Tetrazoles can be degraded upon treatment with the Vilsmeier reagent to the corresponding acyl hydrazone (Equation (138)) ⟨74TL1187⟩.

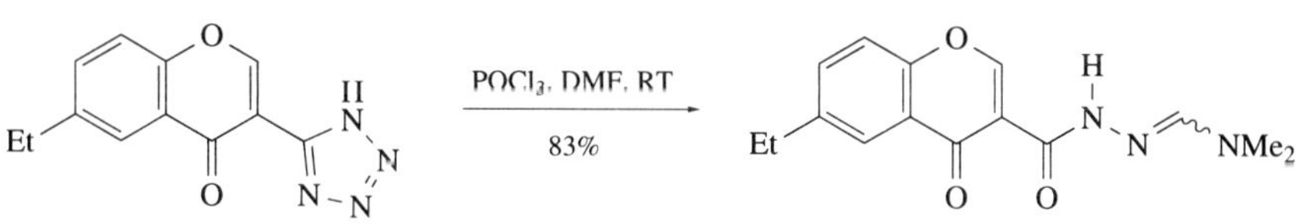

(138)

5.07.4.6.6 From oxidation of conjugated hydrazones

The oxidation of conjugated hydrazones with lead tetraacetate in ether provides the desired acyl hydrazone ⟨68JOC762⟩.

5.07.4.6.7 From azines and enolates

Reaction of 2,3-diazabutadiene with lithium enolates provides *N*-imino-β-lactams (Equation (139)) ⟨85H(23)677⟩. The addition of 2.4 equivalents of HMPA dramatically improves the yield of β-lactam.

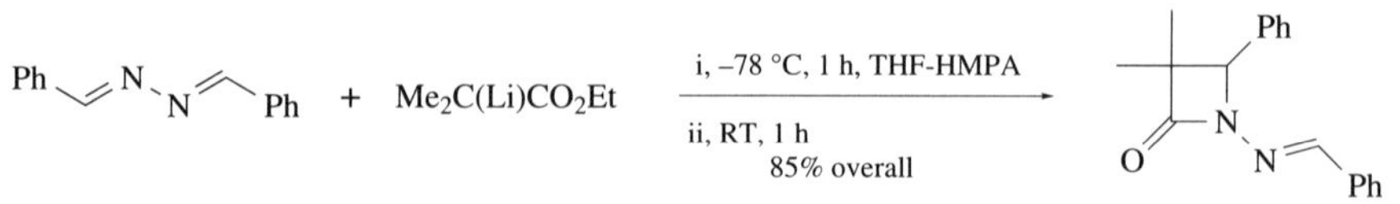

(139)

5.07.4.6.8 From SO_2 extrusion by thiadiazoline

Trans-oriented thiadiazolines (**140**) extrude SO_2 under thermal conditions to form acyl hydrazones (Equation (140)) ⟨81CB802⟩. The reaction of the *trans* isomer is believed to proceed via radical cleavage of sulfur monoxide. This is in contrast to the thermolysis of the *cis*-compound which proceeds in a concerted fashion to the *cis,cis*-diazenes.

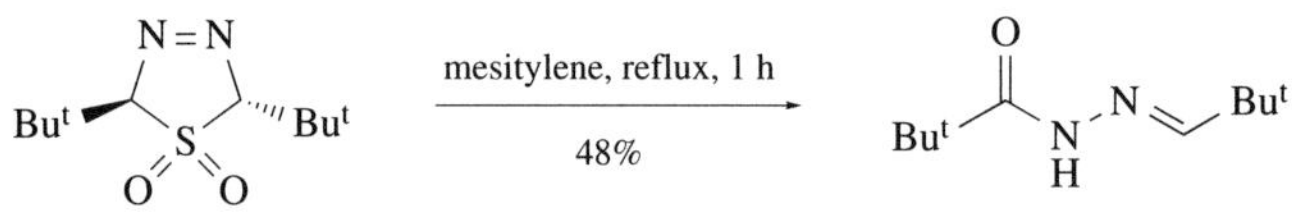

(140)

(140)

5.07.4.7 *N,N*-Diacyl Hydrazones

5.07.4.7.1 From semicarbazones

As mentioned in Section 5.07.4.6.4, heating semicarbazones in acetic anhydride without acetic acid provides the *N,N*-diacyl hydrazone (Equation (141)) ⟨47JA875⟩.

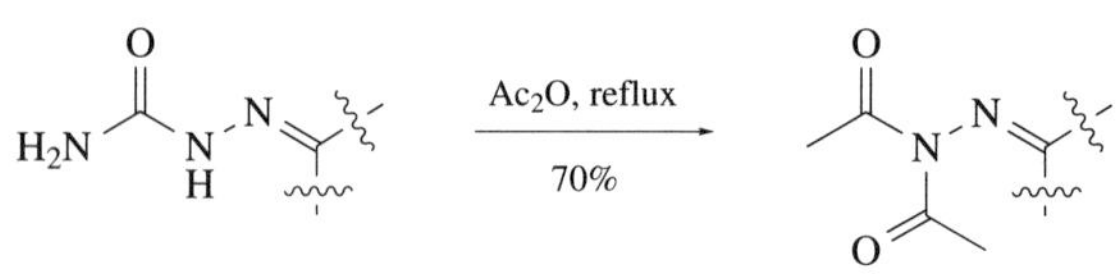

(141)

5.07.4.7.2 From N*-nitrenylphthalimide*

The oxidation of *N*-aminophthalimide (**141**) with lead tetraacetate produces a nitrene intermediate which, when generated in the presence of 2,5-dimethylfuran (**142**), produces phthaloyl hydrazone (**143**) (Equation (142)) ⟨72JCS(P1)2728⟩.

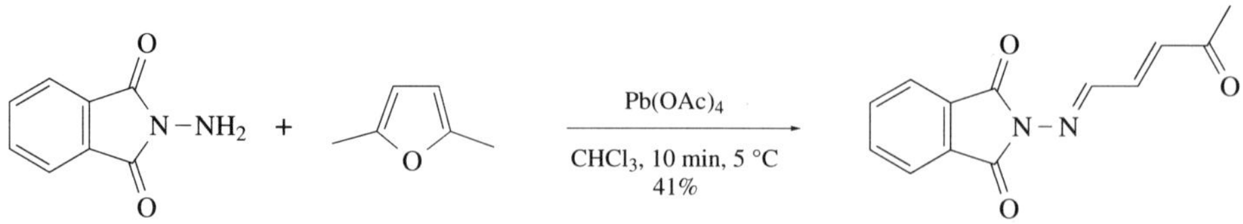

(141) **(142)** **(143)**

(142)

5.07.4.7.3 From ring opening of N*-aziridinylphthalimides*

Treating *N*-aziridinylphthalimides (**144**) with lithium bromide affords the desired hydrazines (Equation (143)) ⟨76BCF1119⟩.

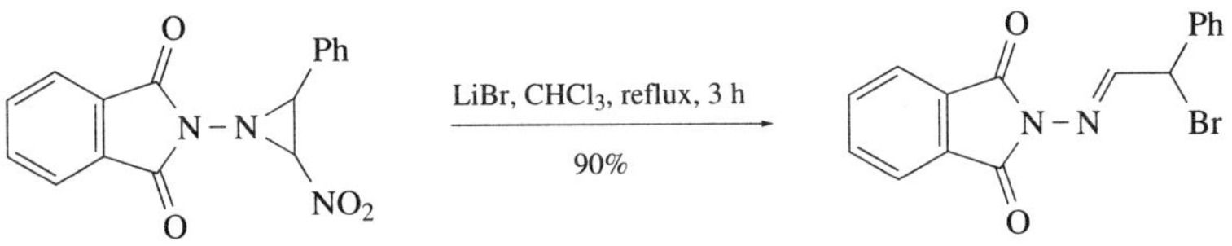

(144)

(143)

5.07.4.7.4 *From rearrangement of 1,2-diacyl diaziridines*

Thermal rearrangement of dibenzoyldiaziridine (**145**) provides the desired hydrazone (**146**) (Equation (144)) ⟨67CB142⟩.

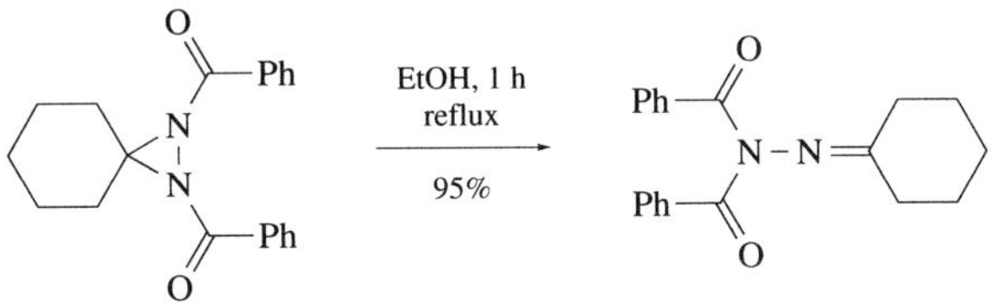

(145) (146)

(144)

5.07.4.8 *N*-Acyl Triazenes

The alkylation of sodium amides with diazonium salts provides a means of producing an unsymmetrical isomer in a regiospecific fashion (Equation (145)) ⟨67JOC1552⟩.

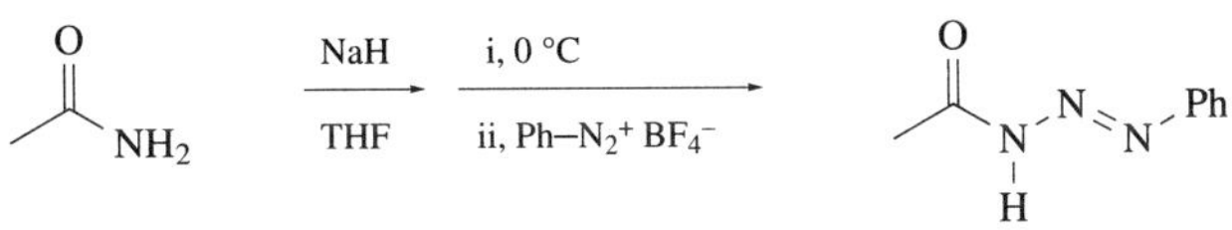

(145)

5.07.4.9 Acyl Tetrazines

5.07.4.9.1 *From oxidation of acyl hydrazines*

The oxidation of 1-acyl hydrazines with lead tetraacetate ⟨76JCS(P1)1150, 87ZC27⟩ or $KMnO_4$ provides the corresponding symmetrical 1,4-diacyl tetrazine (Equation (146)) ⟨56HCA1253⟩.

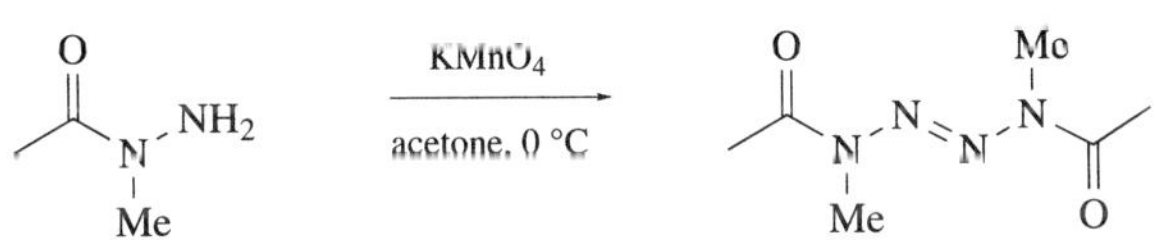

(146)

5.07.4.9.2 *From diazonium salts*

Unsymmetrical acyl tetrazines can be prepared from the reaction of acyl hydrazines with diazonium salts (Equation (147)) ⟨83S478⟩.

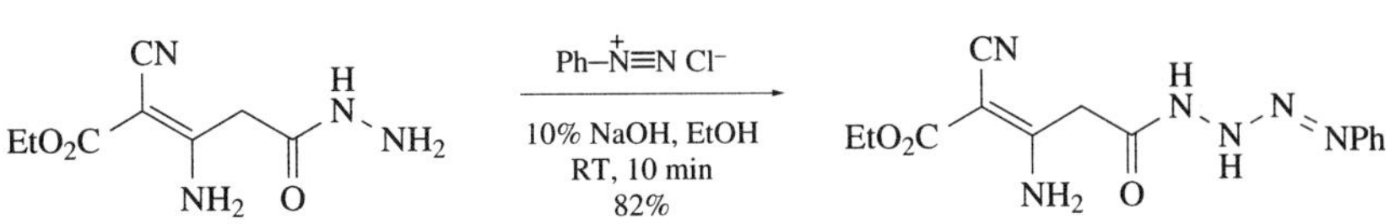

(147)

5.07.4.10 Acyl Diazenes, RC(O)N=NR

The chemistry of diazenes has been reviewed by Kosower ⟨71ACR193⟩.

5.07.4.10.1 From oxidation of aldehyde hydrazones

Hydrazones derived from aldehydes can be oxidized to the corresponding acyl diazene using either mcpba ⟨93PJC711⟩ or benzeneselenic anhydride (Equation (148)) ⟨80JCS(P1)1212⟩. With regard to the latter reagent, it is of interest that neither selenium dioxide nor benzeneselenic acid were able to effect such an oxidation.

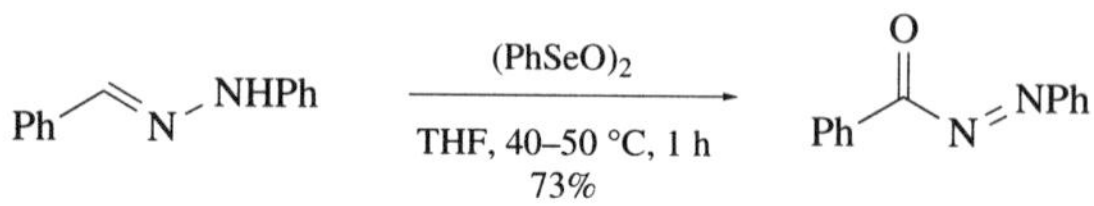

(148)

5.07.4.10.2 From carboxylic acid derivatives

Alkoxydiazenium ions (**147**) react smoothly with carboxylates to provide *N*-acyl-*N′*-alkyldiazenes (Equation (149)) ⟨69CB2889⟩.

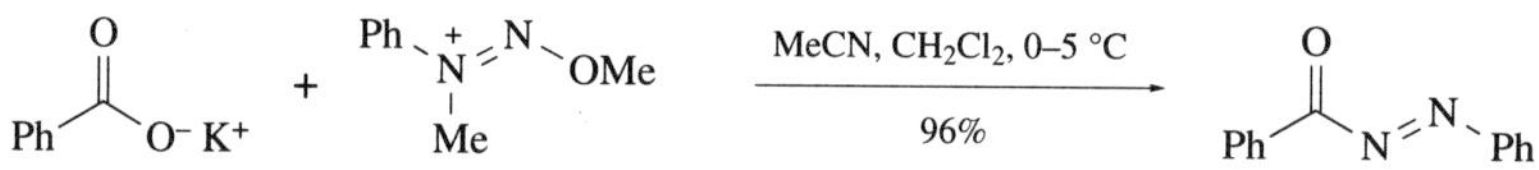

(**147**) (149)

5.07.4.10.3 From dehydrogenation of acyl hydrazines

Oxidants such as lead tetraacetate ⟨90JHC1741, 91JCS(P1)273⟩, NBS ⟨85JOC5604⟩, and iodobenzene diacetate ⟨87SC409⟩, have routinely been used in the field of heterocycle synthesis for the dehydrogenation of urazoles and 1,2-pyridazine-3,6-diones. The oxidation of acyclic acyl hydrazines to acyl diazenes is more precarious owing to the high reactivity/instability of these derivatives. Nevertheless, oxidations such as these have been realized using reagents such as chlorine ⟨54LA(587)93, 79JOC2288⟩, NBS ⟨65JOC64, 69CB2889⟩, iodobenzene diacetate ⟨88JA1638, 89TL2317⟩, lead tetraacetate ⟨84CJC574, 91TL133⟩, mercuric oxide ⟨65JOC1162⟩, mercuric chloride ⟨80CJC1316⟩, nitric acid ⟨82JCED97⟩, and potassium ferricyanide (Equation (150)) ⟨77S339⟩.

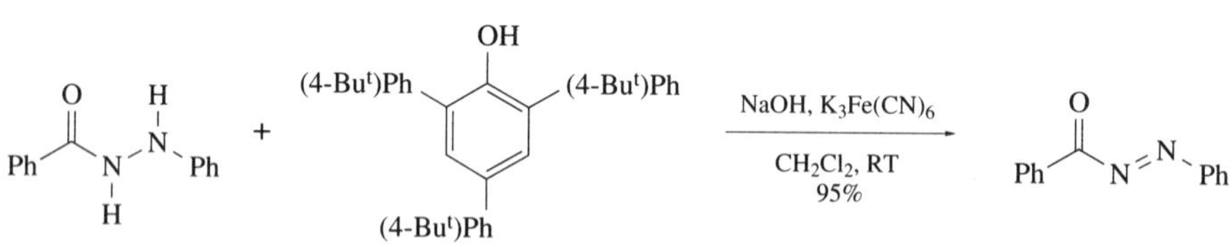

(150)

In the case of 1-acetyl-2-benzylhydrazine (**148**), oxidation with lead tetraacetate provides an equilibrium mixture of the acyl diazene (**149**) and the acyl hydrazone (**150**) (Equation (151)) ⟨72JCS(P1)1692⟩.

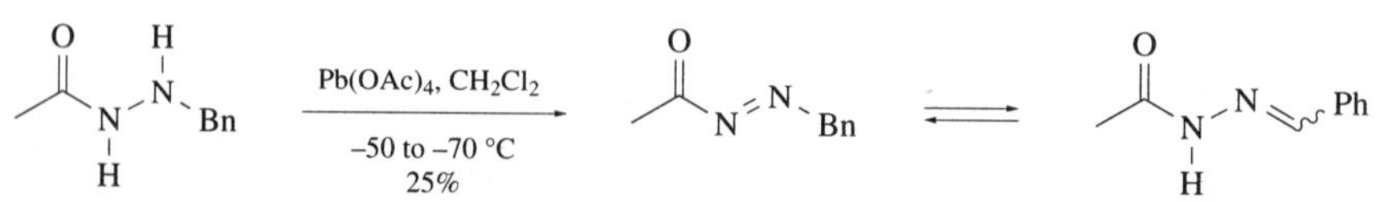

(**148**) (**149**) (**150**) (151)

5.07.4.10.4 From condensation of acyl hydrazines with α-halo ketones

The reaction of acyl hydrazines with α-halo ketones leads to good yields of the corresponding 2-halohydrazone (**151**). Upon treatment with bicarbonate, the corresponding conjugated azoalkenes (**152**) are obtained (Scheme 26) ⟨85OPP385⟩, see also ⟨87JCS(P1)2511⟩.

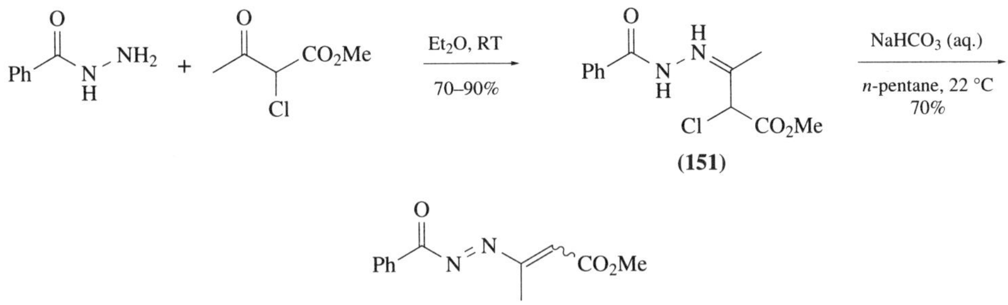

(151)

(152)

Scheme 26

5.07.4.11 Acyl Azides, RCON=N=N

The chemistry of acyl azides has been reviewed ⟨B-71MI 507-01, B-83MI 507-01⟩.

5.07.4.11.1 From acylation of azide

Acyl azides are most commonly prepared by the acylation of sodium azide with acid chlorides ⟨93PS(78)161⟩. Carboxylic acids can be converted to acyl azides following activation of the acyl moiety with chloroformates ⟨90CPB3184, 92BSF360⟩, iminium chlorides (prepared from $SOCl_2$/DMF) ⟨84TL3365⟩, a mixture of NCS/PPh ⟨94PS(89)57⟩, or phenyl dichlorophosphate (Equation (152)) ⟨83SC289⟩. The direct conversion of esters to acyl azides using diethylaluminum azide has been reported ⟨94TL4947⟩.

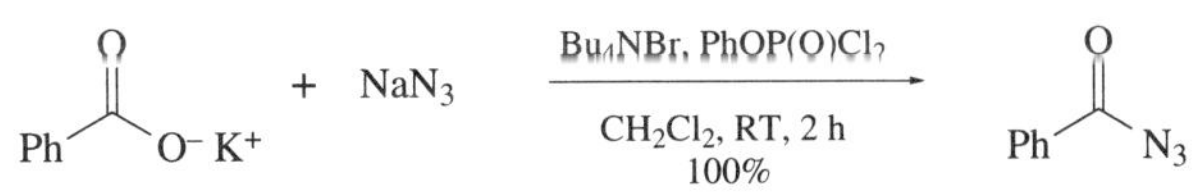

(152)

The low solubility of sodium azide in organic solvents has been overcome by using phase transfer catalysts ⟨90JIC172⟩, polymer-supported azide ⟨89IJC(B)186⟩, or azide reagents such as TMS-azide ⟨83JOC3358, 91SL919⟩ and diphenoxyphosphoryl azide (Equation (153)) ⟨77CPB1651⟩.

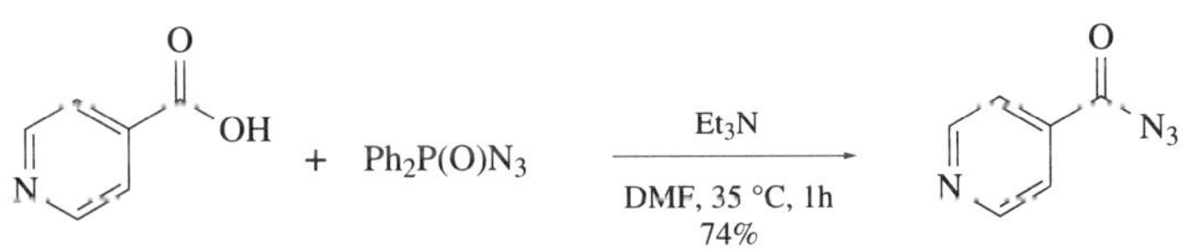

(153)

Acyl azides have also been prepared from the ring-opening of iminium furanones with sodium azide ⟨76CC718⟩.

5.07.4.11.2 From oxidation of acyl hydrazines

Sodium nitrate oxidation of 1-acyl hydrazine provides a reliable method for the preparation of acyl azides. The transformation has been applied to a variety of acyl hydrazines ⟨85JCS(P1)2277, 89JOC145⟩, as well as 1-acyl hydrazones ⟨74TL1187⟩. Various other oxidants, such as isoamyl nitrite

⟨85BCJ352⟩, nitrosyl tetrafluoroborate ⟨87TL5091⟩, dinitrogen tetroxide ⟨86TL4749, 93MI 507-01⟩ and iron(III) nitrate (Equation (154)) ⟨84TL3701⟩, have also been used to effect this transformation.

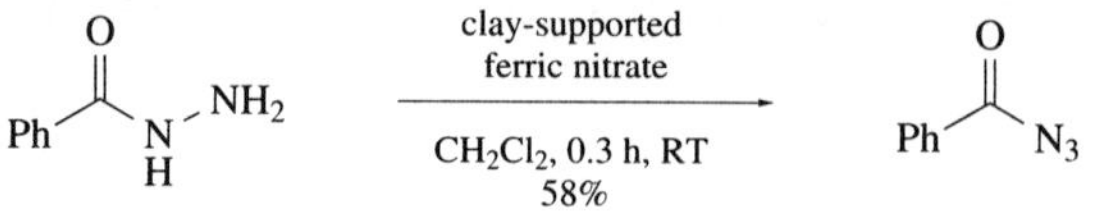

(154)

5.07.4.11.3 From oxidation of aldehydes

The combination of chromic anhydride and azidotrimethylsilane is a mild and useful reagent for the direct conversion of aldehydes to acyl azides (Equation (155)) ⟨92TL3165⟩. A similar oxidation employing pcc instead of chromic anhydride results in mixtures of acyl azides and carbamoyl azides ⟨88SC545⟩.

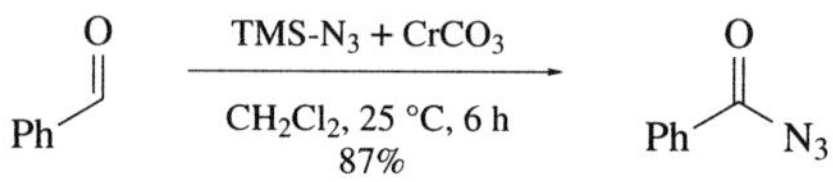

(155)

5.07.4.11.4 From ring opening of N-*aminobenzotriazinone*

N-Aminobenzotriazinone (**153**) can be ring opened under strongly acidic conditions to provide *o*-aminobenzoyl azide (**154**) (Equation (156)) ⟨62JOC1383⟩. The use of milder acid (1 mol l^{-1} HCl) results only in hydrolysis of the imine.

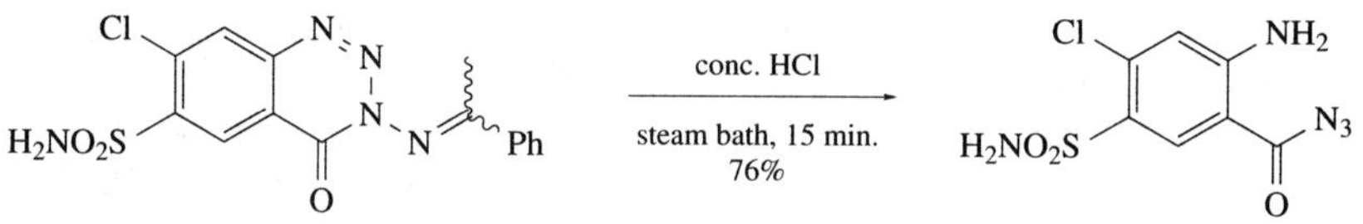

(156)

(153) **(154)**

5.07.5 AMIDES SUBSTITUTED BY PHOSPHORUS, ANTIMONY, ARSENIC, OR BISMUTH

Although these compounds are also located in group 15 of the periodic table, the chemistry of phosphorus, arsenic, antimony, and bismuth is substantially different from that of nitrogen. Phosphorus, like nitrogen, is essentially covalent in all of its chemistry, and exists in both the +3 and +5 oxidation states. The remaining elements, arsenic, antimony, and bismuth, show increasingly ionic character and tend to exist primarily in the +5 state ⟨B-80MI 507-01⟩.

A general review of phosphorus–nitrogen compounds has appeared ⟨B-72MI 507-01⟩.

5.07.5.1 Amide Derivatives: Oxidation State +3, Tricoordinate

5.07.5.1.1 From amides

Amides react with a variety of phosphorus derivatives to provide phosphorylated amide derivatives. Phosphorylation can occur on either the nitrogen or the oxygen of the amide, providing the phosphonamide, the phosphoimidate, or both ⟨85MAC1803⟩.

Acetamide has also been shown to react with diphenylantimony chloride (**155**) to provide *N*-acetyl-As,As-diphenylarsinous amide (**156**) (Equation (157)) ⟨88ZOB829⟩.

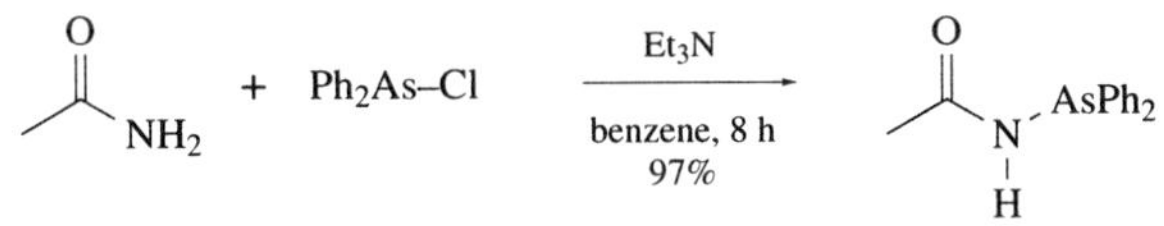

(157)

(liquified) (**155**) (**156**)

5.07.5.1.2 *From silylamides*

The formation of TMS-Cl serves as the driving force in the reaction between silylamides and chlorophosphites. Silylamides react with a variety of chlorophosphites to provide phosphinoamides (Equation (158)) ⟨88ZOB706, 88ZOB2628, 89ZOB2392, 92IZV220⟩.

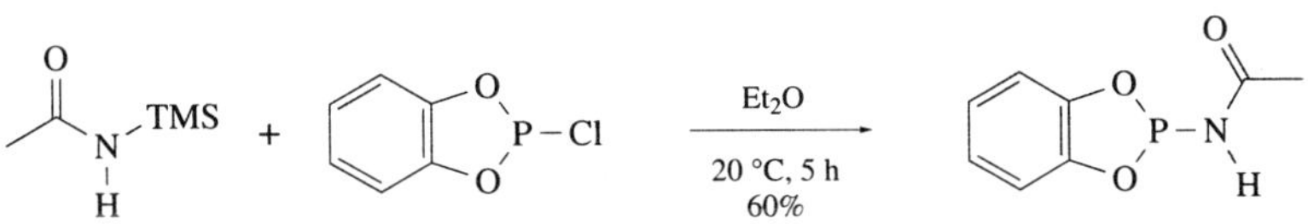

(158)

The deprotection of diethyl *N*-acyl(trimethylsilyl)phosphoramidite with HCl provides a clean route to diethyl *N*-acyl phosphoramidite ⟨78ZOB695⟩.

5.07.5.2 Imide Derivatives: Oxidation State +3, Tricoordinate

The reaction of cyclic imides with phosphorus halides usually provides the *N*-phosphinyl products. Similar reactions with acyclic imides provide a mixture of *N*- and *O*-phosphinylated products ⟨78IZV2412, 90ZOB1187⟩.

Potassium phthalimide reacts with phosphorus trichloride as well as with diethyl chlorophosphite to provide the *N*-phthalimido-bis(diethylamino)phosphite (**157**) ⟨90ZOB1183⟩. Hexaethylphosphorus triamide also reacts with phthalimide to provide tetraethyldiamidophthalimidophosphite (Equation (159)) ⟨85IZV173⟩.

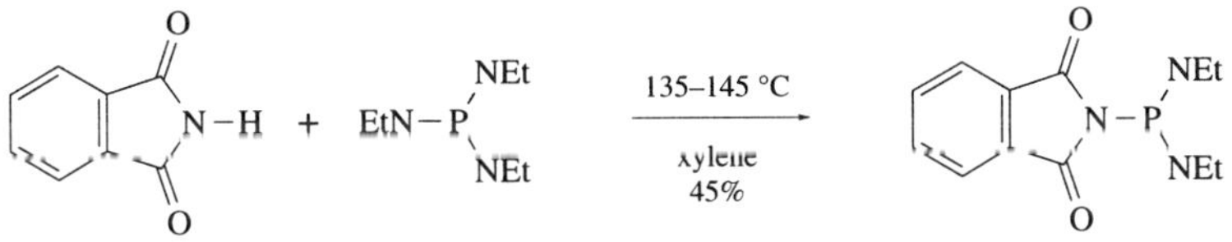

(159)

(**157**)

Succinimide has also been reported to react with diphenylantimony chloride, providing the antimonyimide ⟨89POL581⟩.

5.07.5.3 *N*-Hetero-*N*-phosphoamides (Oxidation State +3)

The reaction of bsa with phosphorus bromodifluoride ⟨80JCS(D)1768⟩, phosphorus trichloride ⟨78ZOB2468⟩, or *S*,*S*-diethyl dithiochlorophosphite ⟨91IZV2103⟩ provides N-(TMS)-*N*-phosphoamides. Monosilylamides can also be used to prepare *N*-hetero-*N*-phosphoamides, providing the silylamide is treated with base prior to addition of the chlorophosphite (Equation (160)) ⟨78ZOB695⟩.

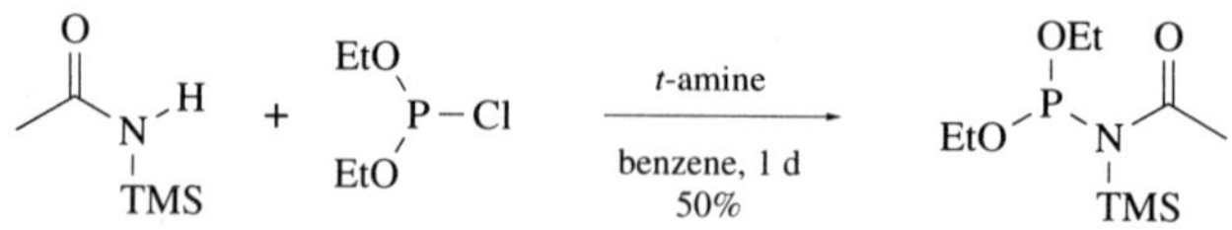

(160)

5.07.5.4 Amide Derivatives: Oxidation State +5: Tetracoordinate

5.07.5.4.1 From phosphorylation of amides

Amides react with phosphoryl chlorides, under basic conditions, to provide *N*-phosphorylamides ⟨70ZOB2464, 82JOC3533, 90JMC2192⟩. The conditions are mild enough to allow for a number of functional groups to be present in the molecule (Equation (161)), ⟨89JMC1886⟩. In cases where the amide contains a *β*-amino ⟨78JHC1169, 81PS(10)261, 89S618⟩, *α*-hydroxy ⟨86BSF101⟩, *β*-hydroxy ⟨74MI 507-01, 92AP325⟩, or *β*-thio ⟨90S661⟩ group, treatment with phosphorus reagents containing two displaceable groups provides cyclic aminophosphates.

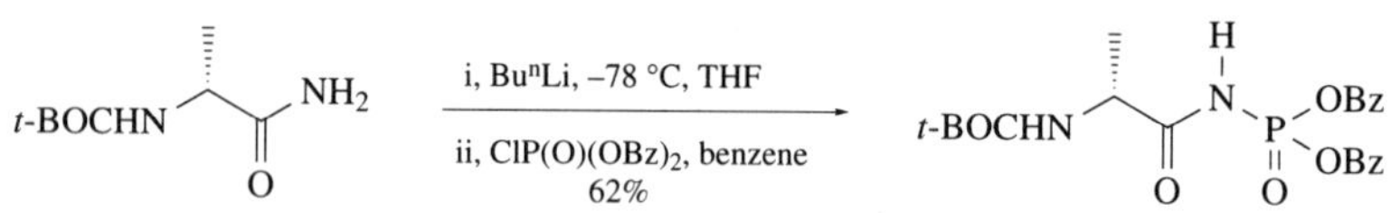

(161)

β-Lactams react with a variety of phosphochloridates to provide *N*-phosphorylated *β*-lactam (**158**) (Equation (162)) ⟨83JA3743⟩.

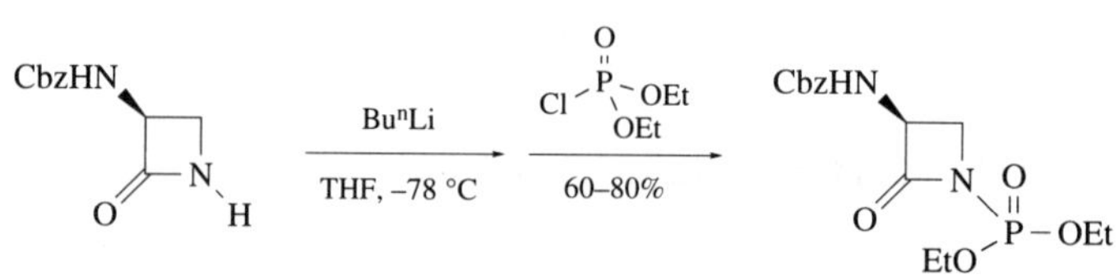

(162)

(**158**)

Cbz = Benzyl carbamate

In a related reaction, ethyldiarylarsinates ($Ar_2As(O)OEt$) have been converted to the corresponding arsenic amides ($Ar_2As(O)NHCOMe$) upon treatment with acetamide ⟨91ZOB450⟩.

5.07.5.4.2 From acylation of phosphoamines

Phosphoramides can be obtained via acylation of phosphoramidate with acid chlorides ⟨70AF790, 80PS(8)27, 87JCS(P2)1489, 88ZOB2424, 90JOC6091, 90PS139⟩. The catalytic use of cobalt(II) chloride has been reported to reduce the side reaction (P—N cleavage) which can occur during the reaction (Equation (163)) ⟨91CI(L)353⟩.

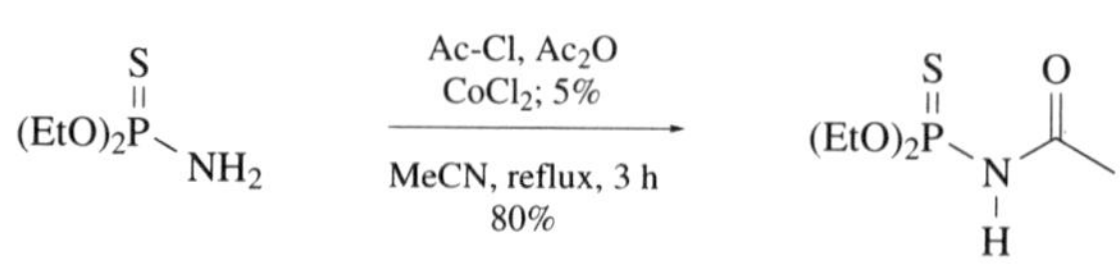

(163)

5.07.5.4.3 From acylation of phosphoaminosilanes

Phosphoaminosilanes (**159**) react with acid chlorides to provide acyl phosphoamides (**160**) and TMS-Cl (Equation (164)) ⟨85BSF1211⟩.

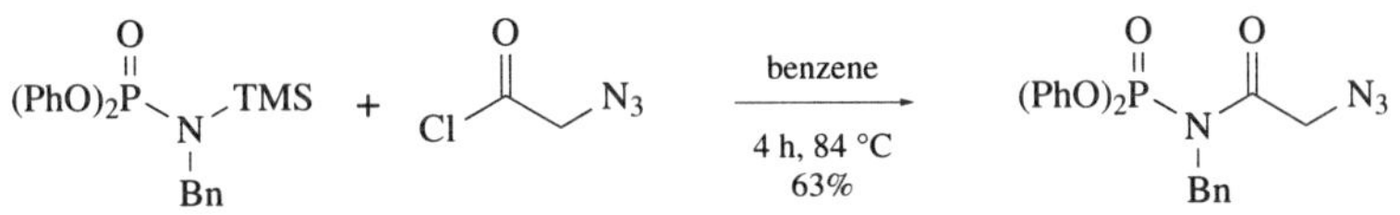

(159) **(160)**

(164)

5.07.5.4.4 From acylation of phosphoimines

The Arbusov reaction of triethoxyphosphinimino derivatives such as (**161**) with acid chlorides provides *N*-phosphoryl-protected amides (Equation (165)) ⟨79TL3851⟩ see also ⟨85ZOB2483, 86BCJ3696⟩. In a related vein, acylation of phosphazenes (N=PR_3) with acid chlorides provides *N*-acyl phosphonamides ⟨78ZOB1594, 86S772, 90JCS(P1)2193, 90TL3497⟩.

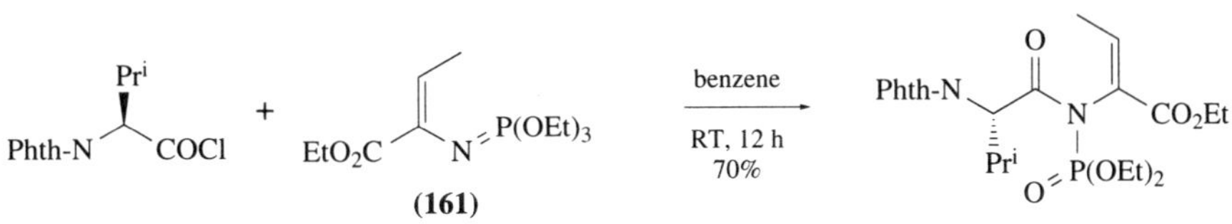

(161)

(165)

5.07.5.4.5 From rearrangement of hydroxyiminophosphinates

α-Hydroxyiminophosphinates and α-hydroxyiminophosphanates undergo Beckmann rearrangement to provide acyl phosphoramidates ⟨89CI(L)389, 92TL2067⟩. It has been shown that only the *anti*-oximes (P and HO on opposite sides of the imine) are converted to amide derivatives (Equation (166)) ⟨90JCS(P1)3263⟩.

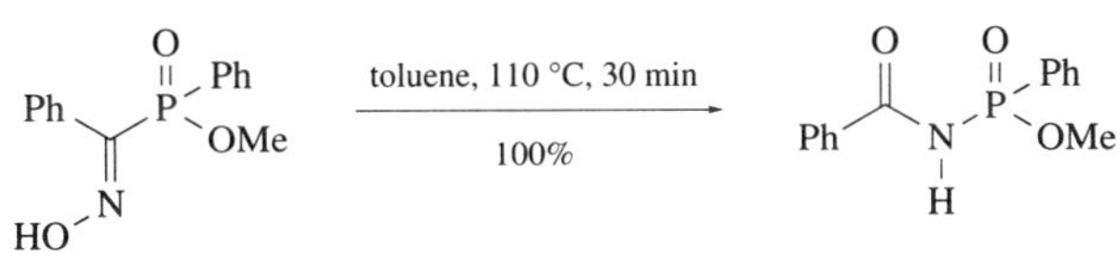

(166)

5.07.5.4.6 From hydrolysis of N*-acyl iminophosphoranes*

N-Acyl iminophosphoranes (**162**) hydrolyze under acidic conditions to *N*-acyl phosphoramidates (**163**) (Equation (167)) ⟨93HAC11, see also 90IZV1139⟩.

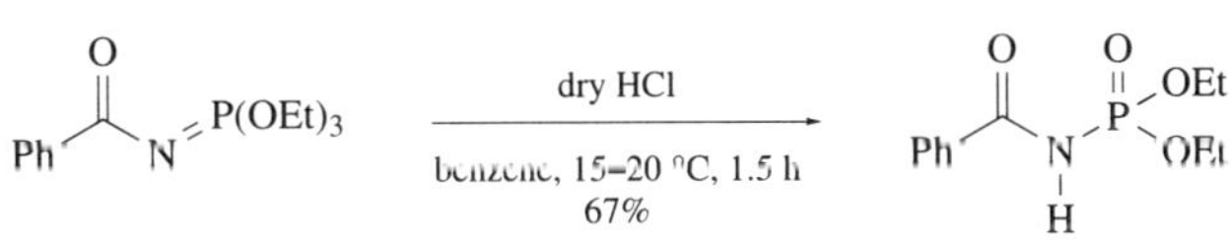

(162) **(163)**

(167)

5.07.5.4.7 Ring opening of phosphoroimides

Ring opening of phosphoroimides with alkyl- or arylamines leads to ring opening and formation of phosphophthalamides (Equation (168)) ⟨82BSF(2)369⟩.

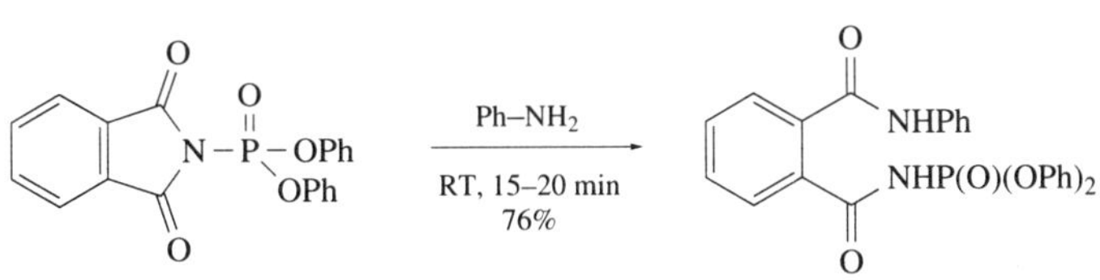

(168)

5.07.5.5 Imide Derivatives: Oxidation State +5, Tetracoordinate

5.07.5.5.1 From imides

Potassium phthalimide reacts with diphenyl phosphochloridate to provide diphenyl phosphorophthalimide (Equation (169)) ⟨81S993⟩.

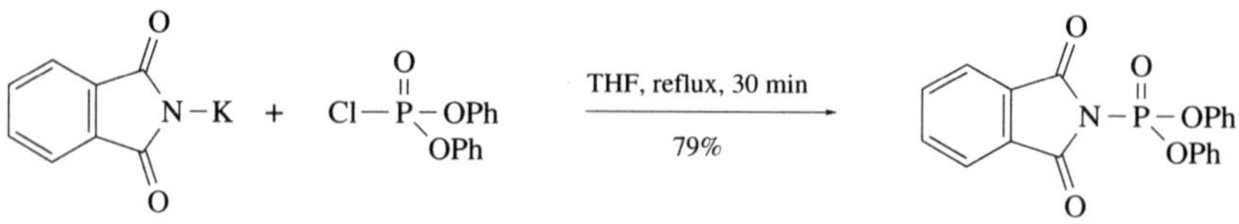

(169)

5.07.5.5.2 From N*-haloimides*

N-Bromoimides undergo an Arbusov reaction with trialkylphosphites to produce *N*-(dialkylphosphonyl)succinimides. The reaction works well for a variety of *N*-alkyl phosphites, but takes place in lower yields with methyl phosphite (Equation (170)) ⟨74JOC922⟩.

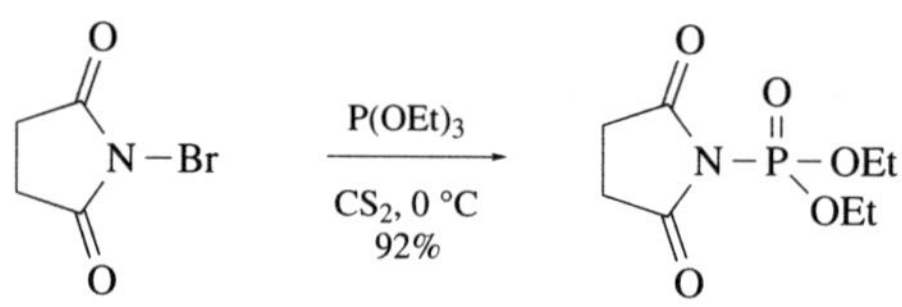

(170)

5.07.5.6 Imide Derivatives: Oxidation State +5, Tetra- and Pentacoordinate

5.07.5.6.1 From imides

Pentacoordinated mono- and diimide derivatives of phosphorus, arsenic, and antimony have been prepared under mild conditions upon treatment of the corresponding metal halides ⟨83IJC(A)120, 87IJC(A)146⟩ or metal alkoxides (Equation (171)) ⟨93SRI1011, see also 84SRI269⟩ with imide.

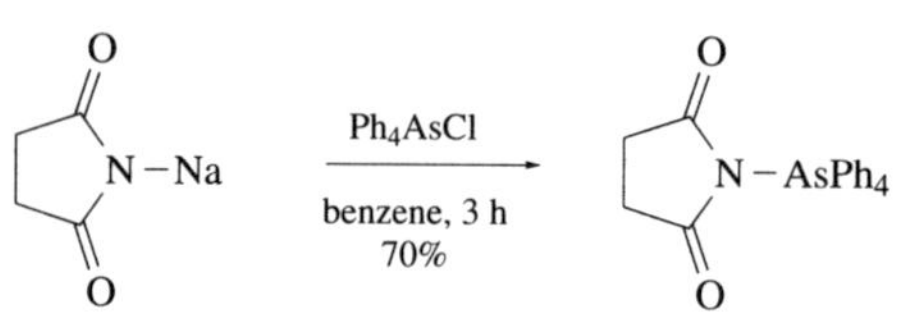

(171)

5.07.5.6.2 From N*-metalated imides*

Silver ⟨83IJC(A)120⟩ and sodium salts of imides react with antimony halides to provide silver halide and the stilbine imide (Equation (172)) ⟨85POL251⟩. The related reaction between sodium succinimide and tetraphenylarsenic chloride has also been reported ⟨93SRI1011⟩.

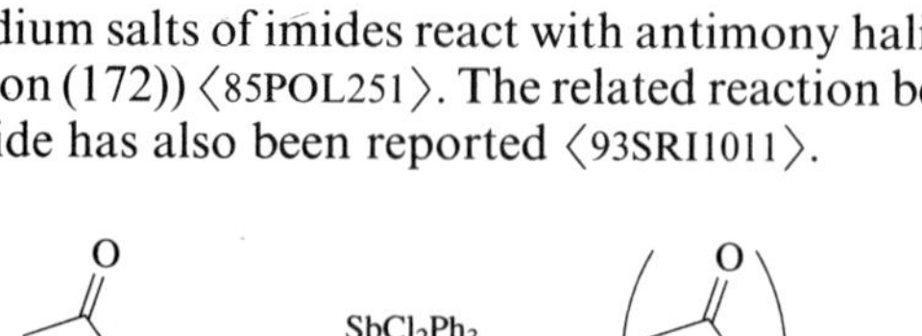

(172)

5.07.5.7 *N*-Acyl Iminophosphoranes, RC(O)N=P

5.07.5.7.1 From acyl azides

The reaction of alkyl azides with phosphines (the Staudinger reaction) provides iminophosphoranes ⟨81T437⟩. When acyl azides are used in place of alkyl azides, *N*-acyl iminophosphoranes are obtained (Equation (173)) ⟨85CC1764, see also 93HAC11, 93PS(78)161⟩.

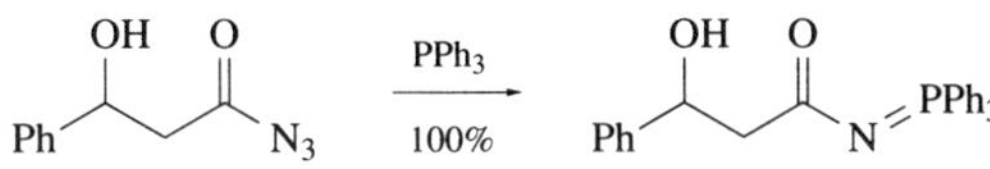

(173)

5.07.5.7.2 From amides

Phosphorus pentachloride (Equation (174)) ⟨70ZOB1949⟩ see also ⟨58JA6442⟩ reacts with primary amides, splitting off HCl and providing the desired trichlorophosphine imides. A similar reaction has been reported for diarylantimony(V) chloride ⟨88ZOB829⟩.

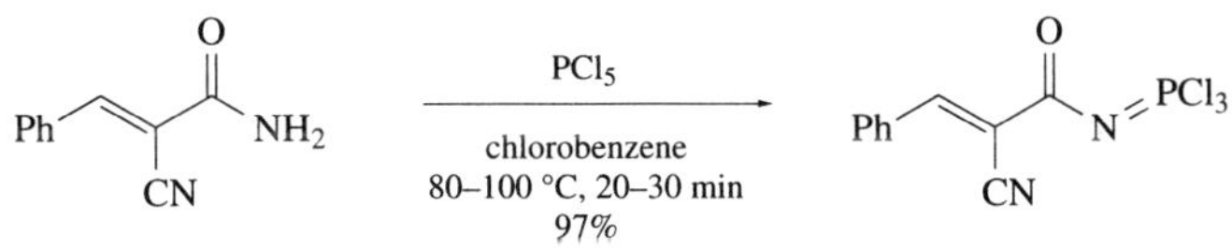

(174)

5.07.5.7.3 From acylation of iminophosphoranes

Triphenylphosphine imide reacts with acid chlorides to produce *N*-acyl iminophosphoranes ⟨69JGU1445⟩. It has also been reported that *N*-lithiated triphenylphosphine imide (**164**) reacts with acyl chlorides and anhydrides to provide the desired *N*-acyl iminophosphoranes (Equation (175)) ⟨91S382⟩.

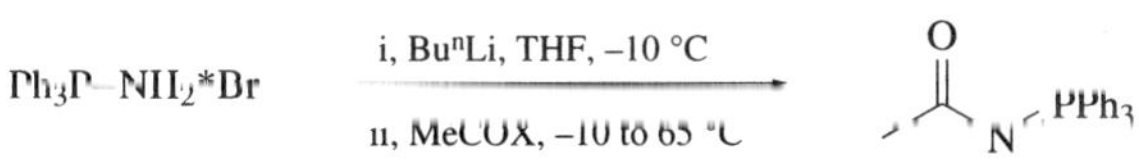

(175)

(164)

X = Cl, 70%
X = OAc, 75%

5.07.5.7.4 From acylation of N-silyliminophosphoranes

Silylated iminophosphoranes react with acid chlorides to produce *N*-acyl iminophosphoranes and TMS-Cl (Equation (176)) ⟨87BSF997, see also 72S695⟩.

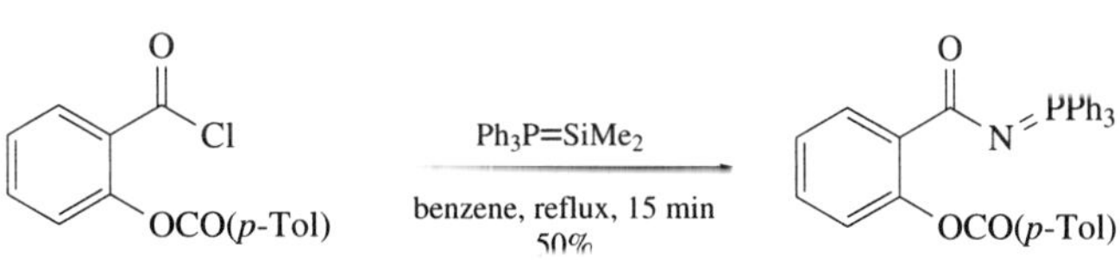

(176)

5.07.5.7.5 From oxidation of acyl phosphoramidites

Dialkyl acyl phosphoramidites react with oxidizing agents such 1,4-benzoquinone, 1,4-naphthoquinone, and chloranil to provide the *N*-acyl phosphoimidates (Equation (177)) ⟨73ZOB1451⟩.

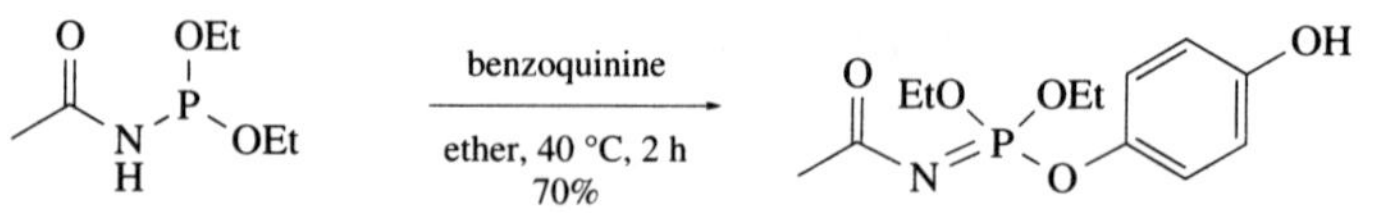

(177)

Amides can also react with the combination dead/PPh_3 and primary amides to give the *N*-acyl iminophosphoranes (Equation (178)) ⟨85JOC1712⟩.

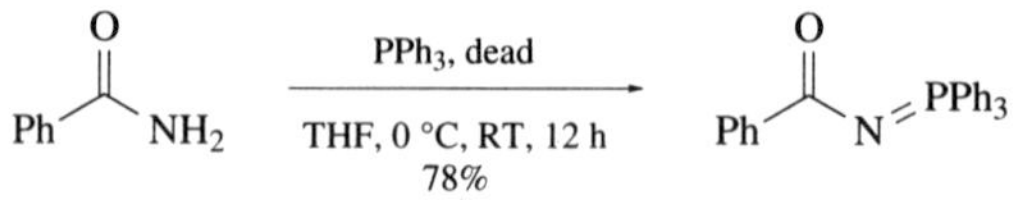

(178)

dead = diethyl azocarboxylate

5.07.6 AMIDES SUBSTITUTED BY SILICON

The chemistry of organosilicon nitrogen compounds has been reviewed by Armitage ⟨B-91MI 507-03⟩.

5.07.6.1 *N*-Silylamide Derivatives

The methods used to prepare *N*-silylamides and lactams depend on the volatility and hydrolytic stability of the resulting *N*-silyl derivatives. TMS-amides are preferred when the product can be distilled or when a labile silicon–amide bond is desired. Larger silyl groups provide greater hydrolytic stability, as well as the opportunity for purification by aqueous workup or chromatography.

5.07.6.1.1 From silylation of amides

The silylation of amides and lactams represents the most common method for the preparation of silylamides. In the lactam field, silylation not only serves a protective role ⟨B-91M 507-01⟩ but also serves to activate the amide nitrogen to alkylation ⟨88ZOB1093⟩ and acylation ⟨60CB2804, 88JHC59⟩. In addition, the silylation of amides has opened areas of amide chemistry that were inaccessible owing to low solubility of the parent amide in organic solvents ⟨84JOC2015⟩.

The TMS group is often introduced by treating a benzene or toluene solution of the amide or lactam with triethylamine followed by TMS-Cl (Equation (179)) ⟨91ZAAC89, see also 88JHC49⟩. These conditions allow for precipitation of the triethylamine hydrochloride and distillation of the TMS-protected product. Such conditions avoid aqueous workup and flash chromatography, conditions known to hydrolyze TMS-amides and lactams ⟨86TL1411⟩.

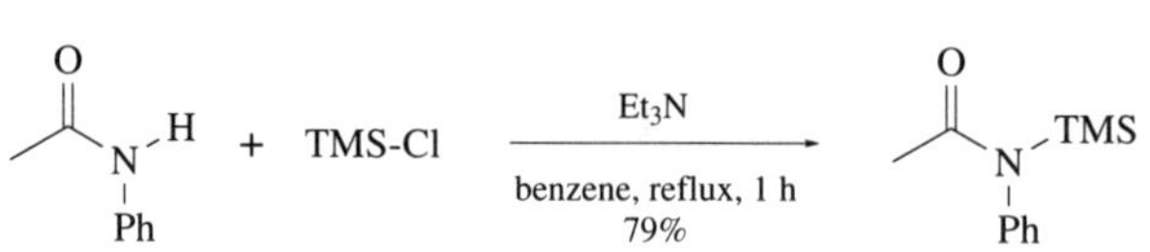

(179)

The introduction of more hindered silyl groups, such as the TBDMS group ⟨80TL31⟩ to amides and lactams, provide more stable silylamides that tolerate a variety of reaction and workup conditions. Because many of the bulky silyl chlorides react more slowly than TMS-Cl, bases such as imidazole ⟨72JA6190⟩, dmap ⟨79TL99⟩, or dbu (Equation (180)) ⟨86TL5285, see also 85TL475⟩ are often used in place of triethylamine to catalyze the reaction. With the silylation of certain lactams, deprotonation with butyllithium prior to silyl chloride addition has been used ⟨86AP(319)203, 86TL5147⟩.

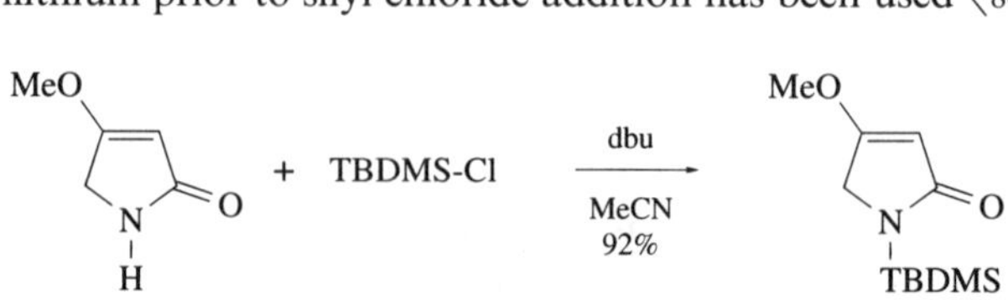

(180)

dbu = 1,5-diazabicyclo[5.4.0]undec-5-ene

Although other silyl chlorides have been used for the silylation of amides and lactams (triethylsilyl ⟨90OM1557⟩, thexyldimethyl ⟨85TL5515⟩, tribenzyl and diphenyl-*t*-butyl (tbdp) ⟨85TL2217⟩), TMS-Cl and TBDMS-Cl remain the two standards. These groups are also introduced to amides using transfer agents other than the silyl chlorides. A summary of more recently developed silyl transfer reagents is found in Table 11.

Table 11 Silylation of amides and lactams.

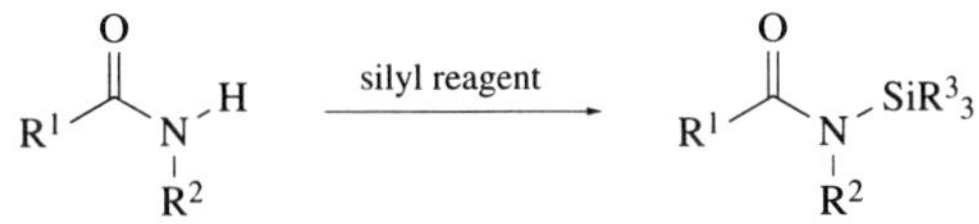

Silyl reagent	*$R^1CON(SiR^3{}_3)R^2$*	*Conditions*	*Ref.*
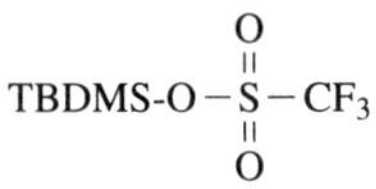	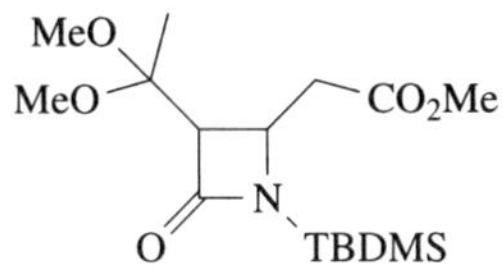	2,6-lutidine CH_2Cl_2 (82%)	89JOC4649
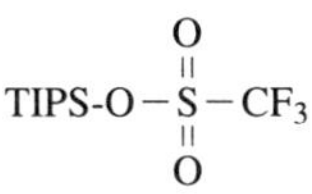	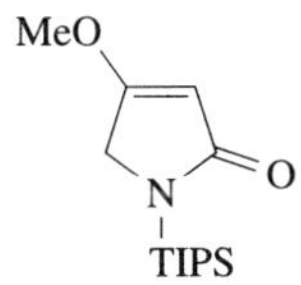	dbu, MeCN (70%)	86TL5285
TBDMS-H	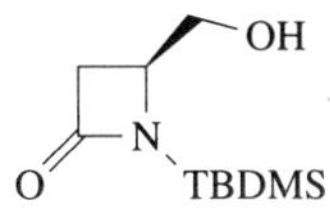	hexane–CH_2Cl_2 10% Pd/C (80%)	89BCJ2111
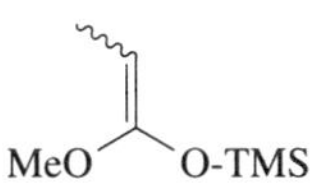	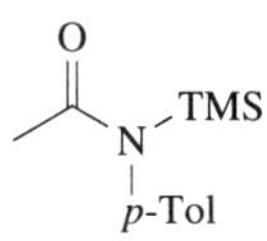	MeCN, 50 °C 10 min (95%)	79TL4311
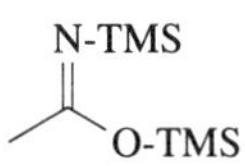 (BSA)	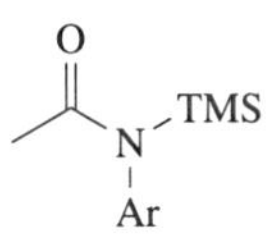	MeCN, reflux 5 min	66JA3390
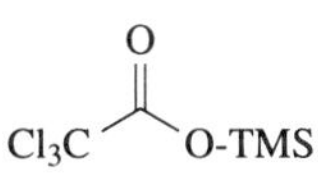	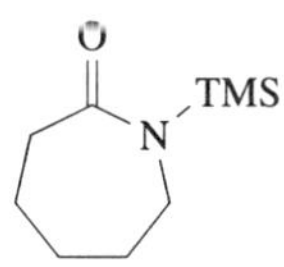	18-crown-6 K_2CO_3 (75%)	85TL1175
Et_2N–TMS	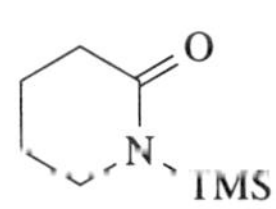	neat, 120 °C 4 h (98%)	65LA(686)226
TMS_2N–H	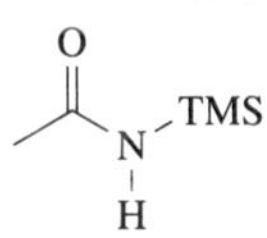	neat, 40 h reflux (80%)	62M352
TMS_2N–H	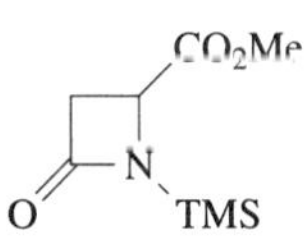	2–5% saccharin $CHCl_3$ (86%)	86JOC558

The use of silyl triflates ⟨81TL3455⟩ allows for the introduction of both TBDMS and triisopropylsilyl (TIPS) groups to *β*-lactams resistant to normal silylating conditions. Many of the remaining methods presented in Table 11 utilize silyl transfer agents that form easy-to-remove by-products. The use of 1,1′,3,3,3′,3′-hexamethyldisilazane (HMDS) as a source of TMS is greatly enhanced when saccharin is added in catalytic amounts.

5.07.6.1.2 *From trans-silylation*

For the introduction of hindered silyl chlorides to amides, trans-silylation of the corresponding TMS-protected amide can also be used. This method has proven especially useful for the preparation of highly substituted silylformanilides (Equation (181)) ⟨90OM1557⟩.

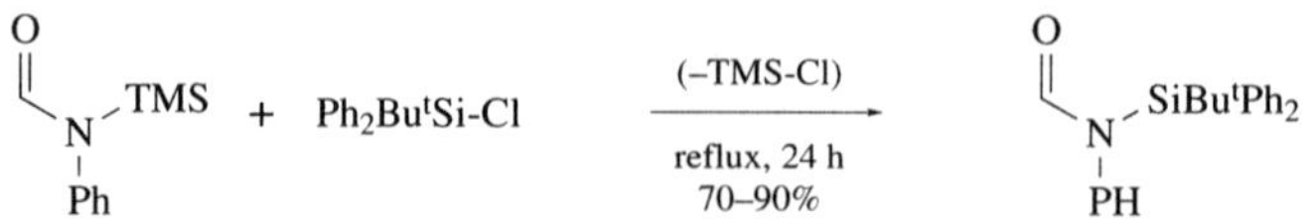

(181)

5.07.6.1.3 *From rearrangement of amido organosilanes*

Amido organosilanes rearrange under thermal conditions to *N*-silylamides (Equation (182)) ⟨73ZOB1753⟩ see also ⟨74ZOB1753, 76ZOB594⟩.

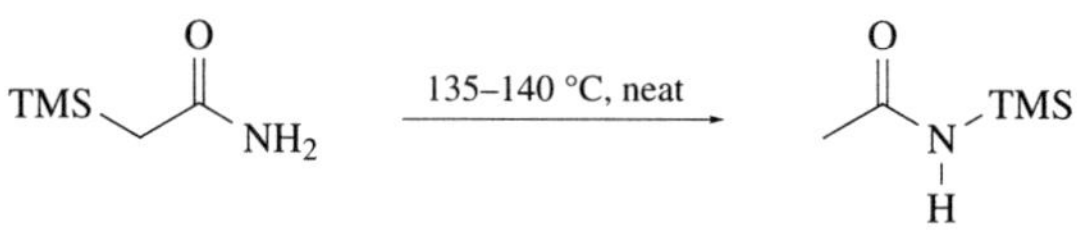

(182)

The migration of silicon from oxygen to nitrogen takes place upon loss of ketene from the *O*-silylvinylacetamide (**165**) (Equation (183)) ⟨69JOM(17)241⟩.

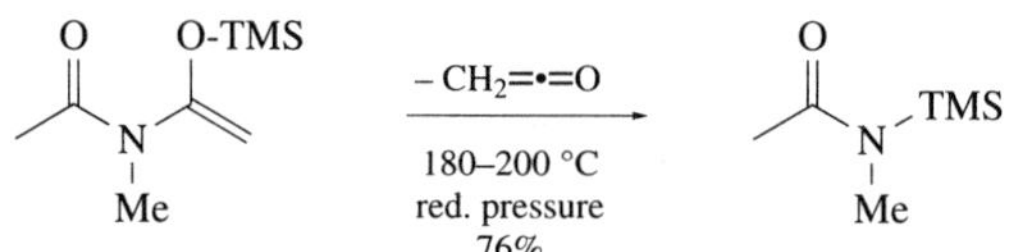

(165)

(183)

5.07.6.1.4 *From acylation of* N*-silylamines*

Trimethylsilylamines react with ketene ⟨73ZOB1753⟩ and activated carboxylic acid derivatives ⟨61CB1263⟩, as well as with acid chlorides ⟨90AF1030⟩ to provide the corresponding *N*-silylamides. *t*-Butyldimethylsilylamines (**166**) containing an appropriately spaced carboxy group cyclize in an intramolecular fashion to provide the corresponding *N*-silyl-β-lactams when treated with *t*-BuMgCl (Equation (184)) ⟨90T4733⟩. The same reaction with trimethylsilylamines yields an unprotected lactam, owing to the instability of the TMS-protected lactam to hydrolysis.

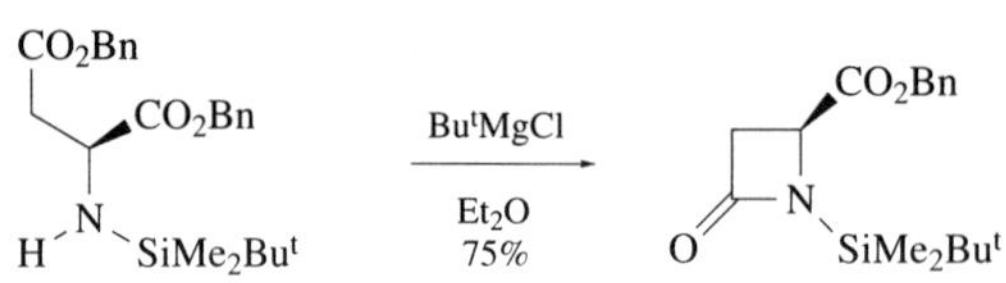

(166)

(184)

TMS-protected amides have been obtained by treating HMDS with carboxylic acids ⟨86ZOB1196, 87ZOB956⟩ or acid chlorides (Equation (185)) ⟨83JOC4111⟩. In the case of acid chlorides, the ratio of acid chloride to amine (1 : > 1) and the order of reagent addition (acid chloride to amine) are important to avoid imide formation from the reaction of silylamides with acid chlorides ⟨91ZAAC89⟩.

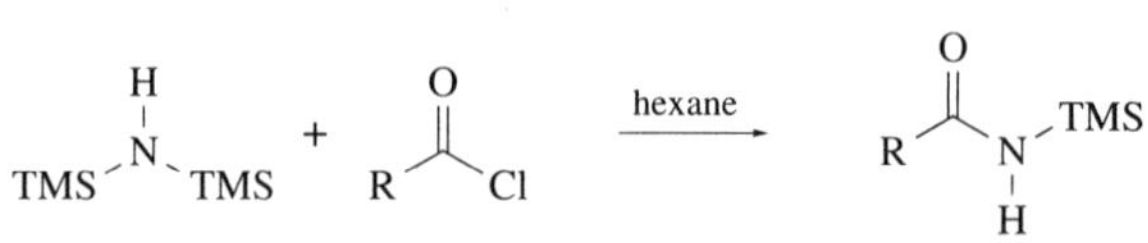

(185)

R = Me, 88%
R = Et, 76%
R = Ph, 84%

In a related process, heptamethylsilazane reacts with trifluoroacetic anhydride to provide the corresponding *N*-methylsilylamide (Equation (186)) ⟨71CB3831⟩.

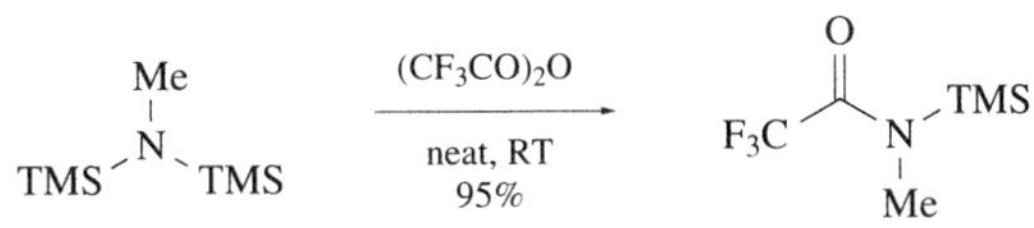

(186)

5.07.6.1.5 *From* N-*silylimines*

The reaction of ester enolates with *N*-silylimines leads to formation of β-lactams ⟨88T4157, 89CRV1447⟩. *N*-(TMS)imines provide the unsubstituted β-lactams (owing to hydrolysis upon workup ⟨91JA5742, 94BMC1865, 94S805⟩, while *N*-TBDMS imines provide the corresponding TBDMS-protected β-lactams ⟨89JCS(P1)1862⟩. Work in this area has revealed that *N*-*t*-butyldiphenylsilyl and *N*-triphenylsilylimines (prepared from the corresponding tin imines) also participate in this reaction, although lower yields of the desired β-lactams (**167a–c**) are obtained compared to TBDMS-protected imines (Equation (187)) ⟨91SL243⟩.

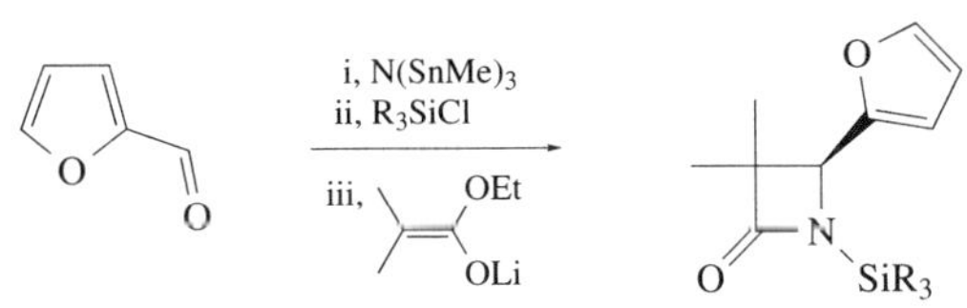

(187)

(167) **a**; SiR_3 = TBDMS, 65%
b; SiR_3 = TBDPS, 29%
c; SiR_3 = TPS, 33%

5.07.6.1.6 *From aldehydes and azides*

The addition of TMS-azide to aliphatic aldehydes leads to formation of α-(trimethylsiloxy)alkyl azides (**168**). These intermediates rearrange to TMS-protected amides upon thermolysis (Scheme (27)) ⟨75LA266⟩.

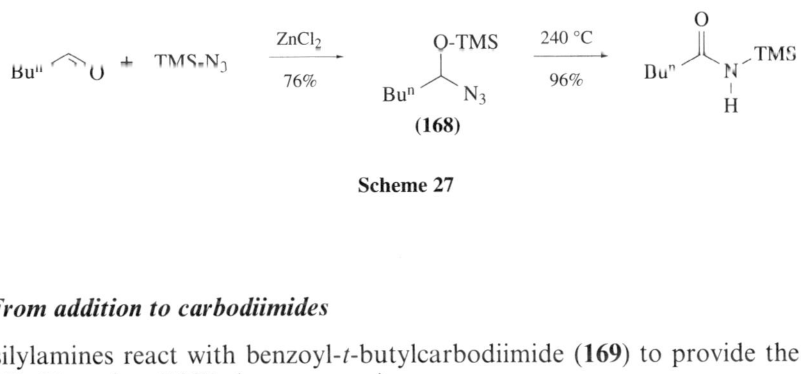

(168)

Scheme 27

5.07.6.1.7 *From addition to carbodiimides*

Trimethylsilylamines react with benzoyl-*t*-butylcarbodiimide (**169**) to provide the desired TMS derivative (**170**) (Equation (188)) ⟨74JOM(69)353⟩.

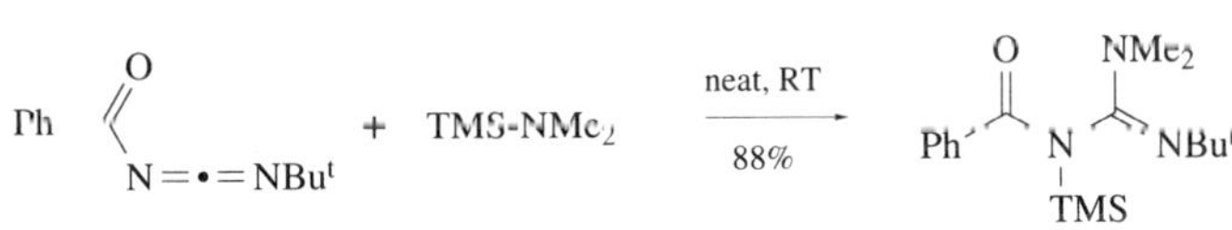

(188)

(169) **(170)**

5.07.6.1.8 *From lactamization of bis(trialkylsilyl)amides*

Knapp has shown that when appropriately substituted bis(trialkylsilyl)amides are treated with iodine, an iodolactonization occurs to provide the corresponding *N*-silylated iodolactam ⟨85TL1803⟩.

With bis(TMS)amides, the unprotected lactam is isolated; with bis(TBDMS)amides, the TBDMS-protected lactam (**171**) is obtained in 88% yield (Equation (189)) ⟨88JOC4006⟩.

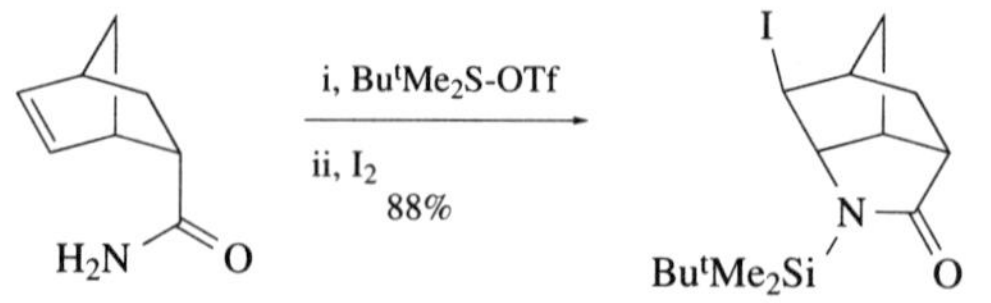

(189)

(171)

5.07.6.2 *N*-Silylimides

5.07.6.2.1 From imides

Imides react with many of the same silyl transfer agents mentioned in Section 5.07.6.1 to provide the corresponding *N*-silylimides. Table 12 lists some of the reagents and conditions used to prepare silylimides from imides.

Table 12 Silylation of imides.

Imide	*Silyl reagent*	*Conditions*	*Product*	*Ref.*
Suc-H	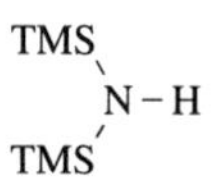	1.5 h (83%)	Suc-TMS	62M352
Phth-K	DMPS-Cl	THF, 60 °C (89%)	Phth-DMPS	71CJC1011
Phth-H	BSA	MeCN, RT, 15 min	Phth-TMS	66JA3390
Phth-H	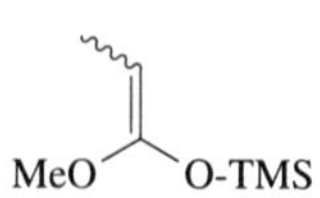	MeCN, 50 °C 10 min (95%)	Phth-TMS	79TL4311
Phth-H	TBDMS-Cl	Et_3N, CH_2Cl_2 reflux, 6 h (71%)	Phth-TBDMS	94S909
Phth-H	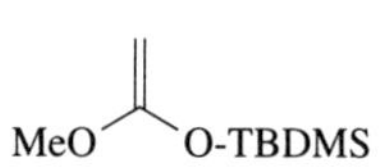	MeCN, 60 °C 15 min (99%)	Phth-TBDMS	81S451

5.07.6.2.2 From N*-haloimides*

N-Bromosuccinimide reacts with a variety of organodisilanes, providing the corresponding *N*-silylimides (Equation (190)) ⟨66BCJ2263⟩. *N*,*O*-Bis(trimethylsilyl)acetamide has also been reported to react with NBS to provide *N*-(TMS)succinimide, although no yield was reported ⟨65TL4007⟩.

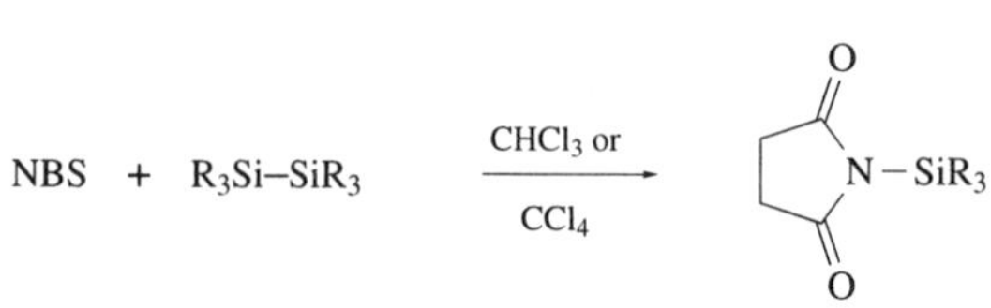

(190)

SiR_3 = TMS, 80%;SiR_3 = $SiMe_2Ph$, 77%

N-Chlorosuccinimide reacts with HMPA in the presence of TMS-Cl to provide TMS-succinimide (Equation (191)) ⟨85ZOB1080⟩.

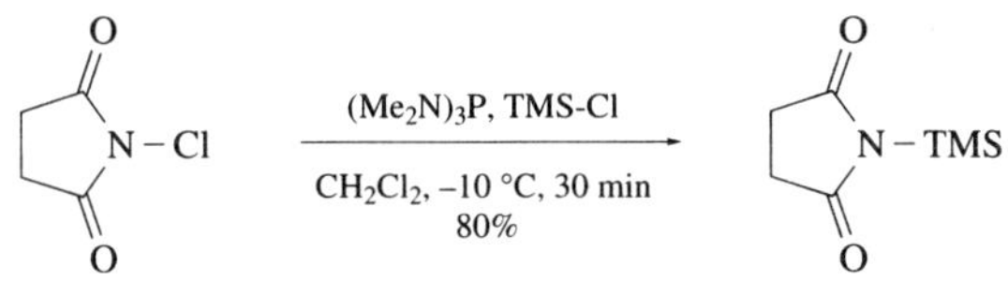

(191)

5.07.6.3 Disilylamides

All bis(trimethylsilyl)amides, except bis(trimethylsilyl)formamide, have been shown to exist in the imidate form ⟨74JA4283⟩. However, depending on the alkyl groups attached to the silicon, some disilylamides do exist in the amide form.

5.07.6.3.1 From silylation of amides

When acetamide ⟨63AG93⟩ or formamide ⟨87JOM(332)69⟩ are treated with two or more equivalents of TMS-Cl, the corresponding bis(TMS)imidates are obtained. This procedure has also been extended to the preparation of dialkylhydrosilanes (**172**) (Equation (192)) ⟨87JOM(332)69⟩.

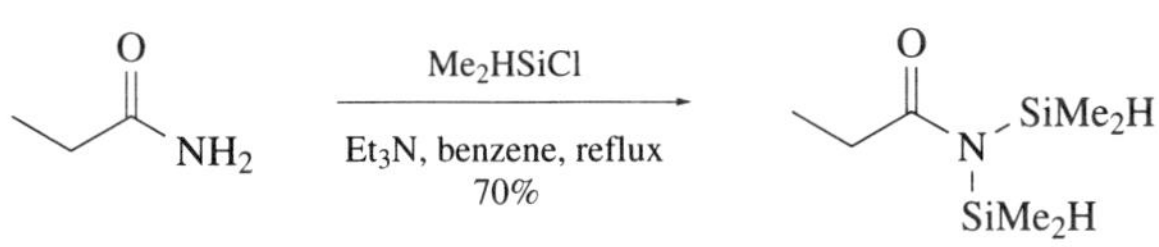

(172)

(192)

For more difficult silylamides, trans-silylation with bsa ⟨87JOM(332)69⟩ or silylation with silyl triflates has also been successful ⟨88JOC4006, 88JOC4773⟩.

5.07.6.3.2 From disilylamines

The reaction of acid chlorides with hexamethyldisilazane leads to cleavage of one TMS group, and formation of an *N*-(TMS)amide (Section 5.07.6.1.2). However, when disilylamines are converted to their lithium ⟨74JA4283, 87JOM(332)69⟩ or sodium ⟨88JOM(341)95⟩ salts prior to acylation, good yields of the corresponding disilylamides are obtained. In a related synthesis, bsa has been prepared by treating hexamethyldisilazane with ketene (Equation (193)) ⟨62RTC430⟩.

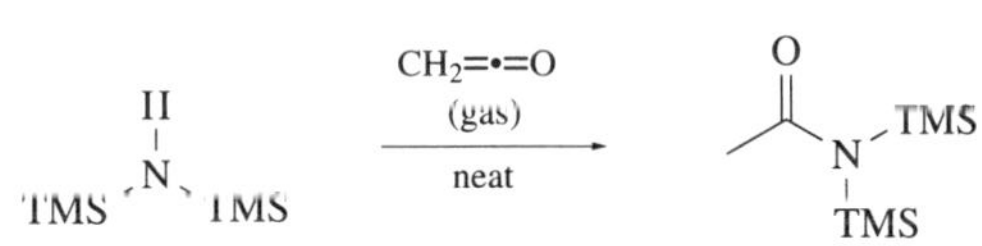

(193)

5.07.6.3.3 From N*-haloamides*

Treatment of *N*,*N*-dichlorobenzamide (**173**) with tris(*N*-ethyl-*N*-phenylamido)phosphite and TMS-Cl provides the *N*,*N*-disilylamide (**174**) (Equation (194)) ⟨85ZOB1080⟩. The steric bulk of the phosphite appears important as the same reaction employing HMPA provides a complex mixture of products.

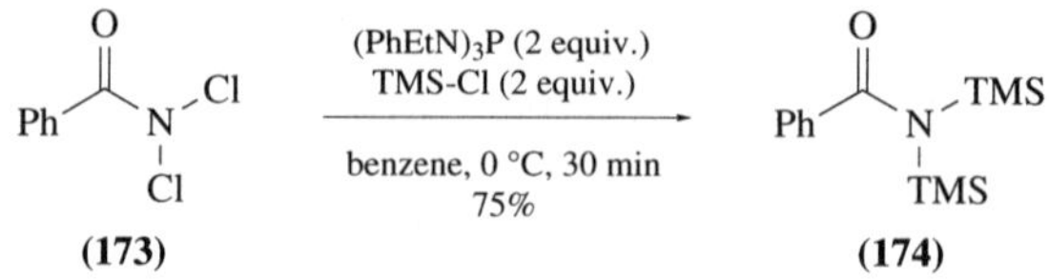

(194)

5.07.6.3.4 *From CO insertion*

Fryzuk has demonstrated that *N*,*N*-disilylamides can be prepared via CO insertion of nickel amine complexes (Equation (194a)) ⟨82OM1540⟩ see also ⟨84JA6993⟩. Similar reactions with palladium require higher pressure (20 atm).

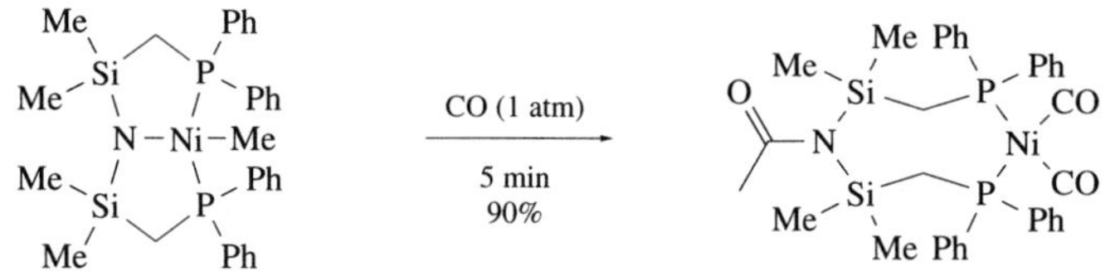

(194a)

5.07.6.4 *N*-Hetero-*N*-silylamides

Like primary amides, hydroxamic acids can be disilylated with hexamethyldisilazane at room temperature to provide the *N*,*O*-bis(trimethylsilyl)hydroxamate (**175**) (Equation (195)) ⟨80TL3367⟩.

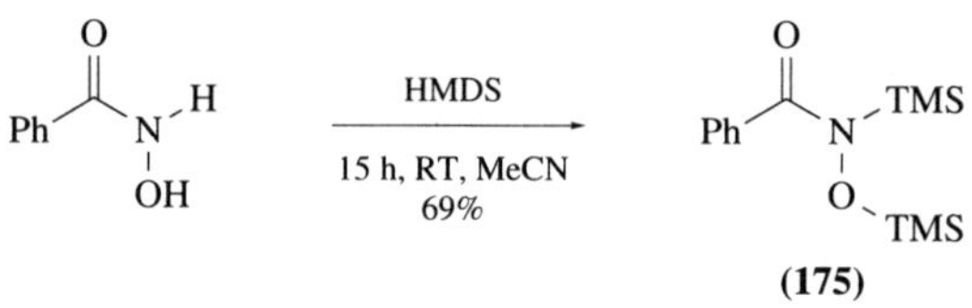

(195)

In a related fashion, 1,2-diacyl hydrazine reacts with two equivalents of TMS-Cl to provide the corresponding acyl silylhydrazine (Equation (196)) ⟨83IZV1413⟩.

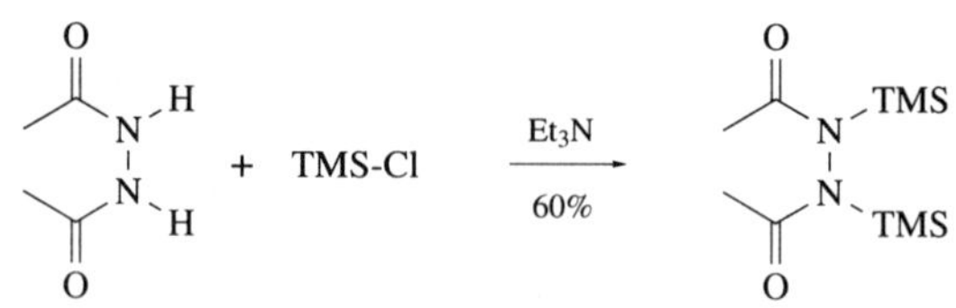

(196)

5.07.7 AMIDES SUBSTITUTED BY BORON OR ALUMINUM

As with amidosilanes, amidoboranes can exist as amides or as imidates. However, unlike silylamides, they can also exist as cyclic or linear oligomers. The substitution on boron has a dramatic effect on the aggregative properties of boronamides. Hence, while amido(amino)boranes exist as monomers ⟨78CB538⟩, amido(halo)boranes exist almost exclusively as dimers ⟨76ZN(B)1546⟩.

5.07.7.1 *N*-Boroamides

Köster has reviewed the methods used to prepare organoboranes as well as boronamides ⟨83HOU(13)87⟩.

5.07.7.1.1 From amides

The reaction between hydroboranes and tertiary amides is known to lead to amide reduction ⟨B-88MI 507-01⟩. In the case of primary or secondary amides, deprotonation occurs, and amidoborane formation takes place. Haloboranes and alkylboranes also react with the active N—H of amides to provide amidoboranes. Some amidoboranes prepared by this method, as well as by other methods, are found in Table 13. Depending on the molar ratio of borane reagent to amide, oxallyldiamides have also been converted to amidoboranes ⟨84CB2112⟩.

Table 13 Reaction of amides with boron reagents.

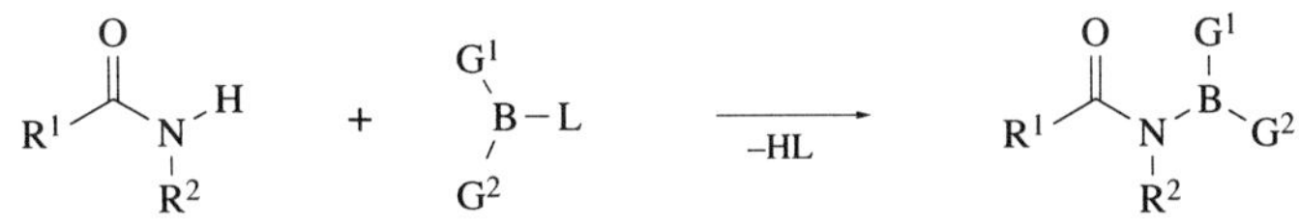

R^1CONHR^2	Conditions	$R^1CONB(G^1G^2)R^2$	Ref.
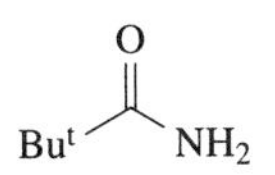	$Pr^i_2BPr^i$, THF, 3 h 50–70 °C(85%)	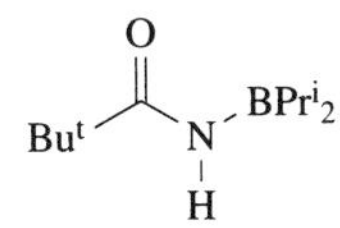	71ZOB2501
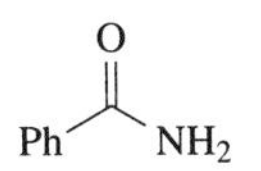	Et_2B–Et, neat, RT (87%)	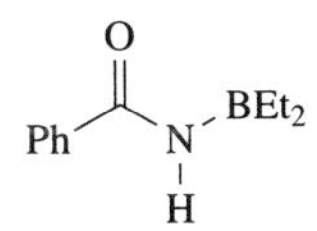	93CB565
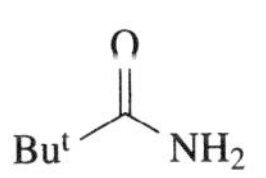	(9-BBN)-H, toluene RT (62%)	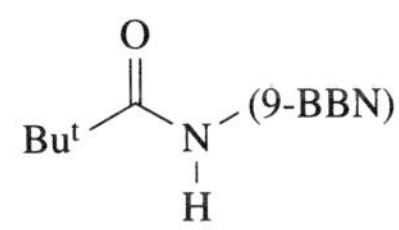	93CB565
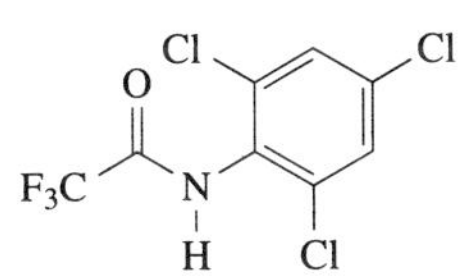	Cl_2B–Cl –30 °C to reflux (90%)	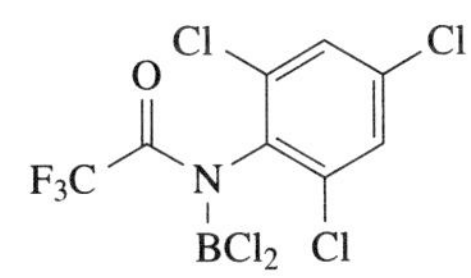	77M1027
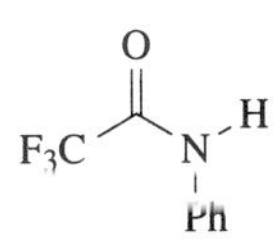	Ph_2B–Br, –30 °C chlorobenzene (44%)	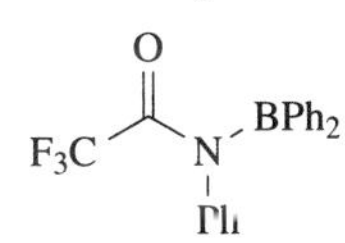	77ZAAC94
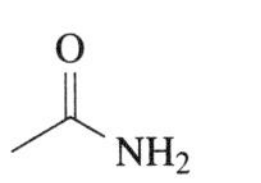	Pr^n_2B–SBu^n vacuum, 60–90 °C 1–1.5 h (70–80%)	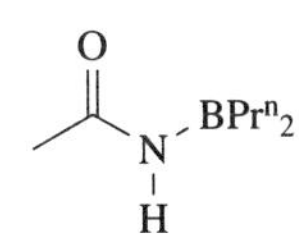	70DOK595
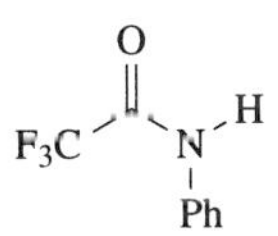	i, Bu^nLi, RT ii, $(Me_2N)_2B$–Cl hexane, reflux (85%)	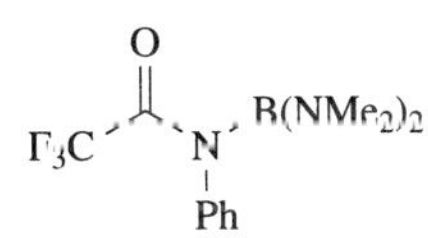	77M1279

5.07.7.1.2 From N*-silylamides*

The reaction of *N*-silylamides with trihaloboranes leads to nearly quantitative formation of amidoboranes. The conditions used to effect this transformation are reported to be milder than those required for simple amides, providing amidoboranes from both primary ⟨79M473⟩ and secondary ⟨77M1279⟩ silylamides. See Table 14.

5.07.7.1.3 From N—H insertion of amino-iminoboranes

Amino-iminoboranes are known to undergo 2+2 cycloadditions with ketones, esters, and alkenes. Similar reactions with amides ⟨93CB1119⟩, lactams ⟨90CB455⟩, or diketopiperazines

Table 14 Reaction of silylamides with boron reagents.

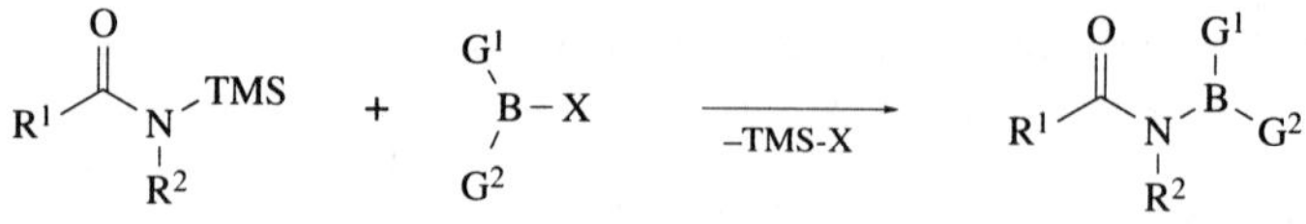

R^1	R^2	G^1G^2B—X	Conditions (Yield)	Ref.
Me	Ph	$(Me_2N)_2B$—Cl	neat, 70°C (90%)	77M1279
Ph	Me	$(Me_2N)_2B$—Cl	neat, 70°C (90%)	77M1279
Me	Me	$(Me_2N)_2B$—Cl	neat, 70°C (90%)	77M1279
CF_3	Ph	Me_2B—Br	−30°C, hexane (90%)	77ZAAC(436)173
CCl_3	Ph	Me_2B—Br	−30°C, hexane (90%)	77ZAAC(436)173
CF_3	Ph	Ph_2B—Br	hexane, RT (90%)	77ZAAC(436)173
CF_3	Ph	$(Bu^n)_2B$—Br	hexane, RT (90%)	77ZAAC(436)173
Me	Me	$(Me_2N)_2B$—NMe_2	neat	78CB538
Me	Me	$Me_2N(Ph)B$—Cl	neat	78CB538
CF_3	H	$(Me_2N)_2B$—Cl	pet. ether, reflux (85%)	79M473
CF_3	H	Me_2B—Br	pet. ether, reflux (65%)	79M473

⟨90CB455⟩ result in N—H insertion and formation of amidoboranes. Depending on the bulk of the alkyl group on nitrogen, varying amounts of amide (**176**) and imidate (**177**) can be obtained (Equation (197)) ⟨90CB63⟩.

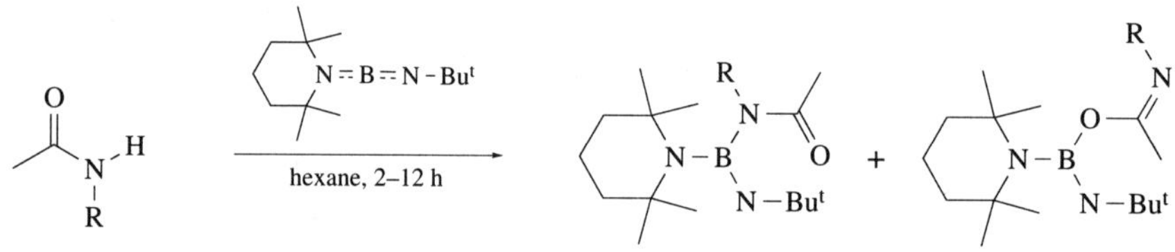

(176) (177)

(197)

Compound	R	Ratio (176):(177)	Yield (%)
a	–Me	100:0	63
b	$–Bu^t$	25:75	75

5.07.7.1.4 *From acylation of aminoborane*

Dialkylboroacetamide (**178**) reacts with a variety of aminoboranes to form a series of amidoboranes (**179**) (Equation (198)) ⟨70DOK(194)595⟩. These compounds exist as "dimers" with boron bound to nitrogen rather than oxygen.

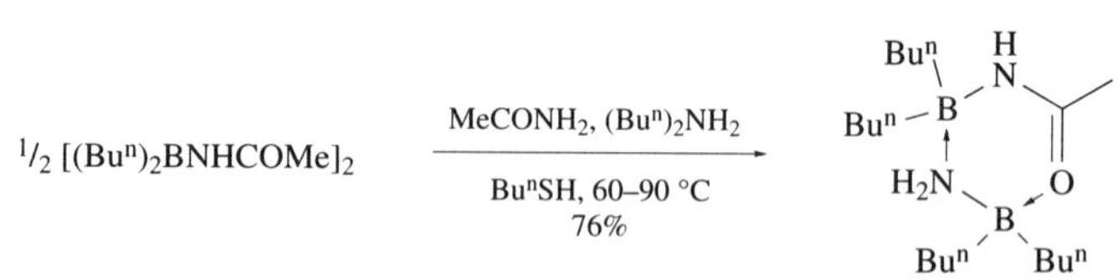

(178) (179)

(198)

The ammonia complex of tris(trifluoromethyl)borane has been reported to be acylated by a variety of anhydrides and acid chlorides. Depending on the acylating agent, deprotonation of the amine was necessary for conversion of the aminoborane complex to an amido complex (Equation (199)) ⟨93JOM(446)25⟩.

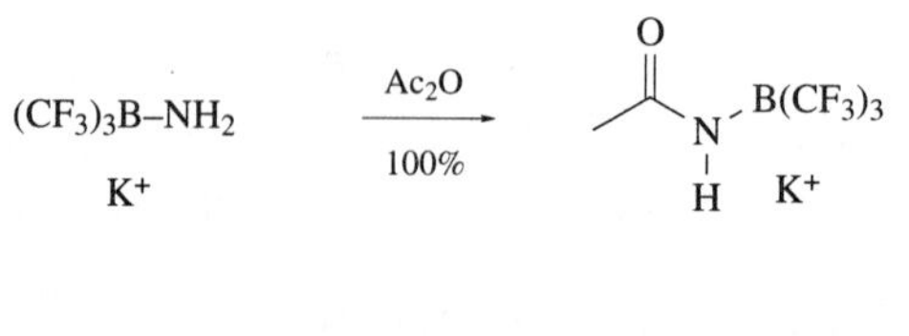

(199)

5.07.7.2 *N*-Hetero-*N*-boronamides

Treatment of 3,6-diphenyl-1,2,4,5-tetraaza-3,6-diborane (**180**) with acetic anhydride provides either the acyclic 1,2-diacyl hydrazine (**181**) or the 1,4-diacylated diborinane (**182**), depending on the reaction conditions (Scheme 28) ⟨70JOM(24)595⟩.

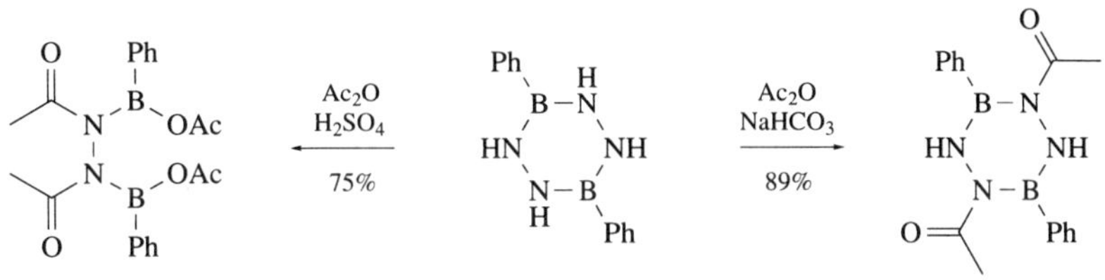

(181) **(180)** **(182)**

Scheme 28

The reaction of 1-trifluoroacyl 2,2-dimethylhydrazine with BF_3 etherate provides the corresponding *N*-aminoboroamide in 74% yield (Equation (200)) ⟨72ZAAC(394)26⟩.

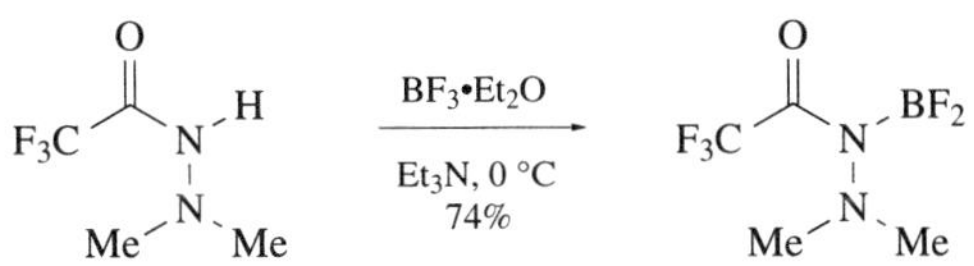

(183) **(184)** (200)

Treatment of *N*,*O*-bis(trimethylsilyl)amide with diorganohaloboranes provides the corresponding silylboroamides (Scheme 29) ⟨78ZAAC(443)148⟩. These compounds were found to be monomeric, existing as a mixture of *N*-silylamidoborane and *N*-silyliminoxyborane.

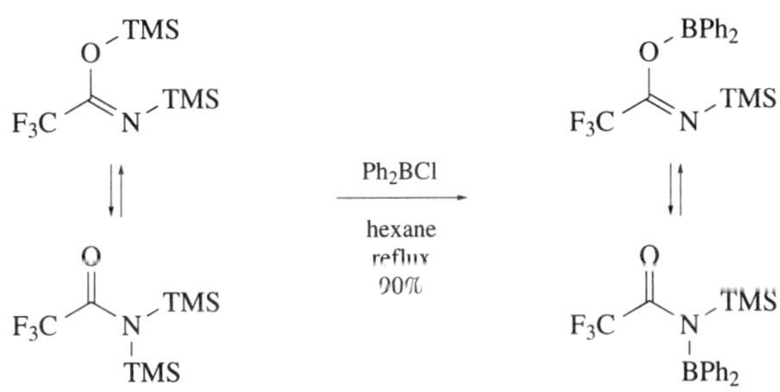

Scheme 29

5.07.7.3 Amidoaluminates

The reaction of trialkylaluminums (R_3Al) with amides leads to the formation of aluminum-substituted amides—metalated amides (Equation (201)) ⟨66MI 507-01⟩ see also ⟨68JCS(A)2535⟩. However, unlike the similar reactions with trialkylboranes, these compounds exist exclusively in the *O*-metalated imidate form.

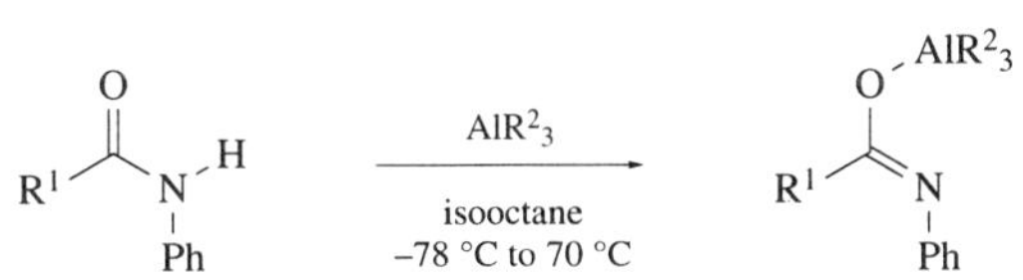

(201)

R^1 = Me, R^2 = Me, >90%
R^1 = Me, R^2 = Ph, >90%
R^1 = Bu^i, R^2 = Me, >90%
R^1 = Bu^i, R^2 = Ph, no reaction

5.07.7.4 Imidoaluminates

Aluminum chloride ($AlCl_3$), reacts with the sodium salt of phthalimide, succinimide, and benzimide to provide the corresponding imidoaluminates ⟨76JIC1230⟩. Depending on the ratio of imide to $AlCl_3$, quantitative yields of the mono-, di-, or triimide derivatives of aluminum could be obtained.

5.07.8 AMIDES SUBSTITUTED BY METALS

The methods used to prepare metalated amides and imides have been reviewed by Döpp ⟨85HOU(E5)1135⟩.

5.07.8.1 Group 1 and 2 Metal Amides

Metalated amides are usually prepared via amide deprotonation with alkyllithiums ⟨86AP(319)216, 91JOM(420)387⟩, metal hydrides ⟨79JOC1999⟩, sodium amides ⟨86AP216⟩, or, in the case of sodium, with the metal itself ⟨79AP822⟩. In cases where the nitrogen is substituted by groups that can delocalize the negative charge (phenyl ⟨91SC1889⟩, phosphate ⟨90ZOB1997⟩ or sulfonate ⟨91T7447⟩), weaker bases such as alkoxide, sodium hydroxide, and potassium carbonate have been used to deprotonate the amide. Metal amides from the group 1 or 2 metals are not generally isolated, rather they are usually prepared *in situ* and transformed further into the desired end product.

5.07.8.2 Group 1 and 2 Metal Imides

Imides are much more acidic than amides and are usually deprotonated by much milder bases such as hydroxide, or alkoxide ⟨76JIC1230⟩.

5.07.8.3 Transition Metal Amides

The field of transition metal amides continues to grow owing to the increasing understanding of their importance in research and in synthesis. A summary of the synthetic methods in this area is hindered by differences in oxidation state, coordination number, and reactivity of metals in this portion of the periodic table. However, at the risk of overgeneralizing, there are some synthetic trends that one can perhaps extrapolate to other metals.

5.07.8.3.1 From amides

The formation of metal amides from amides occurs with varying degrees of difficulty. *In situ* activation of the metal complex/amide mixture by photolysis (palladium oxalates ⟨92OM2024⟩) or thermolysis (diiron nonacarbonyl ⟨90OM560⟩) has resulted in metal amide formation.

Methylcopper derivatives (Equation (202)) ⟨80BCJ1299⟩, as well as mercury(I) carbonate ⟨70CB2388⟩, reacts with the acidic hydrogen on amides, resulting in the formation of the metal–amide bond.

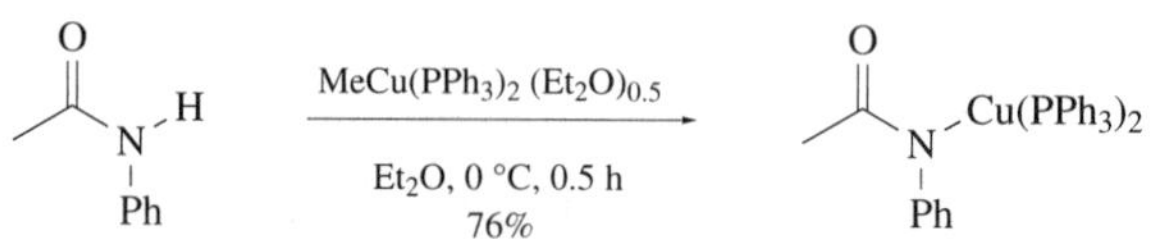

(202)

The ability of amides to displace ligands on metals constitutes one of the most widely applicable methods of metalating amides. In lieu of generalization, selected examples of amide substitution reactions are found in Table 15.

Table 15 Metallation reactions of amides and lactams.

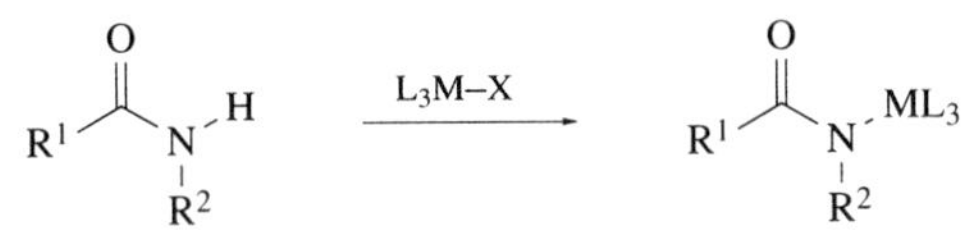

R^1	R^2	$L_3M—X$	*Conditions (Yield)*	*Product*	*Ref.*
Ph	H	$Ti(OPr^i)_3—OPr^i$(2 eq.)	reflux, 10 h	$Ti(OPr^i)_2—(NHCOPh)_2$	91MI 507-04
Ph	H	$Ti(OPr^i)_3—OPr^i$(4 eq.)	reflux, 10 h	$Ti(NHCOPh)_4$	91MI 507-04
Me	H	$Zr(OPr^i)_3—OPr^i$	reflux benzene, 12 h (90%)	$Zr(OPr^i)_3—NHCOMe$	80IJC(A)69
H	H	$[(NH_3)_5Co—OS(Me)_2](ClO_4)_3$	steam bath, 5 h (48%)	$[(NH_3)_5CoNHCOH](ClO_4)_2$	90IC20
CF_3	H	$[(NH_3)_5Co—OS(Me)_2](ClO_4)_3$	DMSO, TMP[a], 70°C 2 h (60–80%)	$[(NH_3)_5CoNHCOCF_3](ClO_4)_2$	90ICA203
Me	H	$[(NH_3)_5Co—OS(Me)_2](ClO_4)_3$	DMSO, TMP[a], 60–80°C (100%)	$[(NH_3)_5CoNHCOMe](ClO_4)_2$	91IC1564
Me	Ph	$Cu(PPh_3)_2—OEt$	Et_2O, 0°C (65%)	$Cu(PPh_3)_2N(Ph)COMe$	78BCJ2909
(2-OH)Ph	H	PdCl—Cl	0.1 mol l^{-1} HCl, MeOH(aq.), 30 min	$Pd(NHCOPh(2\text{-}OH))_2$	84MI 507-01
Bu^t	H	$[(Ph_3PAu)_3O]BF_4$	THF EtOH, K_2CO_3, 20°C (59%)	$Ph_3PAuNHCOBu^t$	88ZOB62
Me	H	$[(3\text{-}F)C_6H_4]Hg\text{-}OH$	warm MeOH (75%)	$[(3\text{-}F)C_6H_4]HgNHCOMe$	72JOM(36)227

[a] TMP = 2,2,6,6-tetramethylpiperidine.

Metals can form cyclic complexes with amides that possess an appropriately spaced nucleophilic moiety. Amides containing pendant phosphine groups form excellent complexes with rhodium ⟨86IC9⟩, iridium ⟨86IC9⟩, palladium ⟨86OM336⟩, and platinum ⟨86OM336⟩. Formation of complexes between α-amino acids and the appropriate nickel ⟨88T5507, 89JOC1584⟩, copper ⟨86JMC199⟩, and palladium ⟨89IC815⟩ metal are also reported. Nickel and copper bind especially well to amino acids containing an attached Schiff base ⟨85CC171, 86BCJ2203, 88JCS(P)305, 93S117⟩. Such compounds are prepared in one pot upon addition of the appropriate ketoamide, amino alcohol, and metal catalyst (Equation (203)) ⟨85JA4252⟩.

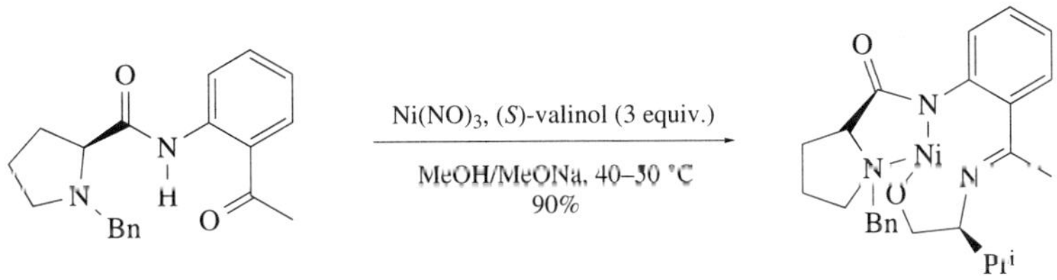

(203)

5.07.8.3.2 *From nitrile hydrolysis*

A convenient route to *N*-bonded metal amide isomers involves the partial hydrolysis of metal-coordinated nitriles. In comparison to the hydrolysis of uncoordinated nitriles, rate accelerations as high as 6–7 orders of magnitude have been reported for nitriles coordinated to transition metal complexes ⟨73JA5649, 74CJC1762⟩. The binding of such complexes usually occurs at the nitrogen of the amide, although an equilibrium between *N*- and *O*-bonded isomers has been observed ⟨92IC2318⟩. With the exception of complexes containing pendant amine groups (which can rearrange to the amine complex ⟨93IC5285⟩), the complexes are relatively stable. A survey of the amide complexes formed by this method is found in Table 16. As the table indicates, a number of functionalized nitriles are tolerated, and a number of metal derivatives can be prepared.

Table 16 Hydrolysis of coordinated nitriles.

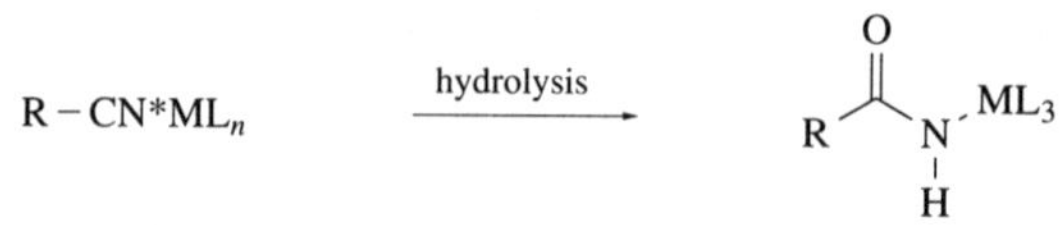

*R—CN*ML$_n$*	*Conditions (Yield)*	*Product*	*Ref.*
$[(NH_3)_5RuNCMe][CF_3SO_3](ClO_4)_2$	1 mol l^{-1} NaOH, 10 min, warm	$[(NH_3)_5RuNHCOMe](ClO_4)_2$	84JA625
$[(NH_3)_5RuNC(3\text{-}PyH)](ClO_4)_3$	$HClO_4$, $Na_2S_2O_8$ 5 h (67%)	$[(NH_3)_5RuNHCO(3\text{-}PyH)](ClO_4)_3$	91IC1862
$[(NH_3)_5RuNCCO_2Et](PF_6)_2$	AgO, NH_4PF_6	$[(NH_3)_5RuNHCOCO_2Et](PF_6)_2$	74CC622
$[(NH_3)_5Co{-}NCPh](ClO_4)_3$	1 eq H_2O, 1 eq NaOH	$[(NH_3)_5CoNHCOPh](ClO_4)_2$	91IC1564
$[(NH_3)_5Co{-}NCMe](ClO_4)_3$	$NaBH_4$, 6 min, 25°C	$[(NH_3)_5CoNHCOMe](ClO_4)_2$	75CC974
$[(NH_3)_5Co{-}NC{-}(2\text{-}CN)Ph](ClO_4)_3$	1 eq NaOH	$[(NH_3)_5Co{-}NHCO{-}(2\text{-}CN)Ph](ClO_4)_2$	79IC937
$[(NH_3)_5Co{-}NCCH{=}CH_2](ClO_4)_3$	H_2CO_3 buffer, 25°C	$[(NH_3)_5Co{-}NHCOCH{=}CH_2](ClO_4)_2$	81JA3559
$[(NH_3)_5Rh{-}NCMe](ClO_4)_3$	NaOH (aq.) (CO_2-free) 25°C	$[(NH_3)_5RhNHCOMe](ClO_4)_2$	75IC42
$[(NH_3)_5IrNCMe][CF_3SO_3](ClO_4)_2$	warm NaOH, 30 min (66%)	$[(NH_3)_5IrNHCOMe](ClO_4)_2$	84JA625
cis-$[Pt(C_6F_5)_2(NCPh)_2]$	20% NBu_4OH acetone, 30 min	$[NBu_4][Pt(C_6F_5)_2(NHCOPh)(H_2O)$	92JCS(D)53
$[PtPh(PMe_2Ph)_2NCPh]BF_4$	KOH, PhCN, H_2O, 2 h (60%)	*trans*-$[PtPh(PMe_2Ph)_2NHCOPh]$	80JOM(202)107
$K_2[Cl_3PtNCMe]$	i, NaOH, 0°C, 1 h; ii, 3 mol l^{-1} HCl (85%)	$K[Cl_3PtNHCOMe]$	93JA5123
CH_3CHFCN	$Hg(OAc)_2$, H_2O 8–10 h, 40–60°C (100%)	$CH_3CHFC(O)NHHgOAc$	74ZOR1428

5.07.8.3.3 *From acyl azides*

The reaction of acyl azides with hydrorhenium ⟨74JOM(76)355, 75JOM(88)237, 76JCS(D)1777⟩, cobalt ⟨70CC100⟩, palladium ⟨75JOM237⟩, or platinum ⟨70CB583⟩ metal complexes leads to formation of the corresponding amide complexes (Equation (204)).

(204)

5.07.8.3.4 *From isocyanates*

The cobalt(I)- and nickel(0)-catalyzed reaction of isocyanates with alkynes ⟨83JOM(252)359, 84JOC4786⟩, dienes ⟨85AG(E)961, 87JOM(328)403, 87JOM(327)429⟩, allenes ⟨85JOM(295)C21⟩, alkenes ⟨88JOM(344)C35⟩, or alkanes ⟨85JOM(288)237, 86JOM(315)245, 88JOM(339)213, 88JOM(344)C35, 89JOM(375)245, 89JOM(378)279, 90JOM(382)C6, 91JOM(412)225, 91JOM(403)401⟩ proceeds through the intermediacy of *N*-metal amides. In certain cases where a stoichiometric zinc complex has been used, it has been possible to characterize the metal–amide complex as an equilibrium mixture of metalocyclic (**185**) and ring-opened (**186**) forms (Scheme 30) ⟨85AG(E)961⟩.

5.07.8.3.5 *From metal oxime*

Under thermolytic conditions, ruthenium oximate (**187**) isomerizes to the corresponding amide complex (**188**) (Equation (205)) ⟨89JA3258⟩.

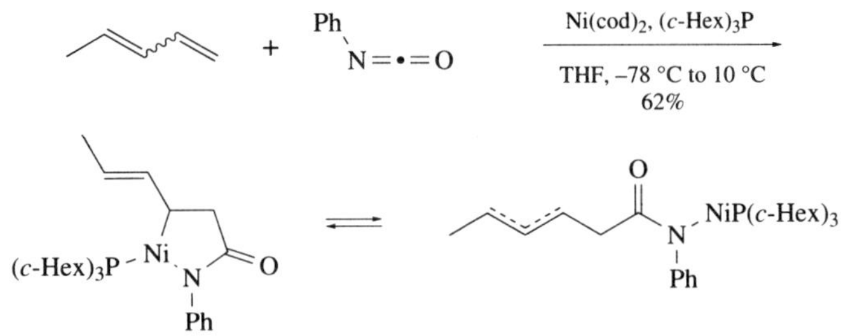

(185) (186)

Scheme 30

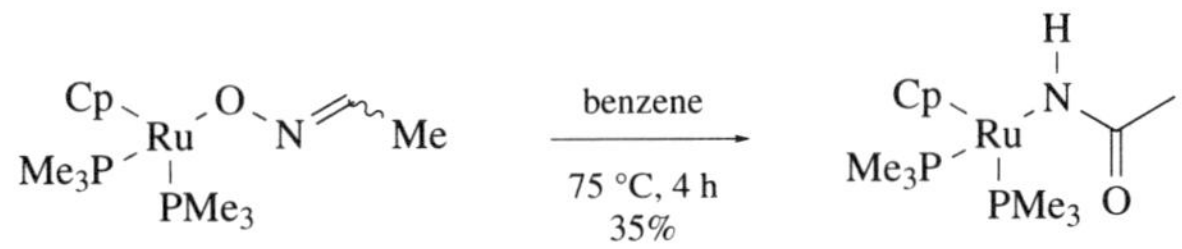

(187) (188) (205)

5.07.8.3.6 From dichloroamides

Reaction of 2,3,4,5-tetrakis(methoxycarbonyl)cyclopentadienebiphenylphosphinoethanepalladium (**189**) with *N*,*N* dichloro-*p*-toluamide (**190**) results in palladium–alkyl bond cleavage and formation of the *N*-chloro-*N*-palladium amide (**191**) (Equation (206)) ⟨75CL197⟩. The reaction of the same dichloroamide with diiron nonacarbonyl provides an *N*-acyl iron nitrene ⟨75CL641⟩.

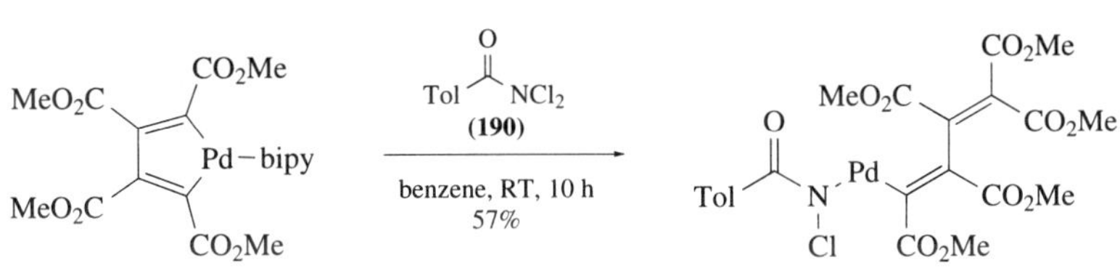

(189) (191) (206)

5.07.8.4 Transition Metal Imides

5.07.8.4.1 From imides

Rhodium(I) and iridium(I) imide complexes of phthalimide and succinimide have been prepared from the corresponding chlorides ⟨86JCS(D)813⟩. Triflate complexes of cobalt also react with succinimide to provide the corresponding imido complexes ⟨91IC4806⟩. The iron complex (**192**) reacts with succinimide and phthalimide under photochemical conditions to provide the corresponding imide derivatives (**193**) (Equation (207)) ⟨89JOM(359)215⟩.

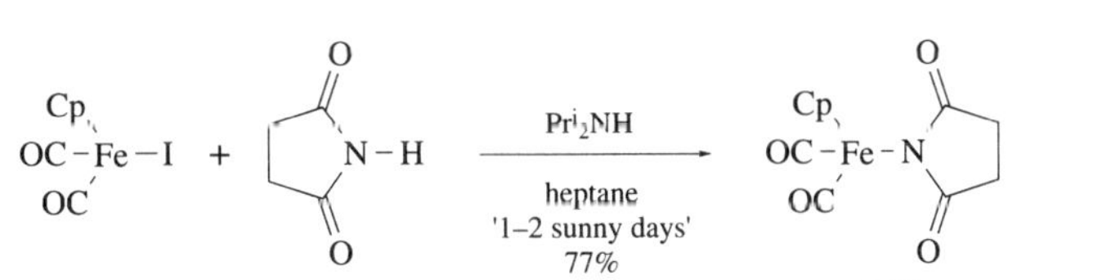

(192) (193) (207)

In addition to displacement reactions, certain dialkylnickel ⟨81BCJ1720⟩, arylpalladium ⟨86JCS(D)813⟩ and alkylcopper complexes ⟨80BCJ1299⟩ react with the acidic NH moiety of imides to provide the corresponding metal imide complexes. In a related fashion, mercury(I) carbonate with imides provideds the mercury imides ⟨70CB2388⟩.

5.07.8.4.2 *From haloimides*

N-Bromosuccinimide reacts with a variety of mercury alkyls, splitting off alkyl halides and producing mercury imides (Equation (208)) 〈62ZN(B)135〉.

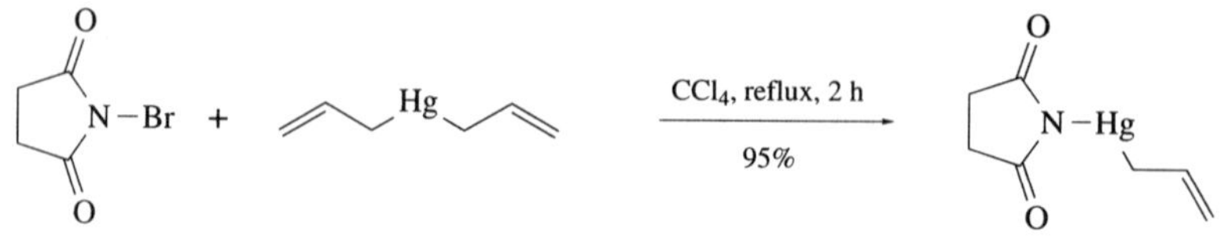

(208)

5.07.8.4.3 *From transmetalation*

Silver succinimide reacts with bis(η-cyclopentadienyl)vanadium (**194**) to provide the corresponding vanadium imide (**195**) (Equation (209)) 〈85ZOB1071〉 and silver metal. The thallium salt of succinimide reacts with bromopentacarbonylmanganese to provide the corresponding metal imide 〈86JCS(D)813〉.

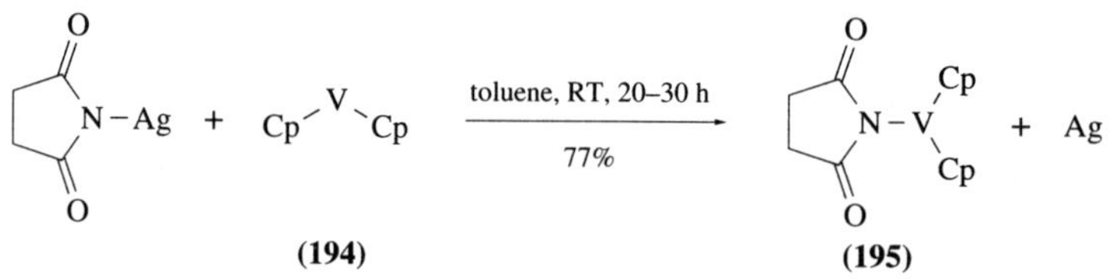

(209)

5.07.8.4.4 *From sulfinylimides*

Platinum π-allyl compounds react with *N*-(alkyl or arylsulfinyl)phthalimides (**196**) to provide the sulfinato complexes (Equation (210)) 〈92ZN(B)1165〉. Such coordinated sulfinic acids do not appear to equilibrate between their *S*-bound and *O*-bound forms.

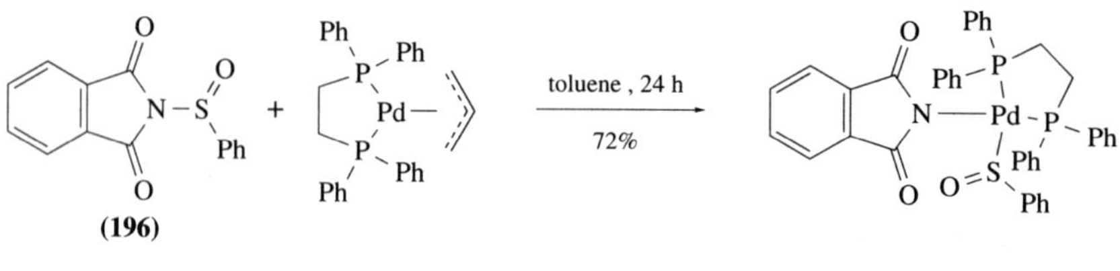

(210)

5.07.8.5 Group 13 Through 16 Metal Amides

The synthesis of N-germylamides has been reviewed 〈92MI 507-01〉.

5.07.8.5.1 *From amides*

The reaction of amides with a variety of trialkyl metal derivatives under basic conditions provides the corresponding metal amides 〈76JOM(107)63〉 (Equation (211)) 〈74JOM(69)353〉.

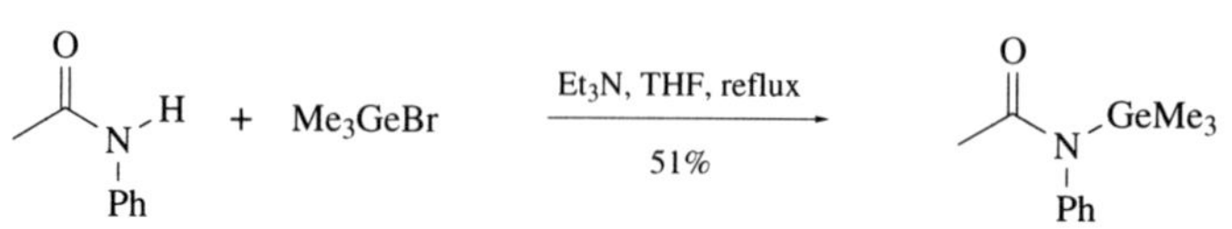

(211)

Amides substituted by trialkylgermanium 〈74JOM(69)353, 88EJM257〉 exist exclusively in their *N*-metalated form. This is in contrast to silylamides, where the difference in bond energy between Si—O and Si—N is large enough to lead to exclusive imidate formation or equilibrium mixtures of

N-silylamides and imidates. Tin amides also show a strong absorption band near 1650 cm^{-1}, indicating that substitution of the amide is at nitrogen rather than at oxygen ⟨77JOM(142)165⟩.

In addition to trialkyl derivatives, dialkyl derivatives of germanium ⟨88EJM257⟩ and tin (Equation (212)) ⟨73JOM(63)221⟩ are converted to the corresponding dialkyldiamide derivatives. Reactions of tetrahalotin compounds with amides have also been reported ⟨69ICA495⟩.

$$R_2Sn(OEt)_2 + MeCONH_2 \xrightarrow[\text{4 h, reflux}]{\text{benzene}} R_2Sn(NCOMe)_2 \quad (212)$$

$$R = Et, 99\%$$
$$R = Bu^n, 97\%$$

5.07.8.5.2 *From transmetalation*

The formation of the strong Si—Cl or Si—O bond serves as a driving force for the reaction of metal chlorides and alkoxides with silylamides. Tin-substituted amides (**198**) are obtained from silylated amides (**197**) and trialkyltin alkoxides ⟨74ZOB1753, 76ZOB594⟩. These compounds can be converted to the corresponding germanium derivatives (**199**) upon treatment with trimethylgermanium bromide (Scheme 31) ⟨73ZOB603⟩.

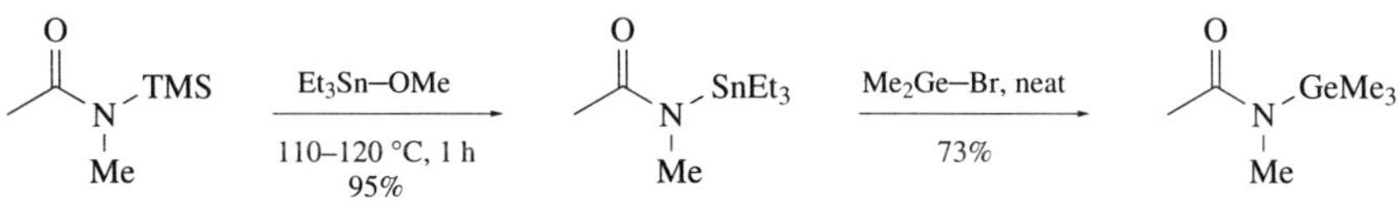

(**197**) (**198**) (**199**)

Scheme 31

5.07.8.5.3 *From nitrile hydrolysis*

Nitriles react with antimony pentachloride and carboxylic acids to provide the corresponding amide and imide complexes ⟨76ZOB75⟩.

5.07.8.5.4 *From acylation of metalated amines*

Triethylgermylphenylamine reacts with ketene (60% yield) or diphenylketene (76%) to provide the corresponding *N*-phenyl derivatives ⟨68BSF4093⟩.

Trimesitylgermylamine (**200**) reacts with acid chlorides to provide the corresponding *N*-germylamides (**201**) (Equation (213)) ⟨92JOM(423)C5, 92PS(70)75⟩.

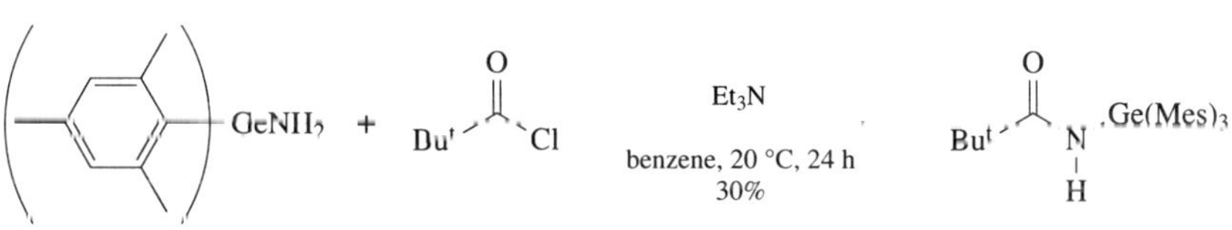

(213)

(**200**) (**201**)

The reaction of anhydrides with bis(trimethylstannyl)pentafluoroaniline (**202**) also provides *N*-stannylated products (Equation (214)) ⟨78IJ132⟩.

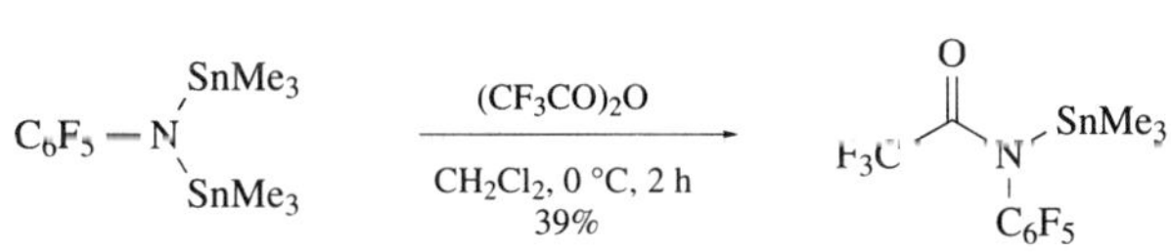

(214)

5.07.8.5.5 *From bis(trialkylstannyl)carbodiimide*

Anhydrides and acid chlorides have been shown to react with bis(trialkylstannyl)carbodiimides such as bis(trimethylstannyl)carbodiimide ⟨76SRI345⟩, bis(tributylstannyl)carbodiimide ⟨78ZOB1146,

88IZV859⟩, and bis(triphenylstannyl)carbodiimide ⟨74JOM(73)319, 75JOM(93)325⟩, to provide the corresponding *N*-substituted *N*-(trialkylstannyl)cyanamides (**202**) (Equation (215)).

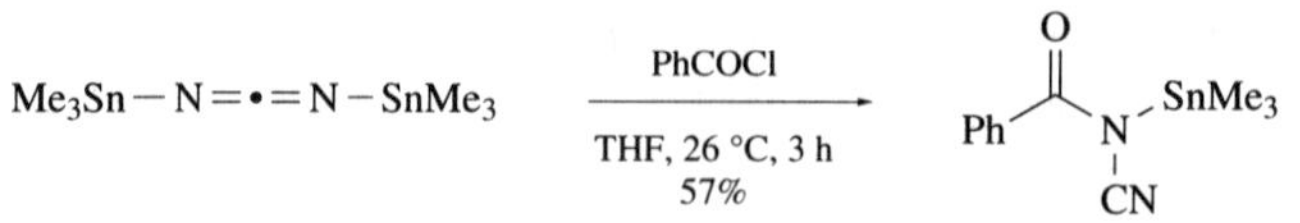

(202)

(215)

5.07.8.5.6 From addition to carbodiimide

In analogy to silicon (see Section 5.07.6.1.7 the addition of trimethylgermanium- or trimethyltindimethylamine to benzoyl *t*-butylcarbodiin. provides the corresponding *N*-metalated amide ⟨71JCS(C)1870, 74JOM(69)353⟩.

5.07.8.5.7 From addition to isocyanates

Triphenylallyltin and triphenylallyllead react with phenylisocyanate to provide *N*-metalated amides ⟨84IJC(A)952⟩. In a related transformation, phosphomethyltriorganostannanes (**203**) add to phenylisocyanate (**204**) to provide *N*-(trialkyltin)amides (**205**) (Equation (216)) ⟨79JOM(182)465⟩.

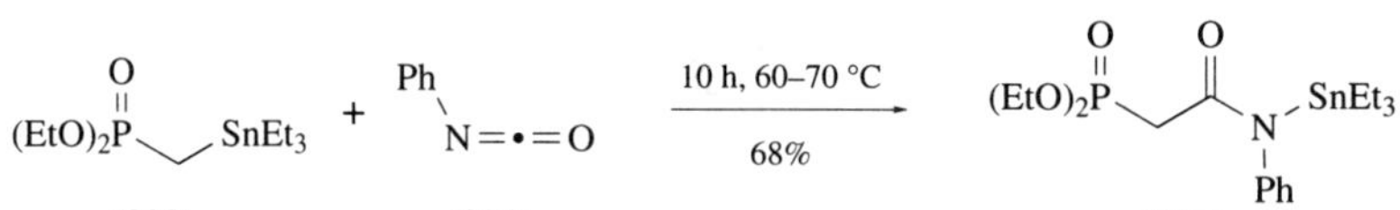

(203) **(204)** **(205)**

(216)

5.07.8.5.8 From ring opening of metal imides

Treatment of certain *N*-(trialkylstannyl)imides (**206**) with amines results in ring opening and formation of *N*-(trialkylstannyl)amides (**207**) (Equation (217)) ⟨80ZOB886⟩.

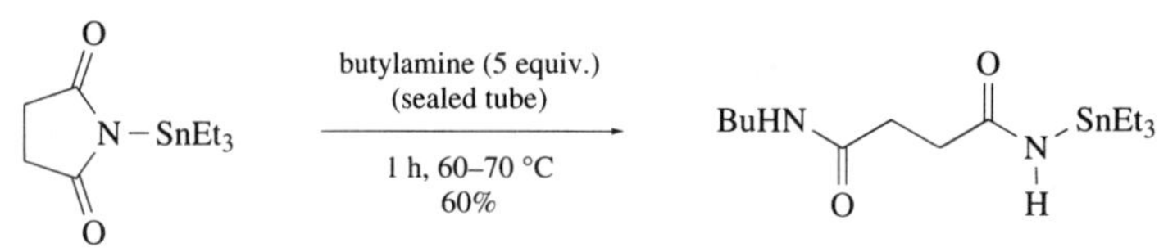

(206) **(207)**

(217)

5.07.8.5.9 From rearrangement of trialkylgermanium amides

As seen earlier with silicon (Section 5.07.6.1.6), trialkylgermaniumacetamides can be thermally converted to *N*-germylamides (Equation (218)) ⟨70ZOB707⟩.

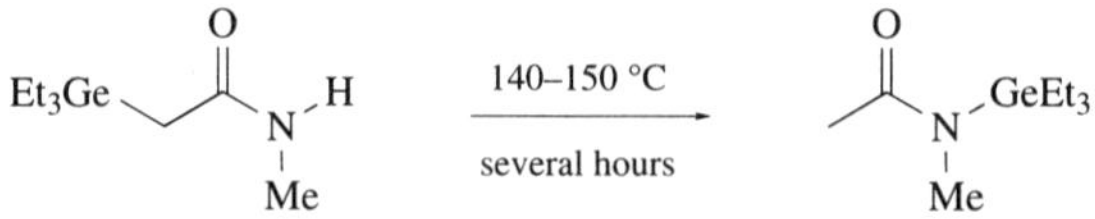

(218)

5.07.8.6 Group 13 through 16 Metal Imides

5.07.8.6.1 From imides

Imides such as phthalimide are converted to the corresponding tributyltin imides upon refluxing with tributyltin oxide with azeotropic removal of water (Equation (219)) ⟨83JOM(255)295⟩. This process has been used to prepare a variety of *N*-trialkyltin imides ⟨77ZOB2294, 81EUP54516, 81ZOB1318⟩.

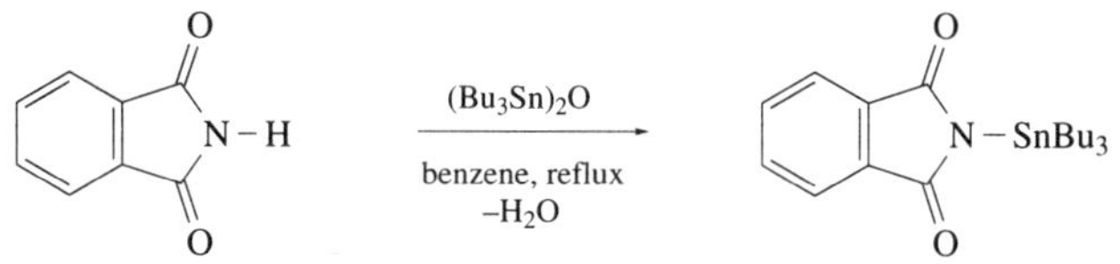

(219)

Imides also react with a variety of other metalated species, such as tributyltin isocyanate ⟨88IZV859⟩, to provide the corresponding tributyltin imide.

5.07.8.6.2 From transmetalation

Treatment of the sodium salts of imides with bromotriethylgermane provides the corresponding *N*-germylated imide ⟨87ZOB1199⟩. *N*-Germylphthalimide can be prepared from the corresponding potassium or silver phthalimide upon treatment with bromotriethylgermane (Equation (220)) ⟨87ZOB1643⟩. These derivatives can be further converted to the corresponding tin or lead imides upon treatment with the appropriate trialkyl metal chloride ⟨87ZOB1643⟩.

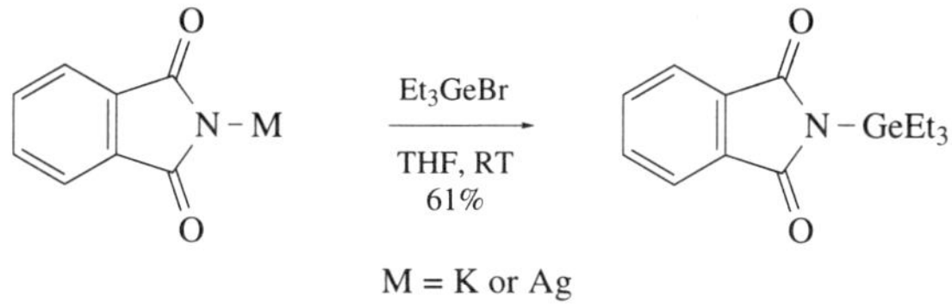

(220)

5.07.8.6.3 From haloimides

N-Chlorosuccinimide reacts with both tetraalkyllead and tetraalkyltin compounds, forming alkyl chlorides and providing trialkylmetal imides (Equation (221)) ⟨87IJC(A)1046⟩.

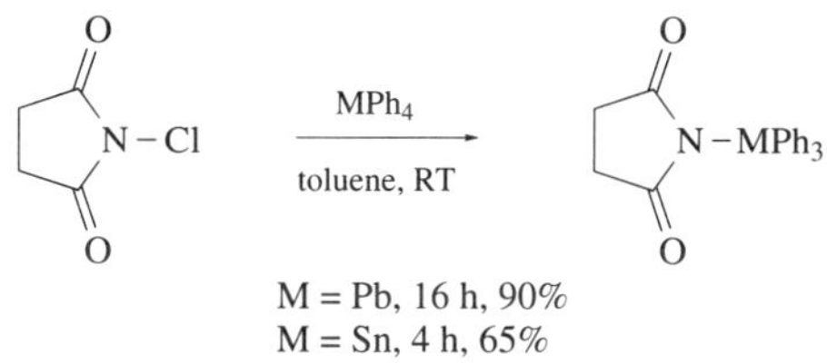

(221)

5.07.8.6.4 From tin isocyanates

Succinic, maleic, and phthalic anhydrides react with tributyltin isocyanate to provide the corresponding *N*-tributylstannyl imides (Equation (222)) ⟨88IZV859⟩. Tributylgermylphthalimide is also prepared by this method from tributylgermanium isocyanate and phthalimide in 45% yield.

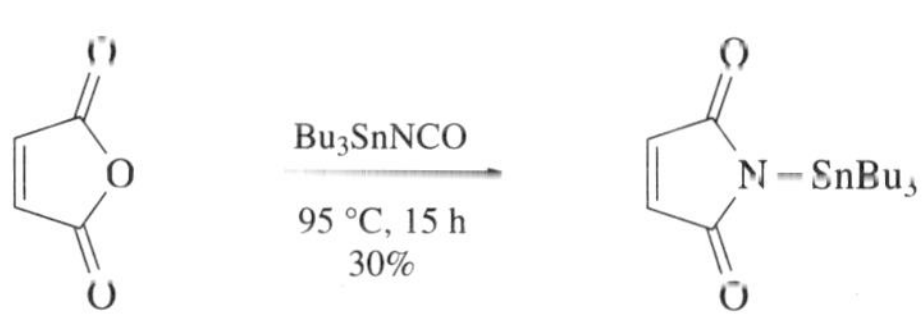

(222)

5.07.8.7 Group 13 through 16 Dimetal Amides

Heating acyl azides with hexabutyldistannane provides *N*,*N*-bis(tributylstannyl)amides in good yield (Equation (223)) ⟨75ZOB2684⟩.

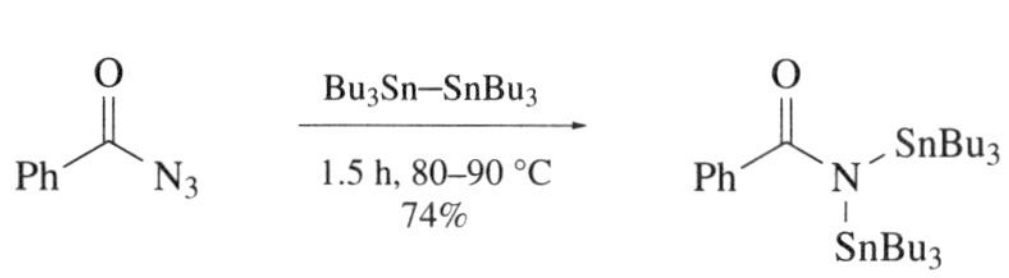

(223)

(208)

5.08
Acylphosphorus, -Arsenic, -Antimony and -Bismuth Functions

KAMYAR AFARINKIA
Kings' College, London, UK

and

M. VICTORIA VINADER
Glaxo-Wellcome Medicines Research Center, Stevenage, UK

5.08.1 INTRODUCTION

The key to understanding the chemical behaviour of acylphosphorus species is the realization that in essence this is a twin functionality comprising a carbonyl and a phosphinyl or phosphonyl group. The chemical behaviour of the functional group as a whole is the sum of the contribution of each of the two constituent moieties. Not only is the chemistry of the carbonyl group within these molecules influenced by the attached heteroatom-bearing functionality, but also, the acyl/aroyl moiety profoundly influences the behaviour of the phosphorus (or another group 15 element) containing functional group.

Acyl- and aroylphosphorus compounds (also referred to as α-ketophosphorus compounds in some of the literature) comprise a group of structurally and chemically diverse entities. The structural diversity arises since phosphorus, as any other group 15 element, can be in different oxidation states or coordinated to by various numbers of ligands, and bear various single or double bonded heteroatom or carbon substituents. These different phosphorus functional groups have varied electronic and steric properties which will influence the adjacent carbonyl function. Thus, the diversity of chemical behaviour is a consequence of the diversity of structure.

However, these differing phosphorus functional groups do have common chemical features. Hence, notwithstanding the structural variations referred to above, a gross chemical analogy can sometimes be made between acylphosphorus species themselves and other phosphorus and nonphosphorus containing functional groups. Most important of these are the parallels that can be drawn with the chemistry of ketones, amides, acid anhydrides or other related functionalities on the one hand, and nonacyl substituted phosphines and phosphonic acids on the other. Such analogies allow us to define the known and expected chemical behaviour of the species as a logical extension of that of other better known functional groups.

Throughout this chapter, phosphorus will feature more frequently than the other group 15 elements. This reflects the relative lack of knowledge about arsenic, antimony and bismuth functionalities. For most parts, it may be assumed that the chemistry of phosphorus containing functionalities is representative of that of the corresponding functional groups of the other elements. Of course, as the chemistry of these other elements is further investigated, the validity of this assumption may come to be questioned.

Various general synthetic routes to acylphosphorus compounds, and their arsenic, antimony and bismuth analogues do exist. Most syntheses, for instance, can be described as the reaction of a nucleophilic phosphorus species with an acylating agent. Indeed, a variety of nucleophilic phosphorus species and many variations on acylating electrophilic agents have been employed. Since both the carbonyl and phosphine/phosphonyl constituents of this functionality have diverse chemistry of their own, functional group transformations or interconversion between functionalities also provide an important synthetic access. This subject has been reviewed ⟨80RCR843⟩ and readers are referred to the references cited therein for further detailed information.

Other 'nonclassical' routes, for example oxidation of α-methylenephosphorus species and related reactions, have also been explored and will be discussed in relevant sections. There appears to be no other established methodology for the synthesis of these species.

Within the scope of this book, various functional groups have been divided according to structure and coordination. To emphasize the chemical trends in this area, in each relevant division synthetic approaches are discussed according to the distinctions mentioned above. The caveat is, however, that because of the inherent differences in the reactivity of each function, a synthetic route for one may not be always applicable to others, or an otherwise generic synthetic approach may fail for one of the functional groups.

Finally, it should be noted that a good number of primary and secondary acylphosphines are only isolated in their enol form with alkyl, boryl or silyl substituents at oxygen. These P=C containing functionalities have a diverse chemistry of their own and are particularly prone to dimerization ⟨92H(34)369⟩.

5.08.2 DICOORDINATED PHOSPHORUS FUNCTIONS

There appears to be no dicoordinated acyl phosphorus species with atoms other than carbon attached to the phosphorus. The reason for this is unclear but may have to do with the inability of the phosphorus atom to take part in second order bonds with highly electronegative atoms whilst attached to an electron withdrawing acyl moiety.

5.08.2.1 Functions With a P=C Bond

Phosphatrifulvenes are isolable molecules prepared from the parent cyclopropenone (Scheme 1). The phosphorus atom in phosphatrifulvenes is highly electron rich and can coordinate to metals, conjugately add to α,β-unsaturated esters, add to ketenes, isocyanates, etc., and be alkylated or acylated with alkyl halides/carboxylic acid halides ⟨91CB2843, 91S1099⟩. This nucleophilic behaviour is unlike that of other phosphaalkenes, where it is expected that the bond is polarized inversely, with a partial negative charge residing on carbon and a partial positive charge residing on phosphorus.

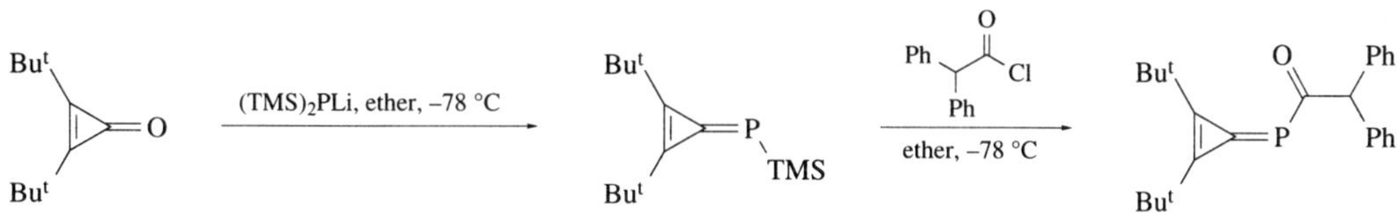

Scheme 1

There is spectroscopic (NMR) evidence that dibenzoylphosphine (**1**) and (**2**) exists in solution exclusively in the enolized form, although methylation of this molecule afforded only the *P*-methylated compound (see Section 5.08.3.1) (Scheme 2). No reports of attempted silylation or boronylation of bisacylphosphines are found, though both the boronylated ⟨88JGU201, 89JGU1817, 90BAU2424, 91PS(55)133⟩ and the silylated ⟨81ZAAC(473)7, 82ZAAC(495)73, 88JGU201, 88JGU2372⟩ enol forms of acylphosphine, for example (**3**), have been synthesized from the parent (Scheme 2).

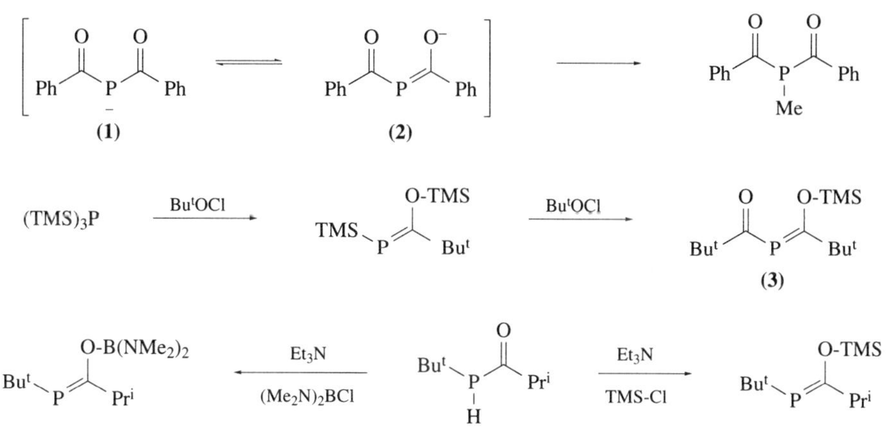

Scheme 2

5.08.3 TRICOORDINATED PHOSPHORUS FUNCTIONS

Tricoordinated acylphosphorus, -arsenic, -antimony and -bismuth functions can be considered as direct analogues of amides. Indeed, many direct analogies can be made between the physical ⟨68T2995⟩ and chemical ⟨66ZAAC(346)241⟩ properties of these functionalities and those of amides. There is strong spectroscopic evidence ⟨68T2995, 84TL1249⟩ that an amide type resonance does not appear to take place in acylphosphides which results in unrestricted C—P bond rotation in this species. This observation is also supported by *ab initio* theoretical calculation ⟨86JST(139)255⟩. The greater reactivity of acylphosphines towards electrophiles also distinguishes them from amides.

5.08.3.1 Functions with a P—H or P—C Bond Only

Primary acetyl, pivaloyl and benzoyl phosphines, for example (**4**), have been synthesized and studied ⟨81ZAAC(473)7, 82ZAAC(495)73, 84TL1249⟩. It appears that in all three, the protons attached to phosphorus are magnetically equivalent, suggesting a lack of amide type resonance. Also, the phosphorus atom is highly nucleophilic and the molecules self-condense to give the corresponding bisacylphosphines, for example (**5**). Methylation of benzoyl phosphine occurs exclusively at the phosphorus, and the secondary acylphosphine thus obtained is also prone to self-condensation, although less severely than the primary analogue (Scheme 3). The choice of reaction conditions and in particular the choice of cation appear to be important for the synthesis of primary acyl phosphines. Under differing conditions, secondary or tertiary acylphosphines are obtained without the isolation of intermediate primary acylphosphine ⟨77ZAAC(430)91⟩.

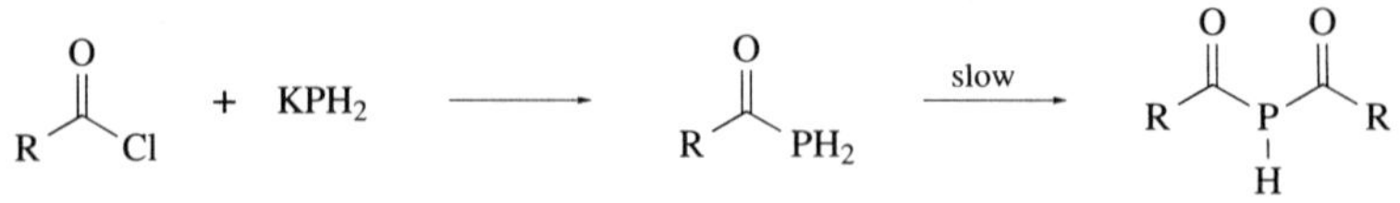

(4) **(5)**

R = Me, Bu^t, Ph

Scheme 3

Secondary acylphosphines can be prepared from reaction of the corresponding primary phosphine and acyl chloride in the presence of potassium carbonate (Scheme 4) ⟨67ZN(B)784⟩ or as discussed before by alkylation of the primary acylphosphine ⟨84TL1249⟩.

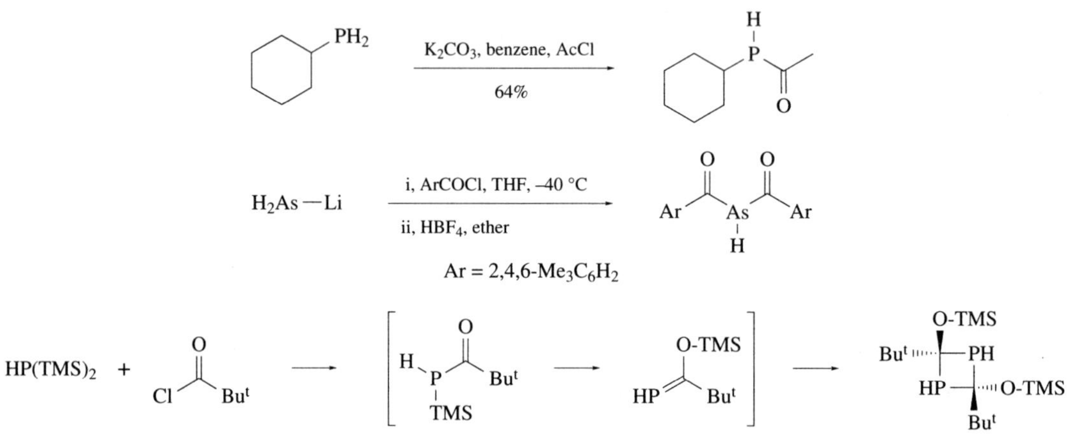

Scheme 4

Bisacylphosphines and arsines are well known and studied. They have been synthesized either directly from phosphine or arsine ⟨84TL1249, 91ZAAC(605)7, 92ZAAC(607)101⟩ or its trimethylsilyl derivative (Scheme 4) ⟨77ZAAC(430)66, 77ZAAC(430)77, 87CC1753, 91ZAAC(605)7, 92ZAAC(608)33⟩. An x-ray crystal structure of dipivaloylphosphine confirms that in the solid state, this molecule exists in the 'enol' form; however, in solution, NMR suggests the presence of some keto as well as enol tautomers (the ratio was found to be solvent dependent) ⟨77ZAAC(430)77⟩. The thermodynamic parameters of keto–enol tautomerization for bisadamantanoyl and biscyclohexanoylphosphines have been calculated from NMR data ⟨92ZAAC(608)33⟩. As already mentioned, the phosphorus atom in these species is highly nucleophilic and reacts with a range of electrophiles ⟨84TL1249, 87CC1753⟩.

As with their primary analogues, treatment of secondary acyl and benzoyl phosphines with an alkylating agent results in alkylation at the phosphorus atom and not at oxygen. This is consistent with the observation that in some instances, when *O*-alkylation/acylation is observed in amides, it proceeds via initial *N*-alkylation/acylation followed by a thermodynamically driven nitrogen to oxygen migration (the Chapman rearrangement). In acylphosphines, a similar phosphorus to oxygen migration is suppressed because the thermodynamic driving force is weaker owing to poorer overlap between phosphorus and carbon *p* orbitals. As we will consider later, a facile phosphorus to oxygen migration is observed if the migrating group is boron or silicon when the thermodynamic gain is enhanced by strong O—B or O—Si bond formations ⟨87DOK114, 88JGU201, 88JGU2372, 89JGU1817⟩.

This ready phosphorus to oxygen migration of silyl groups means that the use of silylated phosphines for the syntheses of primary and secondary acylphosphides cannot always be recommended. In some cases, the required *P*-silylated acylphosphines are never isolable since the molecule rapidly rearranges once it is formed (Scheme 4). Indeed, the resulting silylenol P=C—O—Si species have a diverse chemistry of their own and are not always isolable themselves. For instance, they can dimerize (for a review see ⟨92H(34)369⟩) or be transformed into phosphaalkynes ⟨91PS(55)133⟩.

An interesting example in this series is oxalyl derivative (**6**) prepared as shown in Scheme 5. An unusual feature of this molecule is that deprotonation and treatment with pivaloyl chloride afforded an *O*-acylated product (**7**). This is probably due to steric congestion around phosphorus which prevents the formation of the otherwise expected *P*-acylated species ⟨87AG(E)779⟩ (for comparison see Scheme 8 and related text).

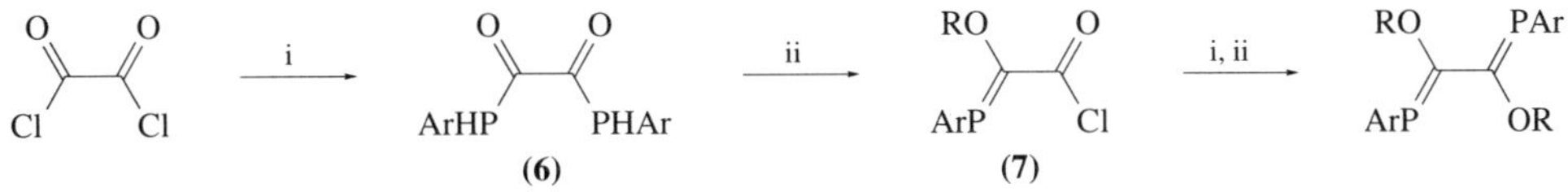

i, $ArPH_2$, –60 °C, *n*-pentane; ii, MeLi, –60 °C, THF then RCl

Ar = $Me_3C_6H_2$; R = Bu^tO

Scheme 5

Tertiary acyl phosphines are more common and have been studied for over four decades ⟨52CB239⟩. The preparation is, as mentioned before, by addition of a phosphorus nucleophile to an acylating agent (Scheme 6). Usually an alkali metal salt of phosphide is employed ⟨66ZAAC(346)241, 68T2995, 78ZN(B)1457⟩. Although no comparative studies could be found, this approach appears to be preferred over the use of a phosphine and an auxiliary base ⟨75BAU816⟩. Lithium phosphides are generated by treatment of the corresponding phosphine with an organometallic base, for example *n*-butyllithium, or through molecular rearrangement ⟨91T6673⟩. Alternatively, a trimethylsilyl derivative of phosphine is added to an acylating agent. This is generally a reliable method ⟨73JOM(49)149, 80CB3268⟩. However, as we mentioned before, this last approach may not always give the desired product (see previous paragraph).

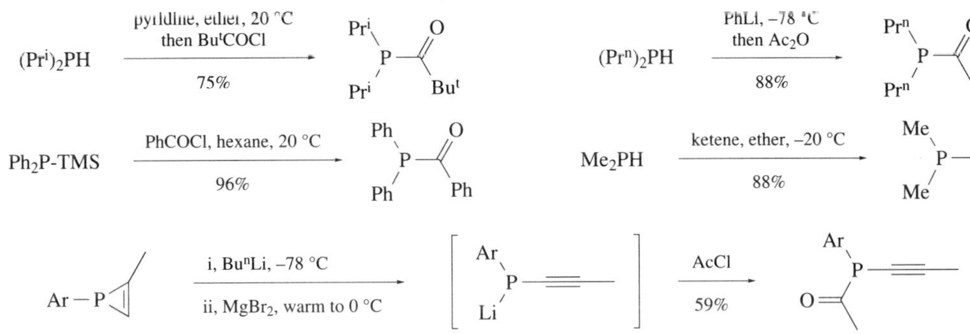

Ar = $Me_3C_6H_2$

Scheme 6

An unusual 1,2-chlorotropic rearrangement has allowed an alternative synthesis of benzoylbis(*t*-butyl)phosphide (**8**) ⟨86JGU2141⟩, though the generality of the reaction is not known (Scheme 7).

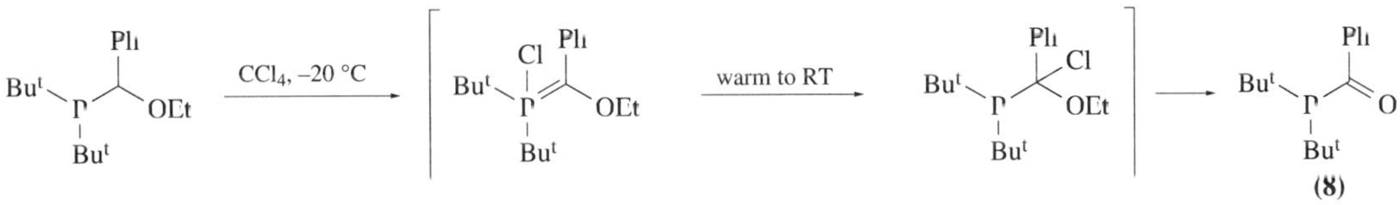

Scheme 7

The chemistry of tertiary benzoylphosphines has been comprehensively studied. It has been concluded that chemically, they closely resemble amides ⟨66ZAAC241, 73JOM(49)149⟩. The exception is their behaviour towards electrophiles where electrophilic attack always occurs through phosphorus and has never been observed through oxygen. This is rationalized in terms of the greater nucleophilicity of phosphorus. In general, acyl phosphines are highly nucleophilic and react readily with alkylating agents to afford quaternary salts ⟨66ZAAC241⟩, and with molecular oxygen to afford the corresponding phosphinyl species (Scheme 8) ⟨77AG(E)269⟩. The bismesityl substituted acylphosphine is stable to air oxidation (but not to hydrogen peroxide) which is attributed to steric congestion around the phosphorus atom (for comparison see Scheme 5 and related text) ⟨80CB3268⟩. The acylphosphonyl species thus generated are labile and hydrolyse readily; hence acylphosphines must be protected from air and moisture.

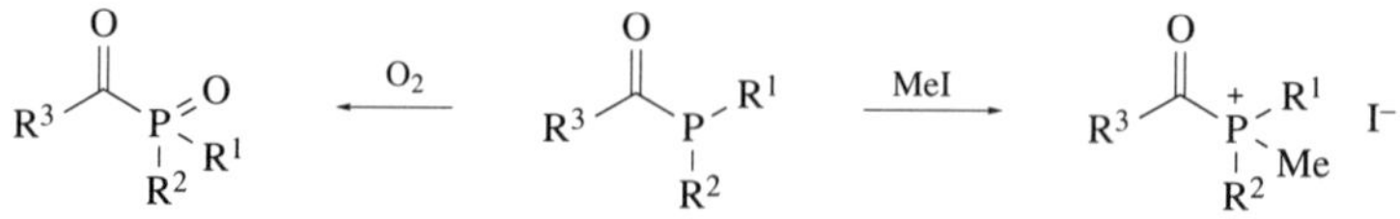

Scheme 8

5.08.3.2 Functions With One P—Heteroatom Bond Only

5.08.3.2.1 P-*Halogen bonded systems*

Only a handful of mono *P*-halogenated acylphosphines are known. The synthesis employs chlorination of the corresponding secondary acylphosphine with carbon tetrachloride in the presence of triethylamine ⟨91JGU114⟩. The reaction has been studied in detail, and it was found that in the absence of bulkier phosphorus substituents, the molecules self-condensed during the course of the reaction to afford diphosphine, for example (**9**) in Equation (1). This self-condensation, which is attributed to the nucleophilicity of phosphorus and is reminiscent of the self-condensation observed with primary acylphosphines (see Scheme 3 and related text), is always observed unless the phosphorus substituent is *t*-butyl. However, with bulkier substituents the rate of reaction also slows and heating becomes necessary. Heating in a sealed tube is recommended since another by-product of the reaction is trichloromethyl substituted acylphosphine (**10**) which is observed if the reaction is carried out under reflux (Equation (1)).

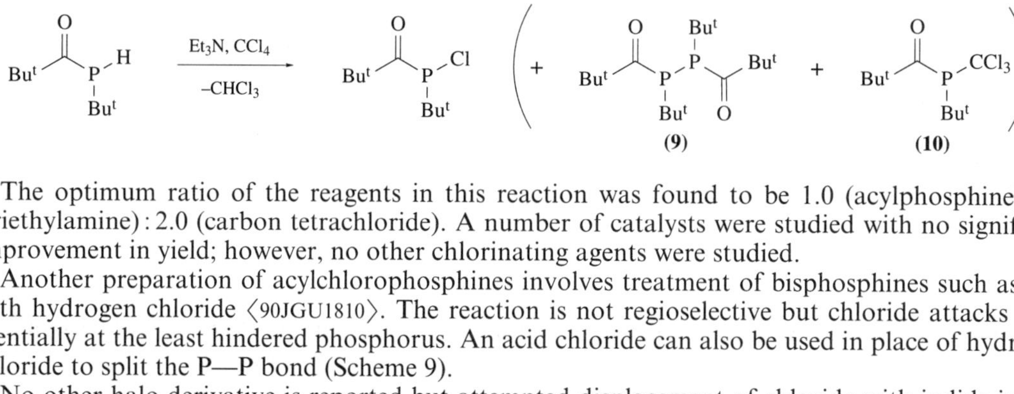

(**9**) (**10**) (1)

The optimum ratio of the reagents in this reaction was found to be 1.0 (acylphosphine) : 1.5 (triethylamine) : 2.0 (carbon tetrachloride). A number of catalysts were studied with no significant improvement in yield; however, no other chlorinating agents were studied.

Another preparation of acylchlorophosphines involves treatment of bisphosphines such as (**11**) with hydrogen chloride ⟨90JGU1810⟩. The reaction is not regioselective but chloride attacks preferentially at the least hindered phosphorus. An acid chloride can also be used in place of hydrogen chloride to split the P—P bond (Scheme 9).

No other halo derivative is reported but attempted displacement of chloride with iodide in (**12**) gave only diphosphine (**13**) (Scheme 9).

5.08.3.2.2 P-*Oxygen, sulfur and nitrogen bonded systems*

The methods of synthesis of these functions are extensions of the general routes described above. No examples of nucleophilic addition of nitrogen substituted phosphines to acylating agents are found, though this probably reflects the inaccessibility of these phosphines. On the other hand, reaction of sulfur substituted phosphide and acyl chlorides has been reported (Equation (2)) ⟨88JGU1965⟩.

R = Me, 34%; CCl_3, 75% (2)

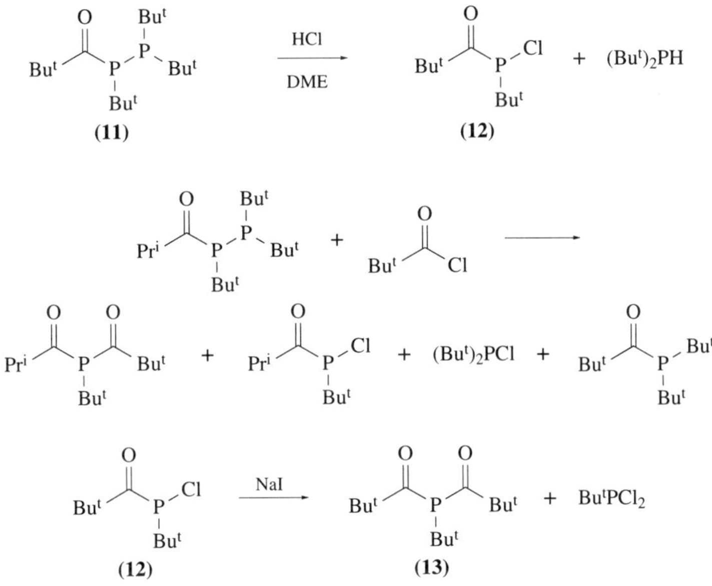

Scheme 9

An interesting alternative is by nucleophilic displacement of chloride from a *P*-chloro substituted phosphine (see Section 5.08.3.2.1). As we saw before, attempted displacement of chloride with iodide in (**12**) gave only diphosphine (**13**) (Scheme 9). However, it has been shown that chloride can be displaced in good to excellent yield by both oxygen and nitrogen nucleophiles (Scheme 10) ⟨91JGU108⟩.

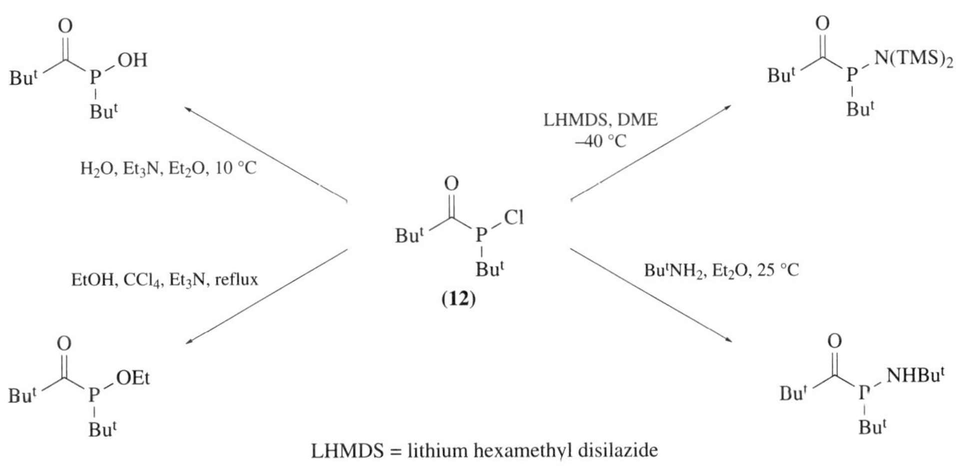

Scheme 10

5.08.3.2.3 P-*Other heteroatom bonded systems*

The tendency of silicon and boron substituted acyl phosphines and acylarsines towards 1,3-migration means that such species are observable at low temperatures spectroscopically, but cannot be isolated in condensed form at ambient temperature ⟨77AG(E)463, 87DOK114, 88JGU201, 88JGU2372, 89JGU1817⟩. The greater tendency of boron towards this rearrangement means that *P*-boronylated species were only observed as transient species in the reaction mixture at $-50^\circ C$ whereas the *P*-silylated species have a short half-life at room temperatures. The use of the lithium salt of the secondary acylphosphine in a nonsolvating medium promotes the formation of *P*-silylated (and boronylated) acylphosphines ⟨88JGU2372⟩, but the use of amine base in a solvating medium promotes the formation of silyl enol species (Scheme 11) ⟨88JGU201⟩.

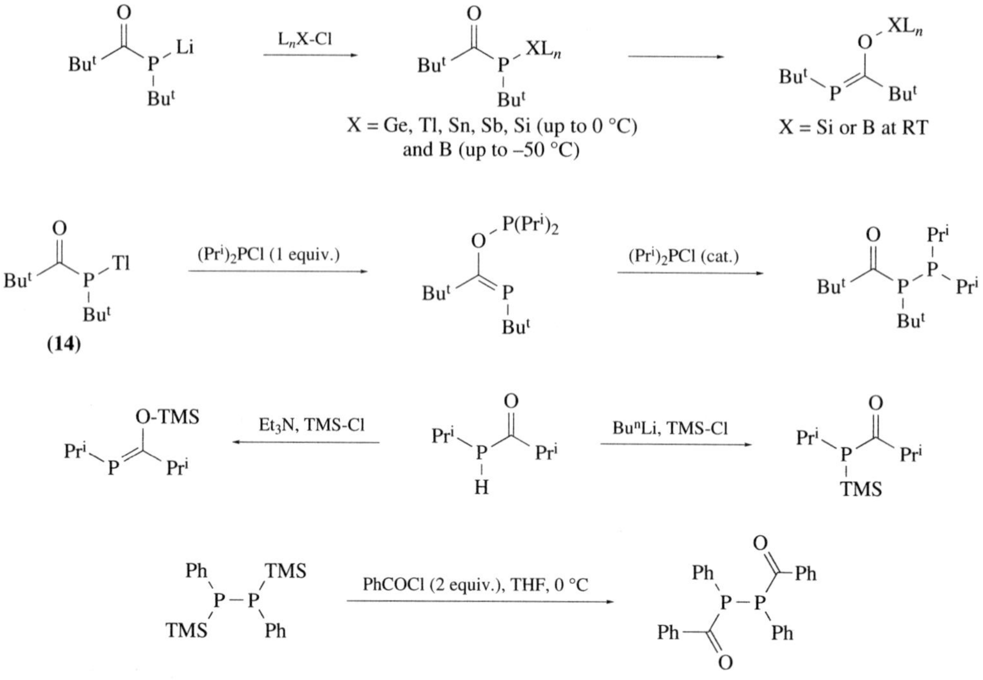

Scheme 11

Germanium ⟨86JGU636⟩, tin ⟨90JGU1822⟩, antimony ⟨90JGU1810⟩, thallium ⟨90JGU1817⟩ and phosphorus derivatives ⟨76CB359, 80ZAAC(468)55, 90JGU1810⟩ of acylphosphines can be isolated in condensed form and it has even been shown that in the case of phosphorus as a substituent, the *P*-substituted isomers are thermodynamically more stable than the *O*-substituted isomer ⟨90JGU1810⟩. The single example of an *O*-substituted isomer is obtained from the reaction of the thallium derivative (**14**) with diisopropylchlorophosphine. The product slowly converts to the *P*-substituted isomer in the presence of catalytic amounts of diisopropylchlorophosphine (analogous to the use of catalytic amounts of alkyl iodide in the Chapman rearrangement).

Preparation of the lithium salt of secondary phosphine and its treatment with X—Cl is the common route applied for the synthesis of P—X derivatives (⟨90JGU1810⟩ and references cited therein). An example has also been reported involving treatment of *vic*-diphosphine with an acylating agent (Scheme 11) ⟨76CB359⟩. The diarsine analogue, however, was prepared by a different route ⟨92ZAAC(606)101⟩.

The reactions of phosphorus substituted acyl phosphine (CO—P—P), which is typical for other semi-metal substituted species (CO—P—X), have been partly discussed above. As a general rule and as can be expected, nucleophiles attack the carbonyl moiety and cause cleavage of carbonyl and P—P(X) units. Electrophiles, on the other hand, cause cleavage of CO—P and P(X) units ⟨90JGU1810⟩.

The phosphorus atom can also be substituted with a transition metal, and a number of such species have been isolated and studied. Iron ⟨86CB1857⟩, osmium ⟨86JOM(311)269, 88JOM(353)355⟩, ruthenium ⟨86ZN(B)1096⟩ and molybdenum ⟨92JOM(426)C28⟩ compounds have been synthesized to date. The chemistry of transition metal bound phosphines closely resembles that of nonmetal substituted compounds. Thus they can be synthesized by treatment of transition metal bound phosphine with an acylating agent (Scheme 12) ⟨86CB1857, 86JOM(311)269, 86ZN(B)1096, 88JOM(353)355, 92JOM(426)C28⟩. A regioisomer, where the metal is attached to a carbonyl instead of to a phosphorus moiety, has been tentatively characterized, but not isolated, as it was found to be stable in solution only at $-78\,^{\circ}C$ (Scheme 12) ⟨88NJC547⟩.

The acylation can also result from a molecular rearrangement, that is it can be intramolecular. The phosphido complex (**15**) can be deprotonated at $-78\,^{\circ}C$ and acylated or alkylated on the phosphorus in the cold. However, if deprotonation is carried out at room temperature, the acyl ligand itself migrates from molybdenum to phosphorus (Scheme 13) ⟨92JOM(426)C28⟩.

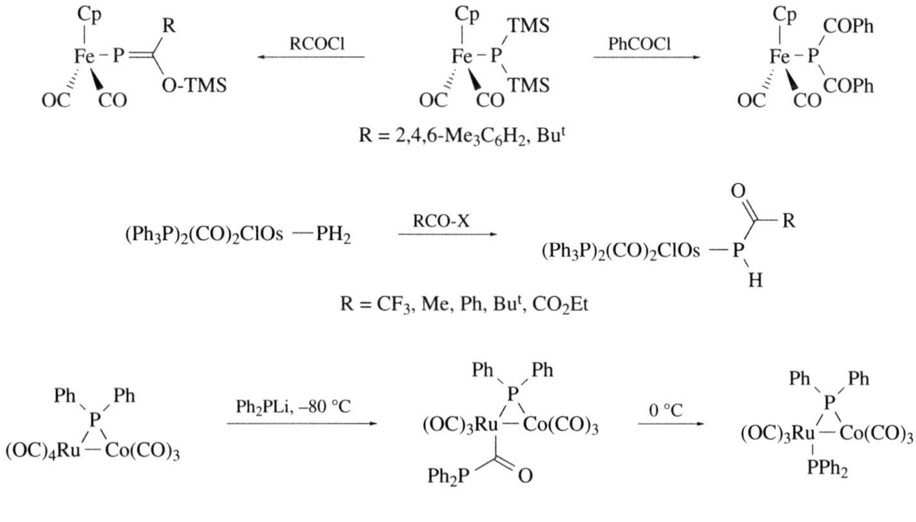

Scheme 12

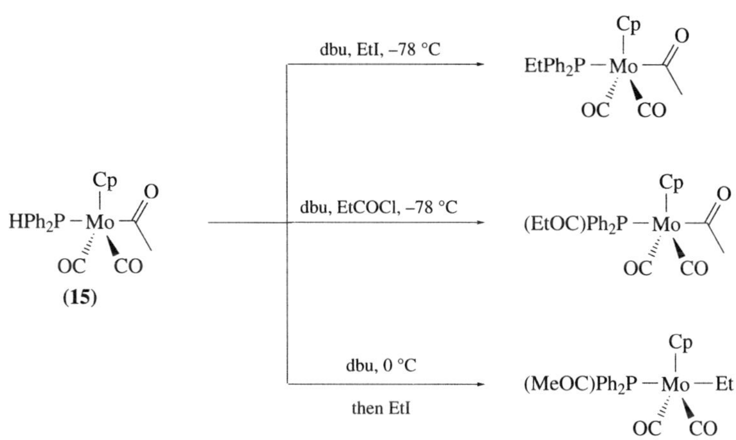

dbu = 1,5-diazabicyclo[5.4.0]undec-5-ene

Scheme 13

5.08.3.3 Functions with Two P—Heteroatom Bonds

Few acylphosphines of this type are reported. Interesting exceptions are shown in Scheme 14 ⟨87JGU1254⟩. The use of dialkoxymethyl as a masked hydrogen substituent for phosphorus is of particular interest (for further examples see ⟨89JCS(P1)1319⟩ and ⟨89JGU1977⟩ and references cited therein).

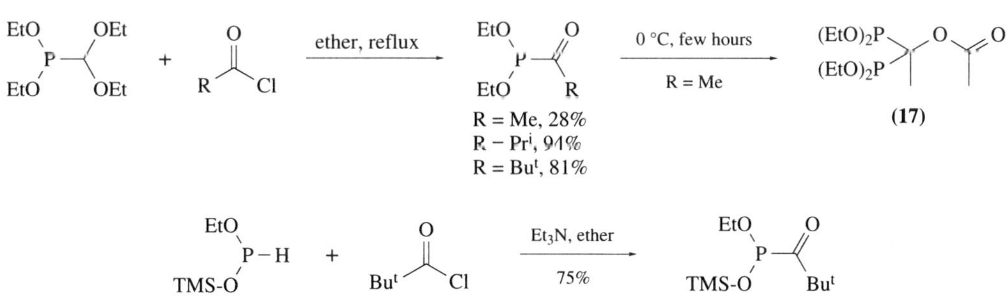

Scheme 14

One interesting feature of this function is the tendency of the molecules to dimerize, depending on the bulkiness of the carbonyl substituent. Thus, although the pivaloyl derivative does not dimerize at all, the acetyl derivative does so quite rapidly even at low temperatures. The dimer has not been fully characterized but is presumed to be (**17**).

5.08.4 TETRACOORDINATED PHOSPHORUS FUNCTIONS

The most common access to this class of compounds is by the addition of a tricoordinated nucleophilic phosphorus species (phosphine or phosphorous acid) to an acylating agent. The choice of both the acylating agent and the phosphine is quite wide and is mainly dependent on the degree of reactivity (nucleophilicity) of the phosphine. Heteroatom substitution in phosphorus(III) species generally increases the nucleophilic aptitude of the phosphorus (the α-effect) though the influence is subject to the stereoelectronic effect ⟨84T3215⟩. Bulkiness of the substituents, however, decreases the rate of nucleophilic attack by phosphorus. The reaction may stop with the formation of a positively charged, tetracoordinated acylphosphonium species. However, when a phosphorus(III) oxyacid is used, this tetracoordinated species breaks down to afford a noncharged tetracoordinated acylphosphorus species (the Arbuzov or Michaelis–Arbuzov reaction). In this two stage reaction, the rate determining step varies with the reactants but as a general rule the first step is fast and exothermic while the second is slower and endothermic (Scheme 15).

Interconversion is an important route for the synthesis of many such compounds and both chemical manipulation of the phosphorus or the carboxyl moiety have been documented.

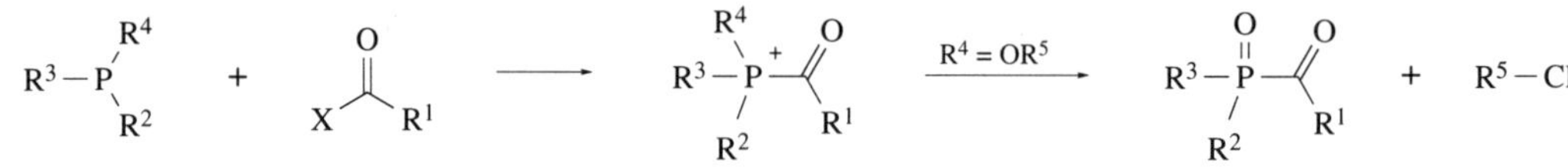

Scheme 15

5.08.4.1 Functions with Single P—C Bonds

Acetylphosphines can be quaternized under mild conditions with iodomethane ⟨59CB3183, 68T2995, 73BCJ1803⟩. Treatment of triphenylphosphine, as well as its antimony and arsenic analogues, with acid chloride also affords quaternary salts ⟨60ANC1896, 89JOM(366)281⟩. Triphenylbismuth, however, afforded only diphenylchlorobismuth and benzophenone (Scheme 16).

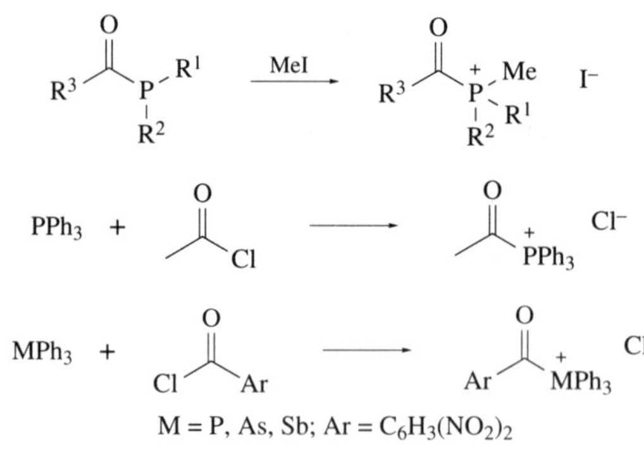

Scheme 16

5.08.4.2 Functions with Singly-bonded P—Heteroatoms Only

The reaction of hexaalkyl phosphorus triamides with benzoyl halide gives different products depending on reaction conditions employed (see Scheme 22 and related text). At low temperatures, benzoyl tris(dialkylamino)phosphonium halides are isolated (Scheme 17) ⟨78JGU414⟩. Although the halo salts themselves are thermally unstable, the hexachloroantimonate salt does not decompose up to its melting point (ca. 110 °C). The phosphorus chemical shift in the NMR of various halo salts is similar suggesting that the species are truly ionic and hence tetracoordinated.

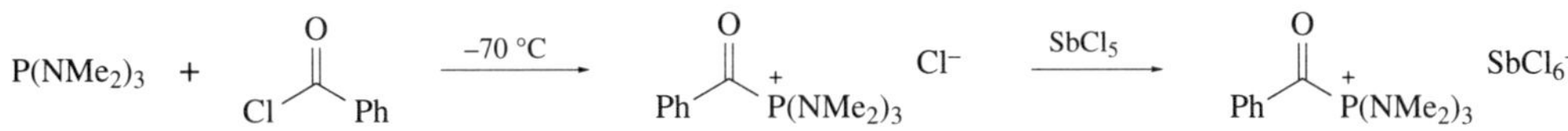

Scheme 17

A number of organometallic derivatives of acylphosphorus species containing tetracoordinated phosphorus have been identified. Since these species contain dative bonds between the phosphorus and the heteroatom, they are discussed in Section 5.08.4.3.3.

5.08.4.3 Functions with P—Heteroatom Double Bonds

These are by far the most common of the acyl phosphorus species. A diverse range of compounds in this family has already been synthesized. As has already been said, nucleophilic addition of phosphorus(III) compounds (the Arbuzov reaction) is the most common approach. Oxygen is the most commonly found doubly-bonded atom. This is a reflection of the ease with which an O—C bond can break in favour of the formation of a P═O bond during the second stage of the Arbuzov reaction (Scheme 15).

The chemistry of these functions is analogous to that of both esters and ketones. For instance, treatment with amine results in cleavage of the C—P(X) bond and formation of an amide and, similarly, hydrolysis affords the corresponding carboxylic acid ⟨80JOC4162, 84JOC2139, 87T3395, 88JCS(P2)2029, 89JCS(P1)765⟩. On the other hand, the carbonyl function is known to undergo a number of ketonic reactions; for instance, it can be reduced ⟨89JCS(P1)765, 90BCJ2728⟩, undergo the Wittig reaction ⟨80BCJ1625⟩, or be transformed to imine ⟨89JCS(P1)765⟩, hydrazone ⟨72CB3357⟩ or oxime ⟨68JOC3090, 90JCS(P1)3263, 92TL2067⟩.

5.08.4.3.1 P═O bonded systems

(i) With no other heteroatom substituent

The most common and reliable route is by reaction of an oxysubstituted, nucleophilic phosphorus(III) species and a carboxylic chloride or anhydride. The first two examples show the range and diversity of the reaction ⟨69JGU597, 70JCS(C)1263⟩, see also ⟨74JGU1685, 78CC285⟩ ((a) and (b), Scheme 18).

An alternative method of synthesis is by oxidation of acyl phosphines (see Scheme 8 and related text). Acylphosphines can be oxidized with both molecular oxygen and hydrogen peroxide ⟨79CB763, 79CB1456, 80CB2769⟩. Overoxidation is a possibility, in particular when stronger oxidizing agents are employed ⟨68TL4405, 79CB793⟩.

Another synthetic route is via rearrangement of acyloxyphosphorus(III) species, as outlined in (c), Scheme 18. Acyloxyphosphorus(III) species are prepared from the reaction of the sodium salt of the corresponding alkanoic acid and halophosphite ((d) and (e), Scheme 18) ⟨86ZN(B)859⟩, or from the reaction of the corresponding acid anhydride and halophosphite ⟨69JGU2541, 79CB763⟩. They are rarely isolable and in most cases spontaneously rearrange to the corresponding phosphine oxide.

(ii) With at least one halogen substituent

Treatment of alkylphosphonous dichlorides with acid chlorides affords an uncharacterized adduct, presumably (**19**), which after alcoholysis affords (**20**) (Scheme 19) ⟨71JGU2513⟩.

Treatment of alkanoylphosphonic acid with a chlorinating agent affords the corresponding chlorophosphonate. Both the bis ⟨92PS(70)109⟩ and mono ⟨91JOC4791, 92PS(70)109⟩ chlorinated acylphosphonates have been prepared in this way (Scheme 19).

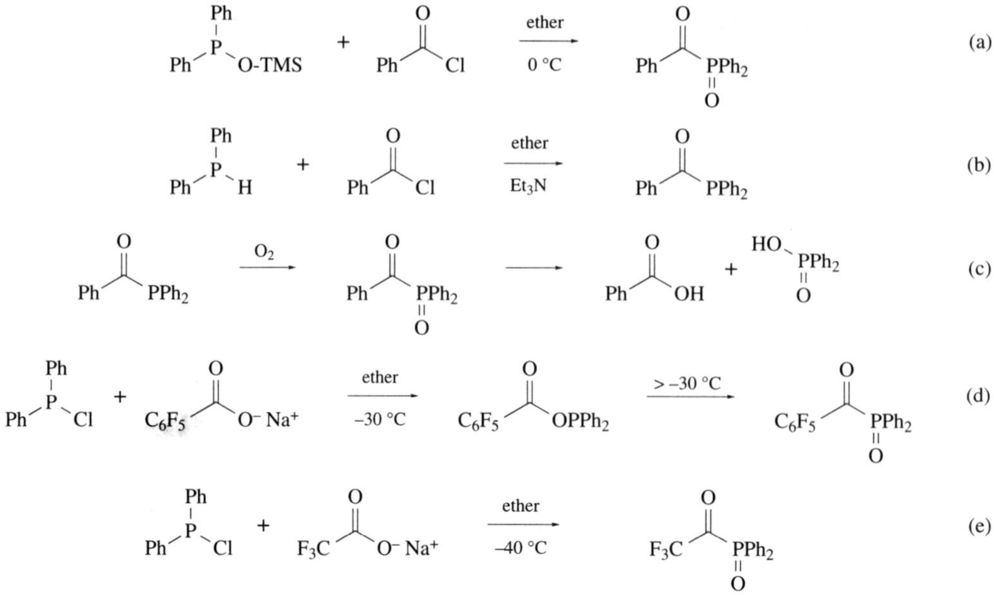

Scheme 18

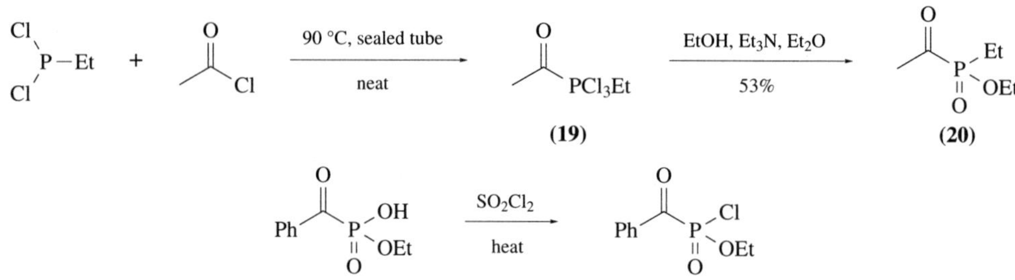

Scheme 19

(iii) With at least one oxygen substituent (but no halogen substituent)

Various examples of preparation of both the mono ⟨67JGU2606, 72JGU1256, 74JGU1685⟩ and bis ⟨67CPB1183, 68JGU1791, 68JOC3090, 80BCJ1625, 80RCR843, 90BCJ2728, 91PS(62)15⟩ oxygen substituted acyl phosphorus compounds of this type are documented. Hydroxy substituted analogues (acylphosphinic acids and acylphosphonic acids) cannot be prepared directly in this way but may be prepared from the corresponding esters by aqueous hydrolysis ⟨89JSC(P1)765, 90BCJ2728⟩ although the yields are often low ⟨80JOC4162, 88JCS(P2)2029, 89JCS(P1)765⟩. *trans*-Esterification of methyl ester with bromotrimethylsilane ⟨77TL155⟩ or iodotrimethylsilane ⟨78CC870, 80JCS(P1)1150⟩ followed by aqueous hydrolysis during workup results in a better yield, and is a cleaner reaction. Alternatively, reaction of tristrimethylsilylphosphorus acid with an acid chloride affords a trimethylsilyl ester which upon aqueous hydrolysis during workup gives the corresponding phosphinic or phosphonic acid (Scheme 20) ⟨68JGU1791⟩.

Surprisingly, little work has been done to investigate the feasibility of preparing acylphosphonates by oxidizing the methylene position adjacent to a phosphonate. α-Hydroxy phosphonates are easily prepared by nucleophilic addition of phosphorus(III) species to aldehydes. Considering the range of mild and selective oxidizing agents suitable for oxidation of alcohols to ketones, this ought to be a fertile area of research. The only reported example is the one outlined in Scheme 21 ⟨92TL4137⟩.

Carbenes generated from α-diazobenzyl phosphonates (themselves prepared from benzoylphosphonates ⟨72CB3357⟩) react with molecular oxygen in a solid argon matrix to afford, amongst other products, the corresponding benzoylphosphonates ⟨92JOC6216⟩.

The monoamide of acylphosphonates can be prepared by the usual nucleophilic addition to acetylchloride ⟨66BAU1432, 67JGU826, 71BAU723⟩ but if bulkier esters are used the yield is poor and nucleophilic addition through nitrogen predominates (Scheme 22). This nucleophilic action of

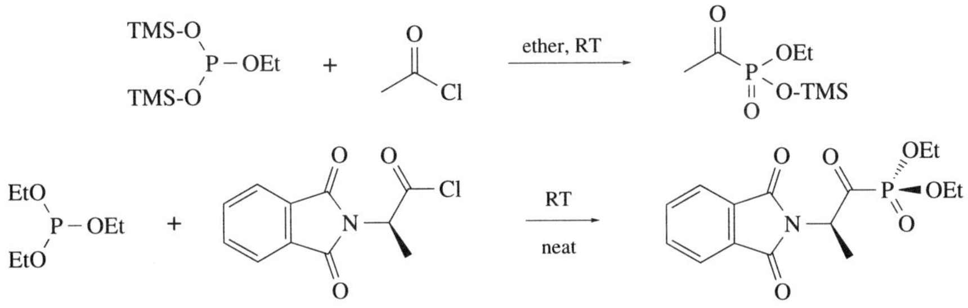

Scheme 20

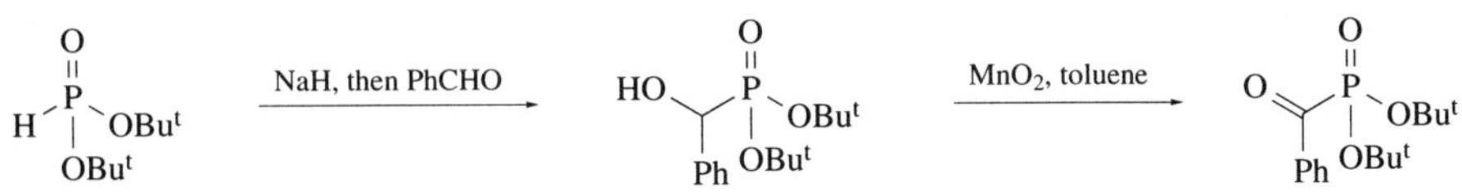

Scheme 21

nitrogen phosphoramidates is well documented and competes well with the nucleophilic action of phosphorus ⟨73PS287, 77JGU505⟩. Substitution at nitrogen with electron withdrawing groups reduces the nucleophilic attack by this atom with little effect on the nucleophilicity of phosphorus ⟨67JGU826, 71BAU723⟩.

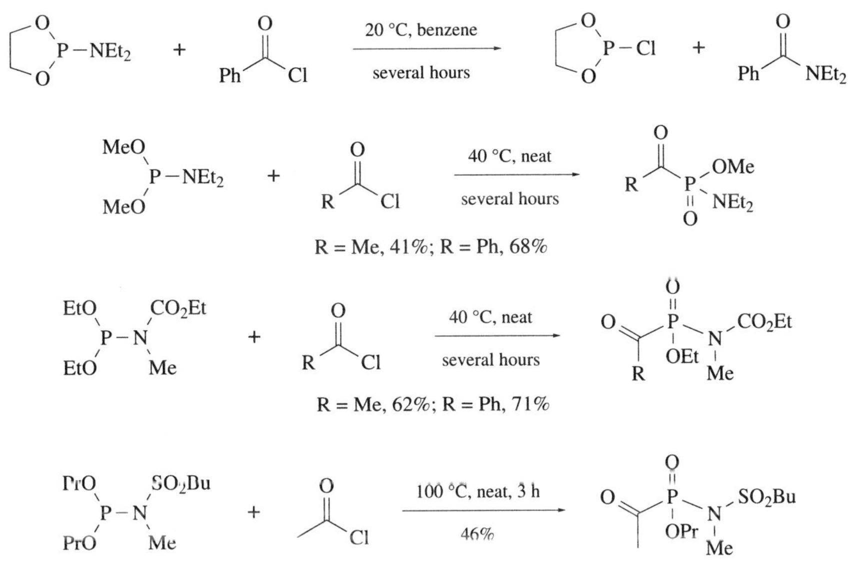

Scheme 22

(iv) With at least one sulfur substituent (but no halogen or oxygen substituent)

The only example appears to be the reaction shown in Equation (3) ⟨89JGU2388⟩.

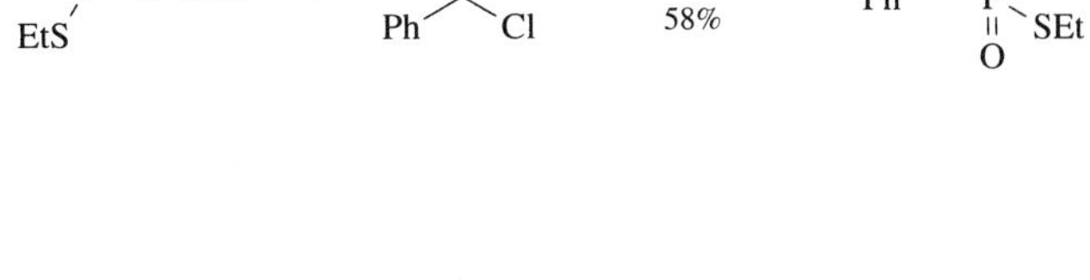

(3)

(v) With at least one nitrogen substituent (but no halogen or chalcogen substituent)

Although hexamethylphosphoramide is considered an inert solvent which can be used at low temperatures without complications with alkyllithium species, at higher temperature it forms the lithiated derivative (**21**), which is then added to electrophiles present in the reaction mixture (Scheme 23) ⟨91JOC4241⟩. Nucleophilic addition of bisamidophosphite also gives the bisamidophosphonate (**22**) (Scheme 23) ⟨90JGU1940⟩.

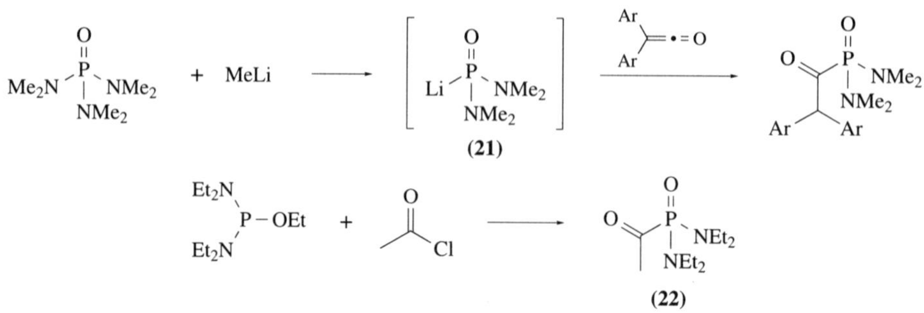

Scheme 23

5.08.4.3.2 P=S bonded systems

By analogy to the oxidation of acyldiphenylphosphine with molecular oxygen to afford the corresponding acylphosphonate, treatment of acylphosphines with sulfur is reported to afford acylthiophosphonate ⟨85JGU1956⟩.

Unlike the oxygen analogue (Scheme 19), *S*-acetyl phosphothioites are stable, isolable species ⟨88JGU1548⟩ which can be easily prepared from the corresponding thioalkanoic acid (Scheme 24). No solid derivative has yet been isolated, however. In solution and presumably in the condensed phase, the compound exists as a mixture of interconverting isomers i.e., in a state of acylotropy. The ratio is based on thermodynamic factors and thus depends on the acyl substituent. Evidence in support of the proposed acylotropy is that treatment with methyl iodide affords a mixture of acetylthiophosphonate and thioacylphosphonate in different ratios, resulting from the kinetic factors (e.g., the more ready rearrangement of phosphothioite).

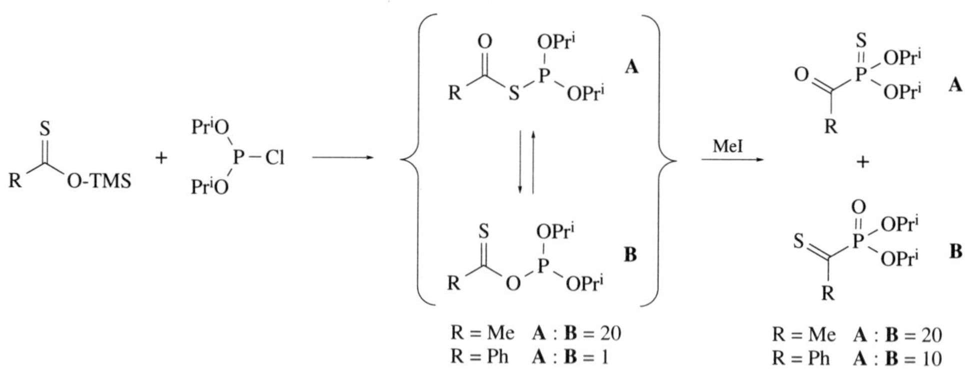

Scheme 24

Treatment of *S*-acetyl cyanophosphorothioite (**23**) with chloral is reported to afford (**24**) (Equation (4)) ⟨91BAU661⟩. No mechanism for this reaction has been proposed.

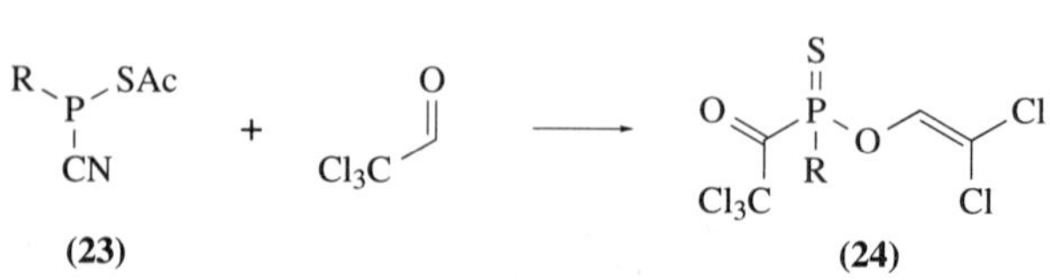 (4)

R = Et, 57%; R = Ph, 64%

5.08.4.3.3 P=Other heteroatom bonded systems

A number of organometallic derivatives of acylphosphorus species with dative bonds between phosphorus and the metal have been identified (Scheme 25) ⟨82OM1488, 84PS(19)311, 91IC1174⟩. The chemical behaviour of the species is very similar to that of the analogous P=O system, as the reactions shown in Scheme 25 demonstrate.

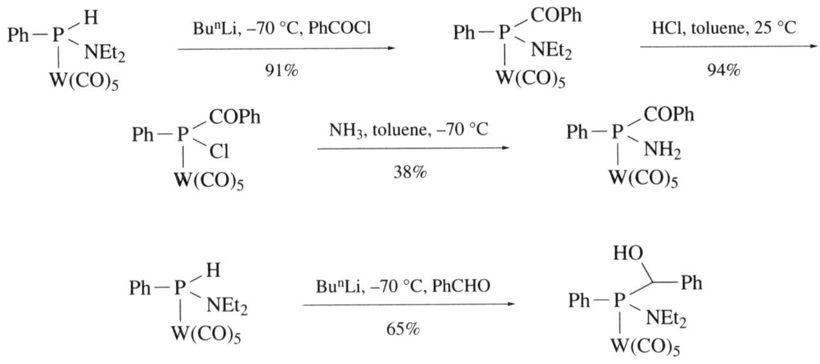

Scheme 25

5.08.5 HIGHER COORDINATED SPECIES

Few five-coordinated acylphosphorus species have been reported (Equation (5)). Fluorodiphenylphosphine reacts with benzoyl fluoride to afford an adduct ⟨70JCS(C)878⟩. The x-ray structure determination of this adduct has unequivocally shown it to contain a pentacoordinated phosphorus ⟨78CC946⟩. Further evidence in support of this comes from a triplet signal in high field ^{31}P NMR suggesting that two magnetically equivalent fluorines are bonded to phosphorus. The reaction of hexamethyl phosphorus triamide with benzoyl chloride and bromide affords adducts which may be considered both as pentacoordinated phosphoranes as tetracoordinated phosphonium salts. However, since the chemical shift for the adduct does not differ depending on the halide, it can only be concluded that the species are ionic and hence tetracoordinated.

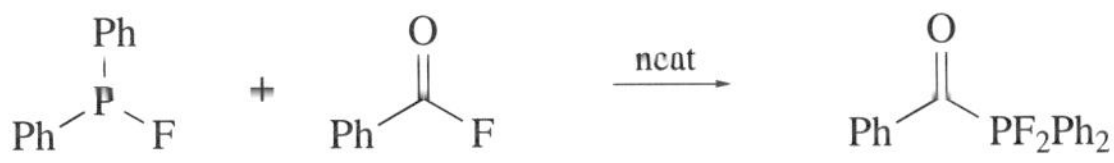

(5)

5.09
Acyl Silicon, Germanium, or Boron Functions

HONGYAN QI and DENNIS P. CURRAN
University of Pittsburgh, PA 15260, USA

5.09.1 ACYL SILICON DERIVATIVES $R^1COSiR^2_3$

The chemistry, synthesis, and properties of acylsilanes have been reviewed previously by Brook ⟨68AOC(7)95⟩, Ricci ⟨89S647⟩, Page ⟨90CSR147⟩, and most recently by Panek ⟨92OPP553⟩. The last review includes reaction methods that were developed during the years 1990–1991, and focuses on stereoselective reactions. The authors have relied on the three recent reviews in preparing this section of the paper, and have tried to condense and organize the most useful methods of acylsilane synthesis. The literature coverage extends through mid-1993.

Many kinds of acylsilanes have been synthesized with various R^1 and R^2 substituents. In this chapter, simple acylsilanes include alkanoyl, aroyl, and heteroaroylsilanes with R^1 = alkyl, aryl, and heteroaryl groups, respectively. If $R^1 = H$, the corresponding acylsilanes are formylsilanes. Functionalized acylsilanes are α-haloacylsilanes (R^1 = α-haloacyl groups), α-ketoacylsilanes (R^1 = α-ketoacyl groups), α,β-unsaturated acylsilanes (R^1 = alkenyl or acetylenyl groups), cyclopropyl acylsilanes (R^1 = cyclopropyl groups), and other functionalized acylsilanes. Although there is no one method that can be used to prepare every type of acylsilane, some methods have good generality, especially the methods for preparation of different simple acylsilanes. These methods are described in Section 5.09.1.1 for simple acylsilanes, and applications to functionalized acylsilanes are briefly mentioned where appropriate. Methods for synthesis of formylsilanes and for each class of functionalized acylsilanes are described in Sections 5.09.1.3 and 5.09.1.4, respectively. Also the applications of some of the general methods to formylsilanes and functionalized acylsilanes are discussed in these sections in detail.

Acylsilanes (and also acylgermanes) are closely related to acyl tin derivatives, and synthetic routes for preparing one class of compounds may often be applicable for preparing the other two. The synthesis of acyl tin compounds is discussed in Chapter 5.10.

5.09.1.1 Simple Aroyl and Alkanoylsilanes

5.09.1.1.1 Hydrolysis of α,α-dihalides and dithianes

The first acylsilanes to be obtained in good yield were a range of benzoyl and substituted benzoyl triphenylsilanes (**3**) by Brook (Scheme 1) ⟨57JA4373⟩. Benzylsilanes (**1**) are converted into the corresponding α,α-dibromobenzylsilanes (**2**) with two equivalents of *N*-bromo-succinimide. Hydrolysis of the dibromides with silver acetate in acetone–ethanol–water ⟨57JA4373⟩ or silica gel ⟨80TL3927⟩ gives the benzoylsilanes (**3**).

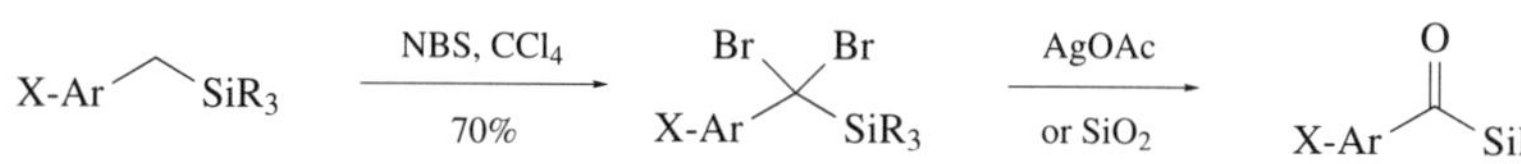

(**1**) (**2**) (**3**)

X = H, Br, Cl, F, NO_2, OMe and Bu^t

Scheme 1

Various aroylsilanes with a number of different groups bonded to the silicon (R) or substituted on the benzene ring (X) have been synthesized by this approach ⟨63CJC2351, 65CJC1175, 70JOM(22)321, 80TL392⟩. But this method is restricted to aroylsilanes where geminal dihalides can be produced easily by NBS bromination. It cannot be applied to alkanoylsilanes because the requisite dibromides cannot be cleanly formed.

An alternative route to the hydrolysis of α,α-dihalide (**2**) is the hydrolysis of a dithiane (**6**) (thioacetal). Originally investigated by Brook ⟨67JA431⟩ and Corey and Seebach ⟨67JA434⟩, this method has proven to be one of the most general ways for the synthesis of acylsilanes. It can be used to prepare aroyl, alkanoyl, and alkenoylsilanes. As shown in Scheme 2, an aldehyde (**4**) is first reacted with 1,3-propanedithiol to obtain the 2-substituted-1,3-dithiane (**5**). Lithiation and silylation then give the α-silyl-1,3-dithiane (**6**). Finally, the acylsilane (**7**) is obtained by unmasking of the latent carbonyl group.

(**4**) (**5**) (**6**) (**7**)

R^1, R^2 = alkyl, aryl

Scheme 2

The first two steps of this method (dithiane formation and silylation) are quite general, but the third step (dithiane hydrolysis) can be problematic. The main side product is the corresponding aldehyde ⟨90T2573⟩. The ease of hydrolysis of the silyl-1,3-dithiane (**6**) depends on the nature of the R^2 substituent. Phenyl substituted dithianes (R^2 = Ph) are much more resistant toward hydrolysis than their methyl analogs ⟨67JA431, 75JOM(87)263⟩. Initially, Brook and Corey ⟨67JA431, 67JA434⟩ used mercury(II) chloride together with either mercury(II) oxide or cadmium carbonate, and the yields of (**7**) varied from 14% to 71%. Other useful reagents for dithiane hydrolysis include chloramine-T in aqueous methanol (81%, R^1 = alkyl, R^2 = Me) ⟨79JA2225⟩, mercury(II) oxide in 35% HBF_4/THF ⟨85TL2509⟩, thallium(III) nitrate trihydrate ⟨86HCA1378⟩ and ceric ammonium nitrate in aqueous acetonitrile ⟨92JOC7010⟩. Anodic oxidation of the 1,3-dithianes (**6**) with a platinum anode in wet acetonitrile generates acylsilanes (**7**) under mild conditions ⟨92TL1355⟩. This electrochemical oxidation provides convenient access to a variety of aryl, saturated, and α,β-unsaturated acylsilanes in good to excellent yields.

Another modification is to use more readily hydrolyzable thioacetals (Scheme 3) ⟨77CB841⟩. Trimethylsilyl lithiobis(methylthio)methane (**8**) undergoes a Michael addition to cyclic enone (**9**) affording adduct (**10**). Hydrolysis of (**10**) by the original Brook–Corey method gives the corresponding acylsilane (**11**) in excellent yield.

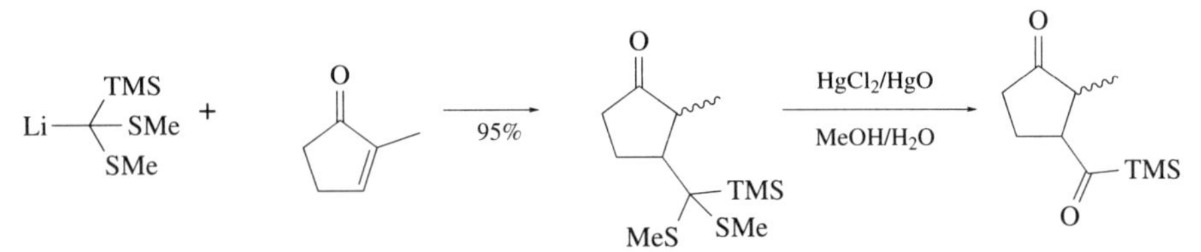

(**8**) (**9**) (**10**) (**11**)

Scheme 3

Oxidative hydrolysis of oxathiolanes also leads to the corresponding ketones ⟨71TL3445, 87JAP6261987⟩. In Scheme 4, oxidation of oxathiolanes (**12**) by sodium periodate in aqueous dioxane gives the acylsilanes in excellent yields ⟨85TL2675, 87JAP6261987⟩.

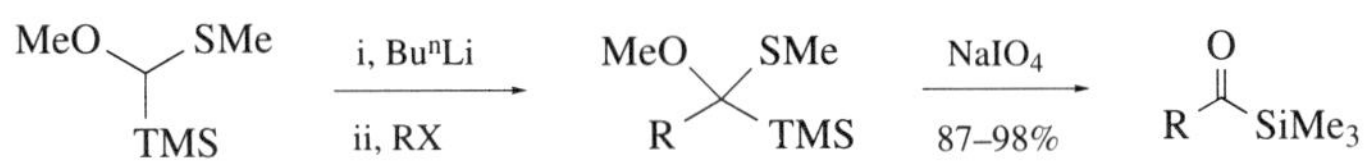

(**12**)

R = primary alkyl

Scheme 4

Many applications of this dithiane methodology have been reported in the literature ⟨92JOC7010, 90JOM(388)47, 90CSR147, 75JOM(87)263⟩. Cyclic acylsilane (**16**) has been synthesized according to Scheme 5 ⟨75JOM(87)263⟩. Hydrosilylation of 4-bromo-1-butene with dimethylchlorosilane provided dihalide (**13**). This was then alkylated by dithiane, first intermolecularly on silicon to give (**14**), and then intramolecularly on carbon to give (**15**). Hydrolysis provided the cyclic acylsilane (**16**). Haloacylsilanes (**17**) were also synthesized via a similar method by reversing the silylation and alkylation processes (Scheme 6) ⟨91TL3515, 92JOC7010⟩.

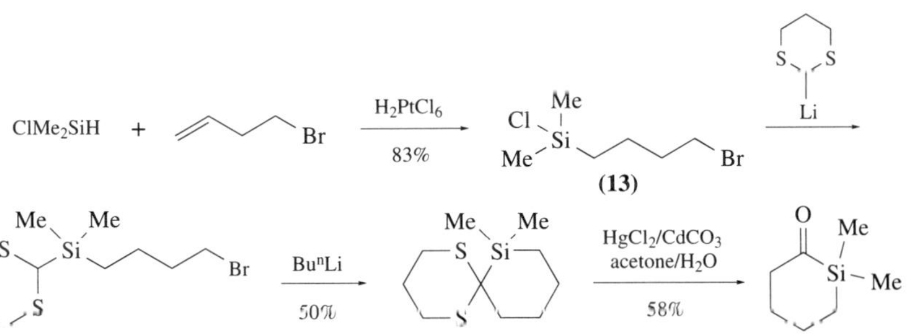

(**13**)

(**14**) (**15**) (**16**)

Scheme 5

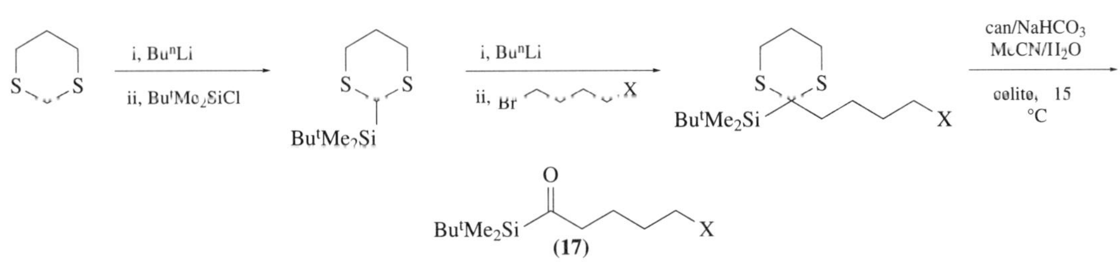

(**17**)

X = Br, Cl

can = ceric ammonium nitrate

Scheme 6

5.09.1.1.2 Oxidation of α-silyl alcohols

The synthesis of acylsilanes by the oxidation of α-hydroxysilanes is attractive because alcohols can be easily oxidized to ketones. However, there are two problems with this ostensibly straightforward method. First, many α-hydroxysilanes have two competing pathways under typical oxidizing conditions. Scheme 7 illustrates that when chromic acid is used as an oxidant, loss of silicon (path b) is often favored over loss of a proton (path a). This leads to the silicon–carbon bond cleavage and results in formation of a silanol and an aldehyde instead of the desired acylsilane ⟨65JOC2566, 68AOC(7)95, 90CSR147⟩. Only acetylsilane, propionyltriphenylsilane ⟨65JOC2566⟩, and cyclopropyl acylsilanes ⟨86HCA44⟩ have been prepared by this method using chromium(VI) oxide in aqueous sulfuric acid–acetone (60–70% yields). However, oxidation of substrates **(18)** and **(19)** with two silicon moieties successfully gives the corresponding acylsilanes by cleavage of only one silicon–carbon bond (Equations (1) and (2)) ⟨76CL993, 77JOC1773⟩.

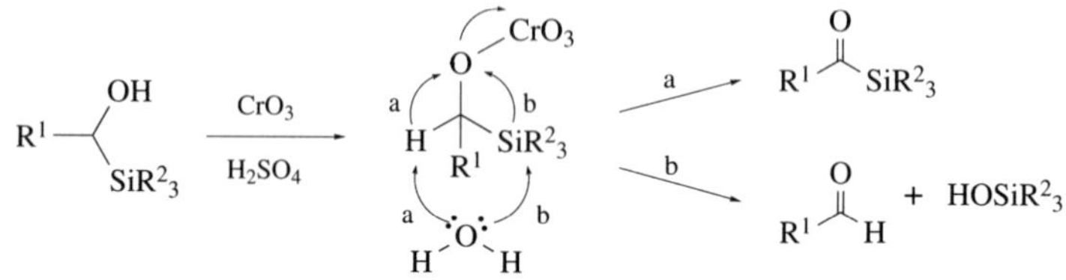

Scheme 7

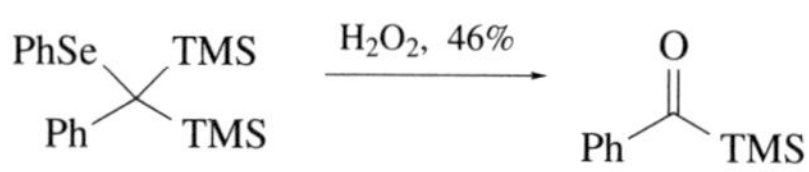

(18) (1)

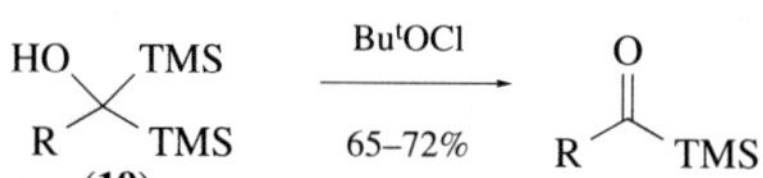

(19) (2)

R = $Ph(CH_2)_2$, n–C_7H_{15}, Ph

Among the several oxidizing systems that have been examined ⟨65JOC2566⟩, the Swern oxidation has emerged as a general method for oxidation of α-hydroxysilanes to acylsilanes ⟨65JOC2566, 85JOC2198, 88JOC1569, 88OS(66)14⟩. Good yields of alkanoyl, aroyl, and acetylenylsilanes have been obtained by this method (Schemes 8 and 9).

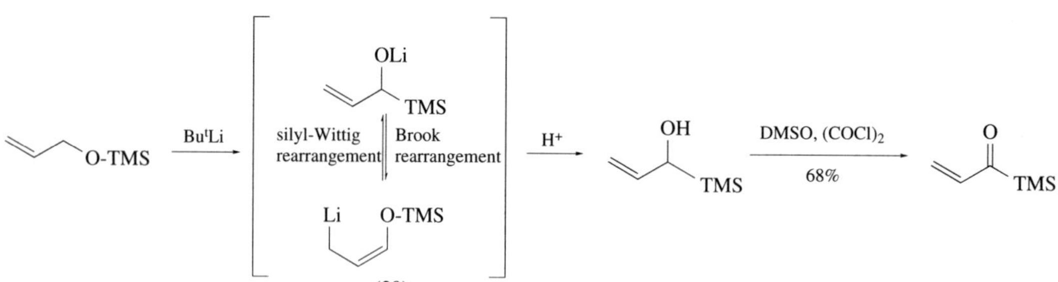

(20)

Scheme 8

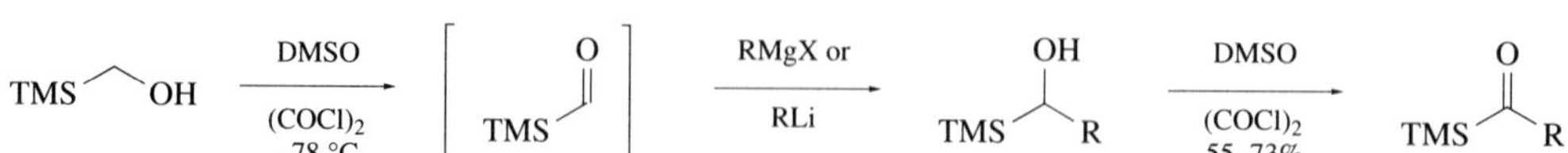

(21) **(22)**

R	*(22)* *Yield* (%)
alkyl	16–46
alkenyl, aryl	46–59
acetylenyl	60–76

Scheme 9

The second difficulty limiting the utility of the oxidation method is that of obtaining the appropriate α-hydroxysilanes. Methods for the direct preparation of (α-hydroxyalkyl)silanes are limited. Addition of a silyllithium reagent to an aldehyde is generally not synthetically useful ⟨82JOC747, 84JOC5282⟩. A silyl-Wittig rearrangement ⟨74ACR77⟩ of trialkylsilyl ether (**20**) has provided an α-hydroxyalkylsilane in good yield (Scheme 8) ⟨85JOC5393⟩. This procedure is limited to allylic ⟨85JOC5393⟩ or benzylic systems ⟨82JOC5051⟩ which are acidic enough to be deprotonated. Linderman ⟨88JOC1569⟩ and Ireland ⟨85JOC2198⟩ have successfully used the direct addition of a nucleophile to *in situ* generated formyl trimethylsilane (**21**) to generate the corresponding α-hydroxysilane (**22**) (Scheme 9). However, good yields of acylsilanes were only obtained with acetylenylcarbinols ⟨88JOC1569⟩. Low yields for alkyl carbinols (16–46%) were attributed to the inherent instability of (**21**) and to the incomplete oxidation of trimethylsilylmethanol.

Synthesis of acylsilanes can also be achieved by the oxidation of phosphonium ylides (**23**) (Scheme 10) ⟨85AG(E)1068⟩, by the photosensitized oxygenation (Scheme 11) ⟨79TL871⟩, or *meta*-chloroperbenzoic acid oxidation (Scheme 12) ⟨86TL2005⟩ of silyl diazo compounds (**24**). These three methods tend to complement each other. While the methods in Schemes 10 and 11 cannot be used for alkanoylsilanes, that in Scheme 12 is only applicable to alkanoylsilanes, particularly primary alkanoylsilanes. The yields of the alkylation from the lithio trimethylsilyl diazomethane reaction drop dramatically for secondary alkanoylsilanes. For the oxidation of (**24**) in Scheme 12, a phosphate buffer solution (pH = 7.6) is used to prevent side reactions. In general, these oxidation routes have limited use.

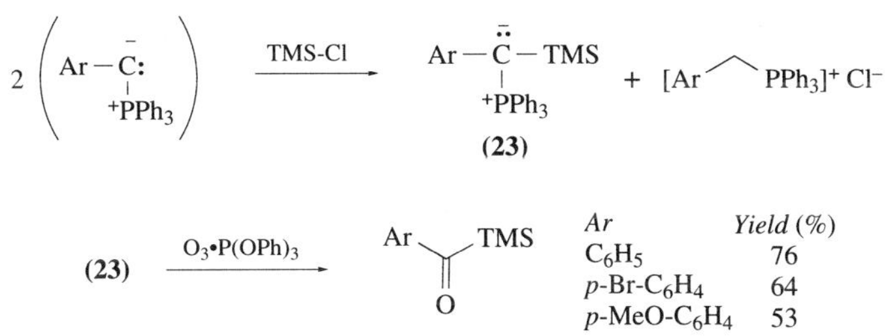

Ar	*Yield* (%)
C_6H_5	76
p-Br-C_6H_4	64
p-MeO-C_6H_4	53

Scheme 10

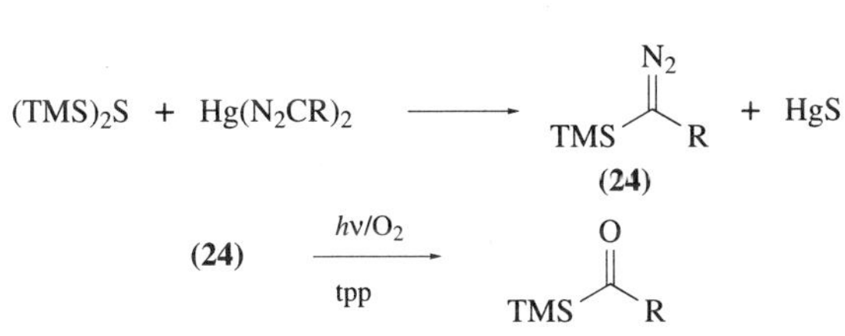

tpp = *meso*-tetraphenylporphyrin

R = Ph, CO_2Me (or CO_2Et) typical yield = 42%

Scheme 11

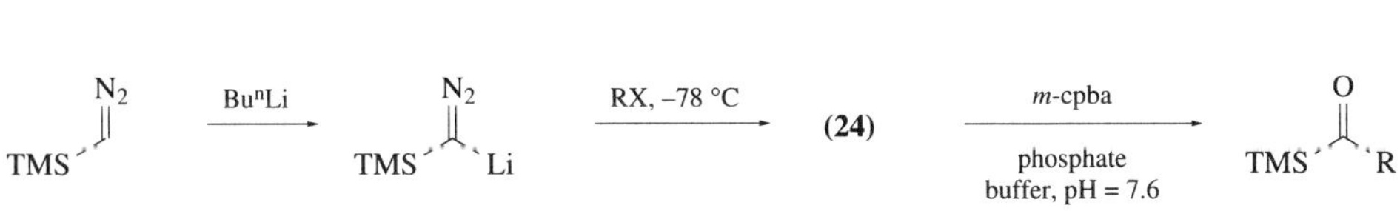

R = primary alkyl, yields = 56–71%

Scheme 12

5.09.1.1.3 Enol ether metalation/hydrolysis

Enol ethers of acylsilanes (**25**) are obvious precursors to acylsilanes since mild hydrolysis of the enol ethers would liberate the corresponding acylsilanes. In addition, enol ethers of acylsilanes may also be reacted with halogens such as chlorine or bromine ⟨78TL259⟩, acetals ⟨77JA5827⟩, and phenylsulphenyl chloride ⟨78JOM(145)C1⟩ to afford α- or β-functionalized acylsilanes. These reactions are illustrated in Scheme 13 for trimethylsilyl enol ethers of acylsilanes (**26**).

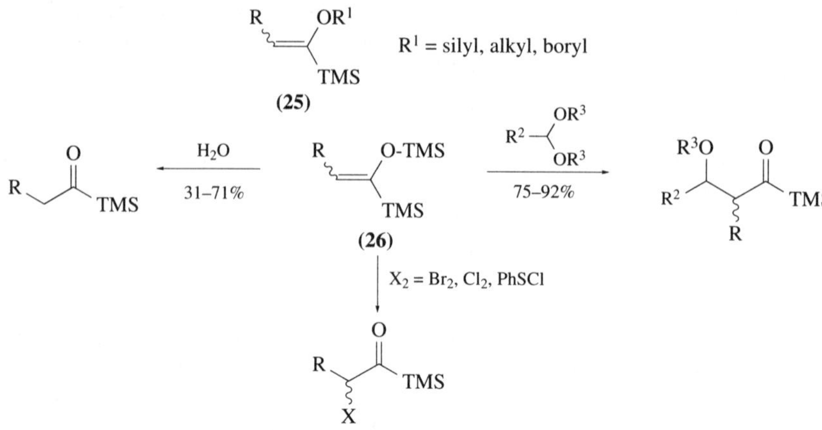

Scheme 13

Much effort has been directed to synthesis of the enol ethers of acylsilanes. These enol ethers of acylsilanes can be classified according to the R^1 group in compound (**25**) as silyl enol ethers (R^1 = TMS), alkyl enol ethers (R^1 = alkyl) and boryl enol ethers of acylsilanes (R^1 = boryl). These individual classes are discussed below.

*(i) Silyl enol ethers of acylsilanes (**26**)*

Silyl enol ethers of acylsilanes may be obtained by a number of different procedures (Scheme 14). Lithium alkoxides (**27**) can be made from the corresponding trimethylsilyl carboxylate esters ⟨76TL1591⟩. Oxidation of (**27**) with benzophenone affords good yields of silyl enol ethers of acylsilanes (**26**) ⟨77JA4181⟩. The mechanism suggested by Kuwajima involves a one-electron-transfer process generating an intermediate radical species (**33**) which disproportionates by loss of a β-hydrogen atom to generate the enol ether (**26**) (Scheme 15).

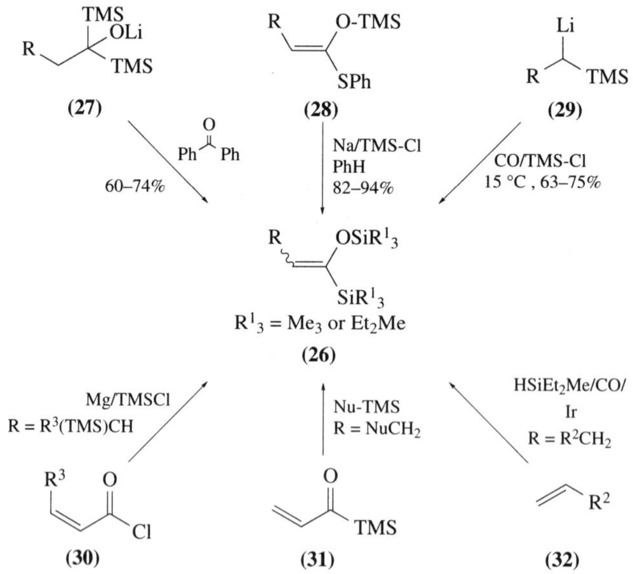

Scheme 14

Treatment of thioether (**28**) with Na/Me_3SiCl in refluxing benzene cleaves the carbon–sulfur bond, and subsequent *in situ* silylation gives the silyl enol ether (**26**). The method for synthesis of the precursor thioether (**28**) from the corresponding ester is detailed in Scheme 16 ⟨78CC478, 80JA6900⟩.

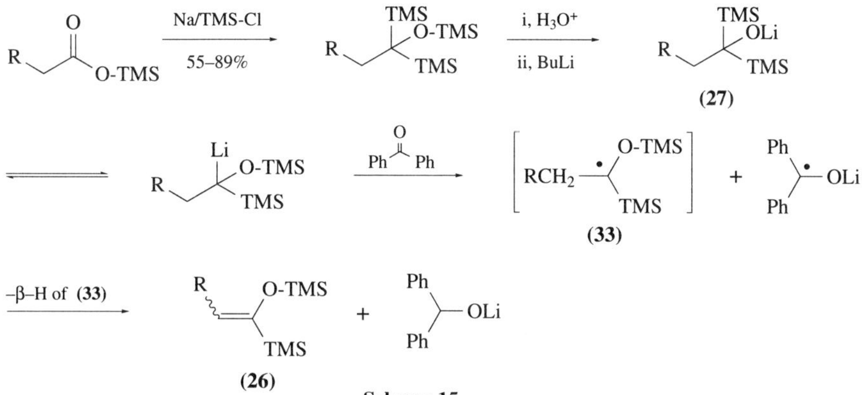

Scheme 15

Reaction of an ester with tris(thiophenyl)aluminum provides a phenylthioester, which is then deprotonated and silylated under standard conditions.

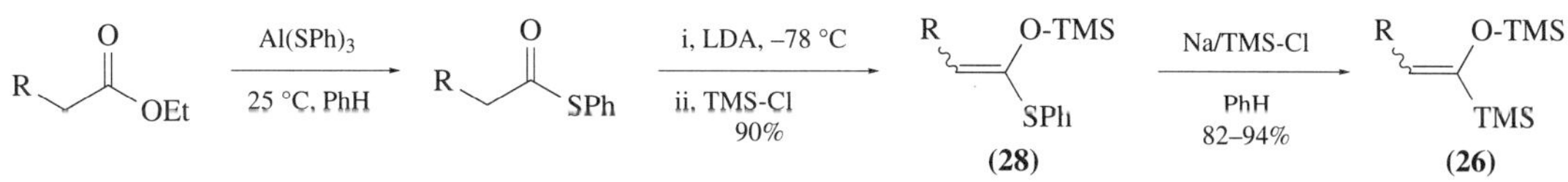

LDA = Lithium diisopropylamide

Scheme 16

α-Lithiosilanes (**29**) add to carbon monoxide (1 atm) at 15 °C to give the α-silyl acyllithium species (**34**) (Scheme 17) ⟨84JA2440⟩. Intramolecular 1,2-silicon shift of (**34**) then gives enolate (**35**). This enolate can be quenched with trimethylsilyl chloride to give the corresponding silyl enol ether (**26**) in good yield with >90% *E* isomeric purity. When the enolate (**35**) is quenched with water, acylsilane (**36**) is produced directly. When the enolate is quenched with benzaldehyde and then with water, syn-β-hydroxyl acylsilane (**37**) is obtained in high diastereoselectivity. Recently, Verlhac *et al.* ⟨92JOM(437)C13⟩ generated the lithium enolate (**35**; R = H) from 1-silyloxyvinyltin (**38**) (Scheme 18). When treated with *n*-BuLi, the tin compound (**38**) underwent rapid Sn–Li exchange and the subsequent 1,2-silicon shift of (**39**) gave the silyl enol ether of enolate (**35**) in 63–73% yield.

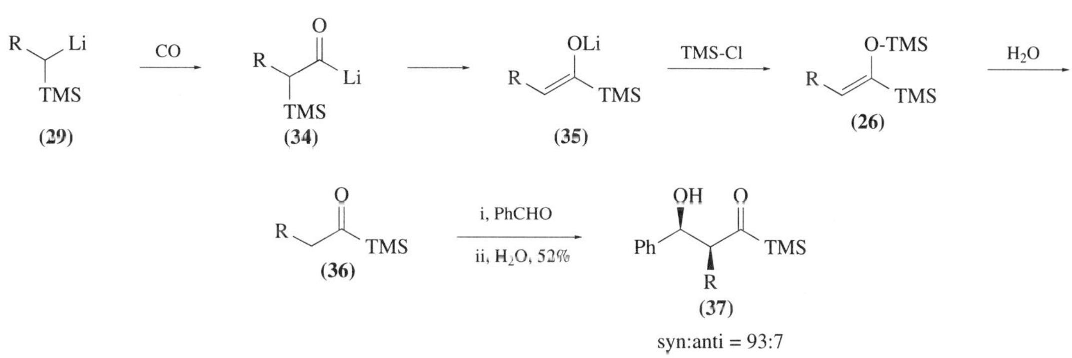

Scheme 17

Scheme 18

Reductive silylation of α,β-unsaturated acyl chlorides (**30**) or acyl imidazoles also gives the corresponding silyl enol ethers (**26**) in moderate yields (Scheme 14) ⟨72JOM(43)157, 74JOM(80)C25⟩.

Propenoyl trimethylsilane (**31**) reacts with a wide variety of silylated nucleophiles ($NuSiMe_3$) to yield the β-functionalized silyl enol ethers of acylsilanes (**26**) (Equation (3)) ⟨87TL4093⟩. The silyl enol ethers obtained through the Michael addition reactions are >90% pure *E* isomers. If these silyl enol ethers are treated with electrophiles other than water (Scheme 13), polyfunctionalized acylsilanes may be obtained easily.

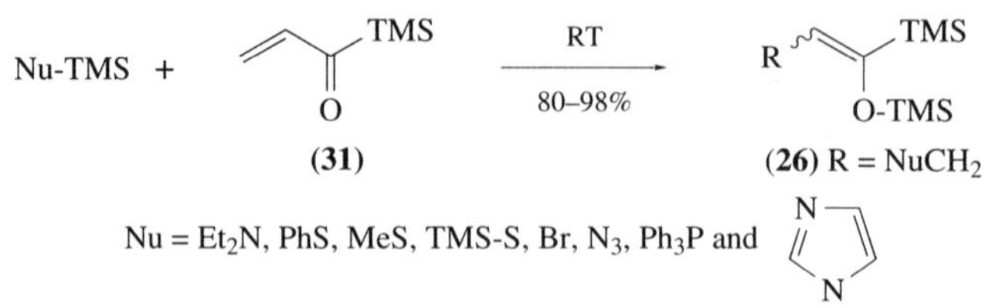

(3)

Murai and co-workers have reported a new method to synthesize silyl enol ethers of acylsilanes ⟨92JA9710⟩. This method is an Ir-catalyzed carbonylation of alkenes (**32**) with a hydrosilane ($HSiEt_2Me$) and carbon monoxide (Equation (4)). The reaction involves the incorporation of a silyloxy(silyl)methylene unit, derived from one molecule of CO and two molecules of $HSiEt_2Me$, into the terminal carbon atom of the alkene (**32**). The reaction is highly regioselective but with only slight preference for the *E*-stereoisomer. A range of substituents can be tolerated in the reaction, including silyl, silyloxy, phenyl, butoxy, acetal, nitrile, and epoxy groups.

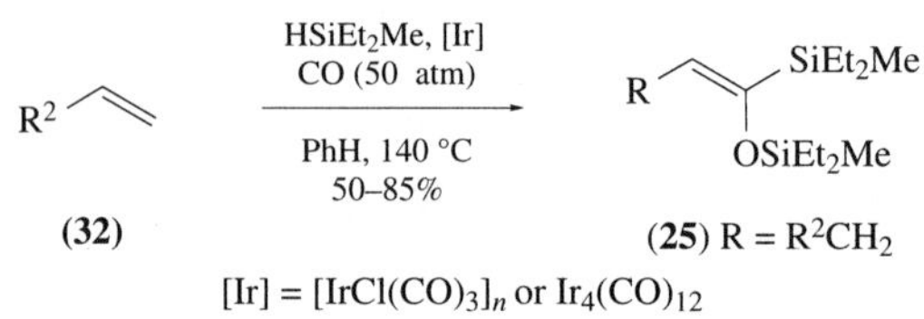

(4)

(ii) Alkyl enol ethers of acylsilanes

Methyl or ethyl enol ethers of acylsilanes have been prepared by methods different from those used for silyl enol ethers of acylsilanes. The first synthesis of these compounds was carried out by Dexheimer's route ⟨76JOM(107)229⟩; however, Soderquist *et al.* ⟨82OM830⟩ modified this route to make it preparatively viable. As shown in Scheme 19, simple deprotonation of methyl enol ethers (**40**) followed by silylation of the intermediate vinyllithium species (**41**) gives the methyl enol ethers of acylsilanes (**42**). Soderquist *et al.* found that the yield of the deprotonation reaction is critically dependent on the reaction conditions. A quantitative yield of vinyllithium (**41**; R = H) was obtained by adding *t*-butyllithium to (**40**) at −78 °C, gradually increasing the temperature to 0 °C, then decreasing again to −78 °C, finally chlorotrimethylsilane was added. Although the reaction conditions are crucial, the reactions themselves are not difficult to execute. Application of this procedure to the synthesis of several different acylsilanes has resulted in excellent yields of (**42**) ⟨90JOM(388)47, 90OS(68)25, 91HCA1477⟩. Good yields were also obtained for acetylsilanes (**43**–**45**) with substituents on silicon ⟨92CB607⟩.

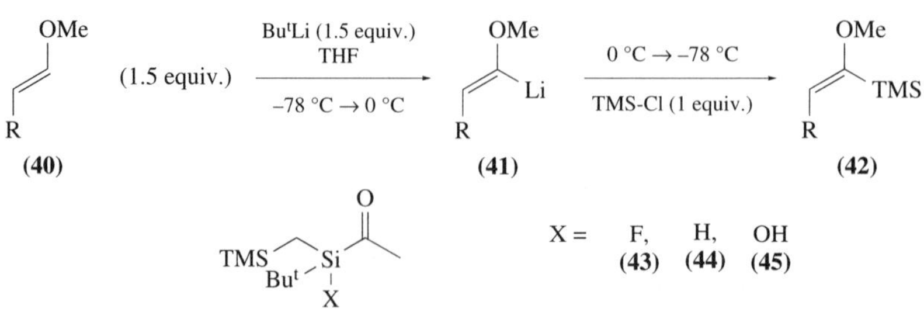

Scheme 19

Yoshida and co-workers ⟨91JOC1307⟩ have developed a new method for the synthesis of methyl enol ethers of acylsilanes by one-carbon homologation of aldehydes. According to this method (Scheme 20), deprotonation of methoxy bis(trimethylsilyl) methane (**46**) with *n*-butyllithium in THF generates the silicon stabilized anion (**47**). This anion reacts with aldehydes to produce the methyl enol ethers (**42**) by a sequence of nucleophilic addition and Peterson elimination. The yields of the acylsilanes after mild acidic hydrolysis range from 51% to 92%.

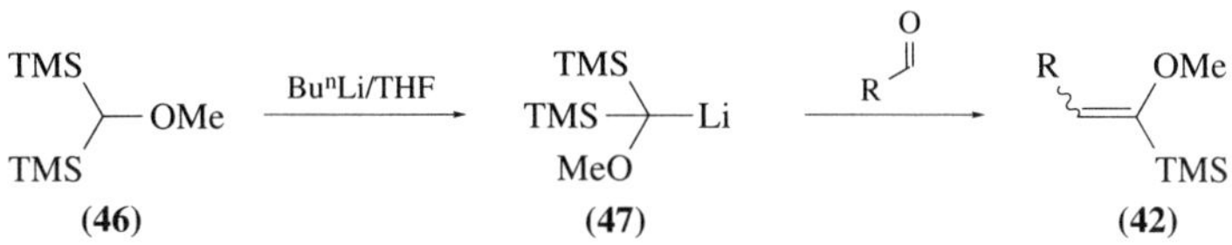

Scheme 20

*(iii) Boryl enol ethers of acylsilanes (**49**)*

Until the early 1990s, boryl enol ethers of acylsilanes had only been generated by hydroboration–oxidation of alkynylsilanes. Until the modification made by Zweifel ⟨81S288⟩, this was not considered a good method for the synthesis of acylsilanes, due to the low yields of boryl enol ethers ⟨77JA3184, 77JOM(131)C1⟩. In Zweifel's modification (Scheme 21), commercially available borane–dimethyl sulfide complex is used as the hydroborating agent instead of the dialkylborane previously used. Oxidation of the resulting trivinylborane (**48**) can be successfully stopped at the boryl enol ether (**49**) stage by using anhydrous trimethylamine oxide. The overall yields from the alkynylsilanes to acylsilanes are 75–91%. Although this procedure cannot be extended to very bulky substituted alkynylsilanes (e.g., R = Bu^t) and cannot be used to synthesize aryl acylsilanes, it is still considered one of the best methods for the preparation of simple acylsilanes ⟨90CSR147⟩.

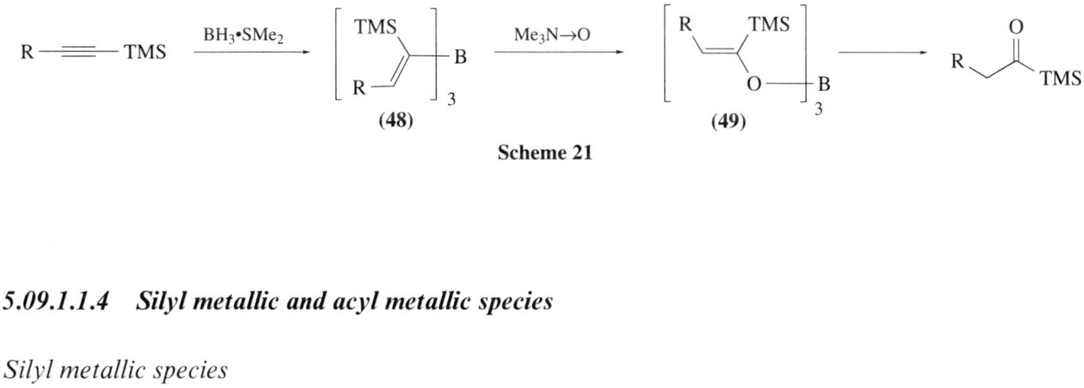

Scheme 21

5.09.1.1.4 Silyl metallic and acyl metallic species

Silyl metallic species

Synthesis of acylsilanes by the direct coupling of silyl metallic species with acyl chlorides is one of the earliest methods attempted for the synthesis of acylsilanes. However, because the reaction of triphenylsilylpotassium with benzoyl chloride (6%) ⟨57JA4373⟩ and the reaction of triphenylsilyllithium with acetyl chloride (Equation (5)) ⟨58JA4529⟩ both occurred in very low yields, this method had been considered to be of little utility.

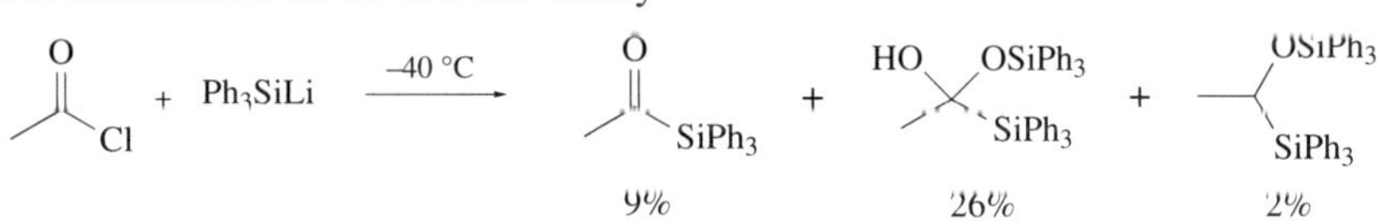

(5)

Improvements since the late 1970s have spurred new interest in this direct coupling method. Less reactive silyl metallic species such as silyl cuprates react smoothly with acyl chlorides to give the corresponding acylsilanes ⟨78JOM(161)C23, 88JOC3612, 89SC275⟩. The silyl cuprates are generated from the silyllithium and CuI or CuCN *in situ*. This synthetic approach may be used with a wide variety of acyl halides leading to a simple and direct synthesis of acylsilanes from commercially available starting materials (Scheme 22).

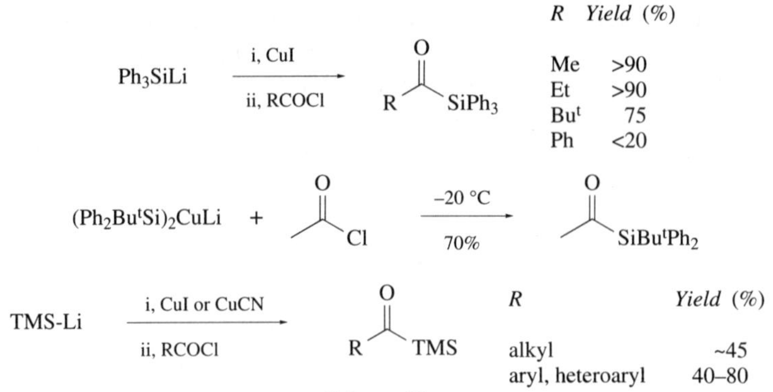

Scheme 22

An even more practical synthesis of acylsilanes utilizes the low ionic character of the aluminum–silicon bond ⟨87TL3261, 91TL4929⟩. In the presence of a catalytic amount of CuCN, lithium tetrakis(trimethylsilyl)aluminum $LiAl(TMS)_4$ or lithium methyl tris(trimethylsilyl)aluminum $LiMeAl(TMS)_3$ reacts with acyl chlorides or anhydrides to yield the corresponding acylsilanes in high yields (Equation (6)). This approach is successful for aliphatic, aromatic, heteroaromatic, and even cyclopropyl acylsilanes. These reaction conditions do not affect ketones, carbamoyl chlorides, esters, nitriles or the acylsilane itself. Therefore even highly functionalized acyl chlorides can be used in the synthesis ⟨87TL3261⟩.

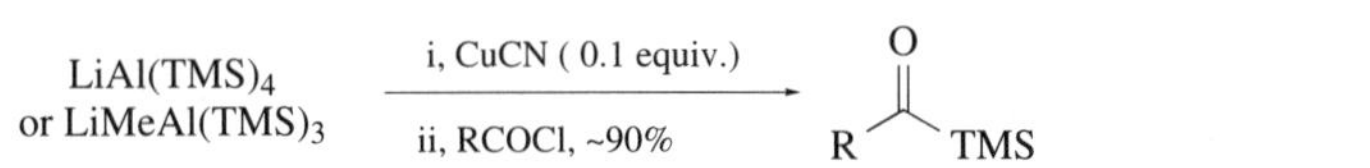

(6)

Although tris(trimethylsilyl)aluminum $Al(TMS)_3$ reacts with acyl chlorides to give only minute quantities of the corresponding acylsilanes ⟨87TL3261⟩, excellent yields have been achieved when it is reacted with *S*-2-pyridyl esters (Equation (7)) ⟨91TL4929⟩. A wide variety of alkyl thioesters with various substituents including an α-alkoxy group can be employed. Problems are encountered with α,β-unsaturated thioesters when using this methodology. These substrates give Michael adducts along with other products.

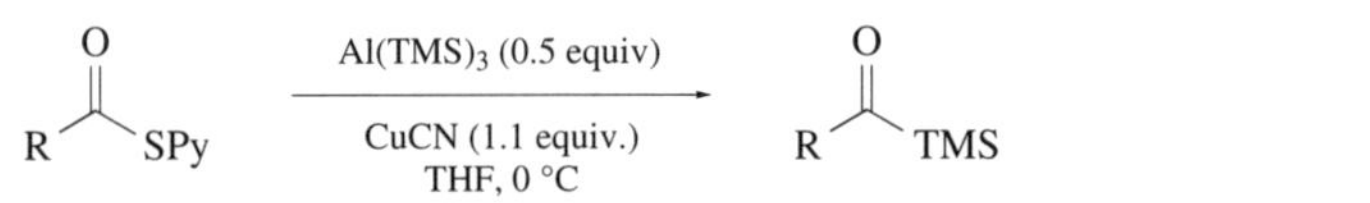

(7)

While simple trialkylsilyllithium reagents are too reactive to be useful for the acylsilane synthesis, methyl bis(trimethylsilyl)silyllithium (**50**) and alkyl bis(trimethylsilyl)silyllithium (**51**) are remarkably stable. These silyl anions react cleanly with acyl chlorides to give the corresponding acylsilanes (**52**) in yields up to 85% (Scheme 23) ⟨82OM994, 86AOC(25)1, 89OM693⟩.

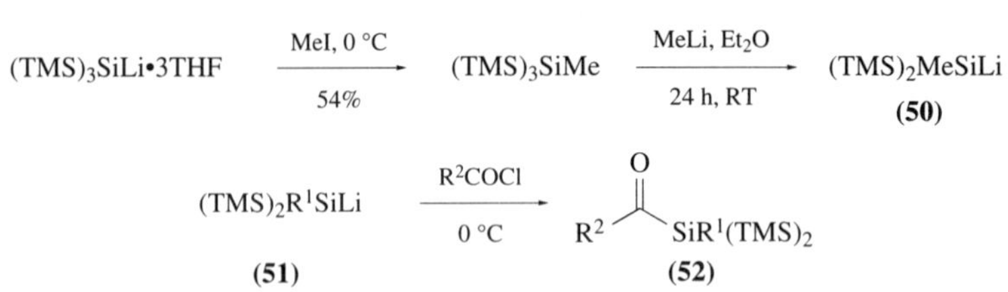

Scheme 23

5.09.1.1.5 *Silylation of acyl metallic species*

There is only one report of the synthesis of acyltrimethylsilanes by the silylation of an acyl lithium species (**53**) (Scheme 24) ⟨82JA5534⟩. Moderate to good isolated yields are obtained only when R is a primary alkyl group. Low temperature must be used to avoid the dimerization of the acyl lithium reagent, which is a prominent side reaction of this method.

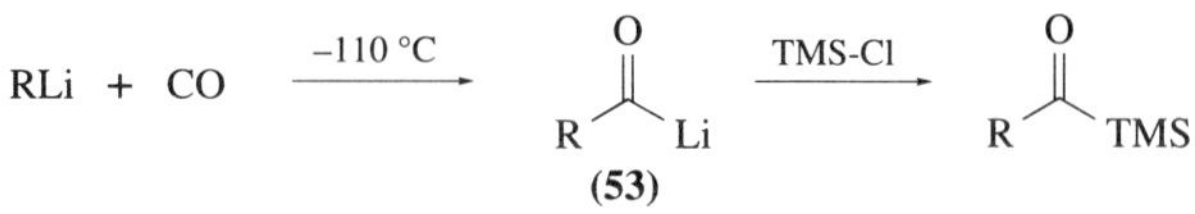

(53)

Scheme 24

5.09.1.1.6 *Transition metal catalyzed synthesis*

(i) Palladium catalyzed coupling of acylchloride and disilanes

The palladium catalyzed synthesis of acylsilanes from suitable acyl chlorides and hexamethyl disilane shown in Equation (8) was first described by Atwell and Bokermann ⟨73USP3772347⟩ and later investigated by Yamamoto *et al.* ⟨80TL1653, 87OM974⟩ and Degl'Innocenti and co-workers ⟨85JOC130⟩. This procedure can be used for the synthesis of a wide variety of substituted aryl and heteroaryl acylsilanes. It can even be used for the synthesis of aroylsilanes with electron-withdrawing substituents, which are often difficult to prepare. The yields of the palladium coupling method are generally lower than the silyl cuprate route, although high yields were occasionally reported with some of the compounds ⟨81JAP81125395, 81JAP81125320, 82JOM(225)331⟩. This method cannot be employed to synthesize alkyl acylsilanes because of very low coupling yield and concomitant formation of several unidentified materials.

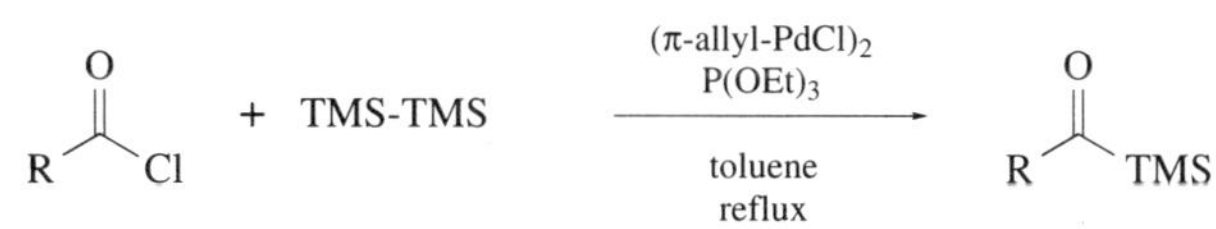

R	*Yield* (%)
o-Ar	24–32
m-Ar	40–57
p-Ar	37–81
hetero-Ar	50–60

(8)

(ii) Isomerization of α-silyloxyallylsilanes

α-Silyloxyallylsilanes (**55**) can be converted into acylsilanes (**56**) by palladium catalyzed isomerization of double bonds under neutral conditions (Scheme 25) ⟨79JOM(175)C1⟩. The yields are generally good (50–80%) for R = H or alkyl group. The allylsilanes (**55**) can be prepared regioselectively in good yield by the reaction of chlorotrimethylsilane with a lithiated derivative of an allyloxysilane (**54**). This isomerization may also be catalyzed by $HRh(PPh_3)_4$ ⟨88JOM(344)71⟩.

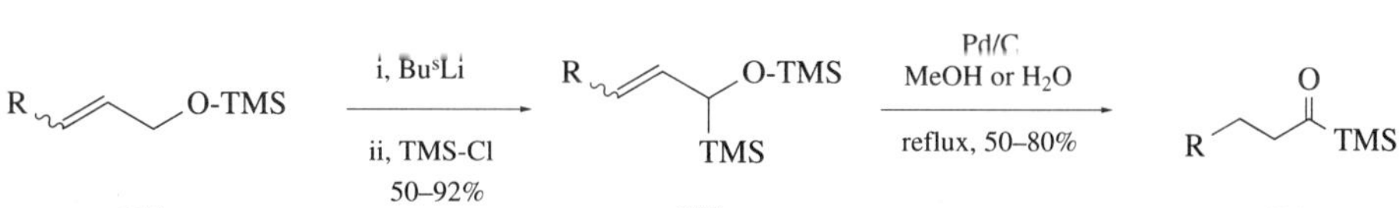

(54) R = H or alkyl **(55)** **(56)**

Scheme 25

5.09.1.1.7 *Reductive silylation*

The reductive silylation of *α,β*-unsaturated acyl chlorides and acyl imidazoles to generate the acylsilanes through the corresponding silyl enol ethers has been discussed in Section 5.09.1.1.3. Reductive silylation of alkyl benzoates (**57**) with Mg/Me_3SiCl/HMPA (hexamethylphosphoramide) also gives acylsilanes, but through the intermediacy of *O*-alkyl-*O*-silyl mixed ketals (**58**) (Scheme 26) ⟨69JOM(19)P33, 71JOM(26)183, 79JOC420⟩. In this reaction, an excess of trimethylchlorosilane relative to HMPA is needed and also basic reagents and high temperatures (> 50 °C) must be avoided during the reaction and workup. A mechanism involving the ketyl (**59**) (Scheme 27) has been proposed for the transformation of the ester to the ketal (**58**; Ar = Ph) ⟨71JOM(26)183⟩. This "magnesium system" has also been reported to be successful when tetrahydrofuran is used as a solvent for the reaction with *N*-phenyl pivalimidoyl chloride (**60**). The acylsilane is formed through the intermediacy of aldimine (**61**) (Scheme 28) ⟨74JOM(76)C1⟩. The hydrolysis of an aldimine like (**62**) often results in the formation of aldehydes and acylsilanes in about 1 : 1 ratio, which limits the generality of this aldimine method (Scheme 29) ⟨74JOC600⟩.

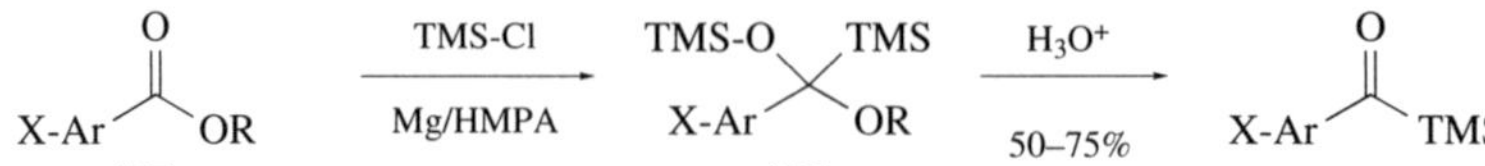

(57) **(58)**

R = Me or Et
X = Me, Me_3C, TMS, Cl, OH

Scheme 26

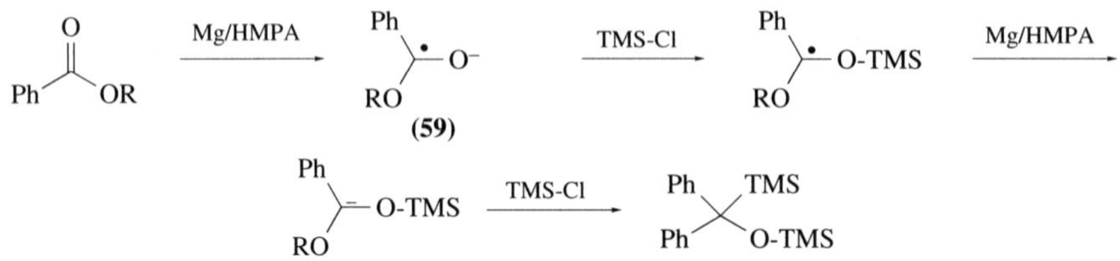

(59) **(58)**

Scheme 27

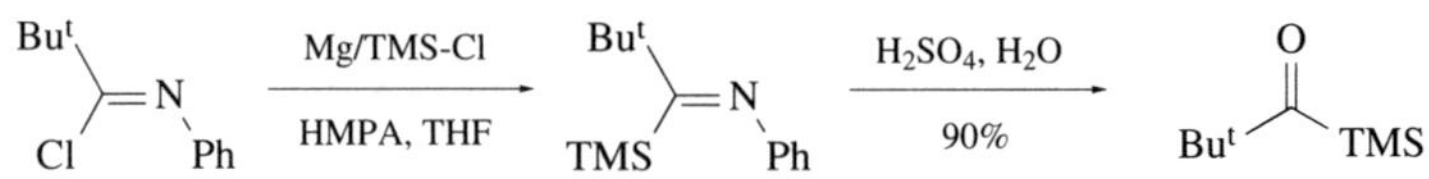

(60) **(61)**

Scheme 28

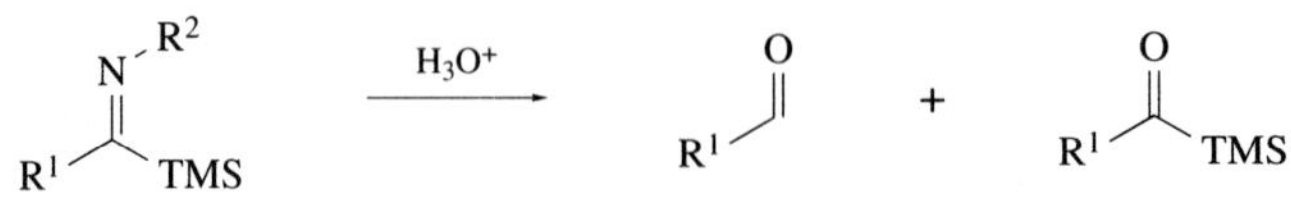

(62) ~1:1

Scheme 29

Comparing the syntheses of benzoyltrimethylsilane, the dihalide method and dithiane method (Section 5.09.1.1.1) gave the product in yields of 40% and 60%, respectively, while the reductive silylation with the Me_3SiCl/Mg/HMPA procedure gave a yield of 70% ⟨71JOM(26)183⟩.

A similar system, Me_3SiCl/Li/THF, can be used in the reductive silylation of aliphatic esters, but the yields are poor and concomitant formation of several other products is observed ⟨75JOM(93)51⟩. Reductive silylation of cyanohydrins (**63**) with this system also gives poor yields of the enamine intermediates (**64**), which, upon hydrolysis, give the corresponding acylsilanes. Simply changing the solvent from THF to HMPA can improve the yields significantly (Scheme 30) ⟨90JOM(391)13⟩. Although this route is experimentally demanding, it produces branched chain acylsilanes (R = 2°-alkyl) in good yields. Most other methods give good yields only for straight chain or β-substituted acylsilanes (R = 1°-alkyl).

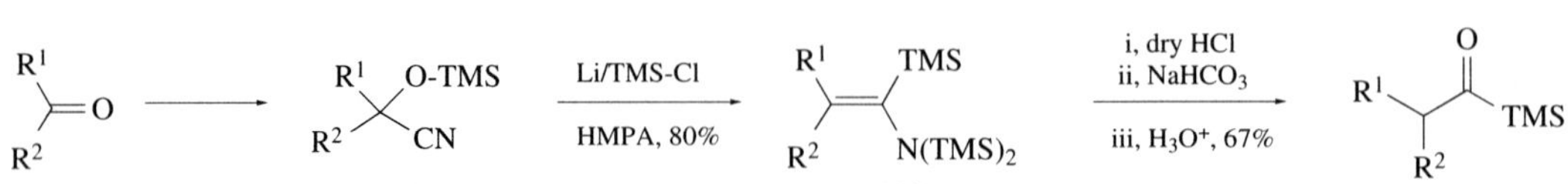

(63) **(64)**

R^1, R^2 = alkyl

Scheme 30

Reductive silylation of carboxylic esters with the "sodium system", Me_3SiCl/Na/THF, does not stop at the acylsilane stage ⟨78BCJ2391⟩. Further reductive silylation gives 1,1-bis(trimethylsilyl)-1-alkanols. Therefore this system is not useful for the synthesis of acylsilanes.

5.09.1.1.8 Rearrangement of silyloxycarbenes

Silyloxycarbenes (**66**) may be expected to insert into the Si—O bond to give acylsilanes. The silyloxycarbenes can be generated either by the pyrolysis of α-ketoesters (**65**) (Scheme 31) ⟨75JOM(99)379⟩, or by the thermal decomposition of acyl cobalt anions (**67**) (Scheme 32) ⟨81JOM(205)31, 77CC73⟩. These two routes are restricted to the preparation of benzoylsilanes and the yields are low (38% from the starting silanes). It is also difficult to prepare the requisite α-ketoesters (**65**) for the pyrolysis. These limitations currently restrict the generality of this silyloxycarbene approach to acylsilanes.

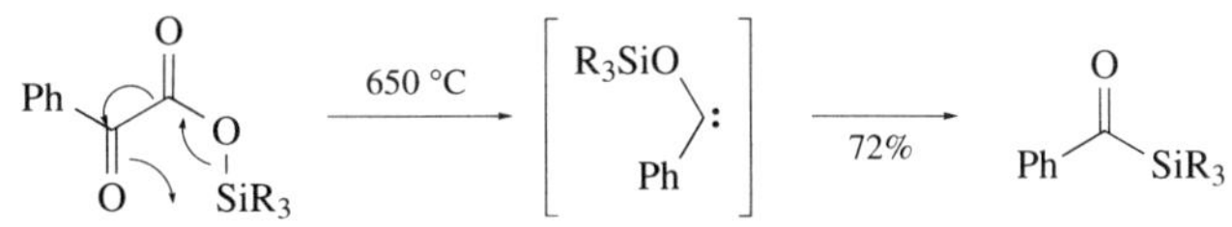

(**65**) (**66**)

$R_3 = Me_2Ph$

Scheme 31

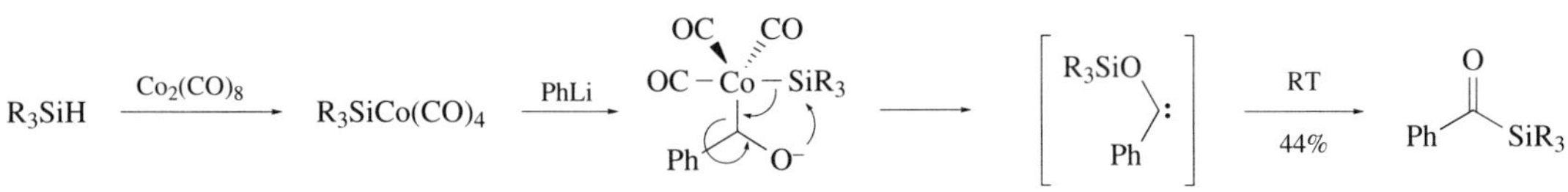

(**67**) (**66**)

$R_3 = MePh_2$

Scheme 32

5.09.1.1.9 Claisen rearrangement

The acylsilanes discussed so far are all achiral. To make chiral acylsilanes, either a chiral starting material must be used and the chirality maintained throughout the synthesis ⟨91JOC1307⟩ or a prochiral substrate may be used with a chiral reagent or catalyst. In the latter case, Yamamoto and co-workers ⟨90JA7791, 91SL793, 91TA647⟩ have used a C2-symmetric organoaluminum reagent (*R*)- or (*S*)-(**68**) to promote the Claisen rearrangement of allylic α-(trimethylsilyl) vinyl ethers (only (*R*)-(**68**)) is shown in Equation (9). Equation (9) shows that the diastereoselectivity of the rearrangement to form (**70a**) and (**70b**) is much higher under the catalyzed conditions than under thermal conditions. With the same catalyst, (*R*)-(**68**), both *trans*- and *cis*-allylic vinyl ethers (**71**) and (**72**) rearrange to acylsilane (**73**) with the same absolute configuration (Scheme 33). The yield and optical purity of (**73**) from the *trans*-ether (**71**), however, are higher than those from the *cis*-ether (**72**). The use of the more bulky *t*-butyldiphenylsilyl group of the catalyst (Ar_3 of **68**) exhibits the highest enantioselectivity for the *trans*-ether (**71**), but the lowest for the *cis*-ether (**72**). Transition states **A, B**, and **C, D** are postulated by Yamamoto for the asymmetric Claisen rearrangement of (**71**) and (**72**), respectively. The orientation of the α-methylene group in chairlike structures **A** and **B** are different. The chiral aluminum catalyst (*R*) (**68**) can discriminate this difference and makes a good match with only structure **A**, leading to the formation of acylsilane (**73**). The rearrangement of *cis*-ether (**72**), however, proceeds through the boatlike structure **C** rather than the crowded chairlike structure **D** (1,3-diaxial interaction). This also leads to the formation of acylsilane (**71**). The more hindered aluminum reagent ($Ar_3 = Bu^tPh_2$) seems to destabilize the boatlike transition state **C**, and it therefore lowers the yield and selectivity of *cis*-ether (**72**).

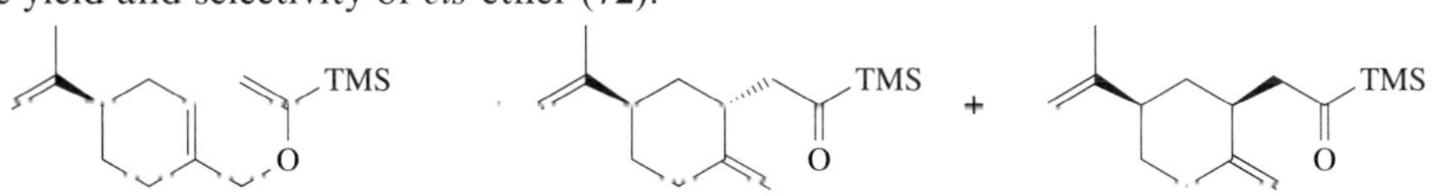

(9)

(**69**) (**70a**) (**70b**)

200 °C 85:15, 91%
(*R*)-(**68**), –20 °C 97:3, 75%

5.09.1.2 Formylsilanes

Formylsilanes are the simplest acylsilanes. Early unsuccessful attempts to prepare them led to the conclusion that they were unstable species ⟨54JA1613, 68AOC(7)95⟩. The first evidence for formyl

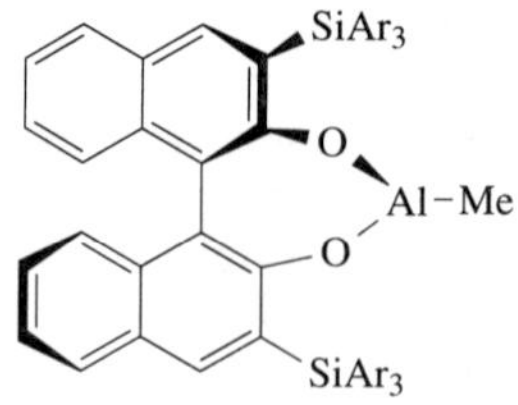

$Ar_3 = Ph_3, Bu^tMe_2, Bu^tPh_2$
(*R*)-(**68**)

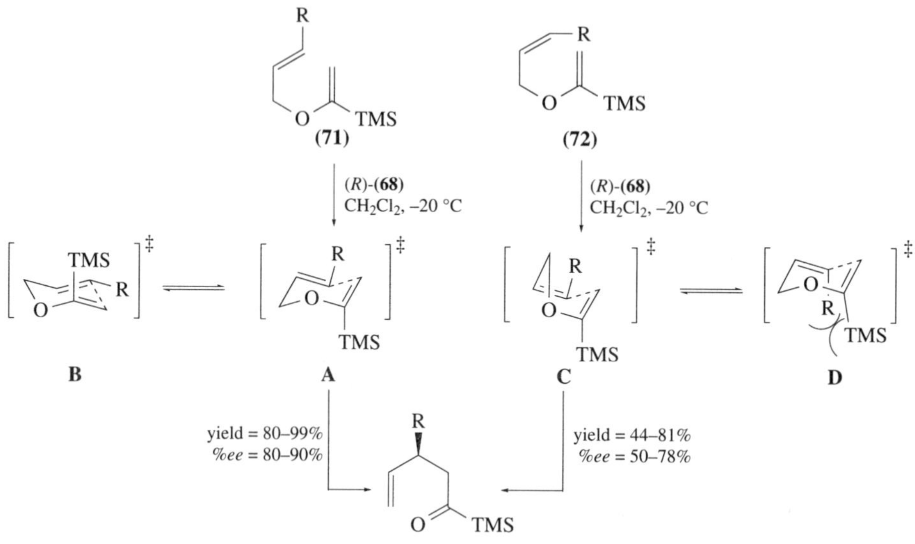

Scheme 33

trimethylsilane Me_3SiCHO was obtained by Ireland and Norbeck ⟨85JOC2198⟩. Ireland and Norbeck generated formyl trimethylsilane *in situ* by Swern oxidation of trimethylsilyl methanol (Me_3SiCH_2OH) and subsequently trapped it with a Wittig reagent. Formyl trimethylsilane is also considered an intermediate in the preparation of acylsilanes from α-hydroxysilanes (Scheme 9, Section 5.09.1.1.2). The intermediacy of formylsilanes was deduced from the structures of the final products, though these species were not actually observed spectroscopically.

Shortly thereafter, Tilley obtained direct evidence for the existence of formyl trimethylsilane Me_3SiCHO by NMR spectroscopy when he treated zirconium η^2-sila-acyl complex (**74**) with hydrogen chloride ⟨87JA2049⟩. Later, he successfully prepared an η^2-$COSi(SiMe_3)_3$ derivative of zirconium (**76**) by the reaction of silylzirconium species (**75**) with carbon monoxide and then treated it with anhydrous HCl to yield the first stable, isolable formylsilane: formyl tris(trimethylsilyl)silane $(Me_3Si)_3SiCHO$ (**77**) (Scheme 34) ⟨88JA313⟩. Formylsilane (**77**) decomposes instantly upon exposure to air, but is thermally stable under nitrogen. In a similar way, another stable formylsilane, formyl triphenylsilane Ph_3SiCHO has also been isolated ⟨92OM2198⟩.

Soderquist ⟨92JA10078⟩ reported a convenient preparation of formyl triisopropylsilane (**80**) Pr^i_3SiCHO from a modified dithiane approach (Scheme 35). Solvolysis of silyl dithiane (**78**) by $HgCl_2/HgO$ in methanol affords the acetal (**79**). Hydrolysis of this acetal in the presence of $LiBF_4$ in aqueous MeCN at reflux provides the formylsilane (**80**) in excellent yield.

5.09.1.3 Functionalized Acylsilanes

5.09.1.3.1 α-Haloacylsilanes

Direct halogenation of simple acylsilanes (acylsilanes as starting material or to be generated *in situ*) affords the α-chloro- or α-bromoacylsilanes in moderate to good yields (Scheme 36) ⟨76CL993,

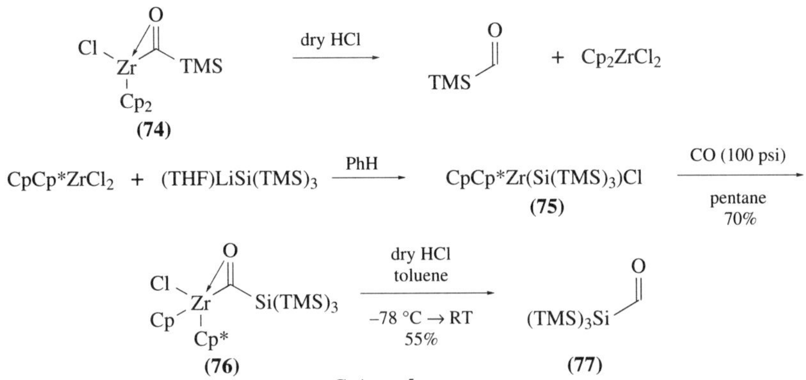

Cp* = η^5-C_5Me_5

Scheme 34

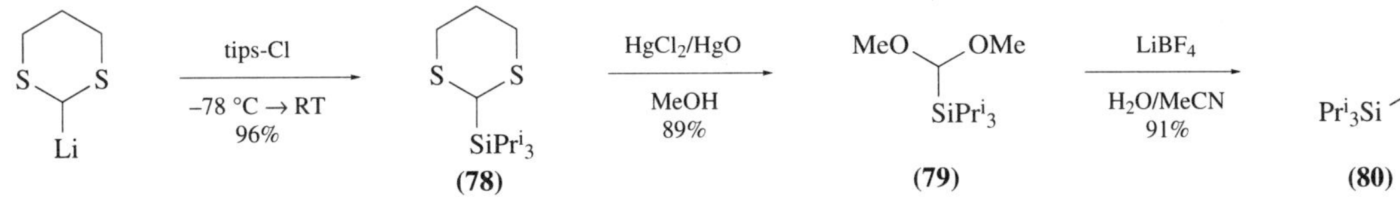

tips-Cl = 1,3-dichloro-1,1,3,3-tetraisopropyldisiloxane

Scheme 35

88T4113⟩. A more general and efficient method involves treatment of enol ethers of acylsilanes with a dihalogen, NCS or NBS (Scheme 13, Section 5.09.1.1.3). High yields (80–95%) in this halogenation are generally obtained from silyl enol ethers ⟨87TL4093, 84BCJ2167⟩ and alkyl enol ethers ⟨88T4113, 89JOC2798, 91JOC1307⟩ of acylsilanes. Other unactivated double bonds in the enol ethers remain unattacked by the halogen. Generation of boryl enol ethers from silyl alkynes (Scheme 21) and the subsequent treatment with NBS can be carried out as a one-pot reaction, but the yield of the resulting α-bromoacylsilanes is lower than the multistep routes (52–61%) ⟨90T2573, 86TL5421⟩.

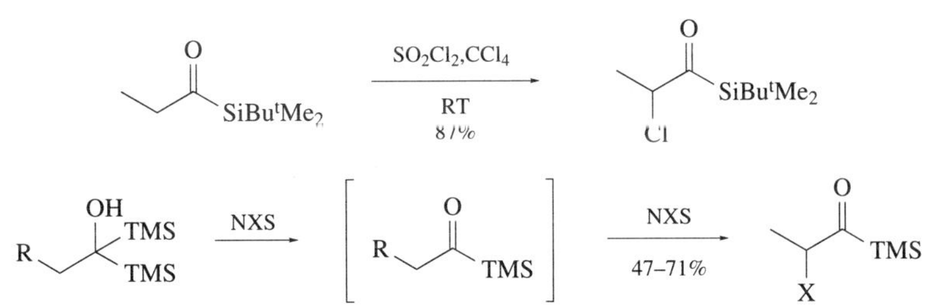

X = Cl, Br
R = $PhCH_2$, n-C_6H_{13}

Scheme 36

In 1991, Jackson and co-workers developed a three-step synthesis of α-bromoacylsilanes by one carbon homologation of aldehydes or ketones (Scheme 37) ⟨91JCS(P1)3103, 91JCS(P1)897⟩. Darzens reaction of chloromethyl phenyl sulfone with aldehydes or ketones affords 2-phenylsulfonyl oxiranes (**81**). Silylation of compound (**81**) followed by treatment with magnesium bromide generates the α-bromoacylsilanes in variable yields.

5.09.1.3.2 α-Ketoacylsilanes

The first α-ketoacylsilanes (**86**) were synthesized by Reich *et al.* (Scheme 38) ⟨82JA1119, 83T949⟩ by using an allene metalation analogous to the alkene metalation method used for simple acylsilanes (Scheme 19). Base catalyzed isomerization of the ethoxyethyl (EE) protected propargyl alcohol (**82**) gives the allene (**83**) in high yield ⟨68MI 509-01⟩. The ethoxyethyl group is used for protection because it is relatively large and polar and therefore makes the protected compounds easier to handle than

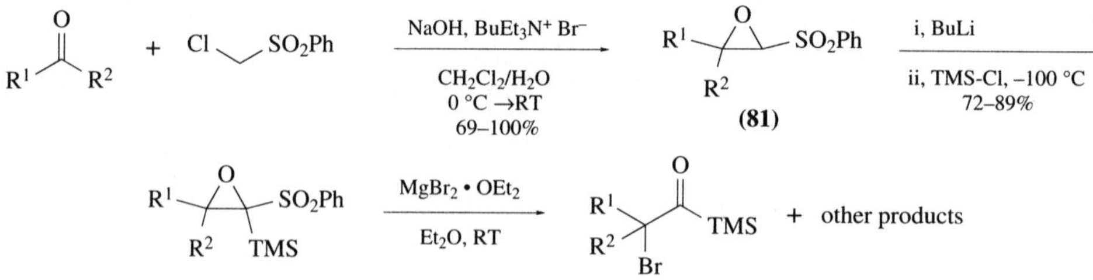

R^1	R^2	*Yield* (%)
Me, Et, Pr, Bu	H	75–80
H	H	50–65
alkyl	alkyl	0–7

Scheme 37

other analogs. Lithiation of the allene (**83**) followed by silylation with a chloroalkylsilane yields the silylated allene (**84**). α-Ketoacylsilanes (**86**) can be obtained after oxidation of allene (**84**) with *m*-chloroperbenzoic acid (mcpba), probably via the allene oxide intermediate (**85**).

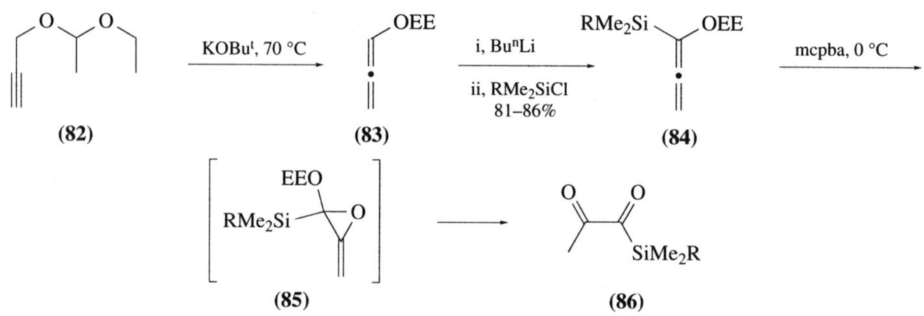

R	*Yield* (%)
Me	42
Bu^t	90

Scheme 38

An alternative route to α-ketoacylsilanes involves the osmium catalyzed oxidation of (*E*)-alkenylsilanes (**87**), followed by Swern oxidation of the resulting *cis*-diols (**88**) (Scheme 39) ⟨90T2573, 86TL2527⟩.

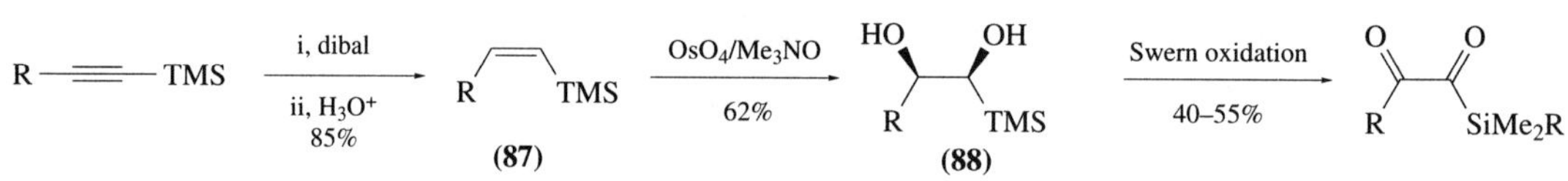

dibal = diisobutylaluminium

R = Ph, n-C_5H_{11}, Bu^n

Scheme 39

5.09.1.3.3 *α,β-Unsaturated acylsilanes*

(i) *Hydrolysis of 1,3-dithianes and 1,3-dioxanes*

Only a few α,β-unsaturated acylsilanes have been synthesized by the dithiane method described in Section 5.09.1.1 ⟨86HCA1378, 85TL2513⟩. The difficulty lies in the final hydrolysis step which is accompanied by partial decomposition of the desired acylsilane products ⟨85TL2513⟩. The analogous 1,3-dioxane route has been successfully used for a few alkynyl acylsilanes ⟨83CC239, 83T3073⟩.

(ii) Oxidation reactions

Oxidation of allylic and acetylenic carbinols has been described in Section 5.09.1.1.2 (Schemes 8 and 9) ⟨85JOC5393, 88JOC1569⟩. The method in Scheme 8 is especially successful for the synthesis of compounds lacking β-alkene substituents such as acryloyl and α-methylacryloylsilanes. Two other oxidation routes to α,β-unsaturated acylsilanes have been reported with mcpba: one involves isoxazoline (**89**) ⟨87TL913⟩ and the other cyclopropene (**90**) (Scheme 40) ⟨91JOC5480⟩; however, these two routes are not general.

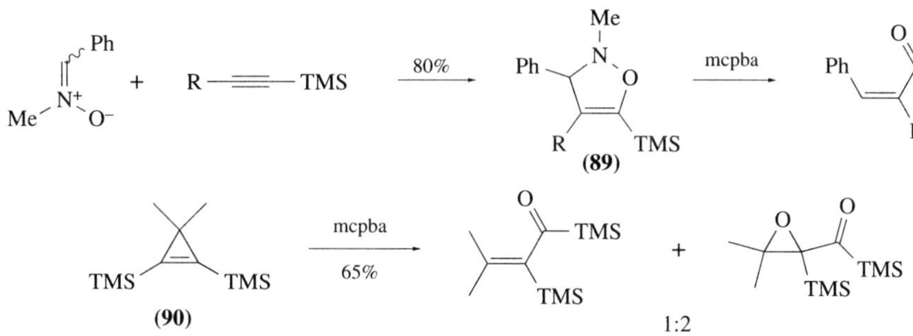

Scheme 40

(iii) Hydroboration–oxidation

Hydroboration of enyne (**91**) (Scheme 41) ⟨77JOM(131)C1, 80JA1577⟩ followed by the oxidative isomerization of vinylboronate (**92**) generates the acryloylsilane (**93**) in a way similar to the hydroboration–oxidation of simple alkynylsilanes. However, compound (**93**) is the only α,β-unsaturated acylsilane that has been synthesized by this method so far.

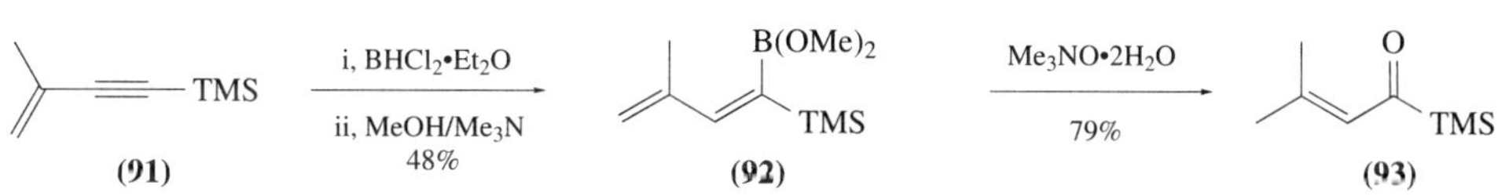

Scheme 41

(iv) Enol ether methodology

Another simple way to prepare α,β-unsaturated acylsilanes is by the hydrolysis of the corresponding methyl enol ethers (**95**) (Scheme 42) ⟨80JA1577⟩. This approach is also not as useful for making more complex α,β-unsaturated acylsilanes due to difficulties in preparation of the appropriate dienes (**94**).

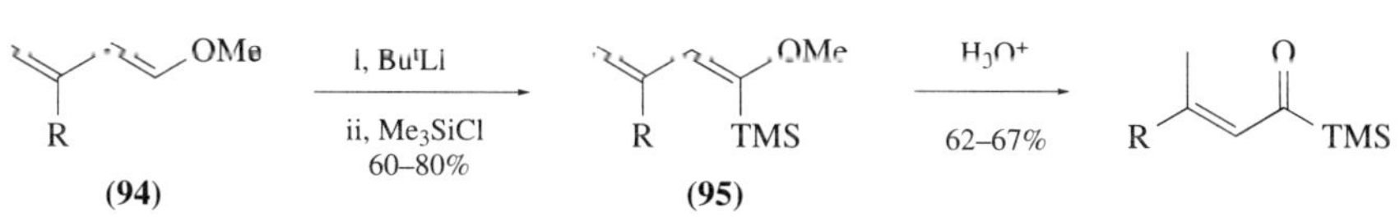

R = H, Me

Scheme 42

Several α,β-unsaturated acylsilanes have been synthesized in good yield from the corresponding silyl enol ethers (Scheme 43) ⟨78JOM(145)C1, 92T7629⟩. Phenyl sulfenylation of the silyl enol ethers

(**96**) gives the α-thioacylsilanes (**97**). Oxidation of the sulfide (**97**) followed by *in situ* thermal decomposition of the resulting α-phenylsulfinylacylsilanes (**98**) yields the α,β-unsaturated acylsilanes.

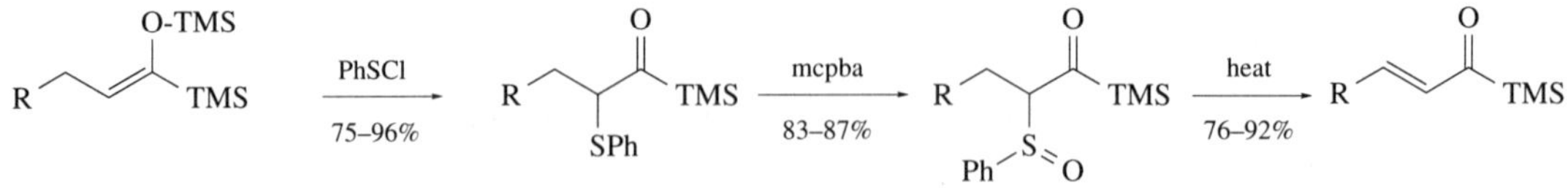

(**96**) (**97**) (**98**)

R = Pr^i, C_5H_{11}, CH_2=CH$(CH_2)_6$, Ph

Scheme 43

A wide variety of α,β-unsaturated acylsilanes have been synthesized from silyl acetylsilane (**99**) by sequential lithiation, alkylation, lithiation; aldol condensation, and Peterson elimination (Scheme 44) ⟨81JA6217⟩. By this route, α,β-unsaturated acylsilanes can be obtained with *E*-configuration in very high yield. Silyl acetylsilane (**99**) can be easily prepared by the hydroboration–oxidation method ⟨81S288⟩. But the acylsilane (**99**) has to be used shortly after preparation as it isomerizes to the silyl enol ether even when stored at low temperatures.

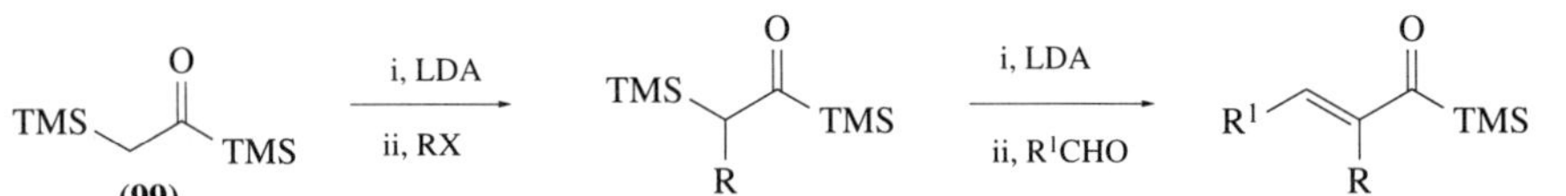

Scheme 44

An alternative alkeneation route to substituted α,β-unsaturated acylsilanes starts from the stable α-phosphonoacylsilane (**100**), which can be prepared in quantitative yield from α-iodoacylsilane via the Arbuzov reaction (Scheme 45) ⟨89JOC2798⟩. Alkylation of (**100**) gives the monosubstituted α-phosphonoacylsilane (**101**). Horner–Wadsworth–Emmons coupling of (**100**) or (**101**) with an aldehyde generates β-substituted or α,β-disubstituted unsaturated acylsilanes (**102**) or (**103**), respectively (74–97%). These substituted alkenoylsilanes are exclusively *E*-isomers.

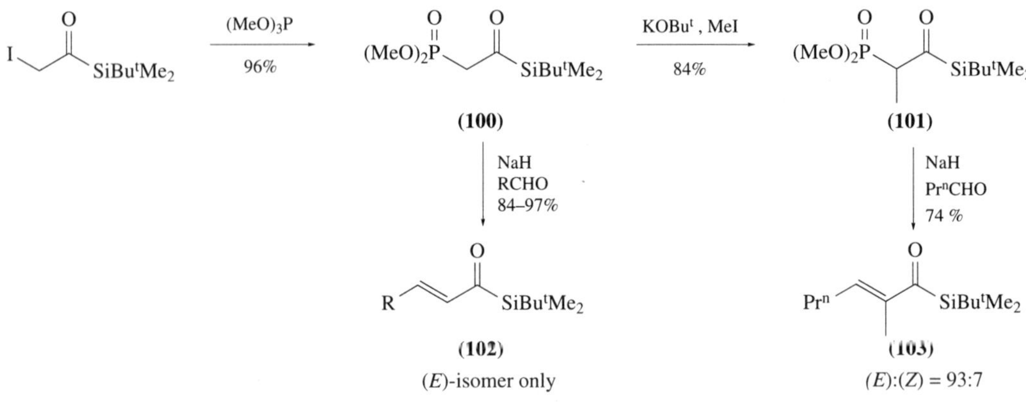

R = Me, Pr^i, Bu^t, cycloproyl, Ph, *p*-NO_2-C_6H_4, etc.

Scheme 45

A new approach to construct conjugated acylsilanes is based on the dichloride (**105**) which is prepared by photolytic addition of bromotrichloromethane to ethyl enol ether (**104**) (Scheme 46) ⟨92TL6751⟩. Treatment of (**105**) with controlled quantities of organocuprates RCu(CN)Li generates the monosubstituted products (**106**) with *E*-selectivity. Further treatment of (**106**) with R_2Cu(CN)Li yields the β,β-disubstituted alkenoylsilanes (**107**). The disubstituted products (**107**) can also be directly obtained by treating the dichloride (**105**) with an equivalent amount of R_2Cu(CN)Li. Sequential treatment of the dichloride (**105**) with two different organocuprates provides mixed disubstituted products with high stereoselectivity.

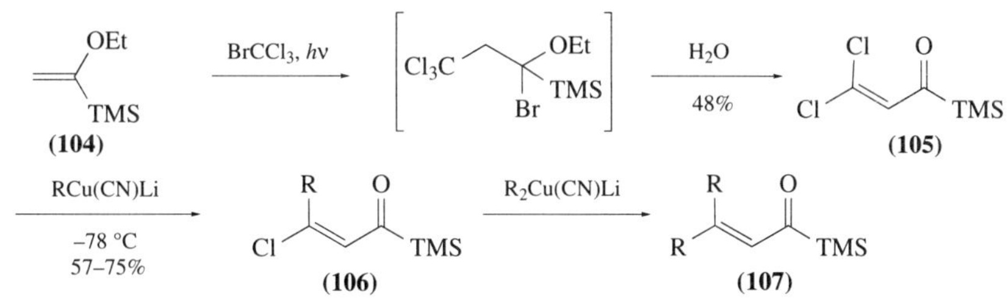

R = Me, Bu^i, Bu^s, Bu^t, Ph

Scheme 46

(v) Hydrolysis of silyl allenes

Allenyl ethers (**84**) are not only precursors of α-ketoacylsilanes (**86**) (Scheme 38, Section 5.09.1.3.2), they are also precursors of fully substituted allenyl ethers (**108**) and (**109**) and of α,β-acetylenic acylsilanes (**113**) (Scheme 47) ⟨71TL591, 73TL2585, 80TL3987, 82TL3101, 82JA1119, 83T949⟩. Sequential alkylation (or silylation, or selenylation) of (**84**) generates the substituted allenyl ethers (**108**) and (**109**). These allenyl ethers can be hydrolyzed to silyl enones (**110**; E = H), or subjected to electrophilic substitution to give α-heterosubstituted silyl enones (**110**; E = SePh, Cl, Br). *m*-Chloroperbenzoic acid oxidation of γ-methylselenyl allenyl ethers (**109**; R^1 = alkyl, R^2 = SeMe) gives the unstable selenoxides (**111**). Upon warming, these unstable intermediates undergo [2,3]-sigmatropic rearrangement to generate the acetals (**112**). Hydrolytic workup of (**112**) yields the acetylenic acylsilanes (**113**) in fair yields. By the same strategy, allenyl acylsilanes (**115**) were synthesized by the hydrolysis of the corresponding 1-silyl-1-silyloxy cumulenes (**114**) (Scheme 48) ⟨81TL2827⟩.

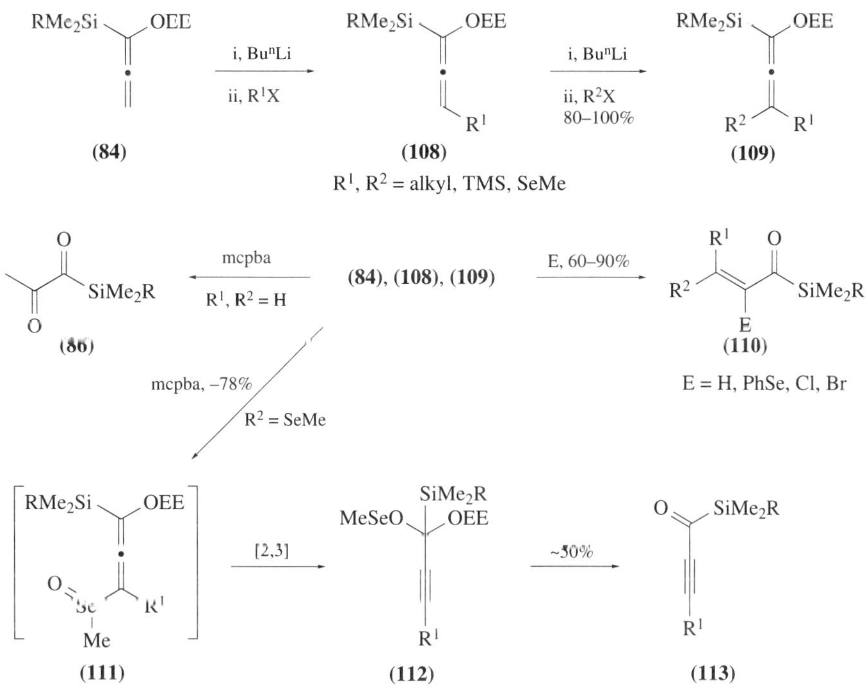

Scheme 47

α,β-Unsaturated acylsilane (**118**) can be obtained from the rearrangement of propargylic alcohol (**116**) (Scheme 49). This rearrangement is catalyzed by tetrabutylammonium perrhenate (Bu_4NReO_4) and *p*-toluenesulfonic acid (*p*-TsOH), and it involves the intermediate allenyl alcohol (**117**) ⟨91CL1413⟩.

(vi) Michael addition reactions

In the early 1990s, Degl'Innocenti and Ricci ⟨92SL332, 92SL329, 92TL1507⟩ developed a new route to α,β-unsaturated acylsilanes based on Michael addition reactions. In this new methodology,

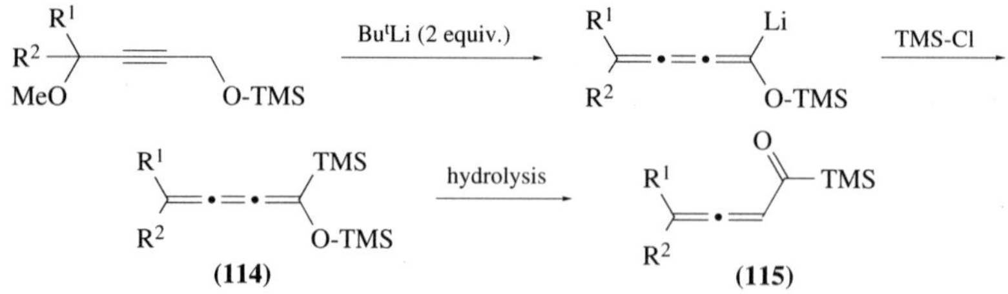

Scheme 48

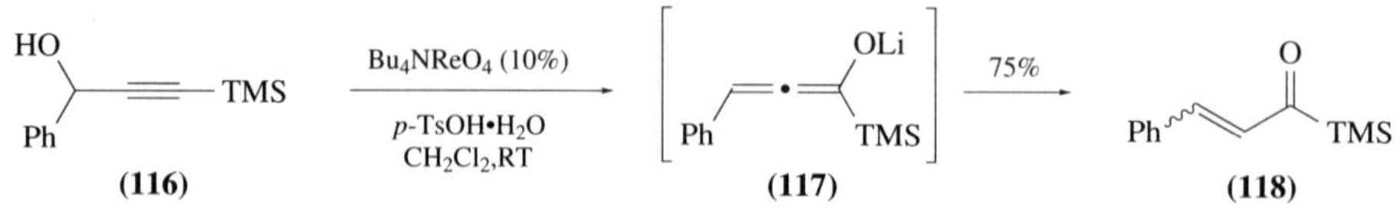

Scheme 49

acetylenic acyltriphenylsilane (**119**) serves as the "lynchpin" precursor for three different synthetic routes (Scheme 50). Nucleophilic 1,4-addition of cuprates or silylated nucleophiles to the acetylenic acylsilane (**119**) affords β-substituted (or β-functionalized) propenoylsilanes (**120, 121**) in good to excellent yields. This addition is regio- and stereoselective; only 1,4-adducts are formed in all cases and usually only the *E*-isomers are obtained. However, when silylated derivatives of ammonia, primary amines or sulfides are used, the *Z*-isomers are formed selectively.

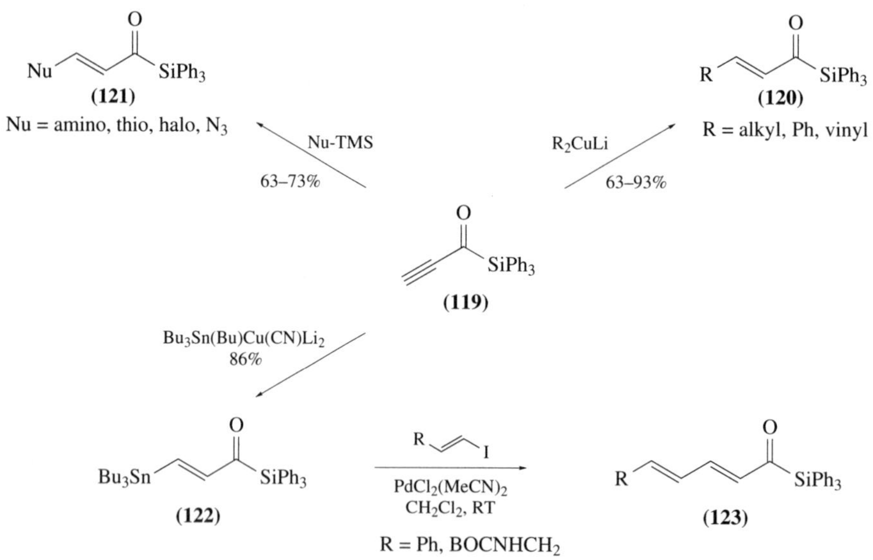

Scheme 50

If acetylenic acylsilane (**119**) reacts with tributylstannyl cuprate, vinyl stannane (**122**) is formed ⟨92SL329⟩. Under palladium catalysis, this vinyl stannane couples with a vinyl iodide providing a stereospecific access to dienoylsilanes (**123**) (Scheme 50). When the reaction intermediate of (**119**) with this specific stannylcuprate is trapped with different electrophiles *in situ* (Scheme 51), such as acetyl chloride, CO_2, and allyl bromide, diketones (**124**) and (**127**) and polyunsaturated ketones (**126**) are synthesized. These enones are otherwise not easily accessible.

5.09.1.3.4 Cyclopropyl acylsilanes

Some convenient syntheses of cyclopropyl acylsilanes have been developed by starting with preformed cyclopropanes. Reaction of a cyclopropyl acyl chloride with $LiAl(TMS)_4$ in the presence of a catalytic amount of CuCN generates the cyclopropyl acylsilane in excellent yield (Equation (6), Section 5.09.1.1.4(i)) ⟨87TL3261⟩. Another approach that includes the cyclopropane ring in the

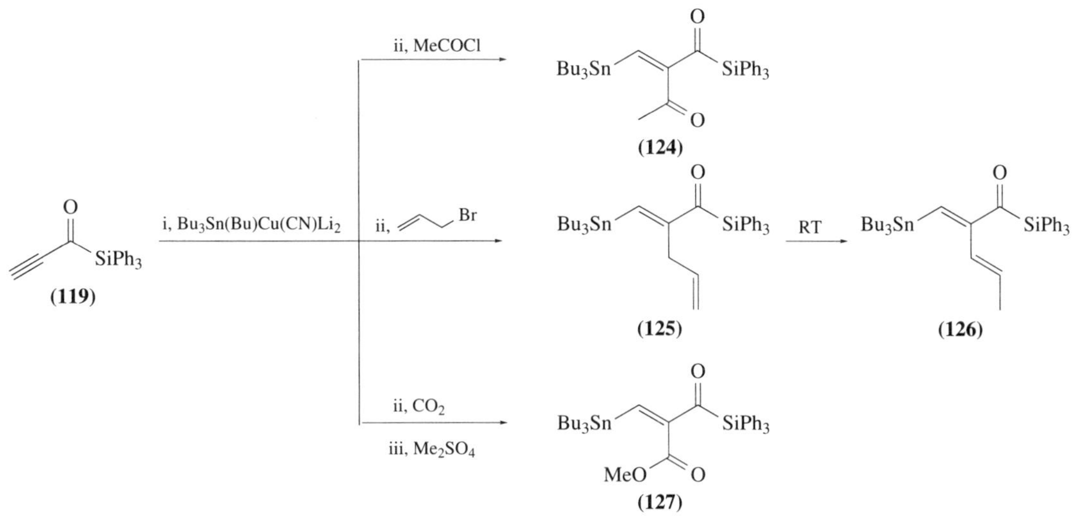

Scheme 51

starting material is the reaction of 1-lithio-1-trimethylsilyl cyclopropanes (**127**) with dichloromethyl methyl ether (Scheme 52) ⟨86CL177⟩. The reaction probably involves a carbene intermediate (**128**) and an intramolecular 1,2-silicon shift occurs to give (**130**) via (**129**).

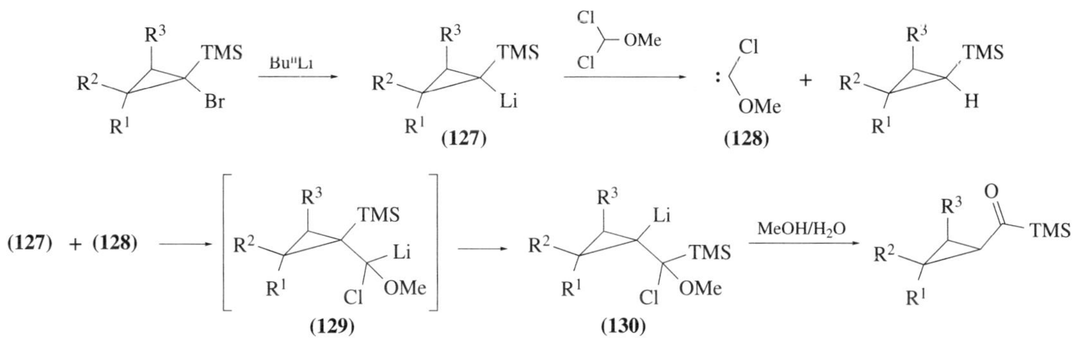

R^1, R^2, R^3 = H, alkyl or Ph
Overall yield = 16–50%

Scheme 52

Other methods involve the synthesis of a cyclopropyl group. To convert an α,β-unsaturated acylsilane to the corresponding cyclopropyl acylsilane, the former was treated with diazomethane followed by vapor phase pyrolysis of the intermediate pyrazoline derivative (Equation (10)) ⟨85TL2513⟩. The cyclopropyl acylsilane (**131**) was the first to be prepared by this method.

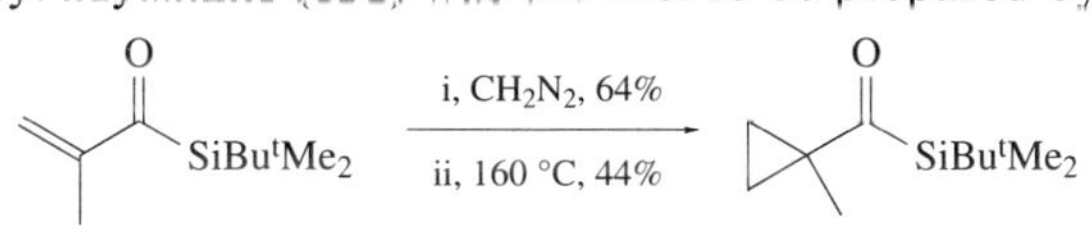

(10)

An alternative way to form the cyclopropane ring is by Simmons–Smith cyclopropanation of an α-silyl allylic alcohol (**132**) followed by oxidation to the acylsilane (Scheme 53) ⟨86HCA44⟩.

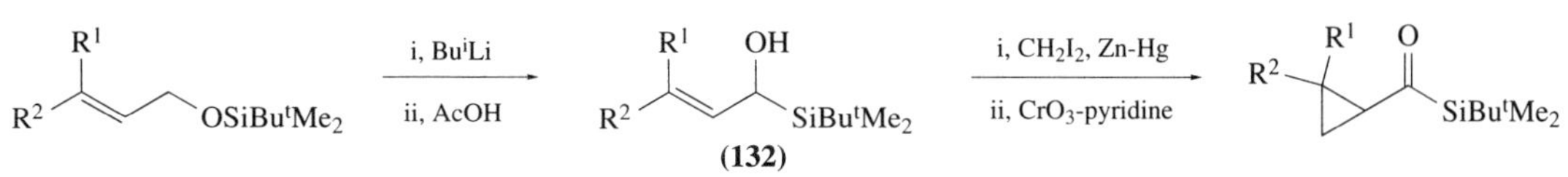

R^1, R^2 = H, alkyl
Yield = 10–85%
Scheme 53

A range of 1-silylketo-2-carbonyl cyclopropanes was prepared by conjugate addition-alkylation of lithium enolates of α-haloacylsilanes (**133**) or sulfur ylides (**134**) with electron deficient alkenes (Scheme 54) ⟨88T4113⟩. This approach can be used on substrates such as α,β-unsaturated ketones, α-carboxyacylsilanes and some nitrile and nitro compounds. Additionally, the sulfur ylide approach is principally used for the cyclopropanation of α,β-unsaturated aldehydes.

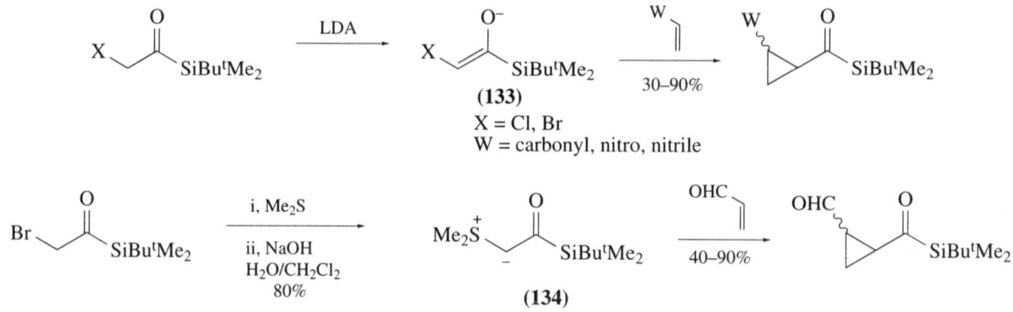

Scheme 54

5.09.1.3.5 *Other functionalized acylsilanes*

α-Carboxyacylsilanes have been reported by photo-oxygenation of silyl diazo compounds (**24**, $R = CO_2Me$ or CO_2Et) (Scheme 11, Section 5.09.1.1.2) ⟨79TL871⟩.

δ-Ketoacylsilanes (**137**) were prepared by the oxidation of 1-silyl cyclopropanol (**135**) with manganese(III) 2-pyridine carboxylate ($Mn(pic)_3$), in the presence of electron-rich alkenes (**136**). This reaction probably involves a radical addition, as shown in Scheme 55 ⟨93BCJ819⟩.

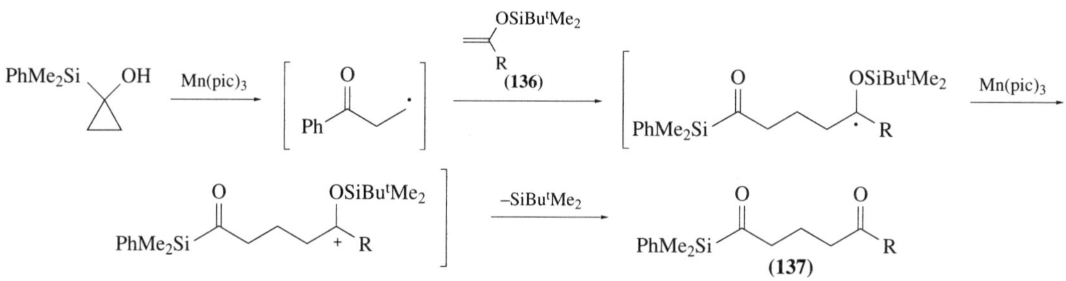

Scheme 55

Two α,β-epoxy silyl ketones (**140**) have been synthesized by the epoxidation of allylic silyl alcohols (**138**) followed by oxidation of the resulting epoxy-silyl alcohols (**139**) (Scheme 56) ⟨89HCA264⟩.

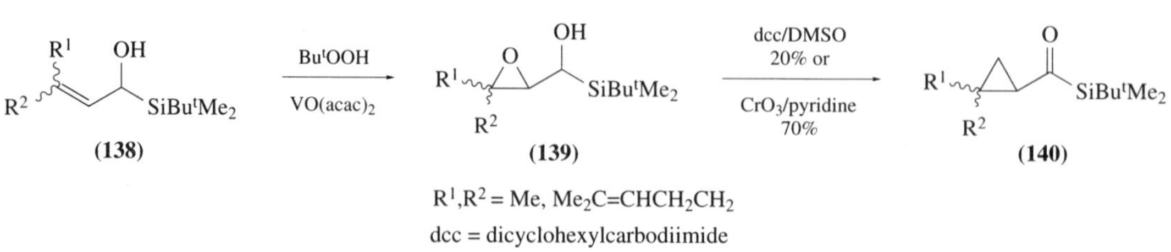

Scheme 56

A number of β-functionalized acylsilanes can be obtained by conjugate addition of appropriate nucleophiles to an α,β-unsaturated acylsilane (Equation (11), NCS = *N*-chlorosuccinimide)

⟨85JCS(P1)409⟩. When the nucleophiles are allylsilanes, 4,5-unsaturated acylsilanes can be isolated in 59–82% yields ⟨85TL2509⟩.

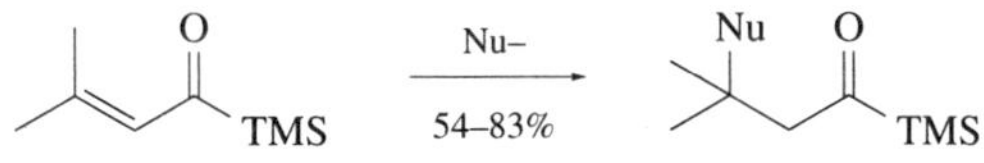

Nu = N_3, $PhSO_2$, NCS (11)

5.09.2 ACYL GERMANIUM DERIVATIVES

Compared to the acylsilanes, acylgermanes are not as widely employed or documented. Many, but not all, of the synthetic methods for acylgermanes are similar to those for acylsilanes.

5.09.2.1 Similar Synthesis to those of Acylsilanes

Some of the synthetic routes of acylsilanes have also been used to prepare the germanium analogs, but the yields are generally lower.

The first acylgermane, benzoyl trimethylgermane, was prepared by the dihalide route ⟨60JA5102⟩. Later, three other acylgermanes ($Et_3GeCOMe$, $Et_3GeCOPh$, $Ph_3GeCOMe$) were obtained by the dithiane method in 54–63% yields ⟨67JA431⟩ (Section 5.09.1.1.1). Like the acylsilanes, the limitation of the dithiane method in acylgermane synthesis lies in the deprotection step. None of the modified hydrolysis techniques for the acylsilanes has yet been applied to acylgermanes.

Acetyl triphenylgermane and benzoyl triphenylgermane were also made by the oxidation of the corresponding substituted methanol (α-hydroxygermane) ⟨66JOM(5)486, 66JOM(5)106⟩. The α-hydroxygermanes (**141**) can be obtained by the reaction of the appropriate aldehyde with a germyllithium reagent ⟨66JOM(5)486⟩ or a hexaorganosilylgermane (**142**) (Scheme 57) ⟨90SC3245⟩. Moderate yields were obtained, except for benzoyl trimethylgermane and benzoyl phenyl dimethylgermane ($PhCOGeMe_3$, $PhCOGeMe_2Ph$, 92%). Both Swern ⟨68AOC(7)95⟩ and DIAD (diisopropyl azodicarboxylate) ⟨92OM2198⟩ oxidation have been used to convert the alcohol into the corresponding acylgermanes in moderate to good yields.

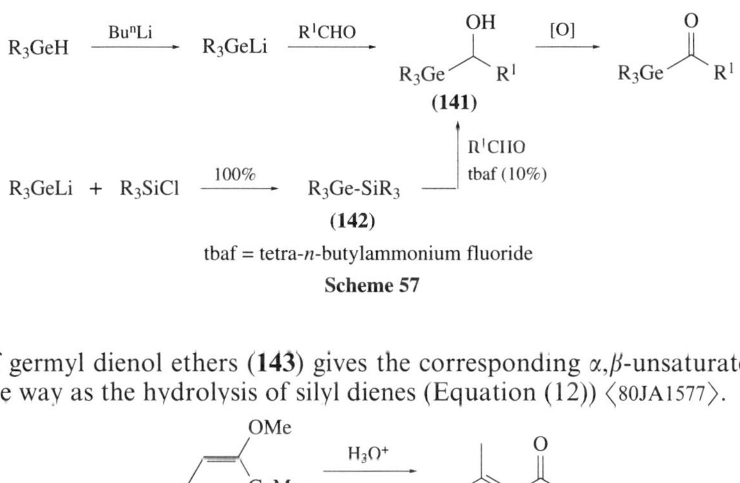

tbaf = tetra-*n*-butylammonium fluoride

Scheme 57

Hydrolysis of germyl dienol ethers (**143**) gives the corresponding α,β-unsaturated acylgermanes (**144**) in the same way as the hydrolysis of silyl dienes (Equation (12)) ⟨80JA1577⟩.

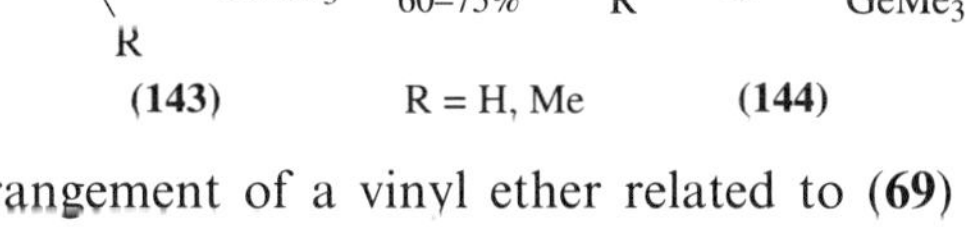

(143) R = H, Me **(144)** (12)

Asymmetric Claisen rearrangement of a vinyl ether related to (**69**) bearing a trimethylgermyl substituent generates the corresponding acylgermane (Scheme 33) ⟨91TA647⟩.

5.09.2.2 Coupling of Germyl Metallic Species and Carboxylic Acid Derivatives

Germyl metallic species, mainly germyllithium reagents, are readily prepared by metalation of the corresponding germane with *n*-butyllithium (Scheme 57). The coupling of germyllithium reagents with acyl chlorides, amides, esters, and lactones is a more general method than the analogous method of preparing acylsilanes.

The addition of triphenylgermyllithium to aromatic acyl chlorides at −78 °C gives the corresponding acylgermanes in 22–84% yields ⟨65IC1747⟩. The yield drops dramatically as the temperature rises. At 25 °C, the bis(triphenylgermyl)carbinol is obtained as the only product. The coupling of triphenylgermyllithium with acetyl chloride and pivaloyl chloride were also reported. Benzoyl trimesitylgermane and triacylgermane (Equation (13)) were also prepared by this method ⟨90JOM(397)7, 92JOM(433)49⟩. Photogermylation of RCOCl with $(Et_3Ge)_2Hg$ gave 65–94% of RCO-$GeEt_3$ (R = Ph, *m*-tolyl, *p*-FC_6H_4) ⟨83IZV1675⟩. While the direct coupling of triethyl germyllithium with pentafluorobenzoyl chloride failed to give any ketone product, the desired pentafluorobenzoyl(triethyl)germane ($C_6F_5COGeEt_3$) was obtained in 54% yield in the presence of CuI ⟨83JOM(248)51⟩.

$$3\ PhH_2GeLi + 3\ MesCOCl \xrightarrow{-LiCl} PhGe(COMes)_3 + 2\ PhGeH_3 \qquad (13)$$

Coupling of triethyl germyllithium (Et_3GeLi) with *N*,*N*-dialkyl carboxamides (Equation (14)) is very dependent on the reaction conditions and the nature of the acyl fragment of the amides ($R^1CONR^2_2$) ⟨83JOM(248)51⟩. When $R^1 = CF_3$, $R^2 = Et$, the reaction was conducted in hexane at −78 °C, trifluoroacetyl(triethyl)germane was obtained in 81% yield. When $R^2 = Me$, $R^1 = Me$ and Me_3C, the coupling was very slow at low temperatures, but at room temperature acetyl triethylgermane and pivaloyl triethylgermane were generated in yields of 50% and 30% respectively. Under the same conditions, benzoyl triethylgermane was prepared in 88% yield ⟨83ZOR1838, 82IZV949, 84MI 509-01⟩. An α,β-unsaturated acylgermane $Me_3SiC{\equiv}CCONR_2$ was also synthesized by this route ⟨84IZV1897⟩.

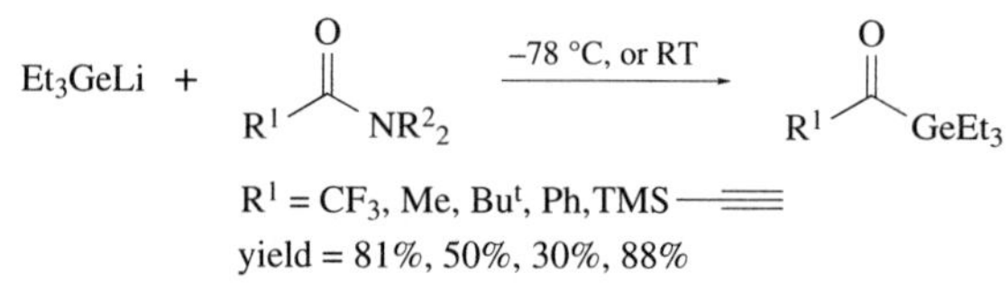

(14)

An easier and more general way to synthesize simple acylgermanes is by the coupling of triphenyl germyllithium with the readily available carboxylic esters and lactones (Scheme 58) ⟨85CL1829⟩. This reaction is conducted in THF at room temperature, and a variety of alkanoyl and aroylgermanes have been synthesized in 76–86% yield. When the reaction was applied to ethyl bromoacetate, triphenylgermylacetyl triphenylgermane (**145**) was obtained in 66% yield ⟨87CL495⟩. (The coupling yield of triphenyl germyllithium with lactones has been improved from 13% ⟨92SL631⟩ to 54% (unpublished results).) β-Hydroxyacylgermanes (**146**) can be synthesized when reagent (**145**) reacts with aldehydes in the presence of boron trifluoride etherate (66–81%).

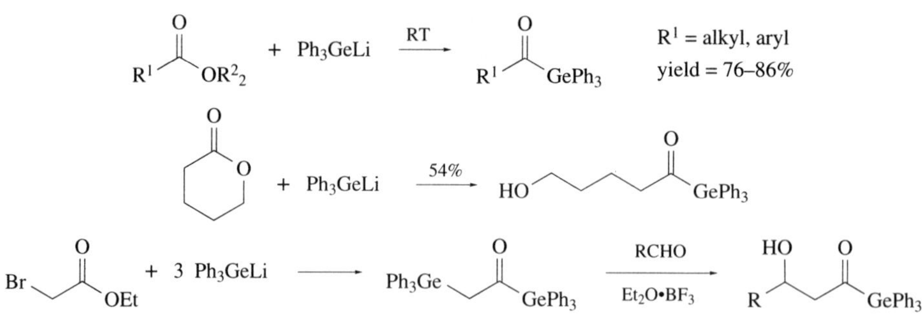

(145) **(146)**

Scheme 58

5.09.3 ACYL BORON DERIVATIVES

A computer substructure search of *Chemical Abstracts* for acylboranes revealed that very little is known about this class of molecules. Acylboranes $R^1_2BCOR^2$ have been postulated as intermediates in the carbonylation of organoboranes ⟨69ACR65⟩. Benzoyldibutylborane Bu_2BCOPh was once claimed to have been prepared and isolated according to Scheme 59 ⟨68CB2502⟩. However, the actual products of Scheme 59 have been proved to be a mixture of benzyloxydibutylborane (Bu_2BOPh) and benzylbenzoate ($PhCO_2CH_2Ph$) ⟨75CC719⟩. Balenkova *et al.* ⟨80MI 509-01⟩ have reported using acetyl and pivaloylborofluorides ($AcBF_2$ and Me_3CCOBF_2) as acylating reagents of (methylene)cyclohexane, (methylene)cyclooctane and (methylene)-cyclodecane. However, the synthesis of these two acyl boranes is not mentioned. *Ab initio* calculations ⟨92MI 509-01⟩ also indicate that acylboranes should be highly reactive and easily rearranged species.

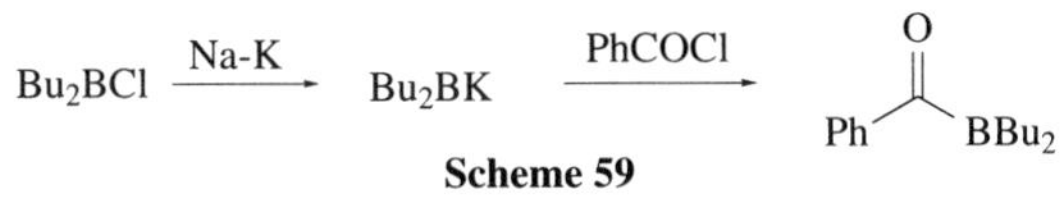

Scheme 59

5.10
Acyl Metal Functions

P. WARNER
ZENECA Pharmaceuticals, Macclesfield, UK

5.10.1 GENERAL INTRODUCTION

Acyl metal derivatives of the main group metals covered in this chapter are rare. Such compounds are usually highly reactive, and their isolation has not been observed at ambient temperatures. In contrast, many transition metals form acyl derivatives which can be readily isolated and characterised. Acyl transition metal complexes have been used in organic synthesis stoichiometrically (e.g., chiral iron acyl compounds) and as key intermediates in catalytic cycles. This chapter focuses primarily on compounds that have been isolated and characterised.

5.10.2 GROUP 1 DERIVATIVES

Four methods for the formation of group 1 acyl metal compounds have been reported:
(i) addition of organolithium reagents to carbon monoxide;
(ii) deprotonation of the formyl group of aldehydes;
(iii) lithium–tellurium exchange of acyltellurides;
(iv) fluoride catalysed desilylation of acylsilanes.

Group 1 acyl metal compounds are relatively unstable and highly reactive and so it has not been possible to characterise such a compound. They have not been isolated, but are generated *in situ*, usually at very low temperature. Streitweiser and Schleyer have reported an *ab initio* study of the stability of main group formyl metal compounds ⟨87JA2553⟩. They found that formyllithium is stable by 3 kcal mol^{-1} with respect to decomposition into lithium hydride and carbon monoxide, but that the formyl complexes of beryllium, sodium, magnesium and aluminum are thermodynamically unstable. The vast majority of group 1 acyl metal derivatives are of lithium. However, Schlosser in a review of organosodium compounds has described the possible formation of benzoylsodium when phenylsodium is treated with carbon monoxide ⟨64AG(E)287⟩.

5.10.2.1 Addition of Organolithiums to Carbon Monoxide

Whitesides and co-workers have published a mechanistic study of the reaction of PhLi with carbon monoxide ⟨73JA8118⟩. No firm evidence was obtained for the production of acyllithium species, but the existence of benzoyllithium (**1**) and other acyllithium compounds (**2**) and (**3**) as transient intermediates was suggested by the formation after acylation, of products (**4**) and (**5**). Compound (**4**) predominates at room temperature, and compound (**5**) is preferred at −78 °C (Scheme 1).

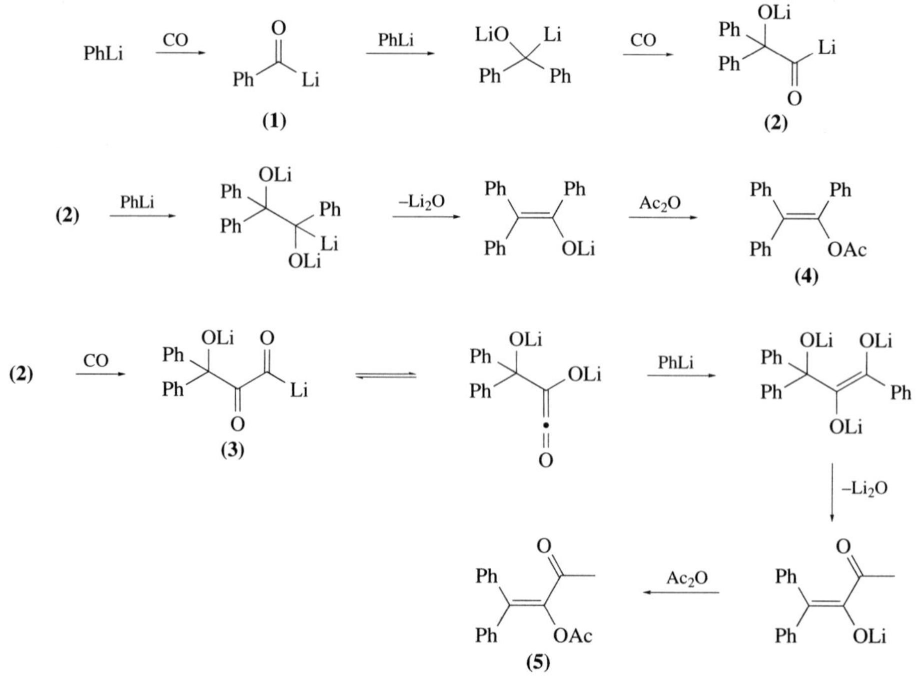

Scheme 1

Seyferth has investigated the controlled formation and subsequent reactions of acyllithium compounds derived from *n*-BuLi, *s*-BuLi and *t*-BuLi with carbon monoxide ⟨83JOC1144⟩. For example, addition of *n*-BuLi and carbon monoxide at $-110°C$ to a solution of 3-pentanone in diethyl ether/pentane gave 3-hydroxy-3-ethyl-4-octanone (**6**) in 67% yield possibly via the formation of *n*-butanoyllithium (**7**) (Equation (1)). Optimal yields are obtained when a 1 : 1 stoichiometry of the organolithium reagent and electrophile is used. Thus, addition of *n*-BuLi to carbon monoxide is favoured over the competitive reaction with the carbonyl group of 3-pentanone. Similarly, the *n*-BuLi/carbon monoxide reaction must be faster than addition of *n*-BuLi to ethyl pentanoate and more rapid than any potential side reactions of *n*-butanoyllithium (**7**).

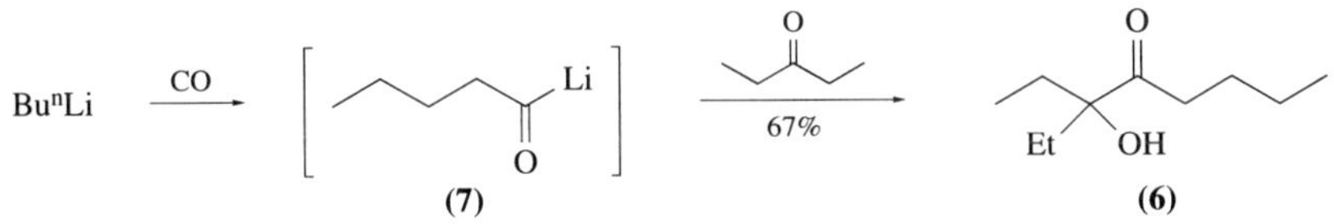

(**7**) (**6**) (1)

5.10.2.2 Deprotonation of the Formyl Group of Aldehydes

Until the mid 1980s the formyl deprotonation of aldehydes to form acyl anions had only been achievable in the gas phase ⟨85RTC74⟩. Shiner has reported that nonenolisable aldehydes and lithium tetramethylpiperidide (LTMP) react at $-78°C$ to give acyloins in high yield (Equation (2)) ⟨88JA957⟩. It is possible that the reaction may involve the formation of an acyllithium intermediate such as (**8**), but alternative mechanisms involving electron transfer can not be excluded.

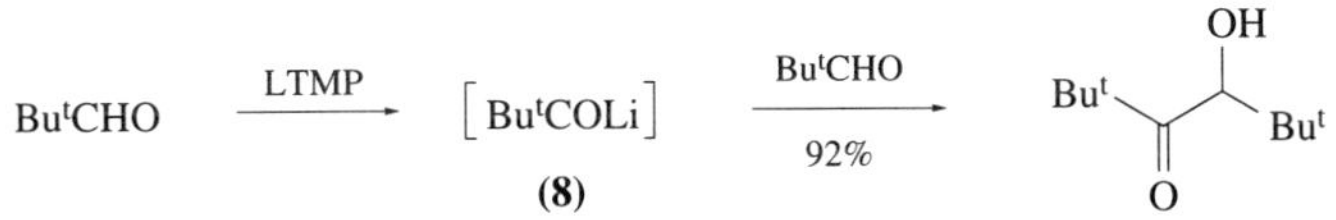

(**8**) (2)

5.10.2.3 Li–Te Exchange of Acyltellurides

Sonoda has described a convenient synthesis of acyllithium compounds based on the efficient lithium–tellurium exchange of tellurolesters ⟨90JA455⟩. For example, when *n*-BuLi was added to *n*-butyl tellurobenzoate (**9**) in THF/diethyl ether at $-105°C$ in the presence of pinacolone, the α-hydroxy ketone (**10**) was obtained in 85% yield (Equation (3)). In contrast, benzoyllithium (**1**) was not observed to be trapped efficiently when formed using the PhLi/carbon monoxide method ⟨84TL1651⟩. If pinacolone is added after 1 min to a preformed solution of benzoyllithium (**1**) at $-105°C$ the yield of (**10**) is reduced (16%) and benzoin (34%) and benzil (21%) are also formed. This demonstrates that benzoyllithium (**1**) is extremely reactive and unstable even at $-105°C$. A selenolester (PhCOSeBu) and acylstanane (*t*-BuCOSnBu$_3$)) were also investigated as potential precursors to carbonyl anions ⟨90JA455⟩, but did not give the desired α-hydroxyketones.

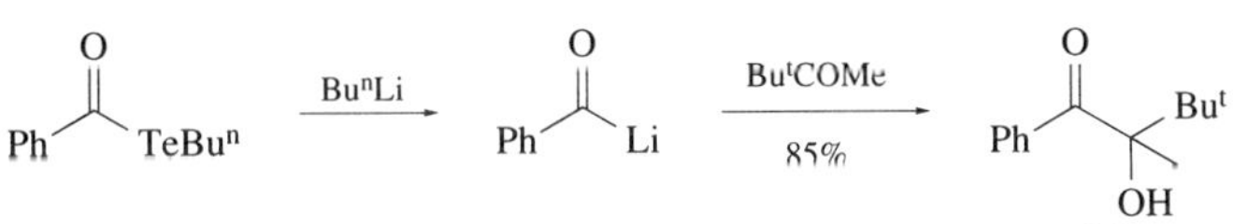

(**9**) (**1**) (**10**) (3)

5.10.2.4 Desilylation of Acylsilanes

Heathcock has reported that benzoyltrimethylsilane (**11**) reacts with potassium fluoride in the presence of methyl iodide to give acetophenone in 55% yield ⟨81TL1881⟩. Similarly, desilylation of (**11**) with sodium methoxide, potassium fluoride, or tbaf in the presence of cyclohexanone gives the hydroxyketone (**12**) (Scheme 2). One plausible mechanism for the formation of these products involves the intermediacy of an acyl anion. However, other mechanisms not involving acyl anion species are conceivable.

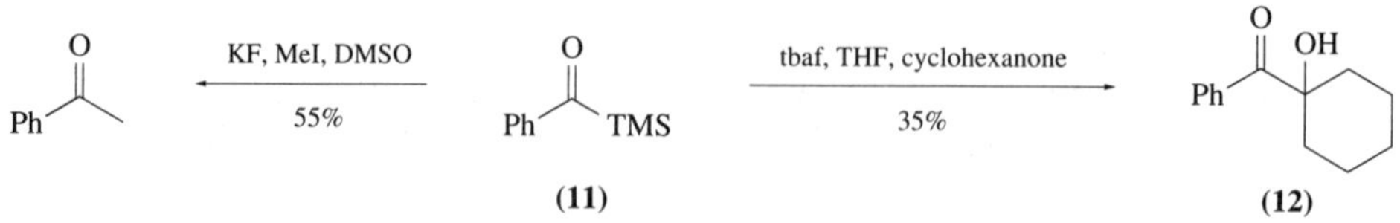

Scheme 2

5.10.2.5 Intramolecular Trapping Reactions

The synthetic utility of acyllithium compounds in intermolecular reactions is relatively limited due to the restricted range of acyllithium species available by current methods. To circumvent the relative instability of acyllithium compounds, intramolecular trapping reactions have been devised. Smith has reported that doubly lithiated *N*-pivaloylaniline (**13**) reacts with carbon monoxide at 0 °C to give the 3-*t*-butyldioxindole (**14**) in good yield (77%) ⟨90AG(E)282⟩. The reaction is likely to involve the generation of a benzoyllithium intermediate (**15**) and a 1,2-migration of the *t*-butyl carbanion (Scheme 3).

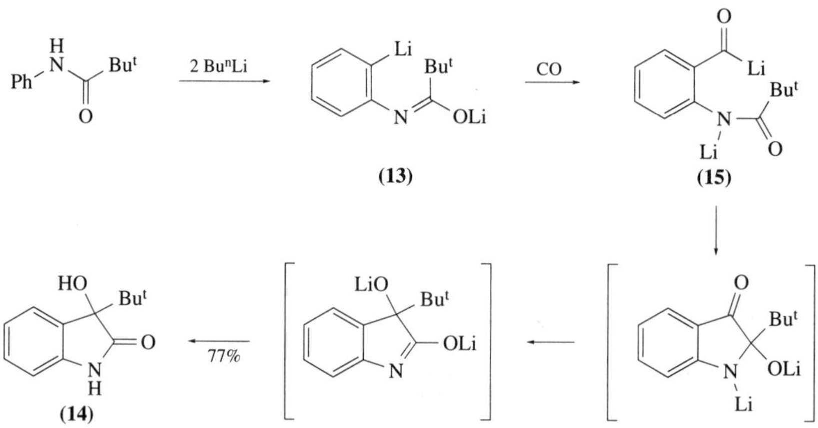

Scheme 3

Murai has described cyclisation reactions of acyllithiums derived from the reaction of carbon monoxide with [1-(silyl)vinyl]lithium (**16**) and [2-phenyl-1-(silyl)vinyl]lithium (**17**) ⟨90JA7061⟩. Compound (**16**) reacts with carbon monoxide to give a cyclopropanone enolate (**18**) which can be isolated as the silylated derivative (**19**) (Scheme 4). Compound (**17**) also reacts with carbon monoxide to give a cyclised product, but in this case it is the indenol (**20**) that predominates. The mechanisms of these remarkable reactions can be rationalised by the formation of intermediate acyllithium species such as (**21**) and (**22**) (Scheme 4).

5.10.3 GROUP 2 DERIVATIVES

Examples of acyl derivatives of group 2 metals are rare. The formation of acylberyllium compounds from beryllium metal and acid chlorides has been claimed ⟨66JGU1945⟩ but the majority of work has involved acylmagnesium compounds.

5.10.3.1 Acylmagnesium Compounds

The reaction of Grignard reagents with carbon monoxide and the characterisation of the resulting products has been described by Sprangers *et al.* ⟨74TL3377, 76JCS(P2)1895, 76C199⟩. These reactions are

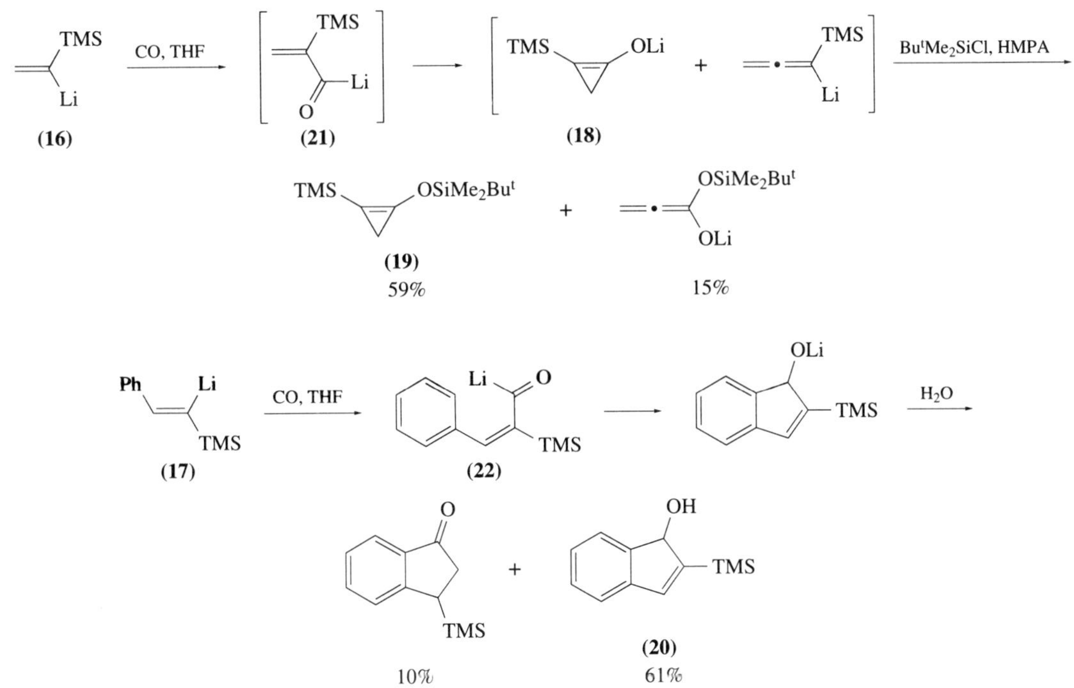

Scheme 4

likely to involve the nucleophilic addition of RMgX to carbon monoxide to give the magnesium acyl compounds RCOMgX. The addition of HMPT to the reaction medium considerably increases the rate of these reactions, possibly by enhancing the polarization of the carbon–magnesium bond resulting in the formation of a more reactive carbanionic species. Addition of water to a reaction which involves the formation of an acylmagnesium species is expected to give an aldehyde product. Carbonylation of PhMgBr at low temperature (−35 °C) under pressure, followed by hydrolysis gave benzaldehyde in 20% yield. No aldehydes were observed for aliphatic systems, and this is presumably due to the high reactivity of the putative acyl species. However, carbonylation of *n*-BuMgBr in HMPT in the presence of $NaBH_4$ gave *n*-pentanol in 6% yield, demonstrating the possible formation of an acylmagnesium species. The reaction pathways observed during the carbonylation of EtMgBr in the presence of ethyl bromide are illustrated in Scheme 5. The acylmagnesium compounds **(23)** and **(24)** are postulated as intermediates.

5.10.4 LANTHANIDE DERIVATIVES

Acyllanthanide compounds are quite rare. Examples of the formation of acylsamarium and acyllutetium compounds are described.

5.10.4.1 Acylsamarium Compounds

The transient formation of the acylsamarium species **(25)** was postulated in the reaction of diphenylethyne with $Cp^*_2Sm(THF)_2$, followed by carbonylation of the intermediate **(27)** at 90 psi in hexane for 24 hours ⟨86JA1722⟩. The formation of product **(28)** can be rationalised as occurring via the acyl species **(25)** (Scheme 6).

The reaction of Cp_2Sm with an acid chloride such as 1-adamantoyl or 1-methylcyclohexanoyl chloride at −20 °C followed by quenching with deuterated water leads to the formation of the corresponding deuterated aldehydes in 63 and 25% yields, respectively ⟨91JOC3118⟩. This reactivity strongly implies that an acylsamarium intermediate is formed. It was not possible, however, to isolate or characterise the postulated acyl intermediate.

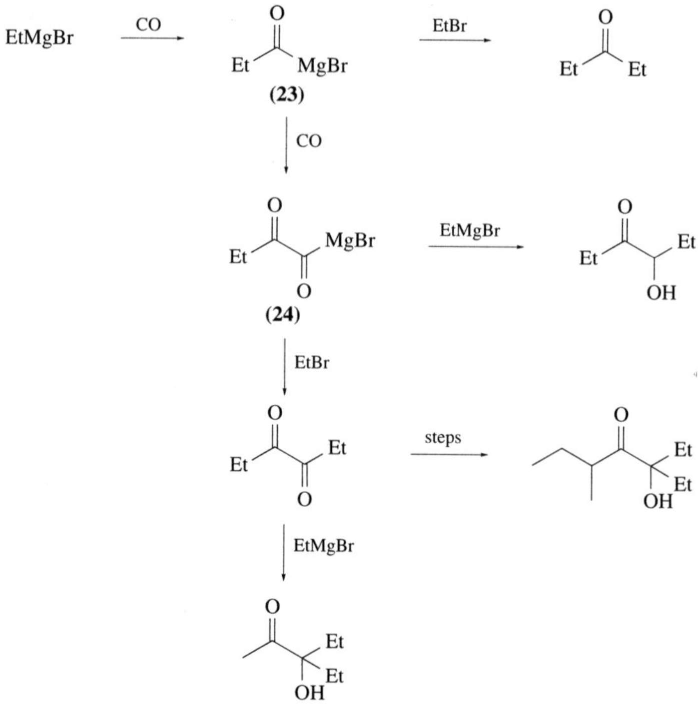

Scheme 5

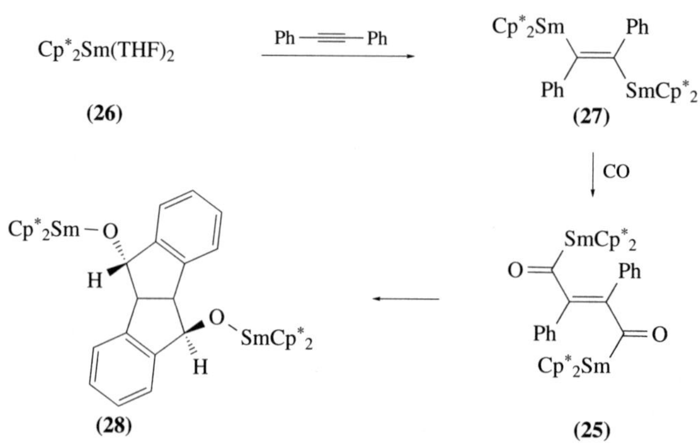

Scheme 6

5.10.4.2 Acyllutetium Compounds

The *t*-butyllutetium compound (**29**) reacts with carbon monoxide in toluene to give the acyl compound (**30**) in 75% yield (Equation (4)) ⟨81CC706⟩. The complex (**30**) was isolated and characterised and found to be stable to heat. The acyl ligand is η^2-coordinated to the metal centre.

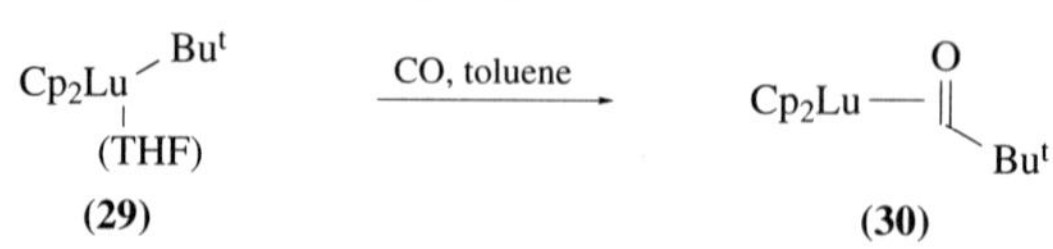

(4)

5.10.5 TRANSITION METAL DERIVATIVES

5.10.5.1 Introduction and General Methods

Most of the transition metals are known to form η^1-acyl metal derivatives. The η^1-acyl ligand is a two-electron donor that can also participate in variable degrees of backbonding to the metal centre. Strong dπ-backbonding leads to short metal–carbon bond lengths and the acyl oxygen is often basic, especially in anionic acyl complexes.

The three main reactions that lead to the formation of acyl metal complexes are:

(i) Migratory insertion of alkyl ligands into a coordinated ligand. This reaction is usually promoted by an external ligand such as carbon monoxide or triphenylphosphine. This reaction has been reviewed by Calderazzo ⟨77AG(E)299⟩.
(ii) Oxidative addition of acid chlorides to transition metal complexes.
(iii) Addition of a metal alkyl reagents (e.g., organolithiums) to a coordinated carbonyl ligand. The binding mode of a carbonyl ligand to a transition metal can give rise to a partial positive charge on the carbon atom (depending on the nature of the metal centre) which facilitates the addition of nucleophilic reagents.

Oxophilic metals at the beginning of the transition series and the actinides and lanthanides can form η^2-coordinated acyl metal derivatives. Hoffmann and co-workers have proposed that these compounds have carbenium ion character rather than the oxycarbene nature that is usually ascribed to them ⟨85JA4440⟩.

5.10.5.2 Acyltitanium Compounds

Titanocene acyl compounds are prepared by reaction of acyl halides (Scheme 7, Method A) or alkyl iodides (Scheme 7, Method B) with $Cp_2Ti(CO)_2$ ⟨77JCS(D)2297⟩. Alternatively, the corresponding monoalkyl compounds react with carbon monoxide in toluene to give acyl products (Scheme 7, Method C) ⟨74JOM(71)C5⟩. These compounds are relatively stable compared with the corresponding alkyl species. In titanium acyl compounds of this type, the acyl ligand is η^2-coordinated to the metal centre. This has been established by x-ray analysis of $Cp_2Ti(COMe)Cl$ and IR spectroscopy ⟨77JCS(D)2297⟩. Attempts to prepare bisacyl complexes of the type $Cp_2Ti(COR)_2$ have failed. For example, both $Cp_2Ti(Bn)_2$ reacting with carbon monoxide and $Cp_2Ti(COMe)Cl$ with MeMgCl in the presence of carbon monoxide give $Cp_2Ti(CO)_2$ ⟨74JOM(71)C5⟩.

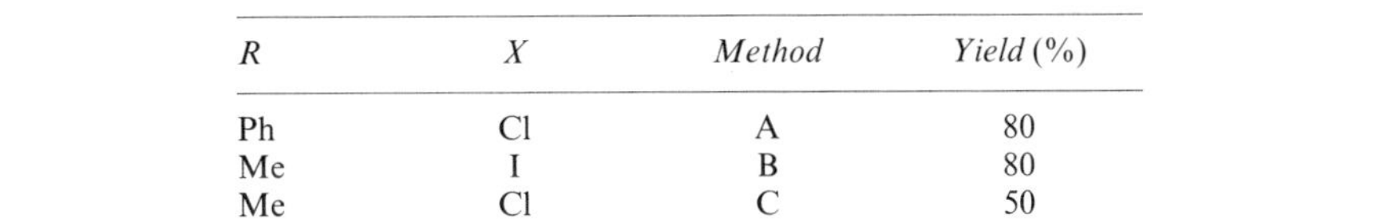

Scheme 7

Table 1 Preparation of acyltitanocene compounds (for R and X see Scheme 7).

R	*X*	*Method*	*Yield* (%)
Ph	Cl	A	80
Me	I	B	80
Me	Cl	C	50
Et	Cl	C	>50
CH_2Ph	Cl	C	>50

The cyclic titanium alkyl compound (**31**) reacts rapidly with carbon monoxide at −55 °C in pentane to give a good yield of the acyl compound (**32**) which decomposes at 0 °C to give cyclopentanone (Scheme 8) ⟨76JA6529⟩. The cyclobutene compound (**33**) reacts with carbon monoxide at −30 °C to give a 5-membered ring cyclic titanium acyl compound (**34**) which can be isolated at room temperature (Scheme 8) ⟨86JA3318⟩. IR studies indicate that the acyl ligand is η^1-coordinated to the metal. Treatment of compound (**34**) with anhydrous HCl at −78 °C under carbon monoxide (100 atm) gives the γ,δ-unsaturated aldehyde (**35**) and $Cp_2Ti(Cl)_2$.

5.10.5.3 Acylzirconium Compounds

The first examples of zirconium acyl complexes were reported by Schwartz and Bertello ⟨75JA228⟩. The biscyclopentadienyl zirconium compound, Cp_2ZrHCl, reacts with alkenes to give the corresponding zirconium alkyl species. This process is known as hydrozirconation. A feature of this

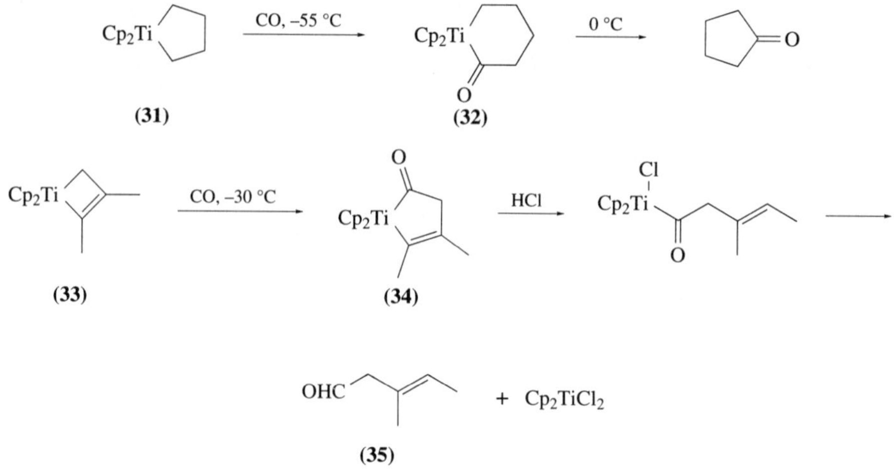

Scheme 8

reaction is that, regardless of the initial position of the alkene double bond in the starting material, the terminal alkyl compound is always formed. The alkyl compounds such as (**36**) react with carbon monoxide with insertion of the carbon monoxide into the zirconium alkyl bond. This process corresponds to a stoichiometric hydroformylation of alkenes and the final zirconium acyl compounds (e.g., (**37**)) can be converted into useful organic materials such as aldehydes, carboxylic acids or esters (Scheme 9). A useful example of this reaction, reported by the same authors, is the overall oxidation of 1,3-dienes to give β,γ-unsaturated aldehydes, such as (**39**), selectively in good yield (Scheme 10) ⟨76JA262⟩. The intermediate alkyl and acylzirconium species (**38**) and (**41**) were isolated and characterised in the reaction between the diene (**40**) and Cp_2ZrHCl to give aldehyde (**39**) (Scheme 10).

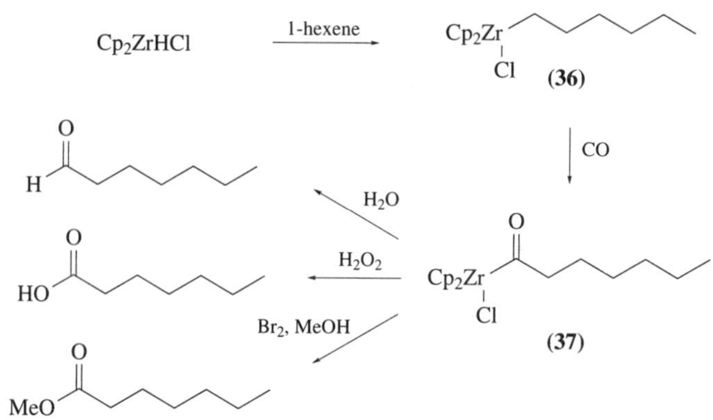

Scheme 9

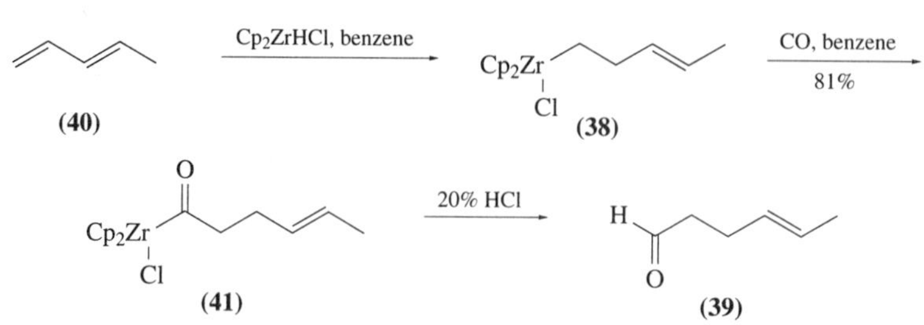

Scheme 10

Zirconium acyl compounds in which an alkyl ligand is also coordinated to the metal centre have been isolated and characterised by Fachinetti *et al.* ⟨76CC522⟩. The bismethyl and bisbenzyl compounds Cp_2ZrMe_2 and Cp_2ZrBn_2 react with carbon monoxide at atmospheric pressure in toluene to produce the monoacyl complexes $Cp_2ZrMe(COMe)$ and $Cp_2ZrBn(COBn)$, respectively. The acyl complexes can be crystallised by addition of heptane and isolated as yellow crystalline solids in 80% yield. The x-ray crystal structure of the acetyl complex $Cp_2ZrMe(COMe)$ shows that both the carbon and oxygen atoms of the acyl group are bonding to the zirconium metal centre. This η^2-type coordination of the acyl group leads to the carbon centre possessing oxycarbenoid reactivity (Equation (5)). Carbonylation of zirconocene complexes has been reviewed by Erker ⟨84ACR103⟩.

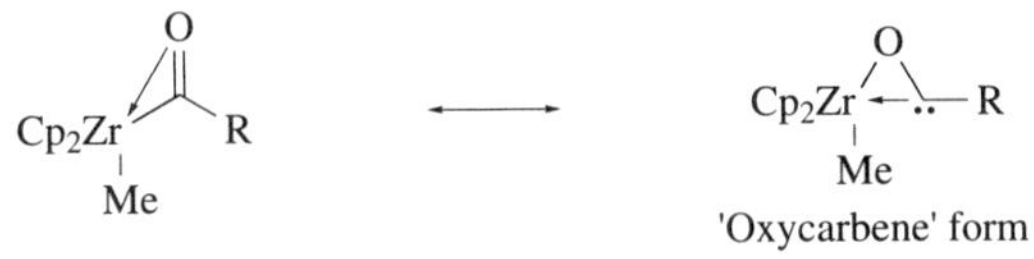

(5)

5.10.5.4 Acylhafnium Compounds

Biscyclopentadienyl bisalkyl hafnium complexes react with carbon monoxide in toluene to give monoacyl compounds ⟨77JCS(D)1946⟩. In the case of the bisphenyl analogue, the benzoyl compound was formed and isolated in 72% yield as a low melting solid (m.p. 22°C) (Scheme 11). The analogous reaction of the bismethyl complex led to an equilibrium mixture of monoacyl compound and starting material. It was not possible to force the reaction to completion. This result, which contrasts with the successful carbonylation of the zirconium and titanium analogues, may be due to the greater hafnium–carbon bond strength compared with the analogous titanium–carbon and zirconium–carbon bond strengths. It is not unusual for the carbonylation of third row transition metal alkyl species to require more forcing conditions than are required for the corresponding first and second row elements. As found for the corresponding zirconium and titanium acyl complexes, the acyl ligand bonds to the metal centre via both the carbon and oxygen atoms.

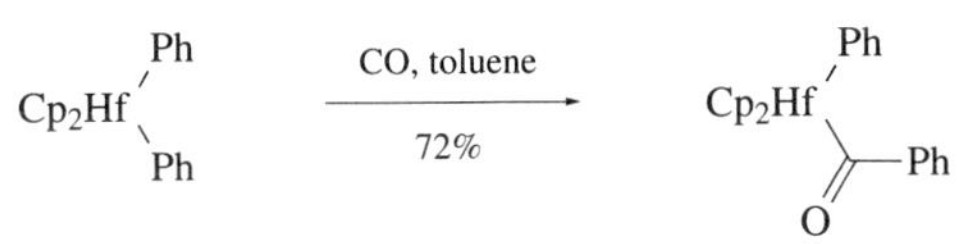

Scheme 11

Bisacyl complexes of early transition metals are relatively rare. Bristow *et al.* have reported the formation of the metallaindane compound (**42**) by reaction of the Grignard reagent (**43**) in THF with the hafnium compound $(\eta^5\text{-}C_5H_4TMS)_2HfCl_2$ ⟨84JCS(D)399⟩. Carbonylation of (**42**) in pentane led to the formation of a stable bisacyl compound (**44**) in which both acyl groups are η^1-bonded through the carbon atoms. The compound was observed to exist as a mixture of two conformers in solution which could be separated and isolated (Scheme 12). In contrast, carbonylation of the corresponding titanium or zirconium metallaindanes gave indan-2-one and $(\eta^5\text{-}C_5H_4TMS)_2M(CO)$ ($M = Ti$ or Zr). No acyl complexes were isolated. This indicates that for these two metals the intermediate monoacyl compounds may eliminate rather than react with a second molecule of carbon monoxide to form a stable bisacyl complex.

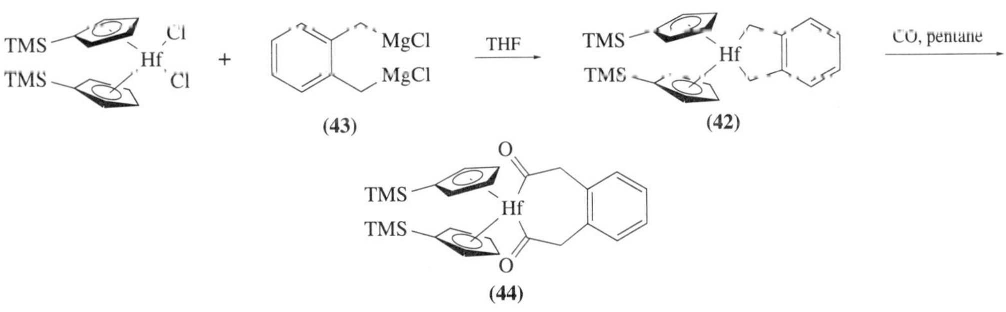

Scheme 12

5.10.5.5 Acylvanadium Compounds

The vanadium acyl species $Cp_2V(COR)CO$ (R = Me, Bn) are formed by reaction of Cp_2VR with carbon monoxide in toluene or heptane at room temperature in about 60% yield as green–maroon crystals ⟨74CC516⟩ (Equation (6)). IR studies indicate that the acyl group is η^2-coordinated to the metal centre. This procedure failed to give the expected acyl product for Cp_2VPh. The addition of carbon monoxide to the metal induced the phenyl group to migrate to the cyclopentadienyl ring leading to the isolation of the dienyl compound (**45**) (Equation (7)) ⟨76JCS(D)203⟩.

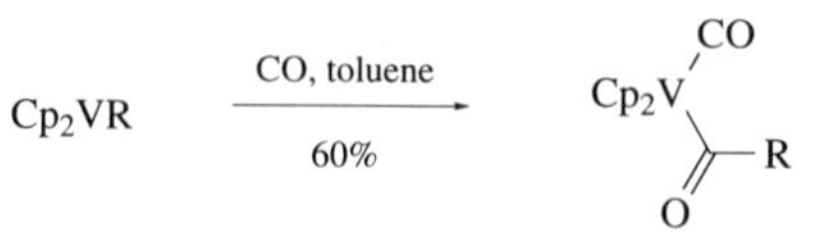

R = Me and Bn (6)

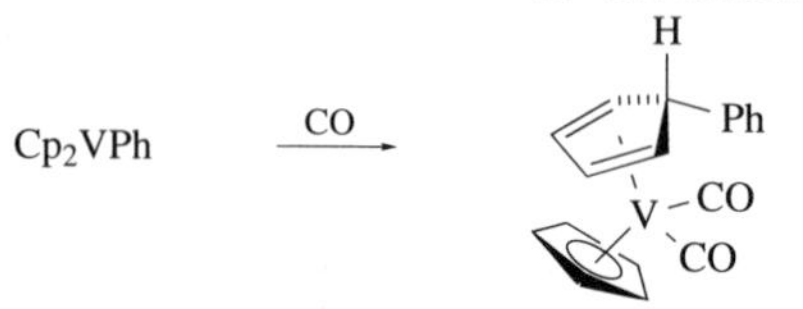

(**45**) (7)

5.10.5.6 Acyltantalum Compounds

Acyl complexes of tantalum are rare. The tantalum chloride complex Cp^*TaCl_4 reacts with 2,2-dimethylpropyl magnesium chloride to give the alkyltantalum compound (**46**) which reacts in pentane with carbon monoxide at 80 psi to give the corresponding acyl species (**47**) (Equation (8)) ⟨89JA149⟩. Compound (**47**), formed in 76% yield, is stable at room temperature as a solid or in solution in hydrocarbon or ethereal solvents. IR spectroscopy indicates that the acyl ligand is η^2-coordinated to tantalum.

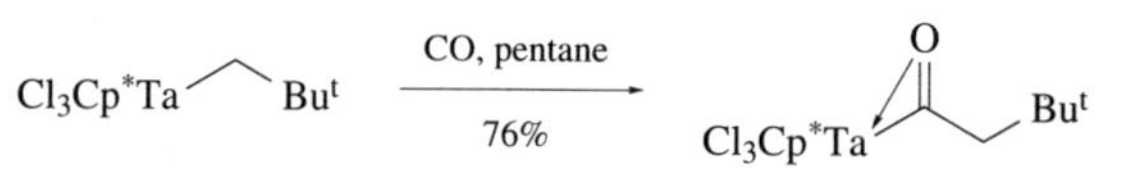

(**46**) (**47**) (8)

5.10.5.7 Acylchromium Compounds

Anionic chromium acyl compounds of the type $[Cr(CO)_5(COR)]NMe_4$ are formed by reaction of alkyl or aryl lithium reagents with $Cr(CO)_6$. For example, PhLi adds to $Cr(CO)_6$ to give the corresponding benzoyl compound (**48**) in 89% and MeLi reacts to give the acetyl analogue (**49**) in 72% yield (Equation (9)) ⟨67CB2445⟩. These acyl salts are relatively unstable and usually prepared *in situ* as intermediates in the synthesis of the well-known chromium carbene compounds ⟨84AG(E)587⟩. Alternative routes to the acyl compounds involve the acylation of the dianion $[Cr(CO)_5]^{2-}$ by acyl chlorides ⟨87OM1839⟩ or the displacement of trimethylamine from $Cr(CO)_5(NMe_3)$ with alkyl or aryllithium reagents and carbonylation (Scheme 13) ⟨93JA4389⟩.

$$Cr(CO)_6 \xrightarrow[\text{ii, } Me_4NBr,\ H_2O]{\text{i, RLi}} [Cr(CO)_5(COR)]NMe_4 \qquad (9)$$

(**48**) R = Ph
(**49**) R = Me

$$Cr(CO)_5NMe_3 \xrightarrow{PhLi} Li[Cr(CO)_5Ph] \xrightarrow{CO} Li[Cr(CO)_5(COPh)]$$

Scheme 13

Carbonylation of the bisaryl chromium compound (**50**) at 80 psi in hexane at room temperature gives the bisacyl complex (**51**) (Scheme 14) ⟨88JCS(D)53⟩. The carbonyl groups in (**51**) may participate in η^2-type bonding with the metal centre. When subjected to heat, compound (**51**) eliminates the amide (**52**) (Scheme 14).

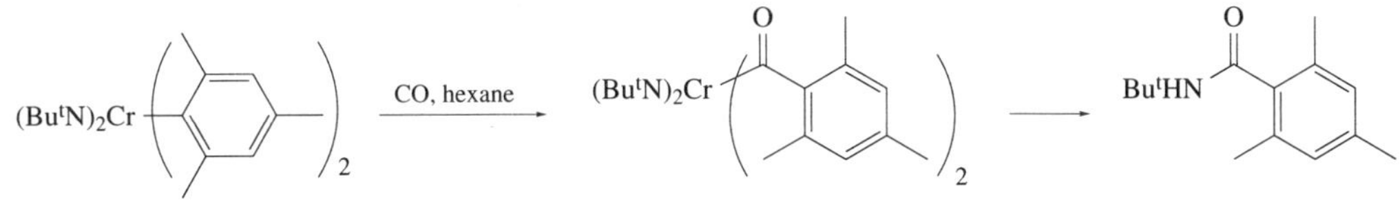

(50) **(51)** **(52)**

Scheme 14

5.10.5.8 Acylmolybdenum Compounds

$Mo(CO)_6$ reacts with alkyl and aryllithium reagents to give anionic acyl complexes ⟨67CB2445⟩. These compounds are readily alkylated with trimethyloxonium tetrafluoroborate or methyl iodide to give the corresponding carbene species (Scheme 15). This reactivity of $Mo(CO)_6$ is similar to that of the chromium and tungsten analogues. η^2-Coordinated acyl complexes of molybdenum such as **(53)** are formed and isolated by addition of Grignard or lithium reagents to the phosphine compound **(54)** in 60–70% yields (Scheme 16) ⟨84JA3214⟩. The structure of **(53)** has been determined by x-ray analysis.

$$Mo(CO)_6 \xrightarrow[\text{ii, } NMe_4Br,\ H_2O]{\text{i, RLi}} [Mo(CO)_5(COR)]NMe_4 \xrightarrow[Me_3OBF_4]{\text{MeI or}} Mo(CO)_5[C(OMe)R]$$

Scheme 15

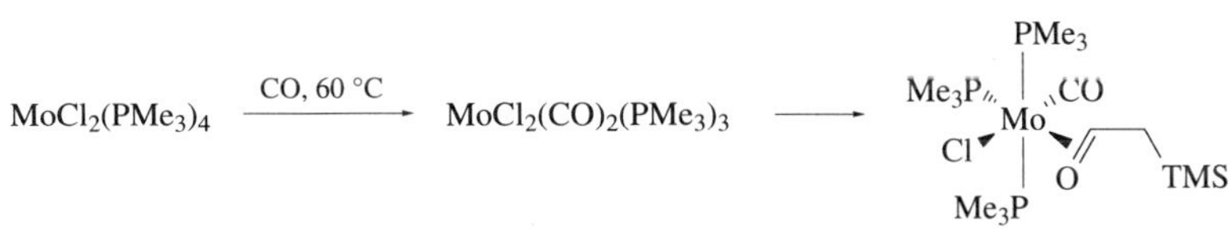

(54) **(53)**

Scheme 16

Barnett has reported the preparation of cyclopentadienyl molybdenum acyl compounds such as **(55)** in poor yield by the addition of acetyl chloride to the anionic species **(56)** (Scheme 17) ⟨72JOM(44)1⟩. The compounds were found to be rather unstable on isolation. Attempts to prepare compound **(55)** by carbonylation of the molybdenum methyl compound **(57)** at 316 atm failed to yield the desired product ⟨78JA1687⟩. However, the desired acyl compound was formed in 51% yield as an $AlBr_3$ adduct **(58)** when **(57)** was carbonylated in the presence of $AlBr_3$ in toluene at 0°C (Scheme 17) ⟨80JA5093⟩. In contrast to the free complex **(55)**, the $AlBr_3$ adduct **(58)** was found to be thermally stable. A range of *para* substituted benzyl analogues of **(57)** react with triphenylphosphine to give the corresponding acyl compounds in which the phosphine complexes to the metal (Equation (10)) ⟨81ICA165⟩. The rate of the reaction correlates with the electron donating ability of the *para* substituent on the aryl ring of the benzyl ligand.

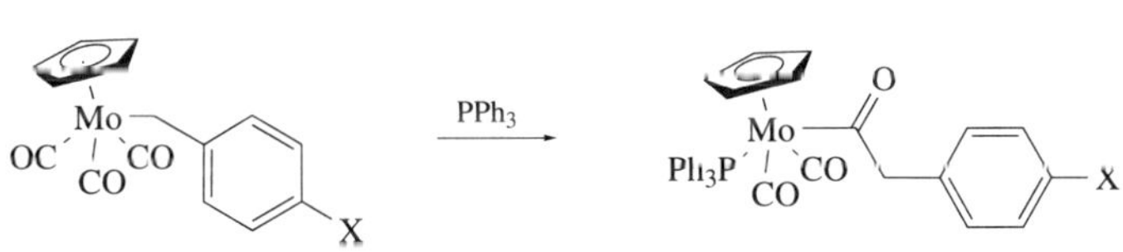

(10)

$$[CpMo(CO)_3]^- \xrightarrow{AcCl} Cp(CO)_3MoCOMe$$

(56) **(55)**

$$CpMo(CO)_3Me \xrightarrow{AlBr_3,\ CO,\ 0\ °C} Cp(CO)_3Mo[C(OAlBr_3)]$$

(57) **(58)**

(**(58)** → **(55)**)

Scheme 17

Metallacyclic acyl complexes of molybdenum such as **(59)** are formed on addition of aziridines to the hydride complex **(60)** (Equation (11)) ⟨73AG(E)77⟩. The corresponding tungsten complexes can also be prepared by this process.

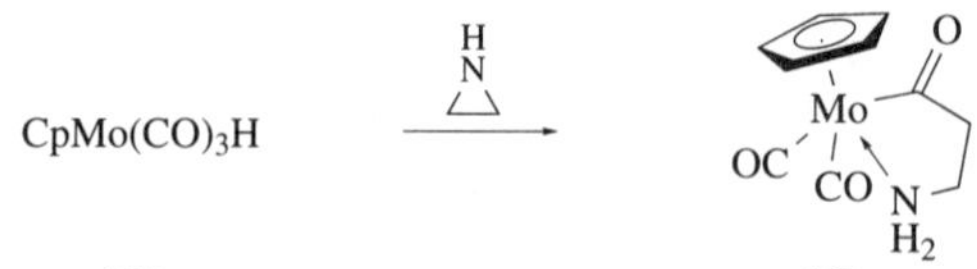

(60) **(59)** (11)

5.10.5.9 Acyltungsten Compounds

$W(CO)_6$ reacts with alkyllithium compounds to give anionic acyl complexes analogous to those of chromium (Equation (12)) ⟨67CB2445⟩. The acetyl and benzoyl compounds **(61)** and **(62)** are obtained in 53% and 71% yield, respectively. The acylalkylidyne compound **(63)** is formed in a similar manner by reaction of **(64)** with MeLi (Equation (13)) ⟨92CB1583⟩. Reaction of the methyl compound $CpW(CO)_3Me$ with $LiEt_3BH$ gives an unusual anionic hydridoacyl complex **(66)** which has been isolated as the lithium salt. The hydride ligand of **(66)** can be replaced by iodine on treatment of **(66)** with iodoform to give a related acyl compound **(67)** (Scheme 18) ⟨88JOM(342)C5⟩. In contrast, the molybdenum analogue of **(66)** can not be isolated and rearranges to Li[CpMo-$(CO)_2$(MeCHO)] ⟨84CC420⟩.

$$W(CO)_6 \xrightarrow[\text{ii, } Me_4NBr,\ H_2O]{\text{i, RLi}} [W(CO)_5(COR)]NMe_4 \qquad (12)$$

(61) R = Me, 53%
(62) R = Ph, 71%

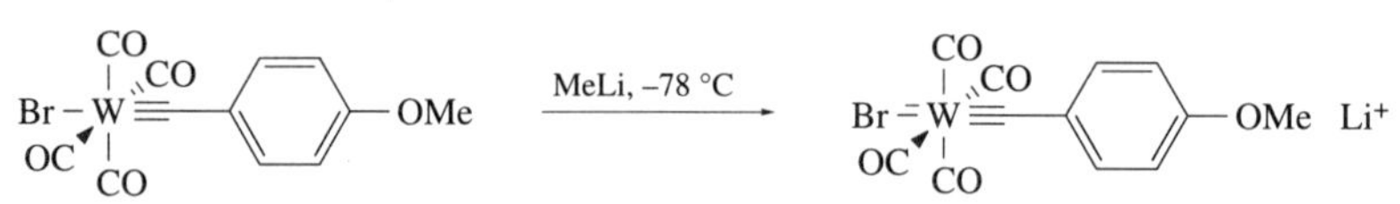

(64) **(63)** (13)

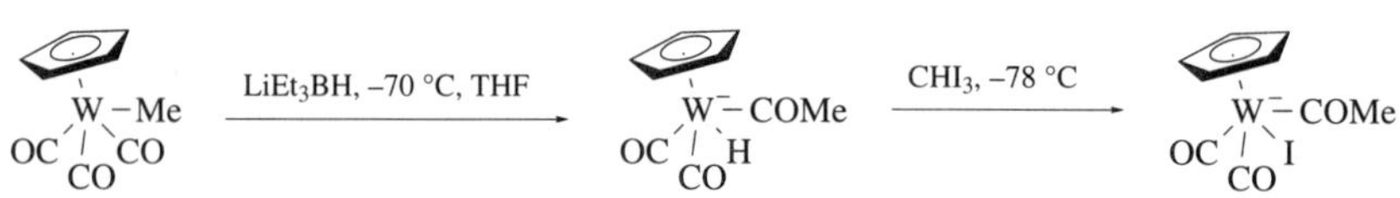

(65) **(66)** **(67)**

Scheme 18

5.10.5.10 Acylmanganese Compounds

The first synthesis of an acylmanganese compound was reported by Coffield *et al.* ⟨57JOC598⟩. $Mn(CO)_6$ in THF was converted into $NaMn(CO)_5$ **(68)** by reaction with sodium amalgam. Addition of acyl chlorides, such as benzoyl chloride, acetyl chloride or propionyl chloride, to **(68)** gave the corresponding acylmanganese pentacarbonyl complexes as white crystalline solids in good yield (Equation (14)). This procedure has been used extensively to prepare a variety of similar acylmanganese compounds (Table 2).

$$Na[Mn(CO)_5] \xrightarrow{RCOCl} (CO)_5MnCOR \qquad (14)$$

(68)

Acetyl manganese pentacarbonyl is formed by insertion of carbon monoxide into the manganese–carbon bond of methyl manganese pentacarbonyl at 500 psi. Acylmanganese compounds have been used to incorporate carbonyl functionality at the anomeric position of sugars ⟨85JA7788⟩. Glycosyl manganese pentacarbonyl compounds such as **(69)** are readily formed by reaction of glycosyl bromides with $NaMn(CO)_5$ **(68)** in THF at $-78\,^{\circ}C$ (Scheme 19). The displacement occurs with inversion of configuration at the anomeric centre. Reaction of the manganese alkyl intermediate **(70)** with carbon monoxide leads to the stereospecific insertion of carbon monoxide into the carbon–manganese bond to give the acyl compound **(69)** with retention of configuration at the carbon

Table 2 Preparation of acylmanganese compounds (for R see Equation 14).

R	*Yield (%)*	*m.p. (°C)*	*Ref.*
CH_2F	50	67–69	66JOM(6)265
CH_2Cl	54	89–91	79JOM(172)177
$(CH_2)_3Br$	56	52.5	90CJC502
CH_2OPh	56	91–92	75JOM(94)62
$COCH_3$	33	76–77	76JA1166
COPh	92	62	76JA1166
$(CH_2)_4COCl$	83	118	68JOM(15)397
C_7H_6	67	83	71JCS(C)1514

centre. The acyl group can be cleaved with a mixture of methanol and sodium carbonate to give the methyl ester (**71**) in 63% yield.

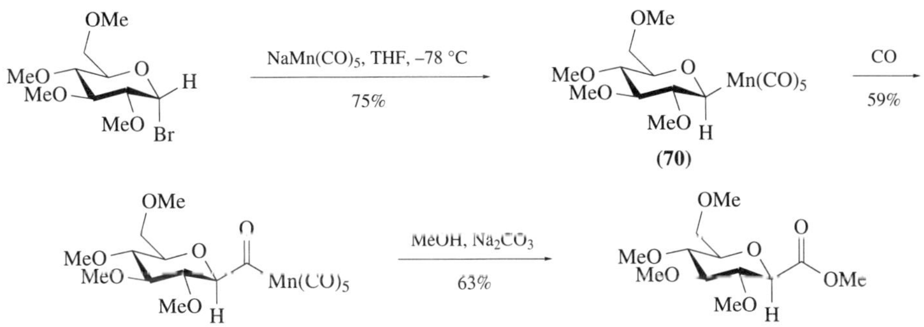

(70) (69) (71)

Scheme 19

The insertion of carbon monoxide into the manganese–carbon bond can be facilitated by strong Lewis acids such as $AlBr_3$ ⟨80JA5093⟩. The Lewis acid increases the rate of alkyl group migration, stabilises the acyl group and provides an electron-rich atom which ligates to the vacant coordination site at the metal centre. Addition of carbon monoxide then occurs which leads to the formation of an acylmanganese complex such as (**72**) after hydrolysis (Scheme 20). In the absence of carbon monoxide gas, complexes such as (**73**) are formed and can be isolated by crystallisation and characterised. The structure of complex (**73**) has been confirmed by x-ray crystallographic analysis. This phenomenon has been observed for other metals (e.g., molybdenum, iron and tungsten). A specific example of the utility of this reaction is the formation of the $Mn(CO)_5(COBn)$ complex from $Mn(CO)_5Bn$ using $AlBr_3$. The initially formed complex coordinated to $AlBr_3$ is obtained in 62% yield and converted into $Mn(CO)_5(COBn)$ by hydrolysis. This compound is not readily prepared by other procedures. The use of BCl_3 or BBr_3 leads to the cleavage of the metal–alkyl bond as the predominant reaction pathway. In contrast, $AlMe_3$, a weaker Lewis acid than $AlCl_3$ lacking an electron-rich ligand to fill the empty coordination site at manganese, does not induce the migration reaction.

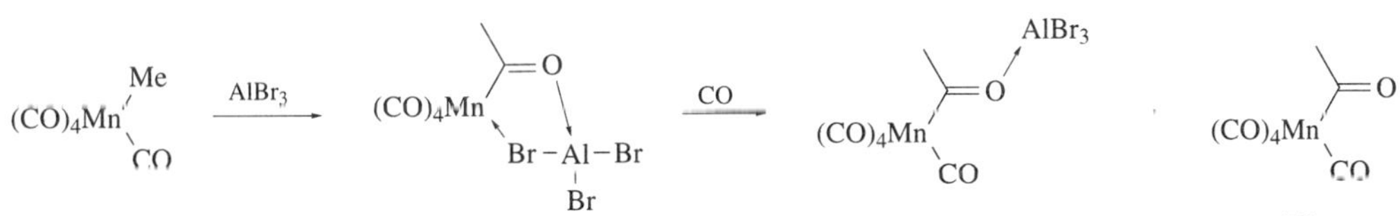

(73) (72)

Scheme 20

Manganese alkyl compounds such as $MeMn(CO)_5$ undergo migratory insertion reactions induced by a variety of donor ligands in addition to carbon monoxide, such as phosphines, arsines, amines, sulphides and halides, to give acyl complexes. Triphenylphosphine reacts with $MeMn(CO)_5$ in THF to give $(PPh_3)Mn(CO)_4(COMe)$ in 72% yield (m.p. 107 °C) ⟨64JA3994⟩. Similarly, amines such as ammonia, aniline and cyclohexylamine react with $MeMn(CO)_5$ to give the corresponding acyl complexes ⟨60JA4204⟩. Aziridines react with $HMn(CO)_5$ to give acyl complexes, e.g. **(74)** (Scheme 21) ⟨73AG(E)77⟩. This reaction can be regarded as a 1,3-insertion of the aziridine into the manganese–hydride bond and alkyl migration to the carbon monoxide ligand with possible intramolecular assistance from the coordinating amine functional group. Analogous sulfur-containing compounds can be formed by reacting $Na[Mn(CO)_5]$ in boiling THF with 2-chloroethyl methyl sulfide ⟨64JA1267⟩. The product **(75)** is isolated in 26% yield as a volatile solid (m.p. 56.5–58.5 °C) (Equation (15)).

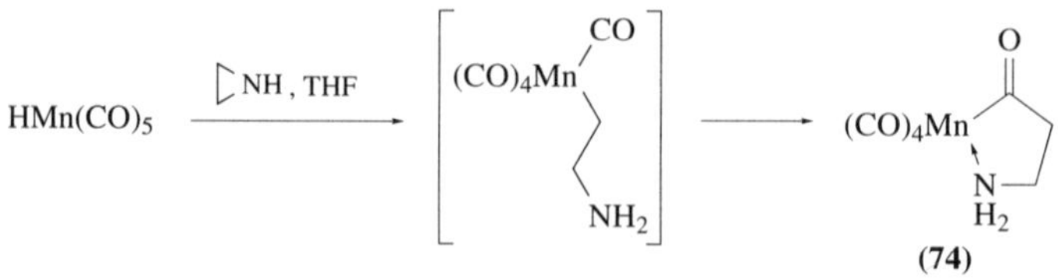

(74)

Scheme 21

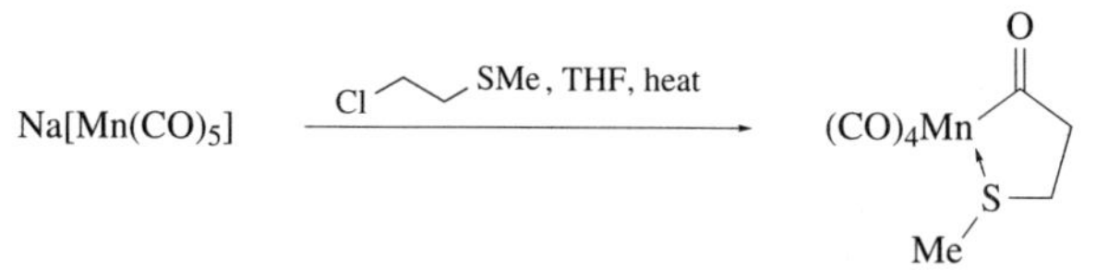

(75)

(15)

The first bisacyl complexes of manganese carrying two different acyl ligands were isolated from the reaction between MeLi and $(CO)_5Mn(COPh)$ at −78 °C in THF after exchange of the counter ion in 54% yield. The product [*cis*-$(PhCO)(MeCO)Mn(CO)_4]NMe_4$ was also formed by addition of PhLi to the analogous $(CO)_5Mn(COMe)$ compound in 46% yield. The structure of the product was confirmed by x-ray crystallographic analysis ⟨76JA437⟩. Cationic arenetricarbonyl manganese complexes such as **(76)** react with lithium cuprates to form acyl complexes (e.g. **(77)**) by addition of the alkyl anion to a coordinated carbonyl group (Equation (16)) ⟨82OM1730⟩. In contrast, the nucleophilic addition of MeLi to $[CpMnNO(PPh_3)CO]PF_6$ gives a mixture of products derived from addition of the methyl anion to the carbonyl and cyclopentadienyl ligands (Equation (17)) ⟨75JOM(87)223⟩. This reaction has been carried out on homochiral $[CpMnNO(PPh_3)CO]PF_6$ **(78)** to give the corresponding homochiral acyl complex **(79)** which is configurationally unstable in solution. The ylide Me_3PCH_2 reacts with a carbon monoxide ligand of the cationic methyl cyclopentadienyl compound **(80)** to give the acyl complex **(81)** which is deprotonated by a second equivalent of the ylide to give the neutral acyl compound **(82)** (Scheme 22) ⟨82CB1488⟩.

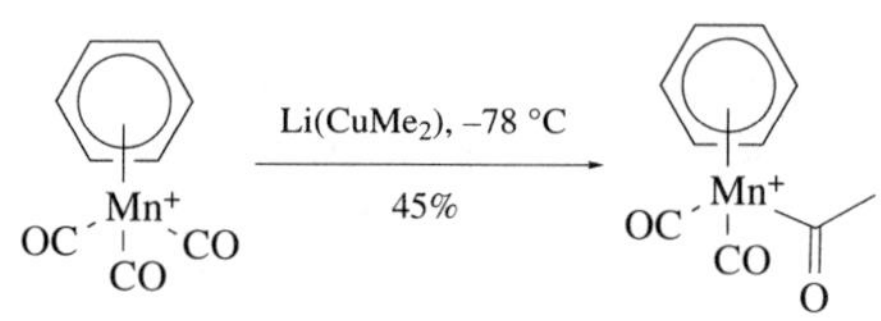

(76) **(77)**

(16)

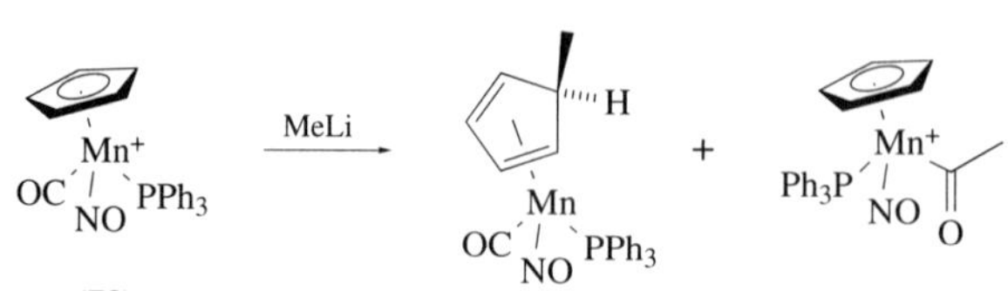

(78) **(79)**

(17)

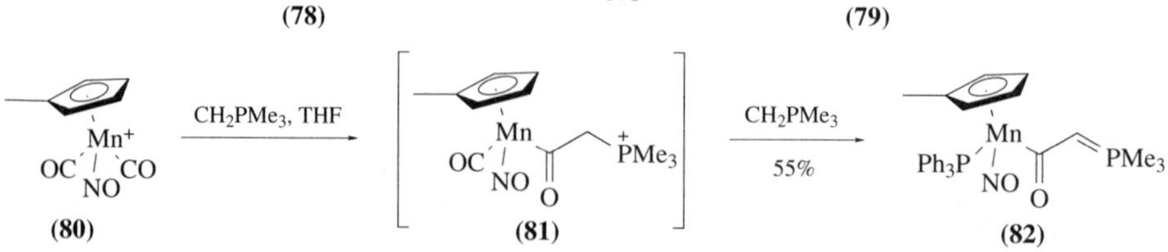

(80) **(81)** **(82)**

Scheme 22

Other methods of forming acylmanganese compounds include the insertion of ketene into the metal–hydride bond of $HMn(CO)_5$ to give $MeCOMn(CO)_5$ ⟨74ZN(B)275⟩ (Equation (18)) and the nucleophilic addition of Grignard reagents such as MeMgBr to coordinated carboxyalkyl groups (Equation (19)) ⟨66JOM(5)341⟩. The manganese silane compound $(CO)_5MnTMS$ (**83**) undergoes a 1,2-addition to aldehydes. Reacting (**83**) with the phosphine-containing aldehyde (**84**) leads to the formation of an acyl complex (**85**) in which the phosphorus adds to the metal to promote the insertion of carbon monoxide into the manganese–carbon bond ⟨86JA1462⟩. The silylated compound (**85**) can be desilylated with KF to give the corresponding α-hydroxyacyl compound (**86**) in 93% yield (Scheme 23).

$$HMn(CO)_5 + H_2C{=}\bullet{=}O \longrightarrow (CO)_5MnCOMe \quad (18)$$

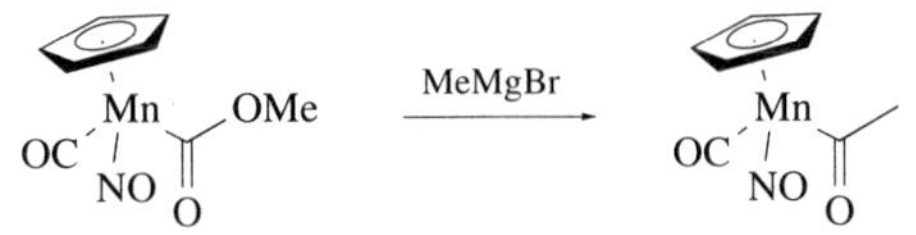

(19)

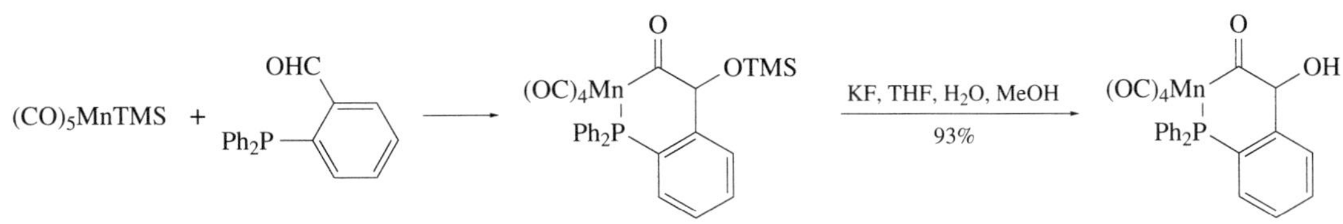

(**83**) (**84**) (**85**) (**86**)

Scheme 23

The formation of α-keto acyl complexes of manganese has been reported and reviewed by Geoffrey *et al.* ⟨89PAC1723⟩. Usually the insertion of carbon monoxide into a metal acyl bond is a very unfavourable process. However, the oxidation of the anionic complex (**87**) with $(Cp_2Fe)PF_6$ gives a 17-electron complex (**88**) which reacts with nitric oxide at low temperature with insertion of the carbon monoxide into the metal acyl bond to give the α-keto acyl complex (**89**) in 75% yield (Scheme 24). An alternative synthesis of α-keto acyl complexes of manganese involves the addition of nitrite anion to cationic carbyne complexes (Scheme 24) ⟨87JA1586⟩. The mechanism of the reaction is believed to involve nitrite addition to the carbyne carbon centre in a compound such as (**90**) to give a carbene intermediate (**91**) and migration of carbon monoxide with migration of nitric oxide to the vacant coordination site.

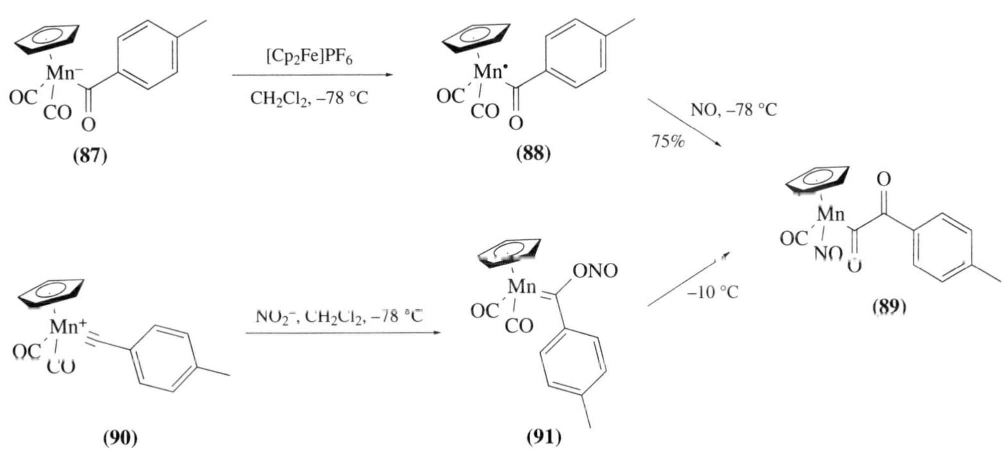

Scheme 24

5.10.5.11 Acylrhenium Compounds

Numerous pentacarbonyl acylrhenium compounds have been synthesised by addition of acid chlorides to $Na[Re(CO)_5]$ which is generated by reduction of $Re_2(CO)_{10}$ with sodium amalgam (Scheme 25) ⟨60CB901⟩. The yields are generally 60–90%. Addition of an organolithium reagent to

$(CO)_5ReCOPh$ leads to the formation of anionic bisacyl complexes such as (**93**). This reaction is improved if an acid such as HCl or CF_3CO_2H is added during the work-up ⟨76JA2365⟩. Generally, these compounds are more stable than the analogous manganese compounds. However, heating (**93**) in deuterated acetone at 68 °C gives the acyl–alkyl compounds (**94**) and (**95**) in the ratio 98 : 2 (Scheme 26) ⟨80JA2723⟩. Compound (**94**) was also synthesised by addition of MeLi to the aryl compound $(CO)_5RePh$ in 53% yield. The acyl–alkyl compounds (**94**) and (**95**) are surprisingly stable to reductive elimination to give ketones despite not being constrained in a metallacycle. This unusual stability may be due to the particularly strong rhenium–carbon bonds.

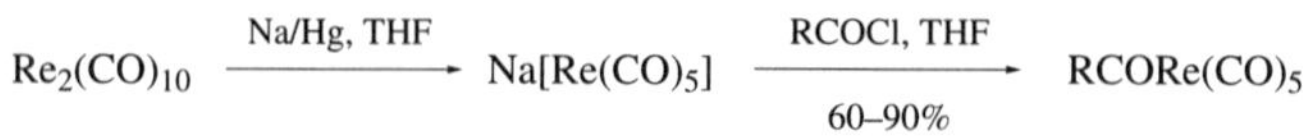

Scheme 25

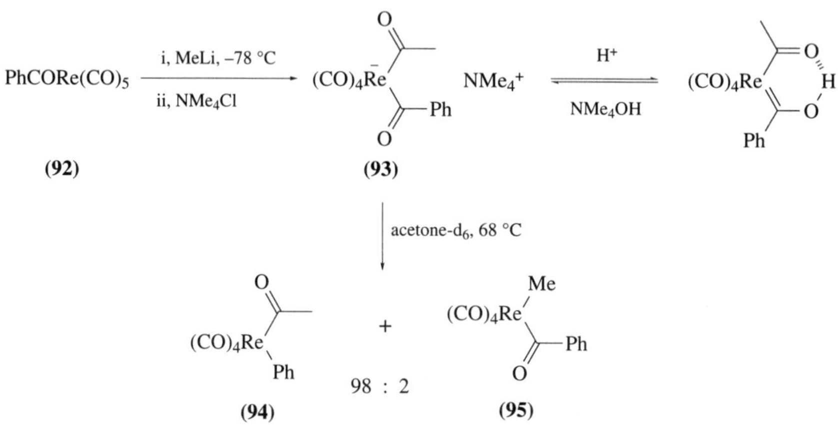

Scheme 26

Addition of MeLi to the cyclopentadienyl compound $CpRe(CO)_3$ (**96**) gives an anionic rhenium acyl compound (**97**) that can be alkylated with methyl iodide to give the acyl–alkyl complex (**98**) in 67% yield (Scheme 27) ⟨89JA1285⟩. An acyl η^3-allylrhenium species (**99**) is a probable intermediate in the reaction of the rhenium vinyl compound (**100**) with allylbromide to give the bisalkene complex (**101**) in which the two alkenic components are added to an initially metal-bound carbonyl ligand (Scheme 28) ⟨90JA4083⟩.

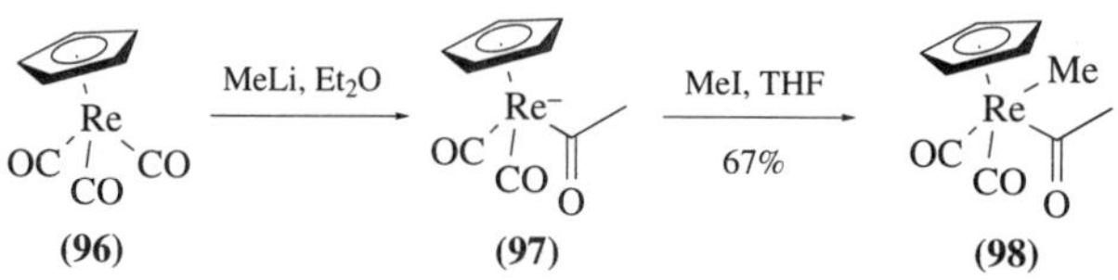

Scheme 27

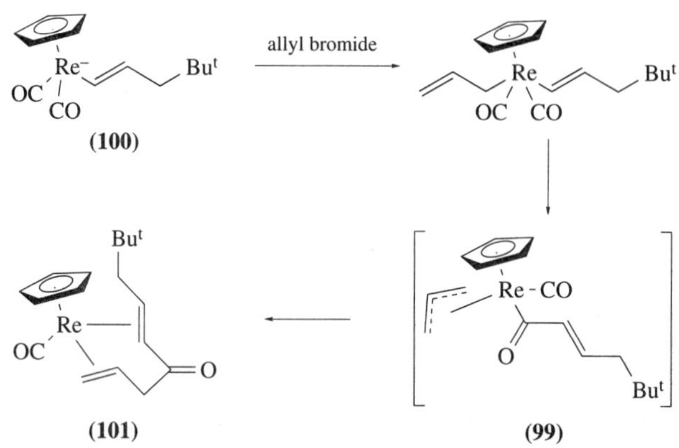

Scheme 28

Deprotonation of the rhenium hydride compound (**102**) by *n*-BuLi at $-78\,^{\circ}C$ generates an anionic compound (**103**) which reacts with benzoic anhydride at $-32\,^{\circ}C$ to give the acyl complex $CpReNO(PPh_3)(COPh)$ (**104**) (Scheme 29) ⟨88JA6110⟩. Similar compounds can be synthesised by addition of Grignard reagents to the carboxymethylrhenium compound (**105**) ⟨83OM1852⟩. For example, the aroyl complex (**106**) is formed from (**105**) in 97% yield (Equation (20)) ⟨87JA4837⟩.

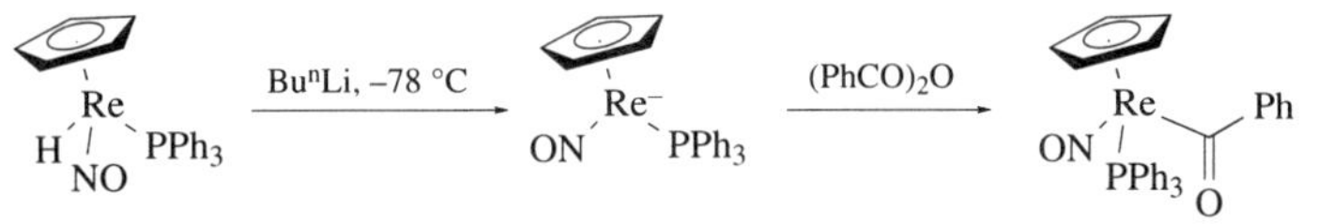

(102) **(103)** **(104)**

Scheme 29

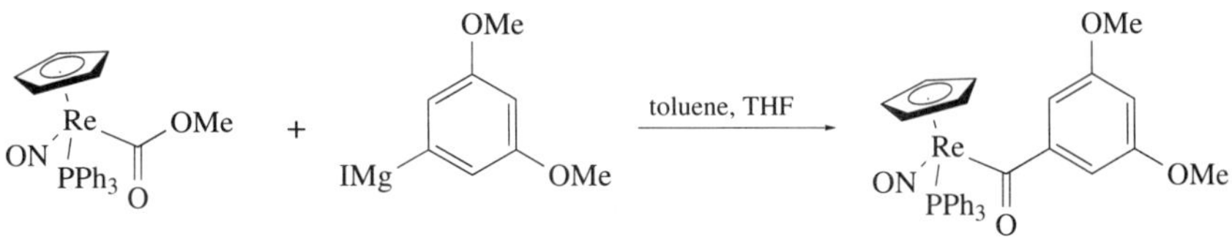

(20)

(105) **(106)**

Rhenium acyl compounds can be synthesised from carbyne compounds such as (**107**) by addition of the nitrite anion in a reaction that formally involves addition of O^- to the carbyne ligand and a replacement of the carbonyl ligand by a nitrosyl ligand (Equation (21)) ⟨87JA1584⟩.

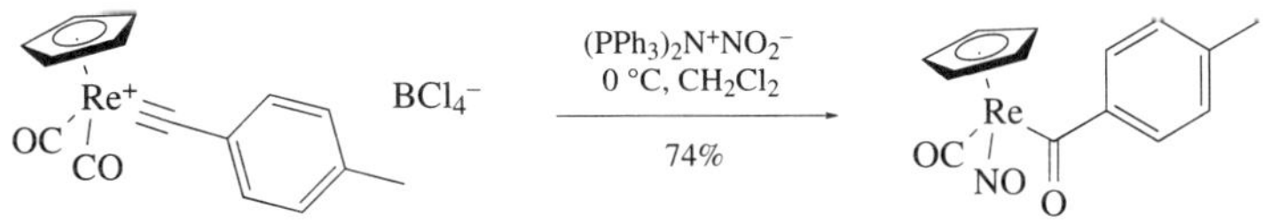

(21)

(107)

The rhenacyclobutenone ring system which occurs in compounds such as (**108**) has been constructed by addition of $Na[Re(CO)_5]$ to dimethyl acetylenedicarboxylate (Scheme 30) ⟨90JOM(383)C1⟩. The mechanism of this process may involve the addition of the rhenium anion to the acetylene to give a vinyl anion which reacts with a metal-coordinated carbonyl group to give the rhenacyclobutenone product (**108**).

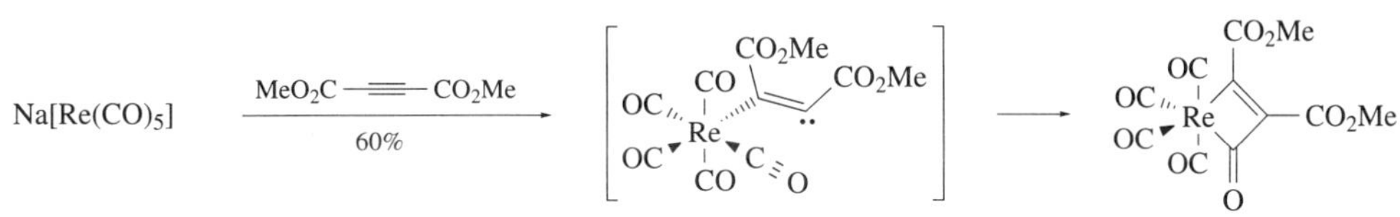

(108)

Scheme 30

5.10.5.12 Acyliron Compounds

Acyl derivatives of iron have found a variety of applications in organic synthesis. The dianionic compound $Na_2Fe(CO)_4$ was developed by Collman as a transition metal analogue of a Grignard reagent ⟨75ACR342⟩. It is formed by reduction of $Fe(CO)_5$ with sodium amalgam ⟨70JA6080⟩ or with molten sodium and an electron carrier, such as benzophenone ketyl, in dioxane ⟨75ACR342⟩. $Na_2Fe(CO)_4$ reacts with primary and secondary alkyl halides by an S_N2 process to give the iron alkyl compounds $NaFeR(CO)_4$ with inversion of stereochemistry at the reacting carbon atom of the alkyl group. These compounds react with carbon monoxide or triphenylphosphine to give the acyl complexes $NaFe(COR)(L)(CO)_3$ which react with alkylating agents to give unsymmetrical ketones. For example, (*R*)-3-methyl-2-nonanone (**109**) was formed from (*S*)-2-octyl tosylate (**110**) in 80% chemical yield and 99% optical yield by reaction with $Na_2Fe(CO)_4$, carbonylation and alkylation with methyl iodide (Scheme 31) ⟨72JA1788⟩. The scope of this process is limited by the poor reactivity of the $NaFe(COR)(L)(CO)_3$ salt. However, aryl iodides will couple with compounds of the type $NaFe(COR)(L)(CO)_3$ in the presence of catalytic quantities of $Pd(PPh_3)_4$ and $ZnCl_2$ to give acyl derivatives (Scheme 32) ⟨88CL1141⟩. The electronic properties of substituents have little effect on the reaction which probably proceeds via an acylpalladium species. The $ZnCl_2$ may act by facilitating the transmetallation. An alternative procedure for forming acyl tetracarbonyl anions is to react $Fe(CO)_5$ with reactive halides under the mild phase transfer conditions shown in Scheme 33 ⟨86CC1754⟩.

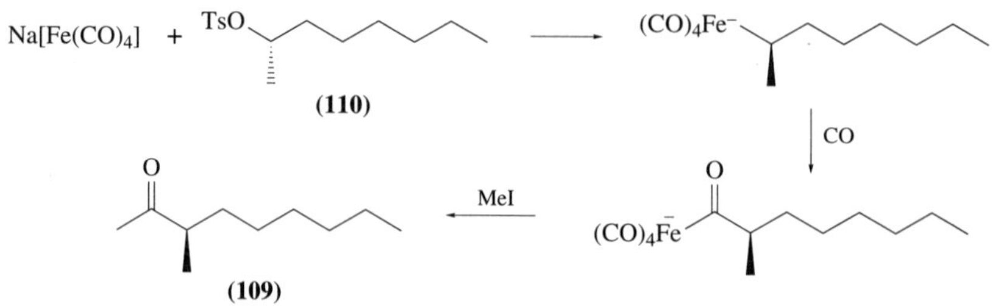

Scheme 31

$RX = Bu^nBr$, $ArX = PhI$, 71%
$RX = EtO_2C(CH_2)_3Br$, $ArX = PhI$, 76%
$RX = Bu^iBr$, $ArX = m\text{-}MeC_6H_4I$, 86%

Scheme 32

$Fe(CO)_5 \xrightarrow{H_2O,\ Bu^n_4NHSO_4,\ NaOH,\ CH_2Cl_2,\ CO} Fe(CO)_4^{2-} \xrightarrow{PhCH_2Cl} PhCH_2Fe(CO)_4^-$

$PhCH_2Fe(CO)_4^- \xrightarrow{H_2O,\ Bu^n_4NHSO_4,\ MeI,\ CH_2Cl_2,\ CO} (PhCH_2CO)FeMe(CO)_4 \xrightarrow{CO} PhCH_2COMe$

Scheme 33

The standard method of preparing iron acyl complexes of the type $CpFe(CO)_2(COR)$ is to acylate, with acyl chlorides, the anion **(111)** formed from $[CpFe(CO)_2]_2$ and sodium amalgam (Scheme 34). For example, the acetyl ⟨63JA1918⟩, trifluoroacetyl ⟨64JOM(2)15⟩ and benzoyl ⟨74JCS(D)2302⟩ compounds **(112)**, **(113)** and **(114)** have been formed by this procedure (Scheme 34). The analogous alkyl compounds such as $CpFe(CO)_2Me$, react with phosphines (e.g., triphenylphosphine, tri-*n*-butylphosphine, tri-*n*-butyl phosphite) in refluxing THF to give phosphine-substituted acyliron compounds (Equation (22) and Table 3) ⟨66IC889⟩. The rate of this reaction is solvent dependent with formation of acyl products three times faster in refluxing acetonitrile than in THF ⟨71JCS(A)367⟩.

$$CpFe(CO)_2R^1 \xrightarrow{PR^2_3,\ MeCN,\ heat} CpFePR^2_3(CO)COR^1 \qquad (22)$$

$$[CpFe(CO)_2]_2 \xrightarrow{Na/Hg} [CpFe(CO)_2]Na \ \textbf{(111)} \xrightarrow{RCOCl} CpFe(CO)_2COR$$

(112) R = Me, 63%
(113) R = CF_3, 23%
(114) R = Bn, 36%

Scheme 34

An alternative route to the compounds in Table 3 is via the carbonylation of iron alkyl complexes. For example, $CpFe(CO)_2Me$ reacts with carbon monoxide at high pressure (325 atm) and tem-

Table 3 Preparation of acyliron compounds (for R^1 and R^2 see Equation (22)).

R^1	R^2	Yield (%)
Et	Ph	85
Et	OPh	90
Et	Bu^n	85
Pr^i	Ph	90

perature (97 °C) in tetradecane to give the corresponding acetyl compound $CpFe(CO)_2(COMe)$ ⟨78JA1687⟩. The analogous phosphine-containing alkyl compound $CpFeCO(PPh_3)(Me)$ reacts with carbon monoxide under much milder conditions in heptane (1 atm, 90 °C) to give the acetyl compound $CpFe(PPh_3)(CO)(COMe)$ (**115**) ⟨66IC1177⟩. Cutler has studied the reaction of compounds of the type CpFeCO(L)(Me) (where L is CO, triphenylphosphine and trimethylphosphite) with carbon monoxide ⟨85OM1247⟩. The reaction was found to be very solvent dependent with solvents such as acetonitrile, nitromethane and trifluoroethanol preferred. Neither DMSO nor DMF led to the significant formation of product. It was proposed that a solvent like trifluoroethanol promotes the reaction by hydrogen bonding to the incipient acetyl ligand. Indeed, the addition of 1% HBF_4 to a solution of $CpFe(CO)_2Me$ or $CpFeCO(PPh_3)(Me)$ in dichloromethane leads to the addition of carbon monoxide at 1 atm. However, neither the use of a hydrogen bonding solvent nor an acid catalyst induces carbonylation of $CpFeCO(PPh_3)(CH_2X)$ where X is OMe, Ph or CO_2Me.

Acyl complexes of the type $CpFeCO(PPh_3)(COR)$ have been studied extensively by Davies and Liebeskind and can be used in the diastereoselective and enantioselective synthesis of various organic compounds. This configurationally stable chiral iron system exerts a powerful stereochemical control in reactions including alkylations ⟨82CC1303, 83CC1202, 83JOM(248)C1, 84JOM(262)49, 86TL623⟩, aldol reactions ⟨84CC956, 84TL2709, 85JOM(285)213, 85TL2125⟩, tandem Michael additions and alkylations ⟨85CC209, 86CC495, 86TL3787⟩ and Diels–Alder reactions ⟨86CC609⟩ to generate a variety of substituted acyl compounds. Examples of these reactions are shown in Scheme 35.

Homochiral menthyl ester iron complexes such as (**116**) react with alkyl lithium species to give the corresponding homochiral acyl compounds (Equation (23)) ⟨72JOM(36)C18⟩. This reaction proceeds with inversion of configuration at the iron centre indicating that the alkyl anion reacts at the carbonyl ligand with elimination of menthylate rather than at the carbonyl group of the menthyl ester ligand.

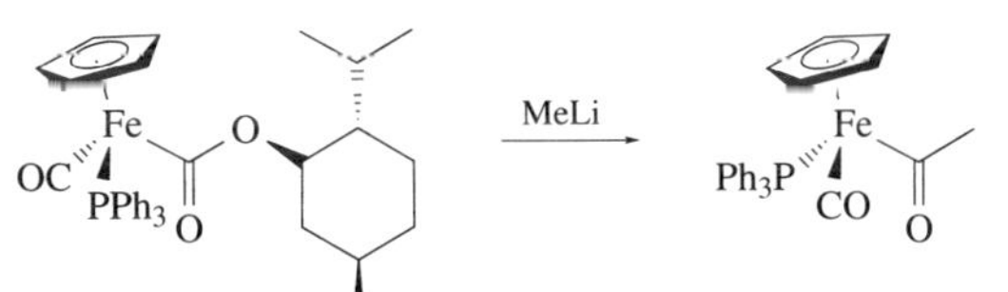

(**116**)

(23)

The absolute stereochemistry of the iron centre of homochiral iron–acyl complexes has been determined by alkylation of the racemic mixture of acetyl compounds (**115**) with homochiral alkyl chloride (**117**), separation of the resulting diastereoisomers and solution of the x-ray crystal structure of the homochiral acyliron compound (**118**) (Scheme 36) ⟨86CC607⟩.

The effect on the stereochemistry at the iron centre of the reaction of homochiral alkyliron compounds such as (**119**) with carbon monoxide has been studied by Flood ⟨84JA2853⟩. The stereochemical outcome of the reaction is solvent dependent. In nitromethane the acyl complex (**120**) is formed highly stereoselectively indicating that the reaction corresponds to formal alkyl group migration (Scheme 37). In DMSO or HMPA the reaction is less stereoselective and gives predominantly the enantiomeric compound (**121**) which corresponds to formal carbon monoxide migration (Scheme 37).

5.10.5.13 Acylruthenium Compounds

Ruthenium acyl complexes of the type CpRuCO(X)(COR) are formed by reactions analogous to those which give the corresponding iron compounds. Thus, the ruthenium salt $Na[CpRu(CO)_2]$

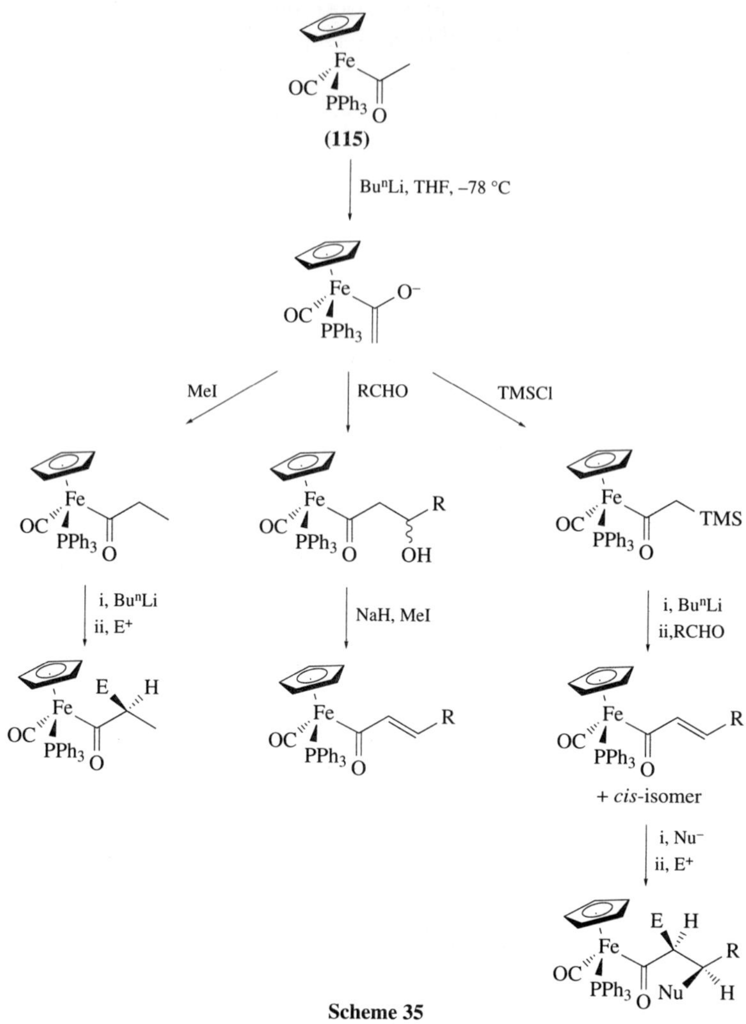

Scheme 35

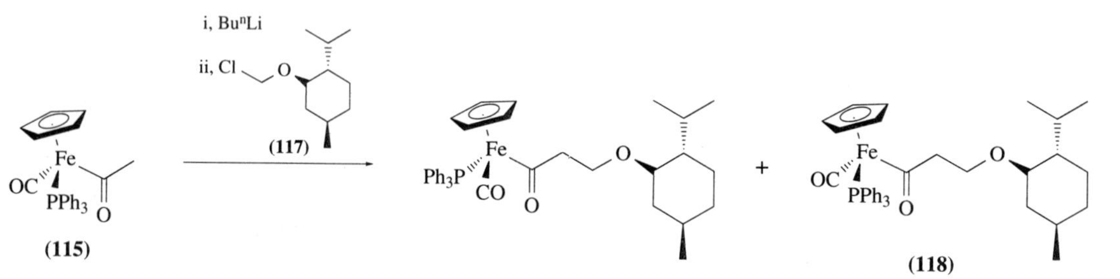

Scheme 36

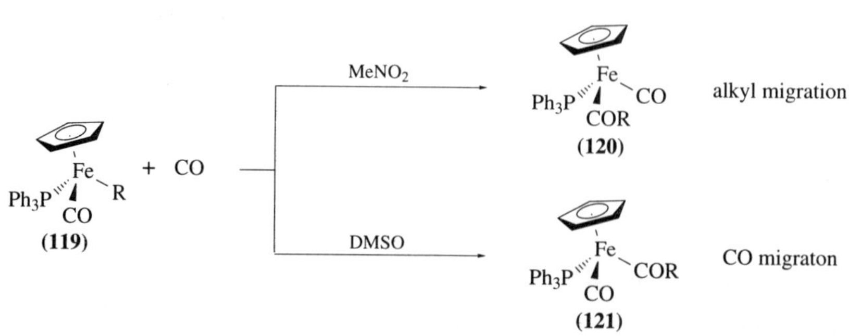

Scheme 37

(**122**) reacts with propionyl chloride to give propionyl compound (**123**) (Scheme 38) ⟨63JCS4096⟩. Alternatively, (**122**) reacts with alkyl iodides to give alkyl intermediates (**124**)–(**126**) which give the corresponding acyl compounds (**127**)–(**129**) on treatment with excess triphenylphosphine in xylene under reflux (Scheme 38) ⟨80JCS(D)1845⟩. The methyl complex (**130**) undergoes a migratory insertion reaction induced by carbon monoxide to give the acetyl compound (**131**) (Equation (24)) ⟨76CC1032⟩.

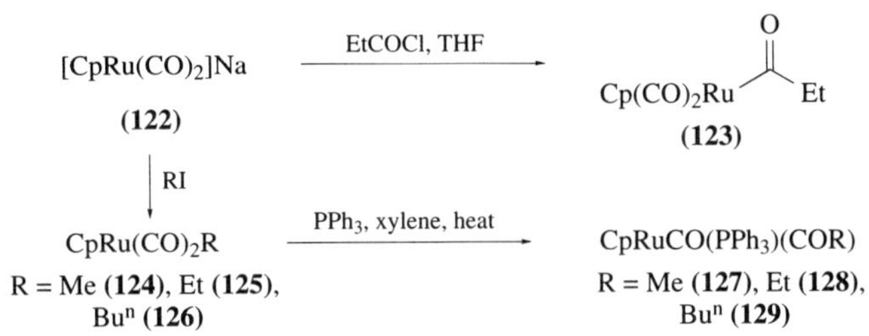

Scheme 38

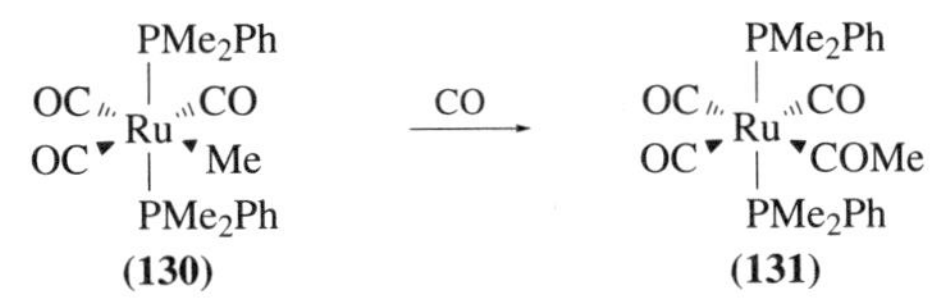

(24)

The 5-coordinate alkenyl ruthenium derivatives (**132**)–(**134**) (route A) and (**135**) and (**136**) (route B) react rapidly with *t*-butylisonitrile at room temperature in dichloromethane to give 6-coordinate cationic acyl compounds (**137**)–(**141**) (Scheme 39) and (Table 4) ⟨90JOM(390)C57⟩.

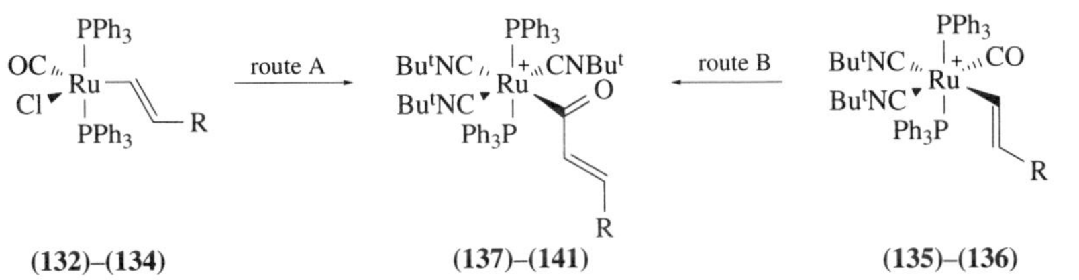

Scheme 39

Table 4 Preparation of acylruthenium compounds (for R and X see Scheme 39).

Route	*R*	*X*	*Yield* (%)
A	TMS	Cl	79
A	*n*-C_8H_{17}	Cl	74
A	Ph	Cl	98
B	*n*-C_8H_{17}	PF_6	86
B	Bu^t	PF_6	84

5.10.5.14 Acylosmium Compounds

There are relatively few reports of acylosmium compounds. The osmium salt $Na_2[Os(CO)_4]$ (**142**) reacts with methyl and ethyl iodide to give the corresponding dialkylosmium compounds (**143**) and (**144**) which react with carbon monoxide to give ethane and $Os(CO)_5$ in the case of the methyl compound (**143**) and the bisacyl osmium species (**145**) for the ethyl compound (**144**) (Scheme 40) ⟨70HCA1091⟩. In both examples, forcing conditions are required for reaction to occur (120 °C and 130 atm). The osmium salt $Na_2[Os(CO)_4]$ (**142**) is prepared by reaction of the osmium cluster $Os_3(CO)_{12}$ with sodium in the presence of bipyridyl.

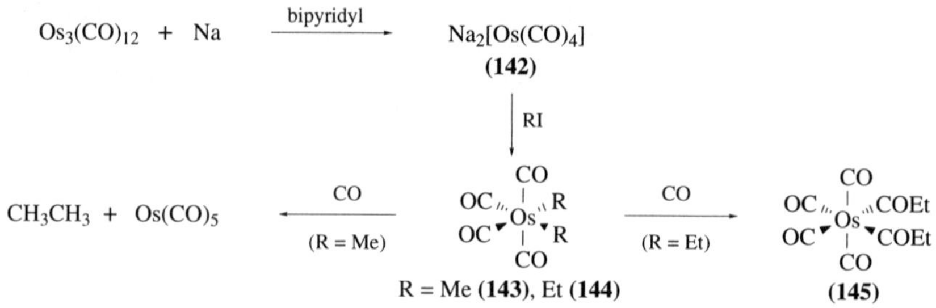

Scheme 40

The acyl complex (**146**) is formed on treatment of the methylimidoosmium complex Os(NMe)(CH_2SiMe_3)$_4$ (**147**) with carbon monoxide at −20 °C (Scheme 41) ⟨90JA8369⟩. The compound (**146**) was assigned the structure shown in Scheme 41 on the basis of NMR and IR spectroscopic data, but was not sufficiently stable to be isolated and fully characterised. Addition of further carbon monoxide to compound (**146**) resulted in the formation of a bisacyl compound (**148**). Attempts to isolate (**148**) resulted in the formation of the ketone (Me_3SiCH_2)$_2$CO and polymeric osmium complexes. The oxo analogue (**149**) (Scheme 41) of (**147**) reacts with carbon monoxide under more forcing conditions but does not give the analogous acyl compound. Presumably in the oxygen-containing compound (**149**) reductive elimination is faster than alkyl migration to coordinated carbon monoxide.

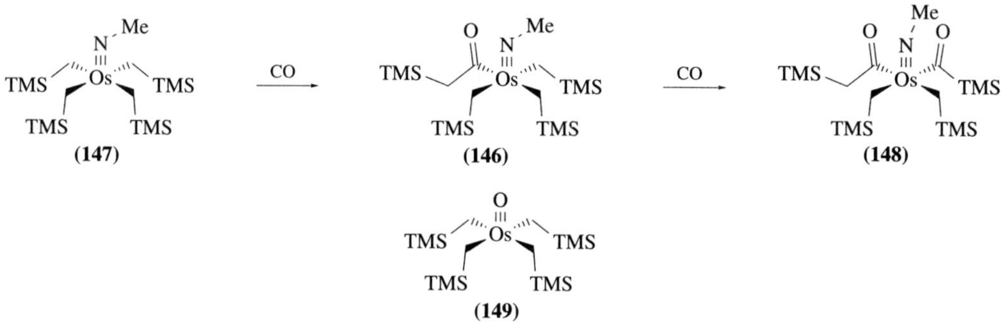

Scheme 41

5.10.5.15 Acylcobalt Compounds

Acylcobalt complexes of the type (RCO)Co(CO)$_4$ have been synthesised by addition of acid chlorides or relatively reactive alkyl halides under an atmosphere of carbon monoxide to Na[Co-(CO)$_4$] (Scheme 42) ⟨62JA2499⟩. Examples of acyl compounds produced by the acid chloride route (Equation (25)) include (RCO)Co(CO)$_4$ where R is Me, Et, Ph and —CH_2CO_2Et; examples from the alkyl halide route (Equation (26)) include R is Me and —CH_2OMe. These types of acylcobalt species are intermediates in the cobalt catalysed hydroformylation of alkenes and are relatively unstable. They have also been prepared by the reaction of CoH(CO)$_4$ with terminal alkenes such as pent-1-ene in the presence of carbon monoxide (Scheme 43) ⟨61JA4023⟩. Addition of triphenylphosphine to the initially formed acyl compounds such as (**150**) results in evolution of carbon monoxide to give stable, crystalline cobalt acyl products such as (**151**) (Scheme 43). The hydride complex CoH(CO)$_4$ reacts with ketenes at −60 °C in pentane to give acyl complexes (e.g., (**152**) and (**153**)) in excellent yield (Scheme 44) ⟨84CC824⟩.

Cobalt acyl compounds such as (**154**) have been used to build up small organic molecules from

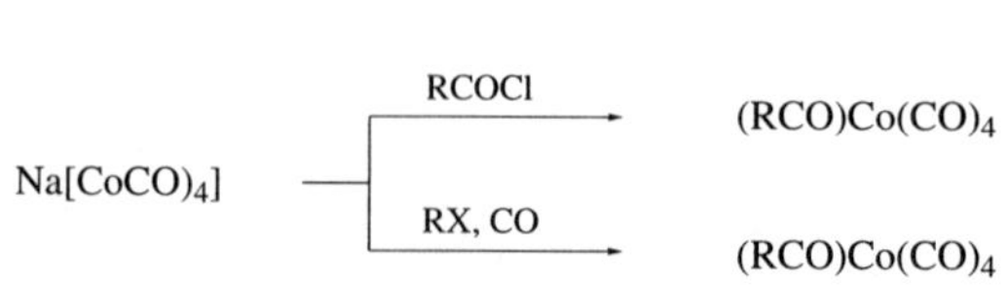

(25)

(26)

Scheme 42

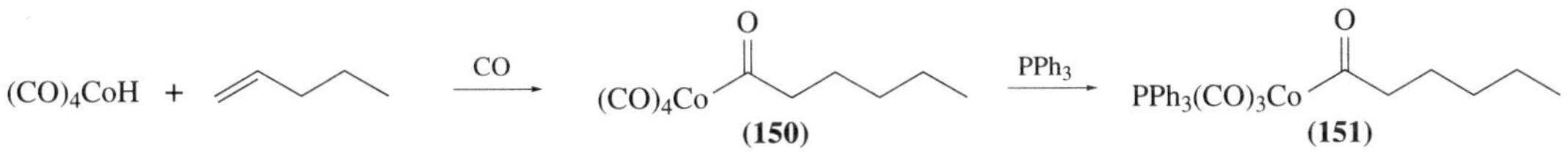

Scheme 43

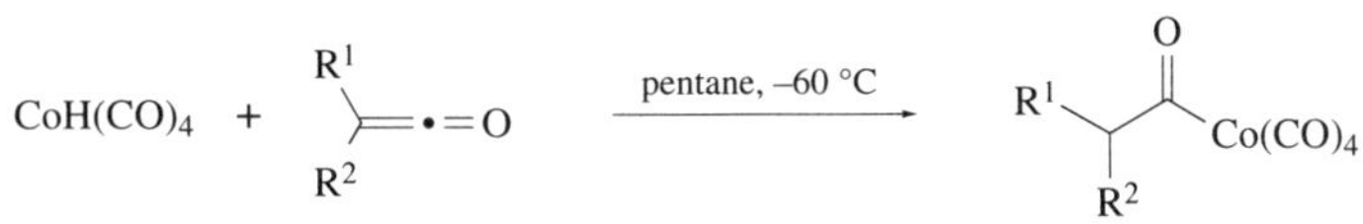

(**152**) R^1 = Me, R^2 = Me, 90%
(**153**) R^1 = Et, R^2 = H, 95%

Scheme 44

carbon monoxide by an iterative process of carbon monoxide insertion, reduction, carbon monoxide insertion, etc. as shown in Scheme 45 ⟨87JA5847⟩. Cutler and co-workers have reported that the cobalt acyl compound (**154**) can be alkylated at oxygen and the intermediate carbene compound (**155**) reduced with Et_3BHLi to give the alkoxycobalt species (**156**), which reacts with carbon monoxide, with insertion of the carbon monoxide into the metal–carbon bond, to form a second acyl compound (**157**). Repetition of this sequence gives a dialkoxy complex (**158**) with essentially complete diastereoselectivity (Scheme 45) ⟨87JA5844⟩.

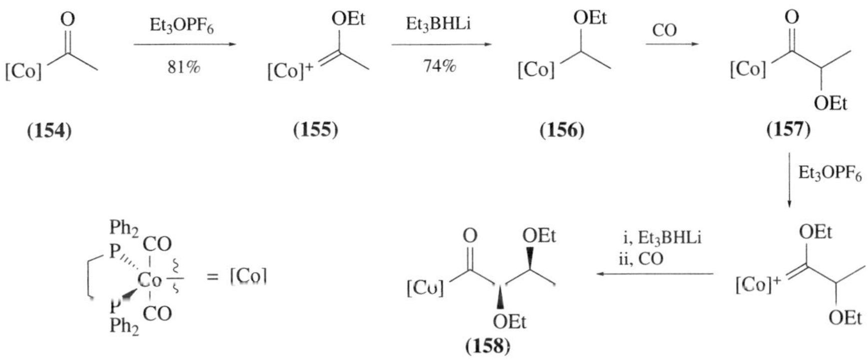

Scheme 45

Acyl cobalt salophen compounds are useful precursors of acyl radicals ⟨90JCS(P1)2721⟩. They are synthesised by reduction of Co(II)salen or Co(III)Br(PPh_3)(salen) with 1% sodium amalgam in THF followed by acylation with acid chlorides (Scheme 46) ⟨67TL1783⟩.

Scheme 46

Maleoyl cobalt complexes such as (**159**) react with alkynes to give quinones (Scheme 47) ⟨84JA6451⟩. Liebeskind has reported the synthesis of complexes such as (**159**) from cyclobutenediones and $ClCo(PPh_3)_3$. The mechanism proposed for the formation of the quinone products is illustrated in Scheme 47 ⟨87JA2759⟩. This methodology has been used in the synthesis of royleanone (Scheme 48) ⟨90TL3723⟩.

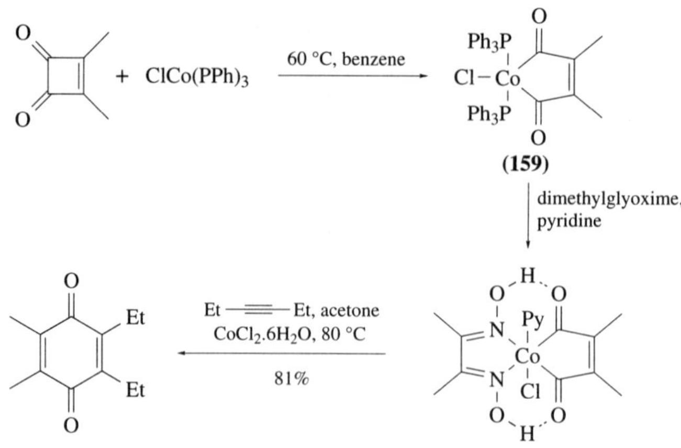

Scheme 47

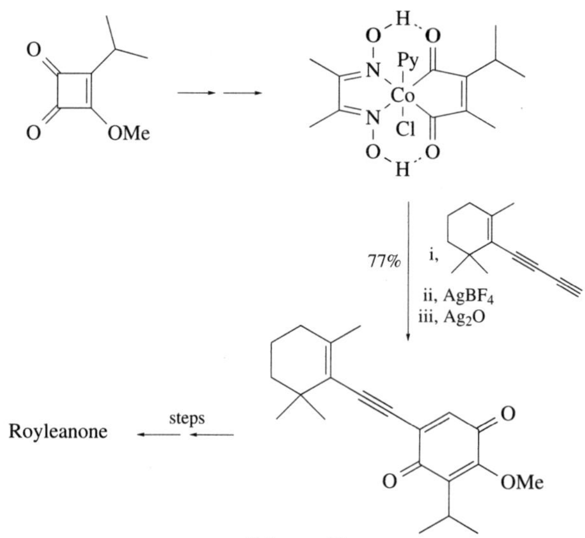

Scheme 48

5.10.5.16 Acylrhodium Compounds

Certain rhodium compounds will undergo reactions with acid chlorides to give acyl species. The phosphine compound (**160**) oxidatively adds to a range of acid chlorides to give the corresponding rhodium acyl complexes in excellent yields on heating at reflux in benzene as shown in Scheme 49 ⟨66JA3452⟩. These acyl compounds will eliminate terminal alkenes in a process which converts acid chlorides to alkenes (Scheme 49) ⟨74JA1518⟩. The cyclopentadienyl rhodium compound (**161**) adds to acid chlorides to give the corresponding acyl compounds (**162**) and (**163**) (Equation (27)) ⟨79CB834⟩.

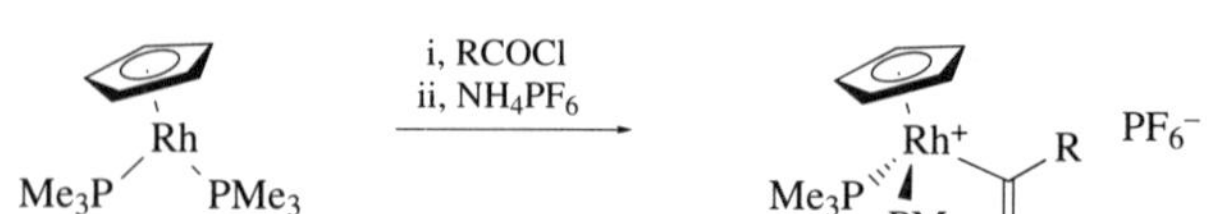

(27)

(**161**)

(**162**) R = Me, 61%
(**163**) R = Ph, 71%

$$\text{ROCCl} \xrightarrow[\text{99\%}]{\text{RhCl(PPh}_3)_3\ \textbf{(160)}} (\text{RCO})\text{RhCl}_2(\text{PPh}_3)_2 \xrightarrow{\text{heat}} \text{Cl(CO)Rh(PPh}_3)_2 + \text{HCl} + \text{alkene}$$

R = Bn, cyclohexyl, *n*-pentyl, $Me(CH_2)_{14}$

Scheme 49

The rhodium complex $RhCl(PPh_3)_3$ can be used synthetically to decarbonylate aldehydes. This process involves the oxidative addition of the aldehyde CH bond to form a coordinatively unsaturated acyl rhodium hydride (**164**) that undergoes an acyl–alkyl rearrangement, with hydrogen transfer to the alkyl group, to give the alkane product and $RhCl(PPh_3)_2(CO)$ (Scheme 50) ⟨69S1⟩. The intermediate acylrhodium hydride (**164**) cannot normally be isolated. However, the quinoline aldehyde (**165**) reacts with $RhCl(PPh_3)_3$ to give the acyl hydride complex (**166**) (Equation (28)). This complex is coordinatively saturated owing to coordination of the quinoline nitrogen to the metal which slows down the acyl-to-alkyl rearrangement and renders the complex isolable ⟨78JA640⟩. The acyl complexes (**167**) and (**168**) were prepared from the quinoline aldehyde (**165**) and the 1,4-pentadiene rhodium species (**169**). The addition of pyridine as a coordinating ligand and reductive elimination with trimethyl phosphite gives a mixture of unsaturated ketones (**170**) and (**171**) in which the pentadiene ligand has added to the aldehyde functional group (Scheme 51) ⟨90JOM(390)361⟩.

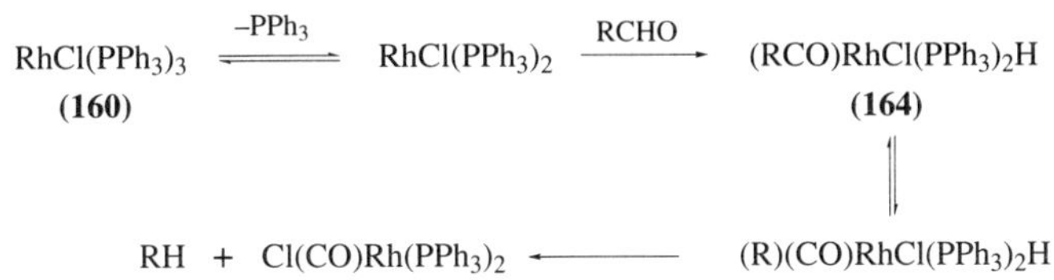

Scheme 50

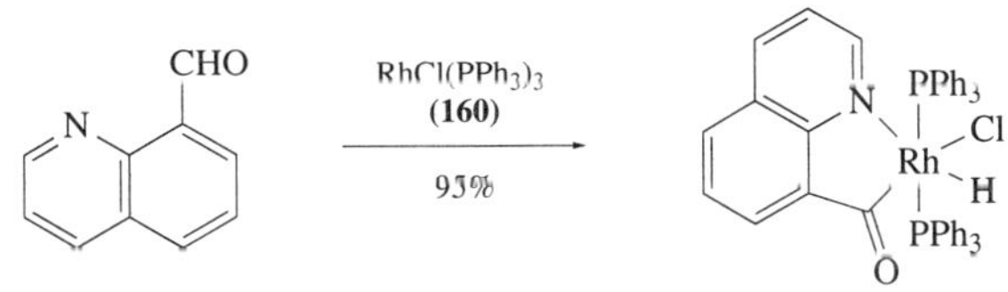

(28)

(**165**) (**166**)

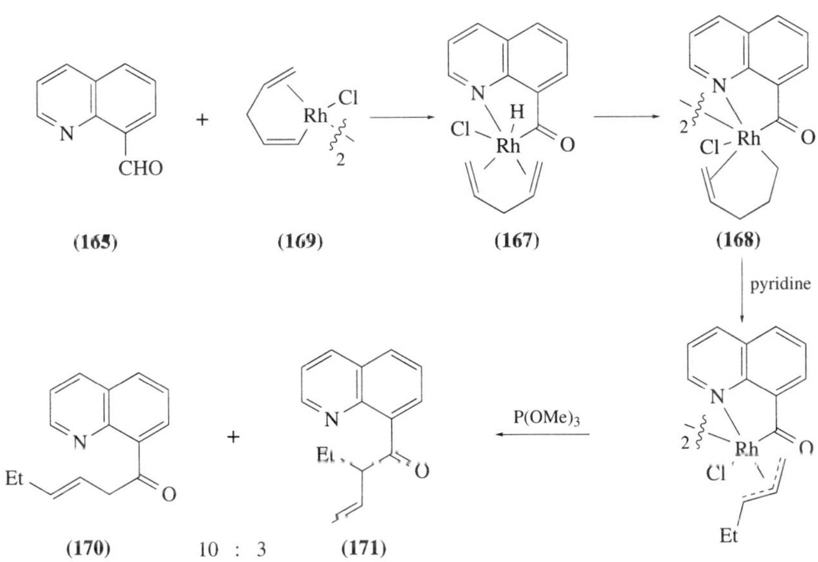

Scheme 51

Rhodium acyl compounds have been prepared in optically pure form by incorporating homochiral phosphine ligands. For example, the complex (**172**) containing an α amino benzylamine-derived phosphine ligand undergoes an addition reaction with methyl iodide to give the acyl complex (**173**) as a 1 : 1 mixture of diastereoisomers which can be separated by fractional crystallisation (Equation (29)) ⟨82JA1096⟩.

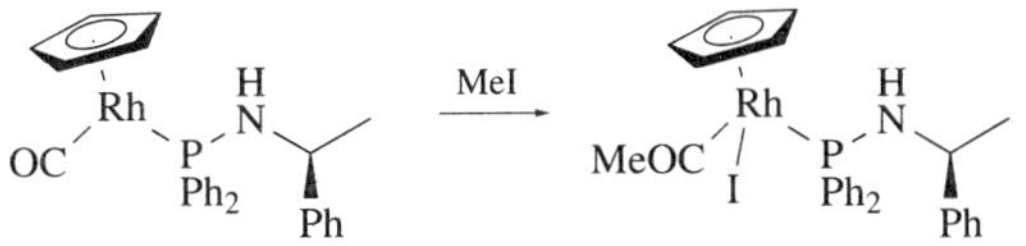

(29)

(**172**) (**173**)

Various strained polycyclic alkanes will react with the dimeric rhodium compound $Rh_2(CO)_4Cl_2$ to give acyl species in which the metal has inserted into a strained carbon–carbon bond to give an alkyl intermediate that inserts into a carbonyl ligand. For example, cubane reacts with $Rh_2(CO)_4Cl_2$ in chloroform to give the acyl compound (**174**) in 90% yield (Scheme 52). This material reacts with triphenylphosphine in dichloromethane to give the cyclic ketone (**175**) in 90% yield ⟨70JA3515⟩. 3,5-Dehydronoriceane also reacts with $Rh_2(CO)_4Cl_2$ in hexane at room temperature to give a stable rhodium acyl compound (**176**) in 83% yield (Equation (30)). However, attempts to react it with triphenylphosphine to generate the expected ketone led to recovery of 3,5-dehydronoriceane ⟨84JOC4460⟩.

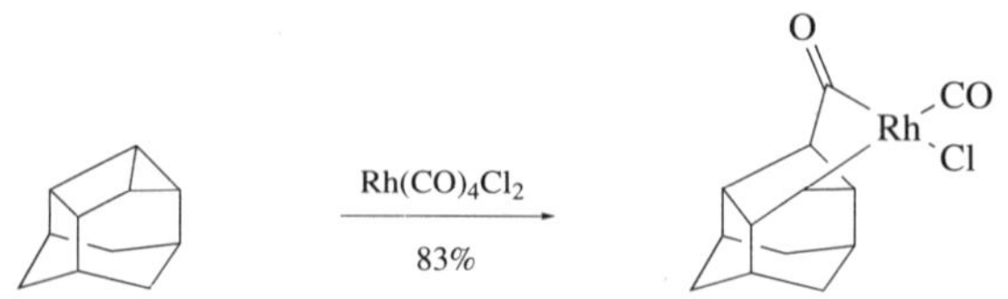

(**176**) (30)

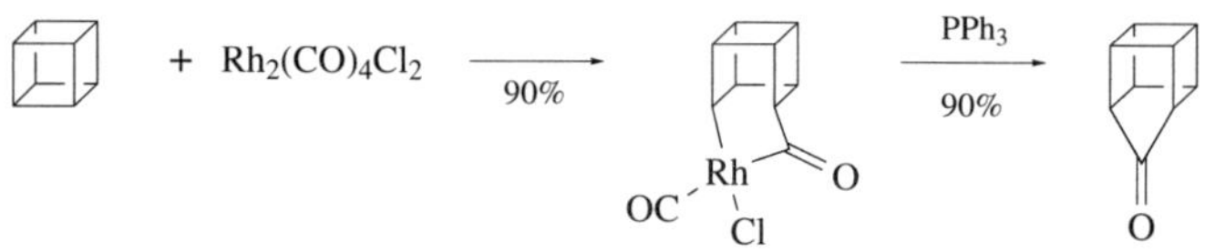

(**174**) (**175**)

Scheme 52

5.10.5.17 Acyliridium Compounds

The iridium acyl compound (**177**) can be prepared by addition of MeLi to $(Ph_3P)_3IrCl$, in diethyl ether at 0 °C, and stepwise carbonylation of the resulting unstable alkyliridium compound (**178**) in diethyl ether, to give the intermediate (**179**), followed by addition of two further moles of carbon monoxide, under pressure in ethanol, to give (**177**) in 70% yield (Scheme 53) ⟨72JA6226⟩. An alternative process used to form the acyl compound (**177**) involves the addition of methyl iodide to the anionic iridium complex (**180**) in the presence of triphenylphosphine followed by carbonylation of the unstable iridium alkyl compound (**181**) (Scheme 54). The overall yield of the reaction sequence is 77% ⟨68JA2282⟩. The anionic iridium complex (**180**) is formed by addition of the chloro complex (**182**) to sodium amalgam.

$$(PPh_3)_3IrCl \xrightarrow{\text{MeLi, 0 °C}} \underset{(\mathbf{178})}{(PPh_3)_3IrMe} \xrightarrow{\text{CO, Et}_2\text{O}} \underset{(\mathbf{179})}{(PPh_3)_3Ir(CO)Me} \xrightarrow{\text{CO, EtOH}} \underset{(\mathbf{177})}{(PPh_3)_2Ir(CO)_2(COMe)}$$

Scheme 53

$$\underset{(\mathbf{182})}{IrCO(PPh_3)_2Cl} \xrightarrow{\text{Na/Hg}} \underset{(\mathbf{180})}{Na[Ir(CO)_3(PPh_3)]} \xrightarrow[\text{THF}]{\text{MeI, PPh}_3} \underset{(\mathbf{181})}{MeIr(CO)_2(PPh_3)_2} \xrightarrow{\text{CO, 40 °C}} \underset{(\mathbf{177})}{Ir(COMe)(CO)_2(PPh_3)_2}$$

Scheme 54

Certain bisphosphine complexes of iridium will undergo oxidative addition reactions with acid halides to give hexa-coordinated iridium acyl compounds (Scheme 55). Examples include the addition of acetyl bromide to the bistriethylphosphine complex (**183**) (Equation (31) in Scheme 55) ⟨68IC27⟩, the addition of various aroyl chlorides to the related compound (**184**) (Equation (32) in Scheme 55) ⟨71JA1368⟩ and the addition of trifluoroacetyl chloride to the bistriphenylphosphine compound (**185**) (Equation (33) in Scheme 55) ⟨74IC1595⟩.

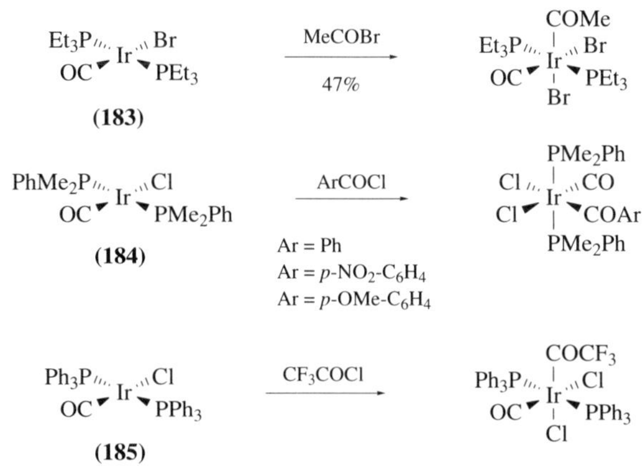

(31)

(32)

(33)

Scheme 55

Allylic complexes of iridium, formed from the addition of butadienes to the hydride complex $IrH(CO)_2(PPh_3)_2$ **(186)**, react with carbon monoxide to give iridium acyl complexes such as **(187)** (Scheme 56) ⟨71JCS(A)850⟩. The reaction occurs to give the acyl product which is formed by insertion of carbon monoxide into the less hindered end of the η^3-allyl ligand.

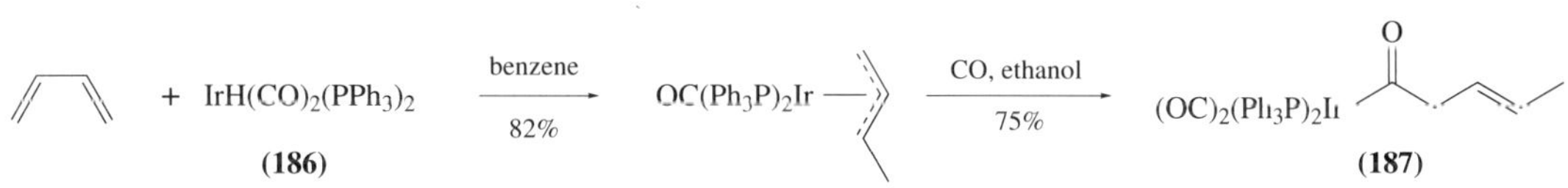

Scheme 56

Alkyliridium complexes such as the methyl compound $MeIr(CO)(PPh_3)_2$ readily add acetylenes such as dimethyl acetylenedicarboxylate (DMAD) in a process that leads to the cyclotrimerisation of the acetylene. The reaction sequence probably involves the formation of iridium acyl species. If the metal bound methyl group in $MeIr(CO)(PPh_3)_2$ is replaced with a bulky 2,2,2-trimethylethyl group and the triphenylphosphine ligands with the *para*-tolyl analogue, the reaction of the resulting compound **(188)** with DMAD gives a stable and isolable acyl compound **(189)** (Equation (34)) ⟨87JA5145⟩. In contrast, the triscarbonyliridium complex $Ir(CO)_3Cl$ reacts with excess diphenylacetylene under carbon monoxide in THF to give a polymeric species which will react with nucleophiles such as triphenylphosphine and bromide ion to give the iridium maleoyl compounds **(190)** and **(191)** (Scheme 57) ⟨75CC267⟩.

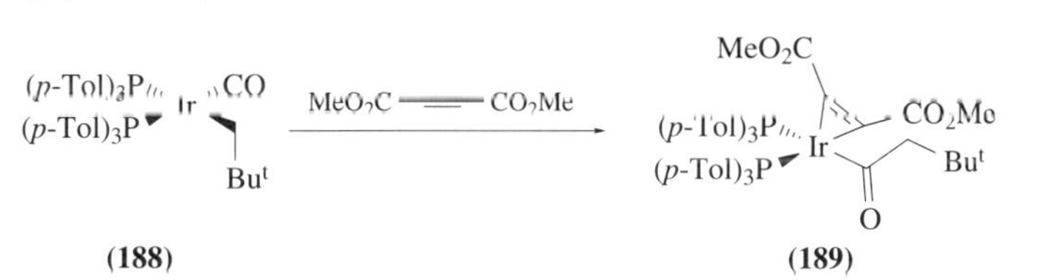

(34)

5.10.5.18 Acylnickel Compounds

The stable acylnickel compound **(192)** was prepared by reaction of the corresponding methylnickel complex **(193)** in pentane with carbon monoxide at room temperature and pressure (Equation (35)) ⟨73AG(E)402⟩. The acyl complex **(192)**, containing the relatively sterically undemanding trimethylphosphine ligand, was formed quantitatively as a crystalline solid (m.p. 98–99 °C). The analogous benzoyl complex is formed in a similar manner in 80% yield and isolated as a solid (m.p. 73–75 °C) ⟨77JA2501⟩. Similarly, the ethylnickel complex **(194)**, in toluene solution, reacts with carbon monoxide at −20 °C to give the corresponding propionyl compound **(195)** in 85% yield (Equation (36)) ⟨76BCJ546⟩.

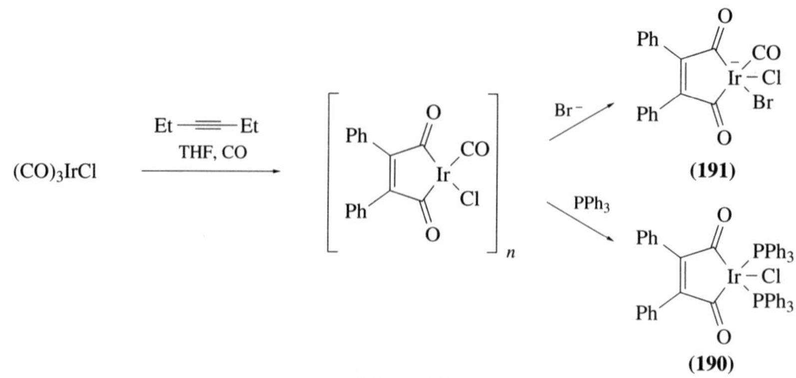

Scheme 57

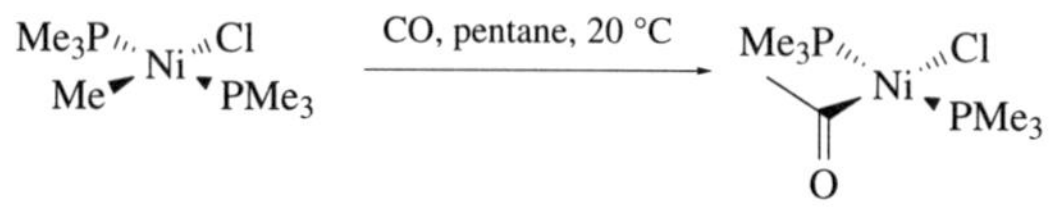 (35)

(193) (192)

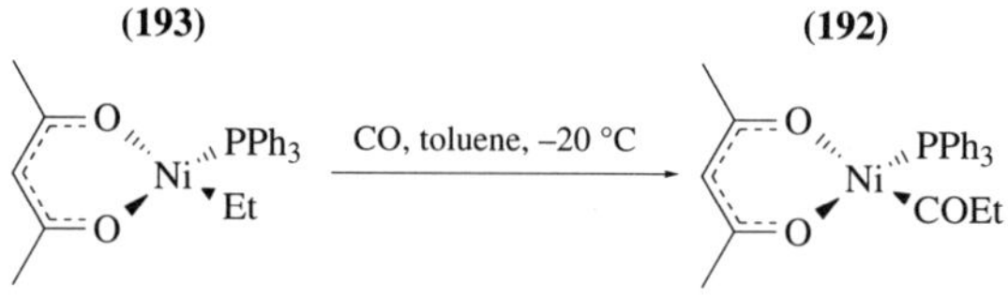 (36)

(194) (195)

Yamamoto *et al.* has studied the synthesis of acylnickel compounds of the type $Ni(COR)YL_2$ from the corresponding alkyl compounds $NiRYL_2$ by carbonylation ⟨81BCJ2161⟩. Examples are listed in Table 5. When the ligand Y makes a moderate to strong bond to the nickel atom the acyl compound is sufficiently stable to be isolated and characterised. Thus compounds (**196**), (**197**) and (**200**) are stable in solution owing to the high strengths of the bonds between nickel and the chloride, succinimide and phthalimide ligands. These compounds react with alcohols and amines to give ester and amide products, respectively. Complexes containing a ligand Y (e.g., alkyl) which forms a relatively weak bond with the nickel atom carbonylate to give acyl compounds which spontaneously reductively eliminate RCOY. The 4-cyanophenoxy substituted complexes (**198**) and (**199**) can be isolated, but readily undergo a reductive coupling of the acyl and 4-cyanophenoxy ligands with elimination of 4-cyanophenyl propionate and acetate, respectively. Dialkyl complexes NiR_2L_2 react with carbon monoxide to give diketone and ketone products. The formation of the diketones can be rationalised by the formation of bisacyl complexes which reductively eliminate (Scheme 58).

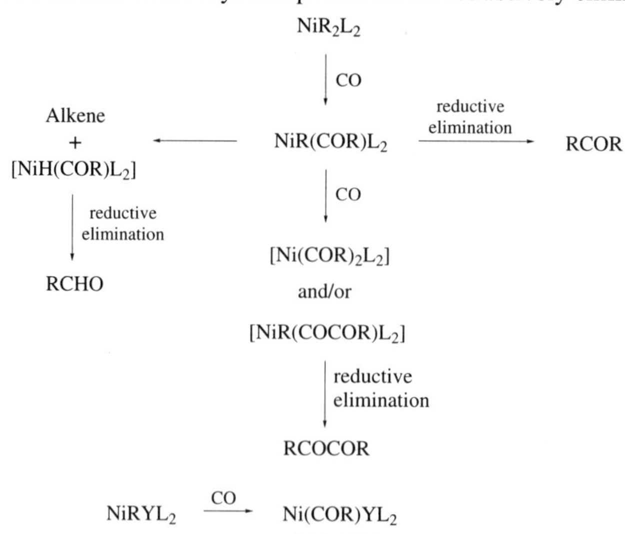

Scheme 58

Table 5 Acylnickel compounds $Ni(COR)YL_2$.

Compound No.	*R*	*Y*	L_2
196	Et	Cl	2,2′-bipyridyl
197	Et	succinimido	2,2′-bipyridyl
198	Et	*p*O-C_6H_4-CN	2,2′-bipyridyl
199	Me	*p*O-C_6H_4-CN	2,2′-bipyridyl
200	Me	phthalimido	2 PEt_3

A penta-coordinate nickel acyl compound (**201**) was formed by oxidative addition of benzoyl chloride to $Ni(Bu^tNC)_4$ (Equation (37)) ⟨69JA7196⟩. The acyl compound (**201**) was isolated as a stable solid (m.p. 83–85 °C) in 74% yield, but decomposed rapidly in solution, undergoing successive insertion reactions. This process may be relevant to the mechanism of nickel catalysed polymerisation of isocyanides.

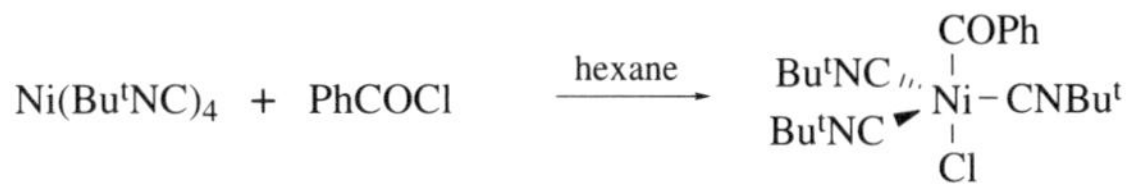

(**201**) (37)

5.10.5.19 Acylpalladium Compounds

The palladium phosphine complex $Pd(PPh_3)_4$ reacts with acetyl chloride in benzene to give the acyl complex $PdCl(PPh_3)_2(COMe)$ ⟨68CC6⟩. Similar acyl compounds, $PdX(PPh_3)_2(COR)$, can be synthesised in 80–95% yield, by the oxidative addition of alkyl halides (RX) such as $CH_2{=}CHCl$, $CH_2{=}CHCH_2Cl$, MeI and PhI to the carbonyl complex $Pd(CO)(PPh_3)_3$ (Scheme 59) ⟨73BCJ2820⟩. Cinnamyl bromide is reported to react with $Pd(CO)(PPh_3)_3$ under 20 atm of carbon monoxide to give an acyl compound (**202**) which is isolated in 74% yield as a yellow crystalline solid (Scheme 60) ⟨88CL377⟩. The triethylphosphine analogues, $PdX(PEt_3)_2(COMe)$ (X = Cl and Br), of $PdCl(PPh_3)_2(COMe)$ are formed from the methyl complexes $PdX(PEt_3)_2Me$ by treatment with carbon monoxide in benzene ⟨66JCS(A)634⟩. The halogen ligand of $PdX(PEt_3)_2(COMe)$ can be replaced with the nitrate, nitrite and thiocyanate ions on treatment with $AgNO_3$, $NaNO_2$ and KSCN, respectively. These palladium acyl compounds are less stable than the analogous platinum compounds.

RX = MeI, PhI, vinyl chloride and allyl chloride

Scheme 59

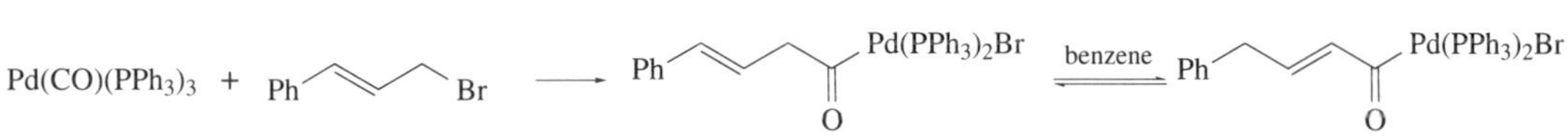

(**202**)

Scheme 60

The stereoselectivity of formation of palladium acyl compounds from chiral alkyl bromides has been studied by Stille *et al.* ⟨74JA4983⟩. The reaction of (*S*)-PhCH(Me)Br with $Pd(PPh_3)_4$ under an atmosphere of carbon monoxide leads to the formation of the acyl compound (**203**) (Scheme 61). The first step in this process involves the oxidative addition of the bromide ion to the palladium metal with inversion of configuration at carbon. This is followed by insertion of carbon monoxide into the palladium–carbon bond of the intermediate alkyl complex with retention of configuration.

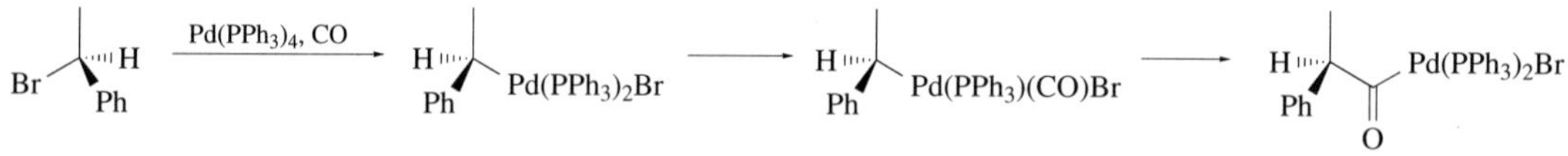

(203)

Scheme 61

Treatment of imines, such as (**204**), and other related systems, with $Pd(OAc)_2$ leads to the formation of *ortho*-palladated intermediates such as (**205**) which react with carbon monoxide to give acyl palladium species (**206**) (Scheme 62). Compound (**206**) rearranges with loss of palladium to give the heterocyclic compound (**207**) (Scheme 62) ⟨75JOC2667⟩. A related process is the carbonylative cyclization of the malonic ester derivative (**208**) to give the cyclic ketone (**209**) (Scheme 63) ⟨89JA8018⟩. The reaction is catalysed by $Cl_2Pd(PPh_3)_2$ and the aryl and acylpalladium intermediates (**210**) and (**211**) have been identified as intermediates in the reaction. This reaction is also effective for the formation of the homologous 6- and 7-membered ring ketones.

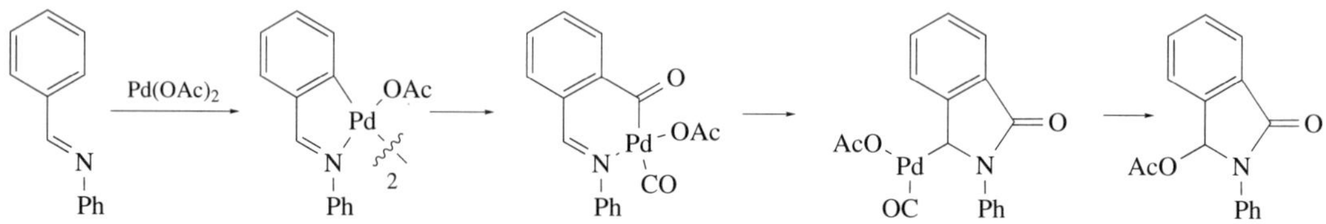

(204) (205) (206) (207)

Scheme 62

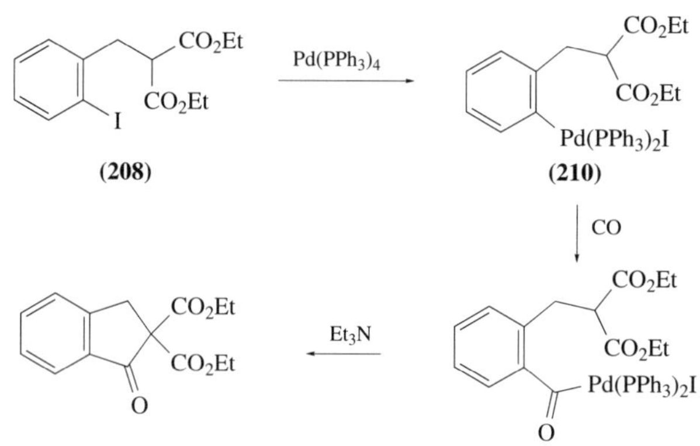

(208) (210) (209) (211)

Scheme 63

Secondary amines react with palladium olefin complexes such as (**212**) to give highly unstable σ-alkyl complexes which cannot be isolated (Scheme 64). Carbonylation *in situ*, however, leads to the formation of isolable and stable acyl complexes such as (**213**) (Scheme 64) ⟨75JA1184⟩. The stability of these acyl species may be related to the ability of the acylamino ligand to chelate to the metal centre to form a favourable 5-membered metallocycle. The acyl products were obtained as single positional isomers.

5.10.5.20 Acylplatinum Compounds

$Pt(PPh_3)_4$ reacts with acid chlorides in benzene at reflux to give the acyl compounds $Pt(PPh_3)_2Cl(COR)$ ($R = Ph$, Me, Et, $CHMe_2$) in high yield ⟨67JCS(A)865, 67CJC301⟩. These acyl compounds react with $AgPF_6$ in acetonitrile to give cationic analogues such as $[Pt(PPh_3)_2(NCMe)(COPh)]PF_6$. Acylplatinum compounds are also prepared by migratory insertion reactions ⟨84ACR67⟩. Booth *et al.* reported the carbonylation reaction of $PtXMe(PEt_3)_2$ ($X = Cl$, Br, I) at 90°C under pressure to give the acyl complexes $PtX(COMe)(PEt_3)_2$ ⟨66JCS(A)634⟩. Mawby and co-workers reported the insertion reaction at a 4-coordinate square planar platinum complex to produce

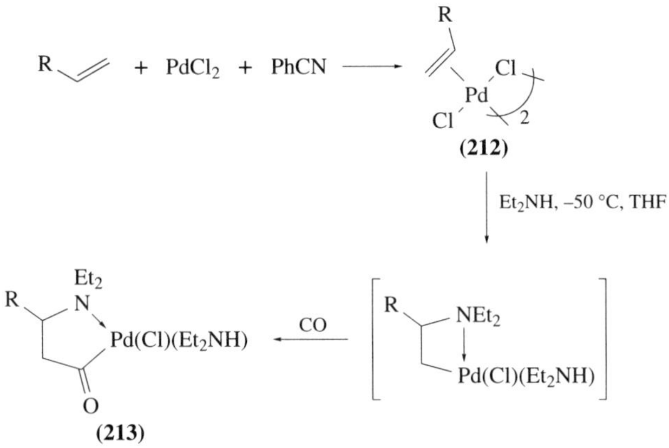

Scheme 64

an acyl derivative in which the insertion to the carbonyl ligand is promoted by the addition of a nucleophile (Scheme 65) ⟨74JCS(D)421⟩. Thus, methyl compound (**214**) reacts with pyridine to give the acetyl compound (**215**). Certain alkyl or aryl complexes, PtXR(CO)L (X = halide, L = tertiary phosphine, R = alkyl), similar to (**214**) will undergo dimerisation reactions in which the metal centres are bridged by halide ligands and the alkyl or aryl group migrates to the carbon monoxide ligand to form an acyl species (Equation (38)). For example, two molecules of the monomer react in chloroform solution to form the dimer in a reversible process ⟨79JCS(D)1246⟩. The position of the equilibrium appears to be relatively insensitive to the solvent, but the dimerisation is promoted by strongly bridging halides in the order I > Br > Cl ⟨80JCS(D)712⟩. The insertion process is promoted by ligands (L) with a strong '*trans*'-effect when the ligand is relatively small. However, more bulky ligands reduce the degree of migration of the R group through steric hindrance ⟨80JCS(D)712⟩. The nature of the R group has a major influence on the migration reaction. The ability to migrate reduces in the order Et > Ph > Me > Bn > CCMe > $CClCCl_2$ ⟨83JCS(D)359⟩.

$$Li_2(PtCl_4) \xrightarrow[\text{ii, PPh}_3]{\text{i, CO, DMF, EtOH}} [Pt_4(CO)_5(PPh_3)_4] \xrightarrow{\text{MeI}} \underset{(\mathbf{214})}{MePt(CO)(PPh_3)I} \xrightarrow{\text{pyridine}} \underset{(\mathbf{215})}{(MeCO)Pt(py)(PPh_3)I}$$

Scheme 65

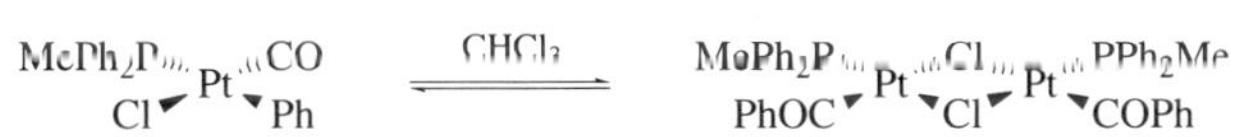 (38)

A variety of cyclic acyl derivatives of platinum have been synthesised. $Pt(PPh_3)_4$ reacts with cyclopropenones to give metallacyclobutenone derivatives, such as (**216**) which is obtained in 60% yield (Equation (39)) ⟨72CC672⟩. A similar process (Equation (40)) is observed on reaction of $Pt(PPh_3)_4$ with benzocyclobutanedione (**217**) to give the metallacycle (**218**) in 65% yield ⟨73CC158⟩ and for the substituted cyclobutadiones (**219**)–(**221**) with the arsine complex $Pt(AsPh_3)_4$ to give the corresponding metallacycles (Scheme 66) ⟨74CC841⟩. A related reaction is the oxidative addition of succinic anhydride to $Pt(COD)_2$ in the presence of a phosphine ⟨83CL115⟩. The initially formed acyl complex (**222**) reacts with the bidentate phosphine ligand dppe in carbon disulfide to form the more stable compound (**223**) (Scheme 66).

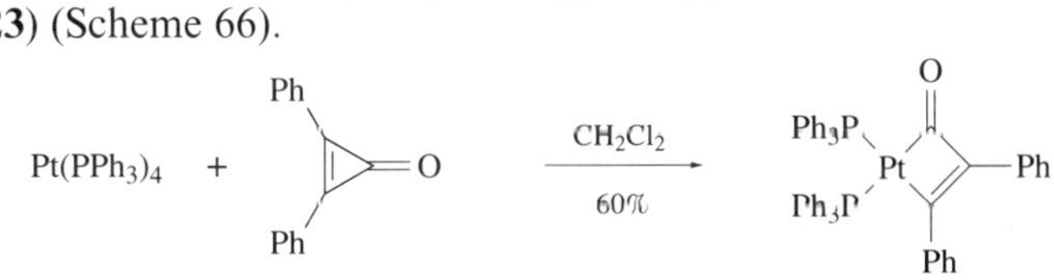 (39)

(**216**)

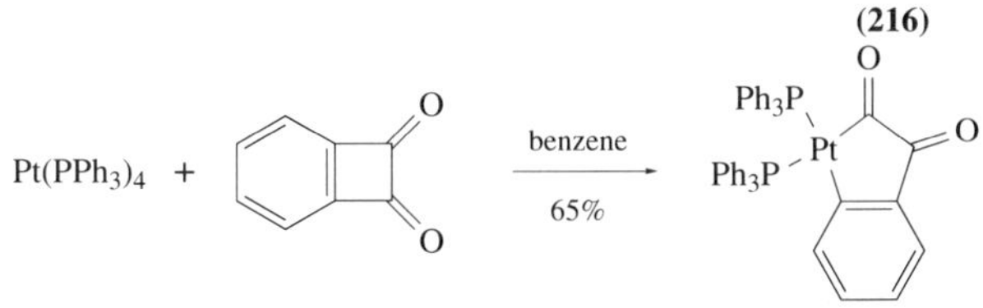 (40)

(**217**) (**218**)

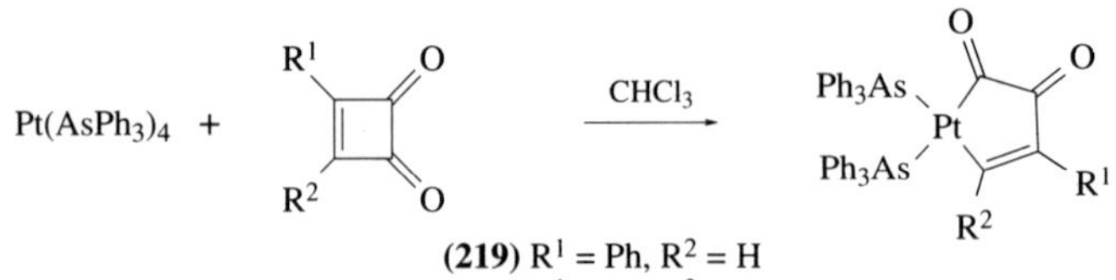

(219) R^1 = Ph, R^2 = H
(220) R^1 = Ph, R^2 = OMe
(221) R^1 = OMe, R^2 = OMe

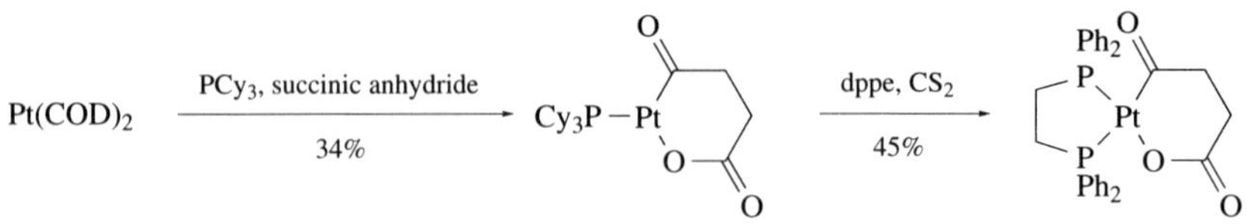

(222) **(223)**

Scheme 66

Other methods that can be used to make platinum acyl compounds include the reaction of $Pt(PPh_3)_4$ with aldehydes in which the carbon–hydrogen bond of the formyl group is cleaved (Equation (41)) ⟨70CC198⟩. This process leads to the formation of diacyl species such as **(224)**–**(226)**. Another method is the reaction of acetylenes with the methyl compound **(227)** to give the acyl **(228)** (Equation (42)) ⟨72JA1532⟩.

$$Pt(PPh_3)_4 + RCHO \longrightarrow Pt(PPh_3)_2(COR)_2 \qquad (41)$$

(224) R = Me
(225) R = Et
(226) R = Ph

(227) **(228)** (42)

5.10.5.21 Acylzinc Compounds

Dialkylzinc compounds do not react with carbon monoxide at atmospheric pressure in the absence of a promoter. However, addition of potassium *t*-butoxide to the reaction of di-*n*-butyl or diisopropylzinc with carbon monoxide gives rise to the formation of acyloin products ⟨72JOC1732⟩. The mechanism postulated for this transformation involves the generation of anionic zinc acyl species (Scheme 67). No direct evidence for the formation of such intermediates has been obtained.

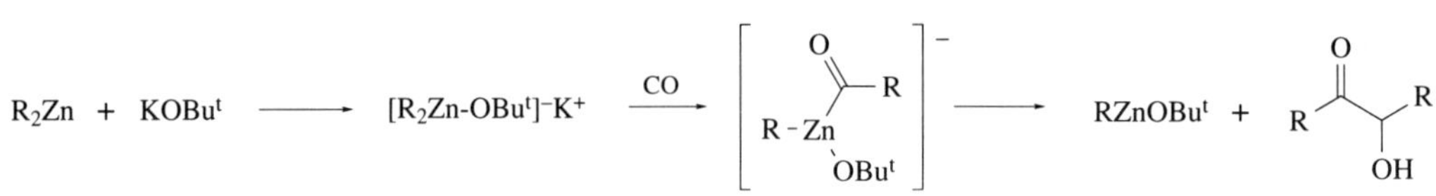

Scheme 67

5.10.6 GROUP 3 DERIVATIVES

5.10.6.1 Acylaluminium Compounds

Acyl derivatives of the main group metals of groups 1, 2 and 3 are not readily available because the acyl metallic species formed can react with the precursor compounds. For example, formation of acylmetal species by deprotonation of an aldehyde with a suitable base leads to a compound that readily reacts with the starting aldehyde. To eliminate this problem, attempts have been made to generate the acylmetal compound in the absence of a substrate that could react with it. This has proved to be successful for the generation of aluminum acyl compounds such as **(229)**. Thus, $Cp_2ZrCl(COPr^n)$ reacts with $AlCl_3$ at $-20\,^{\circ}C$ to give an acylaluminum dichloride compound

(Scheme 68) ⟨79JA3521⟩. The acyl species was observed by low temperature NMR and IR spectroscopy. It is likely that the acyl ligand binds to the aluminum atom through both carbon and oxygen atoms by analogy with early transition metal acyl compounds. In order to prove that the compound is an acyl and not an enolate, it was treated with D_2O. The product obtained was an aldehyde in which the deuterium was incorporated in the aldehydic position (Scheme 68).

(229)

Scheme 68

5.10.7 GROUP 4 DERIVATIVES

5.10.7.1 Acyltin Compounds

Acylstannanes are less well known than the analogous acylsilanes, which are versatile organometallic reagents that undergo many useful synthetic transformations. Acylstannanes can be synthesised *in situ* by reaction of Bu_6Sn_2 with acid chlorides in the presence of $PdCl_2(PPh)_3$ (Scheme 69). The intermediate acyl compounds will react with a second molecule of acid chloride to give diketone products contaminated with ketones (Scheme 69) ⟨85TL6075⟩. The use of Et_6Sn_2 is acceptable, provided that the reaction is carried out under an atmosphere of carbon monoxide to prevent decarbonylation ⟨85JOM(282)421⟩. The reaction of acid chlorides with Me_6Sn_2 has been studied by Mitchell *et al.* In the presence of $(PPh_3)_4Pd$, Me_6Sn_2 reacts with various acid chlorides, RCOCl (R = Me, Et, Ph, 2-furyl, 4-MeC_6H_4), in THF, to give the corresponding acylstannanes which can be isolated in 70–80% yield ⟨90S1001⟩.

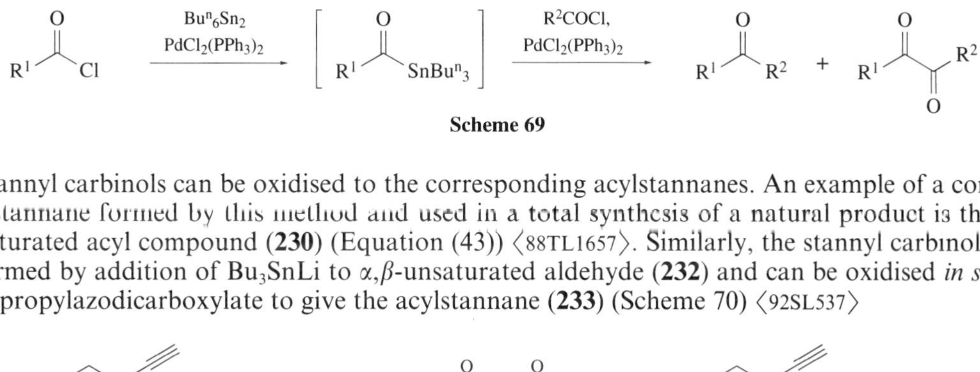

Scheme 69

Stannyl carbinols can be oxidised to the corresponding acylstannanes. An example of a complex acylstannane formed by this method and used in a total synthesis of a natural product is the α,β unsaturated acyl compound (**230**) (Equation (43)) ⟨88TL1657⟩. Similarly, the stannyl carbinol (**231**) is formed by addition of Bu_3SnLi to α,β-unsaturated aldehyde (**232**) and can be oxidised *in situ* by diisopropylazodicarboxylate to give the acylstannane (**233**) (Scheme 70) ⟨92SL537⟩

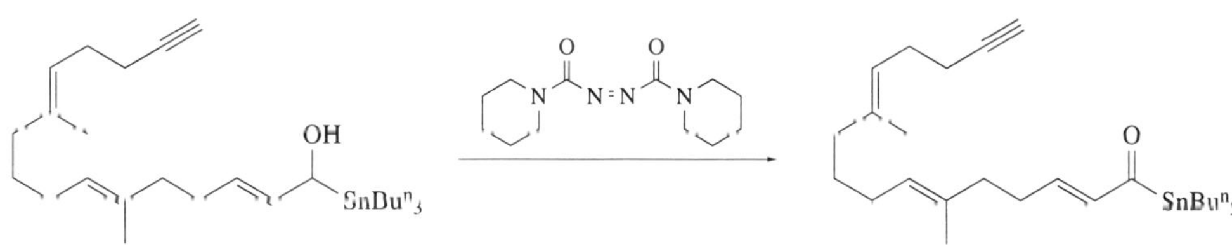

(43)

(**230**)

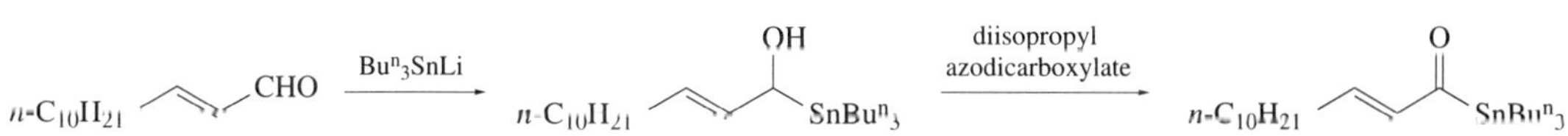

(**232**) (**231**) (**233**)

Scheme 70

Acid chlorides, esters and thioesters may react with trialkylstannyl lithium reagents to give acylstannane products (Scheme 71) ⟨89JOC2966⟩. Generally, the acid chloride route is the least satisfactory giving impure products in poor yields. It is necessary to add a Lewis acid such as $BF_3 \cdot OEt_2$ to the reaction between esters and the stannyllithium to obtain the desired products. The optimal procedure is to add two equivalents of the stannyllithium and 2.4 equivalents of the

$BF_3 \cdot OEt_2$ to the ester. A successful application of this methodology is the synthesis of the furanoyl compound (**234**) in 73% yield at $-78\,^\circ C$ in THF (Equation (44)). The most satisfactory route involves the uncatalysed reaction between the stannyllithium and thioesters. Only one equivalent of the tin reagent is required which simplifies the work-up procedure as stannylated byproducts can be very inconvenient to remove.

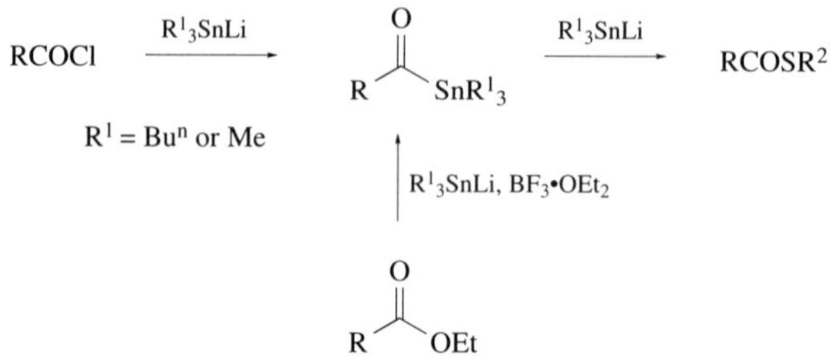

Scheme 71

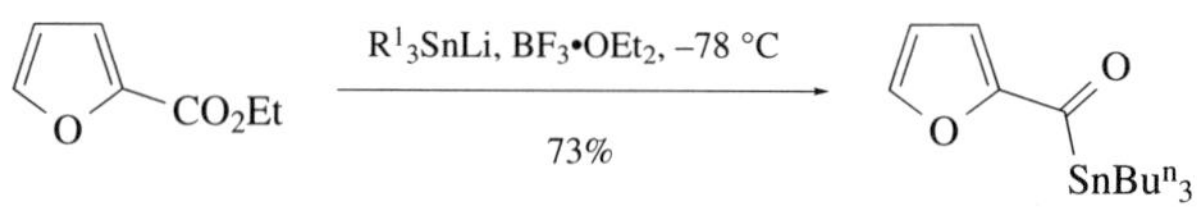

(44)

(234)

An alternative strategy is to form lithiated acyl anion equivalents and transmetallate with trialkylstannyl chloride. For example, the diene (**235**) is lithiated with *t*-BuLi to give the vinyllithium species (**236**). Transmetallation of (**236**) with trimethylstannyl chloride gives the enol ether (**237**) which can be hydrolysed with acid to give the α,β-unsaturated acyl compound (**238**) in which the ketone is unmasked and the terminal double bond brought into conjugation (Scheme 72) ⟨80JA1577⟩. In an analogous process acetylenic tin acyl compounds have been synthesised by reacting the lithium anion of the cyclic acetal (**239**) with trimethylstannyl chloride to give the intermediate compound (**240**) in 76% yield. Hydrolysis of (**240**) gives the deprotected acylstannane (**241**) in 70% yield (Scheme 73) ⟨83CC239⟩.

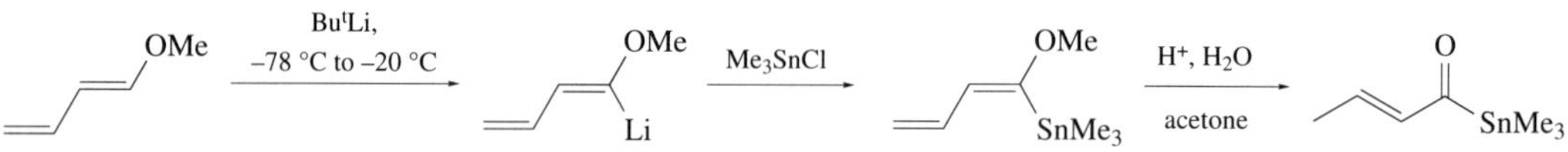

(235) **(236)** **(237)** **(238)**

Scheme 72

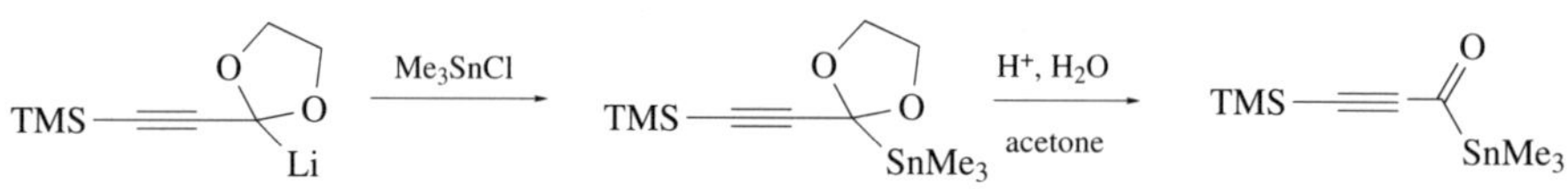

(239) **(240)** **(241)**

Scheme 73

Tin acyl compounds are formed on addition of tin-containing Grignard reagents to an excess of aldehydes in a process related to the Cannizzaro reaction ⟨83JOM(251)175⟩. This reaction has been studied by Kosugi *et al.* Addition of $Bu_3SnMgCl$ to three equivalents of propanal gives the propanoyl tin compound (**242**) in 84% yield (Equation (45)) ⟨87BCJ3462⟩.

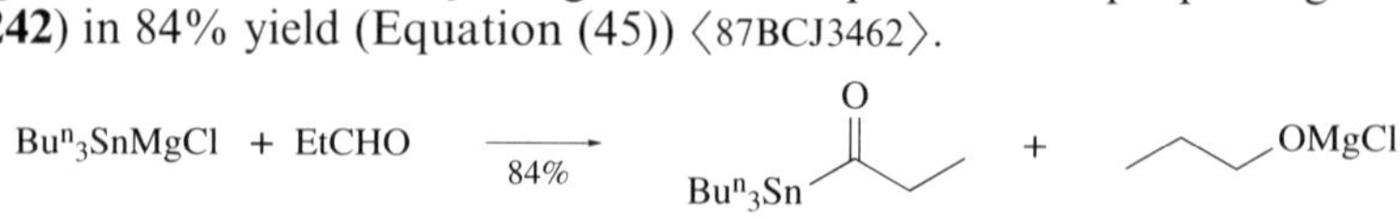

(45)

(242)

5.10.8 ACTINIDE DERIVATIVES

5.10.8.1 Acylthorium Compounds

There are many similarities between organoactinide and early transition metal chemistry. Reaction of $Cp^*_2ThCl(CH_2TMS)$ (**243**) with carbon monoxide in toluene leads to insertion of carbon monoxide into the thorium–carbon bond and a trimethylsilyl migration (Scheme 74) ⟨78JA7112⟩. This process may proceed via an η^2-acyl species (**244**) with oxycarbenoid character as shown in Scheme 74. The 2,2-dimethylpropylthorium compound (**245**) formed by addition of 2,2-dimethylpropyllithium to $Cp^*_2ThCl_2$ reacts with carbon monoxide in toluene solution to give the corresponding acyl complex (**246**) which is isolated as a yellow crystalline solid (Scheme 75) ⟨80JA5393⟩. The molecular structure of (**246**) has been confirmed by x-ray analysis and the acyl ligand is η^2-coordinated to the metal. The η^2 acyl coordination and oxycarbenoid character are related to the high affinity of low valent early transition metals and actinides for oxygen.

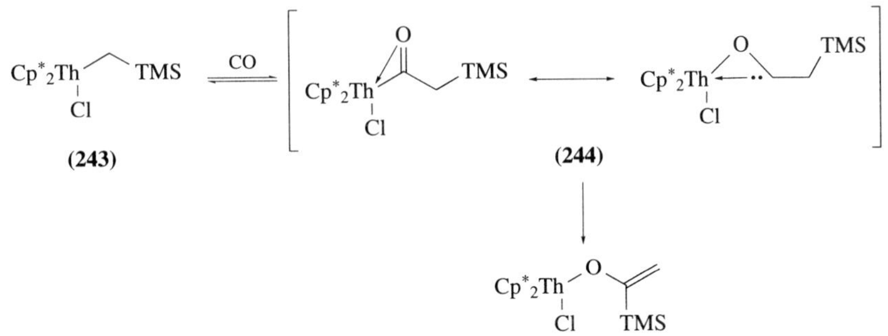

Scheme 74

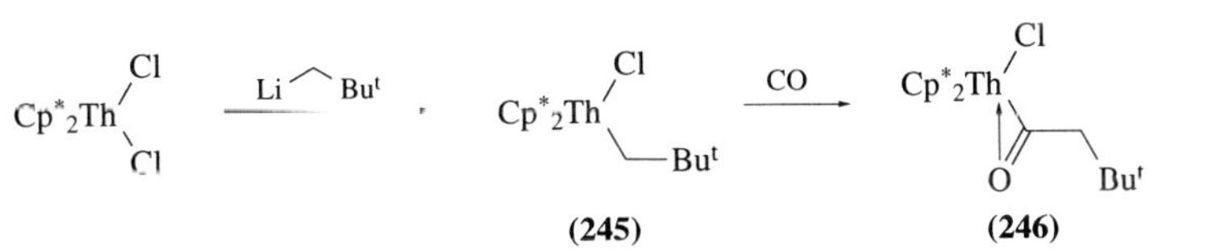

Scheme 75

5.10.8.2 Acyluranium Compounds

Uranium triscyclopentadienyl reacts with acetyl chloride in benzene at 6°C to give $Cp_3UCl(COMe)$ which was characterised by NMR spectroscopy and electrochemical analysis ⟨86ICA107⟩. In addition, the uranium analogue of the thorium compound (**244**) (Scheme 74) has been implicated in the reaction of the uranium analogue of (**243**) with carbon monoxide ⟨78JA7112⟩.

5.11
Thio-, Seleno-, and Telluroacyl Halides

ANDREW C. WILLIAMS
Lilly Research Centre Ltd., Windlesham, UK

5.11.1 INTRODUCTION

The synthesis and chemistry of thioacyl halides (fluorides, chlorides, and bromides) has been reviewed several times, although the comparative discussion of the preparative usefulness and conditions employed in the various methods is often cursory. This chapter aims to collate and expand upon these earlier works as well as describing subsequent new reactions ⟨B-69MI 511-01, B-72MI 511-01, 77SST141, B-79MI 511-01, B-79MI 511-02, 79COC(3)373, 85HOU(E5/1)615, 91COS(6)419⟩.

Although the chemistry of selenium and tellurium analogues of carboxylic acids has been reviewed, no references to the preparation of seleno- or telluroacyl halides were found in the literature up to June 1993 ⟨B-73MI 511-01, B-73MI 511-02, B-86MI 511-01, 86OPP369, 91S793, 91S897⟩. Similarly, no references were found to acyl halides containing higher oxidation states of these chalcogens. The discussion of preparative methods has therefore been confined to thioacyl derivatives alone: perhaps future chemical effort might be directed toward the preparation, isolation, and characterization of higher chalcogen analogues of acid halides.

Halosulfines may formally be thought of as thioacyl halides in which the chalcogen is in a higher(IV) oxidation state. Similarly, halosulfenes may be thought of as sulfur(VI) thioacyl halides. Their preparation is covered in general reviews of sulfine and sulfene chemistry ⟨67AG(E)107, 70SST181, 73SST200, 75ACR10, 75SST219, 77SST136, 77SST141, 79SST133, 79SST139, 79COC(3)373, 81SST160, 81SST165, 82RTC1, 85HOU(E11/1)911, 85HOU(E11/2)1326⟩.

5.11.2 GENERAL METHODS

In contrast to the synthesis of acyl halides, there are few methods for the synthesis of thioacyl halides which could be considered to be truly general. This situation arises from the reactivity and instability of the products. Thioacyl fluorides dimerize or polymerize very readily, as do a number of aliphatic thioacyl chlorides (see Sections 5.11.3.1 and 5.11.4.3). Aliphatic thioacyl chlorides are generally only of sufficient stability to be isolated when they are substituted with electron-withdrawing groups, most commonly halogens, or possess no α-protons (see Sections 5.11.4.2 and 5.11.4.4). Aromatic thioacyl chlorides form the largest subclass of compounds, and their preparation from dithiocarboxylic acids may be considered one of the few truly general synthetic methods described to date (see Section 5.11.4.1). The paucity of thioacyl bromides thus far described does not permit any of the methods known for their synthesis to be considered as general.

Of the preparative methods for halosulfines described to date, that involving the dehydrohalogenation of sulfonyl halides has been applied to the widest range of structural types (see Section 5.11.7.1.1). Approaches to aromatic halosulfines via dichlorosulfenyl chlorides (see Section 5.11.7.1.3), and perhalogenated aliphatic halosulfines (see Section 5.11.7.1.4) are of moderate generality for their subclass of compound.

Since there is only a single method described which is useful for the "preparation" of halosulfenes, namely the dehydrohalogenation of α-halo sulfonyl halides, it constitutes a general method. The limited number of halosulfenes described is probably due to a combination of factors: their high chemical reactivity means that only complexes or cycloadducts are known; no independently existing halosulfenes have been described as of 1995; the preparation of appropriate precursors is not always simple and instability and/or reactivity of the products often makes the unequivocal invocation of the intermediacy of halosulfenes questionable (see Section 5.11.7.2).

5.11.3 THIOACYL FLUORIDES

Thioacyl fluorides are more stable than other thioacyl halides, especially when substituted with fluorine. Their tendency to form polymers has attracted the attention of several research groups, and has resulted in a number of synthetic routes to polyhalo (especially polyfluoro) aliphatic thioacyl fluorides (see Sections 5.11.3.1 and 5.11.3.2). Most thioacyl fluorides described up to 1995 fall into this category.

5.11.3.1 Thioacyl Fluorides via Sulfurization of Polyhaloalkenes and Related Species

Several methods have been described where elemental sulfur reacts with a polyhaloalkene, -alkane or -alkyl derivative to produce a thioacyl fluoride. The trifluoroalkenes (**1**) are converted into thioacyl fluorides (**2**) on treatment with elemental sulfur at elevated temperatures, either in the presence or absence of a catalyst (Equation (1)) ⟨62USP3048629, 62USP3069395, 63USP3097236, 64JCS2944, 65JOC1375⟩. Thioacyl fluorides prepared by this method are summarized in Table 1.

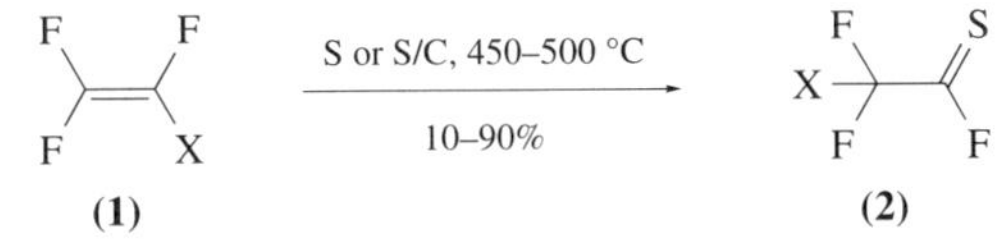

(1)

Table 1 Thioacyl fluorides XCF_2CSF prepared according to Equation (1).

X	*Conditions*	*Yield (%)*	*Ref.*
H	a	10	62USP3048629, 64JCS2944
F	a	20	62USP3048629, 64JCS2944
Cl	a	90	62USP3048629, 64JCS2944
Cl	b	77	62USP3069395, 65JOC1375
Br	b	82	62USP3069395, 65JOC1375

[a] Alkene and sulfur vapours passed over heated charcoal catalyst at 450–500 °C. [b] Alkene vapour passed through refluxing sulfur.

The reaction is thought to proceed via formation of thiirane (**3**), followed by rearrangement to the thioacyl fluoride (**2**), (Scheme 1, path a). If path b were to be followed, a trifluorothioacetyl halide (**4**; X ≠ F) would result. The observation that (**2**; X = Cl) is accompanied by 10% of trifluorothioacetyl chloride (**4**; X = Cl), when prepared by method A, may be an example of this (see Section 5.11.4.3) ⟨64JCS2944⟩.

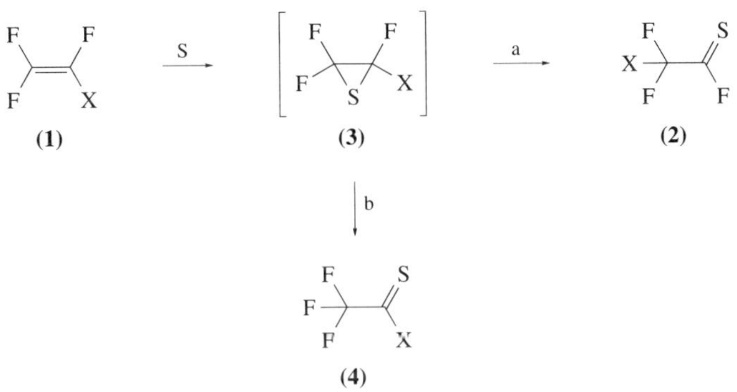

Scheme 1

For iodotrifluoroethene (**5**), the reaction takes a somewhat different course, in that the final product is the dimeric tetrafluorodithiosuccinoyl difluoride (**7**), formed in 61% yield (Scheme 2) ⟨75JOC129⟩. It is thought that the initially formed iododifluorothioacetyl fluoride undergoes homolytic fission to iodine atoms and the resonance-stabilized radicals (**6**), which combine to give the difluoride (**7**) and molecular iodine.

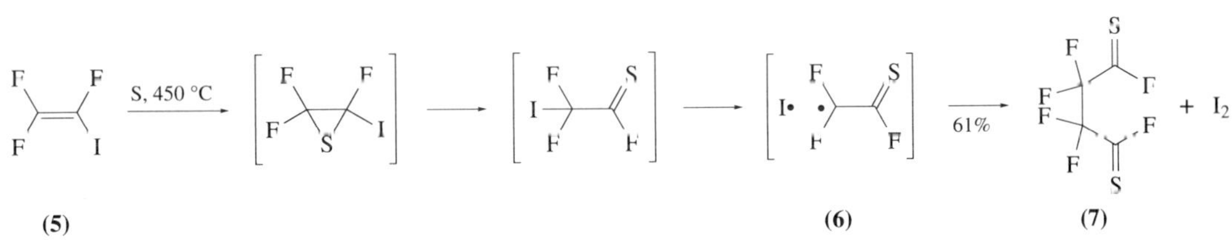

Scheme 2

Iodine has been displaced directly by sulfur in a number of cases. 1,2,2,3,4,4-Hexafluoro-1,1-diiodobut-3-ene (**9**) was converted into 2,2,3,4,4-pentafluorothiobut-3-enoyl fluoride (**10**) in 70% yield by sulfur at reflux. The alternative reaction, at the double bond, potentially to yield

4,4-diiodo-2,2,3,3,4-pentafluorothiobutanoyl fluoride (**8**) was not observed (Scheme 3) ⟨65JOC1375⟩. The perfluorothioacyl fluorides (**12**) and (**14**) were formed in 45% and 55% yields, respectively, when the perfluoroalkyl iodides (**11**) and (**13**) were treated with phosphorus pentasulfide at reflux (Equation (2)) ⟨62USP3069395, 65JOC1375⟩.

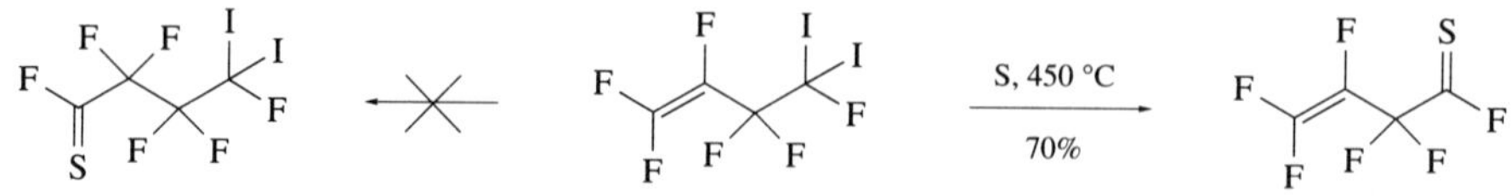

(**8**) (**9**) (**10**)

Scheme 3

$$CF_3(CF_2)_nI \xrightarrow[45\text{–}55\%]{P_4S_{10},\ 550\ °C} CF_3(CF_2)_{n-1}CSF \quad (2)$$

(**11**) $n = 1$ → (**12**) $n = 1$, 45%

(**13**) $n = 2$ → (**14**) $n = 2$, 55%

1-Halo- or 1,1-dihalo-2,2,2-trifluoroethyl mercurials react with boiling sulfur, yielding trifluorothioacetyl fluoride (**18**) (Equation (3)) ⟨61JA2589, 62USP3069395, 65JOC1375⟩.

$$(CF_3CFX)_2Hg \xrightarrow[67\%]{S,\ 450\ °C} CF_3CSF \quad (3)$$

(**15**) X = H
(**16**) X = F
(**17**) X = Cl

(**18**)

Thioacyl fluorides (**19**) dimerize readily, either spontaneously or under the influence of light and/or heat, to dithietanes (**20**). The monomers may be regenerated from these substances by thermolysis, thereby constituting a formal synthesis (Scheme 4) (Table 2) ⟨62USP3069395, 65JOC1375⟩.

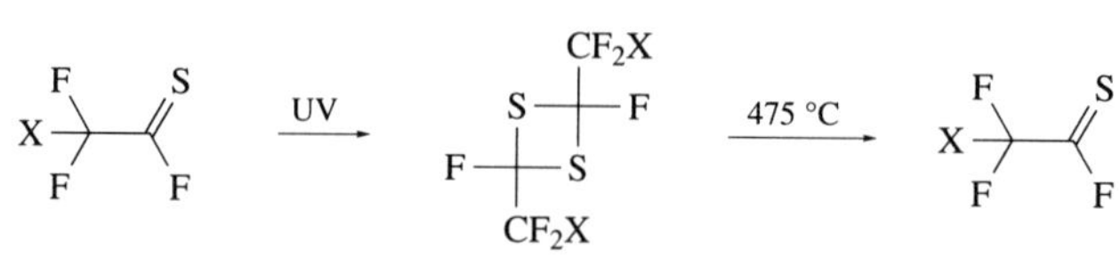

(**19**) (**20**) (**19**)

Scheme 4

Table 2 Dimerization of thioacyl fluorides $XCF_2C(=S)F$ (**19**) to dithietanes (**20**), and regeneration of thioacyl fluorides therefrom.

X	*Yield* (**20**) (%)	*Yield* (**19**) (%)
F	60	90
Cl	83	95
Br	58	

5.11.3.2 Thioacyl Fluorides via Dehydrofluorination of 1,1-Difluorothiols

The treatment of 1,1-difluorothiols (**21**) with a number of agents (most commonly sodium fluoride, but potassium fluoride and secondary amines have also been used) gives thioacyl fluorides (**22**) in fair yields (Equation (4), Table 3) ⟨62USP3069395, 63JA749, 65JOC1375, 67BAU324⟩.

$$\underset{\textbf{(21)}}{RCF_2SH} \xrightarrow[\text{45–62\%}]{\text{NaF, vapor phase}} \underset{\textbf{(22)}}{RCSF} \tag{4}$$

Table 3 Thioacyl fluorides (**22**) by dehydrofluorination of thiols (**21**), Equation (4).

Substituent	*Yield* (%)	*Ref.*
R = CHF_2	59	63JA749
R = CHFCl	62[a]	63JA749
R = CHFOMe	45	63JA749
R = CF_3	60	62USP3069395, 65JOC1375
R = C_2F_5	[b]	62USP3069395, 65JOC1375
R = CF_2Cl	[b]	62USP3069395, 65JOC1375

[a] Also prepared using KF or Et_2NH, not isolated, no yields ⟨67BAU324⟩. [b] Unpurified yield *ca.* 100%, purified yields not quoted.

5.11.3.3 Thioacyl Fluorides via Acidic Cleavage of *S*-Benzyl Thioethers

Thioacyl fluorides (**24**) containing an α-hydrogen atom have been synthesized by the acidic solvolysis of fluorinated alkenyl thioethers (**23**) (Equation (5), Table 4) ⟨90BAU1725⟩.

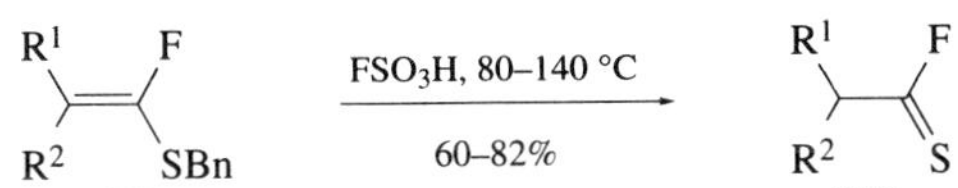

(**23**) (**24**) (5)

Table 4 Thioacyl fluorides (**24**) prepared according to Equation (5).

R^1	R^2	*Yield* (%)	*Ref.*
CF_3	CF_3	82	90BAU1725
CF_3	F	60	90BAU1725
F	Cl	68	90BAU1725

5.11.3.4 Thioacyl Fluorides via Pyrolysis of Thioesters

2,2-Bis(trifluoromethyl)thioacetyl fluoride (**25**) has been obtained in yields of 33–37% by pyrolysis of the thiolbenzoate (**26**) and the dithiophosphate (**27**) (Scheme 5) ⟨77BAU206⟩.

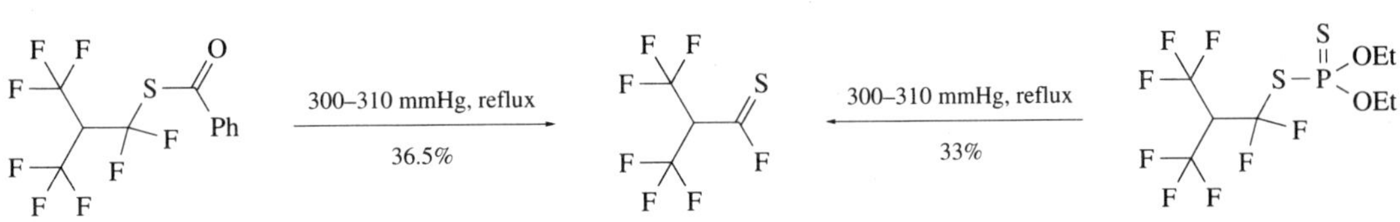

(**26**) (**25**) (**27**)

Scheme 5

5.11.3.5 Miscellaneous Methods

Dichlorofluorothioacetyl fluoride (**29**) was obtained in a yield of 60% by the pyrolysis of dithietane dioxide (**28**) at 450–500 °C and 10^{-2} torr. The reaction is thought to proceed via a similar mechanism to that shown in Scheme 1, after thiirane formation by extrusion of sulfur dioxide (Equation (6)) ⟨64USP3136781, 87CB1499⟩.

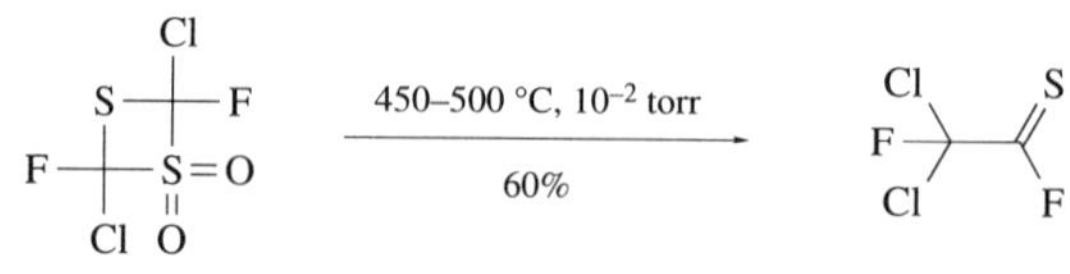

(6)

(28) **(29)**

Trifluorothioacetyl fluoride (**31**) was obtained in a yield of 89% by the pyrolysis of sulfenyl chloride (**30**) at 400 °C and 10^{-2} torr over copper (Equation (7)) ⟨92CB1889⟩.

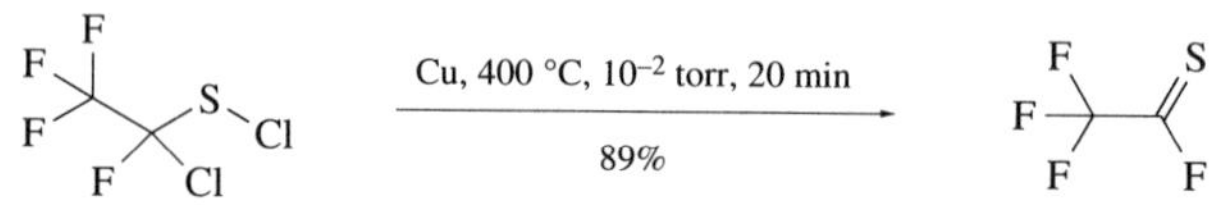

(7)

(30) **(31)**

Treatment of the thioacyl bromide (**32**) (see Section 5.11.5.2) with anhydrous potassium fluoride in acetonitrile at reflux gave the corresponding thioacyl fluoride (**33**) in 86% yield (Equation (8)) ⟨70JOC2002, 72USP3637749⟩.

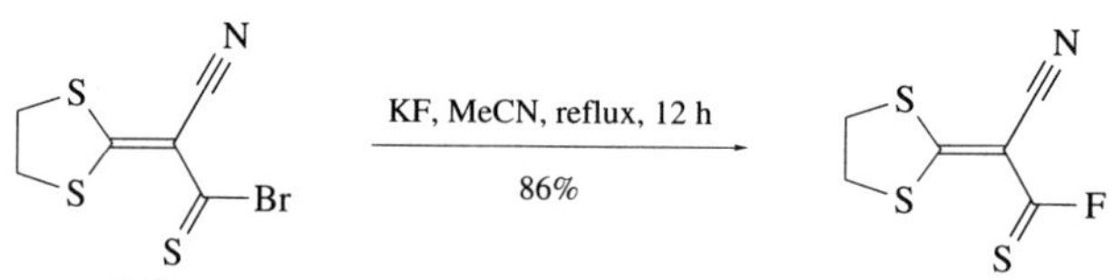

(8)

(32) **(33)**

Dechlorination of 2-chlorotetrafluoroethanesulfenyl chloride (**34**) with tin under acidic conditions gave a modest (16%) yield of chlorodifluorothioacetyl fluoride (**35**) (Equation (9)) ⟨57JGU2301⟩.

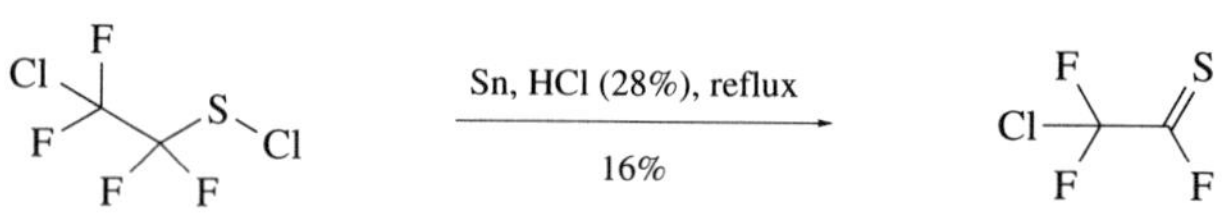

(9)

(34) **(35)**

5.11.4 THIOACYL CHLORIDES

A number of aromatic and heteroaromatic thioacyl chlorides have been described: their aliphatic counterparts are only of sufficient stability to be isolated and fully characterized when they are substituted with electron-withdrawing groups, particularly halogens, or when no α-hydrogens are present (see Section 5.11.2).

5.11.4.1 Thioacyl Chlorides from Dithiocarboxylic Acids

A number of thiobenzoyl chlorides (**37**) have been prepared from dithiocarboxylic acids in yields of up to 80%. The reaction is formally analogous to the classical conversion of carboxylic acids to acyl chlorides. The two reagents most commonly used have been thionyl chloride and phosgene (Equation (10)). The thioacyl chlorides prepared by these methods are summarized in Table 5 ⟨20HCA824, 37CR(204)1342, 43HCA1722, 57MI 511-01, 57MI 511-02, 61BSB519, 61JA4285, 63JPR214, 64TL3363, 65CB829, 65JOC1282, 65JOC2882, 68JOC1783, 68JOC3727, 68ZC142, 75ZC348⟩. Methods A and B have been used to prepare thiophen-2-thiocarbonyl chloride (**39**) from thiophen-2-dithiocarboxylic acid (**38**) in 8% and 1% yield, respectively (Equation (11)) ⟨65CB829, 75ZC348⟩. As discussed previously (see Sections

5.11.2 and 5.11.4) aliphatic thioacyl chlorides are highly unstable species, but thioacetyl chloride (**41**) has been prepared in low yield at low temperature from dithioacetic acid (**40**) (Equation (12)).

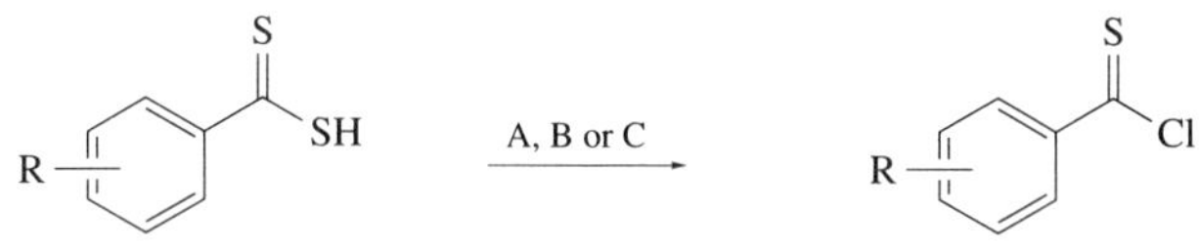

(**36**) (**37**) (10)

Table 5 Thiobenzoyl chlorides prepared according to Equation (10).

R	*Method*	*Yield* (%)	*Ref.*
H	A	60–80	20HCA824, 37CR(204)1342, 43HCA1722, 57MI 511-01, 57MI 511-02, 61JA4285, 63JPR214, 65CB829, 65JOC1282, 68JOC1783, 68JOC3727
H	B	26–48	75ZC348
H	C	46	68ZC142
2-Me	A	50	65CB829
3-Me	A	31	65CB829
4-Me	A	20	65CB829
4-Me	B	53	75ZC348
4-Me	C	43	68ZC142
3-Cl	A	10	65CB829
4-Cl	A	9	65CB829
4-MeO	A	Trace	65CB829
4-MeO	B	36	75ZC348
4-MeO	C	30	68ZC142
4-CF_3	A	20	64TL3363, 65JOC2882

Methods: A, heat under reflux with thionyl chloride; B, treat with phosgene in ether at room temperature; C, treat with (1,2-phenylenedioxy)trichlorophosphorane in ether.

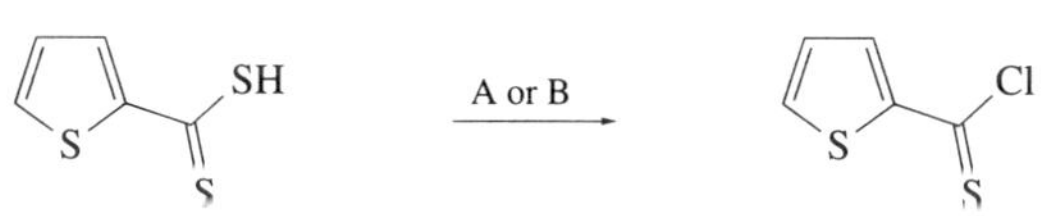

(**38**) (**39**) (11)

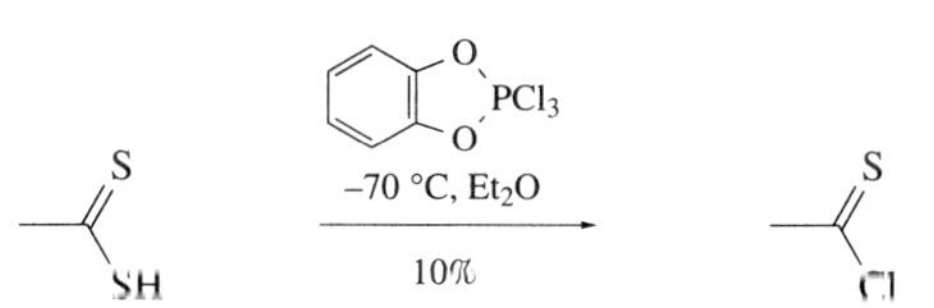

(**40**) (**41**) (12)

The reaction between 1-thiobenzoyl imidazole (**42**; Ar = Ph) and hydrogen chloride constitutes another potential route to thiobenzoyl chloride (**43**) (Scheme 6) ⟨65CB829⟩. This conversion is analogous to a procedure used in the first synthesis of formyl chloride ⟨64AG(E)132⟩. Since the imidazolides (**42**) are available from the dithiocarboxylic acids (**44**) and diimidazolides (**45**) in good yield, this procedure constitutes an indirect conversion of a dithiocarboxylic acid into a thioacyl chloride (Scheme 6) ⟨70LA(739)201⟩.

5.11.4.2 Thioacyl Chlorides via 1,1-Dichlorosulfenyl Chlorides

A range of α-carbamoyl-1,1-dichlorosulfenyl chlorides (**47**) has been prepared from the corresponding *S*-benzyl thioethers (**46**) by treatment with sulfuryl chloride in dichloromethane. Several of these sulfenyl chlorides have been converted into thioacyl chlorides (**48**) by treatment with

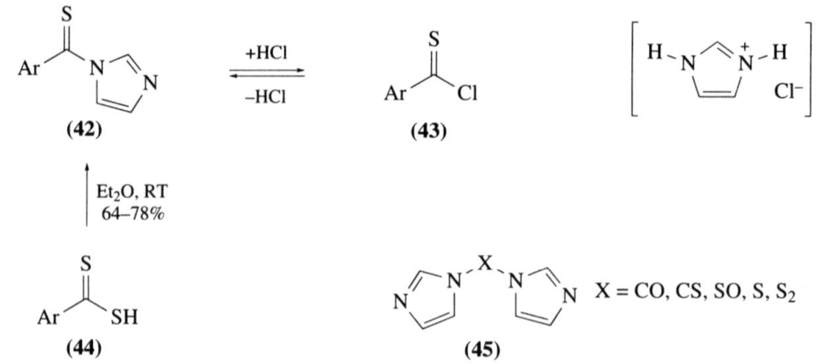

Scheme 6

triphenylphosphine in benzene (Scheme 7) ⟨71JOC3145, 72JOC1526⟩. The thioacyl chlorides prepared by these methods are summarized in Tables 6 and 7.

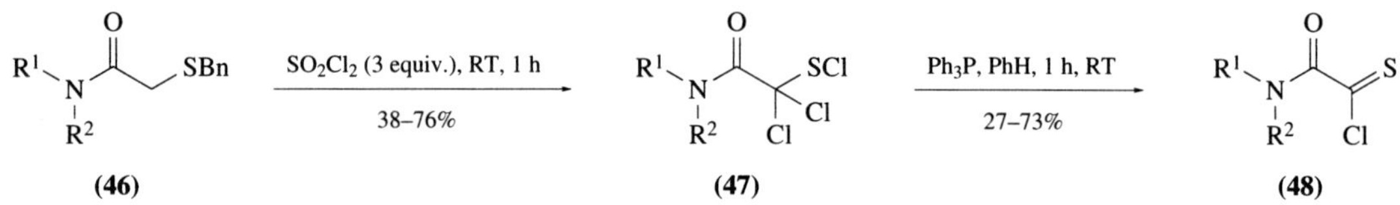

Scheme 7

Table 6 1,1-Dichlorosulfenyl chlorides (**47**) prepared according to Scheme 7.

R^1	R^2	*Yield* (%)	*Ref.*
Ph	Pr^i	68	71JOC3145
Me	Me	37	71JOC3145
Pr^i	Pr^i	65	71JOC3145
CH_2OMe	2,6-$Et_2C_6H_3$	65	71JOC3145
H	2-Bu^t-6-MeC_6H_3	76	71JOC3145

Table 7 Thioacyl chlorides (**48**) prepared according to Scheme 7.

R^1	R^2	*Yield* (%)	*Ref.*
Ph	Pr^i	73	72JOC1526
Me	Me	67	72JOC1526
Pr^i	Pr^i	27	72JOC1526

1,1-Dichlorosulfenyl chlorides (**50**) have also been prepared by the chlorination of dithiocarboxylates (**49**). Subsequent dechlorination with triphenylphosphine gave the thioacyl chlorides (**51**) in good yields (Scheme 8) ⟨77RTC139⟩. In this procedure the 1,1-dichlorosulfenyl chlorides were not isolated and characterized. The thioacyl chlorides prepared by this method are summarized in Table 8.

Although aliphatic thioacyl chlorides are generally unstable (see Sections 5.11.2 and 5.11.4), polyhalogenated derivatives have been isolated and characterized (see Section 5.11.4.3). Trichlorothioacetyl chloride (**53**) has been prepared by reductive dechlorination of pentachloroethanesulfenyl chloride (**52**). Due to its propensity to dimerize to a dithietane (**54**) the thioacyl chloride could not be isolated, but it was characterized as its morpholide (**55**) (Scheme 9) ⟨85SUL177, 87SUL153⟩. It could also be trapped as a cycloadduct with anthracene (see Section 5.11.7.1.4) ⟨91CB1311⟩.

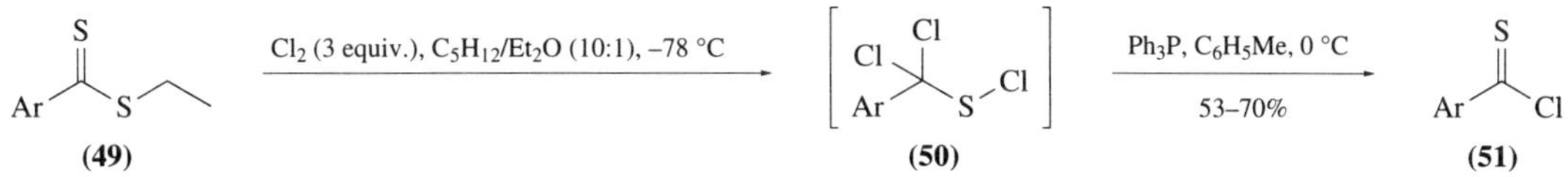

Scheme 8

Table 8 Thiobenzoyl chlorides (**51**) prepared according to Scheme 8.

Ar	*Yield* (%)	*Ref.*
C_6H_5	68	77RTC139
4-ClC_6H_4	53	77RTC139
4-$MeOC_6H_4$	60	77RTC139
2,4,6-$Me_3C_6H_4$	70	77RTC139

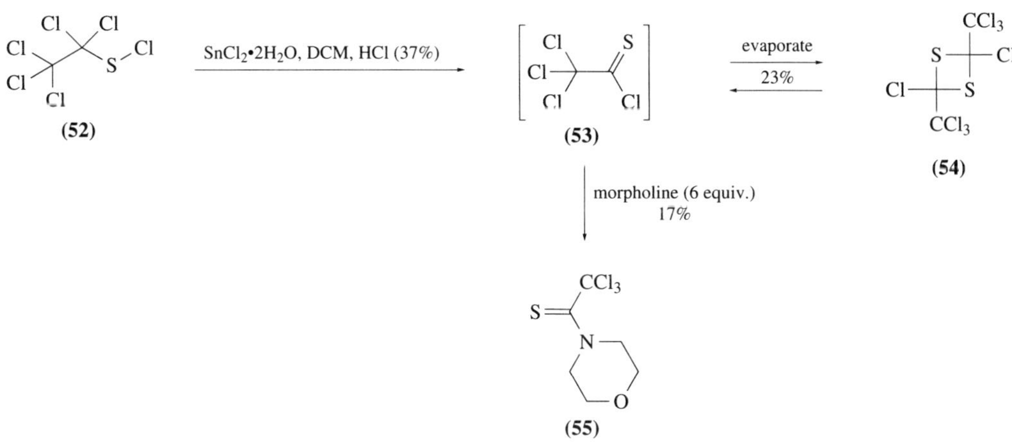

Scheme 9

5.11.4.3 Thioacyl Chlorides via Sulfurization of Polyhaloalkenes, -alkanes and -alkyl Derivatives

A number of the methods described previously for the preparation of thioacyl fluorides have also been used to prepare thioacyl chlorides (see Section 5.11.3.1) ⟨62USP3048629, 62USP3069395, 64JCS2944, 65JOC1375⟩. Thus, when chlorotrifluoroethene (**56**) and sulfur vapour were passed over a heated charcoal catalyst, trifluorothioacetyl chloride (**57**) and chlorodifluorothioacetyl fluoride (**58**) were formed as a 1 : 9 mixture, as determined by ^{19}F NMR (Equation (13)) ⟨64JCS2944⟩. Since the overall conversion of the alkene was only 10%, and separation of the isomeric thioacyl halides difficult, the practical utility of this method is questionable. However, in a closely related procedure, 1,1-dichloro-2,2-difluoroethene (**59**) reacts with boiling sulfur to give a 70% yield of chlorodifluorothioacetyl chloride (**60**) (Equation (14)) ⟨63USP3113936, 65JOC1375⟩. The putative mechanism for these reactions has been discussed (see Section 5.11.3.1).

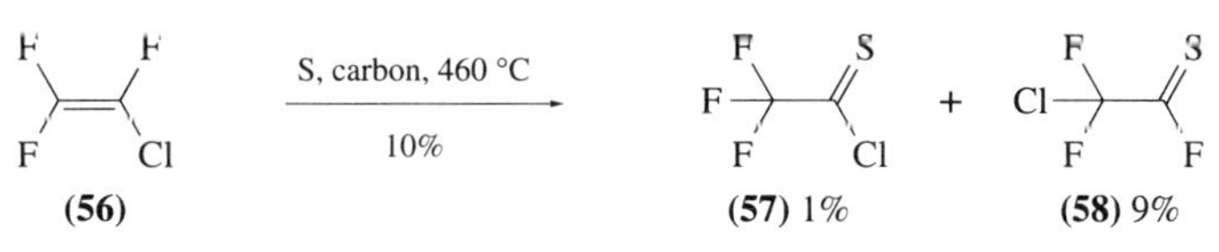 (13)

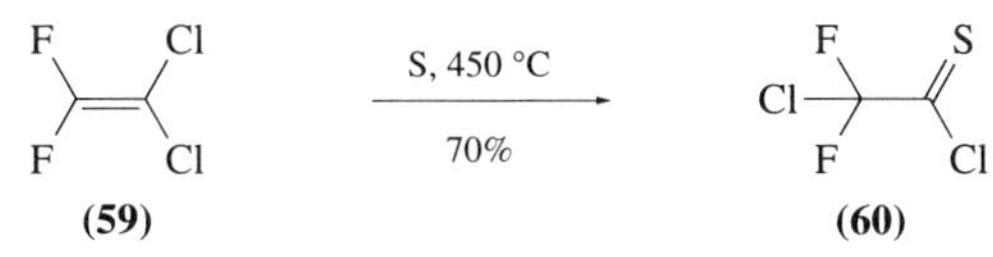 (14)

Bis(2,2,2-trifluoro-1,1-dichloroethyl)mercury (**61**) reacts with sulfur at reflux to give trifluorothioacetyl chloride (**62**) in 37% yield (Equation (15)) ⟨63USP3113936, 65JOC1375⟩.

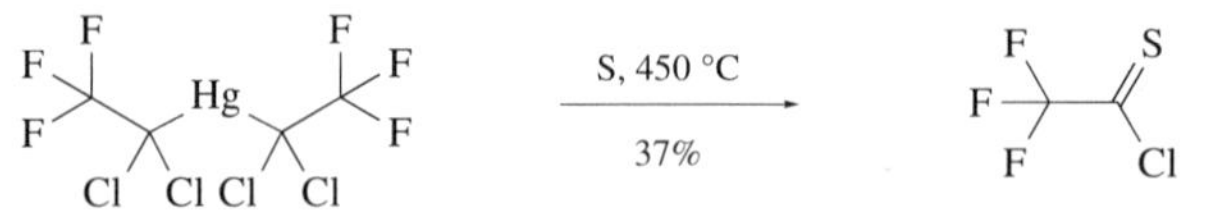

(61) **(62)** (15)

A number of 1,1,1-trihaloalkanes (**63**) have been converted into thioacyl chlorides (**64**) by treatment with sulfur at reflux: related haloalkanes have also been used (Scheme 10) ⟨63USP3113936⟩. The thioacyl chlorides prepared by this method are summarized in Table 9.

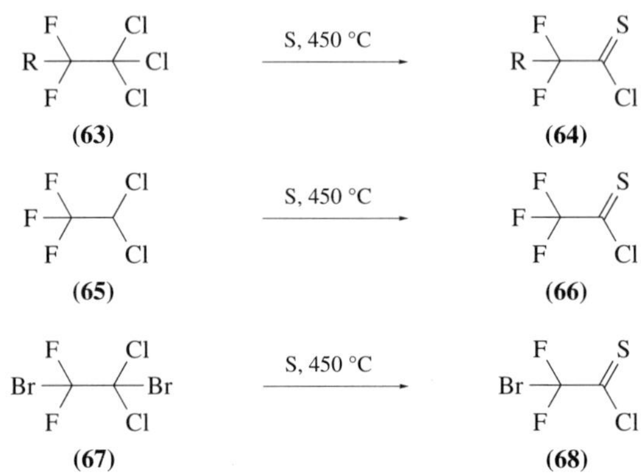

Scheme 10

Table 9 Halogenated thioacyl chlorides RCF_2CSCl prepared according to Scheme 10.

R	*Yield* (%)	*Ref.*
HCF_2		63USP3113936
$ClCF_2$	21	63USP3113936
$Cl(CF_2)_4$		63USP3113936
F		63USP3113936
Br		63USP3113936

5.11.4.4 Thioacyl Chlorides via Thioketenes, Ketenes and Alkenes

Several aliphatic thioacyl chlorides (**72**)–(**74**) have been prepared by the addition of hydrogen chloride to (readily accessible) thioketenes (**69**)–(**71**) (Scheme 11). An analogous α-chlorothioacyl chloride (**75**) was prepared by the addition of chlorine under similar conditions (Scheme 11). Only the highly hindered, *t*-butyl substituted compound (**72**) was of sufficient thermal stability to be isolated (in >90% yield). The products reacted as powerful thioacylating agents, for example, towards amines and thiols ⟨75AG(E)703⟩.

A number of alternative but equivalent routes to (**72**), in which the thioketene is an intermediate, were also described. The first involves the pyrolysis, at 500 °C and 40 torr in a stream of hydrogen chloride and nitrogen, of the thiadiazole (**76**) (Equation (16)) ⟨75AG(E)703⟩. Finally, the thiadiazole (**76**) can be converted into thiolate (**77**) by treatment with potassium ethoxide in dioxan. Further treatment of this thiolate with dry hydrogen chloride in pentane at −80 °C gave (**72**) in 49% yield (Scheme 12) ⟨75AG(E)703⟩.

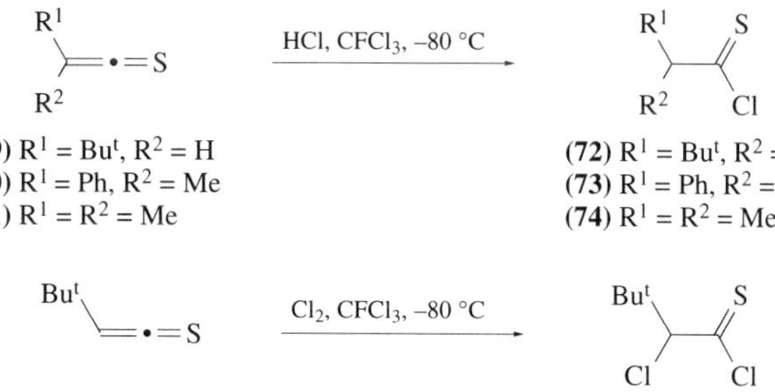

(69) $R^1 = Bu^t$, $R^2 = H$
(70) $R^1 = Ph$, $R^2 = Me$
(71) $R^1 = R^2 = Me$

(72) $R^1 = Bu^t$, $R^2 = H$
(73) $R^1 = Ph$, $R^2 = Me$
(74) $R^1 = R^2 = Me$

(69) **(75)**

Scheme 11

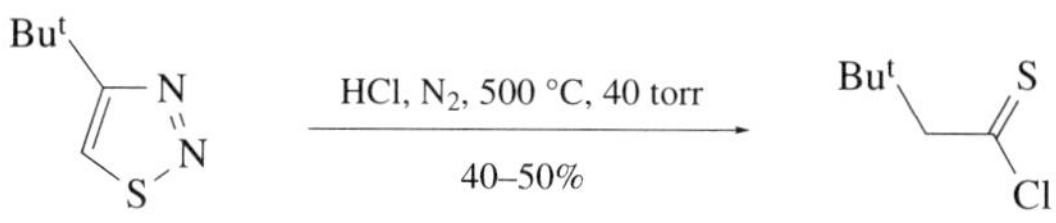

(76) **(72)** (16)

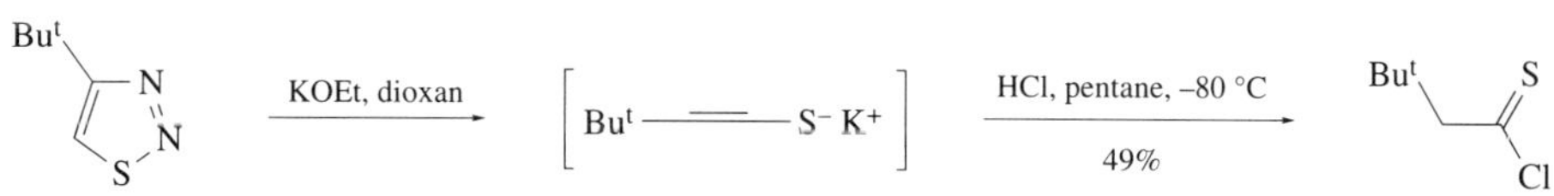

(76) **(77)** **(72)**

Scheme 12

Dimethylketene (**78**) reacts exothermically with thiophosgene (**79**) to produce the thioacyl chloride (**80**) in 89% yield (Equation (17)) ⟨81JOC3911⟩. The high degree of thermal stability of this compound is unusual, and highly atypical of aliphatic thioacyl chlorides in general (see Sections 5.11.2 and 5.11.4). This may be due, at least in part, to the lack of an α-proton in (**80**), since a common decomposition pathway for aliphatic thioacyl chlorides is the elimination of hydrogen chloride to form a thioketene.

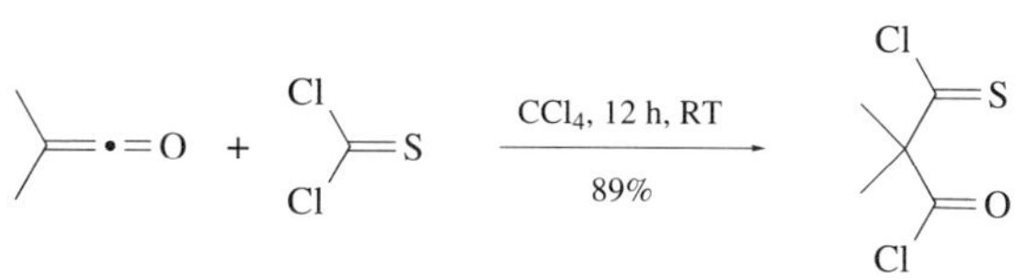

(78) **(79)** **(80)** (17)

Thioacyl chlorides (**85**)–(**88**) have been postulated as intermediates in the formation of a number of thiacumulenes (**89**)–(**92**) from alkenyl precursors (**81**)–(**84**) (Scheme (13)) ⟨74AG(E)273, 75AG(E)53, 75GEP(O)2409357⟩.

5.11.4.5 Thioacyl Chlorides via *S*-Benzyl Thioethers

Several perhalothioalkanoyl chlorides (**94**) have been prepared by the action of Lewis or Brönsted acids on *S*-benzyl thioethers (**93**) (Equation (18)) ⟨87JFC(35)523⟩. The thioacyl chlorides prepared by this method are summarized in Table 10. The reactions of bromofluoro *S*-benzyl thioethers (**95**) produced mixtures of thioacyl bromides (**96**) and thioacyl chlorides (**97**) (Equation (19), Table 11) (see Section 5.11.5.3) ⟨87JFC(35)523⟩.

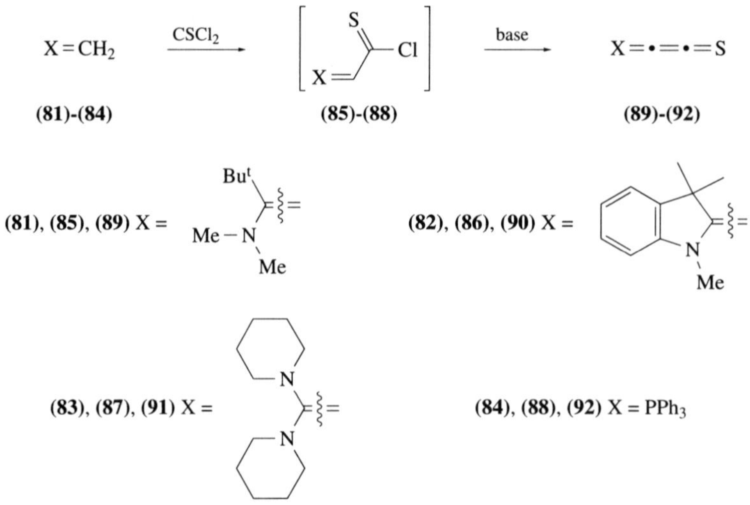

Scheme 13

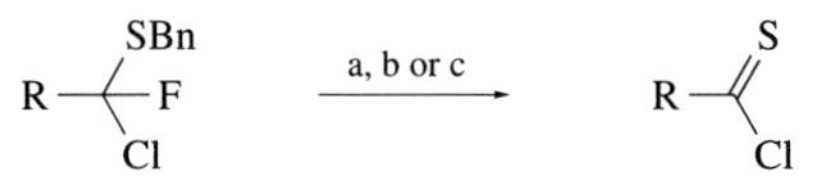

(93) **(94)** (18)

Table 10 Thioacyl chlorides (**94**) prepared according to Equation (18).

R	*Conditions*	*Yield* (%)
CF_3	a	47
CF_3	b	58
CF_3	c	48
CF_2Cl	a	56
CF_2Cl	b	63
CF_2Cl	c	53

[a] TiF_4, heat, no solvent. [b] $TiCl_4$, heat, no solvent. [c] HSO_3Cl, pentane/dichloromethane, 0°C.

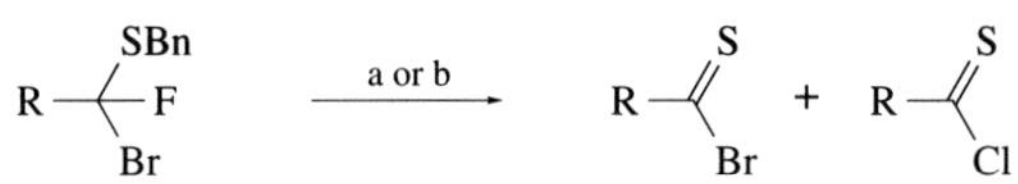

(95) **(96)** **(97)** (19)

Table 11 Thioacyl chlorides and bromides prepared according to Equation (19).

R	*Conditions*	*Yield* (**97**) (%)	*Yield* (**96**) (%)
CF_3	a	35.5	35.5
CF_2Br	a	58	14
CF_2Br	b	55	11

[a] $TiCl_4$, heat, no solvent. [b] HSO_3Cl, pentane/dichloromethane, 0°C.

5.11.4.6 Thioacyl Chlorides via the Reaction between Thiophosgene and Diazoketones

Certain substituted aryl diazoketones (**98**) and (**99**) produce α-chlorothioacyl chlorides (**100**) and (**101**) upon treatment with thiophosgene (Equation (20)) ⟨64LA(673)124⟩. The reaction products were undistillable oils, which were characterized by hydrolysis to the known chloroacetophenones. Other substituents in the diazoketone, such as 4-NO_2, 4-Cl, 4-Me, and 3-NO_2 led to the formation of thiadiazoles (**102**)–(**105**) upon thiophosgene treatment, and the 4-MeO substituted diazoketone gave the chlorodithioformate (**106**). The reaction mechanism, accounting for the formation of all of these products, is discussed.

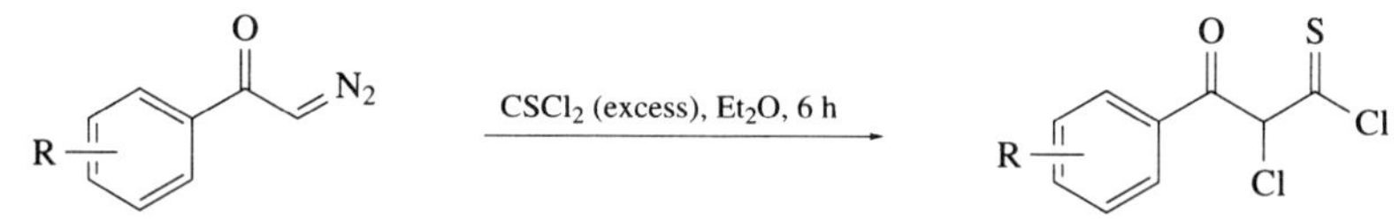

(**98**) R = 3-Me
(**99**) R = 3-MeO

(**100**) R = 3-Me
(**101**) R = 3-MeO

(20)

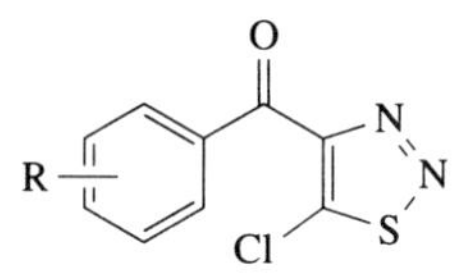

(**102**) R = 4-NO_2
(**103**) R = 4-Cl
(**104**) R = 4-Me
(**105**) R = 3-NO_2

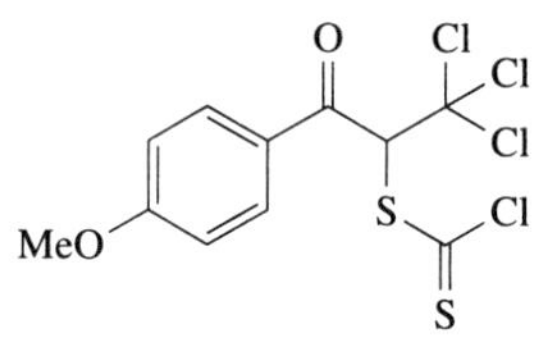

(**106**)

5.11.4.7 Thioacyl Chlorides via *S*-Acetyl 1,1-Dichloroalkyl Disulfides

A range of thioacyl chlorides (**113**)–(**118**) have been prepared and trapped as their morpholides (**119**)–(**124**) by treatment of the *S*-acetyl 1,1-dichloroalkyl disulfides (**107**)–(**112**) with morpholine in ether or benzene (Scheme 14): the thioacyl chlorides prepared by this method are summarized in Table 12. The preparation of the starting materials is also described ⟨86T739⟩.

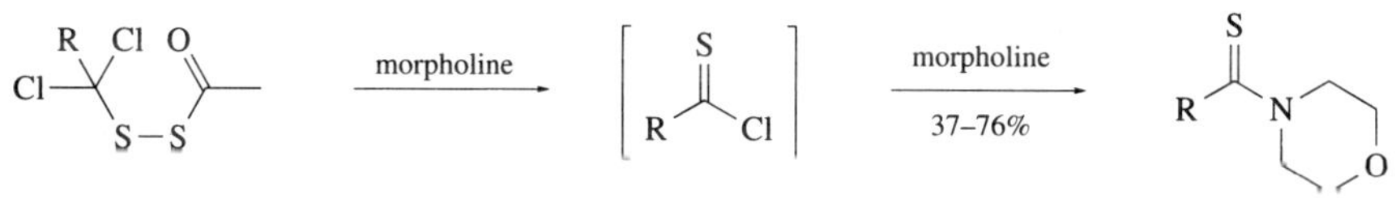

(**107**)–(**112**) (**113**)–(**118**) (**119**)–(**124**)

(**107**), (**113**), (**119**) R = $CHCl_2$
(**108**), (**114**), (**120**) R = CCl_3
(**109**), (**115**), (**121**) R = CCl_2F
(**110**), (**116**), (**122**) R = $CClF_2$
(**111**), (**117**), (**123**) R = EtO_2C
(**112**), (**118**), (**124**) R = PhNHCO

Scheme 14

Table 12 Thioacyl chlorides (**113**)–(**118**) prepared according to Scheme 14.

R	*Yield* (%)[a]
$CHCl_2$	42
CCl_3	42
CCl_2F	37
$CClF_2$	49
EtO_2C	37
PhNHCO	76

[a] Isolated as morpholides (**119**)–(**124**).

5.11.4.8 Thioacyl Chlorides via Acyl Chlorides

The potentially useful conversion of an acyl chloride to a thioacyl chloride has been attempted several times, with little or no success ⟨65CB829, 78BSB293, 80ACS(B)597⟩. However, in one case it has been successful: the thioacyl chloride (**126**) was prepared in 23% yield by treatment of the acyl chloride (**125**) with phosphorus pentasulfide (Equation (21)) ⟨74IZV55⟩.

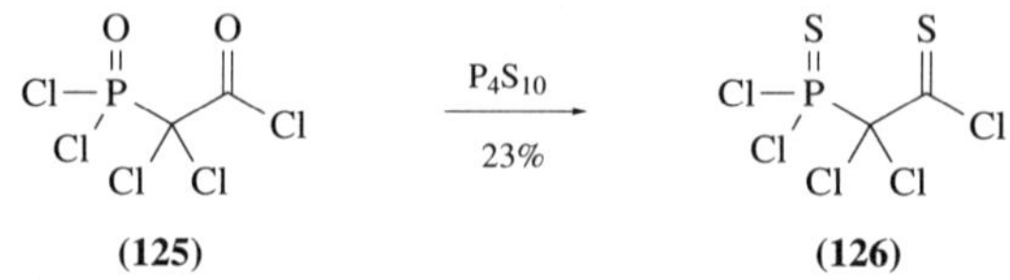

(21)

5.11.4.9 Thioacyl Chlorides via Dehydrofluorination of Thiols

Difluorothioacetyl chloride (**134**) was prepared in low yield by treatment of 1-chloro-1,2,2-trifluoroethanethiol (**129**) with potassium fluoride. The starting material was present as a minor component along with the isomeric 2-chloro-1,1,2-trifluoroethanethiol (**128**). The mixture was formed by the addition of hydrogen sulfide to chlorotrifluoroethene under free-radical conditions (Scheme 15). The acid chloride (**134**) exhibited considerable stability at room temperature. It was postulated as an intermediate in the conversion of thiol (**129**) to thioamide (**135**) when the former was treated with diethylamine (Scheme 16) ⟨67BAU324⟩.

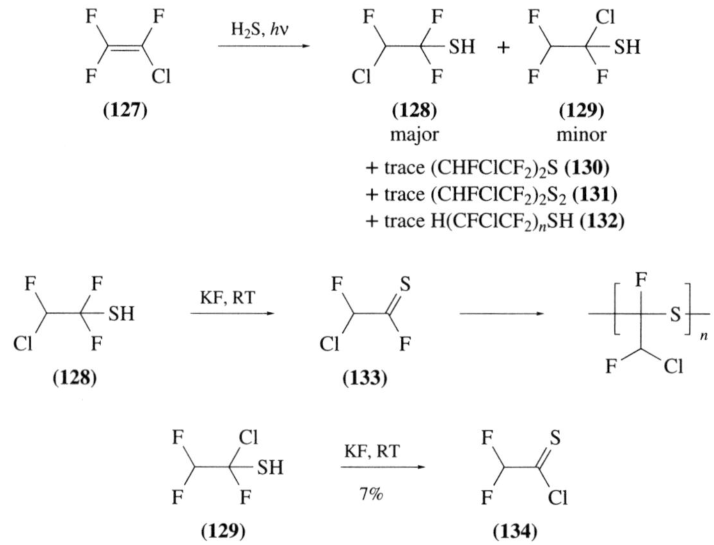

Scheme 15

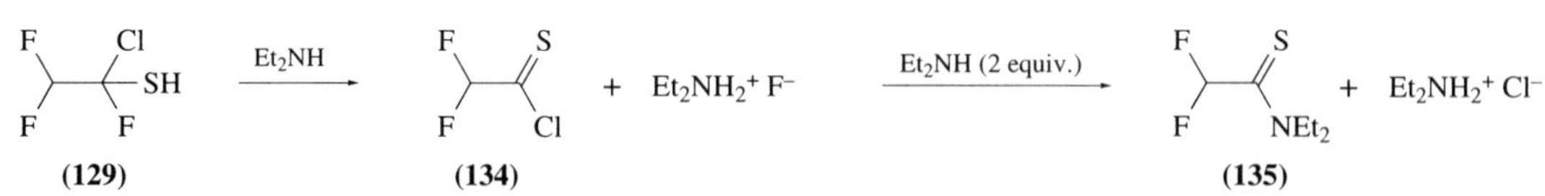

Scheme 16

5.11.4.10 Miscellaneous Reactions

Thioacyl chlorides have been claimed as products upon treatment of a number of active methyl compounds with excess thionyl chloride. Thus, the ketones (**136**)–(**138**) were claimed to produce thioacyl chlorides (**142**)–(**144**) when treated with 10–15 molar equivalents of thionyl chloride and a catalytic amount of pyridine (Scheme 17) ⟨76TL2783⟩. The α-chlorosulfenyl chlorides (**139**)–(**141**) were postulated as intermediates on the basis of NMR data, which showed a mixture of products (**139**) and (**142**), when the reaction was conducted at room temperature. When the reaction was performed at reflux and monitored by NMR over time, signals which the authors ascribed to the sulfenyl chloride (**139**) decreased, and those assigned to the thioacyl chloride (**142**) increased. Similarly, the *t*-butyl ketone (**137**) gave a mixture of products (**140**) and (**143**) at room temperature. The other substrates were all treated with thionyl chloride at reflux, and only the thioacyl chlorides (**144**), (**147**), and (**148**) were claimed as products in these reactions.

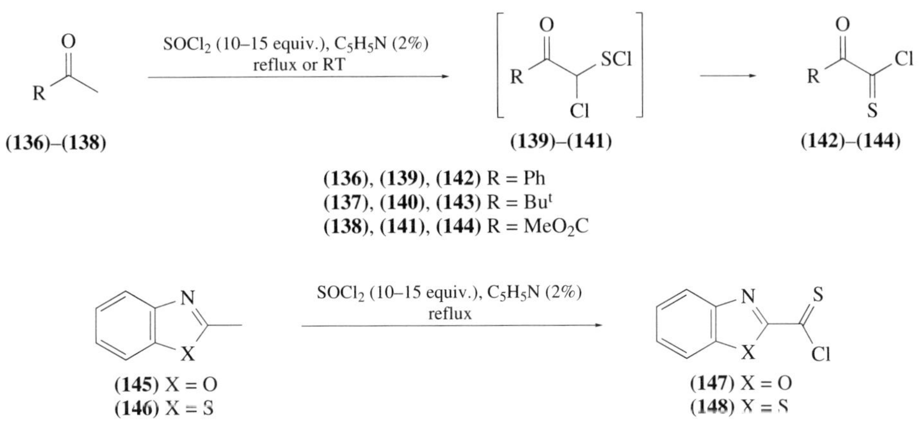

(**136**), (**139**), (**142**) R = Ph
(**137**), (**140**), (**143**) R = Bu^t
(**138**), (**141**), (**144**) R = MeO_2C

Scheme 17

Subsequently, it was rigorously demonstrated by a later group of workers that the thioacyl chlorides were never produced under the reaction conditions described. Treatment of the ketones (**149**)–(**154**) with thionyl chloride in pyridine produced the α-chlorosulfenyl chlorides (**155**)–(**160**) and trisulfanes (**161**)–(**166**) (Scheme 18) ⟨80AG(E)563⟩. Treatment of this mixture with dimethylamine produced α-ketothioamides (**167**)–(**172**) in nearly quantitative yields. Other secondary aliphatic amines, araliphatic amines, and *t*-butylamine reacted similarly. The trisulfanes (**161**)–(**166**) were, in one instance (**161**; R = Bu^t) characterized by x-ray crystallography, ^{1}H and ^{13}C NMR, and in another (**162**; R = Ph) isolated by chromatography, albeit in low yield. The trisulfanes also afforded α-ketothioamides in high yields when treated with secondary amines. The results of these reactions are summarized in Table 13.

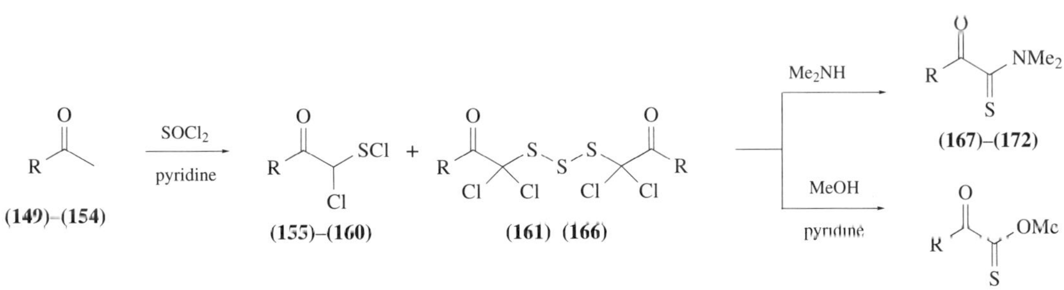

(**149**), (**155**), (**161**), (**167**), (**173**) R = Bu^t
(**150**), (**156**), (**162**), (**168**), (**174**) R = Ph
(**151**), (**157**), (**163**), (**169**), (**175**) R = 4-Bu^tC_6H_4
(**152**), (**158**), (**164**), (**170**), (**176**) R = 3,5-Bu$^t_2$$C_6H_3$
(**153**), (**159**), (**165**), (**171**), (**177**) R = Me
(**154**), (**160**), (**166**), (**172**), (**178**) R = 3,5-$(Ac)_2C_6H_3$

Scheme 18

Again refuting earlier work, a related compound, 4-methoxy-3-nitroacetophenone (**179**) was shown to be converted into the α-ketothioamide (**181**) and α-ketothioester (**182**) under similar conditions. In this case the disulfane (**180**) was shown to be an intermediate (Scheme 19) ⟨80PS(8)1, 81PS(10)213⟩.

In the above reactions there is strong evidence to suggest that α-ketothioacyl chlorides are not intermediates. However, the products, or product mixtures, resulting from the treatment of methyl

Table 13 Preparation of α-ketothioamides $RCOCSNMe_2$ and thioesters RCOCSOMe according to Scheme 18.

R	*Yield* (**155**)–(**160**) (%)	*Yield* (**161**)–(**166**) (%)	*Yield* (**167**)–(**172**) (%)	*Yield* (**173**)–(**178**) (%)
Bu^t	50[a]	40[a]	85[b] (96[c])	79[b]
Ph	45[a]	40[a]	78[b]	56[b]
$4\text{-}Bu^tC_6H_4$			69[b]	40[b]
$3,5\text{-}Bu^t_2C_6H_4$			47[b]	51[b]
Me			5[d]	
$3,5\text{-}(Ac)_2C_6H_4$			3[e]	

[a] Ratio of (**155**):(**161**) and (**156**):(**162**) determined by NMR. [b] Yield referred to ketone; initial mixture treated directly with nucleophiles. [c] Yield from isolated (**161**). [d] Yield of $(Me_2NCS)_2CO$. [e] Yield of $1,3,5\text{-}(Me_2NCSCO)_3C_6H_3$.

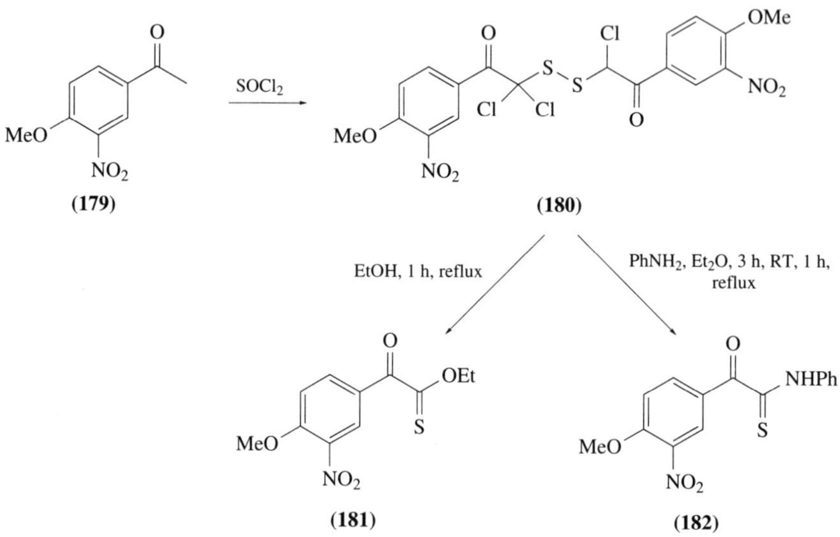

Scheme 19

ketones with thionyl chloride are functionally equivalent to α-ketothioacyl chlorides, in that the products ultimately isolated upon treatment with nucleophiles (amines or alcohols) are those which would be predicted from a reaction of a true α-ketothioacyl chloride.

5.11.5 THIOACYL BROMIDES

Thioacyl bromides are even more unstable and more reactive than the corresponding fluorides or chlorides. Only when special stabilizing features are present, such as aromatic substituents (see Section 5.11.5.1), extensive delocalization (see Section 5.11.5.2) or halogen substitution (see Section 5.11.5.3) are they of sufficient stability to be isolated.

5.11.5.1 Thiobenzoyl Bromides from Dithiobenzoic Acids

The conversion of dithiobenzoic acid (**183**) into thiobenzoyl bromide (**185**) has been accomplished in moderate yield by the use of 1,2-phenylenedioxytribromophosphorane (**184**) in ether (Equation (22)) ⟨68ZC142⟩.

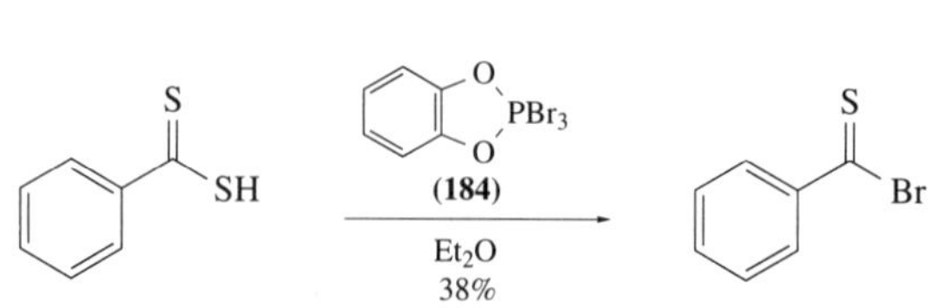

(22)

5.11.5.2 Thioacyl Bromides from Cyclic Trithiocarbonates

The reaction between trithiocarbonate (**186**) and bromocyanoethyne in benzene gave the thioacyl bromide (**187**) in good yield (Equation (23)) ⟨70JOC2002, 72USP3637749⟩. Similarly, the phenyl substituted trithiocarbonate (**188**) gave the thioacyl bromides (**189**) and (**190**) (Equation (24)) ⟨70JOC2002, 72USP3637749⟩.

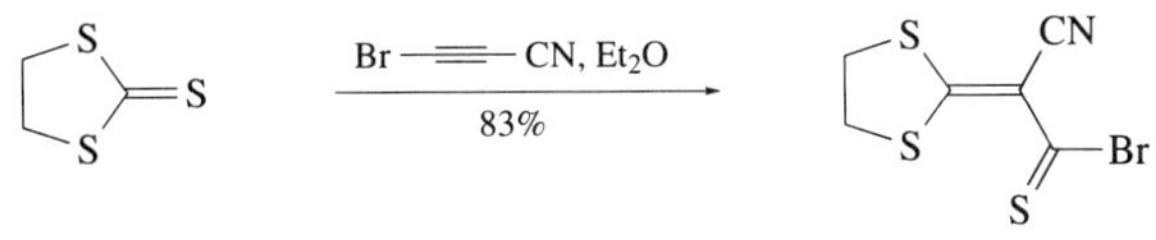

(**186**) (**187**) (23)

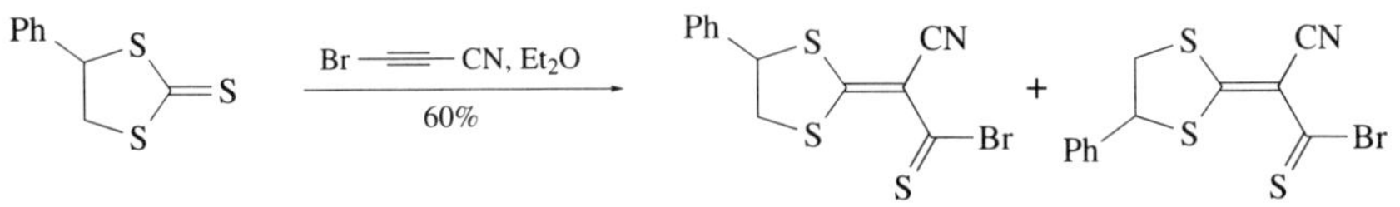

(**188**) (**189**) (**190**) (24)

5.11.5.3 Thioacyl Bromides from Thioethers

As described previously (see Section 5.11.4.5), thioacyl bromides have been prepared (as mixtures with thioacyl chlorides) by the action of acids upon *S*-benzyl thioethers (Equation (25), Table 14) ⟨87JFC(35)523⟩. Using titanium tetrafluoride it was also possible to convert the *S*-methyl thioether (**195**) into trifluorothioacetyl bromide (**196**) in 60% yield (Equation (26)) ⟨87JFC(35)523⟩.

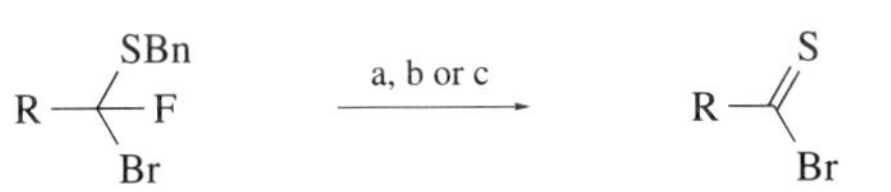

(**191**) R = CF_3
(**192**) R = CF_2Br

(**193**) R = CF_3
(**194**) R = CF_2Br (25)

Table 14 Thioacyl bromides RCSBr prepared according to Equation (25).

R	*Conditions*	*Yield* (%)
CF_3	a	68
CF_3	b	71[d]
CF_3	c	60
CF_2Br	a	80
CF_2Br	b	18[e]
CF_2Br	c	11[f]

[a] TiF_4, heat, no solvent. [b] $TiCl_4$, heat, no solvent. [c] HSO_3Cl, DCM/pentane, 0°C. [d] In a 1:1 mixture with the thioacyl chloride. [e] In a 1:4 mixture with the thioacyl chloride. [f] In a 1:5 mixture with the thioacyl chloride.

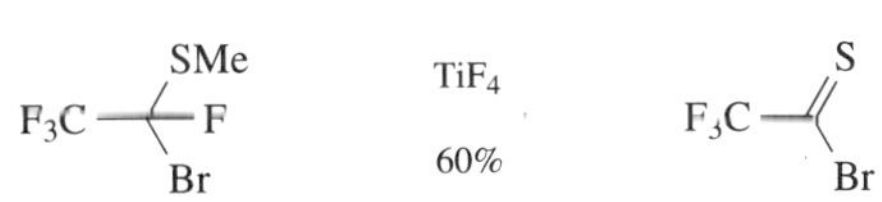

(**195**) (**196**) (26)

5.11.5.4 Thioacyl Bromides from Thioketenes

Bis(trifluoromethyl)thioketene (**197**) adds hydrogen bromide, to produce thioacyl bromide (**198**). This readily eliminates hydrogen bromide upon treatment with a base to form the thioketene dimer, dithietane (**199**) (Scheme 20) ⟨72JOC1347⟩.

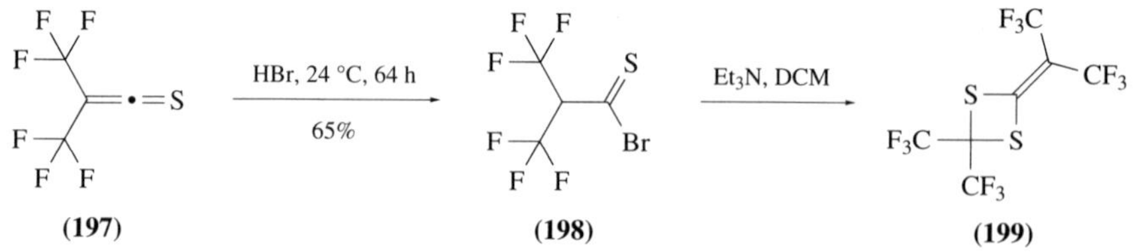

Scheme 20

5.11.6 THIOACYL IODIDES

No references to these compounds were found in the literature up to the end of June 1993.

5.11.7 HIGHER OXIDATION STATES OF SULFUR

5.11.7.1 Sulfur(IV) Derivatives: Halosulfines

The first sulfine (thiocarbonyl-*S*-oxide) to be prepared and characterized was actually the halosulfine (**200**), derived from camphor-10-sulfonyl chloride (see Section 5.11.7.1.1) 〈23CB633〉. Fluorosulfines are too unstable to exist above $-100\,^{\circ}\mathrm{C}$, but generally chlorosulfines are reasonably chemically stable and may be isolated and characterized readily: no examples of bromo- or iodosulfines were found. Halosulfines exhibit *cis/trans* isomerism about the carbon–sulfur double bond, and in many cases the individual isomers have been characterized separately 〈89CB1757, 90JST171〉. It is interesting to note that the first preparation of these sulfur(IV) derivatives involved sulfenes, sulfur(VI) derivatives, as intermediates.

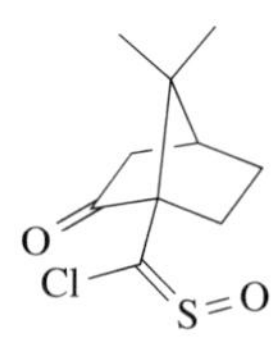

(**200**)

5.11.7.1.1 *Halosulfines via dehydrohalogenation of sulfonyl halides*

As alluded to earlier, the first stable sulfine to be isolated and characterized was the halosulfine (**200**), derived from camphor-10-sulfonyl chloride (**201**) (Equation (27)) 〈23CB633〉. Whilst superficially simple in appearance, the exact mechanism of this reaction is still not completely clear. Much work has shown that it is strongly base and solvent dependent 〈66CJC819, 76CJC2652〉. The reaction is generally favoured by nonpolar solvents; cyclohexane, benzene, and ether are commonly used. The reaction is also favoured by steric hindrance in the base; see the entries in Table 15 comparing trimethylamine, triethylamine, and tributylamine. The isolated yields of halosulfine are greatly increased if an equimolar amount of *p*-toluenesulfonyl chloride is added to the reaction. If an alkyl sulfonyl bromide and *p*-toluenesulfonyl chloride are treated with base together, only the chlorosulfine is isolated, never the bromosulfine 〈64RTC94〉. The currently accepted mechanism involves the elimination of HX from sulfonyl halide (**202**) to give a sulfene (**203**): a second molecule of either the same sulfonyl halide (**204**; $R^2 = R^1CH_2$, $X^1 = X$), or of an external sulfonyl halide (**204**; $R^2 \neq R^1CH_2$, $X^1 \neq X$) reacts with the sulfene, forming a mixed sulfinic–sulfonic anhydride (**205**). This mechanism requires that the halogen in the final product, X^1, is derived from the second sulfonyl halide in the reaction, hence the failure to isolate any bromosulfine from a combination of a sulfonyl bromide and *p*-toluenesulfonyl chloride. Elimination of sulfonate from (**205**) yields the halosulfine (**206**) (Scheme 21) 〈63TL585, 64RTC94, 67AG(E)107, 81S801, 82RTC1, 85HOU(E11/1)911〉. Halosulfines (**208**) prepared from sulfonyl halides (**207**) by this and closely related methods are summarized in Table 15 (Equation (28)).

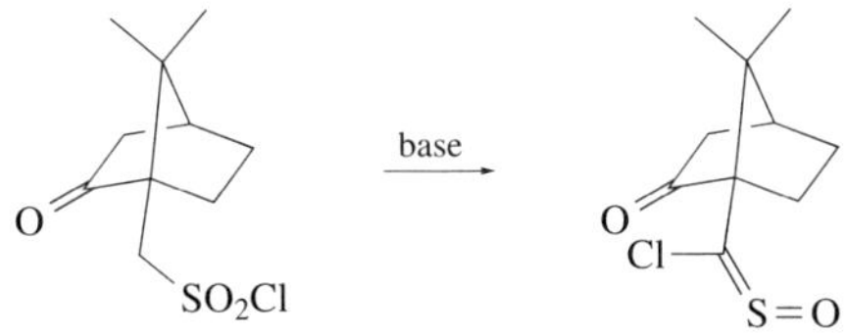

(201) **(200)** (27)

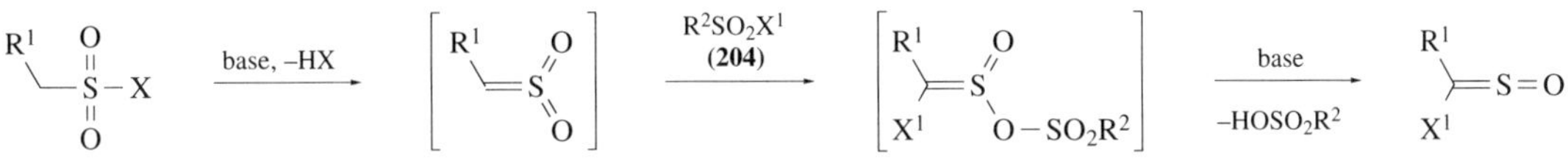

(202) **(203)** **(205)** **(206)**

Scheme 21

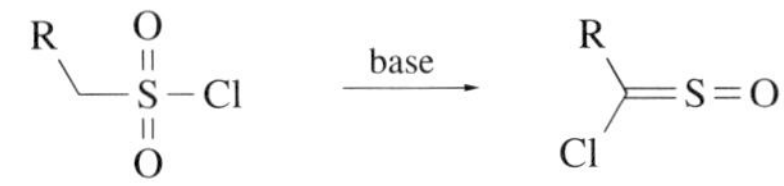

(207) **(208)** (28)

Table 15 Chlorosulfines (**208**) prepared according to Equation (28).

R	*Base*	*Yield* (%)	*Ref.*
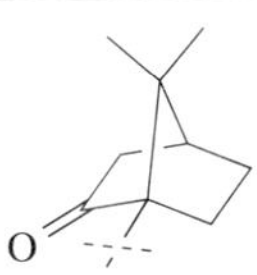	Pyridine	27	23CB633
	Et_3N	38	23CB633[d]
	Pyridine	27	64RTC94
	Pyridine	70[a]	64RTC94
	Pyridine	24[a,b]	64RTC94
	Pyridine	25[c]	63RTC94
	Pyridine	75[a]	80S801
Ph	Et_3N	35	63TL585
	Et_3N	19	63JA2676
	Et_3N	63	66CJC819
	Me_3N	21	76CJC2652
	Et_3N	63	76CJC2652
	Bu_3N	81	76CJC2652
	$EtNPr^i_2$	<50	76CJC2652
Me	1,8-Bis(dimethylamino) na phthalene	64[e]	86TL167
H	1,8-Bis(dimethylamino) naphthalene	47[e]	86TL167
$PhCH_2$	1,8-Bis(dimethylamino) naphthalene	41[e]	86TL167
4-MeC_6H_4	Bu_3N	70	90CB1563
	Bu_3N	70	90CB1989
Bu^t	Et_3N	73	90CB1989
$PhMe_2C$	Et_3N	93	90CB1989

[a] With addition of 4-$MeC_6H_4SO_2Cl$. [b] From camphor-10-sulfonyl bromide. [c] From camphor-10-sulfonyl chloride, with the addition of 4-$MeC_6H_4SO_2I$. [d] See also related references ⟨23CB1557, 25JCS279, 30MI 511-01⟩. [e] Characterized as the 4-alkanesulfinyl-1,8-bis(dimethylamino)naphthalene.

In a related reaction, elimination of hydrogen chloride from an α-chlorosulfinyl chloride (**210**) leads to a chlorosulfine (**211**). The α-chlorosulfinyl chlorides are themselves formed by addition of chlorine to a sulfine (**209**) (Scheme 22) ⟨67RTC641⟩. In the same paper the authors describe another route to phenylchlorosulfine (**215**) (**208**; R = Ph) (Scheme 23) ⟨67RTC641⟩.

Aliphatic analogues, such as methylchlorosulfine (**218**), have also been prepared, although

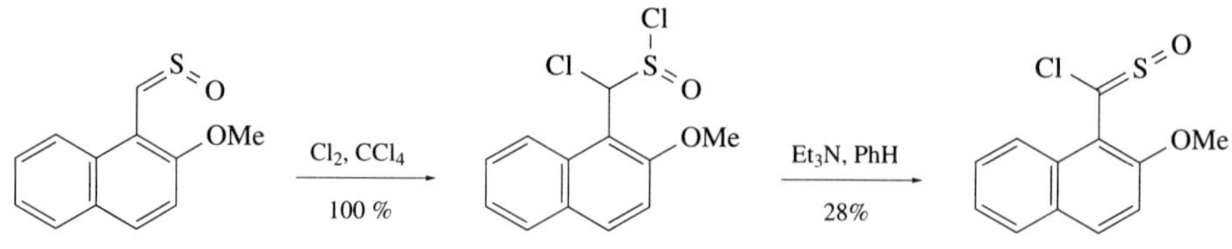

(209) **(210)** **(211)**

Scheme 22

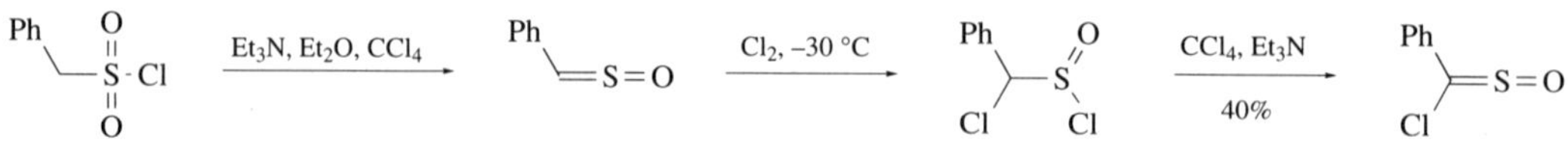

(212) **(213)** **(214)** **(215)**

Scheme 23

isolation was not possible. The compound was identified by conversion into the dichlorosulfinyl chloride (**219**), and subsequent oxidation to the sulfonyl chloride (**220**). This was identical to an authentic sample of sulfonyl chloride prepared by known methods (Scheme 24) ⟨79JOC1177⟩.

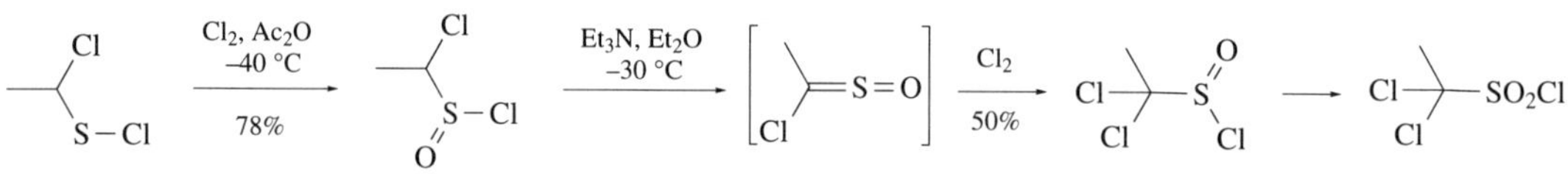

(216) **(217)** **(218)** **(219)** **(220)**

Scheme 24

When the elimination was performed using an excess of 1,8-bis(dimethylamino) naphthalene, the chlorosulfines (**222**) reacted further (Scheme 25) ⟨86TL167⟩.

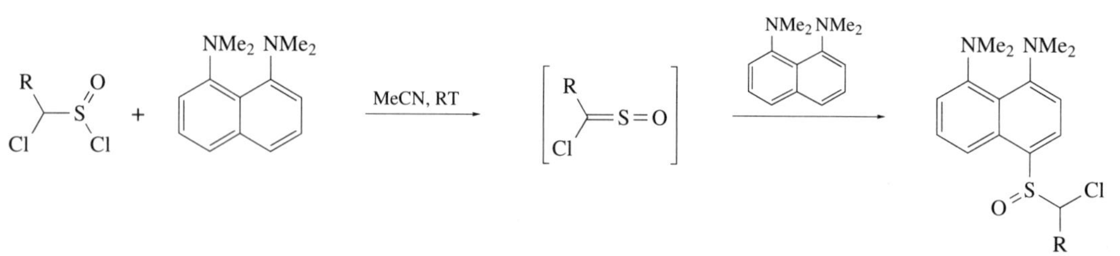

(221)

(222) **a**; R = H
b; R = Me

(223) **a**; R = H, 57%
b; R = Me, 22%

Scheme 25

5.11.7.1.2 Halosulfines via oxidation of thioacyl halides

Phenylchlorosulfine (**225**) has been prepared, as a separable mixture of *cis* and *trans* isomers, by the peracid oxidation of thiobenzoyl chloride (**224**) (Equation (29)) ⟨70RTC687⟩. The preparation of thiobenzoyl chloride has been described above (see Sections 5.11.4.1 and 5.11.4.2). *N,N*-Disubstituted α-carbamoyl chlorosulfines (**229**) may be prepared by dechlorination of dichlorosulfenyl chlorides (**227**), and subsequent oxidation of the resulting thioacyl chloride (**228**) (Scheme 26) ⟨72JOC3818⟩. This sequence of reactions has been discussed in part as a preparative route to the thioacyl chlorides themselves (see Section 5.11.4.2) ⟨71JOC3145, 72JOC1526⟩. A number of simple aryl dichlorosulfenyl chlorides have been prepared by chlorination of the corresponding dithioesters (**230**). The subsequent conversion of the dichlorosulfenyl chlorides into chlorosulfines (**233**), via thioacyl chlorides (**232**), is exactly analogous to that described above (Scheme 27). The sequence of transformations (**230**) to(**232**) has been discussed as a preparative route to thioacyl chlorides (see Section 5.11.4.2) ⟨77RTC139⟩. The chlorosulfines prepared by this route are summarized in Table 16.

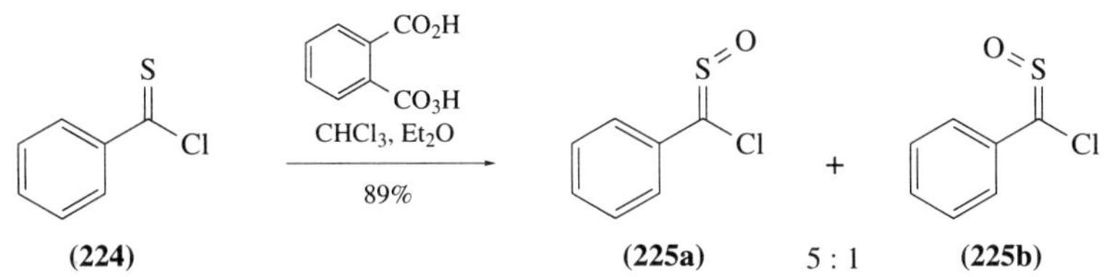

(29)

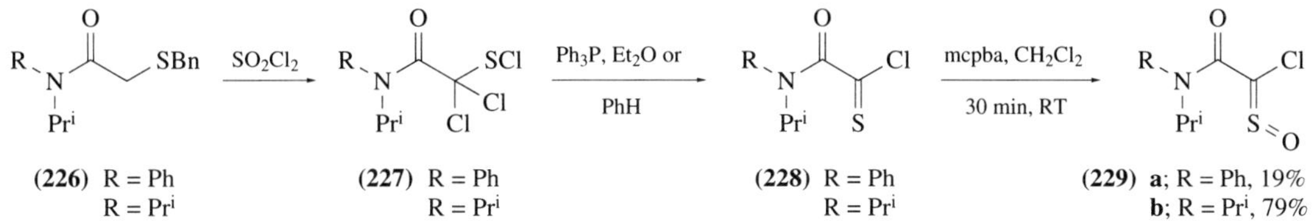

Scheme 26

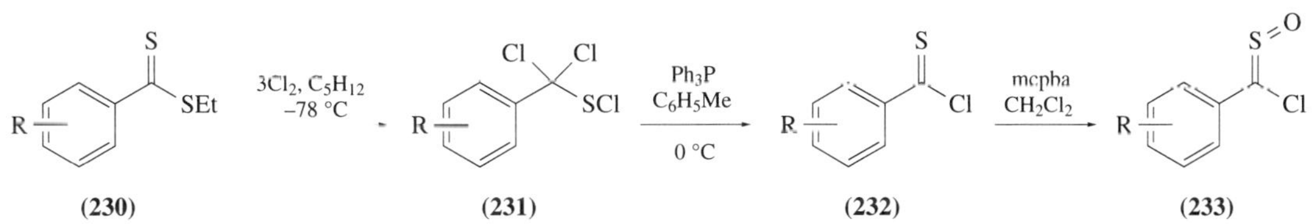

Scheme 27

Table 16 Chlorosulfines (**233**) prepared according to Scheme 27.

R	*Yield* (%)	*Comments*
H	89	(*Z*):(*E*) = 5 : 1
4-Cl	25	(*Z*) only
2,4,6-Me_3	60	(*Z*):(*E*) = 6 : 1

5.11.7.1.3 Halosulfines via dichlorosulfenyl chlorides

In addition to their indirect conversion into chlorosulfines, via thioacyl chlorides as described above (see Section 5.11.7.1.2), α-carbamoyl dichlorosulfenyl chlorides (**234**) may be converted directly into chlorosulfines (**235**) by mild basic hydrolysis. The sulfenyl chlorides (**234**) are prepared from the corresponding acetanilides (**233**) by chlorination with sulfuryl chloride (Scheme 28) ⟨72JOC1526, 72JOC3818⟩. The halosulfines prepared by this method are summarized in Table 17.

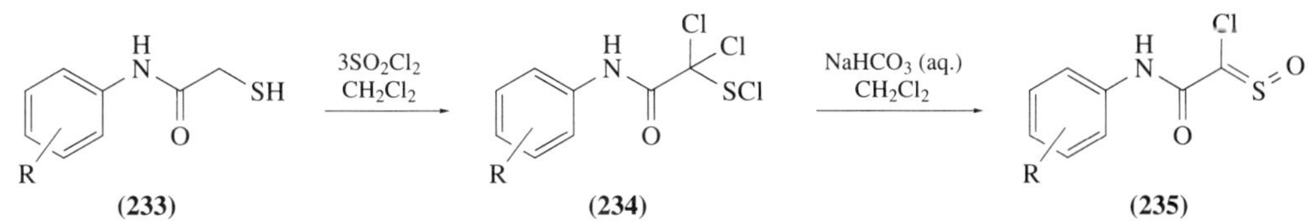

Scheme 28

Table 17 Chlorosulfines (**235**) prepared according to Scheme 28.

R	*Yield* (%)
4-Cl	49
3-CF_3	57
3-Br	52
4-OMe	50
4-Me	26
2,6-Me_2	52
2-Me-6-Bu^t	42

The aromatic 1,1-dichlorosulfenyl chlorides (**237**), the synthesis of which was described in the previous section (5.11.7.1.2), have been converted into chlorosulfines in an analogous procedure (Scheme 29) ⟨77RTC139⟩.

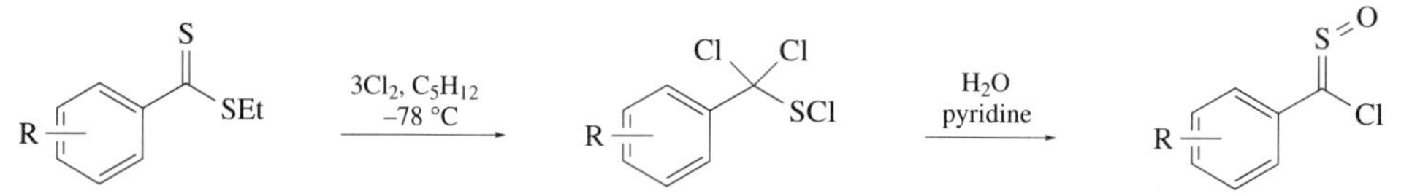

(**236**) **a**; R = H
b; R = 4-Cl

(**237**) **a**; R = H
b; R = 4-Cl

(**238**) **a**; R = H, 48%
b; R = 4-Cl, 39%

Scheme 29

5.11.7.1.4 Preparation of perhalogenated aliphatic halosulfines

In the 1980s and 1990s the preparation of a number of perhalogenated aliphatic halosulfines has been described: the key transformations are depicted in Scheme 30. The pivotal intermediate is a 1,1-dihalosulfenyl chloride (**239**). The most direct route involves hydrolysis of (**239**) under mildly basic conditions; however, the isolated yields of halosulfines are moderate. Greatly improved yields are obtained if the hydrolysis is conducted in the presence of anthracene, when the [4 + 2] cycloadduct is isolated. Pyrolysis of this at low pressure, in an inert atmosphere, gives good yields of the halosulfines (**240**). The oxidized [4 + 2] cycloadduct (**241**) may also be obtained via peracid treatment of the adducts (**243**), themselves obtained from the cycloaddition of thioacyl halides (**242**) and anthracene. The conversion of sulfenyl chlorides to thioacyl halides has been described previously as a preparative method for these latter moieties (see Sections 5.11.3.5 and 5.11.4.2). One further route involves conversion of the sulfenyl halide (**239**) to a sulfinyl halide (**244**), which may then be dechlorinated pyrolytically, over copper at reduced pressure, to the chlorosulfines (**240**), as a mixture of (*E*) and (*Z*) isomers. Whilst not all of these potential routes have been explored for all compounds, the viability of each route has been demonstrated for at least one species. The compounds prepared by these methods are detailed in Table 18 ⟨85TL5505, 88S349, 89CB1757, 91CB1311, 92CB1889⟩.

Interestingly, the chlorotrifluoromethyl sulfine (**240b**), when prepared by direct hydrolysis of (**239b**), or by pyrolysis of the anthracene cycloadduct (**241b**) is formed as a single compound, the more thermodynamically stable (*Z*) isomer: dechlorination of the sulfinyl chloride (**244b**) leads to a mixture of isomers ((*E*/*Z*) = 1 : 1.7). The (*E*) isomer undergoes slow, spontaneous conversion to the (*Z*) at room temperature ⟨89CB1757, 90JST171⟩.

The synthesis, from *S*-benzyl thioethers, of the sulfenyl chlorides required as starting materials is also discussed (Scheme 31). The conversion of (**245**) to (**248**) may be effected in a single pot; however, using less forcing conditions, both the intermediate 1-chloro- and 1,1-dichloro-2,2,2-trifluoroethyl *S*-benzyl thioethers (**246**) and (**247**) may be isolated and characterized. The possibility of controlling the oxidative chlorination process to allow efficient access to intermediates such as (**247**) found utility in the preparation of 1-chloro-1,2,2,2-tetrafluoroethyl *S*-benzyl thioether (**249**), a precursor of sulfenyl chloride **250** (≡**239a**). The conversion of this latter species to fluorotrifluoromethyl sulfine is described above.

The pentachloroethanesulfenyl chloride (**239e**) was prepared as described above (see Section

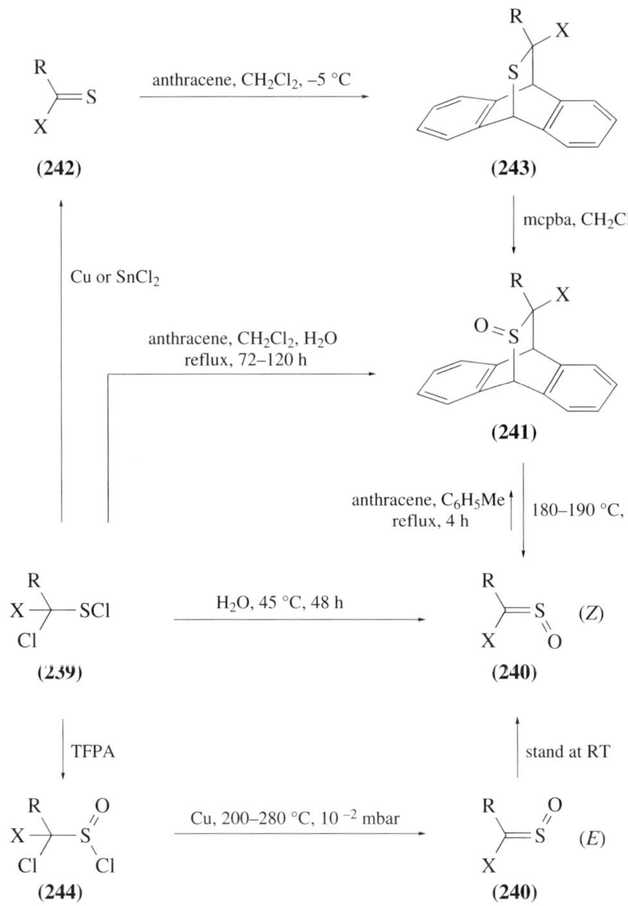

Scheme 30

Table 18 Compounds prepared according to Scheme 30.

Starting material	*Product*	*Yield (%)*	*Ref.*
(239a)	**(241a)**	72	92CB1889
(241a)	**(240a)**	76	92CB1889
(239a)	**(244a)**	69	92CB1889
(239a)	**(242a)**	89	92CB1889
(242a)	**(243a)**	NY	65JOC1390
(239b)	**(240b)**(*Z*)	26	85TL5505, 88S349, 89CB1757
(239b)	**(241b)**	NY	89CB1757
(241b)	**(240b)**(*Z*)	67	89CB1757
(239b)	**(244b)**	72	85TL5505, 88S349, 89CB1757
(244b)	**(240b)**(*E* + *Z*)	62[a]	89CB1757
(242c)	**(243c)**	71	92CB1889
(243c)	**(241c)**	72	92CB1889
(241c)	**(240c)**	86	92CB1889
(242d)	**(243d)**	76	92CB1889
(243d)	**(241d)**	82	92CB1889
(241d)	**(240d)**	6	92CB1889
(239e)	**(243e)**	56	91CB1311
(243e)	**(241e)**	67	91CB1311
(241e)	**(240e)**	78	91CB1311

[a] (*E*):(*Z*) = 1 : 1.7.

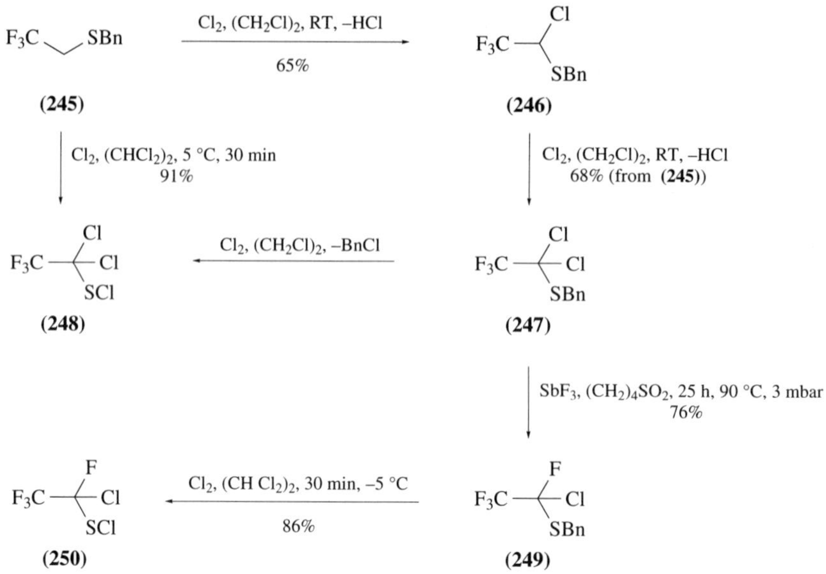

Scheme 31

5.11.4.2) ⟨85SUL177, 87SUL153⟩. The preparation of thioacyl fluorides (**242c**) and (**242d**) used the methods described in Section 5.11.3.1 ⟨65JOC1375⟩.

Prior to the work just described, there was only a single reference to an aliphatic halosulfine in the literature. The fluorotrifluoromethyl sulfine (**257**) (≡**240a**) was obtained, in very low yields, and after a great many steps, as a single isomer, starting ultimately from trifluoroethene (**251**) (Scheme 32) ⟨83MI 511-01, 88S349⟩.

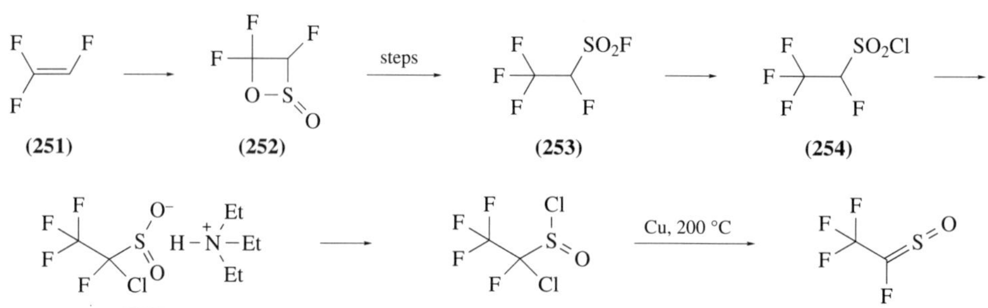

Scheme 32

5.11.7.1.5 Halogenation of stabilized sulfur ylides

Treatment of ethyl (dimethylsulfuranylidene)acetate (**258**) with trichloroacetonitrile in ether at 0–5 °C gave the chloro ylide (**259**) in 78% yield (Equation (30)) ⟨68JOC3517⟩. Later workers described the use of the analogous bromo ylide (**260**), but did not describe its preparation, referring instead to the earlier reported synthesis of (**259**). Presumably the use of tribromo- instead of trichloro-acetonitrile gives the bromo ylide, but since no experimental details are available, no comment can be made regarding conditions, yields, or ease of isolation ⟨75JOC2990⟩.

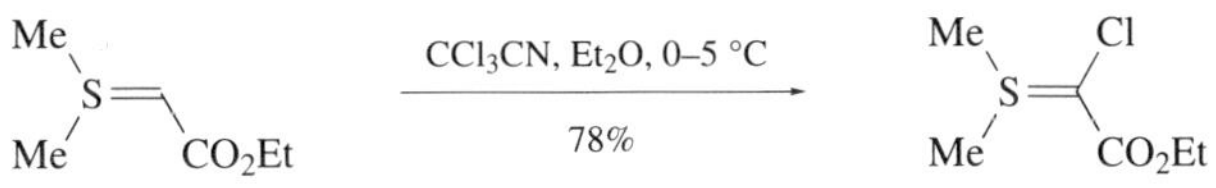

(30)

(258) **(259)**

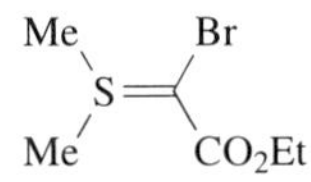

(260)

Related ketone stabilized ylides have also been described. Treatment of 2-(dimethylsulfuranylidene) acetophenone (**261**) with bromine or iodine, followed by sodium hydride in THF gives the corresponding bromo and iodo ylides (**263a**) and (**263b**) in good yields via sulfonium salts (**262**) (Scheme 33) ⟨66JOC1689⟩.

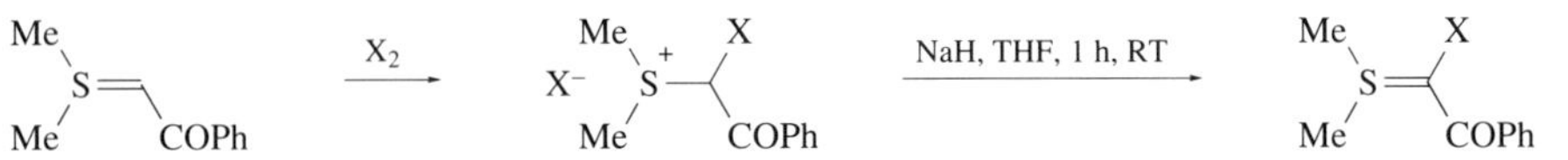

(261) **(262) a**; X = Br, 61% **b**; X = I, 88% **(263) a**; X = Br, 94% **b**; X = I, 69%

Scheme 33

Whilst it will be appreciated that these ylides may formally be thought of as thioacyl halides wherein the sulfur atom is in a higher oxidation state (i.e. IV), their chemistry is more typical of ylides and their preparation is discussed here only for reasons of completeness.

5.11.7.2 Sulfur(VI) Derivatives: Halosulfenes

Despite many attempts in the 1980s to isolate sulfenes as freely existing species in their own right, the high degree of reactivity of the carbon–sulfur double bond has, thus far, permitted only indirect demonstration of their existence, for example by trapping *in situ* as [2 + 2] or [4 + 2] cycloadducts. Their existence has also been demonstrated by the isolation of amine complexes such as (**264**) and (**265**) ⟨66TL5263, 85HOU(E11/2)1326, 88CB1435,89AG(E)221⟩. As a subset of an already "esoteric" group of compounds, halosulfenes are even less numerous and even more poorly described in the literature. The intermediacy of sulfenes has been implicated in the formation of halosulfines from sulfonyl halides, and this may be thought of as an example whereby they are "trapped" by further reaction with an (endogenous or exogenous) sulfonyl halide (see Section 5.11.7.1.1). Most of the halosulfenes described up to 1985 are covered by the few review articles published on sulfenes ⟨67AG(E)107, 75ACR10, 85HOU(E11/2)1326⟩. However, since they are referred to only in passing, and discussed hardly at all, their preparation will be expanded upon here.

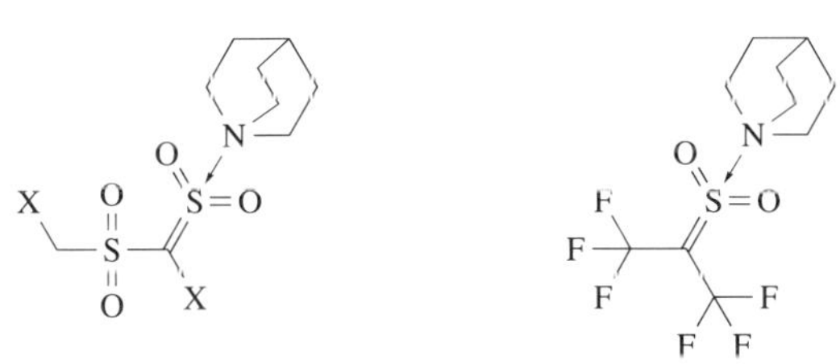

(264) **(265)**
X = H, Cl, Br, Me, CF_3

Since the evidence for the existence of halosulfenes is all, to date, of an indirect, inferential nature, the character of the ensuing section is somewhat different from those preceding it. Essentially, the only synthetic procedure described which is useful for the preparation of halosulfenes is the base-mediated dehydrohalogenation of an appropriately substituted sulfonyl halide. However, since the immediate products of such reactions, the free halosulfenes, are too reactive to permit their isolation, it has been necessary to focus upon those trapping reactions which demonstrate most unequivocally the intermediacy of such transient species.

5.11.7.2.1 Halosulfenes via dehydrohalogenation of sulfonyl halides

The intermediacy of sulfenes in general in the conversion of sulfonyl halides to halosulfines has been discussed above (see Sections 5.11.7.1.1 and 5.11.7.2). If the dehydrohalogenation is performed in the presence of a reagent capable of forming a stable cycloadduct with the sulfene, then it may be intercepted before reaction with further sulfonyl halide occurs, providing more direct evidence of the (transient) existence of a sulfene. By using an appropriately halogenated sulfonyl halide, a halosulfene may be generated and isolated as an analogous, halogenated cycloadduct. Thus, for example, addition of an ethereal solution of chloromethanesulfonyl chloride (**266**) to a solution of equimolar amounts of triethylamine and ketene diethylacetal (**268**) at 25 °C eventually gave a 31% yield of the dithietane-1,1-dioxide (**269**) (Scheme 34) ⟨63JA3231⟩.

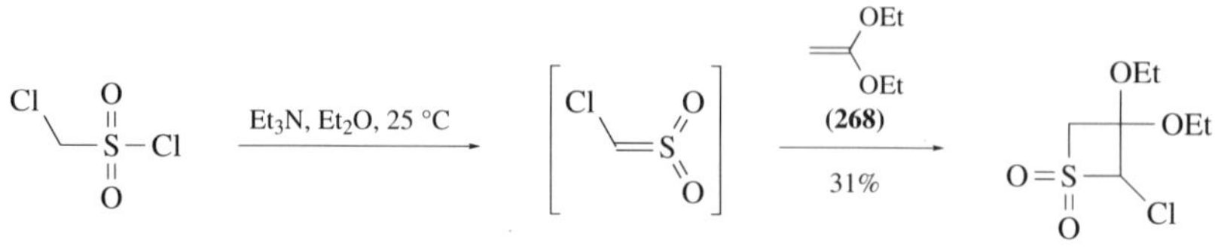

Scheme 34

Chlorosulfene (**267**), generated in a very similar fashion, has also been trapped by reaction with a number of other highly nucleophilic, electron-rich doubly bonded species, such as enamines (Scheme 35) ⟨64JOC2854⟩, (Scheme 36) ⟨72JOC2080⟩, and ketene-*N*,*N*-acetals (Scheme 42) ⟨65LA(684)103⟩. Trapping of *in situ* generated chlorosulfene (**267**) with the morpholine enamines of cyclopentanone, cyclohexanone and cycloheptanone, (**270a**), (**270b**), and (**270c**), respectively, was reported to give moderate yields of single epimers of the corresponding cycloadducts (**271a**)–(**271c**). The reactions of chlorosulfene with the morpholine enamines of 1-methyl-4-piperidone (**272**) and of norborn-5-ene-2-carbaldehyde (**275**) gave reasonable yields of mixtures of epimeric cycloadducts (**273**) + (**274**) and (**276**) + (**277**), respectively (Scheme 35) ⟨64JOC2854⟩. The authors do comment upon the difference between the reactions of chlorosulfene with the carbocyclic enamines (**270**) and the heterocyclic/bicyclic enamines (**272**) and (**275**), discuss the possibility that the failure to obtain both isomers from the former may be an artefact of isolation ⟨64JOC2854⟩.

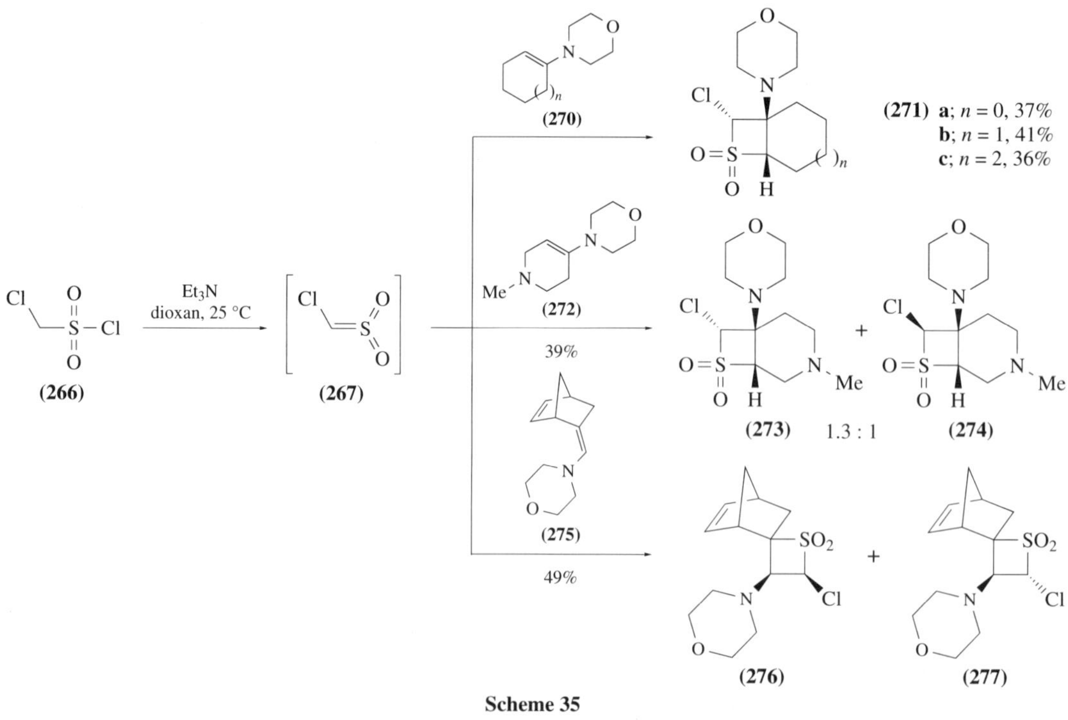

Scheme 35

This explanation may well account correctly for the above observations since, in an almost identical experiment, later workers obtained a mixture of products, rather than a single isomer, from the reaction between chlorosulfene (**267**) and morpholinocyclohexene (**270b**) (Scheme 36) ⟨72JOC2080⟩. The procedure employed benzene as solvent, rather than the dioxan used by the earlier group, but it is difficult to rationalize the difference in observed outcomes as being due to this change alone. Two different crystal forms of (**271b**) were obtained, one of which was identical to that prepared previously by Paquette ⟨64JOC2854⟩. Although the later authors were unable to prepare a pure sample of (**278**) and did not comment upon the ratio of (**271b**) to (**278**), they were able to characterize the latter compound by ^{1}H NMR, having separated a quantity of a mixture highly enriched in it ⟨72JOC2080⟩.

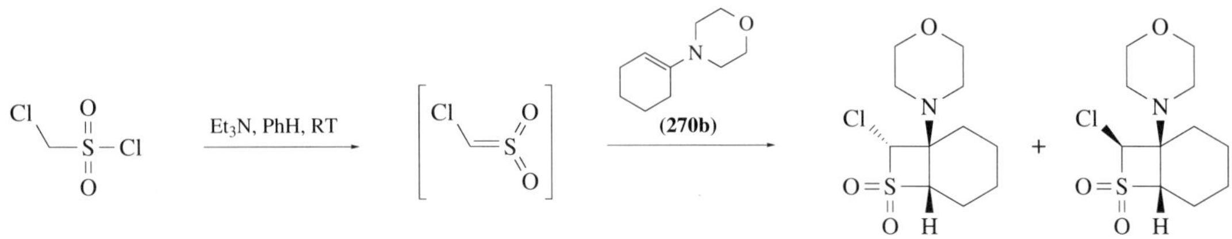

(266) **(267)** **(271b)** **(278)**

Scheme 36

In the same paper, the authors describe an analogous preparation of bromosulfene (**280**) and subsequent *in situ* reaction with the enamine (**270b**) to yield a mixture of epimeric cycloadducts (**281**) and (**282**) (Scheme 37) ⟨72JOC2080⟩. Additionally, two isomeric quaternary ammonium salts (**283**) and (**284**) were obtained. The mechanism of the cycloaddition may be a concerted, one-step, multicenter reaction (a), or a two-step addition, going through a zwitterionic intermediate (**A**) (b) (Scheme 38) ⟨65LA(684)103, 67AG(E)107⟩. The latter is probably to be preferred since it accounts for the observation that only strongly nucleophilic alkenes react with sulfenes, and allows one to correlate the extent of open-chain sulfone (**B**) formation with the degree of charge stabilization in the zwitterionic intermediate (**A**). The observation of a degree of stereoselectivity in product formation can be rationalized if the lifetime of (**A**) is very short. Similar arguments have been applied to the reaction between chlorosulfene and ketene-*N*,*N*-acetals (see below).

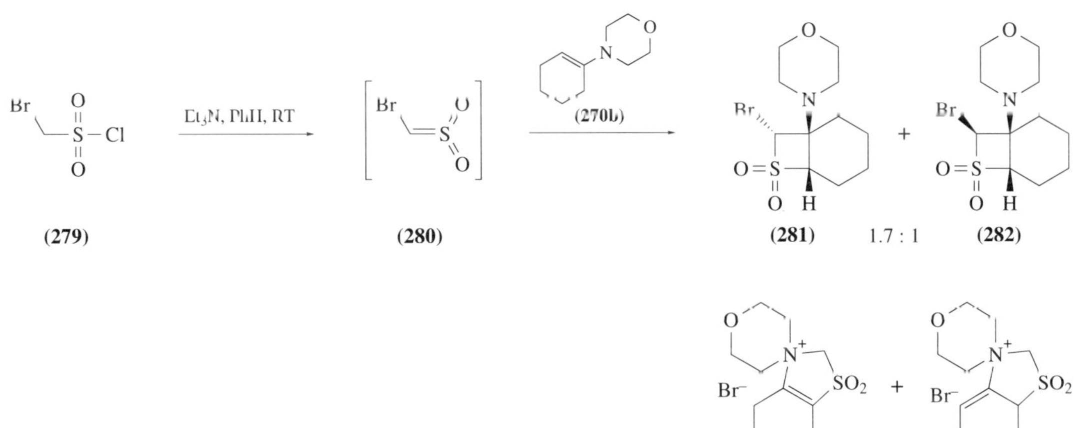

(279) **(280)** **(281)** 1.7 : 1 **(282)**

(283) 1 : 1 **(284)**

(281) + **(282)** + **(283)** + **(284)** – 56%

Scheme 37

Chloro-, bromo-, and iodosulfenes, as well as methylchlorosulfene have been trapped as cycloadducts with enamines of acyclic carbonyl compounds. These reactions are illustrated in Schemes 39 and 40 and Table 19 ⟨73JCS(P1)2540, 74JCS(P2)1483⟩. Two open-chain sulfones, (**288**) and (**289**), were also isolated, which presumably arise from a competing sulfonation of the enamines, since their acyclic analogues, (**300**) and (**294**) + (**299**), were stable under the reaction conditions. The open chain sulfones were formed, indeed can only be formed, when there is at least one hydrogen on the

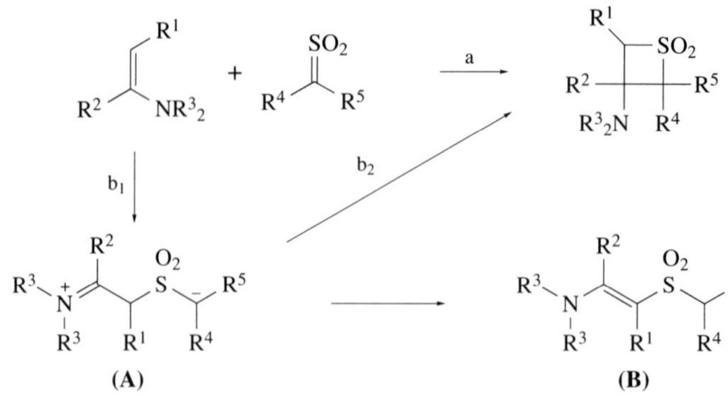

Scheme 38

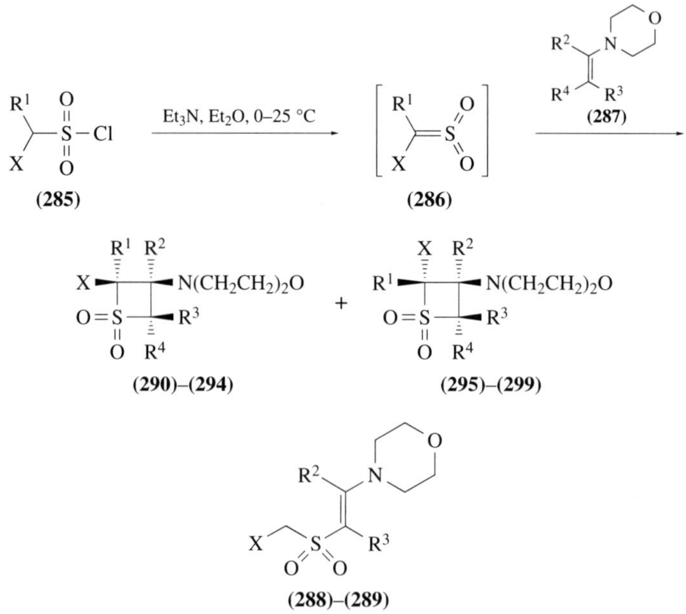

Scheme 39

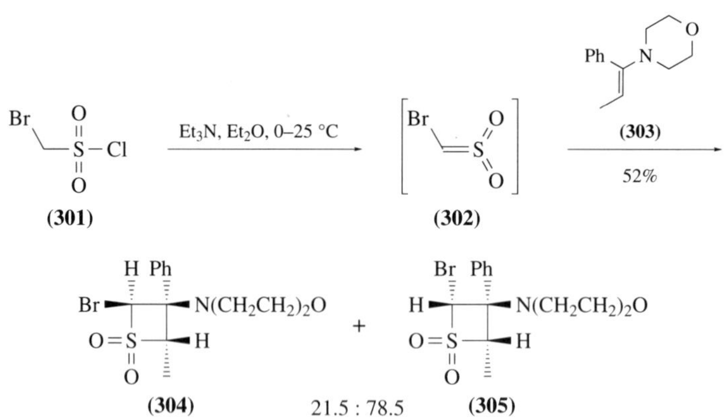

Scheme 40

β-carbon of the enamine. The reaction between chlorosulfene and the morpholine enamine of propiophenone (**287**; $R^2 = Ph$, $R^3 = H$, $R^4 = Me$) produces only two of the possible four diastereoisomers, demonstrating the stereoselectivity frequently observed in these reactions. The prod-

Table 19 Halosulfene adducts prepared according to Scheme 39.

Compound	R^1	R^2	R^3	R^4	*X*	*Yield* (%)
(288)		Ph	H		Cl	8
(289)		Ph	Me		Cl	23
(290)	H	H	Me	Me	Cl	52
(291)	H	H	Me	Me	Br	67.5
(292)	H	H	Me	Me	I	86.4
(293)	Me	H	Me	Me	Cl	27
(294)	H	Ph	H	Me	Cl	38.4
(295)	H	H	Me	Me	Cl	10.5
(296)	H	H	Me	Me	Br	29
(297)	H	H	Me	Me	I	6.8
(298)	Me	H	Me	Me	Cl	6.8
(299)	H	Ph	H	Me	Cl	9.6
(300)	H	Ph	H	H	Cl	70

uct of the reaction between chlorosulfene and the morpholine enamine of acetophenone (**287**; $R^2 = Ph$, $R^3 = R^4 = H$) was assumed to be a mixture of both *cis* and *trans* diastereoisomers, by analogy with the other cycloadducts (**290**)–(**294**) (*cis*) and (**295**)–(**299**) (*trans*). Since later work showed that the *trans* isomers (**295**), (**296**), and (**297**) are more stable than their *cis* counterparts (**290**), (**291**), and (**292**), although the latter are formed in higher amounts, the cycloaddition is clearly under kinetic control. This should be contrasted with the reaction of the bromosulfene (**302**) with enamine (**303**) (≡**287**; $R^2 = Ph$, $R^3 = H$, $R^4 = Me$), which gives more of the more stable *trans* isomer (**305**). In this case it was not possible to obtain the isomer (**304**) in a pure stable state (Scheme 40) ⟨74JCS(P2)1483, 74JCS(P2)1676⟩. Methylchlorosulfene adducts of cyclic enamines (**270a**) and (**270b**) and acyclic enamine (**287**; $R^2 = H$, $R^3 = R^4 = Me$) were later prepared under almost identical conditions to those described above (Scheme 41) ⟨78JCR(M)5501, 78JCR(S)484⟩.

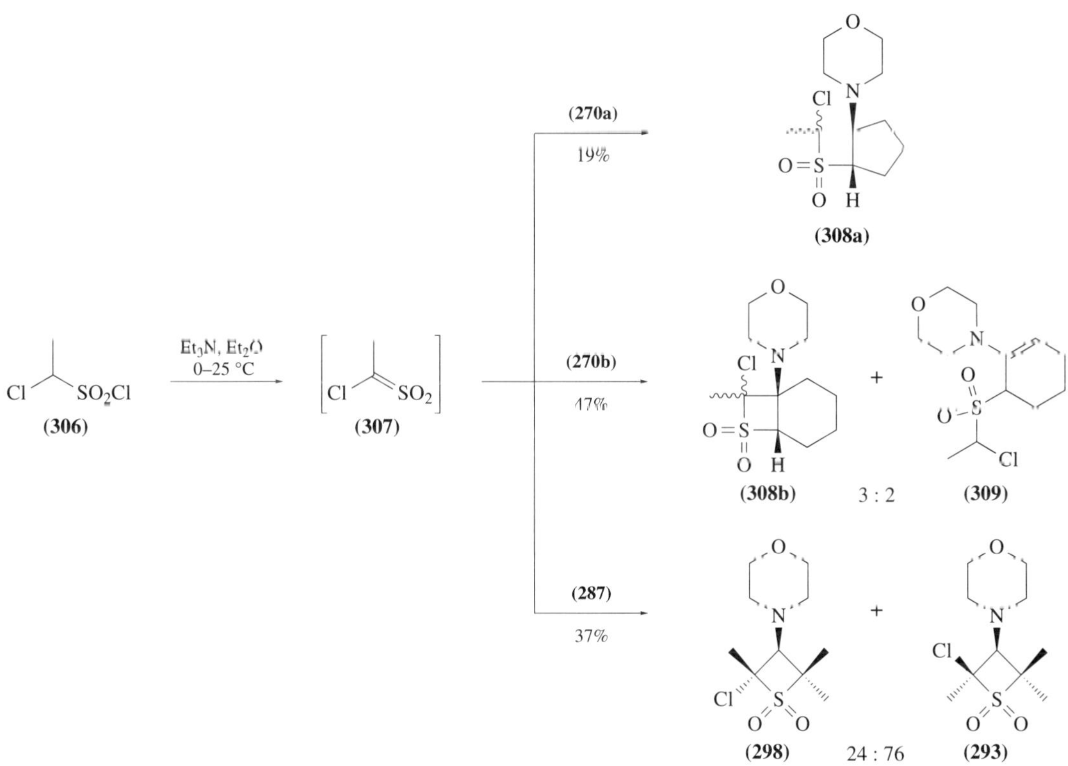

Scheme 41

In contrast to a number of other sulfenes examined, chlorosulfene (**267**) does not form cycloadducts with ketene-*N*,*N*-acetals. Instead, acyclic sulfonation products are obtained (Scheme 42) ⟨65LA(684)103⟩. The reaction is postulated to proceed through a zwitterionic intermediate (**316**), which can close to form the products of a formal cycloaddition, or undergo prototropism to yield the sulfonation product (Scheme 43) ⟨65LA(684)103, 67AG(E)107⟩. The immediate products of sulfonation (**318**) may be hydrolyzed to sulfonyl amides (**321**) or be isolated in their own right. The initially formed cycloadduct (**317**) undergoes spontaneous elimination of one mole of amine to produce the isomeric unsaturated sulfones (**319**) and (**320**; $R^2 \neq R^3$), which, since they are enamines, may be hydrolyzed under mild conditions to 2,4-dialkyl-3-oxothietane-1,1-dioxides (**322**). The formation of open-chain sulfones over cycloaddition products is favoured by unsubstituted ketene-*N*,*N*-acetals ($R^2 = H$), by groups in the sulfonyl chloride which can stabilize the negative charge on the zwitterion (**316**; e.g., $R^3 = Ph$ or Cl), by substitution of the sulfene and by polar solvents which solubilize the triethylammonium hydrochloride. In the preparation of (**313**) there was some evidence from the IR of the mother liquors from recrystallization that the corresponding cycloadducts (**319**) and (**320**; $R^2 = Me$, $R^3 = Cl$) were present ⟨65LA(684)103⟩.

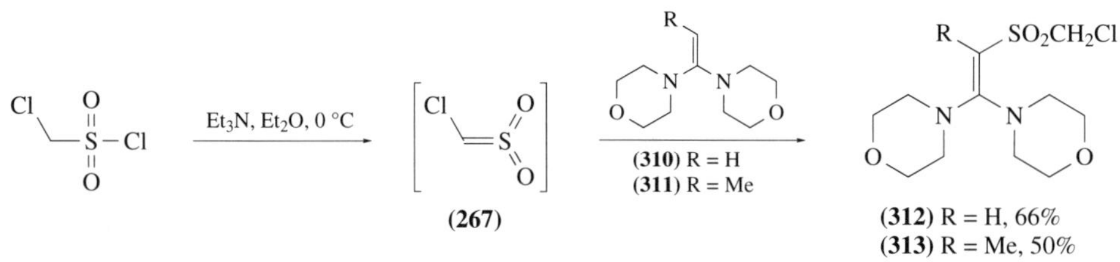

Scheme 42

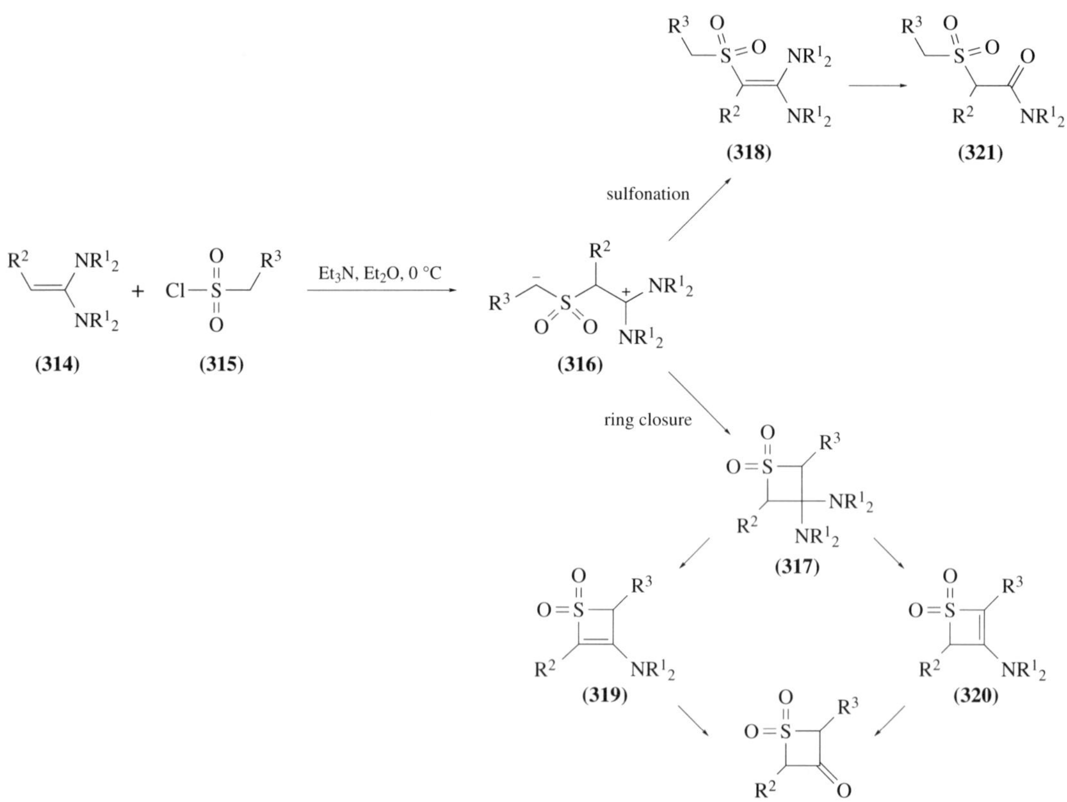

Scheme 43

In addition to their trapping as [2 + 2]-cycloadducts, halosulfenes have been trapped as [4 + 2]-cycloadducts by reaction with aminomethylene ketones of various types (Schemes 44–46, Table 20) ⟨83JHC1549, 86JHC455, 87JHC633⟩. Referring to Scheme 44, with the exception of (**326**), the *cis* cycloadducts could not be isolated in pure form. The existence and spectroscopic properties of (**324**), (**325**), and (**327**) were inferred from the mixtures with their respective *trans* isomers (**328**), (**329**), and (**332**). The *cis* isomers in the remaining cases, corresponding to *trans* isomers (**330**), (**331**), and (**334**), were of insufficient stability to survive even the first step of isolation. The crude mixtures, or indeed pure *trans* isomers, could be dehydrochlorinated in yields of 50–85% by treatment with DBN in refluxing benzene ⟨83JHC1549⟩.

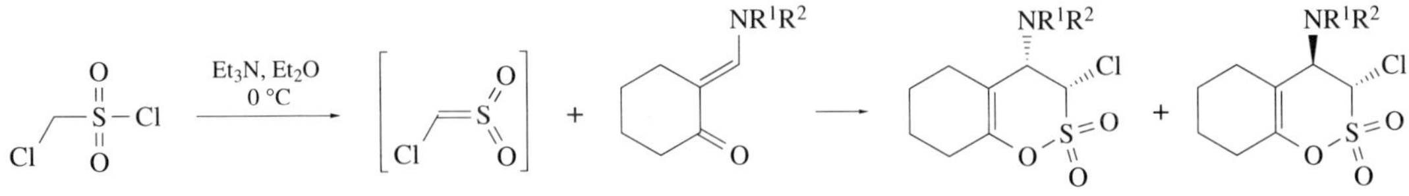

(**266**) (**267**) (**323**) (**324**)–(**327**) (**328**)–(**334**)

Scheme 44

Cycloaddition with the related enaminoketones (**335**) apparently yielded only single, *trans* isomers of products (**336**)–(**340**) in fair to good yields, with the exception of (**336**), where a mixture of both isomers was obtained with a ratio of *trans*:*cis* of 3:2. Only (**336**) could be obtained pure by chromatography. Dehydrochlorination of the immediate products of cycloaddition was effected under similar conditions to those described above in good yields, 75–85%, in all cases (Scheme 45) ⟨86JHC455⟩.

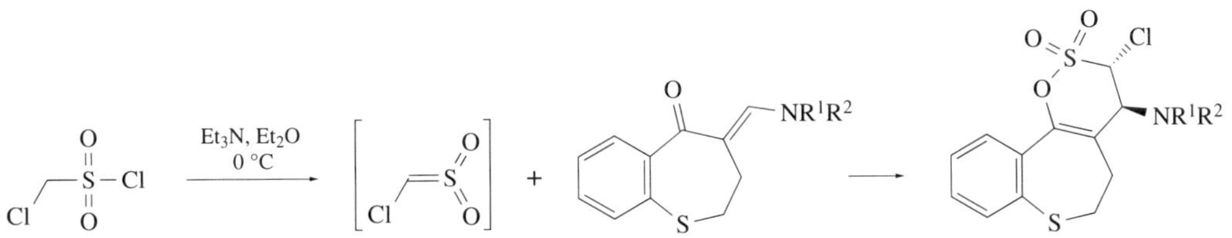

(**266**) (**267**) (**335**) (**336**)–(**340**)

Scheme 45

In the case of enaminones (**341**), the initially formed cycloadducts were not isolated or characterized, but immediately hydrolyzed *in situ*, as described above to produce the unsaturated derivatives (**343**)–(**346**) (Scheme 46) ⟨87JHC633⟩.

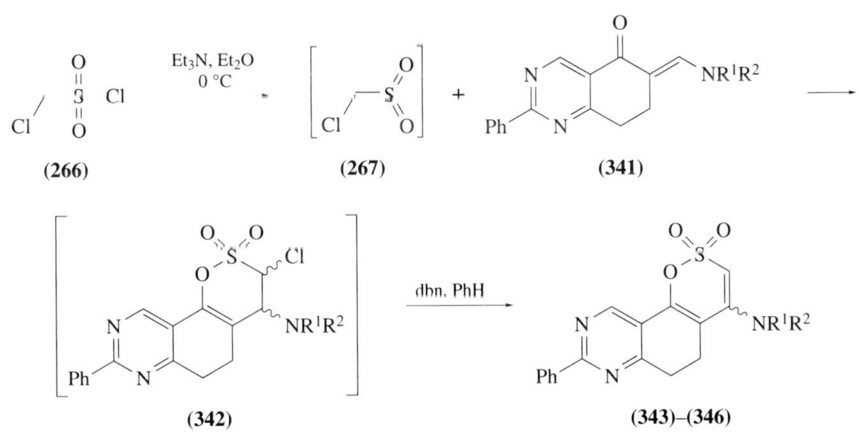

Scheme 46

Table 20 Chlorosulfene adducts prepared according to Schemes 44–46.

Compound	R^1	R^2	*Yield* (%)	*Ref.*
(324)	Me	Me	[a]	83JHC1549
(325)	Et	Et	[a]	83JHC1549
(326)	—$(CH_2)_5$—		16	83JHC1549
(327)	—$(CH_2)_2O(CH_2)_2$—		[a]	83JHC1549
(328)	Me	Me	10	83JHC1549
(329)	Et	Et	25	83JHC1549
(330)	—$(CH_2)_4$—		25	83JHC1549
(331)	—$(CH_2)_5$—		42	83JHC1549
(332)	—$(CH_2)_2O(CH_2)_2$—		33	83JHC1549
(333)	—$(CH_2)_2NMe(CH_2)_2$—		51	83JHC1549
(334)	Me	Ph	33	83JHC1549
(336)	Me	Me	88[b]	86JHC455
(337)	Et	Et	54	86JHC455
(338)	—$(CH_2)_4$—		41	86JHC455
(339)	—$(CH_2)_5$—		83	86JHC455
(340)	—$(CH_2)_2O(CH_2)_2$—		54	86JHC455
(343)	Me	Me	70	87JHC633
(344)	Et	Et	72	87JHC633
(345)	—$(CH_2)_5$—		57	87JHC633
(346)	Me	Ph	40	87JHC633

[a] Not isolated in pure form, see text. [b] As mixture with *cis* isomer: *trans* : *cis* ratio ≅ 3 : 2, see text.

Finally, mention must be made of "amine-stabilized" sulfenes, or "amine–sulfene complexes". The first of these **(349)** was prepared by treatment of 1,2,2,2-tetrafluoroethanesulfenyl fluoride with pyridine at reflux (Scheme 47) ⟨71BAU959, 72AG(E)583⟩. More recently, the triethylamine stabilized fluorosulfene adduct **(351)** was isolated and characterized by ^{19}F NMR, ^1H NMR and x-ray crystallography (Equation (31)) ⟨90AG(E)60⟩. Earlier attempts to prepare the analogous trimethylamine or quinuclidine complexes failed to yield crystalline, isolable products ⟨85CB2208, 88CB1435⟩. Compound **(349)** is considered to be a "normally" stabilized sulfene (Equation (32)), wherein the principal contribution to the stabilization is thought to be a nitrogen–sulfur interaction, *i.e.* **(352)**. Conversely, the triethylamine–fluorosulfene adduct **(351)** is considered to be an "inverse" amine-stabilized sulfene (Equation (33)), wherein the principal stabilizing interaction is thought to be between nitrogen and carbon, *i.e.* **(353)**. These distinctions have been used to account for the differences in the reactions of the two types of amine adduct towards nucleophiles ⟨71BAU959, 90AG(E)60⟩.

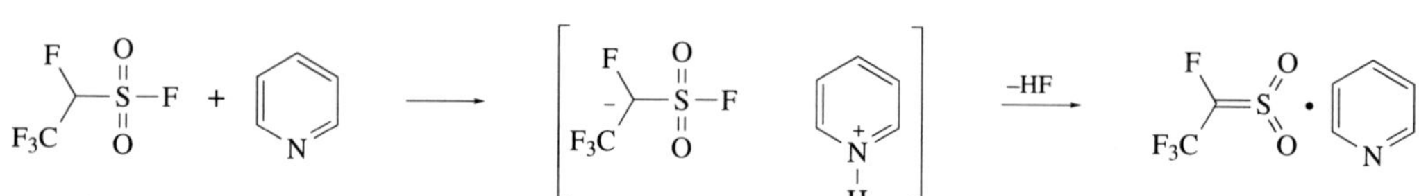

(347) **(348)** **(349)**

Scheme 47

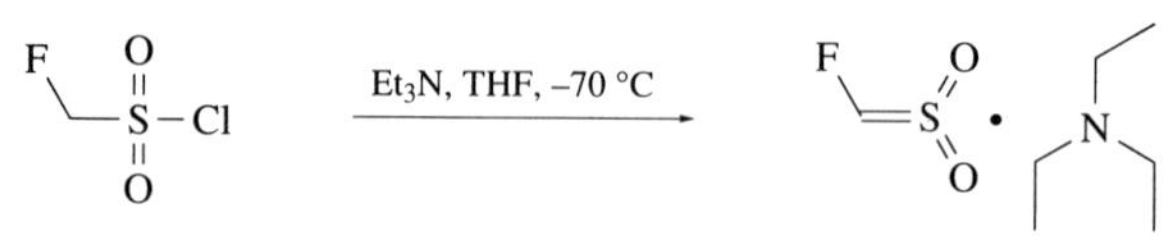

(350) **(351)** (31)

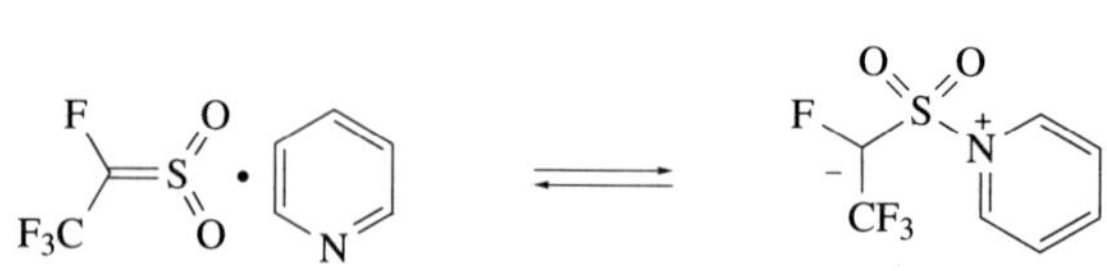

(349) **(352)** (32)

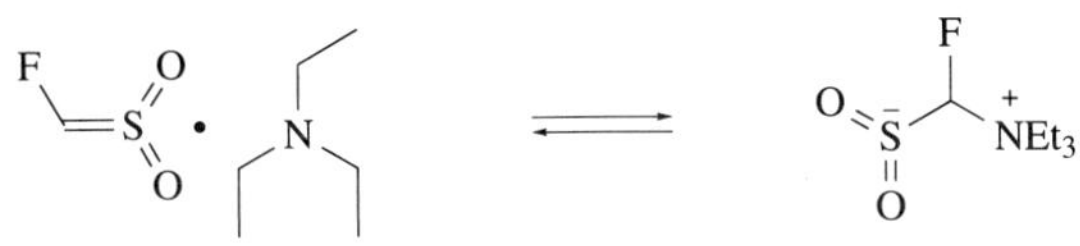

(351) (353)

(33)

5.12
Thio, Seleno, and Telluro Acyloxy Functions, $R^1C(S)OR^2$, $R^1C(Se)OR^2$, $R^1C(Te)OR^2$, etc.

AKIHIKO ISHII and JUZO NAKAYAMA
Saitama University, Japan

5.12.1 CARBOXYLIC ACID ANALOGUES, RC(Y)OH

5.12.1.1 General Aspects

Thiocarboxylic acids can exist as a tautomeric mixture of thiol and thione forms. While the equilibrium generally lies exclusively to the right (Scheme 1) (see Section 5.05.2), the presence of small amounts of the thione forms of thioformic and trichlorothioacetic acids are detectable spectroscopically in an equilibrium mixture ⟨72ZAAC78, 74JST123⟩. *Ab initio* calculations have been carried out for some thiocarboxylic acids to obtain optimized geometries, atomic charges, relative stabilities, and harmonic force fields for s-*cis* and s-*trans* conformers of the thiol and thione tautomers ⟨89JCS(F2)1945, 90JST67⟩.

Scheme 1

There are only two reports on selenocarboxylic acids ⟨72ACS1465, 87NKK1430⟩. However, they have the *Se*-acid structure (**1**) rather than the *O*-acid one (see Section 5.05.2). In the case of tellurocarboxylic acids, no report has appeared up to 1994 on the preparation of the free acids, although their alkali metal or ammonium salts (**2**) have been prepared ⟨87CL475, 89TL1829, 90TL3587, 92JAP(K)04224595, 92MI 512-01⟩.

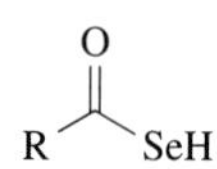

(1)

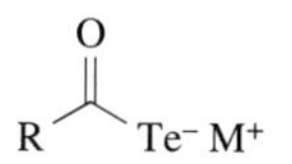

(M = Li, Na, K, $NR^2{}_3$)

(2)

5.12.1.2 Thiocarboxylic *O*-Acids

5.12.1.2.1 Spectroscopic methods

Engler and Gattow prepared thioformic acid (**3**) by reaction of phenyl formate (**4**) with sodium or potassium hydrosulfide, followed by treatment with hydrochloric acid (Scheme 2) ⟨72ZAAC(388)78⟩. The IR spectrum of (**3**) suggested the presence of both thiol (**3a**) and thione (**3b**) forms in the tautomeric equilibrium at room temperature. Thus, absorptions assigned to (**3b**) appeared at 1402, 1235 (COH in-plane bending) and 1160 (C═S stretching) cm^{-1} ⟨72ZAAC(388)78, 74JST123⟩, although later *ab initio* calculations at the 6-31G* level are not in agreement with the assignment (*vide infra*) ⟨89JCS(F2)1945⟩.

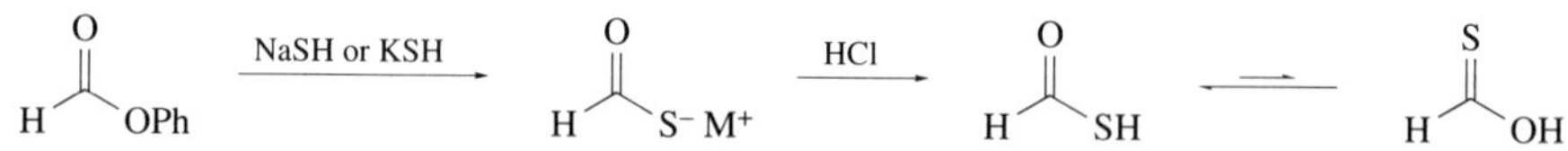

(4) (M = Na or K) **(3a)** **(3b)**

Scheme 2

An IR spectroscopic study of thioacetic acid was reported in the late 1940s ⟨49TFS693, 56CB1110⟩, and IR and ^1H NMR spectra of trichlorothioacetic acid (**5**) were investigated more recently in relation to the thiol–thione equilibrium ⟨77JST25⟩. The IR spectrum of (**5**), prepared by reaction of trichloroacetyl chloride (**6**) with hydrogen sulfide in the presence of $AlCl_3$ at $-20\,^\circ C$ (Scheme 3), showed absorptions due to the thione form (**5b**) at 3570 (OH stretching), 1160 (C═S stretching), 1410 (C—O stretching), 1060 (C—C stretching) and 1240 (C—O—H bending) cm^{-1}. In the ^1H NMR spectrum, determined in the temperature range 75–93 °C, a very weak signal at δ 2.36, which was assigned to the OH proton of (**5b**), was observed. The percentage of (**5b**) was only about 1%.

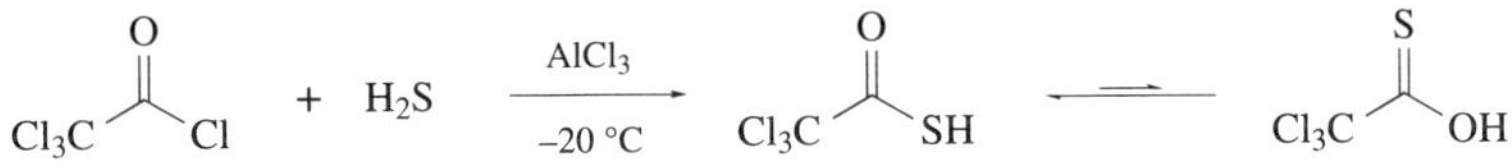

(6) **(5a)** **(5b)**

Scheme 3

5.12.1.2.2 Theoretical methods

Fausto *et al.* reported *ab initio* SCF–MO calculations for s-*cis* and s-*trans* conformers of both the tautomers of thioformic acid (**3**) using the 6-31G* basis set ⟨89JCS(F2)1945⟩. The 6-31G* calculations suggest relative stabilities of the thioformic acid conformers in the order of s-*cis* thiol (***cis*-3a**) (0.0 kJ mol^{-1}) > s-*trans* thiol (***trans*-3a**; 6.3) > s-*cis* thione (***cis*-3b**; 13.0) > s-*trans* thione (***trans*-3b**; 40.2). For both tautomers, the s-*cis* form is more stable than the s-*trans* form, and the energy difference is larger in the thione forms (**3b**) (27.2 kJ mol^{-1}) compared with that in the thiol forms (**3a**; 6.3 kJ mol^{-1}). The effects which make s-*cis* forms more stable than s-*trans* forms have been interpreted in terms of the extended s-*cis* π-electron delocalization, and the occurrence of intramolecular hydrogen bonding in the s-*cis* forms (Scheme 4) ⟨87JST381⟩. In addition, the increase of the energy difference on going from the thiol (**3a**) to thione (**3b**) forms has been ascribed to the more effective conjugating properties of the oxygen 2*p* orbitals in (**3b**; X = S, Y = O) as compared to those of the sulfur 3*p* orbitals in (**3a**; X = O, Y = S), as well as the stronger O—H···S hydrogen bonding interaction in (**3b**) compared with the S—H···O interaction in (**3a**).

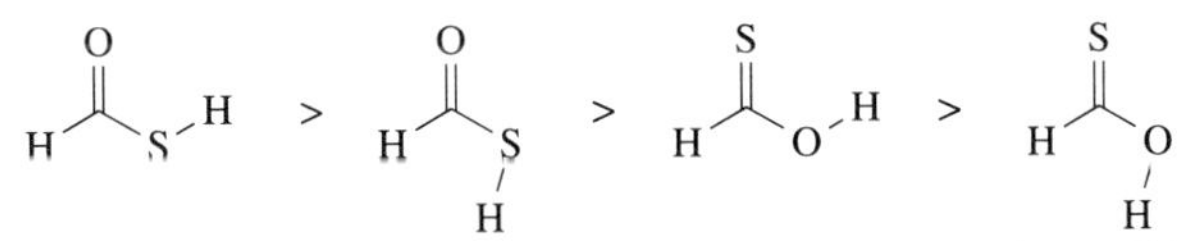

(***cis*-3a**) (***trans*-3a**) (***cis*-3b**) (***trans*-3b**)

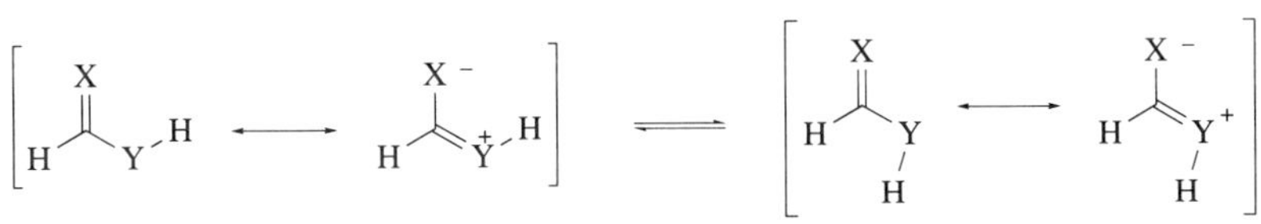

(***cis*-3a**); (X = O; Y = S)
(***cis*-3b**) (X = S; Y = O)

(***trans*-3a**); (X = O; Y = S)
(***trans*-3b**) (X = S; Y = O)

Scheme 4

The fully optimized geometries and calculated vibrational frequencies and the potential energy distribution (PED) of thioformic *O*-acids (**3b**) are given in Figure 1 and Table 1 ⟨89JCS(F2)1945⟩. It is noteworthy that absorptions at 1235 and 1160 cm^{-1}, which were previously assigned to COH in-plane bending and C=S stretching modes, respectively ⟨72ZAAC(388)78, 74JST123⟩, are due to CO stretching and COH bending modes, respectively. The calculated C=S stretching frequency of s-*cis* thioformic *O*-acid (***cis*-3b**) is 935 cm^{-1}.

Ab initio SCF–MO calculations at the 4-31G* level on the relative stability of the tautomers, and

Table 1 Calculated[a] and observed[b] vibrational frequencies (cm^{-1}) and PEDs of s-*cis* thioformic *O*-acid (***cis*-3b**).

Observed	*Ab initio*	*Scaled*	*Symmetry*	*PED*
	4018	3548	1A′	νOH[100]
	3384	2989	2A′	νCH[100]
1402	1627	1460	3A′	δCH[89] + νC=S[12] + δCOH[10] + νCO[8]
1235	1428	1238	4A′	νCO[76] + νC=S[12] + δCOH[10]
1160	1355	1160	5A′	δCOH[66] + δCH[16] + νC=S[10]
	1078	962	1A″	γCH[97]
	1027	935	6A′	νC=S[67] + δCOH[14] + νCO[13] + δSCO[9]
666	731	666	2A″	τCO[121] + γCH[24]
	511	443	7A′	δSCO[10] + δCH[15] + δCOH[6]

[a] 89JCS(F2)1945. [b] 72ZAAC78, 74JST123.

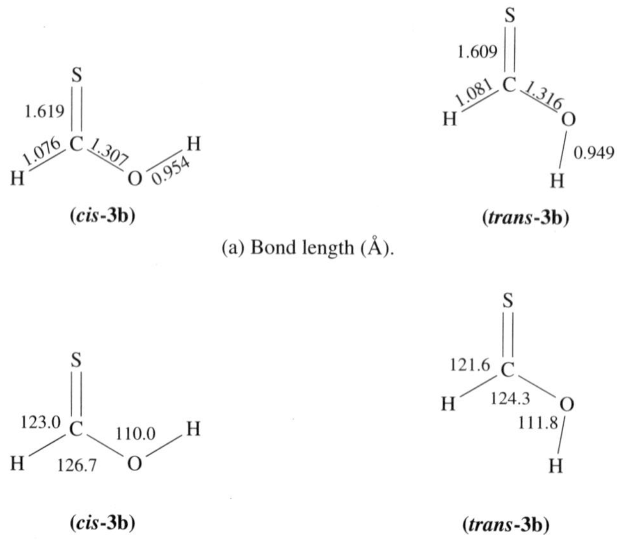

Figure 1 6-31G* calculated geometries of s-*cis* and s-*trans* conformers of thioformic *O*-acid **(3b)**.

their conformers, of thioacetic acid gave similar results to those for thioformic acid at the 6-31G* level ⟨90JST67⟩. Thus, the most stable form is the s-*cis* thiol form (0 kJ mol^{-1}) and the stability decreases in the order of s-*trans* thiol (10.10), s-*cis* thione (12.05) and s-*trans* thione (47.22) forms. Optimized geometries for s-*cis* (***cis*-7**) and s-*trans* (***trans*-7**) thione forms are depicted in Figure 2.

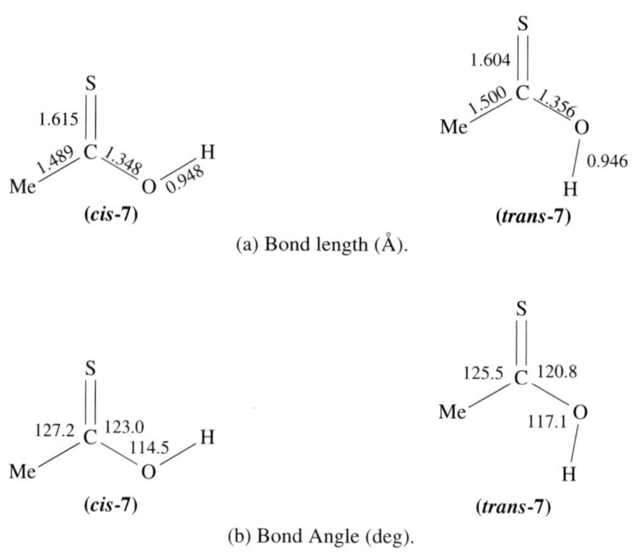

Figure 2 4-31G* calculated geometries of s-*cis* and s-*trans* conformers of thioacetic *O*-acid **(7)**.

5.12.1.3 Selenocarboxylic *O*-Acids

Reaction of benzoyl chloride with hydrogen selenide in pyridine gave selenobenzoic acid **(8)** ⟨72ACS1465⟩. Hirabayashi *et al.* reported that the reaction of sodium 4-biphenylselenocarboxylate or sodium 1-adamantylselenocarboxylate with hydrogen chloride yielded **(9)** and **(10)**, respectively

⟨87NKK1430⟩. These selenocarboxylic acids (**8**)–(**10**) are extremely air-sensitive materials, and their IR and ^{1}H NMR spectra only showed absorptions characteristic of selenocarboxylic *Se*-acids but not *O*-acids.

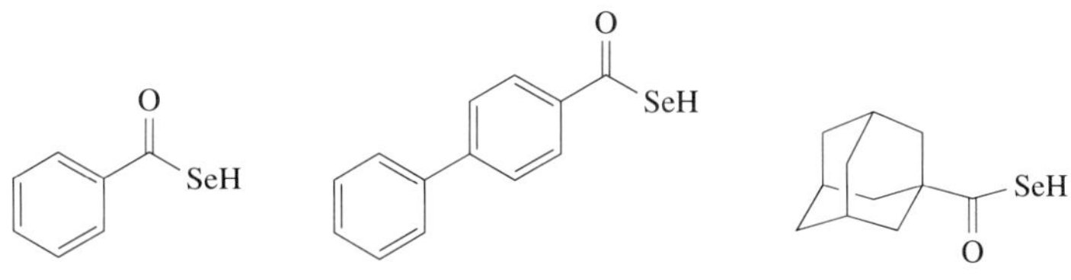

(**8**) (**9**) (**10**)

5.12.1.4 Tellurocarboxylic *O*-Acids

The free acids have not been prepared so far. Only their alkali metal or ammonium salts (**2**) were obtained as highly air-sensitive materials by reaction of diacyl tellurides (**11**) with primary and secondary amines or EtOM (M = Na, K) ⟨87CL475, 90TL3587⟩ and by reaction of acyl chloride with Li_2Te ⟨92JAP(K)04224595, 92MI 512-01⟩ or Na_2Te ⟨89TL1829⟩ (Scheme 5). The salts (**2**) have the unsymmetrical ionic structure in which the negative charge is localized on the tellurium atom ⟨90TL3587⟩.

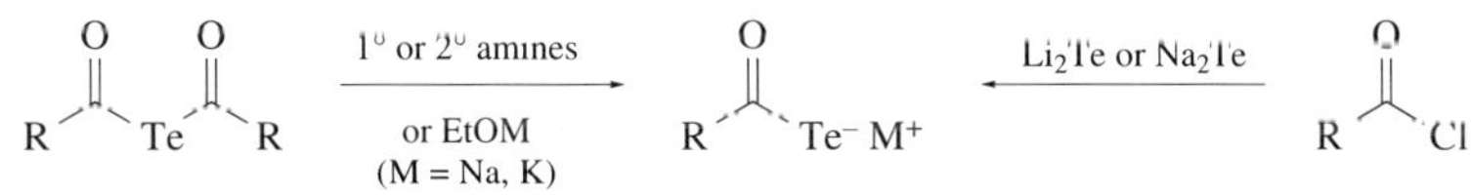

Scheme 5

5.12.1.5 Higher Oxidation States of the Chalcogen

No reports have appeared on compounds bearing higher oxidation states of the chalcogen atoms of thio-, seleno-, and tellurocarboxylic *O*-acids.

5.12.2 CARBOXYLIC ESTERS ANALOGUES, $R^1C(Y)OR^2$

5.12.2.1 General Methods

The preparation of *O*-alkyl or *O*-aryl thio-, seleno-, and tellurocarboxylates (**12**) has been roughly divided into three classes on the basis of the final step of bond formation (Scheme 6) ⟨B-92MI 512-01⟩.

Class A involves, for example, the reaction of thioacyl chlorides, bisthioacyl sulfides, dithio-

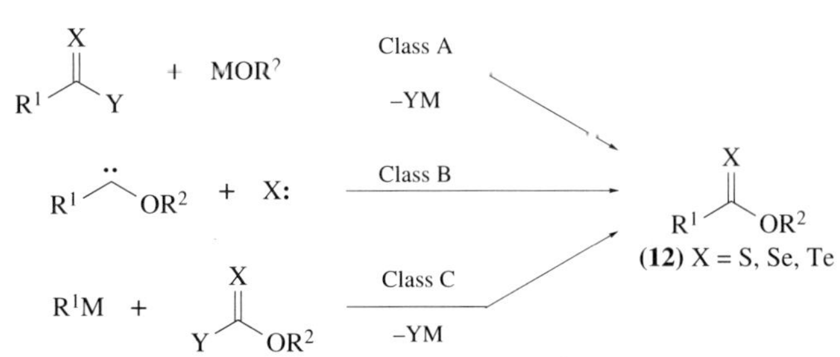

Scheme 6

carboxylates or *O*-alkyl thiocarboxylates (**13**) with alcohols and alkoxides. The reactions of thio- and selenoketenes (**14**) with alcohols are also included in this class (Scheme 7).

(**13**) Y = Cl, SR^3, R^3CS_2 etc. M = Na, K

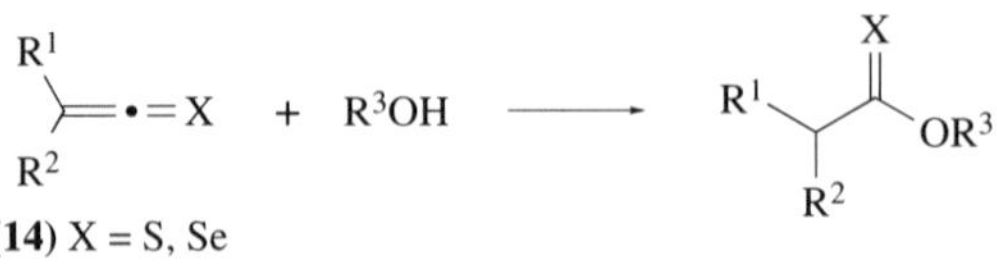

(**14**) X = S, Se

Scheme 7

Thionation, selenation, and telluration of imidate esters or their salts (**15**) with H_2X or NaHX (X = S, Se, Te), and direct thionation or selenation of esters belong to class B (Scheme 8).

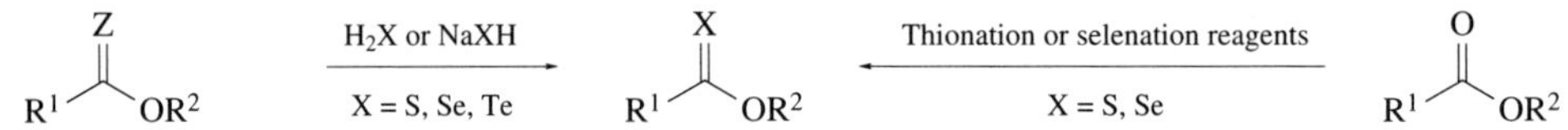

(**15**) Z = $N^+R^3{}_2\,Cl^-$, NH•HCl (**12**)

Scheme 8

A few reactions classified as class C are known. The reaction of alkoxythiocarbonyl chlorides (**16**) with aromatic compounds under the Friedel–Crafts conditions, and the reaction of carbanions with alkoxythiocarbonyl compounds (**17**) belong to this class (Scheme 9) ⟨B-92MI 512-01⟩.

(**16**)

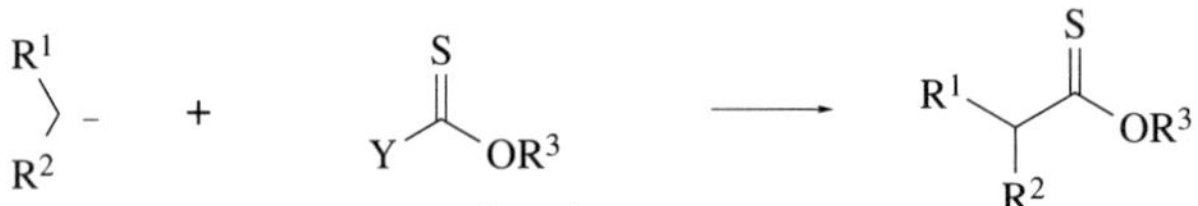

(**17**) Y = Cl, R^4S, R^4OCS_2

Scheme 9

Thiocarboxylates (**12**; X = S) are no longer rare, and their synthetic methods and chemical and physical properties have been well summarized in several review articles ⟨79COC(3)373, 84CRV17, 91COS435, B-92MI 512-01⟩. Therefore, their synthesis is treated briefly in Section 5.12.2.2 according to the above classification with emphasis being placed on results since the beginning of the 1980s.

On the other hand, synthetic methods for seleno- and tellurocarboxylates (**12**; X = Se, Te) have been restricted to class B with only a few exceptions ⟨86OPP369, 91COS(6)461⟩. Sections 5.12.2.3 and 5.12.2.4 cover the synthesis and chemical and physical properties of (**12**; X = Se, Te) as extensively as possible.

5.12.2.2 Thionocarboxylic Esters, $R^1C(S)OR^2$

5.12.2.2.1 Class A

The reaction of thioacyl halides (**18**) with sodium or potassium alkoxides gives the corresponding thiocarboxylates (**19**) (Equation (1)) ⟨63JPR214, 81BRP1589128⟩. Bis(thioacyl) sulfides (**13**) (Y = R^3CS_2) ⟨82LA1229, 90BCJ638⟩ and dithiocarboxylates (**13**; Y = R^3S) ⟨74RTC51⟩ also act as thioacylating reagents ⟨B-92MI 512-01⟩. Voss and coworkers prepared isotope-labeled compounds (**20**) by using a similar methodology (Equation (2)) ⟨79JCR(S)160⟩.

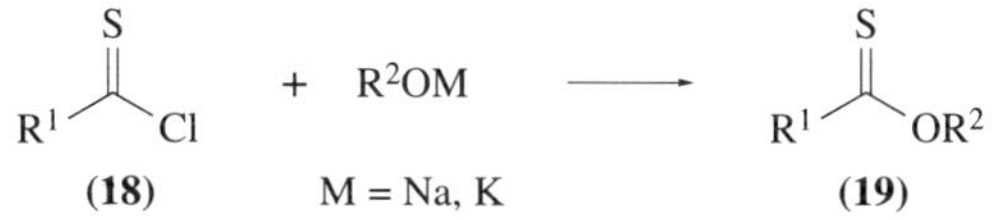

(1)

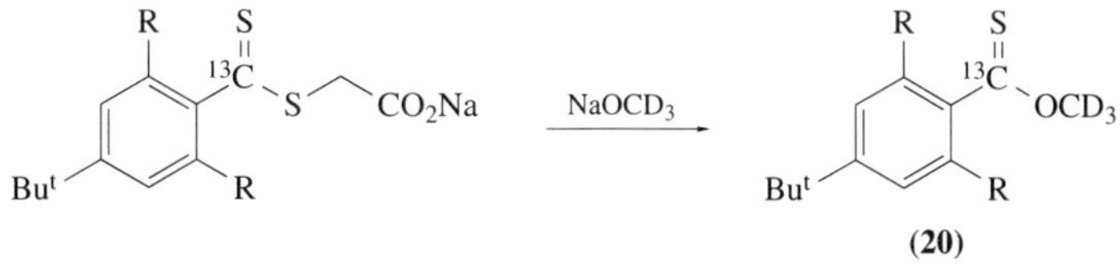

(2)

Transesterification of bis(alkoxythiocarbonyl) compounds with diols gives macrocyclic compounds ⟨83JOC2635⟩. For example, the reaction of *O*,*O*′-dimethyl 2,6-pyridinedicarbothioate (**21**) with glycols (**22**) in benzene using alkali methoxides as catalysts gave dithiono crowns (**23**); (10–71%), which were then converted into crown ethers (**24**) by desulfurization with Raney nickel (Scheme 10).

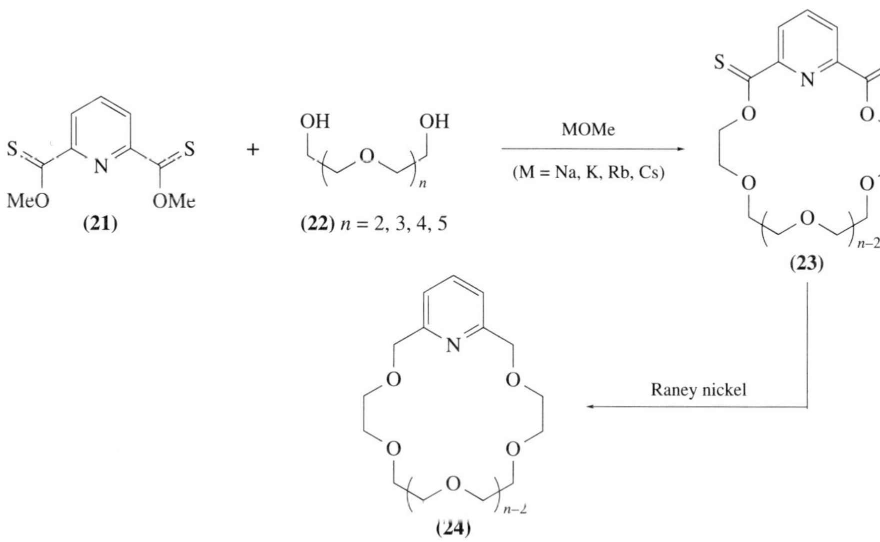

Scheme 10

Addition of alcohols to thioketene intermediates (**25**), generated from 1,2,3-thiadiazoles (**26**) ⟨77CB1225, 77T449, 78JHC697, 79JHC1295, 80CB2519⟩ or 1,3-dithietane derivatives (**27**) ⟨77CB1225⟩, gives the corresponding *O*-alkyl (aryl) thiocarboxylates (**28**) in moderate to high yields (Scheme 11) ⟨84CRV17⟩. Isolable thioketenes such as di-*t*-butyl ⟨68JOC2738⟩ and bis(trifluoromethyl) derivatives ⟨70JOC3470, 72JOC1347⟩, of course, yield (**28**; $R^1 = R^2 = Bu^t$, CF_3), but bis(trimethylsilyl)thioketene (**29**) gave a desilylated product (**30**) (Equation (3)) ⟨76CC1008⟩.

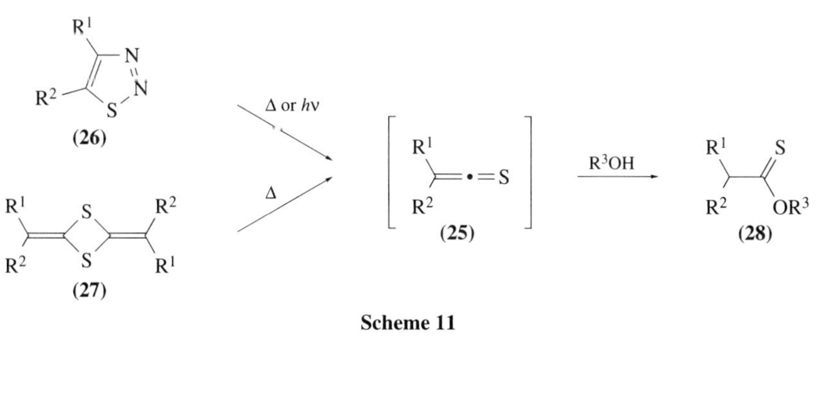

Scheme 11

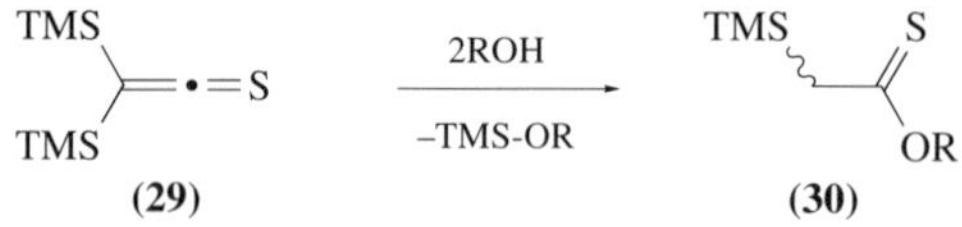

(3)

5.12.2.2.2 Class B

Sulfhydrolysis of salts (**31**) of imidate esters has been widely investigated because the salts are easily prepared from nitriles and amides as starting substrates (Scheme 12) ⟨84CRV17, 91COS435, B-92MI 512-01⟩. Yields are generally high ⟨84CRV17⟩. The reaction of dialkoxycarbenium tetrafluoroborates (**32**) with sodium sulfide or sodium hydrosulfide has also been developed ⟨84CRV17, 91COS435, B-92MI 512-01⟩. The generality and mechanistic features of the reaction have been extensively investigated by Kaloustian and Khouri (Scheme 13) ⟨79JA2249, 80JA7579, 81TL413⟩.

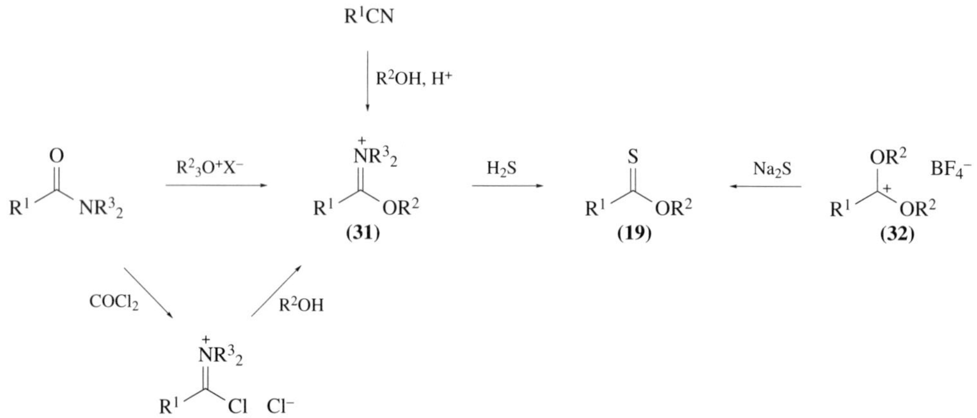

Scheme 12

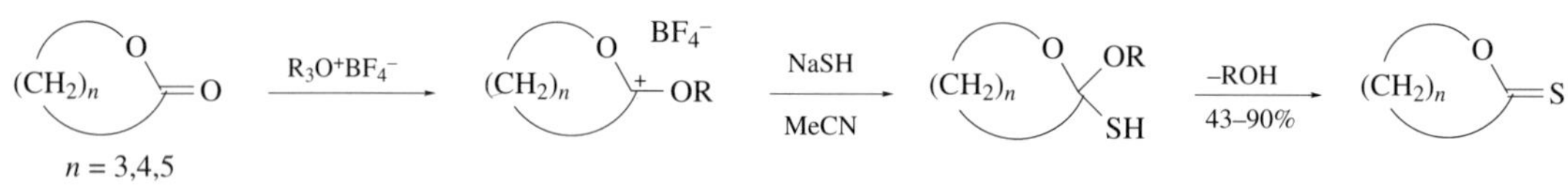

Scheme 13

Direct thionation of esters to thiocarboxylates (**19**) can be achieved by reagents such as P_4S_{10}, 1,3,2,4-dithiadiphosphetane 2,4-disulfides (**33**) (e.g., Lawesson's reagent ⟨85T5061⟩) or freshly prepared B_2S_3 (Equation (4)) ⟨84CRV17, B-92MI 512-01⟩. The yields of the reaction with P_4S_{10} are usually low ⟨73S149⟩. Thiocoumarin (**34**) was prepared by treatment of coumarin with B_2S_3, prepared *in situ* from bis(tricyclohexyltin) sulfide and BCl_3 in refluxing toluene, in 94% yield ⟨82JA3104⟩.

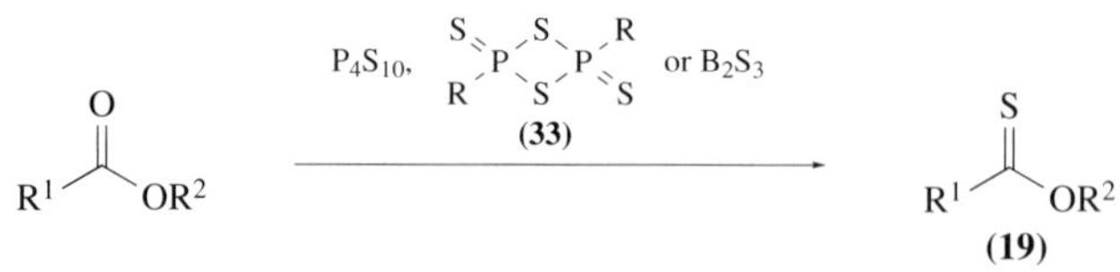

(4)

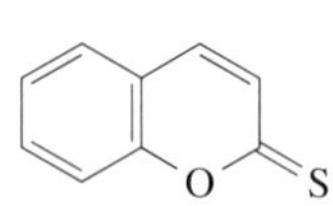

(34)

Nicolaou *et al.* investigated the conversion of lactones to the corresponding thionolactones in the synthesis of medium-sized cyclic ethers. In this study a series of 2,4-diaryl- and 2,4-bis(arylthio)-

1,3,2,4-dithiadiphosphetane 2,4-disulfides (**33**) were shown to be effective in converting 16-membered lactone (**35**) into the corresponding thiolactone (**36**) (Scheme 14) ⟨90JA3029, 90JA3040, 90JA6263⟩.

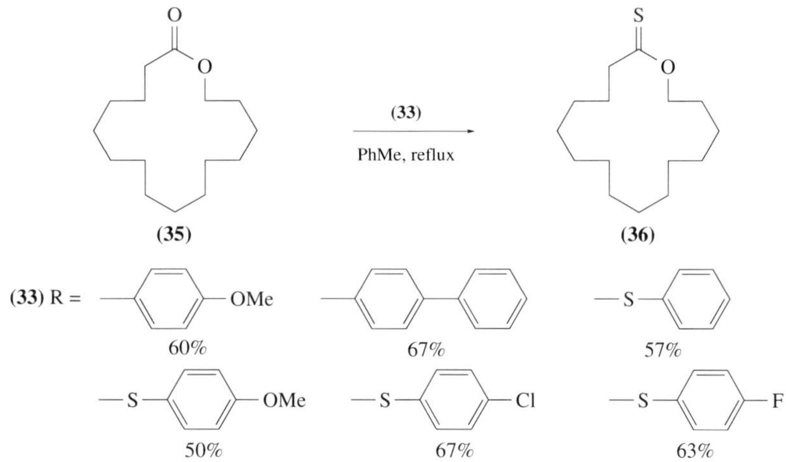

Scheme 14

A mechanistic study concludes that the thionation of carbonyl compounds by Lawesson's reagent (**33**; Ar = 4-MeOC$_6$H$_4$) can be rationalized in terms of nucleophilic attack of carbonyl oxygen on the monomer (**37**) that exists in an equilibrium mixture (Scheme 15) ⟨86TL3445⟩.

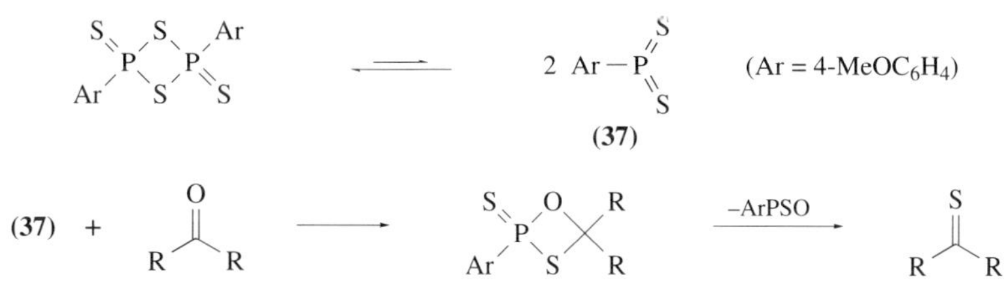

Scheme 15

5.12.2.2.3 *Class C*

Friedel–Crafts thioacylation with *O*-phenyl chlorothioformate (**38**) gives thioesters in moderate yields (Equation (5)) ⟨68CB3517, 91COS435⟩.

(5)

Ar = Ph, 15%; 2,4-(MeO)$_2$C$_6$H$_3$, 22%; *p*-C$_6$H$_5$-C$_6$H$_4$, 44%

The reaction of bis(ethoxythiocarbonyl) sulfide (**39**) with enolates gives highly enolized *O* ethyl *β*-oxothiocarboxylates (**40**) (Scheme 16) ⟨86CL1251⟩. Carbanions prepared from phosphonate (**41**) ⟨85S960⟩, hydrazones (**44**) ⟨88SC323⟩, and acetylenes (**47**) ⟨89TL1073⟩ react with *O*-alkyl chlorothioformates (**42**) or *O*-alkyl dithiocarbonates (**45**) to yield the corresponding functionalized *O*-alkyl thiocarboxylates, (**43**), (**46**), and (**48**) (Schemes 17–19) ⟨91COS435⟩.

Bachi and Bosch reported tri-*n*-butylstannyl radical-initiated radical cyclization of *O*-homoallyl dithiocarboxylates (xanthates) (**49**) leading to *γ*-thionolactones (**50**) (Scheme 20) ⟨89JOC1234, 92JOC6803⟩. The yields are strongly dependent on the substituents in the starting substrates ⟨92JOC6803⟩. The stereochemistry of the radical cyclization was investigated by a Japanese group, and found to be highly stereospecific, giving the *anti* isomer predominantly (9 : 1 ≈ 24 : 1) (Equation (6)) ⟨89CC1265, 91JCS(P1)1173, 91JOC2849⟩.

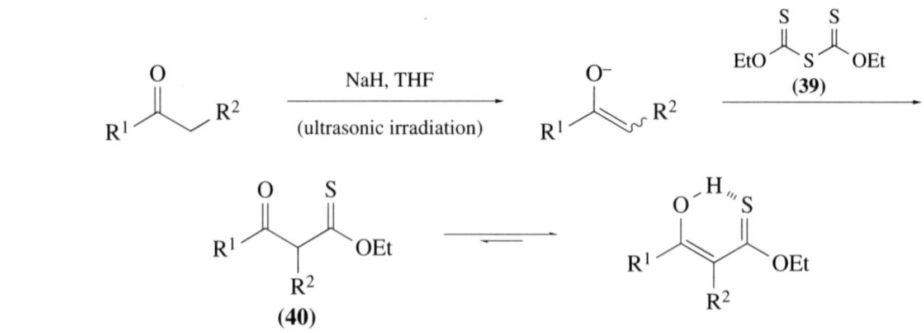

Scheme 16

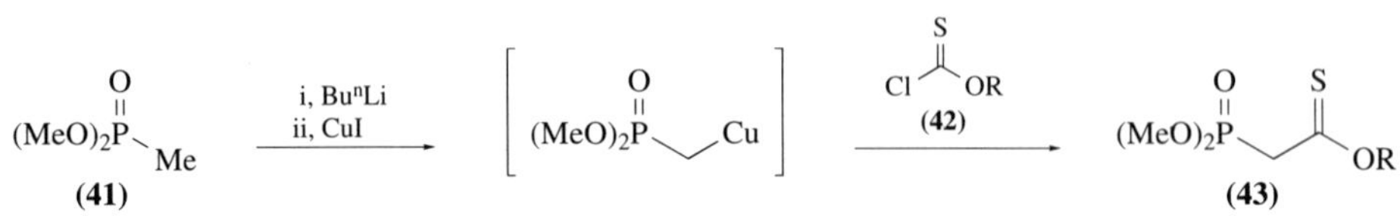

Scheme 17

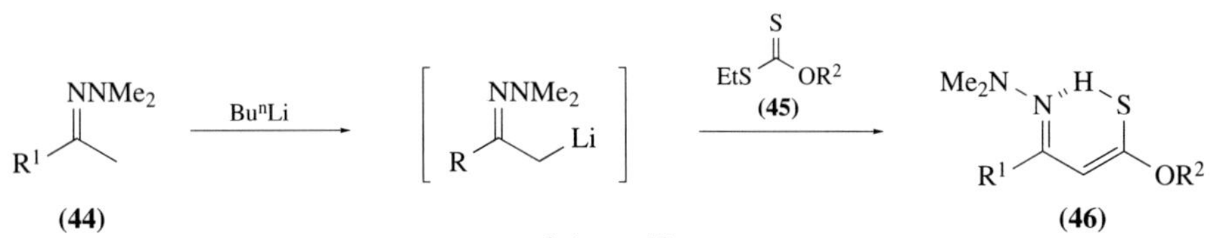

Scheme 18

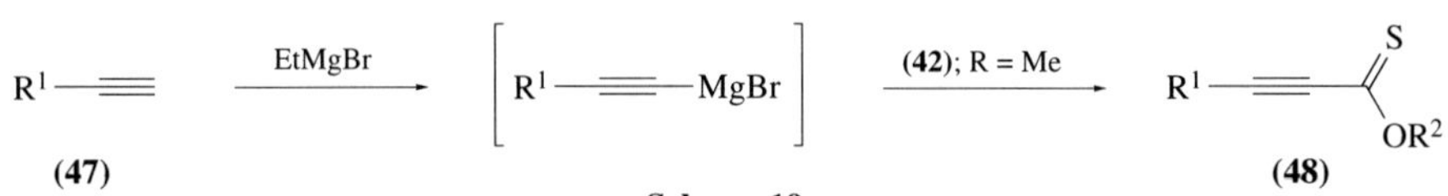

Scheme 19

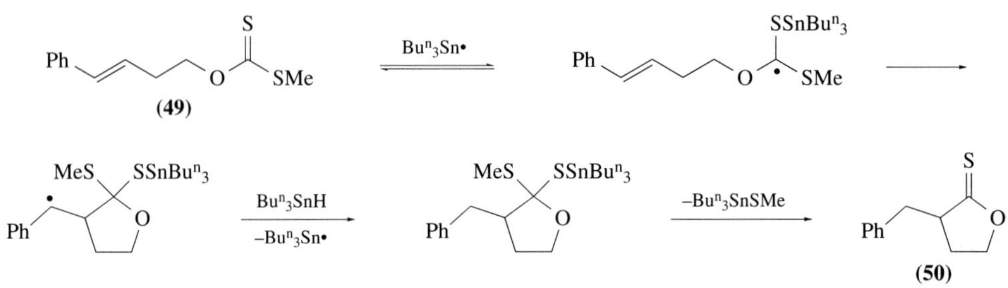

Scheme 20

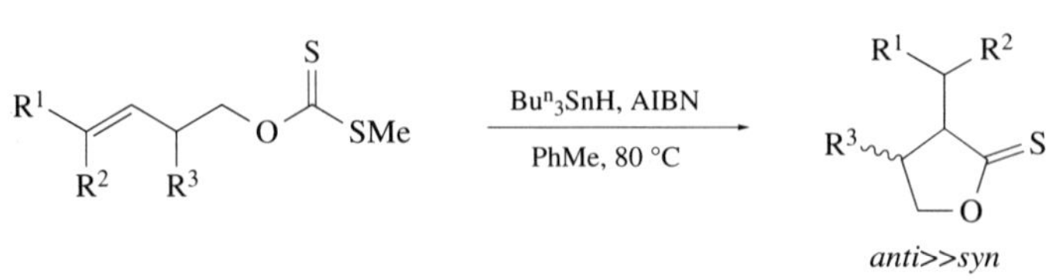 (6)

5.12.2.3 Selenocarboxylic Acid *O*-Esters, $R^1C(Se)OR^2$

5.12.2.3.1 Synthesis

Most synthetic methods for *O*-alkyl or *O*-aryl selenoesters belong to class B and can be classified further as follows:

(a) reaction of imidate esters (**51**), or their salts (**31**), with H_2Se or NaHSe (Scheme 21);

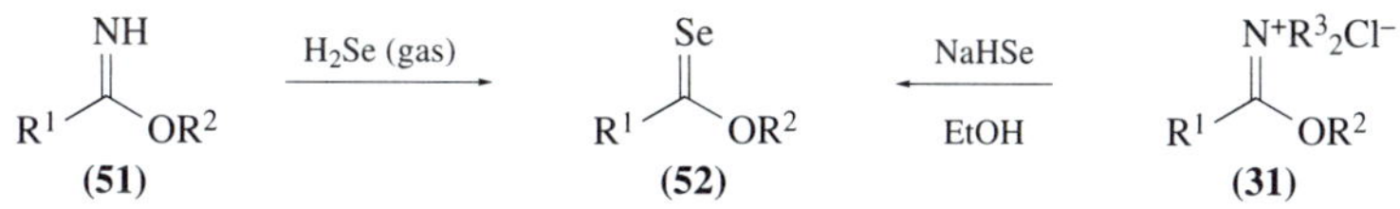

Scheme 21

(b) reaction of alkynylselenolate anions (**53**), or alkynyl selenols generated from acetylselenoalkynes (**55**), with alcohols (Scheme 22);

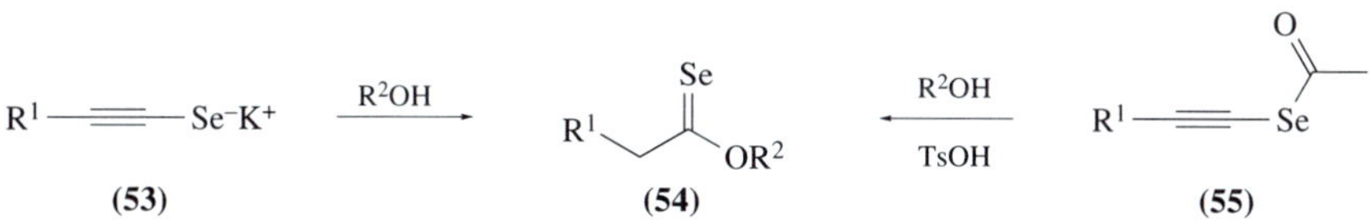

Scheme 22

(c) reaction of alkoxycarbene chromium complexes (**56**) with elemental selenium (Equation (7));

(7)

(d) reaction of esters or orthoesters with selenation reagents (Equation (8));

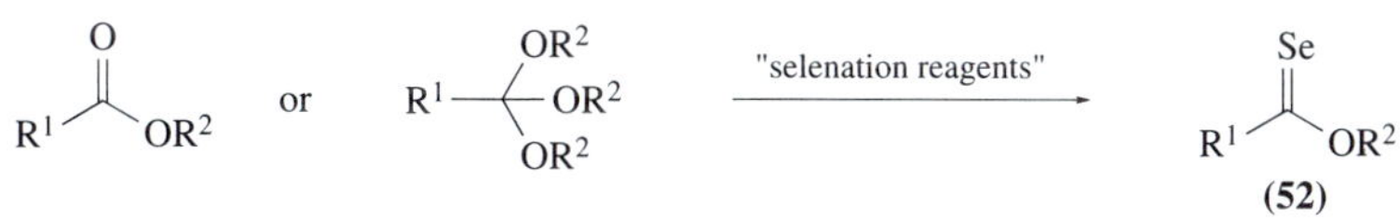

(8)

(e) miscellaneous.

(i) Reaction of imidate esters or their salts with H_2Se or NaHSe

The first preparation of *O*-alkyl selenocarboxylates was reported by Collard-Charon and Renson in 1962 ⟨62BSB563⟩. Thus, the reaction of benzonitrile with methanol and hydrogen chloride gave the imidate ester hydrochloride (**58**), which was then treated with aqueous sodium hydroxide in ether to liberate the free imidate ester (**59**). The dried solution of (**59**) was treated with gaseous hydrogen selenide, generated from aluminum selenide by the addition of water, to yield *O*-methyl selenobenzoate (**60**) as a red liquid in 5.6% yield (Scheme 23). In a similar manner, *O*-methyl 4-methylselenobenzoate was prepared though in a low yield (3%). Selenophthalide (**62**) was prepared by the reaction of the corresponding imido ester hydrochloride (**61**) with hydrogen selenide (Scheme 24) ⟨64BSB419⟩. *O*-Methyl selenobenzoate (**60**) was later directly prepared by the reaction of (**58**) with hydrogen selenide in a higher yield (36%) ⟨66CB1393⟩.

Cohen improved the method to a convenient and general one for *O*-ethyl aliphatic selenoesters (**65**)

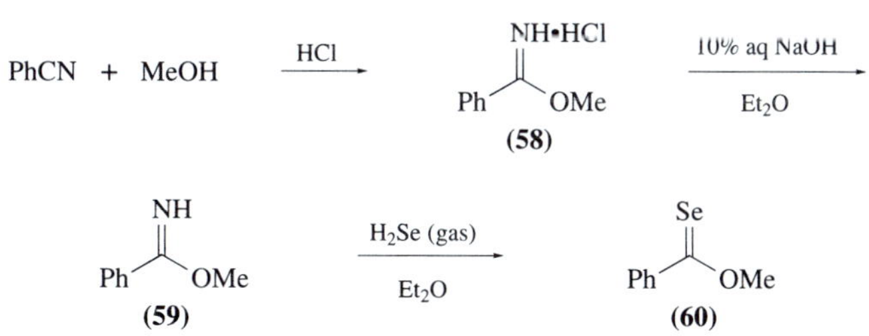

Scheme 23

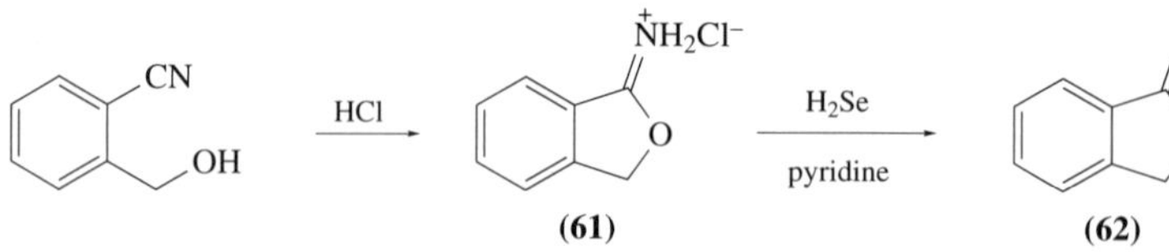

Scheme 24

by treating imidate esters (**64**) with hydrogen selenide in pyridine–triethylamine at low temperatures ($-30 \sim -20\,°C$) (Method A) (Scheme 25) ⟨77JOC2645⟩.

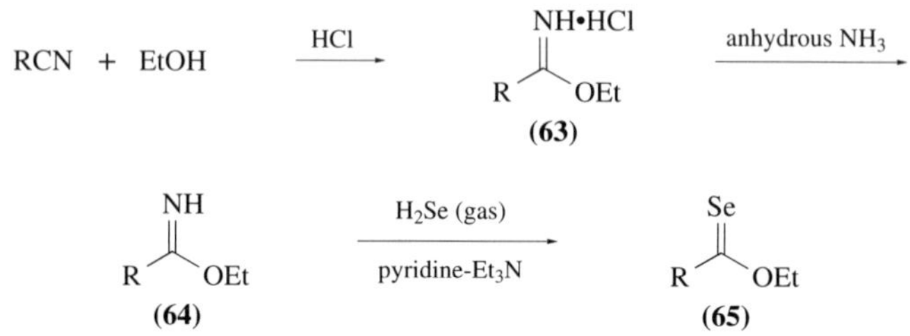

Scheme 25

Since key intermediates (**64**) have been prepared from nitriles and simple alcohols such as methanol and ethanol, the final products are inevitably limited to *O*-methyl or *O*-ethyl selenocarboxylates. The selenoesters thus synthesized are summarized in Table 2 together with those prepared by other methods mentioned below.

A more general synthesis of *O*-alkyl and *O*-aryl selenocarboxylates (**52**) was developed by Barton *et al.* ⟨75JCS(P1)1574, 77JCS(P1)1723⟩. Thus, condensation of alcohols with di-*N*-substituted imidoyl chlorides (**66**), prepared from tertiary amides and phosgene, gave the salt (**67**). Treatment of the salt with ethanolic sodium hydrogen selenide afforded (**52**) in moderate to high yields (Method B) (Scheme 26).

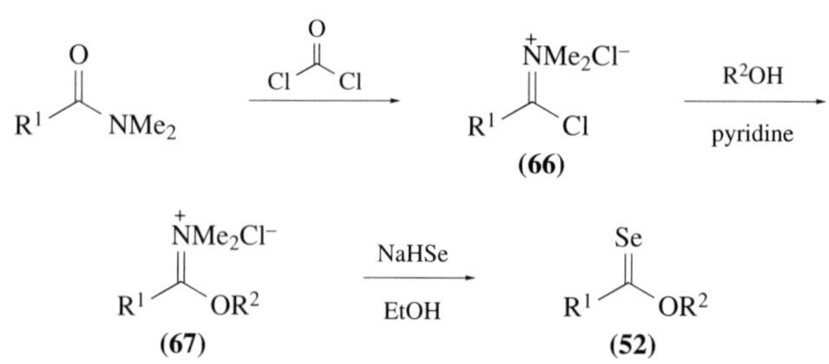

Scheme 26

Various kinds of *O*-alkyl and *O*-aryl selenocarboxylates can be synthesized by this method as shown in Table 2 although it requires the use of highly toxic phosgene gas. *O*-Alkyl selenoformates are also obtained by using *N*,*N*-dimethylformamide as the starting material. However, some limitations exist ⟨77JCS(P1)1723⟩. Thus, the synthesis of *O*-*t*-butyl selenobenzoate failed because of steric retardation of the initial condensation step. Preparation of *O*-benzyl selenoacetate was also unsuccessful because the intermediate *O*-benzyl imidate salt (**68**) (derived from *N*-acetylpiperidine) was rapidly attacked by chloride ion to give benzyl chloride and *N*-acetylpiperidine (Equation (9)). In addition, *O*-phenyl selenoisobutyrate was not obtained because of the undesirable elimination of the phenoxy group (Equation (10)).

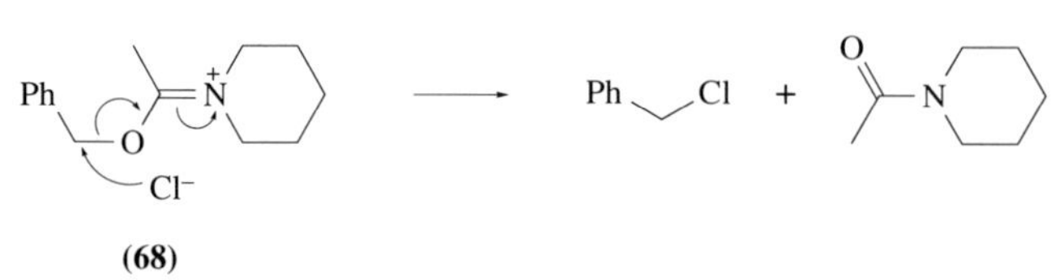

(9)

Table 2 Methods, yields, and melting or boiling points for *O*-alkyl and *O*-aryl selenocarboxylates (**52**).

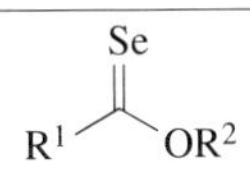

(52)

R^1	R^2	Method[a]	Yield (%)	M.p. (°C) or b.p. (°C) (mmHg)	Ref.
H	Et	B	ca. 15	b	77JPC915
H	Bu^t	B	22	50.5–51.5	86BCJ2529
H	CEt_3	B	6.5	orange oil	86BCJ2529
H	Cholestanyl	B	75	88–90	77JCS(P1)1723
H	Cholesteryl	B	44	126–128	75JCS(P1)1574
		B	69	125–127	77JCS(P1)1723
Me	Me	B	80	108–110 (763)	77JCS(P1)1723
Me	Et	A	26	129 (760)	77JOC2645
		B	82	127–129 (767)	80JCS(P1)1627
Me	Pr^i	B	76	139–141 (772)	77JCS(P1)1723
Me	Ph	B	65	92 (0.7)	77JCS(P1)1723
Me	Cholestanyl	B	74	104–105	77JCS(P1)1723
Me	Cholesteryl	B	83	144–145	77JCS(P1)1723
Et	Et	A	34	147	77JOC2645
Et	Cholesteryl	B	86	151–153	77JCS(P1)1723
EtO_2CCH_2	Et	A	11	57.5–59 (0.2)	68ZC335
Pr^n	Et	A	46	46 (10)	77JOC2645
Pr^i	Me	B	79	58–59 (35)	80JCS(P1)1627
Pr^i	Et	A	48	41 (48)	77JOC2645
		B	69	72–74 (34)	77JCS(P1)1723
Pr^i	Pr^i	B	61	82–84 (33)	77JCS(P1)1723
Bu^n	Me	G	67	c	92TL7865
Bu^n	Et	A	57	60 (9)	77JOC2645
Bu^t	Et	B	46	64 (14)	77JCS(P1)1723
$Cl(CH_2)_4$	Me	D	c	c	93JA5823
n-C_5H_{11}	Me	D	67	yellow liquid	90BCJ835
n-C_5H_{11}	Et	A	60	66 (10)	77JOC2645
		D	60	yellow liquid	90BCJ835
$Me_2CH(CH_2)_2$	Et	A	c	c	78JHC237
n-$C_{17}H_{35}$	Me	B	43	25–26	77JCS(P1)1723
$EtOC(Se)(CH_2)_3$	Et	B	38	d	77JCS(P1)1723
$PhCH_2$	Me	C	ca. 50	c	76JOC729
		B	80	d	77JCS(P1)1723
		D	88	yellow liquid	90BCJ835
$PhCH_2$	Et	C	ca. 50	c	76JOC729
		A	80	105 (2)	77JOC2645
		D	86	yellow liquid	90BCJ835
$PhCH_2$	Pr^i	C	ca. 50	c	76JOC729
		D	83	yellow liquid	90BCJ835
$PhCH_2$	$CH(Me)CHMe_2$	D	77	yellow liquid	90BCJ835
$PhCH_2$	Cyclohexyl	D	80	yellow liquid	90BCJ835
4-$MeC_6H_4CH_2$	Me	D	c	c	93JA5823

Table 2 (continued)

$4\text{-}MeOC_6H_4CH_2$	Me	C	ca. 50	c	76JOC729
		D	c	c	93JA5823
$4\text{-}ClC_6H_4CH_2$	Me	C	ca. 50	c	76JOC729
		D	64	yellow liquid	90BCJ835
$4\text{-}ClC_6H_4CH_2$	Et	D	70	yellow liquid	90BCJ835
$4\text{-}NO_2C_6H_4CH_2$	Me	C	ca. 50	c	76JOC729
$1\text{-}NaphCH_2$[e]	Me	D	c	c	93JA5823
$2\text{-}NaphCH_2$[e]	Me	C	ca. 50	c	76JOC729
Ph	Me	A	5.6	90–91 (1.8)	62BSB563
		A	36	76 (0.4)	66CB1393
		E	10	oil	74CB915
		B	89	c	77JCS(P1)1723
		G	88	c	92TL7865
Ph	Et	A	86	106 (9)	77JOC2645
		B	97	deep red oil	75JCS(P1)1574
		F	14	c	92CL1843
		G	91	c	92TL7865
Ph	Pr^i	B	92	76–78 (0.1)	77JCS(P1)1723
Ph	Cholestanyl	B	82	136–137	76JCS(P1)2112
		B	89	c	77JCS(P1)1723
Ph	Cholesteryl	B	78	160–162	75JCS(P1)1574
		B	87	c	77JCS(P1)1723
Ph	Ph	B	87	164–166	77JCS(P1)1723
Ph	$(CH_2)_2N^+HMe_2Cl^-$	H	56.5	135	72JMC760
Ph	$(CH_2)_2N^+Me_3Br^-$	H	78	158–158.5	72JMC760
Ph	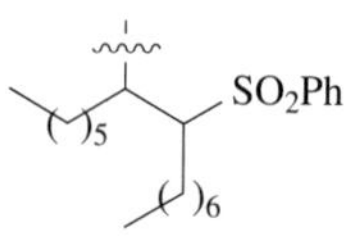	B	35	red oil	91TL2703 92T7109
$4\text{-}MeC_6H_4$	Me	A	3	104 (1.4)	62BSB563
		E	29	oil	74CB915
		G	c	c	93JA5823
$2\text{-}MeC_6H_4$	Me	G	c	c	93JA5823
$4\text{-}MeC_6H_4$	Et	A	90	105 (8)	77JOC2645
		A	quant.	c	87S363
$4\text{-}MeOC_6H_4$	Me	E	27	37	74CB915
		G	c	c	93JA5823
$4\text{-}ClC_6H_4$	Me	E	12	57	74CB915
$4\text{-}BrC_6H_4$	Me	G	c	c	93JA5823
$4\text{-}CF_3C_6H_4$	Me	G	c	c	93JA5823
$4\text{-}Bu^tC_6H_4$	CD_3	b	c	c	79JCR(S)160
2-Furyl	Et	A	quant.	c	87S363
2-Thienyl	Et	A	quant.	c	87S363
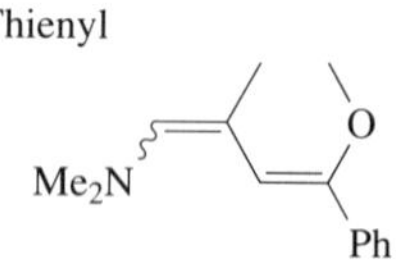		I	68	163	75JOU1077

Table 2 (continued)

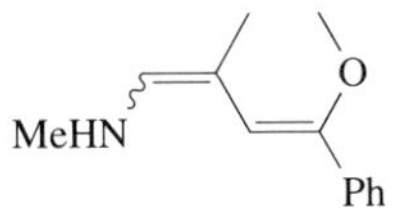	I	86	138	75JOU1077
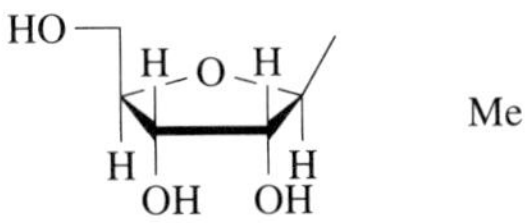	A	ca. 100	yellow syrup	85JOC1741
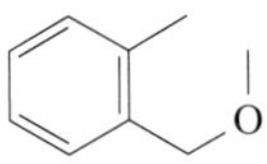	A	62	111	64BSB419

[a] See text. [b] In solution. [c] Not described. [d] Could not be distilled. [e] Naph = Naphthyl.

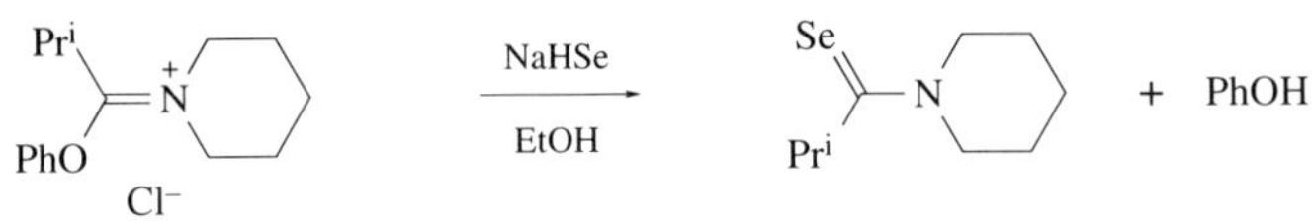

(10)

(ii) Reaction of alkynyl selenolate anions or alkynyl selenols with alcohols

O-Alkyl selenocarboxylates of the type of $R^1CH_2C(=Se)OR^2$ (**54**) were prepared by the reaction of alkynyl selenolate anions or selenols with alcohols. Malek-Yazdi and Yalpani prepared (**54**) by the reaction of the potassium alkynyl selenolates (**70**) with alcohols (R^2 = Me, Et and Pr^i) in about 50% isolated yield ⟨76JOC729⟩. In this synthesis, the selenolates (**70**) were generated from 1,2,3-selenadiazoles (**69**; R^1 = Ph, *p*-$MeOC_6H_4$, *p*-ClC_6H_4, *p*-$O_2NC_6H_4$ and *β*-naphthyl) on treatment with Bu^tOK in Bu^tOH (Method C) (Scheme 27).

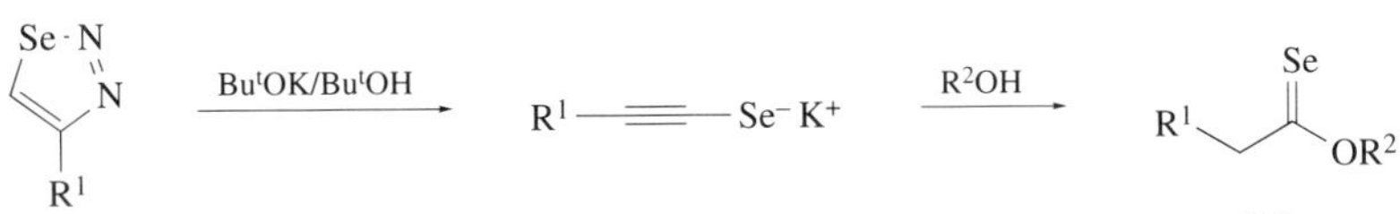

(**69**) (**70**) (**54**)

Scheme 27

On the other hand, Kato *et al.* synthesized a similar type of *O*-alkyl selenocarboxylates (**54**) by the reaction of *Se*-alkynyl selenocarboxylates (**55**; $R^1 = Bu^n$, Ph, and *p*-ClC_6H_4) with alcohols (R^3 = Me, Et, Pr^i, $CH(Me)CH(Me)_2$, and *c*-C_6H_{11}) in the presence of an acid catalyst such as TsOH in 60–88% isolated yields. Selenoketenes (**72**) were proposed as probable intermediates (Method D) (Scheme 28) ⟨90BCJ835⟩.

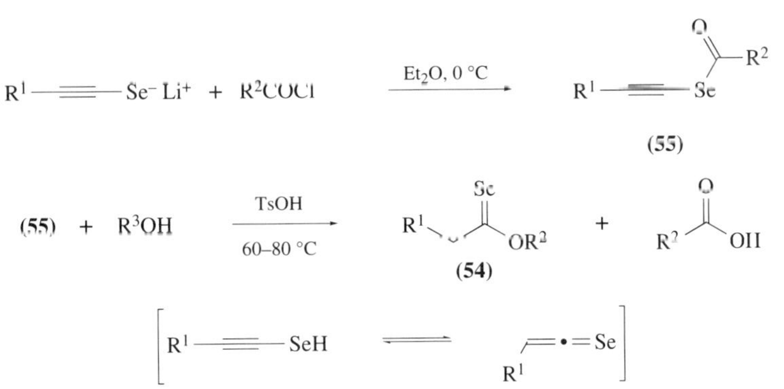

Scheme 28

(iii) Reaction of alkoxycarbene chromium complexes with elemental selenium

The reaction of alkoxycarbene chromium complexes (**56**) ⟨69JOM(16)P29, 71JOM237⟩ with elemental selenium provides another method for the synthesis of *O*-methyl areneselenocarboxylates (**57**) (Method E) (Equation (11)) ⟨74CB915, 75GEP(O)2410157⟩. However, the difficult preparation of (**56**), and low yields of the products seem to be the disadvantage of this method.

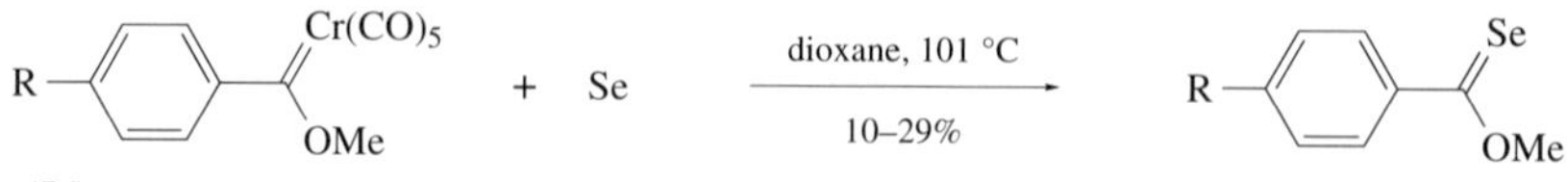

(**56**) R = H, Me, MeO, Cl (**57**) (11)

(iv) Reaction of esters or orthoesters with selenation reagents

In the early 1990s Takikawa and co-workers reported that bis(1,5-cyclooctanediylboryl) selenide (**73**) can directly convert aldehydes, ketones, amides, and an ester into the corresponding selenocarbonyl compounds (Method F) ⟨92CL1843⟩. Thus, the reaction of ethyl benzoate with (**73**) in a sealed tube at 130°C for 10 h gave *O*-ethyl selenobenzoate (**74**) in 14% yield with 81% recovery of ethyl benzoate (Equation (12)).

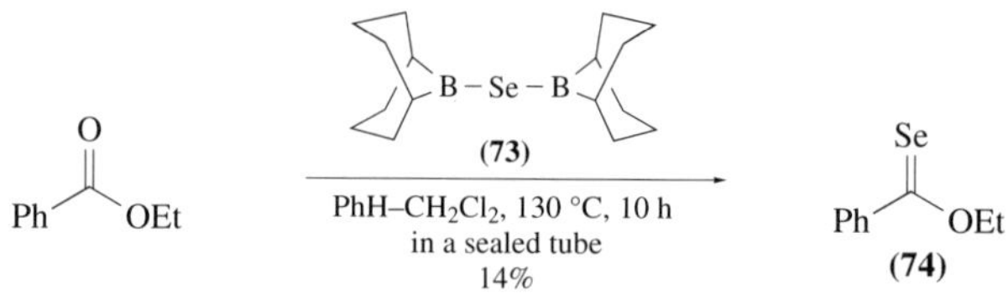

(12)

Segi *et al.* investigated the reaction of bis(dimethylaluminum) selenide (**75**) with acetals and orthoesters and revealed that orthoesters were effectively converted into the corresponding *O*-methyl or *O*-ethyl selenocarboxylates (Method G) (Equation (13)) ⟨92TL7865⟩.

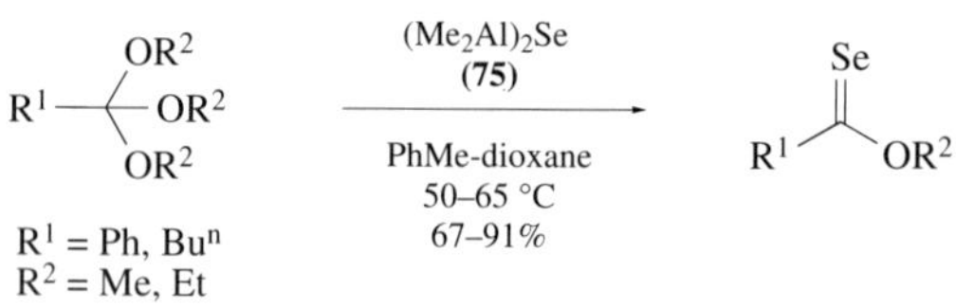

(13)

(v) Miscellaneous

Choline derivatives (**77**) were prepared by transesterification of *O*-methyl selenobenzoate (**60**) with 2-dimethylaminoethanol in toluene at 87°C, followed by treatment with hydrogen chloride or methyl bromide without isolation of the free amine (**76**) (Method H) (Scheme 29) ⟨72JMC760⟩.

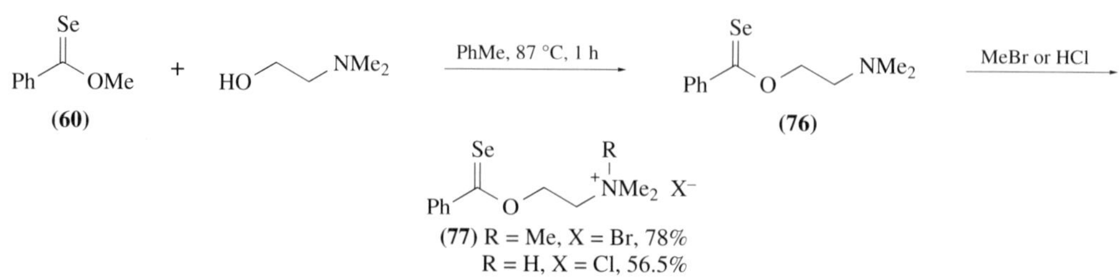

Scheme 29

3-Aminomethylene-5-phenylfuran-2-selones (**79**) and (**80**) were obtained by condensation of 3-formyl-2-hydroseleno-2-phenylfuran (**78**) with methylamine or dimethylamine (Method I) (Scheme 30) ⟨75JOU1077⟩.

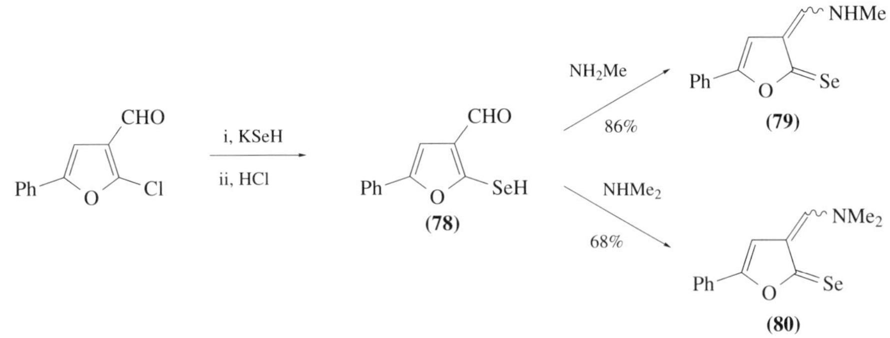

Scheme 30

5.12.2.3.2 Spectroscopies

(i) UV–Vis spectra

O-Alkyl and *O*-aryl selenocarboxylates (**52**) are yellow to red because of a weak absorption around 450 nm (Table 3). The absorptions are due to the lowest electronic transition ($n \rightarrow \pi^*$) of the selenocarbonyl groups ⟨81JA7055, 82JCS(P2)473⟩. A good linear relationship between the $\lambda_{max}^{n \rightarrow \pi^*}$ and δ (^{77}Se) of a series of selenoketones, selenoamides, and a selenocarboxylate was observed ⟨81JA7055⟩.

(ii) NMR spectra

^{13}C, ^{17}O, and ^{77}Se NMR data of *O*-alkyl selenocarboxylates (**52**) are summarized in Tables 4 and 5. Selenocarbonyl carbons of (**52**) resonate at higher field (δ 215–238) than those of selenoketones (ca. δ 290). This ca. 70 ppm of upfield shift can be easily understood by considering resonance interaction between selenocarbonyl and oxygen lone pair electrons which is of significant importance for the selenocarboxylates (**52**) (Scheme 31) ⟨82JCS(P2)473⟩. Similar interpretation can be applied to the upfield shift of ^{77}Se NMR chemical shifts (δ 915–1052) compared to those of selenoketones (δ 1613–2135).⟩

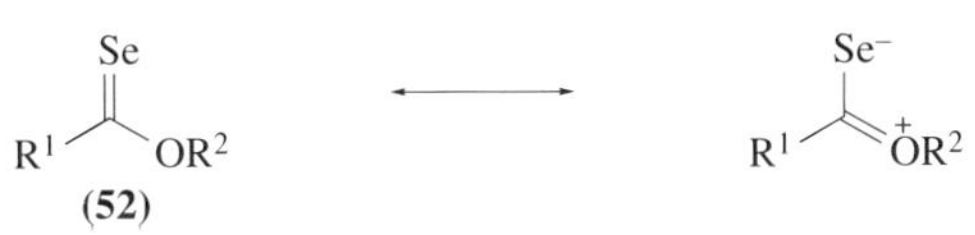

Scheme 31

5.12.2.3.3 Reactions

The fundamental reactions of *O*-alkyl and *O*-aryl selenocarboxylates (**52**) reported by Barton *et al.* are summarized in Scheme 32 ⟨77JCS(P1)1723⟩. Selenocarboxylates (**52**) react with methylenetriphenylphosphorane to give vinyl ethers (**81**) in good yields. Reduction of (**52**) with $NaBH_4$ followed by exposure to air gives diselenides (**82**). While reduction of selenoacetates (**52**; R^1 = Me) with $NaBH_4$–PEt_3 provides the corresponding ethyl ethers (**83**) in high yield, that of selenobenzoates (**52**; R^1 = Ph) and selenoformates (**52**; R^1 = H) under similar conditions gives complex mixtures of products. Later, although the reaction of (**52**; R^2 = alkyl) with PEt_3 was reinvestigated by Hansen, the purple intermediate initially formed could not be fully characterized (Scheme 33) ⟨80JCS(P1)1627⟩. Treatment of *O*-cholesteryl selenoformate with $NaBH_4$ and then with PEt_3 and reduction of (**52**; R^1 = Ph) with W-2 Raney nickel afforded the desired methyl (**84**) and benzyl ethers (**85**), respectively.

Table 3 UV–Vis spectra of *O*-alkyl and *O*-aryl selenocarboxylates (**52**).

R^1*C(=Se)OR*^{2} (**52**)						
R^1	*R*^{2}	λ_{max} (ε)			*Solvent*	*Ref.*
H	Cholestanyl	447 (43)	278 (7700)		CHCl$_3$	77JCS(P1)1723
H	Cholesteryl	448 (60)	277 (8200)		CHCl$_3$	75JCS(P1)1574
Me	Me	443 (50)	270 (7300)		EtOH	77JCS(P1)1723
Me	Et	443 (50)	272 (7200)		EtOH	80JCS(P1)1627
Me	Pr^i	443 (50)	277 (7700)		EtOH	77JCS(P1)1723
Me	Ph	469 (50)	279 (6300)		EtOH	77JCS(P1)1723
Me	Cholestanyl	435 (60)	282 (7400)		CHCl$_3$	77JCS(P1)1723
Me	Cholesteryl	438 (82)	283 (6700)		CHCl$_3$	77JCS(P1)1723
Et	Cholesteryl	434 (60)	283 (8000)		CHCl$_3$	77JCS(P1)1723
Pr^i	Me	443 (50)	271 (7300)		EtOH	80JCS(P1)1627
Pr^i	Et	443 (50)	274 (7100)		EtOH	77JCS(P1)1723
Pr^i	Pr^i	444 (50)	278 (7000)		EtOH	77JCS(P1)1723
Bu^t	Et	450 (50)	274 (7950)		EtOH	77JCS(P1)1723
n-C$_5$H$_{11}$	Me	454 (83)	378 (sh)	255 (24500)	Cyclohexane	90BCJ835
n-C$_5$H$_{11}$	Et	450 (85)	369 (sh)	260 (53700)	Cyclohexane	90BCJ835
n-C$_{17}$H$_{35}$	Me	436 (60)	275 (8200)		CHCl$_3$	77JCS(P1)1723
EtOC(Se)CH$_2$-CH$_2$CH$_2$	Et	446 (110)	274 (15100)		EtOH	77JCS(P1)1723
PhCH$_2$	Me	450 (45)	275 (7800)		EtOH	77JCS(P1)1723
		465 (93)	265 (21400)		Cyclohexane	90BCJ835
PhCH$_2$	Et	457 (93)	278 (12300)		Cyclohexane	90BCJ835
			277 (7700)		EtOH	76JOC729
PhCH$_2$	Pr^i	460 (65)	279 (19000)		Cyclohexane	90BCJ835
			282 (8300)		Pr^{i}OH	76JOC729
PhCH$_2$	CH(CH$_3$)CH(CH$_3$)$_2$	454 (72)	282 (12600)		Cyclohexane	90BCJ835
PhCH$_2$	Cyclohexyl	456 (65)	310 (sh)	260 (36300)	Cyclohexane	90BCJ835
4-MeOC$_6$H$_4$CH$_2$	Me		275 (7600)		MeOH	76JOC729
4-ClC$_6$H$_4$CH$_2$	Me	456 (98)	260 (44700)		Cyclohexane	90BCJ835
			275 (7000)		MeOH	76JOC729
4-ClC$_6$H$_4$CH$_2$	Et	457 (95)	263 (27500)		Cyclohexane	90BCJ835
4-NO$_2$C$_6$H$_4$CH$_2$	Me		275 (16800)		MeOH	76JOC729
2-NaphCH$_2$[a]	Me		276 (13000)		MeOH	76JOC729
Ph	Et	498 (200)	338 (8900)	257 (9600)	CH$_2$Cl$_2$	75JCS(P1)1574
Ph	Pr^i	504 (130)	331 (8800)	254 (7900)	EtOH	77JCS(P1)1723
Ph	Cholestanyl		338 (7700)	254 (9500)	EtOH	76JCS(P1)2112
Ph	Cholesteryl	489 (190)	337 (8800)	257 (9400)	CH$_2$Cl$_2$	75JCS(P1)1574
Ph	Ph	523 (160)	338 (9000)	261 (6300)	EtOH	77JCS(P1)1723
Ph	CH$_2$CH$_2$N$^+$HMe$_2$Cl$^-$		327 (10400)	256 (7020)	EtOH	72JMC760
Ph	CH$_2$CH$_2$N$^+$Me$_3$Br$^-$		328 (10210)	257 (7020)	EtOH	72JMC760

[a] Naph = Naphthyl.

Treatment of (**52**; R^2 = alkyl) with potassium hexamethyldisilazial gives condensation products (**86**) but not normal Claisen-type products.

Cohen *et al.* reported the reaction of *O*-ethyl selenocarboxylates (**65**) with various nitrogen reagents (Scheme 34). The reaction of (**65**) with primary alkylamide magnesium bromide and with secondary amines gave selenoamides (**87**) 〈77JOC2645〉. Condensation of (**65**) with *o*-phenylenediamines, *o*-aminophenols, *o*-aminothiophenols 〈77JHC1321〉, 4-substituted semi- and thiosemicarbazides 〈78JHC237〉, *o*-aminobenzamide, and *o*-aminothiobenzamide 〈78JHC1415〉 gives the corresponding five- and six-membered heterocyclic compounds (**89**)–(**91**). The reaction with hydrazine hydrate yields bis-hydrazones (**92**) 〈79JHC365〉. Two reports describe the condensation leading to selenazole derivatives (**93**) and (**94**) (Equations (14) and (15)) 〈85JOC1741, 87S363〉.

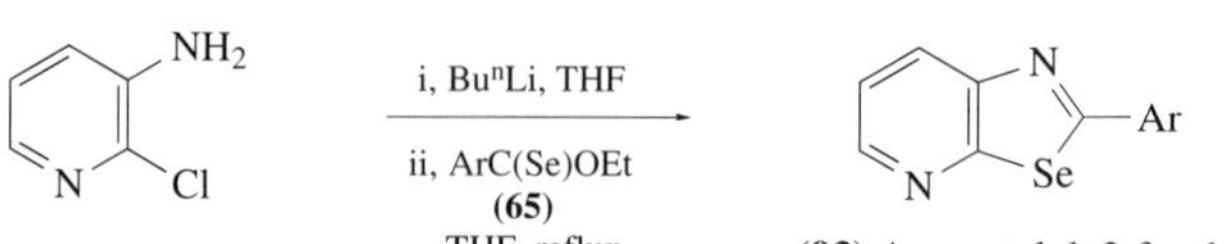

(**93**) Ar = *p*-tolyl, 2-furyl, 2-thienyl

(14)

Table 4 ^{13}C NMR chemical shifts of selenocarbonyl carbons of *O*-alkyl selenocarboxylates (**52**).

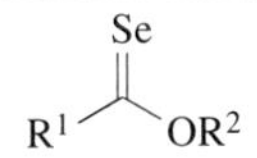

(52)

R^1	R^2	$^{\delta}C=Se^a$	*Ref.*
H	Bu^t	215.0	86BCJ2529
n-C_5H_{11}	Me	237.5	90BCJ835
n-C_5H_{11}	Et	236.8	90BCJ835
$PhCH_2$	Me	232.8	90BCJ835
$PhCH_2$	Et	231.7	90BCJ835
$PhCH_2$	Pr^i	232.1	90BCJ835
$PhCH_2$	$CH(Me)CH(Me)_2$	231.4	90BCJ835
$PhCH_2$	cyclohexyl	230.8	90BCJ835
4-$ClC_6H_4CH_2$	Me	231.9	90BCJ835
4-$ClC_6H_4CH_2$	Et	231.1	90BCJ835
Ph	Et	222.4[b]	82JCS(P2)473
Ph	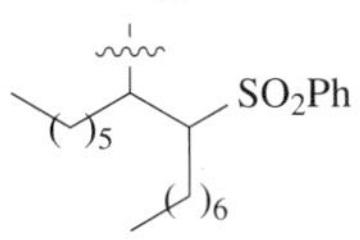	221	92T7109

[a] In $CDCl_3$. [b] $^1J_{C-Se}$ 211.1 Hz.

Table 5 ^{17}O and ^{77}Se NMR chemical shifts of *O*-alkyl selenocarboxylates (**52**).

$R^1C(=Se)OR^2$ (**52**)		^{77}Se NMR[a,b]	^{17}O NMR[c,d]
R^1	R^2	$\delta_{C=Se}$	δ_{OR^2}
H	Bu^t	1052[e]	
H	Cholesteryl	1006[e]	
Ph	Et	915[f]	160.6[g]

[a] In $CDCl_3$. [b] $\delta(Me_2Se) = 0$. [c] In $CDCl_3$—DMSO (2:1 v/v). [d] δ $(H_2O) = 0$. [e] 86BCJ2529. [f] 81JA7055. [g] 83JCS(P2)1471.

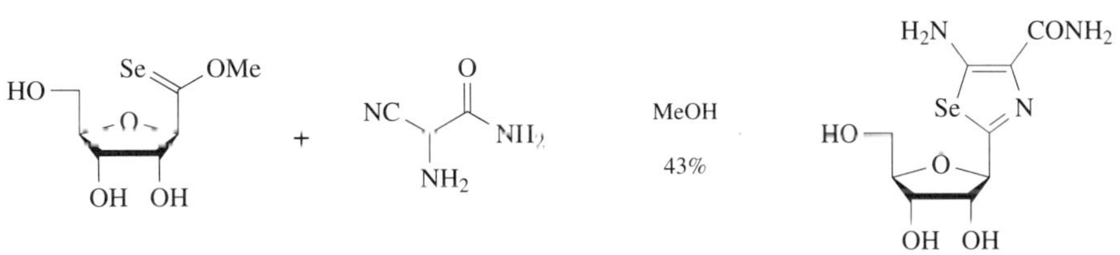

(94) (15)

O-Cholestanyl selenobenzoate (**95**) was converted into the corresponding oxo derivative (**96**) by treatment with benzeneseleninic anhydride (**97**) ⟨78CC393, 80JCS(P1)1650⟩ or bis(p-methoxyphenyl) telluroxide (**98**) ⟨79CC755, 81T(S)213⟩ in a higher yield and in a shorter reaction time than the corresponding sulfur analogue. For the reaction of (**98**) with thiocarbonyl compounds, a tentative reaction mechanism involving cycloaddition and cycloreversion was proposed (Scheme 35).

Selenium atoms of *O*-alkyl selenocarboxylates can be replaced by sulfur on heating the selenoesters with elemental sulfur without solvent (Equations (16) and (17)) ⟨77AG743, 79JCR(S)160⟩. ^{33}S-Labeled thiocarboxylate (**100**) thus prepared, and the parent selenocarboxylate (**99**) were subjected to *in situ* electroreduction to give radical anions, which were investigated by means of ESR spectroscopy ⟨79JCR(S)160⟩. The reaction of *O*-methyl selenocarboxylates (**101**) with $Me_2AlSeMe$ gave the corresponding *Se*-methyl diselenocarboxylates (**102**) in high yields (Equation (18)) ⟨93JA5823⟩.

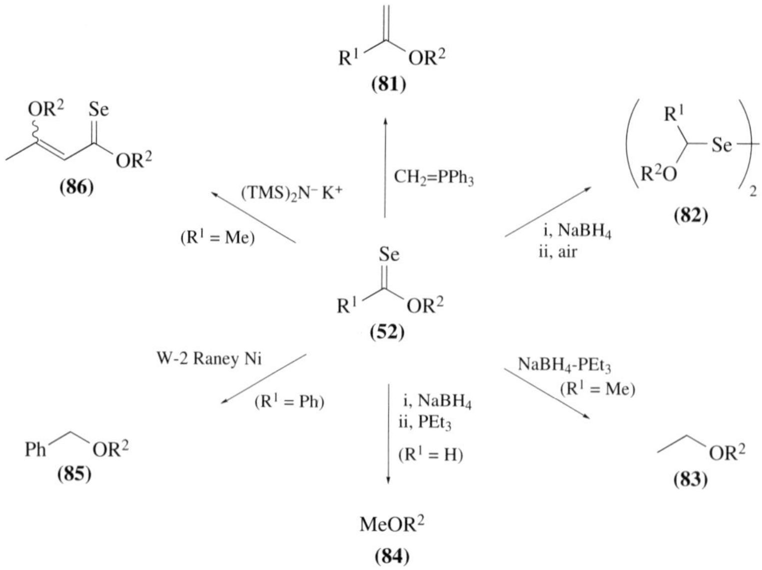

Scheme 32

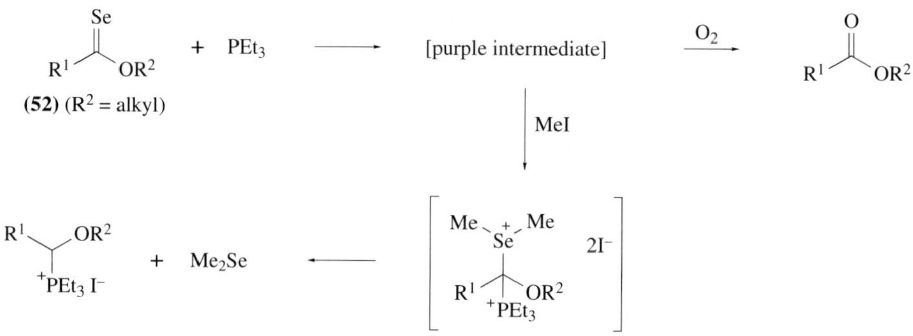

Scheme 33

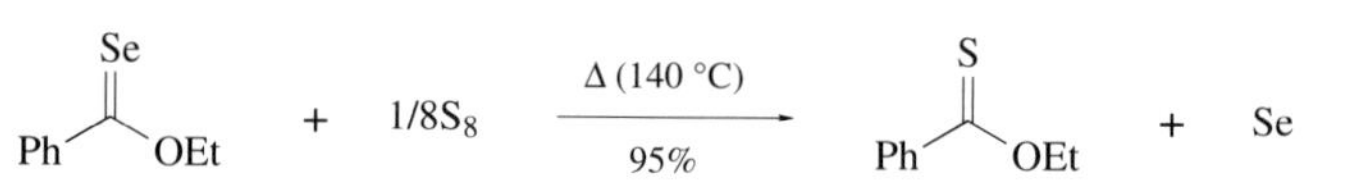

(16)

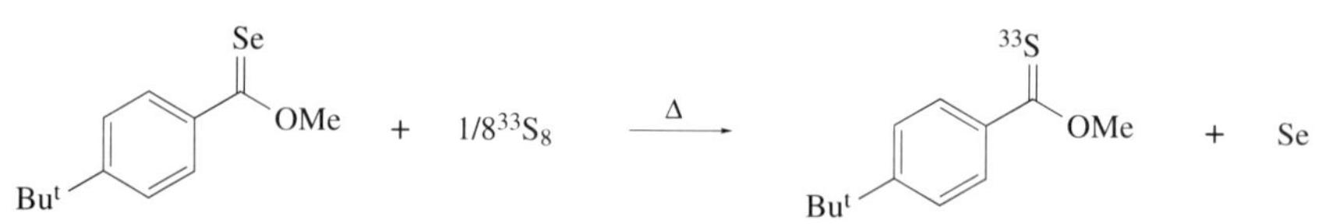

(99) **(100)** (17)

(101) **(102)** (18)

Okazaki and co-workers reported the reaction of 2,4,6-tri-*t*-butylphenyllithium (**103**) with *O*-*t*-butyl and *O*-cholesteryl selenoformates (**104**), where the intermediary formation of selenobenzaldehyde (**105**) was proposed (Equation (19)) ⟨86BCJ2529, 86CC71⟩. The sterically hindered lithium salt (**103**) not only attacked both carbon and selenium atoms of the selenoformyl group but also abstracted the selenoformyl hydrogen ⟨87BCJ1037⟩.

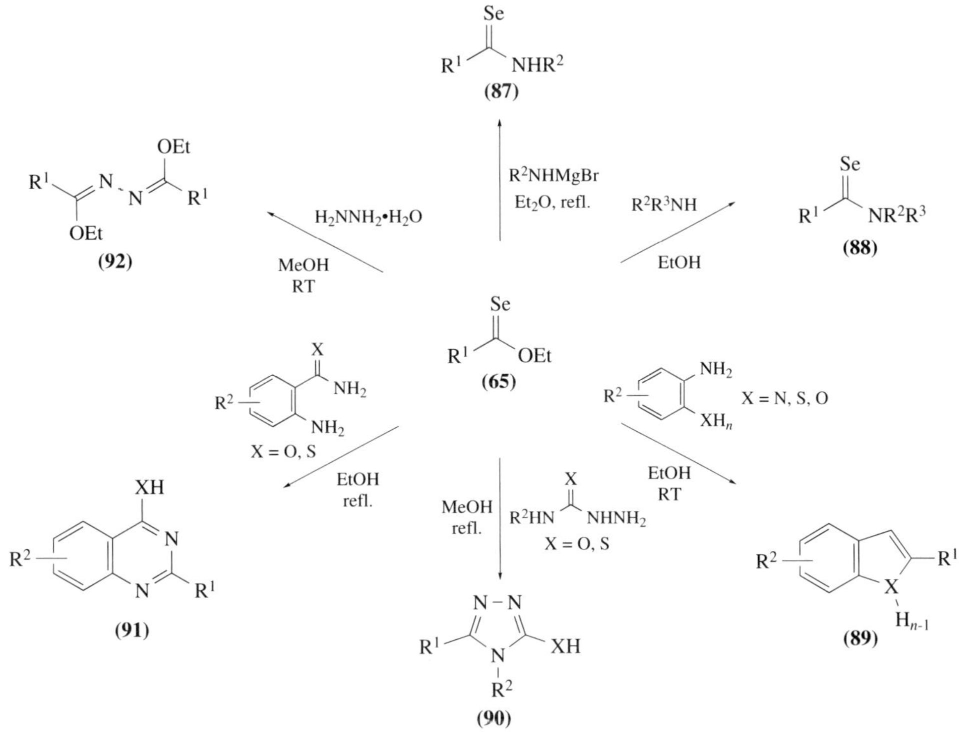

Scheme 34

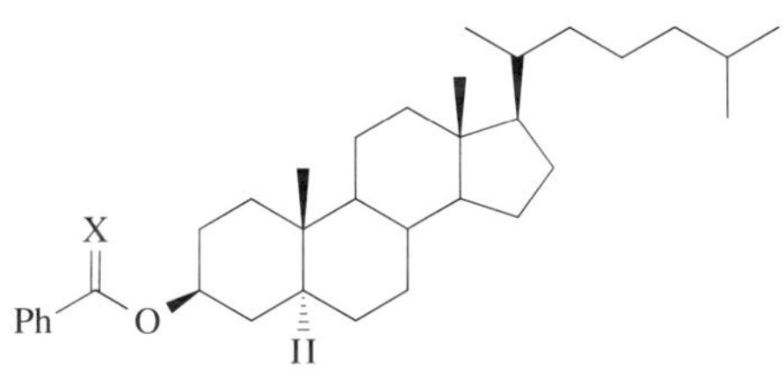

(**95**) X = Se
(**96**) X = O

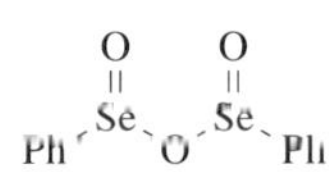

(**97**)

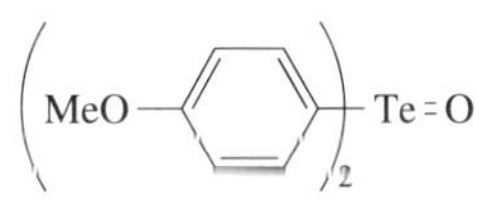

(**98**)

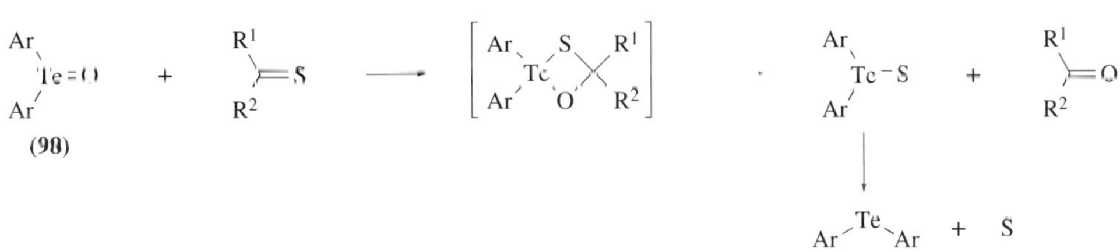

Scheme 35

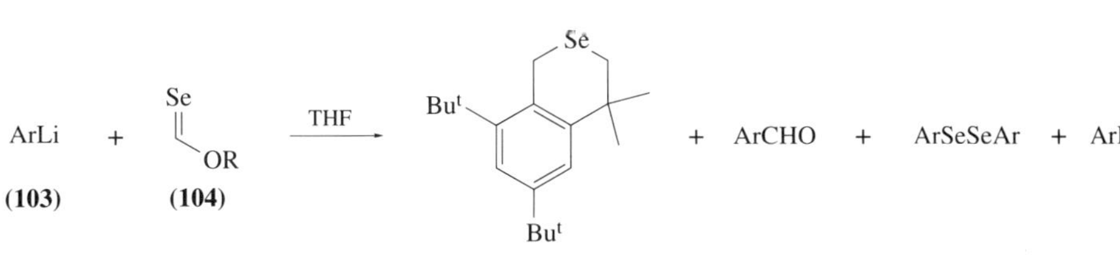

(19)

(Ar = 2, 4, 6-tri-*t*-butylphenyl)

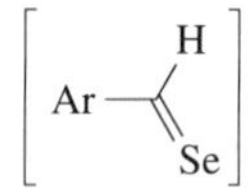

(105)

Reductive alkeneation of a β-hydroxy sulfone was achieved by the reaction of its selenobenzoate derivative (**106**) with *N*-acetoxy-2-thiopyridone (**107**) as the radical source (Equation (20)) ⟨91TL2703⟩. *O*-Cholesteryl selenobenzoate (**108**) was reduced by tributylstannane to yield cholesterine (**109**) in 60% yield ⟨75JCS(P1)1574⟩. In relation to this reduction, ESR studies of adducts of the selenocarboxylate with carbon, tin, and phosphorus radicals were reported ⟨77JPC915⟩. On the other hand, the reaction of *O*-cholestanyl selenobenzoate (**95**) with PhSCl and PhSeCl yielded the corresponding chloride (**110**) and alkene (**111**), where electrophilic addition to the selenocarbonyl selenium occurred initially ⟨76JCS(P1)2112⟩.

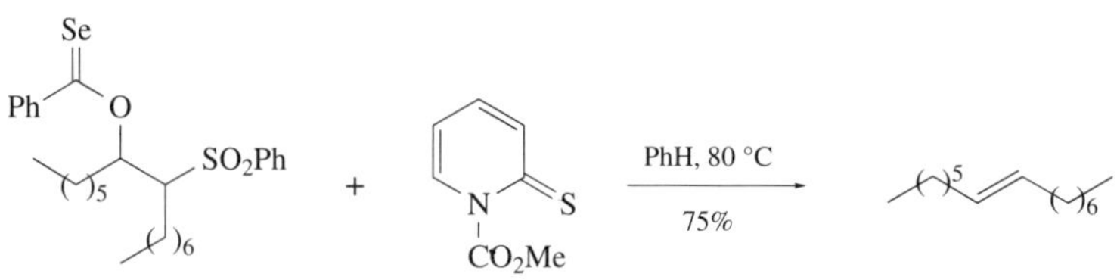

(106) **(107)** (20)

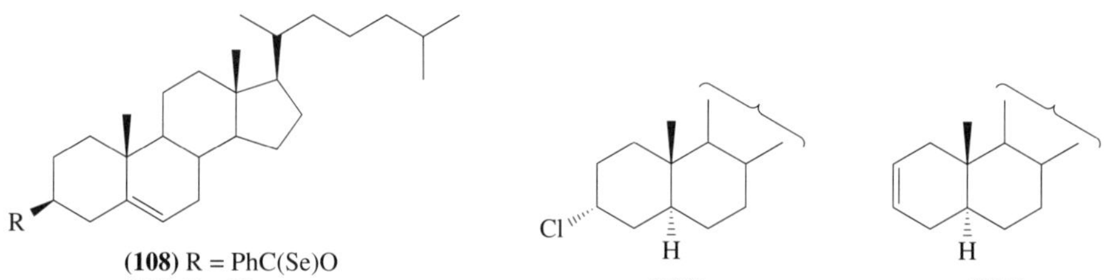

(108) R = PhC(Se)O
(109) R = H
(110) **(111)**

5.12.2.4 Tellurocarboxylic Esters, $R^1C(Te)OR^2$

Barrett, Read and Barton developed the first synthetic method for tellurocarboxylates (**115**), and succeeded in the preparation of sterically protected compounds ⟨79CC645, 80JCS(P1)2191⟩. Thus, *t*-butylchloromethylenedimethylammonium chloride (**113**), prepared from *N*,*N*-dimethylpivalamide (**112**) and phosgene, was treated with alcohols to give alkoxymethylene ammonium chlorides (**114**). Treatment of (**114**) with neutral ethanolic NaHTe (2 equiv.) buffered with acetic acid at a low temperature gave telluropivalates (**115**) (Scheme 36). The reaction of (**114**) with NaHTe and the subsequent purification processes were performed in subdued light, often with the aid of a red photo-filtered lamp. Yields and selected physical and spectroscopic properties of five telluropivalates (**116**)–(**120**) synthesized in this way are summarized in Table 6. These pivalates have a characteristic purple color and are unaffected by water, oxygen in the dark, and anaerobic photolysis (> 500 nm) in benzene solution.

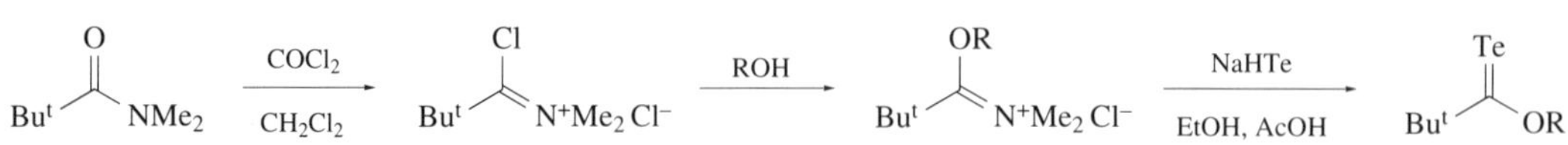

(112) **(113)** **(114)** **(115)**

Scheme 36

On the other hand, preparation of *O*-cholesteryl tellurobenzoate (**121**) was unsuccessful giving only cholesteryl benzyl ether (**85**) in 82% yield ⟨75JCS(P1)1574⟩. A proposed mechanism involved fast electron transfer from NaHTe to intermediary tellurobenzoate (**121**) giving an anion radical (**122**) stabilized by the phenyl group, followed by loss of tellurium after hydrogen capture (Scheme 37). However, this sequence of the reactions provides a mild non-basic procedure for the introduction of the benzyl group ⟨80JCS(P1)2184⟩.

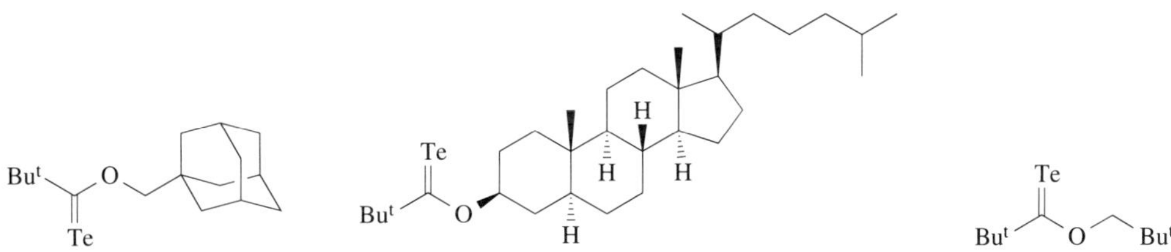

(116) **(117)** **(118)**

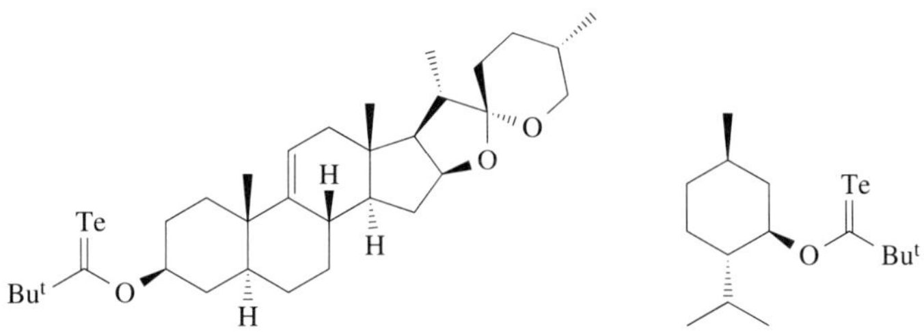

(119) **(120)**

Table 6 Yields and selected physical and spectral properties of *O*-alkyl telluropivalates (**116**)–(**120**)[a].

Compounds	*Yields* (%)	*M.p.* (°C)	^1H *NMR*[b] $\delta_{Bu^tC(=Te)}$	^{13}C *NMR*[b] $\delta_{C=Te}$	*UV–Vis*[c] λ_{max} (*nm*) (ε)
(**116**)	91	<25	1.35	229.20	596 (290) 337 (8400) 266sh (800) 244 (3800)
(**117**)	89	116–120 (decomp.)	1.28	[d]	592 (320) 346 (7600) 266sh (830) 243 (4300)
(**118**)	63	oil	1.37	229.38	584 (170) 336 (5900) 265 (850) 243 (3100)
(**119**)	28	160–165 (decomp.)	1.3	[d]	590 (240) 346 (6800) 265sh (550) 243 (3700)
(**120**)	6	oil	1.27	[d]	589 (240) 346 (6600) 265 (620) 241 (3500)

[a] 79CC645, 80JCS(P1)2191. [b] In $CDCl_3$. [c] In cyclohexane. [d] Not described.

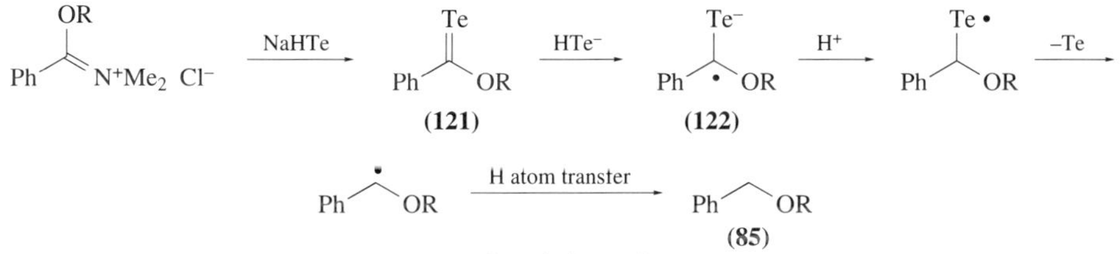

Scheme 37

5.12.2.5 Higher Oxidation States of the Chalcogen

Metzner and co-workers reported oxidation of *O*-methyl thiocarboxylates (**123**) with mcpba ⟨91TL7411⟩. Thus, the formation of both (*E*) and (*Z*) isomers of *O*-methyl thiocarboxylate *S*-oxide (**124**) was observed by ^{1}H and ^{13}C NMR spectroscopy. However, they are thermally unstable and decompose to the corresponding esters at room temperature with precipitation of elemental sulfur (Scheme 38), though their stability is dependent on the bulkiness of the substituents. A mechanism involving oxathiirane intermediates (**125**) was proposed for the decomposition. Their *O*-alkyl (aryl) seleno- and telluro-carboxylate analogues remain unknown.

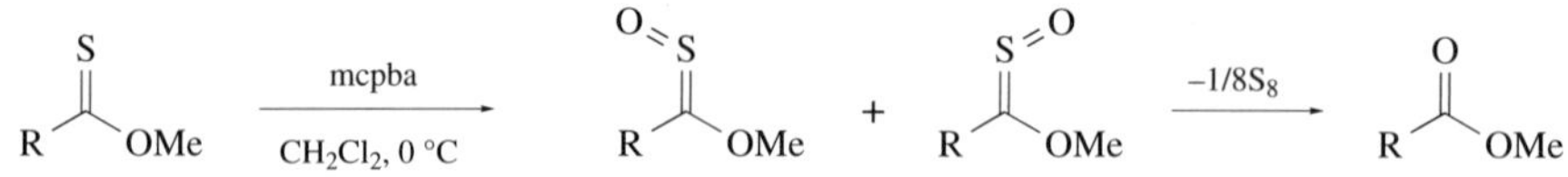

(123) *E*-**(124)** *Z*-**(124)**

Scheme 38

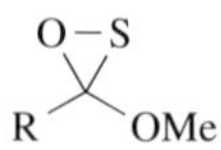

(125)

5.12.3 CARBOXYLIC ANHYDRIDE ANALOGUES, $R^1C(Y)OCOR^2$ AND RC(Y)O(Y′)R

5.12.3.1 Introduction

In 1940 Lewis claimed that reaction of benzoyl chloride with hydrogen sulfide in pyridine gave thionobenzoic anhydride (**126**) ⟨40JCS831⟩. However, Senning and Schank showed that the reported compound is not the anhydride (**126**) but a mixture of bis(benzoylthio)phenylmethane (**127**) and benzoic acid ⟨78ACS(B)780⟩.

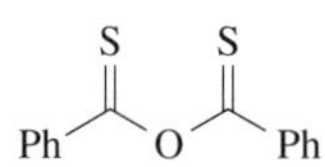

(126)

$PhCH[SC(O)Ph]_2$

(127)

While no acyclic example of the title compounds have been reported up to 1994, a few cyclic thionoanhydrides (**128**)–(**131**) are known though they are still a rare class of compounds ⟨30RZC652, 67JOC3709, 84JA6084, 86S142, 89JOC4746⟩. No reports on seleno- and telluroanhydrides have appeared.

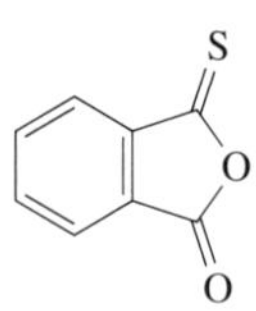

(128)

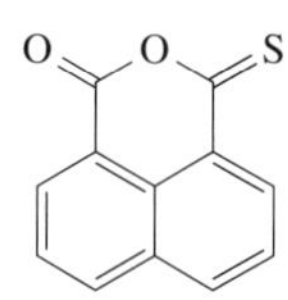

(129)

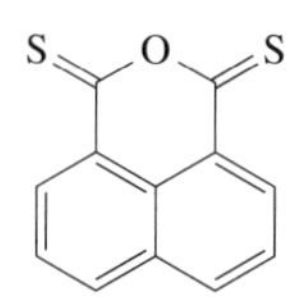

(130)

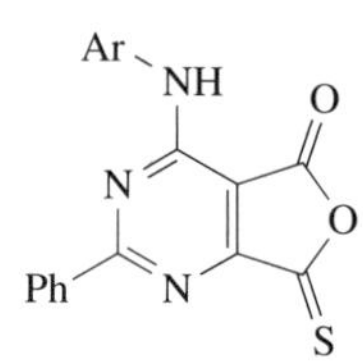

(131)

5.12.3.2 Thionoanhydrides

The five-membered thionoanhydride, thionophthalic anhydride (**128**), was obtained unexpectedly by reaction of phthaloyl chloride with hydrogen disulfide ⟨67JOC3709⟩. However, it is quite unstable thermally, and isomerized readily to the corresponding thioloanhydride (**132**).

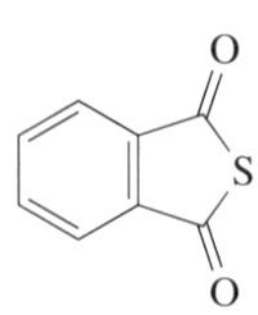

(132)

The stable cyclic thionoanhydride, 1,8-naphthalic dithionoanhydride (**130**), was prepared by Cava and co-workers by the reaction of the pseudo acid chloride (**133**) with Lawesson's reagent (LR) in refluxing chlorobenzene in 57% yield ⟨84JA6084⟩. Later, (**130**) was prepared more practically by direct thionation of 1,8-naphthaloyl chloride (**134**) with LR though in a lower yield (24%) (Scheme 39) ⟨89JOC4746⟩. The dithionoanhydride (**130**) is a red-brown crystalline (needles) solid (m.p. 212 °C, decomp), but rearranged rapidly to the isomeric thionothioloanhydride (**136**) on treatment with a catalytic amount of triethylamine in cold DMF. The isomerization was thought to proceed by intervention of the ring-opened betaine intermediate (**135**) (Equation (21)).

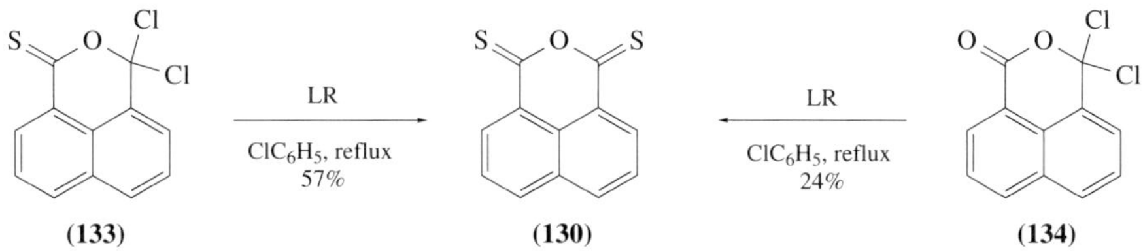

Scheme 39

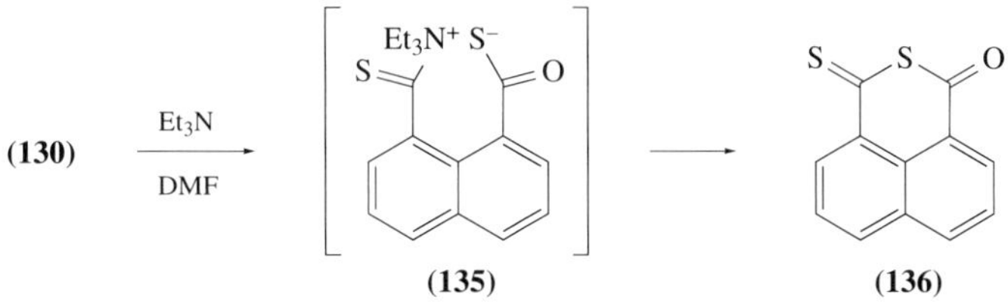

(21)

1,8-Naphthalic monothionoanhydride (**129**) was prepared by reaction of 1,8-naphthaloyl chloride (**134**) with hydrogen sulfide or by a Pedersen thionation reaction upon (**134**) in 81% and 84% yields, respectively (Scheme 40) ⟨30RZC652, 89JOC4746⟩. The monothionoanhydride (**129**) also readily (and cleanly) isomerized to the corresponding thioloanhydride (**137**) on treatment with triethylamine or tributylphosphine.

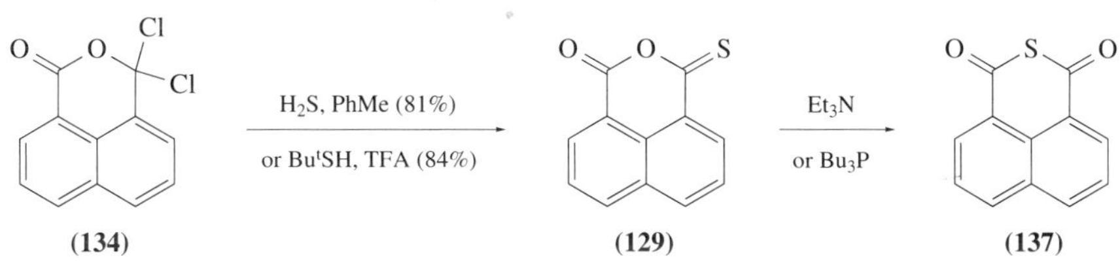

Scheme 40

The reaction of 6-anilino-5-carboxy-4-methyl-2-phenylpyridines (**138**; Ar = C_6H_5, 4-ClC_6H_4, 2-ClC_6H_4, 4-$C_2H_5OC_6H_4$) with thionyl chloride in boiling benzene gave, instead of the expected acid chlorides (**139**), 4-anilino-5-oxo-2-phenyl-7-thioxo-5,7-dihydrofuro[3,4-*d*]pyridines (**131**) in 57–73% yields (Scheme 41) ⟨86S142⟩. The mechanism of the formation of (**131**) was interpreted in terms of an intramolecular cyclization of the thiocarboxylic *O*-acid intermediates (**140**) (Scheme 42).

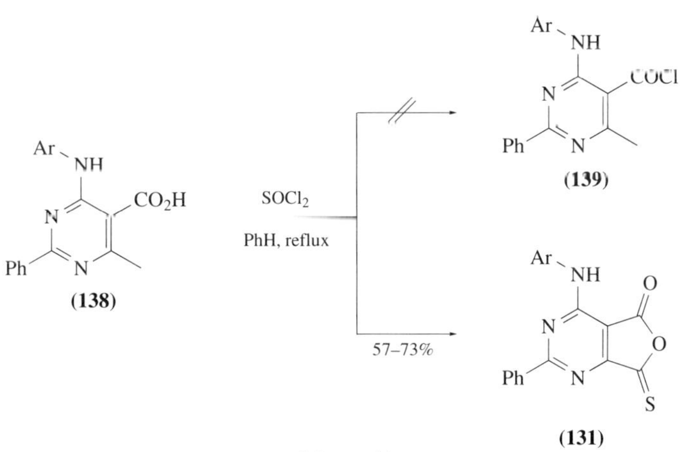

Scheme 41

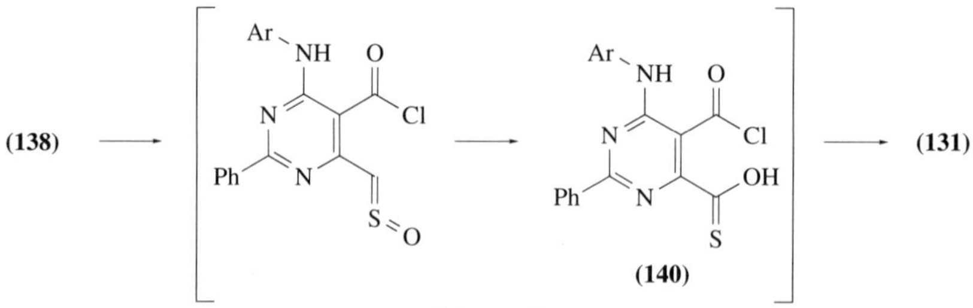

Scheme 42

5.12.3.3 Selenoanhydrides

No example of the selenoanhydrides (**141**) and (**142**) has been reported.

5.12.3.4 Telluroanhydrides

There is no report on telluroanhydrides (**143**) and (**144**).

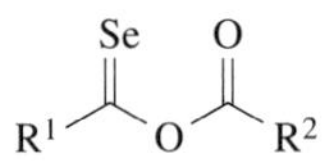

(141)

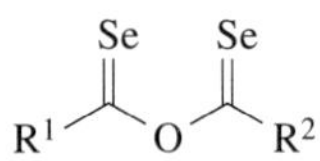

(142)

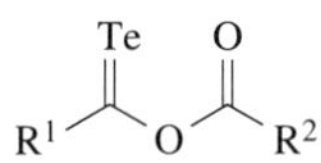

(143)

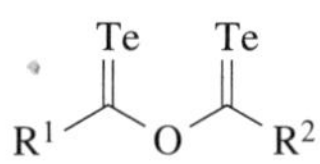

(144)

5.12.4 THIONO, SELENO, AND TELLURO ACYLOXY HETEROATOM FUNCTIONS, RC(Y)OX

5.12.4.1 Halides

There are no reports on *O*-halogeno thio- (**145**), seleno- (**146**), and tellurocarboxylates (**147**).

(145)

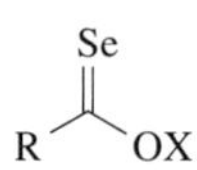

(146)

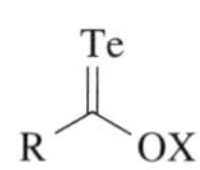

(147) X = F, Cl, Br, I

The reaction of *O*-trimethylsilyl thiopivalate (**148**) with *N*-bromosuccinimide or *t*-butyl hypochlorite resulted in the thiophilic attack of bromonium or chloronium ion giving *S*-bromo or *S*-chloro thiopivalate (**149**), respectively (Equation (22)) ⟨86TL4593⟩.

(148) **(149)** X = Cl, Br (22)

5.12.4.2 Peroxidic Compounds

Russian chemists reported the activation energy of the homolytic decomposition of some peroxides including the peroxythionoester (**150**) ⟨82MI 512-01⟩.

Peroxyseleno- (**151**) and telluroesters (**152**) are unknown.

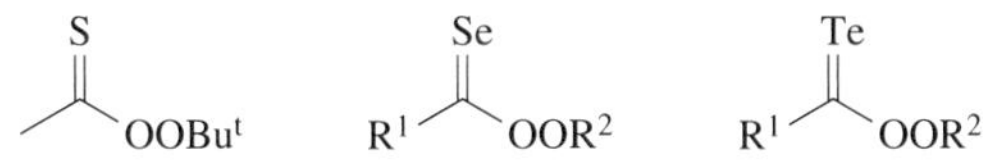

(**150**) (**151**) (**152**)

5.12.4.3 Other Chalcogen-linked Systems

Compounds of types (**153**)–(**155**) have not been reported.

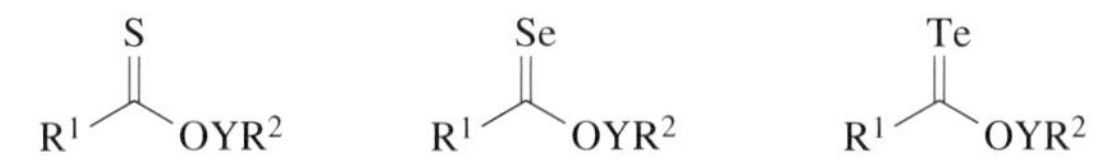

(**153**) Y = Se, Te (**154**) Y = Se, Te (**155**) Y = Se, Te

5.12.4.4 *O*-Thionoacylhydroxylamines, -oximes, and Other *N*-Based Functions and Their Selenium and Tellurium Analogues

5.12.4.4.1 O-*Thionoacylhydroxylamines and -oximes*

Acyclic *O*-thionoacyloximes seem to be thermally unstable ⟨78JCS(P2)1167⟩. It was reported that the reaction of thiobenzoyl chloride with sodium benzaldoximate in ether at −10 °C affords *O*-benzylideneamino thiobenzoate (**156**) in 30% yield as a yellow solid (m.p. 50–53 °C decomp.), though no supporting spectroscopic evidence for (**156**) was given (Scheme 43). The compound (**156**) decomposes to biphenyl, dibenzoyl disulfide, and benzonitrile when heated in boiling chloroform. A mechanism involving free radical intermediates was proposed for this decomposition.

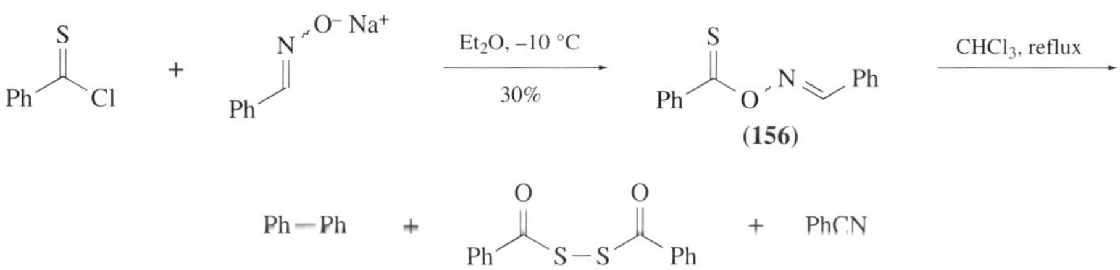

Scheme 43

Although acyclic thionoacylhydroxylamines and -oximes are still a rare class of compound, their cyclic derivatives, i.e., isoxazoline-5-thiones, oxazolidine-5-thiones, and the related heterocyclic compounds, are rather well investigated.

(i) Isoxazoline-5-thiones

2-Isoxazoline-5-thiones (**158**) and 3-isoxazoline-5-thiones (**159**), which are tautomers of 5-mercaptoisoxazole (**157**), can be regarded as five-membered thionoacyloximes and -hydroxylamines, respectively (Scheme 44).

Adembri and Nesi reported the synthesis of 5-mercaptoisoxazole (**157**) by the reaction of 5-

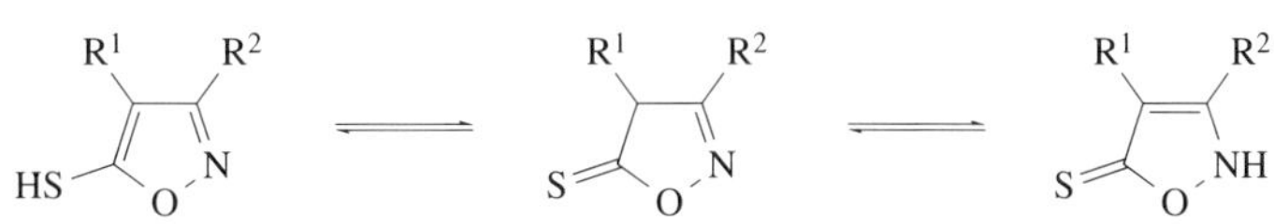

(**157**) (**158**) (**159**)

Scheme 44

chloroisoxazoles (**160**) with potassium hydrosulfide (Equation (23)) ⟨67JHC54⟩. They concluded by inspection of UV and IR spectra that isoxazoles (**157**) exist in the thiolic form exclusively.

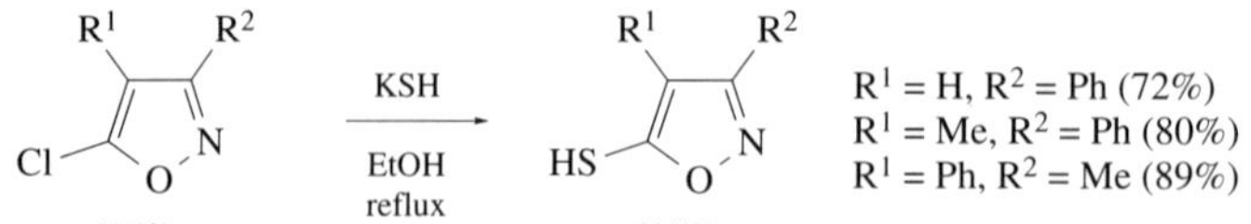

(160) **(157)**

(23)

On the other hand, *N*-methyl-3-isoxazoline-5-thiones (**162**) were prepared by thionation of the corresponding *N*-methylisoxazolin-5-ones (**161**) with P_4S_{10} (Equation (24)) ⟨67JHC54⟩. The longest absorption maximum of (**162**) appears around 350 nm in the UV spectra.

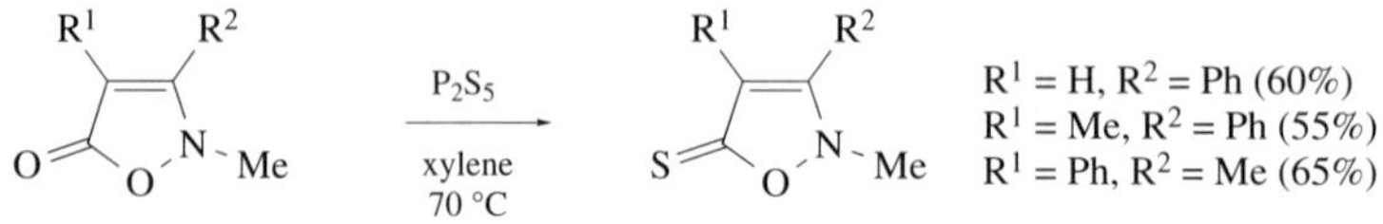

(161) **(162)**

(24)

2-Substituted 3-isoxazoline-5-thiones (**164**) were also prepared by the reaction of 5-chloro-isoxazolium tetrafluoroborates (**163**) with Na_2S at room temperature in good yields (R = Me, 77%; Et, 90%) (Equation (25)) ⟨76ZC270⟩. The isoxazolinethiones (**164**) are stable for a long time in the dark, although they decompose on exposure to light.

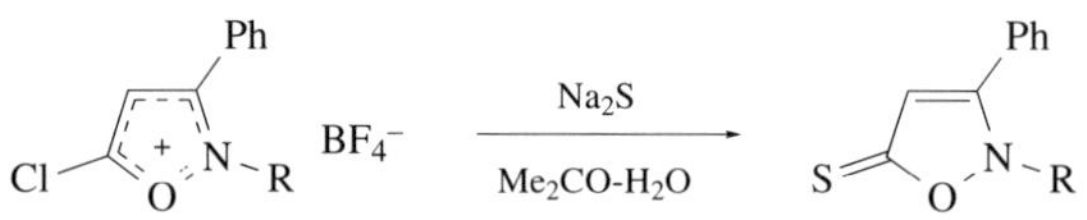

(163) **(164)** R = Me, Et

(25)

4-Arylhydrazono-2-isoxazoline-5-thiones (**166**) were prepared by thionation of the corresponding isoxazolin-5-ones (**165**) with P_4S_{10} (Equation (26)) ⟨65JCS3312, 70GEP(O)2024393, 84JHC253, 89MI 512-01⟩. The reaction of (**166**; R = Ph) with diazomethane results in *N*- and *S*-methylations to give isoxazolinethione (**167**), isoxazole (**168**), and triazole (**169**) derivatives (Equation (27)) ⟨84JHC253⟩. The thiocarbonyl group of 2-isoxazoline-5-thiones (**166**) was readily alkylated by treatment with alkyl halides giving isoxazoles (**170**) or condensed with hydrazines giving 5-hydrazones (**171**) (Scheme 45) ⟨78ZN(B)75, 84JHC253⟩.

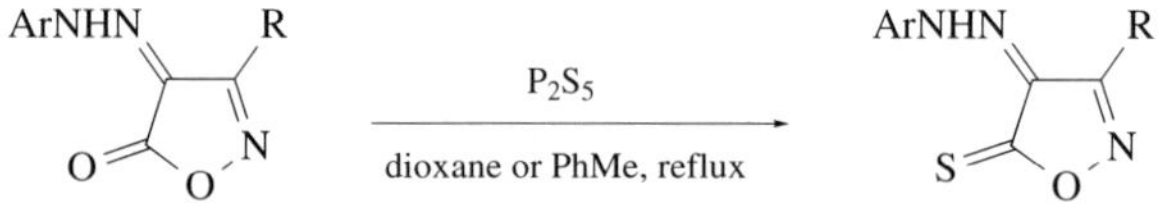

(165) **(166)**

(26)

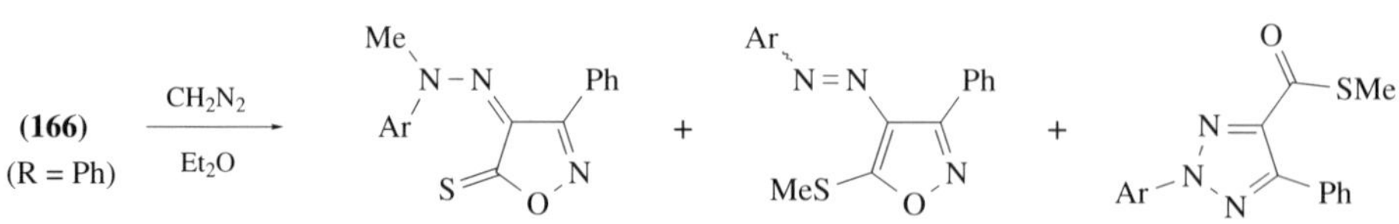

(167) **(168)** **(169)**

(27)

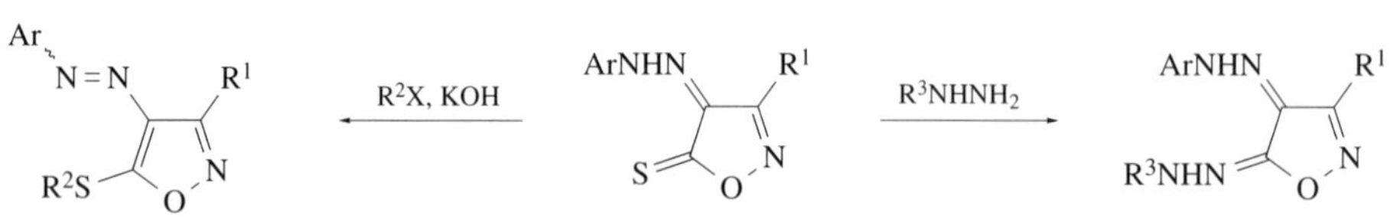

(170) **(166)** **(171)**

Scheme 45

(ii) Oxazolidine-5-thiones

The reaction of thiomalonyl dichlorides (**172**) with *N*-benzylhydroxylamines (**173**) gives 3-oxo-oxazolidine-5-thiones (**174**) (Equation (28)) ⟨79USP4302238⟩.

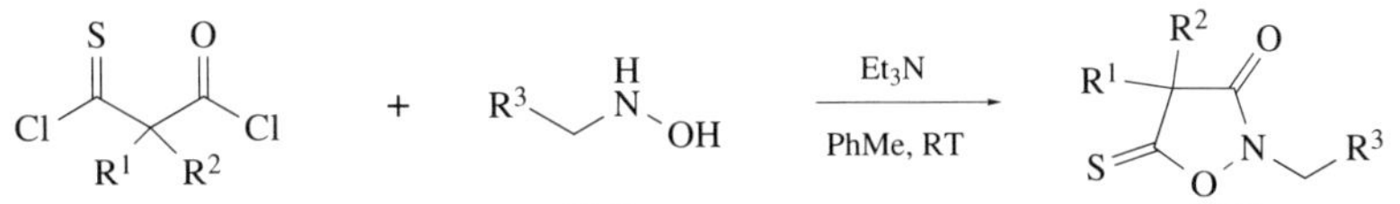

(**172**) (**173**) (**174**)

(28)

(iii) Other heterocyclic systems

6-Oximinocholest-4-ene-4-thiocarboxylic anhydrides (**176**), six-membered *O*-thionoacyloximes, were obtained from the corresponding steroidal nitro compounds (**175**) by reaction with carbon disulfide in the presence of triethylamine in MeCN (Equation (29)) ⟨88JCR(S)376⟩. A rather curious mechanism was proposed for this reaction (Scheme 46). The C═S stretching absorptions of (**176**) appear at 1070–1020 cm^{-1}.

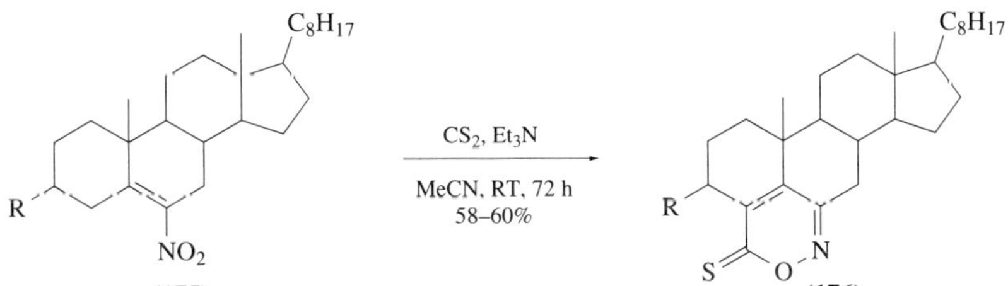

(**175**) (**176**)

(29)

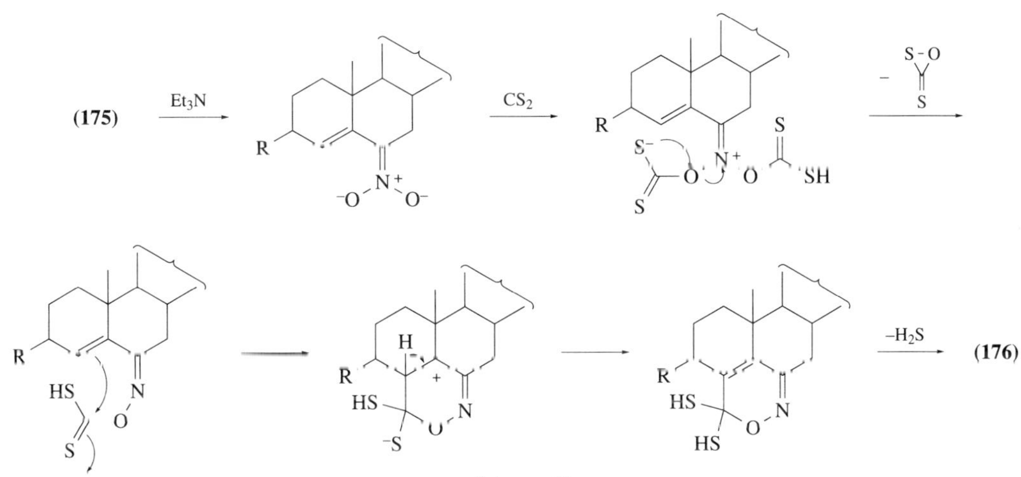

Scheme 46

5.12.4.4.2 O-Selenoacyl and O-telluroacylhydroxyamines and oximes

O-Selenoacyl (**177**) and *O*-telluroacyl (**178**) analogues are unknown.

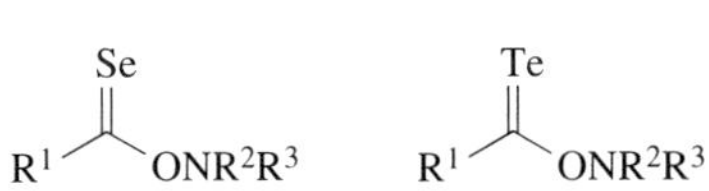

(**177**) (**178**)

5.12.4.5 Other *O*-Heteroatom Derivatives

5.12.4.5.1 B-Thionoacyloxy compounds

In 1993, the first cyclic 1,2-oxaboretane-4-thione (**180**) was obtained by reaction of 9-fluorenylidene(tetramethylpiperidino)borane (**179**) with S=C=O in toluene at 0 °C in 61% yield (Scheme 47) ⟨93CB1119⟩. The oxaboretane thione (**180**), a pale yellow powder (m.p. 195 °C), shows the C=S stretching absorption at 1130 cm^{-1}. Similarly, (**179**) undergoes [2 + 2] cycloaddition with carbonyl and thiocarbonyl compounds to yield the corresponding oxaboretane and thiaboretane derivatives (**181**)–(**183**). The reaction with a substrate bearing both carbonyl and thiocarbonyl groups gives the thermodynamically favored oxaboretane derivative, rather than the thiaboretane.

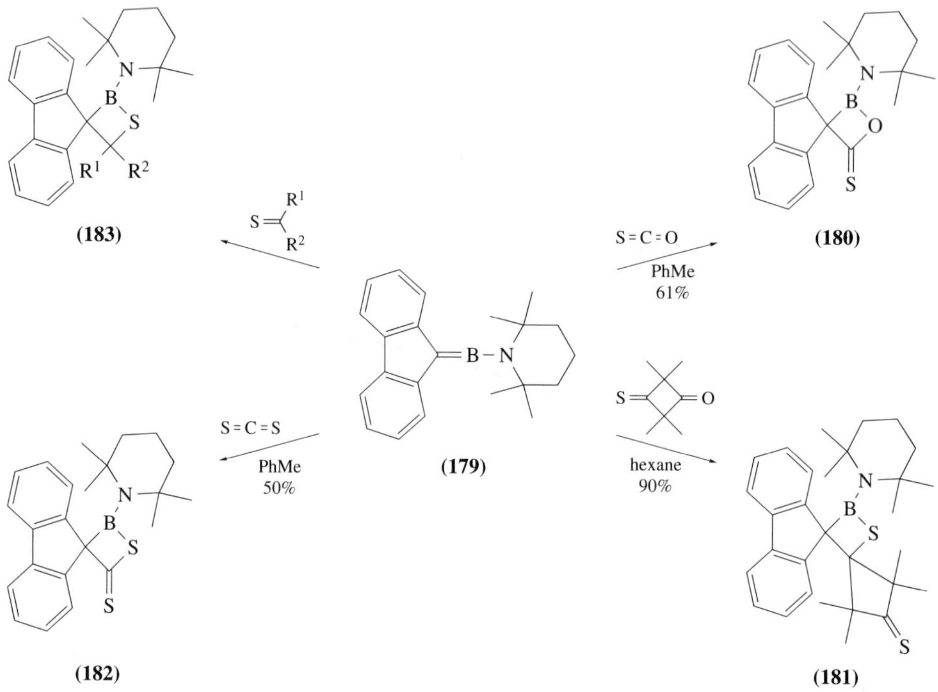

Scheme 47

5.12.4.5.2 P-Thionoacyloxy compounds and their selenium analogues

The reaction of a cyclic phosphinate (**184**) with P_4S_{10} gave a complex mixture from which 1,2-thiaphosphol-3-en-5-one 2-sulfide (**185**) was isolated in 10% yield (Scheme 48) ⟨73JGU2635⟩. The desulfurization of (**185**) with triphenylphosphine at 180 °C gives 1,2-oxaphosphol-3-ene-5-thione (**186**) in 42% yield *via* a carbonyl–thiocarbonyl rearrangement.

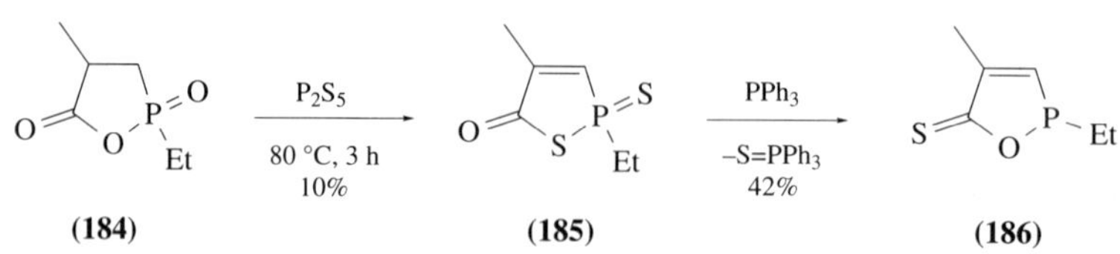

Scheme 48

Alkylaminocrotonates (**187**) react with P_4S_{10} to provide 1,2-thiaphosphetan-4-ones (**188**), which rearrange to 1,2-oxaphosphetane-4-thiones (**189**) when heated in benzene, or subjected to silica-gel chromatography (Scheme 49) ⟨76CB1779⟩. The intramolecular hydrogen bonding has been shown to be considerably stronger in the thione forms (**189**) than in the carbonyl forms (**188**) by their IR and ^{1}H NMR spectra.

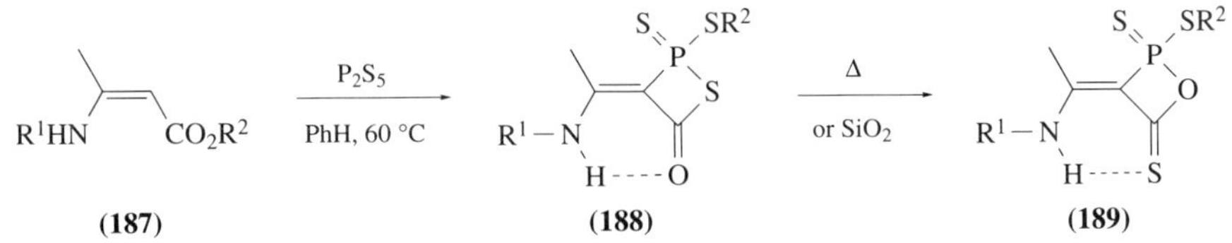

Scheme 49

The equilibrium between *O*-phosphino and *S*-phosphino thio-, and selenoacetates were studied using acyclic substrates ⟨84JCS(D)2301⟩. Thus, the reaction of *S*-tributylstannyl thioacetate (**190**) with $PBrF_2$ at 193K in an NMR tube gave predominantly *O*-difluorophosphino thioacetate (**191**), the formation of which was proved by the ^{31}P (δ_P 114.5), ^{1}H, ^{13}C, and ^{19}F NMR and IR spectra. When the solution was warmed to 243K, a second triplet due to *S*-difluorophosphino thioacetate (**192**) (δ_P 211.5) appeared, and at 308K the ratio of (**191**) to (**192**) reached 4.3 : 1 (Scheme 50). The thermodynamic parameters for the equilibrium between the *O*-bonded (**191**) and *S*-bonded (**192**) isomers were obtained from the relative concentration of the two isomers over the temperature range studied ($\Delta H^\circ = 5.1$ KJ mol^{-1} and $\Delta S^\circ = 4.6$ J K^{-1} mol^{-1}). On the other hand, the reaction of *O*-silyl thioacetate (**193**) with $PBrF_2$ at 193K initially gave the *S*-bonded isomer (**192**) and a second triplet due to the *O*-bonded isomer (**191**) appeared on warming to 228K. A six-membered cyclic intermediate (**194**) was proposed for both reactions. Similar treatment of an equilibrium mixture of *O*- and *Se*-bonded isomers of the silyl selenoacetate (**195**) with $PBrF_2$ at 193K gave a single isomer, assigned as the *Se*-bonded isomer (**196**) (δ_P 244). However, the *O*-bonded isomer (**197**) (δ_P 112.2) was formed at 263K in less than about 10% yield (Scheme 51).

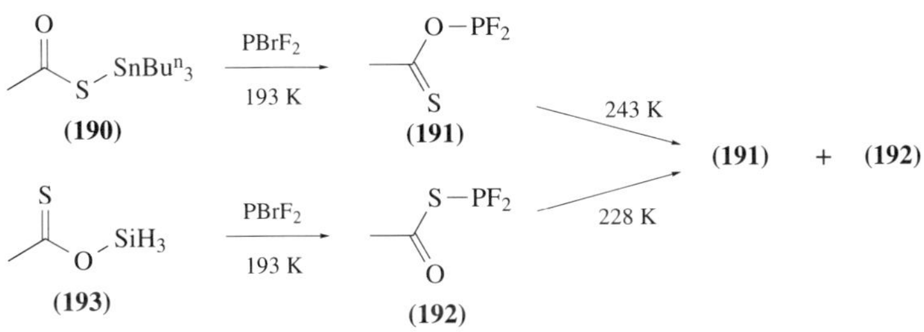

Scheme 50

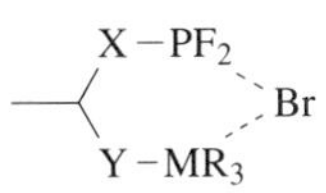

(194)

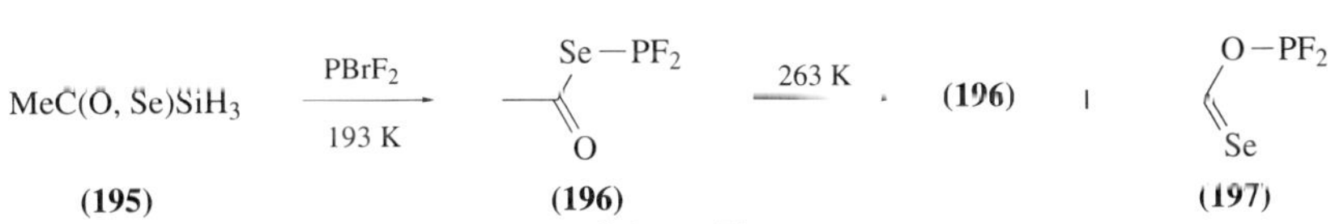

Scheme 51

Phosphorylation of *O*-trimethylsilyl thioacetate (**198**) was reported ⟨89JGU1548⟩. The reaction of (**198**) with dialkoxy phosphorochloridite gave a mixture of thioacetyl phosphite (**199**) and *S*-acetyl phosphorothioite (**200**) in a ratio of 5 : 95 (Equation (30)). Phosphorylation of thioacetic acid (**7**) and its triethylammonium salt (**201**) with diethyl cyanophosphonite and 1-methyl-3-oxo-1-butenyl dipropyl phosphite, respectively, gave similar mixtures though in lower yields (Scheme 52) (54% and 22%, respectively). Interestingly, the replacement of the methyl of the thioacetyl group by phenyl stabilized the thiocarbonyl structure resulting in an increase in the proportion of the *O*-phosphorylated isomers. Thus, the reaction of *O*-trimethylsilyl thiobenzoate (**202**) with diisopropyl phosphorochloridite gave a 1 : 1 mixture of *O*- (**203**) and *S*-phosphorylation (**204**) products (44%).

Treatment of the above mixture with methyl iodide gave a 1 : 9 mixture of (**205**) and (**206**) (65%) (Scheme 53).

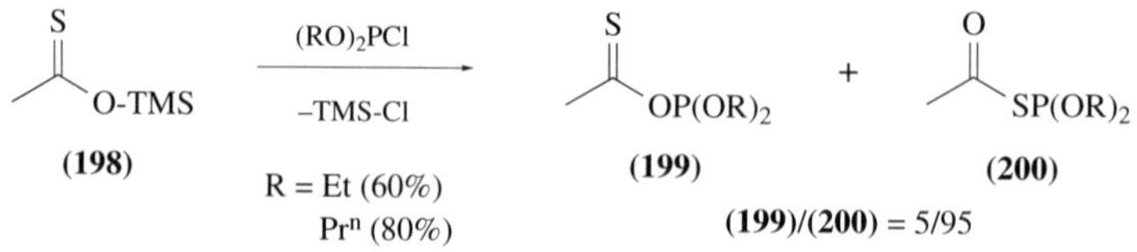

(30)

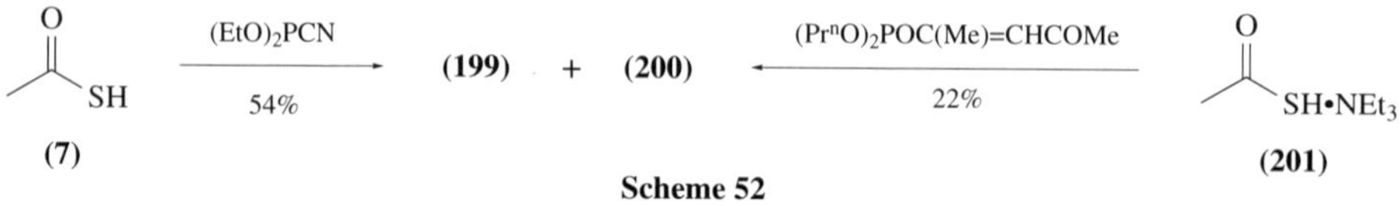

Scheme 52

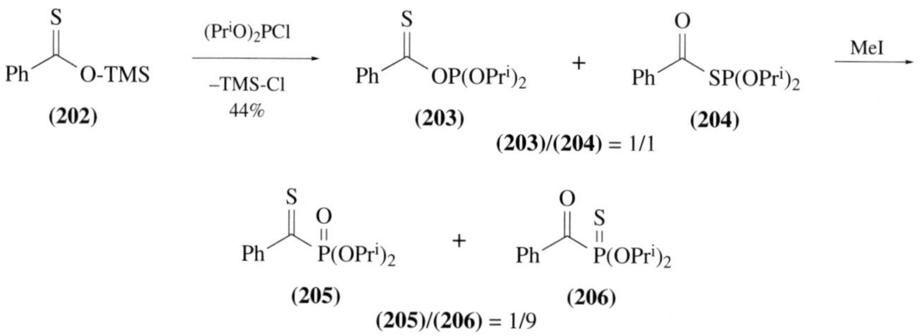

Scheme 53

5.12.4.5.3 Si-Thionoacyloxy compounds and their selenium and tellurium analogues

(i) Si-Thionoacyl compounds

(a) Preparation. Two synthetic routes to *O*-silyl thiocarboxylates (**207**) are available which differ in the final step. Route A involves the reaction of thiocarboxylic acids and their derivatives with silylating agents, and Route B the reaction of carboxylic acid halides or anhydrides with thiosilylating agents (Scheme 54).

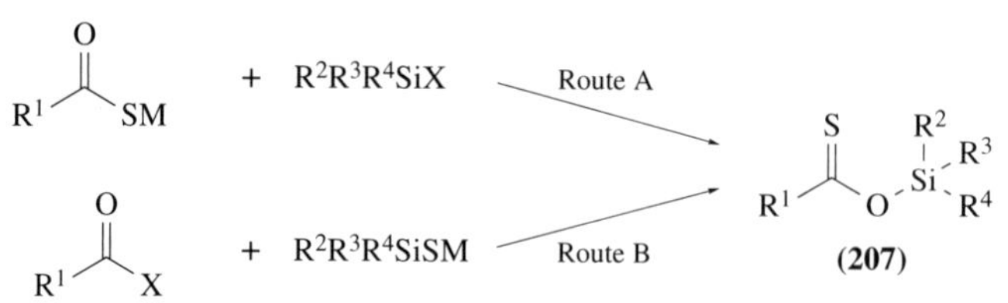

Scheme 54

Route A. Thiocarboxylic acids can be *O*-silylated with disilazanes ⟨66JOC3439, 82JOC3966⟩, silyl chlorides in the presence of a tertiary amine ⟨66MI 512-01, 66JOC3439⟩, or bis(silyl) carbodiimide (Equations (31)–(34)) ⟨76MI 512-01⟩.

(31)

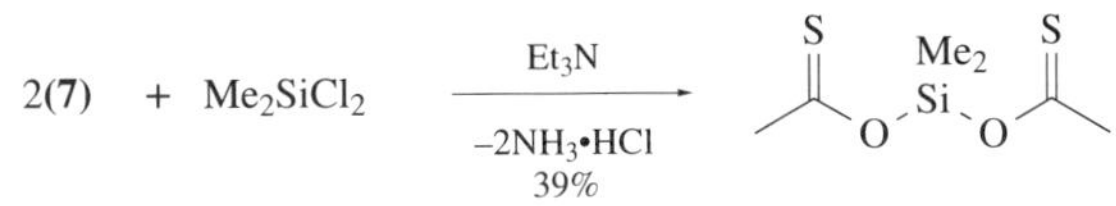

(32)

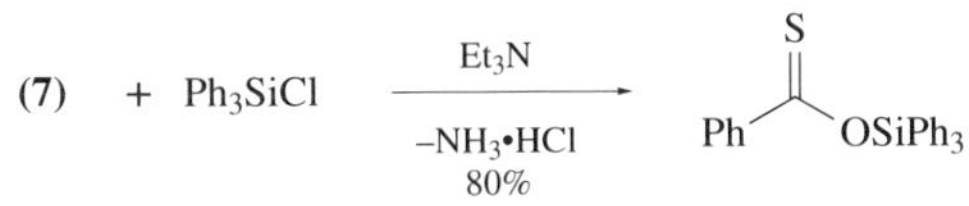

(33)

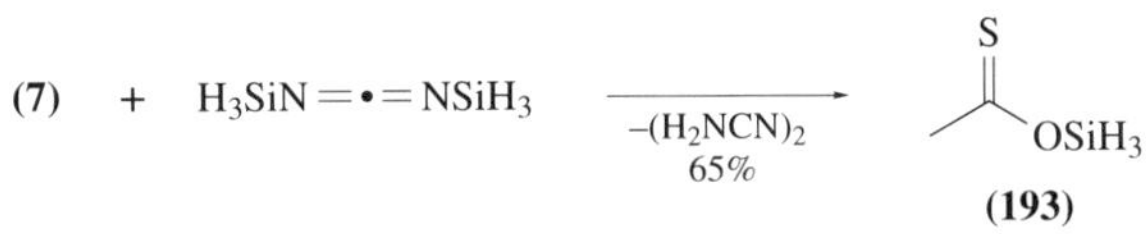

(34)

Potassium or ammonium salts of thiocarboxylic acids react with chlorotrimethylsilane to give the corresponding *O*-trimethylsilyl thiocarboxylate (Equation (35)) ⟨73BCJ244, 72IJS279⟩.

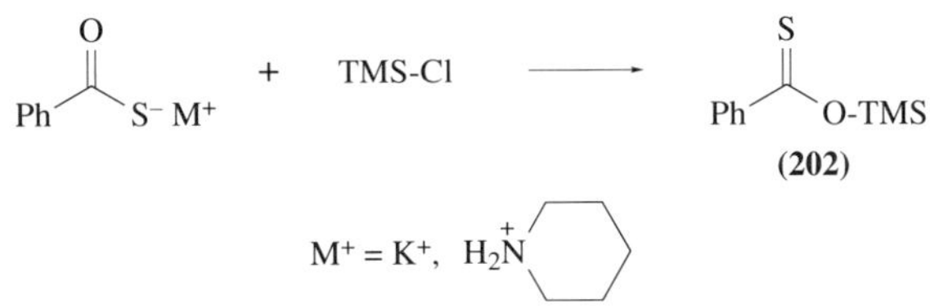

(35)

Kricheldorf and Leppert developed a convenient procedure which does not require isolation of thiocarboxylic acids ⟨71S435⟩. Thus, the reaction of carboxylic acid chlorides with dithiocarbamidate (**208**), prepared *in situ* from propylamine, triethylamine, and carbon disulfide, gave carboxylic thiocarbamic acid thioanhydrides (**209**) which in turn reacted with chlorotrimethylsilane to yield the desired *O*-trimethylsilyl thiocarboxylates (**210**) (Scheme 55). Mixed anhydrides (**211**) and (**212**) were also prepared by the reaction of thiocarboxylic acids with isocyanates or isothiocyanates (Scheme 56) ⟨73MI 512-01⟩.

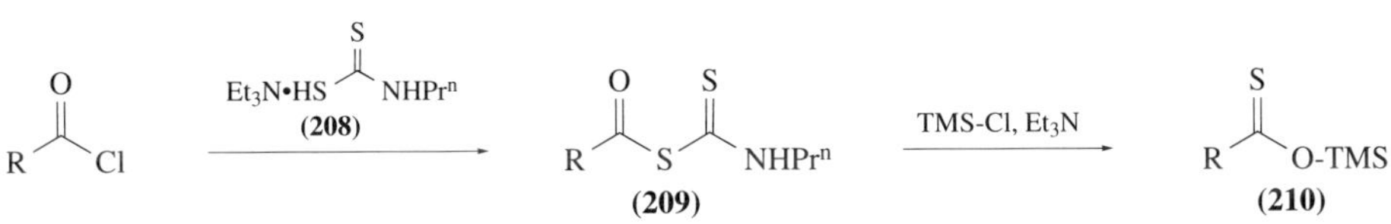

Scheme 55

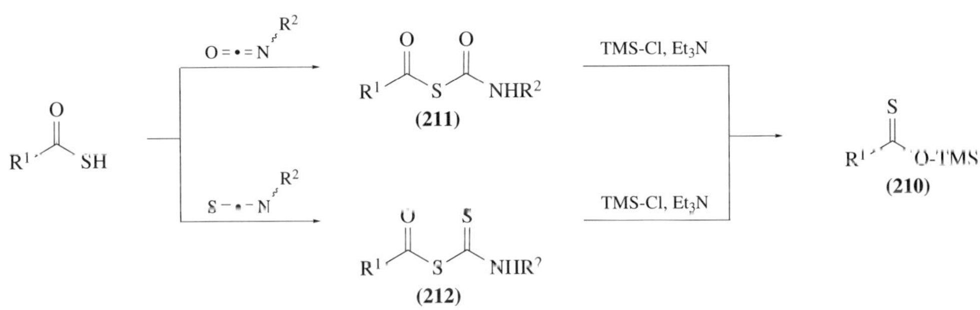

Scheme 56

The reaction of *S*-tributylstannyl thioacetate (**190**) with bromosilane gave *O*-silyl thioacetate (**193**) in 70% yield (Equation (36)) ⟨80M(111)193⟩.

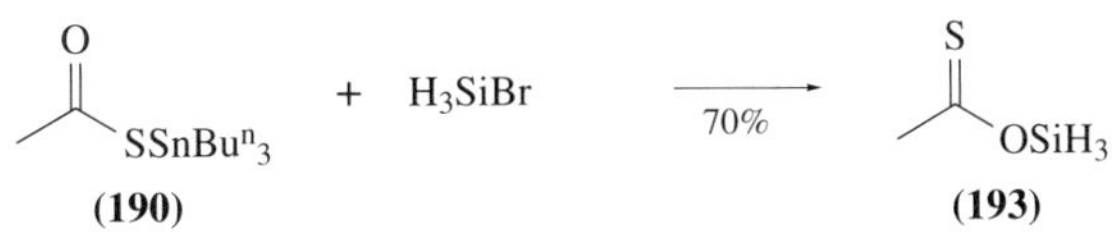

(36)

Route B. The reaction of carboxylic acid chlorides with bis(trimethylsilyl) sulfide (**213**) gave *O*-trimethylsilyl thiocarboxylates (**210**) in high yields (70–90%) (Equation (37)) ⟨72MI 512-01⟩. On the other hand, the reaction of acid anhydrides with (**213**) gives mixtures of (**210**) and trimethylsilyl carboxylates (**214**) whose separation is difficult. Therefore, this reaction is not practical. An exception is the reaction of succinic anhydride with (**213**) which gives the sole product (**215**) (73%) (Equations (38) and (39)) ⟨72MI 512-01⟩.

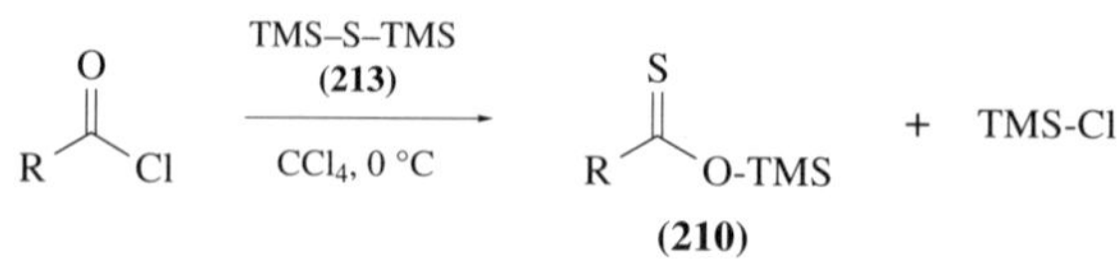

(37)

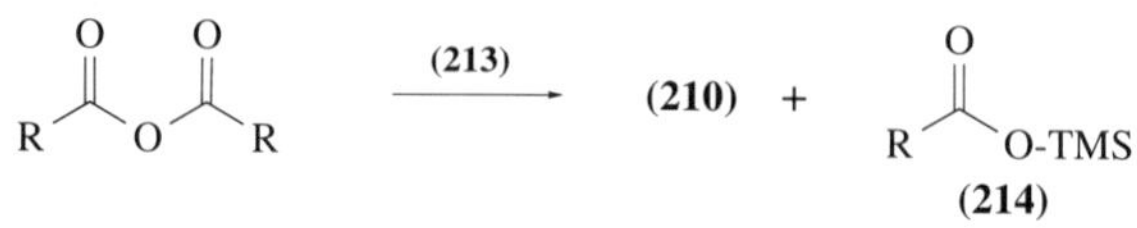

(38)

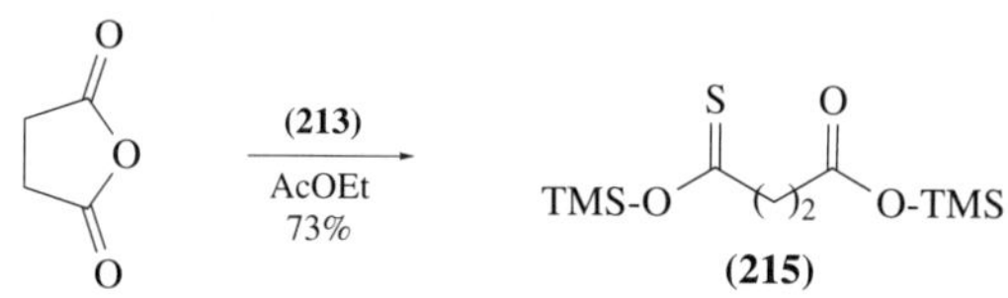

(39)

Later, Levedev *et al.* re-examined the reaction of carboxylic acid chlorides with bis(trimethylsilyl) sulfide (**213**) ⟨79JGU940⟩. The reaction was accelerated by addition of a tertiary amine such as pyridine, and the role of the amine was explained by the formation of an acylammonium ion (**216**) which is a more powerful acylating agent than the acid halide itself (Scheme 57). Interestingly, the reaction of hexyloxydimethylsilanethiol (**217**) with benzoyl chloride in the presence of pyridine initially gave *S*-trimethylsilyl thiocarboxylate (**218**). Thus, the IR spectrum of the reaction mixture (20 °C) showed absorptions due to C═O (1700 cm^{-1}), Si—O—C (1100 cm^{-1}), C═S (1020 cm – 1) and Si—S (460 cm^{-1}) along with S—H (2300 cm^{-1}). When the mixture was heated above 50 °C, the absorptions due to Si—S and C═O vibrations disappeared (Scheme 58). These observations indicate that the rearrangement of (**218**) to *O*-trimethylsilyl thiocarboxylate (**219**) occurred.

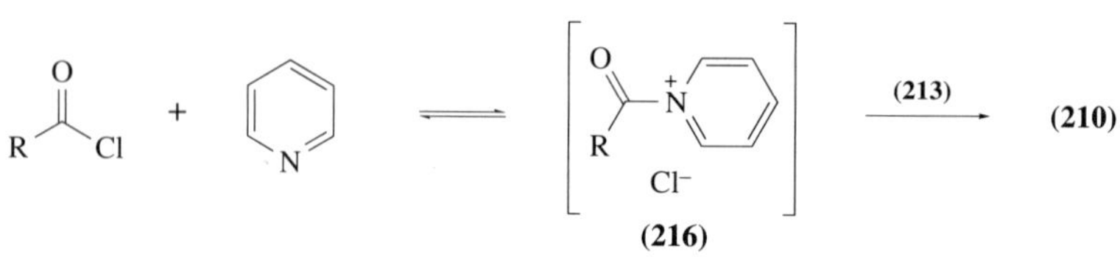

Scheme 57

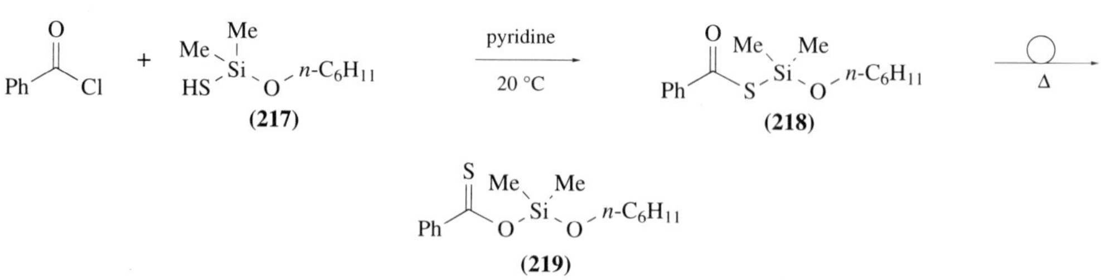

Scheme 58

The reaction of acetyl chloride with lithium silanethiolate (**220**), prepared by treating bis(silyl) sulfide with methyllithium at 210K, gave *O*-silyl thioacetate (**193**) in ca. 80% yield (Equation (40)) ⟨76JCS(D)1661⟩.

(220) **(193)** (40)

Sterically congested *O*-trimethylsilyl thiocarboxylates (**221**) and (**222**) were prepared by the reaction of acid chlorides with anion (**220**) (Equations (41) and (42)) ⟨80JCS(P2)1557⟩. The gas-phase thermolytic decomposition of (**221**) and (**222**) afforded thioketenes (**223**) and (**224**), respectively, though as minor products.

(220) **(221)** (41)

(222) (42)

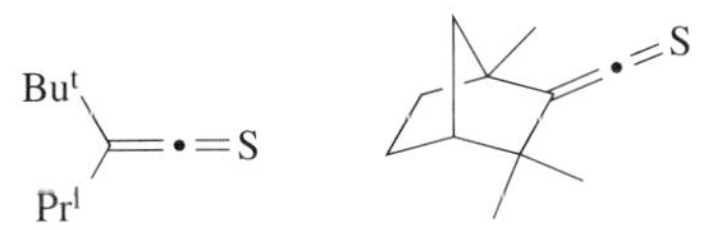

(223) **(224)**

(*b*) *Properties. O*-Silyl thiocarboxylates are extremely sensitive to moisture. *O*-Trimethylsilyl thiocarboxylates (**210**) are readily hydrolyzed by water, or even by moisture in air, to give hexamethyldisiloxane and the corresponding thiocarboxylic acids (Equation (43)) ⟨66JOC3439, 73BCJ244⟩. The color of *O*-silyl thiocarboxylates are generally yellow and the longest absorption maxima, assigned to the $n \rightarrow \pi^*$ transitions of the C=S groups, are observed around 430 nm in their UV–Vis spectra ⟨73BCJ244⟩. In the IR and Raman spectra, absorptions assigned to the C=S stretching are observed in the range 1216–1240 cm^{-1} ⟨73BCJ244, 76MI512-01⟩.

(210) (43)

While formation of an *S*-silyl thiocarboxylate as an intermediate was observed in a certain reaction ⟨79JGU940⟩, most silyl thiocarboxylates were isolated in the *O*-silyl form. The predominance of *O*-silyl form has been interpreted by consideration of bond energies of the both isomers which indicates that *O*-silyl thioacetates are more stable than *S*-silyl thioacetates by approximately 15 kcal mol^{-1} ⟨66JOC3439⟩. Cradock *et al.* who investigated the low-temperature NMR spectra of silyl thioacetate (**193**) reported that the thioester exists as an equilibrium mixture of *O*-silyl (**193**) and *S*-silyl (**225**) isomers in a ratio of 7 : 1 at room temperature (Scheme 59) ⟨72JCS(D)359⟩. On the other hand, the x-ray and electron diffraction analyses of (**193**) confirmed that the silyl group is bonded to oxygen, although the presence of a very small amount of sulfur-bonded isomer in the gas phase can not be excluded ⟨82JCS(D)1131⟩. In addition, the Si and S atoms are in the planar *cis* conformation in both gas and solid phases, and intramolecular Si···S contacts are 3.14 Å in gas and 3.19 Å in solid phases, which are shorter than the sum of van der Waals radii for Si and S (3.9 Å) (Figure 3). Interestingly, the intermolecular Si···S distance and the angle C=S···Si are 3.38 Å and 101°, respectively, indicating that the intermolecular interaction between Si and S is weak but highly directional in character.

(*c*) *Reactions.* Reactions of *O*-trimethylsilyl thiocarboxylates (**210**) are summarized in Scheme 60. The reaction of (**210**) with ethanol was rapid at room temperature, while that with acetic acid was slow but accelerated by addition of sodium acetate ⟨66JOC3439⟩. The reaction with electrophilic

(193)/(225) = 7/1

Scheme 59

(a) (b)

Figure 3 Selected atomic distances (Å) of *O*-silyl thioacetate **(193)**; (a) in a gas phase; (b) in a solid phase.

species such as halonium ions ⟨86TL4593⟩, acyl chlorides ⟨72MI 512-07, 76JCS(P1)564⟩, alkyl halides ⟨72MI 512-01⟩ and triethylethoxystannane ⟨75JGU2421⟩ occurred at the sulfur atom. The reaction with D-glucal derivatives **(226)** involved electrophilic attack of the carbenium ion intermediates on the sulfur atom to give **(227)** ⟨88JOC845⟩. The reaction with aldehydes was explained by the intervention of the salt RC(=O)SH · DBU formed from a trace amount of contaminated RC(=O)SH and DBU ⟨81BCJ790⟩.

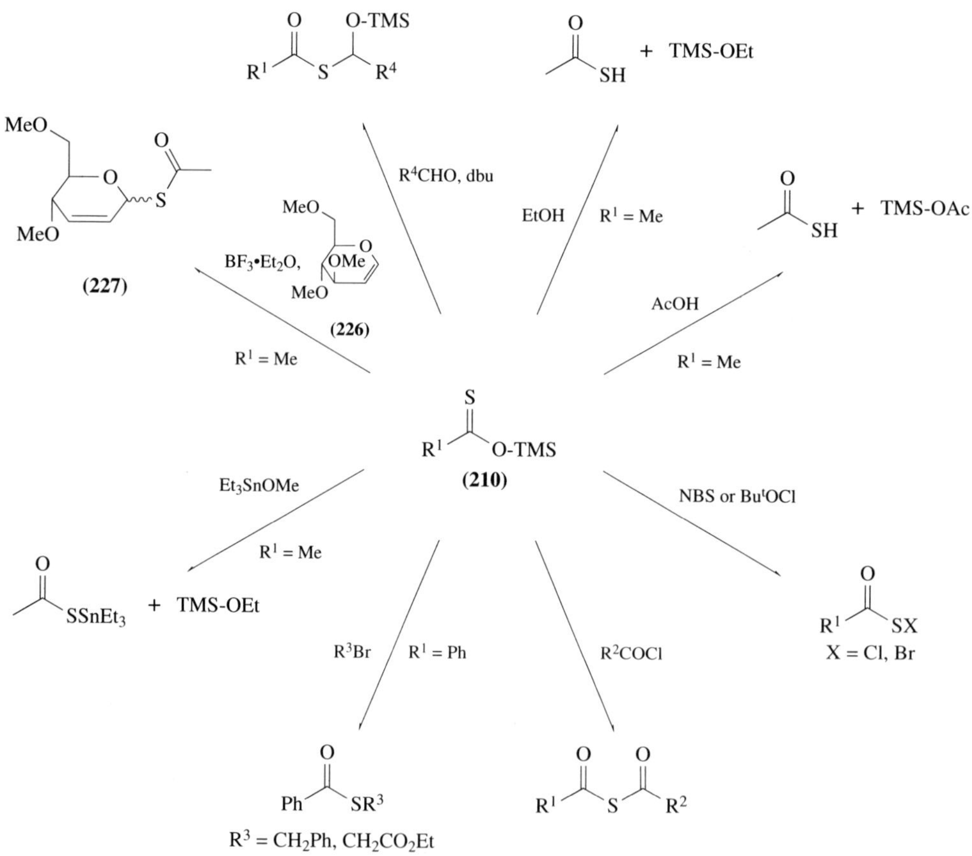

Scheme 60

(d) O*-Stannyl or* O*-germanyl thiocarboxylate.* It has been reported that applications of the synthetic methods for *O*-silyl thiocarboxylates to *O*-stannyl or *O*-germanyl analogues resulted in the formation of *S*-stannyl or *S*-germanyl thiocarboxylates ⟨73BCJ244, 75JGU2421, 76MI 512-01⟩. On

the other hand, Imoto and co-workers reported that the reaction of dichlorodimethyltin with thioacetic acid in water under basic conditions yielded dimethyltin bis(*O*-thioacetate) (**228**) in 83% yield ⟨79NKK1224⟩. The compound (**228**), colorless plates (m.p. 56.5–57.0 °C), show the C=S stretching at 1195 cm^{-1}. In the ^{1}H NMR spectrum, two singlets due to *Me*—Sn and *Me*—C(=S) appear at δ 0.99 and 2.43, respectively. In a similar manner, dimethyltin bis(*O*-thiobenzoate) (**229**) was obtained in 85% yield (Equation (44)).

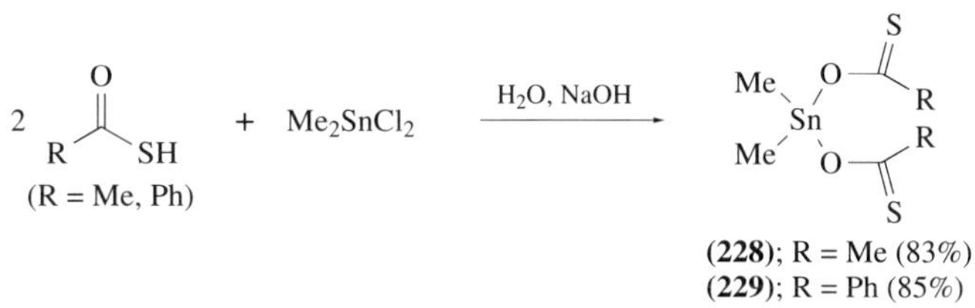

(44)

(ii) Si-Selenoacyl compounds

Silyl selenoacetate (**195**) was first prepared by the reaction of the ammonium salt of silaneselenol with acetyl chloride though as an inseparable mixture contaminated with disilyl selenide (**230**) and diacetyl selenide (**231**) (Equation (45)) ⟨72JCS(D)359⟩. Later, the selenoacetate (**195**) was obtained in a higher purity by the reaction of lithium silaneselenolate (**232**), prepared by treating disilyl selenide (**230**) with methyllithium, with acetyl chloride though still contaminated with ca. 10% of (**230**) (Equation (46)) ⟨76JCS(D)1661⟩. The reaction of (**232**) with acetic anhydride gave a mixture of silyl acetate and (**195**), the ratio depending on the reaction time (Equation (47)). A variable low-temperature ^{1}H NMR study disclosed that (**195**) exists as an equilibrium mixture of *Se*-silyl (major) and *O*-silyl (minor) forms; equilibrium constants (K) in Me_4Si, CCl_4–C_6H_6 and Et_2O are 0.26, 0.30, and 0.45, respectively ⟨72JCS(D)359, 76JCS(D)1661⟩.

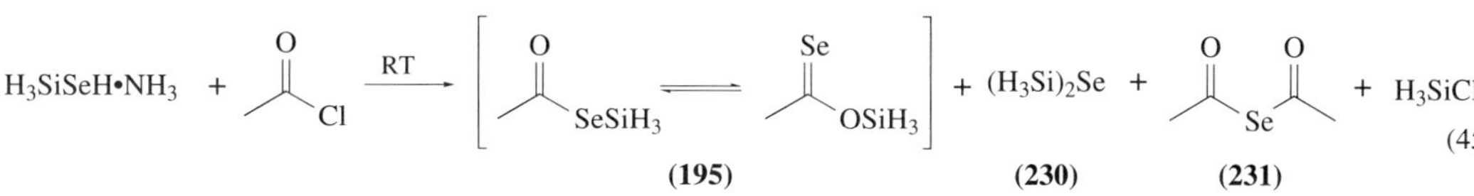

(45)

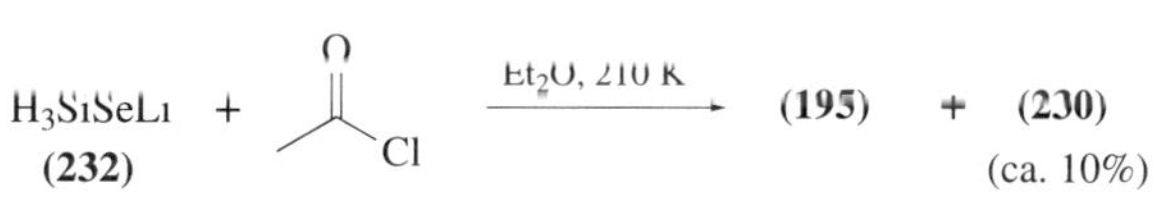

(46)

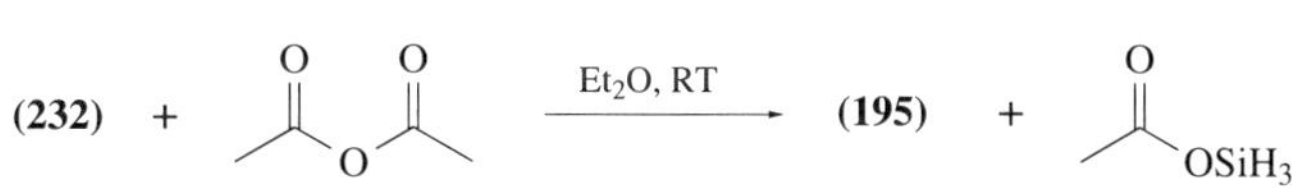

(47)

In contrast to the unsubstituted silyl selenoacetate, triorganosilyl selenocarboxylates exist exclusively in *O*-triorganosilyl forms. Thus, for example, *O*-trimethylsilyl selenobenzoate (**202**) was obtained as a reddish purple liquid by the reaction of potassium selenobenzoate with chlorotrimethylsilane in petroleum ether at room temperature in 68% yield (Equation (48)) ⟨72TL3751, 76CL203⟩. The IR spectrum of (**202**) showed no absorption due to carbonyl in the region between 1720 and 1600 cm^{-1} but a strong absorption at 962 cm^{-1} due to the selenocarbonyl. In 1992, Kato *et al.* synthesized more than twenty *O*-trimethylsilyl or *t*-butyldimethylsilyl selenocarboxylates (**234**) by the reaction of sodium or potassium selenocarboxylates (**233**) with the corresponding chlorosilanes (Equation (49)) ⟨92CB417⟩. These compounds were fully characterized by ^{1}H and ^{13}C NMR, IR, UV–Vis, and MS.

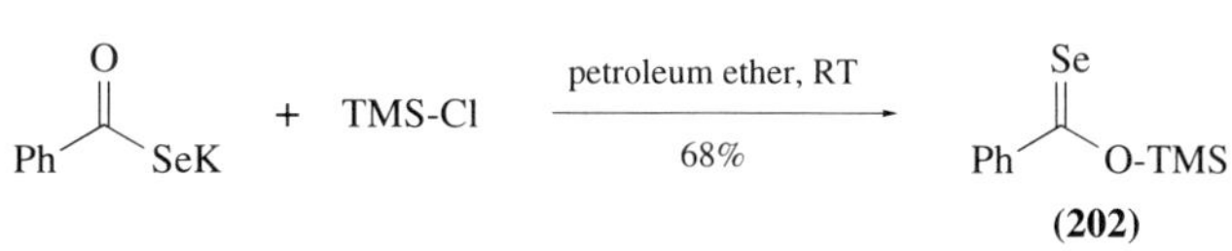

(48)

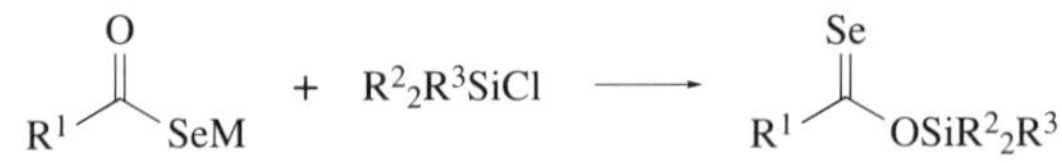

(233) (234) (49)

R^1 = alkyl or aryl; M = Na or K
$R^2 = R^3$ = Me or R^2 = Me, R^3 = Bu[t]

In the above synthesis, chlorotriorganosilanes initially attack on the anionic selenium atom to give *Se*-triorganosilyl selenocarboxylates which quickly rearrange to the corresponding *O*-triorganosilyl selenocarboxylates ⟨92CB417⟩. This intramolecular migration of a silyl group from Se to O was observable by ^{13}C NMR spectroscopy in the reaction of sodium salt (**235**) with *t*-butyldimethylsilyl chloride. Thus, with gradual decrease of the signal at δ 221.2 (C═O) of (**235**), two new signals appeared at δ 200.7 and 223.2, which are assigned to the carbonyl carbon of the *Se*-silyl isomer (**236**) and the selenocarbonyl carbon of the *O*-silyl isomer (**237**), respectively. The intensity of the signal at δ 200.7 quickly decreased and finally the signal at δ 223.2 only remained with increased intensity (Scheme 61).

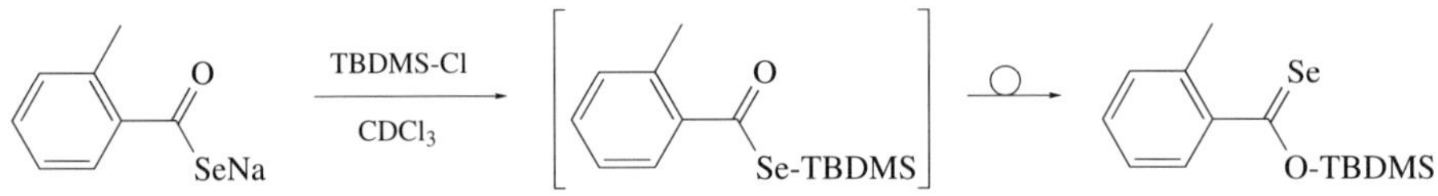

(**235**) $\delta_{C=O}$ 221.2 (**236**) $\delta_{C=O}$ 200.7 (**237**) $\delta_{C=Se}$ 223.2

Scheme 61

O-Triorganosilyl selenocarboxylates are thermally stable but very sensitive to water. For example, *O*-trimethylsilyl 4-methylselenobenzoate (**238**) was quickly hydrolyzed by moisture in the air to afford bis(4-methylbenzoyl) diselenide (**239**) and hexamethyldisiloxane as final products. (Scheme 62) ⟨92CB417⟩. In 1993, the preparation of rubidium and cesium selenocarboxylates (**241**) from *O*-trimethylsilyl selenocarboxylates (**240**) was reported (Equation (50)) ⟨93CC277⟩.

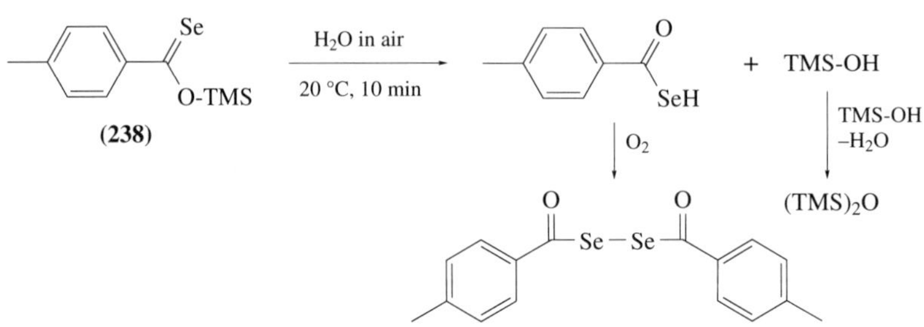

(**238**) (**239**)

Scheme 62

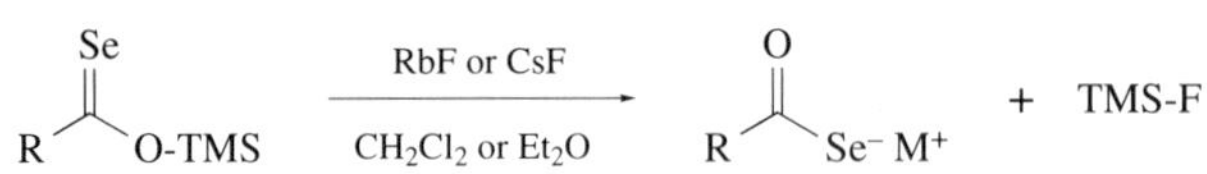

(**240**) (**241**) M = Rb or Cs (50)

(iii) Si-Telluroacyl compounds

The first isolable trimethylsilyl tellurocarboxylate was reported by Severengiz and du Mont in 1987 ⟨87CC820⟩. Thus, the reaction of bis(trimethylsilyl) telluride (**242**) with pivaloyl chloride or dipivaloyl telluride gave a 65 : 35 equilibrium mixture of *O*-trimethylsilyl (**243**) and *Te*-trimethylsilyl (**244**) telluropivalates, which were characterized by ^{13}C, ^{29}Si, and ^{125}Te NMR spectroscopy (Equation (51)). The predominant existence of *O*-trimethylsilyl isomer (**243**) was interpreted in terms of the strong affinity of silicon for oxygen, though it is not sufficient to result in the quantitative formation of a C═Te double bond. The telluropivalate (**243**) has a characteristic blue color and exhibits the longest absorption maximum at 624 nm in the UV–Vis spectrum. The reaction of (**242**) with acetyl chloride or diacetyl telluride gave a blue solution suggesting the formation of *O*-trimethylsilyl telluroacetate ⟨85AG1051⟩.

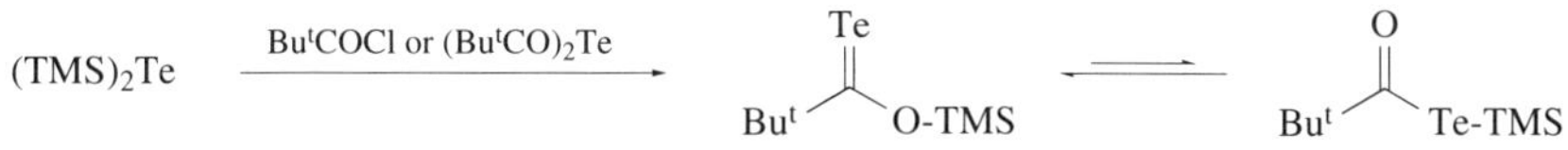

(51)

(242) **(243)** **(244)**

$\delta(^{13}C)$	251.2 (*C*=Te)	207.5 (*C*=O)
$\delta(^{29}Si)$	28.9	3.8
$\delta(^{125}Te)$	1418	229

Interestingly the reaction of sodium 2-methyltellurobenzoate (**245**) with *t*-butyldimethylsilyl chloride gave *Te*-triorganosilyl tellurocarboxylate (**246**) as an intermediate. Its formation and rearrangement to tellurocarboxylate (**247**) were detected by the ^{13}C NMR (Scheme 63) ⟨90TL3587⟩. On the other hand, in reaction of (**245**) with chlorotrimethylsilane and reaction of sodium tellurobenzoate with *t*-butyldimethylsilyl chloride, the rearrangement was too fast to detect the corresponding *Te*-triorganosilyl tellurocarboxylates by NMR spectroscopy.

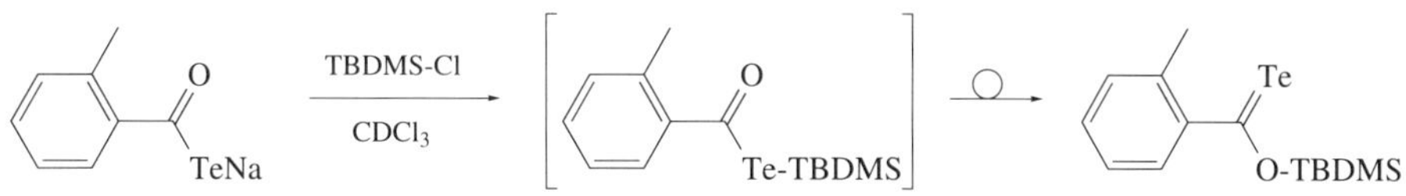

(245) δ_{C-O} 219.9 **(246)** δ_{C-O} 200 **(247)** $\delta_{C=Te}$ 232.1

Scheme 63

These two observations suggest that both steric protection (kinetic stabilization) and thermodynamic stabilization of the tellurocarbonyl group are necessary in order to shift the equilibrium exclusively to the *O*-silyl tellurocarboxylate side.

5.13
Functions with Two Chalcogens other than Oxygen

TOSHIAKI MURAI and SHINZI KATO
Gifu University, Japan

5.13.1 INTRODUCTION

The functionalities belonging to the category of having two chalcogens other than oxygen are shown in Figure 1 in their acid forms, along with their IUPAC names. In this chapter, the preparative methods for each functionality will be described. The most important feature of the compounds in Figure 1 is their stability under aerobic conditions when oxygen atoms of carboxylic acids are replaced with chalcogen atoms, since high-lying π orbitals and low-lying π^* orbitals involving carbon and chalcogen atoms may make the compounds more reactive on going from the first to the third row of Figure 1. In addition, hydrogen bonding stabilization is dramatically decreased, and the chalcogen–hydrogen bond such as S—H, Se—H, and Te—H becomes more susceptible to air oxidation on going from the left to the right column of Figure 1. In fact, only dithioic acids and their esters have so far been studied in detail. Only a few results have been obtained on synthetic efforts towards the selenium analogues. Furthermore, tellurium-containing acid analogues have not yet been reported, despite their potential importance from synthetic and spectroscopic points of view.

5.13.2 DITHIOIC ACIDS AND THEIR Se AND Te ANALOGUES

5.13.2.1 Dithioic Acids

The first example of dithioic acids was reported more than 120 years ago in 1866 ⟨76RCR1035⟩. However, it was in the middle of the 1960s that a large number of derivatives began to be synthesized.

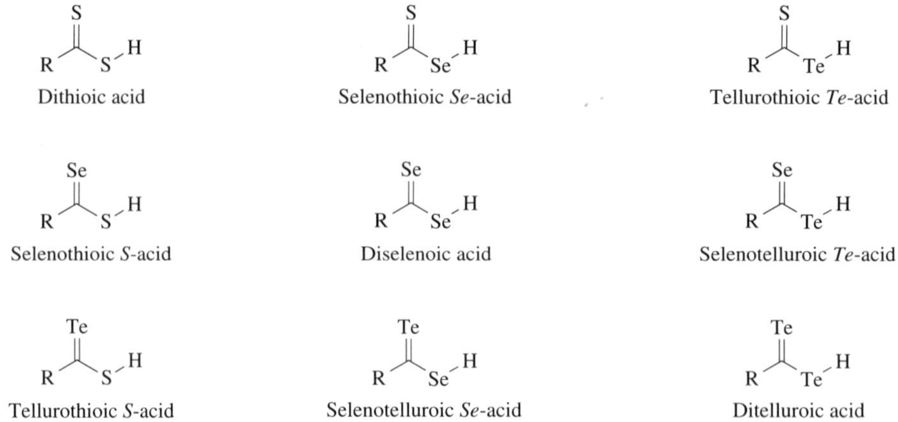

Figure 1 Acid analogues with two chalcogens other than oxygen.

The most important step in constructing the dithiocarboxyl group of dithioic acids was established by the late 1970s ⟨72MI 513-01, 79COC(3)373, B-79MI 513-01, 83S605, 85HOU(E5)891⟩. These methods leading to dithioic acids with simple alkyl and aryl groups have been thoroughly reviewed ⟨B-79MI 513-02⟩. Accordingly, this chapter deals mainly with the later improved methods, and the synthetic examples of dithioic acids with a variety of functional groups.

5.13.2.1.1 Dithioalkanoic acids

In general, the dithioalkanoic acids smell unpleasant and are susceptible to air oxidation to form the corresponding disulfides, or to self-oligomerization. They should therefore be purified immediately prior to use. In contrast, the formation of piperidinium salts of the acids highly enhances their stability. The piperidinium salts are odorless and are easily handled and stored below room temperature ⟨72BCJ3492⟩. The quaternary ammonium salts of dithioic acids have also been used ⟨85BSF891⟩.

The most general methods for the preparation of dithioic acids involve the hydrolysis of acid salts obtained by Grignard reaction using carbon disulfide (CS_2) ⟨B-79MI 513-02, 83S605, 84S797⟩ (Scheme 1). For example, dithiophenylacetic acid was obtained in 30% yield ⟨79JOC569⟩. In this reaction, the use of THF as a solvent efficiently increases the yield of the product. The dithioacetic acid ⟨07CB1303⟩ was prepared from methylmagnesium halide and was used directly for the next reaction ⟨81JOC196⟩. When dithioacetic acid is stored in a refrigerator, isomeric mixtures of its trimer—(**1**), (**2**), and hexathiaadamantane (**3**)—are obtained. The purification of a mixture of (**1**) and (**2**) via chromatography or distillation has been reported to give monomeric dithioacetic acid ⟨78TL3847⟩. Alternatively, the addition of hydrogen sulfide to trifluoroacetonitrile gives trifluorodithioacetic acid in ca. 70% yield (Equation (1)) ⟨69CB3347⟩. Dithioformic acid is formed only in solution and is easily trimerized ⟨72ZAAC(389)145⟩.

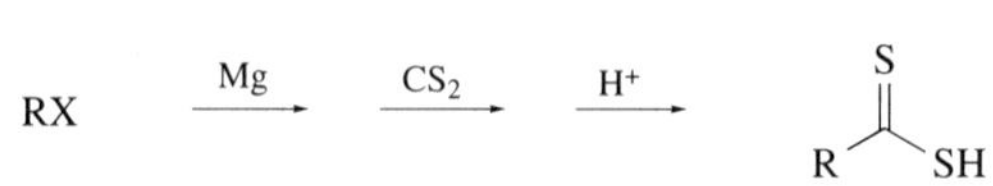

Scheme 1

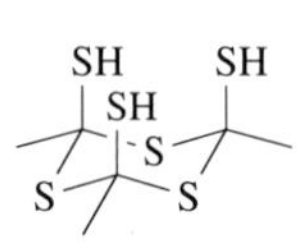

(1)

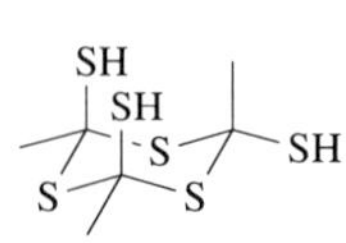

(2)

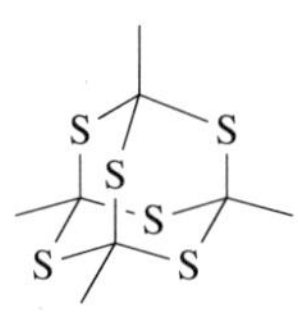

(3)

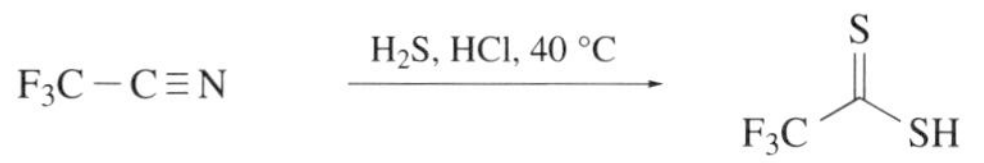

(1)

S-Deuteriodithioalkanoic acids can be obtained by the deuterolysis of dithioic acid *S*-trimethylsilyl esters using an equivalent of D_2O (Equation (2))⟨82S457⟩. This avoids the difficulty in removing H_2O and excess D_2O ⟨76BSF1145⟩.

(2)

γ,δ-Alkenyldithioic acids are prepared via [3,3]-sigmatropic rearrangement of enethiolates, which are derived from the proton abstraction of dithioic acid *S*-allylic esters in high yields using lithium amide ⟨78S439⟩ (Scheme 2).

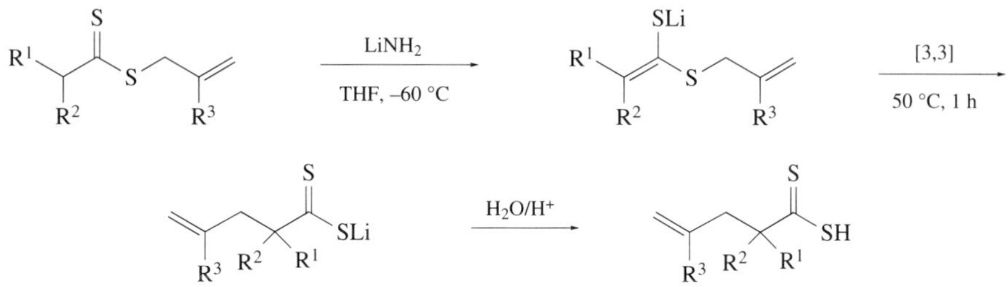

$R^1 = R^2 = R^3 = H$, 90%
$R^1 = R^2 = H$, $R^3 = Me$, 83%
$R^1 = Mc$, $R^2 ■ R^3 = H$, 90%
$R^1 = R^3 = Me$, $R^2 = H$, 83%
$R^1 = R^2 = R^3 = Me$, 80%
$R^1 = R^2 = Me$, $R^3 = H$, 82%

Scheme 2

In contrast to the difficulty of isolating dithioalkanoic acids with only simple alkyl groups, β-oxo- ⟨72T5341, 86T3029⟩ or β-iminodithioalkanoic acids ⟨72JOC1727, 82YGK123⟩ can be obtained as solids and are easily handled. This is mainly due to the formation of the tautomeric isomers involving intramolecular hydrogen bonding, as shown in Scheme 3. The synthesis of β-oxodithioic acids ⟨11CB1693⟩ is effected in 55–93% yields by the reaction of the enolates generated from aromatic ketones with potassium *t*-butoxide, with CS_2 followed by hydrolysis ⟨72T5341, 81OPP81⟩ (Scheme 4). β-Oxodithioic acids derived from aliphatic ketones show lower stability ⟨86T3029⟩, they are therefore formed *in situ* and used for further transformation ⟨93TL3703⟩. Tetrabutylammonium acetylacetone reacts with CS_2 to form ammonium salts, as in Equation (3) ⟨74T93⟩. β-Iminodithioic acids are prepared from cyclic ketones and CS_2 in the presence of ammonium hydroxide (Equation (4)) ⟨69JOC730, 73JCS(P1)1009⟩. The yields are dependent on the ring size and the substitution pattern on the ring. In the reaction of cyclohexanone, a large excess of CS_2 is necessary ⟨79JCR(S)140⟩. *N*-Alkyl-β-iminodithioic acids (**4**) are also formed by either an imido exchange reaction ⟨78JCS(P1)558⟩ or by using primary amines in water ⟨80S566⟩. The reaction of β-iminonitriles gives β-imino-β-cyanodithioic acids (**5**) in 34–62% yields ⟨77JCS(P1)1273⟩.

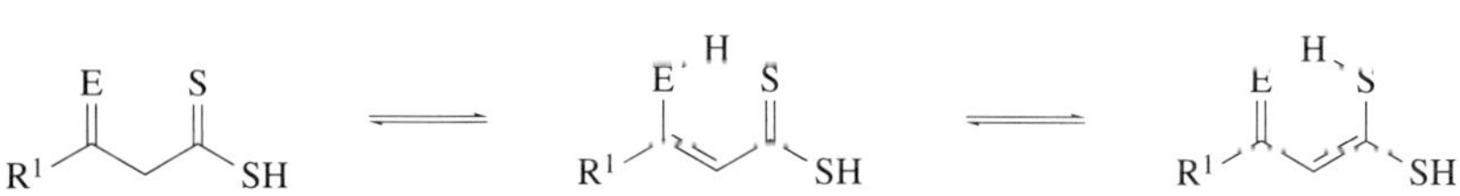

E = O or NR^2

Scheme 3

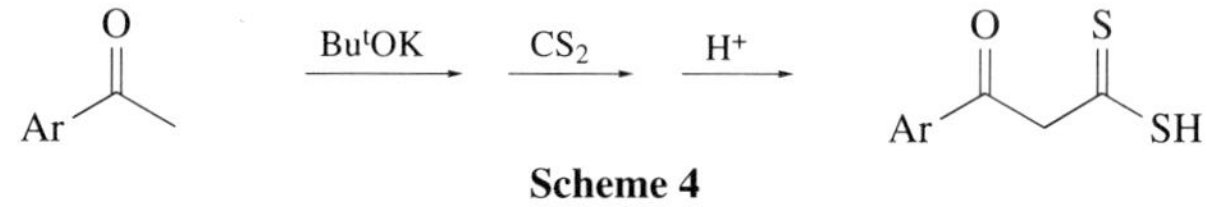

Scheme 4

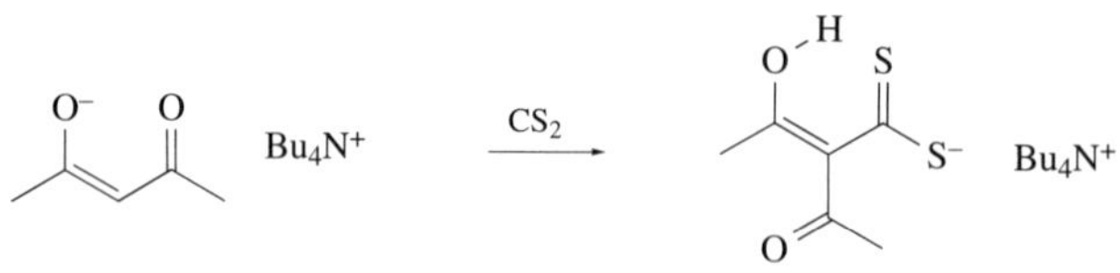
(3)

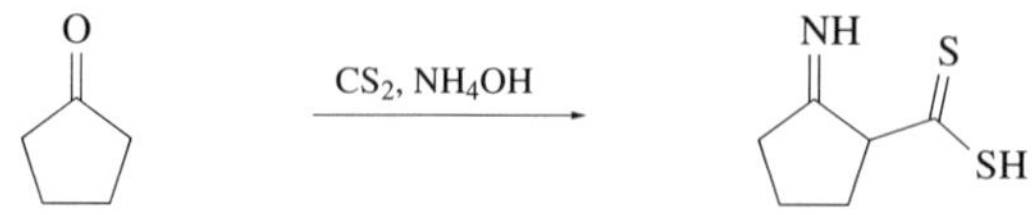
(4)

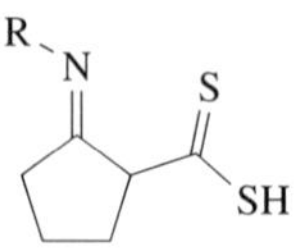
(4)

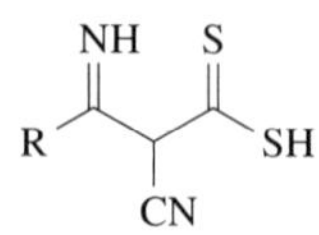
(5)

Attempts to obtain dithiomalonic acid (**6**) by the reaction of enethiolate with CS_2 have failed, resulting instead in the formation of its dimer in 45% yield (Scheme 5) ⟨91TL4679⟩. The addition of dithioic acid dianion to CS_2 initially occurs at the sulfur atom. The 1,3-rearrangement of the dithiocarboxyl group to a carbon atom may then give the tetrathiomalonic acid dianion.

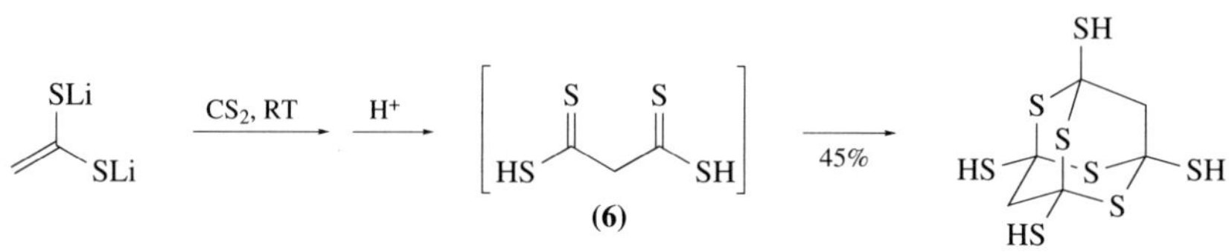

Scheme 5

The inner salts of dithioic acids bearing a diaminoalkyl group have been prepared by the reaction of tetraaminoethylenes and CS_2 (Equation (5)) ⟨65JA2776, 80CB3605, 82LA1451⟩, and by the reaction of diaminoethylene and sulfur (Equation (6)) ⟨92CC1522⟩.

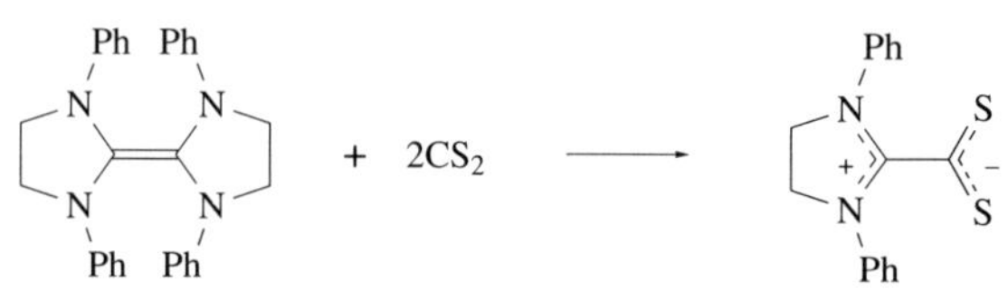
(5)

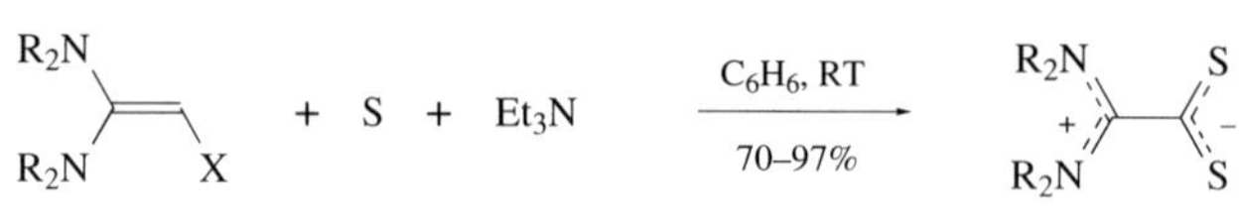
(6)

5.13.2.1.2 α,β-Unsaturated dithioic acids

Dithioalkenoic acids have not been studied to any great extent. The reaction of cinnamaldehyde with $(NH_4)_2S_2$ has been reported to give the dithiocinnamic acid ⟨24G389⟩, although the details are not available. The magnesium salts of dithioalkenoic acids have been prepared by the reaction of vinylmagnesium bromide, as in Equation (7) ⟨80CB919⟩, and are used directly for the synthesis of esters. Alkenyl copper reagents derived from the addition of alkyl copper to alkynes can also react with CS_2 to form copper salts of dithioalkenoic acids as intermediates (Equation (8)) ⟨79S432⟩. The acid hydrolysis of these metal salts has not been reported, probably because of their potential instability.

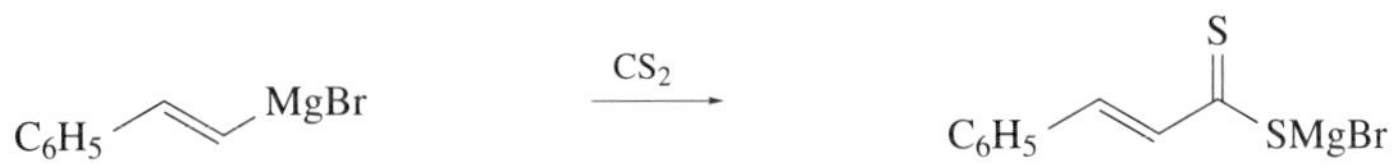
(7)

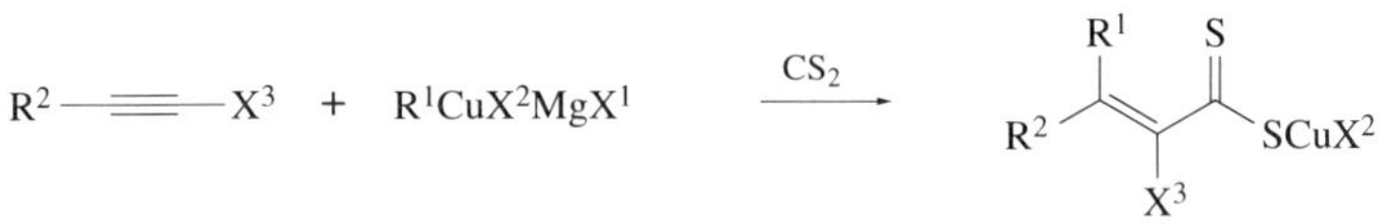
(8)

5.13.2.1.3 *α,β-Acetylenic dithioic acids*

Attempts to prepare α,β-acetylenic dithioic acids have only recently been reported. The reaction of lithium alkynides with CS_2 was reported to give lithium salts (Equation (9)) ⟨89TL1073⟩. The crude salts were obtained as a red powder. However, it is noted that these are not real acid salts, but decomposed polymeric products.

(9)

5.13.2.1.4 *Aromatic and heteroaromatic acids*

By analogy with the dithioalkanoic acids, the most general methods for the preparation of aromatic acids are acid hydrolysis of magnesium salts derived from aryl Grignard reagents with CS_2 ⟨28JA3106, 30JA332, B-79MI 513-02, 83S605, 85HOU(E5)891⟩. Piperidinium salts are suitable for storing in a refrigerator for many years ⟨73IJS359⟩. Ferrocenyl Grignard reagent is reacted with CS_2 to give ferrocenedithioic acid ⟨74JOM(78)405⟩. Alternatively, sulfhydrolysis of trihalomethyl- or dihalomethylbenzene derivatives with an alkali metal salt of H_2S gives the corresponding aromatic acids, as in Equation (10) ⟨B-79MI 513-02⟩. Aromatic aldehydes also react with $(NH_4)_2S_2$ to form dithioic acids in 23–45% yields (Equation (11)) ⟨51JA25⟩. Heteroaromatic compounds can undergo addition to CS_2 to produce the corresponding acids ⟨69LA(723)129⟩ (Scheme 6).

(10)

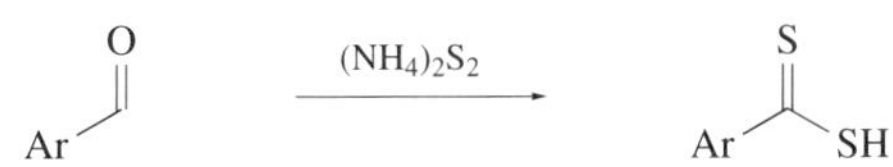
(11)

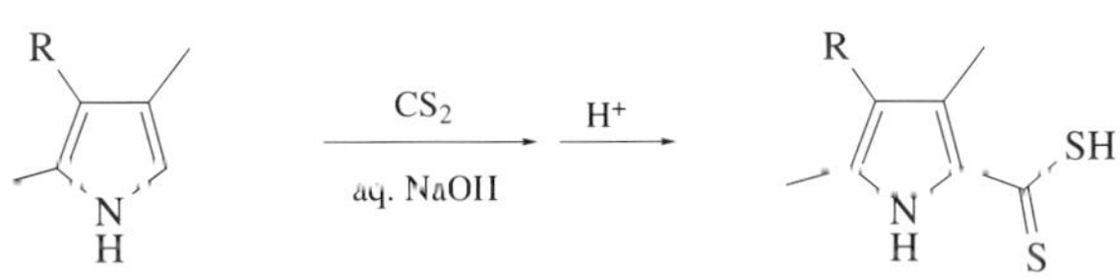

Scheme 6

5.13.2.2 Selenothioic Acids and Diselenoic Acids

The reaction of CSe_2 with $(C_2H_5)_2Zn$ to form zinc salts of diselenoic acid is the only example of the synthesis of selenium isologues of dithioic acids ⟨70MI 513-01⟩. Attempts to hydrolyze this salt have been reported to give the dimer of the acid (Scheme 7).

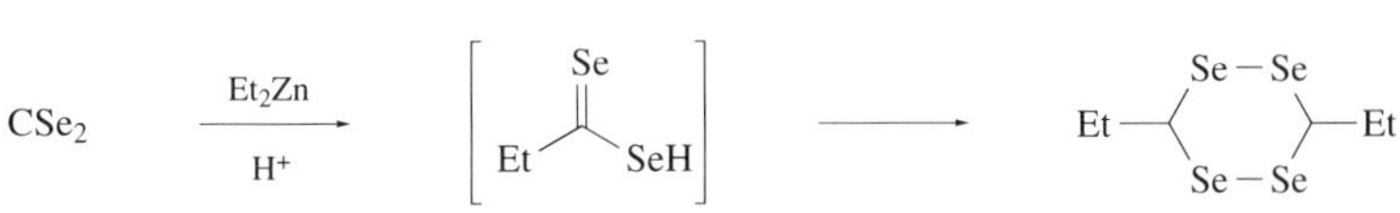

Scheme 7

5.13.3 DITHIOIC ACID ESTERS AND THEIR Se AND Te ANALOGUES

5.13.3.1 Dithioic Acid Esters

In contrast to the acids described above, many more studies on the esters have been reported. The synthetic methods have been thoroughly reviewed ⟨83S605, 85PS(23)253, 88SR155, 89PS(43)129, 89RHA152, 91COC(6)435, 92S1185⟩.

5.13.3.1.1 Dithioalkanoic acid esters

The dithioformic acid esters have been reported to be unstable species ⟨87JOC1156⟩. The formation of dithioformic acid *S*-butyl, *S*-benzyl, or *S*-phenyl ester by the reaction of thiol and carbon monosulfide was suggested as an intermediate leading to bis(alkylthio)methanes (Scheme 8).

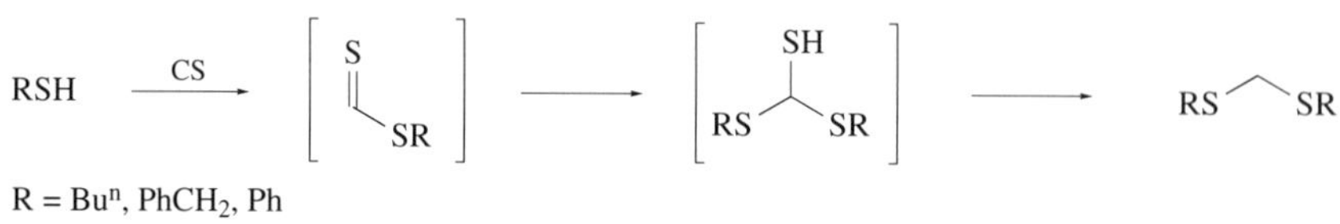

Scheme 8

The standard methods for producing dithioic acid esters are the alkylation of the corresponding acids or their metal or ammonium salts ⟨72BCJ3492, 73RTC601, 83S605, 85BSF891⟩. A variety of alkylating agents, and metal salts or ammonium salts of dithioic acids, can be used for the synthesis of simple dithioalkanoic acid esters ⟨88SR155⟩. Dithioacetic acid *S*-methyl ester is obtained by the methylation of the corresponding acid piperidinium salt ⟨74JOM(82)223⟩. Copper-catalyzed Grignard reaction gives dithioesters in higher yields ⟨79S432⟩. Alkylation using the sodium salt of chloroacetic acid gives the synthetically useful *S*-carboxylmethyl esters ⟨74CI(L)803⟩. Phenylene bis(carboxymethyldibromide) can also be used (Equation (12)) ⟨87CB411⟩. Alkylation with bromochloroalkane selectively proceeds at the carbon attached to bromine ⟨75RTC83⟩. Magnesium salts can react with α,β-unsaturated esters or γ-lactones to produce *S*-carboxyalkyl esters in 60–70% yields ⟨76IJC794⟩. The use of a catalytic amount of copper(I) bromide effects Grignard reaction of CS_2, and even *t*-butyl dithioic acid ester is obtained ⟨79S432⟩. Magnesium salts of dithioic acids react with chlorothioic acid *S*-phenyl ester to give dithioic acid *S*-phenyl esters in 47–73% yields with liberation of COS ⟨87S835⟩ (Scheme 9). Dithiocarboxylation of Grignard reagents with chlorodithioic acid *S*-ethyl ester can also give dithioic acid esters, but in modest yields ⟨76ZC355⟩. Benzyl anions are reacted with *S*-methyl trithiocarbonate to give α-phenyl dithioacetic acid esters ⟨77RTC118⟩.

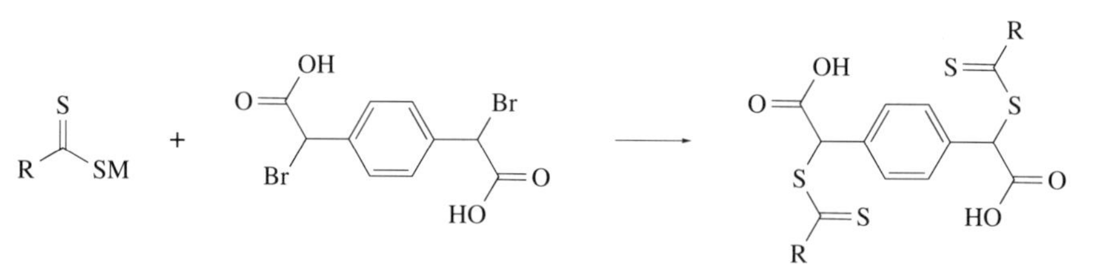

(12)

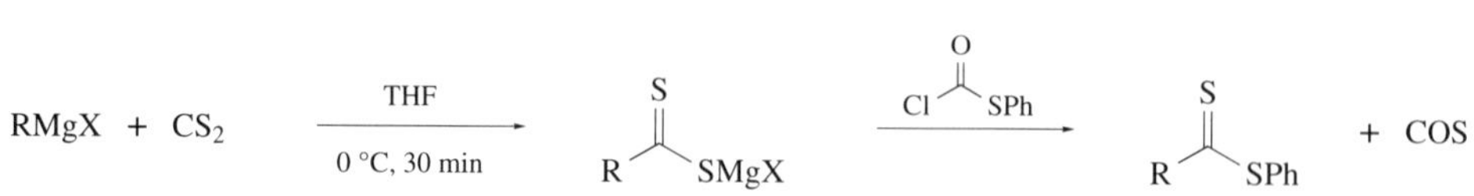

Scheme 9

Addition of dithioic acids to alkenes ⟨72T3203, 76BSF1145⟩ proceeds to give dithioesters. High regioselectivity was observed for the reaction with Michael acceptors ⟨72T3203⟩, whereas isomeric mixtures were obtained from alkyl-substituted alkenes ⟨76BSF1145⟩. By analogy with the synthesis of carboxylic acid esters, thioacylation of thiols using aliphatic thioacyl chlorides are reported ⟨75AG710⟩, although these are not easy to handle. Instead, thioketenes generated *in situ* ⟨66RTC889, 88T1827⟩ (Scheme 10) and dithioic acid thioanhydrides have been used as thioacylating agents. α,α-Bis(trifluoromethyl)dithioesters were prepared by the addition of thiols to the corresponding thioketenes 〈72JOC1347〉. In the reaction using dithioic acid thioanhydrides, *S*-*t*-butyl and *S*-aryl esters, which are not easy to obtain by other methods, are prepared 〈77AG(E)786, 82LA1229〉 (Scheme 11). Ester exchange reaction of dithioic acid esters with thiols has been applied to the synthesis of coenzyme A dithioic acid esters 〈88BBA269〉.

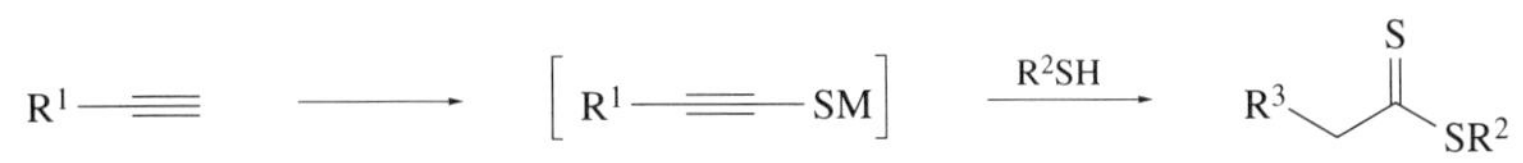

Scheme 10

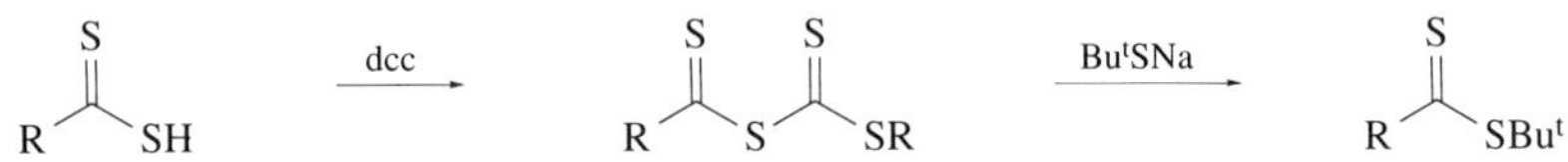

dcc = dicyclohexylcarbodiimide

Scheme 11

Sulfhydrolysis of the corresponding imidoesters with H_2S has been known as a facile preparative method for dithioic acid esters. The starting imidoesters are prepared from the reaction of nitriles with hydrogen chloride and thiols 〈33JA2966, 55JA5997〉 (Scheme 12). Alternatively, iminonium salts derived from thioamides are sulfhydrolyzed 〈79CB1650〉 (Scheme 13). The reaction of Grignard reagents with phenyl isocyanate, followed by alkylation, yields secondary thioamides, which are also used as precursors of imidoesters 〈84T1573〉 (Scheme 14, LDA = lithium diisopropylamide). Alkylation of α,β-unsaturated imidoesters prepared from vinyl Grignard reagents with phenyl isocyanate takes place selectively at the β-carbon. The resulting imidoesters are sulfhydrolyzed 〈84SUL147〉.

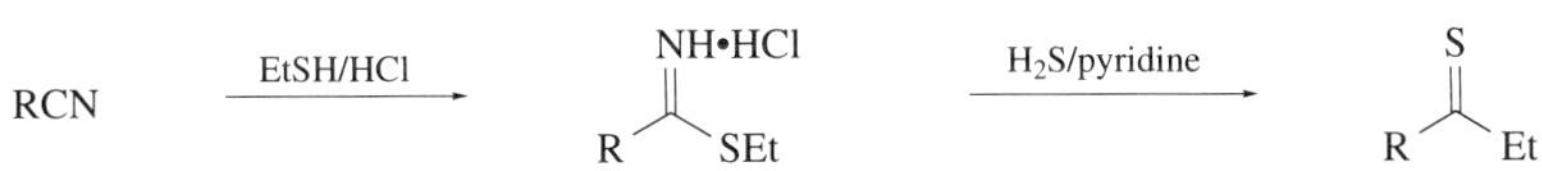

Scheme 12

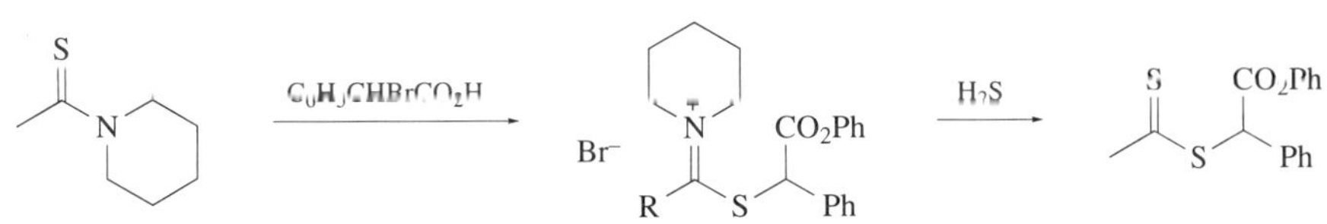

Scheme 13

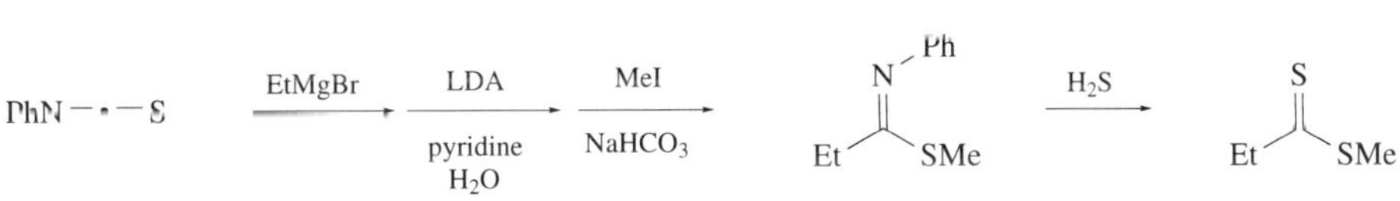

Scheme 14

Lawesson's reagent (**7**) is conveniently used for the conversion of thioic acid esters into dithioic acid esters 〈78BSB293〉, although high temperatures (xylene reflux) and long reaction times (more than 10 h) are necessary. More recently, Davy and Metzner have found that *S*-alkyl phosphorus sulfides (**8**) can effect the direct conversion of carboxylic acids to the corresponding dithioic acid esters in chlorobenzene at 130–140 °C (Equation (13)) 〈82CC457, 85JCR(S)272〉. *S*-Phenyl esters are prepared by using the reagents with an *S*-aryl group similar to (**8**) 〈84S827〉. Dithiophosphoric acids are also used for the similar transformation, but the yields and selectivity of the products are slightly lower 〈77ACS(B)460, 84T2663〉. A one-pot procedure for the conversion of carboxylic acids to dithioic acid esters is reported using P_4S_{10} and MeOH 〈85CI(L)824〉 (Scheme 15). Although careful control of the reaction temperature is necessary, this method appears to be the most convenient for the preparation of *S*-methyl esters.

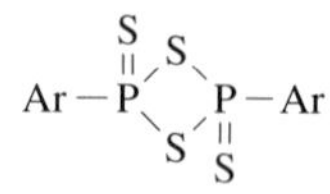

(7)

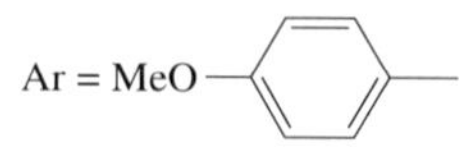

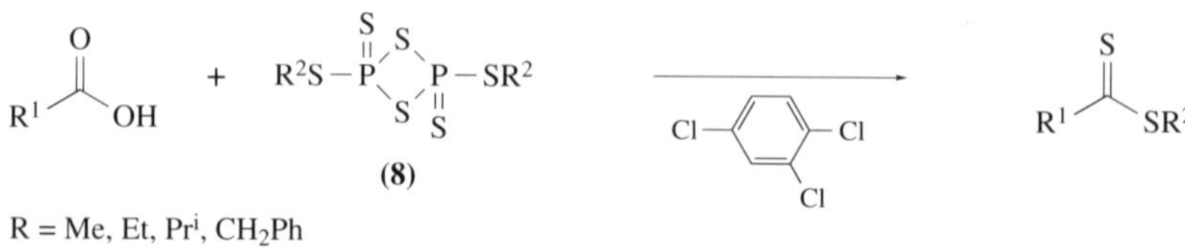

(13)

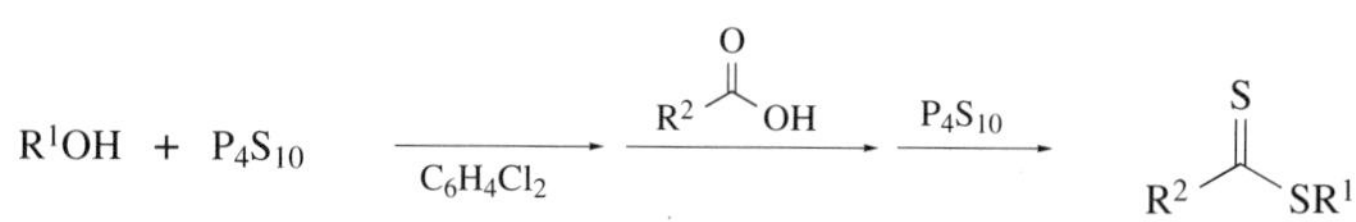

Scheme 15

Alkylation of enethiolates usually takes place at the sulfur atom to form thioacetals. In contrast, protonation of enethiols or their metal or ammonium salts gives the corresponding dithioic acid esters. This method is used for the preparation of α-aryldithioic acid esters, as in Equation (14) ⟨80S736⟩. 1,1-Dichloroethylene is also used as a starting material for dithioesters ⟨80S907⟩ (Scheme 16). *S*-Vinyl esters are obtained by this method ⟨80TL2165⟩. Isomerization of *S*-allyl esters in the presence of an excess amount of potassium *t*-butoxide leads to *S*-vinyl esters ⟨78S439⟩.

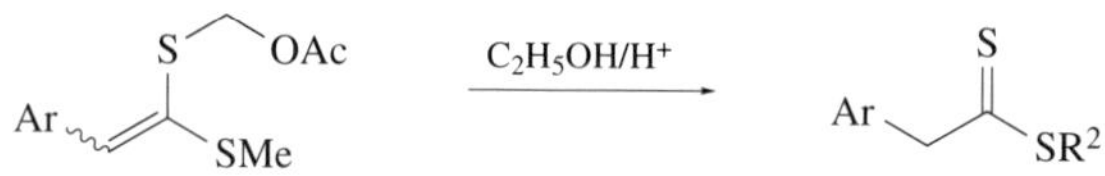

(14)

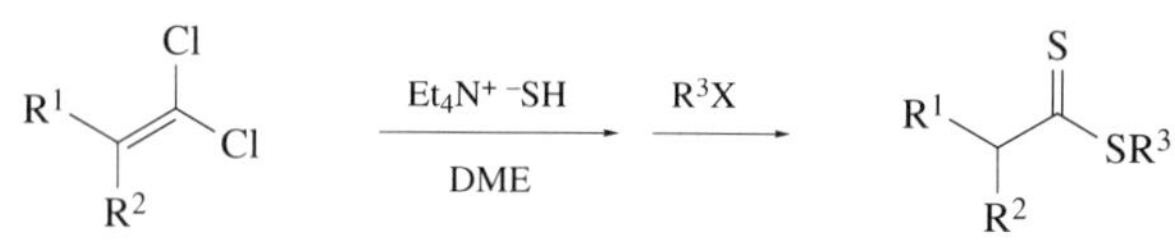

Scheme 16

The reaction of homoallyl magnesium halides with CS_2 followed by alkylation gives γ,δ-unsaturated dithioesters ⟨79TL2271⟩. Sulfhydrolysis of iminonium salts was used in the preparation of γ,δ-unsaturated esters with a norbornyl group ⟨84JCS(P1)865⟩. Brandsma and Schuijl found that addition of alkylthiols to lithium alkynethiolates initially produces allyl vinyl sulfides; thio-Claisen rearrangement then takes place ⟨68RTC929⟩. The introduction of a bulky group at the α-position of a thiocarbonyl group retarded the thio-Claisen rearrangement. Starting from *S*-allyl dithioic esters, the sequential operation—that is, the generation of a dianion, alkylation and rearrangement—gives trisubstituted γ,δ-unsaturated esters stereoselectively ⟨73JA5803⟩. Allylation of enethiolates and the successive thio-Claisen rearrangement ⟨79PS(7)69⟩ leads to γ,δ-unsaturated esters ⟨91PS1⟩. Allylic halides bearing a carbonyl group are also used ⟨80TL917⟩. However, in order to complete the rearrangement, 2 or 3 days are generally needed (Equation (15)). The diallylation of enethiolates formed from active methylene compounds and CS_2 yields *S*-allylic γ,δ-unsaturated esters ⟨73T2077⟩. The stereochemistry of the product depends on the geometry of the starting enethiolates ⟨83TL3617, 85T2691⟩. Allylation of (*E*)-enethiolate leads to the *syn* product via [3,3]-sigmatropic rearrangement at 100 °C, and the (*Z*)-isomer gives the *anti* product (Scheme 17). The allylation of the enethiolates derived from β-substituted dithioic acid esters quantitatively gives allyl vinyl sulfides. The following rearrangement also proceeds stereoselectively ⟨91PS1⟩. The ester having a cyclopropyl group is deprotonated and alkylated with methyl iodide to form the γ,δ-unsaturated ester via ring opening accompanied by [1,5]-hydrogen transfer (Scheme 18) ⟨88LA465⟩. The addition of enethiolates to β-alkenyl-β-propiolactones takes place selectively at the terminal olefinic carbon to give γ,δ-unsaturated esters having a carboxyl group ⟨83CL1901⟩. The stereochemistry of the products is also controlled, although bulky dithioesters give lower yields of the product (Scheme 19).

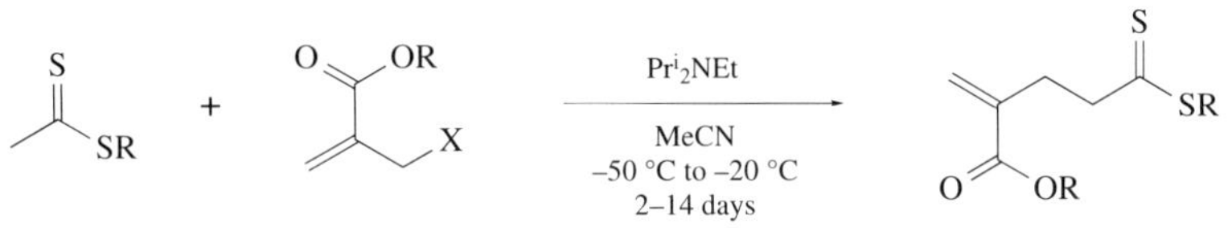

(15)

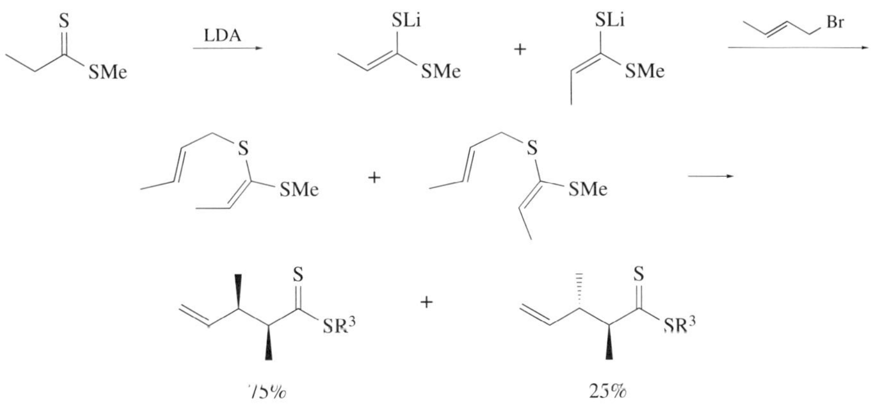

Scheme 17

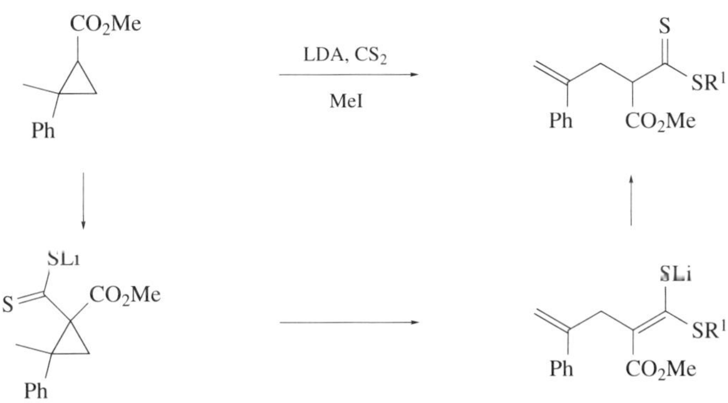

Scheme 18

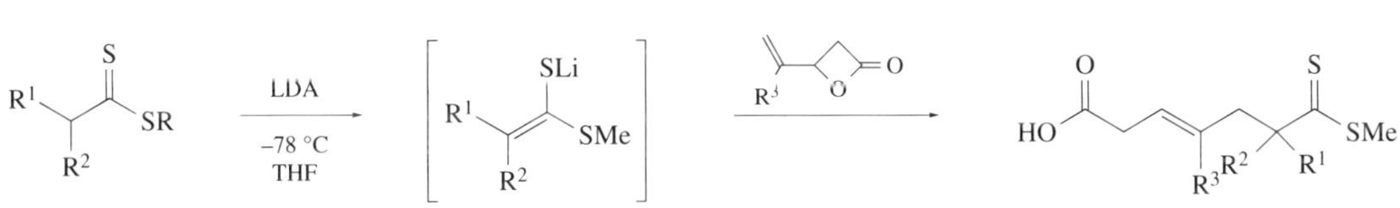

Scheme 19

The synthesis of β,γ-unsaturated esters has also been effected by the reaction of allyl Grignard reagents with CS_2 followed by alkylation. The addition to CS_2 takes place at the allylic carbon away from chlorine ⟨78TL4065⟩. Allylation of phenyl isocyanate, methylation, and sulfhydrolysis also give β,γ-unsaturated esters ⟨78TL2715⟩. Deprotonation of β,γ-unsaturated esters, allylation, and the rearrangement can produce dialkenyl esters (Scheme 20) ⟨86T2025⟩. The [2,3]-sigmatropic rearrangement of the hydrazone derivative is another route to β,γ-unsaturated esters via the carbene (Scheme 21) ⟨72CC354⟩. This procedure was applied to the synthesis of (±)-bakkenolide-A ⟨77JA5453⟩.

[3,3]-Sigmatropic rearrangement via the carbanion generated from allylic trithiane was also carried out ⟨75CR(C)1327⟩.

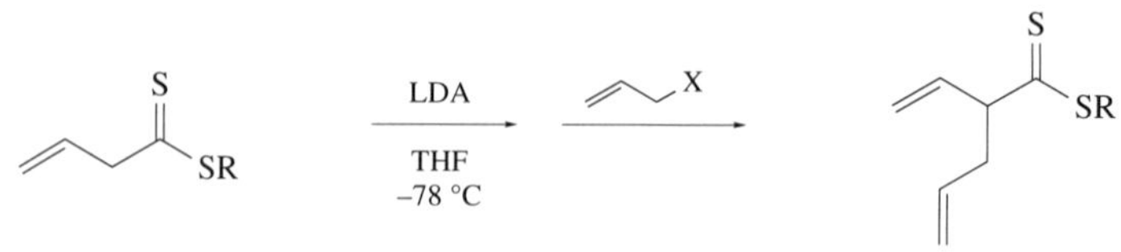

Scheme 20

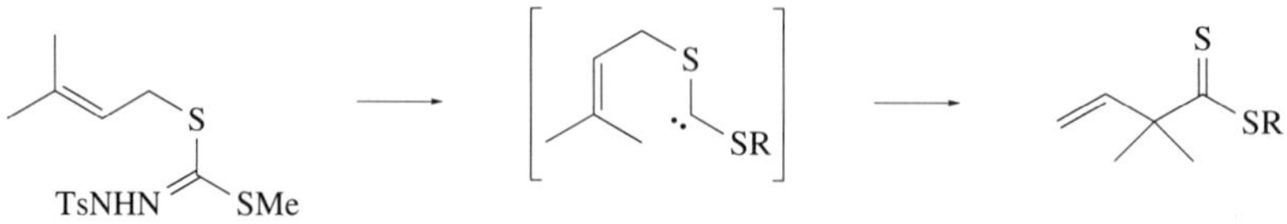

Scheme 21

The alkynylation of enethiolates with propargyl bromide yields α-allenyl esters (**9**) via thio-Claisen rearrangement ⟨69RTC597⟩. The reaction of allenyllithium with CS_2 followed by methylation leads to α-alkynyl esters (**10**) ⟨80TL1771⟩.

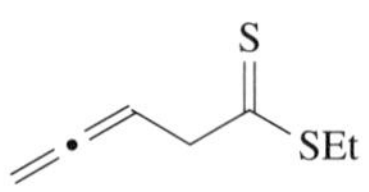

(9)

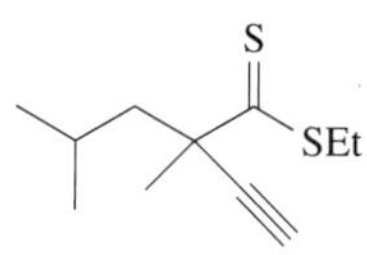

(10)

Synthetic efforts to prepare α-oxo dithioesters have increased since 1980. Alkylation of dithiooxalic acid alkali metal salts gives dialkyl dithiooxalates (**11**) ⟨78S849⟩. Alkylation initially occurs at the sulfur, and then the oxygen atom of oxalates is alkylated. Sulfurization of α-halo esters and amides with sulfur in the presence of amines followed by alkylation leads to α-oxo dithioesters (Scheme 22) ⟨77ZC366, 89JPR243⟩. Halomethyl aryl ketones are employed as a starting material ⟨78ZC90⟩. The use of diazoketones and α-pyridinium ketones also gives α-oxo dithioesters ⟨86S968, 93SL148⟩. α-Sulfonyl ketones prepared from *N*-phenyl glutarimide react with sulfur in the presence of triethylamine in DMSO and DMF at 0°C to give α-oxo dithioester ⟨90T1057⟩. Sulfonium salts (**12**) undergo cycloreversion with base to produce reactive α-oxo dithioesters (Equation (16)) which are trapped with dienes ⟨80JOC2601⟩. Application of this method to the synthesis of complex molecules was reported ⟨82JOC4384⟩. In contrast to the synthetic efforts towards α-oxo dithioesters, much less has been reported about α-thioxo dithioesters. Tetrathiooxalate (**13**) has been suggested as being generated by electrochemical reduction of CS_2 ⟨74JOC511⟩. The formation of the dimethyl ester of (**13**) was confirmed by mass spectrometry (Scheme 23) ⟨82ZC223⟩. Photochemical reaction of 1,3-dithiol-2-one has been reported to give the dimethyl ester of (**13**), as in Equation (17) ⟨76ZC318, 80CB1898⟩. In 1993, a reaction of an ynamine with sulfur under reflux in chlorobenzene was reported to yield an α-thioxo dithioester in 74% yield (Equation (18)) ⟨93BCJ623⟩.

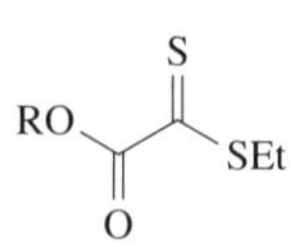

(11)

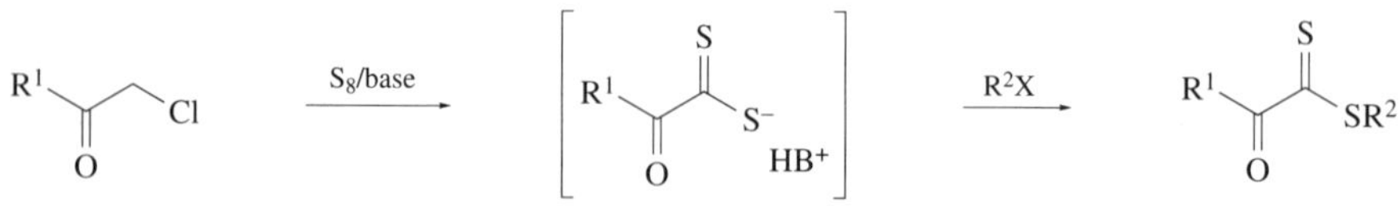

$R^1 = NH_2, NHR^3, OR^3$

Scheme 22

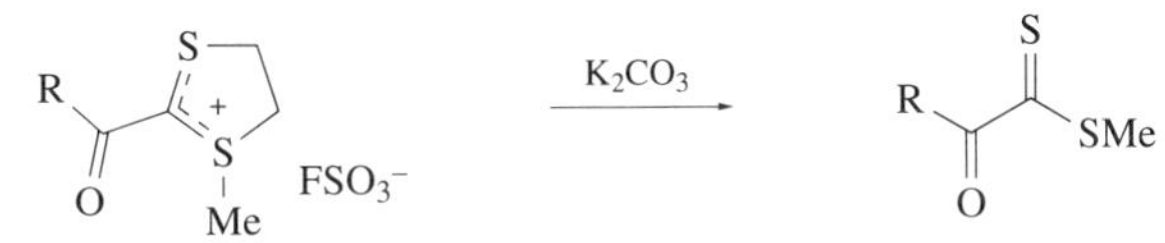

(12)

(16)

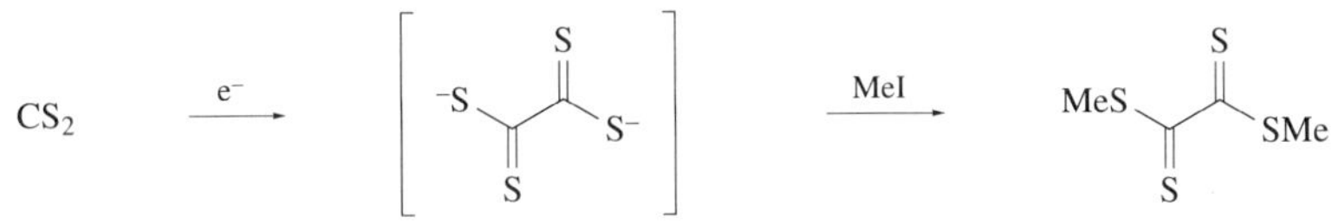

(13)

Scheme 23

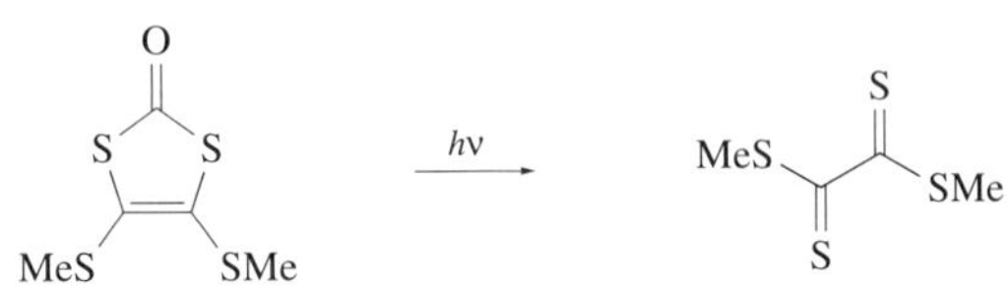

(17)

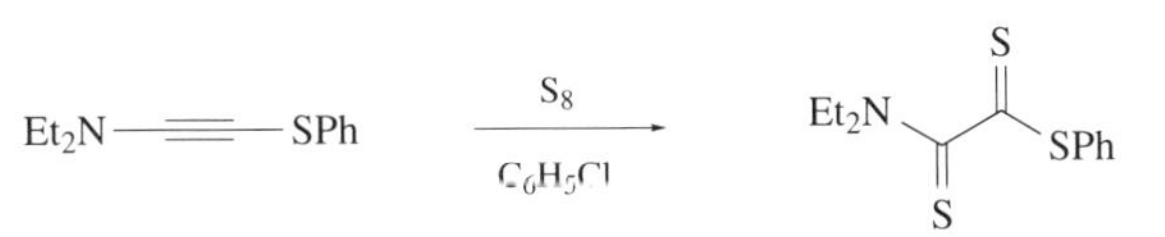

(18)

Alkylation of *β*-oxo dithioic acids usually gives ketenedithioacetals ⟨90T5423⟩ or a mixture with dithioesters ⟨86T3029⟩, except for the reaction of dithiomalonic acid dilithium salts ⟨76T2507⟩. The use of allylic halides as alkylating agents can give *β*-oxo dithioesters ⟨73T2077, 74T93⟩. As mentioned previously, allylation of enethiolates initially takes place at the sulfur atom, and thio-Claisen rearrangement then proceeds at 70–80 °C to give *β*-oxo dithioesters (Scheme 24) ⟨85JCS(P1)641⟩. The reaction of substituted enolate anions (**14**) with CS_2 and methylation produce *β*-oxo dithioesters (Scheme 25) ⟨82SC829⟩. Dimethyl trithiocarbonate (**15**) can be used as an efficient dithiocarboxylating agent for enolates derived from ketones, as in Equation (19) ⟨82S693⟩. The addition reaction of *β*-oxo or *β*-imino dithioic acids leads to the corresponding esters ⟨77JOC3383⟩.

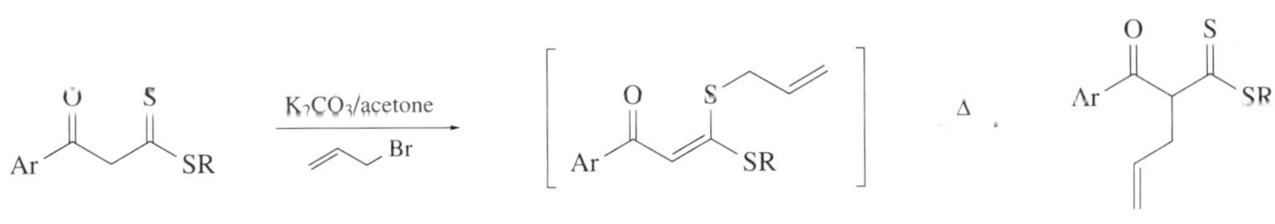

Scheme 24

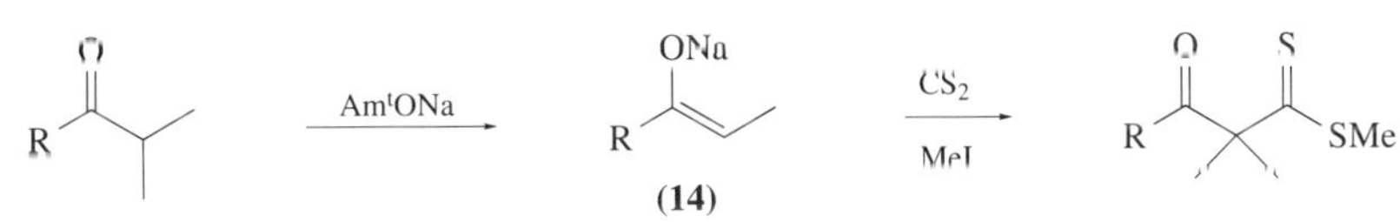

(14)

R = Me, Ph
Am^t = $EtMe_2C$

Scheme 25

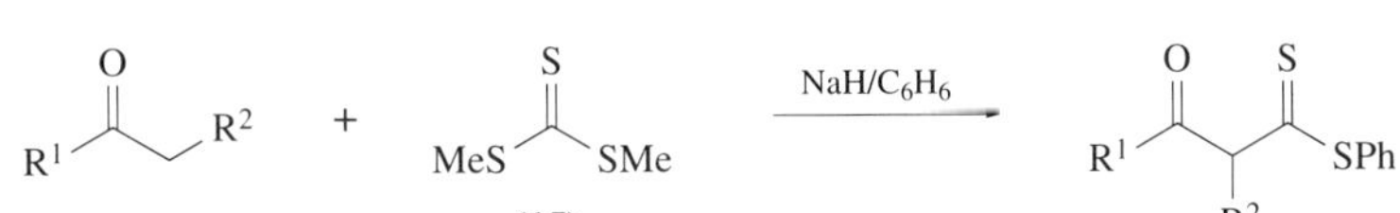

(15)

(19)

α-Carboxy dithioesters are prepared from carboxylic acids, CS_2 and methyl iodide in 77–88% yields ⟨86T3029⟩. The *α*-carboxy dithioesters thus obtained easily undergo decarboxylation (Scheme 26), in sharp contrast to malonic acid derivatives. Tetrathiomalonic acid esters have been prepared by the sulfhydrolysis of dithiomalonic acid dipyrrolidide, as in Equation (20) ⟨80LA483, 88AP(321)903⟩.

Sulfhydrolysis of imidodithioesters (**16**) gave the mixture of dithioesters (**17**) and (**18**) rather than tetrathiomalonic acid esters ⟨76CB3243⟩. γ-Thioxo or δ-thioxo dithioesters are prepared similarly by using the corresponding bisammonium salts ⟨77LA1751⟩. Thermal rearrangement of bis(methylthio) divinyl disulfides leads to tetrathiosuccinic acid esters ⟨74RTC258⟩.

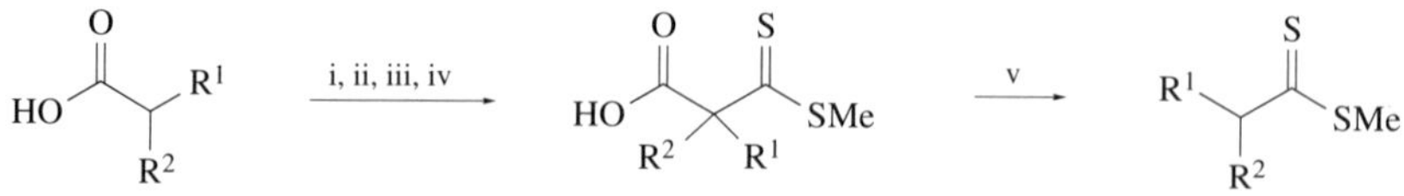

i, LDA (two equiv.), THF, HMPA; ii, Δ, 35 °C, 30 min; iii, CS_2, –30 °C; iv, MeI, –50 °C; v, NaOH or 125 °C, 15 min

HMPA = hexamethylphosphoramide

Scheme 26

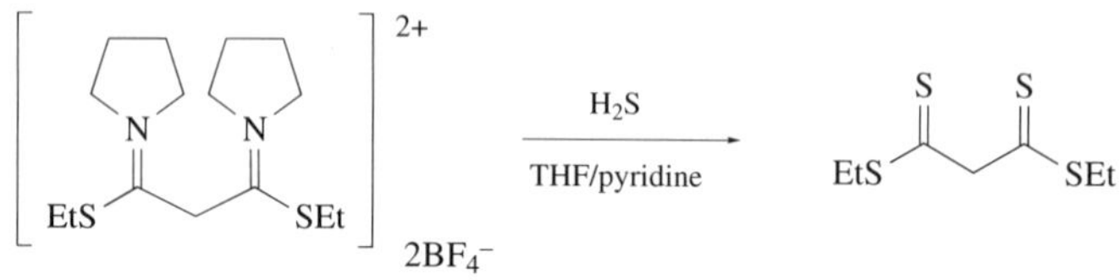

(20)

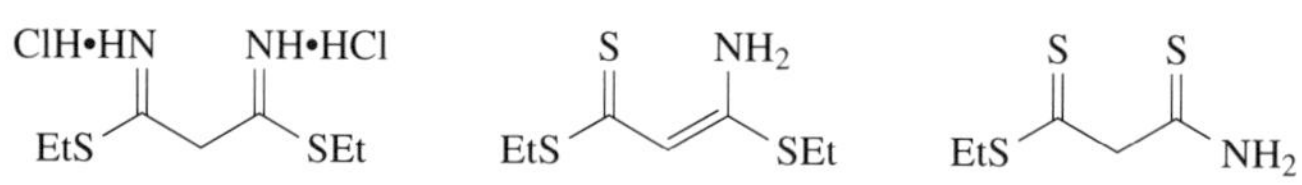

(**16**) (**17**) (**18**)

The reactions of enethiolates with α,β-unsaturated ketones ⟨89RHA152⟩ and esters ⟨90BSF446⟩ have been well studied as an efficient route to δ-oxo dithioesters. For example, the 1,4-addition of enethiolate to cyclohexenone selectively occurs at $-45\,^\circ$C to give the ester (Equation (21)) ⟨82CC335, 85BSF881⟩. The addition of enones to the sulfur atom of enethiolates was observed at $-78\,^\circ$C for disubstituted enethiolates (**19**) ⟨83TL4203⟩. Even in this case the reaction at $-50\,^\circ$C produces the *C*-alkylated product ⟨83CC388⟩. The addition to γ- or δ-methyl enones proceeds stereoselectively to give predominantly *trans* isomers (**20**) and (**21**), whereas the reaction of α-methyl enones exhibits low selectivity ⟨86BSF817⟩. α,β-Disubstituted enones are also reacted with enethiolates to form the *syn* isomers (**22**) exclusively ⟨87TL409, 88T3575⟩. The stereochemistry of the addition of enethiolates to β-substituted enones has disclosed that *cis* and *trans* selectivity depends on the geometry of the starting enethiolates. When (*E*)-enethiolate was selectively formed, the *anti* product predominated ⟨86TL1505, 89T2041⟩ (Scheme 27). A pseudo-eight-membered cyclic transition state is postulated to explain the stereochemical course of the reaction.

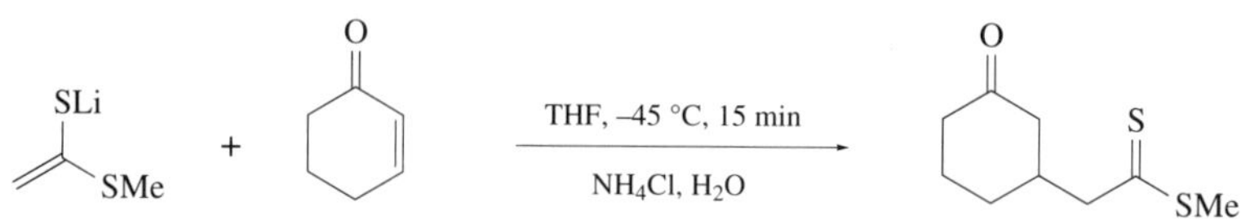

(21)

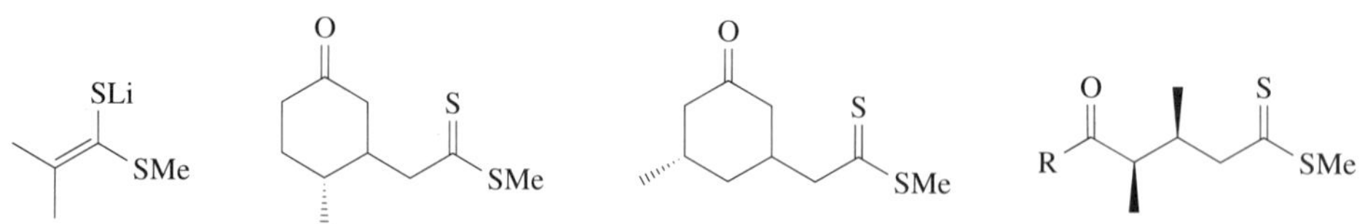

(**19**) (**20**) (**21**) (**22**)

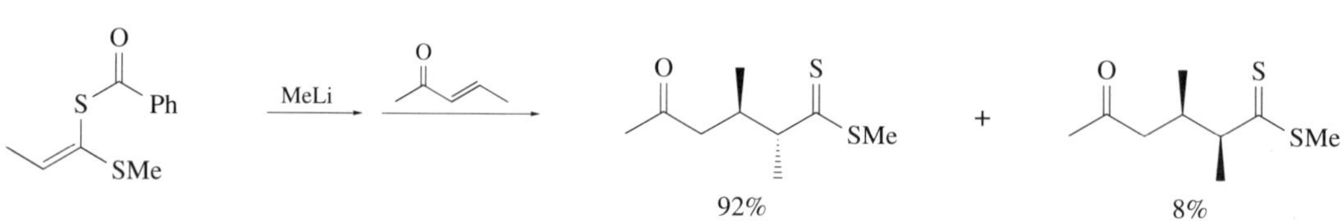

Scheme 27

The reaction of cyclic ketones with lithium alkoxide and CS_2 gives dithioesters having an alkoxycarbonyl group at the terminal carbon ⟨82CL123⟩.

Enantioface differentiation was observed for the 1,4-addition of tin enethiolates to enones in the presence of chiral diamine, although the optical yields ranged from 15% to 70% *ee* (Scheme 28) ⟨88CL1021⟩. The use of a catalytic amount of tin(II) triflate ($Sn(OTf)_2$) and optically active amine showed high enantioselectivity in the reaction of *S*-trimethylsilyl enethiolates with α,β-unsaturated ketones ⟨88CL1025⟩.

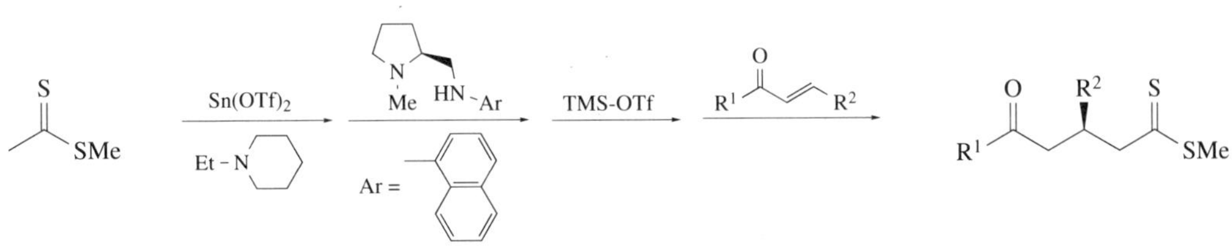

Scheme 28

Meyers *et al.* have developed the synthesis of β-hydroxy dithioesters by the aldol reaction of dithioesters with aldehydes (Scheme 29) ⟨78TL4657⟩. This method has been applied to the total synthesis of natural products ⟨79JA4732, 79JA7104, 81TL3925, 83JA5015⟩. The trityl perchlorate mediated aldol reaction of *S*-trimethylsilyl enethiolate with aldehydes and acetals is developed ⟨89LA321⟩. Double deprotonation of dithiopropanoic acid with LDA followed by the reaction with aldehydes and methyl iodide also gives β-hydroxy dithioesters ⟨89T4445⟩. The stereochemistry of the aldol reaction has also been investigated ⟨85T5089⟩. When enethiolates are reacted with α-methyl aldehydes, the products showing *syn* configuration predominate in 43–78% yields ⟨92T4135⟩, although the ratio of *syn* to *anti* depends on the solvent, temperature, and substituents of the aldehydes. The stereoselective reaction of enethiolates derived from dithiopropanoic acid *S*-ethyl ester with benzaldehyde produces the *syn* isomer ⟨85T5089⟩. The stereochemistry of three contiguous carbon centers is controlled by the reaction of enethiolate with 2-phenylpropanal (Equation (22)) ⟨85T5089⟩. α,β-Unsaturated ketones can be used as the precursor to β-hydroxy dithioesters when titanium enethiolates are generated (Scheme 30) ⟨90TL1853⟩. β-Hydroxy dithioesters having an allylic group are prepared via thio-Claisen rearrangement ⟨89CC414, 91T6275, 92T4135, 93T3131⟩. Although 2 or 3 days are necessary for the completion of the rearrangement, the reaction gives predominantly *syn–syn* isomer irrespective of the geometry of the starting material, as in Equation (23) ⟨93T3131⟩. As additional routes to β-hydroxy dithioesters, alkylation of β-oxo dithioesters with organometallic reagents ⟨82TL4087⟩ or reduction of β-oxo dithioesters with baker's yeast ⟨86TL5405, 87NKK1414⟩ have been used. In both cases, the dithiocarboxyl group is not affected under the reaction conditions. In the latter case, the *syn* isomer is obtained exclusively (Equation (24)).

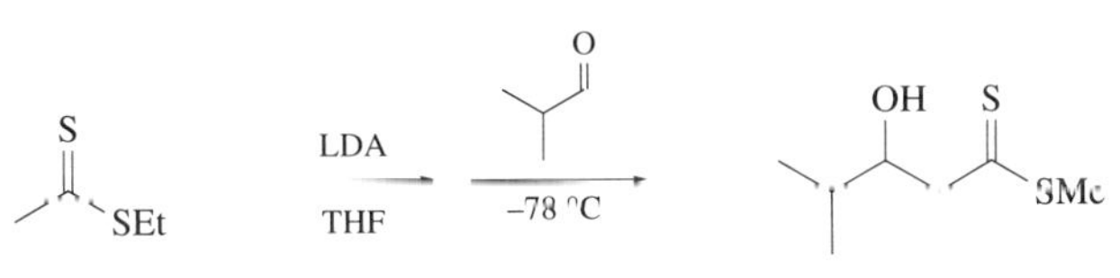

Scheme 29

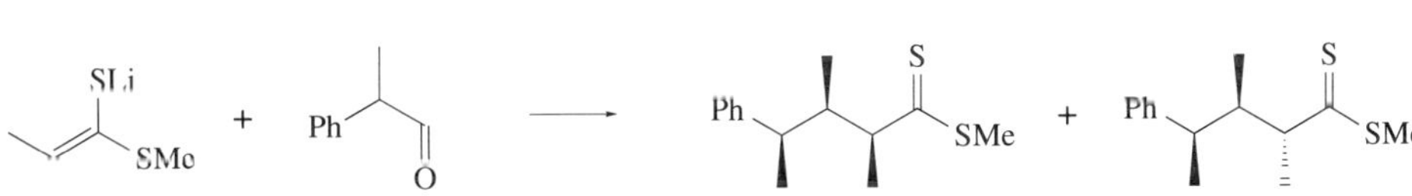

(22)

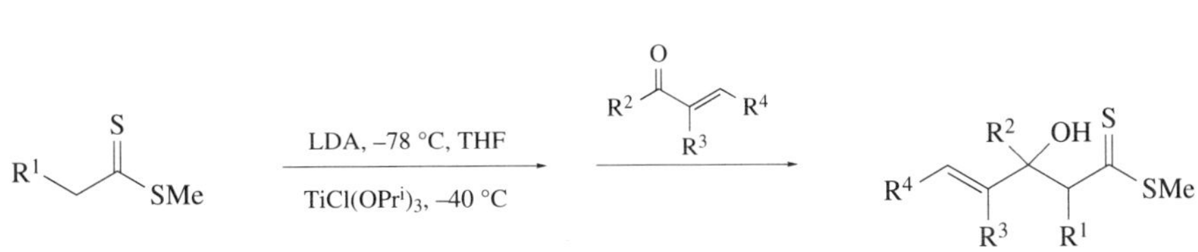

Scheme 30

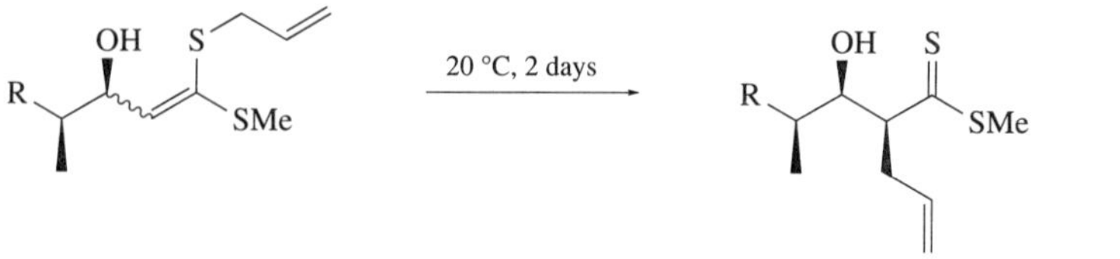

(23)

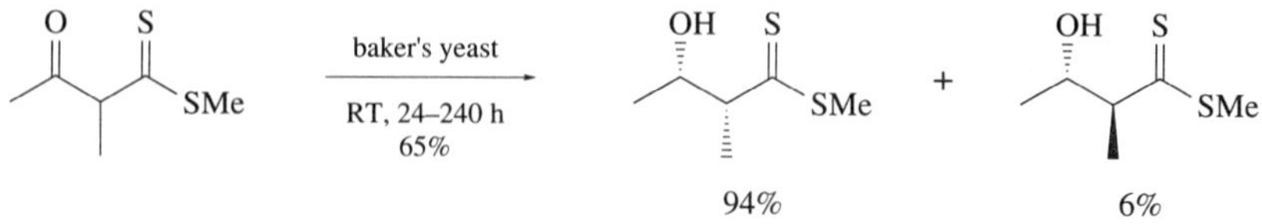

(24)

A variety of α-amido esters (**23**) have been prepared by the sulfhydrolysis of imido thioesters ⟨75BSB61, 82CJC199, 83T3429, 84JCS(P1)785, 92LA595⟩. Similar methods have been applied to the synthesis of α-azido dithioesters (Scheme 31) ⟨92S1199⟩. α-Oxy or α-thio dithioesters, (**24**) ⟨86MI 513-01⟩ and (**25**) ⟨80AP(313)1029⟩, respectively, can also be obtained by sulfhydrolysis of the corresponding imidothioesters derived from nitriles. The synthesis of α-dithio,α-(2-pyridyl) dithioesters has been reported using the reaction of the corresponding carbanion with CS_2 and methyl iodide ⟨87JMC24⟩. α-Sulfinyl and α-sulfonyl dithioesters are prepared from dimethyl sulfoxide or sulfone and trithiocarbonates (Scheme 32) ⟨77RTC254, 84JCS(P1)85⟩. α-Cyano dithioesters are formed via the reaction of the cyanobenzyl anion with CS_2 ⟨73S297, 85JHC1065⟩. The reaction of dithio chloroformic acid esters with copper salts of phosphonic acid esters gives α-dialkyl-phosphono dithioesters (**26**) ⟨83SUL191⟩.

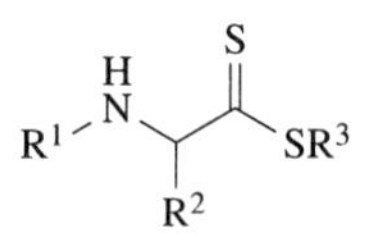

(23)

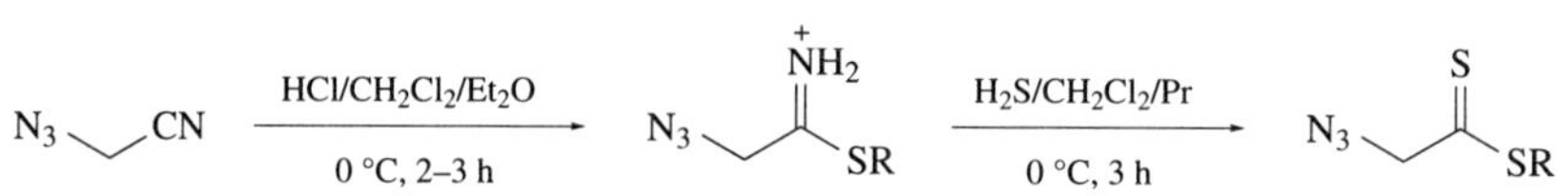

Scheme 31

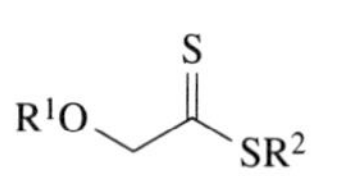

(24)

R^2 = Me, CH_2CO_2H

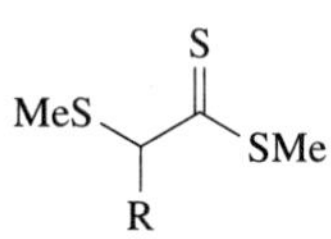

(25)

R = H, Me

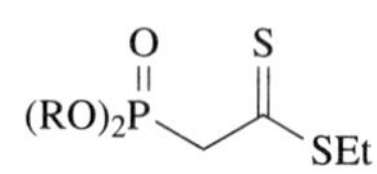

(26)

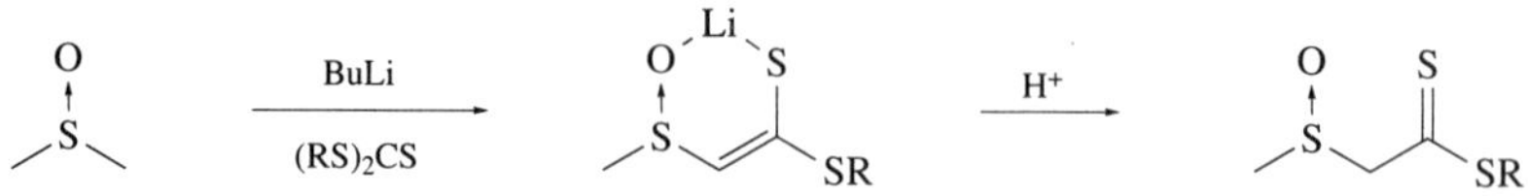

Scheme 32

5.13.3.1.2 α,β-Unsaturated dithioic acid esters

Attempts to prepare dithio acrylic acid methyl ester (**27**) by the flash thermolysis of the dithioester (**28**) ⟨80TL2421⟩, or by dehydration of the β-hydroxy dithioester (**29**) ⟨84JCS(P1)859⟩, have indicated that the ester (**27**) could be trapped at $-196\,^{\circ}C$, but easily underwent dimerization ⟨80TL2421, 84JCS(P1)859, 84JCS(P1)865⟩. The introduction of a methyl group to the ester (**27**) slightly enhances the stability of α,β-unsaturated esters, although they can only be detected below $-40\,^{\circ}C$. Isomerization of β,γ-unsaturated dithioesters as in Equation (25) ⟨78TL2715⟩, dehydration of the corresponding β-hydroxy esters ⟨82TL4087, 84JCS(P1)859⟩, or the reaction of vinylmagnesium bromide with CS_2 ⟨80CB919⟩ leads to alkyl-substituted α,β-unsaturated esters. Among them, β,β-dimethyl dithio acrylic acid ester is stable enough to be handled at room temperature. Methylation of tetrathio malonic acid trianion gives β,β-dimethylthio dithioacrylic acid ester ⟨91TL4679⟩. This ester is also obtained by copper-catalyzed Grignard reaction of vinyl bromide with CS_2 ⟨79S432⟩. The synthesis of dithio cyanamic acid, which shows enhanced stability, is attained by sulfhydrolysis of the corresponding imidoester, as well as Grignard reaction of vinylmagnesium bromide with CS_2 ⟨80CB919⟩. Alternatively, Horner–Emmons reaction of 4-nitrobenzaldehyde with β-phosphinyl dithioic acid esters has been carried out using a two-phase system ⟨85S960⟩. The similar treatment of α-silyl dithioesters with aromatic aldehydes provides α,β-unsaturated esters in 51–84% yields (Equation (26)) ⟨89LA321⟩. Bos and co-workers have observed the formation of methylthio-substituted α,β-unsaturated dithioic acid esters (**30**) by photochemical and thermal reaction of thioketones with 1,2-bis(methylthio)acetylene ⟨76TL209, 78TL4839, 83RTC83, 83RTC91, 83RTC103⟩. Flash vacuum thermolysis has been applied to the generation of dienyl dithioester (**31**) ⟨93TL2605⟩. Trithione was reacted with morpholine and alkylated with methyl iodide or benzyl bromide to give β-amino-α,β-unsaturated dithioesters (Scheme 33) ⟨67T3785⟩. Trithiolium salts were also reacted with amines ⟨64CB1886, 72CR(C)1215⟩. The reaction of uracil with NaOH and CS_2, and alkylation with dimethyl sulfate produced the dithioester (**32**) ⟨79H(12)485⟩. Cyanomethylenephosphorane was deprotonated with triethylamine and reacted with CS_2 and alkyl halides to give the dithioester (**33**) ⟨66JOC3877⟩. Perimidine dithioesters (**34**) are reported to be obtained from chloromethyl-1H-perimidinium chloride ⟨86JPR(328)812⟩

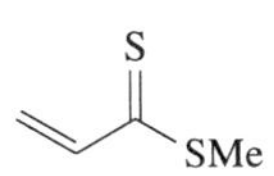

(27)

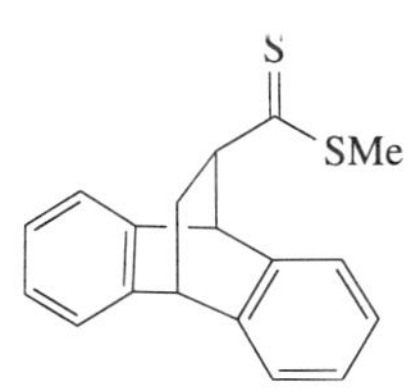

(28)

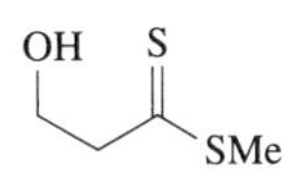

(29)

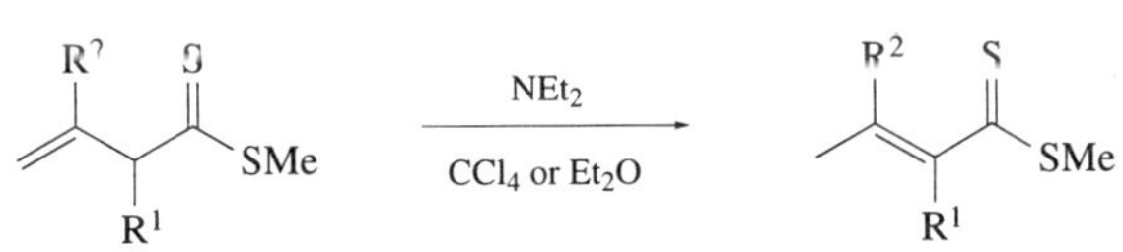

(25)

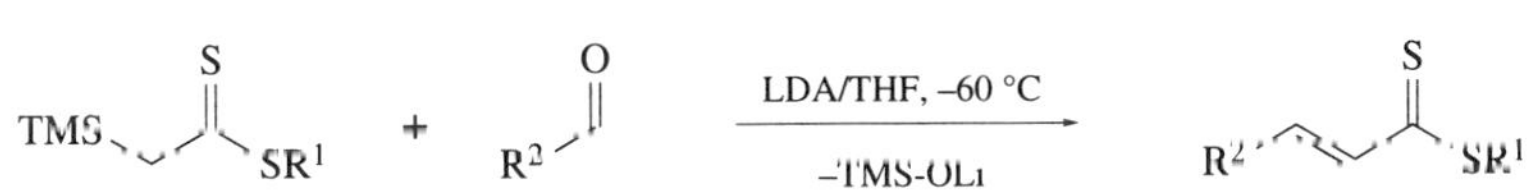

(26)

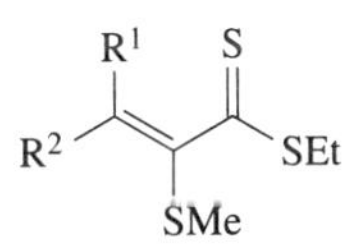

(30)

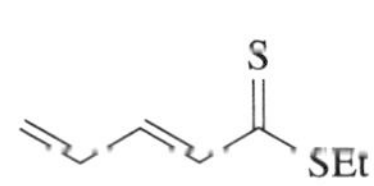

(31)

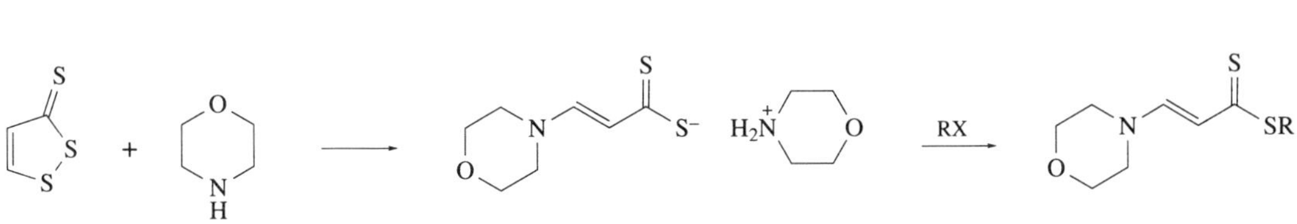

Scheme 33

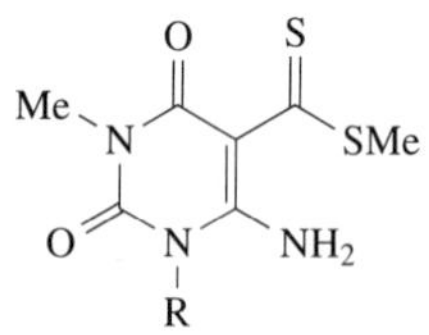

(32)

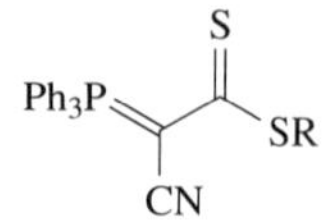

(33)

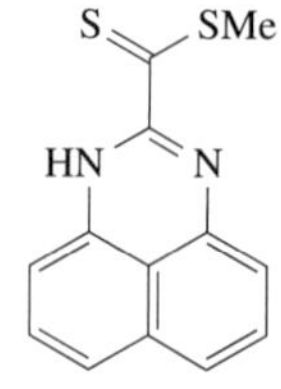

(34)

5.13.3.1.3 *α,β-Acetylenic dithioic acid esters*

In 1989, Hartke and co-workers developed preparative methods for α,β-acetylenic dithioic acid esters based on three different types of reactions ⟨89TL1073, 91LA903⟩. The first is the alkylation of the corresponding dithioic acid lithium salts (Scheme 34). The second method is sulfhydrolysis of the corresponding iminothioesters as in Equation (27). The third is the Pd-catalyzed coupling reaction of terminal acetylenes with chloromethyldithioformate, which leads to α,β-acetylenic dithioic acid esters (Equation (28)). Although the last two cases show higher yields of the products, the isolated products bear aromatic substituents at the terminal carbon of the acetylenic moiety in all cases.

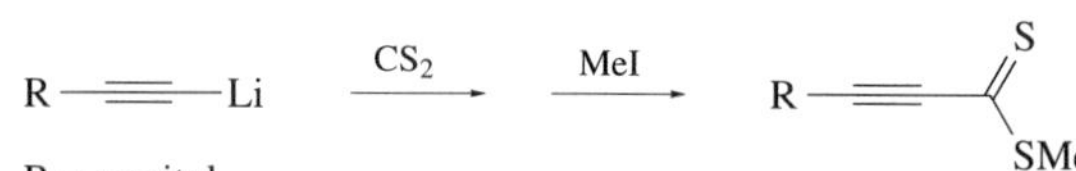

R = mesityl

Scheme 34

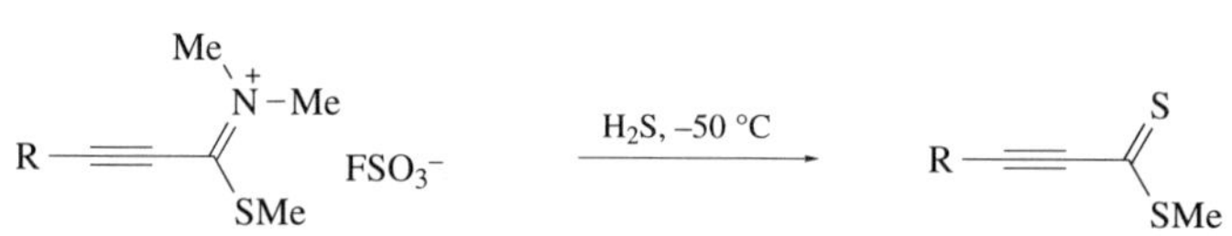

(27)

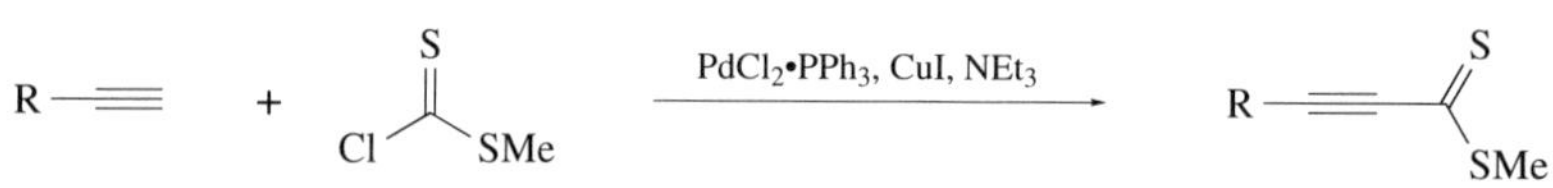

(28)

R = mesityl, 2,4,6-tri-Bu^{t}C$_6$H$_2$

5.13.3.1.4 *Aromatic and heteroaromatic acid esters*

As for the dithioalkanoic acid esters, the general methods for preparing aromatic derivatives are the alkylation of the corresponding acids, sulfhydrolysis of imidothioesters and thioesters with H_2S or Lawesson's reagent ⟨82SUL69⟩, and thiolation of thioacylating agents ⟨88SR155⟩. A one-pot conversion of aromatic acids to the corresponding dithioesters developed by Davy and Metzner is also available ⟨85CI(L)824⟩. Dithiobenzoic acid *S*-phenyl ester is obtained in a similar way ⟨84S827⟩.

Aryl–metal compounds have been believed to undergo selective carbophilic attack on CS_2. However, when a bulky aryl lithium was reacted with CS_2, the product derived from thiophilic attack was obtained ⟨84CC1010⟩. Recently, even with phenylmagnesium bromide the product formed via thiophilic attack was noted to accompany dithiobenzoic acid methyl ester ⟨88ZC269⟩. A copper-catalyzed Grignard reaction is used for the synthesis of 4-dimethylamino or 4-siloxy dithiobenzoic acid esters ⟨79S432, 93JOC3042⟩. A variety of alkylating agents are used for the preparation of aromatic dithioesters. *S*-α-Carboxy ⟨74CI(L)803, 76CB740, 87CB411⟩, *S*-α-phosphonio ⟨86BCJ1403⟩, *S*-β-cyano ⟨73JCS(P1)1571⟩, *S*-β-imino ⟨85PS(22)135⟩, and *S*-β-oxo ⟨77TL4607⟩ esters have been obtained from dithioic acid salts and the corresponding halides. Magnesium salts are alkylated with α,β-unsaturated ketones ⟨81OPP71⟩ and lactones ⟨77JPR169⟩. Indole-substituted dithioester was prepared from indolinone and CS_2 in the presence of sodium hydride ⟨75YZ980⟩. Aryldiazonium salts undergo radical attack on CS_2 at the sulfur atom. The trapping of the generated thioacyl radical with furan forms furan dithioic acid *S*-aryl esters (Scheme 35) ⟨76JOC2639⟩.

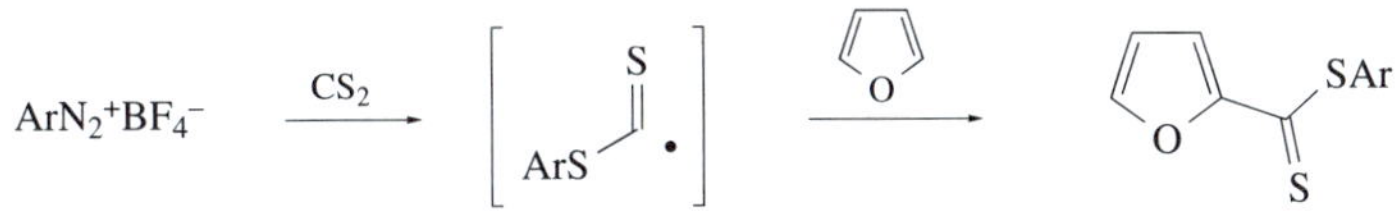

Scheme 35

Tetrathiophthalic acid esters are obtained by the alkylation of the corresponding acids derived from the reaction of 1,4-bis(chloromethylbenzene) with sulfur in the presence of base ⟨87CB61, 88CB313⟩ (Scheme 36). Chlorination of benzyl alkyl sulfides and the subsequent reaction with sulfur in the presence of triethylamine in DMF gives aromatic dithioesters in moderate yields ⟨77ZC92⟩. Sulfhydrolysis of imidothioesters can also lead to tetrathiophthalic acid esters ⟨77JCR(S)144, 81S963⟩. The starting imidoesters are prepared from the reaction of aromatic nitriles with thiols. 3-Cyanofuran is reacted with a thiol, and the imidothioester obtained is sulfhydrolyzed (Equation (29)) ⟨79JOC2807⟩.

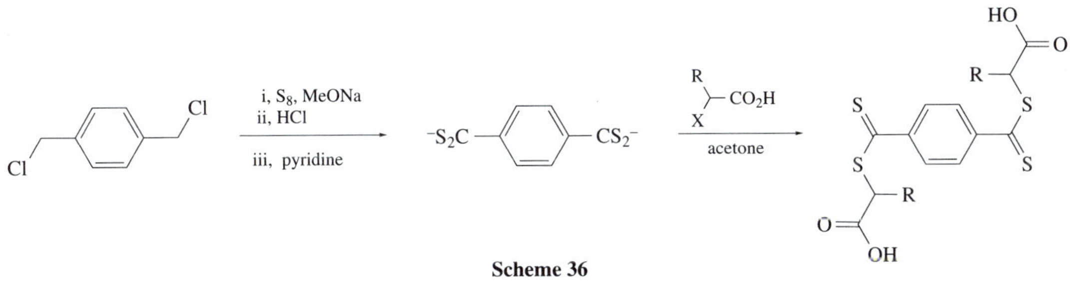

Scheme 36

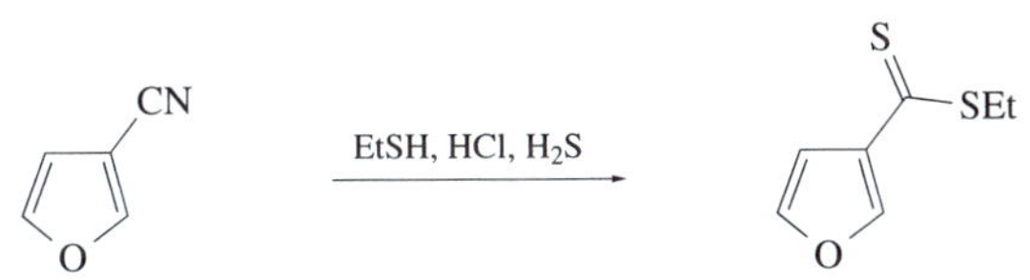

(29)

Pyridinium salts (**35**) have been devised as new thioacylating agents ⟨90ZC67⟩. The salts (**35**) are thermally stable and can react with sodium thiolates at 0–20 °C, as shown in Equation (30). Ferrocenedithioesters are prepared from the corresponding dithioic acid thioanhydride ⟨90BCJ638⟩.

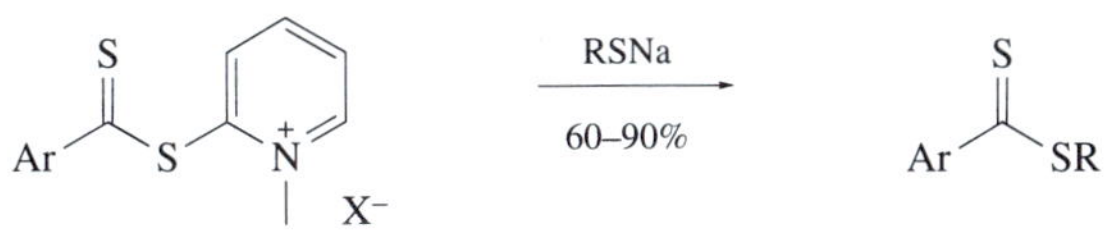

(30)

(35)

The Friedel–Crafts reaction of pyrrole ⟨70LA(737)179⟩ with CS_2, and of anisole in the presence of γ-lactone with CS_2 ⟨72CC345⟩ are also known methods for the synthesis of aromatic dithioesters. A similar Friedel–Crafts reaction is used for the synthesis of 4-hydroxydithiobenzoic acid esters ⟨88S303⟩. The reaction is performed at 40 °C within 1.5 h (Equation (31)), the methoxy group being hydrolyzed in the products. The Friedel–Crafts reaction of *S*-trifluoromethyl fluorodithioformic acid ester with benzene derivatives gives substituted dithiobenzoic acid *S*-trifluoromethyl esters ⟨87CB429⟩.

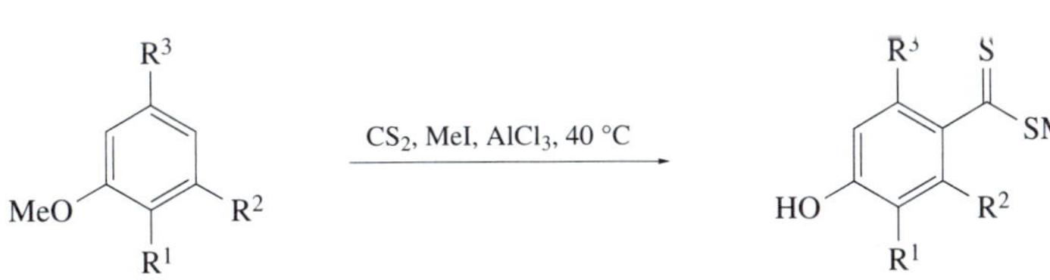

(31)

Polymerization of 4-vinyldithiobenzoic acid ester is also reported ⟨75MI 513-01⟩.

Thioketals prepared from aromatic aldehydes are converted into dithioesters (Scheme 37) ⟨79SC17⟩.

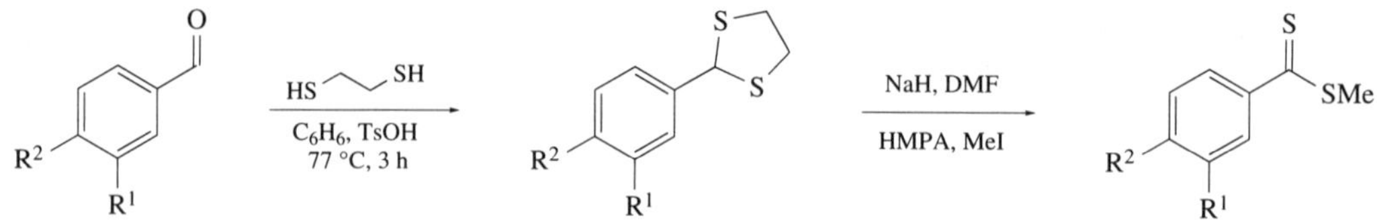

Scheme 37

5.13.3.1.5 *Dithioic acid esters having sulfur in higher oxidation states*

Sulfenyl sulfines (**36**), sulfinyl sulfines (**37**), and sulfonyl sulfines (**38**) are the class of compounds with sulphur in higher oxidation states. Their synthesis has been intensively developed by Zwanenburg and co-workers ⟨82RTC1, 89PS(43)1⟩. The stepwise oxidation of aromatic dithioesters with mcpba initially gives compound (**36**) ⟨68TL2871, 71RTC614⟩. Compounds (**36**) and (**37**) are then formed by further oxidation (Scheme 38). Even dithioesters bearing bulky groups can form sulfines ⟨75JCS(P2)916⟩. The isomeric ratio of sulfenyl sulfines formed by the mcpba oxidation depends on the substituents ⟨67TL3453, 77RTC139⟩. In contrast, (*E*) and (*Z*) isomers are obtained in the same ratio when aromatic dithioesters are treated with chlorine followed by hydrolysis ⟨77RTC139⟩. The oxidation of aliphatic dithioesters can also give sulfenyl sulfines, but they rearrange to form dithioperoxyesters in a few days (Equation (32)) ⟨88CC390⟩. In the oxidation of dithioacetic acid *S*-allyl esters, the olefinic moiety remains intact in the product sulfines ⟨92TL6383⟩.

(**36**) (**37**) (**38**)

Scheme 38

(32)

Alternatively, the Wittig reaction ⟨78TL807⟩ and Peterson olefination ⟨78TL811, 83RTC400⟩ using sulfur dioxide have been applied to the synthesis of sulfenyl sulfines as exemplified by the preparation of phosphoryl substituted sulfines (Scheme 39) ⟨84JOC263⟩. α-Oxo sulfines (Scheme 40) ⟨85JOC2930⟩ can also be generated *in situ*, and trapped with dienes.

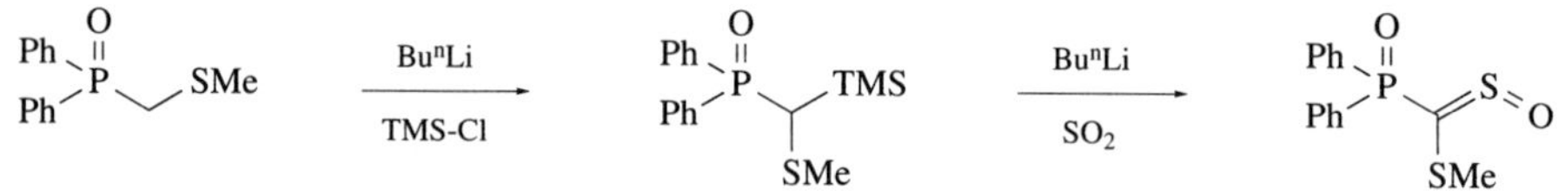

Scheme 39

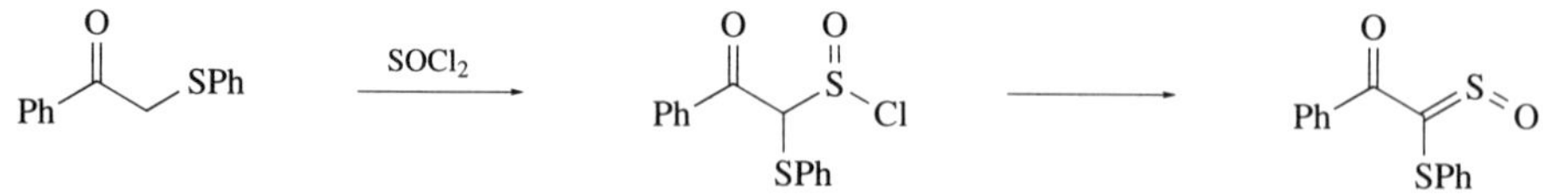

Scheme 40

5.13.3.2 Selenothioic Acid Esters and Diselenoic Acid Esters

Compared with the efforts to develop syntheses of dithioic acid derivatives, only a limited number of examples of the synthesis of selenothioic acid and diselenoic acid esters have been known ⟨8-73MI 513-01⟩. Selenothiobenzoic acid *Se*-aminoethyl ester (**39**) was prepared by the thioacylation of aminoethylselenomagnesium chloride ⟨70JMC214⟩. A variety of aromatic thioacyl chlorides react with sodium benzeneselenolate or trimethylsilylselenide to give the corresponding esters ⟨88BCJ3931⟩. The addition of benzeneselenol to bis(trifluoromethyl)thioketene proceeds similarly to the reaction with thiols to give selenothioic acid *Se*-phenyl ester ⟨72JOC1347⟩. Dithioic acid thioanhydrides are used as a thioacylating agent and a variety of *Se*-aryl esters have been prepared (Equation (33)) ⟨88BCJ3931, 90BCJ638⟩. Pyridinium salts (**35**) also react with sodium selenolates ⟨90ZC67⟩. Alternatively, thioic acid *O*-alkyl esters are reacted with aluminum selenide to produce *Se*-methyl esters in high yields (Equation (34)) ⟨91CC964, 92TL6131⟩. The treatment of alkynyl phenyl selenide in a similar way to Equation (18) gives α-thioxo selenothioic acid *Se*-phenyl ester, although in low yields ⟨93BCJ623⟩. Oxidation of selenothioic acid *Se*-phenyl esters gives selenenyl sulfines along with diphenyl diselenide and selenothioperoxyesters (Equation (35)) ⟨88BCJ3931⟩.

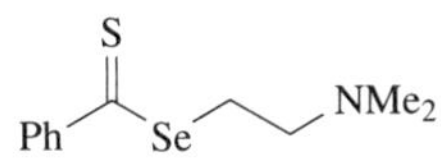

(39)

(33)

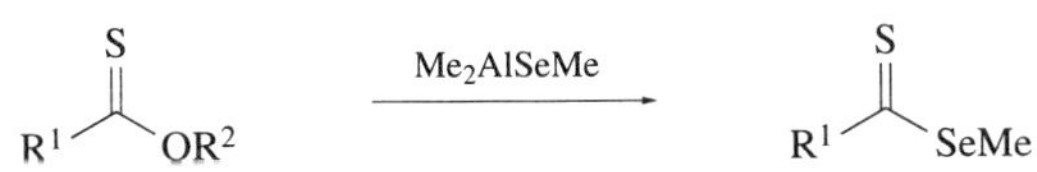

(34)

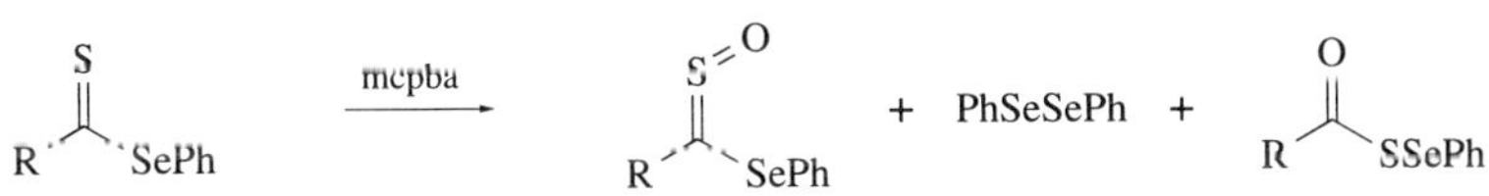

(35)

Although the preparative method for selenothioic acid *S*-alkyl esters has been briefly noted in the literature ⟨66RTC889, B-73MI 513-01⟩, the first example of the isolation of selenothioic acid *S*-alkyl esters derived from selenoacetic acid *Se*-alkynyl esters and thiols was reported in 1993 by Kato *et al.* (Equation (36)) ⟨93JA3000⟩. The facile synthesis of selenothioacetic acid *S*-alkyl esters using trimethylsilylacetylene is also reported as in Equation (37) ⟨93CL1469⟩. These compounds were believed to be too unstable to be isolated, but recent results suggest that they can even be handled under aerobic conditions, although the stability of the compounds is highly dependent on the substituents. For example, in the reaction using benzenethiol, the reaction mixture gradually turns to blue-violet, which is indicative of the formation of the selenothioic acid *S*-phenyl ester, but the color instantly changes to yellow when the mixture is treated with water at room temperature. Selenoacylation of aluminum sulfides using selenoic acid *O*-methyl esters can lead to aromatic selenothioic acid esters ⟨94UP⟩. In a series of selenothioic acid *S*-esters, aromatic esters are less stable than the aliphatic esters. The application of this reaction to aluminum selenide gives diselenoic acid esters in good to high yields as green oils, as in Equation (38) ⟨93JA5823⟩. As an aliphatic diselenoic acid ester, α-naphthyl diselenoacetic acid *Se*-methyl ester was isolated in a pure form.

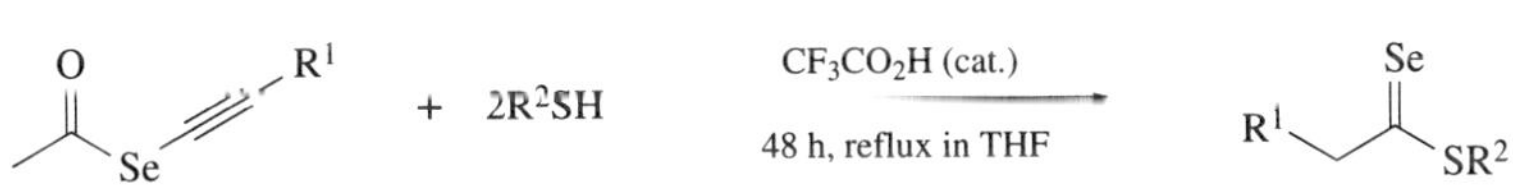

(36)

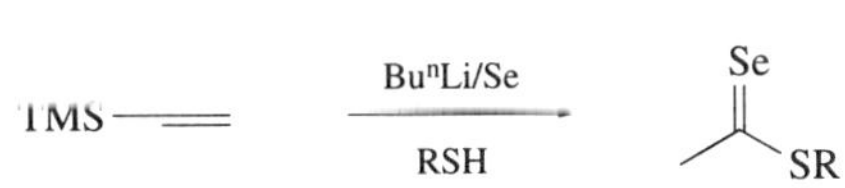

(37)

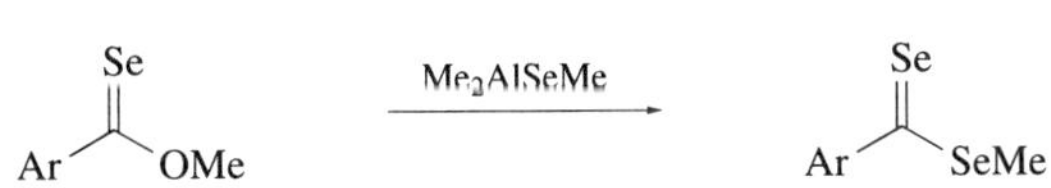

(38)

5.14
Thionoamides and Their Selenium and Tellurium Analogues

COLIN P. DELL
Lilly Research Centre, Windlesham, UK

5.14.1 GENERAL METHODS

A wide variety of methods exists for the preparation of thionoamides and a number of reviews have been published ⟨49BSF272, 60CSR45, 66AG(E)447, 69RCR21⟩. A good starting point is Brillon's review of thionation methods ⟨92SR297⟩. A number of book chapters cover the area thoroughly and serve as excellent sources of material ⟨B-70MI 514-01, 79COC(3)440, 85HOU(E512)1218, 91COS419⟩. Additionally, some specialist periodical reports published by the Royal Society of Chemistry ⟨70SST201, 73SST228, 75SST254, 77SST159, 79SST156, 81SST177⟩ provided reviews on a biannual basis throughout the 1970s. The synthesis and reactions of the slightly more esoteric thionooxamides have also been reviewed ⟨88SR1⟩. Similarly, the synthesis of 3-aminothionoacrylamides has been reviewed ⟨87ZC8⟩. This first section is intended to describe some synthetic methods of general utility.

5.14.1.1 Thionoamides from Amides

Historically, a number of reagents have been used to effect the conversion of an existing amide into its thionated analogue. In terms of popularity, phosphorus pentasulfide (P_4S_{10}) and Lawesson's reagent (the dimer of 4-methoxyphenylthionophosphine sulfide) are the most useful and are described in turn. Other reagents which have received occasional use are described at the end of this section.

5.14.1.1.1 Use of phosphorus pentasulfide

The conversion of amides to thionoamides using phosphorus pentasulfide (Equation (1)) was first described by Hofmann ⟨1878CB338⟩ who obtained thioformanilide from formanilide in 60% yield. The reaction is heterogeneous in nature and is normally carried out at elevated temperatures. A number of solvents have been used, for example aromatic hydrocarbons such as benzene, toluene or xylene, and ethereal solvents such as dioxan or DME. Typical examples are shown in Table 1. Pyridine is reported to be particularly useful.

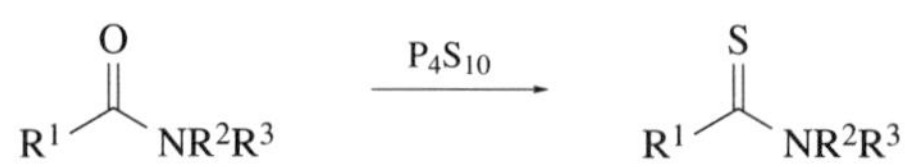
(1)

Various modifications have been used such as a mixture of phosphorus pentasulfide and potassium sulfide ⟨23LA(431)187⟩. Yields from a variety of amides (primary, secondary, tertiary) were in the range 45–93%, with secondary and tertiary amides generally giving the best yields (acetamide did, however, give a useful 75% yield of its thionated analogue). Although in this case the yields were respectable, the isolation of the product was laborious owing to the multiple (4–8) extractions with hot xylene required. Phosphorus trisulfide has been used as an additive ⟨13JCS2272⟩. It is possible to obtain a degree of selectivity in the thionation: thus, thionation of *N*-phenyloxamide reportedly

Table 1 Conversion of amides to thionoamides with P_4S_{10} (Equation (1)).

R[1]	*R*[2]	*R*[3]	*Conditions*	*Yield* (%)	*Ref.*
Ph	Me	Ph	Neat, P_2S_3	not given	13JCS2272
PhCH=CH	H	H	Pyridine, Δ	66	89T6771
PhCH=CH	H	Et	Pyridine, Δ	88	89T6771
$3NO_2—C_6H_4$	H	Ph	Xylene, Δ	70	37HCA691
$4NO_2—C_6H_4$	H	Ph	Xylene, Δ	90	37HCA691
$4NH_2—C_6H_4$	H	Ph	Xylene, K_2S, Δ	5–10	37HCA691
3-Pyridyl	Et	Et	Neat, Na_2S, 125°C	64	51JOC131
4-Pyridyl	H	Pr^i	Neat, 120–150°C	44	55JA4062
2-Furyl	H	H	Pyridine, Δ	84	57CCA319
2-Furyl	H	Me	PhMe, Δ	60	55JA4062
2-Furyl	H	Et	Pyridine, Δ	84	57CCA319
CO_2Et	H	Ph	Xylene, Δ	73[a]	04CB3708
CO_2Et	H	H	Xylene, Δ	63	04CB3708
$CONH_2$	H	Ph	Xylene, Δ	42 (**1**) + 15 (**2**)	04CB3708
$CHCl_2$	H	Ph	PhH, Δ	18	63JOC3492
CH_2Cl	H	Me	PhMe, Δ	49	87PS(31)33
CH_2SMe	Me	Me	Pyridine, Δ	33	77LA1291
CH_2NMe_2	Me	Me	Pyridine, Δ	28	77LA1291
CF_3	H	Ph	Dioxan, Δ	92	76ZOR2213
CF_3	Et	Et	Dioxan, Δ	66	76ZOR2213
Me	H	CH_2Ts	DME	81	76TL143
Me	H	Me	PhMe, 80°C	73	75LA1984
Me	H	Ph	Pyridine, 90–95°C	48	72T199
Bu^t	H	H	Pyridine, 60°C	44	75CB3066
$—CPh_2CHMe—$		Me	PhH, Δ	71	68FES642
$—CEt_2CH_2—$		H	PhH, Et_2O, Al_2S_3, 50°C	17	68FES642
$—[CH_2]_7—$		H	Xylene, K_2S, 80–85°C	50	33HCA1323
H	H	Ph	Neat, melt	60	1878CB338
H	Me	Me	Et_2O, $NaHCO_3$, Δ	85	73S149
H	$—[CH_2]_2O[CH_2]_2—$		PhH, RT	23	47JA1207
H	Ph	Ph	Et_2O, $NaHCO_3$, Δ	49	76CB1309

[a] Isolated yield of the acid following aq. NaOH workup.

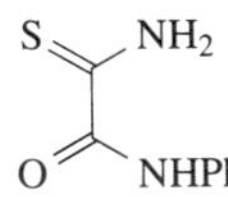

(1)

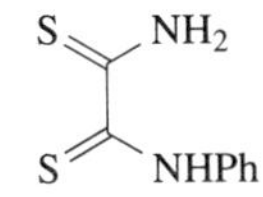

(2)

gives 42% yield of the monothiooxamide (**1**) and 15% of the dithiooxamide (**2**) ⟨04CB3708⟩. Occasionally, the reaction can proceed reasonably successfully in the absence of solvent, for example in the thionation of a pyridinium species (Equation (2)) ⟨74JHC397⟩. Sand has been used as an additive in the thionation of oxindoles ⟨69CPB550⟩. The transformation of a variety of typical functionalised amides, including those bearing indole ⟨76KFZ41⟩, oxindole ⟨93JMC2459⟩, diazo ⟨87KGS845⟩ (note, however, that the diazo group is absent in the product), alkaloid ⟨93JOC6076⟩, steroidal ⟨93SC829⟩, aryl ⟨89JMC2493⟩, benzothiazole ⟨76ZOR2213⟩ and anthranilic acid ⟨71PHA271⟩ portions are shown in Equations (3)–(11).

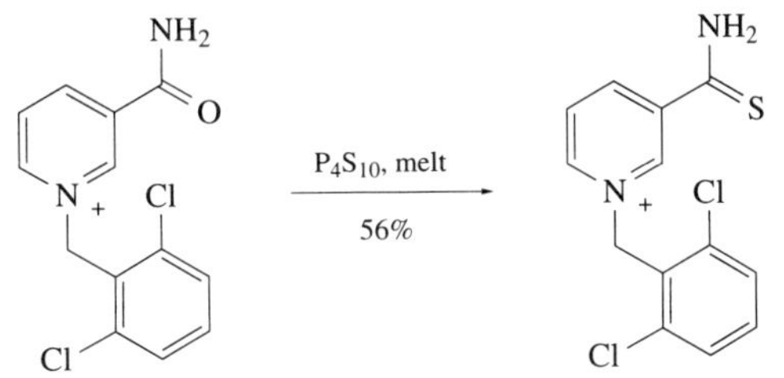

(2)

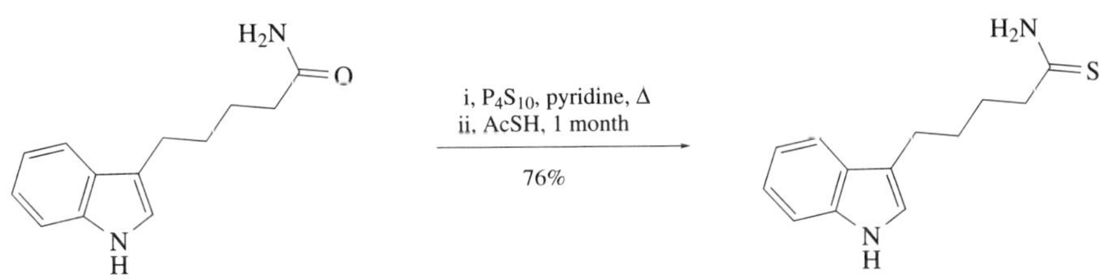

(3)

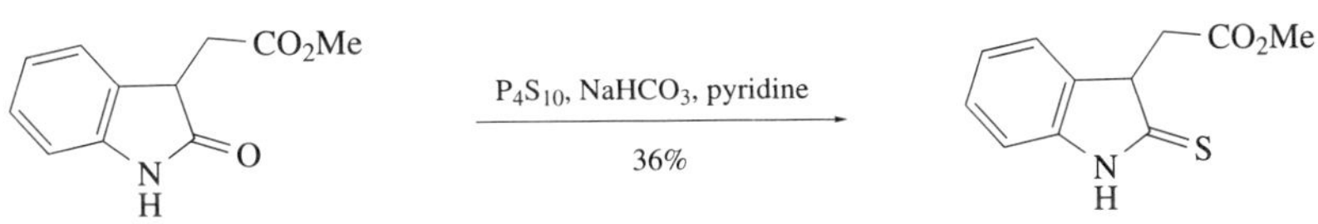

(4)

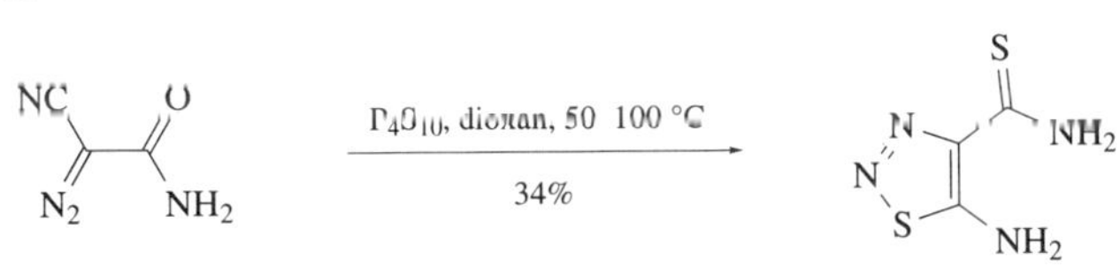

(5)

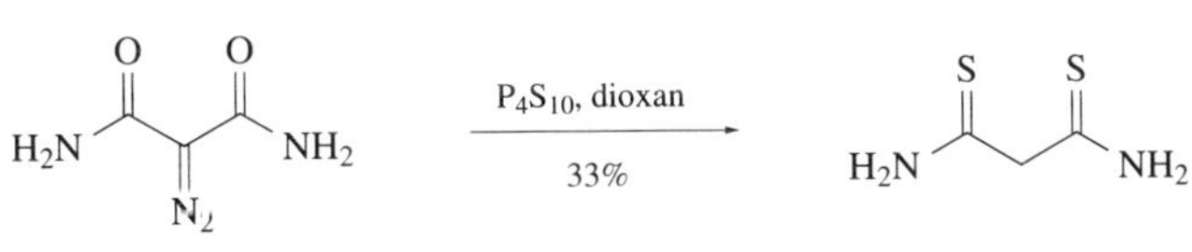

(6)

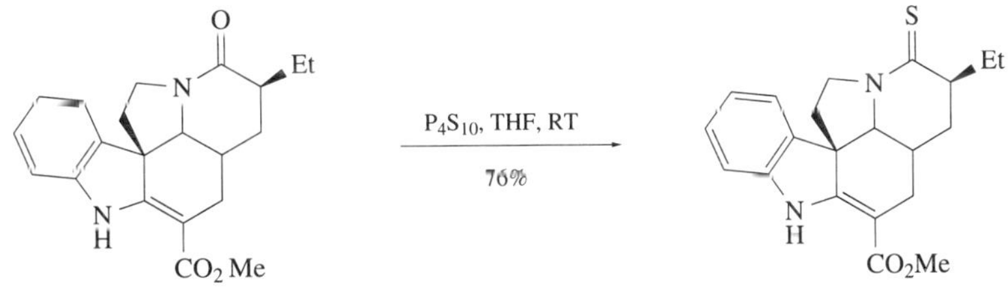

(7)

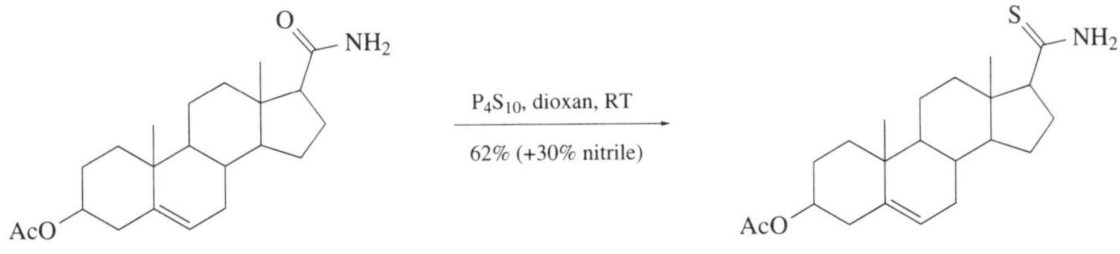

(8)

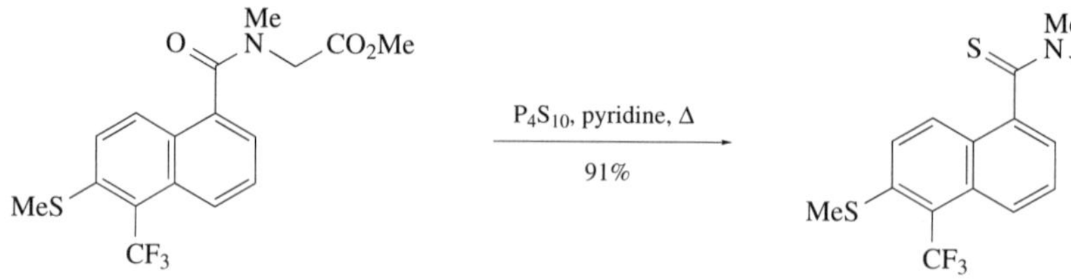

(9)

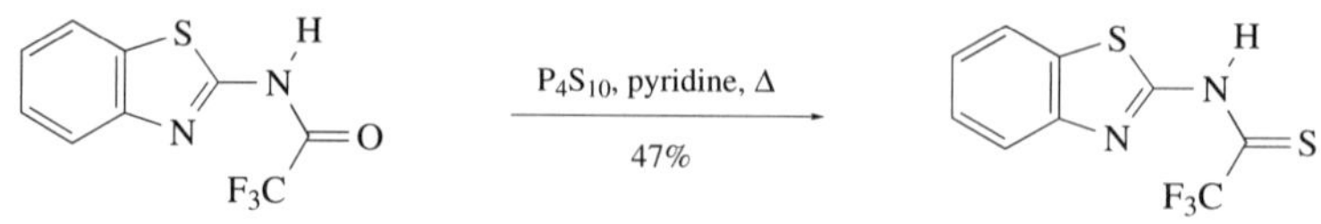

(10)

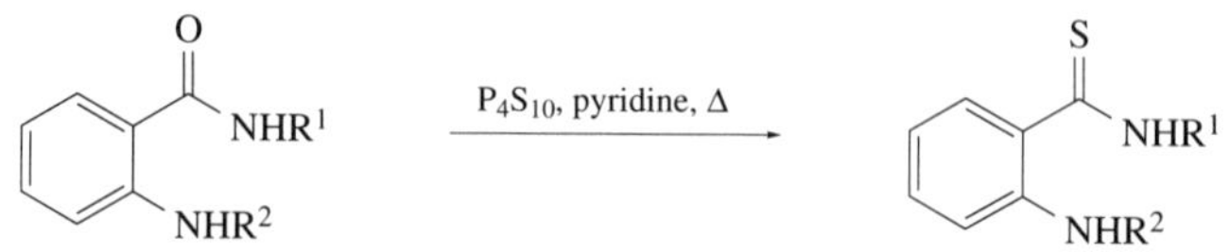

(11)

Thionation reactions can be conducted in THF at 30–40 °C, rather than at the normal elevated temperatures, if ultrasound is used (Equation (12)) ⟨81JOC3558⟩. Reaction times are short and yields of (mainly tertiary) thionoamides are very good (77–97%). Sodium hydrogen carbonate is reportedly a useful additive ⟨73S149⟩, allowing the thionation of *N*,*N*-dimethylformamide under very mild conditions. The mixture is still heterogeneous when ether is used as solvent. The authors suggest that the transformation is accelerated in polar solvents and postulate a thiophosphate nucleophile as the thionating species. One drawback of the use of sodium hydrogen carbonate has been found (Equation (13)), where attempted thionation of secondary acrylamides produced *β*-thiol thionoamides ⟨78AG(E)689⟩ with yields in the 50% region. However, a successful extension of this idea uses an *in situ* reagent generated from a 1 : 1 mixture of phosphorus pentasulfide and sodium carbonate in THF (Equation (14)), apparently giving a homogeneous solution containing a thionating species formulated as $(P_4S_{10}O)^{2-}Na_2{}^{2+}$ by the author ⟨90SC3085⟩. The transformation is applicable to a range of simple amide starting materials and occurs with high yields (80–90%). In addition to these simple substrates, a number of protected amino acids and dipeptides have been thionated chemoselectively at the amide bond ⟨90SC3085⟩. For example, the primary amide and monomethyl amide of *N*-*t*-BOC-L-phenylalanine have been converted to the thionoamides in 80% and 78% yields, respectively ⟨90SC3085⟩. The conditions are sufficiently mild to ensure that the thionoamide (**4**) is generated from amide (**3**) without the competing benzimidazole formation interfering significantly (Equation (15)) ⟨93T10489⟩.

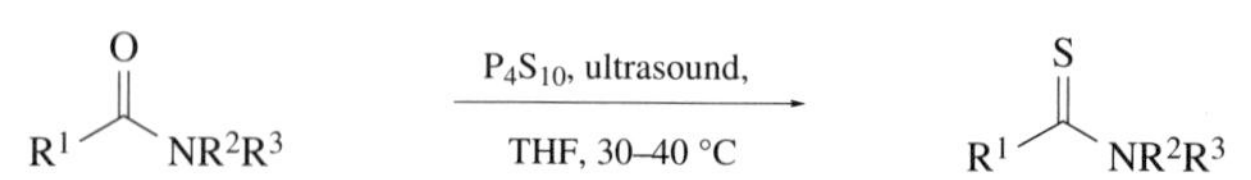

(12)

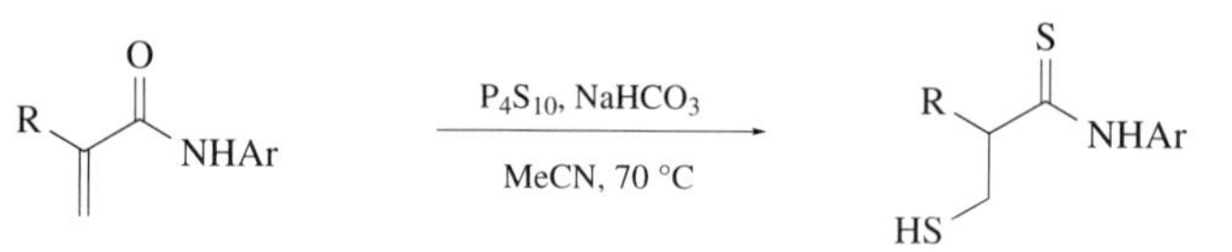

(13)

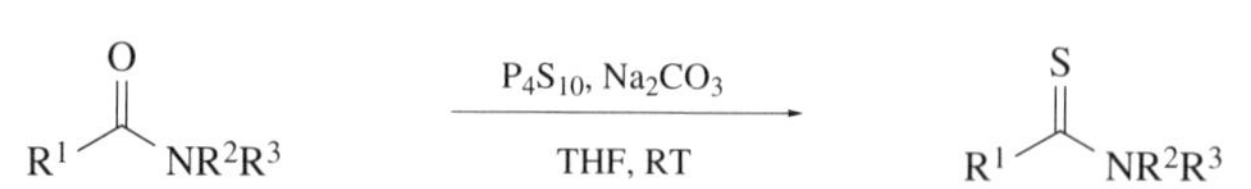

(14)

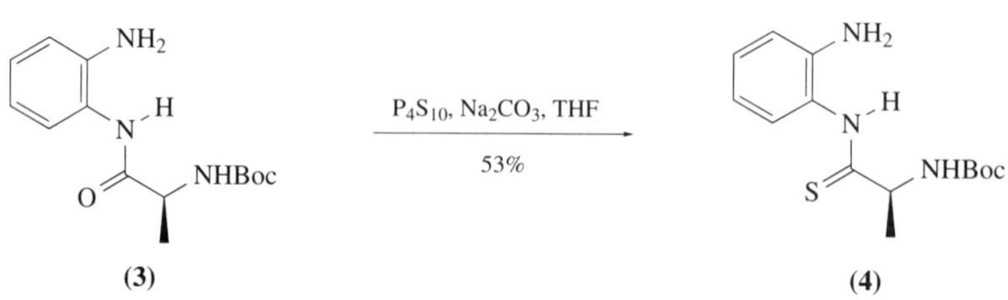

(15)

(3) **(4)**

The yields are not good for the thionation of dipeptides (Table 2) ⟨90SC3085⟩, presumably as a result of steric bulk surrounding the site of reaction. In these cases, a more reactive thionating species was generated by treating the 1 : 1 mixture of phosphorus pentasulfide and sodium carbonate in THF with methyl trifluoromethanesulfonate. This more reactive species was able to some extent to thionate sterically encumbered sites (Table 2). The preparation of this more reactive reagent combination is considered 'risky' on a large scale.

Table 2 Chemoselective monothionation of dipeptides with P_4S_{10}/Na_2CO_3.

Starting material	*Product*	*Yield* (%)	
BocPheGlyOBn	BocPheψ[CSNH]GlyOBn	28	(84)
BocLeuPheOMe	BocLeuψ[CSNH]PheOMe	30	(75)
BocMetLeuOMe	BocMetψ[CSNH]LeuOMe		(72)
BocValPheOMe	BocValψ[CSNH]PheOMe		(28)

Yields in parentheses are those obtained using the reagent generated from $P_4S_{10}/Na_2CO_3/MeOSO_2CF_3$. A blank indicates that no data are given in this chapter.

Another treatment introduced is the use of a combined phosphorus pentasulfide/alkyl lithium reagent ⟨87S162⟩. Here, phosphorus pentasulfide is reacted with four equivalents of *n*-butyllithium, methyllithium or phenyllithium in THF to yield solutions of a new thionating species. Reaction of such complexes with lactams are shown in Equations (16) and (17). Best yields are obtained with the *n*-butyllithium derived reagent. The thionating mixture shows interesting selectivity: *N*-alkylated lactams and lactams in which the carbonyl is conjugated with an aromatic ring, such as **(5)** and **(6)** did not react. On the basis of ^{31}P NMR a structure for the reactive species was proposed.

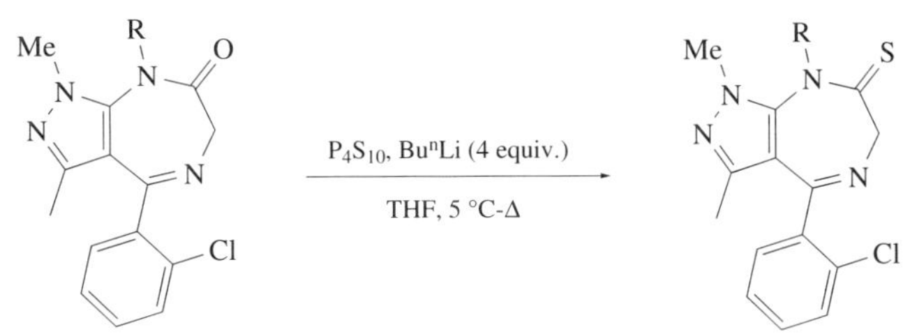

(16)

R = H, 81%
R = Me, no reaction

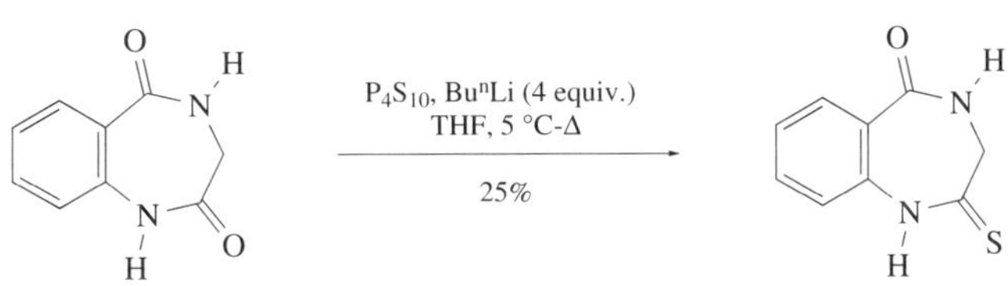

(17)

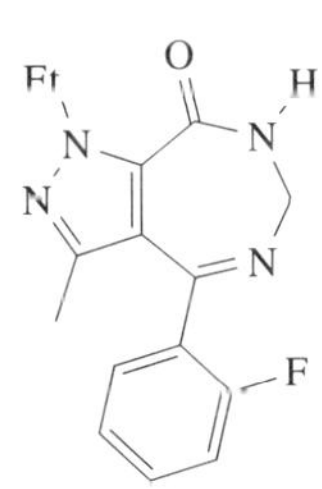

(5)

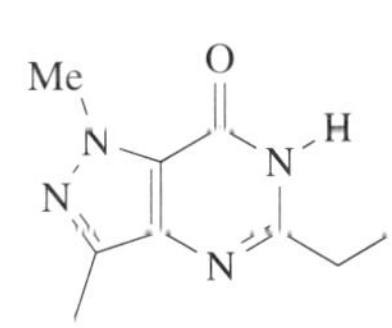

(6)

The heterogeneity of the normal (i.e., without additive) thionation reaction with phosphorus pentasulfide can cause problems since an excess of the reagent and high temperatures are often used. It is therefore a prerequisite that both the starting material and final product are thermally stable. Indeed, one drawback with the use of phosphorus pentasulfide is that primary amides are often directly converted into nitriles (via loss of hydrogen sulfide from the first formed primary thionoamide). This is already apparent in Equation (3), where the thioacetic acid is necessary to reconvert the nitrile into the thionoamide, and in Equation (8). In fact, prior to Hofmann's seminal contribution, Henry had published a novel synthesis of nitriles, prepared by heating a mixture of a primary amide and phosphorus pentasulfide (for example, acetamide gave acetonitrile in about 50% yield and benzamide gave a similar yield of benzonitrile) ⟨1869CB305, 1869CB494⟩.

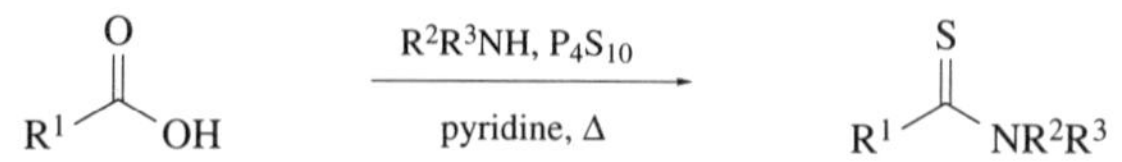

(18)

A number of mechanisms for the transformation have been suggested, among them a proposed attack of the carbonyl oxygen on the phosphorus centre ⟨64JCS1190⟩ and another invoking attack on the carbonyl by a nucleophilic phosphorus species ⟨73S149⟩. The latter hypothesis is supported by the observed catalytic effect of added sodium hydrogen carbonate or organolithiums ⟨87S162⟩.

The direct conversion of a carboxylic acid into a thionoamide by treatment with phosphorus pentasulfide and a primary or secondary amine in pyridine has been reported (Equation (18)) ⟨81JCR(S)58⟩. Yields are poor to moderate for a selection of secondary and tertiary thionoamide products with the amide also being formed, giving a guide to the mechanism involved in this transformation.

5.14.1.1.2 *Use of Lawesson's reagent and related species*

A series of studies by a Danish group developing the use of organophosphorus compounds as thionating agents ⟨77BSB321, 77BSB693⟩ culminated in the discovery of 2,4-bis(4-methoxyphenyl)-1,3,2,4-dithiaphosphetane-2,4-disulfide (**7**) (Figure 1) as an extremely efficient and versatile thionating reagent for a variety of carbonyl containing species ⟨78BSB293⟩, including amides ⟨78BSB229, 78BSB525⟩. The reagent is now referred to as 'Lawesson's reagent' and is available commercially. Its use as a thionating agent has been reviewed ⟨85T5061⟩. It is the reagent of choice for the thionation of aliphatic, unsaturated and aromatic amides and even formamides. Typically substrates are effectively thionated in HMPA at 80–100 °C (Equation (19)) in almost quantitative yields ⟨78BSB229⟩. The use of toluene at 100 °C (Equation (20)) (⟨78BSB525⟩, see also ⟨80TL4061⟩) or benzene at 60–80 °C (Equation (21)) (⟨81ACS(B)239⟩, see also ⟨92JOC2419⟩) frees one from the disadvantages associated with the use of HMPA. These milder conditions also allow thionation of substrates such as *β*-lactams ⟨80NJC47⟩ and *β*,*γ*-unsaturated amides (at room temperature, ⟨87H(26)359⟩).

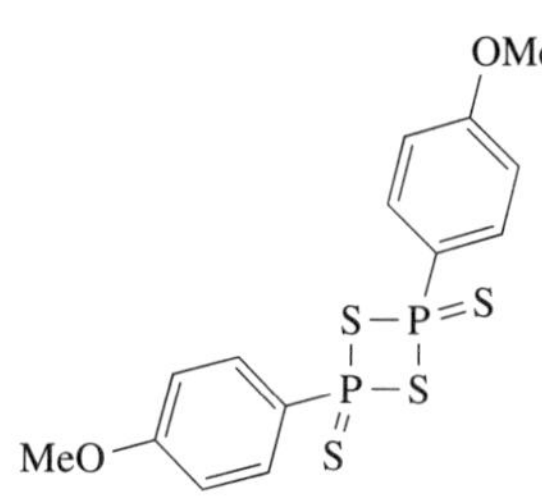

(7)
Lawesson's reagent

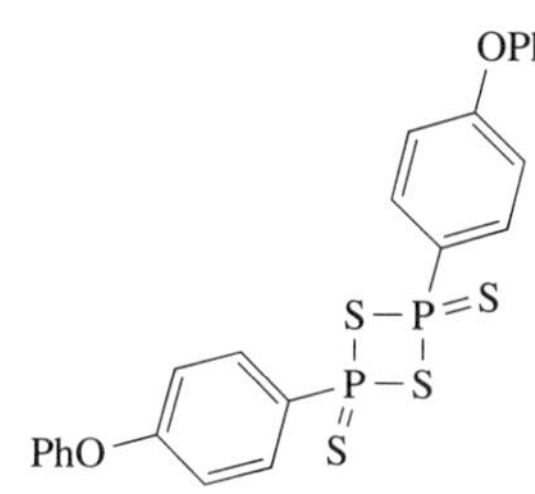

(8)
Belleau's reagent

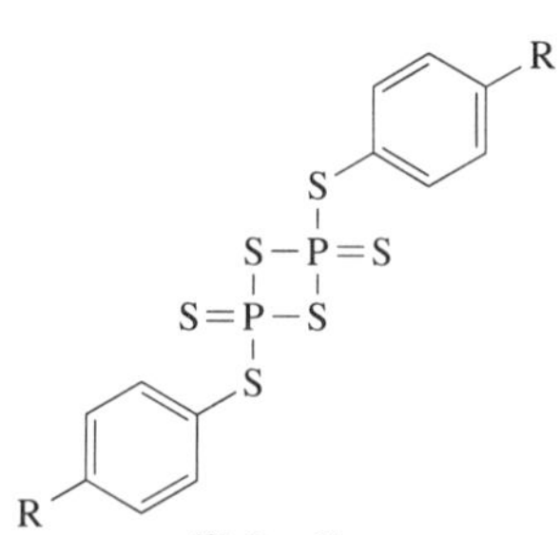

(9) R = H
(10) R = OMe
(11) R = Me

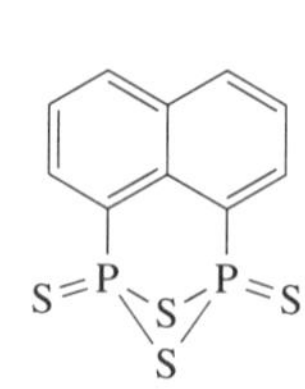

(12)

Figure 1 Lawesson's reagent and related species

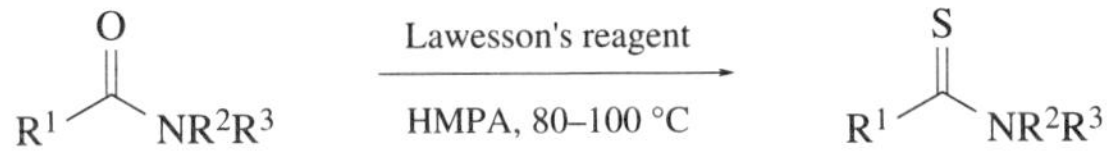

(19)

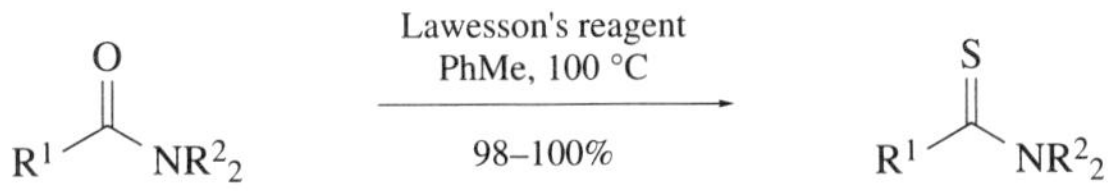

(20)

R^1 = H; R^2 = Me, Et, Pr^n, Bu^n
R^1 = Me; R^2 = Me, Et, Pr^n, Bu^n, C_6H_{13}, $C_{12}H_{25}$
R^1 = Et; R^2 = Et, Pr^n, Bu^n, C_6H_{13}, $C_{12}H_{25}$
R^1 = Pr^n; R^2 = Et, Pr^n, Bu^n, C_6H_{13}, $C_{12}H_{25}$
R^1 = CF_3; R^2 = Et, Pr^n, Bu^n, C_6H_{13}

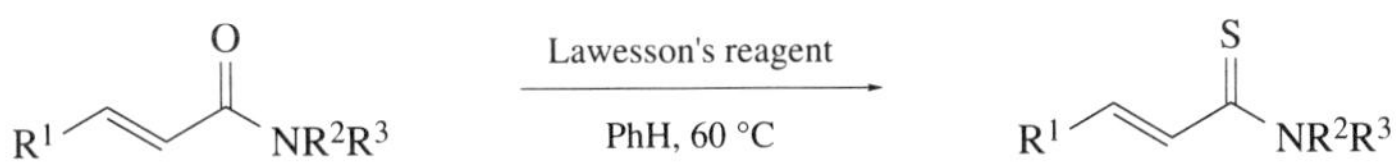

(21)

The reagent is reportedly free from a number of the disadvantages associated with phosphorus pentasulfide, such as the conversion of primary amides to nitriles. One useful facet is that there is no need to protect aromatic hydroxy and amino groups prior to thionating a coexisting amide. The known reaction of Lawesson's reagent with such functionality to form species containing phosphorus–oxygen and phosphorus–nitrogen bonds respectively is in some cases suppressed by using HMPA as a solvent. Lawesson ⟨78BSB229⟩ attributes this to complex formation between HMPA and the reactive functionality. In some cases heterocycles containing phosphorus and sulfur bonds ⟨80NJC47⟩ or phosphorus, sulfur and nitrogen bonds ⟨81ACS(B)239⟩ have been isolated and identified during reactions with this reagent, for example by reaction of Lawesson's reagent with some simple primary acrylamides in HMPA (Equation (22)) ⟨81ACS(B)239⟩ and with some 2-hydroxybenzamides (Equation (23), Table 3) ⟨78BSB229⟩.

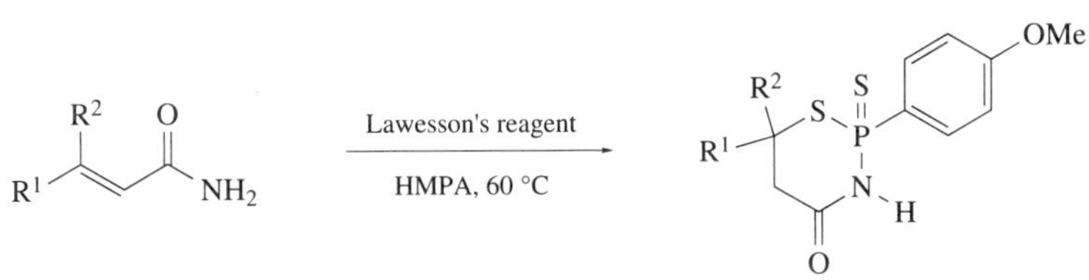

(22)

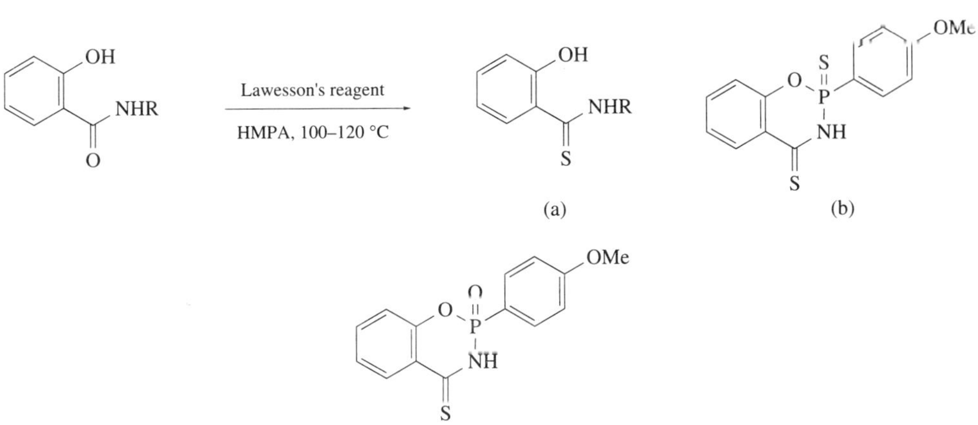

(23)

In some substrates containing particularly reactive functionality, transformation of other groups can occur. For example, diazocyanoacetamide undergoes transformation into a thiadiazole in 42% yield (see Equation (5)) presumably via cyclisation of the diazobisthionoamide ⟨87KGS845⟩. This reaction also occurred with phosphorus pentasulfide, but in slightly lower (34%) yield. Similarly, reaction of amino amide (**3**) with Lawesson's reagent failed to yield the thionoamide (**4**) (Equation (15)), instead producing benzimidazole (**13**) (Equation (24)) ⟨93T10489⟩. When the thionation of some *N*-nitrosoamides was attempted using Lawesson's reagent, the isolated products were the

Table 3 Reactions of 2-hydroxybenzamides (Equation (23)).

R	*Yield* (%) *of product type*		
	(*a*)	(*b*)	(*c*)
Ph	51	8	21
Xylyl	30	24	14
$PhCH_2$	31	23	13
n-C_6H_{13}	15	5	18
Cyclohexyl	36	0	0

thionamides and amides, both lacking the nitroso group ⟨84BSF(2)204⟩. Aromatic nitro substituents are stable to the reagent and dealkylation of ethers is not observed. A range of more interesting substrates is shown in Equations (25)–(30), for example amides with a coexisting highly reactive α-dicarbonyl moiety ⟨90JOC831⟩, sensitive vinyl pyrroles ⟨87TL2265⟩, highly hindered spirocycles ⟨92S778⟩, alkynes ⟨91LA903⟩, cyclopropyl amides ⟨94JOC1129⟩ and enamides ⟨84JOC714⟩ can all be thionated efficiently. Lawesson's reagent has recently been used to thionate Marcfortine A, a fungal metabolite, and the resulting thionoamide used to aid assignment of the absolute stereochemistry of the natural product ⟨94TL1135⟩. Other recent uses include utilisation in syntheses of (±) prosopinine ⟨94TL1669⟩, a polyamine toxin ⟨94TL2057⟩ and a chiral *cis*-dithiooxamide ⟨94TA149⟩. In an approach to the marine natural product sarain A, reaction of a δ-lactam with Lawesson's reagent resulted not in thionation but in epimerisation at a remote stereocentre ⟨94TL3651⟩. A potential mechanistic rationale is given.

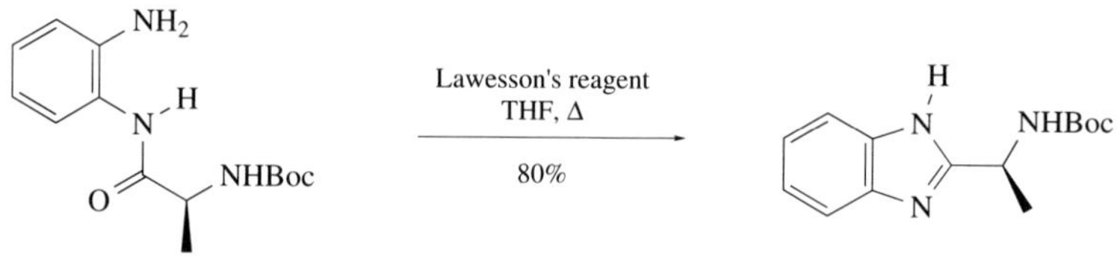

(3) **(13)**

(24)

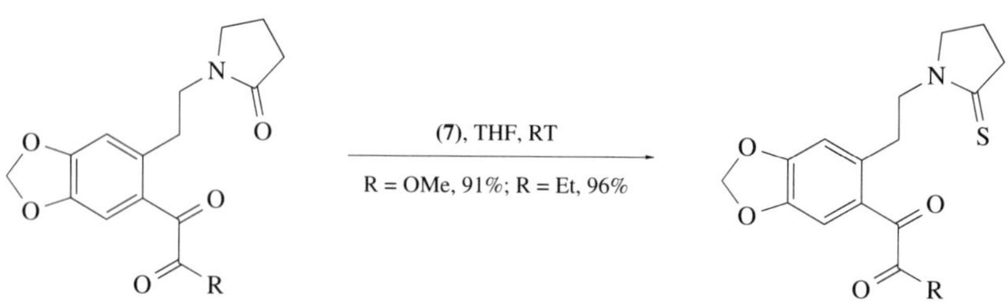

(25)

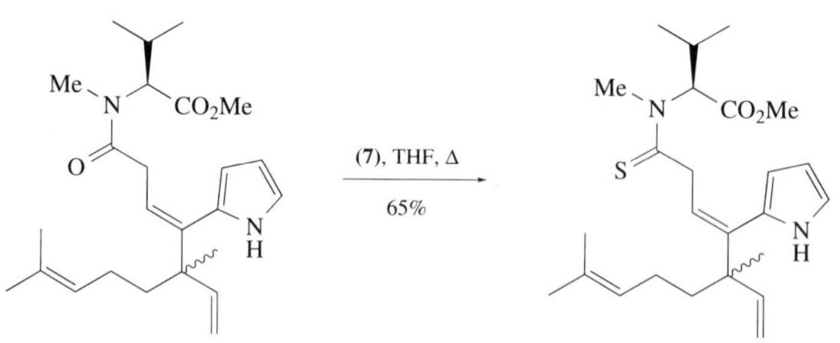

(26)

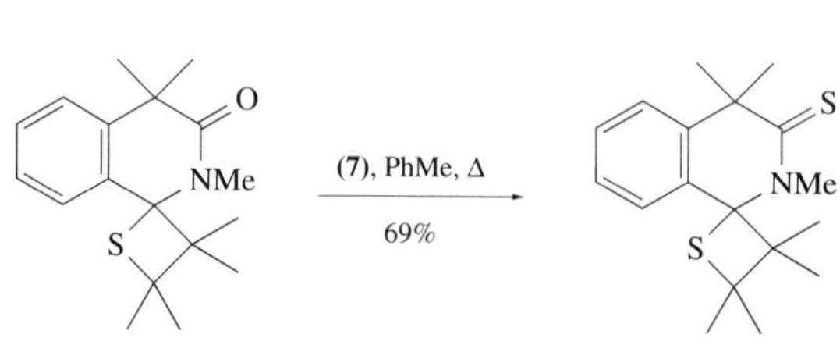

(27)

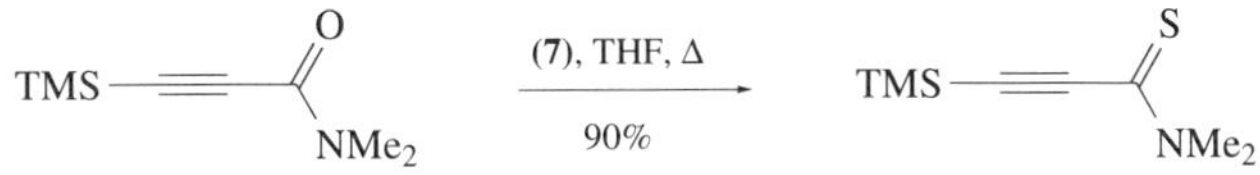

(28)

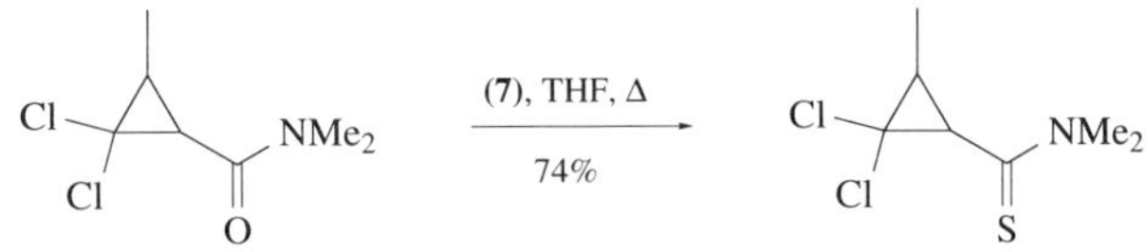

(29)

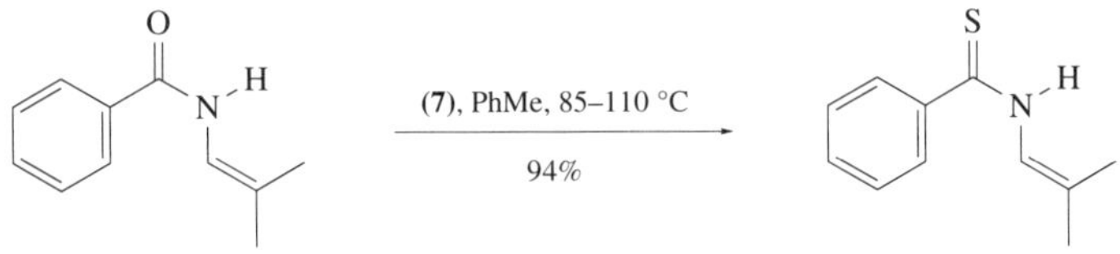

(30)

The normal selectivity noted with Lawesson's reagent (e.g. lactams thionated at 80 °C, ketones and carbamates at 110 °C and acyclic esters at 130 °C) has enabled the extensive, successful use of the reagent for the thionation of amide bonds in suitably protected (*N*-benzyloxycarbonyl or *N*-*t*-BOC) peptide esters (Table 4) ⟨81T3635⟩. This can be carried out without racemisation at the chiral centre ⟨81T3635⟩. The reaction also succeeds with *N*-acyl glycine, alanine, phenylalanine and proline ethyl esters (Equation (31)) ⟨83T3419⟩ in generally excellent yields, with *N*-*t*-BOC glycine, proline, leucine and tyrosine amides ⟨83T3429, 84JCS(P1)785, 92T8601⟩ and with *N*-*t*-BOC-glutamic acid amide 5-benzyl ester ⟨90SC2235⟩. Selectivity can also be achieved as the reagent tends to thionate less hindered amides, particularly glycine ⟨85T5595⟩. In extreme cases of steric hindrance, for example with *N*-*t*-BOCAlaAibOMe (where Aib is 2-amino-2-methylpropanoic acid), the yield of thionoamide can be very poor (16%) and thionation requires forcing conditions (toluene, reflux) in contrast to THF at room temperature for *N*-*t*-BOCGlyAlaOMe ⟨85T5595⟩. Under appropriate conditions, two amide bonds in the tripeptide ZGlyGlyPheOMe can be thionated ⟨87TL2171⟩.

Table 4 Thionation of peptides with Lawesson's reagent.

Starting material	*Product*	*Yield* (%)	*Ref.*
BocGlyGlyOBn	BocGlyψ[CSNH]GlyOBn	89	83T1075
ZGlyGlyOMe	ZGlyψ[CSNH]GlyOMe	92	83T1075
ZGlyGlyOEt	ZGlyψ[CSNH]GlyOEt	78	81TL3635
ZGlyGlyOBn	ZGlyψ[CSNH]GlyOBn	93	81TL3635
ZGlyGlyOPh	ZGlyψ[CSNH]GlyOPh	70	83T1075
BocGlyAlaOMe	BocGlyψ[CSNH]AlaOMe	93	85T5595
ZGlyAlaOBn	ZGlyψ[CSNH]AlaOBn	97	81TL3635
ZGlyAsp(OBn)OBn	ZGlyψ[CSNH]Asp(OBn)OBn	98	81TL3635
ZGlyProOBn	ZGlyψ[CSNH]ProOBn	91	81TL3635
ZAlaGlyOBn	ZAlaψ[CSNH]GlyOBn	95	81TL3635
ZAlaAlaOMe	ZAlaψ[CSNH]AlaOMe	100	83T1075
BocAlaAibOMe	BocAlaψ[CSNH]AibOMe	16	85T5595
ZIleGlyOMe	ZIleψ[CSNH]GlyOMe	86	83T1075
BocLeuGlyOEt	BocLeuψ[CSNH]GlyOEt	83	83T3429
ZPheGlyOMe	ZPheψ[CSNH]GlyOMe	83	83T1075
BocPheLeuOBn	BocPheψ[CSNH]LeuOBn	92	84JCS(P1)785
BocGlyGlyPheOMe	BocGlyψ[CSNH]Glyψ[CSNH]GlyPheOMe	65	87TL2171
BocGlyAlaAibOMe	BocGlyψ[CSNH]AlaAibOMe	85	85T5595

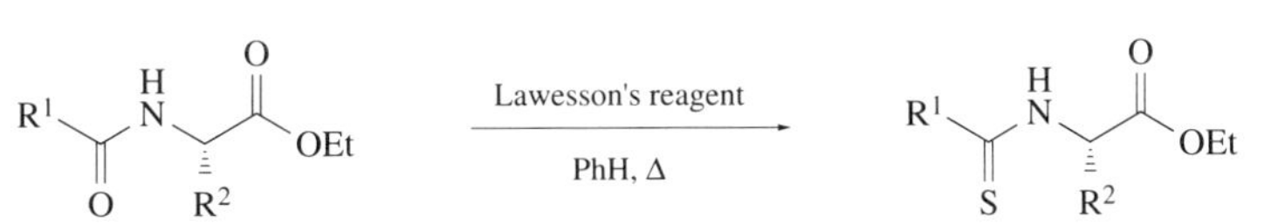

(31)

A number of research groups has attempted to modify Lawesson's reagent (**7**) to improve its selectivity. One problem is chemoselectivity with tri- and higher peptides ⟨87TL2171⟩. The group of Belleau has proposed the use of the phenoxy substituted reagent (**8**). This is reportedly more soluble than Lawesson's reagent and reacts particularly well with peptides in THF, normally at room temperature, with sterically encumbered cases requiring warming to 40 °C and some reactions conducted at 0 °C ⟨83TL3815⟩. Again this reagent is reported to be selective for glycine residues ⟨83TL3815⟩. The reagent has also been used to produce 'retroinverso endothionopeptides' (Equation (32)) ⟨89TL1997⟩ and various thionated protected proline/glutamic acid species ⟨87JPR1039⟩. With other residues, selectivity is not always at a useful level. For example, reaction of Belleau's reagent with *N-t*-BOCMetLeuPheOMe gave a mixture of the two possible mono-endothionated peptides and the dithionated tripeptide (2:2:1 ratio, 80% yield) ⟨85CJC3089⟩ whilst reaction of *f*-Met-LeuPheOMe with the reagent resulted in thionation of the *N*-formyl group ⟨85CJC3089⟩. Typical results using Belleau's reagent (**8**) are shown in Table 5.

Table 5 Thionation of peptides with Belleau's reagent (THF, 0–40 °C).

Starting material	*Product*	*Yield*	*Ref.*
BocGlyGlyOEt	BocGlyψ[CSNH]GlyOEt	81	83TL3815
BocPheLeuOMe	BocPheψ[CSNH]LeuOMe	84	83TL3815
BocGlyPheLeuOMe	BocGlyψ[CSNH]PheLeuOMe	80	83TL3815
BocLeuPheOBu^t	BocLeuψ[CSNH]PheOBu^t	85	85CJC3089
BocMetLeuPheOMe	BocMetLeuψ[CSNH]PheOMe	32	85CJC3089
	+ BocMetψ[CSNH]LeuPheOMe	32	
	+ BocMetψ[CSNH]Leuψ[CSNH]PheOMe	16	
BocPheGlyProOMe	BocPheψ[CSNH]GlyProOMe	64	83TL3815, 86JA182
BocPheAlaProOMe	BocPheψ[CSNH]AlaProOMe	85	83TL3815, 86JA182
O,N-(Boc)$_2$TyrGlyGlyPheLeuOMe	O,N-(Boc)$_2$TyrGlyψ[CSNH]GlyPheLeuOMe	75	83TL3815

(32)

Table 6 Thionation of dipeptides with Lawesson's reagent and reagent (**9**): a comparison.

Starting material	*Product*	*Yield with (7) (%)*	*Yield with (**9**) (%)*
ZGlyGlyOMe	ZGlyψ[CSNH]GlyOMe	69	75
ZGlyGlyOEt	ZGlyψ[CSNH]GlyOEt	69	75
ZGlyPheOEt	ZGlyψ[CSNH]PheOEt	72	94
BocGlyGlyOMe	BocGlyψ[CSNH]GlyOMe	24	93
BocGlyValOMe	BocGlyψ[CSNH]ValOMe	49	59
BocAlaGlyOMe	BocAlaψ[CSNH]GlyOMe	35	61
BocAlaPheOMe	BocAlaψ[CSNH]PheOMe	29	62
BocLeuGlyOMe	BocLeuψ[CSNH]GlyOMe	60	66

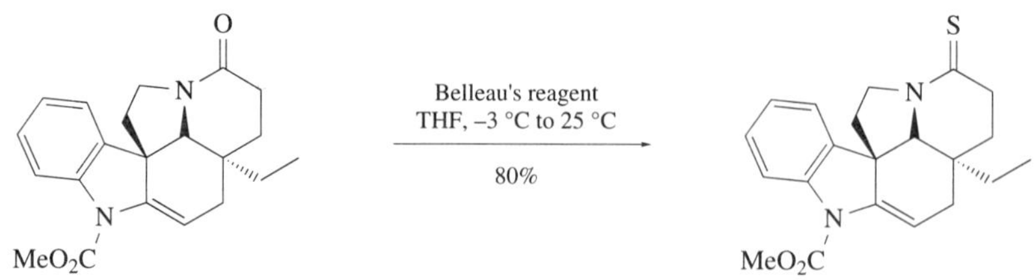

(33)

Belleau's reagent was also used in a recent synthesis of an indole alkaloid (Equation (33)) ⟨86CC1756⟩. A Japanese group has similarly refined the activity of Lawesson's reagent by using the reagents (**9**) and (**10**) ⟨84S827⟩. These two reagents were reported to have good solubility in chloroform, THF, benzene and toluene. Apparently, yields with these two are almost identical, so of the two (**9**) is preferred as the starting materials are cheaper. These react well with cyanoacetamide, caprolactam and a protected dipeptide in THF at room temperature giving high yields of the thionated products ⟨84S827⟩. Under identical conditions, Lawesson's reagent apparently produced only a trace of desired product. A direct comparison of Lawesson's reagent with **(9)** has been carried out ⟨89JCR(S)155⟩ and the results seem to indicate that **(9)** was to be preferred over Lawesson's reagent (Table 6). It was felt that the problems were the relative insolubility of Lawesson's reagent and some competing thionation of the urethane carbonyl. The use of Lawesson's reagent in HMPA at 90–110 °C reduced both of these problems, but **(9)** was deemed superior as reaction could be carried out in THF at room temperature in high yield with no racemisation. A related reagent **(11)** has been used by a Swiss group ⟨87HCA1001⟩. This particular reagent only seems useful for thionating tertiary amides, primary and secondary amides giving poor yields (e.g. 4.5% yield for the thionation of benzamide, 4% for 2-pyrrolidone). A final reagent of this type is **(12)**, derived from bromonaphthalene and phosphorus pentasulfide, which is reported to thionate benzamide in 86% yield ⟨91CC116⟩.

5.14.1.1.3 Miscellaneous reagents

A variety of other reagents has been used to convert amides into thionoamides, but the use of phosphorus pentasulfide and Lawesson's reagent (and the more soluble variants) renders many of the earlier methods redundant. The use of boron sulfide for the thionation of some cephalosporins has been described ⟨75JA5628⟩. Although some regioselectivity was observed, the yields of products were very poor (Equation (34)). This reagent is apparently more reactive than phosphorus pentasulfide for carbonyl–thionocarbonyl transformation ⟨69JCS(C)2192⟩ but seems to have received little attention for the synthesis of thionoamides. One reagent combination purported to generate a form of boron sulfide *in situ* is bis(tricyclohexyltin)sulfide/boron trichloride ⟨82JA3104⟩ which is capable of efficient thionation of β-lactams (Equation (35)) and caprolactam. This mixture is reported to thionate substrates that are inert to Lawesson's reagent and to boron sulfide.

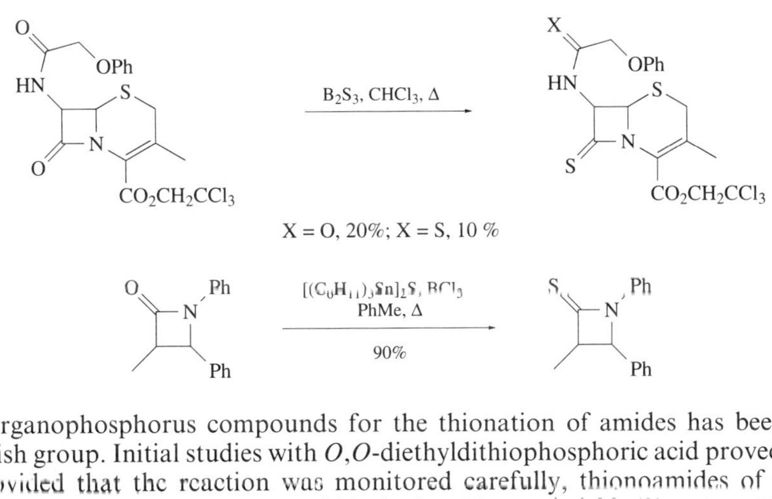

(34)

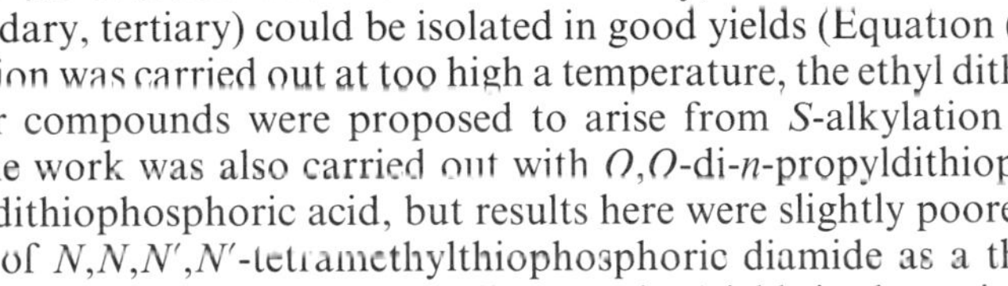

(35)

The use of organophosphorus compounds for the thionation of amides has been explored in depth by a Danish group. Initial studies with O,O-diethyldithiophosphoric acid proved encouraging ⟨77BSB321⟩. Provided that the reaction was monitored carefully, thionoamides of various types (aromatic, aliphatic, secondary, tertiary) could be isolated in good yields (Equation (36)). If heating was prolonged or the reaction was carried out at too high a temperature, the ethyl dithiocarboxylates were produced. The latter compounds were proposed to arise from S-alkylation of the initially formed thionoamide. Some work was also carried out with O,O-di-n-propyldithiophosphoric acid and with O,O-diisopropyldithiophosphoric acid, but results here were slightly poorer than with the original reagent. The use of N,N,N',N'-tetramethylthiophosphoric diamide as a thionating agent was also investigated ⟨77BSB321⟩. This produced excellent results (yields in the region of 90%) with secondary and tertiary amides (Equation (37)), one drawback being the use of HMPA at elevated temperature as the solvent. From formanilide the amidine **(14)** was produced, rather than the expected thionoamide, the product incorporating a dimethylamino group from the solvent. It is noteworthy that under these conditions aromatic primary amides give good (72–87%) yields of nitriles, with the thionoamide being typically formed in small amounts (0–10%). The one exception was the reaction of 3-chlorobenzamide which furnished 18% of the thionoamide and no nitrile. The same research group attempted to refine the selectivity of such reagents by examining the use of a range of reagents of the type $PSCl_x(NMe_2)_{3-x}$ ⟨77BSB693⟩, the results here indicating that $PSCl(NMe_2)_2$ was the best choice of the three (Equations (38) and (39), Table 7). It is interesting that when the reaction is conducted in carbon tetrachloride, chlorinated products are isolated in some cases. Finally, when benzanilide is reacted with an excess of this preferred reagent over an extended time period a sulfide is formed. Another utilised organophosphorus reagent is 2-chloro-2-thiono-5-methyl-1-oxa-2-phospholane, which is reported to thionate *N*,*N*-dimethyl acetamide and *N*,*N*-dimethylbenzamide ⟨91IZV1605⟩.

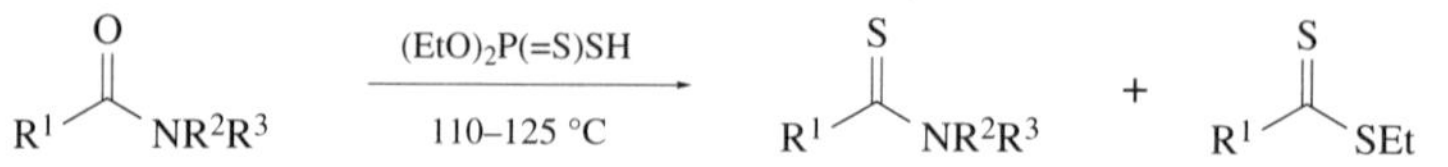

(36)

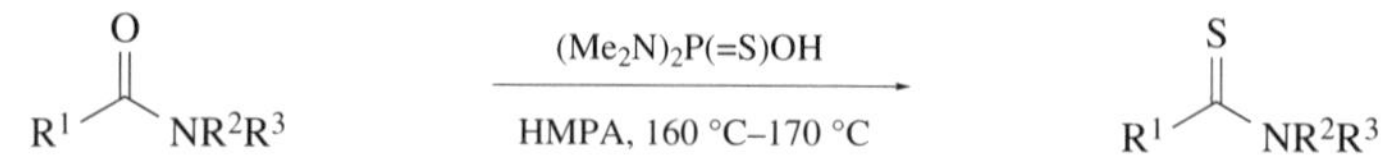

(37)

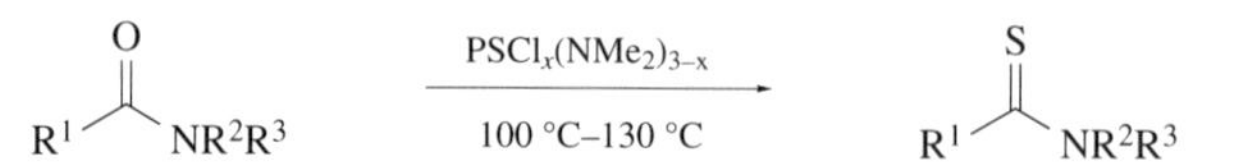

(38)

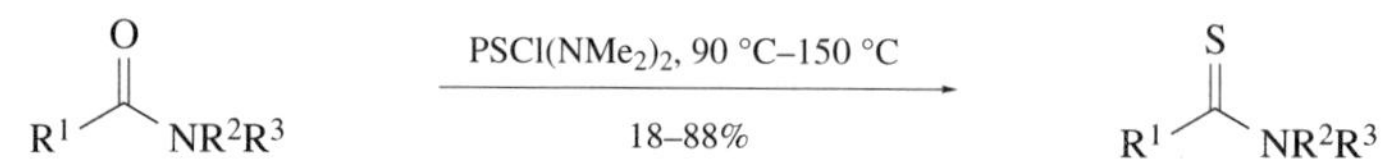

(39)

Table 7 Thionation of amides with $PSCl_x(NMe_2)_{3-x}$ reagents.

R[1]	*R*[2]	*R*[3]	*Yield (%) of thionoamide using:*		
			$PSCl(NMe_2)_2$	$PSCl_2NMe_2$	$PSCl_3$
Ph	Me	Me	90	86	83
Ph	Ph	H	70	23	26
H	Me	Me	69	38	38
Me	Ph	Me	55	28	0

Note: $PS(NMe_2)_3$ failed to give thionoamides.

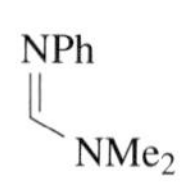

(14)

Some two-step procedures have found application also. For example, treatment of *N*-alkyl lactams with a trialkyl oxonium tetrafluoroborate followed by NaSH in acetone gives reasonable to excellent yields of the thionated lactam (Equation (40), Table 8) ⟨82SC787⟩. The authors compared the yields for this procedure with thionation of the lactams with Lawesson's reagent, with phosphorus pentasulfide in pyridine and with a two-step procedure using trimethyloxonium tetrafluoroborate followed by hydrogen sulfide (Table 8). Broadly speaking their procedure is as effective as using Lawesson's reagent and is better than either of the other two methods. There are some drawbacks, however; the method is inefficient for thionating secondary lactams (competing *N*-deprotonation) and is not applicable to *N*,*N*-disubstituted acyclic amides. A more recent two-step procedure involves activating an amide via treatment with phosphorus oxychloride or oxalyl chloride (generating a chloroiminium salt *in situ*) followed by treatment with bis(trimethylsilyl)sulfide (Equations (41) and (42), Tables 9 and 10) ⟨94JOC348⟩. A number of potential activating agents were tried before deciding on phosphorus oxychloride and oxalyl chloride. Yields are broadly comparable with those obtained using Lawesson's reagent (Tables 9 and 10), although in one case the product *N*-vinyl thionolactam reportedly decomposed.

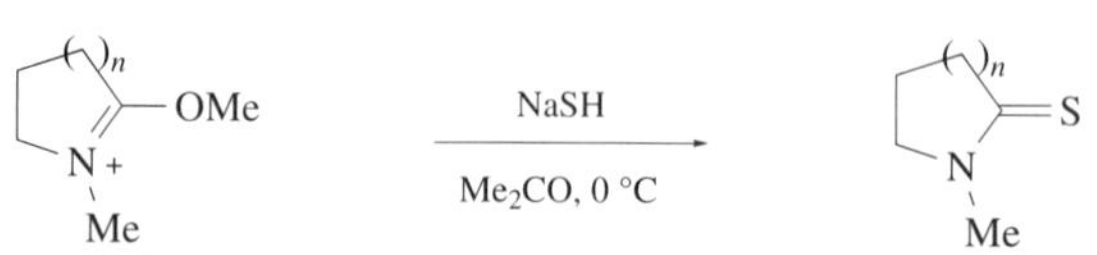

(40)

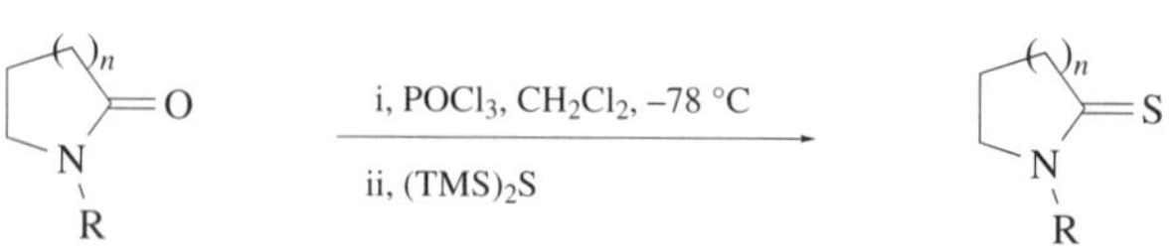

(41)

Table 8 Comparison of thionation methods for lactams.

n	*Yield (%) of thionoamide by:*			
	NaSH	*Me_3OBF_4/H_2S*	*P_4S_{10}/pyridine*	*(7)*
1	100	73	50	95
2	98	63	23	90
3	94	68	29	95

Table 9 Alternative thionation of lactams (Equation (41)).

n	R	*Yield (%) using:*	
		$POCl_3/(TMS)_2S$	*(7)*
1	$PhCH_2$	70	92
2	Me	82	95
2	Et	92	83
2	$CH{=}CH_2$	decomp.	24
2	H	73	75
3	Me	88	95
4	Me	91	94

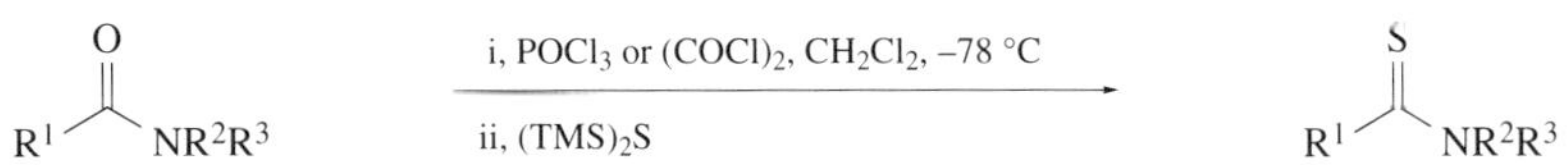
(42)

Table 10 Comparison of amide thionation methods (Equation (42)).

Activator	R^1	R^2	R^3	*Yield (%) using:*	
				$(TMS)_2S$	*(7)*
$(COCl)_2$	Ph	$—[CH_2]_4—$		93	99
$(COCl)_2$	Ph	Me	Me	100	89
$POCl_3$	Ph	H	H	30	92
$POCl_3$	Me	$—[CH_2]_4—$		91	96
$POCl_3$	Me	Me	H	77	89
$(COCl)_2$	Bu^t	$—[CH_2]_4—$		92	86
$(COCl)_2$	H	$—[CH_2]_5—$		93	100
$(COCl)_2$	H	Pr^i	Pr^i	97	86
$POCl_3$	$ArCH_2$[a]	$—[CH_2]_4—$		91	100

[a] $ArCH_2$ = (3,4-methylenedioxy)benzyl.

This is similar to earlier work where *N*,*N*-diethylformamide and *N*-methylformanilide were thionated by treatment with phosphorus oxychloride followed by *N*,*N*-dimethylthionoformamide ⟨79T1087⟩. The authors described this procedure as a 'transcomplexation' via a Vilsmeier-type chloroiminium salt. Carbon disulfide at 100°C and high pressure has also been reported ⟨75T1209⟩ to convert *N*,*N*-dimethylformamide and *N*,*N*-diethylformamide to the thionoformamides in 36% and 23% yields respectively. A route using phosphorus pentachloride followed by hydrogen sulfide to selectively thionate one amide in a penicillin sulfoxide has been reported (Equation (43)) ⟨76TL967⟩. Related reactions to produce various α-chlorothionoamides have also been described (Equations (44) and (45)) ⟨63JOC3492⟩. The combination of phosgene and hydrogen sulfide has also been successful (Equations (46)–(48)) ⟨63CB2671, 85T3331⟩.

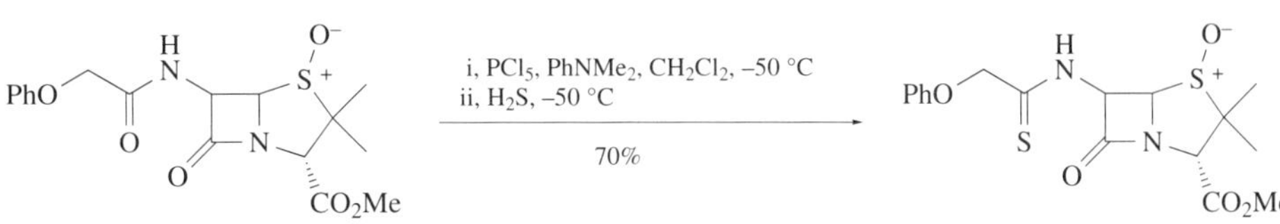
(43)

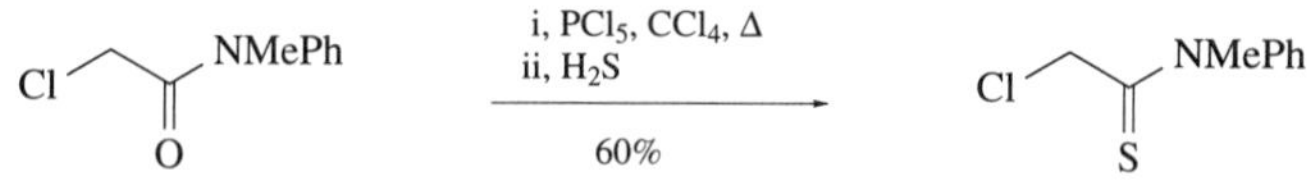

(44)

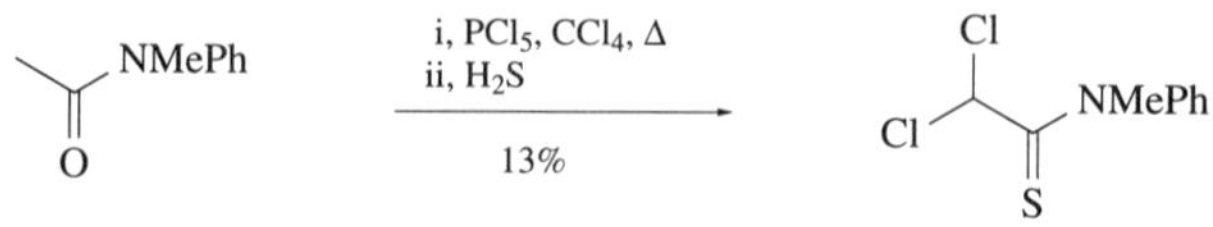

(45)

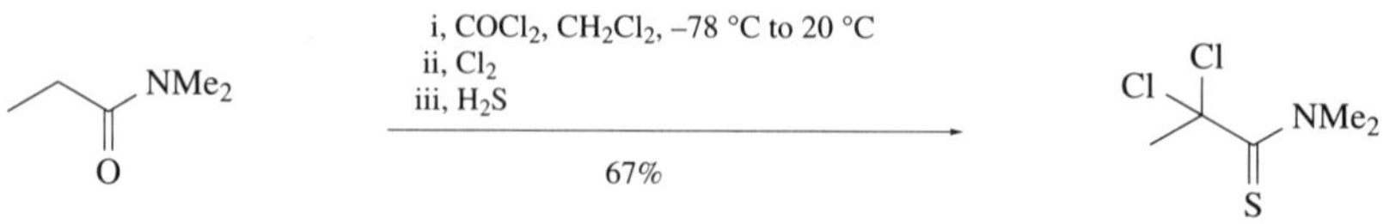

(46)

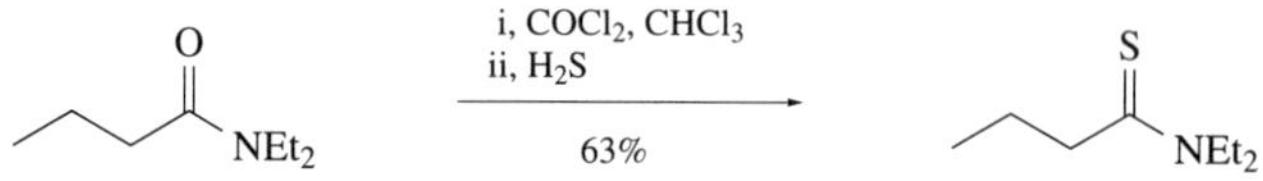

(47)

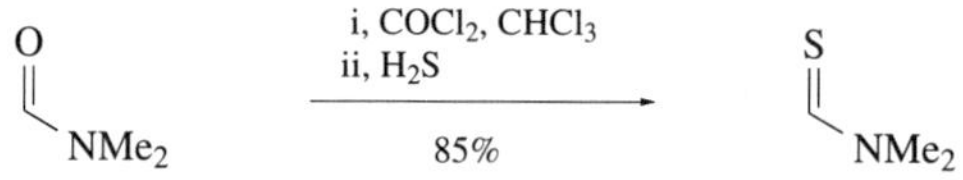

(48)

Ethylaluminum sulfide $(EtAlS)_n$ has been observed to generate thionoamides from tertiary amides (Equation (49)), with yields in the 68–80% range ⟨75JOM(92)139⟩. The reagent $(9\text{-BBN})_2S$ has also been used for the thionation of amides ⟨92LA1081⟩, although its use appears limited to aliphatic amides and *N*-substituted lactams (Equation (50)). Yields are good. *N*,*N*-Dimethylbenzamide reportedly yields thionobenzoic acid 9-BBN ester rather than a thionoamide. Another procedure, although it is technically thioacylation, has been used for the preparation of tertiary thionoamides ⟨78H(11)139⟩. This involves reaction of a tertiary amide with diethylthiocarbamoyl chloride and aluminum chloride (Equations (51)–(54)). The Lewis acid is not necessary for aliphatic cases such as *N*,*N*-dimethylformamide. A reaction mechanism has been proposed. When applied to primary amides, this procedure instead yields nitriles by an alternative pathway. Secondary amides gave no clear products.

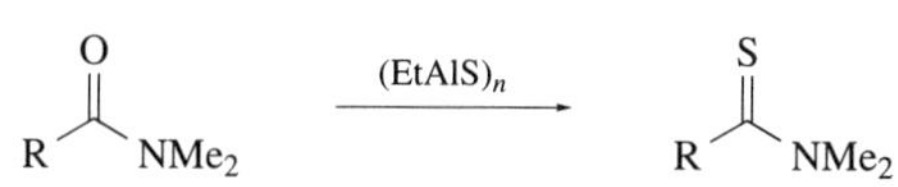

(49)

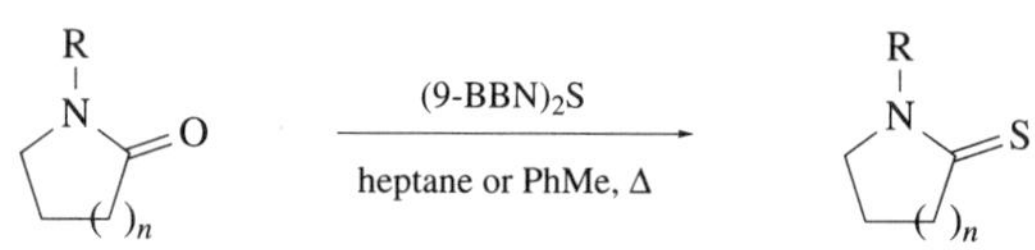

(50)

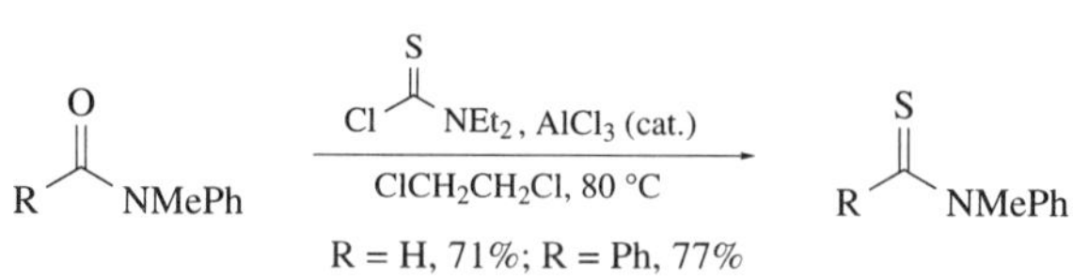

(51)

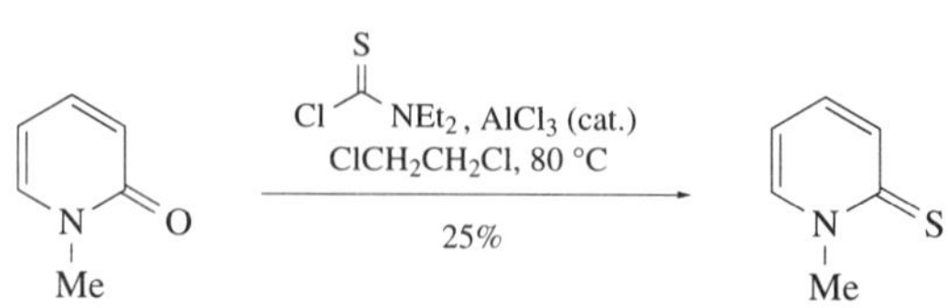

(52)

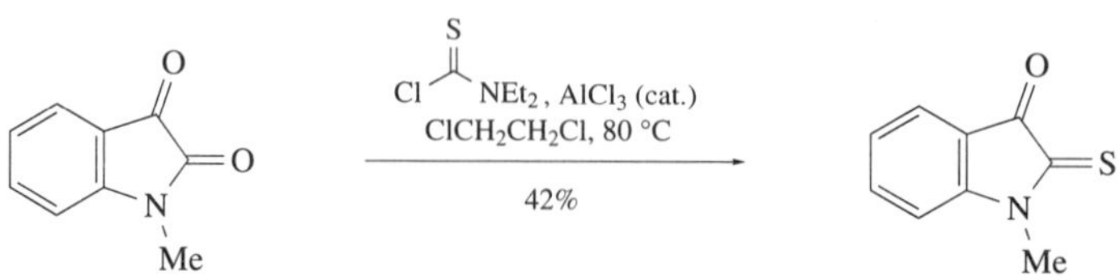

(53)

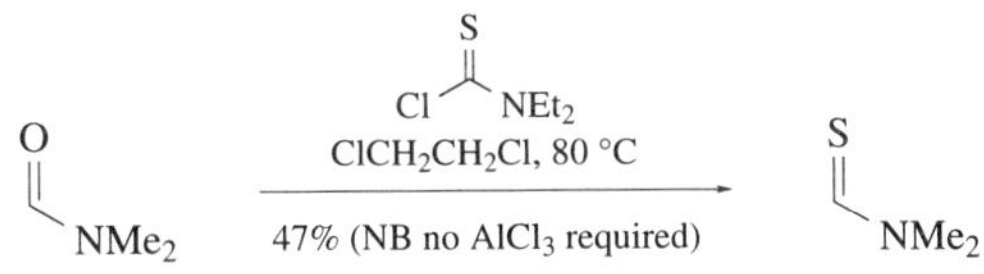

(54)

5.14.1.2 Thionoamides From Nitriles

The earliest reported method for the preparation of thionoamides involved reacting a nitrile with hydrogen sulfide gas. Pertinent references to this transformation can be found in earlier review material ⟨49BSF272, 60CSR45, 66AG(E)447, 69RCR21⟩. The reaction proceeds well under base catalysis, typified by the finding that aromatic nitriles react particularly well with hydrogen sulfide in pyridine solution with one equivalent of triethylamine added ⟨52JCS742⟩. Aliphatic nitriles were reported to react sluggishly under these conditions. Data for typical base-catalysed procedures (Equation (55)) are shown in Table 11.

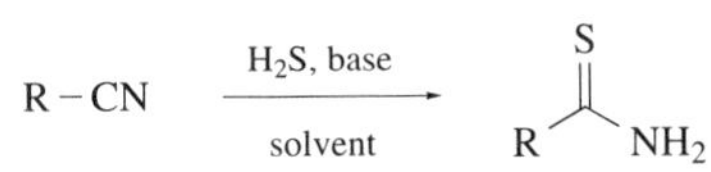

(55)

Table 11 Conversion of nitriles to thionoamides with H_2S.

R	*Base*	*Solvent*	*Yield* (%)	*Ref.*
$AcOCMe_2$	NH_3	EtOH, PhH	50	15CB470
CF_3	None	Pyridine	37	60CB1511
Bu^tCO	NEt_3	Et_2O	35	66LA(698)131
Bu^iCO	NEt_3	Et_2O	84	63AG(E)397, 65M1474
EtO_2C	NEt_3	PhH	86	45JCS601
Bu^tCH_2CO	NEt_3	Et_2O	94	63AG(E)397, 65M1474
$MeNHCH_2$	NEt_3	Pyridine/Dioxan	60	60CB1511
Cyclohexyl	NaOEt	EtOH	6	60CB1511
$PhCH_2$	NEt_3	Pyridine	95	52JCS742
$PhCH_2$	None	EtOH	0	23LA(431)187
$PhCH_2$	NH_3	EtOH	77	23LA(431)187
$PhCH_2$	Me_2NH	EtOH	81	23LA(431)187
$PhCH_2$	NaOEt	EtOH	89	23LA(431)187
$PhCH_2$	KOEt	EtOH	91	23LA(431)187
Ph_2CH	NH_3	EtOH	50	51JOC131
$PhNHCH_2$	NEt_3	Pyridine	83	52JCS742
$PhN{=}NCH_2$	NEt_3	Pyridine	75	52JCS742
2-Furyl	NH_3	EtOH	75	55JA4062
2-Thienyl	NH_3	EtOH	98	55JA4062
2-Pyridyl	NEt_3	Pyridine	66	52JCS742
3-Pyridyl	NEt_3	Pyridine	86	52JCS742
3-Pyridyl	NH_3	EtOH	92	55JA4062
4-Pyridyl	NH_3	EtOH	81	55JA4062
Ph	NEt_3	Pyridine	94	52JCS742
Ph	NH_3	EtOH	78	55JA4062
Ph	Me_2NH	EtOH	100	23LA(431)187
PhCO	NEt_3	Et_2O	66	63AG(E)397, 65M1474
PhCO	NEt_3	Hexane	94	65M1278
4MeO—C_6H_4	NEt_3	Pyridine	90	52JCS742
4MeO—C_6H_4CO	NEt_3	Et_2O	54	66LA(698)131
2Cl—C_6H_4	NEt_3	Pyridine	100	52JCS742
4Cl—C_6H_4	NEt_3	Pyridine	49	52JCS742
$4NH_2$—C_6H_4	NEt_3	Pyridine	82	52JCS742
$4EtSO_2$—C_6H_4	NEt_3	Pyridine	70	52JCS742

This procedure removes the need for high pressure or temperature which have often been used to accelerate the reaction ⟨53JOC748, 69AG(E)278⟩. Some reaction conditions can be quite severe (Equation (56)) and the yields of thionoamides are not always good (e.g., benzonitrile is only converted in 32% yield ⟨53JOC748⟩). Not all nitriles require drastic conditions to react with hydrogen sulfide; thus trifluoroacetonitrile reacts in pyridine at −80°C ⟨60CB1511⟩. One more recent refinement has been to conduct the reaction with added sodium sulfide under phase transfer catalysis (Equation (57)), although the reaction mixture still needs to be heated to 70–80°C ⟨78S917⟩.

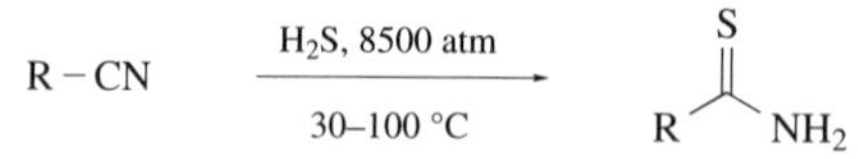

(56)

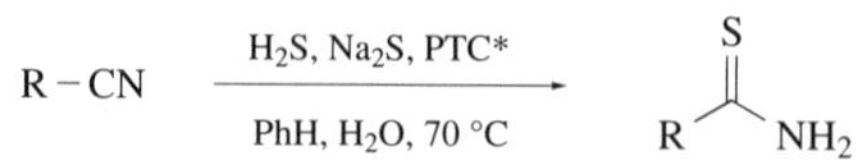

(57)

*PTC is for example, tricaprylyl ammonium chloride

In one interesting piece of work ⟨62JOC2433⟩, the nature of the solvent influenced the course of the reaction (Scheme 1). Reaction of the unsaturated dinitrile (**15**) with hydrogen sulfide and triethylamine in benzene yielded the unsaturated monothionoamide (**16**), yet in ethanol, the isolated product was (**17**). The bisthionoamide (**18**) could be prepared by performing the reaction in DMF at 75–80°C. Transformations of related unsaturated malononitriles have been reported by other workers ⟨70JOC2438, 71JOC2009, 78JCS(P1)629⟩. Treatment of dimethylaminomalononitrile with hydrogen sulfide in a mixture of pyridine and acetic acid reportedly leads to a bisthionoamide ⟨69AG(E)750⟩. Two intermediates en route to this product could also be isolated by running the reaction in benzene at 0°C.

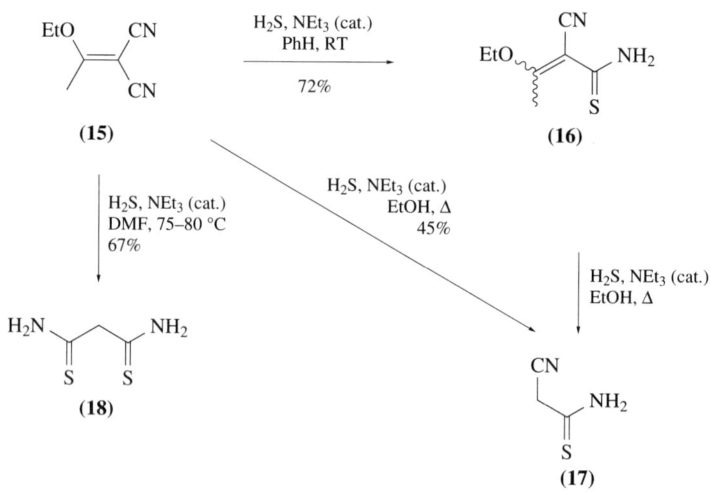

Scheme 1

A route to secondary thionoformamides (Equation (58)) involves the rather unpleasant use of both hydrogen sulfide and hydrogen cyanide ⟨56JOC772⟩. The route is applicable to the synthesis of thionoformamide itself if high pressure is used ⟨69AG(E)278⟩. Hydrogen sulfide in pyridine has also been used in the conversion of interesting α-amino-α-cyano carboxylic acid lactate esters to thionoamides ⟨91BSF760⟩.

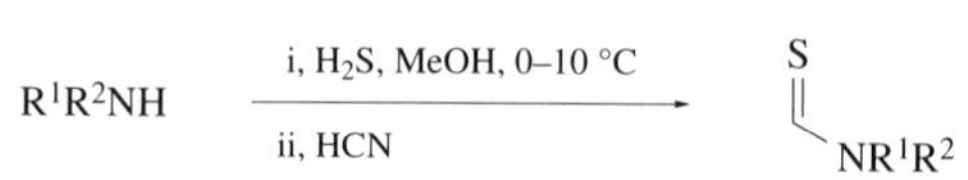

(58)

The use of hydrogen sulfide is not particularly desirable in the laboratory, so a number of reagents have been used which serve as equivalents of hydrogen sulfide. In one particular case thioacetamide acted as the source of sulfur ⟨60JA2656⟩ in reactions with aromatic nitriles or aliphatic dinitriles under acid catalysis (Equation (59)). Ammonium sulfide has also been used (Equation (60)) ⟨24CB981⟩. Cyclohexane 1,1-dithiol (**19**) has served as the source in the preparation of dithionooxamide ⟨70JOC2102⟩. Under appropriate conditions, the monothionooxamide can be isolated (Scheme 2), although the reaction is described as capricious.

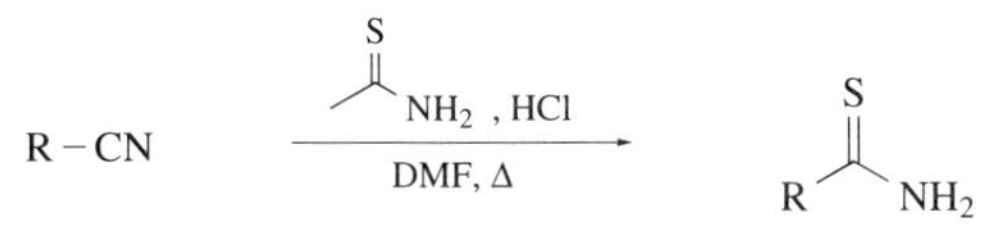

(59)

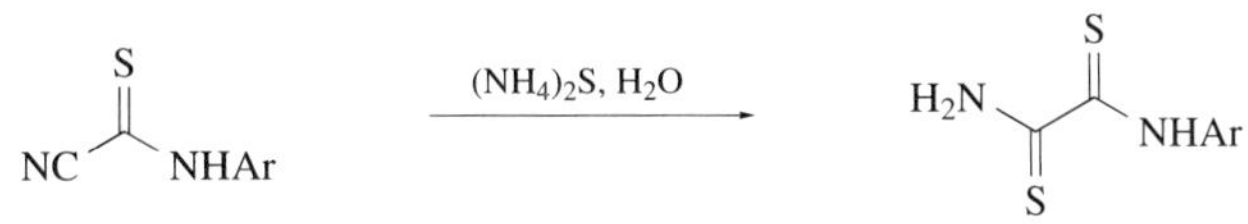

(60)

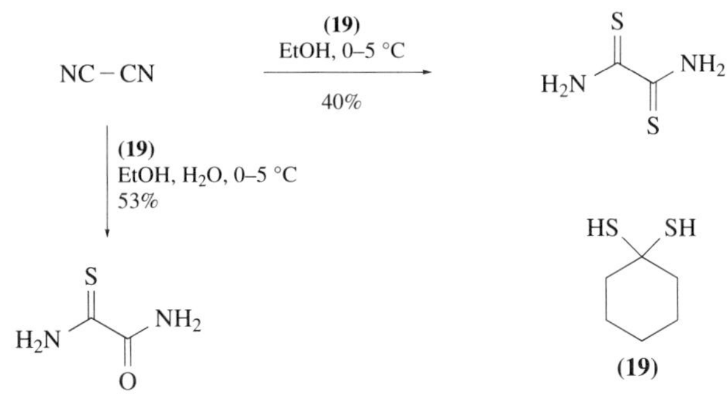

Scheme 2

Tetraethylammonium hydrosulfide has been reported to react with certain unsaturated nitriles (Equation (61)) ⟨84CB2205⟩. Thioacetic acid has been used to convert a series of indole-based nitriles to thionoamides (Equation (62)) ⟨76KFZ41⟩. It can also convert aromatic nitriles ⟨63CB1551⟩ and protected amino acids bearing nitrile functionality ⟨59CB910⟩ to thionoamides. Reactions with thioacetic acid are often very sluggish (days to months). Of more use practically are a group of organophosphorus reagents utilised by a number of research groups. A report has appeared describing the reaction of aromatic nitriles with *O*,*O*-dialkylmonothiophosphoric acids ⟨86ZOB2660⟩. The reactions are sluggish and produce very poor yields of thionoamides, often with a similarly poor amount of the *N*-phosphorylated thionoamide also produced. The reaction of the related *O*,*O*-dialkyldithiophosphoric acids proceeds more effectively giving improved yields of product ⟨66LA(698)131, 79IZV861, 85PS297⟩. The use of either water ⟨85PS(25)297⟩ or an alcohol ⟨92ZOB71⟩ as a co-solvent allows the preparation of a range of thionoamides in good to excellent yields (Equation (63)). It is unfortunate that the work involving water as an additive appears to have been republished virtually identically ⟨89T4599⟩ by a different author from the same institution. Diphenyldithiophosphoric acid has also been used to effect this transformation ⟨81TL1851⟩ and the results are quite impressive (Equation (64)). The author notes that benzophenone, acetone, benzoic acid, benzamide, cyclohexene and some alcohols are not affected by these conditions, although ethyl esters undergo 10–15% transesterification by the isopropanol. Sulfoxides are reportedly deoxygenated and terminal alkynes react.

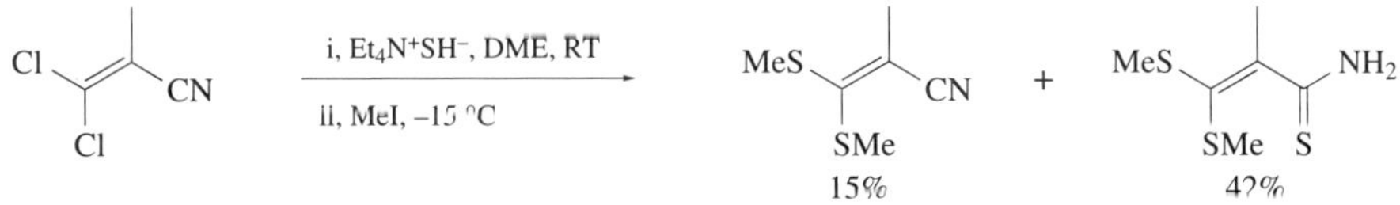

(61)

(62)

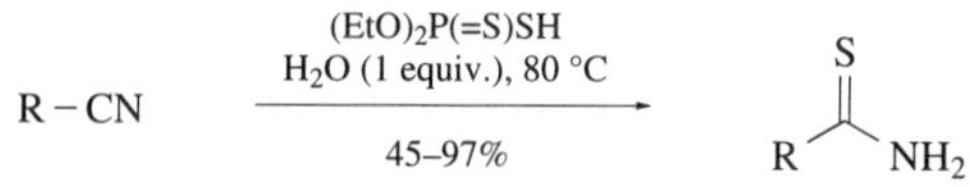

(63)

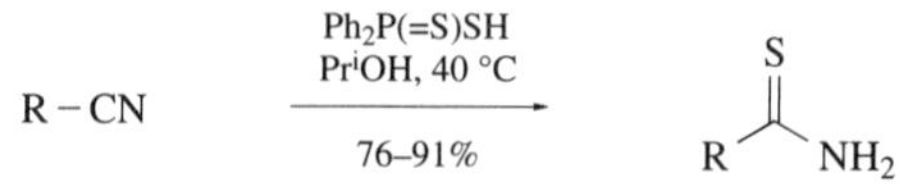

(64)

The use of a mixture of phosphorus pentasulfide and sodium sulfide (1 : 1) in THF at room temperature has recently been described ⟨92SC1397⟩ as a method for the conversion of (mainly aromatic and heteroaromatic) nitriles to the thionoamides in good yields under mild conditions (Equation (65)). Dimethylammonium *N,N*-dimethyldithiocarbamate reportedly converts benzonitrile into *N,N*-dimethylthiobenzamide in 80% yield ⟨67BCJ2209⟩, although the conditions are quite severe (benzene, 150–160 °C, autoclave). Another recent introduction has been the use of a mixture of bis(trimethylsilyl)sulfide and sodium methoxide (generating sodium trimethylsilanethiolate) in a polar aprotic solvent ⟨92S1219⟩. This work focused primarily on aromatic nitriles. Yields were moderate to good. A related procedure published by the same group involves the use of chlorotrimethylsilane and sodium sulfide to generate the sodium trimethylsilanethiolate *in situ* ⟨93JOC4742⟩. This process provided slightly lower yields than the procedure using bis(trimethylsilyl)sulfide (Equation (66), Table 12), but was deemed favourable owing to the expense of the latter reagent.

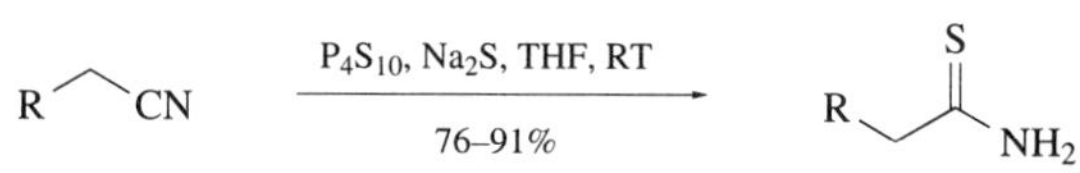

(65)

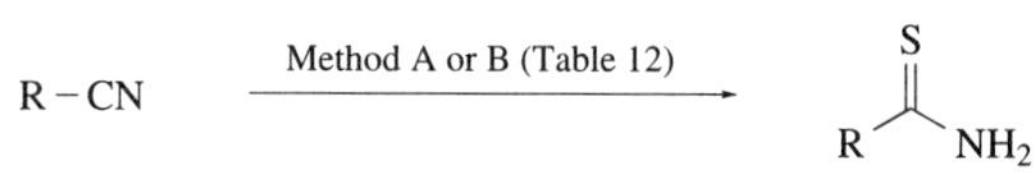

(66)

Table 12 Reaction of nitriles with NaS-TMS.

R	*Yield by Method A* (%)	*Yield by Method B* (%)
n-C_5H_{11}	27	28
$PhCH_2$	57	64
Ph	80	68
4Me-C_6H_4	63	59
3-Pyridyl	71	58

Method A: TMS-S-TMS, NaOMe, DMEU (1,3-dimethyl-2-imidazolinone), 20–35 °C. Method B: TMS-Cl, Na_2S, DMEU, 80 °C.

5.14.1.3 Thioacylation of Amines

A wide variety of reagents exists that enable the conversion of an amine into a thionoamide. Thioacyl halides have only been used to a limited extent, mainly owing to the elusive, unstable nature of simple members of the family. However, thioacyl halides bearing a variety of stabilising groups have been used successfully. The main utility here, however, is in allowing characterisation of the thioacyl halide as a stable derivative. This can be illustrated (Equation (67)) with several examples. Thus trifluorothioacetyl chloride, generated from treatment of an appropriate benzyl haloethyl sulfide with titanium tetrafluoride, has been converted into the thionoamide by reaction with diethylamine ⟨87JFC(35)523⟩. Trichlorothioacetyl chloride, prepared from tin chloride treatment of pentachloroethanesulfenyl chloride, has been reacted successfully with morpholine yielding a mixture of the trichlorothioacetylmorpholide (17%), the dichlorothioacetylmorpholide (27%) and the 1,3-dithietane [2 + 2] dimer of the thioacyl halide ⟨87SUL153⟩. A rationale for the formation of the dichlorothionoamide is given. The functionally complex unsaturated thioacyl bromide (**20**) reacts well with simple primary and secondary amines giving very good yields of the corresponding thionoamides ⟨70JOC2002⟩. On treatment with diethylamine, 2,3,3,3-tetrafluorothiopropanoyl fluoride (obtained by treatment of a benzyl perfluoropropenyl sulfide with fluorosulfonic acid) will also behave as a thioacylating agent, yielding the *N,N*-diethylthionoamide in 50% yield ⟨90IZV1898⟩. Interestingly in this series, a related thioacyl fluoride only yields the heterocycle (**21**) on treatment with diethylamine. This is thought to arise from rapid elimination of hydrogen fluoride (owing to the increased acidity of the α-proton in the hexafluorothioacyl fluoride), generation of the thioketene and dimerisation ⟨90IZV1898⟩. The intermediacy of the α-ketothioacyl chloride (**22**), arising from initial formation of the α-chlorophenacylsulfenyl chloride, has been suggested in the sequential treatment of acetophenone with thionyl chloride and dimethylamine as the isolated product is the *N,N*-dimethylthionoamide (Equation (68)) ⟨76TL2783⟩. *N*-Silylated primary and secondary amines have also been reported to react with thioacyl chlorides ⟨86ZOB1547⟩.

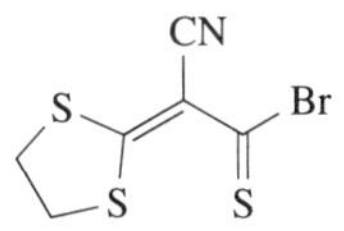

(**20**)

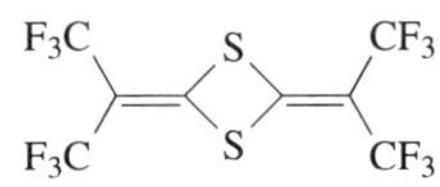

(**21**)

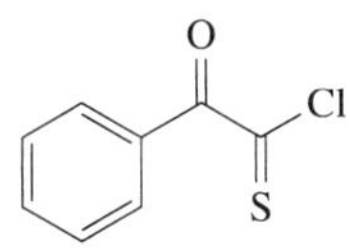

(**22**)

(67)

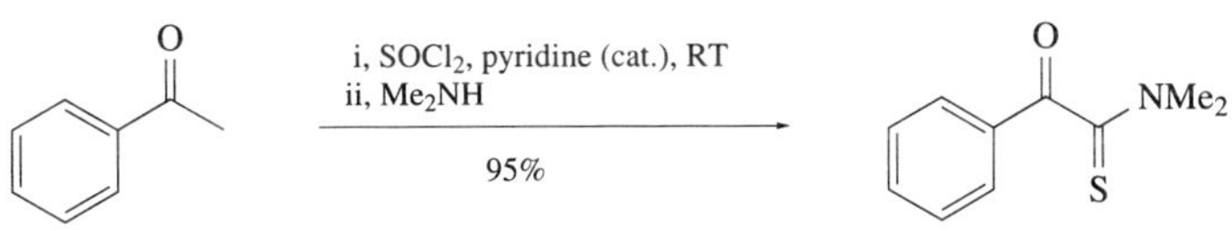

(68)

Trithiocarboxylic acid anhydrides (bisthioacyl sulfides) and related species have received some attention. One useful publication ⟨87ZC90⟩ describes preparative methods and gives a guide to their utility. One early example ⟨59G526⟩ involves the stable heterocyclic compound (**23**) which thioacylates aniline yielding a mixture of the thionoamide (**24**) and the imine (**25**) (Equation (69)). The same group report the preparation of the related trithiocarboxylic anhydride (**26**) ⟨66G125⟩, although its reactions are not described, the authors preferring to use the corresponding dithioacid or methyl dithiocarboxylate. Perhaps of most use in this family of reagents are the aryltrithiocarboxylic anhydrides which react well with a variety of amines providing the thionoamides in moderate yields (Equation (70)) ⟨82LA1229⟩. The other reaction product is the dithioacid amine salt, except in the case of aniline which yields solely the thionoamide. Diphenylamine failed to react ⟨82LA1229⟩. The reactions could not be carried out with aliphatic trithiocarboxylic acid anhydrides, which are reportedly unstable purple oils that dimerise at ambient temperature yielding dithietanes.

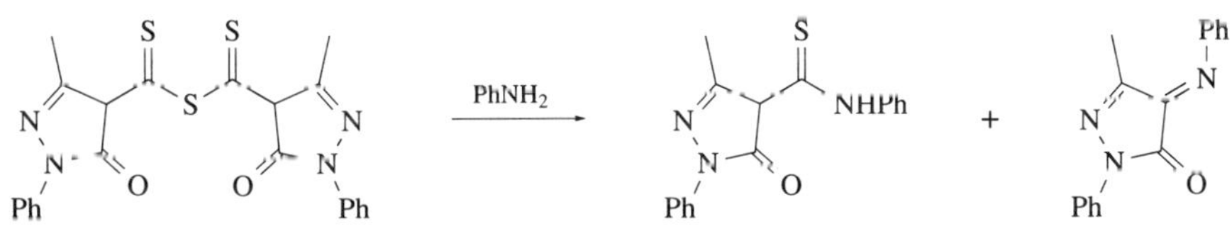

(**23**) (**24**) (**25**) (69)

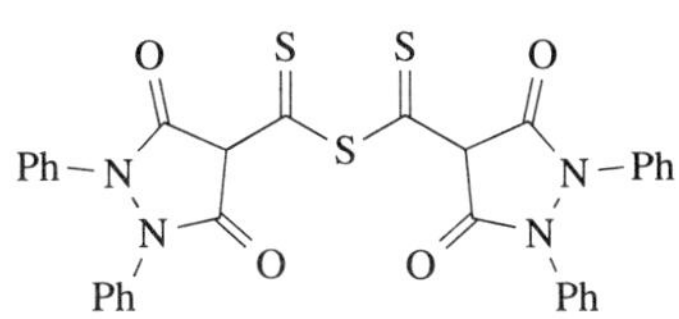

(**26**)

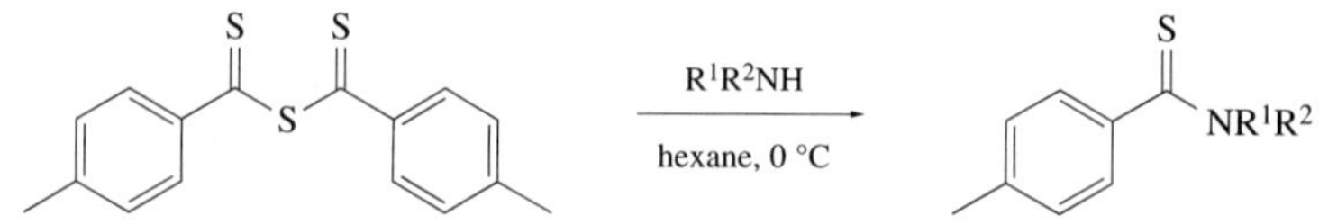

(70)

Cyclic variants have also been used to produce 2-aminothiobenzamides (Equation (71)) ⟨71PHA271⟩. Other efforts involved the mixed reagent (**27**), which on treatment with two equivalents of 4-methylaniline furnished the thionoamide (**28**) in 94% yield, together with the urea (**29**) in 95% yield ⟨72BCJ3653⟩. This appeared to have limited utility as the same reagent (**27**) on reaction with piperidine gave the urea (**30**) (96%) and the piperidinium dithiocarboxylate (**31**) (86%) (Scheme 3). Similar behaviour was noted in the usage of the unsymmetrical acyl thioacyl sulfide (**32**) ⟨81LA1798⟩, yielding a thionoamide only with 4-methylaniline (Equation (72)). The products from reactions with piperidine and *N*-trimethylsilyldiethylamine were the tolyldithiocarboxylic acid piperidinium and diethylammonium salts, respectively. With (diethylamino)trimethylstannane the product was the compound (**33**). A variety of phosphorus-containing reagents, for example thioacyl diphenylthiophosphinoyl sulfides and thioacyl diphenylselenophosphinoyl sulfides, has also been used ⟨85CB1668⟩ producing thionoamides in good yields (Equation (73), Table 13).

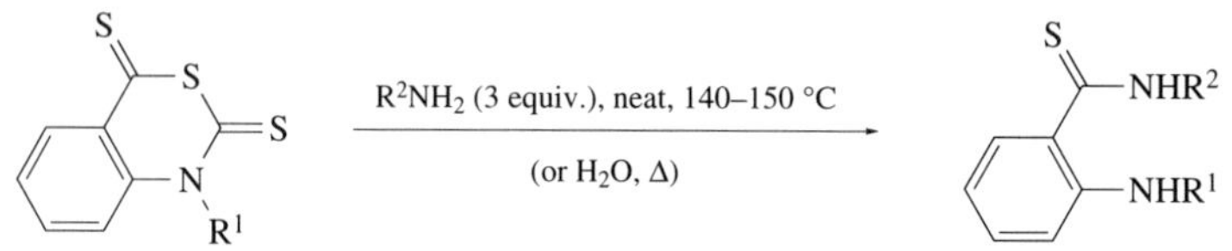

(71)

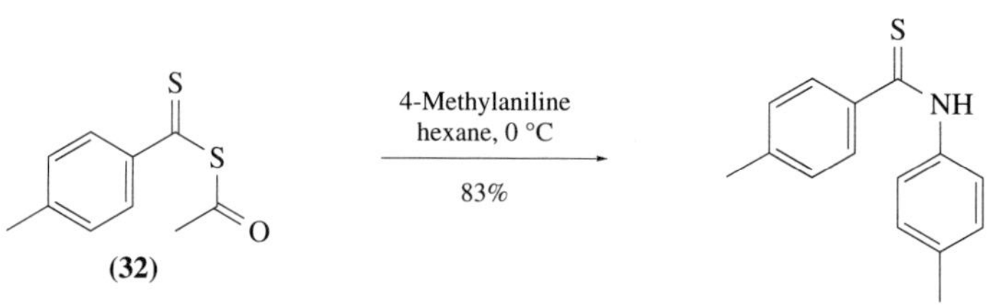

(72)

(73)

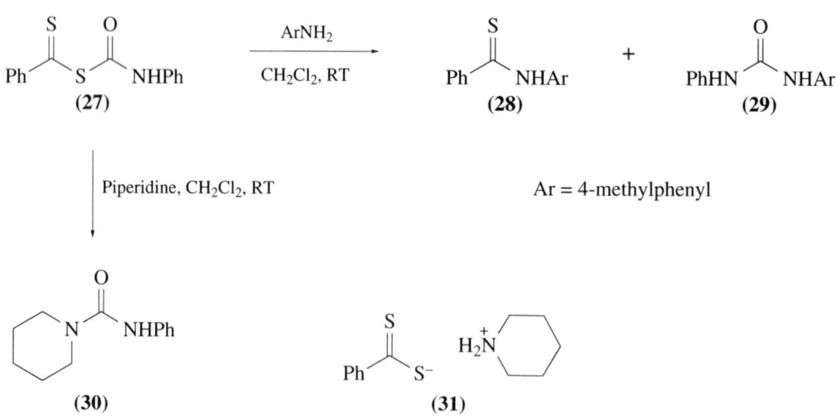

Scheme 3

Thionoesters, dithioesters and *N*-thioacyl azoles (derived from imidazole, triazole, benzimidazole and benzotriazole) have all been used successfully to thioacylate amines. Yields for a range of these transformations (Equation (74)) are shown in Table 14. More specific details on these classes of thioacylating agents follow.

Table 13 Thioacylation using thioacyl diphenylthiophosphinoyl sulfides and related species (Equation (73)).

X	R^1	R^2	*Yield* (%)
S	Cyclohexyl	H	88
S	Et	Et	82
S	Ph	Ph	0
S	Ph	H	78
S	4Me—C_6H_4	H	70
Se	Cyclohexyl	H	70
Se	Et	Et	93
Se	Ph	H	60
	4Me—C_6H_4	H	37
	—$[CH_2]_5$—		26

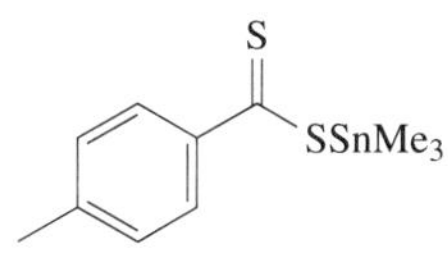

(33)

(74)

Table 14 Thioacylation of amines (Equation (74)).

R^1	*X*	R^2	R^3	*Solvent*	*Yield* (%)	*Ref.*
H	OEt	H	H	None	90	64ZC457
H	OEt	H	Ph	None	95	64ZC457
H	OEt	Me	Bu^t	None	68	76CB1309
H	OEt	Me	$PhCH_2$	None	88	76CB1309
Me	OEt	H	Pr^i	THF	70[a]	65BSF3623
Me	OEt	H	Ph	THF	70[a]	65BSF3623
Me	SCH_2CO_2H	Me	Me	NaOH (aq.)	37	66ACS597
Me	SCH_2CO_2H	—$[CH_2]_2$—		$NaHCO_3$ (aq.)	20–30	75T1245
CN	SMe	H	Bu^n	EtOH	25[b]	85H(23)649
CN	S^-Na^+	H	Et	DMF	53[b]	75S47
Bu^tCOCH_2	SMe	H	Me	PhH	80	73BSF3349
$PhCH_2$	SMe	H	H	MeOH	92	51JOC131
$PhCH_2$	SMe	H	Ph	None	70	51JOC131
$PhCH_2$	SCH_2CO_2H	H	Ph	NaOH (aq.)	96	52ACS327
$PhCH_2$	SMe	Me	Pr^n	None	86	48JA2837
PhCH═CH	OEt	—$[CH_2]_5$—		MeOH	93	80CB919
$PhCONHCH_2$	OMe	—$[CH_2]_5$—		CH_2Cl_2	82	92LA921
Ph	SCH_2CO_2H	c	c	C_5H_5N, NEt_3	52	93SL801
Ph	1-imidazolyl	H	Me	THF	49	70LA(739)201
Ph	1-imidazolyl	H	Ph	C_6H_{12}	71	70LA(739)201
Ph	1-benzimidazolyl	H	Ph	None	73	70LA(739)201
Ph	1-imidazolyl	Et	Et	C_6H_{12}	49	70LA(739)201
Ph	1-1,2,4-triazolyl	Et	Et	C_6H_{12}	88	73LA636

[a] For these cases, the amine was activated by treatment with EtMgBr. [b] In these cases (and others in the references), the products are the dithionoamides. [c] R^2R^3NH is:

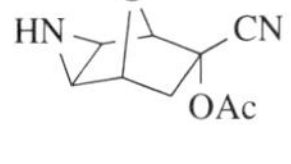

Thionoesters have long been known to thioacylate amines ⟨11CR(153)279⟩. In the case of most secondary nonaromatic amines (e.g. piperidine, morpholine), reactions can proceed in the absence of solvent without the need for any special conditions ⟨65BSF3623⟩. However, diethylamine, primary aliphatic and aromatic amines, and mixed aryl aliphatic amines require treatment with ethyl magnesium bromide in THF or ether in order to react in a satisfactory fashion. For 4-nitroaniline, sodamide in THF at reflux was used to generate the salt of the amine which was then reacted with ethyl thionobenzoate, providing the thionoamide in quantitative yield ⟨65BSF3623⟩. The same workers prepared a wide range of substituted aromatic amides from various substituted anilines, ethyl magnesium bromide and 4-alkoxy substituted ethyl thionobenzoates ⟨65BSF3628⟩. It is noteworthy that the reaction of ethyl thionoformate with primary, secondary and aromatic amines proceeds without the need for treatment with a Grignard reagent ⟨64ZC457, 76CB1309, 82M(113)593⟩. A one-pot procedure for the synthesis of thionoformamides has been published ⟨89JOC1212⟩. This involves the treatment of triethylorthoformate with hydrogen sulfide and a catalytic amount of 70% aqueous perchloric acid, generating the malodorous ethyl thionoformate *in situ*, followed by addition of the primary or secondary amine (Equation (75)). Yields are respectable. Unsaturated thionoesters have been used to prepare α,β-unsaturated thionoamides ⟨80CB919⟩. A kinetic and mechanistic study of the reaction of phenyl thionoacetate with cyclic amines has recently been described ⟨93JOC4908⟩. Thionoesters from protected amino acids (*N*-tosylglycine, *N*-benzyloxycarbonylglycine and alanine, *N*-tosylphenylalanine and β-phthalimidoglycine) have been coupled with glycine and alanine either using aqueous sodium hydroxide/ether ⟨60AG268⟩ or triethylamine in aqueous dioxan ⟨66LA(695)217⟩, thus generating protected thionated dipeptides. Similarly methyl *N*-benzoylthionoglycinate reacts with piperidine to give the thionoamide in 82% yield ⟨92LA921⟩. This publication concentrates on the preparation of *N*-protected chiral amino acid thionesters rather than on their utility. One interesting reaction ⟨71CB2709⟩ is the transformation of ethyl thionopropanoate into the unsaturated thionoamide (**34**) (Equation (76)). Thioisocoumarins have been used as a type of cyclic thionoester providing a variety of *ortho* substituted thionobenzamides on treatment with primary amines (Equation (77)) ⟨66BSF3828⟩.

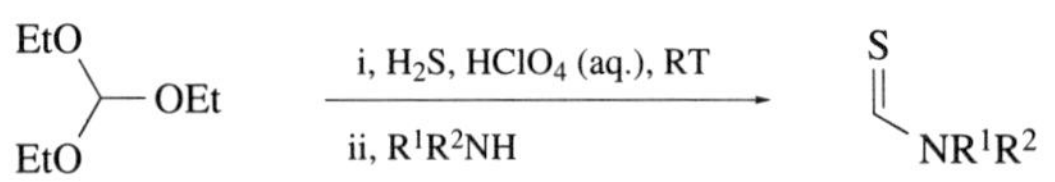

(75)

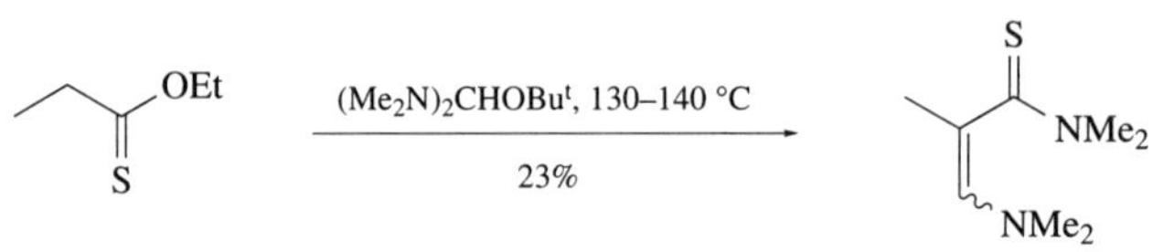

(**34**)

(76)

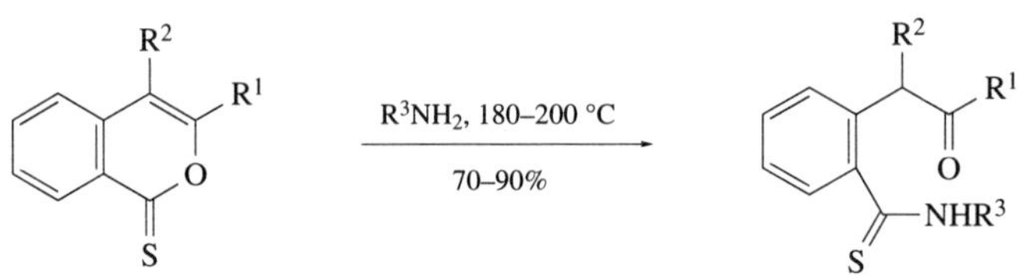

(77)

R^1 = H, aryl; R^2 = H. Ph; R^3 = alkyl, benzyl, aryl

A comparison of the rates of aminolysis of thioesters, thionoesters and dithioesters by piperidine showed that by far the most reactive species were dithioesters ⟨72ZC133⟩. Not surprisingly, therefore, dithioesters have received a large amount of attention as thioacylating agents for amines. Methyl dithiophenylacetate has been successfully reacted with a range of amines ⟨51JOC131⟩ and amino acids ⟨48JA2837⟩. A reagent introduced by Kjaer ⟨52ACS327⟩ is carboxymethyl dithiophenylacetate, allowing the thioacylation reactions to be performed in aqueous sodium hydroxide solution. Carboxymethyl dithioacetate ⟨66ACS597⟩ and carboxymethyl dithiobenzoate ⟨69JCS(C)1117⟩ have also been used. Interestingly, methyl azidodithioacetate fails to provide a thionoamide on reaction with piperidine, complete decomposition occurring instead ⟨92S1199⟩. With β-ketodithioesters, the reaction can provide either the desired β-ketothionoamide or the β-amino-α,β-unsaturated thionoamide ⟨73BSF3349⟩. It is noteworthy that in the case of the methyl dithioacetoacetate reactions other products isolated are the β-amino-α,β-unsaturated dithioesters (Equation (78)). With β-amino-α,β-unsaturated dithioesters amine exchange as well as thionoamide formation occurs (Equation

(79)) 〈84SUL39〉. Aromatic amines apparently do not react with these β-amino-α,β-unsaturated dithioesters.

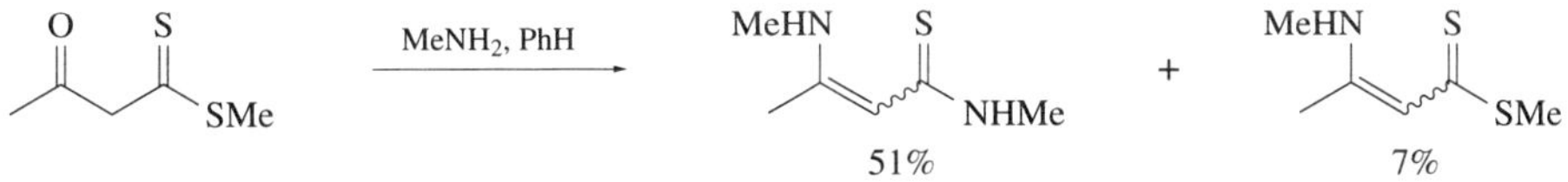

(78)

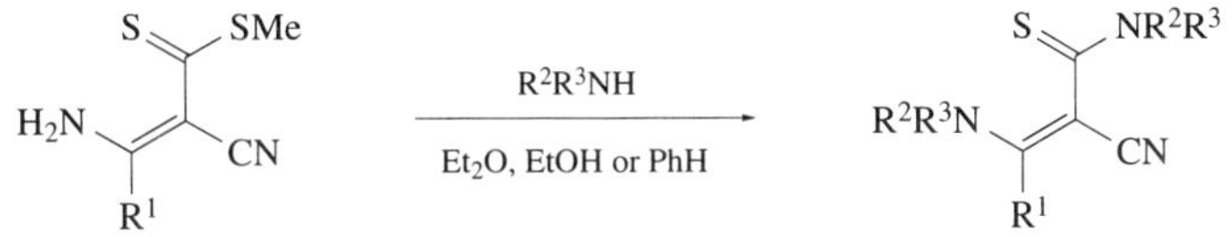

(79)

The presence of adjacent steric bulk, for example quaternary centres (Equations (80) and (81)) does not prevent useful yields being achieved 〈73S297, 93JMC1604〉. Reactions of methyl cyanodithioformate with amines provides a route to dithionooxamides 〈85H(23)649〉 via thioacylation and transformation of the nitrile group. Aniline gives a thiourea rather than a thionoamide. As well as standard dithioesters, their S-oxides (alkylthiosulfines) have been reported to react with amines to provide thionoamides 〈93TL6741〉 (Equation (82)). The oxygen in the starting material is absent from the products. Methyl dithiocarboxylates of *N*-protected amino acids (glycine and tyrosine) have been used in efficient syntheses of monothionated analogues of Leu-enkephalin (H-Tyr-GlyGlyPheLeuOH) (Table 15) 〈84JCS(P1)785〉. It is worth contrasting this efficiency with the much poorer yields or complex mixtures produced when using ethyl *N*-benzyloxycarbonyl thionoglycinate as the thioacylating agent 〈84JCS(P1)785〉. There is one drawback associated with the thioacylation of amino acids with reagents such as (**35**) and that is the potential for racemisation at the amino acid centre. Thus coupling of (**35**) with H-MetOMe gave a 48 : 52 mixture of isomers at the tyrosine centre 〈92T8601〉. A number of conditions and reagents were tried to minimise the proportion of unwanted epimer but this could not be brought below 27%. Some reactions fail to proceed, for example attempted coupling of glycinamide with the methyl dithioester of Boc-leucine resulted in recovery of the thioacylating agent 〈83T3429〉. Other reagents used successfully for preparing *N*-thioacyl amino acids are the carboxymethyl dithioesters introduced by Kjaer 〈52ACS327, 66ACS597〉.

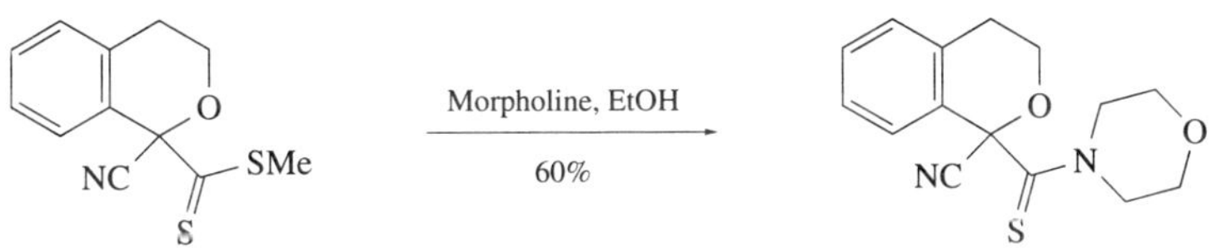

(80)

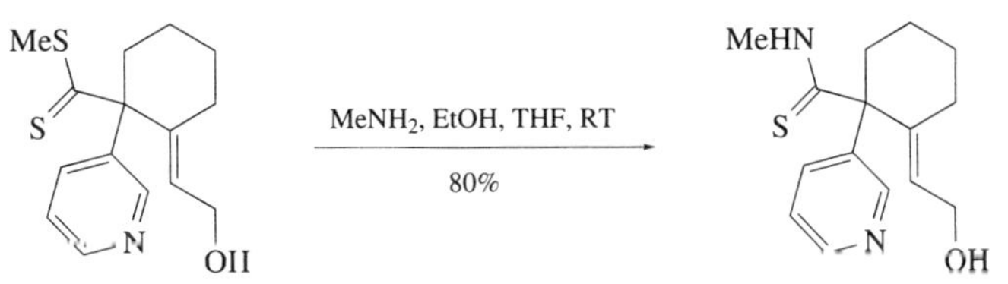

(81)

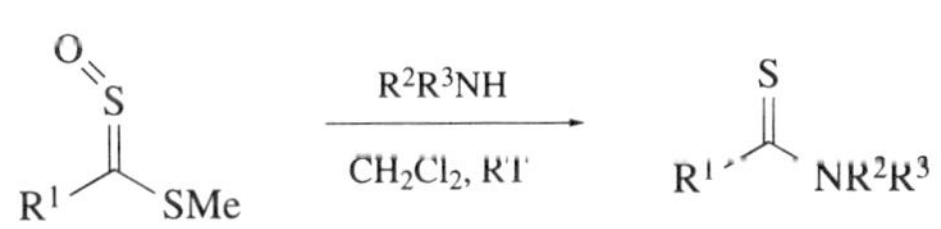

(82)

A range of *N*-thiobenzoyl azoles (incorporating imidazole, triazole, benzimidazole and benzotriazole as the leaving group) have been reacted successfully with a range of amines 〈70LA(739)201, 73LA636〉. Yields are typically only moderate (50–60%). Direct comparison of the reactions of the different *N*-thiobenzoyl azoles with various amines was made in this study. In general, the *N*-thiobenzoyl imidazole gave the best yields. It is worthy of note that coupling of *N*-thiobenzoylimidazole with *N*-methylaniline failed. However, replacement of *N*-thiobenzoylimidazole with *N*-thiobenzoyl 1,2,4-triazole in this reaction gave a 43% yield of the desired thionoamide. Related species such as *N*-thiobenzoylbenzoxazolones are reported 〈68ZC376〉 to react with a variety of primary, secondary and aromatic amines in good yield (Equation (83)). More recently, endothionopeptides have been prepared by coupling *N*-thioacylbenzimidazolones (where the thioacyl

Table 15 Preparation of thionopeptides.

Starting material (HCl salt)	*Thioacylating agent*	*Product*	*Yield (%)*
GlyOEt	$ZNHCH_2C(=S)SMe$	ZGlyψ[CSNH]GlyOEt	97
PheLeuOBn	$ZNHCH_2C(=S)SMe$	ZGlyψ[CSNH]PheLeuOBn	93
GlyPheLeuOBn	$ZNHCH_2C(=S)SMe$	ZGlyψ[CSNH]GlyPheLeuOBn	89
PheLeuOBn	$BocNHCH_2C(=S)SMe$	BocGlyψ[CSNH]PheLeuOBn	89
GlyPheLeuOBn	$BocNHCH_2C(=S)SMe$	BocGlyψ[CSNH]GlyPheLeuOBn	72
GlyGlyPheLeuOBn	(**35**)	BocTyr(OBn)ψ[CSNH]GlyGlyPheLeuOBn	96

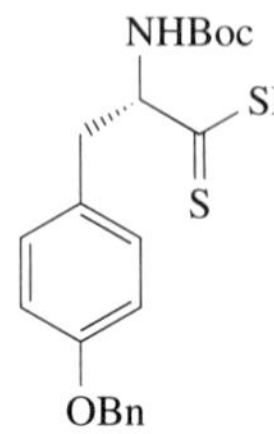

(35)

portion is an *N*-protected amino thionoacid) to amino acid esters ⟨93BMC619⟩. The *N*-thioacylbenzimidazolones of any naturally occurring (or otherwise) suitably protected amino acid could potentially be used in this extremely useful procedure which appears to be free of the racemisation problems discussed earlier. An *N*-aryl *N*-thioacylthiocarbamate has also been reported to react with diethylamine (Equation (84)) ⟨85ZC219⟩. Thioaroyl pyridinium salts also serve as thioacylating agents in the synthesis of aromatic thionoamides (Equation (85)) ⟨90ZC67⟩. A route to pyridinethione thionoamides involves ring opening of dithiolethiones with amines (Equation (86)) ⟨93JHC1079⟩.

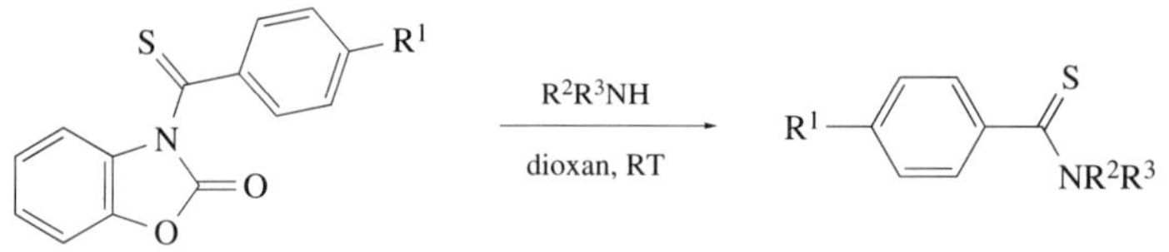

(83)

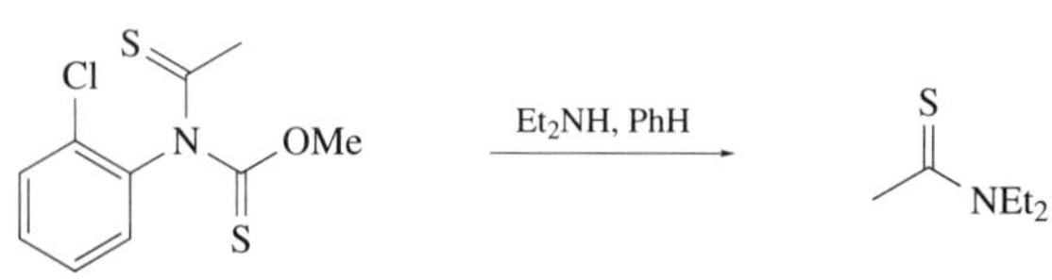

(84)

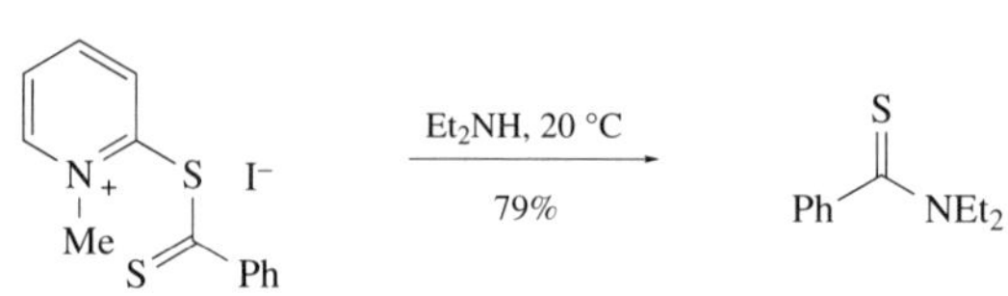

(85)

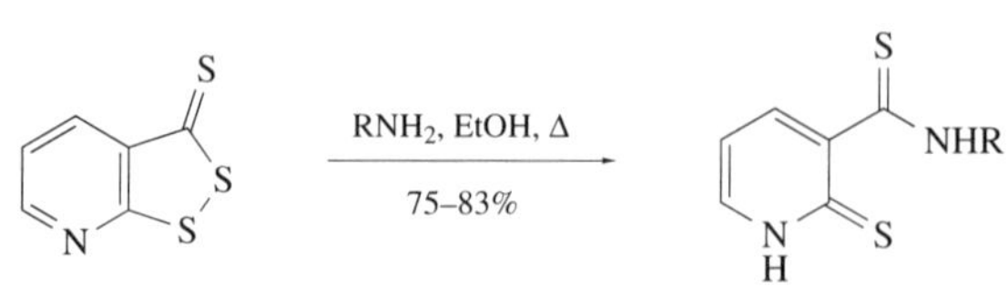

(86)

R = alkyl, benzyl

Dithiocarboxylic acids have been used with the aid of *N,N′*-dicyclohexylcarbodiimide to thioacylate amines derived from the penicillin nucleus (Equation (87)) ⟨75TL3063⟩. A number of early references to the use of dithiocarboxylic acids as thioacylating agents can be found in a review on organic sulfur compounds ⟨48AR198⟩. Not only the free acids but their salts have been used (Equation

(88)) ⟨49JOC962⟩, including one method involving addition of sodium cyanide to carbon disulfide in DMF followed by addition of two equivalents of primary amine providing a route to *N,N′*-disubstituted dithionooxamide preparation ⟨75S47⟩.

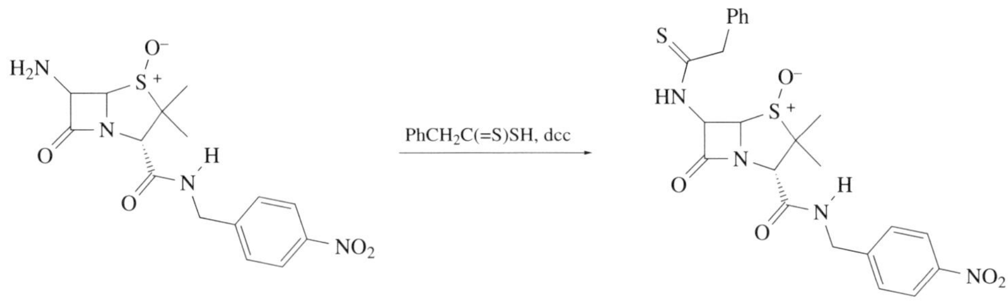

(87)

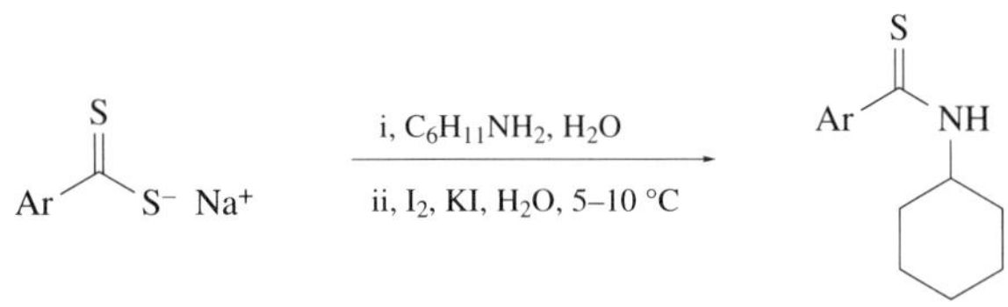

(88)

A recent enhancement involves treating a dithiocarboxylic acid with sodium hydride followed by a catalytic quantity of boron trichloride and finally the primary or secondary amine ⟨93PS(80)31⟩. Related work involves reaction of the dithiocarboxylic acid with one equivalent of catecholborane ⟨90JCR(S)106⟩ or 9-BBN ⟨91PS(56)283⟩ followed by the amine. The procedure using 9-BBN is preferred as the initial reaction of catecholborane with the dithiocarboxylic acid reportedly only proceeds to 75–80% completion. A comparison of yields from these three procedures (Equation (89)) is made in Table 16. Very recently, some studies using monothioacids have been made, in which thioacetic acid is converted to various activated esters using substituted [(benzotriazol-1-yl)oxy] tris(pyrrolidino)phosphonium hexafluorophosphate and then reacted with cyclohexylamine ⟨94JOC1257⟩. Unfortunately, the amide and thionoamide are produced, although the ratio is greatly in favour of the thionoamide.

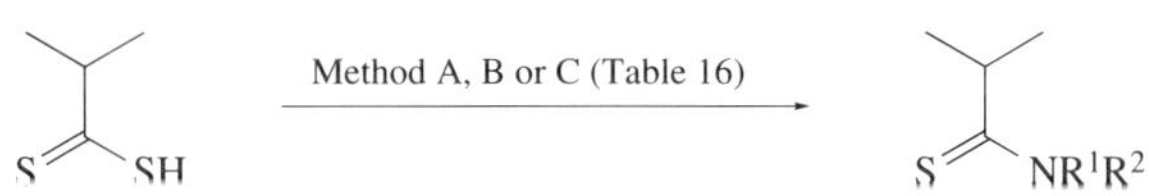

(89)

Table 16 Reactions of dithioacids.

R[1]	*R*[2]	*Yield* (%) *by method*:		
		A	*B*	*C*
Me	Me	92	67	92
—$[CH_2]_4$—		90	55	73

Method A: i, NaH, CH_2Cl_2; ii, cat. BCl_3; iii, R^1R^2NH, RT.
Method B: i, Catecholborane, PhMe; ii, R^1R^2NH, CH_2Cl_2, RT.
Method C: i, 9-BBN, PhMe; ii, R^1R^2NH, PhMe, RT.
See Equation (89).

Thioketenes and thioketenoids generated by [3,3]-sigmatropic rearrangements ⟨68RTC1236, 72RTC578, 77TL4307, 79LA1746, 80CB3024⟩, by thermolysis of heterocycles ⟨75AG(E)248, 78JCS(P1)325, 79LA1715, 84CB2205, 90JHC567⟩ and by other procedures ⟨85TL5269, 76CC1008, 77S328, 79LA1702, 82RTC310⟩ react in variable yields with a range of amines to produce thionoamides. The chemistry of thioketenes has been reviewed by Schaumann, one of the principal workers in the area ⟨88T1827⟩. In many cases, the product thionoamide is only really being used as proof for the intermediacy of the thioketene. Deprotonation of alkynes followed by addition of sulfur and a secondary amine is a useful source of thionoamides ⟨67JPR294, 68RTC38, 71ZOR1120, 77S888⟩. At least two equivalents of the amine are required. In one instance, the addition of one equivalent of *t*-butyl bromide as a proton donor was required ⟨68RTC38⟩. In its absence yields were only 10–20%. The use of primary amines or branched alkynes reportedly give only small to moderate yields of products ⟨68RTC38⟩. Other authors report reasonable yields with branched alkynes ⟨71ZOR1120⟩. The reaction has been suggested to proceed via the thioketene ⟨71ZOR1120, 77S888⟩. Reaction of amines with alkynyl alkyl sulfides at high temperature in a sealed tube also leads to thionoamides ⟨71ZOR1120⟩. Here it is thought that cleavage of the alkyl sulfur bond occurs under the reaction conditions, allowing the generation of an alkynyl thiolate which serves as a thioketene precursor. A range of typical transformations is depicted in Equations (90) to (96).

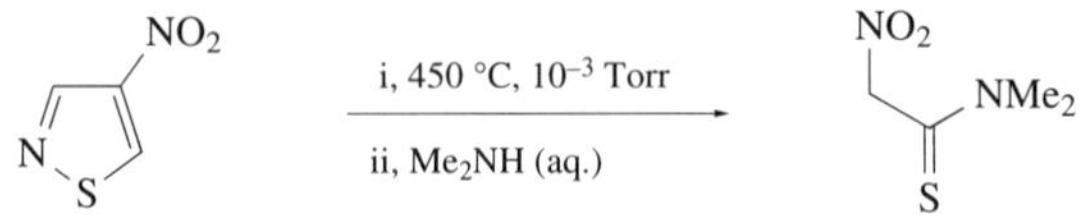

(90)

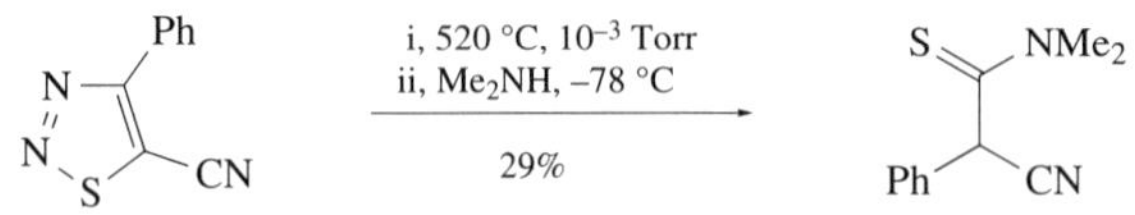

(91)

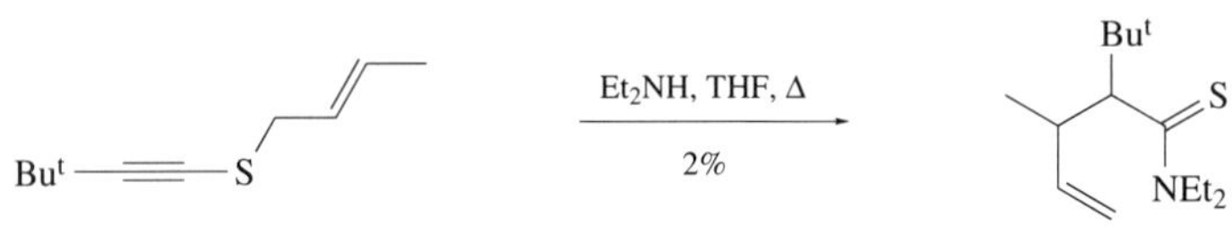

(92)

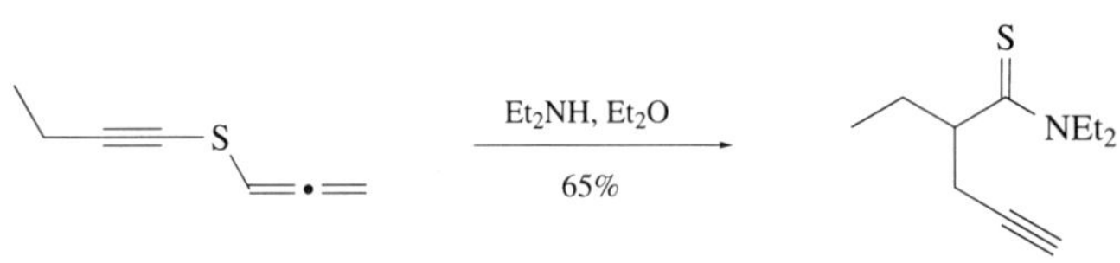

(93)

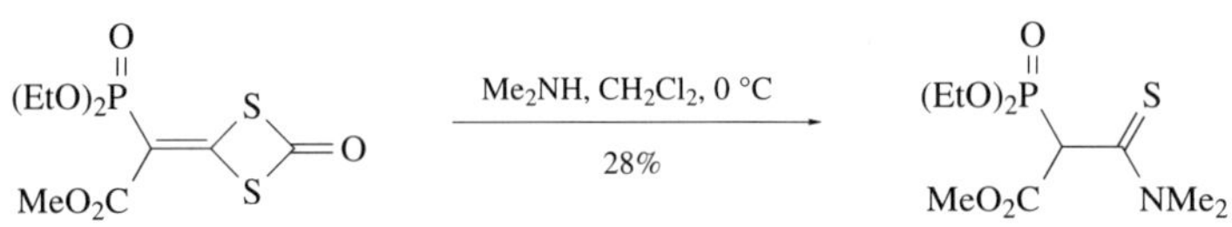

(94)

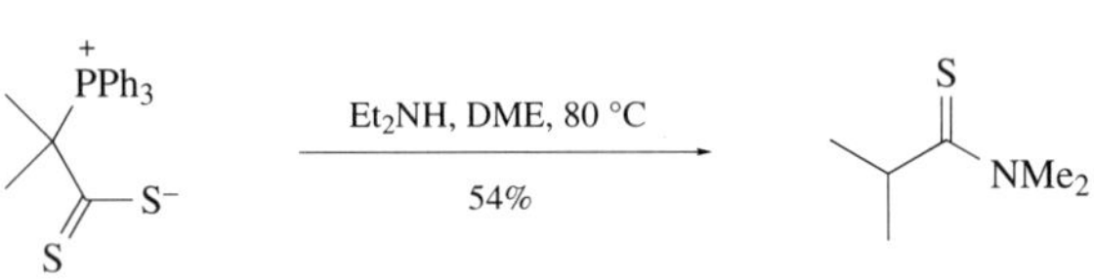

(95)

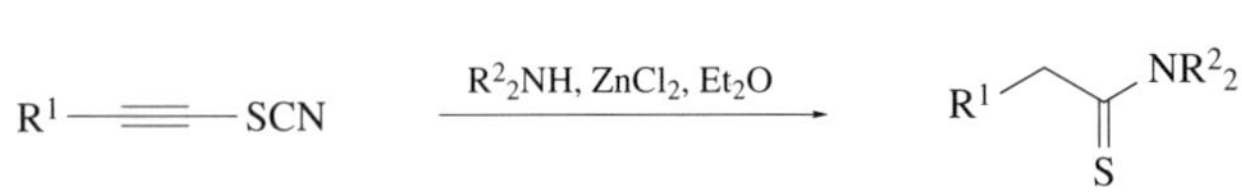

(96)

Acid chlorides have been reacted sequentially with Lawesson's reagent and amines to provide thionoamides (Equation (97)) ⟨87PS(32)51⟩, generally in good yields. Primary amines can be converted into thionoamides by an elaborate multi-step rearrangement sequence (Equation (98)) ⟨90JOC6091⟩ in reasonable yields. Primary and secondary amines can be converted into the thionoformamides by reaction with hydrogen sulfide, sodium ethoxide, ethanol and chloroform (Equation (99)) generally in 60–75% yields ⟨63LA(669)66, 68LA(712)58, 72T199, 76CB1309⟩. It is speculated that this reaction proceeds via the formation of NaSH and dichlorocarbene ⟨63LA(669)66⟩. Acetyl α-chlorodisulfides, available from treatment of α-chlorosulfenyl chlorides with thioacetic acid, react with morpholine to produce interesting halogenated thionoamides (Equation (100)) ⟨86T739⟩ in variable yields. Carbon disulfide reacts with triethylamine under high pressure at elevated temperature to yield *N*,*N*-diethylthionoformamide (Equation (101)) ⟨75T1209⟩. Yields of thionoamides using *N*,*N*-diethylaniline and *N*-ethylmorpholine instead of triethylamine were very low (3% and 6% respectively). A mechanism for the transformation was suggested.

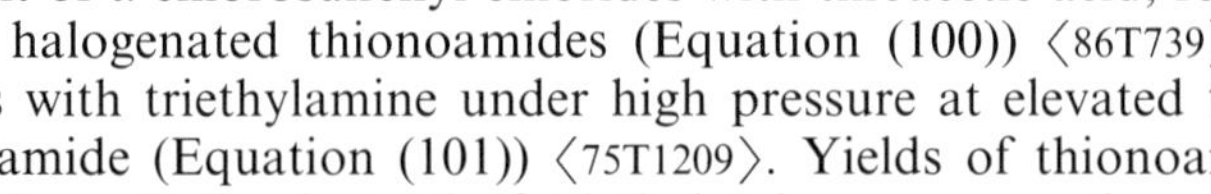

(97)

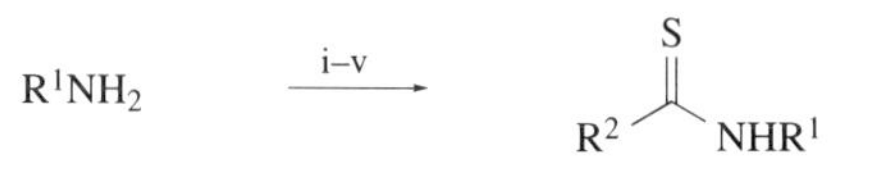

i, $(MeO)_2P(=S)Cl$, Et_2O; ii, NaH, THF, RT; iii, Δ; iv, R^2COCl; v, CCl_4, Δ (98)

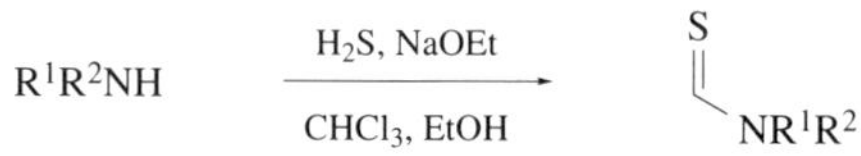

(99)

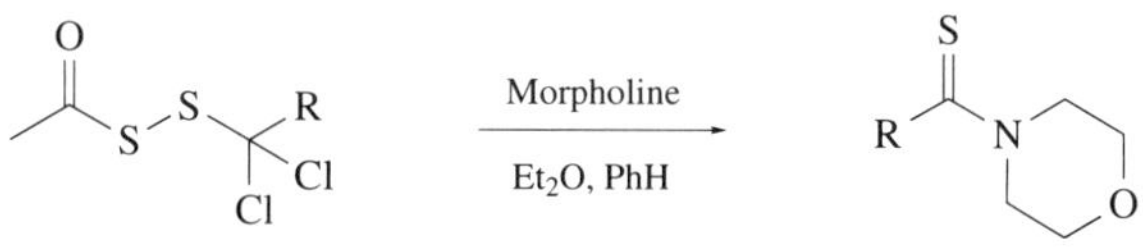

R = $CHCl_2$, CCl_3, $CClF_2$, CO_2Et etc. (100)

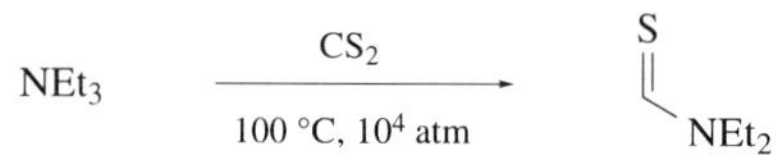

(101)

Finally in this section, thionoamides themselves can act as thioacylating reagents. Benzylamine is reportedly able to transaminate a primary thionoamide ⟨55JA4062⟩ although the yield is modest (Equation (102)). Similarly treatment of monothiooxamide with methylamine hydrochloride yields the monotransaminated product (Equation (103)) where reaction has occurred preferentially at the thionoamide ⟨62LA(660)74⟩. Rather more successful were reactions of dithionooxamide with two equivalents of primary amines ⟨61JOC3980⟩ producing the transaminated products in respectable yields (Equation (104)). Perhaps the most impressive results ⟨86S482⟩ have been obtained by treating *N,N*-dimethylthionoformamide with a range of secondary amines in toluene at reflux (Equation (105)). Yields of the tertiary thionoamides produced were often very good (Table 17).

Table 17 Thionoamides as thioacylating agents (Equation (105)).

R^1R^2NH	*Yield (%)*	*R^1R^2NH*	*Yield (%)*
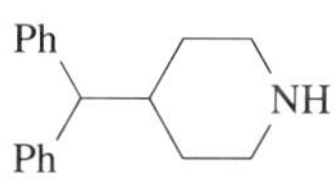	86	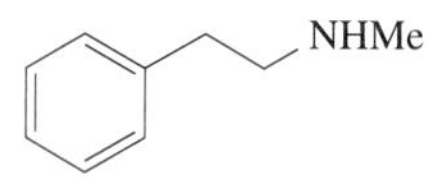	82
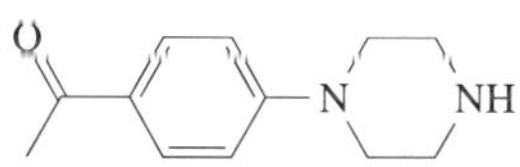	65	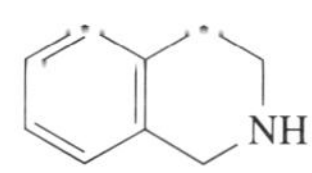	58
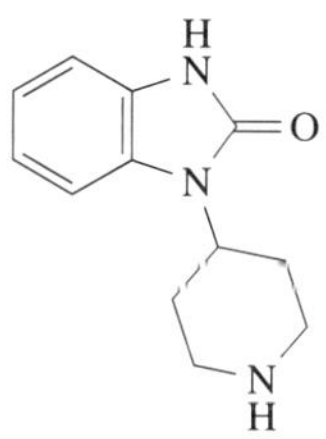	61	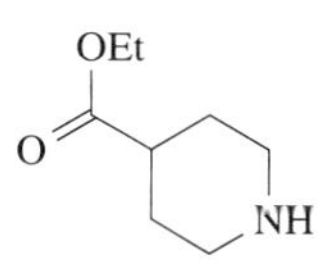	87
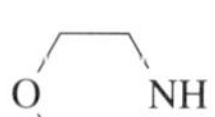	69	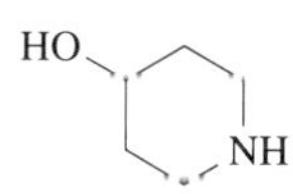	79

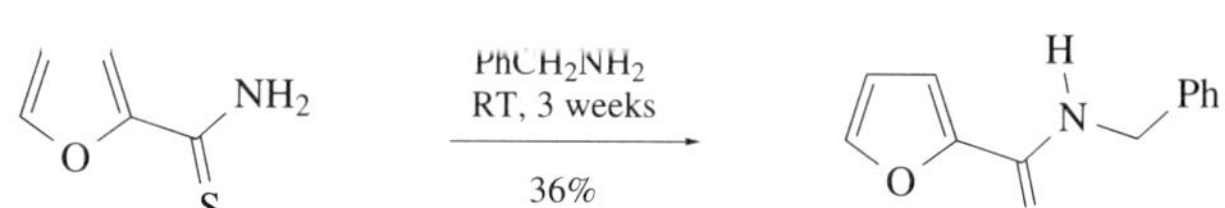

(102)

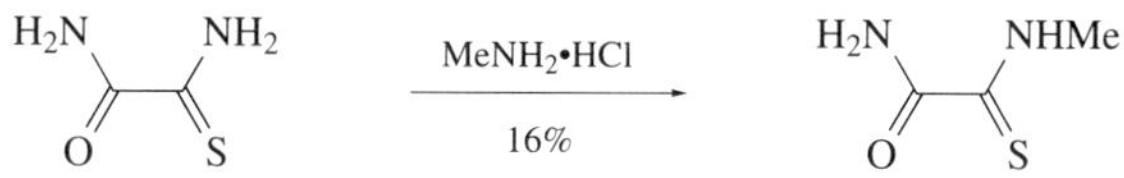

(103)

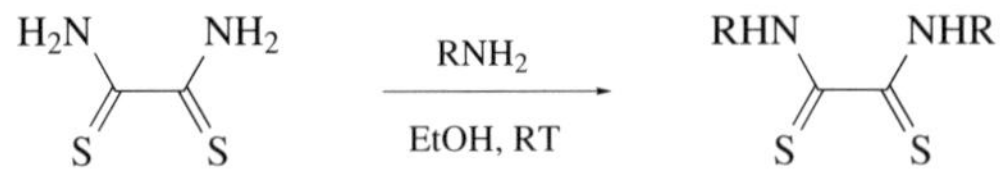

(104)

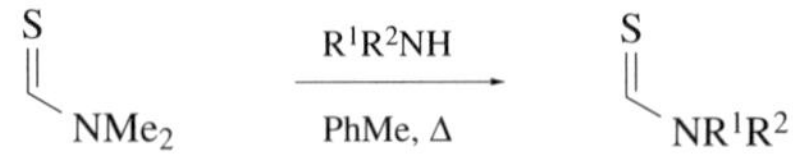

(105)

5.14.1.4 Thiocarbamoylation

The addition of nucleophilic species to isothiocyanates or their synthetic equivalents has proven to be a rich source of thionoamides. The simplest nucleophile, hydride ion, has been reported to add for example to aromatic ⟨63JOC1600, 71LA(753)187⟩ and unsaturated ⟨77TL2145⟩ isothiocyanates (Equation (106)). Aqueous cyanide, with or without an organic cosolvent, has also been used successfully, providing an approach to cyanothionoformamides (Equation (107)) ⟨24CB981, 66LA(698)131, 72ZOR528, 77TL2139⟩. The intermediate cyanothioformamide potassium salts have been alkylated *in situ* to yield a variety of tetrasubstituted secondary enamines ⟨77S607⟩. Addition of cyanide to *C*-sulfonylthionoformamides also provides a route to cyanothioformamides ⟨72S314⟩. Reactions of electron-rich aromatic species with isothiocyanates and a Lewis acid proceed to yield aromatic thionoamides ⟨1894CB1733, 29HCA554, 65LA(681)64⟩. Details of the synthesis of aromatic and heteroaromatic thionoamides (Equation (108)) by means of a variety of thiocarbamoylating agents (all to be covered in this section), including isothiocyanates, are included in Table 18.

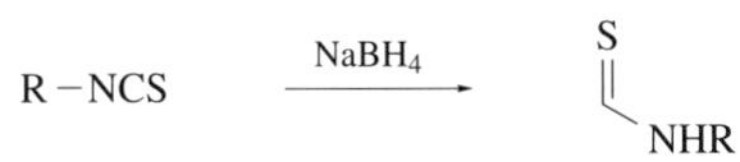

(106)

(107)

(108)

Table 18 Thiocarbamoylation of aromatics.

X	*R*[1]	*R*[2]	*Conditions*	*R*[3]	*R*[4]	*Yield* (%)	*Ref.*
CH=CH	SEt	H	PhNCS, $AlCl_3$, CS_2	H	Ph	a	1894CB1733
CH=CH	OH	H	H_2NCSCl	H	H	25	33HCA999
CH=CH	OH	OH	EtNCS, $AlCl_3$, Et_2O	H	Et	59	29HCA554
CH=CH	Me	H	EtO_2CNCS, $AlCl_3$	H	H	56–58	76JOC962
CH=CH	Me	H	KCNS, HF	H	H	36	76JOC148
CH=CH	Me	H	Me_2NCSCl, $AlCl_3$	Me	Me	53	68CB3517
CH=CH	Me	Me	H_2NCSCl, $AlCl_3$, CS_2	H	H	61	75CB3066
CH=CH	OMe	H	i, EtO_2CNCS, $AlCl_3$ ii, aq. NaOH	H	H	79	76JOC962
CH=CH	OMe	H	KCNS, HF	H	H	61	76JOC148
CH=CH	OMe	H	Me_2NCSCl, $AlCl_3$	Me	Me	77	68CB3517
CH=CH	Cl	H	EtO_2CNCS, $AlCl_3$	H	H	61–64	76JOC962
NH	H	H	i, EtO_2CNCS ii, aq. NaOH, Δ	H	H	84	73JOC667
S	H	H	i, EtO_2NCS, $SnCl_4$ ii, NH_4OH, Δ	H	H	24	74JOC2540
O	H	H	Me_2NCSCl, $SnCl_4$, CH_2Cl_2	Me	Me	10–20	68CB3517

[a] Yield not given in the reference.

Ethoxycarbonylisothiocyanate has been reacted with pyrroles ⟨73JOC667⟩, thiophenes ⟨74JOC2540⟩ and various (mainly *para*-) substituted benzenes ⟨76JOC962⟩. The initially formed *N*-thio(hetero) aroyl carbamate on treatment with base liberates the primary (hetero)aromatic thionoamide. For benzenoid substrates this saponification step is not required if the aromatic entity is the reaction solvent ⟨76JOC962⟩. Recently, reactions of substituted aryltributylstannanes with ethoxycarbonylisothiocyanate and aluminum chloride have been reported ⟨93JOC7022⟩. As these reactions were carried out in dichloromethane, the products were the *N*-thioaroylcarbamates rather than the thionoamides. None the less, the proposed regiospecific *ipso* substitution of the stannane provides an expansion of this methodology into *ortho*-, *meta*- and *para*-substituted aromatic thionoamides. Reactions with isothiocyanates can also proceed in an intramolecular fashion (Equation (109)) ⟨76JCS(P1)33, 77JCS(P1)2357⟩. In these cases, electron-donating substituents on the aromatic ring are not a prerequisite for reasonable yields. Cyclisation to form six- and seven-membered rings fused to the aromatic nucleus proceeded, but the authors had no success in preparing systems containing five- and eight-membered rings fused to the aromatic. Intramolecular tin-mediated radical cyclisations of alkenyl isothiocyanates have been reported (Equation (110)) ⟨90JOC3442⟩. Yields of five- and six-membered thionolactams were generally very good. Reaction of activated aromatics with thiocarbamoyl chlorides has also been described ⟨33HCA999, 68CB3517, 75CB3066⟩. One weakness with the use of isothiocyanates is that they can only provide secondary thionoamides, but this is not so with thiocarbamoyl halides ⟨68CB3517⟩. A variant on this is the reaction of aromatics with potassium thiocyanate in hydrofluoric acid, postulated to proceed via formation of the thiocarbamoyl fluoride ⟨76JOC148⟩. This only proceeds with activated aromatics (toluene and anisole); benzene failed to react.

(109)

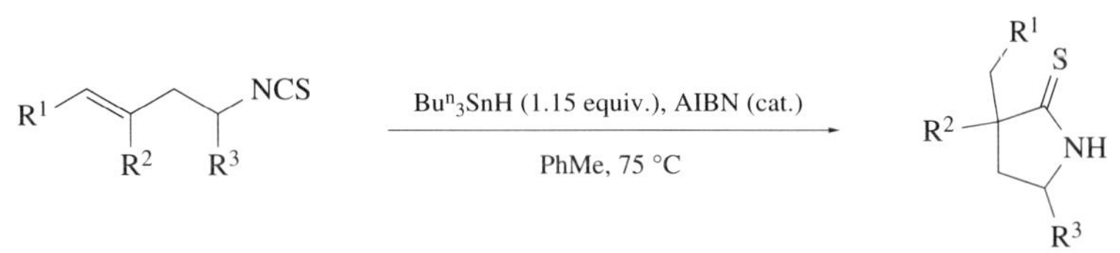

(110)

The reaction of anionic and enolic species derived from active methylene compounds with isothiocyanates is a much-used route to various functionalised thionoamides (Equation (111)). A host of representative examples are detailed in Table 19. Thus treatment of diethyl malonate with base (ethanolic sodium ethoxide or sodium in ether) followed by addition of aryl or alkyl isothiocyanates yields after workup the desired thionoamides ⟨1887JPR449, 08JCS621, 28JA1456, 33JA3672⟩. Yields are greatly reduced if diethyl methyl malonate is used ⟨33JA3672⟩. Similar reactions of ethyl acetoacetate ⟨18JA415, 24JA2834⟩ and acetylacetone ⟨20JA1055⟩ have been described. In some reactions with ethyl acetoacetate, the initial product is hydrolysed and decarboxylated under the reaction or workup conditions, yielding the β-oxothionoamide (this has also been noted in reactions of diethyl malonate with dithiocarbamates ⟨74ZOR1992⟩. Ethyl cyanoacetate has also been used in reactions with isothiocyanates, but the product thionoamide was further reacted with chloroacetonitrile to yield aminothiazolines ⟨76JPR343⟩. Dimedone reacts with phenyl isothiocyanate when triethylamine is used as a catalyst ⟨64CB2209⟩. The sodium (or potassium) enolate of glutacondialdehyde forms pyridine thiones when treated with alkyl or aryl isothiocyanates ⟨76TL3347, 77T341⟩. Simple ketone enolates (derived from acetophenone, cycloalkanones, 1-tetralone and camphor by treatment with sodamide or sodium *t*-pentylate) have also been reacted with alkyl and aryl isothiocyanates to produce thionoamides ⟨67CR(C)99, 75BSF829⟩. The generation of quaternary centres (Equations (112)–(114)) or densely functionalised molecules (Equation (115)) is not a problem using isothiocyanates ⟨73S297, 79LA1715, 93JHC887, 93JMC1604⟩. Active methylene compounds have also been successfully reacted with base and dithiocarbamates as an alternative to isothiocyanates (Equation (116)) by a group of Russian workers ⟨74ZOR1992, 76ZOR69, 76ZOR681⟩, and also with *C*-sulfonylthionoformamides ⟨72S314, 74S433⟩ as a route to tertiary thionoamides (Equation (117)). These *C*-sulfonylthionoformamides are reportedly superior to thiocarbamoyl halides as thiocarbamoylating agents.

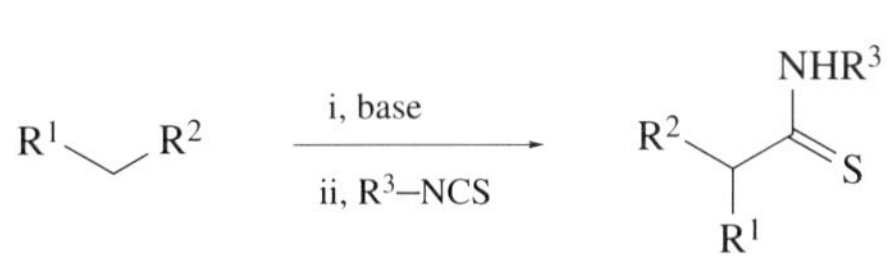

(111)

Table 19 Reactions of active methylenes with isothiocyanates (Equation (111)).

R^1	R^2	R^3	*Base*	*Solvent*	*Yield* (%)	*Ref.*
CO_2Et	CO_2Et	Ph	NaOEt	EtOH	a	1887JPR449, 08JCS621
CO_2Et	CO_2Et	Ph	Na	Et_2O	quant.	33JA3672
CO_2Et	CO_2Et	Me	Na	Et_2O	80	28JA1456
Ac	CO_2Et	Ph	Na	Et_2O	60	18JA415
Ac	Ac	Ph	Na	Et_2O	a	20JA1055
CN	$CONH_2$	Ph	$KOBu^t$	Bu^tOH,THF	64	64CB3106
$CONH_2$	$CONH_2$	Ph	Li/NH_3	DMF	37	64CB3106
CO_2Et	$CONH_2$	Ph	$KOBu^t$	Bu^tOH,THF	88	64CB3106
CO_2Et	$PO(OEt)_2$	Ph	NaH	Et_2O	70	74JPR534
CN	$PO(OEt)_2$	Ph	NaH	Et_2O	93	74JPR534
CHO	PhCO	$PhCH_2$	a	MeCN	88	67LA(710)118
CHO	Ac	Me	a	MeCN	60–65	67LA(710)118
$PhSO_2$	PhCO	Ph	$NaOBu^t$	MeCN	30	73ZOR347
$PhSO_2$	PhCO	$4Br—C_6H_4$	$NaOBu^t$	MeCN	40	73ZOR347
$PhSO_2$	$CONH_2$	Ph	$NaOBu^t$	MeCN	71	71ZOR2352
$PhSO_2$	$CONH_2$	$4EtO—C_6H_4$	$NaOBu^t$	MeCN	67	71ZOR2352
$PhSO_2$	$PhSO_2$	Ph	NaOEt	EtOH/dioxan	82	67JPR302
$PhSO_2$	$PhSO_2$	$4Cl—C_6H_4$	NaOEt	EtOH/dioxan	89	67JPR302
$PhSO_2$	CN	Ph	NaOEt	EtOH/dioxan	96	67JPR302
$PhSO_2$	CN	$4NO_2—C_6H_4$	NaOEt	EtOH/dioxan	96	67JPR302

[a] Not given in the references.

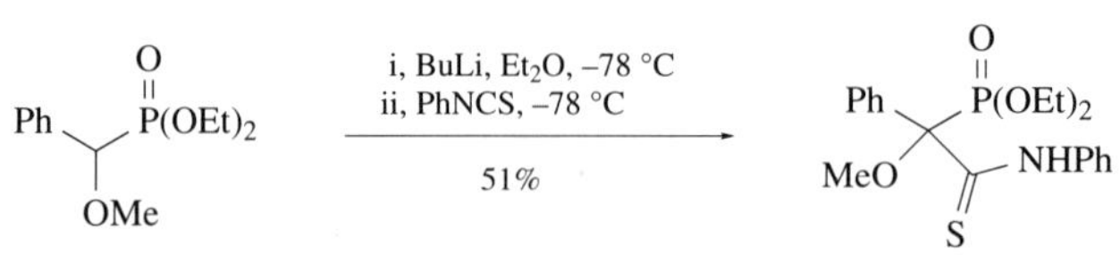

(112)

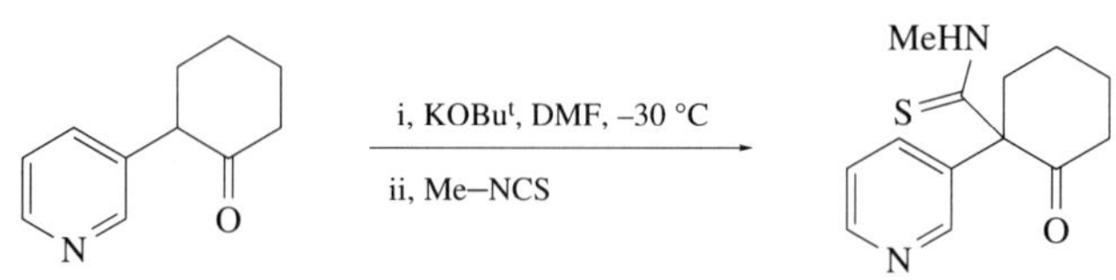

(113)

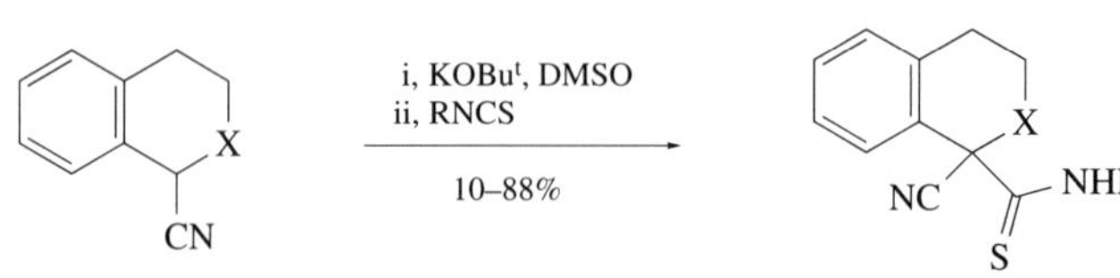

(114)

R = alkyl, $PhCH_2$, Ph; X = O, S

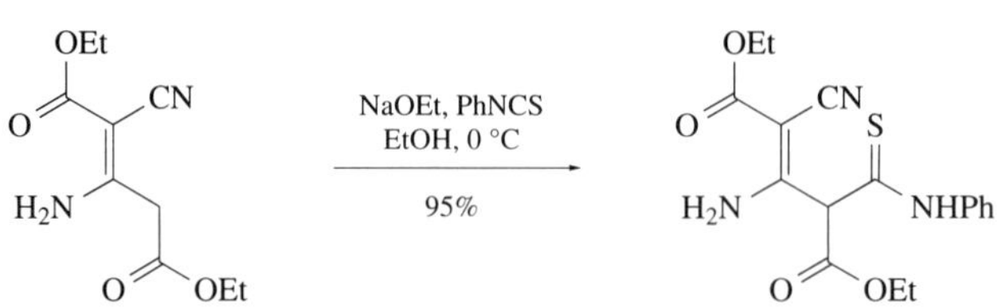

(115)

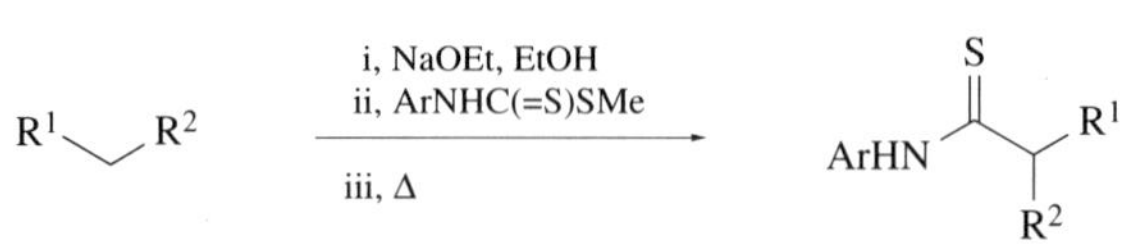

(116)

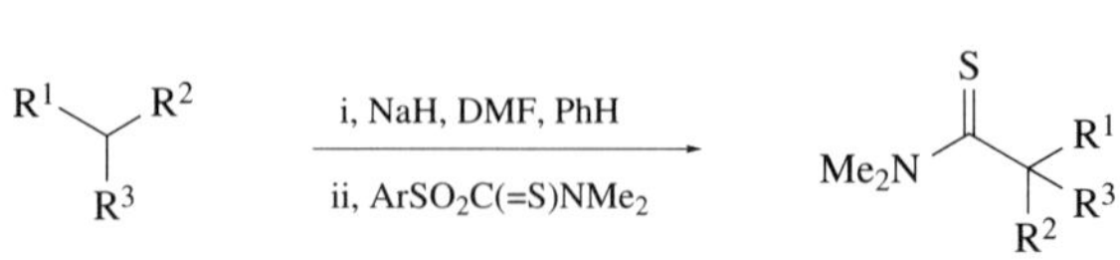

(117)

Grignard reagents have been reported to add to isothiocyanates to form thionoamides ⟨03CB585, 04CB874, 25JA2974⟩. No yields were given in these publications, which all involved addition of alkyl, benzyl and aryl Grignards to aryl isothiocyanates. Mechanistic studies to verify the site of addition of the organometallic to the heterocumulene have been conducted ⟨24JA493⟩. A variety of typical synthetic examples (Equation (118)) are given in Table 20.

(118)

Table 20 Reactions of Grignard reagents with isothiocyanates (Equation (118)).

R[1]	*R*[2]	*Solvent*	*Yield* (%)	*Ref.*
Pr^i	2F—C_6H_4	THF,Et_2O	79	91SC625
Allyl	Ph	THF,Et_2O	87[a]	81TL3409
MeC=CH_2	Me	THF	62	81TL3409
Ph	MeCH=CH	Et_2O	73	60JOC1132
Ph	Cyclohexyl	Et_2O	88	49JOC962
b	$PhCH_2$	Et_2O	40	62JOC4412

[a] Yield is of the α-β-unsaturated thionoamide produced by a base-catalysed conjugation with dbu in THF.
[b] R^1MgBr =

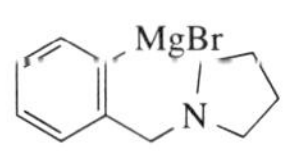

Alkyl and benzyl Grignard reagents also react with triphenylphosphine thiocyanate (Equation (119)), providing primary thionoamides in respectable yields ⟨79S887⟩. The reaction is thought to proceed via a phosphinimine intermediate, which actually proved to be isolable in one case. Alkyl, aryl and alkynyl sodium species reportedly react with isothiocyanates ⟨41JA314, 33JA1262, 37JA1486, 39JA104, 39JA105, 77CB285⟩. The yields are good only in the case of the sodium derivative of phenylacetylene. Triethylaluminum has been successfully reacted with alkyl and aryl isothiocyanates, providing secondary thionopropanoamides ⟨64CB2661, 66CB11⟩. Yields are lower if 'ethylaluminum sesquichloride' ($Et_{1.35}AlCl_{1.45}$) or excess triethylaluminum is used. Organolithiums also react well with isothiocyanates (Equation (120)) ⟨77CZ508⟩. Interestingly, triphenylsilyl isothiocyanate reacts with phenyllithium in ether to provide thiobenzamide in 57% yield ⟨50JA3045⟩, whereas phenyl magnesium bromide gives no thiobenzamide. Recent additions to the repertoire of thiocarbamoylating agents are the thiuram monosulfides (Equation (121)). These react well with alkyl, aryl and heteroaryl organolithium reagents to provide tertiary thionoamides in good yields ⟨93S483⟩. Apparently, Grignard reagents require higher temperature (room temperature), longer reaction times and give poorer yields than the organolithium species.

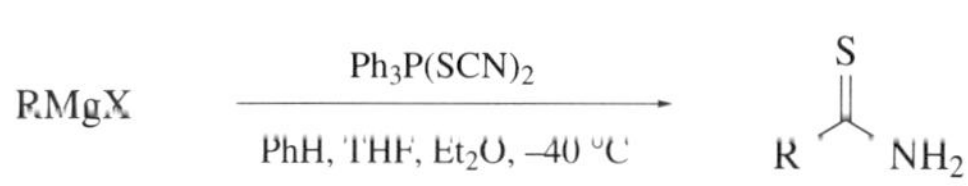

(119)

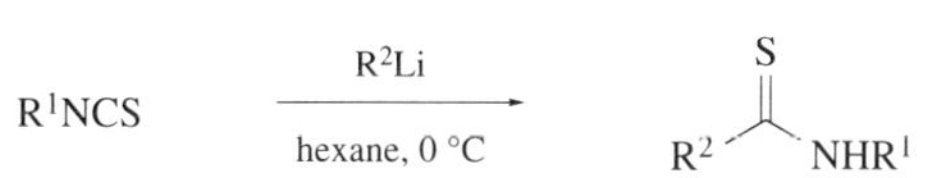

(120)

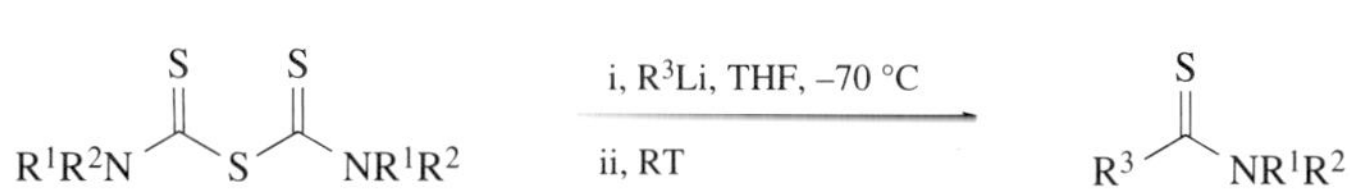

(121)

Deprotonation of alkyl phosphorus ⟨87PS(31)33⟩, tolyl isonitriles ⟨79TL1039⟩, alkylpyridines ⟨93PS(78)97⟩ and tetrahydroquinolines ⟨76JCS(P1)983⟩ with lithium bases followed by quenching with isothiocyanates provides access to a variety of structurally interesting thionoamides (Equations (122)–(125)). Direct metalation of thiazole and benzothiazole followed by isothiocyanate treatment leads to heteroaromatic thionoamides (Equation (126)) ⟨86KGS1405⟩. Cyclic secondary thionoamides can be converted into the dianion and quenched with isothiocyanates (Equation (127)). Lithiated tertiary ynamines derived from trichloroenamines allow access to alkynyl thionoamides bearing a terminal tertiary amine (Equation (128)) ⟨85LA2371⟩. The trialkyl- or triarylstannyl tertiary ynamines also allow the synthesis of such compounds ⟨78TL1951⟩.

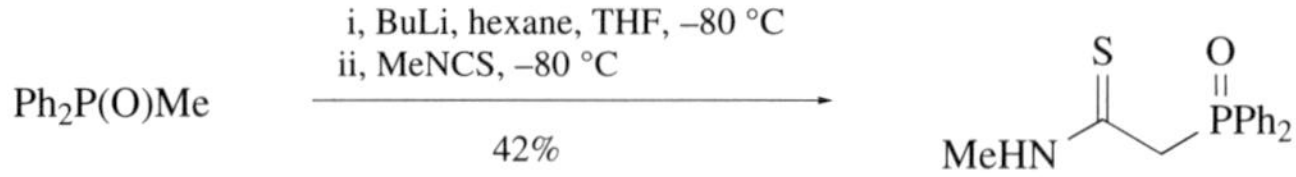

(122)

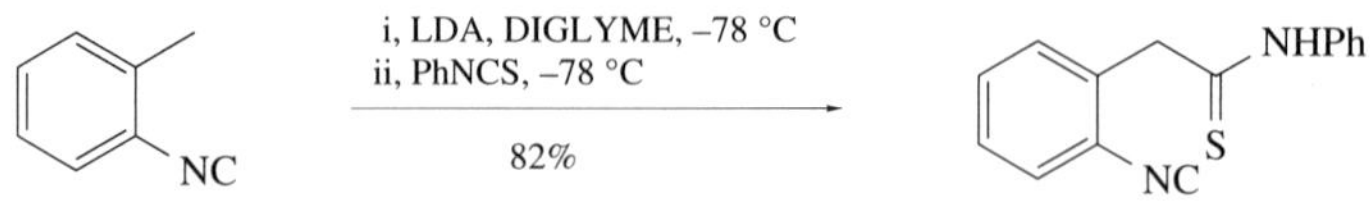

(123)

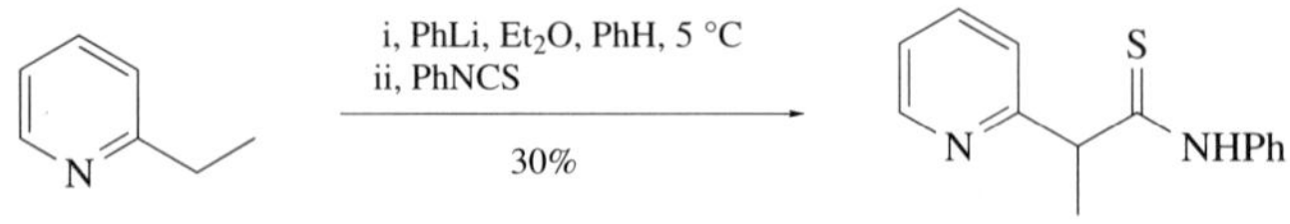

(124)

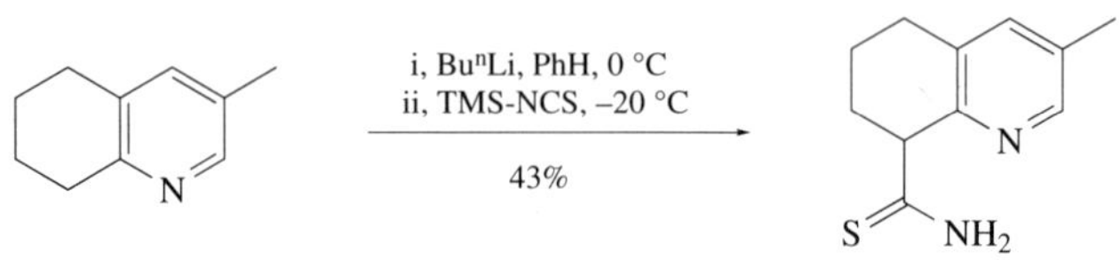

(125)

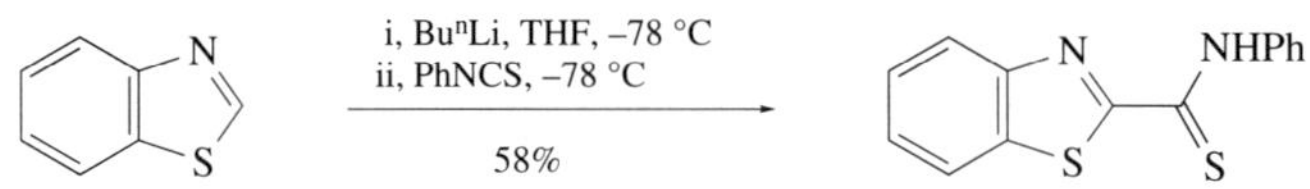

(126)

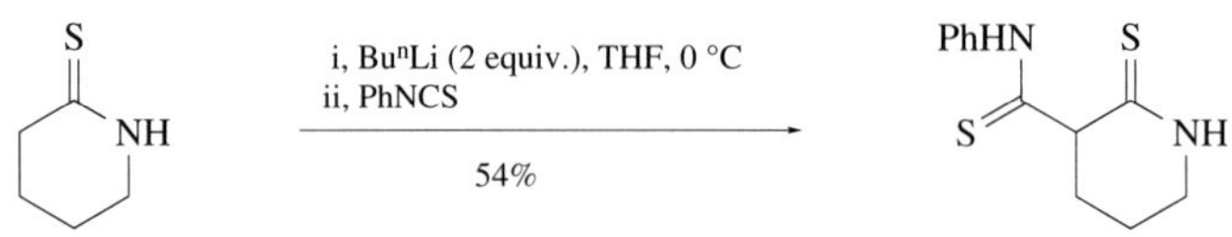

(127)

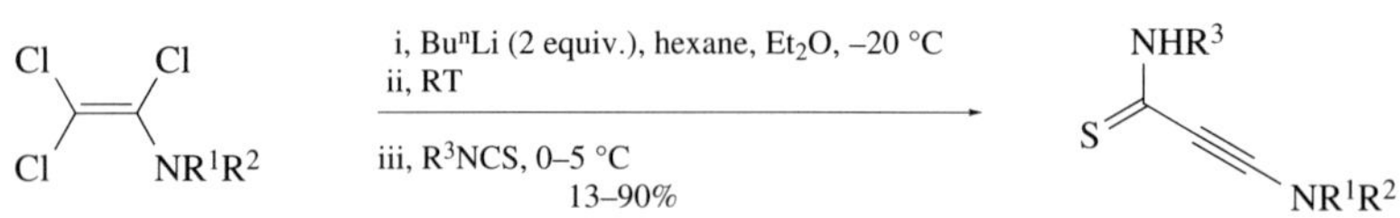

(128)

Reaction of benzyne with phenyl isothiocyanate furnishes a mixture of eight products, two of which are thionoamides but the yields appear to be extremely low ⟨77AP(310)614⟩. Addition of the anions of dimethyl sulfoxide and dimethyl sulfone to alicyclic isothiocyanates has been reported ⟨71JOC1742⟩. Reaction of 1-(carbazol-9-ylmethyl)benzotriazole anion (**36**) with isothiocyanates and removal of the heteroaromatics under acidic conditions followed by derivatisation of the generated aldehyde yields the hydrazone (**37**) (Equation (129)), indicating the use of (**36**) as a formyl anion equivalent ⟨91JOC2143⟩. Acyl anions, generated from carbon monoxide and butyllithium (normal, secondary or tertiary) at very low temperatures, react well with alkyl isothiocyanates to provide α-oxothionoamides in good yields (Equation (130)) ⟨84TL5251⟩. The acyl anions reportedly failed to react with phenyl isothiocyanate. Related work involved the lithium derivatives of tertiary amides (derived from lithium diethylamide and carbon monoxide) being added to isothiocyanates (Equation (131)). This was only successful with phenyl isothiocyanate and the yield was only moderate ⟨93SC2139⟩.

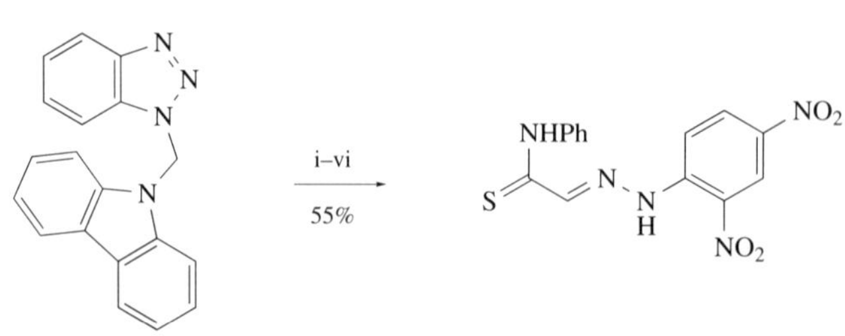

(**36**) (**37**)

(129)

i, Bu^nLi, THF, –78 °C; ii, PhNCS, THF, –78 °C; iii, RT; iv, aqueous work-up; v, conc. H_2SO_4 (cat.), H_2O, THF, RT; vi, 2,4-DNPH, $HClO_4$ (aq.)

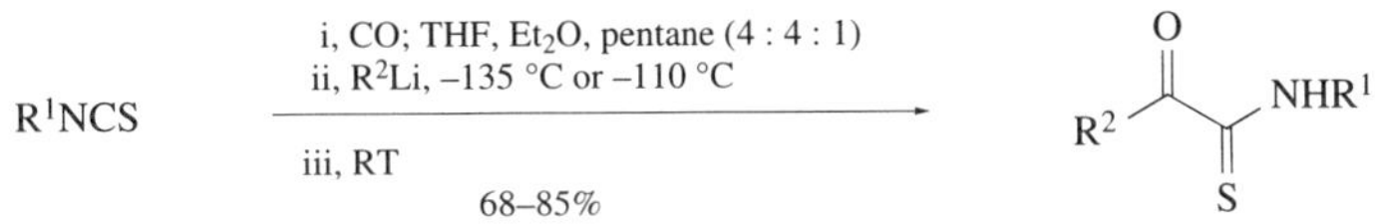

(130)

R^1 = alkyl; R^2 = Bu^n, Bu^s, Bu^t

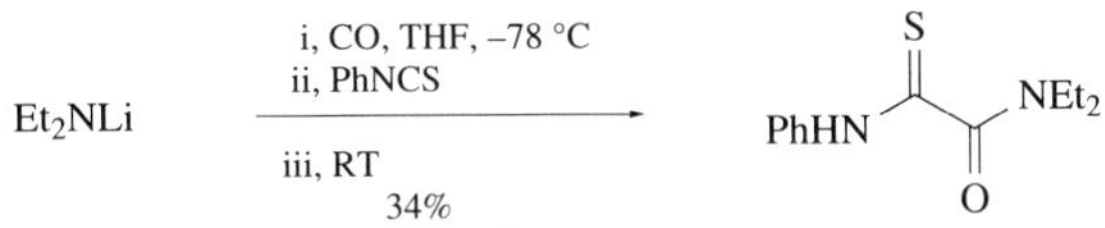

(131)

In the reverse sense, nucleophilic thiocarbamoylation has been used in which deprotonated tertiary thionoformamides add to various ketones and aldehydes to produce the α-hydroxy tertiary thionoamides in generally good yields ⟨73AG(E)1014, 76CB1309, 77CB3733⟩ (Equation (132)). The lithiated thionoformamides can also be alkylated with methyl and ethyl iodide, and acylated with methyl benzoate. Treatment of a mixture of *N*,*N*-dimethylthioformamide and an acylating agent with LDA in THF at low temperature results in the formation of α-oxothionoamides (Equation (133)) ⟨76SC387⟩. Some examples of more neutral reactions are known. Thus alkynes can be reacted with thiocarbamoyl halides under copper and palladium catalysis to produce alkynyl thionoamides (Equation (134)) ⟨91LA903⟩. Stabilised phosphorus ylides are also reported to react with isothiocyanates (Equation (135)) ⟨69ZOB1282, 86CB3812⟩.

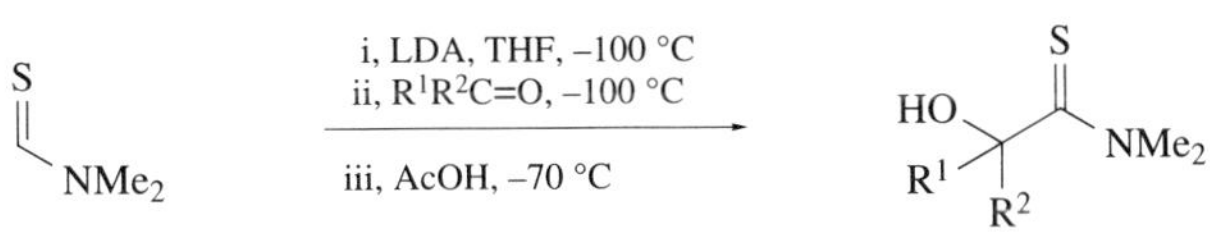

(132)

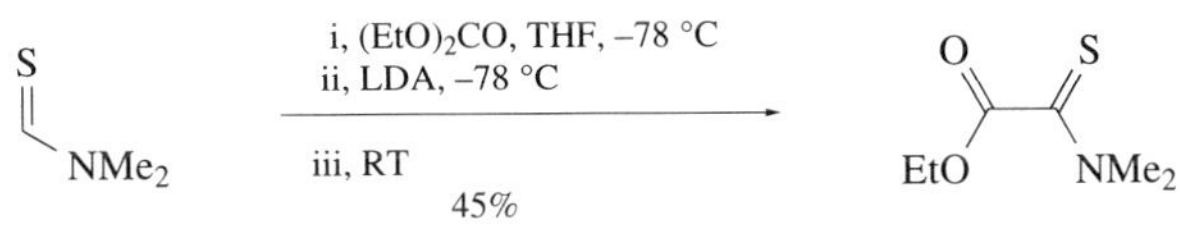

(133)

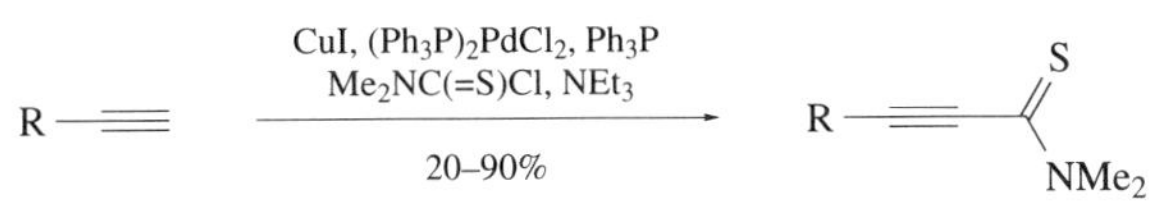

(134)

R = Ph, PhS, Ph_2N, TMS, alkyl

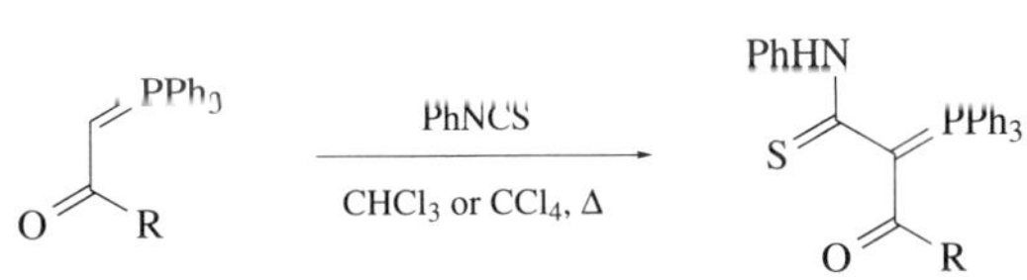

(135)

Simple enamines and those bearing electron-withdrawing substituents on the double bond have been reacted with isothiocyanates (Equation (136)) to yield the expected addition products ⟨00LA(314)200, 03LA(329)341, 62CB926, 63CB526, 63CB1551, 64CB3106, 64CB2209, 64LA(673)132, 65CB1531, 73BSF581, 77CB285, 93M721, 86BCJ3871⟩. Representative yields are shown in Table 21. Treatment of enamine (**38**) with isothiocyanate (**39**) reportedly yields the primary thionoamide (**40**) (Equation (137)) via elimination of 4-methyl-pent-3-en-2-one ⟨79JIC545⟩. Secondary enamines derived from dimedone on treatment with sodium hydride followed by an isothiocyanate yield mixtures of products (Equation (138)). A common spiro intermediate is postulated to account for this ⟨90JCS(P1)1869⟩ Hydrolysis of the adducts of simple enamines with isothiocyanates allows the preparation of *β*-ketothionoamides ⟨62CB926, 64LA(673)132, 79JIC545⟩. Ketone imines have also been reacted with isothiocyanates (Equations (139) and (140)) ⟨68JOC2357, 77CB285⟩. A number of studies have involved reactions of enamines derived from camphor ⟨83FES101, 85FES120, 91FES1033⟩. Ketene *S*,*N*-acetals have also received attention ⟨83S225, 84CPB1658, 84SC1257, 87CPB3139, 88H(27)1953⟩. Some examples of these are included in Table 21.

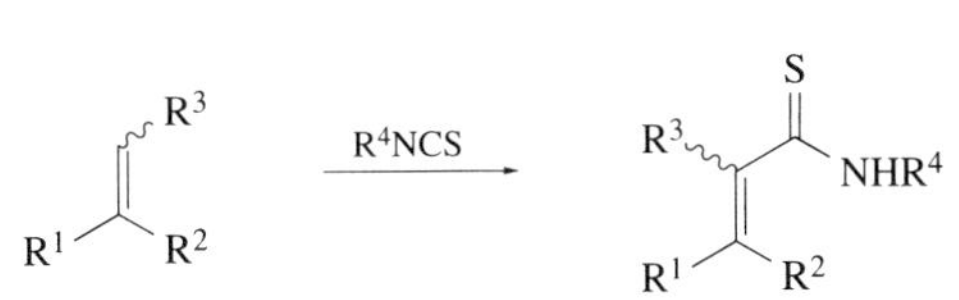

(136)

Table 21 Reactions of enamines with isothiocyanates (Equation (136)).

R[1]	*R*[2]	*R*[3]	*R*[4]	*Yield* (%)	*Ref.*
NH_2	Me	CO_2Et	Me	65	03LA(329)341
NH_2	Me	CO_2Et	Ph	80	00LA(314)200, 03LA(329)341
NH_2	Me	Ac	Me	54	63CB526
NH_2	Me	Ac	Ph	61	63CB526
NH_2	Me	NO_2	Ph	25	86BCJ3871
NH_2	OEt	CO_2Et	$4NO_2—C_6H_4$	85	64CB3106
NH_2	NH_2	CO_2Et	$4NO_2—C_6H_4$	78	64CB3106
NH_2	NH_2	CO_2Et	$PhCH_2$	40	87LA1101
NH_2	Ph	$CONH_2$	$4NO_2—C_6H_4$	63	63CB1551
NHMe	Me	CO_2Et	Me	47	77CB285
NHMe	Me	CO_2Et	$4NO_2—C_6H_4$	93	65CB1531
NHMe	H	Ac	Ph	71	65CB1531
NHPh	Me	Ac	Ph	64	65CB1531
NHPh	Me	CO_2Et	Ph	81	65CB1531
NHPh	Ph	H	Ph	4	73BSF581
NMe_2	SMe	H	Ph	100	84SC1257, 84CPB1658
SMe	$—N(Me)CH_2CH_2—$		Ph	90	83S225, 87CPB3139
Mor	$—[CH_2]_4—$		Ph	76–83	62CB926, 64LA(673)132
Pip	Pip	H	Et	28	64JOC2932
Pip	Pip	Pip	Ph	93	71LA(748)59

Abbreviations: Mor = morpholine; Pip = piperidine, each attached to the double bond via the nitrogen.

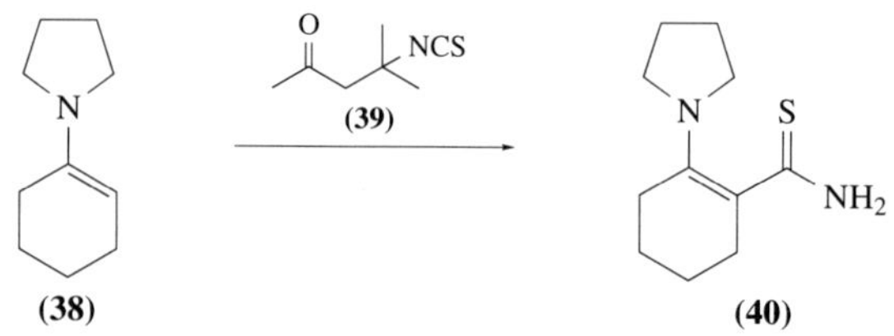

(38) (40)

(137)

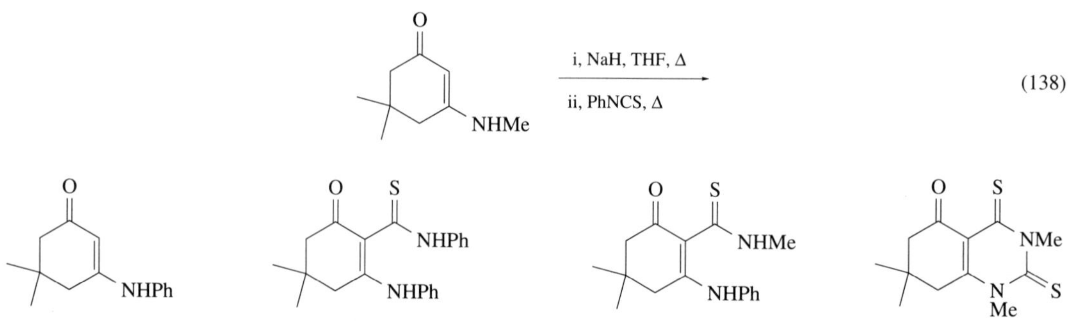

(138)

33% 14% 12% 6%

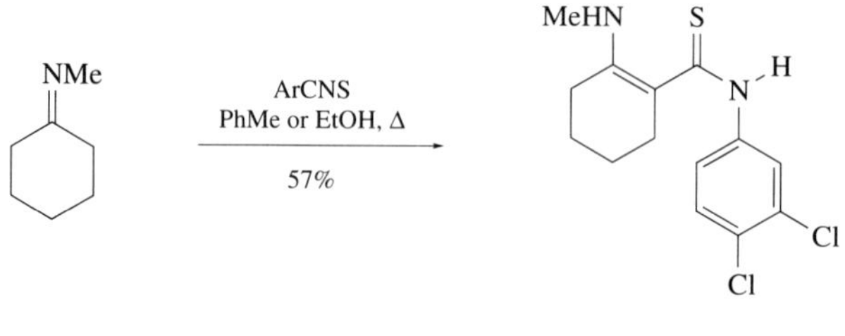

(139)

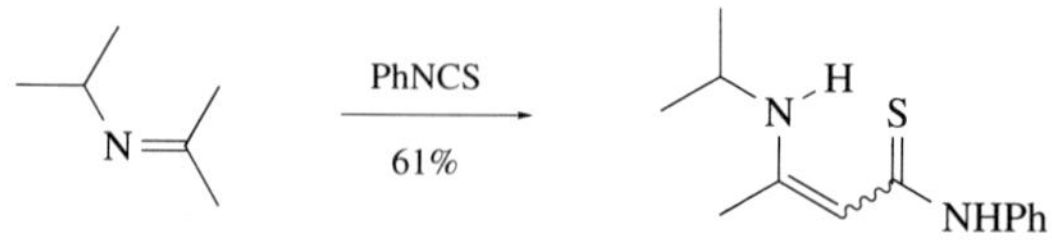

(140)

5.14.1.5 Aminosulfuration/Sulfuration

A powerful reaction for converting a wide variety of aromatic, benzylic, alkenic and alkynic compounds to thionoamides is the aminosulfuration of such substrates. The reaction of compounds, originally aryl alkyl ketones (and more recently many other substrates) with sulfur and an amine is commonly called the Willgerodt–Kindler reaction ⟨46OR(3)83, 75S358⟩. The original Willgerodt reaction involved the conversion of an aryl alkyl ketone to an amide by heating with ammonium sulfide. The Kindler modification involved the use of sulfur and an amine, usually morpholine, and the isolated product was the thionoamide. The most interesting feature of the reaction is the observation that the carbonyl group in the product has moved to the end of the alkyl chain. Yield often diminishes as the chain length increases ⟨56JA4135⟩. A number of authors, notably Asinger and Carmack have addressed the mechanism of the reaction ⟨64AG(E)19, 67AG(E)907, 89JHC1319⟩. A number of papers have been published on the use of multivariate analysis to investigate various parameters of the reaction (e.g., solvent, amine used, ratio of reactants) with respect to the reaction yield and to use this information to predict yields for substrates which had not previously been used ⟨86ACS(B)534, 86ACS(B)694, 87ACS(B)157, 87ACS(B)164⟩. The reactions of a good range of substituted acetophenones with morpholine and sulfur, and of styrene with sulfur and a variety of primary and secondary amines, have been described ⟨48JA2425⟩. It is interesting that acetophenone ⟨47JA1211⟩, the morpholine enamine of acetophenone ⟨64AG(E)705, 67AG(E)294⟩, styrene ⟨46JA2029⟩ and phenylacetylene ⟨46JA2029⟩ can all produce the same thionoamide (Scheme 4). Aminals react with sulfur to provide thionoamides ⟨48JA4143⟩, allowing the authors to postulate such diamines as intermediates in the reaction (Equation (141)). Reaction of acetophenones with dimorpholine disulfide and a catalytic amount of sodium methoxide yields α-oxothionoamides (Equation (142)) ⟨72LA(757)147⟩.

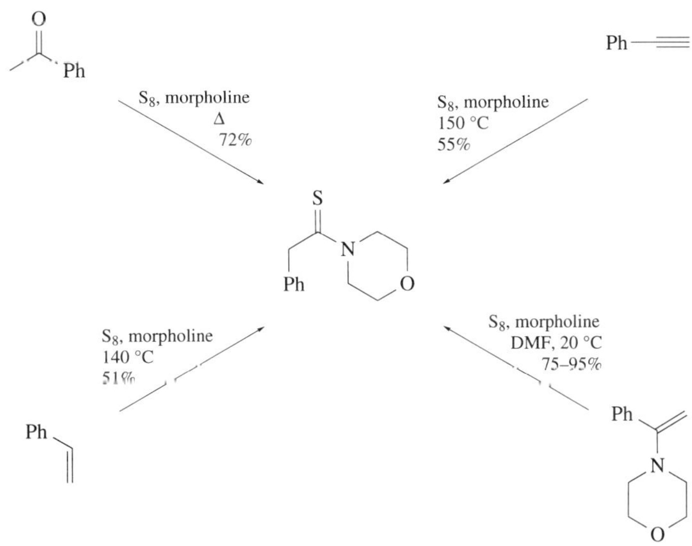

Scheme 4

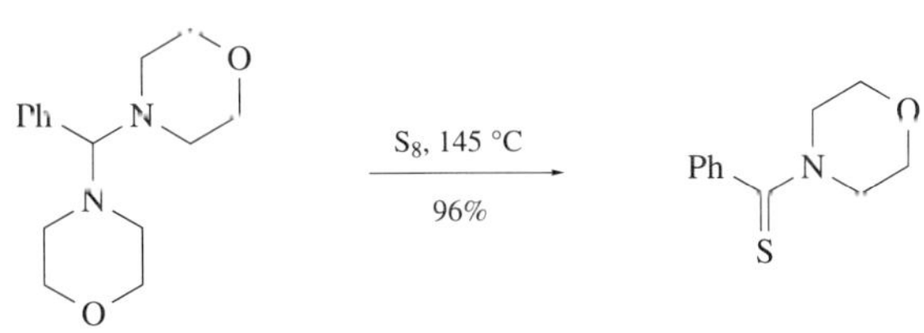

(141)

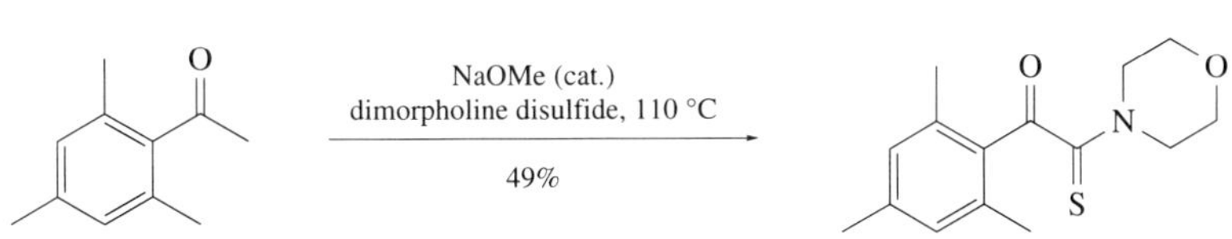

(142)

For the synthesis of primary thionoamides by aminosulfuration, an effective alternative to the use of ammonia under pressure (autoclave or sealed tube) has recently been introduced. This involves reacting an aromatic aldehyde or ketone with sulfur and ammonium acetate in pyridine at reflux ⟨91OPP435⟩. Yields of the thus produced thiobenzamides and phenylthioacetamides are not high and the reaction fails for 4-nitrobenzaldehyde and propiophenone. Apparently if DMF is used as the solvent, *N*,*N*-dimethylthionoamides are produced. An excellent procedure for the synthesis of *N*,*N*-dimethylthiobenzamides and -phenylthioacetamides is the use of sulfur, dimethylamine hydrochloride and sodium acetate in DMF ⟨83S730⟩. Yields are generally very good and the procedure avoids the use of the volatile dimethylamine which would require an autoclave. Representative examples of these two methods (Equation (143)) are shown in Table 22. Another good approach to *N*,*N*-dimethylthionoamides (Equation (144)) is the reaction of a variety of substrates with sulfur in dimethylammonium *N*,*N′*-dimethylcarbamate at reflux (b.p. 61 °C) ⟨89S202⟩. Yields are variable (Table 23). Reaction of various benzylic substrates with sulfur in HMPA at elevated temperature (Equation (145)) allows the preparation of *N*,*N*-dimethylthiobenzamide, often in high yields from a number of different starting materials ⟨75ACS(B)538⟩. Similarly, reaction of esters with sulfur in HMPA (Equation (146)) allows the preparation of *N*,*N*-dimethylthionoamides ⟨75ACS(B)604⟩. Yields can be very good.

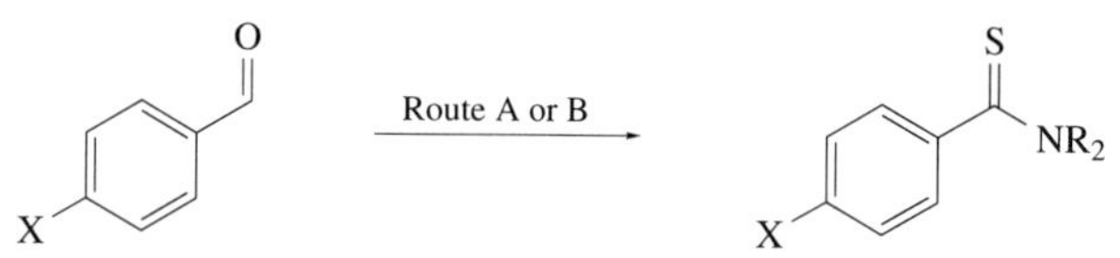

(143)

Table 22 Aminosulfuration of benzaldehydes (Equation (143)).

X	*R*	*Route*	*Yield* (%)	*X*	*R*	*Route*	*Yield* (%)
H	H	A	37	H	Me	B	82
Me	H	A	27				
Cl	H	A	33	Cl	Me	B	92
Br	H	A	51	Br	Me	B	86
CF_3	H	A	54	NO_2	Me	B	83
MeO	H	A	32	MeO	Me	B	91
Me_2N	H	A	25	Me_2N	Me	B	54

Route A: S_8, NH_4OAc, pyridine, Δ.
Route B: S_8, $Me_2NH.HCl$, NaOAc, DMF, 100 °C.

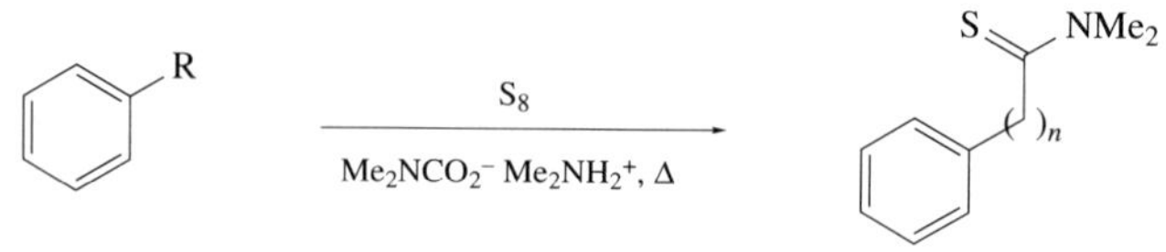

(144)

Table 23 Reactions using dimethylammonium dimethylcarbamate (Equation (144)).

R	*n*	*Yield* (%)	*R*	*n*	*Yield* (%)
CHO	0	80	CH_2Cl	0	40
$CHCl_2$	0	71	CH_2CN	0	59
Ac	1	81	$CH{=}CH_2$	1	51
C≡CH	1	88			

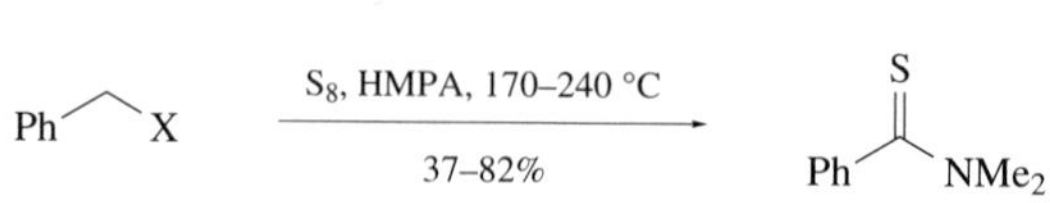

(145)

X = Cl, OH, SH, OEt, NMe_2, CO_2Et

(146)

The aminosulfuration of chloromethyl compounds has been reviewed ⟨69CZ474, 88ZC233⟩. Benzylic chlorides ⟨89IJC(B)673⟩ and bromides ⟨78JPR313⟩ are known to participate (Equation (147)). Steroidal α-chloroketones have been converted to α-oxothionoamides with sulfur and morpholine ⟨79CPB1352⟩.

(147)

Reaction of α-haloketones with sodium thiosulfate produces the so-called Bunte salts (Equation (148)) which also react with amines producing thionoamides ⟨61JCS1194, 62LA(660)74, 66LA(698)131, 76PS75⟩. Yields are variable. The ability to select from a wide range of potential starting materials is a great feature of this family of transformations. Examples from Viehe's group have focused on aminosulfuration of 'capto-dative' methylene species not bearing an aromatic ring (Equation (149)) ⟨85S79, 85T693, 90TL6339⟩. Yields for some examples are shown in Table 24. Other examples from Viehe's group ⟨85S77, 85T693, 90TL6339⟩ are shown in Equations (150)–(155). Allyl alcohol reportedly yields a dithionoamide on treatment with sulfur and morpholine in DMF (Equation (156)) ⟨70ZOR610⟩.

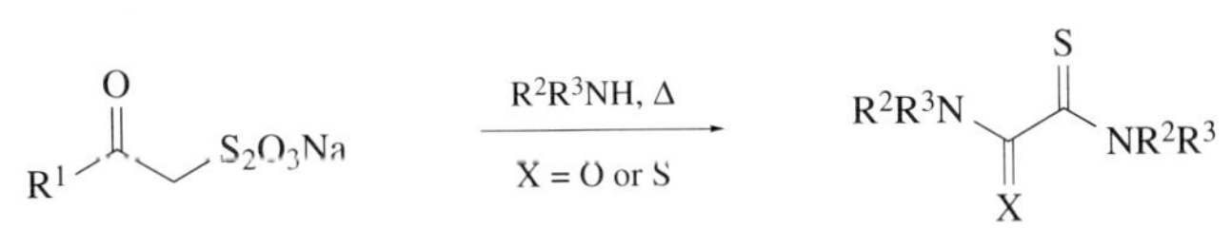

(148)

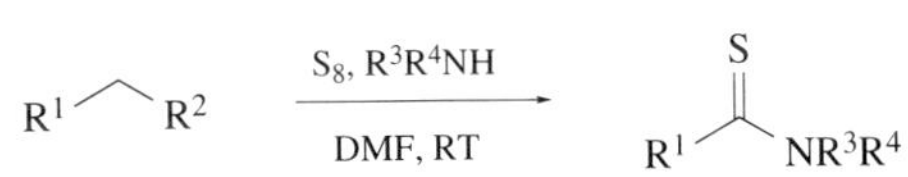

(149)

Table 24 Reactions of 'capto-dative' methylenes (Equation (149)).

R^1	R^2	R^3 R^4	*Yield* (%)	*Ref.*
Ac	SEt	$—[CH_2]_2O[CH_2]_2—$	26	85S79
Ac	Cl	$—[CH_2]_2O[CH_2]_2—$	73	85S79
CN	Cl	$—[CH_2]_2O[CH_2]_2—$	42	85S79
$COCF_3$	Br	$—[CH_2]_4—$	65	90TL6339
$COCF_3$	Br	$—[CH_2]_2S[CH_2]_2—$	25	90TL6339

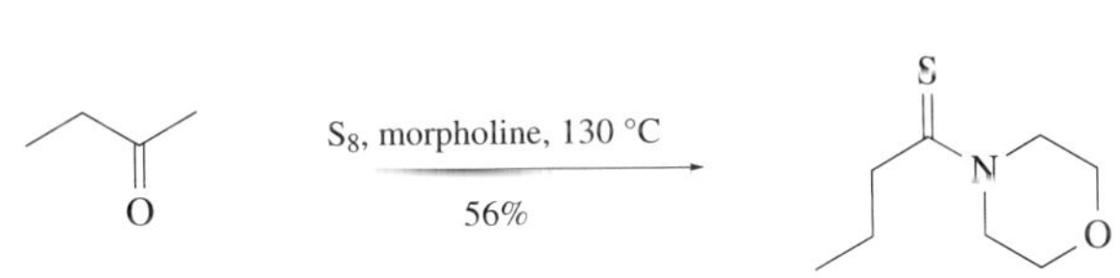

(150)

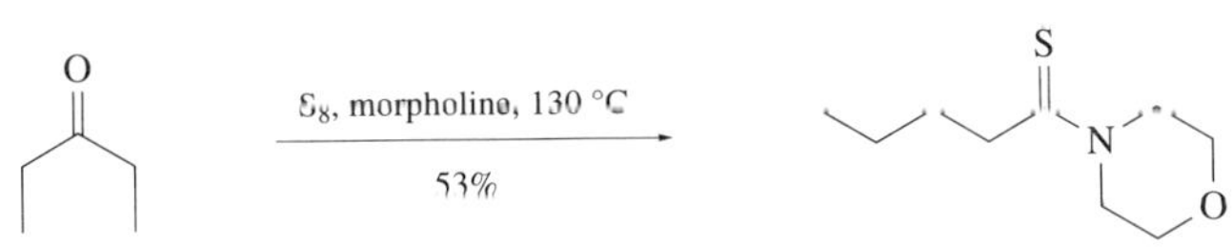

(151)

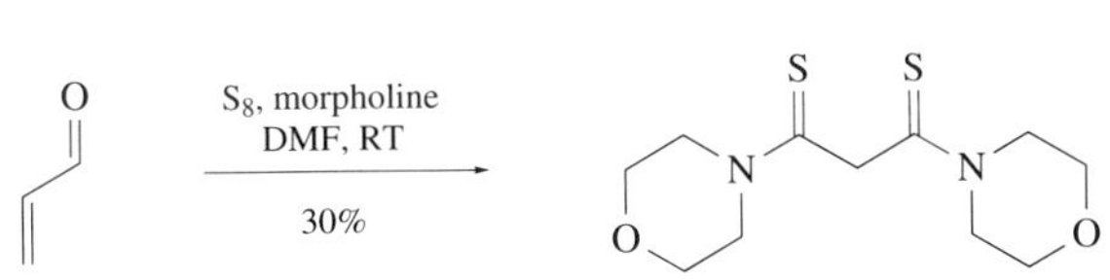

(152)

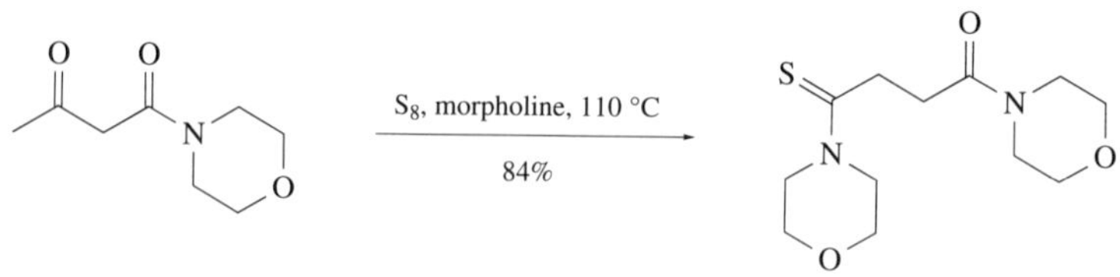

(153)

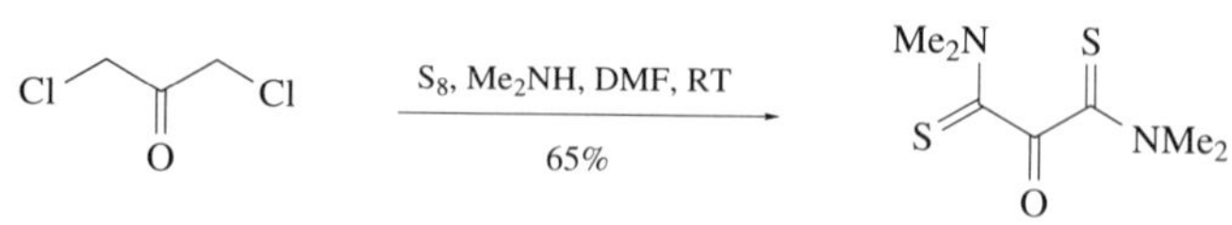

(154)

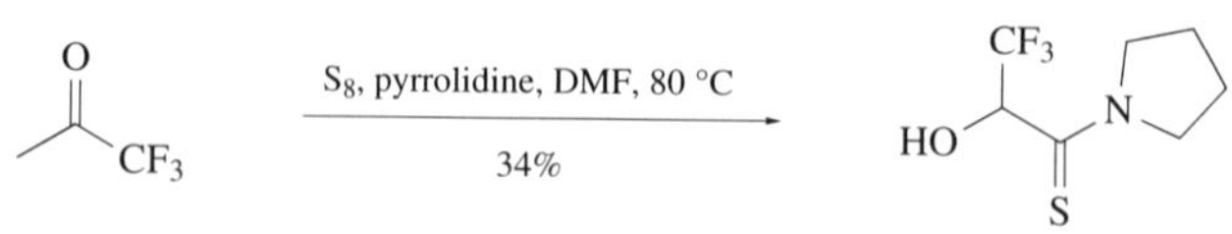

(155)

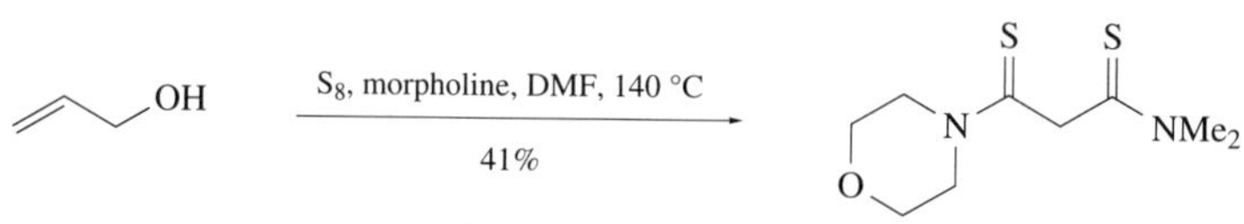

(156)

Dithionooxamides can be prepared from trichloroethylene (Equation (157)). The workers note that 1,1,2,2-tetrachloroethane gives the same product as trichloroethylene ⟨61JCS1194⟩, although the yield is slightly lower (65%). Treatment of α,α,α-trichloroacetamides with sulfur and morpholine (Equation (158)) yields monothionooxamides ⟨61JCS1194⟩. The α-chloro- and α,α-dichloro-acetamides also produce the same product thionoamide ⟨61JCS1194⟩. Reaction of acetophenones with sulfur dichloride followed by treatment with DMF under basic aqueous conditions (Equation (159)) also allows access to *N*,*N*-dimethyl α-oxothionoamides ⟨72IJS89⟩. Substitution of primary and secondary amines for the DMF in this reaction allows the preparation of other thionoamides ⟨76PS75⟩. Sequential treatment of α,α-dichloroketones with NaSH followed by morpholine, reportedly produces α-oxothionoamides (Equation (160)) ⟨65M1278⟩. The addition of sulfur to the reaction increases the yield of the product to 65%.

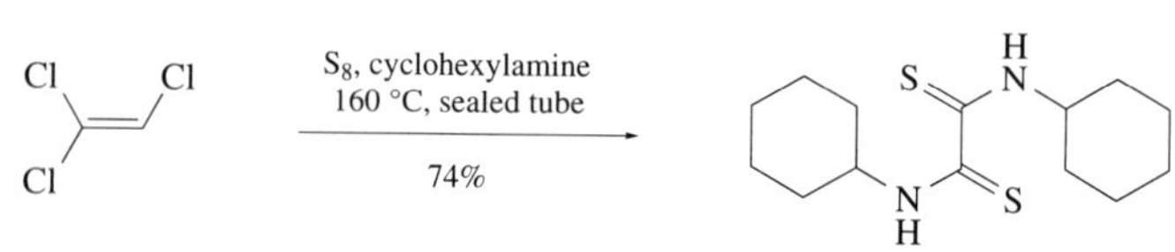

(157)

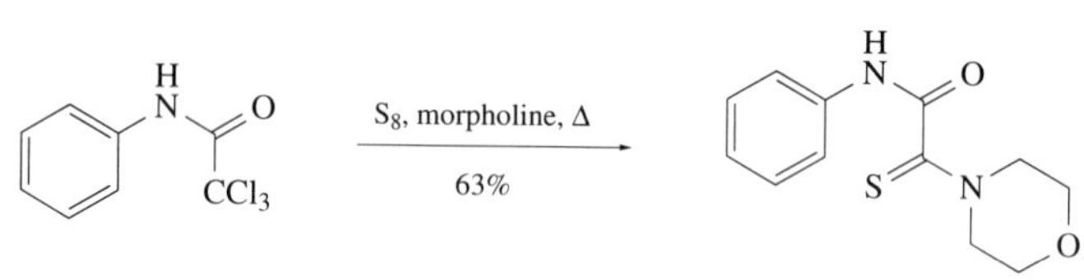

(158)

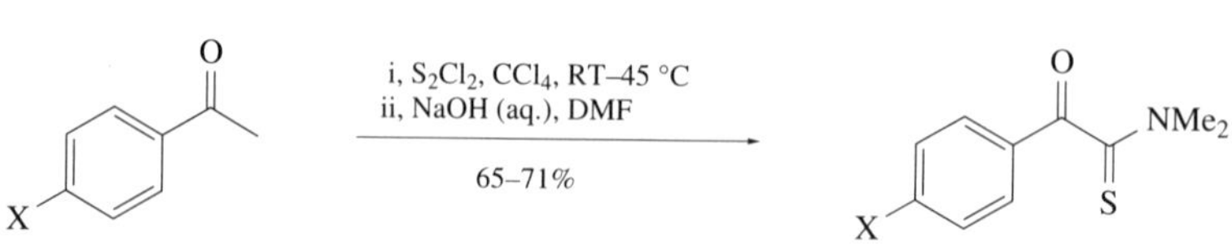

(159)

X = H, Cl, Me, OMe

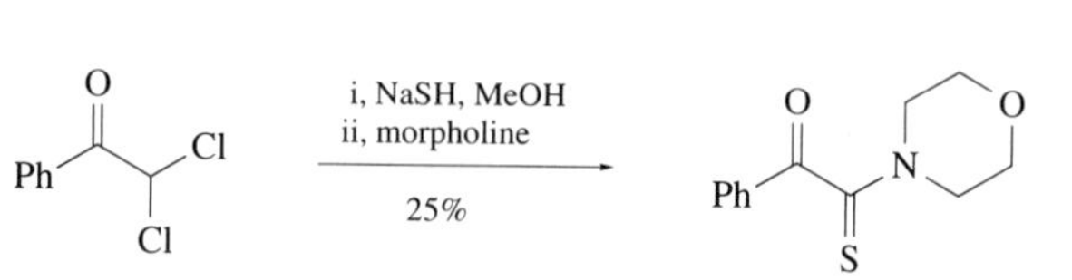

(160)

Sulfuration of 2-nitroamides ⟨89TL3189⟩, 2-nitroketones ⟨89SUL117⟩ and 2-nitroesters ⟨89SUL117⟩ with Lawesson's reagent or phosphorus pentasulfide results in the conversion of the nitromethylene group to a primary thionoamide (Equation (161)). Potential mechanisms to explain the various products are postulated. Reaction of α-iminoacids with sulfur in benzene provides aromatic thionoamides via decarboxylation and reaction of the 1,2-ylide with the sulfur (Equation (162)) ⟨88T7271⟩. In a related transformation described in the same paper, α-oxoacids react with secondary amines and sulfur in benzene providing tertiary thionoamides in good to excellent yields.

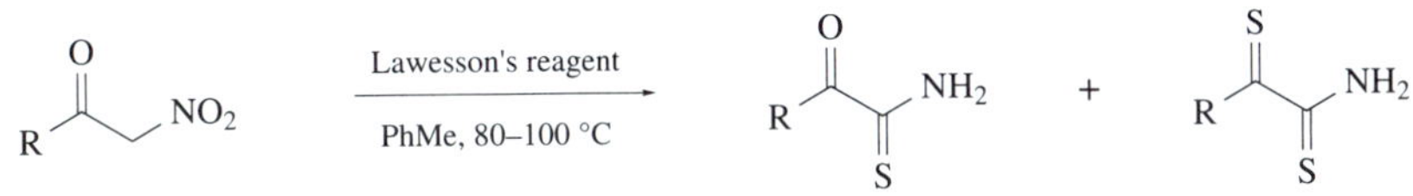
(161)

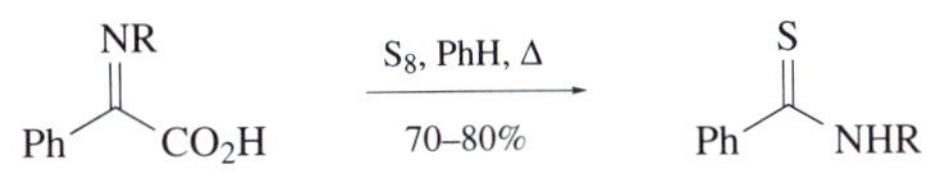
(162)

R = Bu^n, aryl, benzyl

Alkynes react with sulfur in DMF in the presence of catalytic quantities of triphenylphosphine and a cobalt catalyst (Equation (163)) providing *N*,*N*-dimethylthionoamides ⟨89IZV1319⟩. Alkynes also react well, via their sodium ⟨67JPR294⟩ or lithium ⟨68RTC38, 71ZOR1120⟩ salts, with sulfur, secondary amines and sometimes *t*-butyl bromide as a proton source (Equation (164)). Yields are generally only in the 50% region. An alternative procedure ⟨77S888⟩ reportedly produces very good yields of the *N*,*N*-diethylthionoamides (Equation (165)). These reactions have also been described in Section 5.14.1.3 as it is likely that they proceed via the thioketene generated by isomerisation of an alkynethiolate.

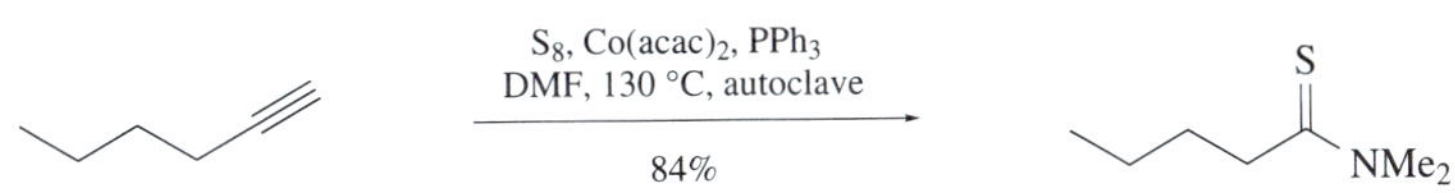
(163)

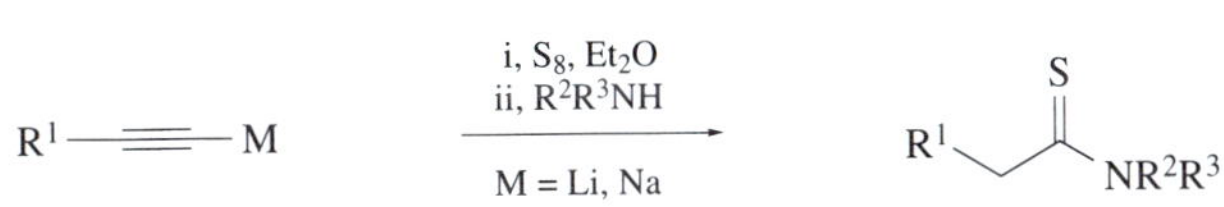
(164)

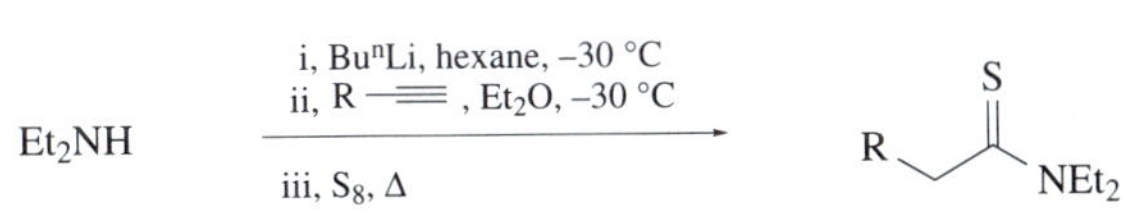
(165)

Stabilised phosphorus ylides ⟨88CL1145, 91BCJ2402⟩ and phosphonate anions ⟨92CL131⟩ have both been reported to react with sulfur and secondary amines to produce thionoamides (Equations (166) and (167)). The reactions are thought to proceed via the thioaldehyde intermediate, support for which can be claimed from the transformation of the α-thiophosphonate (**41**) into the thionoamide (**42**) (Equation (168)) ⟨92CL131⟩. Transformation of the sulfenyl chloride into the thioaldehyde and *in situ* trapping with morpholine (Equation (169)) provides the thionoamide ⟨88CL1145, 91BSJ2402⟩, lending further support to the intermediacy of the thioaldehyde in the phosphorus ylide reaction.

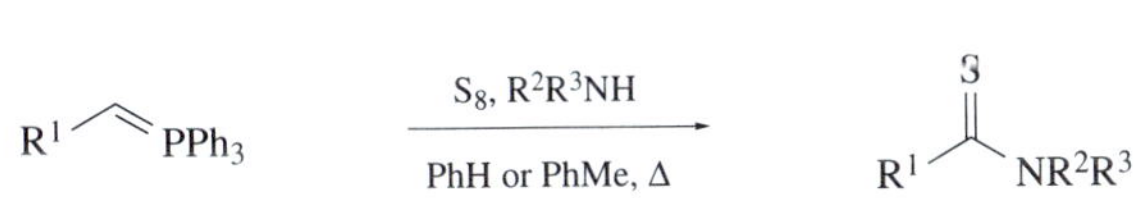
(166)

R^1 = Ph, CO_2Me

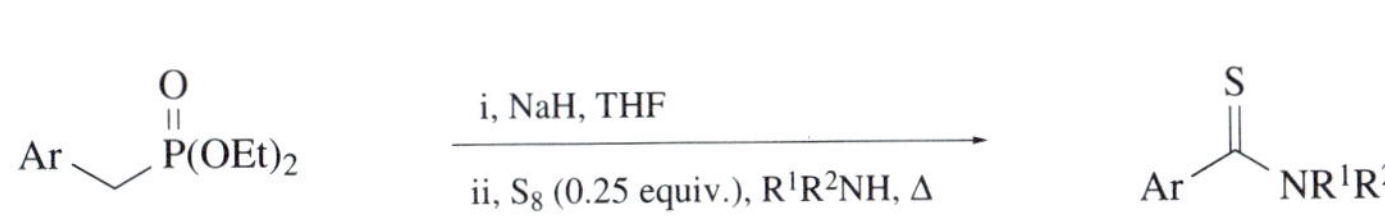
(167)

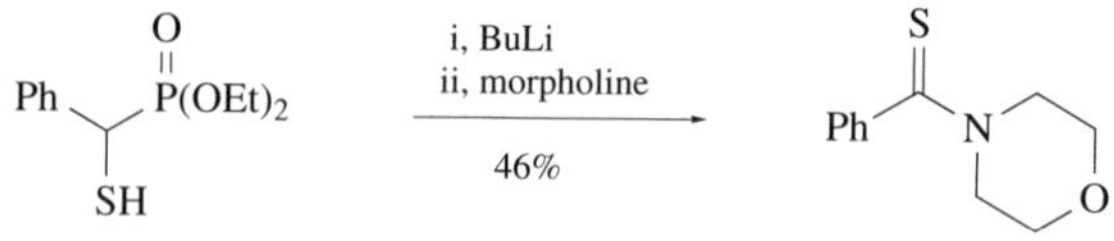

(41) (42) (168)

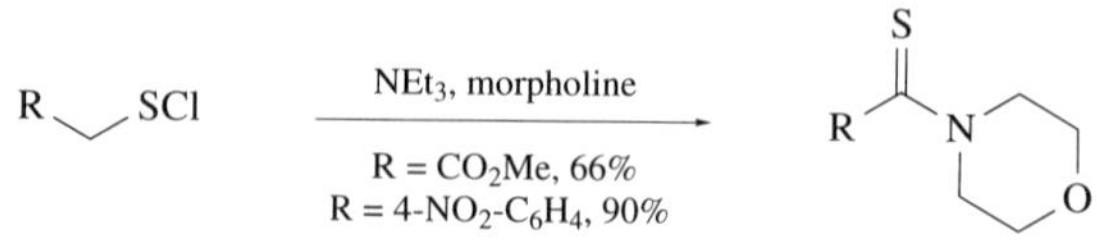

(169)

5.14.1.6 Cycloadditions

As well as being used as thioacylating agents for amines, thioketenes undergo typical [2 + 2]-cycloadditions of the thiocarbonyl bond. The thioketenes can be generated by a number of methods. With imines, the products are thiono β-lactams ⟨76CB906, 79CB1000, 79LA1702, 80CB3024⟩. Typical examples are shown in Equations (170) and (171)). Yields of product can be reasonable (up to about 70%). As well as the thiono β-lactam 1 : 1 adducts, 2 : 1 and 1 : 2 adducts can also be formed ⟨79LA1702, 80CB3010⟩. In some cases, thermolysis of the undesired adducts can lead to the desired thionolactam ⟨84CB2205⟩. Ring opening of adducts to give open chain thionoamides is also reported ⟨79LA1702, 80CB3010, 80CB3024, 84CB2205⟩. The substituents on the thioketene need not be simple alkyls. Other groupings, for example allyl ⟨80CB3010, 80CB3024⟩, trimethylsilyl ⟨80CB3024⟩, cyano ⟨83(AG(E)55, 84CB2205⟩ and benzoyl ⟨83AG(E)55⟩ have also been used. In the latter case, the reaction is a [4 + 2]-cycloaddition (Equation (172)). In one reported case, if the thioketene is substituted with an ester and a cyano group, the regiochemistry of the [2 + 2]-cycloaddition with an imine is reversed and the product is a 1,3-thiazetidine, not a thionoamide ⟨84BSB405⟩.

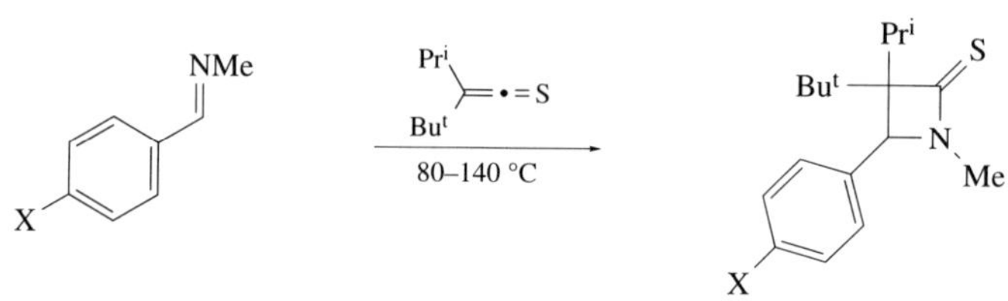

(170)

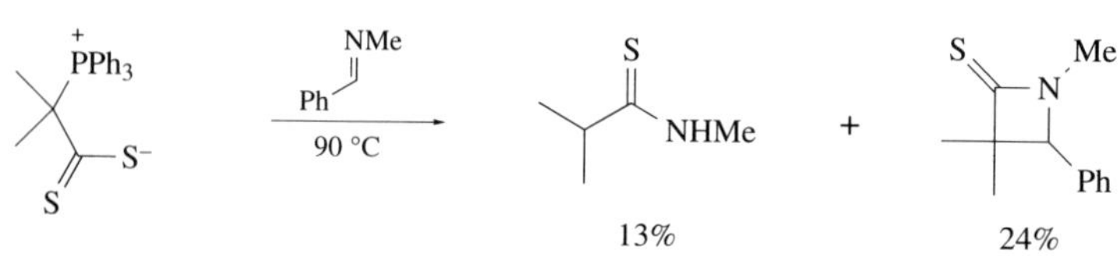

(171)

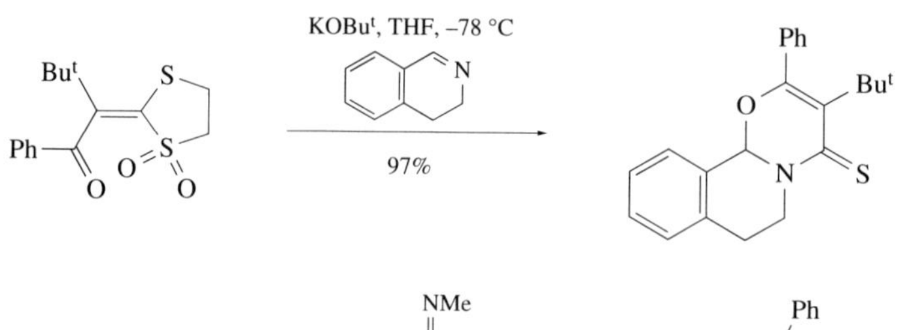

(172)

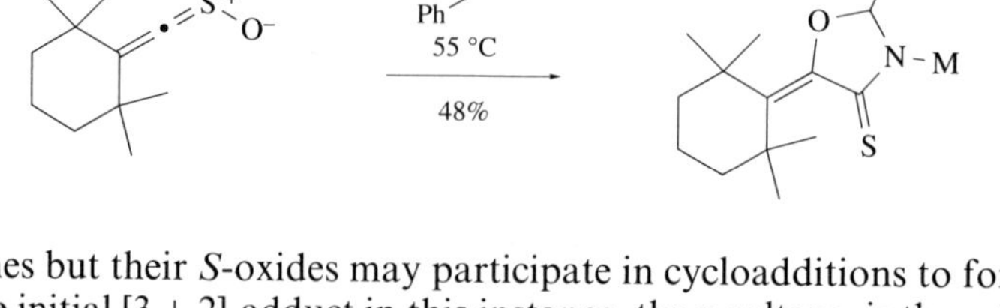

(173)

Not only thioketenes but their *S*-oxides may participate in cycloadditions to form thionolactams (Equation (173)). The initial [3 + 2]-adduct in this instance, the γ-sultene, is thought to rearrange to give the observed product ⟨78AG(E)455⟩. Thioketenes can also react with ynamines, the intermediate cycloadduct rearranging to an allenyl thionoamide ⟨85HOU(E11)251⟩.

A variety of other substrates combine in [2 + 2]-cycloadditions followed by ring opening and rearrangement to yield thionoamides. Thus *S*-silyl alkynylthioethers with imines ⟨80CB3024⟩, ynamines with dithiocarboxylates and trithiocarbonates ⟨81RTC414, 83BSF(2)277⟩, or with thiazoline thiones ⟨86HCA174⟩, thioketones ⟨76TL209⟩, thioketone *S*-oxides ⟨87LA189⟩, thioketone *S*-imides ⟨87CB751⟩, or thioaldehyde pentacarbonyl chromium complexes ⟨87CC559, 88CB2095⟩ and ynamines and alkynes with dithiolones or dithiolonium salts ⟨68CB1428, 83BSF(2)277⟩ all give unsaturated thionoamides as products (Equations (174)–(180)). Even sulfur reacts with ynamines (Equation (181)) to yield α-thioxothionoamides ⟨93TL115, 93BCJ623⟩. Here the reaction is thought to proceed via a 1,2-dithiete.

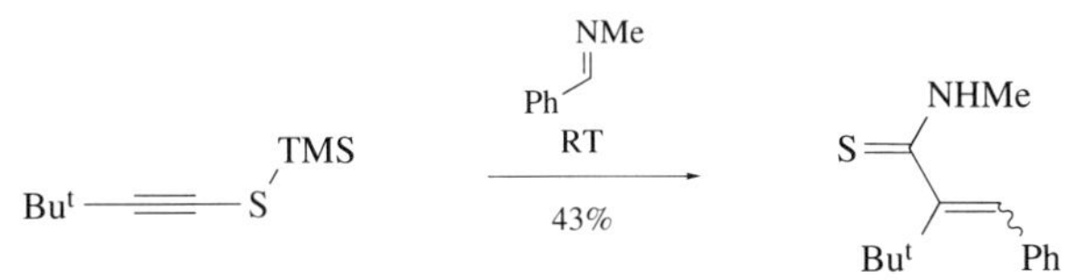

(174)

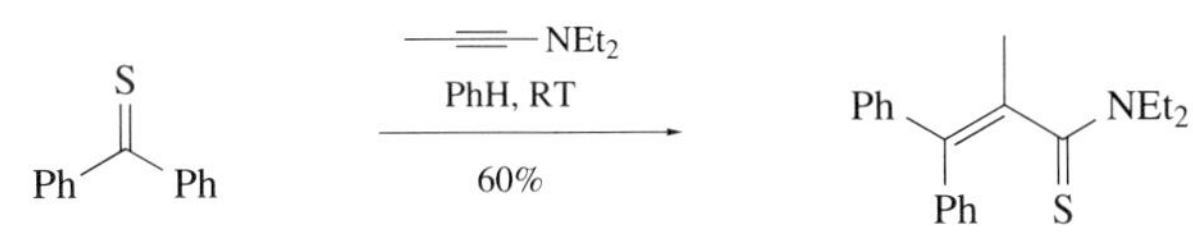

(175)

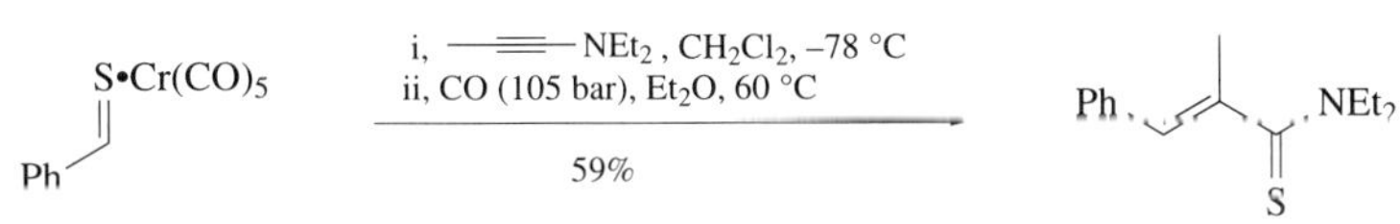

(176)

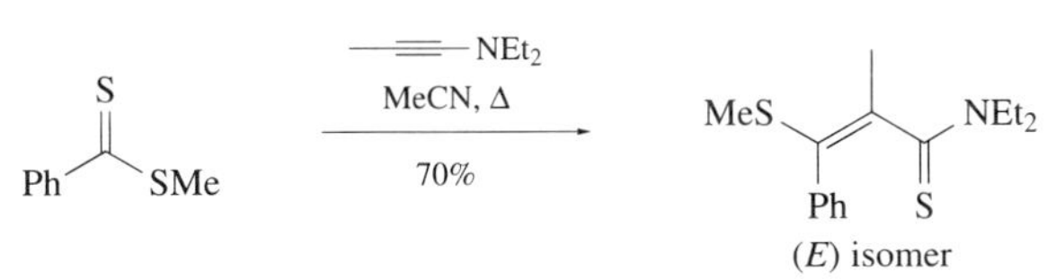

(177)

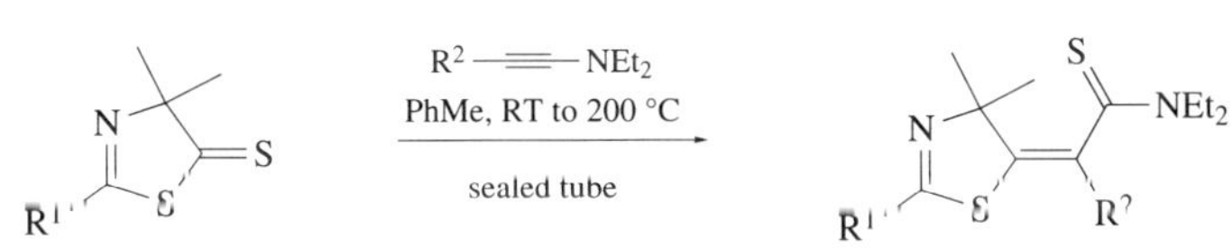

(178)

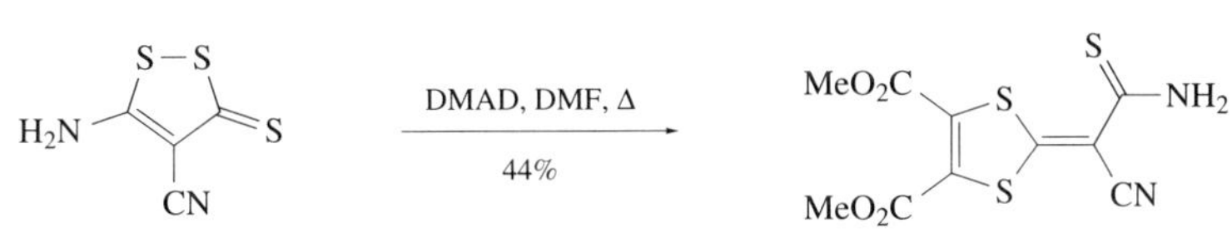

(179)

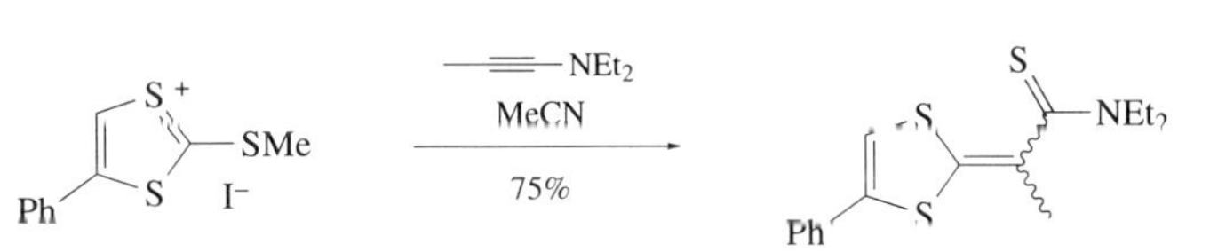

(180)

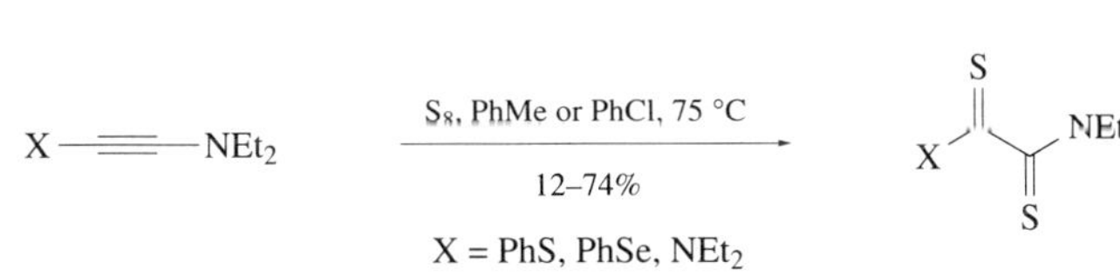

(181)

Aryl isothiocyanates react with cyclic nitrones to yield thionolactams via an initial [3 + 2]-cycloaddition, followed by a cycloreversion eliminating an aryl isocyanate ⟨72TL4191⟩ (Equation (182)). If the nitrone is not sufficiently sterically crowded, the isothiocyanate adds to the C=N bond rather than to the C=S bond. A related approach uses 1,1′-thiocarbonyldiimidazole ⟨83TL4927, 85CJC951⟩ or 1,1′-thiocarbonyldi(1,2,4-triazole) ⟨85CJC951⟩ in reactions with nitrones (Equation

(183)). Isothiocyanates can also react in a photoinitiated chromium(0) promoted [6 + 2] cycloaddition yielding a bicyclic thionolactam (Equation (184)) ⟨93TL5397⟩.

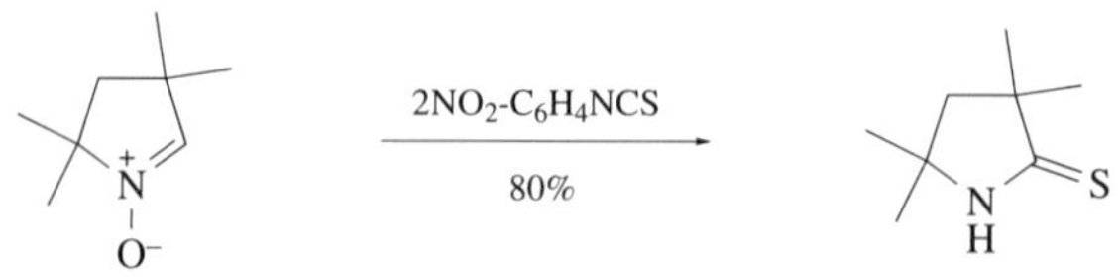

(182)

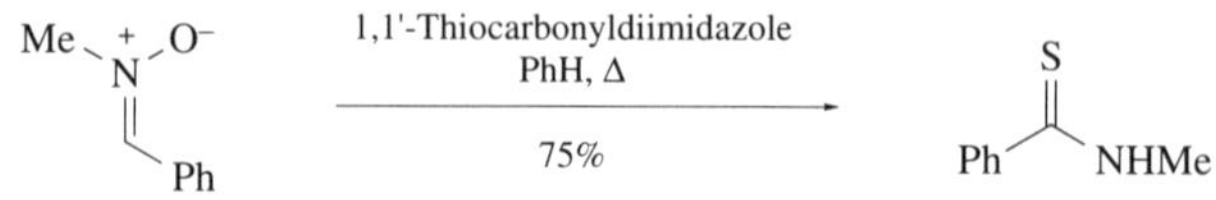

(183)

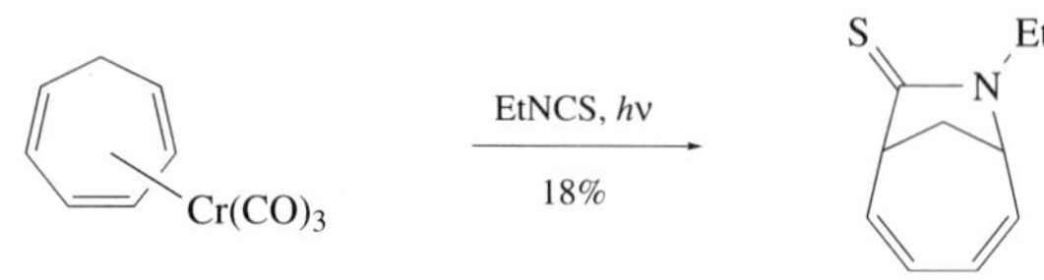

(184)

5.14.1.7 Manipulation of an Existing Thionoamide

Reaction of primary (or secondary) thionoamides with a variety of agents to produce secondary (or tertiary) thionoamides is well described in the literature. Reactions with formaldehyde produce *N*-hydroxymethyl thionoamides ⟨67AP241, 72S34⟩, whilst reaction with formaldehyde and an amine leads to *N*-aminomethyl thionoamides ⟨72T199, 75LA2318⟩. Acetals ⟨77S250⟩, activated halides ⟨69CB3786⟩, ninhydrin ⟨77AP(310)1001⟩, α-ketoaldehydes ⟨74AP(307)801⟩, activated ketones ⟨85ZN(B)124⟩ and imines ⟨85S92⟩ have also been used successfully (Equation (185)). Reaction of thiobenzamide with an aldehyde and benzotriazole, followed by addition of a Grignard reagent, produces secondary thionoamides (Equation (186)) ⟨88TL1755, 88JOC5854⟩. Yields are very good.

(185)

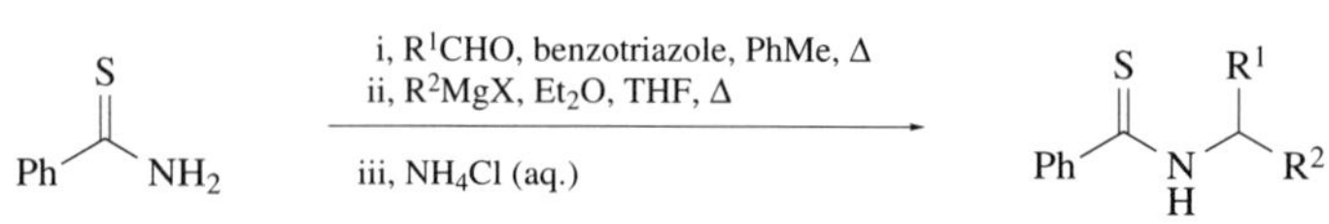

(186)

Treatment of primary thionoamides with a base followed by the addition of a Michael acceptor bearing a *β*-leaving group allows the synthesis of functionalised *N*-alkenyl thionoamides (Equation (187)) ⟨85S86, 92H(34)1573⟩. In the reverse sense, addition of the lithium enolate of *t*-butyl acetate to the amidine (**43**) also yields an *N*-alkenyl thionoamide (Equation (188)) ⟨82PS(12)325⟩. The lithium enolate of *t*-butyl methylketone reacts with the same amidine, but the yield of desired product is only 2%. Treatment of the tertiary thionoamide (**44**) with secondary butyllithium and TMEDA followed by quenching with a variety of electrophiles (alkyl and benzyl halides, aldehydes, ketones) allows preparation of a range of new tertiary thionoamides (Equation (189)) ⟨76AG(E)313⟩. Yields are variable.

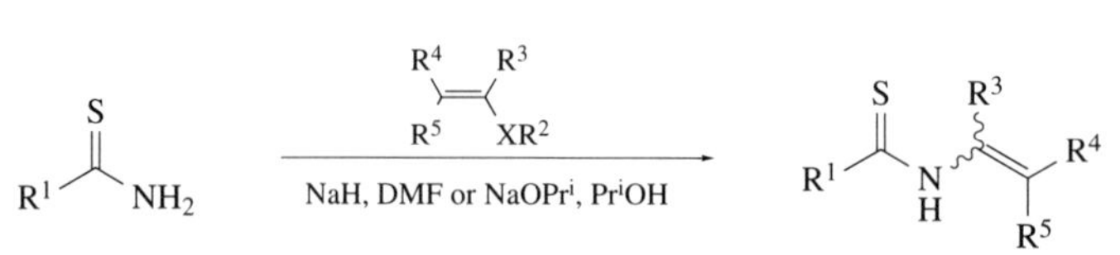

(187)

X = O, S; R^4, R^5 = CN, CO_2R

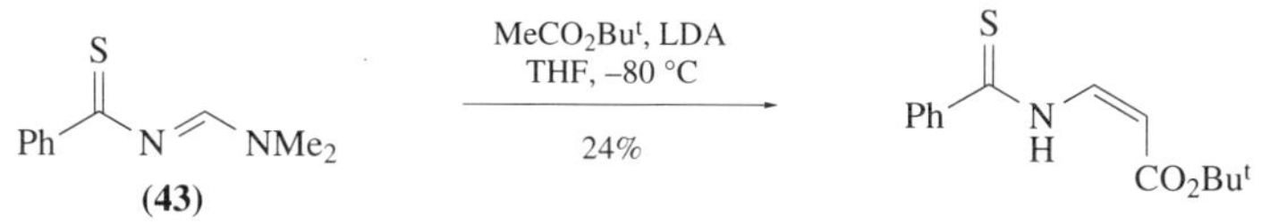

(43)

(188)

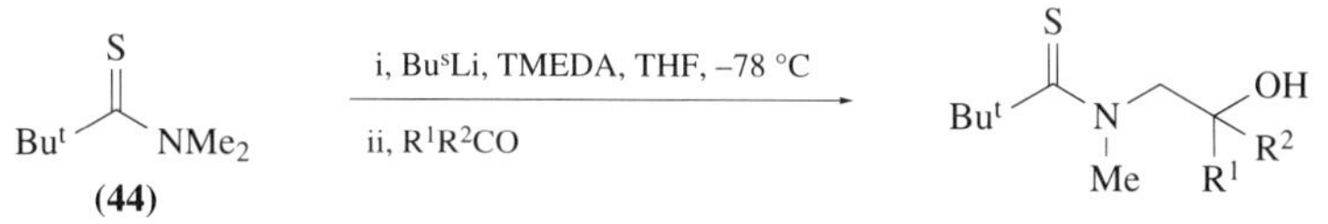

(44)

(189)

Deprotonation of tertiary thionoamides with LDA followed by quenching with a chiral sulfinate allows a precursor for asymmetric aldol reactions to be generated ⟨85T4929⟩. Treatment of this reagent with a Grignard reagent followed by an aldehyde and removal of the chiral auxiliary can lead to aldol products with high enantiomeric excess (Equation (190)). Related diastereoselective procedures via the *S,N*-ketene acetal ⟨83TL4001⟩ and magnesium enolate ⟨84TL5793⟩ are also shown (Equations (191) and (192)). Deprotonation of secondary or tertiary thionoamides with butyllithium followed by quenching with various species allows the generation of various α-substituted thionoamides (Equation (193)) ⟨78JA1923, 81TL3413, 83TL4001, 84TL5793, 85H(23)2213, 85T(24)4929, 86H1247, 86S1063⟩.

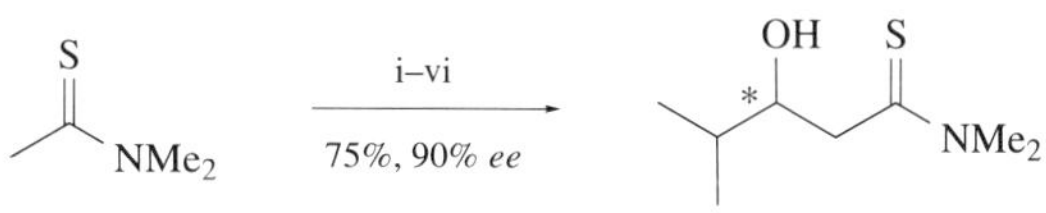

(190)

i, LDA, THF, –78 °C; ii, (–)-(*S*)-menthyl *p*-toluenesulfinate; iii, Bu^tMgBr, THF, –90 °C; iv, Pr^iCHO; v, NH_4Cl (aq.); vi, Na/Hg, MeOH

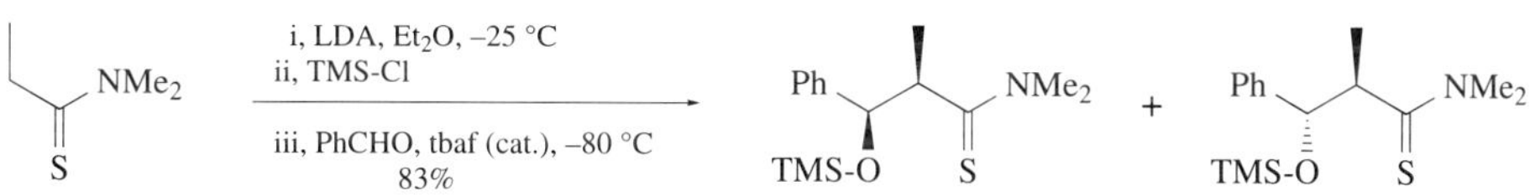

erythro (95) : threo (5)

(191)

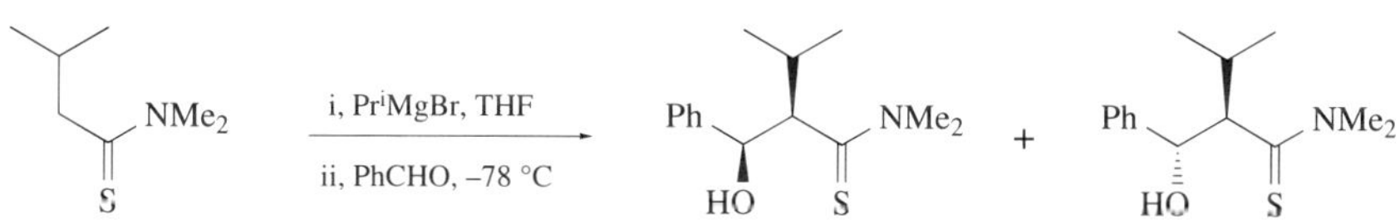

erythro (34) : threo (66)

(192)

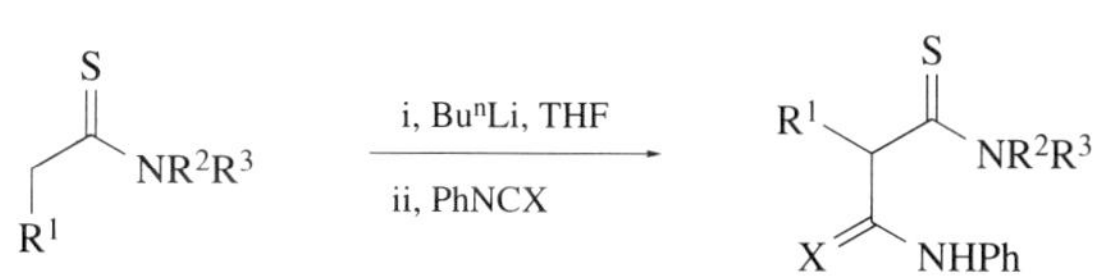

(193)

The stereoselective generation of *Z*-enolates of secondary and tertiary thionoamides and use of such species in diastereoselective aldol reactions and thio-Claisen rearrangements has been reported ⟨80JA7806, 82JA4018⟩. Oxidative dimerisation of deprotonated thionolactams has been observed (Equation (194)). The reaction is thought to proceed via a dithia-Cope rearrangement ⟨78JA1923⟩. Other typical uses of these and related rearrangements have been published ⟨76JA7084, 78TL2167, 81TL4245, 82TL881⟩ and the area has been reviewed ⟨80MI 314-01⟩. Titanium enolates and titanium 'ate' complexes of *N,N*-dimethylphenylthioacetamide have been reacted with cinnamaldehyde and chalcone in a study of both the stereochemistry (*erythro*/*threo* selectivity) and the regiochemistry (1,2- *vs* 1,4-addition) of the addition ⟨94T7193⟩. Reaction of *β*-ketothionoamides with a mild base followed by the addition of acrylonitrile or benzylidenemalononitrile leads to the anticipated conjugate addition products ⟨93M721⟩. In the benzylidenemalononitrile case, the initial adduct cyclises to form an interesting spiro compound (Equation (195)). Reaction of an *S*-allyl thioimidate (from the thionoamide and allyl bromide) with a palladium catalyst results in smooth migration of the allyl group from sulfur to nitrogen (Equation (196)) ⟨81TL4245⟩. Under different conditions, the allyl group migrates from sulfur to carbon. Finally, the secondary thionoamide group can act as a directing group in aromatic and benzylic metalation reactions (Equation (197)) ⟨76JOC4029⟩.

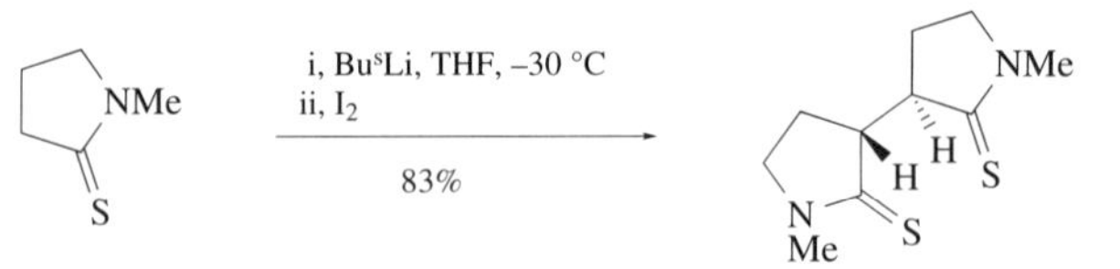
(194)

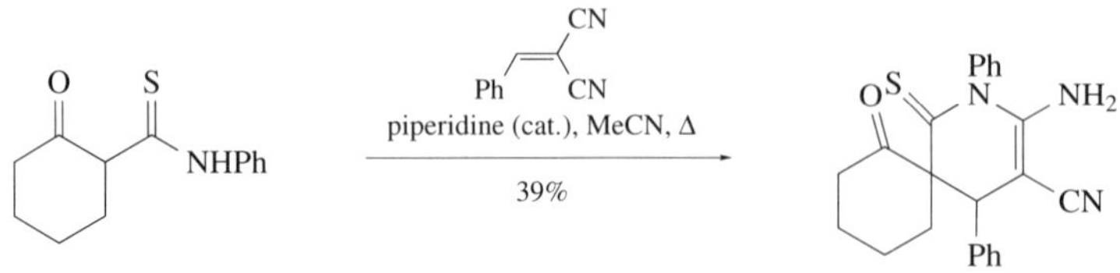
(195)

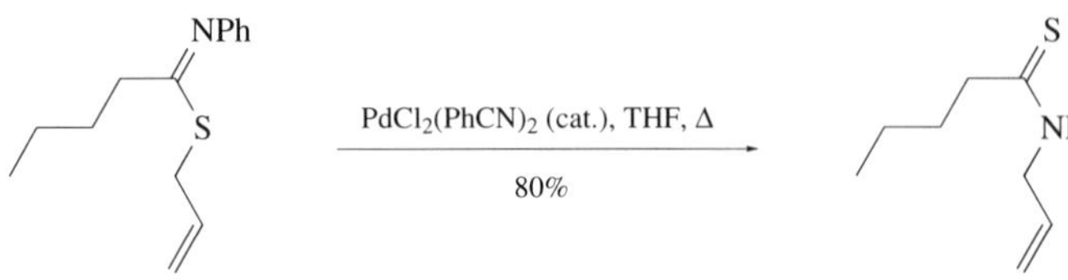
(196)

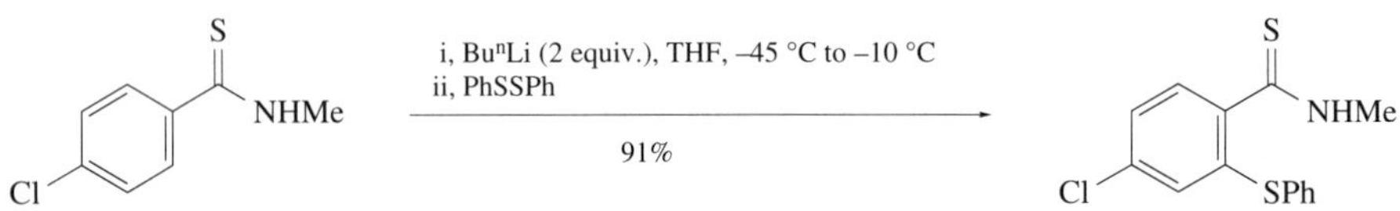
(197)

5.14.2 THIONOFORMAMIDES

Thionoformamides are available by a number of typical procedures, for example treatment of an existing formamide (Equation (198)) with Lawesson's reagent ⟨78BSB229, 78BSB525⟩, phosphorus pentasulfide ⟨47JA1207, 70LA(733)195, 71LA(755)145, 73S149, 76CB1309⟩, ethylaluminum sulfide ⟨75JOM(92)139⟩, carbon disulfide ⟨75T1209⟩ and other reagents ⟨77BSB321, 77BSB693, 78H(11)139⟩. Typical examples are shown in Table 25.

(198)

Two-step procedures via treatment with a chlorinating reagent followed by hydrogen sulfide have

Table 25 Thionation of formamides.

R[1]	*R*[2]	*Reagent*	*Solvent*	*Yield* (%)	*Ref.*
H	Ph	LR[a]	HMPA	85	78BSB229
H	Ph	$(EtO)_2PSSH$	None	28	77BSB321
Me	Me	LR	HMPA	98	78BSB229
Me	Me	$P_4S_{10}/NaHCO_3$	Et_2O	85	73S149
Me	Me	$(EtAlS)_n$	None	80	75JOM(92)139
Me	Me	CS_2	None	36	75T1209
Me	Me	$COCl_2/H_2S$	$CHCl_3$	85	63CB2671
Me	Me	Et_2NCSCl	Not given	47	78H139
Pr[n]	Pr[n]	LR	PhMe	98–100	78BSB525
$—[CH_2]_2O[CH_2]_2—$		P_4S_{10}	PhH	23	47JA1207
Me	Ph	$POCl_3/Me_2NCSH$	None	Not given	79T1087
Me	Ph	$(Me_2N)_2PSCl$	None	72	77BSB693
Et	Ph	P_4S_{10}	PhH	74	71LA(755)145
Ph	Ph	$P_4S_{10}/NaHCO_3$	Et_2O	49	76CB1309

[a] LR = Lawesson's reagent.

also proved popular ⟨63CB2671, 79T1087⟩. Reactions of hydrogen sulfide with isonitriles ⟨1877CB1095⟩, amidines ⟨1877CB1238, 81S299, 82M(113)593⟩, isolated amide chlorides ⟨60AG836⟩ and α-dialkylamino-α-methoxynitriles ⟨82LA1582⟩ are possible (Equations (199)–(204)). Reactions of amines with NaSH generated *in situ* have also received attention (Equation (205)) ⟨63LA(669)66, 68LA(712)58, 72T199, 76CB1309⟩. Thioacylation of amines with thionoesters is also successful (Equation (206)) ⟨64ZC457, 76CB1309, 82M(113)593⟩, with an interesting variant being a one-pot procedure involving *in situ* generation of ethyl thionoformate (Equation (207)) ⟨89JOC1212⟩.

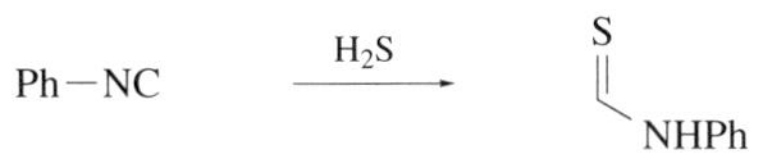
(199)

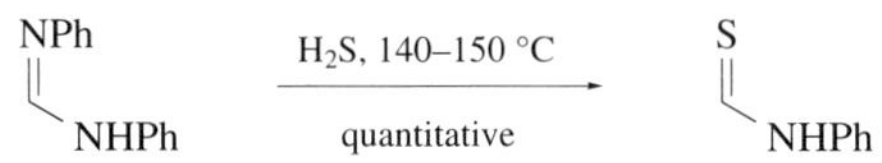
(200)

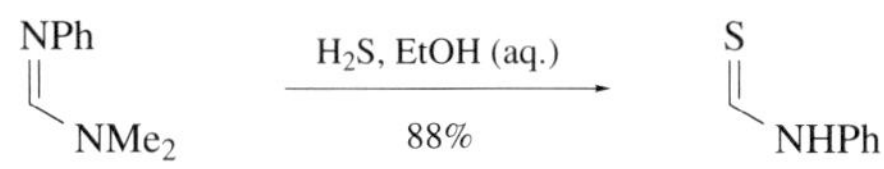
(201)

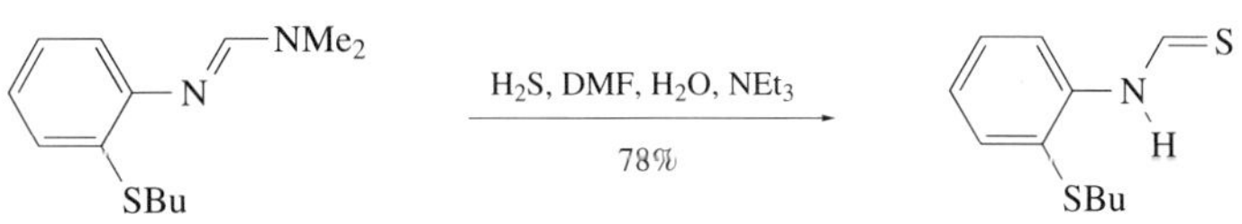
(202)

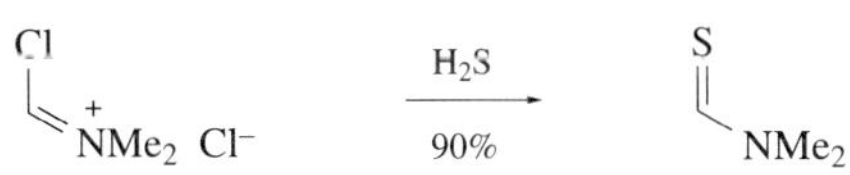
(203)

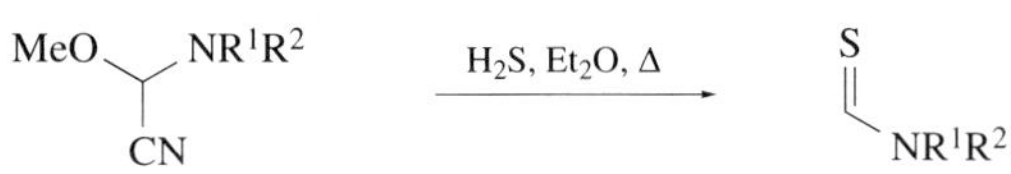
(204)

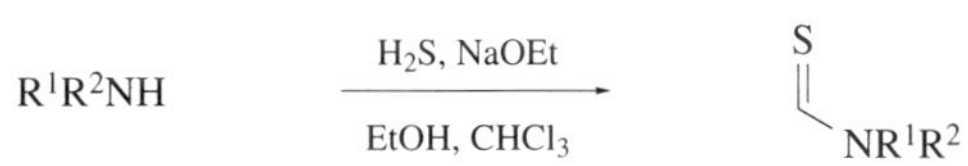
(205)

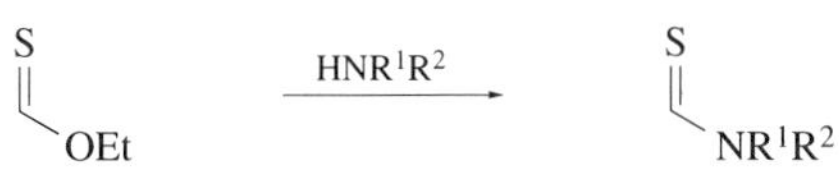
(206)

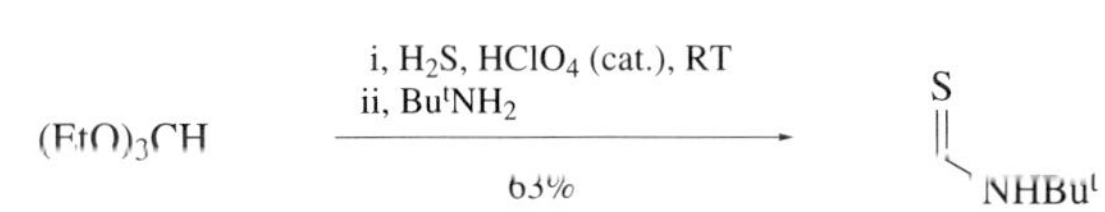
(207)

Transamination of thionoformates is also possible in appropriate cases (Equation (208)) ⟨86S482⟩. Reduction of isothiocyanates with sodium borohydride is another viable method (Equation (209)) ⟨63JOC1600, 71LA(753)187, 77TL2145⟩. Other less general methods include photolysis of monothionoimides ⟨85T5393⟩, reaction of isonitriles with dithioacids ⟨77AG(E)787⟩, reaction of dimethylammonium *N*,*N*′-dimethyldithiocarbamate with carbon disulfide ⟨75T1209⟩ and treatment of thiocarbamoyl chlorides with sodium isopropoxide ⟨69LA(725)234⟩ (Equations (210)–(213)).

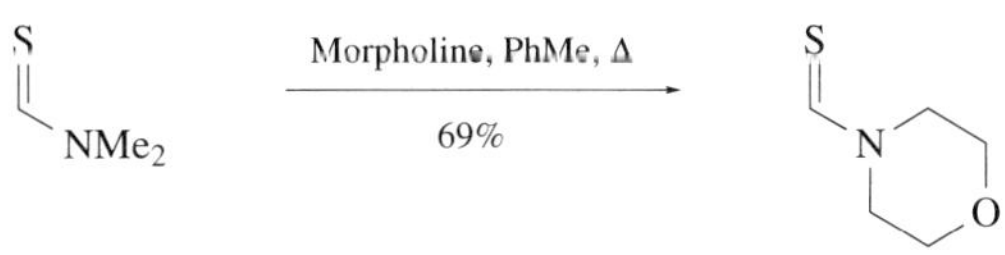
(208)

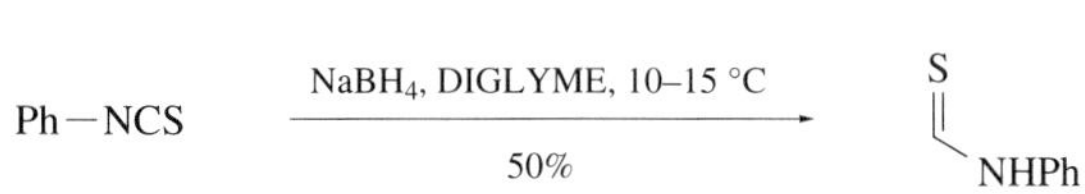
(209)

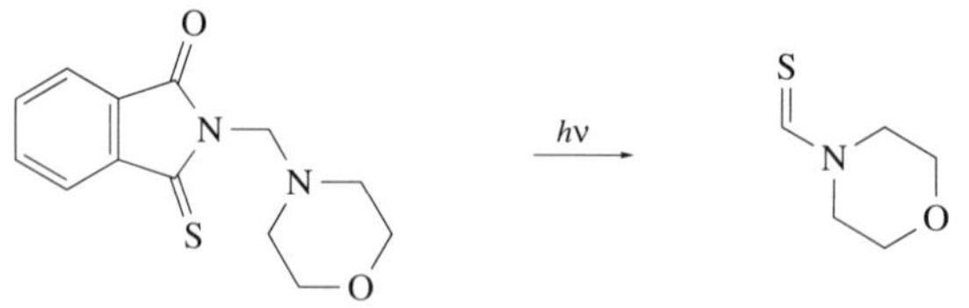

(210)

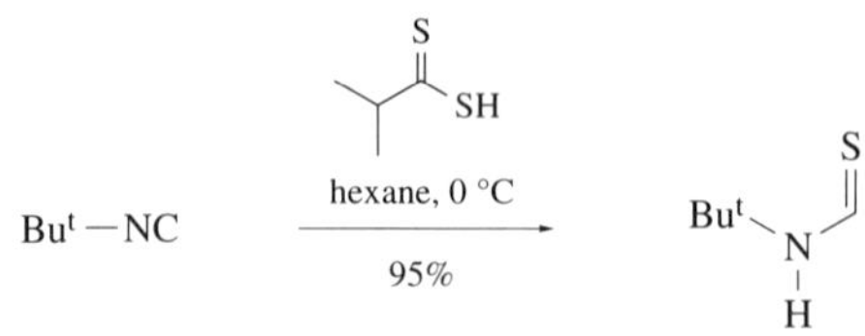

(211)

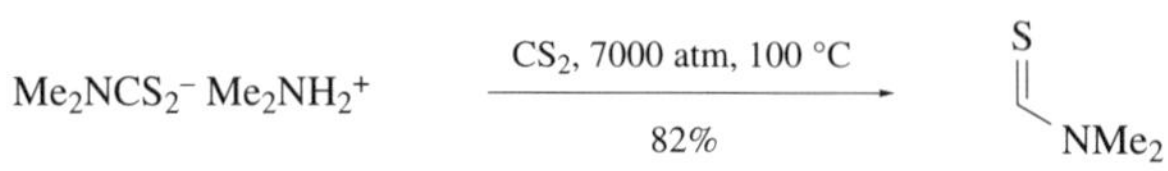

(212)

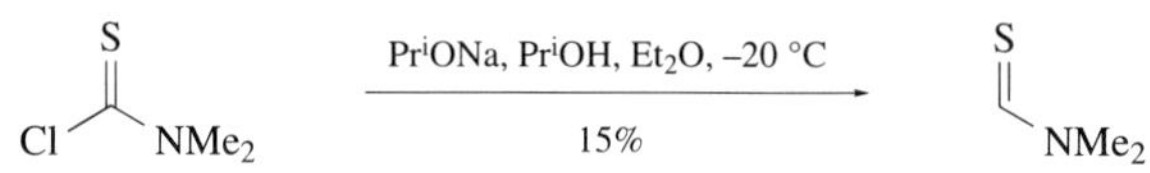

(213)

5.14.3 THIONOALKANOAMIDES

A wide variety of alkanoamides can be converted to the corresponding thionoamides (Equation (214)) by a whole host of reagents. Amongst these are phosphorus pentasulfide, Lawesson's reagent, a variety of other phosphorus containing reagents, ethylaluminum sulfide, 9-BBN sulfide and oxalyl chloride/hydrogen sulfide. Typical examples and references are given in Table 26, with further examples being in Section 5.14.1.1.

(214)

Transformations of nitriles to thionoamides using hydrogen sulfide have been covered in Section

Table 26 Thionation of amides (Equation (214)).

R^1	R^2	R^3	*Reagent*	*Solvent*	*Yield* (%)	*Ref.*
Me	H	H	LR[a]	HMPA	88	78BSB229
Bu^t	H	H	P_4S_{10}	Pyridine	44	75CB3066
CF_3	H	H	LR	THF	53	86PS(26)169
Et	H	$PhCH_2$	$(Me_2N)_2PSCl$	None	67	77BSB693
Me	H	Ph	P_4S_{10}	Pyridine	48	72T199
Me	H	2MeO—C_6H_4	LR	HMPA	98	78BSB229
Me	H	2-Py[b]	LR	PhH	90	83PJC647
$ClCH_2$	H	Me	P_4S_{10}	PhMe	39	87PS(31)33
Me	Me	Me	$(EtAlS)_n$	None	72	75JOM(92)139
Me	Me	Me	$(EtO)_2PSSH$	None	46	77BSB321
CF_3	Et	Et	P_4S_{10}	Dioxan	66	76ZOR2213
Pr^n	Et	Et	LR	PhMe	98	78BSB525
Pr^n	Et	Et	$COCl_2$, H_2S	$CHCl_3$	63	63CB2671
Bu^t	Me	Me	$(EtO)_2PSSH$	None	45	77BSB321
$C_{11}H_{23}$	Et	Et	$(9\text{-BBN})_2S$	Heptane	70	92LA1081
$PhCH_2$	—$[CH_2]_5$—		LR	PhMe	75	80TL4061
$MeSCH_2$	Me	Me	P_4S_{10}	Pyridine	33	77LA1291
Me_2NCH_2	Me	Me	P_4S_{10}	Pyridine	28	77LA1291
Me	—$[CH_2]_2O[CH_2]_2$—		$(Me_2N)_2PSCl$	None	62	77BSB693
—CH_2CHPh—		Ph	LR	PhH	87	80NJC47

[a] LR = Lawesson's reagent. [b] 2-Py = 2-pyridyl.

5.14.1.2. Results using a selection of more interesting reagents for this transformation (Equation (215)) are shown in Table 27. Amongst other substrates, hydrogen sulfide reacts with amidines ⟨66CR(C)788⟩, chloroiminium salts ⟨60AG836⟩, haloenamines ⟨63JOC3492⟩ and thioimidates ⟨84JCS(P1)785⟩ to produce thionoamides (Equations (216)–(219)). Oximes have been treated with benzenesulfonyl chloride followed by hydrogen sulfide yielding rearranged thionoamides (Equation (220)) ⟨58CB972⟩. Sodium sulfide can be added to β-nitroenamines allowing access to α-nitrothionoamides (Equation (221)) ⟨90TL1327⟩.

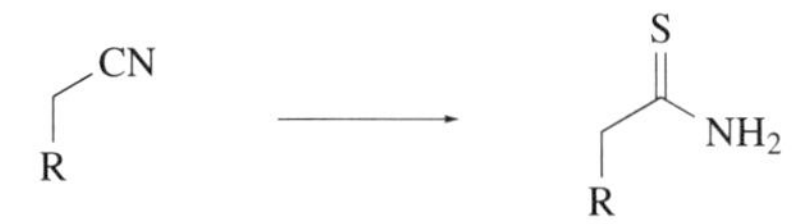

(215)

Table 27 Alternatives to H_2S (Equation (215)).

R	*Conditions*	*Yield* (%)	*Ref.*
Me	$Ph_2P(=S)SH$, Pr^iOH, 40 °C	83	81TL1851
Bu^n	Me_3SiCl, Na_2S, DMEU[a], 80 °C	28	93JOC4742
CH_2Cl	$(EtO)_2P(=S)SH$, H_2O, 80 °C	55	85PS(25)297
CO_2Bu^t	P_4S_{10}, Na_2S, THF, RT	46	92SC1397

[a] DMEU = 1,3-dimethyl-2-imidazolinone.

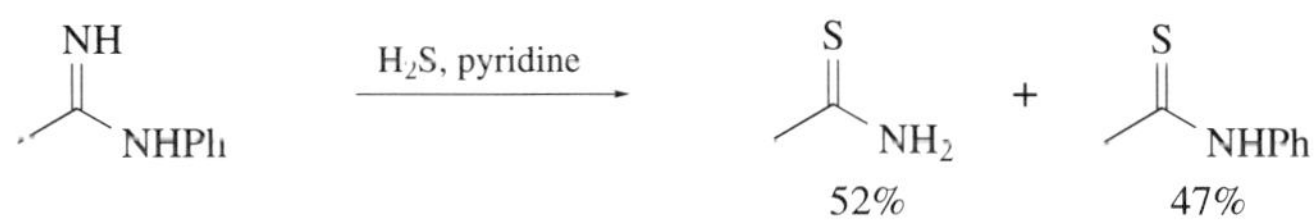

(216)

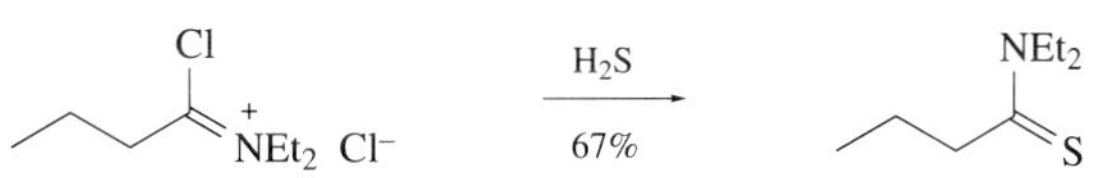

(217)

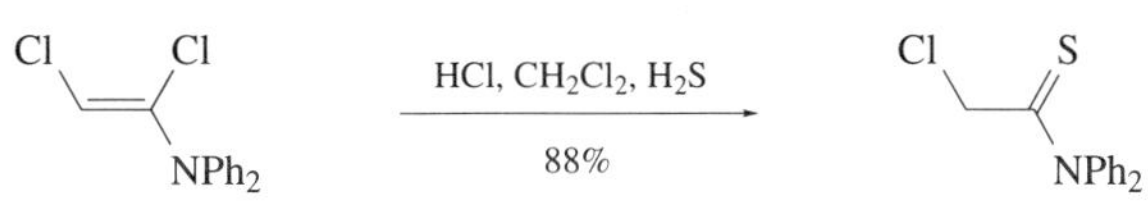

(218)

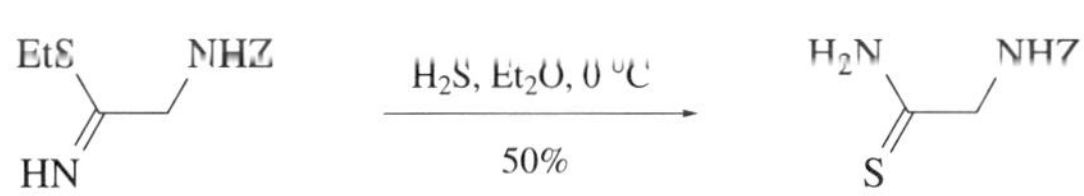

(219)

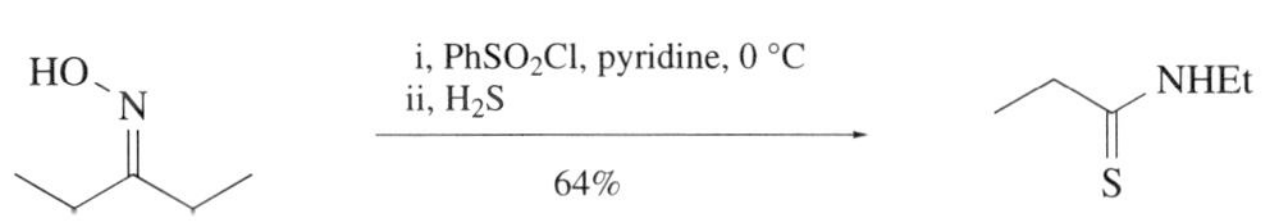

(220)

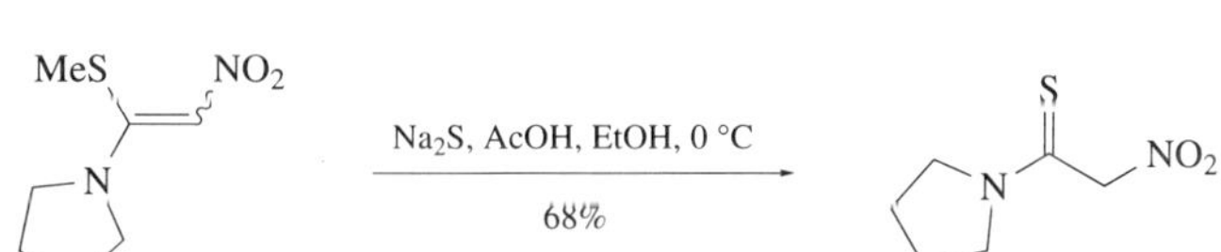

(221)

Thioacylation of amines by thionoesters, dithiocarboxylates and by dithioacids with added boron reagents are all potential routes (Equation (222)). Representative examples are given in Table 28. Even thionoacids have been used ⟨93T7215⟩ although in this case the substrate is an aminoazirine (Equation (223)). Recently, alkylthiosulfines have also been used successfully (Equation (224)) ⟨93TL6741⟩. Reactions thought to proceed via the thioketene, for example reaction of alkynyl thioethers with amines ⟨71ZOR1120⟩ and of alkynes with butyllithium ⟨77S888⟩ or sodium ⟨67JPR294⟩ followed by sulfur and an amine yield thionoamides (Equations (225)–(227)). Other examples are to be found in Section 5.14.1.3. Addition of nucleophiles, for example, alkyllithiums and trialkylaluminums, to isothiocyanates (Equation (228)) is another option. This type of chemistry has already been covered in great detail in Section 5.14.1.4. Willgerodt–Kindler type reactions on dialkyl ketones as demonstrated by Viehe ⟨85S77⟩ are an attractive option (Equation (229)). Other less flexible routes include ring opening of heterocycles ⟨85ZC324⟩ and photolysis of thiazolines ⟨75T1245⟩ (Equations (230) and (231)).

(222)

Table 28 Thioacylation of amines (Equation (222)).

R[1]	*X*	*R*[2]	*R*[3]	*Reagents*	*Yield* (%)	*Ref.*
Pr^i	SH	Me	Me	Catecholborane,Me_2NH	67	90JCR(S)106
Pr^i	SH	Me	Me	9-BBN, Me_2NH	92	91PS(56)283
Me	OEt	H	Bu^n	EtCH=NBu^n	Not given	65CB3819
Et	OEt	Et	Et	Et_2NH, EtMgBr	95	65BSF3623
PhCONH	OMe	—$[CH_2]_5$—		Piperidine, CH_2Cl_2	82	92LA921
$PhCH_2$	SMe	H	Ph	$PhNH_2$	70	51JOC131

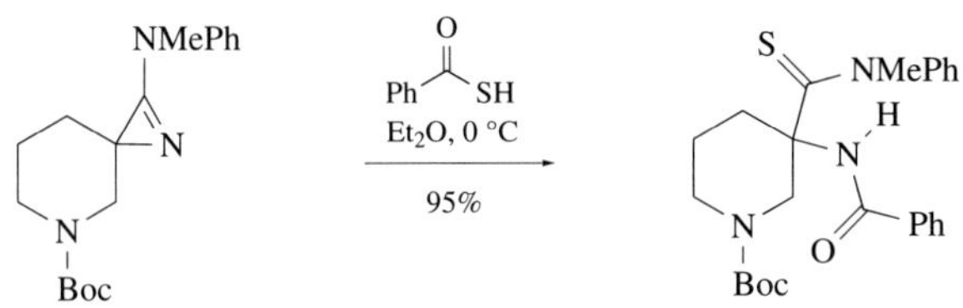

(223)

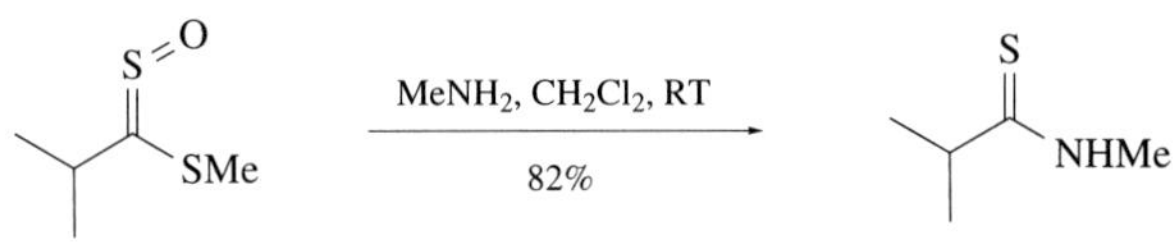

(224)

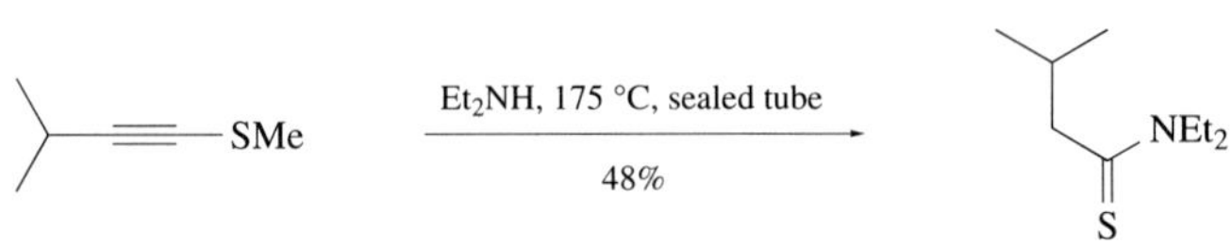

(225)

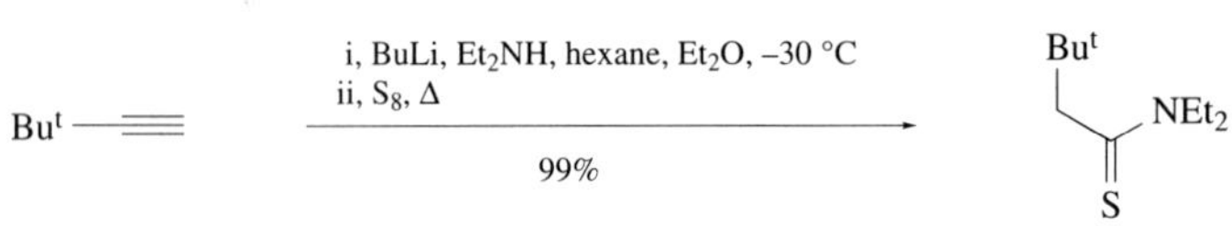

(226)

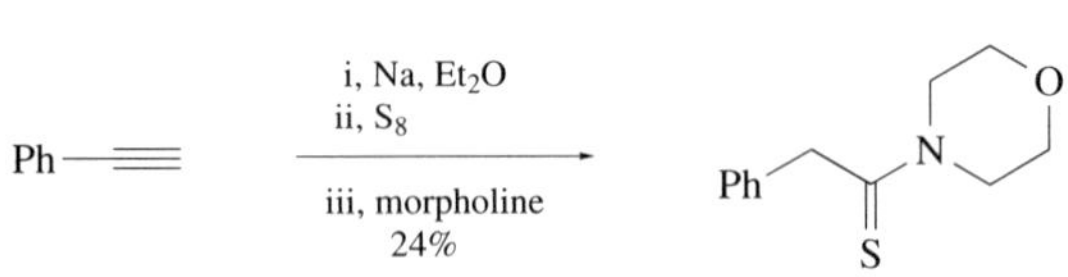

(227)

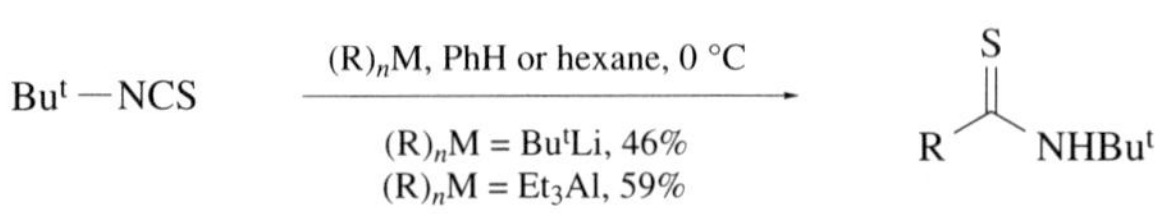

(228)

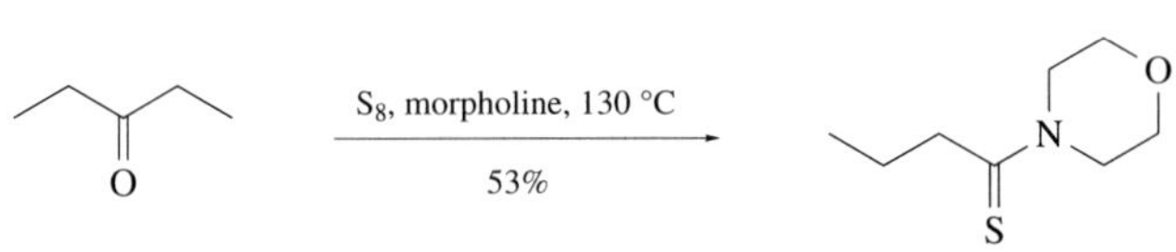

(229)

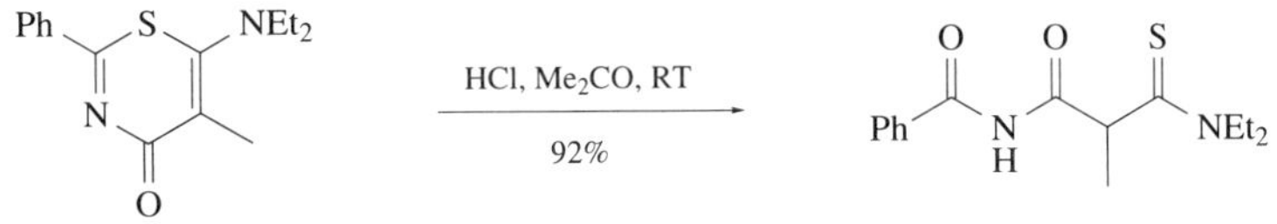

(230)

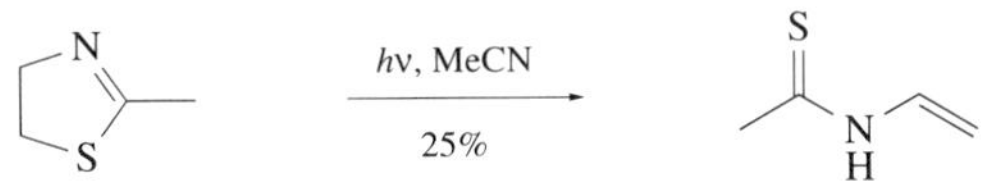

(231)

5.14.4 THIONOALKENOAMIDES

Whilst primary, secondary and tertiary amides derived from cinnamic acid are reportedly converted into the thionoamides in good to excellent yields with Lawesson's reagent in benzene at 60 °C ⟨81ACS(B)239⟩, primary amides derived from acrylic and crotonic acids gave no thionoamide products under these conditions ⟨81ACS(B)239⟩. With Lawesson's reagent in HMPA these latter substrates gave phosphorus containing heterocyclic species ⟨81ACS(B)239⟩. Reaction of secondary amides derived from crotonic, methacrylic and tiglic acid with Lawesson's reagent in benzene provide the thionoamides ⟨92JOC2419⟩. Data for typical examples (Equation (232)) are given in Table 29. Reactions of γ,δ-unsaturated secondary amides with Lawesson's reagent in benzene at room temperature (Equation (233)) proceed normally to provide thionoamides ⟨87H(26)359⟩.

(232)

Table 29 Thionation of amides (Equation (232)).

R^1	R^2	R^3	R^4	*Conditions*	*Yield* (%)	*Ref.*
H	Me	H	Pr^i	LR[a],PhH,Δ	Not given	92JOC2419
Me	H	Me	Me	LR,PhH,60 °C	80	81ACS(B)239
Me	Me	H	$PhCH_2$	LR,PhH,Δ	Not given	92JOC2419
Ph	H	H	H	LR,PhH,60 °C	60	81ACS(B)239
Ph	H	H	H	P_4S_{10},Py[b],Δ	66	89T6771
Ph	H	H	Et	P_4S_{10},Py,Δ	88	89T6771
Ph	H	H	Ph	LR,PhH,60 °C	78	81ACS(B)239
Ph	H	Et	Et	LR,PhH,60 °C	98	81ACS(B)239

[a] LR = Lawesson's reagent. [b] Py = pyridine.

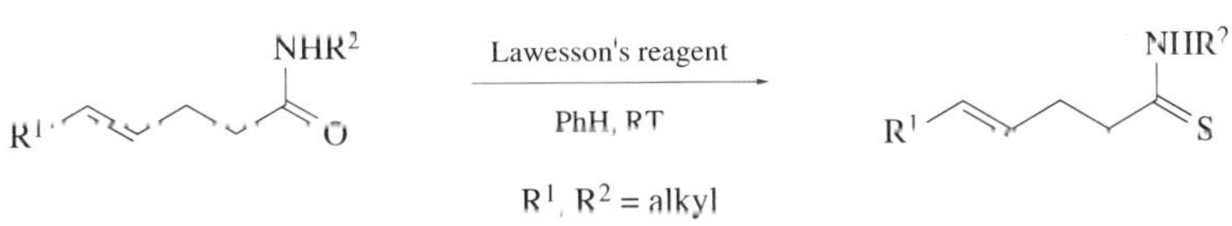

(233)

Primary and secondary amides derived from (ring substituted) cinnamic acids react with phosphorus pentasulfide in pyridine providing thionoamides in reasonable to very good yields ⟨89T6771, 89TL4449, 91TL405⟩. An α,β-unsaturated oxindole has been successfully thionated using phosphorus pentasulfide and sodium hydrogen carbonate in dioxane ⟨93JCS(P1)1835⟩. Secondary aromatic amides derived from acrylic and methacrylic acids on treatment with phosphorus pentasulfide and sodium hydrogen carbonate in acetonitrile give β-thiothionoamides rather than the desired α,β-unsaturated products ⟨78AG(E)689⟩. Thionation of α,β-unsaturated tertiary amides bearing two β-heteroatoms with phosphorus pentasulfide in aromatic solvents has been described ⟨70JOC2002, 83BSF(2)277⟩.

Thioacylation of an α,β-unsaturated thioacyl bromide by a variety of amines has been described ⟨70JOC2002⟩. Dithioacids bearing a β-benzoyl group react readily with two equivalents of amines providing enaminothionoamides (Equation (234)), although the yields are poor ⟨73BSF581⟩. Thionoesters of unsaturated acids react with piperidine providing α,β-unsaturated thionoamides ⟨80CB919⟩. A rather more interesting transformation (Equation (235)) allows access to functionally complex unsaturated thionoamides. Thionoalkenoamides bearing a β-secondary or tertiary amino group have been prepared by a number of means ⟨71CB2709, 73BSF3349, 84SUL39, 94CC1437⟩. One general example is shown in Equation (236). Yields are variable. The synthesis of β-amino-α,β-unsaturated thionoamides has been reviewed ⟨87ZC8⟩. Alternative routes to such compounds involve the addition of enamines to isothiocyanates ⟨62CB926, 64LA(673)132, 64CB3106, 83S225⟩, a topic already covered in depth in Section 5.14.1.4. Reactions of amines with dimethylchlorothiophosphate, followed by sodium hydride and an unsaturated acid chloride provides a novel route to unsaturated thionoamides (Equation (237)) ⟨90JOC6091⟩. Addition of unsaturated Grignard reagents to isothiocyanates also smoothly produces α,β-unsaturated thionoamides (Equation (238)) ⟨81TL3409⟩.

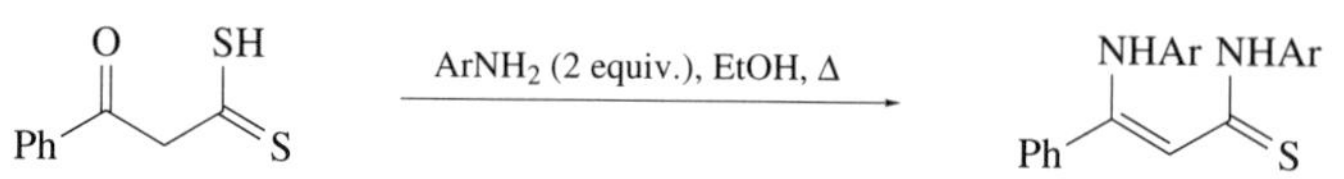

(234)

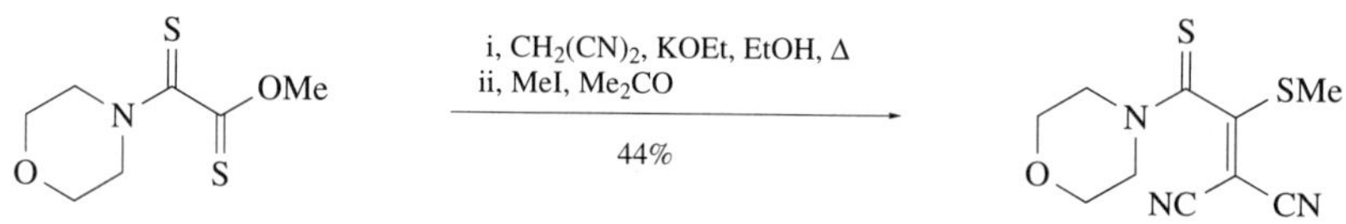

(235)

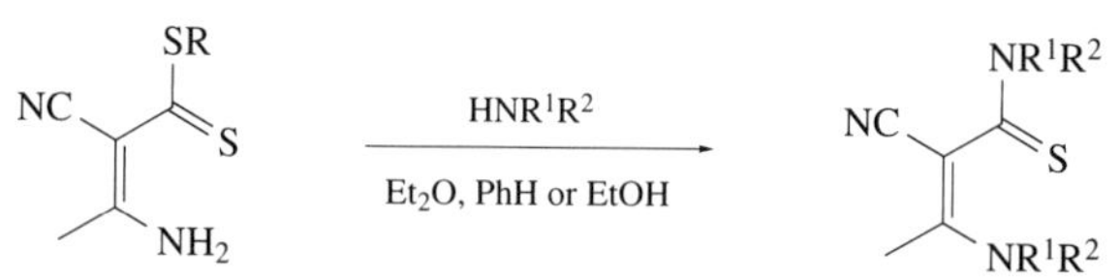

(236)

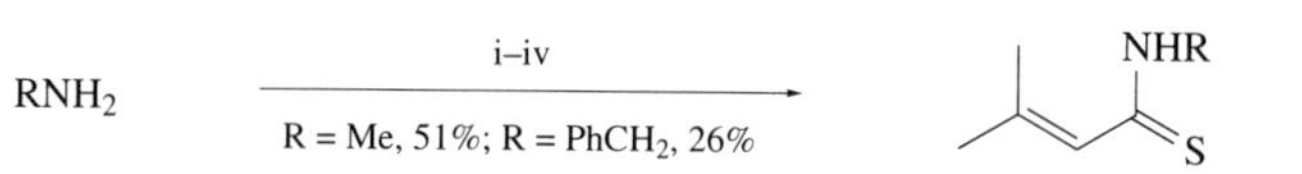

(237)

i, (MeO)$_2$PSCl, Et$_2$O; ii, NaH, THF, RT to Δ; iii, Me$_2$C=CHCOCl; iv, CCl$_4$, Δ

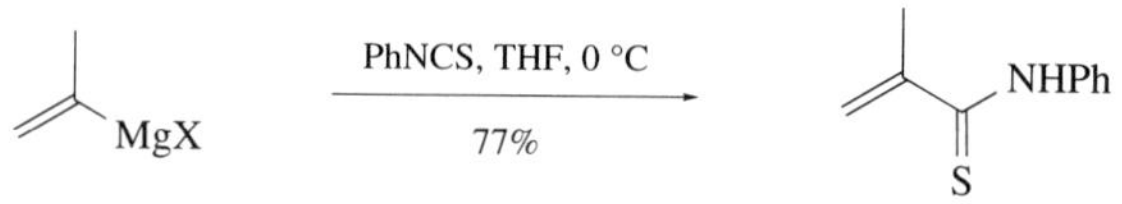

(238)

A powerful route to γ,δ-unsaturated thionoamides involves thio-Claisen rearrangement of alkynyl allyl sulfides, yielding thioketenes, which can be trapped with primary and secondary amines ⟨68RTC1236, 77TL4307, 79LA1746, 80CB3024⟩. Similar rearrangement and trapping for alkynyl propargyl sulfides provides a route to thionoamides of substituted penta-2,3- and penta-2,4-dienoic acids, with conditions available to favour either of these species ⟨72RTC578⟩. Typical examples of these transformations are shown in Equations (239)–(242). Yields are often only moderate (about 40%). In related work, β,γ-unsaturated thionoamides are available by a Wittig-type rearrangement of dithiocarbamates (Equation (243)) ⟨75JA1608⟩. Thermolysis of S-allyl thioimidates produced by reaction of a thionoamide with an allylic halide yields γ,δ-unsaturated thionoamides (Equation (244)) ⟨81TL4245⟩. Under palladium catalysis the allyl group transfers to the nitrogen rather than to the carbon ⟨81TL4245⟩. Already covered in Section 5.14.1.6 are several powerful routes involving reactions of ynamines with dithioesters ⟨81RTC414⟩, with thioketones ⟨76TL209⟩, with thioaldehyde pentacarbonyl chromium complexes ⟨87CC559, 88CB2095⟩ and of imines with S-trimethylsilyl alkynethiols ⟨80CB3024⟩ or thioketenes ⟨80CB3024⟩.

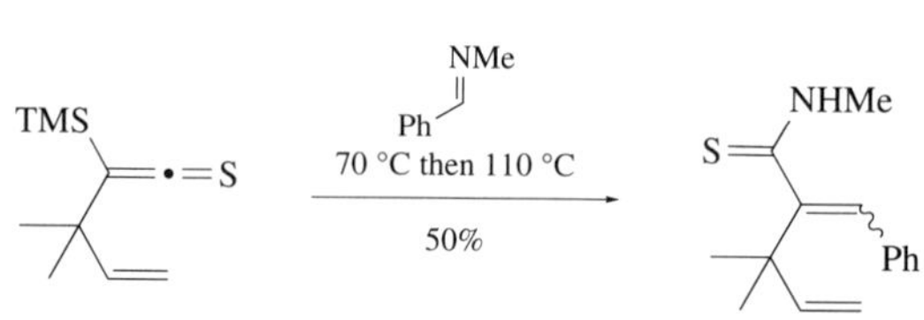

(239)

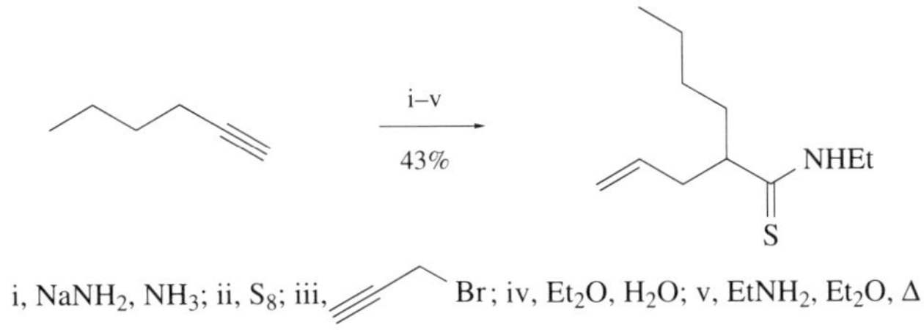

(240)

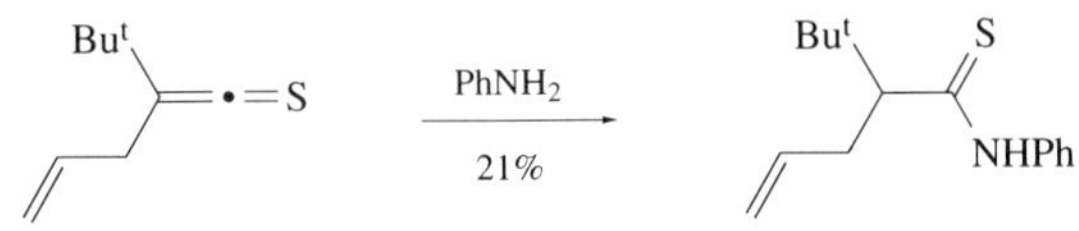

(241)

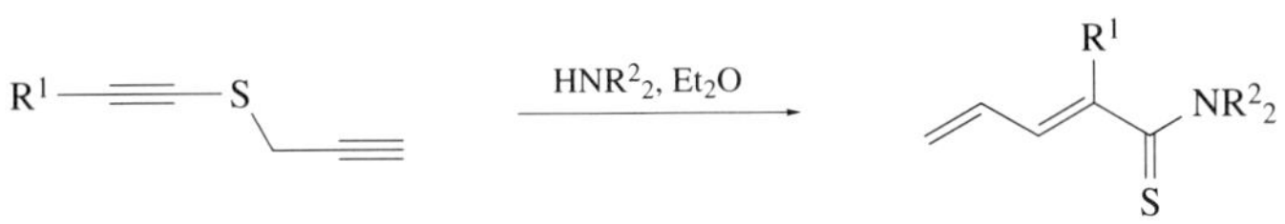

(242)

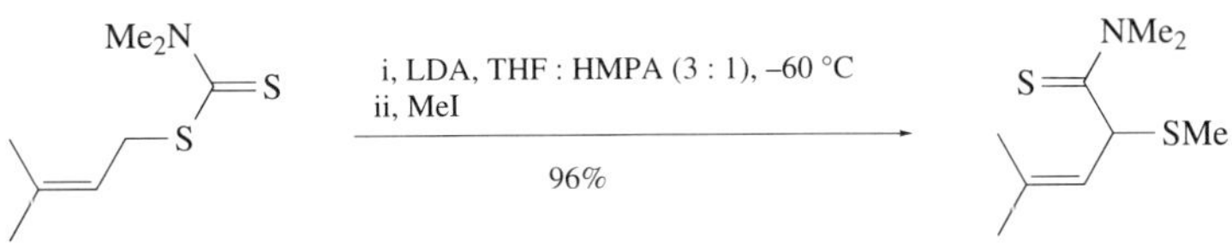

(243)

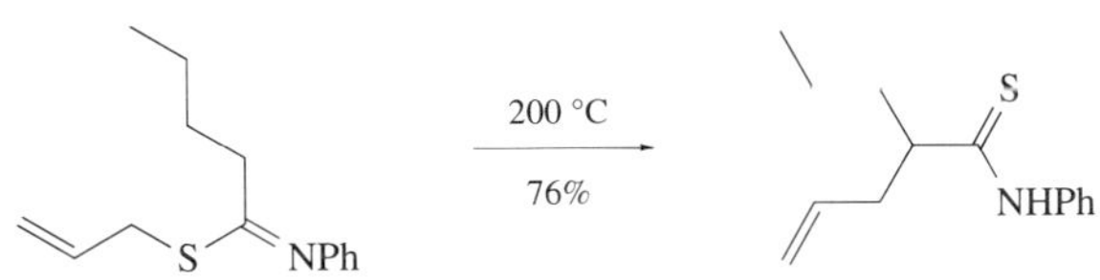

(244)

Routes from existing thionoamides are also popular, particularly condensation of cyanothioacetamide with aromatic and heteroaromatic aldehydes (Equation (245)) yielding unsaturated thionoamides bearing an α-cyano substituent ⟨62JOC2433, 63MI 514-01, 70MI 514-01, 78JCS(P1)629, 87SUL105, 88IJC(B)472, 89ZOR1323, 93H(36)1979, 93SC2259, 94JCS(P1)989⟩. Typical examples are given in Table 30. When the aromatic aldehyde bears certain substituents (*meta*- or *para*- nitro, *para*- halogen) these unsaturated thionoamides undergo [4 + 2]-dimerisation yielding thiopyrans ⟨78JCS(P1)629⟩. In a related observation, 2-thioxoindole reacts with benzaldehyde to yield an unsaturated thionoamide which also undergoes [4 + 2]-dimerisation ⟨93JCS(P1)1835⟩. Cyanothioacetamide also reacts with cyclic ketones to give unsaturated thionoamides ⟨90JCS(P1)1267⟩.

(245)

Table 30 Condensation of cyanothionoacetamide with aldehydes (Equation (245)).

Ar	*Catalyst*	*Solvent*	*Yield* (%)	*Ref.*
4Me—C_6H_4	NEt_3	EtOH or PhH	48	78JCS(P1)629
2Cl—C_6H_4	NEt_3	EtOH or PhH	35	78JCS(P1)629
2Bu^{t}S—C_6H_4	NH_4OAc	PhH	71	93H(36)1979
4Me_2N—C_6H_4	NEt_3	EtOH or PhH	60	78JCS(P1)629
4MeO—C_6H_4	Al_2O_3	MeOH	95–98	93SC2259
3,4-$(MeO)_2C_6H_3$	pyrrolidine	EtOH	90	94JCS(P1)989
2-Thienyl	NEt_3	EtOH or PhH	69	78JCS(P1)629
2-Furyl	NEt_3	EtOH or PhH	58	78JCS(P1)629

Thionoamides have also been successfully condensed with isatins ⟨90PS145⟩, orthoesters ⟨62JOC2433⟩ and related species ⟨71CB2709⟩ (Equations (246)–(248)). Synthesis of γ,δ-unsaturated thionoamides is facile by reaction of a thionoamide with a strong base and an allylic halide ⟨76CC776, 82CL529, 82TL881, 92TL7957⟩. The order of reaction may be reversed (quaternisation followed by deprotonation) as the reaction proceeds via a thio-Claisen rearrangement of an *S*,*N*-ketene acetal. The procedure can be followed more than once to give thionoamides bearing two γ,δ alkenic groupings ⟨76CC776, 82CL529⟩ and has been carried out on α,β-unsaturated thionoamides, too ⟨78TL2167⟩. The synthetic utility of unsaturated thionoamides generated via thio-Claisen and thio-Cope processes has been reviewed ⟨80MI 514-01⟩. Some simple examples are shown in Equations (249)–(251).

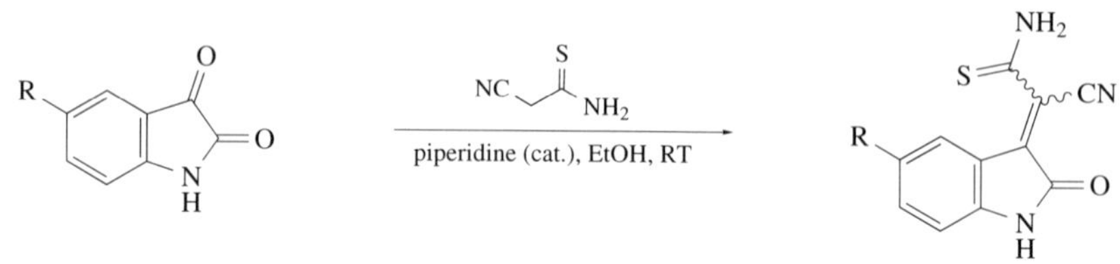

(246)

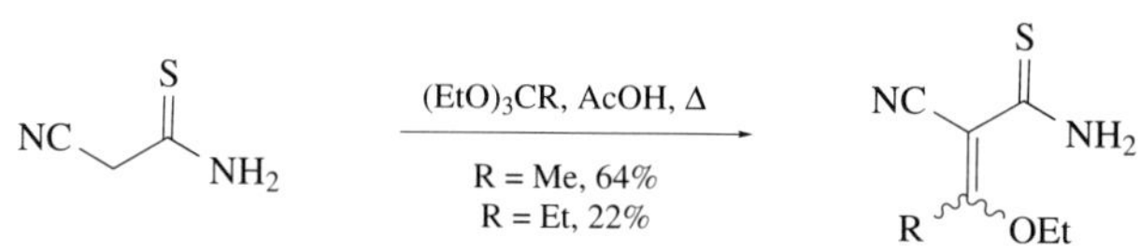

(247)

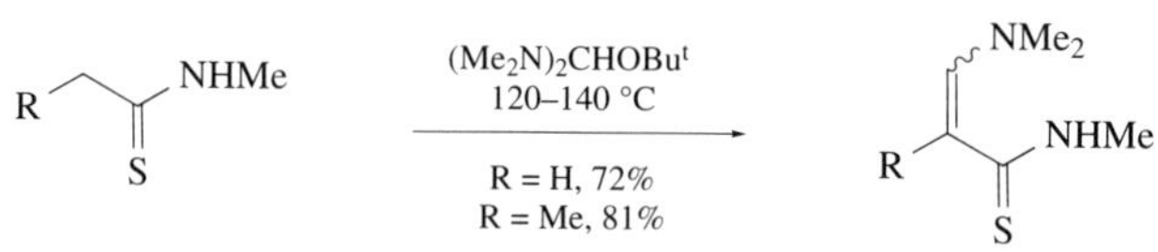

(248)

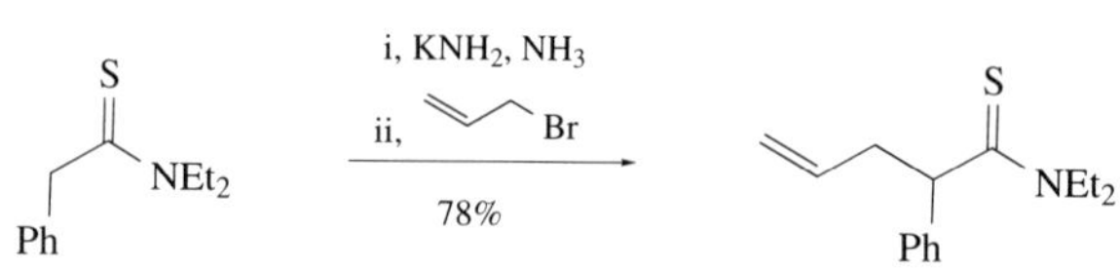

(249)

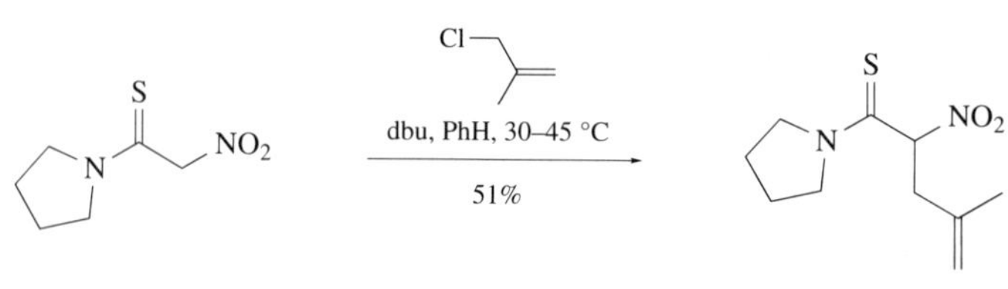

(250)

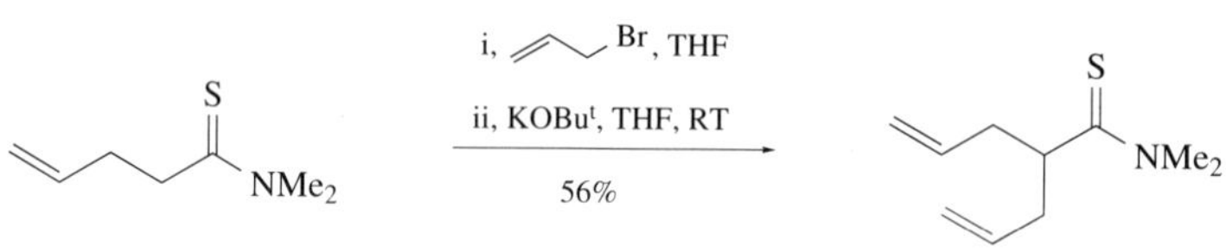

(251)

Deprotonation of *N*,*N*-dimethyl thionoacetamide with LDA followed by addition of α,β-unsaturated ketones provides β-hydroxy γ,δ-unsaturated thionoamides by 1,2 addition (Equation (252)) ⟨85T1289⟩. Oxidation of deprotonated tertiary thionoamides can lead to α,β-unsaturated thionoamides (Equation (253)) ⟨94TL1977⟩. A variety of heterocycles react with amines to produce thionoalkenoamides (Equations (254)–(256)) ⟨68CB1428, 74CB502, 85CB851⟩.

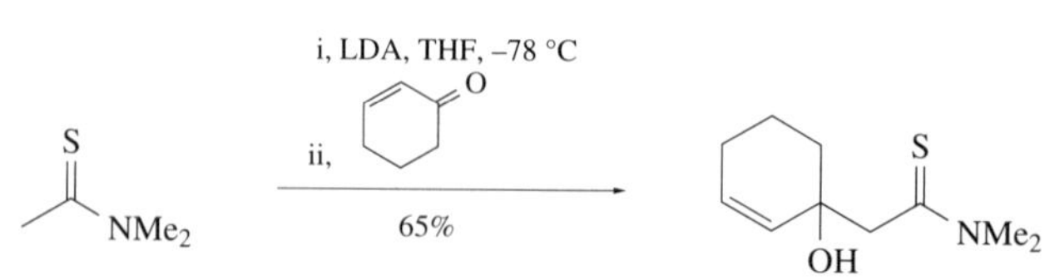

(252)

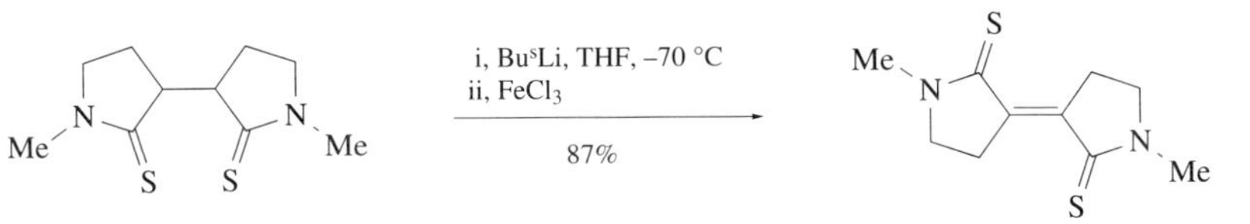

(253)

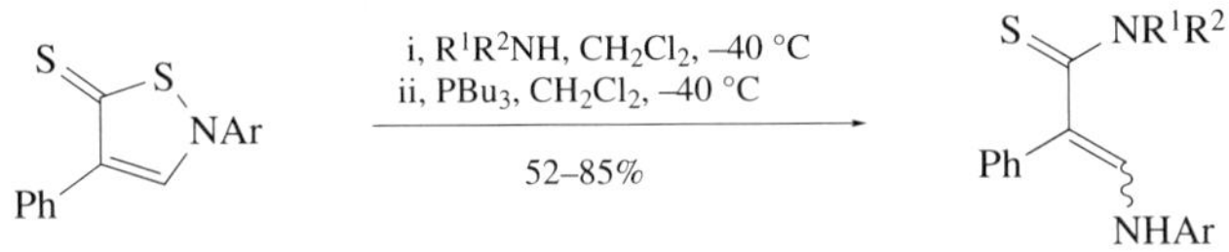

(254)

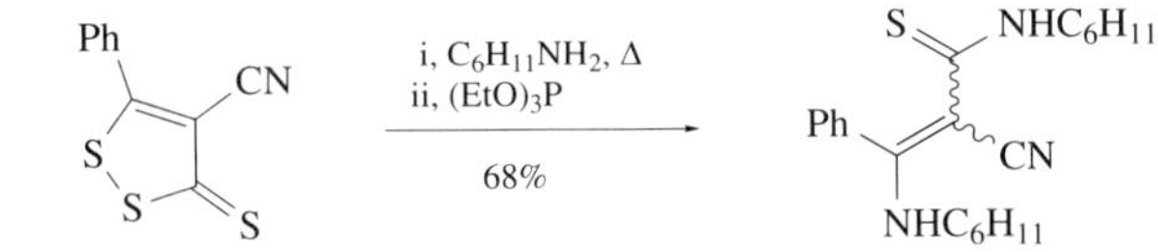

(255)

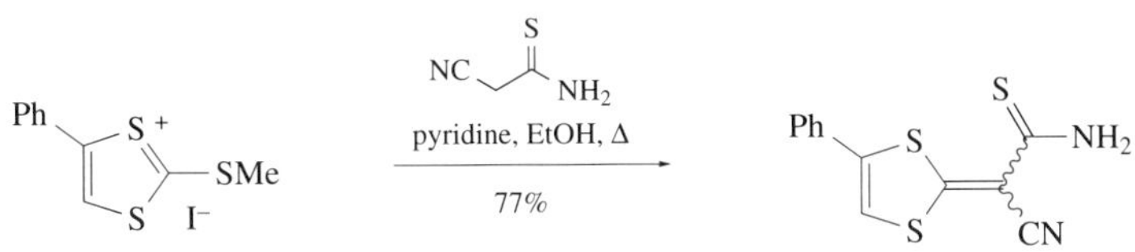

(256)

More classical methods also include elimination of hydrogen chloride from an α,α-dichlorothionoamide (Equation (257)) with an aluminum chloride–triethylamine complex (aluminum chloride alone gives a mixture of the amide and thionoamide ⟨85T3331⟩), and [4 + 2]-cycloaddition of an alkynyl thionoamide to cyclopentadiene (Equation (258)) ⟨91LA903⟩.

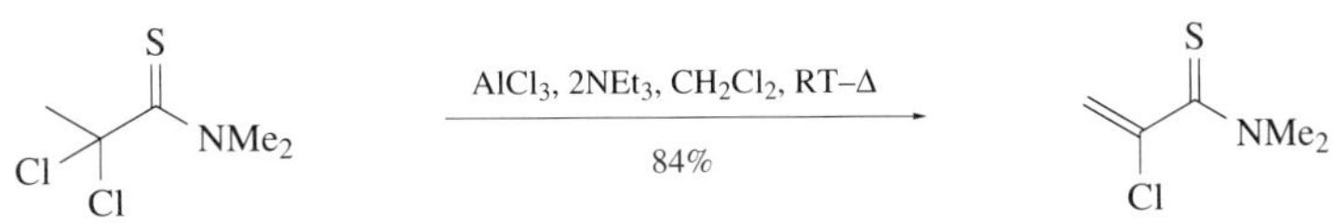

(257)

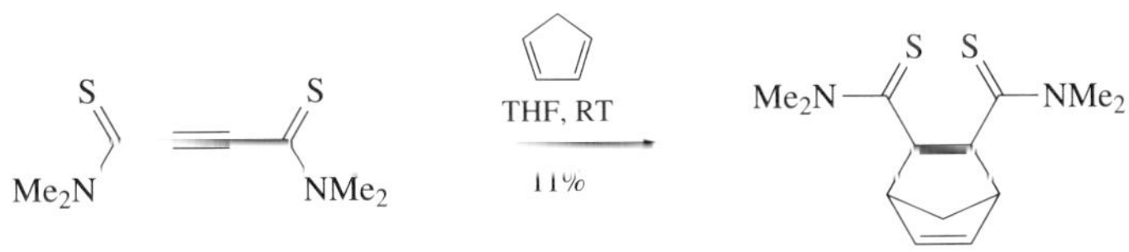

(258)

5.14.5 THIONOALKYNYLOAMIDES

Thionation of tertiary amides of 3-phenyl- and 3 trimethylsilylpropynoic acid with Lawesson's reagent in THF at reflux provides access to the thionoalkynyloamides in good to excellent yields (Equation (259)) ⟨91LA903⟩. As already described in Section 5.14.1.4, phenylacetylene reacts sequentially with sodium and aryl isothiocyanates (Equation (260)) producing thionoalkynyloamides ⟨39JA105⟩.

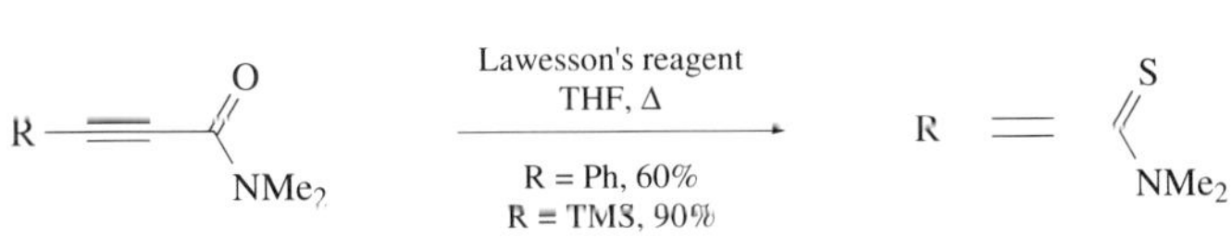

(259)

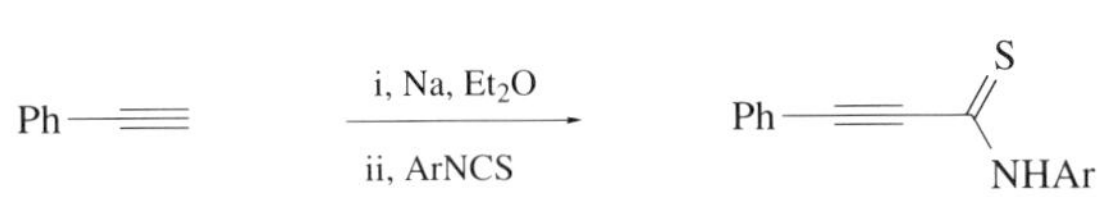

(260)

Phenylacetylene also reacts with dimethylthiocarbamoyl chloride under palladium catalysis producing the desired thionoamides (Equation (261)) ⟨91LA903⟩. Lithiated ⟨85LA2371⟩ and stannylated

⟨78TL1951⟩ ynamines react with aryl isothiocyanates producing ynaminothionoamides (Equations (262) and (263)). Tertiary thionoamides bearing the alkynyl group in the γ,δ-position are available by reaction of secondary amines with thioketenes derived by [3,3]-sigmatropic rearrangement of alkynyl allenyl sulfides (Equation (264)) ⟨72RTC578⟩.

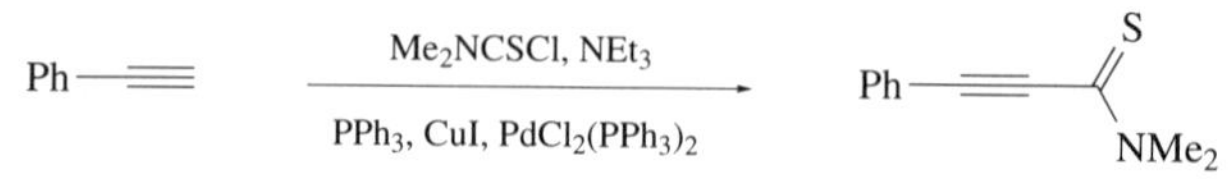

(261)

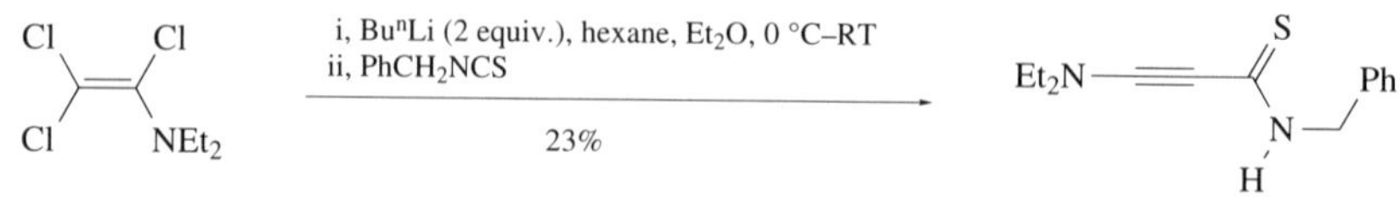

(262)

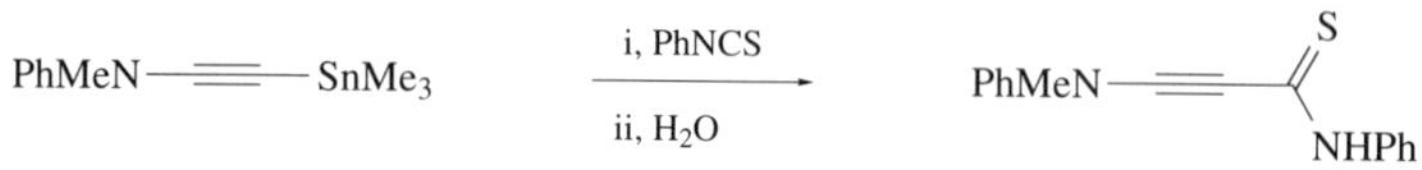

(263)

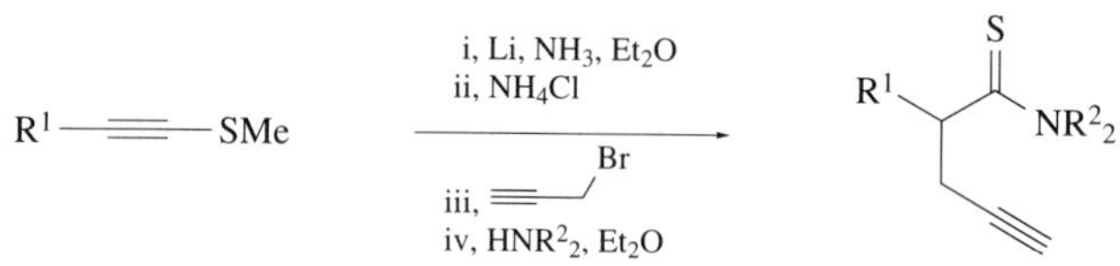

(264)

5.14.6 THIONOARYLAMIDES AND THIONOHETARYLAMIDES

A variety of reagents have been used for the conversion of aromatic amides to thionoamides (Equation (265)) and a few typical examples are shown in Table 31. These reagents have already been discussed at length in Section 5.14.1.1.

(265)

Table 31 Thionation of aromatic and heteroaromatic amides (Equation (265)).

R[1]	*R*[2]	*R*[3]	*Reagent*	*Solvent*	*Yield* (%)	*Ref.*
Ph	H	H	LR[a]	HMPA	92	78BSB229
Ph	H	Me	$(Me_2N)_2PSCl$	None	86	77BSB693
Ph	H	CH_2Ts	P_4S_{10}	DME	97	76TL143
Ph	H	$4NO_2$—C_6H_4	P_4S_{10}	Xylene	70	37HCA691
$2NH_2$—C_6H_4	H	Me	P_4S_{10}	Pyridine	48–78	71PHA271
$2NH_2$—C_6H_4	H	Ph	P_4S_{10}	Pyridine	65	75LA275
Ph	H	Ph	$(Me_2N)_2PSCl$	None	70	77BSB693
2OH—C_6H_4	H	Ph	LR	HMPA	51	78BSB299
Ph	H	CH=CMe_2	LR	PhMe	94	84JOC714
Ph	Me	Me	LR	PhMe	89	80TL4061
Ph	Me	Me	$(Me_2N)_2PSCl$	None	90	77BSB693
Ph	Me	Me	$(EtO)_2PSSH$	None	96	77BSB321
Ph	Me	Me	$(EtAlS)_n$	None	68	75JOM(92)139
2-Furyl	H	H	P_4S_{10}	Pyridine	84	57CCA319
2-Furyl	H	Et	P_4S_{10}	Pyridine	84	57CCA319
2-Furyl	Me	Ph	P_4S_{10}	Pyridine	74	57CCA319
4-Pyridyl	H	Pr^i	P_4S_{10}	None	44	55JA4062
3-Pyridyl	Et	Et	P_4S_{10}/Na_2S	None	64	51JOC131

[a] LR = Lawesson's reagent.

Carboxylic acids have been converted directly into aromatic thioamides (Equation (266)) by the use of phosphorus pentasulfide and morpholine in pyridine ⟨81JCR(S)58⟩. In similar vein the reaction of acid chlorides with Lawesson's reagent followed by an amine (Equation (267)) yields thionoamides ⟨87PS(32)51⟩. The addition of hydrogen sulfide to nitriles is a successful route. Some typical substrates include nitriles of nitroanilines ⟨75LA275⟩, anilines ⟨75LA275⟩, furans ⟨1892CB1311⟩, pyrazoles ⟨70JHC863⟩, pyridines ⟨74JHC397⟩ and pyridinium salts ⟨74JHC397⟩, Equations (268)–(273). A more modern alternative to the use of hydrogen sulfide involves *O*,*O*-diethyldithiophosphoric acid (Equation (274)) ⟨92ZOB71⟩. Early workers found that the reaction of hydrogen sulfide with amidines (Equation (275)) to be successful ⟨1877CB1238⟩, but the reader is directed to earlier reviews if hydrogen sulfide is to be the reagent used.

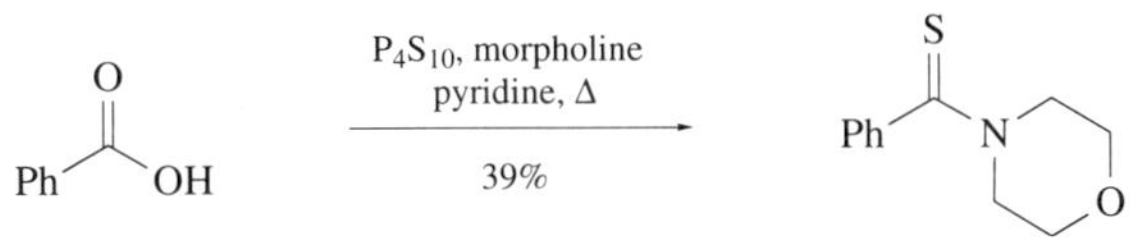

(266)

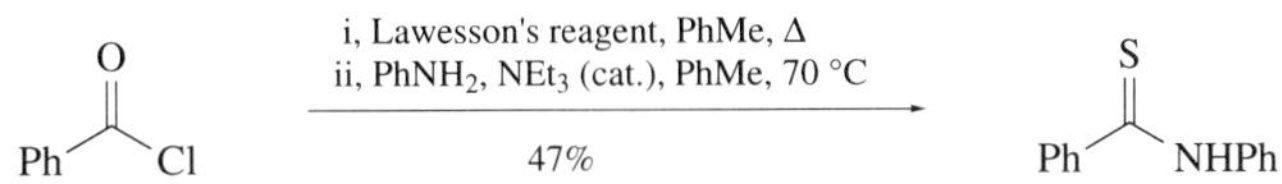

(267)

(268)

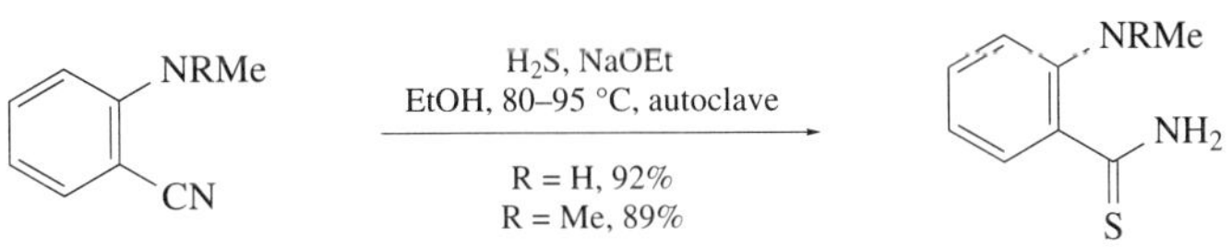

(269)

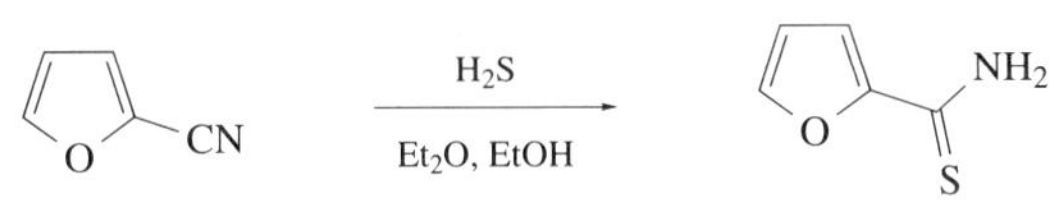

(270)

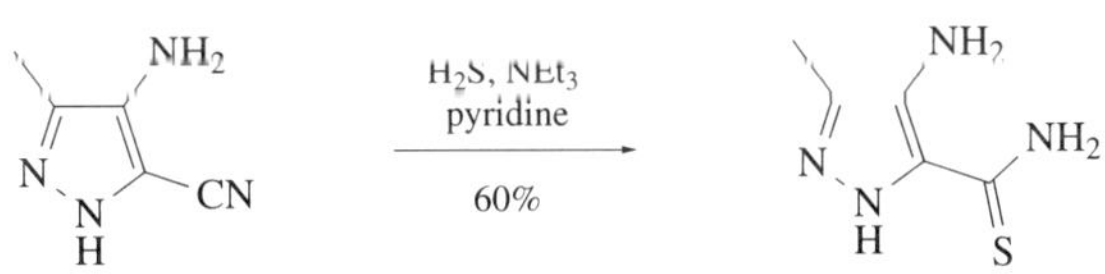

(271)

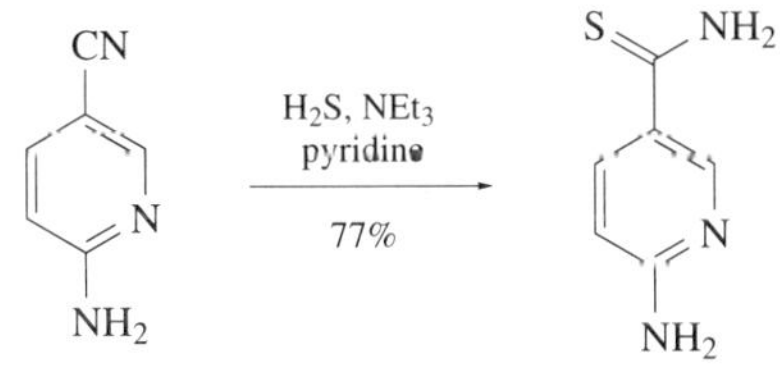

(272)

(273)

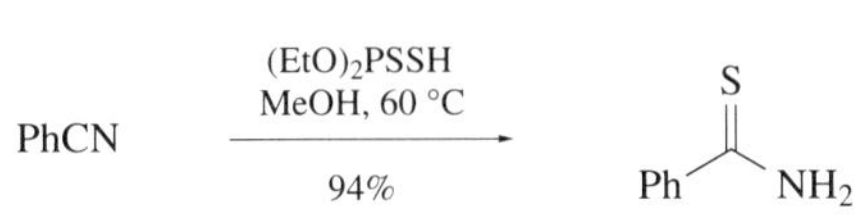

(274)

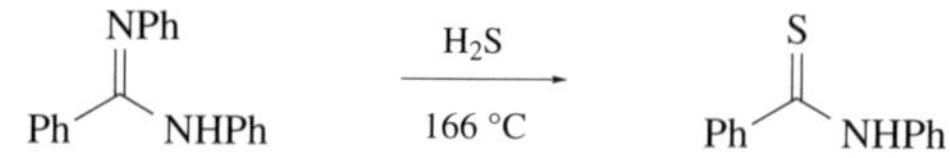

(275)

Thioacylation of amines by a variety of species including carboxymethyl dithiocarboxylates ⟨62LA(660)60⟩, dithiocarboxylic acid salts ⟨49JOC962⟩, thiobenzoylazoles ⟨70LA(739)201⟩, dithioacid anhydrides ⟨82LA1229⟩ and related species ⟨72BCJ3653⟩ are successful approaches. Typical examples are depicted in Equations (276)–(280).

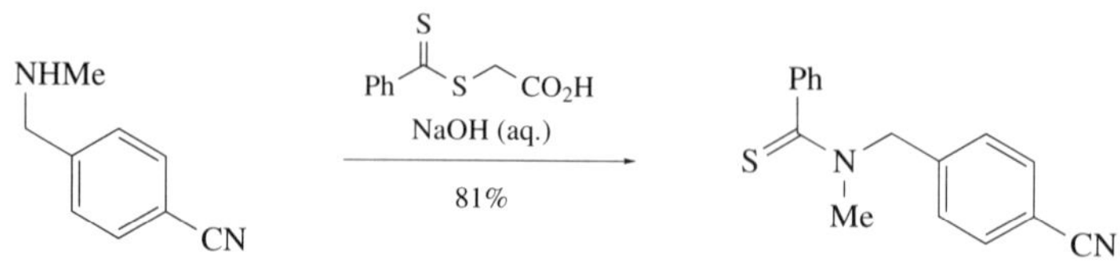

(276)

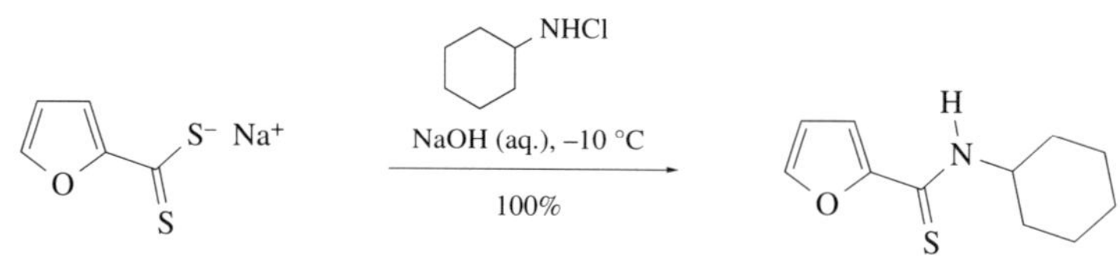

(277)

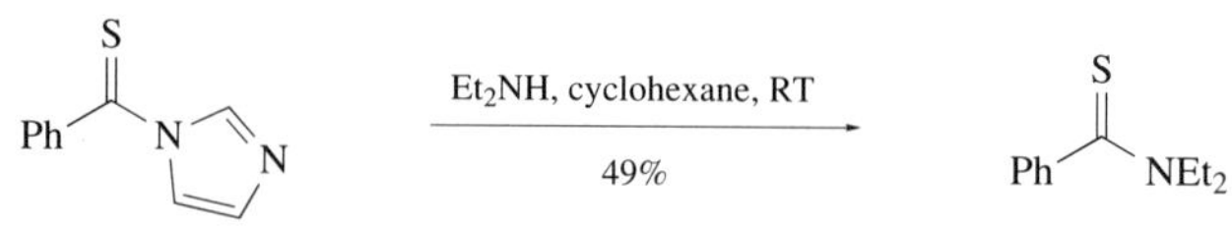

(278)

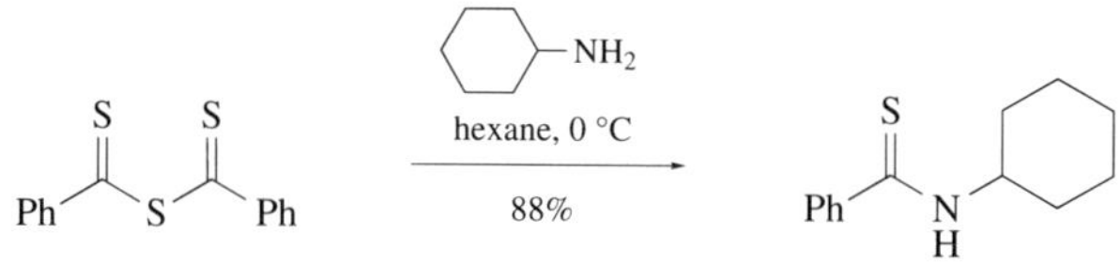

(279)

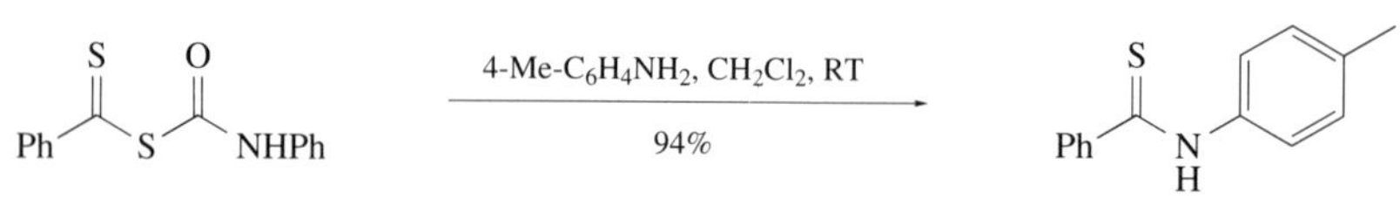

(280)

Thiocarbamoylation by the addition of activated aromatics ⟨85AJC447⟩, aryl lithiums ⟨50JA3045⟩ or Grignard reagents ⟨61JOC1132⟩ to isothiocyanates is popular (Equations (281)–(283)). More references are contained in Section 5.14.1.4. An interesting and useful indole thionoamide synthesis involves treatment of a tolyl isonitrile sequentially with LDA, an isothiocyanate and then more LDA (Equation (284)) ⟨79TL1039⟩. Reaction of aromatics with ethoxycarbonyl isothiocyanate and aluminum chloride proceeds directly to the primary aromatic thionoamide if the aromatic is also the reaction solvent (Equation (285)) ⟨76JOC962⟩. A related approach to tertiary aromatic thionoamides uses aluminum chloride and *N*,*N*-dimethylthiocarbamoyl chloride (Equation (286)) ⟨68CB3517⟩.

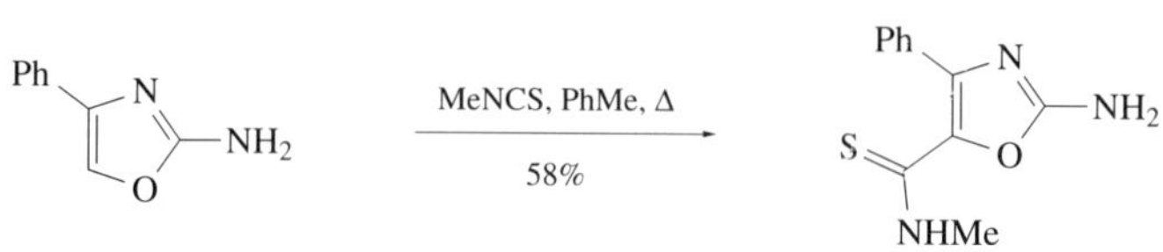

(281)

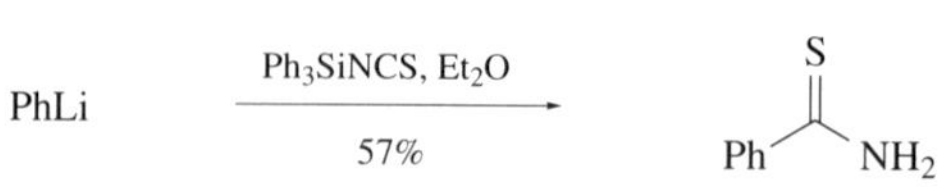

(282)

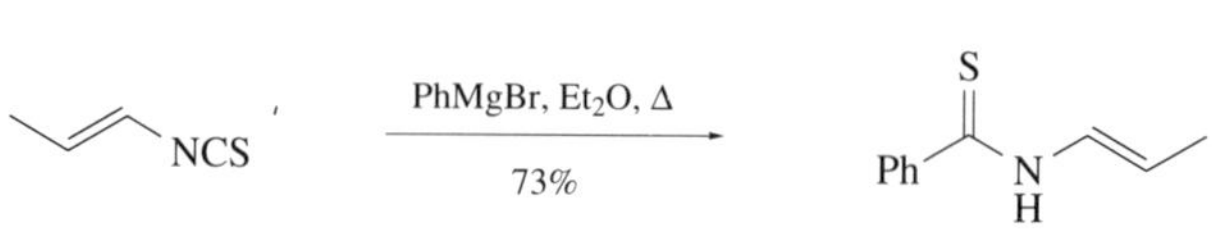

(283)

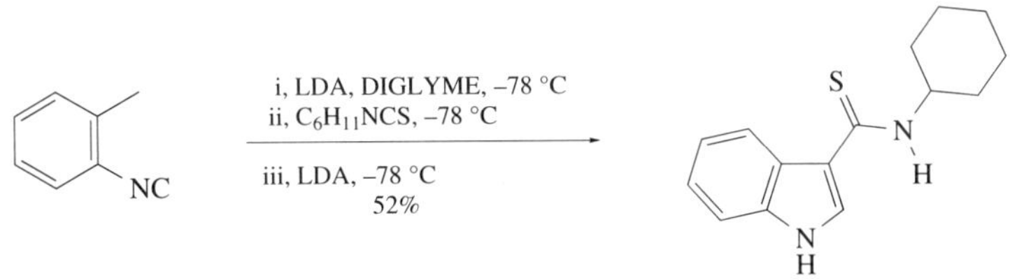

(284)

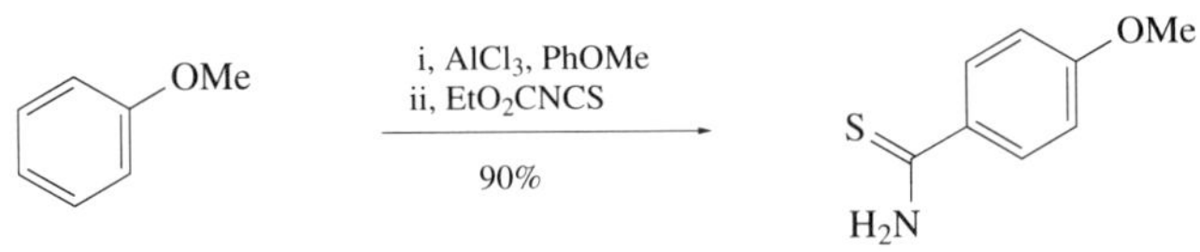

(285)

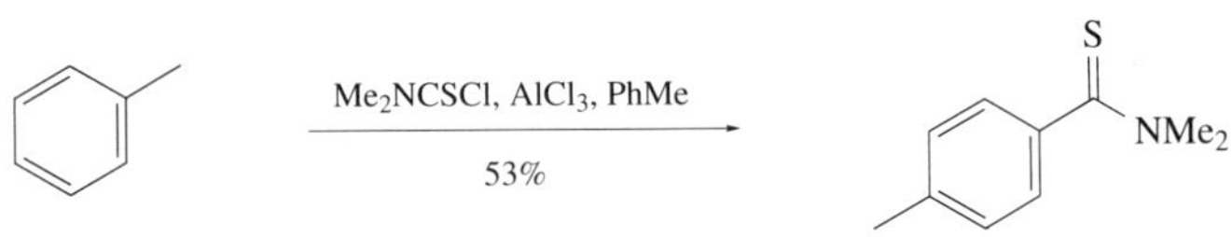

(286)

Aminosulfuration of various species such as benzaldehydes ⟨55JA4062, 83S730, 89S202, 91OPP435⟩, acetophenones ⟨75ACS(B)538⟩, aromatic α-ketoacids ⟨88T7271⟩, esters ⟨75ACS(B)604⟩, imines ⟨75ACS(B)604, 88T7271⟩, benzylic ethers ⟨75ACS(B)538⟩, amines ⟨48JA868, 75ACS(B)538⟩, nitriles ⟨89S202⟩, and mono-, di- and trihalides ⟨61JCS1194, 89IJC(B)673, 89S202⟩ with sulfur and a variety of amines is well described in the literature. A variety of these is shown in Equations (287)–(298). These examples only represent a sample of what could potentially be achieved by variants of the Willgerodt–Kindler reaction.

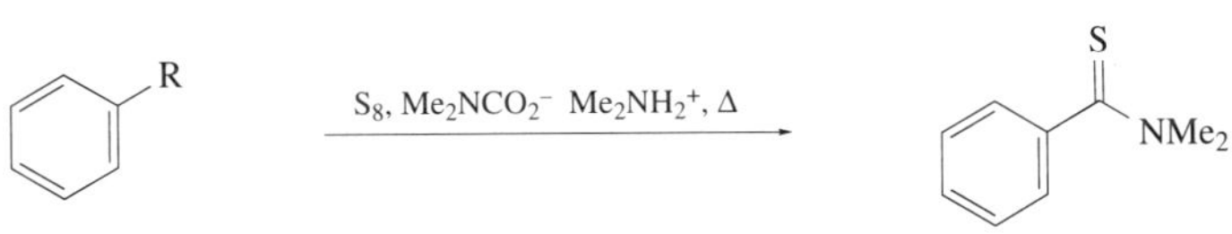

(287)

R = CHO, CH_2Cl, $CHCl_2$, 40–80%

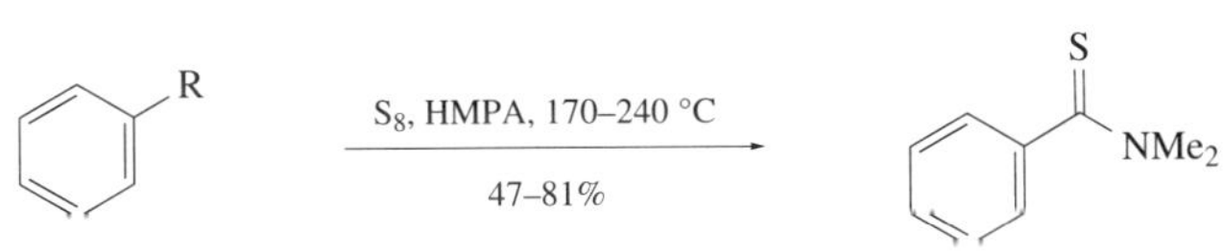

(288)

R = Ac, CH_2OEt, CH_2NMe_2

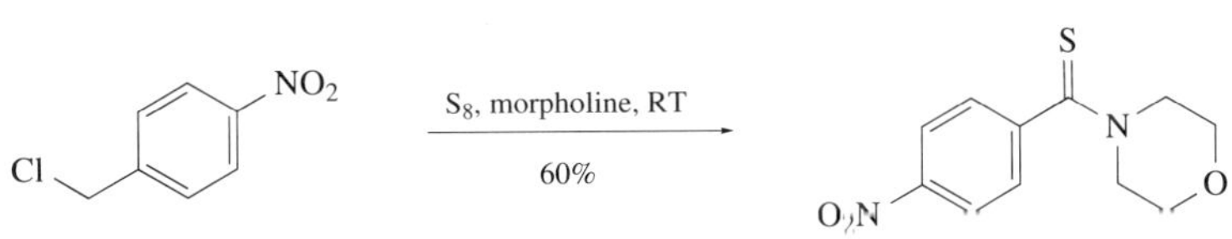

(289)

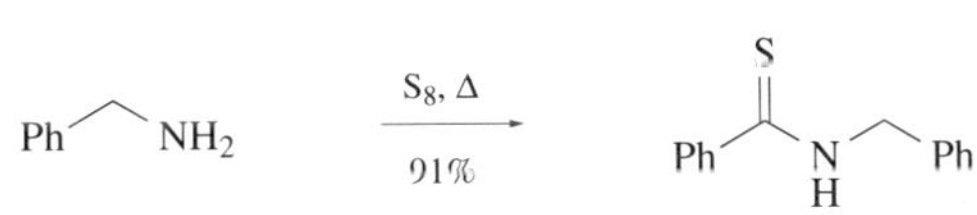

(290)

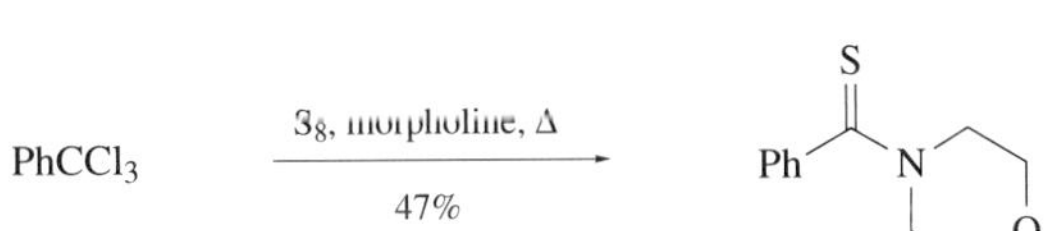

(291)

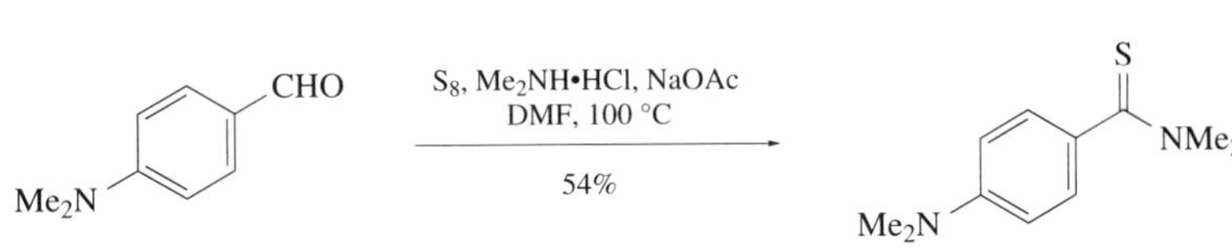

(292)

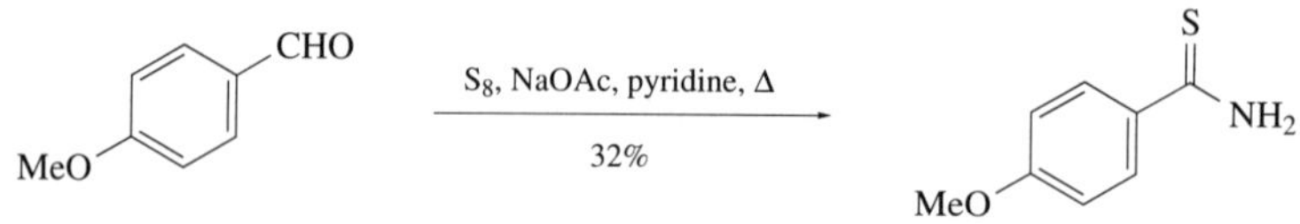

(293)

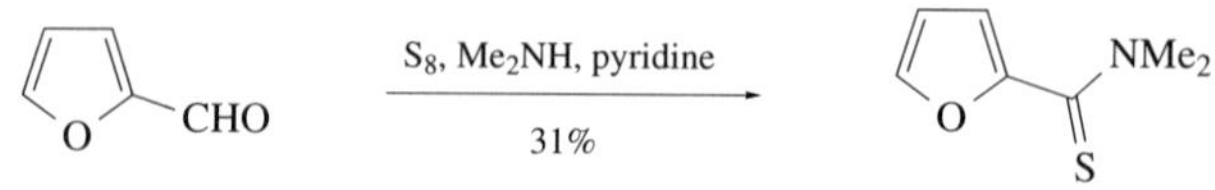

(294)

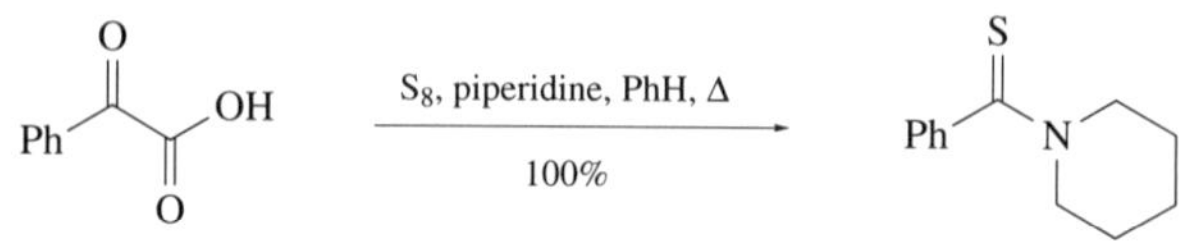

(295)

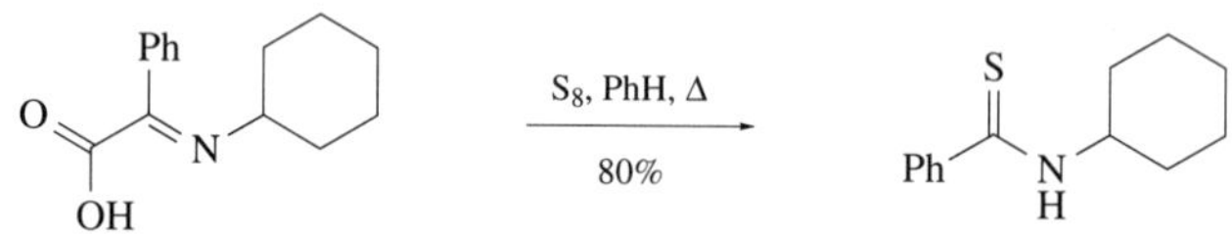

(296)

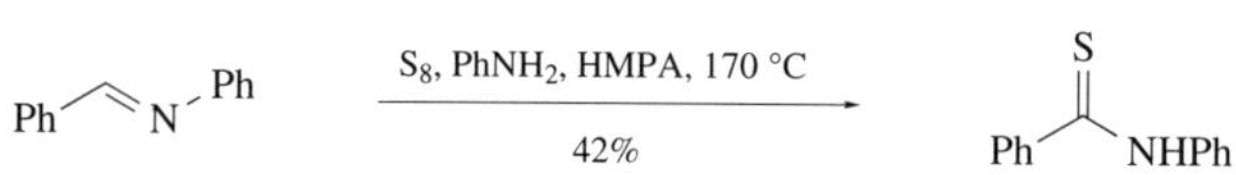

(297)

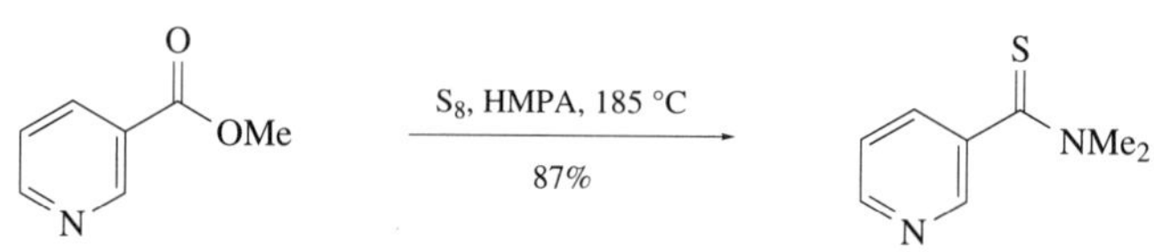

(298)

Other routes have involved phosphonate anions ⟨92CL131⟩, phosphorus ylides ⟨91BCJ2402⟩ and iminophosphoranes ⟨89TL2847⟩ (Equations (299) to (301)). The rearrangement of thiadiazole aldehydes to triazole thionoamides (Equation (302)) is well described ⟨90BSB281, 91JCS(P1)607, 93JCS(P1)1719⟩. Cycloaddition of nitrones to thiocarbonyldiazoles thermally ⟨83TL4927, 85CJC951⟩ or of unsaturated azides with heterocyclic thiones photochemically ⟨84HCA534⟩ also allows access to various aromatic thionoamides (Equations (303) and (304)). Cleavage of oxazolines with hydrogen sulfide ⟨48JCS1919⟩ or of dithiazolium salts with nucleophiles ⟨85ZC364⟩ are other potentially useful procedures (Equations (305) and (306)).

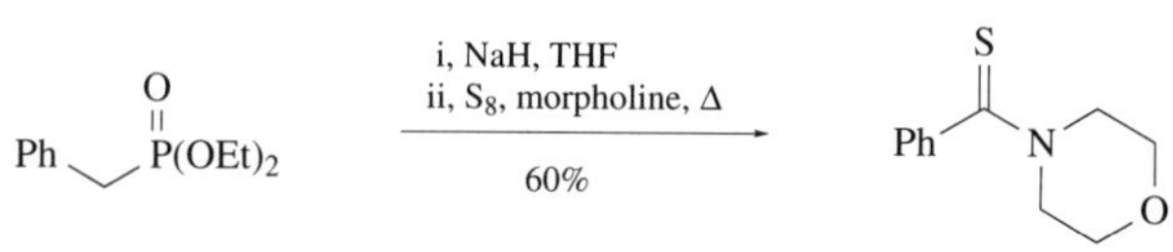

(299)

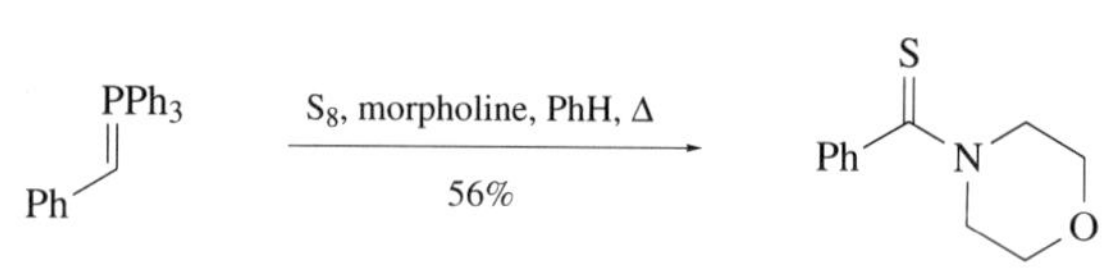

(300)

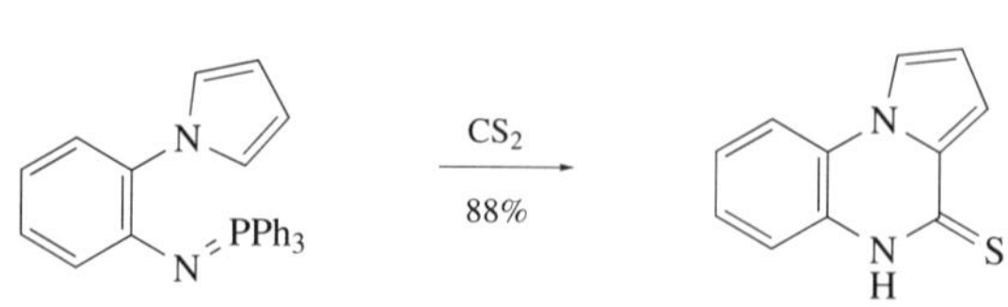

(301)

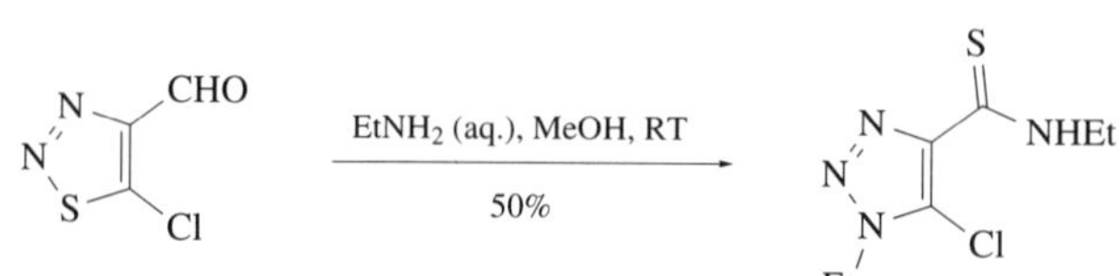

(302)

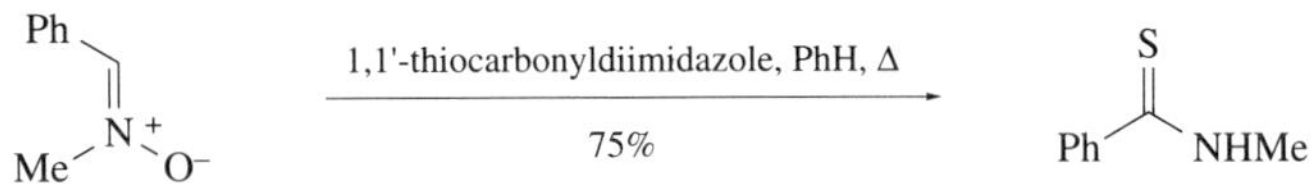

(303)

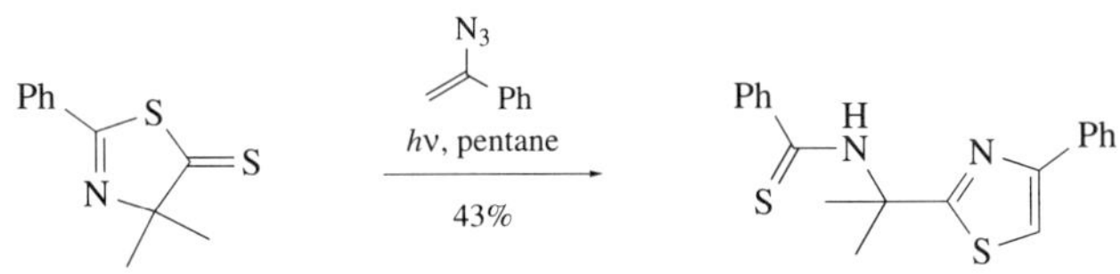

(304)

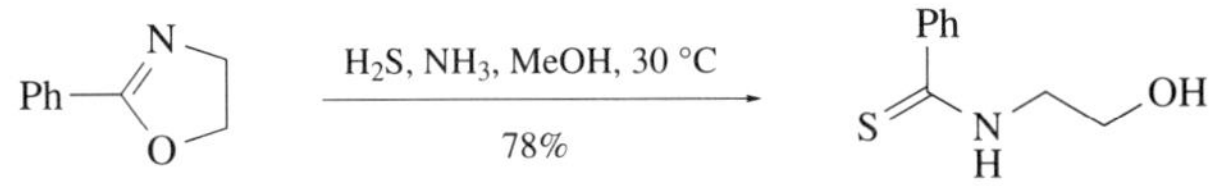

(305)

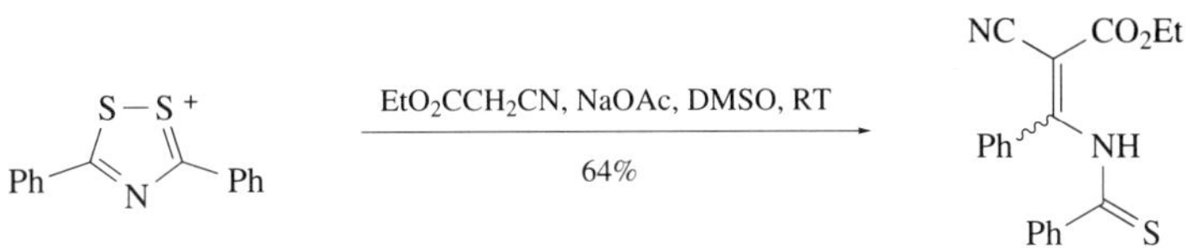

(306)

Finally, reduction of aromatic thionoamide *S*-oxides (Equation (307)) can yield the parent thionoamide ⟨62LA(660)60⟩ whilst treatment of selenoamides with sulfur (Equation (308)) also produces the thionoamide ⟨77AG(E)725⟩.

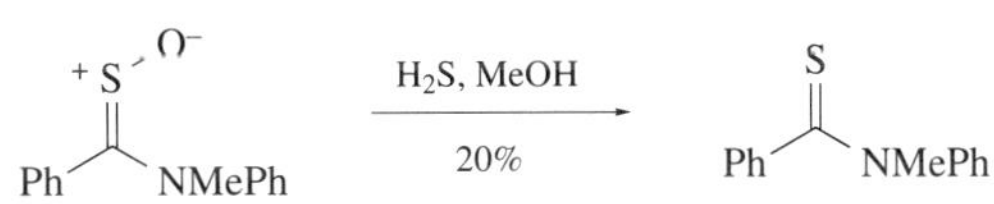

(307)

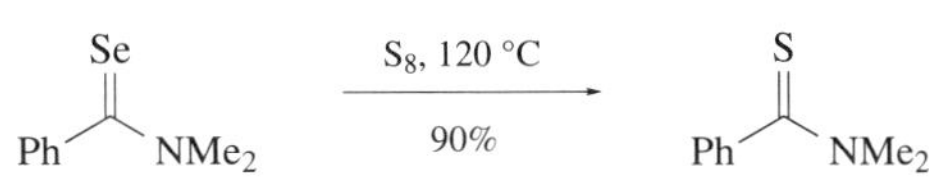

(308)

5.14.7 THIONOAMIDE *S*-OXIDES AND *S,S*-DIOXIDES

A number of methods exist for the preparation of thionoamide *S*-oxides, the most popular being direct oxidation of a thionoamide with hydrogen peroxide (Equation (309)) usually with a buffer present ⟨60CB1511, 60LA(633)35, 62LA(660)60, 62LA(660)74, 65LA(681)64, 65AG(E)1095, 68LA(712)46, 68LA(712)58, 71LA(752)115, 76LA1584, 79CB1000, 86PS(26)169⟩. A wide variety of thionoamides undergo the transformation, primary, secondary, tertiary, aliphatic, aromatic, benzylic, α-oxo, for example. Representative examples are given in Table 32. In many cases the compounds are stable crystalline solids. An alternative preparation is from oxidation of iminomethyldisulfides with perbenzoic acid (Equation (310)) ⟨69LA(727)50⟩, although yields are poor. Reaction of *N*-sulfinylamines and thioketone *S*-oxides with ynamines (Equation (311)) also reportedly yields thionoamide *S*-oxides ⟨87LA189⟩, presumably via an initial [3 + 2]-cycloaddition followed by rearrangement. Treatment of some thionoamide *S*-oxides with diazomethane (Equation (312)) yields the *N*-methylated thionoamide *S*-oxides ⟨62LA(660)60⟩.

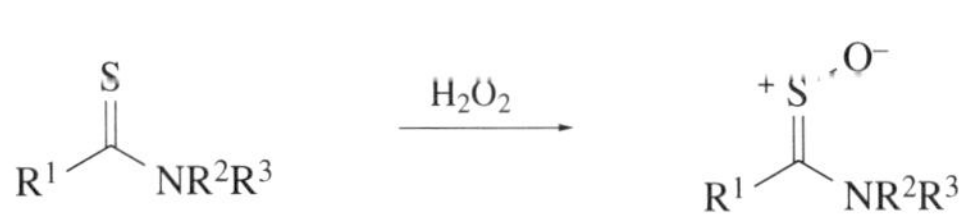

(309)

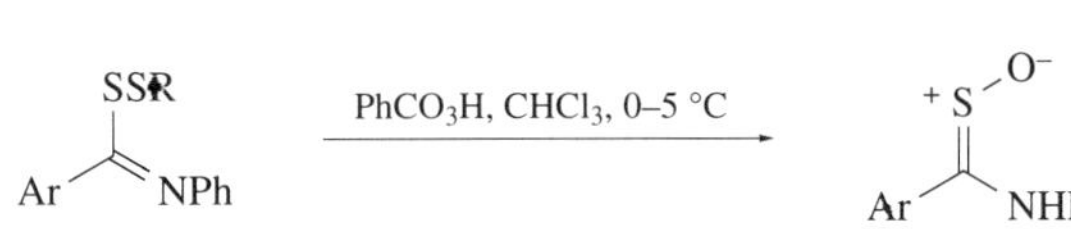

(310)

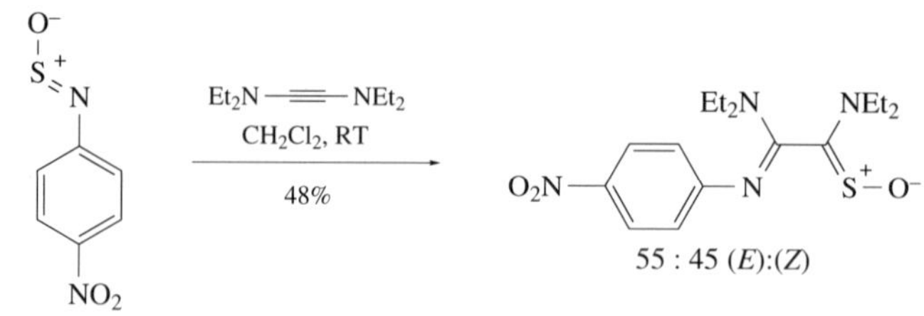

(311)

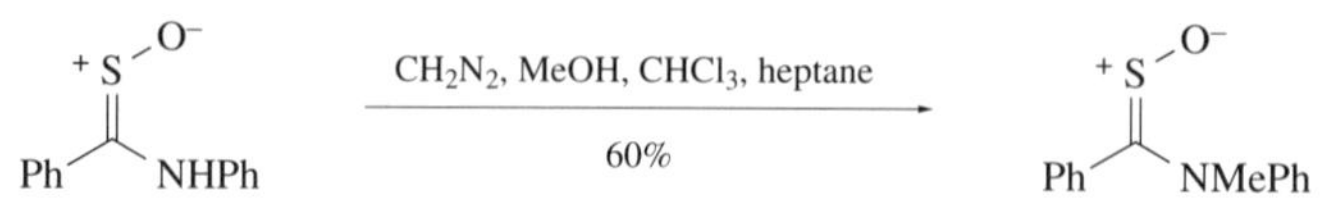

(312)

Table 32 Oxidation of thionoamides (Equation (309)).

R[1]	*R*[2]	*R*[3]	*Solvent*	*Temp.* (°C)	*Yield* (%)	*Ref.*
H	H	H	H_2O	−15	28	60CB1511
H	H	Pr	MeOH	5–20	2	68LA(712)46
H	H	Pr[i]	MeOH	5–20	11	68LA(712)46
H	H	Bu[t]	MeOH	5–20	22	68LA(712)46
H	H	Ph	MeOH	5–20	40	68LA(712)46
H	H	$PhCH_2$	MeOH	5–20	39	68LA(712)46
Me	H	H	H_2O	−15	52	60CB1511
				−30	12–20	60LA(633)35
Bu[t]	H	H	MeCN	0–5	58	86PS(26)169
Bu[t]	Et	Et	AcOH	25	22	86PS(26)169
$CONH_2$	H	H	DMF	40	66	62LA(660)74
Ph	H	H	MeOH	[a]	54	60CB1511
Ph	H	Me	MeOH	55	20	62LA(660)60
Ph	H	2Me-C_6H_4	MeOH, CH_2Cl_2	40	38	71LA(752)115
Ph	H	$PhCH_2$	MeOH	45	83	62LA(660)60
Ph	H	4CN-$C_6H_4CH_2$	MeOH	55	74	62LA(660)60
Ph	Me	Ph	AcOH	30–50	58	76LA1584
Ph	Me	Me	AcOH	30–50	32	76LA1584
Ph	Ph	Ph	AcOH	30–50	68	76LA1584
$PhCH_2$	H	H	MeOH	[a]	55	60CB1511
$PhCH_2$	$PhCH_2$	$PhCH_2$	AcOH	30–50	34	76LA1584
3-Py[b]	H	H	Py	30	41	60CB1511

[a] Temperature not given. [b] Py = pyridine.

Although the intermediacy of thionoamide *S*,*S*-dioxides has been claimed by a number of authors (mainly by product analysis suggesting a mechanism occurring via such species) ⟨75LA305, 76LA1584, 86PS(26)169⟩, to date no one has managed to isolate a member of this class of compounds. Jensen, in a review on thionoamide and thionohydrazide constitution ⟨81MI 514-01⟩, claims it has not been possible to prepare thionoamide *S*,*S*-dioxides by hydrogen peroxide oxidation of thionoamides.

5.14.8 SELENOAMIDES

Unsurprisingly, methods for the preparation of selenoamides parallel those available for thionoamides ⟨79COC(3)522, B-86MI 514-01, B-87MI 514-01, 91COS(6)476⟩. Primary, secondary and tertiary selenoamides are all available from treatment of the corresponding carboxamide with phosphorus pentaselenide (Equation (313)) ⟨63BSB304, 66ACS597, 76CB956, 84ZC99⟩. Whilst yields are typically poor (Table 33), they can often be improved dramatically by the addition of barium carbonate ⟨76IJS519⟩.

(313)

Table 33 Reactions of amides with P_2Se_5 (Equation (313)).

R[1]	*R*[2]	*R*[3]	*Reagent*	*Solvent*	*Yield* (%)	*Ref.*
H	H	H	P_2Se_5	None	34	84ZC99
H	H	Me	P_2Se_5	Not given	2	76CB956
H	Me	Me	P_2Se_5	PhH	12	63BSB304
Me	Me	Me	P_2Se_5	PhH	23	63BSB304
Me	Me	Me	P_2Se_5	PhH	3	76IJS519
Me	Me	Me	P_2Se_5/Ba_2CO_3	Xylene	61	76IJS519
—$[CH_2]_3$—		Me	P_2Se_5/Ba_2CO_3	Xylene	49	76IJS519
—$[CH_2]_4$—		Me	P_2Se_5/Ba_2CO_3	Xylene	48	76IJS519
Ph	Me	Me	P_2Se_5	PhH	5	66ACS597

Hydrogen selenide (generated by a variety of means *in situ*) or sodium hydrogen selenide convert nitriles to selenoamides (Equation (314)) ⟨23LA(431)187, 74JHC397, 78S668, 83JMC445, 85JOC384, 93S870, 94SC1761⟩. Yields are often high (Table 34). Although reaction of 3-cyanopyridine with hydrogen selenide gives the required product, the corresponding methiodide fails to react ⟨74JHC397⟩. Note, however, that reaction of 2-bromopyridine methiodide with sodium hydrogen selenide reportedly provides the *N*-methylpyridineselenone in 58% yield by displacement of the bromine ⟨62JOC3671⟩.

(314)

Table 34 Conversion of nitriles to selenoamides (Equation (314)).

R[1]	*R*[2]	*R*[3]	*Method*	*Yield* (%)	*Ref.*
Me	H	H	E	21	90CL1403
Pr^n	H	H	E	79	90CL1403
Bu^n	H	H	E	50	90CL1403
Ph	H	H	A	76	85JOC384
Ph	H	H	C	92	93S870
Ph	H	H	D	92	78S668
Ph	H	H	E	66	90CL1403
Ph	Pr^i	H	B	41	85BCJ1448
Ph	Bu^t	H	B	0	85BCJ1448
Ph	$PhCH_2$	H	B	87	85BCJ1448
Ph	Me	Me	B	46	85BCJ1448
Ph	—$[CH_2]_5$—		B	45	85BCJ1448
2Me—C_6H_4	H	H	A	22	85JOC384
4Me—C_6H_4	H	H	A	100	85JOC384
4Me—C_6H_4	H	H	C	91	93S870
4Me—C_6H_4	H	H	D	85	78S668
4MeO—C_6H_4	H	H	A	91	85JOC384
4MeO—C_6H_4	H	H	C	74	93S870
4MeO—C_6H_4	H	H	D	70	78S668
$4Me_2N$—C_6H_4	H	H	A	82	85JOC384
4Cl—C_6H_4	H	H	A	99	85JOC384
4Cl—C_6H_4	H	H	C	93	93S870
4Cl—C_6H_4	H	H	D	82	78S668
1-Naphthyl	H	H	A	21	85JOC384
2-Pyridyl	H	H	A	74	85JOC384
2-Pyridyl	H	H	D	72	78S668
3-Pyridyl	H	H	C	67	93S870
2-Thienyl	H	H	D	95	78S668

Conditions: Method A: Se, H_2O, NEt_3, CO, THF, 100 °C, autoclave. Method B: i, Se, H_2O, NEt_3, CO, THF, 100 °C, autoclave, ii, R^2R^3NH. Method C: Se, $NaBH_4$, EtOH, pyridine, aq. HCl, Δ. Method D: Al_2Se_3, NEt_3, pyridine, H_2O, Δ. Method E: $(TMS)_2Se$, $BF_3 \cdot Et_2O$, CH_2Cl_2, $CHCl_3$ or MeCN, Δ.

A combination of bis(trimethylsilyl)selenide and boron trifluoride etherate accomplishes the same transformation as hydrogen selenide for a number of nitrile substrates ⟨90CL1403⟩. For benzonitrile, in the absence of boron trifluoride the yield of the selenoamide is zero. Interestingly, treatment of aromatic or aliphatic nitriles with hydrogen selenide generated *in situ* followed by addition of a primary or secondary amine results in transamination generating secondary and tertiary selenoamides, respectively ⟨85BCJ1448⟩. Reaction of hydrogen selenide with (hetero)aryl formamidines ⟨81S299⟩ or with dimethylformamide azine ⟨70JOC806⟩ provides access to selenoformamides (Equations (315) and (316)). Addition of sodium hydrogen selenide to imines ⟨75CC617⟩ or imminium salts ⟨71ZC60⟩ substituted with leaving groups allows the preparation of various secondary and tertiary selenoamides (Equations (317) and (318)).

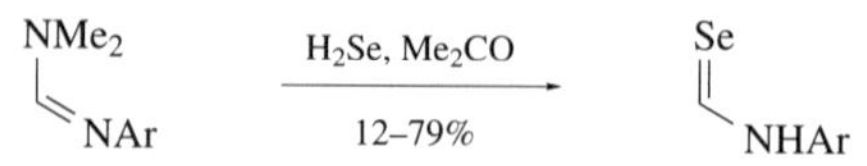

(315)

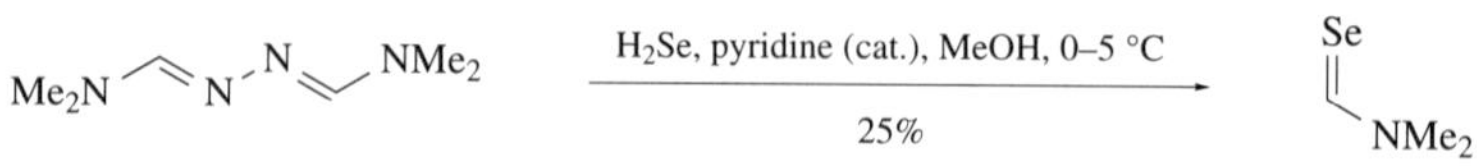

(316)

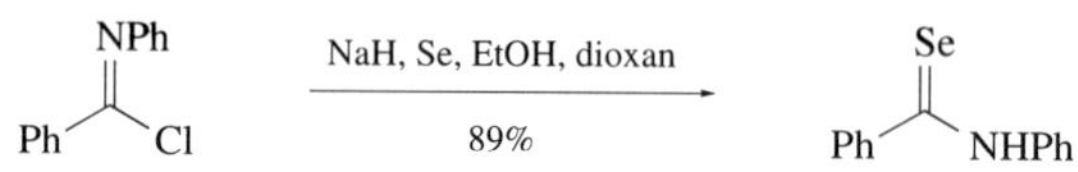

(317)

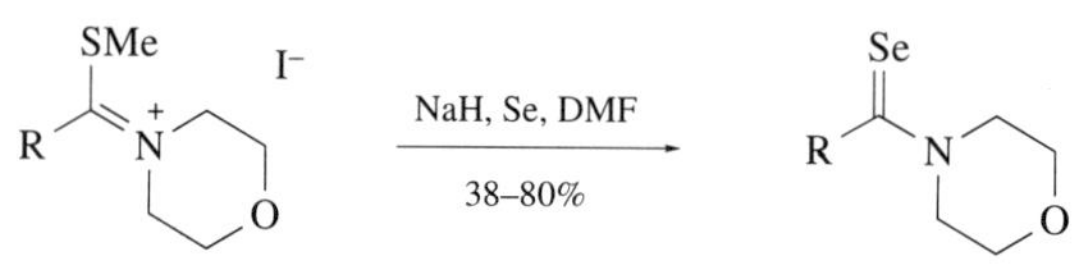

(318)

Procedures for the preparation of aromatic selenoamides via the imidoyl or iminium chloride are reportedly superior to treatment of the amide with phosphorus pentaselenide ⟨76IJS519⟩. Selenoacylations of amines by selenoesters ⟨77JOC2645⟩, selenothioesters ⟨93JA3000⟩, selenoalkynyl carboxylates ⟨90BCJ835⟩, dibenzyltriselenocarbonate ⟨74S501⟩ and selenoketenes or their precursors ⟨77S888, 79RTC55, 80TL4251⟩ have all been described. Selenoacylation with selenoesters (Equation (319)) is poor for primary amines where significant side reactions can occur ⟨77JOC2645⟩. Aromatic amines require treatment with a Grignard reagent (Equation (320)) in order to make them sufficiently reactive to undergo selenoacylation by selenoesters ⟨77JOC2645⟩.

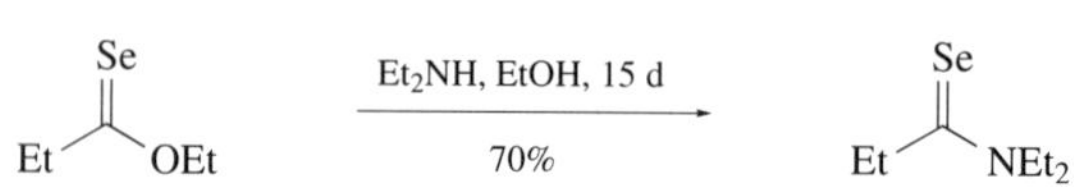

(319)

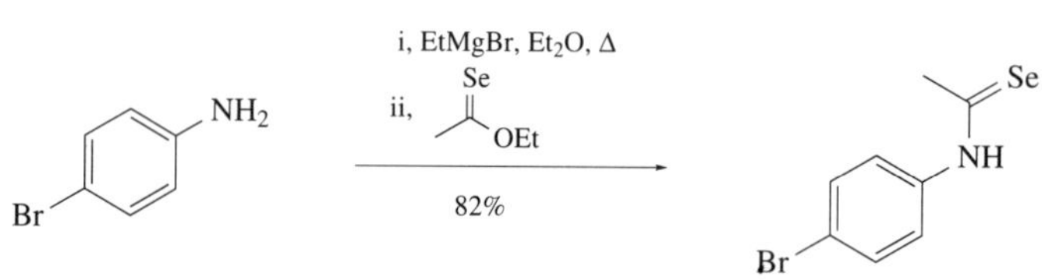

(320)

Selenothioesters react with primary and secondary amines (Equation (321)) although details are scant ⟨93JA3000⟩. Selenoalkynyl carboxylates react well with primary, secondary and aromatic amines (Equation (322)) ⟨90BCJ835⟩. Dibenzyltriselenocarbonate reacts sequentially with benzyl selenol and primary or secondary amines (Equation (323)) providing selenoformamides in moderate yields ⟨74S501⟩. Selenoketenes generated by a variety of means can react well ⟨77S328, 77S888, 79RTC55, 85JCS(P1)907⟩ (Equations (324) to (327)) or poorly ⟨80TL4251⟩ with secondary amines.

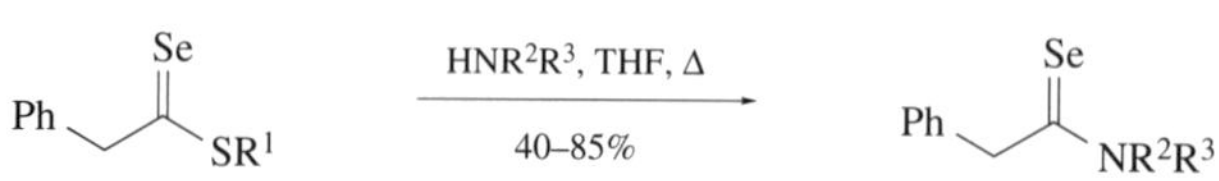

(321)

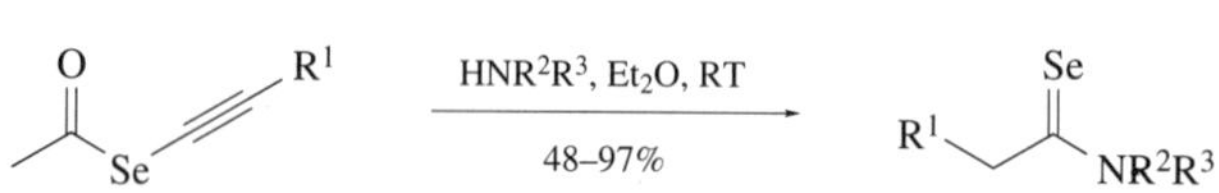

(322)

(323)

i, $PhCH_2SeH$, NEt_3 (cat.), PhMe; ii, Al_2O_3; iii, HNR^1R^2; iv, Al_2O_3

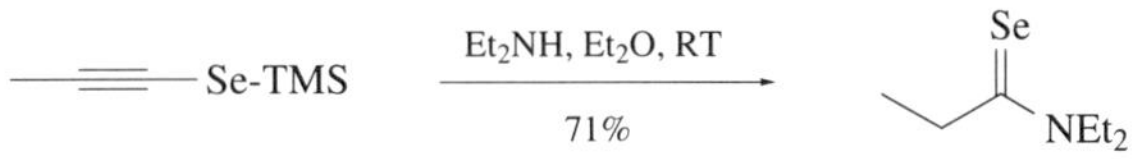

(324)

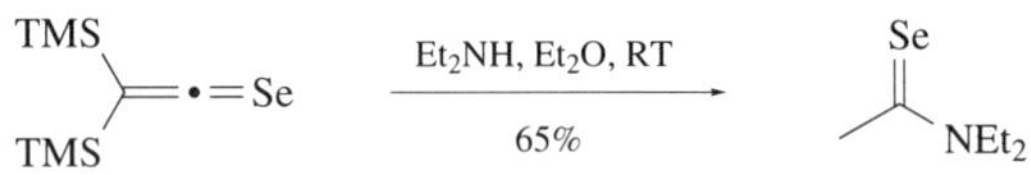

(325)

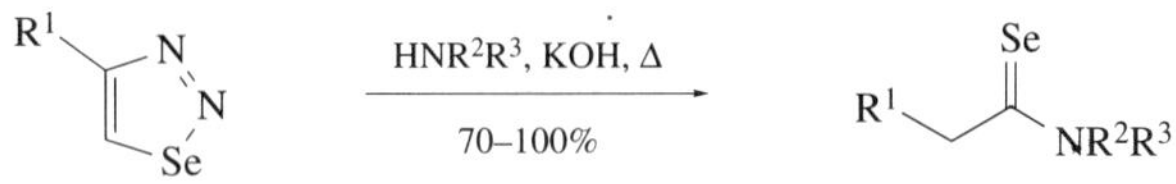

(326)

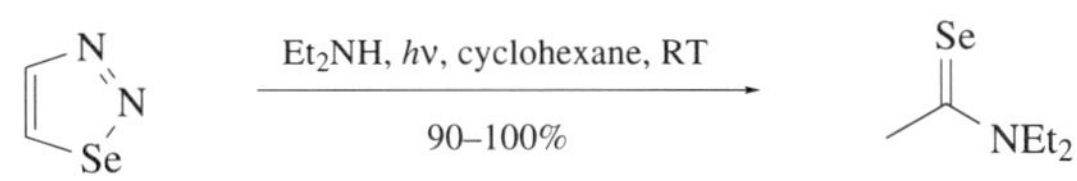

(327)

Reaction of phosphonate anions with elemental selenium (or in much improved yields with an aminoselane derived from morpholine) and a secondary amine provides aromatic tertiary selenoamides (Equation (328)) in low to good yields ⟨92CL131⟩. Ring opening of oxazolium salts ⟨81ZOB1842⟩ with hydrogen selenide (Equation (329)) furnishes selenoamides in good to excellent yields. Addition of selenium ⟨94CL77⟩ or pentacarbonyl chromium or tungsten selenoaldehyde or selenoketone complexes ⟨87CC559, 87JOM(332)105, 88CB2095⟩ to ynamines allows the preparation of diselenooxamides and unsaturated selenoamides, respectively (Equations (330) and (331)).

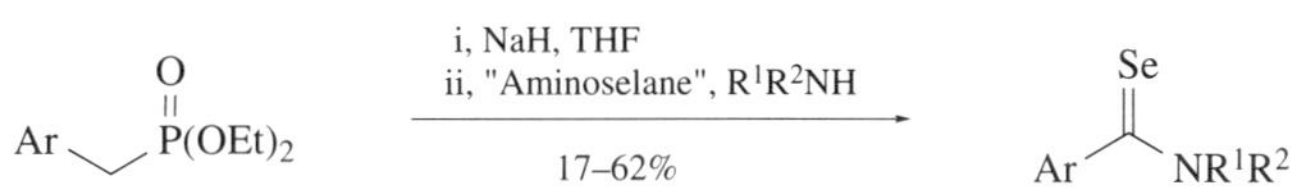

(328)

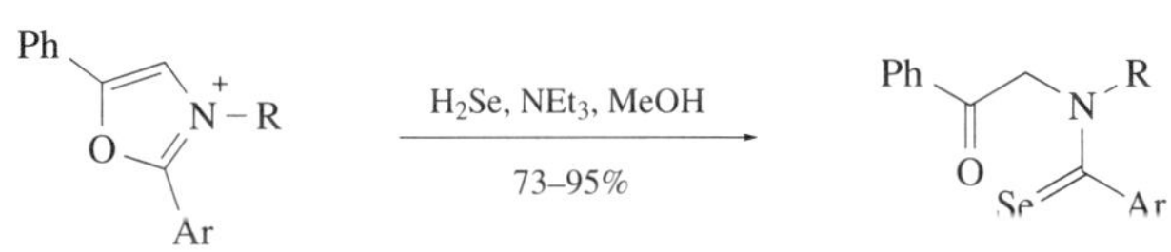

(329)

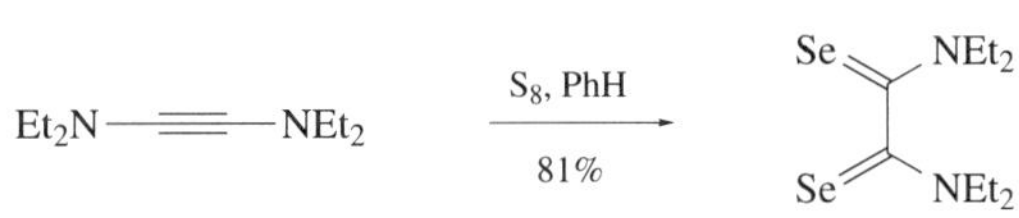

(330)

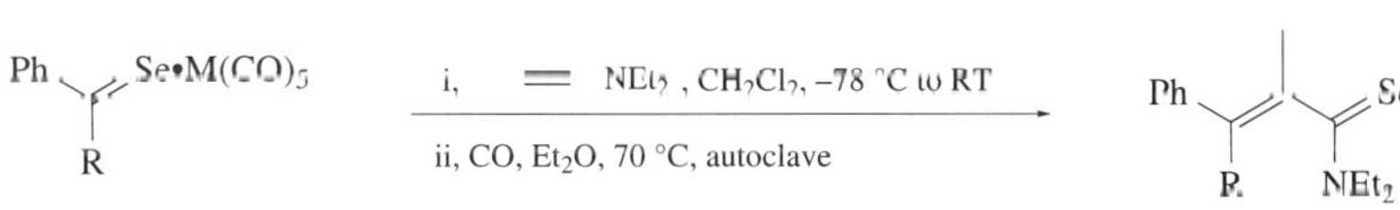

(331)

M = Chromium or tungsten

5.14.9 TELLUROAMIDES

There is scant literature on the preparation of telluroamides. The one reaction resulting in an isolable, identifiable compound is depicted in Equation (332) ⟨79CC1102⟩. What exists suggests that the compounds are unstable. In the atmosphere or in damp solvents decomposition to elemental tellurium and the parent amide is reported ⟨79CC1102⟩. Reaction of a steroidal iminium salt with sodium hydrogen telluride produced an unstable compound suggested to be *N*,*N*-dimethyltelluroformamide ⟨75JCS(P1)1574⟩. A mechanism was proposed to account for the possible formation of this product. Treatment of phenylalkyne with sodium in DMSO followed by the addition of tellurium, *t*-butyl chloride (proton source) and diethylamine produced phenylacetic acid and elemental tellurium ⟨81ZOR2064⟩. The authors argue that these products imply the intermediacy of a telluroketene. On trapping with diethylamine this would produce *N*,*N*-diethylphenyltelluroacetamide, presumably highly unstable and decomposing to produce the observed material.

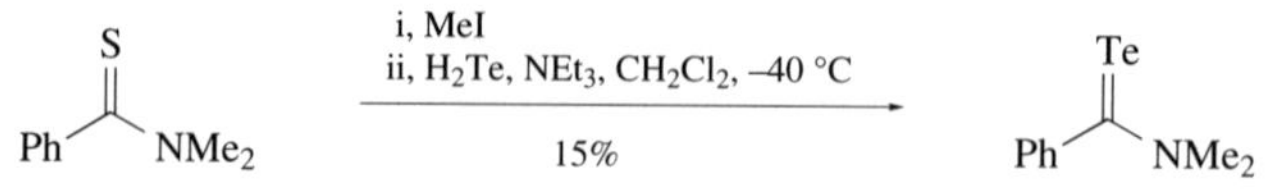

(332)

5.15
N-Substituted Thionoamides and Their Se and Te Analogues

W. MARTIN OWTON
Lilly Research Centre Ltd., Windlesham, UK

5.15.1 OTHER DERIVATIVES OF THIONOAMIDES (AND THEIR Se AND Te ANALOGUES) WITH SINGLY BONDED CARBON ATTACHED TO NITROGEN

5.15.1.1 *N*-Acylthionoamides, -selenoamides and -telluroamides—$R^1C(S)NR^2COR^3$

No recent reviews exist of *N*-acylthionoamides (Equation (1)). A useful introduction, covering the literature up until 1969, is provided by Walter and Voss ⟨B-70MI 515-01⟩; see also ⟨85HOU(E5)1266⟩.

(1)

Primary and secondary thionoamides have been acylated by the following methods.

(i) Acyl halides and anhydrides in neutral solvent in the presence of organic bases ⟨60CB663, 65CB1556, 87CCC2087, 87JOC2442, 88JCS(P1)2203⟩. Thionoamides also react with bisacyl halides in neutral solvent in the presence of organic base. The reaction product(s) is dependent on the chain length of the bisacyl halide (Scheme 1) ⟨60CB671⟩.
(ii) Acyl halides in a liquid–liquid biphasic system with inorganic base ⟨85CPB3479⟩.
(iii) Acyl halides in a solid–liquid biphasic system with inorganic base ⟨77TL2139, 85PS63⟩.
(iv) Ketene in neutral solvent ⟨66C175, 86CB3812⟩.
(v) Acetic acid in the presence of bis(1-naphthyl)bismuthinate ⟨90CL1651⟩.
(vi) Methyl chloroformate in acetone over powdered NaOH ⟨89JMC2493⟩.

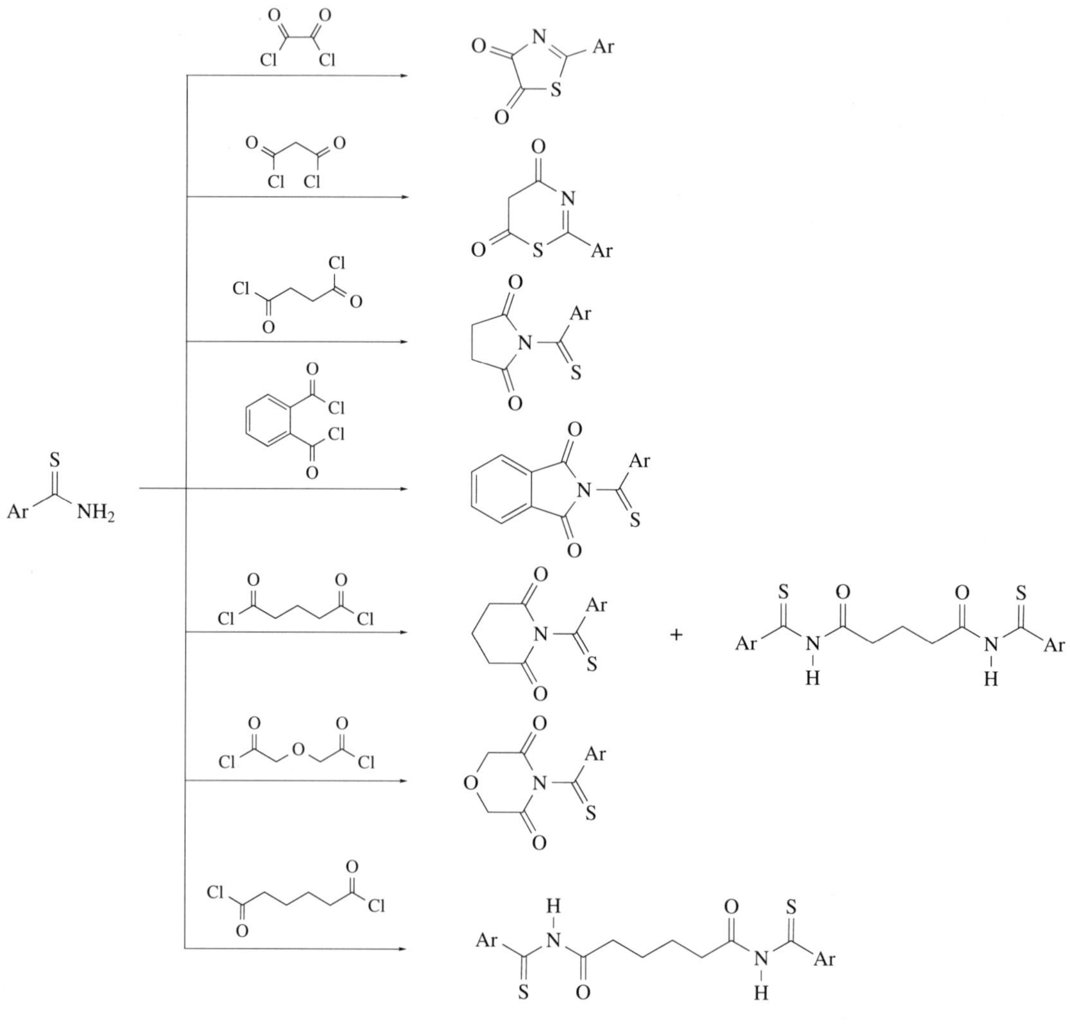

Scheme 1

The acylation of thionoamides with acyl halides proceeds via *S*-acylation followed by rearrangement to the *N*-acylated product. The *S*-acylated intermediate may be isolated if the reaction is run at a low temperature ⟨85PS(25)63⟩.

N-Acylthionoamides have been prepared by a two-step procedure of acid-catalysed condensation of a thionoamide with a nitrile followed by aqueous hydrolysis of the resulting *N*-thioacylamidine hydrochloride (Scheme 2) ⟨61CB2856, 62CB627⟩.

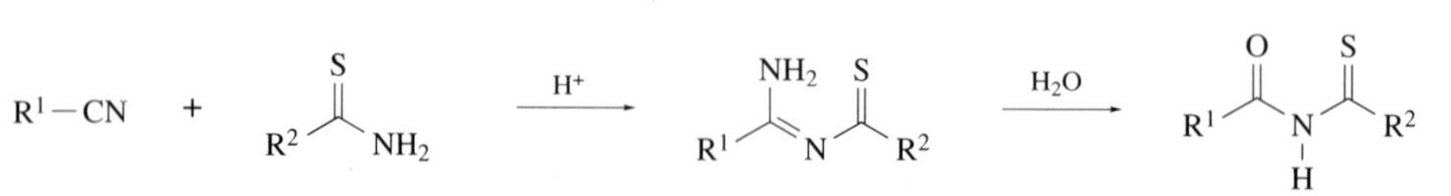

Scheme 2

No reports exist of the isolation of *N*-acylselenoamides and it is reported that acylation of selenoamides occurs in high yield at selenium (Equation (2)) ⟨B-87MI 515-01⟩.

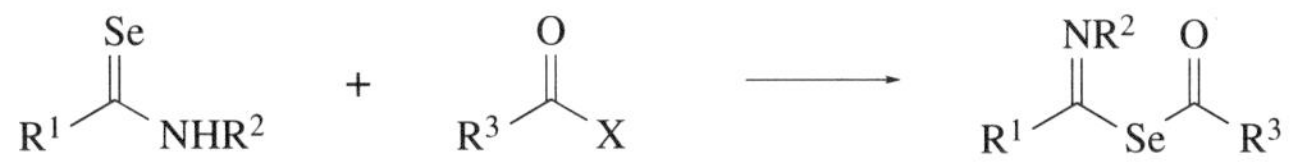

(2)

N-Acylthionoamides have been prepared in very good yield (73–97%) from N_2-acyl-N_1,N_1-dimethylamidines by reaction with hydrogen sulfide in acetic acid (Scheme 3) ⟨84S946⟩.

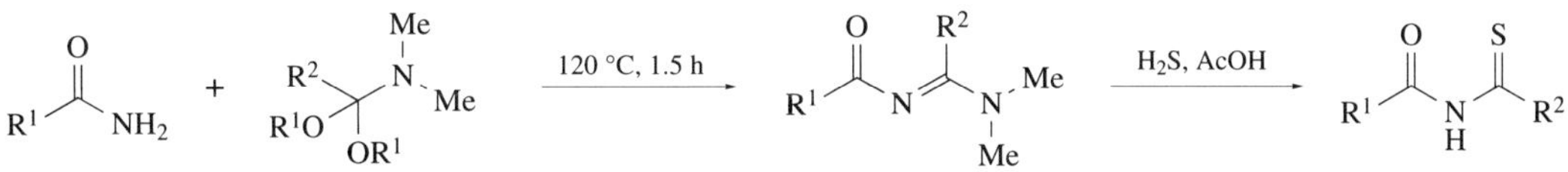

Scheme 3

Acyl isothiocyanates have been reacted with a range of carbon-based nucleophiles to give *N*-acylthionoamides (Equation (3)). The following nucleophiles have been reported: Grignard reagents ⟨73LA476, 76JOC3233⟩, enamines ⟨64JOC1115, 75H(3)1081, 88ZC182⟩, malononitrile, ethyl cyanoacetate, acetylacetone ⟨87S228⟩ and stabilised phosphonium ylides ⟨75ZOB2116⟩.

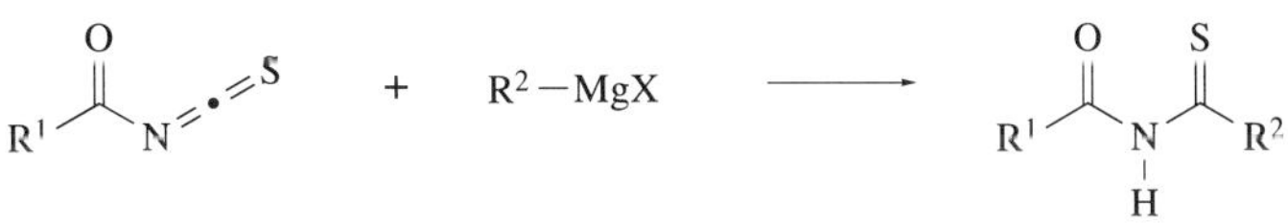

(3)

N-Acylthionoamides also result from the Friedel–Crafts reaction of acyl isothiocyanates with substituted benzenes. Ethoxycarbonyl isothiocyanate has been reacted with a range of substituted benzenes ⟨76JOC962⟩ and phenols ⟨90OPP755⟩ in the presence of aluminum chloride to give good yields (17–90%) of *N*-ethoxycarbonylthionoamides or thionoamides (56–90%), depending on the reaction conditions (Equation (4)).

(4)

Thionation of imides with controlled amounts of Lawesson's reagent in benzene or dioxan gives *N*-acylthionoamides in good yield (Equation (5)) ⟨66ACS689, 85JOC1681, 89H(29)435, 89TL2743⟩. Steric effects appear to be the major factor in determining which carbonyl group is thionated.

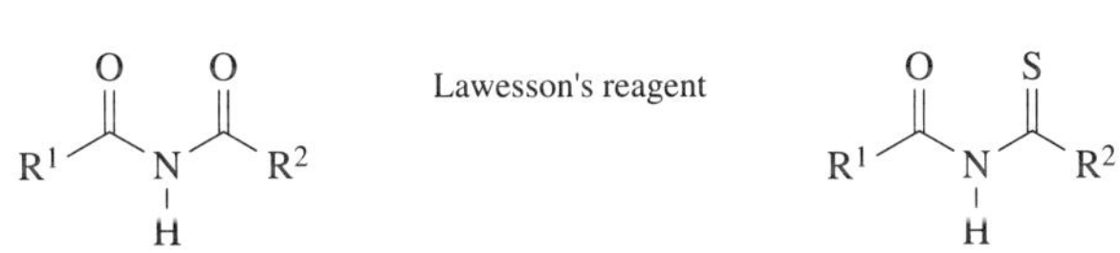

(5)

Thionoamides may be reacted with isocyanates in the presence of Cu_2O to give good yields (48–84%) of substituted thioacylureas (Equation (6)) ⟨74JOC3043, 85IJC(B)977⟩. Without addition of Cu_2O, H_2S is produced along with the symmetrical urea ⟨60CB663⟩. Thionoamides also react with acyl isocyanates ⟨76HCA350⟩ and sulfonyl isocyanates ⟨76S728, 88CZ349⟩ to give acylthioacylureas in high yield (49–95%) and sulfonylthioacylureas in fair yield (19–63%).

(6)

An interesting and singular reaction which is covered within this section is that of isocyanides with thiocarboxylic acids in inert solvent to give the *N*-alkyl-*N*-acylthioformamide accompanied by varying amounts of the *N*-alkylthioformamide (Equation (7)) ⟨75JOC66⟩.

(7)

5.15.1.2 *N*-Thionoacyl-, *N*-Selenoacyl- and *N*-Telluroacylthionoamides (and their Se and Te Analogues)—$R^1C(S)NR^2C(S)R^3$

N-Thionoacylthionoamides (dithioimides) have been prepared from the corresponding imides in satisfactory yields (50–69%) by thionation with phosphorus(V) sulfide or Lawesson's reagent in neutral solvent (Equation (8)) ⟨66ACS689, 79JPR1034, 85JOC1681, 86S768, 89H(29)435⟩. As noted above, it is possible to obtain *N*-acylthionoamides by careful control of the reaction conditions; to obtain *N*-thionoacylthionoamides, excess reagent and reflux temperatures are required. While this reaction may be general for all imides, all the examples quoted in the literature refer to thionation of cyclic imides.

(8)

Thioacyl-substituted thionoamides have also been prepared by reaction of *N*-arylimidoyl chlorides with the *O*-alkyldithiocarbonate potassium salt to give *N*-aryl-*N*-thionoacylthiocarbamates in high yield (Equation (9) and Table 1) ⟨85ZC219⟩. The same products may also be prepared by treatment of a thionocarbamate with sodium hydride and thiobenzoyl chloride, or a thionoamide with sodium hydride and an alkyl chlorothioformate ⟨84ZC438⟩.

Table 1 Reaction of *N*-arylimidoyl chlorides with *O*-alkyldithiocarbonate salts.

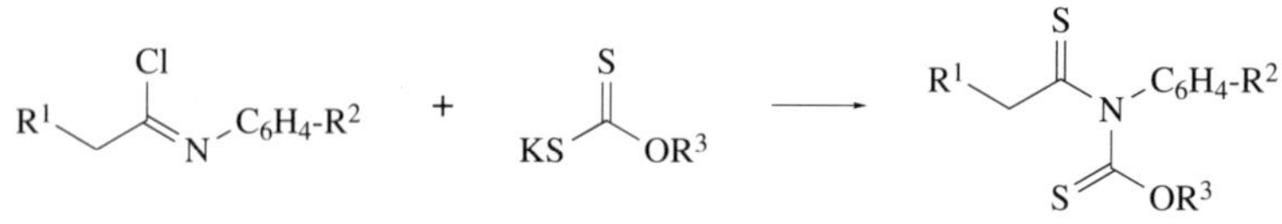

(9)

R^1	*R^2*	*R^3*	*Yield* (%)
H	Me	Me	78
H	Me	Et	93
H	Me	Pr	87
H	Cl	Me	78
H	Cl	Et	84
H	Cl	Pr	86
Me	Me	Me	91
Me	Me	Et	85
Me	Me	Pr	88
Me	Cl	Me	82
Me	Cl	Et	86
Me	Cl	Pr	75

Thionoamides have been reacted with alkyl and aryl isothiocyanates under basic conditions to give thionoacylthioureas in 20–70% yield (Equation (10)) ⟨81ZC271⟩. No references were found to seleno or telluro analogues in this category.

(10)

5.15.1.3 Other Heteroanalogues of *N*-Acylthionoamides (and their Se and Te Analogues)—$R^1C(S)NR^2C(Y)R^3$

As noted above, thionoamides react with nitriles under anhydrous acid conditions to give *N*-thionoacylamidines (Equation (11)) ⟨61CB2856⟩. These compounds are discussed in Chapter 5.19. No references were found to seleno or telluro analogues in this category.

(11)

5.15.2 SINGLY BONDED NITROGEN FUNCTIONS OTHER THAN AMIDES—RC(S)N=Y, RC(S)N≡Z

The condensation of thionoamides with amide dialkyl acetals gives *N*-thionoacylamidines in good to excellent yields (Equation (12)) ⟨85JCS(P1)1875, 85JPR(327)455, 85PS(25)279, 86PS(27)327, 88T139⟩. These compounds are covered in detail in Chapter 5.19.

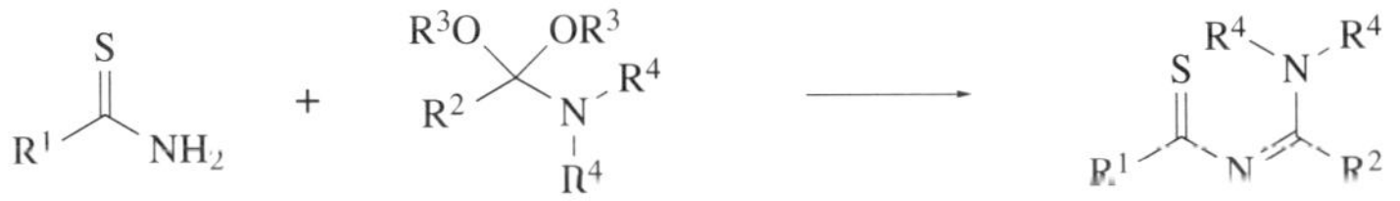

(12)

The reaction of methylenetriphenylphosphorane with *N*-thionoacylformamidines in refluxing THF gives *N*-thionoacyliminoethylidene-triphenylphosphoranes in high yield (Scheme 4) ⟨91BSF574⟩.

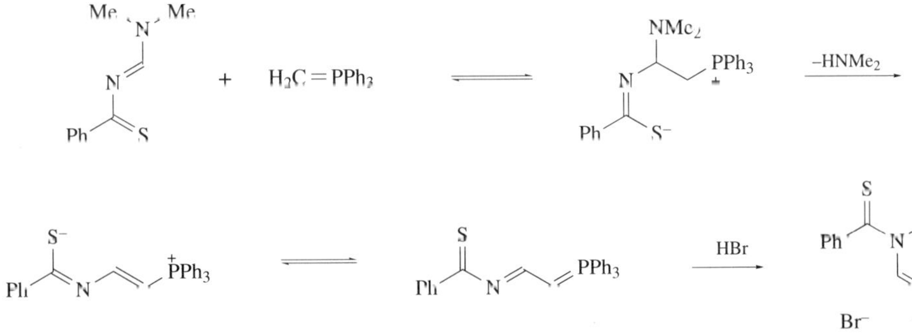

Scheme 4

The reaction of thionoamides with hexafluoroacetone in the presence of trifluoroacetic anhydride (TFAA) gives 6*H*-1,3,5-oxathiazines, which on thermolysis give *N*-thionoacyl perfluoroacetoneimine (Scheme 5). These products are in thermal equilibrium with the 2*H*-1,3-thiazetes ⟨77CB2114⟩.

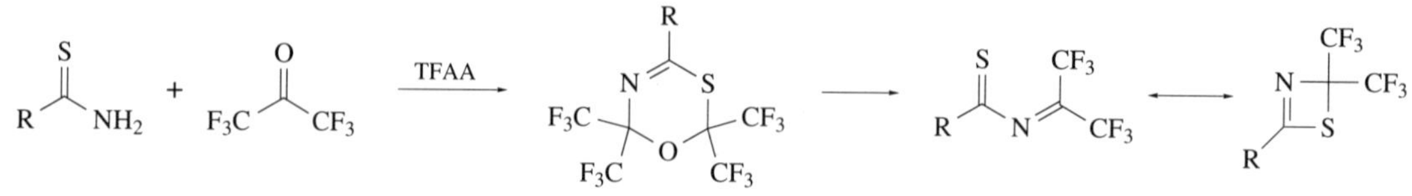

Scheme 5

The alkylation of ethyl 2-isocyanatopropionate and acetate with NaH or KOBu^t and alkyl halides gives functionalised *N*-thionoacylketimines in good yield (50–82%) (Scheme 6) ⟨76TL609⟩. Similar compounds have been prepared by the photocycloaddition of thiones (thioketones) to alkyl nitriles ⟨74TL91⟩, and by reaction of 3-azathiopyrylium salts with 2,6-di-*t*-butyl phenol and 1,3,3-trimethyl-2-methylene-indoline ⟨69CB269⟩.

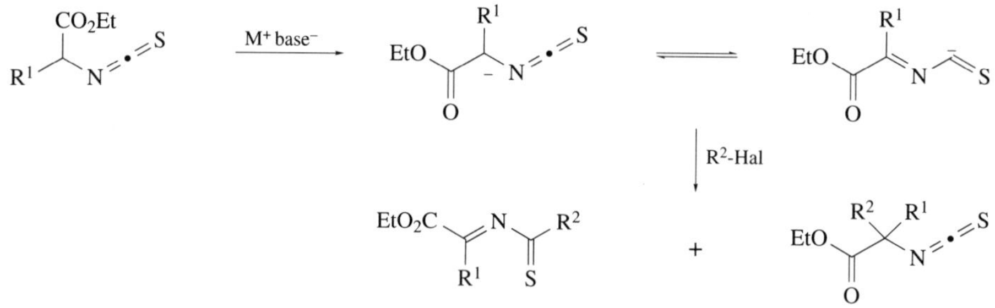

Scheme 6

The syntheses of *N*-thionoacylisocyanates, RC(S)N=C=O, and *N*-thioacylisothiocyanates, RC(S)N=C=S, are discussed in Chapters 5.26 and 5.27. The unstable *N*-thionoacylisocyanates have been trapped by reaction with ethyl 2-(triphenylphosphoranylidene)propionate to give *N*-thionoacylketenimines (Equation (13)) ⟨85LA2305⟩.

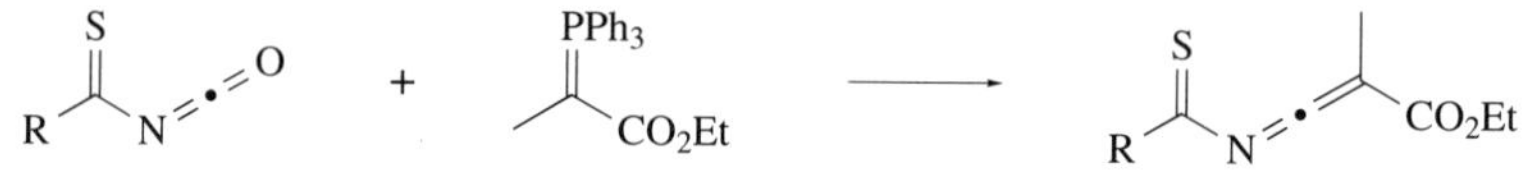

(13)

The reaction of carboxymethyl dithioates with sodium azide might be expected to give thionoacyl azides. Carboxymethyl dithioates do indeed react with sodium azide in aqueous base at room temperature; however, the products have been shown to be thiatriazoles rather than thionoacyl azides (Scheme 7) ⟨61ACS1104⟩. Thiatriazoles may also be prepared by treatment of thionohydrazides with nitrous acid ⟨61ACS1104⟩.

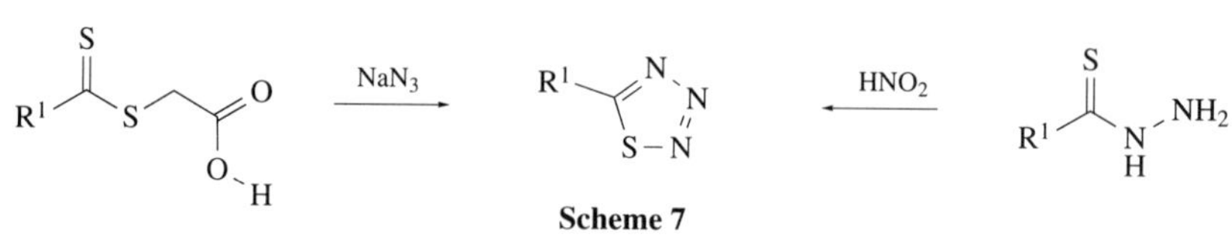

Scheme 7

The photoreaction of 5-phenyl-1,2,4-dithiazole-3-thione with alkenes and alkynes gives 2-thionobenzoylimino-1,3-dithiolans in good yield (40–65%) (Equation (14)) ⟨68CB1428, 70CJC1991, 75JCS(P1)270, 78BCJ301⟩. Similar compounds have also been prepared from the reaction of 5-phenyl-1,2,4-dithiazole-3-thione with benzyne ⟨76BSF120⟩. The chemistry of these products has been explored a little ⟨76CC562⟩.

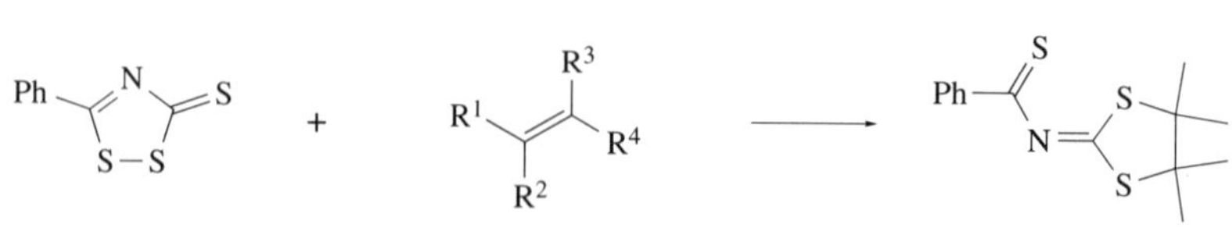

(14)

The 2-thionoacylimino-1,3-dithiolans produced above have also been prepared in good yield (65–84%) by thionation of 2-acylimino-1,3-dithiolans with Lawesson's reagent (Equation (15)) ⟨87H(25)183⟩.

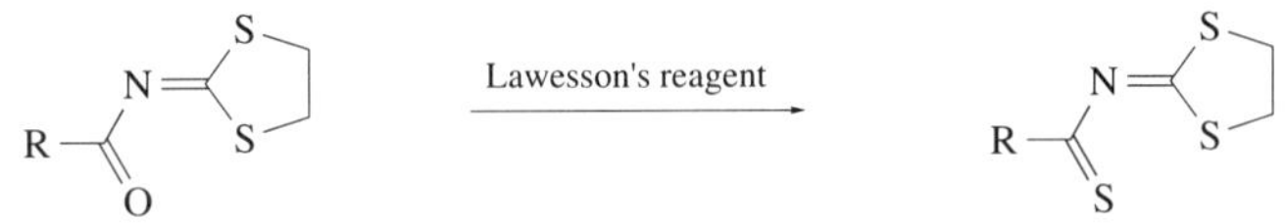

(15)

The reaction of thiobenzamide with chloromethyldichlorophosphane in the presence of a stoichiometric amount of triethylamine in acetonitrile at 0 °C gives a *N*-thionoacyldiphosphane-monoimine. This compound is thermally unstable and heating at 80 °C gives the 1,3,4-thiazaphosphole (Scheme 8) ⟨86PS(27)289⟩. No references were found to seleno or telluro analogues in this category.

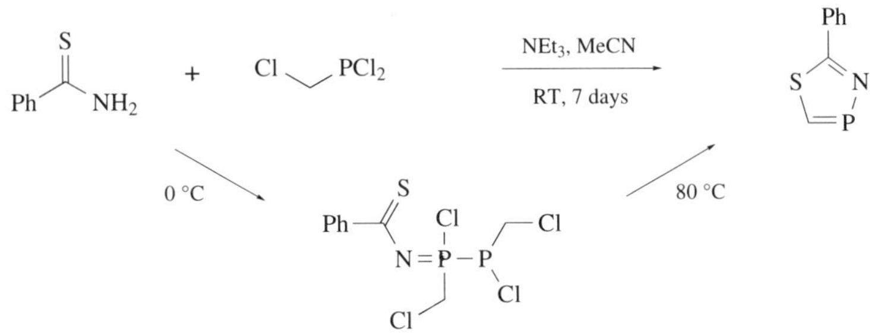

Scheme 8

5.15.3 *N*-HALOTHIONOAMIDES AND THEIR Se AND Te ANALOGUES—RC(S)NHHal

There are no reported examples of this class of compound in the literature. The reason appears to be the susceptibility of thionoamides to oxidation. Reagents which halogenate amides oxidise thionoamides, the principal products being 3,5-disubstituted 1,2,4-thiadiazoles ⟨65AHC(5)123, 68AG(E)856, 85BCJ995⟩. Similar principles apply for the seleno- and telluroamide analogues.

5.15.4 *N*-THIONOACYLHYDROXYLAMINES AND THEIR Se AND Te ANALOGUES—$R^1C(S)NHOR^2$

The synthesis and reactions of *N*-thionoacylhydroxylamines (thiohydroxamic acids) were reviewed in 1971 ⟨71S111⟩ and 1985 ⟨85HOU(E5)1279⟩. Any procedures which give rise to an *N*,*O*-unsubstituted *N*-thionoacylhydroxylamine need to avoid elevated temperatures, at which the compound often decomposes to the nitrile, water and elemental sulfur ⟨67ACS1936, 71S111⟩.

N-Thionoformylhydroxylamines have been prepared from *N*-formylhydroxylamines (hydroxamic acids) by direct thionation with P_4S_{10} ⟨71MI 515-01⟩; however, the yields are only moderate (40%). A general and more modern route which acetylates the free hydroxyl of *N*-acylhydroxylamines before thionation with Lawesson's reagent gives *N*-thionoacylhydroxylamines on hydrolysis with mild aqueous base although the yields are only moderate (10–50%) (Scheme 9) ⟨84S829⟩.

Scheme 9

The most widely used route to *N*-thionoacylhydroxylamines is by thioacylation of hydroxylamine and its *N*-mono- and/or *O*-substituted derivatives with thioacyl compounds (Equation (16)). The thioacyl compounds that have been employed are dithio acids, dithiocarboxylic acid esters, thiocarboxylic acid esters and thioacyl chlorides.

(16)

N-Substituted *N*-thionoacylhydroxylamines have to be prepared from the *N*-substituted hydroxylamine, as *N*-thionoacylhydroxylamines undergo *S*-alkylation with electrophiles ⟨86JOC5198⟩.

The thioacylation with dithio acids uses a metal salt of the acid because the free acids are sensitive to oxidation; potassium salts in aqueous solution have been used ⟨54CB825, 71MI 515-01⟩. However, a more advantageous procedure, which overcomes the general nonavailability of dithio acids, is to prepare the magnesium salt of the dithio acid from carbon disulfide and a Grignard reagent, and react this with the hydroxylamine in one pot (Scheme 10) ⟨57JA1764⟩.

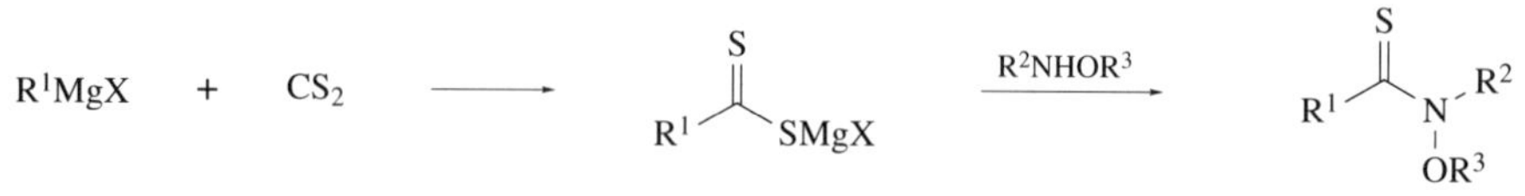

Scheme 10

Dithiocarboxylic acid esters $R^1C(S)SR^2$ have been used extensively for the thioacylation of hydroxylamines ⟨66CPB1249, 68JAP6827842, 71S111⟩. The readily prepared thionoacylglycolic acid esters ⟨61ACS1087⟩, in which the reactivity of an activated ester is coupled with high stability, are particularly useful. This procedure gives good yields (70–80%) and is readily applicable to the preparation of *N*-mono- or *O*,*N*-disubstituted *N*-thionoacylhydroxylamines (Equation (17)) ⟨61LA(643)29, 71LA(743)154, 86JOC5198⟩. The reaction is rapid where R is an alkyl or heteroaryl group; compounds where R is an aryl group usually react more slowly ⟨67ACS1936⟩.

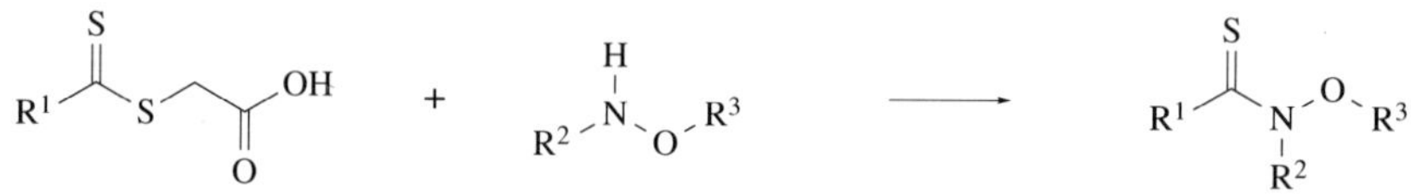

(17)

Thiocarboxylic acid esters $R^1C(S)OR^2$ react with hydroxylamines in a similar fashion to dithiocarboxylic acid esters, although there are rather fewer examples in the literature. Using *O*-ethyl thioformate it is possible to thioformylate *O*,*N*-disubstituted hydroxylamines (Equation (18)) ⟨71LA(743)154, 71S111⟩.

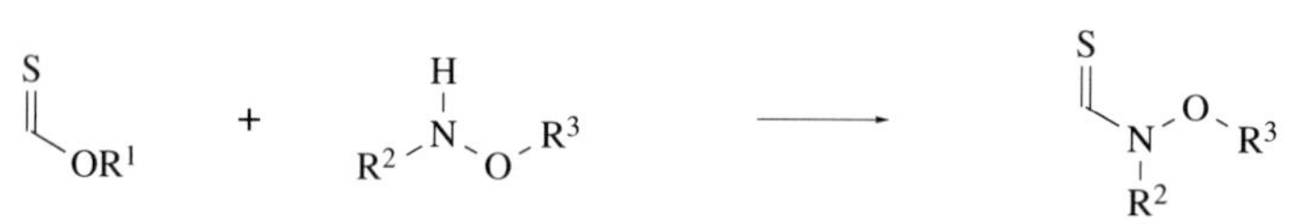

(18)

Attempts to prepare *N*-thioformyl-*N*-substituted-*O*-unsubstituted hydroxylamines have been thwarted by the instability of the products ⟨71S111⟩. In general, short-chain alkyl esters of thiocarboxylic acids (R^1 = Me, Et) have been used to thioacylate hydroxylamines, although the yields appear to be inferior to those obtained with the dithio analogues (Equation (19)) ⟨66CPB1249, 68JAP6827842, 84TL2639, 86JOC5198⟩.

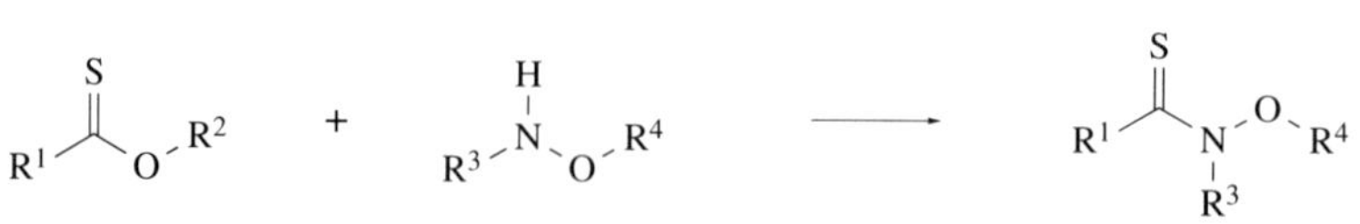

(19)

There are very few examples of the use of thioacyl chlorides for the thioacylation of hydroxylamines, principally due to their instability and inaccessibility (Equation (20)) (see Chapter 5.11); two references are noted in Walter and Schaumann's 1971 review ⟨71S111⟩ but no newer ones appear in a 1985 report ⟨85HOU(E5)1279⟩.

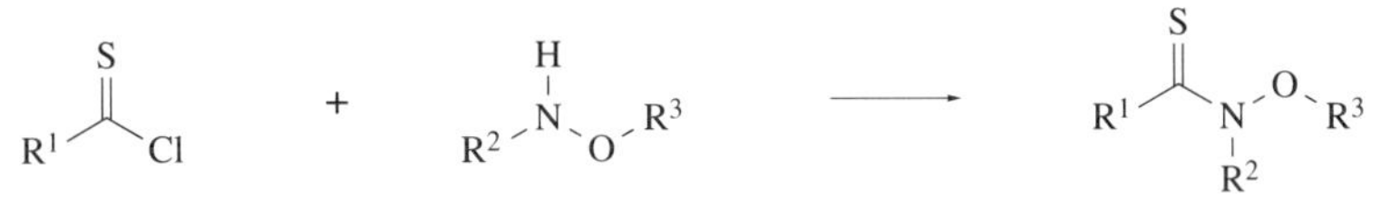

(20)

The addition of hydrosulfide anion to hydroximic acid chlorides, RC(Cl)=NOH, gives *N*-thionoacylhydroxylamines in low yield ⟨71S111⟩ in a manner analogous to the addition of hydrosulfide to imidoyl chorides (see Chapter 5.14). However, at pH > 2 hydroximic acid chlorides begin to form nitrile oxides by loss of HCl ⟨66BSF882⟩, so the reaction will in general proceed via the nitrile oxide. A further complication of this method is the possibility of forming the thionohydroximic acid anhydride by addition of a second equivalent of nitrile oxide; this product results from the addition of hydrogen sulfide or sodium hydrosulfide solutions (pH < 8) to nitrile oxides (Scheme 11). Addition of sodium sulfide solution (pH > 8) to nitrile oxides gives the desired *N*-thionoacylhydroxylamine in high yield ⟨66JOC157⟩.

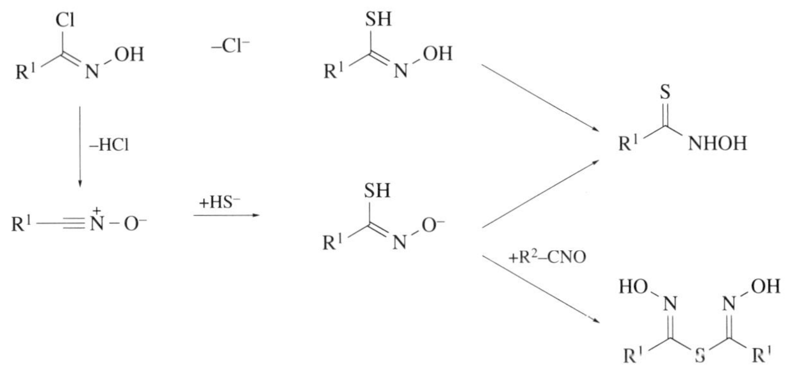

Scheme 11

Primary nitroalkanes have been converted into *N*-thionoacylhydroxylamines in good yield (56–90%) by reaction with KH and hexamethyldisilathiane in THF (Scheme 12) ⟨90T7413⟩. This parallels the reaction of nitro compounds with thiolate anions to give thionohydroximates ⟨77CC597⟩, although this reaction has not been reported for the hydrosulfide anion.

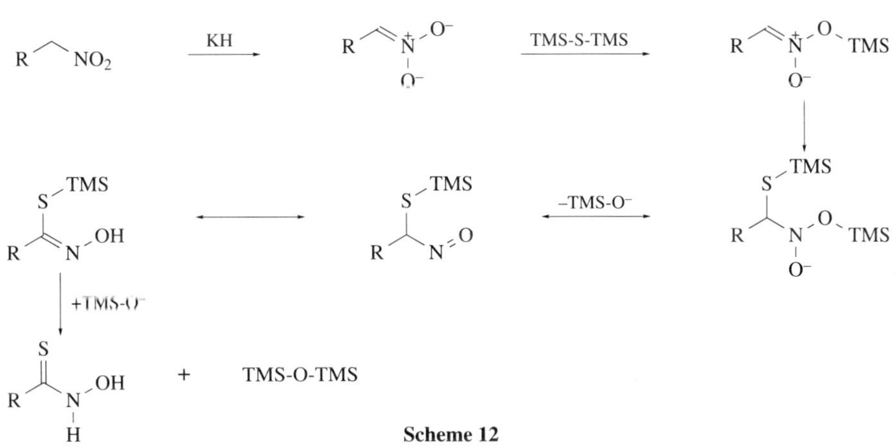

Scheme 12

N-Benzyl-3-aryl-3-phenylpropanothionohydroxamic acids have been prepared in high yield (74–86%) by reaction of 2-benzyl-5-phenylisoxazolidine-3-thiones, prepared by thiation of 2-benzyl-5-phenylisoxazolidin-3-one with P_4S_{10}, with aromatics under Friedel–Crafts conditions (Equation (21)) ⟨91S951⟩. No references to seleno or telluro analogues were found in this category.

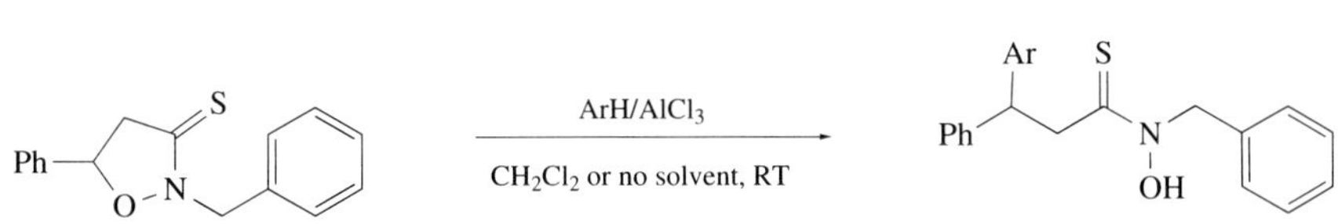

(21)

5.15.5 *N*-THIONOACYLSULFENAMIDES, -SULFINAMIDES AND -SULFONAMIDES (AND THEIR Se AND Te ANALOGUES)—$R^1C(S)NHSR^2$

The reaction of thionoamides with benzenesulfinyl chloride ⟨69LA(727)22⟩ and benzenesulfonyl chloride—as well as with S_2Cl_2, $SOCl_2$ and SO_2Cl_2—gives imidyl sulfides and disulfides as well as *N*-thionoacylamidines, rather than the expected *N*-sulfur compound (Scheme 13) ⟨B-70MI 515-01⟩.

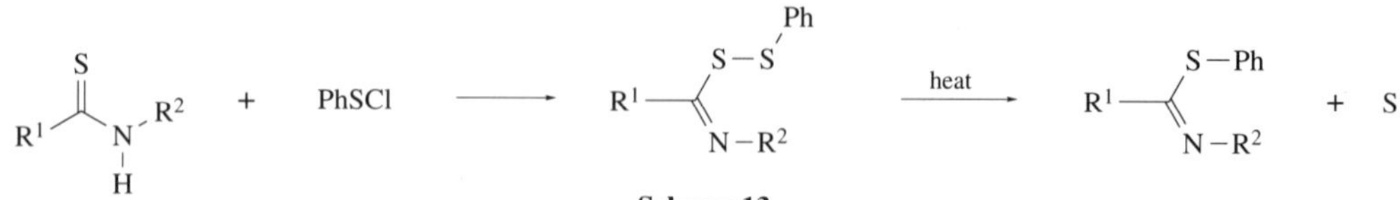

Scheme 13

N-(Arylsulfonyl)thionobenzamides have been prepared by reaction of *N*-arylsulfonylbenzimide chloride with Na_2S solution at elevated temperature (Equation (22)) ⟨66URP181085⟩.

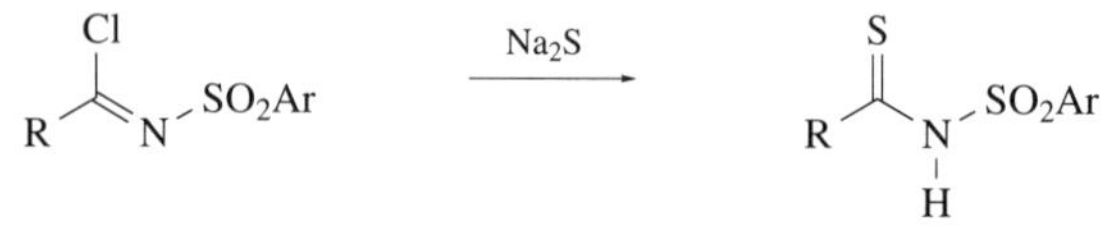

(22)

In addition to the above information, the literature contains a few references to *N*-sulfonylthionoamides where these compounds have arisen under unique circumstances.

Reaction of *p*-toluenesulfonyl isothiocyanate with substituted pyrroles gives *N*-sulfonylpyrrole thionoamides (Equation (23)) ⟨73LA207⟩.

(23)

Reaction of arylsulfonylisothiocyanates with diphenyldiazomethane gives a series of thiiranimines which react with methanol or ethanethiol to give *N*-sulfonyl thionoamides of the structure $Ph_2C(Y)CSNHSO_2Ar$, where Y = OMe, SEt (Scheme 14) ⟨80JOC4366⟩.

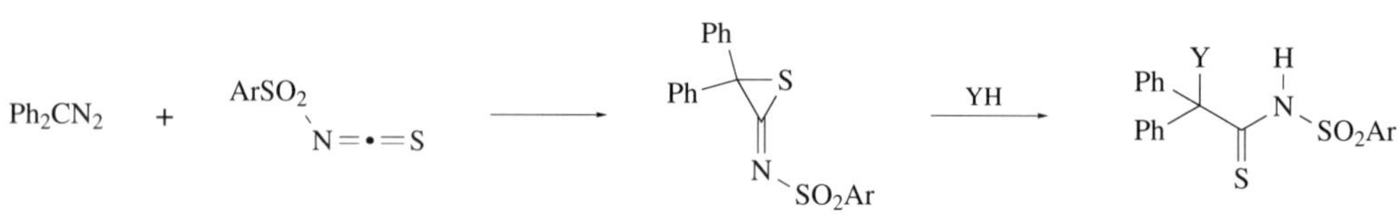

Y = OMe, SEt

Scheme 14

Sulfonyl isothiocyanates also react with β,β-dimethylenamines to give, after protonation with perchloric acid, *N*-sulfonyl-α,α-dimethyl-β-dialkylamino-thionoamides (Scheme 15) ⟨74T4147⟩.

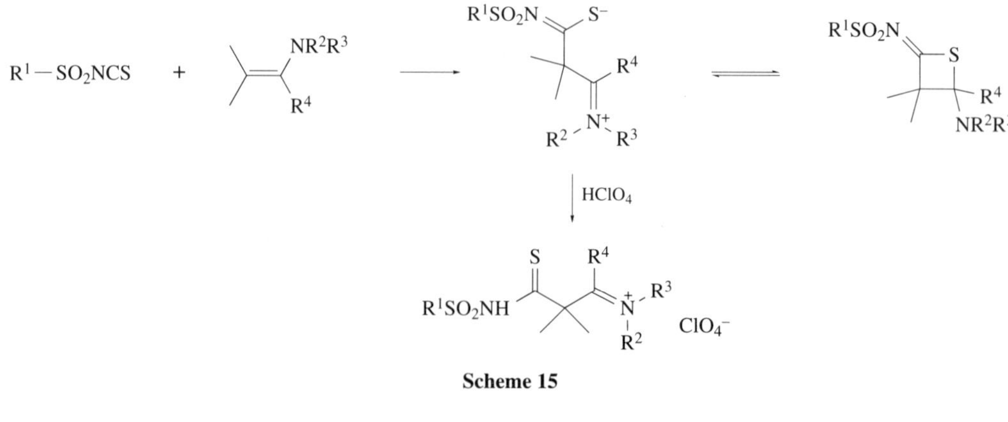

Scheme 15

The [2 + 2] cycloaddition of thioketenes with isocyanates gives 4-thioxo-2-azetidinones in good yield, provided that the thioketene is substituted with bulky groups ⟨80TL4247⟩. These products can be readily hydrolysed to the 4-thioxo-2-azetidinones, which have found some use as intermediates as they are easily alkylated (Scheme 16) ⟨88CB701⟩.

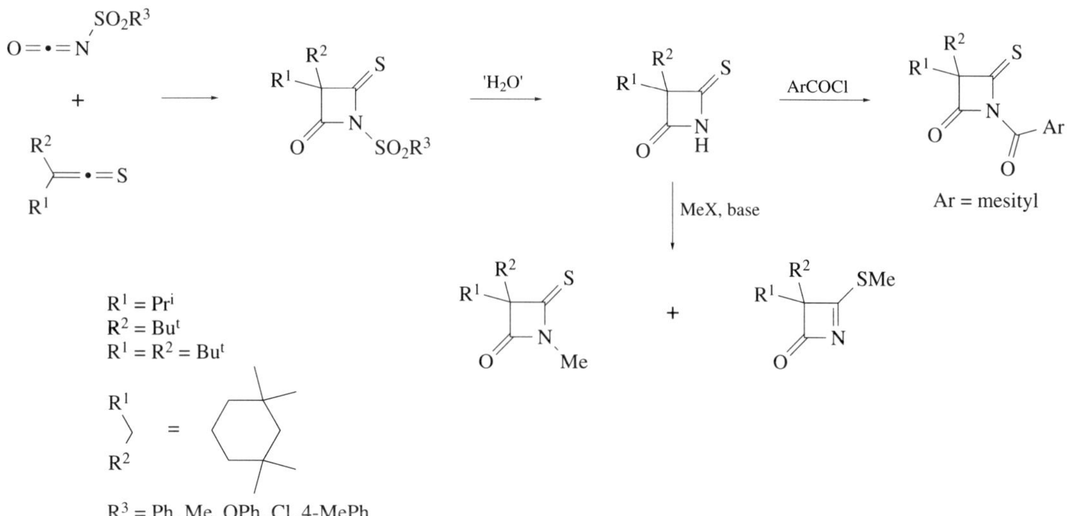

Scheme 16

5.15.6 *N*-THIONOACYLHYDRAZINES AND -HYDRAZONES AND THEIR DERIVATIVES (AND THEIR Se AND Te ANALOGUES)—RC(S)NHNH$_2$, R^{1}C(S)NHN=CR$^2_2$

No reviews exist of the synthesis and chemistry of *N*-thionoacylhydrazines and hydrazones; however, two reports offer good coverage of the general principles ⟨B-70MI 515-01, 85HOU(E5)1291⟩.

The thionation of hydrazides with P_4S_{10} or Lawesson's reagent is not, in general, a useful method, giving thiadiazoles rather than the desired thionohydrazides ⟨61ACS1097⟩. However, N_1,N_2-diacylhydrazides have been successfully thionated to give N_1,N_2-dithionoacylhydrazides with P_4S_{10} ⟨89JHC1341⟩ and Lawesson's reagent ⟨93S463⟩, and in 1991 N_2-phenylbenzoylhydrazide was thionated with Lawesson's reagent in 85% yield ⟨91MI 515-01⟩.

The most common preparation of *N*-thionoacylhydrazides is by thioacylation of hydrazines in a similar fashion to the thioacylation of hydroxylamines discussed above. The work of Jensen and co-workers provides the basis for most methods of thionoacylation currently used ⟨61ACS1097⟩.

The hydrazinium salts of dithioacids are rapidly transformed into hydrazides in dilute aqueous solution ⟨55MI 515-01⟩. This reaction is mainly of importance when the dithioacids can be obtained by the reaction of aldehydes with ammonium polysulfide (Scheme 17) ⟨61ACS1097⟩. This reaction is not straightforward and a number of other products can be produced including hydrazones, thiadiazoles and tetrazines ⟨B-70MI 515-01⟩. Metal salts of dithioacids have also been reacted with hydrazines in aqueous or alcohol solution to give *N*-thionoacylhydrazides ⟨50JCS3389, 54JPJ821, 56BRP758149⟩.

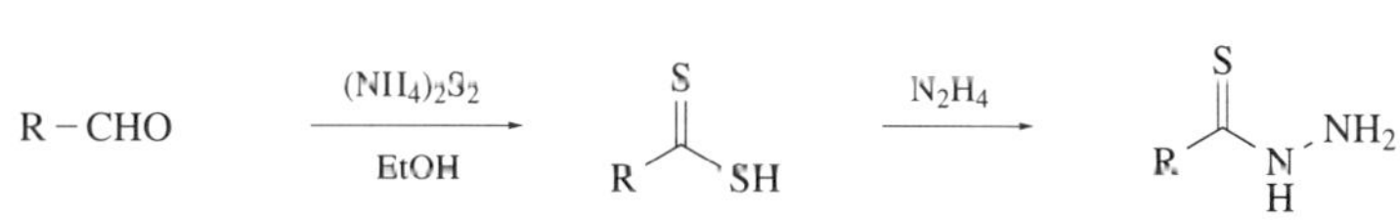

Scheme 17

By far the most common method for the synthesis of *N*-thionoacylhydrazides is via the thioacylation of hydrazines with dithiocarboxylic acid esters, R^{1}C(S)SR^2. Methyl and ethyl esters react with hydrazines to give *N*-thionoacylhydrazides in good yields ⟨86JMC2280, 88CZ239, 88JHC1873⟩. The outcome of the reaction is temperature dependent: the reaction of methyl dithioacetate with hydrazine gives three different products at different temperatures. In ethanol at 50°C the product is

4-amino-3,5-dimethyl-1,2,4-triazole; at 0 °C the product is *N*-thionoacetyl aminoacetamidrazone; and at −78 °C the desired *N*-thionoacetylhydrazine is produced (Scheme 18) ⟨80CPB2116⟩.

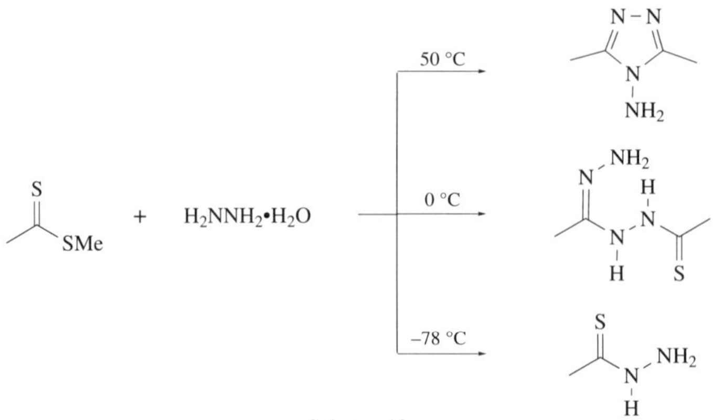

Scheme 18

Higher alkyl esters react slowly or not at all. *O*-Alkyl thioesters react with hydrazine with evolution of hydrogen sulfide; however, no thionohydrazide is formed ⟨61ACS1097⟩. The exceptions to this rule are thionoformate esters which react with hydrazines to give thionoformohydrazides ⟨67AG(E)368, 69CB2117⟩. The outcome of the reaction depends largely on the reaction conditions; in basic media and at moderate temperatures, the thionohydrazide is the major product ⟨B-70MI 515-01⟩. Acylated hydrazines, however, react with aryl *O*-alkylthioesters to form aroyl thionohydrazides ⟨65CB1308⟩. *Se*-Alkyl thioesters have been shown to thioacylate hydrazines readily at room temperature (Equation (24)) ⟨88BCJ3931⟩.

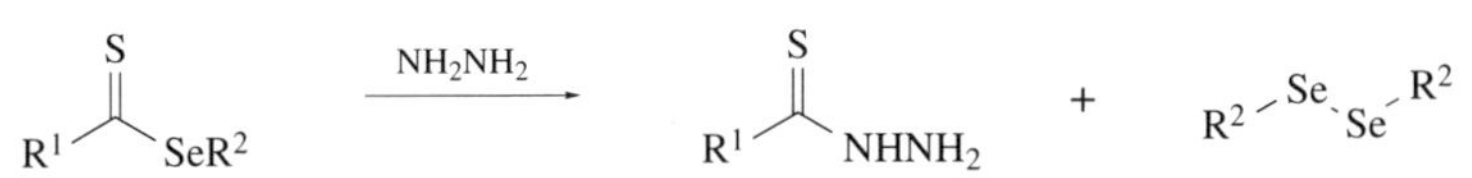 (24)

The most commonly used thioacylating reagents are carboxymethyl dithioates, $R^1C(S)SCH_2CO_2R^2$. These compounds react well with a wide range of hydrazines to give the thionohydrazides in good yields and there are many examples of their use ⟨86JCS(P1)1873, 88JHC1873, 92S637, 93JMC1157⟩. In their definitive study of this chemistry ⟨61ACS1109⟩, Jensen and co-workers showed that monoalkylhydrazines with a primary alkyl group generally yield a mixture of the two possible isomeric thionohydrazides, but the N_2-alkylthionohydrazide is formed almost exclusively with secondary and tertiary alkyl hydrazines (Scheme 19).

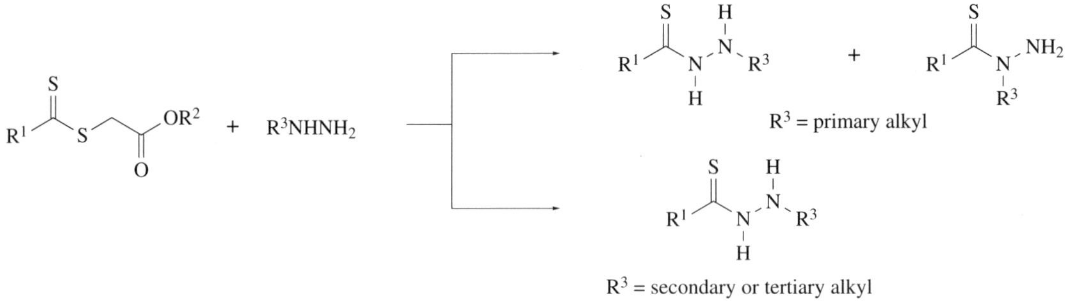

Scheme 19

Symmetrical N_1,N_2-dialkylhydrazines with primary alkyl groups react without difficulty and N_1,N_1-dialkylhydrazines react in all cases. Symmetrical N_1,N_2-dialkylhydrazines with secondary or tertiary alkyl groups do not react. It is not generally possible to prepare thionohydrazides of the type $R^1C(S)NR^2NH_2$ where R^1 is an aryl or secondary or tertiary alkyl group: in these cases the only products are thionohydrazides of the structure $R^1C(S)NHNHR^2$ ⟨61ACS1109⟩. The use of carboxymethyl dithioates thus provides a good route for the synthesis of *N*-thionoacylhydrazides; however, there are some drawbacks. Thionohydrazides prepared from the reaction of aliphatic dithioates bearing at least one hydrogen at the α-carbon atom are unstable with respect to the loss of hydrogen sulfide and concomitant heterocycle formation. For example, from the reaction of carboxymethyl dithiopropionate with hydrazine, 1,3,4-thiadiazole and 2,5-diethyl-1,3,4-thiadiazole could be isolated in good yield. N_1,N_2-Unsubstituted thionohydrazides in most cases react easily with carboxymethyl dithioates, with the formation of thiadiazoles (Scheme 20) ⟨61ACS1124⟩.

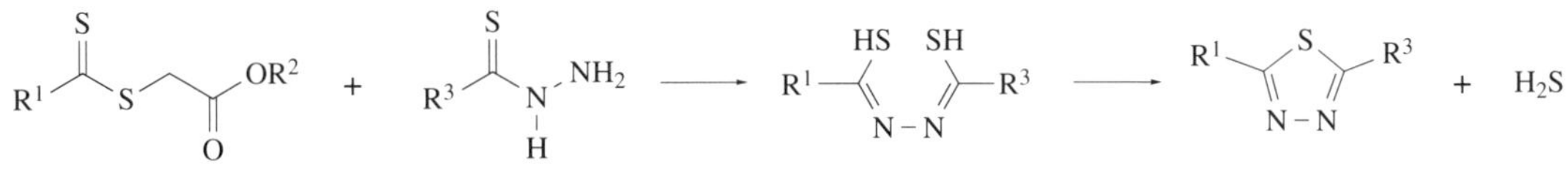

Scheme 20

Isolated references in the literature refer to other species as thioacylating agents for hydrazines. These include *N*-thiobenzoyltriazole and *N*-thiobenzoylbenzotriazole ⟨73LA636⟩, *N*-thiobenzoylimidazole ⟨70LA(739)201⟩ and 2-methylene-1,3-dithietanes ⟨90JHC567⟩. Thioacethydrazide has been prepared from thioacetamide and hydrazine in the presence of Ni(II) ions, but this compound rapidly loses hydrogen sulfide at room temperature (Scheme 21) ⟨79ACS137⟩.

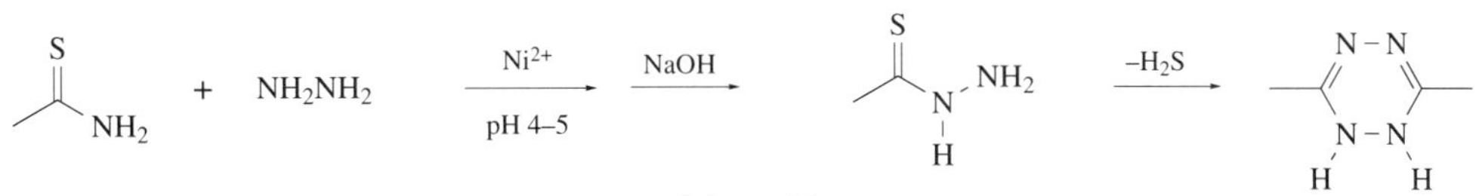

Scheme 21

The other major route for the synthesis of thionohydrazides is by reaction of hydrazonyl halides (prepared from aldehyde hydrazones) with hydrogen sulfide (Equation (25)).

(25)

In these reactions the exact procedure is crucial to minimise the formation of the bishydrazonyl sulfide. The initial report of this route ⟨68TL5973⟩ describes yields of thionoacylhydrazides varying from 83% to 7%; however, a later paper records yields of 50–89% when measures were taken to avoid sulfide formation ⟨74CJC879⟩. Potassium thioacetate has been used as the sulfide source in reactions with hydrazonyl halides in acetonitrile at room temperature to give thionoacylhydrazides in excellent yield (80–90%) (Equation (26)) ⟨70JCS(C)2106⟩; higher reaction temperatures result in the formation of heterocycles and bishydrazonyl sulfides.

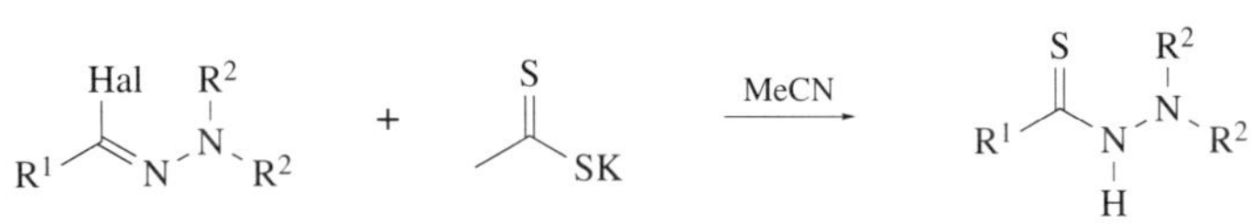

(26)

Thionoacylhydrazides have been isolated from the reaction of hydrazones with SCl_2 in the presence of triethylamine ⟨88JCS(P2)821⟩; however, the complexity of the reaction mixtures and the resultant moderate yields (~20%) prevent this being a synthetically useful method. Hydrogen sulfide has also been reacted with hydrazidines ($RC(NH_2)$—NNR_2) ⟨65CB1308, 65JOC3739⟩ and formazanes ($RC(N{=}NR){=}NNR_2$) ⟨53CB697, 59ZOB1333⟩ to give *N*-thionacylhydrazides, although there do not appear to be any references to the use of this route. Bishydrazonyl disulfides react with potassium cyanide in acetonitrile to give mixtures of equal amounts of iminothiadiazoles and thionoacylhydrazides (Equation (27)) ⟨75CJC3211⟩.

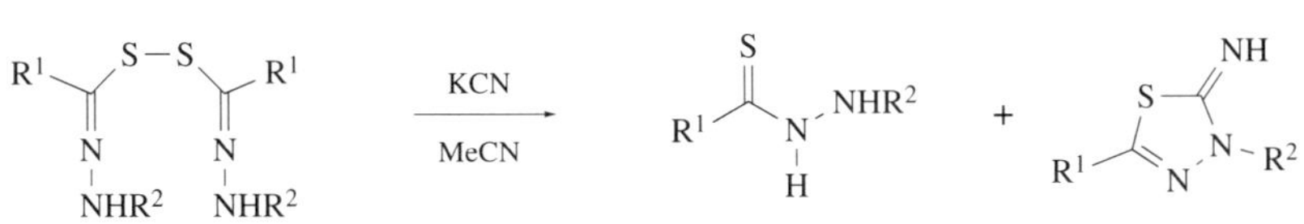

(27)

A further preparation of *N*-thionoacylhydrazides is via the ring opening of isosydnones with hydrogen sulphide (Equation (28)) ⟨68CC499, 74JCS(P1)627⟩.

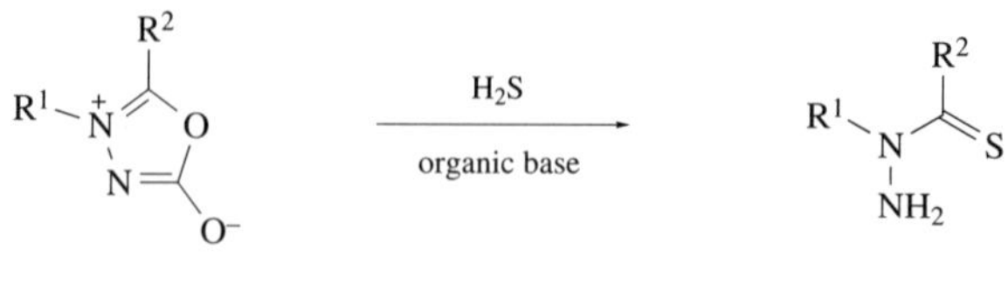

(28)

This route, which has attracted little attention since details of it were first published, allows the preparation of N_1-substituted isomers, $R^1C(S)NR^2NH_2$, where thioacylation yields only the N_2-isomer if R^2 is a secondary or tertiary alkyl or aryl group. N_1-Isomers have also been obtained by thionation of sydnones with P_4S_{10} ⟨73BCJ2921⟩. In a related reaction, N_2-benzoyl-N_1-thiobenzoylphenylhydrazine was prepared by hydrogen sulfide opening of triphenyloxadiazolium perchlorate (Equation (29)) ⟨68CC549⟩.

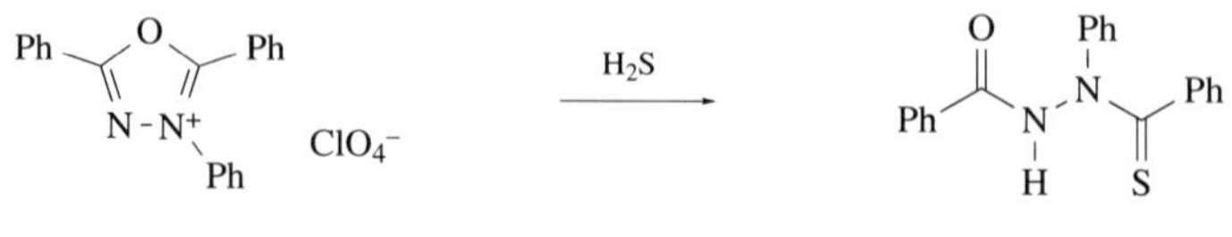

(29)

N-Thionoacylhydrazides form hydrazones, $R^1C(S)NHN{=}CR^2$, with aldehydes by direct reaction in polar solvents ⟨55MI 515-01, 80CPB2116⟩. This reaction has been used as a way of characterising the less stable thionoacylhydrazides.

There is one report in the literature of the synthesis of selenohydrazides. These compounds were prepared in high yield (55–97%) by reaction of 1,3,4-oxadiazolium salts with hydrogen selenide ⟨85ZOB2318⟩, in an analogous manner to the reaction of hydrogen sulfide with triphenyloxadiazolium perchlorate described above (Equation (30)).

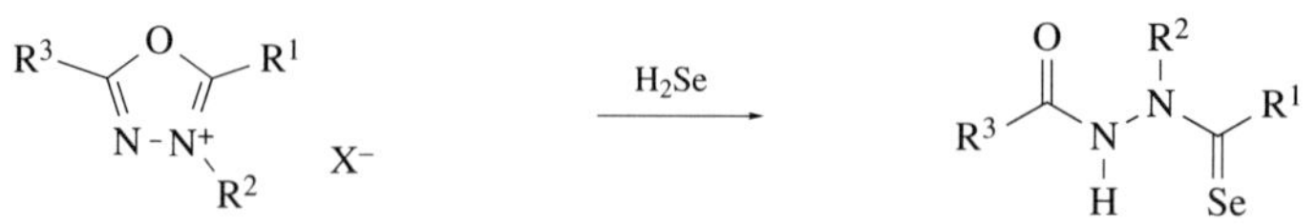

(30)

The selenohydrazides readily undergo acid cyclodehydration to give 1,3,4-selenadiazolium salts in high yield. Reaction of *O*-ethyl selenoacetate with hydrazine gives tetrazines rather than selenohydrazides (Scheme 22) ⟨78JHC1113⟩.

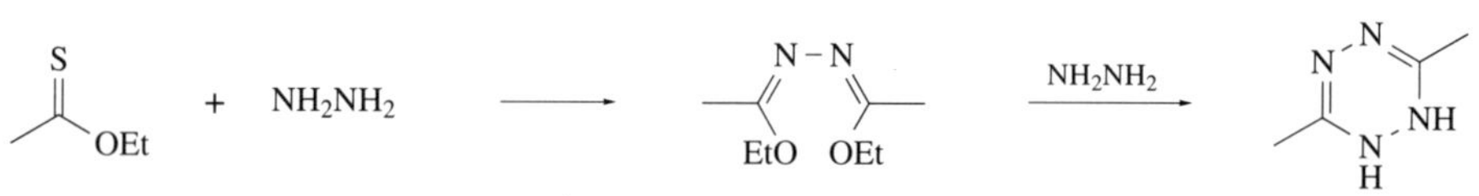

Scheme 22

There is one report in the literature of the synthesis of tellurohydrazides. These compounds were prepared from thionoamides by *S*-methylation with MeI, followed by reaction with hydrogen telluride in the presence of triethylamine in neutral solvent at low temperature (Scheme 23) ⟨79CC1102⟩. The products are stable at room temperature but rapidly decompose in the presence of moisture to the corresponding amides and elemental tellurium.

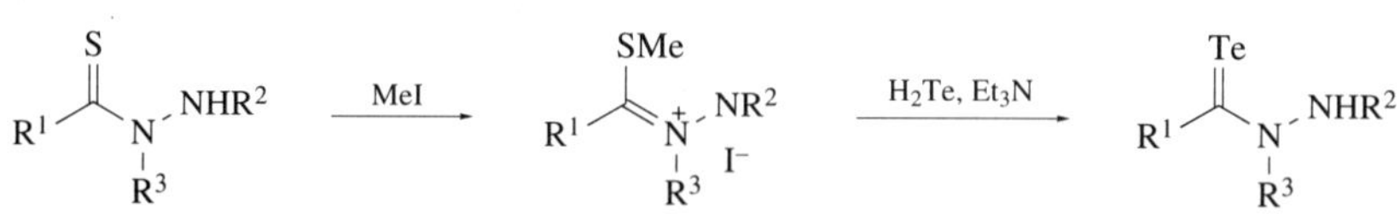

Scheme 23

5.15.7 OTHER *N*-HETEROSUBSTITUTED *N*-THIONOAMIDES (AND THEIR Se AND Te ANALOGUES)

5.15.7.1 *N*-Phosphonothionoamides, -selenoamides and -telluroamides—$R^1C(S)NHPR^2_2$

Much of the reported chemistry of this class of compounds is to be found in the Russian literature and has been reviewed ⟨91RCR1128⟩.

The direct phosphorylation of thionobenzamide in 70% yield has been reported using dialkylthiophosphoryl chloride and potassium *t*-butoxide (Equation (31)) ⟨74ZC56, 79IZV805⟩.

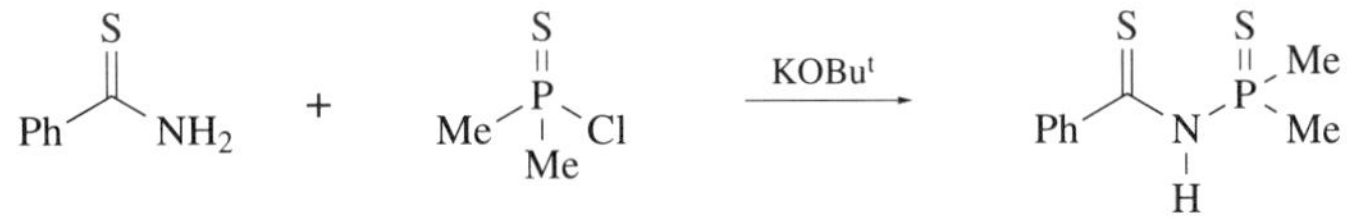

(31)

The same workers reported phosphorylation of phenylthioacetamide with sodium ethoxide/ethanol in 60% yield. More recent work quoted in the review reports good yields of phosphorylated thionoamides under phase transfer and super basic (KOH/anhydrous DMSO) conditions; however, no experimental details have been published as of 1994. It is claimed that only the superbasic conditions are suitable for the preparation of all four possible types of compound, $R^1C(X)NHP(Y)(OR^2)_2$. There is one report of the reaction of thionoamides with phosphorus(III) chlorides. *N*-Methylthiobenzamide was reacted with diethylchlorophosphite to give equal proportions of the *N*- and *S*-phosphorylated products (Scheme 24) ⟨86ZOB1467⟩. The products are thermally unstable with respect to the thionophosphinate.

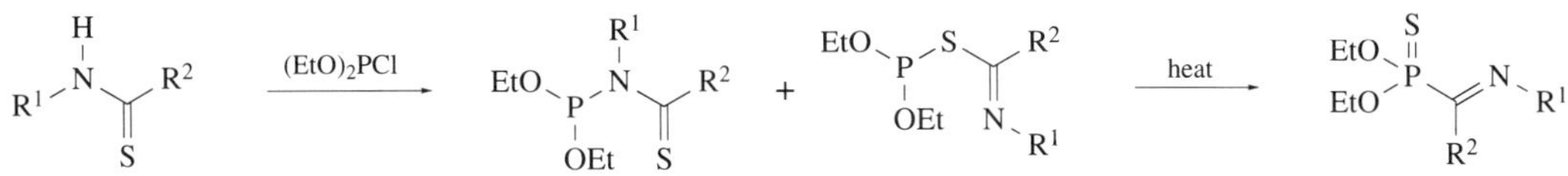

Scheme 24

Thionophosphoryl carbonyl mixed imides have recently been reported as key intermediates in a new synthesis of *N*-monosubstituted thionoamides ⟨90JOC6091⟩. Primary amines are reacted with dimethyl phosphorochloridothionate to give the dimethyl phosphoramidothionate; this is then reacted under basic conditions (NaH, THF) with an acid chloride to generate a thiophosphoryl carbonyl mixed imide in high yield (62–99%). A thermal (ca. 60 °C) rearrangement, which transposes the phosphoryl sulfur atom for the carbonyl oxygen atom, forms the corresponding phosphoryl thiocarbonyl mixed imide. A final hydrolysis (THF/water) cleaves the phosphorus–nitrogen bond to form the thionoamide and dimethyl phosphoric acid (Scheme 25).

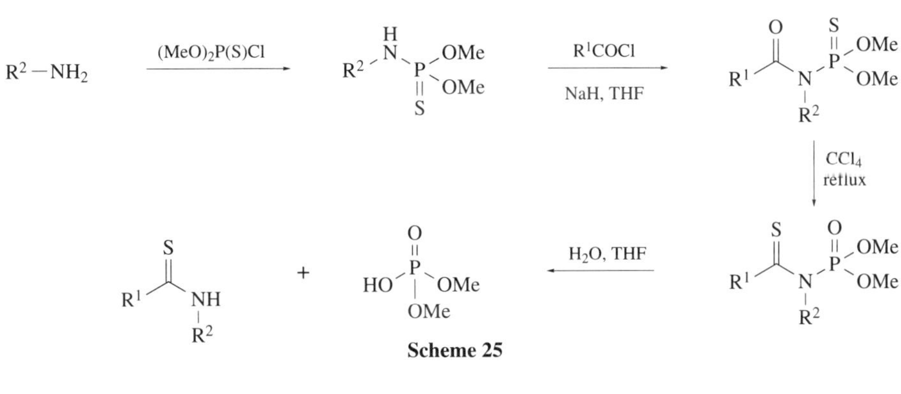

Scheme 25

The reaction of nitriles with monothio- and dithiophosphorus(V) acids, which has been used to synthesise thionoamides ⟨81TL1851⟩, proceeds via *N*-phosphorylated thionoamide intermediates. This reaction proceeds by addition of the thiophosphorus(V) acid to the cyano group with the initial formation of the corresponding *S*-phosphorylthioimidate ⟨78ZOB843⟩. This isomerises to the corresponding *N*-(thio)phosphorylated thionoamide (Scheme 26) ⟨79IZV805⟩. Under anhydrous conditions it is possible to isolate the *N*-thiophosphorylated thionoamide formed between benzonitriles and *O*,*O*-dialkyldithiophosphoric acids ⟨85PS(25)297⟩. This result was confirmed by a subsequent publication from the same laboratories ⟨89T4599⟩.

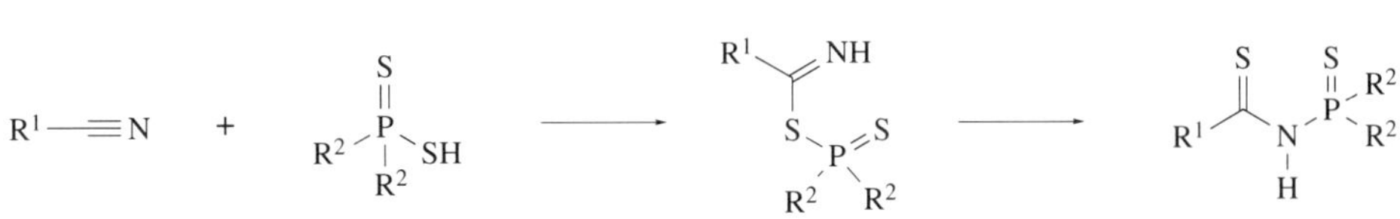

Scheme 26

Aliphatic nitriles either do not react under these conditions or give only poor yields of *N*-thiophosphorylated thionoamides ⟨79IZV805⟩. Interestingly, it was shown that these *N*-thiophosphorylated thionoamides can be quantitatively *N*-methylated with methyl iodide under phase transfer conditions ⟨85PS(25)297⟩, whereas formation of the potassium salt followed by reaction with methyl iodide in acetone at reflux gave the *S*-methylated product (Scheme 27) ⟨90ZOB464⟩.

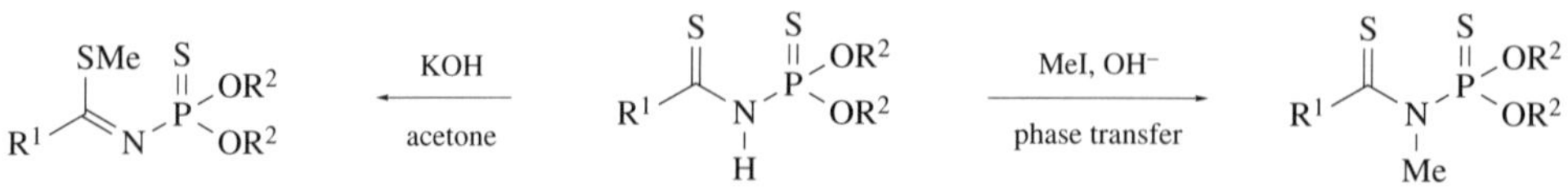

Scheme 27

High yields (70–98%) of *N*-phosphorylated thionoamides may be achieved by cleavage of *N*-dialkyloxyphosphoryl- and *N*-diaryloxyphosphoryliminochlorides with hydrogen sulfide in benzene or dioxan in the presence of tertiary amines (triethylamine, pyridine, dimethylaniline) (Equation (32)) ⟨60ZOB3006⟩. In a related reaction thiophenol was reacted with *N*-phosphoryliminochlorides under basic conditions and the resulting thionoiminoesters hydrolysed to give high yields (60–90%) of *N*-phosphorylated thionoamides ⟨61ZOB582⟩.

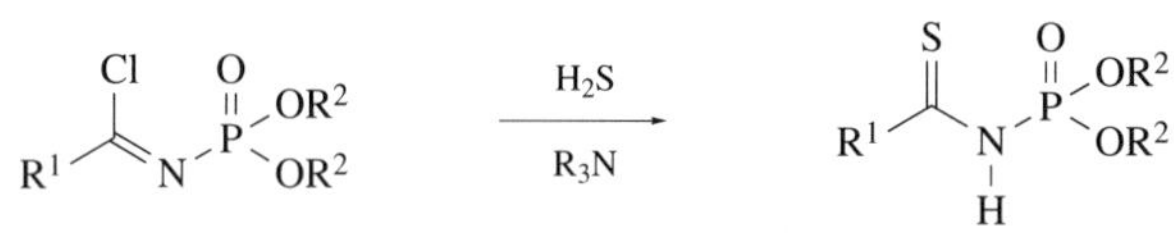

(32)

Tertiary *N*-phosphorylated and *N*-thiophosphorylated thionoamides may also be prepared in high yield by reaction of imidoyl chlorides with salts of monothio- and dithiophosphorus(V) acids (Scheme 28) ⟨79ZOB2323, 86ZOB2359, 89ZOB1917⟩. The products obtained can exist as a tautomeric mixture of the imidoyl dithiophosphate and the phosphorylated thionoamide; the proportions of each product depend on the substitution on the starting imidoyl chloride and on the phosphorus species involved ⟨90ZOB1389⟩.

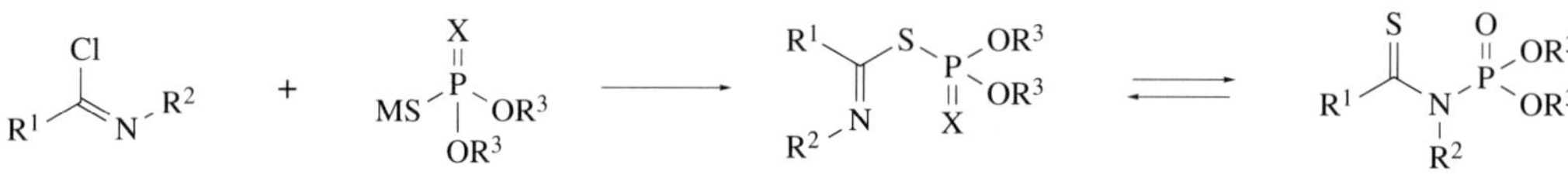

X = O, S
M = Na, K, NH_4

Scheme 28

A similar reaction has been reported with benzohydroximoyl chloride and sodium *O*,*O*-diethyl phosphorodithioate, giving mainly the *O*-phosphorylated thionohydroxylamine and a little *S*-phosphorylation (Scheme 29). The reaction of *O*-methylbenzohydroximoyl chloride with *O*,*O*-dialkyl phosphorodithioates gives high yields (60–90%) of the *N*-phosphorylated *O*-methyl-thionohydroxylamine ⟨85ZOB1526⟩. A low yield (<5%) of *N*-thiophosphorylated thionoamides has also been reported by reaction of iminoesters with dithiophosphoric acids ⟨91RCR1128⟩.

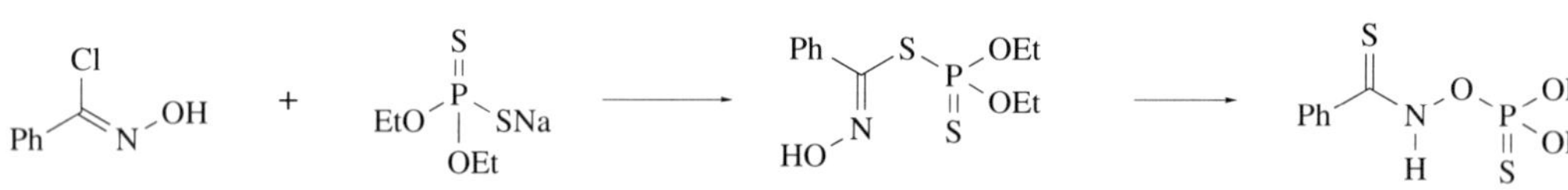

Scheme 29

One reference exists to the reaction between an *N*-phosphorylated isothiocyanate and a carbon nucleophile. The sodium salt of 2-cyanoacetophenone and *O*,*O*-diethyl phosphoroisothiocyanatidate gave a 75% yield of *N*-thiophosphorylated 2-benzoyl-2-cyanothionoacetamide (Equation (33)) ⟨87ZOB685⟩. No references were found to *N*-phosphorylated seleno- or telluroamides.

(33)

5.15.7.2 *N*-Silylthionoamides, -selenoamides and -telluroamides—$R^1C(S)NHSiR^2_3$

Primary thionoamides on treatment with hexamethyldisilazane give *N*-(trimethylsilyl)thionoamides in good yield (67–95%) (Equation (34)) ⟨75LA1808⟩.

(34)

Secondary thionoamides have also been silylated (Bu^nLi, TMS-Cl): those with small substituents on nitrogen and thionoacyl carbon give the expected *N*-silyl product; those having bulkier substituents give a mixture of *N*- and *S*-silylated products ⟨75AG(E)427, 75LA1808⟩.

The reaction of (trimethylsilylmethylene)dimethylphenylphosphorane with isothiocyanates gives *N*-silylated thionoacylmethylenedimethylphenylphosphoranes (Equation (35)) ⟨74JOM(78)339⟩. No references exist for the comparable seleno or telluro analogues

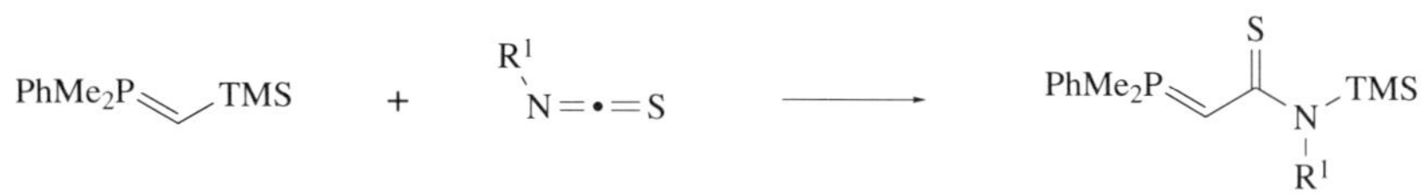

(35)

5.15.8 THIOACYL FUNCTIONS LINKED TO A GROUP 15 ELEMENT OTHER THAN NITROGEN (AND THEIR Se AND Te ANALOGUES)

This is an area of chemistry which has received very little attention and no work has been published establishing general principles. Indeed, a search of the literature reveals only three references to compounds in this category.

Reaction of 1,1,2-trifluoro-2-chloroethanethiol with triethyl phosphite yielded the diethyl ester of fluorochlorothioacetophosphinic acid (Equation (36)) ⟨63ZOB3293⟩.

$$CHFClCF_2SH + (EtO)_3P \longrightarrow CHFClC(S){-}P(O)(OEt)_2 + FC_2H_5 + HF \quad (36)$$

Reaction of α-silylphosphonate anions with sulfur dioxide gives phosphoryl-substituted sulfines in good yield (30–83%)(Scheme 30). These compounds are stable when undergoing chromatography or they may be trapped as [4 + 2] cycloadducts with dienes ⟨84JOC263⟩. There is one example of a similar phosphoryl sulfine being produced in the reaction of an *N*-sulfinylarylsulfonamide with a phosphinyl acetylene ⟨91PS(59)125⟩.

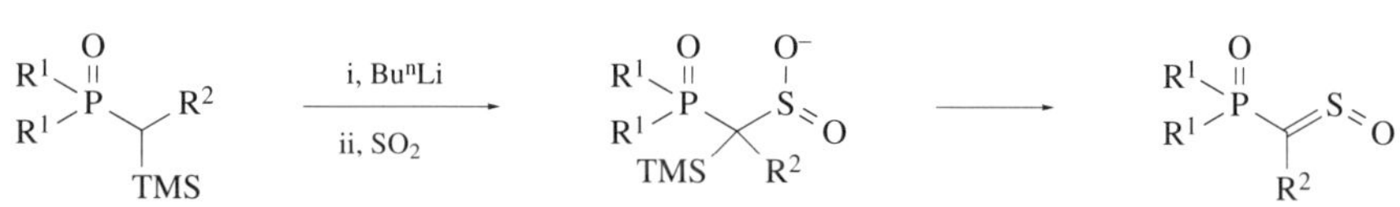

Scheme 30

5.16
Thioacyl Functions Linked to a Metalloid (Si, Ge or B) or Metal; and Their Selenio and Tellurio Analogues

ANDREW D. JONES
ZENECA Pharmaceuticals, Macclesfield, UK

5.16.1 SINGLY BONDED METALLOID DERIVATIVES—RC(S)–METALLOID, RC(Se)–METALLOID, etc.

5.16.1.1 Singly Bonded Silicon Derivatives—$R^1C(S)SiR^2_3$, $R^1C(Se)SiR^2_3$, etc.

It was not until 1981 that the preparation of α-silyl thioketones was reported, in a paper by Bonini *et al.* ⟨81CC822⟩. Not surprisingly, the route chosen to prepare α-silyl thioketones was the most direct, that is from the corresponding α-silyl ketones which are readily accessible via a variety of methods (see Chapter 5.09). The original method for effecting the thiation reaction was to treat the α-silyl ketone with hydrogen sulfide and hydrogen chloride in diethyl ether at −20 °C ⟨86JCS(P1)381⟩. The resulting blue solution was then washed with 5% sodium hydrogen carbonate before being chromatographed under nitrogen to afford the desired α-silyl thioketone as a blue oil or solid (Equation (1)).

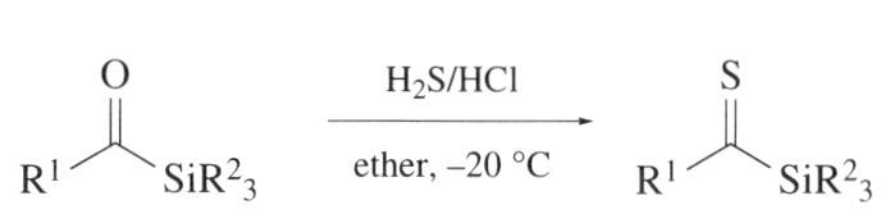

(1)

The most commonly synthesised examples of this organic functional group are the aryl and heteroaryl α-silyl thioketones which tend to be more stable than the corresponding aliphatic analogues. All the reported examples of aryl and heteroaryl α-silyl thioketones are collected in Table 1. It should be noted that, although the attempted preparation of 2,4,6-trimethylphenyl α-trimethylsilyl thioketone from the reaction between the corresponding α-silyl ketone, hydrogen sulfide and hydrogen chloride did not provide any of the desired product, Lawesson's reagent in refluxing toluene provided the anticipated α-silyl thioketone in 68% yield ⟨90JOC3744⟩.

A useful alternative to the above mentioned method of preparing α-silyl thioketones was reported by Ricci *et al.* who showed that the reaction of a variety of α-silyl ketones with bis(trimethylsilyl) sulfide and a catalytic amount of $CoCl_2 \cdot 6H_2O$ afforded the corresponding α-silyl thioketones in good to excellent yield, as shown in Equation (2) and Table 1 ⟨89JOC19⟩. Ricci *et al.* discovered that $CoCl_2 \cdot 6H_2O$ was the best catalyst out of a range that included $BF_3 \cdot Et_2O$ and $FeCl_3$. Although this method was successful for a wide range of substrates, it was shown to be ineffective for the preparation of α-triarylsilyl thioketones. These compounds can, however, be prepared using the hydrogen sulfide and hydrogen chloride method. This result indicates that as well as stabilising the resulting α-triarylsilyl thioketone, the triarylsilyl group also stabilises the starting α-triarylsilyl ketone, thus making the thiation reaction more difficult.

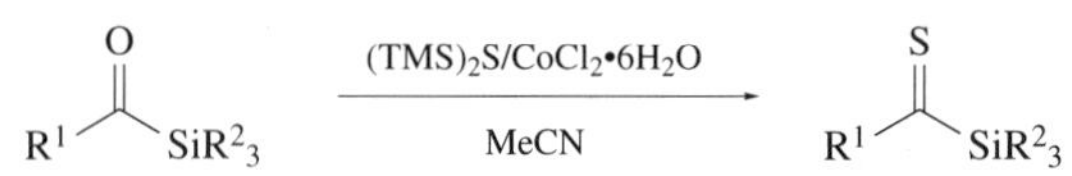

(2)

It has also been demonstrated, as illustrated in Equation (3), that chiral α-silyl thioketones whose chirality is derived from an asymmetric silicon moiety can also be prepared in excellent yield from the corresponding α-silyl ketone, hydrogen sulfide and hydrogen chloride without any racemisation ⟨88CC365⟩.

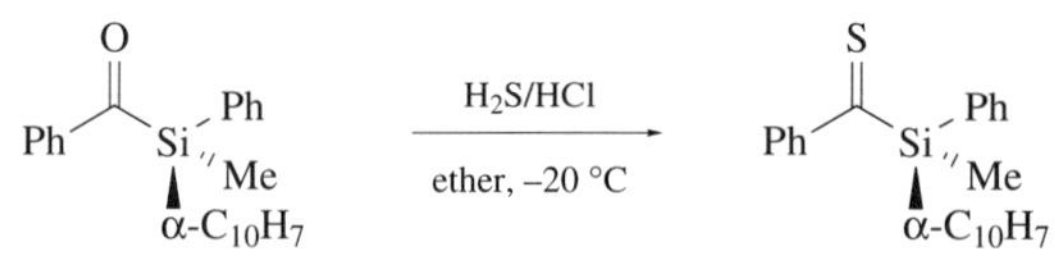

(3)

α-Trialkylsilyl thioketones show a much more varied stability ranging from the very stable *t*-butyl α-trimethylsilyl thioketone—which is isolated as a blue solid—to the very unstable methyl α-trimethylsilyl thioketone, which can be isolated as an unstable oil contaminated with variable quantities of dimeric and trimeric products ⟨89JCS(P1)2083⟩. All of the reported alkyl α-silyl thioketones are summarised in Table 2.

If one chooses to use the hydrogen sulfide and hydrogen chloride method to prepare an α-silyl thioketone, it should be noted that excessively long reaction times lead to the disappearance of the blue colour of the reaction mixture and result in the formation of the stable *gem*-dithiol which can be isolated as a white solid in good yield (Equation (4)). The blue colour of the ethereal solution may be restored if the solution of *gem*-dithiol is washed with aqueous sodium hydrogen carbonate. For example, whereas it is not possible to form methyl α-trimethylsilyl thioketone directly but only via the *gem*-dithiol, *t*-butyl α-trimethylsilyl thioketone does not form a *gem*-dithiol under the reaction conditions and may be isolated directly from the reaction mixture. Care must be taken to ensure that the reaction is performed at temperatures between −15 °C and −20 °C during the addition of the gaseous reagents. If the reaction is performed at 0 °C, significant quantities of trimers are formed which exist in two isomeric forms, as illustrated in Equation (5) ⟨88JCS(P1)1499⟩.

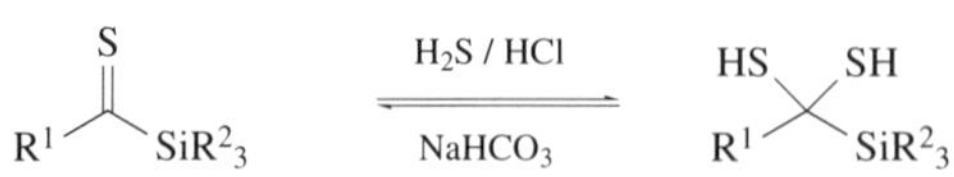

(4)

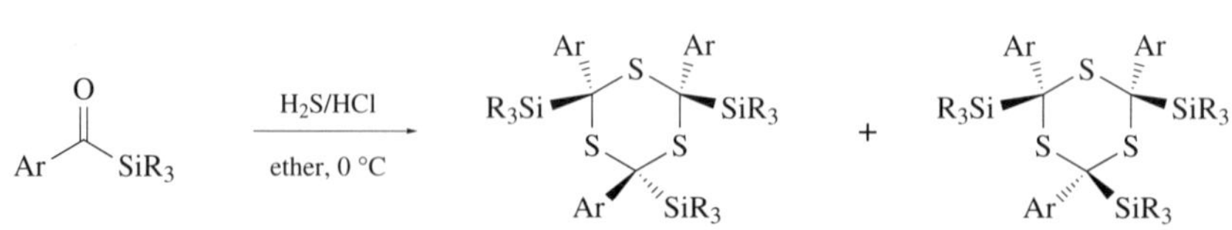

(5)

Up to the mid-1990s no reports of selenio and tellurio analogues of α-silyl thioketones have been published.

Table 1 Preparation of aryl α-silyl thioketones

Product	*Method*	*Yield* (%)	*Ref.*
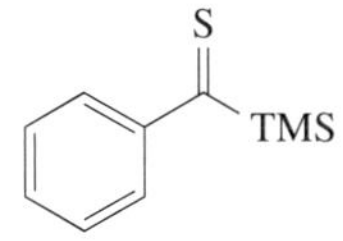	a		91MI 516-01
	a	58	86JCS(P1)381
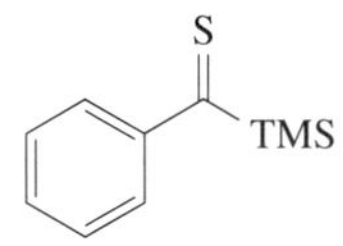	b	92	89JOC19
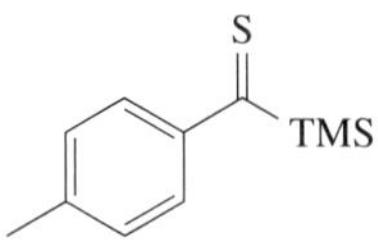	a		86JCS(P1)381
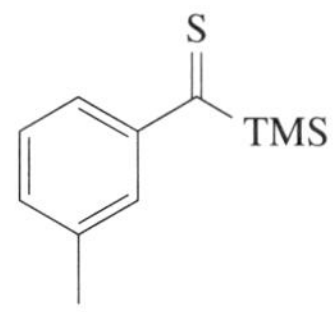	b	74	89JOC19
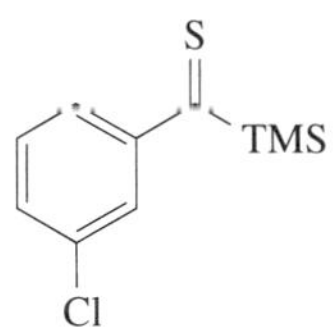	a	55	90JOC3744
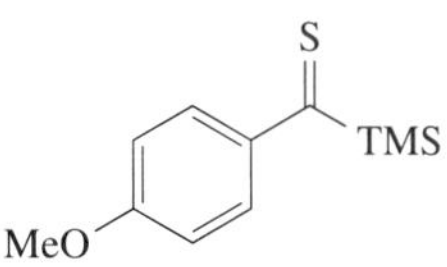	b	66	89JOC19
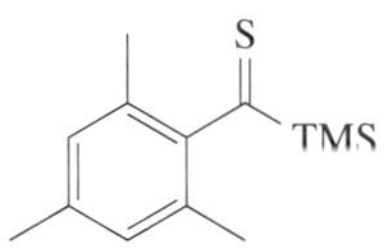	c	68	90JOC3744
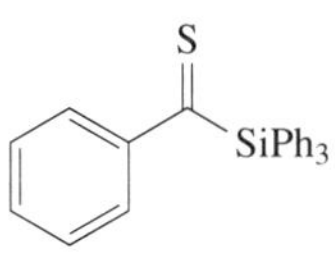	a	85	86JCS(P1)381
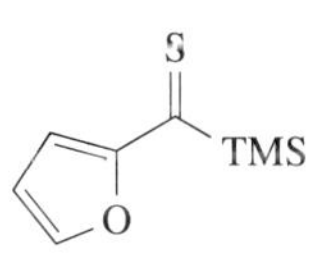	b	68	89JOC19
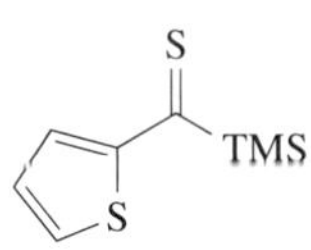	b	59	89JOC19
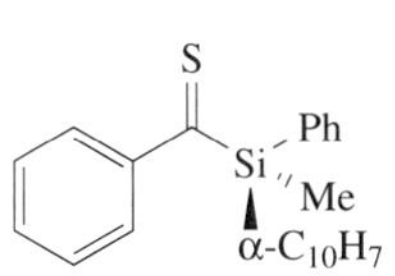	a	95	90JOC3744

[a] i, H_2S/HCl, –30 °C; ii, aq. $NaHCO_3$. [b] $CoCl_2{\bullet}6H_2O/(TMS)_2S$. [c] Lawesson's reagent.

Table 2 Preparation of aliphatic α-silyl thioketones

Product	*Method*	*Yield (%)*	*Ref.*
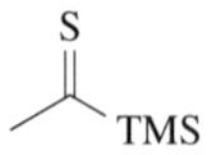	a	d	89JCS(P1)2083
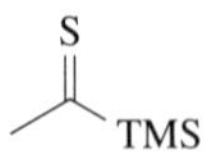	b	30[e]	89JOC19
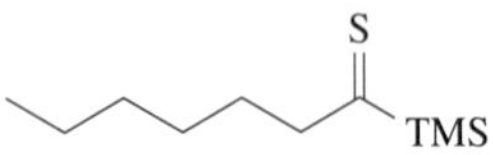	b	64	89JOC19
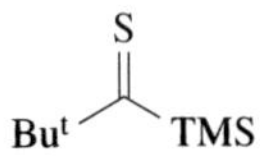	c	90	89JCS(P1)2083

[a] i, H_2S/HCl, –30 °C; ii, aq. $NaHCO_3$. [b] $CoCl_2{\bullet}6H_2O/(TMS)_2S$. [c] H_2S/HCl, –10 °C. [d] unstable, solution only.
[e] Dimers and trimers also formed.

5.16.1.2 Singly Bonded Boron Derivatives—$R^1C(S)BR^2{}_2$, etc.

Up to the mid 1990s no reports of α-boro thioketones or their selenio and tellurio analogues have been published.

5.16.1.3 Singly Bonded Germanium Derivatives—$R^1C(S)GeR^2{}_2$, etc.

Up to the mid 1990s no reports of α-germyl thioketones or their selenio and tellurio analogues have been published.

5.16.2 SINGLY BONDED METAL DERIVATIVES—RC(S)–METAL, RC(Se)–METAL, etc.

5.16.2.1 General Comments and Methods

There are no general methods for the synthesis of metal-bonded thioacyl compounds or their selenio and tellurio analogues.

5.16.2.2 Group 1 and Group 2 Derivatives—RC(S)Li, RC(S)MgX, etc.

No reports of Group 1 and Group 2 bonded thioacyl compounds or their selenio and tellurio analogues have been published.

5.16.2.3 Lanthanide Derivatives

No reports of lanthanide-bonded thioacyl compounds or their selenio and tellurio analogues have been published.

5.16.2.4 Transition Metal Derivatives

There are only two published reports of thioacyl groups bonded to a transition metal. In 1978 Collins and Roper reported the synthesis of osmium complexes containing the thioformyl ligand [Os—CHS] ⟨78JOM(159)73⟩. The complexes were prepared from the corresponding hydrido, thiocarbonyl containing complexes upon treatment with carbon monoxide under pressure, as shown by the example in Equation (6).

(6)

Raubenheimer *et al.* have prepared a thioacyl gold complex, which is also bound to tungsten through the sulfur atom, by reacting the tungsten carbene complex (**1**) with the alkyne thiolate LiSC≡CPh followed by [$AuClPPh_3$], as shown in Scheme 1 ⟨88JOM(355)337⟩. It is interesting to note that the analogous reaction of intermediate (**2**) with the isolobal iron complex [$Fe(CO)_2CpI$] gave the sulfur-bonded product.

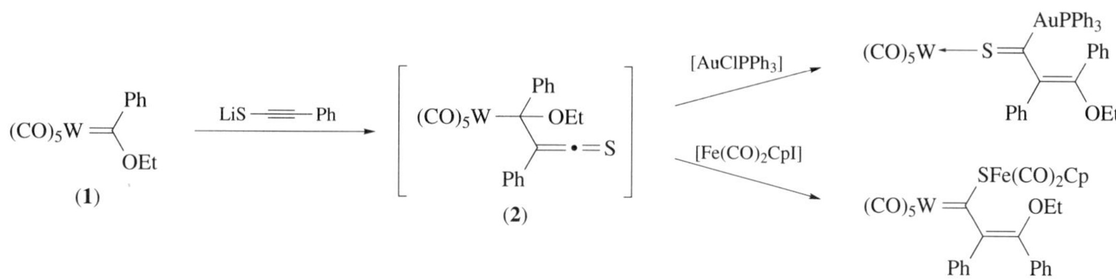

Scheme 1

5.16.2.5 Group 3 and Group 4 Derivatives RC(S)Tl, $R^1C(S)SnR^2_3$, etc.

Up to the mid 1990s no reports of Group 3 and Group 4 bonded thioacyl compounds or their selenio and tellurio analogues have been published.

5.16.2.6 Actinide Derivatives

Again no reports of actinide-bonded thioacyl compounds or their selenio and tellurio analogues have been published.

5.17
Iminoacyl Halides and Oxy Functions

WILLI KANTLEHNER
Universität Stuttgart, Germany

and

WALTER W. MERGEN
Carl Zeiss AG, Aalen, Germany

5.17.1 IMIDOYL HALIDES

5.17.1.1 General Remarks

The activation of amide groups by transformation in the halogenimidoyl group —N═C—Hal has been known since the late nineteenth century ⟨1877LA79, 1892CB2533, 34AG611⟩. Imidoyl halides are of interest because of their high reactivity and there are many procedures known for the formation of these compounds. Monographs ⟨B-68MI 517-01, 85HOU(E5/1)628, 91COS(6)523⟩ and reviews ⟨68AG942, 87RCR1125⟩ describe various methods of preparation up to 1987. The methods published by Holtschmidt deal especially with high temperature chlorination and give references up to 1967 ⟨68AG942⟩.

5.17.1.2 Preparation Procedures

5.17.1.2.1 From amides, cyanohydrins/ureas and halogenating reagents

(i) Reactions with phosgene, thiophosgene and oxalylchloride

The imidoyl chlorides of aliphatic carboxylic acid amides (**1**) are obtained in a general synthesis by reaction with phosgene in dry benzene at room temperature (Equation (1)) ⟨62CB126⟩. The reaction of *N*-phenylpropionamide with phosgene at 0 °C in benzene/pyridine/tetrahydrofuran affords *N*-phenylpropionimidoyl chloride ⟨79JA4245⟩, and *N*-(2-phenoxyethyl)-propionimidoyl chloride (**2**) is obtained from *N*-(2-phenoxyethyl)-propionamide (Equation (2)) ⟨85JHC1121⟩. In a similar way cyclohexane carboximidoyl chlorides and 5-pentene carboximidoyl chlorides, as well as benzimidoyl chlorides with long chain saturated and unsaturated alkyl groups at nitrogen, can be prepared (**3**) (Equation (3)) ⟨91TL1035⟩.

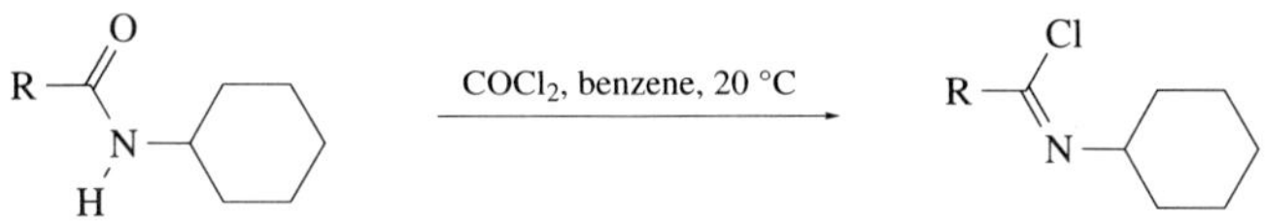

(1)

R = Me, 68%; Et, 60%; Pr^i, 67% **(1)**

(2)

(2)

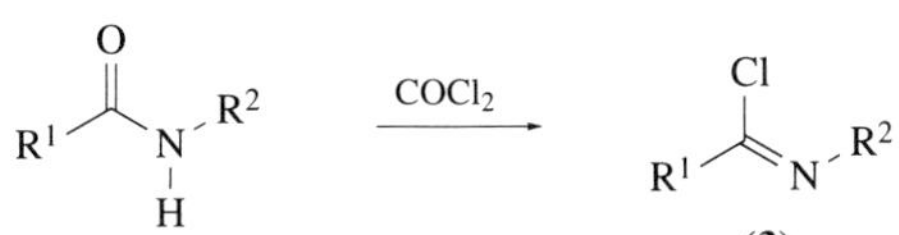

(3)

(3)

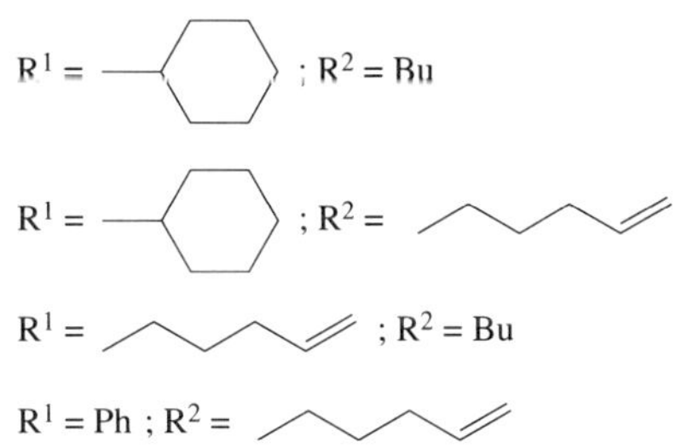

The reaction of *N*-benzoylthioureas or *N*-benzoylselenoureas with thiophosgene yields the corresponding *N*-[thiocarbamoyl]-benzimidoyl chlorides or *N*-[selenocarbamoyl]-benzimidoyl chlorides, respectively ⟨88TL3475⟩.

(ii) Reactions with phosphorus penta- and trihalides

Following a general procedure, *N*-alkylimidoyl chlorides as well as *N*-arylimidoyl chlorides of aromatic carboxylic acids are obtained by treatment of the corresponding amides with PCl_5 ⟨62CB126⟩. For example, *N*-methylbenzamides (**4**) afford *N*-methylbenzimidoyl chlorides (**5**) with PCl_5 (Equation (4)) ⟨78JOU758, 78ZOR820⟩. Using this method the following amides were converted into the corresponding imidoyl chlorides: *N*-methyl-4-nitrobenzamide ⟨89CCC3245⟩, *N*-(3-phenylpropyl)benzamide ⟨88LA833⟩, *N*-(2-phenethyl)benzamides ⟨80JHC1457⟩, *N*-5-pyrazolylbenzamides ⟨87S628⟩, *N*-(1-hydroxy-2,2,2-trihaloethyl)benzamide ⟨76JOU2252, 76ZOR2319⟩, *N*-(2,2-dichloroethenyl)-4-chloro-benzamide ⟨77JOU1471, 77ZOR1594⟩, methyl-2-benzoylamino-3,3-dichloroacrylate ⟨78JOU881, 78ZOR943⟩, *N*-(2-methyl-1-propene-1-yl)acetamide ⟨87JCS(P1)1389⟩, *N*-methyl-trichloroacetamide ⟨78TL1429⟩, *N*-trifluoroacetylbenzamide ⟨87JGU415, 87ZOB475⟩, ethyl-2-[(5-aryl-2-furoyl)-amino]-acetate ⟨84CCC1505⟩, 2-thenoylanilides ⟨87JCS(P1)1153⟩, benzamides with the cephalosporin skeleton at nitrogen ⟨79TL4943⟩, *N*-(arylsulfonylmethyl)benzamides ⟨89JOU2329, 89ZOR2597⟩, phosphoniomethylamides (**6**) (Equation (5)) ⟨84JGU255, 84ZOB288, 86JGU2483, 86ZOB2802, 88JGU2192, 88ZOB2465⟩.

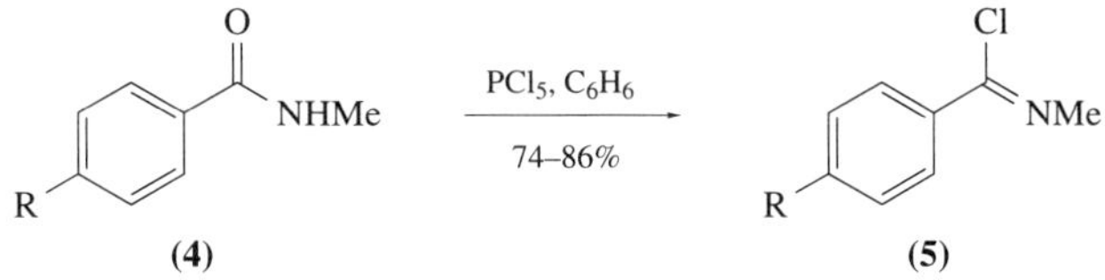

(4)

(4) **(5)**

R = 4-MeO, 4-Me, H, 4-Cl, 3-Cl, 4-NO_2

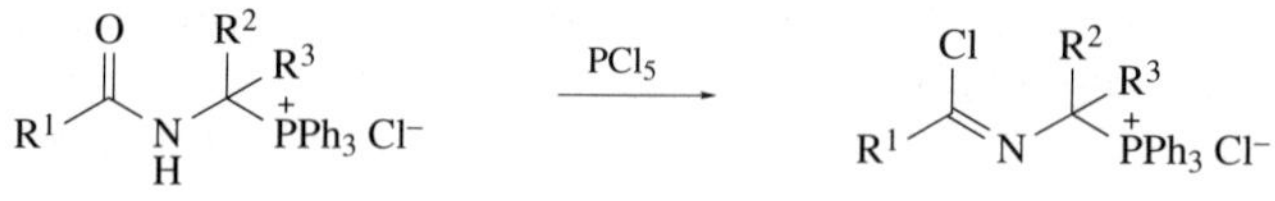
(6)

(5)

Perfluoroglutarimide with PCl_5 yields 2,2,6-trichloro-3,3,4,4,5,5-hexafluoro-1-piperideine (**7**). On treatment with silver fluoride, perfluoro-1-piperideine (**8**) is obtained. Hydrolysis of the trichloro-hexafluoro compound affords 2-chloro-6-oxo-3,3,4,4,5,5-hexafluor-1-piperideine (**9**) (Scheme 1) ⟨62JOC2585⟩. In a similar way perfluorosuccinimide yields 2,5,5-trichloro-3,3,4,4-tetrafluoro-1-pyrroline which on treatment with silver fluoride gives perfluoropyrroline ⟨62JOC2585⟩.

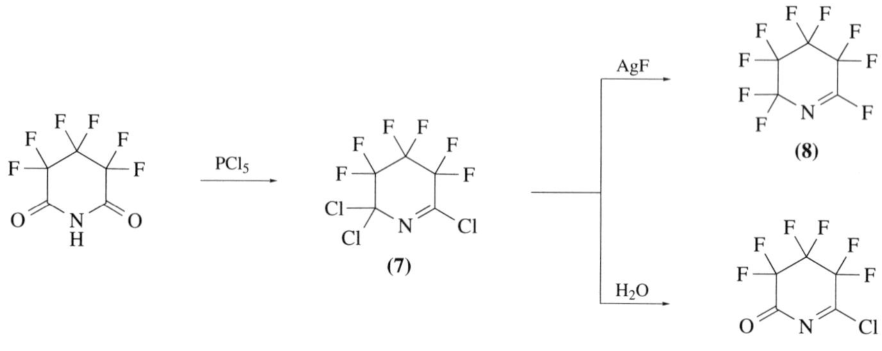
(7) (8) (9)

Scheme 1

From *N*-benzoylurethane and PCl_5 a mixture of *N*-ethoxycarbonyl-benzimidoyl chloride and *N*-benzoyl-ethoxy-formimidoyl chloride is obtained ⟨77JHC765⟩. α-Substituted β-aryl-α,β-unsaturated ketoximes undergo Beckmann rearrangement and form imidoyl chlorides (**10**) (Equation (6)) ⟨80PJC745⟩.

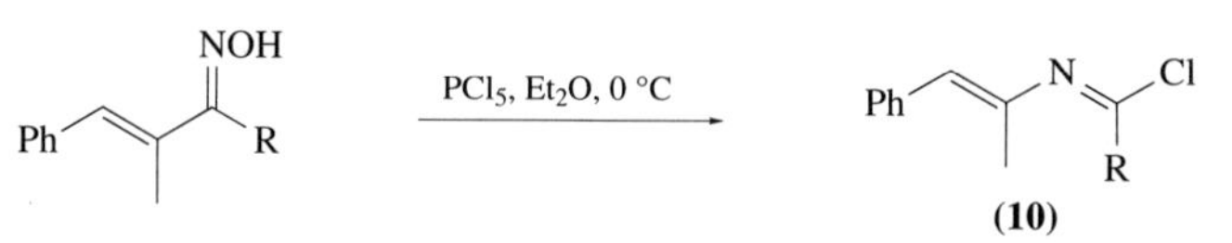
(10)

(6)

Phosphorylated chloroenamines, but not imidoyl chlorides, are the products of the reaction of diarylacetamides with PCl_5 in the presence of SO_2 ⟨81JGU1015, 81ZOB1201⟩. Similar results have been obtained with *N*-methyldiacetamide ⟨87JGU1710, 87ZOB1912⟩. *N*-(Dichloromethylene-diethoxyphosphoryl-methyl)-benzamidines yield *N*-(dichloromethylene-dichlorophosphoryl)-benzamidoyl chlorides (**12**); isolable intermediates are phosphonic analogues of azlactones (**11**) (Equation (7)) ⟨82JGU980, 82ZOB1122⟩.

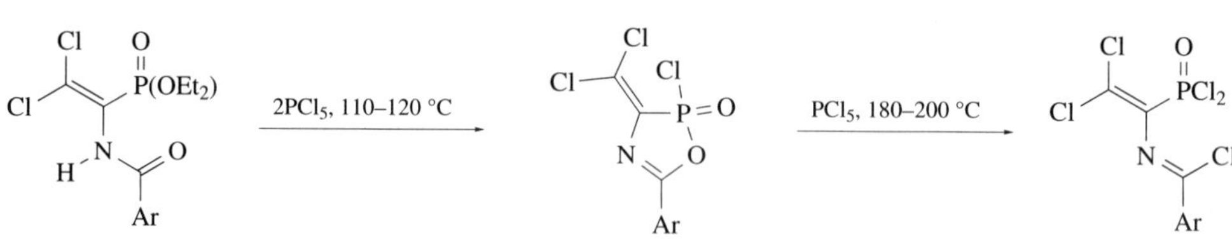
(11) (12)

(7)

N-(1-Arenesulfonyl-2-chloroethenyl)-amides and *N*-(1-aryldiphenylphosphonyl-2-dichloroethenyl)-amides react with mixtures of PCl_5 and $POCl_3$ to give the corresponding imidoyl chlorides (**13**) (Equation (8)) ⟨84JGU255, 84ZOB288, 91UKZ530⟩.

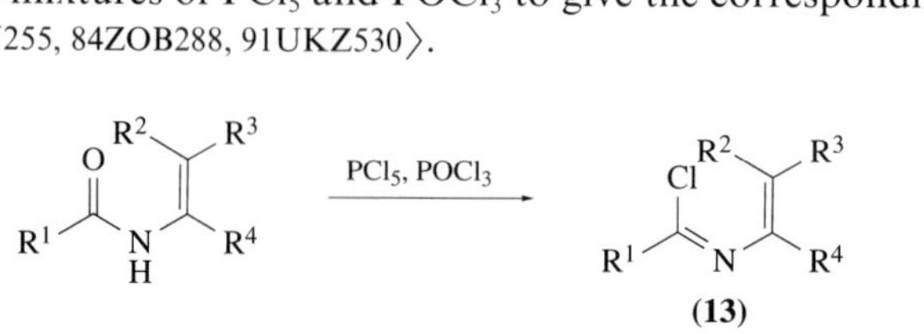
(13)

(8)

Under analogous conditions imidoyl chlorides (**14**) have been prepared from *N*-(2,2,2-trichloro-1-hydroxyethyl)trichloroacetamide, *N*-(1,2,2,2-tetrachloroethyl)acetamides and *N*-(2,2-dichloroethenyl)benzamide (Equation (9)) ⟨76JOU663, 76ZOR673, 76JOU2252, 76ZOR2319, 79JOU32, 79ZOR38⟩.

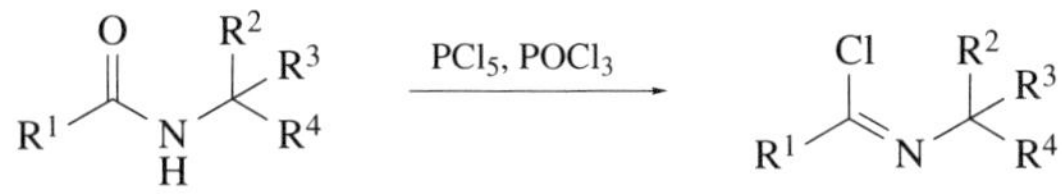

(14) (9)

Bis(benzoylamino)methanes and bis(benzoylamino)arylmethanes react with $PCl_5/POCl_3$ mixtures to yield *N*-chloroalkylbenzimidoyl chlorides ⟨78JOU1457, 78ZOR1559⟩. The reaction of *N*-methylisoinolone with $PCl_5/POCl_3$ affords 2-choroisoinoline ⟨78JHC1513⟩. *N*-Methylbenzamide affords *N*-methylbenzimidoyl bromide upon treatment with phosphorus pentabromide ⟨92JCS(P2)281⟩.

(iii) Reactions with thionyl chloride and sulfuryl chloride

Several papers describe general methods for preparing benzimidoyl chlorides (**15**) from the corresponding benzamides. *N*-Alkylimidoyl chlorides and *N*-arylimidoyl chlorides are mainly obtained in the same way (Equation (10)) ⟨62CB126, 72CB1258, 72JCS(P2)1908, 77CB2996, 78JOC381, 82JOC3658, 84JOC546, 87JCS(P1)1389, 87JGU1104, 87ZOB1233, 88C230⟩. The reaction is usually performed in thionyl chloride as solvent at 80–90 °C. *N*-Benzoyl-*N′*-alkyl- or -*N′*-alkylaryl-thioureas are transferred into nickel complexes which afford on reaction with $SOCl_2$ in benzene or carbon tetrachloride the corresponding *N*-(aminothiocarbonyl)benzimidoyl chlorides (**16**) (Equation (11)) ⟨84T405, 88JPR241, 89ZAAC(576)203⟩.

(15) (10)

(16) (11)

Adamantanecarboximidoyl chlorides are obtained by reaction of the corresponding amide with $SOCl_2$ ⟨79LA74⟩. The reaction of 1,4-diaryl-2-cyano-1,4-diaza-1,3-butadienes with $SOCl_2$ in chloroform leads to 1,4-diaryl-2-cyano-3-chloro-1,4-diaza-1,3-butadienes ⟨88H(27)155⟩.

The disilver salt of *N*-(anilinothiocarbonyl)thiobenzamide reacts with sulfuryl chloride in chloroform to give *N*-[chloro-(iminophenyl)methyl]benzimidoyl chloride. The advantage of this method is that the chlorination of the aromatic nuclei is avoided ⟨81ZOR1720, 82JOU1537⟩.

(iv) Reactions with halogens

N-Phenyltrifluorothioacetamide reacts at 20 °C in $CHCl_3$ with Cl_2 to give *N*-phenyltrifluoroacetimidoyl chloride ⟨76JOU2148, 76ZOR2213⟩.

N-(Anilinothiocarbonyl)thiobenzamides undergo different reactions with chlorine depending on conditions and substituents. In CH_2Cl_2 at 0 °C, 3-aryl-5-arylimino-1,2,4-dithiazoline (**17**) is afforded. Reaction at 40 °C yields the corresponding aryliminochloromethyl benzimidoyl chlorides (**18**). Unsubstituted anilino groups are chlorinated additionally (**19**) (Scheme 2) ⟨81ZOR1720, 81JOU1527⟩.

5.17.1.2.2 From isocyanides, nitriles and isocyanates

N-Methyl-*C*-(trichlorotitanio)formimidoyl chloride is formed in the reaction of methylisocyanide with $TiCl_4$ in CH_2Cl_2 ⟨83HCA1618⟩. Analogously, *N*-phenyl-*C*-(trichlorotitanio)formimidoyl chloride

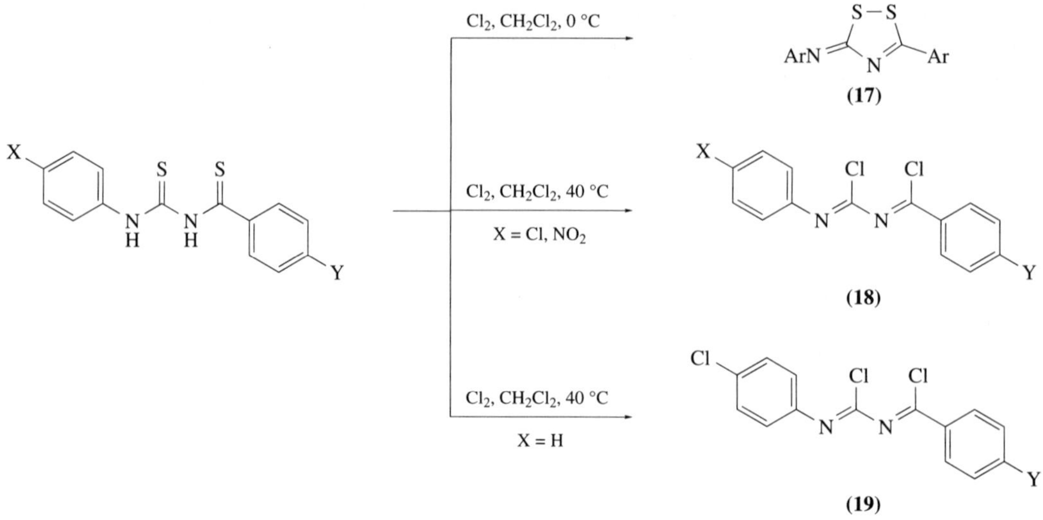

Scheme 2

is accessible from phenylisocyanide and $TiCl_4$ ⟨88CB507⟩. Pyrrolidine adds to the isocyano group of 1,1,2-trichloroethylisocyanide chromium pentacarbonyl complex to give *N*-1,2-dichloroethylidenamino chromium pentacarbonyl complex ⟨89ZN(B)1414⟩.

By a Ritter-type reaction in the presence of HF *N*-cyclohexylacetimidoyl fluoride can be prepared from cyclohexene and acetonitrile; cyclohexene and benzonitrile yield *N*-cyclohexylbenzimidoyl fluoride and 1-pentene gives with benzonitrile a mixture of *N*-(2-pentyl)benzimidoyl fluoride and *N*-(3-pentyl)benzimidoyl fluoride ⟨70JOC1619⟩. Similarly, *N*-(1-chloroisopropyl)benzimidoyl chloride and *N*-(2-chloroethyl)benzimidoyl chloride are obtained in good yield from benzonitrile, Cl_2 and propene or ethene, respectively ⟨81M(112)959⟩.

The reaction of phenylsubstituted benzotrichlorides with trichloroacetonitrile or dichloromalodinitrile in the presence of $AlCl_3$ affords *N*-pentachloroethylbenzimidoyl chlorides or *N*-(1,1,2,2-tetrachloro-2-cyanoethyl)benzimidoyl chlorides, respectively ⟨78S38⟩. The reaction of 2-substituted 3,3-(bisdimethylamino)acrylonitriles with phenylglyoxyloyl chlorides in carbon tetrachloride gives the corresponding 3,3-bis(dimethylamino)-*N*-(phenylglyoxyloyl)acrylimidoyl chlorides (**20**) (Equation (12)) ⟨86LA1997⟩.

(12)

(20)

N-Ethylaminoacetonitrile and *N*-isopropylaminoacetonitrile are transformed into the corresponding *N*-chloroaminoacetonitriles by reaction with $Ca(OCl)_2$. Further treatment with calcium hypochlorite in methylene chloride gives *N*-ethylamino cyanformimidoyl chloride (12% yield) or *N*-isopropylamino cyanformimidoyl chloride (33% yield) ⟨76JA1099⟩. On treatment of acylisocyanates with phosphorus pentachloride, *N*-(α-chloro-alkyliden)-carbamidic acid chlorides (**21**) are accessible. These compounds are reported to be in equilibrium with the corresponding α-dichloroalkyl isocyanates (**22**). The same reaction leads to *N*-chlorobenzylcarbamide chloride (Scheme 3) ⟨66CB239⟩. The photochlorination of benzylisocyanate at 100 °C in carbon tetrachloride yields *N*-(chlorocarbonyl)benzimidoyl chloride as well ⟨89JOU1365, 89ZOR1572⟩.

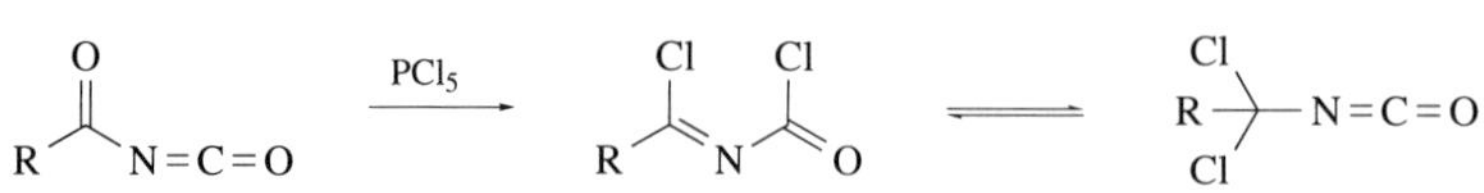

(21) **(22)**

R = Ph, CH_2Cl, $CHCl_2$, CCl_3

Scheme 3

5.17.1.2.3 From imines

The reaction of benzylidenealkylamines with either fluorine ⟨67JOC416⟩ or CF_3OF (Equation (13)) ⟨73TL2771⟩ leads to *N*-alkylbenzimidoyl fluorides such as (**23**), a mixture of *syn*- and *anti*-isomers. By-products are *N*,α,α-trifluorobenzylamines (**24**). The *anti*-isomer of *N*-*t*-butylbenzimidoyl fluoride decomposes rapidly at 25 °C to benzonitrile, isobutylene, *t*-butyl fluoride and, presumably, HF ⟨67JOC416⟩. The *syn*-isomer is more stable and does not decompose at room temperature.

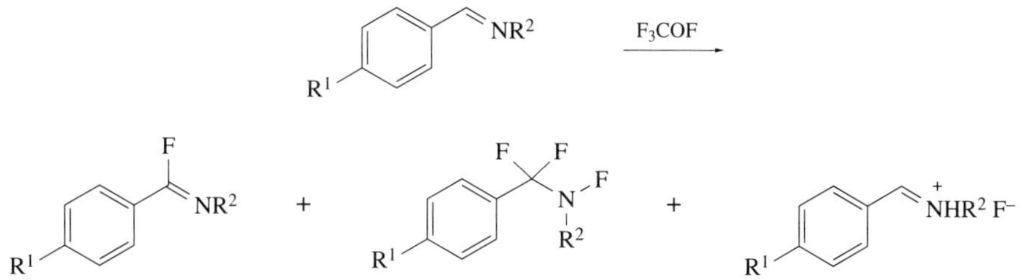

(23) (24) (13)

5.17.1.2.4 From amines with haloalkanes or haloalkenes

Trifluoroacetic acid is transformed into *N*-alkyl- or *N*-aryltrifluoroacetimidoyl chlorides or bromides in good yields by reaction with alkyl- or arylamines in the presence of triphenylphosphine, triethylamine and carbon tetrachloride or carbon tetrabromide, respectively. No reaction takes place with Cl_4, neither with perfluoropropanoic acid nor perfluorooctanoic acid ⟨93JOC32⟩.

The reaction of *N*-dichloroperfluoropropylamine with perfluoroethene or trifluorochloroethene gives *N*-chloro-*N*-perfluoropropyl perfluoroethylamine or *N*-chloro-*N*-perfluoropropyl-(2-chloro-tetrafluoroethyl)amine, respectively, from which the corresponding *N*-perhaloethylpropylimidoyl fluorides are accessible on reaction with triphenylphosphine or CsCl ⟨92IC4917⟩.

5.17.1.2.5 From imidoyl halides

N-(1,2,2-Trichloroethyl)benzimidoyl chlorides and *N*-(2,2,2-tribromo-1-chloroethyl)- or *N*-(2-bromo-1,2,2-trichloro-ethyl)benzimidoyl chlorides undergo elimination of HX to form the corresponding *N*-(1-chloro-2,2-dihalogenoethenyl)benzimidoyl chlorides on reaction with triethylamine ⟨76JOU2252, 76ZOR2319⟩. The chlorination of these compounds leads to *N*-(1,1-dichloro-2,2,2-trihalo-ethyl)benzimidoyl chlorides.

The photochlorination of *N*-(trichloroethenyl)trichloroacetimidoyl chloride proceeds with addition of chlorine to give *N*-(pentachloroethyl)trichloracetimidoyl chloride (**25**) (Equation (14) ⟨77JOU1471, 77ZOR1594⟩.

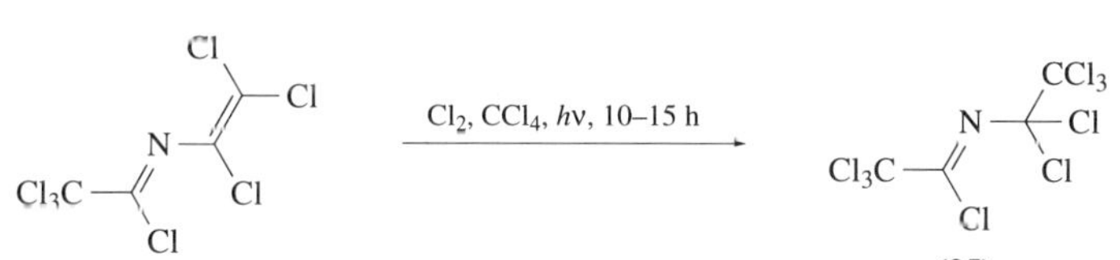

(25) (14)

The result of the photochlorination of *N*-(benzyl)trichloroacetimidoyl chloride depends on the reaction temperature. At room temperature, a mixture of *N*-(pentachloroethyl)benzimidoyl chloride (**27**) and *N*-(α-chlorobenzyl)trichloroacetimidoyl chloride (**26**) results. At 70–80 °C, *N*-(pentachloro-ethyl)benzimidoyl chloride (**27**) is formed exclusively (Scheme 4) ⟨79JOU32, 79ZOR38⟩.

N-Benzyltrifluoroacetimidoyl chloride is isomerized by triethylamine to give *N*-(2,2,2-trifluoro-ethyl)benzimidoyl chloride ⟨86JGU2372, 86ZOB2681⟩. Perhalogenated imidoyl chlorides are dehalogenated with activated zinc in sulfolane in a temperature range of −196 °C to 25 °C to form perfluorinated *N*-1-chloroalkenyl-alkanecarboximidoyl chlorides (**28**) (Equation (15)) ⟨91JFC(55)47⟩.

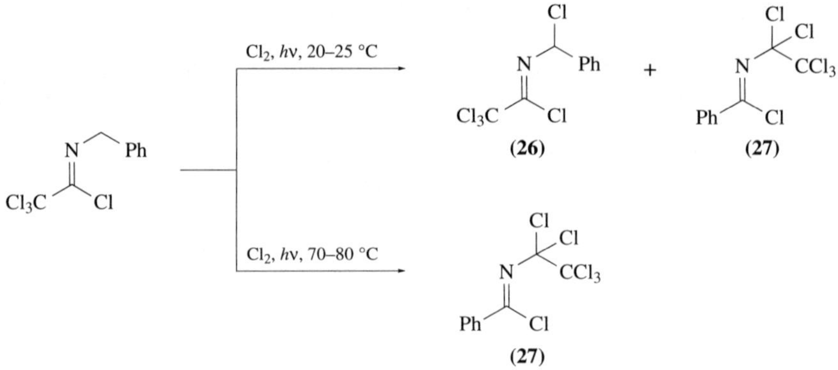

Scheme 4

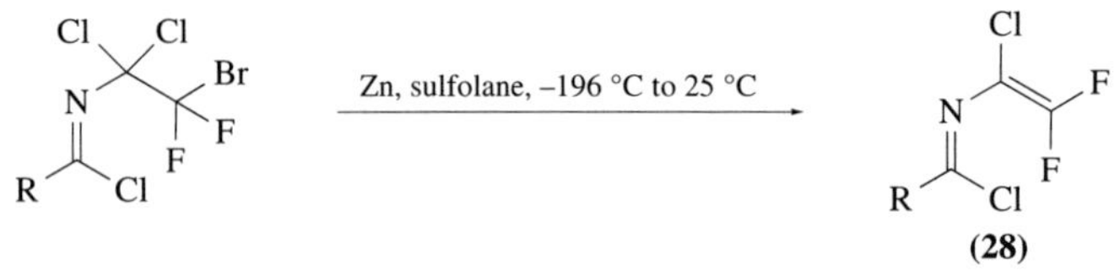

(15)

Imidoyl chlorides are transformed to the corresponding fluorides by treatment of imidoylchlorides with KF in acetonitrile in the presence of crown ethers (18-crown-6) ⟨78TL1429⟩. The reverse reaction can be achieved with $AlCl_3$ (**29**). Fluorine present in the *N*-α-position is also exchanged (**29**), (**30**) (Equation (16)) ⟨91JFC(55)47⟩. The chlorine of the α-chlorobenzyl group is very reactive. It can be replaced by various nucleophiles without affecting the imidoyl group. Examples are given in Scheme 5 (**32**–**34**) ⟨87JGU1349, 87ZOB1514⟩. *N*,*N*-Dialkylchloromethyleneiminium chlorides transform aromatic nitriles, amides or aldoximes to *N*-(dialkylaminomethylio)benzimidoyl chlorides (**31**) (Equation (17)) ⟨84JPR311⟩.

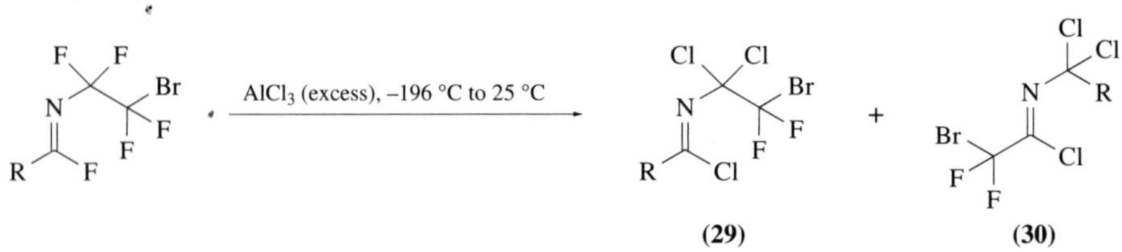

(16)

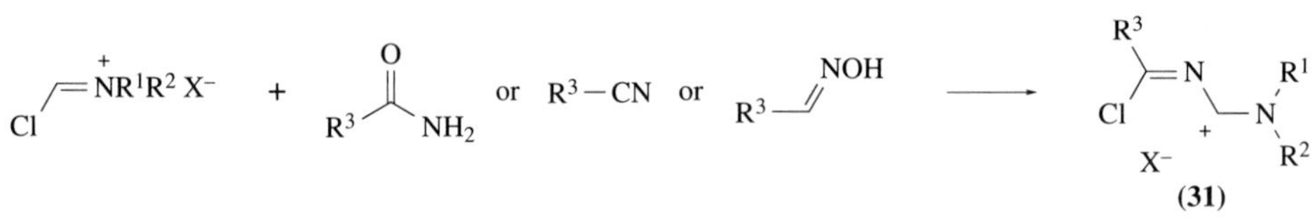

(17)

5.17.1.2.6 Miscellaneous methods

The reaction of *N*-alkyl- and *N*-arylhalogenacetamides with boron trihalides or dimethylboron bromide yields the corresponding imidoyl halides (**35**) (Equation (18)). If diphenylboron bromide is used, the reaction products are usually amidoboranes or carbiminoboranes ⟨77ZAAC(433)94⟩. With boron tribromide the pentacarbonyl(dimethylamino-[methoxy (phenyl)methyleneamino]carbene)-chromium(0) reacts to give ([bromo(phenyl)methyleneamino]dimethylaminocarbene)pentacarbonyl-chromium(0) and in a similar procedure ([bromo(phenyl)methyleneamino]dimethylaminocarbene)-pentacarbonyltungsten(0) can be obtained ⟨84JOM(267)277⟩.

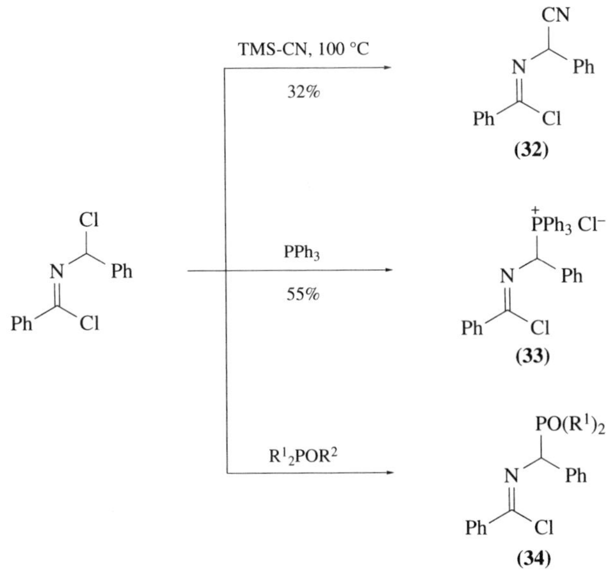

(32) **(33)** **(34)**

Scheme 5

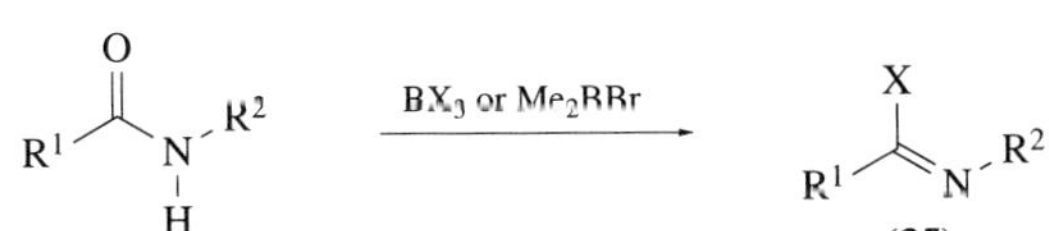

(35) (18)

The Beckmann rearrangement of acetophenone-oximeethyl carbonate in the presence of trimethylsilyl iodide affords *N*-phenylacetimidoyl iodide 〈83TL3255〉.

The reaction of 5-chloro-2*H*-1,4-oxazine-2-ones with alkenes leads to 6-chloro-2-oxa-5-aza-bicyclo[2.2.2]oct-5-en-3-ones 〈92S705〉. From 3-diethylamino-5-chloro-2*H*-1,4-oxazin-2-one and styrene the corresponding 6-chloro-2-oxa-4-diethylamino-5-aza-bicyclo[2.2.2.]oct-5-en-3-one (**36**) is obtained (Equation (19)) 〈92TL2049〉. Treatment of ethyl-2[Δ^2-1,3-oxazolinyl]-acetate in HCl and chloroform affords *N*-(2-hydroxyethyl)ethoxycarbonylacetimidoyl chloride (**37**) (Equation (20)) 〈91KGS1205, 92CHE579〉.

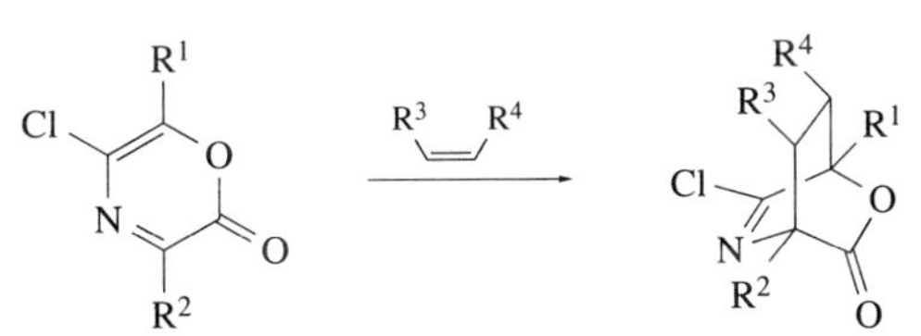

(36) (19)

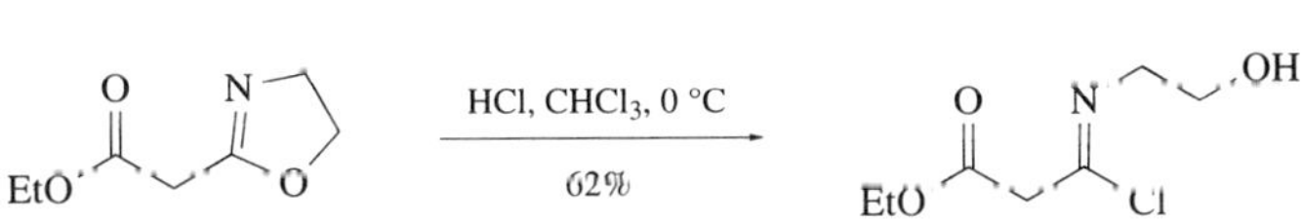

(37) (20)

The reaction of 4-chlorobenzotrichloride or perchloropropene with $SbCl_5$ and benzonitrile in 1,2-dichloroethane at -30°C affords the corresponding azonia allene salts (**38**) (Equation (21)) which can be transformed into other benzimidoyl chlorides (**39**) (**40**) by treatment with quaternary ammonium chlorides and subsequent hydrolysis (Scheme 6) 〈92T8271〉.

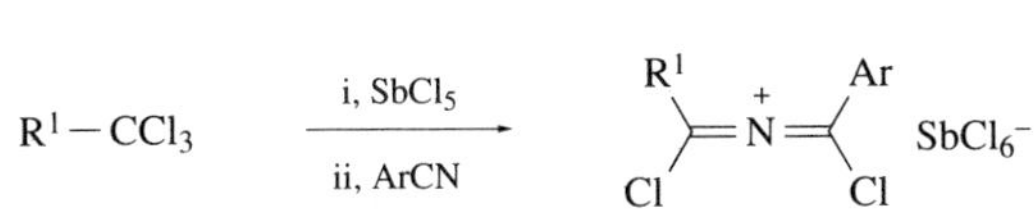

(38) (21)

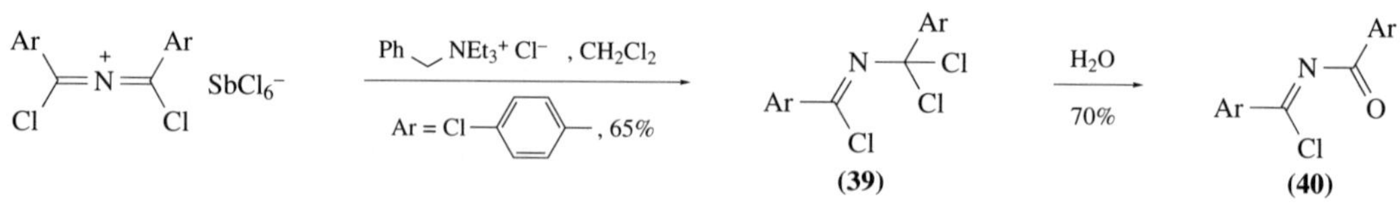

Scheme 6

Perfluoropiperidine is defluorinated by 2-substituted malonic acid esters/NaH in THF to give perfluoro-1-piperideine (**8**) (Equation (22)) ⟨91JFC(52)89⟩. α,α-Dihalogenated nitriles and trihalogenated acetonitriles react with bromine in the presence of CsF to give the corresponding *N*-bromoimidoyl fluorides (**41**) (Scheme 7). Further reaction with 1,1-difluoroethylenes yields *N*-(2-bromo-1,1-difluoroalkyl)haloalkaneimidoyl fluorides (**42**) (Scheme 7) ⟨86MI 517-01⟩. Hydrolysis of some of these compounds affords *N*-(2-bromo-1,1,2,2-tetrafluoroethyl)perfluorocarboxylic acid amides, which eliminate HF on heating with KF to give *N*-acylimidoyl fluorides (**43**) (Scheme 8) ⟨91JOC5256⟩. The reaction of perfluoropropyloxy *N*-perfluorobutylimino-fluoroformimidate with CsF yields *N*-(trifluoromethyl)-perfluorobutane-imidoyl fluoride (**44**) (Equation (23)) ⟨93JCS(P1)505⟩.

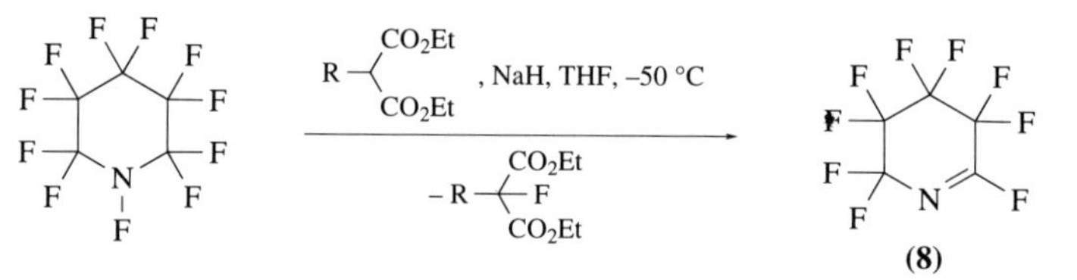

(22)

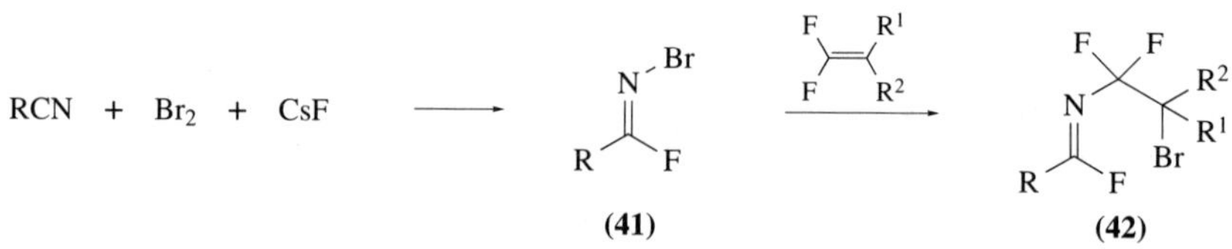

Scheme 7

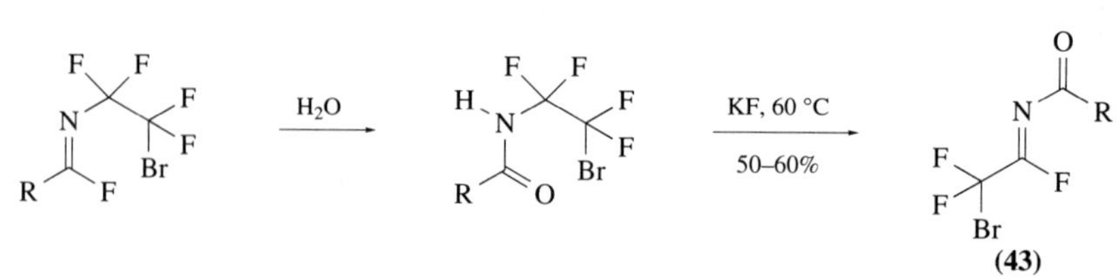

Scheme 8

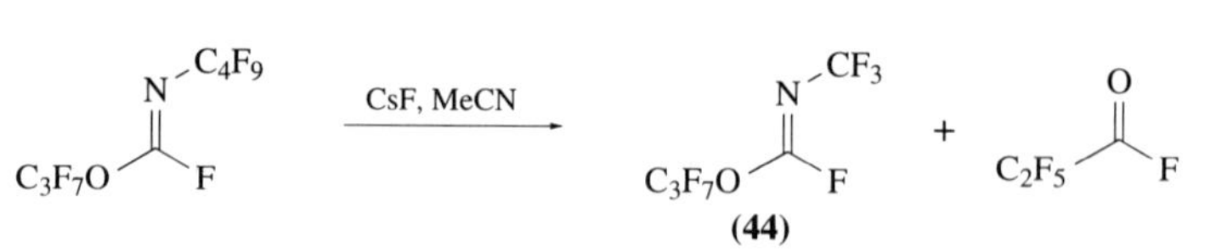

(23)

5.17.2 *N*-HALO-IMIDOYL HALIDES

Several *N*-chloroimidoyl chlorides have been prepared by addition of chlorine to the cyano groups of trichloroacetonitrile, cyanogen and malononitrile (Equations (24) and (25) and Scheme 9) ⟨71RTC458, 71RTC463⟩. *N*-Halogenated imidoyl fluorides are accessible by addition of chlorine or bromine to α-perfluorinated nitriles or trichloroacetonitrile in the presence of HgF_2 ⟨82CC839, 85CB4459⟩ or CsF ⟨84JOC1467, 86MI 517-01⟩ (Scheme 10). When these additions were applied to cyanogen, it turned out that the nature of the resulting *N*-halogenated imidoyl fluorides was strongly dependent on the stoichiometric amounts of the starting materials and the reaction conditions

(Scheme 11) ⟨82CC839, 83CB724, 91IC535⟩. Similar addition reactions have been performed with *N*-fluoro-cyanoformimidoyl fluoride (Equation (26)) ⟨90JFC(48)367⟩.

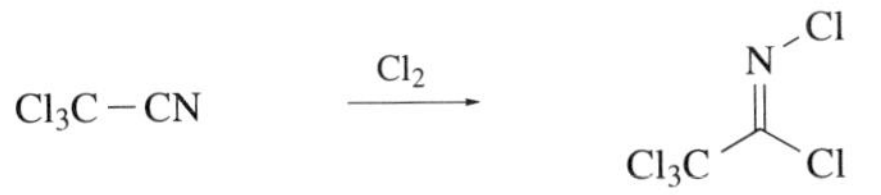

(24)

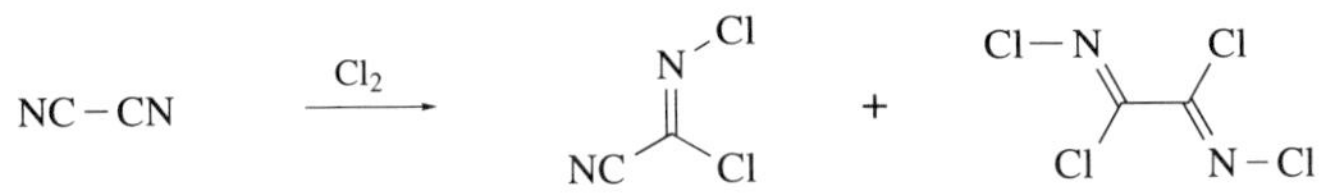

(25)

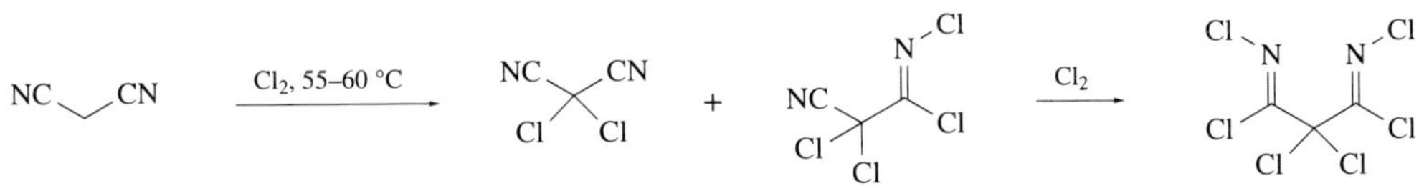

Scheme 9

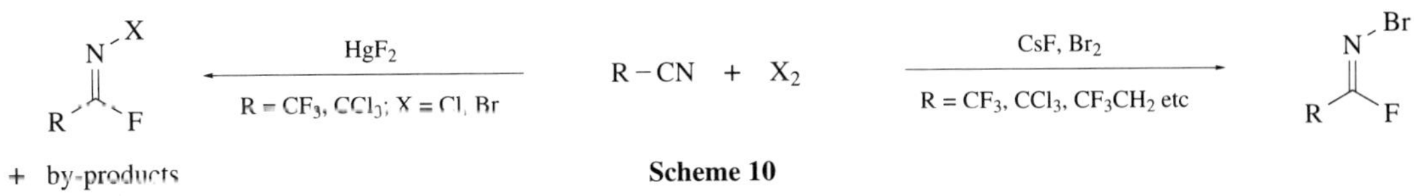

Scheme 10

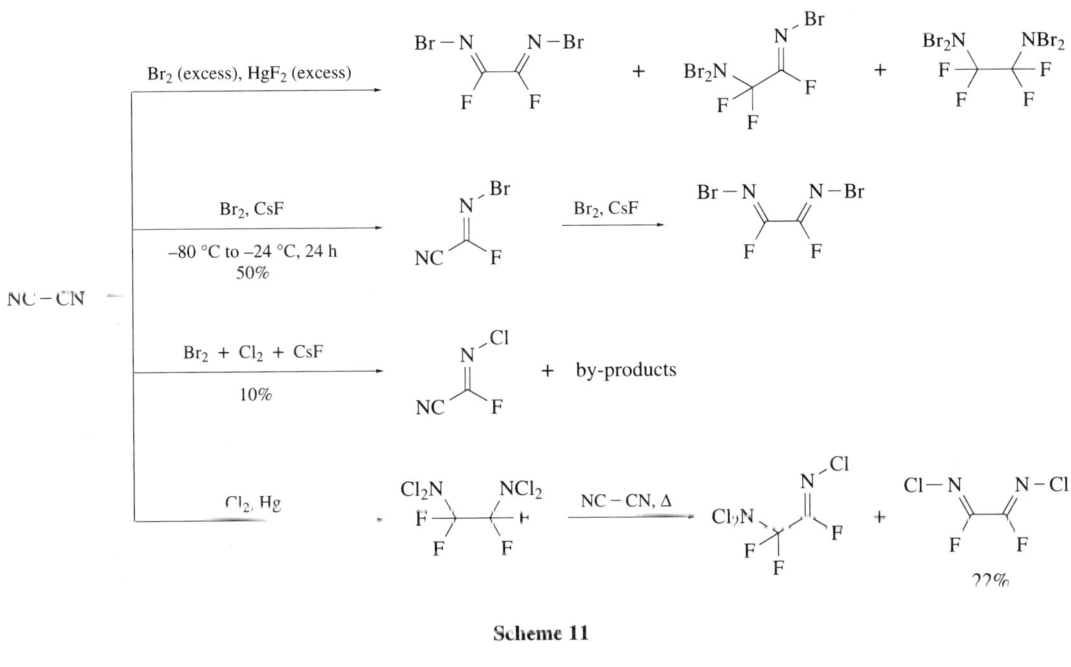

Scheme 11

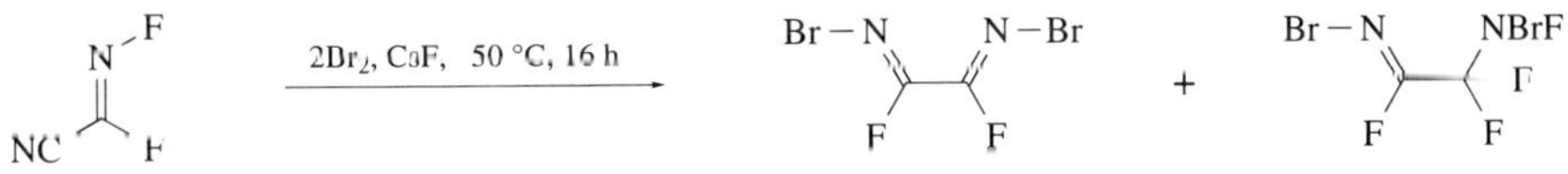

(26)

Perfluorohydrazine reacts with trifluoroethene in the presence of KF to give perfluoroamino acetimidoyl fluoride. Under similar conditions, starting from 1-chloro-2,2-difluoro-ethene, a mixture of *N*-fluoro-cyanoformimidoyl chloride and *N*-fluoro-cyanoformimidoyl fluoride results (Scheme 12) ⟨71T51⟩. On heating with CsF, perfluoroalkoxy *N*-perfluoroalkyl fluoroformimidates are cleaved to give *N*-fluorinated carboximidoyl fluorides (Equation (27)) ⟨93JCS(P1)505⟩.

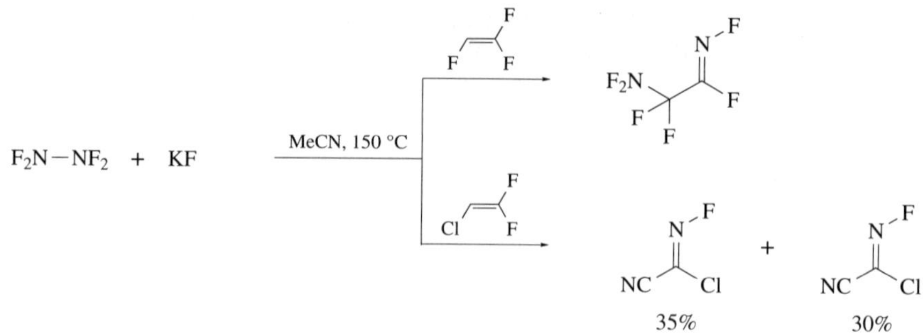

Scheme 12

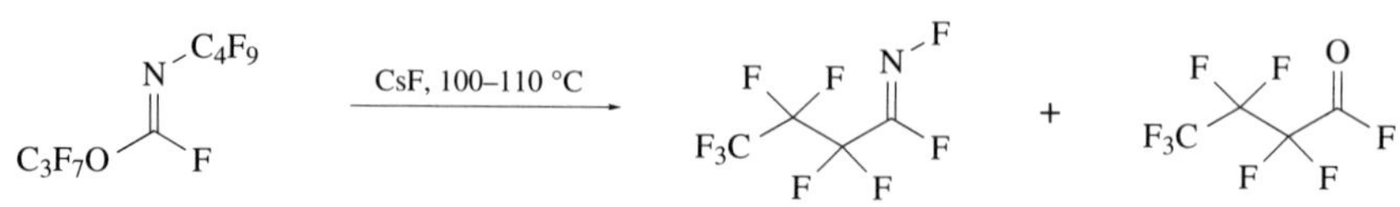

(27)

5.17.3 HYDROXAMOYL HALIDES

5.17.3.1 General Remarks

Hydroxamoyl halides are a well known class of compounds with high reactivity. Nitrile oxides, which are easily accessible from them, are important compounds for cycloaddition reactions. As a consequence of the increased interest in this type of reaction, many hydroxamoyl chlorides with great structural variety have been synthesized. Reviews on these compounds are available—some with detailed preparation procedures ⟨B-68MI 517-02, 85HOU(E5/1)631⟩.

5.17.3.2 Preparation Procedures

5.17.3.2.1 By halogenation of aldoximes

The halogenation of aldoximes is a standard method for the preparation of hydroxamoyl chlorides ⟨B-68MI 517-02, 85HOU(E5/1)631⟩; suitable solvents for such reactions are halogenated solvents, for example chloroform ⟨90SRI437⟩ or carbon tetrachloride. In many cases the chlorination can be achieved advantageously in concentrated hydrochloric acid at temperatures of about 0 °C ⟨81JCR(S)129⟩; the chlorine can be produced *in situ* from NaOCl ⟨84G133⟩ or from potassium hydrogenperoxomonosulfate(oxone) ⟨92JOC6649⟩ and HCl. Photochlorination of oximes also affords hydroxamoyl chlorides ⟨89OPP517, 92SRI851⟩, for example terephthaloylhydroxamoyl chloride ⟨89OPP517⟩. *N*-Chlorosuccinimide (NCS) is a suitable reagent for the conversion of substituted benzaldoximes ⟨80JOC3916⟩ and heterocyclic aldoximes ⟨84JMC1201⟩ to the corresponding hydroxamoyl chlorides. Benzhydroxamoyl chlorides, cinnamhydroxamoyl chlorides and thienoylhydroxamoyl chlorides are available from the corresponding oximes and *t*-butyl hypochlorite in CCl_4 at 20 °C ⟨86SC763⟩ (Scheme 13). With the aid of NBS in DMF, trifluoroacethydroxamoyl bromide was prepared from trifluoroacetaldoxime ⟨84BCJ2184⟩.

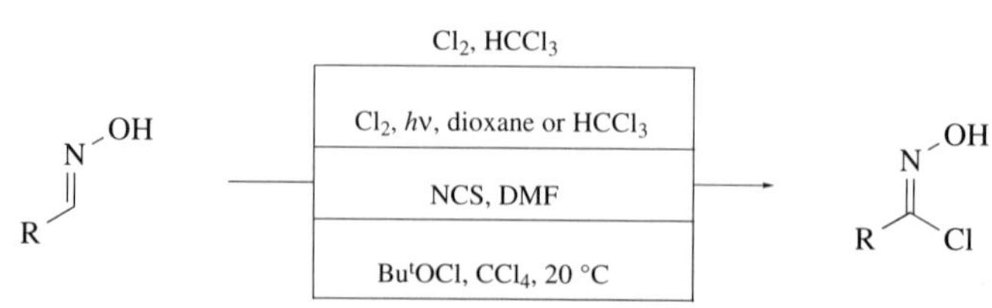

Scheme 13

Aldoximes are transformed to hydroxamoyl chlorides at temperatures of about $-20\,^{\circ}C$ by nitrosyl chloride in ether ⟨B-68MI 517-02, 72JCS(P2)1911⟩. Furancarboxhydroxamoyl chlorides are also accessible by this procedure (Equation (28)) ⟨76CCC3085⟩.

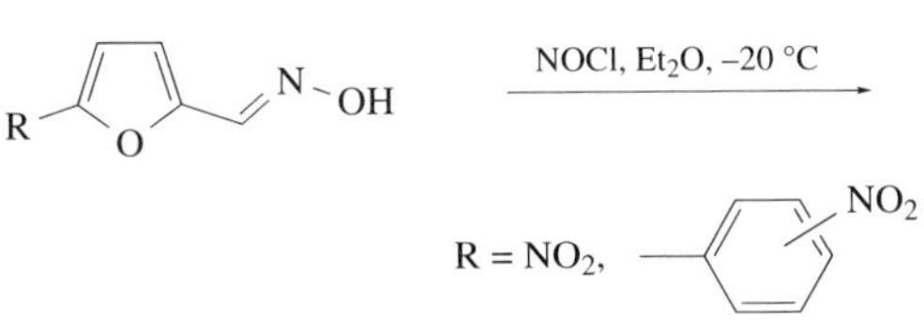

(28)

5.17.3.2.2 *By halogenation of acyl- and heteroaryl nitroalkanes*

When 3-(2-nitroethyl)indole is treated with sodium methoxide in methanol/ether and subsequently with $SOCl_2$ in DME at $-7\,^{\circ}C$, the hydroxamoyl chloride is obtained **(45)** (Equation (29)) ⟨90TL1417⟩. On heating in chloroform, benzoylnitromethane and $TeCl_4$ give the hydroxamoyl chloride **(46)** in good yield (Equation (30)) ⟨90CL559⟩.

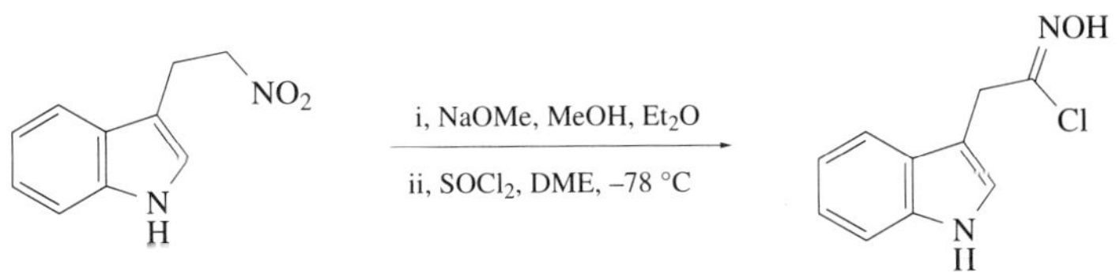

(45)

(29)

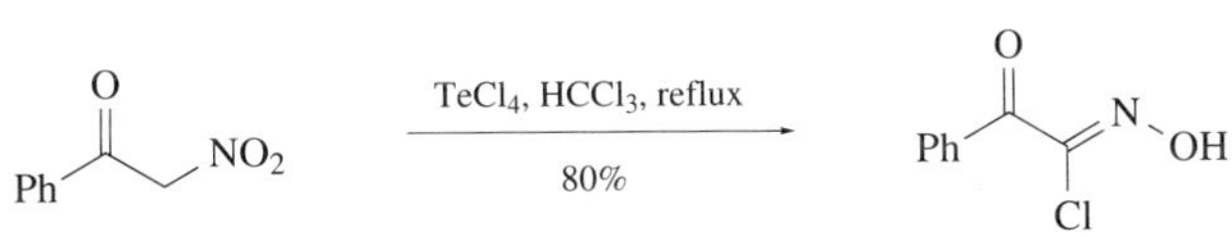

(46)

(30)

5.17.3.2.3 *From hydroxamic acids or hydroxamic acid derivatives and halogen carriers*

N-Alkoxy-carboxylic acid amides react with thionyl chloride ⟨80LA791, 84IJC(B)728, 85JOC993⟩ or phosphorus pentachloride ⟨B-68MI 517-02, 80LA791, 84IJC(B)728, 84JOC919⟩ or $POCl_3/PCl_5$ ⟨91S750, 92JOC3245⟩ to give the corresponding *N*-alkoxycarboximidoyl chlorides (Equation (31)). The PCl_5 route has been applied to prepare the trifluoroacetyl compound **(47)** (Equation (32)); a by-product is the 1,3,4-dioxazole **(48)**. The homologue 1,3,4-dioxazole **(50)** (Equation (33)) is the sole product, if the fluorinated amide **(49)** is treated with PCl_5 ⟨84JOC919⟩.

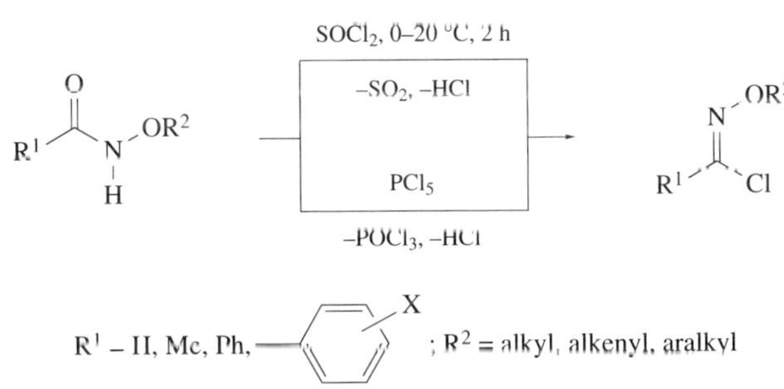

(31)

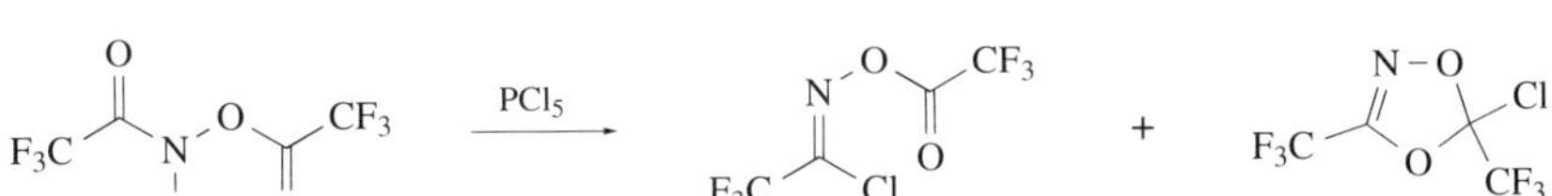

(47) 58% **(48)** 11%

(32)

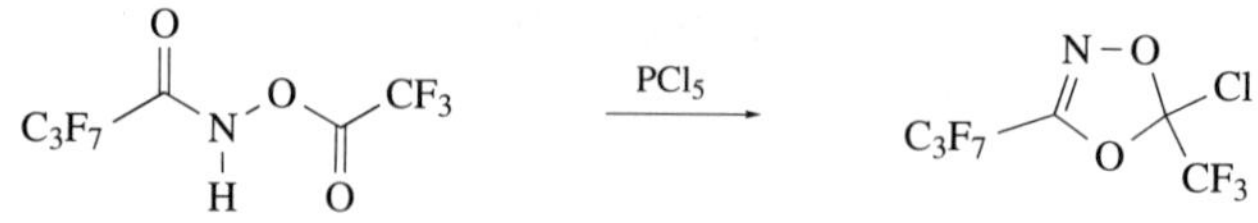

(49) **(50)** (33)

Thionyl chloride reacts with the *N*-hydroxyamide (**51**) (Scheme 14) via bis-chlorosulfites (**52**) to give the *N*-chlorosulfonyloxybenzimidoyl chlorides (**53**), which on alcoholysis affords the aromatic hydroxamoyl chlorides (**54**) ⟨84BCJ2011⟩. The action of $SOCl_2$ on benzhydroxamic acid gives the 1,3,2,4-dioxathiazol-*S*-oxide (**55**) (Equation (34)); this finding suggests that these methods are in general only applicable for *O*-protected hydroxamic acids. *N*-Alkoxyamides of aromatic and aliphatic carboxylic acids are converted by CBr_4/PPh_3 in acetonitrile to *N*-alkoxyimidoyl bromides (**56**) (Equation (35)) ⟨91S750⟩, and a route to *N*-alkoxy carboxylic acid imidoyl bromides by this method, starting from the carboxylic acids, has also been described (Equation (36)) ⟨93SC1061⟩.

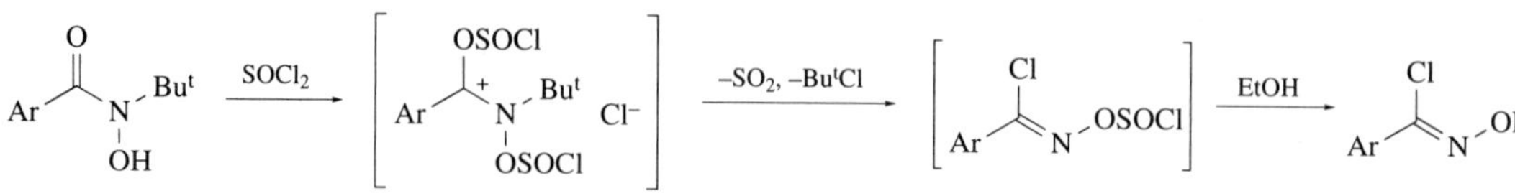

(51) **(52)** **(53)** **(54)**

Scheme 14

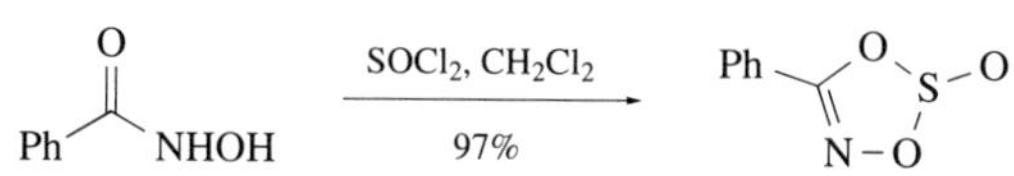

(55) (34)

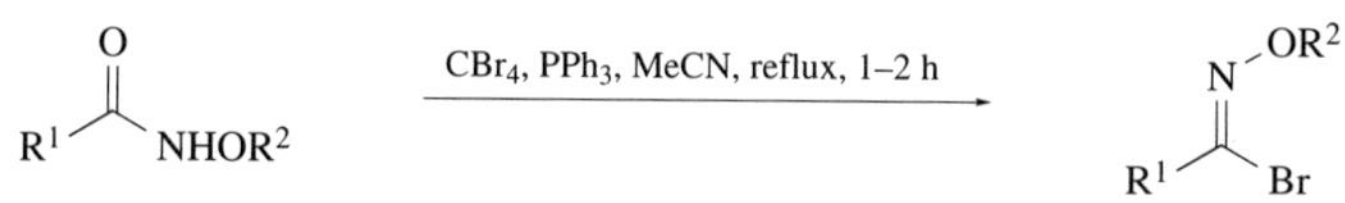

(56) (35)

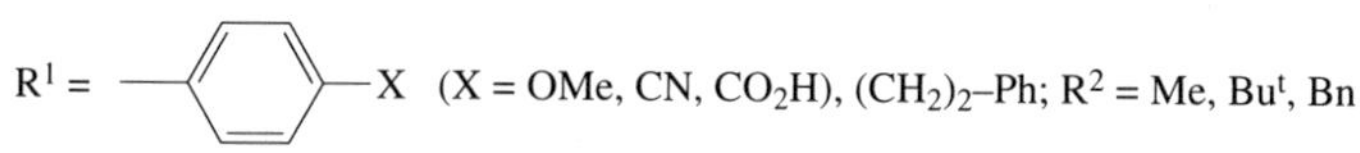

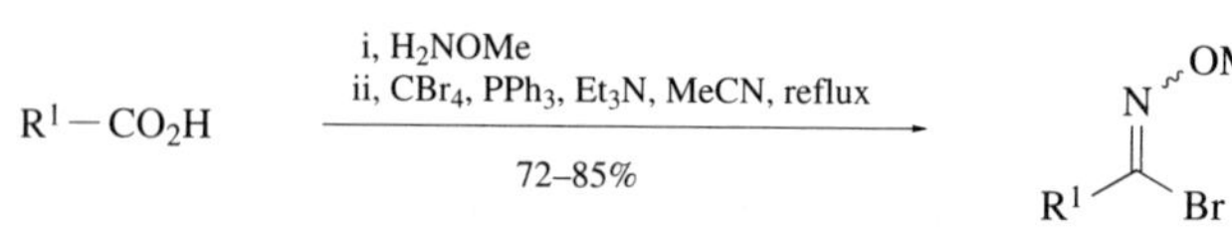

(36)

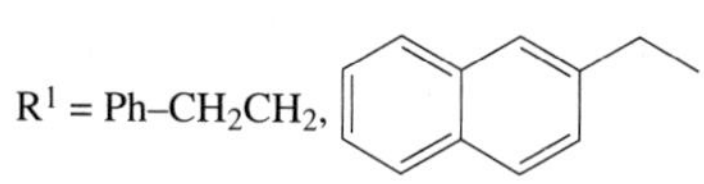

5.17.3.2.4 *From trichloromethyl compounds or activated 1,1-difluoroalkenes*

2,4,6-Trimethylbenzotrichloride reacts with excessive hydroxylamine in ethanol, with replacement of two chlorines by the hydroxyimino group, to give the hydroxamoyl chloride (**57**) (Scheme 15). The corresponding hydroxamic acid esters (**58**) are formed with sterically less hindered benzotrichlorides ⟨90BAU1538, 90IZV1692⟩. From chloral and hydroxylamine the acethydroxamoyl chloride (**59**) can be obtained (Equation (37)) ⟨83JOC366⟩. Perfluoromethacrylic acid derivatives react with hydroxylamine to yield hydroxamoyl fluorides (**60**) (Equation (38)) ⟨86MI 517-02⟩.

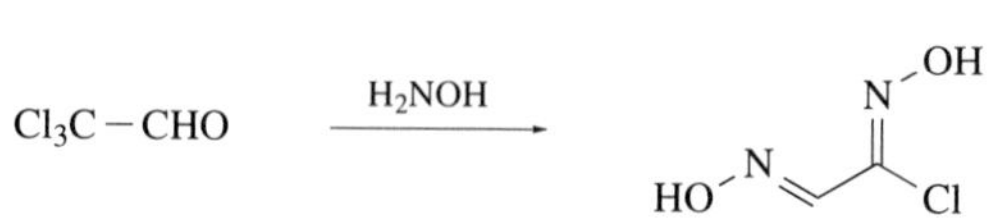

(59) (37)

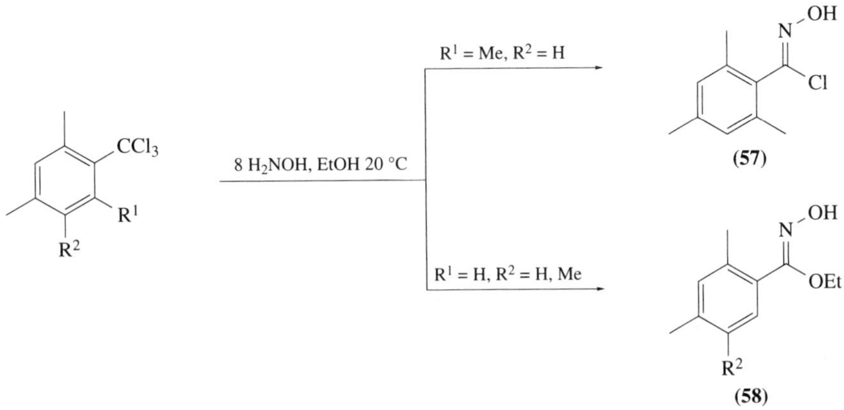

Scheme 15

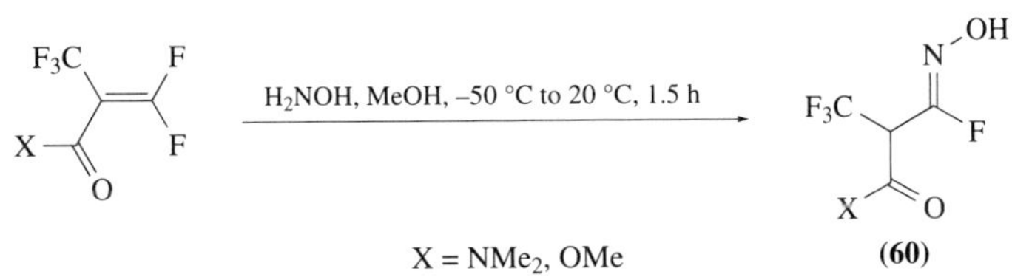

(38)

5.17.3.2.5 *By nitrosation of α-halogenated ketones, chloromethyl heterocycles, arylmethylketones and amidoximes*

The nitrosation of 1,3-dibromoacetone affords a mixture of the two hydroxyimidoyl bromides (**61**) and (**62**) (Scheme 16) together with the pyrazolone-1,2-dioxide (**63**). In contrast to this, the bis-hydroxamoyl chloride (**64**) as well as the mono-hydroxamoyl chloride (**65**) can be obtained from 1,3-dichloroacetone by variation of the nitrosation conditions. The nitrosation of 3-bromo-2-oxopropionhydroxamoyl halides yields bis-hydroxamoyl halides (**66**) (Equation (39)) ⟨91BAU438, 91IZV506⟩.

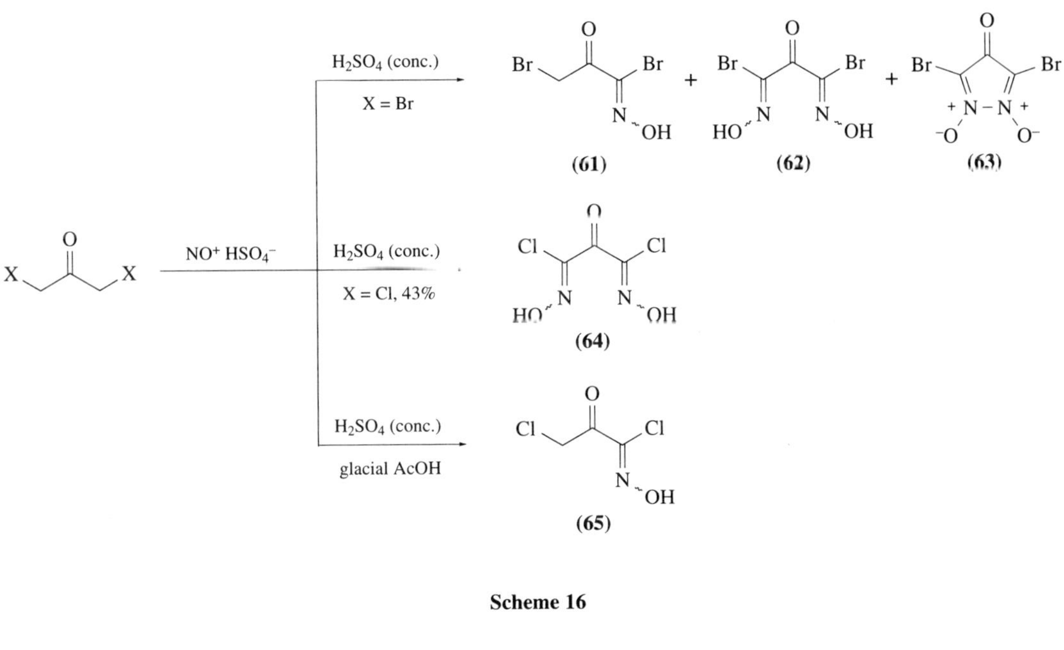

Scheme 16

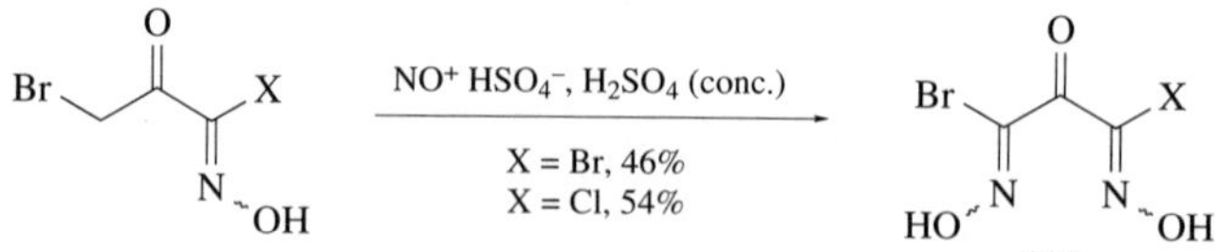

(66)

(39)

Chloromethyl heterocycles are transformed to the hydroxamoyl chlorides (**67**) (Equation (40)) by isopentyl nitrite in the presence of HCl in dioxane ⟨79S102⟩. By a similar procedure arylmethyl ketones are converted into 2-aryl-2-oxo-acethydroxamoyl chlorides (**68**) (Equation (41)) ⟨86JMC1689⟩. The nitrosation of amidoximes (**69**), (**72**) and (**74**) derived from malonic acid, affords the corresponding α-hydroxyimino-substituted imidoyl halides (**70**) and (**71**) (Scheme 17); (**73**) and (**75**) (Equations (42) and (43)) ⟨92SC453⟩.

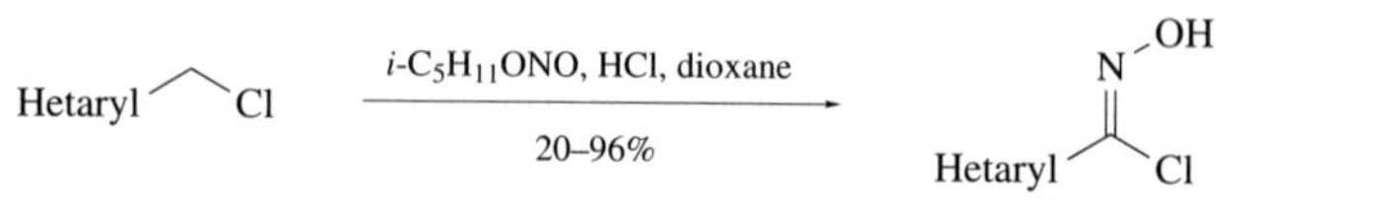

(67)

(40)

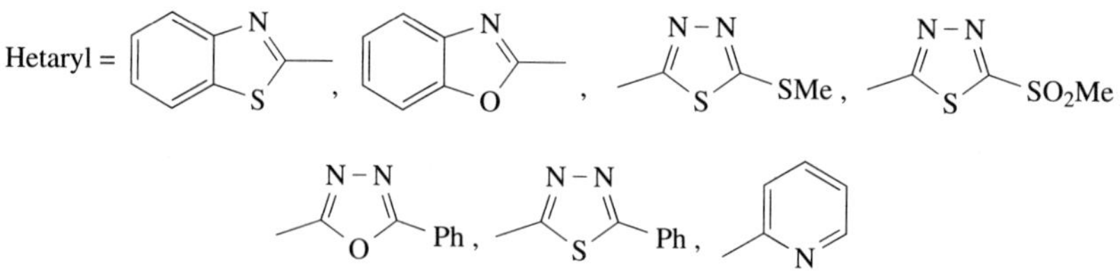

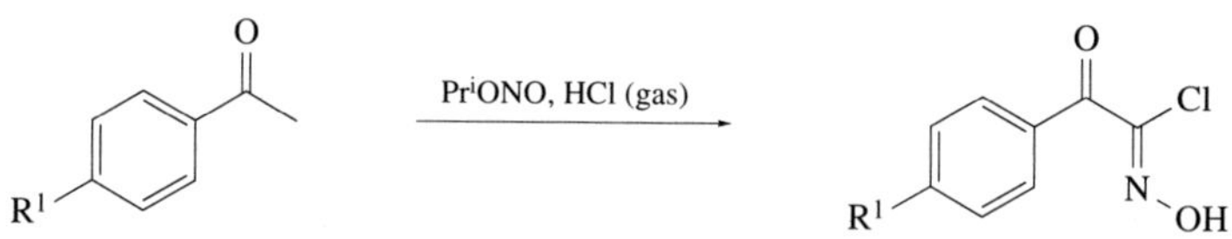

(68)

(41)

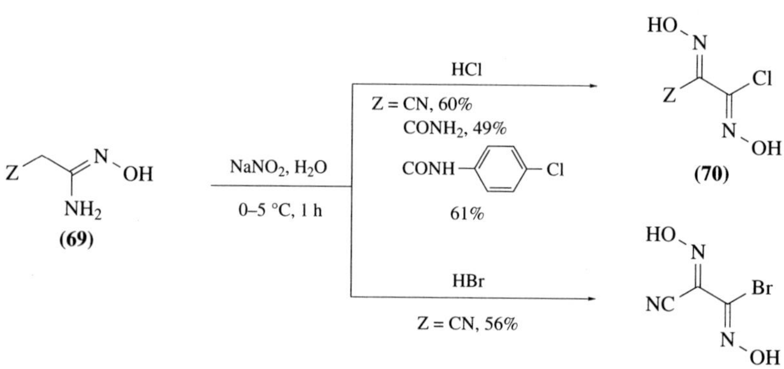

(69) (70) (71)

Scheme 17

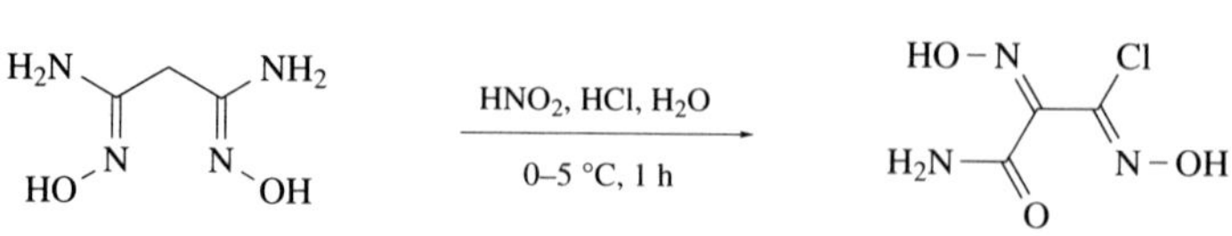

(72) (73)

(42)

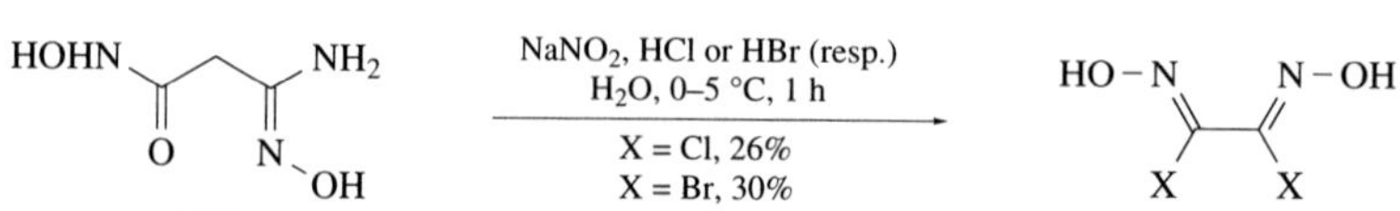

(74) (75)

(43)

5.17.3.2.6 *From dichloro nitrosoalkanes and phosphorus compounds, carboxylic acid chlorides or chlorosilicon compounds*

Various halogenated phosphorus derivatives have been reacted with dichloronitrosoalkanes in the presence of zinc ⟨88BAU1506, 88IVZ1691⟩, SO_2 ⟨87JGU1479, 87ZOB1659, 91PS(60)85⟩ or alcohols ⟨93PS(78)153⟩ to give *O*-phosphorylated hydroxamoyl chlorides **(76)**–**(79)**, (Scheme 18). Under the same conditions, carboxylic acid chlorides ⟨89KFZ1317, 90BAU415, 90IZV477⟩ or trimethylchlorosilane ⟨89BAU672, 89BAU2612, 89IZV745, 89IZV2849⟩ react analogously with dichloronitrosoalkanes to give the compounds **(80)** and **(81)** (Scheme 18). Similar reactions have been performed with phosphinic acid esters ⟨87BAU204, 87IZV226⟩, chlorophosphinic acid derivatives ⟨87BAU1000, 87IZV1086⟩, diphenylphosphineous acid ⟨87JGU1478, 87ZOB1658⟩ dialkoxychlorophosphines ⟨90BAU1007, 90IZV1122⟩, and alkyl dichlorophosphines ⟨89BAU2628, 89IZV2865⟩ yielding the hydroxamoyl chloride derivatives **(82)**–**(85)** (Scheme 19).

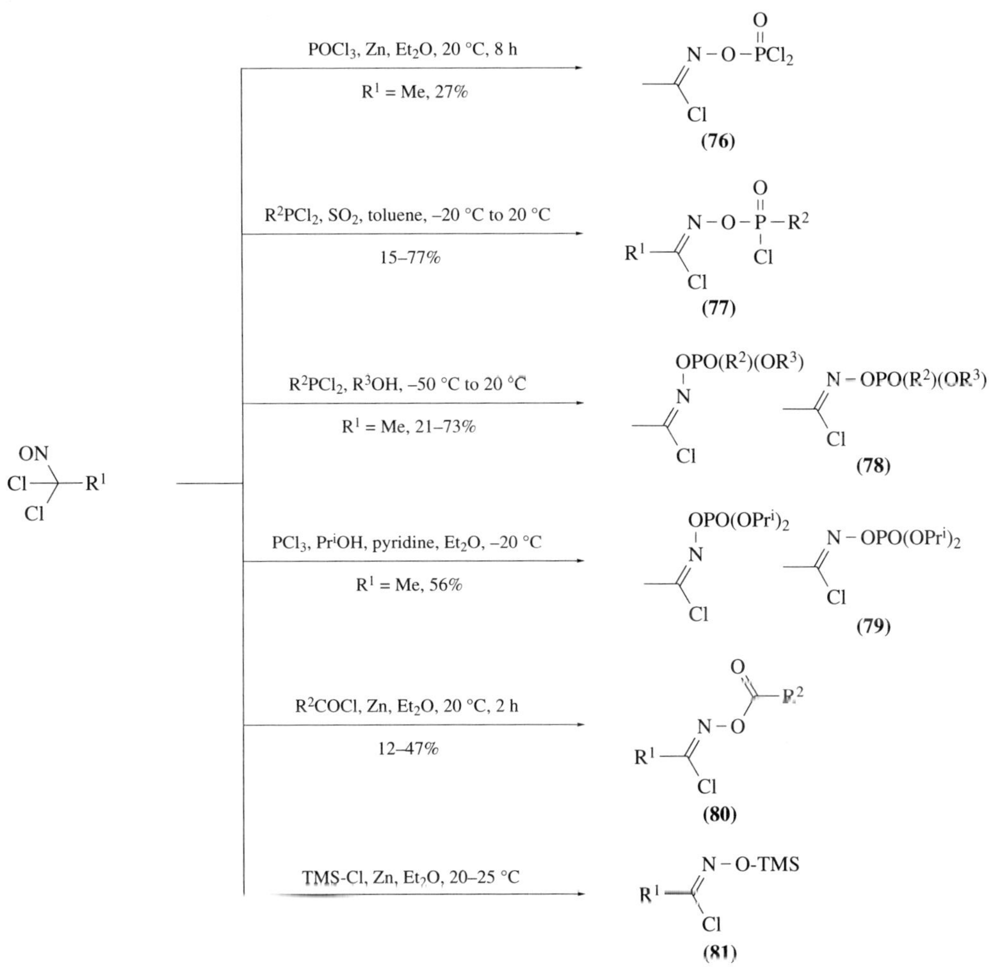

Scheme 18

5.17.3.2.7 *From nitrile oxides and from 1-nitroso-1,2,2-trichloroethene*

Hydrogen chloride can be added to nitrile oxides to give hydroxamoyl chlorides (Equation (44)) ⟨B-68MI 517-02⟩. This preparation method is of little interest, since nitrile oxides are usually prepared from hydroxamoyl halides. The addition of carboxylic acid chlorides ⟨91BAU382, 91IZV444⟩ or chlorocarbonyl isocyanate ⟨90JOU287, 90ZOR339⟩ to nitrile oxides is well suited to prepare *O*-acylated hydroxamoyl chlorides, for example **(86)** and **(87)** (Scheme 20 and Equation (45)). Similar reactions of aromatic nitrile oxides and perhalogenated isocyanates afford reactive hydroxamoyl chlorides, for example **(88)** (Equation (46)) ⟨89UKZ1071, 91UKZ202⟩.

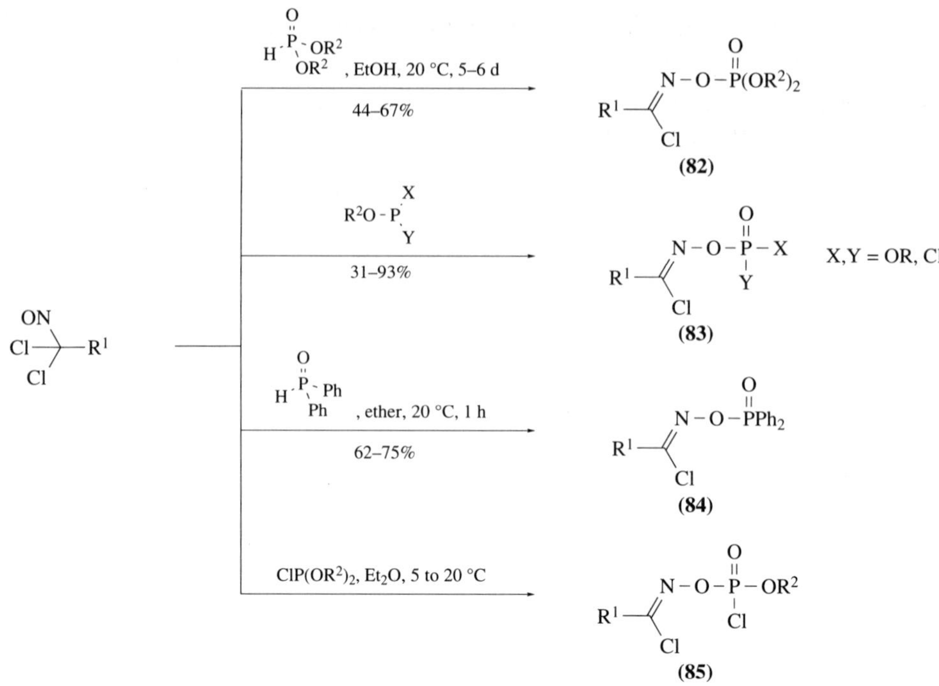

Scheme 19

N-Methanesulfonyloxy imidoylchlorides (**89**) (Equation (47)) can be obtained from nitrile oxides and methanesulfonyl chloride under catalysis of triethylamine ⟨66CJC297⟩. Treatment of trichloro-acethydroxamoyl chloride with sodium carbonate in CH_2Cl_2 forms 1-nitroso-1,2,2-trichloroethene (**90**) (Equation (48)) which undergoes cycloadditions with various alkenes and thiobenzophenone to give 'heterocyclic' hydroxamoyl chlorides ⟨87BSB51⟩.

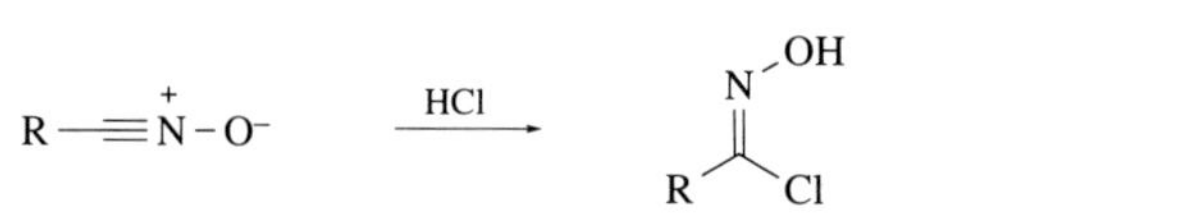

(44)

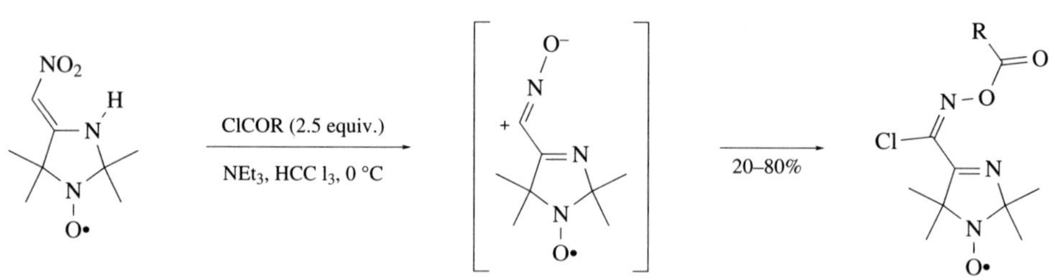

(86)

Scheme 20

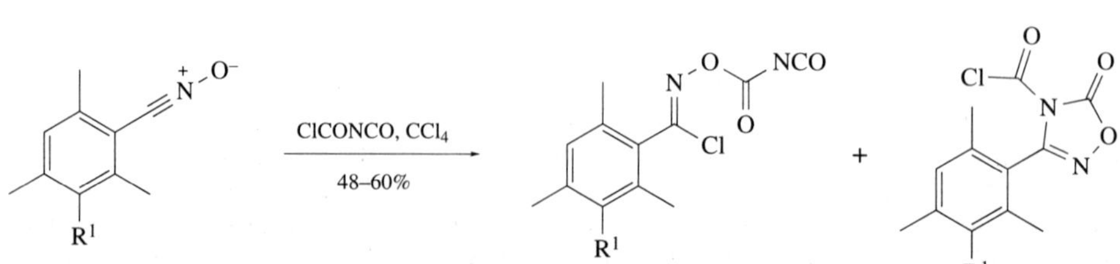

(87)

(45)

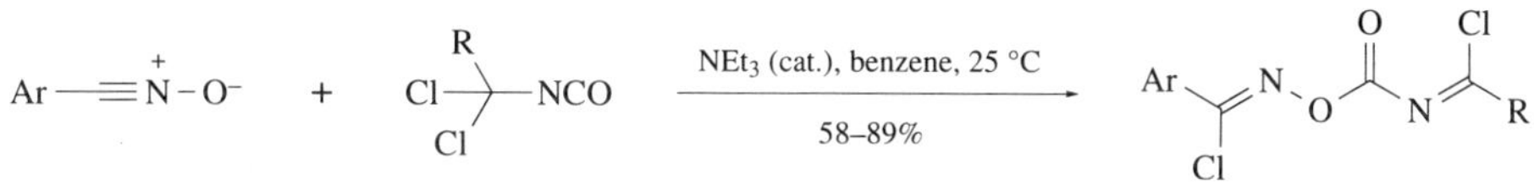

(88) (46)

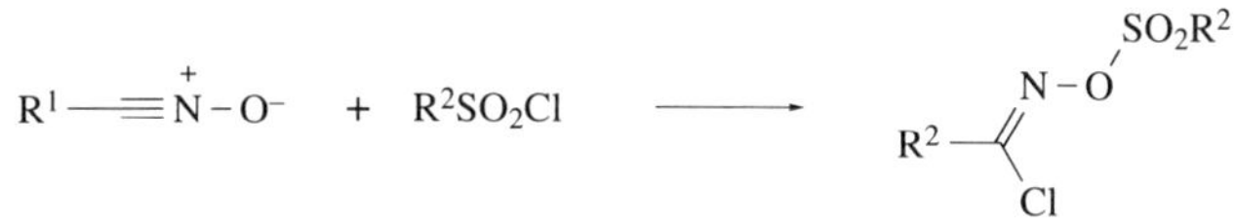

(89) (47)

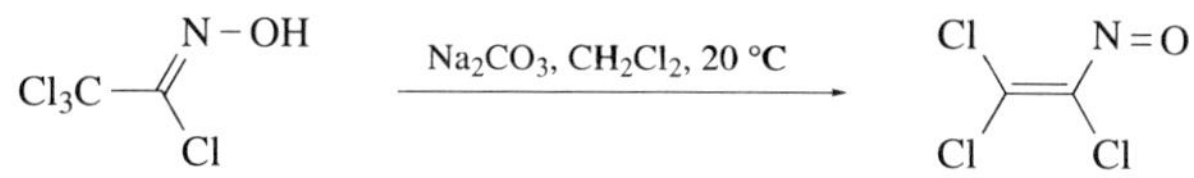

(90) (48)

5.17.3.2.8 From hydroxamoyl halides

The skeleton of hydroxamoyl halide molecules can be varied by many standard procedures without affecting the imidoyl halide function. Thus, brominations of compounds of the **(91)** type (Equation (49)) ⟨91BAU438, 91IZV506⟩ as well as transformations of an aldoxime group in **(92)** (Equation (50)) to a nitrile function ⟨83JOC366⟩, or the formation of carbamates **(94)** (Equation (51)) from the hydroxamoyl chloride **(93)** and isocyanates ⟨90JPR731⟩ are possible. Even the hydrolysis of an activated dichloromethylene group in the heterocyclic hydroxamoyl chloride **(95)** (Equation (52)) has been reported ⟨87BSB51⟩. Perfluorohydroxamoyl fluorides can be converted to the corresponding imidoyl halides by treatment with hydrogen chloride or hydrogen bromide (Equation (53)) ⟨66JGU1340, 66ZOB1326⟩. Hydroxamoyl chlorides (e.g. **(96)**–**(100)** can be functionalized at the hydroxy groups by treatment with thionyl chloride ⟨84BJC2011⟩, acetic anhydride ⟨84JHC725⟩, isocyanates ⟨81PJC655⟩, ethylvinyl ether ⟨87BSB51⟩ and hexafluoroacetone ⟨86MI 517-02⟩ (Equations (54)–(58)).

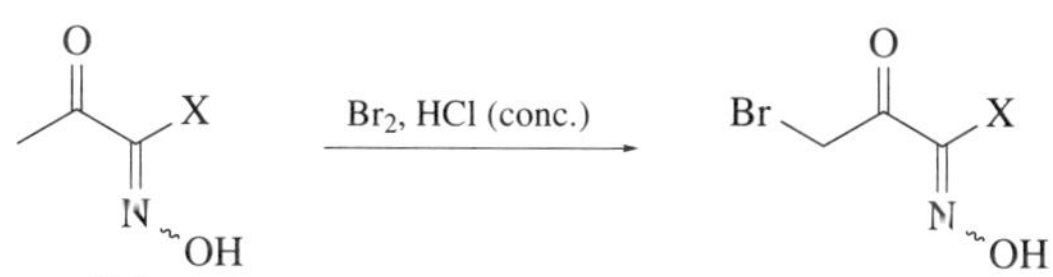

(91) (49)

(92) (50)

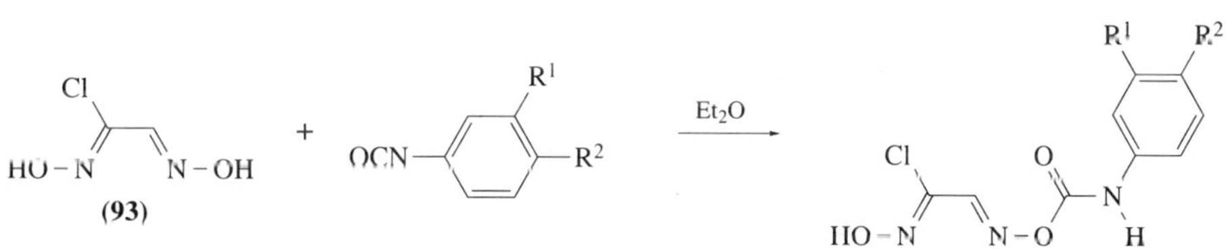

(93) (94) (51)

(95) (52)

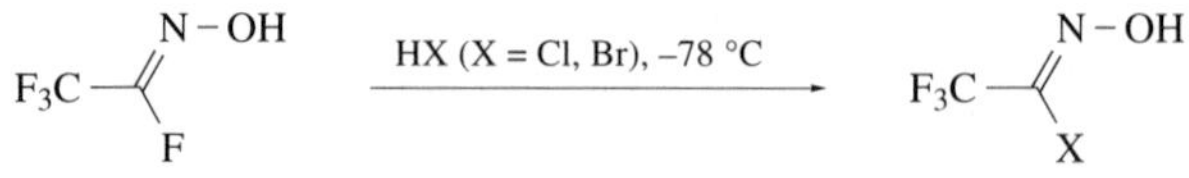

(53)

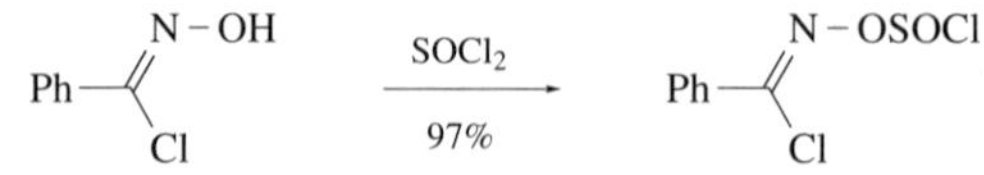

(96)

(54)

(97)

(55)

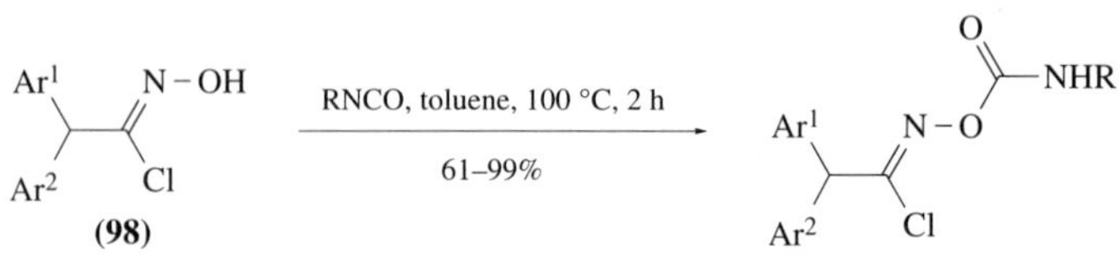

(98)

(56)

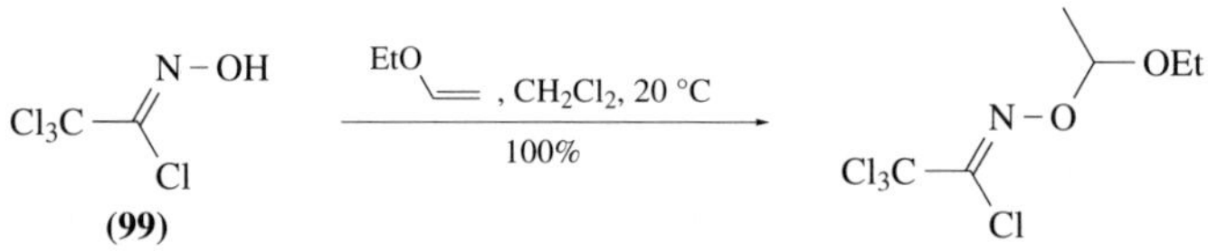

(99)

(57)

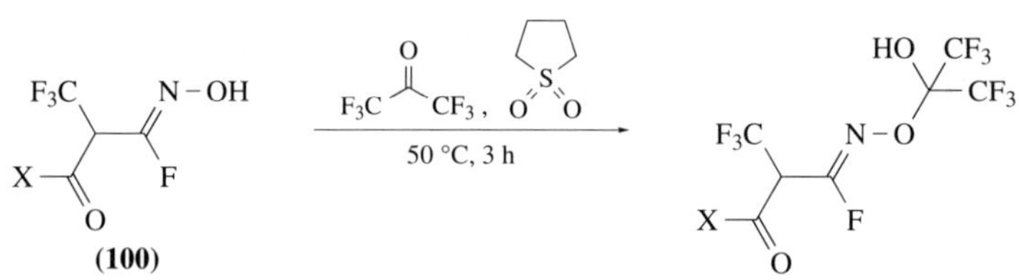

(100)

(58)

On heating with the halogenated isocyanate (**102**), the *O*-silylated hydroxamoyl chloride (**101**) gives the hydroxamoyl chloride (**103**) (Equation (59)) ⟨89BAU2201, 89IZV2390⟩. Hydroxamoyl chloride derivatives with reactive functional groups attached to the oxygen of the oximino function can be derived by nucleophiles at these electrophilic centers without affecting the hydroxamoyl chloride function. Examples of such hydroxamoyl chlorides in which additional electrophilic groups are present include: trifluoroacetyl, for example in (**104**) (Equation (60)) ⟨84JOC919⟩; chlorocarbonyl, for example in (**105**) (Scheme 21) ⟨89BAU2612, 89IVZ2849, 91JGU1152, 91KFZ33, 91ZOB1268, 92BAU236, 92IZV315⟩; isocyanatocarbonyl, for example in (**106**) (Equation (61)) ⟨90JGU1718, 90ZOR339⟩; alkyltrichlorophosphoranyl, for example in (**107**) (Scheme 22) ⟨89BAU2628, 89IZV2865, 90JGU1718, 90ZOB1923, 91PS(60)85⟩, chloroalkoxyphosphono, for example in (**108**) (Scheme 23) ⟨87BAU1000, 87IZV1086, 90BAU1007, 90IZV1122⟩, chlorophenylphosphono, for example in (**109**) (Equation (62)) ⟨91PS(60)85⟩.

Z-Isomers of *O*-alkylated and *O*-acylated hydroxamoyl chlorides ⟨85JOC993, 86JCS(P2)995⟩ and bromides ⟨92JOC3245⟩ are isomerized to the *E*-isomers on irradiation with UV light in benzene or toluene (Equation (63)).

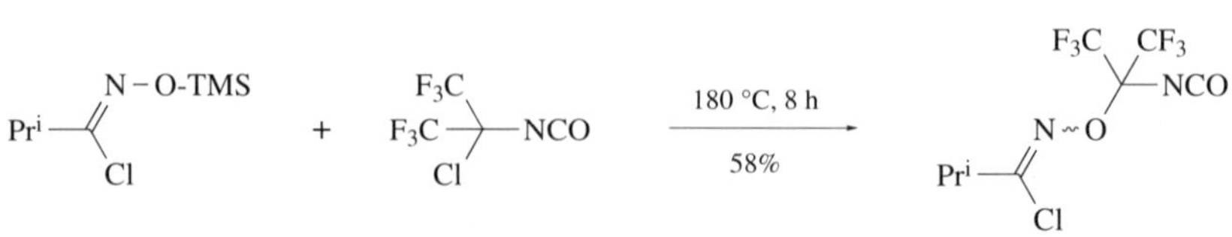

(59)

(101) **(102)** **(103)**

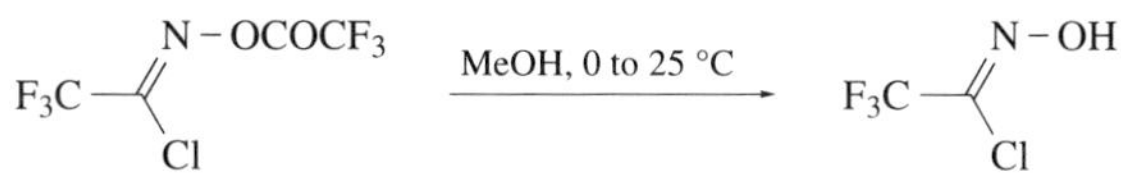

(104) (60)

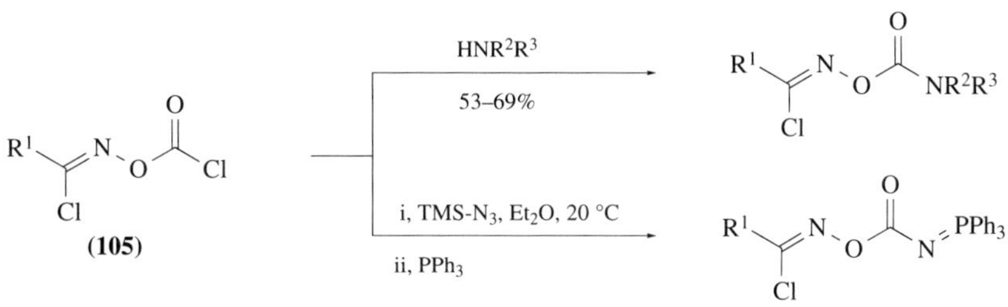

(105)

Scheme 21

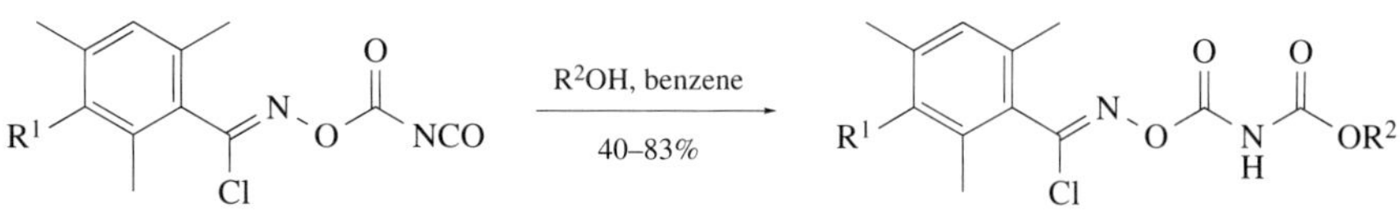

(106) (61)

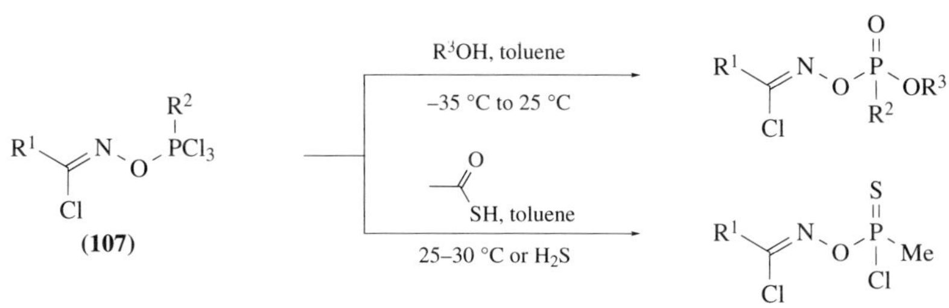

(107)

Scheme 22

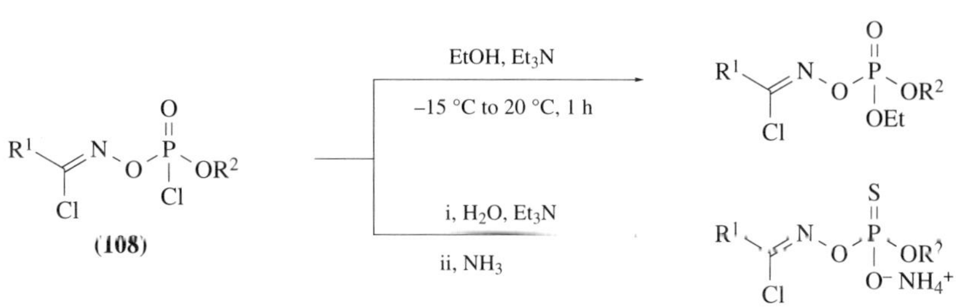

(108)

Scheme 23

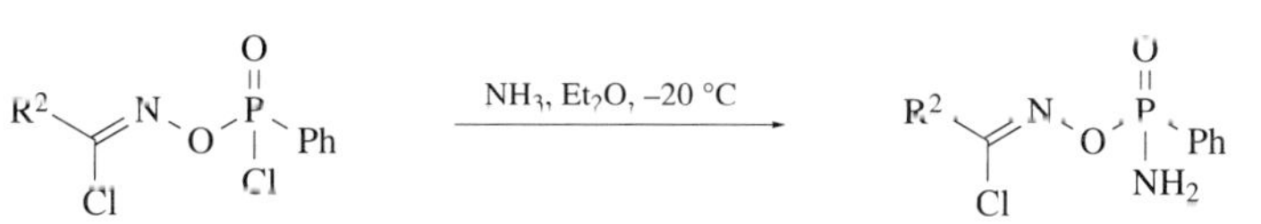

(109) (62)

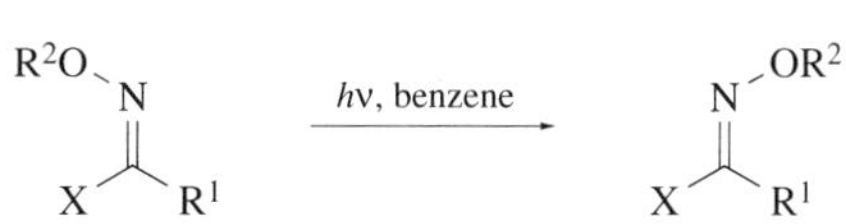

(63)

5.17.3.2.9 *Miscellaneous methods*

3-Fluoro-oxaziridines are ring-opened on heating or by treatment with ketones and CsF, or SbF_5, to give *N*-perfluoroalkylimidoyl fluorides or *N*-alkoxyimidoyl fluorides (Scheme 24) ⟨93JCS(P1)505⟩. The action of amines on 2-chloro-1-methoxy-3-nitropyridinium perchlorate affords 5-dialkylamino-2-nitro-2,4-pentadienhydroxamoyl chlorides (Equation (64)) ⟨83AP(316)21⟩.

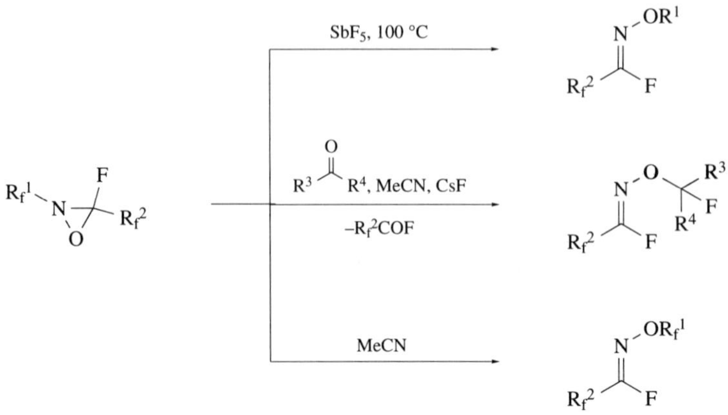

Scheme 24

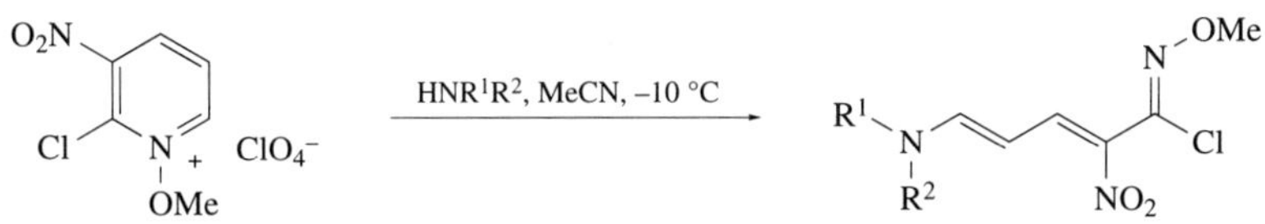

(64)

5.17.4 *N*-ORGANOTHIO-IMIDOYL HALIDES

With the aid of PCl_5, *N*-benzenesulfonyl carboximidoyl chlorides can be prepared from *N*-sulfonylated carboxylic acid amides (Equations (65) and (66)) ⟨34CB1762, 80ZC97⟩. *N*-arenesulfonyl benzimidoyl chlorides can also be obtained by heating sodium *N*-arenesulfonyl benzamides with $SOCl_2$ in toluene at 100–110 °C ⟨76URP534448⟩. Arylsulfonamides form complexes with aluminum trichloride, which on treatment with trichloromethylarenes in 1,2-dichloroethane at room temperature afford *N*-arenesulfonylbenzimidoyl chlorides (Equation (67)) ⟨65ZOR1841, 66JOU1809⟩.

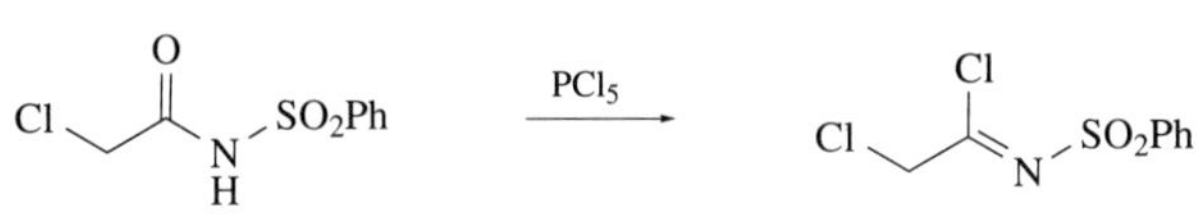

(65)

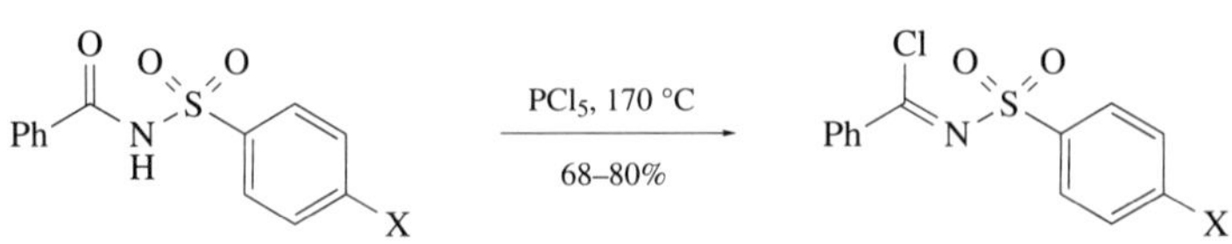

(66)

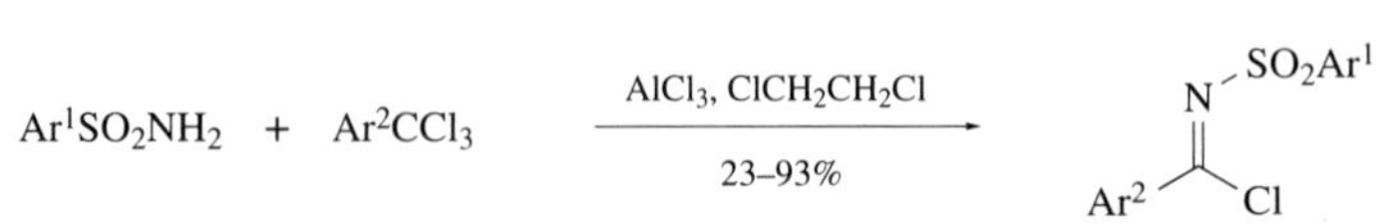

(67)

4,5-Dichloro-1,2,3-dithiazolium chloride is readily accessible from chloroacetonitrile and S_2Cl_2. Chlorination of the 1,2,3-dithiazolium chloride affords *N*-chlorosulfeno- or *N*-(2-chlorodisulfanyl)trichloro acetimidoyl chloride (Scheme 25). By chlorination of 2-chloro-5*H*-1,3,4-dithiol-5-one, 2-chloro-2-(chlorosulfenoimino)acetyl chloride can be prepared (Equation (68)) ⟨85CB1632⟩.

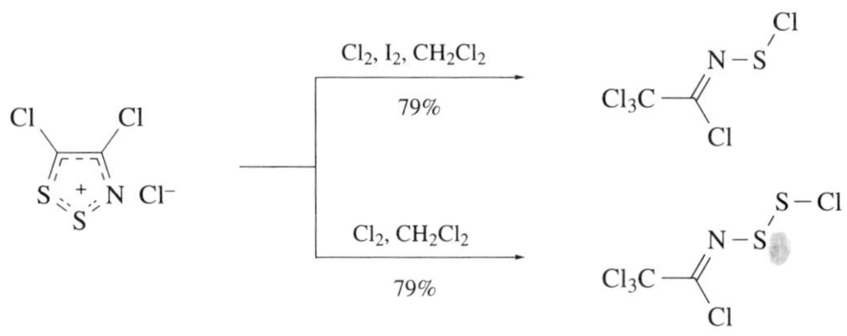

Scheme 25

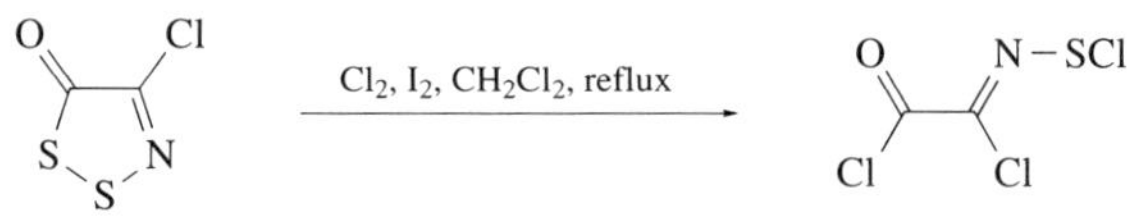

(68)

By means of NaF ⟨64JA357⟩ or HgF_2 ⟨90JFC(48)407⟩, *N*-pentafluorothio-perfluorinated imidoyl chlorides can be converted into the corresponding imidoyl fluorides (Equation (69)). On UV irradiation, sulfurpentafluoride chloride adds to the nitrile function of perfluorinated alkylnitriles ⟨64JA357, 90JFC(48)407⟩ and *N*,*N*,2,2-tetrafluoroaminoacetonitrile ⟨93IC287⟩ to yield α-fluorinated imidoyl chlorides (Equation (70)).

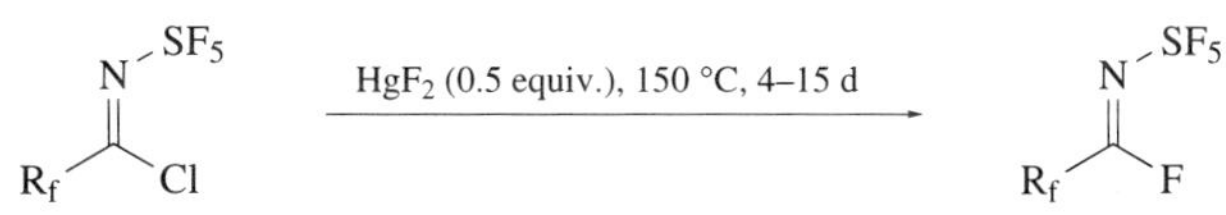

(69)

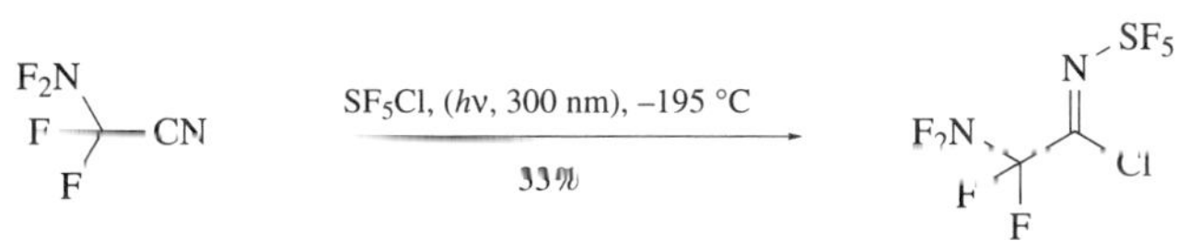

(70)

5.17.5 HYDRAZONYL HALIDES (HYDRAZIDOYL HALIDES)

5.17.5.1 Preparation Methods

5.17.5.1.1 From carboxylic acid hydrazides and halogen carriers

The conversion of arylcarboxylic acid hydrazides into the corresponding hydrazidoyl chlorides with phosphorus pentachloride is a standard procedure ⟨B-68MI 517-03, 85HOU(E5(1)632⟩, which can be performed either without solvent or in chlorinated solvents such as methylene chloride, chloroform or carbon tetrachloride. The following examples (Equations (71)–(73))—which have been taken from the literature ⟨74TL195, 81JMC532, 84JOC3083, 84S146, 85CB1008, 86CC1196⟩—will demonstrate the scope of the method. The compounds thus prepared can often be used as crude products without further purification ⟨81CC1121⟩. The preparation of hydrazidoyl bromides from hydrazides and a mixture of PBr_3 and PBr_5 has been reported ⟨B-68MI 517-03⟩. In a mild procedure for the synthesis of hydrazidoyl chlorides and bromides from carboxylic acid hydrazides, triphenylphosphane/tetrachloromethane ⟨84JOC3083⟩ and triphenylphosphane/tetrabromomethane ⟨90H(31)247⟩ in acetonitrile can be used, respectively, as halogenating reagents (Equation (74)).

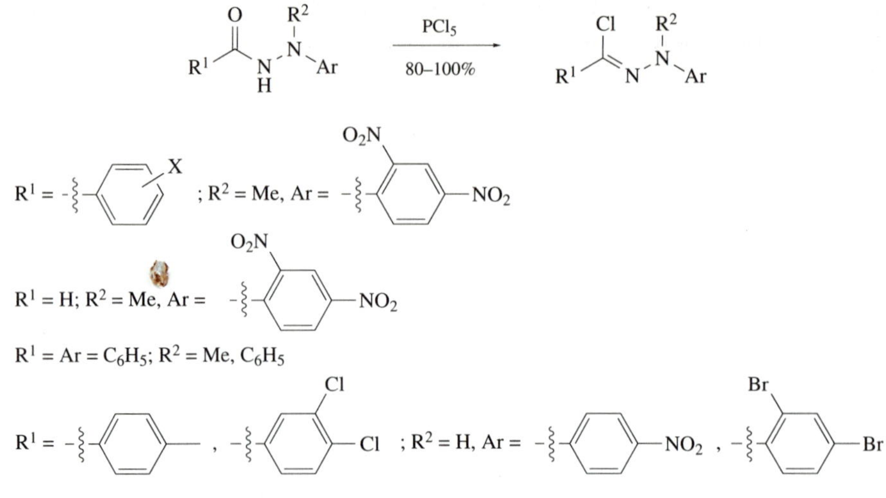

(71)

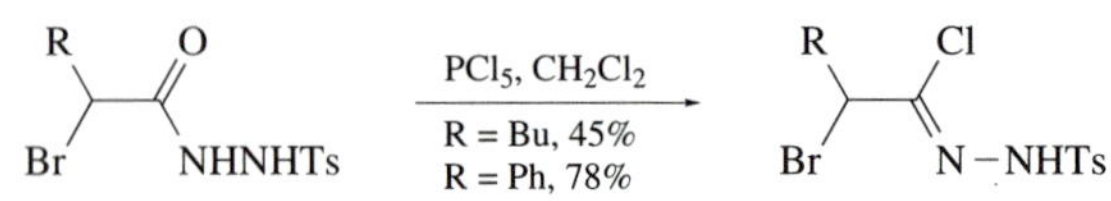

(72)

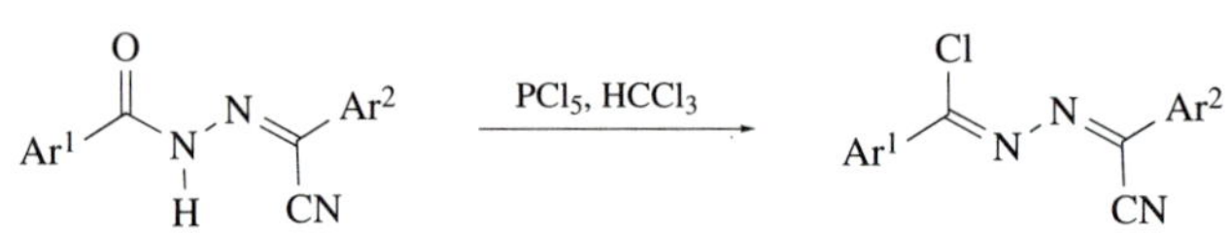

(73)

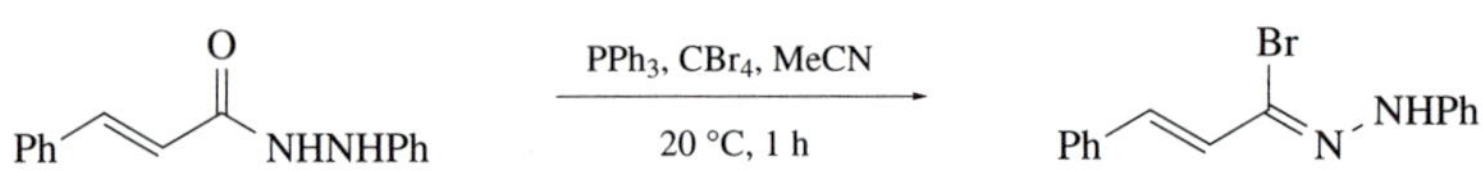

(74)

1-Benzoyl-2-arenesulfonylhydrazines (**110**) (Equation (75)) react with thionyl chloride to give hydrazidoyl chlorides ⟨84JHC797⟩. Phosphorus oxychloride/*N*,*N*-dimethylaniline has been used to convert *N*,*N*′-bis(trifluoroacetyl)hydrazine to the corresponding bis-imidoylchloride (**111**) (Equation (76)) ⟨83JFC293⟩.

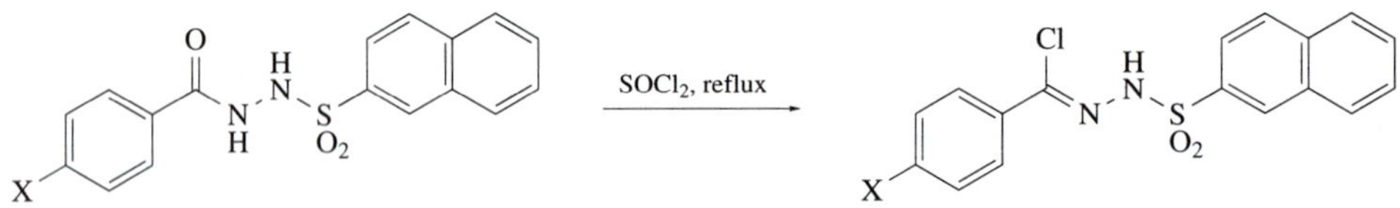

(75)

(110)

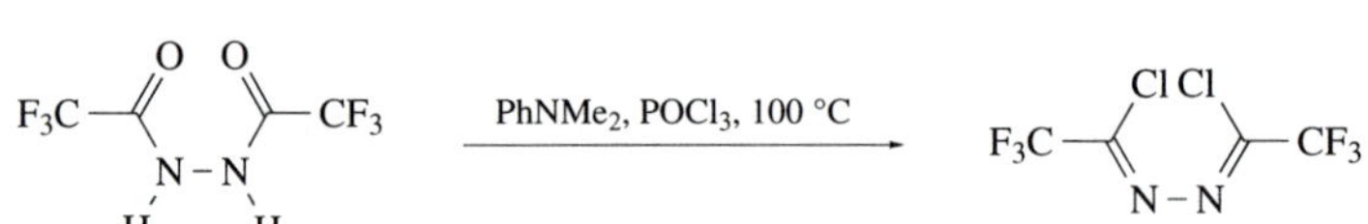

(76)

(111)

5.17.5.1.2 From diazonium halides

Diazonium halides couple with halogenated CH-acidic compounds. In these compounds, the CH acidity usually arises from the presence of two additional electron-withdrawing groups; typical examples are 2-chloroacetic acid esters, chloromalonic acid esters, chloromalonic acid, chloromalonoaldehyde, chloroacetone (Equation (77)) and methyl-phenacyl-phenylsulfoniumbromide ⟨B-68MI 517-03⟩. The reaction has been extended to 3-chloro-2,4-pentanedione ⟨82H(19)319, 82JCED103, 85T2503⟩, 2-chloro-1-phenyl-1,3-butanedione ⟨82JHC73⟩, 2-chloro-1-(2-furyl)-1,3-butanedione ⟨82H(19)57⟩, 1-chloro-1-(4-toluene-sulfonyl)acetone ⟨85T2503⟩, 1-chloro-1-(4-nitrophenyl)acetone ⟨84JCS(P1)2535⟩ and dimethyl-[2-oxo-2-thienyl-(2)]ethyl sulfonium bromide ⟨88OPP521⟩. Additionally the range of diazonium salts that can be used in the coupling reaction has been extended to benzenediazonium salts with reactive groups in the 2-position, such as vinyl ⟨81JOC1402⟩, vinyloxy ⟨93JHC559⟩, (alkylthio)methyl ⟨92OPP352⟩, alkylthio ⟨81JCS(P1)2245, 84JCS(P1)2535⟩, benzylthio ⟨81JCS(P1)2245⟩, formyl ⟨85T2503⟩, hydroxycarbonyl ⟨83JHC719⟩, alkoxycarbonyl ⟨82JHC73, 83JHC719⟩ and *N*-allyl-*N*-alkylaminocarbonyl ⟨89T7427⟩. Alternatively, diazonium salts derived from heterocyclic compounds, such as 3-diazonium-benzo[c]pyrazole hydrogensulfate ⟨82JCED103⟩, 5-diazonium-3-methyl-5-phenylpyrazole chloride ⟨87AP(320)850⟩ and 2-diazonium-5-ethoxycarbonyl-4-methyl-1,3-thiazole hydrogensulfate can also be used ⟨91PS(60)119⟩. Several examples are shown in (Equations (78)–(82)). Diazonium salts couple with diazo compounds to afford hydrazidoyl chlorides (Equation (83)) ⟨55LA(591)218, 55LA(591)226⟩.

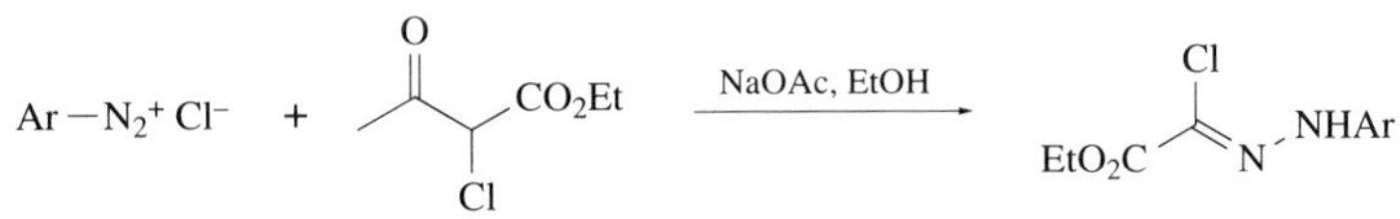

(77)

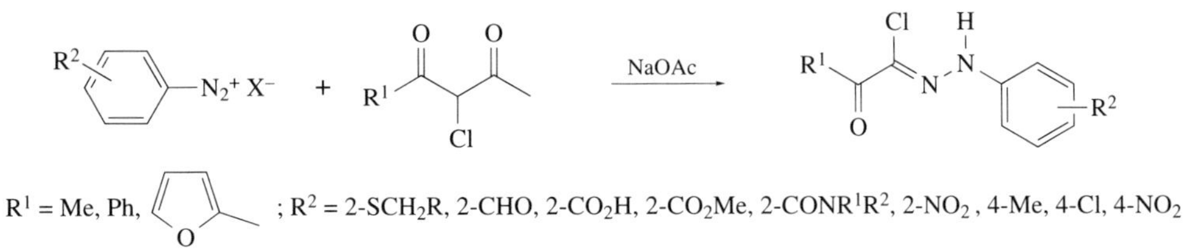

(78)

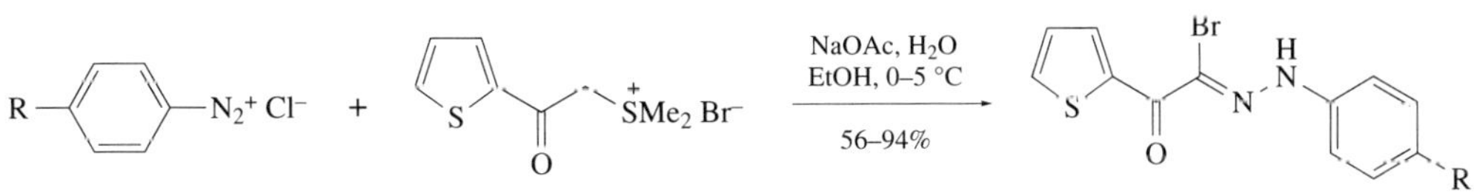

(79)

R = H, Me, NH_2, OMe, Cl

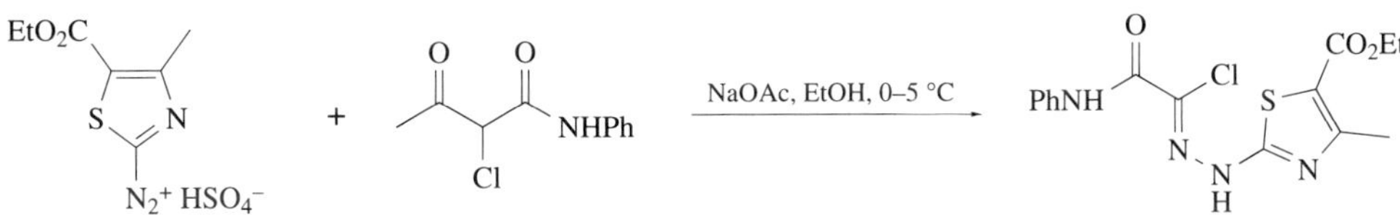

(80)

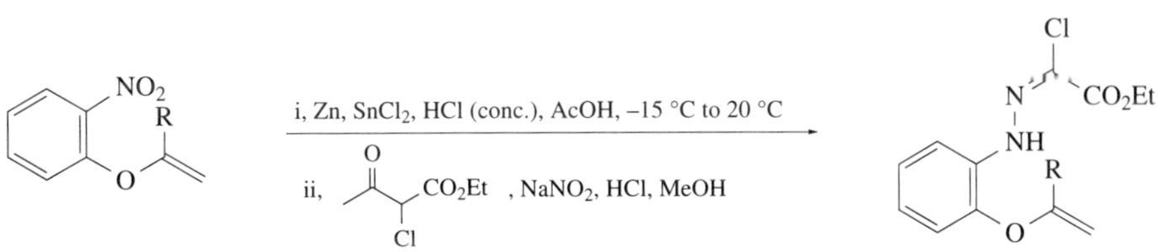

(81)

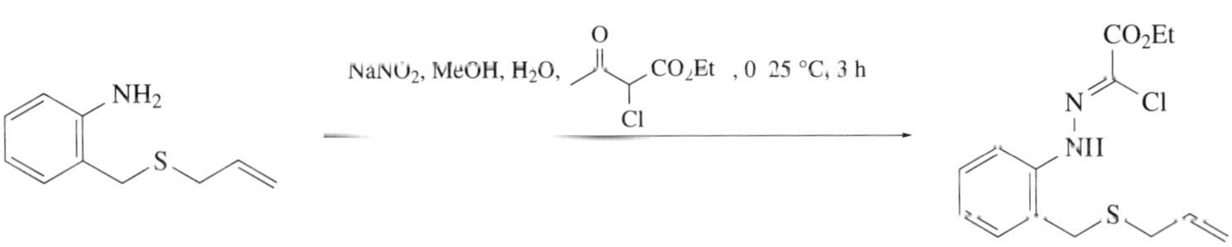

(82)

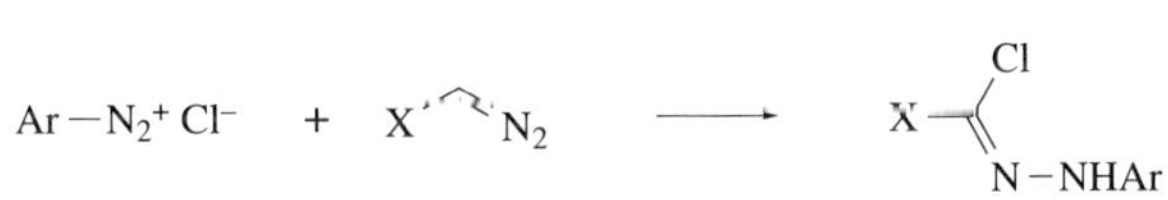

(83)

X = H, CO_2Et

5.17.5.1.3 *By halogenation of hydrazones and azines*

The halogenation of aromatic aldehyde arylhydrazones produces hydrazidoyl halides. Usually, the reactions are performed in glacial acetic acid. Unless the aromatic nucleus of the arylhydrazine part is not strongly deactivated against electrophilic attack (e.g., by an NO_2 group), the desired hydrazidoyl halide formation is accompanied by halogenation of the aromatic ring. In the preparation of hydrazidoyl chlorides, *N*-chlorination can also occur. The *N*-chloro compound thus formed can rearrange on heating if suitable substituents are present (Scheme 26) ⟨B-68MI 517-03⟩.

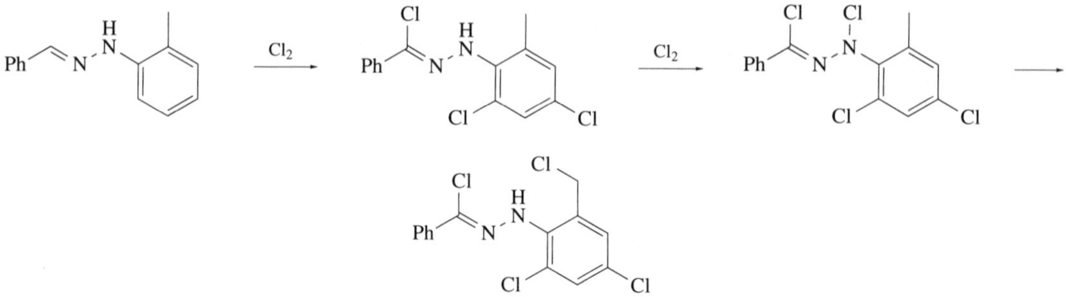

Scheme 26

2-Arylhydrazino-3-oxobutyric acid esters can also be transformed to hydrazidoyl halides. In the case of the bromination of 2-phenylhydrazino-3-oxobutyric acid ethyl esters, the type of final product depends on the reaction conditions (Scheme 27) ⟨B-68MI 517-03⟩. As the following examples show, this reaction has been extended to the halogenation of arylhydrazones of substituted benzaldehydes (**112**) ⟨81JMC532⟩, 2,4-dinitrophenylhydrazones of aliphatic and aromatic aldehydes (**113**) ⟨81JCED418⟩, arylhydrazones of 2-pyridine carbaldehyde (**114**) ⟨81H(16)717⟩ and heteroarylhydrazones of benzaldehyde (**115**) (Equations (84)–(87)) ⟨84CPB4437⟩.

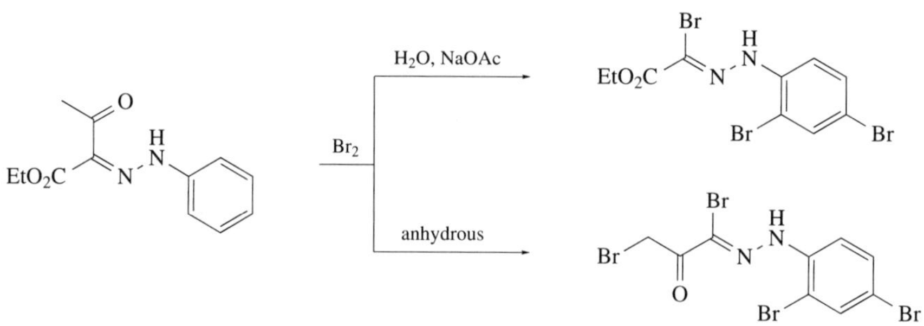

Scheme 27

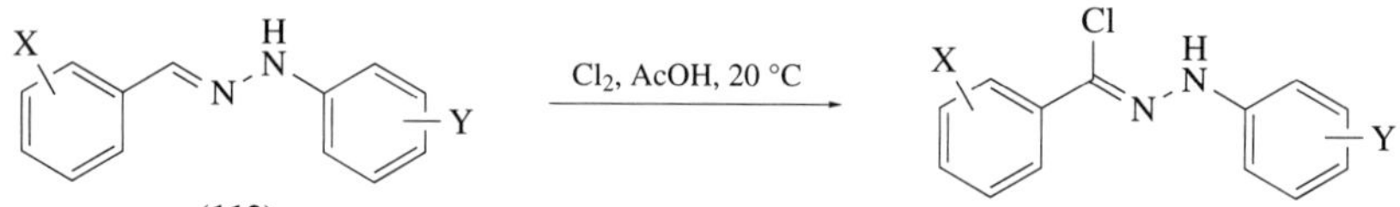 (84)

(112)

$X = H, 4\text{-}Me, 3,4\text{-}Cl_2;\ Y = 4\text{-}NO_2, 2,4\text{-}Br_2$

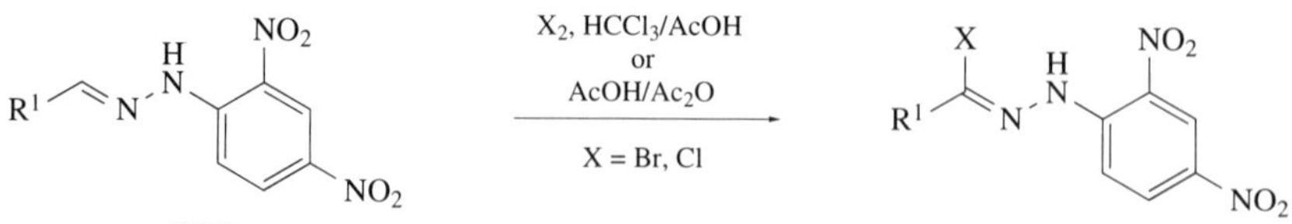 (85)

(113)

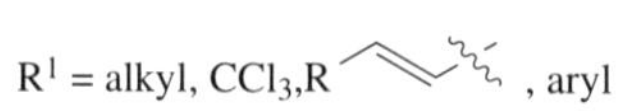

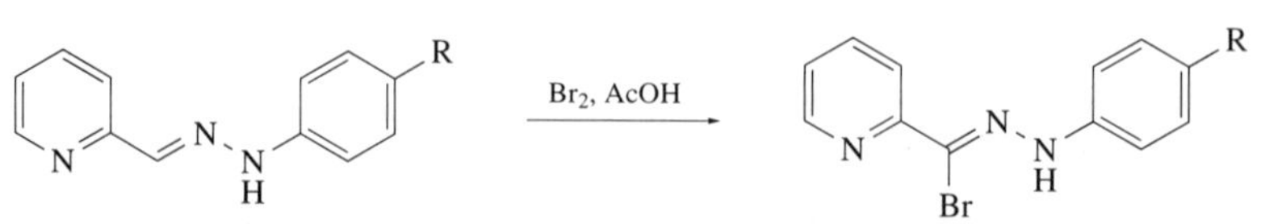 (86)

(114)

$R = H, OMe, NO_2$

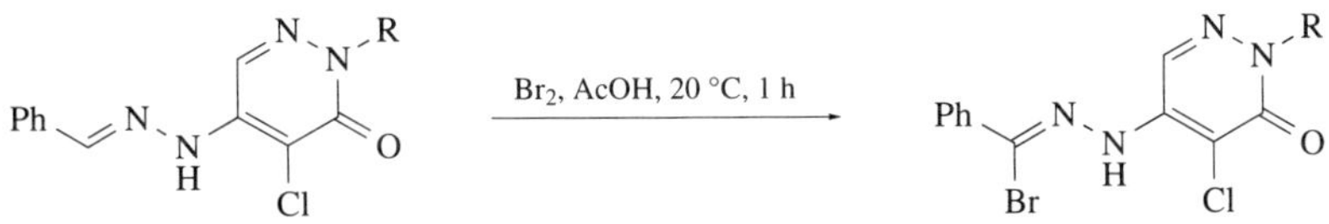

(115) (87)

By chlorination of benzaldazines 'dibenzhydrazidoyl chlorides' (**116**) are accessible (Equation (88)) ⟨B-68MI 517-03⟩. The action of bromine on ethyl-2-arylhydrazono-3-oxobutanoates (**117**) and 3-arylhydrazono-2,4-pentanediones (**118**) in glacial acetic acid and sodium acetate proceeds with deacylation to give hydrazidoyl bromides (Equation (89)) ⟨81JCED418⟩. The bromination of glyoxal-bis(dialkoxyphosphoryl hydrazones) affords the corresponding diimidoyl bromides (**119**) (Equation (90)) ⟨91JGU2372, 91ZOB2257⟩.

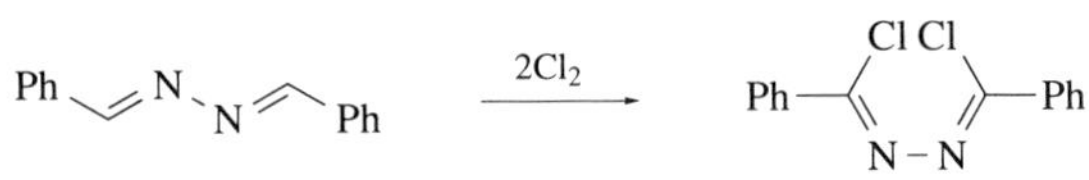

(116) (88)

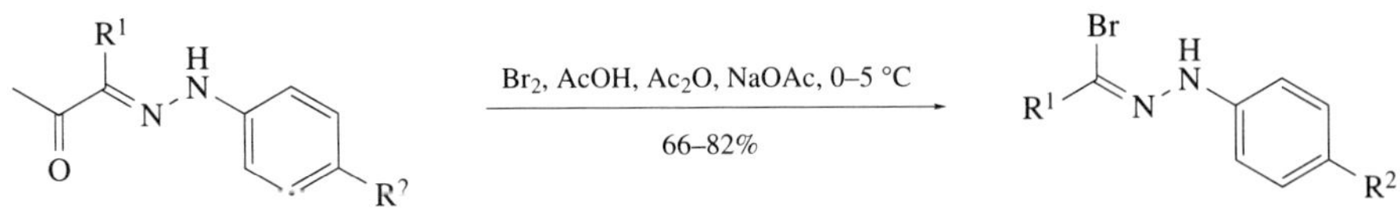

(89)

(**117**) $R^1 = CO_2Et$
(**118**) $R^1 = MeCO$

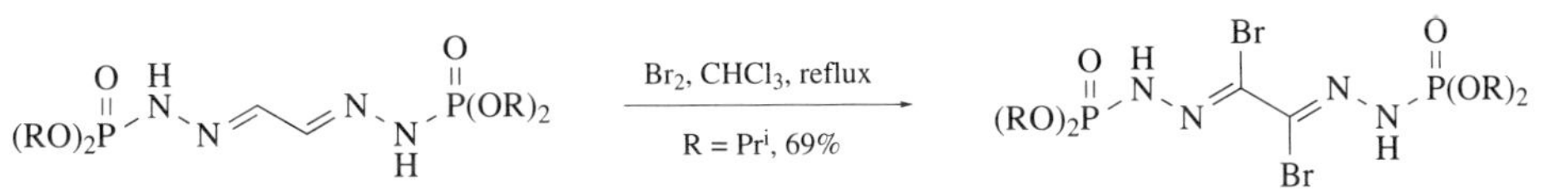

(119) (90)

5.17.5.1.4 *From* N*-chlorinated perfluoroalkylamines,* N*-chloro perfluoroalkyl hydrazines and trifluoroacetonitrile*

On irradiation, *N*-chlorinated perfluoroalkylamines add to the CN bond of trifluoroacetonitrile to give perfluoroalkylated hydrazidoyl chlorides (Scheme 28) ⟨90JFC(48)395, 92ICA527⟩. In the analogous reaction of *N*-chloroperfluoroalkylhydrazine (**120**) with trifluoroacetonitrile, the imidoyl compound (**121**) is formed only in minor amounts, the main product being the perfluoroalkyl tetrazane (**122**) (Equation (91)) ⟨90IC4255⟩.

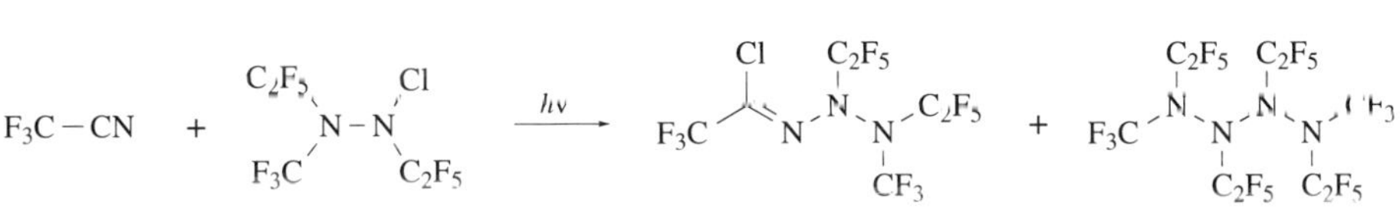

(120) (121) (122) (91)

5.17.5.1.5 *From perfluorinated alkenes*

Perfluoro-1-hexene reacts with phenylhydrazine to yield the hydrazidoyl chloride (**123**) (Equation (92)) ⟨91IZV1918⟩.

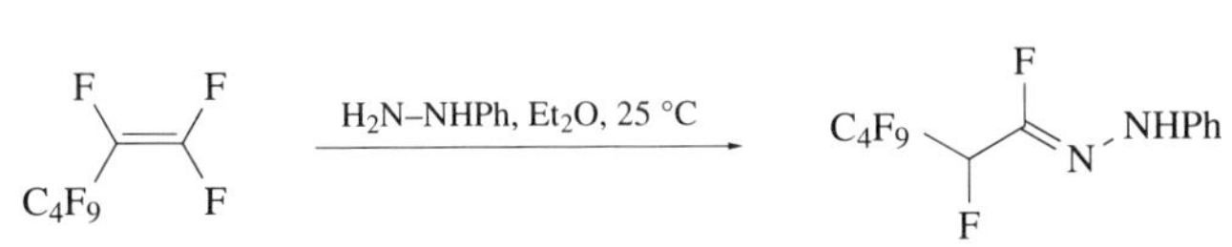

(123) (92)

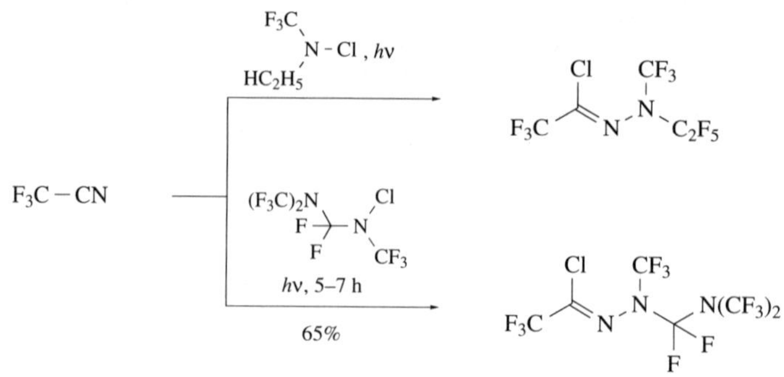

Scheme 28

5.17.5.1.6 *From triphenylphosphoranylidene hydrazine derivatives*

2-Methylthio-4-phenyl-1-triphenylphosphoranylidenaminoimidazole (**124**) reacts with carboxylic acid chlorides on heating to yield the hydrazidoyl chlorides (**125**) (Equation (93)) ⟨88H(27)1935, 89JCR(S)262, 89S843⟩.

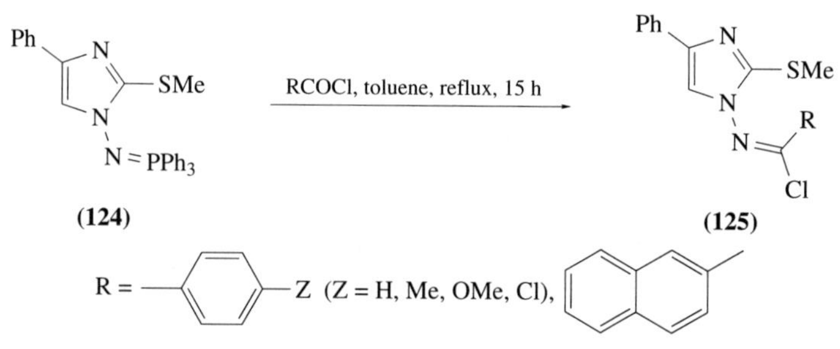

(93)

5.17.5.1.7 *From diazene-N-oxides*

Treatment of the diazene-*N*-oxide (**126**) with HCl in acetic acid yields the hydrazidoyl chloride (**127**) (Equation (94)) ⟨90BAU294, 90IZV352⟩. Ethylvinyl ether reacts with the 2-(tribromophenyl) diazene-2-oxide (**128**) to give the formhydrazidoyl bromide (**129**) (Equation (95)) ⟨91MI 517-01⟩. Mixtures of the hydrazidoyl bromides (**131**) and (**132**) are produced in the decomposition of the diazene oxides (**130**) in chloroform (Equation (96)) ⟨91MI 517-01⟩.

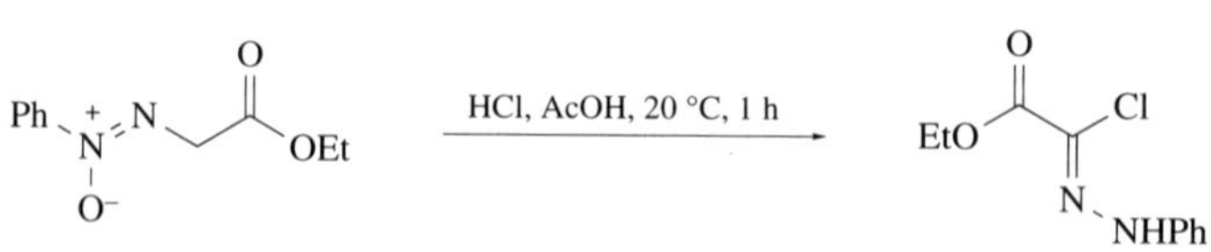

(94)

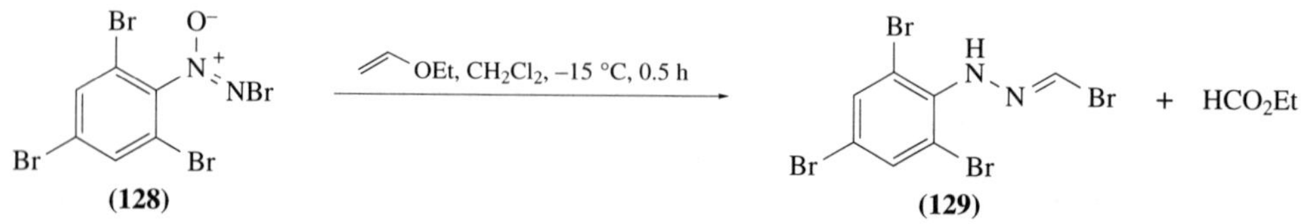

(95)

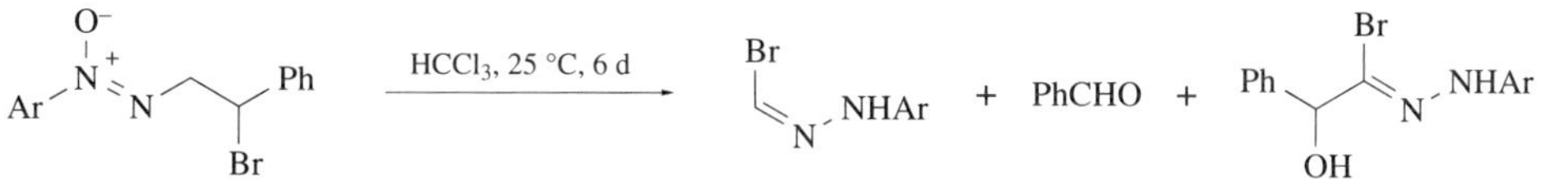

(130) (131) (132) (96)

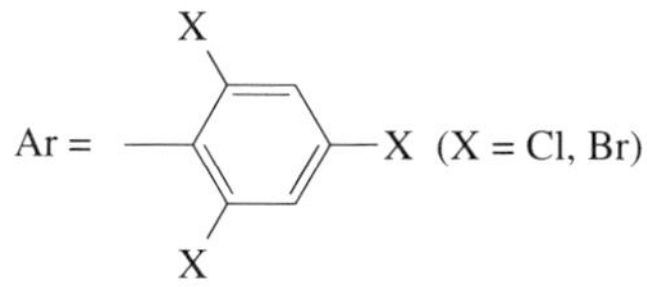

5.17.5.1.8 *From hydrazidoyl halides*

In bishydrazidoyl chlorides, for example (**133**) (Scheme 29), the chlorine can be exchanged for fluoride ⟨83JFC(23)293⟩ or iodide ⟨83JFC(23)293, 93JFC(65)101⟩ by treatment with the appropriate alkali salts. The *N*-methylation of N^2-benzenesulfonylhydrazidoyl chlorides (**134**) (Equation (97)) by means of diazomethane has been described ⟨83BCJ545, 84BCJ539⟩. The hydrazidoyl chloride (**135**) is deprotonated by triethylamine; the nitrilimine thus formed can be carboxylated at the nitrogen with methyl isocyanate to afford the hydrazidoylchloride (**136**) (Equation (98)) ⟨87JHC1391⟩.

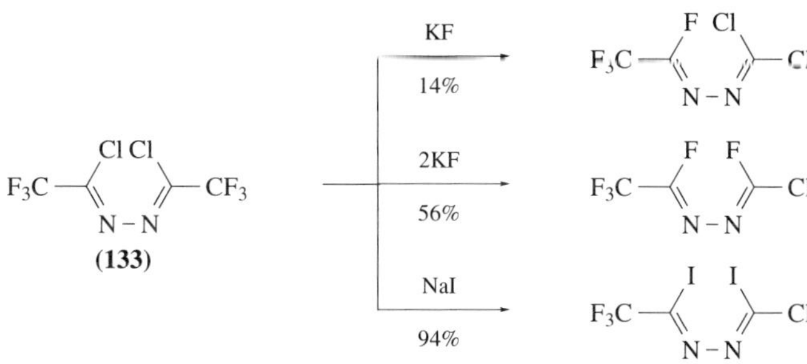

Scheme 29

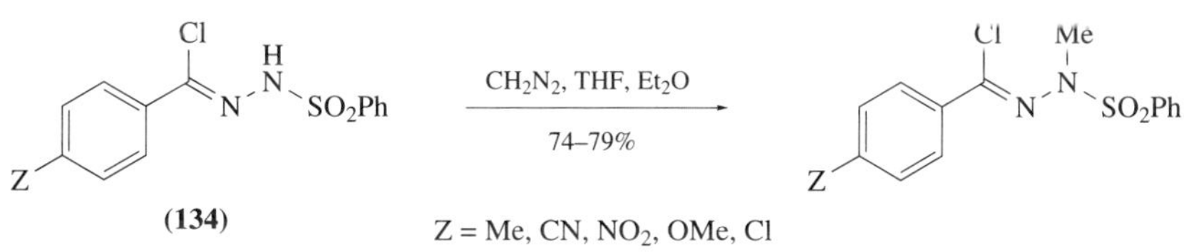

(134) (97)

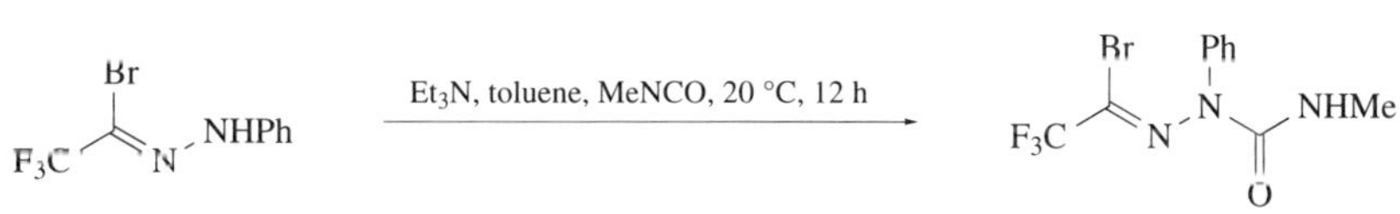

(135) (136) (98)

In N^2-(2-alkylthio-phenyl)hydrazidoyl chlorides (**137**) the alkylthio groups can be oxidized by hydrogen peroxide to give sulfonyl groups; any alkenic functions are epoxidized under these conditions (Equation (99)) ⟨86JCR(S)210⟩. The electrochemical oxidation of benzhydrazidoyl chlorides occurs via radicals which dimerize to give various diphenyl derivatives, among them the hydrazidoyl compounds, for example (**138**) (Equation (100)) ⟨77IZV393⟩. The hydrazidoyl chloride (**139**) rearranges on treatment with polyphosphoric acid at 90 °C to give the hydrazidoyl chloride (**140**) (Equation (101)) ⟨83JHC225⟩. In the reaction of bishydrazidoyl halides (**141**) with α,ω-diamino alkanes, one or two halides are replaced (depending on the number of CH_2 groups between the two amino groups) giving rise to the formation of (**142**) and (**143**) (Equation (102)) ⟨93JFC(65)157⟩. By treatment of the bishydrazonoyl halides (**144**) with tributyltin hydride or thiophenolate, mixtures of imidoyl compounds are produced (Scheme 30) ⟨93JFC(65)101⟩. *N*-Methylsulfonylmethylated compounds (**145**) have been obtained from *N*-acylhydrazidoyl halides and DMSO (Equation (103)) ⟨84S146⟩.

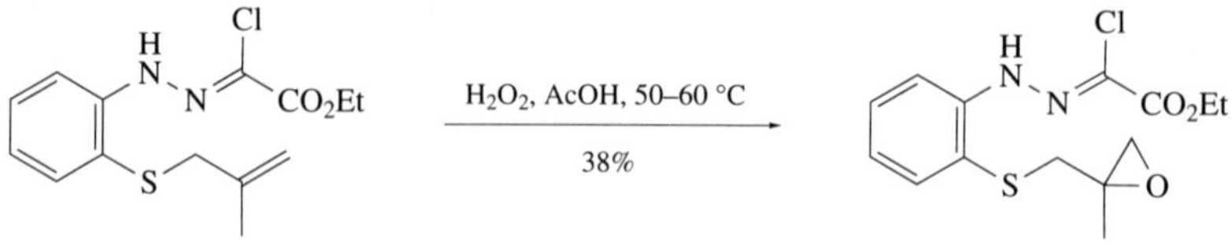

(**137**)

(99)

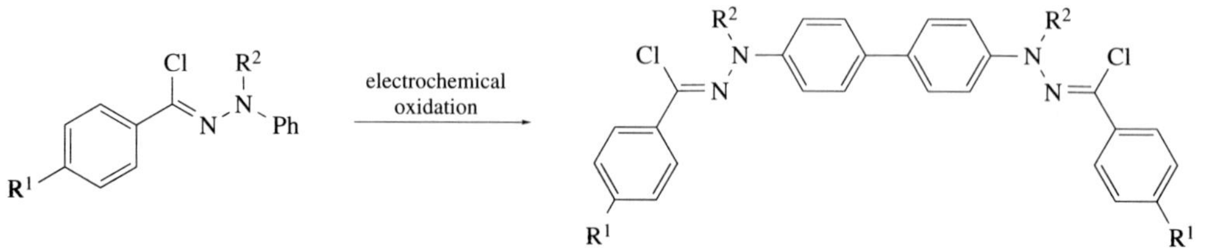

(**138**)

(100)

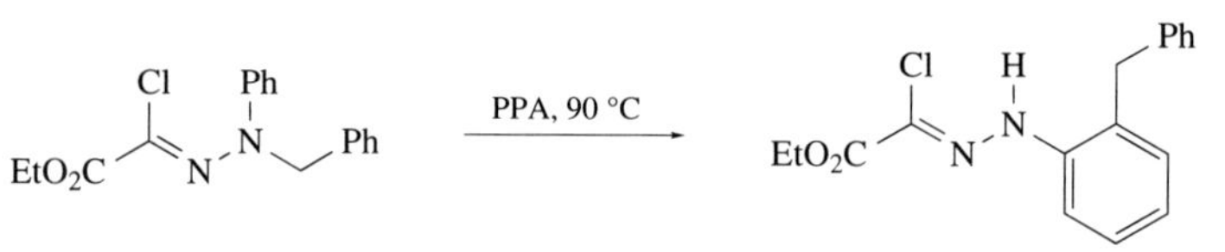

(**139**) (**140**)

(101)

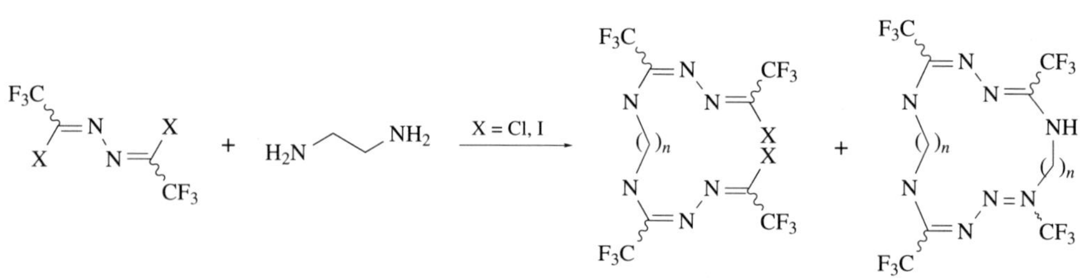

(**141**) (**142**) (**143**)

(102)

		Yields	
X	*n*	*(142)*	*(143)*
Cl	2	0	52
I	4	75	21
I	6	99	0

(**145**)

(103)

5.17.5.1.9 *Miscellaneous methods*

1-Phenylazo-1,2,2-trichloro ethene reacts with chlorine to yield the hydrazidoyl chloride (**146**) (Equation (104)) ⟨B-68MI 517-03⟩.

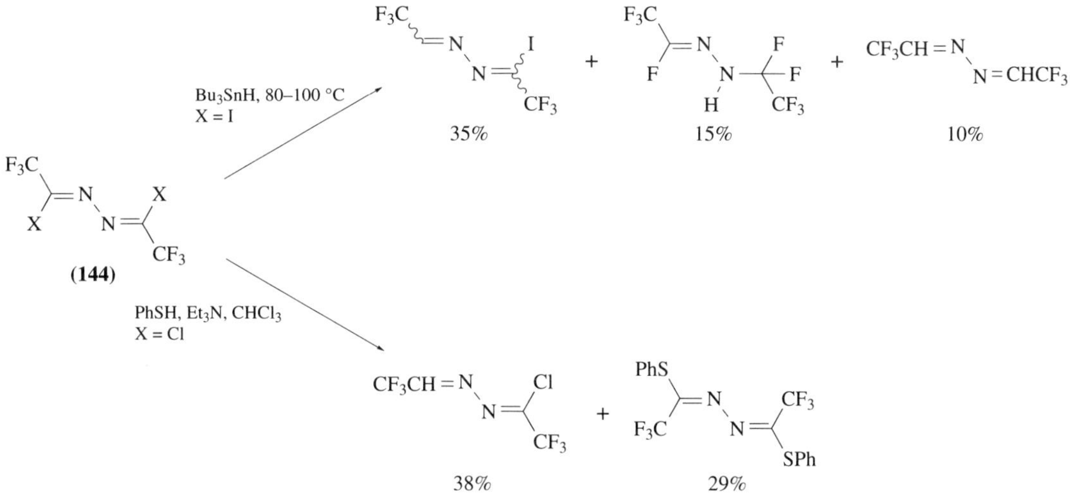

Scheme 30

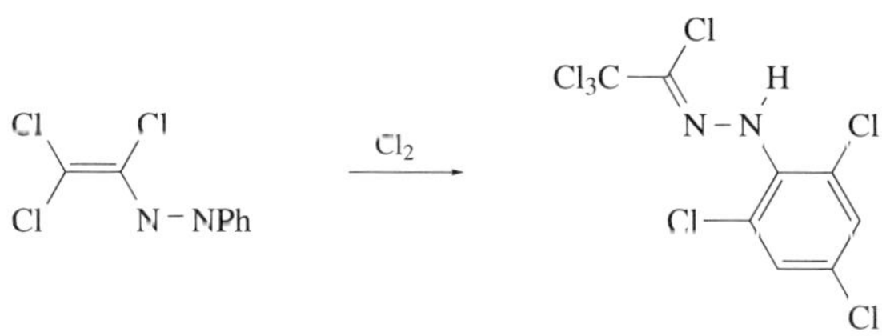

(146)

(104)

5.17.6 *N*-PHOSPHORYLATED IMIDOYL HALIDES

N-Phosphorylated carboxylic acid amides react with PCl_5 to give the corresponding imidoyl chlorides ⟨56JGU2239, 56JGU2935, 56ZOB2009, 56ZOB2631, 57JGU3284, 57ZOB3248, 58JGU1930, 58ZOB1887, 59JGU224, 59ZOB241, 65JGU1019, 65JGU1206, 65ZOB1014, 65ZOB1202⟩; some examples are given in Equations (105)–(108).

1,1,2-Trichloronitrosoethane reacts with dichloro-(2,2,3,3-tetrafluoropropyloxy)phosphine in ether to yield *N*-(dichlorophosphono)chloroacetimidoylchloride; in liquid SO_2 *N*-(chloro-tetrafluoropropyloxy-phosphonoxido)chloroacetimidoyl chloride is formed (Scheme 31) ⟨87BAU1314, 87IZV1422⟩. From long chain dinitriles and PCl_5, *N*-phosphorylated 'amidochlorides' are produced which on treatment with acetic acid in benzene are transformed to bisimidoyl chlorides (Scheme 32) ⟨66JGU1276, 66ZOB1260⟩. *N*-Dichlorophosphorylimidoyl chlorides can be converted into *N*-dichlorothiophosphorylimidoyl chlorides by heating with P_2S_5 (Equation (109)) ⟨87UKZ397⟩. In the reaction of 1-trichlorophosphazo-pentachloroethane with *p*-methoxybenzaldehyde oxime, the products include a phosphorylated imidoyl chloride (Equation (110)) ⟨76JGU239, 76ZOB243⟩.

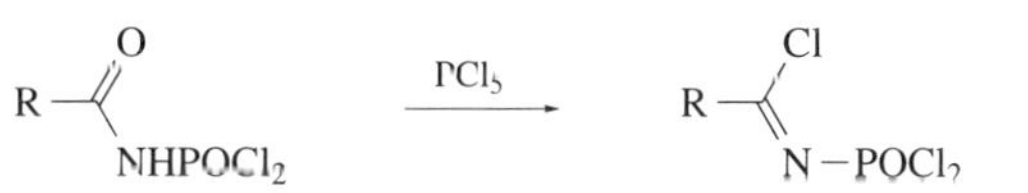

(105)

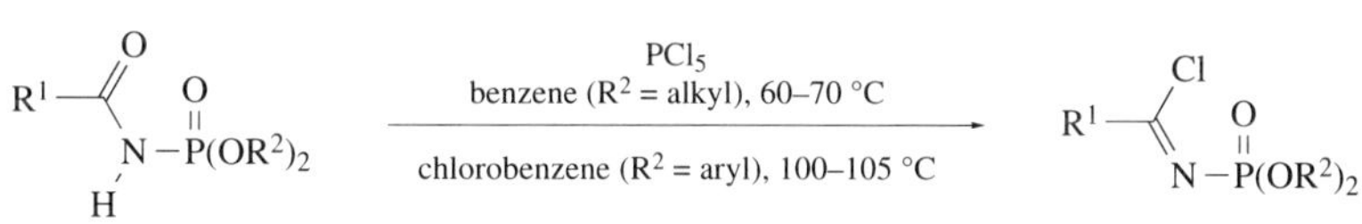

(106)

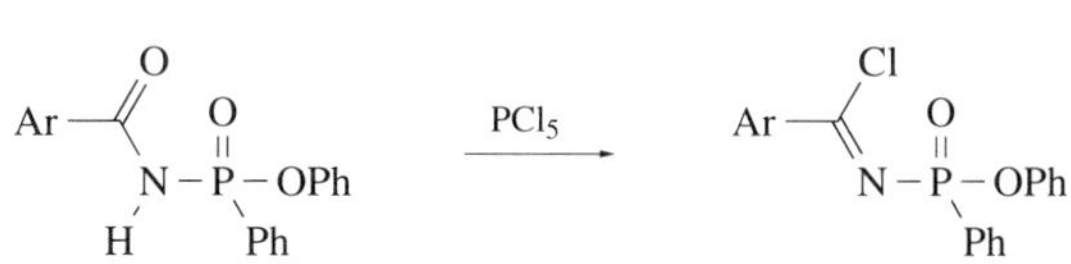

(107)

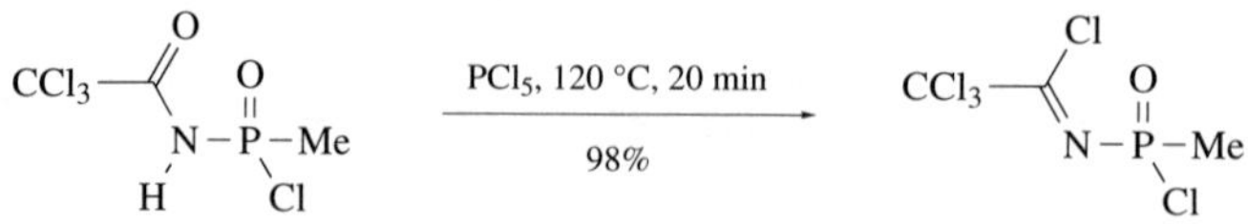

(108)

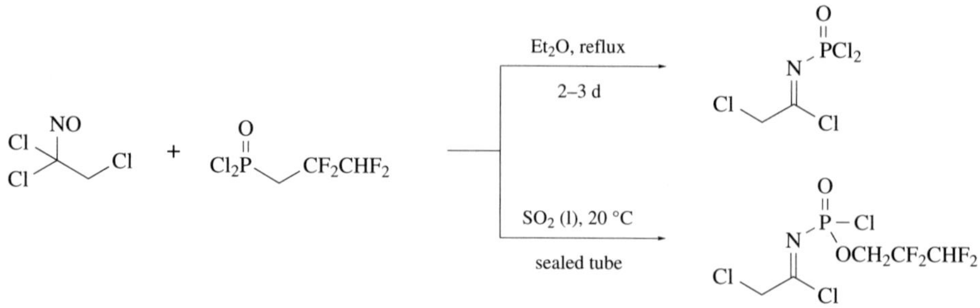

Scheme 31

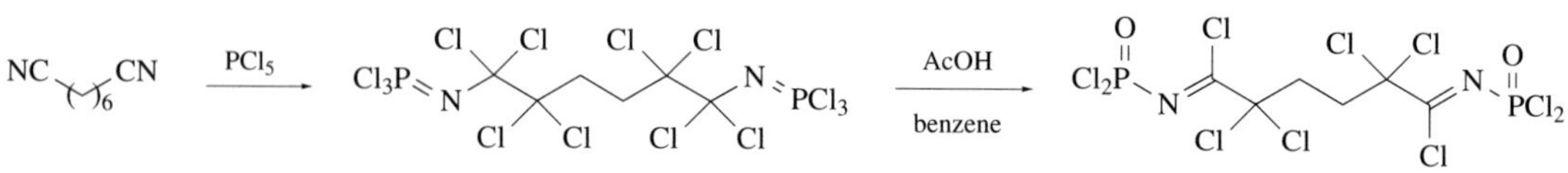

Scheme 32

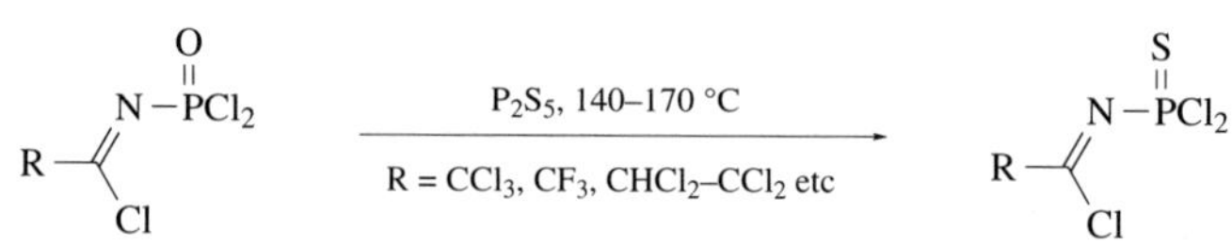

(109)

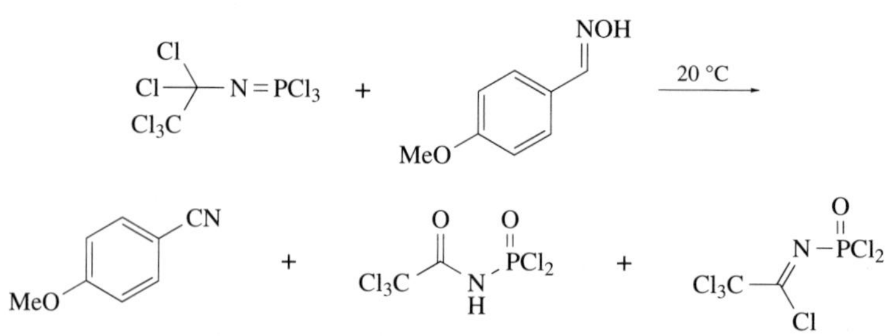

(110)

5.17.7 *N*-ORGANOELEMENTAL SUBSTITUTED IMIDOYL HALIDES

Boron trichloride adds to α-fluorinated nitriles and cyano-imidoyl compounds to give dimeric *N*-borylated imidoyl halides (Equations (111) and (112)) 〈92JFC(56)335〉.

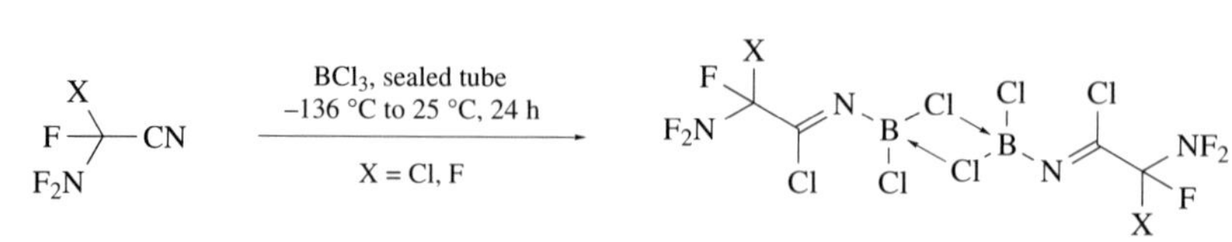

(111)

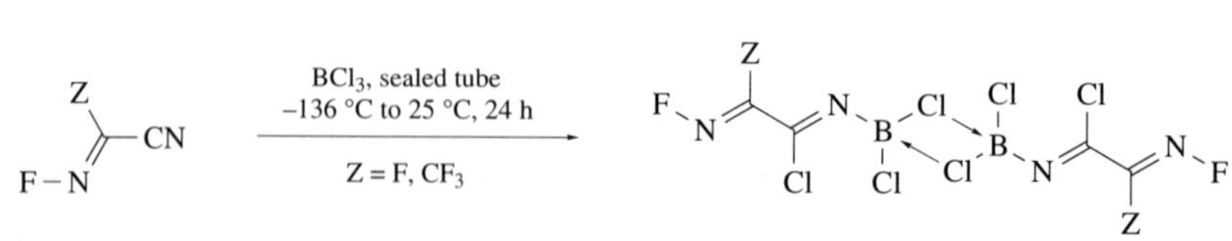

(112)

5.17.8 IMIDIC ACIDS

The tautomeric forms ofamides are imidic acids (Equation (113)). In several reviews ⟨B-70MI 517-01, B-75MI 517-03⟩ dealing with imidoyl compounds the question of whether imidic acids exist in general is discussed. It is clearly stated that they are unstable compounds which may be in equilibrium with the stable amide molecule. This is confirmed by theoretical investigations using semiempirical methods, which demonstrate that the amide form is about 11 kcal/mol (46 Kj/mol) more stable than the tautomeric imidic acid ⟨89JCR(S)56, 91MI 517-02⟩. Nevertheless, in some papers it is claimed that with a special substitution pattern, the amide form is dominated by the imidic acid form. For example, ethereal solutions of compounds which seem to be *N*-fluoroimino acids rather than *N*-fluoro amides are accessible by hydrolysis of perfluorinated imidoyl fluorides. Methylation of the *N*-fluoroimino acids gives the corresponding iminoesters (Scheme 33) ⟨71T51⟩.

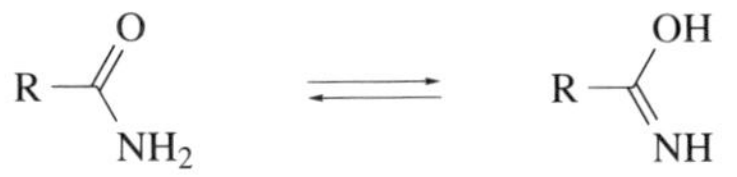

(113)

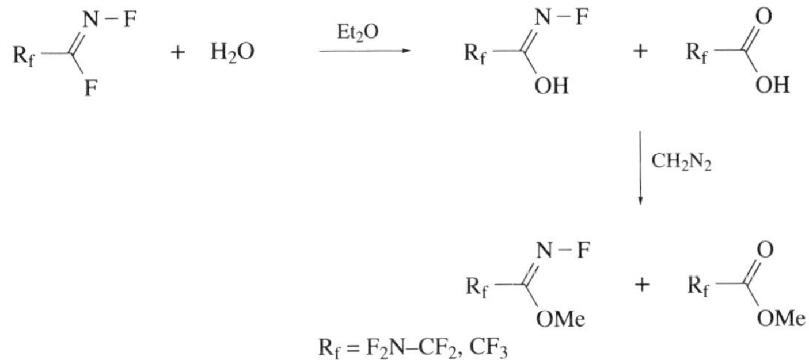

Scheme 33

5.17.8.1 Imidic Acid Esters

5.17.8.1.1 Carboximidic acid esters (in which R^2 and R^3 are carbon based)—general remarks

The chemistry of iminoesters has been reviewed several times ⟨61CRV179, B-72MI 517-01, B-75MI 517-02, 85HOU(E5/1)812, 91COS(6)485, B-91MI 517-03⟩.

5.17.8.1.2 From imidoyl halides, nitrilium salts and related compounds

The action of alcohols or phenols on imidoyl chlorides in the presence of suitable bases (e.g., alcoholates, tertiary amines) is a standard procedure for the synthesis of imidates and has been reviewed several times ⟨61CRV179, B-72MI 517-01, B-75MI 517-02, 85HOU(E5/1)812, 91COS(6)485, B-91MI 517-03⟩. In this way arylimidates (e.g., (**147**), Equation (114)) ⟨93MI 517-01⟩, as well as imidates with a great variety of substituents at the imidoyl nitrogen (e.g., (**148**), (**149**), Scheme 34), can be prepared advantageously ⟨75JOU1926, 75ZOR1913, 77JOU1474, 77ZOR1597⟩. A phase transfer catalyzed procedure has also been described ⟨80S114⟩. Bisimidates (**150**) (Equation (115)) ⟨90JOU1408, 90ZOR1629⟩ and alkenylimidates (**151**)–(**155**) (Equations (116) and (117)) Scheme 35 are also accessible by this method ⟨86TL2743, 88T6841, 89CB657, 89CB669, 89TL4443⟩. Under the catalysis of $ZnCl_2$, the imidoyl chloride (**156**) and paraformaldehyde react to give the *N*-chloromethyl imidester (**157**) (Equation (118)) ⟨78ZN(B)636⟩. On treatment with steroidal alcohols in the presence of NH_4Cl/NEt_3, the iminium chloride (**158**) is converted to the iminoester (**159**); if the ammonium chloride is replaced by hydrazine hydrochloride azinodi-alkoxyphenylmethanes (**160**) and hydrazono alkoxyphenylmethanes (**161**) are obtained (Scheme 36) ⟨80JCS(P1)2184⟩.

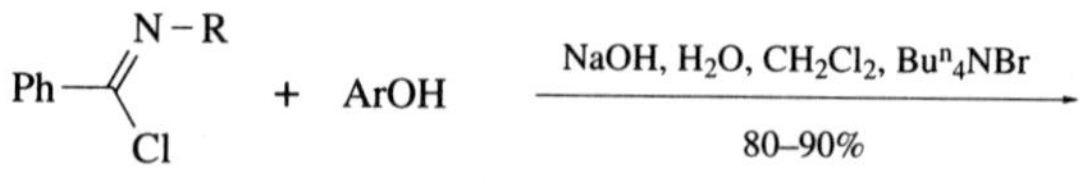

(147)

(114)

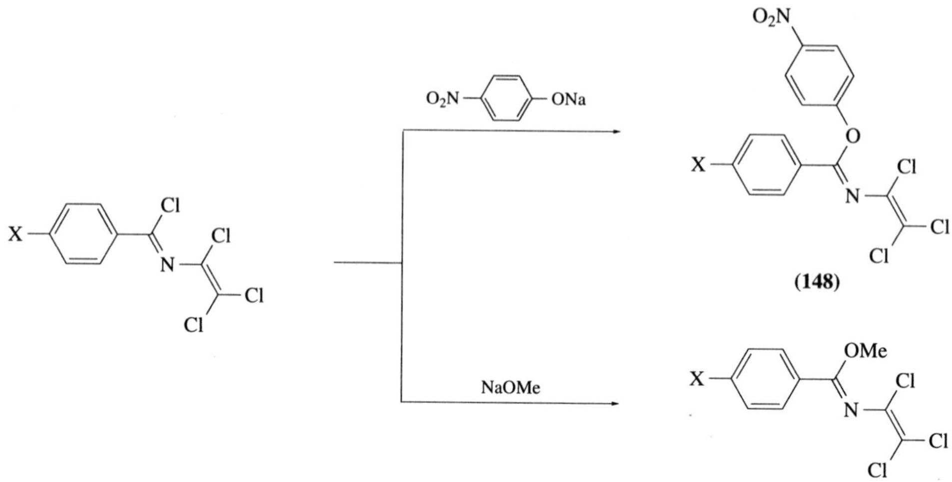

(148)

(149)

Scheme 34

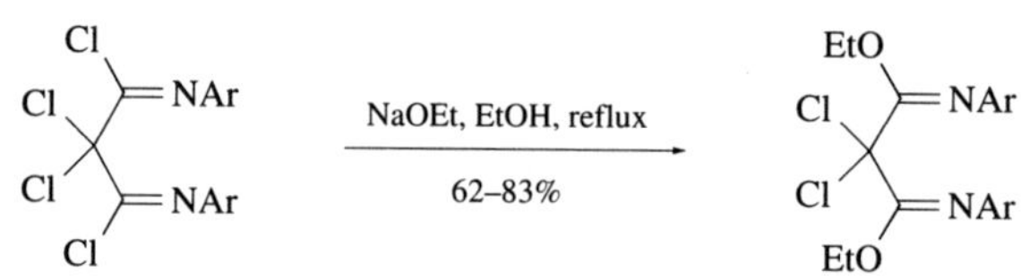

(150)

(115)

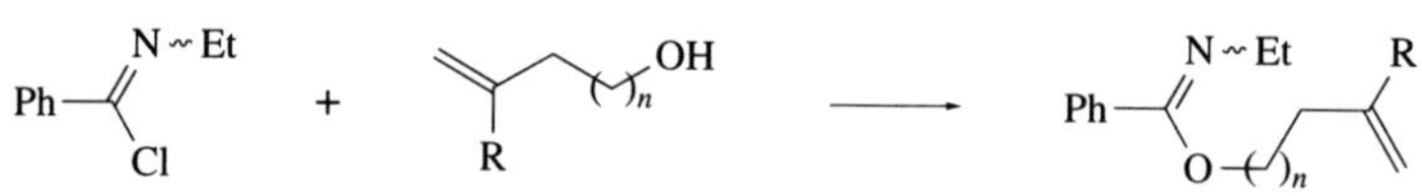

(151)

(116)

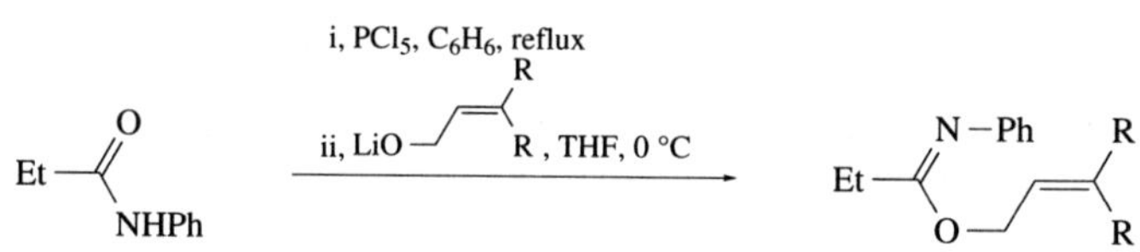

(152)

(117)

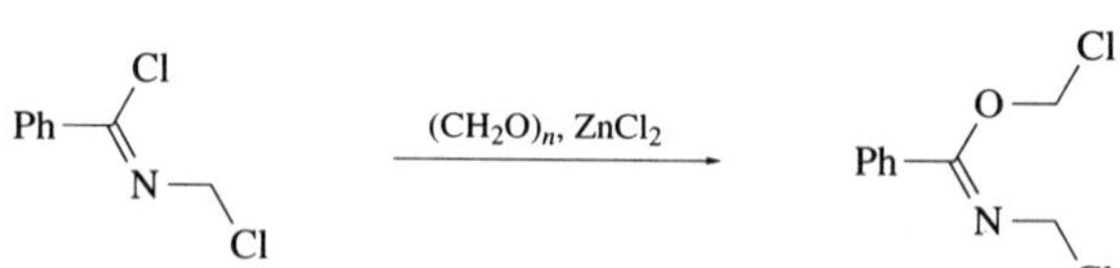

(156) (157)

(118)

5.17.8.1.3 From alkoxymethyleniminium salts and related compounds

The deprotonation of *N*-unsubstituted and *N*-monosubstituted alkoxymethyleniminium salts can be achieved by tertiary amines, carbonates or alcoholates, etc. The final products are the cor-

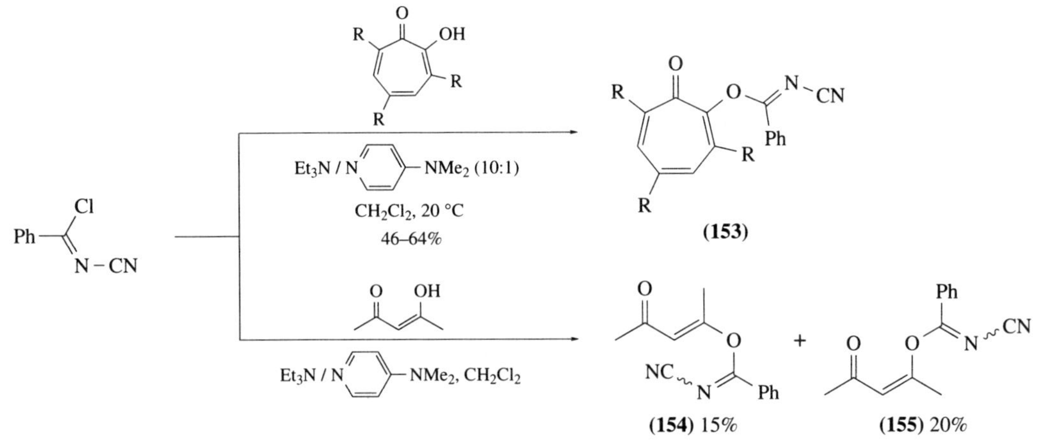

Scheme 35

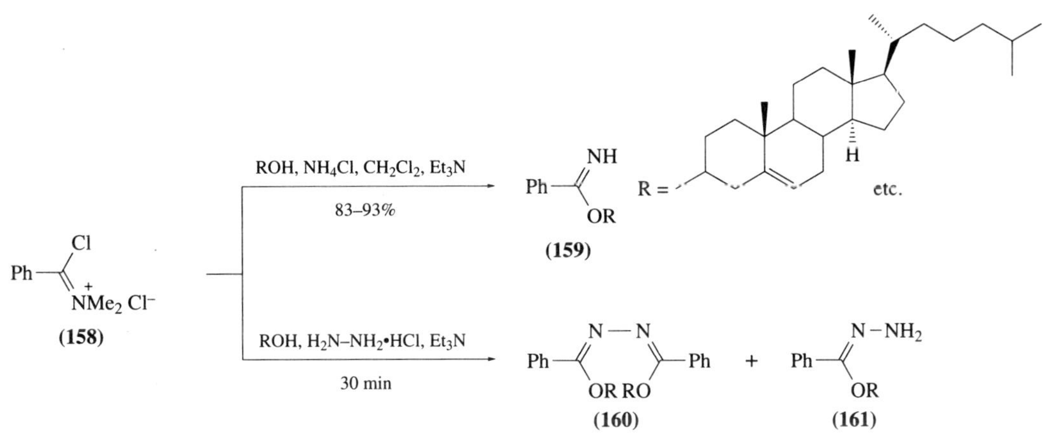

Scheme 36

responding imidates (**162**) (Equation (119)) ⟨61CRV179, B-72MI 517-01, B-75MI 517-02, 85HOU(E5/1)812, 91COS(6)485, B-91MI 517-03⟩.

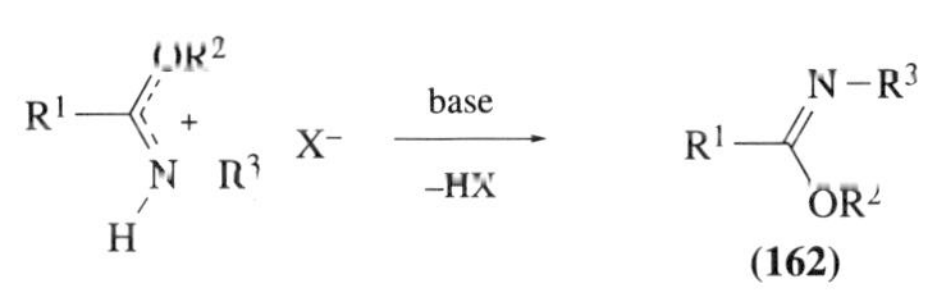

(**162**) (119)

The *N*-unsubstituted salts are accessible by Pinner synthesis, which has been reviewed ⟨65HOU(6/3)295⟩. A compilation of applications of this reaction is available ⟨91COS(6)485⟩. The syntheses of the compounds (**163**) and (**164**) are actual examples of the synthesis of imino esters via iminoester hydrochlorides (Equations (120) and (121)) ⟨87JHC789, 91CHE299, 91KGS369⟩. In 1993, iminoesters were obtained from unsaturated nitriles and unsaturated alcohols by this procedure ⟨93AJC1213⟩.

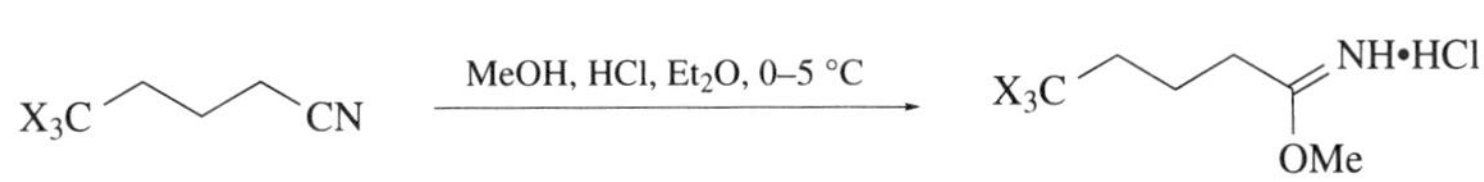

(**163**) (120)

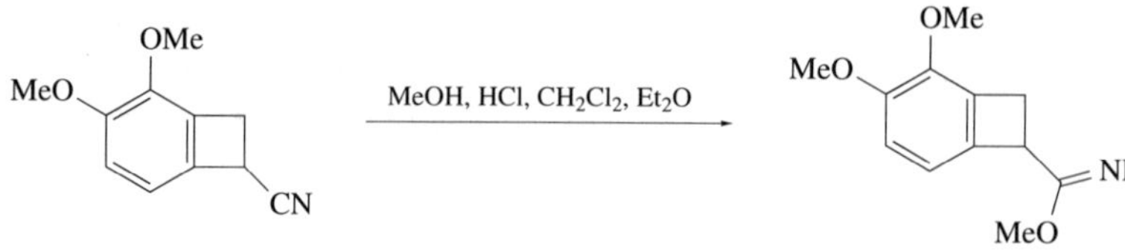

(164)

(121)

Thealkoxymethyleniminium salts can also be prepared from amides, thioamides, lactams and chloroformic acid esters; additionally they are accessible in wide structural variety by alkylation of *N*-unsubstituted and *N*-monosubstituted carboxylic acid amides with dialkyl sulfates, trialkyloxonium tetrafluoroborate, methyl trifluoromethanesulfonates (for reviews see ⟨B-79MI 517-01, 91COS(6)485⟩) and diethoxycarbenium perchlorate (*in situ* from triethyl *ortho*-formate and $HClO_4$ ⟨81JOU1986, 81ZOR2225⟩. The synthesis of the iminoesters **(165)**–**(168)** demonstrates the scope of the reaction (Equations (122)–(125)) ⟨87S262, 89JOC4335, 90IC3218, 90JA6439⟩. As can be seen from the preparation of the compounds **(170)**–**(172)** (Equations (126) and (127) and Scheme 37) the reaction is also useful for the transformation of lactams (e.g. **(169)**) to lactim ethers ⟨B-75MI 517-02, 81JOU1986, 81ZOR2225, 85HOU(E5/1)812, 89PJC265, 90TL123, 91COS(6)485⟩.

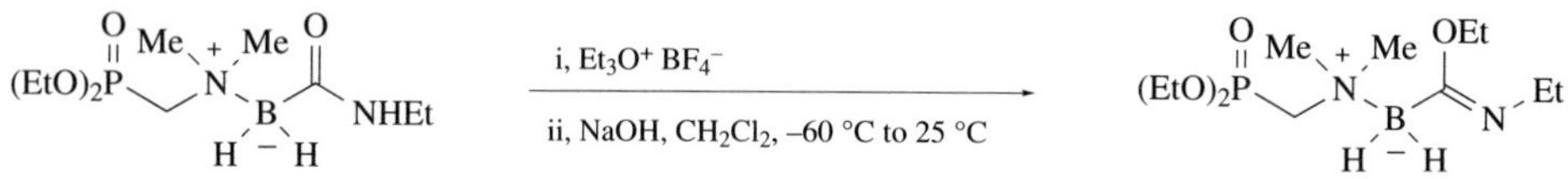

(165)

(122)

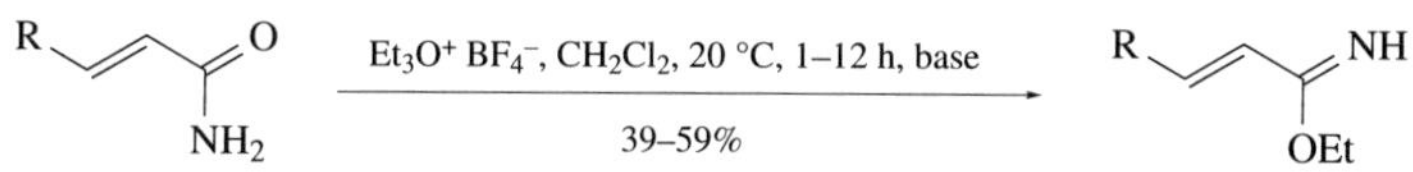

(166)

(123)

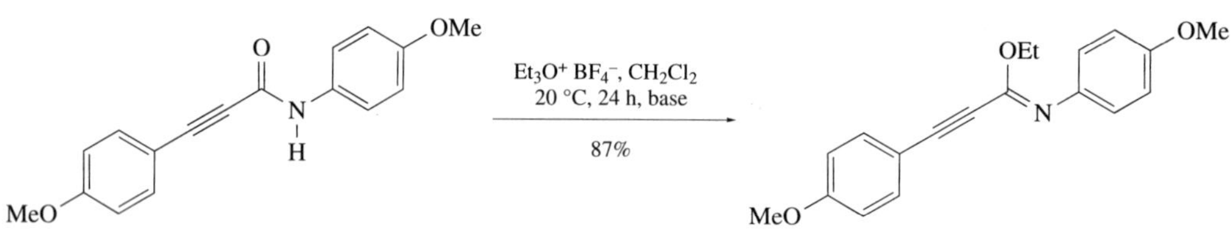

(167)

(124)

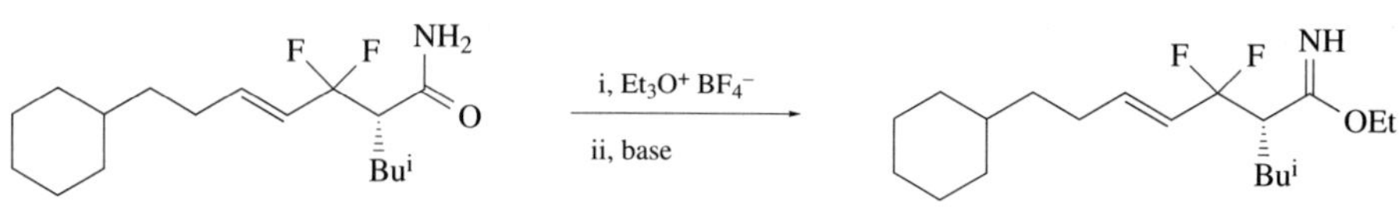

(168)

(125)

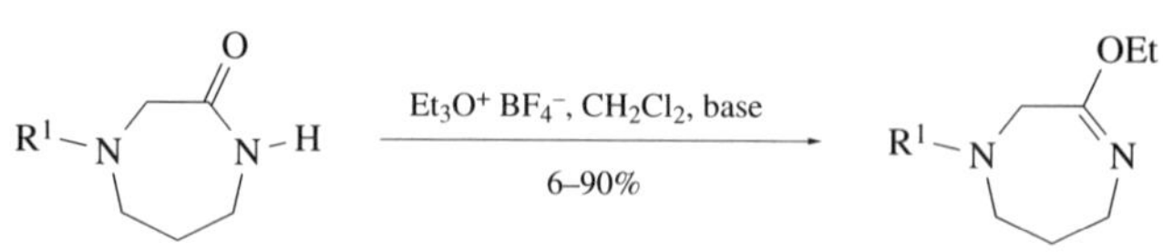

(169) **(170)**

(126)

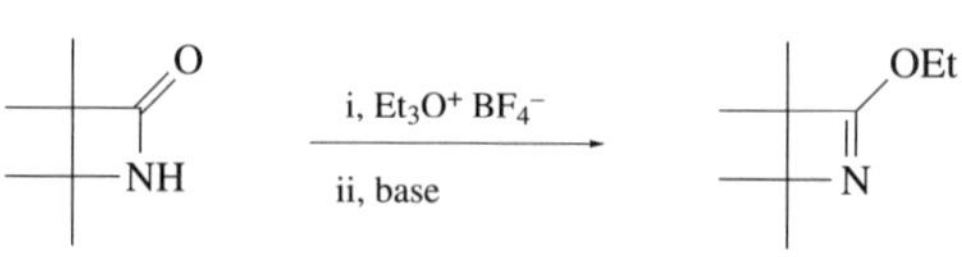

(171)

(127)

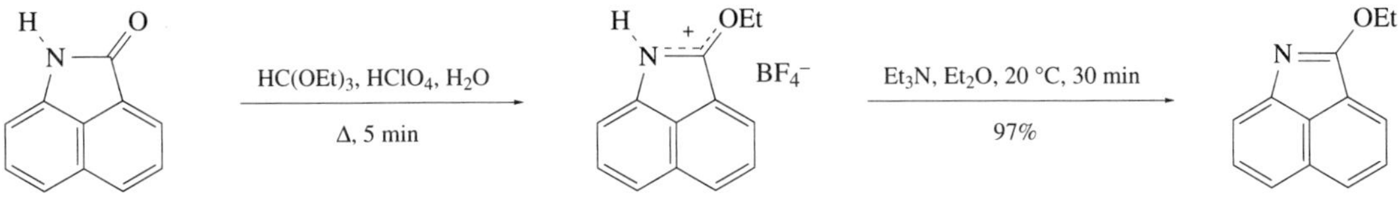

(172)

Scheme 37

Iminoesters have been synthesized via alkoxymethyleniminium salts, which were obtained by addition of alcohols to nitrilium salts (Equation (128)) ⟨69ACS81, 85HOU(E5/1)812, 91COS(6)485⟩.

(128)

O-Arylsulfonyloximes rearrange on heating with phenols in dry toluene to give iminium sulfonates (**173**) (Scheme 38); the iminoesters are set free from them by addition of sodium hydroxide ⟨48JCS1514⟩. Similar results have been achieved by rearrangement of oxime carbonates ⟨76CJC647⟩.

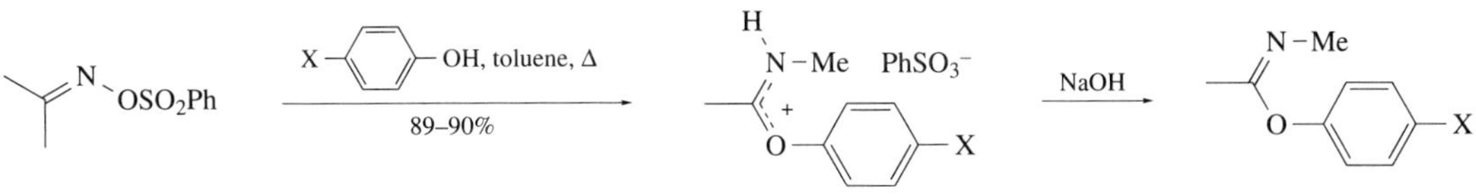

(173)

Scheme 38

N-Monosubstituted carboxylic acid amides are converted into iminoesters by dialkoxy diphenylsulfuranes (Equation (129)) ⟨73JA2017⟩. *N*-Acylcarbamates undergo the Mitsunobo reaction to give mixtures of *O*- and *N*-alkylated compounds (Equation (130)) ⟨93JCS(P2)655⟩.

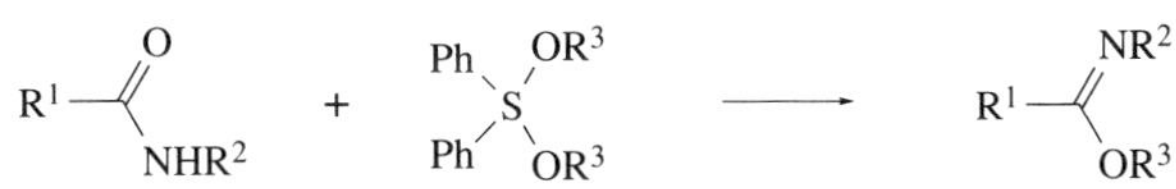

(129)

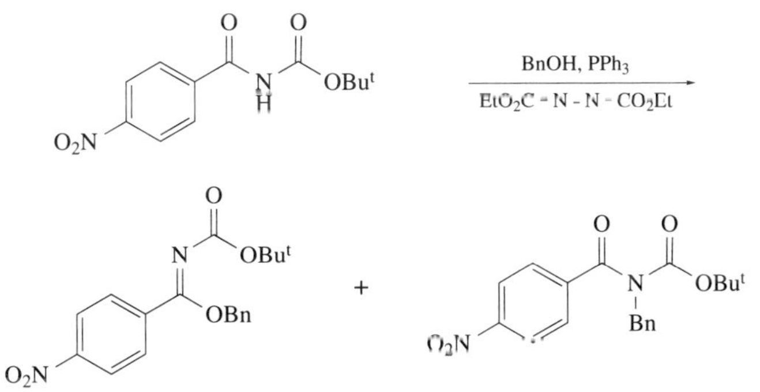

(130)

N-Chloroamides react with dimethylsulfide via sulfonium salts (**174**) to give methylthiomethoxy imidates (**175**) (Scheme 39) ⟨76S46⟩. In the reaction of the nitrilium salt (**176**) with DMSO the sulfonium salt (**178**) is formed via the iminium salt (**177**); (**178**) is converted to the iminoester (**179**) on treatment with base (Scheme 40) ⟨80LA1064⟩. The salts (**181**), obtained from the sulfonio-imide (**180**) and carboxylic acid chlorides, react with triethylamine to yield mixtures of amides (**182**) and imidates (**183**) (Scheme 41) ⟨76S46, 80LA1075⟩.

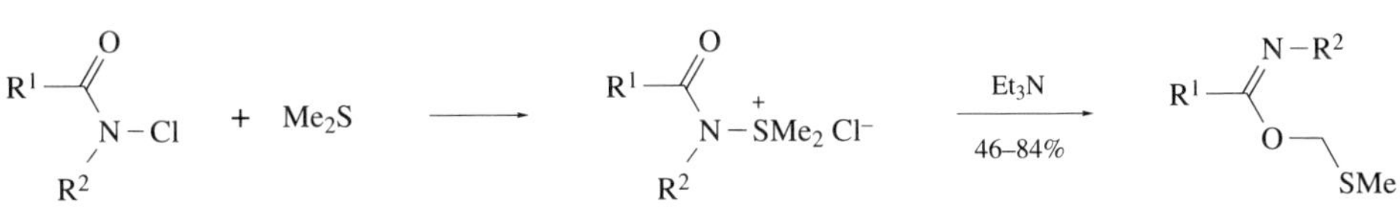

(174) **(175)**

Scheme 39

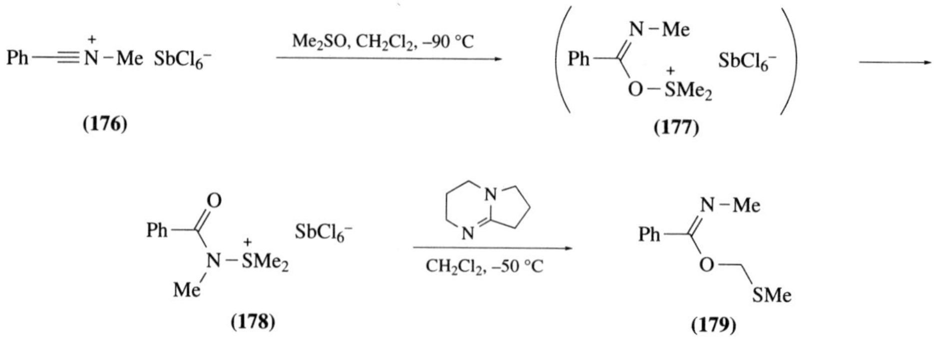

Scheme 40

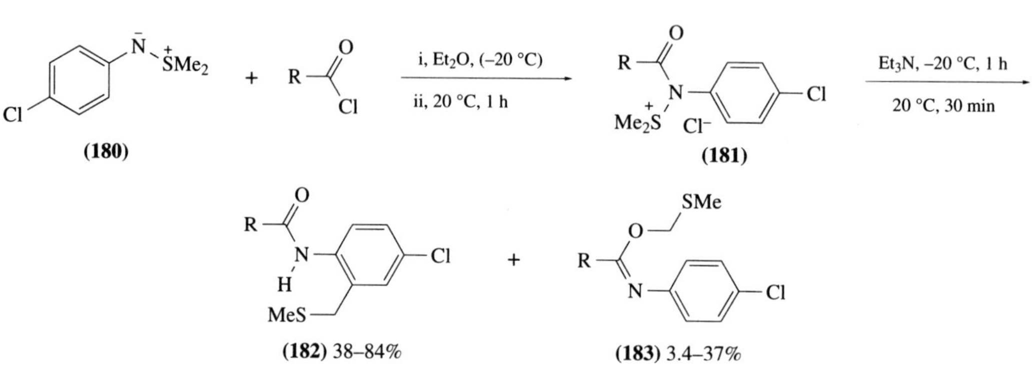

Scheme 41

The alkylation of anions derived from carboxylic acid amides affords imidates and/or amides. The course of the reaction is strongly influenced by the nature of the alkylating reagent and the metal cation used ⟨B-75MI 517-02, 85HOU(E5/1)812, 91COS(6)485⟩. An application of this reaction is given in Equation (131) ⟨90ZAAC(588)59⟩.

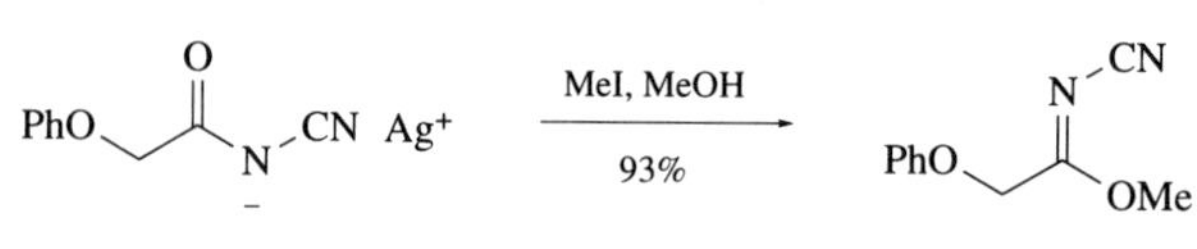 (131)

5.17.8.1.4 *From carboxylic* ortho-*esters, carboxylic* ortho-*amide derivatives and related compounds*

The acid-catalyzed reaction of *ortho*-esters with NH_2-containing compounds has often been used to prepare the corresponding iminoesters (Equation (132)) ⟨61CRV179, B-72MI 517-01, B-75MI 517-02, 85HOU(E5/1)812⟩. A compilation of examples is available ⟨91COS(6)485⟩. In addition, anilines ⟨79JOC633, 80LA1216, 88JMC1031, 88JOC3852⟩, aminopyrazoles ⟨82CHE1302, 82IJC(B)989, 82KGS1682, 88S242, 93MI 517-02⟩, aminomalonodinitriles ⟨83JOC3, 88JOC382⟩, *cis*-2,3-diamino-butenedinitrile ⟨90JCS(P1)1705⟩, aminotriazoles ⟨76JCS(P1)291, 88LA1107⟩, aminopyrano[2,3-*c*]pyrazoles ⟨90S704⟩, aminoimidazoles ⟨85LA1910, 88S242, 93T11029⟩, aminothieno[2,3-*d*]pyrimidines ⟨87CHE907, 87KGS1131, 88CHE1297, 88KGS1559⟩, aminothiophenes ⟨86JHC1757, 92BSB445⟩, aminothieno[2,3-*b*]quinolines ⟨90JIC260⟩, aminouracils ⟨86JHC1411⟩, aminothieno[2,3-*c*]pyran and aminothieno[2,3-*c*]thiopyran ⟨77KFZ38⟩, aminofuro[2,3-*c*]thiopyran ⟨90CHE1170, 90KGS1403⟩, aminocarbonylfuran ⟨86CS347⟩, aminoacetonitrile ⟨84JOC1212⟩, aminopyridine ⟨88JOC5309⟩ and 5-amino-6-cyano-2,3-dimethylimidazo[4,5-*b*] pyridine ⟨92JCS(P1)2789⟩ have been transformed to the corresponding iminoesters with the aid of various types of *ortho*-esters, for example trialkyl *ortho*-formates, trialkyl *ortho*-acetates, trialkyl *ortho*-benzoates, triethyl *ortho*-propionate, diethoxymethyl acetate and 2,5-bis(trimethoxymethyl)-1,3,4-thiadiazole. Imines (**184**), derived from alkyl-aryl or alkyl-heteroaryl ketones react with *ortho*-

formates on heating with catalytic amounts of toluenesulfonic acid to give *N*-alkenyliminoesters (**185**) (Equation (133)) ⟨91BCJ766, 92BCJ674⟩.

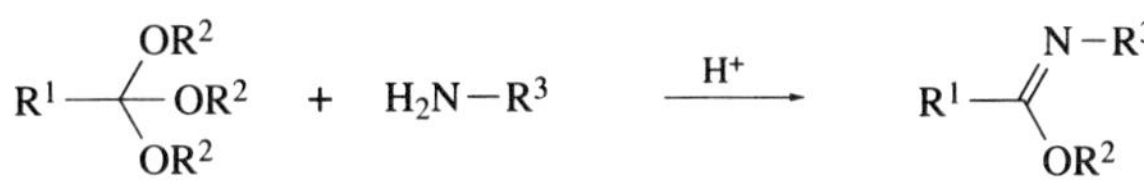

(132)

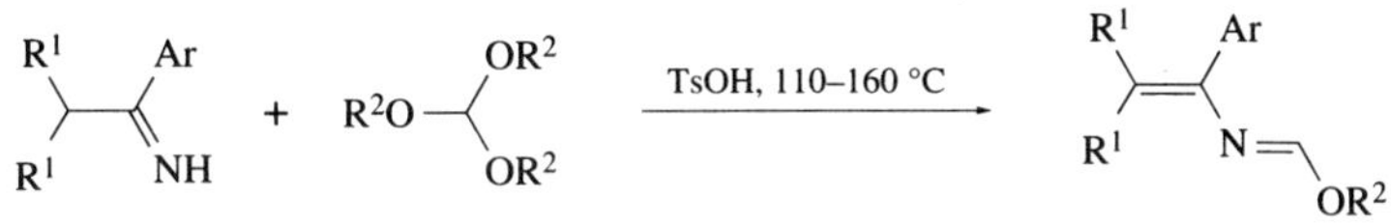

(133)

(184) **(185)**

Iminoesters can be prepared from ketene acetals and primary amines (Equation (134)) ⟨91COS(6)485⟩. The reaction of ynamines with aminoalcohols affords heterocyclic imidates (Scheme 42) ⟨88JOU1049, 88ZOR1165, 91CHE424, 91KGS532⟩. Imidates are also accessible from ketenimines and alcohols (Equation (135)) ⟨B-75MI 517-02, 85HOU(E5(1)812, 91COS(6)485, 91TL4279⟩.

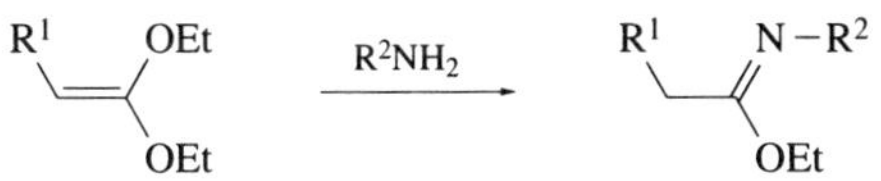

(134)

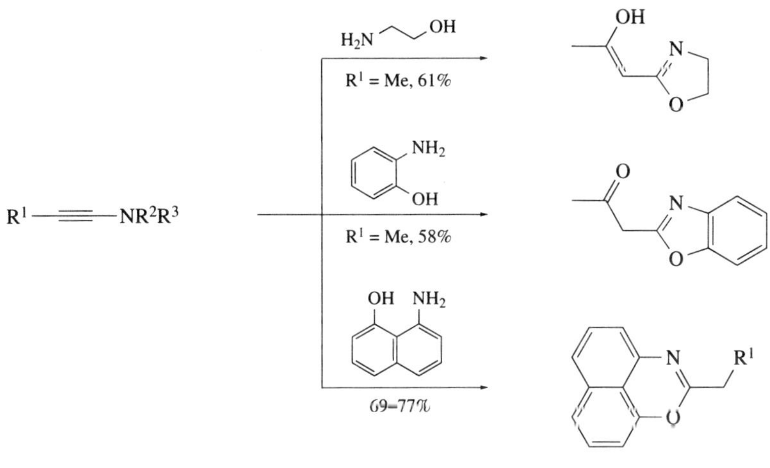

Scheme 42

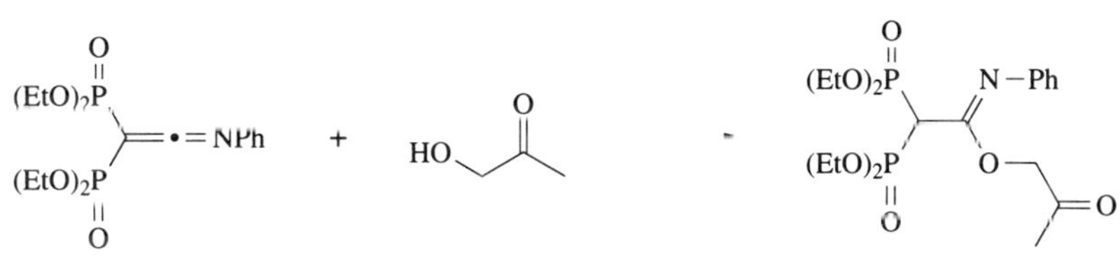

(135)

5.17.8.1.5 *From nitriles and isonitriles*

In nitriles with electron-withdrawing groups in the α-position, the basicity of the nitrile nitrogen is decreased; as a consequence, the Pinner synthesis does not work well. However, in such cases, the base-catalyzed addition of alcohols to the nitrile function proceeds well. From 5-substituted-2-cyanofuranes the corresponding iminoesters have been obtained by both methods ⟨92JOU1399, 92ZOR1750⟩. A range of aromatic and heterocyclic imidic acid esters, as well as α-arylthioimidic acid esters, have been prepared ⟨91COS(6)485⟩. In 1992, a procedure for the addition of ethanol to nicotinic acid nitrile with the aid of basic ion exchange resin was described ⟨92ZPK1430⟩. The base-catalyzed addition of alcohols to trichloro- and trifluoroacetonitrile has gained much attention, since the iminoesters thus formed (e.g., Equation (136)) are useful in the synthesis of natural products ⟨86AG213, 86AG(E)212, 87MI 517-01, 91COS(6)33, 91COS(6)485⟩. Some more recent applications are shown in Equations (137)–(140) (Structures **(186)**–**(188)**) ⟨86MI 517-02, 88ACS(B)605, 89S256, 93TL5983⟩.

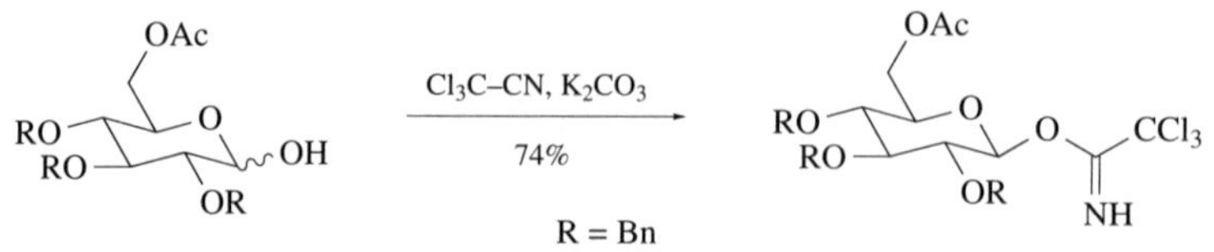

(136)

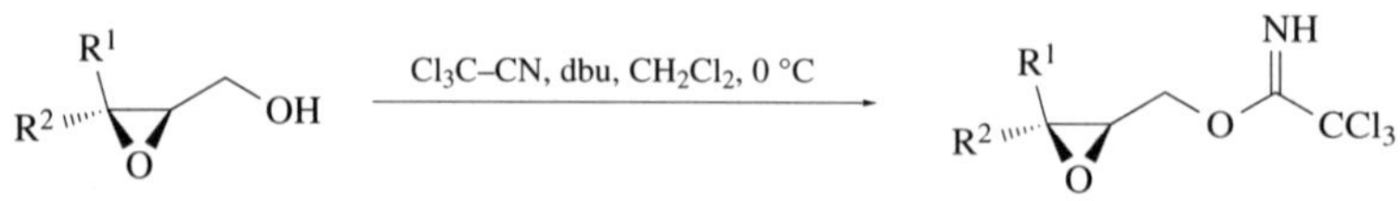

(137)

(186)

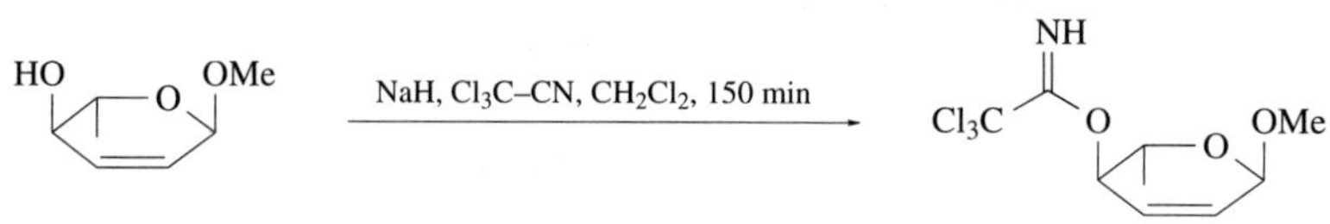

(138)

(187)

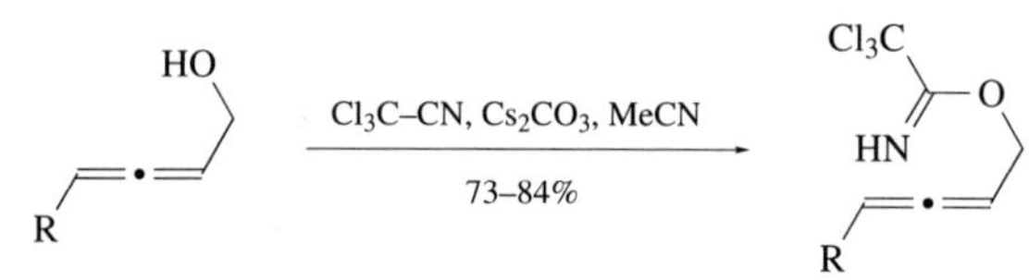

(139)

(188)

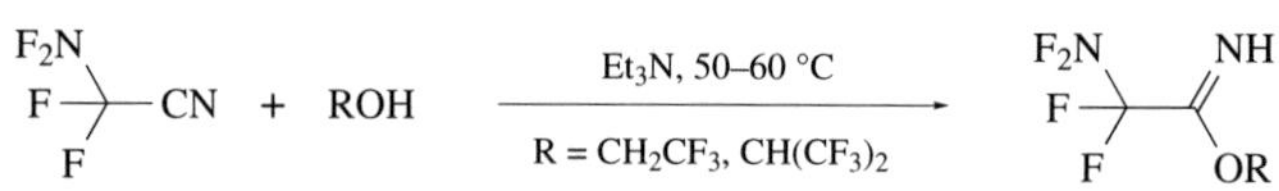

(140)

(189)

With triethylamine as a catalyst, iminoesters **(189)** (Equation (140)) have been obtained from *N*,*N*,2,2-tetrafluoro-aminoacetonitrile and fluorinated alcohols ⟨88IC3100⟩. Alcohols add to α-nitro-nitriles catalyzed by NaCN to give the corresponding iminoesters **(190)** (Equation (141)) ⟨88BAP139⟩. In some cases iminoester formation was observed as the main reaction or as a consecutive reaction **(191)**, **(192)** (Equations (142) and (143)) ⟨84CB1935, 88BCJ3731⟩.

Oxiranyliminoesters **(193)** are formed from dichloroacetonitrile and ketones in the presence of isopropoxide (Equation (144)) ⟨74BSF1965⟩; in a related reaction, an oxetanylimidic acid ester **(195)** was prepared from the aldehyde **(194)** (Equation (145)) ⟨89JCR(S)337⟩. The copper-catalyzed addition of alcohols to isonitriles affords formiminoesters (Equation (146)) ⟨B-75MI 517-02, 75S291, 85HOU(E5/1)812, 91COS(6)485⟩. Iminoesters can also be prepared from isonitriles and aldehyde-*O*,*N*-acetals and α,β-unsaturated ketones (Scheme 43) ⟨91COS(6)485⟩.

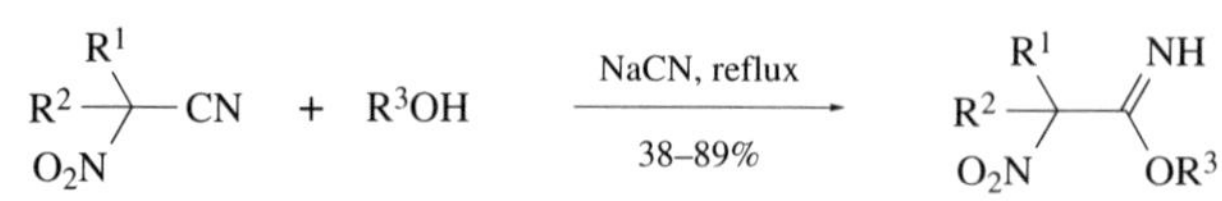

(141)

(190)

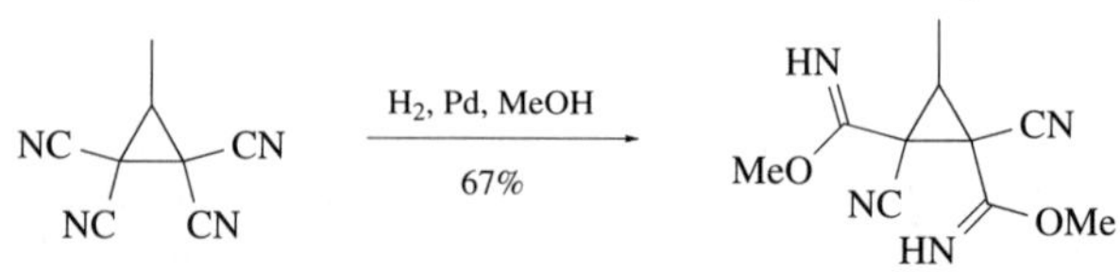

(142)

(191)

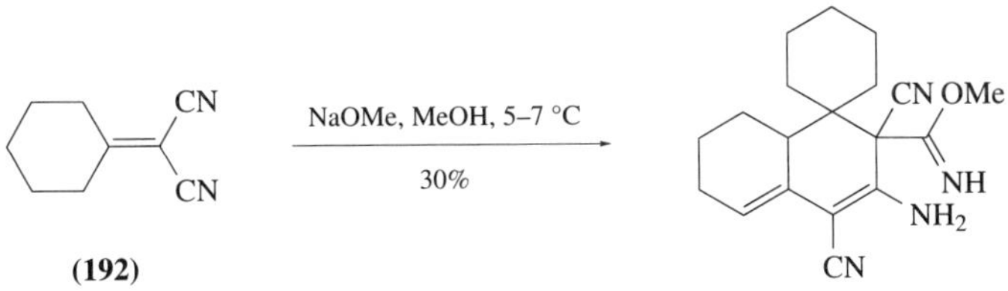

(192) (143)

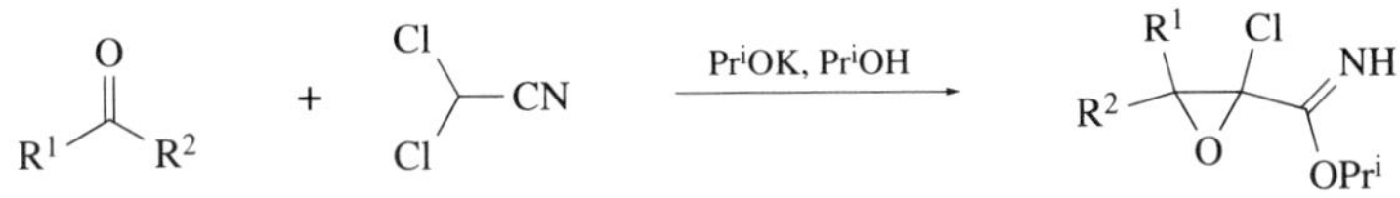

(193) (144)

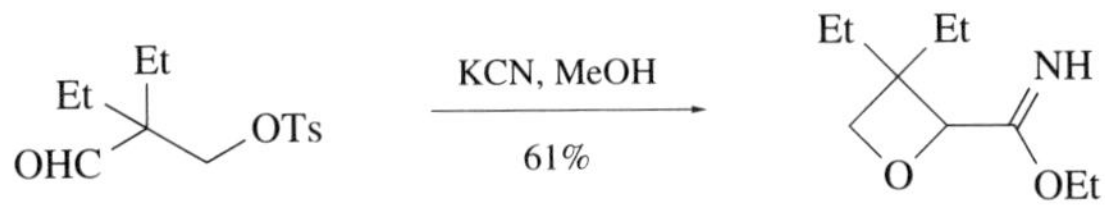

(194) (195) (145)

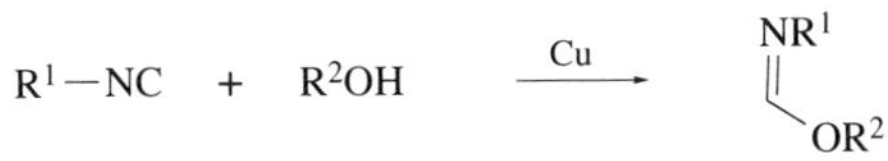

(146)

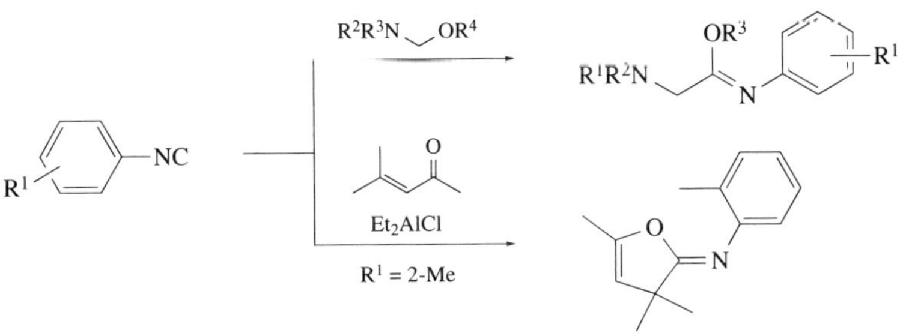

Scheme 43

5.17.8.1.6 From heterocyclic compounds (ringopening reactions)

The action of alkoxides on aziridines gives rise to the formation of iminoesters by opening the aziridine ring (Equation (147)) ⟨91COS(6)485⟩. Similar reactions have been performed with derivatives of pyrimidone (**196**) (Equation (148)) ⟨81JOC1769⟩ 1,2,4-triazin-5-one (**197**) (Equation (149)) ⟨88CBP3354⟩, 1,2,4-thiadiazolium fluorosulfate (**198**) (Equation (150)) ⟨77JCS(P1)1791⟩ and 1,4-diazepine (**199**) (Equation (151)) ⟨84CPB4694⟩. Some of these reactions proceed under photochemical conditions.

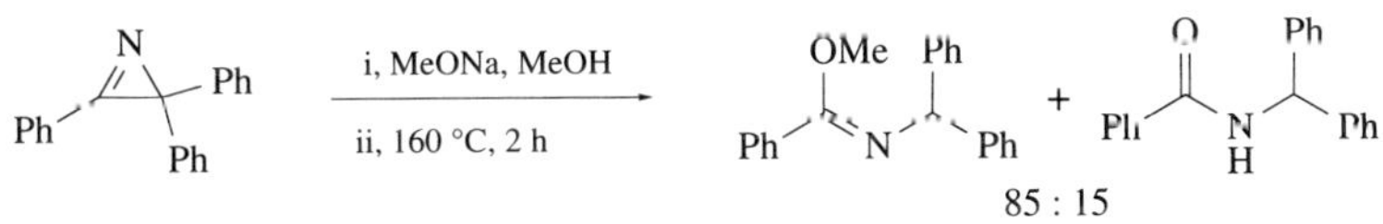

(147)

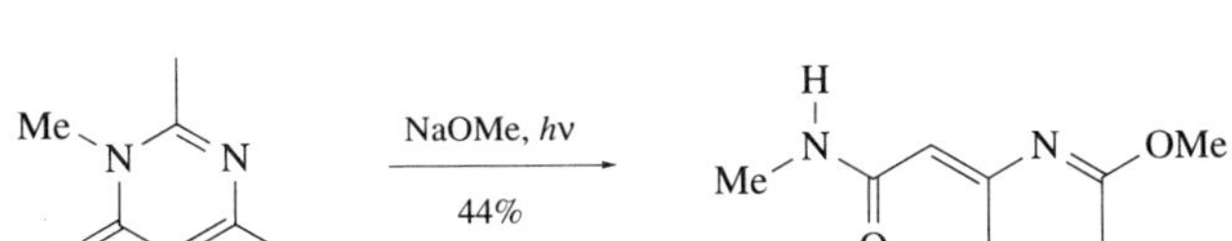

(196) (148)

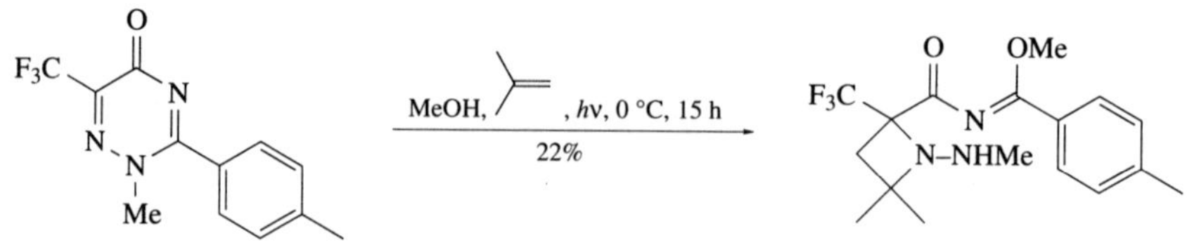

(197)

(149)

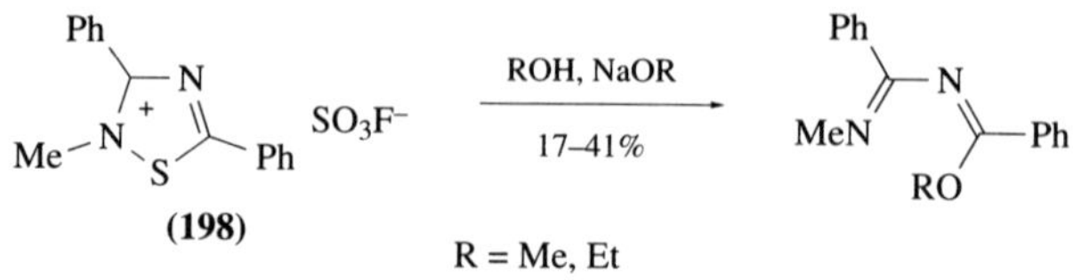

(150)

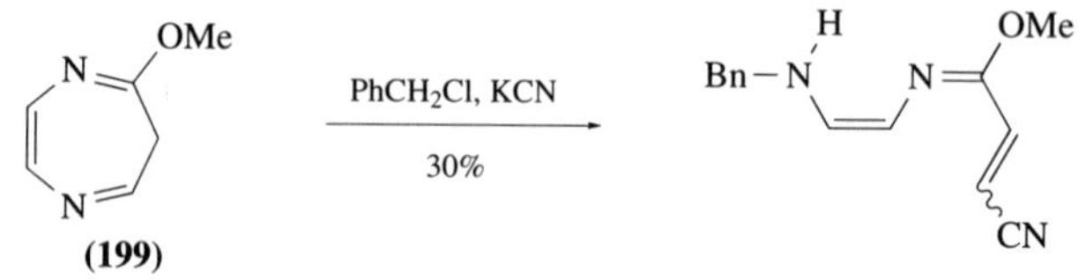

(199)

(151)

5.17.8.1.7 *From imidates, thioimidates, amidines and related compounds*

Upon alkoxide catalyzed transesterification, the alkoxy group in imidates can be exchanged ⟨B-75MI 517-02⟩; however, there are known examples where this reaction was also performed under acid catalysis (Equation (152)) ⟨89LA665⟩. *N*-Aryl substituted imidates have been prepared from *N*-unsubstituted iminoester hydrochlorides and aromatic amines (Equation (153)) ⟨87JHC325⟩. *N*-Unsubstituted imidates can be carboxylated at the nitrogen with the aid of isocyanates, isothiocyanates or chloroformic esters (Scheme 44) ⟨B-75MI 517-02, 91COS(6)485⟩; analogous *N*-acylations have been performed with carboxylic acid chlorides ⟨B-75MI 517-02, 85CB3089, 89JOC4335, 89TL3069, 91COS(6)485⟩, carboxylic acid anhydrides ⟨80JCS(P1)2184 (e.g., Equation (154)) and imidoyl chlorides (Equation (155)) ⟨76JOU60, 76ZOR64⟩. Reactive functional groups which are bound to the nitrogen of the imidate group can be modified with several reagents without degradation of the imidate function (Equations (156) and (157) and Scheme 45) ⟨75JOU1926, 75ZOR1913, 77JOU1474, 77ZOR1597, 89TL3069⟩. The *N*-phosphorylated benzamide **(200)** ⟨91TL4193⟩, thioimidates (e.g., **(201)** ⟨B-75MI 517-02, 87NKK1304⟩ and the amidine **(202)** ⟨76JOU1789, 76ZOR1826⟩ yield imidates on alcoholysis (Equations (158)–(160)).

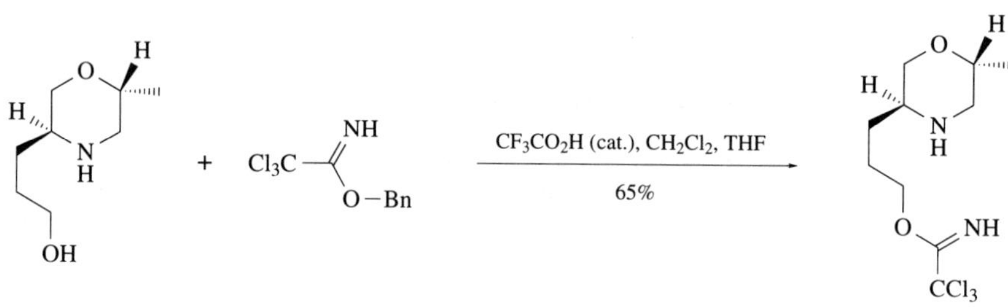

(152)

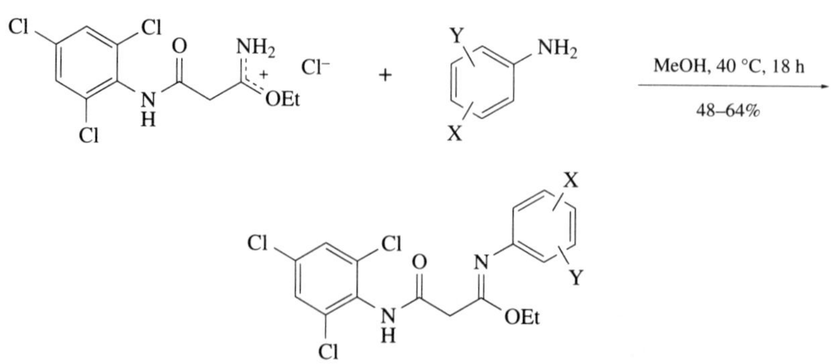

(153)

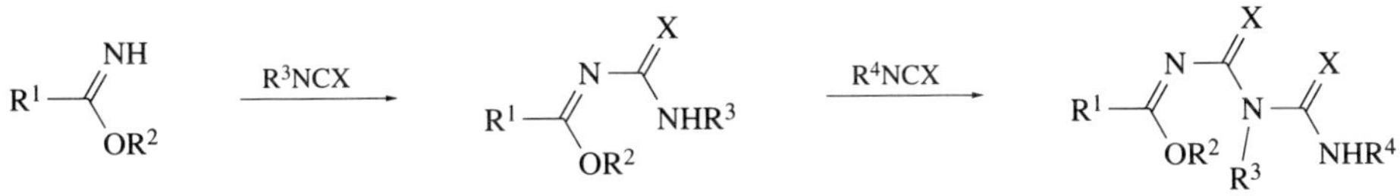

Scheme 44

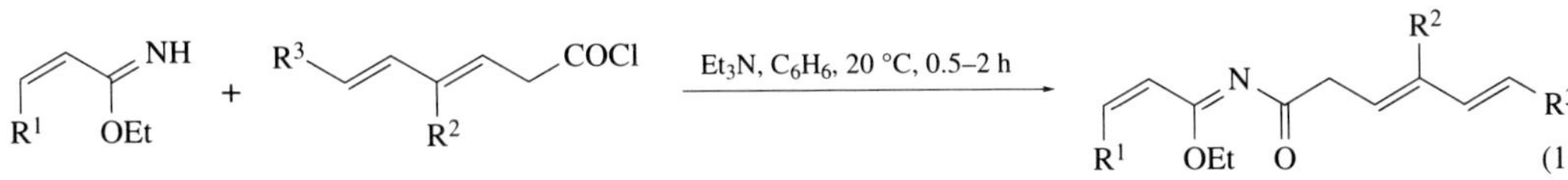

(154)

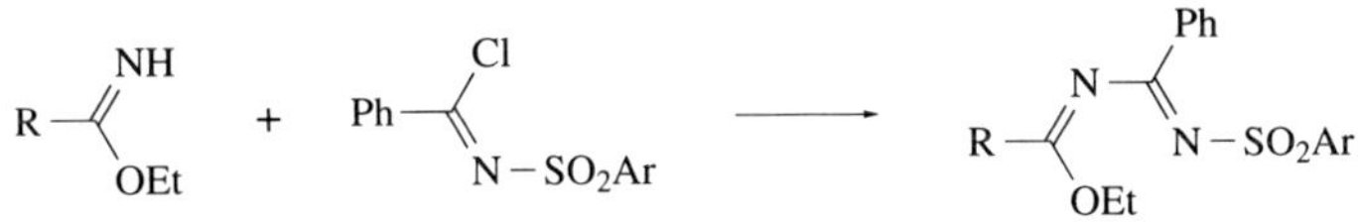

(155)

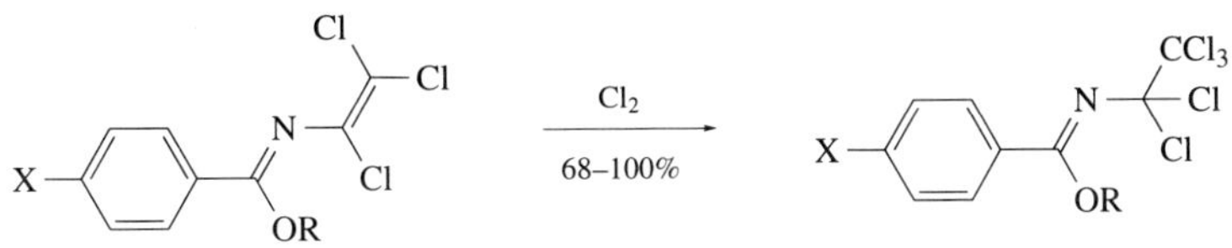

(156)

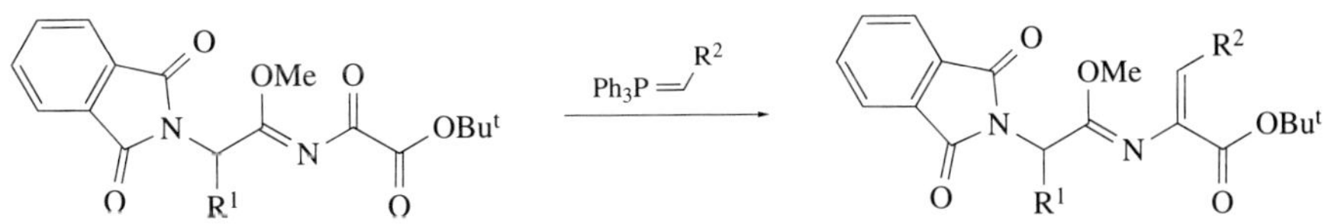

(157)

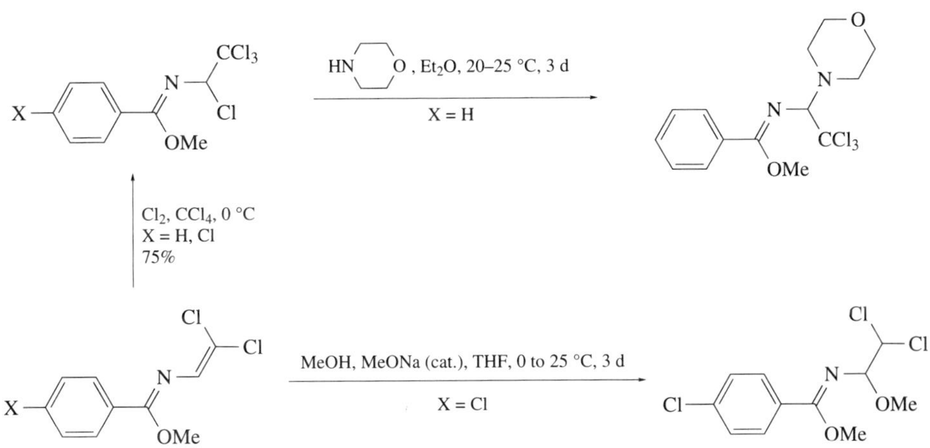

Scheme 45

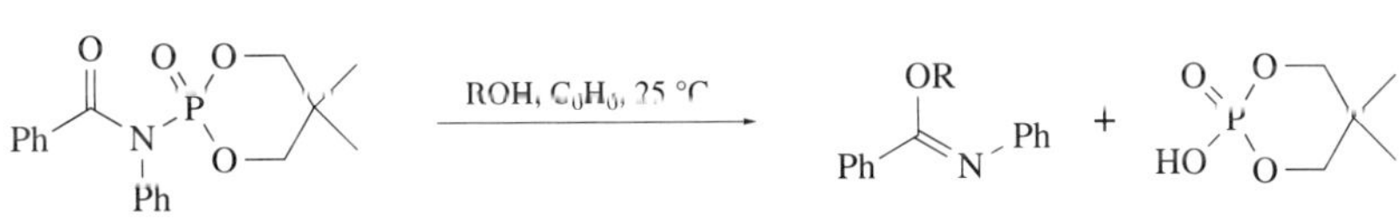

(200)

R = Me, Bu^t, Ph

(158)

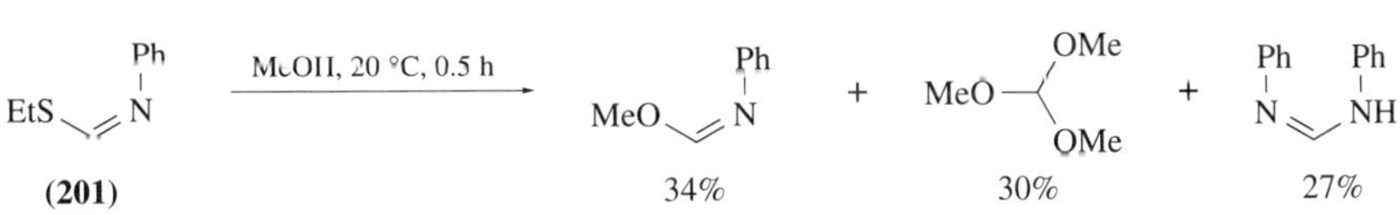

(201)

(159)

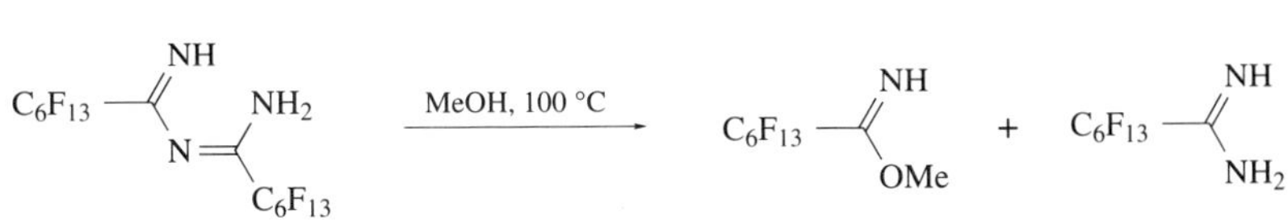

(202)

(160)

N-Benzyl- and *N*-(methoxycarbonyl)methyl-benzimidates (**203**) are metalated by lithium diethylamide at the *N*-methylene group; the anions thus formed can be reacted with electrophiles to yield the corresponding *N*-alkylated imidates (Scheme 46) ⟨81HCA1322⟩.

5.17.8.1.8 *From metal complexes*

On irradiation or on heating, alkoxycarbene-pentacarbonyl chromium reacts with nitriles ⟨88JA307⟩, sulfilimines ⟨91OM11, 93JOC3886⟩ and azirines ⟨85OM1749⟩ to give iminoesters (Scheme 47). Similar reactions have been performed with Pt-nitrile complexes (Equation (161)) ⟨89JCS(D)947⟩. Iminoesters can also be prepared from alkylamino methoxycarbene gold complexes ⟨74JOM(71)453⟩.

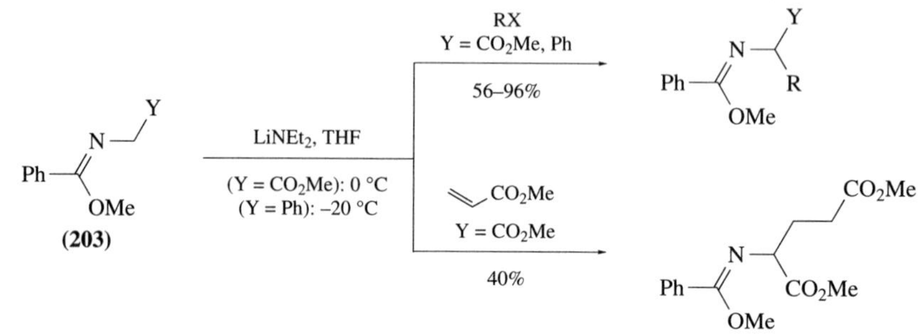

Scheme 46

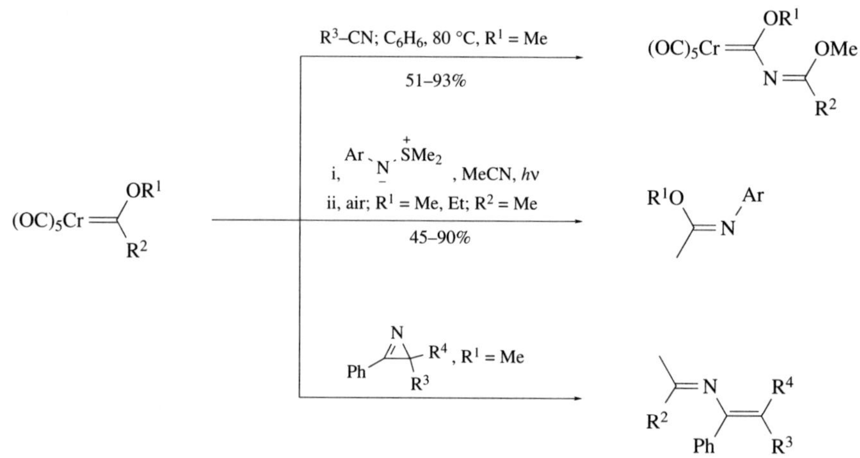

Scheme 47

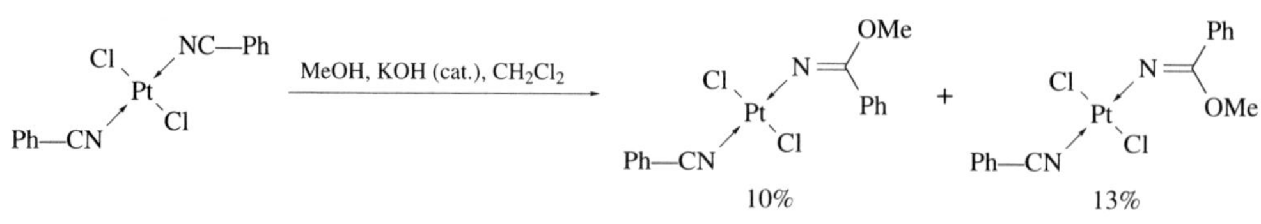 (161)

5.17.8.1.9 *Miscellaneous methods*

When the *O*,*N*-acetals (**204**) are treated with strong bases, iminoesters are formed (Scheme 48) ⟨81JCS(P1)2435⟩. Aldehydes or ketones and hydrazoic acids give α-azidoethers (**205**), (**206**), which on photolysis or thermolysis afford iminoesters (Schemes 49 and 50) ⟨88JOC22⟩. Ethylvinyl ether reacts with the azidocephalosporine (**207**) to give the iminoester (**208**) (Equation (162)) in low yield ⟨84H(22)2487⟩. Iminoesters can also be synthesized from α-chloro-imines and various reagents (Equation (163) and Scheme 51) ⟨90T6753, 92T7345⟩.

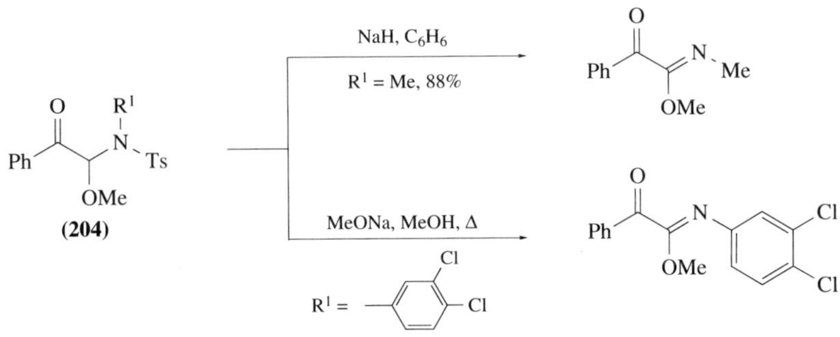

Scheme 48

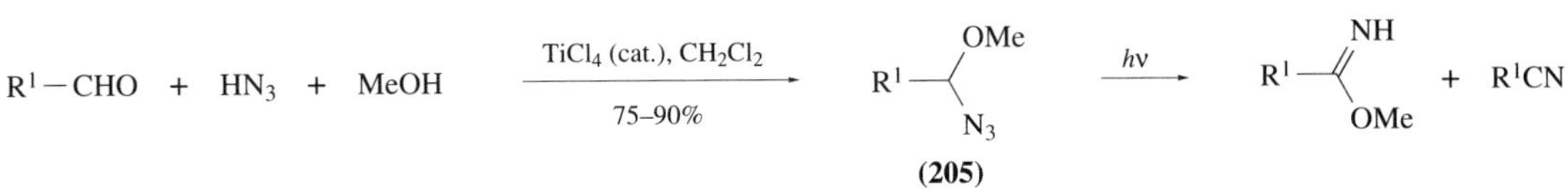

Scheme 49

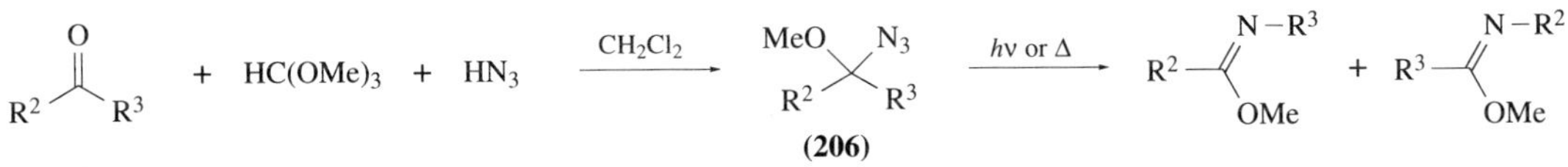

Scheme 50

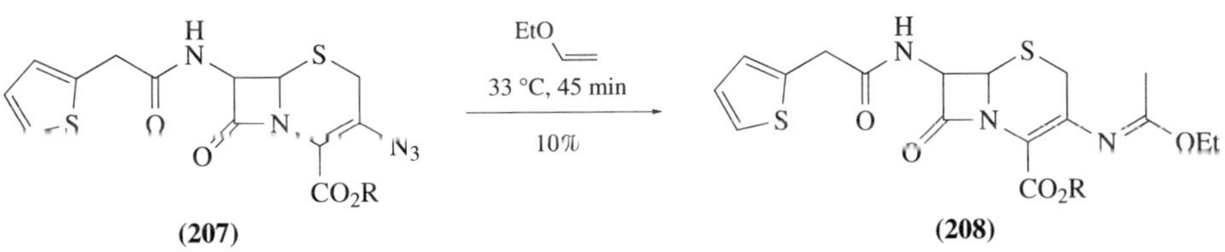

(162)

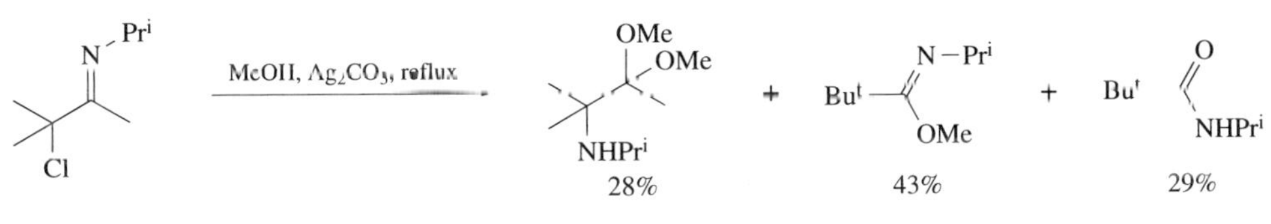

(163)

Scheme 51

Oxidation of *N*-phenyl-thiobenzamide-*S*-oxide (**209**) in the presence of alcohols gives iminoesters and *N*-phenylbenzamide (Equation (164)) ⟨75LA305⟩.

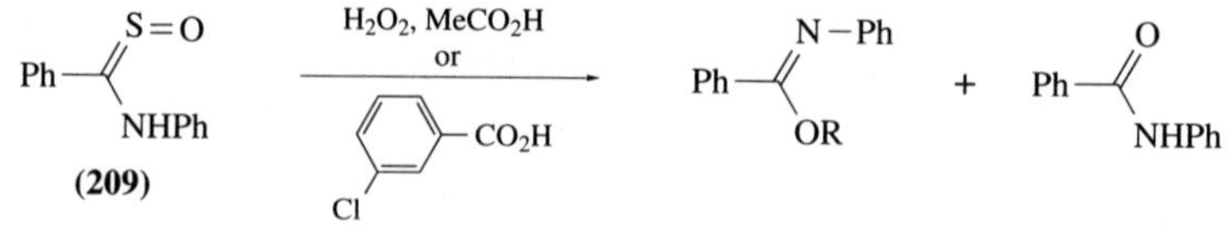

(209) (164)

The electrochemical oxidation of cyanide ions in methanol in a divided cell affords dimethyl oxaldiimidate; in an undivided cell, dimethyl carbimidate is produced (Scheme 52) ⟨93CL521⟩.

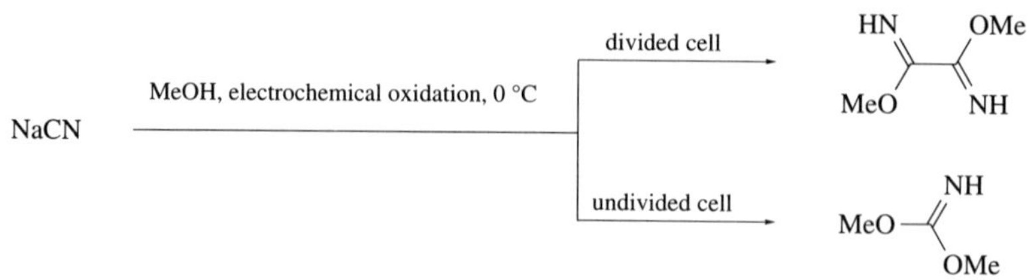

Scheme 52

The malonoamide-imidoester (**211**) is accessible by flash vacuum pyrolysis of the dimedone derivative (**210**) (Equation (165)) ⟨92CC1571⟩. Benzylidencyanalkylamino oxides (**212**) and alcoholic alcoholates give iminoesters (**213**) (Equation (166)) ⟨77CZ154⟩.

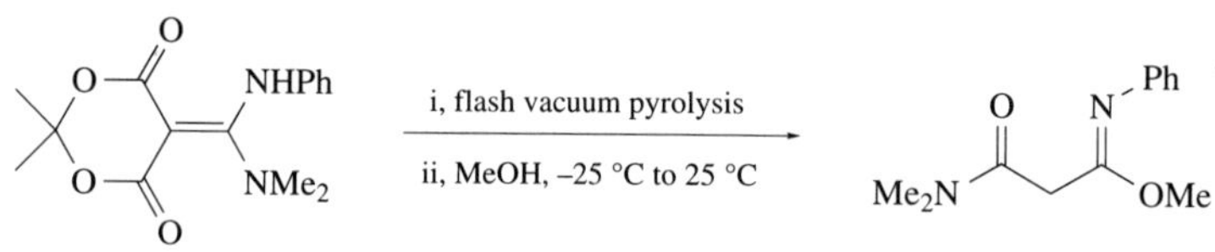

(210) (211) (165)

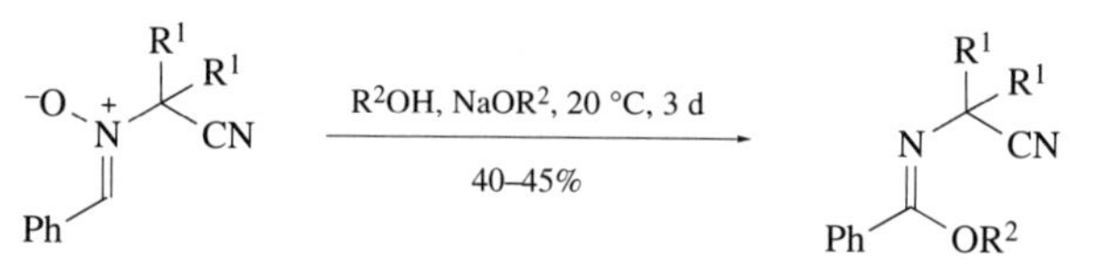

(212) (213) (166)

5.17.8.2 *N*-Haloimidates

The chemistry of *N*-haloimidates is included in a general review of imidates ⟨B-75MI 517-03, B-91MI 517-04⟩.

Alkoxides transform *N*-fluorinated imidoyl fluorides to *N*-fluorinated imidates (Equation (167)) ⟨66JOC3689, 71T51⟩. In a related reaction *N*-fluoroimidates were obtained from *N*,*N*,*N′*-trifluoro-amidines and alkoxides (Equation (168)) ⟨78IZV2802⟩. Reaction of hypochlorite with *N*-unsubstituted imidoates gives *N*-chloro-⟨1896JA751, 01CB1615, 13CB3616, 66JOC3708⟩, *N*-bromo- and *N*-iodoimidoates ⟨13CB3616⟩. The *N*-chloro compounds can also be obtained (sometimes in lower yield) using *t*-butyl hypochlorite ⟨66JOC3708, 87AJC701⟩. *N*-Bromoimidoates have been prepared using *t*-butyl hypobromite (Equation (169)) ⟨93AJC1213⟩. *N*-Fluorinated imidoates are accessible by fluorination of *N*-unsubstituted imidoates with mixtures of fluorine and nitrogen (Equation (170)) ⟨82IZV937⟩.

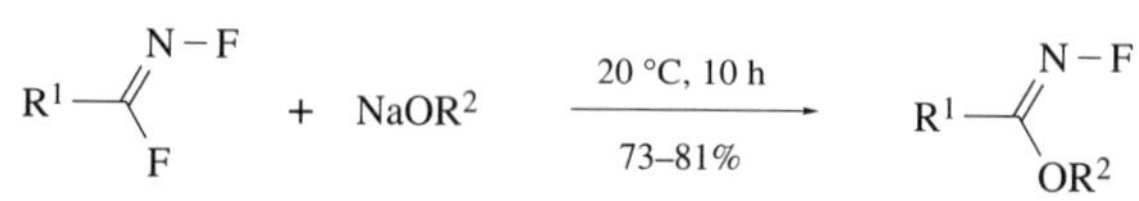

(167)

$R^1 = CN, F_2N–CF_2$

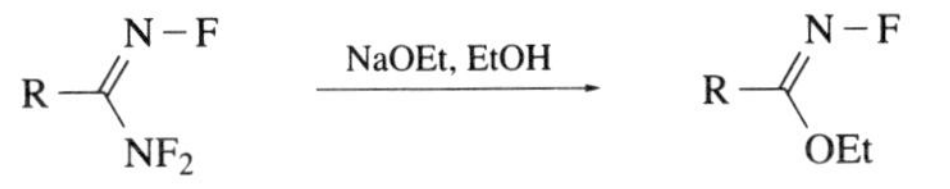

(168)

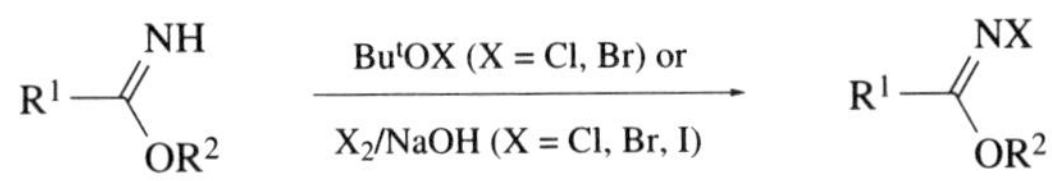

(169)

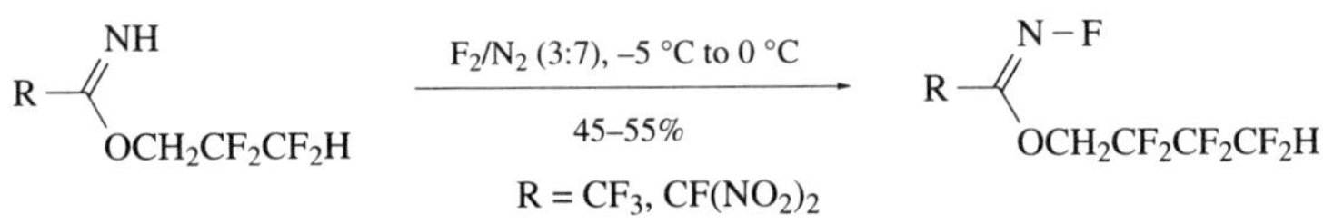

(170)

2,2,3,3,4,4,5,5-Octafluoropentyl hypochlorite adds to the nitrile group of perfluorobutyronitrile to give the corresponding *N*-chlorobutyrimidic acid ester (Equation (171)) ⟨81JFC(18)553⟩.

N-Fluorinated imidic acids react with diazomethane to give the corresponding methylimidates (Equation (172)) ⟨71T51⟩.

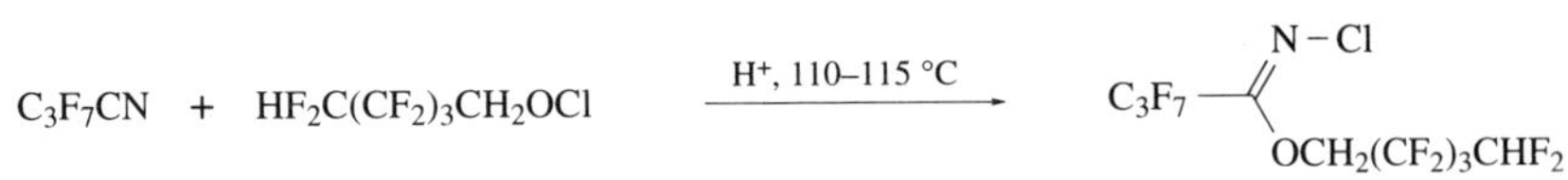

(171)

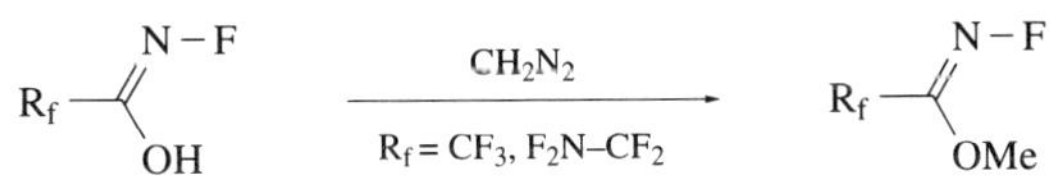

(172)

5.17.8.3 *N*-Hydroxy- and *N*-Alkoxyimidates (Hydroxamic Acid Esters)—General Remarks

The chemistry of these compounds is included in several reviews ⟨B-75MI 517-04, 85HOU(E5/1)812, 85HOU(E5/2)1591, B-91MI 517-05⟩.

*5.17.8.3.1 From hydroxamoyl chlorides (*N*-Hydroxy- and *N*-alkoxyimidoylhalides), nitrile oxides and iminoesters*

Hydroxamoyl chlorides are attacked by alkoxides to give alkyl-*N*-hydroxyimidoates ⟨B-68MI 517-04⟩; analogous reactions can also be performed with *N*-hydroxy-imidoyl bromides and sodium and thallium alkoxides (Equation (173)) ⟨B-75MI 517-04, 85HOU(E5/1)812⟩. The alkoxide catalyzed addition of alcohols to nitrile oxides is a well known method of synthesis for *N*-hydroxyimino esters (Equation (174)) ⟨78JOC388, 85HOU(E5/1)812, 85HOU(E5/2)1591⟩. Aromatic nitrile oxides react with methanol and BF_3-etherate to give methyl *N*-hydroxyimidoates provided that the reactions are stopped after a short time (Equation (175)) ⟨81BCJ3863⟩. The methoxyaminolysis of benzimidates affords a mixture of *N*-methoxy-benzamidines (**214**) and *N*-methoxy-benziminoesters (**215**) (Equation (176)) ⟨82JA6769⟩.

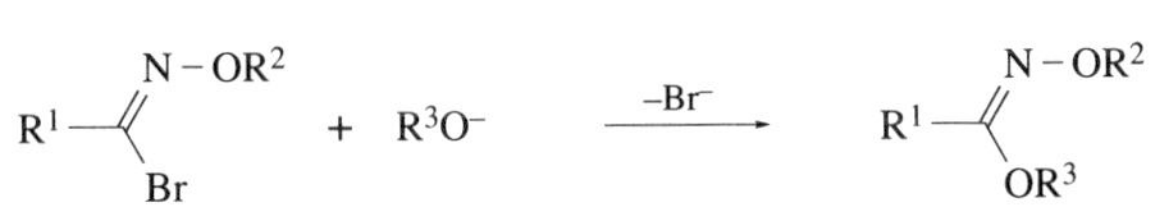

(173)

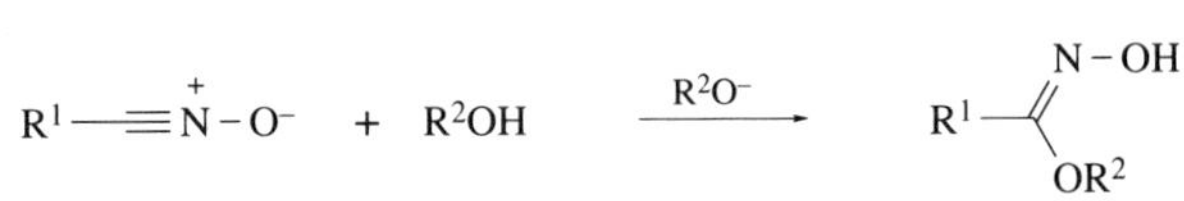

(174)

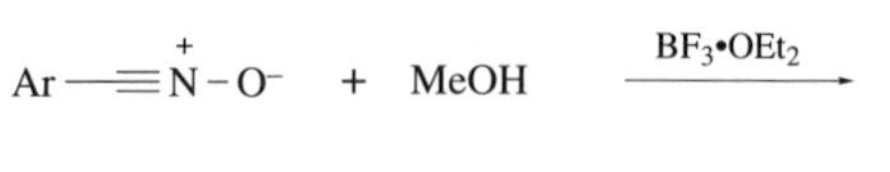

(175)

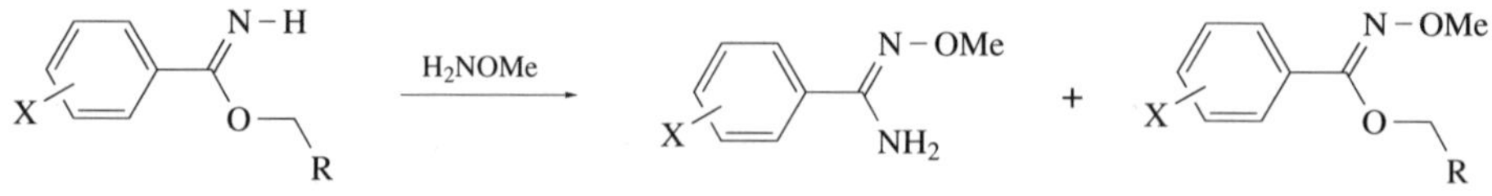

(176)

(214) **(215)**

5.17.8.3.2 *By alkylation of* N-*alkoxyamides (hydroxamic acids)*

Alkylation of *N*-alkoxy carboxylic acid amides gives mixtures of *N*-alkoxyiminoesters and *N*-alkyl-*N′*-alkoxyamides, because the substrate molecules can be attacked by the alkylation reagent either at the carbonyl oxygen or at the nitrogen of the *N*-alkoxy group (Equation (177)) ⟨85HOU(E5/1)812⟩. Salts of *N*-alkoxyamides are alkylated to give *N*-alkoxyiminoesters. Interestingly, the stereochemistry of the reaction products is strongly influenced by the counterion; thus from potassium salts, *N*-alkoxyiminoesters are formed with *Z*-configuration exclusively, whereas from silver salts a mixture of *E*/*Z*-isomers is obtained ⟨85HOU(E5/1)812⟩. On irradiation or heating, the *Z*-isomers can be transformed into the *E*-isomers ⟨85HOU(E5/1)812⟩. Methylation of *N*-acyloxyamides with diazomethane results in *syn* (i.e., *Z*) products (Equation (178)) ⟨69HCA569⟩.

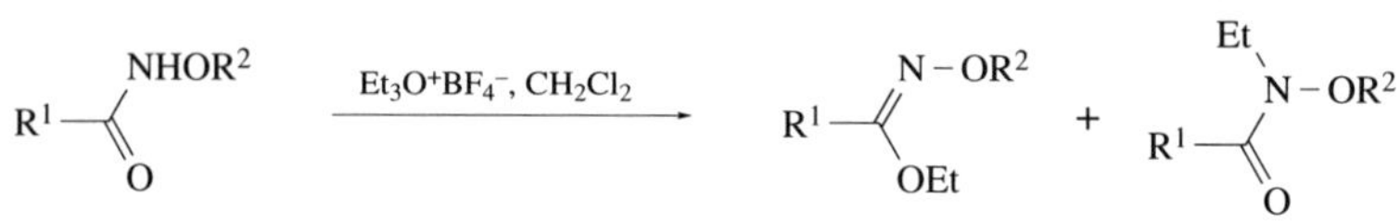

(177)

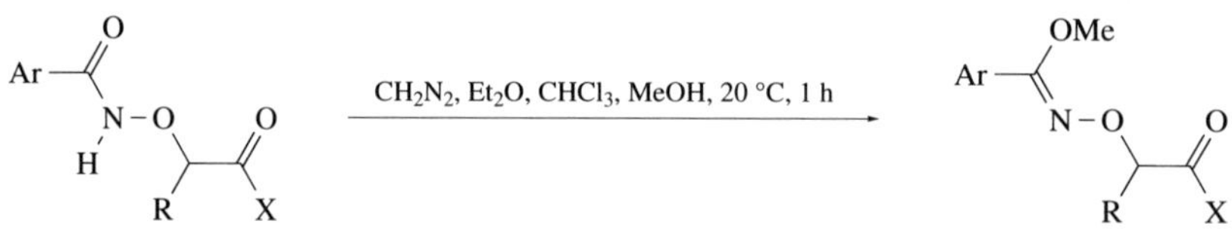

(178)

5.17.8.3.3 *From hydroxylamine, hydroxylamine derivatives and thiocarboxylic acid esters*

N-Hydroxy- and *N*-alkoxyimino esters can be prepared in good yield from thiocarboxylic acid esters and the appropriate hydroxylamine derivative at room temperature (Equation (179)) ⟨85HOU(E5/1)812⟩. The necessary thionoesters can be prepared from Lawesson's reagent and esters (Equation (180)) ⟨92JCS(P1)2127⟩ or from carboxylic *ortho*-esters and H_2S (Equation (181)) ⟨93ACS1004⟩.

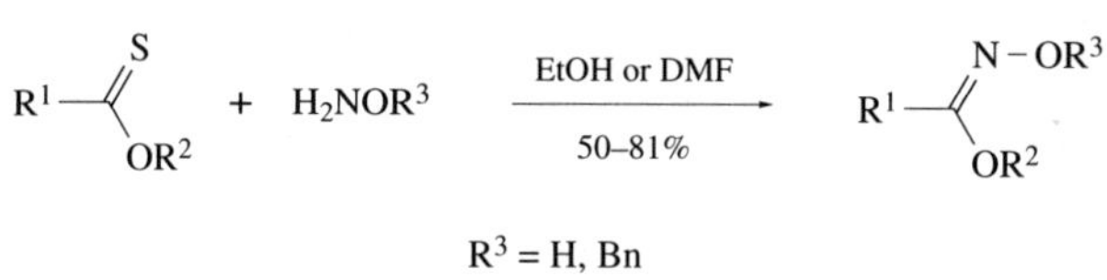

(179)

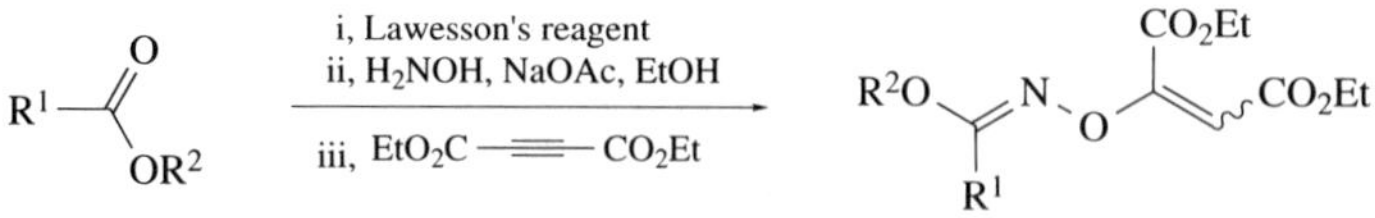

(180)

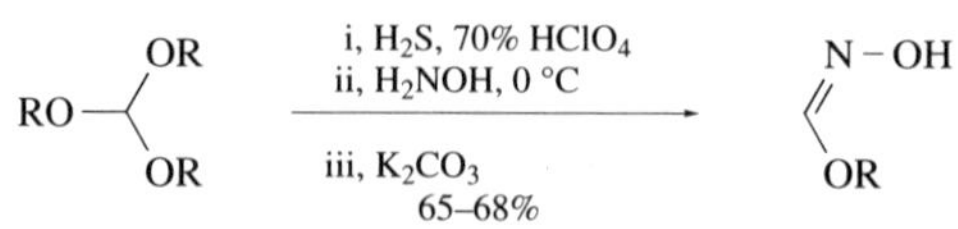

(181)

5.17.8.3.4 *From* N-*hydroxyimidic acid esters*

N-Hydroxyimidic acid esters have been alkylated at the hydroxy group with various reagents, such as α,ω-dihaloalkanes ⟨90MI 517-01⟩, α-bromocarboxylic acid esters ⟨92SC1985⟩, α,β-unsaturated aldehydes ⟨77OPP57⟩, alkyl propiolates and dialkyl alkenedicarboxylates ⟨92JCS(P1)2127⟩. Some of the products thus formed have additional reactive groups, which can be used to introduce further functional groups ⟨90MI 517-01, 92SC1985⟩. Examples are given in Schemes 53 and 54, and Equation (182). By double alkylation of hydroxyamic acids with ethyl bromide or triethyloxonium tetrafluoroborate, ethyl-*N*-ethoxyimidoates have been prepared which were α-brominated with NBS. From these α-bromo-*N*-ethoxyimidoates the α-fluoro- as well as the α-aminoimidoates could be obtained (Scheme 55) ⟨92LA997⟩.

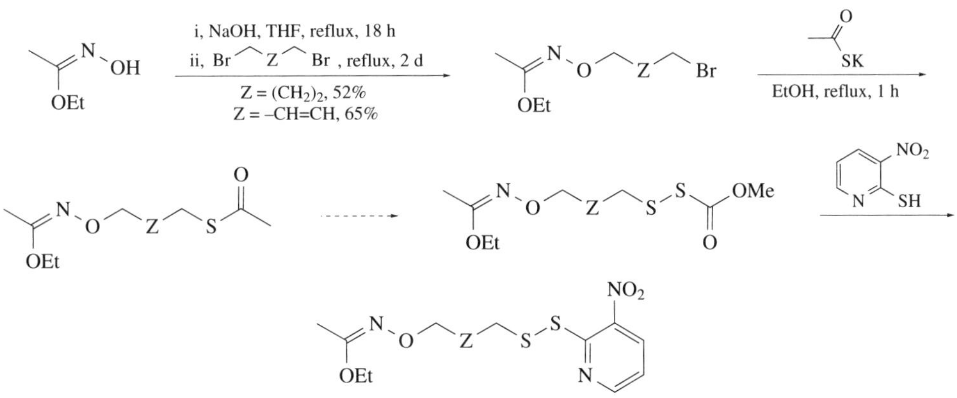

Scheme 53

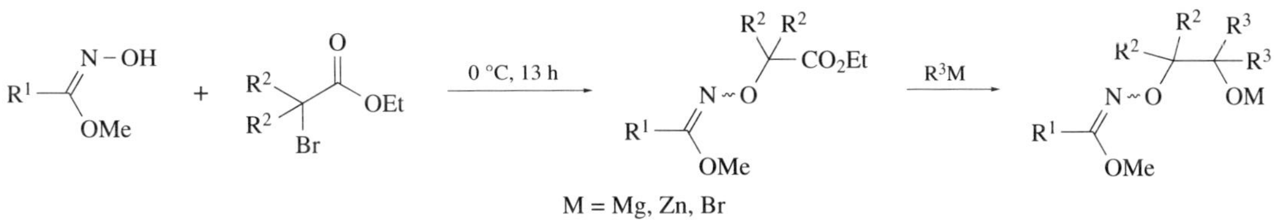

Scheme 54

 (182)

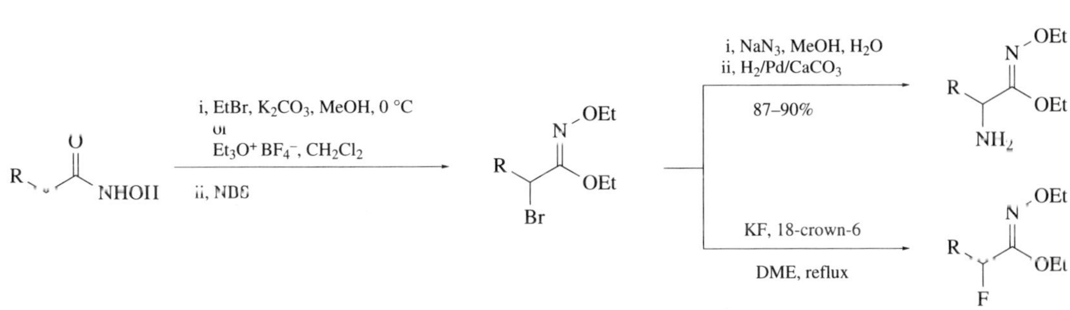

Scheme 55

5.17.8.4 *N*-Organothioiminoesters and Related Compounds

In the reaction of *N*,*N*-dibromobenzenesulfonamide with cyclohexene and formamide (2-bromocyclohexyl) *N*-benzenesulfonyl formimidate is formed (Equation (183)) ⟨71CPB1606⟩ among other products. The condensation of carboxylic *ortho*-esters with benzenesulfonamide represents a valuable method for the preparation of *N*-benzenesulfonyl carboximidates (Equation (184)) ⟨69CB1641⟩.

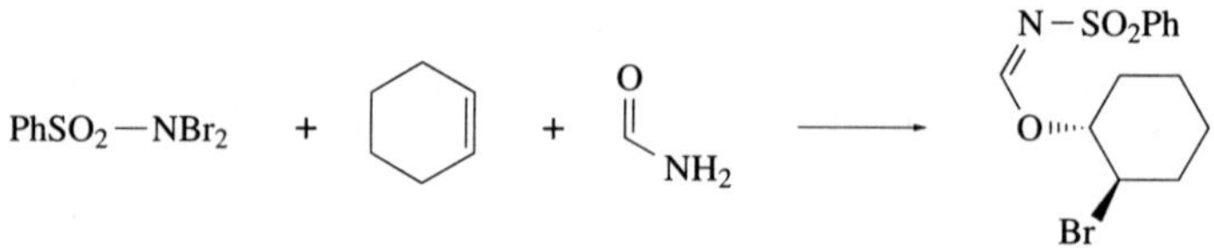
(183)

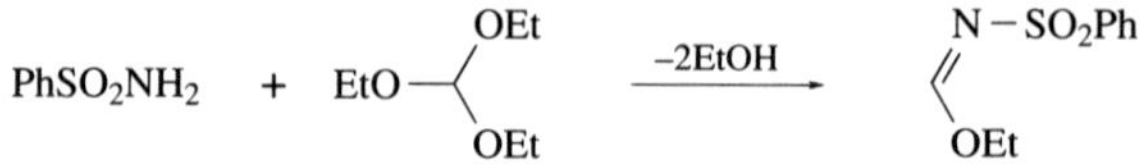
(184)

N-Acyl-*N*-toluenesulfonylaminosulfonium salts, which are accessible from dialkylsulfonio-*N*-toluenesulfonamides and acylium hexachloroantimonates, are deprotonated by triethylamine to give mixtures of *N*-toluolsulfonyl carboximidic acid esters and *N*-toluolsulfonyl carboxylic acid amides (Scheme 56) ⟨80LA1055⟩. 1-Alkoxy-1,4-cyclohexadienes ⟨86TL4701, 91TL2457⟩, 1-alkoxy-cyclohexenes ⟨88JOU2154, 88ZOR2389, 91TL2457⟩ and noncyclic enol ethers ⟨88JOU2149, 88JOU2154, 88ZOR2384, 88ZOR2389⟩ react with arenesulfonyl azides to give *N*-arenesulfonyl-substituted iminoesters (Equations (185)–(187)). On heating, 2-ethoxy-1-methanesulfonyl-4-trichloromethyl-3-trimethylsilyl-2-azetine undergoes ring opening to give an *N*-sulfonylated α,β-unsaturated imidoate (Equation (188)) ⟨91JGU1268, 91ZOB1389⟩.

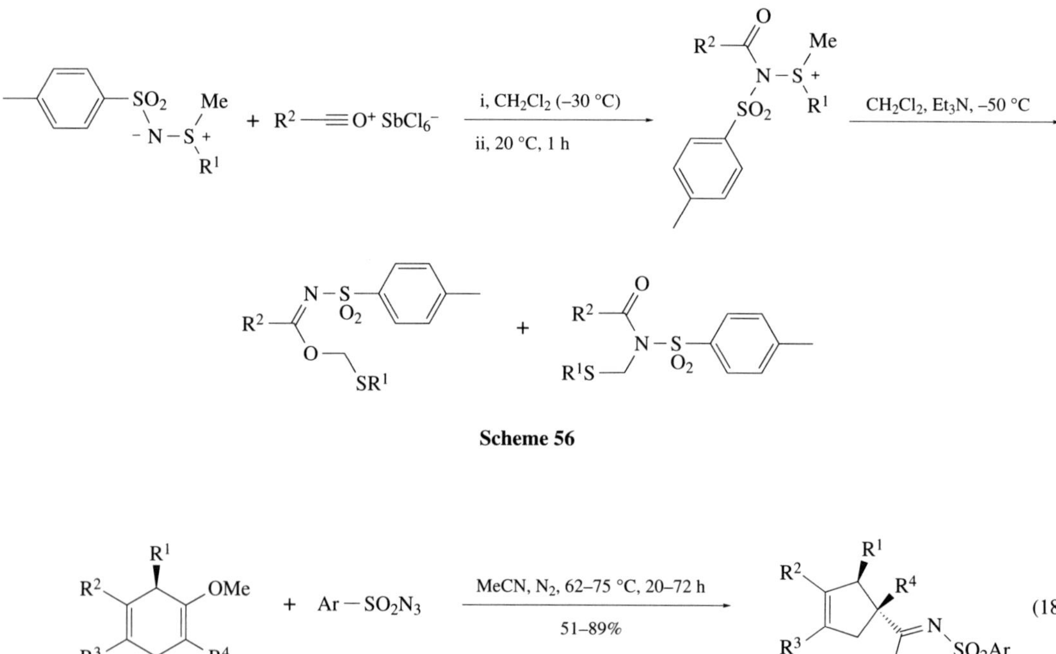

Scheme 56

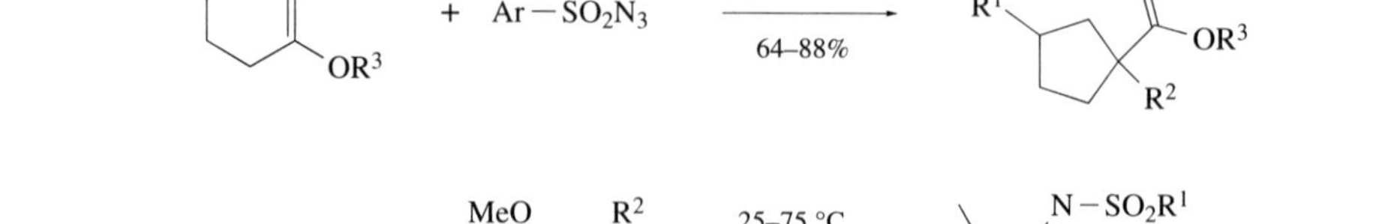
(185)

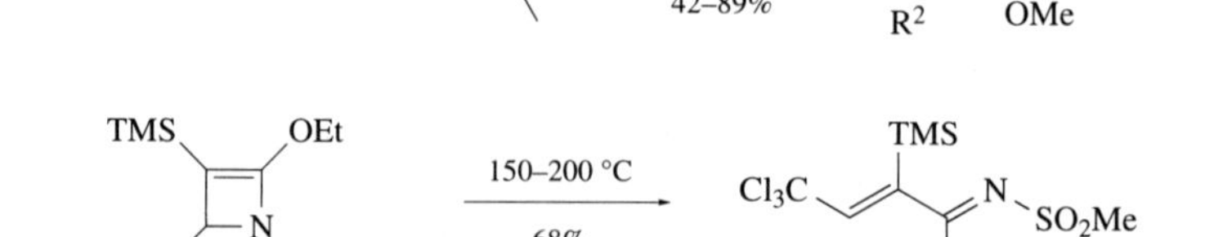
(186)

(187)

(188)

Iminoesters are attacked by arenesulfonyl halides to give *N*-arenesulfonyliminoesters (Equation (189)) 〈43JCS101〉. Analogous reactions can be performed with arylsulfenyl chlorides (Equation (190)) 〈87JOU75, 87ZOR86〉. The reactions of *N*-chloroimidates with sulfenyl chlorides 〈85JOU1166, 85ZOR1281〉 and diarylsulfides 〈87JOU75, 87ZOR86〉 yield *N*-(aryl-dichlorothio)imidates and *N*-(arylthio)imidates (Scheme 57). *N*-(Aryldichlorothio)imidates, which seem to be in equilibrium with *N*-(arylchlorosulfoniono)imidate chlorides, are also accessible by chlorination of *N*-arylthioimidates (Scheme 58) 〈85JOU1166, 85ZOR1281〉.

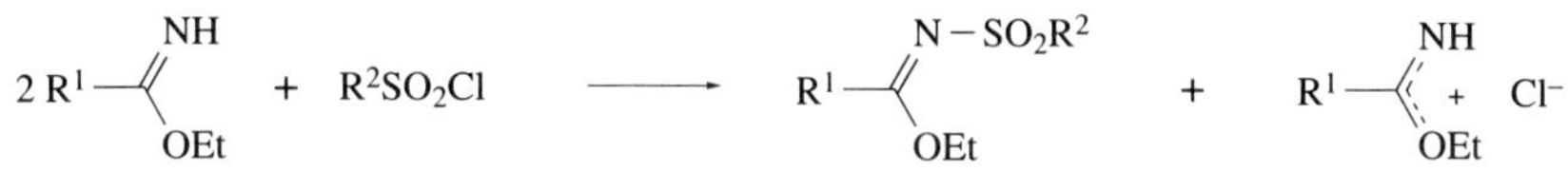
(189)

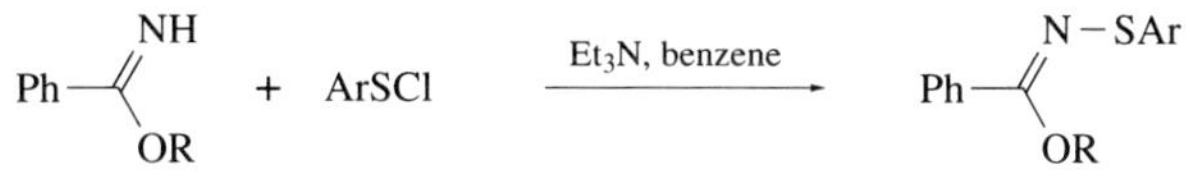
(190)

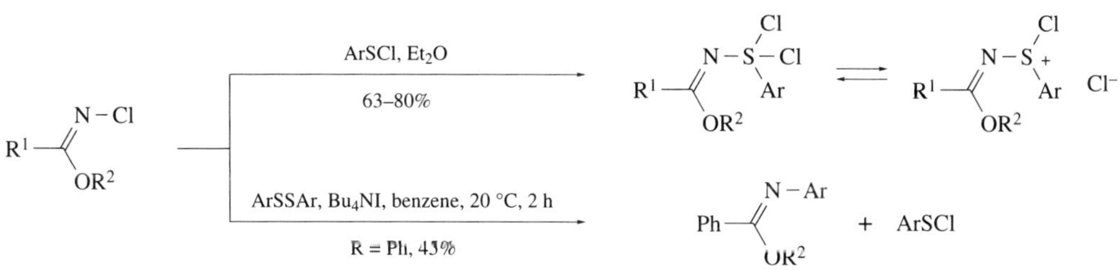

Scheme 57

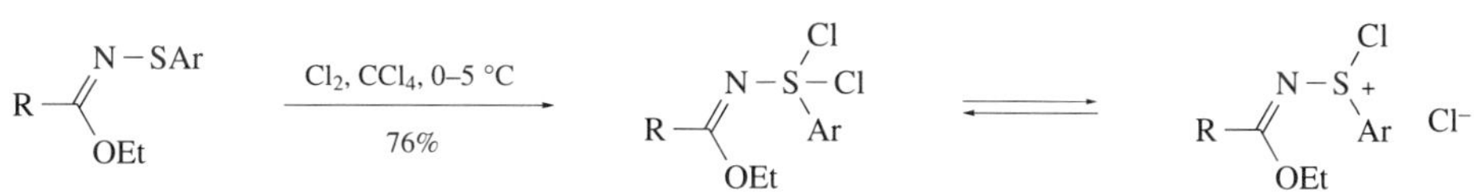

Scheme 58

Alkoxides and phenoxides react with *N*-arenesulfonylimidoyl chlorides to give *N*-arenesulfonyl imidates (Equation (191)) 〈69JOU252, 69ZOR263, 88JMC2066〉.

(191)

5.17.8.5 Hydrazonic Acid Esters (Hydrazonoates, *N*-Aminoimidic Acid Esters

5.17.8.5.1 From hydrazonoyl halides (hydrazidoyl halides), α-arylhydrazono-nitroalkanes and benzotrichlorides

Hydrazonoyl chlorides 〈65TL841, 71CC689, 77JCS(P2)1136, 84JOC3083, 87JHC577, 88H(27)695, 90IC4255, 92ICA527〉 and bromides 〈77JCS(P2)1136, 88H(27)695, 88OPP521〉 have been reacted with sodium alkoxides 〈77JCS(P2)1136, 84JOC3083〉, lithium alkoxides 〈90IC4255, 92ICA527〉, and sodium phenoxides in ethanol 〈87JHC577, 88H(27)695, 88OPP521〉 or with phenols in dioxane/water 〈71CC689〉, in diethylether 〈93JFC(65)101〉, in the presence of triethylamine or in alcohols buffered with sodium acetate 〈65TL841〉 to give the corresponding alkyl- or arylhydrazonoic acid esters (Equations (192)–(194)). The *O*-arylesters thus prepared form mixtures of *E*/*Z*-isomers, which can be separated by HPLC. In some cases, the *E*- and *Z*-isomers of the *O*-methylesters can be prepared in a pure state by recrystallization (Scheme 59) 〈84JOC3083〉. When choosing the reaction conditions, it should be kept in mind that hydrazonic acid esters can rearrange at temperatures above 100°C to give the corresponding carboxylic acid hydrazides (Equations (195) and (196)) ⟨71CC689⟩.

Nitration of aromatic aldehyde arylhydrazones gives aryl-hydrazono-aryl-nitromethanes, which on treatment with phenoxide afford phenyl-hydrazonoates (Scheme 60) ⟨80JHC1745⟩. When hydrazides of aliphatic carboxylic acids are treated with benzotrichloride and primary alcohols in the presence of sodium carbonate, hydrazonic acid esters are formed; the analogous reaction with aromatic carboxylic acid hydrazides affords 1,3,4-oxdiazoles (Scheme 61) ⟨79S946, 82H(19)1477⟩.

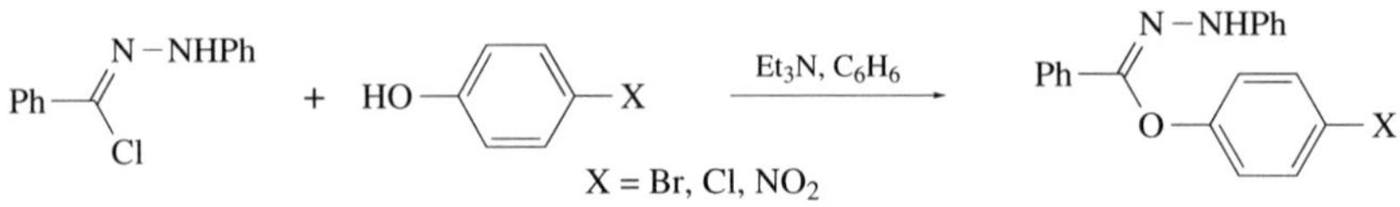

(192)

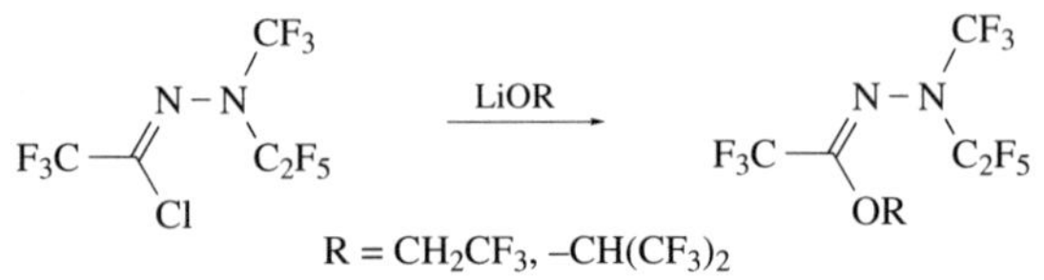

(193)

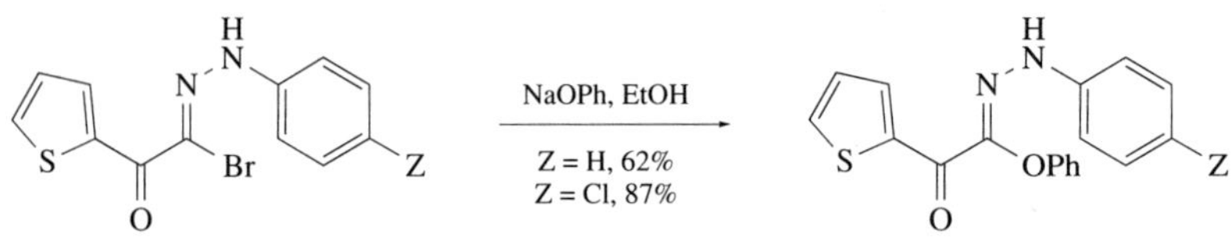

(194)

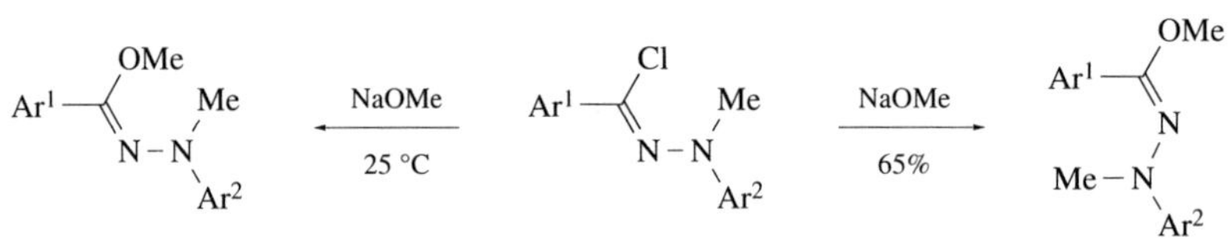

Scheme 59

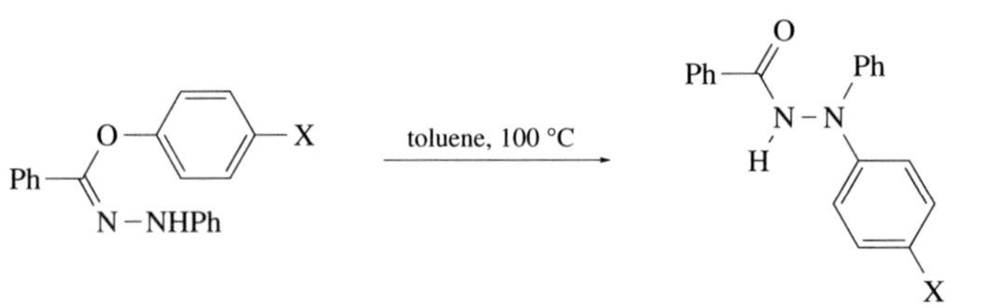

(195)

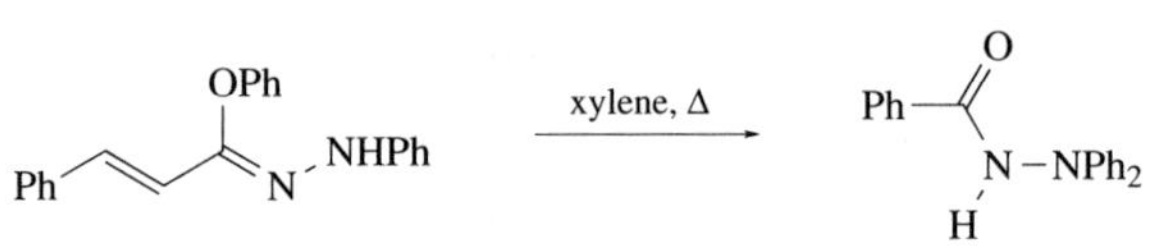

(196)

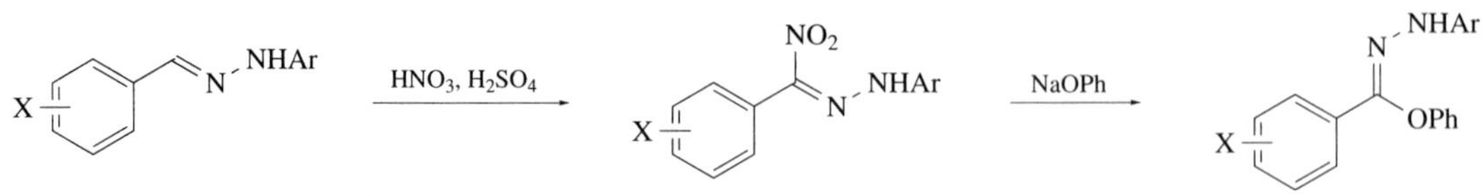

Scheme 60

5.17.8.5.2 From selenoesters or iminoesters and hydrazine derivatives

Hydrazonic acid esters are formed in the reactions of selenoesters with hydrazine ⟨78JHC1113⟩, semicarbazides and thiosemicarbazides (Scheme 62) ⟨78JHC237⟩. Similar results have been obtained

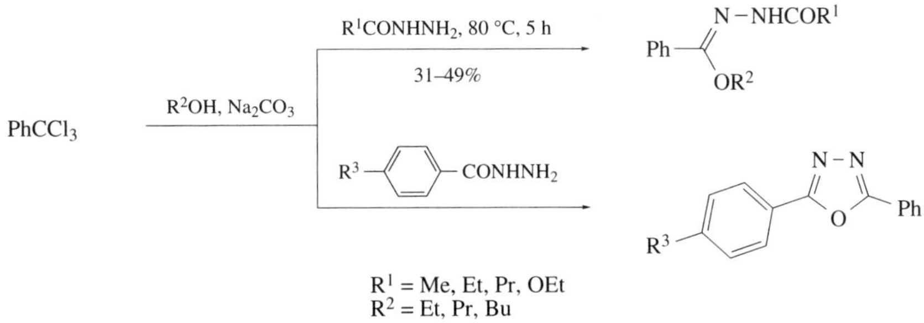

Scheme 61

from the reactions of iminoesters with phenylhydrazine hydrochloride (Equation (197)) ⟨78CR(C)69⟩, and iminoester hydrochlorides with thiosemicarbazide (Equation (198)) ⟨84JHC1689⟩.

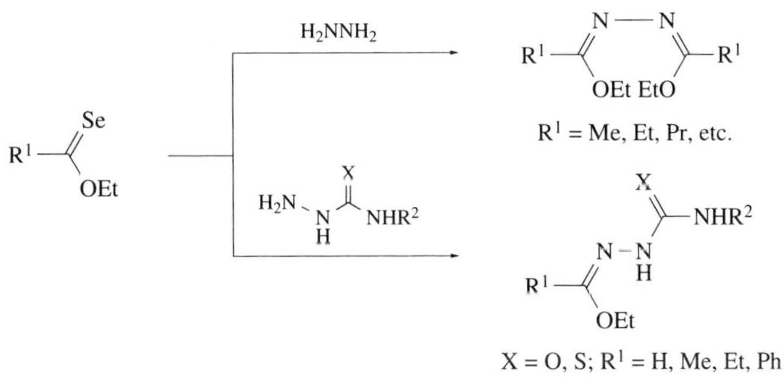

Scheme 62

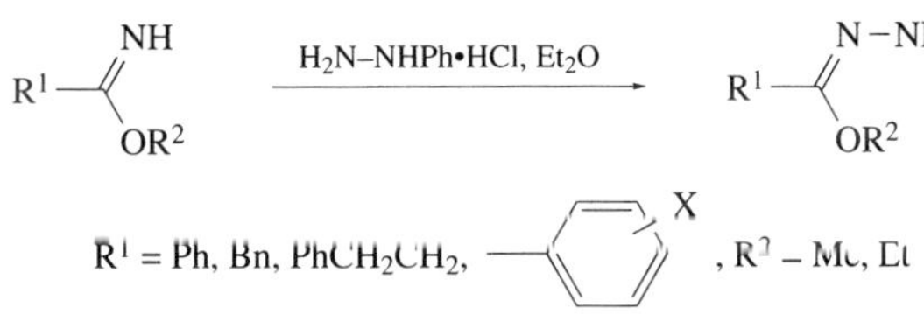

(197)

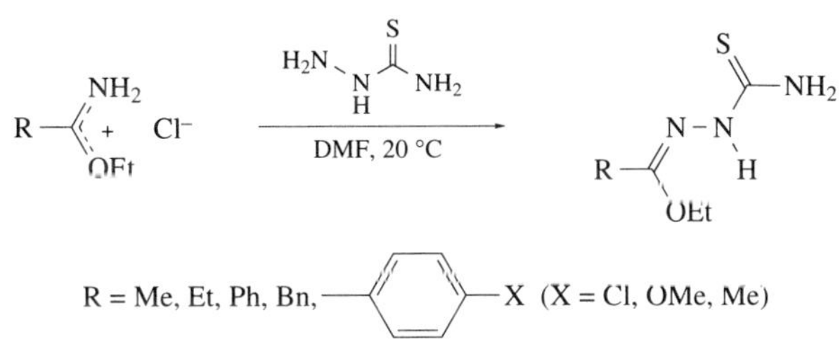

(198)

5.17.8.5.3 *From carboxylic* **ortho**-*esters and related compounds and hydrazine derivatives*

Carboxylic *ortho*-esters react with *N*,*N*-disubstituted hydrazines to give the corresponding hydrazonic acid esters. This synthesis is of wide applicability, usually proceeds in good yields (65–98%), and has therefore often been used. Triethyl *ortho*-formate ⟨77JHC749⟩, triethyl *ortho*-acetate ⟨77JHC749, 78CR(C)69, 81TL2731⟩, triethyl *ortho*-propionate ⟨78CR(C)69⟩, triethyl *ortho*-benzoate ⟨78CR(C)69, 78PS(4)317⟩ with phenylhydrazine, *N*-methyl-*N*-phenylhydrazine, benzaldehyde hydrazone, acetone hydrazone ⟨78CR(C)69⟩, arenesulfonyl hydrazides ⟨81TL2731⟩, chlorophosphorus ester hydrazides, *N*-methyl-chlorothiophosphorus ester hydrazides ⟨78PS(4)317⟩, thiophosphonic dihydrazides, 2,4-dimethyl-3-(1-methylhydrazino)-1,2,4,5,3-tetraza-phospha-5-cyclohexen-3-thiones

⟨77JHC749⟩ and benzo[c]-pyrazole-3-carbohydrazide ⟨84JHC91⟩ have been treated in this way. Examples are given in Schemes 63 and 64, and Equation (199).

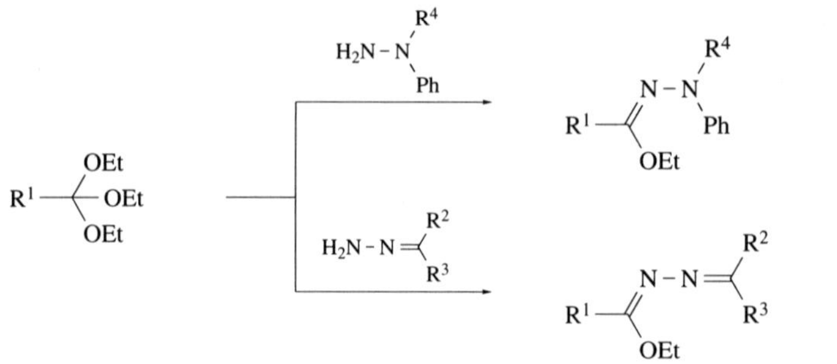

Scheme 63

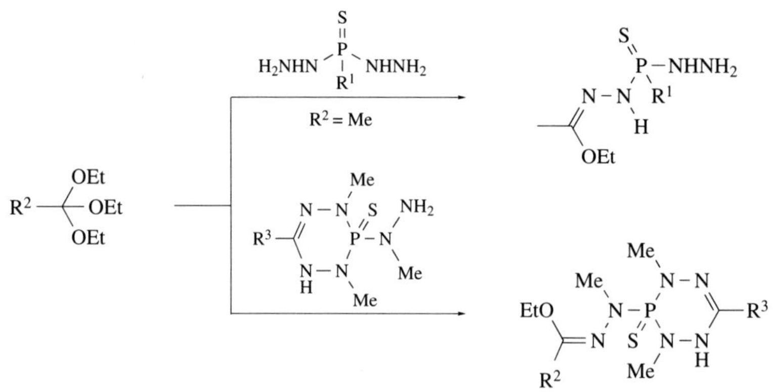

Scheme 64

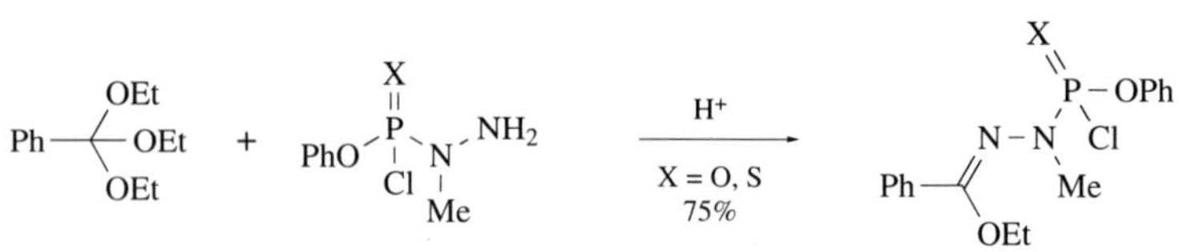

(199)

Ortho-esters and hydrazine derivatives in which one nitrogen is a part of a heterocyclic ring system give hydrazonic acid esters (Equation (200)) ⟨85H(23)281, 85JHC1435, 90SC23⟩. Hydrazine derivatives react with ketene-*O*,*O*-acetals to give hydrazonic acid esters ⟨55JA4571⟩ (Equation (201)).

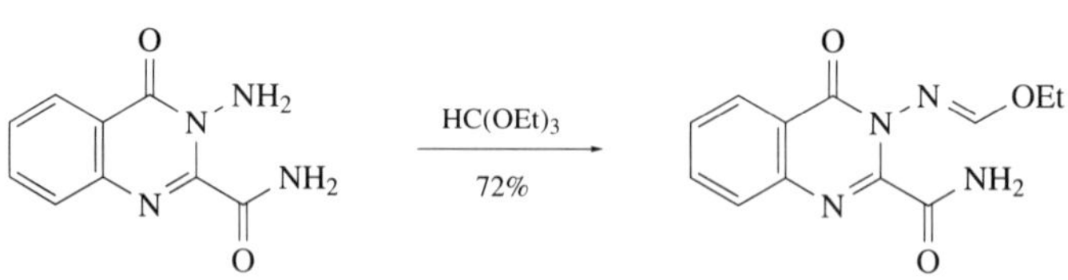

(200)

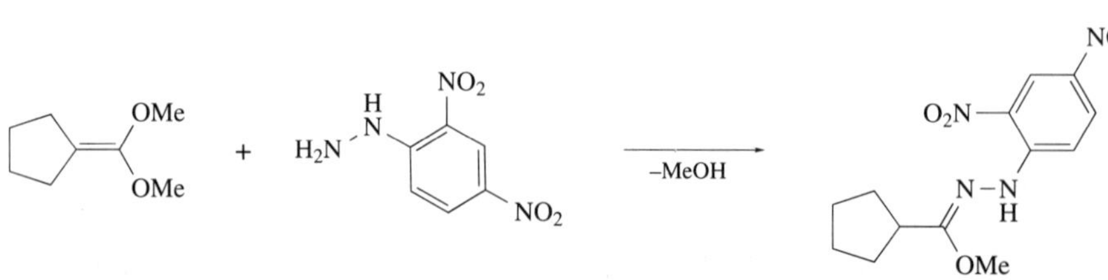

(201)

5.17.8.5.4 Miscellaneous methods

The photolysis of 2,5-diphenyltetrazoles in the presence of phenols affords aryl-hydrazonoates (Equation (202)) ⟨88ZC24⟩. Mixtures of hydrazonic acid esters are accessible by the thermolysis of oxadiazolines (Equation (203)) ⟨84JOC3466⟩.

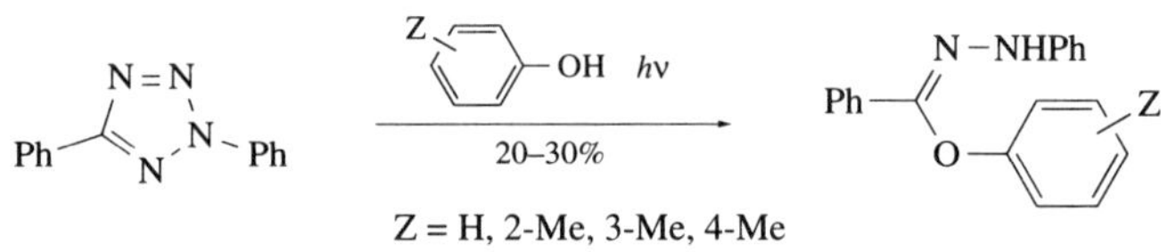
(202)

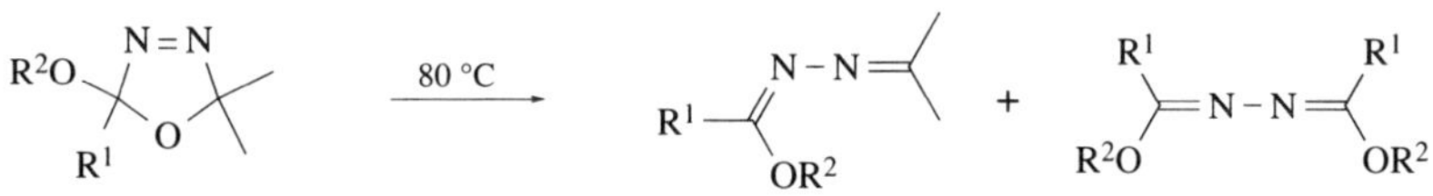
(203)

5.17.8.6 *N*-Phosphorylated Iminoesters

5.17.8.6.1 From imidoyl halides

On treatment with alkoxides and phenoxides, *N*-(dichlorophosphinoyl)- and *N*-(dichlorophosphinthioyl)imidoyl chlorides are converted into alkyl- and aryl-*N*-(diorganophosphin)thioyl imidates (Equation (204)) ⟨56JGU2935, 56ZOB2631, 57JGU1065, 57JGU3284, 57ZOB1080, 57ZOB3248, 58JGU2283, 58ZOB2247, 59JGU596, 59ZOB600, 61JGU3192, 61ZOB3424, 87UKZ397⟩. Similarly, in *N*-(chloromethyl-phosphinoyl)imidoyl chlorides both chlorines are replaced by aryloxy groups on treatment with sodium nitrophenoxide giving rise to the formation of the corresponding 4-nitrophenyl imidates (Equation (205)) ⟨68JGU1815, 68ZOB1867⟩.

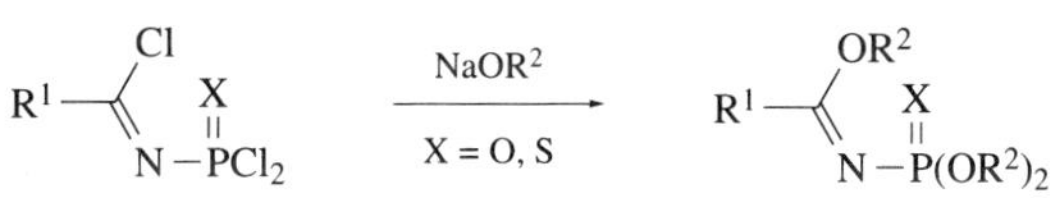
(204)

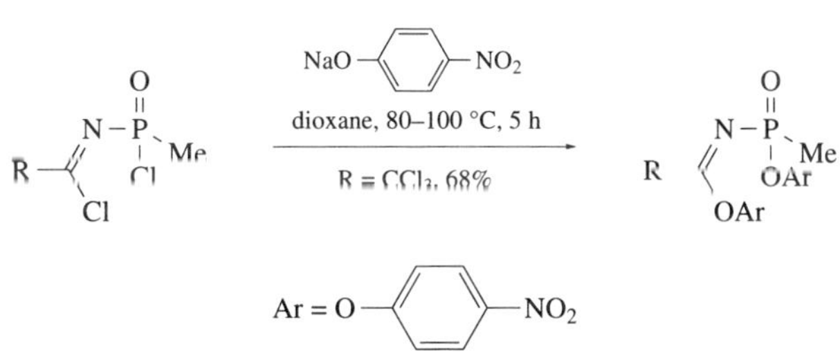
(205)

5.17.8.6.2 From diphenylphosphinic acid amide, diethyl phosphoramidate and carboxylic ortho-*esters*

Diphenylphosphinic acid amide ⟨79S691⟩ and diethyl phosphoramidate ⟨68JGU2501, 68ZOB2586, 88PS(40)263, 93HAC11⟩ react with carboxylic *ortho*-esters to give *N*-phosphinoyl imidates and *N*-diethoxyphosphoryl imidates respectively (Scheme 65).

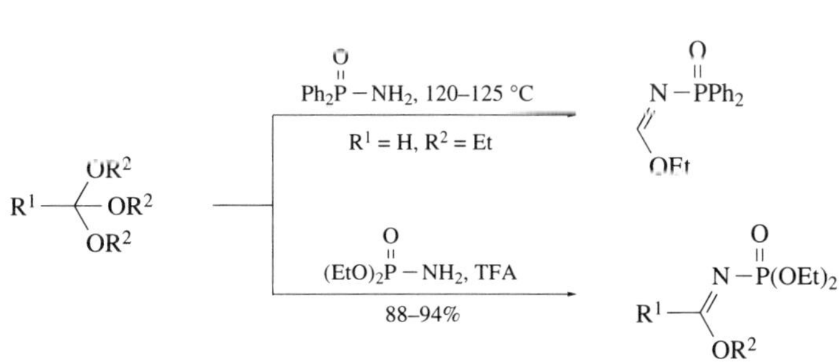

Scheme 65

5.17.8.6.3 From imidates and phosphorus compounds

N-Phosphorylated imidates can be prepared from *N*-chloroimidates and trialkoxyphosphines ⟨61JGU473, 61ZOB516, 62JGU167, 62ZOB171, 63JGU551, 63ZOB557, 65JGU1014, 65ZOB1009⟩ phosphorus trichloride or phosphorus pentachloride ⟨64JGU1152, 64ZOB1161⟩, dialkoxyhalophosphines, dialkoxycyanophosphines ⟨61JGU473, 61ZOB516⟩, and alkoxydihalophosphines (Equations (206) and (207) and Scheme 66) ⟨61JGU473, 61ZOB516, 65JGU1014, 65ZOB1009⟩. Ethyl *N*-chlorobenzimidate reacts with aryl-dialkylphosphines to give benzoliminophosphoranes (Equation (208)) ⟨63JGU1006, 63ZOB1017⟩. Other *N*-phosphorylated imidates have been obtained by treatment of imidates with diethoxychlorophosphine, bis(dimethylamino)chlorophosphine, tris(dimethylamino)phosphine ⟨70BSF1363⟩, chloro-bis(2,2,3,3-tetrafluoro)propylphosphine ⟨86JGU1096, 86ZOB1245⟩ and chlorodiphenylphosphine ⟨71CB1199⟩ (Scheme 67 and Equation (209)).

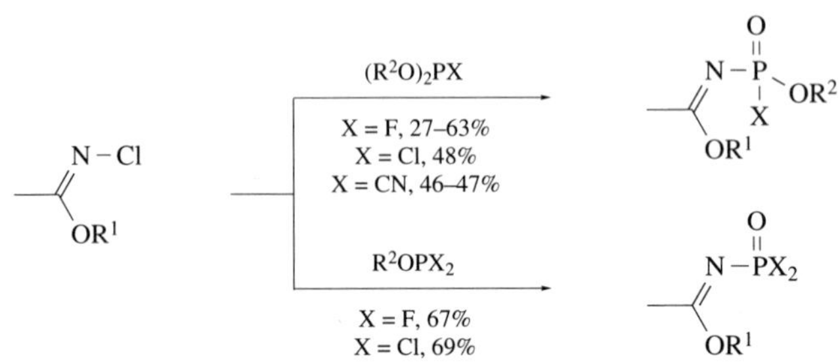

Scheme 66

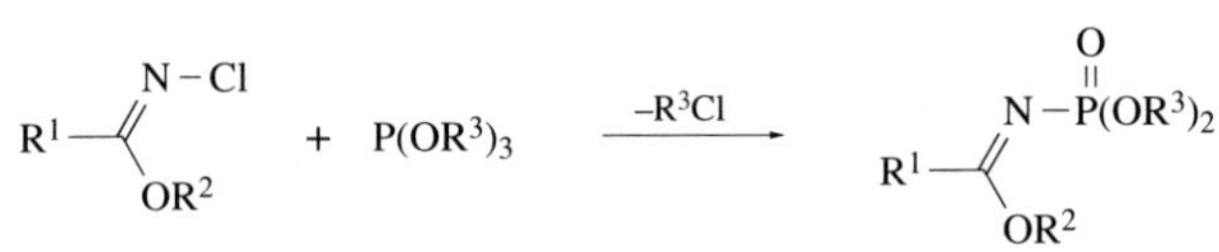

(206)

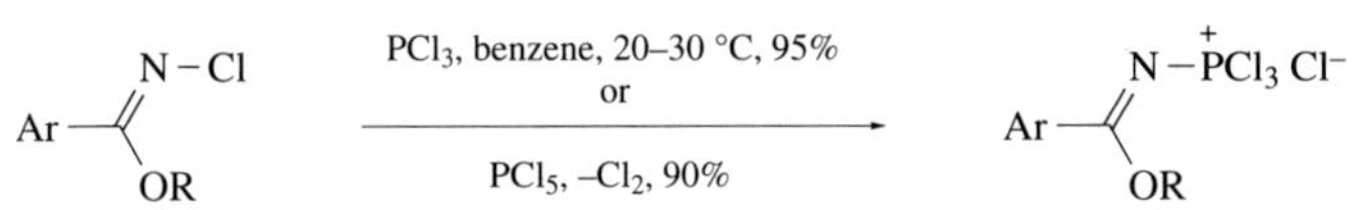

(207)

(208)

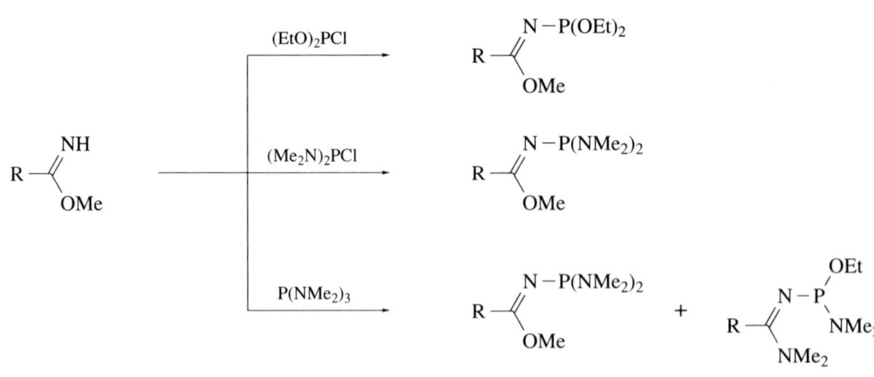

Scheme 67

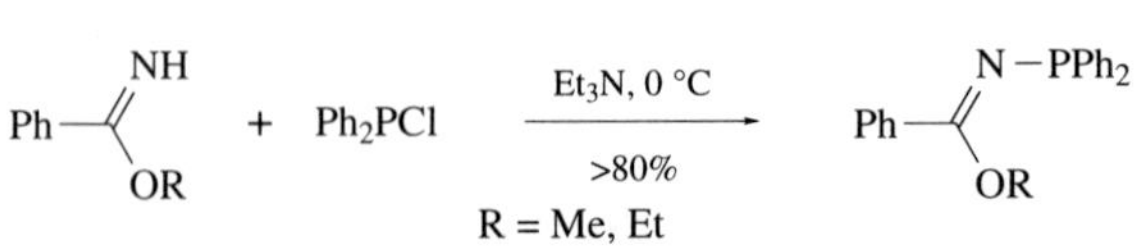

(209)

5.17.8.6.4 Miscellaneous methods

N-Phosphinoyl imidates are formed in the reactions of dialkylphosphinoylazides with vinylethers (Equation (210)) ⟨66T975⟩. Ethyl *N*-(dimethyloxyphosphoryl)benzimidate reacts with chloral to give ethyl *N*-(methoxy-vinyloxy-phosphoryl)benzimidate (Equation (211)) ⟨84JGU1969, 84ZOB2203⟩. Other iminoesters have been prepared from *N*-phosphorylated imidoates by manipulations (substitutions or addition reactions) at the phosphorus atom. ⟨65JGU1014, 65ZOB1009, 71CB1199, 84JGU1969, 84ZOB2203⟩. Some examples are given in Equation (212) and Scheme 68.

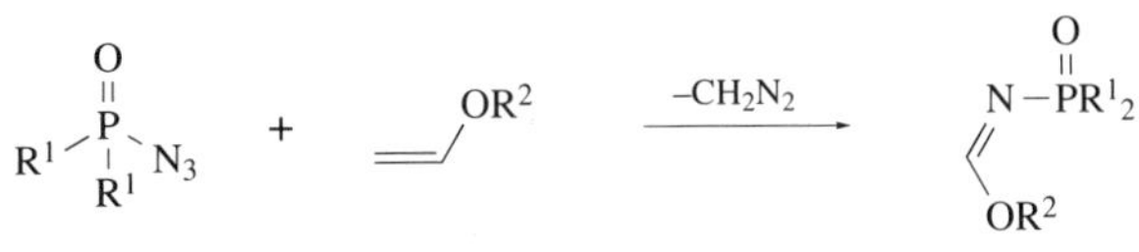

(210)

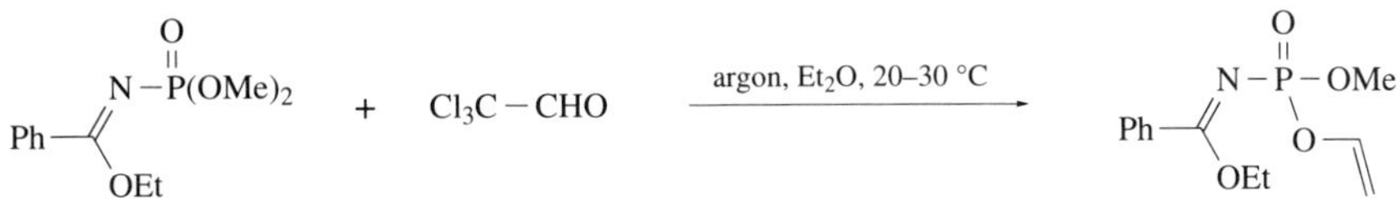

(211)

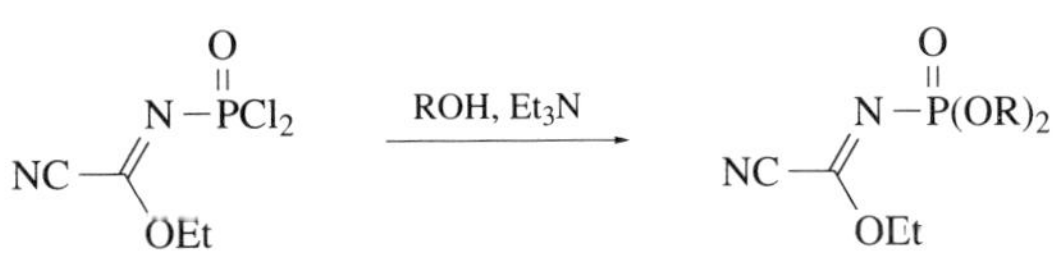

(212)

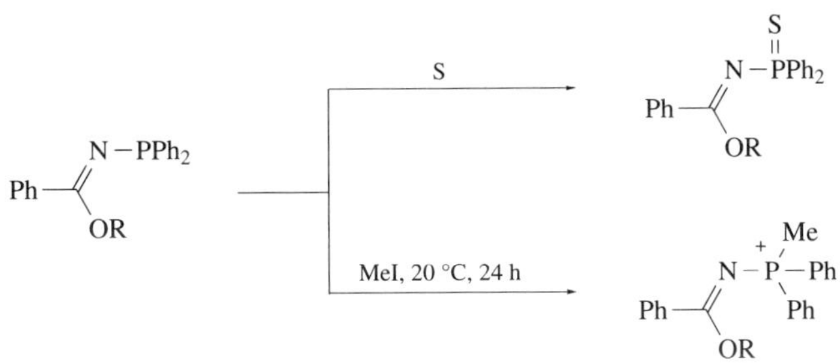

Scheme 68

5.17.8.7 Mixed Anhydrides From Carboximidic Acids and Inorganic Acids or Organic Acids

5.17.8.7.1 Imidoyl compounds of the type RC(NY)OX

These compounds are shown as structure (**216**). From kinetic studies of the hydrolysis of *N*-chloro-*N*-methylbenzamide, it was concluded that 'imino hypochlorites' (**217**) might be intermediates (Scheme 69) ⟨67JCS(B)1151⟩. The saponification of nitriles with H_2O_2 in weakly alkaline medium is a well known reaction ⟨1884CB1289⟩. Hydroperoxy imidates (**218**) are assumed to be intermediates (Scheme 70) ⟨53JA3961⟩. Hydroperoxy imidates have also been discussed as intermediates in the epoxidation of acrylonitrile ⟨61JOC651, 61JOC659⟩; and as the oxidizing species in mixtures of H_2O_2, $NaHCO_3$ and acetonitrile ⟨64JCS1189⟩ since it was found that yields of epoxidation reactions are improved when this mixture is used as epoxidation reagent ⟨77GEP751220, 82MI 517-01⟩. In accordance with this assumption, the system trichloroacetonitrile/H_2O_2 has been recommended for epoxidation reactions ⟨83JOC888, 87MI 517-02⟩.

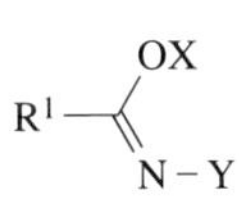

(216)

X = Hal, OR^2, $–S(O)_n–R^2$, $P(O)_nR^2R^3R^3$, $N(O)_nR^2R^3$, -COR, SnR_3, PbR_3, BR^2R^3, AlR^2R^3

Y = alkyl, aryl, Hal, OR, $S(O)_nR^2$, NR^1R^2, $P(O)_nR^2R^3R^4$, SnR^2_3, BR^2R^3, AlR^2R^3

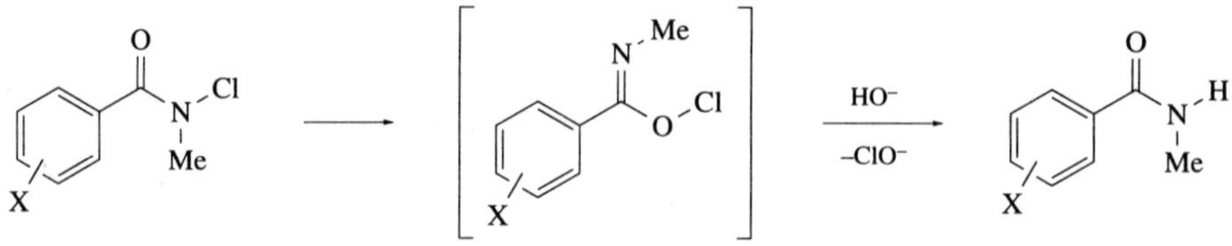

(217)

Scheme 69

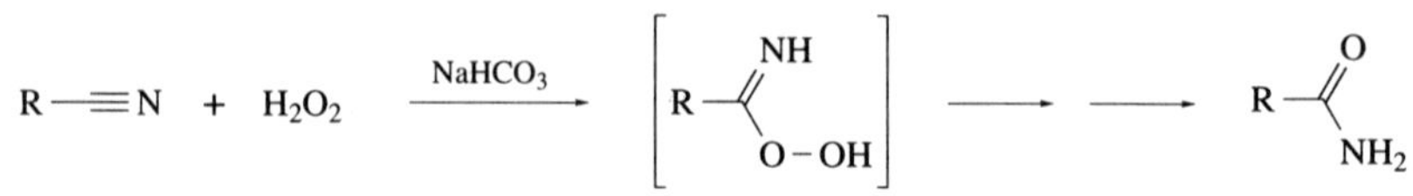

(218)

Scheme 70

In connection with investigations on the Ritter reaction of isobutene and acrylonitrile in the presence of sulfuric acid, a product was isolated in good yields, which was assumed to be a mixed anhydride of *N*-*t*-butylacrylimidic and sulfuric acid (Equation (213)) ⟨66BSF1376⟩. Secondary carboxylic acid amides and benzenesulfonyl chloride with pyridine as a base give imidic acid benzenesulfonic acid anhydrides (Scheme 71), which have not been isolated but reacted further with various nucleophiles to give iminoesters, amidines or thioamides ⟨48JCS1514, 48JCS1618, 58CB972, 58CB1129⟩. Trifluoromethanesulfonic acid converts secondary carboxylic acid amides to mixed anhydrides of carboximidic and trifluoromethane sulfonic acid (Equation (214)) ⟨81H(15)165⟩. Analogous products have been obtained from *N*-alkoxybenzamides, sulfonyl chlorides and pyridine in benzene (Equation (215)) ⟨89CCC2738, 92CCC540⟩. Mixed anhydrides from *N*-acyloxy imides and arylsulfonic acids can be prepared by the same procedure ⟨93CCC2150⟩. At low temperatures perfluorinated nitriles react with chlorine fluorosulfate to give *N*-chlorocarboximidic fluorosulfonic acid anhydrides (Equation (216)) ⟨81IZV2370, 81JFC(18)553⟩. Addition of toluenesulfonic acid to isonitriles results in mixed anhydrides from formimidic and toluenesulfonic acid (Equation (217)) ⟨82JPR(324)667⟩.

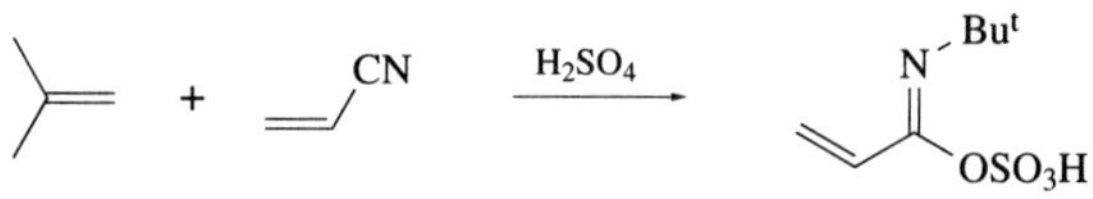

(213)

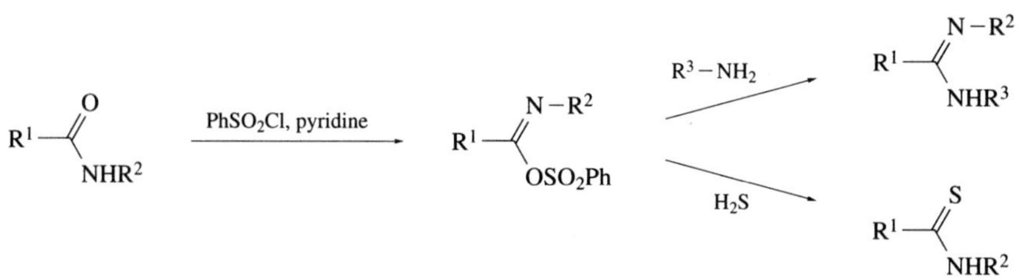

Scheme 71

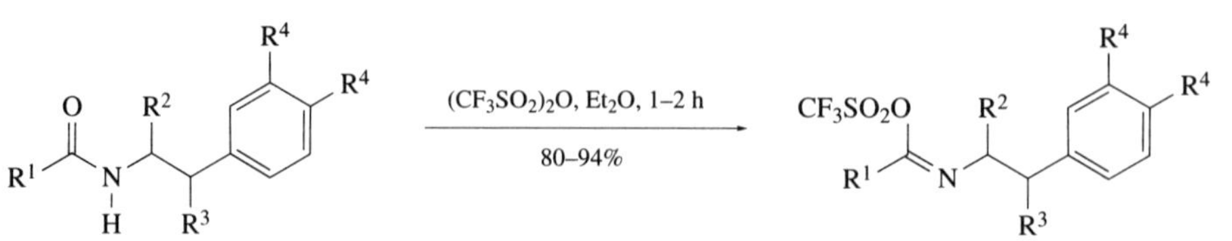

(214)

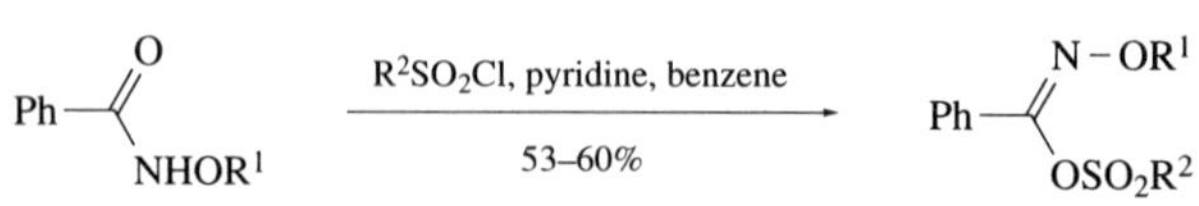

(215)

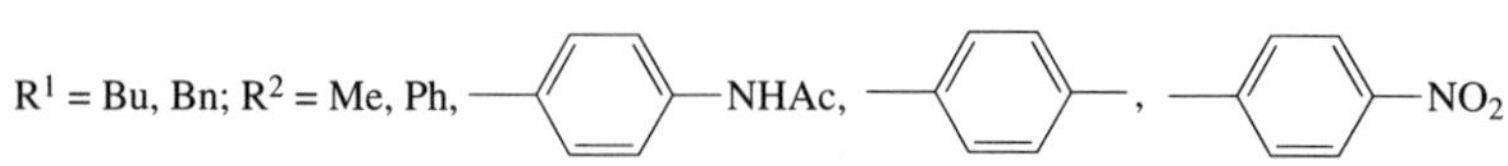

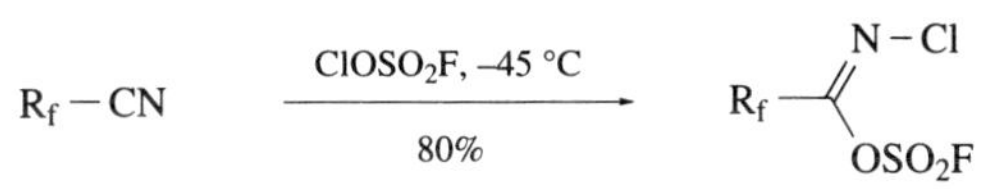

$R_f = CF_3, C_3F_7, C_4F_9, CF_3\text{–}OCF_2, O_2N\text{–}CF_2$

(216)

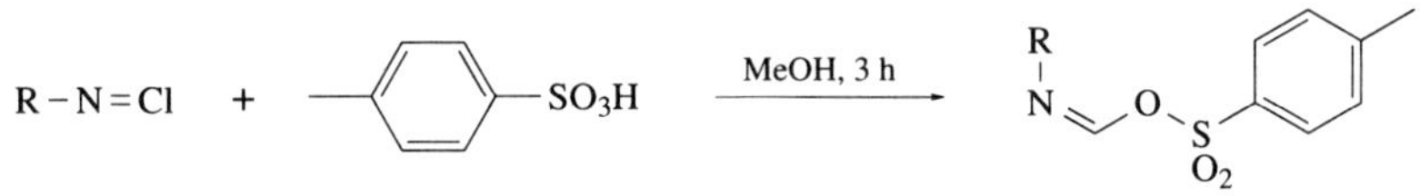

$R = C_6H_{11}$, Ph

(217)

N-Arylbenzimidoyl chlorides have been reacted with silver nitrate ⟨88CC1249, 92JCS(P2)281⟩. If the 2,4,6-positions of the aryl group are blocked by substituents, *N*-nitro amides are the final products, otherwise nitration of the aromatic nucleus occurs in the *o*- and *p*-positions. The mechanism of this reaction was carefully investigated, and it was concluded that ionic and radical reactions are involved, in which unstable imidoyl nitrates play an important role (Scheme 72) ⟨92JCS(P2)281⟩. In the synthesis of *N*-nitroso amides from imidoyl chlorides and silver nitrite, imidoyl nitrites are considered to be the intermediates (Scheme 73) ⟨85JCR(S)132⟩.

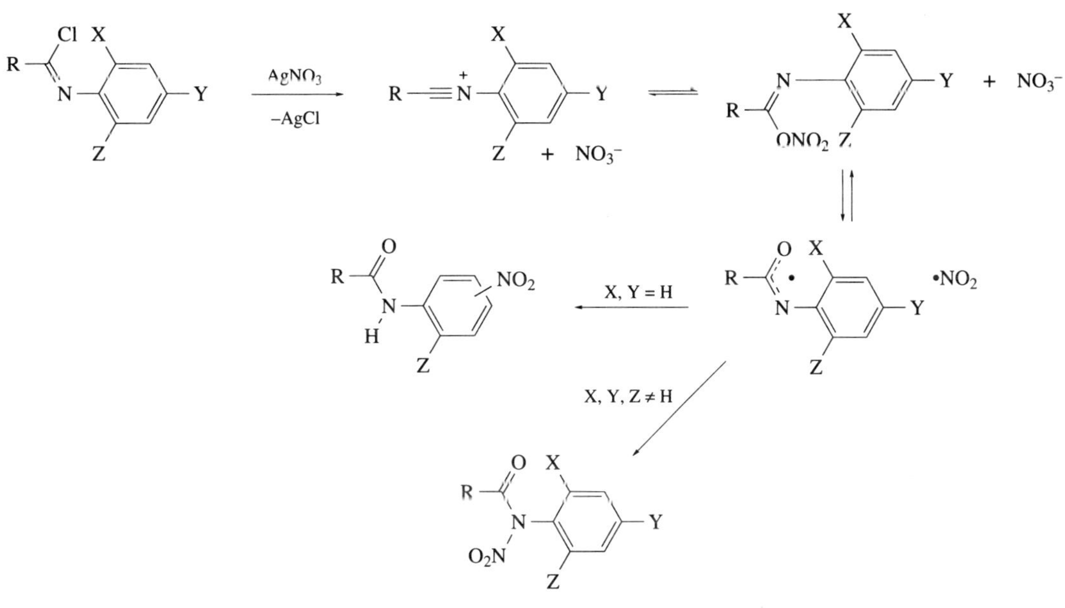

Scheme 72

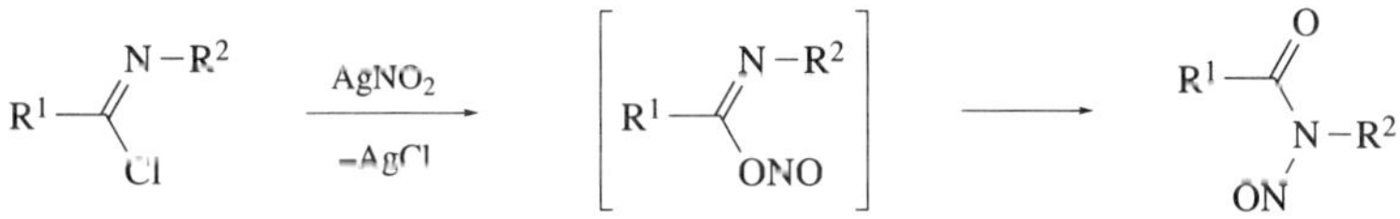

Scheme 73

Acetoxime adds to perfluorinated nitriles affording *O*-perfluoro alkyl imino acetaldoximes (Equation (218)) ⟨65JOC3724⟩. The iminoacylation of sodium-*N*-aryl hydroxamates proceeds via nonisolable *O*-carboximidoyl-*N*-aryl hydroxamic acids to give *o*-phenylene diamine derivatives (Scheme 74) ⟨82AG375, 85TL5281⟩. Analogous behavior was observed in the reaction between the sodium salts of *N*-hydroxypyrimidine derivatives and benzimidoyl chlorides (Scheme 75) ⟨85SC643⟩.

The corresponding imidoyl compounds were prepared from sodium pyridine-2-thione-1-olate and imidoyl chlorides (Equation (219)) ⟨91TL1035⟩. Analogous products can be obtained from *N*-sulfonylated benzimidoyl chlorides, carboxylic acid anhydrides and benzhydroxamic acids (Equation (220)) ⟨75JOU1719, 75ZOR1724⟩. The reactions of pyridine-*N*-oxides with imidoyl halides have been reviewed ⟨88RCR234, 88UK405⟩.

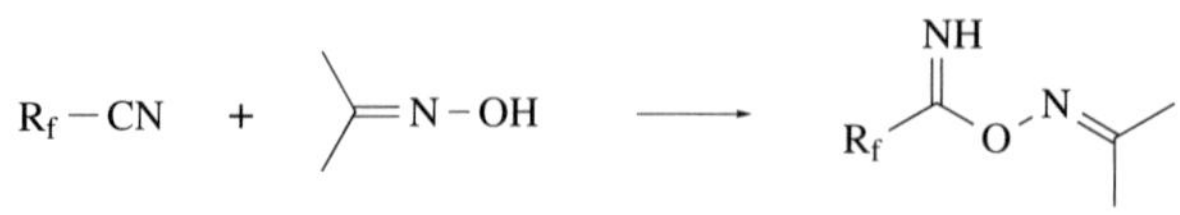

(218)

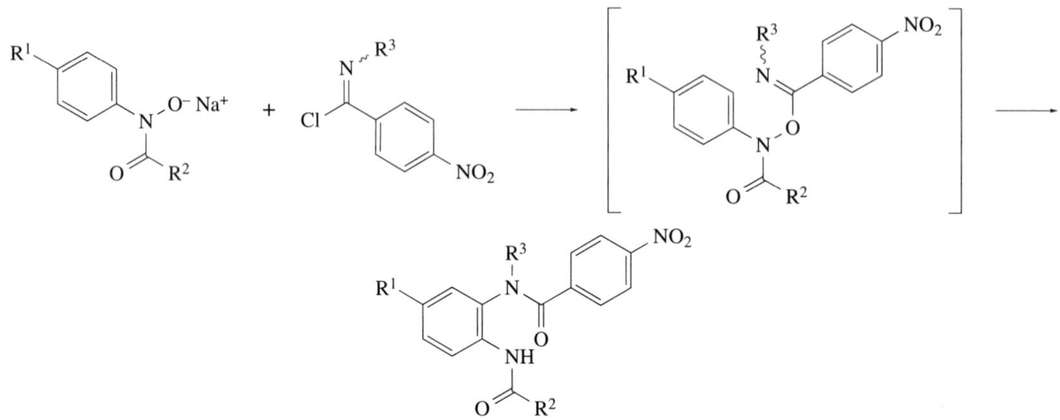

Scheme 74

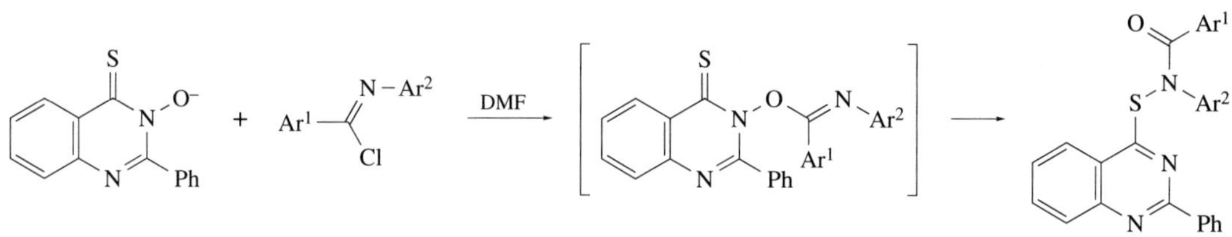

Scheme 75

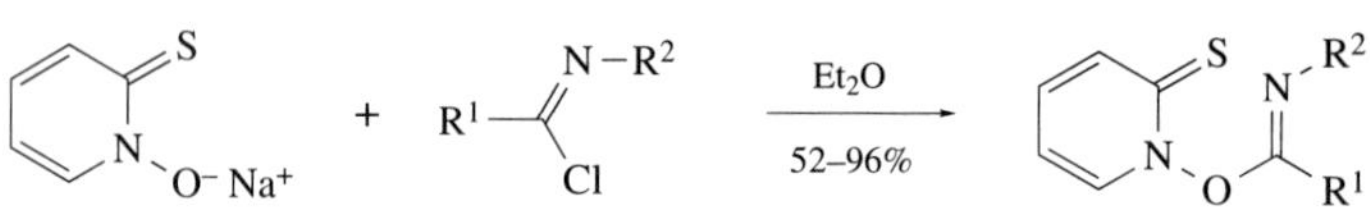

(219)

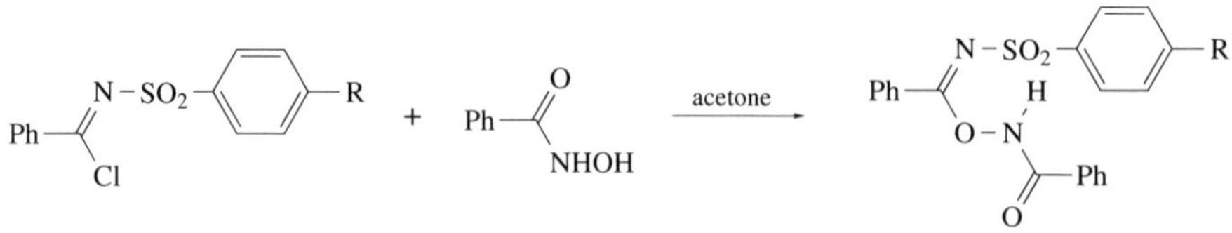

(220)

The addition of phosphonic acids to isonitriles in methanol yields mixed anhydrides of formimidic and phosphonic acid (Equation (221)); methyl formimidates are by-products ⟨82JPR(324)667⟩. Imidoyl chlorides have been reacted with silver dibenzylphosphate to give mixed anhydrides ⟨55CI(L)1183, 56JCS1231⟩. Compounds of this type can also be obtained from *O*-sulfonated oximes and dialkylphosphates (Scheme 76) ⟨56JCS1231⟩; in the absence of nucleophiles the products readily rearrange to *N*-phosphorylated amides, whereas in the presence of triethylammonium dibenzylphosphate, pyrophosphates are formed ⟨55CI(L)1183, 56JCS1231⟩.

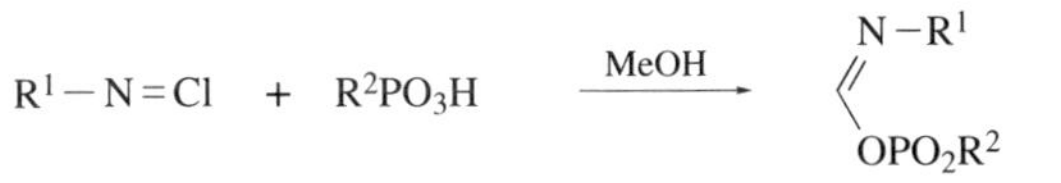

(221)

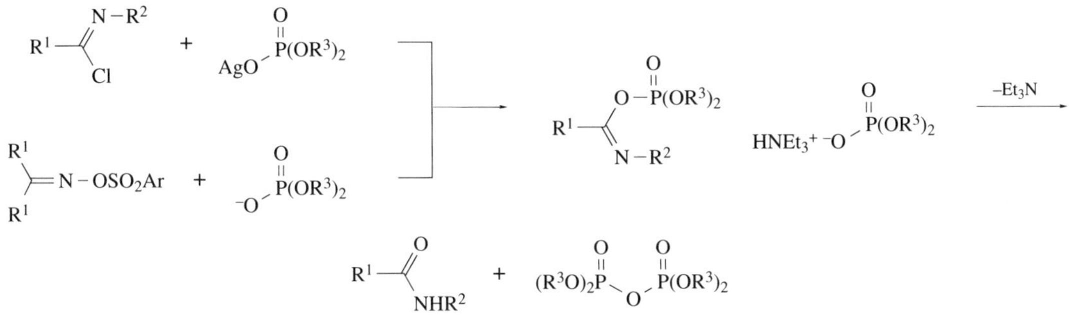

Scheme 76

N-Alkoxybenzimidoyl phosphates and thiophosphates are accessible from the corresponding N-alkoxyimidoyl chlorides and sodium dialkyl thiophosphates, or by heating N-alkoxybenzoic and amides with dialkylphosphonates in the presence of CCl_4/triethylamine; the products thus formed can rearrange to N-phosphorylated N-alkoxy-amides (Scheme 77) ⟨85JGU1526, 85ZOB1718⟩.

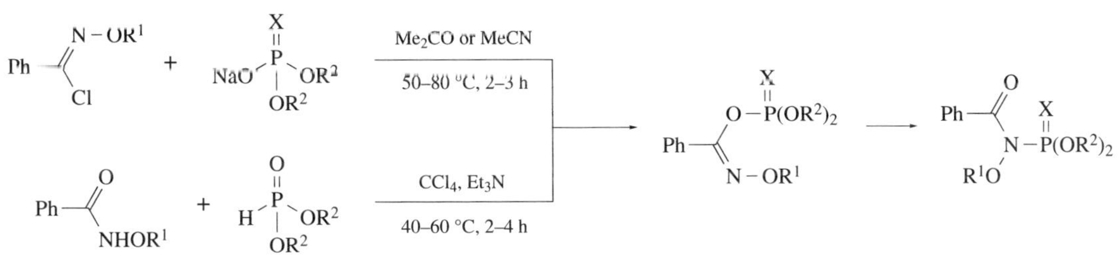

Scheme 77

Mixed anhydrides of N-methylbenzimidic and phosphorus acid diesters are formed in the reaction of dialkylphosphonates with N-methylbenzimidoyl chloride; they rearrange on standing with triethylamine hydrochloride within a few days or on distillation *in vacuo* to give N-methylimino benzylphosphonic acid esters, which can also be prepared from the imidoyl chloride and trialkyl phosphites (Scheme 78) ⟨84JGU1925, 84ZOB2154⟩.

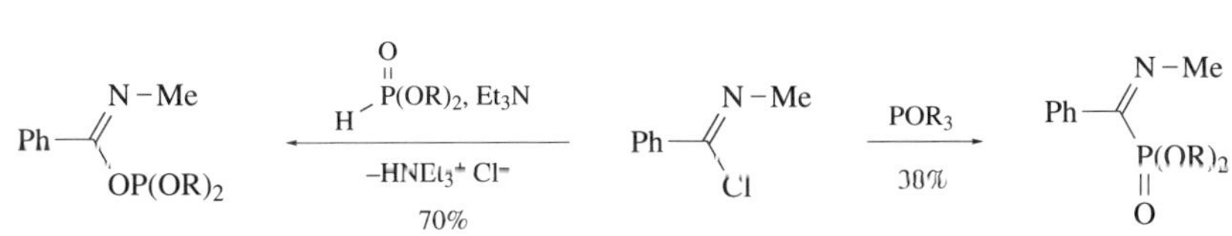

Scheme 78

Carboxylic acid amides react with dialkoxychlorophosphine in the presence of triethylamine to give a mixture of unstable N- and O-phosphorylated compounds which rearrange on prolonged standing or on heating to give various phosphorus compounds, for example N-alkylbenzamido dialkoxyphosphines, 1-(N-alkylamino)alkylphosphonic acid esters or phosphorenamines (Scheme 79). The stability of the primary products is strongly influenced by the nature of the substituents, for example the reaction products from bis(trifluoroacetyl)amine rearrange so rapidly, that the process could not be monitored by spectroscopic methods ⟨86JGU1113, 86ZOB1262, 89JGU2228, 89ZOB2492, 90JGU849, 90JGU1057, 90JGU1059, 90ZOB963, 90ZOB1186, 90ZOB1187⟩.

Bis(trimethylsilyl)acetamide reacts with bromo difluorophosphine at temperatures below −80 °C to give O-trimethylsilyl-N-difluorophosphino acetimidate or N,N-bis(difluorophosphino) acetamide depending on the stoichiometric relations used (Scheme 80) ⟨80JCS(D)1768⟩. The silylation of N-diethoxyphosphoryl formamide produces an unstable trimethylsilyl formamidate which reacts

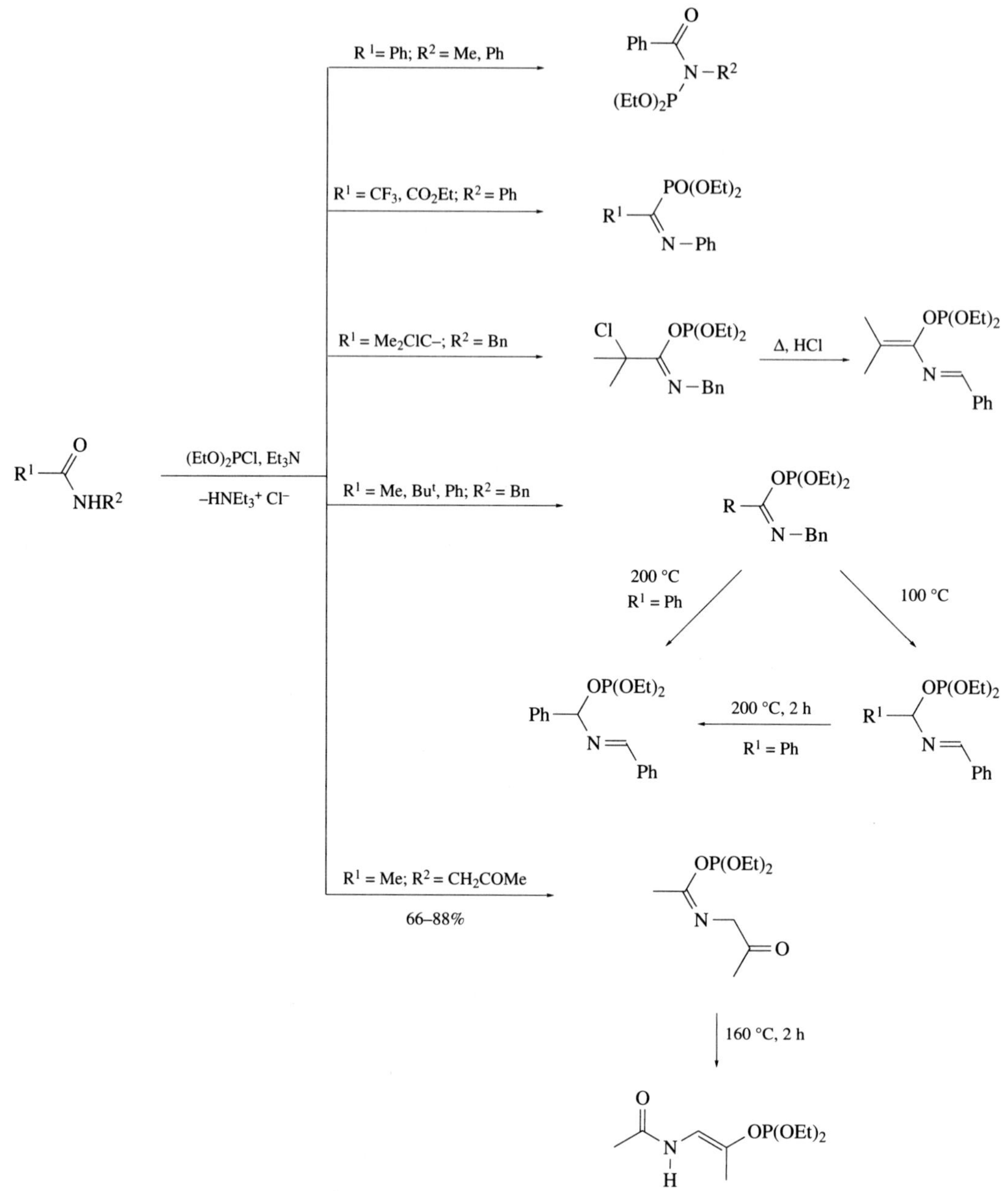

Scheme 79

with HCl to give a mixture of *N*-diethoxyphonylformamide and diethyl-iminomethylphosphate (Scheme 81) ⟨92PS(72)201⟩. Mixed anhydrides from nitrobenzoic and formimidic acids can be obtained from isonitriles and nitrobenzoic acids in methanol ⟨68ZC219, 82JPR667⟩. When acetic or benzoic acid are reacted under similar conditions methyl formimidates are formed (Scheme 82) ⟨82JPR(324)667⟩. By the action of imidoyl chlorides on the silver salts of benzoic acids benzimidic carboxylic acid anhydrides were prepared (Equation (222)) ⟨67JA637⟩. Analogous reactions have been performed with the silver salts of hydroxamic acid derivatives and carboxylic acid chlorides (Equation (223)) ⟨77JCS(P2)1080, 77JCS(P2)1085⟩. Reaction of *N*-sulfonylated-benzimidoyl chlorides with the sodium salts of carboxylic or thiocarboxylic acids in acetone yields [*N*-(sulfonylimino)benzyl]carboxylates and thiocarboxylates, respectively (Scheme 83) ⟨75JOU1719, 75ZOR1724⟩. The acylation of *N*-alkoxy- or *N*-acyloxycarboxylic acid amides with acid chlorides/pyridine ⟨89CCC2738, 92CCC540, 93CCC2150⟩ or acetic acid anhydride/pyridine ⟨69AJC161, 79JCS(P2)634⟩ has been reported (Scheme 84) to give mixed anhydrides from carboxylic and *N*-alkoxyimidic acids.

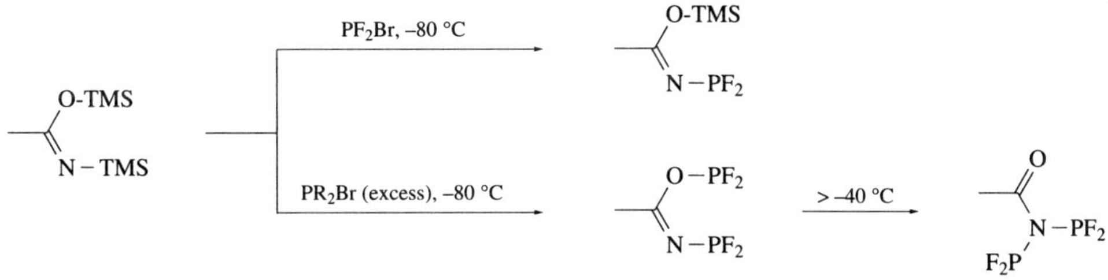

Scheme 80

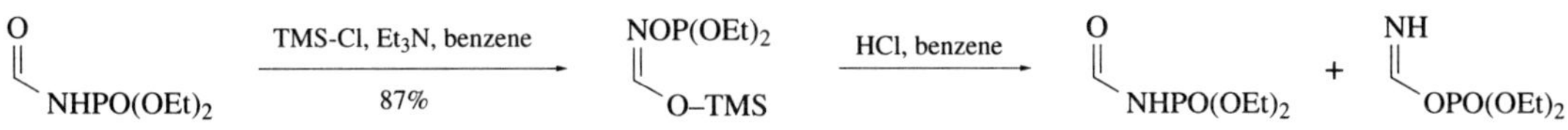

Scheme 81

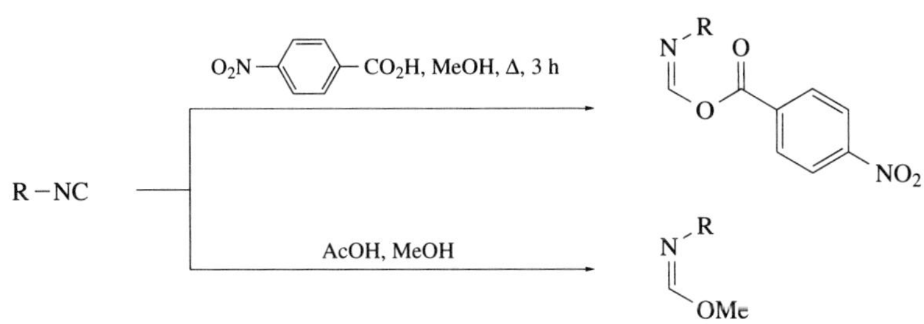

$R = C_6H_{11}$, Ph

Scheme 82

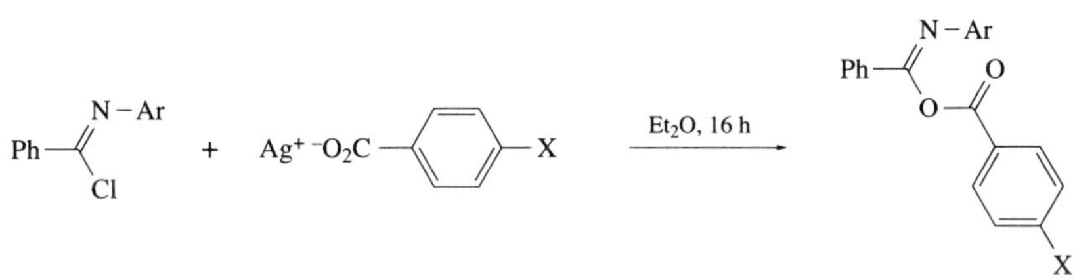

(222)

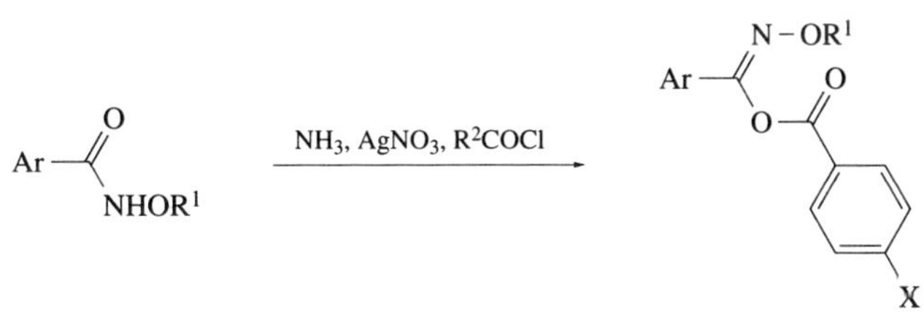

(223)

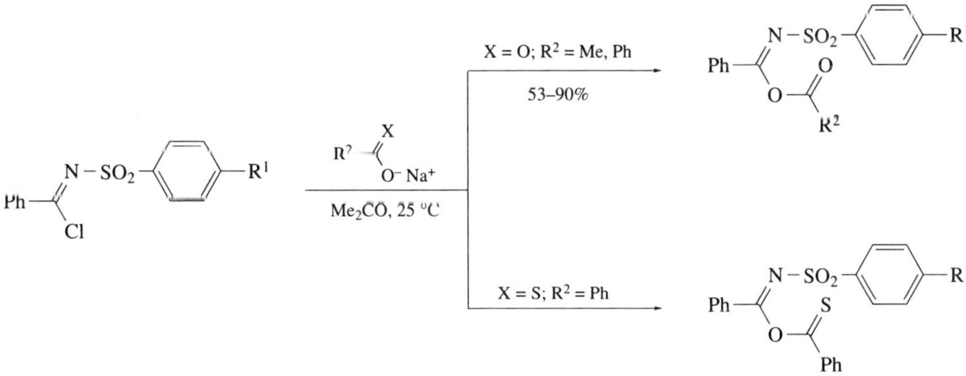

Scheme 83

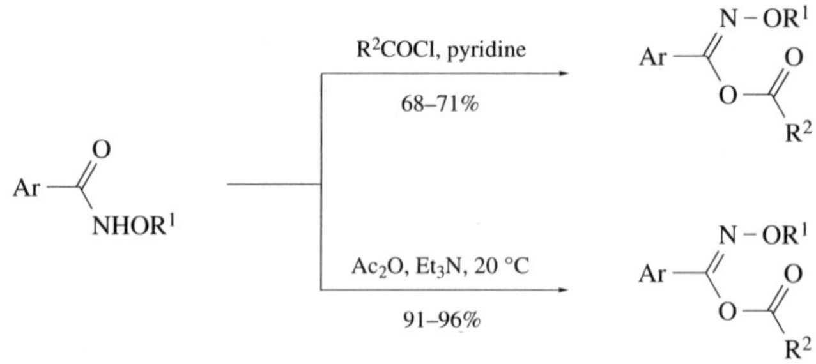

Scheme 84

Similar products have been obtained from carboxamides and trifluoroacetic acid anhydride (Equation (224)) ⟨81H(15)165⟩. The final products from the reactions of imidoyl chlorides with the triethylammonium salts of carboxylic acids are diacylamines; intermediates are mixed anhydrides from carboxylic and imidic acids (Scheme 85) ⟨79ACS(B)544, 80ACS(B)313, 80JCS(P2)121, 87LA483⟩. Cyclic mixed anhydrides of carboxylic and imidic acids can be prepared by treating the monoamides of dicarboxylic acids with acetic anhydride/$HClO_4$. The salts thus formed are deprotonated by triethylamine to give the mixed anhydrides (Scheme 86) ⟨78JCS(P1)1338⟩.

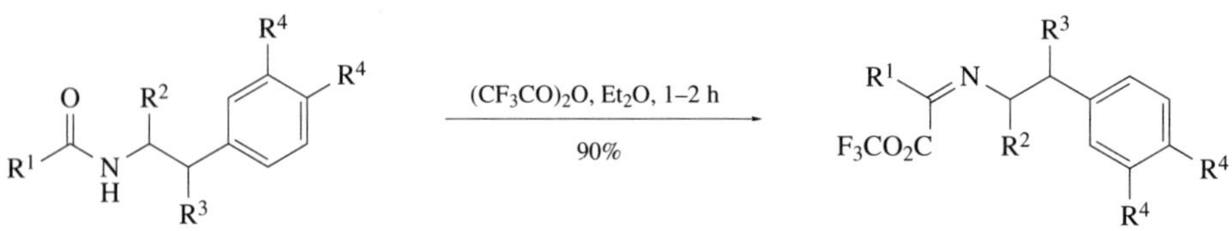

(224)

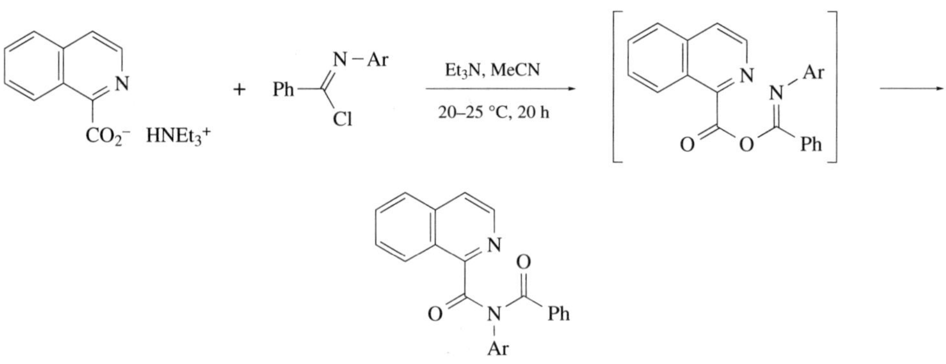

Scheme 85

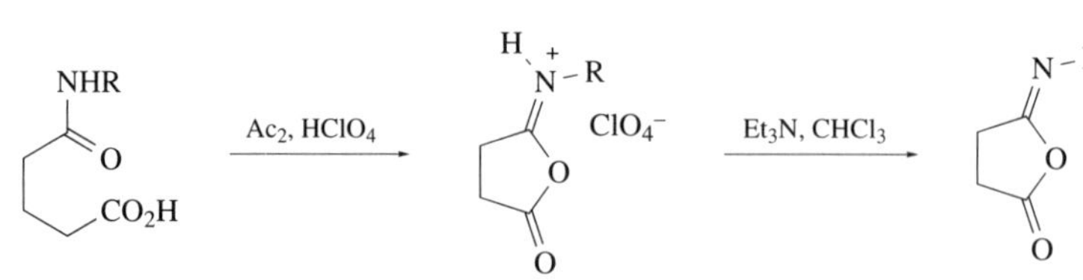

Scheme 86

Addition of trialkyltin alkoxides to trichloroacetonitrile gives stannylated imidates (Scheme 87) ⟨66JOM(5)590, 67JCS(C)1309⟩. Methoxytripropylstannane adds to acetyl isocyanate to give an *O*-stannylated imidate (Equation (225)) ⟨72JGU1857, 72ZOB1868⟩. The acylation of bis(tributylstannyl)carbodiimide with acetyl chloride or benzoyl chloride proceeds with the formation of the corresponding tributylstannyl *N*-cyanoacetimidate and benzimidate respectively (Equation (226)) ⟨76JGU714, 76ZOB714⟩. Compounds of this type seem to exist as a mixture of isomers, since it was found that the reaction products bis(tributylstannyl)carbodiimide and acetic acid anhydride form an equilibrium mixture containing three compounds (Scheme 88) ⟨78JGU1046, 78ZOB1146⟩.

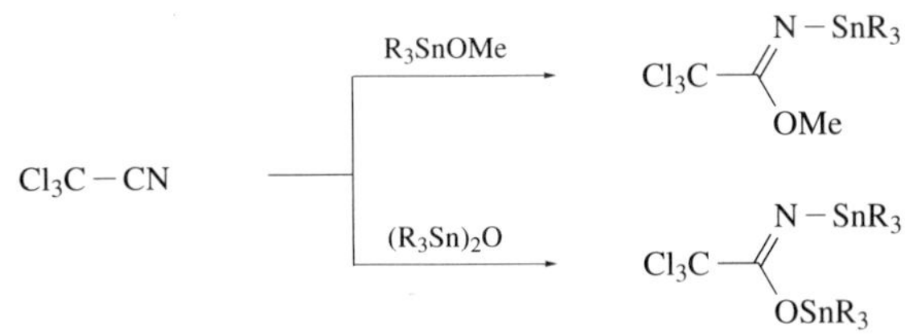

Scheme 87

(225)

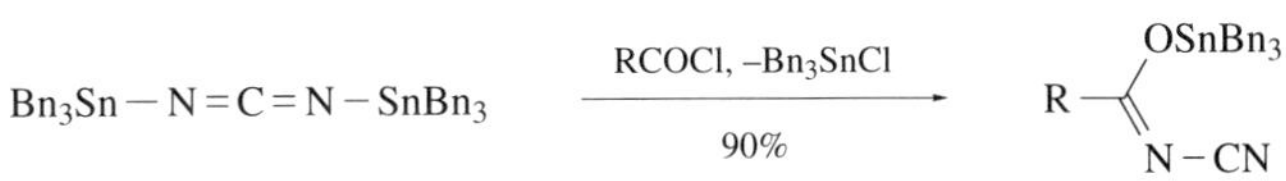

(226)

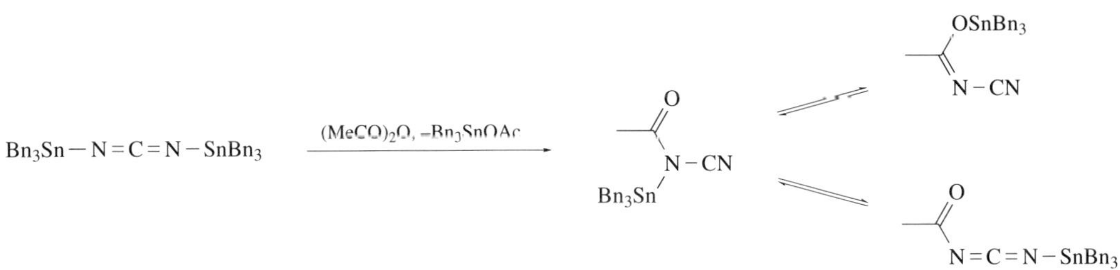

Scheme 88

Trialkylmethoxyplumbanes and trichloroacetonitrile react to yield *N*-(trialkylplumbyl)- and *O*,*N*-bis(trialkylplumbyl)imidates (Equation (227)) ⟨66JOM(5)590⟩. From the reaction of *N*-aryltrifluoroacetamides and bromodimethylborane, mixtures of *N*-borylated amides and *O*-borylated imidates are formed in low yields. Better yields are obtained when dibromomethylborane is used (Scheme 89) ⟨77ZAAC(433)94⟩.

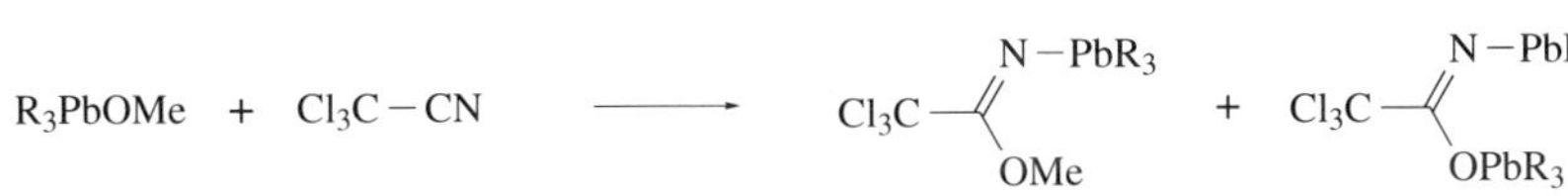

(227)

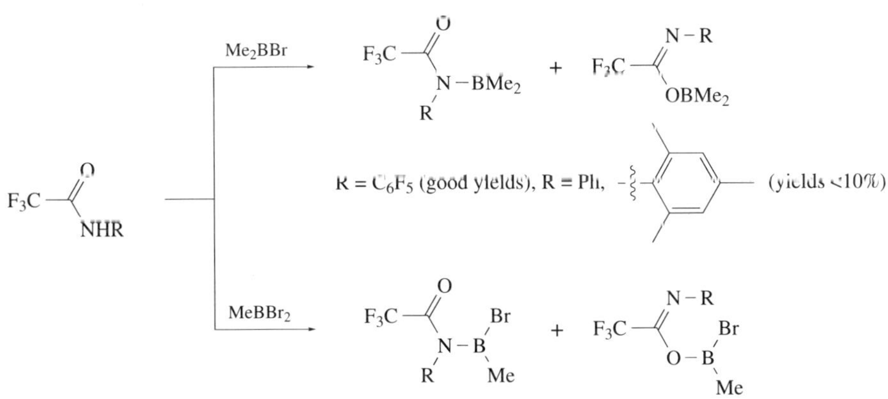

$R = Me, C_4H_9$

Scheme 89

Reaction of trimethylaluminum with *N*-phenylbenzamide gives a dimeric product with an imidate structure (Equation (228)) ⟨68JPS(B)389, 69JPS(B)17⟩.

(228)

5.17.8.8 Trialkylsilyl Imidates [*N*-Trialkylsilyl-Imidic Acid Esters, *O*,*N*-Bis(Trimethylsilyl)-Imidates]

5.17.8.8.1 From carboxylic acid amides, salts of carboxylic acid amides and silylating reagents

The mono- and disilylation of carboxamides can be achieved by treating the amides with trimethylchlorosilane ⟨63AG96, 63CB1473, 64CB627, 66JA3390, 70CB553, 73JCS(P1)2614, 84JGU426, 84JOU1963, 84ZOB477, 84ZOR2153, 86JHC183, 87JGU1073, 87ZOB1199, 88BAU378, 88IZV460, 91ZAAC(603)89⟩ in the presence of base. Solvents for these types of reactions are ether, THF, hexane or benzene. Thus acetamide ⟨63CB1473⟩, diacetylamine ⟨87JGU1073, 87ZOB1199⟩, *N*-ethylacetamide ⟨73JCS(P1)2614⟩, *N*-phenylacetamide, *N*-phenylpropionamide, *N*-phenyl-phenylacetamide ⟨91ZAAC(603)89⟩, dipeptide derivatives ⟨73JCS(P1)-2614⟩, trifluoroacetyl hydrazides ⟨84JGU426, 84ZOB477⟩, benzhydrazides ⟨88BAU378, 88IZV460⟩, 2-aminocarbonyl-5-oxo-pyrrolidine (pyroglutaminic acid amide) ⟨86JHC183⟩ and *N*,*N*-dibenzoyl sulfurdiamide ⟨84JOU1963, 84ZOR2153⟩ have been used. Examples are given in Equations (229)–(233).

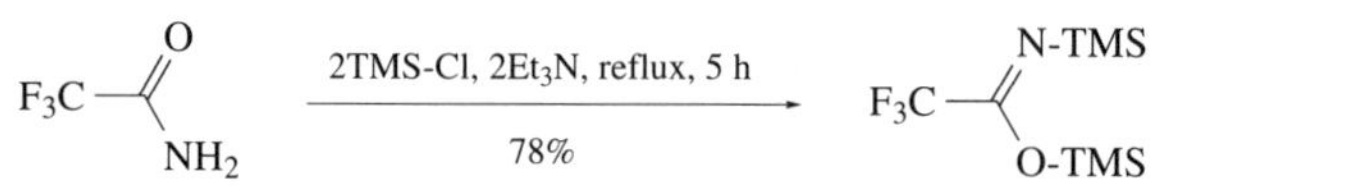

(229)

(230)

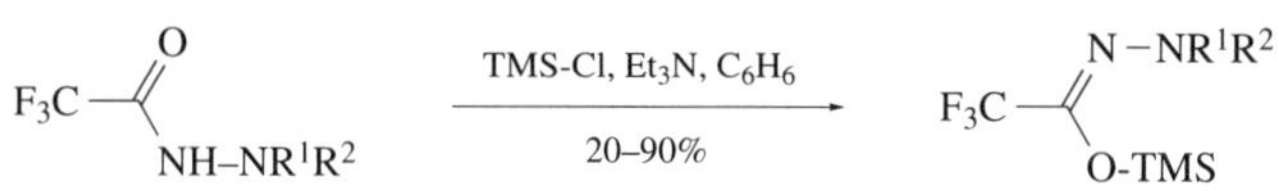

(231)

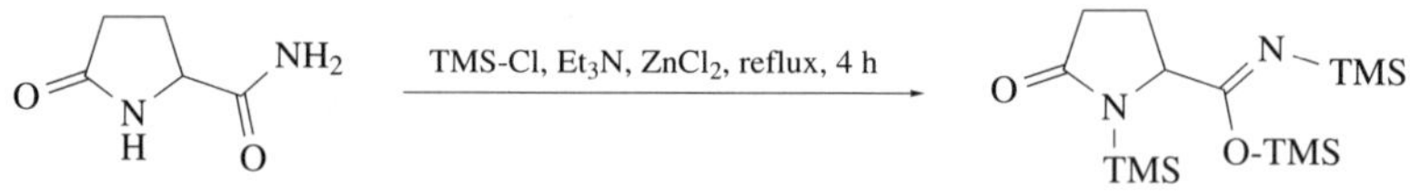

(232)

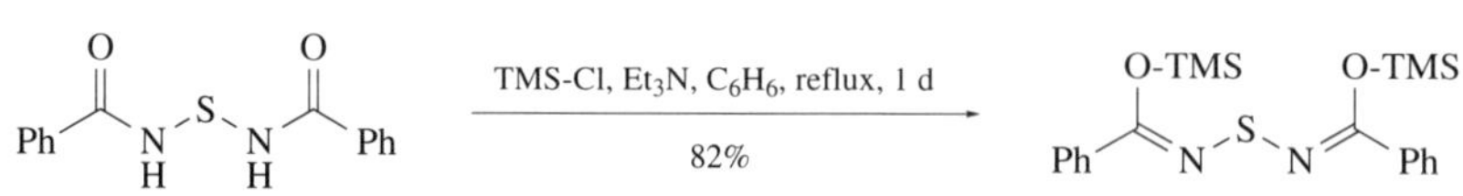

(233)

With the aid of trialkylsilyl triflates/triethylamine, diacylamines ⟨82JA1428, 87SC657⟩, *N*,*N*′-diformylhydrazine ⟨76S259⟩ and *N*-trifluoroacetylaminocarboxylic acid esters ⟨85S403, 87LA687, 90S525⟩ have also been converted into trialkylsilyl imidates. Similar reactions are known for cyclic imides, for example glutarimide ⟨87CB657⟩. Examples are given in Equations (234)–(237).

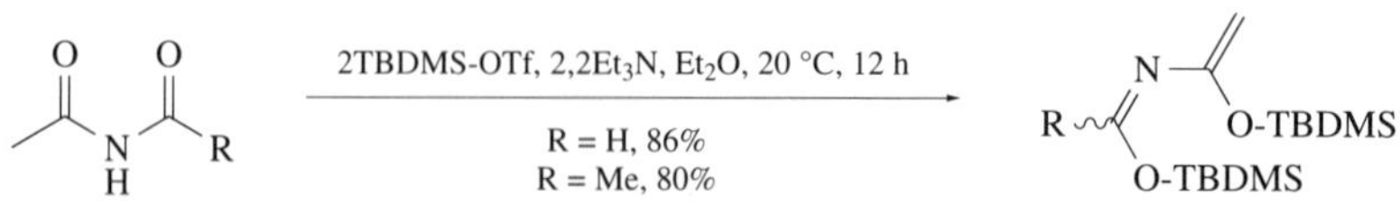

(234)

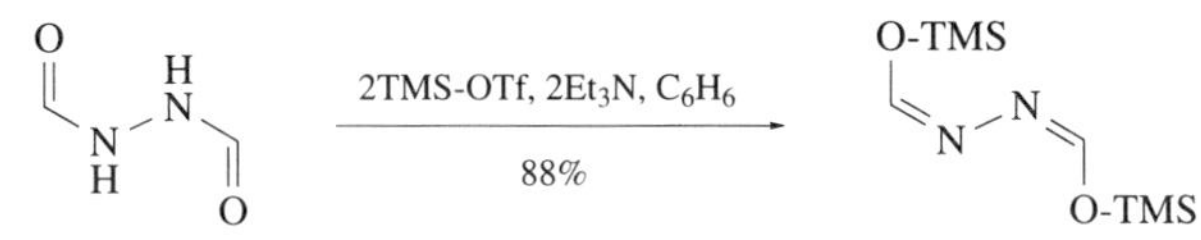

(235)

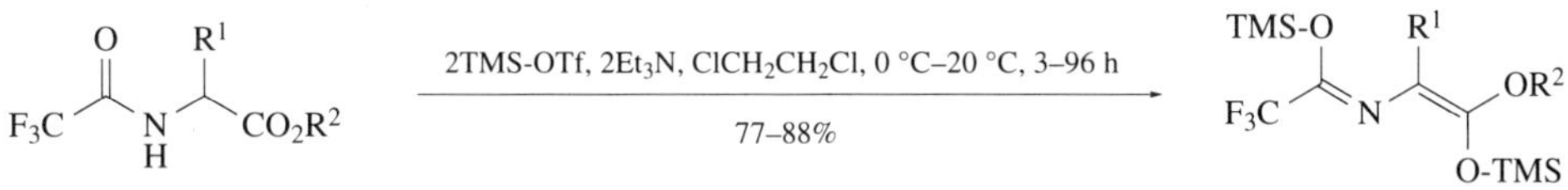

(236)

R^1 = H, Me

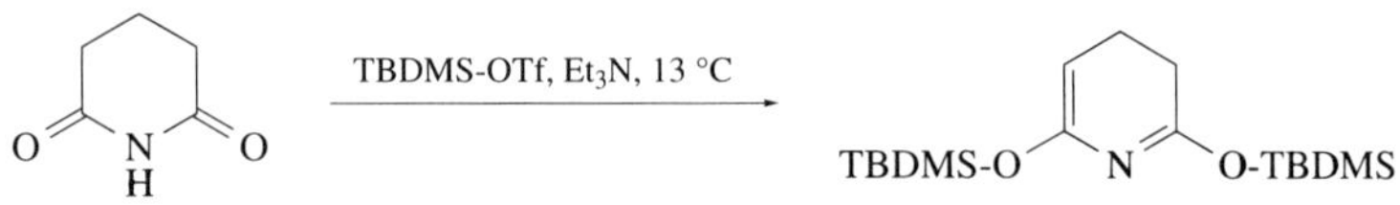

(237)

The alkali metal salts (Na, Li) of *N*-substituted acetamides, and homologous amides, react with trimethylchlorosilane to give monosilylated products, which seem to exist as an equilibrium mixture of the *O*- and *N*-silylated compounds (Scheme 90) ⟨72JCS(P2)1043⟩. Analogous silylations have been achieved when the sodium salts of the *N*-methylamides of formic, acetic and trifluoroacetic acid were treated with *t*-butyldimethylchlorosilane; it is also possible to prepare *O*,*N*-bis(*t*-butyldimethylsilyl)acetamide via the sodium salt of acetamide ⟨82JOC3336⟩.

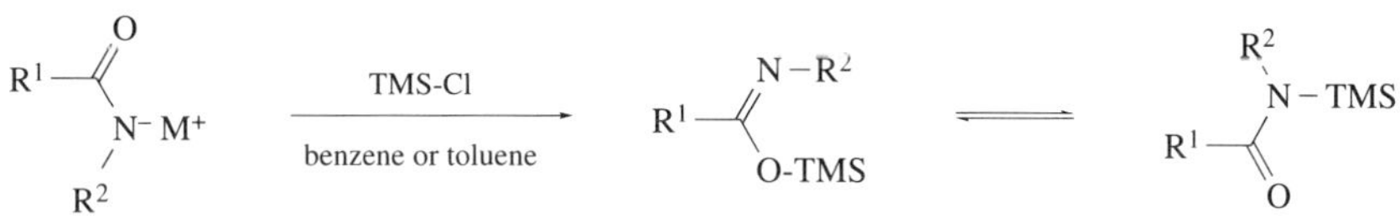

R^1 = CF_3, CCl_3, $CHCl_2$, Me, Et, Pr^i, Bu^t, Ph; R^2= Ph
R^1 = Ph; R^2 = Me

Scheme 90

The lithium salt of *N*-ethylpyruvic acid amide, which is obtained by "nucleophilic acylation" of ethyl isocyanate, is converted at low temperatures by trimethylchlorosilane to trimethylsilyl *N*-ethyl-2-oxo-propionimidate (Scheme 91) ⟨84TL5251⟩.

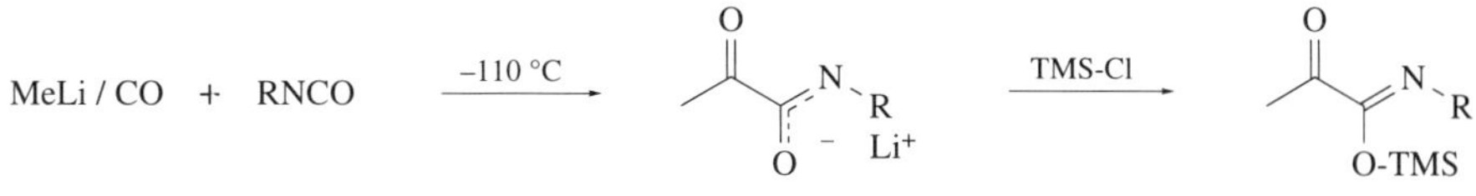

Scheme 91

In the presence of KCN and 18-crown-6, trimethylpentafluorophenylsilane silylates *N*-acetylacetamide at the carbonyl oxygens once or twice depending on the amount of the silylating reagent, and the nature of the solvent (Scheme 92) ⟨83BAU1373, 83IZV1515, 83JGU200, 83JGU1851, 83ZOB229, 83ZOB2051⟩. Disilylation is also observed in the reaction of *N*-acetylacetamide with dimethyl-bis(pentafluorophenyl)silane (Scheme 92) ⟨83IZV1515, 83JGU1851, 83ZOB2051⟩. Fluorinated hydroxamic acids are converted by hexamethyldisiloxane to the corresponding *N*-trimethylsilyl hydroxamic acid trimethylsilyl esters (Equations (238) and (239)) ⟨84JOC4541⟩. Similar reactions have been performed with phenyltrichlorosilane; the disilylated compounds thus formed decompose on heating to give nitriles (Scheme 93) ⟨83JGU1513, 83ZOB1683⟩.

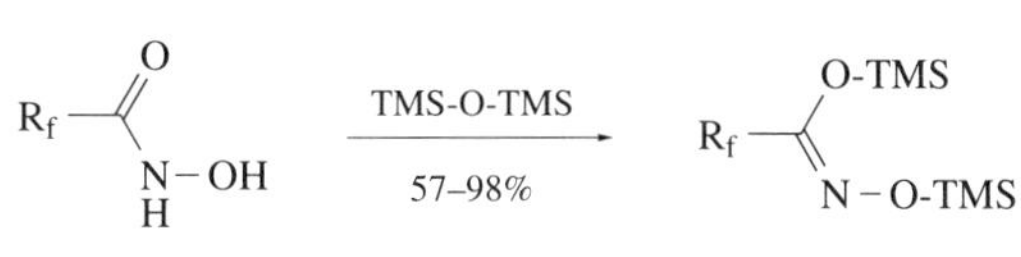

(238)

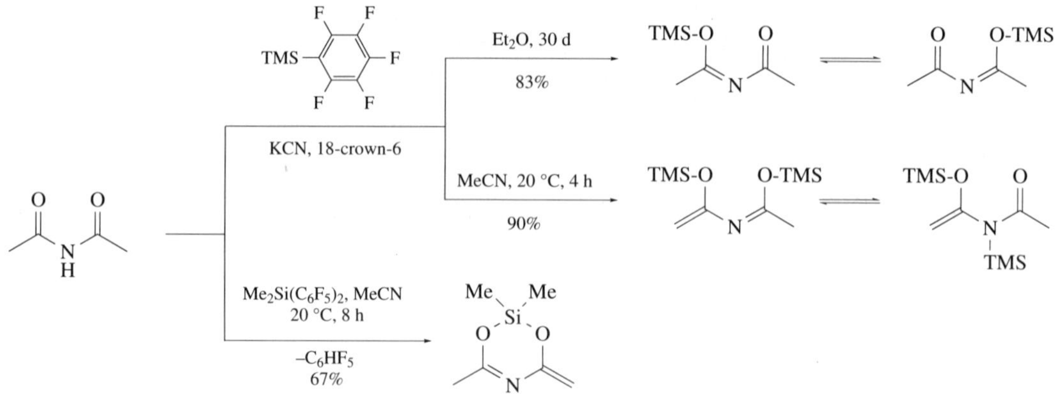

Scheme 92

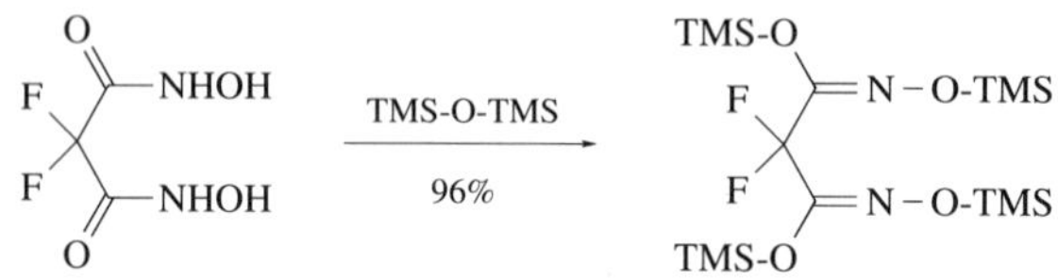

(239)

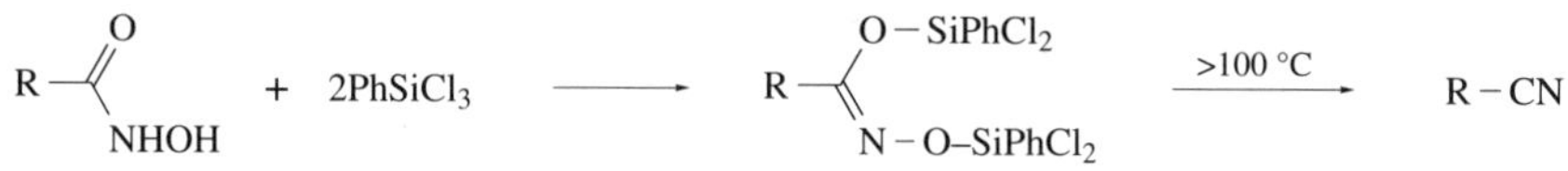

Scheme 93

Under the influence of a nickel catalyst, carboxylic acid amides are silylated by triethylsilane (Scheme 94) ⟨74CR(C)213⟩. Diethyl-trimethylsilylphosphine adds to *N*-acetyl-benzophenoneimine via a 1,4-addition, giving rise to the formation of an *O*-silylated acetamide derivative (Equation (240)) ⟨75HCA1316⟩.

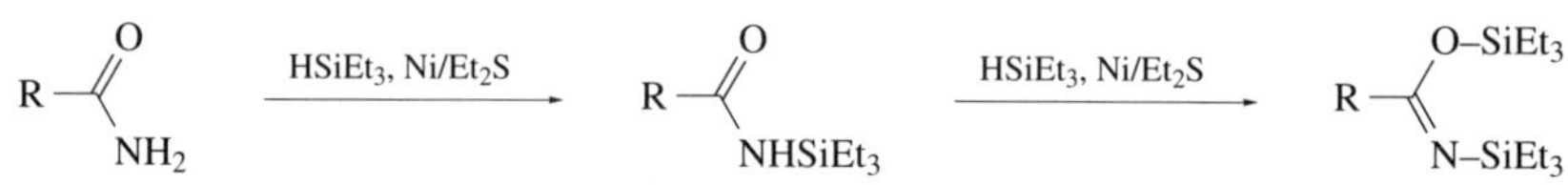

Scheme 94

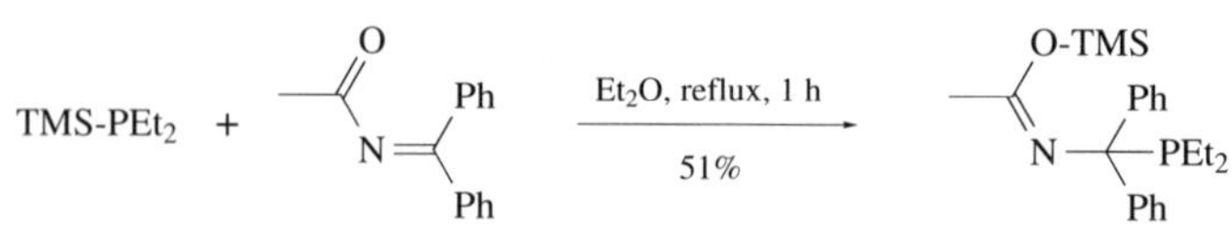

(240)

It should be noted that monosilylated amides usually exist as equilibrium mixtures of *O*- and *N*-silylated compounds ⟨72JCS(P2)1043, 77JOM(134)21, 79JCS(P1)2478, 79JOM(175)273, 83BAU1373, 83IZV1515, 83ZOB2051⟩ (cf. Equation (241)); this is true also for *N*-aryl-(*t*-butyldimethyl)silyl formamides, although the *N*-methyl compound, which exists exclusively in the amide form ⟨78JOM(161)313⟩, is an exception. In contrast to the behavior of carboxylic amides most lactams are monosilylated at the nitrogen ⟨70IZV1157, 72IZV1150, 79JCS(P1)2478, 90JOC3682⟩. With the exception of bis(trimethylsilyl)formamide—which exists as the *N*,*N*-disilylated amide—the other homologous bis-trimethylsilylated amides seem to exist as *O*,*N*-disilylated iminoesters ⟨72JA5285, 79JCS(P1)2478, 79JOM(175)273, 87JOM(339)69⟩. In contrast to this, in bis(dimethylsilyl)acetamide and -propionamide, both dimethylsilyl groups are attached at the amide nitrogen ⟨87JOM(339)69⟩.

(241)

5.17.8.8.2 *From carboxylic acid derivatives, carbonic acid derivatives and* N*-silylated amines and amine derivatives*

The reaction of sodium bis(trimethylsilyl) ⟨63CB2138⟩ or lithium bis(trimethylsilyl)amide ⟨87JOM(339)69⟩ with benzoic acid esters, benzoic acid anhydride or benzamide gives *O*,*N*-bis(trimethylsilyl)benzimidate (Equation (242)). By the same procedure *O*,*N*-bis(trimethylsilyl)trifluoroacetimidate has been synthesized ⟨76CB2542⟩. Trifluoroacetyl chloride has been reacted with silylated amines and amine derivatives to give the corresponding *O*-silylated imidic acid derivatives (Scheme 95). Silylated amines and benzoyl isocyanate give monosilylated compounds, which seem to exist as a mixture of *O*- and *N*-silylated isomers (Scheme 96) ⟨71JCS(C)1870⟩. Trimethylsilyl cyanide adds to trifluoromethyl isocyanate to give trifluoromethylimino trimethylsilyloxy acetonitrile (Equation (243)) ⟨79CB2158⟩. The reaction of carboxylic acid chlorides with bis(trimethylsilyl)carbodiimide affords an equilibrium mixture which contains *O*-silylated imino acid derivatives (Scheme 97) ⟨78JGU1313, 78ZOB1429⟩.

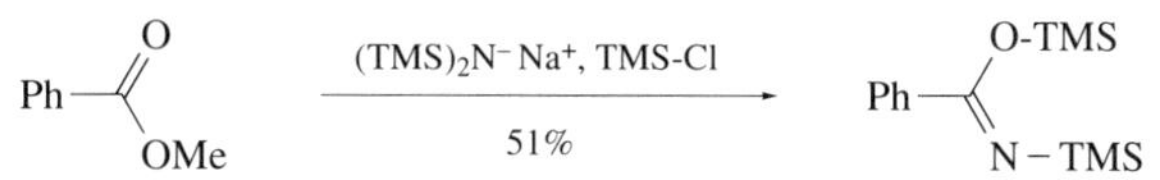

(242)

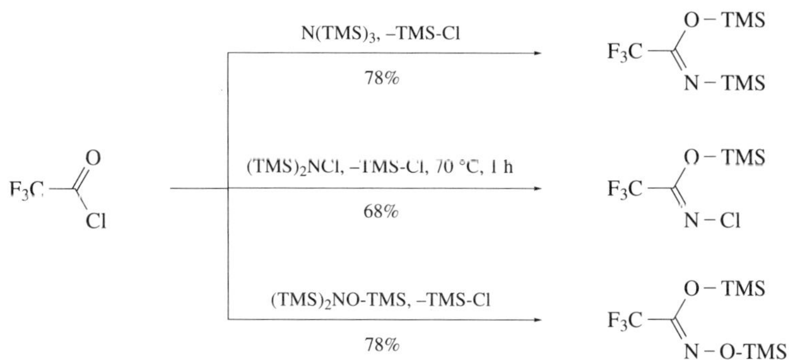

Scheme 95

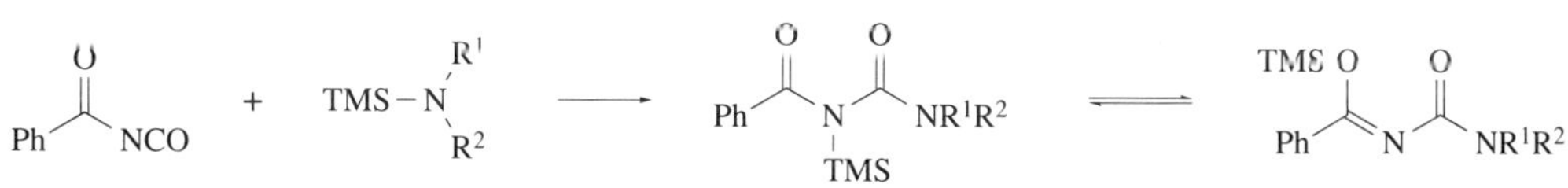

Scheme 96

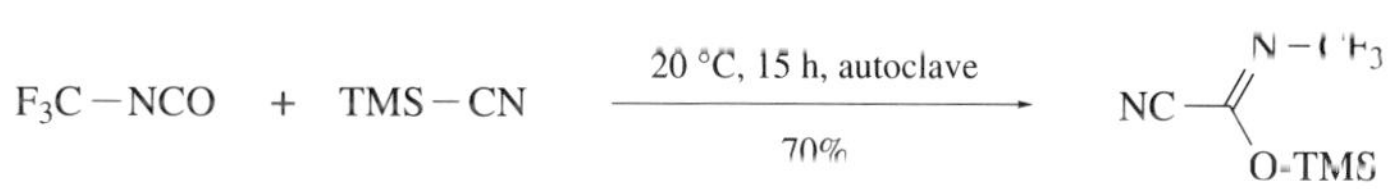

(243)

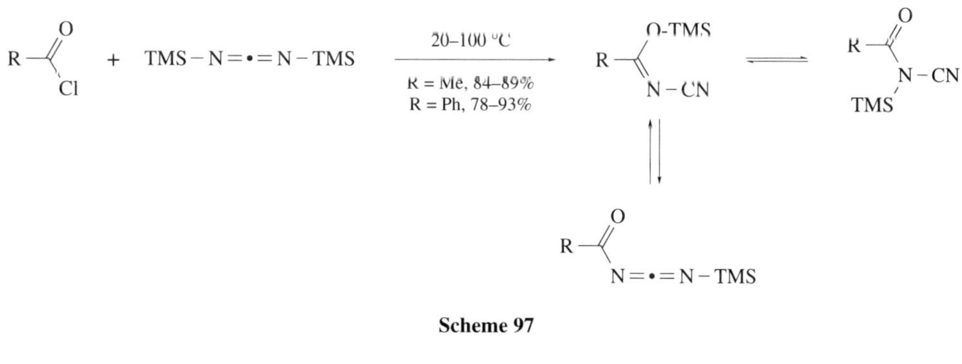

Scheme 97

5.17.8.8.3 From imidic acid derivatives

N-Unsubstituted imidic acid esters are silylated by trialkylchlorosilanes/triethylamine to give *N*-silylated imidates (Equations (244) and (245)) ⟨71DOK(199)103, 75JGU935, 75ZOB949⟩. Electrophilic reagents such as methyl chloroformate ⟨72JGU1859, 72ZOB1870⟩, phosgene ⟨83JGU1944, 83ZOB2155⟩, trifluoroacetyl chloride ⟨76CB2542⟩, cyanogen chloride ⟨78JGU1313, 78ZOB1429⟩, formaldehyde ⟨81JGU2055, 81ZOB2382, 82JGU1222, 82ZOB1386⟩, *N*-halosuccinimides, *N*-chloro carboxylic acid amides ⟨68CB2585⟩, *t*-butyl hypochlorite ⟨87AJC701⟩ and organoboron halides ⟨78ZAAC(443)148⟩ usually attack the imidic group giving rise to the formation of *O*-silylated imidic acid derivatives; some reaction products are equilibrium mixtures of *O*- and *N*-silylated isomeric compounds. Examples are given in Schemes 98 and 99.

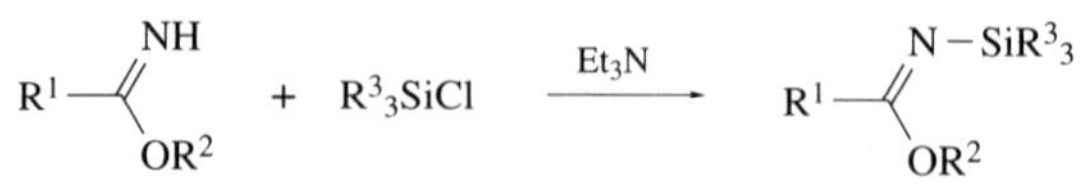

(244)

R^1 = Me, Ph; R^2, R^3 = Me, Et

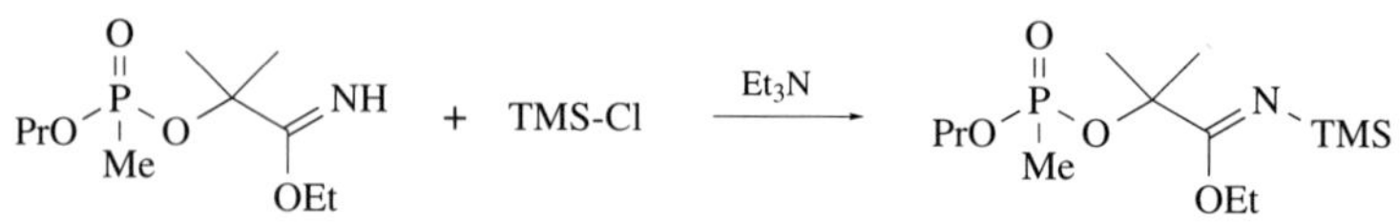

(245)

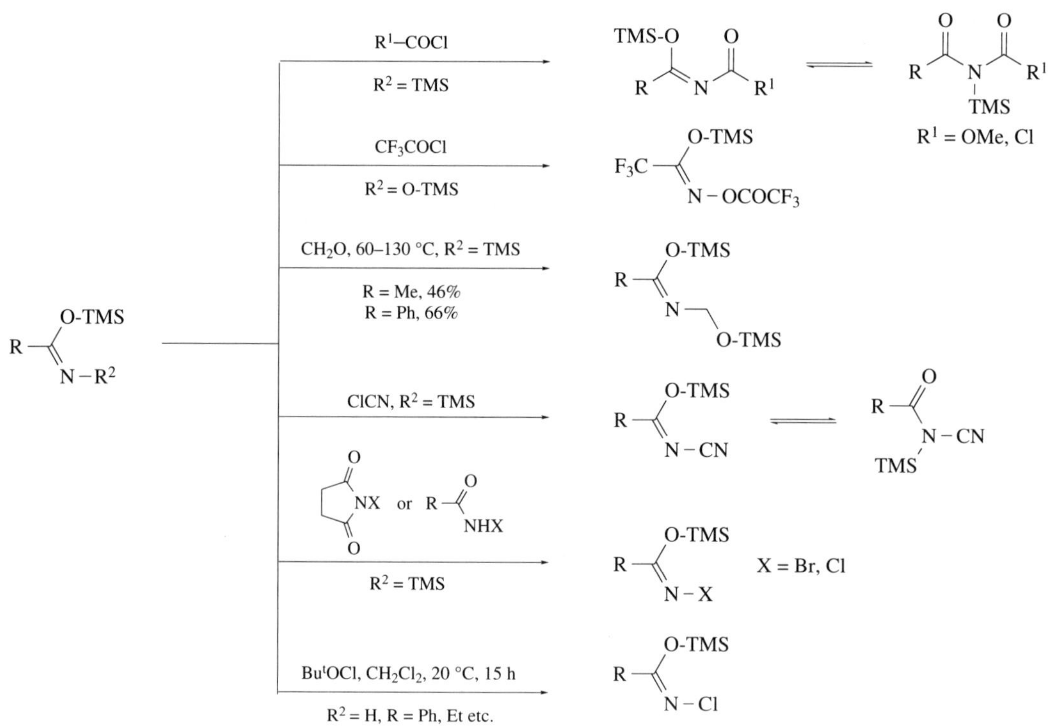

Scheme 98

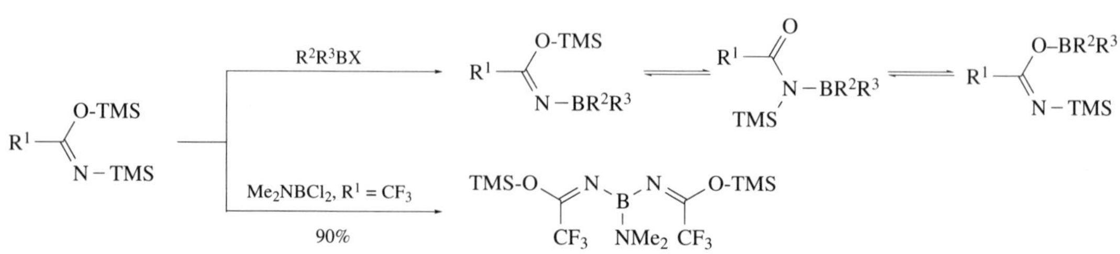

Scheme 99

5.17.8.8.4 *Miscellaneous methods*

1,1-Dialkoxy-2-trimethylsilylethenes can be aminoalkylated with the aid of *N*-acetyltrichloroacetaldoxime; the final products are trimethylsilyl imidates (Equation (246)) ⟨89JGU2055, 89ZOB2293⟩.

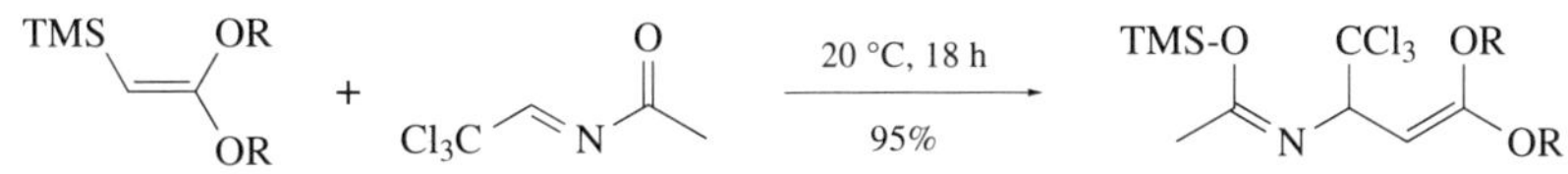

(246)

5.17.8.8.5 *Imidoyloxymetals*

Carboxylic acid amides and the corresponding imidic acids are tautomeric compounds; thus the anions obtained by *N*-deprotonation of amides and by *O*-deprotonation of hypothetical imidic acids are identical (Scheme 100).

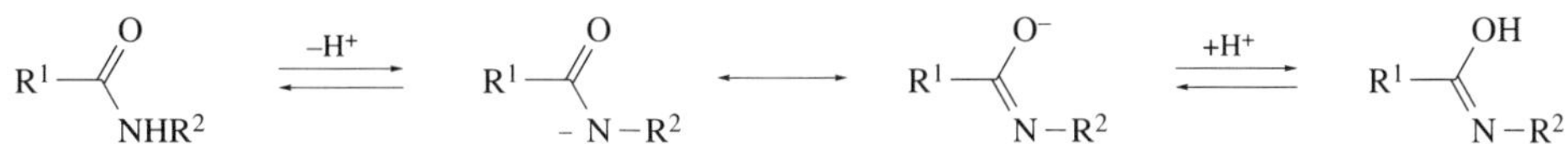

Scheme 100

Since primary and secondary carboxylic acids can act as weak NH acids ⟨B-70MI 517-02⟩, they are deprotonated even by sodium hydroxide in protonic solvents (H_2O, ROH). The anions from secondary amides can exist as *E*/*Z*-isomers (Scheme 101). The *E*-isomer is generally more stable than the *Z*, but the *E*/*Z* equilibrium constant is strongly solvent-dependent ⟨85JOC1405⟩.

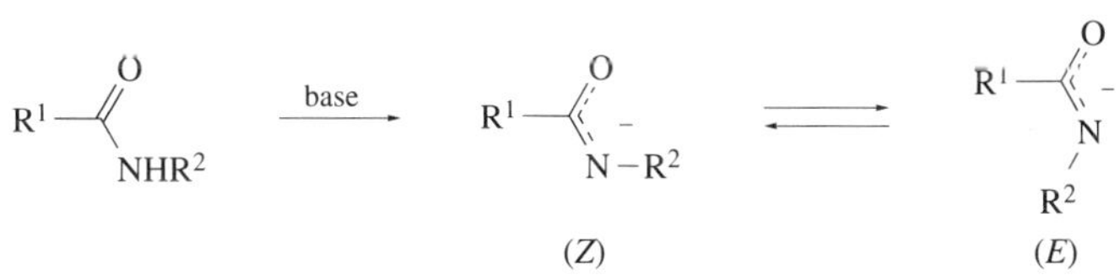

Scheme 101

On a preparative scale the alkali salts of carboxylic acid amides (e.g., formanilides) can be prepared by treating the amides with NaH or KH in DMSO, THF, glycol or cyclohexanol/dioxane 1 : 1 ⟨85JOC1405⟩. The lithium salt of formanilide was prepared by heating LiH with formanilide in DMF ⟨74CJC4050⟩.

In another preparation, the sodium *N*-acylamides are obtained by treating the amides (e.g., acetanilide) with sodium wire in refluxing benzene ⟨35CB751⟩, or by dropping the molten amide (e.g., formanilide) on to molten sodium in toluene at about 100 °C ⟨77JOC1786⟩. The potassium salts are obtained by treating the amides with potassium in benzene ⟨74CJC4050⟩. In most cases the salts have not been isolated; usually they have been reacted further immediately after their generation. Many such salts (including silver salts) have been reacted with alkylating reagents in order to determine the site of alkylation (*O* versus *N*) in these ambident anions ⟨B-70MI 517-03, 74CJC4050⟩.

Various metal salts from *N*-cyano phenoxyacetic acid amide ⟨90ZAAC(588)59⟩ have been prepared in analytically pure state; examples for the procedures used are given in Scheme 102. Potassium ⟨69AJC161⟩, silver ⟨1890JA499, 71JOC284⟩ and thallium salts ⟨77JCS(P2)1080⟩ have also been prepared from hydroxamic acids.

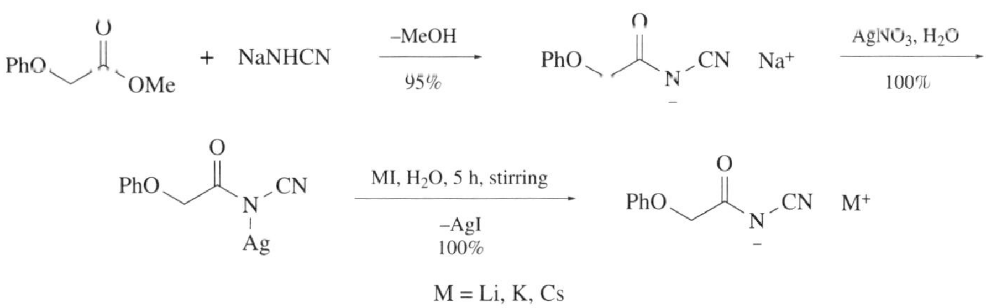

Scheme 102

5.18
Iminoacyl Functions Linked to Chalcogens Other Than Oxygen

STEPHEN CHALLENGER
Pfizer Central Research, Sandwich, UK

5.18.1 IMIDOYLSULFUR, -SELENIUM AND -TELLURIUM DERIVATIVES

5.18.1.1 Thiolimidic Acids and their Selenium and Tellurium Derivatives, $R^1C(NR^2)SH$, $R^1C(NR^2)SeH$ and $R^1C(NR^2)TeH$

Thiolimidic acids can be considered as tautomers of primary and secondary thioamides. Spectroscopic methods, including x-ray diffraction ⟨60JCS997⟩, NMR ⟨67CJC1⟩ and IR ⟨65JCS5230⟩, have clearly differentiated *cis–trans* isomerism from thioamide–thiolimidic acid tautomerism and demonstrated that negligible concentrations of the imidic tautomer (**2**) exist in solutions of thioamides (**1**) or in the solid state. This subject has been reviewed by Walter and Voss ⟨B-70MI 518-01⟩. A quantitative estimation of amide–imidic acid tautomerism in the corresponding selenium and tellurium derivatives has not been reported. The IR spectra of selenoamides ⟨66ACS597⟩ and telluroamides ⟨79CC1102⟩ has been shown to be very similar to thioamides suggesting that the amide tautomer is favoured in these systems also.

5.18.1.2 Thioimidic Esters and Anhydrides, $R^1C(NH)SR^2$, $R^1C(NR^2)SR^3$ and $R^1C(NR^2)SC(NR^2)R^1$

Thioimidic esters are most frequently referred to in the literature as thioimidates. Cyclic thioimidic esters have also been called thiolactim ethers. Methods of synthesis of this class of compounds have been reviewed previously, the most recent of which was by Kantlehner ⟨91COS(6)536⟩. Reviews by Roger and Neilson ⟨61CRV179⟩, by Sandler and Karo ⟨B-72MI 518-01⟩ and by Neilson ⟨B-91MI 518-01⟩ on the chemistry of imidates include examples of the preparation of thioimidic esters. The five principal methods of preparation of this class of compound are exemplified in the following sections.

5.18.1.2.1 *Thioimidic esters and anhydrides from thioamides, thiolactams and related compounds*

The most general and widely used method of preparation of compounds of this class is the reaction of thioamides (**3**) or thiolactams (**4**) with alkylating agents (Scheme 1).

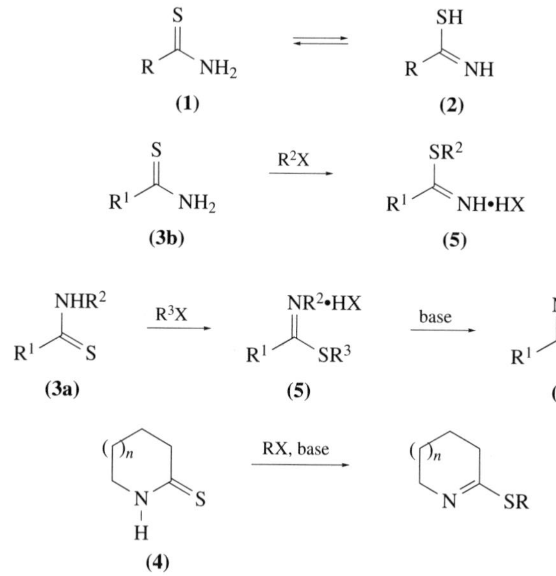

Scheme 1

This reaction is applicable to both primary and secondary thioamides and in the absence of added base leads to the formation of salts of thioimidic esters (**5**), also named alkylthiolmethylene-iminium salts. Several examples of this reaction are tabulated in a review by Kantlehner ⟨B-79MI 518-01⟩. Deprotonation of these salts or the inclusion of a base in the alkylation reaction leads to thioimidic esters (**6**), usually as a mixture of (*E*) and (*Z*) isomers. The alkylation of tertiary thioamides leads to the formation of salts of *N*,*N*-disubstituted thioimidic esters. A variety of alkylating reagents have been employed including alkyl halides ⟨B-79MI 518-01⟩, dialkyl sulfates ⟨92CPB602⟩, trifluoromethanesulfonic acid esters ⟨89TL4443⟩, sultones ⟨64LA(676)114⟩, alkyl fluorosulfonates ⟨77LA1291⟩ and oxonium salts ⟨83SC753⟩. In the majority of cases the product of *S*-alkylation is observed, however there are examples particularly when the alkylating agent can form a stable carbenium ion in which *N*-alkylation occurs ⟨B-70MI 518-01⟩.

Some useful specific preparations of compounds of this class involving the alkylation of thioamides have been described. The methylation of thioamides and thiolactams with diazomethane in the presence of silica gel has been reported to give high yields of thioimidic esters ⟨79TL4671⟩. Thioacetanilide reacts with 2-allyl-1,3-dicyclohexylisourea (**7**) in the presence of a palladium catalyst to give the *S*-allylated thioimidic ester (**8**) in 60% yield (Equation (1)) ⟨85BCJ2721⟩. The thioimidic ester (**10**) has been obtained from thioamide (**9**) (70%) in an intramolecular displacement reaction under Mitsunobu conditions (diethyl azodicarboxylate (dead), triphenylphosphine) (Equation (2))

⟨92TL5117⟩. A related, less general preparation of thioimidic esters involves the alkylation of thioamides with ethyl thiochloroformate (**11**) (Equation (3)) ⟨73JOC2242⟩. The mechanism of this reaction has been studied and a steric effect on product yield observed. Introduction of bulkier substituents in the thioamide or alkyl thiochloroformate has been shown to decrease yields of thioimidic ester ⟨83JOC1544⟩. A specific preparation based on the alkylation of a thiolactam with a chloroformate is the formation of thiolactim ether (**13**) (67%) by reaction of thiocaprolactam (**12**) with benzyl chloroformate (Equation (4)) ⟨84SC1159⟩. The thioimidic anhydride (**15**) has been prepared in 30% yield by the reaction of thioamide (**14**) with phenylsulfenyl chloride (Equation (5)) ⟨69LA50⟩.

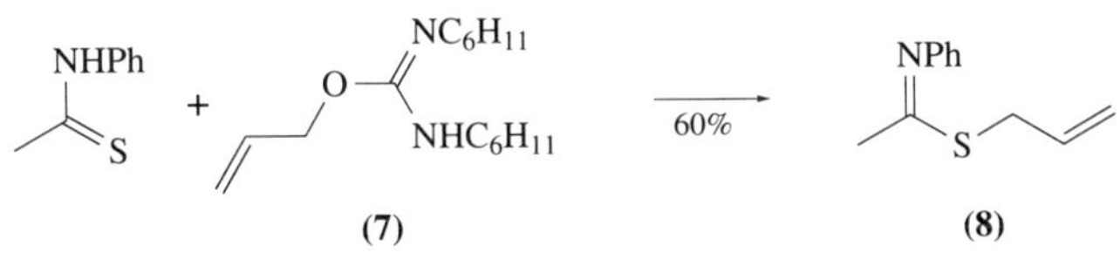

(1)

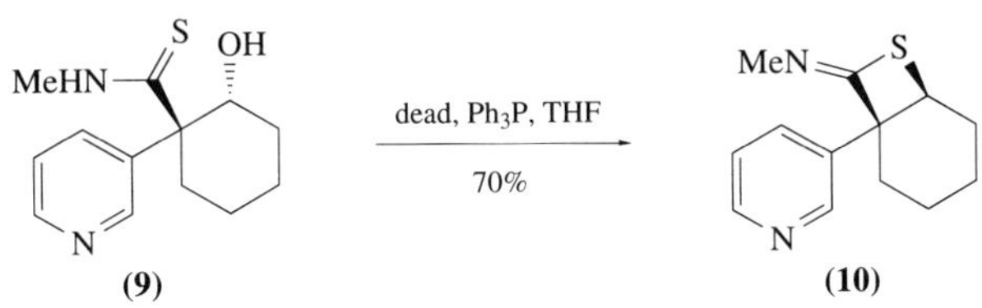

(2)

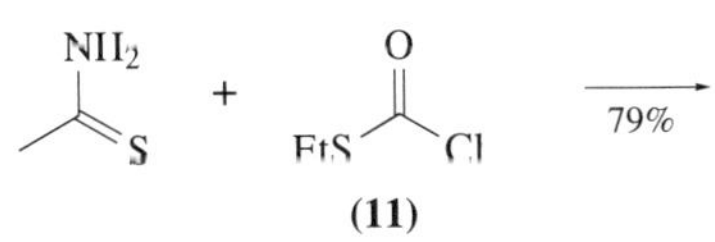

(3)

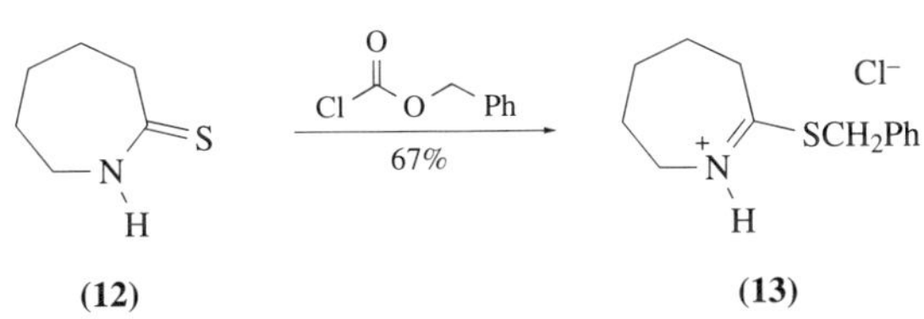

(4)

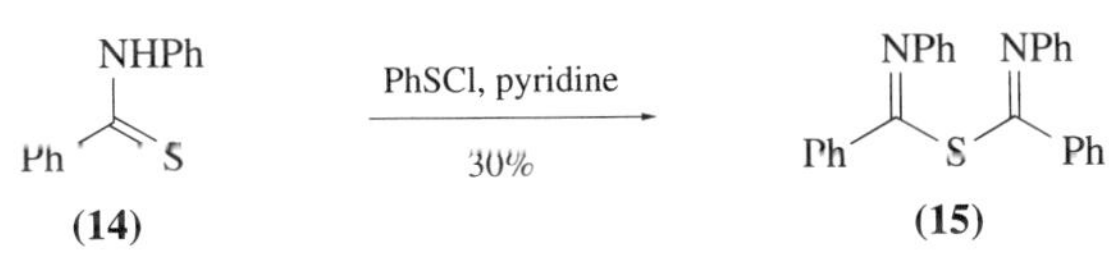

(5)

Radical arylation of *N*-substituted thioamides with nitrosoacetanilide or phenylazotriphenylmethane ⟨82IZV2121⟩ and aryldiazonium salts ⟨70ZOR805⟩ has been reported to give thioimidic esters. A related method involving carbon radicals has provided a route to thioimidic esters. On heating to 170 °C thiohydroxamic ester (**16**) undergoes decarboxylative rearrangement to give (**17**) in 60% yield (Equation (6)) ⟨83TL5889, 86JCS(P1)39⟩.

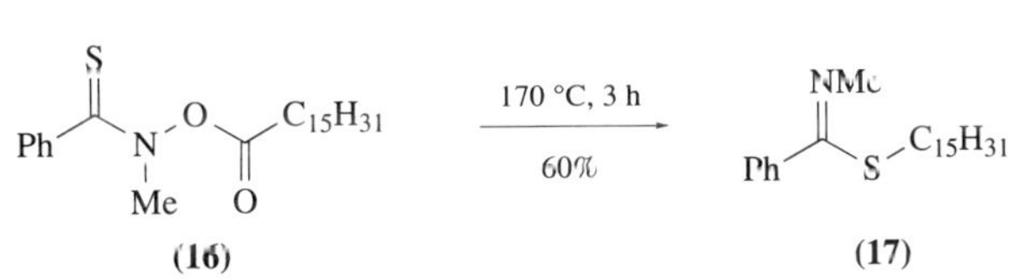

(6)

5.18.1.2.2 *Thioimidic esters from nitriles, isonitriles and thiocyanates*

The addition of thiols to alkyl or aryl nitriles in the presence of hydrogen halides is an established method for the preparation of salts of *N*-unsubstituted thioimidic esters, which was first described by Pinner in 1878 ⟨1878CB1825⟩. The method has been applied to a variety of nitriles, including α-amino ⟨72S622⟩, α-azido ⟨92S1199⟩ and α-hydroxy nitriles ⟨70CJC3662⟩, with few limitations. *Ortho*-substituted aromatic nitriles tend to fail to undergo the reaction. In a related preparation the

N-alkyl thioimidate salts (**19**) and (**21**) have been isolated from the reaction of nitrilium ions (**18**) and (**20**) with thiols (Scheme 2) ⟨83JCS(P1)1067, 85JOC4415⟩.

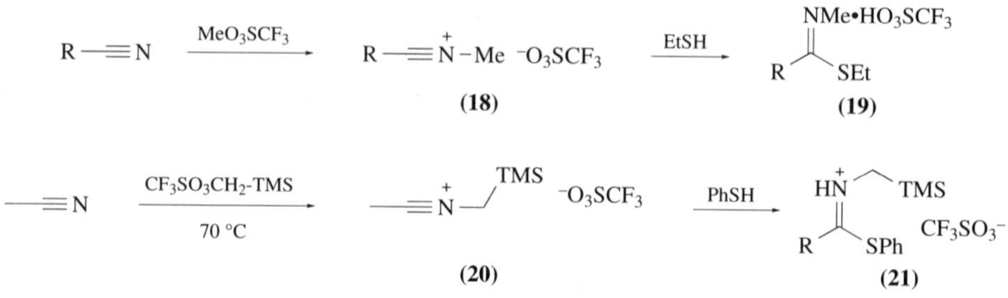

Scheme 2

The base-catalysed addition of thiols to the nitrile function has been reported to give thioimidic esters ⟨76BSF667⟩. The reaction is facilitated if electron-withdrawing groups are present in the α-position to the nitrile group. The addition of thiols or thiophenols and other divalent sulfur compounds to isonitriles has been used as a method of preparation of compounds of this class ⟨82TL2023, 83CC210, 84T1075, 88T3501⟩: some examples are shown in Scheme 3. The condensation of aliphatic and aromatic thiocyanates with substituted phenols in the presence of HCl and zinc chloride has been reported to give salts of thioimidic esters (Equation (7)) ⟨23JA1744⟩.

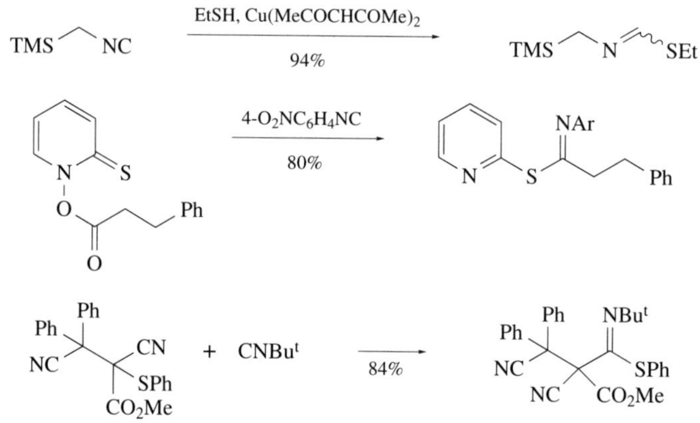

Scheme 3

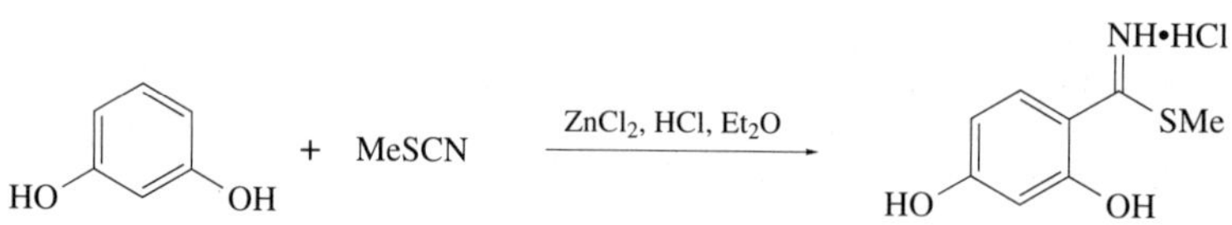

(7)

5.18.1.2.3 *Thioimidic esters from imidoyl halides*

The displacement of chloride from imidoyl chlorides by thiols has been reported as a method of preparation of compounds of this class ⟨26JCS2296, 63CB2671⟩, although this reaction is not frequently used. This is the only method which has been used to prepare all three chalcogenoimidic esters. For example the thioimidic ester (**23**) was isolated (45%) from the reaction of the corresponding imidoyl chloride (**22**) with 4-chlorothiophenol (Equation (8)) ⟨80S114⟩. Related reactions include the preparation of thioimidic ester (**25**) from the imidoyl iodide (**24**) and lithium ethanethiolate (Scheme 4) ⟨83TL3255⟩, and the reaction of imidoyl chloride (**26**) with thallium(I) thiophenoxide to give (**27**) in 91% yield (Equation (9)) ⟨80JOC80⟩. An alternative method is the conversion of thioimidoyl chlorides into thioimidic esters on reaction with carbon nucleophiles ⟨77S626, 85JOC771, 89CL1261⟩. For example, *N*-phenyl phenylthioimidoyl chloride (**28**) was converted into thioimidic esters (**30**) with alkynyltin reagents (**29**) and a palladium catalyst (Equation (10)) ⟨89CL1261⟩. The α,β-unsaturated thioimidic ester (**33**) is obtained from thioimidoyl chloride (**31**) and triphenylphosphonium methylide, followed by Wittig reaction of the intermediate phosphorane (**32**) with benzaldehyde (Scheme 5) ⟨77S626⟩.

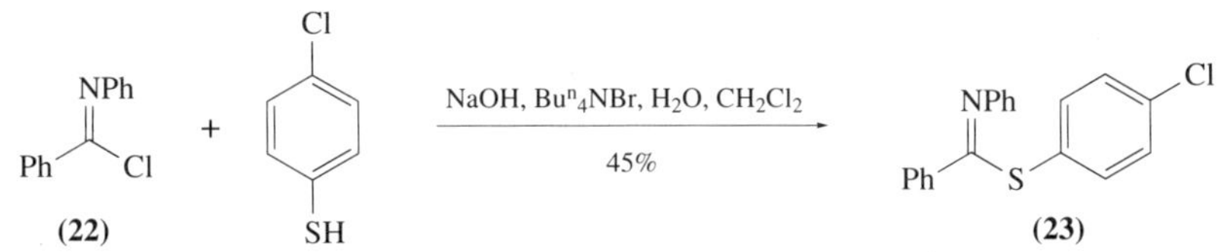

(**22**) (**23**) (8)

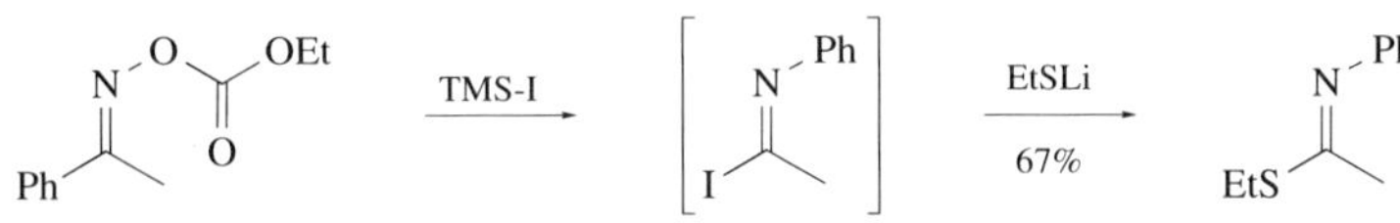

(**24**) (**25**)

Scheme 4

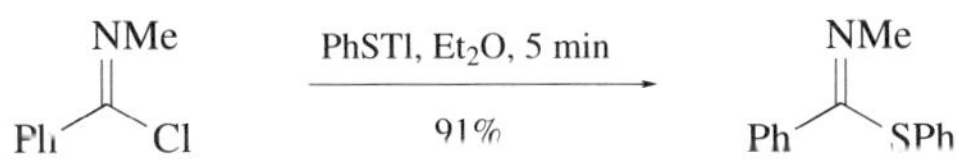

(**26**) (**27**) (9)

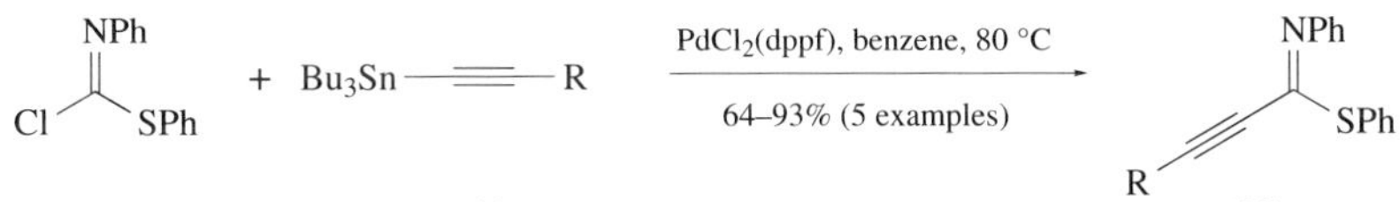

(**28**) (**29**) (**30**) (10)

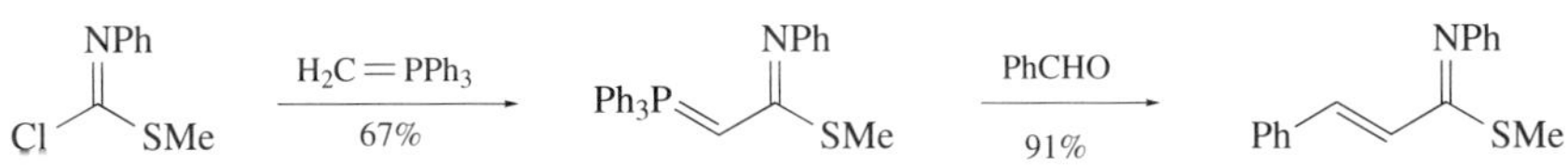

(**31**) (**32**) (**33**)

Scheme 5

5.18.1.2.4 Thioimidic esters from heterocumulenes, isothiocyanates and ketenimines

The conversion of phenyl isothiocyanate into *N*-phenyl thioimidic esters has been achieved by reaction with carbon nucleophiles, including Grignard reagents ⟨78TL2715, 83CI(L)391, 87JOC2523⟩, alkyl lithiums ⟨84T1573⟩, ester enolates ⟨88LA471⟩, ketene *O,N*-acetals ⟨74CB3589⟩ and phosphorus stabilised carbanions ⟨85S172⟩, followed by *S*-alkylation; for example (**35**) is obtained in 67% yield from phenyl isothiocyanate (**34**) and phenylmagnesium bromide followed by alkylation with dimethyl sulfate (Equation (11)) ⟨24IA493⟩. Intramolecular variants of this reaction have been used to prepare cyclic thioimidic esters. Lithium–halogen exchange of the iodoisothiocyanate (**36**) with *n*-butyllithium followed by *S*-alkylation, of the intermediate arising from intramolecular addition to the isothiocyanate group, with iodomethane gave (**37**) in 92% yield (Equation (12)) ⟨81JCS(P1)40⟩. Reaction of the lithium anion of 2-oxetanone (**38**) with phenyl isothiocyanate gave, after decarboxylation of intermediate (**39**), the thioimidic ester (**40**) (Scheme 6) ⟨80AG(E)466⟩. The reaction of *t*-butyllithium with carbon monoxide in the presence of alkyl isothiocyanates gives, after alkylation with iodomethane, α-oxothioimidates (**41**) (Equation (13)) ⟨84TL5251⟩.

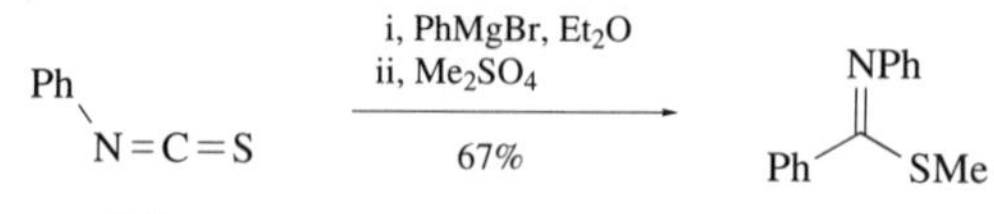

(34) (35) (11)

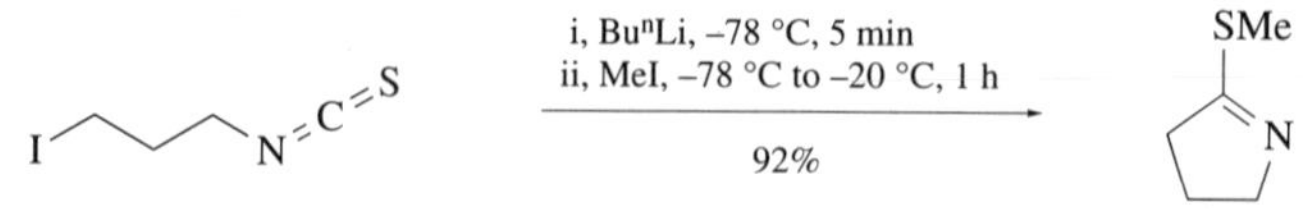

(36) (37) (12)

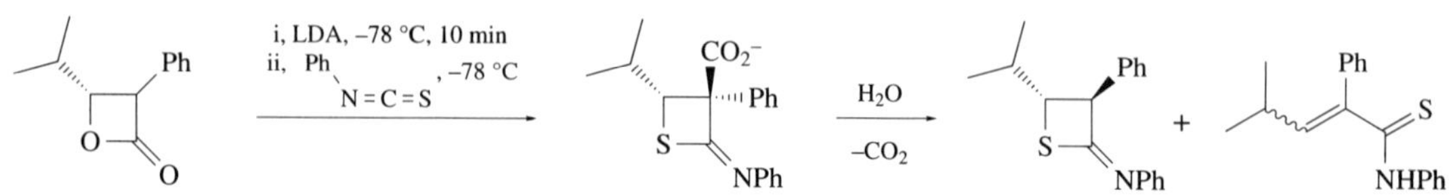

(38) (39) (40) 41% 36%

Scheme 6

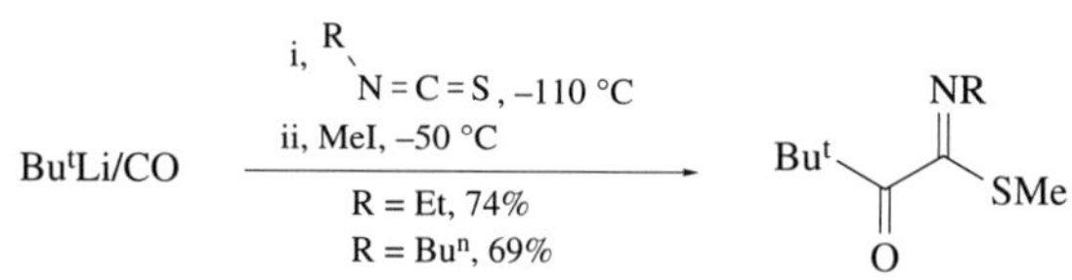

(41) (13)

N-Aryl thioimidic esters have been obtained in 44–90% yields by the reaction of alkyl and aryl thiols with ketenimines (**42**) ⟨73JOC3951⟩ or phosphacumulene ylides (**43**) (Scheme 7) ⟨80CB912⟩. The preparation is limited by the availability of ketenimines. On warming above 10 °C the lithium anion of 1,4-diphenyl-1,2,3-triazole (**44**) undergoes fragmentation with loss of nitrogen to give the lithium anion of ketenimine (**45**) which has been trapped with ethanethiol to give thioimidic ester (**46**) in 80% yield (Scheme 8) ⟨71CJC1792⟩.

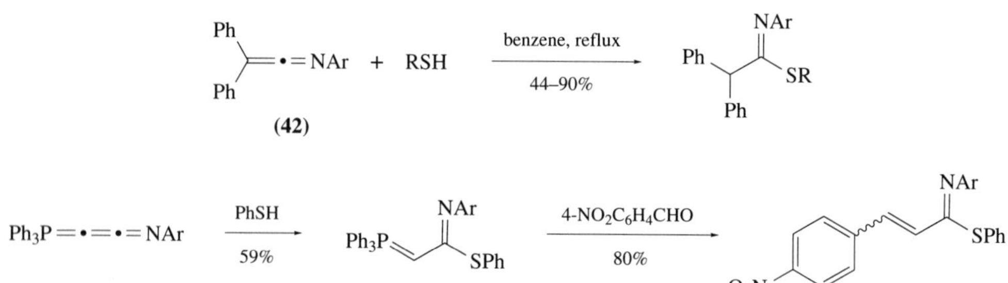

(42) (43)

Scheme 7

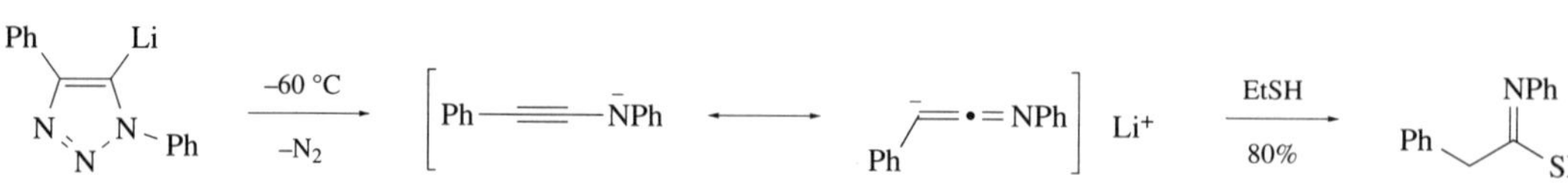

(44) (45) (46)

Scheme 8

5.18.1.2.5 *Thioimidic esters from thioimidic esters*

The metallation and alkylation of thioimidic esters (**47**), containing at least one hydrogen on the α-carbon atom, has provided a route to several substituted derivatives (**48**). At low temperatures alkylation occurs regioselectively at nitrogen (R^1 is alkyl) or at the α-carbon atom (R^1 is phenyl, vinyl, trimethylsilyl or hydrogen) (Equation (14)) ⟨84T1573, 91JCS(P1)3389, 91PS(59)141⟩. Kinetic metallation of thiolactim ether (**49**) occurs at the *S*-methyl group to give (**50**) after alkylation with iodomethane. Alternatively, thermodynamic metallation of (**51**) at the α-carbon leads to (**52**) in 82% yield after alkylation (Scheme 9) ⟨80JA7929⟩. Thioimidic ester (**54**) has been prepared in 78% yield in an exchange reaction on heating (**53**) with methyl 3-thiopropionate (Equation (15)) ⟨72JCS(P1)2326⟩.

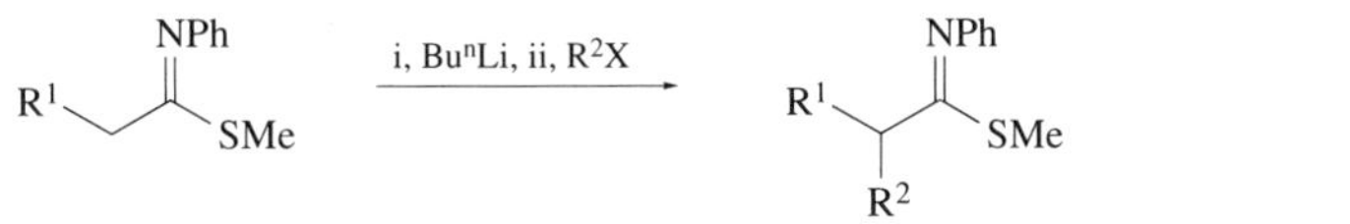

(**47**) (**48**) (14)

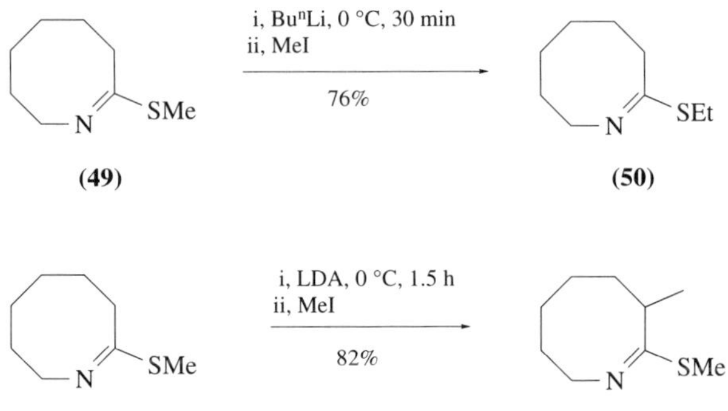

(**49**) (**50**)

(**51**) (**52**)

Scheme 9

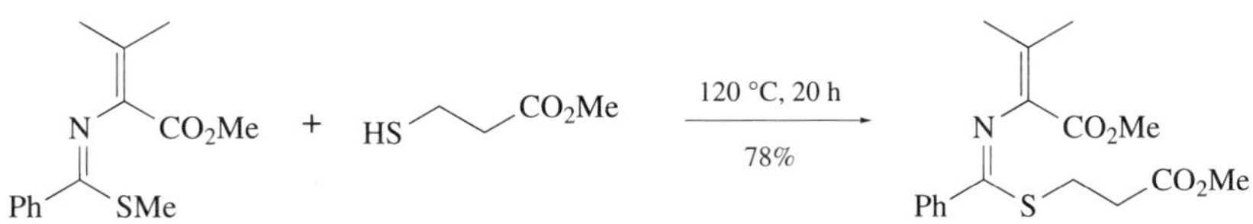

(**53**) (**54**) (15)

5.18.1.2.6 *Thioimidic esters by miscellaneous methods*

Some specific preparations of thioimidic esters have been described. The Beckmann rearrangement of oxime sulfonates (**55**) or (**56**) in the presence of benzylthiol or ethanethiol and triethylamine gives high yields of *N*-phenyl thioimidic esters (Scheme 10) ⟨67BCJ2705⟩. In a similar reaction, Beckmann rearrangement of oxime sulfonates in the presence of dialkylaluminum thiolates provides thioimidic esters in good yields ⟨83JA2831⟩. Some examples of this reaction are included in Scheme 10.

Cyclic thioimidic esters (**58**) have been prepared from the rearrangement of 1-azido-1-methylthiocycloalkanes (**57**), which are themselves derived from thioketals (Equation (16)) ⟨80JA7929⟩. Salts of *N*,*N*-disubstituted thioimidic esters (**60**) and (**63**) have been prepared by the electrophilic substitution of activated aromatic compounds with chlorothioformamidinium salts (**59**) ⟨74AJC2635⟩ and the reaction of methyl dithiobenzoate (**61**) with *N*-sulfinyldimethylammonium tetrafluoroborate (**62**) (Scheme 11) ⟨81LA65⟩. Thioimidic esters (**66**) have been prepared by the reaction of benzil dianils (**64**) with 2,4-bis(phenylthio)-1,3-dithia-2,4-diphosphetane-2,4-disulfide (**65**) (Equation (17)) ⟨92PS(66)87⟩, and by the oxidation of *S*,*N*-acetals with lead tetracetate ⟨88JCS(P1)169, 88T1667⟩ or nitrosyl chloride ⟨87S547⟩.

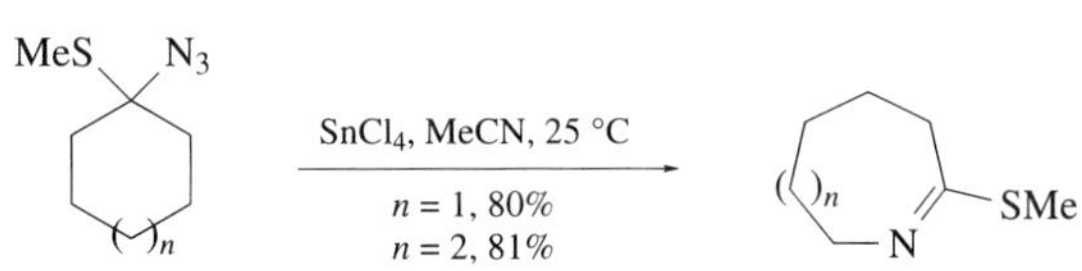

(**57**) (**58**) (16)

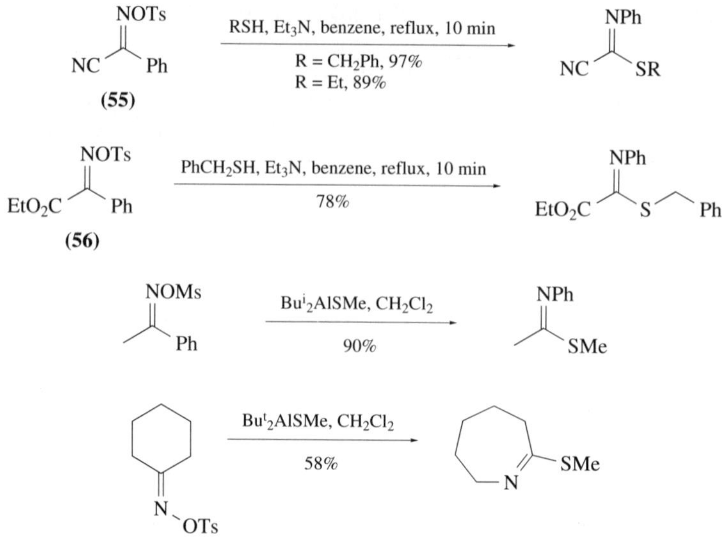

(55) (56)

Scheme 10

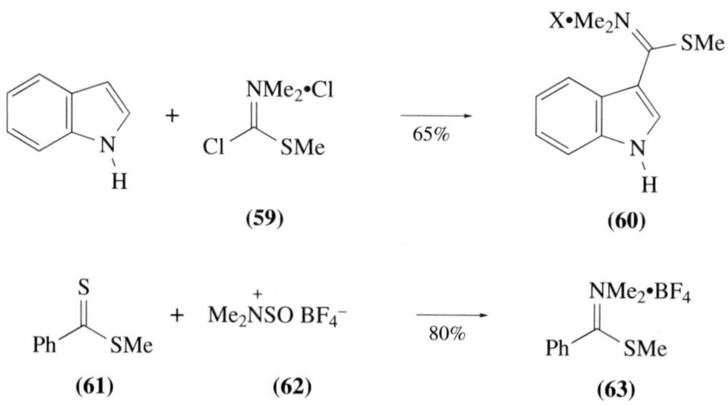

(59) (60) (61) (62) (63)

Scheme 11

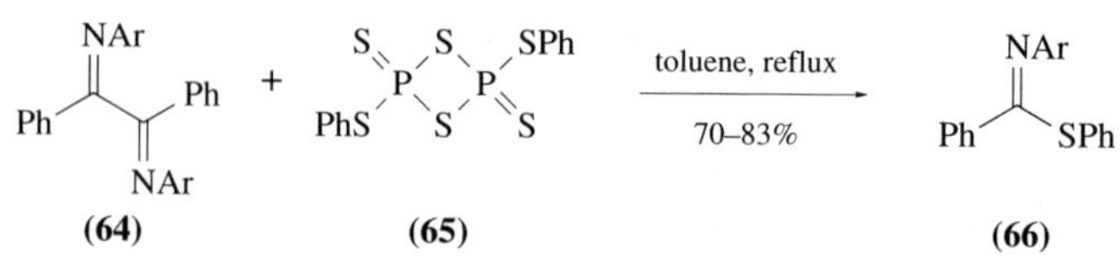

(64) (65) (66)

(17)

5.18.1.3 Selenoimidic and Telluroimidic Esters, $R^1C(NR^2)SeR^3$ and $R^1C(NR^2)TeCR^3$

The preparation of selenoimidic esters, also called selenoimidates, has been reviewed by Paulmier ⟨B-86MI 518-01⟩. The methods available for the synthesis of this class of compound are analogous to those used for the corresponding thioimidic esters, but with fewer examples reflecting their greater reactivity. There are examples of the preparation of selenoimidic esters by the alkylation of primary and secondary selenoamides with alkyl halides (Scheme 12) ⟨66ACS597, 75CC617, 76CB956⟩. The displacement of chloride from imidoyl chlorides by selenium nucleophiles has been used as a method of preparation for a few types of compounds of this class ⟨76CB956⟩. For example, the selenoimidic ester **(68)** was isolated (67%) from the reaction of imidoyl chloride **(67)** with sodium benzeneselenate (Equation (18)) ⟨76CB956⟩. A related reaction is the preparation of selenoimidic ester **(69)** from imidoyl chloride **(67)** and thallium(I) phenylselenide (Equation (19)) ⟨80JOC80⟩. Other methods for the formation of selenoimidic esters which are analogous to those used for thioimidic esters are illustrated in Scheme 13.

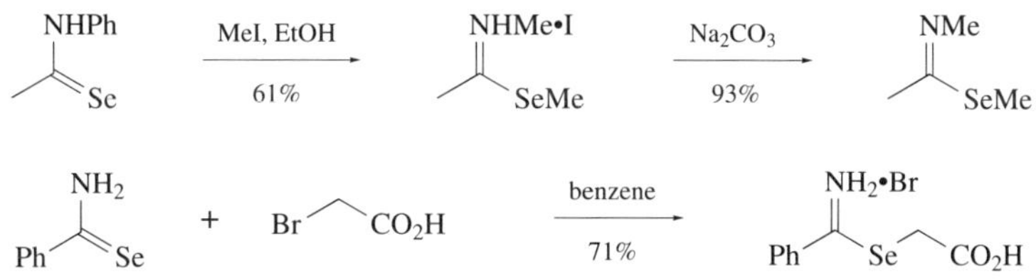

Scheme 12

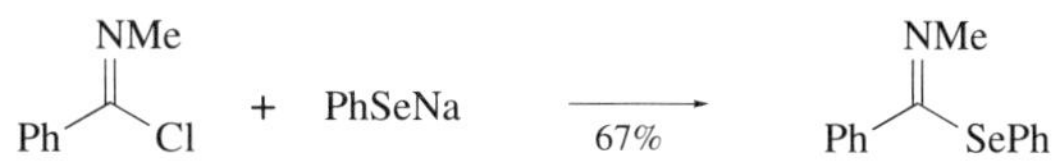

(66) (67) (18)

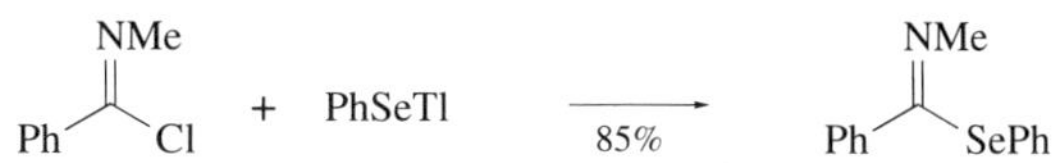

(68) (69) (19)

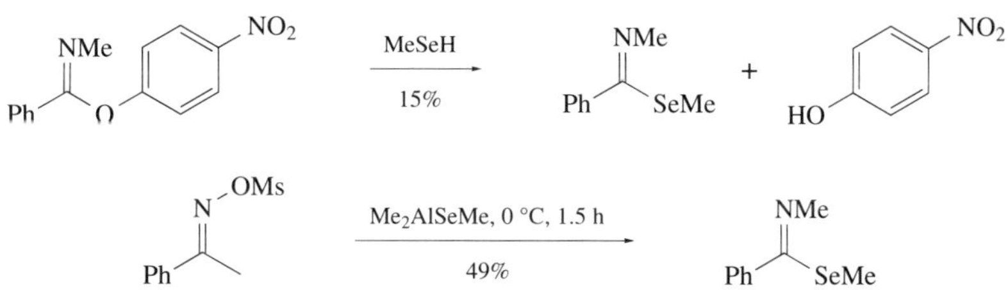

Scheme 13

A few telluroimidic esters (**71**) have been prepared by the displacement of chloride from imidoyl chlorides (**70**) with sodium benzenetellurolate prepared from diphenyl ditelluride and sodium borohydride (Equation (20)) ⟨93TL7933⟩.

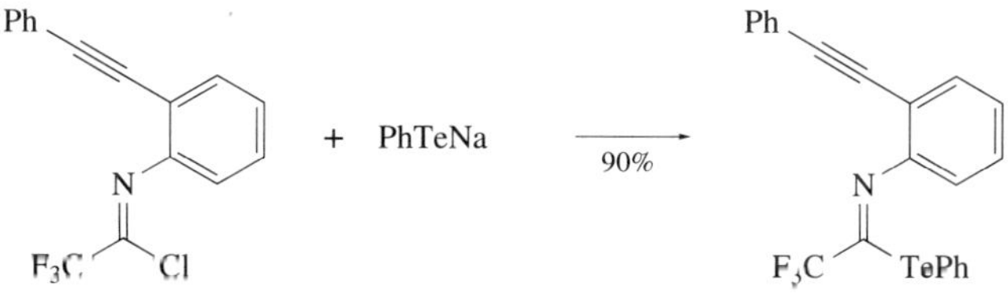

(70) (71) (20)

5.18.1.4 Other Thioimidic Derivatives Where R^2 is Carbon Based, $R^1C(NR)SR^2$ (R^2 = -COR^3, For Example)

The acylation of primary and secondary thioamides with carboxylic acid chlorides in general leads to the formation of the more stable *N*-acylthioamides ⟨B-70MI 518-01⟩. *S*-Acyl derivatives (**72**) have been obtained from the reaction of secondary thioamides with acid chlorides in the presence of sodium hydroxide at low temperature (Equation (21)) ⟨85PS(25)63⟩.

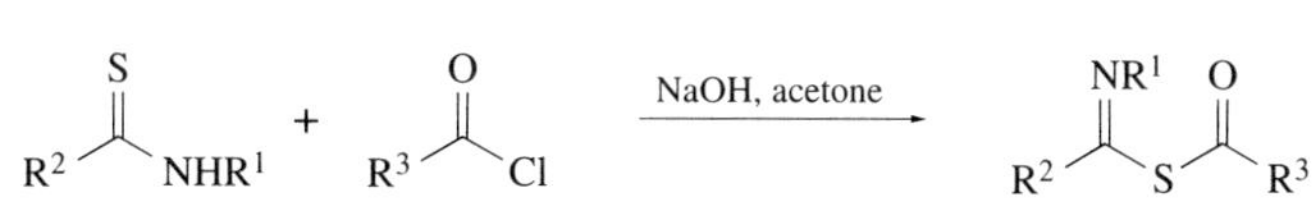

(72) (21)

5.18.1.5 Imidoyl Sulfenyl Halides and their Selenium and Tellurium Analogues, $R^1C(NR^2)SX$, $R^1C(NR^2)SeX$ and $R^1C(NR^2)TeX$

Isolable examples of these classes of compound are virtually unknown. *N,N*-Disubstituted thiobenzamides have been found to react with bromine in anhydrous solvents to give air and moisture sensitive 1 : 1 adducts of low stability (Equation (22)) ⟨82SC865, 84JCR(S)404, 84JCS(P1)897⟩. The adducts were assigned the sulfurane structure (**73**) on the basis of ^{13}C NMR spectra ⟨84OMR724⟩. The higher valent tellurium diiodides (**75**) have been prepared by the reaction of dibenzoyltellurium diiodide (**74**) with diamines (Equation (23)) ⟨86IJC57⟩.

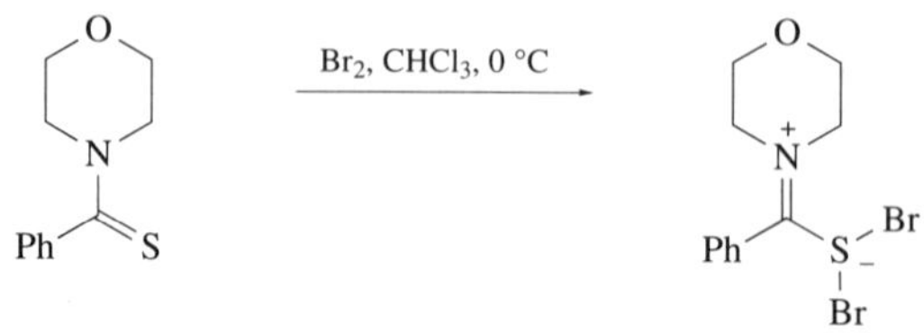

(**73**)

(22)

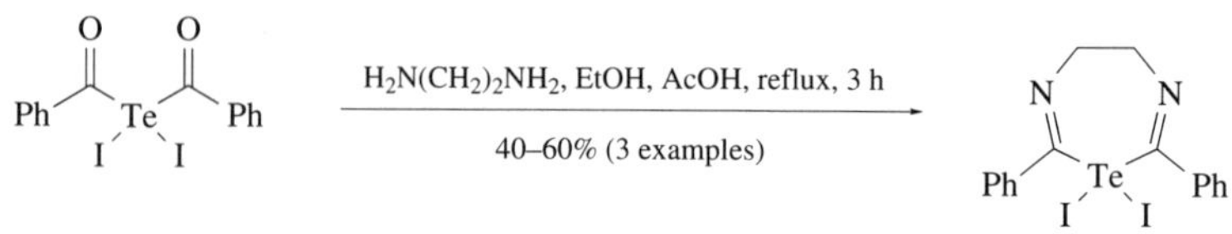

(**74**) (**75**)

(23)

5.18.1.6 *S*-Imidoyl Sulfenates and *S*-Imidoyl Sulfoxides, $R^1C(NR^2)SOR^3$ and $R^1C(NR^2)S(O)R^3$

A few *S*-imidoyl sulfenates (**77**) have been prepared by the *O*-alkylation of secondary thioamide *S*-oxides (**76**) with triethyloxonium tetrafluoroborate followed by aqueous sodium carbonate (Scheme 14) ⟨84RTC342⟩. The ethyl *S*-imidoyl sulfenates are obtained as a mixture of (*E*)- and (*Z*)-isomers. The isomeric *S*-imidoyl sulfoxides (**78**) have been obtained by deprotonation of thioamide *S*-oxides with sodium hydride followed by *S*-alkylation with an alkyl halide (Equation (24)) ⟨84RTC342⟩.

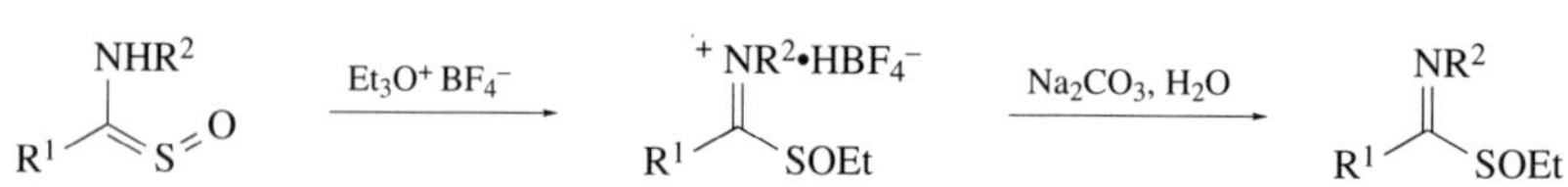

(**76**) (**77**)

Scheme 14

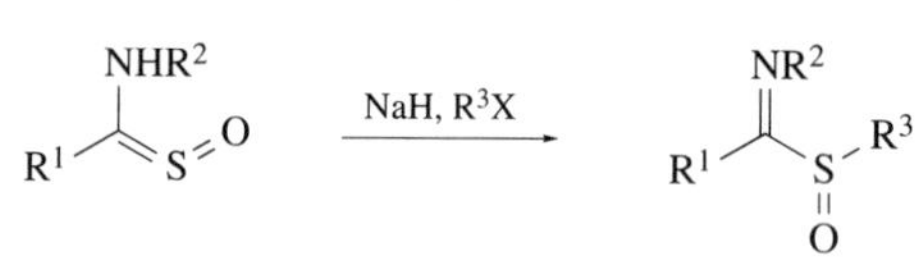

(**76**) (**78**)

(24)

5.18.1.7 Imidoyl Disulfides, $R^1C(NR^2)SSR^3$

There are two principal methods for the preparation of compounds of this type:

5.18.1.7.1 Imidoyl disulfides from the reaction of thioamides with sulfenylating reagents

The addition of sulfenyl chlorides to secondary thioamides provides a general route to imidoyl disulfides (**79**) (Scheme 15). The reaction is usually carried out, by adding the sulfenyl chloride to a solution of the thioamide in ether containing pyridine, and the products are isolated as a mixture of (*E*)- and (*Z*)-isomers in good yields ⟨69LA(727)22, 69LA(727)50, 73LA462⟩. Hydrochloride salts of imidoyl disulfides (**80**) obtained from the reaction of primary thioamides and sulfenyl chlorides have been isolated (Scheme 15) ⟨69LA(727)22⟩. In general, the free bases are unstable and decompose to nitriles and sulfur containing by-products ⟨69LA(727)22, 73LA462⟩. Other sulfenylating reagents also react with thioamides to give imidoyl disulfides: examples of reactions with *N*-thiosuccinimides ⟨77JAP73023033⟩, *N*-alkylpyridyl disulfides ⟨91JOC6697⟩ and tribenzenesulfenamide ⟨74JCS(P1)853⟩ have been reported.

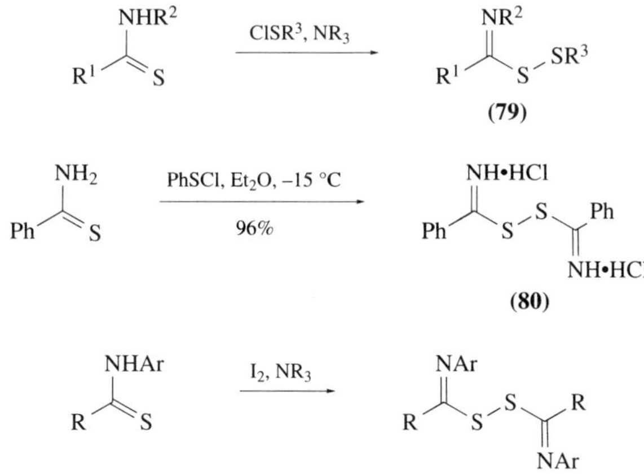

(**79**)

(**80**)

(**81**)

Scheme 15

5.18.1.7.2 Imidoyl disulfides from the reaction of thioamides with oxidising agents

The second method, which is applicable to the preparation of symmetrical imidoyl disulfides, involves oxidation of the corresponding thioamides under carefully controlled conditions. The products obtained on oxidation of thioamides have been reviewed by Hurd and DeLaMater ⟨61CRV45⟩. Imidoyl disulfides can be obtained depending on the reagent employed, the reaction conditions and the nature of the thioamide. Iodine is used most often as the oxidising agent in the presence of a tertiary amine (Scheme 15) ⟨67JOC392⟩. The reaction is a useful method for the preparation of bis(*N* arylimidoyl) disulfides (**81**). Electron-withdrawing substituents in the *para* position of the benzene ring enhance the stability of the disulfide to aerial oxidation ⟨67JOC392⟩. The *N*-alkyl thioamides *N*-methylthioacetamide (**82**) and piperidine-2-thione (**83**) gave tar with iodine and triethylamine ⟨80JCS(P1)665⟩. Oxidation of primary thioamides with iodine in general leads to the formation of1,2,4-thiadiazoles (**84**) (Equation (25)) ⟨61CRV45⟩, however there are reports of the isolation of disulfides under acidic conditions ⟨52AP(285)375, 69JIC1096⟩. Alternative oxidising agents have been used for this transformation including potassium hexacyanoferrate(III) ⟨37HCA691⟩, dibenzoyl peroxide ⟨57BSF633, 77RRC1111⟩, dead ⟨71BCJ1362⟩ and phenyl iodosodiacetate ⟨81MI 518-01⟩.

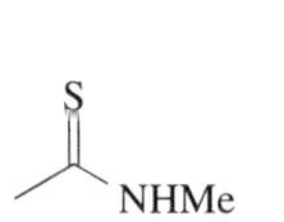

(**82**)

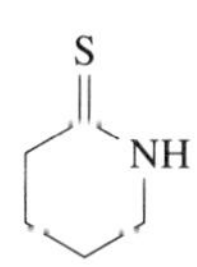

(**83**)

(**84**)

(25)

Some specific preparations of compounds of this class have been described. On heating to 200 °C in a sealed tube the bis(thiobenzoyl)hydrazine (**85**) undergoes Cope rearrangement to give imidoyl disulfide (**86**) in 33% yield (Equation (26)) ⟨90ZOR214⟩. Azetidinone imidoyl disulfides (**88**) have been prepared by heating penicillin sulfoxides (**87**) with thioamides (Equation (27)) ⟨79CP1051008⟩. 5-Arylimino-4-chloro-5*H*-1,2,3-dithiazoles (**89**) react with primary and secondary alkylamines in dichloromethane at room temperature to give imidoyl disulfides (**90**) (Equation (28)) ⟨93JOC7001⟩. Imidoyl disulfides (**91**) containing at least one hydrogen on the α-carbon atom exist in tautomeric equilibrium with aminovinyl disulfides (**92**) (Equation (29)) ⟨73LA462, 74LA776⟩. The position of the equilibrium depends on the substituents, when R^3 is an electron-withdrawing group (**91**) is stabilised, whereas (**92**) is stabilised when R^1 or R^2 are electron-withdrawing.

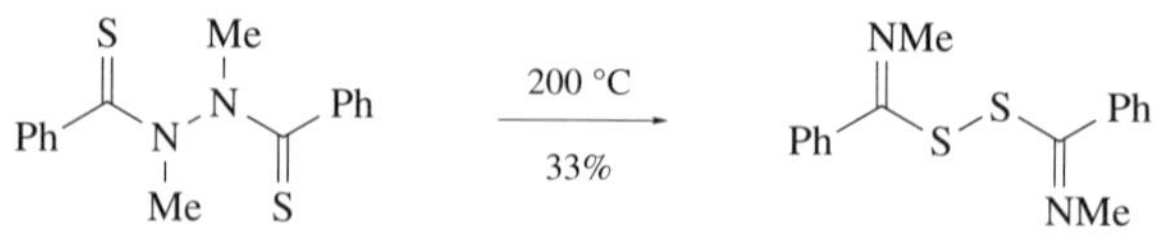

(**85**) (**86**) (26)

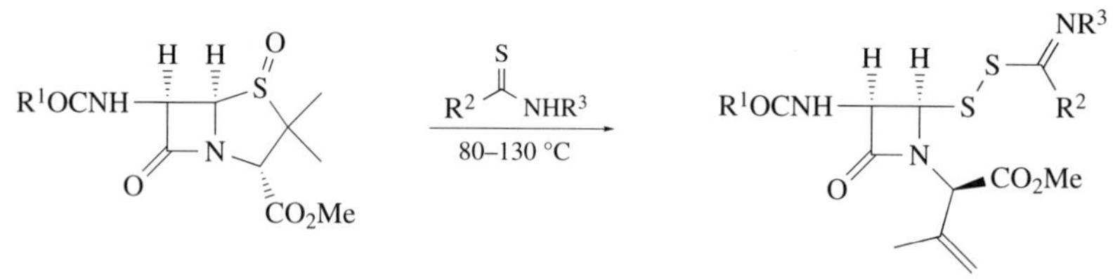

(**87**) (**88**) (27)

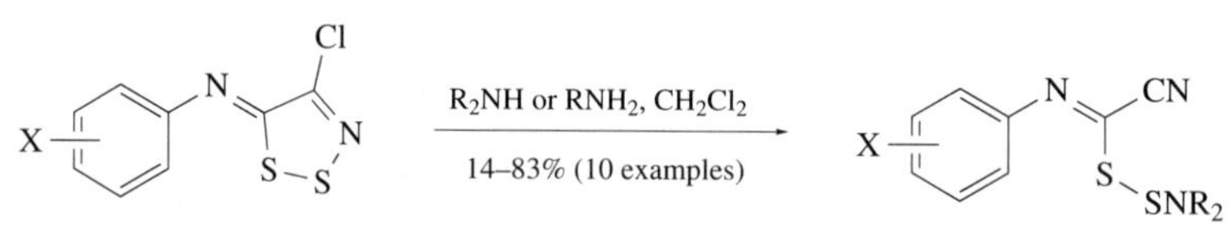

(**89**) (**90**) (28)

(**91**) (**92**) (29)

5.18.1.8 *S*-Imidoyl Sulfenamides and Related Structures, $R^1C(NR^2)SNR^3R^4$ and $R^1C(NR^2)SN{=}CR^3$

The parent compounds in this class ($R^3 = R^4 = H$) have also been called *S*-imidoyl hydrosulfamines. The principal method for the preparation of compounds of this type is the *S*-amination of secondary thioamides, a method which was first described in 1971 ⟨71LA(753)169, 71USP3631071⟩. The amination of thiobenzanilide with hydroxylamine-*O*-sulfonic acid in alkaline solution leads to the formation of *S*-imidoyl sulfenamide (**93**) in 94% yield ⟨71USP3631071, 72JOC3820⟩. This compound can be recrystallised from cyclohexane and stored at 4 °C. It decomposes to sulfur and benzamidine on storage at 24 °C. Experimental details have been given for the preparation of (**93**) and the thermally unstable analogue (**94**) (88%) by this method (Scheme 16). These *S*-imidoyl sulfenamides react with isocyanates to form stable ureas (**95**) and with salicylaldehyde to form the Schiff base (**96**) (Scheme 16) ⟨72JOC3820⟩. In an alternative procedure several secondary thioamides have been treated with chloroamine to give *S*-imidoyl sulfenamides ⟨71LA(753)169⟩. The reaction of imidoyl sulfenates (**97**) with primary or secondary amines in carbon tetrachloride at room temperature has been used as a method of preparation for a few compounds of this class (Scheme 16) ⟨84RTC342⟩.

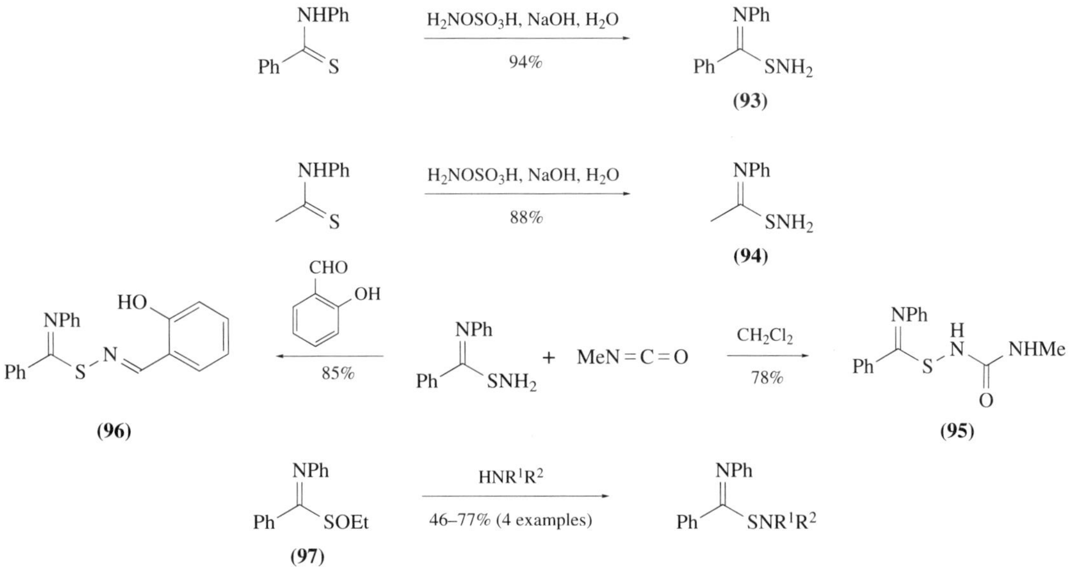

Scheme 16

5.18.1.9 Imidoylthiophosphorus, -Arsenic, -Antimony and -Bismuth Derivatives, $R^1C(NR^2)SP$, $R^1C(NR^2)SAs$, $R^1C(NR^2)SSb$ and $R^1C(NR^2)SBi$

5.18.1.9.1 Imidoyldithiophosphates, thiophosphates and dithiophosphinates

The three most important methods available for the synthesis of imidoylthiophosphorus derivatives are analogous to methods used for the preparation of thioimidic esters (see 5.18.1.2).

(i) S-Imidoylthiophosphites from the reaction of thioamides with phosphorylating reagents

The reaction of primary or secondary thioamides with phosphorylating reagents in the presence of a tertiary amine has been used to prepare *S*-imidoylthiophosphites (**98**) ⟨85ZOR559⟩ and *S*-imidoylthiophosphoramidates (**99**) ⟨86ZOR1654⟩ (Equations (30) and (31)).

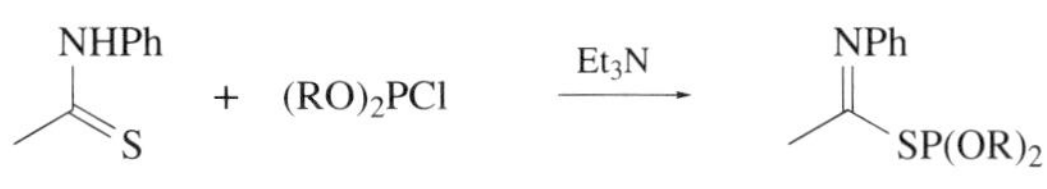

(30)

(98)

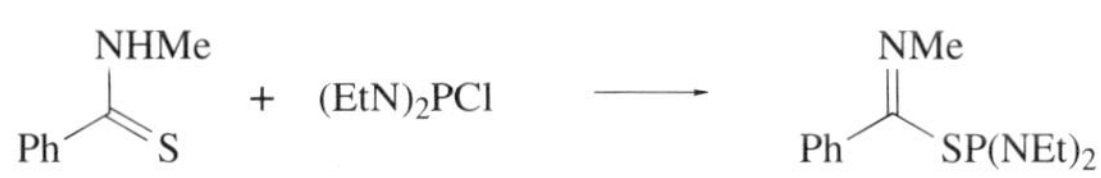

(31)

(99)

(ii) S-Imidoyldithiophosphates from the reaction of nitriles and isonitriles with dialkyl dithiophosphates

The imidoyldithiophosphate (**100**) has been isolated in 46% yield from the reaction of diisopropyldithiophosphate with benzonitrile (Equation (32)) ⟨73DOK(211)113⟩. Aryl isonitriles react exothermically with dialkyl dithiophosphates to give crystalline imidoyldithiophosphate derivatives (**102**) in good yields (Scheme 17) ⟨77JGU916⟩. The reaction is believed to proceed through intermediate (**101**), which has not been isolated and reacts with excess isonitrile to give the product.

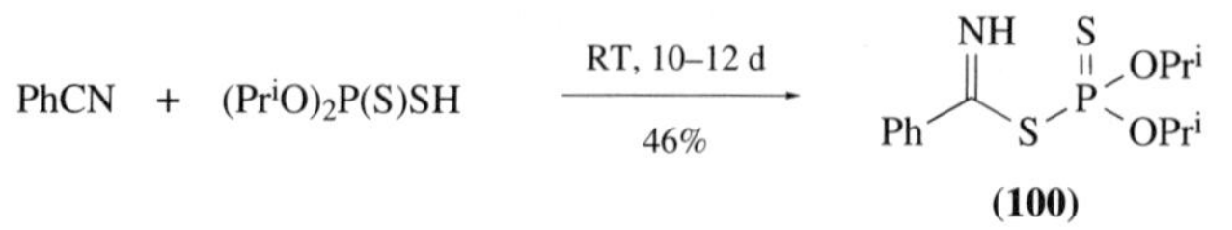 (32)

(100)

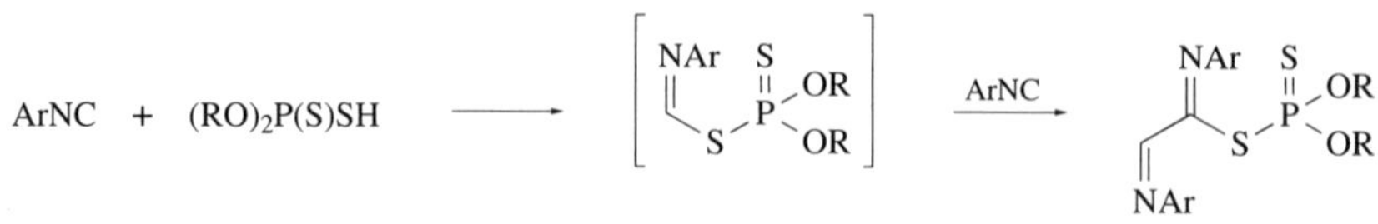

(101) **(102)**

Scheme 17

(iii) S-*Imidoyldithiophosphates, thiophosphates and dithiophosphinates from imidoyl chlorides*

There are several examples of the formation of imidoyl thiophosphorus compounds by the displacement of chloride from imidoyl chlorides with salts of dithiophosphoric or dithiophosphinic acids ⟨59USP3053876, 79JGU2323, 87EUP243970, 90JGU1389⟩. Reaction of imidoyl chloride (**103**) with sodium diethyl dithiophosphate (**104**) leads to the formation of a tautomeric mixture of the imidoyl dithiophosphate (**105**) and the *N*-phosphorylated thioamide (**106**) (Scheme 18) ⟨79JGU2323⟩. The equilibrium is shifted to the imidoyl tautomer (**105**) with increased temperature and decreased solvent polarity ⟨79JGU2323⟩. The reaction of the imidoyl chloride (**103**) with sodium diethylthiophoshate (**107**) occurs at sulfur to give a tautomeric mixture of the imidoylthiophosphate (**108**) and *N*-phosphorylated thioamide (**109**) in which the thioamide tautomer predominates (Scheme 19) ⟨79JGU2323⟩. Imidoyldithiophosphinate (**112**), prepared from potassium diphenyl dithiophosphinate (**111**) and imidoyl chloride (**110**), is stable and does not undergo migration of the phosphinyl group from sulfur to nitrogen (Equation (33)) ⟨90JGU1389⟩.

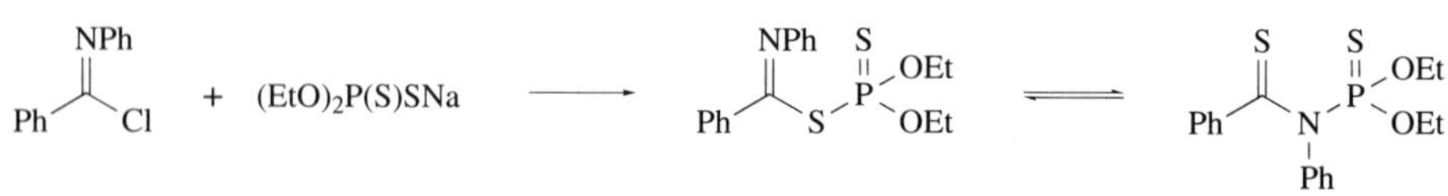

(103) **(104)** **(105)** **(106)**

Scheme 18

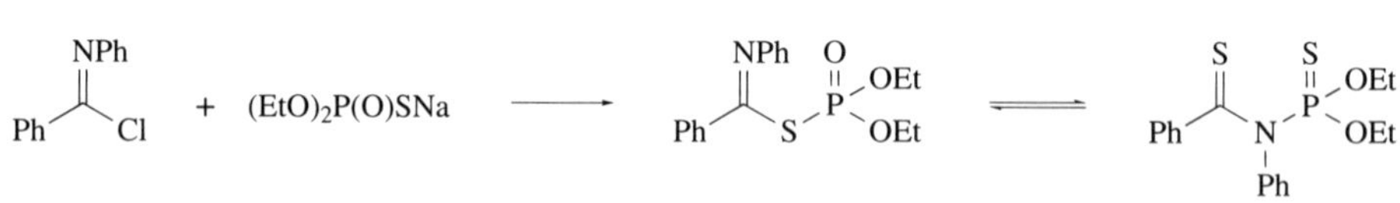

(103) **(107)** **(108)** **(109)**

Scheme 19

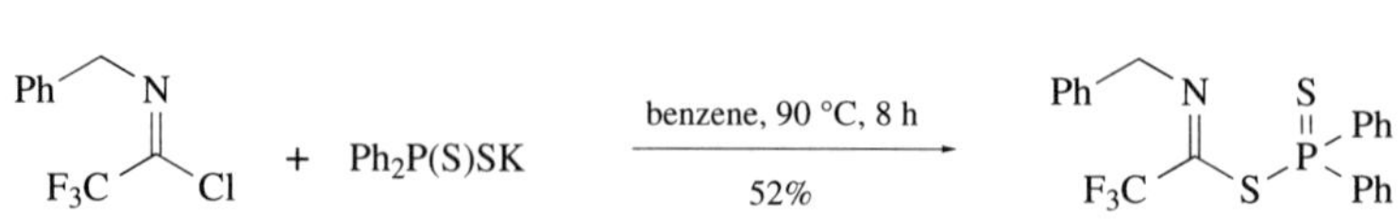 (33)

(110) **(111)** **(112)**

5.18.1.9.2 Imidoylthioarsenic, -antimony and -bismuth derivatives, $R^1C(NR^2)SAs$, $R^1C(NR^2)SSb$ and $R^1C(NR^2)SBi$

The preparation of these compounds has not been reported in the literature.

5.18.1.10 Imidoylthiometalloid Derivatives, $R^1C(NR^2)SSi$, $R^1C(NR^2)SB$ and $R^1C(NR^2)SGe$

Imidoylthiometalloid compounds have been reported where the metalloid is silicon or boron. Silylation of metallated secondary thioamides with chlorotrimethylsilane leads to an equilibrium mixture of *N*-(trimethylsilyl)thioamides (**113**) and *S*-(trimethylsilyl)thioimidic esters (**114**) (Scheme 20) ⟨75LA1808⟩. An analogous equilibrium has been found, where the metalloid is boron, with the reaction of *N*-trimethylsilylthiobenzanilide (**115**) with dibromophenylborane (**116**) (Scheme 21) ⟨77JOM(141)241⟩.

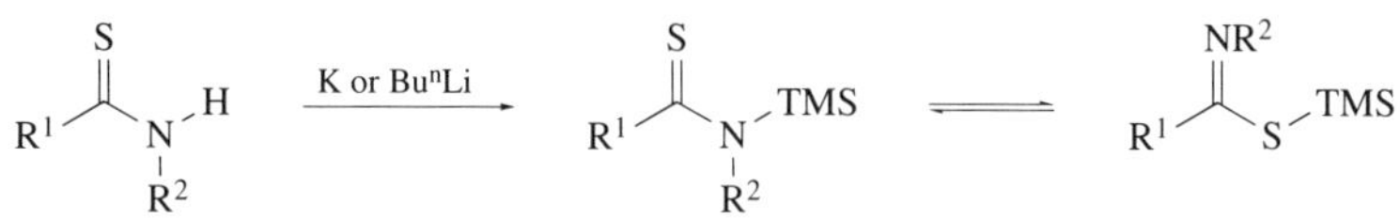

(**113**) (**114**)

Scheme 20

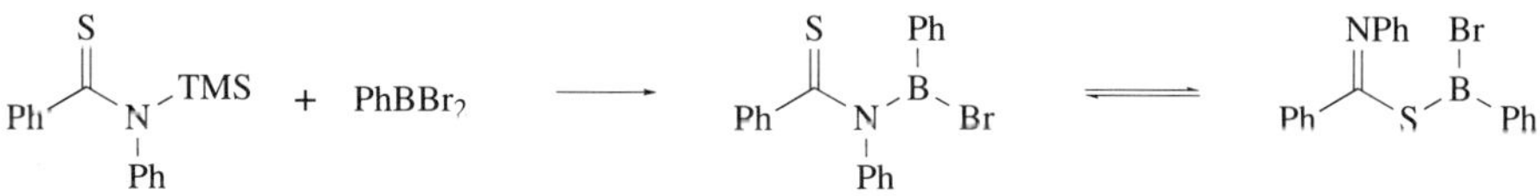

(**115**) (**116**)

Scheme 21

5.18.1.11 Imidoylthiometals, $R^1C(NR^2)SMetal$

Imidoylthiometals have been reported were the metal is mercury or tin.

5.18.1.11.1 S-Imidoyl mercury derivatives $R^1C(NR^2)SHg$

The symmetrical thiolide (**118**) has been prepared in 70% yield by the reaction of mercury-bis(phenylalkyne) (**117**) with thiobenzanilide (Equation (34)) ⟨70CB32⟩.

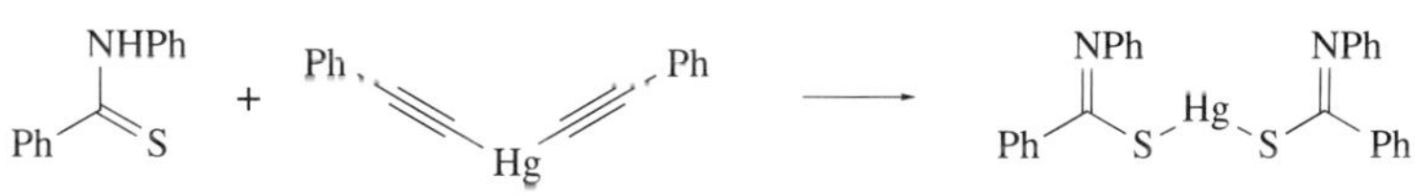

(**117**) (**118**) (34)

5.18.1.11.2 S-Imidoyl tin derivatives $R^1C(NR^2)SSn$

Several *N*-arylthioamides have been converted into *S*-(triphenylstannyl)thioimidic esters by reaction with an equimolar amount of triphenyltin iodide and a slight excess of triethylamine in refluxing acetonitrile, generally in good yield ⟨75JOM(97)39⟩. The method is exemplified by the preparation of the thioimidic ester (**119**) (96%) from thioacetanilide and triphenyltin iodide (Equation (35)). In a similar reaction the *S*-(tri-*n*-butylstannyl)thioimidic ester (**121**) has been prepared from thioamide (**120**), tri-*n*-butyltin chloride and sodium hydride (Equation (36)) ⟨89JA1886⟩. The reaction of (stannylethynyl)amines (**122**) with aryl isothiocyanates has been used to prepare *S*-stannylthioimidic esters (**123**) (Equation (37)) ⟨78TL1951, 85LA1669⟩. The *S*-(trimethylstannyl)thioimidic ester (**124**) has been prepared in 60% yield by the reaction of phenylisothiocyanate with methylmagnesium chloride followed by stannylation with trimethyltin chloride (Equation (38)) ⟨76JOM253⟩.

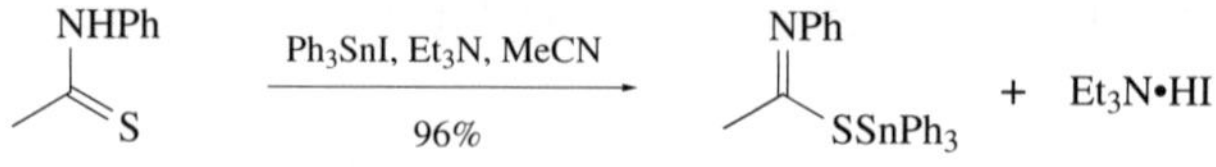

(119)

(35)

(120) **(121)**

(36)

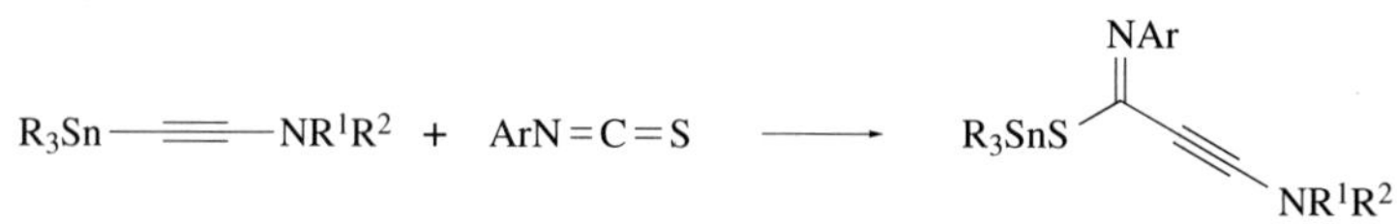

(122) **(123)**

(37)

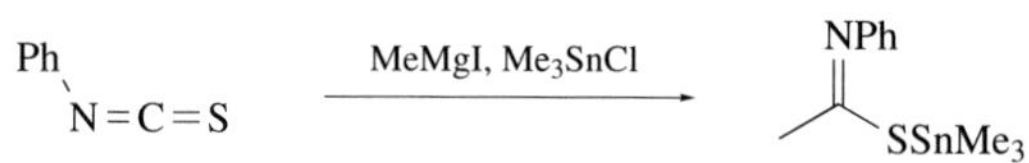

(124)

(38)

5.19
Amidines and *N*-Substituted Amidines

PETER J. DUNN
Pfizer Central Research, Sandwich, UK

5.19.1 AMIDINES

5.19.1.1 Introduction and General Methods

5.19.1.1.1 Introduction

In this chapter the synthesis of amidines is reviewed. This review is restricted to compounds of structural formula (**1**) where R^1 is hydrogen or a carbon radical. Thus guanidines ($R^1 = NR^5_2$, etc.) and haloamidines are excluded along with cyclic amidines such as (**2**), which are covered in *Comprehensive Heterocyclic Chemistry* ⟨84CHEC⟩. Some cyclic amidines may be included when the method of synthesis is likely to be applicable to acyclic compounds.

(**1**) (**2**)

There have been five previous reviews of amidine synthesis ⟨44CRV351, 58HOU(9)39, B-75MI 519-01, 83RCR377, B-91MI 519-01⟩. (The review by Boyd includes haloamidines.) A smaller chapter in *Houben-Weyl* ⟨85HOU(E5)1305⟩ and related reviews on imidic ester chemistry ⟨61CRV179, B-75MI 519-02⟩ also give useful information.

5.19.1.1.2 General methods

The most common methods of preparing amidines are from nitriles, amides or thioamides. All of these syntheses involve disconnection of the product to an iminium cation synthon and a nitrogen nucleophile.

Nitriles are most commonly activated using the classical Pinner reaction to form the intermediate imidic ester salt ⟨61CRV179⟩. The nitrile is treated with an alcohol under anhydrous conditions in the presence of hydrogen chloride or hydrogen bromide. The reaction is normally carried out in a chlorinated, ethereal or aromatic solvent. Subsequent reaction of the imidic ester salt with primary or secondary amine or with an alcoholic solution of ammonia furnishes the amidine, normally in excellent yield. The Pinner reaction has been very widely used to prepare primary amidines.

A modification of the Pinner synthesis involves the formation of thioimidic ester (**3**) (Scheme 1). Reaction of the isolable thioimidic ester with aromatic amines or the acetate salts of aliphatic amines or ammonium acetate reliably forms amidines. The thioimidic ester method can be used with acid sensitive substrates, such as those with an *N*-BOC group ⟨81JOC2455⟩. This method also works for hindered aryl cyanides where the Pinner procedure fails.

In addition to activating the nitrile, it is also possible to activate the nitrogen nucleophile as its alkali metal salt, for example, benzyl cyanide and the lithium derivative of aromatic amines yield *N*-aryl amidines ⟨72ZOR1246⟩.

A method which has been very widely used to prepare *N*-substituted or *N*,*N*-disubstituted amidines

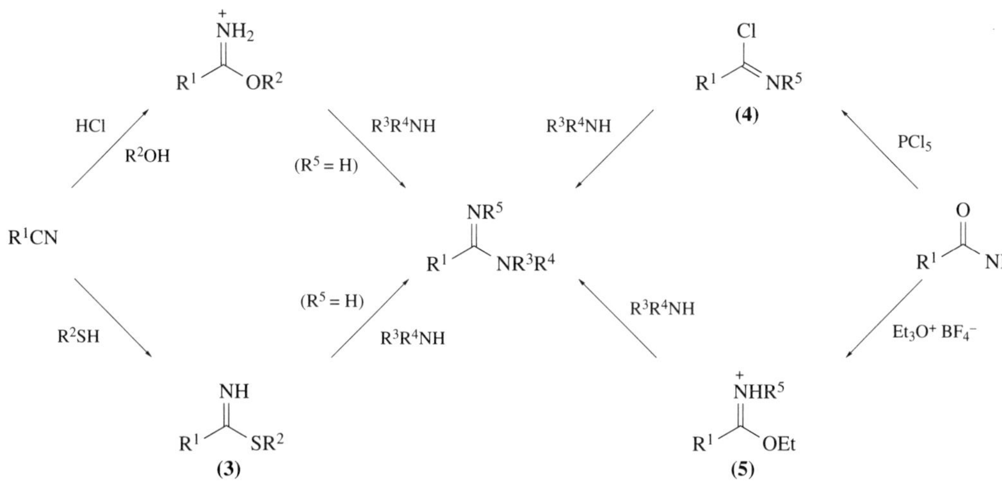

Scheme 1

is to heat the nitrile with a primary or secondary amine in the presence of aluminum chloride ⟨58HOU(9)39⟩.

More highly substituted amidines are prepared from amides or thioamides and these are often the best starting materials from which to prepare trisubstituted amidines. Activation of the mono-substituted amide to give the imidoyl chloride (**4**) is best achieved with phosphorus pentachloride or with oxalyl chloride though a variety of chlorinating agents such as thionyl chloride, phosphorus oxychloride and phosgene can be used. The imidoyl chloride is reacted with a wide variety of mono- or disubstituted alkyl or aryl amines to produce amidines. In many cases, excellent yields of amidines can simply be obtained by heating the monosubstituted amide, amine and phosphorus pentachloride in an aromatic solvent. Though in some cases where the amine is subject to attack by the phosphorus chloride it is best to isolate the imidoyl chloride.

An alternative activation of the amide involves alkylation and then displacement of the resulting imidic ester salt (**5**).

Methods of preparation in which *both* carbon–nitrogen bonds are formed include the synthesis of amidines from *ortho*-esters, acetals or thioesters. The reaction of *ortho*-esters (**6**) with primary aromatic and aliphatic amines gives symmetrically disubstituted amidines (Scheme 2) ⟨54JA2411, 63JOC1108⟩. Symmetrically disubstituted amidines are also the products formed when thioesters (**7**) are reacted with primary amines under acid catalysed conditions ⟨76BSF301⟩.

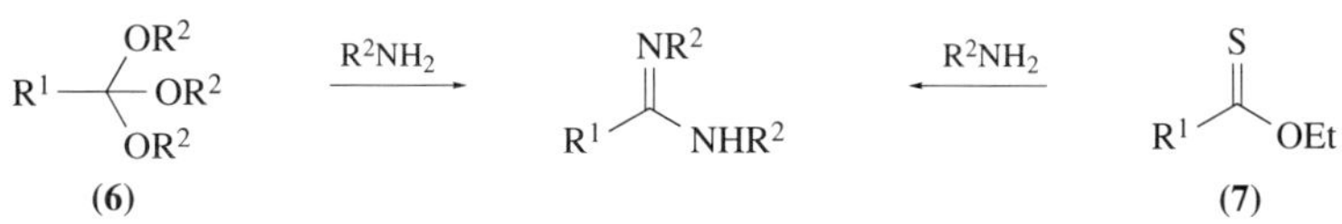

Scheme 2

The best method of preparing an *N*,*N*-disubstituted or a trisubstituted formamidine is usually by the high yielding reaction of a formamide acetal, such as (**8**) with a primary amine or ammonia. The formamide acetal is also an excellent precursor to *N*-heteroatom substituted formamidines.

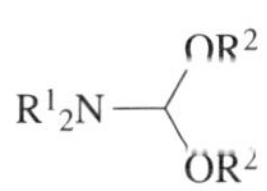

(8)

The synthesis of amidines via carbon–carbon bond formation is achieved by adding Grignard or organolithium reagents to carbodiimides or to cyanoamides followed by hydrolysis of the resulting metallated intermediate.

Substituted amidines can be prepared by *N*-alkylation of simpler amidines. Best results are obtained by exhaustive alkylation of primary amidines (RC(NH)NH_2) to give a trialkylated amidine ⟨23JCS3359⟩ or alkylation of a symmetrical *N*,*N*′-disubstituted amidine. Primary amidines can also be selectively monoalkylated with care ⟨1878CB4, 47JCS738⟩. Less successful are alkylation reactions of monosubstituted amidines and unsymmetrical *N,N′*-disubstituted amidines as these often lead to mixtures of alkylated products ⟨23JCS3359⟩. An exception to the rule are *N*-arylamidines which are alkylated selectively.

5.19.1.2 Formamidines ($HC(NR^1)NR^2{}_2$)

5.19.1.2.1 Preparation from formamides and thioformamides

(i) Formamidines from monosubstituted formamides

Di- or trisubstituted formamidines are readily prepared from monosubstituted formamides. A series of symmetrical and unsymmetrical diaryl substituted formamidines were prepared by Mandel using PCl_5 as chlorinating agent (Equation (1)) ⟨54JA3978⟩.

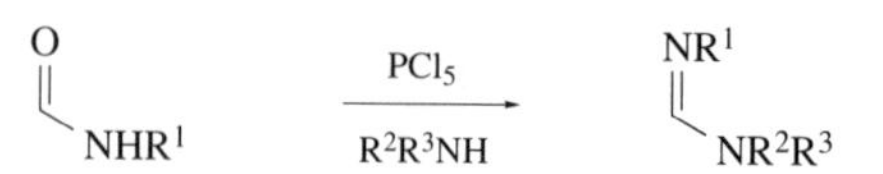

(1)

Over 40 different trialkyl substituted formamidines have been prepared by Gaetzi and Fischer using either PCl_5 or dimethyl sulfate as the activating agent ⟨72SZP563109⟩. If phosgene is used as the chlorinating agent then *N*-(dichloromethyl)formamidines (**9**) are formed (Equation (2)) ⟨64CB1361⟩.

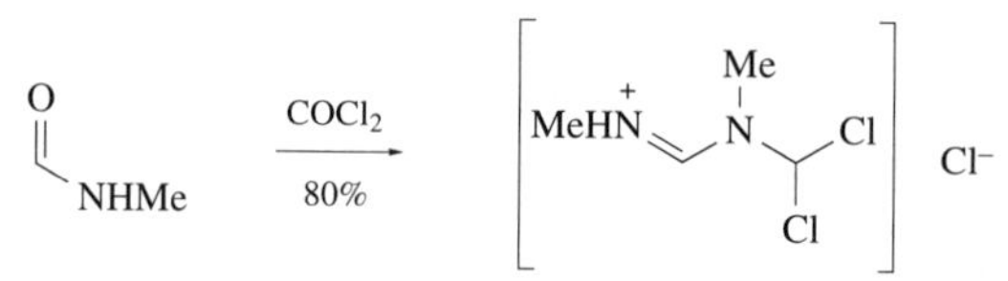

(2)

(9)

A variant of this method is the reaction between a monosubstituted formamide with a carbamoyl chloride in toluene at about 80 °C. The formamidine is isolated as its hydrochloride salt after loss of carbon dioxide. Examples of formamidines prepared by this method are found in Table 1 ⟨83S35⟩.

Table 1 The reaction of formamides with carbamoyl chlorides.

Starting formamide	*Starting carbamoyl chloride*	*Product*	*Yield (%)*
$HCONHMe$	$(Bu^n)_2NCOCl$	$HC(NMe)N(Bu^n)_2$	83
$HCONHPr^n$	$(Bu^s)_2NCOCl$	$HC(NPr^n)N(Bu^s)_2$	80
$HCONHC_6H_{11}$	Et_2NCOCl	$HC(NC_6H_{11})NEt_2$	71
$HCONH(n\text{-}C_{18}H_{37})$	$(Bu^n)_2NCOCl$	$HC[N(n\text{-}C_{18}H_{37})]N(Bu^n)_2$	68
$HCONH(n\text{-}C_{18}H_{37})$	Me_2NCOCl	$HC[N(n\text{-}C_{18}H_{37})]NMe_2$	73

Symmetrical diaryl formamidines can be prepared by heating the arylformamide with dimethyl sulfate in benzene or THF. The product is initially isolated as the methosulfonate salt as shown in Equation (3) ⟨59CB837⟩.

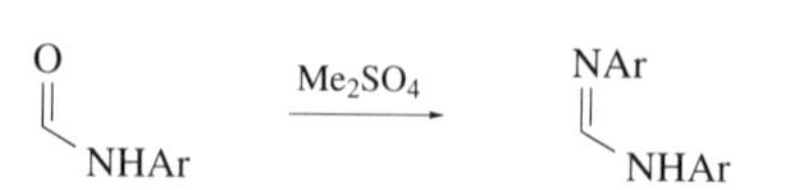

(3)

(ii) Formamidines from disubstituted formamides, Vilsmeier reagents and disubstituted thioformamides

Trisubstituted formamidines have been widely prepared from Vilsmeier reagents, most commonly from dimethylformamide and phosphorus oxychloride. The reaction of the Vilsmeier salt, derived from DMF, with both aromatic and aliphatic amines generally proceeds in good yield ⟨59CB837⟩. Examples of trisubstituted formamidines prepared by this method are given in Table 2. In addition to these examples some 20 further formamidines have been prepared using Vilsmeier intermediates ⟨77MI 519-01⟩.

Table 2 Preparation of formamidines from Vilsmeier-type reagents.

Formamide	*Product*	*Yield* (%)	*Ref.*
$HCONMe_2$	$HC(NPh)NMe_2$	85	59CB837
$HCONMe_2$	$HC(NTol)NMe_2$	71	59CB837
HCON(Me)Ph	HC(NPh)N(Me)Ph	56	59CB837
$HCONMe_2$	$HC(NC_6H_{11})NMe_2$	63	59CB837
$HCONEt_2$	$HC[N(n\text{-}C_5H_{11})]NEt_2$	40	77MI 519-01
$HCONPh_2$	$HC(NPh)NPh_2$	25	77MI 519-01
$HCONBu_2$	$HC(NPh)NBu_2$	54	54JA3978

Vilsmeier salts will also react directly with acetylated arylamines to give a trisubstituted formamidine and acetyl chloride ⟨77ACS(B)261, 79S546⟩. A very convenient procedure is to dissolve the acetanilide in DMF at 100 °C. One equivalent of phosphorus oxychloride is added dropwise to the solution resulting in simultaneous distillation of acetyl chloride from the reaction ⟨80S883⟩.

In the preparation of trisubstituted formamidines, it is reported to be more favourable to start with a disubstituted formamide and a primary amine as opposed to a monosubstituted formamide and a secondary amine (Scheme 3). Hence reactions A and B are much higher yielding than the alternative routes C and D where *N,N'*-diphenylformamidine is formed as a major side product ⟨54JA3978⟩.

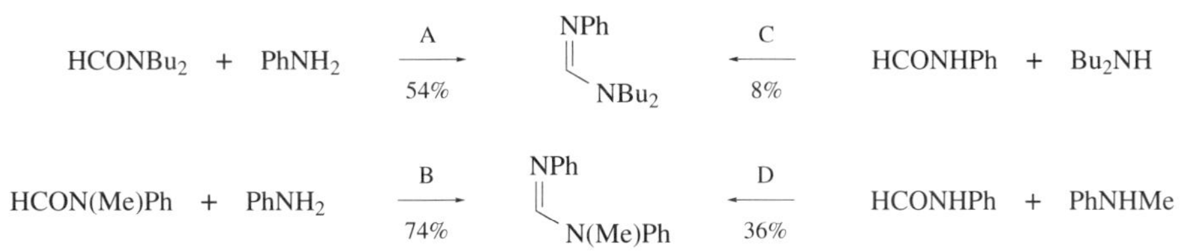

Reaction conditions: (A to D) Formamide (1 equiv.), amine (1 equiv.), PCl_5 (1.2 equiv.) in boiling chloroform (3 h)

Scheme 3

In addition to activating the formamide it is also possible to activate the nitrogen nucleophile; hence tetrakis(dimethylamino)titanium ($(Me_2N)_4Ti$) undergoes reaction with monosubstituted formamides to give *N,N*-dimethyl-*N'*-aryl or *N,N*-dimethyl-*N'*-alkylformamidines ⟨71JOC1613⟩.

The reaction of dimethylformamide with amines in the presence of sodium methoxide gives *N*-formylated amines ⟨61JOC2563⟩, however dimethylthioformamide reacts directly with 1-naphthylamine to give the formamidine (**10**) in greater than 90% yield in the presence or absence of sodium methoxide (Scheme 4) ⟨65CJC2640⟩.

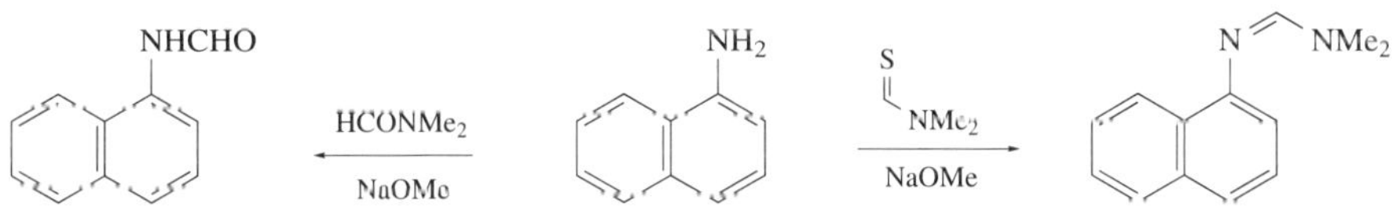

(10)

Scheme 4

Dimethylformamide undergoes reaction with both phenyl isocyanate and 1-naphthyl isocyanate to give trisubstituted formamidines ⟨60JOC2245, 65CJC2640⟩.

5.19.1.2.2 Formamidines, prepared by reduction of carbodiimides and ureas

1,3-Dialkyl ureas are reduced to formamidines by the action of sodium borohydride in pyridine ⟨69TL699⟩; yields of 50–80% were obtained but these conditions were not successful with

1,1,3-trialkyl ureas. In work done in the early 1960s, lithium aluminum hydride in benzene–ether or toluene–*n*-butyl ether was used to reduce 1-aryl-3,3-dialkyl ureas (16 examples) and 1,1,3-trialkyl ureas (nine examples) to the corresponding formamidines in moderate yield ⟨64JOC3697⟩. The preparation of formamidines by reduction of ureas is complicated by overreduction leading to amine formation. A much milder and higher yielding synthesis of formamidines by reduction, is from carbodiimides. These compounds are reduced by sodium borohydride in warm alcohols ⟨78CPB2246⟩ or by hydrogenation ⟨65CB2128⟩. In both of these methods nearly quantitative yields are obtained (Scheme 5).

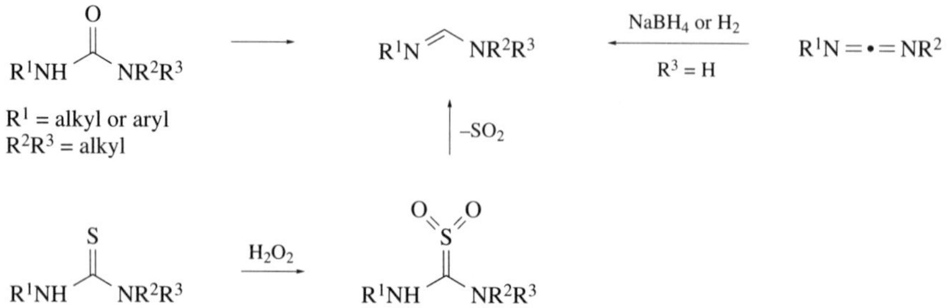

(11)

Scheme 5

Trisubstituted formamidines can be prepared from trisubstituted thioureas. The overall reaction is a reduction though the reagent is hydrogen peroxide. The reaction proceeds via the *S*,*S*-dioxide (**11**) followed by expulsion of sulfur dioxide ⟨69CB2640⟩.

5.19.1.2.3 *Formamidines from* ortho-*formates, acetals and aminals*

Symmetrical disubstituted formamidines can be prepared by heating ethyl *ortho*-formate with 2 moles of either an aliphatic ⟨63JOC1108⟩ or aromatic primary amine ⟨49JOC277⟩. In the case of an aliphatic amines a catalytic amount of acid (usually acetic acid) is required. In both cases, very high yields are obtained, though diarylformamidines have been shown to react with ethyl *ortho*-formate in the presence of acetic acid to give ethyl *N*-arylformimidates ⟨49JA3848⟩.

Since the early 1970s an excellent and very widely used method of preparing formamidines has been by the reaction of an amine with a formamide acetal such as (**12**). The chemistry of formamide acetals, including their use in preparing formamidines, has been reviewed by Abdulla and Brinkmeyer ⟨79T1675⟩. The method gives very high yields under mild reaction conditions and even works well for weakly basic amines; hence the room temperature reaction of 2-amino-3-cyano-5-methylpyridazine (**13**) with (**12**) in DMF proceeded in greater than 90% yield (Equation (4)) ⟨77JOC1523⟩, see also ⟨80JOC2485⟩.

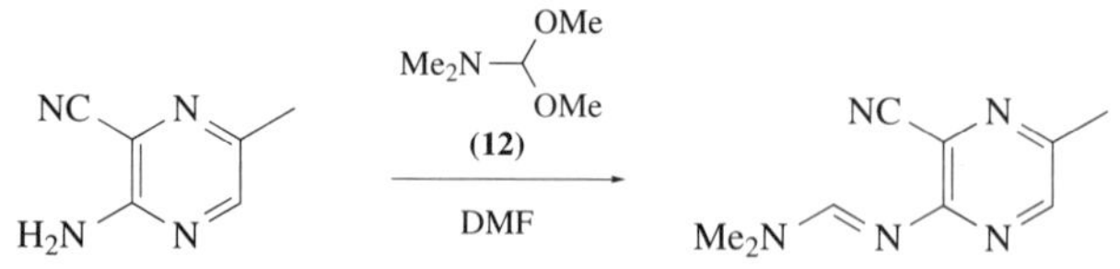

(13)

(4)

There has been some synthetic interest in preparing formamidines from α-amino acids (Scheme 6). In the original procedure, the free amino acid was heated at reflux in dimethylformamide dimethyl acetal to give the formamidine ester (**14**) directly ⟨77JOC2639⟩. Subsequently it was discovered that this method leads to complete loss of stereochemical integrity of the α-centre. This problem is overcome by the initial preparation of the amino acid ester followed by reaction with dimethylformamide dimethylacetal under mild conditions and at room temperature ⟨84TL3651⟩. This procedure also gives enantiomerically pure formamidine from the reaction of stereochemically labile phenylglycine methyl ester with (**12**). Chiral formamidines derived from isoquinoline have been prepared by formamidine exchange as shown in Equation (5); hence treatment of tetrahydroisoquinoline with chiral formamidine (**15**) resulted in an exchange to optically pure (**16**) with concomitant loss of dimethylamine ⟨87T5095⟩.

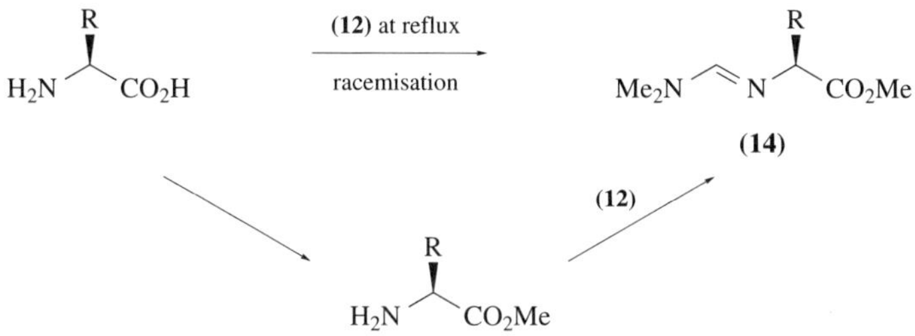

Scheme 6

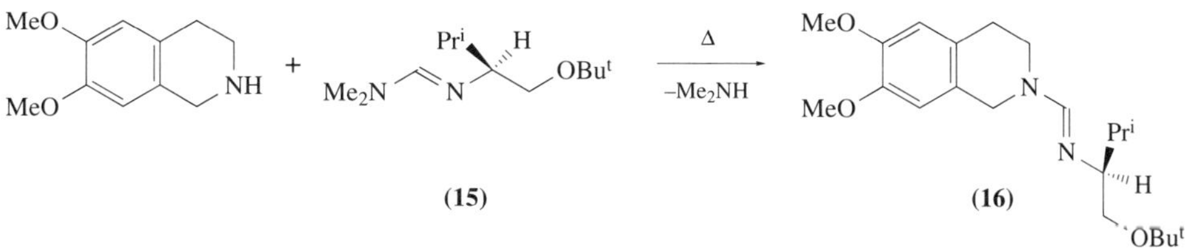

(5)

In addition to the acetals, another form of activated amide is the ester aminal ($HC(OBu^t)(NMe_2)_2$). This compound can react directly with *N* formyl secondary alkyl or aryl amines to give formamidines with DMF as the by-product although high temperatures (160 °C) are required ⟨80LA344⟩.

5.19.1.2.4 Formamidines from 1,3,5-triazine

A high yielding method of preparing symmetrically *N,N'*-disubstituted formamidines is to heat 6 moles of primary aromatic amine with 1 mole of 1,3,5-triazine. The reaction is successful in a variety of solvents or under neat conditions. With primary aliphatic amines the reaction is extremely facile ⟨55JA6559⟩. In the early 1970s, it was demonstrated that three component reactions are possible to form *N*-cyanoformamidines; hence a variety of secondary amines (**17**) undergo reaction with cyanamide and 1,3,5-triazine (**18**) to give cyanoformamidines (**19**). In the absence of a secondary amine the parent cyanoformamide (**19**; $R^1 = H$, $R^2 = H$) is formed in 91% yield (Equation (6)) ⟨72T4877⟩.

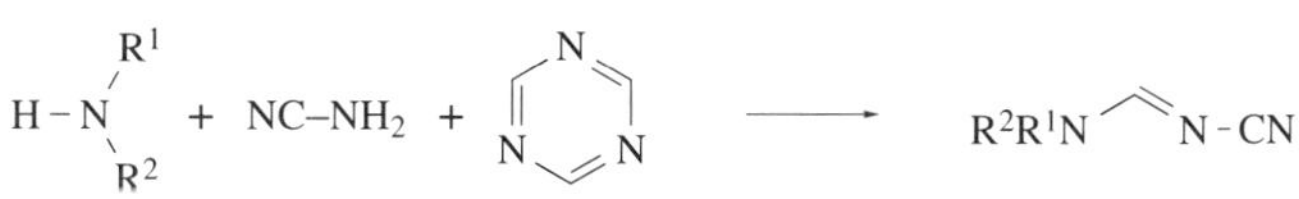

(6)

(**17**) (**18**) (**19**)

$R^1, R^2 = Me, Et$
$R^1 = Me, R^2 = Ph$
$R^1 = Me, R^2 = Tol$

5.19.1.2.5 Formamidines from isonitriles

Formamidines can be prepared by the reaction of isonitriles with primary or secondary amines (Equation (7)). The reaction is catalysed by Ag(I), Cu(I), Zn(II) or Cd(II) salts. For aliphatic isonitriles, a catalyst is essential to form any formamidine ⟨69BCJ3310, 80TL1459⟩. It has been known since the beginning of the twentieth century that aryl isonitriles undergo reaction with amines without a catalyst ⟨1892LA(270)267⟩; however, the use of a catalyst permits the reaction under very mild conditions. Thus the reaction of phenyl isonitrile with piperidine, catalysed by less than 1 mol% of CuCl, gives a 96% yield of formamidine after just 0.3 h at 0 °C ⟨69BCJ3310⟩. The use of a catalyst also allows relatively unstable (*Z*)-formamidines to be prepared. A series of aryl isonitriles

(**20**; Ar = Ph; *o*-, *m*- and *p*-ClC_6H_4; *o*- and *p*-$NO_2C_6H_4$) were treated with catalytic silver chloride and one mole of a secondary amine at −15°C to give the (*Z*)-formamidines (**21**). This product, characterised by infrared and NMR spectroscopy can be converted to the more stable (*E*)-isomer (**22**) by heating in boiling chloroform for 6 h or by treatment with acid at room temperature (Scheme 7) ⟨80CC130, 80TL885⟩.

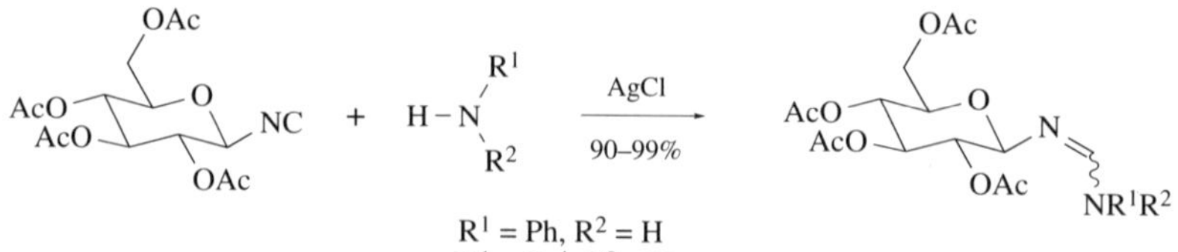

(7)

R^1 = Ph, R^2 = H
R^1 = Bu^i, R^2 = H
R^1R^2 = Et
HNR^1R^2 = piperidine

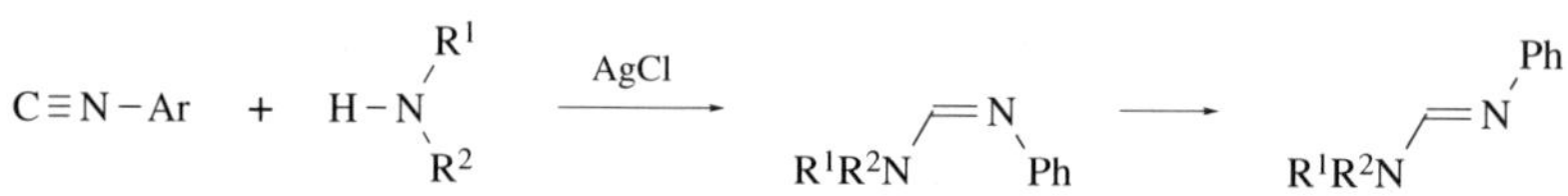

(**20**) (**21**) (*Z*)-Formamidine (**22**) (*E*)-Formamidine

R^1, R^2 = Me, Et
or $-(CH_2)_n-$, *n* = 2 to 5

Scheme 7

5.19.1.3 Aliphatic Amidines $R^1C(NR^2)NR^3_2$ (R^1 = alkyl, allyl, propargyl, etc.)

Nearly all of the methods described in this section are equally applicable to aromatic amidines. Only when there are important differences will the syntheses of aromatic amidines be discussed separately in Section 5.19.1.4.

5.19.1.3.1 Aliphatic amidines from nitriles

In 1877, Pinner described the synthesis of amidines from nitriles via the imidic ester (Scheme 8) ⟨1877CB1889⟩. This classical procedure is a reliable method of preparing a wide variety of amidines and is used extensively. The Pinner procedure is the most widely used method of preparing primary amidines ⟨41OSC(1)5, 59JOC1309, 67JMC1123⟩.

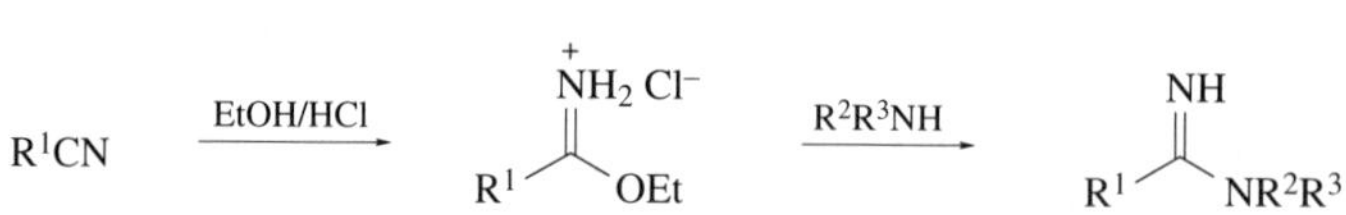

R^1 = alkyl or aryl

Scheme 8

The preparations of monosubstituted and *N*,*N*-disubstituted amidines via the imidic ester are tabulated in *Houben-Weyl* ⟨58HOU(9)39⟩. Reactions of imidic ester with weakly basic amines, such as 2-aminopyridine successfully give amidines though in lower yields ⟨53MI 519-01⟩. The nitrile is usually dissolved or suspended in anhydrous alcohol, cooled and treated with an excess of HCl to form the imidic ester hydrochloride. Subsequent reaction with a primary or secondary amine or with an alcoholic solution of ammonia or ammonium chloride forms the amidine hydrochloride.

The reaction can be performed with a reduced amount of ethanol (1–3 equiv.) in an inert solvent such as dioxan, ether, benzene or chloroform and this is sometimes advantageous; for example, a side reaction to form *ortho*-esters is reduced ⟨55MI 519-01⟩. Another side reaction associated with the Pinner method is the formation of *N*,*N*′-disubstituted amidines ⟨26JA732, B-75MI 519-01⟩.

If the starting substrate contains an ester, a possible side reaction is ammonolysis; this is especially prevalent when aqueous ammonia is used (Equation (8)) ⟨54CLY1210⟩. One method of avoiding this side reaction is to hydrolyse the ester-nitrile to an acid-nitrile and then perform the Pinner reaction with a limited amount of ethanol in an inert solvent such as diethyl ether. This procedure has given the respective amidino-acid in excellent yield ⟨49MI 519-01⟩.

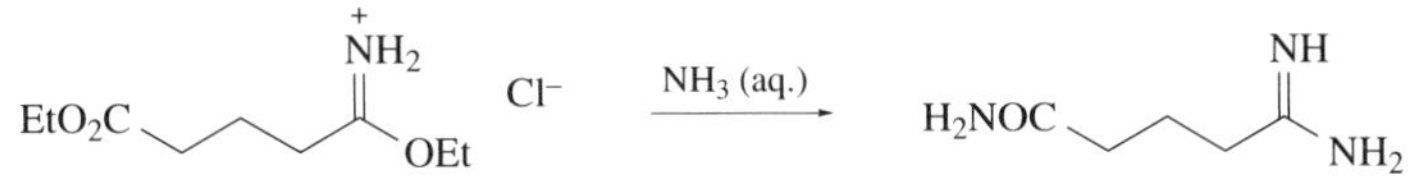

(8)

A less common method of forming the intermediate imidic ester is by 'base-catalysed reaction'. An electron-withdrawing substituent is required in the α-position of the nitrile in order to achieve a favourable equilibrium concentration of the imidic ester. Good yields of amidines are obtained on subsequent treatment with ammonium chloride as shown in (Scheme 9) ⟨61JOC412, 87MI 519-01⟩.

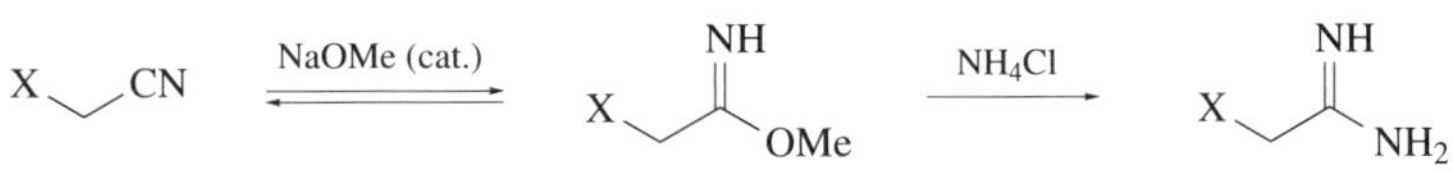

X = Cl, 2-step yield, 96%
X = Bu^{t}O, 2-step yield, 78%
X = PhNH, 2-step yield, 72%

Scheme 9

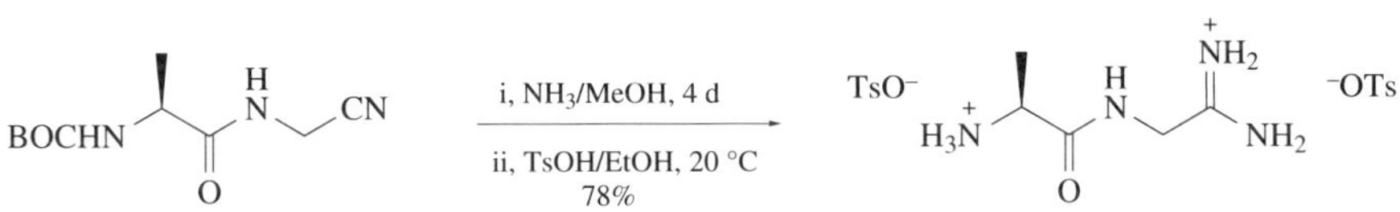

(23)

(9)

A similar method was used by Eschenmoser and co-workers to prepare peptide-like amidinium salts such as (**23**) (Equation (9)). The amidine formation can be catalysed by the addition of L-cysteine ⟨86HCA1224⟩.

Highly electrophilic nitriles such as trichloroacetonitrile will react directly with amines to give amidines ⟨51RTC638, 61CJC761⟩.

Aliphatic and aromatic nitriles react with primary or secondary amines in the presence of aluminum chloride. This is a good general method of making monosubstituted and *N*,*N*-disubstituted amidines. Representative examples are given in Table 3.

A more modern preparation of amidines from nitriles involving aluminum activation has been reported by Garigipati. Aluminum amides undergo reaction with alkyl or aryl nitriles in toluene at 80 °C to give primary or substituted amidines (Equation (10)) ⟨90TL1969⟩.

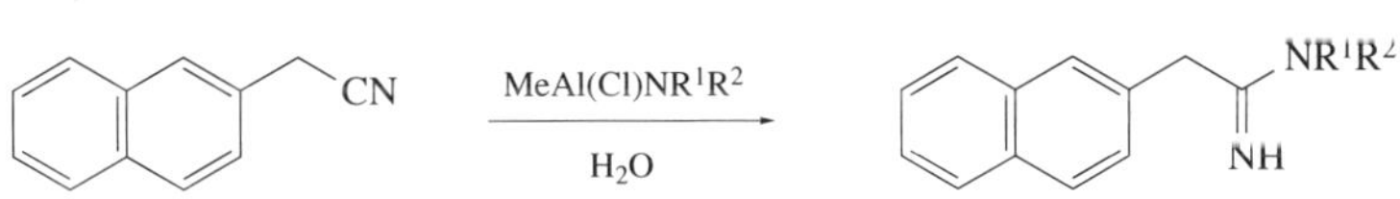

(10)

$R^1 = R^2 = H$, 95%
$R^1 = H$, $R^2 = Me$, 94%
$R^1 = R^2 = Me$, 60%

Heating a molten mixture of alkyl or aryl nitriles with ammonium alkylsulfonates or ammonium arylsulfonates gives amidines (initially as the sulfonic acid salt) in variable yield. The reaction can also be applied to amine salts of sulfonic acids; in both cases very high reaction temperatures (180–300 °C) are required (Equation (11)) ⟨46JCS147, 51JCS485, 58HOU(9)39⟩.

(11)

R = alkyl or aryl

Table 3 Preparation of amidines by aluminum chloride-induced reaction of nitriles with primary and secondary amines.

Amidine Product	*Temperature*	*Time*	*Yield* (%)	*Ref.*
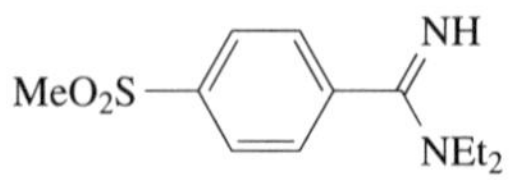	100 °C	1.5 h	100	47JCS1110
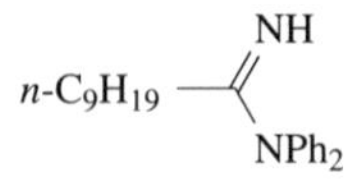	160–200 °C	0.3–1 h	63	51JCS485
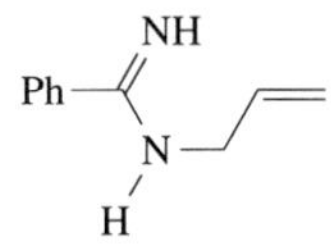	reflux		66	73JCS(P1)453
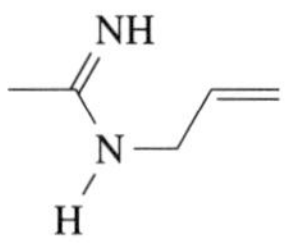	reflux		62	73JCS(P1)453
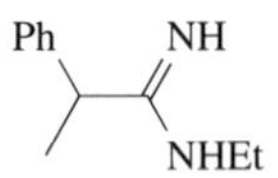	160 °C		76	61BCF2067
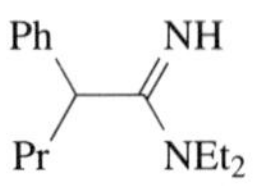	170 °C		54	61BCF2067

Slightly lower reaction temperatures (180 °C) are possible using ammonium or alkylammonium thiocyanate, and this can lead to good yields in some cases ⟨47JCS390⟩. Schaefer and Krapcho decreased the reaction temperature to 125–150 °C by using simple ammonium salts in liquid ammonia as solvent. Under these high-pressure conditions, reasonable yields of amidines are obtained ⟨62JOC1255⟩.

A new method which allows direct formation of the amidines under mild conditions is the stoichiometric reaction of aliphatic amines with acetonitrile or benzonitrile promoted by Cu(I) chloride (Equation (12)) ⟨93TL6395⟩.

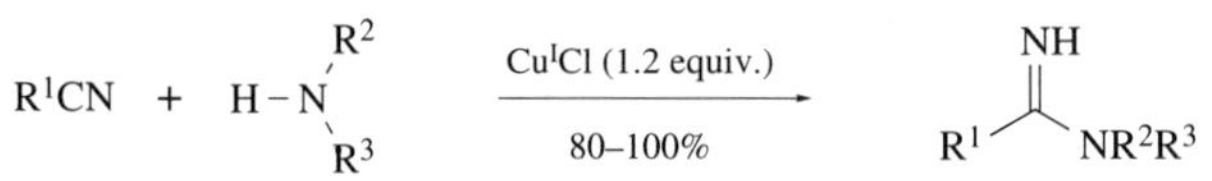

(12)

R^1 = Me or Ph
R^2, R^3 = H or alkyl

With secondary amines, the reaction can be run in methanol, ethanol or DMSO (80 °C) as solvent. With primary amines, formation of the *N*,*N*′-disubstituted amidines is a major side product and best yields are obtained when the nitrile is the solvent. If the *N*,*N*′-disubstituted amidine is the desired product excellent yields can be obtained using four equivalents of the primary amine in ethanol or preferably DMSO as solvent.

Amidines are formed from aliphatic nitriles by reaction with alkali amide bases (Equation (13)). Thus propane-, butane- and pentanenitrile react with potassium amide in liquid ammonia to give a 30–50% yield of the potassium salt of the respective primary amidine. Yields of 85–90% are obtained from branched-chain nitriles such as 2-ethylpropanenitrile and a large number of branched-chain nitriles were converted to amidines by reaction with sodamide in refluxing benzene ⟨47JCS738⟩. The reaction also works for highly hindered aliphatic nitriles such as **(24)** and this can be useful as such nitriles do not undergo successful reactions under the Pinner conditions ⟨47JCS738⟩.

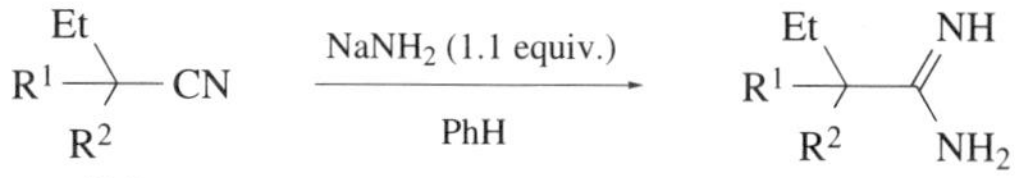

(24) (13)

R^1 = Et, R^2 = Bu^n, 80%
R^1 = Bu^n, R^2 = n-C_6H_{13}, 75%
R^1 = Pr^i, R^2 = n-C_6H_{13}, 40%
R^1 = Bu^n, R^2 = s-octyl, 32%

Reaction with alkali amides fails completely with nitriles containing reactive methylene groups such as phenylacetonitrile. Phenylacetonitrile is reported to react with lithium derivatives of aromatic amines to yield *N*-arylamidines ⟨72ZOR1246⟩.

Other methods of activating aliphatic nitriles which lead to amidine formation include the formation of *N*-substituted nitrilium salts with alkyl halides and iron(III) chloride ⟨73T2147⟩ or with powerful alkylating agents such as triethyloxonium tetrafluoroborate ⟨76BEP831934, 77BRP1466047, 78BSB717⟩, or activation with zinc chloride ⟨47JCS1110⟩ or tributylborane (see section 5.19.2.6.2) or by forming a pentaamminocobalt complex (see section 5.19.2.7.3).

5.19.1.3.2 Aliphatic amidines from amides

Scheme 10 shows a method which is generally poor for preparing primary amidines ⟨44CRV351, B-75MI 519-01⟩ but is an excellent and general method of preparing di- and trisubstituted amidines. The secondary or tertiary amide can be activated by a number of methods but most commonly as the imidoyl chloride, subsequent reaction with primary or secondary amines yield amidines. Examples of preparations starting from monosubstituted amides are given in Table 4.

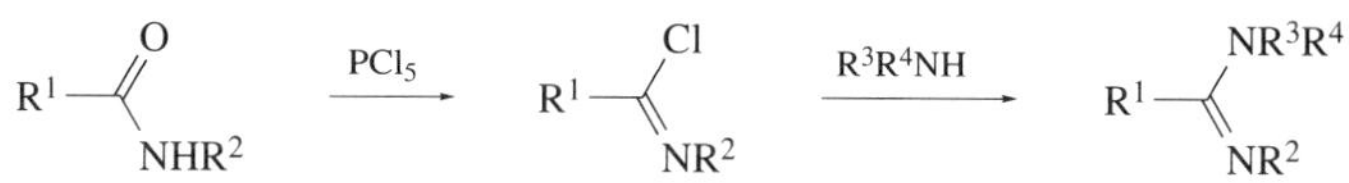

Scheme 10

A reliable procedure is to heat the aliphatic amide with the amine in the presence of phosphorus pentachloride for several hours in a suitable solvent (usually benzene or chloroform at reflux) ⟨49JCS1308, 53JA1761, 73JCS(P1)453⟩. Isolation of the imino chloride has given poor results in the preparation of aliphatic amidines due to the formation of *N*-(chlorovinyl)amidines (**25**) (Scheme 11) ⟨27LA(151)153, 31AG611⟩; however, this procedure is used successfully in the synthesis of formamidines (see section 5.19.1.2.1) and aromatic amidines (see section 5.19.1.4.2).

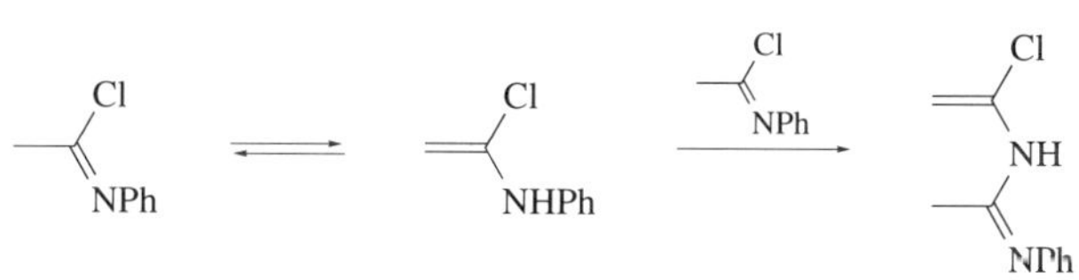

(25)

Scheme 11

Other chlorinating agents can be used to form the imino chloride. Phosphorus oxychloride, trichloride or thionyl chloride all give amidines although generally in poorer yield ⟨B-75MI 519-01⟩.

One problem with the preparation of unsymmetrically substituted amidines from monosubstituted amides is that symmetrical amidines ((**26**) and (**27**)) can be formed as side reactions. Delaby proposes this occurs according to Scheme 12 ⟨56MI 519-01, 56MI 519-02⟩.

Tertiary amides react under similar conditions to secondary amides to give trisubstituted amidines ⟨59CB837⟩.

An alternative method of activating the amide is by alkylation and this has the advantage that it can be used to prepare primary amidines via the alkylation of a primary amide as well as more highly substituted amidines (Scheme 13). Hence treatment of an amide (**28**) with triethyloxonium tetrafluoroborate at room temperature gave the isolable imidic ester fluoroborate. Subsequent room temperature reaction with an amine or an alcoholic solution of ammonia gives amidines in good yield.

Table 4 Preparation of aliphatic amidines from monosubstituted amides.

Starting amide	*Amidine product*	*Reaction conditions*	*Yield (%)*	*Ref.*
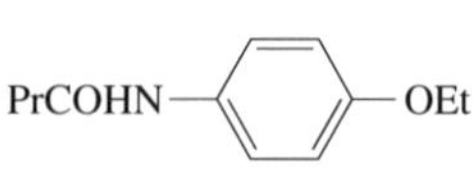	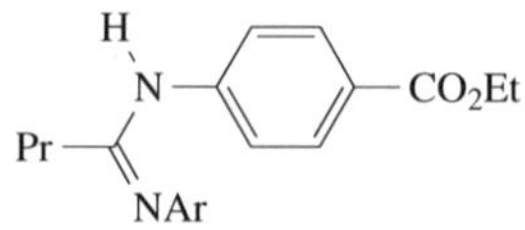	PCl_5, Δ, C_6H_6, 3 h	51	26JACS3214
	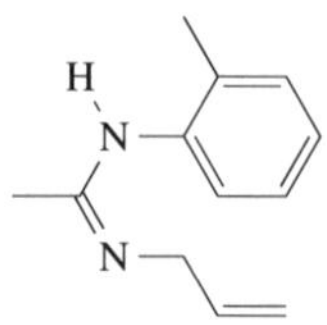	PCl_5, Δ, C_6H_6, 3 h	80	73JCS(P1)453
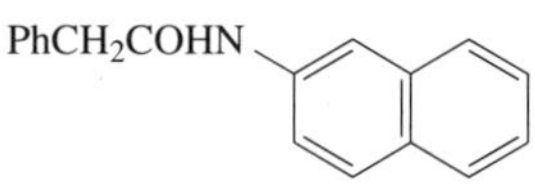	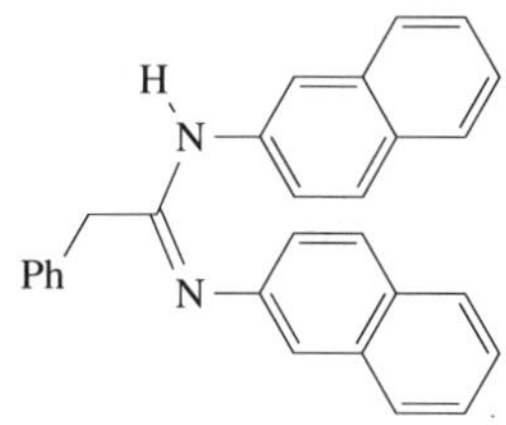	$POCl_3$, Δ, C_6H_6, 5 h	50	56MI 519-01
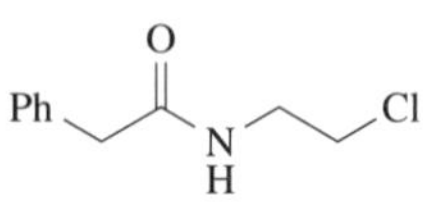	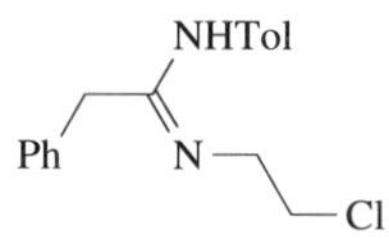	PCl_5, Δ, C_6H_6, 10 min add amine, Δ 3 h	97	49JCS1308
MeCONHPh	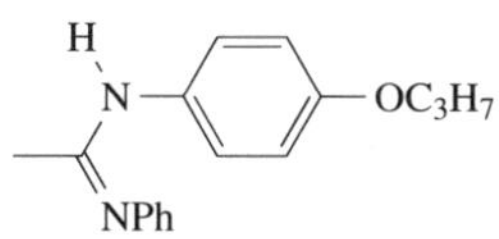	$POCl_3$, steam bath, 2 h	92	56MI 519-02
$Et_2CHCONHAr$	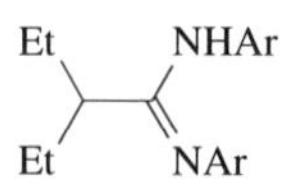 Ar = 4-$EtOC_6H_4$	PCl_5, Δ, C_6H_6	75	42CB1585

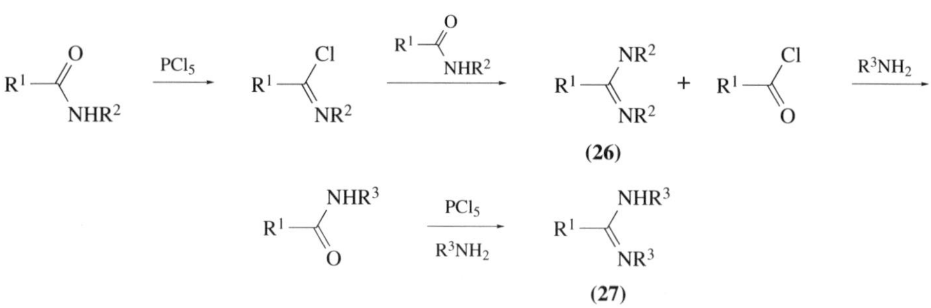

Scheme 12

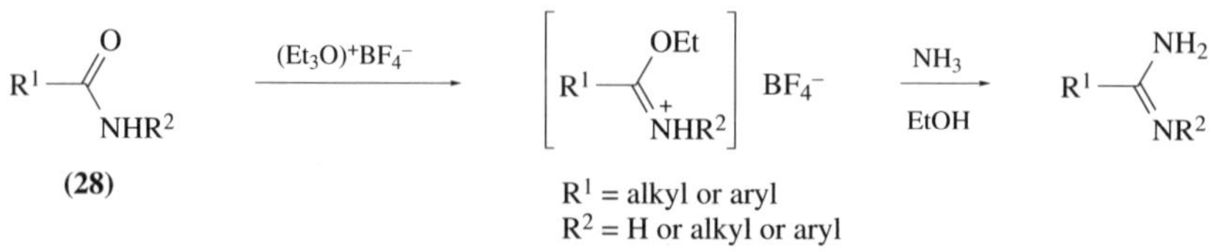

Scheme 13

Activation of the nitrogen nucleophile allows direct reaction with unactivated amides. Reaction of tetrakis(dimethylamino)titanium with various secondary amides in THF (mostly at room temperature) gives a variety of trisubstituted amidines as shown in Equation (14) ⟨71JOC1613⟩.

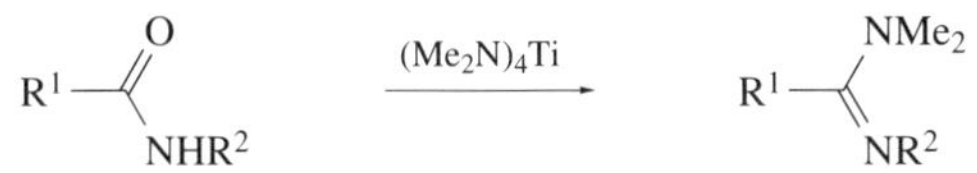

(14)

R^1 = H, alkyl or aryl
R^2 = Me, Ph, Bu^t

5.19.1.3.3 Aliphatic amidines from thioamides and thioimidic esters

The condensation of ammonia or an amine with a thioamide gives the H_2S salt of the amidine ⟨52LA(577)234⟩. The reaction may be improved by the addition of a mercury(II) salt as a sulfide scavenger (Equation (15)).

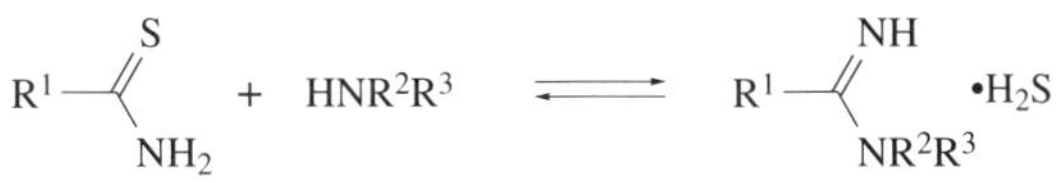

(15)

R^1 = alkyl or aryl

Alkylation of thioamides with alkyl iodides or triethyloxonium tetrafluoroborate gives thiomidic esters which react more readily with ammonia or amines to form amidines ⟨57CB1837, 64CR4067⟩ (Scheme 14). This procedure was used to convert cyanoformates via the thioamide and thioimidic ester into amidinoformates (**29**). The reaction of the thioimidic ester (**30**) with a variety of amines proceeds almost instantaneously at room temperature or 0°C and in nearly quantitative yield ⟨79TL2517⟩. It is reported that the reaction of thioimidic esters with aliphatic amines is improved by using pH buffer control (see section 5.19.1.4.3).

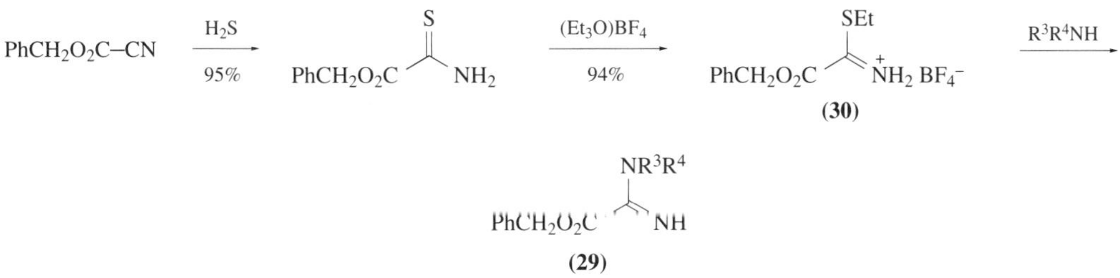

Scheme 14

5.19.1.3.4 Aliphatic amidines from ortho-*esters*

Ethyl *ortho*-acetate (**31**) undergoes reaction with two equivalents of an aliphatic or aromatic amine in the presence of 1 mole of acetic acid to give good yields of the symmetrical *N,N′*-disubstituted acetamidine (**32**) (Equation (16)). Much lower yields (25–30%) have been observed for electron-poor aromatic amines ⟨63JOC1108⟩.

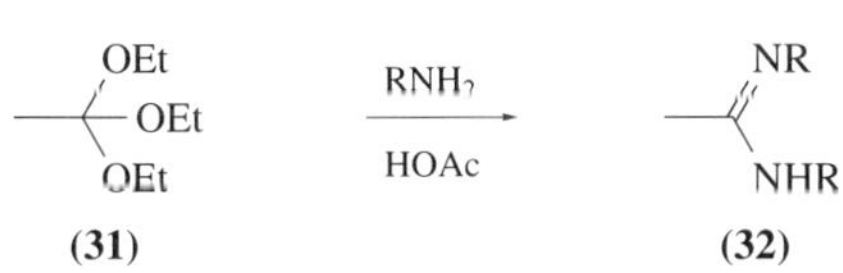

(16)

5.19.1.3.5 Aliphatic amidines from compounds with cumulated double bonds

Amidines can be prepared from the addition of amines to ketenimines (Equation (17)). This chemistry has been used to prepare polyamidines ⟨77JPS2163⟩.

(17)

The reaction of isocyanates with tertiary amides gives amidines. The reaction proceeds via unstable four-membered ring compounds which decompose spontaneously to give carbon dioxide and amidines (Scheme 15) ⟨49JA3746, 60JOC2245⟩.

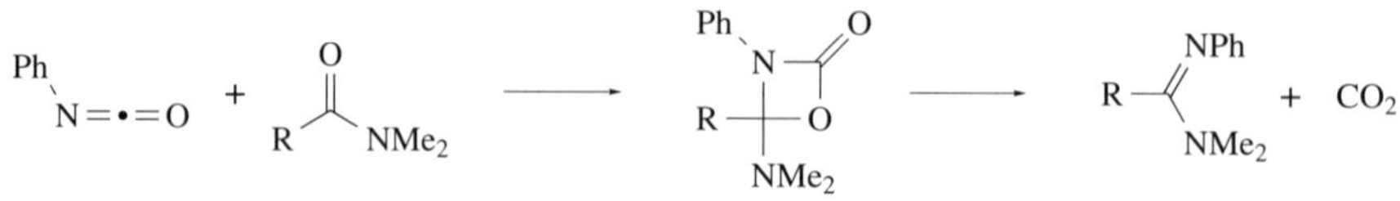

R = alkyl or aryl

Scheme 15

With secondary amides different products are possible (Scheme 16), benzanilide (**33**) yields the amidine whereas with acetanilide, (**34**) the major product is the urea ⟨49JA3746⟩.

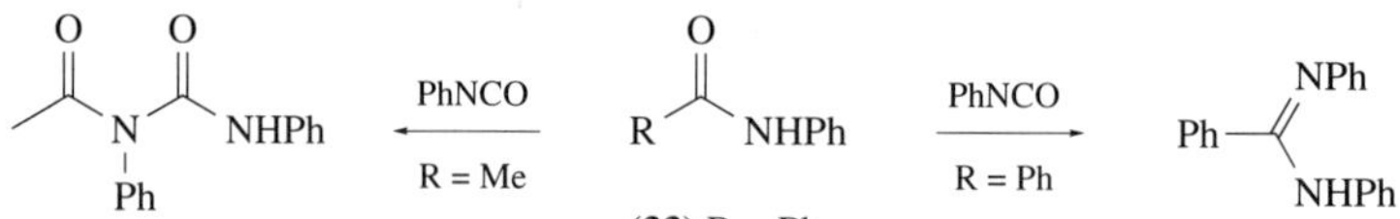

(**33**) R = Ph
(**34**) R = Me

Scheme 16

The addition of organometallic compounds to carbodiimides is a rare example of an amidine synthesis via carbon–carbon bond formation. Organolithiums, Grignard reagents and in some cases organozinc reagents have all been utilised in high yields (Equation (18)) ⟨74BCF994⟩.

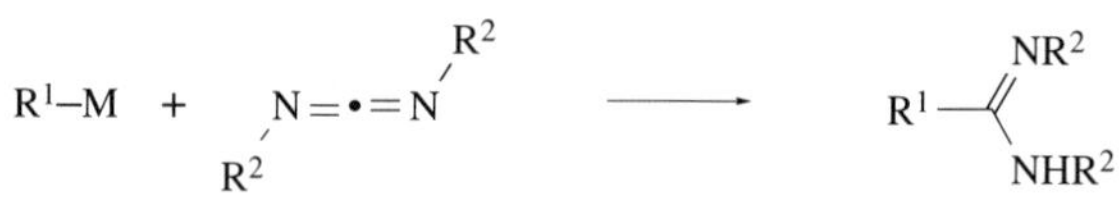

(18)

R^1 = alkyl, aryl, allyl; R^2 = Ph, cyclohexyl, Pr^i

5.19.1.3.6 Amidines, prepared by N*-alkylation of simpler amidines*

This method is most effective for alkylation of symmetrical *N,N′*-disubstituted amidines ⟨69URP256779, 69URP256781⟩, alkylation of *N,N*-disubstituted amidines or monoalkylation of a primary amidine. Exhaustive alkylation of primary or monosubstituted amidines to give a trisubstituted amidine is also possible. Primary aliphatic amidines are monoalkylated by preparing the monosodio derivative with sodamide in benzene, then heating with the appropriate alkylbromide ⟨47JCS738⟩.

Attempts to monoalkylate monosubstituted amidines have attracted little synthetic interest. Such alkylations are complicated by poor regioselectivity and lead to a mixture of *N,N*- and *N,N′*-substituted amidines ⟨B-75MI 519-01⟩. The most useful results have been observed in the alkylation of *N*-aryl amidines which are alkylated predominantly on the nitrogen atom bearing the aryl group. Hence Gautier *et al.* alkylated a series of *N*-aryl substituted amidines in 32–57% yield as shown in Equation (19) ⟨71BCF478⟩.

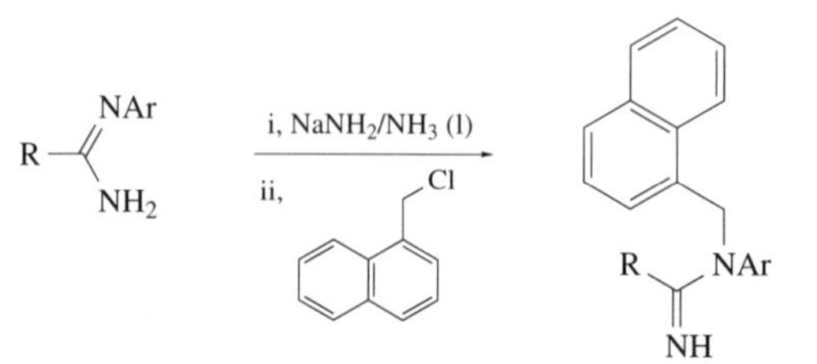

(19)

In his work in the 1920s, Pyman showed that *N*-phenylbenzamidine was methylated on the substituted nitrogen in good yield (83%) and with a regioselectivity estimated to be greater than 100 : 1 ⟨23JCS3359⟩. *N*,*N*′-Unsymmetrically disubstituted amidines are normally alkylated with poor regioselectivity but *N*-aryl-*N*′-alkylamidines such as (**35**) and (**36**) are an exception which again are alkylated predominantly on the nitrogen atom bearing the aryl group (Equation (20)) ⟨23JCS3359⟩.

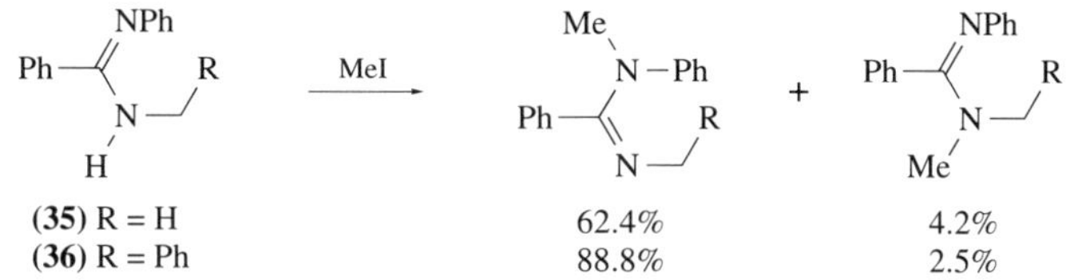

(20)

5.19.1.3.7 *Amidines, prepared by miscellaneous methods*

Acetylenic Grignard reagents react with chloroformamidines to give alkynylamidines. These rare examples of amidines prepared by carbon–carbon bond formation are shown in Equation (21) ⟨71CB3329⟩.

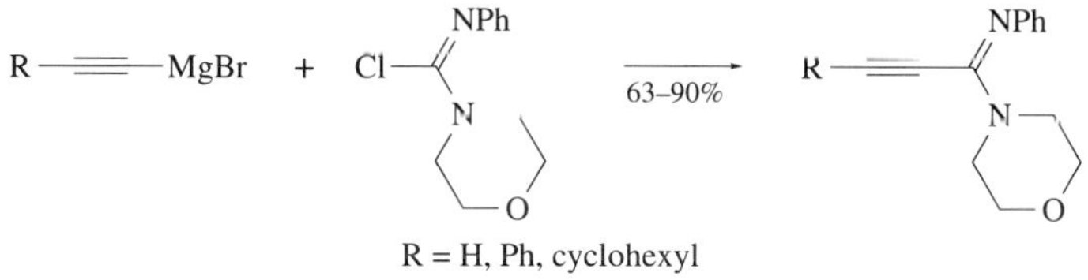

(21)

Amidines have been prepared by a type of Beckmann rearrangement when ketoximes are heated with hexamethylphosphortriamide (Equation (22)) ⟨83SC1083⟩.

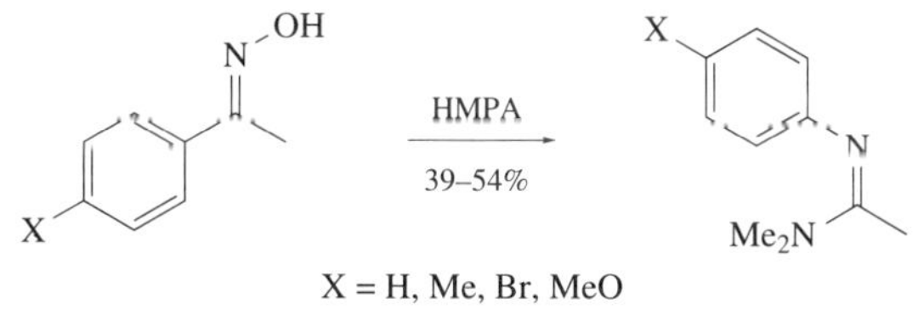

(22)

The preparation of amidines from imines, hydrazones, halogenated compounds, aldoximes, by ammonolysis and by addition reactions to enynamines are covered in previous reviews ⟨B-75MI 519-01, B-91MI 519-01⟩.

5.19.1.4 Aromatic Amidines $ArC(NR^1)NR^2_2$

5.19.1.4.1 *Aromatic amidines from nitriles*

Nearly all of the synthetic methods described in section 5.19.1.3.1 also apply to aromatic amidines. An important exception is the synthesis of aryl amidines from *ortho*-substituted aryl cyanides (**37**; R = Me, Cl, SO_2NH_2) (**38**; R = NO_2, NH_2) and 1-naphthonitrile which exhibit a 'proximity effect' and do not form imidic esters when treated with ethanol and anhydrous HCl ⟨1890CB2917, 03JCS766, 47BCF406⟩. Positional isomers where the positions *ortho*- to the nitrile are unoccupied behave normally, hence 4-methylbenzonitrile ⟨44CRV351⟩ and 2-naphthonitrile ⟨78PHA568⟩ form imidic esters, which are then converted into amidines. Furans, thiophens and selenophens with carboxamidine substituents in either the 2- or 3-position have also been prepared by this method ⟨78JCR(S)132⟩.

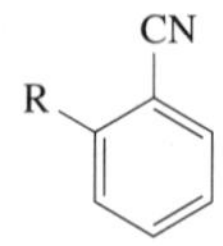

(37)

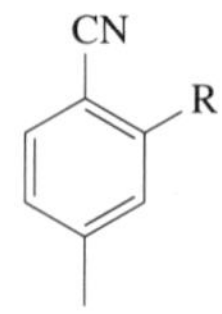

(38)

The extent of the limitations of this proximity effect have not been fully investigated but some benzonitriles with oxygenated substituents in the *ortho*-position (**37**; R = OH or OEt) will form imidic esters. The proximity effect has some synthetic use in differentiating between vicinal dinitriles; hence the symmetrical dinitriles ((**39**)–(**41**)) all form salts of mono-imidic esters. Presumably the cylindrical, sterically undemanding *ortho*-cyano group does not inhibit the formation of the first imidic ester group. Mono-imidic esters of (**41**) have been converted into amidines; 1,3- and 1,4-benzonitriles have been converted into dicarboxamidines ⟨1884CB1428⟩. A number of methods of overcoming the proximity effect have been developed, two of which are outlined in Scheme 17.

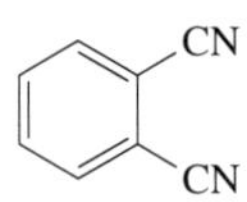

(39)

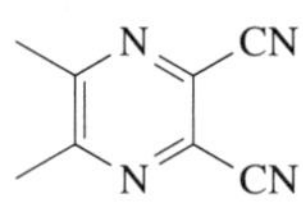

(40)

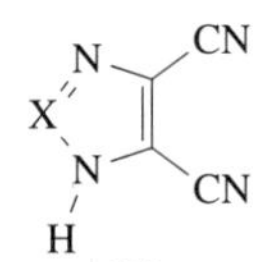

(41)

X = N or CH

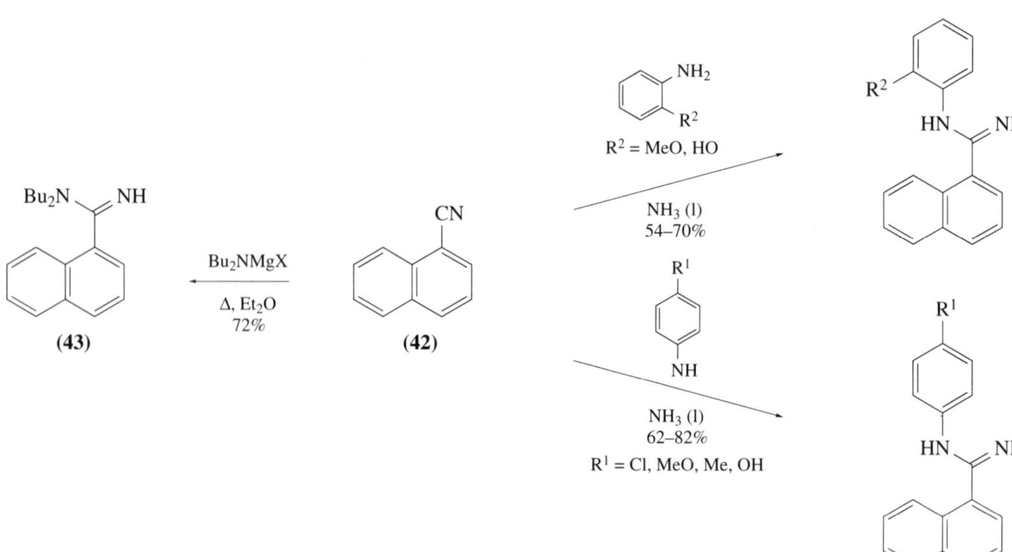

Scheme 17

1-Naphthonitrile (**42**) can be condensed with aryl amines in liquid ammonia ⟨70BCF200⟩. 1-Naphthonitrile also reacts with the magnesium salts of secondary amines to give the amidine (**43**). This method also works for sterically hindered nitriles such as 2-methyl-1-naphthonitrile and 2,4,6-trimethylbenzonitrile ⟨48JA1904⟩. Both of the methods outlined in Scheme 17 are poor when used with aliphatic nitriles which have an α-hydrogen atom but work well for aryl nitriles including 2-chloro- and 2-methylbenzonitriles.

Ortho-substituted benzonitriles undergo room temperature reaction in DMSO with anilines which have been metallated by sodium hydride ⟨71CC498⟩. In an older procedure the reaction of aniline, 2-methylbenzonitrile and sodamide in boiling benzene gave *N*-phenyl 2-methylbenzamidine ⟨53JCS255⟩. When 1-naphthonitrile is heated with the ammonium salt of a sulfonic acid at 260 °C for 2 h, 1-naphthamidine is formed in 30% yield ⟨46JCS147⟩. A better method of forming the primary 1-naphthamidine and its derivatives is to convert the nitrile into the thioimidic ester (see section 5.19.1.4.3).

5.19.1.4.2 Aromatic amidines from amides

Aromatic amides (**44**) are converted into aromatic amidines via the imino chloride (normally by reaction with PCl_5). Two procedures have been widely used; the first involves heating the aryl amide, amine and phosphorus chloride in an inert solvent such as chloroform or benzene ⟨27JCS2318, 49JCS3043, 54JA920⟩. However, this procedure is unsuitable for amines which react with phosphorus chlorides. A second method, which can be used for arylamides and also for other amides which do not possess a β-hydrogen, is to isolate the imino chloride. This procedure is tolerant of amines which react with phosphorus chlorides, hence aminophenols can be used directly without phenolic protection (Scheme 18) ⟨47JCS160, 49JCS3319, 53JA1761⟩.

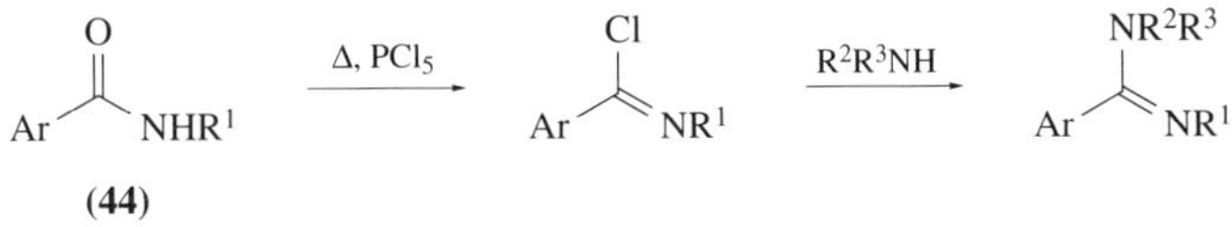

(44)

Scheme 18

5.19.1.4.3 Aromatic amidines from thioamides and thioimidic esters

The reaction of aromatic thioimidic esters with aromatic amines give amidines in very good yield but with more basic aliphatic amines, retroreaction to generate the starting nitrile occurs (Scheme 19). Hence the reaction of thioimidic ester (**45**) with aniline leads to amidine formation in 88% recrystallised yield. The reaction of (**45**) with diisopropylamine gave a nitrile recovery of 90% with only a trace of amidine formation. The problem is overcome by adding acetic acid–sodium acetate buffer to the reaction whereby the room temperature reaction of (**45**) with methylamine or diisopropylamine gives 90% and 55% yields of the respective amidines ⟨79JOC3726⟩.

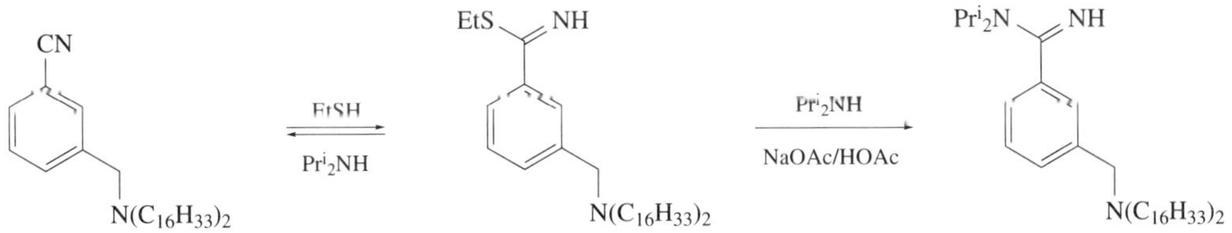

(45)

Scheme 19

The unwanted regeneration of the starting nitrile was also observed by Baker and Erickson when thioimidic ester (**46**) was reacted with alcoholic ammonia. Amidine formation was obtained by reaction with ammonium acetate in boiling ethanol (Scheme 20) ⟨67JMC1123⟩.

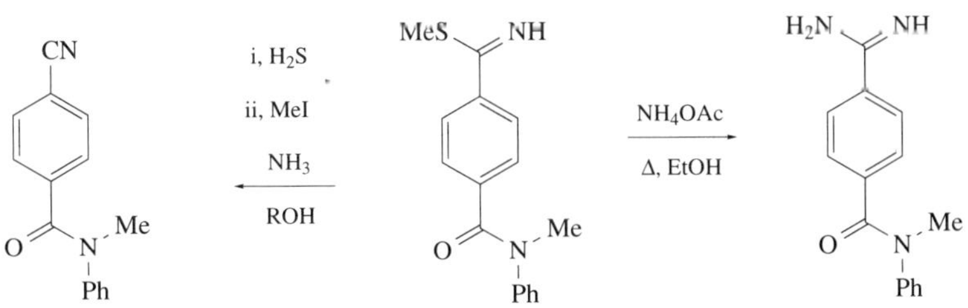

(46)

Scheme 20

1-Naphthamidines (**47**) can be prepared by reaction of thioimidic esters (**48**) with ammonium acetate in ethanol at 60°C (Scheme 21). Attempts to prepare these 1-naphthamidines via the imidic ester failed. The thioimidic esters can be prepared from 1-naphthonitriles (**49**). Hence these procedures represent a way of overcoming the 'proximity effect' in the preparation of primary amidines ⟨78PHA568⟩.

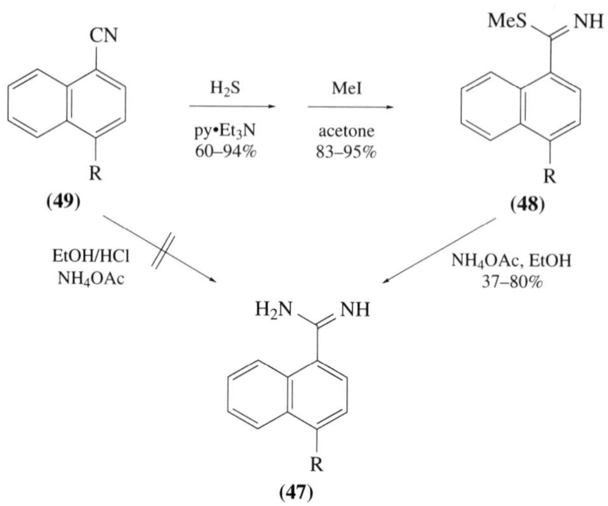

R = Me, COPh, CH_2COPh, CH_2CH_2COPh

Scheme 21

5.19.1.5 *N*-Acyl and *N*-Heteroacyclamidines

5.19.1.5.1 N-*Acylamidines* $R^1C(NR^2)NR^3COR^4$

(i) By direct acylation of amidines

This is the most common method of preparing acylamidines. Acetamidine hydrochloride and benzamidine hydrochloride are benzoylated using benzoyl chloride in aqueous acetone in 96% and 94% yields, respectively ⟨74JCS(P2)546, 76BCJ3607⟩. Phenyl benzoate and benzoic anhydride have also been utilised for monoacylation ⟨11JCS1493, 11JCS1507, 77JCS(P1)1791⟩. Acetylation and benzoylation of amino amidines such as (**50**) were reported by Rapoport and co-workers ⟨81JOC2455⟩. The reactions were performed either neat or in methylene chloride solution (Equation 23). A large number of primary amidines have been monoacylated under two-phase reaction conditions as shown in Equation (24) ⟨83GEP3134144⟩.

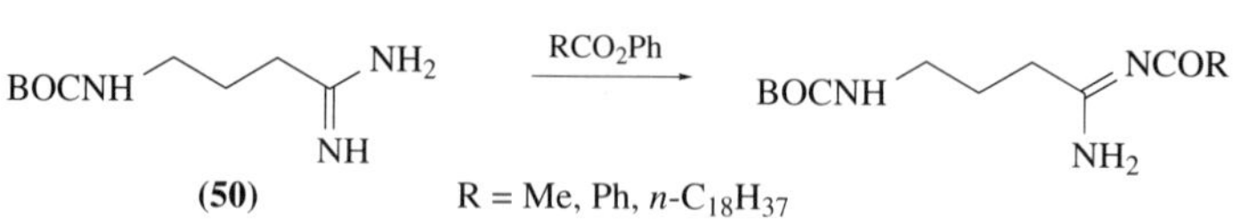

(23)

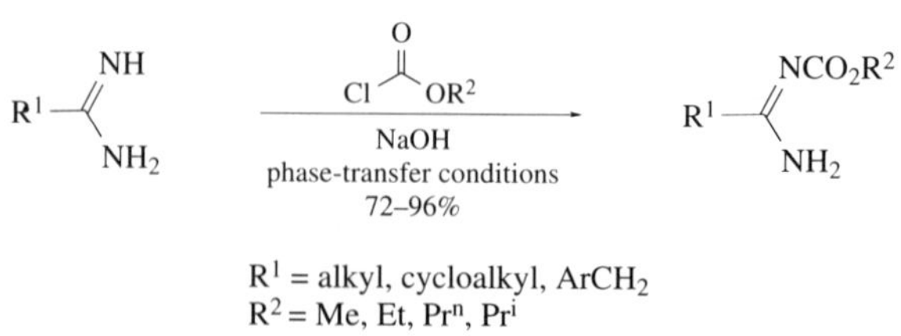

(24)

R^1 = alkyl, cycloalkyl, $ArCH_2$
R^2 = Me, Et, Pr^n, Pr^i

Mono-substituted amidines such as (**51**) are acylated on the nitrogen bearing the substituent. If that substituent is aryl, then the regioselectivity of the acylation reaction is excellent and a wide variety of benzamidines and naphthamidines have been prepared in very good yield by this method a few of which are outlined in Scheme 22 ⟨72FRP2081556⟩. Subsequently, it was reported that acylation products of this type for example (**52**; $R^1 = Ph$, $R^2 = H$) rearrange when heated in ethanol containing hydrochloric or sulfuric acid to give *N*-aryl-*N*′-acylamidines ⟨82MI 519-01⟩.

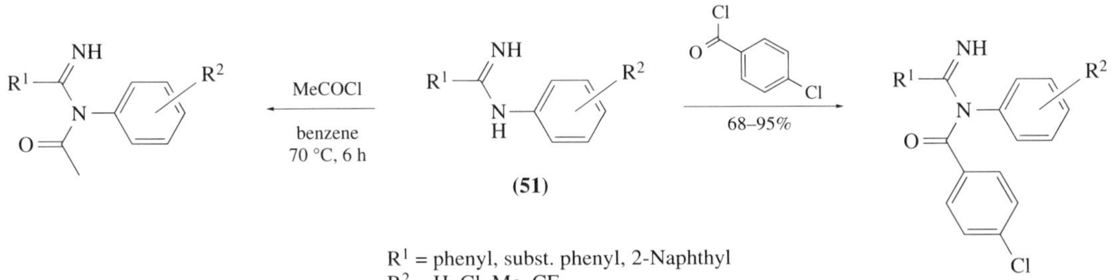

(51) (52)

R^1 = phenyl, subst. phenyl, 2-Naphthyl
R^2 = H, Cl, Me, CF_3

Scheme 22

N-Methylbenzamidine (**53**) is also preferentially acylated on the substituted nitrogen though the regioselectivity and yield are poor (Equation (25)) ⟨77JCS(P1)1791⟩.

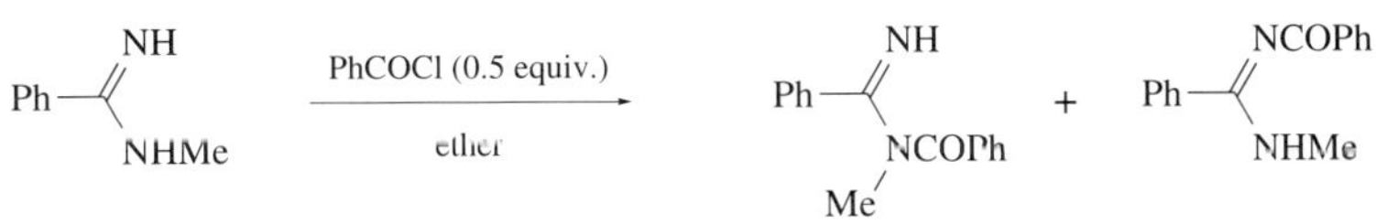

(25)

(53) 20% 10%

N-Silylformamidines react with acid chlorides to give *N*-acylformamidines with N—Si bond cleavage. The *N*-silylformamidine can be prepared by hydrosilylation of a carbodiimide (see section 5.19.2.6.1) ⟨77JOM(140)97, 91S954⟩.

(ii) By miscellaneous methods

Amides can be converted in high yield into either *N*′-acyl-*N*,*N*-dimethylformamidines or N′-acyl-*N*,*N*-dimethylacetamidines by reaction with the appropriate amide acetal (**54**) (Equation (26)) ⟨79JOC4160⟩.

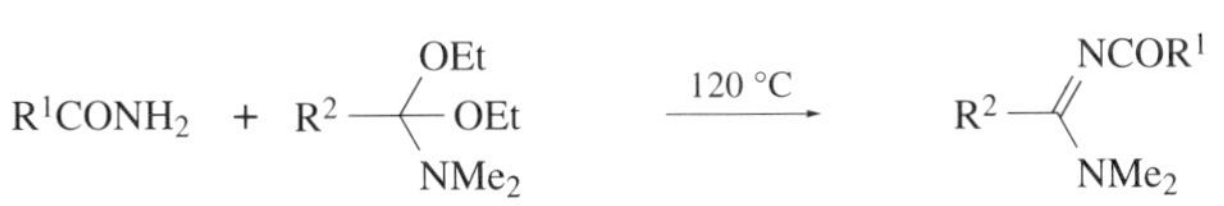

(26)

(54)

R^2 = H, 94%
R^2 = Me, 90%

N,*N*′-Dibenzoylamidines have been prepared in high yield by the reaction of imidazoles with singlet oxygen ⟨68TL3277⟩. Monobenzoylamidines have been prepared by heating 1,2,4-thiadiazolium salts (**55**) and (**56**) with sodium hydroxide (Equation (27)) ⟨77JCS(P1)1791⟩.

(27A)

(55)

(27B)

(56)

N-Allyl isocyanide (**57**) undergoes Cu(I)-catalysed reaction with the formamide (**58**) to give the *N*-formyl amidine (**59**) (Equation (28)) ⟨71T3795⟩.

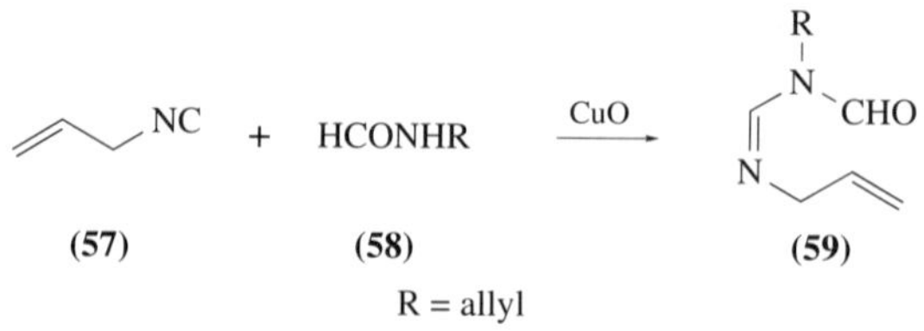

(28)

5.19.1.5.2 N-*Thioacylamidines*

There are a number of routes to thioacylamidines. Amidines themselves can be thioacylated by reaction with thiochloroformates (**60**) under phase-transfer conditions ⟨83GEP3134144⟩ or by room-temperature reaction with isothiocyanates such as (**61**) (Scheme 23) ⟨69T5437, 79JCS(P1)1762⟩.

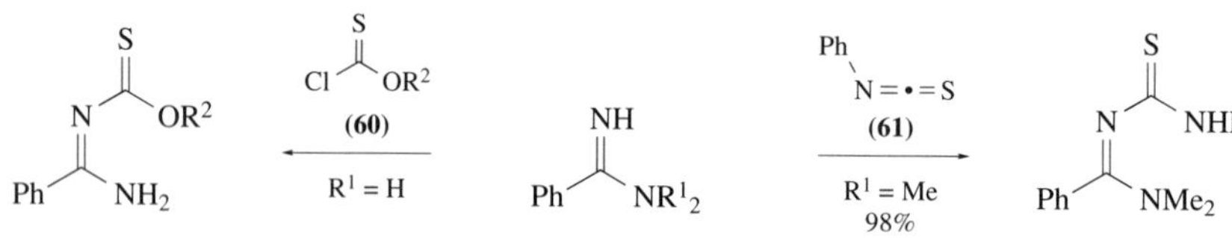

Scheme 23

Thiobenzamide undergoes reaction with protonated acetonitrile to give thiobenzoylacetamidine (**62**). The reaction of thiobenzamide with dimethylformamide dimethyl acetal (**63**) forms the thiobenzoylformamide (**64**) in excellent yield and under mild conditions ⟨80JOC3750⟩. The reaction of (**63**) and other formamide acetals with a variety of thioamides and thioureas has been a widely used, reliable method of preparing thioacylformamidines (Scheme 24) ⟨80JHC767, 83PS(15)143, 84CC157, 84S51⟩.

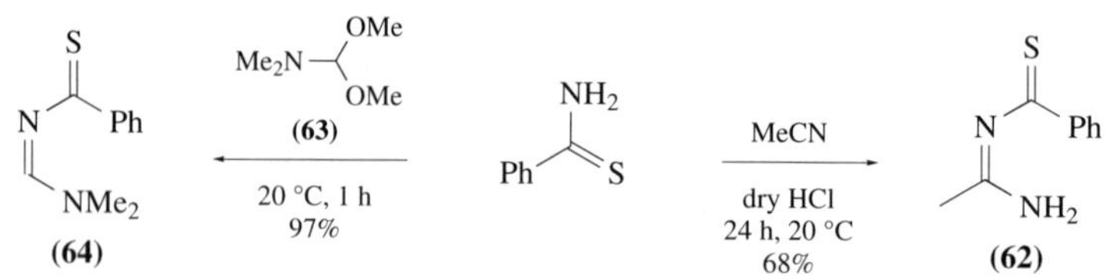

Scheme 24

The reaction of thiobenzamide with a thioamide–bromine adduct such as (**65**) forms thiobenzoylamidines along with some minor products (Scheme 25). The reaction is thought to proceed via formation of the disulfide cation (**66**), which undergoes 1,5-electrocyclic ring closure and loss of a proton to give (**67**). Ring opening and sulfur extrusion gives the amidine (**68**) ⟨84JCS(P1)897⟩.

N-Thioacylbenzimino chlorides such as (**69**) are prepared from the nickel chelate (**70**). In the

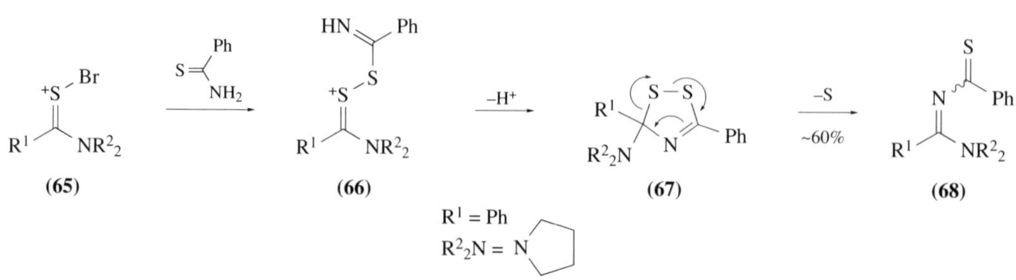

Scheme 25

absence of a transition metal, cyclisation occurs. Fourteen thiocarbonylbenzamidines were formed by the reaction of **(69)** with ammonia, primary or secondary aliphatic amines, or primary aromatic amines (Scheme 26) ⟨84TL405⟩.

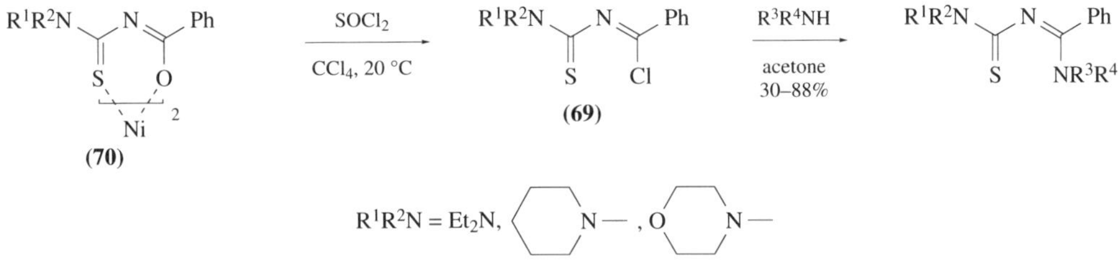

(70) **(69)**

Scheme 26

5.19.1.5.3 *N-Selenoacylamidines*

Isoselenocyanates **(71)** have been prepared from their respective iminochlorides. Subsequent addition of aniline gave the crystalline selenoacylamidine derivatives **(72)** (Scheme 27) ⟨70UKZ372⟩.

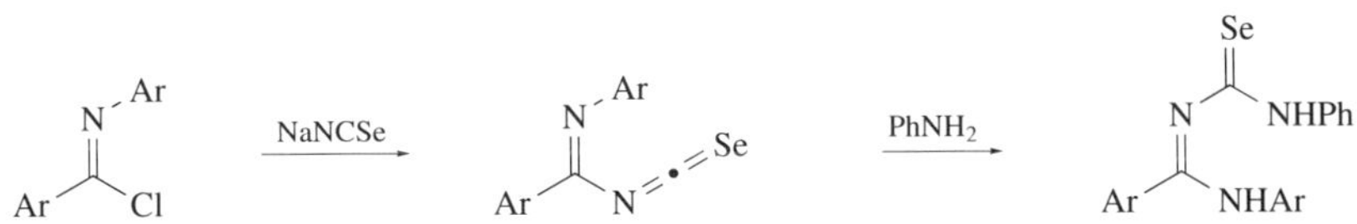

(71) **(72)**

Ar = Ph or 4-ClC_6H_4–

Scheme 27

5.19.2 AMIDINE DERIVED STRUCTURES WITH AN *N*-HETEROATOM BOND

5.19.2.1 *N*-Haloamidines

5.19.2.1.1 *N-Fluoroamidines*

CAUTION: fluorine–nitrogen compounds are potentially explosive; please read the primary literature carefully and take appropriate precautions.

Fluoroiminonitriles, such as **(73)** and **(74)**, react with amines and ammonia to give *N*-fluoroamidines. The (*Z*)-fluoroimine **(73)** reacts with diethylamine to selectively form the (*Z*)-amidine **(75)**. The (*E*)-isomer **(74)** likewise gives the (*E*)-amidine **(76)**. Reaction of either **(73)** or **(74)** with ammonia or 4-chloroaniline yields amidines **(77)** or **(78)** as a single geometric isomer. An explanation for this difference is that **(77)** and **(78)** have the possibility of forming a hydrogen bonded dimer, thus favouring the (*E*)-isomer. Molecular-weight determination by boiling-point elevation gives support to this explanation (Scheme 28) ⟨66JOC3689⟩.

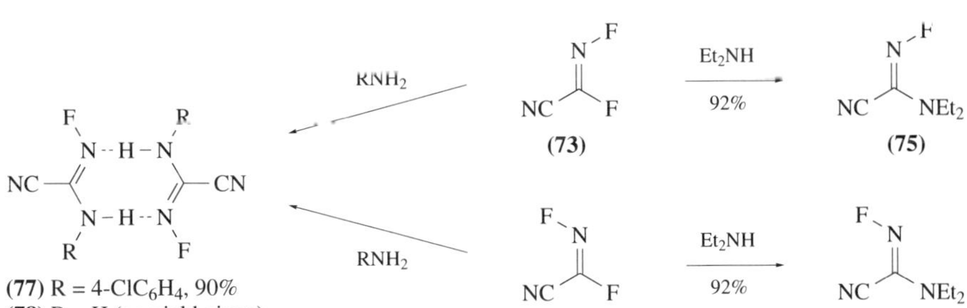

(73) **(75)** **(74)** **(76)**

(77) R = 4-ClC_6H_4, 90%
(78) R = H (no yield given)

Scheme 28

N-Fluoroamidines are formed in high yield by the action of ammonia or dimethylamine on the fluoroimine (**79**) (Equation (29)) ⟨73JOC1075⟩.

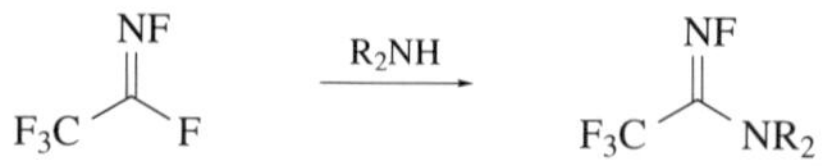

(**79**) R = H, 95% R = Me, 90% (29)

N,N,N'-Trifluoroacetamidine and *N,N,N'*-trifluorobutanamidine have been reported ⟨78IZV2802⟩. (*Z*)-*N,N,N'*-Trifluorobutanamidine undergoes room-temperature, acid-catalysed isomerisation to give (*E*)-*N,N,N'*-trifluorobutanamidine in high yield ⟨73IZV2159⟩.

5.19.2.1.2 N-*Chloroamidines*

N-Chloroamidines are relatively stable compounds which can be crystalline solids with sharp melting points (see Table 5). *N*-Chloroamidines are prepared by direct chlorination of amidines using *N*-chlorosuccinimide, sodium hypochlorite or *t*-butyl hypochlorite. The chlorination proceeds under mild conditions and normally in very high yield. Monosubstituted amidines are chlorinated on the unsubstituted nitrogen. Representative examples of monochloroamidines are given in Table 5.

Table 5 *N*-Chloroamidines.

Starting amidine	*Chlorinated amidine*	*Chlorinating agent*	*Yield (%)*	*Melting point*	*Ref.*
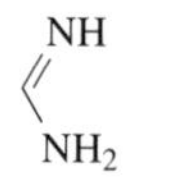	 [a]	NaOCl		−17 °C	53CB400
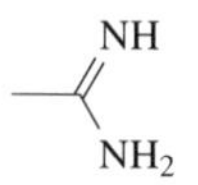	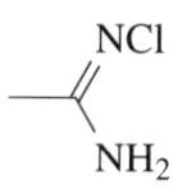	NaOCl	80	31 °C	53CB400
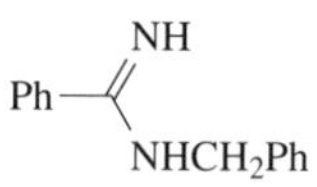	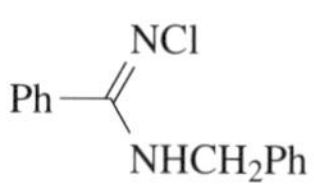	NCS	not given	72 °C	78JCS(P1)309
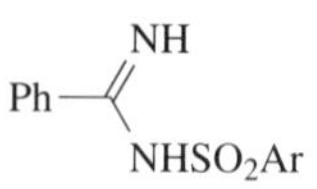 Ar = 4-$O_2NC_6H_4$	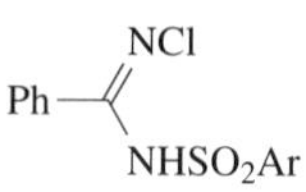	Bu^{t}OCl	99	155 °C	87T4785
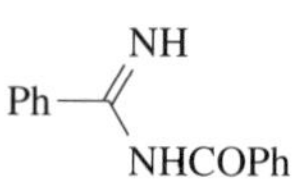	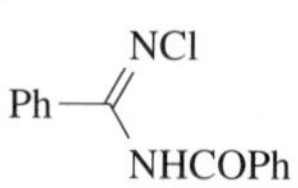	Bu^{t}OCl NaOCl	not given not given	98 °C 110 °C	76BCJ3607 78TL2175
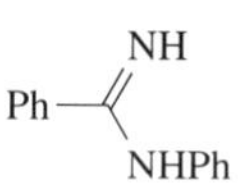	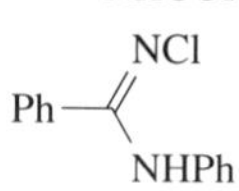	NCS	almost quantitative	128 °C	76TL1839 78JCS(P1)309
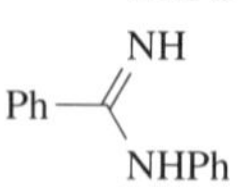	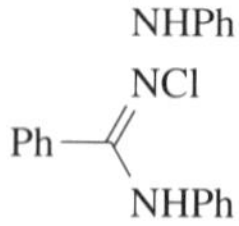	NaOCl	85	130–131 °C	65JOC259
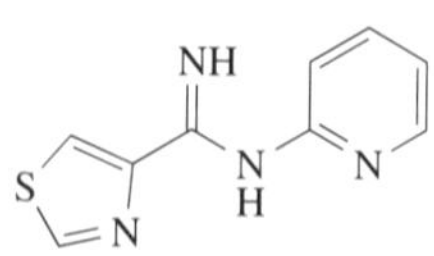	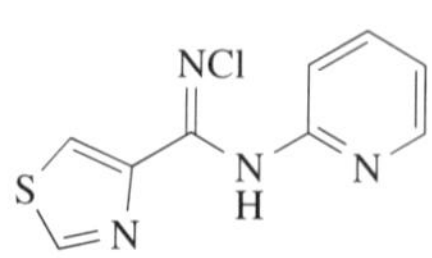	NaOCl	98	93–96.5 °C	65JOC259
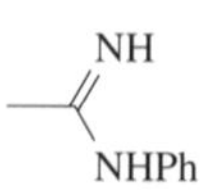	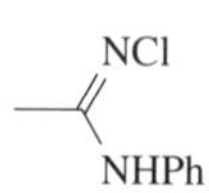	NCS	not given	88 °C	78JCS(P1)309

[a] Thermally unstable.

Further chlorination in an attempt to form a dichloroamidine leads to diazirine formation ⟨65JA4396⟩. *N,N,N'*-Trichloroamidines have been prepared by dissolving acetamidine or butanamidine hydrochlorides in water containing excess potassium chloride; treatment with diluted fluorine gas then gave good yields of the potentially explosive *N,N,N'*-trichloroamidines as shown in Equation (30). When heptafluorobutanamidine hydrochloride was subjected to the same reaction conditions only monochlorination was observed ⟨66TL3537⟩.

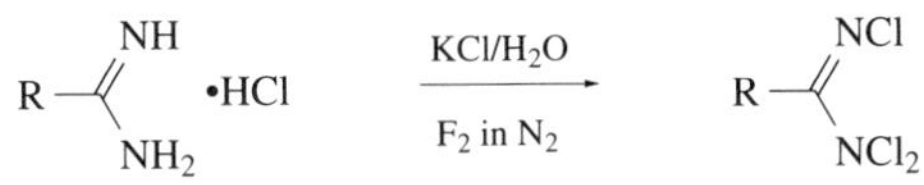

R = Me, Pr^n R = Pr^n, 90% R = Me, 65% (30)

5.19.2.1.3 N-*Bromoamidines*

N-Bromoamidines are prepared by bromination of the parent amidine with sodium hypobromite. Equation (31) shows some of the examples which have been reported ⟨53CB890, 60CB397⟩. *N*-Methylbenzamidine is brominated on the unsubstituted nitrogen to give *N*-bromo-*N'*-methyl-benzamidine ⟨53CB890⟩. When an excess of sodium hypobromite is used under certain reaction conditions, diazirine formation is observed ⟨65JA4396⟩.

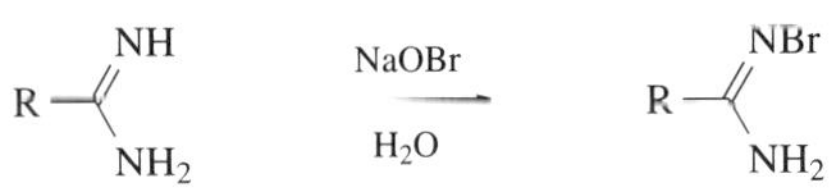

R = 3-$O_2NC_6H_4$–, 80% (31)

R = Ph, $PhCH_2$, 2-$O_2NC_6H_4$–, 3-$O_2NC_6H_4$, 4-$O_2NC_6H_4$

5.19.2.1.4 N-*Iodoamidines*

N-Iodobenzamidine and *N*-iodo-4-methoxybenzamidine are prepared by treating the appropriate benzamidine with an aqueous solution of sodium hydroxide, iodine and potassium iodide ⟨20CR(171)38, 21CR(172)452⟩.

5.19.2.2 *N*-Imidoylhydroxylamines and Related Structures

N-Imidoylhydroxylamines are also known as amide oximes, hydroxyamidines and most commonly as amidoximes. The synthesis and chemistry of *N*-imidoylhydroxylamines has been previously reviewed by Eloy and Lenaers ⟨62CRV155⟩.

5.19.2.2.1 N-*Imidoylhydroxylamines from hydroxylamine*

Hydroxylamine is sufficiently nucleophilic to undergo an addition reaction to a nitrile without the preformation of an imidic ester. This is the most widely used method of preparing *N*-imidoylhydroxylamines, for example the reaction of *p*-bromobenzonitrile or pyridine-3-carbonitrile with hydroxylamine in boiling aqueous ethanol furnishes the desired imidoyl hydroxylamines which can be readily *O*-alkylated or acylated if required (Scheme 29) ⟨54JCS4251⟩. When (*E*)-3-phenylpropenitrile derivatives (**80**) are treated with hydroxylamine, formation of the imidoylhydroxylamines (**81**) occurs in preference to any competing Michael addition (Equation (32)).

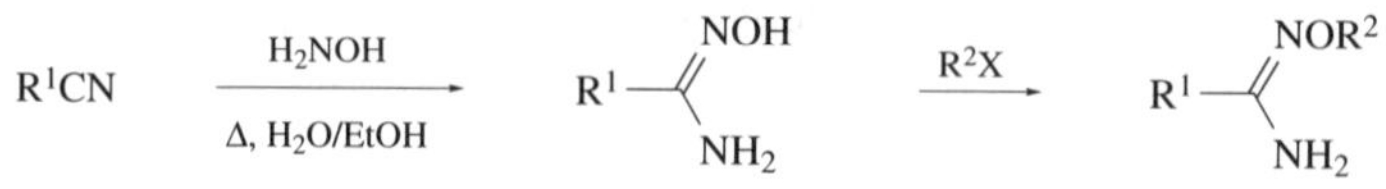

R = 4-BrC_6H_4, 89% 3-pyridyl

Scheme 29

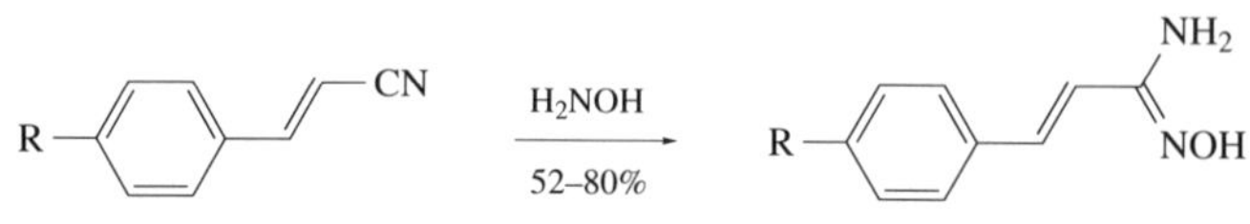

(32)

(80) R = MeS, NO_2, Cl, MeSO– **(81)**

Hydroxylamine is also sufficiently nucleophilic to react directly with thioamides without the need for thioimidic ester formation. The reaction of (**82**) with hydroxylamine in boiling ethanol gave the imidoylhydroxylamine in 50% two-step yield from the starting amide (**83**) (Scheme 30) ⟨76JCS(P1)2166⟩.

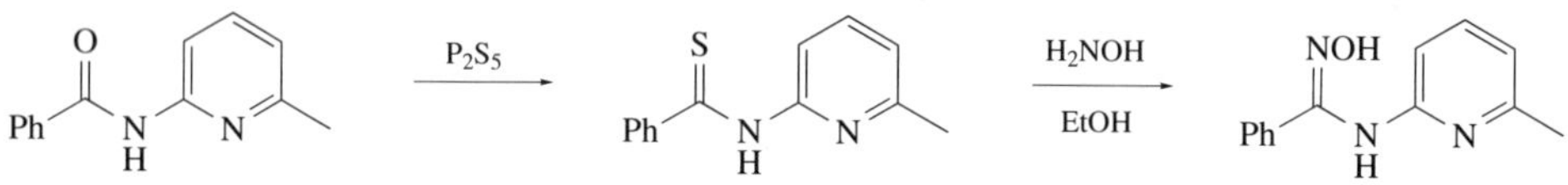

(83) **(82)**

Scheme 30

More highly substituted imidoylhydroxylamines have been prepared by the facile, room temperature reaction of imidoyl chlorides with hydroxylamine ⟨76AJC357⟩. *N*-Phenylhydroxylamine also undergoes reaction with imidoyl chlorides to produce *N*-imidoyl-*N*-phenylhydroxylamines (**84**) (Equation (33)) ⟨76AJC357⟩.

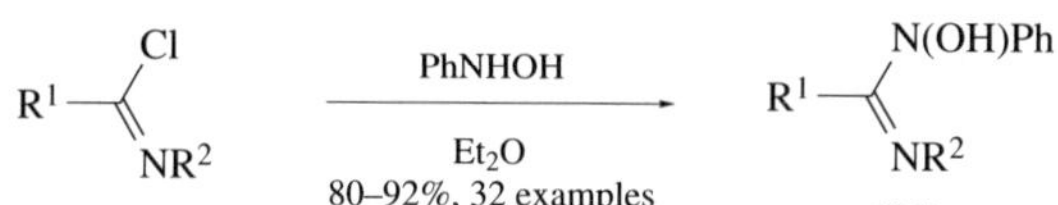

(33)

(84)

R^1, R^2 = alkyl or aryl

N-Formidoylhydroxylamines are often best prepared from formamide acetals. This can be by the reaction of hydroxylamine with a formamidine acetal ⟨89JCS(P1)2207⟩. A second strategy involves converting an amine, such as (**85**), into the formamidine (**86**). Subsequent reaction with hydroxylamine gives the desired imidoylhydroxylamine (**87**) even in the presence of a readily substituted chlorine atom (Scheme 31) ⟨76JHC487⟩.

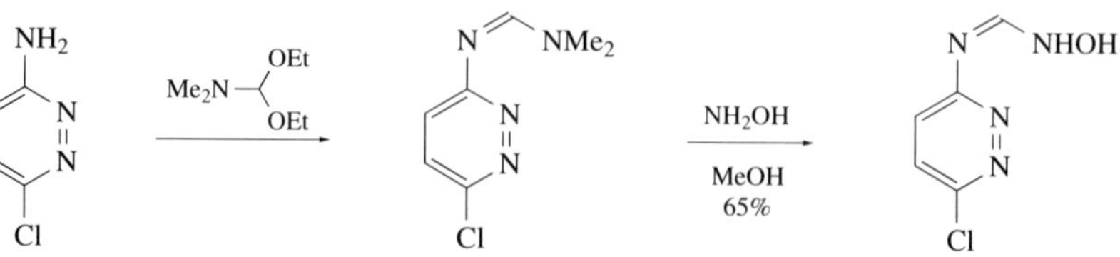

(85) **(86)** **(87)**

Scheme 31

5.19.2.2.2 N-*Imidoylhydroxylamines from amines and ammonia*

Chlorination of aldoximes gives oxyimidic chlorides (**88**) which react readily with ammonia to give imidoylhydroxylamines ⟨62CRV155⟩. Reaction of (**88**) with benzamide gives the benzoylated product (**89**) (Scheme 32) ⟨83CCC596⟩.

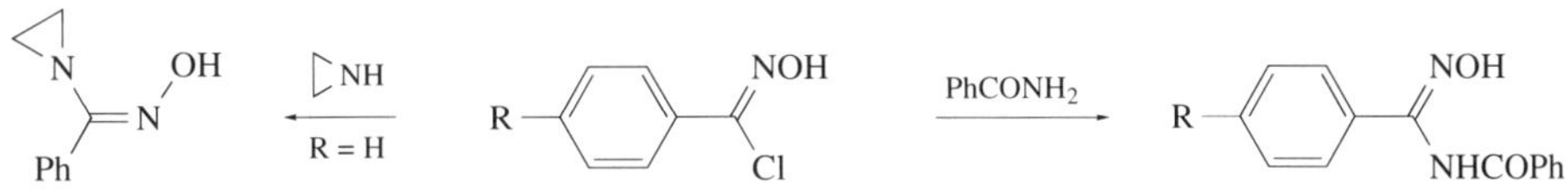

(90) **(88)** **(89)**

Scheme 32

Oxyimidic chloride (**88**; R = H) undergoes reaction with aziridine to give the thermodynamically stable (*Z*)-imidoylhydroxylamine (**90**) ⟨86JHC1861⟩. The reaction is thought to proceed via stereospecific addition of aziridine to benzonitrile oxide. The relative thermodynamic stability of (*E*)- and (*Z*)-imidoylhydroxylamines is finely balanced and depends on the exact nature of the substituents ⟨86JHC1861⟩.

N-Trimethylsilyldiethylamine (**91**) undergoes facile addition to acetonitrile oxide to give the kinetically favoured *O*-silylated (*Z*)-imidoylhydroxylamine. Over several days at room temperature the (*Z*)-isomer rearranges to the thermodynamically more stable (*E*)-isomer (Scheme 33) ⟨85JOM(281)135⟩.

Et₂N-TMS +

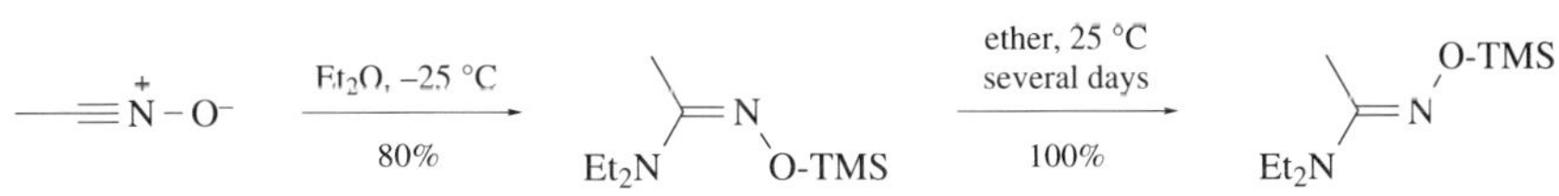

(91)

Scheme 33

5.19.2.2.3 N-*imidoylhydroxylamines by reduction*

Oxidation of cyclohexanol with concentrated nitric acid gives the nitrolic acid (**92**). Hydrogenation of this product using 5% platinum on carbon in acetic acid gave the imidoylhydroxylamine (**93**) (Scheme 34) ⟨56JA1461⟩.

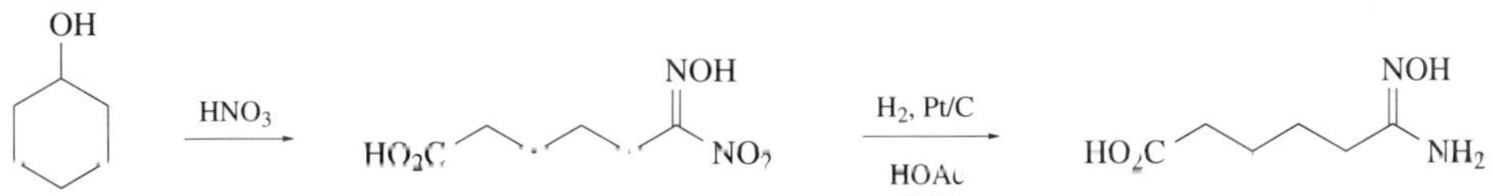

(92) **(93)**

Scheme 34

5.19.2.3 *N*-Imidoylsulfenamides, -sulfimides, -sulfinamides and -sulfonamides

5.19.2.3.1 N-*Imidoylsulfenamides* $R^1C(NR^2)\ NR^3\ SR^4$

N-Imidoylsulfenamides, also known as sulfenylamidines, are stable materials which are often isolated as crystalline, sharp melting solids; they are slowly hydrolysed by aqueous alcohol ⟨57CB1638⟩. *N*-Imidoylsulfenamides are prepared by the reaction of amidines with sulfenyl chlorides, with *N*-(phenylthio)phthalimide ⟨93JA7584⟩ or with sulfenate esters. Over 30 imidoylsulfenamides are reported by Goerdeler *et al.* ⟨57CB1638, 59CB2563⟩. Representative examples of preparations by Goerdeler and from other authors are given in Table 6.

Table 6 *N*-Imidoylsulfenamides.

Imidoylsulfenamide	*Sulfenating reagent*	*Yield (%)*	*Ref.*
1. 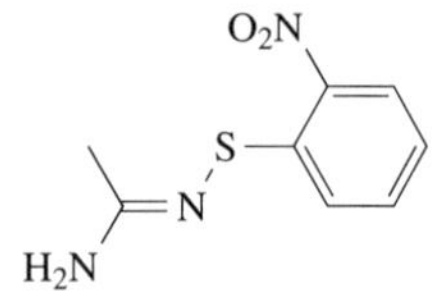	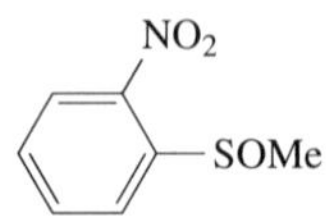	90	57CB1638
2. 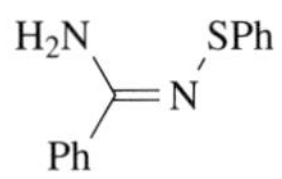	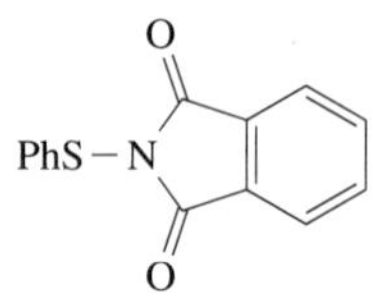	99	93JA7584
3. 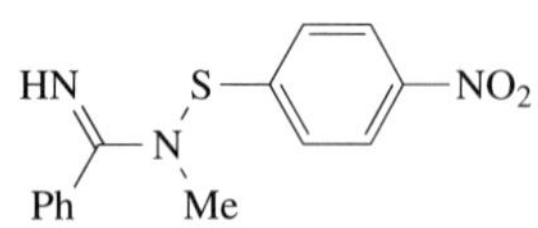	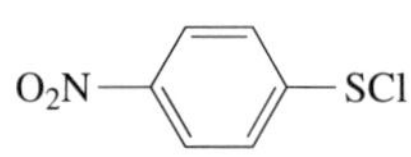	59	57CB1638
4. 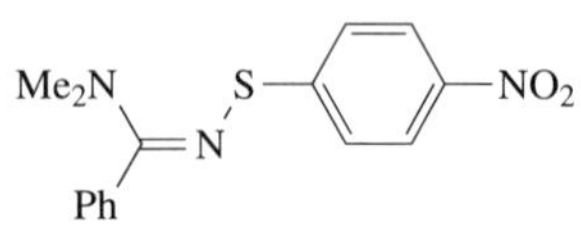	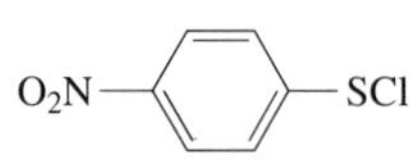	79	57CB1638
5. 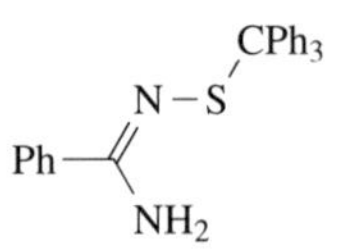	$Ph_3C–SOC_6H_4NO_2$	80–90	73JCS(P2)534
6. 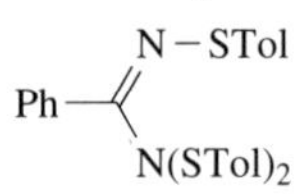	TolSCl	52	83CL885
7. 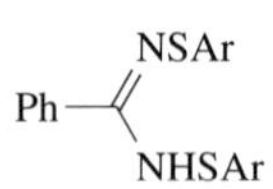	ArSCl	73	83CL885

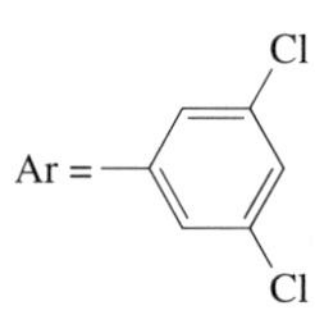

N-Alkylamidines are preferentially sulfenated on the substituted nitrogen (Table 6, entry 3). Interestingly benzamidine reacts with 3,5-dichlorobenzenesulphenyl chloride to give the disubstituted amidine (entry 7), but reacts with benzenesulfenyl chloride or toluenesulfenyl chloride to give trisubstituted amidines (entry 6).

5.19.2.3.2 N-*Imidoylsulfimides*

N-Imidoylsulfimides (**94**) are prepared by three methods which are shown in Scheme 35. Method 1 involves the reaction of *N*-chloroamidines with dialkylsulfides ⟨74CL247, 75JCS(P1)1964⟩. In method 2, the imidoylsulfimide is prepared by the reaction of an amidine with a sulfonium salt (**95**) ⟨74TL3319, 75JCS(PI)1964, 83JCS(PI)49⟩. *S*,*S*-Diphenylsulfimides (**94**; R = Ph) are prepared by method 3, in which a benzimidoyl chloride undergoes reaction with the sulfimide $Ph_2S{=}NH$ ⟨75JCS(PI)1964⟩. The chemistry and preparation of sulfimides, including imidoylsulfimides are reviewed by Gilchrist and Moody ⟨77CRV409⟩.

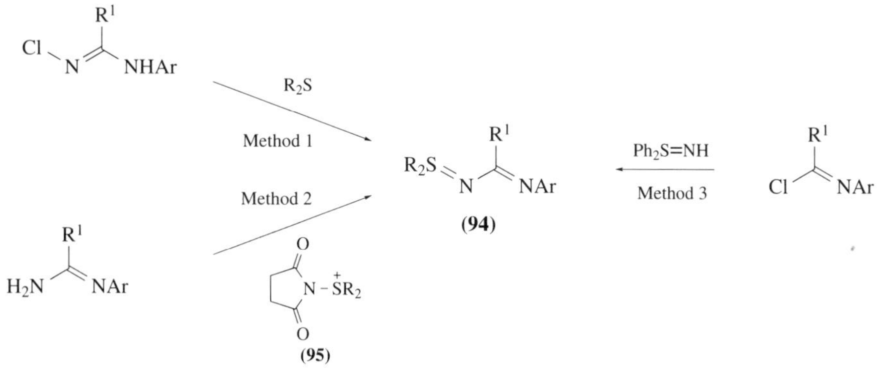

Scheme 35

5.19.2.3.3 N-*Imidoylsulfinamides*

No examples of *N*-imidoylsulfinamides have been reported.

5.19.2.3.4 N-*Imidoylsulfonamides*

N-Imidoylsulfonamides, also known as sulfonylamidines, have been reported widely. They are most commonly prepared by direct sulfonation of amidines or by converting an *N*-sulfonylcarboxyamide into its iminochloride followed by reaction with an amine. Amidines undergo reaction with aryl or alkylsulfonyl chlorides to give *N*-imidoylsulfonamides ⟨43JCS101, 71JCS(C)1704, 74JCS(P2)546⟩. The analogous preparations of imidoylsulfonamides from vinyl, styryl or substituted vinylsulfonyl chlorides have also been reported ⟨74JMC549, 74JOC3080⟩.

N-Sulfonylcarboxyamides are converted into their imino chlorides by treatment with PCl_5 in boiling benzene. Subsequent reaction with an amine or ammonia forms the imidoylsulfonamide (Scheme 36) ⟨34CB1762, 56JA3825, 70JCS(C)1429⟩.

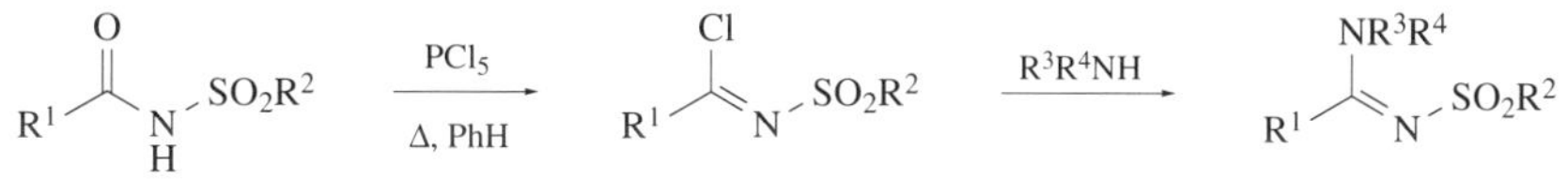

R^2 = Me or Ph

Scheme 36

Less common methods of preparing *N*-imidoylsulfonamides include the reaction of arylsulfonamides with isocyanates or imidic esters; the reaction of *N*-sulfinyl sulfonamides with ketenimines, ynamines or amides; the reaction of formamides or acetamides with *N*-sulfonyl isocyanates. All of these approaches are outlined in Scheme 37.

5.19.2.3.5 *Amidine derivatives with an* N-*selenium or* N-*tellurium bond*

(*N*,*N*-Dimethylmethanimamidato-*N'*)pentafluorotellurium (**95**) is prepared by the reaction of the isonitrile (**96**) with dimethylformamide accompanied by the loss of carbon dioxide (see section 5.19.1.3.5) ⟨84IC2217⟩. The selenium compound (**97**) is less reactive than (**96**) and does not react with dimethylformamide; indeed no amidine with an *N*-selenium bond has been reported (Scheme 38).

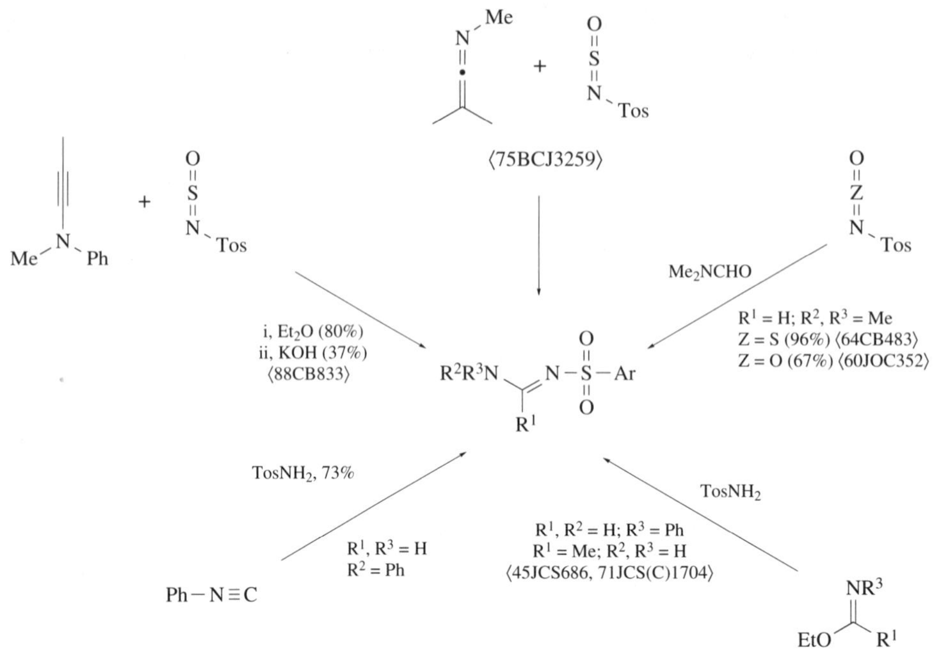

Scheme 37

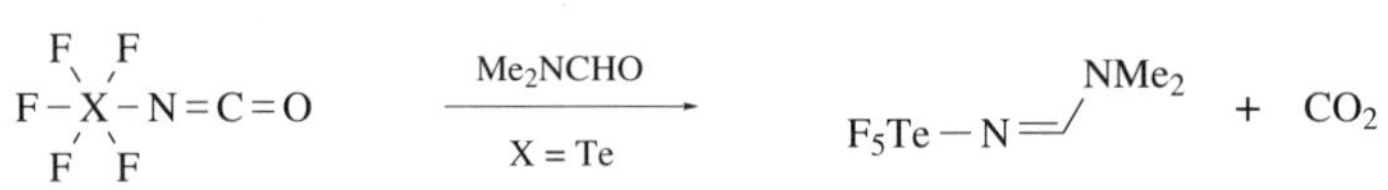

(96) X = Te
(97) X = Se

(95)

Scheme 38

5.19.2.4 Amidrazones and Related Structures

5.19.2.4.1 Introduction and nomenclature

Amidrazones are also known as hydrazidines, amide hydrazines or *N*-imidoylhydrazines. In this review the name hydrazidines will be used for structures such as (**98**). There have been two previous reviews on amidrazones by Neilson *et al.* ⟨70CRV151, B-75MI 519-03⟩. This present review in common with those previous will number the nitrogen atoms as shown for compound (**99**) which is therefore named N^1-phenyl-N^1,N^3,N^3-trimethylpropanamidrazone. Hydrazidines (**98**) are not covered in this section but are included in the two reviews by Neilson.

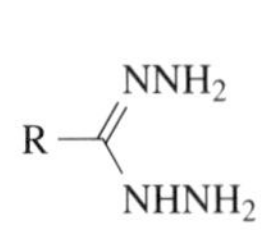

(98)

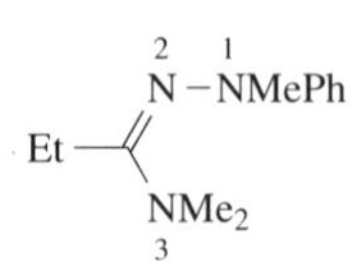

(99)

5.19.2.4.2 Primary amidrazones RC(NH)NHNH₂

Hydrazine undergoes nucleophilic addition reactions with electron deficient nitriles (Scheme 39). Thus the heterocyclic nitrile (**100**) adds hydrazine in good yield to give a primary amidrazone

⟨82CPB4521⟩. The addition of hydrazine to other heterocyclic nitriles including 2-pyridinecarbonitrile has been reported ⟨65JOC931, 68JHC223, 70MI 519-01⟩. A side product is the formation of the triazole (**101**) and this is the predominant product for nitriles which are not attached to an electron-withdrawing substituent. It has been thought since the mid 1950s that 4-pyridinecarboxyamidrazone could not be formed directly from 4-pyridinecarbonitrile ⟨56JCS2253⟩; however, more recently it has been shown that this preparation is successful if the reaction is performed at room temperature ⟨89MI 519-01⟩.

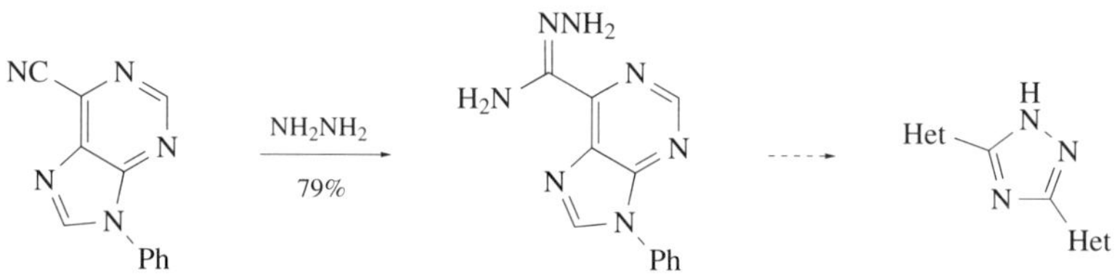

(100) **(101)**

Scheme 39

The sodium salt of hydrazine will react with nitriles which are electron-neutral. Hence benzonitrile and aliphatic nitriles form primary amidrazones when treated with sodium hydrazide in an inert solvent ⟨64CB3436, 66CB2600⟩.

The most widely used method of preparing primary amidrazones is by the substitution reaction of imidic esters (**102**; X = O) with hydrazine (Scheme 40) ⟨61CRV179, 70CRV151, B-75MI 519-02⟩. The best reaction conditions for aliphatic imidic esters are reported to be the use of one equivalent of anhydrous hydrazine with the imidic ester (free base) at 0 °C in ether, ethanol or ethanol–ether as solvent ⟨31M106, 33M285⟩. These conditions have also been successfully applied to aromatic imidic esters ⟨72JOC3958⟩. Using a higher reaction temperature, more than one equivalent of hydrazine or using hydrazine hydrate leads to greater formation of the hydrazidine (**103**) as well as heterocyclic by-products ⟨B-75MI 519-03, 76T1031⟩. The imidic ester method has also been used for the preparation of heterocyclic primary amidrazones ⟨56JCS2253⟩.

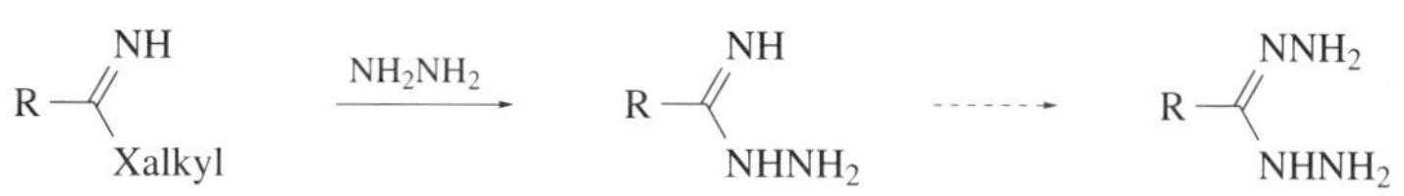

(102)
X = O or S
(103)

Scheme 40

Thioimidic esters (**102**; X = S) have not been widely used to prepare amidrazones but Kurzer and Doyle prepared both benzamidrazone and phenylacetamidrazone in about 80% yield by the action of anhydrous hydrazine on a thioimidic ester at 0 °C ⟨74S583⟩.

Primary amidrazones are also obtained by the rapid reaction of hydrazine with aryl or heteroaryl thiocarboxyamides. The preparation of 2- and 4-pyridinecarboxyamidrazones by this method gives 66% and 42% recrystallised yields, respectively, which are comparable with other procedures ⟨55HCA1560, 55RTC257⟩.

5.19.2.4.3 N-*Alkyl-, aryl- or alkenyl-substituted amidrazones*

(i) N-*Substituted amidrazones from hydrazines*

The room-temperature reaction of imidic esters with monosubstituted, and especially with *N*,*N*-disubstituted hydrazines, give amidrazones with fewer side products than the equivalent reaction with hydrazine itself. When substituted hydrazines react with *N*-substituted imidic esters such as (**104**), N^1,N^3-disubstituted or N^1,N^1,N^3-trisubstituted amidrazones (Equation (34)) are formed ⟨68JIC729, 72LA(758)162⟩.

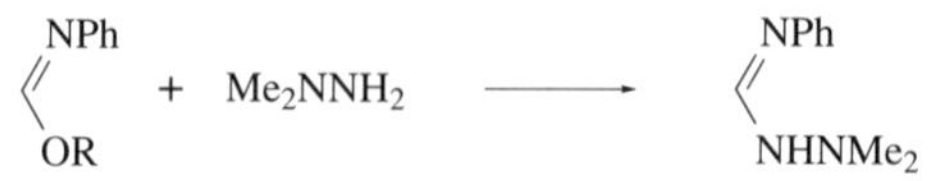

(**104**)

(34)

Imidoyl chlorides react with mono-, 1,1-di-, 1,2-di- and trisubstituted hydrazines to give amidrazones. The high reactivity of the imidoyl chloride leads to some loss of selectivity. Hence (**105**) undergoes reaction with phenylhydrazine to give a mixture of N^1- and N^2-substituted isomers ⟨70CRV151⟩. *N*-Phenylbenzimidoyl chloride (**105**) cleanly forms the trisubstituted amidrazone (**106**) with a large excess of 1,1-dimethylhydrazine, but with just 2 moles of 1,1-dimethylhydrazine a mixture of (**106**) and (**107**) is obtained (Scheme 41).

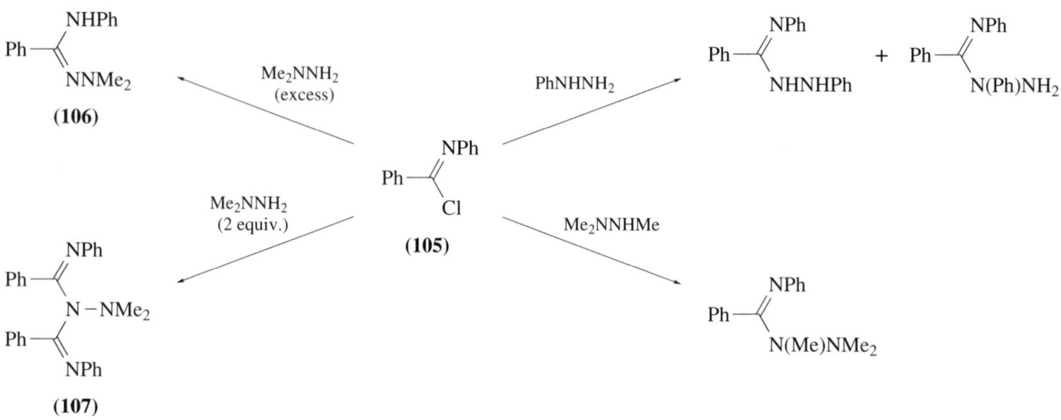

Scheme 41

Amidrazones are formed by the addition of hydrazines to nitriles which have been activated by complexation with a Lewis acid, such as aluminum chloride ⟨72LA(758)162⟩. One procedure involves complexation of acrylonitrile with iron(III) chloride; treatment *in situ* with an alkylating agent and phenylhydrazine to give propenamidrazones (**108**) (Scheme 42) ⟨92MI 519-01⟩. Further elaboration is achieved by the 1,4-addition of a thiol to (**108**) in up to 94% yield ⟨92MI 519-02⟩.

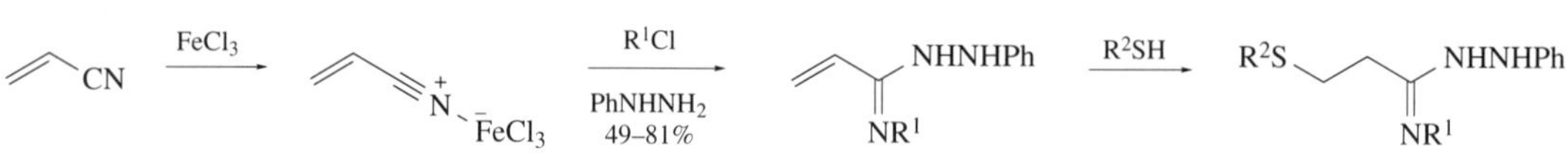

(**108**)

R^1 = Bn, Bu^t, Et

Scheme 42

Hydrazine undergoes an extremely facile reaction with the ketenimine (**109**) to give an N^3-substituted amidrazone ⟨65JOC3718⟩. The reaction of 1,3,5-triazine with most hydrazines gives complex mixtures but with 1,1-disubstituted hydrazines, formamidrazones can be prepared ⟨63AG(E)309⟩. The formamidrazone (**110**) was prepared by treating the appropriate arylhydrazine with boiling dimethylformamide dimethyl acetal ⟨84JCS(P1)993⟩. Condensation of the *N*-aryl thioamide (**111**) with hydrazine hydrate below room temperature gives the N^3-substituted amidrazone (**112**) with evolution of hydrogen sulfide ⟨62JGU3330⟩. The last four amidrazone preparations are outlined in Scheme 43.

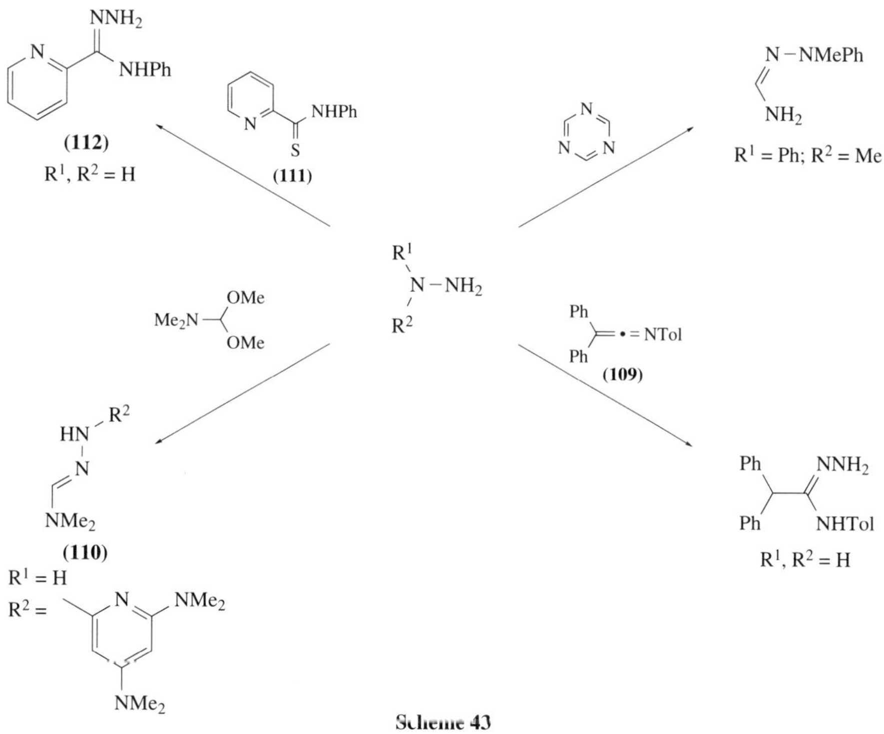

Scheme 43

(ii) N-*Substituted amidrazones from amines and ammonia*

The very reactive halogen atom of hydrazanoyl halides is readily displaced with ammonia or amines. This method has been very widely used to prepare amidrazones particularly where R^1 is aryl, acyl or alkoxycarbonyl (Equation (35)) ⟨72GEP2146192, 80JHC877, 80LA1376, 82JCS(P1)2663, 83JHC225, 84BCJ1650⟩. The reaction works well for weakly basic heterocyclic amines such as given in Scheme 44 ⟨83JHC225, 84BCJ1650⟩. The hydrazanoyl halide can be prepared by chlorination of the appropriate acyl hydrazine as in Scheme 44 or by bromination of a hydrazone (Scheme 45) ⟨72LA(758)162⟩.

(35)

R^1 = aryl, acyl, CO_2R; X = Cl, Br

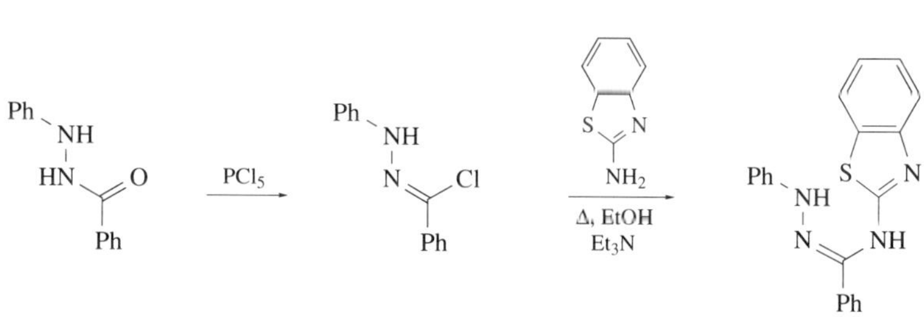

Scheme 44

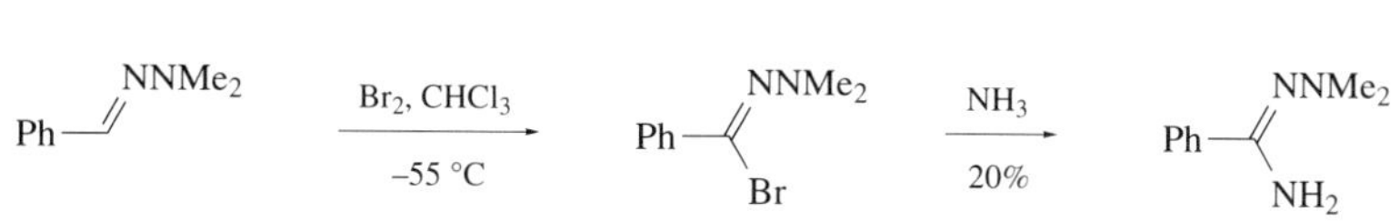

Scheme 45

The condensation of hydrazines with amides is promoted by phosphorus oxychloride or *p*-toluenesulfonyl chloride ⟨67JCS(C)690, 71JOC1155⟩. The intermediate (**113**) reacts with the hydrazone (**114**) to give the amidrazone (**115**) (Scheme 46).

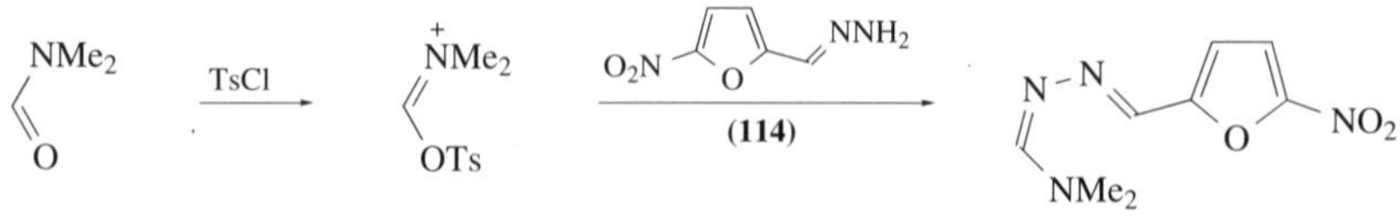

(113) **(115)**

Scheme 46

Amidrazones of this type can also be prepared by condensing a primary amidrazone with an aldehyde ⟨83EJM413⟩.

The reaction of ammonia or amines with hydrazonate esters $R^1C(OR^2)NNR^3_2$ is a relatively poor method of preparing amidrazones ⟨B-75MI 519-03⟩. A hydrazone with a nitro group in the α-position (**116**; R = Me) can be hydrogenated over Raney nickel to give an N^3-unsubstituted amidrazone ⟨51LA(574)85, 55CB1284⟩. The nitro group of (**116**, R = H) is displaced by morpholine to give an amidrazone (Scheme 47) ⟨77JOC2091⟩.

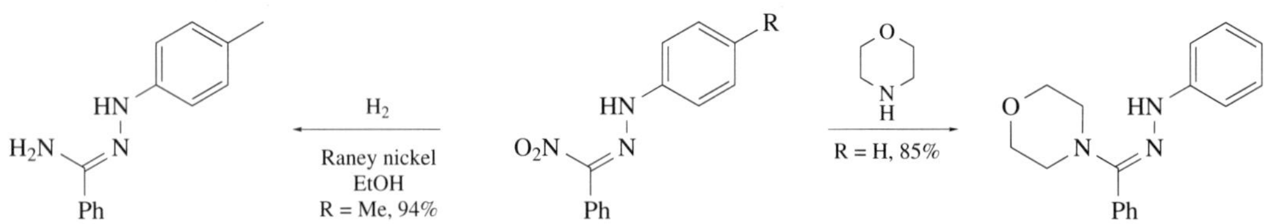

(116)

Scheme 47

(*iii*) N-*Substituted amidrazones by* N-*alkylation of simpler amidrazones*

The alkylation of amidrazones has been studied by Smith *et al.* ⟨71JOC1155, 73JOC1344, 77JOC1862⟩. In most cases, methylation at N^2 or N^3 occurs so that an 'amidinium like' delocalised cation is formed. N^1,N^1-Dimethylacetamidrazone (**117**) and the N^1,N^1,N^3-substituted amidrazones (**118**)–(**120**) are all alkylated at N^2 (Schemes 48 and 49). Amidrazones (**121**) and (**122**) are also methylated to give a delocalised cation, but with these substrates, this is accomplished by alkylation at the N^3-atom (Equation (36)).

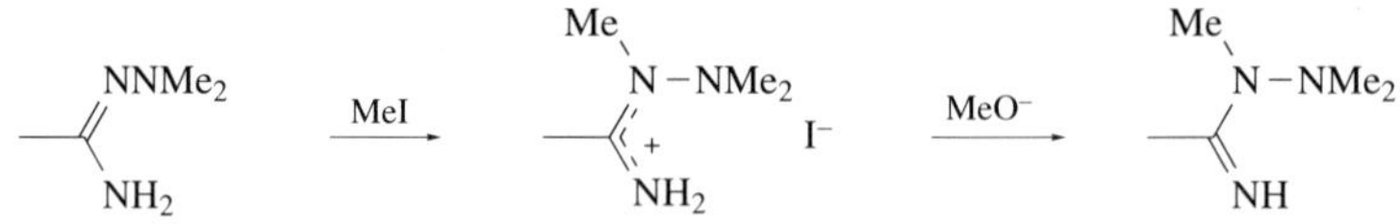

(117)

Scheme 48

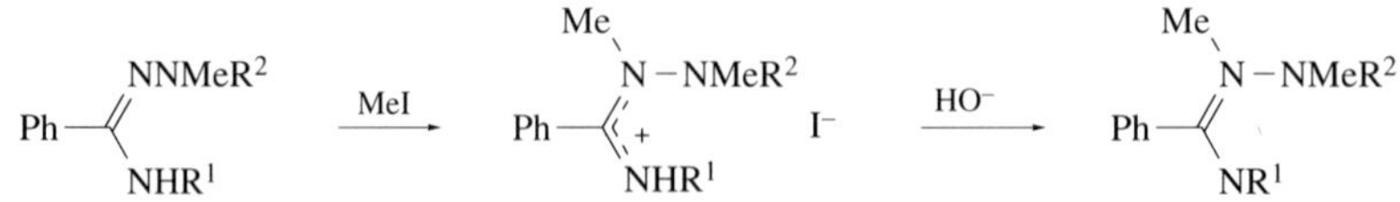

(**118**) R^1 = Me; R^2 = Me
(**119**) R^1 = Ph; R^2 = Me
(**120**) R^1 = Me; R^2 = Ph

Scheme 49

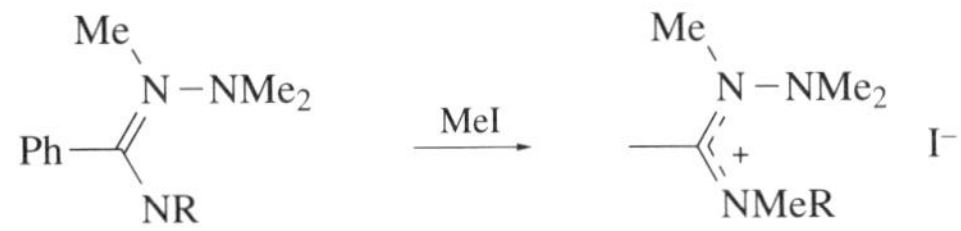

(36)

(121) R = Ph
(122) R = Me

With N^3-disubstituted amidrazones, the reactions are more finely balanced. Amidrazones **(123)**–**(125)** are quaternised at the N^1-atom by methyl iodide. In contrast, N^1-phenyl-N^1-methyl-N^3-dimethylformamidrazone **(126)** and propanamidrazone **(127)** are methylated at the N^2-position.

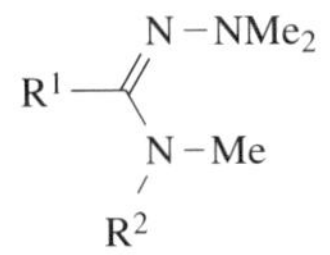

(123) R^1 = Ph; R^2 = Ph
(124) R^1 = Ph; R^2 = Me
(125) R^1 = Et; R^2 = Me

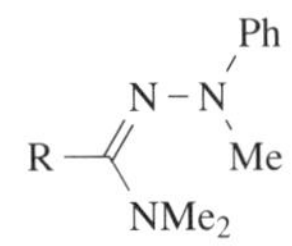

(126) R = H
(127) R = Et

5.19.2.4.4 N-*Acylamidrazones*

(i) N^1-*Acylamidrazones*

A number of routes to N^1-acylamidrazones have been described. Primary amidrazones are acylated directly at N^1 in good yield ⟨71CB639, B-75MI 519-03⟩. Acylhydrazines react with imidic esters to give acylamidrazones ⟨70JPR601, 82TL3357⟩. Not surprisingly, the reaction is slower but much cleaner than with hydrazine. Some care must be taken to avoid cyclisation to a triazole; thus **(128)** can be isolated in good yield but on heating with 1 mole of potassium carbonate, cyclisation to the triazole **(129)** is observed ⟨83JMC1187⟩. N^1-Formylamidrazones have also been prepared by this method ⟨68CB2028⟩ (Scheme 50).

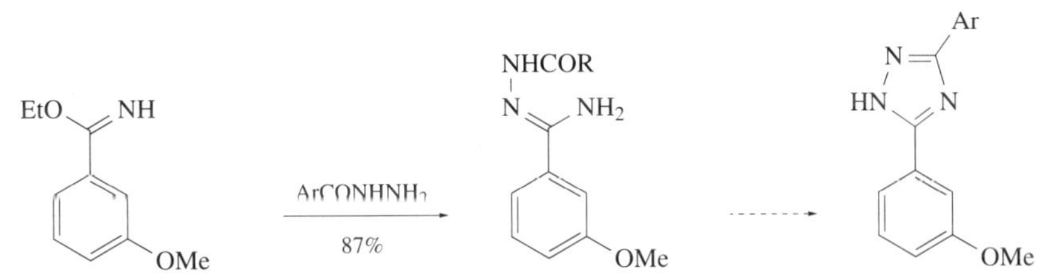

Ar = 2-ethylphenyl **(128)** **(129)**

Scheme 50

Acylhydrazines will also react directly with nitriles if catalysed by triethylaluminum ⟨81CB3684⟩. Reaction of acylhydrazines with dimethylformamide dimethylacetal is an excellent way of preparing *N*-acylformamidrazones. The reaction can be extended to acetamidrazones, hence the acyl hydrazine **(130)** reacts with the acetal **(131)** to give the acetamidrazones **(132)** ⟨87T4185⟩. Reaction of the hydrazonate ester **(133)** with a boiling aqueous solution of dimethylamine is an alternative method of preparing **(132)** (Scheme 51).

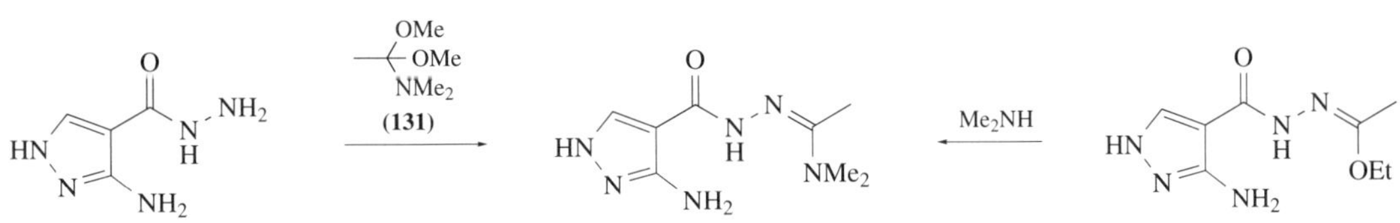

(130) **(132)** **(133)**

Scheme 51

A more unusual way of preparing N^1-acylformamidrazones is by treating a tetrazole with phosphorus oxychloride and DMF at room temperature (Equation (37)) ⟨74TL1187⟩.

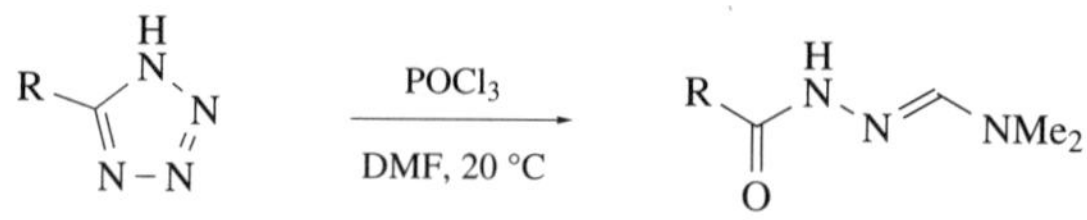

(37)

(*ii*) N^3-*Acylamidrazones*

N^1,N^1-Disubstituted amidrazones are acylated on the N^3-atom (Equation (38)) ⟨68TL3569, 83JHC69⟩. N^1,N^3-Diphenylbenzamidrazone (**134**) is benzoylated at N^3 by treating with benzoyl chloride followed by perchloric acid to give the triazolium salt (**135**). Subsequent ring opening with hydroxide ion gives the N^3-benzoylamidrazone (**136**) (Scheme 52) ⟨69G69⟩.

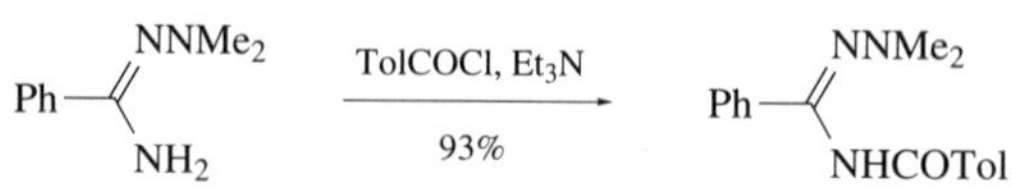

(38)

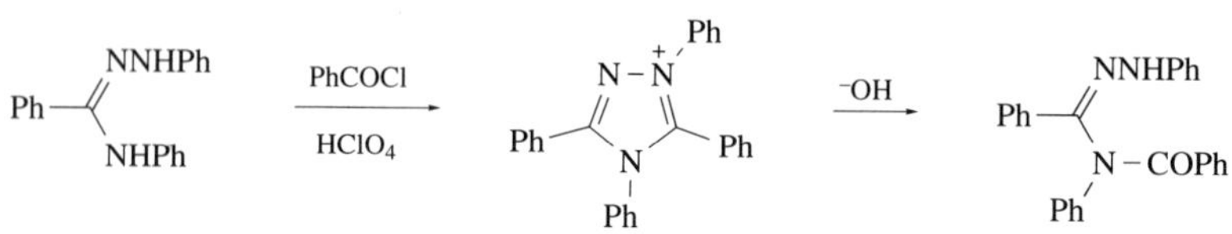

(**134**) (**135**) (**136**)

Scheme 52

5.19.2.5 Amidine Derivatives with an *N—P, N—As or N—Sb Bond*

5.19.2.5.1 N-*Phosphorylamidine derivatives*

N-Phosphorylamidine derivatives have been prepared by several routes. *N*-Chlorobenzamidines undergo an Arbusov-type reaction with tribenzyl phosphite to give *N*-phosphorylbenzamidines (Equation (39)) ⟨92NKK1344⟩.

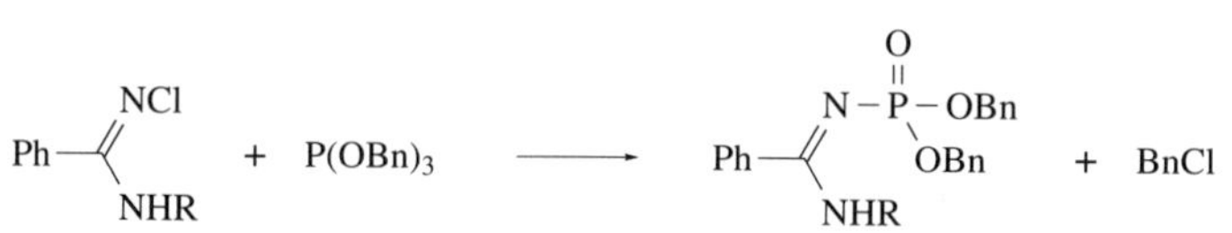

(39)

Phosphoryl- and thiophosphorylformamidines have been widely prepared by treating the imidate esters (**137**)–(**139**) with a wide variety of amines ⟨78ZOB78, 78GEP2701309, 78GEP2758173, 89JPR149, 90MI 519-01⟩. The iminochloride derivatives (**140**) undergo similar reaction with amines or ammonia (Equation (40)) ⟨71MI 519-01⟩.

(40)

(**137**) R^1 = H; Z = O; X and Y = alkoxy
(**138**) R^1 = H; Z = S; X = alkoxy; Y = SR^2
(**139**) R^1 = H; Z = O; X = alkoxy; Y = SR^2
(**140**) R^1 = CH_2Cl, CF_3; Z = O; X = Cl; Y = alkoxy

Direct thiophosphorylation of primary amidines proceeds in excellent yield in aqueous sodium hydroxide (Equation (41)) ⟨78ZOB2465⟩.

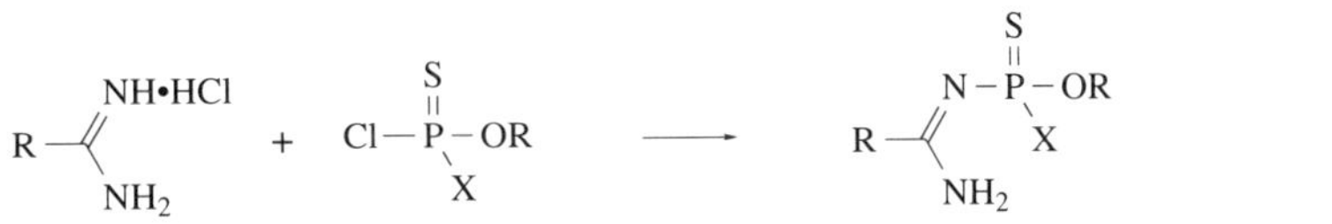 (41)

X = alkoxy, PhO, Me_2N, Et_2N

The enamine (**141**) undergoes a 1,3-dipolar cycloaddition reaction with diphenylphosphorylazide to give the intermediate (**142**); a 1,2-phenyl migration and evolution of nitrogen gives the phosphorylated amidine (**143**) (Scheme 53) ⟨83CPB2564, 83CPB3139⟩.

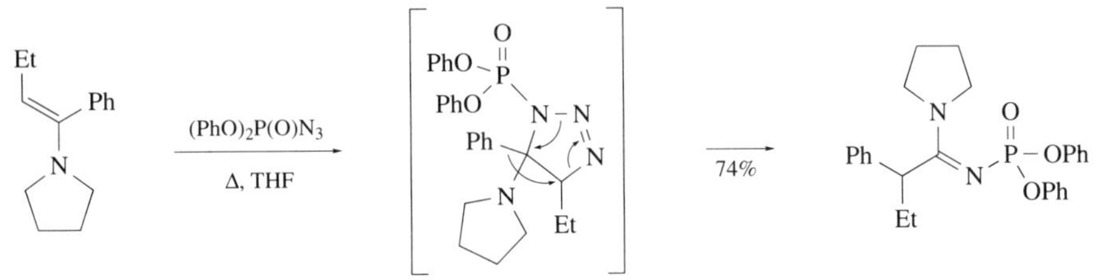

(**141**) (**142**) (**143**)

Scheme 53

The condensation of *N*-imidoylhydroxylamines (**144**) with tris(dimethylamino)phosphine gives the N^2-[bis(dimethylamino)phosphinoyl]amidines (**145**) along with the cyclic co-products (**146**) (Equation (42)) ⟨77JCS(P1)1806⟩.

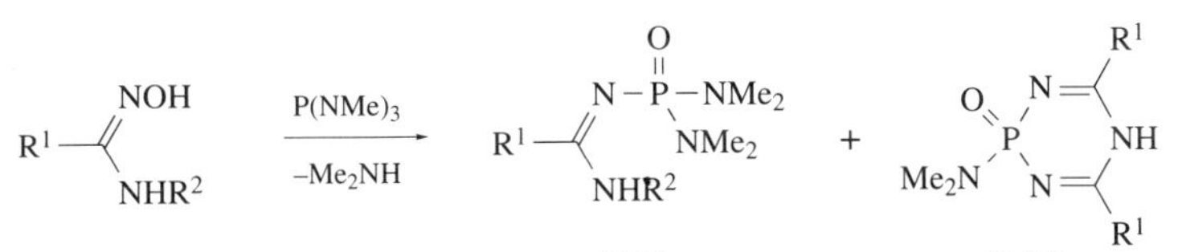 (42)

(**144**) (**145**) (**146**)

R^1 = Me; R^2 = H
R^1 = Me; R^2 = Ph
R^1 = Ph; R^2 = H
R^1 = Ph; R^2 = Me

5.19.2.5.2 N-*Phosphorus amidines (excluding* N-*phosphorylamidines)*

The dithiophosphonylformamidine (**147**) is prepared from the appropriate dithiophosphonylamine and dimethylformamide dimethyl acetal ⟨85GEP3501452⟩. The reaction of *N*-chlorobenzamidines with triphenylphosphine gives the ylide (**148**) in moderate yield. The ylides are readily protonated by HCl to form a salt (**149**) ⟨79BCJ1541⟩. *N*-Chlorobenzamidines will also undergo a high yielding reaction with phosphorus trichloride in the presence of pyridine at 0–5°C to give the ylide (**150**) ⟨84ZOR1784⟩. The phosphinoamidine (**151**) has been prepared by the reaction of diisopropylphosphorus chloride with the trimethylsilylamidine (**152**) (Equation (43)) ⟨91ZOB1581⟩.

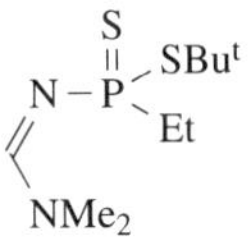

(**147**)

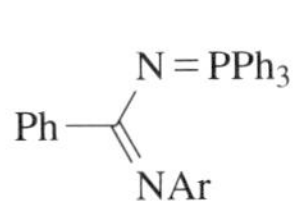

(**148**)

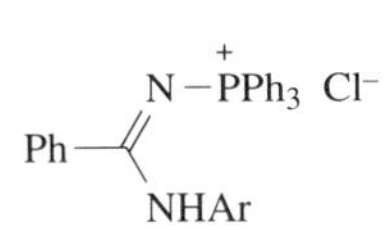

(**149**)

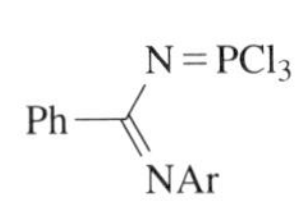

(**150**)

(152) **(151)** (43)

Condensation of iminochlorides with phosphoramides in the presence of triethylamine gives *N*-phosphorylbenzamidine derivatives (Equation (44)) ⟨85ZOB776⟩.

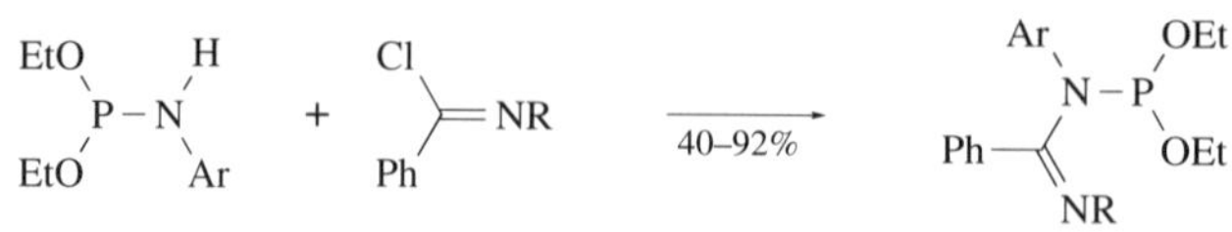

(44)

R = Me, Ph

5.19.2.5.3 Amidines with a N*-arsenic bond*

The arsoranylbenzamidine derivative (**153**) is prepared by the reaction of *N*-chloro-*N*,*N′*-dimethylbenzamidine (**154**) with triphenylarsenic. Variable temperature NMR studies show that there is 1,3-migration of Ph_3As between the two nitrogen atoms (Scheme 54) ⟨80CB1394⟩. The reaction of (**154**) with arsenic trichloride gives the octahedral arsenic complex (**155**) for which x-ray data is available ⟨86ZAAC(533)55⟩.

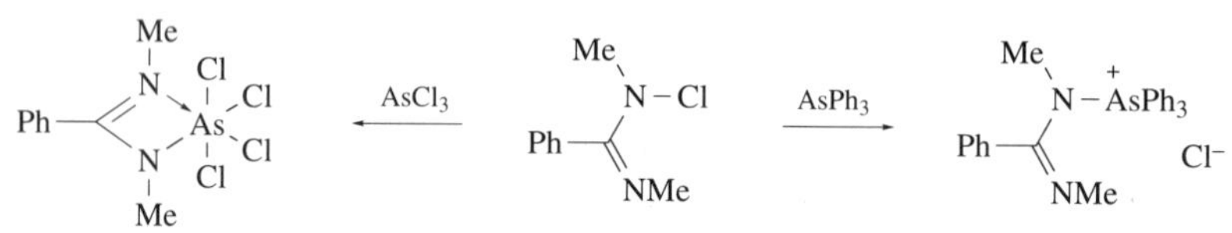

(155) **(154)** **(153)**

Scheme 54

5.19.2.5.4 Amidines with an N*-antimony bond*

The antimony–benzamidine complex (156) is prepared from *N*-chloro-*N*,*N′*-dimethylbenzamidine (**154**) and triphenylantimony. Unlike the arsenic derivative (**153**) described above, the antimony complex prefers the chelated structure (**156**) ⟨80CB1394⟩. The related antimony complexes (**157**)–(**159**) have also been prepared from (**154**). The x-ray structures of (**157**) and (**158**) are available. Structures (**156**) and (**159**) are inferred from Mössbauer spectroscopy ⟨80CB1394, 82ZAAC(486)61, 86ZAAC(533)55⟩.

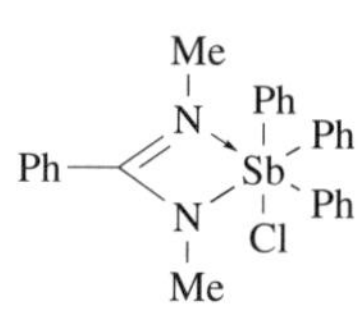

(156)

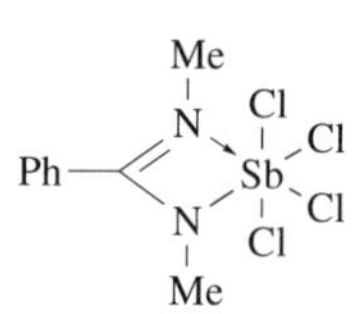

(157)

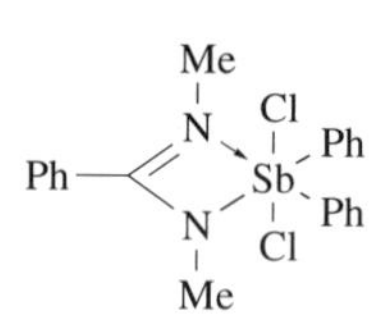

(158)

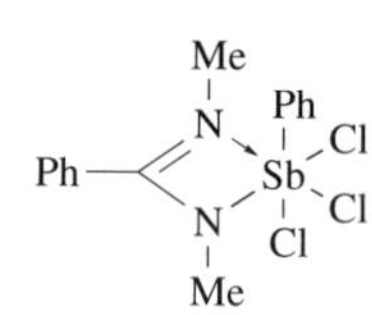

(159)

Arylamidoximes react with antimony pentachloride to give stable, octahedral antimony complexes (**160**). Sodium arylamidoximes react with Ph_3SbBr_2 to give the complex (**161**) in which one amidoxime molecule is a bidendate ligand and the other a monodentate ligand ⟨90SRI1413⟩.

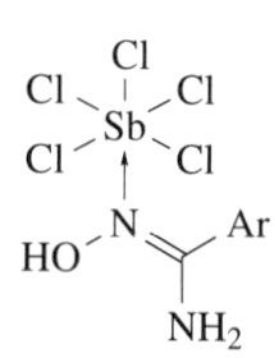

(160)

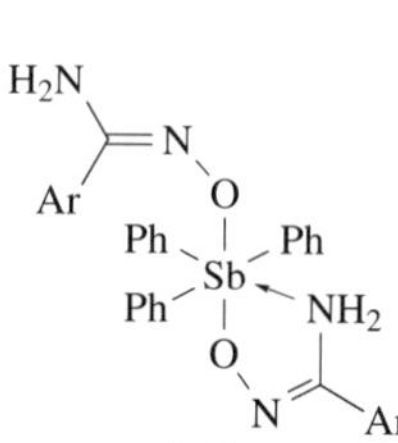

(161)

5.19.2.6 Amidine Derivatives with an *N′*—Metalloid Bond

5.19.2.6.1 N-*Silylamidines*

(i) Monosilylamidines

Reactions of the lithium salt of *N*-(trimethylsilyl)aniline (**162**) with the appropriate iminochlorides give the mono-*N*-silylamidines (**163**) ⟨78ZAAC(445)107⟩. *N*-Methyl-*N′*-trimethylsilylbenzamidine (**164**) is formed by the initial reaction of lithium methylamide with benzonitrile and quenching the resulting lithium anion with trimethylsilyl chloride ⟨68CB2533⟩. Direct monosilylation of *N,N′*-diphenylformamidine is accomplished with trimethylsilyl chloride and triethylamine ⟨78ZAAC(445)107⟩. The diamidine (**165**) is silylated with an excess of bis(trimethylsilyl)amide to give (**166**) and ammonia (Equation (45)) ⟨81JA4186⟩.

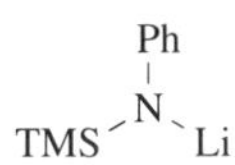

(162)

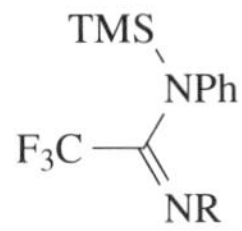

(163)
R = Ph or 3-$CF_3C_6H_4$

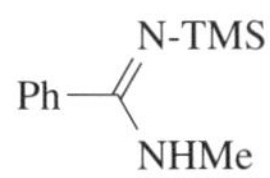

(164)

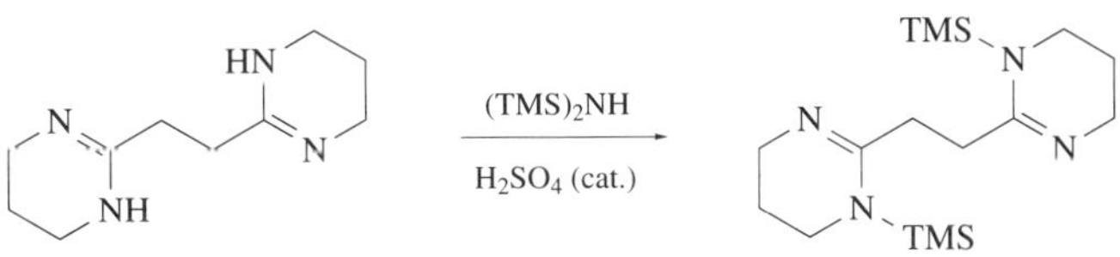

(165) **(166)** (45)

Monosilylformamidines are prepared by hydrosilylation of a symmetrical carbodiimide with a variety of trialkyl or aryldialkylsilanes (Equation (46); $R^2 = H$). The reaction is catalysed by palladium(II) chloride or tris(triphenylphosphine)rhodium(I) chloride ⟨74JOM(72)C11, 77JOM(140)97, 91S954⟩. Trimethylsilyl cyanide will react with *N,N′*-diisopropylcarbodiimide without a catalyst to give *N,N′*-diisopropyl-*N′*-trimethylsilyl-1-cyanoformamidine though a high reaction temperature (190 °C) is required (Equation (46); $R^1 = Me$, $R^2 = CN$). If the reaction is catalysed by 5 mol% of aluminum trichloride, the cyanoformamidine (**167**, $R^2 = CN$) is formed at room temperature (Equation (46)) ⟨75JOM(99)C5⟩.

(46)

$R^1 = Et$; $R^2 = H$, 82%
$R^1 = Me$, $R^2 = CN$, 84%

(167)

(ii) Disilylamidines

N,N′-Bis(trimethylsilyl)benzamidine is prepared by the high yielding reaction of lithium or calcium bis(trimethylsilyl)amide with benzonitrile. The product is isolated as either the lithium or calcium salt ⟨87JOM(331)161, 92ZN(B)453⟩. The reactions are best performed in ether or THF, respectively, as this reduces the competing base-catalysed trimerisation of the benzonitrile. *N,N′*-Bis(trimethylsilyl)benzamidine can also be prepared by the addition of phenyllithium to N^1,N^3-bis(trimethylsilyl)carbodiimide. Again the product is isolated as the lithium salt (Equation (47)) ⟨90CZ295⟩. If an additional *N*-alkyl substituent is required then a sequence such as that outlined in Scheme 55 can be used ⟨68CB2533⟩.

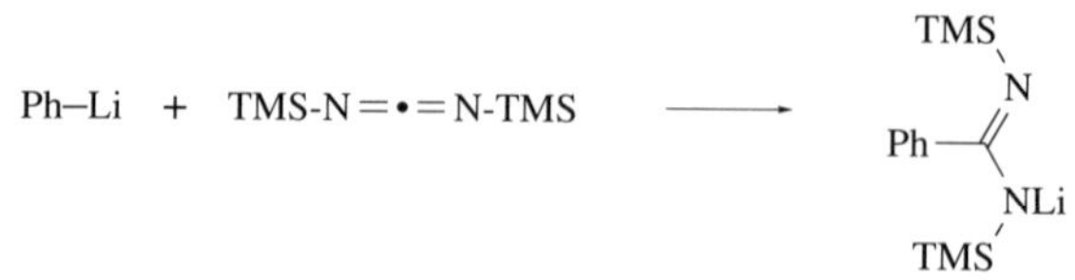

(47)

(168)

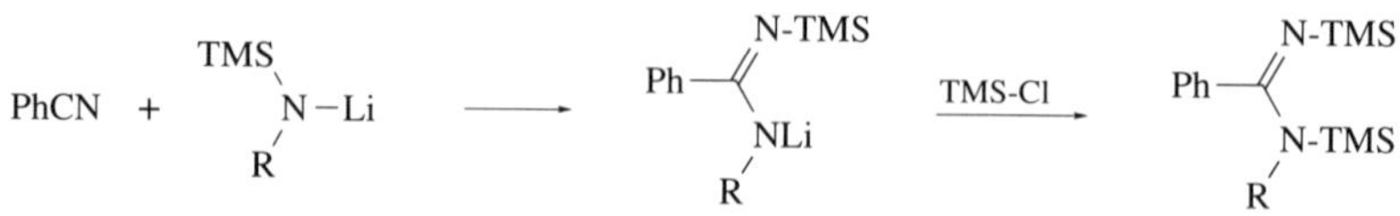

R = Me, Et

Scheme 55

(iii) Trisilylamidines

N,*N*,*N*′-Tris(trimethylsilyl)benzamidine and its derivatives are prepared by the addition of lithium bis(trimethylsilyl)amide to benzonitriles (**169**) followed by quenching with trimethylsilyl chloride (Equation (48)) ⟨87JOM(331)161⟩.

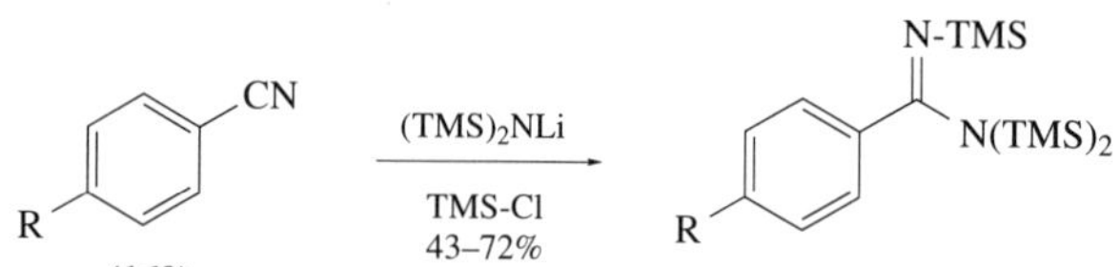

(48)

(169)

R = H, Me, Cl, MeO, Me_2N, CF_3, CN, NO_2, Ph

5.19.2.6.2 N-*borylamidines*

(i) N-*(Trialkylboryl)amidines*

Acetamidine forms an isolable complex with tripropylborane, which exists as a low melting crystalline solid ⟨71DOK(199)1328⟩.

(ii) N-*(Dialkylboryl)amidines and* N-*(diarylboryl)amidines*

A number of dialkylborylamidines have been prepared in good yield by heating *N*,*N*′-disubstituted amidines and trialkylboranes in THF at between 70 and 90 °C. The reaction proceeds via a trialkylborylamidine complex which eliminates a molecule of a saturated hydrocarbon ⟨78BAU2304⟩. An alternate procedure involves heating the *N*,*N*′-disubstituted amidine with a dialkylthioborinate at a slightly higher temperature ⟨74BAU1593⟩. The dialkyborylamidines (**170**) to (**172**) are reported to exist in solution as the intercoordinate compounds, whereas some dialkylborylamidines including (**173**) have been shown by ^{1}H and ^{11}B NMR to exist as a mixture of cyclic and open chain structures (Equation (49)) ⟨77PAC749, 78BAU2304⟩. Over 30 borylamidines derivatives have been prepared by Maringelle and Meller by the reaction of silylamidines with boron halogen compounds (Equation (50)) ⟨78ZAAC(445)107⟩. *N*-Borylamidines can also be made directly from a nitrile, a trialkylborane and an amine, if that amine contains an additional coordinating group. Thus heating a mixture of 3-amino-1,2,4-triazole, tripropyl- or tributylborane and an arylcarbonitrile gives the dialkylborylamidine complex as shown in Equation (51) ⟨87IZV165⟩. Such chelates can be hydrolysed to the parent amidines by treating with a mixture of aqueous HCl, butanol and benzene ⟨89IZV2353⟩.

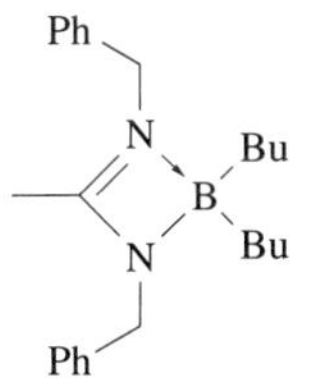

(170)

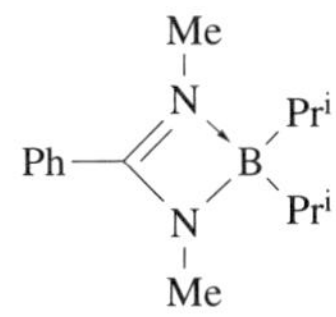

(171)

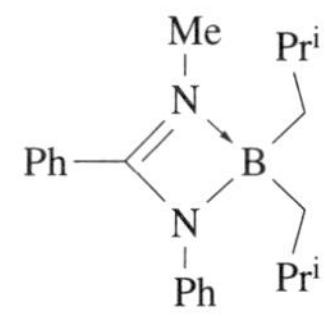

(172)

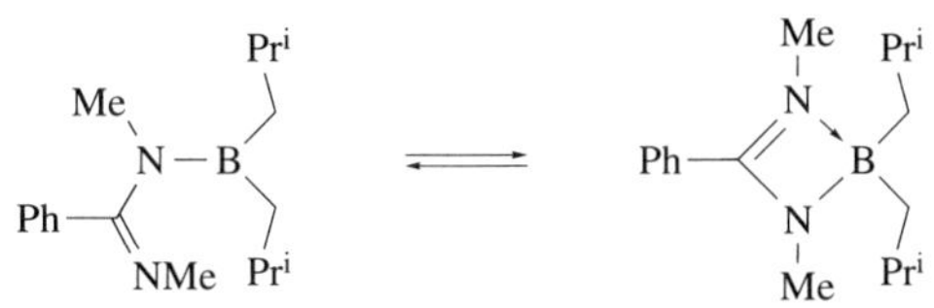

(173) **(173)**

(49)

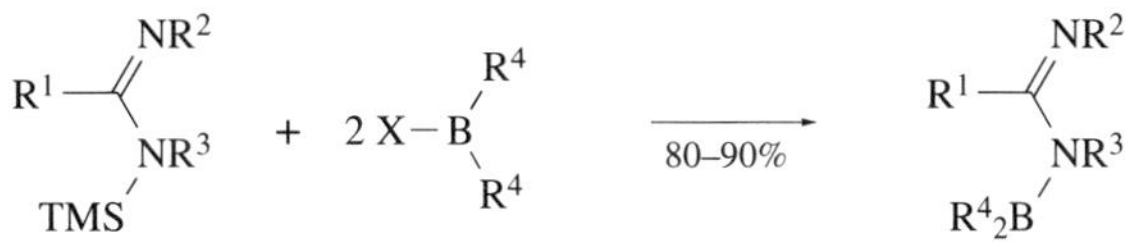

(50)

R^1 = H or CF_3; R^2, R^3 = aryl; R^4 = Ph, Me, NMe_2; X = Cl or Br

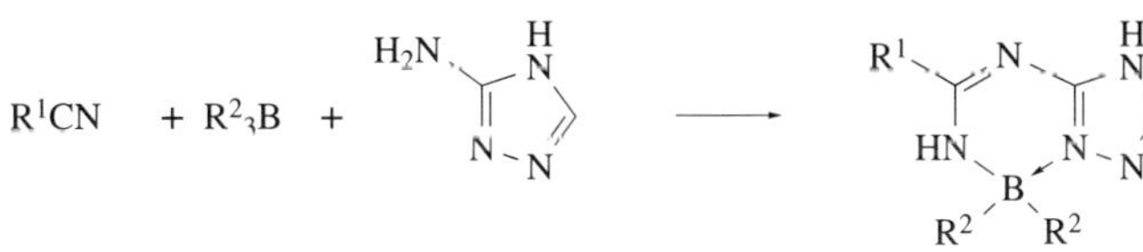

(51)

R^1 = Ph, Tol, 2-MeC_6H_4; R^2 = Pr, Bu

(iii) Miscellaneous borylamidines

Oxamidines (**174**) have the possibility of forming either four-membered or five-membered inter-coordinated complexes with trisubstituted boranes. In practice, heating 1 mole of the oxamidine with 2 moles of the boron compound in either boiling toluene, chlorobenzene or dichlorobenzene give *N,N′*-diborylamidines (**175**) as colourless crystalline solids which are thermally stable at above 300 °C (Equation (52)) ⟨67JA7014⟩.

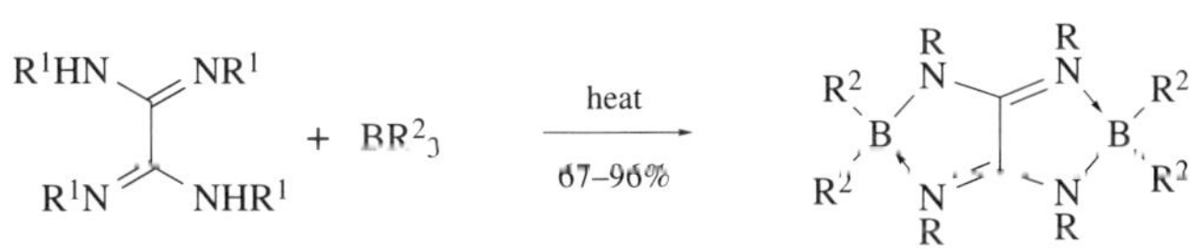

(52)

(174) R^1 = Me, Et, Pr, Ph; R^2 = Et, Ph, Br **(175)**

Borylamidines are also prepared by phenyl boration of *N,N′*-ditolylcarbodiimide with either triphenylborane or chlorodiphenylborane to give the borylamidines (**176**). In the reaction of chlorodiphenylborane, insertion into the boron—phenyl bond occurs in preference to insertion into a boron—chlorine bond (which occurs when boron trichloride reacts with a carbodiimide) (Equation (53)) ⟨66JCS(A)1584⟩.

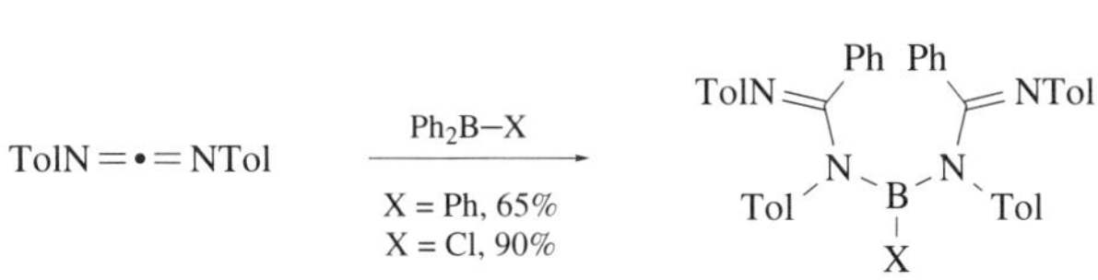

(53)

(176)

5.19.2.7 Amidine Derivatives with an *N*-metal bond $R^1C(NR^2)NR^3$-M

Amidines form salts with the alkali metals and alkaline earth metals and complexes with most other metals. The amidinato ligand can be either η_1 or η_2 with π-type or N—M σ-bonding. Since 1988, a large number of complexes formed from *N*-silylated amidines have been reported.

5.19.2.7.1 *Amidines with an* N-*metal bond, where M is a Group 13 metal*

The interaction of *N,N'*-dimethylacetamidine with trialkylaluminum, gallium or indium forms the puckered eight-membered ring structure (**177**) ⟨78ZN(B)24⟩. With *N,N',N'',N'''*-tetramethyloxamidine, the trimethylmetals give monomeric bis(dimethylmetal)oxamidines (**178**) ⟨79JOM(175)33⟩. Thallium(III) chloride reacts with *N,N,N'*-tris(trimethylsilyl)benzamidine in THF solution to give (**179**). Analysis by x-ray crystallography shows the structure consists of an amidinium cation and an anion in which the thallium is pentacoordinate ⟨91ZAAC(596)47⟩.

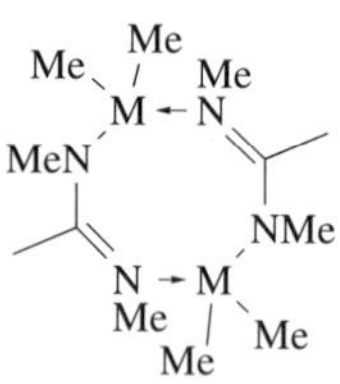

(177)
M = Al, Ga, In

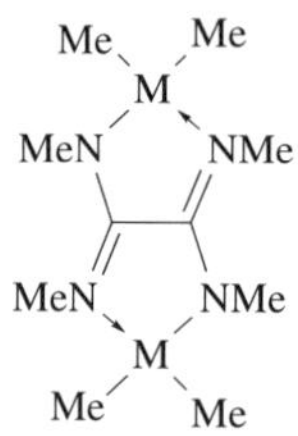

(178)
M = Al, Ga, In

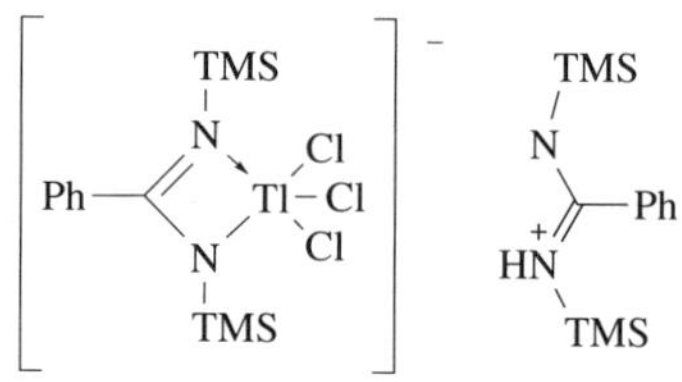

(179)

5.19.2.7.2 *Amidines with an* N-*metal bond, where M is a Group 14 metal*

The stable, crystalline lithium salt (**180**) is *trans*-metallated with germanium, tin or lead at room temperature to give the cyclic compounds (**181**) formed with a 1,3-migration of the trimethylsilyl group ⟨68CB2533⟩. The *N*-(triethyltin)formamidine derivative (**182**), or its intercoordinate form, is formed in 73% yield by hydrostannylation of *N,N'*-dicyclohexylcarbodiimide in the presence of 2,2'-azobisisobutyronitrile (AIBN) (Equation (54)) ⟨65LA(688)14⟩.

(180) **(181)** (54)

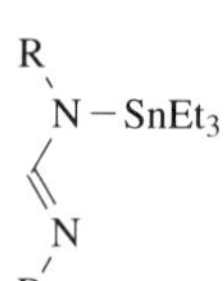

(182)
R = cyclohexyl

5.19.2.7.3 *Amidines with an* N-*metal bond, where M is a transition metal*

Amidines form complexes with most of the transition metals and with all the first-row transition elements. When a nitrile is coordinated to pentaaminocobalt, it undergoes extremely facile reaction with liquid ammonia to give the hexacoordinate cobalt(III) complex (**183**) in quantitative yield. The complex can be broken to generate the primary amidine by treatment with aqueous acid (Equation (55)) ⟨90IC140⟩. In related reactions, amines have been added to nitrile–tantalum complexes to give amidine–tantalum complexes ⟨87MI 519-02⟩. *N,N′*-Bis(trimethylsilyl)benzamidine has been widely used as a ligand for transition metals. Two-to-one complexes are formed with chromium, manganese, zinc, mercury and tetrachloroferrate(III) (**184**) ⟨i, 91ZN(B)1328; ii, 91ZAAC47; iii, 89ZAAC(575)10⟩. The reaction of tris(trimethylsilyl)benzamidine with either copper(I) chloride, silver acetate or ClAuCO in acetonitrile gives complexes (**185**) which have been shown by x-ray analysis to have a short metal–metal bond distance ⟨i, 88ZN(B)1628; ii, 90ZN(B)1273⟩. If tris(trimethylsilyl)benzamidine is treated with copper(II) chloride in acetonitrile, then a 2 : 1 amidinato copper(II) complex is formed in which one amidine molecule acts as a bidendate ligand and the other amidine molecule is monocoordinate (Equation (55)) ⟨88ZN(B)1679⟩.

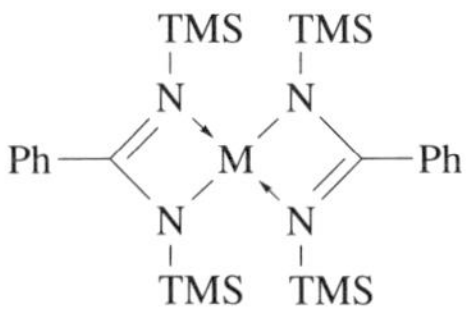

(184)

i, M = Cr, Mn, Zn
ii, M = Hg
iii, M = $FeCl_4$

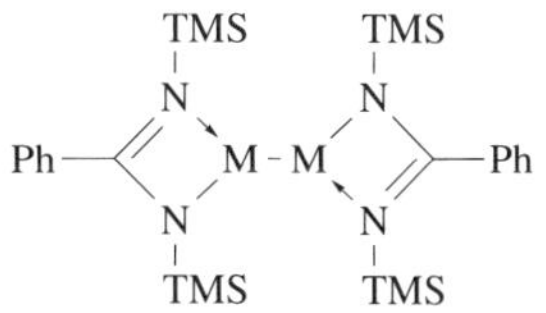

(185)

i, M = Cu
ii, M = Ag, Au

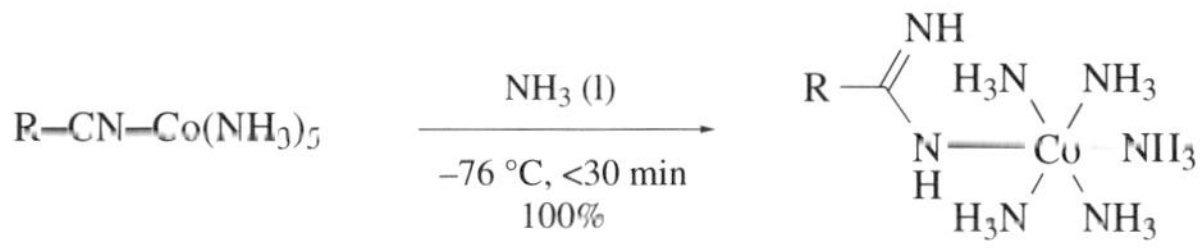

(183) (55)

R = Me, CH_2=CH, Ph, 2-$O_2NC_6H_4$, 4-FC_6H_4

Certain transition metals will also form 2 : 1 complexes with amidine ligands which have not been silylated. Simple nickel(II), palladium(II) and platinum(II) complexes have been reported (**186**) ⟨86JCS(D)1359, 87JCS(D)2687⟩, along with ruthenium(II), ruthenium(III) and osmium(IV) complexes (**187**) ⟨92JCS(D)3199, 93JCS(D)2827⟩. Amidinato complexes with a 3 : 1 stoichiometry, $[MeC(NPh)_2]_3M$, have been reported for rhodium(III) and iron(III) ⟨84ICA85, 93POL563⟩.

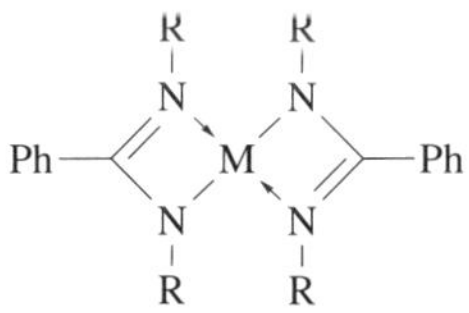

(186)

M = Pt, Ni, Pd
R = Me or Ph

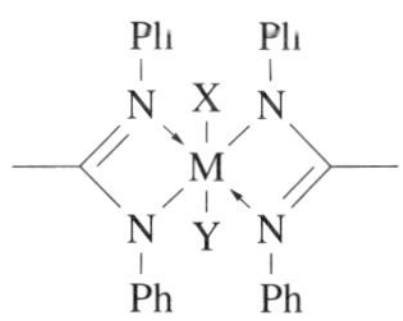

(187)

M = Ru^{II}; X, Y = PPh_3
M = Ru^{II}, X = PPh_3, Y = CO
M = Ru^{III}; X = PPh_3; Y = Cl
M = Os^{IV}; X, Y = Cl

Acetamidine derivatives with an *N*-niobium or *N*-tantalum bond are prepared by C—N insertion from a carbodiimide into a metal—carbon bond. One to one complexes (**188**) are formed when an excess of Me_2MCl_3 or Me_3MCl_2 are reacted with the carbodiimide. When an excess of the carbodiimide is used Me_3MCl_2 forms 2 : 1 complexes (**189**) in which two methyl groups have been transferred. Transfer of the third and final methyl group is not observed even when a very large excess of carbodiimide is used ⟨74JOM(80)349⟩.

Tetrakis(formamidinato)dimolybdenum and dichromium complexes (**190**) are formed from the reaction of molybdenum and chromium hexacarbonyl with formamidines. In the formation of the molybdenum complex the reaction proceeds via formation of pentacarbonylformamidine molybdenum and tetracarbonyldiformamidinemolybdenum complexes followed by formation of a metal—metal bond ⟨77JOM(135)183, 77MI 519-02⟩. Preparation of amidine complexes containing an *N*-nickel or *N*-titanium bond is also possible ⟨75JCS(D)930, 91ZN(B)1328⟩.

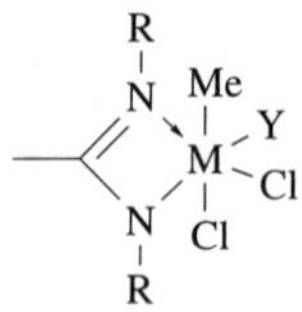

(188)

M = Ta; R = Tol; Y = Cl
M = Nb; R = Tol; Y = Cl or Me
M = Nb; R = cyclohexyl; Y = Cl or Me
M = Ta; R = cyclohexyl; Y = Cl

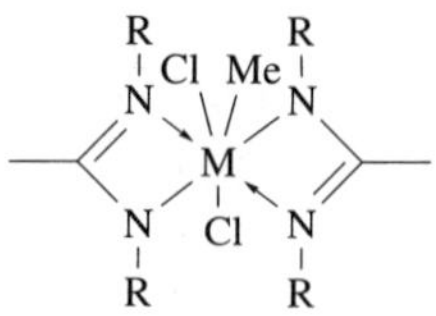

(189)

M = Ta; R = Tol
M = Ta; R = cyclohexyl
M = Nb; R = cyclohexyl
M = Ta; R = isopropyl

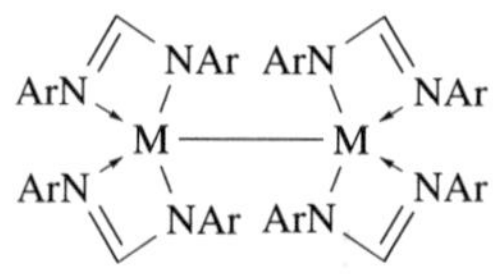

(190)

M = Mo; Ar = 3,5-$Me_2C_6H_3$, 4-ClC_6H_4, 4-MeC_6H_4
M = Cr; Ar = 3,5-$Me_2C_6H_3$

5.19.2.7.4 Amidines with an N*-metal bond, where M is a lanthanide or actinide metal*

Anhydrous $LnCl_3$ (Ln = lanthanide or scandium) react with the sodium salts of *N,N′*-bis(trimethylsilyl)benzamidine and its derivatives to give monomeric lanthanide(III) complexes, Ln[4-RC_6H_4(N-TMS)$_2$]$_3$ 〈92CB2171〉. Amidinato complexes of the actinide elements, uranium and thorium have been prepared for metals in their +4 oxidation state by the reaction of MCl_4 with the sodium or lithium salt of the appropriate *N,N′*-bis(trimethylsilyl)amidine. With sterically more demanding amidinato ligands, complexes such as **(191)** to **(194)** are formed exclusively 〈90JOM(388)21〉. With less sterically demanding amidinato ligands, complexes of different stoichiometries, such as [4-$CF_3C_6H_4$C(N-TMS)$_2$]$_3$MCl (M = U, Th), can be isolated 〈88JOM(345)C1, 90JOM(338)21〉.

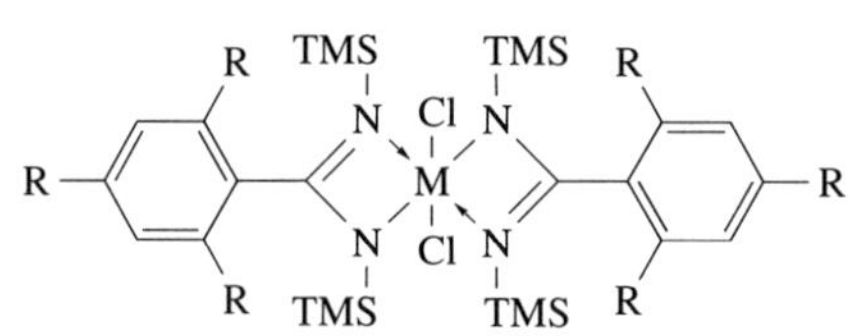

(191) M = U; R = CF_3
(192) M = U; R = Me
(193) M = Th; R = CF_3
(194) M = Th; R = Me

5.20
Iminoacyl Functions Linked to Any Heteroatom Other Than Halogen, Chalcogen or Nitrogen

MICHAEL BUTTERS

Pfizer Central Research, Sandwich, UK

5.20.1 IMIDOYLPHOSPHORUS, -ARSENIC, -ANTIMONY AND -BISMUTH FUNCTIONS—$R^1C(NR^2)PR^3_2$, etc.

5.20.1.1 Dicoordinate Phosphorus Derivatives

No reports of this structural class have been made.

5.20.1.2 Tricoordinate Phosphorus Derivatives (Phosphaamidines)

One of the earliest reported examples (see below) of a tricoordinate imidoylphosphorus compound was a derivative of the *N*-phenylimide of benzoic acid. Eight different synthetic methods for making the tricoordinate phosphorus derivatives have appeared in the literature. In most of the functional group transformations employed, the imidoyl carbon to phosphorus bond is newly formed. In some cases, the reactions involve reagents containing a nucleophilic phosphorus centre, whilst others involve an electrophilic source of phosphorus.

5.20.1.2.1 Imidoyl-tricoordinate phosphorus functions from imidoyl chlorides or ethers

Owing to the nucleophilic character of certain organometallic phosphorus compounds, imidoylphosphorus derivatives are often readily prepared by displacement of chloride or methoxide from the corresponding imidoyl chlorides or ethers (Equation (1); Scheme 1). This presents an attractive synthetic method considering the accessibility of imidoyl chlorides from the parent amides. Both $LiPEt_2$ and $KPPh_2$ have been reacted with *N*-phenylbenzimidoyl chloride (**1**; $R^1 = R^2 = Ph$) in ether to give the phosphaamidine (**2**) ⟨66ZAAC(346)241⟩. Likewise, the monosilylated tertiary phosphine Ph_2P-TMS gives the same product ⟨78JOM(160)47⟩ but in 89% yield (cf., only 35% using $KPPh_2$). In cases where the imidoyl chloride starting material carries a strongly electronegative alkyl radical (e.g., perfluoroalkyl), it is not necessary to use metallated phosphines, and simple bases (e.g., trimethylamine) will suffice ⟨86JFC(30)439⟩. Heterocyclic imidoylphosphorus compounds have also been synthesized from imidoyl chlorides (**3**) (Scheme 1). In one example, where the nitrogen was functionalized with a 2-chloroethyl group, dilithio phenylphosphine gave a 1,3-azaphospholine (**4**) ⟨86ZC407⟩. Alternatively, reaction of a secondary 2-phosphinoethylamine (**5**) with an iminoester hydrochloride liberates ammonium chloride in the presence of hot aqueous hydroxide to give the same 1,3-azaphospholines ⟨86ZC407⟩.

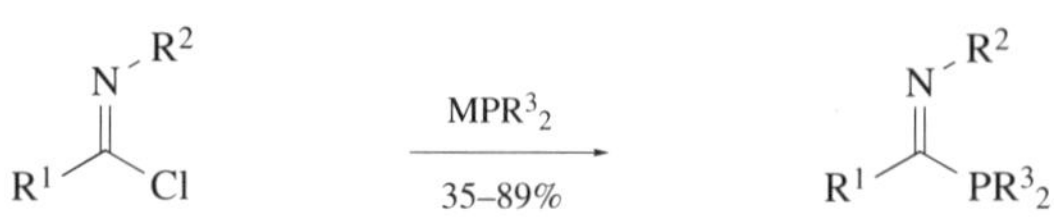

(1) (2) (1)

$R^1 = Ph$; $R^2 = Ph$; $R^3 = Et, Ph$; M = Li, K, TMS

5.20.1.2.2 Imidoyl-tricoordinate phosphorus functionalities from isonitriles

1,3-Azaphospholines (Scheme 1) bearing the desired imidoylphosphorus (tricoordinate) grouping have also been prepared by the reaction of vinylisonitrile (**6**) with a primary phosphine in the

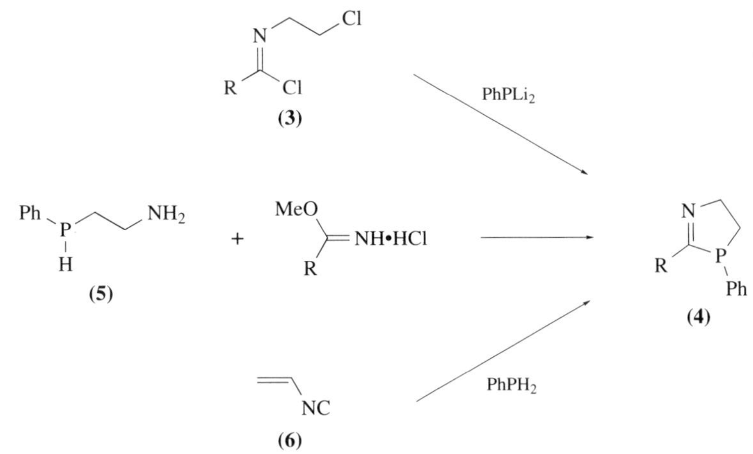

Scheme 1

presence of KOBu^t base ⟨86ZC407⟩. Disubstituted phosphines (Equation (2)), however, react with isonitriles in a radical-mediated process. Thus, reaction of diethylphosphine with *n*- or cyclohexylisonitrile in benzene with 2,2′-azobisisobutyronitrile (AIBN) radical initiator (70°C) gave moderate yields of the corresponding diethylformimidoyl phosphines ⟨70JOC4238⟩. Better yields (81%) of imidoylphosphines are observed when copper(I) oxide is used as the catalyst ⟨68TL935⟩. Moreover, the copper-catalysed process may be extended to a wide range of isonitriles.

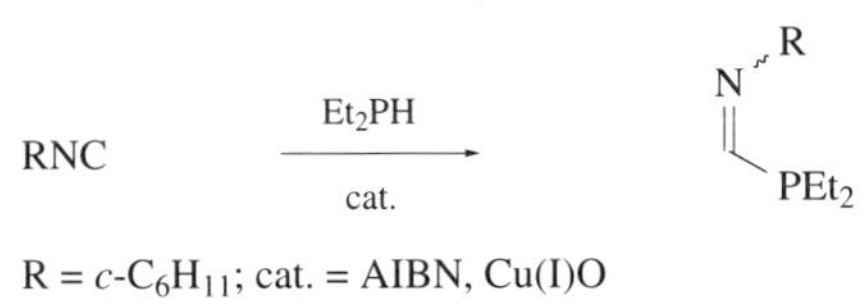

(2)

5.20.1.2.3 *Imidoyl-tricoordinate phosphorus functionalities from imines*

The reaction of Schiff bases (e.g., benzylideneaniline) with PCl_3 in acetone/pyridine (Equation (3)) is facile and imidoylphosphorus compounds can be prepared in which the tricoordinate P is bonded to two chlorines ⟨85POL1151⟩. Although not exploited, this reaction provides a simple entry to reactive imidoylphosphorus dichloride compounds and offers much potential for further functionalization by virtue of the reactive -PCl_2 moiety.

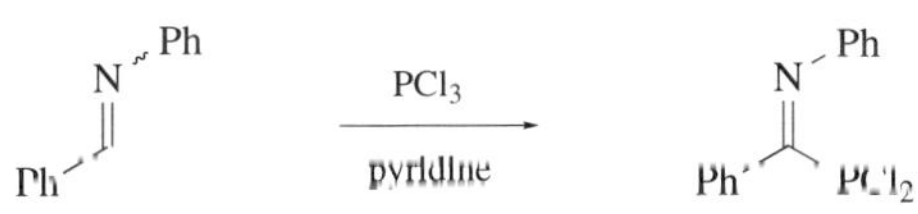

(3)

5.20.1.2.4 *Imidoyl-tricoordinate phosphorus functionalities from nitrile ylides*

The 1,3-azaphospholine system (referred to in Sections 5.20.1.2.1 and 5.20.1.2.2) has been prepared (Equation (4)) from nitrile ylides ⟨83CJC1105⟩ and diazaphospholes. Photochemical ring-opening of azirines gives a reactive nitrile ylide (**7**) that undergoes a cycloaddition reaction with diazaphospholes to give a new heterocyclic bicycle (**8**) as a diastereoisomeric mixture. This adduct is one of few examples in which the imidoylphosphorus derivative has phosphorus functionalized with nitrogen.

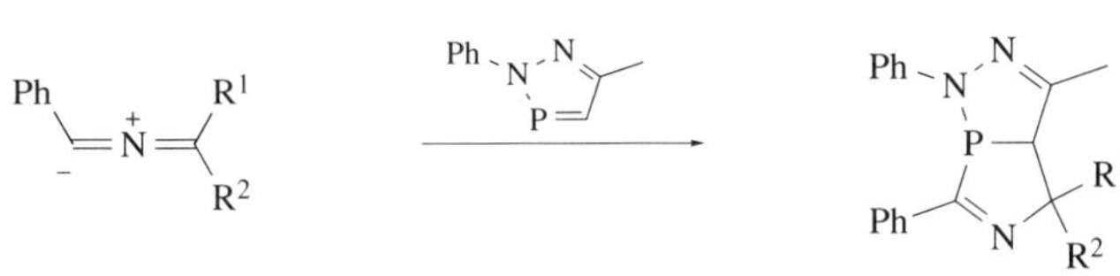

(**7**) (**8**)

(4)

5.20.1.2.5 Imidoyl-tricoordinate phosphorus functionalities from perfluoroalkylnitriles

Perfluoroalkylnitrile compounds readily react with mono- and diphenylphosphine ⟨85JFC(30)269, 85JFC(30)289⟩ to give the corresponding bis- or monoimidoylphosphine adducts (Scheme 2). Although complex mixtures of products were formed it was found that a 10-fold excess of perfluorooctanonitrile reacts with phenylphosphine to give the bisimidoylphosphorus compound (**9**) in 28% yield ⟨85JFC(30)269⟩. Similarly, the imidoylphosphorus compound (**10**) was formed in 27% yield from diphenylphosphine.

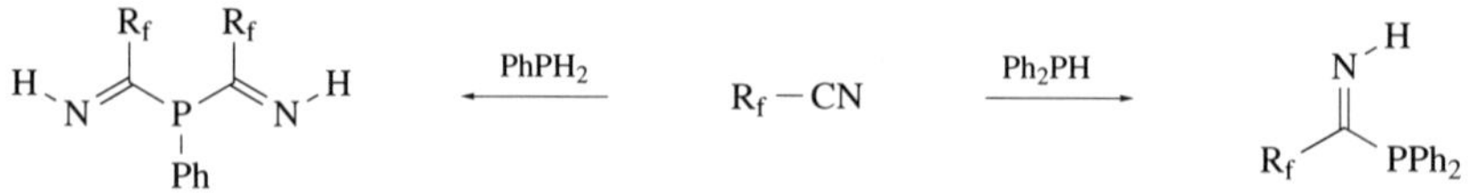

(**9**) (**10**)

R_f = perfluoroalkyl

Scheme 2

5.20.1.2.6 Imidoyl-tricoordinate phosphorus functionalities from primary amines and perfluorophosphines

$(F_3C)_2PH$ reacts with excess isopropylamine in dichloromethane (−78 °C) initially to give a perfluoro-2-phosphapropene $F_3CP{=}CF_2$. Subsequently another two moles of HF are eliminated as one mole of isopropylamine becomes condensed to form a new C=N bond. Reaction of a second mole of isopropylamine then saturates the P=C bond to give a novel formimidoyl phosphine (**11**) ⟨91ZN(B)978⟩.

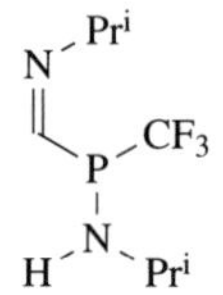

(**11**)

5.20.1.3 Tetracoordinate Phosphorus Derivatives

A great number of imidoyl-tetracoordinate phosphorus compounds have been prepared and at least 10 different synthetic methods are known. The most prevalent examples within this class are those with a phosphonate grouping (-P(O)(OR)_2) or a dialkylphosphinyl grouping (-P(O)R_2). No doubt in most cases these tetracoordinate phosphorus derivatives have a tetrahedral arrangement of substituents around an sp^3 hybridized phosphorus atom. A number of texts have appeared which make only limited reference to imidoylphosphorus compounds ⟨B-67MI 520-01, B-70MI 520-01, B-72MI 520-01⟩.

5.20.1.3.1 Imidoyl-tetracoordinate phosphorus functionalities from nitriles

The reaction of dialkylphosphinous acids with nitriles occurs readily when the nitrile carries a neighbouring electronegative substituent. Pudovik and Sudakova have published details of this chemistry in which iminophosphine oxides are the main products. This chemistry is also successful (Equation (5)) for phosphinothious acids and acceptable phosphorus substituents include alkyl or aryl moieties ⟨72JGU1637,1715, 74JGU2371, 75JGU2512⟩. In general, the reactions are promoted by alkoxide bases and the rate is dependent upon the temperature and the extent of electron withdrawal from the nitrile group. Thus, a mixture of diphenylphosphinous acid and 4-nitrobenzonitrile in the presence of sodium ethoxide and nitrobenzene at 160 °C gave the α-iminophosphine oxide (**12**) in 52% yield. When acidic protons are available α to the parent nitrile group, the product imidoyl phosphinate compound exists in equilibrium with the tautomeric β-aminoacrylonitrile system (Scheme 3).

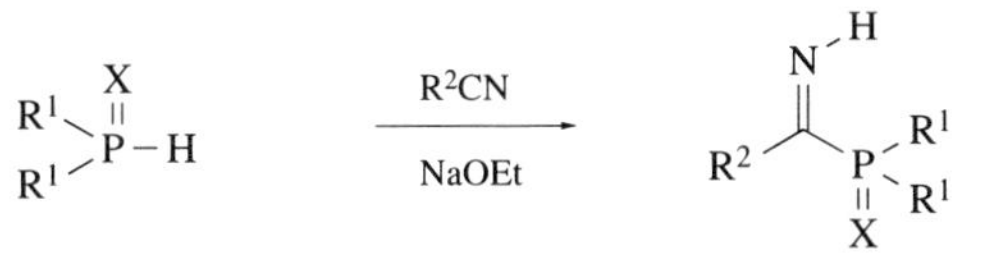

(5)

R^1 = alkyl, aryl;
R^2 = $-CH_2CN$, $-CH_2P(O)R_2$, $-CH_2CO_2Et$, $-CH(Me)OP(O)R_2$, –Ph
X = S, O

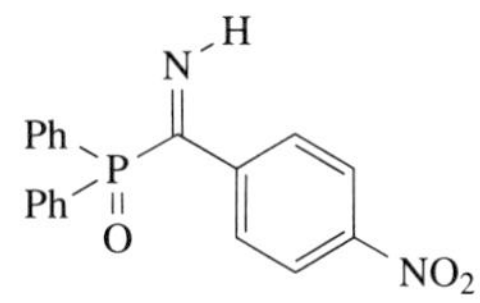

(12)

Scheme 3

5.20.1.3.2 Imidoyl-tetracoordinate phosphorus functionalities from α-chloroalkyl isocyanates

Dichloromethyl isocyanate reacts with triethyl phosphite to give chloro(diethoxyphosphinyl) methyl isocyanate (**13**) ⟨73JGU10⟩; reaction of this with aniline provides an α-phosphonyl-α-chloroalkyl urea that loses HCl in the presence of triethylamine giving a formimidoyl phosphonate (**14**) (Equation (6)). A similar reaction sequence can be applied to α,α-dichlorobenzyl isocyanate, ⟨90JGU864⟩ although in this system further reaction of the imidoyl phosphonate with aniline occurs to give the anilide of *N*-[α-(bis(phenylamino)phosphonyl)benzylidene]carbamic acid.

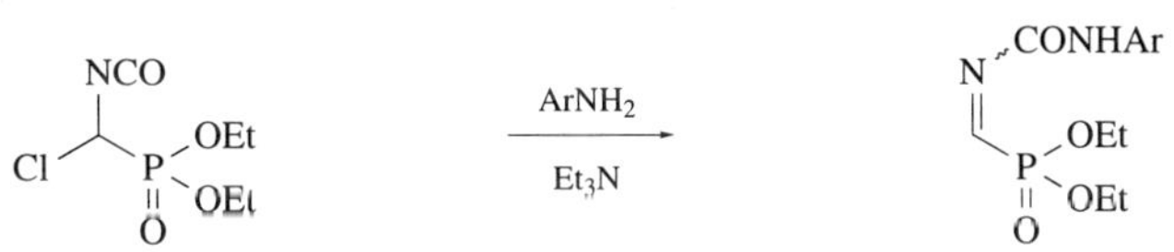

(6)

(13) **(14)**

5.20.1.3.3 Imidoyl-tetracoordinate phosphorus functionalities from imidoyl chlorides or ethers

Trialkyl phosphites ($P(OEt)_3$, $P(OMe)_3$, $P(OPr)_3$) readily react with imidoyl chlorides bearing an electron-withdrawing substituent on carbon (e.g., aryl, trifluoromethyl, cyano and ethoxycarbonyl) to give imidoyl phosphonates as shown in Equation (7) and Table 1. Very little has been reported in connection with cyclic imidoyl phosphonates ⟨77JGU441⟩. However, when the imidoylnitrogen carries a 2,2-dichlorovinyl moiety, it is possible to achieve ring closure after chlorination (with Cl_2) and thermal elimination of chloroethane (Scheme 4). Although most of the work in the area of imidoylphosphorus derivatives has employed trialkyl phosphites, a number of different phosphorus reagents have also been applied including tris(trimethylsilyl) phosphite ⟨89JGU1496⟩, dialkyl hydrogen phosphites ⟨86JGU2372⟩, trialkyl phosphinites ⟨87JGU1349⟩ and diethyl phenylphosphoramidite ⟨89JGU1129⟩.

(7)

Table 1 Imidoyl-tetracoordinate phosphorus derivatives from imidoyl chlorides (Equation (7)).

R[1]	*R*[2]	*Yield* (%)	*Ref.*
CN	Ph	64	76JPR272
Bu^t	CH_2Ph	77	89JGU1129
Ph	Ph		76JPR272
CF_3	CH(Ph)Me	59	90JGU1165, 86JGU2372
CO_2Et	Ph	72	76JPR272, 79UKZ624
Ph	CH_2CF_3	64	86JGU2372
Ph	$CHCCl_2$	87	77JGU441
Ph	Me		85JGU402
CF_3	COPh	30	87JGU415
Ph	$CH(Ph)PO(Ph)_2$		87JGU1349
$CCl(CH_3)_2$	Me	50	89JGU2228
$CCl(CH_3)_2$	CH(Ph)Me	62	89JGU2228

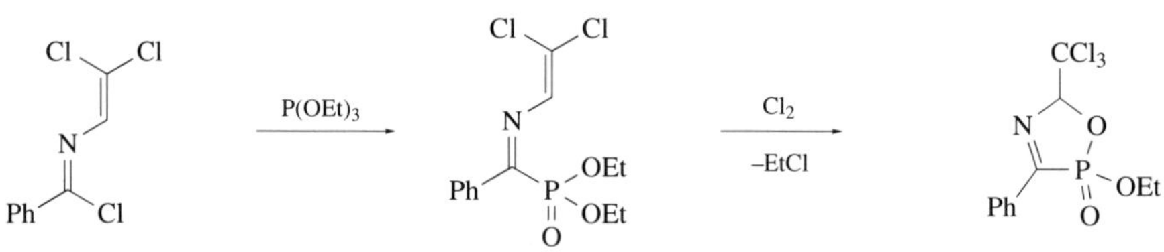

Scheme 4

5.20.1.3.4 Imidoyl-tetracoordinate phosphorus functionalities by rearrangement of phosphites

Dialkylchlorophosphites react with secondary amines to give a mixture of phosphonamidites and imidoyl phosphites. These materials are stable at room temperature, but undergo rearrangement to an imidoyl phosphonate on heating ⟨90JGU1059⟩. Evidence has been published to indicate that imidoyl phosphites (Scheme 5; X = O, (**15**)) are more readily rearranged to phosphonates than the phosphonamidite counterparts (X = O, (**16**)) ⟨86JGU1113⟩. A similar rearrangement process has been observed in the imidoyl thiophosphite series (Scheme 5; X = S) ⟨86JGU1467⟩.

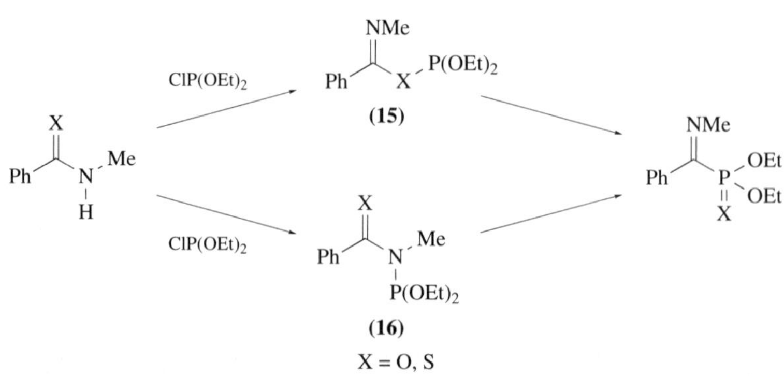

Scheme 5

5.20.1.3.5 *Imidoyl-tetracoordinate phosphorus derivatives from nitrile ylides*

Thermolysis of the oxazaphospholidine (**17**) gives an unstable nitrile ylide which reacts with trimethyl phosphite to give a mixture of two imidoyl phosphonates (**18**) in 57% combined yield (Equation (8)) ⟨76CZ340⟩.

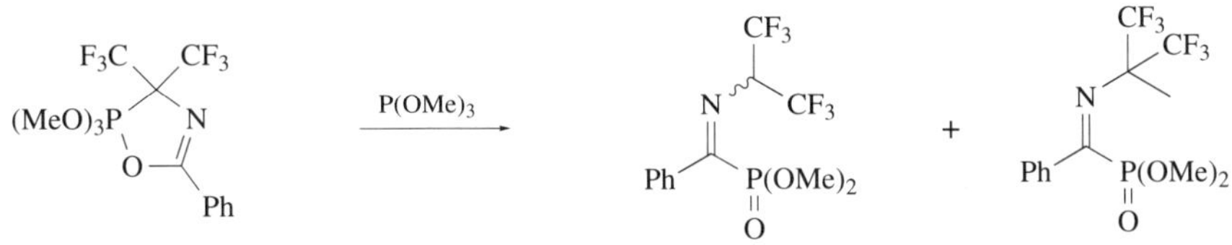

(8)

(17) **(18)**

5.20.1.3.6 *Imidoyl-tetracoordinate phosphorus derivatives from nitrones*

Trialkyl phosphites can react with nitrones via three different pathways depending on the relative pH of the medium ⟨79T43⟩. Studies on a steroidal nitrone (**19**) have illustrated that in the presence of acetic acid, the dominant pathway (Scheme 6) involves phosphonylation of the nitrone carbon followed by dehydration to the imidoyl phosphonate (**20**).

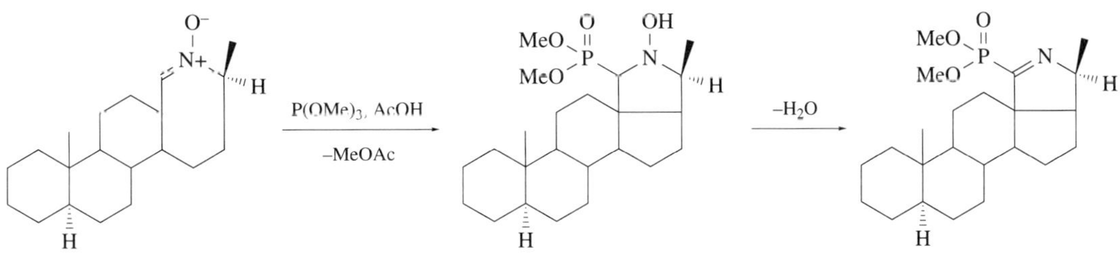

(19) **(20)**

Scheme 6

5.20.1.3.7 *Imidoyl-tetracoordinate phosphorus derivatives from enamines or imines*

Diphosphonylalkynes react with alkylamines to give enamines (Scheme 7). The lithiated enamines (**21**) undergo *β*-alkylation with carbon electrophiles such as alkyl halides or aldehydes with concomitant migration of the double bond to give an imidoyl phosphonate (**22**) in good yield (70–94%) ⟨83TL1679⟩. Alternatively, imidoyl phosphonates can be prepared by double bond migration (prototropic isomerization) in which *N*-trifluoroethylidenylbenzylamine undergoes a rearrangement ⟨86JGU2372, 88JGU30, 89JGU1123⟩. For example, trimethyl phosphite was reacted with an equimolar amount of triethylamine and *α*-chloroimine (**23**) in ether. The product phosphonate (**24**) was readily isomerized at room temperature to the desired imidoyl phosphonate (**25**), driven by the energy of conjugation of the phenyl substituent with the C=N bond (see Scheme 8). A full account of this chemistry has been published in which the nature of the phosphorus functionality has been extended to include phosphonofluoridate ⟨89JGU1123⟩.

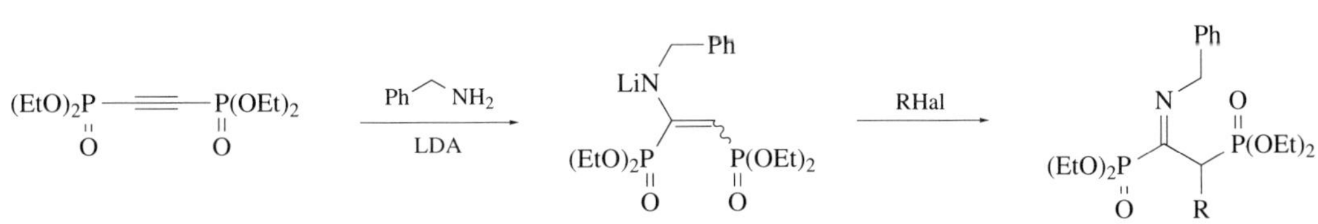

(21) **(22)**

Scheme 7

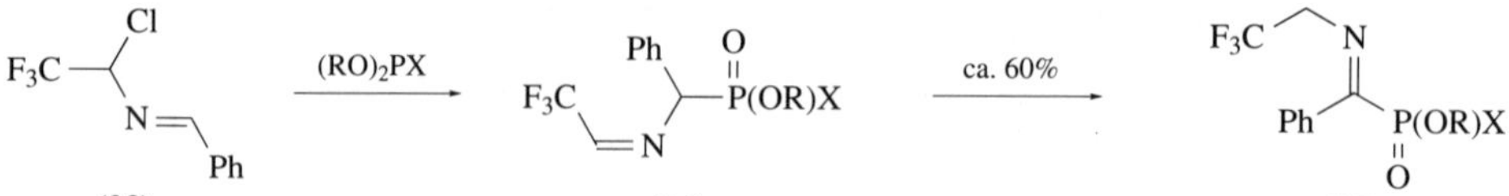

(23) (24) (25)

R = Me, Et, Pr; X = OMe, OEt, OPr, F

Scheme 8

5.20.1.3.8 Imidoyl-tetracoordinate phosphorus derivatives from α-halo-α-aminophosphinates (and phosphonates)

Reaction of α-aminophosphinates (**26**) with NBS provides the α-bromo-α-aminophosphinates (**27**). Treatment of the bromo derivatives (**27**) with Hunig's base at −78 °C affords the formimidoyl phosphinate series (**28**) in good yield (Scheme 9) ⟨89S97⟩. Similarly, chlorination of 1-formamido-2,2,2-trichloroethanephosphonic esters in an excess of pyridine leads to imidoyl phosphonates in 90% yield ⟨91PS(63)95⟩.

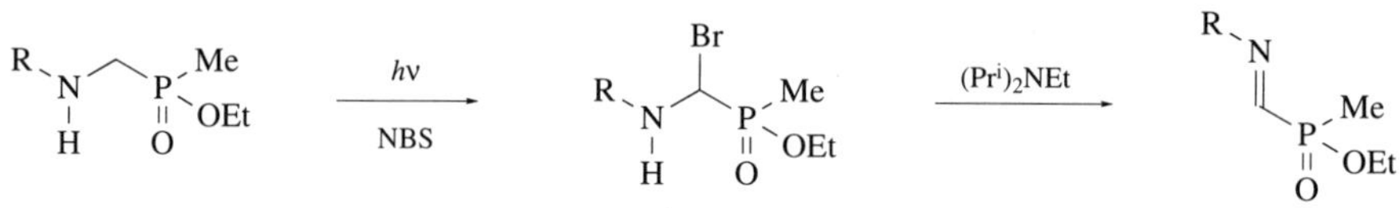

(26) (27) (28)

R = $Cl_3CH_2O_2C$–, PhCO–, $PhSO_2$–

Scheme 9

5.20.1.4 Higher Coordinate Phosphorus Derivatives

In Section 5.20.1.3.3, a number of examples of imidoyl phosphonates were described. Most of these compounds were prepared from imidoyl chlorides and trialkyl phosphites and involved the liberation of an alkyl chloride. In many cases, the mechanism of reaction is thought to involve initial formation of an imidoyl phosphorane. Little has been published on these compounds because of their limited lifetime ⟨87JGU415⟩.

5.20.1.5 Imidoylarsenic Derivatives

All reports of imidoylarsenic derivatives refer to tricoordinate arsenic compounds and the synthetic methods to such compounds are relatively underdeveloped. Existing synthetic routes to imidoylarsenic derivatives can readily be grouped into three types.

5.20.1.5.1 Imidoylarsenic derivatives from imidoyl ethers

2-Aminoalkylarsines (**29**) react with iminoester hydrochlorides (**30**) to give the corresponding 1,3-azaarsolines (**31**) with the desired imidoylarsenic grouping (Equation (9)) ⟨76JPR409, 89ZC71⟩. Thus, phenyl-2-aminoethylarsine reacted with methyl acetimidate hydrochloride to give an acetamidine intermediate, which upon treatment with sodium carbonate as base followed by thermal (150 °C) elimination of ammonia gave the azaarsoline (**31**; R^1 = Ph, R^2 = CH_3).

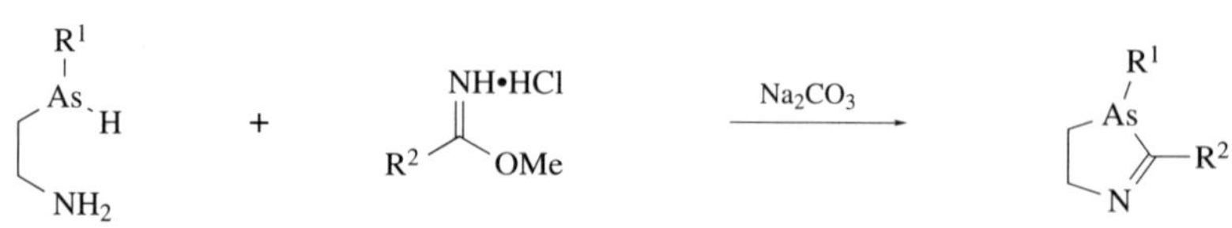

(29) (30) (31) (9)

R^1 = Ph, Bn; R^2 = Ph, Me

5.20.1.5.2 Imidoylarsenic derivatives from imidoyl chlorides

Reactions of alkylarsines with imidoyl chlorides require activation of the arsenic centre by forming a lithium arsenide function. The reaction of a range of lithium arsenides (**32**) with imidoyl chlorides (**33**) has been studied ⟨78ZC452⟩. For example, phenylethylarsine was treated with one equivalent of *n*-butyllithium (in hexane/diethylether), followed by *N*-phenylbenzimidoyl chloride at 0 °C to give a 73% yield of the imidoylarsenic derivative (**34**) as shown in Equation (10). Reaction of *N*-(2-chloroethyl)benzimidoylchloride (**36**) with the dilithio anion of phenylarsine (**35**) at −50 °C (diethylether) provides an alternative synthesis of 1,3-azaarsolines (**37**) (Equation (11)) that is complementary to the route described above (Section 5.20.1.5.1) ⟨86ZC407⟩. Imidoyl chlorides are also reactive towards Si–As bonds. Thus, reaction of phenylbis(trimethylsilyl)arsine with pivaloylimidoyl chloride generates an imidoylsilylarsine (**40**) which rearranges to the *N*-silylated As=C isomer (**41**) (Scheme 10) according to a temperature-dependent equilibrium ⟨79JOM(166)175⟩. Reaction of *o*-arsinophenols (**42**) with imidoyl chlorides (for example (**43**), *N*-phenylbenzimidoyl chloride) proceeds via a primary attack of the arsino group to give the hydrochloride salt of an imidoylarsine (**44**) ⟨80JOM(186)39⟩. Such compounds are unstable, however, and after a period of a few days react intramolecularly to give benzoxaarsoles (see Equation (12)).

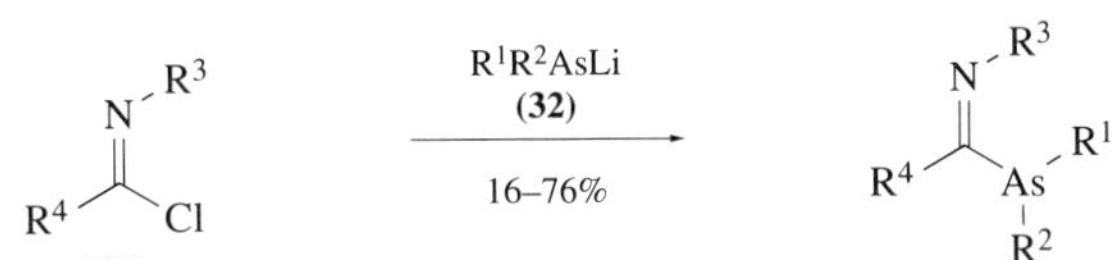

(**33**) (**34**) (10)

R^1 = Ph, Et; R^2 = Et, Bu^s, H, CH_2CH_2NHEt
R^3 = Ph, Bu^n; R^4 = Ph, Bu^t

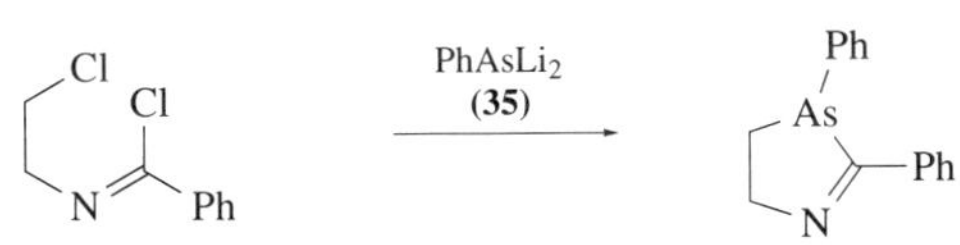

(**36**) (**37**) (11)

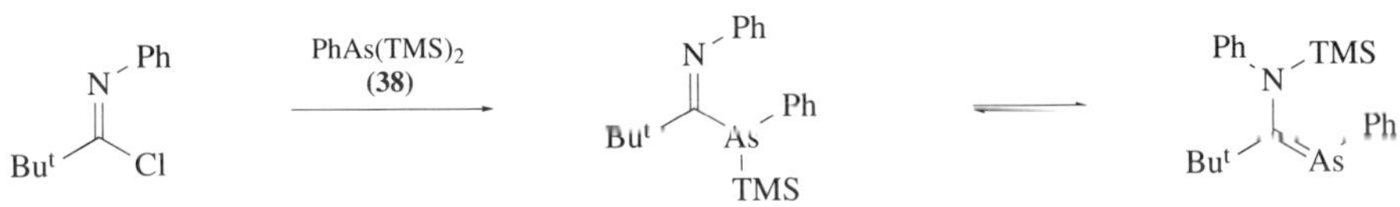

(**39**) (**40**) (**41**)

Scheme 10

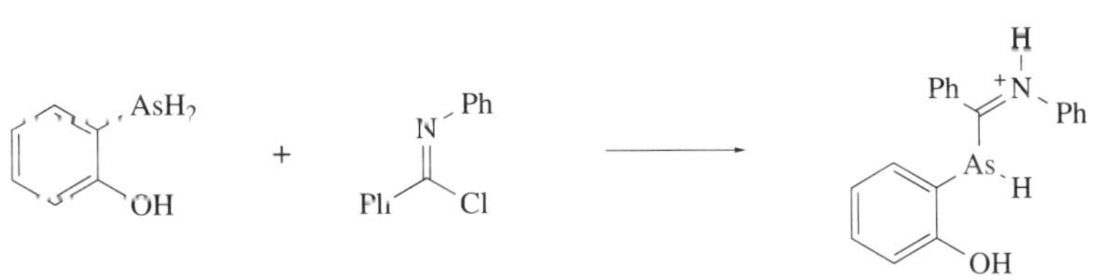

(**42**) (**43**) (**44**) (12)

5.20.1.5.3 Imidoylarsenic derivatives by cycloaddition of nitrile ylides across diazaarsoles

A nitrile ylide cycloaddition process involving diazaphospholes has been described in Section 5.20.1.2.4. A similar reaction occurs with diazaarsoles (**45**) and in superior yield (43%; cf. 18% with its phosphorus counterpart). Thus, a benzene solution of diazaarsole (**45**) together with phenylazirine produced nitrile ylide (**46**) and was irradiated to give the bicyclic product (**47**) possessing the imidoyl-arsenic unit (Equation (13)) ⟨83CJC1105⟩.

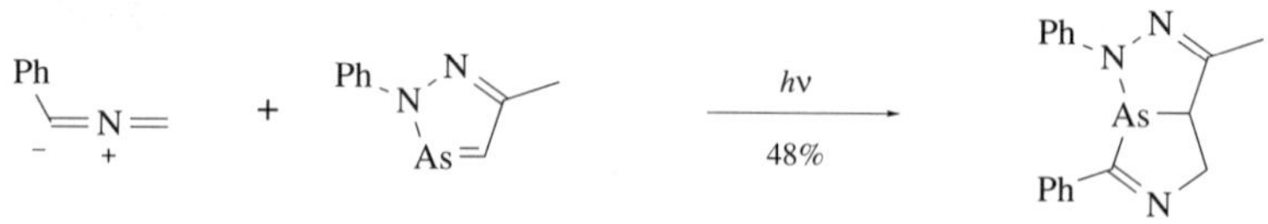

(13)

(**46**) (**45**) (**47**)

5.20.1.6 Imidoylantimony Derivatives and Imidoylbismuth Derivatives

Although no examples of imidoylantimony derivatives were found, it is well known that alkali metal derivatives of organoantimony compounds may be formed by hydrogen–metal exchange or halogen–metal exchange and the so formed Sb–metal (Na, Li) bond is reactive towards a range of aliphatic and aromatic halides providing access to new antimony derivatives ⟨70MI 520-02⟩. Similar chemistry has been established for organobismuth compounds.

5.20.2 IMIDOYL METALLOID FUNCTIONS—$R^1C(NR^2)SiR^3_3$, $R^1C(NR^2)GeR^3_3$, $R^1C(NR^2)BR^3_2$

5.20.2.1 Tetracoordinate Silicon Derivatives

Several compounds of this type have been described with either carbon or hydrogen attached to the imidoyl carbon (R = H,C). Groups on nitrogen or silicon tend to be alkyl, although a few cases of hydrogen or silicon also appear. Only a handful of publications describe these imidoyl silicon derivatives and there are three main methods of synthesis, delineated below. In all cases the silicon is present as a tetracoordinate derivative.

5.20.2.1.1 Imidoylsilicon derivatives directly from isonitriles

Isonitriles are readily converted into formimidoylsilanes (Equation (14)) by a copper-catalysed 1,1-hydrosilation ⟨67JA2240, 68TL935⟩. For example, cyclohexylisonitrile (**48**) has been converted into the corresponding *N*-cyclohexylformimidoyltrimethylsilane (**50**) (in 86% yield) by reaction with trimethylsilane (**49**) in the presence of a copper(II) acetylacetone catalyst at 100°C. This method is only useful for formimidoylsilanes and consequently it has not been investigated in any depth.

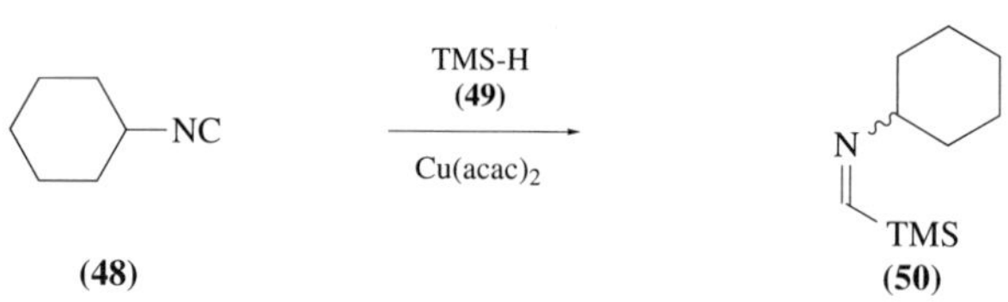

(14)

(**48**) (**50**)

5.20.2.1.2 Imidoylsilicon derivatives from metalloaldimines

Another method for preparing imidoylsilanes involves the reaction of a metalloaldimine (see Section 5.20.3.1.) with a chlorotrialkylsilane (Equation (15)). In most cases, the intermediate metalloaldimine can be prepared from an isonitrile and an alkyllithium or alkyl Grignard reagent. This method was first introduced by Walborsky ⟨70JA6675, 74JOC600⟩ and was more fully exploited by Matern and co-workers ⟨78CJC2286⟩. For example *t*-butylisonitrile reacts with *n*-butyllithium to give the lithioaldimine (**51**) which can subsequently be converted to the corresponding imidoylsilane (**53**) (31%) by reaction with chlorotrimethylsilane (**52**). A range of alkyl- or aryllithium reagents can be used, although a major limitation of the methodology is that only isonitriles without an α-hydrogen (*t*-octyl, *t*-butyl, 2,6-dimethylphenyl) are successfully employed. Imidoylsilanes have been employed to facilitate fluoride induced 1,4-elimination for the generation of *o*-quinodimethanes ⟨84TL5139⟩. Thus *o*-lithiobenzyldimethylamine is added to *t*-butylisonitrile followed by silylation (using chlorotrimethylsilane) to give the synthetically useful *N*-*t*-butyl-*o*-(dimethylaminomethyl)benzimidoylsilane (**54**).

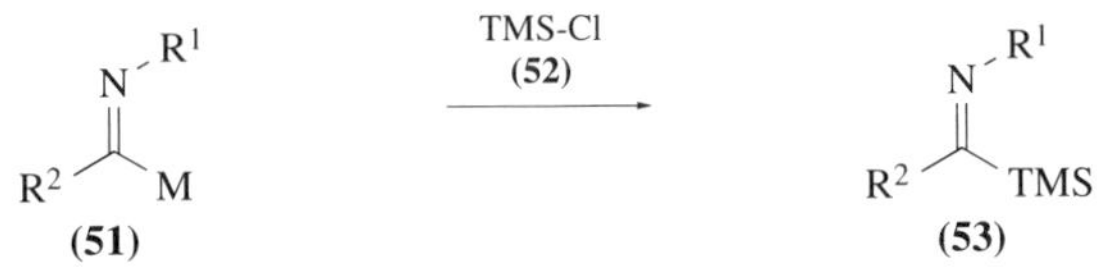

(15)

$R^1 = Bu^t$, 2,6-dimethylphenyl, octyl; $R^2 = Bu^n$; M = Li, MgBr

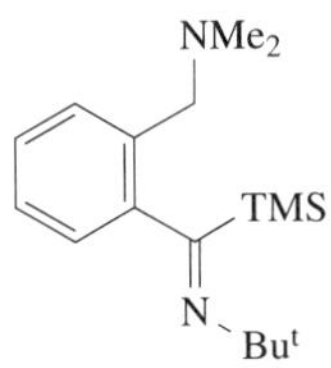

(54)

Metalloaldimines bearing a silicon linked to carbon can be alkylated (Scheme 11) with a range of carbon electrophiles (ketones, aldehydes and alkyl bromides) or protonated using methanol. For example, ((2,6-dimethylphenylimino)(trimethylsilyl)methyl), trimethylstannane (**55**) (prepared from the parent isonitrile and an organosilylstannane) is selectively transmetallated with *n*-butyllithium to give the intermediate lithioaldimine (**56**) which when treated with ethyl bromide forms the corresponding imidoylsilane (**57**) in 75% yield ⟨87JA7888⟩. When (**56**) is treated with pivaldehyde the resulting adduct (**58**) undergoes the Brook-type migration (the silyl group moves from imino carbon to oxygen) providing a new lithioaldimine intermediate (**59**). Reaction of this lithioaldimine with chlorotrimethylsilane provides a new imidoylsilane (**60**) in 62% yield.

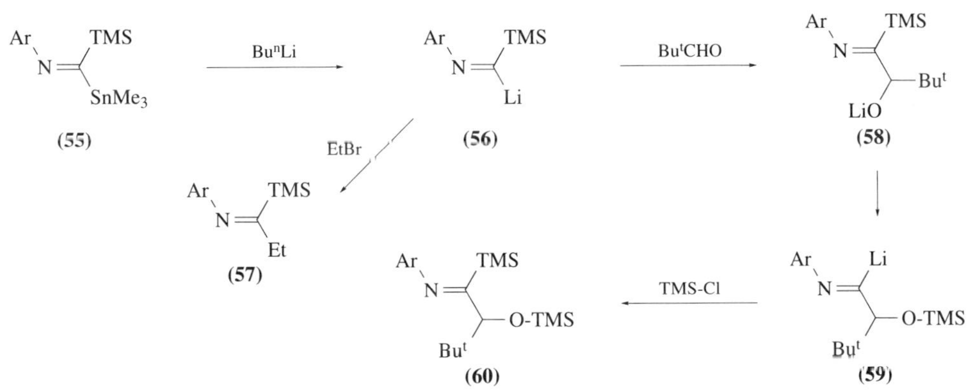

Scheme 11

5.20.2.1.3 Imidoylsilicon derivatives from imidoyl chlorides

The reaction of imidoyl chlorides (**61**) with chlorotrimethylsilane in the presence of magnesium and hexamethylphosphoramide (HMPA) leads to imidoylsilanes (**62**) (Equation (16)). Very little has been published on this process and the mechanism is not clear. One possible mechanism involves an organosilyl Grignard intermediate ⟨76ZN(B)281⟩ and may occur via a radical anion species. *N*-Phenylpivalimidoyl chloride ⟨74JOM(76)C1⟩ and a series of *C*-chlorobenzylideneanilines ⟨81JCS(P2)1043⟩ have been converted into the corresponding imidoylsilanes (with yields in the region of 65%). A disadvantage of this method is contamination by desilylated counterparts which are difficult to remove without causing decomposition of the desired imidoylsilanes. However, when appropriate isonitrile starting materials are not available for the synthesis of a given imidoylsilane, this method would appear to offer some advantages.

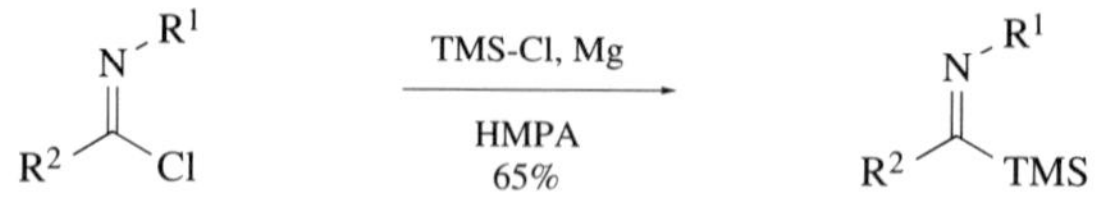

(61) (62) (16)

R1 = Ph, Ar; R2 = Bu^t, Bn

5.20.2.1.4 *Imidoylsilicon derivatives from enamines of acylsilanes*

The reductive silylation of acetonecyanohydrin using TMS-Cl/lithium/HMPA affords the *N*-silylenamine of 2-methylpropanoylsilane ⟨90JOM(391)13⟩. Protodesilylation using dry HCl then gives the enamine (**63**) which slowly isomerizes to the corresponding imidoylsilane (**64**) (imine–enamine isomerization). The limitations of this method are associated with the structure of the ketone precursor (cyanohydrin formation) and the fact that only NH groups in the final imidoylsilane are possible. Clearly the relative thermodynamic stability of the enamine versus imine isomers will also affect the ease of synthesis.

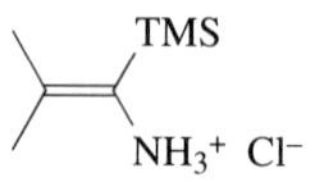

(63)

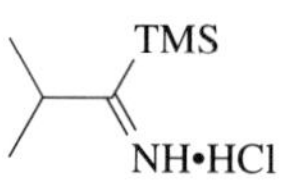

(64)

5.20.2.1.5 *Imidoylsilicon derivatives from trimethylazidosilane*

Matrix photolysis of trimethylazidosilane provides a dimethylsilylimine. However, the isomeric form is ambiguous and the structure might in fact be an *N*-silylimine ⟨78CC762⟩.

5.20.2.2 Tetracoordinate Germanium Derivatives

A few examples of this class of imidoyl metalloid have been reported from imidoyl chlorides, lactones and isonitrile intermediates. The first publication on imidoylgermanium compounds described the preparation of *N*-phenylbenzimidoyltriphenylgermanium (**66**) from *N*-phenylbenzimidoyl chloride (**65**) and triphenylgermyllithium in boiling THF (Scheme 12) ⟨71MI 520-02⟩.

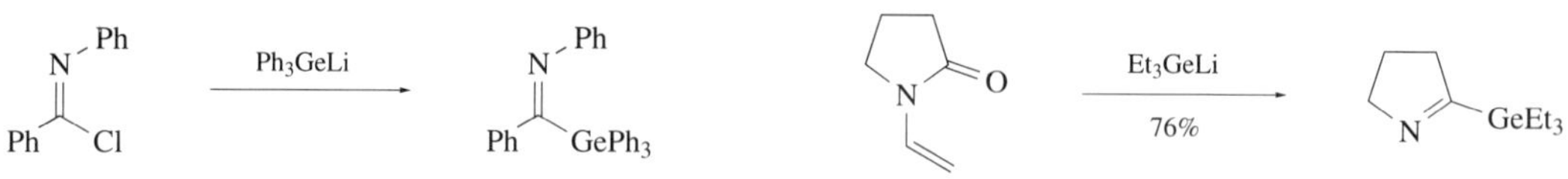

(65) **(66)** **(67)**

Scheme 12

An unusual reaction between triethylgermyllithium and *N*-vinylpyrrolidinone (or *N*-vinylcaprolactam) affords the corresponding cyclic imidoylgermanium compound (**67**) (Scheme 12). This nucleophilic process only takes place in nonpolar solvents like hexane (THF promotes enolization of the lactone). The mechanism is unclear but it is noteworthy that acetaldehyde is produced after hydrolytic workup ⟨89MI 520-01⟩. A number of organogermyl metallic compounds are known and have been reacted with a range of organic functional groups ⟨70MI 520-02⟩, it would appear therefore that a potential synthesis of imidoylgermanes would involve reaction of organogermyl metallics with isonitriles.

5.20.2.3 Tricoordinate Boron Derivatives

The most general method (see also ⟨B-71MI 520-01⟩) for preparing imidoylboron derivatives (Equation (17)) involves the reaction of metalloaldimines (**68**) with aminohaloboranes or dialkylhaloboranes (**69**). Both alkyl- and phenylisonitriles have been converted into the corresponding lithium aldimines (Section 5.20.3.1) by reaction with alkyllithiums. These lithium aldimines ⟨75BCJ3682, 75TL2689, 82JOM(231)191⟩ afford fully substituted imidoylboron compounds (**70**) by reaction with dialkylhaloboranes. For example, an ethereal solution of *t*-butylisonitrile reacted with *n*-butyllithium to give a lithium aldimine intermediate which then reacted with di(*n*-hexyl)chloroborane yielding a di(*n*-hexyl)imidoylborane. Alternatively, the *t*-butylisonitrile derived lithium aldimine was successfully reacted with chlorobis(dimethylamino)borane to give the imidoylborane (**70**) (Equation (17)).

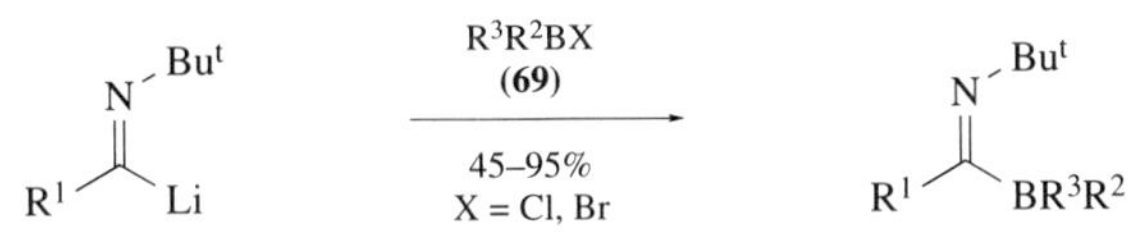

(68) **(70)** (17)

$R^1 = Bu^n, Pr^i, Me; R^3, R^2 = n\text{-}C_6H_{11}, Bu^n$
$R^1 = Bu^n; R^2 = NMe_2; R^3 = Me, Ph, NMe_2$

The ethylation of aminecyanoboranes with triethyloxonium tetrafluoroborate leads to cationic boron compounds of the nitrilium ion type. For example, the TMEDA adduct of cyanoborane (TMEDA.2BH$_2$CN) reacts with Et_3OBF_4 to give the tetrafluoroborate nitrilium salt (TMEDA.2BH$_2$CNEt)$^{2+}$ (BF_4^-)$_2$ in 93% yield. This salt readily reacts with anionic reagents like aqueous cyanide to give the *N*-ethyliminocyanomethylborane adduct (**71**) in 95% yield ⟨88JOM(344)29, 88MI 520-01⟩.

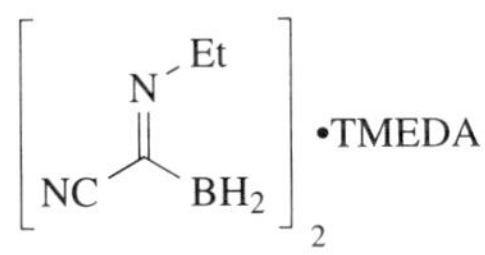

(71)

5.20.2.4 Imidoyl Borate Derivatives

The reaction of *n*-hexyl-9-borabicyclo[3.3.1]nonane with the lithium aldimine from *t*-butylisonitrile and butyllithium generates a lithium borate complex (**72**) in which the imidoyl borate carries a negative charge. This borate complex is not isolable and readily undergoes rearrangement to a ketone derivative ⟨75BCJ3682⟩.

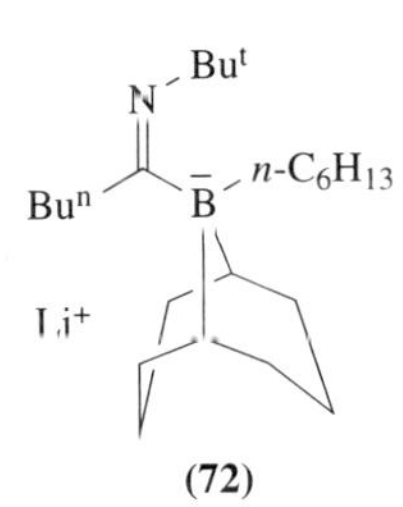

(72)

5.20.3 IMIDOYL METAL FUNCTIONS—$R^1C(NR^2)$ METAL

5.20.3.1 Imidoyl Derivatives of Group 1 Metals—Li, Na, K, Rb and Cs

Imidoyl functions bearing group 1 metals clearly offer much potential for the synthesis of new carbon–carbon bonds and for accessing other imidoyl–heteroatom systems. Despite this there is a very limited choice of methods for preparing them. Indeed, only imidoyllithium intermediates (often referred to as 'lithium aldimines') have received significant attention, and the other group 1 metals (Na, K, Rb and Cs) have not been investigated. Two reported methods for preparing imidoyllithium intermediates involve either additions to isonitriles or lithium–tin exchange ⟨87JA7888⟩. Routes via lithium–halogen exchange or deprotonation have apparently not been investigated.

5.20.3.1.1 Imidoyllithium intermediates from isonitriles

The addition of organolithium reagents to isonitriles lacking α-hydrogens proceeds by an α-addition to produce imidoyllithium compounds (lithium aldimines). Early attempts to add organolithium reagents to isonitriles involved simple compounds like ethyl isonitrile in which the α-hydrogens were subject to deprotonation ⟨74JOC600⟩. Later it was found that aryl or *t*-alkyl isonitriles provide imidoyllithium compounds (Equation (18)) in good yield (45–100%) ⟨74JOC600⟩. In general (see Table 2) it was found that poor yields were obtained in cases where the desired lithium aldimine could equilibrate with the starting materials. It has been suggested that conjugate bases of acids with $pK_a < 37$ will not add appreciably to 1,1,3,3-tetramethylbutylisonitrile (TMBI) ⟨74JOC600⟩. Representative reaction conditions for the reaction of an isonitrile (TMBI) with an alkyllithium (*s*-BuLi) have appeared in *Organic Synthesis* ⟨71OS(51)31⟩. Thus, equimolar equivalents of TMBI and *s*-BuLi were reacted in THF at −5 °C before quenching the imidoyllithium intermediate with D_2O. α-Addition and *ortho*-metallation of phenylisonitrile can be achieved using *t*-butyllithium and TMEDA in diethyl ether (−78 °C) ⟨78JOC731, 89JOM(364)C17⟩.

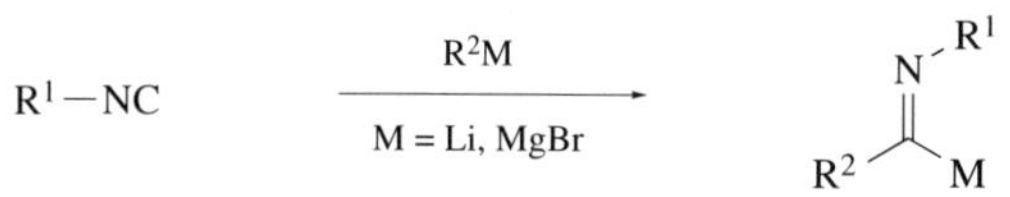

(18)

5.20.3.2 Imidoyl Derivatives of Group 2 Metals—Be, Mg, Ca, Sr, Ba

There are very few reports of imidoylmagnesium compounds in the literature ⟨70JA6675, B-71MI 520-01, 71OS(51)31, 74JOC600⟩ and no significant chemistry has been done with the remaining group 2 metals (Be, Ca, Sr and Ba) in this respect. Very much like their organolithium counterparts (see Section 5.20.3.1), organomagnesium compounds successfully undergo α,α-addition to isonitriles in

Table 2 Preparation of iminoacyl lithium intermediates from isonitriles (Equation (18)).

R¹NC	*R²Li*	*Yield* (%)	*Ref.*
TMBI[a]	Bu^sLi, nBuLi	100	69JA7778, 71OS(51)31
TMBI[a]	PhLi, EtLi	67–81	69JA7778
TMBI[a]	Bu^tLi	93	74JOC600, 82JOC52
TMBI[a]	$CH_2{=}CHLi$	0	74JOC600
TMBI[a]	$PhC{\equiv}CLi$	0	74JOC600
TMBI[a]	MeLi	28	82MI 520-01
TBI[b]	Bu^nLi, iso PrLi	77–92	70JA6675, 75TL2689
TBI[b]	$CH_2{=}CH(CH_2)_3Li$, MeLi	46	75TL2689
TBI[b]	$CH_3CH(CH_3)CH_2CH_2Li$	92	75BCJ3682
TBI[b]	EtLi	86	75BCJ3682
TBI[b]	Bu^tLi, Bu^sLi	62–95	78CJC2286, 84TL3091
isoPrNC	Bu^nLi	45	82JOM(231)191
Bu^nNC	Bu^nLi	43	82JOM(231)191
Pr^nNC	Bu^nLi	31	82JOM(231)191
$c\text{-}C_6H_{11}NC$	Bu^nLi	37	82JOM(231)191
DMPI[c]	EtLi, isoPrLi	50	70JA6675, 78CJC2286
DMPI[c]	Bu^tLi	73	78CJC2286
DMPI[c]	MeLi	41	78CJC2286

[a] 1,1,3,3-Tetramethylbutylisonitrile (TMBI) [b] Tertiary-butylisonitrile (TBI) [c] 2,6-Dimethylphenylisonitrile (DMPI).

good yield (see Equation (18) and Table 3). Unfortunately, due to an unfavourable equilibrium, phenylmagnesium bromide does not add to isonitriles in a viable yield (ca. 2%).

Table 3 Addition of alkyl Grignard reagents to TMBI (Equation (18)).

R^2MgBr	*Yield* (%)	*Ref.*
Bu^s	67	71OS(51)31, 70JA6675
Bu^t	48	71OS(51)31, 70JA6675
n-Hexyl	62	71OS(51)31, 70JA6675
2-Phenylethyl	63	71OS(51)31, 70JA6675
Cyclopentyl	66	71OS(51)31, 70JA6675
Bu^n	>34	74JOC600

5.20.3.3 Transition Metals

Isonitriles have attracted an enormous interest in the study of transition metal chemistry, particularly as a replacement ligand for CO. In many cases, however, the isonitrile moiety can undergo insertion reactions to give novel imidoyl transition metal derivatives. Other methods for preparing such compounds include nucleophilic attack on transition metal–isonitrile ligands; oxidative addition of imidoyl chloride compounds to metal complexes; and the reaction of metal carbonyl anions with imidoyl chlorides. The synthesis of different groups of transition metal derivatives are discussed below.

5.20.3.3.1 Imidoyl-Ti, Zr and Hf derivatives

All reported preparations of imidoyl-linked group 4 metals (Ti, Zr, Hf) have been prepared via insertion of an isonitrile reagent into either a metal–carbon or a metal–hydrogen bond. Most examples involve Zr or Ti complexes with unsaturated ligands like cyclopentadienes or styrenes. In general (see Equation (19)), four coordinate Zr or Ti complexes with two cyclopentadiene type ligands and two other groups will react with aryl or alkyl isonitriles. These reactions are typically run in diethyl ether at −20 °C and yields tend to be in the range 80–95% ⟨81JCS(D)2088, 85JOM(280)355, 91OM1201, 91OM2665⟩.

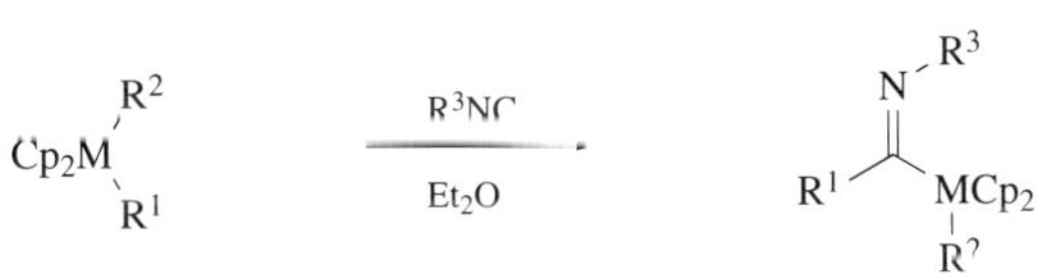

(19)

R^1, R^2 = H, alkyl; R^3 = Me, Bu, Ar; M = Zr, Ti

Dialkylbis[bis(trimethylsilyl)amido]hafnium(IV) compounds readily react with alkyl isonitriles with double insertion ⟨79IC2928⟩.

5.20.3.3.2 Imidoyl-V, Nb, and Ta derivatives

Although imidoylvanadium chemistry is underdeveloped there are some reports of imidoylniobium and imidoyltantalum compounds ⟨74JOM(67)269, 81JCS(D)2088⟩. As in Section 5.20.3.3.1, the chemistry is dominated by isonitrile insertion reactions. The stoichiometry of isonitrile insertion reactions with alkyltantalum complexes is determined by the character of the isonitrile. Thus, reaction of dichlorotrimethyltantalum (**73**) with excess *t*-butylisonitrile gives (**74**) with maximum insertion. However, reaction with cyclohexyl- or *p*-tolylisonitrile gives (**75**) in which only two of the Ta–Me bonds have reacted (Scheme 13).

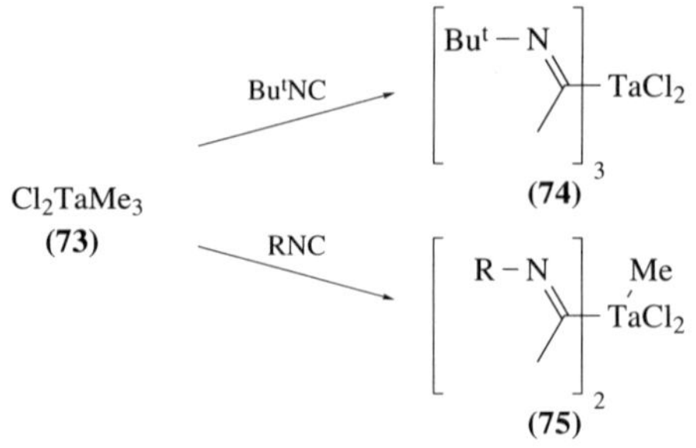

R = c-C_6H_{11}, Tol

Scheme 13

5.20.3.3.3 *Imidoyl-Cr, Mo and W derivatives*

Both Cr and W have been bonded to η^1-imidoyl groups and two general methods of synthesis have been employed.

(i) Imidoylchromium and -tungsten derivatives from isonitriles

Insertion of *t*-butylisonitrile with the trimesitylchromium complex (**76**) is promoted by the presence of trimethylphosphine ⟨92JCS(D)19⟩ and provides the trisimidoylchromium complex (**77**) in 48% yield (Equation (20)). Similar isonitrile insertions have been observed for tungsten ⟨88JCS(D)27⟩.

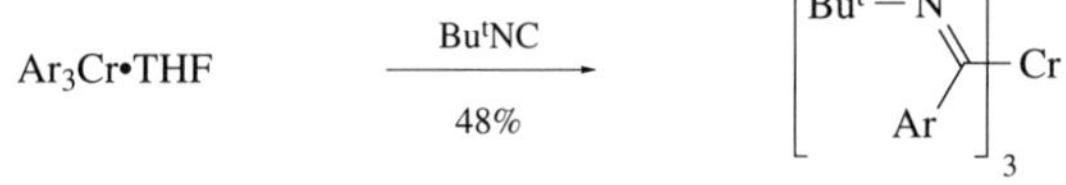

(20)

(**76**) (**77**)

(ii) Imidoyltungsten derivatives by displacement of imidoyl halides

A number of metal carbonyl anions react with imidoyl halides ⟨79JOM(172)251⟩ in moderate to good yield (see also Section 5.20.3.3.5). Thus, the reaction of the cyclopentadienyltricarbonyl tungstonate anion (**78**) with the benzimidoyl chlorides (**79**) at −78°C to 25°C (THF solvent) provides the desired imidoyltungsten derivatives (**80**) in 44–62% yield (Equation (21)) ⟨79JOM(172)251⟩.

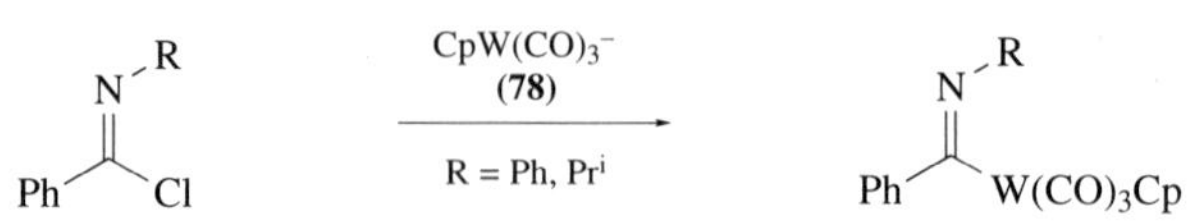

(21)

(**79**) (**80**)

(iii) Imidoyltungsten and -chromium derivatives from metalloalkoxycarbenes

The C═N bond of imidoyltungsten ⟨87JOM(327)339, 87JOM(333)245⟩ and -chromium ⟨89OM2253, 90JOM(381)191⟩ derivatives can be formed by the reaction of allylamine with metalloalkoxycarbenes.

The imidoylnitrogen of these products is protonated, with a negative charge residing on the metal centre.

5.20.3.3.4 Imidoyl-Mn derivatives

The reaction of isonitriles (**81**) with alkyl- ⟨89OM2589⟩ and aryl- ⟨89OM1523⟩ manganese pentacarbonyl complexes (**82**) involves PdO-catalysed monodecarbonylation with two consecutive isonitrile insertions to give a diazabutadiene system bonded to Mn via the imidoyl carbon (**83**). These compounds (Equation (22)) are formed in ca. 50% yields based on the input of (**82**).

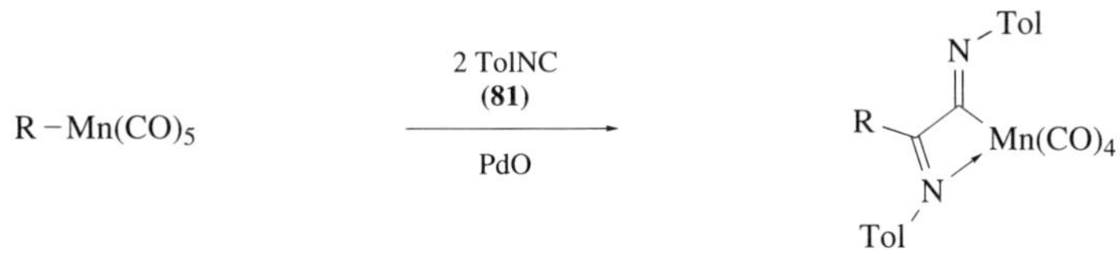

(**82**) R = CH_2Tol, C_6H_4OMe (**83**) (22)

5.20.3.3.5 Imidoyl-Fe, Co and Ni derivatives

(i) Imidoyliron derivatives from iron carbonyl anions

Reaction of iron carbonyl anions (**84**) with imidoyl chlorides (**85**) occurs through nucleophilic displacement of the chloride atom to produce imidoyliron derivatives (**86**) in which the imidoyl group is bonded in a monodentate fashion ⟨79JOM(172)251⟩. A detailed mechanistic study of this process has been published ⟨91G101⟩. Similar chemistry has been observed in the reaction of iron carbonyl anions with ketenimines (**87**) ⟨82JOM(224)153⟩. Here, monoprotonation of a ferrazetidine intermediate (using HCl in diethyl ether) affords imidoyliron derivatives (**88**) (Scheme 14).

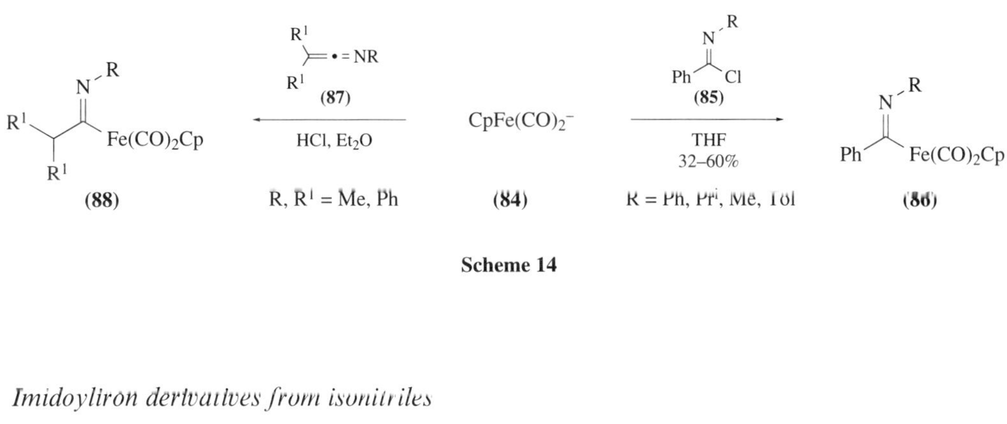

(**88**) R, R^1 = Me, Ph (**84**) R = Ph, Pr^i, Me, Tol (**86**)

Scheme 14

(ii) Imidoyliron derivatives from isonitriles

A wide variety of imidoyliron derivatives can be accessed using isonitrile-insertion chemistry. Pentakis(*t*-butylisonitrile)iron (**89**) undergoes oxidative addition with a range of alkyl halides and subsequent intramolecular isonitrile insertion can then take place (Equation (23)) to provide new imidoyliron derivatives (**90**). Many of these derivatives are unstable and rapidly rearrange to give the η^2-iminoacyl derivatives. When the iron complex carries a cyclopentadiene ligand (**91**), stable monoimidoyliron derivatives (**92**) can be formed in the presence of CO (Equation (24)). However, in the presence of excess isonitrile reagent, the second insertion is rapidly followed by polymerization ⟨88JCS(D)1591⟩.

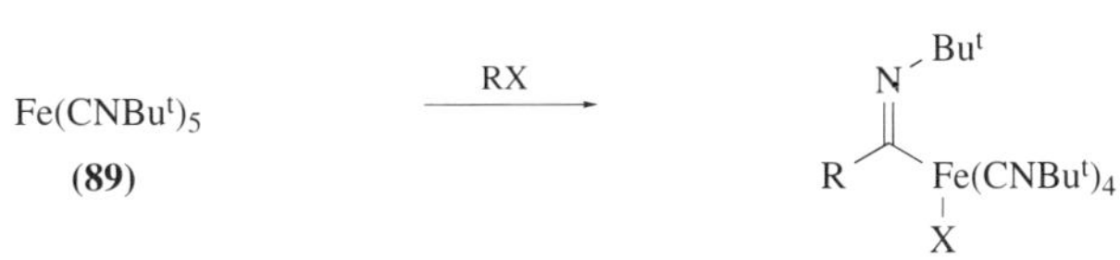

(**89**) (**90**) (23)

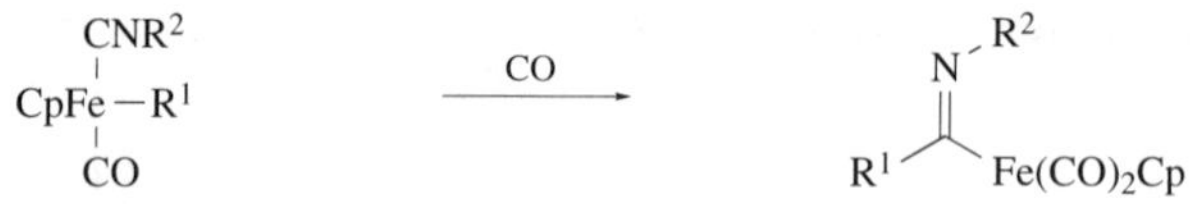

(24)

(91) **(92)**

R^1 = Me, Bn, C_6H_4Cl; R^2 = *c*-C_6H_{11}, Bu^t

Although mechanistically implicated ⟨79JOM(172)251⟩, imidoylcobalt derivatives are generally unstable and very few examples have been cited ⟨85CB1517⟩.

(iii) Imidoylnickel derivatives from nickel carbonyl anions

Work on nickel carbonyl anions (**93**) has demonstrated that, like iron (Equation (25)), benzimidoyl chlorides (**94**) react to give imidoylnickel derivatives (**95**) in 29% yield ⟨79JOM(172)251⟩.

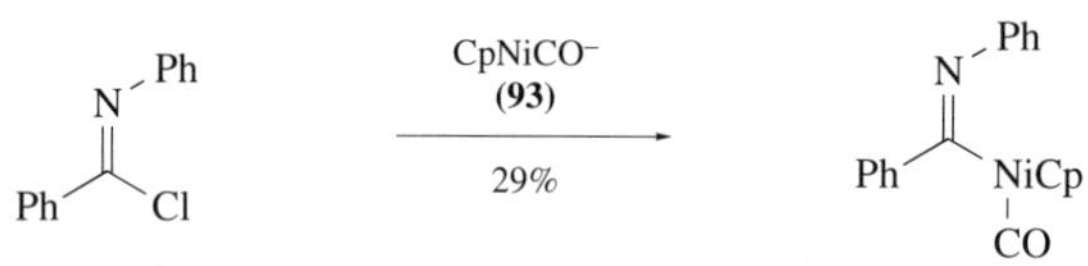

(25)

(94) **(95)**

(iv) Imidoylnickel derivatives from isonitriles

Much of the nickel–isonitrile chemistry is dominated by multiple insertion reactions ⟨69JA7196, 87JOM(332)371, 89JOM(377)157, 90OM583⟩. A stepwise isonitrile insertion process is possible when phosphine ligands are present on the nickle centre ⟨87JOM(332)371, 89JOM(377)157, 90OM583⟩. Treatment of the nickel complex (**96**) with isonitrile reagents has provided good yields of mono-insertion products (Equation (26)).

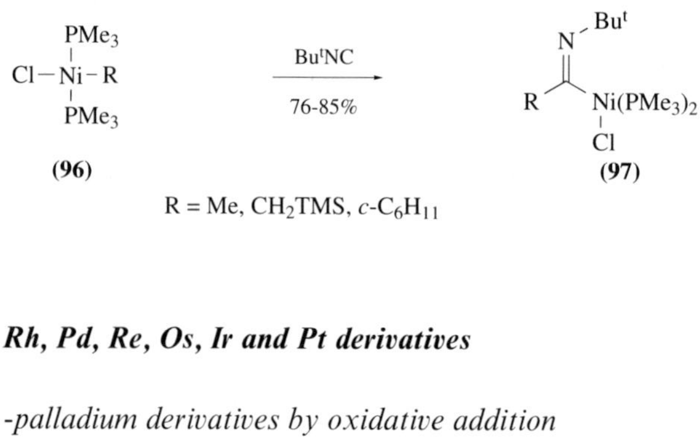

(26)

(96) **(97)**

R = Me, CH_2TMS, *c*-C_6H_{11}

5.20.3.3.6 Imidoyl-Ru, Rh, Pd, Re, Os, Ir and Pt derivatives

(i) Imidoylrhodium and -palladium derivatives by oxidative addition

Isonitrile insertion reactions have received little attention as a method of forming imidoylruthenium or -rhodium complexes. As part of a study of rhodium catalyst hydroxylation, imidoylrhodium complexes (**99**) were prepared from 3-methyl-2-aminopyridyl aldimines (**98**) by oxidative addition with Wilkinson's catalyst in THF at 55°C in greater than 90% yield (Equation (27)). The desired imidoylrhodium hydride (**99**) has been further reacted with alkenes ⟨79JA489, 90MI 520-01⟩ to provide an interesting route to unsymmetrical ketones.

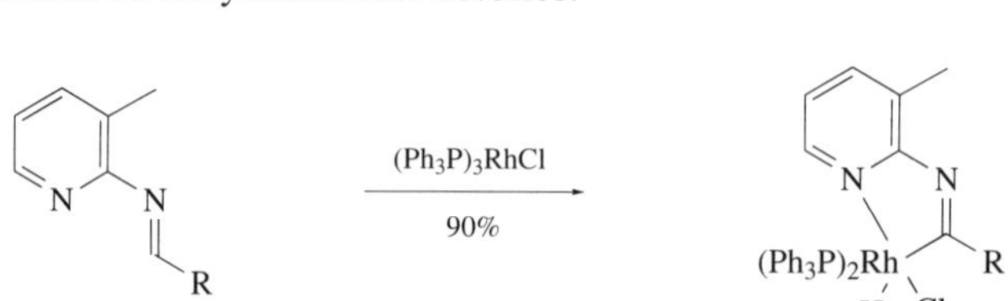

(27)

(98) **(99)**

Imidoylpalladium(II) complexes (**101**) have been prepared by oxidative addition of imidoyl chlorides with palladium(0) complexes like $Pd(PPh_3)_4$ ⟨79JOM(168)97, 80BCJ1171, 81JOM(206)C11⟩. In general, a slight excess of the imidoyl chloride (**100**) in benzene is added to $Pd(PPh_3)_4$ at room temperature (Equation (28)). Crystallization of the product imidoylpalladium(II) derivatives then gives yields in the range 70–90%.

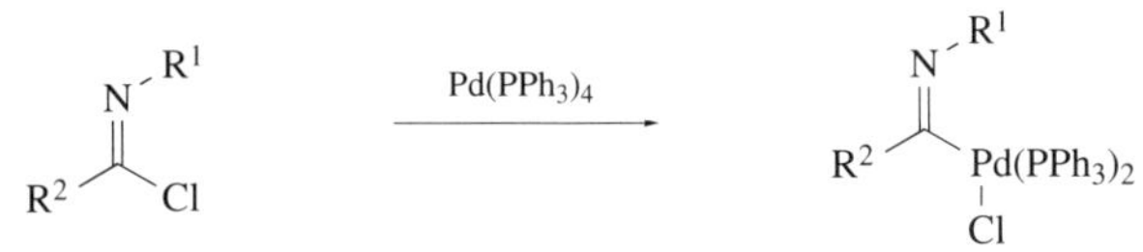

(**100**) (**101**) (28)

R^1 = Bu^t, *c*-C_6H_{11}, C_6H_4OMe, Tol, $C_6H_4NO_2$; R^2 = Ph, Tol, COMe

(ii) Imidoylpalladium derivatives from isonitriles

Isonitrile insertion reactions with Pd—C bonds (Equation (29)) are very prominent and exceed the number of examples of oxidative addition routes to imidoylpalladium derivatives (**103**). With few exceptions ⟨76JCS(D)327⟩, the palladium centre (**102**) generally carries a mixture of halogen and phosphine ligands and the site of isonitrile insertion involves benzylic ⟨76JCS(D)327⟩, vinylic ⟨76JCS(D)327⟩, acetylenic ⟨88CC561, 91JCS(D)1531, 92AG(E)851⟩, aromatic ⟨90JCS(D)3193⟩, heteroaromatic ⟨82JOM(236)C37, 86OM1369⟩ or simple methyl groups ⟨69JA7196, 74IC438⟩. A variety of isonitriles have been employed in this transformation.

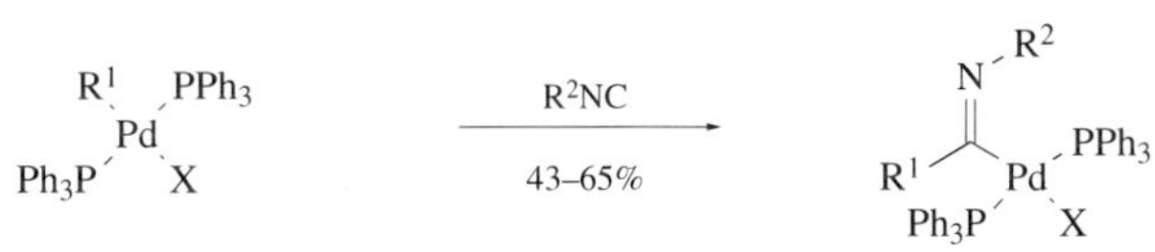

(**102**) (**103**) (29)

X = Cl, Br, I; R^1 = Bn, Me C_5H_4N, $CH{=}CHCO_2Me$
R^2 = C_6H_4OMe, Tol, $C_6H_4NO_2$, Ph, Bu^t, *c*-C_6H_{11}

(iii) Imidoylrhenium derivatives from functionalized amines

With the exception of one imidoylrhenium derivative prepared via isonitrile insertion chemistry ⟨81JCS(D)2088⟩, most of the imidoylrhenium derivatives (**105**) are protonated on nitrogen and have been derived from rhena-*β*-diketones (**104**) using the Schiff-base condensation (Equation (30)) with a variety of functionalized primary amines ⟨78IC2369, 81JA1467, 82IC1278, 83IC3954, 86ICAL15, 86ICA67⟩.

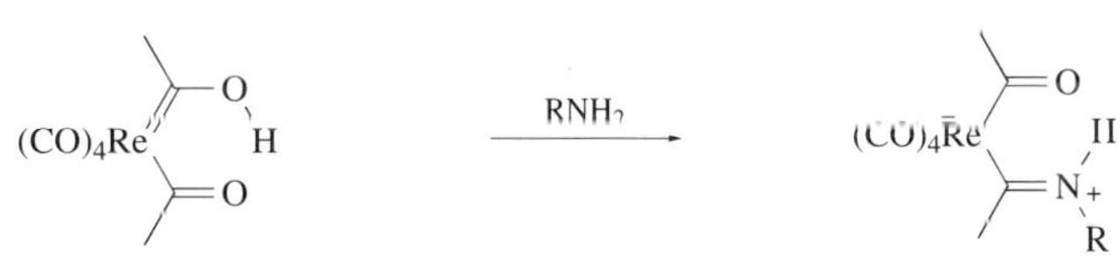

(**104**) (**105**) (30)

(iv) Imidoylosmium and -platinum derivatives from isonitriles

A mono-insertion product is formed in the reaction of an osmium methylimide complex with *t*-butylisonitrile ⟨90JA8369⟩.

The reactions of a number of *cis*-dialkyl- or *cis*-diaryl-triphenylphosphineplatinum complexes with isonitriles typically result in substitution only. However, the presence of a smaller and more basic phosphine (e.g., PEt_3) does result in isonitrile insertion. Thus, in general (Equation (31)), platinum complexes (**106**) carrying two phosphine ligands, one halogen and an alkyl, alkenyl, aryl or hydride group react with alkyl or aryl isonitriles to give imidoylplatinum derivatives (**107**) ⟨73JOM(61)415, 76JCS(D)327, 76JOM(112)209⟩. Ring expanded zwitterionic complexes, containing the imidoylplatinum unit, have also been prepared ⟨89JCS(D)345, 90JCS(D)781⟩.

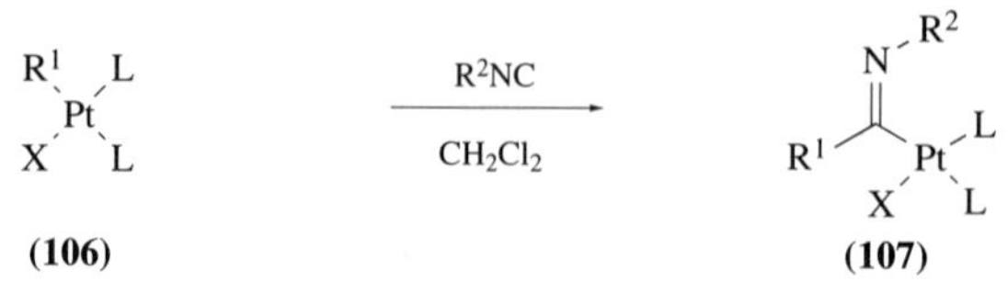

(31)

$X = Cl, Br, I; R^2 = Me, C_6H_4Cl$, Tol;
$R^1 = Me, Ph; L = Et_3P, PPhMe_2$
$R^1 = CH{=}CHCO_2Me; L = PPh_3$
$R^1 = H; L = Et_3P, PPhMe_2$

(v) Imidoylplatinium derivatives by oxidative addition

Oxidative addition of α-ketoimidoyl chlorides (**108**) with platinum(0) complexes yields α-keto-imidoylplatinum derivatives (Equation (32)). Thus, platinum tetrakis(triphenylphosphine) (**109**, $L = PPh_3$) or platinum tetrakis(phenyldimethylphosphine) (**109**, $L = PPhMe_2$) react with (**108**) in benzene at 60 °C to give the corresponding imidoylplatinum derivatives (**110**) in 55–75% yield ⟨81JOM(206)C11⟩.

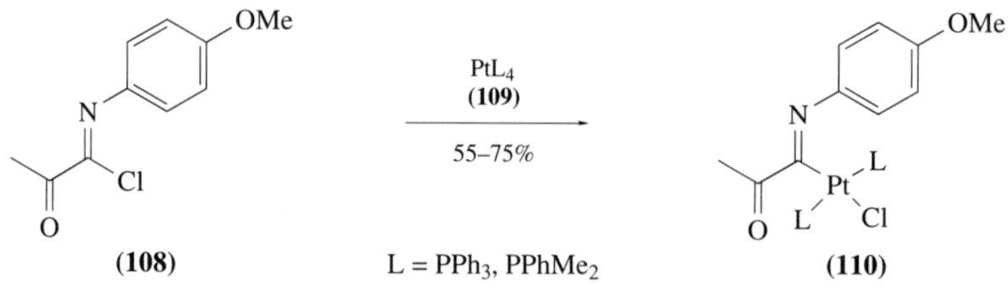

(32)

$L = PPh_3, PPhMe_2$

5.20.3.3.7 *Imidoyl-Cu, Ag and Au derivatives*

Imidoylcopper derivatives (Equation (33)) have been prepared by transmetallation of (**111**), the parent imidoyllithium compounds (see Section 5.20.3.3.1). A number of copper(I) salts have been employed ⟨78JOC731, 84TL3091⟩ with CuI being most effective. The reactive imidoylcopper products (**112**) have not been characterized and have found some synthetic utility in the conjugate addition of imidoyl anions to enones providing 1,4-diketones. They are not stable at 25 °C and hydrolysis results in oxidation of the imidoylcarbon to give an amide function together with liberation of copper metal.

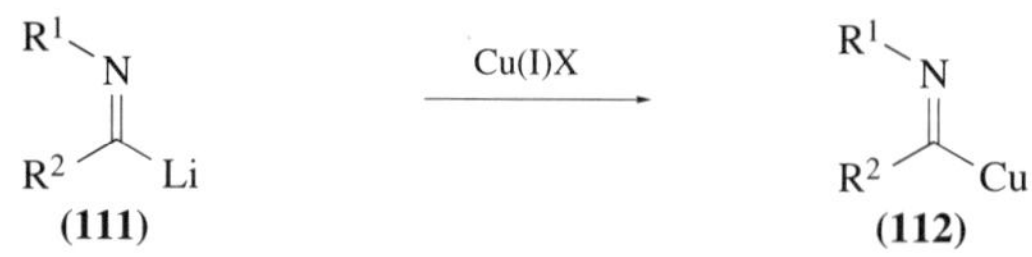

(33)

$R^1 = Ph; R^2\ Bu^t; X = Cl$
$R^1 = Bu^t; R^2 = Bu^n, Bu^s; X = I$

There appears to be no synthetic interest in imidoylsilver derivatives, and only one synthesis of an imidoylgold complex has been reported ⟨88BCJ3649⟩.

5.20.3.3.8 *Imidoyl-Zn, Cd and Hg derivatives*

A number of organozinc reagents (**113**) have been reacted with isonitriles ⟨88JOC4158, 89CL1603⟩ to give imidoylzinc derivatives (Equation (34)). With the exception of divinylzinc (11% yield of insertion product) the yields are reliable (60–74%). Most success has been experienced with aromatic isonitriles; α-addition (i.e., insertion) is accelerated by electron-withdrawing groups *ortho* to the isonitrile function and aliphatic isonitriles do not react at all. These imidoylzinc derivatives (**114**) have proved quite valuable as acyl anion equivalents ⟨89CL1603⟩ and can be coupled to aromatic halides under Pd catalysis. There have been no reports of imidoylcadmium or -mercury derivatives.

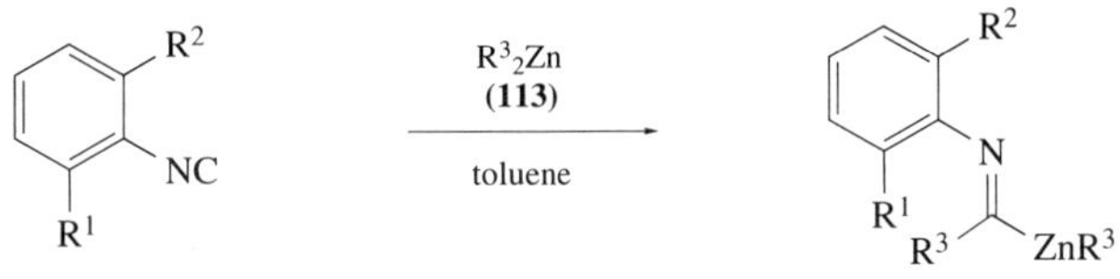

(34)

(**114**)

$R^1 = R^2 = H, Me, CF_3$
$R^3 = Et, Pr^i, CH_2CH{=}CH_2, Ph$

5.20.3.4 Imidoyl Derivatives of Group 13 and Group 14 Metals: Al, Ga, In, Tl, Sn and Pb

The chemistry of imidoyl group 13 metal derivatives is largely unknown. However, formimidoylaluminum derivatives (**116**) have been prepared using the insertion reaction of isonitriles (**115**) with the Al—H bond in diisobutylaluminum hydride (dibal-H) (Equation (35)) ⟨75LA1124⟩.

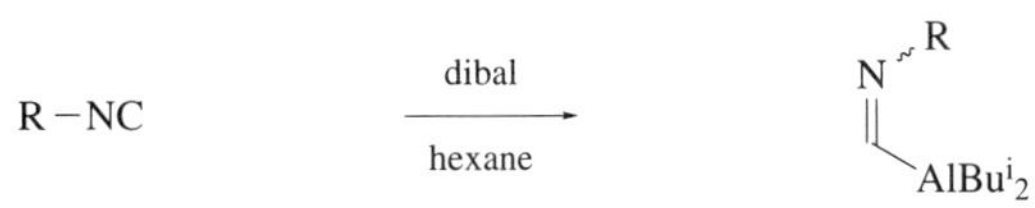

(35)

(**115**) (**116**)

$R = Bu^t$, 93%; c-C_6H_{11}, 43%; Bn, 27%

Imidoyltin and imidoyllead derivatives can be prepared by displacement of chloride from the corresponding imidoyl chlorides ⟨71MI 520-02, 81JCS(P2)1043⟩. Thus trimethylstannyllithium and triphenylstannyllithium (**117**) react in boiling THF with a series of benzimidoyl chlorides (**118**) to give imidoyltin derivatives (**119**) (Equation (36)), in poor yield (20–38%). Similar results are obtained in the preparation of imidoyllead compounds.

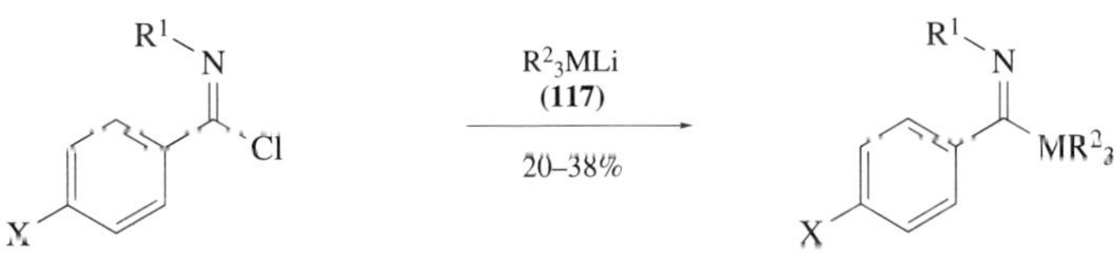

(36)

(**118**) (**119**)

$R^1 = Ar$, c-C_6H_{11}; $R^2_3M = Me_3Sn, Ph_3Sn$; X = H, Me, Cl, OMe; $R^1 = Ph$; $R^2_3M = Ph_3Pb$; X = H

5.20.3.5 Imidoyl Derivatives of Lanthanides and Actinides

The chemistry of imidoyllanthanides and -actinides has not been developed. Only imidoylsamarium derivatives (**122**) have attracted interest (Equation (37)) because of their utility in the addition of acyl anion equivalents to carbonyl functionalities ⟨90JA2437⟩. The imidoylsamarium derivative (**122**) has been prepared by the reaction of samarium iodide, benzyl chloromethyl ether

(**120**) and 2,6-xylylisonitrile (**121**) in THF at -15°C. The yield is nearly quantitative based upon the subsequent reactions with a variety of ketones ⟨90JA2437⟩.

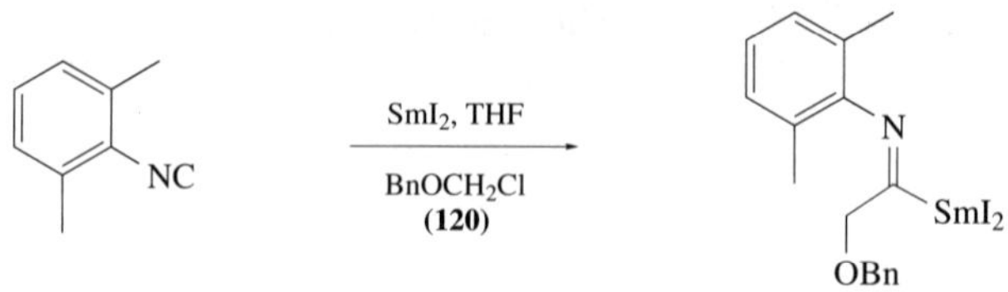

(**121**) (**122**) (37)

5.21
N-Hetero-substituted Iminoacyl Functions

BRANKO STANOVNIK
University of Ljubljana, Slovenia

5.21.1 HALOIMIDIC DERIVATIVES

CAUTION. Many *N*-halo compounds are known to be powerful explosives. Reactions should be carried out on a small scale to minimize laboratory handling hazards of the products. Adequate protective equipment should be used during all phases of the work, including the manipulations during spectroscopic and elemental analysis. For quantities greater than 1 g, remote handling is recommended. Liquid samples confined in metal containers are even more hazardous.

5.21.1.1 *N*-Fluoroimino Fluorides

Highly fluorinated organonitrogen compounds encompass a broad range of materials, whose synthesis, properties, and chemical reactions are of great interest.

Compounds with an *N*-fluoroimino group may be prepared in different ways. $CF_2{=}NF$ (**2**) was prepared, along with other compounds (**3**)–(**6**) and the highly fluorinated stable polymeric material (**7**), by the action of fluorine on acetonitrile (**1**), using helium or nitrogen as a diluent, in 7 : 1 : 21 molar ratio, under vigorous conditions at 60–140 °C (Equation (1)) ⟨52JA710⟩.

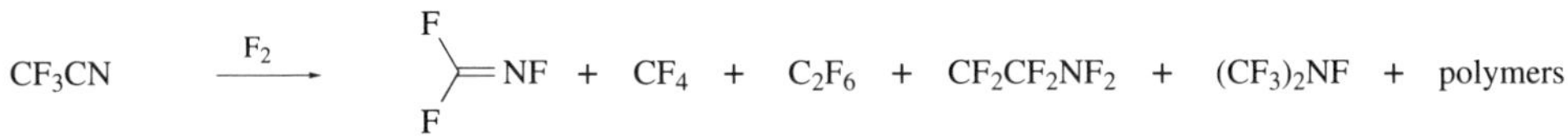

(1) (2) (3) (4) (5) (6) (7) (1)

The two best methods for the preparation of *N*-fluoroimines are as follows. First, *N*-fluoroimines are prepared by 1,2-elimination of Cl_2 or F_2 from compounds such as $F_2ClCNClF$ (**8**), by mercury in polar solvents, for example trifluoracetic acid (TFA) or trifluoroacetic anhydride (TFAA) (Equation (2)). The yields are good and no CF_3Cl is formed as a side product ⟨71IC1635⟩. When TFAA is used as a solvent, the yields are >90% and the product is easily purified, while with TFA at 0 °C two products are formed, $CF_2{=}NF$ and ClF_2CNHF, due to competition between the loss of Cl^- and the abstraction of a proton from the solvent by ClF_2CNF ⟨81JOC1277⟩.

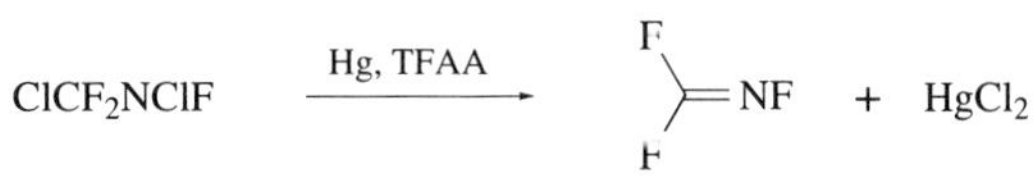

(8) (9) (2)

The following compounds have been obtained according to this method: $CF_2{=}NF$ (64%), $CF_3CF{=}NF$ (59%), $CF_3CF_2CF{=}NF$ (54%), and $CF_3CF{=}NCl$ (5%). The reaction in which $CF_3CF_2NCl_2$ in TFA produces $CF_3CF{=}NCl$ may be a useful synthetic method for the preparation of $R_fCF{=}NF$ type compounds. Perfluoro-1-amino-2-imino ethane (**11**) is obtained from 1,2-bis(difluoroamino)trifluoroethane (**10**) with caesium fluoride at 100 °C in 85% yield (Equation (3)) ⟨68ZOB1408⟩.

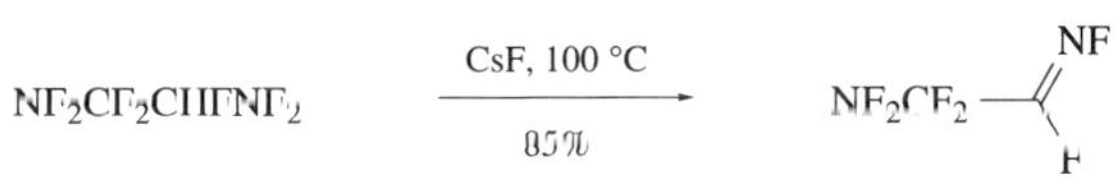

(10) (11) (3)

In the second method, *N*-fluoroimines are prepared by 1,2-elimination of ClF from compounds such as $NCCCl_2NF_2$ ⟨71IC407⟩, Cl_2FCNF_2 and Cl_3CNF_2 ⟨71IC1635⟩ by mercury. With TFA as the solvent $ClCF_2NClF$ is dehalogenated with mercury to give excellent yields of $CF_2{=}NF$. Other *N*-chloro-*N*-fluoroamines (**12**) undergo dehalogenation with mercury to form the corresponding *N*-fluoroimidic fluorides (**13**) in good yields (Equation (4)). Only *syn*-isomers are formed in this reaction ⟨81JOC1277⟩. *N*-Chlorodifluoromethyleneimine (**15**) is formed by thermolysis of the corresponding *N*,*N*-dichloroperhaloalkylamine (**14**) (Equation (5)) ⟨66IC1795, 67ZOB1413⟩.

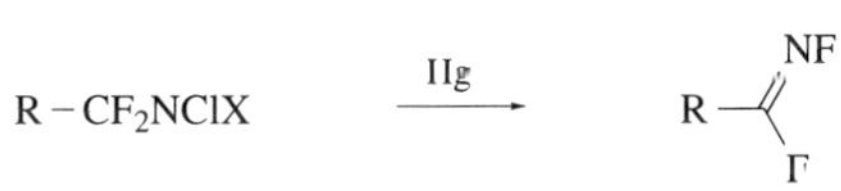

(12) (13) (4)

R	*X*	*Solvent*	*Yield* (%)
Cl	F	TFA	90
Cl	F	TFAA	88
CF_3	F	TFAA	4
C_2F_5	F	TFAA	84
Cl	F	none	64
CF_3	F	none	59
C_2F_5	F	none	

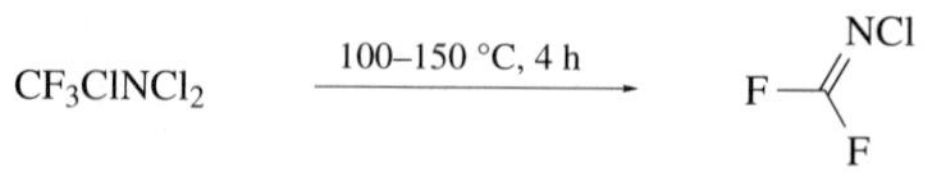

(14) (15) (5)

Direct fluorination of *N*-methylformamide (**16**) and *N*,*N*-dimethylformamide (**17**) with fluorine at 275 °C gives $CF_2{=}NF$ (**19**), among other products in 6% and 22% yield, respectively (Equation (6)). The same product has also been obtained by fluorination of methylthiocyanate (**18**) ⟨59JA3599⟩.

(16) (17) (18) (19) (6)

The reductive defluorination reaction of perfluoroamines, in which dicyclopentadienyliron (ferrocene) or dicumenechromium act primarily as reducing agents, serves to introduce a —CF=NF moiety into the fluorocarbon structure in high yield. The use of these reagents is especially convenient since they permit the use of aprotic hydrocarbon solvents at room temperature. According to this procedure a difluoroamino compound, such as perfluoro-1-azapentane (**20**; $R = C_3F_7$), is treated with 2 mol of ferrocene in carbon tetrachloride at room temperature, producing perfluoro-1-azapentene-1 (**21**; $R = C_3F_7$) in high yield (Equation (7)). Perfluoro-1,5-diazapenta-1,4-diene (**23**) can be prepared in an analogous manner from (**22**) in 57% yield (Equation (8)) ⟨65JA328⟩. Many compounds of the general formulae R_f—CF=NF ⟨72USP3674785⟩ and X—CF=NF (X = perfluoroalkyl, NF_2, F) ⟨73USP3751312⟩ have been prepared accordingly.

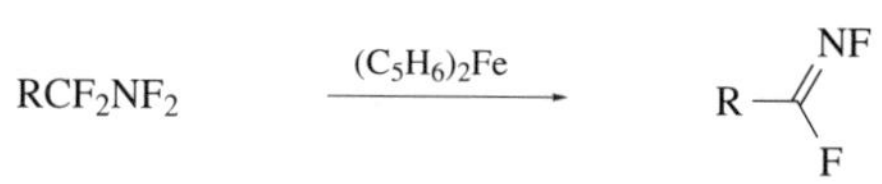

(20) $R = C_3F_7, CFCl_2$ (21) (7)

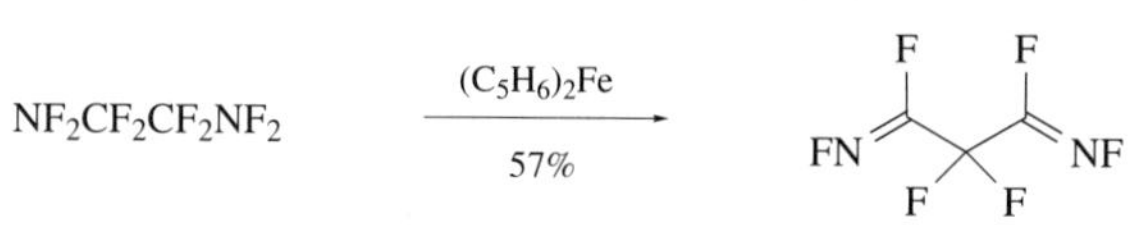

(22) (23) (8)

5.21.1.2 From Nitriles

The general synthesis of perfluoro-*N*-chloroimidic fluorides (**25**) and *N*-bromoimidic fluorides is represented by the reaction of perfluoronitriles (**24**) with chlorine or bromine in the presence of metal fluorides, such as NaF, KF, PbF_2, CuF_2, and HgF_2 (Equation (9)). Sometimes the more reactive AgF is used. The reactions have been extensively studied in the temperature range 50–200 °C ⟨60JOC2016, 62JA2337, 62USP3023208, 63USP3076843⟩. The method can be applied also to carbonitriles of chlorocarbons. Azo compounds (**26**) are formed as by-products in this reaction ⟨62JA2337⟩.

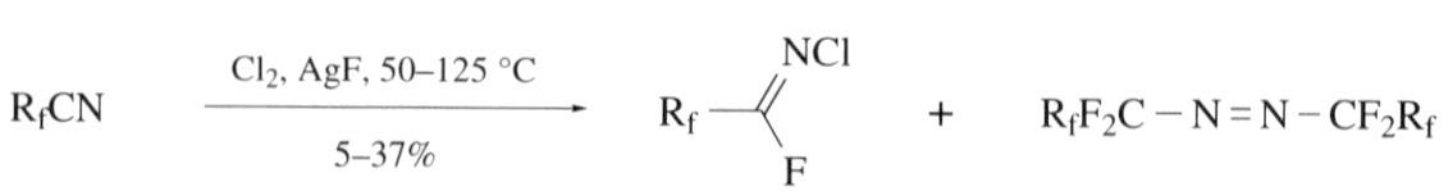

(24) (25) (26) (9)

Perhalogenated nitriles (**27**) react readily with bromine and active caesium fluoride to afford *N*-bromoperhalo-1-alkanimines (**28**) (Equation (10)). Acetonitrile and pentafluorobenzonitrile do not react under these conditions ⟨84JOC1467, 86MI 521-01⟩.

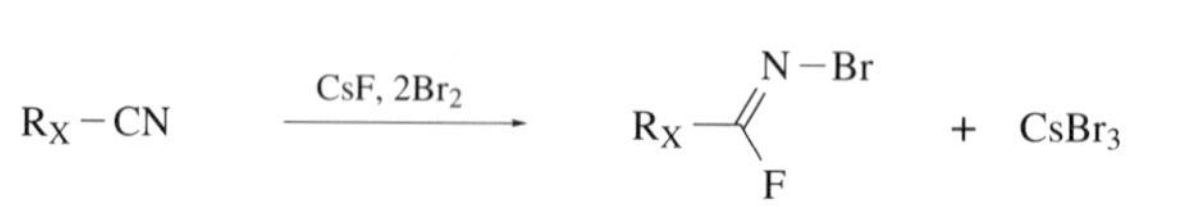

(27) (28) (10)

$R_X = CF_3, C_2F_5, n\text{-}C_3F_7, CClF_2, CCl_3$

The nitriles (**29**) react with chlorine and bromine to form *N*,*N*-dihalo-1,1-difluoroamines as intermediates, from which the corresponding haloimidic fluorides (**30**) are obtained (Equation (11)) ⟨85CB4459⟩. Trifluoroacetonitrile and trichloroacetonitrile react with a stoichiometric amount of HgF_2 in the presence of halogen, while with acetonitrile only traces of the corresponding haloimidic fluorides are formed ⟨85CB4459⟩.

The chlorination of sulfonyl cyanides (**31**) produced *N*-chloroformimidoyl chlorides (**32**) (Equation (12)) ⟨70TL3763⟩.

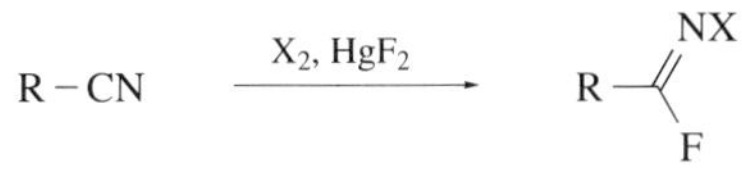

(11)

(**29**) (**30**)

R = Cl, CF_3, CCl_3, Me; X = Cl, Br

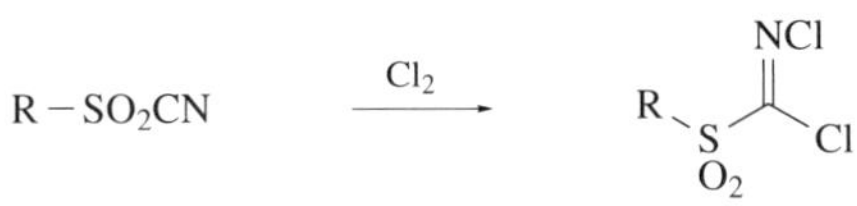

(12)

(**31**) (**32**)

R = Me, Ph, 4-Me-C_6H_4, 4-Cl-C_6H_4

The chlorination of cyanogen (**33**) under pressure at temperatures below 100 °C affords a mixture of two products in 80% yield: *N*-chlorocyanoformimidoyl chloride (**34**) and *N*,*N′*-dichloro-oxalimidoyl dichloride (**35**) (Equation (13)) ⟨71RTC458⟩. The addition of chlorine to cyanogen chloride (**36**), in 2 : 3 molar ratio, under pressure gives *N*-chlorochloroformimidoyl chloride (**37**) in 77% yield (Equation (14)) ⟨69AG572⟩ and addition of bromine to cyanogen fluoride (**38**) in the presence of potassium fluoride or other alkali metal fluorides gives the corresponding *N*-bromo-fluoroformimidoyl fluoride (**39**) in around 60% yield (Equation (15)) ⟨84JA4266, 88JOC4443⟩.

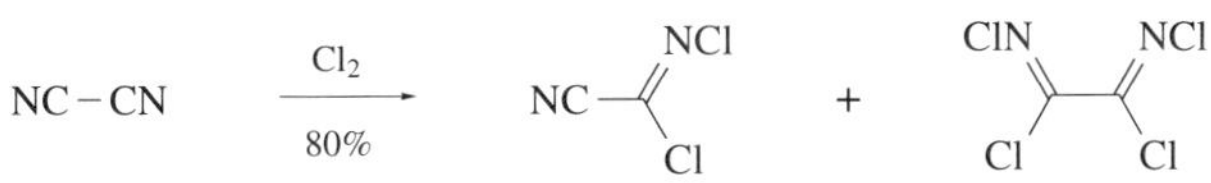

(13)

(**33**) (**34**) (**35**)

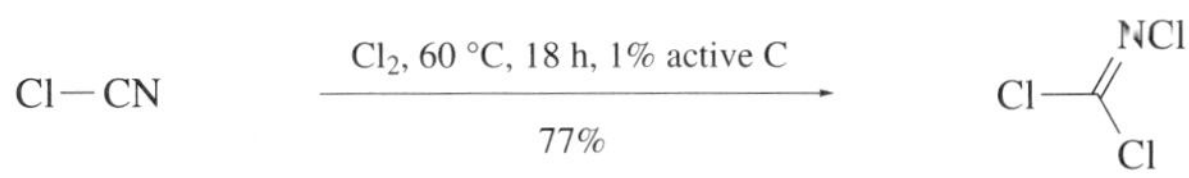

(14)

(**36**) (**37**)

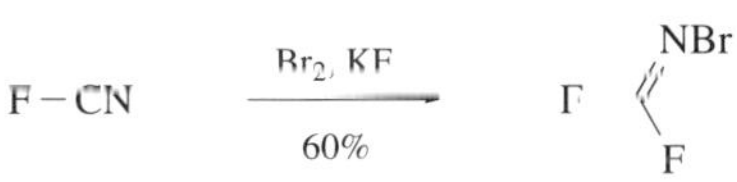

(15)

(**38**) (**39**)

On chlorination of malononitrile (**40**) at elevated temperatures with free chlorine at normal pressure, *N*,2,2-trichlorocyanoacetimidoyl chloride (**41**) is formed in 45–60% yield (Equation (16)). Chlorination of (**42**) under pressure affords tetrachloromalonimidoyl dichloride (**43**) in 60% yield (Equation (17)) ⟨71RTC458⟩. *N*,2,2-Trichlorocyanoacetimidoyl chloride has been isolated as a by-product by chlorination of malononitrile ⟨70TL3763⟩, and *N*-chlorothiochloroformimidoyl chloride (**45**) by chlorination of thiocyanogen (**44**) (Equation (18)) ⟨66AG908⟩.

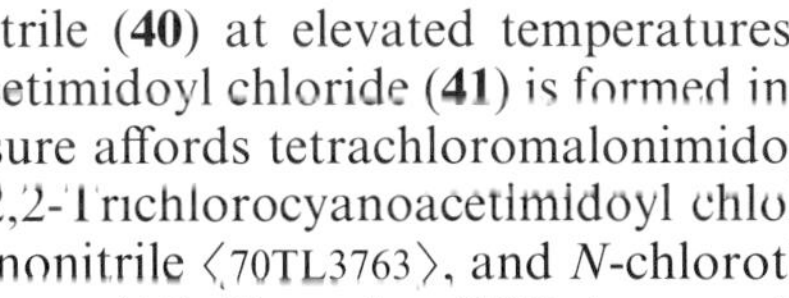

(16)

(**40**) (**41**)

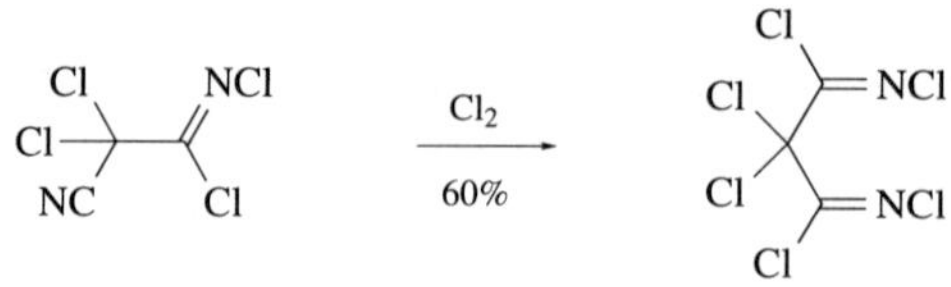

(42) (43) (17)

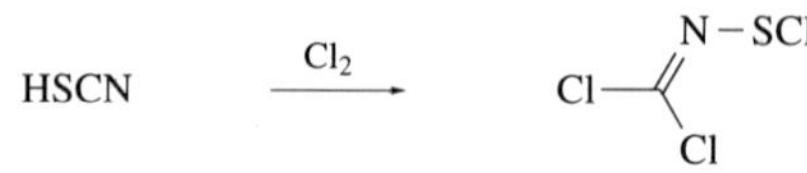

(44) (45) (18)

The chlorination of trichloroacetonitrile (**46**) with chlorine in the presence of catalytic amounts of tetramethylammonium chloride affords, in low yield, *N*-chlorotrichloroacetimidoyl chloride (**47**) (Equation (19)) ⟨71RTC458⟩.

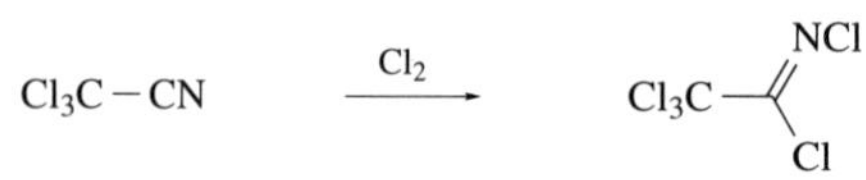

(46) (47) (19)

The chlorination of sodium salt of tricyanomethane (**48**) gives, dependent on the molar ratio between the substrate and chlorine, the following compounds (**49**), (**51**), and (**52**) (Scheme 1) ⟨72ZOR695⟩, while in the reaction of potassium salt of tricyanomethane (**54**) with thionyl chloride tricyanochloromethane (**55**) is formed as the first product, which is transformed further into 2-chloro-2,2-dicyano-*N*-chloroiminoacetic acid chloride (**56**) (Scheme 2) ⟨75ZOR1981⟩. Reaction of sodium salt of dicyanamide (**57**) with thionyl chloride in the presence of tetramethylammonium chloride as a catalyst gives *N*-chloro-*N'*-chlorothio-*N'*-cyanoaminoformimidoyl chloride (**58**) in 65% yield (Equation (20)) ⟨62HOU(5/3)503, 70JCS(C)875⟩.

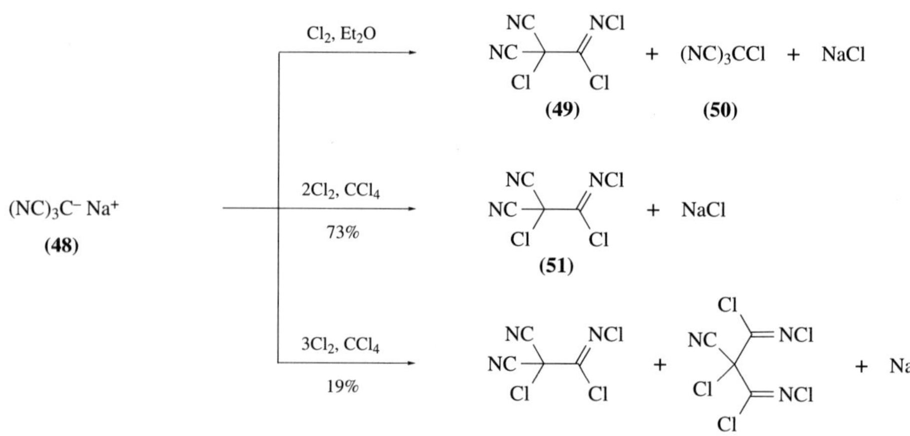

(48) (49) (50) (51) (52) (53)

Scheme 1

(54) (55) (56)

Scheme 2

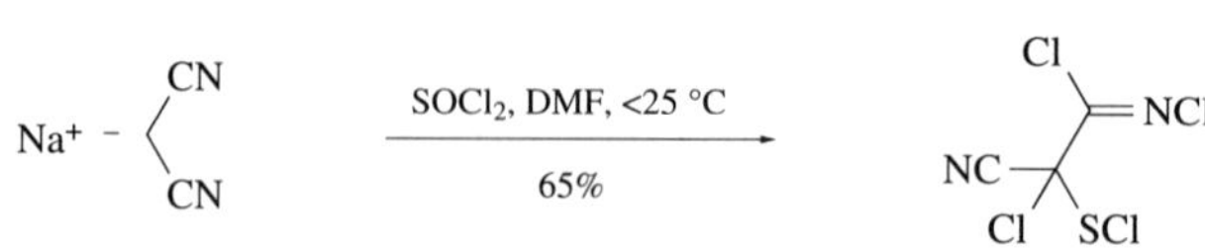

(57) (58) (20)

5.21.1.3 From Halo Alkenes

In the reaction of hydrogen-containing halo alkenes with tetrafluorohydrazine in the presence of sodium fluoride a number of perfluorinated *N*-fluoroimidic fluorides are formed ⟨68IZV1134⟩. For example, fluorination of 1,1,2-trifluoroethene **(59)** in the presence of potassium fluoride gives the corresponding *N*-fluoroimidic fluoride of difluoroaminodifluoroacetic acetic acid **(60)** (Equation (21)) ⟨68IZV657⟩, while with tetrabromoethylene **(61)** a mixture of two compounds **(62)** and **(63)** in 67% and 14% yield are isolated, respectively (Equation (22)) ⟨68JOC2330⟩.

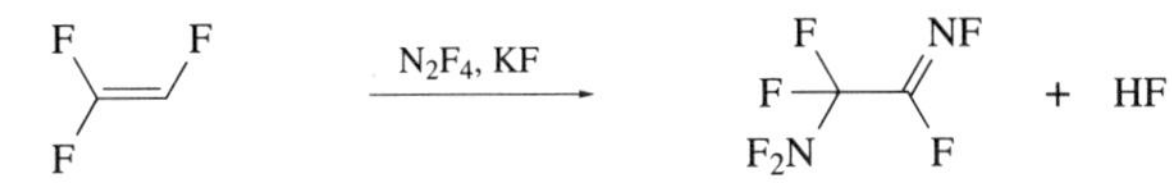

(59) **(60)** (21)

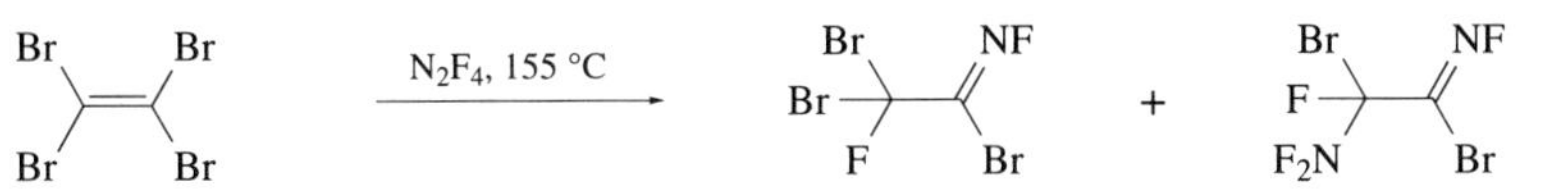

(61) **(62)** **(63)** (22)

5.21.1.4 Other Methods

Direct fluorination of guanylurea sulfate **(64)** with fluorine diluted with nitrogen produces a mixture of compounds **(65)**–**(70)**, from which the *N*-fluoroimidic fluoride derivative **(67)** can be isolated in low yield (Equation (23)) ⟨67JOC3859⟩.

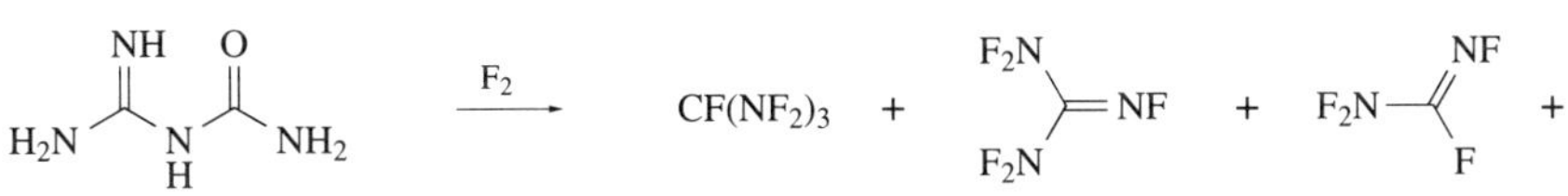

(64) **(65)** **(66)** **(67)** (23)

$(F_2N)_2CFNFCOF$ + $(F_2N)_2CFNCO$ + $CF_2(NF_2)_2$

(68) **(69)** **(70)**

In the reaction of semicarbazide hydrochloride with pentafluoroguanidine **(71)** four products **(72)**–**(75)** are formed, among them *N*-fluorofluoroformimidoyl chloride **(73)** and *N*-fluoro-difluoroaminoformimidoyl fluoride in low yield (Equation (24)) ⟨73JOC1075⟩.

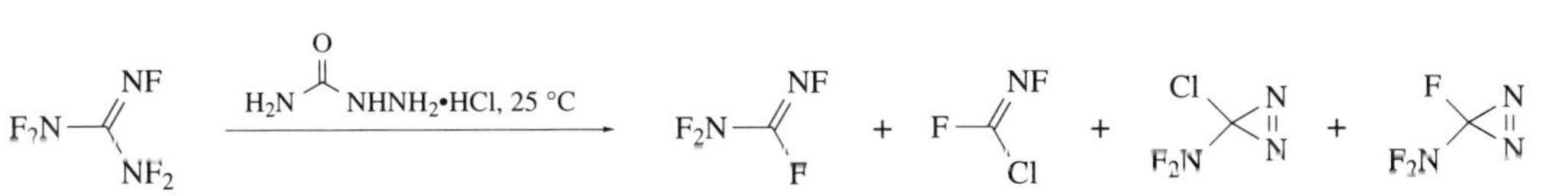

(71) **(72)** **(73)** **(74)** **(75)** (24)

Some difluoroaminoperfluorocarboxylic acids **(77)** decarboxylate spontaneously to give, after elimination of hydrogen fluoride, the corresponding *N*-fluoroimidoyl fluorides **(78)** (Scheme 3) ⟨67IZV1060⟩.

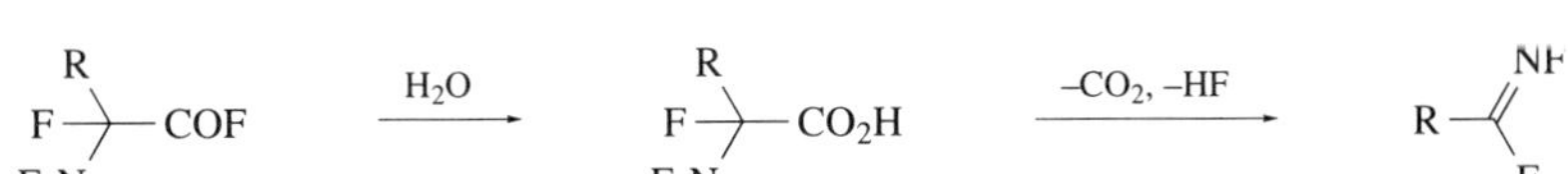

(76) **(77)** **(78)**

R = Cl, CF_3

Scheme 3

5.21.1.5 Pyrolysis of 2,2-bis(difluoroamino)propane and other compounds

The major product formed by the pyrolysis of 2,3-bis(difluoroamino)propane (**79**) at 450–550 °C is a 50:50 mixture of *syn*-(**80**) and *anti*-(**81**) of *N*-fluoriminoacetyl fluoride (62%) and 2-(*N*-fluorimino)propane (**82**) (21%) (Equation (25)) ⟨72JA8776⟩. Another method is pyrolysis of perfluoro-2-alkyl-1,2-oxazetines (**83**) or the corresponding copolymer (**85**) formed from R_fNO and $CF_2{=}CF_2$ (Scheme 4) ⟨B-81MI 521-01⟩.

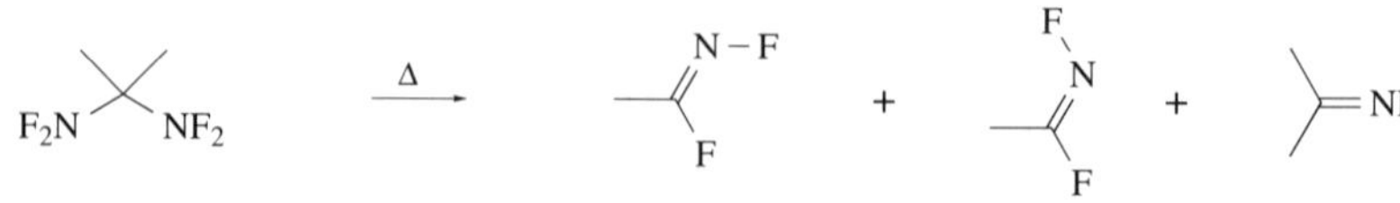

(25)

syn 1 : 1 *anti*

(**79**) (**80**) (**81**) (**82**)

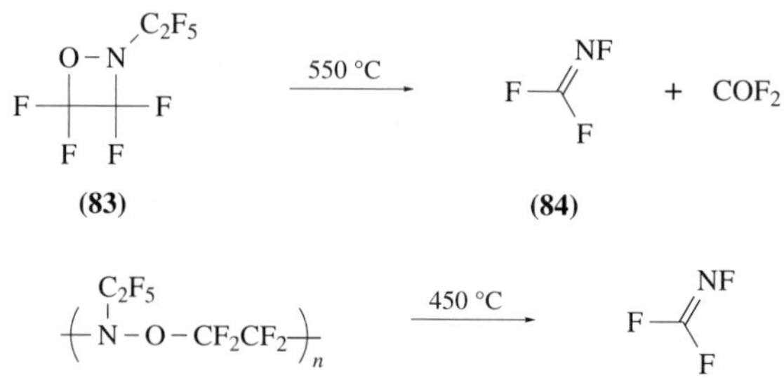

(**83**) (**84**) (**85**) (**86**)

Scheme 4

5.21.1.6 From Fluoroethenes

In the reaction of fluoroethene (**87**) with N_2F_4 in *o*-dichlorobenzene in the presence of powdered sodium fluoride, addition of N_2F_4 to the double bond occurs first, followed by elimination of hydrogen fluoride to give 1-cyano-*N*-fluoroformimidoyl fluoride in 35% yield (Scheme 5). The structures *syn*-(**89**) and *anti*-(**90**) are stable enough to be separated by gas chromatography ⟨66JOC3689⟩.

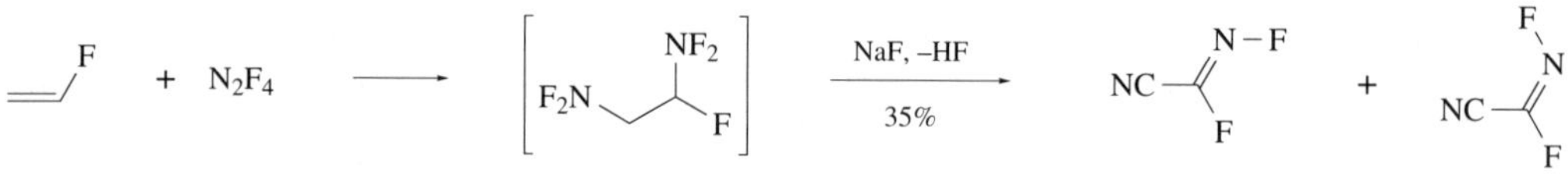

syn *anti*

(**87**) (**88**) (**89**) (**90**)

Scheme 5

5.21.1.7 Other *N*-Haloimidic Derivatives

5.21.1.7.1 N-*Haloimidic acids*

N-Haloimidic acids are prepared easily by hydrolysis of other *N*-haloimidic derivatives. For example, hydrolysis of N^1-fluoroimidoyl-N^2,N^2-difluoroaminodifluoroacetyl fluoride takes place in diethyl ether solution in the presence of water to give the corresponding N^1-fluoroimino-N^2,N^2-

difluoroaminodifluoroacetic acid ⟨69IZV958⟩. Perfluoro *N*-fluoroimidic fluorides (**91**) react similarly and have been shown to exist in *N*-fluoroimino form (**92**) (Equation (26)) ⟨71T51⟩.

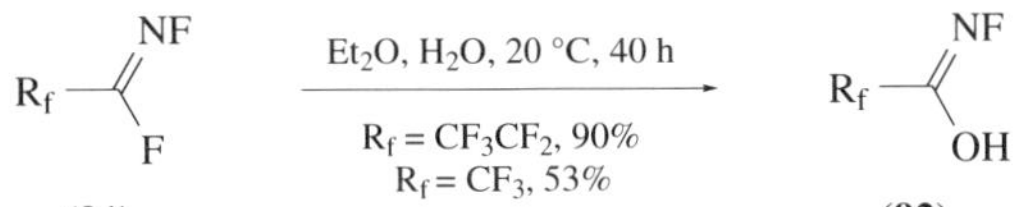

(**91**) (**92**) (26)

5.21.1.7.2 N-*Haloimidic esters (*N-*haloimidates)*

(i) From imidic esters

N-Haloimidic esters (**94**) are prepared from the corresponding imidates (**93**) with sodium hypochlorite ⟨43JCS101⟩, hypobromite ⟨13CB3616, 43JCS101⟩, *t*-butyl hypochlorite ⟨66JOC3708⟩, bromine, and iodine ⟨1897JA129⟩ (Equation (27)). Accordingly, *N*-fluoroimidic esters (**96**) can also be prepared from (**95**) (Equation (28)) ⟨70JOC3093, 71T51⟩. *N*-Bromoimidic esters (**98**) are formed as intermediates by the bromination of alkenes (**97**) with NBS (Equation (29)) ⟨71CJC1384⟩. However, all of these methods are inferior to substitution reactions as far as the yields are concerned ⟨71T51⟩.

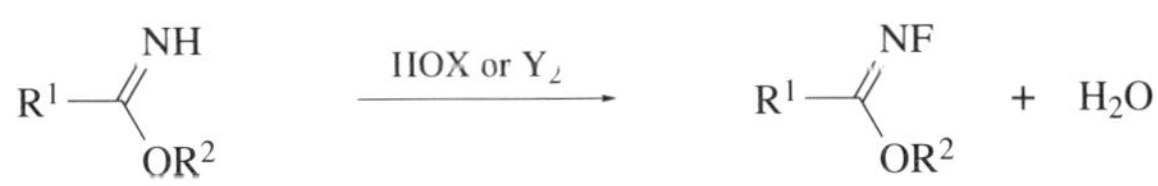

(**93**) (**94**) (27)

$X = Cl, Br; Y_2 = Br_2, I_2$

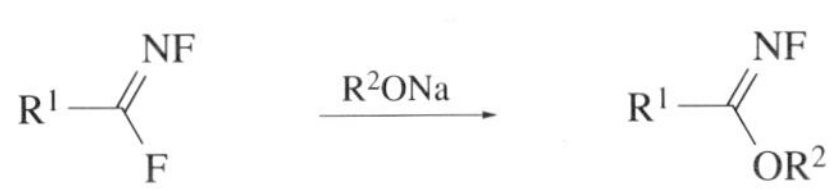

(**95**) $R^1 = CN, R_f$ (**96**) (28)

(**97**) (**98**) (29)

(ii) From nitriles

In the reaction of *n*-perfluoropropionitrile (**99**) with 1,1,5 trihydroperfluoroamyl hypochlorite at elevated temperatures the corresponding 1,1,5-trihydroperfluoroamyl-*N*-chloroperfluorobutanimidate (**100**) is formed (Equation (30)) ⟨80IZV2623⟩.

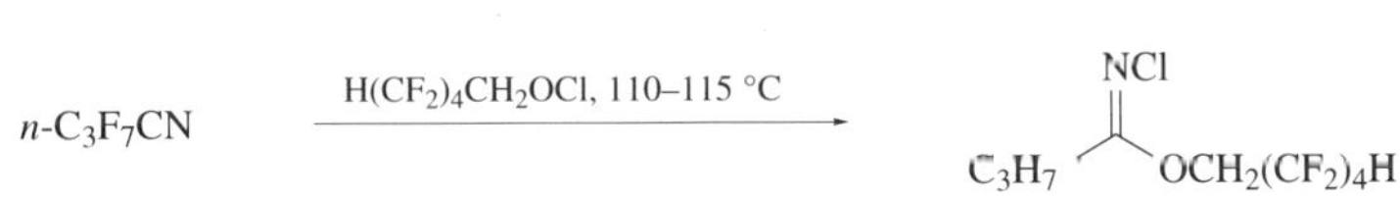

(**99**) (**100**) (30)

(iii) From N,N-*perfluoroamidines*

N,*N*-Perfluoroformamidines (**101**) and other amidines form *N*-fluorimidates by treatment with alkoxides in an alcohol solution at $-90\,^{\circ}C$ to $-80\,^{\circ}C$ to give the corresponding *N*-fluoroimidates (**102**) (Equation (31)) ⟨78IZV2802⟩.

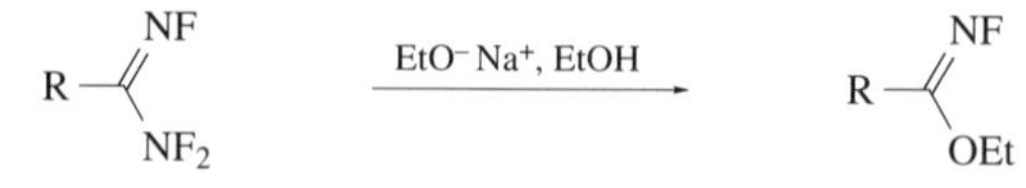

(101) (102) (31)

Perfluoroamidines (**103**) add methanol without catalyst. The intermediate difluoroaminoalkyl methyl esters (**104**) could not be isolated from the reaction mixture without decomposition. However, further fluorination of the methanol adducts gives three products (**105**)–(**107**) in equal amounts (Scheme 6) ⟨70JOC3093⟩.

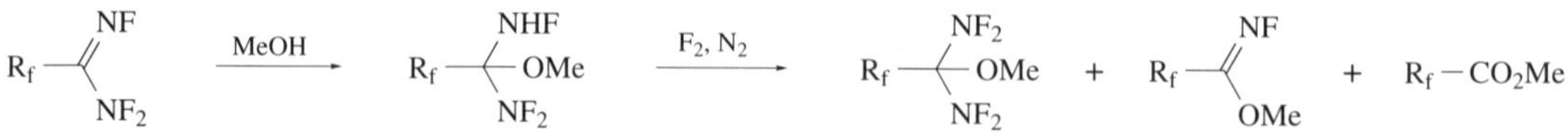

(103) (104) (105) (106) (107)

Scheme 6

Other *N*-fluoroimines, such as perfluoroguanidine (**108**), react similarly with hydroxy group containing compounds, such as alcohols and oximes to give the corresponding *N*-fluoroimidic esters (**110**) and derivatives. The same compounds can be obtained by treatment of tetrafluoroformamidine (**111**) with alcohols, followed by elimination of hydrogen fluoride from the unstable intermediate adduct (Scheme 7) ⟨73JOC1065⟩.

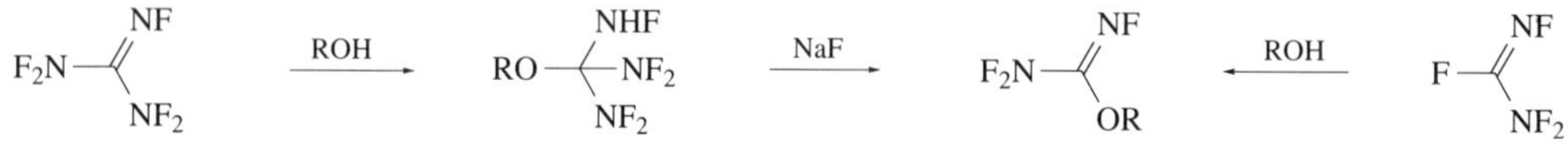

(108) (109) (110) (111)

R = Me, Bu^n, Pr^i, $PhCH_2$, $HOCH_2CH_2$, CF_3CH_2, 4-NO_2-C_6H_4, HC≡C–CH_2, H_2C=CH–CH_2, $H_2NCH_2CH_2$, Me_2C=N, Cl_2C=N; yields 70–90%

Scheme 7

N-Halo-*O*-silyl imidates (**113**) are prepared from *N*-trimethylsilylacetimidic silyl esters (**112**) and NCS or NBS, or by reaction of *N*-halocarboxamides (**114**) with *N*-trimethylsilylsuccinimide by gentle heating for a short time (Scheme 8) ⟨68CB2585⟩.

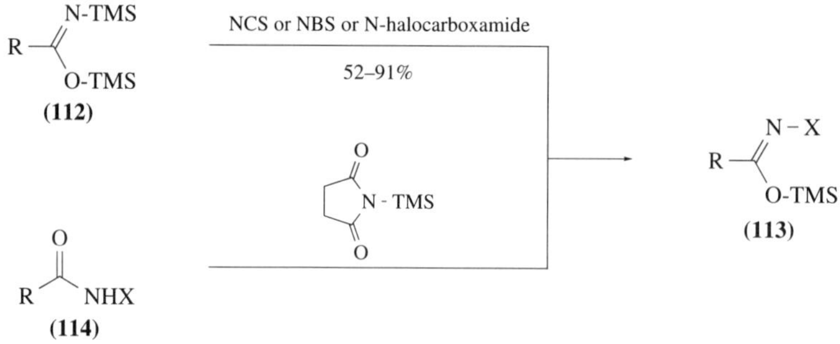

Scheme 8

(iv) From vinylsulfur pentafluoride

Vinylsulfur pentafluoride (**115**) reacts with hydrazine tetrafluoride in the presence of sodium fluoride in *o*-dichlorobenzene at 130 °C to give cyano(fluoroiminosulfur pentafluoride) (**116**) (Equation (32)) ⟨66JOC3689⟩.

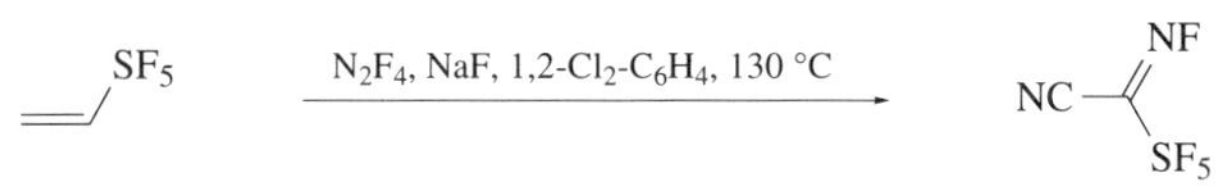

(32)

(115) **(116)**

5.21.1.7.3 N-*Haloamidines*

The synthesis of *N*-haloamidines has been studied extensively ⟨52JA4876, 53CB400, 66TL3537⟩.

(*i*) *From amidines*

By direct fluorination of tetrafluoroformamidine, pentafluoroguanidine ⟨67JOC3859⟩, 1-[bis(difluoramino)fluoromethyl]-1,2,3,3-tetrafluoroguanidine and some other compounds containing the trifluoroguanyl group ⟨68USP3361815⟩ are obtained from the corresponding amidines and guanidines. Heptafluorobutyramidine hydrochloride (**117**) and some other perfluoro amidines treated in the solid phase with fluorine diluted with nitrogen gives perfluoroamidine derivatives (**118**) (Equation (33)) ⟨56JA6032⟩. The use of alkali metal fluorides as a diluent has been described ⟨66JOC4232⟩.

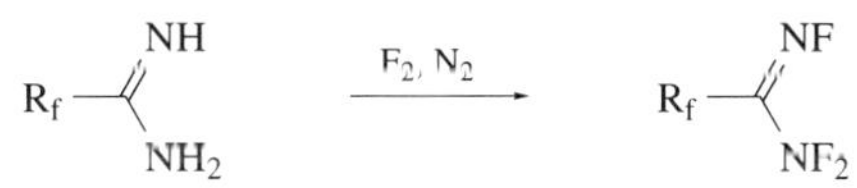

(33)

(117) $R_f = C_3F_7, O_2NCF_2$ **(118)**

By dehydrofluorination of 1,1-bis(difluoroamino)alkanes (**119**), a series of *N,N,N'*-trifluoroamidines (**120**) are prepared (Equation (34)). The reaction is carried out in dichloromethane in the presence of pyridine and the products obtained are 1 : 1 mixtures of the *syn*- and *anti*-isomers. *N,N,N'*-Trifluoropropionamidine (**120a**), *N,N,N'*-trifluorohexanamidine (**120b**) and 2-chloro-*N,N,N'*-trifluoropropionamidine (**120c**) can be prepared in this manner. Other bases, such as piperidine, quinoline, or aqueous sodium hydroxide can be used ⟨70JOC3093⟩.

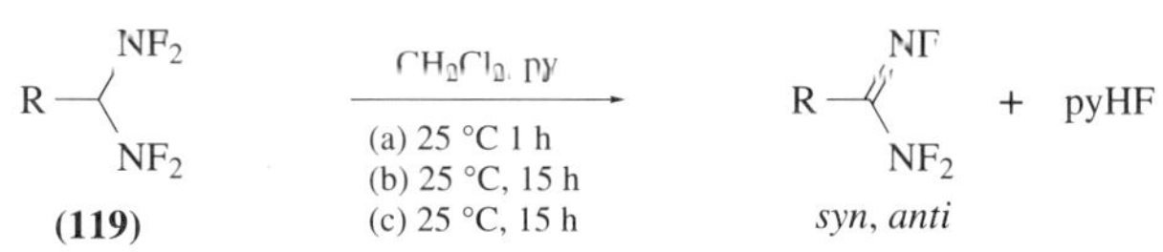

(34)

(119)

(120) **a**; R = Et, 90%
b; $R = Me(CH_2)_4$, 62%
c; $R = ClCH_2CH_2$, 64%

The synthesis of *N,N,N'*-trifluoroamidines (**121**), (**122**) and (**124**) in which other structural elements are attached to the amidine carbon atom, such as hydrogen, nitro, acyl, and the like, which can be cleaved and replaced by fluorine, is achieved by direct fluorination (Equation (35)). The method is also suitable for the preparation of the compounds in which Y is a group with nitrogen as a central atom or halogen. The fluorination with elemental fluorine in pure or diluted form is conducted between −100 °C and 40 °C. Guanylureas, biguanide, cyanoguanidine, and their derivatives can also be fluorinated according to this procedure. An interesting example is ammeline (**125**), which produces by fluorination a mixture of products (**126**)–(**129**) (Equation (36)) ⟨73USP3726903⟩.

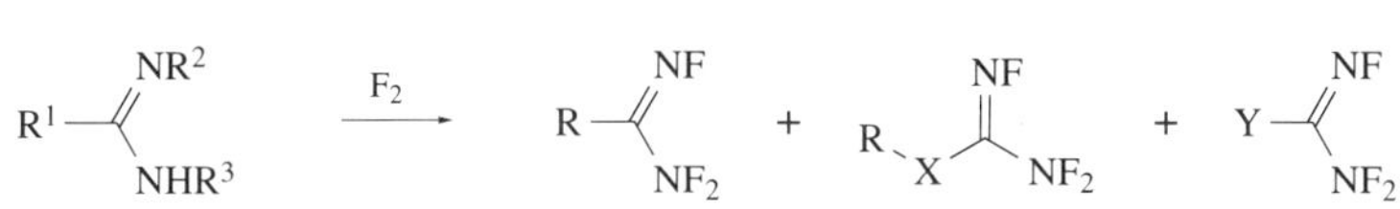

(35)

(121) **(122)** **(123)** **(124)**

X = O–O, S, O, NH, NR, CH=N, CH=N-O
$Y = NH_2, N_3$, NCO, F, Cl, Br, CN, NCS

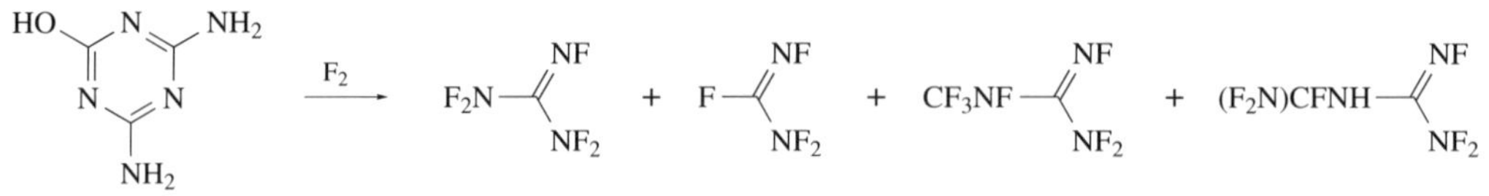 (36)

(125) **(126)** **(127)** **(128)** **(129)**

The mono *N*-halogenation of amidines (**130**) takes place with alkoxy hypohalides to give *N*-halo compounds (**131**) in good yields (Equation (37)) ⟨84USP4446068⟩. *N*-Chloro-*N′*-arylamidines (**133**) are obtained from *N*-arylamidines (**132**) with *N*-chlorosuccinimide (5% excess) in dry dichloromethane at room temperature (Equation (38)) ⟨78JCS(P1)309⟩, and some heteroaryl-substituted *N*-chloroamidines, such as *N*-chloro-*N′*-phenyl-4-thiazolecarboxamidine (**134a**) and *N*-chloro-*N′*-(3-pyridyl)4-thiazolecarboxamidine (**134b**) are obtained from the corresponding amidines by the action of sodium hypochlorite in 98% and 89% yield, respectively ⟨65JOC259⟩.

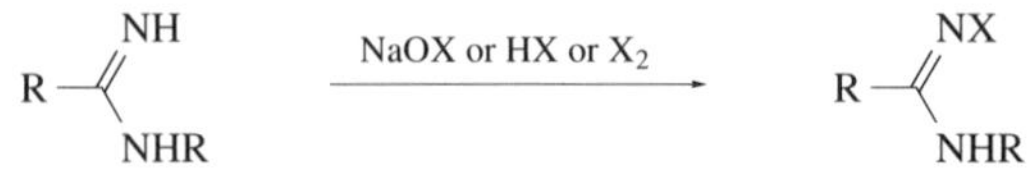 (37)

(130) **(131)**

R = alkyl, perfluoroalkyl, OAr, NO_2, halogen; X = F, Cl, Br

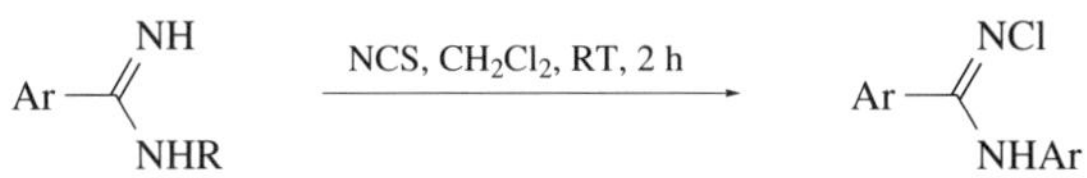 (38)

(132) **(133)**

Ar = Ph, 4-Br-C_6H_4; R = $PhCH_2$, 4-NO_2-C_6H_4, 4-Br-C_6H_4, 4-MeO-C_6H_4, 4-Me-C_6H_4

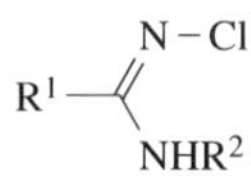

(134)

(134a) R^1 = 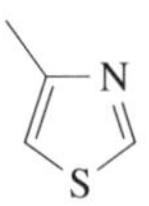 , R^2 = Ph

(134b) R^1 = 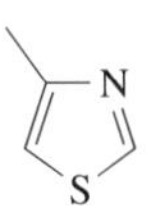 , R^2 = 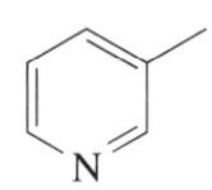

(ii) From N-*haloimidic halides*

In the reaction of *N*-haloimidic halides with nucleophiles the substitution of carbon-bonded halogen is typical.

Various *N*-chloroamidines (**137**) and *N*,*N′*-dichloroamidines (**138**) are prepared from *N*-chlorocyanoformimide chloride (**135**) or *N*,*N′*-dichlorooxalimide chloride (**136**) by nucleophilic substitution of chlorine with amino compounds in the presence of an organic base, such as trialkylamine, dialkylaniline, paridine, or inorganic base, such as alkali hydroxide, carbonate or hydrogen carbonate, in diethyl ether, dioxane, dimethoxyethane, tetrahydrofuran, benzene, toluene, xylene, chloroform, acetonitrile, dimethyl sulfoxide, acetone, or methyl ethyl ketone (Equation (39)) ⟨73GEP2242772⟩.

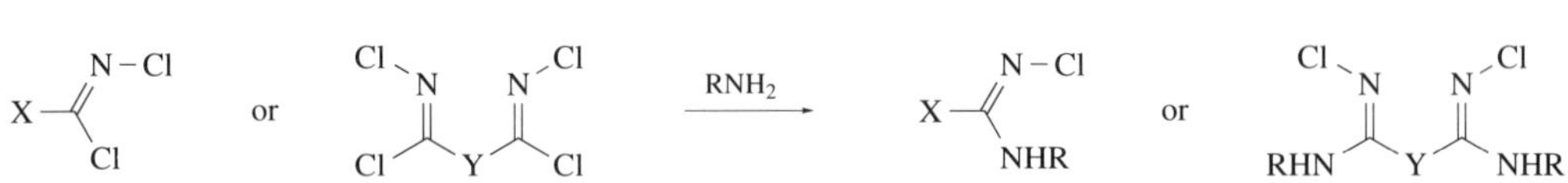 (39)

(135) **(136)** **(137)** **(138)**

R = H, alkyl, cycloalkyl; X = CN, CCl_3, CCl_2CN; Y = CCl_2

In the reaction of either *syn*-(**139**) or *anti*-(**141**) of 1,1-cyano-*N*-fluoroformimidoyl fluoride with diethylamine both *syn*-(**140**) and *anti*-1-cyano-*N*,*N*-diethyl-*N′*-fluoroformamidine (**142**) are formed. However, when ammonia or *p*-chloroaniline are used only the *anti*-isomer of the product (**142**) is formed from both isomers, due to the equilibrium with a dimeric structure (**143**) (Scheme 9) ⟨66JOC3689⟩.

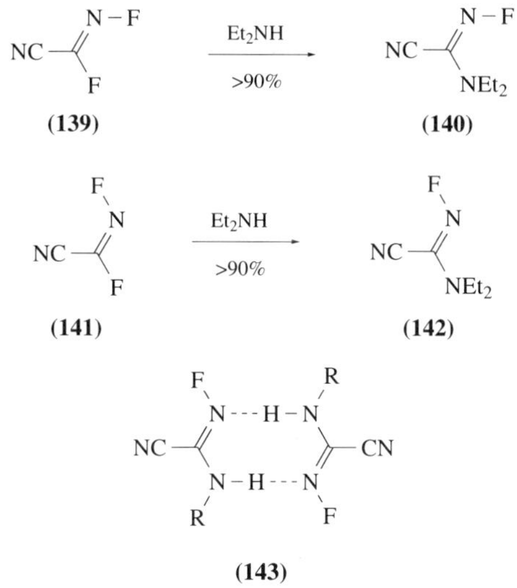

Scheme 9

(iii) From N*-fluoroimines*

Stepwise introduction of —NF_2 group into fluoroimino compounds by reaction with ammonia and amines followed by elimination of a suitable group is a convenient method for preparation of *N*-fluoroamidines. Further fluorination of these compounds, usually with elemental fluorine gives perfluoroamidines (Scheme 10) ⟨73USP3726903⟩.

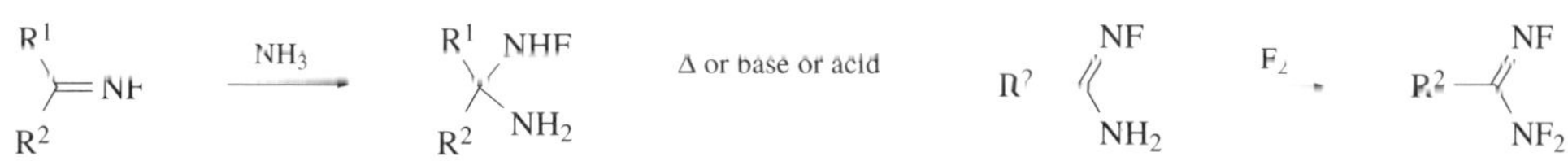

Scheme 10

Derivatives of pentafluoroguanidine (**144**) react with ammonia and amines to give bis(difluoro-amino)fluoroaminomethylamines (**145**) as intermediates, which are transformed by elimination of HNF_2 into the corresponding substituted 1,1,2-trifluoroguanidines (**146**) (Scheme 11). The addition of ammonia ($R^1 = R^2 = H$) takes place rapidly in dimethyl ether at $-110\,^{\circ}C$ to afford a high yield of the saturated compound, which, on warming to room temperature, eliminates HNF_2. More basic amines, such as *n*-butylamine ($R^1 = R^2 = Bu^n$), dimethylamine ($R^1 = R^2 = Me$) react rapidly at $-78\,^{\circ}C$ to yield exclusively the unsaturated compounds. Amino compounds of very low nucleo-philicity, such as succinimide, require a basic catalyst for the addition reaction to take place. Polar solvents, such as acetonitrile, dimethyl ether, diethyl ether, and tetrahydrofuran facilitate the reaction and prevent explosions ⟨73JOC1075⟩.

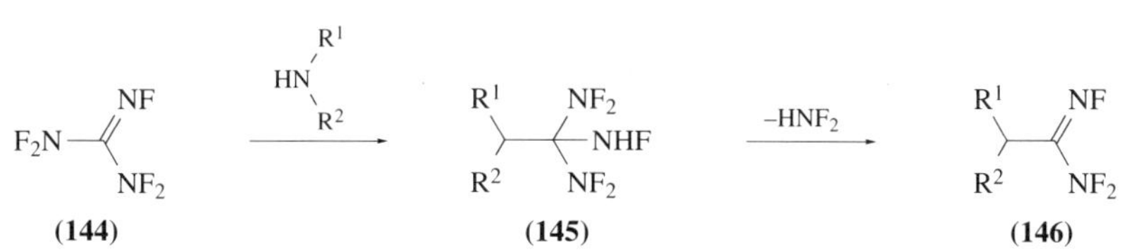

Scheme 11

Perfluoroamidines (**147**) with an electron-withdrawing group attached to amidine carbon atom are versatile starting compounds for preparation of various perfluoroamidines (**148**), since the electron-withdrawing group can be easily substituted with active hydrogen-containing compounds. The reaction is general and involves two steps: nucleophilic addition of R^2XH compound to an electron-deficient double bond, sometimes in the presence of a basic catalyst, such as urea, *N,N*-dimethylurea, triethylamine, and sodium cyanide, which promote the addition reaction. The reaction takes place in polar solvents, such as acetonitrile, trifluoroethanol, nitromethane, ethyl acetate, and sulfolane. In the second step, the adducts (**148**) are heated, treated with sodium or potassium fluoride, or triethylamine, in order to eliminate R^1H (Scheme 12) ⟨73USP3726903⟩.

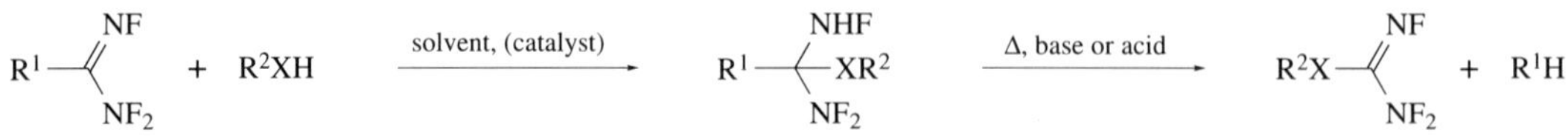

(147) **(148)**

X = O, O–O, CH=N, CH=N–O, S, N, etc
$R^1 = NF_2$, F, Cl
R^2 = alkyl, aryl, heteroaryl, perfluoroalkyl
$R^2XH = NH_2$, NHCO, HN_3, HF, HCl, HBr, HCN

Scheme 12

Perfluoroamidines (**150**) are obtained by the pyrolysis of saturated NF containing compounds, such as tris(difluoroamino)methyl derivatives (**149**) (Equation (40)) ⟨73USP3726903⟩.

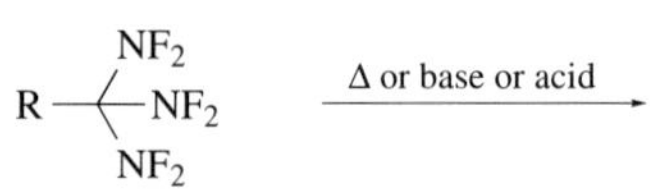

(149) **(150)** (40)

Trifluoroamidines (**151**) react with free radicals to form new trifluoroamidines (**152**) (Equation (41)). The free radicals may be generated by heating peroxides or azo compounds or by photolysis. The method is especially useful when photolysis at low temperatures can be employed. Higher yields are usually obtained when $R^1\cdot$ is more stable than $R^2\cdot$. Therefore, perfluoroamidine is a useful starting material, since NF· is a stable radical ⟨73USP3726903⟩.

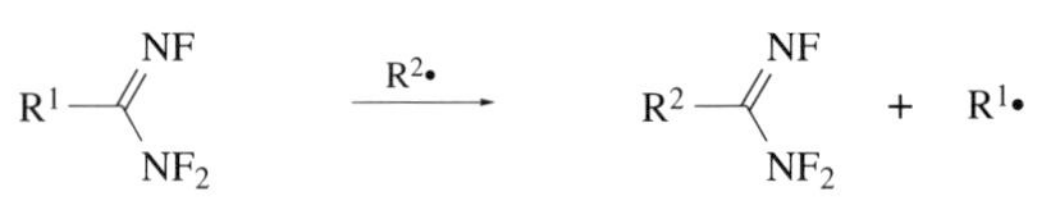

(151) **(152)** (41)

(iv) From N-*chloroimines*

An interesting example is transformation of diiminosuccinonitrile (**153**) by chlorine into *N,N'*-dichlorodiiminosuccinonitrile (**154**), which gives by treatment with *n*-propylamine *N-n*-propyl-*N'*-chloro-1-cyanoformamidine (**155**) in 45% yield (Scheme 13) ⟨72JOC4136⟩.

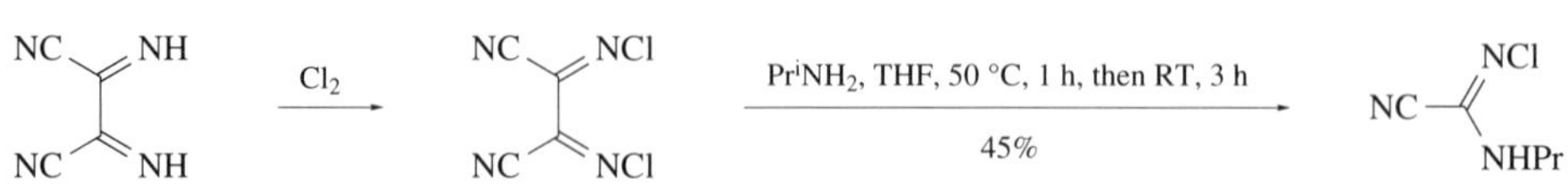

(153) **(154)** **(155)**
R = Me, Pr^i

Scheme 13

5.21.1.7.4 N-*Haloimidoyl azides*

N-Fluoroimidoyl fluorides react with sodium azide to give the corresponding azides. For example, 1-cyano-*N*-fluoroformimidoyl fluoride, *syn*- and *anti*-form gives the *syn*- and *anti*-form of 1-cyano-*n*-fluoroformimidoyl azide. However, attempts to isolate the compound in the pure form led to explosion (Equation (42)) ⟨66JOC3689⟩.

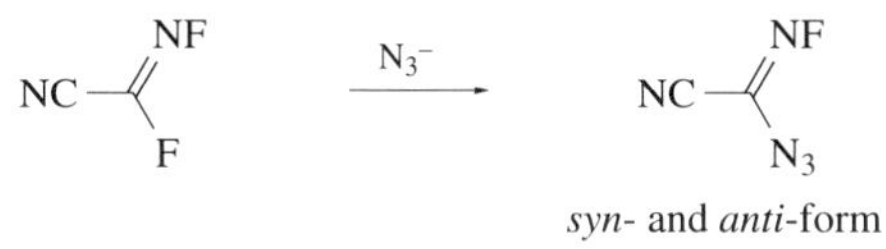

(42)

5.21.2 OXYIMIDIC DERIVATIVES

5.21.2.1 Oxyimidic Halides

5.21.2.1.1 *From nitrile oxides*

Fulminic acid ⟨1894CB2816, 1894LA(280)305⟩, silver fulminate ⟨30LA(484)236⟩ and nitrile oxides ⟨31G561, 36G116⟩ are converted with hydrochloric acid into the corresponding oximidic chlorides (Scheme 14) ⟨68HOU(10/4)1⟩.

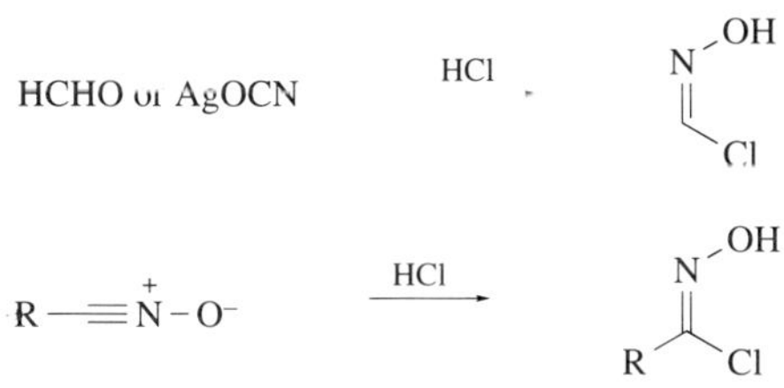

Scheme 14

The addition to nitrile oxides is especially suitable for the preparation of the corresponding bromides and iodides ⟨65JOC2809, 66JOC157, B-71MI 521-01⟩. With sulfonyl chlorides the corresponding *O*-sulfonates are obtained (Scheme 15) ⟨66CJC297⟩.

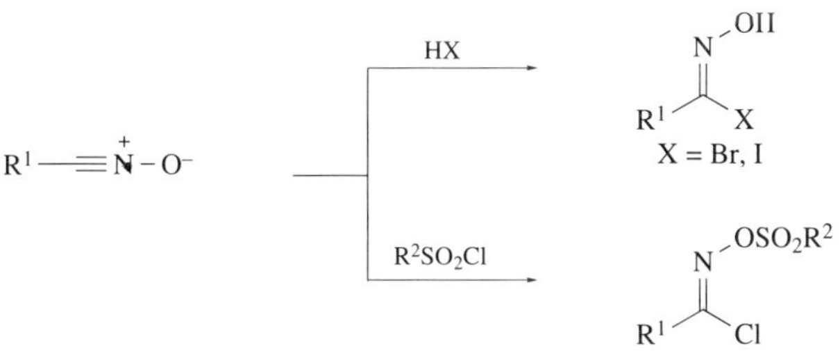

Scheme 15

5.21.2.1.2 *From* aci-*nitro compounds and nitro alkenes*

Treatment of primary *aci*-nitro compounds with hydrogen chloride gives *gem* chloronitroso derivatives, which tautomerize in solution on standing, or by heating into the corresponding oxyimidic chlorides. For example, *aci*-phenylnitromethane (**156**) gives in ether saturated with hydrogen chloride oxyimidic acid (**157**) (Equation (43)) ⟨11JPR687⟩. Ethyl *aci*-nitroacetate (**158**) is transformed in dry ether with hydrogen chloride into ethoxycarbonyl formoxyimidic chloride (**159**) in 60% yield (Equation (44)) ⟨11JPR453⟩. The *aci*-nitro group of ammonium salt of *aci*-nitroacetamide (**160**) can be transformed into the corresponding oxyimidic chloride (**161**) with thionyl chloride (Equation (45)) ⟨11JPR453⟩.

(156) **(157)** (43)

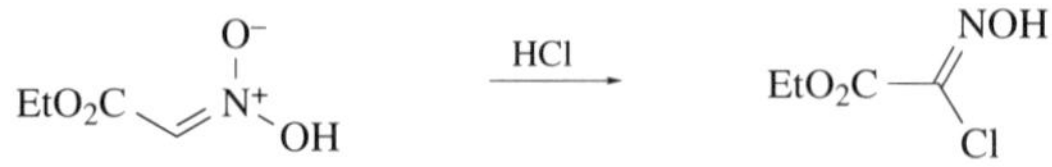

(158) **(159)** (44)

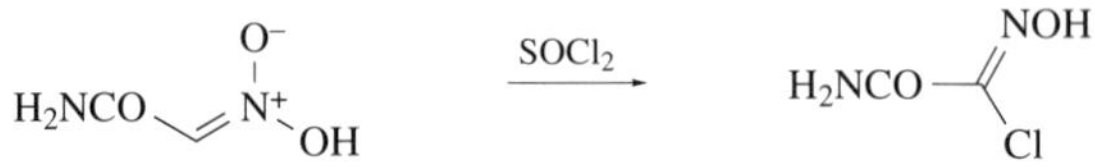

(160) **(161)** (45)

Oxyimidic chlorides with another chlorine at α-position can be obtained from nitroalkenes with hydrogen chloride by 1,4-addition followed by tautomerization into the final products. For example, 1-nitro-2-nitropropene (**162**) gives, at 0 °C in ether saturated with hydrogen chloride, 2-chloro-2-methylpropaneoxyimidic chloride (**163**) in 80% yield (Equation (46)) ⟨10JPR216⟩.

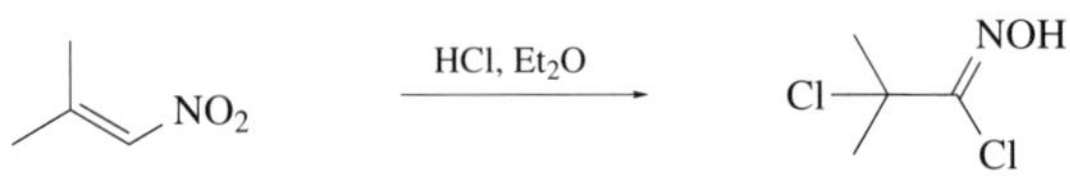

(162) **(163)** (46)

Nitroalkenes (**162**) and (**165**) react with benzene and other aromatic compounds in the presence of aluminum chloride to give the corresponding oxyimidic chlorides (**164**) and (**166**) (Scheme 16) ⟨49JCS42, 50JA4697⟩.

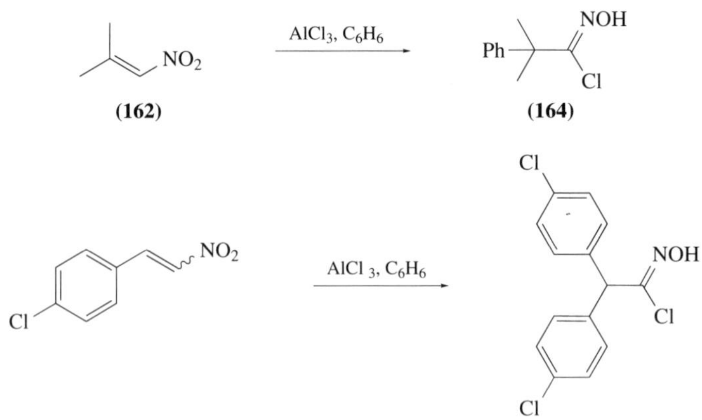

(165) **(166)**

Scheme 16

When nitroalkenes (**167**) are treated with acetyl, propanoyl, or butanoyl bromide, with or without a catalyst, such as zinc sulfate or sulfuric acid, the corresponding *O*-acyl oxyimidic bromides (**168**) are obtained (Equation (47)) ⟨60ZOB1937⟩.

(167) **(168)** (47)

5.21.2.1.3 From aldoximes

Addition of nitrosyl chloride (X = NO), chlorine (X = Cl), hypochloric acid (X = OH), or *t*-butyl hypochlorite (X = Bu^{t}O) to C=N bond in aldoximes, followed by elimination, is an excellent reaction and the most frequently employed method for preparation of oxyimidic chlorides. The reaction is carried out in ether, tetrachloromethane, or in chloroform at around 0 °C. The reaction can be applied to a variety of compounds with the R group being alkyl ⟨25LA(444)133⟩, phenyl ⟨1894CB2193⟩, substituted phenyl ⟨1893LA(274)95, 1894CB2193, 1894CB2846, 02CB3101⟩, haloalkyl ⟨11JPR453, 52CB344, 59T24⟩, 2-furyl ⟨27LA(451)161⟩, 2-pyridyl ⟨63JOC2260⟩, and others ⟨67TL327, B-68MI 521-01, 86JCS(P2)995⟩. Aldoximes (**169**) can be chlorinated in aqueous hydrochloric acid to give 1-chloro-1-nitroso derivatives (**170**), which rearrange into the corresponding oxyimidic chlorides (**171**) by standing at room temperature in ether solutions, while ketoximes produce chloronitro compounds (Scheme 17) ⟨02CB3101, 62HOU(5/3)1⟩. Aldoximes react with bromine to give the corresponding oximidic bromides. In this manner, oxyiminodibromomethane ⟨30G700⟩, and other derivatives have been prepared ⟨30G700, 32CB65, 58JIC893⟩.

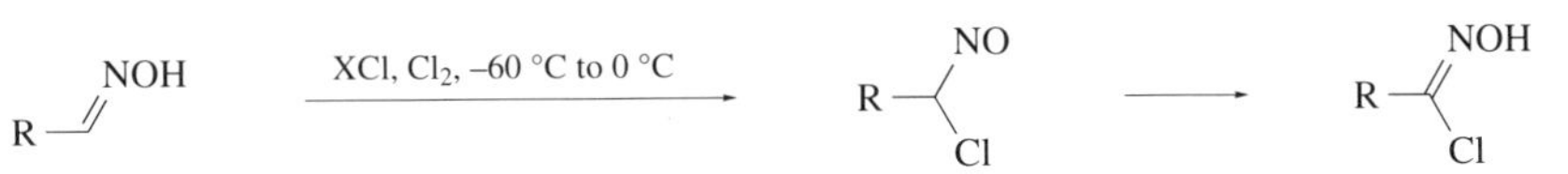

(169) **(170)** **(171)**

R = H, Cl, NO_2, Me, CH_2Cl, $CHCl_2$, CF_3, Ph

Scheme 17

Terephthalohydroxymoyl chloride (**173**) has been prepared from terephthalaldoxime (**172**) by chlorination with chlorine under UV irradiation (λ = 366 nm) (Equation (48)) ⟨89OPP517⟩.

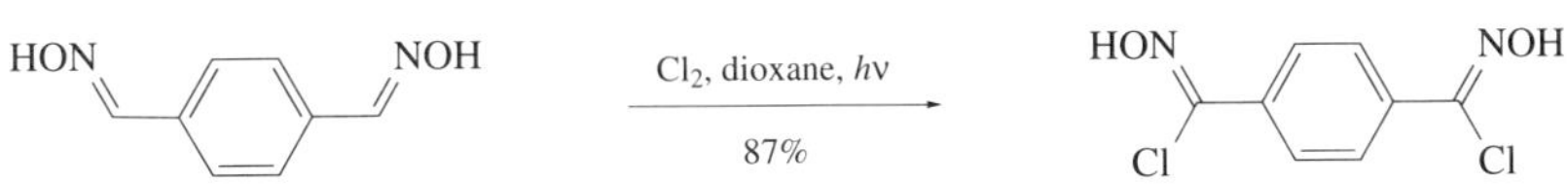

(172) **(173)** (48)

Aliphatic and aromatic hydroxymoyl chlorides (**175**; X = Cl) and bromides (**175**; X = Br) are prepared from the corresponding aldoximes (**174**) with *N*-chloro- ⟨80JOC3916⟩ or *N*-bromo-succinimide ⟨76T1599⟩ in DMF in high yields (Equation (49)). The method is especially suitable for those compounds that chlorinate in the aromatic ring when chlorine gas is employed.

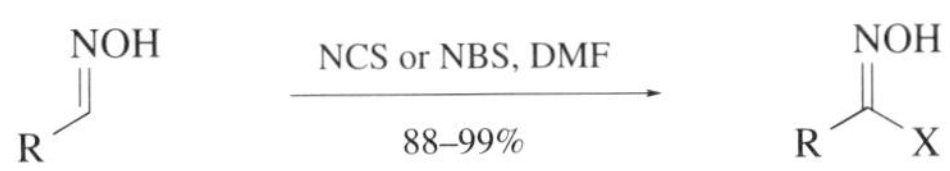

(174) **(175)** (49)

R = alkyl, aryl; X = Cl, Br

Chlorination of *N*-[2-(diethoxyphosphoryl)ethylidene]hydroxylamine (**176**), easily available from bromoacetaldehyde diethyl acetal and triethyl phosphite, with NCS give *N*-[1-chloro-2-(diethoxy-phosphoryl)ethylidene]hydroxylamine (**177**) (Equation (50)) ⟨87BCJ2463⟩.

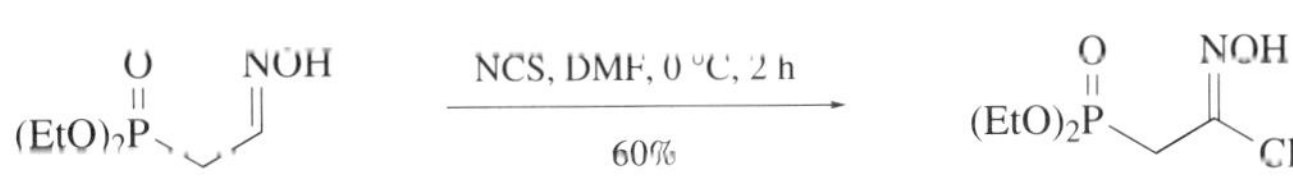

(176) **(177)** (50)

5.21.2.1.4 From activated alkenes and other activated compounds

The addition of hydroxylamine to a C=C double bond, followed by elimination of hydrogen fluoride requires activation. Fluoroalkenes are suitable substrates for this type of transformation. For example, hexafluoropropene (**178**) reacts in ether at 0 °C to give 2,3,3,3-tetrafluorooxyimidic fluoride (**180**) in 67% yield (Scheme 18) ⟨59DOK(127)337⟩.

(178) **(179)** **(180)**

Scheme 18

In chloral hydrate (**181**) two chloro atoms are substituted with oxyimidic group, when the compound is treated with hydroxylamine in 1 : 3 molar ratio in the presence of sodium hydrogen carbonate, to give the corresponding oxyimidic chloride (**182**) (Equation (51)) ⟨1892CB705, 13G58⟩. Trichloroacetonitrile (**183**) gives, with hydroxylamine in the presence of sodium carbonate in 1 : 5 : 3 molar ratio in water at 65 °C, 2-chloro-1-aminodioximino ethane (**184**) in moderate yield (Equation (52)) ⟨08JPR198⟩.

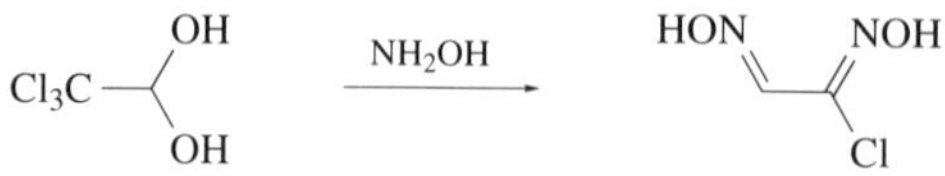

(181) **(182)** (51)

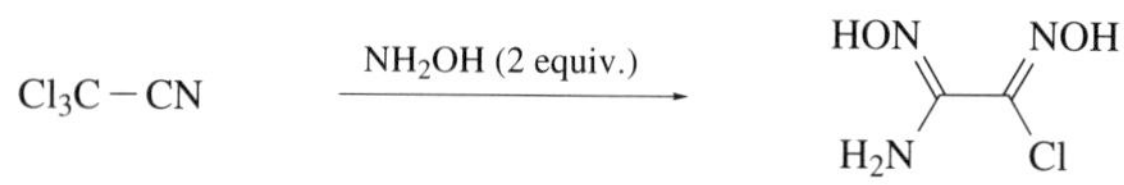

(183) **(184)** (52)

5.21.2.1.5 *From nitroso and nitro compounds*

Reduction of perfluorochloro- or bromonitro alkanes with hydrogen iodide gives the corresponding oxyimidic fluorides. For example, 1,2,2-trifluoro-1,2-dichloro-1-nitroethane (**185**) is transformed into difluorochloroacetoximidic fluoride (**186**) (Equation (53)) ⟨60ZOB2409⟩. Nitrotrichloromethane (**187**) is reduced with tin in concentrated hydrochloric acid in ether at 0–5 °C to give phosgene oxime (**188**) in 75% yield (Equation (54)) ⟨48BSF597⟩. The same compound can be obtained also by hydrolysis of chloronitroacetyl chloride (**189**) followed by elimination of carbon dioxide ⟨65ZOB248⟩ or by the reaction of mercury fulminate with chlorine ⟨32CB65⟩, while the corresponding dibromo analogue is obtained in the reaction of the mercury fulminate with bromine ⟨30G703⟩.

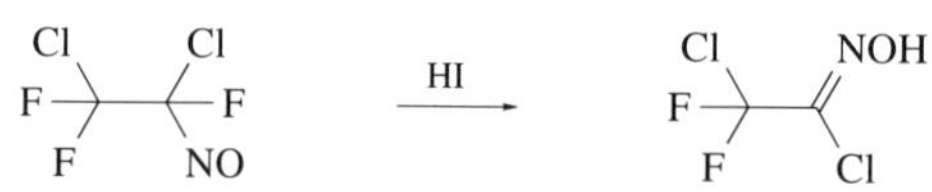

(185) **(186)** (53)

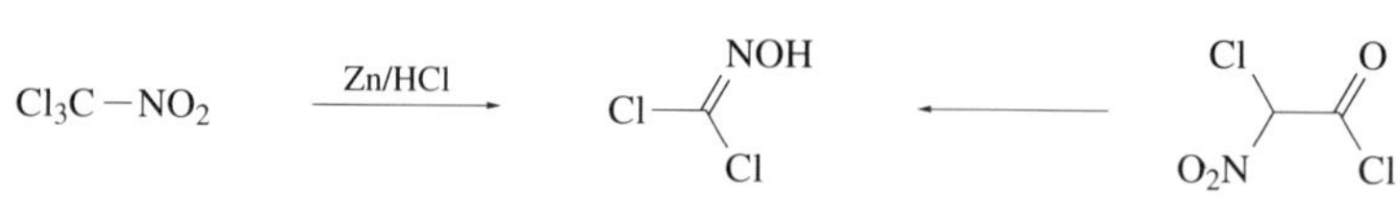

(187) **(189)** **(188)** (54)

2-Nitroacetaldehyde oxime (**190**) gives with hydrochloric acid the corresponding glyoxylic amidoxime oxyimidic chloride (**191**) (Equation (55)) ⟨10JPR224, 11JPR453⟩. ω-Nitroacetophenone (**192**) reacts with acetyl chloride or bromide in the presence of aluminum chloride to produce the corresponding oxyimidic halides (**193**) ⟨55LA(594)185⟩. The oxyimidic iodides are prepared by exchange of chlorine with iodine by treatment with potassium iodide in acetone ⟨11JPR453⟩.

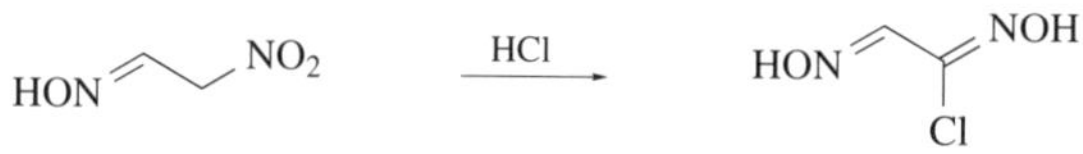

(190) **(191)** (55)

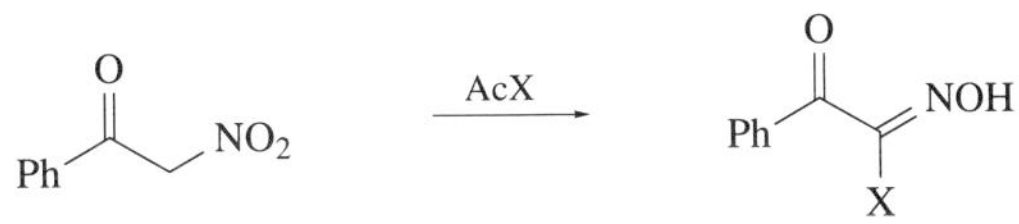

(192) X = Cl, Br (193)

(56)

The reaction has been extended also to heteroaryl-substituted nitroalkanes. For example, nitromethylquinoline (**194**) and other derivatives react with benzoyl chloride to form *O*-benzoyl oxyimidic chloride (**195**) derivatives (Scheme 19). The same products can be obtained also from 2-(nitrobromo)ethyl quinoline (**196**) ⟨60ZOB506⟩.

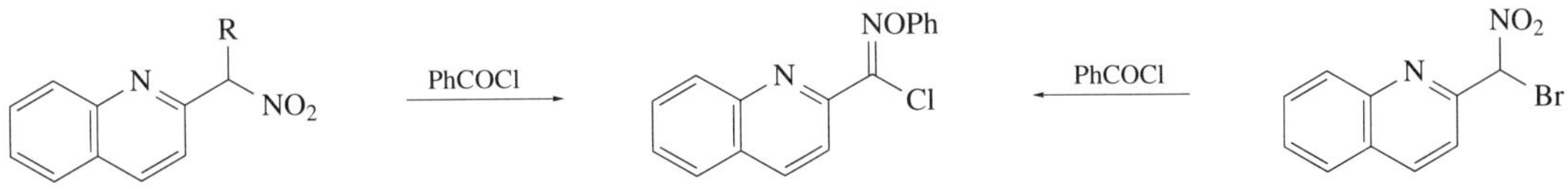

(194) (195) (196)

Scheme 19

5.21.2.1.6 From hydroxy and carbonyl compounds

Some secondary alcohols, such as 2-propanol (**197**), react with nitrosyl chloride to give the corresponding oxyimidic chloride (**198**) (Equation (57)) ⟨63JOC1673⟩.

(197) (198)

(57)

Methyl ketones (**199**) react with nitrosyl chloride in excess in tetrachloromethane or ether at low temperatures (around 0°C) to give the corresponding oxyimidic chlorides (**200**) (Equation (58)) ⟨25LA(444)113, 28CB32⟩. Acetophenone reacts with amyl nitrite in the presence of hydrogen chloride ⟨1887CB252⟩.

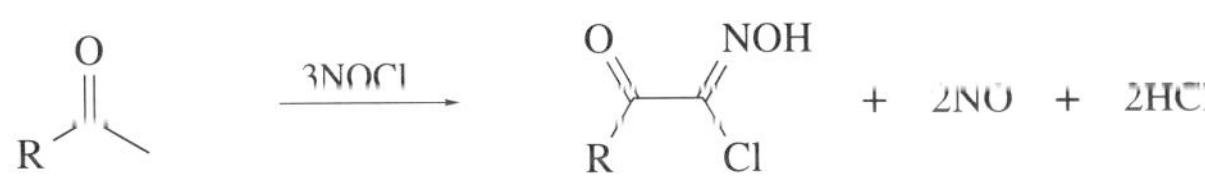

(199) (200)

R = alkyl, aryl

(58)

α-Chloro ketones, such as chloroacetone (**201**), are converted using nitric acid, while chloromethyl aryl ketones (**203**) are transformed by nitrosation into oxyimidic chlorides (**202**) and (**204**) with alkyl nitrites, such as butyl or amyl nitrite in chloroform or ether in the presence of anhydrous hydrogen chloride or concentrated hydrochloric acid at low temperature (Scheme 20) ⟨29JA2262, 61MAC62⟩. The yields of the corresponding oxyimidic chlorides are above 80% with few exceptions ⟨42JOC408, 55CB130⟩.

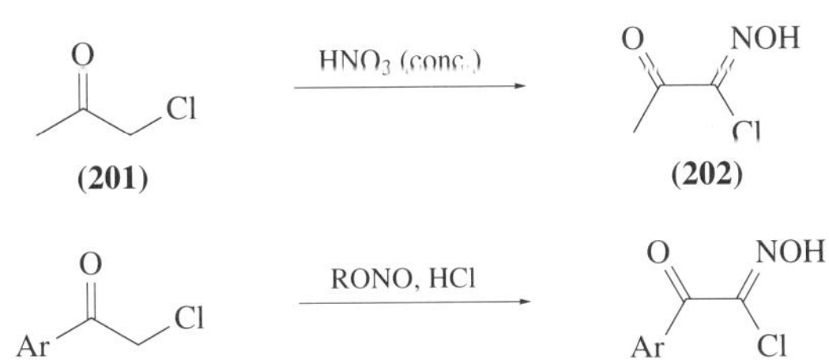

(201) (202)

(203) (204)

Scheme 20

Conversion of α-chloroacetoacetates (**205**) takes place with isoamyl nitrite in hydrochloric acid. The nitrosation is followed by elimination of the acetyl group to give the oxyimidic chloride derivative (**206**) (Equation (59)) ⟨55CB130⟩.

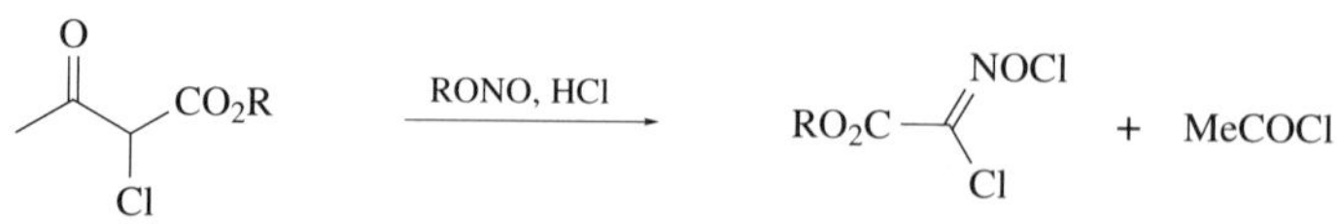

(205) **(206)** (59)

5.21.2.1.7 From N-*alkoxyamides*

O-Methoxy-, *O*-benzyloxy, and *O*-*t*-butoxyimidoyl halides (**208**) are prepared from *N*-alkoxyamides (**207**) according to two methods. *O*-Methoxyamides are treated with a mixture of phosphorus pentachloride and phosphorus oxychloride at reflux temperature, while other *N*-alkoxyamides are treated with a mixture of triphenylphosphine and tetrachloro- or tetrabromomethane in acetonitrile ⟨76JOC252, 91S750⟩, phosphorus tribromide ⟨76JOC252⟩, or thionyl chloride ⟨92JOC3245⟩ (Scheme 21).

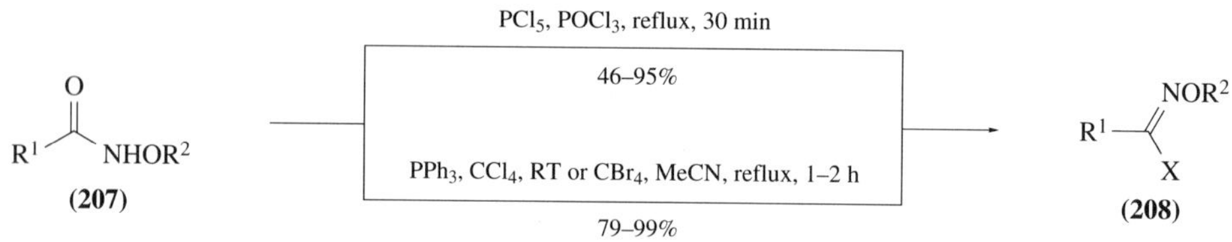

(207) **(208)**

R^1 = 4-Me-C_6H_4, 4-$MeCO_2C_6H_4$, 4-CN-C_6H_4, $PhCH_2CH_2$, 4-MeO-$C_6H_4CH_2$, $MeO_2C(CH_2)_4$, $NC(CH_2)_{10}$
R^2 = Me, Bu^n, Bu^t
X = Cl, Br

Scheme 21

Oxyimidic amide derivatives (**209**) and (**211**) can be converted by nitrosative deamination in the presence of chloride ion into the corresponding oxyimidic chlorides (**210**) ⟨86JHC1861⟩ and (**212**) ⟨82JHC1397⟩ (Scheme 22).

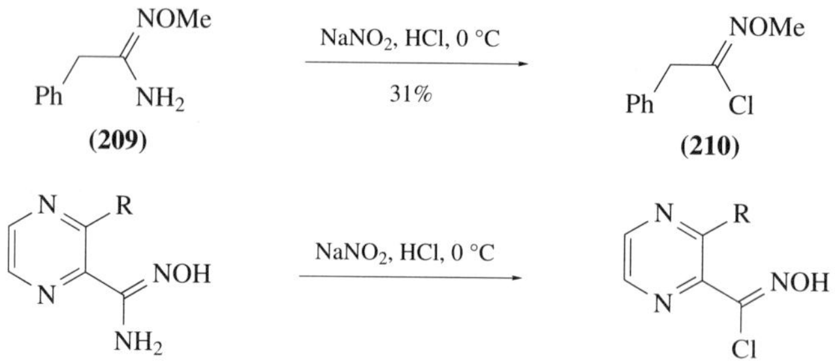

(209) **(210)**

(211) **(212)**

Scheme 22

5.21.2.1.8 From oxaziridines

Perfluoro-2,3-dialkyloxaziridines (**213**) and (**215**) rearrange in the presence of CsF at room temperature or with SbF_5 at elevated temperature into oxyimidic fluoride derivatives as the sole products (Scheme 23) ⟨93JCS(P1)505⟩. Similarly, they react with fluorinated carbonyl compounds, such as hexafluoro- or chloropentafluoroacetone to give *O*-perfluoroalkyl oxyimidic fluorides **(214)** and **(216)** in high yields. 1,3-Dichlorotetrafluoroacetone and difluorophosgene react analogously ⟨93JCS(P1)505⟩.

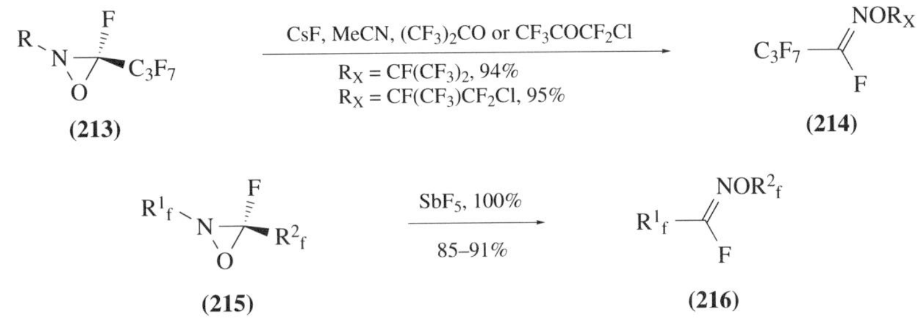

(213) **(214)** **(215)** **(216)**

$R^1{}_f = C_3F_7, C_4F_9, C_6F_{13}$; $R^2{}_f = C_2F_5, C_3F_7, C_5F_{11}$

Scheme 23

5.21.2.1.9 *From oxyimidic amides*

Some oxyimidic amides can be transformed into oxyimidic chlorides. For example, when ethyl oxalic acid oxyimidic amide is treated with nitrosyl chloride in slight excess in methylene chloride for 18 h at 0°C, the corresponding oxyimidic chloride is obtained in approximately 50% yield (Equation (60)) ⟨58JOC1517⟩. Several *O*-substituted oxyimidic acid and amide derivatives can serve as starting compounds for oxyimidic chorides. For example, formic acid *O*-benzyl oxime is transformed with phosphorus pentachloride into formic acid oxyimidic chloride ⟨1900LA(310)11⟩ and benzoic acid *O*-ethyl oxyimidic amide in hydrochloric acid gives with sodium nitrite the corresponding oxyimidic chloride ⟨1884CB1685, 1885CB727, 1891CB3453⟩.

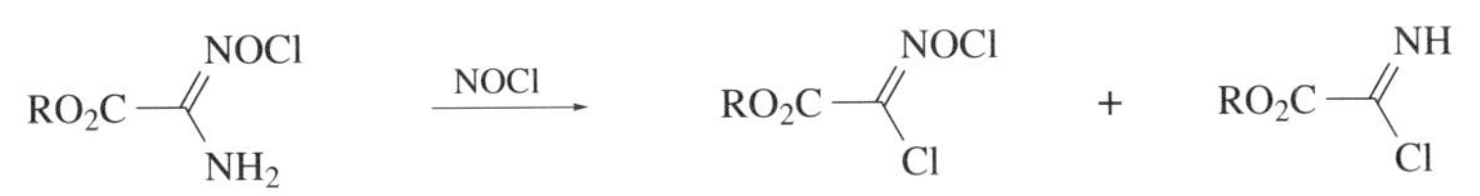

(60)

5.21.2.1.10 *Isomerization*

Oxyimidic halides prepared according to the methods described above usually exist in (*Z*)-form. By UV irradiation in hexane a mixture of (*E*)- and (*Z*)-forms is obtained, which can be separated (Equation (61)) ⟨92JOC3245⟩. *O*-Acetyl and *O*-benzoyl esters, prepared from oxyimidic chlorides and acyl chlorides in benzene and dry pyridine isomerize under photochemical conditions into (*E*)-isomers (Scheme 24) ⟨86JCS(P2)995⟩.

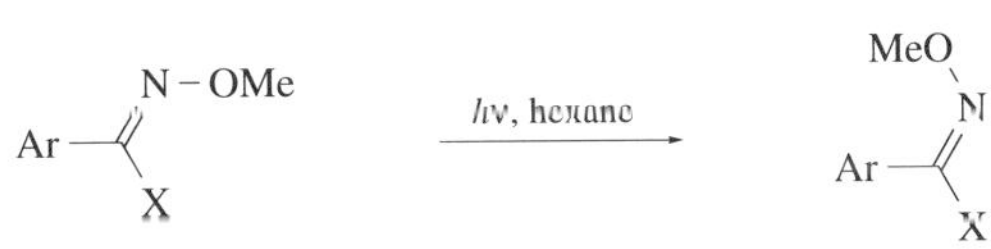

(61)

X = Cl, Br

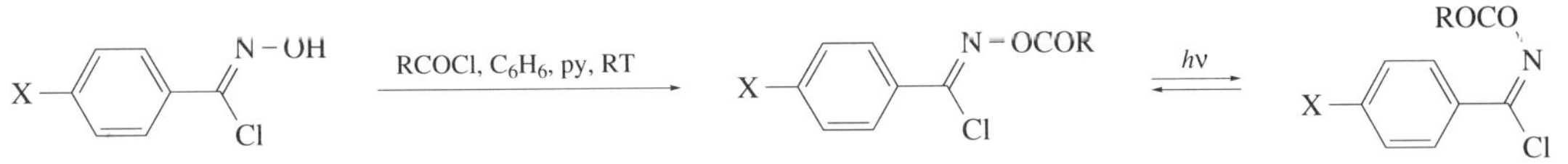

X = H, Cl, NO_2; R = Me, CH_2Cl, $CHCl_2$, CF_3, Ph

Scheme 24

5.21.2.2 Oxyimidic Esters and Related Structures

5.21.2.2.1 From imidic esters

The replacement of the imino group in imidic esters (**217**) with hydroxylamine is the most widely used reaction for the preparation of oxymidic esters (**218**) (Equation (62)) ⟨67IZV1820, B-75MI 521-01⟩. In the same manner imidic thioesters (**219**) react to give the corresponding oxyimidic thioesters (**220**) (Equation (63)) ⟨B-75MI 521-01⟩. *O*-Alkylhydroxylamine can also be used to obtain the corresponding *O*-alkyl substituted oxyimidic esters ⟨77JOC1610⟩. Some other compounds, such as *O*-acyl oxyimidic thioesters are obtained analogously ⟨83IJC(B)51, 87JOC4570⟩.

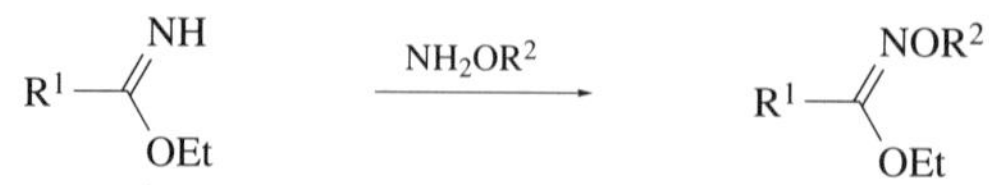

(**217**) (**218**) (62)

R^1 = alkyl, aryl; R^2 = H, Me

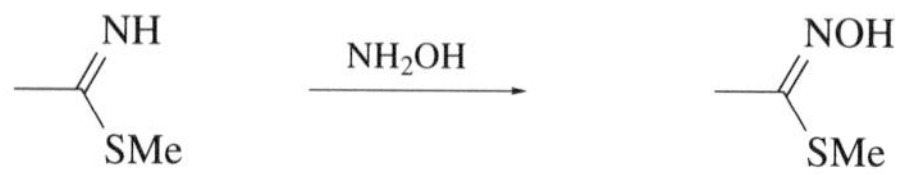

(**219**) (**220**) (63)

5.21.2.2.2 From oxyimidic chlorides and derivatives

O-Alkyl derivatives of oxyimidic chlorides (**221**) react with alcoholates to give derivatives of *O*-alkyl oxyimidic esters (**222**) (Equation (64)) ⟨1895LA(281)169, 1896CB1161⟩.

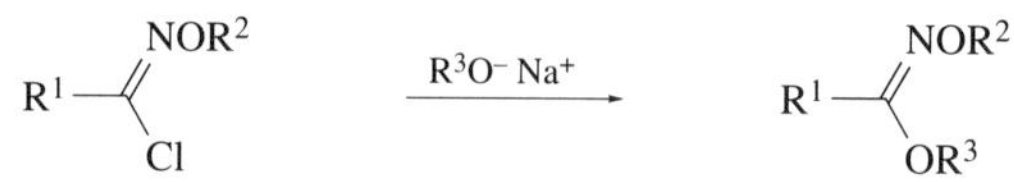

(**221**) (**222**) (64)

5.21.2.2.3 From nitrile oxides

Nitrile oxides (**223**) react with alcohols in the presence of a strong base or a diluted acid to give *O*-alkyl oxyimidic esters (**224**) (Equation (65)) ⟨66FCF62⟩, while thiols react with aromatic and aliphatic nitrile oxides without any catalyst or in the presence of triethylamine to form the corresponding oxyimidic thioesters ⟨66FCF62⟩.

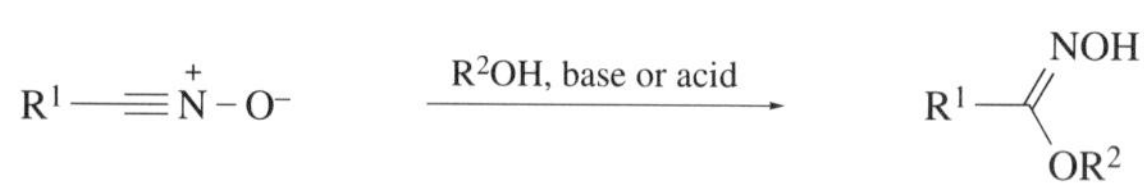

(**223**) (**224**) (65)

R^1 = alkyl, aryl; R^2 = alkyl

5.21.2.2.4 From tosylated malononitrile oxime

One of the cyano groups in tosylated malononitrile oxime (**225**) can be exchanged with one equivalent of ethoxide ion as a base to give the corresponding *O*-tosyl oxyimidic ester (**226**) (Equation (66)) ⟨74T999⟩.

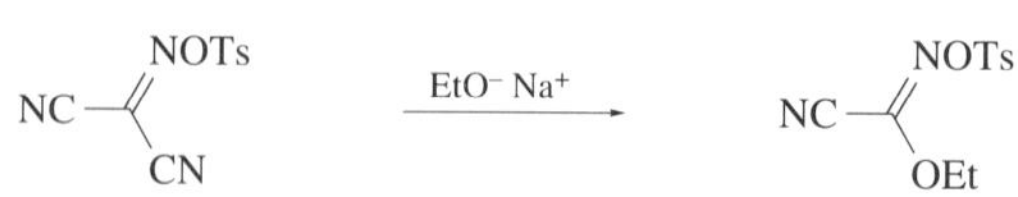

(**225**) (**226**) (66)

5.21.2.2.5 *From thiocarboxylic acid O-esters*

Thiocarboxylic acid *O*-esters (**227**) react with hydroxylamine or *O*-alkylhydroxylamine to give *O*-alkyl oxyimidic esters (**228**) (Equation (67)) ⟨83S220⟩.

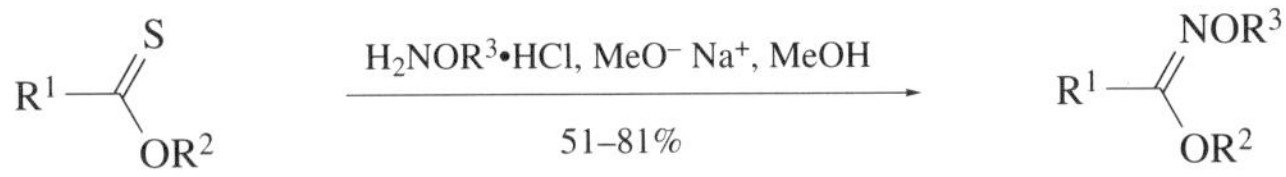

(67)

(227) **(228)** (*E*) or (*E*/*Z*)

R^1 = Me, Et, $PhCH_2$, Ph; R^2 = Et, $(CH_2)_2NMe_2$; R^3 = H, CH_2, Ph

5.21.2.2.6 *From carboxylic acid alkoxy amides and acyloxy amides*

Alkylation of carboxylic acid alkoxamides (**229**) with trialkyloxonium tetrafluoroborate takes place on the oxygen of the carbonyl group to give *O*-alkyl oxyimidic esters (**230**) together with *N*-alkylated products (**231**) (Equation (68)) ⟨71JOC284⟩. *O*-Alkyl oxyimidic bromides, which can be easily obtained from *O*-alkyl aldoximes by bromination, are transformed into *O*-alkyl oxyimidic esters with sodium or thallium alcoholates (Equation (69)) ⟨70MI 521-01⟩. Another method is represented by the alkylation of the silver salt of benzoyloxycarboxamides with alkyl halides in ether at room temperature ⟨71JOC284⟩.

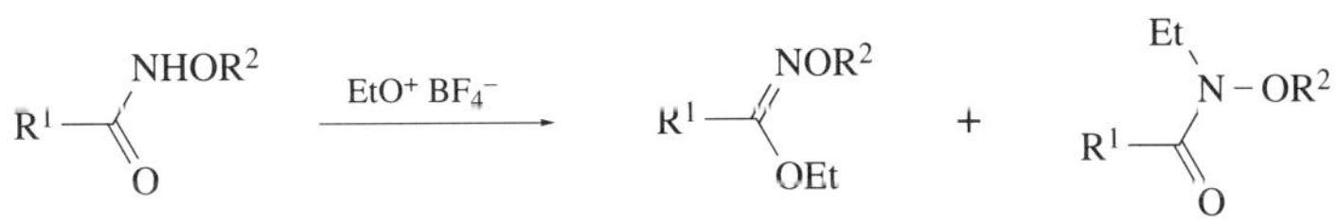

(68)

(229) **(230)** **(231)**

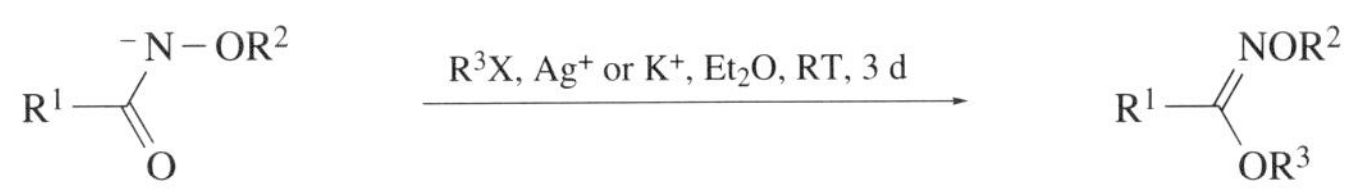

(69)

R^1 = Ph
R^2 = Et, Pr^n, Bu^n, COPh
R^3 = Me, Et, Pr^i, alkyl, benzyl

yields (from potassium salts) 5–99%
mostly above 60%

(from silver salts) 22–100%
mostly above 60%

5.21.2.2.7 *From oxyimidic esters*

Oxyimidic esters (**232**) can be alkylated with benzyl chloride or substituted benzyl chloride and analogous compounds in the presence of sodium alkoxides to give (**233**) (Equation (70)) ⟨67ZOR1207⟩.

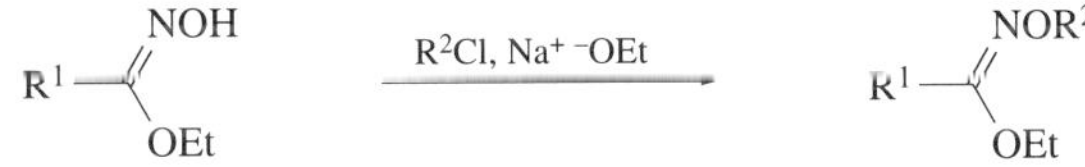

(70)

(232) **(233)**

R^2 = $PhCH_2$, 4-MeO-C_6H_4-CH_2, 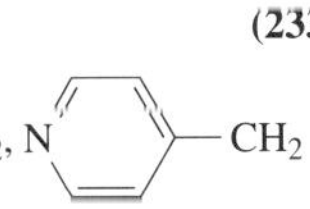

5.21.2.2.8 *Thioimidic esters*

A good route to prepare sulfonyl derivatives (**235**) of imidates is treatment of a sulfonamide (**234**) with an *ortho* ester (Scheme 25) ⟨69CB1641⟩. Addition of sulfonyl azides to ethoxyacetylene results in the formation of triazole derivatives (**236**), which exist in equilibrium with *N*-sulfonylimidates (**237**) with a diazo function attached to the side chain (Scheme 25) ⟨72CB2975⟩.

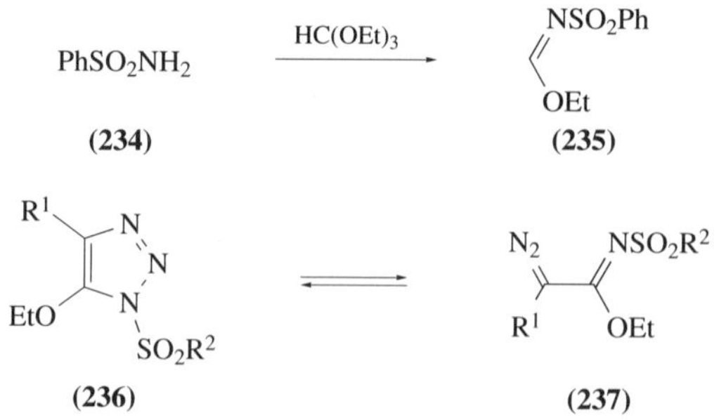

Scheme 25

5.21.2.3 Oxyimidic Sulfur, Selenium, and Tellurium Derivatives

5.21.2.3.1 From nitrile oxides

Nitrile oxides (**239**), usually prepared *in situ* from oxyimidic acid chlorides (**238**), react with thiols and thiophenols to form the corresponding *S*-aryl oxyimidic thioesters (**240**) (Scheme 26) ⟨66FCF62⟩. Aliphatic nitrile oxides (**241**; R^1 = alkyl) react with thiols without any catalyst or in the presence of triethylamine to give *S*-alkyl esters (**240**; R^1 = alkyl) (Equation (71)) ⟨66FCF62⟩. When cyanogen *N,N*-dioxide (**242**) reacts with sulfide ion, the corresponding thiooxalic bis(hydroxamic acid) (**243**) and tetraoximino-1,4-dithiane (**244**) are obtained (Equation (72)) ⟨65LA(687)191⟩. Oxyimidic thioesters can be prepared also from fulminates ⟨70ACS1367⟩.

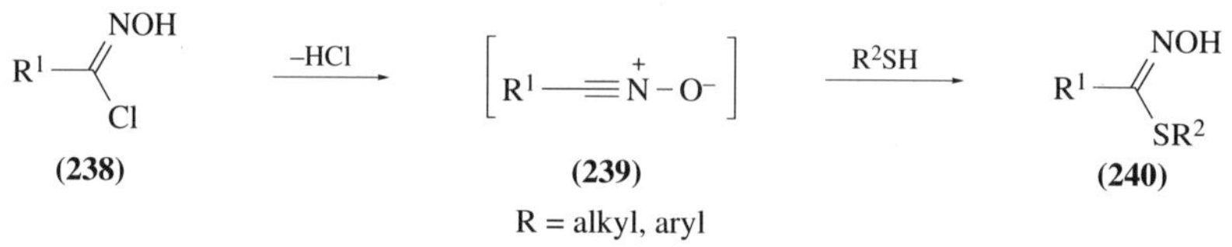

Scheme 26

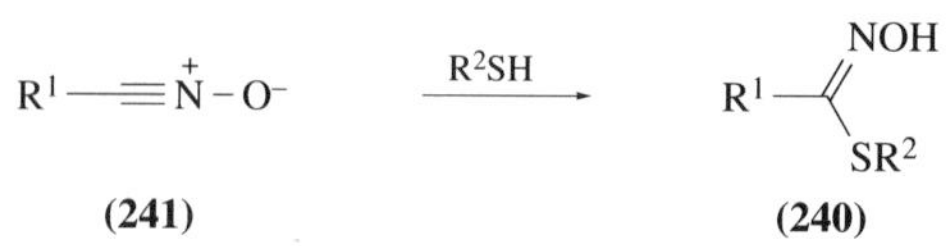

(71)

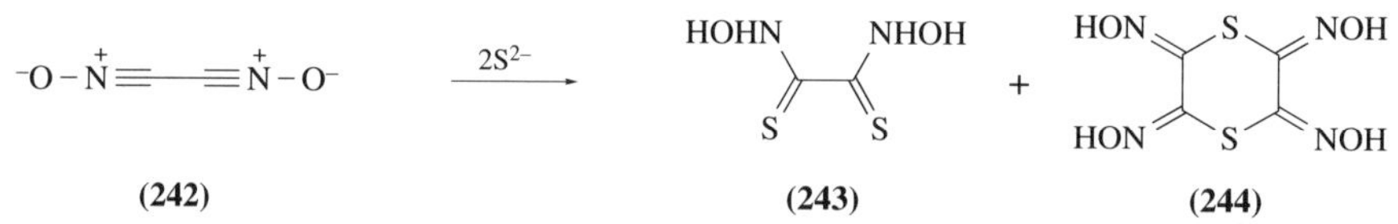

(72)

5.21.2.3.2 From oxyimidic halides

Thiol ions react with oxyimidic chlorides (**245**) to give *S*-alkyl or *S*-aryl oxyimidic thioesters (**246**) (Scheme 27) ⟨66FCF62, 68JCS(C)431⟩. Oxyimidic iodides (**247**) react analogously to give (**246**) ⟨70ACS1367⟩. Some *S*-2-hydroxyethyl esters and many *S*-aryl esters have been synthesized by reacting oxyimidic chlorides, prepared from aryl and diarylacetonitriles, with various thiol derivatives ⟨72MI 521-01⟩. Oxyimidic derivatives with SO_2Me group (**249**) are obtained in the reaction of oxyimidic chlorides (**248**) with methanesulfinate ion (Equation (73)) ⟨68CJC1873⟩.

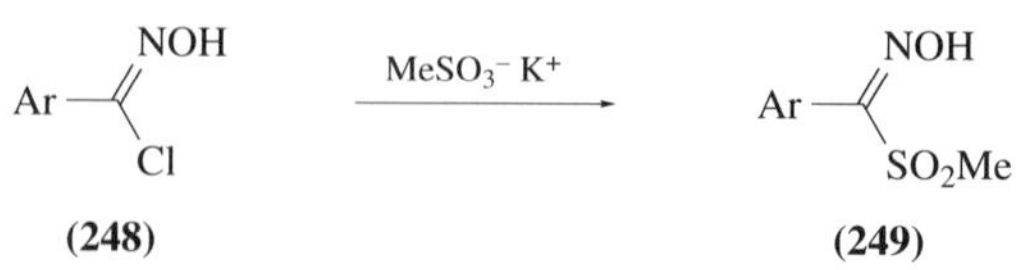

(73)

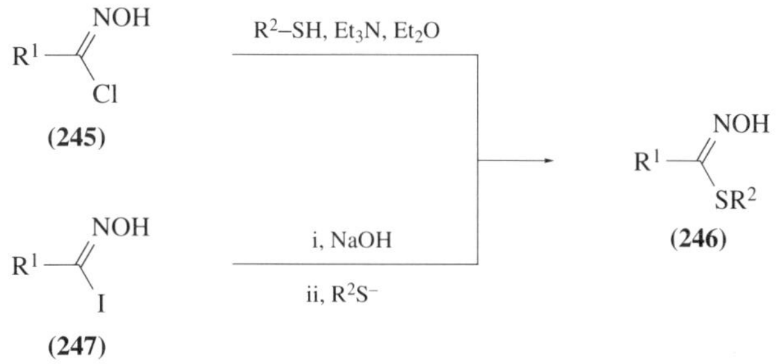

R^1 = Ph, Ph_2CH, *o*-(*m*-, *p*-)-C_6H_4, *o*-(*m*-, *p*-)-X-C_6H_4, 5-NO_2-C_6H_4, furyl-2
R^2 = CH_2CH_2OH, Ph, 2,4,6-$Cl_3C_6H_4$, C_6Cl_5

Scheme 27

5.21.2.3.3 From nitro compounds

Oxyimidic thioesters (**253**) and (**255**) are synthesized by reaction of primary nitro compounds (**250**) and (**254**) with aliphatic and aromatic thiols in the presence of tributylamine ⟨54USP2786865⟩ or alkoxide under pressure ⟨69FRP1573102⟩ (Scheme 28). A one-step synthesis of α-acyl oxyimidic thioesters (**256**) is represented by the nucleophilic dehydration of α nitro ketones, prepared by reaction of sodium salt of nitromethane with acylimidazoles, followed by treatment with thiophenols and titanium tetrachloride in the presence of triethylamine (Scheme 29). It has been shown by x-ray analysis that the *N*-hydroxy group is oriented *syn* to the phenylthio group ⟨87JOC4570⟩.

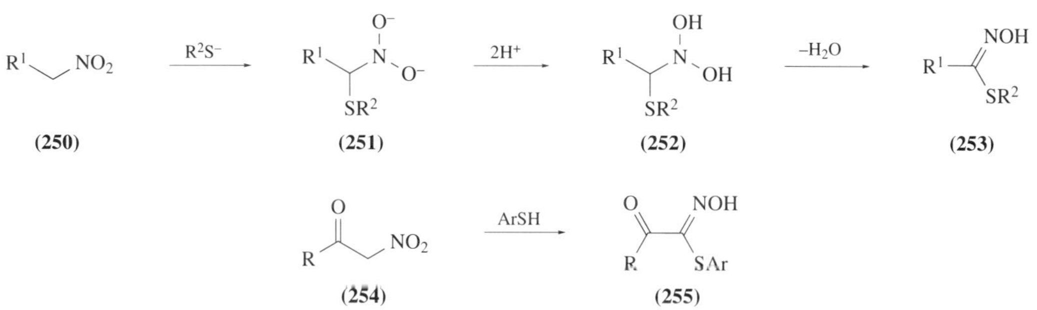

Scheme 28

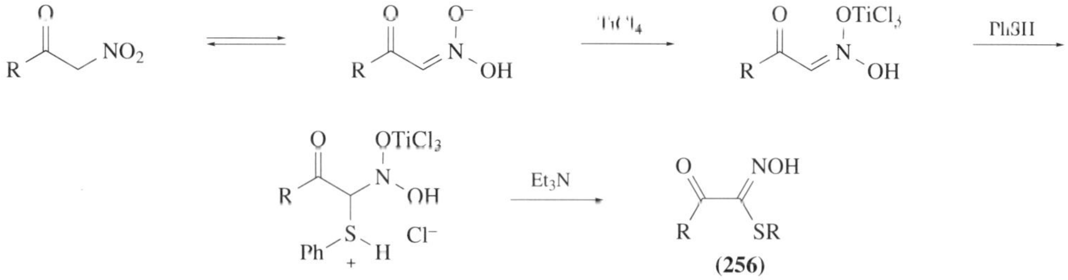

R = Ph, 60%; Et, 52%; Me, 31%

Scheme 29

Primary nitro compounds (**257**) are converted into mixtures of (*E*)- and (*Z*)-isomers of oxyimidic thioesters (**258**) by reaction with *n*-butyllithium and thiosilanes, such as ethylthiotrimethylsilane and phenylthiotrimethylsilane in high yields (Equation (74)) ⟨90T7413⟩.

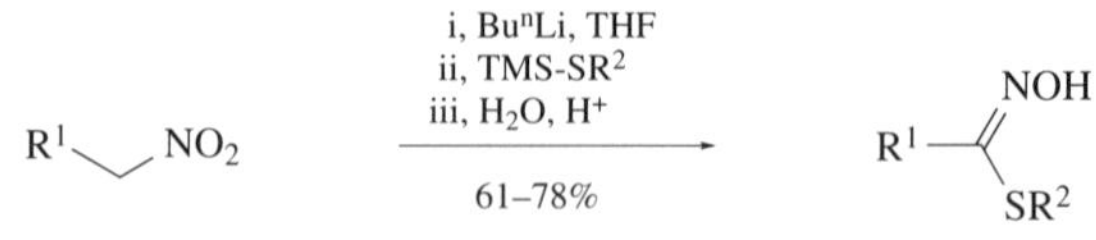

(257) (258) (74)

5.21.2.3.4 *From carbonyl compounds*

α-Keto oxyimidic thioesters (**261**) are prepared according to two methods. (a) Acetophenones (**259**) are converted first into oxyimidic chlorides (**260**) ⟨66ZC313⟩ followed by treatment with *N*,*N*-dialkylaminoalkanethiols (Scheme 30) ⟨81JMC1124⟩. (b) An alternative method is bromination of acetophenone to give (**263**) and subsequent thiolation with *N*,*N*-dialkylaminoalkanethiols to give (**262**) followed by nitrosation (Scheme 31) ⟨86JMC1689⟩.

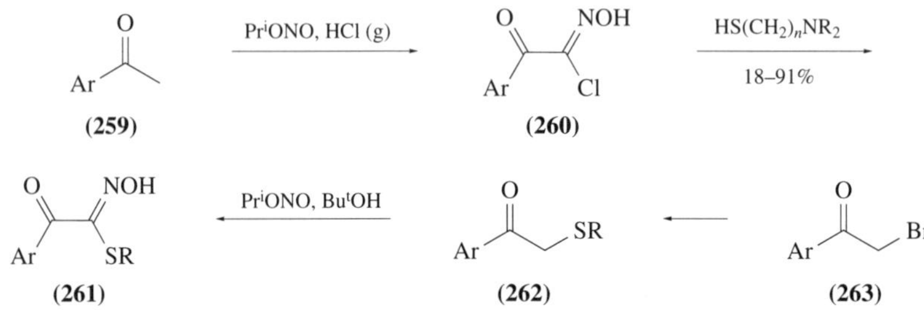

Scheme 30

A one-step synthesis of α-keto oxyimidic thioesters (**265**) represents a nucleophilic dehydration of α-nitro ketones (**264**) with thiophenol and titanium tetrachloride in the presence of triethylamine (Equation (75)) ⟨87JOC4570⟩.

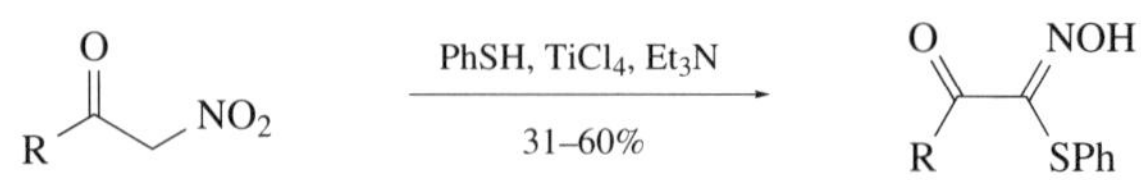

(264) (265) (75)

5.21.2.3.5 *Oxidation and reduction of oxyimidic thioacid derivatives*

A mild oxidation of oxyimidic thioacids (**266**), their sodium salts or esters leads to the formation of the corresponding disulfides (**267**) (Scheme 31). The oxidation agents used for this purpose are hydrogen peroxide ⟨66CPB1263⟩, iron(III) or copper(II) salts ⟨67CPB61⟩. The disulfides (**267**) react with methyl iodide by cleavage of the disulfide bridge to form a mixture of *N*,*O*-dimethyl (**268**) and *S*-methyl (**269**) derivatives (Scheme 31) ⟨66CPB1263⟩.

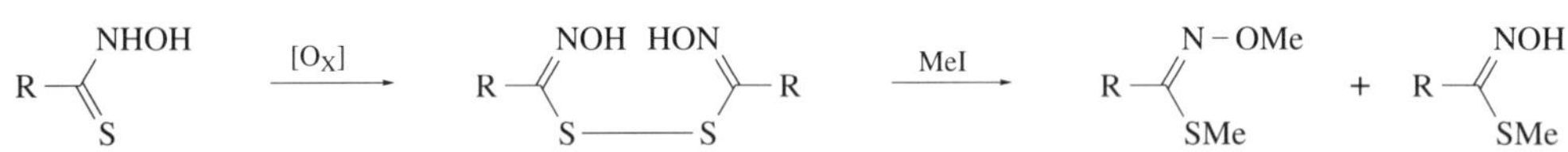

(266) (267) (268) (269)

Scheme 31

5.21.2.3.6 *From alkyl thioimidates*

The imino group in thioimidates (**270**) can be replaced by the oximino group to give the corresponding oxyimidic thioesters (Equation (76)) ⟨36CB2352, B-75MI 521-02⟩. The method can be extended to thioimidates of higher acids ⟨68USP3374260⟩ and unsaturated acids ⟨1886CB1668⟩. An exception is *S*-benzyl imidate of thioformic acid (**271**), from which the final product (**272**) is formed by reaction of two molecules of thioimidate with one molecule of hydroxylamine (Scheme 32) ⟨70ACS1367⟩.

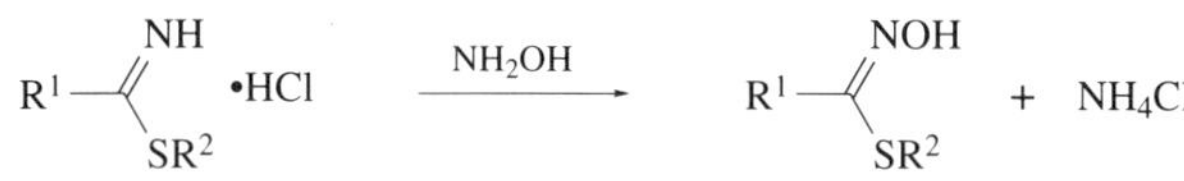

(**270**)

(76)

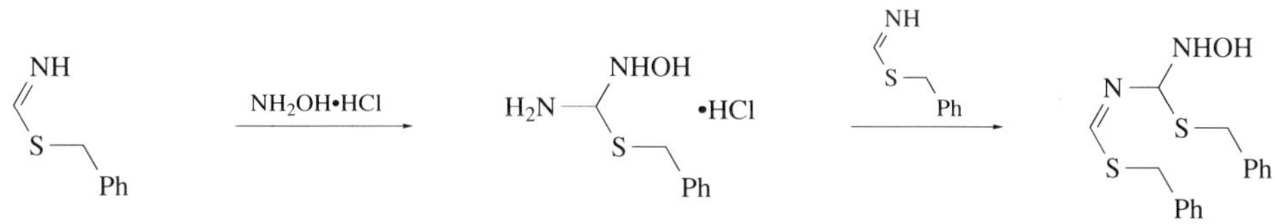

(**271**) (**272**)

Scheme 32

5.21.2.3.7 *From oxyimidic thioacids*

Electrophilic attack of alkylating agents on the sodium or potassium salts of oxyimidic thioacids (**273**) in an alcohol ⟨09MI 521-01, 11G166, 57IA1764, 58MI 521-01, 62JOC304, 66CPB1263⟩, acetone or 1,2-dimethoxyethane takes place to give (**274**) (Equation (77)) ⟨62JOC304⟩. With *para*-substituted thiobenzhydroxamic acids (**275**) and benzyl chloride the *O*,*S*-dibenzyl compounds (**276**) are formed (Equation (78)) ⟨58MI 521-01⟩. The nickel complexes, which can be employed instead of alkali salts, give with methyl iodide or dimethyl sulfate in aqueous sodium hydroxide solution the *S*-alkyl and *O*,*S*-dialkyl oxyimidic acid derivatives ⟨69CPB653⟩. *N*-Hydroxy thioureas (**277**) react with alkyl halides to give the corresponding *S*-alkyl-*N*-hydroxyisothioureas (**278**) (Equation (79)) ⟨1897LA(289)117⟩, while with diazomethane the corresponding *N*-methoxythioureas are formed ⟨70CB82⟩.

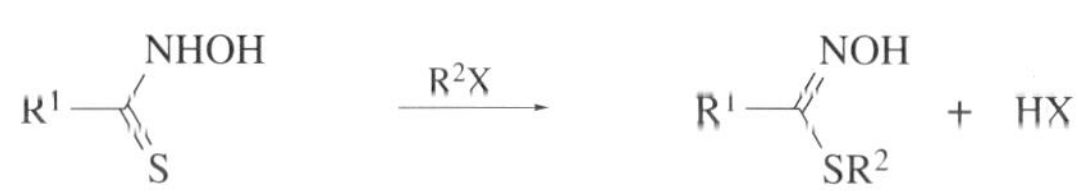

(**273**) (**274**)

(77)

(**275**) (**276**)

(78)

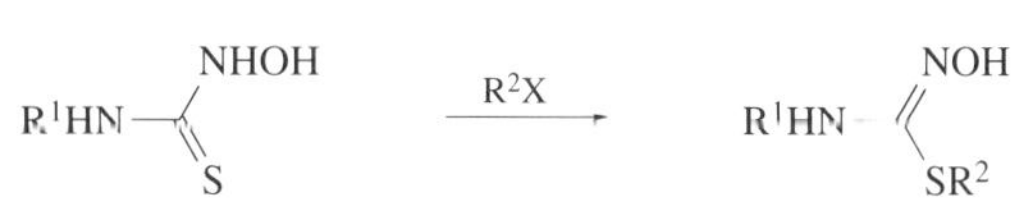

(**277**) (**278**)

(79)

Acylation of oxyimidic thioacids (**279**) with acyl halides produces only the *O*,*S*-diacyl derivatives (**280**) (Equation (80)) ⟨09MI 521-01, 66CPB1263⟩. Attempts to prepare the *S*- or *N*-monoacyl derivatives according to this method failed. However, if the nickel complex of the oxyimidic thio acid reacts with benzoyl chloride, the corresponding *N*-monobenzoyl derivative was isolated as a nickel complex ⟨69CPB653⟩.

(279) **(280)** (80)

5.21.2.3.8 Other derivatives of oxyimidic acid

Oxyimidic thioesters (**281**) are oxidized to their *S*-oxides (**282**) with hydrogen peroxide (Scheme 33) ⟨61AG656, 68CJC1873⟩. The same compounds can be obtained from benzonitrile (**283**) and dimethyl sulfoxide in the presence of sodium hydride ⟨67G1664⟩.

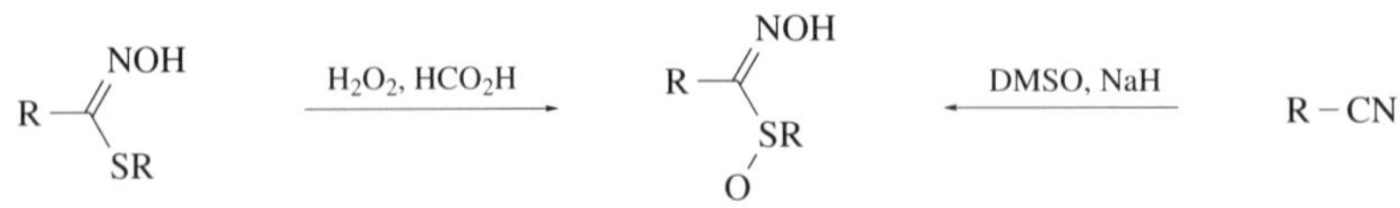

(281) **(282)** **(283)**

R = Ph

Scheme 33

Chlorination of 2-cyanothionacetamide (**284**) in tetrachloromethane produces two products: 1,2,2-trichlorocyanoacetimidoyl chloride (**285**) preferentially formed at room temperature and *N*-2,2-trichlorocyanoacetimidoylsulfenyl chloride (**286**), which is formed at low temperatures (Equation (81)) ⟨70CI(L)238⟩.

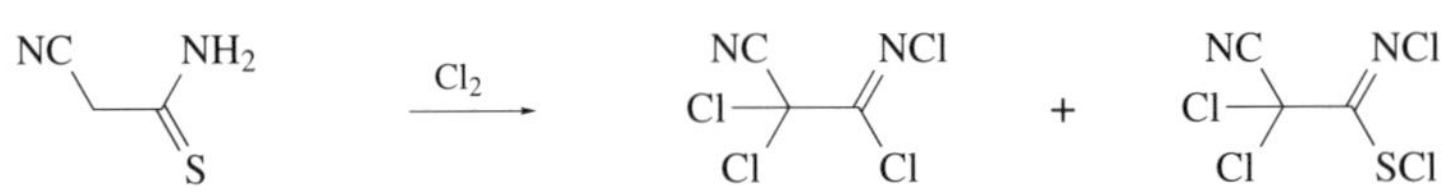

(284) **(285)** **(286)** (81)

A Beckmann rearrangement of substituted arylthiohydroxamic acid anhydrides (**287**), obtained by reaction of 2,3,5,6-tetramethylbenzonitrile oxide and 2,5-dichlorobenzonitrile oxide with hydrosulfide ion, takes place at temperatures around 200 °C to give the rearranged product (**288**) in the first instance, which further decomposes into an isothiocyanate and urea derivatives (Equation (82)) ⟨66JOC157⟩. On the other hand, 2,4,6-trimethoxybenzonitrile oxide is readily reduced to 2,4,6-trimethoxybenzonitrile under the same conditions.

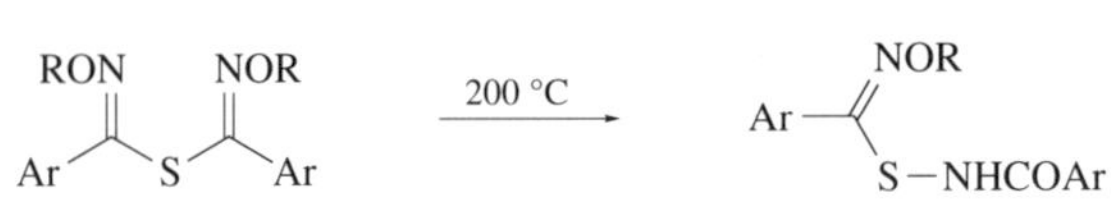

(287) R = H, Ac **(288)** (82)

5.21.2.3.9 Oxyimidic selenium derivatives

The reaction of oxyimidic chloride (**289**) with selenophenol produces the corresponding oxyimidic selenoester (**290**) (Equation (83)) ⟨64CJC2393⟩. Selenoglucosinates (**292**) are prepared from tetra-*O*-acetyl-1-seleno-*β*-D-glucopyranose anion (**291**) with oxyimidic chlorides (Equation (84)) ⟨87ACS(B)29⟩.

(289) **(290)** (83)

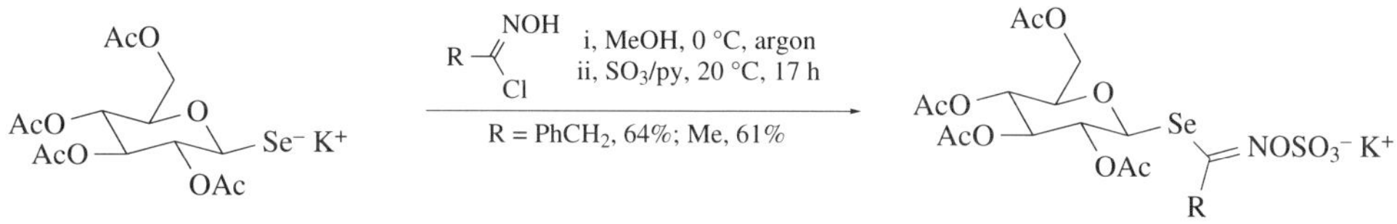

(291) (292) (84)

5.21.2.4 Oxyimidic Amides and Other Related Derivatives

Oxyimidic amides are also known as amidoximes or oxamidines. According to *Chemical Abstracts* they are named hydroxyamidines. Review articles have been published ⟨62CRV155, B-70MI 521-02⟩.

5.21.2.4.1 From nitriles

Nitriles (**293**) are the most frequently employed starting materials for this class of compounds. They react with hydroxylamine liberated usually from its hydrochloride with sodium carbonate, sodium or potassium hydroxide, or sodium ethoxide and the reaction mixture is kept for several hours at 60–80°C to give (**294**) (Equation (85)) ⟨1884CB126⟩. To avoid the separation of the product from sodium or potassium chloride a solution of free hydroxylamine in anhydrous methanol or ethanol is used ⟨53JOC64⟩. The highest yields are obtained when 15% excess of hydroxylamine in butanol is used and the mixture is left for 48 h at 60°C. The product separates as a practically pure crystalline material ⟨62HCA441⟩.

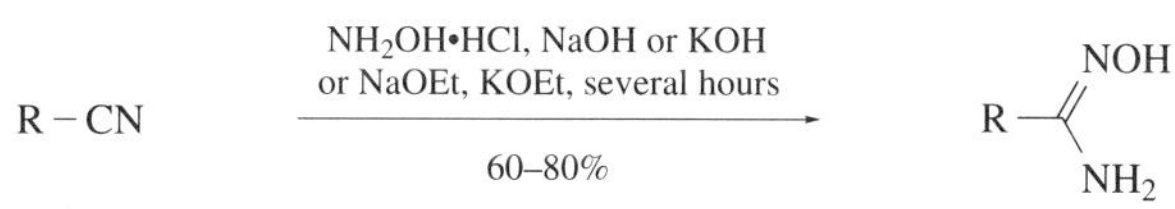

(293) (294) (85)

Diamidoximes (**297**) can, in principle, be prepared according to the same method from dinitriles, such as cyanogen (**295**) with hydroxylamine or alternatively, from its addition compounds with aniline, for example, diphenyloxamidine (**296**), which is treated with hydroxylamine hydrochloride (Scheme 34) ⟨1889CB1936, 1889CB2946, 1891CB803⟩. The best yields and the purest products are obtained if gaseous cyanogen is led directly in an aqueous hydroxylamine solution at 0°C ⟨62CVR155⟩. Polyoxyimidic amides ⟨59USP2902514⟩ and polymers containing the oxyimidic amide function have been prepared from polyacrylonitrile of low molecular weight with a slight excess of hydroxylamine ⟨57MAC25, 58MAC246⟩.

(295) (296) (297)

Scheme 34

In the formation of di(oxyimidic amides) intramolecular cyclization can occur and mono- or dioximes of the corresponding cyclic imides are formed. For example, when succinic dinitrile (**298**) reacts with hydroxylamine in 1:2 molar ratio in aqueous ethanol, succinamide dioxime (**299**) is formed as the major product accompanied with a small amount of succinimide dioxime (Scheme 35) ⟨1889CB2958⟩. When the molar ratio is 1:1, either the diimide oxime (**300**) or cyclic imide oxime (**301**) is formed, depending on the reaction conditions ⟨1891CB3426⟩.

Similar reactions occur with glutaronitrile ⟨1886CB1491, 1889CB2973, 1891CB3420⟩, phthalonitrile ⟨54JCS4251⟩, and others ⟨62CRV155⟩.

α-Aminooxyimidic amides (**303**) are obtained from the corresponding α-aminonitriles (**302**) with hydroxylamine hydrochloride (Equation (86)) ⟨88CB723⟩.

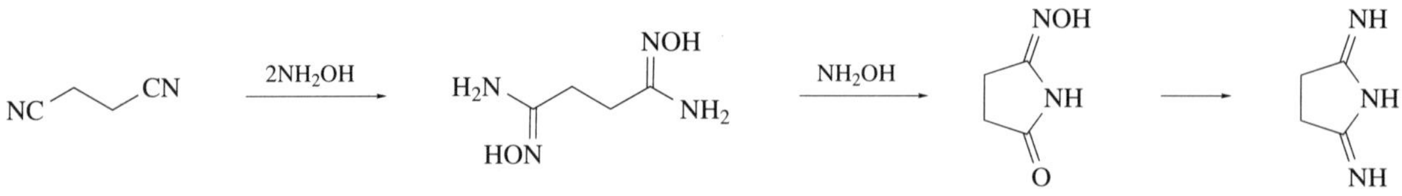

(298) (299) (300) (301)

Scheme 35

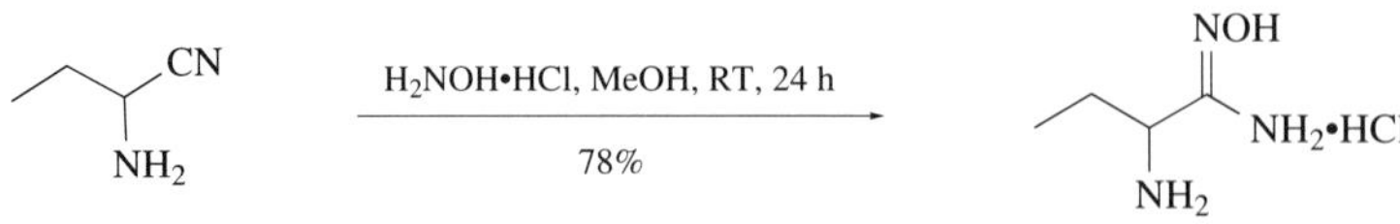

(302) (303)

(86)

C-Heteroaryl-substituted oxyimidic amides (**305**) are prepared from heterocyclic nitriles (**304**) by addition of hydroxylamine in ethanol at room temperature (Scheme 36). The yields in pyridine, pyridazine, pyrimidine, and pyrazine series are very high. The compounds (**305**) can be methylated with methyl iodide to give *O*-methylated products (**306**; R = Me) or acetylated with acetic anhydride to give *O*-acetyloxime derivatives (**306**; R = Ac) ⟨78H(9)1327, 79JOC1695, 82JHC1397, 86S807⟩.

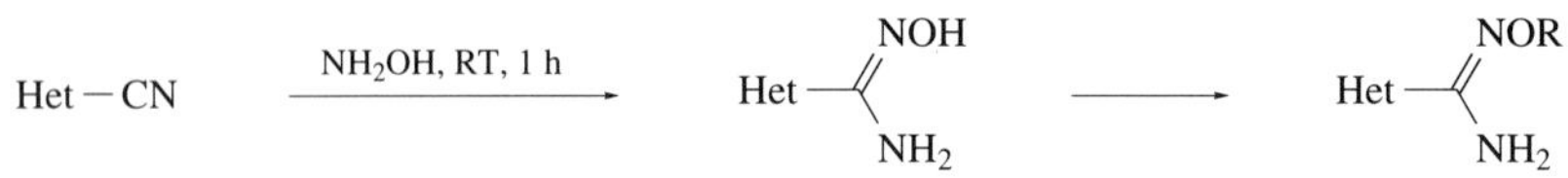

(304) (305) (306)

R = Me, COMe

Scheme 36

5.21.2.4.2 *From amides and thioamides*

Generally, amides do not react with hydroxylamine, while thioamides (**307**) are more reactive and they condense with hydroxylamine to form the corresponding oxyimidic amides (**308**). The reactions are usually carried out in ethanolic solutions (Equation (87)) ⟨1891CB3658⟩.

(307) (308)

(87)

5.21.2.4.3 *From oxyimidic halides*

Oxyimidic halides (**309**) react easily with ammonia to yield the corresponding oxyimidic amides (**310**) (Equation (88)). Several aryl-substituted derivatives have been prepared by this method ⟨1894CB2193, 1899CB1975⟩.

(309) (310)

(88)

The mechanism and stereochemistry of substitution of chlorine in (*Z*)-*O*-methyl-4-nitrophenyl-oxyimidic chloride (**311**) with morpholine, piperidine, pyrrolidine, and azetidine has been studied in detail, showing that (*Z*)-isomers of (**312**) are formed exclusively in benzene solutions, while in diphenyl ether at 100°C an equilibrium mixture exists between the (*Z*)- and (*E*)-isomer with (*Z*)-isomer predominating (Equation (89)) ⟨92JOC4648⟩.

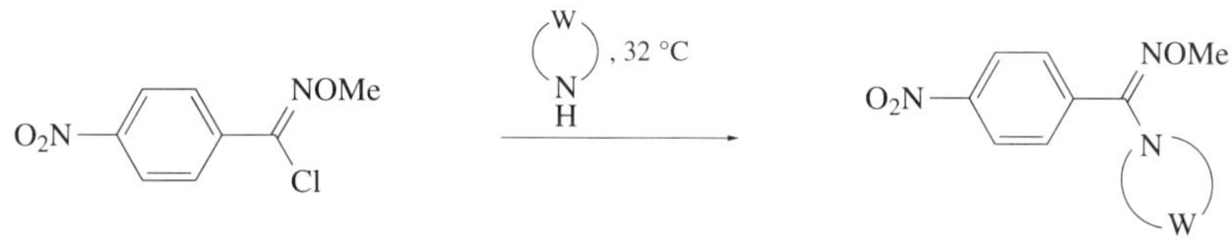

(89)

(311) W = $(CH_2)_3$, $(CH_2)_5$, $(CH_2)_2O(CH_2)_2$ **(312)**

5.21.2.4.4 *From oxyimidic esters*

Oxymidic esters (**313**) react with ammonia in ethanol by heating for several hours at 175°C to give (**314**) (Equation (90)) ⟨1889LA(252)214⟩.

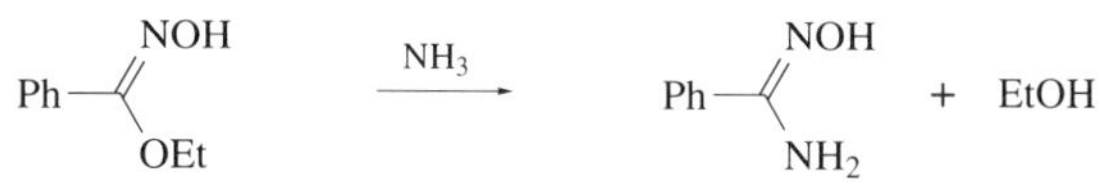

(90)

(313) **(314)**

Oxyimidic amides (**316**) with aromatic substituents are prepared by treating oxyimidic esters (**315**) with hydroxylamine (Equation (91)) ⟨1884CB183, 1884CB1587⟩. The method is of limited practical value, since oxyimidic esters are usually prepared from the corresponding nitriles.

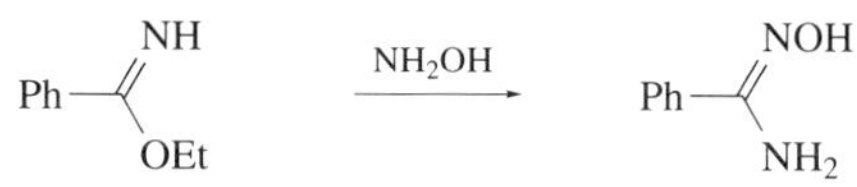

(91)

(315) **(316)**

5.21.2.4.5 *From amidines*

(i) Aromatic oxyimidic amides

Aromatic oxyimidic amides (**318**) can be prepared from benzamidine (**317**) and other aryl-substituted amidine hydrochlorides and hydroxylamine (Equation (92)) ⟨1884CB183⟩. However, this method is of no practical value, since amidines are obtained usually from nitriles, thioamides or oxyimidic esters, which can be directly transformed into amidoximes.

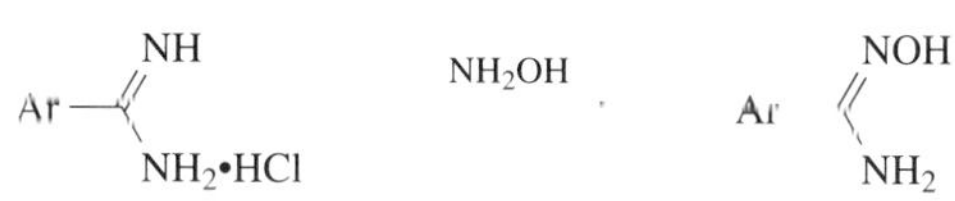

(92)

(317) **(318)**

(ii) N-*Heteroaryl-substituted oxyimidic amides*

A great number of *N*-heteroaryl-substituted formamide oximes and acetamide oximes have been prepared as versatile intermediates in the synthesis of various heterocyclic systems ⟨82MI 521-01, 83H(20)1591, B-87MI 521-01, 87MI 521-02⟩.

N-Heteroaryl-substituted oxyimidic amides (**321**) can be prepared in several ways. From heterocyclic amines (**319**) and *N*,*N*-dimethylformamide dimethyl acetal (DMFDMA) to form the corresponding amidines, followed by treatment with free hydroxylamine or its hydrochloride in ethanol to give *N*-heteroarylformamide oximes (**321**). The intermediate amidines (**320**) can be isolated and then separately converted into the final products, but the yields, in most cases above 80%, are generally better if the amidines are transformed directly into formamide oximes (Scheme

37) ⟨73TL1677, 74JOC2143, 86M(117)221, 87M601⟩. They exist in solution exclusively in the hydroxyimino form, as shown by comparison of NMR spectra of ^{14}N and ^{15}N labelled compounds ⟨74JHC103⟩. In order to avoid the methylation of OH, SH, and secondary NH groups, if present in the heterocyclic moiety, *N*,*N*-dimethylchloroformiminium chloride (DFCC) was selected as an alternative reagent for preparation of formamidines ⟨84H(22)1545⟩.

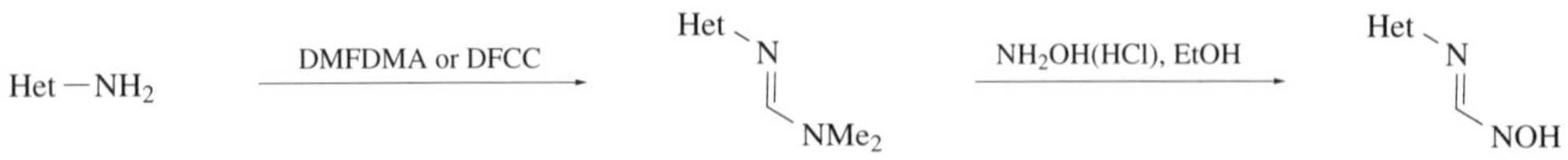

(319) **(320)** **(321)**

Scheme 37

Another preparation of (**321b**) by a synthetic procedure is represented by the transformation of heterocyclic amines (**319**) with trisformamidomethane into (**321a**), followed by treatment with hydroxylamine hydrochloride in methanol for 12 h at room temperature to give (**321**) (Scheme 38). The yields are in the range 54–95% ⟨81H(16)2173⟩.

$$\underset{\textbf{(319)}}{\text{Het}-\text{NH}_2} \xrightarrow{\text{HC(NHCHO)}_3} \underset{\textbf{(321a)}}{\text{Het}-\text{NHCHO}} \xrightarrow{\text{NH}_2\text{OH, MeOH, 12 h}} \underset{\textbf{(321b)}}{\text{HetNHCH}=\text{NOH}}$$

Scheme 38

Heterocyclic amines could be transformed via amidine derivatives into thioformyl derivatives (**322**) with hydrogen sulfide, followed by treatment with hydroxylamine hydrochloride in methanol at room temperature to give *N*-formamide oximes (**323**) in around 50% yield (Equation (93)) ⟨81S299⟩.

$$\underset{\textbf{(322)}}{\text{HetNHCHS}} \xrightarrow{\text{NH}_2\text{OH}} \underset{\textbf{(323)}}{\text{HetNHCH}=\text{NOH}} \qquad (93)$$

The corresponding *N*-heteroarylacetamide oximes (**326**; R = H) are analogously obtained from heterocyclic amines (**324**) and *N*,*N*-dimethylacetamide dimethyl acetal (DMADMA), followed by reaction with hydroxalamine hydrochloride in methanol or ethanol, usually without isolation of the corresponding intermediate amidine (**325**), in 74–98% yield (Scheme 39). When *O*-methyl-hydroxylamine hydrochloride is used, the corresponding *O*-methyl derivatives (**326**; R = Me) are obtained in high yields ⟨82JHC577⟩.

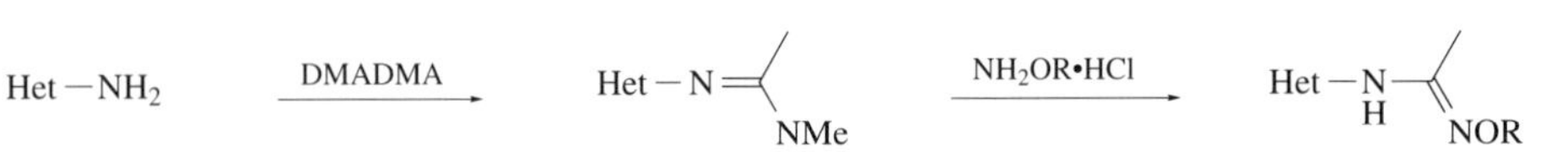

(324) **(325)** **(326)**

R = H, Me

Scheme 39

Methylation of *N*-heteroarylacetamide *O*-methyl oximes (**327**) with DMFDMA takes place at nitrogen to give *N*-heteroaryl-*N*-methylacetamide *O*-methyl oxime (**328**) in 80% yield or better (Equation (94)) ⟨82JHC577⟩.

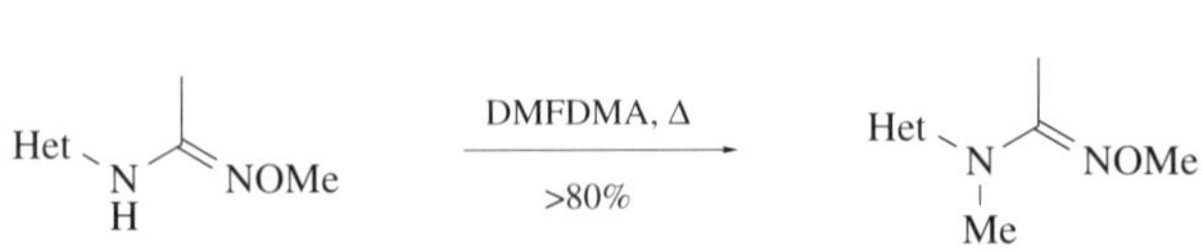

(94)

(327) **(328)**

With heterocyclic amines (**329**) having the cyano group attached in the heteroaromatic ring the addition of hydroxylamine to the cyano group takes place besides the exchange of the dimethylamino group in the formamidine part of the molecule to give the corresponding *C*,*N*-bisamide oximes (**330**) (Equation (95)). The reaction with free hydroxylamine takes place readily at 0°C to give the products in high yields ⟨81H(15)293, 84CS209, 85H(23)2819⟩.

(95)

(**329**) (**330**)

5.21.2.4.6 *From formamide oxime*

Formamide oxime (**331**) reacts with aromatic aldehydes to give the corresponding oxyimidic amides (**332**) (Equation (96)) ⟨08AC(P)84⟩.

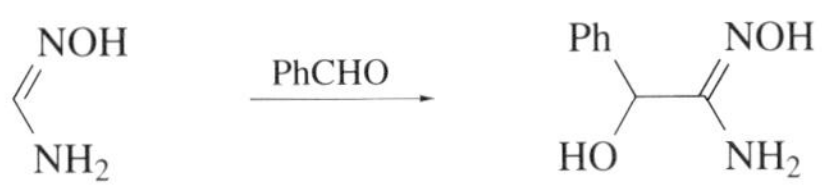

(96)

(**331**) (**332**)

5.21.2.4.7 *From nitrile oxide oxime derivatives*

Nitrile oxide oxime (**333**) react with ammonia to give the corresponding oxyimidic amides (**334**) (Equation (97)) ⟨31G561, 66JOC157⟩. The addition of various other nucleophiles, such as primary ⟨64AG440⟩ and secondary amines ⟨64AG440⟩, and hydrazine derivatives ⟨66JOC157⟩ to nitrile oxides (**335**), is a general and mild reaction for the preparation of substituted oxyimidic amides (**336**)–(**339**) (Scheme 40). Dinitrile oxides (**340**) react in the same manner (Equation (98)) ⟨65LA(687)191⟩.

(97)

(**333**) (**334**)

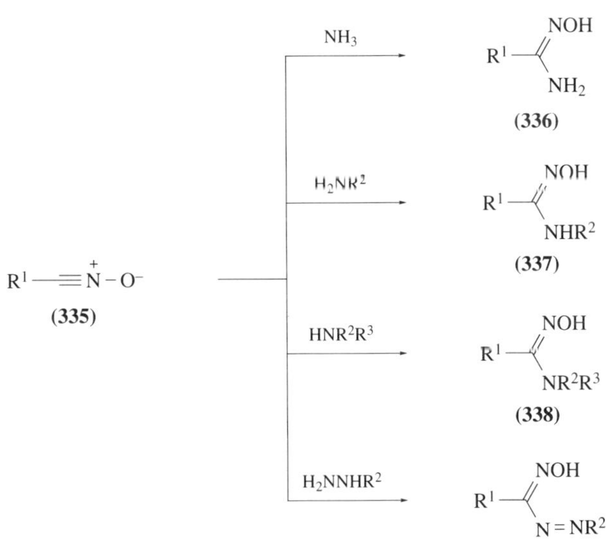

(**335**) (**336**) (**337**) (**338**) (**339**)

Scheme 40

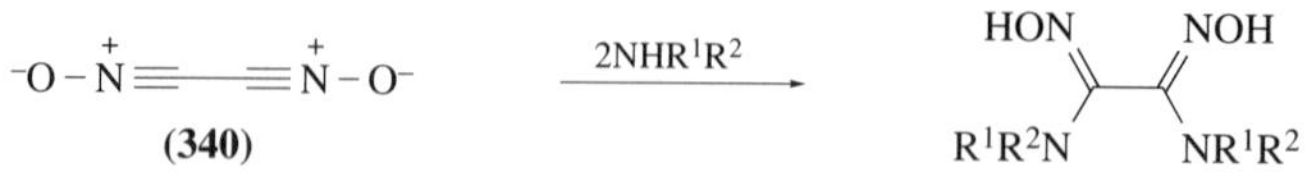

(340) (98)

5.21.2.4.8 *From aliphatic nitro compounds*

Nitrosation of the methylene group activated by a nitro group, such as primary nitro compounds (**341**), with alkyl nitrites produces nitrolic acids (**342**) (Equation (99)) ⟨1873CB1492, 1874CB425, 1874CB670, 1874CB790, 1875CB114, 1875(175)LA93, 1882(214)LA329, 1898CB2854, 10JPR209⟩, which can be reduced to oxyimidic amides (see below).

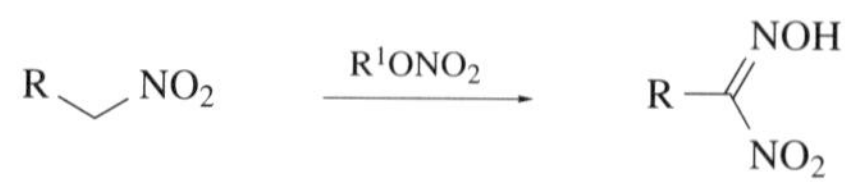

(99)

(**341**) R = alkyl, subat.alkyl (**342**)

5.21.2.4.9 *From oxyamide oximes*

Oxyamide oximes (**343**), prepared usually from oxyimidic halides and hydroxylamine, are reduced by sulfur dioxide to the corresponding oxyimidic amides (**344**) (Equation (100)) ⟨14CB2938⟩.

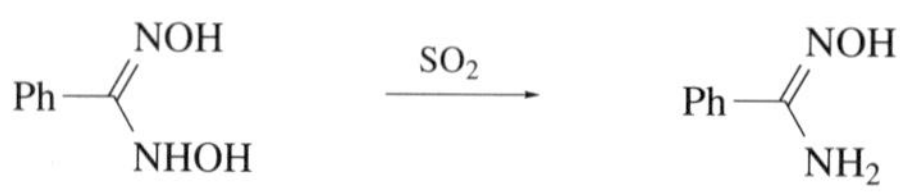

(100)

(**343**) (**344**)

5.21.2.4.10 *From nitrosolic and nitrolic acids*

The reduction of nitrosolic acids (**345**) with hydrogen sulfide gives the corresponding oxyimidic amides (**346**) (Equation (101)) ⟨06CB1480⟩. Monooxyimidic amides (**348**) of dicarboxylic acids have been prepared by reduction of mononitrolic derivatives (**347**) of dicarboxylic acids (Equation (102)) ⟨58USP2851489⟩.

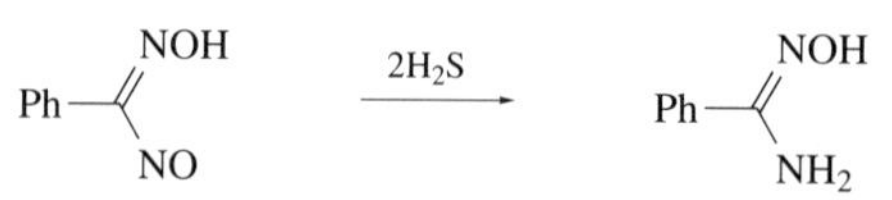

(101)

(**345**) (**346**)

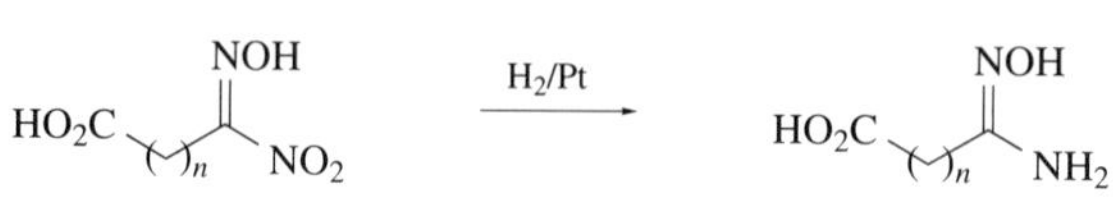

(102)

(**347**) (**348**)

5.21.2.4.11 *Preparation of nitrolic acids*

Nitrolic acids (**350**) are prepared by addition of dinitrogen tetraoxide to oximes (**349**) ⟨02CB3884, 61IZV477⟩, nitronic acids ⟨52DOK(83)243, 60IZV1783⟩ or nitronate salts ⟨58ZOB2305⟩ (Equation (103)). The reaction has been investigated in detail for benzaldehyde oxime ⟨1890CB3490⟩, acetaldehyde oxime, and propionaldehyde oxime ⟨1899CB3137⟩. All aryl-substituted nitrolic acids are formed as an isomeric mixture of (*E*)- and (*Z*)-isomers, which can be separated ⟨91JCS(P2)249⟩. The (*E*)-isomer of acetonitrolic acid (**352**) is prepared from nitroethane (**351**) and sodium nitrite in the presence of a base at low temperatures. The (*E*)-isomer of the *O*-acetyl derivative (**353**) can be completely converted by UV irradiation into the (*Z*)-isomer (**354**) (Scheme 41) ⟨91JCS(P2)249⟩.

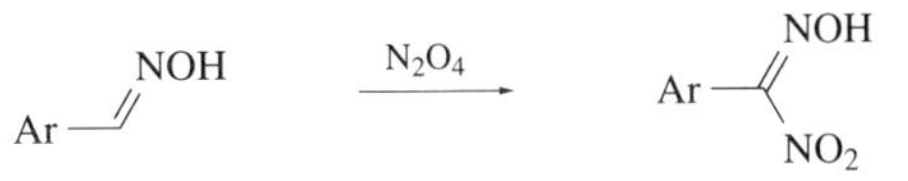

(103)

(349) **(350)**

Ar = Ph, 4-$NO_2C_6H_4$, 2-$NO_2C_6H_4$, 3-$NO_2C_6H_4$, 4-ClC_6H_4

(351) **(352)** **(353)** (*E*) **(354)** (*Z*)

Scheme 41

5.21.2.5 Oxyimidic Phosphorus Derivatives

5.21.2.5.1 From phosphanes

The oxyimidic phosphorus structural element is known only in cyclic structures, formed usually by the cycloaddition reaction. For example, (2,2 dimethylpropylidyne)phosphane (**355**) undergoes a regiospecific [3 + 2]cycloaddition reaction with nitrile oxides to form 1,2,4-oxazaphospholes (**356**) and the diadduct (**357**) (Equation (104)) ⟨86S31⟩. Related cycloadditions have also been described ⟨85ZN(B)1077, 87S689⟩.

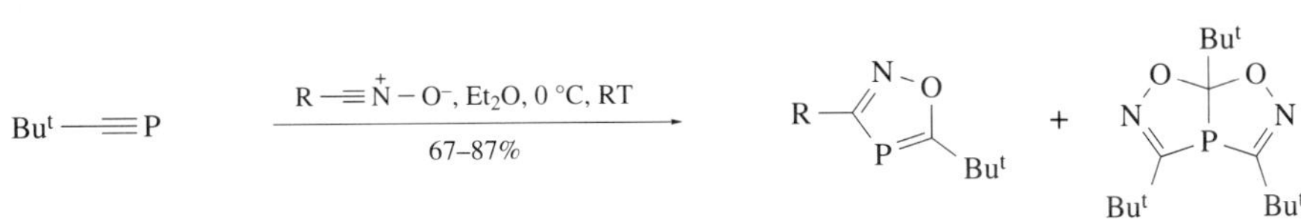

(104)

(355) **(356)** **(357)**

5.21.3 HYDRAZONOYL DERIVATIVES

5.21.3.1 Hydrazonoyl Halides

5.21.3.1.1 From fluoroalkenes

Fluoroalkenes (**358**) and (**360**) with a difluoromethylene terminal group react with hydrazine hydrate in a nonpolar solvent, such as ether or dioxane, by heating in an autoclave for 8 h at 60°C to form the corresponding hydrazonoyl fluorides (**359**) and (**361**) (Scheme 42) ⟨56USP2817662, 58JA5793⟩. However, they are transformed during the reaction or by standing at room temperature, into 1,2-dihydro-1,2,4,5-tetrazine derivatives (**362**). The stable hydrazonoyl fluorides (**363**) and (**364**) are obtained from fluoroalkenes with substituted hydrazines, such as phenyl hydrazine or semicarbazide (Scheme 42) ⟨62HOU(5/3)1⟩.

5.21.3.1.2 From hydrazones

Hydrazones (**365**) are readily transformed into hydrazonoyl halides with a variety of halogenating agents. By halogenation with bromine in chloroform at low temperature the corresponding hydrazonyl bromides (**366**) are obtained (Equation (105)) ⟨69G69, 72LA(758)162⟩.

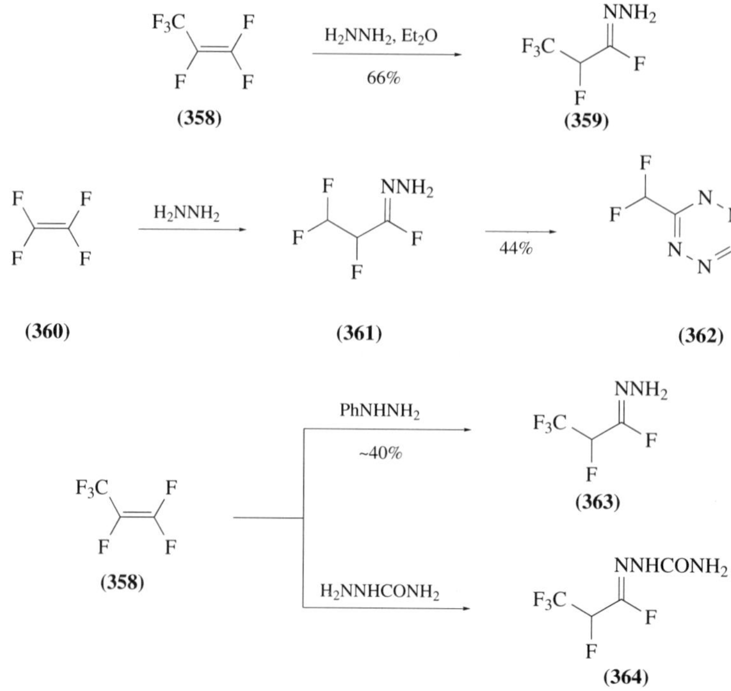

Scheme 42

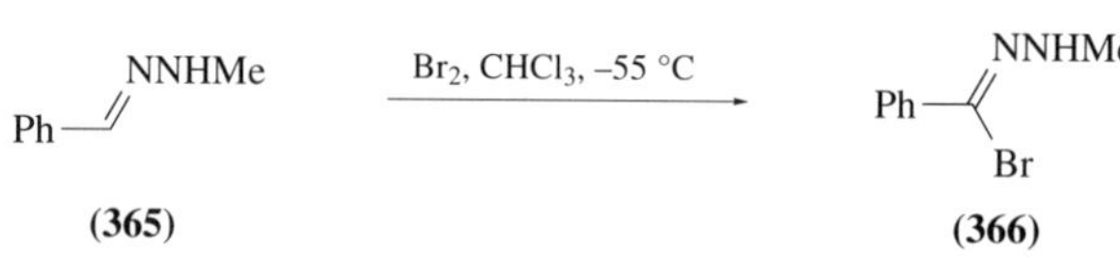

(105)

Chlorination of benzaldehyde phenylhydrazone (**367**) in acetic acid or in tetrachloromethane also gives products with chlorine incorporated into the aromatic ring, dependent on the ratio between the substrate and chlorine; 2,4-dichloro- (**368**) and 2,4,6-trichlorohydrazonoyl chloride (**369**) derivatives are formed. The 2,4-dibromo derivative (**370**) is obtained by bromination when the ratio between the substrate and bromine is 1 : 3 (Scheme 43) ⟨25JCS975, 25JCS1304, 65TL3801⟩.

Glyoxal diphenylosazone (**371**) is transformed by chlorination into the bishydrazonoyl derivative

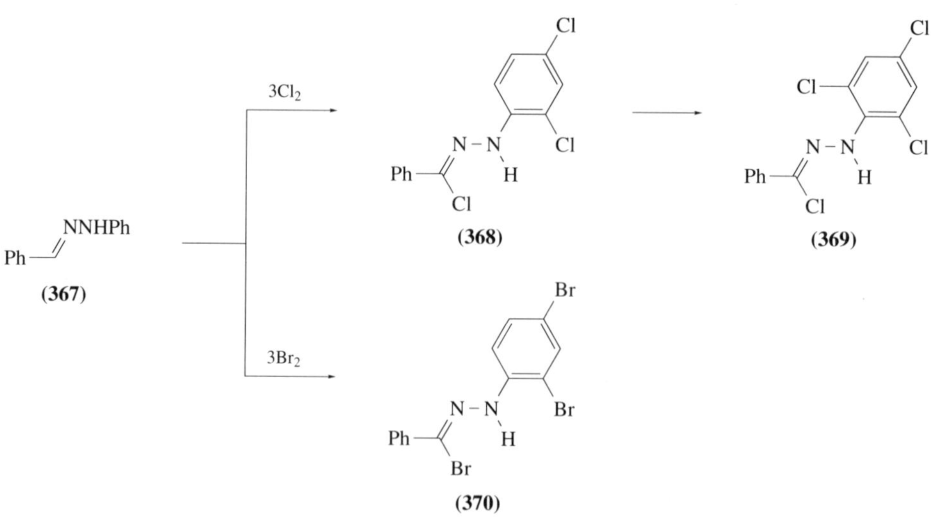

Scheme 43

(**372**) containing chlorine at position 2 and 4 in both aromatic rings (Equation (106)) ⟨30JCS94⟩. In some instances, addition of chlorine occurs also in unsaturated side chains. For example, when a solution of α,β-dichlorocrotonaldehyde 2,4,6-trichlorophenylhydrazone (**373**) is left for 24 h at room temperature in acetic acid saturated with chlorine, polychlorinated hydrazonoyl chloride (**374**) is formed (Equation (107)) ⟨32JA263⟩.

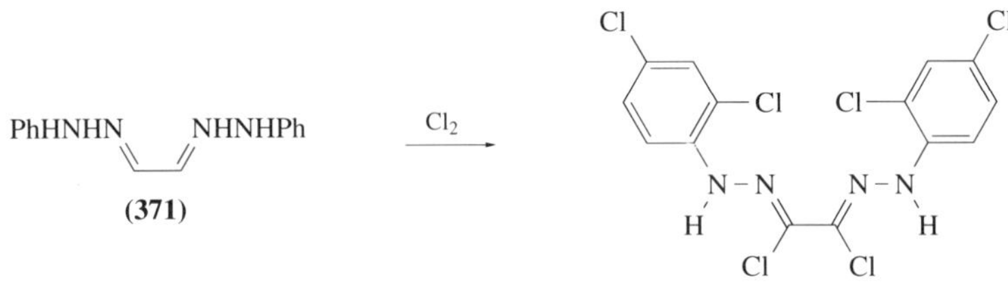

(**371**) (**372**) (106)

(**373**) (**374**) (107)

Chlorination of hydrazone (**375**) with hypochlorite results in the formation of isolable *N*-chloro hydrazonoyl chloride (**376**), which can be transformed by heating into hydrazonoyl chlorides (**377**) substituted either in the aromatic ring or side chain (Equation (108)) ⟨25JCS1687⟩.

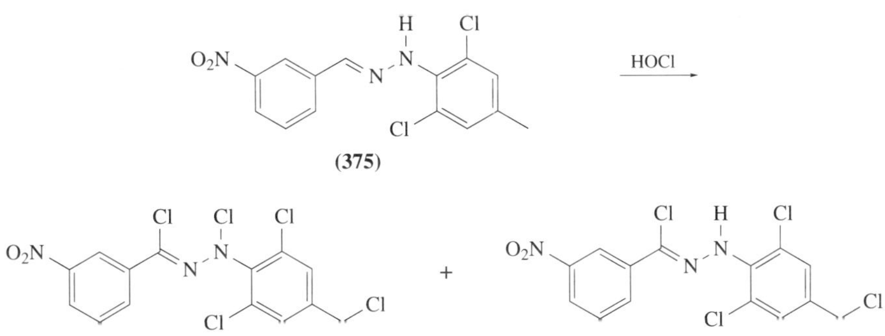

(**375**) (**376**) (**377**) (108)

Benzaldehyde hydrazones, prepared from heterocyclic hydrazino compounds and benzaldehyde, react easily with bromine in acetic acid in 1 : 1 molar ratio. For example, benzaldehyde tetrazolyl-5-hydrazone (**378**) gives the corresponding hydrazonoyl bromide (**379**) in 72% yield (Equation (109)) ⟨57JOC692⟩.

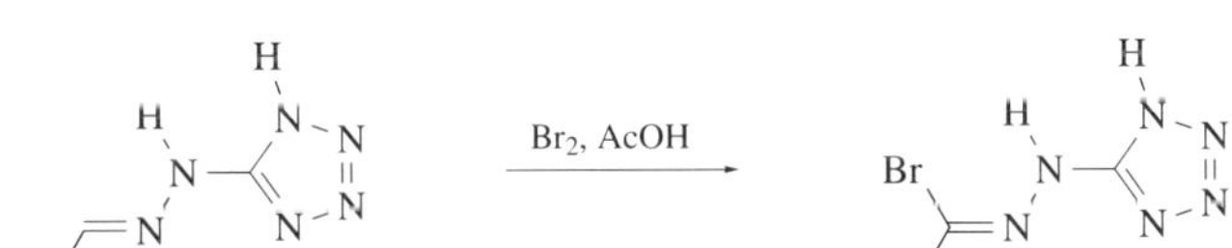

(**378**) (**379**) (109)

Chlorination of 2-hydrazones of 1,2,3-tricarbonyl compounds (**380**), usually prepared from 1,3-dicarbonyl compounds or β-keto esters and diazonium salts, in acetic acid or ethanol, gives the corresponding hydrazonoyl chlorides of oxalic acid derivatives (**381**) by substitution of the acyl group with chlorine. At the same time, one or two chlorine atoms in position 4 or in positions 2 and 4 are introduced into the aromatic nucleus, dependent on the ratio between hydrazone and chlorine (Equation (110)) ⟨12CB3732, 13CB2032, 17CB1478⟩. Chlorination of 2-phenylhydrazone (**382**) in free acetoacetic acid is followed by decarboxylation to give hydrazonoyl chloride of pyruvic acid (**383**) (Equation (111)) ⟨13CB2370⟩. The hydrazones of dioxobutyric esters (**384**) and dioxosuccinic esters (**385**) lose the acyl function by conversion into the corresponding hydrazonoyl bromides (**386**) and (**387**) with bromine in boiling acetic acid (Equations (112) and (113)) ⟨33JCS1389, 46JA588⟩.

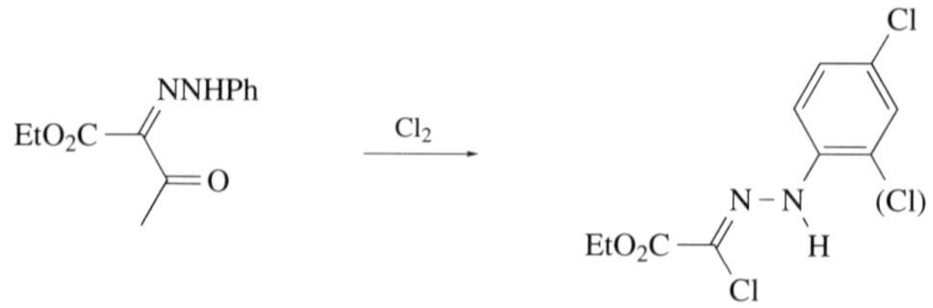

(**380**) (**381**) (110)

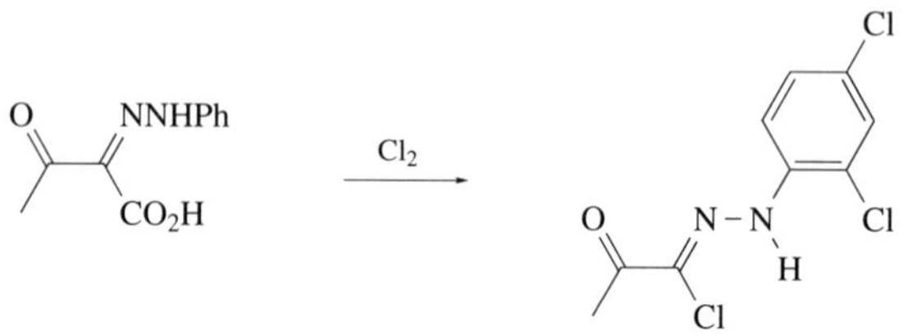

(**382**) (**383**) (111)

(**384**) (**385**) (112)

(**386**) (**387**) (113)

Bromination of acetophenone phenylhydrazone (**388**) with excess bromine results in elimination of the methyl group to give the hydrazonoyl bromide (**389**), brominated in both aromatic rings (Equation (114)) ⟨33JCS1389⟩.

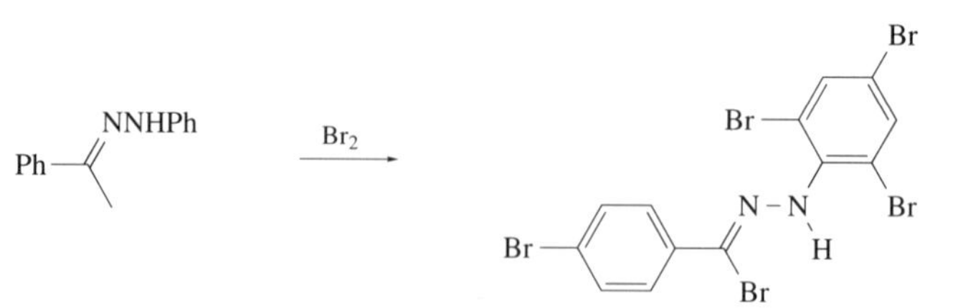

(**388**) (**389**) (114)

5.21.3.1.3 *From hydrazides*

Hydrazonoyl halides (**392**) are easily obtained from hydrazides (acylhydrazines) (**390**) only in aromatic series by treatment with phosphorus pentachloride in ether. At room temperature a phosphorus-containing intermediate (**391**) can be isolated, which gives after cleavage with phenol the products in 58–85% yield (Scheme 44) ⟨1894CB320, 1894CB2121, 62T3, 66JHC155, 67CB71, 67HOU(10/2)378⟩.

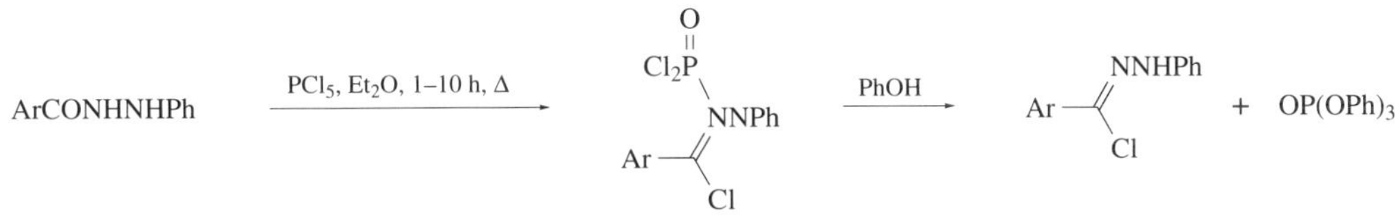

(390) **(391)** **(392)**

Scheme 44

5.21.3.1.4 From 2-halo-1,3-dicarbonyl compounds

2-Chloro-(or bromo)-1,3-diketones or 2-chloro-(or bromo)-β-keto esters (**393**) react with diazonium salts to give hydrazonoyl halides (**394**). In this manner, chloroacetone ⟨27BSF1494⟩, ω-chloroacetophenone ⟨63ZOB3917⟩, chloromalonic acid ⟨64UKZ68⟩, ethyl 2-chloroacetoacetate ⟨02CR1312, 65HOU(10/3)538⟩, 2-chloro acetoacetanilide ⟨24LA(439)211⟩, chloromalondialdehyde ⟨05CB2986⟩, and 3-chloroacetylacetone ⟨27BSF1603⟩ react with aryldiazonium salts to give after the cleavage of the acyl group the corresponding hydrazonoyl chlorides or bromides, respectively, in 36% to practically quantitative yields (Equation (115)).

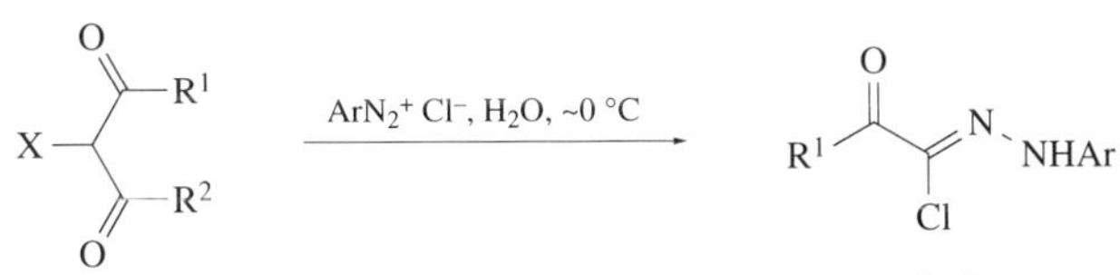

(115)

(393) **(394)**

R^1 = H, Me, OEt, OH Ar = Ph, 2-NO_2-C_6H_4
R^2 = H, Me, Ph, OH X = Cl, Br

The coupling of chloromalonic dialdehyde (**395**) with aryldiazonium salts gives hydrazonyl chloride of glyoxylic acid (**396**). The coupling of chloromalonic acid (**397**) with aryldiazonium salts in acetic acid affords normally the corresponding formazanes. However, in sulfuric acid a mixture of hydrazonoyl halide derivative (**398**) in 50% yield together with the formazane derivative is formed. In analogous manner a large number of oxalic acid hydrazonoyl chlorides can be prepared (Scheme 45) ⟨61ZOB1621⟩.

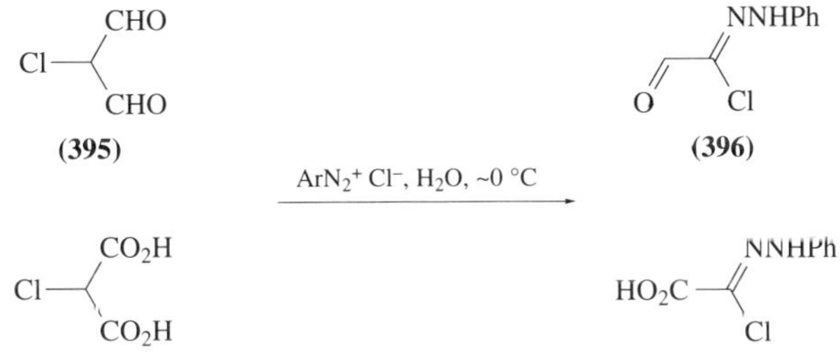

Scheme 45

5.21.3.1.5 From aliphatic diazo compounds

Aliphatic diazo compounds, such as diazo ketones and diazo esters, react with aryl diazonium salts to form hydrazonoyl halides. For example, ethyl diazoacetate (**399**), diazomethane (**401**) and diazoacetone (**403**) react with 4-nitrobenzenediazonium chloride in methanol to give the corresponding hydrazonoyl chlorides (**400**), (**402**) and (**404**) in 64%, 81%, and 32% yield, respectively (Scheme 46) ⟨55LA(591)190, 67HOU(10/2)467⟩.

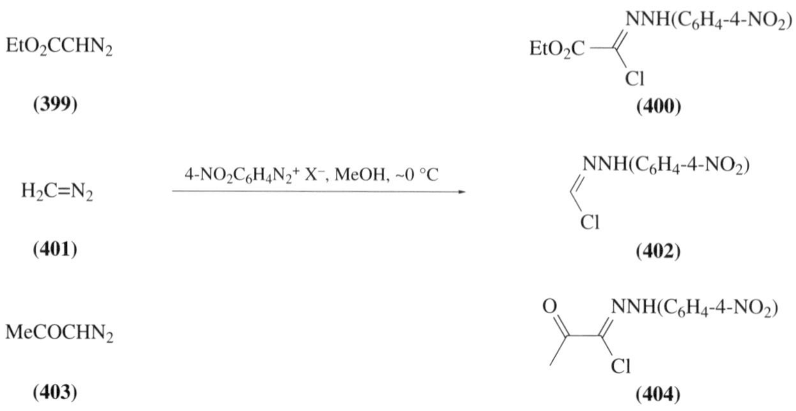

Scheme 46

5.21.3.1.6 Preparation of 1-diazo-1-haloalkanes

Chloro- or bromodiazomethane is prepared in the reaction of diazomethane with *t*-butyl hypochlorite or hypobromite ⟨62JA4350, 65JA4270⟩. Ethyl bromo- (**405**), chloro- (**406**) and iododiazoacetates (**407**) are obtained by treatment of silver or mercury derivatives of ethyl diazoacetate with iodine, cyano bromide, sulfuryl chloride, or bromine (Scheme 47) ⟨68LA(716)204, 69CB488, B-77MI 521-01⟩.

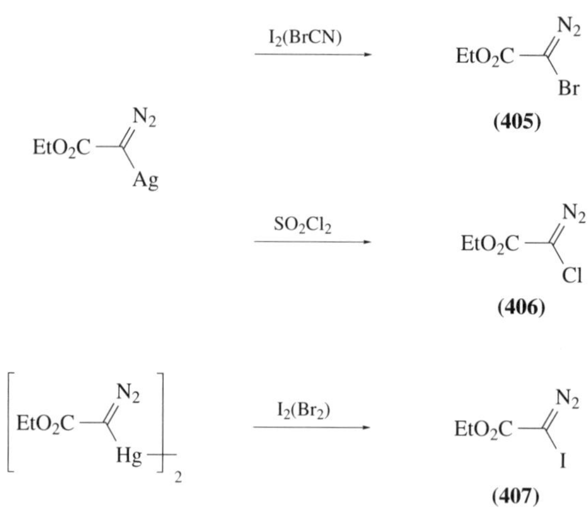

Scheme 47

5.21.3.2 Hydrazonic Acid Derivatives

5.21.3.2.1 From aromatic hydrazonoyl halides

Hydrazonates can be readily prepared from the corresponding hydrazonoyl halides (**408**) only when an electron-withdrawing group is attached to the aromatic ring (Equation (116)). Reactions in sodium acetate-buffered solutions of ethanol give high yields of the hydrazonates and are least complicated by side reactions. Although the use of triethylamine provided a rapid method for the synthesis of these compounds, workup is hindered by the presence of quantities of tar ⟨65TL841, 70JCS(C)968⟩. The analogues (N^1-4-tolylsulfonyl)hydrazonates can be prepared in the same manner ⟨65CJC126⟩.

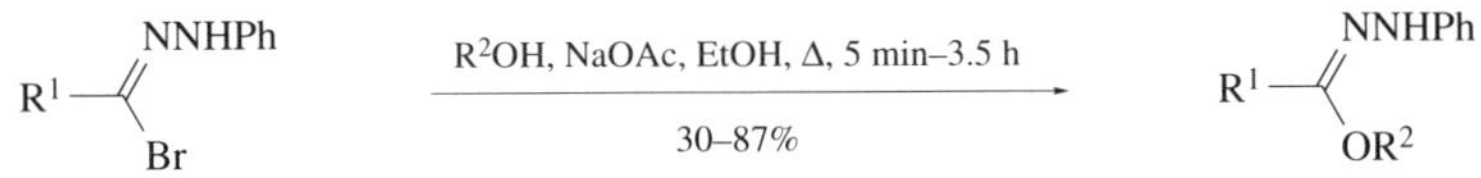

(408) (116)

R^1 = Ph, 4-Pr^i-C_6H_4, 4-Cl-C_6H_4, 4-NO_2-C_6H_4; R^2 = Me, Et, Pr

5.21.3.2.2 *From heteroarylhydrazides*

Heteroaryl-substituted hydrazonates (**410**) are prepared from heteroarylhydrazides (**409**) and triethyl *ortho*-formate (Equation (117)) ⟨86H(24)1899⟩.

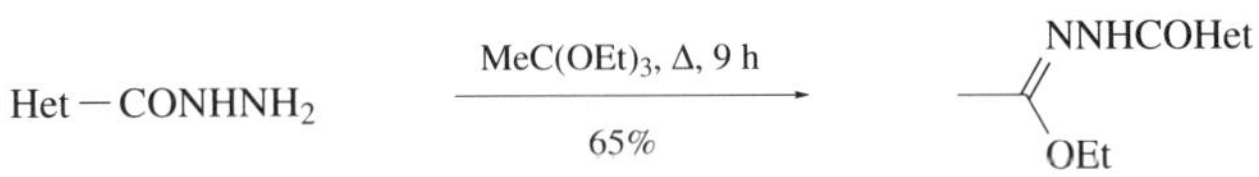

(409) **(410)** (117)

5.21.3.2.3 *From nitrene precursors*

Aliphatic hydrazonates, which cannot be obtained by the above methods, can be prepared by another method. Ethoxycarbonyl nitrene (**412**), prepared from (**411**) in the presence of a base reacts with the NH bond of imidates to form hydrazonates (**413**) (Scheme 48). The base can be the imidate itself, but triethylamine is more useful because of its greater basicity and the insolubility of its 4-nitrobenzenesulfonate salt. The best yields are obtained in concentrated dichloromethane solutions and a 5–8 times excess of the imidate. (*E*)- and (*Z*) isomers can be separated by crystallization ⟨73CC932⟩.

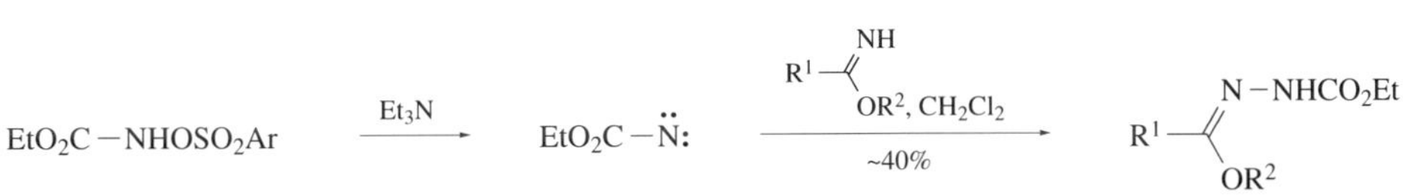

(411) **(412)** **(413)**

Ar = 4-NO_2-C_6H_4; R^1 = Me; R^2 = Me, Et

Scheme 48

5.21.3.2.4 *From ketene acetals*

Ketene acetals (**414**) react with aromatic hydrazines in the presence of fused zinc chloride to give the corresponding hydrazonates (**415**) (Equation (118)) ⟨55JA4571⟩. Dicyanoketene acetal (**416**) reacts similarly with 1,1-disubstituted hydrazines to give (**417**), while monosubstituted or unsubstituted hydrazine react further to give cyclic products (Equation (119)) ⟨58JA2829⟩.

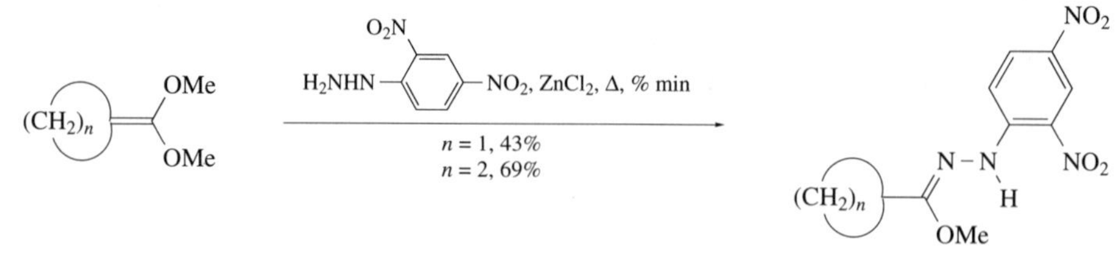

(414) (415) (118)

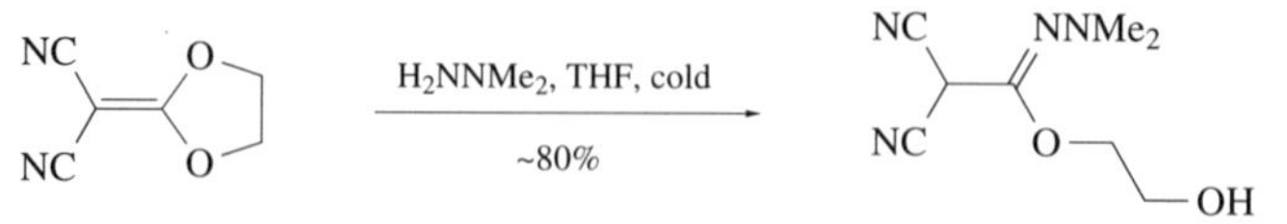

(416) (417) (119)

5.21.3.3 Thio- and Selenohydrazonic Acid Derivatives

For a review see ⟨85HOU(E/5)1300⟩.

5.21.3.3.1 From thiohydrazides

Alkylation of thiohydrazides (**418**) with alkyl halides takes place at the sulfur atom to give N[1]-alkyl or aryl hydrazonide esters of thiocarboxylic acids (**419**) (Equation (120)). The products of unsubstituted hydrazides are rarely isolable, since they cyclize under the reaction conditions into 1,4-dihydrotetrazine derivatives ⟨54BCJ624, 54JPJ821, 63ACS95, 65ACS2395, 68TL2463⟩.

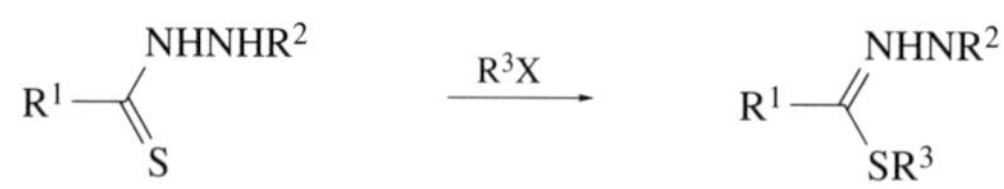

(418) (419) (120)

Methylation of the (*Z*)- and (*E*)-forms of the diethyl hydrazide of thioformic acid (**420**) and (**422**) gives the (*Z*)- and (*E*)-forms of diethyl hydrazonic thioester (**421**) and (**423**) (Scheme 49) ⟨69CB2117⟩.

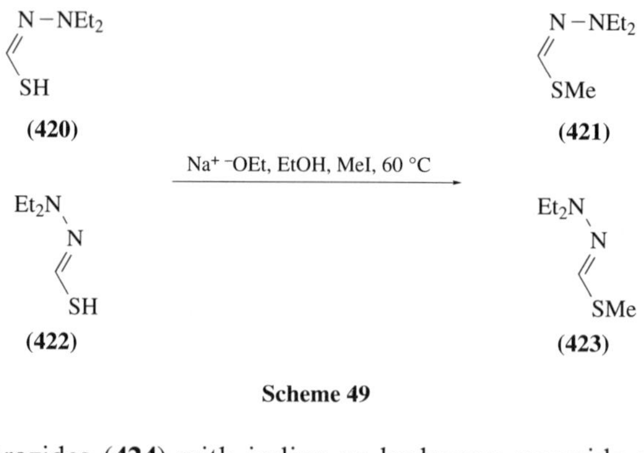

Scheme 49

Oxidation of thiohydrazides (**424**) with iodine or hydrogen peroxide gives the corresponding disulfides (**425**) ⟨76JCS(P1)38⟩, which, when *N*-unsubstituted, cyclize in acidic media at elevated temperatures into 1,3,4-thiadiazole derivatives (**426**) (Scheme 50) ⟨65JOC3739⟩.

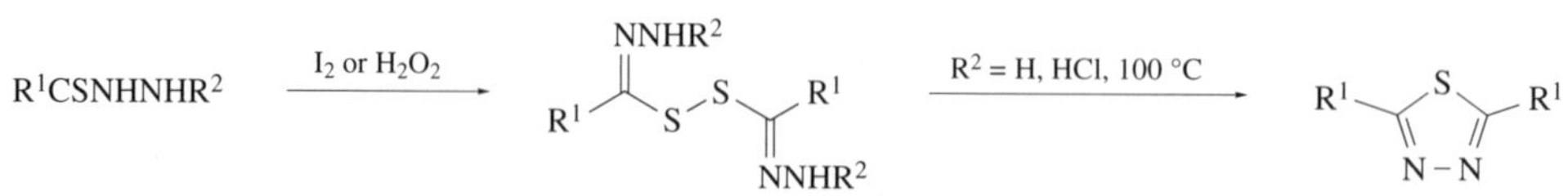

(424) (425) (426)

Scheme 50

5.21.3.4 Hydrazonamides and Related Structures

There are several comprehensive reviews about this type of compound. The most important of these reviews are by Watson and Neilson 〈B-75MI 521-02〉, Perrin 〈B-91MI 521-01〉, and Neilson *et al.* 〈70CRV151〉.

There are two general types of methods for preparing these types of compounds. (a) Addition reactions in which a hydrazine or its derivative is added to a cyano group or to a carbon of the multiple bond, such as in a ketimine or carbodiimide. (b) Substitution reactions in which either a suitable leaving group, attached to an imino carbon, is substituted by a hydrazine residue, or reactions in which ammonia or an amine reacts as a nucleophile attacking a hydrazone with a suitable leaving group.

Besides these two general principles, hydrazonoamides are obtained by the ring opening of various heterocyclic systems, usually after quaternization, or from the compounds with a suitable structural element, as demonstrated by the reduction of formazans.

5.21.3.4.1 From nitriles

(i) Reactions of nitriles with hydrazine

The nucleophilic attack of hydrazine, mono- and disubstituted hydrazine on a nitrile (**427**) with an electron-withdrawing group results in the formation of hydrazonamides (**428**) as the primary products (Equation (121)). The electron-withdrawing groups are heterocyclic 〈65JOC931, 66JOC2398〉, such as 2-cyanothiazole 〈56JCS2253〉, 2-cyanopyridine 〈65JOC931, 66JOC2398〉, 3-cyanoisoquinoline 〈56JCS2253, 67JHC483〉, and perfluoroalkyl groups 〈60JA4700〉, including the synthesis of fluorinated polymers containing hydrazonamide groups 〈62USP3061590〉, while 4-cyanopyridine does not react under the same conditions 〈56JCS2253〉.

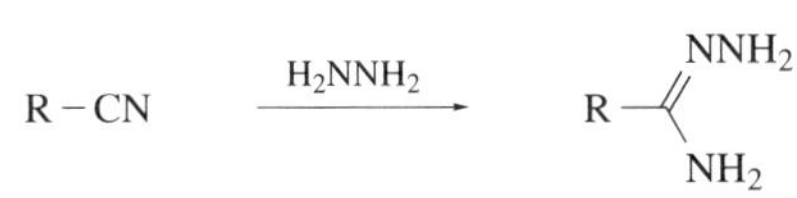

(121)

(427) **(428)**

Anhydrous hydrazine 〈70JPB1696〉 and hydrazine hydrate 〈69JPS(A)945〉, in an alcohol usually at room temperature, have been used in all these reactions. Accordingly, aminoguanidine (**429**) can be prepared from cyanamide (**430**) (Equation (122)) 〈14G72〉, while cyanogen (**431**) reacts with hydrazine in aqueous media to give, dependent on the reaction conditions, either cyanohydrazoamides (**432**) or oxaldihydrazonamides (**433**) (Scheme 51) 〈59CRV841〉.

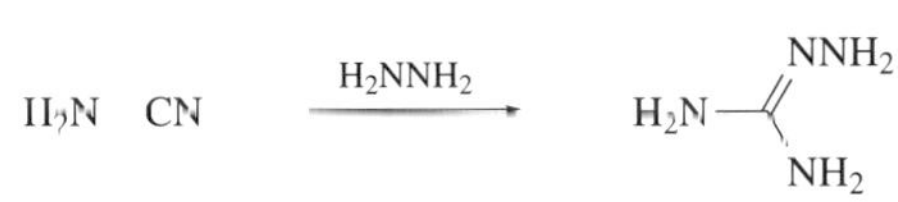

(122)

(429) **(430)**

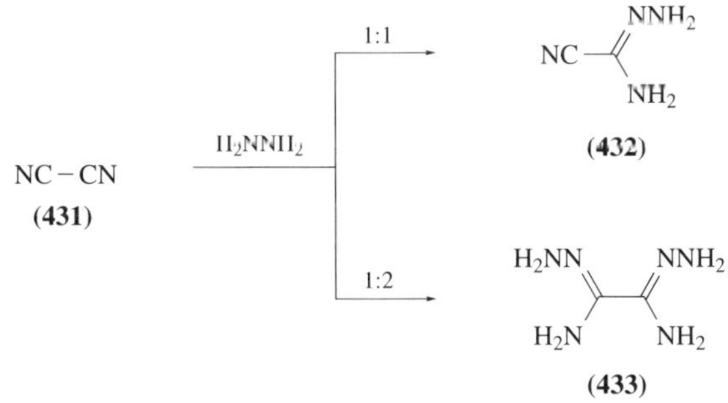

Scheme 51

In this type of reaction some secondary products are formed, for example dihydrazidines (**435**), by the addition of hydrazonamide to the starting nitrile (**434**). These can cyclize by loss of ammonia into triazole derivatives (**436**) (Scheme 52) ⟨56JCS2253, 65JOC318, 72JPJ275⟩.

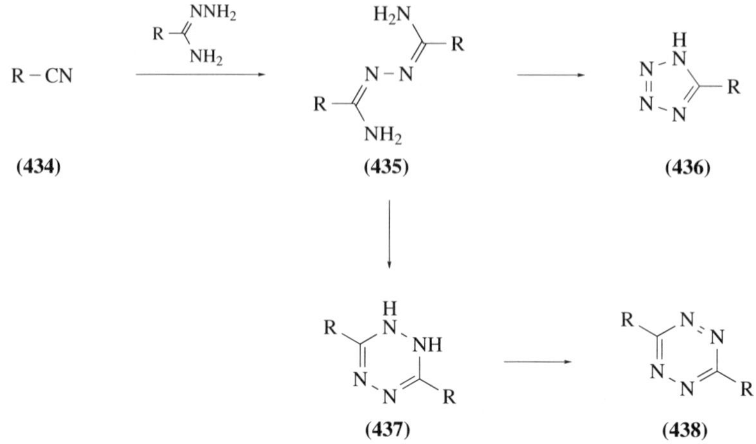

Scheme 52

Hydrazonamides, in the absence of hydrazine, can undergo self-condensation to form dihydrazidines (**435**), dihydrotetrazines (**437**), and tetrazines (**438**) by further oxidation. These secondary products are formed in particular, on prolonged reaction times, higher temperatures, excess of hydrazine, and on the nature of group R. Monosubstituted hydrazines give two types of products, N^1- (**439**) and N^2-substituted compounds (**440**), depending the nature of the group R^2 (Equation (123)). In most cases, N^1-substituted hydrazonamides are formed. 1,1-Disubstituted hydrazines also react similarly ⟨56JCS2253, 65JOC318, 72JPJ275⟩.

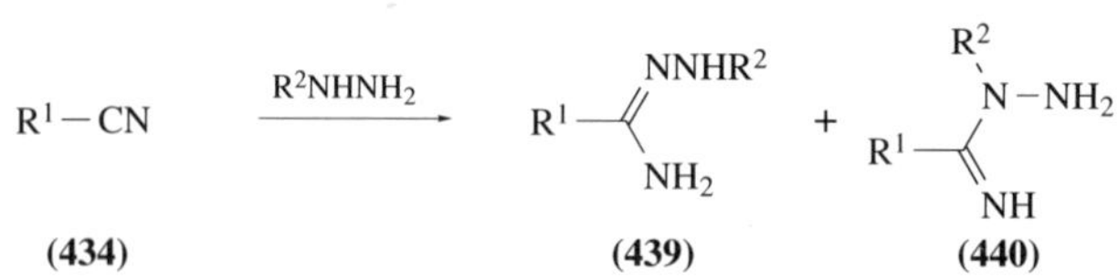 (123)

(ii) *Reactions of nitriles with hydrazine in the presence of alkali*

A modification to the above mentioned procedure is the reaction of nitriles with hydrazine in benzene in the presence of sodium. For example, benzonitrile and *N*-methyl-*N*-phenylhydrazine give *N*-methyl-*N*-phenylbenzohydrazonamide (Equation (124)).

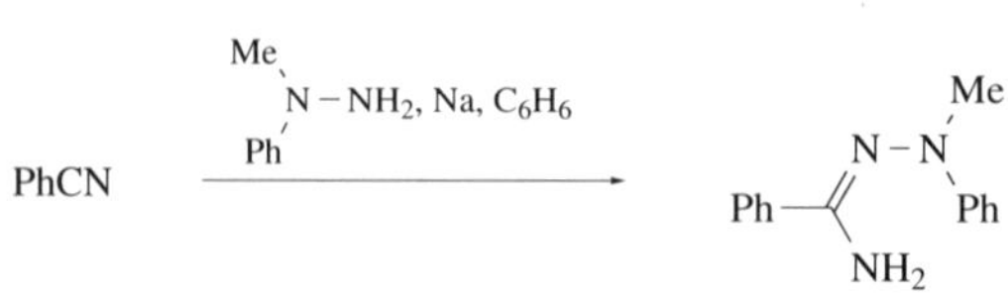 (124)

With equimolar quantities of phenylhydrazine the corresponding N^1-phenylhydrazonamide is obtained, while with two equivalents of phenylhydrazine a triazole derivative is formed ⟨1896JPR54⟩.

The reaction of nitriles with hydrazine in the presence of alkali is general, and many aliphatic and aromatic nitriles and dinitriles react with sodium hydrazide in inert solvents to give the corresponding hydrazonamides in good yields ⟨63AG(E)217, 64CB3436, 66CB2600⟩. The method is advantageous, since the less reactive nitriles, which form by direct reaction side products, can be employed. Lithium ⟨71ZN(B)296⟩ and barium hydrazide ⟨72UKZ171⟩ have been used instead of sodium hydrazide. In these cases, benzonitrile affords diphenylhydrazine, while phenylacetonitrile gives phenylacetamidrazone.

5.21.3.4.2 From carbodiimides and related compounds

Ketimines (**441**) form hydrazonamides (**442**), by addition of hydrazine, in greater than 90% yield (Equation (125)) ⟨65JOC3718⟩. Carbodiimides (**443**) react with hydrazine in either 1 : 1 or 2 : 1 molar ratio to form the corresponding hydrazonamides (**444**) and (**445**) (Scheme 53) ⟨67CRV107, 68JCS(C)1375⟩.

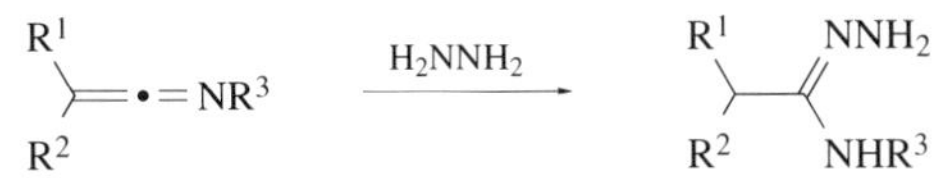

(125)

(**441**) (**442**)

$R^1 = R^2 = Ph$, R^3 = 4-Me-C_6H_4

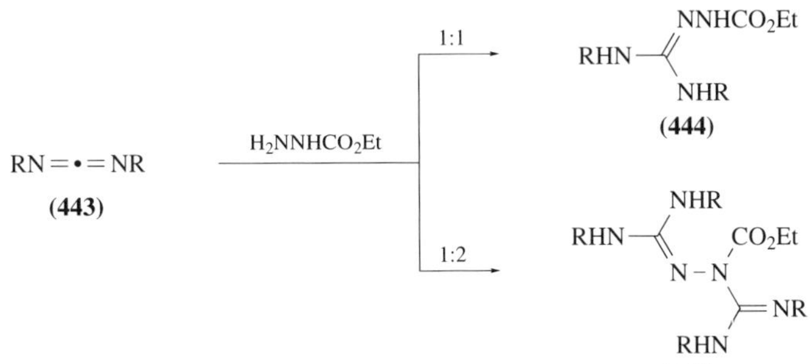

(**443**) (**444**) (**445**)

Scheme 53

5.21.3.4.3 From imidic esters

Imidic esters (**446**) are more reactive than nitriles, and therefore the reaction, in which the alkoxy group is substituted with hydrazine, has found wide application for the preparation of hydrazonamides (**447**) (Equation (126)) ⟨67CRV107, 70CRV151⟩.

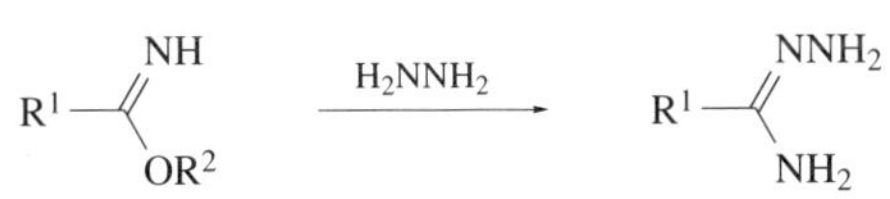

(126)

(**446**) (**447**)

However, in this reaction, besides the hydrazonamide, dihydrazides, triazole derivatives, *s*-tetrazines, and/or 4-amino-1,2,4-triazoles are formed, depending on the reaction conditions. Hydrazonamides are formed as major products, when an equimolar amount of imidic ester or its salt and hydrazine react at temperatures at 0 °C or below in an ethanol/ether solution. According to this method many unsubstituted hydrazonamides have been prepared from aliphatic, aromatic, and heterocyclic imidic esters ⟨61CRV179, 72JOC3958, 72LA(760)102⟩. *N*-Substituted imidic esters give, with hydrazine, *N*-substituted hydrazonamides ⟨72LA(755)101⟩, monosubstituted hydrazines react with equimolar quantities of imidic esters or their salts ⟨54JCS3319, 70CRV151⟩ and acylhydrazines form *N*-acylhydrazonamides (**448**) (Equation (127)) ⟨72JCS(P1)2927⟩.

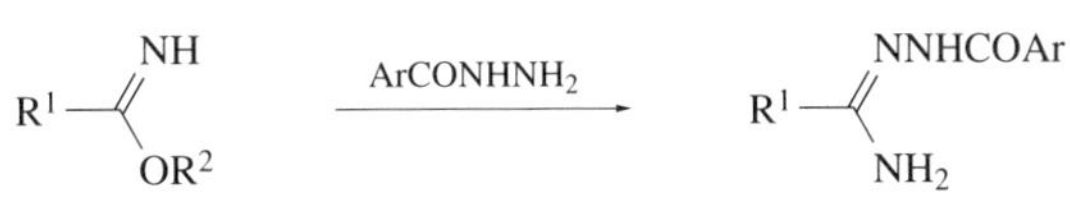

(127)

(**446**) (**448**)

Formylhydrazonamides (**449**) can be converted into the corresponding unsubstituted hydrazonamides (**450**) by acid hydrolysis of the formyl group (Equation (128)) ⟨68CB2028, 70JPS105⟩.

(449) **(450)**

(128)

Imidic esters (**451**) produce with 1,1-disubstituted hydrazines at room temperature the corresponding N^1,N^1-disubstituted hydrazonamides (**452**), whilst at higher temperatures in the presence of ammonium salts, dihydroformazans are formed (Equation (129)) ⟨69ZOR1404, 70CRV151⟩. Accordingly, a series of silylhydrazonamides and acrylhydrazonamides have been prepared ⟨73JGU317⟩.

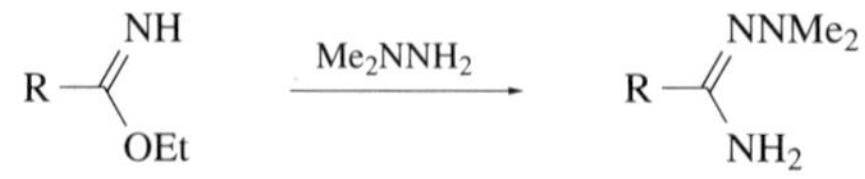

(451) **(452)**

(129)

N-Substituted imidic esters (**453**) give with substituted hydrazine N^1,N^3-disubstituted and N^1,N^1,N^3-trisubstituted hydrazonamides (**454**) (Equation (130)) ⟨68JIC729, 72LA(758)162⟩.

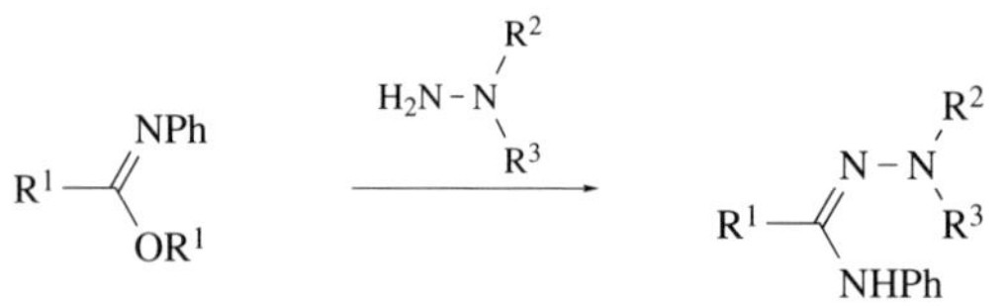

(453) **(454)**

(130)

5.21.3.4.4 From thioimidic esters

Cyclic thioimidic esters (**455**) form with hydrazine the corresponding hydrazonamides (**456**) (Equation (131)) ⟨68TL3569⟩. For further examples see ⟨84CHEC⟩.

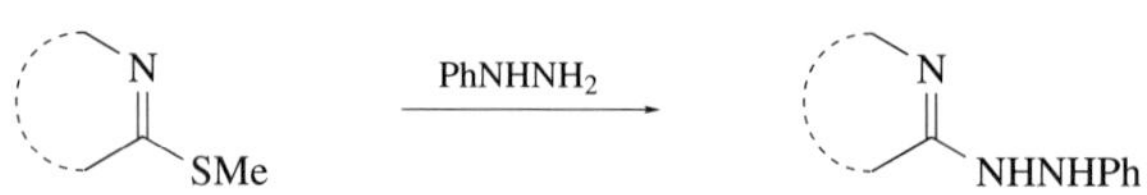

(455) **(456)**

(131)

5.21.3.4.5 From imidoyl halides

Monosubstituted hydrazines, such as arylhydrazines, give with *N*-phenylbenzimidoyl chloride (**457**) two products, N^1- (**458**) and N^2-substituted hydrazonamides (**459**) (Scheme 54) ⟨70CRV151⟩. *N*-phenylbenzimidoyl chloride (**457**) yields with 1,1-dimethylhydrazine three products (**460**)–(**462**), dependent on the molar ratio, while with a large excess of methylhydrazine only N^2-substituted hydrazonamide (**460**) is formed (Scheme 54) ⟨71JOC1155, 73JOC1344⟩.

5.21.3.4.6 From amidines

Amidines are less useful than imidic esters and imidoyl chlorides for the preparation of hydrazonamides. Frequently, amidines or their salts are used for this purpose. Accordingly, amidines (**463**) and (**465**) are converted into hydrazonamides (**464**) and (**466**) (Scheme 55) ⟨60JA4700, 70CRV151, 85HOU(E/5)1304⟩.

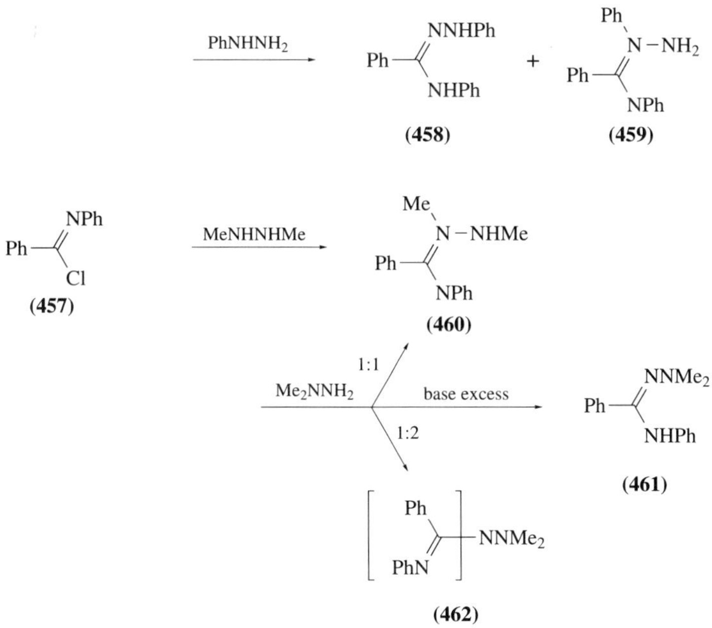

Scheme 54

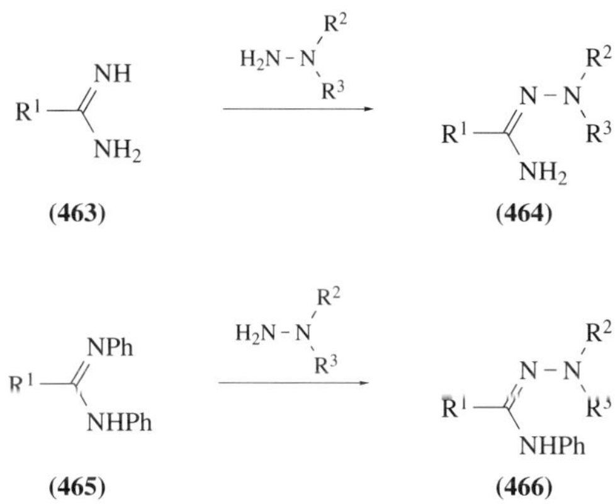

Scheme 55

N,*N*-Disubstituted amidines (**467**) react with phenylhydrazine at about 100°C to give N^1,N^3-disubstituted hydrazonamides (**468**) (Equation (132)) ⟨1898JPR223, 08JPR489⟩.

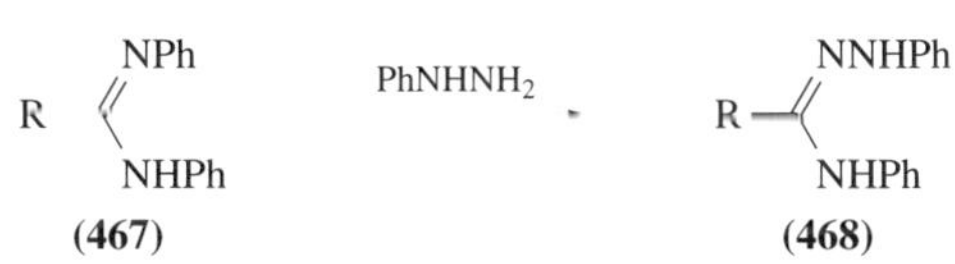

(132)

5.21.3.4.7 *From amides*

The transformation of amides such as (**469**) and (**472**) into hydrazonamides (**470**) and (**476**) requires the formation of activated intermediates such as (**471**) and (**473**) (Scheme 56) ⟨67JCS(C)690,

71JOC1155⟩, or the intermediate formation of imidic ester tetrafluoroborate (**474**) and amide acetal (**475**) ⟨61LA(641)1⟩. N^1,N^3-Disubstituted and N^1,N^2,N^3-trisubstituted hydrazonamides can be prepared according to this procedure ⟨68JOC1679⟩.

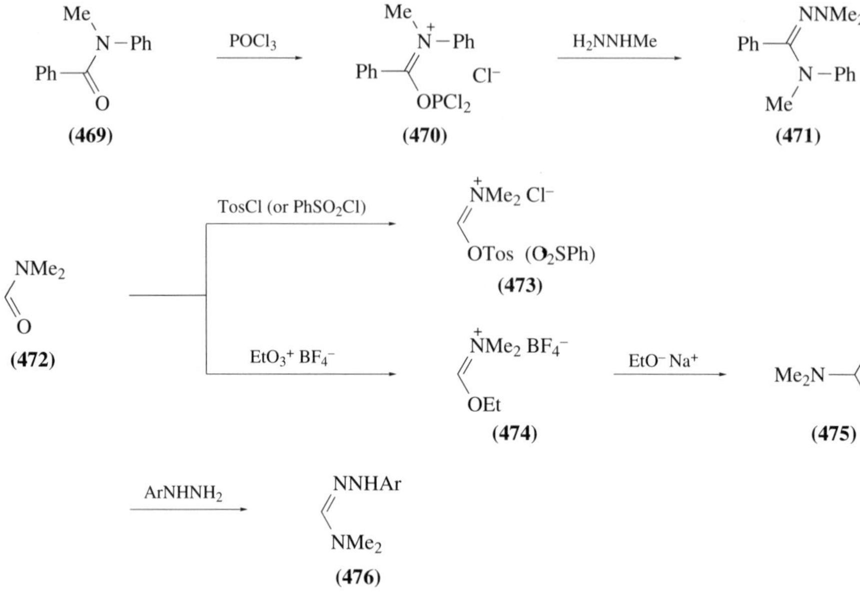

Scheme 56

5.21.3.4.8 *From thioamides*

Thioamides are used for the preparation of unsubstituted and substituted hydrazonamides. The reaction is dependent on the conditions. For example, isonicotinic thioamide (**477**) gives the corresponding hydrazonamide (**478**) by heating with hydrazine for a short time, while on prolonged treatment at room temperature the thiadiazole derivative (**479**) is formed (Scheme 57) ⟨55RTC257⟩. Arylthiocarboxanilides (**480**) yield hydrazonamides (**481**) in the cold, while on heating dehydrotetrazines (**482**) are obtained (Scheme 58) ⟨62JGU3330, 66CB3734⟩.

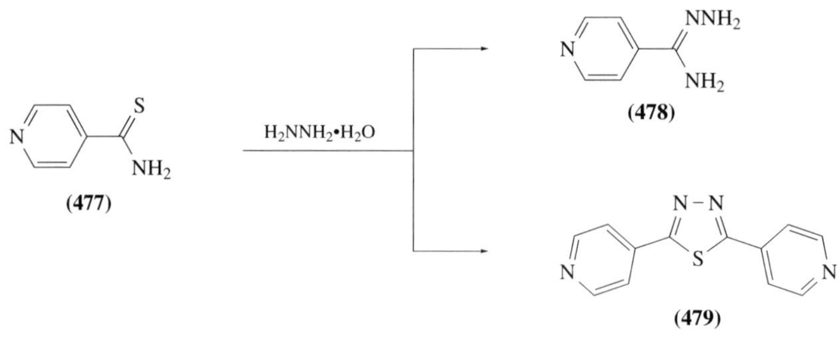

Scheme 57

Various oxalic acid derivatives containing the thioamide group such as (**483**), (**485**) and (**487**) have been converted into the corresponding hydrazonamides (**484**), (**486**), and (**488**) (Scheme 59) ⟨1893CB2385, 55HCA1560, 58JOC1931⟩.

5.21.3.4.9 *From hydrazonic esters and thioesters*

Hydrazonic esters (**489**) react with ammonia or amines to give hydrazonamides (**490**) (Equation (133)) ⟨68CB3947, 70LA(736)166⟩. However, amidines can be formed as side products ⟨66CB850⟩.

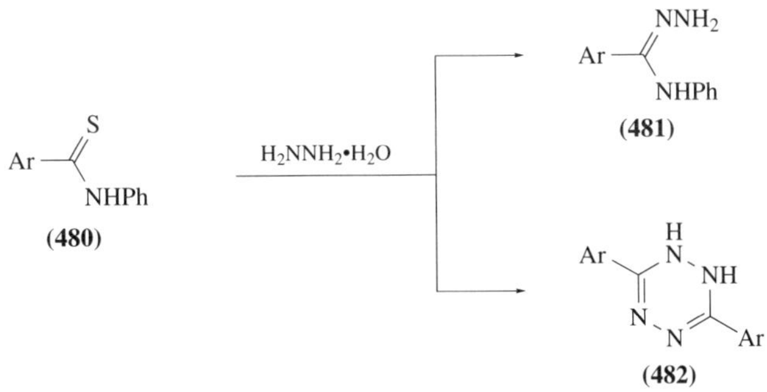

Scheme 58

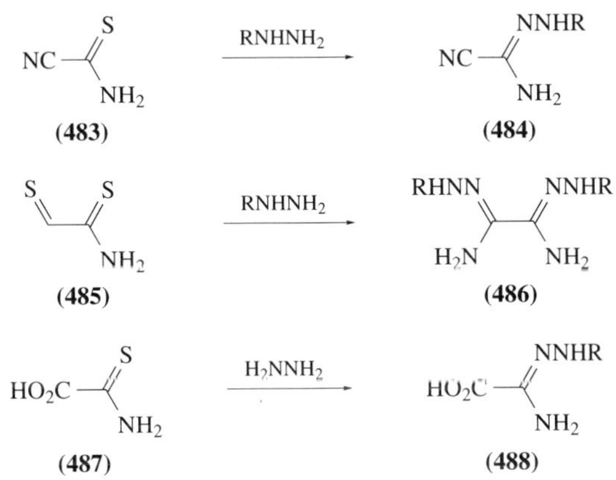

Scheme 59

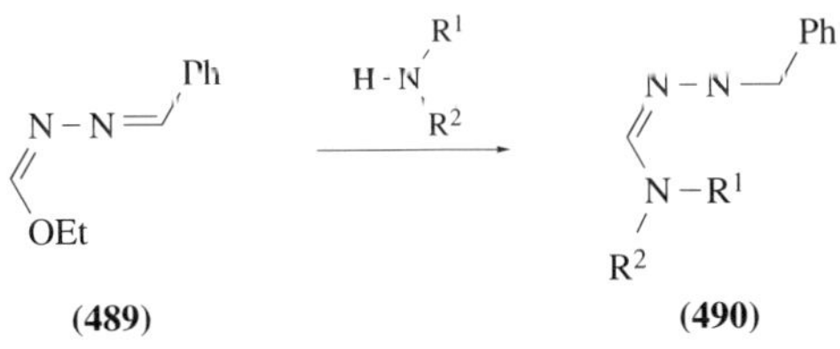

(133)

Ethyl thioformate (**491**) reacts with 1 methyl-1-phenylhydrazine in ethanol at 0°C to give hydrazonic ester (**492**). At higher temperatures, either in ethanol or in the absence of solvent a mixture of the corresponding thiohydrazide (**493**), hydrazonic ester (**492**) and hydrazonamides (**494**) and (**495**) are produced (Scheme 60) 〈70LA(736)166〉.

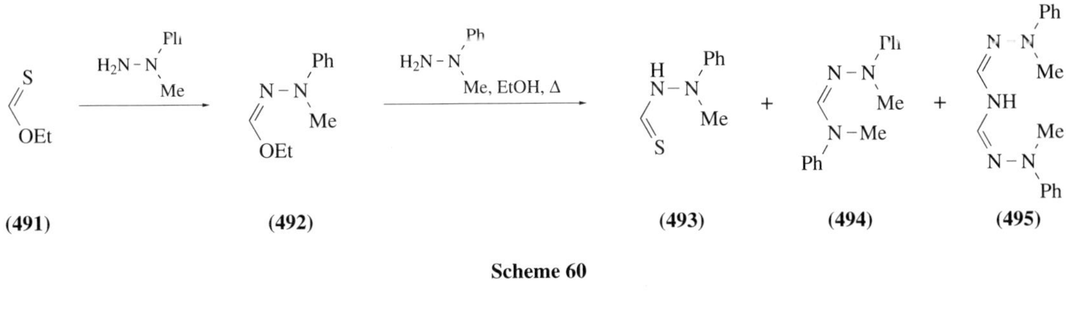

Scheme 60

5.21.3.4.10 From hydrazonoyl halides

Hydrazonoyl halides (**496**) react with ammonia and amines to give the corresponding hydrazonamides (**497**) (Equation (134)) ⟨69G69, 72GEP2146192⟩.

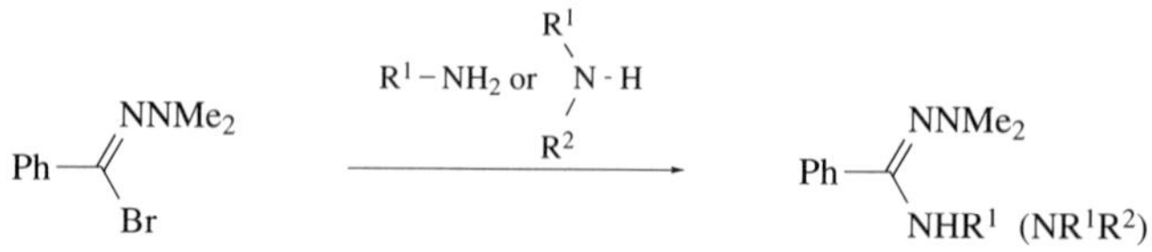

(496) **(497)** (134)

5.21.3.4.11 From 1-nitroaldehyde hydrazones

Substitution of the nitro group in 1-nitroaldehyde hydrazones (**498**) with ammonia results in the formation of hydrazonamides (**499**) (Equation (135)) ⟨10G77⟩.

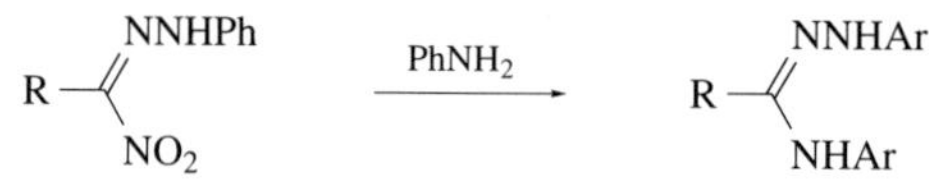

(498) **(499)** (135)

5.21.3.4.12 From N^3-unsubstituted hydrazonamides

Hydrazonamides (**500**) with unsubstituted amino group react with primary or secondary amines in the presence of ammonium salts as catalysts to give N^3-substituted derivatives (**501**) (Equation (136)) ⟨69ZOR1411⟩.

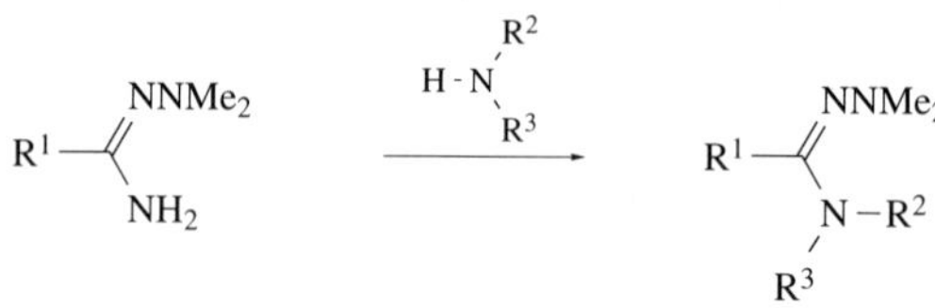

(500) **(501)** (136)

5.21.3.4.13 From acylaminomalonates

Hydrazonamides (**503**) are formed by coupling of aryldiazonium salts with acylaminomalonates (**502**) (Equation (137)) ⟨51LA(574)85, 61CB1868⟩.

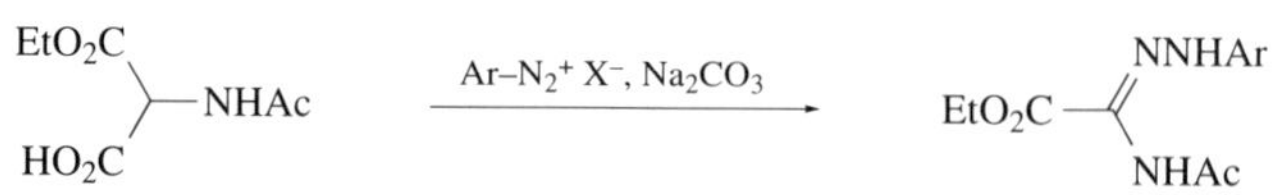

(502) **(503)** (137)

5.21.3.4.14 From heterocyclic precursors

Nucleophilic ring opening of 1,3,4-oxadiazole derivatives (**504**), possessing strong electron-withdrawing substituents, such as perfluoroalkyl groups, with ammonia produces hydrazonamides (**505**) (Equation (138)). The reactivity of the oxadiazole ring can be increased by converting it into a quaternary salt. The hydrazonamides frequently cyclize into triazoles or triazolium salts. Primary amines give the corresponding dihydrazidines ⟨62JOC3240, 71JCS(C)409⟩.

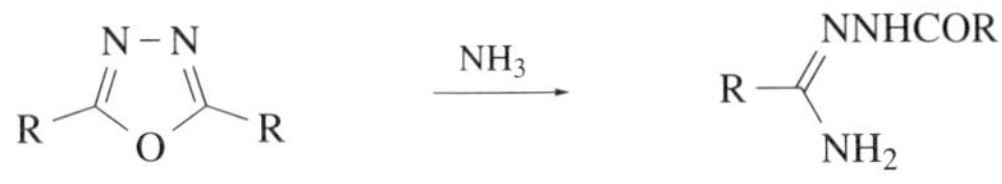

(504) (505) (138)

The 1,2,4-triazolium salts (**506**) give the corresponding hydrazonamides (**507**) under alkaline conditions (Equation (139)). However, the equilibrium between the open and cyclic structures in acidic media is shifted towards the cyclic structure ⟨69G69⟩.

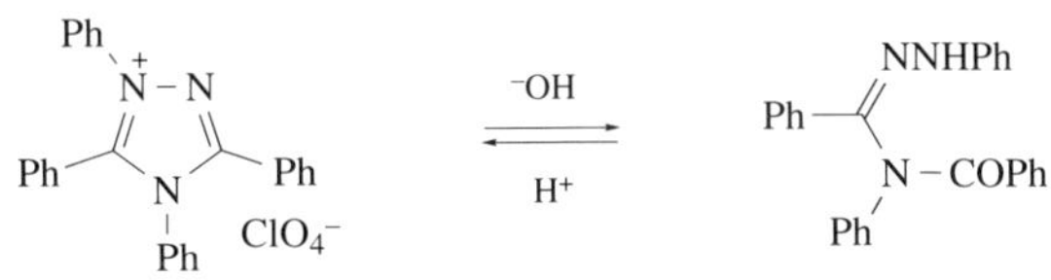

(506) (507) (139)

Quaternary salts of 1,4-dimethyl-1,4-dihydro-*s*-tetrazine (**508**) undergo ring opening in the presence of alkali in a short period to form hydrazonamides (**509**) (Scheme 61) ⟨72JOC3504⟩.

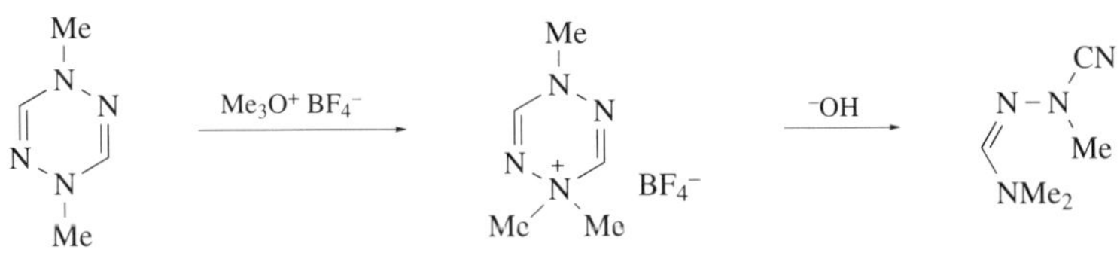

(508) (509)

Scheme 61

s-Triazine (**510**) reacts with 1,1-disubstituted hydrazines to form the corresponding hydrazonamides (**511**) (Equation (140)) ⟨63AG(E)309⟩.

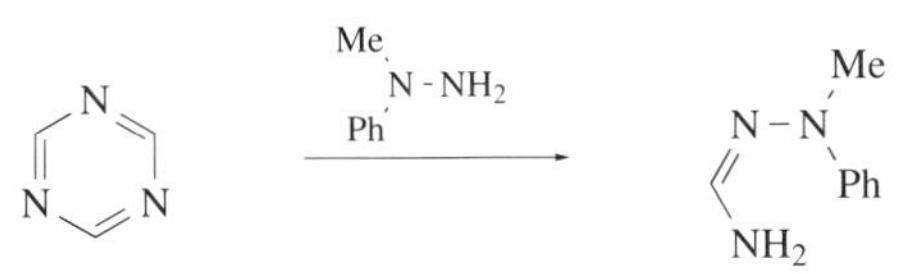

(510) (511) (140)

5.21.3.4.15 By reduction

Hydrazonamide acid derivatives (**513**), (**515**), (**517**), and (**519**) can be prepared by reduction of nitrazones (**512**) ⟨02CB1084, 51LA(574)85, 55CRV355⟩, formazanes (**514**) ⟨52CB470, 62BSF186, 63CB3120⟩, tetrazolium salts (**516**) ⟨50LA(568)185, 51LA(574)85, 57LA(605)191⟩, amidrazones (**518**) and many other imide derivatives ⟨14CB2938, 70CRV151⟩ (Scheme 62). The most frequently employed reducing agents are Raney nickel, palladium, tin and its salts, sulfur dioxide, and lithium aluminum ⟨55CRV355, 70CRV151⟩.

5.21.3.4.16 N^2- *and* N^3-*alkylated hydrazonamides*

Hydrazonamides can be methylated either at N^2 or N^3 to give first the corresponding salts, from which free hydrazonamides are obtained by basification. The compounds (**520**) undergo methylation at N^1 to give (**521**) (Equation (141)) ⟨71JOC1155, 73JOC1344⟩.

(520) (521) (141)

$R^1 = R^2 = Ph$; $R^1 = Ph$, $R^2 = Me$; $R^1 = Et$, $R^2 = Me$

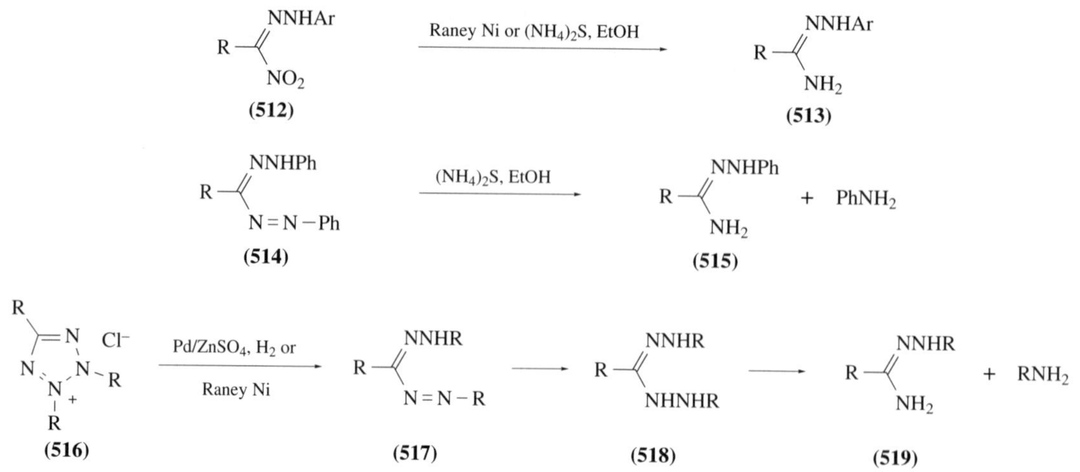

Scheme 62

5.21.3.4.17 *From hydrazonamides*

Hydrazonamides (**522**) react under mild conditions with hydroxylamine to give N^3-hydroxy-hydrazonamides (**523**) (Equation (142)) ⟨70JPR610, 71LA(750)12⟩.

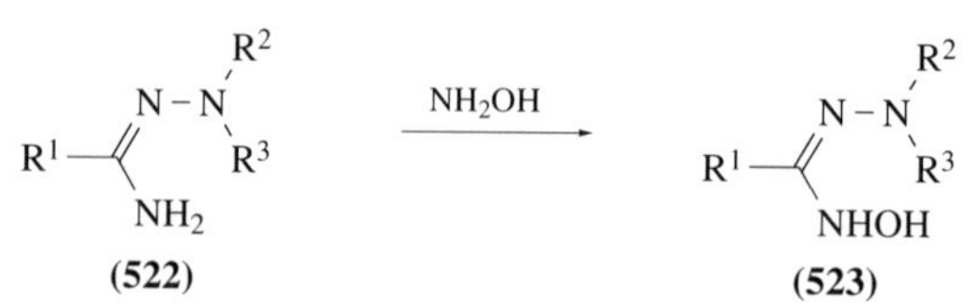

(142)

5.21.3.4.18 *Formazans*

(i) *From hydrazones*

Arylhydrazones of aldehydes (**524**) react with aryldiazonium salts at $pH > 8$ to form formazans (**525**) (Equation (143)) ⟨26CB1162⟩. At a lower pH an unstable intermediate tetrazene is formed, which is converted by treatment with alkali or pyridine into formazane. Besides formazans some other products are also formed ⟨56CB1081⟩. The coupling is carried out in solvents, miscible with water, such as methanol, ethanol, acetone, dioxane, or tetrahydrofurane, to which an aqueous solution of aryldiazonium salt is added, followed by addition of alkali or sodium acetate ⟨1900CB2741, 42CB75⟩.

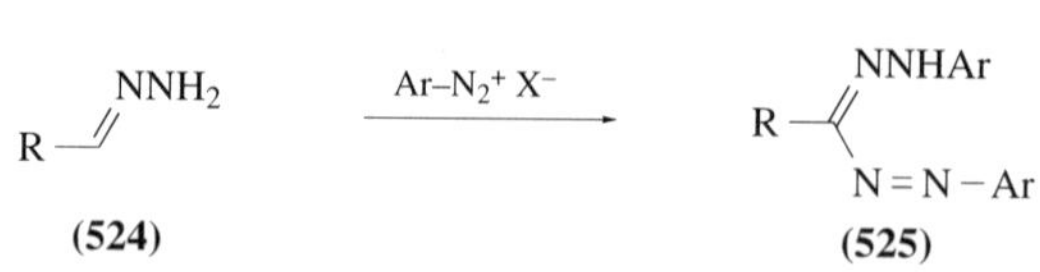

(143)

Arylhydrazones of aliphatic dialdehydes with up to eight carbon atoms give the corresponding bisformazanes in yields above 65% ⟨64LA(678)132⟩. Formazans unsubstituted at *C*-atom (**527**) could not be prepared from arylhydrazones of formaldehyde. They are obtained more easily from glyoxylic acid arylhydrazones (**526**) (Equation (144)). The coupling of diazonium salts is followed by elimination of carbon dioxide ⟨15JPR1⟩.

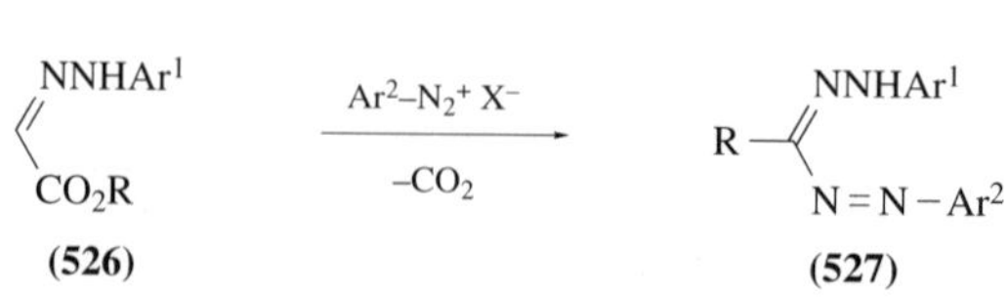

(144)

The coupling of benzenediazonium salts to phenylhydrazones of carbohydrates give also the corresponding formazans. However, the osazone of D-glucose (**528**) reacts with benzenediazonium salts to give phenylhydrazone of the corresponding formazan (**529**) (Equation (145)) ⟨55JA4297, 55JA4301⟩.

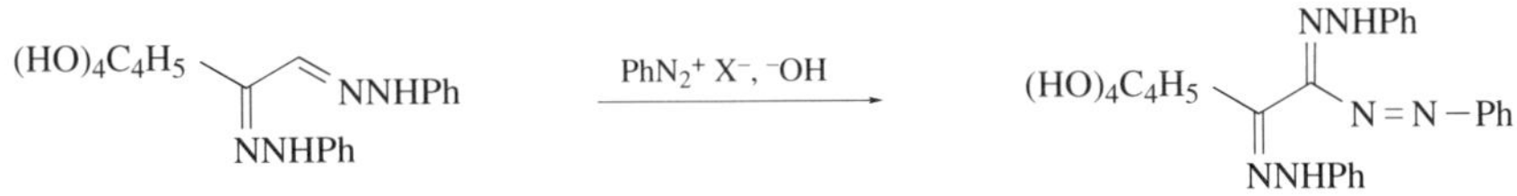

(528) (529) (145)

α- and β-Naphthyldiazonium salts ⟨1900CB751⟩ and bisdiazonium salts can be employed in all these reactions ⟨50MI 521-01⟩. Heterocyclic benzaldehyde hydrazones react normally ⟨53AG442, 53JA5309, 56CB563⟩, while acylhydrazones of aldehydes, such as benzoyl- or thiobenzoylhydrazone, do not form the corresponding formazans ⟨53JA5309, 53LA(581)44⟩. Arylhydrazones of α-keto acids (**530**) react very easily with aryldiazonium salts by elimination of carbon dioxide ⟨01JPR199⟩ to give (**531**) (Equation (146)). Bisphenylhydrazone of dioxosuccinic acid gives the corresponding bisformazane (**532**) ⟨1893CB2978⟩. This method is frequently employed instead of the reaction with hydrazones of aldehydes, since the yields are usually much higher.

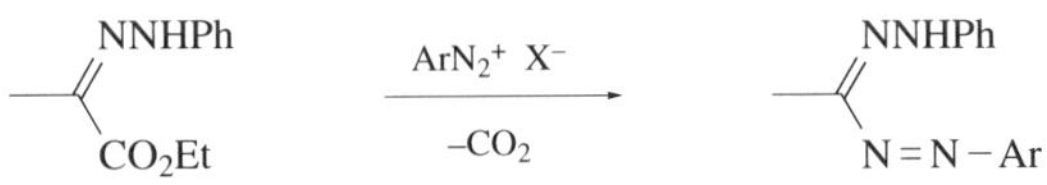

(530) (531) (146)

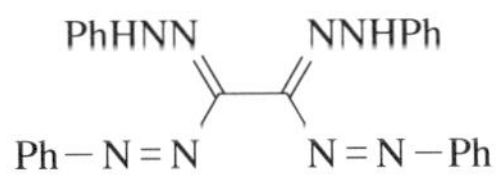

(532)

(ii) From active methylene compounds

The coupling of diazonium salts to active methylene compounds such as (**533**), gives in the first step the corresponding hydrazones (**534**), which react further to form formazans (**535**) (Scheme 63). The second step is frequently faster, especially, when the reaction is carried out in alkaline solutions. In some cases unsymmetrical formazans can be prepared by using two different diazonium salts. The electron-withdrawing groups, such as a carbonyl, nitro, or heterocyclic ring can activate the adjacent, ethyl, or methylene group. When only alkoxycarbonyl or carboxy group is present, the reaction does not take place ⟨53LA(579)28⟩. The Japp–Klingemann reaction, carried out in neutral solutions, involves substitution in the methylene group, followed by tautomerization to form an arylhydrazone. In alkaline solution, the second molecule of diazonium ion couples to the arylhydrazone followed by elimination of other groups. In this reaction a variety of aryldiazonium salts with *o*-, *m*- and/or *p*-halo-, nitro-, alkyl-, and a great variety of active methylene or methine compounds, including aldehydes, ketones, α keto aldehydes, α keto acids, esters and their derivatives, malonic acid and its derivatives, nitroalkanes, unsaturated acids, α-halo diacids, and others can be used. The yields vary from below 10% to quantitative ⟨55CRV355⟩.

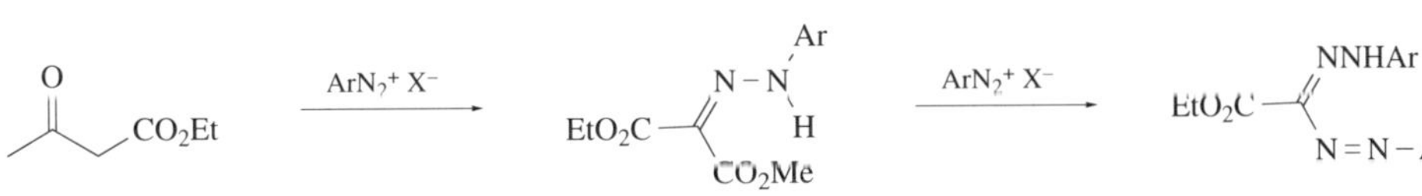

(533) (534) (535)

Scheme 63

In this reaction the substitution of a variety of groups, such as hydrogen, acyl and carboxy, ethoxycarbonyl and benzoyl group, can be replaced by aryldiazonium ion ⟨1892CB3175⟩. Formazan, formed in the first step from acetoacetic acid (**536**), reacts further with another molecule of diazonium ion to form the formarylazobenzene derivative (**537**), obtained also from nitroformazan (**538**) and benzenediazonium salt (Scheme 64) ⟨1892CB3547⟩.

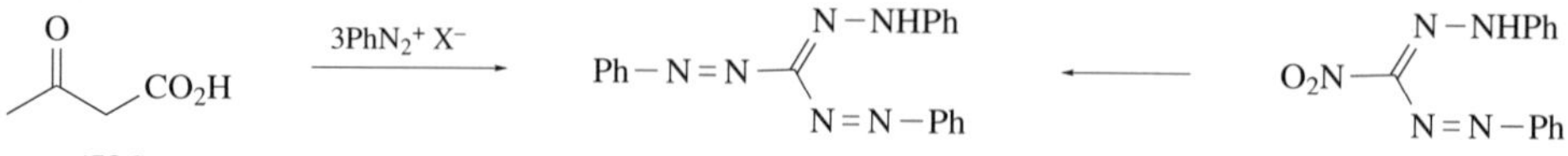

(**536**) (**537**) (**538**)

Scheme 64

Diethyl malonate and ethyl benzoylacetate also behave differently from the corresponding free acids ⟨1892CB3175, 01CB2001⟩. In some instances, the intermediate hydrazone can be isolated, which can react with different diazonum salts to give formazans with two different groups. The formazans are not formed in this second step, when diazotized *p*-anisidine or sulfanilic acid are used on one hand, or mono- or diesters of chloromalonic acid on the other ⟨48G332⟩. By coupling of diazonium salts with malonic acid, three products are formed: formazans, glyoxylic acid arylhydrazones, and azooximes ⟨05JPR366⟩.

A detailed study of the relative ability of various methylene compounds (**539**), to which different groups X and Y are attached to form formazans (**540**), has been published ⟨53LA(579)28⟩. The groups X and Y are: NO_2, CHO, COMe, CN, CO_2R, $CONH_2$, CO_2H, SO_2Me, COPh, SO_2Ph, and others. The coupling with *p*-nitrobenzenediazonium tetrafluoroborate over a pH range 2–10 shows that the lowest pH at which a compound reacts is indicative of its reactivity. Only the most active methylene compounds react in acid solutions. The order of substituents is roughly the order of their decreasing activation. The displacement of groups depends on the pH and can occur for the following X,Y pairs:

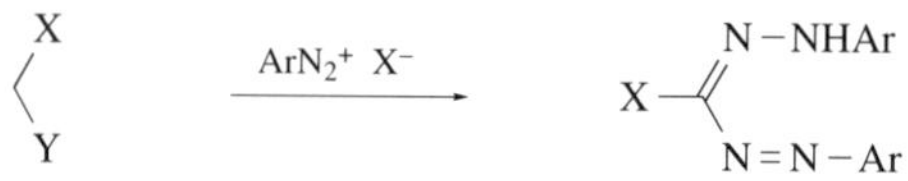

(**539**) (**540**) (147)

X	*Y*	*X*	*Y*	*X*	*Y*
NO_2	$CONH_2$	CO_2R	$CONH_2$	CO_2H	CO_2H
NO_2	CO_2H	CN	CO_2H	$CONH_2$	SO_2Me
COMe	CO_2H	CO_2R	CO_2H	SO_2Me	SOMe
CN	$CONH_2$	$CONH_2$	$CONH_2$	CN	SOMe
		$CONH_2$	CO_2H		

The reactivity is dependent on four factors: the acidity of the CH form, the rate of dissociation of the CH form, the amount of the enol form present, and the acidity of the enol form ⟨53LA(579)28⟩.

A variation of this general method is especially suitable for preparation of *C*-heterocyclic formazans (**544**), when heterocyclic aldehydes are not easily accessible. In such cases, heterocyclic pyruvic esters (**541**), obtained from heterocyclic compounds with an active methylene group by condensing with diethyl oxalate, are coupled with diazonium salts in ammonia solution to give (**544**) via (**542**) and (**543**) (Scheme 65). In some instances, heteroarylacetonitriles (**545**) can be converted into the corresponding formazans (**546**) in ammonia solutions. The yields in all of these reactions are moderate, usually between 30% and 60%, rarely above 80% (Scheme 65) ⟨53LA(581)23, 54JCS2968⟩.

The formazans (**548**) are formed from oxalocrotonic acid (**547**; $n = 1$) ⟨05LA(338)360⟩ and oxalosorbic ester (**547**; $n = 2$) (Equation (148)) ⟨34LA(512)97⟩.

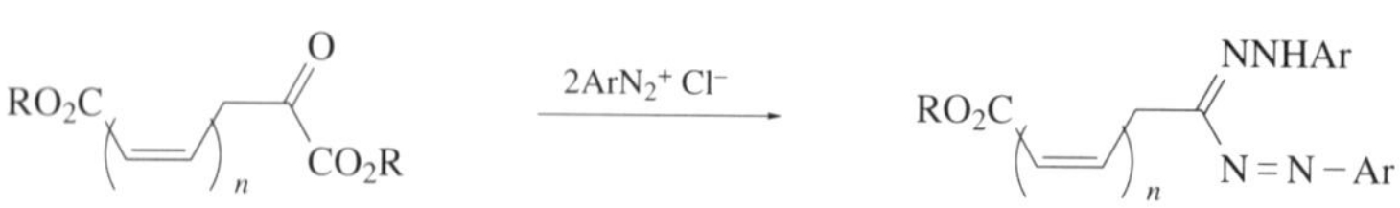

(**547**) $n = 1, 2$ (**548**) (148)

(*iii*) *From aliphatic azo compounds*

Diazonium salts couple with aliphatic azo compounds ⟨01JPR222⟩, sulfonic acid derivatives of ethyl dizoacetate ⟨1896CB2166⟩, and benzenesulfonyl acetic acid (**549**), (**551**), (**553**) and (**555**) ⟨21JPR1⟩ to give (**550**), (**552**), (**554**) and (**556**), respectively (Scheme 66).

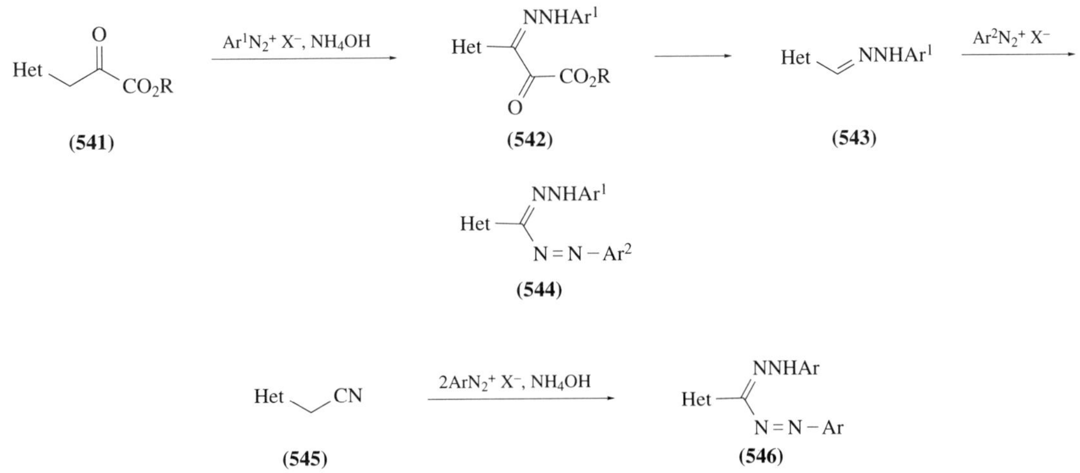

Scheme 65

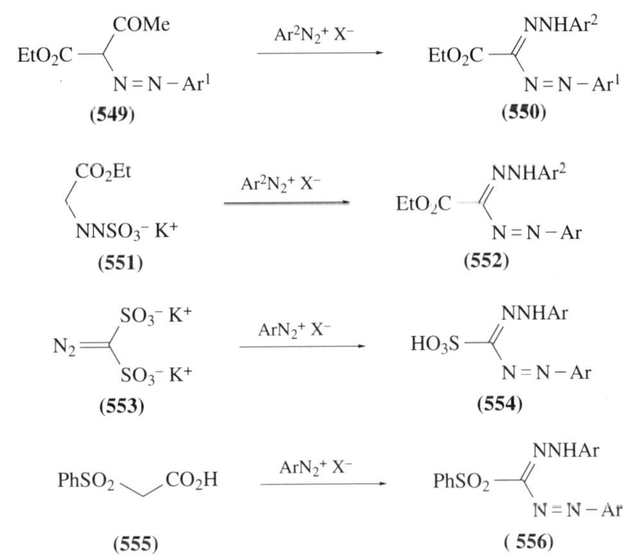

Scheme 66

(iv) From arylhydrazines

Arylhydrazines (**557**) react with a variety of compounds to form formazans (**561**), via (**558**) or (**559**) and (**560**) (Scheme 67). Examples include ethyl formate ⟨1892CB3175, 1894CB320⟩ and triethyl *ortho* formate ⟨1896JPR475⟩.

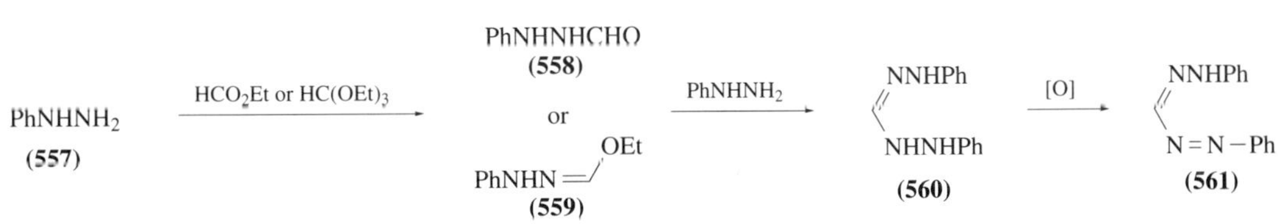

Scheme 67

Arylhydrazines react with imino ethers (**562**) to form formazans (**563**) in good yield, usually above 60% (Scheme 68). However, the method is in some cases disadvantageous, since amidrazones are produced at the same time ⟨1884CB183, 03CB2484, 51LA(574)85⟩. Formazans (**566**) are also obtained from hydrazonoyl halides (**564**) by substitution of halogen with a hydrazine derivative, followed by oxidation ⟨25JCS975, 74ZOR2200⟩ and dichloroacetanilide (**565**) (Scheme 68) ⟨15JPR1⟩. Some formazans with heterocyclic substituents have also been prepared ⟨78KGS991⟩.

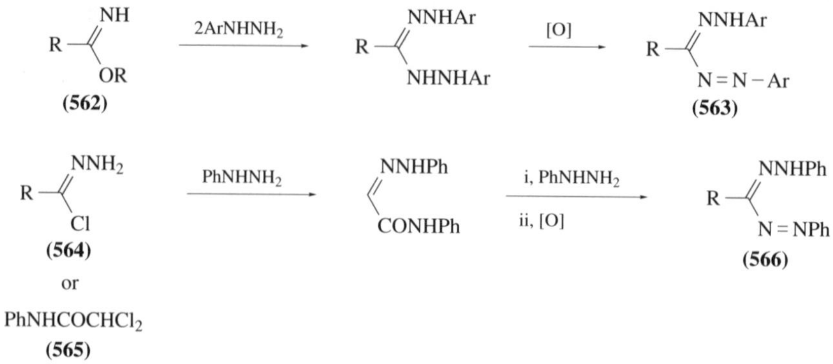

Scheme 68

5.21.4 IMIDOYL DERIVATIVES WITH AN *N*-PHOSPHORUS BOND

5.21.4.1 Imidophosphorane Chlorides

Imidophosphorane chlorides (**570**) are prepared from chlorodiphenylacetonitrile (**567**) and triphenyl phosphine via (**568**) and (**569**) (Scheme 69). The reaction is to be carried out in an aprotic solvent, since in protic solvents the phosphonium salt can decompose into diphenylacetonitrile and triphenylphosphine ⟨65JA5068, 66JA4996⟩.

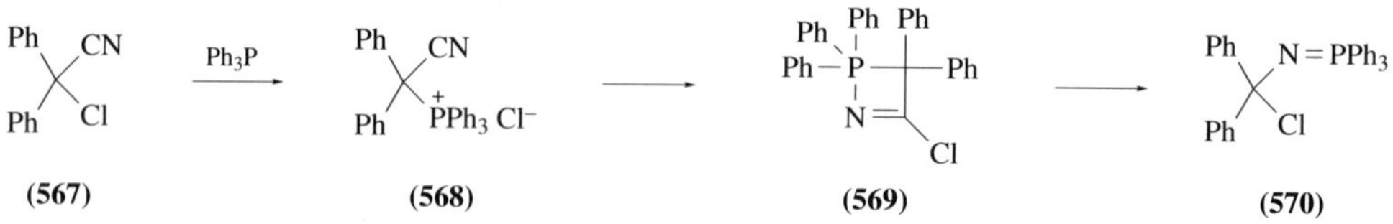

Scheme 69

5.21.4.2 Imidophosphorane Esters

N-Chloroimidic (**571**) esters react with tertiary phosphines to form phosphonium salts (**572**), which are transformed into imidophosphorane esters (Equation (149)) ⟨62ZOB1873, 63ZOB1017⟩.

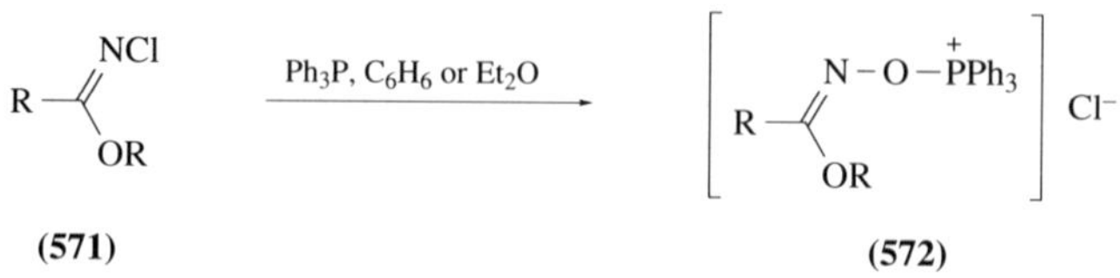 (149)

R = Me, Ph, 4-BrC_6H_4, 4-$NO_2C_6H_4$, OEt

Some phosphorus derivatives (**574**), (**576**), (**578**), and (**579**) are obtained by direct interaction of imidates (**573**), (**575**), and (**577**) with phosphorus pentachloride (Scheme 70) ⟨62JGU3689⟩. Other phosphorylating agents for this purpose are chlorodiphenylphosphine ⟨71CB1199⟩ and other chlorophosphines ⟨63JGU2880, 66JGU1101, 70BSF1363⟩, and triaminophosphines ⟨68CR(C)1241⟩.

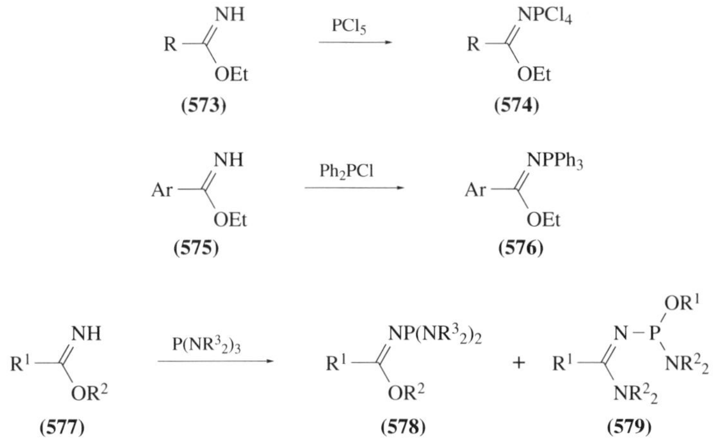

Scheme 70

Phosphorazidates (**580**) react with vinyl ethers by an aza-transfer reaction to give the corresponding imidophosphorane esters (**581**) (Equation (150)) ⟨61ZOB3424, 63JGU2984, 66T975⟩.

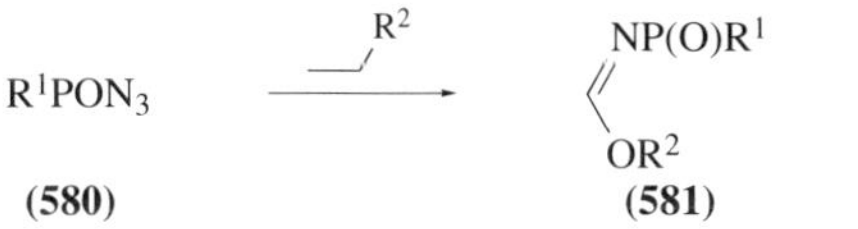

(150)

5.21.4.3 Imidophosphorane Thioesters

Imidophosphorane thioesters (**583**) and (**585**) are prepared by alkylation of the corresponding thioamides (**582**) ⟨68JGU1815⟩ or by the reaction of phosphorus pentachloride with alkyl and aryl thiocyanates (**584**) (Scheme 71) ⟨61ZOB3424, 63JGU2984, 68ZOB326, 68ZOB850⟩.

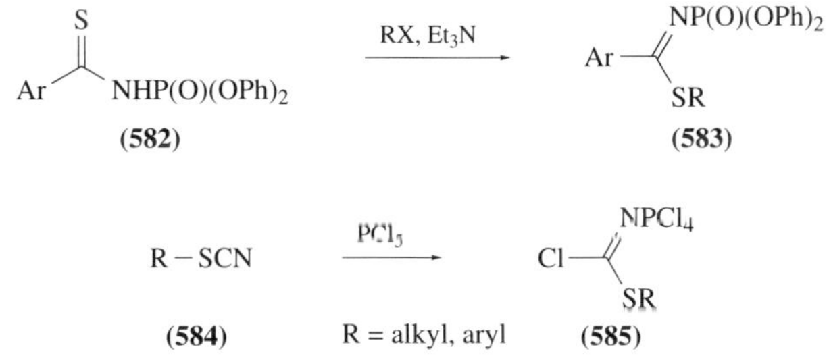

Scheme 71

5.21.4.4 Imidophosphorane Amides

N-Chloroarylamidines (**586**) react with triphenylphosphine to give phosphonium salts (**587**) as intermediates, which are transformed with a base into the corresponding imidophosphoranes (**588**) (Scheme 72) ⟨62ZOB2245⟩. Another method is by substitution of the chlorine atom in 1-chlorovinyliminophosphorane (**589**) with aromatic amines (Scheme 73) ⟨62ZOB2245⟩.

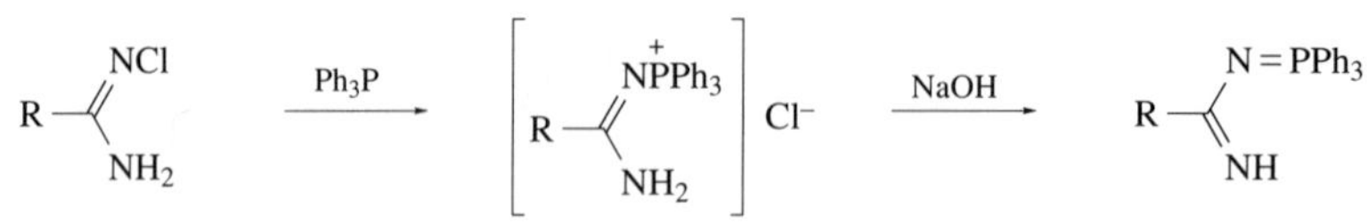

(586) **(587)** **(588)**

R = Ph, 4-BrC_6H_4, 4-$NO_2C_6H_4$

Scheme 72

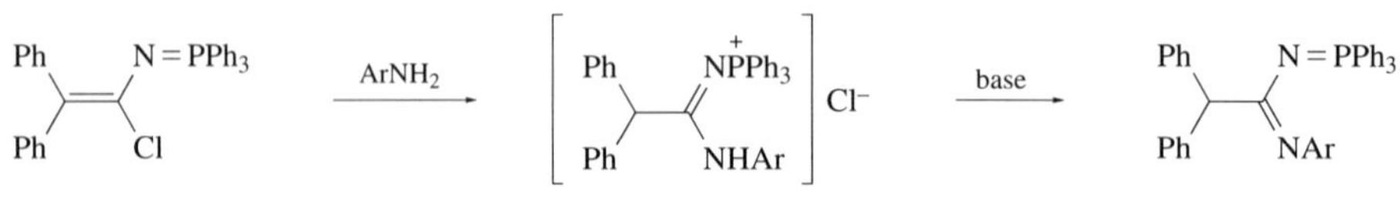

(589)

Scheme 73

5.21.4.5 Oxyimidic Imidophosphoranes

O-Aroyl-substituted oxyimidic azides (**591**), obtained from (**590**) react with triphenylphosphine in dry dichloromethane to give the corresponding imidophosphoranes (**592**) in high yields (Scheme 74) ⟨86S843⟩.

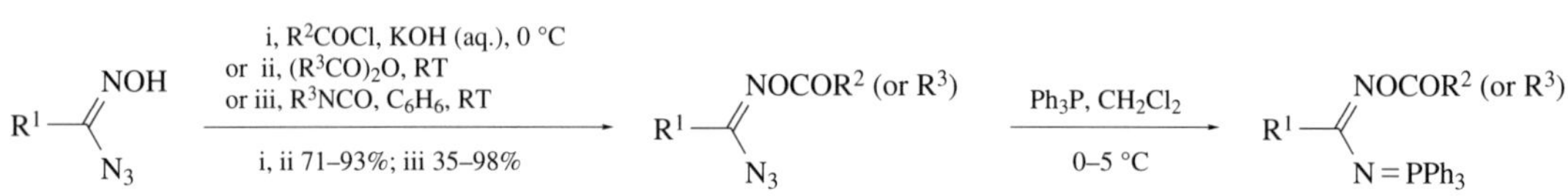

(590) **(591)** **(592)**

R^1 = Ph, 4-ClC_6H_4; R^2 = Me, Ph, 4-MeC_6H_4, 4-ClC_6H_4, Et_2O, MeNH, PhNH; R^3 = Me, Ph

Scheme 74

5.21.4.6 Phosphoranes derivatives of hydrazonamides

The Staudinger reaction provides a simple and efficient method for the preparation of iminophosphoranes ⟨81T437⟩. According to this reaction, hydrazonazides (**593**) give the corresponding hydrazonimidophosphoranes (**594**) (Equation (151)) ⟨86JCS(P1)2177⟩.

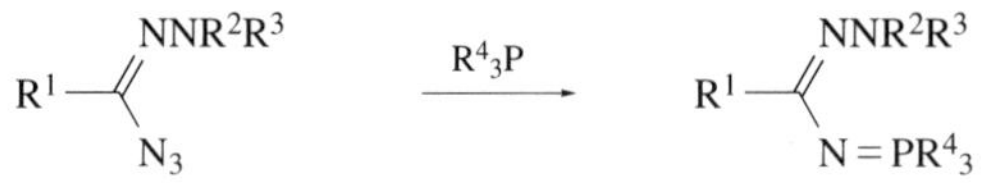

(593) **(594)** (151)

Phosphorane derivatives of hydrazonamides (**596**) are easily prepared by the reaction of a mixture of triphenylphosphine and iodine with benzophenone 1-aminoethylidene hydrazone, readily obtainable from *S*-methyl thioacetimidate hydroiodide (**595**) and benzophenone hydrazone (Equation (152)) ⟨94S1057⟩.

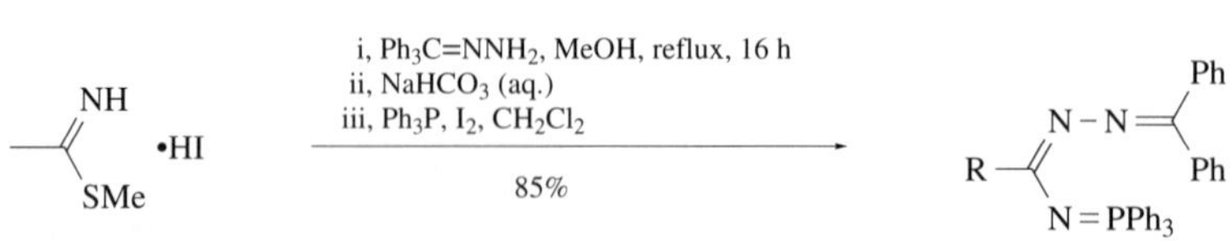

(595) **(596)** (152)

The amidines (**597**) react with tris(diethylamino)phosphine to five N^1,N^1-diethyl-N^2-tetra-ethyldiamidophosphitoamidines (**598**) ⟨80ZOB2413⟩.

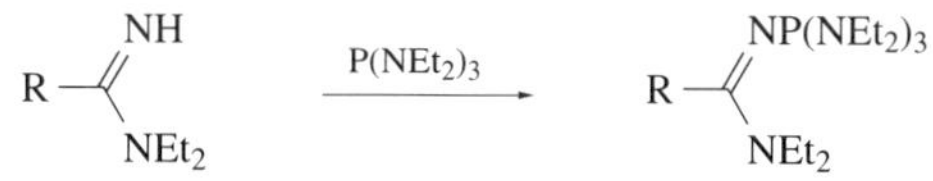

(**597**) (**598**) (153)

5.21.5 IMIDOYL DERIVATIVES WITH AN *N*-METALLOID BOND

5.21.5.1 *N*-Silylimidoyl Derivatives

N-Silylimidoyl compounds are usually obtained by silylation of the imido group.

N-Silylimidoyl esters (**600**) and (**602**) are prepared from imidates (**599**) and (**601**) by the action of halosilanes (Scheme 75) ⟨70GEP2011937⟩. Silylation of imidates with trialkylchlorosilanes takes place in the presence of triethylamine ⟨71DOK(199)103⟩. Silylation of imidates of β-dialkylaminopropionic acid with trimethylchlorosilane in anhydrous benzene at $>5\,^{\circ}C$ in the presence of triethylamine gives the corresponding trimethylsilyl derivatives. Instead of trimethylchlorosilane, trimethyldiethylsilane can be employed ⟨72ZOB868⟩.

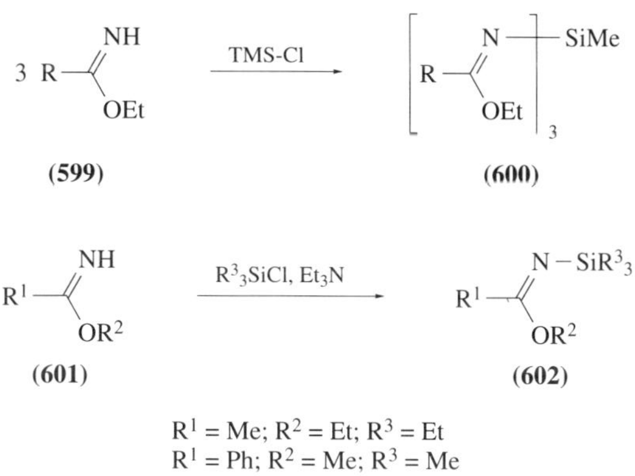

(**599**) (**600**)

(**601**) (**602**)

$R^1 = Me$; $R^2 = Et$; $R^3 = Et$
$R^1 = Ph$; $R^2 = Me$; $R^3 = Me$

Scheme 75

5.21.5.2 *N*-Borylimidoyl Derivatives

Alkylthioiminoborane hydrobromides (**604**) are formed by reaction of monomeric bromo substituted iminoboranes (**603**) with alkylthiols (Scheme 76). Thermal dehydrobromination yields free iminoboranes (**605**), while dehydrobromination with triethylamine leads to unstable imidic thioesters. Under the same conditions, thiophenol also replaces the bromine bonded to boron in (**606**) producing iminoboranes (**607**) with a phenylthio group at both boron and carbon (Equation (154)) ⟨71M121⟩.

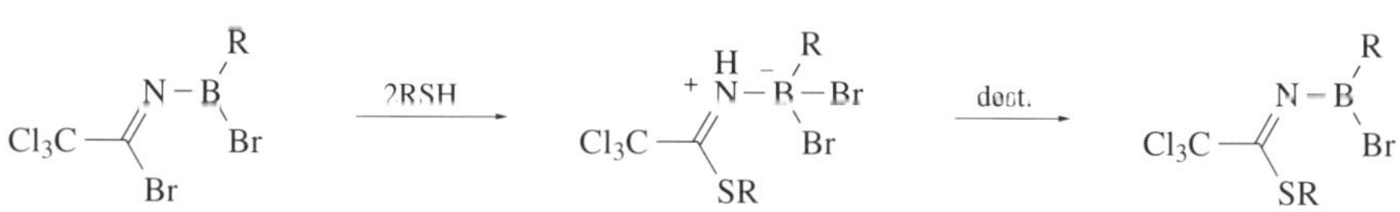

(**603**) (**604**) (**605**)

R = Me, Bu^n, Bu^s

Scheme 76

(606) **(607)**

(154)

Isothiocyanato substituted iminoboranes (**609**) are obtained from the corresponding imidoyl bromides (**608**) by treatment with methylthiocyanate (Equation (155)) ⟨71M121⟩.

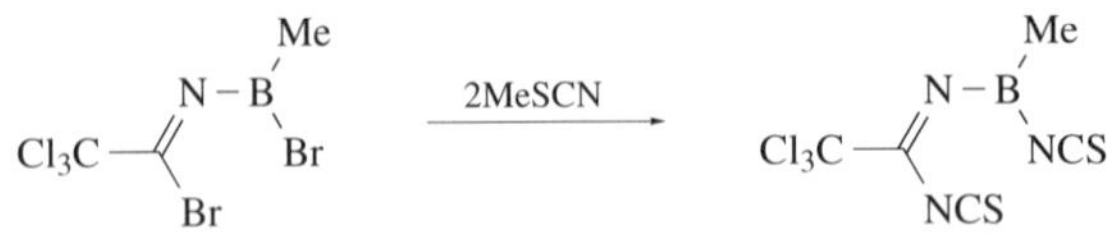

(608) **(609)**

(155)

5.21.6 IMIDOYL DERIVATIVES WITH AN *N*-METAL BOND

Salts of oxyimidic derivatives have been prepared with nearly all elements of groups 1, 2, and 13 of the periodic table, as well as zirconium, thorium, cobalt, nickel, and iron ⟨43CRV209⟩. The derivatives with a free hydroxy group, such as oxyimidic esters and oxyimidic halides, give the color reaction with iron(III) chloride. This test is very sensitive. It detects the presence of 2γ of the material ⟨34MI 521-01, B-37MI 521-01⟩. The formation of the complexes of oxyimidic amides with various metals has been extensively studied due to their practical applications, especially in analytical chemistry ⟨62CRV155, 69MI 521-01, B-75MI 521-03, B-91MI 521-02⟩ including their applications for spectrometric determinations of Hg(I), Hg(II), Cu(II), Mo(VI), Fe(III), Fe(II), Co(II), and Ni(II) ions ⟨82AQ(8)327, 86POL1277⟩. The oxyimidic amide group is binding as either metal ions as such, or in the anion form, and can behave as a bidentate ligand through the amide nitrogen and oxyimidic nitrogen atom. When the oxyimidic amide ligand contains an additional dialkylamino donor site, the second coordinating atom can be either the oxyimidic nitrogen (**610**) or the amide nitrogen (**611**).

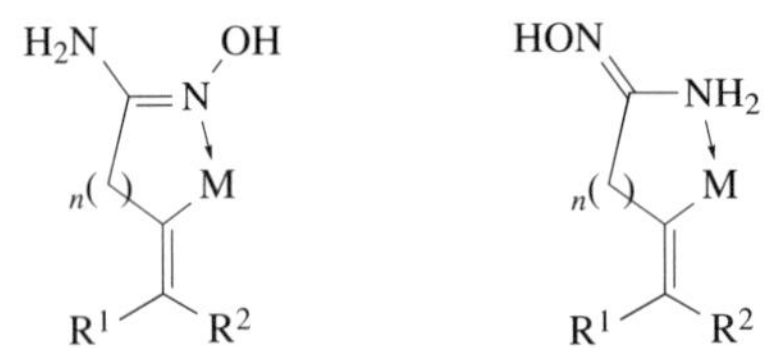

(611) **(612)**

$n = 1, 2$

Complex formation equilibria between Ni(II) and 2-aminoacetamidoxime and its methylated derivatives have been studied, showing that mono-, bis-, and tris-complexes are formed stepwise, and that stability constants decrease with increasing *N*-substitution ⟨83ACS(A)631⟩. Some poly-amidoximes easily form complexes with Cu(II), Co(II), Ni(II), Fe(II), Fe(III), Pd(II), Mo(VI), and U(VI) ions ⟨86POL1277⟩. Divinylbenzene cross-linked poly(acrylamidoxime) resins have been prepared and they are used for concentration of trace metals from aqueous solutions, including sea water, since they exhibit no affinity to alkali or alkaline earth metals. The order of decreasing selectivity is Cu(II), Ni(II), Co(II), Zn(II), and Mn(II) ⟨80ANC967⟩. Some other resins of this type have been prepared ⟨86MI 521-01, 87MI 521-03⟩ and used for the recovery of rare earth metals from aqueous solutions by extraction with organic solvents ⟨71GEP2044574⟩. Complexes of Cu(II), Ni(II), Co(II), Fe(II), and Mn(II) with pyridine-2-amidoxime ⟨83IJC(A)307⟩ and pyrazine-2-amidoxime have been prepared as potential biologically active compounds ⟨86SRI113⟩.

5.22
Diazo Functions with an α-Heteroatom (RC(X)N_2)

KAMYAR AFARINKIA
King's College, London, UK

and

M. VICTORIA VINADER
Glaxo-Wellcome Medicines Research Center, Stevenage, UK

5.22.1 DIAZO FUNCTIONS WITH AN α-HETEROATOM

Diazo compounds have a wide and fascinating chemistry. They are useful reagents in modern organic chemistry, mainly as carbene precursors or as 1,3-dipoles in cycloaddition reactions. This subject has been twice reviewed and both reviews give excellent accounts of both preparation and reactions of this functionality ⟨B-78MI 522-01, B-86MI 522-01⟩.

Due to their ylide nature, aliphatic diazo compounds are most stable when substituted by at least one group that can stabilise negative charge. In general, aliphatic diazo compounds bearing only alkyl- or electron-releasing groups are unstable, though this is not to say that they cannot be prepared. In this chapter, the preparation of diazoalkanes with one α-heteroatom substituent will be discussed. For most compounds, the other substituent is hydrogen or carbon but occasionally references are made to bisheteroatom-substituted diazomethanes.

Access to α-diazo compounds is synthetically limited to routes outlined in Scheme 1. The routes are general though exceptions are found when the chemistry of a route is incompatible with functional groups within the molecule. Broadly speaking, it is difficult to distinguish between routes according to the heteroatom preference. Indeed, it appears from the literature that the routes are not restricted by heteroatom X but rather by the type of functionality that incorporates X, or in effect the electronic nature of X. Thus, parallel synthetic routes can be identified for synthesis of each different class of compound.

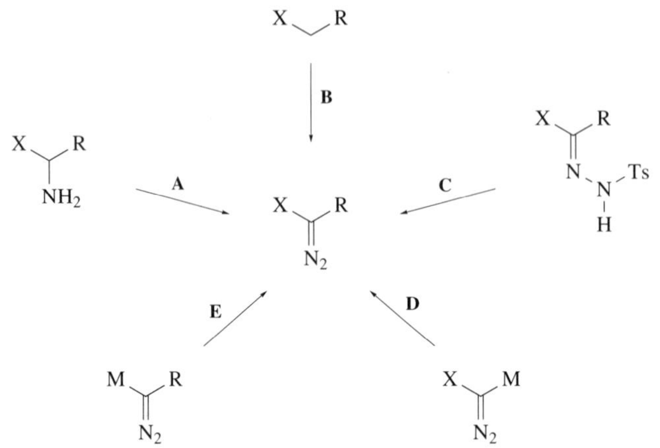

Scheme 1

This chapter will start with α-diazo halo compounds and will then progress to other α-diazo-heteroatom compounds. However, it is recommended that for a general understanding of this subject, the reader should first consult the section on α-diazo phosphorus compounds (5.22.5) for they have been the most comprehensively studied functional group.

5.22.2 α-DIAZO HALO COMPOUNDS

Halogenated α-diazo compounds are unusual in the sense that their synthesis does not always conform to the rational classifications outlined previously. Also, many are best characterised as halodiazirines, structural isomers of the corresponding diazo form.

The best preparation from nondiazo compounds is via halogenation–oxidation of amidines ⟨65JA4396⟩. For further experimental details see Equation (1) ⟨76JA1278, 81JA6164⟩. This reaction, sometimes referred to as the Graham reaction, is reported to be applicable for both electron-rich (e.g., R = OMe) and electron-poor (e.g., R = C≡N, CF_3, etc.) substituents.

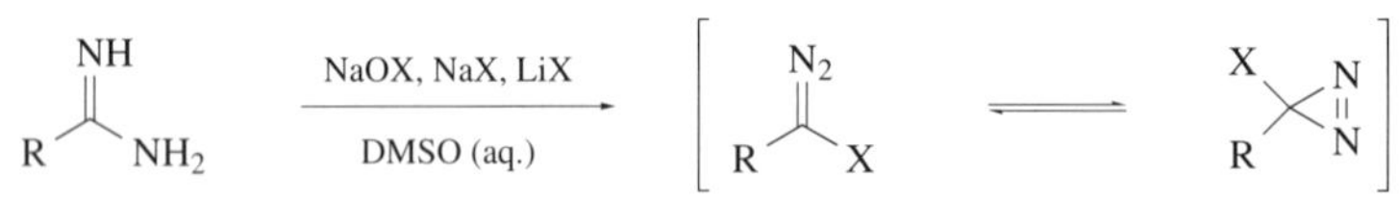

(1)

Owing to aromatic stability of carbocation (**1**), halodiazo (halodiazirine) compounds undergo rapid SN_1-type displacement reactions with nucleophiles. Thus, it is shown that a fluorodiazo (fluorodiazirine) is prepared from the corresponding bromodiazo (bromodiazirine) compound by treatment with fluoride anion ⟨87TL5801⟩. This interconversion of diazo compounds is unusual, and hence rather interesting, since in a typical interconversion of type E or D (Scheme 1), the starting diazo molecule acts as a nucleophile and not an electrophile. Unfortunately, in the published literature, the scope of the reaction has remained restricted to halogen exchange.

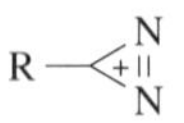

(1)

A more typical preparation of halosubstituted diazo compounds by interconversion is by electrophilic halogenation (Scheme 1, route E and Scheme 2) ⟨62JA4350, 68LA(716)204⟩.

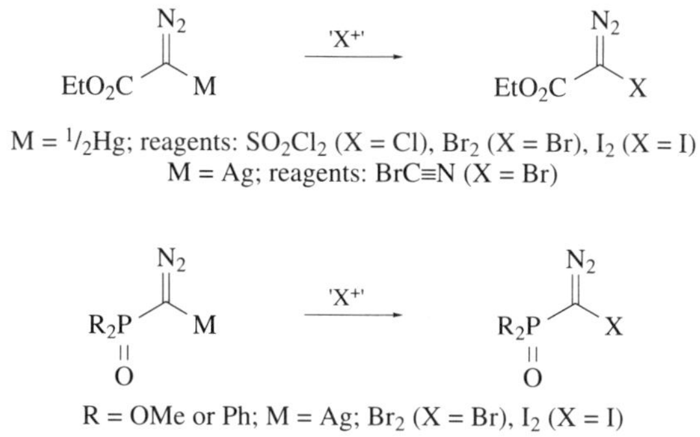

Scheme 2

5.22.3 α-DIAZO OXYGEN AND SULFUR COMPOUNDS

Broadly speaking, it can be expected that preparative routes to α-diazoalkanesulfonates (and sulfones) are similar to those for α-diazoalkanephosphonates (and phosphines). Traditionally, however, access to α-diazoalkanesulfur compounds has proved to be synthetically nontrivial. For instance, the first example of an α-diazoalkanesulfonate was only reported in 1989 ⟨89CB1147⟩.

As with phosphorus compounds, the most reliable synthesis of simple sulfonyldiazomethanes is by route A (Scheme 1), that is from the corresponding *N*-nitrosamine (for comparison see Equation (6) and related text). Thus, nitrosocarbamates are used for the preparation of a range of compounds ⟨73TL5207, 75T597⟩. On the other hand, diazo transfer to a methylene adjacent to sulfoxide, sulfone or sulfonate is only recommended if it is further substituted by an electron-withdrawing group such as a carboxylate or ketone ⟨68JSC(C)1602, 77JCS(P1)2298, 89CB1147⟩. An alternative is a two-step process involving formylation followed by diazo transfer ⟨75T597, 89CB1147⟩. Thus, deprotonation of alkyl alkanesulfonate (**2**) with *n*-butyllithium and subsequent reaction with ethyl formate afforded alkyl 1-formylalkanesulfonate (**3**) which was then treated with 4-carboxybenzenesulfonyl azide or tosyl azide to afford α-diazoalkenesulfonate (**4**) (Scheme 3). An alternative employing an α-ketoethyl ester in place of the formyl group is reported ⟨80JOC1717⟩.

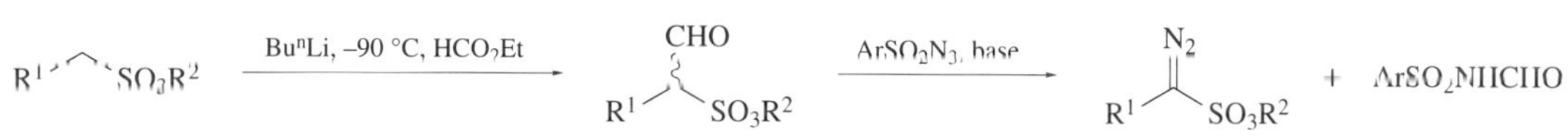

(**2**) (**3**) (**4**)

Scheme 3

Interconversion is also applicable and as in the case of phosphonyldiazomethane (see Scheme 9 and related text), sulfonyldiazomethane is easily deprotonated with organic amine base and the anion reacts with electrophiles such as acid chloride ⟨82CPB899⟩. α-Diazomethylsulfoxide (**5**) is prepared by reaction of diazomethane and phenylsulfinyl chloride (Equation (2)) ⟨75JA3238, 84AG(E)717⟩. As hydrogen chloride is generated during the reaction, chloromethyl phenyl sulfoxide is a by-product (for analogy see Equation (4) and related text.)

(**5**) (2)

The Forster reaction has been applied to the synthesis of α-diazoalkanesulfones but the yields are unacceptably low ⟨75T597⟩. A nongeneral route restricted to the preparation of diazomethane derivatives is by the base-induced cleavage of the corresponding α-diazo-β-ketosulfone compound (Equation (3)) ⟨68JSC(C)2201⟩.

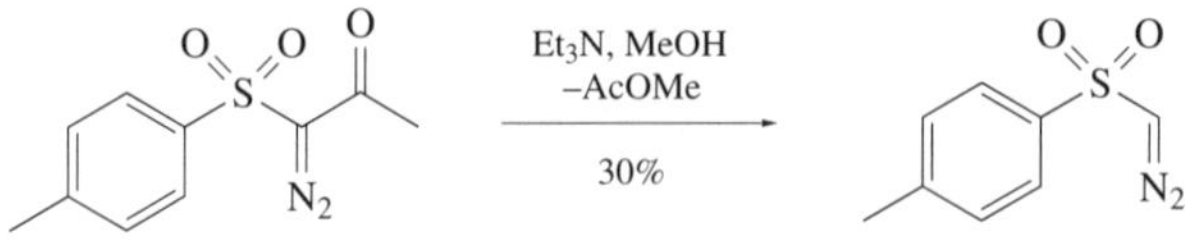

(3)

5.22.4 α-DIAZO NITROGEN COMPOUNDS

Nitrodiazo compounds are prepared by careful nitration of a diazo compound (Equation (4)) ⟨66TL6199, 90TL7341, 91JOC2258⟩. The reaction is limited to systems bearing strongly electron-withdrawing groups with no hydrogens on the β-position. The reaction requires two equivalents of diazo compound since the second mole acts as a base to deprotonate the intermediate diazonium ion.

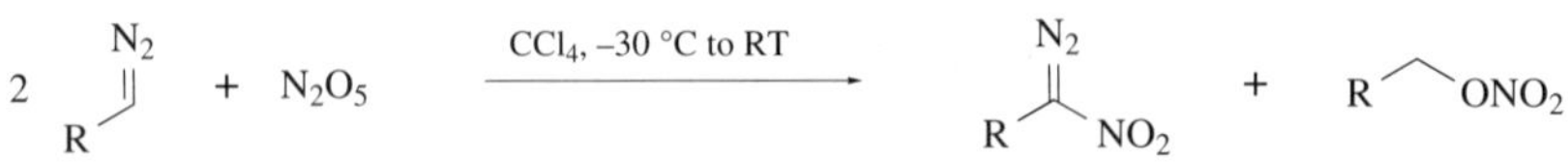

(4)

R = CN, CO_2Et, CO_2Bu^t, COPh, SO_2Tol, CF_3

5.22.5 α-DIAZO PHOSPHORUS COMPOUNDS

5.22.5.1 Diazotisation of α-amino compounds (Scheme 1, route A)

As with other primary amines, diazotisation of α-amino phosphonates and phosphinates (via an *N*-nitroso intermediate) leads to the formation of the corresponding α-diazo compounds (Equation (5)). This route requires acid catalysis but since the phosphonyl moiety is not as good an electron-withdrawing group as the carbonyl, protonation at the α-position resulting in the decomposition of the α-diazo compound is facile. Hence to obtain better yields, the use of strong aqueous mineral acids should be avoided in favour of acetic acid ⟨B-86MI 522-01⟩.

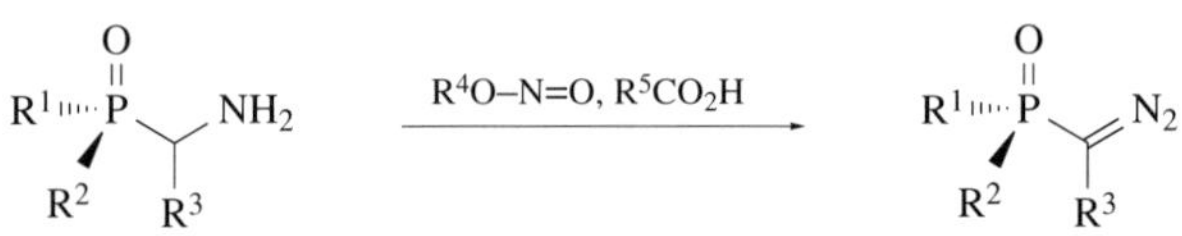

(5)

This route is particularly useful for the synthesis of diazomethylphosphonates and phosphinates, for example **(6)** ⟨67JA4811, 70TL2493, 71JOC1379, 92T1465⟩, which cannot be easily prepared by other means (Equation (6)). It should be emphasised that there is no known limitation to the range of substituents, R^1, R^2 and R^3. However, since the synthesis of α-amino phosphonates and phosphinates themselves is not so easy, this route has found little favour.

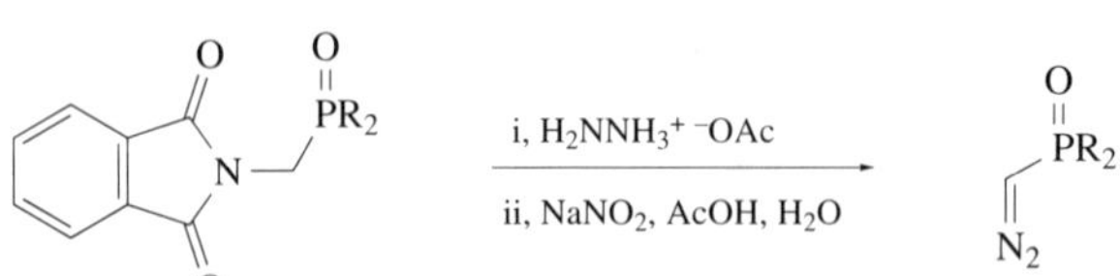

(6)

(6)

5.22.5.2 Diazo transfer (Scheme 1, route B)

Diazo-group transfer reactions have provided preparative routes to a diverse range of α-diazo phosphonyl compounds. The inductive electron-withdrawing power of the phosphonyl moiety is expected to be sufficiently strong to activate the adjacent methylene protons towards substitution. However, the number of α-diazo phosphonates and phosphinates prepared by this route which do not bear a further activating group (R^3) is small ⟨B-86MI 522-04⟩. This is due to the fact that most easily available α-unsubstituted phosphonates and phosphinates are prepared by the Arbuzov reaction which is facilitated when the R^3 group is electron withdrawing (Scheme 4).

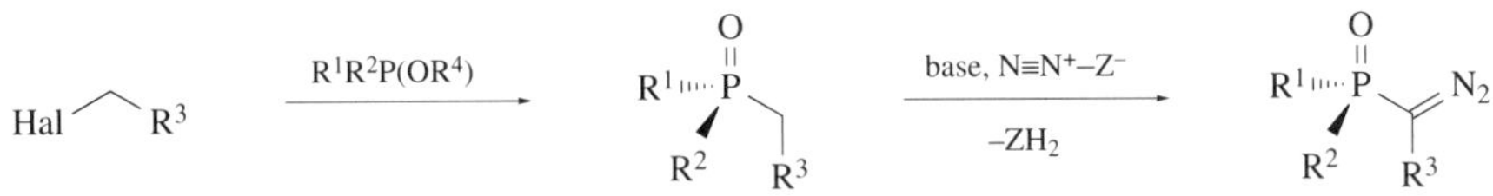

Scheme 4

The choice of base largely depends on this other group (R^3). With carbonyl groups which activate the adjacent methylene protons highly, organic amine bases suffice. With less strongly activating groups such as phenyl, potassium *t*-butoxide may be used. The use of organometallic bases such as butyllithium and phenyllithium is not recommended as ligand exchange on phosphorus may occur (Equation (7)) ⟨68CB3734⟩.

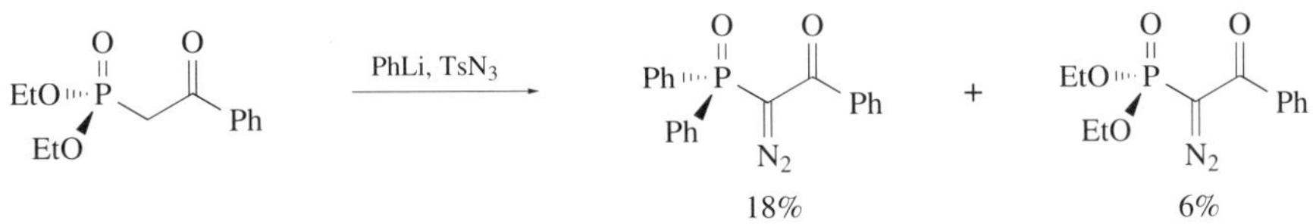

(7)

The diazo transfer reagent is commonly tosyl azide (**8**) ⟨67AG(E)733⟩, although on occasions other reagents such as mesyl azide (**9**) ⟨86JOC4077⟩, diphenyl phosphorazidate (DPPA) (**7**) ⟨84OS(62)187⟩, 4-nitrophenyl azide (**10**) ⟨73LA578, 76CB2039⟩ and (**11**) ⟨61LA(647)1, 61LA(647)11⟩ as well as polymer bound reagents ⟨74TL1391⟩, are also used.

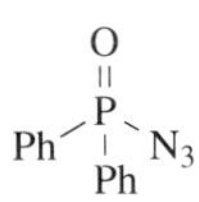

(**7**)

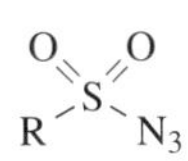

(**8**) R = 4-MeC_6H_4
(**9**) R = Me

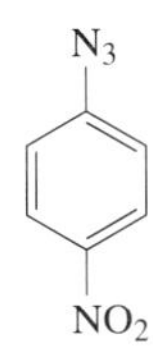

(**10**)

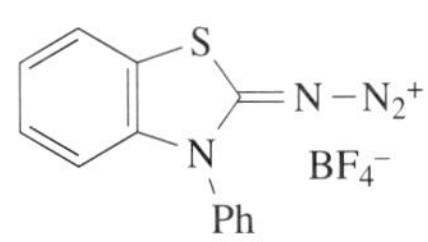

(**11**)

Diazo group transfer sometimes fails to give the expected α-diazo phosphono compound. Such behaviour is often due to the group R^3 and can therefore be predicted from the chemistry of the other functional groups present in the molecule. Thus, in the following examples (Scheme 5), the presence of nucleophilic groups in R^3 which interact with the diazo functions leads to the formation of 1,2,3-triazoles ⟨69CB2216⟩.

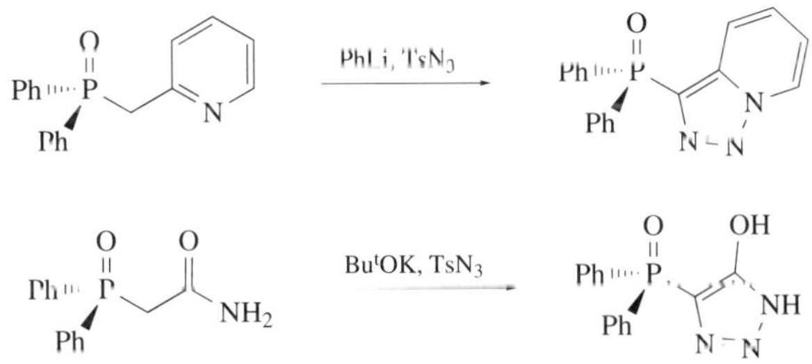

Scheme 5

Occasionally, such cyclisations are reversible and are put to good use for synthesis of α-diazo phosphorus compounds from precursors which do not have an adjacent methylene group. Ynamine phosphonate (**12**) and cyclopropene phosphonate (**14**) are starting materials for the synthesis of diazo phosphonates (**13**) (the major isomer in solution and the sole isomer in the crystalline state) ⟨B-78MI 522-02⟩ and (**15**) ⟨B-78MI 522-03⟩ respectively (Scheme 6).

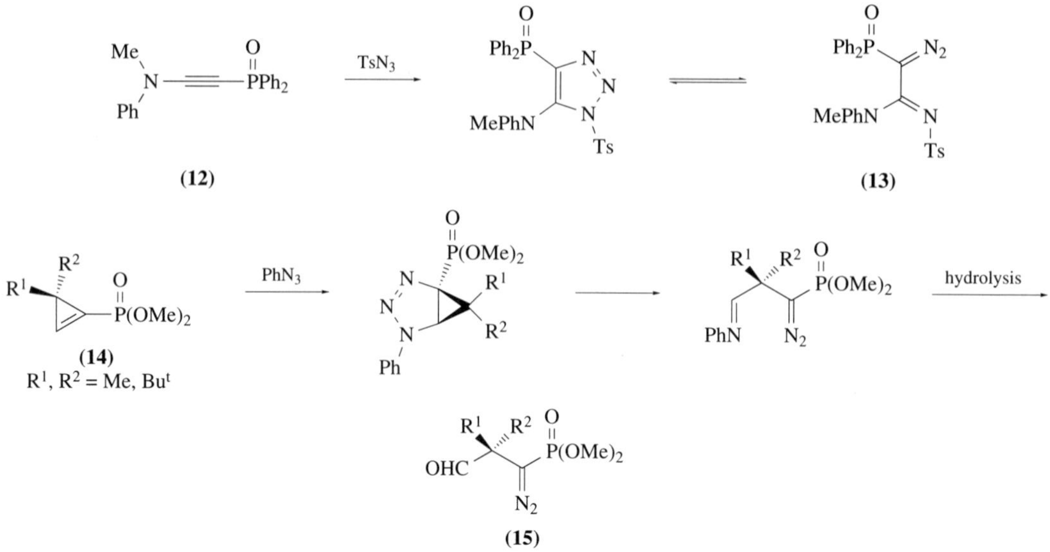

Scheme 6

5.22.5.3 Bamford–Stevens reaction (Scheme 1, route C)

Synthesis of α-diazo phosphonates and phosphinates by the Bamford–Stevens reaction (Scheme 7) is, like the other two methods already discussed, not limited in the range of R^1, R^2 and R^3, though, as before, the availability of starting materials has to be taken into account ⟨71JOC128⟩. Thus, treatment of phosphorus(III) esters with acid chlorides gives α-oxo phosphono compounds (**16**). Dehydrative condensation with tosyl hydrazide affords tosyl hydrazone (**17**) which upon addition of a base decomposes to yield the diazo compound. Bases used for this step vary in strength, from sodium hydride to aqueous sodium carbonate, depending on substituents ⟨71JOC128, 85T2625⟩.

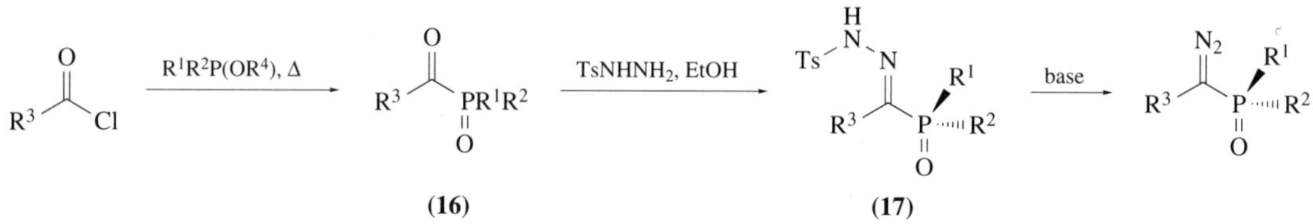

Scheme 7

Two closely related reactions, the Forster reaction (Scheme 8) ⟨B-86MI 522-02⟩ and oxidation of hydrazones are both used for the synthesis of α-diazoalkyl carbonyl ⟨B-86MI 522-03⟩ and sulfonyl compounds (Scheme 8) ⟨75T597⟩, but have not yet been reported for the synthesis of α-diazo phosphono compounds.

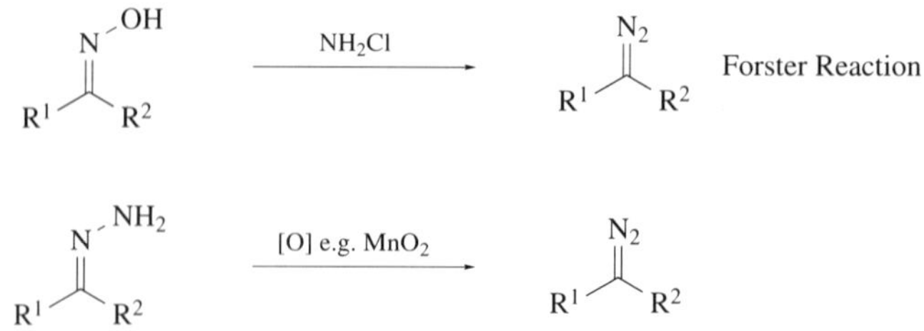

Scheme 8

5.22.5.4 From other α-diazo compounds (interconversions) (Scheme 1, routes D and E)

This route is by far the most versatile approach to the synthesis of these species. The diazomethyl proton is acidic, owing to the diazo group's strong anion stabilising effect arising from its electron-withdrawing property. Further stabilising effects of the conjugate anion by the X group can make this proton even more acidic. For instance, α-diazomethylphosphonates and phosphinates can be easily deprotonated and the anion thus formed can react with a range of heteroatom or carbon electrophiles to afford new substituted α-diazophosphonates. Triethylamine is often a strong enough base, and the reactions usually proceed under mild conditions (Scheme 9) ⟨67AG(E)261, 69CB488, 71LA(748)207, 72TL3799, 76CB546, 76CB2039⟩. As an extension to this preparation, it has been shown that some α-diazophosphonates can directly react with potent (or positively charged) electrophiles without prior metallation ⟨85T811, 85T825, 86T1815, 89CB565⟩.

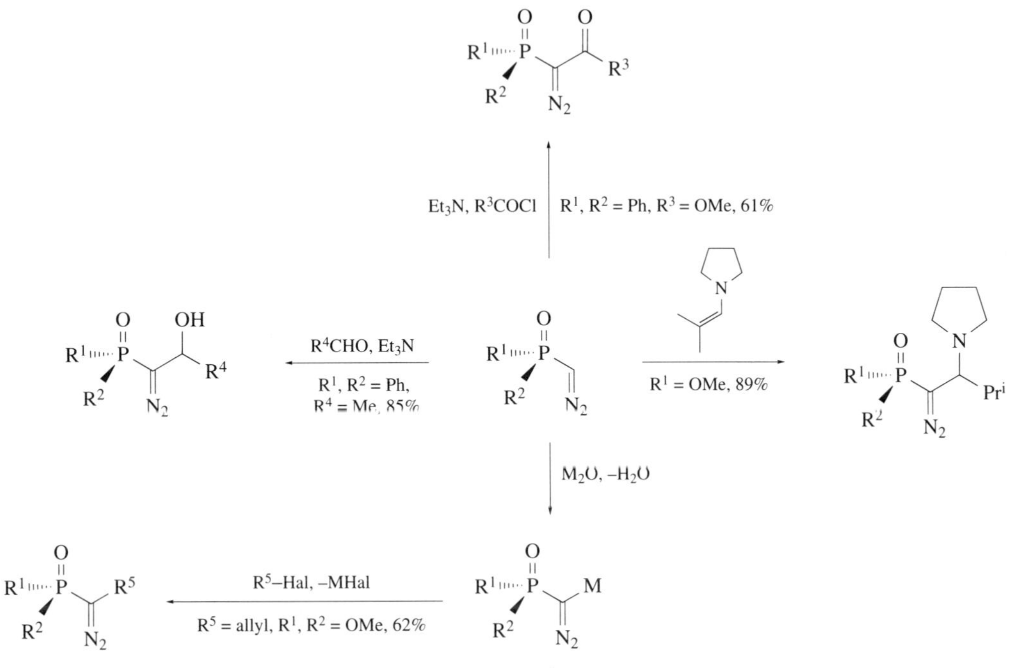

(18) M = ½ Hg
(19) M = Ag

Scheme 9

Metallated species are easily generated and some, for instance mercury (**18**) ⟨71LA(748)207⟩ and silver (**19**) ⟨67AG(E)261, 69CB488⟩ derivatives, are stable and isolable.

Alternatively, a mono-substituted diazoalkane or diazomethane itself can react with a phosphorus or arsenic electrophile ⟨88JA2663, 88JA6463, 90JA6277, 91JOC1801, 92JA6059⟩. The synthesis of trimethylarsinediazomethane, under carefully controlled reaction conditions, from the corresponding arsinamide is shown (Scheme 10) ⟨76JOM(110)195⟩. The trimethylsilyl substituted derivative is also known ⟨80JOM(191)371⟩. No reports of preparation of nonsubstituted or C-substituted antimony or bismuth diazomethanes exist, but the trimethylsilyl substituted analogues are prepared in a similar way ⟨80JOM(191)371⟩.

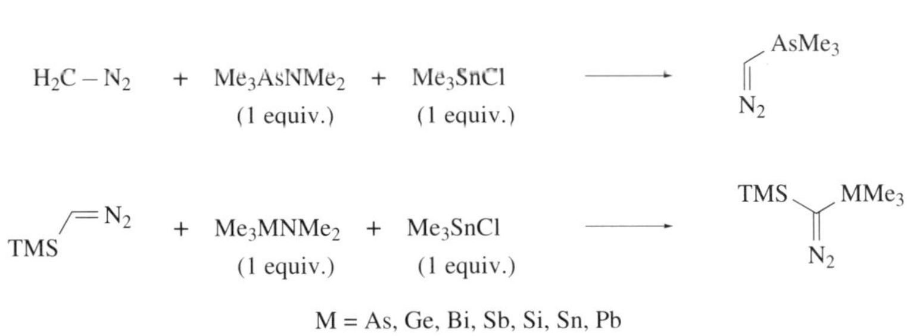

Scheme 10

Finally, an alternative approach, the oxidation of an α-diazophosphorus(III) to an α-diazophosphorus(IV) is demonstrated in Scheme 11 ⟨88JA2663, 88JA6463, 90JA6277, 91JOC1801, 92JA6059⟩.

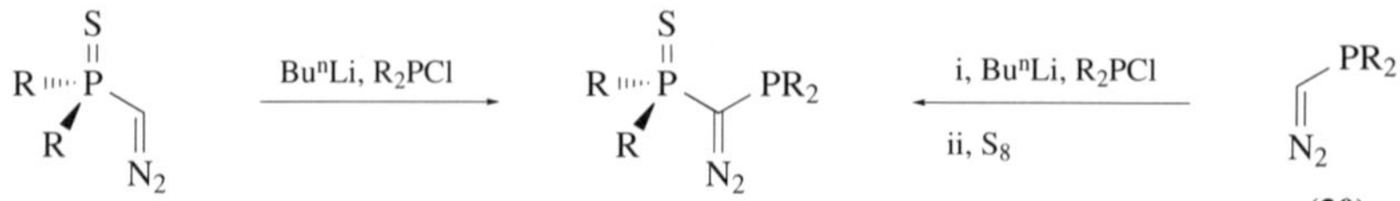

(20)

$R = NPr^i_2$

Scheme 11

It should be noted that base-induced cleavage (Equation (8); see also Equation (3) and related text) as applied to the synthesis of α-diazomethylcarbonyl ⟨73OSC(5)179⟩ and α-diazomethylsulfonyl compounds ⟨68JCS(C)2201⟩ is also reported for the synthesis of α-diazomethylphosphorus compounds ⟨89SC561⟩.

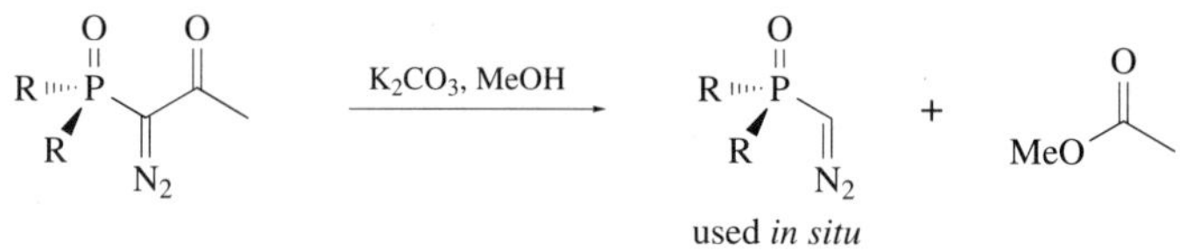

(8)

5.22.6 α-DIAZO BORON, SILICON AND GERMANIUM COMPOUNDS

5.22.6.1 α-Diazo boron compounds

The two reported diazomethylboron compounds, (**21**) and (**22**), are prepared from the corresponding chloroboranes (Scheme 12) ⟨74LA1767, 91JA5856⟩. Unlike the analogous α-diazomethylphosphorus compound, for example (**20**), deprotonation and reaction of (**22**) with electrophilic reagents affords only a nitrilimine, which will not rearrange to the α-diazomethylboron compound ⟨91JA5856, 91OM3205⟩.

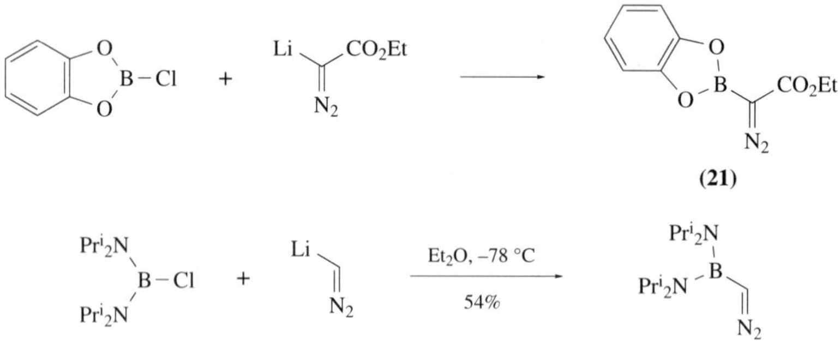

(21)

(22)

Scheme 12

Addition of borane (Y = H) or boron trifluoride (Y = F) to diazomethylenephosphorane (**23**) ⟨92BSF367⟩, furnished the adducts (**24**) and (**25**), as coloured oils (Equation (9)) ⟨91AG(E)1154⟩.

(9)

(**23**) $R = Pr^i_2N$

(**24**) Y = H, 87%
(**25**) Y = F, 83%

The Bamford–Stevens reaction (Scheme 1, route C) has also been applied to the synthesis of a diazomethylcarborane ⟨92JA1094⟩.

5.22.6.2 α-Diazo silicon compounds

The best known example of this family is diazomethyltrimethylsilane (**27**). Although a synthesis of this compound from the corresponding *N*-nitrosourea is reported ⟨81CPB3249⟩, the most convenient and efficient general preparation route is a modified diazo transfer reaction (Scheme 13) ⟨82CPB3380, 83CL871, 83TL4245, 90OS(68)1⟩. Diazosilane (**27**) is now commercially available.

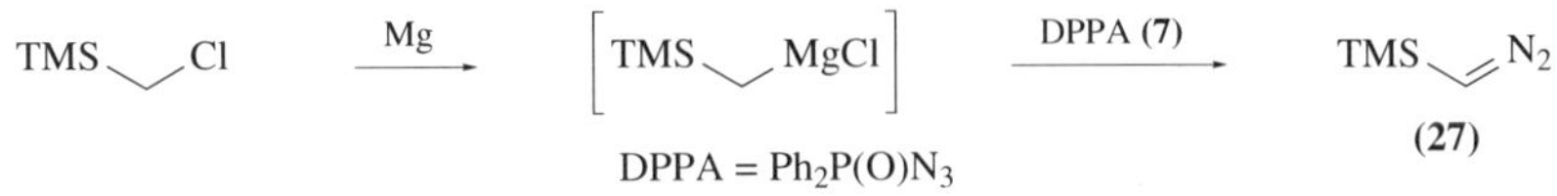

Scheme 13

A number of nongeneral syntheses of α-diazosilyl compounds exist. Silylation of both the lithium ⟨70AG(E)301, 72LA(761)137⟩ and mercury ⟨67AG(E)884⟩ salts of ethyl diazoacetate proceed smoothly (Scheme 14). α-Diazosilyl compounds can also be prepared by Bamford–Stevens reaction (Scheme 1, route C) ⟨77JA6995⟩ or by oxidation of hydrazones ⟨68TL4973⟩ (for comparison see Scheme 8). It is also reported that intramolecular rearrangement of pyrazoline (**28**) under thermal conditions afforded α-diazosilyl compound (**29**) (Scheme 14) ⟨87T215⟩. This reaction is analogous to the one observed previously for phosphorus (see Scheme 6).

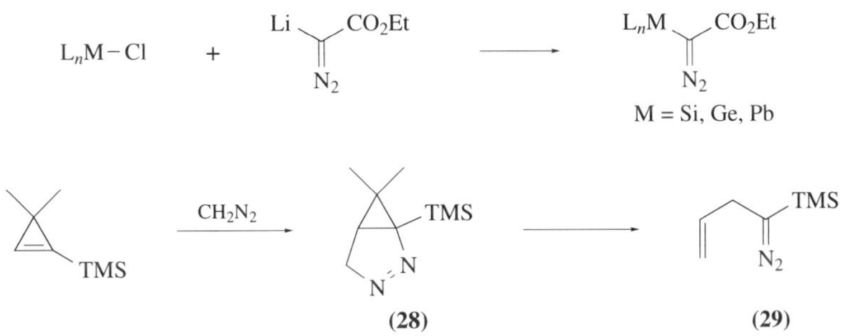

Scheme 14

Diazomethyltrimethylsilane (**27**) is a stable and safe substitute for diazomethane which can be efficiently used in its place for the Arndt–Eistert reaction ⟨80TL4461, 81CPB3249⟩, homologation of ketones ⟨80TL4619, 82CPB119⟩ and aldehydes ⟨81H(15)975⟩ and methyl esterification of carboxylic acids ⟨81CPB1475⟩.

Other α-trimethylsilyldiazoalkanes, for example (**30**), can be prepared by deprotonation of (**27**) with strong base such as *n*-BuLi, and quenching the resulting anion with an alkyl halide (Scheme 1, route D) ⟨86TL2005⟩ or other electrophiles (Scheme 15) ⟨80JOM(191)371⟩. Conversion of the resulting compound to acylsilanes is easily achieved by oxidation with mcpba ⟨86TL2005⟩.

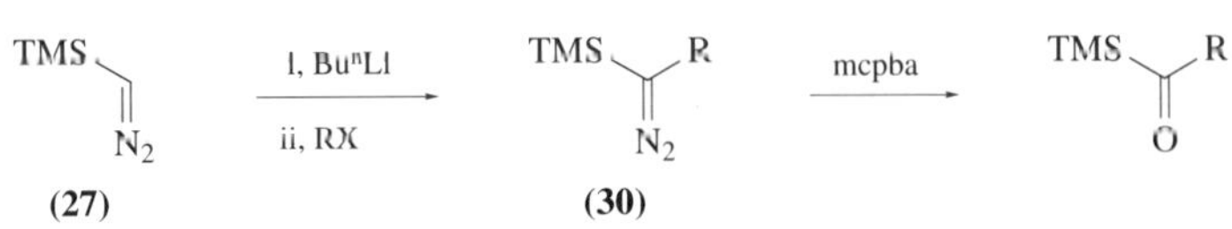

Scheme 15

The trimethylsilyl substituted diazomethane derivative of phosphorus ⟨88JA2663, 88JA6463, 90JA6277, 91JOC1801⟩, arsenic, bismuth, antimony, germanium, tin and lead ⟨80JOM(191)371⟩ are all known, and are prepared in a similar manner (Scheme 10).

The lithium salt (**31**) is also reported to react with a number of other electrophiles including cyclopropenylium cation ⟨85TL1845, 86AG(E)1093⟩ and ketenimines to give a triazole (**32**) or, if bearing electron-withdrawing groups at the C2 position, a 4-amino-3-trimethylsilylpyrazole (**33**) (Scheme 16) ⟨91S1163⟩.

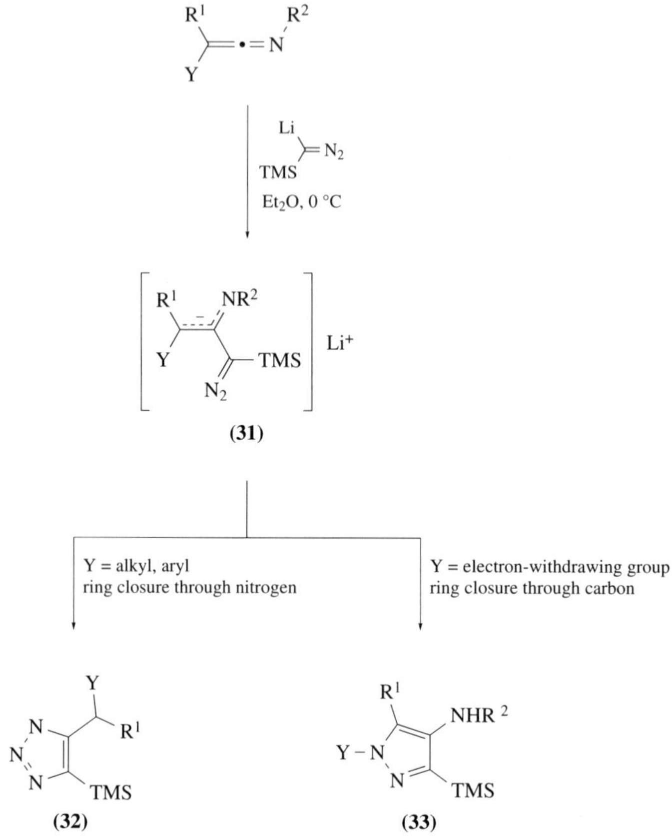

Scheme 16

5.22.6.3 α-Diazo germanium compounds

As with silicon derivatives, α-diazo germanium compounds are prepared by germylation of the lithiated diazo compounds (Scheme 14) ⟨72LA(761)137⟩. The only other reported example of a germanium substituted diazoalkane appears to be that in which the diazomethane is further substituted with a silicon group (Scheme 10) ⟨80JOM(191)371⟩.

5.22.7 α-DIAZO METAL COMPOUNDS

The subject of α-diazo metal derivatives has been reviewed ⟨80RCR357⟩. Metallated diazo species are easily generated ⟨B-86MI 522-01⟩, and some, in particular mercury ⟨68LA(716)204, 68TL5237, 71LA(748)207, 73JOC3937⟩ and silver ⟨67AG(E)261, 69CB488, 74CC466⟩ derivatives are stable and isolable. Tin derivatives are also known and studied ⟨67CC836, 70AG(E)301⟩. Lead-substituted diazoalkanes are known ⟨72LA(761)137, 80JOM(191)371⟩.

5.23
Phosphoacyl Functions and their As, Sb, and Bi Analogues

LOTHAR WEBER
University of Bielefeld, Germany

5.23.1 DICOORDINATE PHOSPHORUS, ARSENIC, ANTIMONY, AND BISMUTH FUNCTIONS—$R^1C(PR^2)X$, $R^1C(AsR^2)X$, $R^1C(SbR^2)X$ AND $R^1C(BiR^2)X$

5.23.1.1 Dicoordinate Phosphorus Functions—$R^1C(PR^2)X$

5.23.1.1.1 General comments

Phosphoacyl compounds of the type $R^1C(=PR^2)X$ are more commonly addressed as methylene phosphanes or phosphaalkenes, which expresses their close relationship to alkenes rather than to imines $R^1_2C(=NR^2)$. In keeping with this, the formation of the P=C double bond resembles in many respects well-known strategies of alkene syntheses. According to the relative importance of the respective method, eliminations, condensations, rearrangements, additions, and reactions at peripheral substituents at the P=C moiety are discussed. According to the so-called "double-bond rule", multiple bonding involving *p*-orbitals is a privilege only to the elements of the second period. Thus the rule excludes stable molecules with P=C, As=C, Sb=C, and Bi=C double bonds. In spite of this, the preparation of compounds with "forbidden multiple bonds" became possible by kinetic stabilization with bulky substituents, or by thermodynamic stabilization by means of effective electron delocalization.

Since the first report on acyclic compounds with carbon–phosphorus multiple bonds ⟨64AG(E)384⟩, this area of research has been rapidly expanding. Accounts on the progress in phosphaalkene chemistry are given in a number of excellent reviews ⟨81AG(E)731, 82HOU(E1)27, 89MI 523-01, 89T6019, B-90MI 523-02⟩. Heteroarenes such as phosphinines and phospholes, or heterophosphinines and heterophospholes are not considered in this chapter.

5.23.1.1.2 Halogen derivatives—$R^1C(PR^2)X(X = F, Cl, Br, I)$

(i) Fluorine derivatives—$R^1C(PR^2)F$

By analogy to the well-known isonitrile reaction, the stable 2,4,6-tri-*t*-butylphenylphosphane (supermesitylphosphane, Mes^*PH_2 (**1**)) reacts smoothly with dichlorofluoromethane in the presence of potassium hydroxide to afford the phosphaalkene (**2**) in 38% yield as a mixture of (*E*/*Z*)-isomers. The pure *E*-isomer of (**2**) could be separated by means of column chromatography and crystallization (Equation (1)) ⟨87ZAAC(553)7⟩.

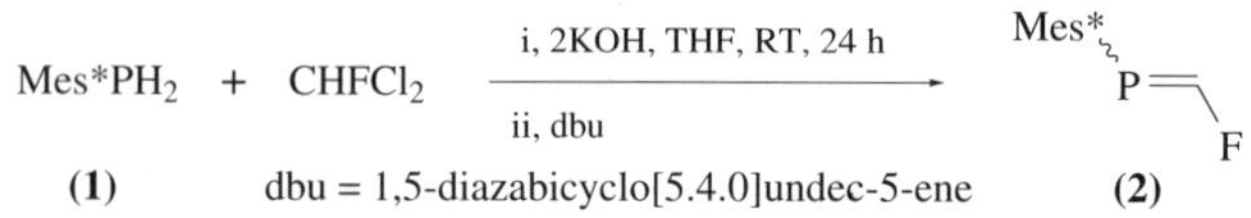

(1)

(**1**) dbu = 1,5-diazabicyclo[5.4.0]undec-5-ene (**2**)

(ii) Chlorine derivatives—$R^1C(PR^2)Cl$

(a) From chloromethyldichlorophosphane. The phosphane Cl_2PCH_2Cl is readily converted into the (*Z*)-configurated phosphaalkene (**3**) by treatment with two equivalents of sodium hexamethyldisilazide in ether at −70°C (80% yield). Initially the (*E*)-isomer of (**4**) was formed, but this rearranged to the final product after the reaction mixture was warmed to 20°C (Equation (2)) ⟨84ZOB1520⟩.

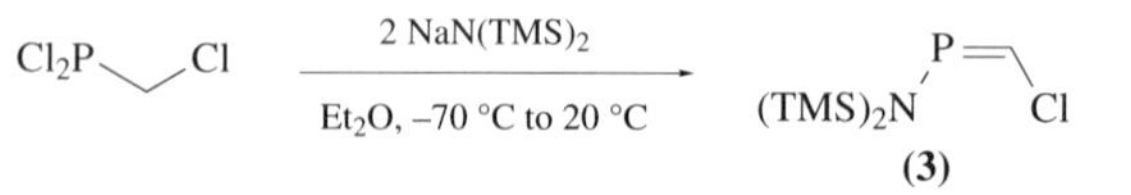

(2)

(**3**)

(b) From primary phosphanes, chloroform, and base. An analogous procedure as described for (**2**) was employed for the synthesis of an (*E*/*Z*) mixture of the chloro derivative (**4**) from (**1**) and chloroform (31% yield). The pure (*E*)-isomer was obtained by chromatography of the crude product and subsequent crystallization.

This reaction involves the insertion of *in situ* formed dichlorocarbene into a PH bond of (**1**). The intermediate phosphane Mes*P(H)—$CHCl_2$ was quantitatively converted into (**4**) by addition of 1,5-diazabicyclo[5.4.0]undec-5-ene (dbu) (Scheme 1) ⟨87ZAAC(553)7⟩.

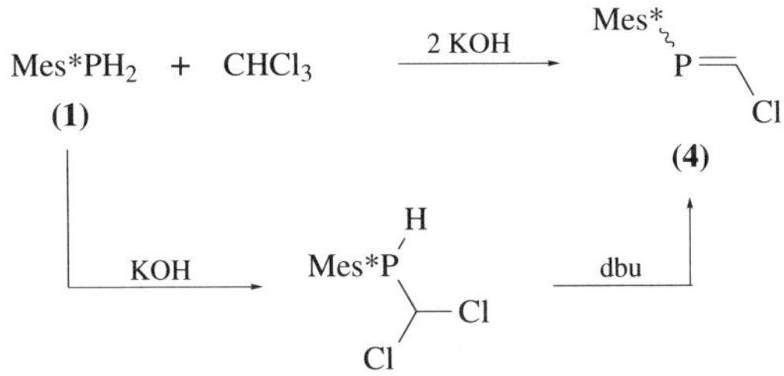

Scheme 1

(c) From chloromethylene phosphanes. Another synthetic approach to (**4**) makes use of the lithiation of the dichloromethylene phosphane (**5**) by *n*-butyllithium in an ether–toluene mixture (1 : 4) at −80 °C. The thermolabile lithium derivative (**6**) was quenched either with methanol to give (**4**) ⟨85TL3551⟩ or with methyl iodide to afford cleanly the (*Z*)-isomer of (**7**) (Scheme 2) ⟨85TL3551, 88CL1733⟩.

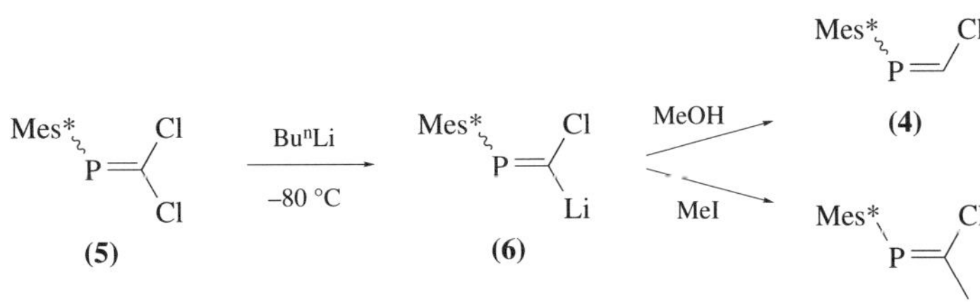

Scheme 2

Pure (*E*)-(**7**) was synthesized by deprotonation of (*E*)-(**4**) with two equivalents of *t*-butyllithium and the subsequent quench of the intermediate lithium derivative (*E*)-(**6**) with methyl iodide at −78 °C (Equation (3)) ⟨88CL1733⟩.

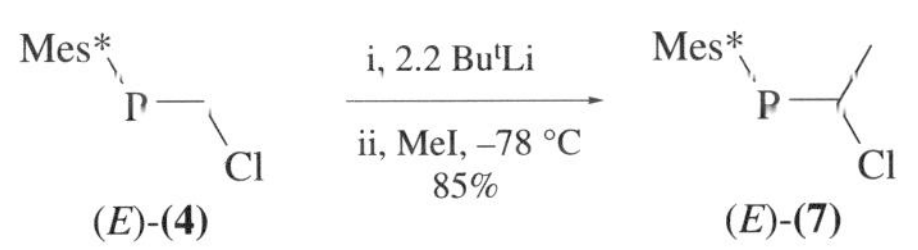

(3)

(iii) Bromine derivatives—$R^1C(PR^2)Br$

(a) From bromomethyldibromophosphane. By analogy with the preparation of (**3**), the (*E*)-configurated bromomethylene derivative (**8**) was obtained in 65% yield from Br_2PCH_2Br and two equivalents of $NaN(TMS)_2$ ⟨84ZOB1520⟩.

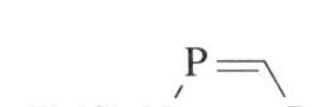

(**8**)

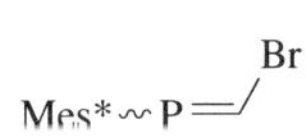

(**9**)

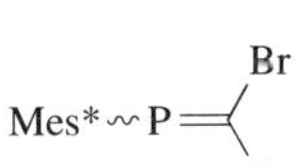

(**10**)

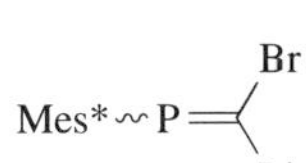

(**11**)

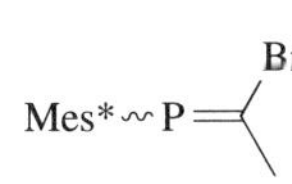

(**12**)

(b) From primary phosphanes, bromoform, and base. The synthesis of an (*E*/*Z*) mixture of Mes*P=CHBr (**9**) from $CHBr_3$, (**1**) and KOH followed the principles described above for the

chlorine analogue (**4**). Again it was possible to separate the pure (*E*)-isomer from the crude product ⟨87ZAAC(553)7⟩.

(c) From bromomethylene phosphanes. By analogy to Scheme 2, compound (**10**) (see Chapter 6.22) was lithiated by *n*-butyllithium at −80 °C to give unstable (**11**), which upon protonation with methanol afforded (**9**). Similarly, (**11**) was converted into a 1 : 1 mixture of (*E*)- and (*Z*)-(**12**) ⟨85TL3551⟩.

The C—Si bond of phosphaalkene (**13**) is subject to facile cleavage by NaOMe in boiling methanol with the result of the quantitative formation of an (*E*/*Z*) mixture of (**9**) from which the pure (*Z*)-isomer could be isolated (Equation (4)) ⟨84TL4109, 85TH86⟩.

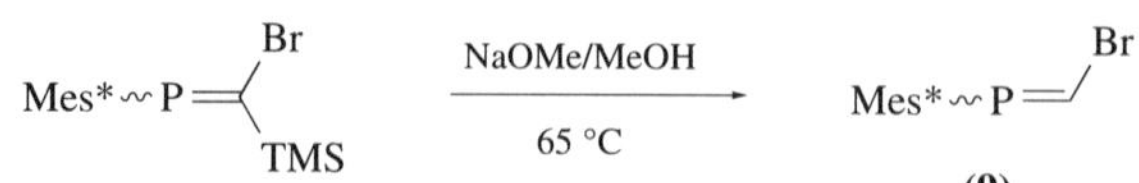

(13) **(9)** (4)

(iv) Iodine derivatives—$R^1C(PR^2)I$

(a) From iodoform. Two synthetic approaches to iodine-functionalized phosphaalkenes based on iodoform are available. The first parallels the preparation of (**4**) and (**9**) utilizing phosphane (**1**), CHI_3, and solid KOH in THF. An (*E*/*Z*) mixture of Mes* ~ P═CHI (**14**) was obtained as a dark brown oil (22%) ⟨87ZAAC(553)7⟩. Alternatively, the treatment of supermesityl dichlorophosphane (**15**) with CHI_3 and two equivalents of *n*-butyllithium in THF at −100 °C results in an (*E*/*Z*) mixture (4 : 1) of (**14**) in 90% yield. In the latter transformation, a metal/halogen exchange affords $LiCHI_2$, which serves as nucleophile and base (Scheme 3) ⟨91CB2677⟩.

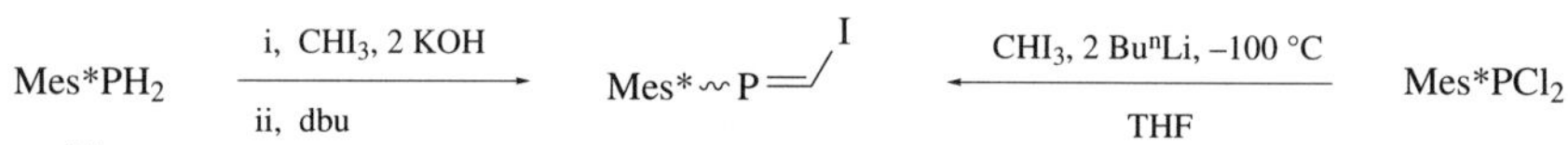

(1) **(14)** **(15)**

Scheme 3

(b) From diiodomethylene phosphanes. Compound (**16**) shows high reactivity towards *n*-butyllithium and undergoes halogen–metal exchange even at −120 °C to furnish (*Z*)-(**17**) exclusively; this was followed by methanolysis to give pure (*Z*)-(**14**) (75%) (Scheme 4) ⟨91CB2677⟩.

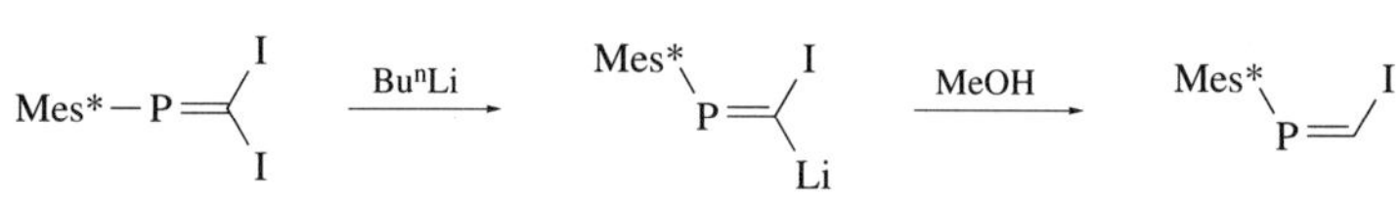

(16) (*Z*)-(**17**) (*Z*)-(**14**)

Scheme 4

5.23.1.1.3 Chalcogen derivatives—$R^1C(PR^2)ER^3{}_n$ (E = O, S, Se, Te)

(i) Oxygen derivatives—$R^1C(PR^2)OR^3$

(a) From silylphosphanes and acid chlorides. The most important and versatile synthesis of oxygen-functionalized phosphaalkenes is based on the observation by Becker, that silylated acylphosphanes $R^2(TMS)PCOBu^t$ (**18**; R^2 = Me, Bu^t, *c*-C_6H_{11}, Ph, TMS) underwent a facile 1,3-silatropic rearrangement to afford the phosphaalkenes (**19**). The driving force of this process is the formation of a stable Si—O bond. The precursors (**18**) are conveniently accessible by the reaction of equimolar amounts of pivaloyl chloride and bis(trimethylsilyl)phosphanes $R^2P(TMS)_2$ ⟨76ZAAC(423)242⟩ or $P(TMS)_3$ ⟨77ZAAC(430)66⟩. Other acid chlorides R^1COCl with bulky substituents have also been employed successfully (R^1 = Pr^i, Bu^tCH_2, 1-MeC_6H_{10} ⟨86JOM(306)39⟩, R^1 = 1-adamantyl ⟨86S31⟩. Alternatively the more nucleophilic $LiP(TMS)_2 \cdot 2THF$ may be used instead of $P(TMS)_3$ (R^1 = Bu^t ⟨87CC980⟩, R^1 = 1-MeC_5H_8 ⟨86JOM(306)39⟩, R^1 = 1-adamantyl ⟨86S31⟩. In the case of benzoyl chloride the formation of monomeric phosphaalkenes is possible with bulky

disilylphosphanes $(TMS)_2PR^2$ ($R^2 = Bu^t$, Mes ⟨80TH158⟩) or with $LiP(Mes^*)(SiBu^tMe_2)$ (Scheme 5) ⟨83CL1653⟩.

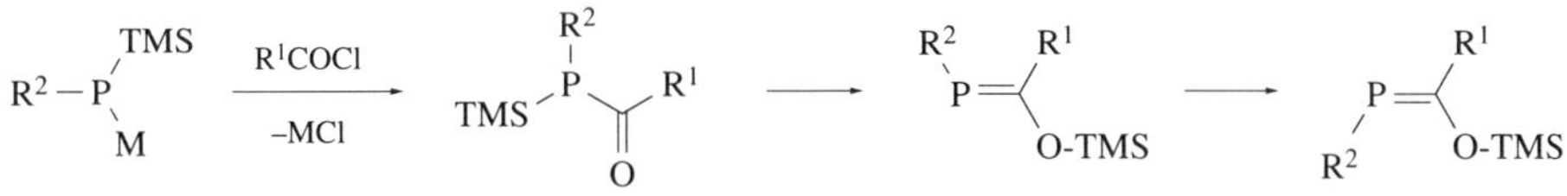

M = Li, TMS (**18**) (*E*)-(**19**) (*Z*)-(**19**)

Scheme 5

Compounds (**18**) cannot be isolated because of their pronounced tendency for silatropy. In some cases, however, they were detected by ^{31}P-NMR spectroscopy. The ease of the 1,3-silyl migration is governed by the steric requirements of the substituent R^2 at phosphorus. Usually the (*E*)-phosphaalkenes are initially formed and are rearranged subsequently into the thermodynamically more stable (*Z*)-isomers.

The size of the substituent R^2 at the phosphorus atom is also responsible for the kinetic stabilization of (**19**). In keeping with this at 20 °C, phosphaalkene $H—P=C(Bu^t)OTMS$ slowly underwent a head-to-tail dimerization to the corresponding 1,3-diphosphetane ⟨81ZAAC(473)7⟩. The conditions of the synthesis of a few selected phosphaalkenes of the type (**19**) are compiled in Table 1.

Table 1 Selected phosphaalkenes $R^2—P=C(R^1)OTMS$ from carboxylic acid chlorides and silylphosphanes.

*R*¹	*R*²	*Reagents and conditions*	*Yield* (%)	*Ref.*
Pr^i	TMS	Pr^iCOCl, $(TMS)_3P$, n-C_5H_{12}, 20 °C	84	86JOM(306)39
Bu^t	H	Bu^tCOCl, $(TMS)_2PH$, c-C_5H_{10}, 20 °C		81ZAAC(473)7
Bu^t	Me	Bu^tCOCl, $(TMS)_2PMe$, c-C_5H_{10}, 0–20 °C	>95	76ZAAC(423)242
Bu^t	Bu^t	Bu^tCOCl, $(TMS)_2PBu^t$, c-C_5H_{10}, 0–20 °C	>95	76ZAAC(423)242
Bu^t	c-C_6H_{11}	Bu^tCOCl, $(TMS)_2P$-c-C_6H_{11}, c-C_5H_{10}, 0–20 °C	>95	76ZAAC(423)242
Bu^t	Ph	Bu^tCOCl, $(TMS)_2PPh$, c-C_5H_{10}, 0–20 °C	>95	76ZAAC(423)242
Ph_2CH	Bu^t	$Ph_2CHCOCl$, $(TMS)_2PBu^t$, Et_2O, −40 °C to 20 °C	80	81JGU1883
Bu^t	TMS	Bu^tCOCl, $(TMS)_3P$, c-C_5H_{10}, 20 °C	91	77ZAAC(430)66
Bu^tCH_2	TMS	Bu^tCOCl, $(TMS)_3P$, n-C_5H_{12}, 20 °C	91	86JOM(306)39
1-MeC_5H_8	TMS	1-MeC_5H_8COCl, $(TMS)_2PLi \cdot 2THF$, n-C_5H_{12}, −40 °C	75	86JOM(306)39
1-MeC_6H_{10}	TMS	1-$MeC_6H_{10}COCl$, $(TMS)_3P$, n-C_5H_{12}, 20 °C	81	86JOM(306)39
1-Ad	TMS	1-AdCOCl, $(TMS)_2PLi \cdot 2THF$, c-C_5H_{10}, −40 °C	96	86S31
Ph	Bu^t	PhCOCl, $(TMS)_2PBu^t$, c-C_5H_{10}, −60 °C to 20 °C	74	80TH158
Ph	Mes	PhCOCl, $(TMS)_2PMes$, DME, −50 °C to 20 °C	72	80TH158

The reaction of $P(TMS)_3$ or $LiP(TMS)_2 \cdot 2THF$ with acyl chlorides R^1COCl is very sensitive to the nature of the substituent R^1 and to the stoichiometry of the reactants. Thus monoacetyl- and diacetylsilylphosphanes could not be synthesized from acetyl chloride and $P(TMS)_3$. These species are prone to facile disproportionation to give PAc_3. The latter compound was obtained in 83% yield, when the reaction was conducted with an excess of the acid chloride ⟨81ZAAC(480)21⟩. On the other hand, a 2:3 mixture of mesitoyl chloride and $LiP(TMS)_2 \cdot 2THF$ in THF gave rise to the formation of chelate (**20**) (Equation (5)) ⟨91ZAAC(605)7⟩.

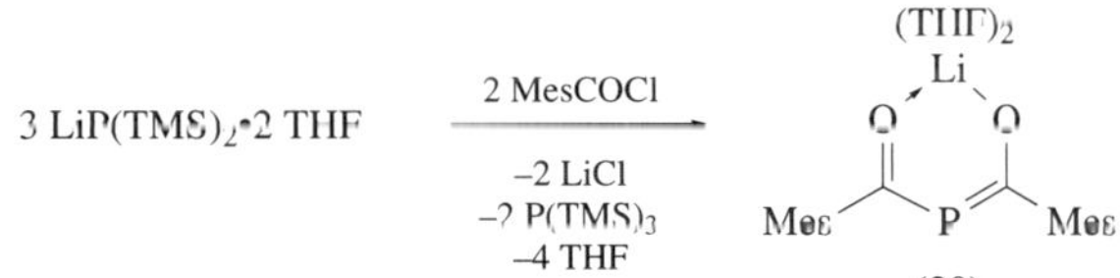 (5)

(**20**)

(b) From lithium phosphides and carboxylic acid derivatives. Aromatic acid chlorides and lithium dihydrogen phosphide (**21**) in a molar ratio of 1:2 undergo reaction to the dimeric lithiated phosphaalkenes (**22**) (Equation (6)) ⟨85MI 523-01⟩.

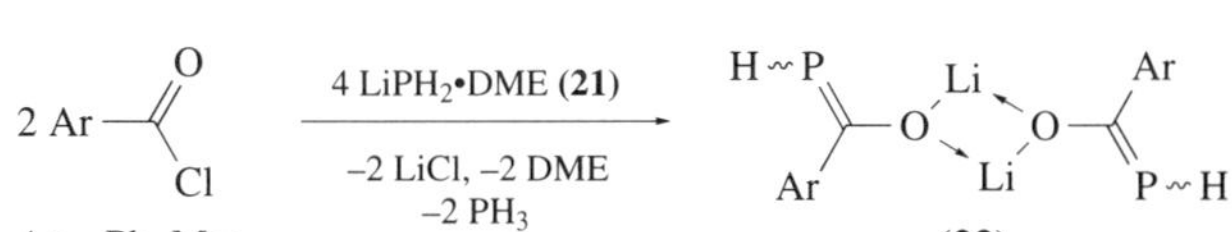 (6)

Ar = Ph, Mes (**22**)

With aliphatic carboxylic acid chlorides, the reaction was complicated by a double acylation of (**21**). This problem was circumvented by the employment of carboxylic esters, where no further acylation took place (**23**) (Scheme 6) ⟨85MI 523-01, 92ZAAC(612)72⟩.

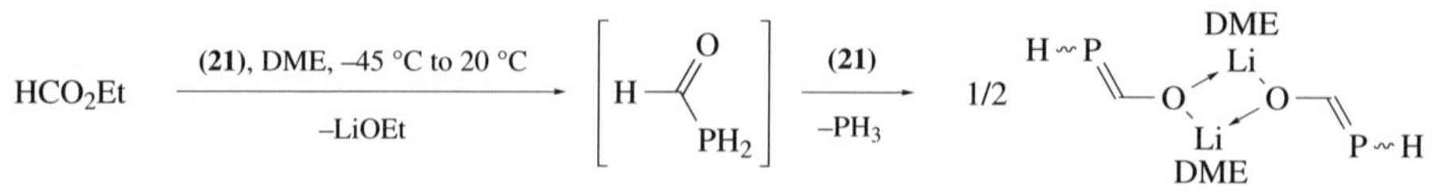

(**23**)

Scheme 6

(c) From lithiated acyl phosphides. In some cases the lithiation of stable and isolable acyl phosphides was possible. The lithium derivatives were subsequently converted into the corresponding phosphaalkenes by quenching with electrophiles (Scheme 7) ⟨87AG(E)779⟩. In 1,4-diphosphabuta-1,3-diene (**24**), the presence of the bulky supermesityl group is crucial to avoid intramolecular cyclization.

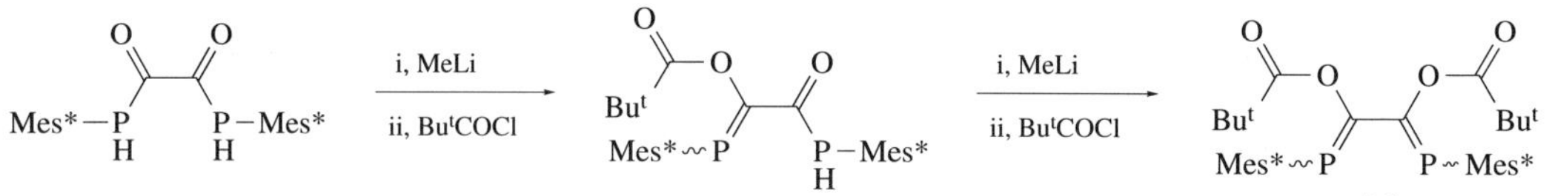

(**24**)

Scheme 7

(d) From imidoyl chlorides and α,ω-hydroxyalkyl phosphanes. A different synthetic approach to oxygen derivatives of phosphaalkenes is based upon the condensation of *N*-*p*-tolylpivalimidoyl chloride and α,ω-hydroxyalkylphosphanes in ether at 20 °C (Scheme 8) ⟨91CB493⟩.

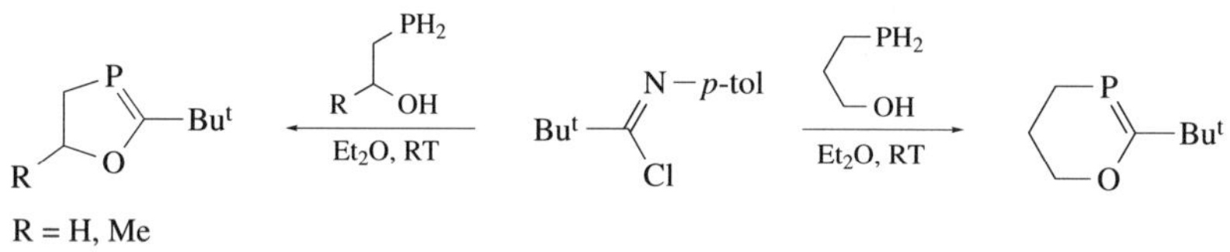

Scheme 8

(e) From peripheral reactions. The silyl group at the phosphorus atom in TMS-P=C(Bu^t)O-TMS (**25**) can conveniently be replaced by other functionalities leading to further phosphaalkenes. Chlorination by hexachloroethane furnished the isolable species (**26**) (65%), which is further converted into (**27**) by an excess of (**25**) (75%) ⟨83CB938⟩, whereas treatment of (**25**) with an excess of neat pivaloyl chloride at 70–80 °C gave compound (**28**) in 68% yield (Scheme 9) ⟨77ZAAC(430)66⟩.

With a number of diazoalkanes $RCHN_2$ (R = H, Me, CO_2Bu^t) a formal insertion of carbene RCH into the PSi bond of (**25**) with formation of (**30**) was observed. It was believed that this process was initiated by a 1,3-dipolar cycloaddition to give transient (**29**). This route to phosphaalkenes (**30**) is not generally applicable. With more bulky diazoalkanes (R = Bu^t, TMS) (**25**) was transformed into 1,2,4-diazaphospholes ⟨85ZN(B)1077⟩.

(ii) Derivatives—$R^1C(PR^2)ER^3$

Acyclic derivatives $R^1C(PR^2)ER^3$ with sulfur, selenium, and tellurium are not described in the literature.

5.23.1.1.4 Nitrogen derivatives—$R^1C(PR^2)NR^3_2$

The major synthetic pathway to nitrogen-functionalized phosphaalkenes utilizes the condensation of functionalized phosphanes and phosphides with carboxylic amides, imidoyl chlorides, and other amino-functionalized electrophiles.

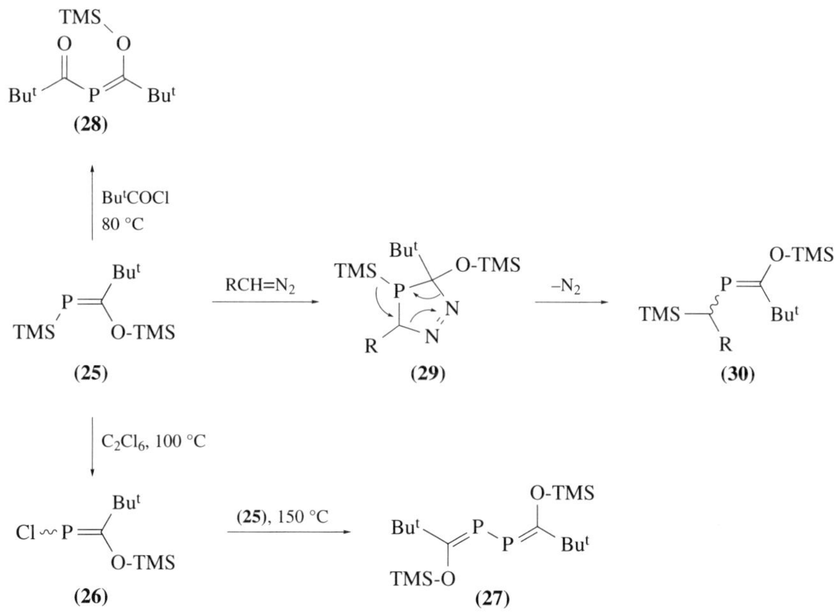

R = H, Me, CO_2Bu^t

Scheme 9

(i) From imidoyl chlorides

The transformation of disilylphosphanes and imidoyl chlorides to amino-substituted phosphaalkenes (**32**) parallels the synthesis of the oxygen-functionalized analogues (**19**). Thus, heating a 1 : 1 mixture of phenyl bis(trimethylsilyl)phosphane with imidoyl chlorides at 150–160°C led to (**32**) in 23–85% yield. Intermediates such as (**31**) could not be detected by ^{31}P-NMR spectroscopy (Scheme 10). According to preliminary and informative experiments, this reaction was also operative with alkyl disilylphosphanes ⟨78JOM(160)47⟩.

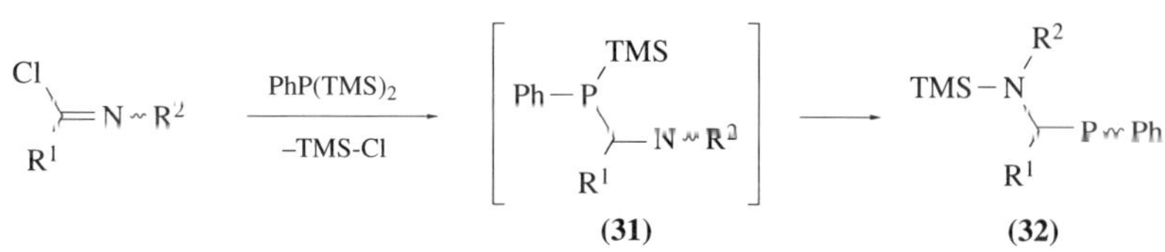

R^1 = Ph, Mes, Bu^t; R^2 = Ph, 2,6-$Et_2C_6H_3$, Me, c-C_6H_{11}

Scheme 10

(ii) From chloromethylene immonium salts

Chloromethylene immonium salts $[Me_2N{=}C(Cl)R]Cl$ have been successfully converted into amino derivatives of phosphaalkenes.

The reaction of lithium (triphenylstannyl)phenyl phosphide with chloromethylene dimethylimmonium chloride gave phosphaalkene (**33**) in 50% yield. This synthetic principle cannot be generalized as shown by the reaction of $Li(Ph_3SnPPh)$ with $[Me_2N{=}C(Cl)Bu^t]^+Cl^-$ and $[Me_2N{=}C(Cl)Ph]^+Cl^-$ which led to tetraphospholanes (Scheme 11) ⟨85MI 523-02⟩.

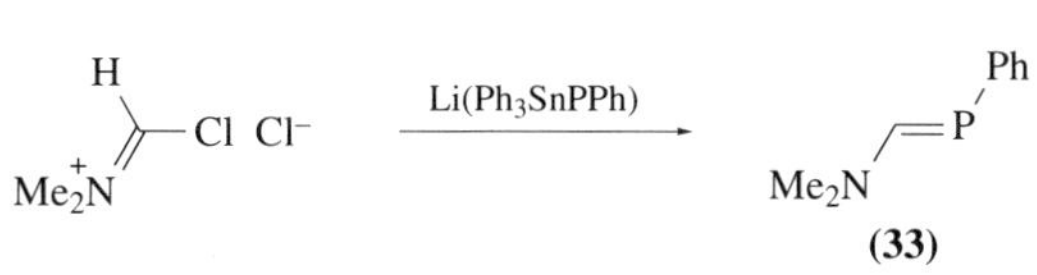

Scheme 11

2-Phosphaallyl chlorides (**34**), with the structural feature of a nitrogen-derived C═P bond, are accessible by the reaction of $P(TMS)_3$ with a twofold excess of chloromethylene immonium salts in acetonitrile (80–96% yield) (Equation (7)) ⟨84AG(E)903⟩.

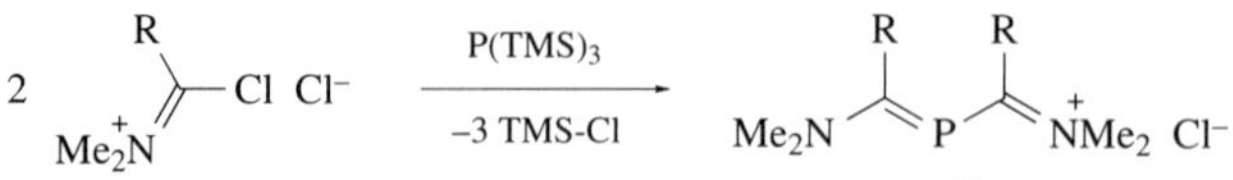

(7)

(34)

R = H, Bu^t, c-C_6H_{11}, Ph, 4-$O_2NC_6H_4$

In principle, the conversion of $[Me_2N{=}C(Ph)Cl]Cl$ to (**34**) may also be achieved with white phosphorus. This variation, however, has not become preparatively useful ⟨B-90MI 523-01⟩. Phosphaallylic cations and chloromethylene immonium salts take part in an exchange reaction by which unsymmetrical examples such as (**35**) are obtained (Equation (8)) ⟨B-90MI 523-01⟩.

(8)

(34) **(35)**

R^1 = Ph, R^2 = H; R^1 = Ph, R^2 = NMe_2; R^1 = H, R^2 = NMe_2

The reaction of the 2-chloroquinolinium salts (**36**) with trishydroxymethylphosphane ⟨64AG(E)384, 66CB1325⟩, or alternatively $P(TMS)_3$ ⟨76TL3489⟩, furnished phosphamethine cyanines (**37**), which were the first compounds with a PC multiple bond to be prepared (Equation (9)).

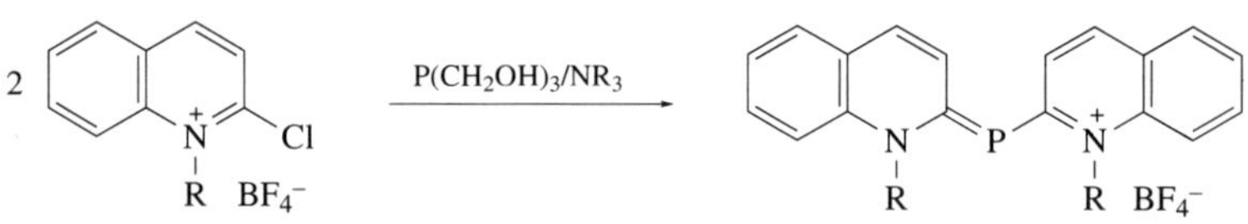

(9)

(36) **(37)**

(iii) From amides

The very sluggish condensation between neat organodisilylphosphanes and an excess of DMF can be markedly accelerated by solid sodium hydroxide. For example, the time for quantitative conversion of $Bu^tP(TMS)_2$ to $Bu^tP{=}CHNMe_2$ is shortened from one year to only three days. Aryl phosphanes are more reactive towards DMF as indicated by the condensation with $MesP(TMS)_2$ to give $MesP{=}CHNMe_2$ within the period of 24 h (85% yield). The catalytic influence of solid NaOH is rationalized by the formation of reactive NaP(TMS)R at the surface of the pellet. In keeping with this, the reaction of the formamide with an ethereal solution of LiP(TMS)Mes to give nitrogen-functionalized phosphaalkenes is straightforward (90% yield) (Equations (10) and (11)). The generation of monomeric phosphaalkenes of this type is crucially dependent on the size of the substituents at phosphorus.

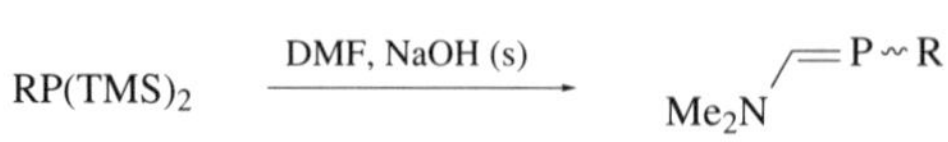

(10)

R = Bu^t, Ph, Mes

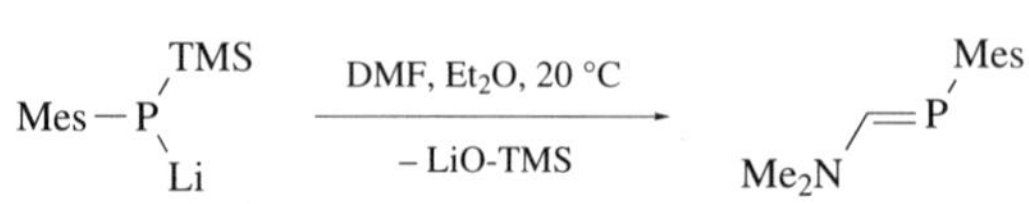

(11)

(38)

The methyl derivative, though detectable by ^{1}H-NMR, rapidly dimerized to the corresponding 1,3-diphosphetane. Upon warming, the phenyl derivative (**33**) also underwent cyclodimerization ⟨81ZAAC(479)41⟩.

(iv) From carbenium salts

The phosphaalkenes (**33**) and (**38**) are also available from the reaction of the carbenium salt $[EtO(Me_2N)CH]^+BF_4^-$ (**39**) with sodium phenylphosphide or lithium mesitylphosphide, respectively. This synthetic approach is restricted to arylphosphides and KPH_2. The more basic alkylphosphides give rise to deprotonation of the cation with the formation of ethene, DMF, and primary phosphane. The PH-functionalized species (**41**) and (**42**) are obtained in low yield (23% and 27%, respectively) from the interaction of KPH_2 with the carbenium salts (**39**) and (**40**) in THF at ambient temperature (Scheme 12) ⟨83ZC99⟩.

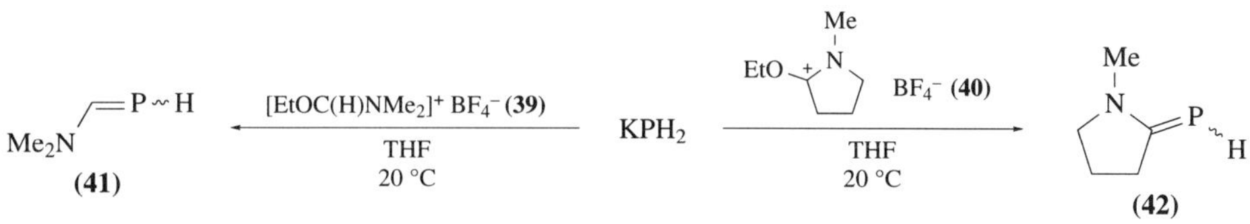

Scheme 12

(v) From carboxylic amide acetals

A further synthetic route to the compounds under discussion is based on the condensation of primary arylphosphanes with amide acetals at 80 °C (Equations (12) and (13)) ⟨80TL1141, 83TL5885⟩. The transformation is catalyzed by acid, but cannot be extended to alkyl phosphanes, mainly because of the reduced acidity of these species.

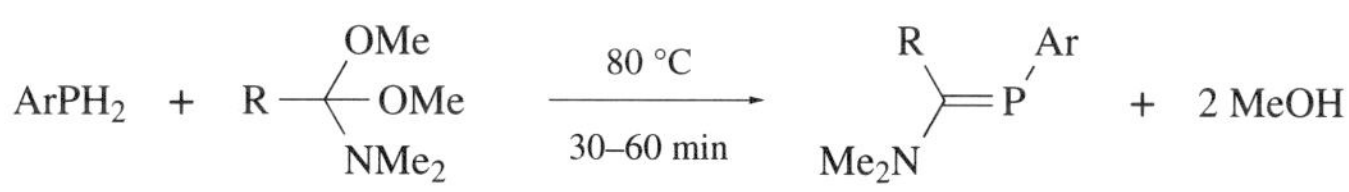

(12)

Ar = Ph, Mes, R = H; Ar = Ph, R = Me; Ar = Mes*, R = H, Me

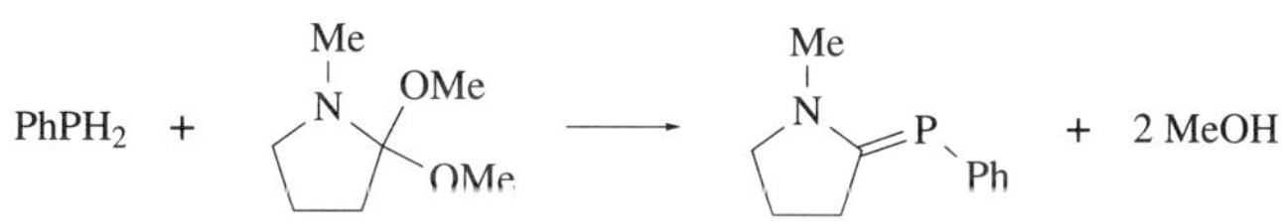

(13)

(vi) From phosphaalkynes and azetes

Phosphaalkynes Bu^t—C≡P and Bu^{t}CH$_2$C≡P undergo a Diels–Alder reaction with tri-*t*-butylazete to give the 1 Dewar 1,3-azaphosphinines (**43**) in high yield (Equation (14)) ⟨88AG(E)272, 88AG(E)1484⟩.

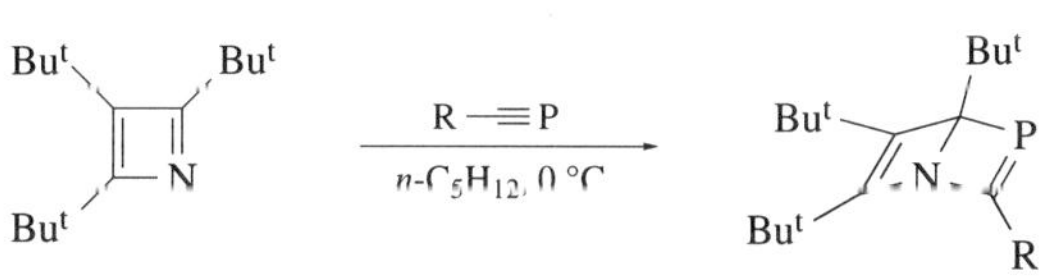

(14)

R = Bu^t, Bu^{t}CH$_2$

(**43**)

5.23.1.1.5 *Group 15 element derivatives—R^{1}C(PR^2)ER^3_n (E = P, As, Sb, Bi)*

Phosphaalkenes with substituents based on the heavier group 15 elements are restricted to phosphorus up to early 1995.

(i) Derivatives with dicoordinate phosphorus—$R^1C(PR^2)PR^3$

The reaction of phosphaalkyne Mes*C≡P with methyllithium in 2:1 stoichiometry produced the 1,3-diphosphabutadienyl lithium derivative (**44**) which was intercepted by isopropyl chloride to afford compound (**45**) (Scheme 13) ⟨88CC171⟩.

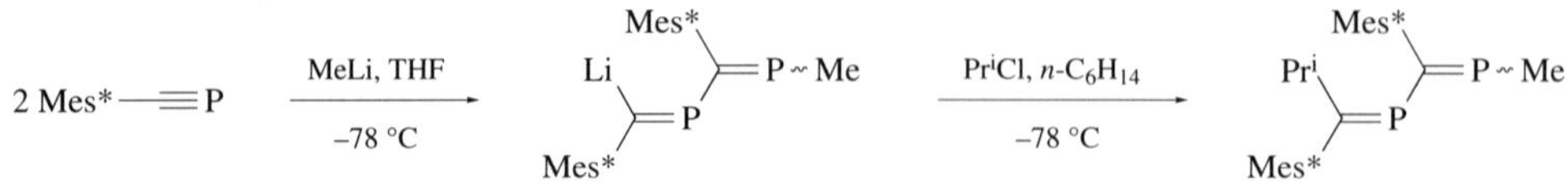

Scheme 13

(ii) Derivatives with tricoordinate phosphorus—$R^1C(PR^2)PR^3_2$

(a) From 1,3-dichloro-1,3-diphosphapropanes. Dehydrohalogenation of the 1,3-dichloro-1,3-diphosphapropane (**46**), by one equiv. of dbu furnished the phosphino-substituted phosphaalkene (**47**). The P—Cl function of (**47**) is subject to facile nucleophilic substitution by LiR (R = I, Bu^t, Mes*, H) to give the derivatives (**48**) (Scheme 14) ⟨84TL3687⟩.

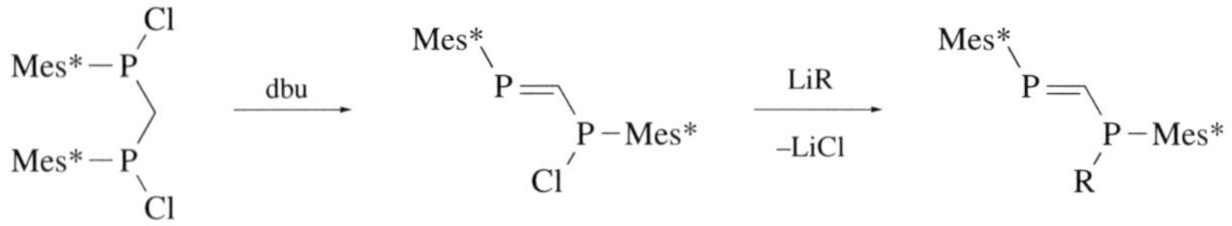

Scheme 14

When bis(dichlorophosphino)methane (**49**) was allowed to react with four equivs of sodium bis(trimethylsilyl)amide in ether in the range −40 °C to 20 °C, phosphaalkene (**51**) was isolated in 94% yield. The process proceeds via intermediate (**50**) which experiences subsequent nucleophilic substitution and dehydrochlorination by the amide. This mechanism was supported by the conversion of (**50**) into (**51**) in a separate experiment (Scheme 15) ⟨81DOK(256)1401, 81ZOB482⟩. However, a different situation was encountered in the reaction of (**52**) with two equivs of the amide whereupon phosphaalkene, (**54**), was produced. Obviously the initially formed phosphaalkene (**53**) suffered from a rapid 1,3-shift of the *t*-butyl group from the tricoordinate to the dicoordinate phosphorus atom (Scheme 16) ⟨81DOK(256)1401, 81ZOB482⟩.

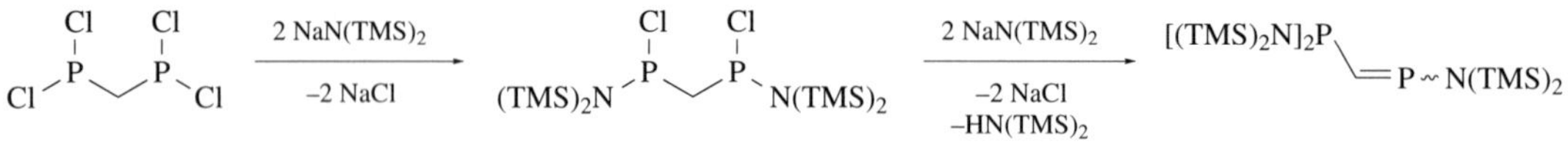

Scheme 15

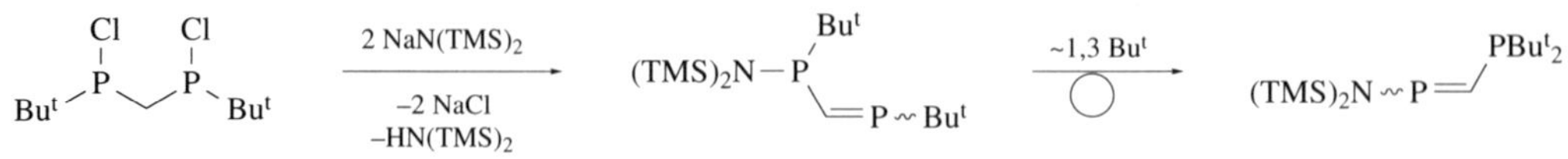

Scheme 16

(b) From a 1,3-diphosphaallene. The reduction of the 1,3-diphosphaallene Mes*P=C=PMes* (**55**) with Red-Al gave phosphaalkene (*E*)-Mes*P=CHPHMes*. Thermolysis of (**55**) in refluxing toluene furnished (**56**) as the product of an intramolecular cyclization (Equation (15)) ⟨88TL333⟩.

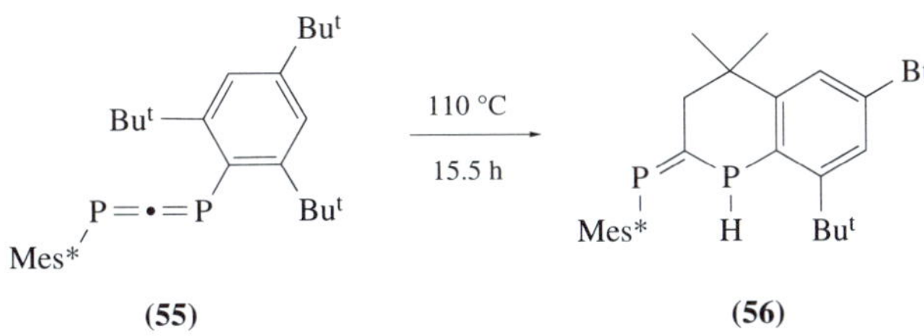

(15)

(c) From phosphaalkynes. The zwitterionic phosphaalkene (**58**) resulted from a spirocyclotrimerization between phosphaalkyne (**57**) and aluminum chloride in a 3 : 1 molar ratio (95% yield) (Scheme 17). When (**58**) was freed from $AlCl_3$ by the addition of an excess of DMSO in the presence of (**57**) in CH_2Cl_2 ($-45\,°C$), the cyclotetramer (**60**) was isolated in 37% yield. This process involves the generation of transient (**59**) which was intercepted by the phosphaalkyne in a homo-Diels–Alder reaction.

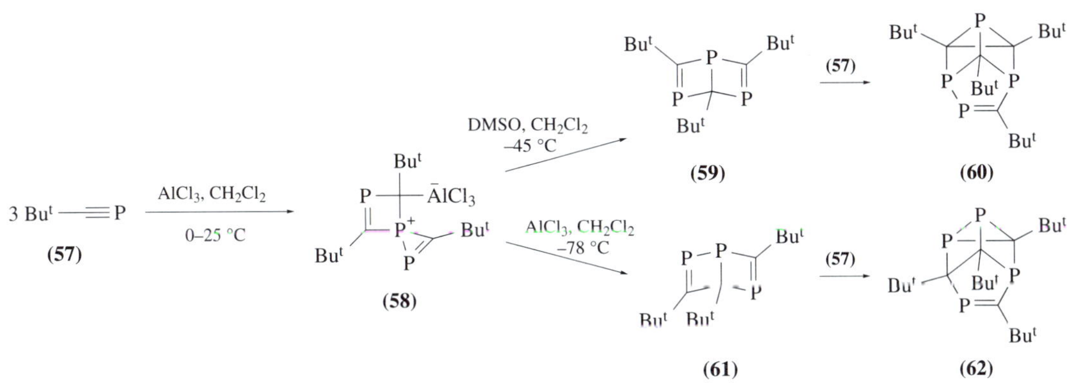

Scheme 17

The cyclotetramerization took a different course when (**58**) was exposed to an excess of $AlCl_3$ in CH_2Cl_2 at $-78\,°C$. In the presence of (**57**) the transient 1,2,5-triphospha-Dewar-benzene (**61**) was trapped to give the isomeric tetraphosphabishomoprismane (**62**) in 38% yield. Both processes are highly selective 〈92AG(E)1055〉.

When a suspension of the benzene-1-diazonium-2-carboxylate (**63**) was thermolyzed at 80–90 °C in the presence of an excess of phosphaalkynes R—C≡P (R = Bu^t, CMe_2Et, 1-adamantyl), the dihydrodiphosphetes (**66**) with the structural features of a *C*-phosphino phosphaalkene were isolated in low yield (3–16%). The key steps of the reaction are the [4 + 2] cycloaddition of betaine (**64**) with R—C≡P to give transient (**65**), which undergoes a [2 + 2] cycloaddition with the phosphaalkyne educts (Scheme 18) 〈93CB1945〉.

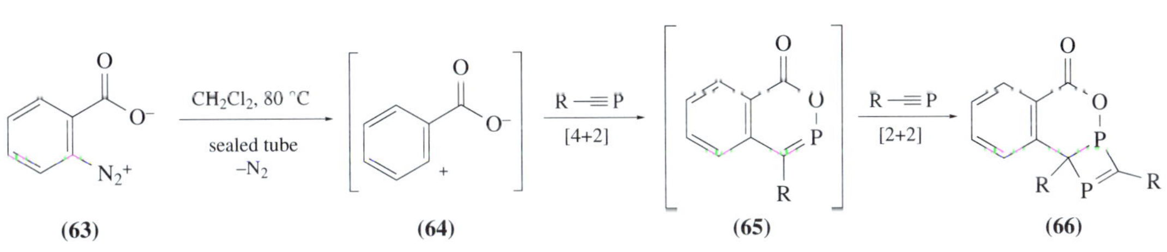

Scheme 18

The transition metal-mediated cooligomerization between (**57**) and tolane led to complex (**67**) in 76% yield (Equation (16)) 〈91AG(E)310〉. Treatment of (**57**) with one equiv. of the iminophosphanes (**68**) and (**69**) produced the 1,2,4-azadiphosphetines (**71**). In the case of (**68**) the unstable [2 + 1] cycloadduct (**70**) has been detected as an intermediate of this transformation (Equation (17)) 〈89TL459〉. The rate of the ring-opening reaction is influenced by the steric bulk of the iminophosphane. Whereas with (**68**) the rearrangement of (**70**) to give (**71**) took several hours at 20 °C, with (**69**) no product of the type (**70**) was observed. In contrast, the more sterically demanding

Et_3C—P=N—Mes* and (**57**) afforded a three-membered ring product, which completely resisted ring opening.

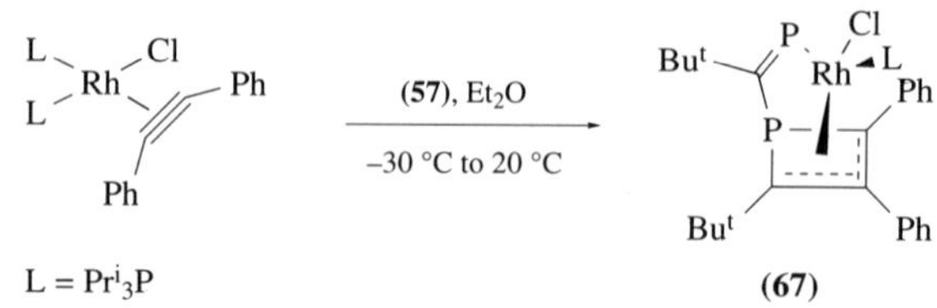

(16)

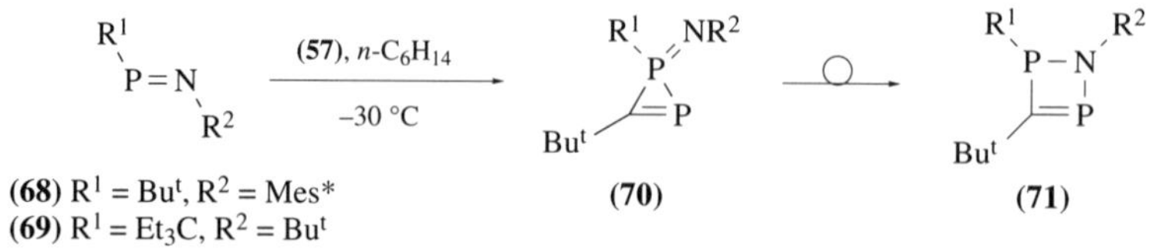

(17)

(d) By miscellaneous methods. The phosphaalkenes (**73**) were obtained as products of the reductive ring opening of the halogenated diphosphiranes (**72**) by Na[Cp$(CO)_2$Fe] in boiling toluene (Equation (18)). This reaction is of no general applicability since (**72**) is usually converted by carbonyl metalates such as $Na^+[C_5H_5(CO)_3M]$ (M = Mo, W) or Na[Co$(CO)_4$] into phosphenium- or η^3-1,3-diphosphaallyl complexes ⟨93PS(76)41⟩.

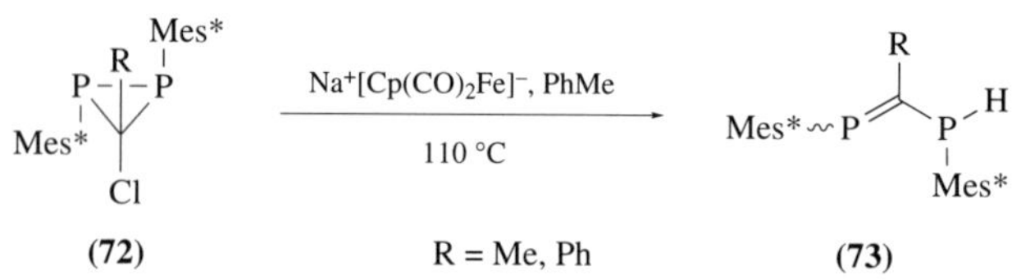

(18)

Usually the interaction of 1,3-diphospholides with transition-metal halides gives rise to the formation of η^5-diphospholyl transition-metal complexes. In contrast to this, the treatment of a solution of Na[$P_3C_2Bu^t_2$] and Na[$P_2C_3Bu^t_3$] with $[RuCl_2(COD)]_n$ in DME at −30 °C produced orange crystals of (**74**) in 37% yield. Careful control of the reaction temperature is necessary, since, at 20 °C, the reaction proceeded differently to afford complex [Ru(η^5-$P_2C_3Bu^t_3$)(η^5-C_8H_{11})] (Equation (19)) ⟨93CC267⟩.

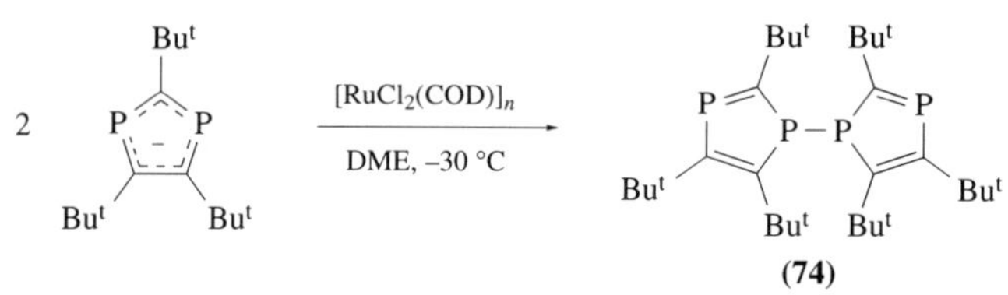

(19)

(iii) Derivatives with tetracoordinate phosphorus—$R^1C(PR^2)PR^3_3$

(a) From ylides. Phosphonio-substituted phosphaalkenes (**76**) are readily available by the reaction of the ylides (**75**) with two equivalents of $BF_3 \cdot OEt_2$ or $AlCl_3$ in CH_2Cl_2 in yields >90% (Equation (20)). From an x-ray analysis of (**76**), it is clear, that the multiple bond is mainly localized between the carbon atom and the dicoordinate phosphorus center. The stability of ions such as (**76**) is highly dependent on the presence of the amino function. In keeping with this, the attempted synthesis of *P*-alkyl derivatives of (**76**) was thwarted by dimerization of the expected phosphaalkene ⟨89AG(E)740⟩.

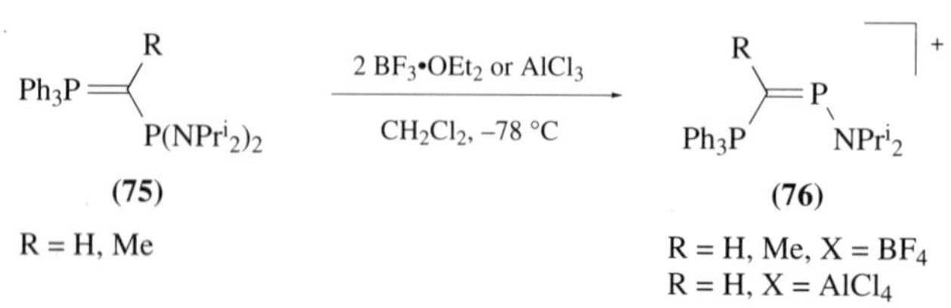

(20)

(b) By miscellaneous methods. The phosphaalkene (**78**) can be synthesized either by the protonation of the 1,3-diphosphaallene (**55**) with HBF_4 or by a chloride abstraction from (**47**) with $AlCl_3$. The initially formed 1,3-diphosphaallyl cation (**77**) is highly reactive and collapses to (**78**) via the insertion of the phosphenium center into the CH bond of an *ortho-t*-butyl group (Scheme 19) ⟨86AG(E)454⟩.

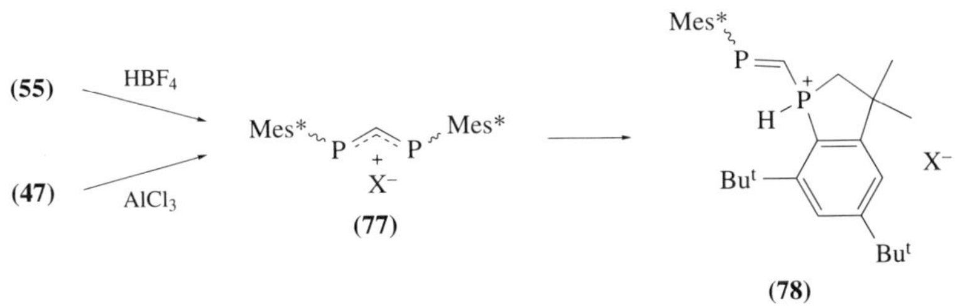

Scheme 19

In a few cases, addition processes to the P≡C triple bond of phosphaalkynes provide a suitable pathway to the systems under discussion. Unstable (**79**) was trapped by di-*iso*-propylamine to yield the phosphonio-functionalized phosphaalkene (**80**) quantitatively (Equation (21)) ⟨91CC302⟩.

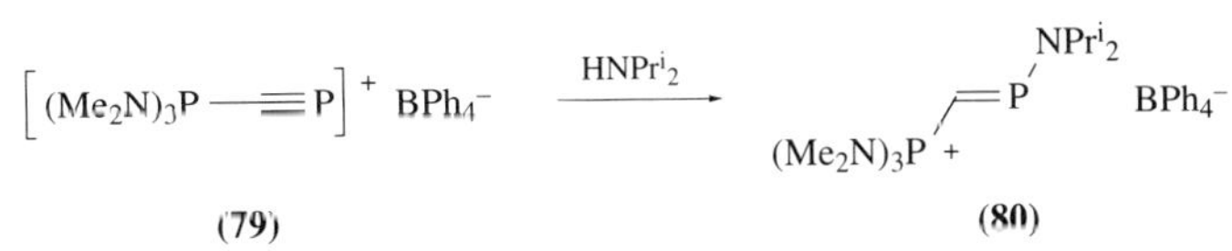

(21)

5.23.1.1.6 *Metalloid derivatives—$R^1C(PR^2)ER^3_n$ (E = Si, Ge, B)*

(i) Silicon derivatives—$R^1C(PR^2)SiR^3_3$

(a) From P-*halo(silylmethyl)phosphanes*. Elimination processes provide an important entry into silylated phosphaalkenes. Usually dehydrohalogenations are base-mediated, whereas thermolysis of the precursors leads to the elimination of TMS-Cl. Thus compound (**81**) was readily converted into phosphaalkene (**82**) by treatment with 1,4-diazabicyclo[2.2.2]octane (dabco) in ether at room temperature (54% yield) (Equation (22)) ⟨80AG(E)556⟩. Dehydrochlorination of (**83**) to give (**84**) was effected by dbu (Equation (23)) ⟨87PS(30)463⟩. The methylene phosphane (**86**) was prepared in 73% yield from chlorophosphane (**85**) in THF using $LiN(TMS)_2$ as a base (Equation (24)) ⟨81IC1679⟩. Compound (**86**) was also obtained by pyrolysis of precursor (**87**) at 150 °C *in vacuo* (Equation (25)) ⟨81AG(E)131⟩. Heterocycle (**88**) was conveniently metallated by *n*-butyllithium to a lithio derivative, which was subsequently condensed with $F_2PN(R)TMS$ to afford phosphaalkenes (**89**) in ca. 50% yield (Equation (26)) ⟨83ZAAC(501)57⟩.

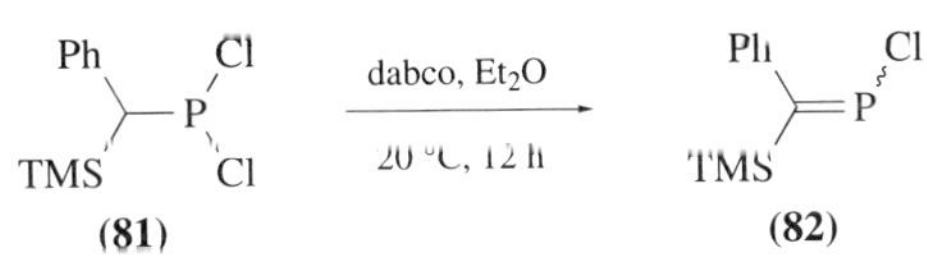

(22)

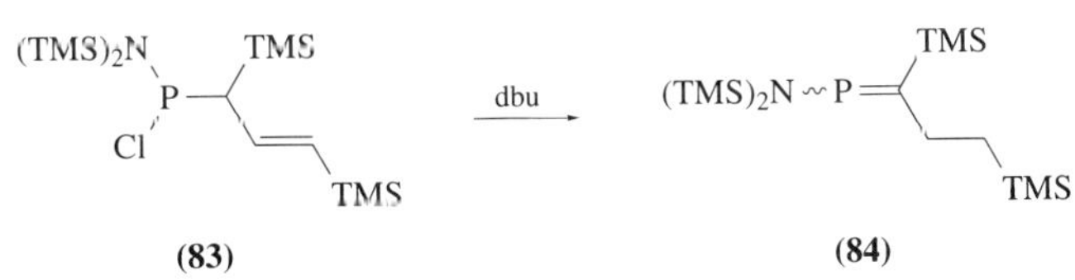

(23)

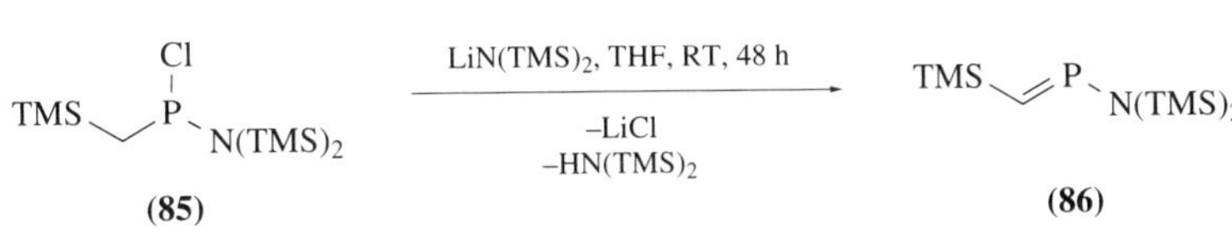

(24)

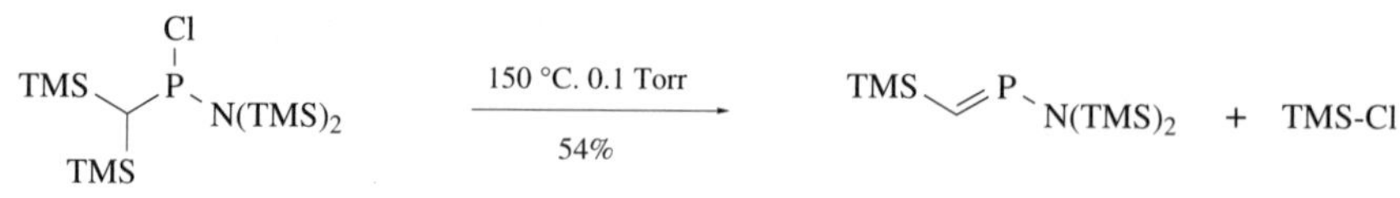

(87) **(86)** (25)

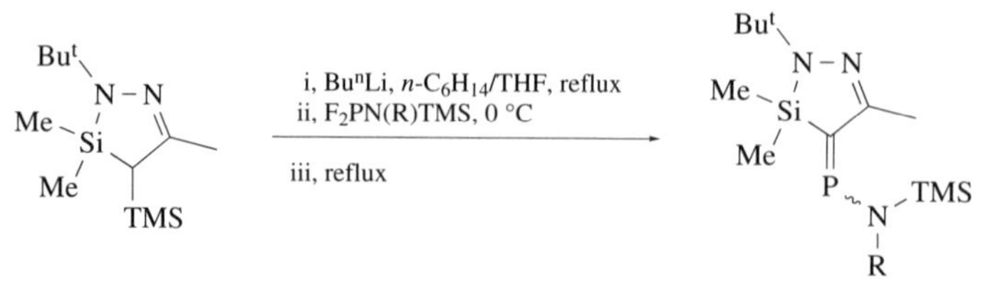

(88) **(89)** R = TMS, Bu^t (26)

(b) From ylides. Phosphenium ylides (**90**) were subjected to reaction with BEt_3 in toluene at $-80\,^{\circ}C$ to form the thermolabile adducts (**91**), which decomposed to phosphaalkenes (**92**) and the aminoboranes R_2NBEt_2 (Scheme 20). The steric demands of the amino functions as well as of the boranes markedly influenced the course of this transformation. The dicyclohexylamino derivative (**91**) is significantly more stable than the diisopropylamino analogue. No reaction occurred between (**90**) and tricyclohexylborane or dimesityl(methyl)borane, whereas with $B(OMe)_3$ ylide $(R_2N)_2(MeO)P{=}C\,(TMS)[B(OMe)_2]$ was isolated ⟨93CC1354⟩.

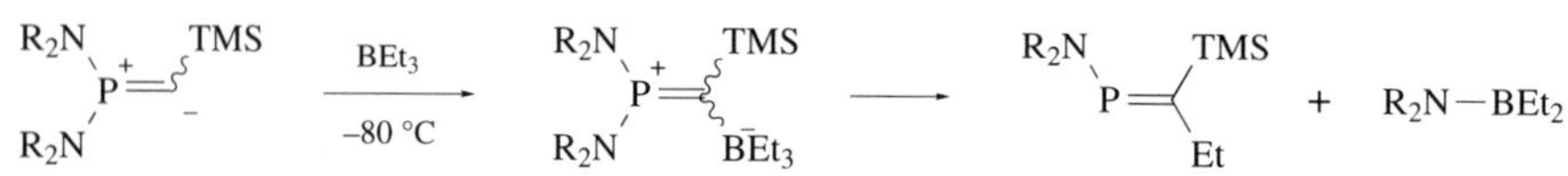

(90) **(91)** **(92)**

R = Pr^i, c-C_6H_{11}

Scheme 20

Thermolysis of *P*-chloro(methylene)phosphorane (**93**) at 120 °C resulted mainly in the formation of the phosphaalkene (**95**). This conversion was explained by invoking the intermediacy of phosphenium ylide (**94**). The general applicability of this method, based upon the 1,2-(*P*,*C*) shift of an organic substituent is restricted so far to *P*-arylated ylides (Equation (27)) ⟨81ZAAC(475)18⟩. Halide abstraction from the *P*-chlorinated ylide (**96**) with a catalytic amount of $SnCl_2$ afforded the (*E*)-configurated phosphaalkene (**97**) in 75% yield (Equation (28)) ⟨93CC673⟩.

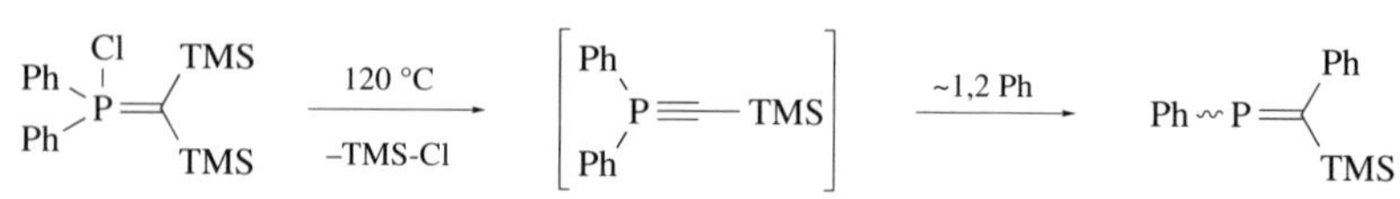

(93) **(94)** **(95)** (27)

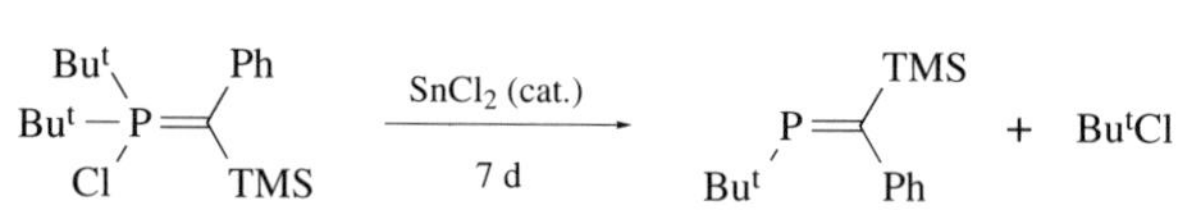

(96) **(97)** (28)

(c) From iminophosphanes. The interaction of the iminophosphane (**98**) with bis(trimethylsilyl)methyl lithium in THF at $-78\,^{\circ}C$ furnished phosphaalkene (**86**) via a sequence of nucleophilic substitution and rearrangement processes in 55% yield (Equation (29)) ⟨84PS(19)189⟩.

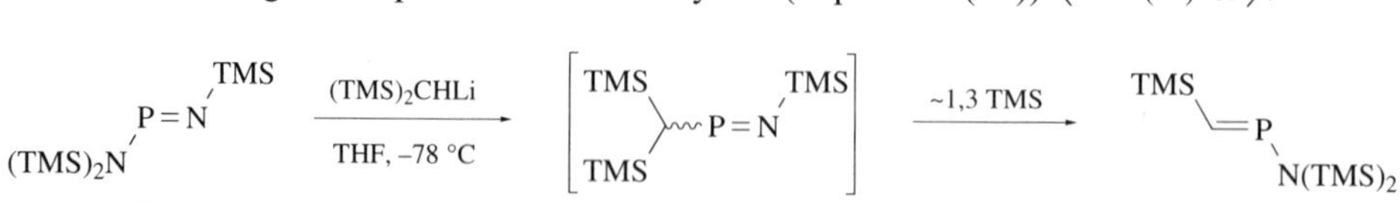

(98) **(86)** (29)

(d) From phosphaalkynes. Irradiation of a 1 : 2 mixture of the cyclotrisilane (**99**) in the presence of phosphaalkynes gave rise to the generation of phosphasilirenes (**100**). This transformation is rationalized by the [2 + 1] cycloaddition of transient silylene $[SiBu^t_2]$ to the P≡C triple bond (Scheme

21) ⟨87AG(E)776⟩. In common with other three-membered rings containing heavier elements, the stability of (**100**) is crucially associated with the presence of bulky substituents.

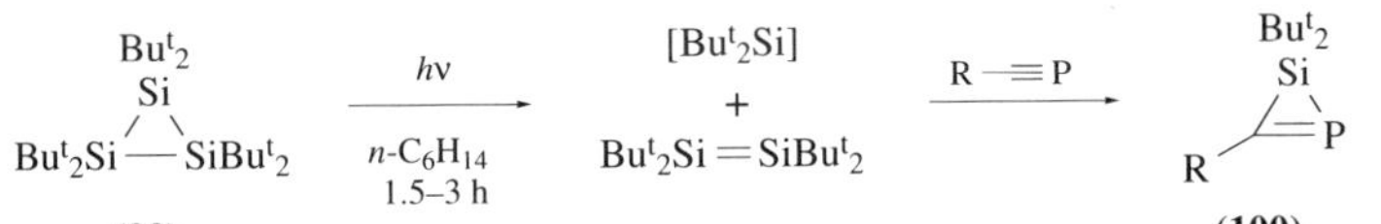

(**99**) (**100**)

R = Bu^t, 1-adamantyl

Scheme 21

(e) From peripheral reactions. Phosphaalkene (**14**) can be lithiated by means of *n*-butyllithium at −100 °C in ether to a *C*-lithiophosphaalkene, which with TMS-Cl readily underwent silylation (Equation (30)). Nucleophilic substitution at the phosphorus atom in *P*-halophosphaalkenes provides an elegant entry into numerous compounds of the general type R ~ P═C(Ph)TMS (Scheme 22) ⟨85CB1352⟩.

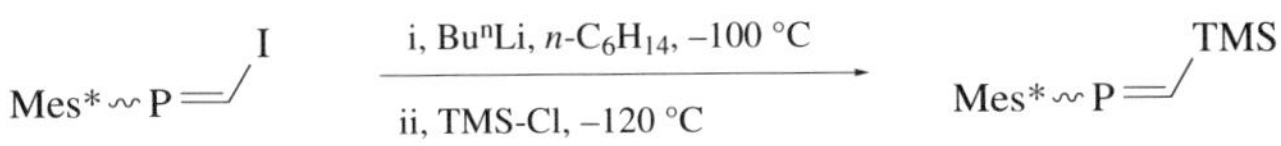

(**14**) (**101**) (30)

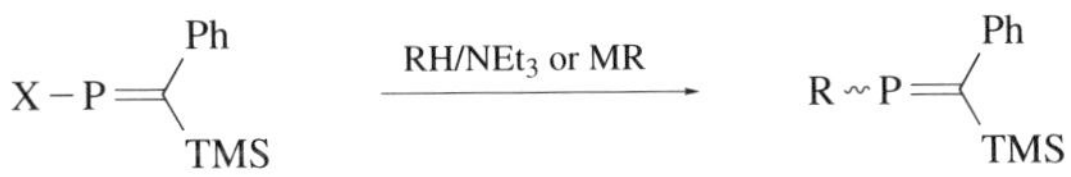

RH = HNR^{1}R^2, HOR^1, HSR^1, HPR^{1}R^2
MR = Li alkyl, Li alkenyl, Li aryl, Li alkynyl
X = Cl, Br

Scheme 22

(ii) Germanium derivatives—$R^1C(PR^2)GeR^3_3$

(a) From phosphaalkynes. Treatment of phosphaalkyne (**57**) with germylene Ge[CH(TMS)$_2$]$_2$ in a mixture of *n*-hexane and THF at 25 °C furnished the phosphagermirene (**102**) as a yellow crystalline solid (80%) (Equation (31)) ⟨88CC753⟩.

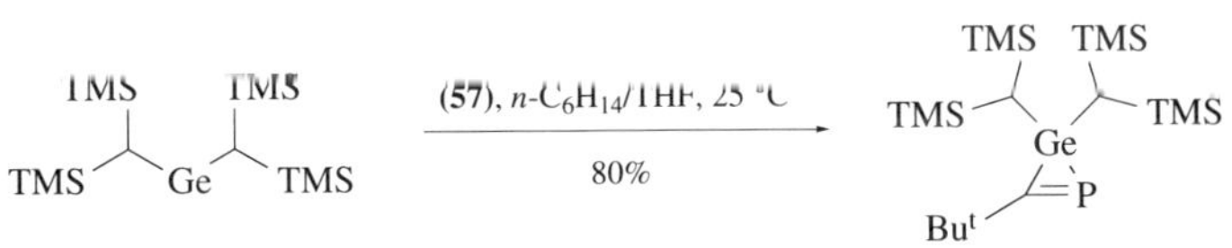

(**102**) (31)

(b) From peripheral reactions. Quenching the transient lithium species obtained from the metal/iodine exchange of (**14**) with Me$_3$GeCl at −120 °C yielded the germylated phosphaalkene (**103**) as a yellowish oil (98%). The (*E*/*Z*) ratio was determined to 80 : 20 ⟨91CB2677⟩.

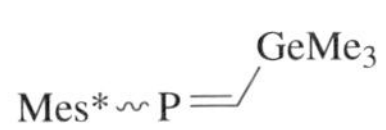

(**103**)

(iii) Boron derivatives—$R^1C(PR^2)BR^3_2$

No stable C-borylated phosphaalkenes of this type have been reported as of early 1995.

5.23.1.1.7 *Metal derivatives—$R^1C(PR^2)M$*

No isolable phosphaalkenes which are functionalized by alkali or earth alkali metals have been described as of early 1995.

(i) Transition metal derivatives

(a) Titanium and zirconium derivatives. A series of phosphaalkenes with titanium and zirconium functions are available from phosphaalkyne (**57**) and/or suitable metal complexes thereof. 3,3-Dimethyl cyclopropene displaced the trimethylphosphane ligand from complexes (**104**) with incorporation of the cycloalkene into bicyclic systems (**105**) in high yields (Equation (32)) ⟨88TH17, 90CB1617⟩. Alkynes or aldehydes may, though less reactive, also be used in this type of cycloaddition, provided that the reaction mixture is freed from the trimethylphosphane by BEt_3 below room temperature (Scheme 23) ⟨90CB1617, 90PS(49/50)337, 90TH29⟩.

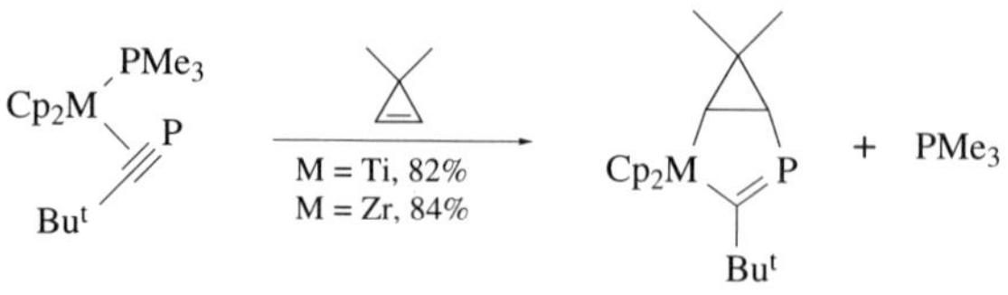

(104) **(105)** (32)

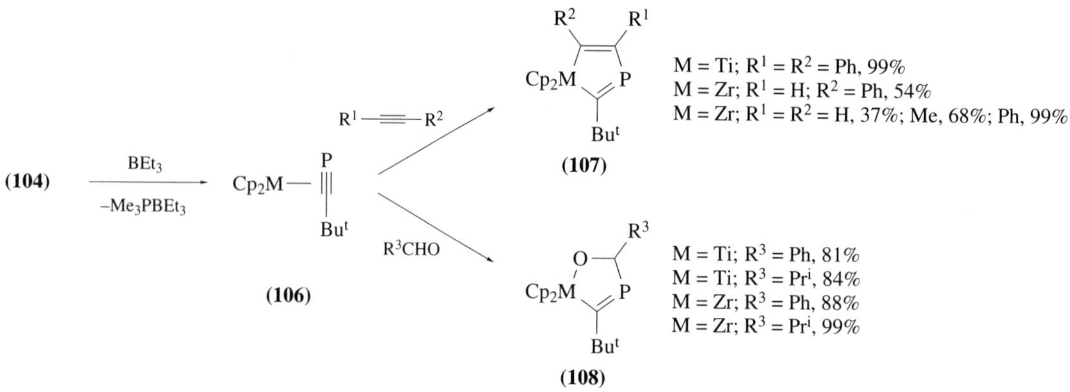

Scheme 23

(b) Mercury derivatives—$R^1C(PR^2{}_2)HgR^3$. Reaction of (*E*/*Z*)-(**14**) with *n*-butyllithium followed by a $HgCl_2$-quench gave the mercurated phosphaalkenes Mes* ~ P=C(H)HgCl and [Mes* ~ P=C(H)]$_2$Hg depending upon the stoichiometry of the reactants ⟨92OM3844, 93PS(77)252⟩.

(c) Derivatives with Group 13 metals. No phosphaalkenes functionalized by the Group 13 metals (Al, Ga, In, and Tl) at the methylene carbon atom are known as of early 1995.

(d) Derivatives with Group 14 metals. Stannylated phosphaalkenes such as (**109**) or (**110**) were either obtained by the addition of two equivs of the stannylene $[(TMS)_2CH]_2Sn$ to phosphaalkyne (**57**) ⟨88AG(E)838⟩ or by the *in situ* stannylation of the lithiated phosphaalkene Mes*P=C(H)Li with Me_3SnCl in ether at $-120\,^{\circ}C$ (80%), respectively (Equation (33)) ⟨91CB2677⟩. Lead derivatives are unknown as of early 1995.

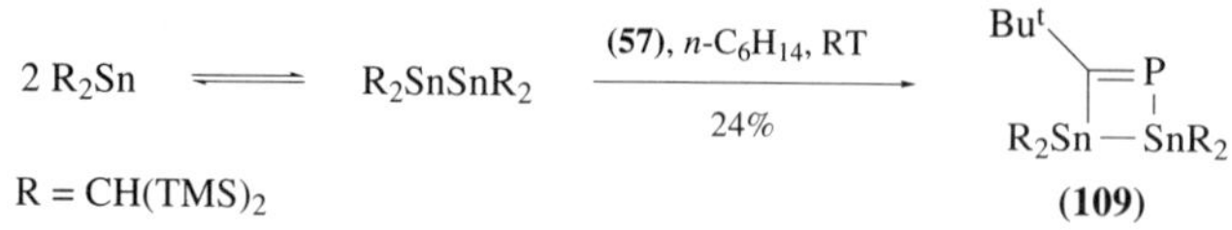

(109) (33)

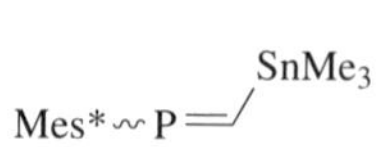

(110)

5.23.1.2 Dicoordinate Arsenic Functions—$R^1C(=AsR^2)X$

5.23.1.2.1 General methods

Compounds of the type $R^1C(=AsR^2)X$ are usually described as methylene arsanes or arsaalkenes. Their chemistry has been developed to a much lesser extent than that of the related phosphaalkenes.

Synthetic routes to such compounds mainly parallel those affording the phosphorus analogues. Thus organodisilylarsanes, their lithium derivatives as well as lithium arsenide, are commonly used in condensations and rearrangements. Moreover, dehydrohalogenations are important in the synthesis of carbon-substituted arsaalkenes.

Heteroarenes such as arsenines arsoles, or azaarsoles are not discussed in this chapter.

5.23.1.2.2 *Halogen derivatives—$R^1C(AsR^2)X$ (X = F, Cl, Br, I)*

Acyclic species of this type are unknown.

5.23.1.2.3 *Chalcogen derivatives—$R^1C(AsR^2)ER^3$ (E = O, S, Se, Te)*

(i) Oxygen derivatives—$R^1C(AsR^2)OR^3$

Pivaloyl chloride and equimolar amounts of alkyl- or arylbis(trimethylsilyl)arsanes undergo a reaction resulting in the quantitative formation of pivaloyl(silyl)arsanes (**111**). This transformation proceeds in the temperature range of −40 °C to 20 °C and takes about 12 h. In the case of $Bu^tC(O)As(Bu^t)TMS$, the reaction is very sluggish and after 14 days there is still starting material present. This can be circumvented by condensing $LiAs(Bu^t)TMS \cdot 2DME$ with the acid chloride. The oily compounds (**111**) underwent a clean and quantitative rearrangement to the *Z*-configurated arsaalkenes (**112**), when heated *in vacuo* at 70–80 °C for several hours (Scheme 24).

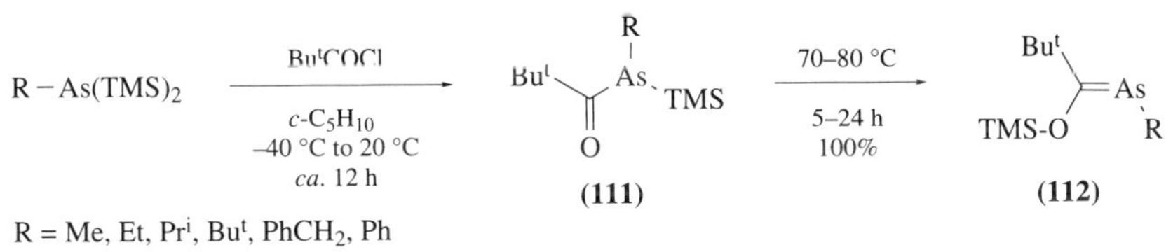

(**111**) (**112**)

Scheme 24

Compounds (**112**) are stable at ambient temperature under anaerobic conditions. With small substituents R(Me, Et), arsaalkenes dimerize slowly when exposed to daylight ⟨80ZAAC(470)131, 80ZAAC(470)144⟩.

The acylation of lithium dihydrogen arsenide · 2THF with mesitoyl chloride in a molar ratio of 3:2 at −50 °C in THF afforded the chelate (**113**) as orange–yellow crystals (24% yield). Chelate (**113**) was converted into the enol tautomer of dimesitoylarsane (**114**) by treatment with ethereal HBF_4 in THF at −40 °C (40%) (Scheme 25) ⟨91ZAAC(605)7⟩.

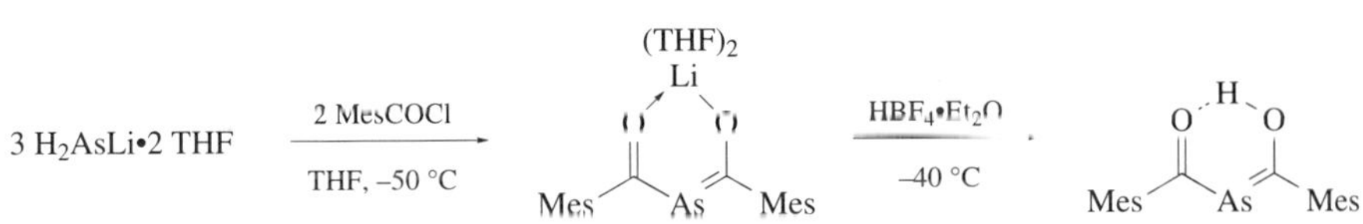

(**113**) (**114**)

Scheme 25

5.23.1.2.4 *Nitrogen derivatives—$R^1C(AsR^2)NR^2_2$*

(i) From disilylarsanes and imidoyl chlorides

N-Aryl pivaloylimide chlorides and $PhAs(TMS)_2$ interact upon prolonged stirring at room temperature (4 weeks) or heating on a water bath (20 h) with the generation of an equilibrium mixture of the isomers (**115**) and (**116**). They were obtained as green–brown oils in 50% (Ar = *p*-Tol) and

70% yield (Ar = Ph), respectively. Crystals of pure arsaalkene (**116**) slowly separated from this mixture (Scheme 26). This method seems to be limited to the *t*-butyl substituent at the As=C bond, since a similar experiment with benzoylimide chlorides afforded complex mixtures ⟨79JOM(166)175⟩.

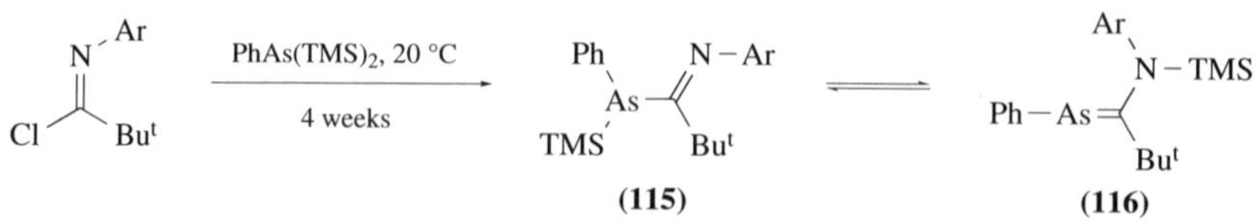

Ar = Ph, 70%
Ar = *p*-Tol, 50%

Scheme 26

(ii) From DMF and silylarsanes

The treatment of $PhAs(TMS)_2$ with an excess of DMF in the presence of catalytic amounts of solid NaOH resulted in the formation of (**117**). In the absence of the catalyst, this conversion was very sluggish. An alternative approach to (**117**) made use of the condensation of Ph(TMS)As-Li(DME) with the amide (Scheme 27) ⟨81ZN(B)1080⟩.

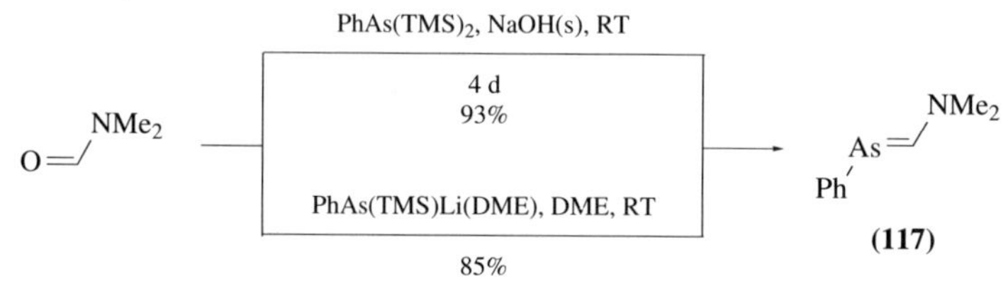

Scheme 27

5.23.1.2.5 *Metalloid derivatives—$R^1C(AsR^2)ER^3_n$ (E = Si, Ge, B)*

Only one acyclic arsaalkene with a metalloid substituent has been described in the literature up to the early 1990s. In a pseudo Wittig-reaction arsasilene (**118**) and benzophenone gave rise to compound (**119**) (83%), which upon heating (160°C, 10 h) suffered from a cycloreversion to afford the isomeric arsaalkenes (**120**) and (**121**) as well as a 1,3-dioxo-2,4-disiletane derivative (Scheme 28) ⟨92AG(E)316⟩. Pure (**120**) was independently synthesized from benzophenone and $Is_2Si(F)$—$As(Li)SiPr^i_3$ (Is = 2,4,6-$Pr^i_3C_6H_2$). The thermal rearrangement of (**120**) furnished (**121**).

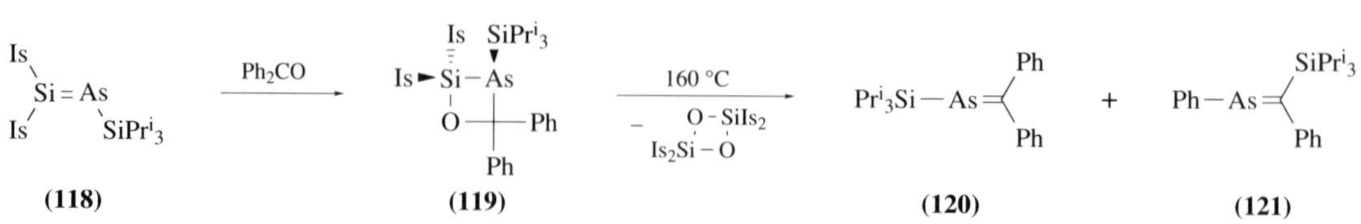

Scheme 28

5.23.1.3 Dicoordinate Antimony and Bismuth Functions—$R^1C(=ER^2)X$ (E = Sb, Bi)

Compounds of this type have not been described in the literature as of 1995.

5.23.2 TRICOORDINATE PHOSPHORUS, ARSENIC, ANTIMONY, AND BISMUTH FUNCTIONS—$RC(PX^1Y)X^2$, $RC(AsX^1Y)X^2$, $RC(SbX^1Y)X^2$, AND $RC(BiX^1Y)X^2$

5.23.2.1 Tricoordinate Phosphorus Functions—$RC(PX^1Y)X^2$

Compounds of the above type are quite rare if simple transition metal coordination compounds of phosphaalkenes are neglected. Basically, cationic species $RC(P^+X^1Y)X^2$ may be derived from *P*-halogenated ylides by halide abstraction or alternatively from electron-rich phosphaalkenes by *P*-alkylation. The reaction at the carbon atom of phosphenium ylides $RC{\equiv}PX^1Y$ with electrophiles $(X^2)^+$ may also lead to $[RC(PX^1Y)X^2]^+$ ions.

5.23.2.1.1 Methylene phosphonium ions

Chloride abstraction from ylide (**122**) was effected by $AlCl_3$ in CH_2Cl_2 at −78 °C to afford the sensitive methylene phosphonium ion (**123**) ⟨93PS(76)21⟩. The corresponding ion $[Bu^t_2P{=}CPh(TMS)]^+$ was obtained similarly from ylide (**96**) (Equation (34)) ⟨93CC673⟩.

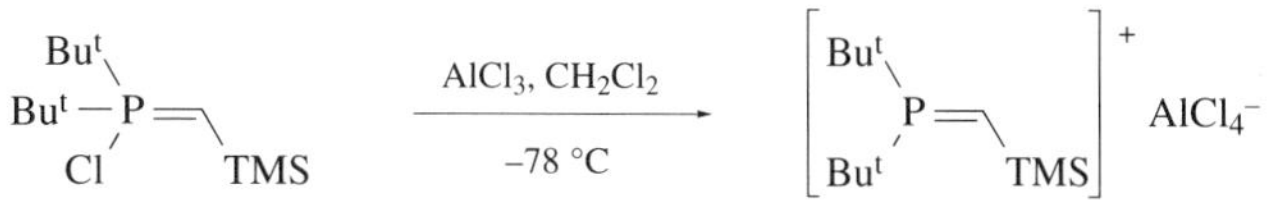

(34)

(122) **(123)**

5.23.2.1.2 η^1-Phosphaalkene complexes—$RC[PX^1(M)]X^2$

Compounds of this type are generated by treatment of phosphaalkenes with reactive transition metal complexes. An example is given by the synthesis of (**124**) from (**86**) and $Fe_2(CO)_9$ in pentane (42%) (Equation (35)) ⟨85IC1993⟩. In another example of a Fe—C=P linkage (**126**) results from the rearrangement of the η^3-phosphaallyl complex (**125**), which is effected by an equimolar amount of $Fe_2(CO)_9$ in toluene (71%) (Equation (36)) ⟨91CB989⟩.

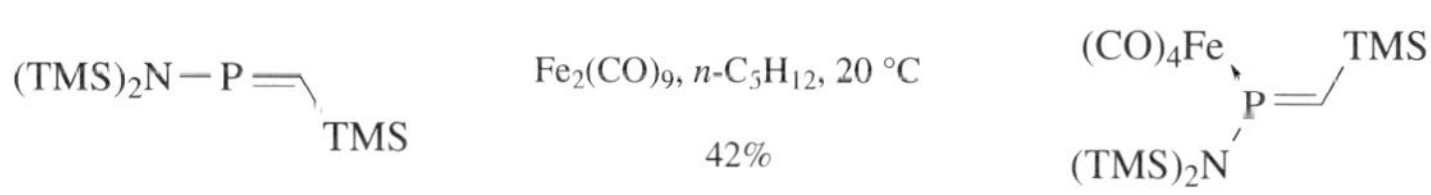

(35)

(86) **(124)**

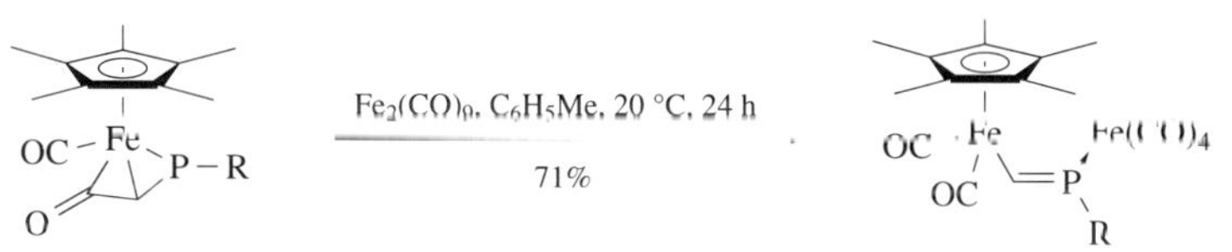

(36)

(125) **(126)**

R = $CH(TMS)_2$

Phosphaalkyne (**57**) was reduced by a samarium(II) species to give red crystalline (**127**) in 41% (Equation (37)) ⟨89AG(E)445⟩.

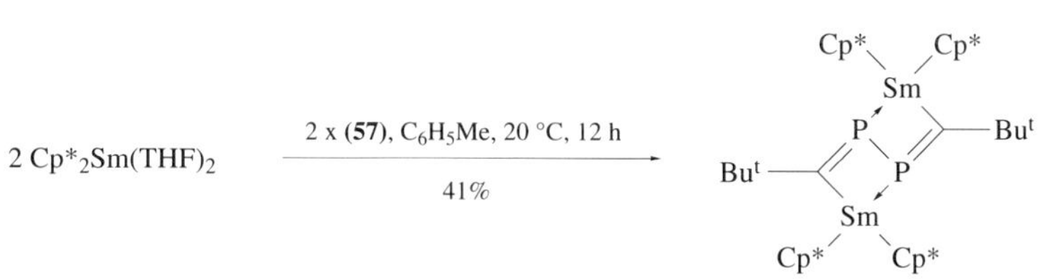

(37)

(127)

5.23.2.2 Tricoordinate Arsenic, Antimony, and Bismuth Functions—$RC(EX^1Y)X^2$ (E = As, Sb, Bi)

Representatives of such types of compounds are unknown as of early 1995.

5.23.3 TETRACOORDINATE PHOSPHORUS, ARSENIC, ANTIMONY, AND BISMUTH FUNCTIONS—$R^1C(PR^2_3)X$, $R^1C(AsR^2_3)X$, $R^1C(SbR^2_3)X$, AND $R^1C(BiR^2_3)X$

5.23.3.1 Tetracoordinate Phosphorus Functions—$R^1C(PR^2_3)X$

5.23.3.1.1 General methods

Phosphorus ylides or methylene phosphoranes have been prepared by several methods, the most useful being the proton abstraction from phosphonium salts. Phosphonium salts are also intermediates in the reactions of ylides with electrophiles, although they suffer from rapid deprotonation by the parent ylide to afford novel functionalized ylides (transylidation). The conversion of tertiary phosphanes into *P*-halo ylides by carbon tetrahalides involves the deprotonation of a transient phosphonium ion by the trihalomethanide anion. Several reviews and monographs on ylide chemistry are available ⟨B-66MI 523-01, 72MI 523-07, 75ACR62, 78PAC19, 80PAC1057, 82HOU(E1)616, B-82MI 523-01, 83AG(E)907, 89ZC396, B-93MI 523-01⟩.

5.23.3.1.2 Halogen derivatives—$R^1C(PR^2_3)X$ (X = F, Cl, Br, I)

(i) Fluorine derivatives—$R^1C(PR^2_3)F$

In the literature on α-fluoroylides no isolable species with fluoromethylene functions are described. Instead the *in situ* generation of such species and their subsequent conversion into fluoroalkenes by the Wittig reaction is the subject of interest of several papers ⟨69S75, 75JOC2796, 90JOC2311, 93TL1925⟩.

(ii) Chlorine derivatives—$R^1C(PR^2_3)Cl$

(a) From phosphonium salts. Whereas a number of chemists have described the preparation of $Ph_3P{=}CHCl$ (**129**) and its *in situ* reaction with carbonyl compounds, only one paper is concerned with the isolation of this ylide. Ylide (**129**) is the product of the reaction between phosphonium salt (**128**) and $Ph_3P{=}CH_2$, which serves as a base (Equation (38)) ⟨77AG(E)180⟩.

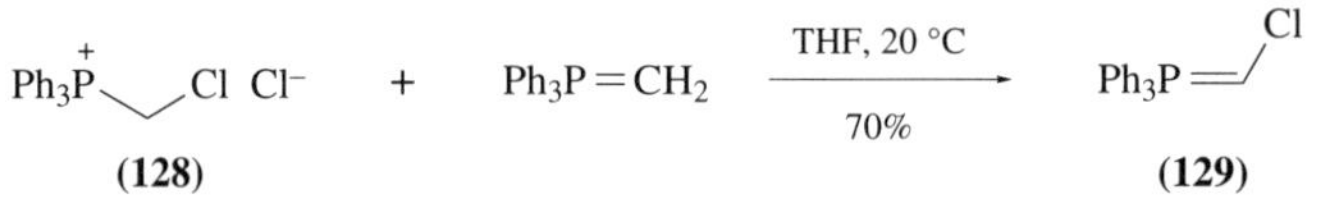

(38)

(**128**) (**129**)

Phosphonium salts with anion-stabilizing groups have been shown to interact with a variety of chlorinating agents, providing direct access to α-chloroylides. High yields of the chloroylides $Ph_3P{=}C(Cl)(EWG)$ have been obtained by the treatment of the salts $[Ph_3PCH_2(EWG)]Cl$ with Chloramine T in aqueous hydrochloric acid ⟨70S590⟩ or with chlorinated lime in ethanolic HCl (Table 2) ⟨93T1053⟩.

Table 2 Preparation of α-chloroylides from phosphonium salts.

Entry	*Substrate*	*Conditions*	*Product*	*Yield (%)*
1	$(Ph_3PCH_2CO_2Me)Cl$	a	$Ph_3P{=}C(Cl)CO_2Me$	91
2	$(Ph_3PCH_2CO_2Et)Cl$	a	$Ph_3P{=}C(Cl)CO_2Et$	71
3	$(Ph_3PCH_2COMe)Cl$	a	$Ph_3P{=}C(Cl)COMe$	86
4	$(Ph_3PCH_2COPh)Cl$	a	$Ph_3P{=}C(Cl)COPh$	73
5	$(Ph_3PCH_2COMe)Cl$	b	$Ph_3P{=}C(Cl)COMe$	100
6	$(Ph_3PCH_2CN)Cl$	b	$Ph_3P{=}C(Cl)CN$	74
7	$(Ph_3PCH_2CO_2Et)Cl$	b	$Ph_3P{=}C(Cl)CO_2Et$	80

[a] $4\text{-}MeC_6H_4SO_2N(Na)Cl$, HCl/H_2O, 70–80 °C. [b] (i) $3Ca(OCl)Cl \cdot Ca(OH)_2 \cdot 5H_2O$; (ii) 2 M HCl in EtOH, 20 °C.

(b) From ylides. In a number of instances the free stabilized ylides have been subjected to chlorination. The treatment of ylides $Ph_3P{=}CH(EWG)$ (EWG = COPh, CO_2Et, $CONPh_2$) with elemental chlorine gave rise to the formation of α-chlorophosphonium salts which were subsequently deprotonated by aqueous base (Table 3) ⟨62JOC998, 63JOC465⟩.

Table 3 Preparation of α-chloroylides from stabilized ylides.

Entry	*Substrate*	*Conditions*	*Product*	*Yield (%)*
1	$Ph_3P{=}CHCOPh$	a	$Ph_3P{=}C(Cl)COPh$	87
2	$Ph_3P{=}CHCOPh$	b	$Ph_3P{=}C(Cl)COPh$	76
3	$Ph_3P{=}CHCOPh$	c	$Ph_3P{=}C(Cl)COPh$	93
4	$Ph_3P{=}CHCOMe$	b	$Ph_3P{=}C(Cl)COMe$	46
5	$Ph_3P{=}CHCONPh_2$	d	$Ph_3P{=}C(Cl)CONPh_2$	67
6	$Ph_3P{=}CHCO_2Me$	e	$Ph_3P{=}C(Cl)CO_2Me$	85–93
7	$Ph_3P{=}CHCO_2Me$	f	$Ph_3P{=}C(Cl)CO_2Me$	95
8	$Ph_3P{=}CHCO_2Et$	b	$Ph_3P{=}C(Cl)CO_2Et$	80
9	$Ph_3P{=}CHCO_2Et$	d	$Ph_3P{=}C(Cl)CO_2Et$	50
10	$Ph_3P{=}CHCN$	c	$Ph_3P{=}C(Cl)CN$	90

[a] (i) $Cl_2/CH_2Cl_2/CCl_4$, −70 °C, (ii) $NaOH/H_2O$; [b] $Cl_2/CH_2Cl_2/CCl_4$/py or NEt_3, −70 °C; [c] Bu^tOCl, CH_2Cl_2, −70 °C; [d] (i) $Cl_2/CHCl_3$, 0–5 °C, (ii) $NaOH/H_2O$; [e] 0.5 Cl_2, C_6H_6, 10 °C; [f] 0.5 $PhICl_2$, C_6H_6.

When the halogenation was performed in the presence of pyridine or triethylamine, the α-chloro product was obtained directly ⟨62JOC998⟩. The same is true for the action of the ylides $Ph_3P{=}CHCOPh$ ⟨62JOC998⟩ and $Ph_3P{=}CHCN$ ⟨79HCA1401⟩ with *t*-butyl hypochlorite (Scheme 29).

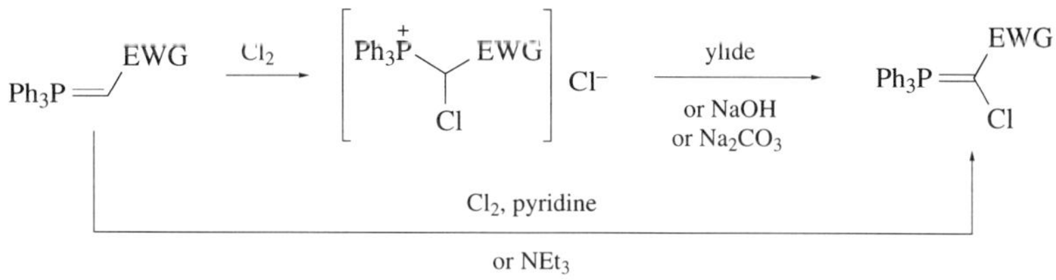

EWG = electron withdrawing group

Scheme 29

Reaction of $Ph_3P{=}CHCO_2Me$ with chlorine or iodobenzene dichloride in benzene solution afforded an equimolar mixture of $Ph_3P{=}C(Cl)CO_2Me$ and phosphonium salt $(Ph_3\overset{+}{P}CH_2CO_2Me)Cl$. This result invokes a spontaneous deprotonation of the transient chlorophosphonium salt by the more basic unconverted ylide. A disadvantage of this transylidation process is the loss of one-half of the starting ylide through its action as a base ⟨61CB2996⟩.

Low yields (17–24%) of the ylides $Ph_3P{=}C(Cl)CONPh_2$ and $Ph_3P{=}C(Cl)CO_2Et$ resulted from the acylation of the dichloromethylene phosphorane $Ph_3P{=}CCl_2$ with $ClCONPh_2$ or $ClCO_2Et$ and the subsequent basic workup ⟨63JOC465⟩.

In a completely different approach, the chlorination of ylide (**130**) to give (**131**) was accomplished by reaction with carbon tetrachloride. In the case of methylene phosphoranes and CCl_4, both hydrogen atoms at the carbanionic center are exchanged by chlorine (Equation (39)) ⟨88ZOB491⟩.

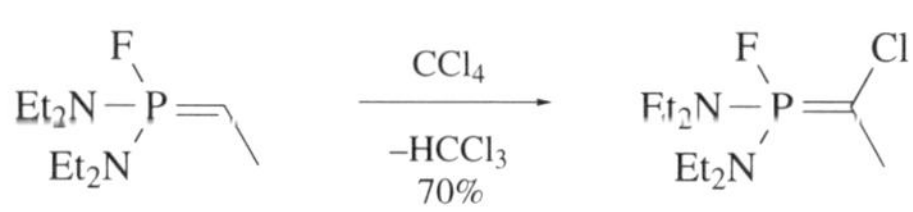

(**130**) (**131**) (39)

(iii) Bromine derivatives—$R^1C(PR^2_3)Br$

The ylides $Ph_3P{=}C(R)Br$ (R = H, Et, C_5H_{11}) serve as important bromomethylene transfer reagents in Wittig reactions, where they are usually pregenerated from the corresponding α-bromophosphonium salts and organolithiums, and quenched without prior isolation. The same applies for the *in situ* preparation of such ylides from PPh_3 and organomercurials ⟨66JOM(5)267, 85TL1623, 86TL1995, 93TL1925⟩.

Isolable α-bromoylides usually experience stabilization from electron-withdrawing groups at the carbanionic center, and possible approaches to α-bromophosphonium ylides $Ph_3P{=}C(Br)(EWG)$ involve the bromination of either the parent ylides $Ph_3P{=}C(H)(EWG)$ with the element and subsequent deprotonation of the α-bromophosphonium salt with aqueous base ⟨62JOC998, 63JOC465⟩. Alternatively, the bromination of the phosphonium salts such as $(Ph_3PCH_2CN)Cl$ ⟨79HCA1401⟩ or $(Ph_3PCH_2CO_2Me)Br$ ⟨61CB2996⟩ with either bromine in the presence of an amine in CH_2Cl_2 solution or with a bromine/bromate mixture in water. In the latter case, the resulting α-bromophosphonium salt was converted into $Ph_3P{=}C(Br)CO_2Me$ by means of aqueous sodium hydroxide (Scheme 30).

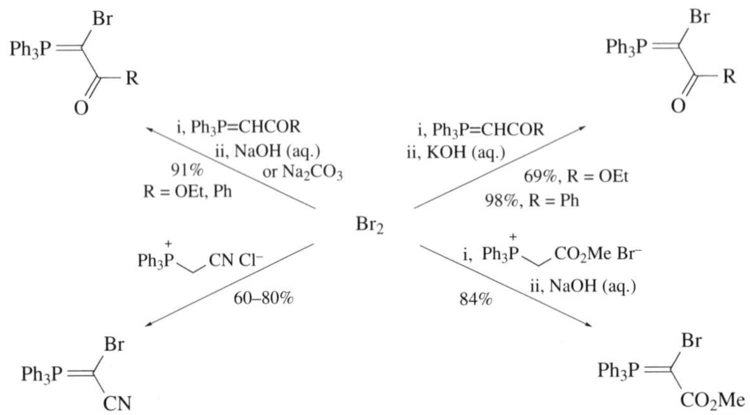

Scheme 30

(iv) Iodine derivatives—$R^1C(PR^2_3)I$

The preparation of stable α-iodo-functionalized ylides has the same limitations encountered with the brominated analogues.

Ylide $Ph_3P{=}CHI$ was prepared by several routes from either $(Ph_3PCH_2I)X$ employing various bases ⟨93TL1925⟩. Best results were obtained with $NaN(TMS)_2$ ⟨89TL2173, 89TL5261⟩. In a different approach, the ylide $Ph_3P{=}C(I)TMS$ was converted into $Ph_3P{=}CHI$ by *t*-butanol and caesium fluoride.

No reports on the successful isolation of this ylide are given; instead the interception of $Ph_3P{=}CHI$ in the Wittig reaction to produce iodoalkenes was the matter of interest. The isolation of stable α-iodoylides is only feasible when anion-stabilizing functionalities are present in the molecule. Thus phosphonium salt (**132**) was smoothly converted into ylide (**133**) by treatment with iodine in the presence of two equivs of triethylamine (Equation (40)) ⟨79HCA1401⟩.

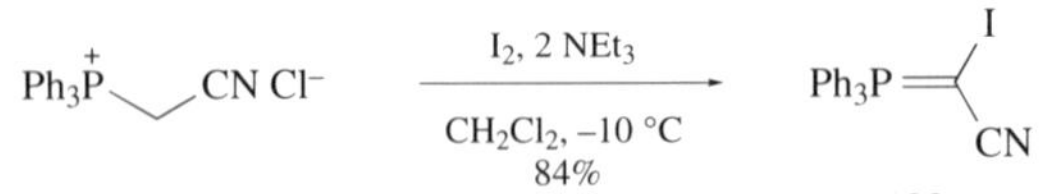

(**132**) (**133**) (40)

The iodination of the free ylides $Ph_3P{=}CHCOPh$ with I_2 in $CHCl_3$ caused problems due to the incomplete conversion into salt $[Ph_3PCHI(CO)Ph]^+I^-$. Thus the yield of $Ph_3P{=}CIC(O)Ph$ after deprotonation with aqueous base was only 44% ⟨63JOC465⟩.

Difficulties were also encountered in the preparation of $Ph_3P{=}C(I)CO_2Me$ from $Ph_3P{=}CHCO_2Me$ and I_2 in MeOH, where the loss of one-half of the starting ylide is due to its function as a base ⟨61CB2996⟩. The nucleophilic substitution of bromide by iodide in aqueous methanol circumvented this hindrance, converting (**134**) into the α-iodoylide (**135**) in 90% yield (Equation (41)) ⟨61CB2996⟩.

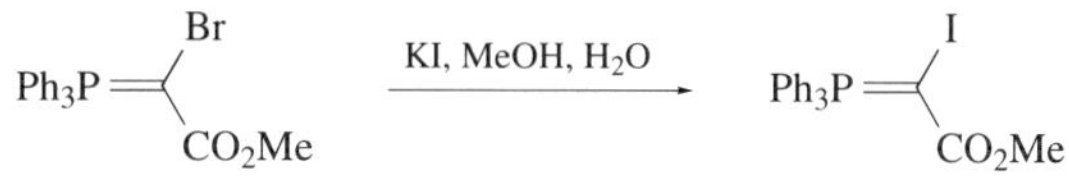

(134) **(135)** (41)

Phosphorus ylides such as (**136**) with an α-aryliodonium substituent were synthesized by the reaction of the appropriate ylide $Ph_3P{=}CHCOR$ with ArI = O, $ArI(OAc)_2$ or $ArIF_2$ in a mixture of methanol and aqueous HBF_4 at 0 °C (Equation (42)) ⟨84JA6082⟩. In contrast to this iodoarene dichlorides furnished α-chlorophosphonium salts.

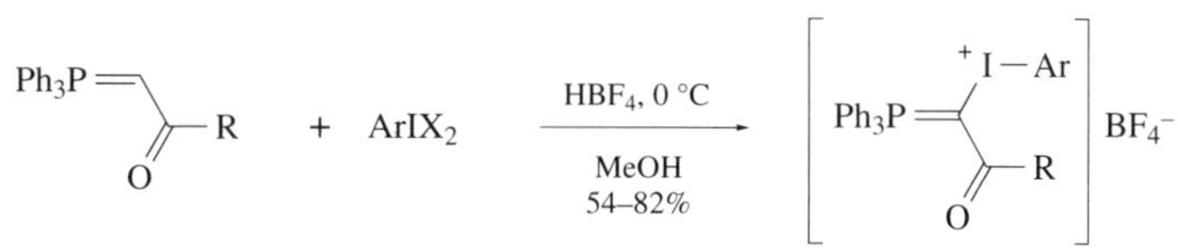

(42)

(136)

X_2 = O, F_2, $(OAc)_2$
R = Ph, OEt, 4-MeC_6H_4, 4-$MeOC_6H_4$, 4-PhC_6H_4
Ar = Ph, 2-MeC_6H_4

5.23.3.1.3 *Chalcogen derivatives—$R^1C(PR^2{}_3)ER^3{}_n$ (E = O, S, Se, Te)*

(i) Oxygen derivatives—$R^1C(PR^2{}_3)OR^3$

The oxygen-functionalized ylide $Ph_3P{=}CHOMe$ (**137**) has been extensively employed for the homologation of carbonyl compounds. It has been generated from the corresponding phosphonium salt by deprotonation with various bases. Due to its thermolability, the freshly prepared ylide was used *in situ* for further transformations ⟨61CB1373, 92TL6915⟩. Similarly, α-methoxybenzylidene- and α-methoxyethylidene phosphoranes proved to be non-isolable compounds ⟨73CB1062⟩. The presence of an α-acyl substituent, however, makes feasible the isolation of stable α-methoxyylides (**138**) ⟨66M1797⟩. Their generation is based on (**137**) which is prepared from $(Ph_3PCH_2OMe)Br$ by deprotonation with phenyllithium in ether. Acylation and transylidation occur upon treatment of (**137**) with half an equiv. of an acyl chloride or an ethyl carboxylate. For the success of this transformation, the use of an ester in place of the acyl chloride is recommended in some cases. Moreover acylation of (**137**) with esters demands the pregeneration of the ylide by means of an organolithium reagent (Scheme 31).

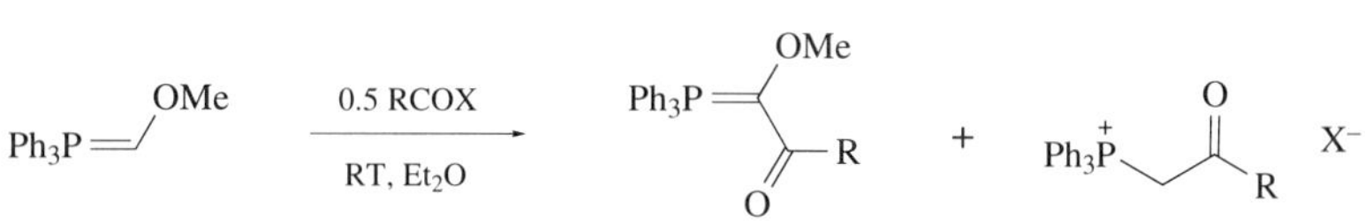

(137) **(138)**

i, c-C_4H_7COCl, 51%; ii, c C_3H_5COCl, 57%; iii, Pr^iCOCl, 49%; iv, Bu^nCOCl, 27%; PhCH=CHCOCl, 17%,
v, 2-$C_5H_4NCO_2Et$, 51%; vi, Ph–C≡C–CO_2Et, 17%; vii, sorbinic acid chloride, 14%; viii, 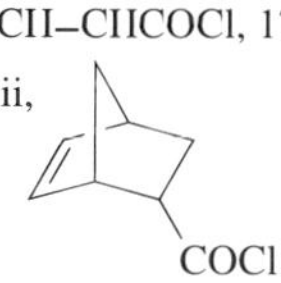 , 47%

Scheme 31

(ii) Derivatives with dicoordinate sulfur—$R^1C(PR^2{}_3)SR^3$

(a) From phosphonium salts. Frequently organothiomethylphosphonium salts $[R^2_3PCH(R^1)SR^3]X$ were deprotonated by organolithium reagents to give the ylides under discussion. However, no efforts were made to isolate these species. Instead, the products were subsequently involved in reactions with carbonyl compounds ⟨61CB1373⟩, acyl halides ⟨66M1797⟩, or PhSeCl ⟨83CB1955⟩.

(b) From ylides. A variety of useful methods are available to functionalize phosphorus ylides

by sulfur-containing groups. Transylidation was observed when ylides such as $Ph_3P{=}CHMe$ or $Ph_3P{=}CH_2$ were allowed to react with half an equiv. of benzene- or *n*-butane sulfenyl chloride ⟨68TL3787⟩. Nevertheless, this method is accompanied by the loss of one-half of the starting ylide. This disadvantage was overcome when the stabilized phosphonium ylides $Ph_3P{=}CHR$ were allowed to react with sulfonium chloride (**139**) in CH_2Cl_2. The resulting salts (**140**) were heated in boiling THF or dioxane to give (**141**) with release of methyl chloride (Scheme 32) ⟨71S431⟩.

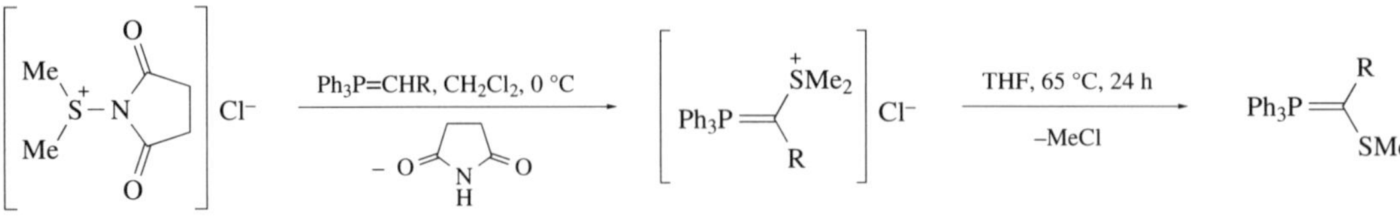

(139) **(140)** **(141)**

i, R = CO_2Et, 50%; ii, R = CN, 29%; iii, R = COMe, 47%; iv, R = COPh, 52%

Scheme 32

The dealkylation of similar ylidic cations was also achieved by PPh_3, by fusing the reactants together in a vacuum (Equation (43)) ⟨79JGU1740⟩. The phenylthio group can also be introduced into the α-position of an ylide by means of *N*-methyl-*N*-phenyl-thioacetamide (Scheme 33) ⟨70BCJ2870⟩. Other methods of arylthiolation of stabilized ylides made use of either arenesulfonyl chlorides in benzene at 20 °C for 10 h ⟨68YZ1039⟩ or *p*-toluene thiosulfonates (Scheme 34) ⟨72RTC37⟩. This procedure is not of general utility as more basic ylides were chlorinated by sulfonyl chlorides ⟨72RTC37⟩. Thiocyanogen reacts with an equimolar amount of $Ph_3P{=}CHCOMe$ in benzene at 20 °C to give a 69% yield of ylide (**142**) which is smoothly converted into ylide (**143**) by dissolving in a saturated ethanolic NH_3 solution (Scheme 35) ⟨69LA(721)121⟩. Another synthetic approach to ylide (**143**) involves the oxidation of the respective phosphonium thiocyanate with $Pb(OAc)_4$ in $CHCl_3$ (80%) ⟨66M180⟩.

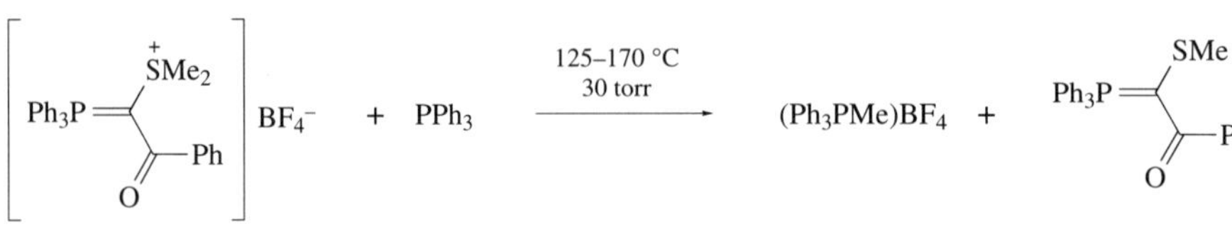

(43)

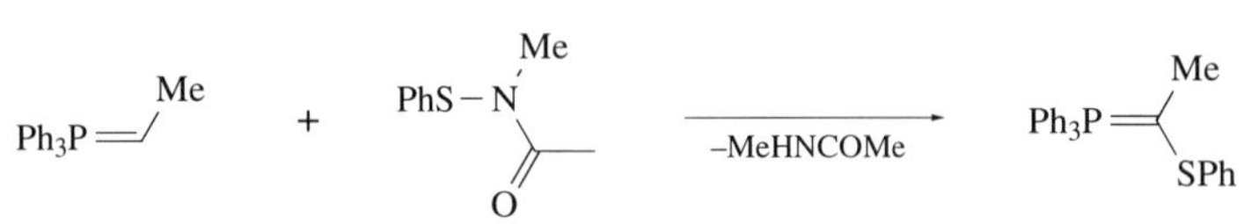

Scheme 33

It is also conceivable to further functionalize simple organothiomethylene phosphoranes to more complex molecules. In keeping with this, the *in situ* reaction of ylide $Ph_3P{=}CHSMe$ with a series of acyl halides in the molar ratio of 2 : 1 led to the formation of α-methylthio-β-keto phosphoranes (Scheme 36) ⟨66M1797⟩.

(c) From organothiomethylene phosphanes. Sterically hindered phosphanes such as (**144**) smoothly react with an equimolar amount of carbon tetrachloride in pentane in the temperature range of −80 °C to 20 °C to afford ylide (**146**) as an unpleasant smelling colourless oil (70%) (Scheme 37). The steric screening of the phosphorus atom by the bulky *t*-butyl groups has a decisive influence on the course of the reaction and on the character of the products. Sterically unhindered phosphanes fail to produce *P*-haloylides in favor of trichloromethylphosphonium salts and derivatives thereof ⟨81JGU2125⟩. It is believed that the phosphane attacks the carbon tetrachloride at a chlorine atom to afford the ion pair (**145**). Deprotonation of the α-methylene group by the CCl_3^- anion completes the formation of (**146**). Electron-acceptor substituents at the α-carbon atom facilitate abstraction of the proton and also prevent the produced halogenated ylides from rearrangement to α-halomethylphosphanes.

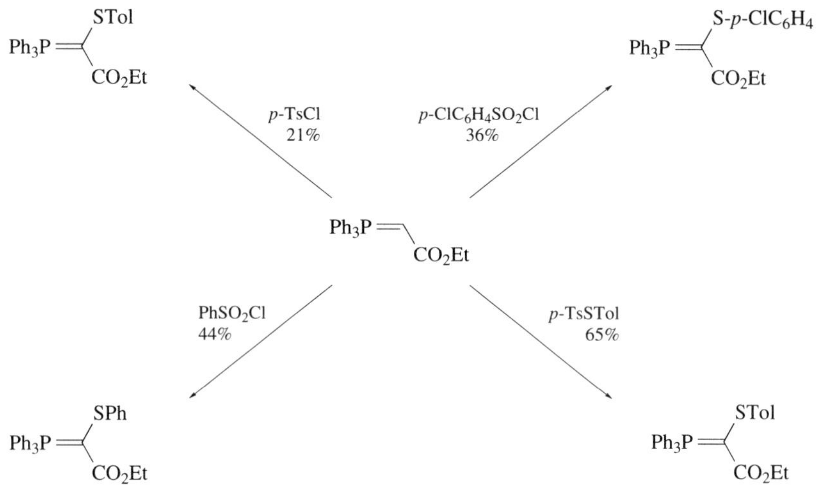

Scheme 34

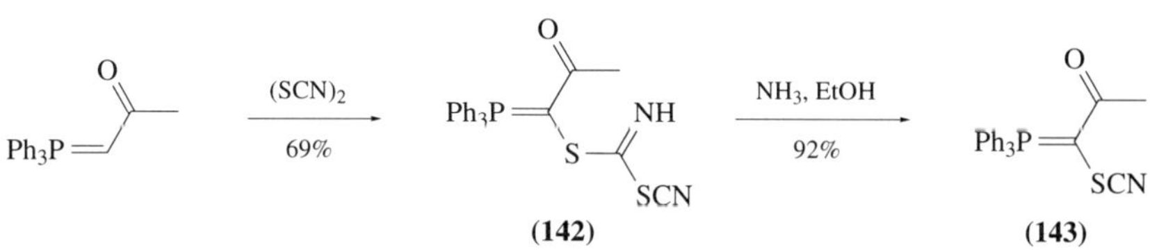

Scheme 35

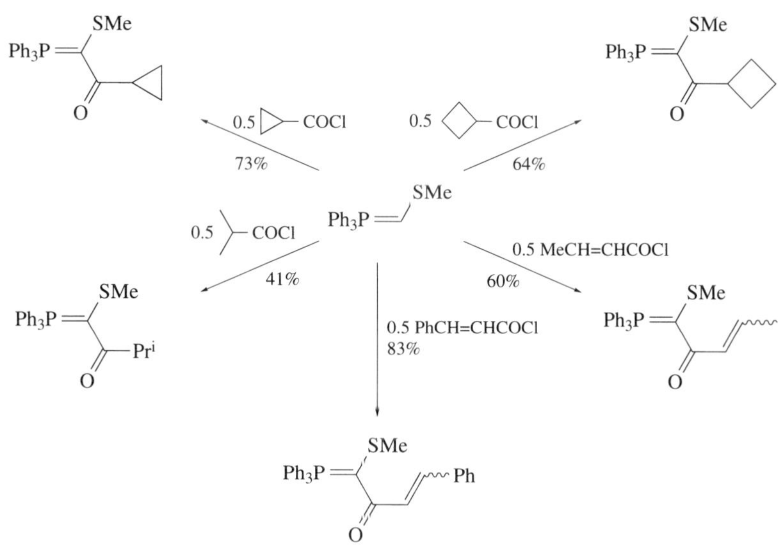

Scheme 36

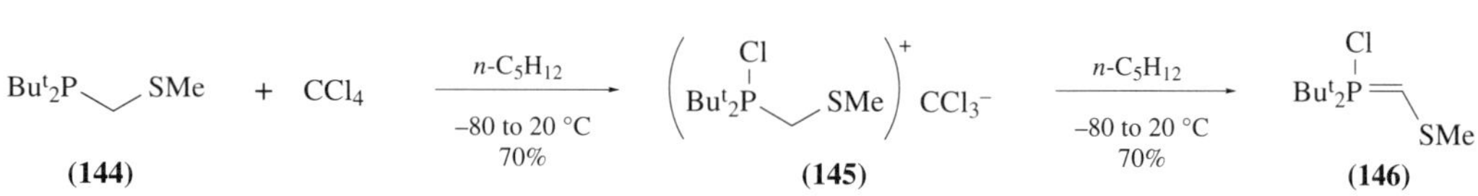

Scheme 37

(d) Miscellaneous methods. Slow addition of $Fe(CO)_4CS$ to an excess of $P(NMe_2)_3$ in *n*-pentane at 0°C furnished the complexes (**147**) and (**148**) in yields less than 15%. $(Me_2N)_3P{=}S$ is formed as a by-product (Scheme 38). As yet the desulfurization of the terminal CS ligand of $Fe(CO)_4CS$ seems to be restricted to $P(NMe_2)_3$ and phosphites $P(OR)_3$ ⟨93OM4056⟩.

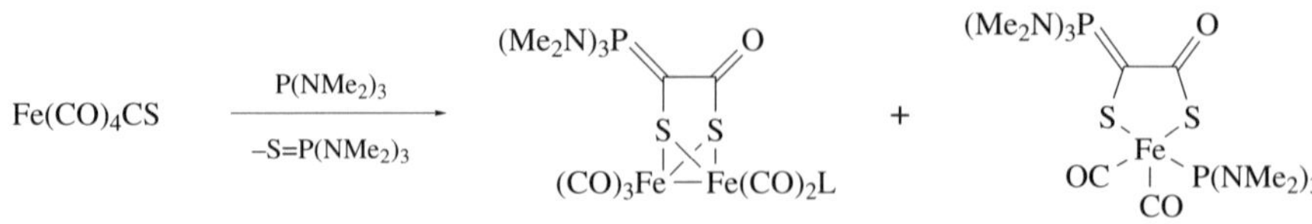

(**147**) **a**; L = CO
b; L = CS

(**148**)

Scheme 38

(iii) Derivatives with tricoordinate sulfur—$R^1C(PR^2{}_3)SR^3R^4$

(a) Sulfonium-substituted ylides. Basically, three different routes to phosphonium ylides functionalized by an α-sulfonium group are known. Deprotonation of the salts (**149**) suspended in cold anhydrous ethanol by equimolar amounts of sodium ethanolate afforded the colourless ylide salts (**150**) (70–80%) (Equation (44)) ⟨67AG(E)249⟩. The condensation of $Ph_3P{=}CHR$ with the amino functionalized sulfonium chloride (**139**) to give (**140**) has already been mentioned (Scheme 32).

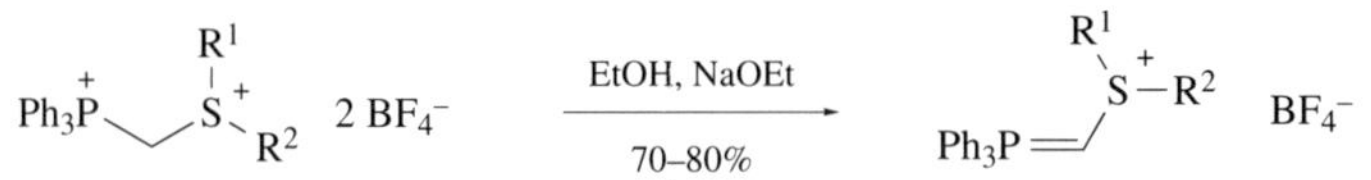

(**149**) (**150**) (44)

$R^1 = R^2 = Et$ or Me
$R^1 = Me, R^2 = Et$

The acetic anhydride-promoted condensation of dimethyl sulfoxide and tetrahydrothiophene-*S*-oxide with phosphonium salt $(Ph_3PCH_2CO_2Et)^+BF_4^-$ in boiling methylene chloride provides a useful entry to the ylide salts (**151**) (Equation (45)) ⟨79ZOR2396⟩.

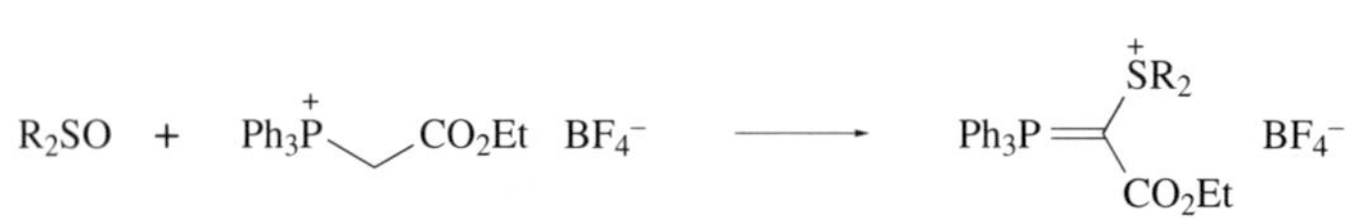

(**151**) (45)

$R_2 = Me_2$: Ac_2O, CH_2Cl_2, 40 °C, 30 h, 90%
$R_2 = C_4H_8$: Ac_2O, CH_2Cl_2, 40 °C, 25 h, 94%

(b) Sulfinyl-substituted ylides. An α-sulfinylylide (**152**) was prepared from equimolar amounts of $Ph_3P{=}CHCO_2Me$ and *p*-toluenesulfinyl chloride in the presence of triethylamine (80.5%) (Equation (46)). The related compounds $Ph_3P{=}C(EWG)S(O)R$ (EWG = CO_2Me, R = Ph, *p*-BrC_6H_4, Me, Et; EWG = CO_2Et, R = Ph, *p*-Tol, *p*-ClC_6H_4, *p*-BrC_6H_4, Et; EWG = CN, R = Ph, *p*-Tol) were synthesized analogously ⟨72USP3647856, 94CC805⟩. The reaction of $Ph_3P{=}CHCO_2Et$ with phenylmethanesulfinyl chloride under similar conditions led to (**153**) via the intermediacy of sulfine $PhCH{=}S{=}O$ (Equation (47)) ⟨68JCS(C)1612⟩.

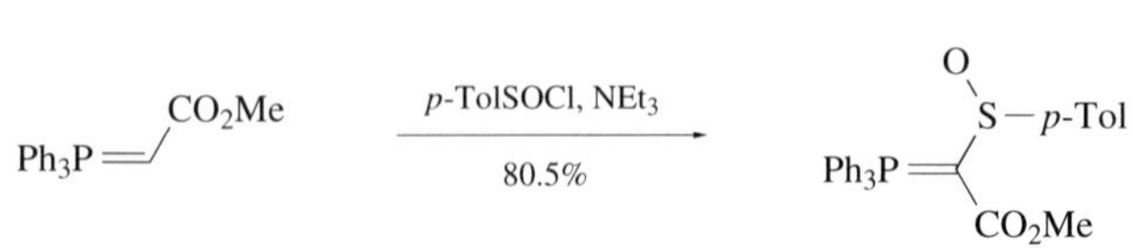

(46)

(**152**)

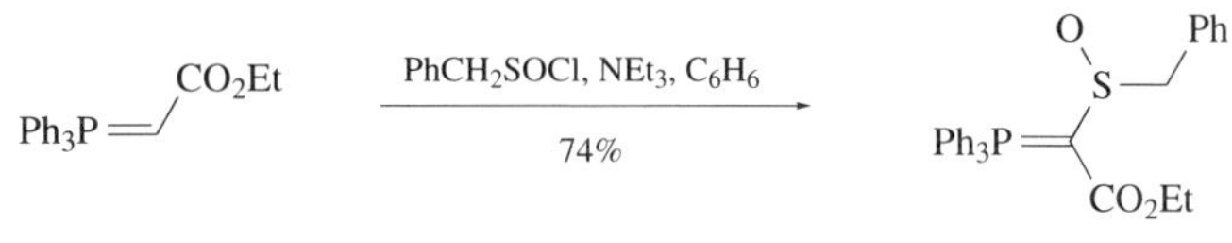

(153)

(47)

(iv) *Derivatives with tetracoordinate sulfur—$R^1C(PR^2_3)SO_2R^3$*

(a) From phosphonium salts. α-Sulfonylphosphonium ylides (**155**) were obtained by the treatment of phosphonium salts (**154**) with aqueous base (Equation (48)). This method, however, is restricted by the availability of the respective phosphonium salts ⟨61TL724, 65JA5603, 72RTC37⟩.

(154) (155)

(48)

R = Me; X not defined, NaOH/H_2O, 65%
R = *p*-Tol; X = Br, NaOH, H_2O, 76%
R = Ph, *p*-Tol, *p*-ClC_6H_4; X = Br, Na_2CO_3, H_2O (80, 42 and 51% resp.)

(b) From ylides. A more promising approach involves the sulfonylation of ylides by arenesulfonyl fluorides in a THF/ether mixture at 20°C (Scheme 39) ⟨72RTC37⟩. Whereas the sulfonylation of ethylidenephosphorane with tosyl fluoride provides $Ph_3P{=}C(Me)SO_2$*p*-Tol (50%), no reaction with the more bulky neopentylidenephosphorane or 1-adamantylmethylidenephosphorane occurred. Benzylidenephosphorane was converted into $Ph_3P{=}C(Ph)SO_2$*p*-Tol in only 12–15% yield.

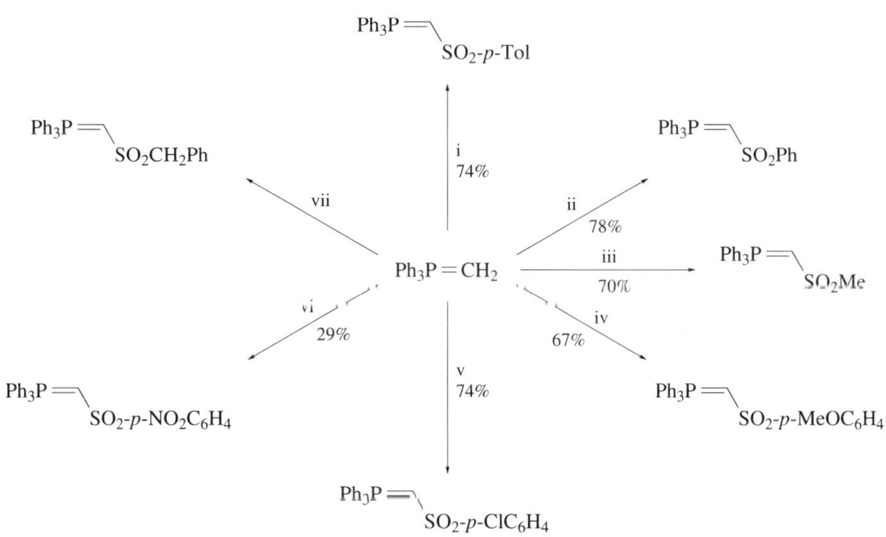

i, 0.5 *p*-$TolSO_2F$, THF, Et_2O, 20 °C, 80 min; ii, 0.5 $PhSO_2F$; iii, 0.5 $MeSO_2F$; iv, 0.5 4-$MeOC_6H_4SO_2F$; v, 0.5 4-$ClC_6H_4SO_2F$; vi, 0.5 4-$O_2NC_6H_4SO_2F$; vii, 0.5 $PhCH_2SO_2F$

Scheme 39

In addition, rearranged products were formed when the ylides $Ph_3P{=}CR^1H$ (R^1 = Ph, Bu^t, 1-adamantyl) were allowed to react with alkanesulfonyl fluorides $R^2CH_2SO_2F$ or anhydrides $(R^2CH_2SO_2)_2O$ containing relatively small groups R^2 such as H, Me or Ph (Table 4) (Equation (49)).

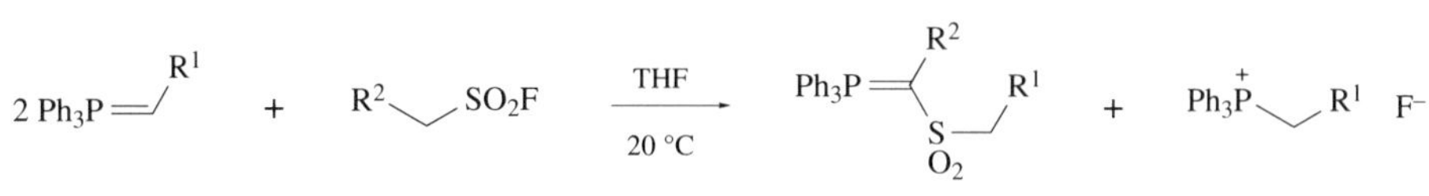

(49)

Table 4 Sulfonylation of ylides $Ph_3P{=}C(R^1)H$ by alkane sulfonyl fluorides $R^2CH_2SO_2F$[a].

Entry	R^1	R^2	*Reaction time* (h)	*Product*	*Yield* (%)
1	Ph	H	0.5	$Ph_3P{=}C(H)SO_2CH_2Ph$	95
2	Ph	Me	0.5	$Ph_3P{=}C(Me)SO_2CH_2Ph$	94
3	Ph	Ph	0.5	$Ph_3P{=}C(Ph)SO_2CH_2Ph$	91
4	*p*-$MeOC_6H_4$	Ph	1	$Ph_3P{=}C(Ph)SO_2CH_2C_6H_4$-*p*-OMe	88
5	Ph	*p*-$O_2NC_6H_4$	1.5	$Ph_3P{=}C$(*p*-$O_2NC_6H_4)SO_2CH_2Ph$	48
6	*p*-$O_2NC_6H_4$	Ph	3	$Ph_3P{=}C(Ph)SO_2CH_2C_6H_4$-*p*-$NO_2$	82
7	Bu^t	H	0.5	$Ph_3P{=}C(H)SO_2CH_2Bu^t$	69
8	Bu^t	Me	1	$Ph_3P{=}C(Me)SO_2CH_2Bu^t$	65
9	1-adamantyl	H[b]	0.5	$Ph_3P{=}C(H)SO_2CH_2$-1-adamantyl	51
10	H	H	1	$Ph_3P{=}C(H)SO_2Me$	70
11	Me	Bu^t	1	$Ph_3P{=}C(Me)SO_2CH_2Bu^t$	50
12	Me	Me	0.5	$Ph_3P{=}C(Me)SO_2Et$	75
13	H	Ph	1	$Ph_3P{=}C(H)SO_2CH_2Ph$	63

[a] THF, 20 °C; [b] $(MeSO_2)_2O$ was employed here.

Conversely, only unrearranged ylides were produced when bulky sulfonyl fluorides ($R^2 = Bu^t$) were allowed to react with sterically nondemanding ylides. The course of the reactions between alkanesulfonyl fluorides or anhydrides and ylides was rationalized by invoking sulfenes as intermediates ⟨74JOC2728⟩. Similar results were obtained when phosphorus ylides without α-hydrogens were submitted to reaction with alkanesulfonyl chlorides in the presence of an excess of NEt_3 in dioxane (Equation (50)) ⟨67T2137⟩. Derivatization of α-sulfonylylides with an α-hydrogen is conceivable by a lithiation/acylation sequence (Equation (51)) ⟨72RTC37⟩. One report was concerned with the sulfonation of stabilized phosphonium ylides to give the sulfobetaines (**156**), which were readily deprotonated to give salts of sulfonic acids with the ylidic features (**157**) (Scheme 40) ⟨63DOK(151)856⟩. The dimethylsulfoxide fragment in sulfonium ylide (**158**) was readily replaced by Et_2PCl in the presence of triethylamine to yield phosphonium ylide (**159**) (45%) (Equation (52)) ⟨77JGU658⟩.

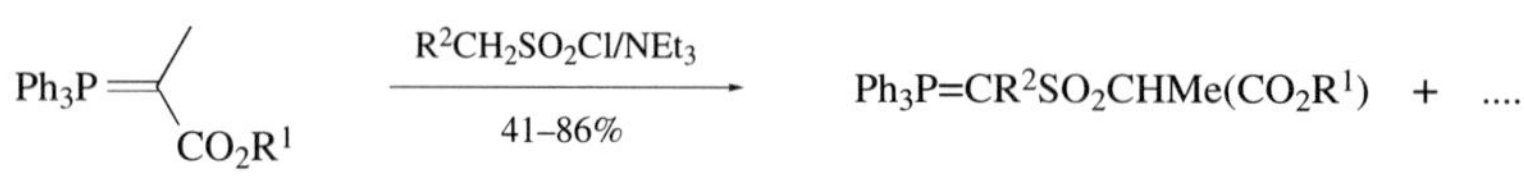

(50)

R^1 = Et, Bz, $(CH_2)_2Ph$, *c*-C_6H_{11}, Ph; R^2 = H, Ph

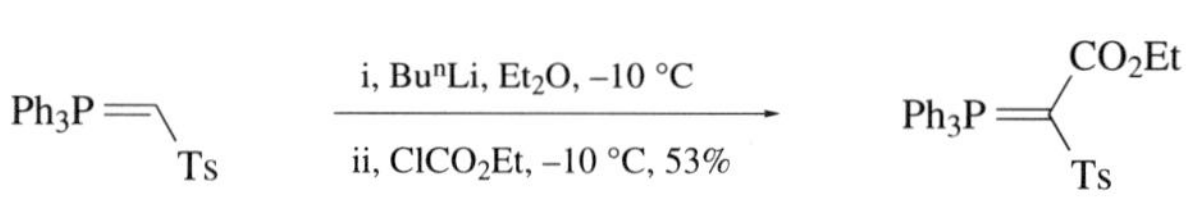

(51)

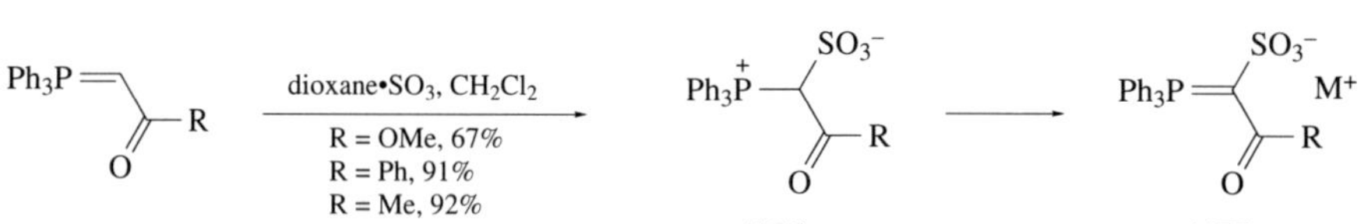

(**156**) (**157**)

R = OMe: 5% Na_2CO_3 (aq.), M = $Na{\bullet}H_2O$, 100%
R = Ph: NH_3 (liq.), $PhCH_2SC(NH_2)_2^+\ Cl^-$, H_2O, M = $PhCH_2SC(NH_2)_2^+$, 80%
R = Me: NH_3, $CHCl_3$, $PhCH_2SC(NH_2)_2^+\ Cl^-$, H_2O, M = $PhCH_2SC(NH_2)_2^+$

Scheme 40

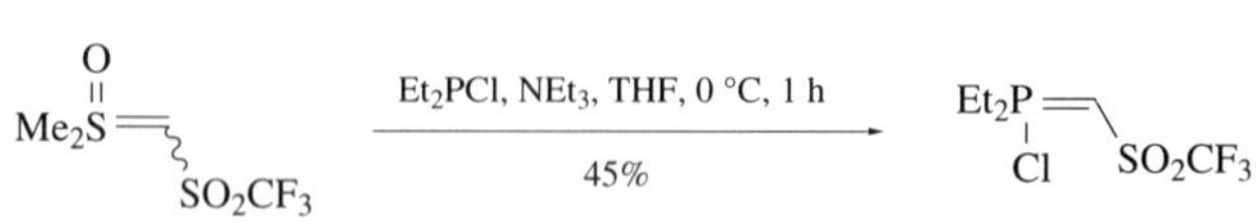

(52)

(**158**) (**159**)

(c) From phosphanes. Phosphanes with mobile α-hydrogen atoms such as (**160**) tend to disproportionate when distilled *in vacuo* with the generation of ylide (**161**) (50%) (Equation (53)) ⟨79JGU88⟩. Ylide (**161**) was obtained directly from $MeSO_2CF_3$ by treatment with two equivs of

Et_2PCl in the presence of an excess of NEt_3 in Et_2O (Equation (54)) ⟨75JGU2509⟩. Without triethylamine the reaction is very sluggish and does not furnish the expected product. Presumably the amine favors the formation of the anion of (**160**), in which the nucleophilicity of the P atom is increased. On the other hand, phosphane (**160**) was readily converted into ylide (**159**) by reaction with carbon tetrachloride and triethylamine in ether (Equation (55)) ⟨79JGU88⟩.

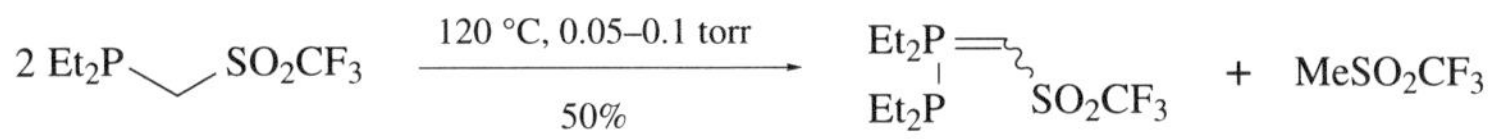

(**160**) (**161**) (53)

$$MeSO_2CF_3 \xrightarrow[50\%]{2\ Et_2PCl,\ NEt_3,\ Et_2O,\ 30\text{–}60\ min} \mathbf{(161)} \quad (54)$$

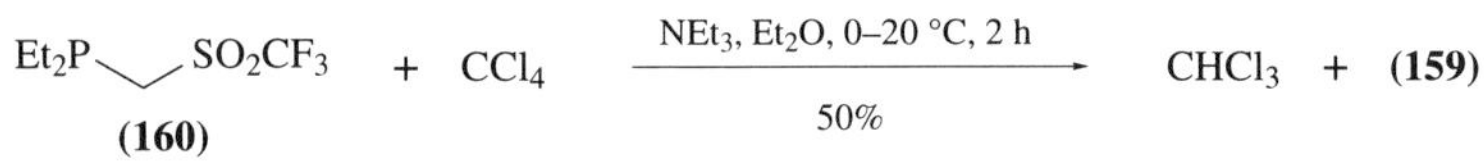

(**160**) (55)

(v) *Derivatives with dicoordinate selenium—$R^1C(PR^2_3)SeR^3$*

(a) From phosphonium salts. The deprotonation of $(Ph_3PCHRSePh)Br$ by *n*-butyl lithium in THF gave the selenium-functionalized ylides which were further reacted with carbonyl compounds without isolation ⟨76JOM(114)281, 77JOM(124)1⟩. In a somewhat different approach, the oxidation of phosphonium selenocyanate (**162**) furnished ylide (**163**) in 50–60% yield (Equation (56)) ⟨66M180⟩.

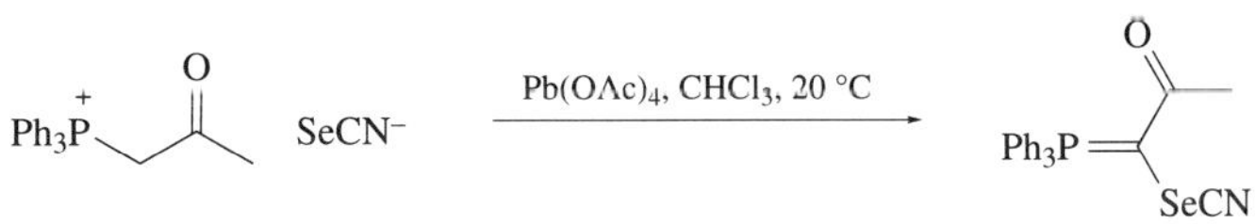

(**162**) (**163**) (56)

(b) From ylides. α-Selenylylides were also obtained from ylides $Ph_3P{=}CHR$ (R = H, Me) and PhSeBr via selenation. Again no efforts were made to isolate the products before treatment with carbonyls ⟨76JOM(114)281, 77JOM(124)1⟩.

Ylides $Ph_3P{=}CHCO_2R$ (R = Me, Et) underwent reaction with either PhSeCl, $Ph(Me)SeCl_2$, or $PhCH_2(Ph)SeCl_2$ in THF to give selenium derivatives (**164**) (Scheme 41) ⟨74JGU1677⟩. Melting an equimolar mixture of triphenylphosphane and the salts (**165**) at 125–170 °C at 30 torr for 5 min resulted in the formation of (**166**) in yields of 52% (R = Me) and 68% (R = Ph) (Equation (57)) ⟨79JGU1740⟩.

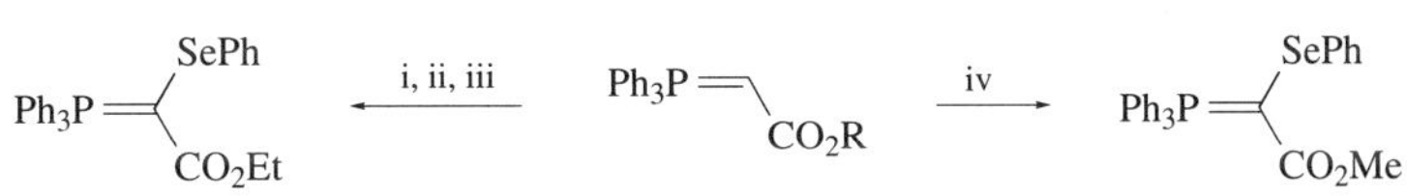

(**164a**) (**164b**)

i, $PhMeSeCl_2$, 20 °C, 90%; ii, $PhCH_2(Ph)SeCl_2$, –25 °C, 31%; iii, PhSeCl, 20 °C, 66%; iv, $PhMeSeCl_2$, 20 °C, 60%

Scheme 41

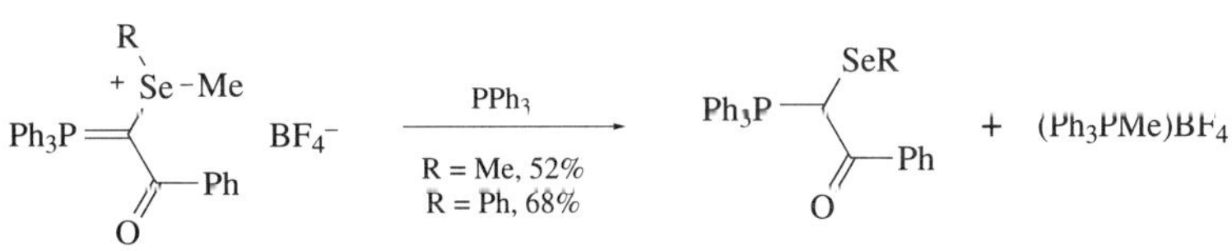

(**165**) (**166**) (57)

(vi) *Derivatives with tricoordinate selenium—$R^1C(PR^2_3)SeR^3_2$*

Phosphonium ylides with α-selenonium functions result from condensation of ylides or phosphonium salts with tetravalent selenium species. Thus, ylides $Ph_3P{=}CHCO_2R$ (R = Me, Et) were converted into (**167**) by reaction with tetrahydroselenophene 1,1-dichloride (Equation (58)) ⟨74JGU1677⟩.

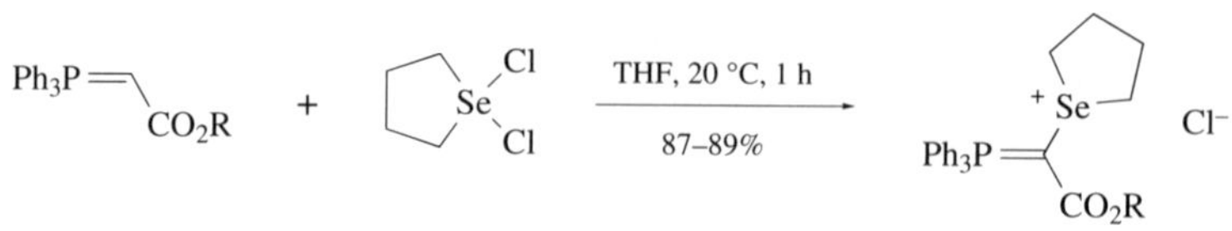

(**167**)

(58)

Condensation of selenium oxides with $(Ph_3PCH_2CO_2Et)BF_4$ to give selenium-substituted ylides (**167**) and (**168**) was achieved by acetic anhydride in boiling methylene chloride (Scheme 42) ⟨79ZOR2396⟩.

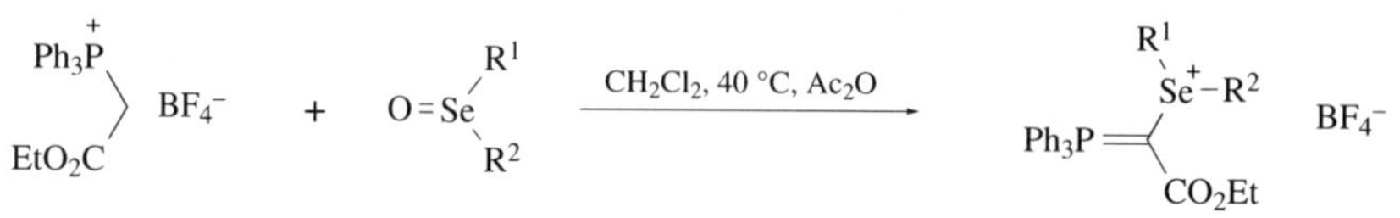

(**167**) 92%
(**168**) $R^1 = R^2 = Ph$, 93.5%
$R^1 = Bn$; $R^2 = Ph$, 47%

Scheme 42

5.23.3.1.4 *Nitrogen derivatives—$R^1C(PR^2_3)NR^3_n$*

(i) Derivatives with dicoordinate nitrogen—$R^1C(PR^2_3)NR^3$

(a) From phosphonium salts and base. Phosphonium salt $(Ph_3PCH_2NC)Cl$ and a threefold excess of sodamide were vigorously stirred in THF at −78 °C to furnish the yellow crystalline isocyanomethylene phosphorane $Ph_3P{=}CHNC$ (93%) (Equation (59)) ⟨85AG(E)979⟩.

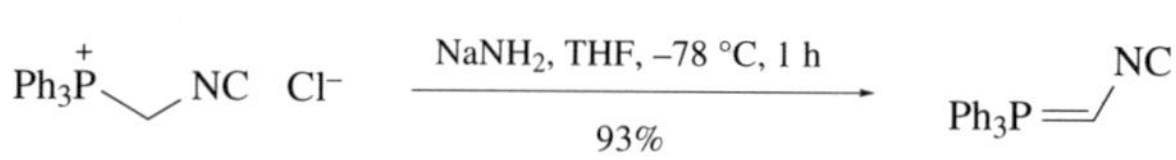

(59)

Ethyl nitroso(triphenylphosphoranylidene)acetate was the product of the deprotonation of the corresponding phosphonium salt by sodium ethoxide in anhydrous ethanol. This procedure is of no general value since usually the treatment of α-nitrosophosphonium salts with base furnishes nitriles (Equation (60)) ⟨71JGU2019⟩.

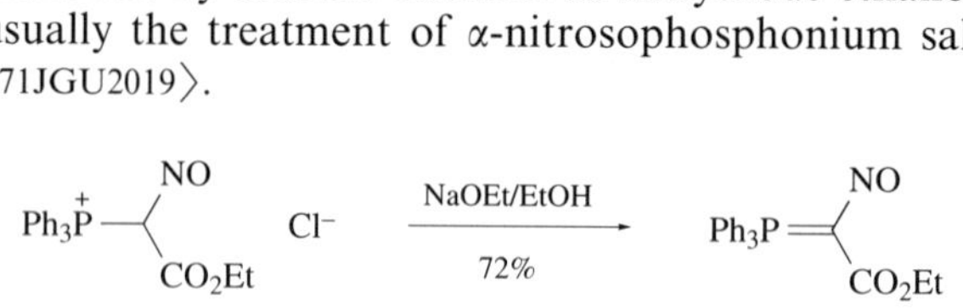

(60)

Arylazo-functionalized phosphonium salts were converted in high yields into the corresponding ylides, when treated with aqueous NaOH. Since the educts result from the reaction of stabilized ylides and diazonium salts in water, it is obvious that this method is restricted to such ylides, which resist hydrolysis (Equation (61)) ⟨61TL807⟩.

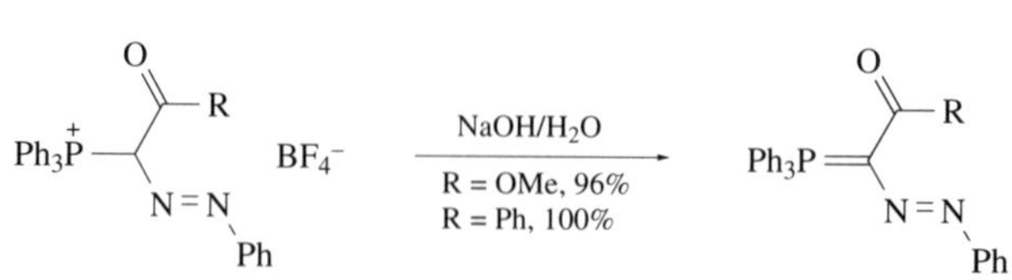

(61)

(b) From ylides. Arylazo-functionalized ylides resulted in high yields from the addition of 2N NaOH to an aqueous or methanolic solution of phosphonium and diazonium salts. Alternatively, the

azo-functionality may be introduced into the stabilized ylide by reaction with *N*-nitrosoacetanilide in methanol (Table 5) ⟨61TL807⟩.

Table 5 Preparation of azomethylenephosphoranes from ylides or phosphonium salts.

Entry	*Substrate*	*Conditions*	*Product*	*Yield* (%)
1	$(Ph_3PCH_2CO_2Me)X$	a	$Ph_3P{=}C(CO_2Me)N{=}NPh$	98
2	$Ph_3P{=}CHCO_2Me$	b	$Ph_3P{=}C(CO_2Me)N{=}NPh$	91
3	$(Ph_3PCH_2CO_2Me)X$	a	$Ph_3P{=}C(CO_2Me)N{=}NC_6H_4NO_2\text{-}p$	100
4	$[Ph_3PCH_2(CH{=}CH)_2CO_2Me]X$	a	$Ph_3P{=}C[(CH{=}CH)_2CO_2Me]N{=}NC_6H_4NO_2\text{-}p$	78
5	$Ph_3P{=}CHCOPh$	b	$Ph_3P{=}C(COPh)N{=}NPh$	94
6	$(Ph_3PCH_2C_6H_4NO_2\text{-}p)X$	a	$Ph_3P{=}C(C_6H_4NO_2\text{-}p)N{=}NPh$	89

[a] $ArN{=}N^+BF_4^-/NaOH/H_2O/MeOH$; [b] *N*-nitrosoanilide/MeOH.

(c) From phosphanes. Arylazomethylenetriphenylphosphoranes (**170**) are conveniently synthesized in high yield when hydrazonoyl chlorides (**169**) were allowed to react with PPh_3 in the presence of triethylamine in acetonitrile at 20 °C. As this conversion failed in the absence of base it was suggested that transient nitrilimines were added in a carbenoid form to the phosphane (Equation (62)) (Table 6, (**170a–i**)) ⟨79S299⟩. This reaction is also useful for the transformation of α-alkenyl-substituted arylhydrazonyl chlorides into the corresponding ylides. However, in a few cases, this reaction was complicated by the intramolecular formation of cyclopropacinnolines and benzodiazepines. These transformations are favored by high temperatures and polar solvents, whereas ylide generation is the major pathway at low temperature and in apolar solvents ⟨81G285, 84T2165⟩.

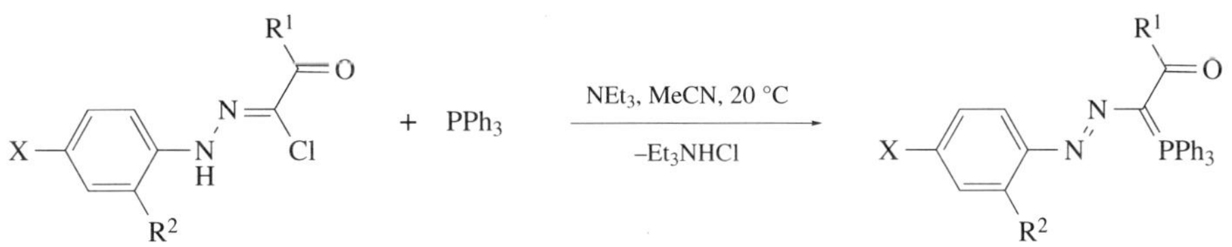

(169) (170) (62)

Table 6 Preparation of arylazomethylenephosphoranes from hydrazonoyl chlorides.

(170)	R^1	R^2	X	*Yield* (%)
a	OBu^t	CHO	H	85
b	OBu^t	COMe	H	57
c	OBu^t	COPh	Cl	95
d	OMe	CHO	H	45
e	Me	H	OMe	86
f	OMe	H	OMe	83
g	OMe	H	Cl	76
h	Ph	$C({=}CH_2)Ph$	Cl	75
i	OEt	$CH{=}C(Me)CO_2Et$	H	88

(ii) Derivatives with tricoordinate nitrogen—$R^1C(PR^2_3)NR^3_2$

Woodward and co-workers developed an elegant synthesis of 6-acylaminopenem-3-carboxylic acids, which involved a three-step transformation of the *N*-substituted azetidinones (**171**) into the ylides (**173**). The latter were obtained from the epimeric chlorides (**172**) by mild heating (50 °C) with PPh_3 in dioxane in the presence of polymeric Hünig base (45–60%) (Scheme 43) ⟨78JA8214⟩.

The ylides (**174**) arose from nucleophilic attack of ylides $Ph_3P{=}CHCOR^1$ (R^1 = Ph, Me) on triazolinediones (Scheme 44). The presence of an α-proton in the ylidic educts is essential for the addition ⟨76JOC2102⟩.

Pentacarbonyl(isocyanomethylenetriphenylphosphorane)chromium and methyl isocyanate

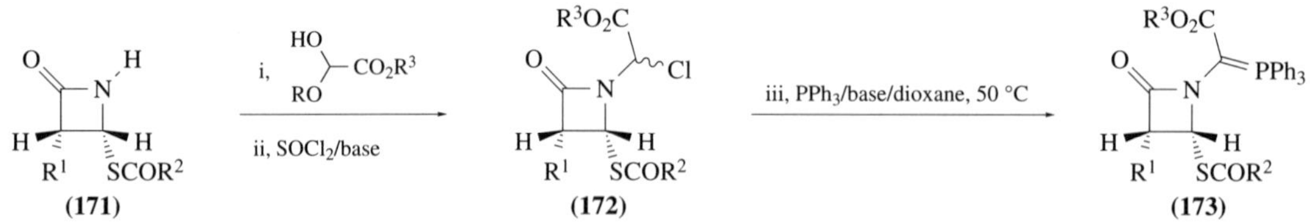

R^1 = $PhOCH_2ONH$; R^2 = Me, Pr^i, Ph, p-$O_2NC_6H_4$; R^3 = Me, Bu^t, CH_2CCl_3, p-O_2NBn

Scheme 43

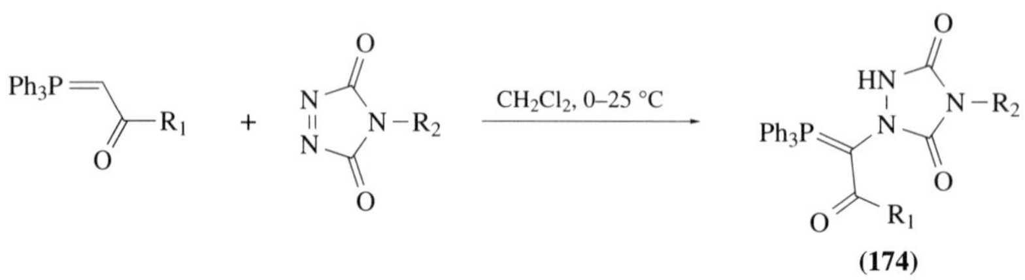

R^1 = R^2 = Ph, 82%; R^1 = Ph, R^2 = Me, 78%; R^1 = R^2 = Me, 85%

Scheme 44

underwent a [3 + 2] dipolar cycloaddition to give complex (**175**), with the features of an amino-functionalized ylide (Equation (63)) ⟨85AG(E)979⟩.

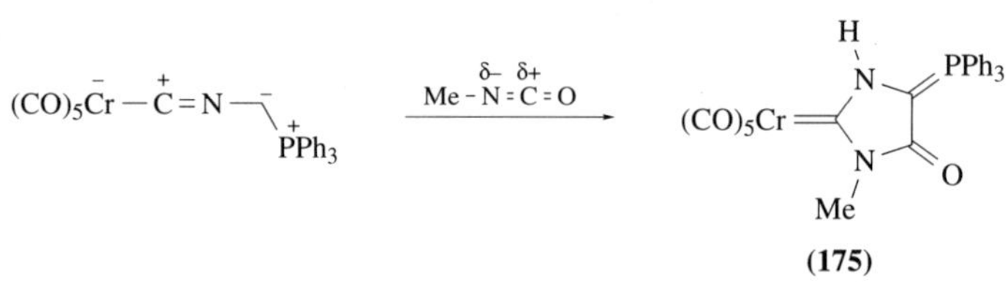

(63)

Upon heating *o*-formylarylazomethylenephosphoranes of the type (**176**) undergo intramolecular cyclization with generation of 3-oxoindazolinylmethylenephosphoranes (**177**) (Scheme 45). The scope of this isomerization is limited, since with azo-functionalized ylides (**176**; R = SO_2Ph or Ts) only 4-oxodihydroquinazoline derivatives were obtained. Nevertheless, the latter compounds are the products of prolonged heating of (**176**) or (**177**). Ylide (**177**; R = CS_2Me) was synthesized directly from *o*-aminobenzaldehyde by diazotation with HNO_2 and subsequent quenching of the diazonium salt with methylthio(thiocarbonyl)methylenetriphenylphosphorane and aqueous NaOH (42%) ⟨85T3321⟩.

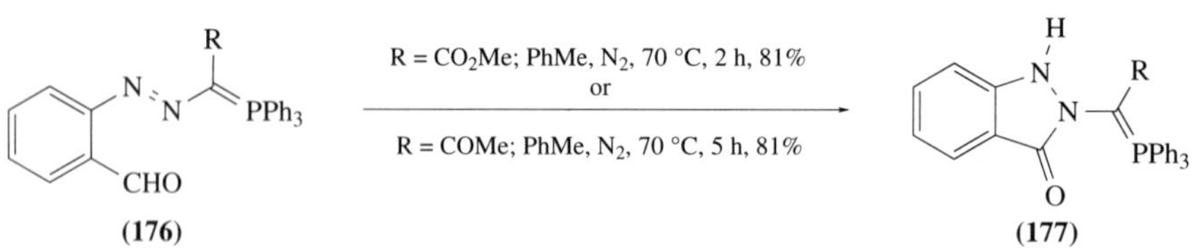

Scheme 45

An alternative route to (**177**; R = CO_2Me, X = Br) was based on 5-bromoisatin, which was subsequently hydrolyzed, diazotized, and coupled with $Ph_3P{=}CHCO_2Me$ (Scheme 46) ⟨83CC337⟩.

Nitrogen-functionalized ylides (**179**) are the products of the thermolysis of arylazophosphoranes with *o*-aldonitronyl substituents in toluene or DMSO (Scheme 47). This transformation is sensitive to the nature of the substituents R^1 and R^2 and the solvent. When the thermolysis of (**178a**) was carried out in neat toluene at 110 °C an imidazo [1,2-b] indazole derivative was generated in 63% yield, whereas the same procedure in *o*-dichlorobenzene furnished a mixture of the latter system and (**179a**). In the more polar DMSO, this complication was absent. The preparation of compounds such as (**179**) from (**178**) with R^1 = Ts, R^2 = Me failed, whereas heating (**178d**) in toluene at 110 °C or in DMSO at 80 °C gave intractable mixtures ⟨85T2503⟩.

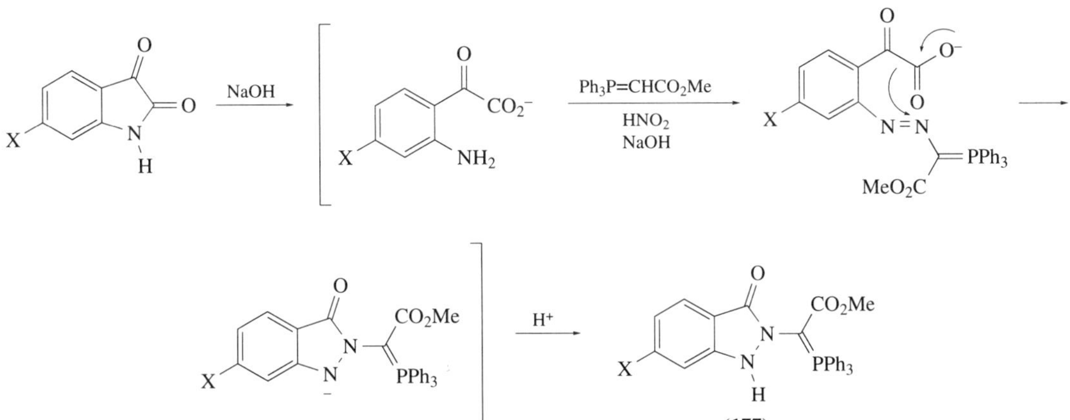

(177)

Scheme 46

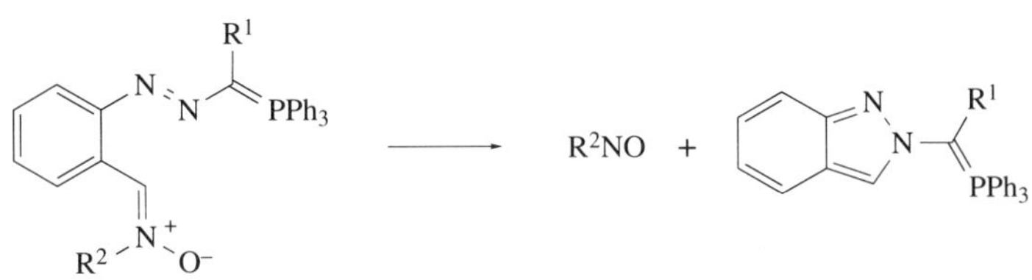

(178) **(179)**

a; $R^1 = CO_2Me$, $R^2 = Me$, PhMe/NEt_3, 110 °C, 6 h, 80%
b; R^1 COMe, $R^2 = Me$, PhMe or DMSO, 110 °C, 6 or 8 h, 47 or 35% resp.
c; $R^1 = COMe$, $R^2 = Ph$, PhMe or DMSO, 110 °C, 7 or 4 h, 54 or 40%, resp.
d; $R^1 = Ts$, $R^2 = Ph$, PhMe, 80 °C, 10 h, 60%

Scheme 47

5.23.3.1.5 *Group 15 element derivatives—$R^1C(PR^2_3)ER^3_n$ (E = P, As, Sb, Bi)*

(i) Derivatives with dicoordinate phosphorus—$R^1C(PR^2_3)PR^3$

Ionic species (**180**), with the features of ylides functionalized by a dicoordinate phosphorus substituent, precipitated when two equivs of a silylated ylide were treated with a benzene solution of PCl_3 (Equation (64)) ⟨92TL471⟩. Orange crystalline ylides $Ph_3P{=}C(R)P{=}S$ with P=S functions were isolated from the thiolysis of $Ph_3P{=}C(R)PCl_2$ (R = Me, Et) with sodium sulfide in THF. Aryl-substituted ylides (R = Ph, *m*-Tol) were similarly formed. Here, however, the reaction was not straightforward, and the products were unstable (Equation (65)) ⟨93AG(E)1089⟩.

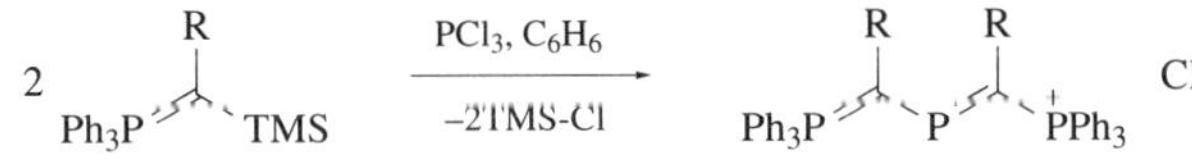

R = Me, Et, Ph **(180)** (64)

(65)

(ii) Derivatives with tricoordinate phosphorus—$R^1C(PR^2_3)PR^3_2$

(a) From phosphonium salts. Phosphinomethylphosphonium salts which are conveniently accessible by monoalkylation of the corresponding bisphosphinomethanes $(R_2P)_2CH_2$ were deprotonated to afford phosphino-functionalized ylides of the type $R^1_2R^2P{=}CHPR^1_2$. The employment of

equimolar amounts of base ($NaNH_2$, NaH, organolithium reagents) is recommended to avoid proton abstraction from the ylides (Scheme 48) ⟨79ZN(B)31, 80CB902, 82ZN(B)950, 83CB1386⟩.

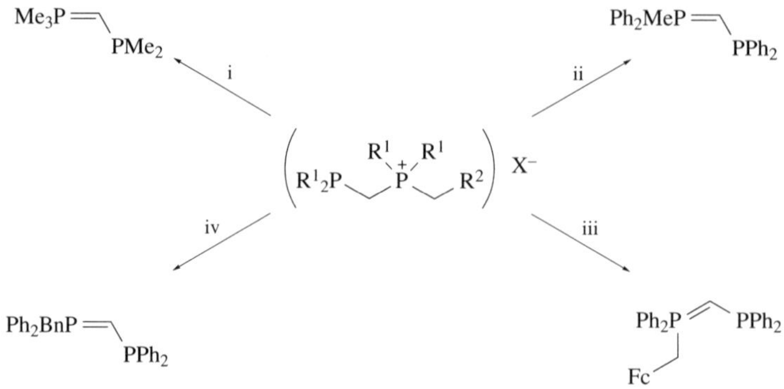

i, ($Me_3PCH_2PMe_2$)Br, $NaNH_2$, THF, 20 °C, 1 d
ii, ($MePh_2PCH_2PPh_2$)I, $Me_3P{=}CH_2$, THF, 25 °C, 3 h, 96.5%
iii, [(FerrocenylCH_2)$Ph_2PCH_2PPh_2$]I, $Me_3P{=}CH_2$, THF, 88%
iv, ($BnPh_2PCH_2PPh_2$)Br, 1 equiv. $NaNH_2$, THF, 20 °C, 3.5 h, 87%

Scheme 48

Whenever group R^2 in the phosphonium species efficiently delocalizes negative charge, the P═C bond is essentially generated at the periphery of the ylide ⟨81CB2491⟩. Skeletal rearrangements were observed upon base treatment of the phosphonium salts (**181**) and (**183**) resulting in the formation of ylides (**182**) and (**184**) respectively (Equations (66) and (67)) ⟨79CB510⟩.

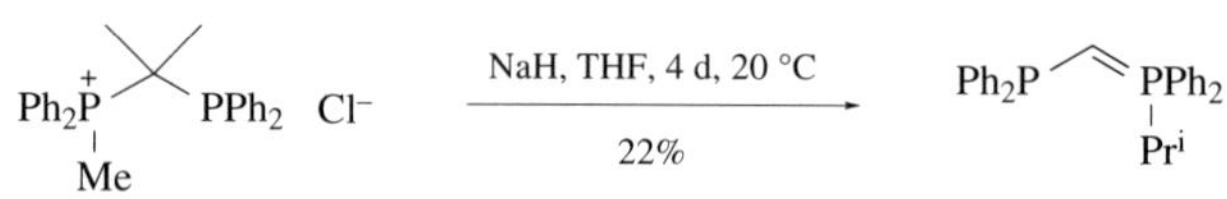

(181) (182) (66)

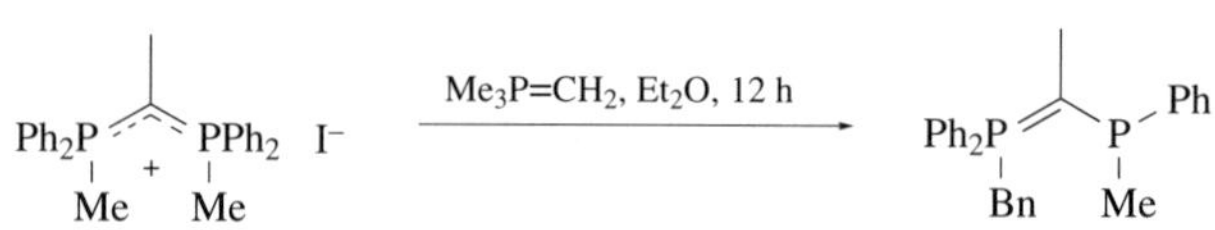

(183) (184) (67)

(*b*) *From ylides.* A powerful method for the preparation of a series of phosphino-substituted ylides is based upon the reaction of monochlorophosphanes with two equivs of an ylide which has at least one α-hydrogen atom via transylidation. Usually the chlorophosphane solution is added to the solution of the ylide to guarantee an excess of base throughout the process. Ylides derived from triarylphosphanes as well as trialkylphosphanes were successfully employed (Scheme 49) ⟨66LA(699)40, 68CB3545⟩.

An α-phosphino function may also be introduced into an ylide by treating α-silylated ylides with halophosphanes (Equation (68)). These ylides are usually not accessible from $Ph_3P{=}CRH$ via transylidation ⟨92TL471⟩. Ylides with an α-hydrogen and an α-silyl group react with organochlorophosphanes preferentially via transylidation and not via elimination of TMS-Cl ⟨68CB3545⟩.

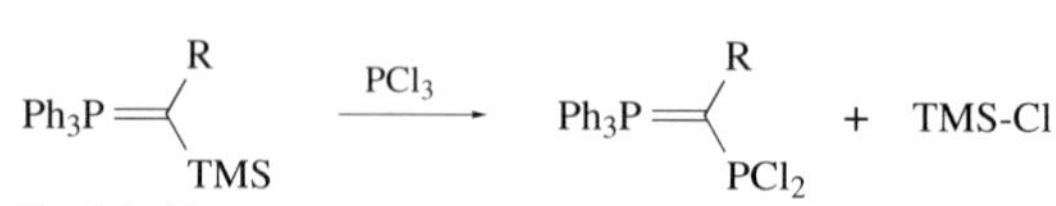

R = Me, Et (68)

The nucleophilic attack of methylene phosphoranes at the phosphorus atom of 2-vinyl phosphiranes selectively led to ring opening and transprotonation to afford the α-phosphino ylides (**185**–**187**) in 85–90% yield (Scheme 50) ⟨84T3273⟩.

The sulfurization of $Ph_3P{=}C(Et)P{=}S$ gave $Ph_3{=}C(Et)PS_2$. Ylides of the latter type were also available from $Ph_3P{=}C(R)PCl_2$ by the nucleophilic substitution of chloride with Na_2S_2 (Equations (69) and (70)) ⟨93AG(E)1089⟩.

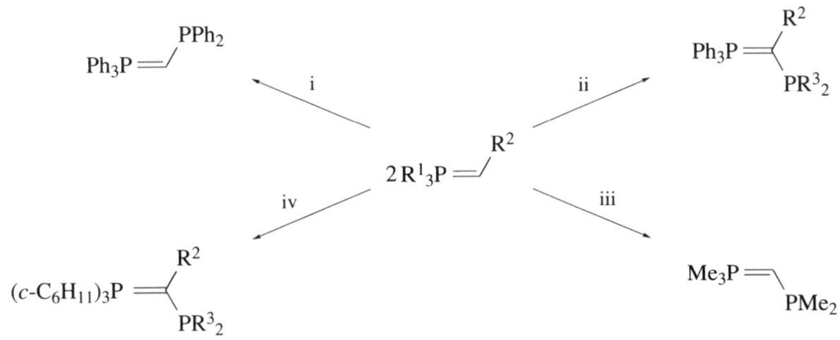

i, $Ph_3P{=}CH_2$, Ph_2PCl, C_6H_6, 50–60 °C, 10 h, 81%
ii, $Ph_3P{=}CHR^2$ (R^2 = Me, Ph, CO_2Me, CO_2Et), Ph_2PCl, C_6H_6, 80 °C; $Ph_3P{=}CHMe$, $R^3{}_2PCl$ (R^3 = Et, *c*-C_6H_{11}), C_6H_6, 80 °C
iii, $Me_3P{=}CH_2$, Me_2PCl, Et_2O, 10 °C, 12 h, 58%
iv, (*c*-$C_6H_{11})_3P{=}CHR^2$ (R^2 = H, Me, Ph), Ph_2PCl, C_6H_6; (*c*-$C_6H_{11})_3P{=}CHMe$, (*c*-$C_6H_{11})_2PCl$, C_6H_6

Scheme 49

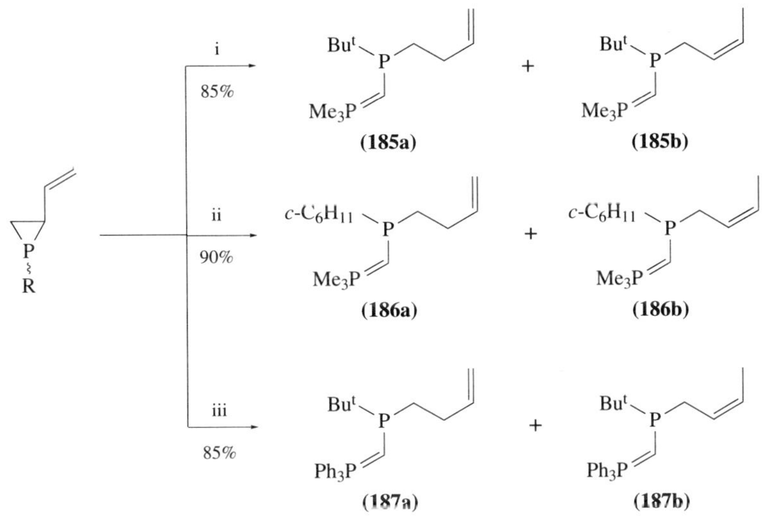

i, R = Bu^t: $Me_3P{=}CH_2$, *n*-C_5H_{12}, 0–20 °C, 16 h (**a** : **b** = 2 : 3)
ii, R = *c*-C_6H_{11}: $Me_3P{=}CH_2$, *n*-C_5H_{12}, (**a** : **b** = 3 : 2)
iii, R = Bu^t: $Ph_3P{=}CH_2$, *n*-C_5H_{12}, 20 °C, 6 d (**a** : **b** = 8 : 5)

Scheme 50

(69)

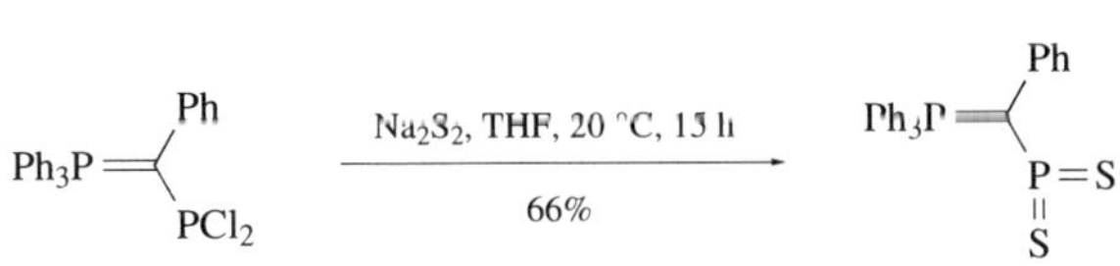

(70)

(c) From lithium diphosphinomethanides. Lithium diphosphinomethanides (**188**) are difunctional nucleophiles which undergo P—P bond formation with bulky diorganochlorophosphanes to give C,P-diphosphino-functionalized ylides (**189**). With less bulky chlorophosphanes $R^2{}_2PCl$ (R^2 = Me,

Ph) usually the carbon center of (**188**) is attacked. If the reaction is carried out at $-78\,^{\circ}\mathrm{C}$, mixtures of trisphosphinomethanes and ylides are formed. Solutions of these ylides completely rearrange at $20\,^{\circ}\mathrm{C}$ to the trisphosphinomethanes (Scheme 51) ⟨79CB1093, 79ZN(B)1171⟩.

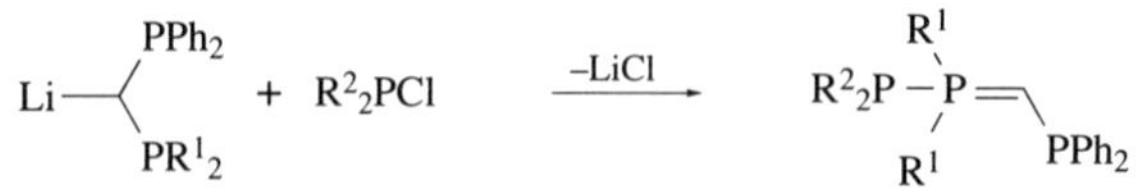

(**188**) (**189**)

$LiCH(PPh_2)_2$, TMEDA, $Bu^t{}_2PCl$, C_6H_6, 20 °C, 2 h, 63%
$LiCH(PPh_2)_2$, TMEDA, $(c\text{-}C_6H_{11})_2PCl$, C_6H_6, 20 °C, 2 h, 70%
$LiCH(PPh_2)(PMe_2)$, Et_2O, –78 °C, 2 h

Scheme 51

(iii) Derivatives with tetracoordinate phosphorus—$R^1C(PR^2{}_3)PR^3{}_3$

(a) From phosphonium salts. Conjugated double ylides such as (**192**) and (**194**) having two phosphorus atoms in 1,3-positions are only accessible with substituents which are capable of stabilizing negative charge. Otherwise cumulated isomers $R_3P{=}C{=}PR_3$ are generated instead. The deprotonation of salts (**190a,b**) with $Me_3P{=}CH_2$ or KH afforded (**192a,b**) in fair yields via the intermediacy of salts (**191a,b**) (Scheme 52) ⟨81CB1428⟩.

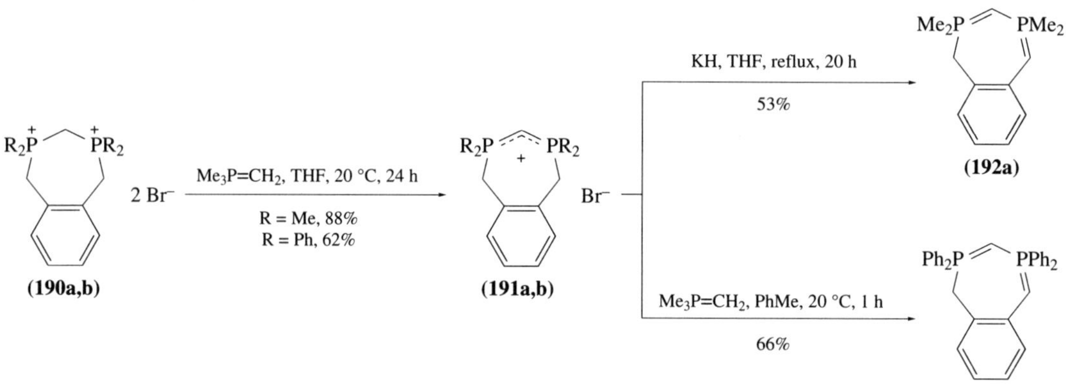

(**190a,b**) (**191a,b**) (**192a**) (**192b**)

Scheme 52

Similarly, the bis(phosphonium)salt (**193**) gave rise to the formation of (**194**) (81%) when treated with two equivs of $Me_3P{=}CH_2$ in THF. This deprotonation (Equation (71)) was also carried out with two equivs of $NaNH_2$ albeit in only 44% yield ⟨80CB902⟩.

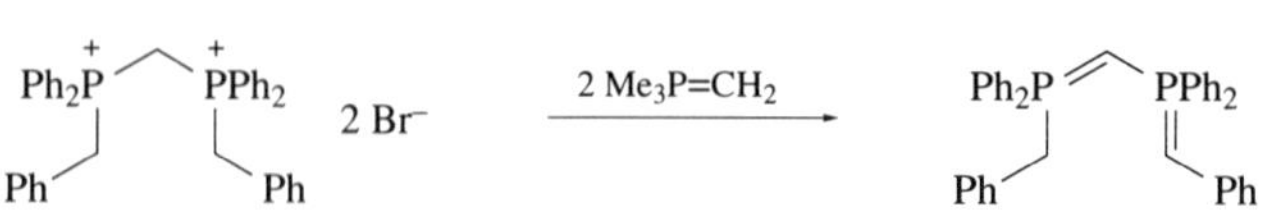

(**193**) (**194**) (71)

(b) From ylides. Diphenylphosphinic chloride and diphenylthiophosphinic chloride react with ylides $Ph_3P{=}CHR$ (R = H, Me, Ph, but not COMe) resulting in transylidation to yield *P*-functionalized ylides (**195**) (Equation (72)) ⟨67LA(707)112⟩. A similar process accounts for the formation of compounds $Ph_3P{=}CHP(O)(OPh)_2$ from $Ph_3P{=}CH_2$ and the respective phosphonic diester chloride (60%) ⟨68TL5731⟩.

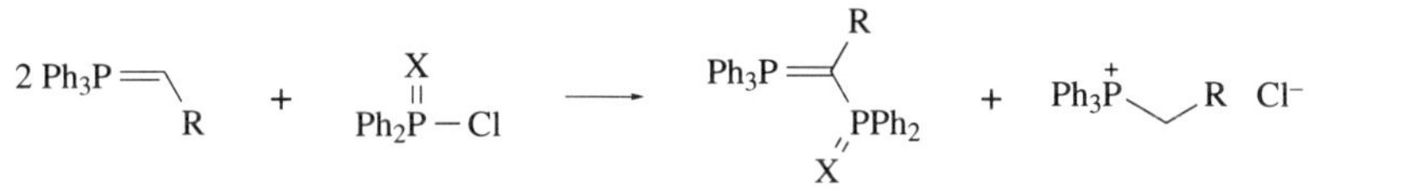

(72)

(195) **a**; R = H, X = O, C_6H_6, 60 °C, 3-4 h, 73%
b; R = Me, X = O, C_6H_6, 80 °C, 1 h
c; R = Ph, X = O, C_6H_6, 80 °C, 10 h
d; R = Ph, X = S, C_6H_6, 80 °C, 40 h

(c) From double ylides. When crystalline (**196**) was exposed for 20 h to a stream of wet nitrogen, a nearly complete conversion to (**195a**) took place (Equation (73)) ⟨61JA3539⟩.

(73)

(196) **(195a)**

Similarly, hydrolysis of (**192b**) and (**194**) in moist THF or ether led to the formation of ylides (**197**) ⟨81CB1428⟩ and (**198**) (Equations (74) and (75)) ⟨80CB902⟩.

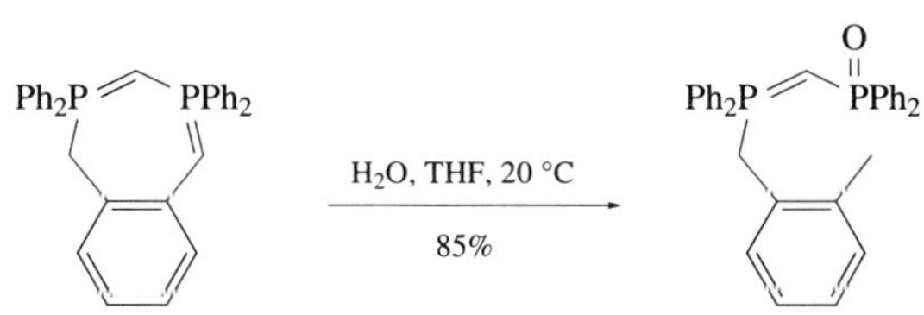

(74)

(192b) **(197)**

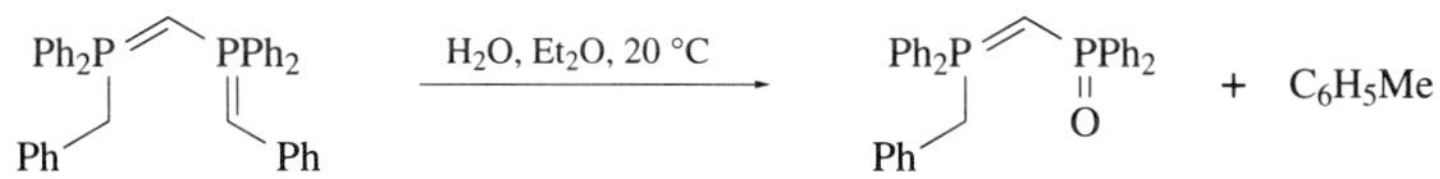

(75)

(194) **(198)**

(iv) Derivatives with pentacoordinate phosphorus—$R^1C(PR^2_3)PR^3_4$

Ylide (**199**) resulted from the exposure of $Me_3P{=}CH\text{-}TMS$ to Me_3PF_2 in boiling benzene or alternatively from the transylidation of $Me_3P{=}CH_2$ and Me_3PF_2 (Scheme 53) ⟨77CB3501⟩.

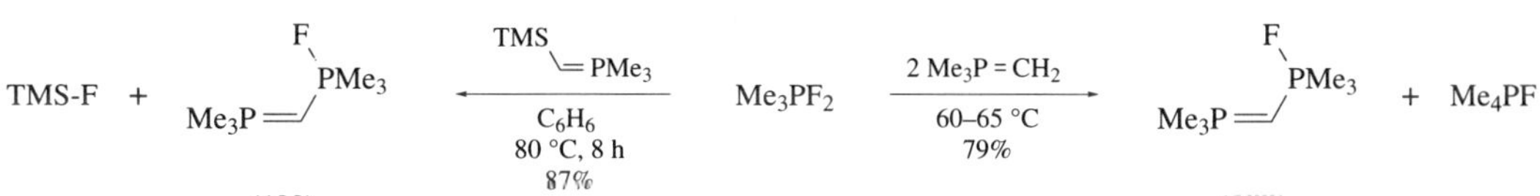

(199) **(199)**

Scheme 53

(v) Arsenic derivatives—$R^1C(PR^2_3)AsR^2_2$

Triphenylphosphonium-diphenylarsinomethylide (**200**) was easily prepared via transylidation using $Ph_3P{=}CH_2$ and Ph_2AsCl as the starting materials in a 2 : 1 molar ratio (50%) (Equation (76)) ⟨84ZN(B)1456⟩. Transylidation processes were also operative in the heterolytic cleavage of the M—As bond of metalloarsanes by $Me_3P{=}CH_2$ (Equation (77)) ⟨80CB3318, 82JOM(236)C1⟩. An alternative synthesis of (**201a**) was based upon the ligand exchange in $Me_3P{=}CH\text{-}TMS$ by means of the arsino-functionalized siloxane $TMSOAsMe_2$. The driving force for this conversion is the formation of stable hexamethyldisiloxane (Equation (78)) ⟨68CB3545⟩.

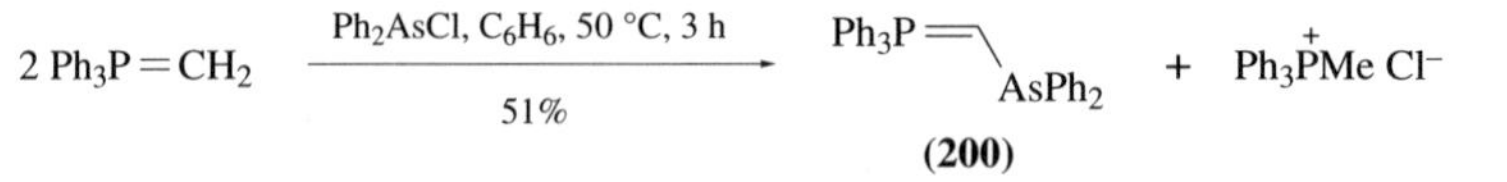

(76)

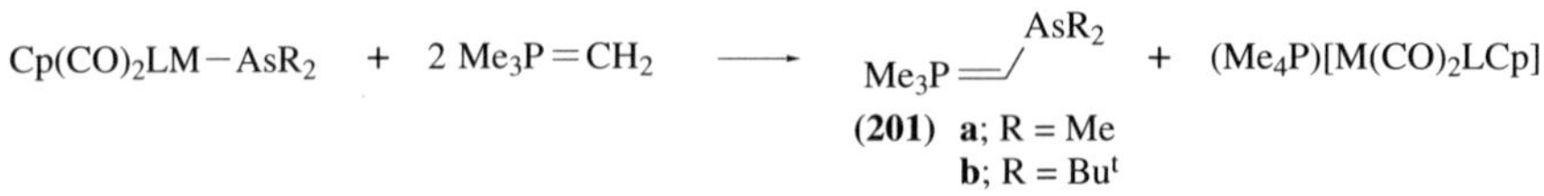

(77)

(201a): $CpMo(CO)_3AsMe_2$, Et_2O, 20 °C, 30 min, 91% or $CpW(CO)_2(CNMe)AsMe_2$, C_6H_6, 20 °C, 12 h
(201b): $CpW(CO)_2(CNBu^t)AsBu^t{}_2$, C6H6, 20 °C, 12 h

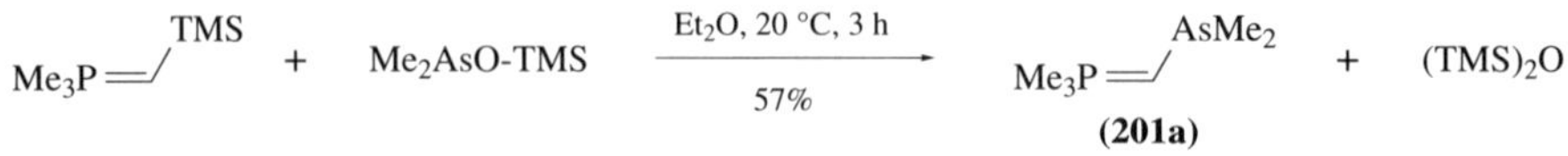

(78)

(vi) Antimony derivatives—$R^1C(PR^2{}_3)SbR^3{}_2$

Similar to the preparation of arsino-substituted phosphorus ylides, the heterolytic cleavage of the M—Sb bond in metallostibanes led to the transfer of the Me_2Sb group onto the ylide (Equation (79)) ⟨75JOM(99)421⟩. Compound **(202)** was first synthesized from $Me_3P{=}CH_2$ via lithiation and the subsequent quench of the lithium derivative by Me_2SbCl in ether (40%) (Equation (80)) ⟨68CB3545⟩.

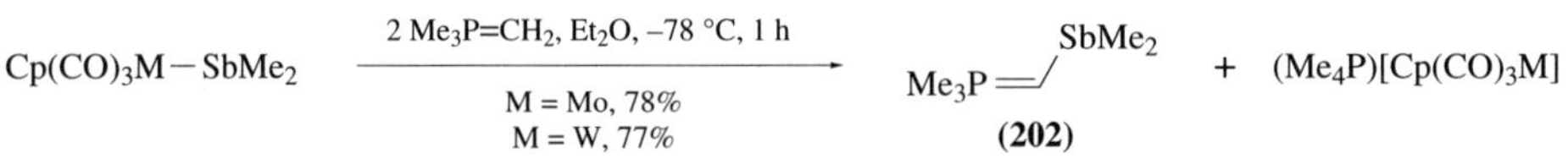

(79)

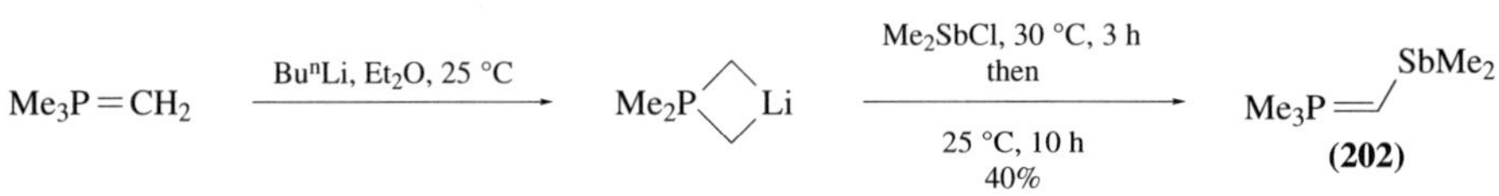

(80)

5.23.3.1.6 *Metalloid derivatives—$R^1C(PR^2{}_3)ER^3{}_n$ ($E = Si, Ge, B$)*

(i) Silicon derivatives—$R^1C(PR^2{}_3)SiR^3{}_3$

(a) From phosphonium salts. The silylated ylide $Me_3P{=}CHTMS$ results from the reaction of $(Me_3PCH_2TMS)Cl$ with an equimolar amount of *n*-butyllithium in dry ether at 20 °C. After the removal of all volatiles, the colorless residue was thermolyzed at 100–170 °C *in vacuo*. Redistillation of the resulting volatiles, which have been collected in a trap at −196 °C, afforded the pure ylide (99%) ⟨65IC1458⟩. Similarly, SiH-functionalized ylides are available (Equation (81)) ⟨70CB3007⟩.

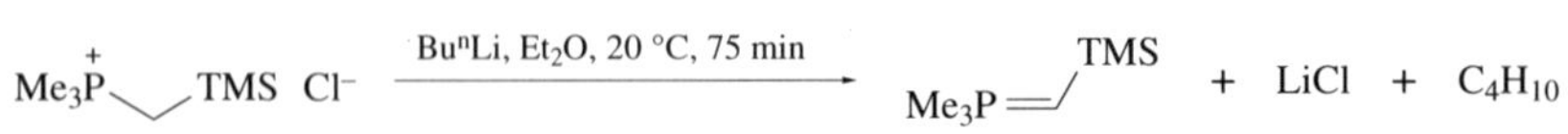

(81)

Due to the increased acidity of the hydrogen atoms of the methylene group bridging the two heteroatoms, usually only those ylides were formed, where the silyl group stabilizes the carbanionic function. In cases where no methylene hydrogens are present in the phosphonium salt, the primarily formed ylides undergo rearrangements with silyl shifts to produce isomers where the silyl groups are associated with the ylidic function (Equation (82)) ⟨70CB3448⟩. The same holds for salts such as **(203)** where the bridging carbon atom is substituted by an α-hydrogen and an alkyl group. Because of the destabilizing effect of the alkyl group towards an ylidic center, deprotonation at a *P*-methyl group with subsequent rearrangement to **(204)** was observed (Scheme 54) ⟨68AG(E)220⟩.

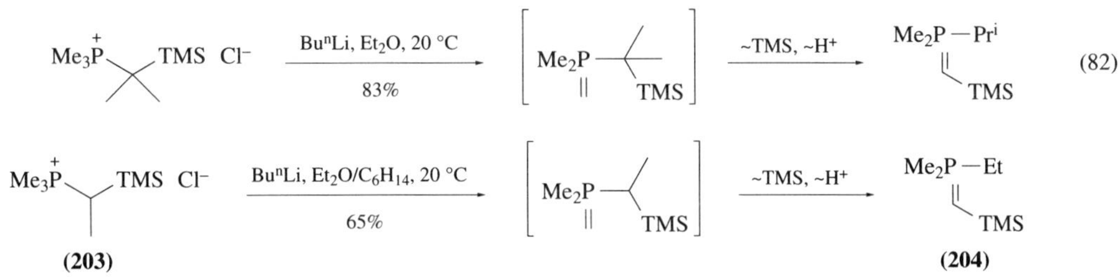

Scheme 54

The pyrolysis of salt $(Ph_3PCH_2\text{-TMS})Cl$ at 200 °C and 1 torr afforded ylide $Ph_3P{=}CH\text{-TMS}$ (**205**) (98%) ⟨67CB1032⟩. This process, which also applies for the synthesis of $Me_3P{=}CH\text{-TMS}$, invokes the dissociation of the salt into $Ph_3P{=}CH_2$ and TMS-Cl and a subsequent transylidation step (Equation (83)) ⟨65IC1458, 65JA390⟩.

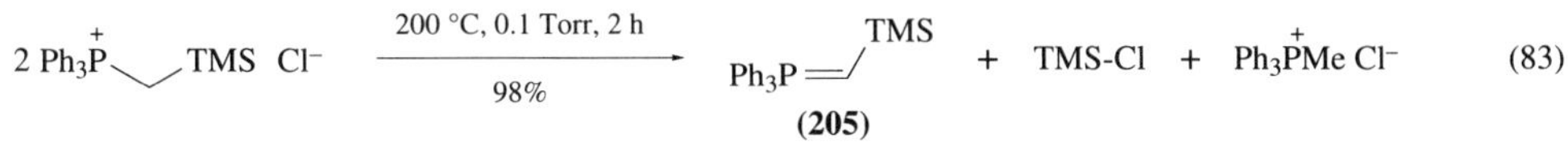

(b) From ylides. A very efficient and widely applicable approach for the synthesis of silylated ylides is based upon the treatment of α-H functionalized ylides with half an equiv. of a triorganosilyl halide. Thus $Ph_3P{=}CH_2$ was converted by TMS-Cl into ylide (**205**) and $(Ph_3PMe)Cl$ (67%) (Equation (84)) ⟨67CB1032⟩. Due to the relatively low basicity of (**205**), a second silylation was not observed. Such a process, however, was reported for $Me_3P{=}CH\text{-TMS}$ ⟨67CB1032⟩ and $Me_3P{=}CHSiH_2Me$ ⟨77AG(E)639⟩ as well as for $Ph_3P{=}CH\text{-TMS}$ when an excess of TMSBr was employed ⟨72CB1084⟩. The scope of silylating reagents was extended to 1-chlorosiletanes, 2-chloroethylsilane, chloromethylsilane, disilanyl chlorides, metallosilanes, and $(TMS)_2C{=}PMe_3$ (Scheme 55) ⟨67CB1032, 72CB3173, 75CB2554, 75CB2851, 77AG(E)639, 80CB1637⟩. In a variant, the synthesis of (**205**) was achieved from equimolar amounts of $(Ph_3PMe)Br$ and $Me_3P{=}CH\text{-TMS}$ in boiling ether as the result of transylidation and synchronous transsilylation steps ⟨72CB1084⟩.

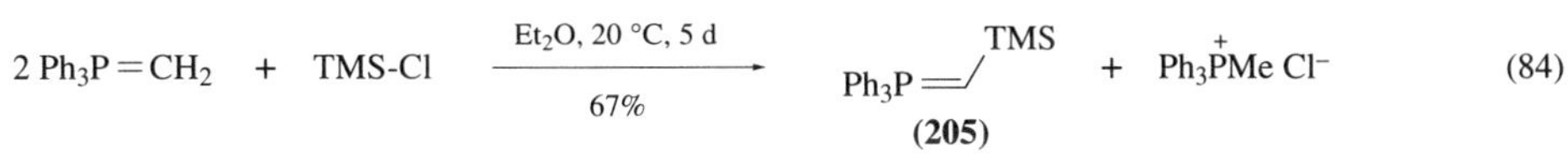

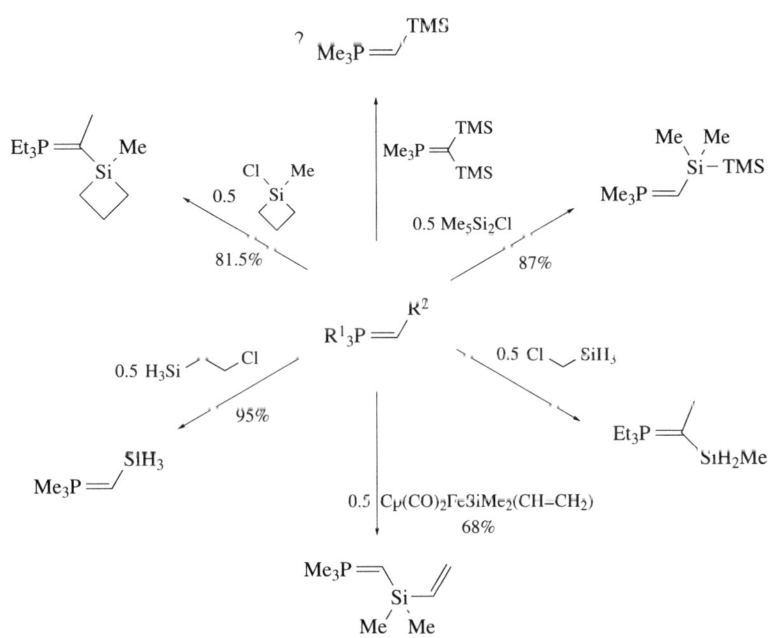

Scheme 55

Polyhalogenated organosilanes or disilanes were reported to give rise to cyclic species when subjected to the reaction with phosphorus ylides. The addition of four molar equivs of $Me_3P{=}CH_2$ to an ethereal solution of Me_2SiCl_2 at $-10\,^{\circ}C$ resulted in the formation of (**206**) (61%). However, when the solution of the ylide was added to the dichlorosilane in a molar ratio of 3 : 1, the cyclic ylide (**207**) was isolated (55%) ⟨70CB97⟩. Upon treatment with an excess of ylide, (**207**) isomerized to afford ylide (**208**) (Scheme 56) ⟨74AG(E)540⟩.

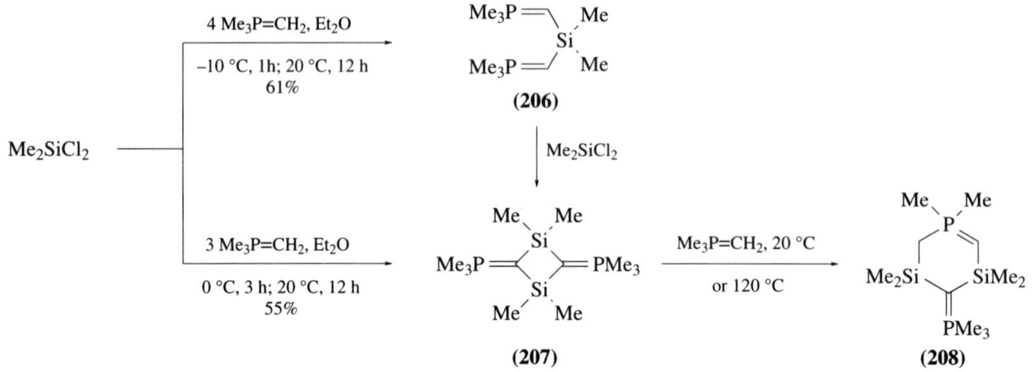

Scheme 56

Cyclic derivatives (**210**) and (**211**) were also obtained from quenching the dilithiated ylide (**209**) with bis(chlorodimethylsilyl)methane. The separation of (**211**) was accomplished via its HCl adduct and repeated deprotonation (69%) (Equation (85)) ⟨78CB2696⟩.

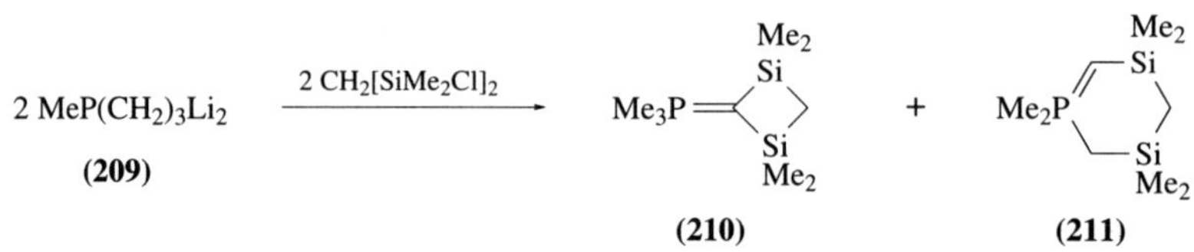

(85)

For the preparation of the acyclic ylide (**212**) from $Me_3P{=}CH_2$ and 1,2-dichlorotetramethyldisilane, a twelve-fold excess of $Me_3P{=}CH_2$ was crucial. The employment of only eight equivs of the ylide afforded a mixture of the heterocycles (**213**) and (**214**) instead of (**212**) (Scheme 57) ⟨72CB3173⟩. Ring opening of hexamethylsilirane with the result of (**215**) occurred with $Me_3P{=}CH_2$ (Equation (86)) ⟨78JOM(159)137⟩.

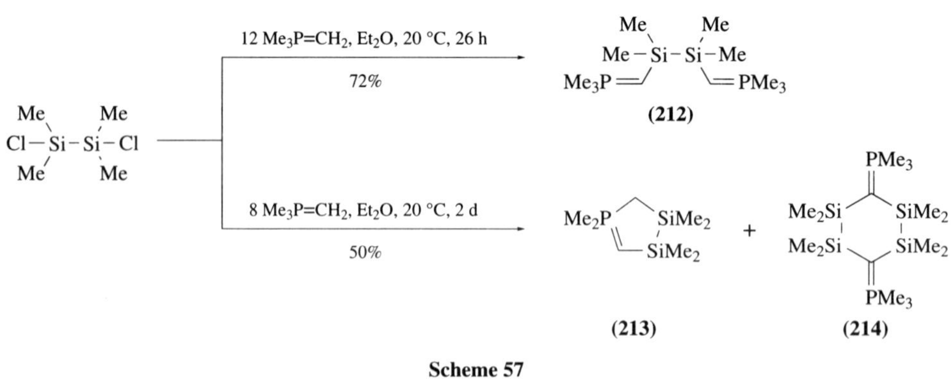

Scheme 57

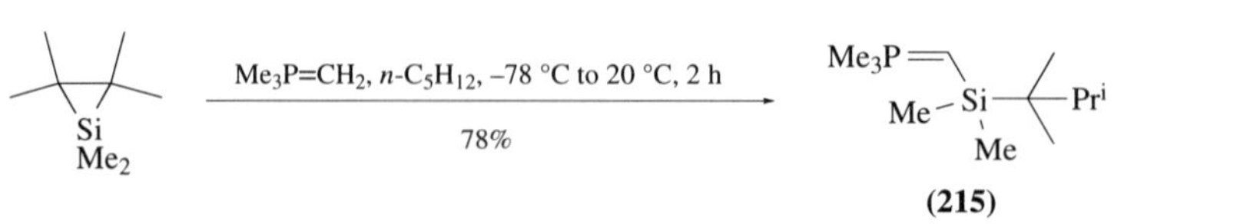

(86)

SiH-functionalized siletanes are cleaved by strongly nucleophilic ylides such as $Me_3P{=}CH_2$,

$Et_3P{=}CH_2$, and $Et_3P{=}CHMe$. After subsequent recyclization, the cyclic ylides (**216**) were isolated (Scheme 58) (Table 7) ⟨75CB2842⟩.

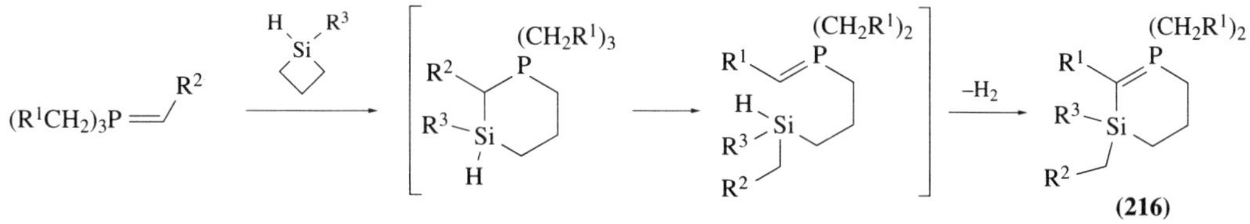

(216)

Scheme 58

Table 7 Ylides (**216**) from silacyclobutanes $H(R^3)\overline{Si(CH_2)_2CH_2}$ and $(R^1CH_2)_3P{=}CHR^2$.

R^1	R^2	R^3	*Reagents and conditions*	*Yield* (%)
H	H	H	$Me_3P{=}CH_2$, Et_2O, −78 °C to 25 °C	78
Me	H	H	$Et_3P{=}CH_2$, Et_2O, −78 °C to 25 °C	54
Me	Me	H	$Et_3P{=}CHMe$, Et_2O, −78 °C to 25 °C	62
H	H	Me	$Me_3P{=}CH_2$, Et_2O, −78 °C to 25 °C	63
Me	H	Me	$Et_3P{=}CH_2$, Et_2O, −78 °C to 25 °C	62
Me	Me	Me	$Et_3P{=}CHMe$, Et_2O, −78 °C to 25 °C	63

1,1-Difluoro-1-silacyclobutane was converted by two equivs of ylide into the analogous six-membered heterocycles (**217**). Here the last step of the transformation is the base-induced HF elimination (Equation (87)). Interestingly, with the less reactive ylide $Pr^i_3P{=}CH_2$ no ring cleavage of 1-silacyclobutane was observed under similar conditions. Ylide (**218**) was isolated instead (57%).

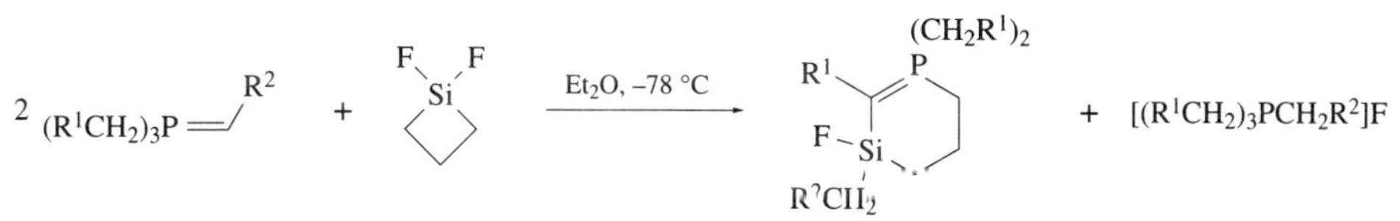 (87)

(217) $R^1 = R^2 = H$, 82%
$R^1 = Me$, $R^2 = H$, 45%
$R^1 = R^2 = Me$, 47%

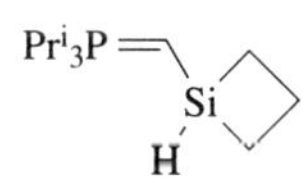

(218)

The smooth transformation of 1,3-disiletanes (**219**; $R^2 = H$, Me) into the silyl functionalized ylides (**220**) was effected by one equiv. of the ylides $R^1_3P{=}CH_2$(R^1 = Me, Et, Pr^i) (Equation (88)) (Table 8) ⟨75CB2834⟩.

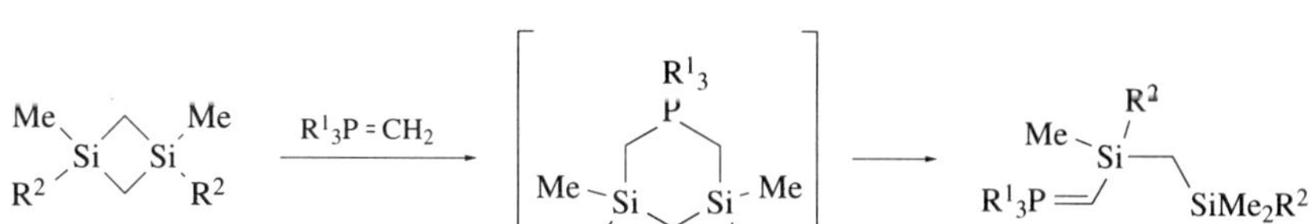 (88)

(219) **(220)**

(c) From phosphanes. Bulky tertiary phosphanes containing α-hydrogen functions such as $Bu^t_2PCH_2$-TMS were readily converted into *P*-chloroylides such as $Bu^t_2ClP{=}CH$-TMS by treatment with CCl_4 in *n*-pentane in the temperature range between −80 °C and 25 °C (80%) ⟨81JGU2125⟩.

Table 8 Ylides (**220**) from 1,3-disilacyclobutanes $[Me(R^2)Si\text{-}CH_2]_2$ (**219**) and $R^1_3P{=}CH_2$.

R[1]	*R*[2]	*Reagents and conditions*	*Yield* (%)
Me	Me	$Me_3P{=}CH_2$, THF, 20 °C	85
Et	Me	$Et_3P{=}CH_2$, THF, 60 °C, 24 h	61
Pr^i	Me	$Pr^i_3P{=}CH_2$, 100 °C, 7 days	48
Me	H	$Me_3P{=}CH_2$, Et_2O, 20 °C	89
Et	H	$Et_3P{=}CH_2$, Et_2O, 20 °C	86
Pr^i	H	$Pr^i_3P{=}CH_2$, Et_2O, 25 °C, 3 h	72

(ii) Germanium derivatives—$R^1C(PR^2_3)GeR^3_3$

Reaction of *n*-butyllithium with an equimolar amount of $(Me_3PCH_2GeMe_3)Cl$, slurried at 20 °C in ether, gave off $Me_3P{=}CHGeMe_3$ (**221**) (86%). The same ylide (**221**) resulted from the exposure of $Me_3P{=}CH\text{-}TMS$ to hexamethylgermanosiloxane (Equation (89)) ⟨68CB3545⟩.

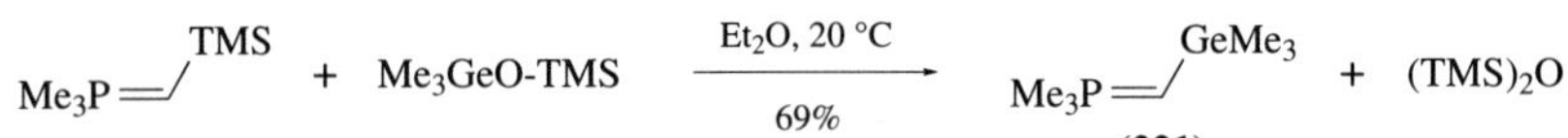

(**221**) (89)

Monogermylated triphenylphosphonium ylides were not available from $Ph_3P{=}CH_2$ and Me_3GeCl; instead, the bis-germylated ylide $Ph_3P{=}C(GeMe_3)_2$ was invariably formed ⟨67CB1032⟩.

(iii) Boron derivatives—$R^1C(PR_3)BR^3_2$

Crystalline dialkylboryl derivatives of triphenylphosphonium ylides were readily prepared by the transylidation of $Ph_3P{=}CHR^1$ with bulky dialkylchloroboranes in boiling benzene (Equation (90)) ⟨84AG(E)381⟩. Isolable adducts obtained from a 1 : 1 mixture of an ylide $Ph_3P{=}CHR^1$ and a sterically demanding dichloroalkylborane $RBCl_2$ underwent transylidation to give the boranediyl-bridged species (**222**). If the transylidation was conducted with a different ylide $Ph_3P{=}CHR^3$, unsymmetrical bis-ylides (**223**) were formed (Scheme 59) ⟨86AG(E)559⟩.

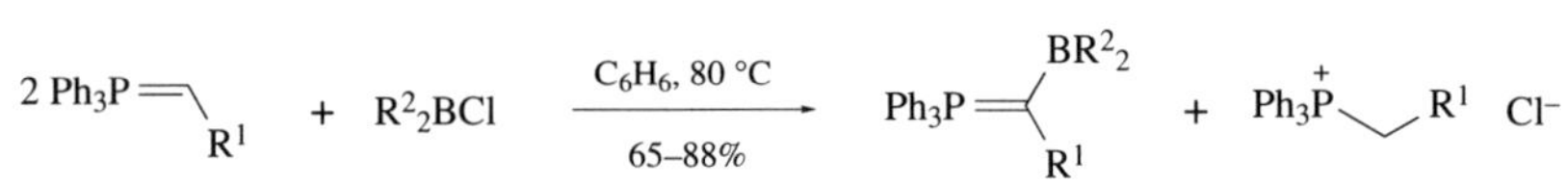

(90)

R^1 = Me, Et, Ph; R^2 = *n*-C_5H_{12}, *c*-C_5H_9, *c*-C_6H_{11}

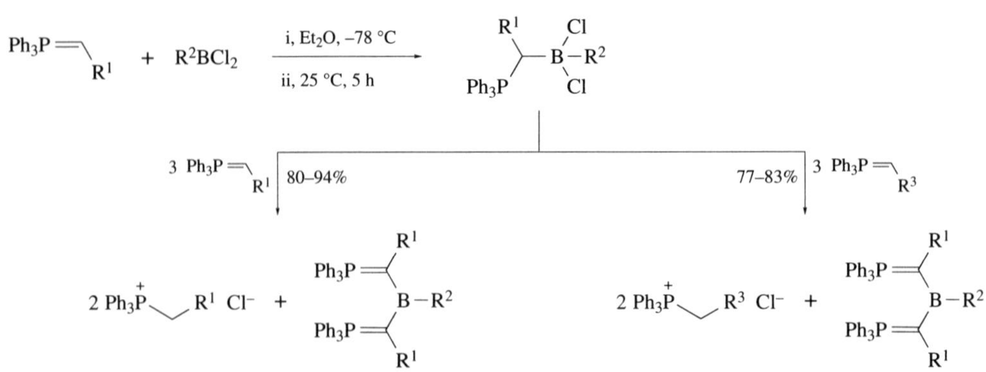

(**222**) (**223**)

R^1 = H; R^2 = *c*-C_6H_{11}, 2,2,6,6-$Me_4C_5H_6N$; CMe_2–CMe_2H

R^1 = H; R^2 = *c*-C_6H_{11}; R^3 = Me

Scheme 59

5.23.3.1.7 *Metal derivatives*

(i) Lithium, sodium, potassium, rubidium, and caesium derivatives—$R^1C(PR^2_3)M$

No isolable and stable representatives of this class of compounds have been reported as of early 1995.

(ii) Beryllium, magnesium, calcium, strontium, and barium derivatives—$R^1C(PR^2_3)MR^3$

No compounds of this type have been reported as of early 1995.

(iii) Transition metal derivatives—$R^1C(PR^2_3)ML_n$

(a) Titanium, zirconium, and hafnium derivatives. Phosphorus ylides with one α-titanium, zirconium, or hafnium substituent of the general type $R_3P{=}CHM(Cl)Cp_2$ resulted from Cp_2MCl_2 and an excess of ylide $R_3P{=}CH_2$. This transformation occurs via transylidation of an initially formed transition metal ylide complex (Scheme 60) ⟨80ZN(B)1289, 87CB89⟩.

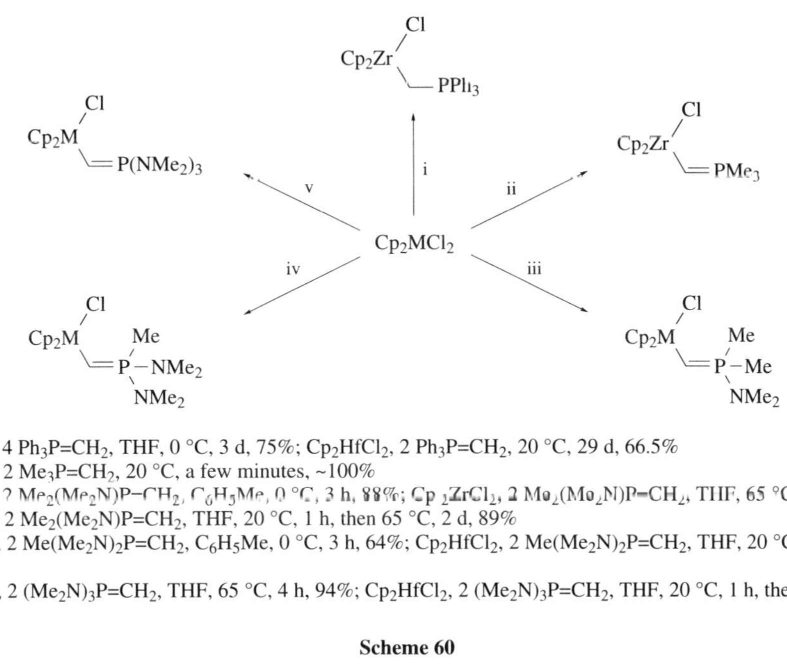

i, Cp_2ZrCl_2, 4 $Ph_3P{=}CH_2$, THF, 0 °C, 3 d, 75%; Cp_2HfCl_2, 2 $Ph_3P{=}CH_2$, 20 °C, 29 d, 66.5%
ii, Cp_2ZrCl_2, 2 $Me_3P{=}CH_2$, 20 °C, a few minutes, ~100%
iii, Cp_2TiCl_2, 2 $Me_2(Me_2N)P{=}CH_2$, C_6H_5Me, 0 °C, 3 h, 88%; Cp_2ZrCl_2, 2 $Me_2(Me_2N)P{=}CH_2$, THF, 65 °C, 1 h, 78%; Cp_2HfCl_2, 2 $Me_2(Me_2N)P{=}CH_2$, THF, 20 °C, 1 h, then 65 °C, 2 d, 89%
iv, Cp_2TiCl_2, 2 $Me(Me_2N)_2P{=}CH_2$, C_6H_5Me, 0 °C, 3 h, 64%; Cp_2HfCl_2, 2 $Me(Me_2N)_2P{=}CH_2$, THF, 20 °C, 1 h, then 65 °C, 2 d, 91%
v, Cp_2ZrCl_2, 2 $(Me_2N)_3P{=}CH_2$, THF, 65 °C, 4 h, 94%; Cp_2HfCl_2, 2 $(Me_2N)_3P{=}CH_2$, THF, 20 °C, 1 h, then 65 °C, 2 d, 96%

Scheme 60

The complex $Cp_2Zr(Cl)CH{=}PMe(NMe_2)_2$ was always obtained as a mixture with an isomeric bisylide chelate complex. An alternative access to complex $Cp_2Zr(Cl)CH{=}PMe_3$ involved chlorination of $Cp_2Zr(H)[(CH_2)_2PMe_2]$ with methyl chloride and the heat-induced rearrangement of intermediate $Cp_2Zr(Cl)[(CH_2)_2PMe_2]$ (76%) ⟨80IC3207⟩.

The deprotonation of initially formed ylide complexes was accomplished by ancillary ligands such as η^2-arynes or butadiene. Thus heating equimolar amounts of the diaryl complexes (**224**) and $Ph_3P{=}CH_2$ in heptane solution at 80 °C for 5 h furnished complexes (**225**) (90%) or an isomeric mixture (3:2) of (**226a**) and (**226b**) (71%), respectively (Scheme 61) ⟨85OM1310⟩.

Butadiene complex (**227**) and $Ph_3P{=}CH_2$ underwent reaction in toluene at −50 °C to give the zirconium-substituted ylide (**228a**). At −20 °C isomerization to a mixture of (**228a**) and (**228b**) occurred. When this transformation was conducted in the range 0 °C to 20 °C, a 54:46 mixture of (**228a**) and (**228b**) was isolated as a yellow powder (79%) (Scheme 62) ⟨85ZN(B)1177⟩. Similarly a mixture of the corresponding crotylhafnium ylide $Cp_2Hf(CH_2CH{=}CHMe)CH{=}PPh_3$ (76%) was synthesized ⟨85ZN(B)1177⟩.

(b) Tantalum derivatives. One report deals with the tantalum-substituted phosphonium ylide

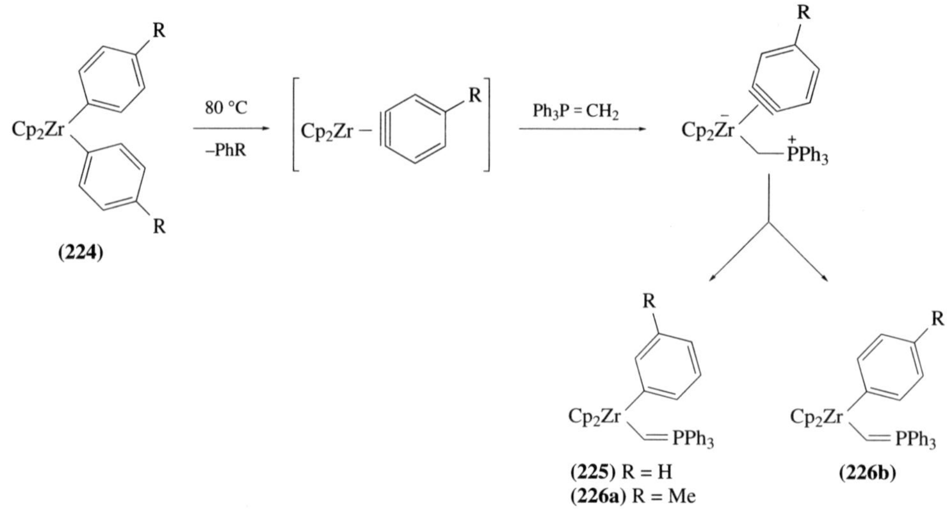

Scheme 61

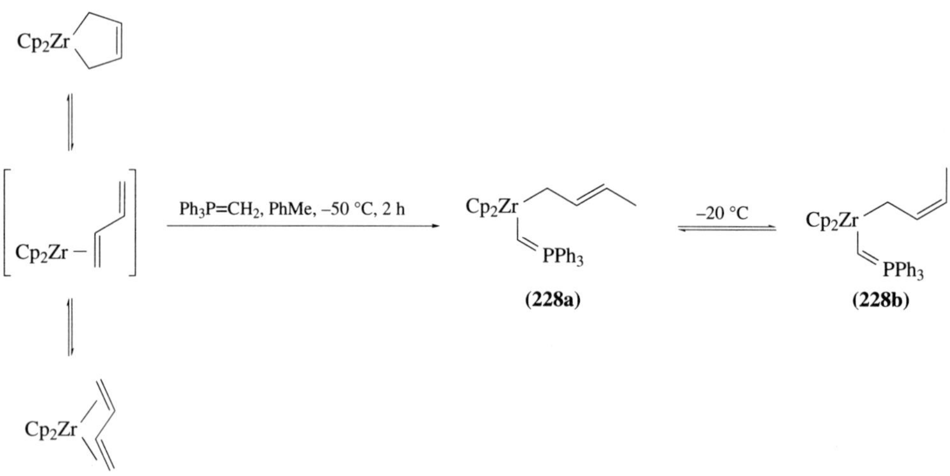

Scheme 62

Cp*Ta(CHPh) (CH=PPh_3)Cl which resulted from the attempted HCl elimination of Cp*Ta(CHPh) (CH_2Ph)Cl by one equiv. of Ph_3P=CH_2 (64%) ⟨80JA6744⟩.

(c) Chromium, molybdenum, and tungsten derivatives. Transylidation reactions of Mo(NBu^t)$_2$$Cl_2$ or W(NBu^t)$_2$$Cl_2(py)_2$ with four equivs of Ph_3P=CH_2 in toluene smoothly afforded the metal-functionalized derivatives (**229**) and (**230**) in 81% and 86% yield, respectively. The use of only two equivs of the parent ylide led to the formation of monoylidic species (Bu^tN)$_2$(Cl)W—CH=PPh_3 (89%) (Equation (91)) ⟨92CC1631⟩. The corresponding dark-violet chromium derivatives (Bu^tN)$_2$(Cl)Cr—CH=PPh_3 resulted from Cr(NBu^t)$_2$$Cl_2$ and Ph_3P=CH_2 under comparable conditions (93%) ⟨93AG(E)731⟩.

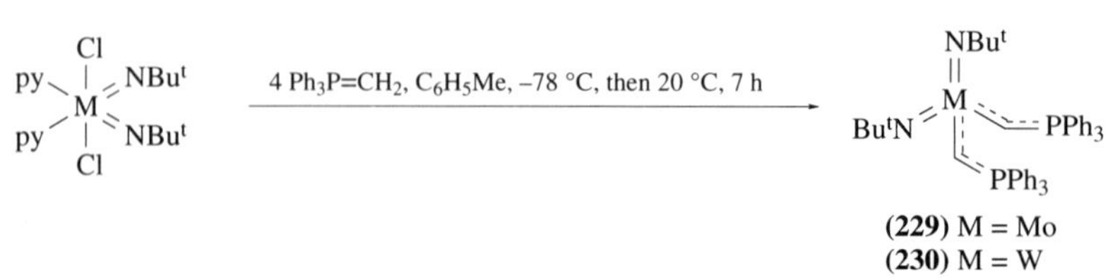

(91)

When Mo(O)$_2$(Mes)$_2$ was treated with two equivs of Bu_3P=CH_2 at 20 °C in pyridine, yellow–brown crystalline (Mes)(O)$_2$Mo—C(Mes)—PBu_3 was obtained (10–15%) ⟨85CC1151⟩.

A different approach to molybdenum- and tungsten-substituted ylides was based upon the treatment of carbyne complexes such as *trans*-$X(CO)_4M{\equiv}C{-}R$ (M = Mo, W, X = Cl, R = *p*-Tol; M = W; X = Br; R = Ph, *p*-Tol, $SiPh_3$) with PMe_3 ⟨78CB3283⟩. At low temperatures in CH_2Cl_2, two molecules of phosphane were incorporated to give (**231**), which for M = W at 40°C experienced another CO/PMe_3 exchange to finally yield (**232**) ⟨77CB805⟩. The related chromium species $X(CO)_4Cr{-}C(Ar){=}PMe_3$ are thermolabile. This type of reaction is restricted to PMe_3 since sterically more demanding and less basic phosphanes do not add to the carbyne center (Scheme 63).

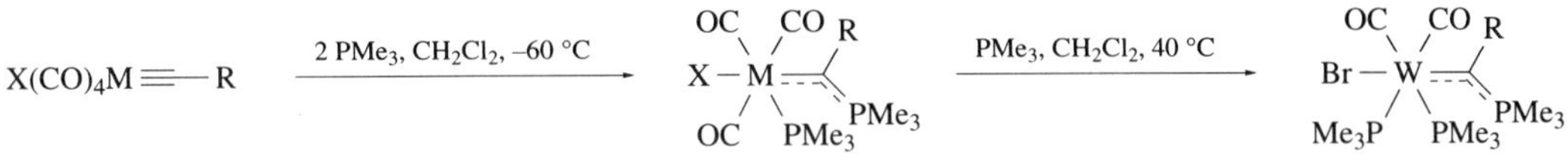

(**231**)

M = Mo, X = Cl, R = *p*-Tol, 85%
M = W, X = Cl, R = *p*-Tol, 82%
M = W, X = Br, R = *p*-Tol, 86%
M = W, X = Br, R = Ph, 49%
M = W, X = Br, R = $SiPh_3$, 76%

(**232**) R = Ph, 46%
R = *p*-Tol, 70%

Scheme 63

(d) Manganese derivatives. The addition of PMe_3 to a cationic manganese carbyne complex provides an entry into the class of manganese-functionalized ylides (Equation (92)) ⟨77CB3040⟩.

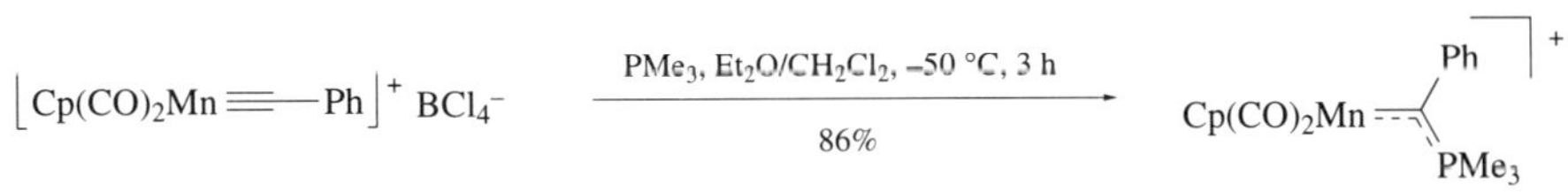

(92)

(e) Rhenium derivatives. Two different synthetic routes to isolable and stable ylidic rhenium species have been reported in the literature. The transylidation between $Ph_3P{=}CH_2$ and $Re(NBu^t)_3Cl$ provided an 89% yield of orange–yellow crystalline $(Bu^tN)_3ReCH{=}PPh_3$ ⟨92CC1631⟩. Transylidation was also operative in the preparation of $(CO)_5ReCH{=}PPh_3$ from $(CO)_5ReBr$ and $Ph_3P{=}CH_2$ in THF (41%) ⟨74JA2847⟩. Analogous to the reaction described in Scheme 63, $[Cp(CO)_2Re{\equiv}C{-}Ph]BCl_4$ and PMe_3 afforded scarlet crystals of $Cp(CO)_2ReC(Ph)CHPMe_3$ (95%) ⟨77CB3040⟩.

(f) Mercury derivatives. Mercury dichloride and $Ph_3P{=}CH{-}COPh$ form a mercurated phosphonium salt, which upon deprotonation with NaOMe in DMF at 60°C affords ylide (**233**) (90%) (Equation (93)). The $HgCl_2$ adducts of $Ph_3P{=}CHCOR$ (R = CO_2Me, COMe) were not prone to ylide formation ⟨65JOM(4)202⟩. However, base treatment of the adducts $[(Ph_3PCHR)HgOAc]OAc$ (R = CO_2Me, CN) afforded the mercury-bridged bisylides (**234**) in 80% yield (Equation (94)) ⟨70DOK(195)98⟩.

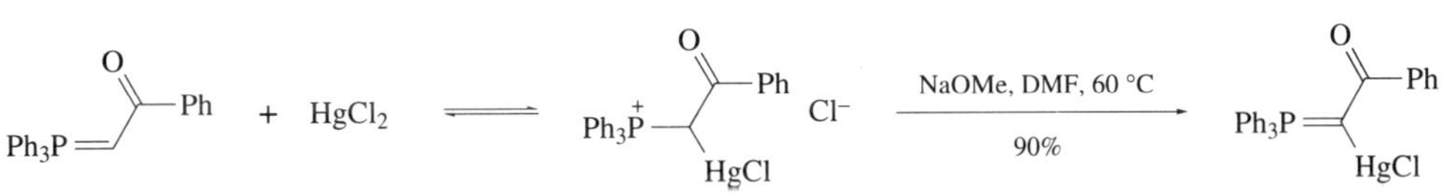

(93)

(**233**)

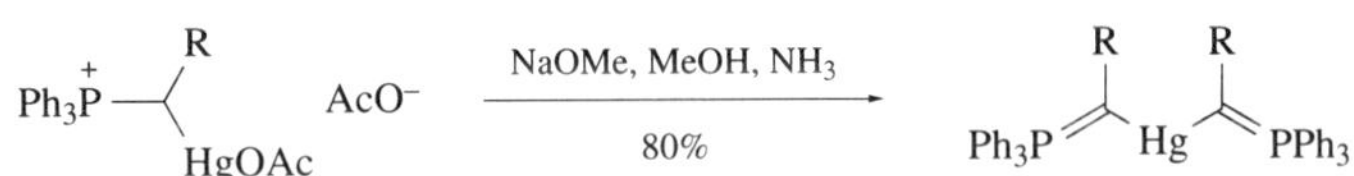

(94)

(**234**)

(iv) Group 13 metal derivatives

Such ylidic species have not been reported as of early 1995.

(v) Group 14 metal derivatives

Stannanediyl SnR_2 and ketene ylide $Ph_3P{=}C{=}C{=}O$ formed a thermolabile adduct (**235**), which slowly rearranged to (**236**) by 1,3-migration of an aryl group (Equation (95)) ⟨94AG(E)456⟩. The cleavage of the transition metal–tin bond in metallostannanes by ylides is the best method for the synthesis of the tin-functionalized ylides $Me_3P{=}CHSnMe_3$ and $Et_3P{=}C(Me)TMS$ (Equation (96)) ⟨73JOM(61)C15⟩. The attempted preparation of $Ph_3P{=}CHSnMe_3$ from $Ph_3P{=}CH_2$ and Me_3SnCl invariably afforded the bis-stannylated ylide $Ph_3P{=}C(SnMe_3)_2$ underlining the high propensity of monostannylated ylides to disproportionation ⟨67CB1032⟩. Similarly, the synthesis of $Me_3P{=}CHPbMe_3$ failed in favor of the disproportionation products $Me_3P{=}C(PbMe_3)_2$ and $Me_3P{=}CH_2$ ⟨77CB677⟩.

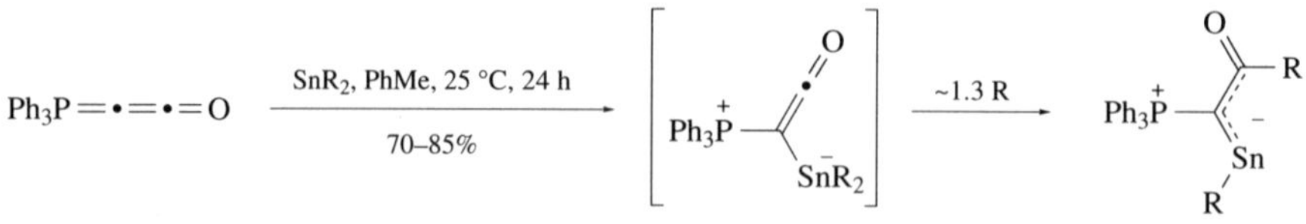
(95)

R = 2,4,6-$(CF_3)_3C_6H_2$ **(235)** **(236)**

(96)

R^1 = Me, R^2 = H; [M] = $Cp(CO)_3Mo$, $Cp(CO)_3W$, $Co(CO)_4$, $Cp(CO)_2Fe$
R^1 = Et, R^2 = Me; [M] = $Cp(CO)_3Mo$, $Cp(CO)_3W$, $Cp(CO)_2Fe$

(vi) Actinide derivatives

Phosphorus ylides with one α-actinide function are only known from uranium. Thus the reaction between Cp_3UCl and $Li(CH_2)_2PPh_2$ or $Li(CH_2)_2P(Me)Ph$ in 1:1 molar ratio in THF at −50°C produced dark-green crystalline $Cp_3UCH{=}P(Me)Ph_2$ (57%) or $Cp_3UCH{=}PMe_2Ph$ (62%). An excess of the lithium reagent led to dinuclear complexes with double ylide bridges ⟨81IC2466⟩.

5.23.3.2 Tetracoordinate Arsenic Functions—$R^1C(AsR^2_3)X$

5.23.3.2.1 Chalcogen derivatives—$R^1C(AsR^2_3)ER^3_n$ (E = O, S, Se, Te)

The synthesis of arsonium ylides or methylene arsoranes in most cases follows those principles employed successfully for the phosphorus analogues.

A facile method for the effective preparation of a variety of stabilized arsonium ylides has been developed by the action of active methylene compounds with tertiary arsenic oxide. Thus, the condensation of triphenylarsane oxide with dimethylphenacylsulfonium tetraphenylborate in the presence of acetic anhydride in $CHCl_3$ slowly afforded arsonium ylide (**237**) which is functionalized by a dimethylsulfonium group (Equation (97)) ⟨81JOU1676⟩. *p*-Bromobenzoylmethylene triphenylarsorane (**238**) reacted with phenylsulfine to give the sulfinated product (**239**), whereas carbethoxymethylene triphenylarsorane reacted with phenylsulfene under formation of benzylsulfonyl derivative (**240**) (21.5%) accompanied with ethyl cinnamate (44%) and triphenylarsane (Scheme 64) ⟨71JCS(C)1114⟩.

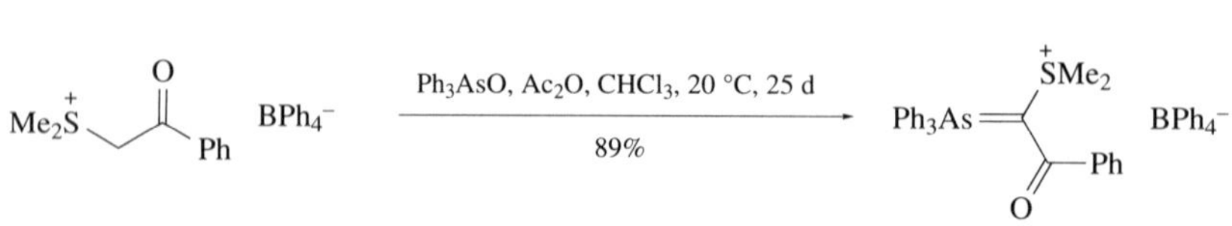
(97)

(237)

The synthesis of selenium derivative (**242**) by an analogous procedure as depicted in Equation (97) failed. Compound (**242**), however, resulted from the dicyclohexylcarbodiimide (dcc) mediated condensation of salt (**241**) and Ph_2SeO under the catalysis of trifluoracetic acid (Equation (98)) ⟨81JOU1676⟩.

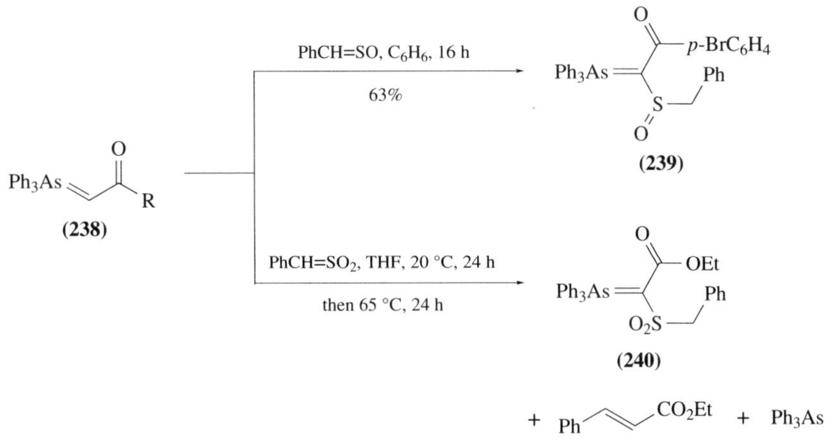

Scheme 64

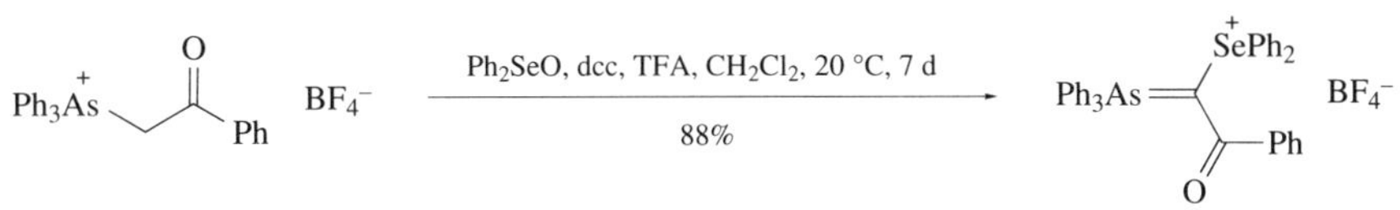 (98)

(241) dcc = dicyclohexylcarbodiimide **(242)**

5.23.3.2.2 *Nitrogen derivatives—$R^1C(AsR^2_3)NR^3_2$*

An arsonium ylide functionalized by a nitro group was the product of condensing Ph_3AsCl_2 with phenylnitromethane in the presence of triethylamine (Equation (99)) ⟨58CB437⟩.

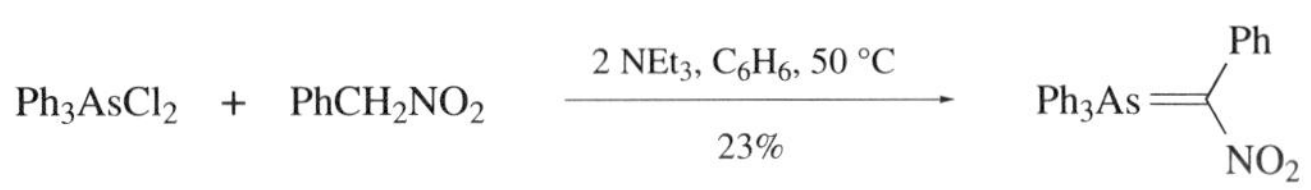 (99)

5.23.3.2.3 *Group 15 element derivatives—$R^1C(AsR^2_3)ER^3_n$ ($E = P$, As, Sb, Bi)*

The only stable arsino-substituted arsonium ylide reported as of 1995 (**243**) was obtained by the deprotonation of the corresponding arsonium salt with an excess of sodium amide (Equation (100)) ⟨85OM344⟩. A similar treatment of $(Me_3AsCH_2AsMe_2)I$ or $(Et_3AsCH_2AsEt_2)I$ furnished the thermolabile and non-isolable alkyl analogues ⟨87CB1281⟩.

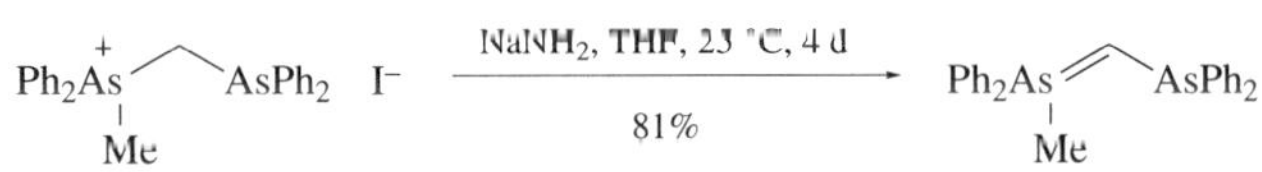 (100)

(243)

5.23.3.2.4 *Metalloid derivatives—$R^1C(AsR^2_3)ER^3_n$ ($E = Si$, Ge, B)*

Deprotonation of salt $(Me_3AsCH_2\text{-TMS})Cl$ in dry ether with an equimolar amount of *n*-butyllithium produced a colorless solid adduct $Me_3AsCH{\cdot}TMS\text{-}LiCl$ which upon thermolysis *in vacuo* liberated $Me_3As{=}CH\text{-}TMS$ (**244**) (54%) (Equation (101)) ⟨65IC1458⟩. Analogously, $Et_3As{=}CH\text{-}TMS$ was obtained in 87% yield. Methylation of this compound with MeBr in ether at $-50\,^\circ C$ gave the salt [$Et_3AsCH(Me)$-TMS]Br (**245**) (76%), which was converted into ylide (**246**) by means of *n*-butyllithium (Equation (102)) ⟨77CB1312⟩.

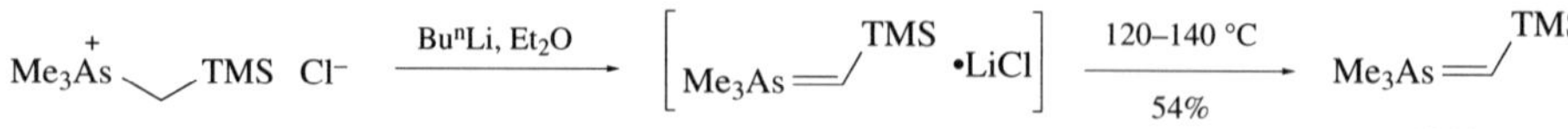

(244) (101)

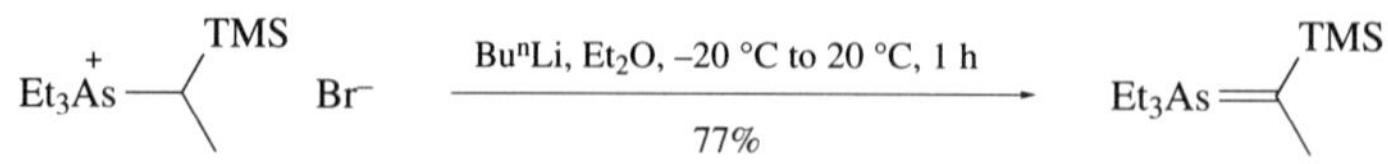

(245) (246) (102)

A germanium derivative $Me_3As{=}CHGeMe_3$ was observed as an unstable intermediate during the reaction of $Me_3As{=}CHTMS$ with $Me_3GeOTMS$. It, however, suffered from facile disproportionation into $Me_3As{=}CH_2$ and $Me_3As{=}C(GeMe_3)_2$ ⟨77CB677⟩.

No other stable arsonium ylides with the structural features under discussion have been reported.

5.23.3.3 Tetracoordinate Antimony Function—$R^1C(SbR^2_3)X$

5.23.3.3.1 Metalloid derivatives—$R^1C(SbR^2_3)EMe_3$

Stable and isolable antimony ylides or methylene stiboranes are extremely rare. Dehydrochlorination of [(TMS-$CH_2)_2]_4SbCl$ with $Et_3P{=}CHMe$ in *n*-pentane in the temperature range of −15 °C to 20 °C gave ylide (TMS-$CH_2)_3Sb{=}CH$(TMS) as a yellow oil (65%), which upon distillation partially rearranged into Me(TMS-$CH_2)_2Sb{=}C(TMS)_2$ (30%) ⟨78CB2702⟩.

5.23.3.4 Tetracoordinate Bismuth Functions—$R^1C(BiR^2_3)X$

Such compounds are unknown.

5.24
Doubly Bonded Metalloid Functions, $R^1C(X)=SiR^2_2$ $R^1C(X)=BR^2$, $R^1C(X)=GeR^2_2$

BARRY P. CLARK
Lilly Research Centre Ltd., Windlesham, UK

5.24.1 SILICON DERIVATIVES $R^1C(X)=SiR^2_2$

Reviews on silenes Si=C are included in articles entitled multiply bonded silicon ⟨85CRV419⟩ and stable silenes ⟨86AOC(25)1⟩. Examples of the atom X (not including carbon) in the general structure $R^1C(X)=SiR^2_2$ were found in *Chemical Abstracts* only for oxygen, silicon and germanium. Silenes are generally reactive species that are not isolated but rather trapped *in situ*. It was not until 1981 that the first solid silene was isolated ⟨81CC191⟩ and the x-ray structure determined ⟨85AX(C)1632⟩. More recently stable double bond isomers have been characterised but not isolated ⟨92OM3088⟩

5.24.1.1 Oxasilenes, $R^2_2Si=CR^1—OR^3$

5.24.1.1.1 From acylsilanes

There are many examples of the well-established method to prepare oxasilenes from acylsilanes, largely due to Brook and co-workers. References with full experimental details on the photochemical and thermal rearrangements to silenes are included ⟨79JA83, 82JA5667, 89OM693⟩. The 1,3–shift of a trimethylsilyl group from silicon to oxygen during the irradiation of acylsilanes under mild conditions ($\lambda > 360$ nm) produced silenes in high yield (Scheme 1). The subsequent chemistry of the silenes (**1**) was dependent on the steric size of the R^1 and R^2 groups and on their facile photochemical excitation by UV light. Similarly, thermolysis at 160 °C in a sealed tube produced silenes that were trapped with methanol (Scheme 2) ⟨79JA83⟩.

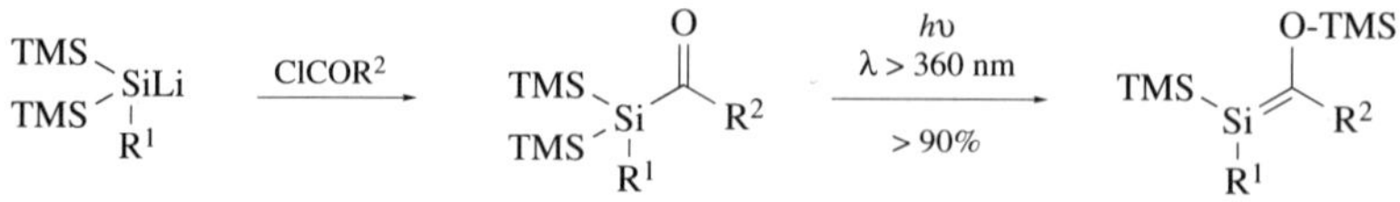

(**1**)

Scheme 1

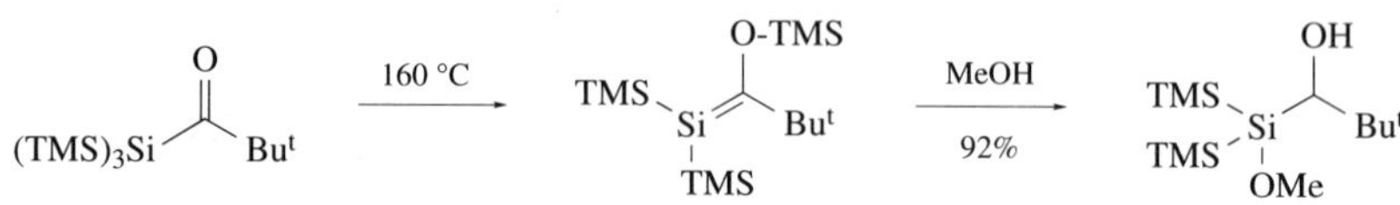

Scheme 2

The utility of oxasilanes as reactive intermediates has been demonstrated by their reaction with ketones ⟨83TL1477, 87OM1246⟩, dienes ⟨87OM2128, 92OM2317⟩, phosphonium ylides ⟨88JOM(341)C9⟩, isonitriles ⟨88OM2245, 89OM850⟩, enones ⟨91OM2752⟩, and α,β-unsaturated esters ⟨91OM2758⟩. In the absence of trapping agents, silenes undergo self-dimerisation to produce head to head dimers. However, if sterically bulky groups (such as 1-adamantyl or triethylmethyl) were attached on the silene carbon then dimers were disfavoured ⟨79JA6750, 89OM693⟩ and a stable silene can be isolated. These crystalline solids were unchanged after nine months' storage, sealed under argon but reacted immediately on exposure to air producing white smoke. Solutions of oxasilenes reverted to starting acylsilanes after several days ⟨81CC191⟩.

The x-ray structure of the silene (**1**; $R^1 = $ TMS $R^2 = $ 1-adamantyl) showed planar sp^2-hybridised silicon and carbon ⟨84CC768, 85AX(C)1632⟩. The double bond character of silenes has further been demonstrated by the characterisation of stable geometric isomers by NMR spectroscopy ⟨82OM994⟩. The ratio of isomers shown in Scheme 3 was unchanged after 16 hours at 65 °C but the silenes decomposed after one week at 100 °C ⟨92OM3088⟩. There is one report of the formation of lithium silaenolate anions with triethylgermanyl lithium followed by alkylation on silicon (Scheme 4) ⟨88IZV1947, 89JOM(368)163⟩.

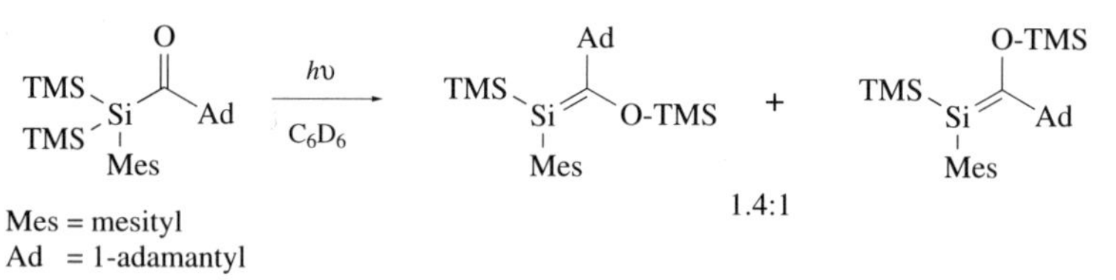

Scheme 3

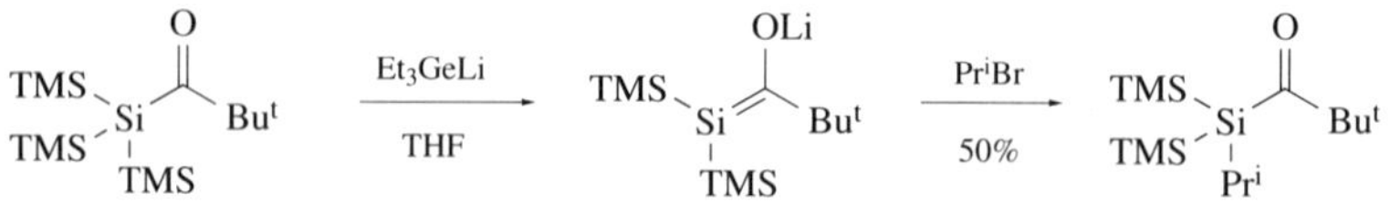

Scheme 4

5.24.1.1.2 *Retro dimerisation*

The reversibility of head-to-head dimerisation of silenes was shown in 1976 ⟨76JA3381⟩. These dimers are stable solids and form more convenient reactants compared to the reactive silenes. Thermolysis of the dimer (**2**) at 90 °C generated the silene (**3**) (Scheme 5) which was subsequently trapped with an azete to form the pyrrole (**4**) ⟨88CB1363⟩.

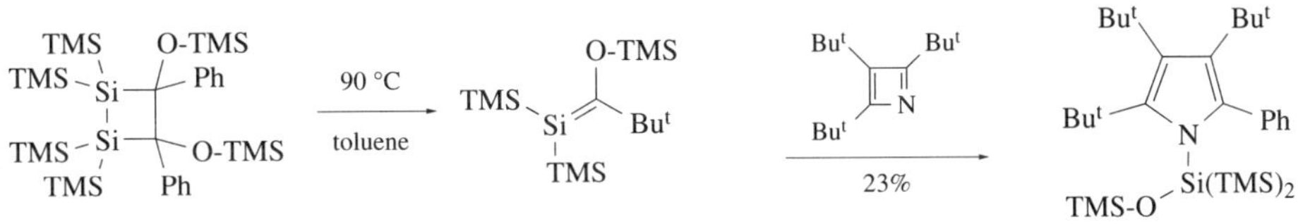

(2) **(3)** **(4)**

Scheme 5

5.24.1.2 Silasilenes, $R^2{}_2Si{=}CR^1{-}SiR^3{}_3$

5.24.1.2.1 From diazo derivatives

The photolysis or thermolysis of silyl substituted diazo derivatives is a general route to silenes and usually proceeds in high yield. The reaction of diazo compound (**5**) occurred via an α-silylcarbene to give the silene (**6**), which at room temperature rearranged to the silylketene (**7**) ⟨81JA5573, 87OM2337⟩ as shown in Scheme 6. The silene (**6**) was stable at 10 K to photolysis in an argon matrix but was converted by brief irradiation at shorter wavelength $\lambda > 240$ nm to the silylketene (**7**) ⟨85TL2337⟩. Various silenes (**8**) were isolated in a 3-methylpentane matrix at 77 K by photolysis of the corresponding diazo derivatives, and their UV spectra measured ⟨86CL2025⟩.

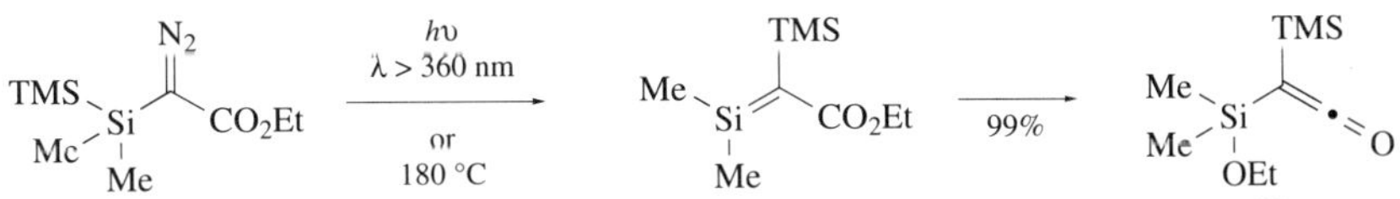

(5) **(6)** **(7)**

Scheme 6

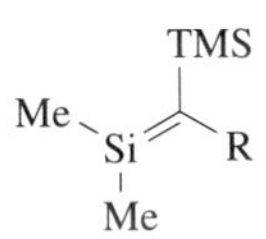

R = Me, CO-1-adamantyl

(8)

5.24.1.2.2 From silenes

There are no general methods to prepare silenes from silenes, but two unusual examples are known. A remarkable silene photochemical rearrangement was observed (Scheme 7) when one of the silyl groups on silicon in (**9**) was replaced by *t*-butyl ⟨85JA4338⟩. A complex pathway of rearrangements was proposed for this clean first-order reaction ⟨89OM693⟩. The silasilene (**10**) was characterised as the methanol adduct and by x-ray structure determination of the head to tail dimer ⟨89OM709⟩. The other example involved [1,3]-sigmatropic migration of a *t*-butoxy group that occurred spontaneously ⟨92JA4910⟩. The expected dichlorosilene (**11**) could not be trapped but isomerised to the dimethylsilene (**12**) shown in Scheme 8. The silene (**12**) was characterised by *in situ* reaction with methoxytrimethylsilane or with quadricyclane to produce the exocyclic adduct.

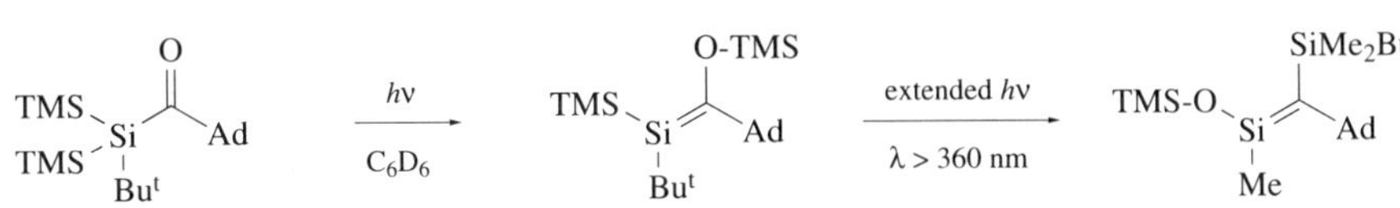

(9) **(10)**

Scheme 7

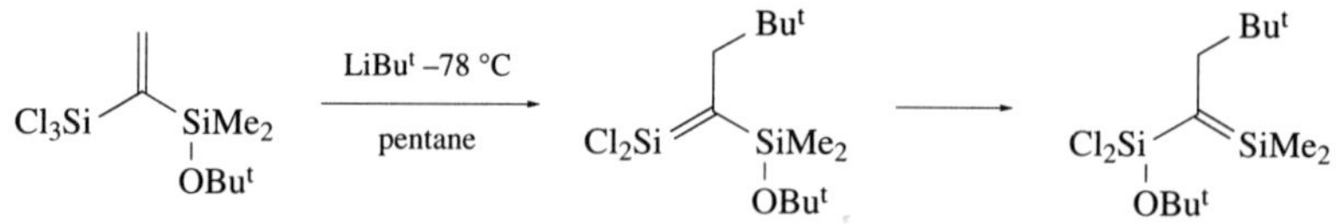

(11) (12)

Scheme 8

5.24.1.2.3 *From alkenes*

There are several examples of producing silenes from alkenes. Neopentyl silenes were formed in the reaction of chlorovinyl silanes with *t*-butyl lithium in nonpolar solvents ⟨92OM2494⟩. The α-lithiated species (**13**) shown in Scheme 9 rapidly eliminated lithium chloride to form the silenes (**14**) which, without isolation, readily participated in cycloaddition reactions with dienes. Dichlorosilenes prepared by this method can undergo rearrangement as described in Scheme 8. A 3 : 1 mixture of geometric isomers about the double bond of (**14**) where R,R = Ph,Me was formed ⟨87JA913⟩.

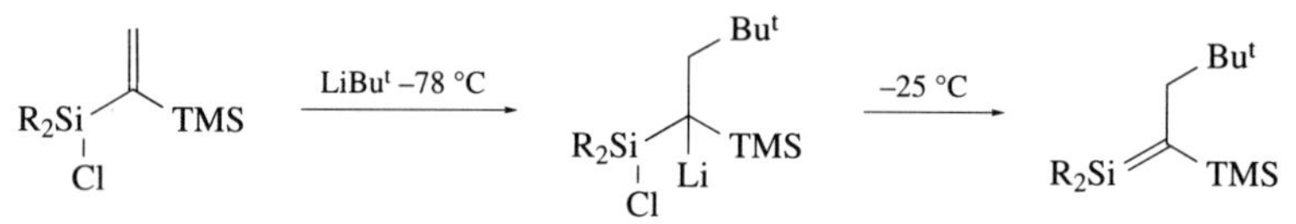

$R_2 = Cl_2, Pr^i_2$

R, R = Ph, Me

(13) (14)

Scheme 9

5.24.1.2.4 *Retro Diels–Alder*

The thermal retro Diels–Alder reaction of anthracene adducts stereospecifically generated silenes in high yield ⟨87JA913⟩. Thermolysis and methanol trapping of the pure crystalline adduct (**15**) resulted in a single isomer of (**16**), whereas a 69 : 31 mixture of isomers resulted in a 69 : 31 ratio of products (Scheme 10). The methanol addition was a stereospecific *syn* reaction.

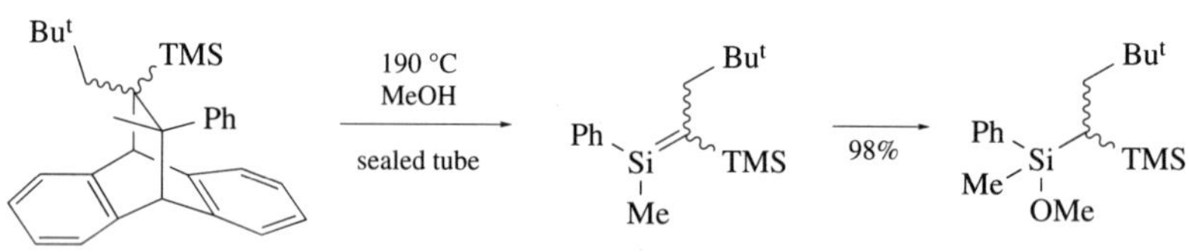

(15) (16)

Scheme 10

5.24.1.3 Germasilenes, $R^2_2Si{=}CR^1{-}GeR^3_3$

There are no general synthetic methods for germasilenes and only one example is known ⟨80JA1584⟩. The germasilene was formed by pyrolysis of the trimethylgermyldiazomethane in the presence of benzaldehyde, or by photolysis in the presence of methanol as a trapping agent (Scheme 11). Interestingly, both silene and germene products can be formed. The carbene produced from N_2 extrusion underwent methyl migration from silicon to produce 79% silene (**17**) and methyl migration from germanium to produce 21% germene (**18**).

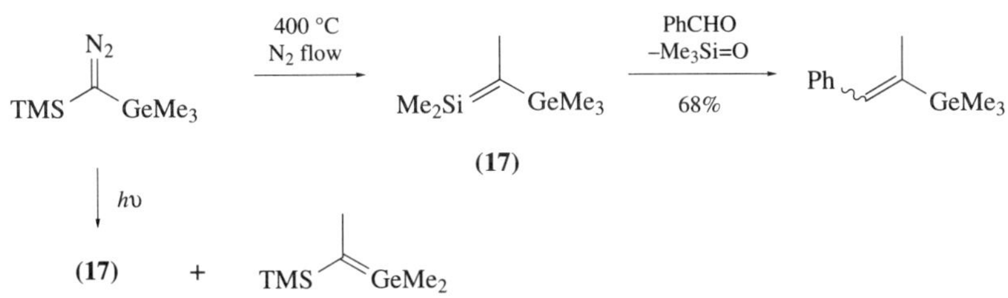

(17)

(18) 21%

Scheme 11

5.24.2 GERMANIUM DERIVATIVES—R[1]C(X)═GeR[2]$_2$

Reviews on germenes Ge═C are included in articles entitled multiply bonded germanium species ⟨90CRV283⟩ and doubly bonded germanium derivatives ⟨82AOC(21)241⟩. The literature describing germenes of the type Ge═C—X relevant to this section is sparse. Stable germenes of this structural class have not been isolated but were either trapped *in situ* or allowed to form dimers ⟨80JA1584⟩.

5.24.2.1 Silagermenes, R[2]$_2$Ge═CR[1]—SiR[3]$_3$

Photolysis of a diazo derivative produced the silagermene (**18**) as described in Section 5.24.1.3 (Scheme 11) and was reacted with methanol ⟨80JA1584⟩.

5.24.2.2 Germagermenes, R[2]$_2$Ge═CR[1]—GeR[3]$_3$

The dimeric germane (**20**) was formed from bis(trimethylgermyl)diazomethane by either pyrolysis at 400 °C using a nitrogen flow system in 32% yield or, more efficiently, by photolysis in benzene solution in 89% yield (Scheme 12). The germagermene (**19**) can be trapped by methanol or benzaldehyde ⟨80JA1584⟩.

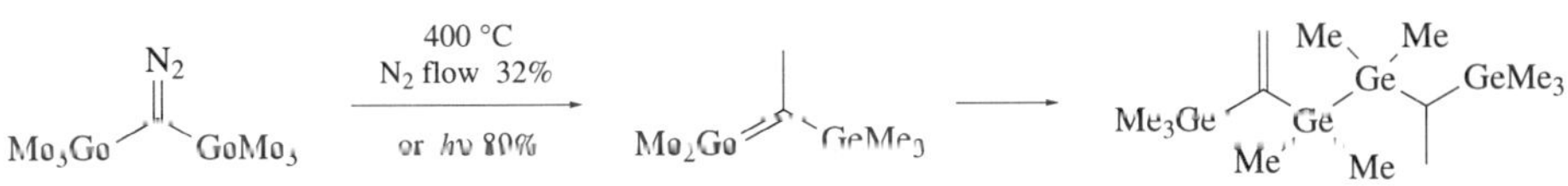

(19) (20)

Scheme 12

5.24.3 BORON DERIVATIVES—R[1]C(X)═BR[2]

Despite the vast literature on boron chemistry, the subject of boron carbon double bonds B═C does not appear to have been reviewed as a single topic. Compounds containing B═C—B are described below. A search of *Chemical Abstracts* also revealed compounds containing B═C—M where M = Ti, Cr, Mn, Fe, Co and W ⟨87JOM(325)185, 88AG(E)1331, 90AG(E)199, 92AG(E)1603, 92CB73⟩ but these organometallic complexes are not included here.

5.24.3.1 Boryl Derivatives, R[2]B═CR—BR[1]$_2$

The first example of a stable compound containing a borene B═C was reported by Berndt and co-workers in 1983 ⟨83AG(E)877⟩. The structure (**21**) was subsequently revised ⟨85AG(E)788⟩ in view of theoretical calculations which showed a lower energy isomer (**22**) (Scheme 13) ⟨84AG(E)825,

84CPL521, 85JA2773, 90MI 524-01⟩. This was later confirmed by x-ray crystallography ⟨90AG(E)398⟩. (*Z*)- and (*E*)-isomers about the B═C bond have been observed by NMR spectroscopy and in solution these interconvert at room temperature ⟨88AG(E)960⟩.

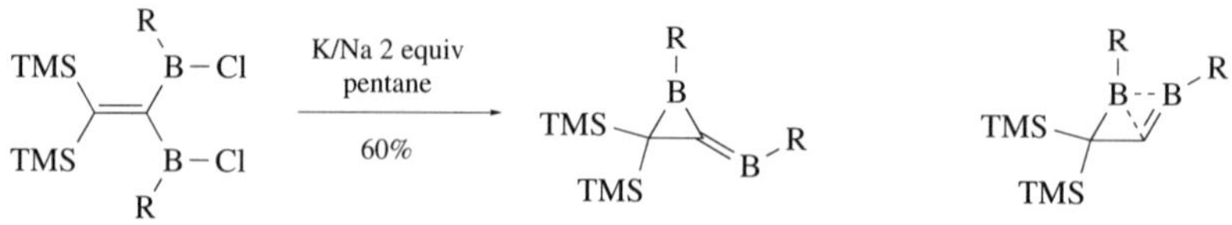

(21) **(22)**

R = Bu^t, 2,3,5,6-tetramethylphenyl

Scheme 13

5.24.3.1.1 *From dichlorides*

The reduction of dichlorides is the only known general method to prepare borenes. The choice of magnesium in THF as reducing agent gave a higher yield (88%) ⟨88ZN(B)801⟩ than the potassium/sodium alloy in pentane shown in Scheme 14 ⟨83AG(E)877, 84AG(E)371, 84AG(E)372⟩. The borene (**22**; R = Bu^t) is a distillable liquid that is stable for months in the absence of oxygen and water. The four bulky substituents are essential for chemical stability and replacing one of the trimethylsilyl groups by methyl resulted in dimerisation ⟨83AG(E)877⟩. The borene (**22**; R = Bu^t) reacted with alkynes ⟨84AG(E)369⟩, and tin and germanium derivatives ⟨87AG(E)546, 87AG(E)798⟩. The structures (**21**) and (**23**) drawn with a formal B═C are better represented by the delocalised structures (**22**) and (**24**) in which the two bond lengths of B—C—B are nearly equal as shown by x-ray analysis ⟨84AG(E)371, 90AG(E)398⟩.

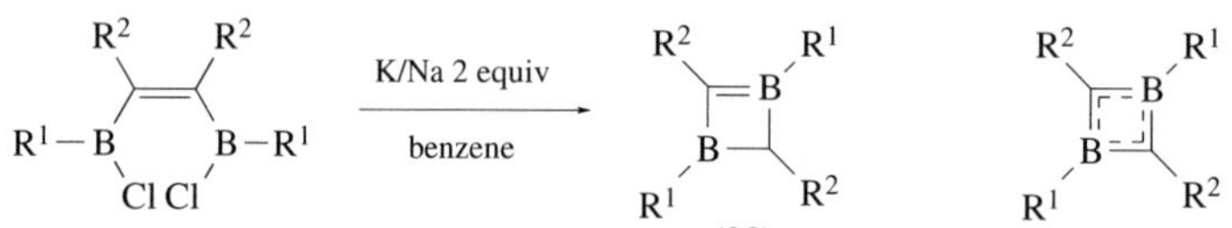

(23) **(24)**

R^1 = NMe_2, R^2 = Bu^t, 65%

R^1 = Bu^t, R^2 = Et, 30%

R^1 = Bu^t, R^2 = Me, 9%

Scheme 14

Reduction of dichlorides (**25**; R = Bu^t) using four equivalents of potassium/sodium alloy resulted in a diborirane dianion that can be protonated to the borene (**26**) shown in Scheme 15 ⟨85AG(E)788⟩. The diborirane dianion was converted into various metal salts (Li^+, Na^+, Cs^{2+} and Mg^{2+}) ⟨88ZN(B)801⟩ and was oxidised with cadmium chloride to the borene (**22**; R = Bu^t). When four equivalents of lithium metal were used to reduce the dichloride (**25**; R = 2,4,6-trimethylphenyl) (Scheme 15), the resulting dianion was protonated to the anion (**28**) containing a linear B═C^-—B bond, indicative of triple bond character, although no yields were quoted ⟨88AG(E)961⟩. An unusual thermal rearrangement of the dichloride (**25**; R = 2,3,5,6-tetramethylphenyl) at 135 °C resulted in the borene (**27**) via migrations of chloro groups from boron to silicon and methyl groups from silicon ⟨92ZN(B)983⟩.

5.24.3.1.2 *From borenes*

There are no general methods to prepare borenes from borenes, although there are miscellaneous examples involving migration of groups between boron and carbon, reduction of dianions and adduct formation. The thermal migration of a *t*-butyl group at 170 °C formed the 1,3-diborete (**29**) as the thermodynamically more stable product (Scheme 16) ⟨84AG(E)372⟩. The diborete (**29**; R = TMS) was also formed using triphenylphosphine at 20 °C ⟨84AG(E)826⟩.

The addition of phenyllithium to a borene produced the borenyl anion (**30**) as a yellow solid (Scheme 17) ⟨92AG(E)1238⟩. Heating a toluene solution of the anion (**30**) resulted in a mixture of

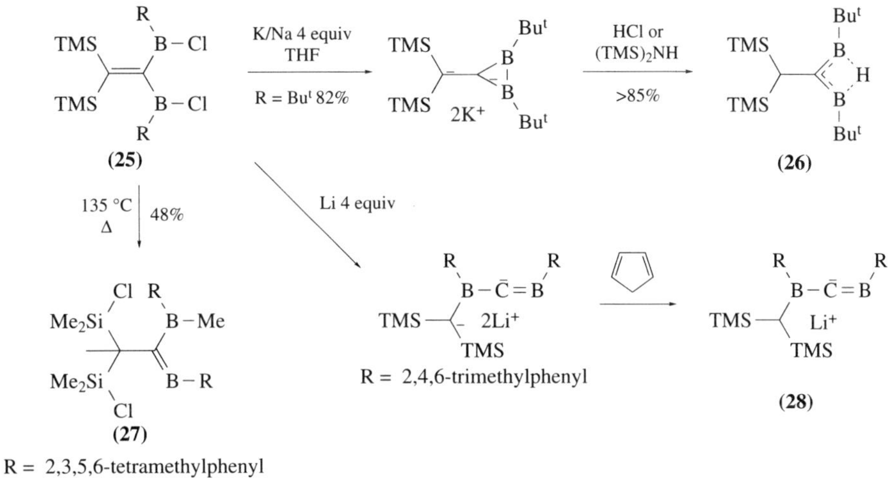

Scheme 15

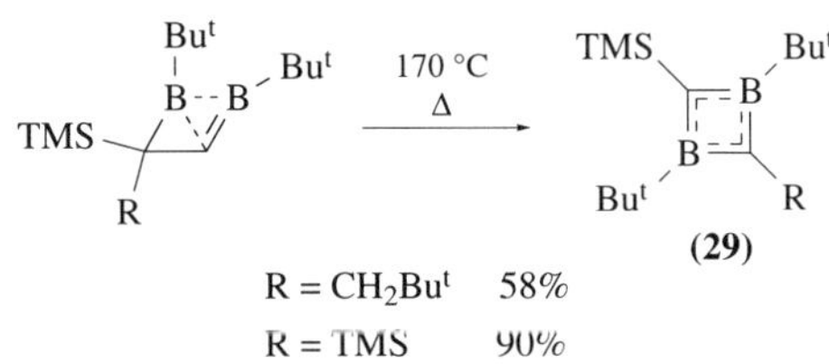

Scheme 16

anions (**31**) and (**32**) probably via a carbene intermediate. When 2,3,5,6-tetramethylphenyllithium was used a single product analogous to (**31**) was formed in 52% yield ⟨92AG(E)1238⟩.

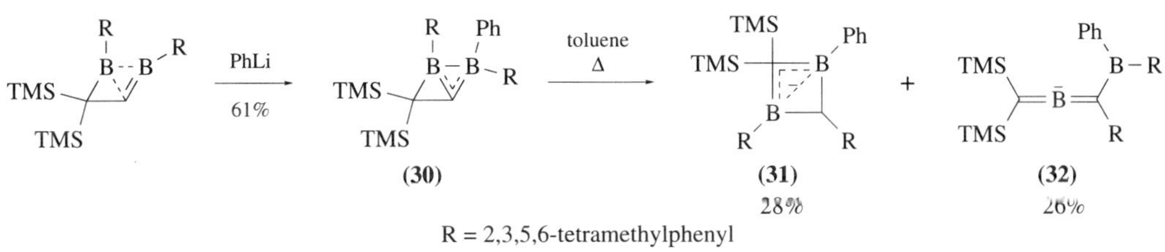

Scheme 17

An unusual mesityl-bridged borene (**34**) was formed by oxidation of the dilithioborene (**33**) with tin(II) chloride as shown in Scheme 18 ⟨90AG(E)399⟩. The x-ray structure of the borene (**34**) showed the bridging bonds from the boron atoms to the aryl ring were comparable in length. The aryl bridge can stabilize the electron deficient boron atoms and the borene (**34**) reacted quantitatively with 4-*t*-butylpyridine to form the adduct (**35**) ⟨90AG(E)399⟩.

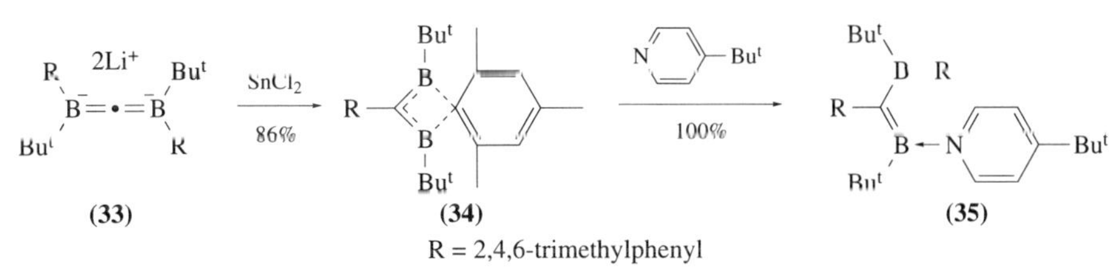

Scheme 18

The borenes (**36**) reacted with a variety of nucleophiles to form differing products depending on the nucleophile (Scheme 19). The boriranes (**37**) were obtained from the additions of hexamethyldisilazane, hydrogen chloride or methyl lithium to the boron carbon double bond of the borenes (**36**) ⟨92AG(E)1384⟩. The borene (**38**) was formed by a thermal rearrangement involving a

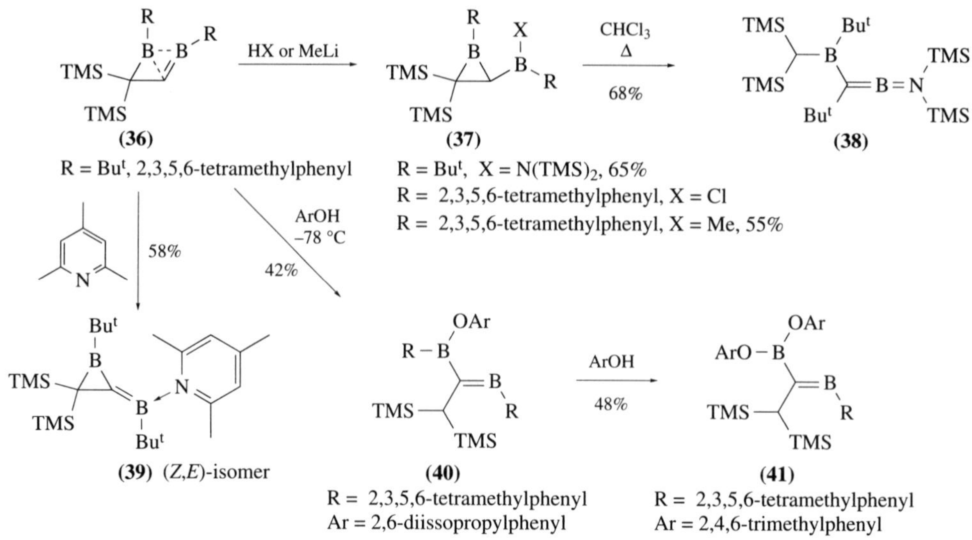

Scheme 19

carbene and 1,2-migration of the *t*-butyl group. The presence of B=C double bond isomers was demonstrated for the 2,4,6-trimethylpyridine adduct (**39**) but the identity of (*Z*)- or (*E*)-stereochemistry was not determined ⟨88AG(E)960⟩. The major isomer of the adduct (**39**) partly crystallised at $-40\,^{\circ}C$ leaving a 1 : 1 isomer ratio in solution as detected by [^{13}C]-NMR (Scheme 19). Borenes with one or two O-aryl substituents on the boryl group (**40**) and (**41**) were obtained by cleavage of the boron carbon single bonds of the borene (**36**) with sterically hindered phenols ⟨92ZN(B)983⟩. The x-ray structure of (**41**) showed a nearly normal sp^2-hybridised carbon atom for B—C=B with a bond angle of 118°, this is comparable to an alkene carbon atom.

5.25
Functions Doubly Bonded to a Metal

PETER QUAYLE
University of Manchester, UK

5.25.1 INTRODUCTION

Transition metal carbene complexes possess a formal carbon–metal double bond which has attracted countless investigations from theoretical, physical, synthetic organometallic and latterly organic chemists ⟨72CRV545, 74AG651, 84AG(E)587, 88AG(E)1456, 89AG(E)397, 90NJC433, 91SL381⟩.

From the perspective of the synthetic chemist, the ability to 'tame' reactive intermediates via coordination to a metal centre has been a much expounded paradigm although, in many instances, carbene complexes exhibit reactivity patterns which bear little similarity to that of the uncomplexed all-carbon, six-electron 'parent' system. The rapid growth in the number of publications in this area stems from their novel reactivity patterns with simple organic molecules leading to a variety of organic and organometallic products with much increased molecular complexity or potential synthetic utility. The use of organometallics in organic synthesis continues to be a fertile area of investigation, which will no doubt stimulate further interest in delineating the reactivity trends of this class of compounds and ultimately spawn new synthetic applications ⟨94AG(E)303⟩.

Most transition metal carbene complexes can be classified as being either 'Fischer-type' or 'Schrock-type' ⟨84AG(E)587⟩. Fischer complexes are usually 18-electron species derived from transition metals in groups 6 to 8 in which the metal is in a low (0 or 1+) formal oxidation state. The presence of a phenyl or heteroatom attached to the carbene carbon is usually required in order to confer stability. Schrock complexes on the other hand are usually electron deficient (10- to 16-electron) complexes in which the metal is in a high formal oxidation state and are derived from an early transition metal (Figure 1). The two different classes of complexes exhibit widely different chemical reactivity—Fischer complexes being electrophilic at the carbene carbon whilst Schrock complexes are nucleophilic at the carbon. In many respects Fischer carbene complexes resemble carboxylic acid derivatives in terms of their chemical reactivity, whereas Schrock complexes in many ways can be regarded as reacting as transition metal ylides. It is the chemistry of Fischer carbene complexes which will be discussed in this review.

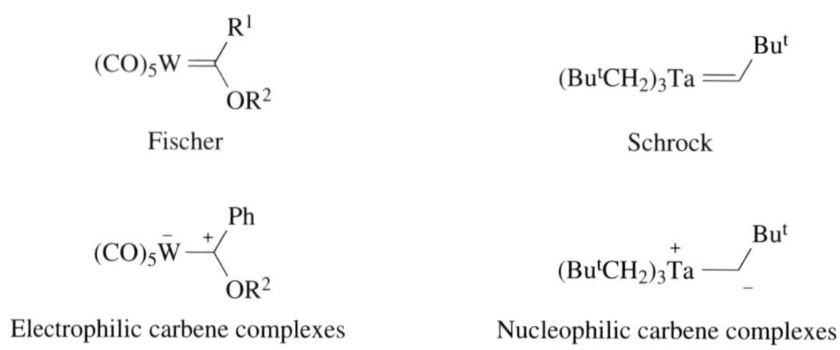

Figure 1

5.25.2 STRUCTURE AND BONDING

The preparation of the first transition metal carbene complex (**1**) was reported in 1964 by Fischer and Maasböl ⟨64AG(E)580⟩ with a definitive single crystal x-ray structure determination appearing the following year ⟨65AG(E)1082⟩. The x-ray structure determination reported by Mills and Redhouse ⟨65AG(E)1082, B-75MI 525-01⟩ clearly demonstrates that in complex (**1**) the carbene carbon is indeed bound to chromium with the chromium atom, carbene carbon and alkoxy substituent adopting a coplanar arrangement (Figure 2). The carbene–oxygen and methyl–oxygen bond lengths (131 ± 2 pm and 143 ± 3 pm respectively) are similar to those found in esters, whereas the carbene phenyl distance (149 ± 3 pm) corresponds to bonding between two sp^2 carbon atoms; in the solid state complex (**1**) adopts an *s-trans* conformation about the carbene carbon-oxygen bond. The x-ray data suggests considerable double bond character in the carbene carbon–oxygen bond (bond order of 1.2) which results in an appreciable barrier to rotation about this bond. Variable temperature ^1H NMR experiments in $CDCl_3$ ⟨68JOM(13)209⟩ indicate that complex (**2**) exists predominantly in the *s-trans* conformation over the temperature range $-50\,^{\circ}C$ to $120\,^{\circ}C$, while in d_6-acetone ⟨69AG(E)761⟩ coalescence between the *s-trans* and *s-cis* conformers occurs at $-40\,^{\circ}C$. This translates to an activation energy of 12.4 kcal mol^{-1} for the *s-trans* to *s-cis* interconversion which is similar to the theoretical value of 18.3 kcal mol^{-1} ⟨85JA1993⟩. The analogous aminocarbene complexes (**3**) have an appreciably higher barrier for *s-trans* to *s-cis* interconversion which is in excess of 25 kcal mol^{-1} for chromium carbene complexes ⟨90T1235⟩. Multinuclear NMR techniques have also been utilized in order to correlate chemical reactivity with bonding parameters ⟨88JA8413⟩. From a theoretical standpoint ⟨76JA441, 82JA3879, 84JA1576, 84JA5892, 86JA2173, 86JA2180, 86JA4746, 87JA3953, 92JA539⟩, Fischer carbene complexes may be regarded as being derived from a singlet state carbene in which electron density is donated from a filled 'sp^2 hybrid' HOMO on the carbene carbon to an empty $4s4p3d$ hybrid (LUMO) on the metal. The π-bond is formed by back transfer from degenerate dπ MOs of the $Cr(CO)_5$ fragment with the vacant LUMO of the carbene (Figure 3). Theoretical studies ⟨83JA426⟩ suggest that there is essentially free rotation about the carbene carbon–metal bond (the barrier to rotation may be as small as 0.41 kcal mol^{-1} in the case of chromium carbene complexes), a situation which appears to have experimental grounding ⟨83JA426⟩.

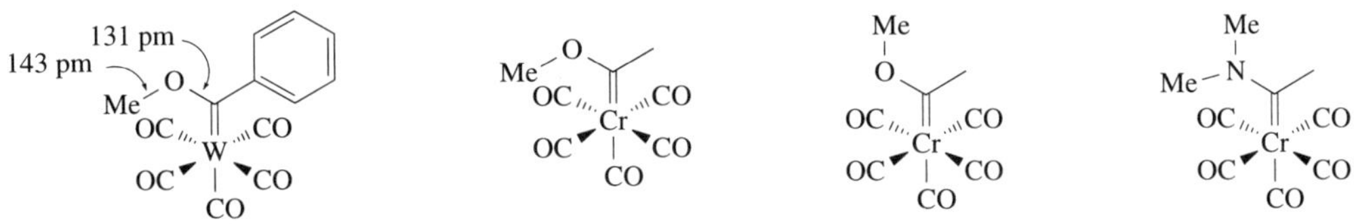

Figure 2

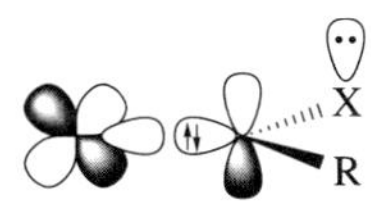

Figure 3

5.25.3 PREPARATION FROM NONCARBENE STARTING MATERIALS

5.25.3.1 Acyclic Mononuclear Alkoxycarbene Complexes

Alkoxycarbene complexes are in many cases intensely coloured red crystalline compounds or low melting solids. Purification may be achieved either by low temperature recrystallization from nonpolar solvents or by flash chromatography on silica using degassed solvents. Sometimes low temperature chromatography is required in order to purify the less stable complexes ⟨77IS164, 79IS95⟩. They usually have good stability when stored under nitrogen or argon in the cold, but may decompose relatively rapidly in solution at or above ambient temperature. Alkoxycarbene complexes are relatively stable towards oxidation in the bulk state, but will oxidize readily to carboxylic esters in the presence of even the mildest of oxidizing agents (e.g., DMSO or dissolved molecular oxygen) when in solution (Equations (1) and (2)). Alkoxycarbene complexes themselves prove to be valuable starting materials for the preparation of a variety of more functionalized complexes as outlined in Section 5.25.3.6.

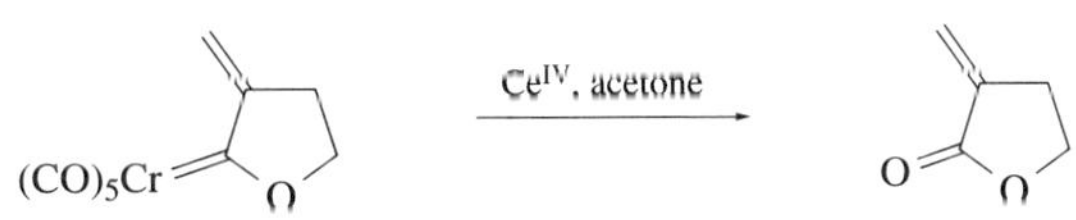

(1)

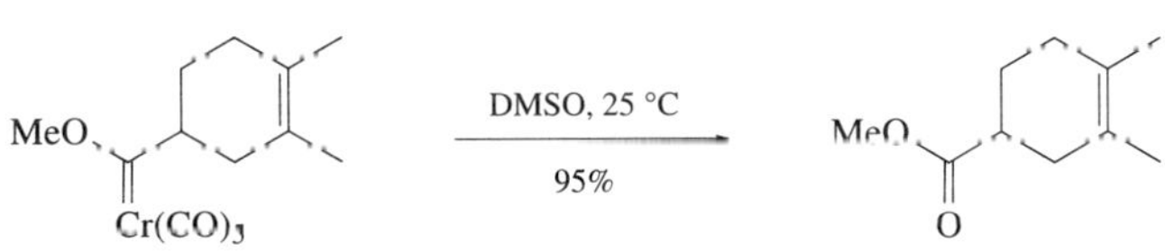

(2)

5.25.3.1.1 The Fischer reaction

In their landmark paper Fischer and Maasböl ⟨64AG(E)580⟩ showed that reaction of tungsten hexacarbonyl with phenyllithium proceeds by nucleophilic attack at a carbon monoxide ligand to afford an acylate complex (**4**) (Scheme 1). In their original procedure, the acylate (**4**) was precipitated out of aqueous solution as its tetramethylammonium salt (**5**), which upon protonation to the hydroxycarbene complex (**6**) and subsequent reaction with diazomethane afforded the carbene complex (**1**) as a yellow crystalline solid (m.p. 59 °C). Attempts to isolate the hydroxycomplex (**6**) on this occasion were unsuccessful, leading to the isolation of benzaldehyde upon workup. Since their original report ⟨64AG(E)580⟩, an operationally simpler and more direct procedure has been developed ⟨77IS164, 79IS95⟩ which avoids the preparation of intermediate hydroxycarbene complexes, but rather involves direct capping of the acylate complex with a powerful alkylating agent such as a trialkyloxonium salt (Equation (3)). This method of carbene synthesis is general for a number of transition metals (e.g., Cr, Mo, W, Mn, Tc, Re and Ni) and affords ready access to literally hundreds of carbene complexes ⟨80MI 525-01⟩, the majority of which are derived from the group 6 transition metals. This method enables the preparation of alkyl ⟨86JA520⟩, aryl ⟨80CB(113)1010, 93CB2573, 94TL561⟩, heteroaromatic ⟨89JOC3249, 89TL1633⟩, vinyl ⟨84JA434, 88JOC1203, 91TL7759, 94T2003⟩ and acetylenic carbene complexes ⟨91AG(E)1658, 95CB645⟩ (Equations (4)–(15)).

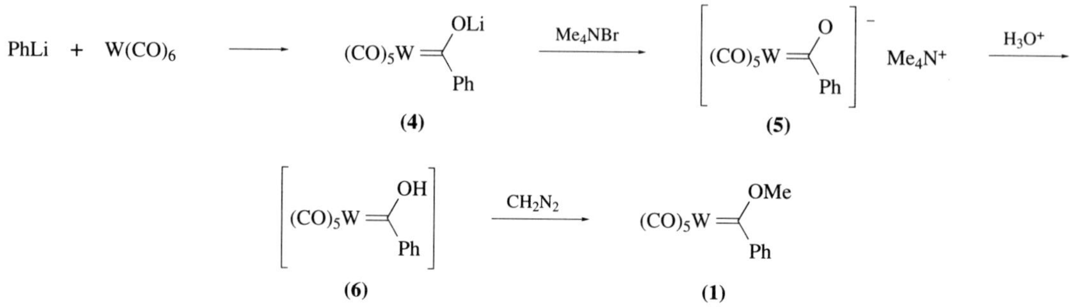

Scheme 1

(3)

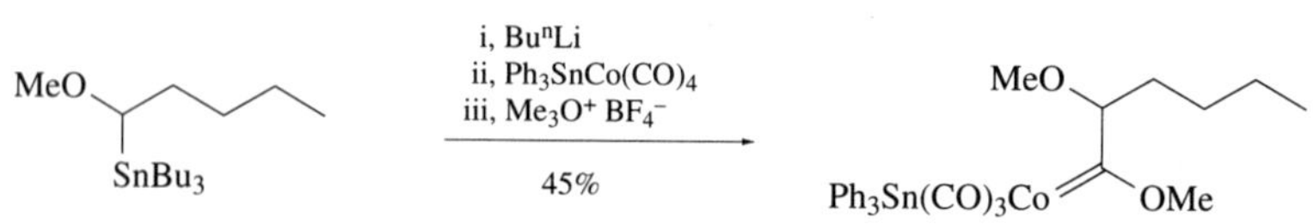

(4)

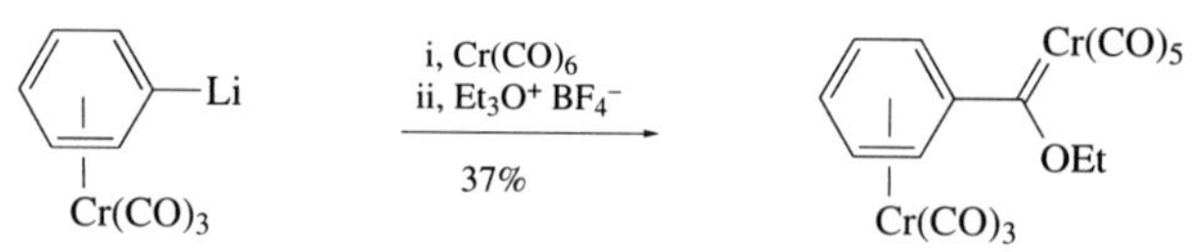

(5)

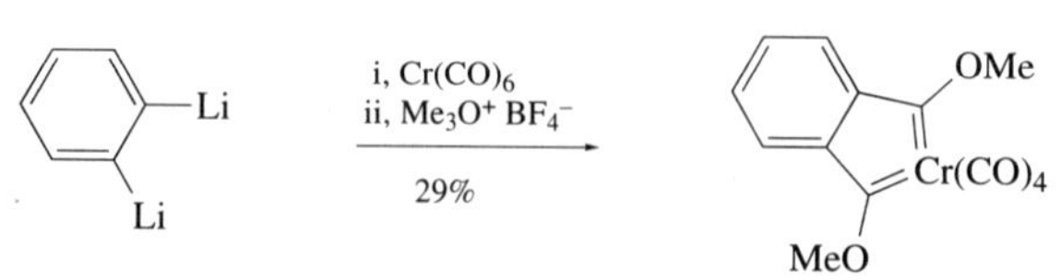

(6)

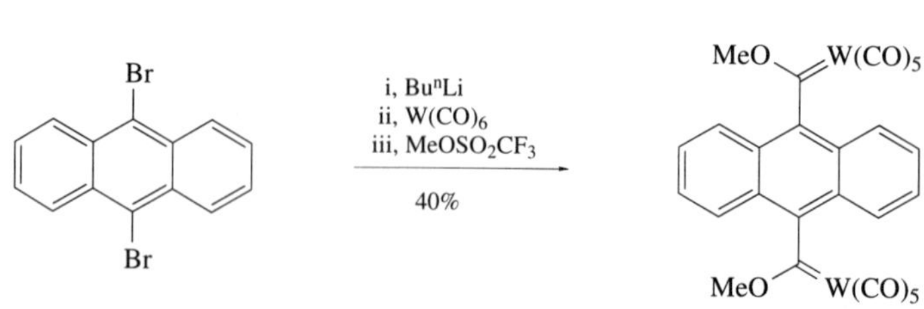

(7)

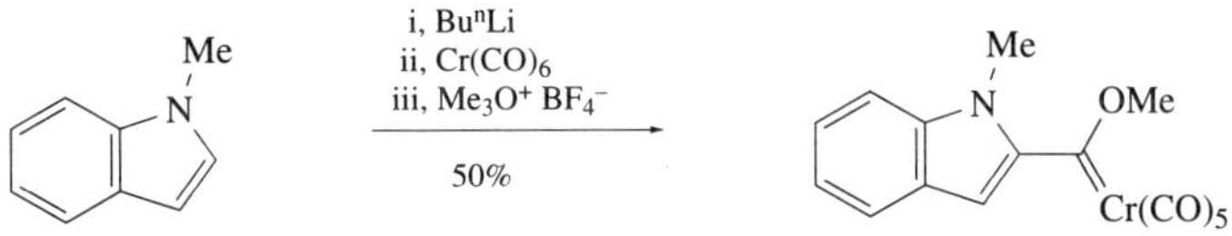

(8)

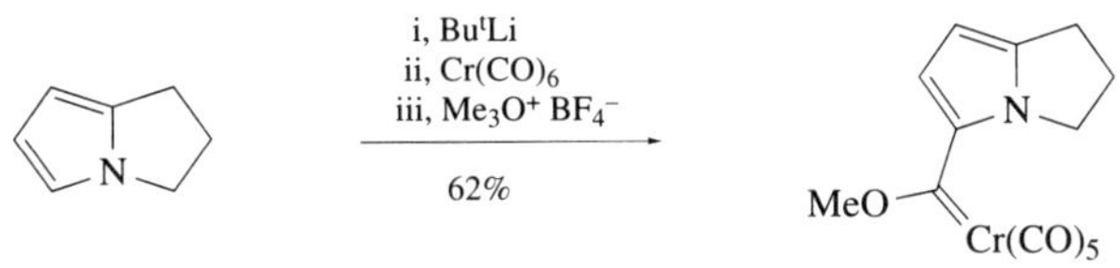

(9)

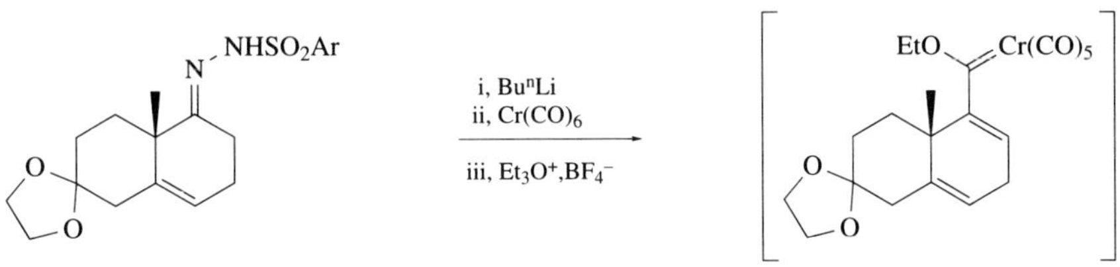

(10)

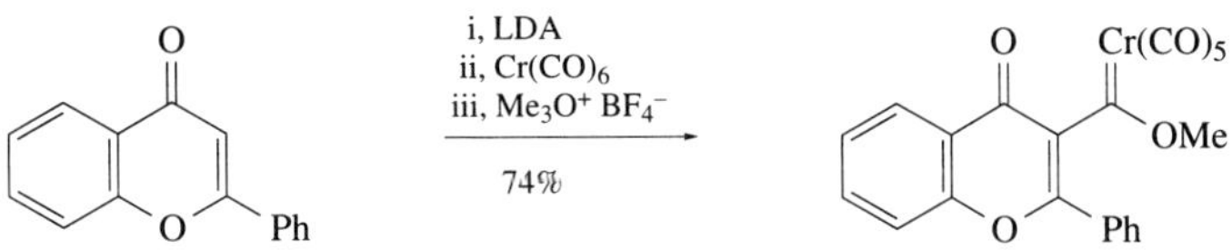

(11)

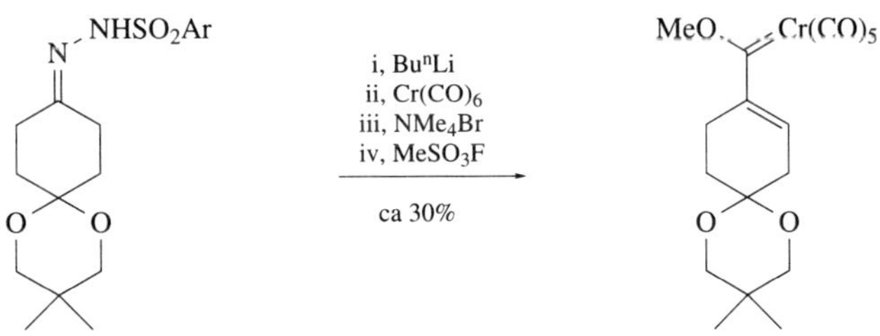

(12)

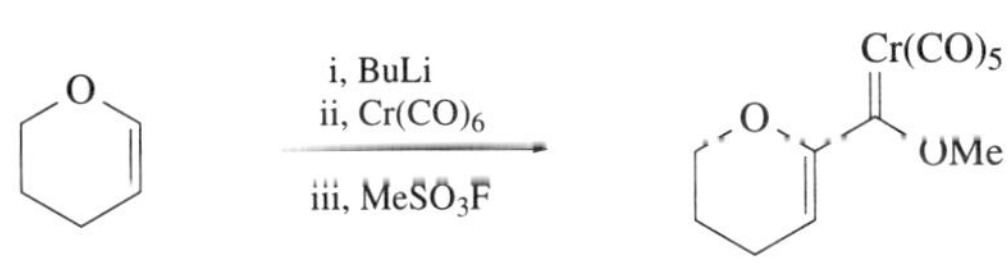

(13)

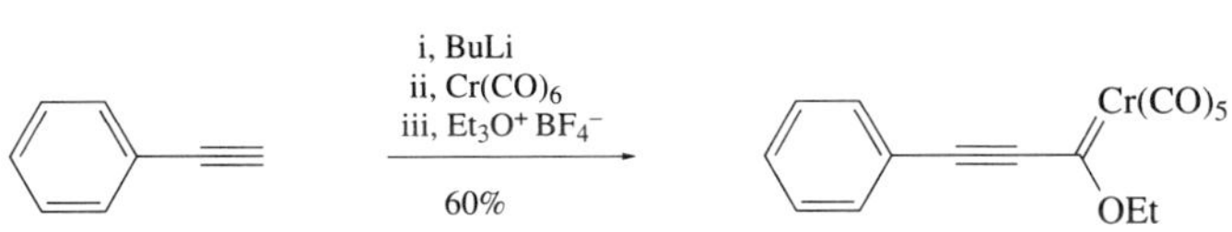

(14)

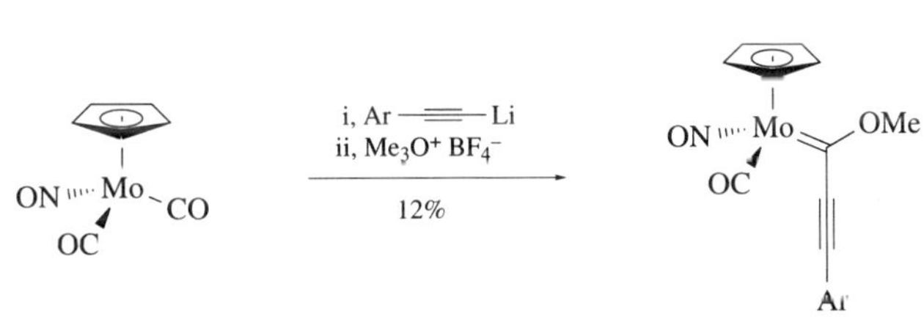

(15)

The acylate complexes used in this procedure are relatively non-nucleophilic and therefore require the use of powerful alkylating agents in the capping step. Although trialkyloxonium salts ⟨77IS164⟩ were initially used for this purpose, a number of workers have advocated the use of the more reactive fluorosulfonate ⟨90JA7806, 90TL2529⟩ or triflate esters ⟨91TL5223, 94OM4189⟩ in order to generate optimal yields of the desired carbene complexes. In certain cases ⟨83JA4099, 89JA1285⟩ the 'hardness' of the alkylating reagent may also determine the site of alkylation in the acylate complex (i.e., at either the metal centre or at oxygen) (Equation (16)). In a related sequence, the alkylation of an acylate complex with benzenediazonium tetrafluoroborate affords a phenoxycarbene complex in 60% overall yield (Equation (17)) ⟨72JOM(46)C15⟩. The generality of this methodology has not been investigated thus far. In 1994 ⟨94CL1537⟩, the use of alkyldiphenylsulfonium salts has been advocated as a useful alternative to oxonium salts and sulfonate esters in the alkylation of acylate complexes for the preparation of functionalized alkoxycarbene complexes (Equation (18)). This procedure works well (yields ca. 65–95%) for the preparation of chromium, tungsten and even molybdenum carbene complexes, which some authors claim to be somewhat unstable ⟨67CB2445, 90JA7806⟩. A variety of organometallic reagents have also been utilized in the acylate capping reaction as illustrated below (Equations (19)–(23)) ⟨72JOM(40)159, 77CB2574, 79CB708, 92JA8735, 92JOM(426)343, 92OM1759, 94JA10921⟩.

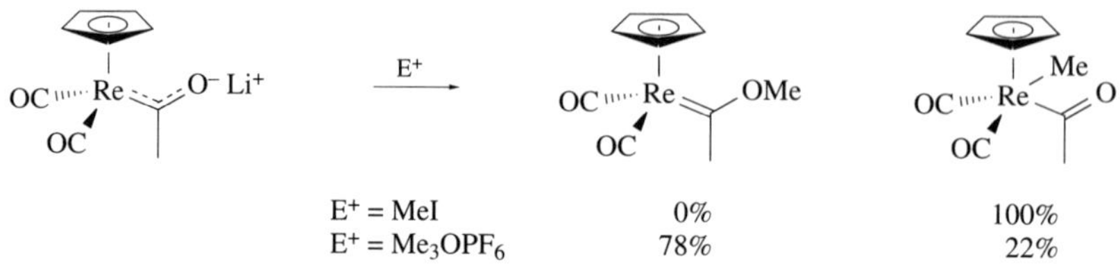

(16)

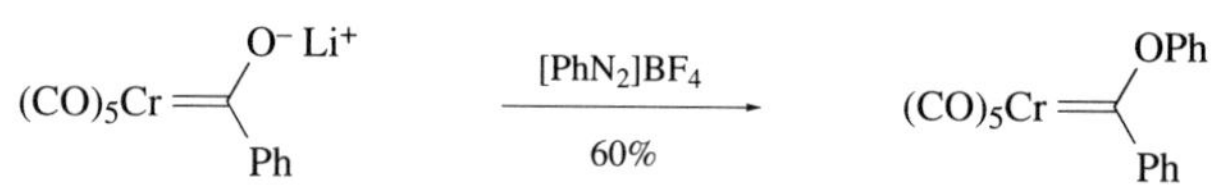

(17)

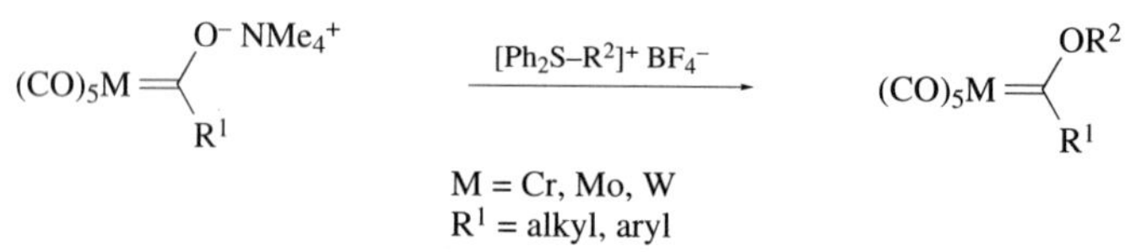

(18)

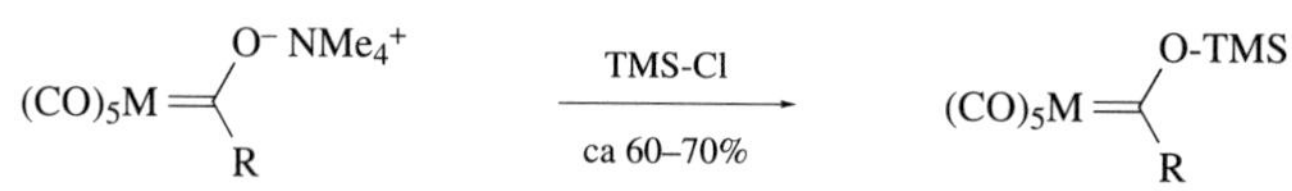

(19)

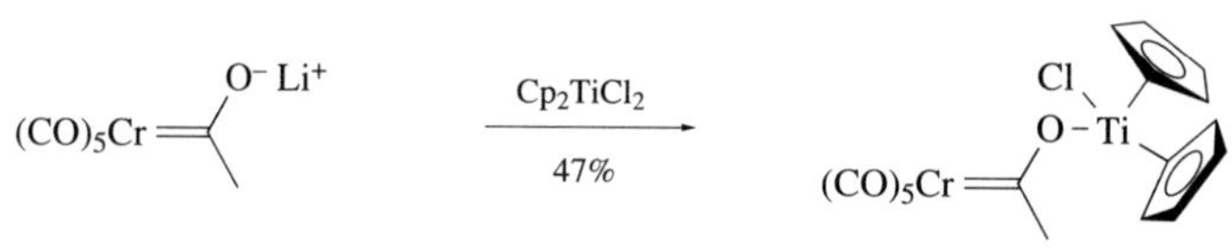

(20)

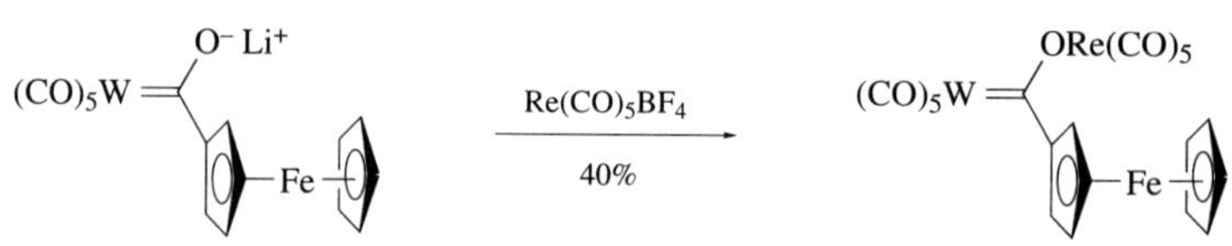

(21)

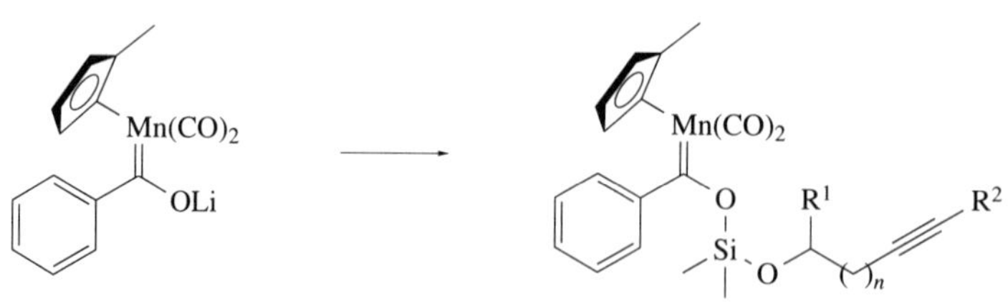

(22)

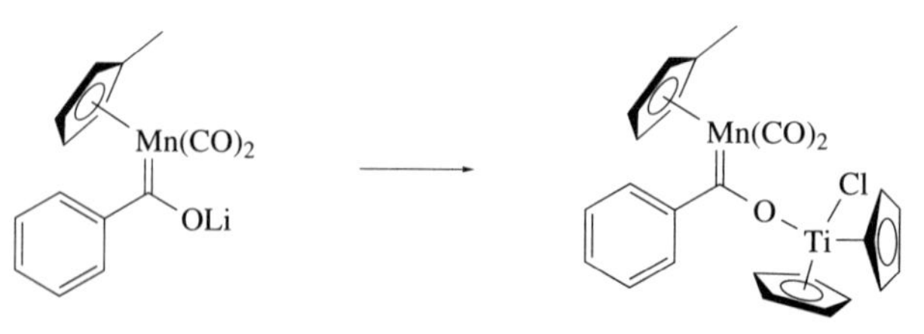

(23)

A possible limitation of the classical Fischer carbene synthesis resides in the use of highly reactive organometallic reagents (typically alkyllithium and magnesium reagents) in the initial generation of the acylate complex. Clearly this severely limits the range of functional groups which can be incorporated into the nucleophilic partner if protecting group strategies are to be avoided ⟨95OM3163⟩. A complementary route to acylate complex generation has been developed by Semmelhack ⟨87OM1839⟩, although its synthetic utility awaits further scrutiny.

5.25.3.1.2 The Semmelhack reaction

The alkali metal salts (**7**) and (**8**) are readily prepared by reduction of the amino complex (**9**) with lithium or sodium naphthalenide in THF under strictly anaerobic conditions. A stock solution of the salt (**7**) or (**8**) in THF is then reacted with a carboxylic acid chloride in THF at −78 °C to afford the acylate complex (**10**) which may be capped in the normal manner in order to generate the carbene complex. The salts (**7**) and (**8**) are unreactive towards a variety of functional groups (e.g., esters, benzylic halides, etc.) thereby allowing the introduction of the carbene moiety in a chemoselective fashion (Scheme 2) ⟨87OM1839⟩. Overall yields are comparable to those reported by Fischer in the cases where direct comparisons can be made.

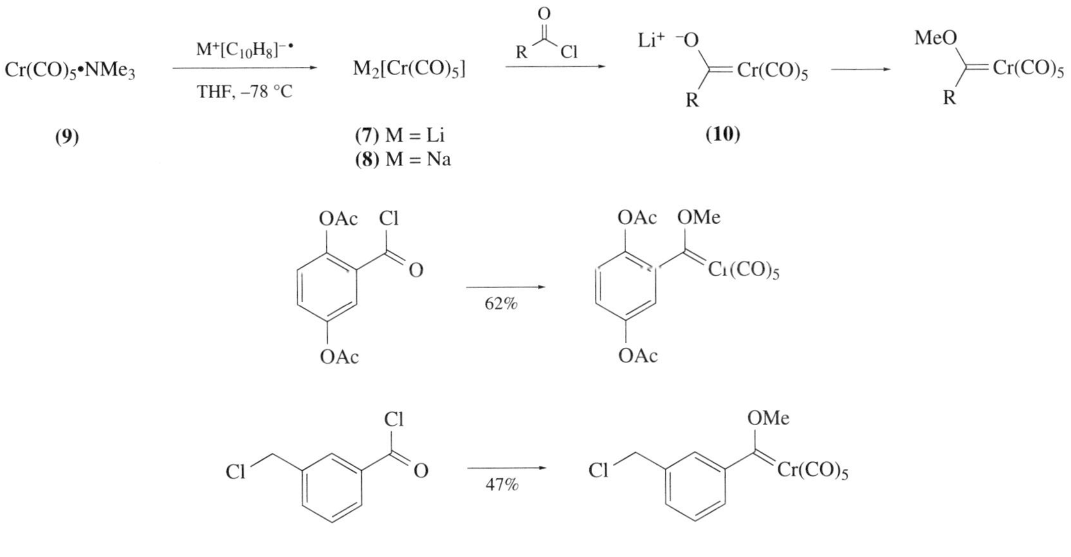

Scheme 2

5.25.3.1.3 Preparation from acyl complexes

Acyl complexes are readily prepared and are ubiquitous throughout transition metal chemistry ⟨82COMC-I⟩. Their chemistry has been extensively investigated, and they serve as useful intermediates for the preparation of carbene complexes. For example, Green ⟨71JCS(A)794⟩ has demonstrated that the acyl complexes (**11**; M = Fe, Rh, Ru, Mo) can be converted into the thermally stable, cationic complexes (**13**) either via the hydroxy carbene complexes (**12**) or directly ⟨81JOM(213)C31⟩ by reaction with trialkyloxonium salts. Usually the yields are high for this sequence (ca. 80%). This protocol is quite general and enables the synthesis of relatively unstable species such as cyclopropyl carbenes (Scheme 3) ⟨85OM943⟩. Related reaction sequences have also been performed on chiral rhenium and iron acyl complexes (Scheme 4) ⟨83JA258, 83OM1852, 84OM1764, 91JA927⟩. Acyl complexes undergo acylation ⟨87OM2249⟩ with triflic anhydride generating unstable intermediates such as (**15**). Facile elimination of triflic acid from (**15**) generates the electrophilic vinylidene complex (**16**) which can be intercepted by alcohols to generate the carbene complexes (**17**) (Scheme 5).

Jones has utilized the α-elimination reactions (1,2-rearrangement reactions) of a variety of metal acyl complexes in the synthesis of cyclic carbenes (Scheme 6) ⟨86OM178⟩. Hence the cyclobutyl complexes (**18**) rearrange upon photolysis to the corresponding ferracyclopentene (**19**) whereas the cyclopropyl derivative (**20**) undergoes a 1,2-shift followed by insertion of carbon monoxide to afford the ferracyclopentenone (**21**) ⟨85OM944⟩. In 1992 ⟨92OM622⟩, the same group demonstrated that the

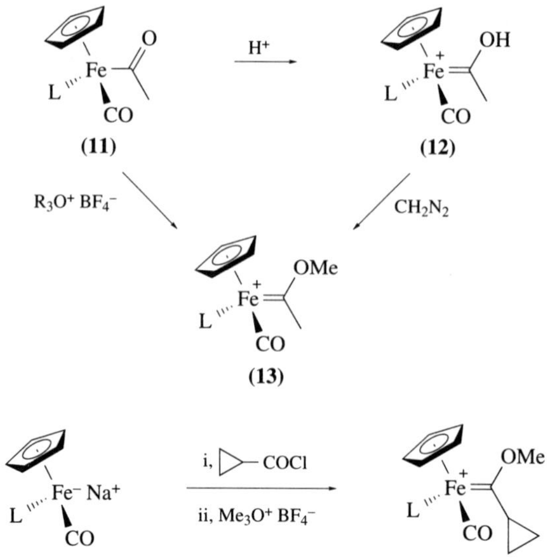

Scheme 3

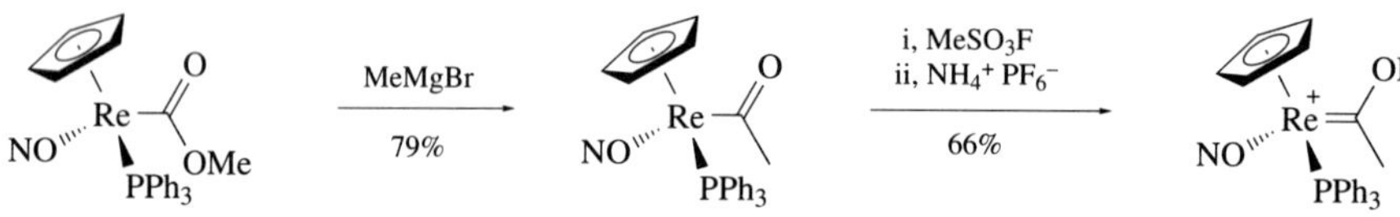

Scheme 4

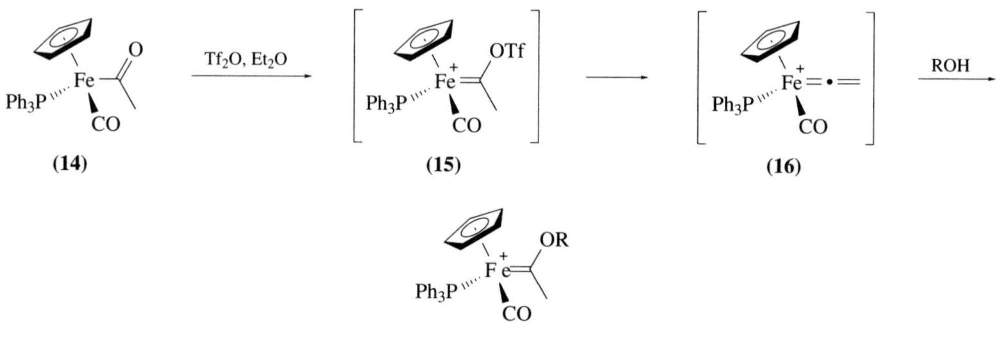

Scheme 5

bicyclic acyl complex (**22**) undergoes ring expansion to the corresponding carbene complex (**23**) upon mild thermolysis in d_6-benzene. Photolysis of the iron complex (**24**) occurs in a similar manner to generate a 3.5 : 1 mixture of the *endo*- and *exo*-ferracyclopentenones (**25**) in 45% yield (Scheme 6). Late transition metal systems such as the ruthenium acyl complex (**26**) did not rearrange to the ruthenium carbene complex (**27**) under similar conditions but rather afforded the alternate ring-expanded complex (**28**) ⟨92OM1442⟩. Fortunately, an alternative synthesis of the ruthenium carbene complex (**27**) was accomplished starting from the alkyl complex (**29**) (Scheme 7). The metalloindenes (**27**) are thermally stable and show no signs of interconversion into the isomeric structure (**30**) upon thermolysis, which would proceed via a 1,2-shift to the cyclobutane (**31**) which is presumably a thermodynamically unfavoured process (Scheme 7) ⟨94OM971⟩.

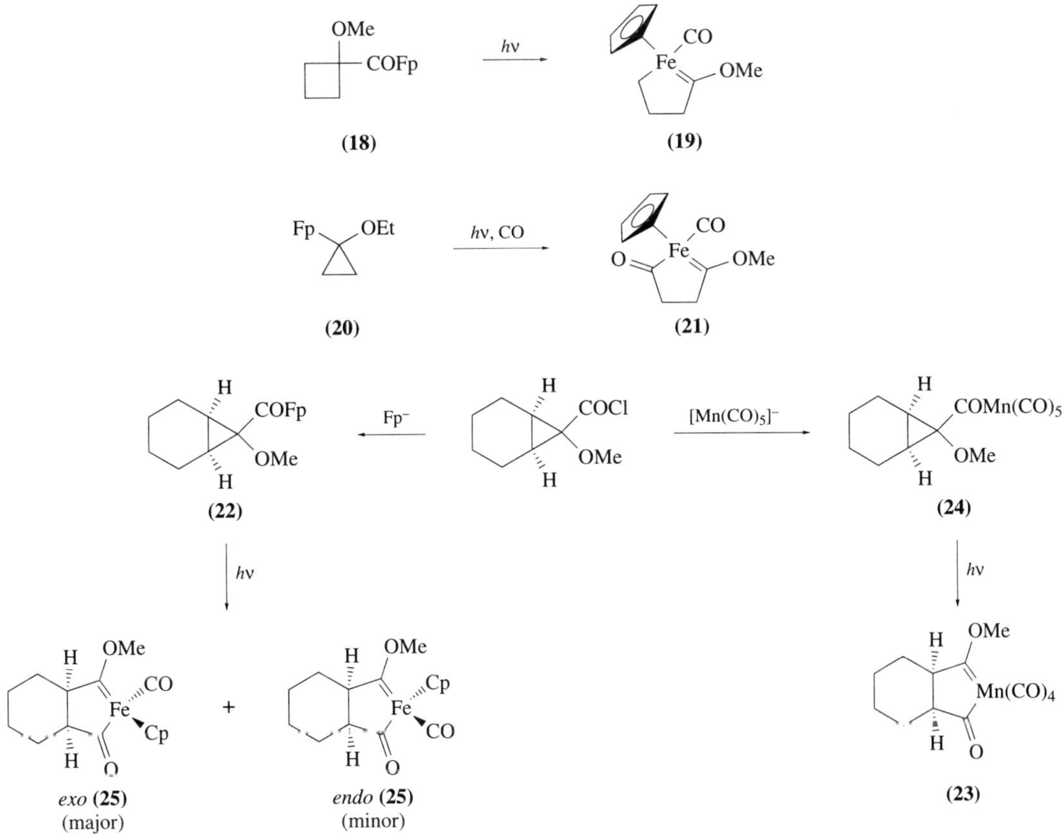

Scheme 6

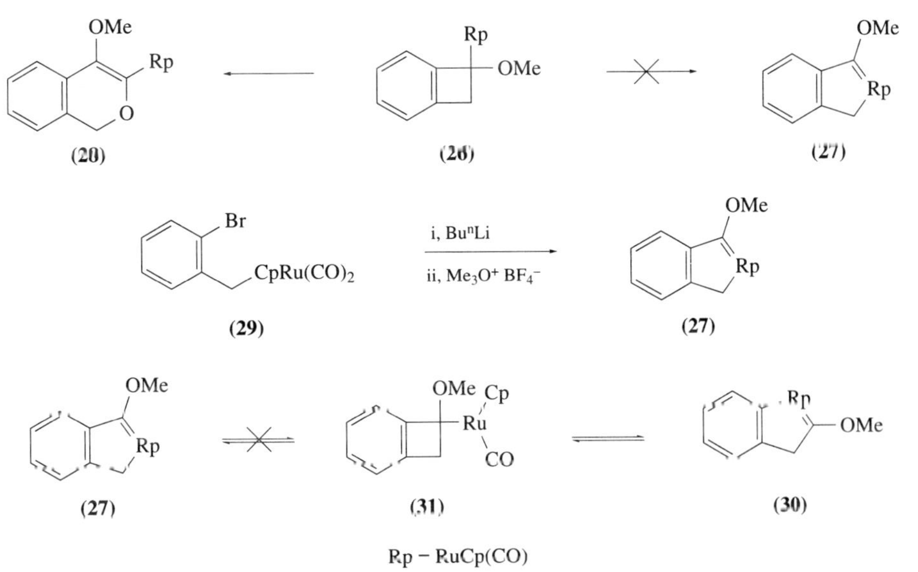

Scheme 7

5.25.3.1.4 *Alkyne–vinylidene and related rearrangements*

In 1970 Chisholm and Clark reported that alkynes were rapidly converted to cationic platinum carbene complexes (**32**) upon reaction with platinum(II) salts ⟨70CC763, 73ACR202⟩. A mechanism involving π-complexation, intramolecular hydride transfer generating the carbonium ion (**33**) followed by solvent capture is in keeping with experimental findings (Scheme 8) ⟨71IC1711, 72CRV545, 72IC503, 72JA1532, 77IC677⟩. Since these initial observations this reaction sequence has been found to be quite general for a number of the transition metals (Equations (24) and (25)) ⟨73JOM(47)C17, 90CC641, 95JOM(488)C9⟩. Preformed alkynide σ-complexes such as (**34**) and (**35**) have also been utilized in carbene generation (Schemes 9 and 10) ⟨78JOM(159)417, 86OM1976, 91OM2768⟩.

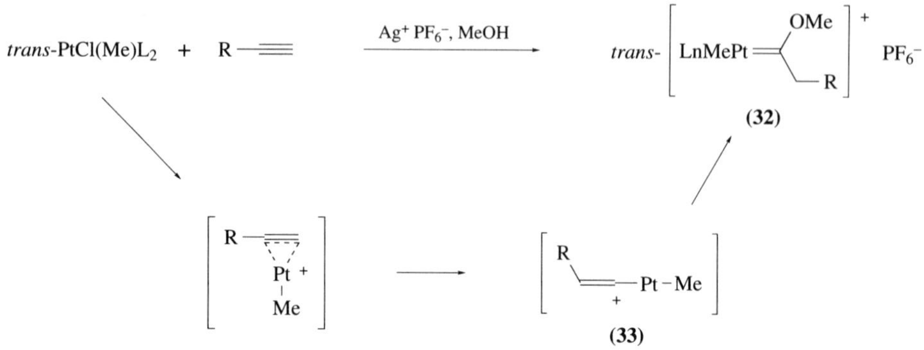

Scheme 8

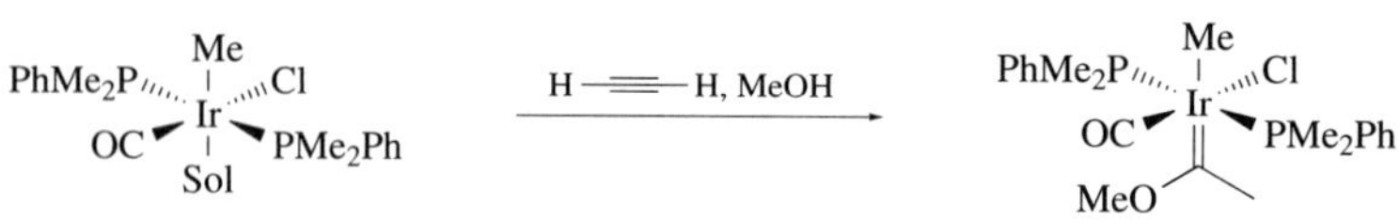

(24)

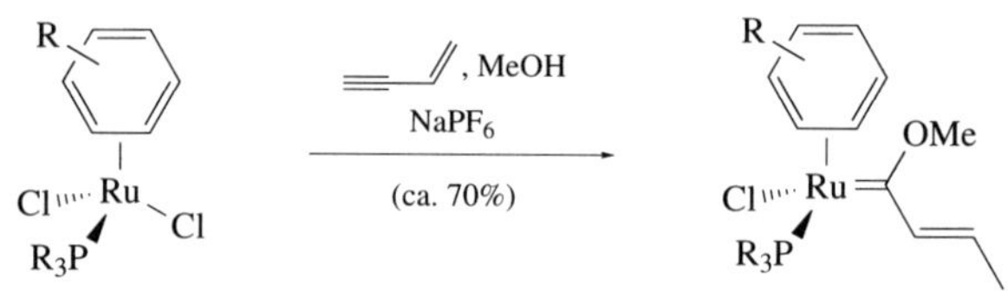

(25)

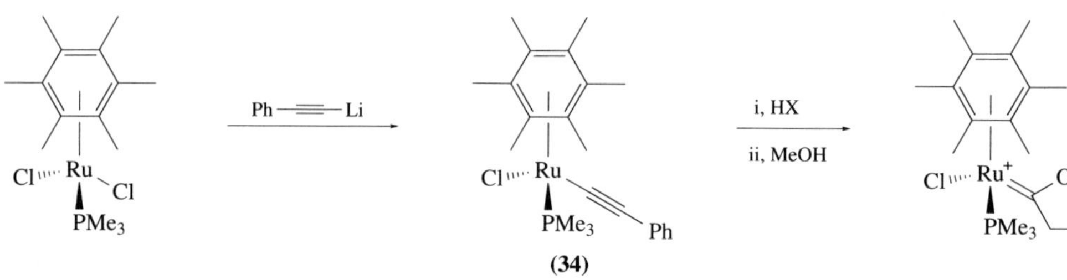

Scheme 9

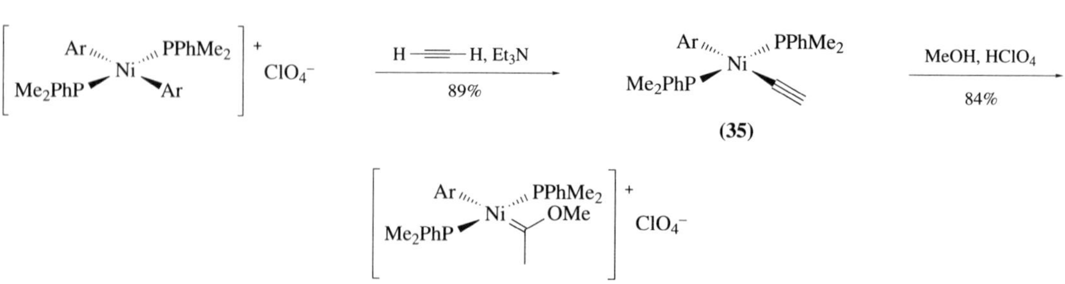

Scheme 10

Interaction of a terminal alkyne with a variety of transition metal complexes often results in a rapid, low activation rearrangement reaction leading to the generation of vinylidene complexes (Scheme 11) ⟨85JA4474⟩. This is to be compared with the isomerization of ethyne itself into vinylidene which is a high energy process (ca. 37 kcal mol^{-1}) ⟨91CRV197⟩. This so-called alkyne–vinylidene rearrangement has been extensively studied from both preparative and mechanistic perspectives and has been reviewed ⟨83AOC(22)59, 89AOC(29)163, 90AOC(30)1, 91CRV197⟩. Irrespective of mechanistic niceties this process allows access to organometallic intermediates such as (**36**) which are akin to ketenes in terms of their reactivity patterns. Hence generation of the vinylidene complex (**36**) in the presence of a nucleophilic reagent (e.g., an alcohol) leads directly to the carbene complex (**37**) without recourse to any additional synthetic manipulations (Scheme 11) ⟨90CC641, 95JOM(488)C9⟩.

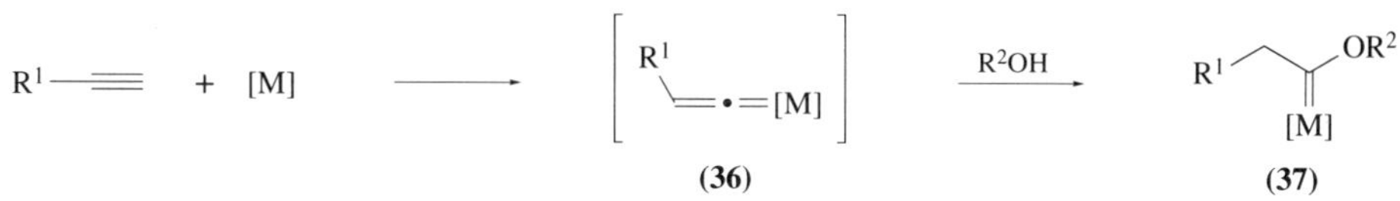

Scheme 11

In a related series of investigations, French workers ⟨89AOC(29)163, 91CC881, 92CC1220, 92OM809, 95OM1938⟩ have shown that propargylic alcohols undergo alkyne–vinylidene rearrangements in the presence of (arene)Ru^{II}, zerovalent group 6 hexacarbonyl complexes to generate unstable vinylidene complexes, for example (**38**). Complexes such as (**38**) readily dehydrate under the reaction conditions to allylidene complexes (**39**) which then undergo rapid *in situ* capture with the alcohol solvent to afford the α,β-unsaturated carbene complexes (**40**) (Scheme 12). This sequence may also be carried out with a variety of more functionalized propargylic alcohols and represents the most effective synthesis to date of highly unsaturated systems such as (**41**) (Equation (26)) ⟨95OM1938⟩.

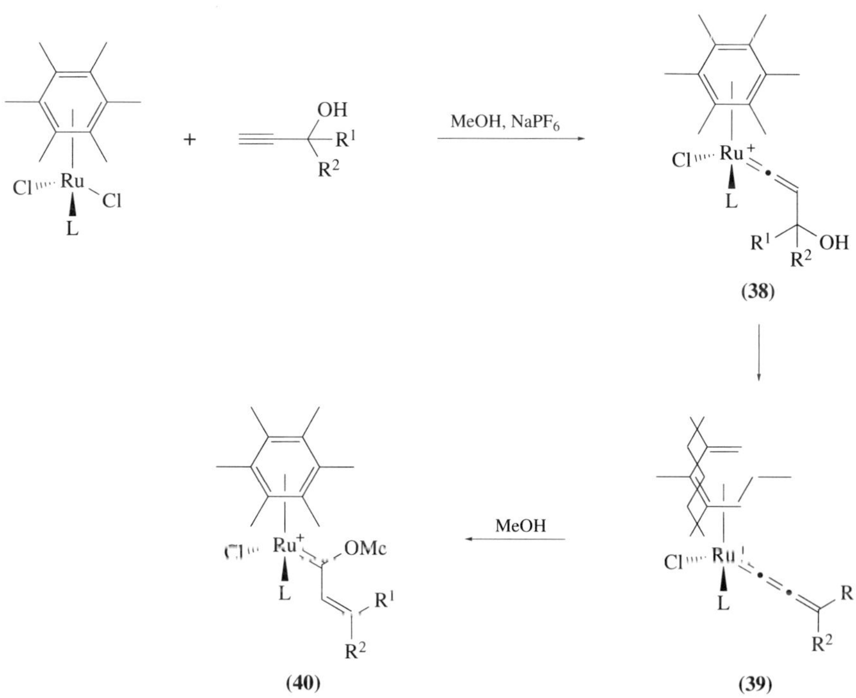

Scheme 12

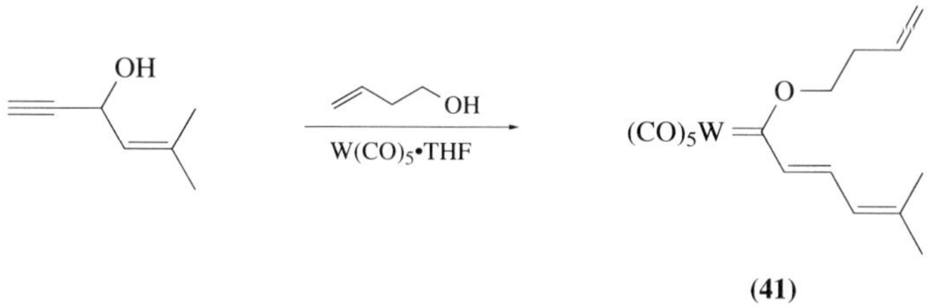

(26)

(41)

5.25.3.1.5 Miscellaneous methods

Alkoxycarbene complexes are similar in many respects to carboxylic acid esters and as such undergo 'addition–elimination' reactions with a variety of nucleophiles effecting an overall displacement of the alkoxy group. In certain cases this reactivity has been put to good effect in the preparation of homologated alkoxycarbene complexes. For example, treatment of arylmethoxycarbene complex (**42**) with an excess of but-3-en-1-ol in the presence of a catalytic quantity of alkoxide base affords the carbene complex (**43**) in 68% yield (Scheme 13) ⟨85OM736⟩. Care must be taken when preparing unsaturated carbene complexes in this manner as they may undergo relatively rapid intramolecular cyclopropanation reactions upon mild heating (Scheme 13) ⟨85OM736, 87CRV411⟩.

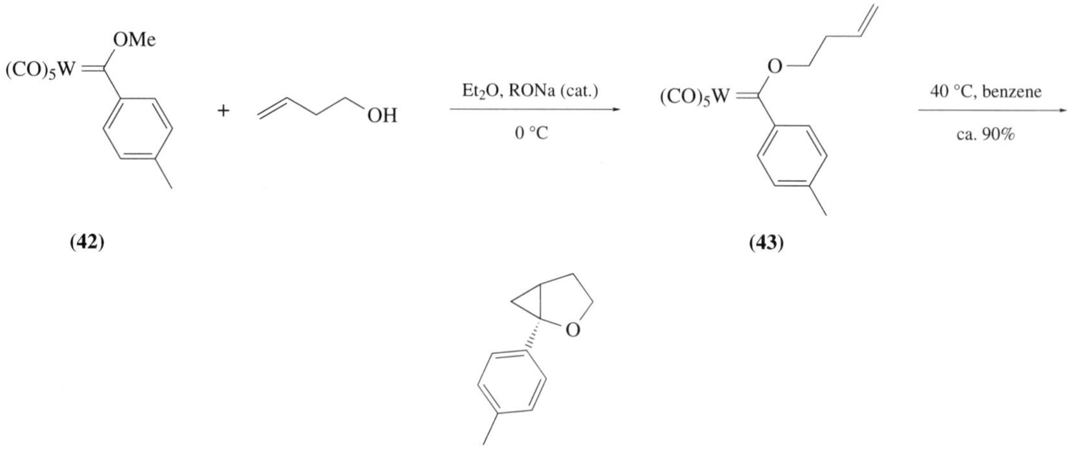

Scheme 13

Readily available ⟨86AOC(25)121, 87AOC(27)51, 91AOC(32)227⟩ transition metal carbyne complexes such as (**44**) react with alcoholates to afford carbene complexes (**45**) in high yields (Equation (27)) ⟨77CB1140⟩. This reaction has been put to good effect in the synthesis of the first carbohydratocarbene complexes, for example (**46**), of manganese and rhenium ⟨95CB373⟩.

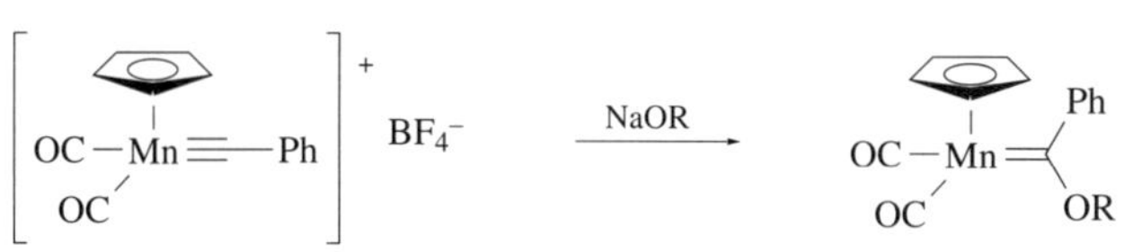

(27)

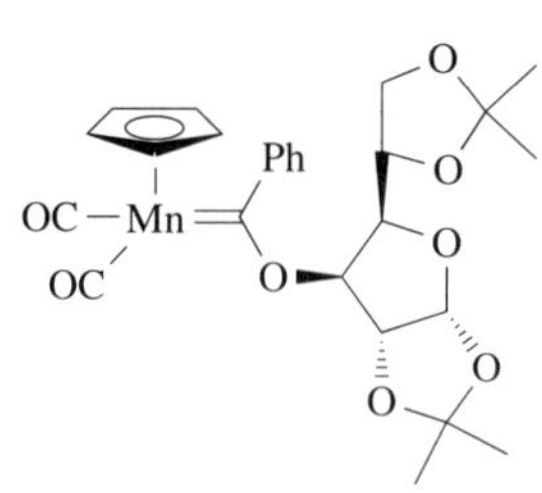

(46)

It was discovered ⟨93JOM(445)143⟩ that iron alkyls react with triethyloxonium tetrafluoroborate under an atmosphere of carbon monoxide to afford iron ethoxycarbene complexes (Equation (28)). Although the mechanism of this reaction has yet to be established, its ease of operation will doubtless attract much attention from synthetic chemists.

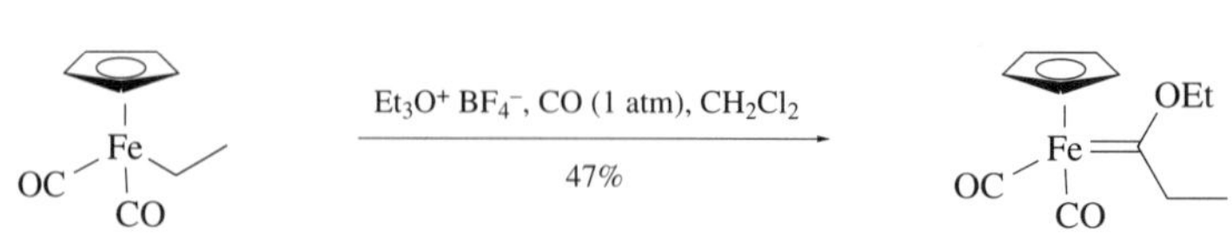

(28)

By analogy to the direct conversion of diazo compounds into transition metal carbene complexes ⟨74AG(E)599⟩, reaction of the diazirine (**47**) with chromium pentacarbonyl afforded the Fischer carbene complex (**48**), albeit in low overall yield (Equation (29)). Experimental evidence points to a pathway involving insertion of the metal into the diazirine ring followed by loss of nitrogen rather than via generation of free (methoxy)phenylcarbene ⟨83CC880⟩.

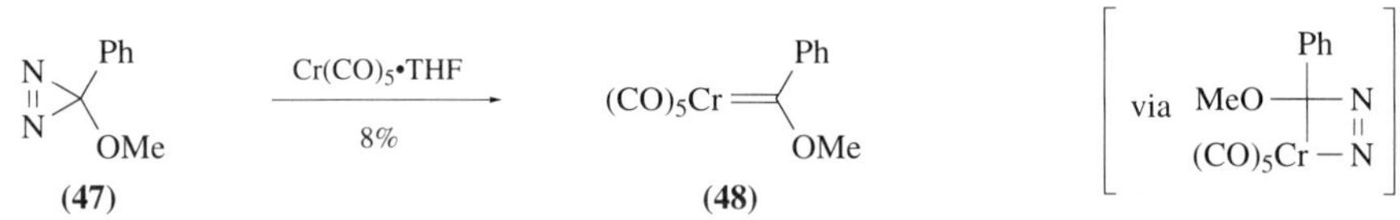

(29)

Nucleophilic addition of $Na^+[Re(CO)_5]^-$ to activated alkynes followed by alkylation with $Et_3O^+PF^-_6$ affords the novel cyclic Fischer carbene complexes (**49**) in good overall yields (ca. 60%). The chemistry of these complexes has been briefly explored and closely resembles that of more conventional Fischer complexes (Equation (30)) ⟨93JA9986⟩.

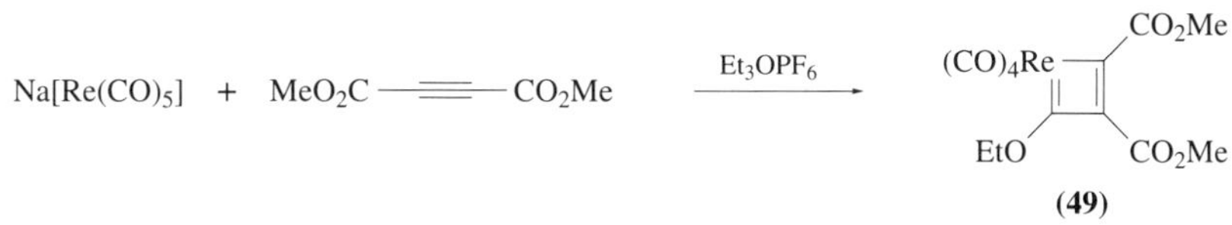

(30)

5.25.3.1.6 *Cyclic alkoxycarbene complexes*

(i) The Fischer reaction

Maiorana and co-workers ⟨92CC1623⟩ have described a direct approach to cyclic complexes such as (**50**) utilizing Cohen's methodology for the generation of γ-lithioalkoxides from oxetanes. Although yields are variable (Scheme 14) the directness of this approach from readily available starting materials suggests that it may well be adopted as a facile entry into this class of complex.

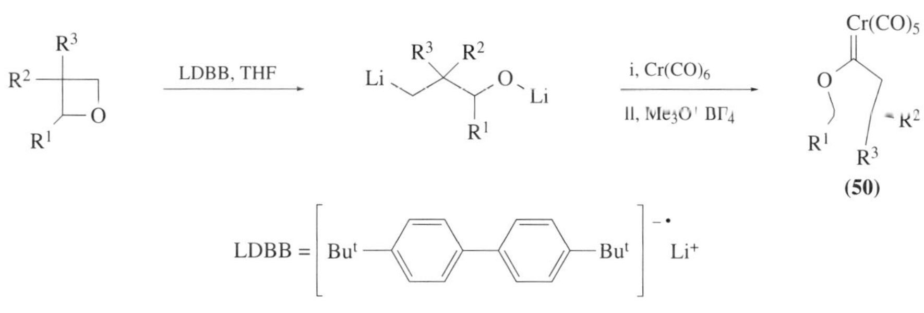

Scheme 14

(ii) From α,ω-dihaloalkanes and ω-halocarboxylic acid chlorides

In 1963 King ⟨63JA1922, 95AOC(37)169⟩ reported that the reaction of 1,3-dibromopropane with $Na^+[(CO)_5Mn]^-$ afforded the bimetallic, acyclic compound (**51**) in 28% yield. However, it was not until 1970 that Casey reformulated the structure of this compound as the cyclic carbene complex (**52**) ⟨70CC1220⟩ and suggested a mechanism for its formation (Scheme 15) ⟨71JA3554⟩. The sequence as depicted in Scheme 16—that is alkylation, promotion of alkyl migration by nucleophilic attack at the metal centre followed by cyclization—is now known to be relatively general ⟨71JA2672, 82CC215, 84CC747, 90OM2621, 92JOM(441)63⟩. The initial displacement of halide by the nucleophilic metal centre takes place preferentially at the sterically least hindered carbon ⟨90OM2836⟩. With longer chain bishalides, the ω-bimetallic species may indeed be formed in preference to the cyclic carbene complex

⟨83JCS(D)2397⟩. In 1994, this methodology was applied to the synthesis of functionalized oxacyclopentylidene and oxacyclohexylidene complexes in which ω-halo epoxides rather than dihalides were used in the initial alkylation sequence. Chemoselective alkylation at the halogen bearing carbon is observed, with cyclization of the metal acyl complex being effected by the addition of a strong Lewis acid such as $SnCl_4$. The overall yield is approximately 50% for the whole reaction sequence. This reaction sequence also bears much promise for the synthesis of functionalized lactones, given the ease of oxidative removal (pyridine-*N*-oxide) of the metal centre (Scheme 17) ⟨94TL6717⟩.

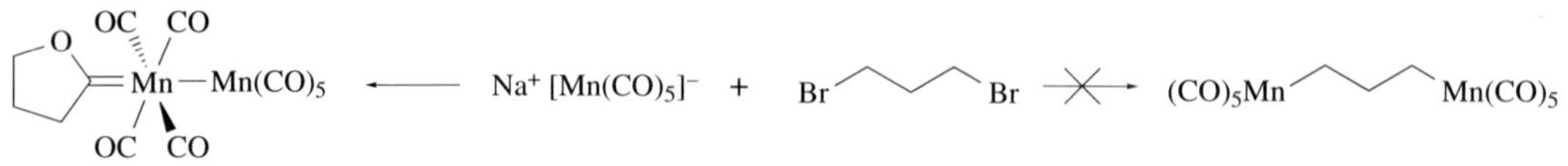

(52) **(51)**

Scheme 15

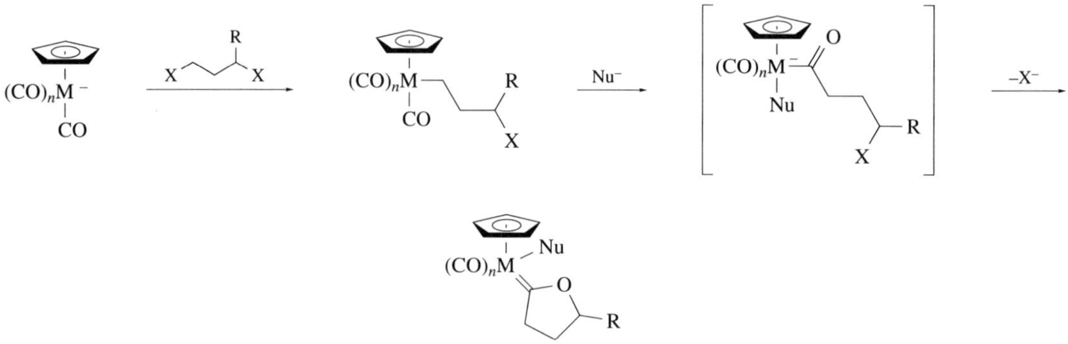

Scheme 16

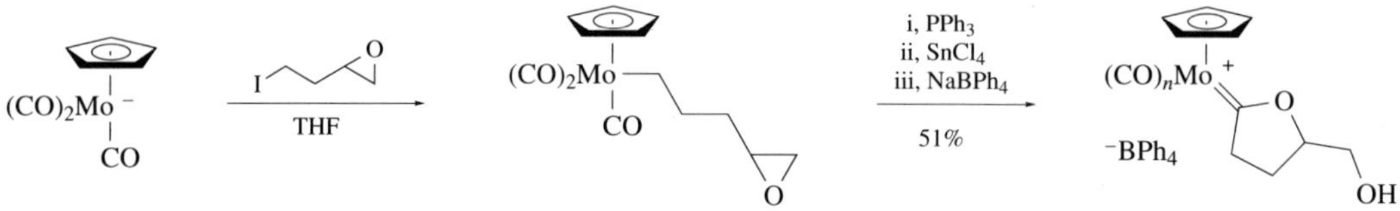

Scheme 17

(iii) From alkynes

In their original investigations ⟨70CC763⟩ Chisholm and Clark noted that but-1-yn-4-ol reacted with platinum salts to afford the oxacyclopentylidene complex (**53**) (Equation (31)). Since this observation a number of workers have reported similar reactions of alkynic alcohols with a range of transition metal centres ⟨80AJC2605, 80CC341, 84CC747, 86CC514, 87JA7578, 87OM1424, 89JA4129, 90JA464, 90JA6232, 91CB1121, 92JOM(431)247, 93JOM(460)C25, 93OM3510, 94JCS(D)1671, 94TL3801, 95OM2102⟩. The generally accepted mechanism which has been adopted for this transformation involves the initial formation of the vinylidene complex (**54**) followed by intramolecular capture of the carbene complex (**55**) (Scheme 18). The Dötz cyclization ⟨87OM1424⟩ of alkynic alcohols with group 6 metal pentacarbonyl·THF complexes has been put to good effect in the preparation of a variety of functionalized carbene complexes (Scheme 19) ⟨94TL8883, 95UP 525-01⟩. In the case of the molybdenum complexes, facile removal of the metal fragment to the enol ether occurs ⟨93JOC6952, 95AG(E)350⟩. In certain cases, cyclization of the intermediate hydroxyvinylidene complexes is unfavourable and dehydration to the corresponding allenylidene complexes (**56**) may be observed (Equation (32)) ⟨95CB49⟩. In 1995, the direct conversion of enynols to unsaturated ruthenium carbene complexes has been disclosed ⟨95OM1095⟩ as has the preparation of oxacycloalkenylidene complexes from lithiated propargylic ethers (Scheme 20) ⟨95CB157⟩.

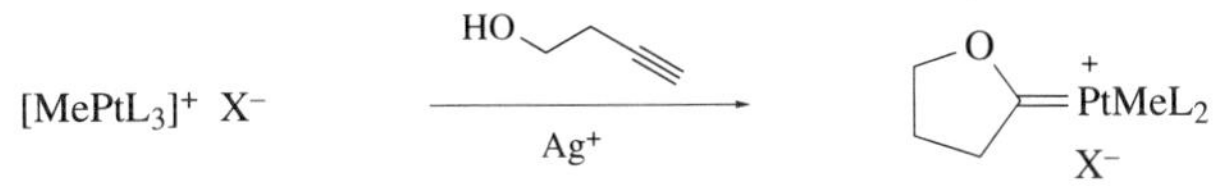

(53)

(31)

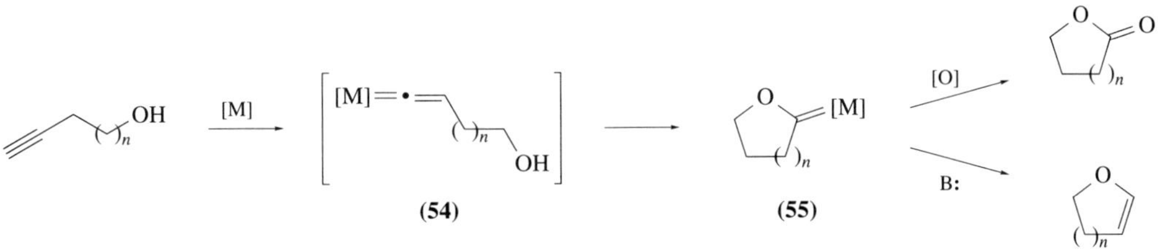

[M] = complexes of Fe, Os, Ru, Cr, W, Mo, Ir, Re, Pt

Scheme 18

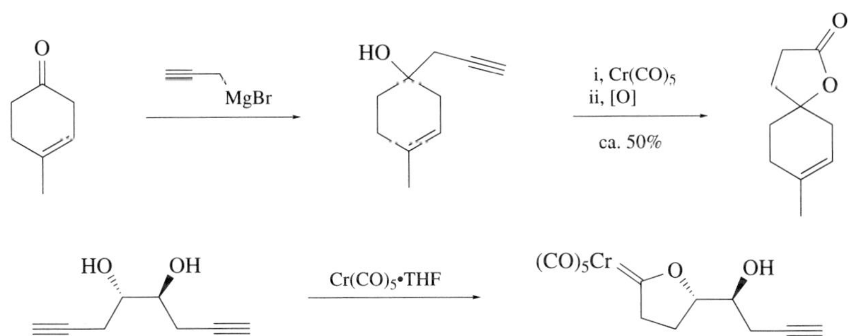

Scheme 19

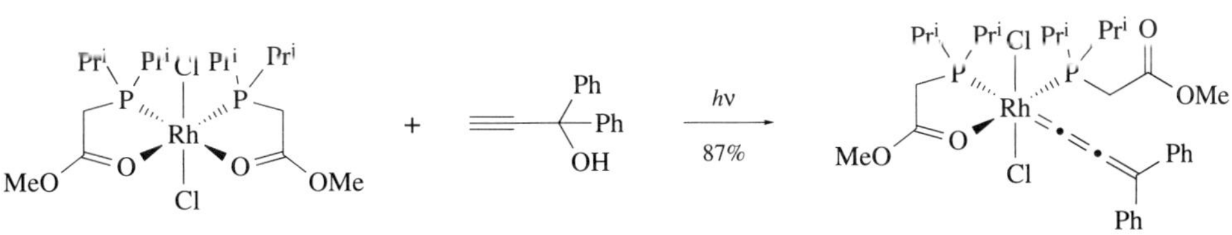

(56)

(32)

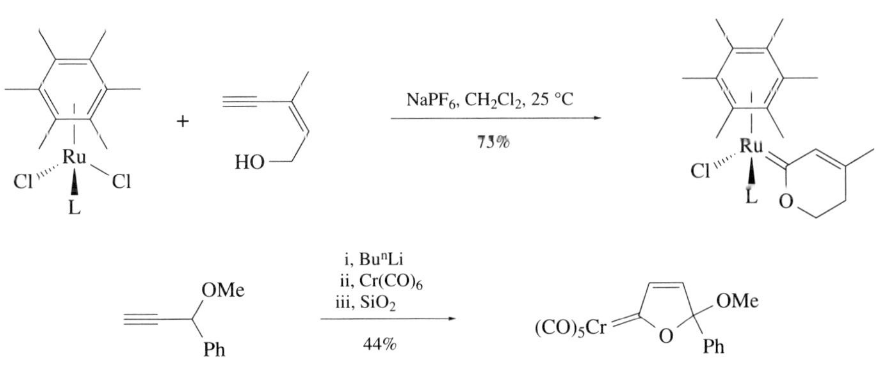

Scheme 20

(iv) From carbene complexes

As pointed out in Section 5.25.3.1.1 the classical Fischer carbene synthesis has a number of limitations and much attention has focused upon the manipulation of readily available complexes in order to generate more complex structures of potential interest to the synthetic organic chemist. In 1974 Casey and Anderson ⟨74JOM(73)C28⟩ reported that treatment of the carbene complexes **(57)** with Bu^nLi in ether afforded nucleophilic anions **(58)** which react with epoxides at the least hindered position to afford oxacyclopentylidene complexes in moderate overall yields (18–50%) (Scheme 21). It was not until 1991 that this basic reaction was optimized by Licandro and co-workers ⟨91OM807⟩ into a general preparative process, the key modification being the addition of $BF_3 \cdot OEt_2$ which greatly facilitates the epoxide ring opening reaction. Again nucleophilic attack of the epoxide is highly regioselective, except in the case of styrene oxide. Homochiral complexes are readily available using this methodology starting from optically pure epoxides. Of interest is the observation that the reaction sequence can be performed on alumina, without the addition of any other base, affording the unsaturated complex **(60)** in 50% yield based on recovered starting material (Scheme 22). In a similar vein, treatment of the manganese carbene complexes **(61)** with two equivalents of Bu^nLi generates the highly nucleophilic alkynide complexes **(62)** which also undergo regiospecific alkylation reactions with epoxides and oxetanes in the presence of a Lewis acid ($BF_3 \cdot OEt_2$) to afford, after an acidic workup, the cyclic carbene complexes **(63)**–**(65)** in high yields (Scheme 23) ⟨90JA3220, 92JA6735⟩.

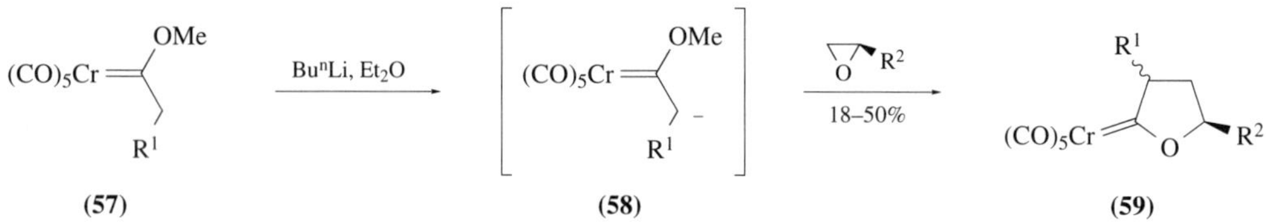

Scheme 21

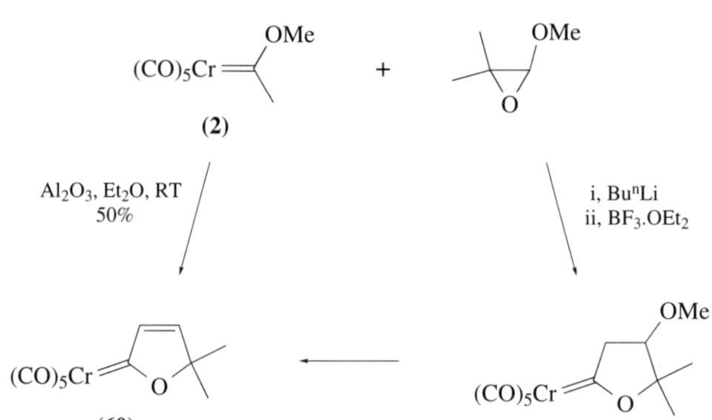

Scheme 22

Alkylation/acylation of preformed carbenes followed by cyclization have been reported by a number of workers to provide rapid access to functionalized carbene complexes (Equations (33)–(40)) ⟨87CB537, 92JA6735, 93CC1068, 93JA3806, 93SL486, 93TL227, 94ICA233, 94OM3510, 95TL3699⟩. The examples cited in Equations (38) and (39) are particularly noteworthy as they indicate the proclivity of α,β-unsaturated carbene complexes to behave as Michael acceptors towards carbon-centred anions and radicals, a much underdeveloped aspect of the chemistry of this class of complex.

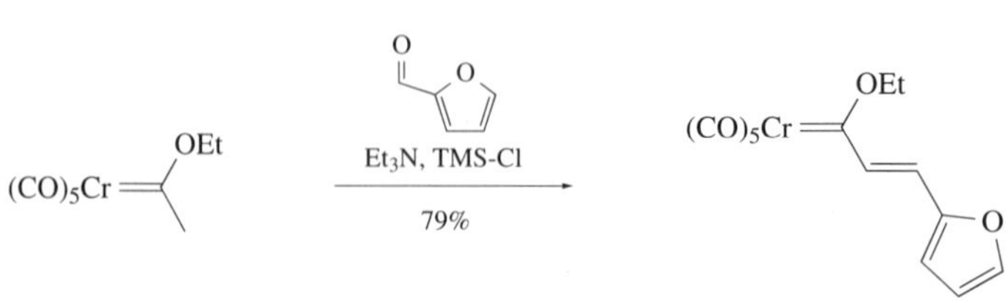

(33)

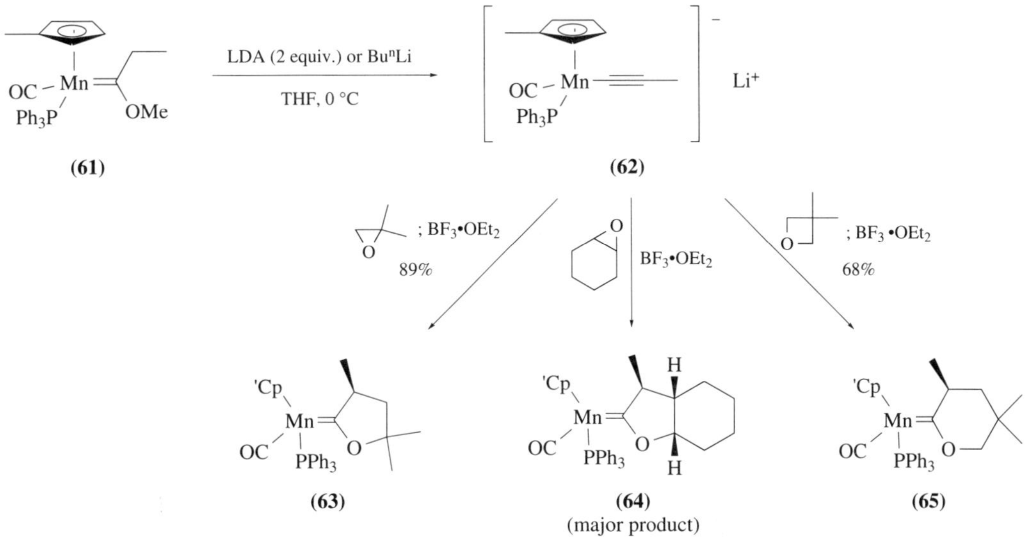

Scheme 23

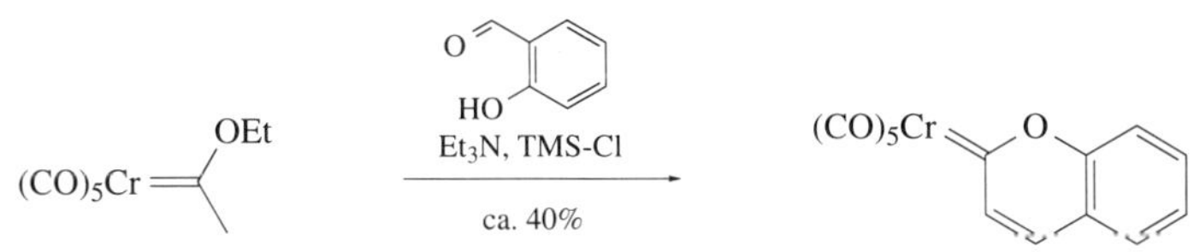

(34)

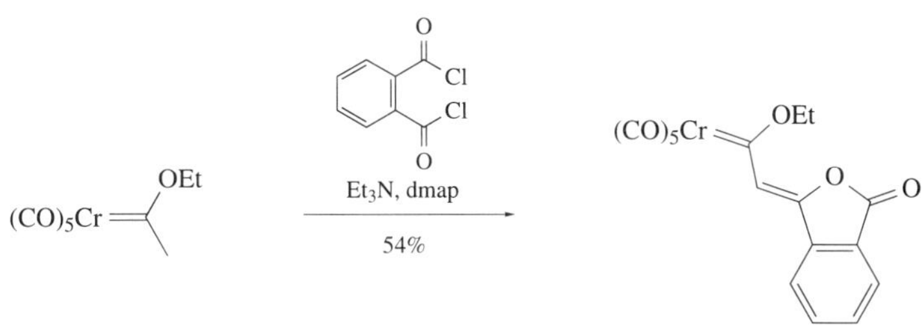

(35)

(36)

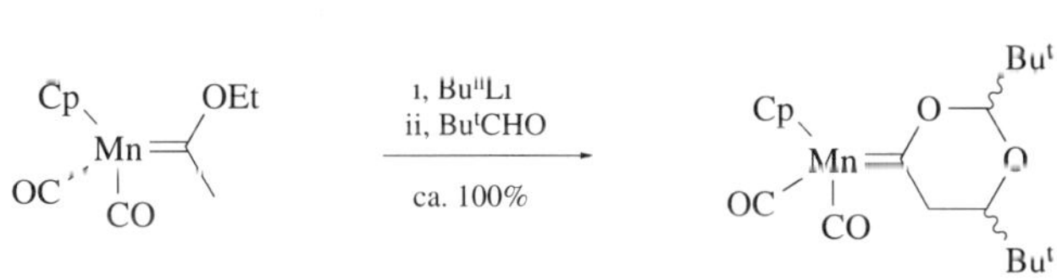

(37)

(single diastereoisomer)

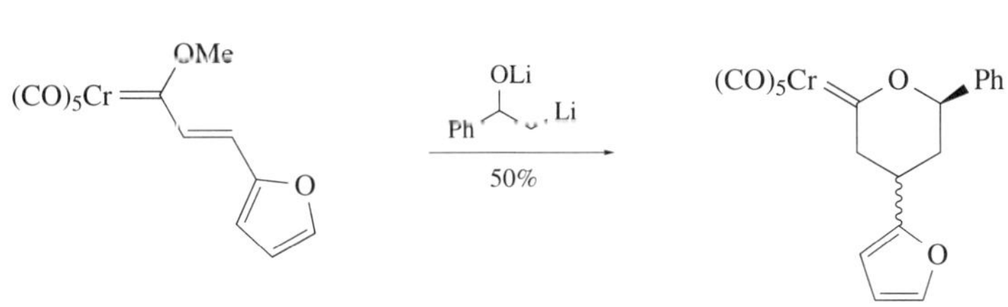

(38)

(1:1 mixture of diastereoisomers)

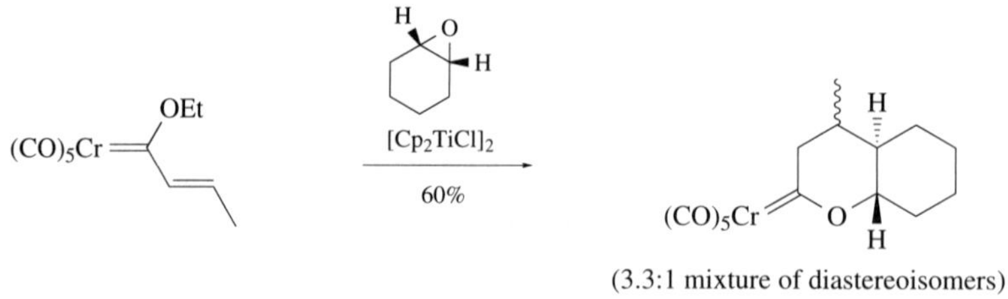

(39)

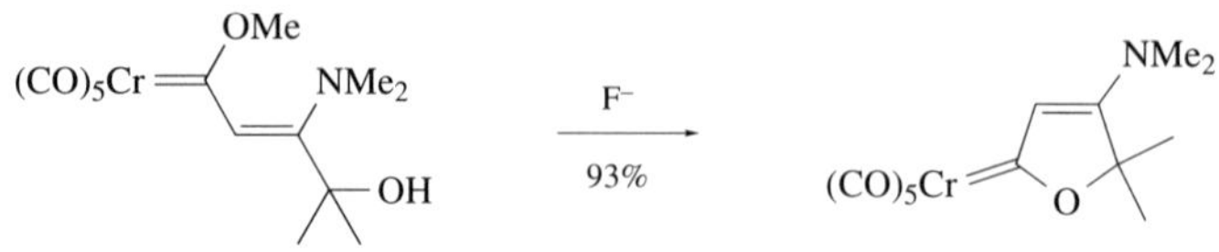

(40)

(v) *From σ-vinyl complexes*

σ-Vinyl complexes undergo protonation to afford carbene complexes as illustrated in Scheme 24 for the preparation of cyclic nickel and iridium complexes ⟨82CC1178, 82JCS(D)793, 89OM2075⟩. In the case of the iridium complex (**66**), purification was hampered by virtue of its hydrolytic instability with respect to the open chain acyl complex (**67**) (Scheme 24).

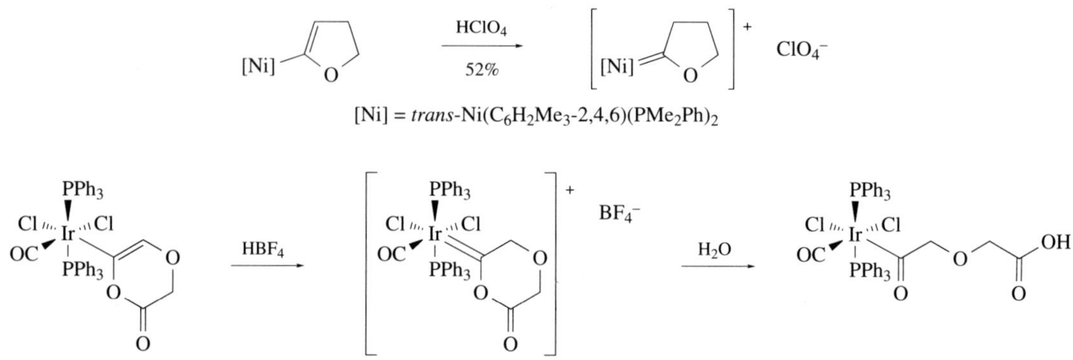

(**66**) (**67**)

Scheme 24

(vi) *C—H activation*

There are three reports which detail the direct formation of cyclic Fischer complexes from the corresponding saturated cyclic ethers as depicted in Equation (41) ⟨92JA7288, 94JA791, 94JA11584⟩.

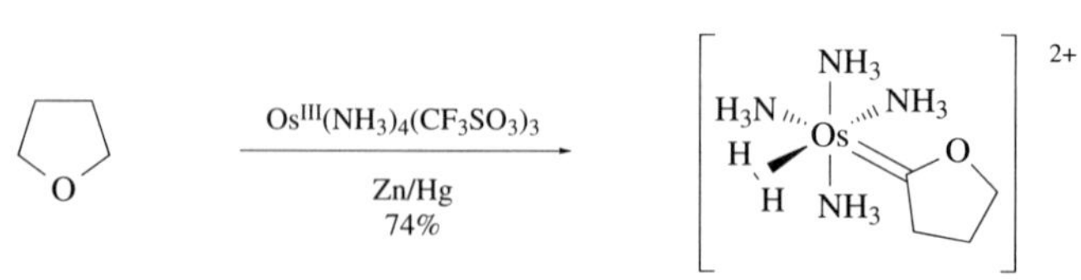

(41)

5.25.3.2 Hydroxycarbene Complexes

In their original paper, Fischer and Massböl ⟨64AG(E)580⟩ suggested that hydroxycarbene complexes were somewhat unstable and could not be isolated in a pure state. In certain cases however it was discovered that protonation of the initial acylate complexes with HBr afforded the hydroxycarbene complexes (**68**), which could be purified by recrystallization and column chromatography (Equation (42)) ⟨73JCS(D)975, 73JOM(56)C37⟩. Similarly, protonation of the acylate complexes

(**69**; X = I or Br) with phosphoric acid afforded the crystalline hydroxycarbene complexes (**70**) in which the hydroxy group was hydrogen-bonded to the halogen. Notably, complex (**70**; X = Cl) was unstable and could not be purified (Equation (43)) ⟨73JCS(D)975⟩.

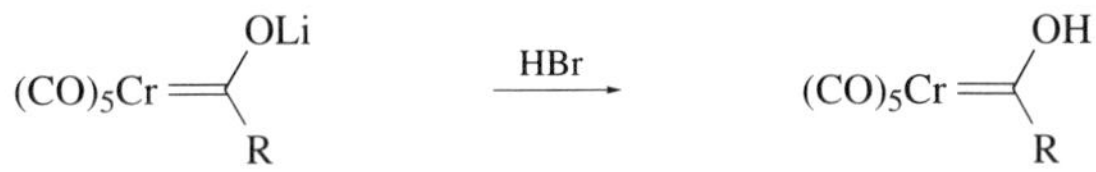

(10) **(68)**

R = Ph, Me

(42)

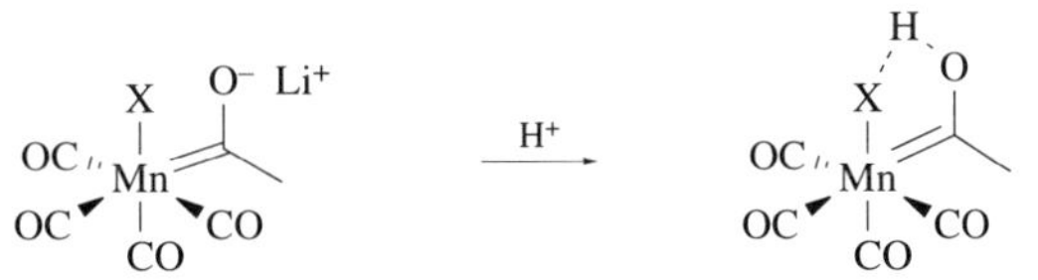

(69) X = I, Br, Cl **(70)**

(43)

It has been shown that treatment of the alkyl complex (**71**) with HCl at low temperatures leads to the reversible formation of the hydroxycarbene (**72**) which has been characterized spectroscopically. If a solution of the carbene (**72**) is allowed to warm up from −80 °C, then the equilibrium is displaced back to the starting material (Equation (44)) ⟨91JA6673⟩.

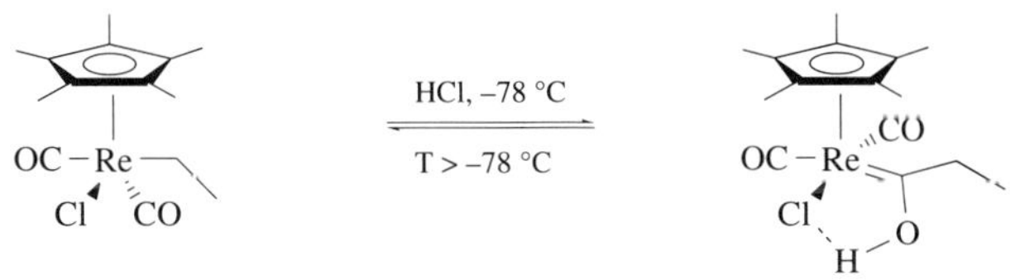

(71) **(72)**

(44)

Hydroxy complexes such as (**68**) may be regarded as being the synthetic equivalents of carboxylic acids. As such, they are observed to undergo dehydration reactions ⟨74CB3554, 76CB1128⟩ on treatment with dicyclohexylcarbodiimide (dcc) to afford highly electrophilic 'anhydride' complexes (**73**). These complexes are readily transposed into amino complexes upon reaction with amines (Scheme 25) ⟨76CB1120⟩.

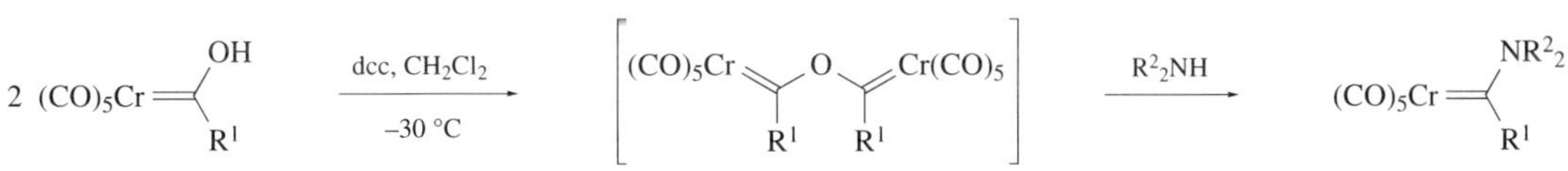

(68) **(73)**

R^1 = Ph

Scheme 25

5.25.3.3 Acyloxycarbene Complexes

Tetraalkylammonium pentacarbonylacylmetalates of chromium, tungsten and molybdenum react rapidly with acyl halides between −78 °C and −20 °C to afford acyclate complexes (**74**) (Scheme 26) ⟨71CC570, 77CB2947⟩. These complexes are usually thermally labile, although there are specific cases in which such a complex has been isolated as a stable crystalline solid ⟨71JCS(A)3368⟩. Chromium and molybdenum acyloxycarbenes rapidly decompose at ambient temperature, although the analogous tungsten complexes appear to be somewhat more stable, decomposing at temperatures in excess of 50 °C ⟨91OM3951⟩. Despite their intrinsic instability acyloxycarbene complexes are valuable synthetic intermediates, as they mirror mixed anhydrides in terms of their chemistry and have been used to prepare amino-, thio- ⟨71CC570⟩ and alkoxycarbene ⟨92JA10665⟩ complexes as illustrated in Scheme 26.

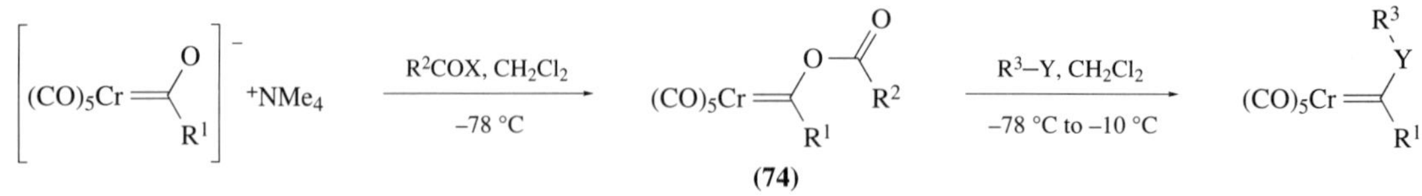

Scheme 26

Treatment of the iron acyl complex (**14**) with triflic anhydride generates the highly unstable carbene complex (**15**) which rapidly eliminates triflic acid to afford the vinylidene complex (**16**) ⟨82OM628⟩. Complex (**16**) is itself a useful intermediate in that it undergoes reactions with thiols, amines and HCl to afford the thio- (**75**), amino- (**76**) and chlorocarbene complexes (**77**) (Scheme 27) ⟨82OM635⟩.

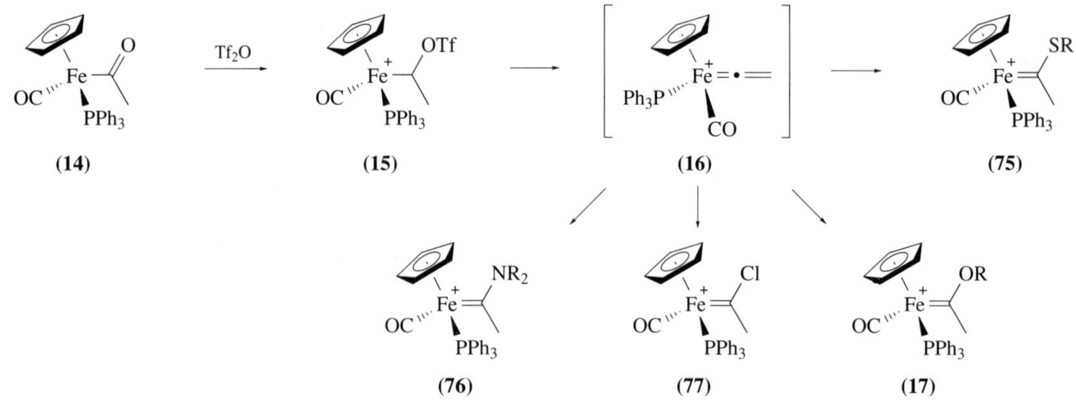

Scheme 27

5.25.3.4 Thioether Complexes

Since the first preparation of a thiocarbene complex was documented in 1972 ⟨72CB150⟩ a variety of methods have been developed for their preparation. Their chemistry has not been as extensively investigated as in the case of amino- or alkoxycarbene complexes. Generally these group 6 complexes are more stable towards thermal or oxidative decomposition than the alkoxycarbenes (stability follows the pattern Cr ~ Mo < W ⟨74JOM(70)273⟩), but not as stable as the analogous aminocarbenes ⟨90CB2053⟩. The barrier to rotation ⟨72CB150⟩ about the carbene C—S bond in (**78**) is in the region of 15–18 kcal mol^{-1}, which is appreciably more than in the case of (**2**) (12.4 kcal mol^{-1}). This is manifested in bond length data derived from the x-ray structure ⟨72JCS(D)653⟩ of (**78**) in which the C—S bond length has been determined to be 169 pm (cf. C—S ~ 182 pm; C═S ~ 155 pm) (Figure 4).

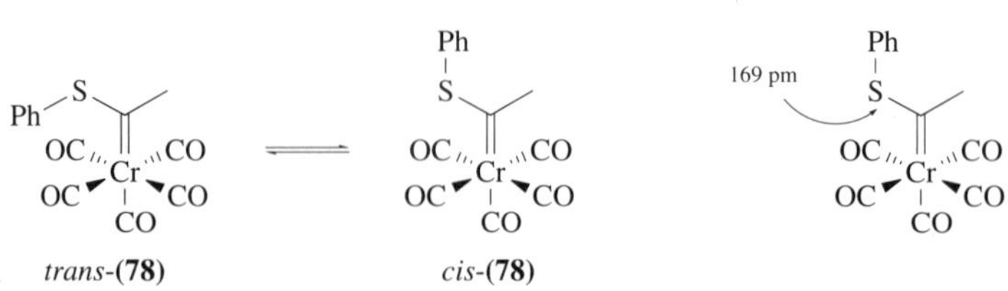

Figure 4

5.25.3.4.1 Preparation from carbene complexes

The original method of preparation developed by Fischer ⟨67JA7141, 72CB150⟩ utilized a 'trans-acylation' between an alkoxycarbene complex which was dissolved in an excess of thiol at room temperature for 4–48 h, the exchange reaction thereby generating the more stable thiocarbene. Removal of excess thiol and purification by chromatography on silica afforded the highly coloured complexes as viscous oils or low melting solids in 25–68% yield. Not unsurprisingly, aryl thiols gave the lowest yields in these exchange reactions.

Subsequently, a number of modifications have been developed for the preparation of these complexes from Fischer complexes ⟨74JOM(70)273, 89JOC4481, 90CB2053, 93CB1849⟩. The most general of which include exchange reactions of acyloxycarbene complexes ⟨89JOC4481⟩ (see Section 5.25.3.3) or the modification developed by Aumann and co-workers where sodium carbonate in methanol has been found to be an ideal base for the promotion of the exchange reaction between alkoxycarbene complexes and thiols (Equation (45)) ⟨90CB2053, 93CB1849⟩.

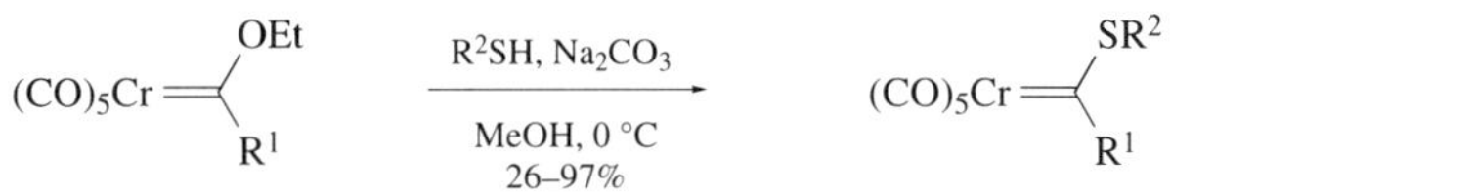

(45)

Treatment of alkoxycarbene complexes, for example (**79**), with lithiodithianes at −60 °C initiates an unexpected rearrangement reaction generating unusual fragmentation products including (**80**) ⟨81AG(E)1055, 85OM275⟩ as depicted in Equation (46).

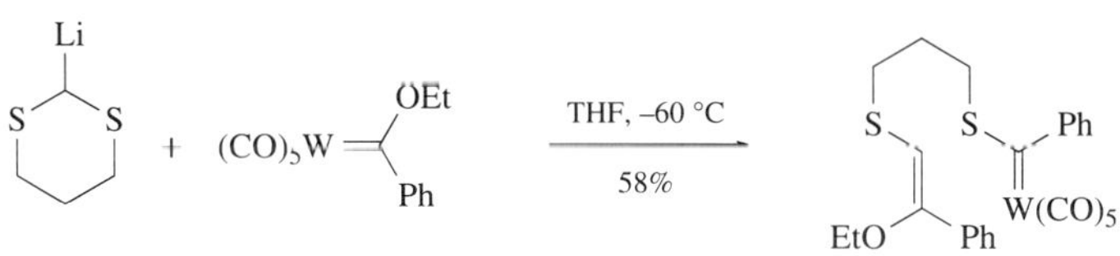

(**79**) (**80**)

(46)

5.25.3.4.2 Miscellaneous methods of preparation

Helquist and co-workers have developed a general method for the preparation of iron thiocarbene complexes starting from iron acyl complexes ⟨87OM988⟩ (cf. Sections 5.25.3.1.3 and 5.25.3.3).

Treatment of the iron complex (**81**) with the anion (**82**) followed by alkylation with either TFA or CF_3SO_3Me constitutes a rapid entry into the thiocarbene complexes (**83**) ⟨94JOM(479)C12⟩. Casey and co-workers ⟨77IC3059⟩ have reported the preparation of the cyclic thiocarbene complex (**84**) from (**2**) in modest overall yield (Scheme 28).

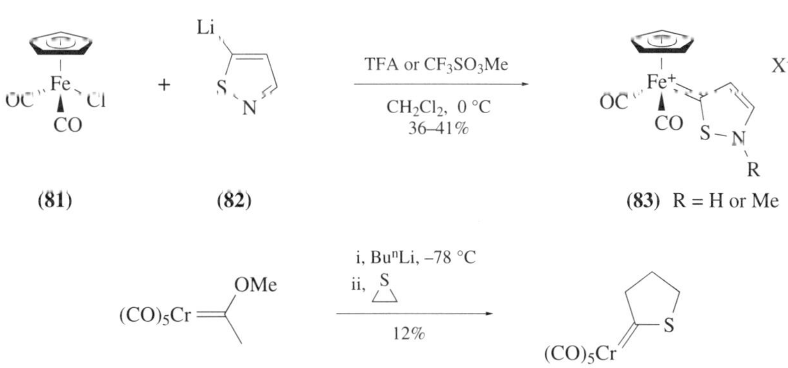

Scheme 28

5.25.3.5 Selenocarbene Complexes

The preparation of selenocarbene complexes has received scant attention ⟨67AG(E)961, 73CB3910⟩. Treatment of the tungsten or chromium Fischer carbene complexes with methyl selenol afforded the corresponding selenocomplexes (**85**) in low yields (6–39%). This reaction is somewhat substrate dependent, as attempted reaction of the same complexes with selenophenol afforded the σ-complex (**86**), the product of 'carbene' insertion into the Se—H bond (Scheme 29) ⟨67AG(E)961⟩.

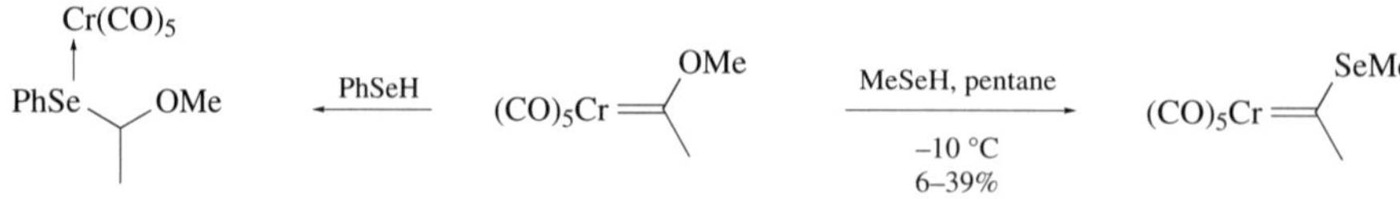

Scheme 29

5.25.3.6 Phosphinocarbene Complexes

Attempts to displace alcohols from alkoxycarbene complexes lead either to the formation of phosphine complexes with liberation of the organic ligand as an enol ether ⟨68JOM(14)P9⟩, or to the generation of phosphonium salts ⟨73JOM(51)C11⟩. The carbyne complex (**87**) when treated with CF_3I undergoes an Arbusov-type reaction followed by migration of the phosphonate group to the alkylidyne carbon generating the chelated carbene complex (**88**) in 35% yield (Scheme 30) ⟨80JA7811⟩. The related phosphinocarbene complex (**89**) has been prepared from the carbyne complex (**90**) (Scheme 30) ⟨95CB289⟩.

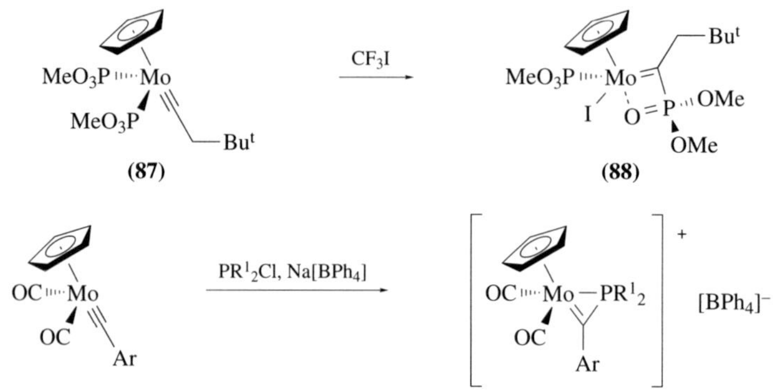

Scheme 30

5.25.3.7 Aminocarbene Complexes

Aminocarbene complexes of the group 6 metals have been the subject of much interest and a review of their properties and chemical transformations has appeared ⟨91CSR503⟩. Aminocarbene complexes of chromium, molybdenum and tungsten are often bright yellow crystalline solids which are appreciably more stable than the corresponding alkoxycarbene complexes having in certain cases been left in the open air for protracted periods of time without showing any significant deterioration in purity ⟨93JA4602⟩. Aminocarbene complexes are similar in many respects to carboxylic acid amides, a consequence of the substantial delocalization of charge from nitrogen into the carbene system. This results in a high activation barrier for rotation about the carbene C—N bond ⟨85OM2102, 91CSR503⟩. Indeed this barrier is sufficiently large to permit isolation of isomeric *syn* and *anti* complexes by chromatography or chemical manipulation ⟨87JOM(333)245⟩. Thermal equilibration of the *syn* and *anti* complexes is usually not possible as thermolysis at temperatures in excess of 120°C are required. Base-mediated equilibration of *syn* and *anti* complexes has been observed ⟨68JOM(15)147⟩.

5.25.3.7.1 Preparation from alkoxycarbene, acyloxycarbene and carbyne complexes

The first example of an aminocarbene complex was reported by Klabunde and Fischer ⟨67JA7141⟩ who found that alkoxycarbene complexes undergo rapid exchange with a variety of amines (e.g., Equation (47)). An in-depth mechanistic study of these exchange reactions has been reported ⟨93JA1341⟩. This reaction works well for ammonia and relatively unhindered primary and secondary amines ⟨67CC1024, 68JOM(13)387, 71JOM(28)359, 74CB3326, 88JA2122, 91CB1635, 92CB2137, 95JOM(486)279, 95JOM(494)43⟩ with the lithium or potassium salt of pyrrole ⟨95OM2522⟩, with formamidines ⟨93JOM(459)169⟩, hydrazines ⟨94CL777, 95OM2447⟩ and imines ⟨71JOM(31)C68⟩. In the case of more sterically congested amines reactions may become sluggish or produce amino complexes which have suffered dealkylation ⟨67CC1024⟩. Studies have shown that the addition of a catalytic quantity of an alkoxide base greatly accelerates such reactions (Equation (48)) ⟨95JOM(486)279⟩. Acyloxycarbene complexes have been used extensively for the preparation of functionalized aminocarbene complexes **(91)** (Equation (49)) ⟨90JA6255, 94JOC6882⟩. Aminocarbene complexes derived from amino acids have been used in peptide synthesis ⟨73CB1277, 76CB1868⟩. Reaction of Fischer carbene complexes with pyridine leads to disruption of the complex affording enol ethers and (pyridinato)pentacarbonyl complexes ⟨71JOM(28)359, 74CB3326, 92JA2985, 92JA8088⟩ whereas reaction with tertiary amines affords isolable nitrogen ylide pentacarbonyl metal complexes ⟨74CB183⟩. Reaction of carbyne complexes with amines provides a facile method for the preparation of aminocarbene complexes ⟨85OM2102⟩. α,β-Unsaturated alkoxy chromium carbene complexes undergo tandem insertion–rearrangement reactions to afford the thermolabile dihydropyrrole complexes **(92)** (Equation (50)) ⟨86CB3801⟩.

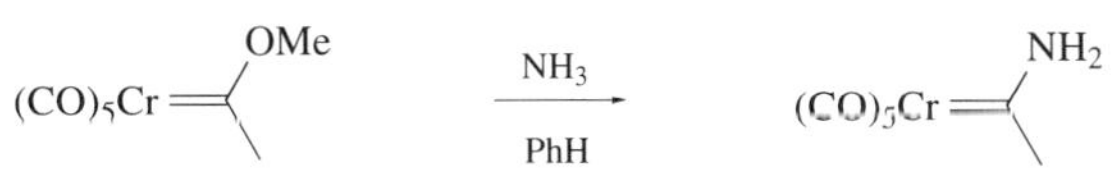

(2)

(47)

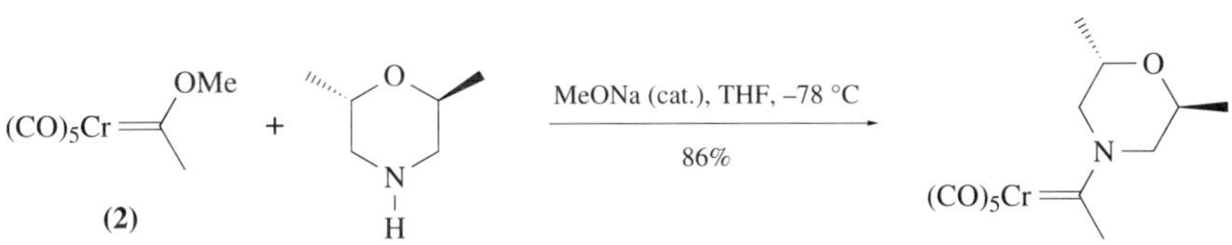

(2)

(48)

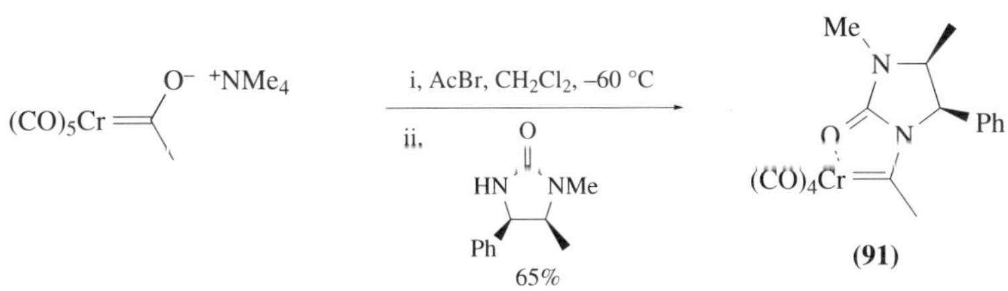

(91)

(49)

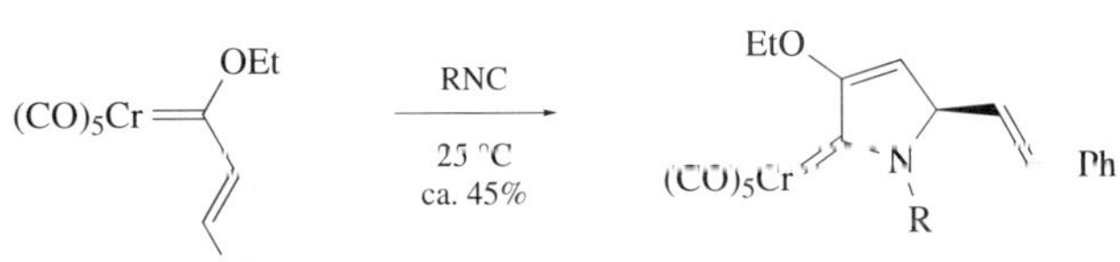

(92)

(50)

5.25.3.7.2 Preparation from carboxylic acid amides and related substrates

The preparation of chromium ⟨88OM702⟩ and iron aminocarbene complexes ⟨95OM570⟩ can be accomplished directly from tertiary carboxylic acid amides upon reaction with $Na_2[Cr(CO)_5]$ or $Na_2[Fe(CO)_4]$ in the presence of TMS-Cl. In the case of the preparation of iron complexes in this manner, care must be taken to avoid solvents such as dioxane which tend to generate unreactive adducts with the nucleophilic iron complex (Scheme 31). A limitation yet to be addressed in the case of the iron system appears to be the incompatability of carboxamides possessing enolizable α-hydrogens with the reaction conditions. In the case of the chromium system a wide variety of carbene complexes are available using this methodology including substrates possessing enolizable hydrogens. In a related series of investigations, it has been shown that anionic silylstannylmanganese ⟨91JA2228⟩, -chromium ⟨93CB657⟩ and -rhodium ⟨94JOM(472)C15⟩ complexes undergo nucleophilic addition–elimination reactions with the chloropyridinium salt (**93**) to generate cyclic aminocarbene complexes (Scheme 32).

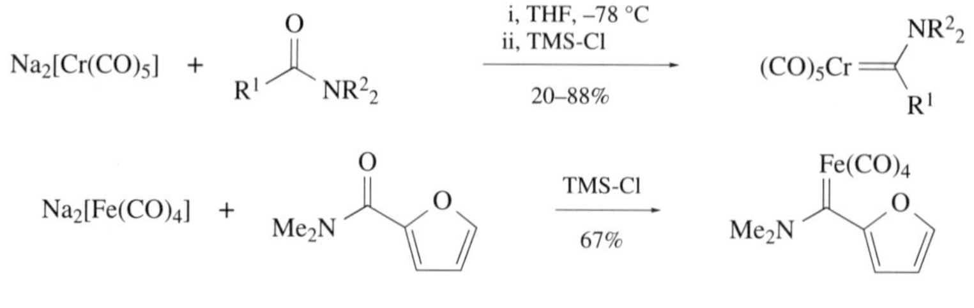

Scheme 31

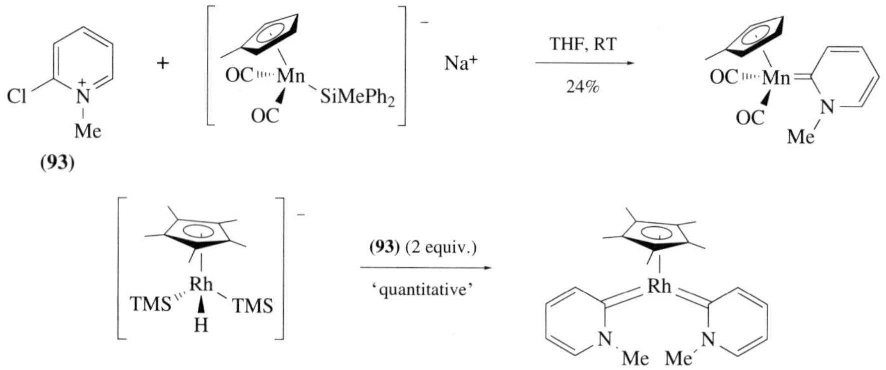

Scheme 32

5.25.3.7.3 *Alkyne–vinylidene rearrangements*

Treatment of propargylic alcohols with group 6 metal pentacarbonyl complexes followed by dehydration affords allenylidene complexes which undergo reaction with amines to afford aminocarbene complexes (Scheme 33) ⟨93JOM(454)133⟩.

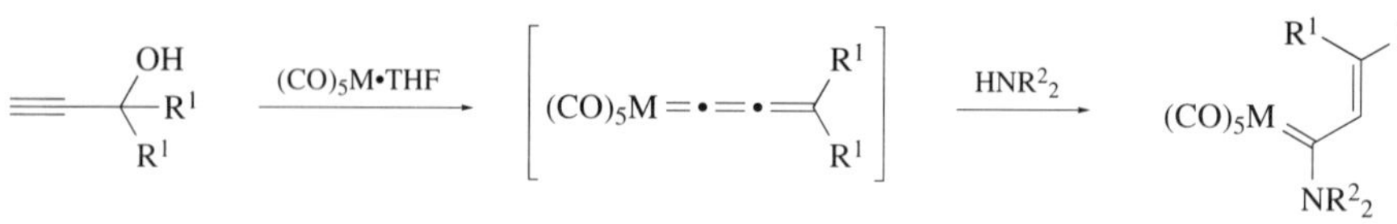

Scheme 33

5.25.3.7.4 *Cycloaddition strategies*

Iron alkoxycarbene complexes undergo formal [1 + 1 + 2] cycloaddition reactions with isocyanides to afford iminoazetidinylidene complexes (**94**) in excellent yields ⟨87CB1297⟩. The related complexes (**95**) are also available by way of a formal [2 + 2] cycloaddition of cationic vinylidene complexes (**96**) with imines ⟨88OM2553⟩. Molybdenum and tungsten carbyne complexes undergo formal [3 + 2] cycloaddition reactions with azidoformates to afford triazametallole complexes (**97**) in near quantitative yields ⟨92CB1571⟩.

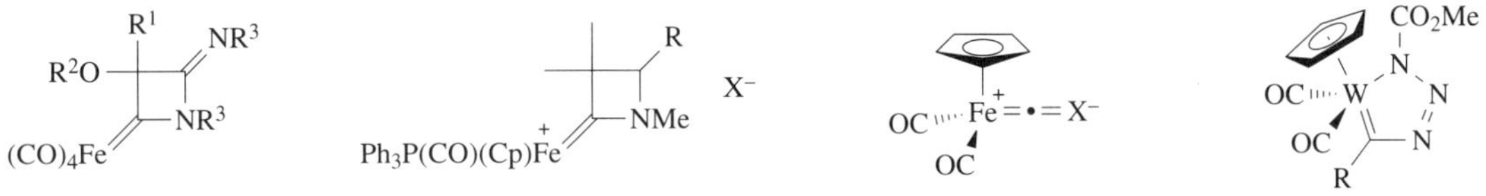

(94) **(95)** **(96)** **(97)**

5.25.3.8 Preparation from Alkyl σ-Complexes

Winter and co-workers have developed an elegant route to functionalized cyclic aminocarbene complexes starting from the readily available anionic complex (**98**). Treatment of (**98**) with an iodoepoxide generates the intermediate (**99**) which cyclizes via a 5-*exo-tet* pathway to the carbene complex (**100**) in 38% overall yield (Scheme 34) ⟨95JOM(494)C12⟩.

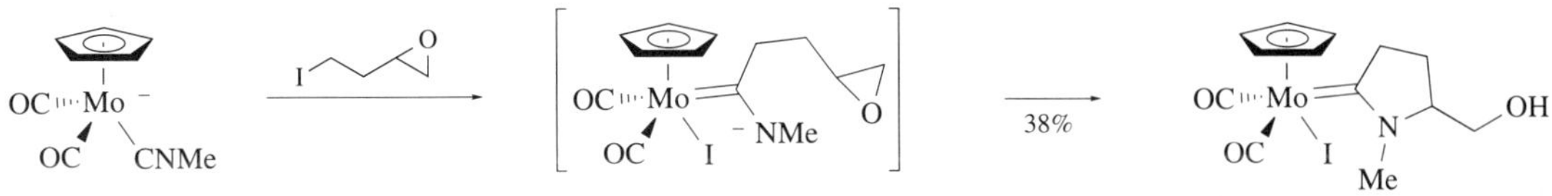

(98) **(99)** **(100)**

Scheme 34

5.25.4 FUNCTIONALIZATION OF CARBENE COMPLEXES WITH RETENTION OF THE CARBENE MOIETY

Although there are now several fairly general and reliable methods for the preparation of Fischer carbene complexes from noncarbene precursors (see above) their exploitation as synthetic reagents depends upon the ability to manipulate the chemistry of these intermediates in a practical sense. Since 1985 a wealth of methodological chemistry has been developed in this area, much of which has been discussed in Wulff's excellent review ⟨91COS(5)1065⟩.

5.25.4.1 Alkylation Reactions

5.25.4.1.1 Alkylation at the α-position

Kreiter was the first to report that protons attached to the α-position of transition metal carbene complexes undergo rapid base-catalysed hydrogen–deuterium exchange in dilute NaOMe/CH_3OD solution ⟨68AG(E)390⟩. Casey and Anderson subsequently showed that the carbene complex (**2**) has approximately the same acidity as *p*-cyanophenol in THF ⟨74JA1230⟩. Stopped-flow studies have shown that complex (**2**) has a pK_a of about 12.3 in aqueous acetonitrile ⟨89OM2282, 93JA12526⟩.

Treatment of Fischer complexes with bases such as Bu^nLi in THF rapidly generates the anions (**58**) which undergo alkylation reactions with a variety of electrophiles (Scheme 35) ⟨87JOC3263, 89OM1275, 91CB2431, 92CB1141, 94CC2457, 94JOM(476)C25, 94OM4676⟩. One consequence of the thermodynamic stability of the anions (**58**) derived from alkoxycarbene complexes is that they are somewhat unreactive and only undergo alkylation reactions with reactive electrophiles such as methyl fluorosulfonate ⟨72JA8947⟩, epoxides, aldehydes and α-chloromethyl ethers ⟨75JOM(102)175⟩, α-bromoesters and allylic halides (Scheme 36) ⟨76JOM(118)309⟩. A severe limitation of this procedure is the problem associated with polyalkylation ⟨76JOM(118)309⟩. Amin and Sarbir ⟨95OM547⟩ have presented a partial solution of this problem by carrying out the alkylation reactions of alkoxycarbene complexes under phase-transfer conditions. Aminocarbene complexes are unreactive under similar conditions. Wulff

⟨87JOC3263, 94ICA215⟩ has demonstrated that aminocarbene complexes, having a $pK_a \sim 20.4$ (in DMSO), generate conjugate bases with much higher nucleophilicities. Importantly, mono functionalized products are usually obtained in alkylation reactions of these intermediates ⟨89TL4061⟩. The allyl anion (**101**) undergoes γ-selective functionalization with less reactive carbon electrophiles such as CH_3I (Scheme 37) ⟨94TL6377⟩.

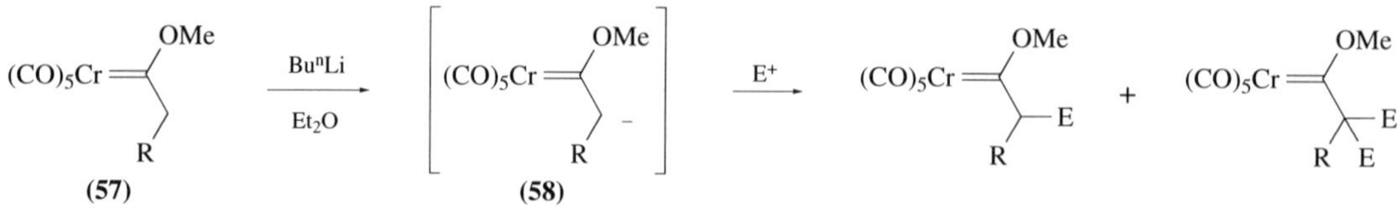

Scheme 35

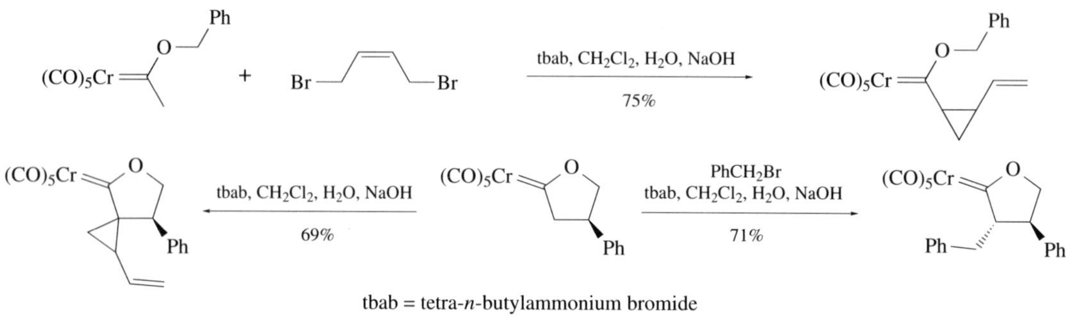

Scheme 36

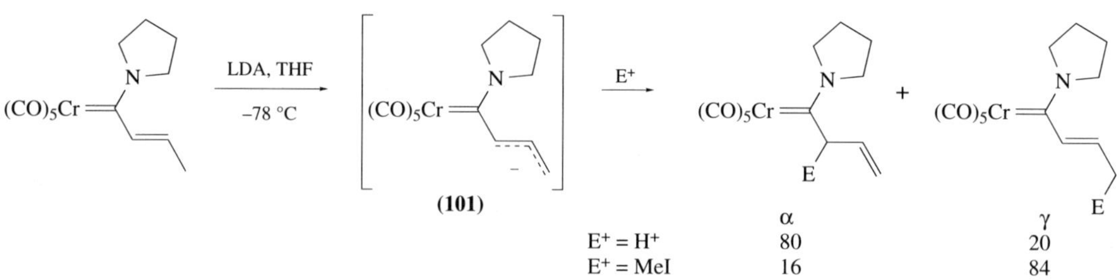

Scheme 37

5.25.4.1.2 *Remote alkylation*

Two papers have demonstrated that carbene complexes may be deprotonated at centres distal to the carbene centre and that the carbanion so formed may undergo synthetically useful alkylation reactions without disruption of the carbene moiety. Clearly the generality of these processes should be pursued in order to evaluate their synthetic potential (Scheme 38) ⟨92JA2985, 93OM597⟩.

5.25.4.2 'Aldol' Reactions

Due to the unreactivity of the 'enolates' derived from alkoxycarbene complexes, the addition of Lewis acids (e.g., $BF_3 \cdot OEt_2$ or $TiCl_4$) was required in order to effect 'aldol' reactions with ketones and some aldehydes. In certain cases, useful levels of simple asymmetric induction were observed ⟨85JA503⟩. Aminocarbene complexes bearing chiral auxiliaries have been shown to undergo aldol reactions with high levels of asymmetric induction (Scheme 39) ⟨94ICA(220)215, 94JOC6882⟩.

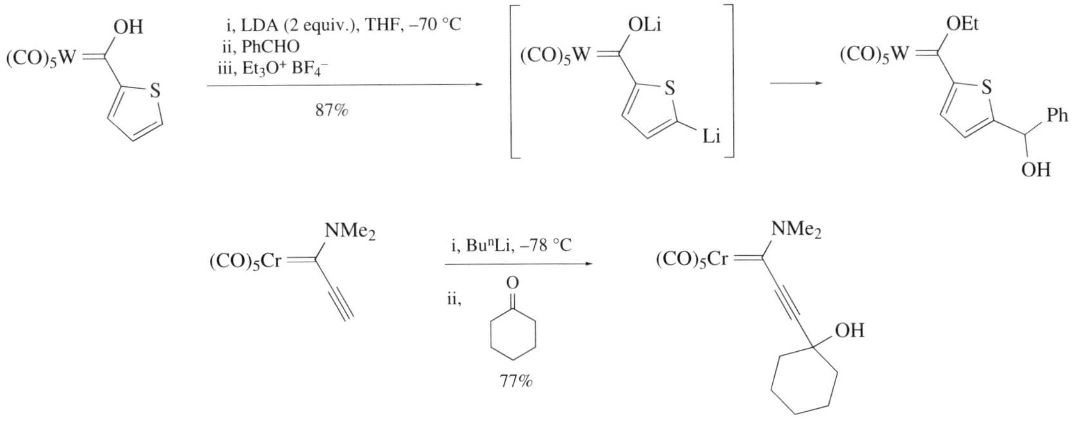

Scheme 38

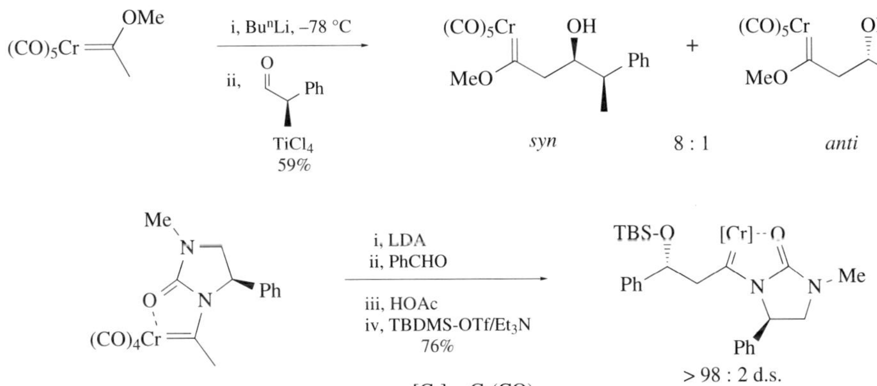

[Cr] = $Cr(CO)_4$
TBDMS = *t*-butyldimethylsilyl; OTf = triflate

Scheme 39

In certain cases the β-hydroxycarbene complexes formed during the course of these reactions undergo dehydration to the corresponding α,β-unsaturated complexes (cf. Equation (33)) ⟨93JA3806, 94JA9363⟩. Indeed, the alkylation reactions with carbonyl acceptors when conducted with an amine base in the presence of TMS-Cl produce high yields of the α,β-unsaturated carbene complexes and appear to have general synthetic utility (Scheme 40) ⟨87CB537, 91CB957⟩. Amides undergo a similar alkenation reaction using $Et_3N/POCl_3$ ⟨89CB365, 90CB611, 90CB2047, 92CB2765⟩, although one should be aware of the possibility of competing rearrangement processes during the course of these alkylation reactions ⟨89CB365⟩. Acylation reactions of carbene complexes have been discussed (cf. Equation (35)) ⟨94OM3510⟩.

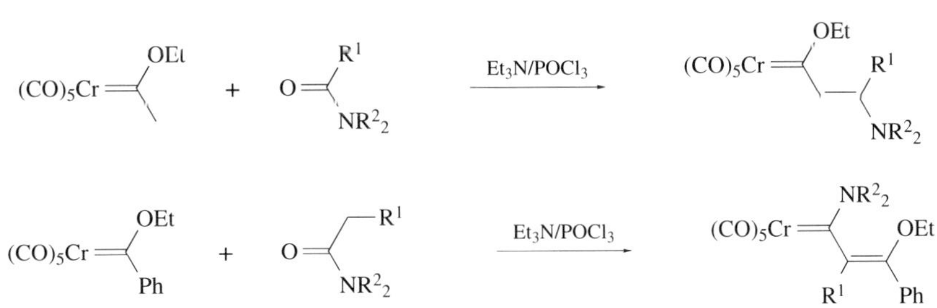

Scheme 40

5.25.4.3 Conjugate Reactions Involving Carbene Complexes

5.25.4.3.1 *Carbene as a Michael donor*

Fischer carbene complexes, especially the aminocarbene complexes, undergo conjugate addition reactions with a variety of Michael acceptors in the presence of either stoichiometric or catalytic quantities of base to afford Michael adducts in good isolated yields ⟨77IC3059⟩. Mono alkylation can be achieved by judicious choice of reaction conditions or by changing the electronic nature of the carbene complex ⟨77IC3059, 91OM2121, 93JA4602, 95JOM(486)279, 95TL3699⟩. Good levels of induction have been observed in those cases which incorporate a chiral auxiliary in the carbene partner (Scheme 41) ⟨93JA4602, 95JOM(486)279⟩.

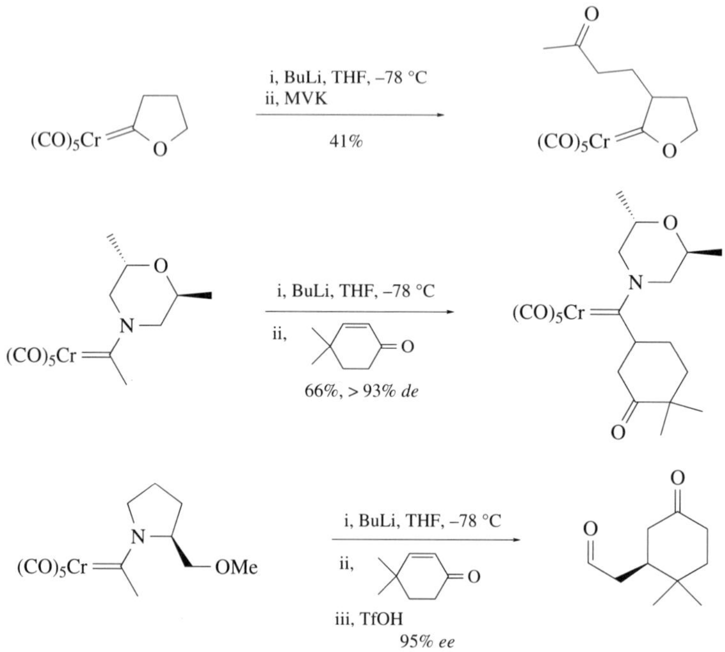

MVK = methyl vinyl ketone

Scheme 41

5.25.4.3.2 *Carbene as a Michael acceptor*

(i) Carbon-centred donors

α,β-Unsaturated carbene complexes behave as Michael acceptors towards a variety of nucleophilic reagents. The use of carbon-centred nucleophiles in such reactions has so far been restricted to a few examples (Scheme 42) ⟨74JOM(77)345, 91OM737, 91OM2121, 93CC1068, 95TL3937⟩. Enolate anions undergo Michael addition reactions with acyclic α,β-unsaturated alkoxycarbene complexes in a *syn* selective manner ⟨77IC391, 92JA9715, 93JA9015⟩. Chelated aminocarbene complexes also undergo conjugate addition reactions with enolates in a *syn* selective manner but with a higher intrinsic bias (Scheme 42) ⟨94JOC5122⟩. Carbon-centred radicals also undergo conjugate addition to unsaturated carbene complexes as discussed earlier (see Section 5.25.3.1.5(iv)).

5.25.4.4 Heterocentred Donors

β-Functionalized α,β-unsaturated carbene complexes are readily available from the Michael addition between alkynic carbene complexes and a variety of O-, S-, N- and P-centred reagents (Scheme 43) ⟨92CB2051, 94CB725, 94CB2475, 94OM3502⟩.

5.25.4.5 Cycloaddition Reactions

The cycloaddition chemistry of unsaturated carbene complexes has been discussed in some detail in Wulff's review ⟨91COS(5)1065⟩, see also ⟨94ICA233⟩.

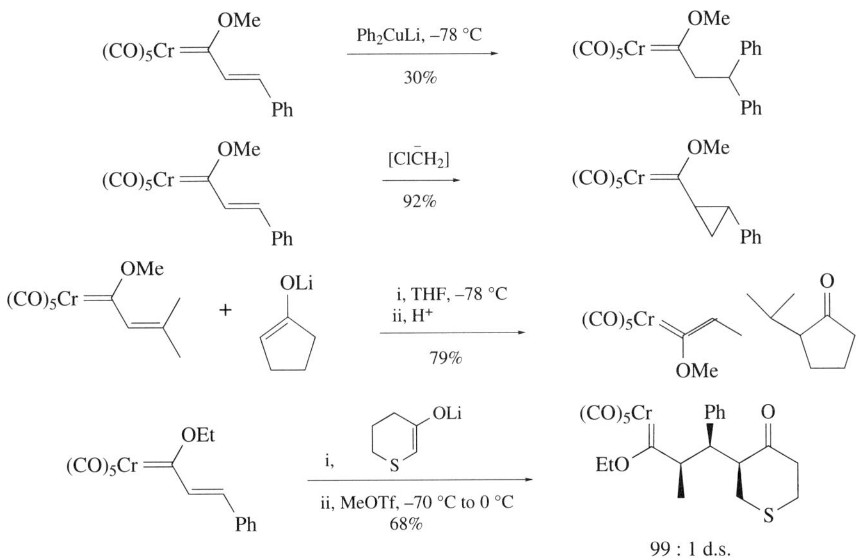

Scheme 42

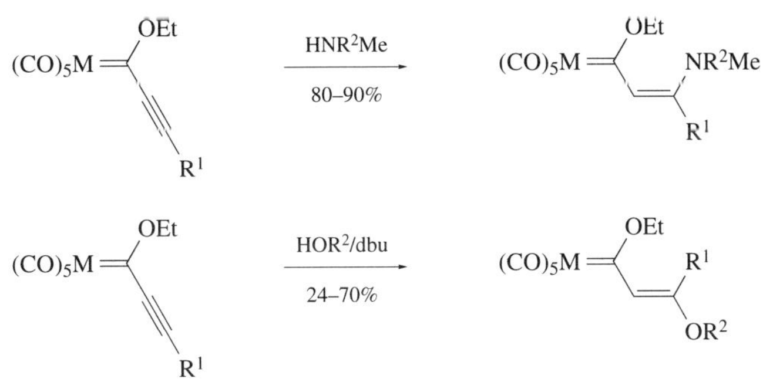

dbu = 1,5-diazabicyclo[5.4.0]undec-5-ene

Scheme 43

5.26 Functions With at Least One Oxygen, Y═C═O

PEDRO MOLINA and ALBERTO TARRAGA

Universidad de Murcia, Spain

5.26.1 INTRODUCTION

Although the formation and spectroscopic behaviour of RP=C=O, R_2Si=C=O and RB=C=O has been reported, of all the functions in this section, the most relevant one is the RN=C=O which, in contrast to the above, has been widely studied and reviewed ⟨57CRV47, 72CRV457, B-77MI 526-01, 83HOU(E4)738, 84JOM207, 87S525, 89CRV1927⟩. No examples of compounds such as R_2S=C=O, R_2Se=C=O and related sulfur and selenium derivatives have been reported.

Isocyanates are esters of isocyanic acid and the first member of this class of compounds was prepared by Wurtz in 1848. Organic isocyanates are a fascinating class of compounds which have found wide-ranging applications in organic synthesis. In this context, stable isocyanates, having the isocyanate group attached to elements other than carbon, have become available since the mid 1960s, and silicon, phosphorus and sulfur isocyanates are important chemical intermediates. In addition, complexes with coordinated isocyanate ligands are thought to be intermediates in some interesting transition-metal catalyzed reactions.

Although for many, the name "isocyanate" is unfortunately associated with the tragic accident in Union Carbides' Bhopal plant, organic isocyanates are used industrially, in particular in polymer chemistry, where they have found their widest application in the manufacture of polyurethanes by addition of dihydroxy compounds. They are also employed as useful reagents or sometimes involved as intermediates in the synthesis of urea and carbamate derivatives, which are used in the production of herbicides, crop protection agents, as well as in the manufacture of antidiabetic drugs.

5.26.2 NITROGEN FUNCTIONS: ISOCYANATES, RN=C=O

5.26.2.1 From Alkyl Halides

5.26.2.1.1 Reaction with cyanogen chloride

Alkylation of cyanogen chloride with alkyl chlorides in the presence of $FeCl_3$, as Lewis acid, gives some extremely hygroscopic alkyl isocyanide dichloride—$FeCl_3$ complexes (Equation (1)), which are converted into alkyl isocyanates with various reagents, such as zinc oxide, salts of carboxylic acids, trifluoroacetic acid or dimethyl sulfoxide ⟨73T297⟩.

$$ClCN + RCl + FeCl_3 \longrightarrow (RN{=}CCl_2)_n(FeCl_3)_m \qquad (1)$$

$$R = Pr^i, n = 3, m = 2; R = Bu^t, n = m = 1$$

In this way, the nonisolated isopropylisocyanide dichloride complex (**1**) was transformed, in 85% yield, into isopropyl isocyanate by a smooth reaction with zinc oxide. The yields of isocyanate are, however, much lower when the reaction is carried out starting from the isolated complex, owing to its high sensitivity to moisture. Successful results have also been obtained with various salts of carboxylic acids such as sodium or mercury(II) acetate or sodium oxalate (Scheme 1).

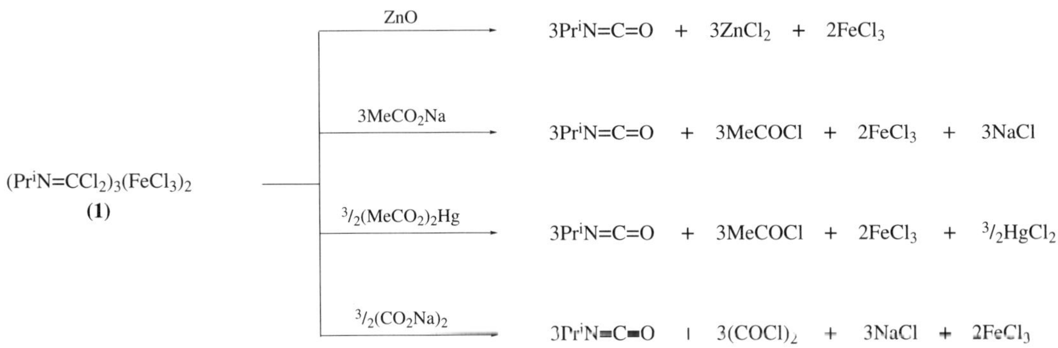

Scheme 1

In contrast, *t*-butylisocyanide dichloride complex (**2**) did not react with zinc oxide, under the same reaction conditions. However, *t*-butyl isocyanate has been prepared starting from the corresponding complex (**2**), by treatment either with trifluoroacetic acid in the presence of tributylamine or with dimethyl sulfoxide (Scheme 2).

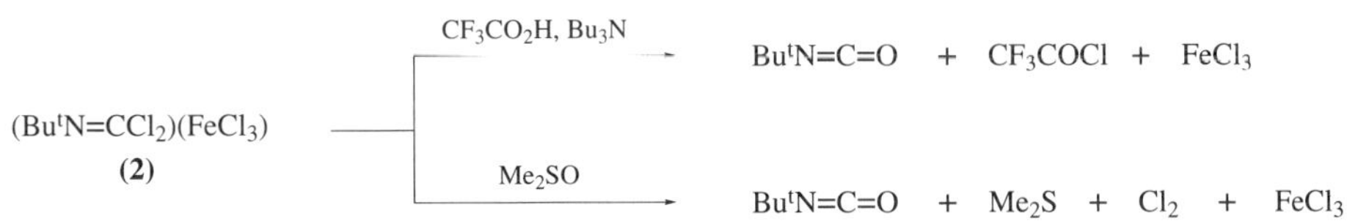

Scheme 2

5.26.2.1.2 Reaction with metal nitrocyanamides

On the basis of the results obtained by studying the reaction between trityl chloride and silver phenylcyanomethylene nitronate or silver nitrocyanamide ⟨86H(24)2813, 87JCS(P1)2163, 87S907⟩ which gave rise to the formation of trityl isocyanate, among other compounds, Boyer *et al.* have reported a general synthetic method for the preparation of isocyanates by alkylation of silver nitrocyanamide with a wide variety of alkyl halides (primary, secondary and tertiary aliphatics, including alicyclic, aralkyl and heteroalkyl systems) and certain nonvicinal dihalides ⟨88JCS(P1)2137⟩. In these processes, the intermediate alkylnitrocyanamide (**4**) was thermolysed (−20 to −80 °C) to the isocyanate in yields ranging from 63 to 95%. In certain examples, silver nitrocyanamide is generated *in situ* from sodium nitrocyanamide and silver nitrate. The assumption that each alkylnitrocyanamide underwent thermal isomerization to an undetected *N*-alkyl-*N*′-nitrocarbodiimide

(**5**) followed by ring closure and dissociation, provided a straightforward explanation for the formation of the isocyanate (Scheme 3).

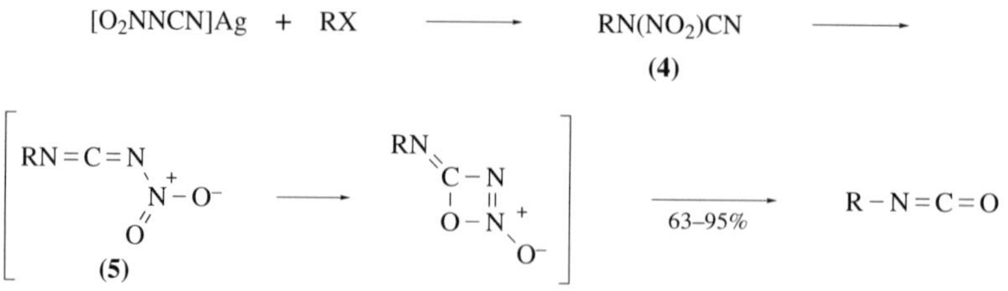

Scheme 3

However, under the same reaction conditions, silver nitrocyanamide does not react with cyclopropyl bromide, acetyl chloride, tosyl chloride or phenacyl bromide, and the results with vicinal dihalides are very erratic, without detectable formation of vicinal diisocyanates.

It is also reported that *m*-chloroperbenzoic acid oxidatively assisted a reaction at 25 °C between a straight-chain primary alkyl iodide and sodium nitrocyanamide to give the corresponding alkyl nitrocyanamide, isolated as the major product, together with the *O*-alkylation product, as the minor product. Alkyl nitrocyanamides, so obtained, were thermolysed (neat or in benzene at 80 °C) to give alkyl isocyanates in nearly quantitative yield (Scheme 4) ⟨89JCR(S)331⟩. In the absence of the oxidant, sodium nitrocyanamide suspended in an inert solvent did not convert these alkyl halides into the corresponding isocyanates. This method has the advantage of using sodium nitrocyanamide (nonexplosive) instead of silver nitrocyanamide (explosive).

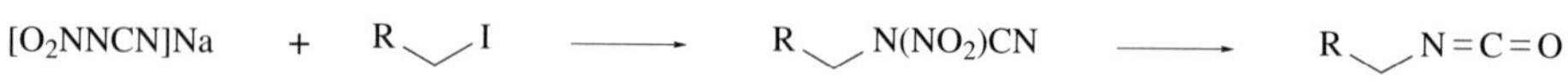

Scheme 4

5.26.2.1.3 *Reaction with isocyanic acid and alkali cyanates*

Alkyl isocyanates are obtained by treatment of alkyl chlorides with isocyanic acid in the presence of NiO, under rigorous conditions. Following this procedure, the preparation of alkylene diisocyanates has been reported ⟨72CRV457⟩.

On the other hand, alkali metal cyanates react with alkyl halides to give either isocyanates or isocyanurates, depending on the reaction conditions and solvent. Thus, a patent has reported that organic isocyanates can be prepared by the reaction of alkali or alkaline-earth metal cyanates, and ammonium cyanates with an alkyl halide in either dimethylformamide or alkyl sulfone solvents. In this way, preparation of ethyl- (95%) and butyl isocyanate (70%) and di-, hexamethylene- (80%) and *p*-xylylene (50%) diisocyanates have been reported (Equation (2)) ⟨62USP3017420⟩.

$$\text{RX} + \text{KNCO} \longrightarrow \text{RN=C=O} + \text{KCl} \qquad (2)$$

Other patents claim that methylene dibromide ⟨70JAP026894⟩, allyl chloride, alkenyl chlorides ⟨70USP3558684⟩ and chloromethyl methyl ether ⟨64GEP1205087, 69MI 526-01⟩, react with alkali cyanates to give the corresponding isocyanate or diisocyanate. It is emphasized that the solvent used in these transformations is very important because in aprotic polar solvents, such as dimethylformamide, the isocyanate produced usually trimerizes so readily that it is isolated with difficulty.

Under similar conditions, α- and ω-halo- as well as α,ω-dihalocarboxylic alkyl esters react with potassium cyanate, in dipolar aprotic solvents, to yield the corresponding isocyanate esters which, on treatment with hydrochloric acid, lead to the amino acid hydrochlorides in nearly quantitative yields (Equation (3)) ⟨81CB173⟩.

(3)

R = Me: X = Br, Y = H
X = H, Y = Br
R = Et: X = Br, Y = H
X = H, Y = Br

R = Me: X = NCO, Y = H
X = H, Y = NCO
R = Et: X = NCO, Y = H
X = H, Y = NCO

Silver cyanate has also been reported to be a useful reagent for the preparation of isocyanates starting from the corresponding halides ⟨69AP335, 69AP617, 70ZOB1933, 76USP3954441⟩. As an example, treatment of α,α-dihaloalkyl ether (**6**) with silver cyanate in diethyl ether leads to the formation of the diisocyanate (**7**), in good yield (Equation (4)) ⟨69AP617⟩.

(4)

(6) **(7)**

5.26.2.2 From Amines

5.26.2.2.1 Reaction with phosgene, diphosgene and triphosgene

Despite the availability of numerous methods for the preparation of isocyanates, the most widely used route to these important compounds is the treatment of primary alkyl and aryl amines, or their salts, with phosgene (Scheme 5) ⟨72CRV457, 73CRV75⟩.

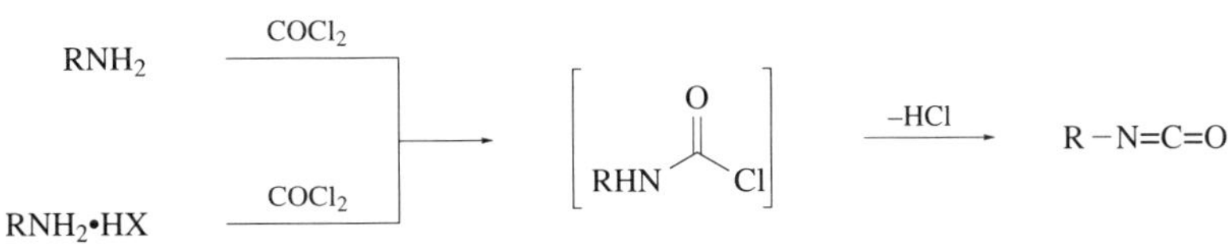

Scheme 5

The most common procedure for the preparation of isocyanates using gaseous phosgene, involves either the addition of the amine to a cold solution of phosgene in an inert solvent, to yield a mixture of the corresponding carbamoyl chloride and amine hydrochloride, or direct addition of phosgene to the amine hydrochloride. The reaction mixture is then heated in the presence of a large excess of phosgene, and reaction of the amine hydrochloride, as well as the dehydrohalogenation of carbamoyl chloride, takes place. In the presence of catalysts such as morpholine, boron trifluoride, aluminum chloride or activated carbon, the reaction of an amine hydrochloride with phosgene is particularly effective ⟨44USP2362648, 56USP2733254, 73CRV75⟩. Phosgenation of primary amines has also been carried out in the presence of fully *N*-substituted amides, amidines and guanidines ⟨70USP3492331⟩.

Highly halogenated aryl isocyanates or amino aryl isocyanates (**8**)–(**13**) have also been obtained by phosgenation of the adequate aromatic amine ⟨59USP2884435⟩. It is worth noting that the azido function is not destroyed on transforming the amino group into an isocyanato group by phosgene. Thus, aromatic mono- and polyamines containing azido groups are converted into the corresponding azidoaryl isocyanates by treatment with phosgene (Equation (5)) ⟨66NEP6606456⟩.

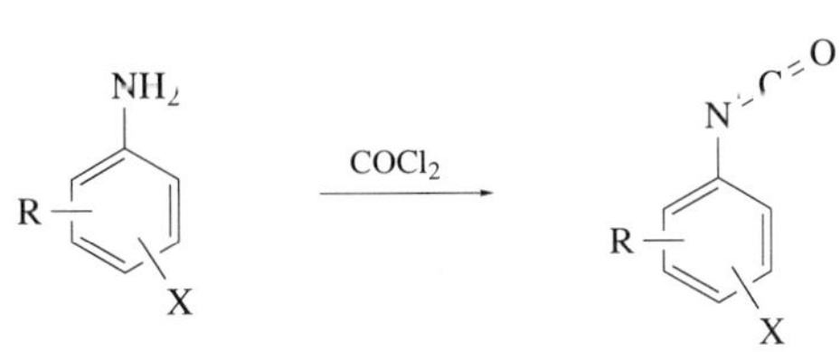

(5)

R = 4-Me, X = 2-N_3; R = 2-Me, X = 5-N_3; R = H, X = 4-SO_2N_3
R = H, X = 3-SO_2N_3; R = 2-Me, X = 4-SO_2N_3; R = 2-Cl, X = 4-SO_2N_3

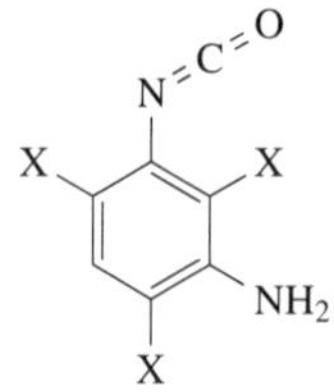
(8)
X = Cl, Br

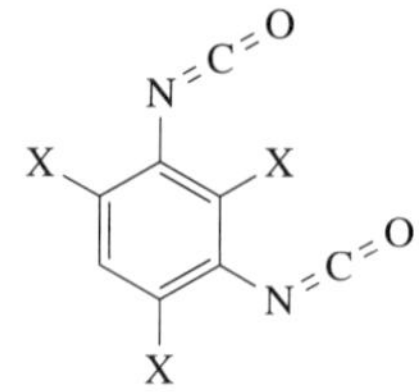
(9)

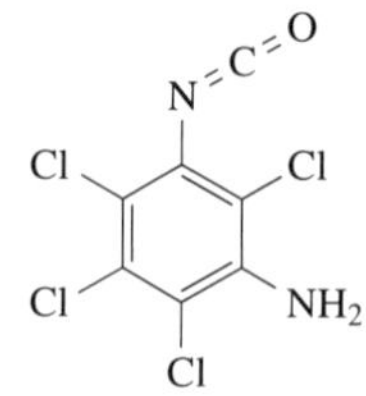
(10)

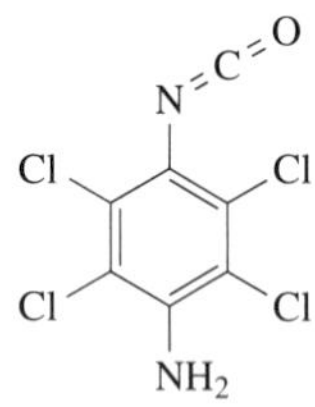
(11)

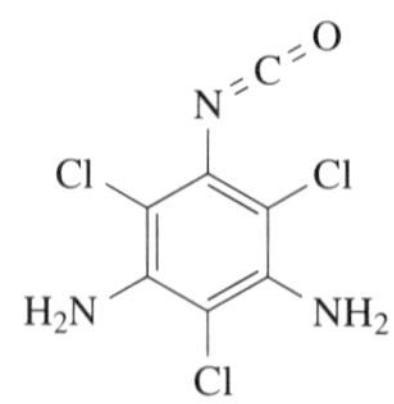
(12)

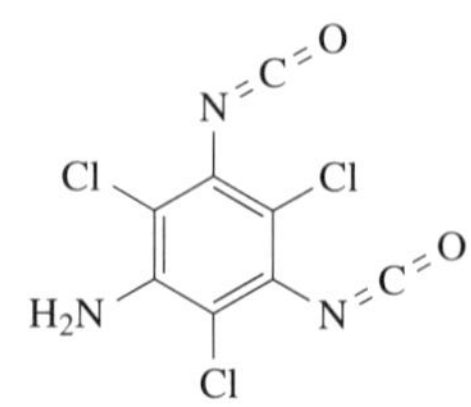
(13)

Phosgenation of thiocyanatoanilines in chlorobenzene also gives the corresponding thiocyanato aryl isocyanates in high yields (70–80%) (Equation (6)) ⟨66ZOB1707⟩.

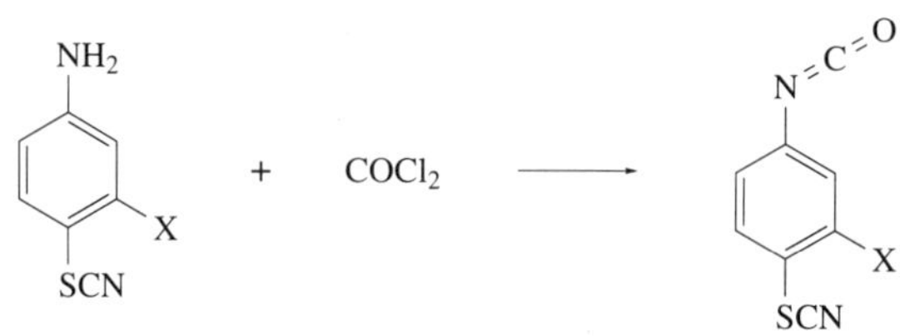
(6)

X = H, Cl, *m*-Me-Ph

Isocyanates containing an hydroxyl group have also been prepared by treating aminophenols having *sec*- or *tert*-alkyl, cyclohexyl, aryl or halogen substituents in both *o*-positions with phosgene (Equation (7)) ⟨64GEP1170398⟩. On the other hand, *m*- and *p*-isocyanatophenol have been prepared from the adequate aminophenol with phosgene in a highly polar aprotic solvent in almost quantitative yields ⟨69BRP1153261⟩. A patent has also claimed that chloroformic acid esters of aryl isocyanates were obtained by treatment of aminophenols with phosgene and *N*,*N*-dimethylaniline in chlorobenzene ⟨66GEP1228249⟩.

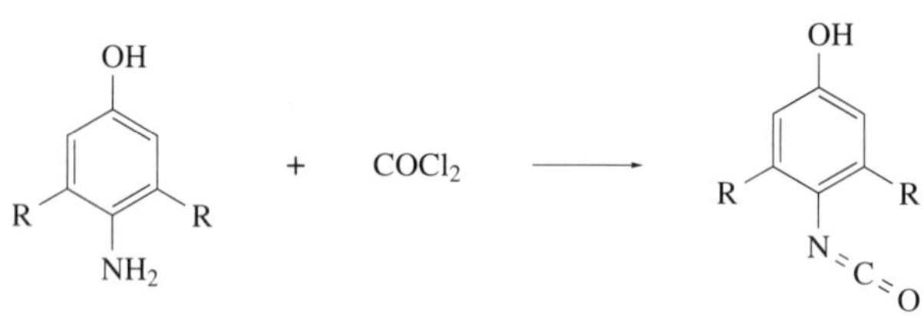
(7)

R = *t*-alkyl, cyclohexyl, aryl, halogen

Preparation of isocyanates and diisocyanates through a phosgenation process has also been carried out by addition of a cooled solution of the amine and NaOH to a solution of phosgene in toluene kept at −7°C ⟨69BRP1152877⟩. This method is also useful for the preparation of isocyanates which are difficult to obtain under normal phosgenation conditions ($COCl_2$/HCl), and it also applies to the preparation of diisocyanates.

Prolonged phosgenation of *o*-tolylenediamines at an elevated temperature affords the corresponding *o*-tolylene diisocyanates in moderate yields ⟨69JOC1162⟩. The reaction seems to proceed via the tolyleneaminocarbamoyl chlorides (**14**), the methylbenzimidazolones (**15**) and the tolyleneallophanoyl chlorides (**16**), as is set out in Scheme 6. *o*-Tolylene diisocyanates are very unstable, a fact which impedes their preparation in high yields and makes their isolation difficult; they polymerize readily to form solids, but can be depolymerized and isolated by heating and distillation. Repeated polymerization–distillation cycles tend to increase the stability of the diisocyanates slightly, indicating that trace amounts of impurities, contained in the original samples, may act as a polymerization catalyst.

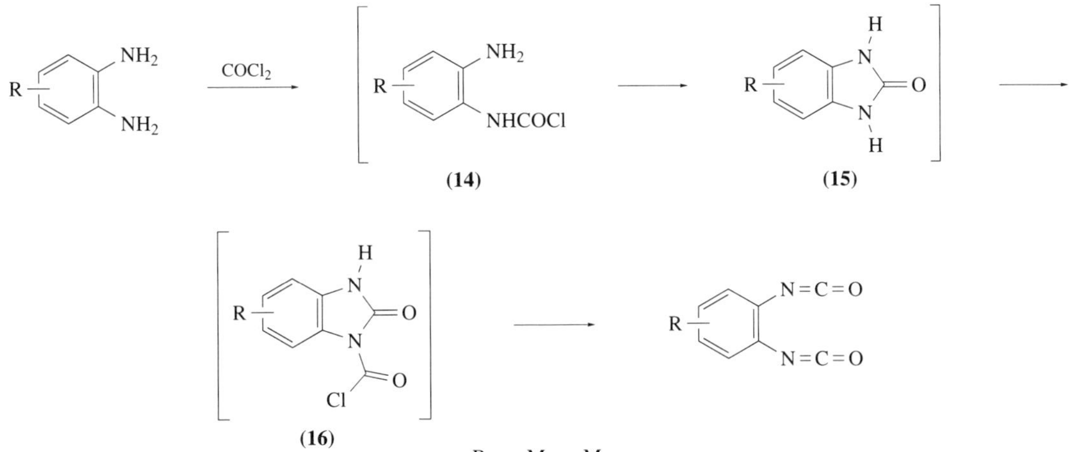

R = *o*-Me, *p*-Me

Scheme 6

The treatment of a primary amino function with phosgene has been extended to various types of molecules such as iminoacyl isocyanates (**17**) ⟨69ZOR119⟩, isocyanato derivatives of aliphatic ethers (**18**) ⟨69JAP4427365⟩, diisocyanatophenyl ethers (**19**) ⟨70USP3509199⟩, *p*-styrylisocyanate (**20**) ⟨69ZOR1888⟩, and fluoroalkyl or aryl isocyanates (**21**)–(**24**) ⟨67JPS(A)2757, 67PP1149, 69URP241428, 69USP3468924⟩.

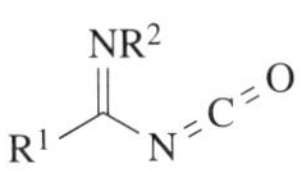

(17)

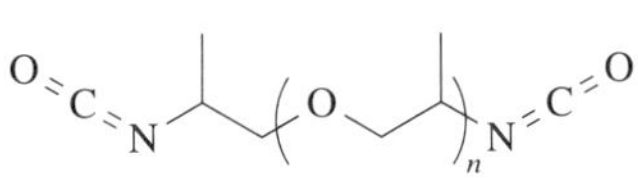

(18)

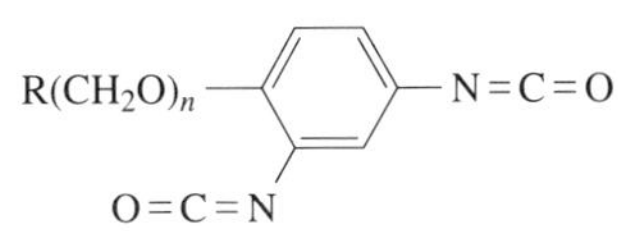

(19)

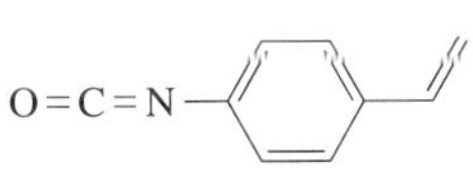

(20)

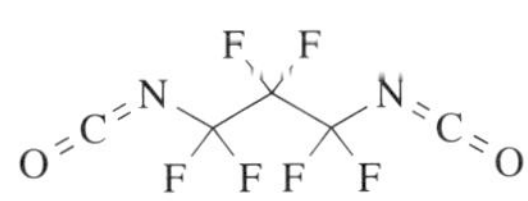

(21)

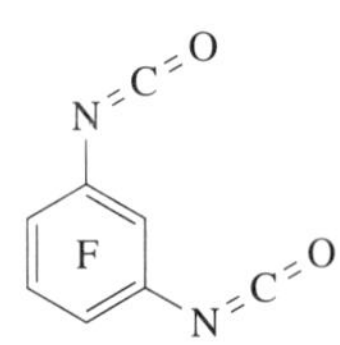

(22)

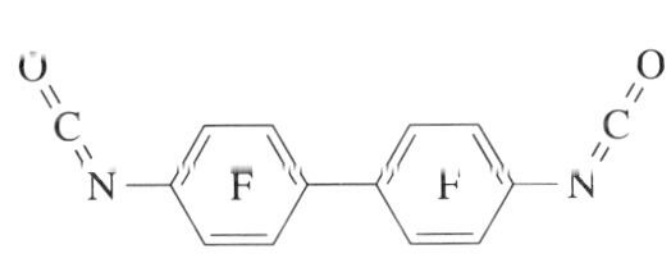

(23)

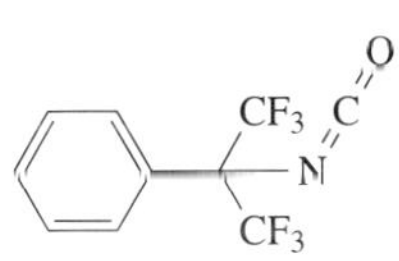

(24)

The preparation of fluorinated diisocyanates with two intervening methylene groups has been achieved from α,ω-diiodoperfluoroalkanes ⟨80JOC347⟩ by insertion of ethylene into the CF_2–I bond ⟨82JOC2251⟩, displacement of the iodo groups with azide, reduction of the resulting azide groups to amines and phosgenation of the diamine (Scheme 7) ⟨91JOC3043⟩

N-Alkylamidines undergo phosgenation in hot inert solvents yielding iminoalkylisocyanates ⟨67AG(E)864⟩. These products are relatively stable, although they react with amines or alcohols to give substituted ureas or urethanes, respectively. However, they react very slowly with water.

When enaminonitriles with a primary nitrogen atom are allowed to react with phosgene in refluxing ethyl acetate, *N*-acylation occurs to give the corresponding β-cyano-α,β-unsaturated isocyanates (**25**) in 37–79% yield (Equation (8)) ⟨72JOC3030⟩.

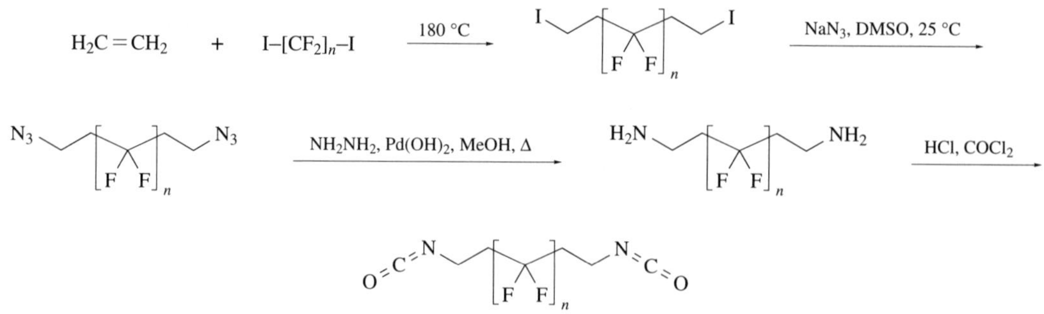

$n = 4, 6, 8$

Scheme 7

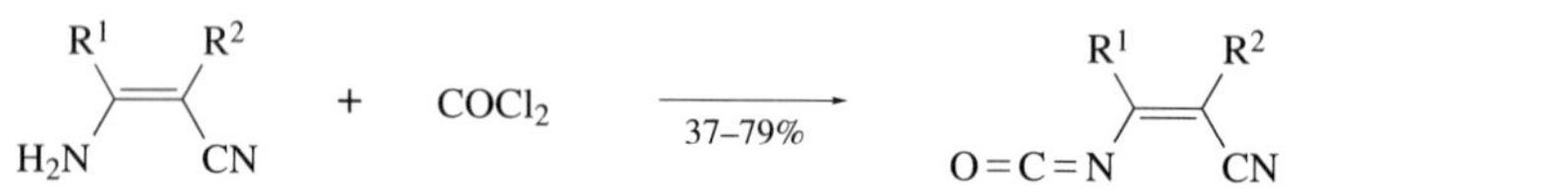

(8)

$R^1 = Et, R^2 = Me; R^1 = Pr^n, R^2 = Et; R^1 = Ph, R^2 = Me; R^1, R^2 = (CH_2)_4$

The facile reaction of amines with thionyl chloride gives *N*-sulfinyl derivatives (**26**) which undergo subsequent phosgenation to produce the corresponding isocyanate with regeneration of thionyl chloride (Scheme 8). The reaction is general, and aliphatic and aromatic amines can be converted into the corresponding isocyanates ⟨69JOC3200⟩.

$$R{-}NH_2 + SOCl_2 \xrightarrow{-2HCl} \underset{(\mathbf{26})}{R{-}N{=}S{=}O} \xrightarrow{COCl_2} R{-}N{=}C{=}O + SOCl_2$$

Scheme 8

Aminoaryl esters of silicic acid, readily available from silicon halides and the corresponding aminophenol, can be easily converted into isocyanates (**27**)–(**30**) in high yields starting either from the hydrochloride or from the free base ⟨62AG(E)617⟩.

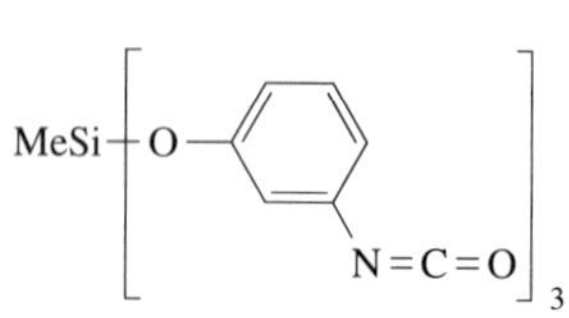

(**27**) 86%

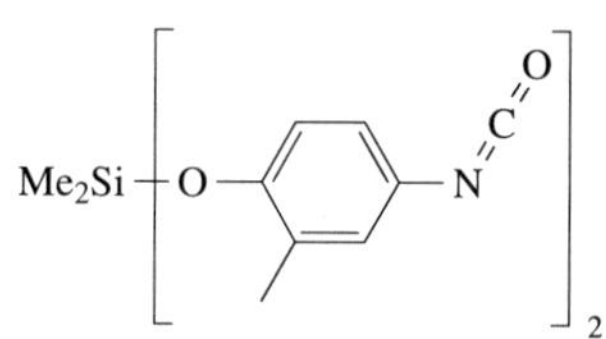

(**28**) 93%

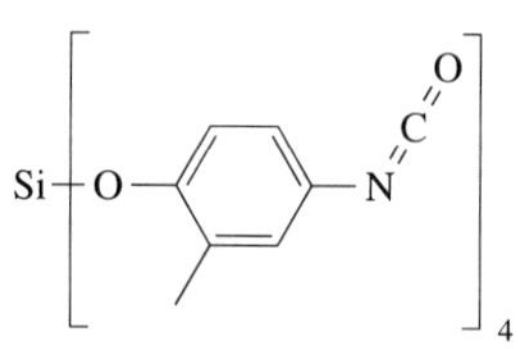

(**29**) 80%

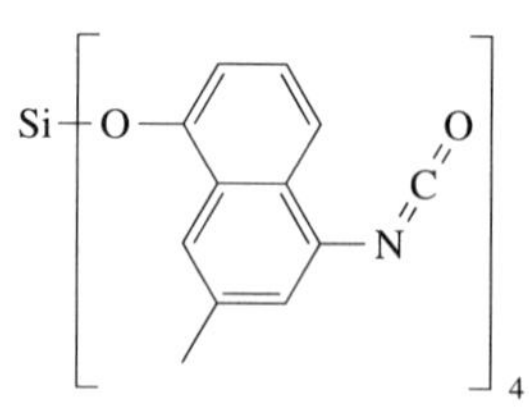

(**30**) 85%

On the other hand, Mironov *et al.* found a method for the preparation of isocyanates and diisocyanates by treating silyl primary amines with phosgene at low temperatures (Scheme 9) which has many advantages over the application of free amines in this reaction ⟨69FRP1563380, 69ZOB2598, 70DOK(190)110, 71BRP1235156, 73ZOB623, 73ZOB1502, 75ZOB2025, 79RCR473, 84JOM207⟩. The high yields of isocyanate obtained are accounted for firstly by the fact that the intermediate *N*-silylcarbaminoyl chlorides contain a σ,σ-conjugated Si—N—C—Cl bonding system, and therefore they can easily undergo an intramolecular β-elimination reaction, and secondly, the isocyanate and trimethylchlorosilane products do not react with each other ⟨84JOM207⟩. This method has also been applied to the preparation of isocyanates from *N,N*-disilylamines (Equation (9)), while phosgenation of *N,N′*-disilylated diamines resulted in the formation of diisocyanates of the aliphatic and aromatic series (Equation (10)) ⟨73ZOB1502, 75ZOB2025⟩.

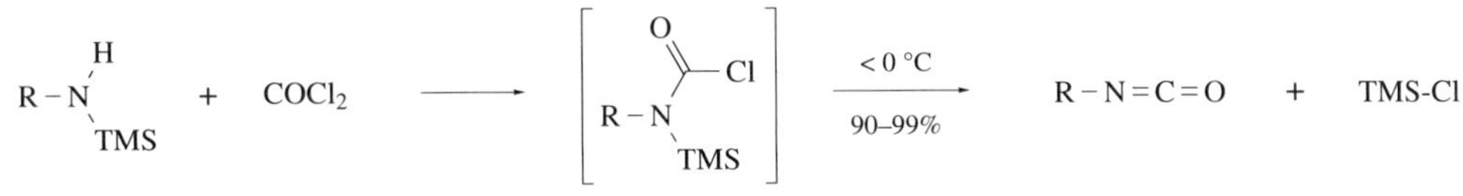

Scheme 9

$$R^1N(SiR^2_3)_2 + COCl_2 \xrightarrow{0\,°C} R^1{-}N{=}C{=}O + 2R^2_3SiCl \qquad (9)$$

$$(TMS)(H)N{-}R{-}N(H)(TMS) + 2COCl_2 \longrightarrow O{=}C{=}N{-}R{-}N{=}C{=}O + 2TMS\text{-}Cl \qquad (10)$$

R = alkyl, aryl

On phosgenating those compounds containing, in addition to amino functions, other functional groups sensitive to the action of phosgene or HCl (amino alcohols, amino acids, etc.), as well as thermally weakly-stable compounds, the change from free amines to silyl derivatives makes it possible to achieve not only quantitatively better results, but also qualitatively superior ones. For instance, phosgene treatment of silyl derivatives of aminophenol and anthranilic acid results in the formation of the corresponding isocyanates in good yields, while direct phosgenation of the free amine gives rise to the formation of benzoxazolone and isatoic anhydride, respectively (Scheme 10) ⟨79ZOB784, 80ZOB620⟩. Adamantyl isocyanate ⟨82ZOB2654⟩, as well as a number of carbon-functional isocyanates from Group 14 elements, have been obtained in the same way (Scheme 11) ⟨70DOK(190)110, 72ZOB2501, 73ZOB623, 73ZOB1502, 75ZOB2185⟩.

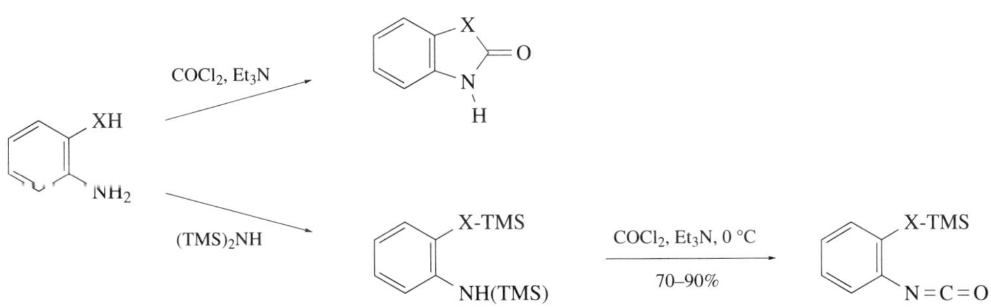

Scheme 10

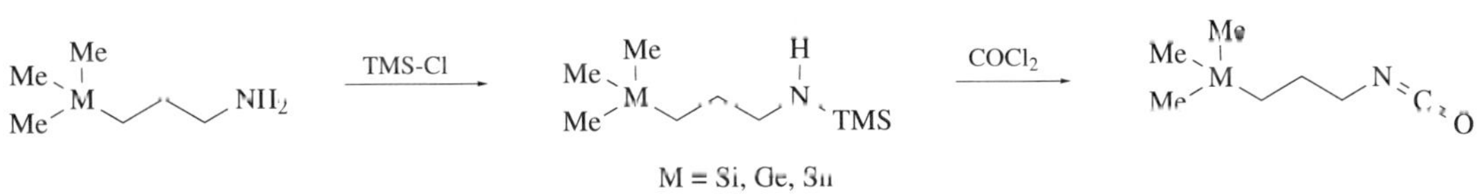

Scheme 11

Aminoaryl esters of phosphoric and thiophosphoric acids, prepared by the reaction of phosphorus pentahalides with nitrophenols followed by catalytic hydrogenation or by treating aminophenols with phosphorus trihalides and oxidation to pentavalent phosphorus, have been converted into isocyanato esters (**31**)–(**35**) by the action of phosgene. The same principle permits the preparation of isocyanates of phosphonic and phosphinic esters, when applied to phosphorus halides in which one or two hydrocarbon groupings are linked to the phosphorus. Thus, bis(3-isocyanato-4-methylphenyl) styrylphosphonate (**36**) has been synthesized in 72% yield from the corresponding styrylphosphonic dichloride and 3-amino-4-methylphenol followed by treatment with phosgene ⟨62AG(E)617⟩.

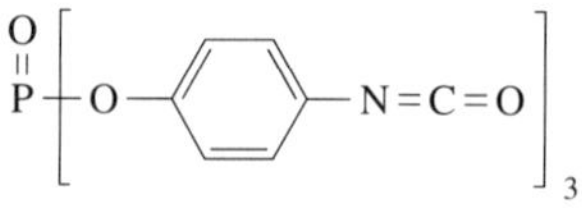

(31) 87%

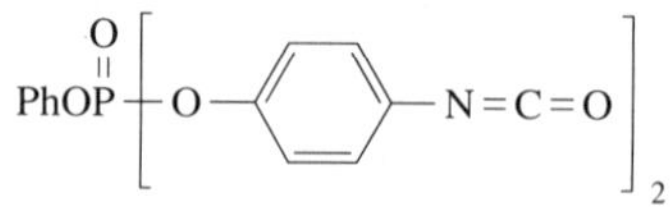

(32) 97%

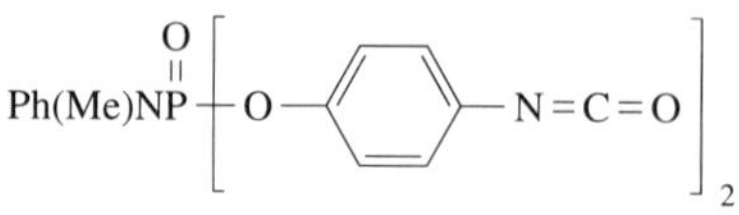

(33) 88%

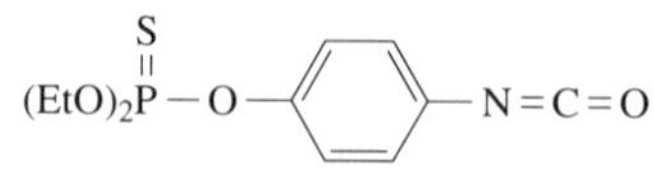

(34) 82%

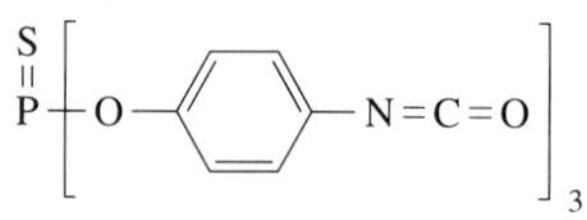

(35) 90%

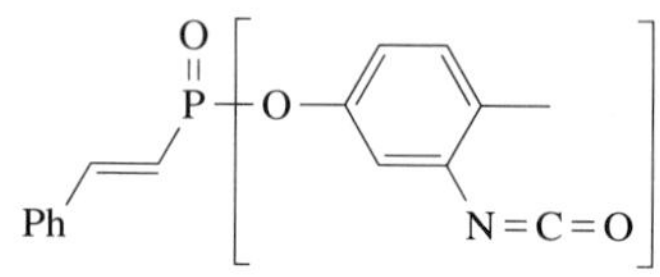

(36) 72%

It has been well established that the reaction of α-amino acids with phosgene, in an inert solvent such as dioxane, gives 2,5-oxazolidinediones in good yields under suitable conditions. However, on prolonged reaction, (*N*-chloroformylamino)acyl chlorides are formed, which are converted into 2-isocyanatoacyl chlorides (**37**), with evolution of hydrogen chloride, when distilled (Scheme 12) ⟨65JOC1158⟩. The reaction temperature had little effect on the yield, although addition of hydrogen chloride to the reaction mixture considerably increased the yield of the isocyanatoacyl chloride.

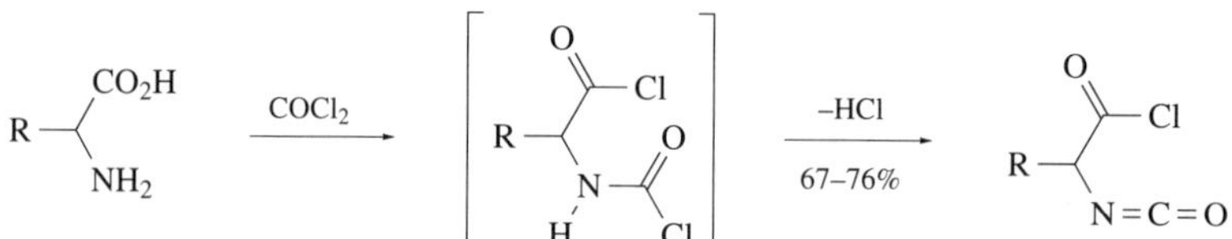

(37)

Scheme 12

By a similar reaction, ω-isocyanatoalkanecarboxylic acid chlorides and isocyanatobenzoyl chloride have been prepared by treatment of the corresponding amino acids or their hydrochlorides with phosgene and hydrogen chloride ⟨65JOC1158, 66JOC142⟩. With aliphatic compounds, 3-aminopropanoic acid gave 3-isocyanatopropanoyl chloride in 36% yield without added hydrogen chloride, and 92% yield with added hydrogen chloride. Similarly, 6-aminohexanoic acid without added hydrogen chloride gave only a trace of 6-isocyanatohexanoyl chloride, but the yield was increased to 91% when hydrogen chloride was added. 4-Aminobutanoic acid, however, with hydrogen chloride gave the cyclic product 1-chloroformyl-5-chloro-2,3-dihydropyrrole in 40% yield, while 5-aminopentanoic acid gave principally the isocyanatopentanoyl chloride, in 65% yield, along with a small amount of 1-chloroformyl-2-chloro-2,3-dehydropiperidine (Equation (11)). With aromatic compounds, *p*- and *m*-aminobenzoic acid reacted with phosgene, in the presence of hydrogen chloride, to give isocyanatobenzoyl chlorides in 45 and 19% yield, respectively. With the *ortho*-isomer, however, the reaction with phosgene gave isatoic anhydride in 96% yield with or without added hydrogen chloride. Nevertheless, the latter was subsequently converted into *o*-isocyanatobenzoyl chloride with thionyl chloride (90% yield) or with phosphorus pentachloride (47% yield) in the presence of a trace of pyridine (Scheme 13).

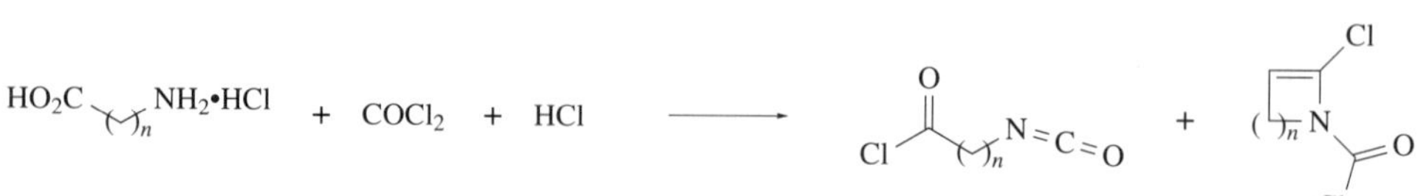

(11)

$n = 2, 4, 5$ $n = 2, 3$

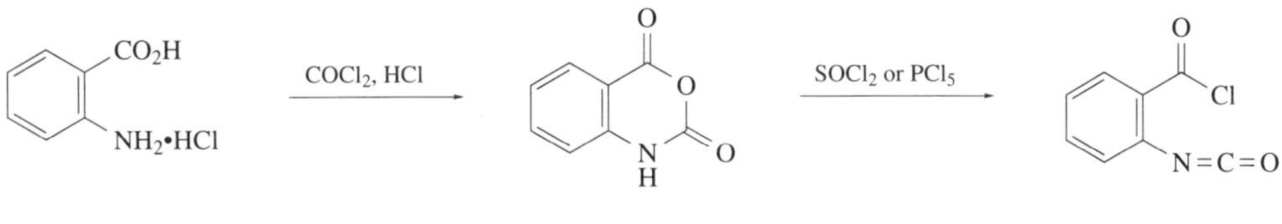

Scheme 13

In the early 1990s, a method has been found for the preparation of enantiomerically pure aminoacid ester isocyanates that avoids the use of gaseous phosgene, elevated temperatures and strongly acidic conditions. The method consists of the addition of a commercially available solution of phosgene in toluene to a mixture of an amino acid ester hydrochloride and pyridine (Equation (12)). This method is compatible with suitably protected functionality in the amino acid side chains, and the mild conditions used allow the preparation of derivatives containing acid-sensitive functionality ⟨92JOC7364⟩.

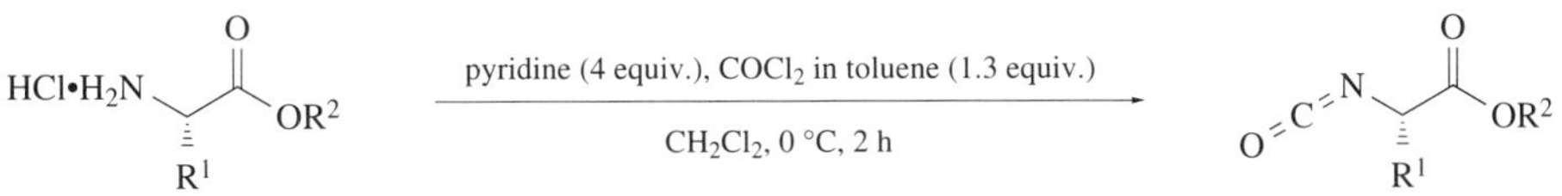 (12)

R^1 = Me, R^2 = Et; R^1 = Pr^i, Bu^i, Bu^s, Bn, $MeS(CH_2)_2$, R^2 = Me

ε-Aminocaprohydroxamic acid hydrochloride, obtained by reaction of ε-caprolactam and hydroxylamine hydrochloride, undergoes phosgenation to give pentamethylene diisocyanate in 32% yield ⟨70S589⟩.

Secondary amines react with phosgene to give disubstituted carbamoyl chlorides which decompose readily at low temperatures, with or without catalysis, to give isocyanates. Thus, pyrolysis of *N*-*t*-butyl-*N*-alkyl carbamoyl chloride (**38**) to the primary carbamoyl chloride (**39**) or even to the corresponding alkyl isocyanate (**40**) provides a synthetic approach to mono- and diisocyanates (Scheme 14) ⟨63JOC2076⟩. Similarly, the reaction of *N*,*N'*-diphenylmethylenediamine (**41**) with excess phosgene is accompanied by the cleavage of a carbon–nitrogen bond to give *N*-chloromethyl-*N*-phenylcarbamoyl chloride (**42**) and phenyl isocyanate (Equation (13)) ⟨74JOC2897⟩. In the same way, 4-aminobenzylaniline, upon reaction with phosgene, produces *N*-phenyl-*N*-4-isocyanatobenzylcarbamoyl chloride in high yield, which on reaction with hydrogen chloride, undergoes a carbon–nitrogen bond cleavage to give phenyl isocyanate and 4-isocyanatobenzyl chloride ⟨74JOC2897⟩.

(**38**) (**39**) (**40**)

n = 1, R = Bu, Bu^n, $H_2C{-}CHCH_2$
n = 2, R = $CH_2CH{=}CHCH_2$, *p*-xylylene, $(CH_2)_2$

Scheme 14

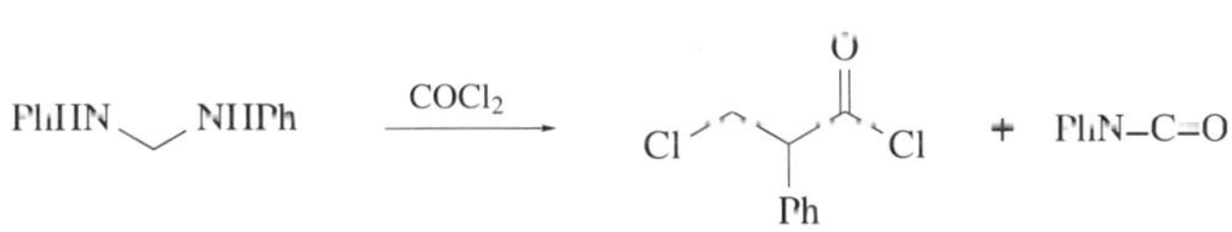 (13)

(**41**) (**42**)

Although all the methods described above involving phosgene are very useful and some are utilized on an industrial scale, the toxicity of the reagent represents a drawback in the laboratory synthesis of isocyanates.

A synthetic route to isocyanates which uses diphosgene (trichloromethyl chloroformate, TCF) as a phosgene precursor has been described to afford these compounds in excellent yields ⟨75JAP89344,

77BCJ2406, 83JAP08052⟩. The procedure consists of mixing the appropriate amines with diphosgene and heating the reaction mixture in an inert solvent, in the presence of active carbon.

Diphosgene is prepared by the photochlorination of methyl chloroformate or methyl formate ⟨80OS(59)195⟩ and although it is also toxic, it is a dense liquid (b.p. 128°C) with a vapour pressure of only 10 mmHg at 20°C. This compound is stable at room temperature and, consequently, it is easier and safer to handle than phosgene, but at elevated temperatures or in contact with iron(III) oxide or charcoal, it decomposes to gaseous phosgene and thereby 1 mole of diphosgene corresponds in its reactivity to 2 moles of phosgene.

In contrast to the preparation of alkyl isocyanato acid chlorides, the diphosgene method with aromatic amino acids gave similar results to those with phosgene ⟨76JOC2071⟩. Also, reactions of amino alcohols with diphosgene proceed similarly to those with phosgene ⟨76JOC2071⟩. Thus, diphosgene is far superior to phosgene in the alkyl isocyanate acid chlorides syntheses, but it is comparable to phosgene in the preparations of aromatic isocyanato acid chlorides and alkyl isocyanato chloroformates.

In relation to the previously reported reactions it is significant to note that formation of *N*-carboxy anhydrides (NCAs) of α-amino acids, which are very important compounds in peptide chemistry, because they are used for the preparation of poly(α-amino acids) ⟨B-56MI 526-01⟩ and peptides ⟨66JA3163, 69JOC1631⟩, takes place in a similar way. Although the conventional procedure for its preparation involves the phosgenation of an amino acid ⟨50JCS3213, 76MI 526-01⟩, an alternative method for the preparation of the NCAs which involves the decomposition of diphosgene to phosgene in a warm solvent, followed by reaction of the resulting phosgene with the amino acid, has been reported ⟨73CL1143⟩. However, the advantage of using a calculated amount of liquid diphosgene is diminished by the very slow decomposition of diphosgene to phosgene.

Katakai and Iizuka have developed a simple method for the preparation of the NCAs using diphosgene ⟨85JOC715⟩. This improved method involves the rapid decomposition of diphosgene with activated charcoal and the simultaneous reaction of the resulting phosgene with the amino acids in the same reaction vessel.

Diphosgene has also been used as a convenient reagent in the preparation of diisocyanates with benzene or furan rings by treating a suspension of the appropriate diamino derivative in toluene with diphosgene at elevated temperature, in the presence of active charcoal ⟨92JPR187⟩. The reactions can be carried out with an equimolecular amount of diphosgene but the yields are increased if an excess of reagent is used (Equations (14a) and (14b)).

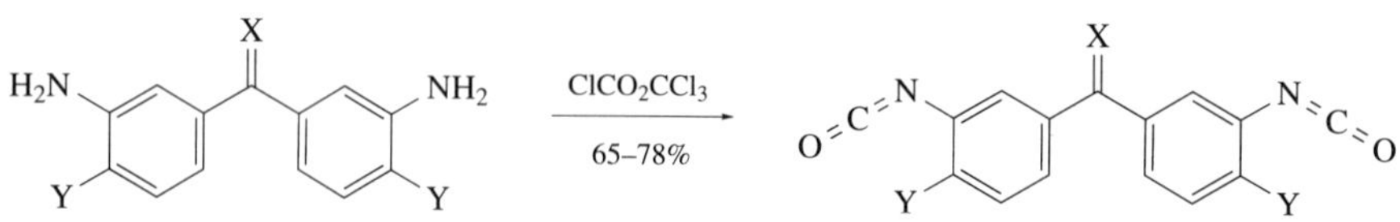

(14a)

X = CCl_2, Y = Cl, OMe; X = O, Y = Cl, OMe

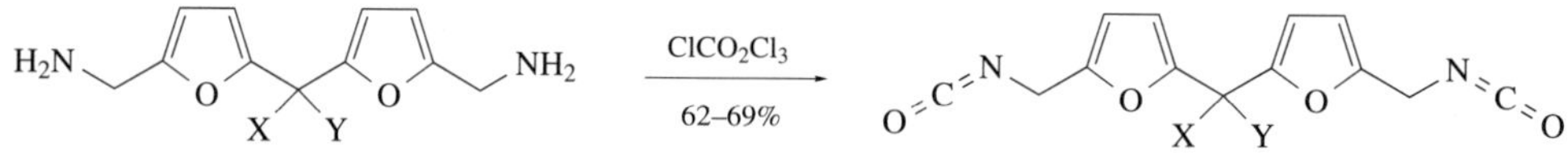

(14b)

X = Y = H or Me; X = H, Y = Me

Even safer, especially during transport and storage, is the stable crystalline triphosgene (bis(trichloromethyl)carbonate) (m.p. 79°C; b.p. 206°C) ⟨87AG(E)894⟩. Although an excess of triphosgene, which can be measured accurately, is not necessary, the yields are mostly higher than those from the corresponding reactions with phosgene.

5.26.2.2.2 *Reaction with* N,N′*-carbonyldiimidazole*

In reactions with amines and alcohols, *N,N′*-carbonyldiimidazole shows about the same reactivity as phosgene. The fact that it can be prepared in such good yields from phosgene and imidazole can be ascribed to the insolubility of imidazolium chloride in tetrahydrofuran, benzene and chloroform, as for the preparation of imidazolides from acid chlorides and imidazoles. Thus, the preparative use of *N,N′*-carbonyldiimidazole in reactions in which phosgene is normally used is of practical interest only in special cases where the following advantages are important: (i) carbonyldiimidazole reactions take place in neutral medium without formation of HCl — the reaction product is accompanied only by imidazole which is usually inert and can be removed readily; (ii) once prepared, *N,N'*-carbonyldiimidazole is a nontoxic, crystalline substance and is more convenient to handle and easier to measure than toxic, gaseous phosgene ⟨62AG(E)361⟩.

By using primary amines in a 1 : 1 molar ratio, primary imidazole-*N*-carboxamides (**43**) are obtained. These compounds are interesting for their ability to dissociate into isocyanates and imidazole, even at room temperature. This dissociation forms the basis for a simple method of preparing isocyanates from aliphatic, alicyclic, and aromatic amines (Scheme 15) ⟨61LA(648)72⟩.

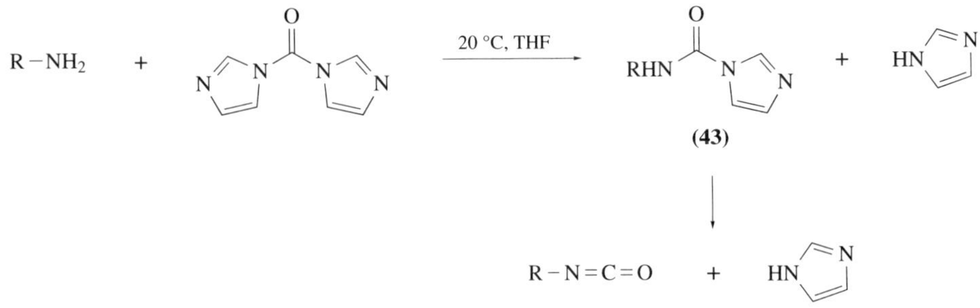

Scheme 15

The dissociation of imidazole *N*-carboxamides into isocyanates was also attempted for the preparation of alkoxyisocyanates. However, their tendency to trimerize is apparently so great that the only product isolated was the corresponding trialkoxyisocyanuric acid ⟨61AG657⟩.

5.26.2.2.3 *Reaction with oxalyl chloride*

Quinones carrying an isocyanato substituent in the quinone ring have been obtained by reactions of the corresponding amines with oxalyl chloride (Equation (15)) ⟨71AG(E)403⟩.

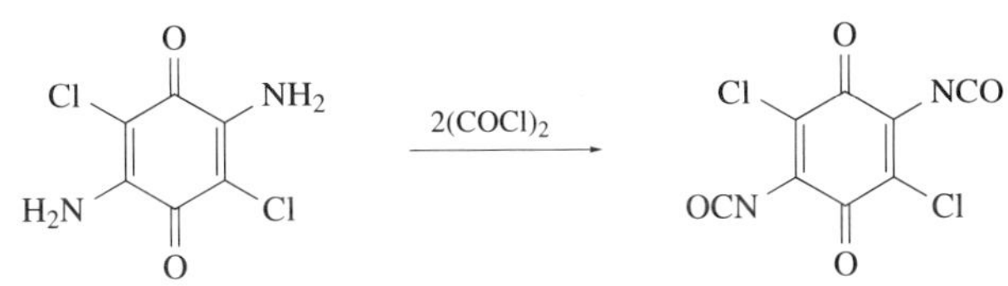

(15)

Pyridine (**44**), pyrimidine (**45**) or pyrazine (**46**) derivatives in which an isocyanato group occurs in the vicinal position to nitrogen, have also been isolated in very good yields when the aminoheterocycle is treated with oxalyl chloride ⟨71AG(E)402⟩. Mono- and diisocyanato derivatives of *s*-triazine (**47**) and (**48**) can be readily prepared in good yield by reaction of the appropriate *s*-triazinamine with oxalyl chloride ⟨71AG(E)403⟩. The reaction proceeds better the more electronegative the substituent R is. This method also applies to the preparation of pyrazole isocyanate derivatives such as (**49**) ⟨66NEP6603215⟩.

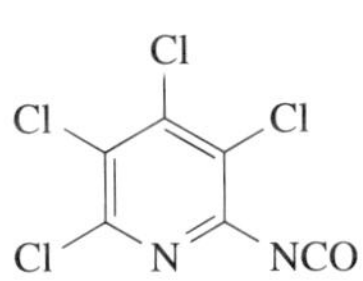

(**44**) 85%

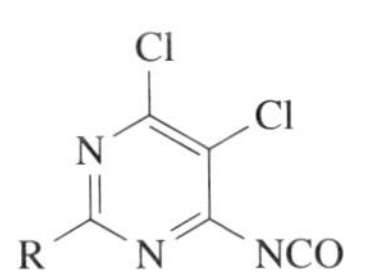

(**45**) 72–90%
R = Cl, CCl_3, C_2Cl_5, Ph

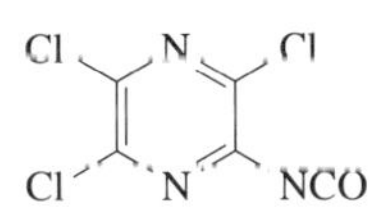

(**46**) 70%

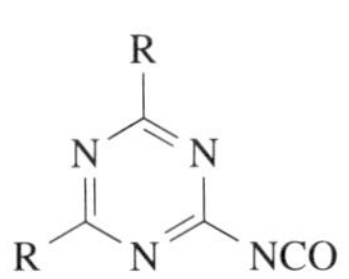

(**47**)
R = Cl, F, CCl_3 70–93%

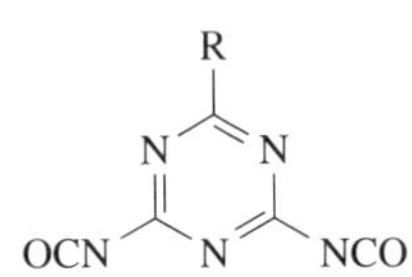

(**48**)
R = Cl, CCl_3, Ph 75–85%

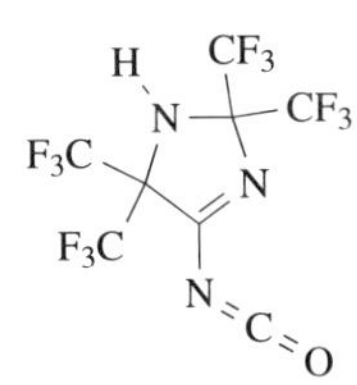

(**49**)

On the other hand, some heterocycles which are formally dimers of isocyanates can be prepared by treatment of the corresponding amino precursor with oxalyl chloride under standard conditions. This is the case of the 2-(2-benzothiazolyl)benzothiazolo[3,2-a][1,3,5]-triazine-1,3-dione (**50**) which was prepared by reaction of 2-aminobenzothiazole with oxalyl chloride as shown in Scheme 16 ⟨68AG(E)381⟩.

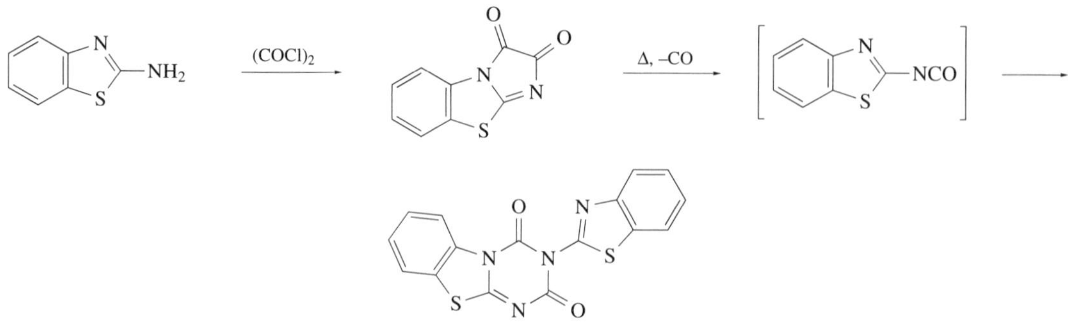

(50)

Scheme 16

5.26.2.2.4 *Reaction with carbon monoxide*

A more direct preparation of isocyanates involves combining primary aliphatic or aromatic amines with carbon monoxide in the presence of $PdCl_2$. The reaction proceeds readily under mild conditions, and is accompanied by reduction of $PdCl_2$ to palladium metal (Equation (16)) ⟨66JOC596⟩.

$$R{-}NH_2 + CO + PdCl_2 \longrightarrow R{-}N{=}C{=}O + Pd + 2HCl \qquad (16)$$

It is clear that isocyanate formation is not due to *in situ* phosgene formation from $PdCl_2$ and CO. In the absence of amine, no reduction of $PdCl_2$ by CO was observed during extended periods of time at temperatures ranging from 40 °C to 200 °C. The reaction occurs irrespective of the order in which the amine or CO are added to $PdCl_2$ and, moreover, has been observed when amines are added to $(PdCOCl_2)_2$, or CO to $Pd(RNH_2)_2Cl_2$.

The reaction of primary amines, carbon monoxide, allyl chloride and palladium chloride has also been investigated and it was found that both π-allylpalladium chloride and isocyanate were formed together ⟨66CC828⟩.

A patent claims that polyisocyanates can be prepared by treating aromatic polyamines with a thermally labile Ni or Ti carbonyl and oxygen at 100 °C to 350 °C; this causes decomposition of the metal carbonyl, releasing CO and metal, which catalyzes the reaction with formation of the polyisocyanate and water ⟨62USP3070618⟩.

5.26.2.2.5 *Reaction of phosphoramidate anions and iminophosphoranes with carbon dioxide*

Iminophosphoranes were discovered in 1921 by Staudinger and Hauser and are extremely interesting intermediates for the synthesis of a wide variety of unsaturated nitrogen compounds ⟨21HCA861⟩. Their utility, like that of the Wittig reagents to which they bear a formal analogy, is based on the fact that they react with a number of carbonyl-containing molecules to give phosphine oxides, and an unsaturated nitrogen compound. In this context, anilines and alkylamines can be converted into isocyanates by the reaction, under mild conditions, of carbon dioxide with their iminophosphorane derivatives ⟨82S596⟩. The iminophosphoranes were in turn prepared from the corresponding amine and triphenylphosphine dibromide (Scheme 17) ⟨59LA(627)142⟩.

$$\text{R–NH}_2 + \text{Ph}_3\text{PBr}_2 \xrightarrow{\text{Et}_3\text{N}} \text{R–N=PPh}_3 \xrightarrow[\text{82–99\%}]{\text{CO}_2} \text{R–N=C=O} + \text{Ph}_3\text{P=O}$$

R = alkyl, aryl

Scheme 17

In a similar way formation of 3′-isocyanato-3′-deoxythymidine was obtained in 15% yield from the iminophosphorane prepared by reaction of 3′-azido-3′-deoxythymidine (AZT) and triphenylphosphine ⟨92S358⟩.

The phosphoramidate anions show a reactivity toward carbonyl groups which is analogous to the one shown by iminophosphoranes and therefore they also constitute some very useful reagents for laboratory preparations of similar unsaturated organic nitrogen compounds. Dialkyl *N*-alkylphosphoramidates may be conveniently prepared by the classical reaction of an amine with the commercially available dialkyl phosphorochloridates. An alternate and frequently superior synthesis is based on the reaction of a dialkyl phosphite with amines in the presence of carbon tetrachloride ⟨45JCS660⟩. Phosphoramidates react smoothly with sodium hydride to form stable anions in either benzene or GLYME solvents. Reaction of these compounds with carbon dioxide at room temperature and atmospheric pressure give a carbamate (**51**) which, when heated to an appropriate temperature (usually 80 °C), yields an isocyanate in good yield (Scheme 18) ⟨62JA1316, 64JOC2816, 67JOC1279⟩. If the reaction of carbon dioxide with a branched *N*-alkylphosphoramidate is initially carried out at a temperature of 80 °C, the carbamate fragments, possibly concurrently with its formation, and the liberated isocyanate reacts with a second mole of the anion to give carbodiimide as the major product.

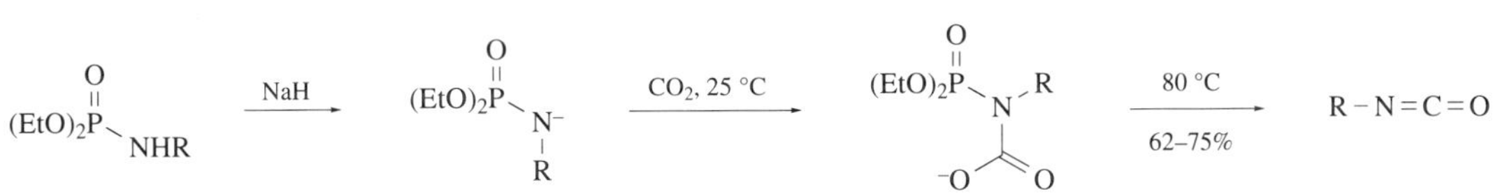

(51)

Scheme 18

5.26.2.2.6 *Other methods*

The first isocyanate derived from sulfur hexafluoride was prepared by reaction of the secondary amine (**52**) and sodium fluoride with benzoic acid (Equation (17)) ⟨64JA357⟩. The starting amine was prepared by addition of HF to the appropriate perfluoroazomethine which, in turn, was prepared by an ultraviolet light-mediated addition of sulfur chloride pentafluoride to cyanogen chloride, followed by an exchange reaction of the chlorine atoms when treated with fluoride ion

$$\underset{(\mathbf{52})}{\text{F}_5\text{SNHCF}_3} + 2\text{NaF} + \text{PhCO}_2\text{H} \longrightarrow \text{F}_5\text{S}-\text{N}=\text{C}=\text{O} + 2\text{NaHF}_2 + \text{PhCOF} \qquad (17)$$

Hydrazine reacts with the cyclopentadienyliron tricarbonyl cation to form, first, a carbazoyl intermediate (**53**) which then loses ammonia to give an isocyanate complex (**54**). The isocyanate complex may also be prepared by reaction of the same cation with azide and isocyanate anions, respectively (Scheme 19) ⟨69JA3197⟩.

Metal *N* alkylcarbamato complexes ($M(O_2CNHR)_n$) (M = Na, $n = 1$; M = Mn, Co, $n = 2$), prepared by treating carbon dioxide with a primary alkyl or aryl amine and various complex precursors, readily decompose in hydrocarbon media when treated with a stoichiometric amount of acyl halides. The overall process, from carbon dioxide to alkyl isocyanates, corresponds to the deoxygenation of carbon dioxide ⟨88CB1891⟩. From Equation (18) it follows that the metal chloride is not involved in the mass balance, so that the total transformation may be described as a metal-assisted reaction. In a pictorial way, Calderazzo *et al.* have shown that the thermodynamically unfavorable conversion of carbon dioxide and primary alkyl amines to alkyl isocyanates has been accomplished by the

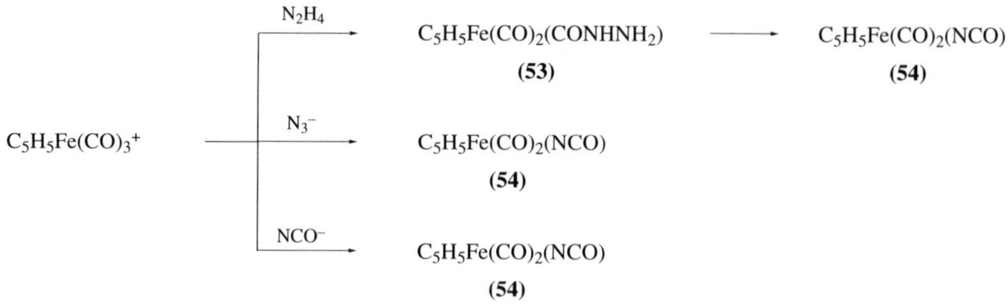

Scheme 19

simultaneous conversion of amine and acyl chloride to ammonium salt and carboxylic anhydride (Scheme 20) ⟨88CB1891⟩.

$$2CO_2 + 4RNH_2 + 2MeCOCl \longrightarrow 2RNCO + Ac_2O + 2RNH_3Cl + H_2O \qquad (18)$$

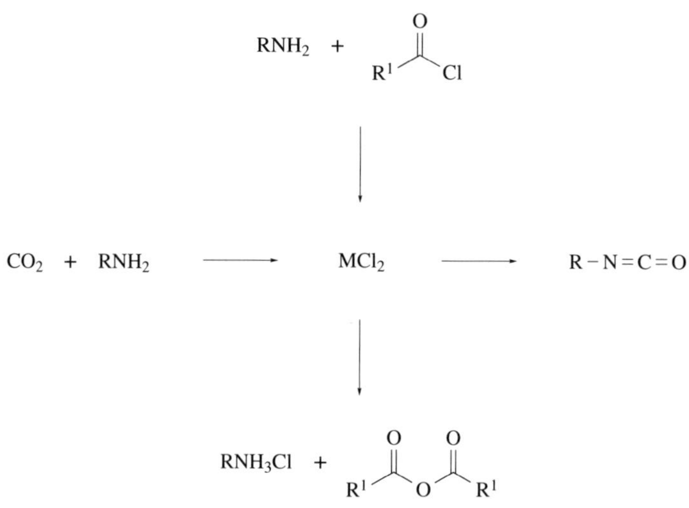

Scheme 20

Isocyanates are also prepared by the reaction of an aliphatic or aromatic primary monoamine or a long-chain aliphatic diamine with diphenyl carbonate at 50 °C to 350 °C ⟨68USP3366662⟩.

5.26.2.3 From Carbamic Acid Derivatives

5.26.2.3.1 From carbamates and phosphorus halides

The synthesis of isocyanates from carbamates by means of phosphorus pentachloride ⟨1897MI 526-01⟩ or pyrocatechol–phosphorus trichloride ⟨63CB1387⟩ has been reported. However, it often involves difficulties in separating the isocyanate from the phosphorus oxide chlorides that are also formed. Nevertheless, it was shown that the reaction of *N*-β-hydroxy- or chloroethyl carbamate with phosphorus pentachloride yielded β-chloroethyl isocyanate in 49%, irrespective of the route used ⟨36JA2608⟩.

In the late 1980s it was demonstrated that aryl isocyanates could be obtained in good yield

(60–90%) when the corresponding carbamate was treated with PCl_5 or $POCl_3$ at 120°C to 140°C ⟨89ZOR885⟩ (Equation (19)).

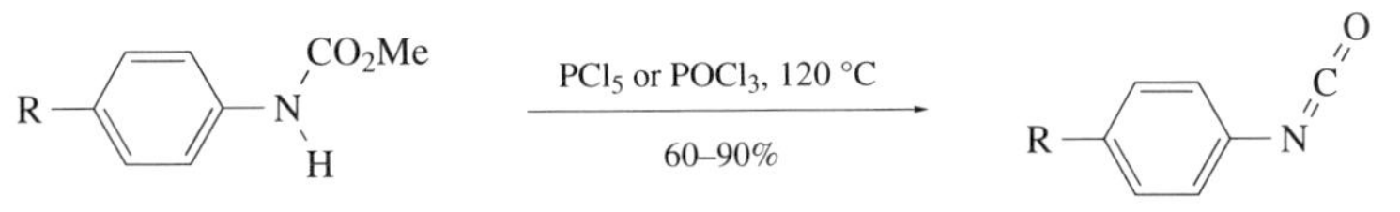

(19)

R = H, Me, Br, OMe, NO_2

5.26.2.3.2 *From carbamates and phosgene*

Reaction of carbamates in an inert solvent containing *N*,*N*-dimethylformamide as catalyst gives isocyanates and alkyl chlorides, even at low temperatures. This process seems to proceed by thermal decomposition of the intermediate chloroformimidates (**54**), formed by reaction of the enolic form of the carbamates with the formimidoyl chloride derived from phosgene and DMF (Scheme 21) ⟨67AG(E)636⟩.

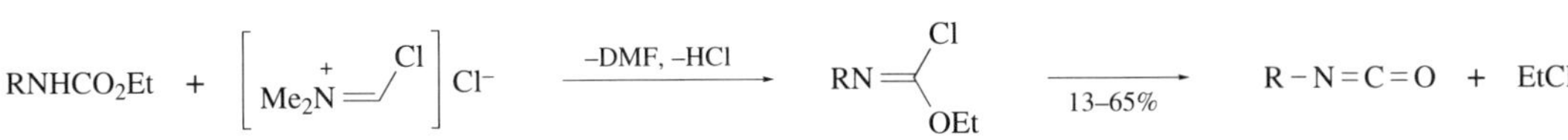

Scheme 21

Following this method, 1,2,2,2-tetrachloroethyl isocyanate has been obtained, in 48% yield. However, the chloro group adjacent to nitrogen is quite labile, and further treatment with triethyl phosphite results in the formation of the corresponding isocyanatophosphonate ⟨68JOC2887⟩.

5.26.2.3.3 *From carbamates and oxalyl chloride*

It has been reported that treatment of primary amides with oxalyl chloride constitutes a very general method for the preparation of acyl isocyanates. According to this general procedure, benzyloxycarbonyl isocyanate was obtained in 53% yield by reaction of benzyl carbamate and oxalyl chloride (Equation (20)) ⟨88S992⟩.

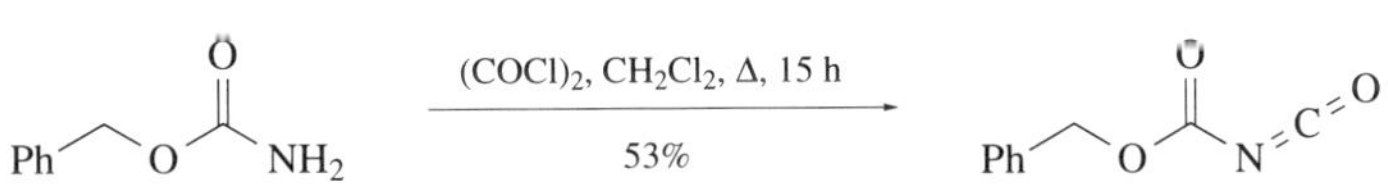

(20)

5.26.2.3.4 *By thermolysis of carbamates*

Thermal cleavage of carbamoyl chlorides and carbamates usually gives only poor yields of isocyanates because the high temperatures (200°C to 300°C) favor side reactions, and the decomposition products undergo reverse reaction at slightly lower temperatures. However, it is reported that heating of carbamates in the presence of phosphorus pentoxide produces isocyanates ⟨36JA2608⟩. Also, a patent describes the preparation of alkyl- and arylisocyanates by heating the appropriate *N*-alkyl- or aryl-substituted alkylthiocarbamate, at 250°C to 400°C, under reduced pressure ⟨63BEP631961⟩.

In contrast to the thermal behavior of carbamates, it was found that *N*-silylated carbamic acid derivatives (**55**), readily available by treatment of acid chlorides, anhydrides and esters either with chlorotrimethylsilane/triethylamine or with mono- or bis(trimethylsilyl)-acetamide, decompose to isocyanate and the corresponding silicon compound (**56**) between −20°C and 120°C (Equation (21)) ⟨68AG(E)941, 69DOK(181)115⟩. Since the isocyanate and (**56**) undergo neither the reverse reaction nor, to an appreciable extent, side reactions, even sensitive or very reactive isocyanates can be obtained in yields between 70% and 95%. For urethanes R^1—NH—$COOR^3$ the temperature for cleavage depends on the nature of R^3; if R^1 remains the same, it decreases rapidly with increasing stability of the anion R^3O^-.

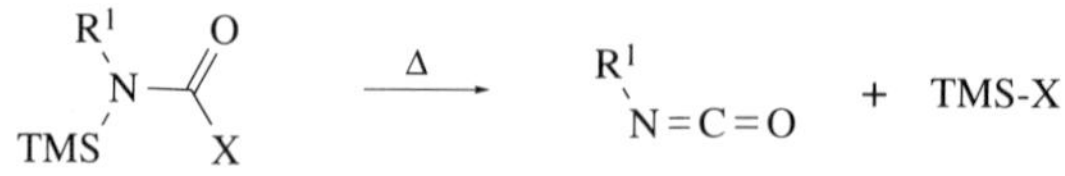

(55) (56) (21)

$X = Cl, O_2CR^2, OR^2; R^1, R^2$ = alkyl, aryl or cycloalkyl

Following the same methodology, Kricheldorf has achieved the preparation of isocyanatoalkane-, alkene- or arenecarboxylic acid trimethylsilyl esters (**58**), from the corresponding functionalized carbamates (**57**) (Scheme 22) ⟨70S649, 72AG(E)128⟩.

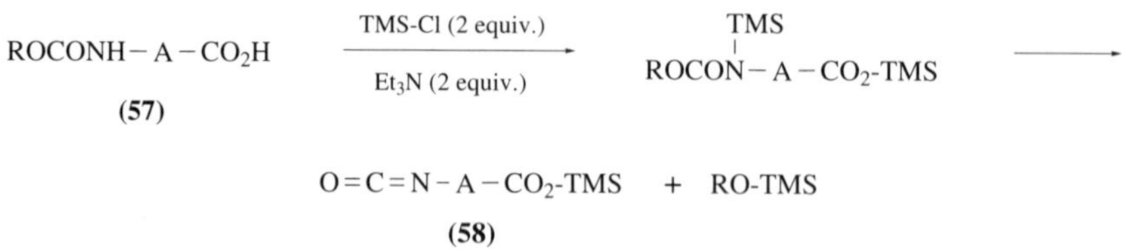

R = alkyl, aryl; A = aliphatic or aromatic

Scheme 22

Mironov *et al.* have shown that, from the preparative point of view, a more favorable synthetic variant is the direct reaction of *N*-alkyl- and *N*-aryl carbamates with polychlorosilanes in the absence of HCl acceptors (Scheme 23) (Table 1) ⟨81ZOB1814⟩. The intermediate *N*-chlorosilylurethanes are less stable thermally than the *N*-trimethylsilylurethanes, and fragment with elimination of isocyanate at the moment of their formation ⟨84JOM207⟩.

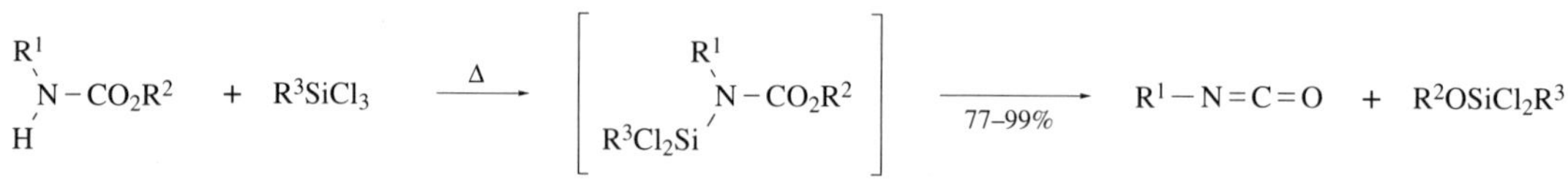

R^1 = alkyl, cycloalkyl, aryl

Scheme 23

Table 1 Preparation of isocyanates from carbamates and chlorosilanes.

R^1	*R^2*	*Chlorosilane*	*Yield (%)*
Me	Me	$PhSiCl_3$	95
Me	Me	$(Cl_3Si)_2O$	68
Me	Me	$Cl_3SiCHCl_2$	51
Me	Bu^n	$PhSiCl_3$	64
Et	Ph	$PhSiCl_3$	73
Bu^t	Bu^t	$PhSiCl_3$	97
c-C_6H_{11}	c-C_6H_{11}	$PhSiCl_3$	97
c-C_6H_{11}	c-C_6H_{11}	$Cl_3SiCHCl_2$	87
Ph	Ph	$PhSiCl_3$	79

The method for the synthesis of alkyl and aryl isocyanates by thermolysis of *N*-silyl carbamates has also been extended to acyl carbamates, which can be formed by thermolysis of *N*-acyl-*N*-silyl carbamates ⟨72ZOB1870⟩.

On the other hand, *O*-silyl carbamates, upon heating, decompose to the corresponding isocyanate and trimethylsilanol ⟨73ZOB2089, 75ZOB2007⟩. However, due to the subsequent dehydration of trimethylsilanol, a large part of the isocyanate is hydrolyzed by the released water, and the main reaction products may be hexamethyldisiloxane and the *sym*-dialkylurea. Generally, if the thermolysis is carried out in the presence of polychlorosilanes, the organic isocyanate can appear in high yield, the formation of *O*-chlorosilyl carbamates which are thermally more unstable than the *O*-trimethylsilyl derivatives being the role of the polychlorosilane (Scheme 24).

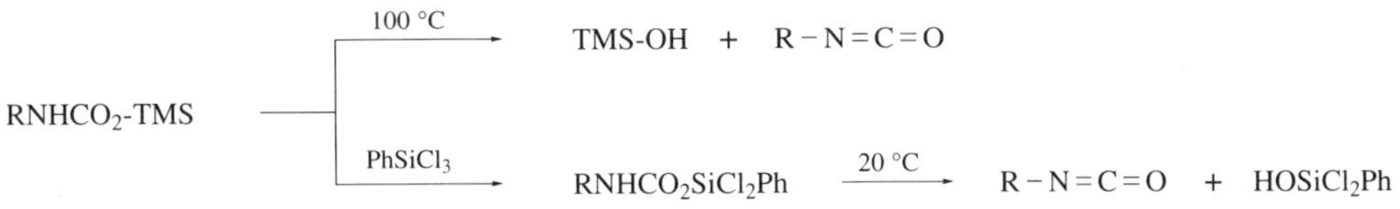

Scheme 24

Similarly, both aryl- and alkylsulfonyl carbamates can be easily silylated, and subsequent thermolysis of the *N*- and *O*-trimethylsilyl intermediates produces the corresponding sulfonyl isocyanates in excellent yields. This suggests that a rapid equilibration between the *N*- and *O*-silyl isomers is occurring under the thermolysis conditions (Scheme 25) ⟨74JOC1597⟩.

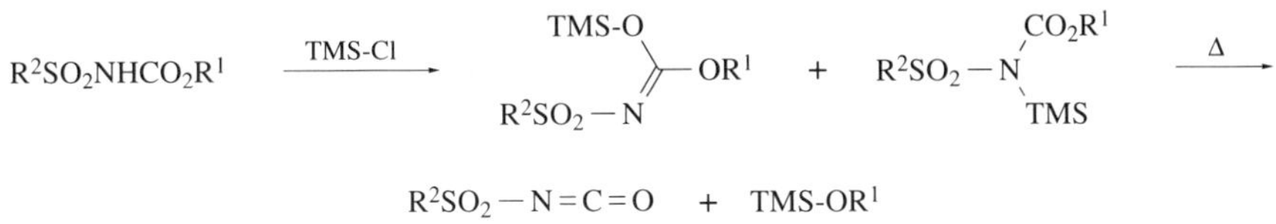

Scheme 25

Moreover, reaction of some alkenes with ethyl carbamate or ethyl *N*-chloro carbamate give rise to the formation of *N*-substituted carbamates which upon reaction with a high boiling isocyanate yield the lower boiling isocyanate, which is separated from the reaction mixture by distillation. Following this method alkyl-, cycloalkyl-, and aryl isocyanates have been prepared in yields ranging from 34% to 87% ⟨65CB1097⟩.

5.26.2.3.5 *From thiocarbamates*

Methyl thiocarbamates are easily converted into the corresponding isocyanates in good yields on treatment with 2-chloro-1-ethylbenzoxazolium tetrafluoroborate in the presence of an equimolecular amount of triethylamine (Scheme 26) ⟨77CL1345⟩. The method can be used for the synthesis of both aliphatic and aromatic isocyanates at room temperature, under neutral conditions and in yields ranging from 64% to 83%. It also permits the preparation of allyl isocyanates which are accessible only in low yield by conventional methods ⟨79AG(E)707⟩.

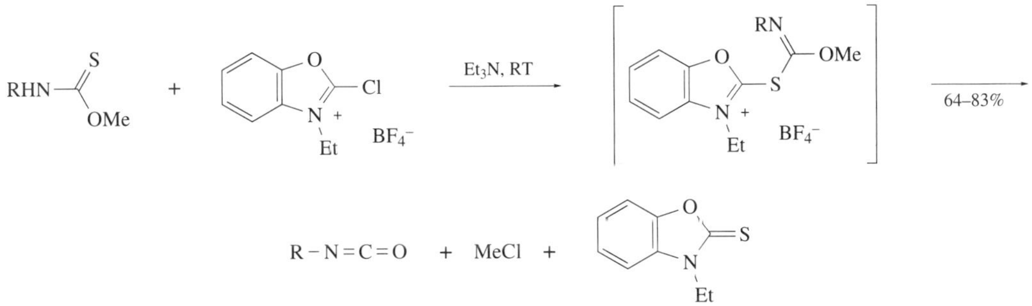

Scheme 26

5.26.2.4 From Ureas

5.26.2.4.1 *By treatment with Lewis or mineral acids*

Alkyl or aryl isocyanates can be obtained by the reaction of gaseous boron trifluoride with the corresponding monosubstituted urea, below 40°C and under reduced pressure ⟨61USP3013045⟩. However, this reaction does not occur with all urea derivatives.

The preparation of several alkyl, cycloalkyl, and aryl isocyanates, in yields ranging from 30% to 82%, by treatment of the corresponding substituted ureas with sodium nitrite in the presence of HCl, is also reported ⟨60CI(M)1243⟩.

On the other hand, a patent describes the preparation, in 72–92% yield, of aryl isocyanates and diisocyanates OCN—X—NCO (X = alkylene, alkylenearylene or arylene) by cleavage of the appropriate trisubstituted urea with HCl ⟨73GEP(O)2225365⟩.

5.26.2.4.2 By treatment with phosgene

Although Hentschel was the first to describe the preparation of phenyl isocyanate by phosgenation of 1,3-diphenylurea, Ulrich and Sayigh have later developed a simple reaction principle (Equations (22) and (23)) which permits the preparation of a number of mono- and diisocyanates that had not previously been obtained by direct phosgenation of the corresponding amines. The method is based on the preparation of a urea derivative, by treatment of a primary amine or sulfonamide with a readily available alkyl or aryl isocyanate (Equation (22)), which in a second step is phosgenated (Equation (23)) to yield the new isocyanate plus the starting one. Thus, the overall reaction is equivalent to the phosgenation of an inert amine, or of an amine that is not accessible to direct phosgenation ⟨66AG(E)704⟩.

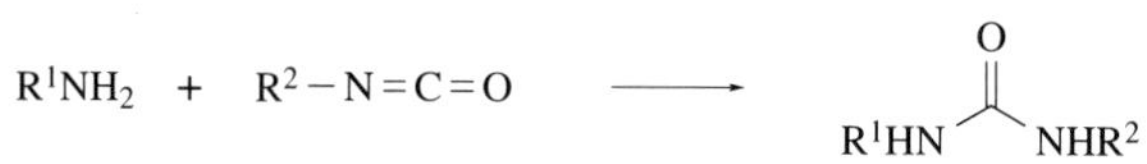

(22)

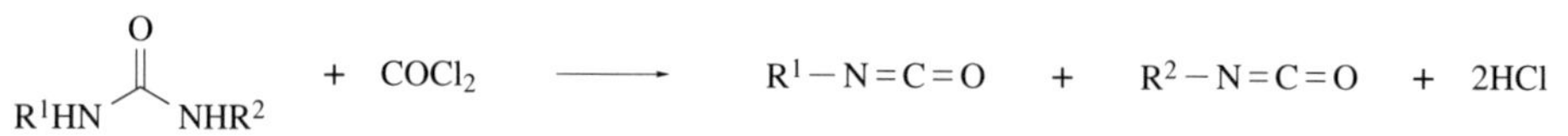

(23)

R^1, R^2 = aryl, alkyl, arylsulfonyl

The more basic 1,3-dialkylureas react with phosgene, even at room temperature, leading to 2,4-dialkylallophanoyl chlorides (**59**) and/or 1,3-dialkylchloroformamidinium chlorides (**60**). The allophanoyl chlorides decompose into isocyanate, almost in quantitative yields, when heated above 100 °C in an inert solvent. However, if the solution contains more than 10% of (**59**), or if the heating is carried out in the absence of a solvent, the products include not only isocyanates but also isocyanurates (**61**), which are formed by the reaction of the allophanoyl chlorides with the isocyanate formed (Scheme 27) ⟨64JOC3344⟩.

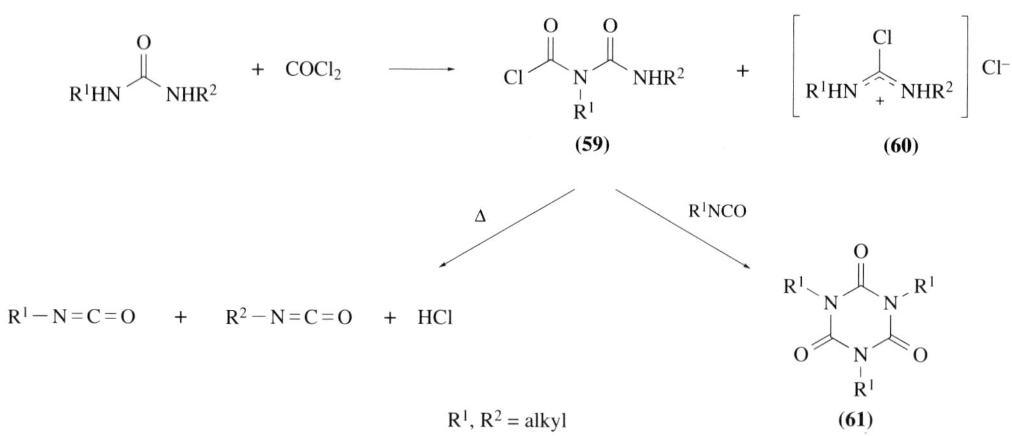

Scheme 27

Alternatively, the reaction of 1-arylsulfonyl-3-alkylureas with phosgene leads to attack exclusively on the alkyl-substituted nitrogen owing to the acidity of the proton in α-position to the sulfonyl group. The resulting allophanoyl chlorides decompose immediately to give the corresponding isocyanates and hydrogen chloride (Equation (24)). Since alkyl isocyanates have much lower boiling points than sulfonyl isocyanates, they can be easily separated from the latter by fractional distillation ⟨66AG(E)704⟩.

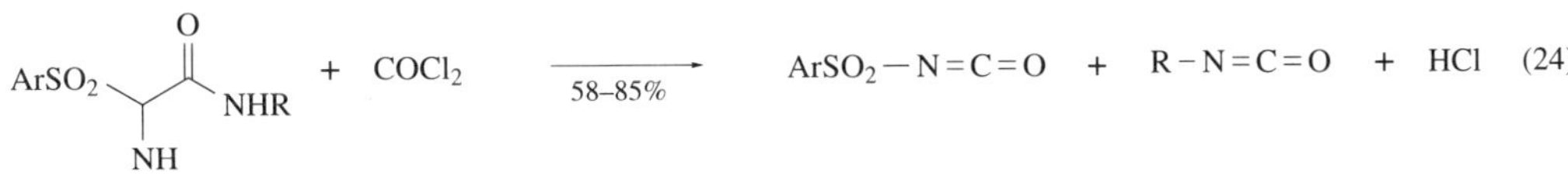

In order to obtain arylsulfonyl isocyanates, it is even simpler not to use the ureas, but to add catalytic amounts of an isocyanate to sulfonamides (Equation (25)) ⟨66AG(E)704⟩.

$$R^2SO_2NH_2 + COCl_2 + R^1{-}N{=}C{=}O \xrightarrow{32\text{–}93\%} R^2SO_2{-}N{=}C{=}O + 2HCl \quad (25)$$

In the case of phosgenation of alkylenediamines to obtain the corresponding diisocyanates, difficulties are encountered, owing to the low solubility of alkylenediamine dihydrochlorides. Nevertheless, alkylenediamines and sulfonyl isocyanates give 1,1′-alkylenebis(3-arylsulfonylureas) **(62)** in almost quantitative yields, which, unlike the free amines, react readily with only a small excess of phosgene leading to the corresponding diisocyanates (Scheme 28) ⟨66JOC2658⟩. Similarly, arylbis(sulfonylalkylureas) were phosgenated to give aryldisulfonyl diisocyanates.

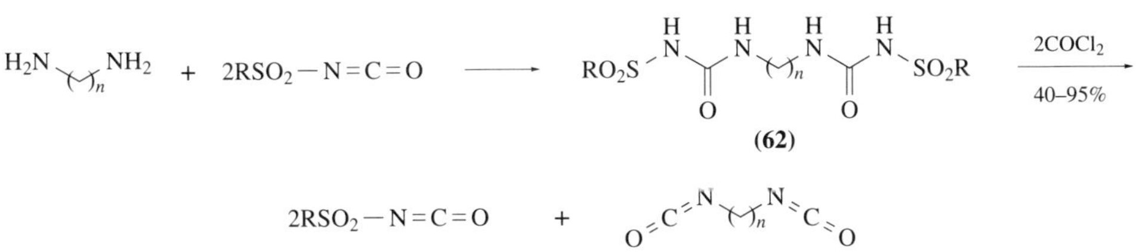

R = *p*-tolyl

Scheme 28

Another possibility for the preparation of interesting compounds is offered by the direct phosgenation of sulfanilylureas, which leads to difunctional compounds in which the sulfonyl isocyanate group is much more reactive than the aryl isocyanate group (Equation (26)) ⟨66JOC2658⟩.

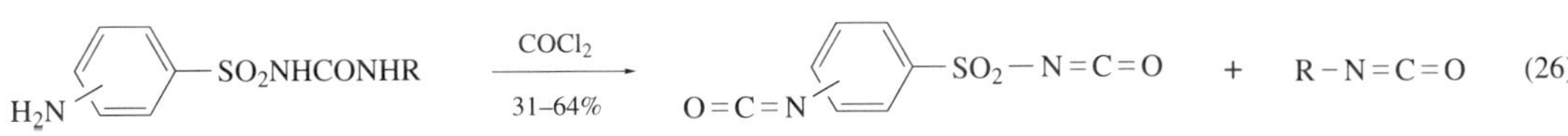

5.26.2.4.3 *By treatment with diaryl carbonates*

Instead of phosgene, diaryl carbonates may be used to prepare isocyanates from ureas (Equation (27)), although this method has the disadvantage that the products readily undergo further reaction with one another. Nevertheless, several patents have claimed that this method can be used for the preparation of both mono- and diisocyanates ⟨64GEP1154090, 73JAP7301653, 73JAP7320534⟩.

R = alkyl

5.26.2.4.4 *By treatment with thionyl chloride*

The reaction of 1,3-disubstituted ureas with thionyl chloride follows the same pattern as the reaction with phosgene (Scheme 29). Ureas containing primary alkyl groups are preferentially attacked on the nitrogen, whereas those containing secondary alkyl groups undergo reaction on the oxygen atom. The intermediates (**63**) are dehydrochlorinated even at room temperature leading to a mixture of isocyanate and sulfinylamine although the yields of the latter are somewhat lower, since part of this product reacts with the liberated hydrogen chloride ⟨66AG(E)704⟩.

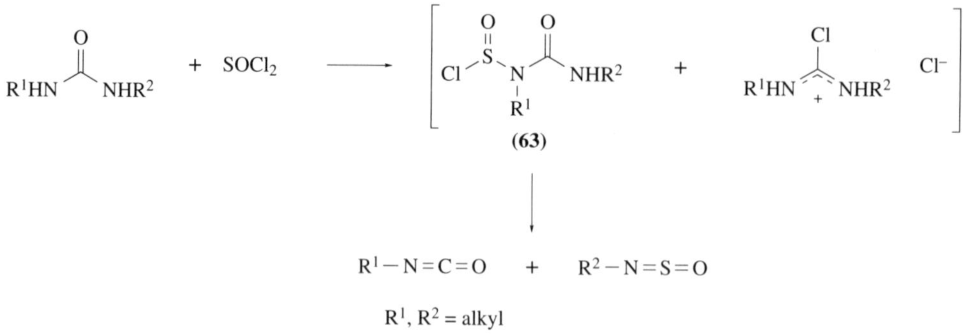

Scheme 29

5.26.2.4.5 *By treatment with phosphorus pentachloride*

In the reaction of 1,3-disubstituted ureas containing primary alkyl substituents with phosphorus pentachloride, *N*-attack is also observed leading to the formation of 1,2,4-phosphadiazetidin-3-ones (**64**) which decompose, on heating, into trichlorophosphazenes (**65**) and isocyanates (Scheme 30). However, this method has the disadvantage that the two products formed in the process readily undergo further reaction with one another to give carbodiimides ⟨62AG(E)595, 63JCS5558⟩. Compound (**64**) could not be isolated when 1-arylsulfonyl-3-alkylureas reacted with phosphorus pentachloride. However, it decomposed to give the corresponding isocyanates which did not react with the sulfonyltrichlorophosphazene (**65**) which was also formed ⟨65JOC2779⟩.

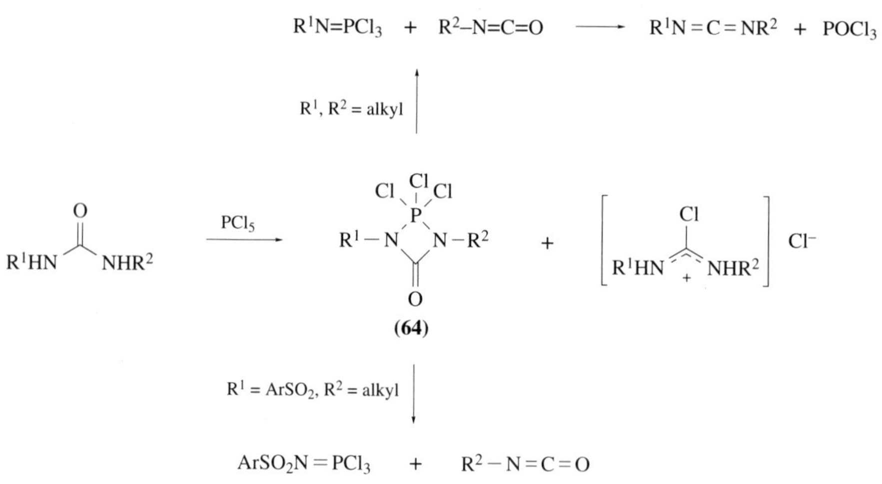

Scheme 30

5.26.2.4.6 *By thermolysis*

Although it is reported that *N*-acyl-*N*,*N'*-diarylureas decompose at 80°C giving 64–77% yields of isocyanate ⟨90UKZ1313⟩, it is very well known that *sym*-diorganylureas dissociate by thermolysis to give a mixture of amine and isocyanate, which is very difficult to separate due to the reversibility of the reaction (Equation (28)).

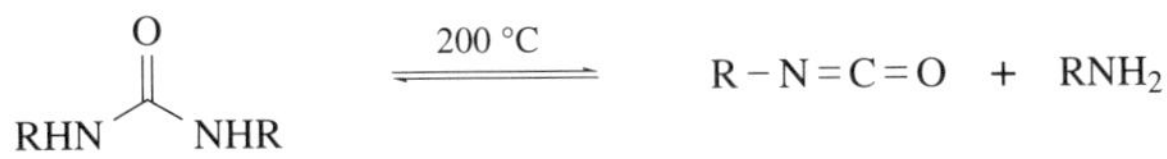

(28)

However Mironov and co-workers have found that aliphatic and aromatic isocyanates can be formed in high yield when these ureas are heated with high-boiling chlorosilanes (Equation (29)) (Table 2) ⟨80ZOB960, 81ZOB2017⟩. An increase in the reactivity of the chlorosilane leads to a decrease in the starting temperature of the reactions and to a rise in the reaction rate.

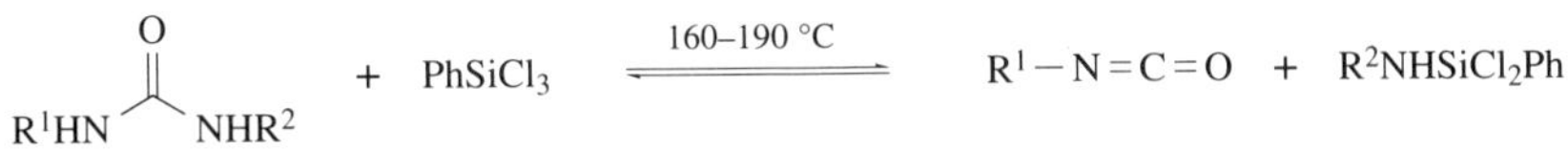

(29)

Table 2 Preparation of isocyanates from ureas and chlorosilanes.

R^1	*R^2*	*Chlorosilane*	*Yield (%)*
Me	Et	$PhSiCl_3$	77
Et	Me	$PhSiCl_3$	87
Bu^n	Et	Ph_2SiCl_2	85
Bu^n	Et	$PhSiCl_3$	89
$i\text{-}C_6H_{13}$	Me	$PhSiCl_3$	84
$c\text{-}C_6H_{11}$	Me	$PhSiCl_3$	80
$c\text{-}C_6H_{11}$	Me	$SiCl_4$	98
Ph	Me	$PhSiCl_3$	85
Ph	Et	Ph_2SiCl_2	69
Ph	Et	$PhSiCl_3$	99

5.26.2.5 From Azides

Collman *et al.* have demonstrated that the reaction of acyl or aryl azides with iridium(I) or rhodium(I) carbonyl complexes leads to the formation of organic isocyanates ⟨67JA4809⟩ (Equation (30)). Addition of CO regenerates the original carbonyl complex which can then react with more organic azide so that the overall process is catalytic. The activation energies are small and these reactions are rapid at low temperatures.

$$ArN_3 + CO \xrightarrow[M = Rh \text{ or } Ir;\ L = PPh_3]{MCl(CO)L_2} Ar{-}N{=}C{=}O + N_2 \qquad (30)$$

A related process is the stoichiometric reactions of the iridium(I) and rhodium(I) carbonyl azides (**66**) with CO to yield the corresponding isocyanates (**67**) (Equation (31)). The bisazido complexes of palladium and platinum (**68**) undergo a similar transformation (Equation (32)) ⟨67JA4809, 67AG(E)169⟩.

$$\underset{\mathbf{(66)}}{MN_3(CO)L_2} + CO \longrightarrow \underset{\mathbf{(67)}}{MNCO(CO)L_2} + N_2 \qquad (31)$$

M = Rh or Ir

$$\underset{\mathbf{(68)}}{M(N_3)_2L_2} + CO \longrightarrow \underset{\mathbf{(69)}}{M(NCO)_2L_2} + 2N_2 \qquad (32)$$

M = Pd or Pt

In contrast to the above results, the synthesis of free isocyanates from aromatic azides and uncomplexed CO, through an uncatalyzed reaction ⟨68JA3295⟩ has been reported. However, this reaction is carried out in an autoclave under high CO pressure (200–300 atm.) (Equation (33)).

$$ArN_3 + CO \xrightarrow{\text{pressure}} Ar\text{–}N{=}C{=}O + N_2 \qquad (33)$$

5.26.2.6 From Nitro and Nitroso Compounds

The one-step conversion of aromatic nitro compounds into isocyanates can be carried out by treating a solution of the nitro compound with carbon monoxide under pressure (200–1350 atm.) in the presence of a catalyst and at temperatures below 200 °C (Equation (34)). Benzene, chlorobenzene, cyclohexane, and 1,1,2-trichloro-1,2,2-trifluoroethane have been used as solvents, while the catalyst system consists of a noble metal and a Lewis acid. The noble metal can be supported, unsupported or in the form of a salt, oxide or complex and iron(III) chloride or bromide, iron(II) chloride or bromide, aluminum chloride or bromide, tin(IV) chloride, copper(II) chloride, and anhydrous hydrogen chloride are among the Lewis acid co-catalysts used successfully ⟨67TL961, 73JOC1316⟩.

$$RNO_2 + 3CO \xrightarrow[\text{R = alkyl, aryl}]{\text{catalyst}} R\text{–}N{=}C{=}O + 2CO_2 \qquad (34)$$

Similarly, the preparation of organic mono- and diisocyanates following the same scheme is also reported, although the catalyst system used in this case is formed by heteroaromatic sulfur compounds and a noble metal halide or oxide. The catalyst system may also contain a third component selected from oxides of the group 5 or 6 metals ⟨69GEP1910303⟩.

Aromatic nitro compounds are also reductively carbonylated to form aryl isocyanates in the presence of rhodium carbonyl complexes and metal halides of groups 5 and 6 of the periodic table. Substituents in *o*-position have proved to increase the reaction rate ⟨79JPR928⟩. The same methodology can also be applied to the carbonylation of aromatic nitroso compounds (Equation (35)) by using rhodium and iridium carbonyls ⟨77JPR841⟩.

$$RNO + 2CO \xrightarrow{\text{catalyst}} R\text{–}N{=}C{=}O + CO_2 \qquad (35)$$

When the catalytically-activated reductive carbonylation of nitrobenzene is carried out in the presence of alcohols, the corresponding *N*-phenylcarbamates are formed which are important precursors for the preparation of both mono- and diisocyanates ⟨92CCC1505⟩.

5.26.2.7 From Alkenes and Arenes

5.26.2.7.1 Reaction with iodine isocyanate

The addition of iodine isocyanate (INCO) to a wide variety of alkenes proceeds smoothly even at room temperatures, constituting a useful approach to the synthesis of isocyanates ⟨60CB519, 63AG(E)213⟩. It has been demonstrated that the additions generally occur in a stereospecific manner and that the iodine and isocyanate functions are introduced *trans* to each other, and diaxially in rigid, fused cyclohexanes (Equation (36)) ⟨64TL1125, 65AG(E)370, 65AG(E)449, 65JOC1748⟩. These results have been explained by the formation of a three-membered ring iodonium ion intermediate which is opened from the backside by the isocyanate ion. Consistent with this interpretation is the fact that the open chain *cis*- and *trans*-2-butenes give cleanly and in almost quantitative yield *threo*- and *erythro*-iodo isocyanate adducts, respectively ⟨70JA1326⟩.

(36)

The reagent was first prepared by Birckenbach and Lindhard ⟨30CB2528, 30CB2544⟩, who proved that it existed as a monomer although it was never isolated as such. Nevertheless, an ethereal solution of INCO can be prepared from iodine and freshly prepared silver cyanate in ether. Iodine reacts only slowly with silver cyanate at 0 °C in ether but, if an alkene is present in the mixture, the INCO generated *in situ* will react rapidly with the alkene, shifting the equilibrium of the reaction of iodine and silver cyanate toward INCO. In the latter case, the overall reaction with alkene is often complete within 30 min. Alternatively, iodine can be allowed to react with a large excess of silver cyanate in a variety of solvents, and the precipitated silver iodide can be filtered off. Alkenes can then be added to the preformed solution of INCO, and reaction usually takes place within minutes. It is also possible to generate INCO in solution by treating iodine or iodine monochloride with other metal cyanates but the yields are usually lower than those obtained from the silver cyanate–iodine reaction ⟨67JOC540⟩.

The reaction has general applicability and the rate of addition of INCO decreases in the following order: cyclopentene > cycloheptene > cyclohexene > cyclooctene > cyclododecene. Methylene cycloalkanes react readily as do styrenes. In these cases the direction of the addition is such that the isocyanate function appears at the more substituted carbon. In simple terminal alkenes such as 1-hexene, addition takes place readily but a mixture of products is obtained with the isocyanate at the terminal as well as at the secondary carbon. Although INCO addition has all the markings of electrophilic additions to alkenes, and proceeds via a three-membered ring iodonium-ion intermediate, the reaction seems to have definite steric requirements ⟨67JOC540, 69JOC1677⟩.

The addition of INCO to conjugated and nonconjugated aliphatic and alicyclic dienes, alkenes, and methylallene can also be achieved, while conjugated unsaturated carbonyl compounds are recovered unchanged on exposure to INCO ⟨67JOC3665⟩. Alternatively, reactions of INCO with allenes at −78 °C gave addition preferentially to the substituted double bond of alkyl allenes, while addition at 0 °C gave a 1 : 1 mixture of isomers from addition to both double bonds, except in the case of 1,1-dimethylallene, which gave preferentially addition to the unsubstituted double bond. Phenylallene in both cases furnished a 1 : 1 mixture of isomers ⟨73ACS3368⟩.

5.26.2.7.2 *Reaction with isocyanic acid*

In general, an alkene reacts with isocyanic acid with difficulty, but certain alkenes with electron-releasing substituents react readily. As an example, α-alkoxy isocyanates could be prepared through the addition of isocyanic acid to α,β-unsaturated ethers, such as ethyl and butyl vinyl ether, vinyloxyethyl methacrylate and butyl isopropenyl ether, in the presence of *p*-toluenesulfonic acid (Equation (37)). Solvents, such as diethyl ether or benzene, are necessary to moderate the reaction and to stabilize the isocyanic acid against trimerization. Polymerization of the unsaturated ether is the principal competitive reaction and is minimized by using an excess of isocyanic acid ⟨63JOC2082, 63CI(L)1280⟩.

(37)

The same method applies to the addition of isocyanic acid to *m*- and *p*-diisopropenylbenzenes, isoprene, styrene, α-methylstyrene, and other alkenes (Equation (38)). In general, the most reactive alkenes were those having terminal double bonds with at least one electron-releasing group, such as methyl or phenyl, on C2. Isopropenylbenzenes were considerably more reactive than the other alkenes studied ⟨64JOC143⟩.

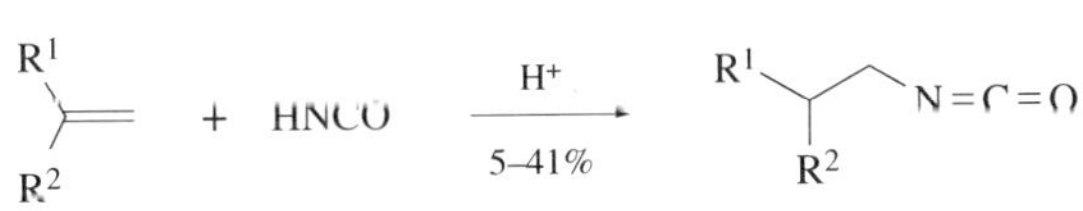

(38)

Certain cyclic and bicyclic alkenes such as norbornene, 2-ethylidenenorbornene, norbornadiene, and dicyclopentadiene, react with isocyanic acid to give the corresponding isocyanate ⟨66FRP1440591, 72CRV457, 73JAP7300455⟩.

A patent describes the preparation of organic isocyanates by treating hydrocarbons containing at least one alkenic double bond with isocyanate ion under anhydrous conditions, and in the presence of a salt or coordination compound of Pd or Pt, but yields are not given ⟨69BRP1142991⟩.

5.26.2.7.3 *Electrochemical substitution by cyanate ion*

When methanolic solutions of anisole containing potassium cyanate was electrolysed at an anode potential of +1.8 V, substitution of hydrogen by a methylcarbamyl group along with the formation of methyl carbamate was observed. Although an intermediate isocyanate was postulated, it was not isolated ⟨68TL3341⟩.

Related to the above results, it has also been reported that when alkenes or aromatic hydrocarbons are bubbled, in the presence of alcohols or amines, through a melt obtained by fusing cyanate and ureas at 135°C through which an electric current is passed, carbamates or ureas, respectively, are produced ⟨72CRV457⟩.

5.26.2.8 From Aldehydes and Ketones and their Derivatives

A number of α-hydroxy isocyanates were synthesized by addition of isocyanic acid to certain carbonyl compounds such as formaldehyde, chloral, *s*-dichlorotetrafluoroacetone, trifluoroacetaldehyde, and perfluorocyclobutanone (Equation (39)). It appears that most carbonyl compounds capable of forming hydrates undergo this reaction and, in general, the carbonyl compounds should have electronegative substituents in the α-position. The reaction takes place at temperatures of 0°C or below ⟨63JOC1825⟩.

(39)

Due to the high reactivity of hydroxy isocyanates, this approach has been successfully utilized to prepare a wide variety of other isocyanates. For example, hydroxymethyl isocyanate forms adducts with α,β-unsaturated ethers such as dihydropyran and butyl vinyl ether (Equation (40)). This reaction proceeds readily at room temperature with an acidic catalyst such as *p*-toluenesulfonic acid.

(40)

Hydroxymethyl isocyanate, formed from the reaction of formaldehyde, reacts in the cold with thionyl chloride to give first the chlorosulfite, which loses sulfur dioxide on warming to room temperature leading to the formation of chloromethylcarbamoyl chloride (Scheme 31) which, in turn, can be used to prepare other isocyanates. For example, it can be converted into chloromethyl isocyanate by treatment with α-pinene, D,L-limonene, acrylonitrile or pyridine, or it can be added to styrene in the presence of small amounts of mercury(II) chloride to give cinnamyl isocyanate, or it can be reacted with *m*-xylene, in the presence of zinc chloride as a catalyst, to give the corresponding isocyanate (Scheme 32) ⟨63JOC1825⟩.

The formation of phenylazomethyl isocyanates (**71**) can be achieved, in 30–71% yield, by the

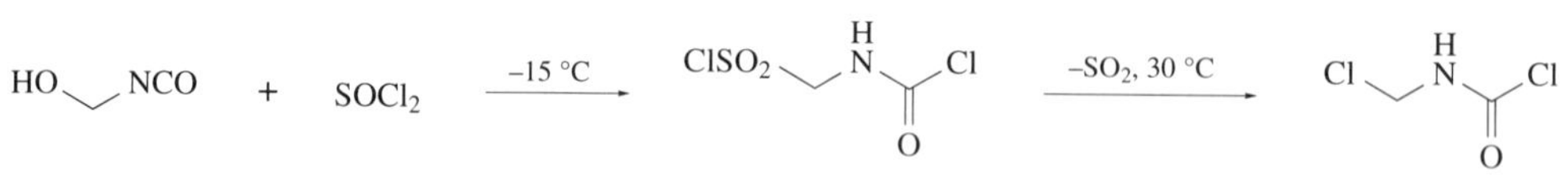

Scheme 31

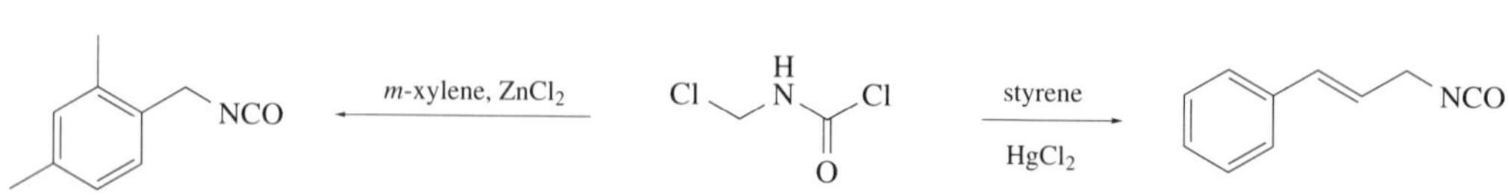

Scheme 32

oxidation of 2-phenylsemicarbazones (**69**) with chromyl acetate (Scheme 33). Presumably, the initially formed nitrene (**70**) rearranges to the isocyanate via a five-membered ring ⟨68AG(E)293⟩.

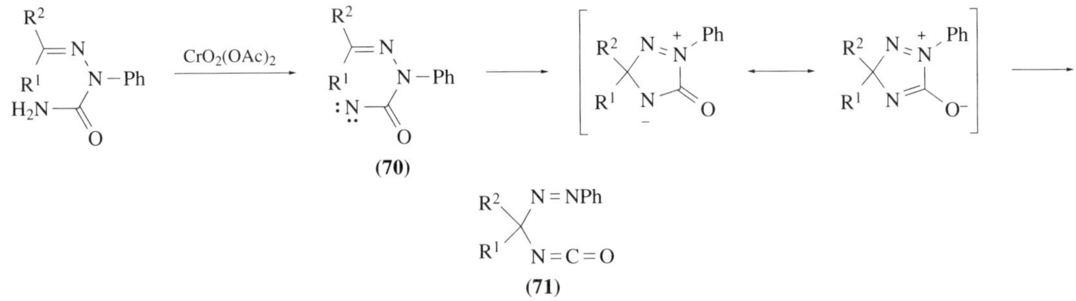

Scheme 33

A patent has reported the preparation of perfluoroisopropyl isocyanate by reaction of perfluoroacetone with potassium fluoride, followed by treatment of the intermediate alkoxide with ClCN ⟨72BRP1298049⟩.

5.26.2.9 From C=N-Containing Functionalities

5.26.2.9.1 From amidines

N-Arylamidines (**72**; R = aryl) react with phosgene in hot inert solvents (benzene, toluene, xylene or chlorobenzene) (80–110 °C) yielding 4-quinazolones (**74**) in almost quantitative yields, which are formed by way of the iminoalkyl isocyanate (**73**) as a nonisolable intermediate. However, *N*-alkylamidines (**72**; R = alkyl) react with phosgene under the same reaction conditions to give iminoalkyl isocyanates (**73**), as relatively stable colorless liquids, in yields ranging from 30–63% (Scheme 34) ⟨67AG(E)864⟩.

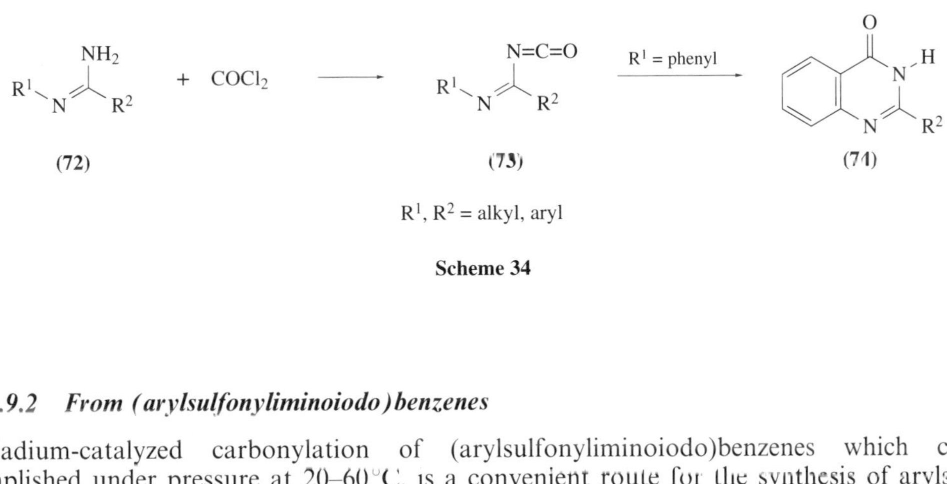

Scheme 34

5.26.2.9.2 From (arylsulfonyliminoiodo)benzenes

Palladium-catalyzed carbonylation of (arylsulfonyliminoiodo)benzenes which can be accomplished under pressure at 20–60 °C, is a convenient route for the synthesis of arylsulfonyl isocyanates in yields ranging from 52–80% (Equation (41)) ⟨93TL2839⟩. Suitable catalysts for the transformation are chloro-complexes of palladium(II), such as $PdCl_2(PhCN)_2$, $PdCl_2(CH_3CN)_2$, or $PdCl_2$ with CH_3CN or PhCN additive. It is noteworthy that carbonylation is restricted to the formation of sulfonyl isocyanates, since even in the case of aromatic rings bearing nitro substituents, this function remains untouched.

$$ArSO_2N{=}IPh \ + \ CO \xrightarrow{\text{Pd-catalyst, solvent}} ArSO_2{-}N{=}C{=}O \ + \ PhI \quad (41)$$

The fact that carbonylation of (arylsulfonyliminoiodo)benzenes takes place under identical reaction conditions to carbonylation of *N*-chloroarylsulfonamidates ⟨90AG(E)1147⟩, and in both cases the reactions afford arylsulfonyl isocyanates, makes it plausible that these reactions occur via

CO insertion into the palladium—nitrogen bond of coordinated arylsulfonylnitrenes, which is in agreement with the behavior of some imidocarbonylruthenium clusters which underwent carbonylation to give phenyl isocyanate and other phenyl isocyanate clusters ⟨92OM1242⟩.

5.26.2.9.3 *From alkoxycarbonylimino phosphoranes*

Kirsanov found in 1954 that phosphorus pentachloride reacts very readily with alkyl carbamates to give alkoxycarbonyliminotrichlorophosphoranes which lose alkyl chloride, when heated, to form phosphoryl dichloride isocyanate (Scheme 35). The mixture must be heated very slowly since the reaction is strongly exothermic, so much so that explosions may occur if heating is fast. The higher the molecular weight of the alkyl groups, the more stable the intermediate phosphoranes, and the less vigorous their decomposition. The intermediate in the reaction, which proceeds almost quantitatively, need not be isolated ⟨54ZOB1033, 59ZOB2256, 61ZOB1607, 65ZOB923⟩.

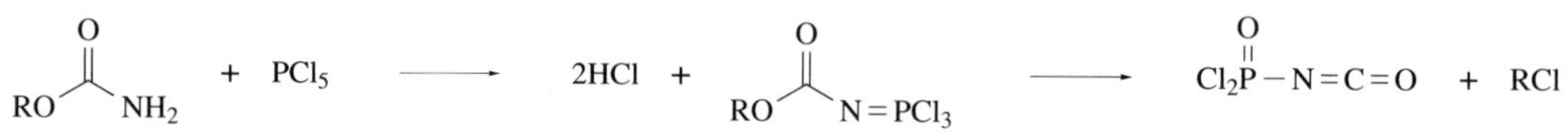

Scheme 35

Chlorophosphazo compounds ROCON=PX_2Cl with at least one chlorine atom attached to the phosphorus can also decompose to give the isocyanate (Equation (42)) ⟨61ZOB3062, 62ZOB3759, 63ZOB1587⟩.

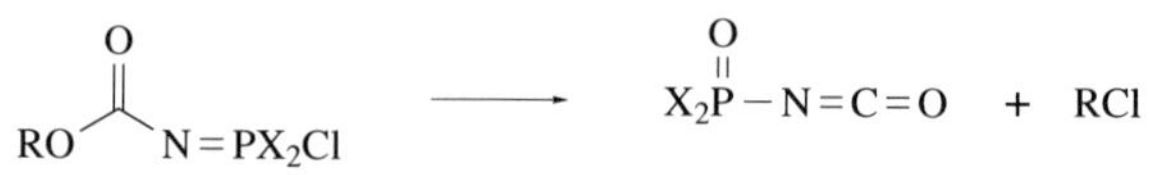

(42)

R = alkyl, X = alkyl, aryl, OAr, Cl

Derkach found a general and convenient method for the preparation of similar compounds by reaction of chloroiminocarbonates with phosphorus(III) halides (Scheme 36). However, this method is not suitable for the preparation of alkylphosphonic chloride isocyanates [R(Cl)P(O)NCO] ⟨69AG(E)421⟩. On the other hand, phosphoryl dichloride isocyanate [$Cl_2P(O)NCO$] is obtained only in low yields from diethyl *N*-chloroiminocarbonate ⟨64ZOB1161⟩.

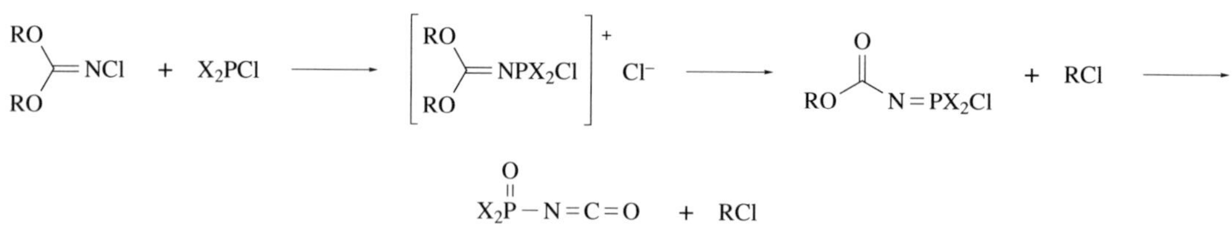

Scheme 36

5.26.2.9.4 *From hydrolysis of perfluoroazaalkenes*

Haszeldine and co-workers found that perfluoro-2-azapropene could be converted into trifluoromethyl isocyanate by reaction with water, although the conditions described are very difficult to apply on a preparative scale ⟨56JCS3428⟩.

A more convenient method of achieving the same result is to heat the perfluoro-2-azapropene with a hydrated fluoride. Thus, an estimated 83% yield of isocyanate was obtained. The most important disadvantage of this method is the large amount of azapropene which has to be used, since it acts as an HF (high-frequency) acceptor (Equation (43)) ⟨59JA1587⟩.

(43)

5.26.2.9.5 From fluoroazaalkanes

The preparation of fluoroalkyl isocyanates has been achieved by carbonylation of fluoroazaalkanes with carbon monoxide, which suggests that the corresponding nitrene may be an intermediate. For example, trifluoromethyl isocyanate was obtained in 88% yield by reaction of hexafluoroazomethane with carbon monoxide at 325 °C under 650 atm. ⟨62JA2337⟩.

5.26.2.10 From Carboxylic Acid Derivatives

5.26.2.10.1 From acyl azides (Curtius rearrangement)

The decomposition of acyl azides to isocyanates and nitrogen is known as the Curtius rearrangement (Equation (44)). This is a useful preparative method for isocyanates, and for compounds derived from isocyanates. When coupled with a hydrolytic step, the Curtius rearrangement becomes a practical procedure for replacing a carboxyl group by an amino group. Curtius degradation has been successfully carried out on aliphatic, alicyclic, aromatic, and heterocyclic acids, on saturated and unsaturated acids and on acids containing various functional groups ⟨46OR(3)337, B-71MI 526-01⟩, using different solvents, especially mixtures of nonpolar and polar aprotic solvents ⟨89MI 526-01⟩.

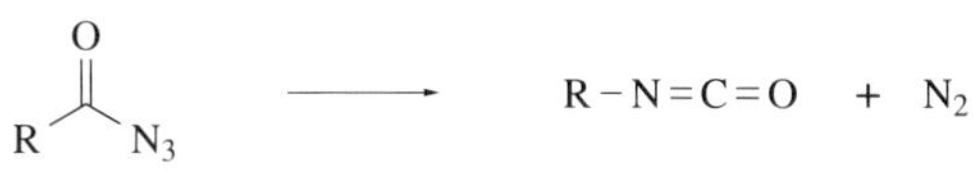
(44)

The carbonyl azides, discovered by Curtius ⟨1890CB3023⟩, can be obtained either by reaction of carboxylic acid chlorides with sodium azide, or by treatment of carboxylic acid hydrazides with nitrous acid. These compounds can undergo rearrangement to isocyanates on thermal decomposition, although the temperatures required to bring about decomposition of acyl azides are quite low, usually being about 100 °C ⟨1894CB778⟩. The mechanism of the Curtius rearrangement has been debated since Stieglitz proposed a carbonyl nitrene intermediate ⟨1896MI 526-01⟩. Some authors have adopted this mechanism while others prefer a concerted one. Nevertheless, the problem has been critically discussed by Smith, who considers the possibility of a spectrum of mechanism between the two extreme ones ⟨B-63MI 526-01, B-66MI 526-01⟩. The Curtius rearrangement can be carried out thermally or photolytically, and it is one of the best methods for preparing isocyanates.

Following the above method, even functionalized isocyanates have been prepared. For example, 4-difluoroaminobutyric acid chloride was transformed into the corresponding acyl azide, which decomposes on heating in benzene to give the isocyanate in 85% yield ⟨70JOC1545⟩. Moreover, the synthesis and applications in heterocyclic synthesis of isocyanates having alkyl- or arylthiomethyl substituents has also been described to proceed through the corresponding acyl azide ⟨85S276⟩. Similarly, this reaction has also been described as the key step in the synthesis of 1-aminocyclopropenecarboxylic acid, which is the unsaturated analogue of 1-aminocyclo-propanecarboxylic acid (ACC), the biosynthetic precursor of ethylene in plants ⟨87JOC4875⟩.

Vinyl and isopropenyl isocyanates are obtained in 86% and 84% yield, respectively, by reacting a toluene solution of acrylic or methacrylic acid with an aqueous solution of sodium azide ⟨56BSB291⟩. Similarly, by the thermal Curtius reaction, each *trans*-cinnamoyl azide (**76**), obtained by treating *trans*-cinnamoyl chloride (**75**) with sodium azide, gave the corresponding *trans*-β-styryl isocyanate (**77**) in overall yield ranging from moderate to good. In an inert solvent at 250 °C, *trans*-β-styryl isocyanate (**77**) and its β-methyl and β-phenyl derivatives were efficiently isomerized into the corresponding isocarbostyril (**78**) (Scheme 37) ⟨72JOC724⟩. A highly convergent construction of polycyclic 2-pyridone systems has been described by reaction of vinyl isocyanates, prepared directly from the corresponding α,β-unsaturated acids by the Curtius procedure, with enamines ⟨89JOC224⟩.

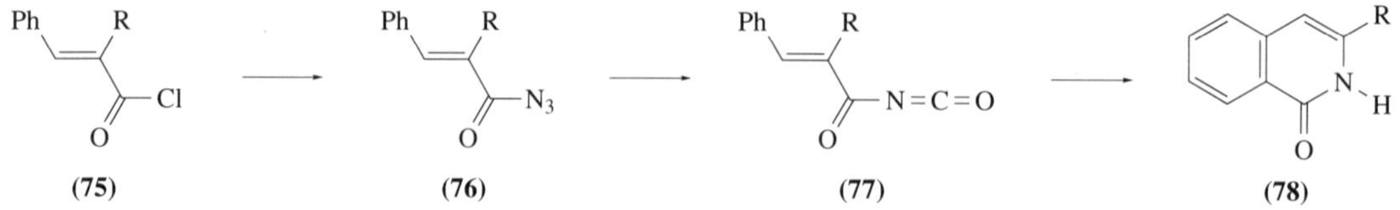

(75) **(76)** **(77)** **(78)**

R = H, Me, Ph

Scheme 37

Thermolysis of heterocarbonyl azides to isocyanates by Curtius rearrangement, appears to be one of the most convenient routes to heteroaromatic isocyanates. For example, β-(5-nitro-2-furyl)vinyl isocyanate (**79**) could be isolated in quantitative yield from the Curtius rearrangement of β-(5-nitro-2-furyl)acryloyl azide ⟨68BCJ1258⟩. However, whereas β-thienylvinyl isocyanate (**80**) readily undergoes cyclization giving rise to the thienopyridone isomer ⟨70BSB301⟩, thienyl isocyanate (**81**) was formed by thermal rearrangement of the corresponding thienoyl azide in boiling carbon tetrachloride ⟨77JOC1508⟩. Also, pyridine-3-isocyanate (**82**) has been successfully synthesized in 64% yield by decomposition of nicotinyl azide in benzene ⟨67CI(L)1406⟩, and 5-isocyanato derivatives of uracil (**83**) and of 1,3-dimethyluracil (**84**) were similarly prepared in 98% and 95% yields, respectively, by heating the corresponding uracil-5-carbonylazides in xylene, mesitylene or toluene. In contrast to these results, all attempts to isolate the isocyanates derived from uracil-6-carbonylazide or 1,3-dimethyluracil-6-carbonylazide were unsuccessful. This behavior may be explained by the fact that the 5-position of uracil is less electron-deficient than the 6-position ⟨72JHC1267⟩.

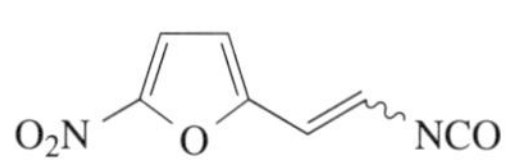

(79)

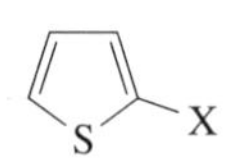

(80) X = CH=CH–NCO
(81) X = NCO

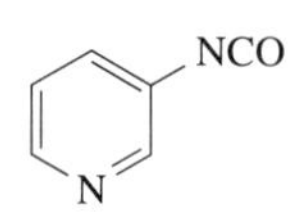

(82)

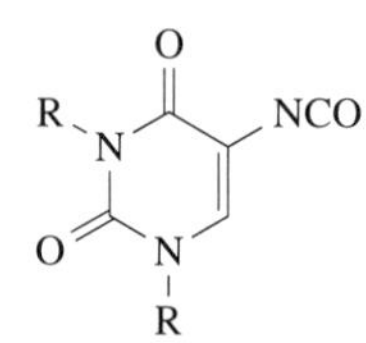

(83) R = H
(84) R = Me

It is also reported that 2-chloro-4-(trifluoromethyl)-5-isocyanatothiazole (**85**) and 2,5-dichloro-4-isocyanatothiazole (**86**) are obtained, in 87% and 73% yields, respectively, by thermal decomposition in dry refluxing carbon tetrachloride of the corresponding acyl azide (Scheme 38) ⟨91JHC1003⟩.

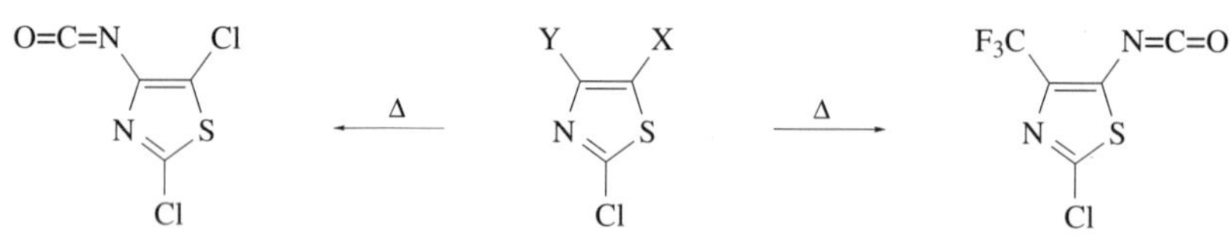

(86) X = CON_3, Y = CF_3; X = Cl, Y = CON_3 **(85)**

Scheme 38

Δ^2-Isoxazoline 5-isocyanates are used for the preparation of isoxazole derivatives through a process which involves elimination of isocyanic acid from the starting isocyanates, which in turn were prepared by Curtius rearrangement of the corresponding acid azide ⟨72MI 526-01⟩.

The Curtius rearrangement has also been used to synthesize diisocyanates. For example, diisocyanates (**87**) have been synthesized in quantitative yields by refluxing the corresponding diacylazides in dry benzene ⟨92MI 526-01⟩. Similarly, ether-diisocyanates (**88**) or sulfido-diisocyanates (**89**) and (**90**) were prepared *in situ* in 51–60% yield from the corresponding acyl diazides, which in turn were prepared by sequential treatment of the corresponding aliphatic diesters with hydrazine hydrate, followed by transformation of the dihydrazides into acyl diazides by means of acidified water solution of sodium nitrite ⟨93JPR294⟩. Moreover, following this route, Lesiak and Nielek have described the preparation of 2,5-diisocyanatofuran in 84% yield by heating 2,5-dicarbazidofuran in anhydrous toluene ⟨88JPR825⟩; using a closely related process, pyridine-2,6-diisocyanate was also obtained in 31% yield ⟨67CI(L)1406⟩.

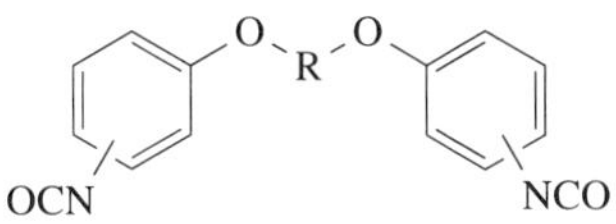

(87)
R = CH_2, $(CH_2)_2OCH_2O(CH_2)_2$

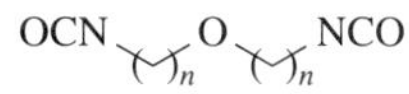

(88)
n = 2, 3, 4

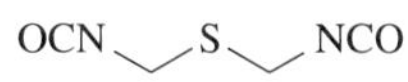

(89)

(90)
n = 1, 2

The usual procedure for obtaining acid azides from acid chlorides and sodium azide cannot be applied to the industrial synthesis of isocyanates since it suffers from the following disadvantages: (i) it is often difficult to obtain pure acid chlorides in good yields from acids; (ii) the insolubility of sodium azide in common organic solvents makes it desirable to use a water-containing solvent mixture; (iii) it is very hazardous to handle the acyl azides on a large scale. For these reasons, modifications to this reaction have been developed, which involve both variation of the acyl component and the azido reagent used in the preparation of the starting carbonyl azide.

Thus, mixed anhydrides react with sodium azide under very mild conditions to form acid azides, in excellent yields, which can be rearranged to the isocyanate ⟨61JOC3511⟩. A variety of *trans*-1-*N*-acylamino-1,3-dienes can be conveniently prepared, on large scales, from dienoic acids using the previously mentioned Curtius modification: dienoic acids (**91**) were converted, via their mixed anhydrides, into the azide derivatives (**92**) which were not isolated. The diene isocyanate produced after heating in toluene was either trapped as it was formed or cooled to room temperature before the trapping agent was added. This procedure provides a general route to the reactive dienamides (**93**) which have heteroatom acyl substituents (Scheme 39) ⟨78JOC2164⟩.

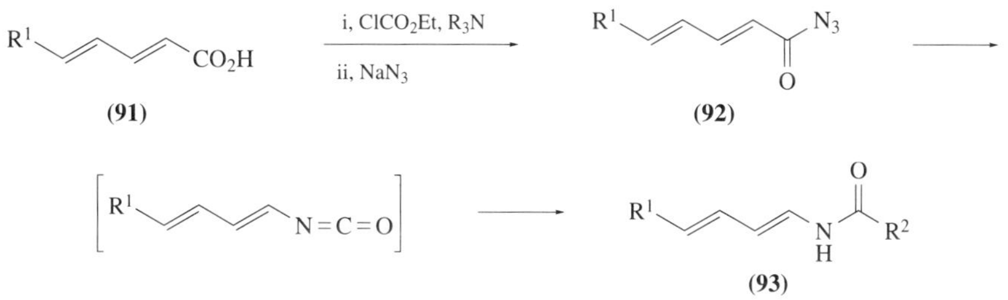

R^2 = OR^3, SR^3, NR^3_2

Scheme 39

Formation of acyl azides has also been achieved by intermolecular trapping of presumed acyl iron(III) complexes with azide ion. Curtius rearrangement of these compounds provides substituted 7-*syn*-amino-2-norbornenes ⟨86JOC5123⟩.

Acyl azides can be prepared under mild and homogeneous conditions by using hydrazoic acid and pyridine ⟨73S413⟩, or tetramethylguanidinium azide ⟨71JOC2387⟩. However, these methods require the handling of hydrazoic acid with its concomitant hazards. In contrast, tetrabutylammonium azide, which has been obtained in pure crystalline form, is soluble in a number of organic solvents, and the reaction with carboxylic acid chlorides can thus be carried out in a one-phase system. The resulting acyl azides can be transformed into isocyanates by heating, and it is possible to perform reactions with the isocyanates in the same solution that was used in the Curtius rearrangement. Final separation from the tetrabutylammonium chloride should ordinarily cause no problems, since the salt is readily removed from organic solvents by water ⟨74ACS(B)699⟩.

Another modification of this reaction is the method developed by Kricheldorf, which involves the reaction of alkanoyl, and substituted-alkanoyl, chlorides with the stable and nonhazardous trimethylsilyl azide. The alkanoyl azides so formed undergo Curtius rearrangement under moderate thermal conditions to the corresponding isocyanates in high yields (Scheme 40) ⟨72S551⟩. This procedure affords mainly primary alkyl isocyanates, but has been further applied to the preparation of some other functionalized primary alkyl isocyanates such as 2-fluoro-alkylethyl isocyanates ⟨92JFC(56)85⟩. This isocyanate synthesis via silyl azides offers a route, under nearly neutral conditions, to diene isocyanates which undergo electrocyclic ring closure to give 3-substituted-2(1H)-pyridones ⟨73JOC2982⟩.

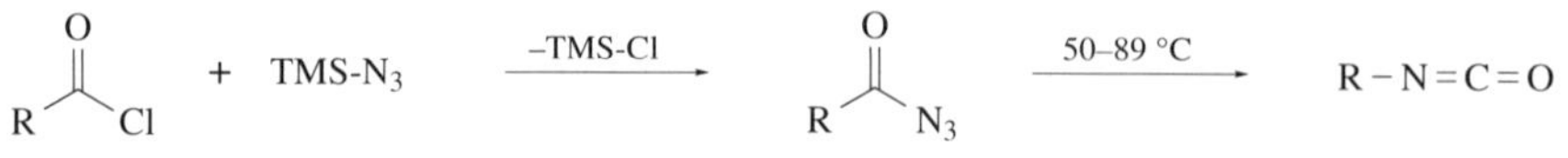

Scheme 40

Heterocyclic isocyanates such as 2-thienyl isocyanate, 2-benzo[*b*]thienyl isocyanate and 3-thienyl isocyanate have also been prepared in 52%, 79% and 70% yields, respectively, by reaction of the corresponding thienoyl chloride with trimethylsilyl azide, even at room temperature ⟨92JCS(P1)1101⟩.

Although trimethylsilyl azide reacts with acid chlorides under reflux temperature to give the corresponding isocyanate directly, Olah and co-workers have reported that under zinc iodide catalysis, aroyl chlorides react with trimethylsilyl azide at room temperature in dichloromethane solution, providing the corresponding aroyl azides which can be isolated in high yields (85–96%). The reaction does not proceed without the catalyst, and it works well with electron-releasing as well as electron-withdrawing substituents, although the reaction is slower with the latter ⟨83JOC3358⟩. Similarly, tertiary cyclo(polycyclo)alkyl isocyanates have been prepared in near quantitative yield by the zinc iodide catalyzed reaction of tertiary alkanoyl chlorides and trimethylsilyl azide (Equation (45)) ⟨90JOC4282⟩.

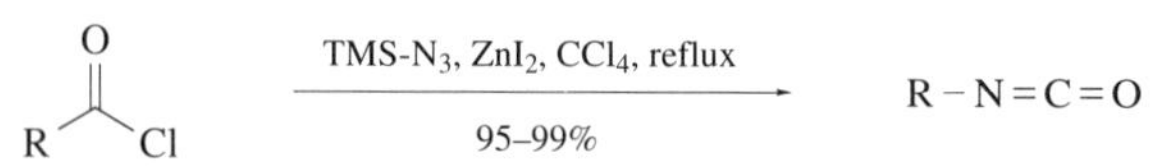

(45)

This procedure has also been applied to the preparation of similar isocyanates from anhydrides. Thus, aliphatic or aromatic carboxylic acid anhydrides (**94**) react with silyl azides on heating to form carboxylic acid silyl esters and isocyanates, while cyclic carboxylic acid anhydrides (**95**) lead, in the same way, to isocyanatocarboxylic acid silyl esters (**96**) (Scheme 41 and Equation (46)) ⟨72CB3958⟩. In the case of maleic anhydride, formation of 1-trimethylsilyl-1,3-oxazine-2,4-(6H)-dione is reported, which could be explained through a cyclization process of the intermediate trimethylsilyl 2-isocyanatoacrylate which arises from a Curtius rearrangement of a maleic anhydride-silyl azide adduct ⟨72JOC1738⟩.

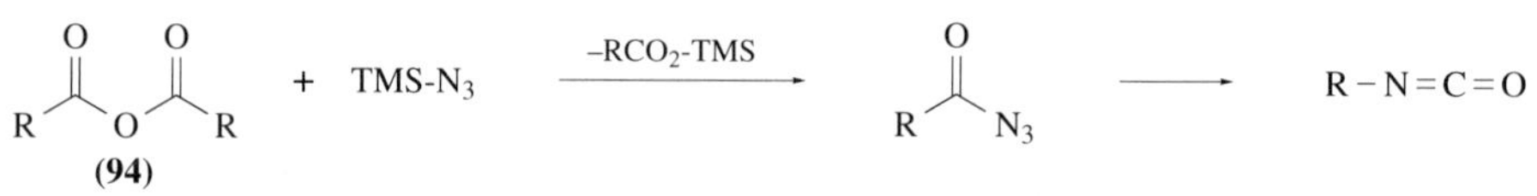

Scheme 41

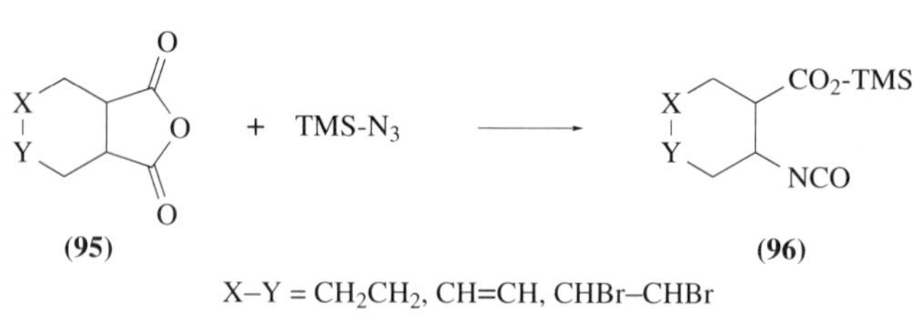

(46)

On the other hand, while under standard conditions 1,2-dimethyl-1,2-cyclopropanedicarboxylic anhydride (**97**) reacts with trimethylsilyl azide to give the expected trimethylsilyl 2-isocyanato-1-cyclopropanecarboxylates (**98**), in the presence of pyridine as catalyst, the initially formed isocyanate (**98**) undergoes decarboxylation and ring expansion with formation of 3,5-dimethyl-2-(trimethylsilyloxy)pyrrole (**99**), which in turn is reversibly isomerized to 3,5-dimethyl-1-trimethylsilyl-3-pyrrolin-2-one (**100**) (Scheme 42) ⟨73CB3753⟩.

Another viable modification for the synthesis of isocyanates is the reaction of carbonyl chlorides with diphenyldiazidosilane. For example, cyclopropanecarbonyl chloride reacts with this reagent to yield cyclopropyl isocyanate ⟨73CB3753⟩.

Acyl chlorides as well as carboxylic anhydrides and thioanhydrides can all be converted into isocyanates with stannyl azides (Scheme 43) ⟨76S329⟩. The preferred reagent is tri-*n*-butylstannyl azide because it is not explosive, not sensitive to hydrolysis, not particularly toxic, stable on storage,

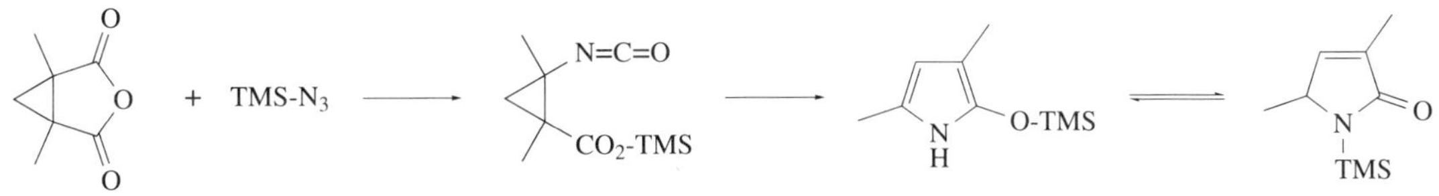

(97) **(98)** **(99)** **(100)**

Scheme 42

and readily accessible in larger quantities by reaction of stannyl chlorides with sodium azide. However, stannyl azides also add to isocyanates to form readily decomposable *N*-stannylcarbamoyl azides. This side reaction can be avoided by first acylating the stannyl azide, for example, with equimolecular amounts of dicarboxylic anhydride, at 25–35 °C, and subjecting the resulting acyl azides to Curtius degradation. In this way, the stannyl esters of ω-isocyanatocarboxylic acids are formed almost quantitatively. Since stannyl esters are resistant to alcoholysis, in contrast to silyl esters, alcohols can be added to the isocyanate group without side reactions; hydrolysis of the resulting *N*-alkoxycarbonyl-ω-amino stannyl ester gives *N*-protected ω-amino acids ⟨77AG(E)550⟩.

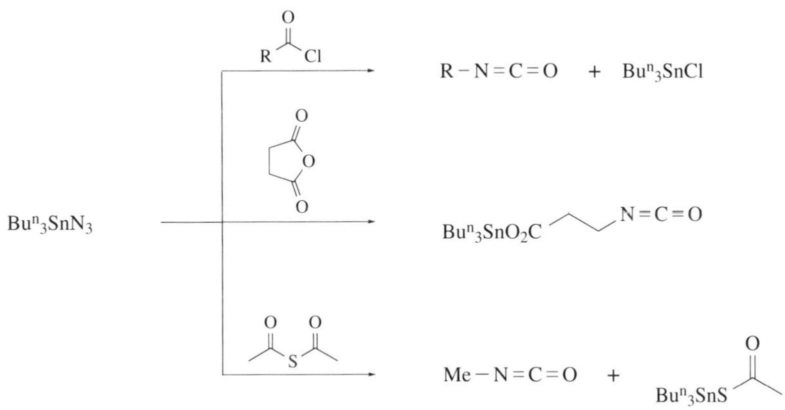

Scheme 43

Yamada and co-workers reported a simple one-step conversion of carboxylic acids to urethanes by refluxing an equimolecular mixture of a carboxylic acid, diphenyl phosphorazidate (dppa) and triethylamine in the presence of an alcohol (Scheme 44) ⟨72JA6203⟩. This reaction is general for aliphatic, aromatic, and heterocyclic carboxylic acids, and it may be considered as a modified Curtius reaction ⟨74T2151⟩. This modified Curtius reaction is much simpler and less laborious than the classical one; it proceeds under more or less neutral and nonoxidizing conditions, and neither strong alkali, as in the Hofmann reaction, nor strong acid, as in the Schmidt reaction, are required.

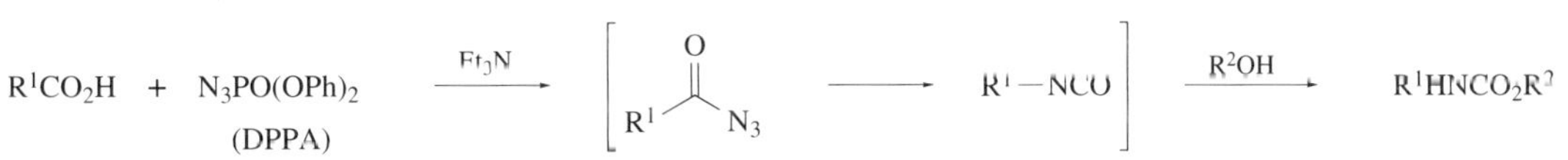

Scheme 44

This procedure, minus the *in situ* functionalization with alcohol, has been applied to prepare 1,3-diisocyanatoadamantane ⟨79JAP54/46762⟩, *n*-heptadecyl isocyanate ⟨83JAP53/35194⟩, and to synthesize vinyl isocyanates, used as aza diene equivalents in reactions with enamines ⟨84JOC4569⟩, benzyne addends ⟨89JOC4019⟩, or ester enolates ⟨86JOC1374⟩.

The preparation of *N*-benzyloxymethyl isocyanate and *t*-butyldiphenylsiloxymethyl isocyanate, which have been used in the synthesis of the herbicide (**101**), have also been obtained via the diphenylphosphoryl azide modified Curtius rearrangement (Scheme 45) ⟨92SC411⟩.

In 1993, a simple method for the synthesis of high-purity isocyanates from carboxylic acids was developed using diphenylphosphoryl azide and 1,8-bis(dimethylamino)naphthalene (proton

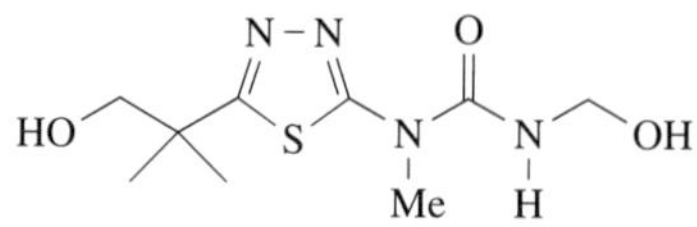

(101)

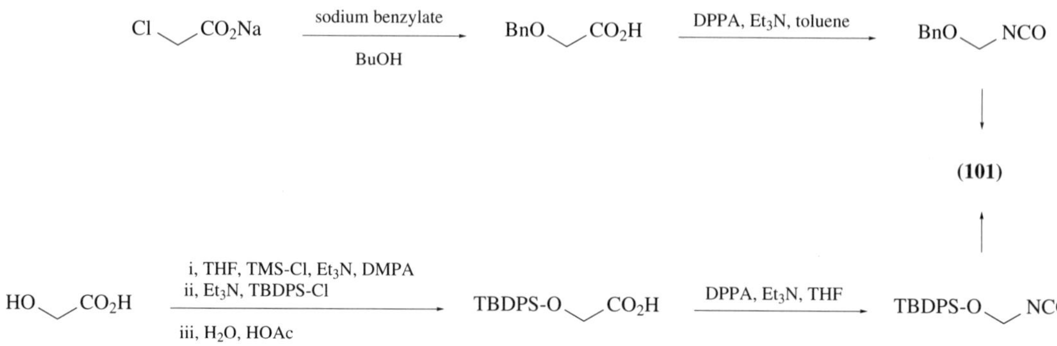

Scheme 45

sponge). Yields for the aromatic and aliphatic mono-isocyanates prepared ranged from 60–81%. Out of the di- and tricarboxylic acids evaluated, only hexamethylene diisocyanate was successfully synthesized ⟨93SC335⟩.

5.26.2.10.2 Reaction of carbonyl chlorides with isocyanic acid or metal cyanates

Isocyanic acid, obtained by thermal decomposition of cyanuric acid, reacts with carboxylic acid chlorides, in the presence of a proton acceptor B (pyridine, triethylamine or *N*-methylmorpholine) at low temperatures to give the corresponding isocyanates (Equation (47)). The best results were achieved by using THF as solvent and pyridine as base ⟨63JOC586, 64USP3155700⟩.

(47)

Acyl isocyanates were first made by reaction of the corresponding acyl chlorides with silver cyanate (Equation (48)) ⟨03CB3213⟩. This reaction was later used by several groups to prepare acyl isocyanates ⟨40JA1595, 54JCS4018, 57JCS1091⟩. For instance, different dialkylacetyl isocyanates were prepared in yields ranging from 40–90% when the appropriate dialkylacetyl chloride was treated with freshly prepared silver cyanate suspended in dry ether ⟨40JA1595⟩. A modification of those methods consists of treating the acyl chloride with potassium cyanate in a lithium chloride–potassium chloride melt at 480 °C ⟨76CB1491⟩.

(48)

Although aroyl isocyanates can also be prepared by the condensation of aroyl chlorides with another metal cyanate such as sodium cyanate, it is reported that, under these conditions, special catalysts and solvents must be used for achieving satisfactory yields. Among the catalysts studied, $ZnCl_2$ and $SnCl_4$ appear to be those which give the best results, while out of the solvents used, 1,2-dichlorobenzene is an excellent reaction medium. This method has also been applied to some nonaromatic acyl chlorides ⟨88T6079⟩.

The above methodology can also be applied to alkyl or aryl sulfonyl chlorides which react with silver cyanate to give the corresponding monosulfonyl isocyanates although the yields achieved are generally low (5–38%) (Equation (49)) ⟨05CB2013⟩. Moreover, sulfonyl diisocyanates **(102)** and **(103)** were prepared by the reactions of chlorosulfonyl isocyanate with silver cyanate ⟨58CB1200⟩ and potassium cyanate with sulfur trioxide (Equation (50)) ⟨58AG271⟩.

$$ClSO_2NCO + AgNCO \longrightarrow \underset{\textbf{(102)}}{O{=}C{=}N{-}\underset{O_2}{S}{-}N{=}C{=}O} \tag{49}$$

$$4SO_3 + 2KNCO \longrightarrow K_2S_2O_7 + \underset{\textbf{(103)}}{S_2O_5(NCO)_2} \tag{50}$$

In contrast to other aliphatic sulfinyl chlorides, trichloromethanesulfinyl chloride is converted by silver cyanate in boiling benzene into the corresponding isocyanate, which was the first sulfinyl isocyanate to be reported ⟨66AG(E)1047⟩. Perfluoromethane- and perfluorobutanesulfinyl isocyanates were similarly obtained, by reacting the corresponding perfluoroalkanesulfinic amides with chlorosulfonyl isocyanate ⟨74CB508⟩. Analogously, interaction of trifluoromethylsulfenyl chloride with silver cyanate in a sealed tube at room temperature, yielded the corresponding trifluoromethylsulfenyl isocyanate in addition to *N*,*N*-bis(trifluoromethylsulfenyl)carbamoyl isocyanate ⟨63AG(E)552⟩. Related to this reaction is the *in situ* preparation of arylsulfenylisocyanates by treatment of arylsulfenyl chlorides with silver or sodium cyanate ⟨88CL1385⟩.

A modification consists of the alkylation of cyanates with alkyl phosphates. Thus, triethyl phosphate is capable of ethylating potassium cyanate affording ethyl isocyanate in 63% yield. It should be pointed out that the ratio of phosphoric ester to potassium cyanate used in this reaction is considerably greater than unity. Butyl isocyanate was similarly prepared, only in a smaller yield (29%), by the reaction of potassium cyanate with tributyl phosphate ⟨52JA4700⟩.

5.26.2.10.3 *From anhydrides*

Preparation of 2-isocyanatobenzoyl chloride has been achieved by phosgenation of isatoic anhydride in the presence of a catalytic amount of *N*,*N*-dimethylformamide. In the absence of catalyst, no reaction took place ⟨67JOC4052⟩.

A patent claims that perfluoroacyl isocyanates, such as trifluoroacetyl, perfluorooctanoyl, and perfluorooctadecanoyl isocyanates, can be prepared by mixing equimolecular amounts of isocyanic acid and the corresponding perfluorocarboxylic anhydride at low temperatures ⟨70USP3541120⟩. Similarly, trifluoroacetyl isocyanate was prepared in 37% yield by the reaction of isocyanic acid and trifluoroacetic anhydride in a 2:1 mole ratio, but without solvent ⟨68JOC441⟩.

A closely related process consists of the preparation of monosulfonyl isocyanates which has been achieved by reaction of sulfonic anhydrides with silver cyanate ⟨52JA394⟩.

5.26.2.10.4 *From amides*

(i) Reaction of amides with oxalyl chloride

Speziale and Smith have reported the reaction of amides with oxalyl chloride as a very convenient and general method for the preparation of acyl isocyanates ⟨62JOC3742, 63JOC1805⟩ (Equation (51)) which appears to be far superior in yield and scope to the previously described silver cyanate method ⟨40JA1595⟩, especially when compounds sensitive to silver salts are involved (i.e., α-halo carbonyl compounds). Also, this method is an alternative to the phosgene method, used in the preparation of alkyl and aryl isocyanates from amines, since the reaction of phosgene with primary amides leads to nitriles or complex mixtures instead of the corresponding acyl isocyanates.

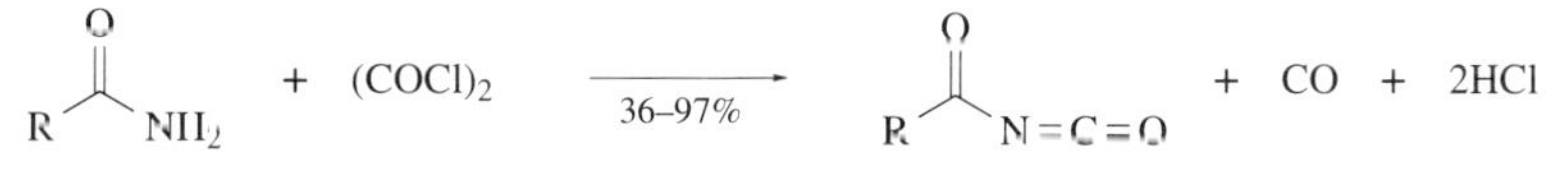

(51)

This reaction has been applied to a variety of amides, clearly showing that aromatic amides give excellent results regardless of substitution, but yields with aliphatic amides are not very satisfactory unless an electron-withdrawing group is present on the α-carbon atom or there are no α-hydrogen atoms ⟨65JOC4306⟩. Moreover, reaction of cyclopropanecarboxamide with oxalyl chloride gave the γ-chlorobutyroyl isocyanate instead of the expected cyclopropanecarbonyl isocyanate ⟨65JOC4306⟩.

Speziale's method can also be applied to the synthesis of aliphatic and aromatic diacyl diisocyanates (Equation (52)) ⟨67CB698, 68T2583⟩. For example, reactions of sebac-, phthal-, isophthal-, and terephthalamide with oxalyl chloride progressed smoothly, forming the corresponding diacyl diisocyanate. However, in the reaction of succinamide or adipamide, a large amount of oxalyl chloride is required and, in addition to the expected isocyanate, 1,3-bis(2-(1,3-oxazolidine-4,5-dione)) ethanediylidene or 1,3-adipoylimidazoline-2,4,5-trione is obtained. In contrast to the above results, when this method is applied to *N,N*-disubstituted acetamides, the reaction leads to 5-disubstituted amino-2,2,4-trichloro-3(2H)-furanones ⟨62JOC4361⟩.

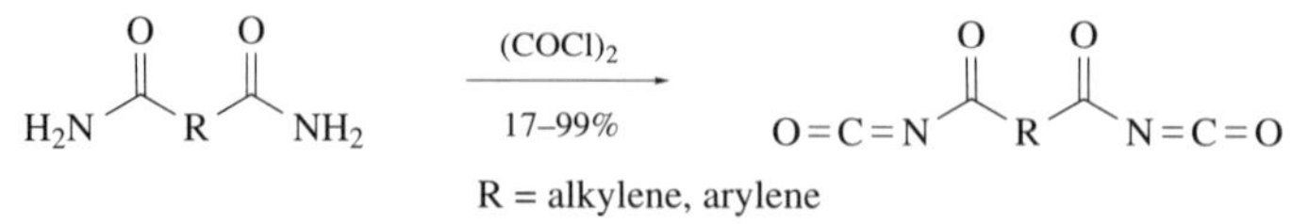

(52)

This process has also been successfully used in the case of aromatic thioamides which react with oxalyl chloride giving rise to the formation of thiazolinediones (**104**) which undergo thermolysis, with elimination of carbon monoxide, to form highly reactive thioacyl isocyanates (**105**) (Scheme 46) ⟨63AG(E)552, 65CB2954⟩, some of which can undergo thermal isomerization to the corresponding acyl isothiocyanates. This reaction seems to be general, and the 2-functionalized thiadiazolinediones, prepared from the appropriate functionalized thioamide and oxalyl chloride, behave similarly and undergo thermal CO elimination to give the corresponding functionalized thiocarbonyl isocyanates ⟨66CB1258, 66CB3572⟩.

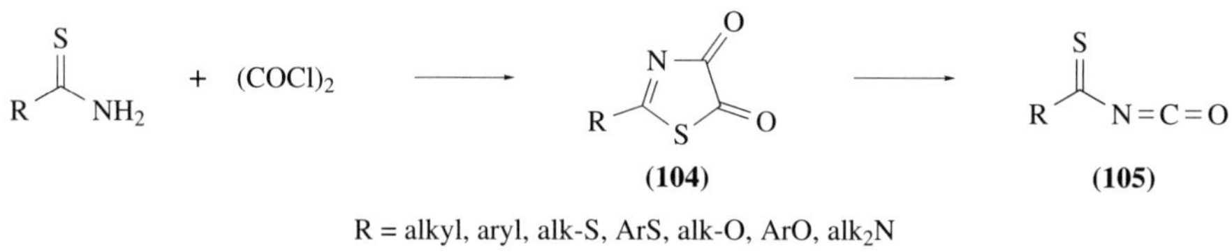

Scheme 46

The reactions of aryl and alkyl sulfonamides with oxalyl chloride have also been studied, and it was shown that the uncatalyzed reaction produced sulfonyloxamides or sulfonyloxamoyl chlorides depending on the mole ratio of reactants. However, in the presence of a tertiary amine catalyst the major products were sulfonylparabanates or mixtures of sulfonyl isocyanate and sulfonyl chloride (Scheme 47) ⟨64JOC2592⟩. Nevertheless, pyrolyses of parabanates yielded benzenesulfonyl isocyanate in 45% yield.

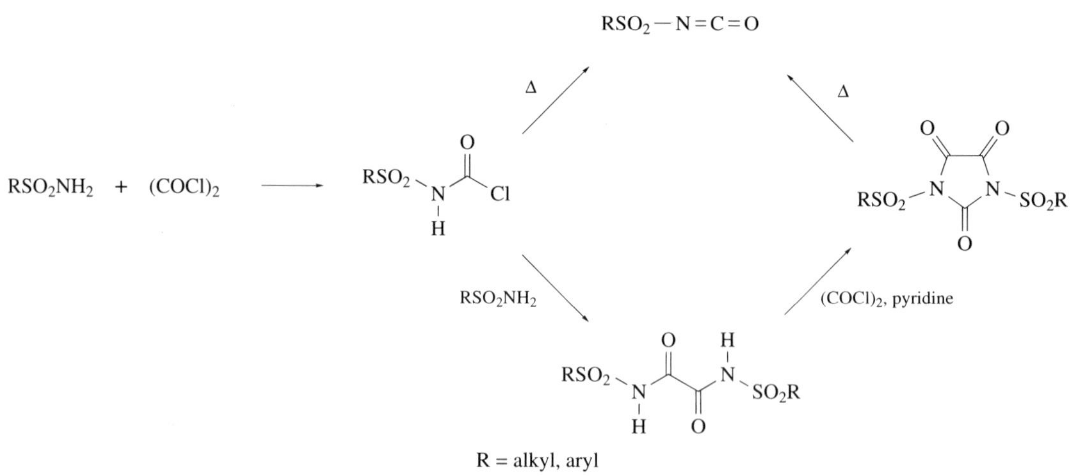

Scheme 47

(ii) Reaction of amides with phosgene

The reaction of amides with phosgene, widely used for preparing alkyl and aryl isocyanates is unsuitable for the synthesis of acyl isocyanates from amides. However, it was found that bis(trimethylsilyl)amides of carboxylic acids, in contrast to nonsilylated initial amides, react more easily with phosgene, giving the corresponding acyl isocyanates in a monomeric form in almost quantitative yields (Scheme 48) ⟨83ZOB2155⟩. This is a comprehensive method for the preparation of acyl isocyanates, including the lower members of the aliphatic series which are formed via Speziale's method in extremely low yields.

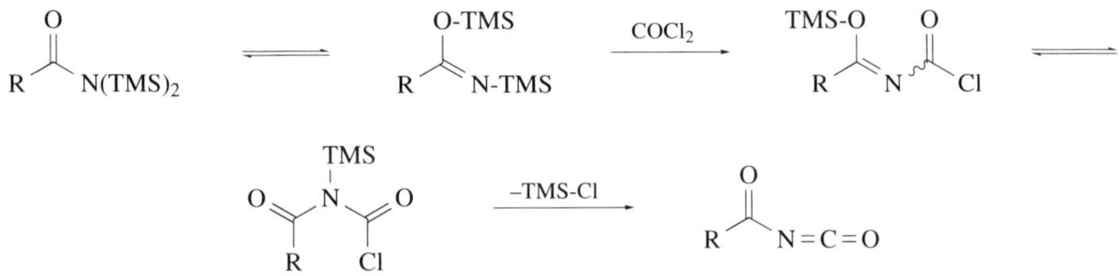

Scheme 48

In contrast to the behavior of amides derived from carboxylic acids, direct phosgenation of *p*-aminobenzenesulfonamide gave rise to the formation of *p*-isocyanatobenzenesulfonyl isocyanate in 87% yield. Similarly, organodisulfonamides (**106**) yielded the corresponding organodisulfonyl isocyanates (**107**) in yields ranging from 54% to 62% (Equation (53)) ⟨65JOC1260⟩. A patent has reported that arenesulfonyl isocyanates can also be obtained by thermal decomposition, in the presence of P_2O_5, of the adduct formed by reaction of sulfonamides with *N*,*N*′-carbonyldiimidazole ⟨70JAP7019893⟩.

$$\underset{\textbf{(106)}}{H_2NSO_2-R-SO_2NH_2} + COCl_2 \xrightarrow[54\text{–}62\%]{} \underset{\textbf{(107)}}{O{=}C{=}N{-}SO_2-R-SO_2{-}N{=}C{=}O} \qquad (53)$$

Isocyanates derived from phosphorus acids can similarly be obtained, in 49–51% yield, by phosgenation of the corresponding phosphonamide in toluene ⟨69ZOB1712⟩.

(iii) By thermal treatment of substituted amides

The pyrolysis of cyanoacetamides at elevated temperatures gives isocyanates and acetonitrile, the amounts of these compounds being increased as the temperature rises, but decreasing markedly above 700 °C, because of side reactions. These results indicate that, for this transformation, a six-membered cyclic intermediate is involved ⟨63BCJ95⟩.

Pyrolysis of trialkylaminimides (**108**), prepared by alkylating hydrazides or by treating an ester with an unsymmetrically trisubstituted hydrazinium quaternary salt and base in an appropriate solvent, gives isocyanates (Scheme 49). As an example, an excellent yield of isopropenyl isocyanate can be obtained by pyrolysis of trimethylamine methacrylimide either at atmospheric or at reduced pressure. This means that, starting from the readily available methyl methacrylate, it is possible to synthesize this isocyanate in two steps by a safe method in an overall yield of 78% ⟨67CJC2625⟩.

Also, a patent claims the preparation of isocyanates by thermolysis of amides and diamides of α,α-disubstituted β-ketoacids ⟨73CZP148911⟩.

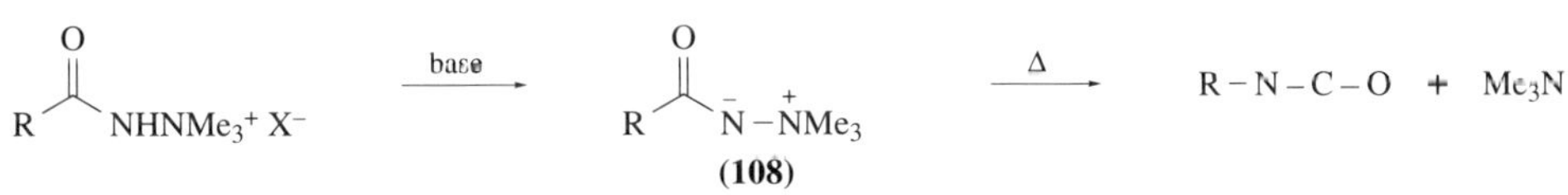

Scheme 49

(iv) Reaction of amides with hypobromite and lead tetraacetate (Hofmann rearrangement)

Another approach to the preparation of isocyanates is the Hofmann rearrangement which is also a very useful reaction for the conversion of amides into primary amines uncontaminated with secondary amines. The course of this reaction has already been established, as is set out in Equations (54)–(56), by isolating reaction intermediates. The rearrangement step, where the conjugate base of *N*-haloamide is converted into an isocyanate is the most important stage in the course of the Hofmann reaction. Although this rearrangement is well known to be intramolecular ⟨46OR(3)267, 57JA2544⟩ and to be accelerated by electron-releasing substituents in the phenyl group of *N*-bromobenzamides and retarded by electron-attracting substituents ⟨37JA121, 50JA4284⟩, other kinetic studies to elucidate the reaction mechanism more precisely have been carried out ⟨66JIC481, 68MI 526-01, 71BCJ1632, 71BCJ1639, 71BCJ1644⟩.

(54)

(55)

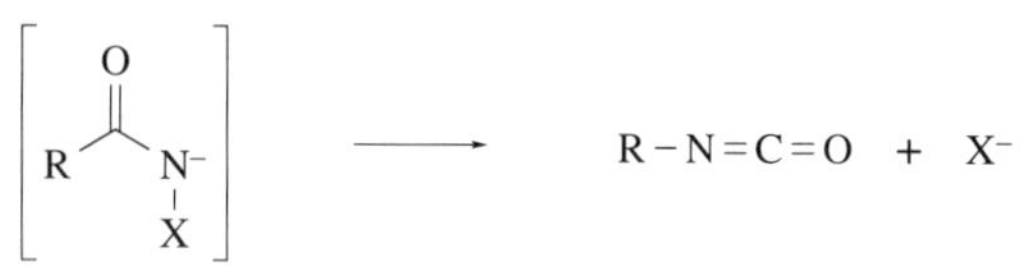
(56)

Under the usual conditions of Hofmann reaction, the isocyanate formed is readily hydrolyzed to the amine. Consequently, it can only be used to synthesize hydrolytically stable isocyanates. However, it is also possible to trap the reactive isocyanate intermediate in an organic phase by performing the reaction in a two-phase system and by using a phase-transfer catalyst to facilitate the transfer of the ionic reactants between phases. In this way, the syntheses of several aliphatic and alicyclic isocyanates have been achieved, in yields ranging from 16% to 87%, by treating the corresponding amide with two equivalents of bromine and aqueous sodium hydroxide in methylene chloride and tetra-*n*-butylammonium bisulfate at room temperature (Equation (57)) ⟨80TL2223⟩.

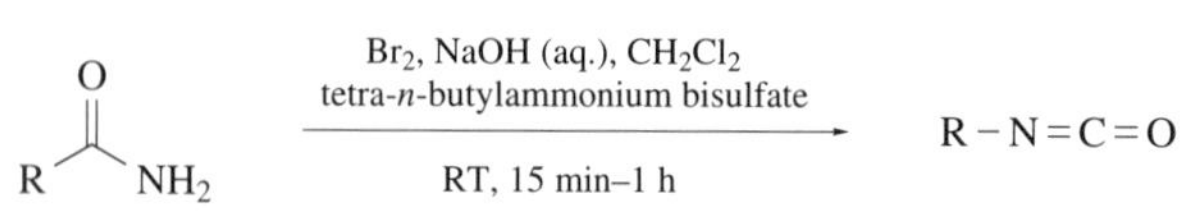

(57)

Anhydrous potassium fluoride has also been shown to behave as a base in the Hofmann reaction, converting *N*-chlorobenzamide into phenyl isocyanate. However, the latter undergoes addition of a second mole of the amide followed by hydrolysis of the resulting product to yield *N*-benzoyl-*N'*-phenylurea ⟨60JOC1837, 65JOC48⟩.

A patent claims that monofunctional isocyanates can also be prepared by decomposition of an *N*-haloamide in a moving inert gas stream at 150–300 °C and using a catalyst such as NaOH, CaO, Na_2CO_3 or Al_2O_3 plus a solvent such as nitrobenzene or chloroform ⟨70USP3483242⟩.

Related to the above reaction is the rearrangement of *N*-bromo- or *N*-chloro-*β*-lactams (**109**) in the presence of alkenes or alkynes, to give *β*-bromo- and *β*-chloroalkyl isocyanates (**110**). This rearrangement is a radical reaction, which can be promoted by irradiation, by radical donors or by heating (Equation (58)) ⟨69TL117⟩. From the pure enantiomers of the *β*-lactams, the optically active isocyanates are obtained ⟨71LA(752)142⟩.

(58)

(**109**) X = Cl, Br (**110**)

ω-Isocyanatoalkanoates, useful for the preparation of cellular plastics, polymers, pharmaceuticals, and plant-protection agents, have also been prepared in high yields by passing phosgene into a solution of *O*-methyl-4-butyrolactim containing sodium bicarbonate and NaOH at $-5\,^{\circ}C$ to $0\,^{\circ}C$ ⟨70GEP1913273⟩.

Oxidative rearrangement of simple amides using lead tetraacetate yields products similar to those obtained from the previously described reaction and, because of that, it is a useful alternative to Hofmann rearrangement and can therefore be considered as a Hofmann-like reaction. Thus, under suitable reaction conditions, a number of aliphatic and aromatic amides have been quickly converted into the corresponding isocyanates in moderate to excellent yields (Equation (59)) ⟨65JA1141, 71QR411⟩. Simple aliphatic amides did not react appreciably with lead tetraacetate in acetic acid, and reacted only sluggishly in nonpolar solvents such as benzene, chloroform or methylene chloride. Aromatic amides were less reactive and reacted most satisfactorily in dimethylformamide with added triethylamine. The rearrangement can be run very rapidly in dimethylformamide solution in such a way as to permit isolation of the intermediate isocyanate or to proceed directly via acid hydrolysis to the amine hydrochloride ⟨75JOC3554⟩. Similarly, formation of the corresponding carbamates was observed when the reactions of primary amides and lead tetraacetate were carried out in alcoholic solvents ⟨65TL4039⟩.

(59)

When the molecule contains a suitably situated carbonyl or carbamyl group, intramolecular cyclization with the isocyanate group can occur, giving six-membered heterocyclic rings. Thus, phthalaminic acid gives isatoic anhydride, whilst pyridine 2,3-dicarboxamide gives the corresponding pyrido–pyrimidine derivative ⟨68AJC2756⟩.

Similarly, lead tetraacetate in pyridine has been found to provide a mild procedure for effecting a rapid, high-yield, Hofmann-like rearrangement of *β*-hydroxy primary amides to 2-oxazolidinones, via the corresponding *β*-hydroxyisocyanate. These products, in turn, give the corresponding *β*-hydroxyamines, so the reaction can also be used to transform primary amides into amines in high yield, with retention of configuration ⟨73JOC414⟩.

The reagent bis(trifluoroacetoxy)iodo benzene (PIFA) also brings about the simple oxidative rearrangement of aliphatic amides to amines in mildly acidic mixed aqueous–organic solvents. Aromatic amines are further oxidized by the reagent and therefore cannot be prepared by this method. The rearrangement, which is in effect an acidic Hofmann rearrangement, occurs with complete retention of configuration in the migrating group, and the rate of the reaction approximately follows the migratory aptitudes of the migrating groups determined for other similar reactions. Isocyanates are intermediates in the rearrangement and survive long enough to be isolated because they hydrolyze somewhat more slowly than they form. Nevertheless, they are rapidly hydrolyzed to the amines under the mildly acidic conditions. The acidic conditions protect the amines from reacting with the isocyanate and, consequently, avoid the formation of the corresponding ureas. This reaction formally resembles the previously described reaction of lead tetraacetate with amides; however, PIFA is noteworthy for its mildness as an oxidizing reagent ⟨84JOC4272, 84JOC4277⟩.

The conversion of primary aliphatic amides to alkylammonium tosylates by (hydroxy(tosyloxy)iodo)benzene seems also to proceed through the formation of a nonisolated alkyl isocyanate as in the classic Hofmann reaction ⟨86JOC2669, 88JOC5158⟩.

Formation of carbamates in near quantitative yield by treatment of primary aliphatic and aromatic carboxamides with NBS–$Hg(OAc)_2$–ROH, dibromantin–$Hg(OAc)_2$–ROH, NBS–AgOAc–ROH or dibromantin–AgOAc–ROH in DMF under argon, is a new method for Hofmann rearrangement which proceeds without isolation of the intermediate isocyanates ⟨90TL1559⟩.

5.26.2.10.5 From hydroxamic acid derivatives (Lossen rearrangement)

In 1872, Lossen discovered the rearrangement bearing his name when he observed that pyrolysis of mixed anhydrides of hydroxamic and carboxylic acids afforded the corresponding isocyanates ⟨1872LA(161)347, 1877LA(186)1, 1889LA(252)170, 1894LA(280)169⟩. This rearrangement is a useful method for the preparation of organic isocyanates from hydroxamic acids. *O*-Acylation is essential for a smooth rearrangement; even then, such acyl derivatives only undergo pyrolytic rearrangement under fairly stringent conditions. Although the Lossen rearrangement of hydroxamic acid derivatives to isocyanates usually has the same limitations as the Hofmann rearrangement, under certain conditions and by using aprotic solvents it can be used for the synthesis of isocyanates. The usual route is to convert compounds (**111**) by means of a strong base, preferably in nonhydroxylic solvents, at low temperature into the salt (**112**). Rearrangement of the anion of (**112**) to the isocyanate involves loss of the carboxylate ion with concomitant migration of the group R from carbon to nitrogen (Scheme 50). The rate of this base-catalyzed rearrangement depends on the electronic nature of the groups R and R^1: the rate increases as the leaving group $R^1CO_2^-$ becomes more electron-attracting, and as the electron-releasing power of R in (**111**) increases ⟨74AG(E)376⟩. The stereospecificity of the Lossen rearrangement is also proved since the configuration of R in (**111**) remains unchanged during the rearrangement ⟨41JA1674, 41JCS263, 59JOC1293⟩.

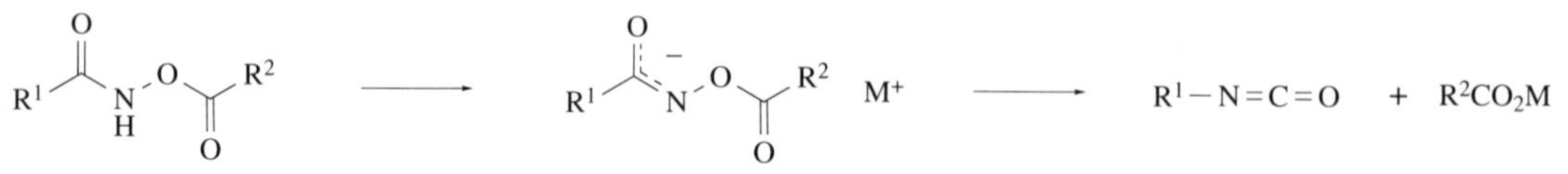

(111) **(112)**

Scheme 50

A variation to obtain the corresponding *O*-acylhydroxamic acid consists of the reaction of aromatic hydroxamic acids with triphenylphosphine, ethanol, and diethyl azodicarboxylate which gives rise to the formation of *O*-(*N*-phenylcarbamyl) hydroxamates through a Lossen rearrangement, which involves the formation of a nonisolated isocyanate intermediate ⟨74TL1965⟩.

The reaction of hydroxamic acids with sulfur trioxide–tertiary amine complexes proceeds by *O*-sulfation to give crystalline water-soluble *t*-ammonium *N*-acylhydroxylamine-*O*-sulfonates, which have been found to undergo a simple Lossen rearrangement to form isocyanates when they are treated with a tertiary amine in aqueous solution, and in the presence of an inert solvent such as petroleum ether or carbon tetrachloride (Scheme 51) ⟨69JOC2908⟩. As the isocyanate is formed, it is extracted into the organic layer. Yields of aliphatic isocyanates prepared by this procedure range from 63% to 75%. Nevertheless, attempts to prepare aliphatic diisocyanates by this method were unsuccessful.

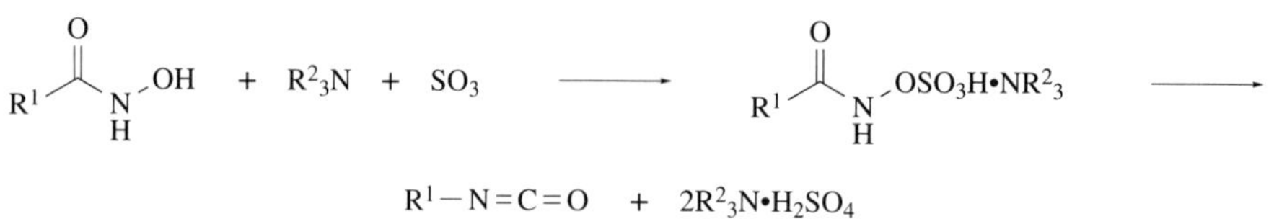

R^2 = Me, py

Scheme 51

Silylated hydroxamic acids (**113**), readily prepared from acid chlorides and TMS-ONH-TMS or TMS-ON$(TMS)_2$, fragment cleanly and rapidly to give the corresponding isocyanates in virtually quantitative yields (Scheme 52) ⟨78CC351⟩. Each one of the initial steps has its particular merit: (i) because TMS-ONH-TMS is prepared directly from TMS-Cl and hydroxylamine; (ii) it is useful for low-boiling acid chlorides; (iii) it is attractive in that the whole sequence (acid chloride to isocyanate) can be carried out in homogeneous solution without isolation of intermediates.

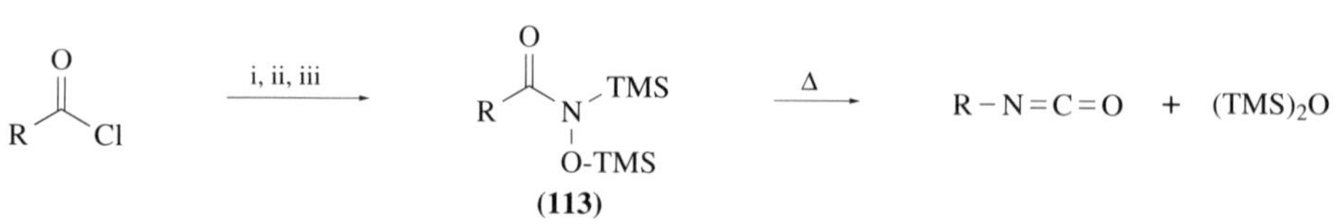

(113)

i, TMS-ONH-TMS, Et_3N (excess), light petroleum, R = alkyl, aryl
ii, TMS-OLi-TMS, ether, R = ethynyl
iii, TMS-ON$(TMS)_2$, mesitylene, R = aryl, styryl

Scheme 52

It is also reported that bis(trimethylsilyl) derivatives of hydroxamic acids, when heated at 150°C, are converted into isocyanates by the Lossen rearrangement ⟨73UK1451⟩. However, it was found that a more convenient method from the preparative point of view, which does not require the special synthesis of bis(trimethylsilyl) derivatives of hydroxamic acids, is by heating at 100 °C the hydroxamic acids or their salts in the presence of a polychlorosilane (Scheme 53) ⟨84ZOB876⟩. Such a modification to the Lossen rearrangement is distinguished favorably from the classical variant by an almost complete absence of side reactions, which allows the preparation of isocyanates in close to quantitative yields.

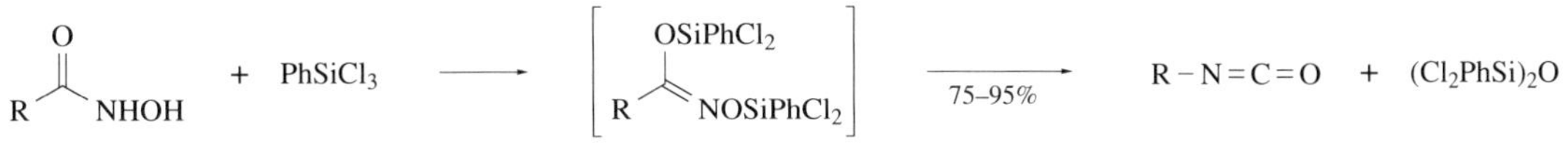

Scheme 53

An attempt to combine the preparation of hydroxamic acids, their silylation, and their subsequent thermolysis to isocyanate into one process gave good results only in the aromatic series (Equation (60)) ⟨84ZOB878⟩.

$$(PhCO)_2O + HONH_2{\bullet}HCl \xrightarrow{PhSiCl_3} Ph{-}N{=}C{=}O + PhCO_2SiPhCl_2 + HCl + (Cl_2PhSi)_2O \qquad (60)$$

Mukaiyama *et al.* have reported a method for the conversion of hydroxamic acids into isocyanates which involves pyrolysis, at 350–400 °C under reduced pressure, of the adducts **(114)** formed by treatment of the corresponding hydroxamic acid with diketene (Scheme 54) ⟨61JOC782, 61JOC4381⟩.

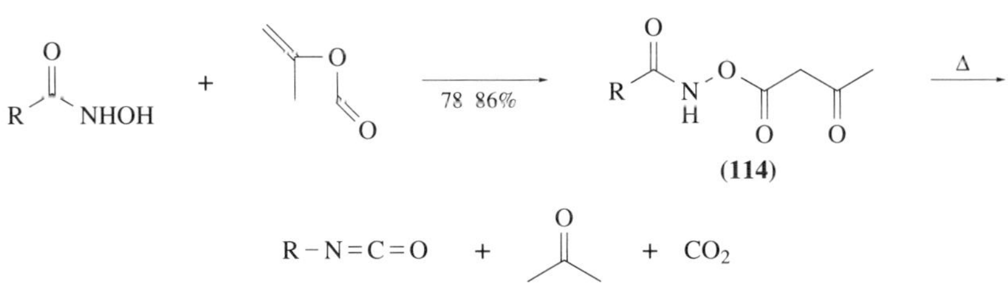

R = alkyl, aryl

Scheme 54

Also, a patent claims that mono- and diisocyanates can be prepared by thermal decomposition of acetates of hydroxamic acids when they are heated under nitrogen to 150 °C ⟨69USP3465024⟩. On the other hand, phenyl isocyanate and 3-chloro-4-methoxyphenyl isocyanate were obtained in 70% and 18% yields, respectively, when benzhydroxamic chloride and 3-chloro-4-methoxybenzhydroxamic chloride were pyrolyzed at 180 °C ⟨71JOC2155⟩.

5.26.2.10.6 Reaction of carboxylic acids with hydrazoic acid (Schmidt rearrangement)

The reaction between equimolecular amounts of hydrazoic acid and carbonyl compounds in the presence of strong mineral acid in an inert solvent at about 40 °C has become known as the Schmidt reaction. The key step is a rearrangement and the reaction has close similarities to, and indeed some overlap with, the Curtius rearrangement. The latter differs in that acyl azides form a distinct stage either as starting materials or as isolable intermediates, and acid catalysis is not obligatory. Although the Schmidt reaction affords an additional method to the Hofmann and Curtius degradations, its scope is somewhat smaller than that of the Curtius rearrangement, owing both to the forcing conditions which can isomerize sensitive substrates and products, and to the multistep nature of the reaction that gives scope for the formation of by-products ⟨46OR(3)307, 71MI 526-02⟩. The Curtius, Hofmann, Lossen, and Schmidt reactions have advantages and disadvantages which have been discussed in detail by Wolff ⟨46OR(3)307⟩.

Nevertheless, although this approach does not constitute an excellent method for the preparation of isocyanates, in some cases the isocyanates can be isolated when the corresponding carboxylic acids are treated under the Schmidt conditions. This is the case of 4-phenanthrene isocyanate, isolated in 95% yield, when a solution held at 0–5 °C of 4-phenanthrene carboxylic acid in a mixture of trifluoroacetic acid and trifluoroacetic anhydride was treated with sodium azide ⟨57JA213⟩. It is noteworthy that no mineral acid is required, and also that both trifluoroacetic acid and the anhydride were needed as the reaction did not proceed when either the acid or the anhydride was used alone. This was the first isocyanate isolated in the Schmidt reaction with organic acids, which supports the postulate that a protonated isocyanate is an intermediate in the Schmidt rearrangement ⟨48JA317⟩.

5.26.2.10.7 *Thermal decomposition of* N*-acylamino pyridinium salts and pyridone derivatives*

Katritzky and co-workers have investigated reactions of *N*-substituted pyridinium salts and found that the pyridine ring can act as a leaving group in a variety of synthetically useful transformations. Following this methodology they have described the preparation of isocyanates, in yields ranging from 76% to 95%, by pyrolysis of the *N*-acylimines (**115**), prepared either from acid hydrazides and 2,4,6-triphenylpyrylium salt, or from acid chlorides and 1-amino-2,4,6-triphenylpyridinium salt (Scheme 55) ⟨76TL2691, 79JCS(P1)446⟩.

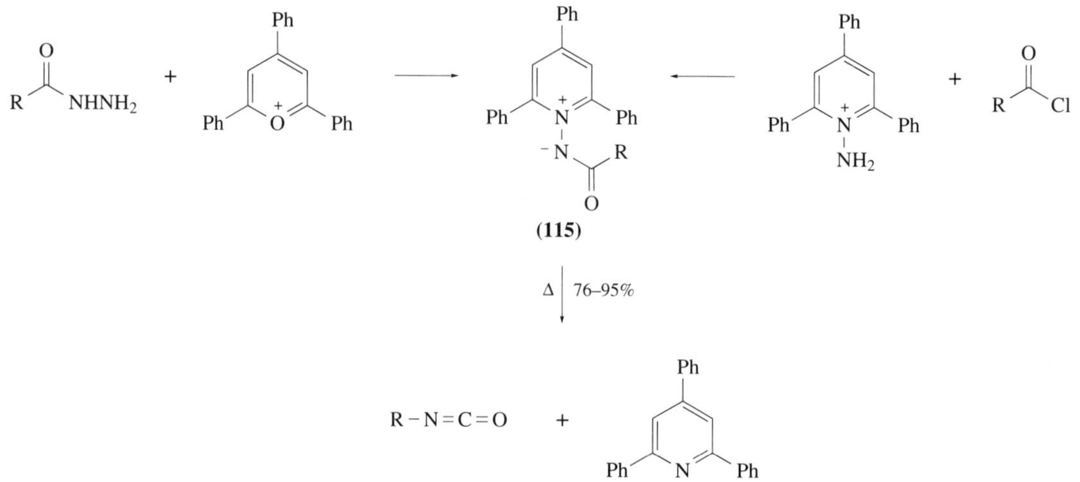

R = alkyl, aryl

Scheme 55

Similarly, they have shown that monoacyl derivatives of 1-amino-4,6-diphenyl-2-pyridone can undergo pyrolysis, in the presence of 4,6-diphenyl-2-pyridone, giving rise to the formation of isocyanates in yields ranging from 85% to 95%. (Equation (61)) ⟨79JCS(P1)1957⟩.

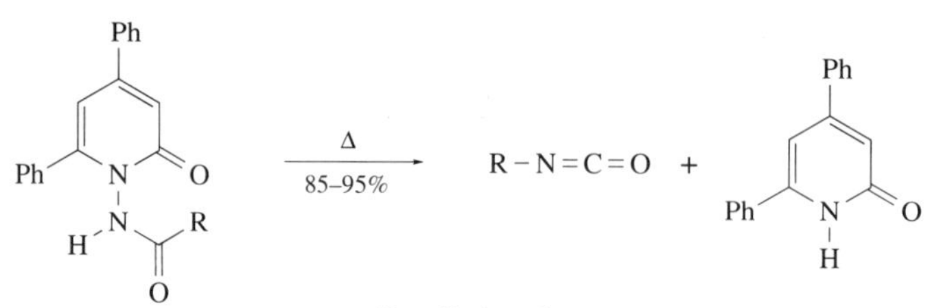

(61)

R = alkyl, aryl

5.26.2.11 From Nitriles

5.26.2.11.1 *By reaction with phosgene*

The reaction of arylacetonitriles, bearing at least two methyl groups at the 2- and 6-positions of the benzene ring, with phosgene in the presence of hydrogen chloride in a sealed glass tube, gives rise to isomeric mixtures of *cis*- and *trans*-α,β-unsaturated isocyanates. However, isolation of pure isocyanates from the reaction mixtures has been reported to be difficult since large amounts of unreacted nitriles were present (Equation (62)) ⟨71JOC3542, 73BCJ1275⟩.

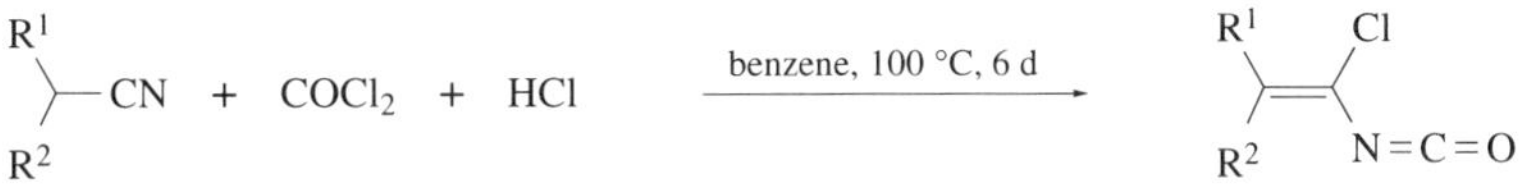

cis, trans

(62)

5.26.2.11.2 *By reaction with carbonyl fluoride*

Carbonyl fluoride adds to the triple bond of the nitrile function to give isocyanates. Thus, trifluoroacetonitrile at 200–300 °C gave pentafluoroethyl isocyanate although in very low yield (Equation (63)) ⟨62JA4275⟩. A patent also claims that CH_3CF_2NCO and F_5SNCO were prepared by treating acetonitrile or NSF_3 with carbonyl fluoride ⟨72USP3666784⟩.

(63)

5.26.2.11.3 *From cyanomethylene amines*

A patent has reported that poly(isocyanatoethyl)amines can be prepared by hydrogenation of cyanomethyleneamines, such as $N(CH_2CN)_3$, in the presence of phosgene, a solvent such as chlorobenzene or xylene, an acid acceptor such as pyridine, *N,N′*-diethylpiperazine or triethylenediamine, and a catalyst such as Pd—C and Pd or Pt oxides ⟨65USP3218345⟩.

5.26.2.11.4 *From nitrile oxides*

It is known that aromatic nitrile oxides (**116**) are unstable and dimerize within hours or days to the corresponding furoxans (**117**). However, it has been described that aromatic nitrile oxides bearing substituents at the *o,o′*-positions undergo slow rearrangement to the corresponding isocyanates instead of giving furoxans (Scheme 56) ⟨64AG(E)585⟩. This rearrangement is of particular interest because of the simultaneous shift of two ligands along a CN unit. It has also been achieved by using a suitable catalyst, such as sulfur dioxide, which participates in the reaction as a dipolarophile. In this way, the rearrangement of nitrile oxides to the isocyanates takes place even in the absence of substituents at the *o,o′*-positions of the aromatic ring ⟨77AG(E)555⟩.

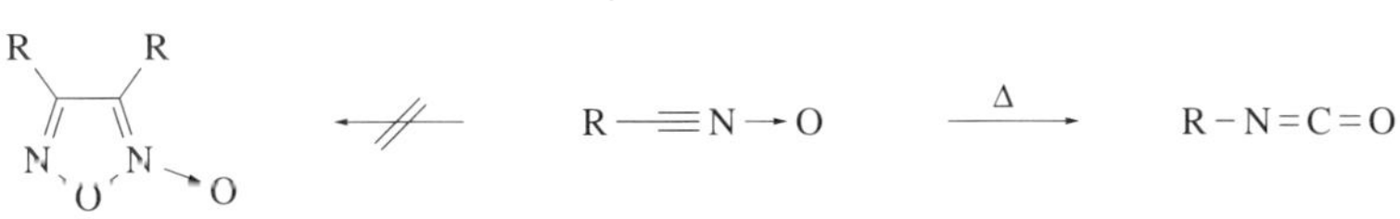

(**117**) (**116**)

Scheme 56

5.26.2.12 From Isonitriles and Derivatives

5.26.2.12.1 *By oxidation of isonitriles*

Oxidation of isonitriles to isocyanates was first carried out using ozone as an oxidizing agent ⟨58JOC1107⟩, but since then, several methods have been developed. Another oxidation of isonitriles to isocyanates is based on the treatment with dimethyl sulfoxide in the presence of a catalytic amount of bromine ⟨64JOC246⟩. Presumably this reaction involves the initial formation of an isocyanide dihalide, which is oxidized by dimethyl sulfoxide in a second step (Equations (64) and

(65)). Isopropyl and phenyl isocyanates were prepared in 80% and 82% yield following this procedure.

$$\text{R–NC} + \text{Br}_2 \longrightarrow \text{RN}{=}\text{CBr}_2 \qquad (64)$$

$$\text{RN}{=}\text{CX}_2 + \text{Me}_2\text{SO} \longrightarrow \text{R–N=C=O} + \text{Me}_2\text{S} + \text{Br}_2 \qquad (65)$$

It has been found that in oxidation of isocyanide with dimethyl sulfoxide, catalysis with Brönsted acids (*p*-toluenesulfonic acid or anhydrous HCl) or triphenylmethyl perchlorate is more convenient. The exothermic reaction requires only a short time for completion at 50–80 °C in the absence of a solvent and it gives the isocyanates in 50–90% yields ⟨67AG(E)168⟩. Similarly, some aliphatic and aromatic isonitriles have been oxidized to the corresponding isocyanate by pyridine *N*-oxide in the presence of iodine (Scheme 57) ⟨67JOC1939⟩.

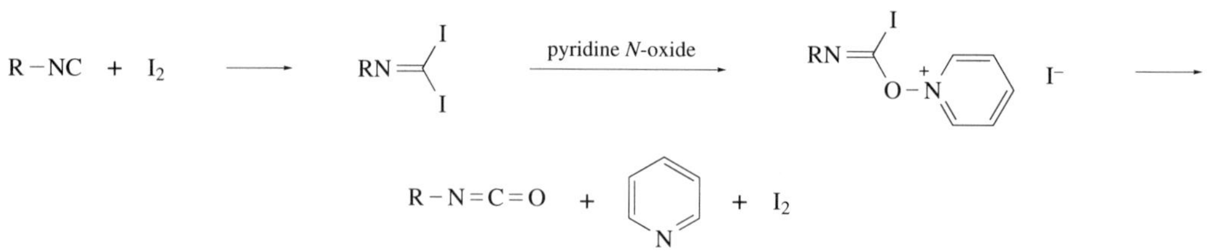

Scheme 57

Nitrile oxides are also known to be easily reduced to nitriles by treatment with isonitriles to form isocyanates and benzonitrile (Equation (66)), through a process which implies an oxygen transfer ⟨65TL4645, 80NJC127⟩.

$$\text{ArCNO} + \text{R–NC} \longrightarrow \text{ArCN} + \text{R–N=C=O} \qquad (66)$$

Reactions of nitrogen oxides with isocyanides are also reported to produce isocyanates, although it was shown that nitrogen oxide is reduced more readily than dinitrogen oxide (Equation (67)).

$$\text{R–NC} + \text{NO} \xrightarrow[\text{50–77\%}]{} \text{R–N=C=O} + {}^{1}\!/_{2}\,\text{N}_2 \qquad (67)$$

5.26.2.12.2 From isonitrile dichlorides

Isonitrile dichlorides, readily available through a wide variety of processes ⟨71MI 526-03⟩ can undergo replacement of the two chlorine atoms by oxygen, giving rise to isocyanates. To achieve this conversion, several methods have been developed.

The first conversion of an isocyanide dichloride into an isocyanate was described in 1874 by treatment of phenyl isocyanide dichloride with silver oxide ⟨1874CB1228⟩. Later, the preparation of dichlorophenyl isocyanate, in 55% yield, by reaction of dichlorophenyl isocyanide dichloride with phosphorus pentoxide at 240–250 °C was reported ⟨69AG(E)20⟩.

Other approaches, which constitute very simple and productive methods for the preparation of aliphatic and aromatic isocyanates, have been described by Arlt. One method consists of the reaction, at room temperature, of isocyanide dichlorides with very strong acids, such as chlorosulfuric or dichlorophosphoric acid, and subsequent distillation (Scheme 58). Similarly, phosphorus oxychloride in the presence of water or methanesulfonic acid can also be used ⟨70S20⟩. Related to the latter process, is the preparation of several acyl isocyanates (**119**) by treatment of the appropriate *N*-(chlorocarbonyl)imidoyl chloride (**118**) with methanesulfonic acid (Equation (68)) ⟨73AG(E)999⟩.

Previously it had already been found that some acyl isocyanates could be obtained by partial hydrolysis of the readily accessible imidoyl chlorides ⟨62AG(E)632, 69AG(E)607⟩.

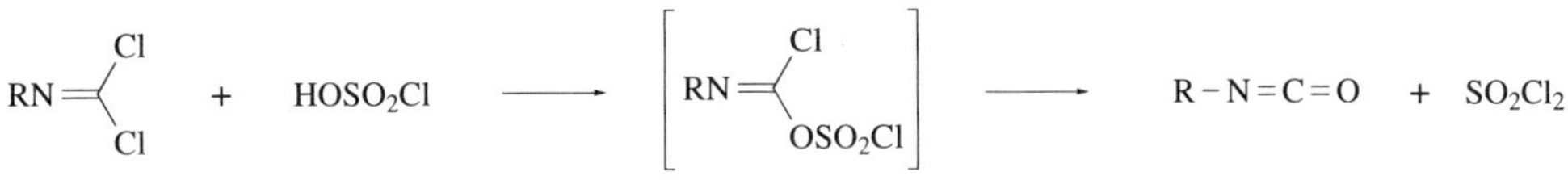

Scheme 58

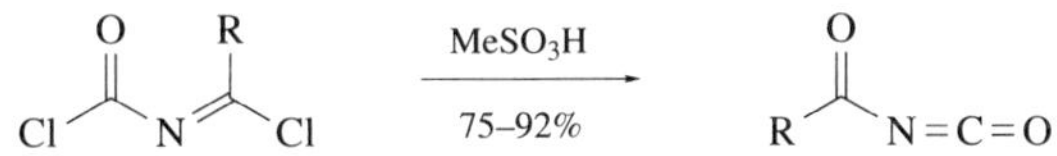

(118) **(119)**

R = Ph, OAr, SAr (68)

Alternatively, Mukaiyama has used phenyl isocyanide dichloride as a dehydration agent to prepare phenyl isocyanate from benzaldoxime (Equation (69)) ⟨62BCJ1104⟩.

(69)

Moreover, a phosphorylated isocyanate has been prepared by reaction of a dialkoxyphosphoryl isocyanide dichloride with anhydrous formic acid. This method has the advantage that all by-products are gaseous (Equation (70)) ⟨64ZOB2812⟩.

(70)

5.26.2.12.3 From other isocyanates

(i) By using chlorosulfonyl isocyanate (csi)

Chlorosulfonyl isocyanate has been reported to be a useful reagent to synthesize sulfonyl iso cyanates by reaction with different types of functional groups. By using this reagent, aryloxy- and alkoxysulfonyl isocyanates (**121**) and (**122**) have been prepared. The former were obtained by reaction of phenols with chlorosulfonyl isocyanate in refluxing toluene, in yields ranging from 36% to 92% ⟨72CB2791⟩, while the latter involved the thermal decomposition of carbamates (**120**), prepared by sequential treatment of chlorosulfonyl isocyanate and phenols, at room temperature, followed by reaction with an alcohol (Scheme 59) ⟨72CB2800⟩. Following this method, other aryloxysulfonyl isocyanates, which have been proved to be useful in the synthesis of β-lactams, have been prepared ⟨85JOC169⟩. Several perfluoroalkanesulfonyl isocyanates have been prepared in a similar way, in 75–91% yield, by reaction of the corresponding sulfonamide with csi ⟨73GEP(O)2133795⟩.

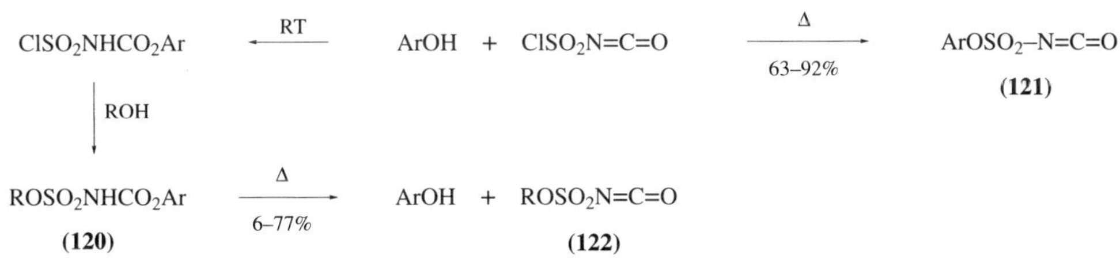

Scheme 59

Arenesulfonyl isocyanates can also be prepared by an electrophilic substitution reaction between csi and the corresponding arene. Following this process, 2-naphthalenesulfonyl isocyanate has been prepared in 55% yield ⟨69GEP1289526⟩. In the early 1990s, a reaction between csi and trialkylstannyl substituted arenes was described. The trialkylstannyl group, as an efficient leaving group, allows mild reaction conditions to be applied and unusual substituted patterns to be obtained, normally not accessible by electrophilic aromatic substitution (Scheme 60) ⟨91CB1997⟩. This method can also be applied to polycyclic ring systems such as naphthalene derivatives.

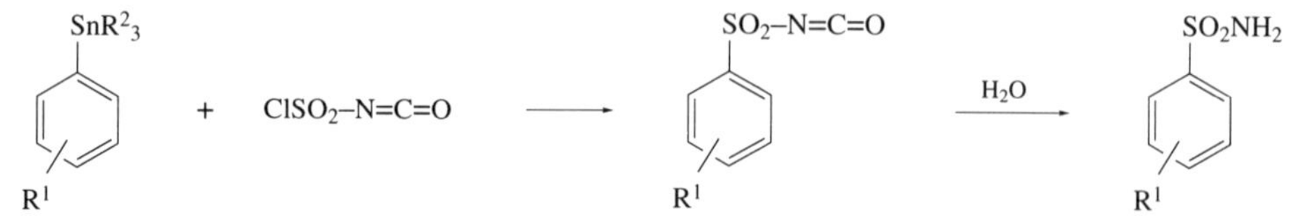

Scheme 60

5.26.2.12.4 By using organosilicon and organotin derivatives

Isocyanatosilanes (**123**), which can be available, among other suitable methods, by reaction of urea with hexaalkyldisilazanes ⟨76ZOB1531, 83ZOB1096⟩, are extremely stable and highly soluble in organic solvents and behave as sources of the isocyanate group in organic synthesis, giving rise to many convenient ways for preparing a great variety of isocyanates ⟨84JOM207⟩.

In the presence of catalytic amounts of halides from Group 2–8 elements, isocyanatosilanes were found to undergo exchange reaction with acyl and sulfonyl halides to form acyl and sulfonyl isocyanates in high yield (Equation (71)) ⟨75ZOB477, 77ZOB2259⟩. Triethyltin isocyanate behaves similarly and its reaction with acetyl bromide leads to the formation of acetylisocyanate in 58% yield ⟨72ZOB1868⟩.

$$\text{R–X–Y} + \underset{(\mathbf{123})}{\text{TMS–N=C=O}} \xrightarrow[32\text{–}98\%]{\text{catalyst, 90–150 °C}} \text{R–X–N=C=O} + \text{TMS-Y} \quad (71)$$

$X = CO, CO_2, SO_2$; $Y = Cl, Br$

Related to the above processes, the high-yielding preparation of diisocyanatoesters and -diesters has been reported by treatment of trimethylsiloxyalkyl and -aryl isocyanates with alkanedioic acid dichlorides (oxalyl-, adipoyl-, suberoyl dichloride), or with isocyanatoalkanoyl chlorides (3-isocyanatopropanoyl-6-isocyanatohexanoyl chloride) ⟨89JPS(A)4341⟩.

The same reaction with dialkylcarbamoyl chlorides leads to dialkylcarbamoyl isocyanates ⟨83ZOB1434⟩ in much better yields than those obtained in the reaction of *N,N*-dialkylureas with oxalyl chloride (Scheme 61).

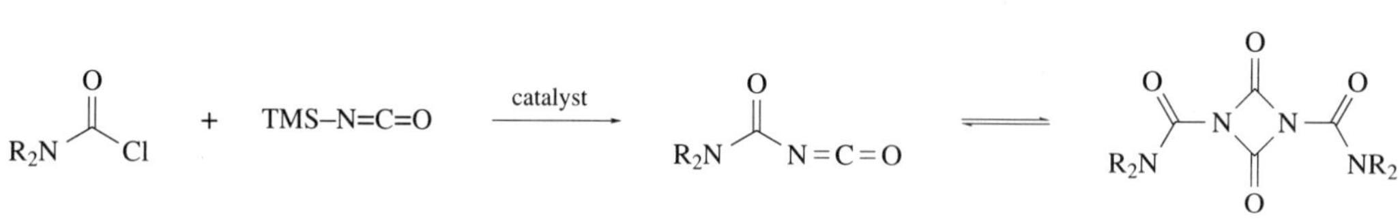

Scheme 61

The exchange reaction of trimethylsilyl isocyanate with chlorides of silylcarboxylic acids leads to organosilicon-substituted acyl isocyanates through two different routes, depending on the catalyst involved ⟨77ZOB2259⟩. Of the catalysts used in these reactions, the most effective was tin tetrachloride. On the other hand, carboxylic acid anhydrides react with isocyanatosilanes, at low temperatures, with great ease leading to the corresponding acyl isocyanates in high yields (Equation (72)) ⟨77ZOB2259⟩.

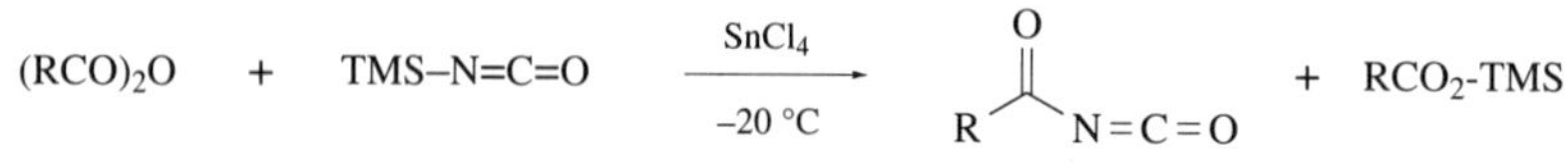

(72)

Acyl isocyanates are also formed in satisfactory yield, as a result of the acylation of *N,O*-bis(trimethylsilyl)carbamates with acyl chlorides, and subsequent treatment of the reaction product with silicon tetrachloride or with phenyltrichlorosilane (Scheme 62) ⟨81ZOB239⟩.

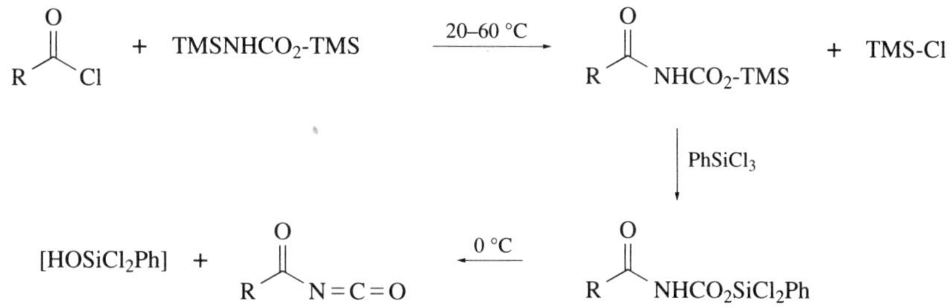

Scheme 62

Trimethylsilyl isocyanate and $Si(NCO)_4$ have been used for the preparation of fluorocarbonyl isocyanate and difluorophosphoryl isocyanate ⟨67CB1082⟩.

A method for the preparation of polyfunctional aliphatic isocyanates in high purity and quantitative yields, by the hydrosilylation of *m*-isopropenyl-α,α-dimethylbenzyl isocyanate (*m*-TMI) with cyclic and acyclic hydrogenmethylsiloxanes has also been reported ⟨91JPS(A)1097⟩.

5.26.2.12.5 By heating with high boiling isocyanates

A simple and versatile synthesis of trimethylsiloxy-substituted isocyanates consists of the reaction of the *O*-trimethylsilyl derivatives of aminoalcohols, aminophenols, 4-aminobenzoic acid, and hydroxylamine with a high boiling isocyanate (Scheme 63) ⟨88S990⟩. Regioselective reaction of trimethylsiloxy isocyanates with isocyanato alkanoyl chlorides or alkanoyl chlorides gives rise to the formation of diisocyanatoesters and diisocyanatodiesters ⟨89JPS(A)4341⟩.

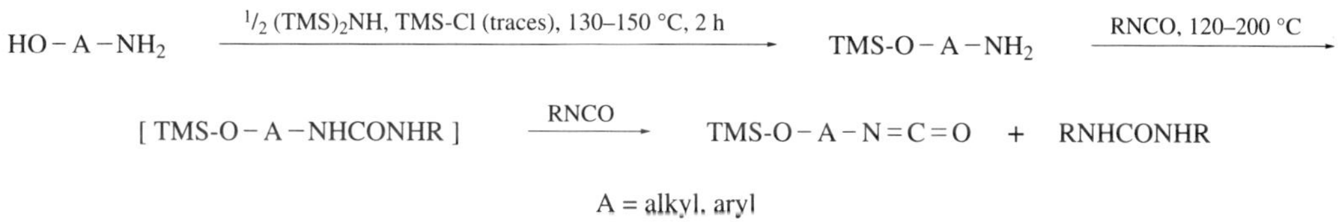

Scheme 63

5.26.2.12.6 By using N*-chlorocarbonyl isocyanate*

N-Chlorocarbonyl isocyanate, easily prepared in 90% yield, by reaction of phosgene and cyanogen chloride followed by partial hydrolysis by methanesulfonic acid, has proved to be an interesting reagent to prepare isocyanates. Nucleophiles such as alcohols, thiols, phenols, thiophenols, secondary amines, and sulfonamides react with this reagent to give *N*-carbonyl-substituted carbamoyl chlorides (**124**) which lose hydrogen chloride at 70–125 °C, thus being transformed into the relevant isocyanates (**125**) in yields ranging from 43% to 90% (Scheme 64) ⟨77AG(E)743⟩. 1,6-Hexanediol, triethylene glycol, and α,α-dimethyl-4,4′-methylenediphenol lead, on reaction with *N*-chlorocarbonyl isocyanate, to the corresponding diisocyanate. Similarly, this reaction proceeds very smoothly even with polyethers containing several hydroxyl groups, which becomes a simple route to high-molecular polyisocyanates.

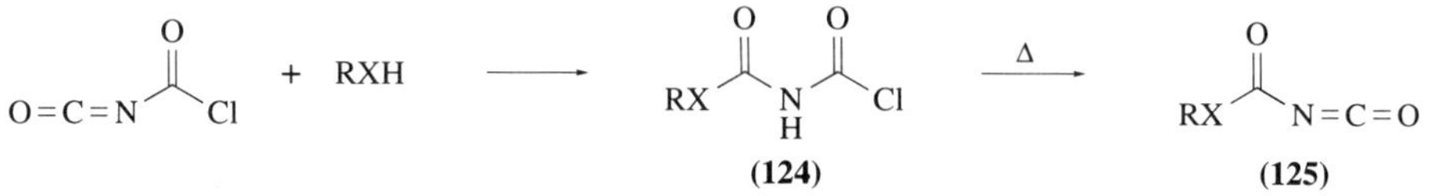

(**124**) (**125**)

Scheme 64

The chlorocarbonyl group of this reagent can undergo a selective reaction with polar π-bonds and with the C—O bond of epoxides. So, from azomethines it leads to the formation of carbamoyl isocyanates with an *N*-(chloroalkyl)amide partial structure (Equation (73)) ⟨72AG(E)1011⟩. Isocyanates (**126**) can be prepared in moderate to high yields (25–80%) starting from the reactions of aldehydes or ketones with *N*-chlorocarbonyl isocyanate ⟨76ZOR1963⟩. Moreover, preparation of 3-alkoxyacryloyl isocyanates (**127**) (36–43% yield) can be achieved by reaction of vinyl ethers with *N*-chlorocarbonyl isocyanate in the presence of triethylamine ⟨75ZOR2227⟩.

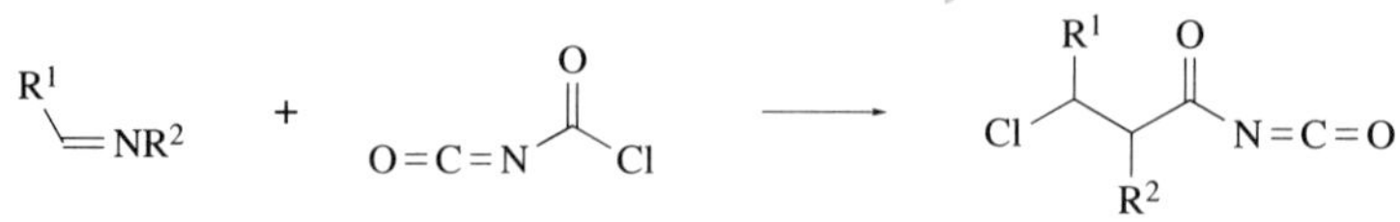

R = alkyl, aryl (73)

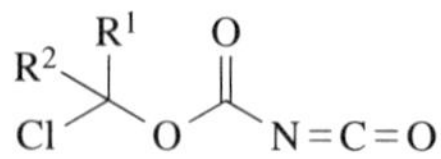

(126)

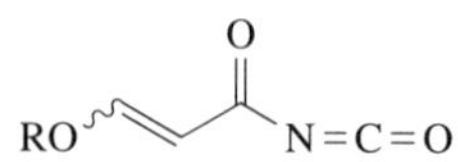

(127)

Another method is the addition of the acid chloride group of *N*-chlorocarbonyl isocyanate to the C≡N triple bond of aromatic cyanates which gives rise to the formation of *N*-(isocyanatocarbonyl)-chloroformimidic acids (**128**), in yields from 55% to 60% ⟨76ZOR2473⟩.

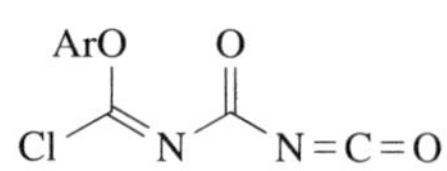

(128)

Epoxides undergo ring-opening addition when treated with this reagent, and formation of the corresponding 1-(chloromethyl)alkoxycarbonyl isocyanates (**129**) takes place in 55–60% yield (Equation (74)) ⟨77AG(E)743⟩.

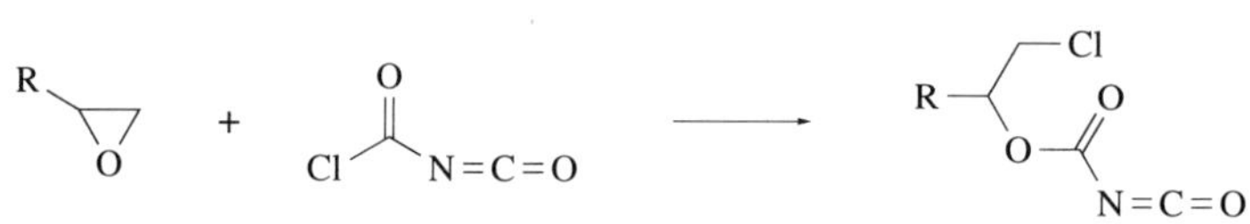

(129) (74)

5.26.2.12.7 *By using phosphorus isocyanates*

Trialkyl and triaryl phosphites react with phosphorus triisocyanate and phosphoryl triisocyanate (Equations (75) and (76)). Moreover, while the reaction of alkoxydiisocyanatophosphine with chlorine or bromine leads to phosphoryl halide diisocyanate in good yields (Scheme 65), dialkoxyisocyanatophosphine and chlorine lead to the formation of alkyl chloroisocyanatophosphates (Equation (77)) ⟨66ZOB1639⟩.

$$(RO)_3P + P(NCO)_3 \longrightarrow ROP(NCO)_2 + (RO)_2PNCO \qquad (75)$$

$$(RO)_3P + P(O)(NCO)_3 \longrightarrow ROP(O)(NCO)_2 + (RO)_2(PO)NCO \qquad (76)$$

$$ROP(NCO)_2 + Cl_2 \longrightarrow [ROP(NCO)Cl_2] \longrightarrow ClP(O)(NCO)_2 + RCl$$

Scheme 65

$$(RO)_2PNCO + Cl_2 \longrightarrow RO(Cl)P(O)NCO \qquad (77)$$

5.26.2.12.8 *By using α-haloalkylaryl isocyanates*

Isocyanates containing phosphonic ester groupings were prepared, in a very simple way, from phosphorus-free polyisocyanates and trialkyl phosphites by using the Michaelis–Arbusov reaction.

For instance, although all attempts to obtain the isocyanatophosphonic esters (**130**) by reacting 2,4-diisocyanatobenzyl chloride with trialkyl phosphites failed, it was found that by simultaneous action of equimolar amounts of trialkyl phosphite and alcohols on the diisocyanatobenzyl chloride, the reaction led to the formation of phosphorylated monoisocyanates (**131**), and phosphorylated polyisocyanates, e.g., (**132**), when using di- or polyols (Scheme 66) ⟨62AG(E)617⟩.

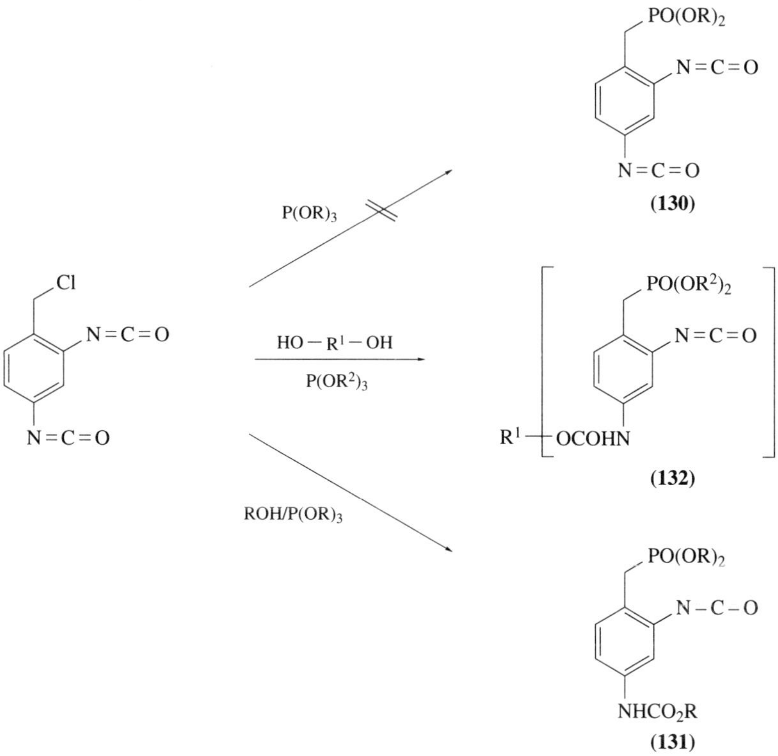

Scheme 66

5.26.2.13 From Isothiocyanates

Isocyanates and isothiocyanates have been found to undergo an exchange reaction simply when heated together at temperatures in the range of 190–230 °C ⟨59NAT675⟩. Since isocyanates boil at appreciably lower temperatures than the corresponding isothiocyanates, separation of the desired isocyanates was conveniently achieved by conducting the reaction in the stillpot of a fractionating column. In this way, *n*-hexyl isocyanate was prepared, in 61% yield, from *n*-hexyl isothiocyanate and *m*-chlorophenyl isocyanate; allyl isocyanate was obtained in yields of 22% and 23% from allyl isothiocyanate and toluene diisocyanate and 2-methylallyl isocyanate was obtained in 31% yield from 2-methylallyl isothiocyanate and toluene diisocyanate ⟨64JOC2091⟩. It was proposed that this exchange reaction proceeds through the intermediate formation of uretidinethione ketones (Scheme 67).

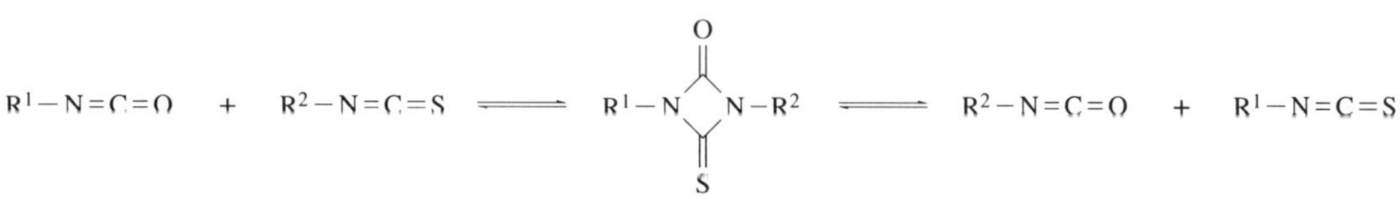

Scheme 67

Phosphorus(V) isothiocyanates (**133**) are also converted into the corresponding phosphorus(V) isocyanates (**135**) through a process which implies the initial formation of the very reactive *N*-dichloromethylenephosphoric diester amides (**134**) which, in a second step, react with formic acid to give the final isocyanate (Scheme 68) ⟨69AG(E)421⟩. On the other hand, *N*-dichloromethylenealkylphosphonic amide chloride and *N*,*N*-bis(dichloromethylene)alkylphosphonamide, react similarly with formic or acetic acid to give phosphorus(V) isocyanates (Equation (78)) ⟨64ZOB2812, 66ZOB461⟩.

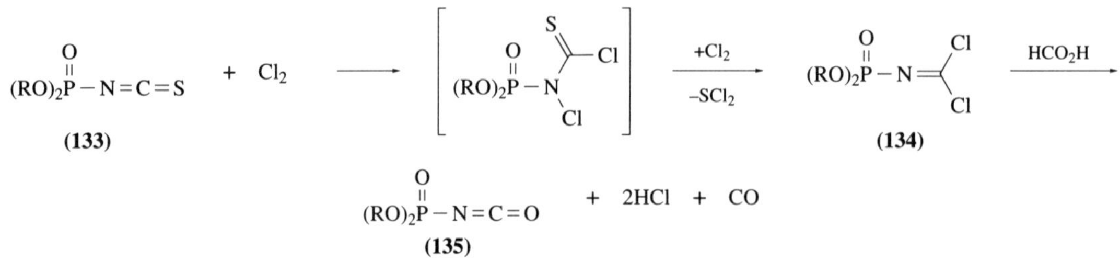

Scheme 68

(78)

5.26.2.14 By Ring Opening of Nitrogen Heterocycles

5.26.2.14.1 From three-membered nitrogen heterocycles

The reaction of aziridines with an equimolecular amount of 1 : 1 complex of phosgene and a tertiary amine (triethylamine or pyridine) in ether, at or below room temperature, gives fair to good yields of the corresponding 2-chloroalkylisocyanates. As an example, isocyanates (**136**) and (**137**) were obtained in 53% and 63% yield. The 1 : 1 complexes should be prepared *in situ* immediately before use, by adding the tertiary amine to a slight excess of phosgene in ether (Equations (79a) and (79b)) ⟨67JHC178, 67JOC1508⟩. In contrast to these results, the reaction of aziridines with phosgene in the absence of base yields β,β'-dichloroalkylureas ⟨50LA(566)210⟩.

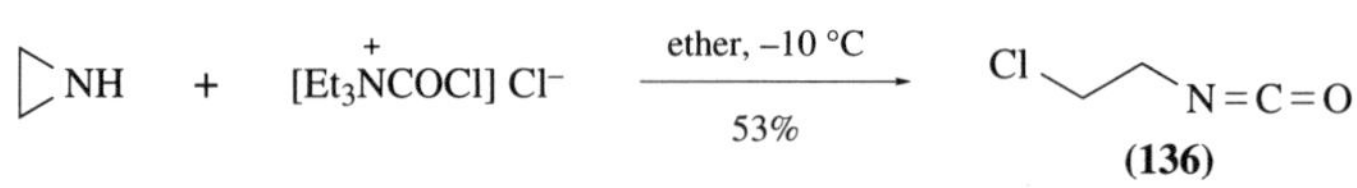

(79a)

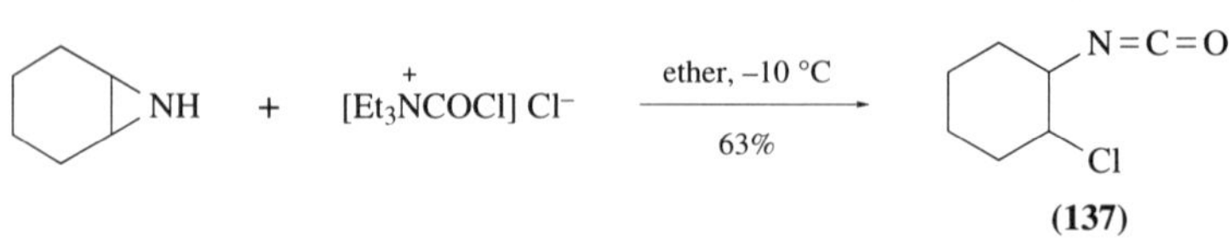

(79b)

Thermal rearrangement of 1-phenylthiocarbonylaziridine leads to the isomeric 2-phenylthioethyl isocyanate (**138**) (Equation (80)) ⟨63JA2743⟩. It was shown that when the isomerization of 1-(arylthiocarbonyl) aziridines to 2-(arylthioalkyl) isocyanates is catalyzed by triethylamine or thiophenolate, the process may occur under relatively mild conditions (25–77°C) ⟨70JOC47⟩.

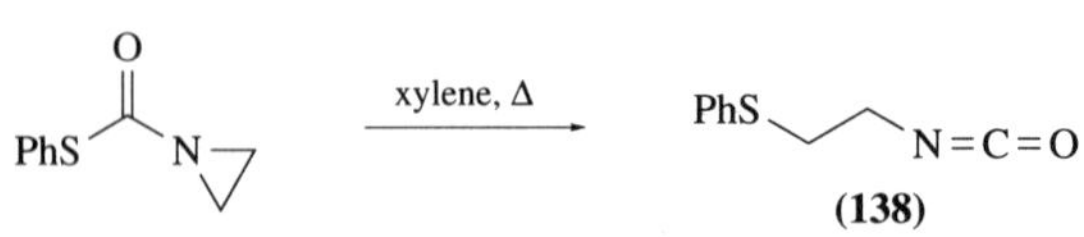

(80)

2-Isopropyl-4,6-di-*t*-butylphenyl isocyanate (**141**) has also been prepared by ring opening of the oxaziridine ring formed by irradiation of 5,7-di-*t*-butyl-3,3-dimethyl-3*H*-indole-1-oxide (**139**) or by oxidation of 5,7-di-*t*-butyl-3,3-dimethyl-3*H*-indole (**140**) with peracid (Scheme 69) ⟨74TL2445⟩.

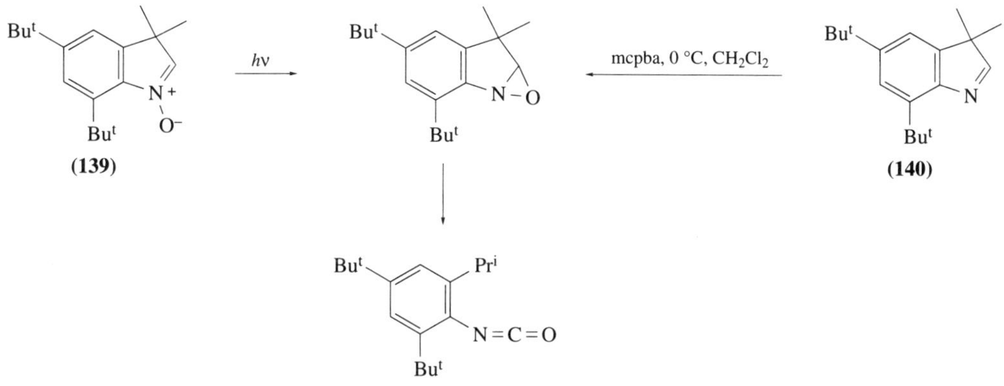

(139) **(140)** **(141)**

Scheme 69

It is also worth noting the thermal rearrangement of 2,2-bis(trifluoromethyl)-3-methoxy-2*H*-azirine (**142**) to the isocyanate (**143**), which was isolated in 20% yield (Equation (81)) ⟨69JOC1278⟩.

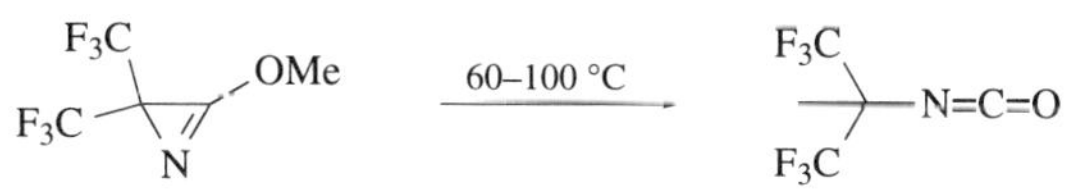

(142) **(143)** (81)

5.26.2.14.2 *From four-membered nitrogen heterocycles*

Tetraaryl-1,2-diazetidinone (**144**) undergo decomposition in refluxing benzene to give the corresponding isocyanate and the ketimine (**145**) (Equation (82)). This reaction is promoted by strong electron-donating groups in the *para* positions of the *N*-1 and *N*-2 phenyl groups ⟨66JOC1079⟩.

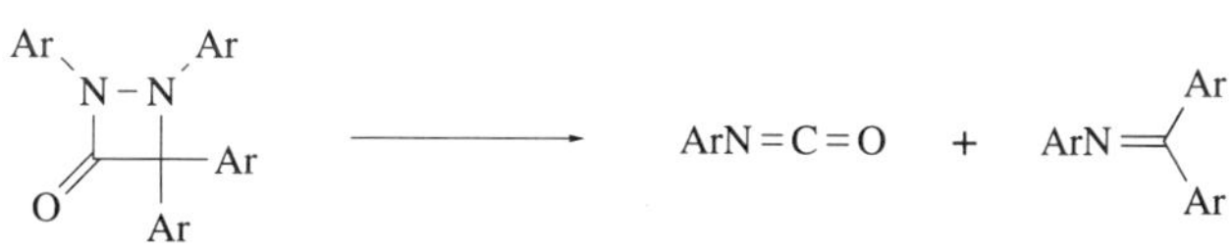

(144) **(145)** (82)

Alkyl- and aryluretidinones also decompose by thermal treatment to give the monomeric isocyanates, while trifluoromethyl isocyanate has been obtained, as a special case, by pyrolysis of 2-fluoroformyloxazetidine (**146**) (Equation (83)) ⟨69JOC1088⟩.

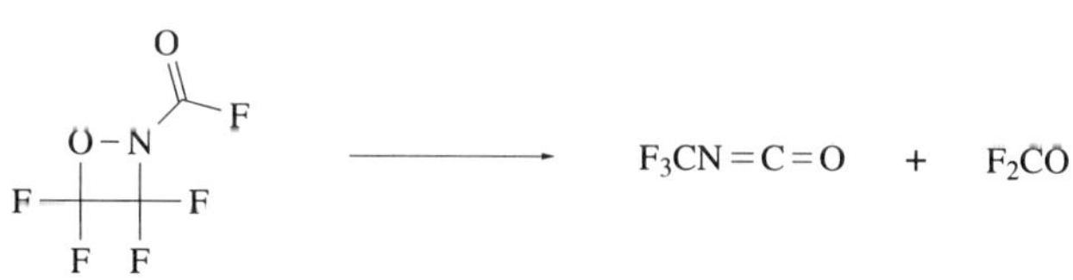

(146) (83)

5.26.2.14.3 *From five-membered nitrogen heterocycles*

Ethoxycyclopropyl isocyanate (**148**) has been isolated in 78% yield by irradiation of 2-ethoxypyrrolin-5-one (**147**) in a nonhydroxylic solvent. When this photochemical ring contraction was carried out in alcoholic solutions, the corresponding *N*(ethoxycyclopropyl)carbamate was formed (Equation (84)) ⟨73JA3957⟩.

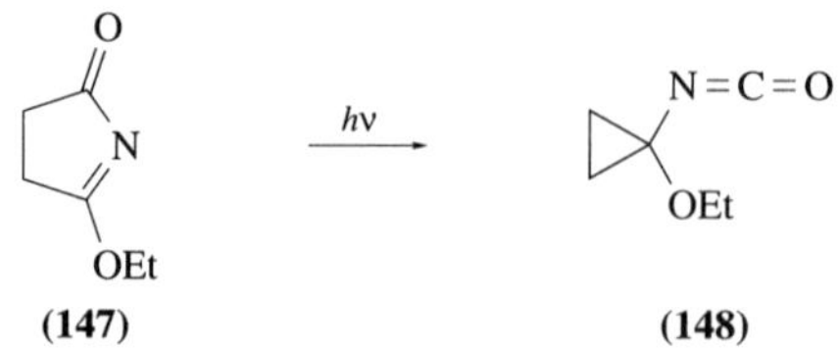

(84)

N-Silylated aliphatic or aromatic oxazolidin-2-ones are also converted on heating, reversibly in some cases and irreversibly in others, into the isomeric 2-siloxyalkyl and 2-siloxyphenyl isocyanates ⟨73LA772⟩. For example, *N*-silyl-substituted benzoxazolone (**149**) undergoes rearrangement, when heated, to give the corresponding isocyanate (**150**) (Scheme 70) ⟨84JOM207⟩.

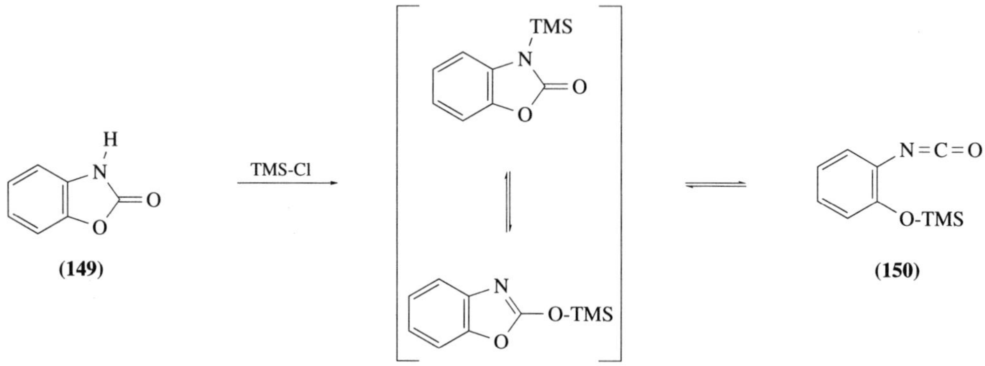

Scheme 70

Pyrolysis of *N*-substituted 4-hydroxy-5,5-dimethyl-2-oxazolidinones and bis(2-oxazolidinones) has been proved to be a general method for obtaining both mono- and diisocyanates in yields ranging from 17% to 76% ⟨89NKK1881⟩. In this way, *β*-chloroethyl isocyanate has been prepared in 46% yield by treatment of oxazolidinone (**151**) with tosyl chloride and pyridine (Equation (85)) ⟨75TL695⟩.

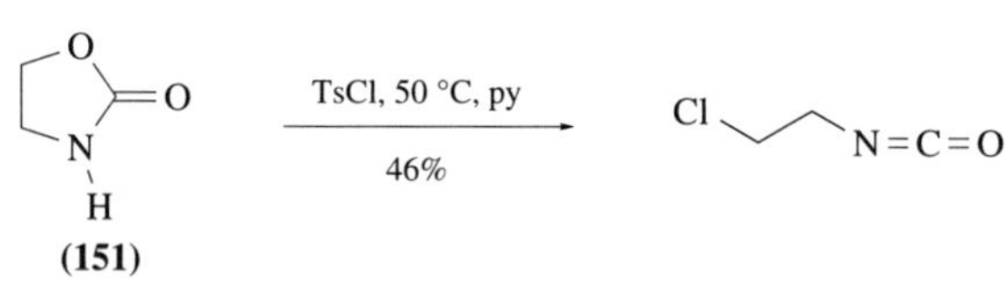

(85)

Oxazolidine-2,5-diones may be silylated with trimethylchlorosilane and the resulting *N*-(trimethylsilyl)oxazolidine-2,5-diones (**152**) are in equilibrium with the isomeric *α*-isocyanatocarboxylic acid trimethylsilyl esters (**153**), the ratio at 0°C being about 1 : 1 (Scheme 71) ⟨71CB3131⟩.

Scheme 71

Linear isocyanate esters have been prepared by ring opening of cyclic lactim ethers or thioethers when treated with phosgene ⟨70GEP(O)1913273, 71BRP1252099, 73GEP(O)2145514⟩. *β*-Isocyanatoethyl thiocarboxylate (**154**) was prepared in this way (Equation (86)).

(154)

(86)

As in the case of the 1,3-dialkylureas, the reaction of 2-imidazolidinones (**155**) with phosgene takes place almost exclusively on the nitrogen, with formation of 2-imidazolidinone-1-carbonyl chlorides (**156**) which undergo dehydrochlorination at room temperature in the presence of an HCl acceptor to give diisocyanates (**157**) (Scheme 72) ⟨66AG(E)704⟩. Since the 2-imidazolidinones are readily available from 1,2-alkylenediamines and carbon dioxide, this provides a simple method for the industrial production of alkylene diisocyanates. Similarly, phosgenation of benzimidazolones gives rise to the formation of *o*-diisocyanates ⟨69JOC1162⟩.

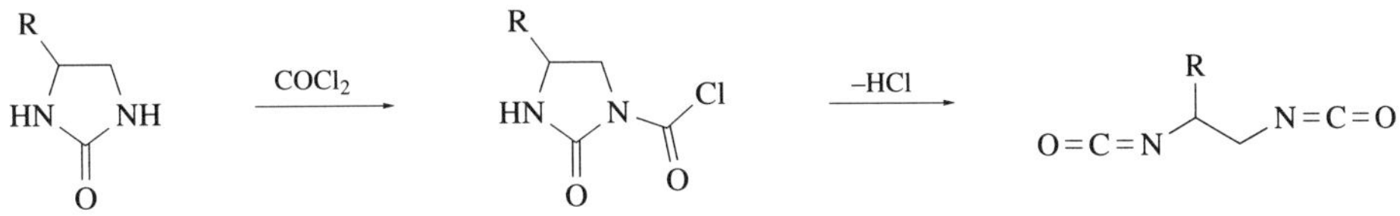

(155) (156) (157)

Scheme 72

Thermal ring opening of furazan *N*-oxides in the absence of dipolarophiles results in the formation of isocyanates through the intermediate nitrile oxides (Equation (87)) ⟨76JCS(CC)240⟩. On the other hand, pyrolysis of both 3,5-diphenyl-1,2,4-oxadiazole (**158**) and 2,5-diphenyl-1,3,4-oxadiazole (**159**) leads to the formation of phenyl isocyanate and benzonitrile (Scheme 73) ⟨66CC336⟩.

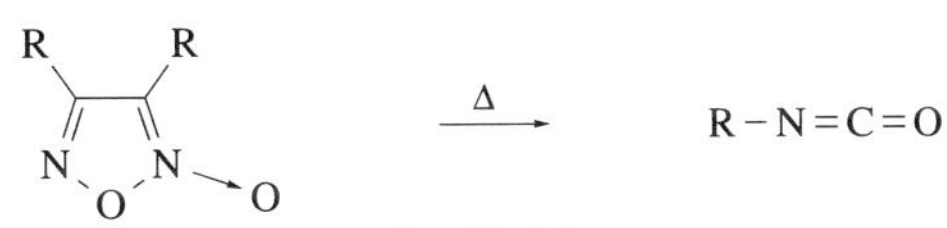

(87)

R = Ph, Me

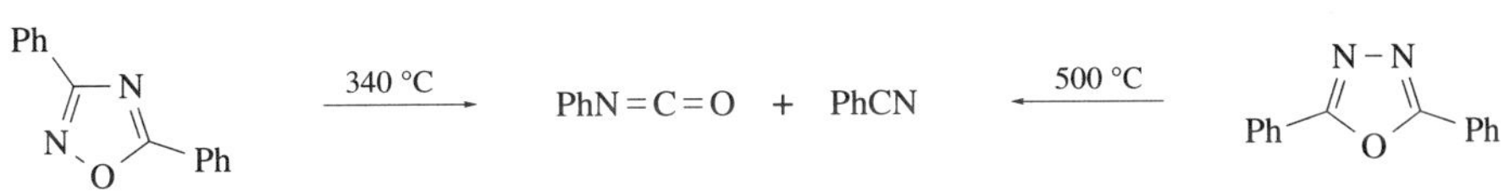

(158) (159)

Scheme 73

The thermal rearrangement of dioxazolones (**160**) to aliphatic and aromatic isocyanates has been well documented, and has also been applied to the preparation of functionalized isocyanates such as methoxymethyl isocyanate (Equation (88)) ⟨68TL319, 70JHC177, 74JOC2472⟩.

(88)

(160)

R = alkyl, aryl

The thermal decomposition of 5-alkoxy- and aryloxy-1,2,3,4-thiatriazoles (**161**) leads to the formation of the corresponding alkyl- or aryl cyanates (**162**) which isomerize rapidly to the corresponding isocyanates (Scheme 74) ⟨64TL2829, 65ACS438⟩. In contrast to these results, 5-(2-alkenyloxy)-1,2,3,4-thiatriazoles decompose in solution to give 2-alkenyl isocyanates without isolation of the precursor cyanates ⟨70ACS1512, 70ACS1852⟩.

An attractive method for the synthesis of mono- and diisocyanates in high yields lies in the thermal rearrangement of 1,3-dioxa-2,4-thiazole 2-oxides (**163**), readily available from reaction of hydroxamic acids and thionyl chloride (Equation (89)) ⟨70JHC177⟩, or by 1,3-cycloaddition of nitrile oxides with SO_2 ⟨77AG(E)555⟩.

(161) **(162)**

R = alkyl, aryl

Scheme 74

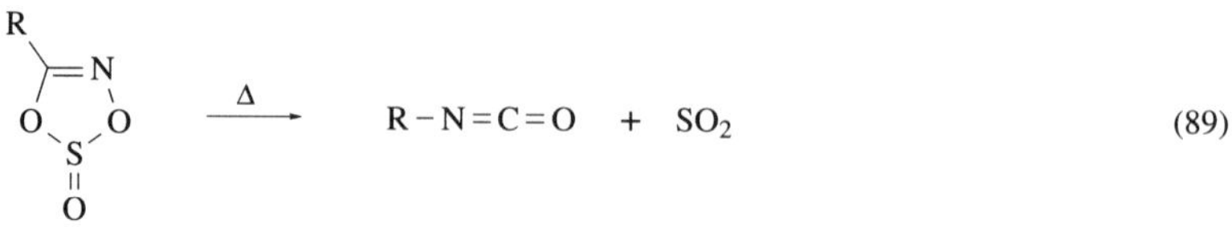

(163)

Arenesulfonyl isocyanates (**165**) can be prepared by thermal cleavage of tetrazolin-5-ones (**164**) (Equation (90)). However, because of the fact that the latter are formed by cycloaddition reactions between isocyanates and azides, this method is of no preparative importance ⟨73JOC675⟩.

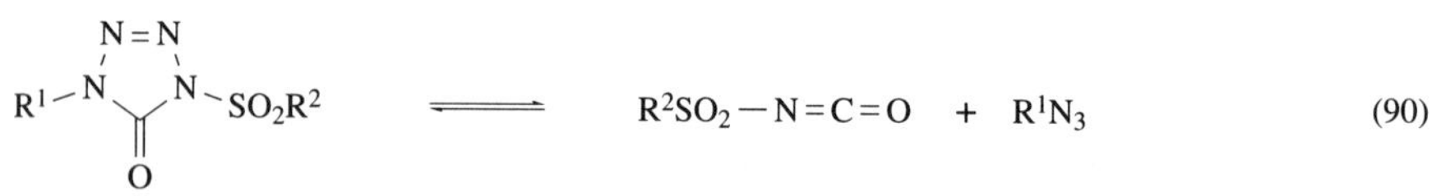

(164) **(165)**

However, the thermolytic reaction of cyclic silylurethanes proceeds along different pathways, depending on the *exo*- or *endo*-arrangement of the Si—N—C—OR′ bond system. For example, in compound (**166**) the *β*-elimination proceeds with ring opening giving an isocyanate containing silicon (**167**), while for compound (**168**), the reaction proceeds with ring contraction to give methyl isocyanate and the cyclic compound (**169**) (Equations (91a) and (91b) ⟨68DOK(181)115, 74ZOB553⟩.

(166) **(167)**

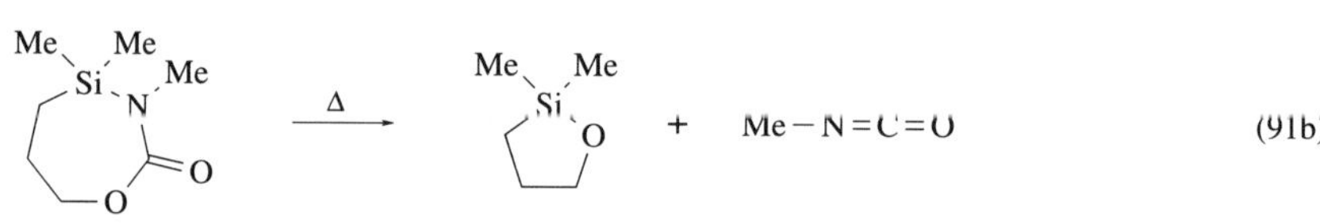

(168) **(169)**

5.26.2.14.4 From six-membered nitrogen heterocycles

Dihydro-1,3-oxazine (**170**) undergoes thermal rearrangement giving the *ω*-bromoisocyanate (**171**), presumably through a dipolar intermediate which undergoes bond rotation and recombination (Equation (92)) ⟨69TL5151⟩.

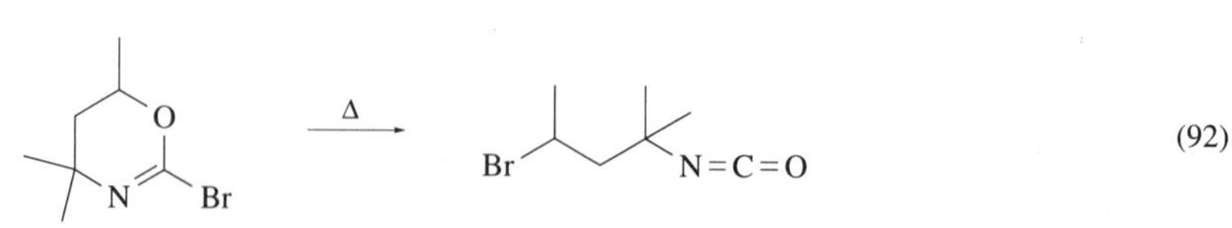

(170) **(171)**

Functionalized isocyanates of the general structure (**174**) can be prepared in almost quantitative yields, by treatment of the oxazinone (**172**) with tosyl chloride in pyridine, to give a crystalline intermediate pyridinium tosylate (**173**), which undergoes ring opening by the action of pyridinium halides (Scheme 75) ⟨75TL695⟩.

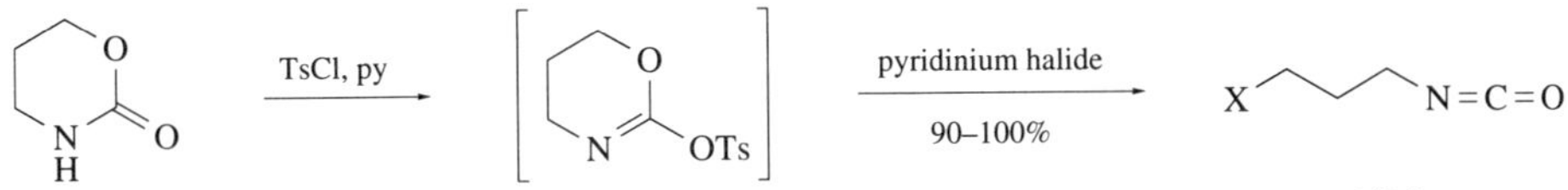

(172) **(173)** **(174)**
X = Cl, Br, I

Scheme 75

Thermolytic cleavage of the six-membered cyclic silylurethane (**175**) leads to the formation of two molecules of isocyanate containing a silicon atom (**176**) (Equation (93)) ⟨71ZOB2340, 72ZOB879, 73ZOB1044, 84JOM207⟩.

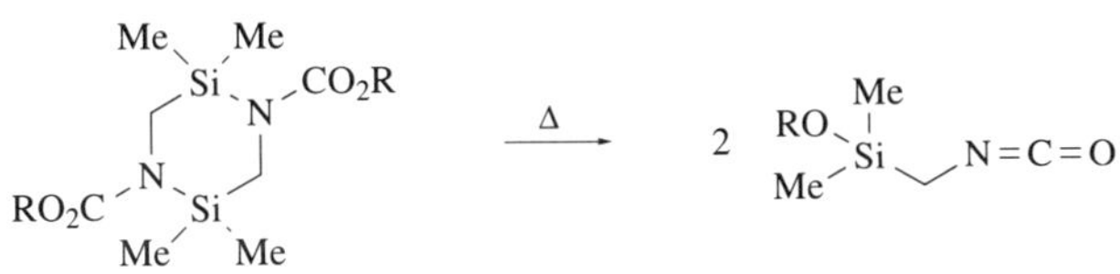 (93)

(175) **(176)**

Thermal decomposition of trichloroisocyanuric acid, under vacuum, leads to the formation of chloroisocyanate while, under normal pressure, carbonyl isocyanate is formed, which is due to the disproportion of the chloroisocyanate initially formed (Scheme 76). Triiodoisocyanuric acid also forms carbonyl isocyanate on thermolysis ⟨66M115, 66M361, 70M655, 71M264⟩. Similarly, a patent claims that pyrolysis of trivinyl-, triallyl-, and triphenylisocyanurate, at 500–1000 °C under reduced pressure, gives the corresponding pure isocyanates, and states that this method is especially useful for the preparation of isocyanates whose corresponding amines are unstable or unreactive ⟨69USP3470228⟩.

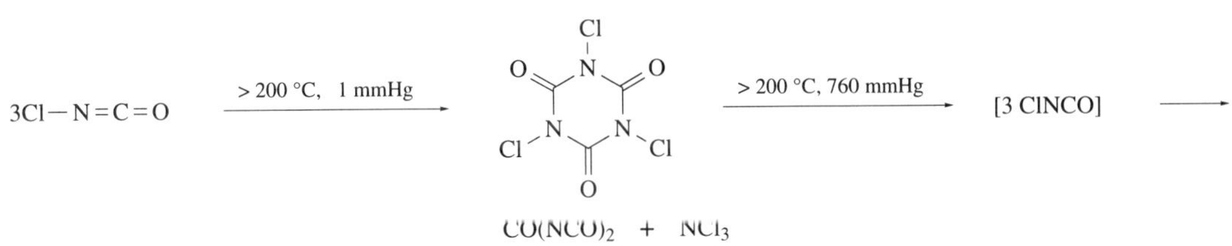

Scheme 76

5.26.2.15 From Inorganic Halides and Isocyanic Acid and Metal Cyanates

In general, the most important methods for the preparation of isocyanates bearing the NCO group attached to an atom other than carbon consists of the reaction of the corresponding halides with isocyanic acid or cyanate salts, though metal oxides can also be used instead of the halide. This method also applies for the preparation of bromocarbonyl isocyanate, by reaction of gaseous bromine with silver cyanate ⟨71AG(E)416⟩, and fluorocarbonyl isocyanate by reaction of carbonyl fluoride with molten potassium cyanate ⟨73CB1752⟩.

5.26.2.15.1 Boron isocyanates

Although mononuclear three-coordinate isocyanatoboranes such as the tris-compounds $B(NCO)_3$ ⟨40JA761⟩, dimethylisocyanatoborane Me_2BNCO ⟨60CB1379⟩, and 2-isocyanato-1,3,2-dioxaborinane $(CH_2)_3O_2BNCO$ ⟨63SA1621⟩ has been reported, a general method for the preparation of three-coordinate mononuclear isocyanatoboranes, of general structures (**177**), (**178**) and (**179**), is based on the reaction of the corresponding chlorides or bromides with silver, sodium or lead cyanate in solvents such as benzene, methylene chloride, and acetonitrile ⟨65JCS4256⟩. The yields reported ranged from 53% to 95%.

$(RO)_2B-N=C=O$ **(177)**

$(R_2N)_2B-N=C=O$ **(178)**

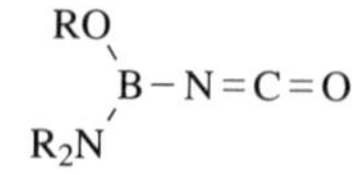

(179)

5.26.2.15.2 *Silicon isocyanates*

Silicon isocyanates have been prepared from the reaction of silicon chloride with silver cyanate ⟨40JA761⟩. The same general technique was used for the preparation of a variety of isocyanatosilanes from chlorosilanes and silver cyanate ⟨47JA3048, 48JA1043, 48JA1222, 50JA3045⟩. Silicon tetraisocyanate can be used to prepare fluorosilicon isocyanates by reacting with antimony trifluoride in the absence of a catalyst ⟨47JA1241⟩.

Isocyanatosilanes can also be prepared by reaction of the corresponding chlorides with isocyanic acid in the presence of tertiary amines (Equation (94)). Unlike the reactions of acid chlorides with metal cyanates, these fast, exothermic (usually cooling is necessary) processes are carried out in homogeneous solutions. The isocyanosilanes so obtained are easy to isolate, do not polymerize or undergo side reactions during their distillation, are stable on storage at room temperature and are prepared in moderate yields ⟨63JOC586⟩.

$$R_nSiCl_{4-n} + (4-n)\,HNCO + (4-n)\,B \xrightarrow[57\text{–}60\%]{} R_nSi(NCO)_{4-n} + (4-n)\,BH^+\,Cl^- \quad (94)$$

Other approaches to the synthesis of silicon isocyanates are the thermal treatment of silyl ureas (Equation (95)) ⟨68ZOB1179⟩, the exchange reactions between trimethyl chlorosilane and other isocyanates, such as chloroisocyanate or tri-*n*-propyltin isocyanate ⟨66ZOB952, 71JOM(31)C3⟩, and the ring opening of ethylene oxide by silicon tetraisocyanate (Equation (96)) ⟨63BRP923581⟩.

(95)

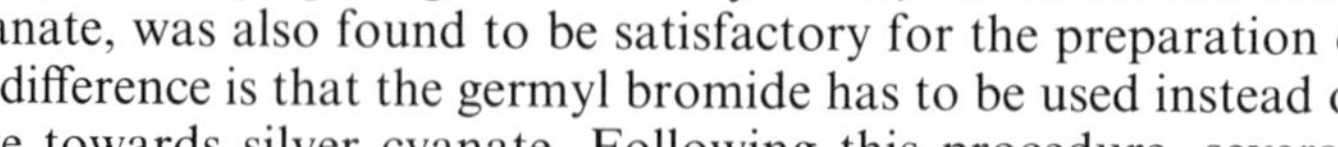

(96)

5.26.2.15.3 *Germanium isocyanates*

The method developed for preparing silicon isocyanates, based on the reaction of a silicon halide with silver cyanate, was also found to be satisfactory for the preparation of the germanium analogues. The only difference is that the germyl bromide has to be used instead of germyl chloride as it is more reactive towards silver cyanate. Following this procedure, several germanium isocyanates have been prepared ⟨43JA1783, 62CJC739, 66JOM(5)383, 68JA6247⟩.

5.26.2.15.4 *Tin, antimony, and arsenic isocyanates*

The heterogeneous reaction between the appropriate organotin chloride and silver cyanate in ether or light petroleum, gave triphenyltin isocyanate, tributyltin isocyanate, and dibutyltin diisocyanate.

Treatment of triphenyltin or tributyltin chloride with a suspension of the sodium salt of urethane in xylene also furnished the isocyanates in high yields (Scheme 77) ⟨65JCS5055⟩.

$$R_3SnCl + NaNHCO_2Et \longrightarrow [R_3SnNHCO_2Et] \longrightarrow R_3Sn{-}N{=}C{=}O + EtOH$$

Scheme 77

Although triethyltin oxide does not react with free isocyanic acid to form triethyltin isocyanate ⟨54JOC1300⟩, the reaction takes place rapidly and almost quantitatively when isocyanic acid is heated to 70–80 °C with organotin oxides. In a similar way, trialkylantimony diisocyanates could be isolated by treating trialkylantimony oxides with isocyanic acid ⟨65JOC693⟩.

When organotin oxides were fused with urea at 130 °C, organotin isocyanates were obtained, in yields from 85% to 95% (Equations (97) and (98)). In an analogous reaction, trialkylantimony and triphenylarsenic diisocyanates were obtained by fusion of trialkylantimony oxides (or bishydroxides) with urea (Equation (99)) ⟨65JOC693⟩.

$$R_3SnOH + (H_2N)_2CO \longrightarrow R_3Sn{-}N{=}C{=}O + NH_3 + H_2O \quad (97)$$

$$(R_3Sn)_2O + 2(H_2N)_2CO \longrightarrow 2R_3Sn{-}N{=}C{=}O + 2NH_3 + H_2O \quad (98)$$

$$R = Bu^n, Bu^i, Et$$

$$R_3MO + 2(H_2N)_2CO \longrightarrow R_3M(N{=}C{=}O)_2 + 2NH_3 + H_2O \quad (99)$$

$$M = Sb, R = Bu^n, Bu^i; M = As, R = Ph$$

Fused urea also reacted with organotin and organoantimony chlorides. However, yields of isocyanates were only of the order of 20–30%, and the separation of the isocyanates from unreacted chlorides by fractional distillation was difficult ⟨65JOC693⟩.

5.26.2.15.5 *Lead isocyanates*

The reaction of trimethyllead chloride and silver cyanate in refluxing benzene allows the preparation of trimethyllead isocyanate, obtained in 10% yield ⟨66JOM(5)383⟩.

5.26.2.15.6 *Phosphorus isocyanates*

Preparation of phosphorus isocyanate $P(NCO)_3$ and phosphoryl isocyanate $PO(NCO)_3$ can be achieved by using different synthetic pathways ⟨69AG(E)421, B-77MI 526-01⟩, some of which are reported in previous sections related to the synthesis of C—NCO because of their similarity, for example, thermal decomposition of alkyl phosphorylcarbamates or reaction of oxalyl chloride with phosphoric ester amides. However, valuable approaches based on the use of phosphorus halides are described in this case.

The first phosphorus(III) isocyanates were prepared by a reaction between phosphorus halides and silver cyanate in boiling benzene (Equation (100)) ⟨40JA761⟩. This compound was also formed from silver cyanate and phosphorus triiodide in nitromethane ⟨50JA193, 50JA2761⟩. However, when silver cyanate is slowly introduced into an excess of phosphorus trichloride Cl_2PNCO and $ClP(NCO)_2$ were obtained ⟨45JA2176⟩. Alkyl- and aryldichlorophosphines and methyldibromophosphine react similarly (Equation (101)) ⟨56JA842⟩.

$$PCl_3 + 3AgOCN \longrightarrow P(N{=}C{=}O)_3 + 3AgCl \quad (100)$$

$$RPCl_2 + 2AgOCN \longrightarrow RP(N{=}C{=}O)_2 + 2AgCl \quad (101)$$

Phosphorus(III) isocyanates have also been obtained from phosphorus chlorides and potassium, sodium, and lithium cyanates in benzene or sulfur dioxide ⟨59USP2873171, 62BEP612359, 62BRP907029⟩. The action of silver cyanate on phosphoryl trichloride yields mixtures of phosphoryl triisocyanate and phosphoryl tricyanate (Equation (102)). In contrast to that result, alkyl- and arylphosphonic dichlorides react with silver cyanate to give only isocyanates (Equation (103)) ⟨56JA842, 64ZOB3609⟩.

$$POCl_3 + 3AgOCN \longrightarrow O{=}P(NCO)_3 + O{=}P(OCN)_3 + 3AgCl \qquad (102)$$

$$RPOCl_2 + 2AgOCN \longrightarrow OCN{-}\overset{O}{\overset{\|}{\underset{R}{\underset{|}{P}}}}{-}NCO + 2AgCl \qquad (103)$$

In the presence of triethylamine or pyridine, alkoxydichlorophosphines react very readily with cyanic acid to form alkoxydicyanatophosphines in yields of about 50–60% (Equation (104)).

$$ROPCl_2 + 2HNCO + 2Et_3N \longrightarrow ROP(N{=}C{=}O)_2 + 2Et_3\overset{+}{N}H\ Cl^- \qquad (104)$$

The reaction of tetraisocyanatosilane with phosphoryl fluoride is also a method for the preparation of phosphoryl difluoride isocyanate ⟨67CB1082⟩.

5.26.2.15.7 *Sulfur isocyanates*

As for phosphorus isocyanates, there are many different methods for the preparation of isocyanates in which the NCO moiety is directly linked to a sulfur atom ⟨65CRV369, 72IJS(B)319, B-77MI 526-01⟩, and some of them are already described in previous sections. However, methods which use sulfur halides are also commonly followed.

Preparation of sulfuryl diisocyanate following a route based on the reaction of chlorosulfonyl isocyanate and silver cyanate was reported ⟨58CB1200⟩. The above-mentioned diisocyanate undergoes further reaction with fluorosulfonic acid to give fluorosulfonylisocyanate (Scheme 78) ⟨64CB849⟩.

$$ClSO_2{-}N{=}C{=}O + AgOCN \longrightarrow SO_2(NCO)_2 \xrightarrow{FSO_3H} FSO_2NCO + OCN{-}SO_3H$$

Scheme 78

Reaction of trifluoromethanesulfenyl chloride with silver cyanate gave trifluoromethanesulfenyl isocyanate (CF_3SNCO) as the major product (75% yield) but, in addition, the dimer $(CF_3SNCO)_2$ was also formed ⟨63AG(E)552, 63JCS1272⟩. In a similar way, dichlorofluoro- and chlorodifluoromethanesulfenyl isocyanates were prepared (Equation (105)) ⟨67CB480⟩. Preparations of fluorocarbonylsulfenyl and trichloromethylthiosulfenyl isocyanates have also been described through an analogous route (Equation (106)) ⟨69CB2718, 71CB2588⟩.

$$Cl_{3-n}F_nCSCl + AgOCN \longrightarrow Cl_{3-n}F_nCS{-}N{=}C{=}O + AgCl \qquad (105)$$

$$FCOSCl + AgOCN \longrightarrow FCOS{-}N{=}C{=}O \qquad (106)$$

5.26.3 PHOSPHORUS FUNCTION: RP═C═O

Only a few systems with cumulative double bonds and containing P═C sequences are known. These compounds, moreover, have a tendency to undergo self-addition. Thus, *t*-butylphosphaketene (**182**), the first phosphaketene to be unequivocally detected, is stable as a monomer only below −60 °C. At higher temperatures, it dimerizes to the diphosphetanedione (**183**). This phosphaketene was prepared by phosgenation of the silylphosphanes (**180**) at −100 °C, which leads to the formation of the corresponding phospha analogues of the carbamic acid chlorides (**181**), which lose trimethylchlorosilane or HCl to form the phosphaketene (Scheme 79) ⟨83TL2639⟩.

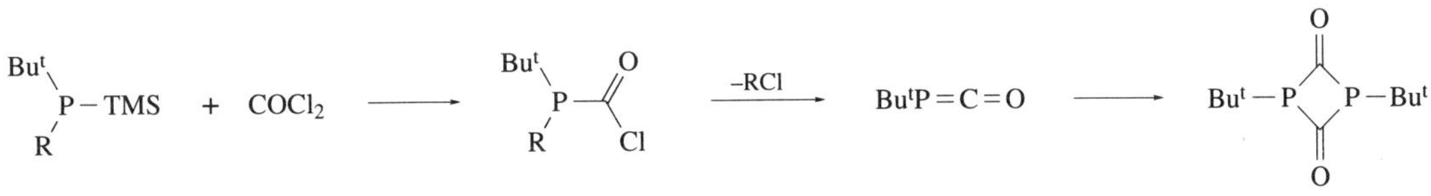

(**180**) (**181**) (**182**) (**183**)

R = TMS, H

Scheme 79

The preparation of the first phosphaketene (**186**) which is stable at room temperature, in monomeric form as orange prisms, was achieved by reaction of the disilylated phosphane (**184**) with phosgene (Scheme 80) ⟨83AG(E)785⟩. It is characterised by the ^{31}P-NMR signal which appears in the high field region at $\delta = -207.4$ p.p.m. Likewise, the infrared band for the unsymmetric stretching vibration of the PCO group at 1953 cm^{-1} is comparable to that of other cumulative double-bond systems. The phosphaketene is no doubt stabilized by the presence of the bulky tri-*t*-butylphenyl moiety. The monocondensation product (**185**), which, in the case of the *t*-butyl derivative could be detected by ^{31}P-NMR spectroscopy in the reaction solution at $-90\,^\circ$C, cannot be observed during the synthesis of (**186**), because of the drastic reaction conditions.

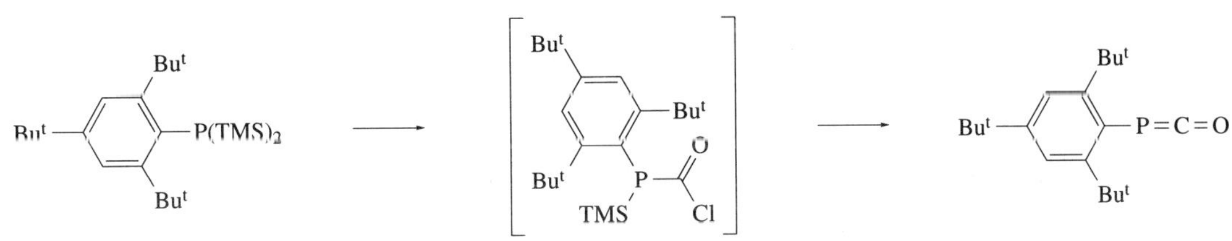

(**184**) (**185**) (**186**)

Scheme 80

5.26.4 METALLOID FUNCTIONS: R_2Si—C—O AND RB=C=O

5.26.4.1 R_2Si=C=O

Arrington and co-workers have reported some evidence for the reaction of dimethylsilylene with carbon monoxide, resulting in the formation of a species in which the CO is bonded to the silicon atom. The initial expectation was that the reaction product represented a Lewis-base adduct with the carbon lone pair of CO forming a coordinate covalent bond with the empty p-orbital on silicon, which would assume a pyramidal geometry. Formation of the silaketene would formally require loss of one C—O π bond to form the Si—C π bond. However, MNDO (Modified Neglect of Diatomic Overlap)/AM1 (Austin Model 1) and *ab initio* calculations give quite different structures for optimized geometries. Both AM1 and MNDO calculations have a minimum energy for the silaketene geometry. Calculated geometries have an Si—C bond length of 1.625 Å (MNDO) and 1.644 Å (AM1), clearly in the range of Si=C double bond lengths. *Ab initio* calculations with a 3-21G basis set give a lowest energy structure which is pyramidal at Si with an Si—C bond length of 2.89 Å and a C Si CO angle of 89°, while the silaketene structure is 84 kJ/mol higher in energy than the pyramidal structure. The CO bond lengths are 1.185 Å (MNDO) and 1.125 Å (3-21 G) for the adduct compared with 1.63 Å (MNDO), 1.129 Å (3-21 G), and 1.128 (experimental) for CO. In the light of this experimental evidence the structure of the adduct could not be determined with certainty ⟨88JA6240⟩.

Independently, another report studied the interactions of diorganosilylenes with carbon monoxide in frozen hydrocarbon matrices at 77 K (Scheme 81) ⟨88JA7228⟩. These interactions were monitored by observing the UV–vis spectrum of the reaction mixture, both at 77 K and upon warming. In 3-methylpentane (3-MP), (**187a**) and (**187b**) gave silylenes (**188a**) and (**188b**), identified by their characteristic visible absorption bands. Annealing of the matrix resulted in the formation of the silylene—CO complex (**189a**) or (**189b**). Upon further warming the disilenes (**190a**) and (**190b**) were formed. With precursors (**187c**) and (**187d**), photolysis in 3-MP gave not only the absorption bands for silylenes (**188c**) and (**188d**), but also an absorption attributed to the complex (**189c**) and (**189d**). On standing for several hours, the complex band grew more intense, while the silylene band decreased. Annealing of the matrix led to complete disappearance of the silylene band and to a simultaneous great increase of the absorption band attributed to the complex. Further warming gave the stable disilene (**190c**) from (**189c**).

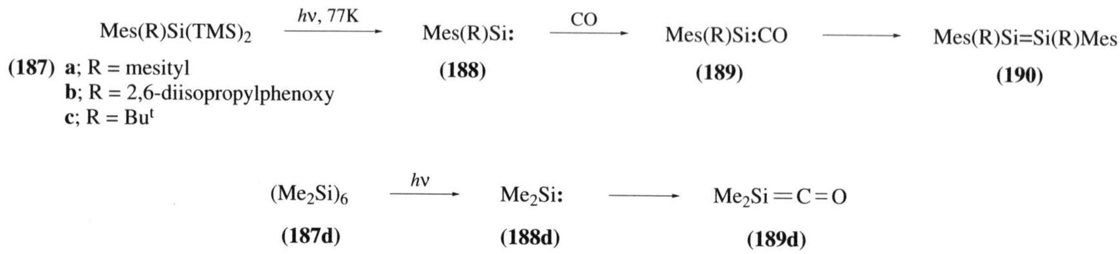

Scheme 81

Photolysis of (**187a**)–(**187d**) was also carried out in a 4 : 1 2-methylbutane : 3-methylpentane mixture, which forms a soft matrix at 77 K. Under these conditions only the absorption bands attributed to the silylene—CO complex were observed. Annealing of the matrix led again to the disilenes (**190a**)–(**190c**) but not (**190d**).

In the light of these results, it was assumed that carbon monoxide might therefore react with silylenes to give a nonplanar complex or might combine to form silaketene. Silylene—CO complexes (**189a**)–(**189c**) behave similarly to other acid–base complexes of silylenes (**188a**)–(**188c**) with donor molecules since they produce disilenes (**190a**)–(**190c**) when the matrix is melted ⟨87OM2270, 87OM2617, 88OM558⟩. These results are consistent with formulation of the CO adducts as acid–base complexes rather than silaketenes, which would not be expected to dissociate readily. Theoretical calculations also indicate that the nonplanar complexes are more stable than silaketenes ⟨89JCP1031⟩.

Atomic charges in silaketene and dimethylsilaketene (for both the pyramidal minimum and the planar transition state) have also been calculated within the GAPT (Generalized Atomic Polar Tensor) population analysis ⟨90JA4183⟩, and it concludes that unlike the ketene molecule, silaketene has a nonplanar geometry whereas the planar arrangement of atoms corresponds to a transition state ⟨89JCP1031⟩. The C═O bond in the nonplanar geometry is less polarized than in the ketene molecule ⟨89JA8333⟩, and as expected, the polarity decreases, although slightly, upon the inclusion of electron correlation. Similar observations are relevant to the dimethylsilaketene system.

5.26.4.2 RB═C═O

The infrared spectra for Ar/CO samples codeposited with laser-ablated boron atoms on a 12 ± 1 K salt window have been reported. BCO was identified as the primary reaction product, based on isotopic substitution at all atomic positions and *ab initio* structure and frequency calculations. The dicarbonyl, $B(CO)_2$, and the linear dimer, $(BCO)_2$, were also identified from the infrared spectra ⟨92JPC10195⟩.

5.27
Functions With at Least One Chalcogen Other Than Oxygen

JEREMY GILMORE and PETER T. GALLAGHER
Lilly Research Centre Ltd., Windlesham, UK

5.27.1 CHALCOGEN FUNCTIONS

Compounds covered in this section would be of the general Structure (**1**). A computer search of *Chemical Abstracts* revealed no examples of such structures.

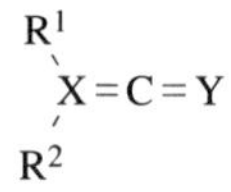

(1)

X, Y = S, Se, Te

5.27.2 NITROGEN FUNCTIONS

5.27.2.1 Functions Based on Nitrogen and Sulfur

This section describes synthetic approaches to organic isothiocyanates where the —NCS functionality is bonded directly to a carbon atom. The carbon atom may form part of an alkyl, alkenyl, alkynyl, aryl or heteroaryl grouping (Structure (**2**))—syntheses of these compounds have been categorised according to the functionality undergoing transformation to the isothiocyanate. Alternatively, the carbon atom bonded to the isothiocyanate may be sp^2-hybridised to a heteroatom (Structure (**3**))—syntheses of such acyl, thioacyl and imidoyl isothiocyanates are discussed separately at the end of the section.

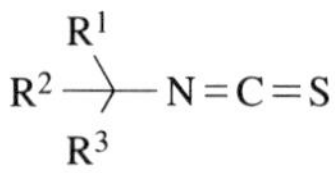

(2)

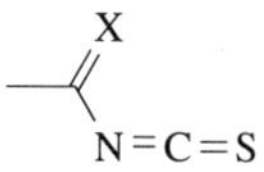

(3)

X = O, S, NR

The preparation of isothiocyanates has been reviewed previously. Schroeder's review of thioureas includes a synopsis of the most frequently employed syntheses of isothiocyanates ⟨55CRV181⟩. Smith describes common routes in brief, including methods for preparing acyl isothiocyanates ⟨B-65MI 527-01⟩. Assony's review of 'The chemistry of isothiocyanates' similarly includes the synthesis of isothiocyanates and their acyl derivatives ⟨B-61MI 527-01⟩. For the period from 1969 to 1980, the literature of isothiocyanates was regularly reviewed in a series of specialist periodical reports but this was unfortunately discontinued ⟨70SST83, 73SST73, 75SST65, 77SST59, 79SST52, 81SST66⟩.

The most comprehensive of reviews are to be found in the appropriate volumes of the standard texts on functional groups by Patai and by *Houben-Weyl*. In the former, Drobnica *et al.* cover the chemistry and synthesis of all —NCS-containing compounds with a particularly strong emphasis on the reaction of thiophosgene with amines ⟨B-77MI 527-01⟩; while Tsuge has written a specialist chapter on the synthesis of acyl and thioacyl isothiocyanates ⟨B-77MI 527-02⟩. In *Houben-Weyl*, Hartmann reviews the literature on all classes of isothiocyanates and provides typical experimental procedures ⟨83HOU(E4)834⟩. For a more brief, general overview of the subject, the appropriate sections of *Comprehensive Organic Chemistry* are to be recommended ⟨79COC(2)513, 79COC(3)305, 79COC(3)461⟩.

A number of specialist reports on specific classes of isothiocyanates are also available, covering six-membered heterocyclic isothiocyanates ⟨87S525⟩, acyl isothiocyanates ⟨68RCR363⟩ and 2-polyfluoroalkylethyl isothiocyanates ⟨91JFC(53)1⟩.

5.27.2.1.1 Isothiocyanates from amines

The reaction of amines with thiocarbonyl reagents provides the most widely employed and versatile procedure for the synthesis of isothiocyanates. This section is subdivided according to the three classes of thiocarbonyl reagents available for this transformation:

(i) carbon disulfide;
(ii) thiophosgene; and
(iii) other thiocarbonyl derivatives.

Close derivatives of amines (e.g., amides, imines, etc.) which may be transformed directly to isothiocyanates with such thiocarbonyl reagents, or precursors which undergo *in situ* conversion to amines (e.g., azides or nitro compounds), are included in a separate section (see Section 5.27.2.1.1.(iv)); as many of these derivatives may be readily prepared from amines, reference should also be made to this latter section.

(i) Carbon disulfide

Amines may react with carbon disulfide in the presence of an organic or inorganic base to yield a dithiocarbamate salt, which in turn, with the assistance of a suitable activating reagent, may decompose to yield the isothiocyanate (Scheme 1). The general procedure is frequently referred to as the Kaluza reaction after the latter's early reports on the synthesis of isothiocyanates from *N*-substituted dithiocarbamates ⟨09M717, 12M363⟩. This approach is dependent upon the basicity of the amine allowing initial formation of the dithiocarbamate, and so can be limited in its application.

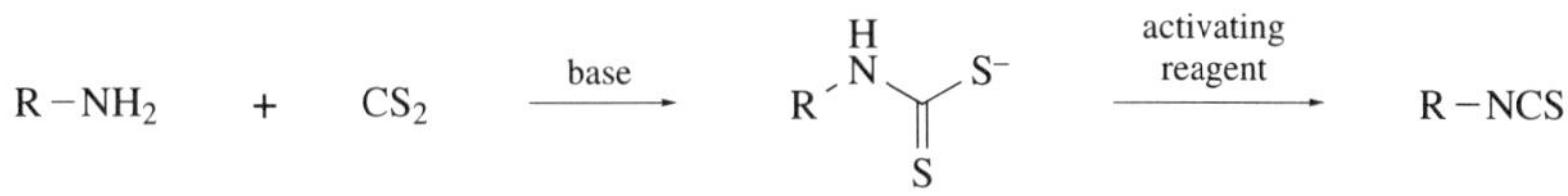

Scheme 1

One of the early methods of achieving this reaction employed heavy metal salts to enhance the decomposition of the intermediate dithiocarbamate. Thus an *Organic Syntheses* procedure uses lead nitrate to give a good yield of phenyl isothiocyanate ⟨32OSC(1)447⟩, but substituted aromatics give only modest yields at best. Similarly, mercury ⟨07CR(144)1125, 09LA(371)201, 66JCS(C)950⟩, lead ⟨07CR(144)1125, 22MI 527-01⟩, copper ⟨1891CB3021⟩ and iron salts ⟨38BRP495510⟩ have been used, but without general applicability.

The decomposition of dithiocarbamates may be enhanced by formation of an unstable derivative. Thus reaction of aryl dithiocarbamates with phosgene can give good yields of the isothiocyanate (Equation (1), R = H, 4-MeO, 4-EtO, 4-OH) ⟨30CB888⟩—this method even allows the preparation of 4-hydroxyphenyl isothiocyanate in 55% yield. In the mid 1960s, phosphorus oxychloride was similarly employed for the decomposition of both alkyl and aryl dithiocarbamates formed *in situ* (Equation (2), R = Et, Bu^n, allyl, Bn, Ph, 4-ClC_6H_4) ⟨65CB2425⟩. Although the yield for phenyl isothiocyanate is not optimal (68%), this method does allow the formation of alkyl isothiocyanates in moderate yield. Use of the catechol derivative appears to enhance yields giving, for instance, 84% benzyl isothiocyanate.

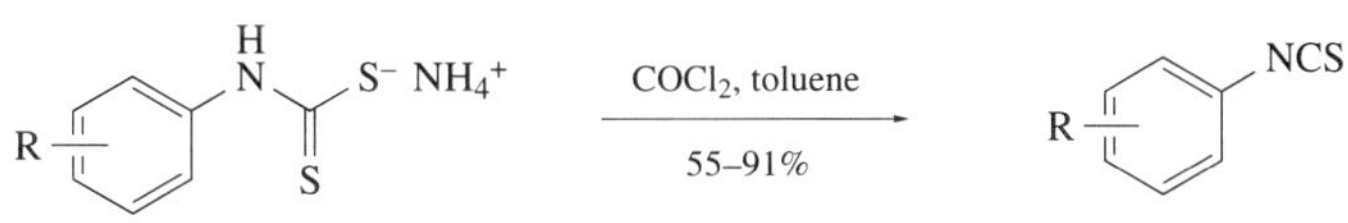

(1)

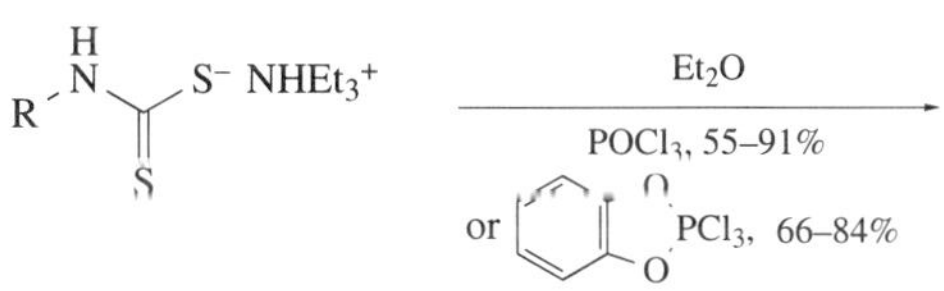

(2)

Two of the more popular reagents originally utilised in the synthesis of isothiocyanates from dithiocarbamates were methyl or ethyl chloroformate. First demonstrated by Andreasch, a reaction takes place by formation of the unstable carbomethoxy or carbethoxy dithiocarbamate with subsequent base-catalysed decomposition ⟨06M1221⟩. Typical examples of this procedure are shown in Table 1.

In the late 1960s, new activating agents were introduced which give reliable decomposition of both alkyl and aryl dithiocarbamates generated *in situ*. Thus Jochims and Seeliger demonstrated the utility of dicyclohexylcarbodiimide (dcc), first for the formation of alkyl isothiocyanates (Equation (3); e.g., R = Bn, $(EtO)_2CHCH_2$) ⟨67AG(E)174⟩ and then for the aryl isothiocyanates (Equation (4); e.g., Ar = Ph, 4-HO-C_6H_4, 3-pyridyl) ⟨68CB1746⟩.

$$\text{R}-\text{NH}_2 \;+\; \text{CS}_2 \;+\; \text{dcc} \xrightarrow[\text{87–90\%}]{\text{Et}_2\text{O or THF}} \text{R}-\text{NCS} \;+\; N,N'\text{-dicyclohexylthiourea} \qquad (3)$$

$$\text{Ar}-\text{NH}_2 \xrightarrow[\text{75–87\%}]{\text{CS}_2\text{, dcc, pyridine}} \text{Ar}-\text{NCS} \qquad (4)$$

Yamamoto *et al.* proposed that as cyanamide is a tautomeric form of carbodiimide it might prove useful as an agent for the dehydrosulfurisation of dithiocarbamates ⟨92OPP346⟩. They reported that this reagent could be used in a one-pot synthesis of isothiocyanates in good yield. The method

Table 1 Isothiocyanates from amines via dithiocarbamates with ethyl chloroformate.

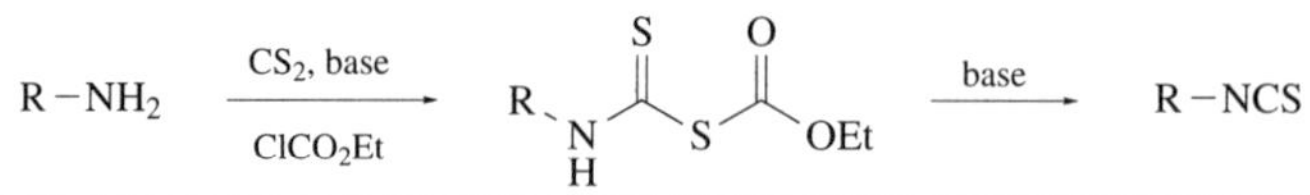

Reactant	*Base*	*Yield* (%)	*Ref.*
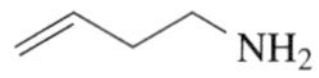	KOH	75	55JA1831
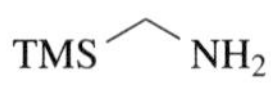	NMe_3	50	55JA3149
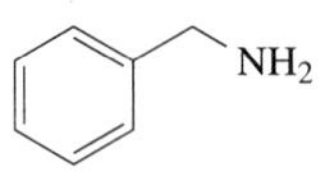	KOH	80	56JOC204
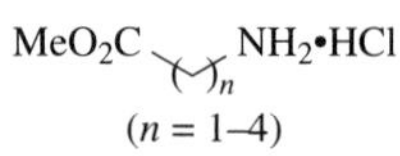	NEt_3	37–81	58JA3332
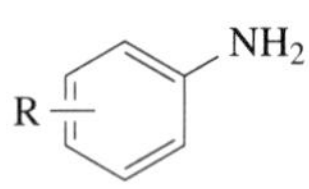	NEt_3	57–73	64JOC3098
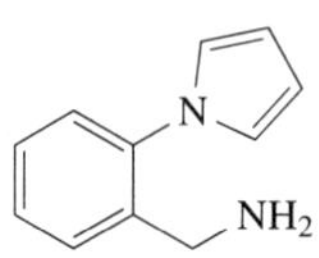	NEt_3	73	84JHC1685
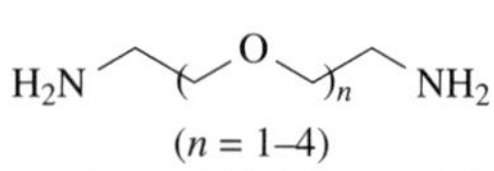	NEt_3	65–78	85ZOB2100

works particularly well with alkylamines (giving yields of 62–95% for primary and secondary amines, but only 27% for Bu^tNH_2), when only a few drops of triethylamine are required to catalyse the reaction, but yields proved poorer (28–56%) with arylamines even in the presence of one equivalent of triethylamine and were 20–30% lower than those achieved with dcc. Alternatively, dicyandiamide may be used in a similar way ⟨91JAP(K)03193760⟩—thus *n*-propylamine is initially reacted with carbon disulfide at 0°C for 3 h, then dicyandiamide added and stirred at 40°C for a further 3 h to yield *n*-propyl isothiocyanate in 95% yield.

Other reagents employed include 2-chloro-1-methylpyridinium iodide, acetic anhydride, *S,S′*-bis(1-phenyl-1*H*-tetrazol-5-yl)dithiocarbonate and triphenylphosphine/carbon tetrachloride (Table 2). For the latter reagent mixture, a one-pot method is preferred for reaction of alkylamines with the increased nucleophilicity of their amino function, giving high yields (70–91%) of the desired products.

Sakai *et al.* have demonstrated that direct decomposition of dithiocarbamates is possible in strong base. Lithium dithiocarbamates, formed from amines with *n*-butyllithium and carbon disulfide, were treated with an additional equivalent of Bu^nLi and carbon disulfide, giving the dilithio derivative (**4**) (Scheme 2) ⟨74JOC1970⟩. When R = alkyl, decomposition of this dilithio species occurred at room temperature (55–85%), while the corresponding aryl-substituted derivatives required heating at reflux in THF (71–99%). The reaction is obviously restricted to amines containing no groups sensitive to Bu^nLi. The author has extended the procedure to the use of ethylmagnesium bromide as base, where the addition of a second equivalent of carbon disulfide is unnecessary ⟨75BCJ2981⟩. Although easier to perform, the yields tend to be somewhat lower than those achieved with Bu^nLi. The reaction is considered to involve the formation of a dimagnesio dithiocarbamate (where R = aryl, this is formed directly by addition of two equivalents of EtMgBr to the amine, followed by carbon disulfide; while when R = alkyl, because of the lower acidity of the N—H bond, a stepwise method must be employed (Scheme 3)).

Table 2 Isothiocyanates from dithiocarbamate salts with miscellaneous reagents.

Reactant	*Reagent*	*Yield* (%)	*Ref.*
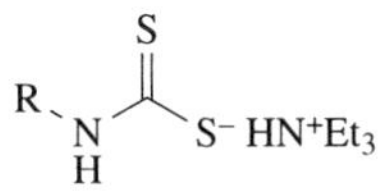	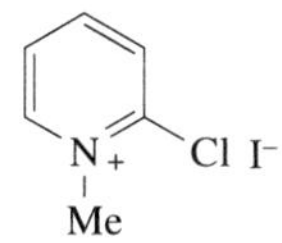	84–100	77CL573
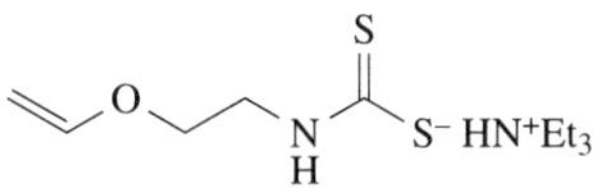	Ac_2O	95	88ZOR1771
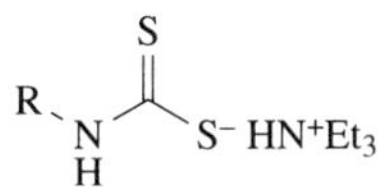	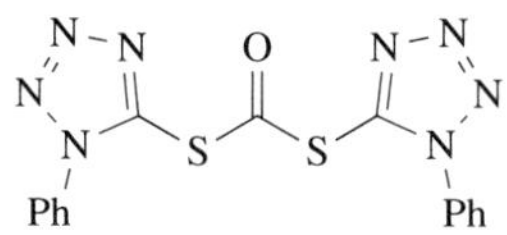	58–95	89CPB2334
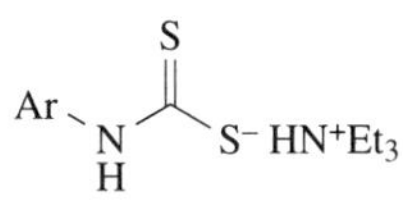	PPh_3/CCl_4	49–80	88CE215
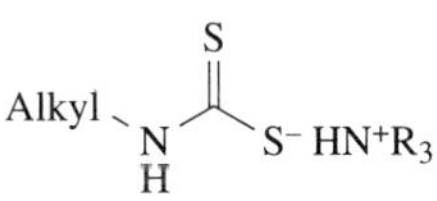	PPh_3/CCl_4	70–90	89NKK822

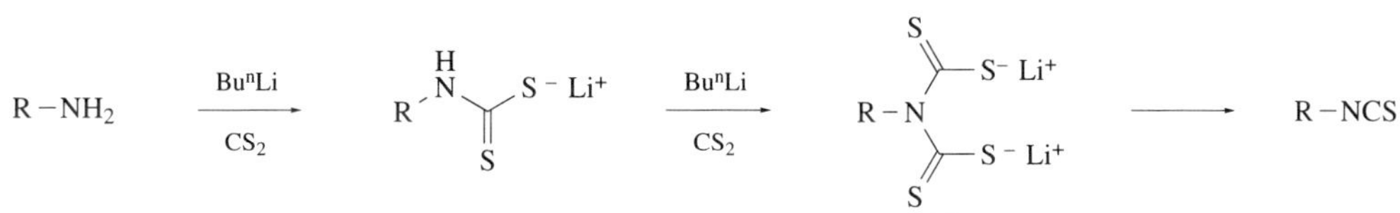

(4)

Scheme 2

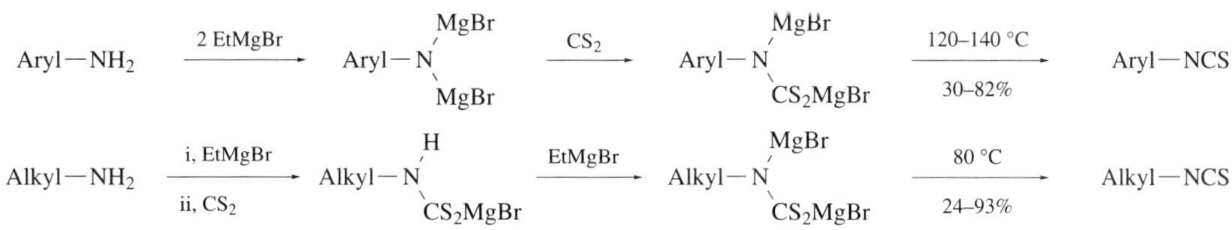

Scheme 3

3-Isothiocyanatoindoles have been prepared from the 3-aminoindole by decomposition of the intermediate dithiocarbamate with excess carbon disulfide and sodium ethoxide in 40–60% yield ⟨78KGS1497⟩. Alkyl isothiocyanates may be formed in high yield from their amines with carbon disulfide in the presence of arsenic trioxide, ideally in the ratio 3:3:1 for amine:CS_2:As_2O_3 ⟨78YGK1090⟩.

Dithiocarbamates may also be cleaved under oxidative conditions. An early approach involved initial oxidation with iodine to the thiouram disulfide, followed by cleavage with a second equivalent of iodine and sodium to yield the isothiocyanate (70% yield of phenethyl isothiocyanate) ⟨12CB2188⟩. Alternatively, a solution or suspension of the preformed ammonium dithiocarbamate in a biphasic, water–dichloromethane system may be treated with an aqueous solution of sodium hypochlorite ⟨54LA(585)230⟩ or sodium chlorite ⟨55LA(594)233⟩ with the resultant isothiocyanate partitioning into the organic phase for ease of isolation. The method has been applied to a number of alkyl, substituted phenyl, ω-dialkylaminoalkyl and 3-alkoxypropyl isothiocyanates in yields of 71–91%; yields appear to be generally higher with sodium chlorite as oxidant. The addition of emulsifying agents allows the preparation of long-chain alkyl isothiocyanates (C_{12}–C_{18}) by the chlorite oxidation method ⟨59LA(621)1⟩. Other oxidising agents employed have included hydrogen peroxide ⟨70IJC759⟩ and iron(III) chloride ⟨77S582⟩.

A more lengthy procedure for the decomposition of dithiocarbamates involves an initial conversion to the corresponding alkyl ester. It was first demonstrated by Hofmann that such esters could be decomposed thermally to yield isothiocyanates ⟨1869CB120⟩ but more comprehensive studies of this pyrolysis have been carried out since the beginning of the 1970s. In 1975, Moharir showed that a range of alkyl and aryl isothiocyanates were formed (in 62–87% yield) on heating methyl or ethyl esters of the corresponding dithiocarbamates at temperatures of 145–180 °C and steam distilling the product ⟨75JIC148⟩. A similar method was employed by Blotny to give primary, secondary and tertiary alkyl isothiocyanates in excellent yield from the ethyl dithiocarbamates (Scheme 4) ⟨82LA1927⟩.

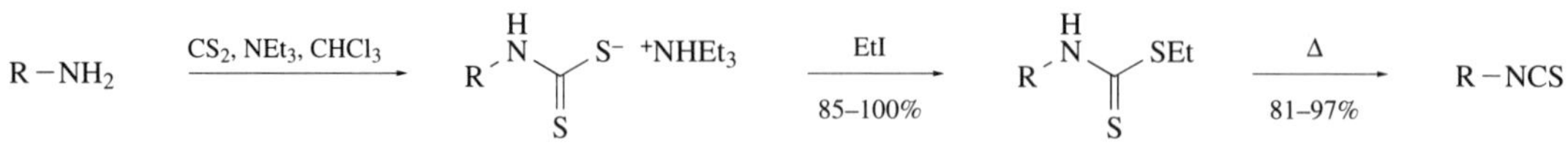

Scheme 4

The decomposition of dithiocarbamate esters has been shown to be base-catalysed. Aromatic diisothiocyanates are synthesised in good yield by the action of mercury(II) oxide and a catalytic amount of aqueous ammonium hydroxide in DMF (Equation (5)) ⟨89BSB289⟩. The authors also report that alkyl diisothiocyanates may be prepared by this methodology, though yields are not quoted. Formation of 2-(vinyloxy)ethyl isothiocyanate (Equation (6)) from its methyl dithiocarbamate requires the use of catalytic amounts of potassium hydroxide to ensure good yields ⟨90ZOR2059⟩.

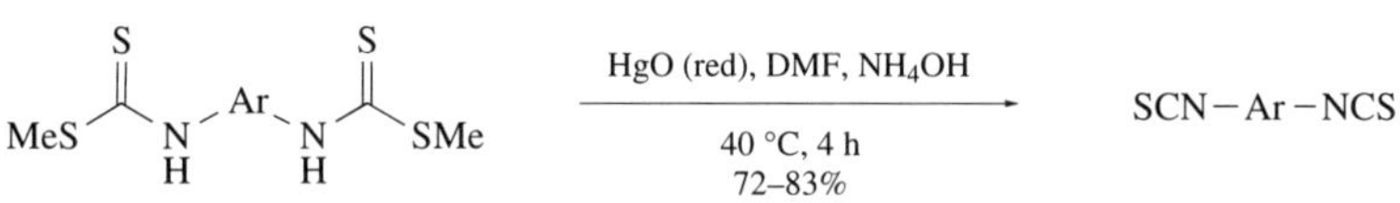

(5)

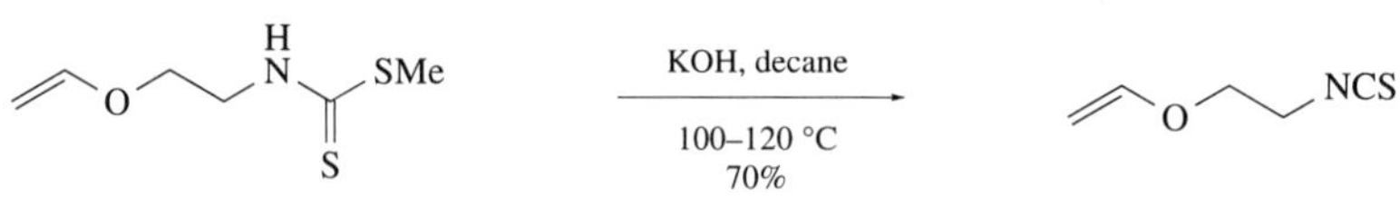

(6)

Van der Kerk *et al.* have reported an interesting variation whereby the carboxymethyl esters of dithiocarbamates are prepared, and their zinc salts precipitated at pH 4–5 and isolated ⟨55RTC1262⟩. Resuspension in water and adjustment to pH ~ 7 causes spontaneous decomposition and formation of the isothiocyanate (Equation (7); R = H, NCS, AcNH, EtO, Cl, Br).

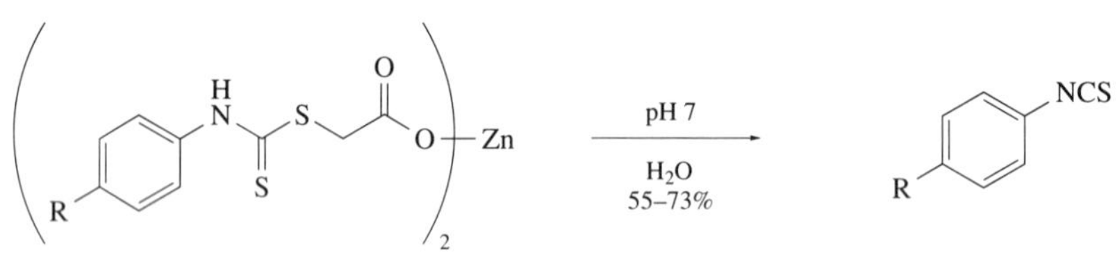

(7)

Kricheldorf has shown that ethyl or benzyl esters of the dithiocarbamates derived from amino acids or their homologues are converted to the isothiocyanates with trimethylsilyl chloride in refluxing toluene (Equation (8)) ⟨70AG(E)526⟩.

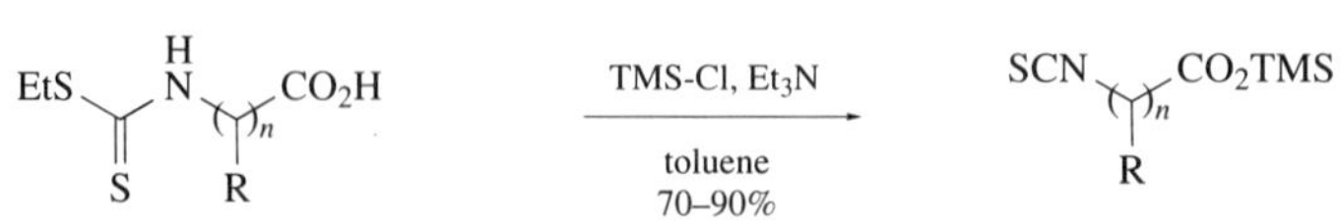

(8)

(ii) Thiophosgene

The ability of primary amines to react with thiophosgene has been recognised since the 1870s ⟨1872CB799⟩ and, despite the unpleasant nature of this compound, it retains its popularity as a versatile reagent for the formation of isothiocyanates. The reaction of thiophosgene with amines has been described by Sharma as part of two general reviews on the chemistry of thiophosgene ⟨78S803, 86SR1⟩.

Thiophosgene, compared with its oxygen analogue, is relatively stable in water and reactions are frequently conducted in aqueous medium. The reaction proceeds through the intermediacy of a thiocarbamoyl chloride which then loses a molecule of hydrogen chloride to give the isothiocyanate (Scheme 5) ⟨24JCS1702⟩. As this intermediate or the final product may react with free amine to give symmetrical thioureas, the thiophosgene must be kept in excess, except when the amine is of sufficiently weak basicity.

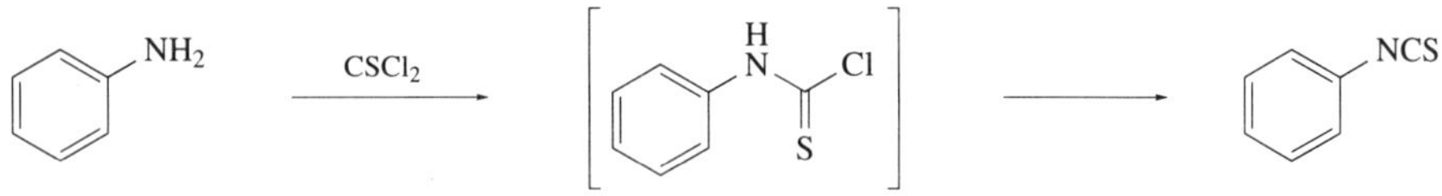

Scheme 5

Many of the early examples of this reaction were undertaken on arylamines, added either as the free base or as the soluble hydrochloride salt to thiophosgene, which were then stirred vigorously in water at room temperature. The variety of substituents tolerated under these reaction conditions demonstrates the versatility of this methodology (Table 3), respectable yields being achieved even in the presence of electron-withdrawing groups.

Table 3 Isothiocyanates from amines with thiophosgene.

Reactant	*R*	*Solvent*	*Yield* (%)	*Ref.*
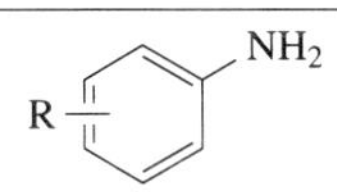	Me, Hal, NO_2, OH, CO_2Et, CN	Water	Good	24JCS1702 26JCS3041 27JCS436
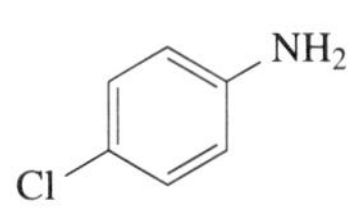		Water	81	32OSC(1)165
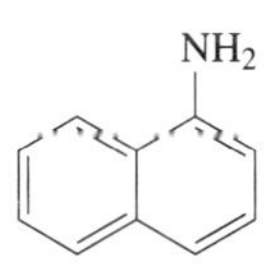		Water/$CHCl_3$	53	42JCS374
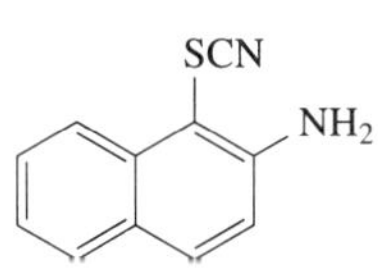		Acetone	95	65AP51
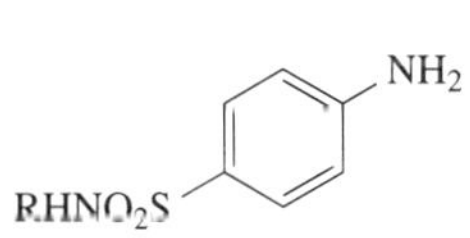	H, Ac, 2-pyridyl, 2-thiazolyl	Aq. HCl	76–98	46JA2506
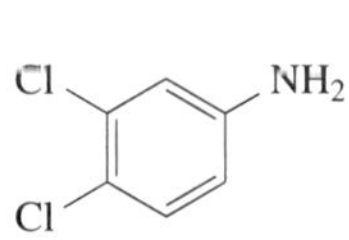		Aq. HCl/toluene	95	57JA1236

The presence of hydrogen chloride, liberated during the course of the reaction, does not appear to inhibit the reactivity of the amine to thiophosgene, but could prove detrimental to other acid-labile functionality. This problem may be overcome by the addition of an organic base—thus, the presence of three equivalents of triethylamine allows the conversion of amino acid trimethylsilyl esters to the corresponding isothiocyanates with retention of the acid-sensitive silyl protecting group (Equation (9), X = $(CH_2)_3$, 1,3-C_6H_4, 1,4-C_6H_4, Pr^iCH, Bu^iCH) ⟨71LA(748)101⟩.

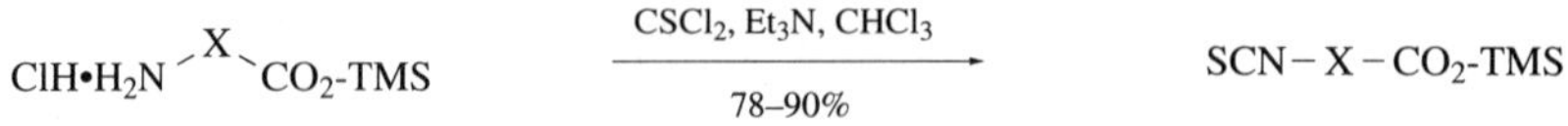

(9)

An inorganic base is more frequently employed, with the reaction performed in a biphasic system of water and a halogenated organic solvent. Calcium carbonate has found particular favour and has been employed in the synthesis of a wide range of alkyl, aryl and heteroaryl isothiocyanates (Table 4). A number of alternative alkalis have been similarly employed, as exemplified in Table 5.

Table 4 Isothiocyanates from amines with thiophosgene and calcium carbonate.

Reactant	*Solvent*	*Yield* (%)	*Ref.*
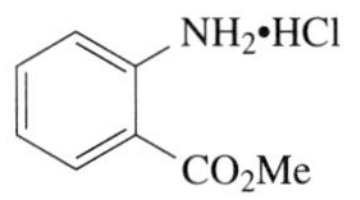	Water/$(CH_2Cl)_2$	63	62JOC3701
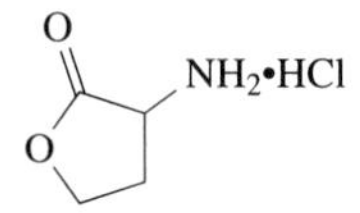	Water/$CHCl_3$	72	76MI 527-01
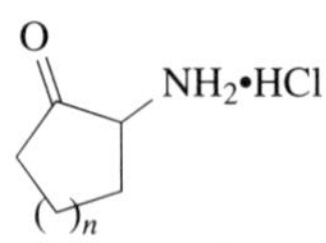 (*n* = 1–3)	Water/$CHCl_3$	58–81	83CCC3421
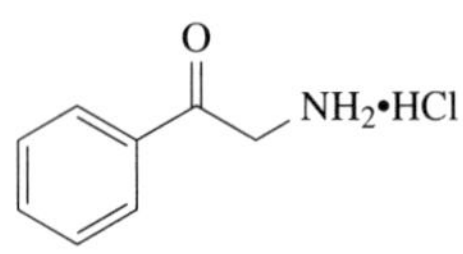	Water/$CHCl_3$	72	83CCC3421
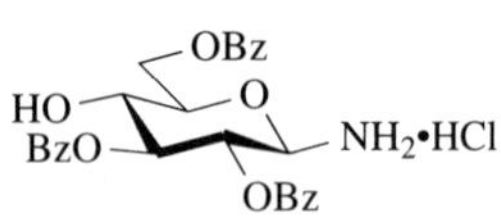	Water/$CHCl_3$	78	89CAR35
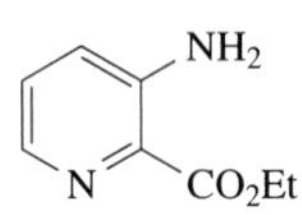	Aq. HCl/CH_2Cl_2	70	90JHC407
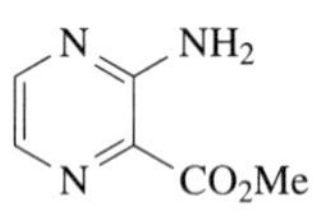	Water/CH_2Cl_2	53	90JHC433
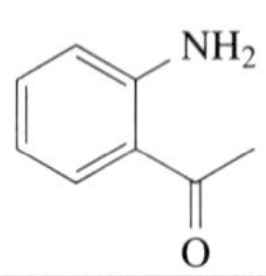	Water/$CHCl_3$	70	91JHC1091

Connolly and Dyson demonstrated that thiocarbonyl tetrachloride ($CSCl_4$) (prepared from chlorine and carbon disulfide) could form isothiocyanates in modest yields with arylamines ⟨34JCS822⟩. If the reaction was carried out in the presence of tin(II) chloride, thiophosgene was formed *in situ*, giving substituted aryl isothiocyanates in moderate to good yields (47–99%) ⟨35JCS679⟩.

(iii) Other thiocarbonyl derivatives

A number of derivatives of thiophosgene have been developed where one or both chlorines have been replaced by an alternative leaving group. These frequently have advantages of ease of handling, due to increased stability or reduced toxicity, or of improved yields.

Table 5 Isothiocyanates from amines with thiophosgene and various inorganic bases.

Reactant	*Solvent*	*Base*	*Yield* (%)	*Ref.*
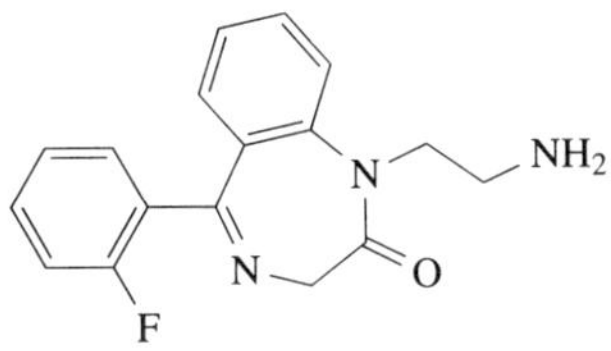	Water/$CHCl_3$	$NaHCO_3$	80	79NAT854
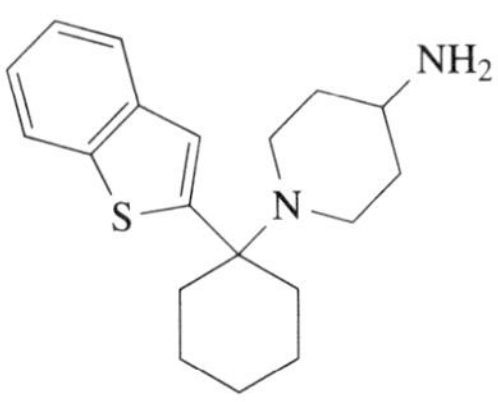	Water/$CHCl_3$	$NaHCO_3$	81	92JCS(P1)1671
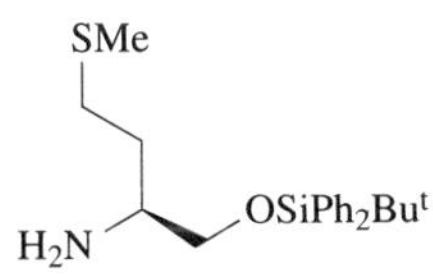	Water/CH_2Cl_2	Na_2CO_3	87	90JOC3749
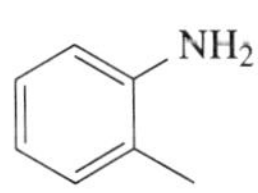	Water/$CHCl_3$	K_2CO_3	84	89JHC1735
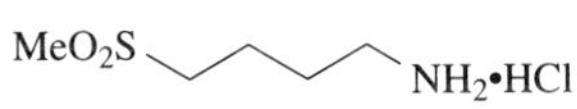	Water	NaOH	78	54ACS295

Bis(trichloromethyl)pentathiodiperoxycarbonate (**5**), a stable red liquid of low volatility, provides a convenient and less toxic alternative to thiophosgene, giving comparable yields when reacted with a number of amines in aqueous suspension ⟨66AG(E)963⟩.

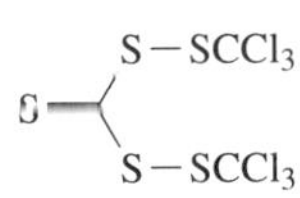

(5)

Diethylthiocarbamoyl chloride (**6**) reacts preferentially with arylamines substituted with electron-withdrawing groups ⟨65JOC2465⟩—in fact, the more basic unsubstituted aniline is unreactive towards this reagent. The related bisdiethylthiocarbamoyl sulfide (**7**) and disulfide (**8**) also react with the electron-deficient 4-nitroaniline in refluxing chlorobenzene saturated with hydrogen chloride, giving 4-nitrophenyl isothiocyanate in 53% and 73% yield, respectively ⟨66HCA1716⟩.

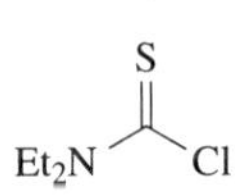

(6)

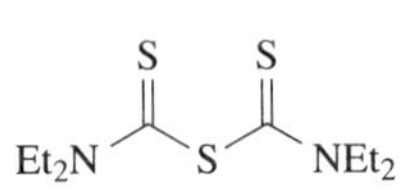

(7)

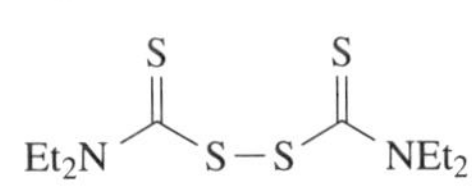

(8)

Perhaps the most widely used thiophosgene derivative is *N,N'*-thiocarbonyl diimidazole, a hygroscopic crystalline solid which reacts with both alkyl and aryl amines at room temperature in good yield (Equation (10); R = Bu^n, *c*-hexyl, Ph) ⟨62LA(657)104⟩.

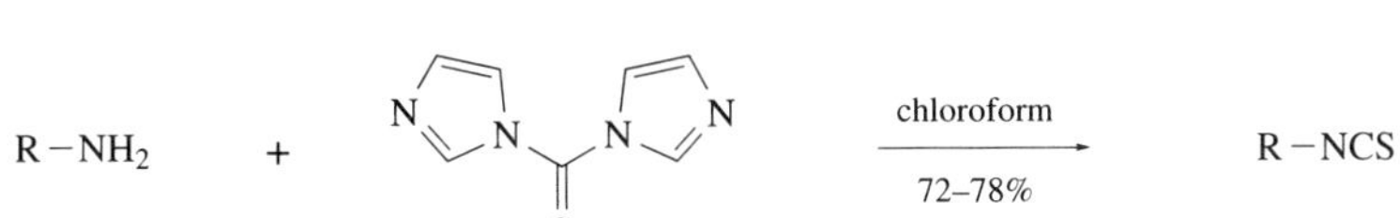

(10)

Kim and Yi have developed two 2-pyridyl-derived thiocarbonyl reagents which react with amines at room temperature under essentially neutral conditions. Thus, di-2-pyridyl thiocarbonate (**9**) ⟨85TL1661⟩ and 1,1′-thiocarbonyl-2,2′-pyridone (**10**) ⟨86JOC2613⟩ react with a range of aryl and alkyl amines in similarly high yields (85–95%). The latter, which is commercially available, is formed by heating the former in refluxing toluene (Equation (11)). The only by-product from reaction with an amine is the water-soluble 2-hydroxypyridine. With the easy workup, high yields and neutral reaction conditions, Structure (**10**) would seem to be preferable to thiophosgene or other thiocarbonyl-derived reagents.

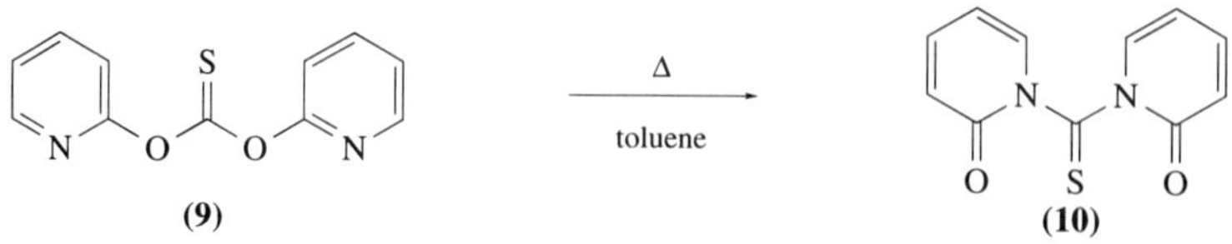

(11)

Thiophosgene reacts with 1-trimethylsilyloxybenzotriazole (**11**) to yield the crystalline 1,1′-(thiocarbonyldioxy)dibenzotriazole (**12**) (Equation (12)) which may then react very rapidly with alkyl and aryl amines at room temperature in the presence of triethylamine ⟨86MI 527-01⟩. The intermediate benzotriazole (**12**) explodes on drying at 114°C, so the authors recommend its generation *in situ* and immediate use. However, with the availability of alternative reagents, its use seems inadvisable.

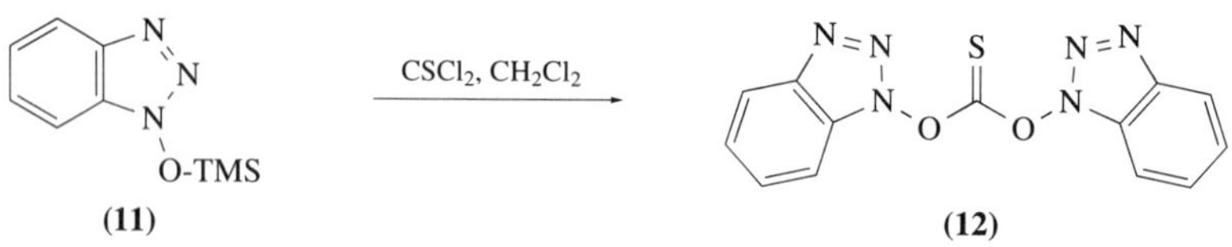

(12)

(iv) From amine derivatives

A wide range of amine derivatives may be transformed directly into isothiocyanates, frequently employing similar conditions to those described earlier in this section. The conversion of the following derivatives is described: *N*-silylamines, amides, thioamides, phosphoramidates, thiophosphoramidates, iminophosphoranes, azides, imines, carbonimidoyl dichlorides, oxaziridines, *N*-sulfinylamines and aromatic nitro compounds.

N-Trimethylsilylamines react readily with carbon disulfide to yield the intermediate trimethylsilyl ester of the dithiocarbamate (**13**), which may be decomposed to the isothiocyanate via the unstable disilylated compound (**14**) (Scheme 6). This approach has been applied to the synthesis of alkyl isothiocyanates, and in particular allows the formation of 2-isothiocyanatoalkanoic acid trimethylsilyl esters by reaction of amino acids ⟨70S539⟩.

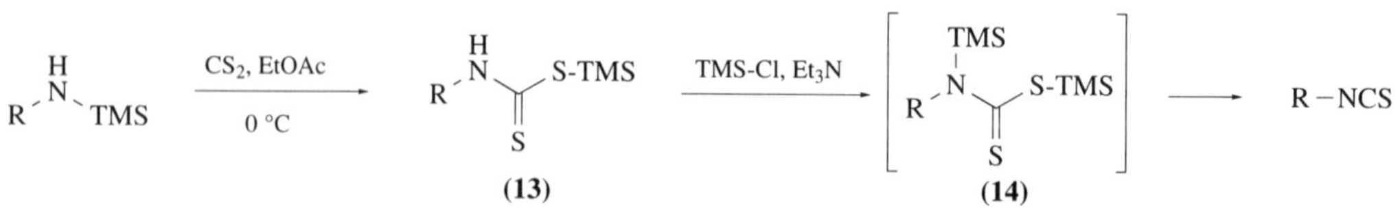

Scheme 6

Shahak and Sasson discovered a novel method of cleaving secondary amides under anhydrous conditions to yield the isothiocyanate and a thiol acid by reaction of the amide anion with carbon disulfide (Scheme 7) ⟨73JA3440⟩. Good to excellent yields were reported for arylamines (R = Ph, 4-MeC_6H_4, 1-naphthyl) with the acyl function having a marked effect (from the limited examples quoted, R^1 = Ph appears most advantageous). For alkylamides (R = Me, Et), the isothiocyanate products were not isolated. Albanese and Penso have adapted this methodology, employing a mixture of anhydrous potassium carbonate and sodium hydroxide in acetonitrile as the base, and shown its applicability to both aryl and alkyl trifluoroacetamides ⟨91S1001⟩. Yields were mostly in the range 55–82%; only R = Bu^t and cinnamyl gave poor yields (Scheme 8). Sakai and co-workers have shown that thioamides decompose directly to isothiocyanates by reaction with *n*-butyllithium (cf. method described in Section 5.27.2.1.1.(i)) ⟨78NKK265⟩.

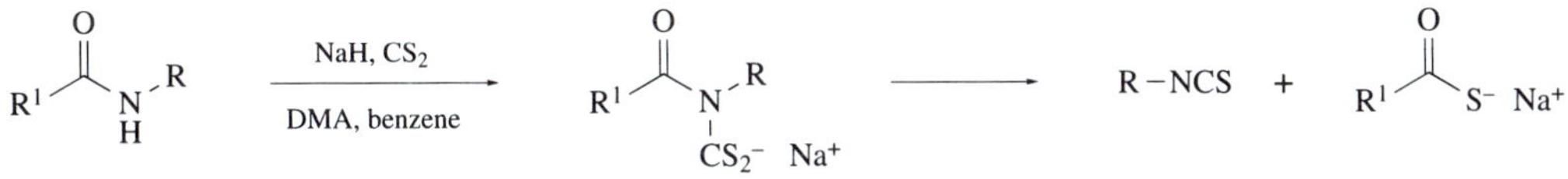

Scheme 7

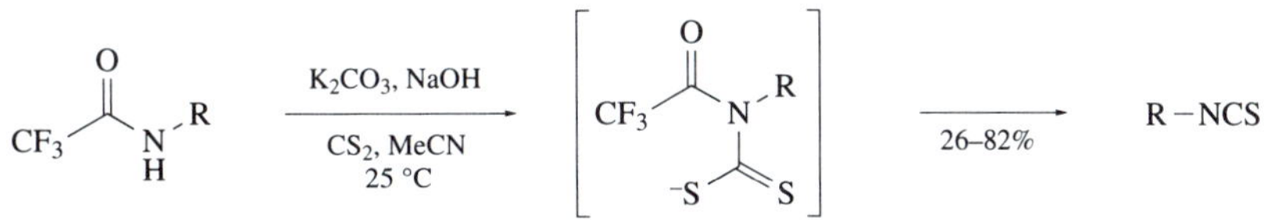

Scheme 8

Lactams, as their *N*-trimethylsilyl derivatives, react exothermically with thiophosgene at −10 °C, giving ring-opening to the ω-isothiocyanatoacyl chlorides (Equation (13)) ⟨75AG(E)502⟩. The yield of 7-isothiocyanatoheptanoyl chloride is low due to decomposition during high temperature vacuum distillation.

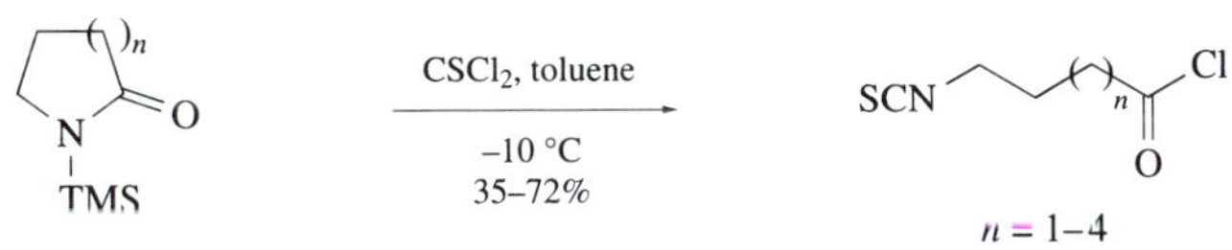

(13)

Phosphoramidate anions react with carbon disulfide in a similar fashion to amides to yield the isothiocyanates. Wadsworth and Emmons demonstrated that moderately good yields of alkyl isothiocyanates could be achieved (Equation (14), R = Bu^n, *c*-hexyl, Bu^t) but phenyl isothiocyanate was formed in only 30% ⟨64JOC2816⟩. During the late 1980s, Olejniczak and Zwierzak have improved this approach by addition of catalytic amounts of tetrabutylammonium bromide and using benzene as solvent, achieving better and more reproducible yields from a range of *N*-alkyl and *N*-aryl phosphoramidates (74–91%) ⟨89S300⟩. They point out that a general method for transforming alkyl bromides into diethyl alkylphosphoramidates via their azides makes this route particularly attractive ⟨88JPR473⟩. This interconversion may also be carried out with the thiophosphoramidates (Equation (15)) ⟨75ZC56⟩—yields are poor for unbranched alkyl isothiocyanates (R = Me, Et, Bu^n) but are good for aryl isothiocyanates (R = Ph, 4-$MeOC_6H_4$, 4-ClC_6H_4).

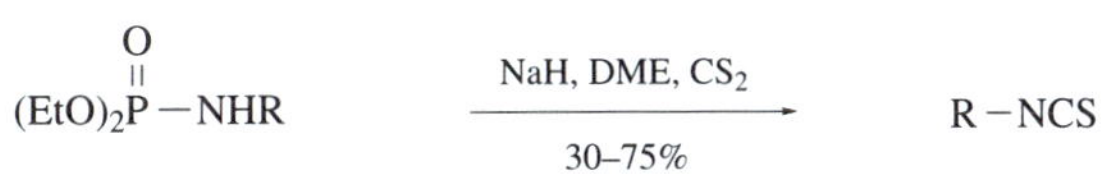

(14)

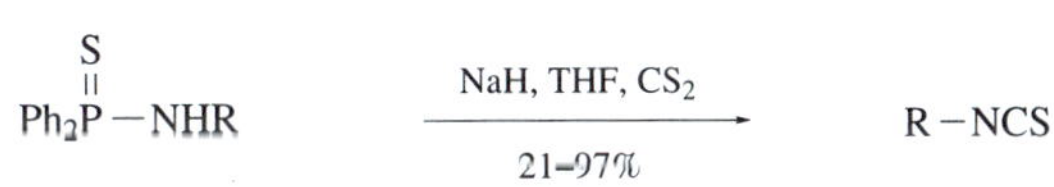

(15)

Iminophosphoranes react directly with carbon disulfide to give high yields of both alkyl and aryl isothiocyanates ⟨82S596⟩. The reactants may themselves be prepared from the corresponding amines and triphenylphosphine dibromide (Scheme 9). The method differs from other preparations with carbon disulfide in that it may be used with aromatics substituted with electron-withdrawing groups (R = 4-$NO_2C_6H_4$, yield 88%). Katritzky *et al.* have prepared benzyl isothiocyanate by a novel, one-carbon homologation of the phenyl Grignard reagent with a benzotriazolylmethyl iminophosphorane, followed by reaction with carbon disulfide (Scheme 10) ⟨90S565⟩. Tsuge *et al.* used a one-pot procedure to form an isothiocyanate from trimethylsilylmethyl azide by *in situ* generation of the iminophosphorane with triphenylphosphine ⟨84JOC2688⟩.

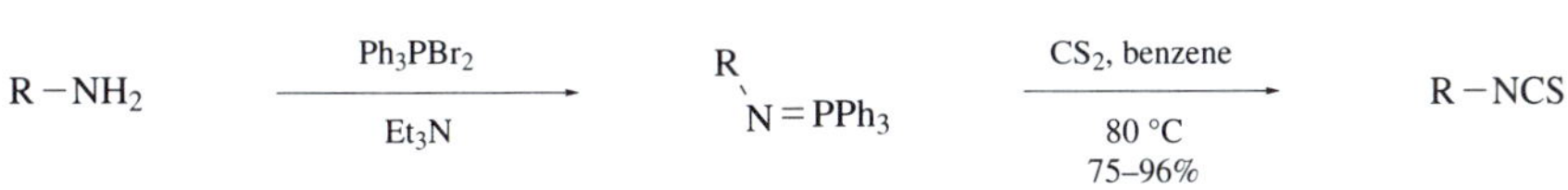

Scheme 9

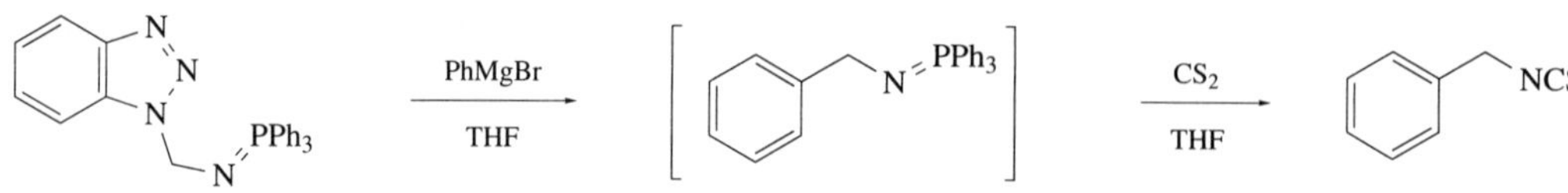

Scheme 10

Ketimines react with thiophosgene in refluxing toluene ⟨73AG(E)842⟩. Where there is no β-hydrogen, an α-chloroalkyl isothiocyanate is formed (Equation (16); R^1, R^2 = Bu^t, Ph, 1-naphthyl); but when dehydrohalogenation is possible, an alkenyl isothiocyanate is formed (Equation (17); R^1 = H, Me; R^2 = Me, Pr^n, Pr^i).

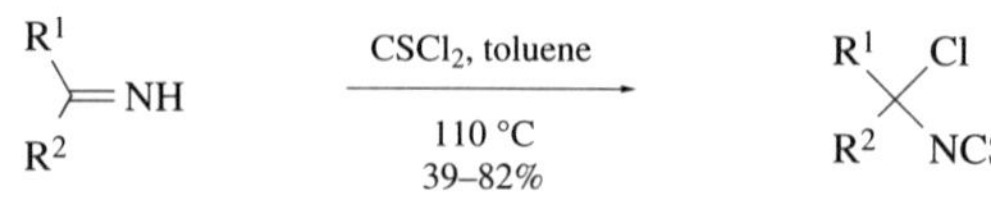

(16)

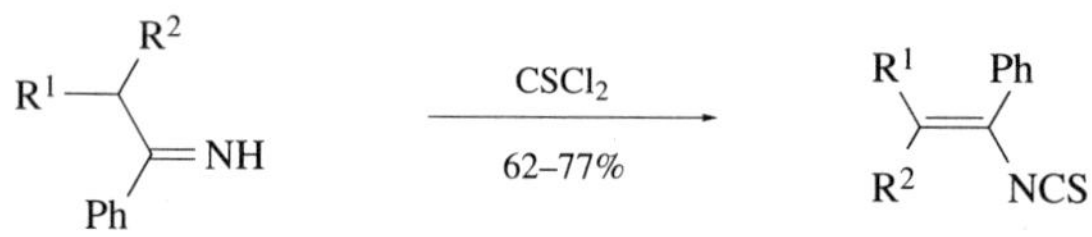

(17)

The chemistry of carbonimidoyl dichlorides (isocyanide dichlorides) has been reviewed by Kühle ⟨69AG(E)20⟩. He reported that aliphatic and aromatic carbonimidoyl dichlorides react with sodium or ammonium sulfide in aqueous acetone to give high yields (> 80%) of the isothiocyanate (Equation (18)). Alternatively, the transformation can be carried out under anhydrous conditions with phosphorus pentasulfide in boiling toluene or chlorobenzene. Use of such anhydrous media allows reaction of carbonimidoyl dichlorides containing reactive groups such as isocyanates or acyl chlorides. Again yields are greater than 80%. Alternatively, bis(trimethylsilyl)sulfide has been reacted with the carbonimidoyl dichloride to prepare phenyl isothiocyanate in 91% yield ⟨72ZOR1822⟩. Isothiocyanates have also been prepared from carbonimidoyl dichlorides by reaction with two equivalents of potassium thioacetate (Equation (19)) in THF at room temperature ⟨77BCJ722⟩. Use of thallium(I) thioacetate required boiling THF, while potassium thiocyanate in THF or ethanol gave only modest yields of phenyl isothiocyanate (< 45%).

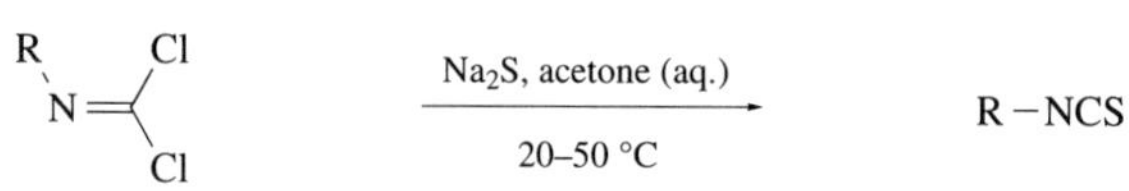

(18)

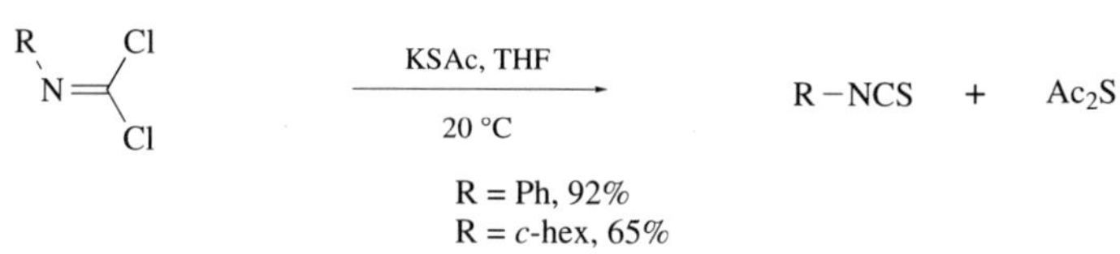

(19)

2-Alkyloxaziridines may be converted to isothiocyanates in virtually quantitative yield (Scheme 11; R^1 = Me, Bu^n, *c*-hexyl; R^2 = H or Me) by reaction with excess carbon disulfide at reflux ⟨74JOC957⟩. The reaction is assumed to proceed via the intermediate thiaziridine-3-thione (**15**), which decomposes to the isothiocyanate and sulfur.

N-Arylsulfinylamines have been shown by Minami and Agawa to react with carbon disulfide at

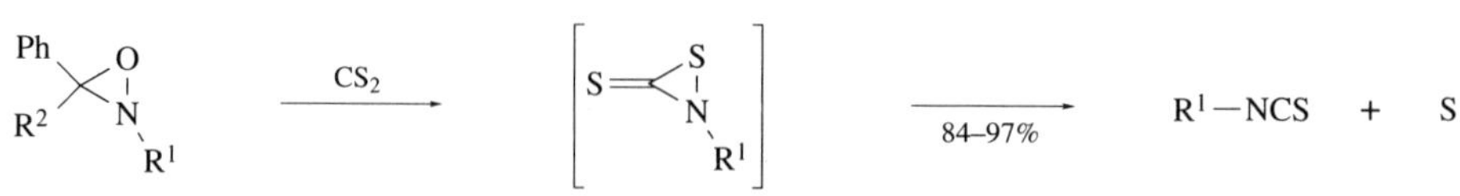

(15)

Scheme 11

200°C in a sealed tube to give quantitative yields of the aryl isothiocyanates (Scheme 12; R = H, 2-Me, 3-Me, 2-Cl) via the intermediate (**16**) ⟨68TL2651⟩.

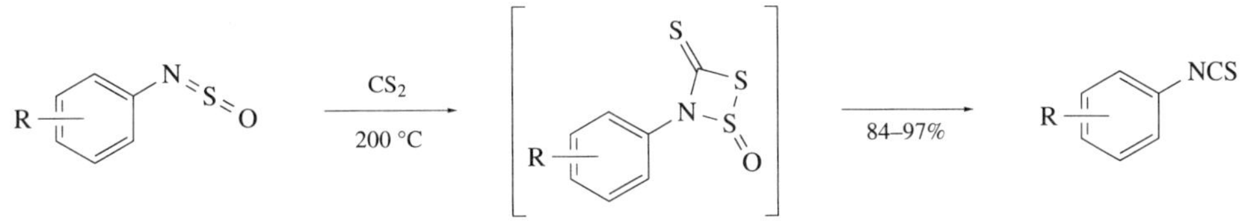

Scheme 12

Aromatic nitro compounds may be converted directly into the corresponding isothiocyanates by reaction with carbon disulfide or carbonyl sulfide in the presence of a base such as an alkali metal alkoxide or phenoxide, or an alkyl or aryl sulfide ⟨69AG(E)760⟩. The best percentage conversions and yields were reported using the sodium salt of 3-methylthiophenol and carbon disulfide. Reactions were performed in an autoclave in neat carbon disulfide at 150–170°C (Equation (20)).

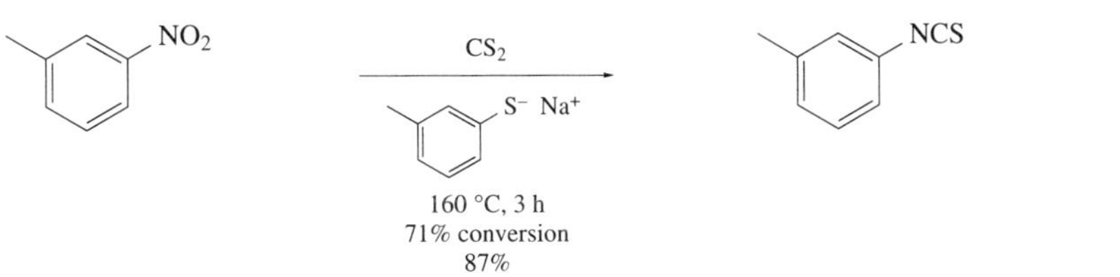

(20)

5.27.2.1.2 *Isothiocyanates by cleavage of nitrogen heterocycles*

A number of examples exist where thiophosgene reacts with nitrogen heterocycles with ring-opening to yield an ω-substituted isothiocyanate. Hull, during studies on the formation of Reissert compounds with various thiocarbonyl chlorides, observed that thiophosgene reacted with quinoline in the presence of potassium cyanide to give a low yield (4%) of 2-isothiocyanato-*trans*-cinnamaldehyde ⟨68JCS(C)1777⟩. The yield was increased using sodium hydroxide as base, while employing barium carbonate yielded the thermally labile *cis* isomer (Scheme 13).

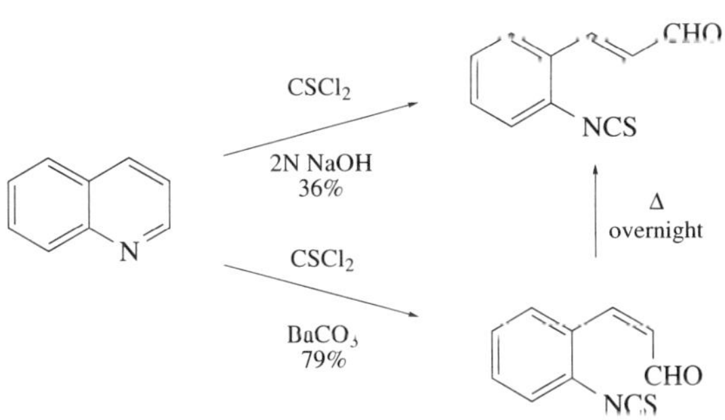

Scheme 13

Pyridine, isoquinoline, 4,7-dichloroquinoline and 2,7-diazabiphenylene were similarly shown to undergo ring-opening to the corresponding isothiocyanatoaldehydes. Hull extended this reaction to benzoxazoles, benzimidazole and imidazole (Table 6). A modification of Hull's procedure has been applied to the opening of 2-aryloxythiazolines, 2-alkyloxazolines and to the cyclic iminoether derived from methyl 6-aminohexanoate (Table 7).

Aziridine was originally reported as reacting with thiophosgene in the presence of triethylamine to yield a stable thiocarbamoyl chloride based only on an elemental analysis ⟨54LA(590)123⟩. It has now been demonstrated that ring-opening occurs to yield the 2-chloroethyl isothiocyanate (Scheme 14; R = H or Me), although the thiocarbamoyl intermediate (**17**) can be detected in solution at −15°C by ^{1}H-NMR ⟨66JHC384⟩. In the absence of triethylamine, the chloroethyl isothiocyanates are also formed in 45–50% yield accompanied by the chloroethylamine hydrochloride.

Table 6 Isothiocyanates from heteroaryls with thiophosgene.

Reactant	*Product*	*Yield (%)*	*Ref.*
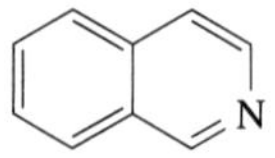	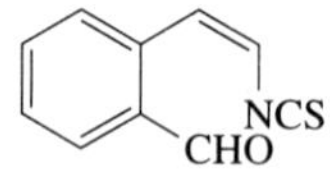	42	74JCS(P1)1541
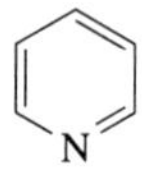	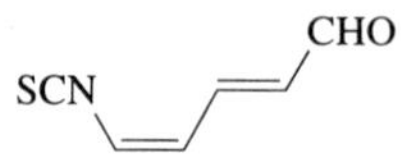	8	74JCS(P1)1541
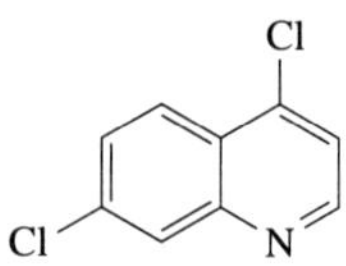	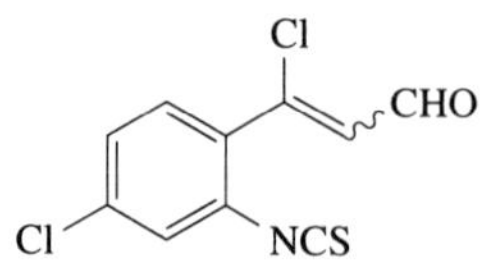	58	75JCS(P1)922
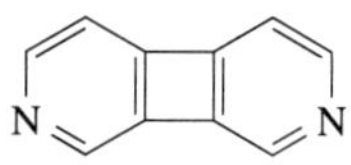	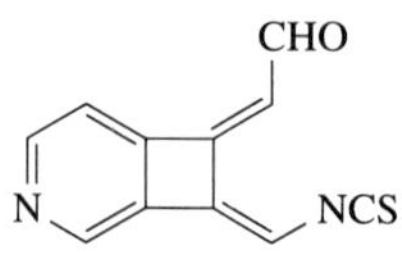	60	83CC74
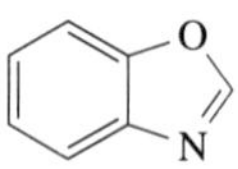	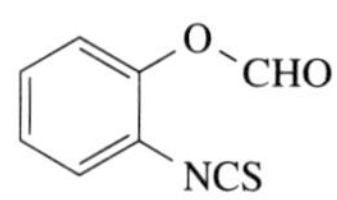	72	79JCR(S)148
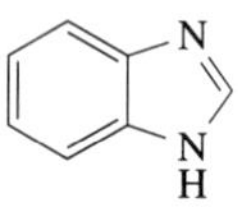	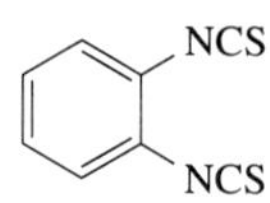	53	79SC477
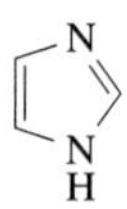	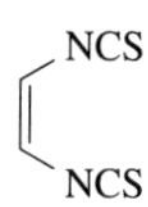	46	80SC489

Table 7 Isothiocyanates from miscellaneous heterocycles with thiophosgene.

Reactant	*Product*	*Yield (%)*	*Ref.*
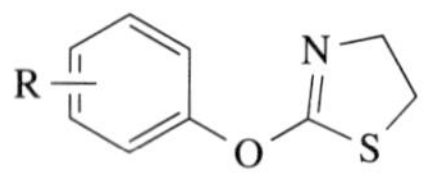	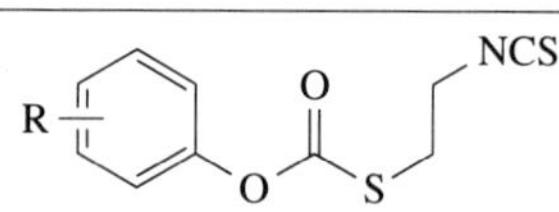	61–81	84CCC295
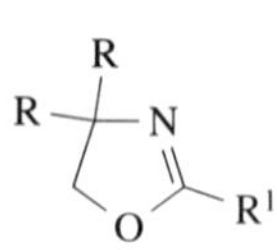	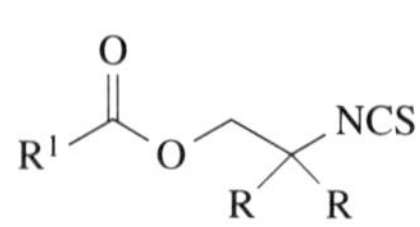	30–68	86CCC112
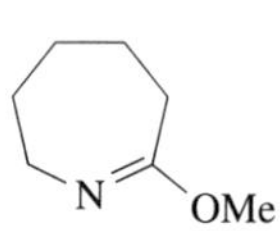	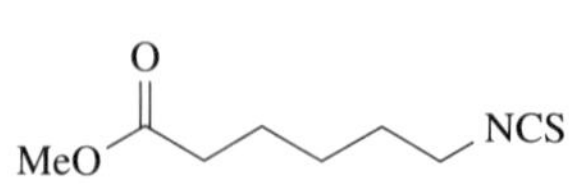	56	86CCC112

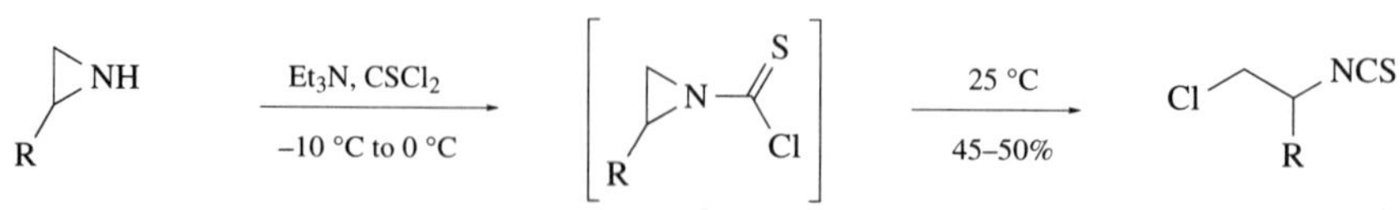

(17)

Scheme 14

5.27.2.1.3 *Isothiocyanates from thioureas*

Heating of arylthioureas in an inert solvent causes their fission into ammonia and the aryl isothiocyanate in moderate to high yield (Equation (21)) ⟨54CI(L)785⟩. The authors have investigated the mechanism and kinetics of the reaction using the decomposition of α-naphthylthiourea as a model ⟨56JCS659⟩. The rate of reaction and yield of products could be explained by the involvement of two simultaneous, slow first-order reactions, followed by two fast reactions leading, in addition to the required isothiocyanate, to the formation of low yields of the *N*,*N*′-diarylthiourea (Scheme 15; $k_1 = 0.629\ h^{-1}$, $k_2 = 0.053\ h^{-1}$). The experimental method for the high yielding preparation of α-naphthyl isothiocyanate by this approach has been reported in detail ⟨63OSC(4)700⟩. A recent patent shows the utility of the method in the preparation of substituted phenoxyphenyl isothiocyanates in nearly quantitative yield ⟨91USP4997967⟩.

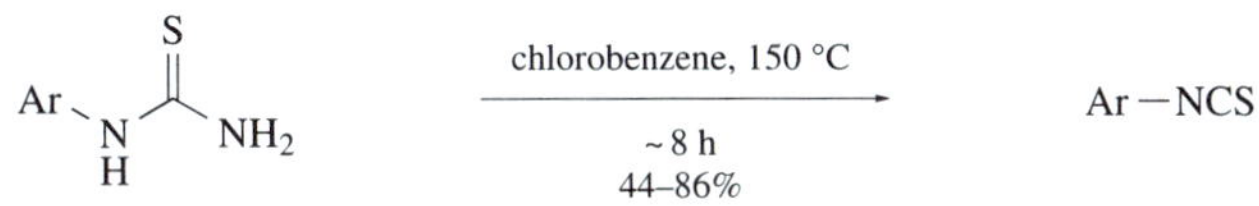

(21)

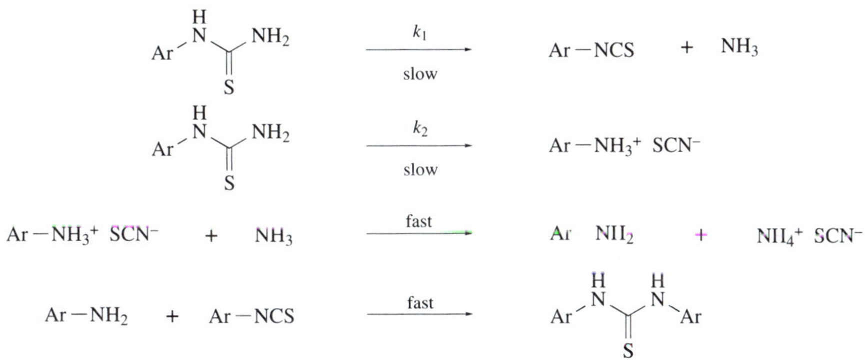

Scheme 15

A related approach involves pyrolysis of the *N*-aryl-*N*′-benzoylthiourea in 1,2-dichlorobenzene, used to prepare 4-isothiocyanato-4′-nitrodiphenylamine (Equation (22)) ⟨79JCS(P1)2001⟩.

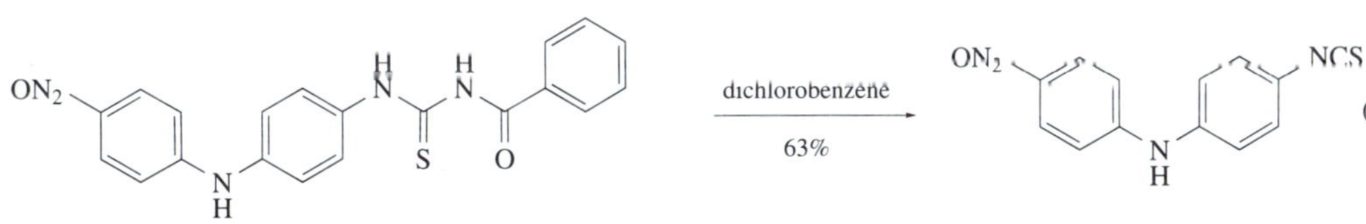

(22)

Fission of a *sym*-diarylthiourea with acetic anhydride also gives the isothiocyanate but involves loss of half the arylamine ⟨1891JCS396⟩. Alternative acidic reagents have been employed—for example, concentrated hydrochloric acid ⟨49JA2240⟩ and phosphoric acid ⟨1882CB985⟩. A mild method has been developed in the early 1990s to achieve this cleavage in the presence of *N*,*N*-dimethyl dichloromethaniminium chloride ⟨91UKZ431⟩. Reaction occurs with an organic base at room temperature to give good yields of the isothiocyanates (Equation (23); R = Pr^i, Bu^n, Bu^t, *c*-hexyl, Ph, 2,4,6-$Me_3C_6H_2$). Cleavage by metallation with *n*-butyllithium or ethylmagnesium bromide followed by addition of carbon disulfide affords 2 mol isothiocyanate from 1 mol thiourea (Equation (24)) ⟨77BCJ425⟩. As well as giving high yields (73–93%) of aryl isothiocyanates, this methodology also gives modest yields (40–43%) of the isothiocyanate with dialkyl thioureas.

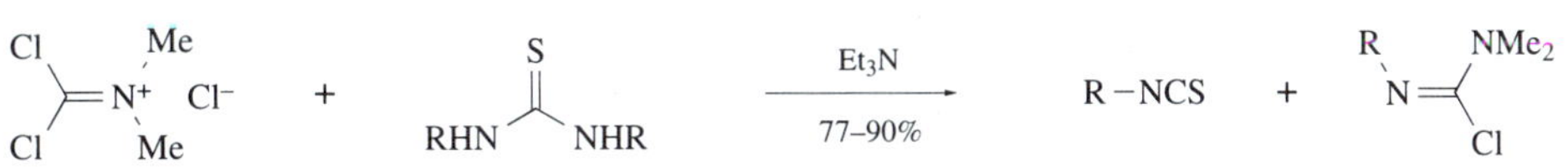

(23)

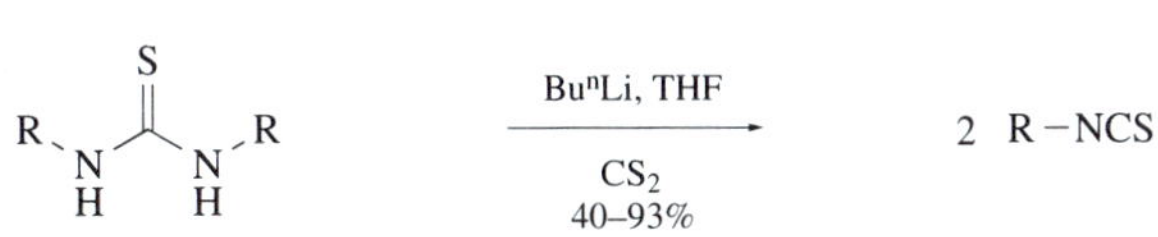

(24)

The related thiosemicarbazides may be oxidised with iron(III) chloride to the isothiocyanate ⟨50JCS1582⟩—this method has been applied to various alkyl and phenyl substituents, but the authors only quoted a yield for the oxidation of benzyl semicarbazide (Equation (25)).

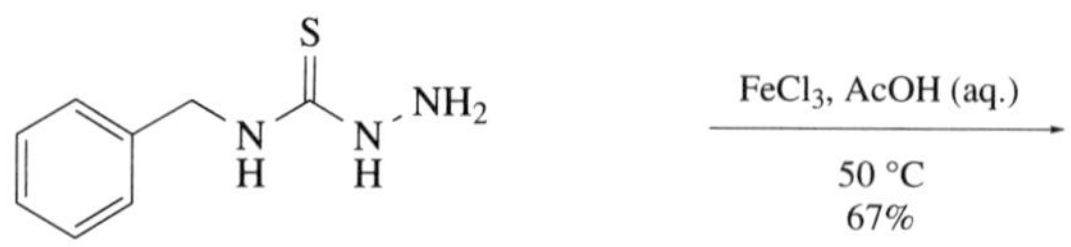

(25)

5.27.2.1.4 *Isothiocyanates from isocyanides*

Although it was recognised at the end of the nineteenth century that isocyanides could react with sulfur ⟨1873CB210⟩, the first detailed studies were only undertaken in the 1970s when it was demonstrated that aryl isocyanides reacted with elemental sulfur in benzene at reflux ⟨72JOC1360⟩ (Equation (26); R = MeO, NO_2, Ph). Ramakrishnan and Boyer have also shown that the reaction is promoted by irradiation at 254 nm, although the yields are lower ⟨72IJS139⟩.

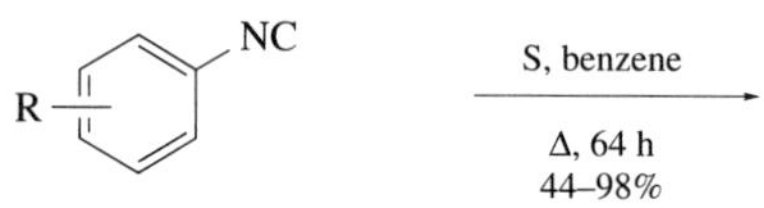

(26)

The yields are markedly improved by selenium catalysis in the presence of triethylamine, proceeding via the isoselenocyanate, with subsequent Se–S exchange as the rate-determining step in the catalytic cycle ⟨91TL3503⟩. This approach gives high yields of both alkyl and aryl isothiocyanates (Table 8).

Table 8 Isothiocyanates from isocyanides with sulfur and selenium catalysis.

$$\mathrm{R{-}NC} + \mathrm{S} \xrightarrow[\text{THF, reflux}]{\text{Se, Et}_3\text{N}} \mathrm{R{-}NCS}$$

R	*Reaction time* (h)	*Isolated Yield* (%)
Bu^n	2	73
Bu^i	3	71
Bu^t	5	58
c-Hex	2	89
Ph	2	91
4-MeO-C_6H_4	0.5	82
4-Cl-C_6H_4	1	83

Other sulfur-containing compounds may also react with isocyanides to yield isothiocyanates. Thus, isocyanides react with diacyl disulfides in chloroform at reflux in the presence of thallium(I) acetate or lead(II) acetate to give both alkyl and aryl isothiocyanates, frequently in virtually quantitative yield ⟨77BCJ2785⟩. In the absence of the metal salts, cyclohexyl isocyanide reacts with dibenzoyl disulfide only in toluene at reflux over 24 h to give 49% of the isothiocyanate ⟨78BCJ659⟩. The lower yield is thought to be due to the competing reaction of the isocyanide with dibenzoyl sulfide formed during the course of the reaction. Cyclohexyl isocyanide also reacts with chlorocarbonylsulfenyl chloride (ClC(O)SCl) to yield the isothiocyanate with elimination of phosgene ⟨70AG(E)54⟩.

In the presence of a radical initiator, isocyanides may react with thiols to form isothiocyanates

⟨70JOC2118⟩. The mechanism is considered to involve the thiyl radical and formation of an intermediate imidoyl radical (**18**). β-Scission occurs with a tertiary thiol ($R = Bu^t$) to give the isothiocyanate and an alkyl radical (Scheme 16).

$$R{-}SH \xrightarrow{AIBN} R{-}S\bullet \xrightarrow{PhNC} \left[R{-}S{-}\dot{C}{=}N{-}Ph\right] \xrightarrow[97\%]{} Ph{-}NCS + R\bullet$$

(18)

Scheme 16

Aryl isocyanides also participate in the transfer of sulfur from an isothiocyanate but, being a reversible process (Equation (27)), the reaction is of limited synthetic utility ⟨72JOC1360⟩.

$$Ar{-}NC + Ar^1{-}NCS \rightleftharpoons Ar{-}NCS + Ar^1{-}NC \qquad (27)$$

5.27.2.1.5 *Isothiocyanates from thioamides*

In the course of a study of the oxidation of chlorinated thiobenzamides with nitrous acid ⟨79ZN(B)768⟩, it was observed that, while most of the thioamides yielded a mixture of nitriles and 1,2,4-thiadiazoles, 2,6-dichlorothiobenzamide gave the isothiocyanate (Equation (28)). A more detailed investigation of 2,6-disubstituted thiobenzamides revealed that this method appears to be restricted to the 2,6-dichloro- and 2,6-dibromo- derivatives ⟨80CZ365⟩.

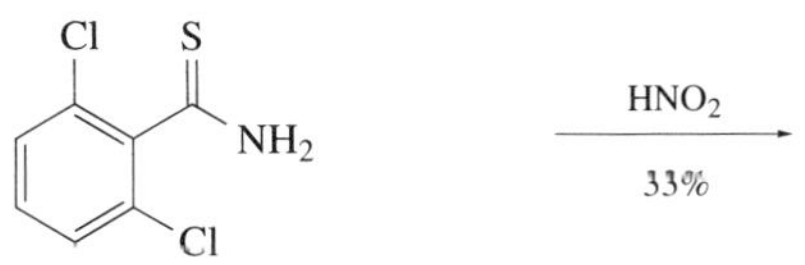

(28)

5.27.2.1.6 *Isothiocyanates from cyanates or isocyanates*

Alkyl and aryl isocyanates can be converted into isothiocyanates by reaction with *O*,*O*′-diethyl hydrogen dithiophosphate and thermal decomposition of the intermediate *S*-carbamoyl *O*,*O*′-diethyl dithiophosphate ⟨66AG(E)725⟩. The latter intermediate is usually formed virtually quantitatively and on heating at 130 °C gives the isothiocyanate in 50–75% yield. This approach has been applied to the synthesis of 2- and 3-thienyl isothiocyanates, avoiding the use of the unstable aminothiophenes (Scheme 17) ⟨77JOC1508⟩. Isocyanates and isothiocyanates undergo an exchange reaction when heated together at 190–230 °C through the intermediacy of the uretidinethione ketone (**19**) (Scheme 18). However, the method appears to be most practicable for the preparation of the lower boiling isocyanate rather than the isothiocyanate ⟨64JOC2091⟩.

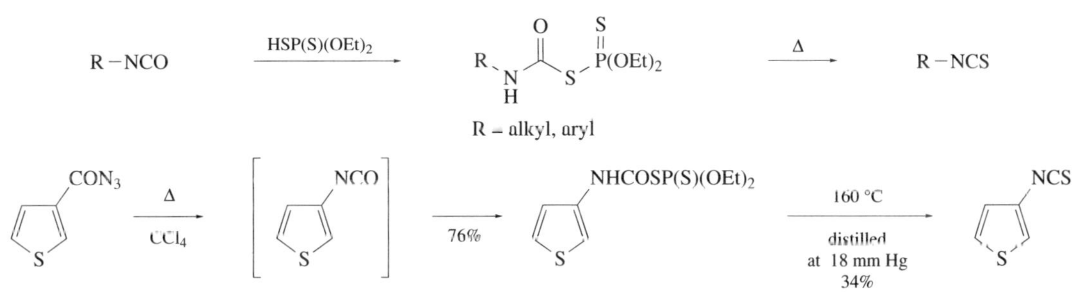

Scheme 17

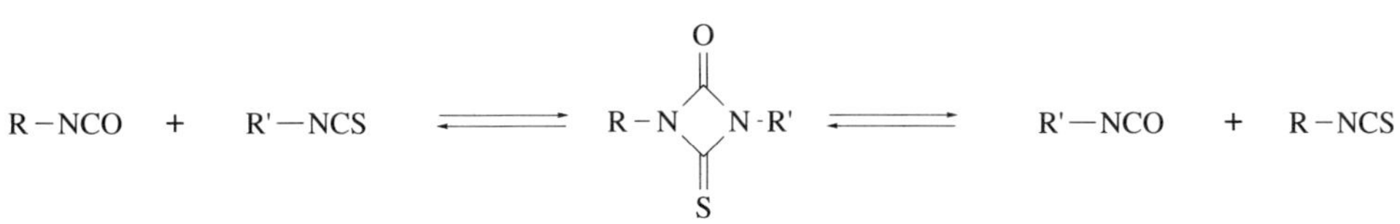

(19)

Scheme 18

Aryl cyanates (or nitrile oxides) undergo a cycloaddition reaction with thiobenzophenones to afford 3,5,5-triaryl-1,4,2-oxathiazoles (**20**), which undergo thermal decomposition to aryl isothiocyanates and benzophenones in high yield (Scheme 19) ⟨71S378⟩.

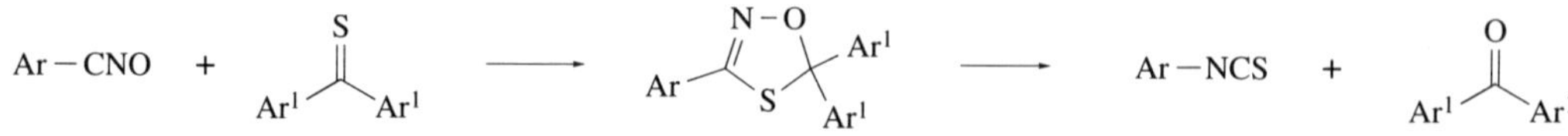

(20)

Scheme 19

5.27.2.1.7 *Isothiocyanates from isothiocyanates*

[^{35}S]-Phenyl isothiocyanate has been prepared by sulfur exchange with [^{35}S]-labelled ammonium hydrosulfide, catalysed by lead(II) nitrate (Equation (29)) ⟨58CB683⟩.

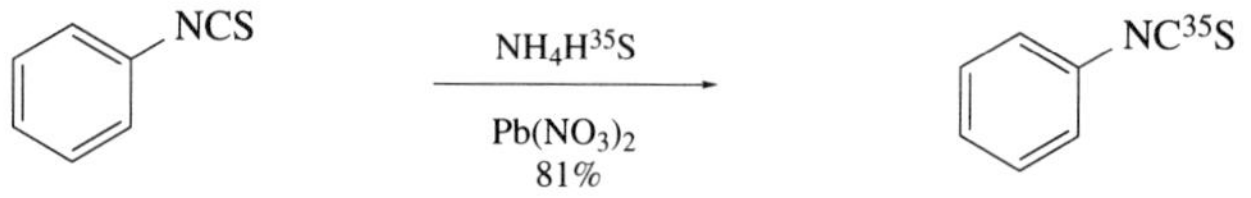

(29)

Phenyl isothiocyanate undergoes an exchange reaction with 2-*t*-butylamino-1,3,2-benzodioxaborole to yield *t*-butyl isothiocyanate and the phenylaminobenzodioxaborole as its dimer (Equation (30)) ⟨64JCS2108⟩.

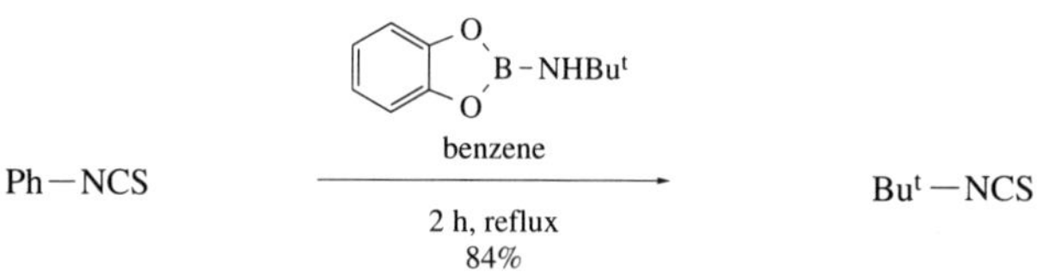

(30)

5.27.2.1.8 *Isothiocyanates from halides and thiocyanates*

The reaction of alkyl and aryl halides with ammonium or metal thiocyanates is one of the oldest methods of preparing isothiocyanates (Equation (31)). Due to the ambidentate character of the thiocyanate anion as a nucleophile, mixtures of the organic thiocyanate and isothiocyanate are usually formed. However, as the isothiocyanate is the more thermally stable species, rearrangement to the required product may frequently be achieved by heating, catalysed by anhydrous zinc chloride, or during distillation ⟨B-65MI 527-01⟩. The rearrangement is considered to involve ion-pair formation and therefore ease of isomerisation is dependent on the stability of the carbocation. Thus tertiary alkyl and benzyl thiocyanates rearrange with ease, with the rate increasing in a more polar solvent. A detailed study of the isomerisation of benzhydryl thiocyanates showed the reaction to follow first-order kinetics ⟨61JA2729⟩. It has been demonstrated that (−)-4-chlorobenzhydryl thiocyanate isomerises to the isothiocyanate with net retention of configuration ⟨69JOC4032⟩, in agreement with the proposed involvement of an ion pair. Alternatively, an organic thiocyanate may isomerise in the presence of inorganic thiocyanate, despite the lower nucleophilicity of the nitrogen versus sulfur ⟨65JA4791⟩. A more detailed discussion of the rearrangement of thiocyanates to isothiocyanates is to be found in Drobnica *et al.*'s review ⟨B-77MI 527-01⟩.

$$\text{R—Hal} \xrightarrow{M^+\ SCN^-} \text{R—NCS} \qquad (31)$$

A number of different inorganic thiocyanate salts have been employed for either direct formation of an isothiocyanate or for initial formation of the mixed thiocyanate/isothiocyanate prior to rearrangement.

t-Butyl isothiocyanate was prepared in 98% yield by reaction of *t*-butyl chloride with ammonium thiocyanate in the presence of zinc chloride in water, followed by rearrangement of the isolated thiocyanate/isothiocyanate mixture with powdered anhydrous zinc chloride ⟨69JCED110⟩.

Alkali metal thiocyanates may react with halides to yield the isothiocyanates directly (Table 9, entry 1), or on distillation of the crude reaction product (entry 2), or by initial isolation of the thiocyanate followed by thermal rearrangement (entry 3).

Table 9 Isothiocyanates from halides with alkali metal thiocyanates.

$$R-Hal \xrightarrow{M^+SCN^-} R-NCS$$

Entry	*R–Hal*	*M^+*	*Solvent*	*Yield (%)*	*Ref.*
1	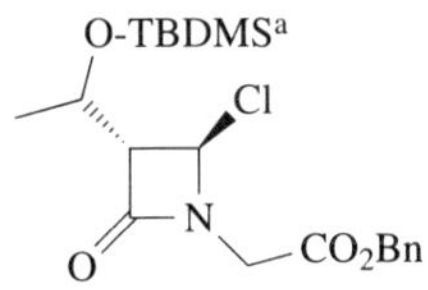	Na	Acetone	69	85H3069
2	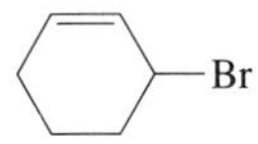	K	MeOH	76	51JA4495
3	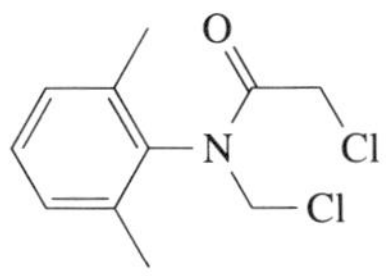	K	Acetone, then 2-butanone	70	86S817

[a] TBDMS = *t*-butyldimethylsilyl

Benzyl halides have been reported to give high yields of the isothiocyanates on treatment with potassium thiocyanate in DMF at 30 °C ⟨62MI 527-01⟩, but Muthusamy and Ramakrishnan only isolated the benzyl thiocyanate in 80% yield on repetition of this procedure. These authors have shown that reacting benzyl halides with potassium thiocyanate in refluxing 1,2-dichlorobenzene in the presence of 18-crown-6 gives moderate yields of the isothiocyanates (Scheme 20) ⟨89OPP228⟩. Rearrangement of benzyl thiocyanates in the presence of quaternary ammonium salts (hexamethylenetetraammonium iodide or laurylpyridinium bromide) similarly gives high yields (80–90%) of the isothiocyanates ⟨69CZP131547⟩.

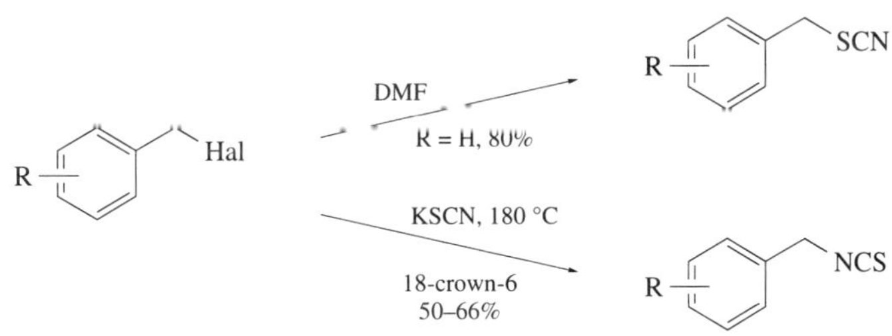

Scheme 20

A study of inorganic solid-supported alkaline earth metal thiocyanates shows a dramatic change in the sulfur–nitrogen selectivity of the —SCN substitution of *t*-butyl bromide from that observed with unsupported thiocyanates (Table 10) ⟨90CC1213⟩.

Phase-transfer catalysis has also been employed in the formation of glycosyl isothiocyanates using potassium thiocyanate (Equation (32); R = OAc, NHAc) ⟨84S509⟩.

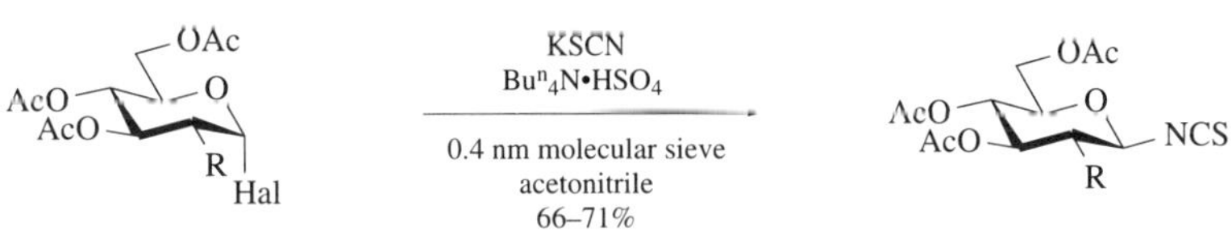

(32)

The more frequent approach to the formation of glycosyl isothiocyanates has been the reaction of halides with silver thiocyanate (Table 11).

A specialist example of the preparation of isothiocyanates from halides is the aromatic nucleophilic substitution of halide in electron-deficient aromatic species. Kristian showed that 2- and

Table 10 *t*-Butyl isothiocyanate from bromide with solid-supported thiocyanates.

$$Bu^t{-}Br \xrightarrow[\text{Support}]{M(SCN)_2} \underset{(N)}{Bu^t{-}NCS} + \underset{(S)}{Bu^t{-}SCN}$$

M	*Support*	*Yield* (%)	*N:S ratio*
Ba		6	9:91
	$CaCO_3$	57	5:95
	CaF_2	79	99:1
Ca		19	5:95
	BaF_2	36	89:11
	CaF_2	100	17:83

Table 11 Glycosyl isothiocyanates from halides with silver thiocyanate.

Reactant	*Solvent*	*Yield* (%)	*Ref.*
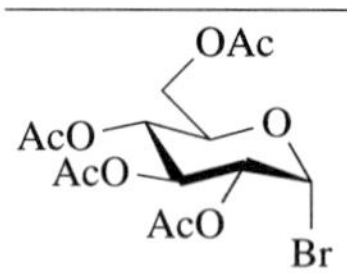	Toluene	74	56CB133
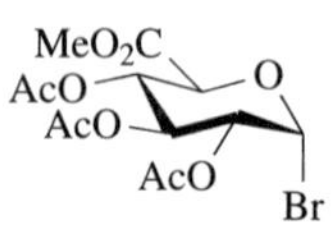	Xylene	85	57CB1590
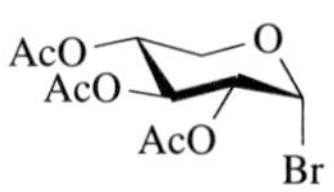	Xylene	93	75MI 527-01

4-substituted 9-chloroacridines could be reacted with either silver thiocyanate in xylene or with potassium thiocyanate in DMF to yield the 9-isothiocyanato derivatives in moderate yield (Equation (33); R = H, 2-Me, 4-Me, 2-MeO, 2-EtO, 2-Cl, 2-Br) ⟨69MI 527-01⟩. Picryl fluoride reacts with potassium thiocyanate to yield the isothiocyanate ⟨73AJC273⟩, but an improved synthesis has been reported whereby picryl chloride is first reacted with pyridine and the resultant pyridinium chloride treated with potassium thiocyanate (Scheme 21) ⟨90JHC1059⟩.

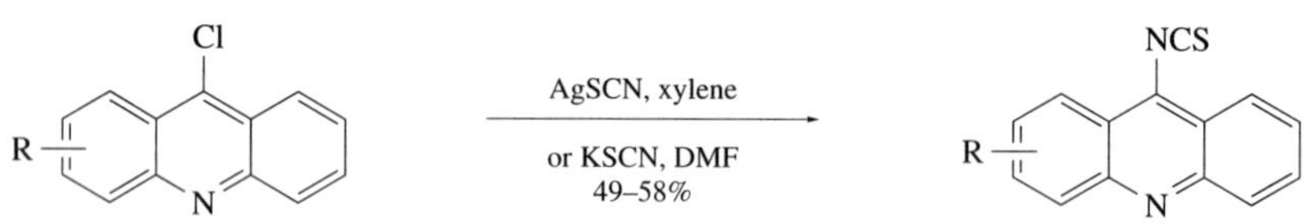

(33)

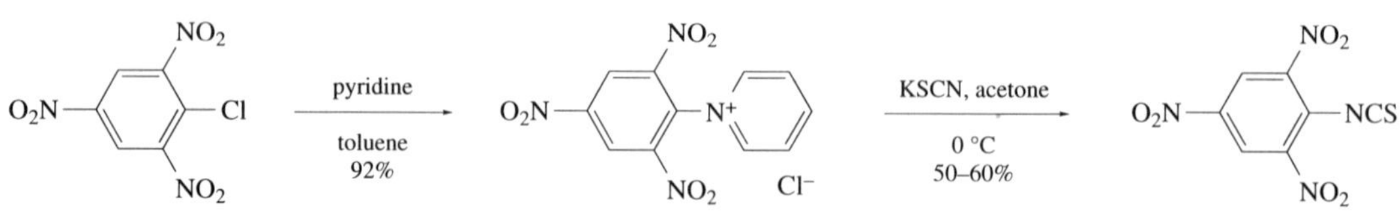

Scheme 21

The conversion of allyl thiocyanates to allyl isothiocyanates takes place by a facile thermal [3,3] sigmatropic rearrangement, as demonstrated by the observation that crotyl thiocyanate gives α-methylallyl isothiocyanate rather than the crotyl isothiocyanate on distillation at atmospheric pressure (Equation (34)) ⟨40CB843⟩. The exception to this rule is that cinnamyl thiocyanates give only the cinnamyl isothiocyanates ⟨60JA3076⟩. This methodology has been used to prepare a number of α-isothiocyanatoacrylic esters (Scheme 22), although only the ethyl 2-isothiocyanatocrotonate was formed in good yield (76%; R = Et, R^1 = H) ⟨78SC541⟩. The authors also observed an unexpected reaction with the α,β-unsaturated cyclic ketone (**21**, Scheme 23). The isolated products result from the reaction of 1-thiocyanatobuta-1,3-dienes with reactive dienophiles when the initial [4 + 2] cycloaddition is followed by a [3,3] sigmatropic rearrangement (Scheme 24) ⟨85CC533⟩.

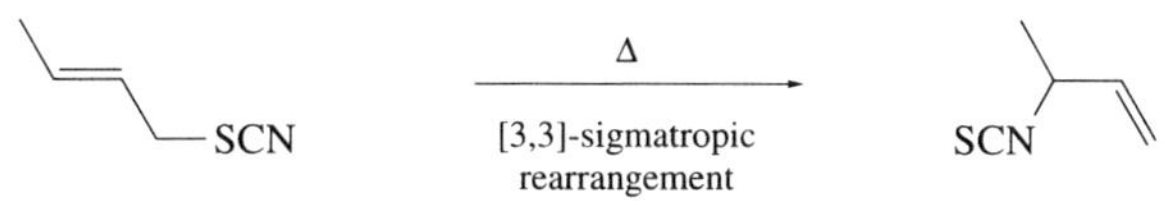

(34)

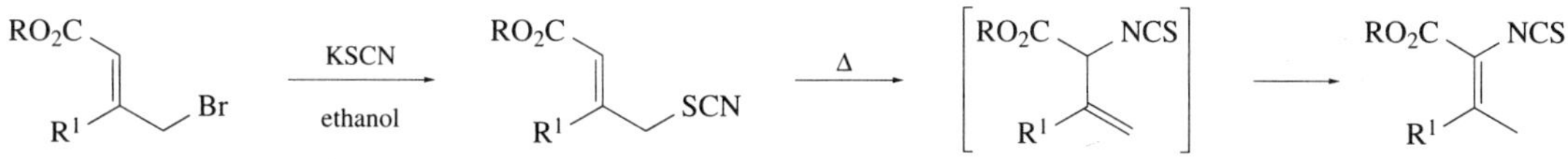

Scheme 22

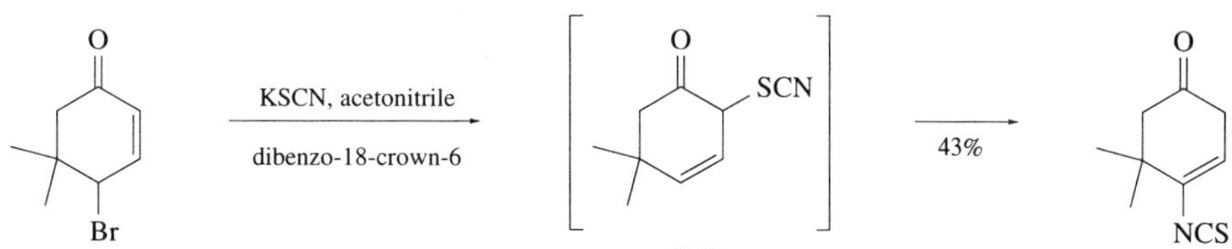

(21)

Scheme 23

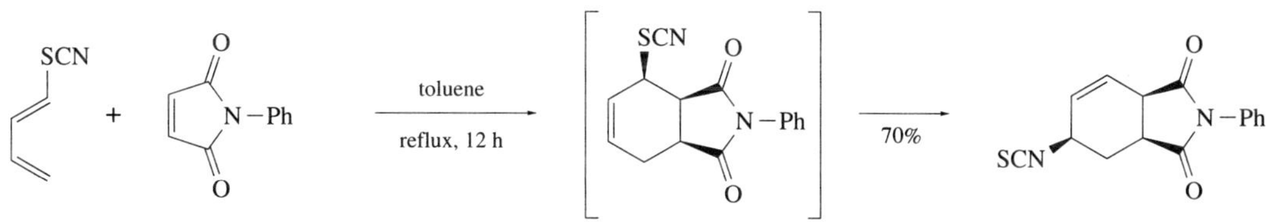

Scheme 24

Isothiocyanate-substituted allenes and 1,3 butadienes have been prepared by [3,3]-sigmatropic rearrangements of propargyl and 2,3-butadienyl thiocyanates ⟨92AG(E)90⟩. The products are extremely reactive and have been used to form thiazoles, and to act as dienes in cycloaddition reactions (Scheme 25).

5.27.2.1.9 Isothiocyanates from alcohols or derivatives

The use of organophosphorus isothiocyanate reagents allows tertiary alcohols to be converted directly into isothiocyanates without prior conversion to the halide. Initial studies with triphenylphosphine isothiocyanate ($Ph_3P(NCS)_2$), formed by reaction of triphenylphosphine with thiocyanogen, gave mixed success, with the yield of isothiocyanate very dependent upon substitution at the tertiary centre ⟨77TL4417⟩. Thus, while α,α-dimethylbenzyl alcohol yielded 80% of the isothiocyanate, *t*-butyl alcohol gave only 8% of product. However, use of phosphoryl isothiocyanate (synthesised from phosphoryl chloride and potassium thiocyanate in benzene with 18-crown-6) appears to provide a reliable route for this interconversion, giving good yields with a number of different alkyl- and aryl-substituted tertiary alcohols (Equation (35); R^1, R^2, R^3 = Me, Et, $MeOCH_2$, Ph) ⟨90SC509⟩. The reaction is carried out in neat alcohol, with the exception of solid trityl alcohol where diethyl ether is used as solvent.

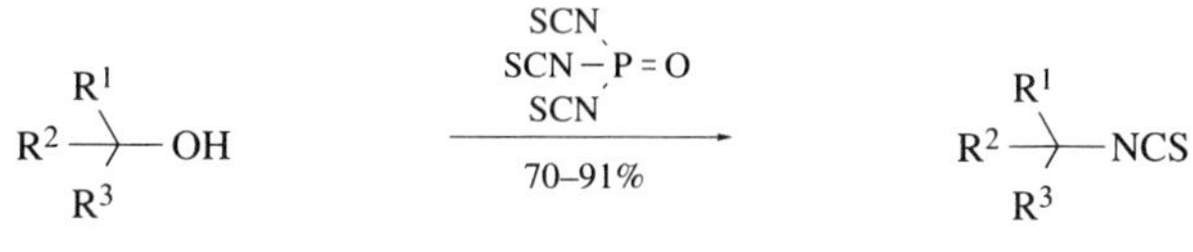

(35)

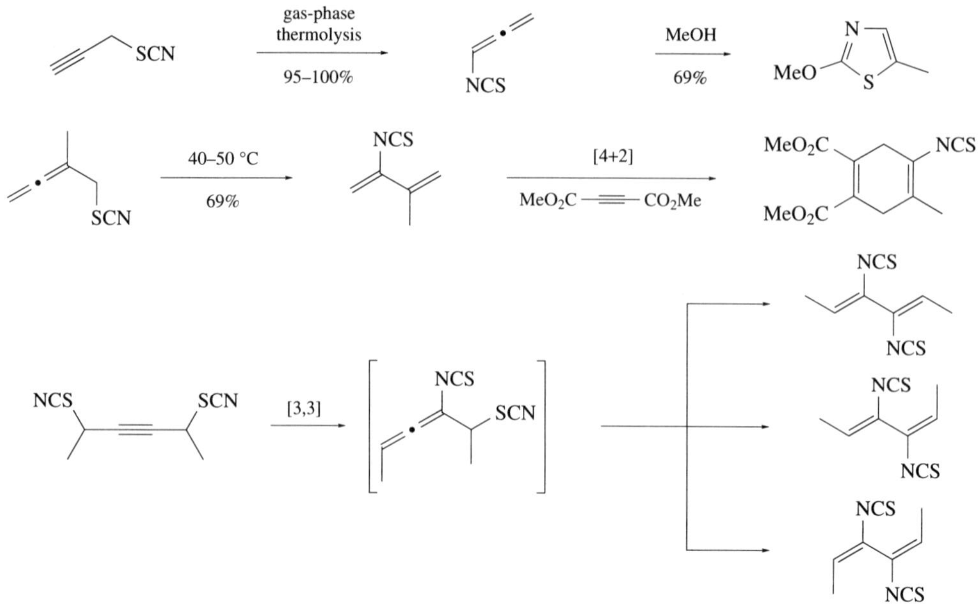

Scheme 25

The anomeric hydroxyl of a β-mannosyl-chitobiose trisaccharide, protected as a dihydrooxazole, reacts with potassium thiocyanate and tetrafluoroboric acid in the presence of 18-crown-6 to form the isothiocyanate via rearrangement of the intermediate thiocyanate (Equation (36); R = disaccharide) ⟨90AG(E)1050⟩.

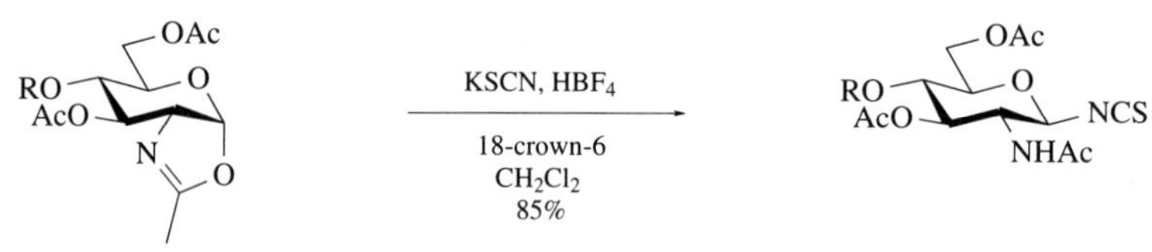

(36)

Cyclic ethers may undergo ring-opening with mercury(II) thiocyanate and a primary or secondary alkyl halide to yield alkoxyalkyl isothiocyanates. THF gives approximately a 1 : 1 ratio of isothiocyanate to thiocyanate ⟨75BCJ3205⟩; tetrahydropyran gives a similar ratio in low yield (7%). However, substituted oxetanes give predominantly the isothiocyanate (Equation (37); R^1 = Et, Pr^i) (ratio of isothiocyanate to thiocyanate between 82 : 18 and 94 : 4), probably through the intermediacy of an oxonium ion ⟨79BCJ3611⟩.

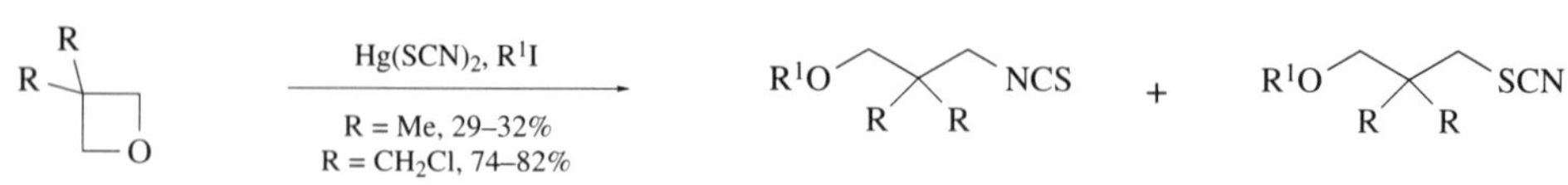

(37)

5.27.2.1.10 *Isothiocyanates from aldehydes or ketones*

Alkyl aldehydes react with trimethylsilyl isothiocyanate in the presence of a catalytic amount of zinc chloride to yield the symmetrical disubstituted ether, via the trioxane derivative (Scheme 26; R = Et, Pr^n, Pr^i, Bu^t) ⟨87BCJ2289⟩. Benzaldehyde was unreactive under these conditions. The authors also showed that the corresponding acetals and ketals reacted under similar conditions to give the α-isothiocyanatoether (Equation (38); R = alkyl, aryl; R^1 = H, alkyl; R^2 = Me, Et). Isolated yields were frequently low due to the moisture sensitivity of the products.

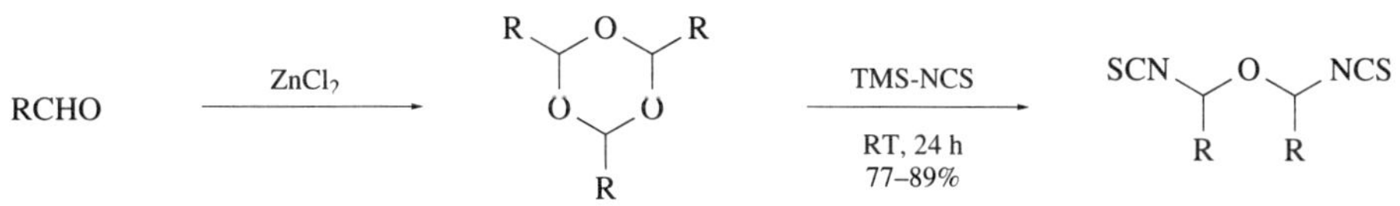

Scheme 26

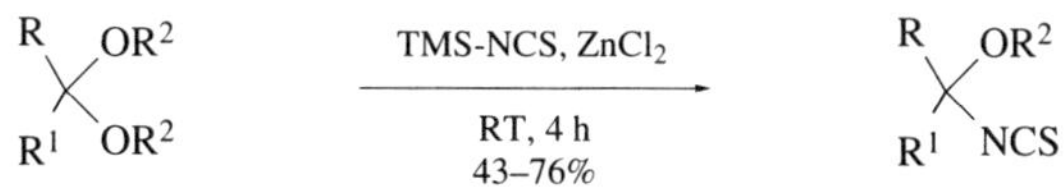

(38)

If aldol reactions are carried out with acetaldehyde or acetone in the presence of phosphorus oxychloride and potassium thiocyanate, the intermediate aldol undergoes substitution to yield the isothiocyanate (Scheme 27) ⟨88ZC141⟩.

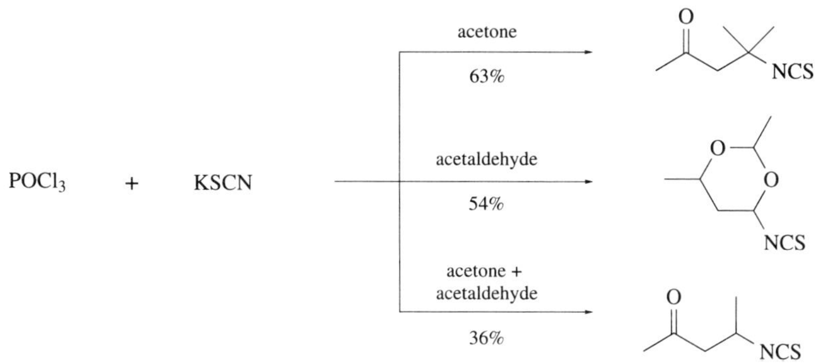

Scheme 27

Cycloalkanones react with silicon tetraisothiocyanate in the presence of bis(trimethylsilyl) sulfate and tri-*n*-butylstannyl fluoride to give 1-cycloalkenyl isothiocyanates in high yield (Equation (39)) ⟨84TL4683⟩.

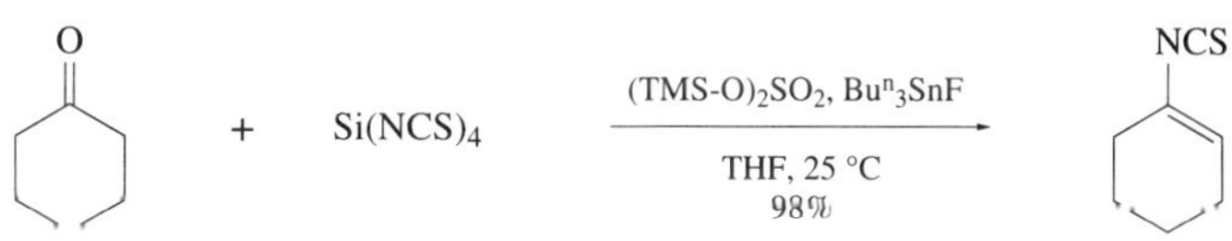

(39)

5.27.2.1.11 *Isothiocyanates from alkenes and alkynes*

Isothiocyanic acid, generated *in situ* from sodium thiocyanate and aqueous sulfuric acid, adds to alkenes to give predominantly the isothiocyanate ⟨56JA4965⟩. Early studies of the addition of isothiocyanic acid to *endo*-dicyclopentadiene suggested that the ring system underwent a typical Wagner–Meerwein rearrangement ⟨45JA1178, 59ZN(B)814⟩. However, a detailed study of the addition of isothiocyanic acid to bicyclo[2.2.1]heptene systems, undertaken by Diveley *et al.*, clearly demonstrated that isothiocyanic acid adds predominantly *cis–exo* without rearrangement of the bicyclic ring structure (Equations (40) and (41)) ⟨69JOC616⟩.

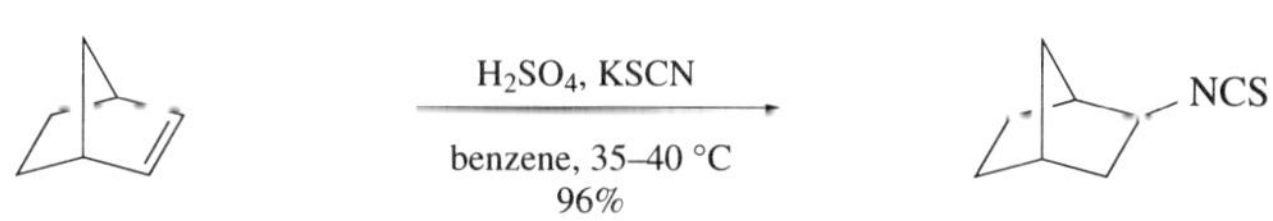

(40)

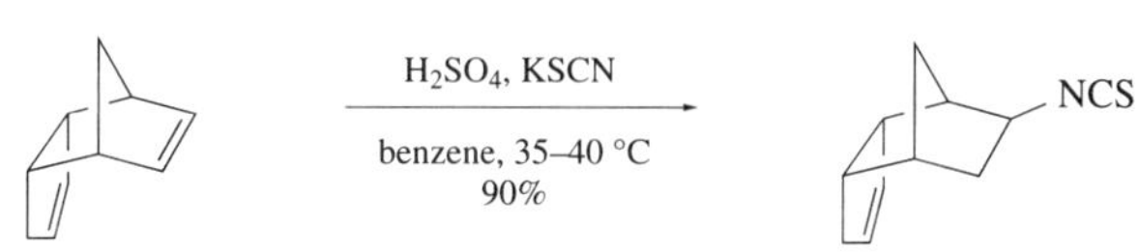

(41)

The similar addition of isothiocyanic acid to camphene was shown to give 2-isothiocyanatoisocamphane (Equation (42)) ⟨56JOC1430⟩. This methodology has been successfully applied to the synthesis of a number of marine sesquiterpene isothiocyanates ⟨93TL6717⟩. Reaction times of between 2 and 7 days were required and usually a mixture of the thiocyanate and isothiocyanate was formed. The presence of a hydroxyl function in the reactant sesquiterpene appeared to slow down the reaction rate and lead to higher percentages of thiocyanate.

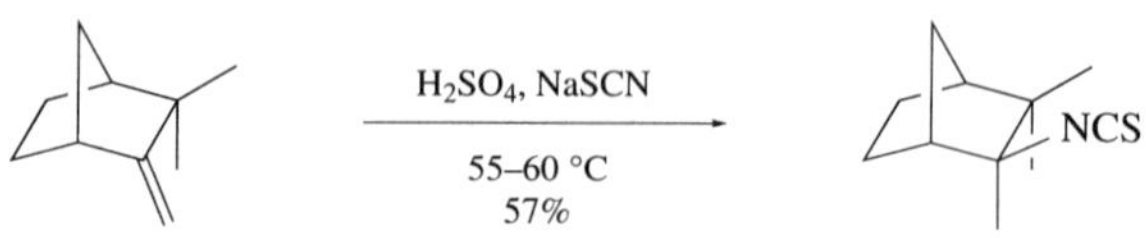

(42)

If benzonorbornadiene is reacted with thiocyanogen, addition occurs with exclusive formation of the product from a Wagner–Meerwein rearrangement (Equation (43)) ⟨91ZOR340⟩. The ratio of isothiocyanate (**22**) to thiocyanate (**23**) is dependent on the solvent (49 : 51 in chloroform; 25 : 75 in acetic acid). A study in the 1960s investigated the reaction of thiocyanogen with simple alkyl- and aryl-substituted alkenes and showed that while the former gave exclusively *trans* addition products, the latter gave non-stereospecific addition ⟨69CI(L)1702⟩.

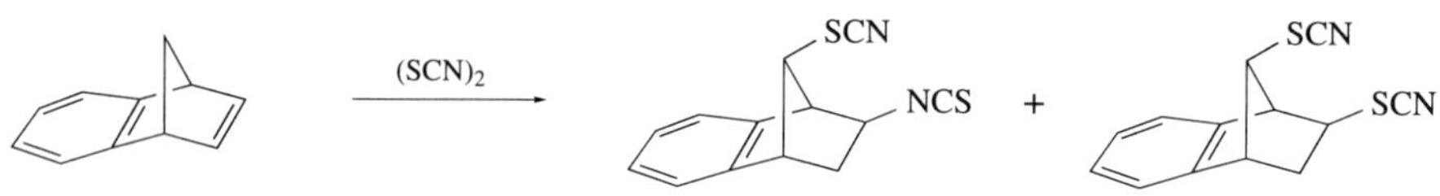

(**22**) (**23**)

(43)

Dienes may react with aryldiazonium tetrafluoroborates and alkali thiocyanates with catalytic quantities of copper salts to give 2-isothiocyanato-1-aryl-3-butenes in yields of 45–63% (Equation (44); R^1 = H, 4-Me, 4-OMe; R^2 = H, Me) ⟨91UKZ89⟩. This should be compared with the authors' previous publication where the products were identified as 4-isothiocyanato-1-aryl-2-butenes ⟨88ZOB616⟩.

(44)

Alkynes may undergo addition of thiocyanic acid. Thus acetylenedicarboxylates initially form a mixture of the thiocyanatomaleate and fumarate, which on thermal isomerisation yields predominantly the isothiocyanatofumarate (Scheme 28) ⟨78LA227⟩. Reaction of an alkyne with potassium thiocyanate and iodine in the presence of copper(II) tetrafluoroborate yielded the 1,2-iodofunctionalised alkene (Equation (45)) ⟨87CC1491⟩. The related addition of iodine(I) thiocyanate to alkenes gives predominantly the *vic*-iodothiocyanates, although these may be rearranged to the thermodynamically more stable *vic*-iodoisothiocyanates in the presence of boron trifluoride etherate ⟨79JCS(P1)757⟩. A report from the mid 1990s on the synthesis of analogues of sulforaphane, isolated from broccoli, used the addition of iodothiocyanate followed by radical dehalogenation to form *cis*-1,4-substituted isothiocyanatocyclohexanes (Scheme 29) ⟨94JMC170⟩.

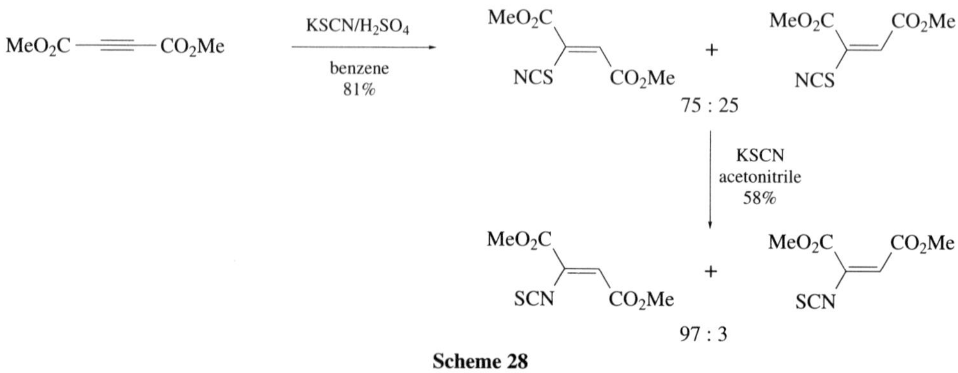

Scheme 28

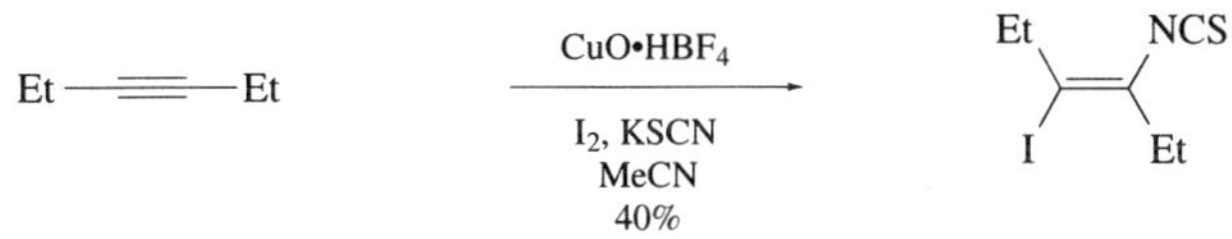

(45)

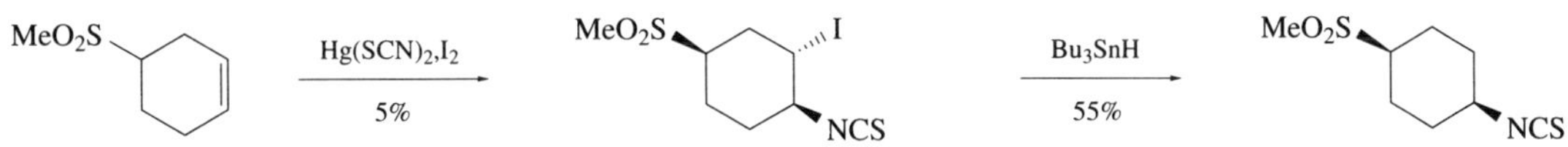

Scheme 29

A 1,3-difunctionalised adamantane has been prepared by transannular cyclisation with trimethylsilyl isothiocyanate in the presence of a Lewis acid, giving a 1 : 1 mixture of the isothiocyanate and thiocyanate (Scheme 30) ⟨90S646⟩.

Scheme 30

5.27.2.1.12 Isothiocyanates by aromatic electrophilic substitution

Reaction of electron-rich aromatic species with thiocyanogen usually yields the thiocyanate, but there are examples where the isothiocyanate is formed in preference. Thus, when thiocyanogen is prepared by electrochemical oxidation (probably present as trithiocyanate $(SCN)_3^-$), it yields isothiocyanate derivatives with substituted anilines (Equation (46)) ⟨92SC2741⟩.

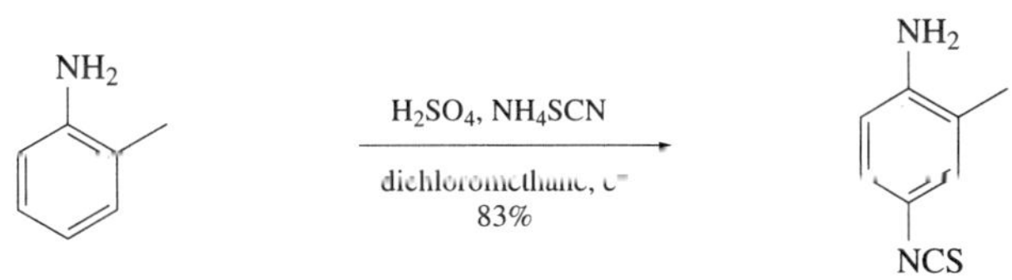

(46)

3-Alkylindoles similarly form isothiocyanates during electrochemical thiocyanation ⟨90TL7229⟩—for example, methyl indole-3-acetate gives 67% of the 2-isothiocyanate and only 7% of the 2-thiocyanate (Equation (47)). Using 'nascent' thiocyanogen generated *in situ* from potassium thiocyanate and bromine also gives predominantly the isothiocyanate although in lower yield (44%). Substitution is considered to proceed initially by an *ipso* attack at the 3-position, followed by an unprecedented [3,5]-sigmatropic shift of this 3-thiocyano-3*H*-indolium cation intermediate.

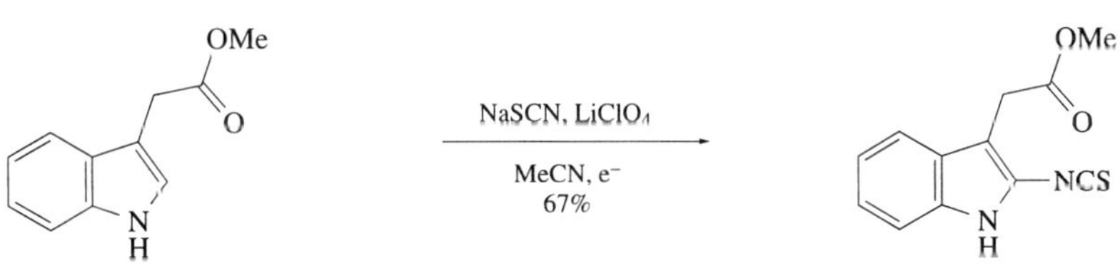

(47)

5.27.2.1.13 Isothiocyanates by cleavage of nitrogen–sulfur containing heterocycles

Isothiocyanates may be formed where the three atoms of the —NCS functionality are derived from cleavage of an appropriate nitrogen–sulfur-containing heterocycle.

4-Alkyl or 4-aryl thiosemicarbazides react with nitrous acid to yield substituted 5-amino-1,2,3,4-thiatriazoles (also formed by the action of hydrazoic acid on isothiocyanates). The alkyl-substituted heterocycles are reported to decompose in the presence of aqueous base to yield the isothiocyanate (no yields are quoted) (Scheme 31) ⟨57CJC832⟩.

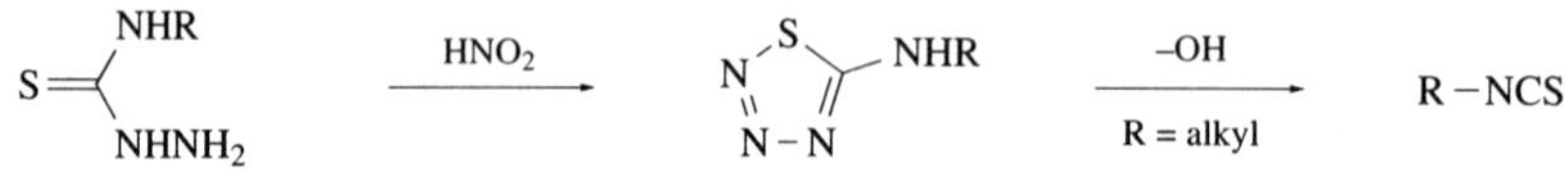

Scheme 31

Carbon disulfide reacts with azide ion to yield the salt of 5-thiol-1,2,3,4-thiatriazole, which on alkylation with benzhydryl or trityl halides forms the 4-substituted 1,2,3,4-thiatriazoline-5-thiones (Scheme 32; R = Ph_2CH or Ph_3C). These decompose to the isothiocyanates on refluxing in benzene (no yields given) ⟨63JOC194⟩.

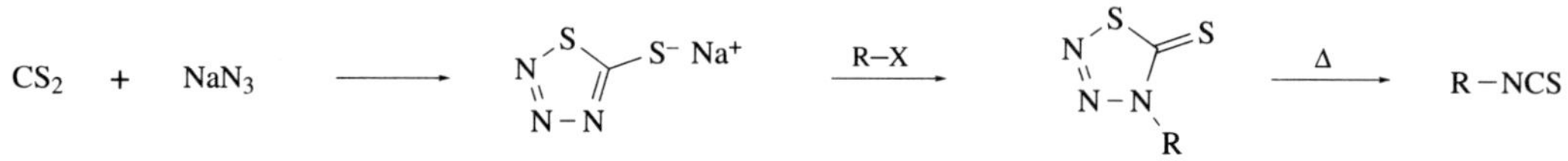

Scheme 32

1,4,2-Oxathiazoles have been mentioned previously (see Section 5.27.2.1.6) as intermediates in the conversion of aryl cyanates to isothiocyanates with thiobenzophenones. They have also been formed by reaction of thiohydroxamic acids with diaryldichloromethanes, and are decomposed by heating at 80–140°C. Thus, the 3-(4-nitrophenyl)-1,4,2-oxathiazole gives 4-nitrophenyl isothiocyanate (Equation (48)) in good yield on heating at 120°C ⟨72CB2815⟩.

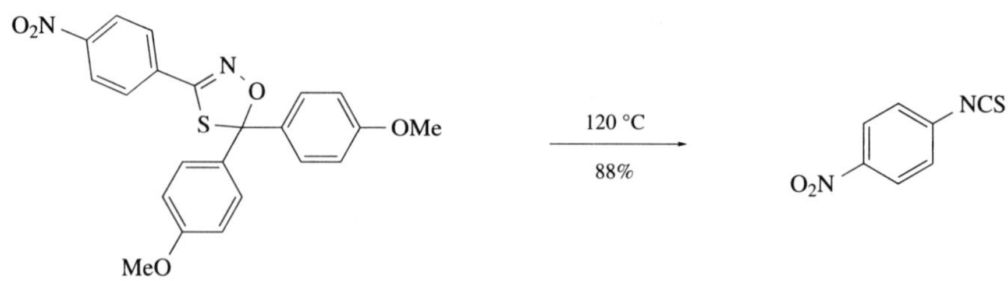

(48)

2-Keto isothiocyanates may be formed by cleavage of 5-hydroxythiazolidine-2-thiones with dicyclohexylcarbodiimide (dcc) or acetic anhydride (Equation (49); R^1, R^2 = Me, Ph) ⟨76CB154⟩. If one of the substituents α to the nitrogen is a hydrogen, the resultant isothiocyanate undergoes ring closure to the 4-oxazoline-2-thione. The related Δ^3-thiazoline-2-thiones eliminate sulfur either thermally, photolytically or with trimethyl phosphite to give the vinyl isothiocyanates (Equation (50); R = Me, Ph) ⟨78CB3750⟩. 2-Isothiocyanatoacrylic esters are formed by a base-catalysed ring-opening of the *N*-phosphorylated oxazolidine-2-thiones (Scheme 33; R^1 = H, R^2 = Pr^i, Ph, PhCH = CH; R^1 = Me, R^2 = Me, Ph; R^1,R^2 = -$(CH_2)_4$-, -$(CH_2)_5$-) ⟨73AG(E)923⟩.

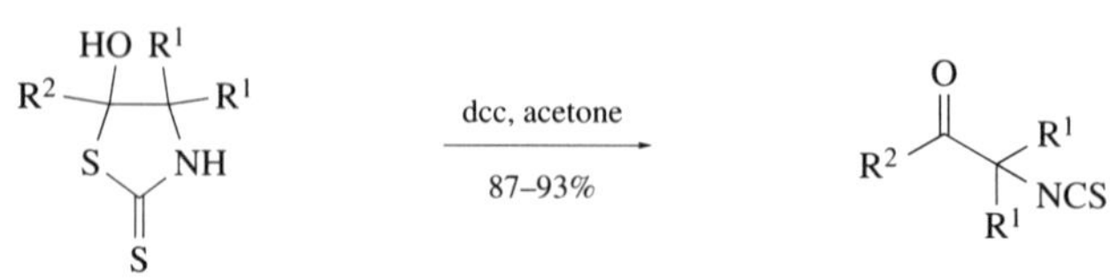

(49)

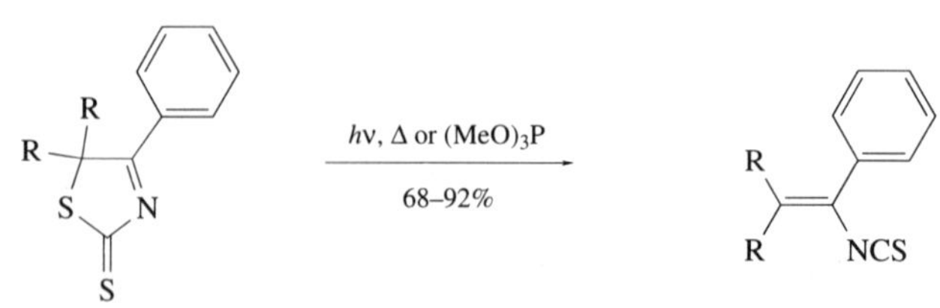

(50)

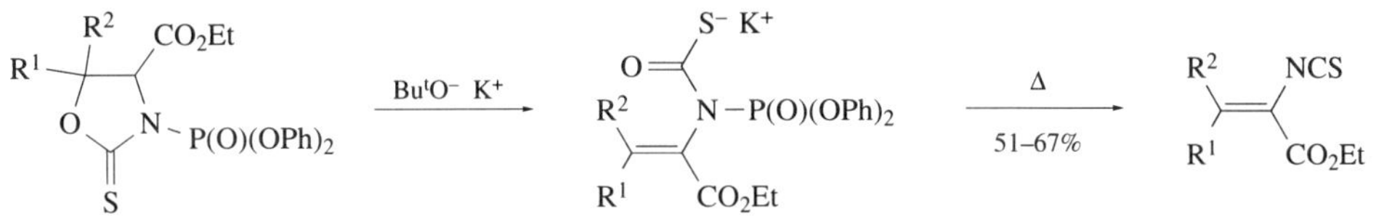

Scheme 33

5.27.2.1.14 *Acyl isothiocyanates*

Acyl isothiocyanates are more reactive than the corresponding alkyl or aryl isothiocyanates, readily undergoing addition, substitution and cyclisation reactions. In most cases, they are colourless or pale yellow liquids, or low melting solids, and have frequently been purified by distillation under high vacuum. On standing, acyl isothiocyanates polymerise and thus are often used immediately for further reaction.

Known since the late nineteenth century, the standard preparation of acyl isothiocyanates remains as treatment of the corresponding acyl halide with an inorganic thiocyanate. The preferred reagent has been lead(II) thiocyanate in solvents such as benzene, toluene or xylene at reflux. Dixon, following the earlier preparation of acetyl and benzoyl isothiocyanate by Miquel ⟨1877MI 527-01⟩, showed this method to be applicable to a number of simple alkanoyl and aroyl isothiocyanates (Table 12).

Table 12 Acyl isothiocyanates from acyl chlorides with lead thiocyanate.

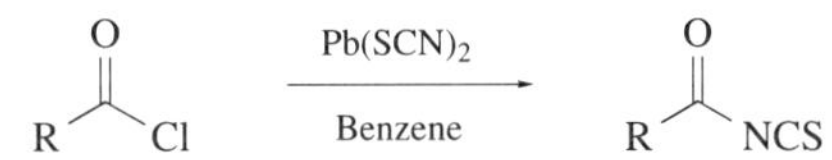

Reactant	*Product*	*Ref.*
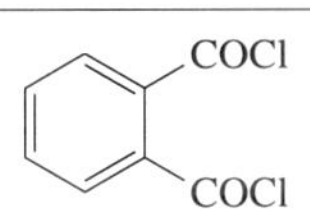	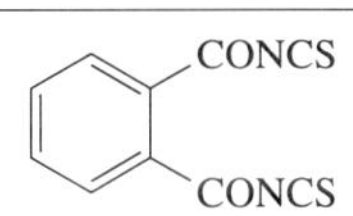	1895JCS565
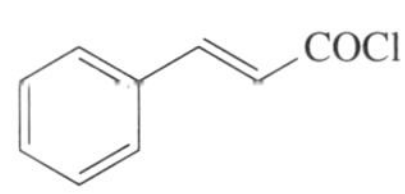	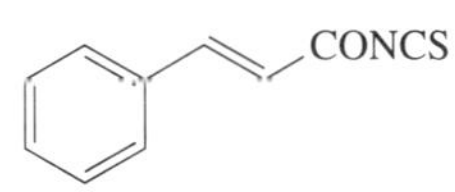	1895JCS1041
EtCOCl	EtCONCS	1896JCS855
$C_{17}H_{35}COCl$	$C_{17}H_{35}CONCS$	1896JCS1593

These early references gave no report of yields but Dixon's methodology has been extended to a wide range of alkanoyl, alkenoyl, aroyl and heteroaroyl substrates. Lipp *et al.* reported the synthesis (in 29–98% yield) and physical properties of 47 diverse acyl isothiocyanates as intermediates in the synthesis of bacteriostatic agents ⟨58CB1660⟩; while Smith and Kan prepared 26 acyl isothiocyanates (in 22–82% yield) for participation in Friedel–Crafts acylation reactions ⟨64JOC2261⟩. Pohloudek-Fabini and Schropl examined the formation of benzoyl isothiocyanate from benzoyl chloride and lead thiocyanate in different solvents and demonstrated the highest yield in benzene ⟨68MI 527-01⟩. They also prepared acetyl, propanoyl, butanoyl, pentanoyl, heptanoyl and halogenated alkanoyl isothiocyanates. Potassium thiocyanate similarly appears to react well with a variety of acid chlorides in a number of different solvents (Table 13). Other inorganic thiocyanates have also been employed with success, as exemplified in Table 14.

The use of phase transfer catalysis avoids the need for long reaction times and for anhydrous reaction conditions. The method has been employed with aroyl and long-chain alkanoyl isothiocyanates (Equation (51), e.g., R = substituted Ph, 2-naphthyl, furanyl, C_5H_{11}), using tributylammonium bromide (TBAB) as catalyst ⟨81SC781⟩. This approach is unsuccessful with lower alkyl homologues due to competing hydrolysis of the acid chlorides.

Table 13 Acyl isothiocyanates from acyl chlorides with potassium thiocyanate.

Reactant	*Solvent*	*Yield (%)*	*Ref.*
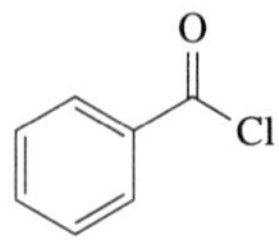	Benzene	64	39JA632
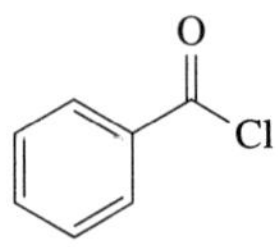	Acetone	61	58JCS1141
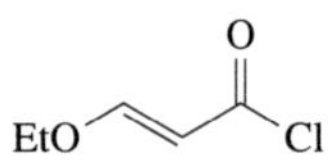	Acetonitrile	57	58JCS153
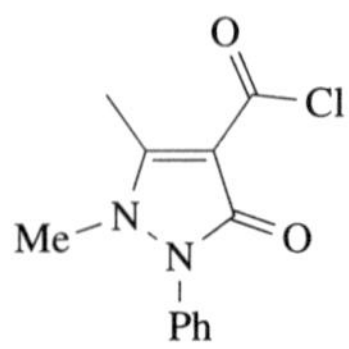	Xylene	75	60AP150
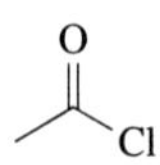	Acetone	20	63BCJ1214
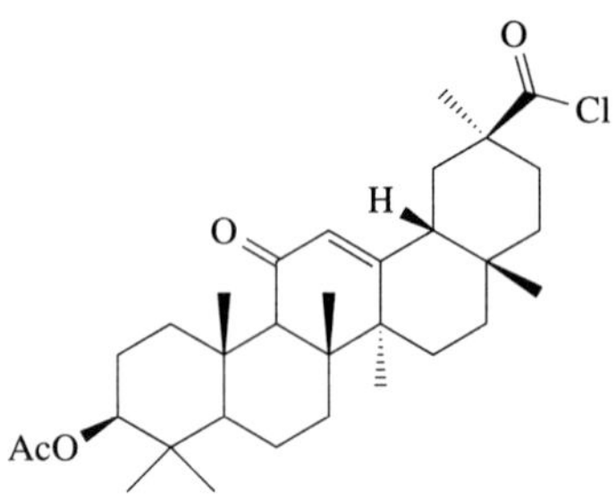	Acetone	85	70AP905

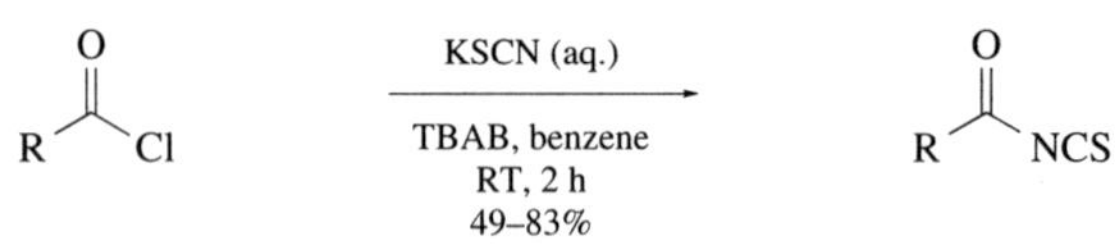

(51)

An alternative to inorganic thiocyanates is the use of silyl isothiocyanates which react with acyl halides in the absence of solvent (Equation (52)) ⟨66GEP1215144⟩. This methodology seems particularly appropriate for large scale synthesis, where the liberated silyl chloride may be distilled off and reacted with inorganic thiocyanate for recycling.

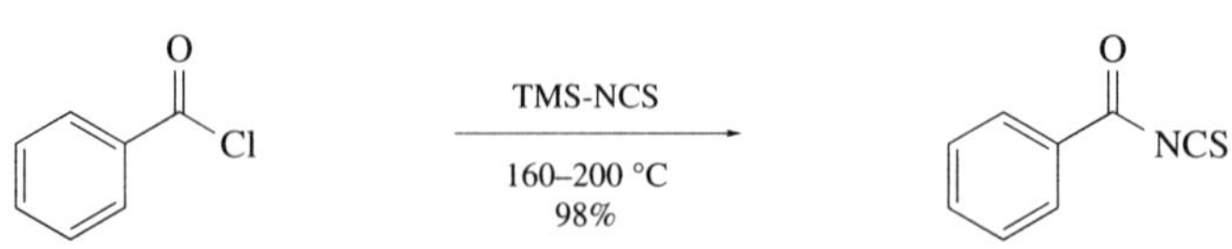

(52)

In a manner similar to that employed with alcohols (see Section 5.27.2.1.9), organophosphorus bis(isothiocyanates) may be used to convert carboxylic acids directly into their acyl isothiocyanates. Thus, triphenylphosphine isothiocyanate ($PPh_3(NCS)_2$) has been utilised with some success ⟨78CI(L)806⟩, but phosphoryl isothiocyanate ($P(O)(NCS)_3$) gives better yields (generally >60%)

Table 14 Acyl isothiocyanates from acyl chlorides.

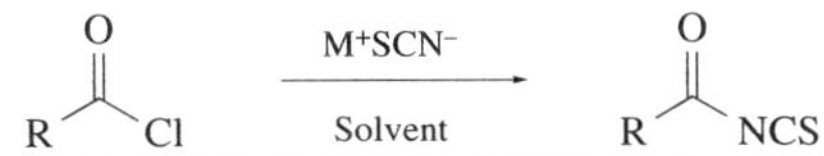

Reactant	M	Solvent	Yield (%)	Ref.
Oleoyl chloride	Mg	Benzene	73	58BRP800807
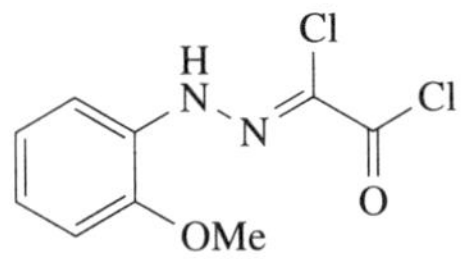	Na	Acetone	97	65ZOR1415
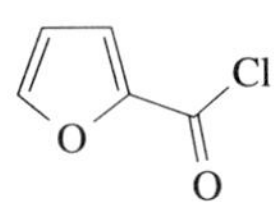	NH_4	Acetone	Used without isolation	34JA719
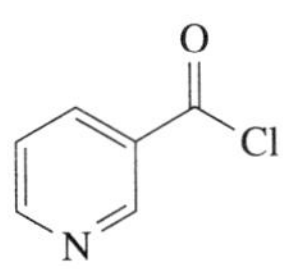	NH_4	Toluene	Low yield on isolation	34JA1609
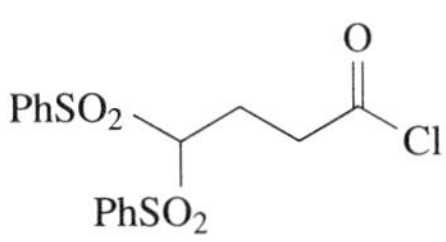	Ag	Benzene	80	68ZOR658
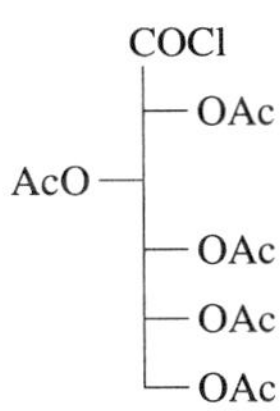	Ag	Xylene	83	79CPB1137
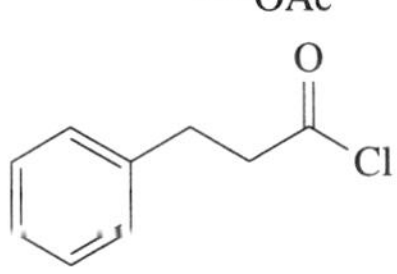	Ag	Diethyl ether	Unstable	76CCC1388

⟨90SC509⟩. As this approach appears to be applicable to a wide range of substrates (Equation (53); e.g., R = substituted Ph, Bn, 2-thienyl, $C_{17}H_{35}$) and avoids the need to synthesise the intermediate acid chloride, it provides a good general method for the synthesis of acyl isothiocyanates.

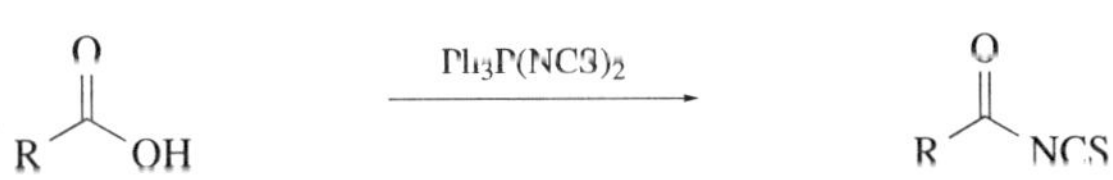

(53)

5.27.2.1.15 *Thioacyl isothiocyanates*

Syntheses of thioacyl isothiocyanates are limited to only a few examples in the literature. Behringer and Deichmann suggested the intermediacy of a thioacyl isothiocyanate in the formation of 5-aryl-3*H*-1,2,4-dithiazol-3-thiones (**24**) from the aroyl isothiocyanates (Scheme 34; R = H, MeO, Cl) ⟨67TL1013⟩. The same authors reported that an isomer of (**24**) was formed by reaction of thiobenzamide with thiophosgene in carbon disulfide, and suggested that it could be the 1,3,4-dithiazol-2-thione (**25**) (Equation (54)).

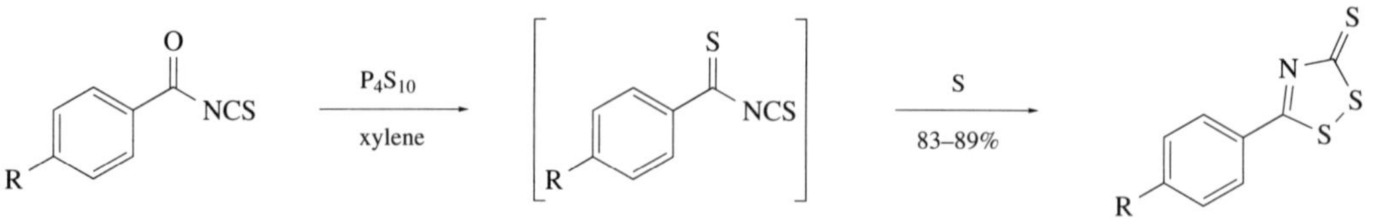

(24)

Scheme 34

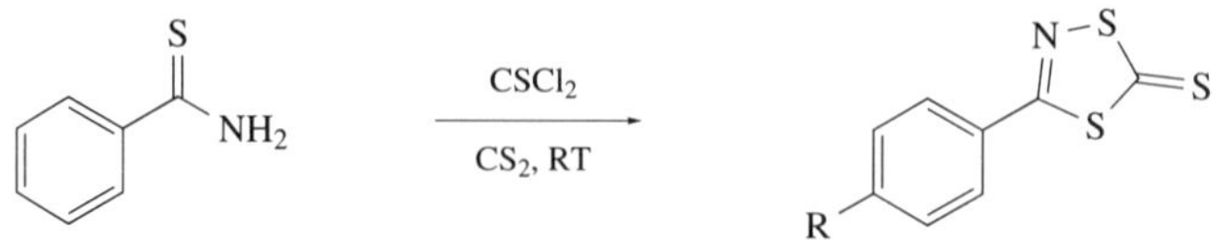

(25)

(54)

A paper by Barnikow and Gabrio implied that these latter conditions gave the thiobenzoyl isothiocyanate ⟨68ZC142⟩. They proposed that alternatively the thioaroyl isothiocyanates could be prepared by reaction of thioaroyl chlorides with sodium thiocyanate in ethyl acetate. The IR spectrum was interpreted as showing an NCS absorption at the anomalously high position of 2220–2230 cm^{-1}. The situation was resolved by Goerdeler and Teller, who demonstrated that thioaroyl chlorides reacted with sodium thiocyanate in ethyl acetate to yield initially the thioaroyl thiocyanate (IR —SCN absorption 2165 cm^{-1}), which on standing in solution rearranged to the isothiocyanate (IR NCS absorption 1980–2000 cm^{-1}), the rate of isomerisation depending on the aryl substitution ⟨72TL1513⟩. The same isothiocyanates could be prepared by desulfurisation of the corresponding 5-aryl-1,2,4-dithiazol-3-thiones (mentioned above) with triphenylphosphine (Scheme 35). The related desulfurisation of 5-amino-1,2,4-dithiazol-3-thiones to yield thiocarbamoyl isothiocyanates has been reported in more detail by the same authors ⟨74CB502⟩.

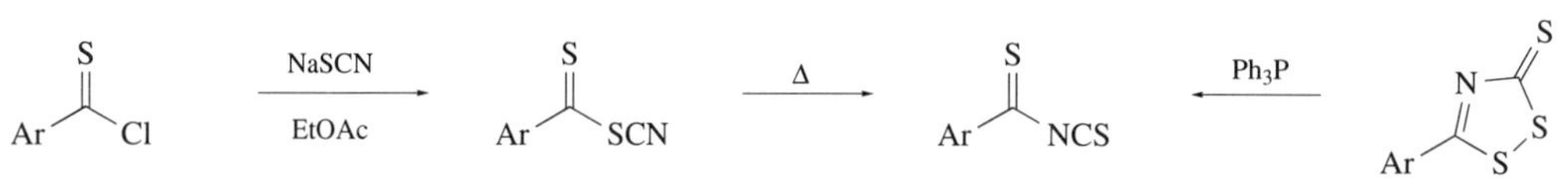

Scheme 35

5.27.2.1.16 Imidoyl isothiocyanates

This section describes compounds of the general structure (**26**), where R may be alkyl or aryl; alternatively, R may take the form of a protecting group of the free imino function (e.g., *N*-diaryloxyphosphinyl, *N*-sulfonyl or *N*-thiocarbamoyl derivatives)—these isothiocyanates will be discussed separately at the end of the section. In 1994, the majority of the imidoyl isothiocyanates described contain the arylimidoyl moiety.

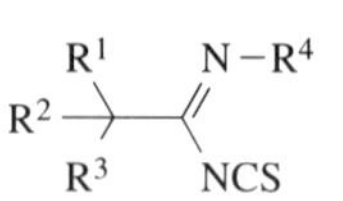

(26)

Crystalline imidoyl isothiocyanates appear to be stable at room temperature but when oils, dimerisation can occur either in the presence of sunlight or at elevated temperatures to give the *sym*-triazine derivatives (**27**) ⟨64TL799, 64TL1087⟩. *N*-arylimidoyl isothiocyanates may also undergo ring closure in refluxing benzene or toluene (Scheme 36) to give the 4(3*H*)-quinazoline thiones (**28**) ⟨64TL855⟩.

The preferred method of synthesis is by displacement of the imidoyl halide with an inorganic thiocyanate, closely analogous to that employed for acyl isothiocyanates (see Section 5.27.2.1.14). As with the formation of acyl isothiocyanates, lead, potassium, sodium or ammonium thiocyanates are the salts of choice, and the reactions may be carried out in solvents such as benzene, acetone or acetonitrile (Table 15).

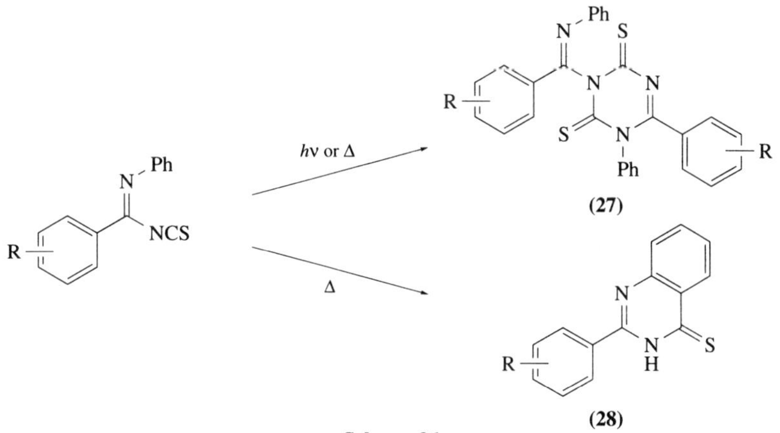

Scheme 36

Barnikow and Ebeling have demonstrated the formation of *N*-methylbenzimidoyl isothiocyanate by treatment of the bis(trimethylsilyl)-*N*-methylbenzamidine (**29**) with thiophosgene in diethyl ether (Scheme 37) ⟨71ZC420⟩. The resultant red oil could not be purified by distillation without decomposition.

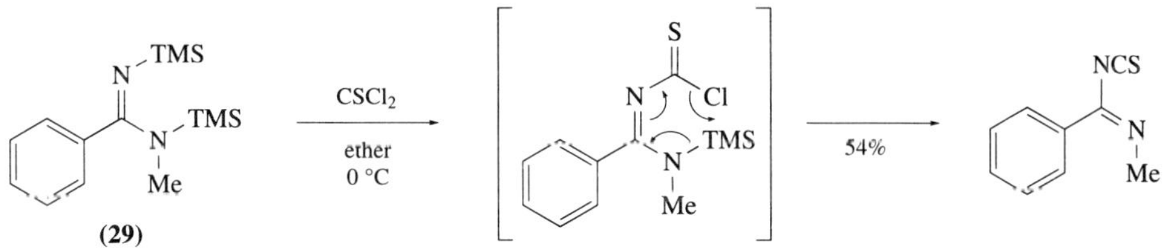

Scheme 37

A number of derivatives of benzimidoyl chlorides of general formula (**30**) (Table 16) (where X = phosphoryl, sulfonyl or thiocarbamoyl protecting groups) have been converted into the corresponding isothiocyanates by reaction with ammonium or alkali metal thiocyanates.

5.27.2.2 Functions Based on Nitrogen and Selenium or Tellurium

This section describes synthetic approaches to organic isoseleno- and isotellurocyanates, where the —NCSe or —NCTe functionality is bonded directly to a carbon atom. Isoselenocyanates of alkanes, alkenes, arenes and heteroarenes are described according to the precursor function undergoing transformation, while acyl and imidoyl isoselenocyanates are described separately at the end of the section. Up to 1994, all attempted approaches to synthesise isotellurocyanates have proved unsuccessful.

The preparation of isoselenocyanates has been reviewed previously. In the course of a study of selenosemicarbazides, Collard-Charon and Renson researched the possible approaches to the precursor isoselenocyanates, describing routes analogous to those employed for the preparation of isothiocyanates, though a number of these failed with the corresponding selenium compounds ⟨62BSB531⟩. Some general experimental procedures are described. Short reviews in *Comprehensive Organic Chemistry* ⟨79COC(3)522⟩ and by Paulmier ⟨B-86MI 527-01⟩ reiterated these methods. More extensive reviews are to be found in the appropriate volumes of Patai ⟨B-77MI 527-03, B-87MI 527-01, B-87MI 527-02⟩ and of *Houben-Weyl* ⟨55HOU(9)951⟩.

5.27.2.2.1 Isoselenocyanates from isocyanides

The reaction of isocyanides with elemental selenium provides the most widely employed method of synthesising isoselenocyanates. Jensen and Frederiksen demonstrated in 1936 that phenyl isocyanide heated at reflux with selenium in chloroform gave low yields (29%) of the isoselenocyanate on vacuum distillation ⟨36ZAAC(230)31⟩. Chloroform has remained the solvent of choice, though various alkanes, THF and pyridine have found utility. The methodology has been successfully applied to

Table 15 Imidoyl isothiocyanates from imidoyl chlorides.

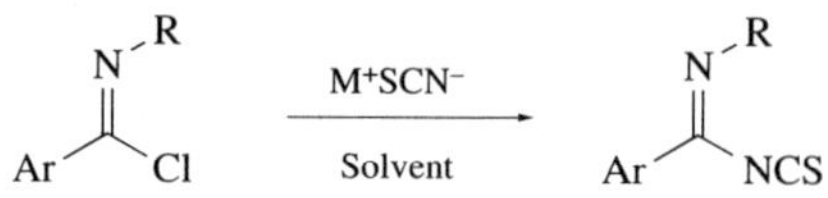

Reactant	*X*	*M*	*Solvent*	*Yield (%)*	*Ref.*
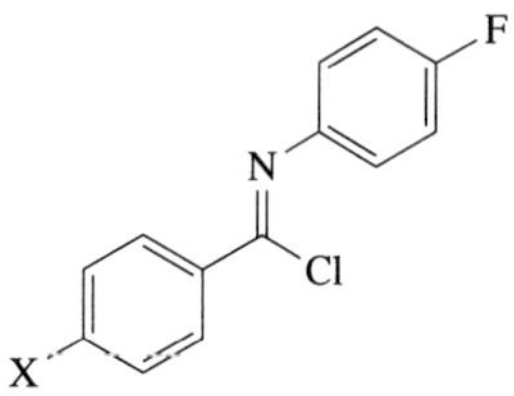	H, F, Me, OMe	Pb	Benzene	61–80	66JOC722
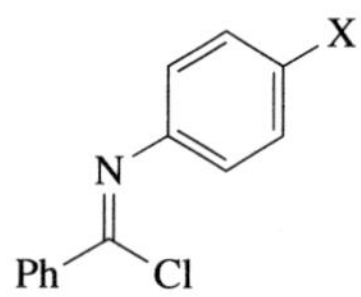	H, Cl, NO_2	Na	Acetone or Acetonitrile	94–96	68CB3475
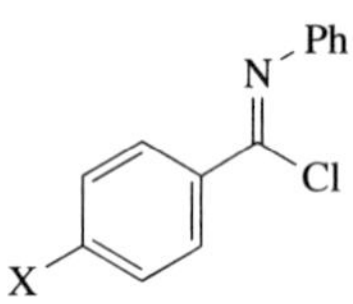	H, OMe, NO_2	Na	Acetone or Acetonitrile	90–95	68CB3475
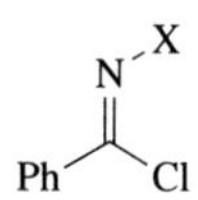	Et, Bu^n	K	Acetonitrile	20–30	69ZOR1285
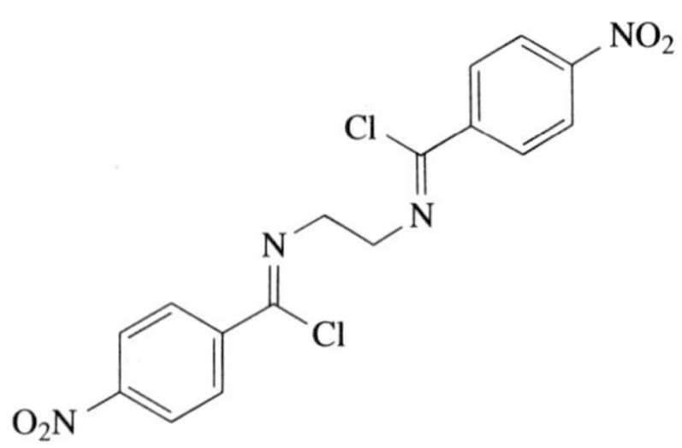		NH_4	Acetone	52	74ZC356
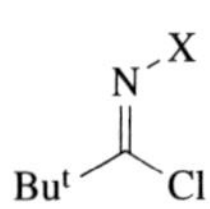	*c*-hex, Ph, 2,6-$Me_2C_6H_3$	Na	Acetone	85–89	75CB3071
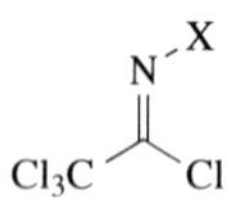	Me, Et, *c*-hex	Na	Acetone	70–78	75CB3071

alkyl, aryl and heteroaryl substrates, the crude yields being generally high, though appreciable losses may occur on purification by vacuum distillation or recrystallisation (Table 17). A synthesis of an isoselenocyanato derivative of *N*-[1-(2-thienyl)cyclohexyl]ethylamine reports the use of ultrasound to give a finely divided suspension of selenium to aid reactivity with the isocyanide ⟨91SL654⟩.

The reports from the early 1970s of the reaction of alkyl isocyanides in petroleum ether showed only slow reaction rates and low yields, but the presence of an equivalent of triethylamine in THF allows the formation of high yields of alkyl isoselenocyanates with much shorter reaction times ⟨72BCJ2937⟩ (Table 18). This method has been successfully applied by Witczak to yield a number of

Table 16 Imidoyl isothiocyanates from imidoyl chlorides.

(30)

Reactant	*R*	*M*	*Solvent*	*Yield (%)*	*Ref.*
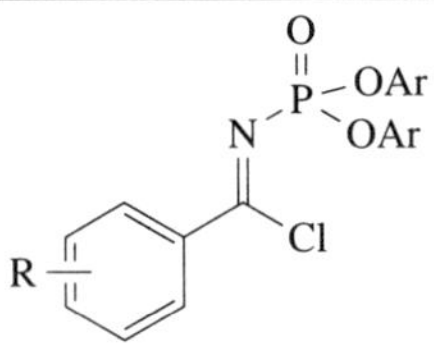	H, 4-Cl, 3-NO_2, 3,5-$(NO_2)_2$	K	Acetone	Almost quantitative	61ZOB2275
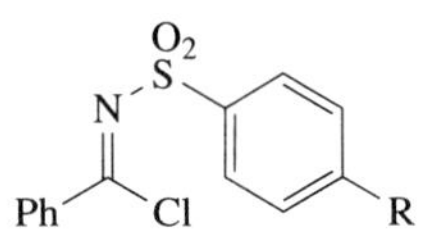	H, NO_2	NH_4	Acetone or THF	86–89	80ZC97
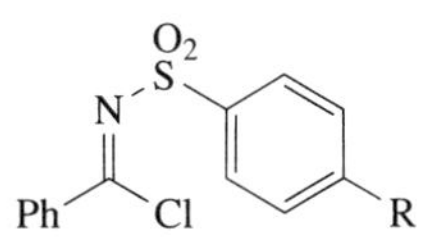	H, Me	Na	Acetonitrile	Not isolated	89ZOR2597
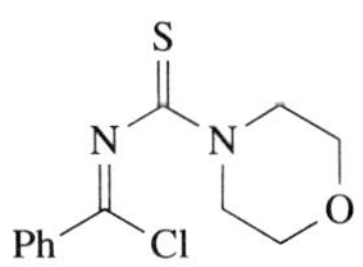		K	Acetone	35	88JPR241

O-acetylglycosyl isoselenocyanates ⟨85T4781⟩. He showed that yields were noticeably increased and reaction rates quadrupled by use of THF rather than chloroform as solvent. Barton and co-workers have reported the reaction of an isocyanide with selenium in the presence of *n*-butyllithium at −80 °C (Equation (55)) ⟨76JCS(P1)2079⟩.

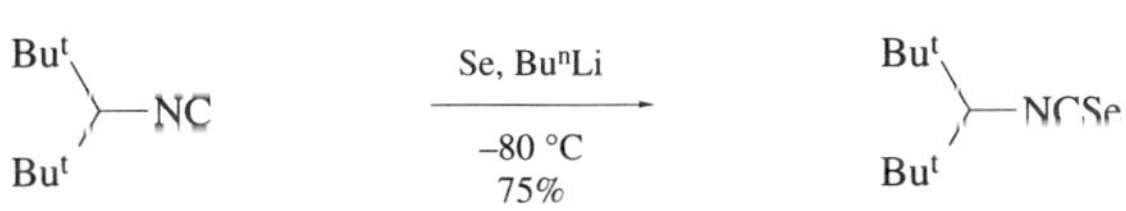

(55)

5.27.2.2.2 *Isoselenocyanates from halides, selenocyanates and alcohols*

Reaction of organic halides with inorganic selenocyanates has been little used as an approach to the synthesis of isoselenocyanates as the strong nucleophilicity of the selenium atom leads predominantly to the formation of the selenocyanate rather than the required isoselenocyanate. Thus, direct reaction of trityl chloride with potassium selenocyanate yields the isoselenocyanate (Equation (56)) ⟨62AC(R)79, 63ACS1459⟩, but reaction of diphenylmethyl bromide gave only the selenocyanate—attempted isomerisation with zinc chloride in refluxing acetone was unsuccessful.

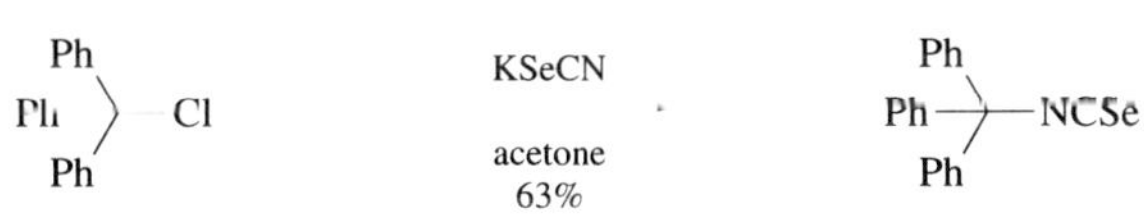

(56)

Uemura *et al.* examined the reaction of a number of primary, secondary and tertiary alkyl halides with mercury(II) and potassium selenocyanates, and again observed that only trityl chloride gave a good yield of the isoselenocyanate ⟨78BCJ1818⟩. Use of mercury(II) selenocyanate does, however, allow the formation of low yields of certain secondary and tertiary isoselenocyanates (isopropyl, 11%; *t*-butyl, 8%).

In a limited number of cases, it has been possible to isomerise selenocyanates to isoselenocyanates

Table 17 Isoselenocyanates from isocyanides.

$$R{-}NC \; + \; Se \longrightarrow R{-}NCSe$$

Reactant	*Conditions*	*Yield* (%)	*Ref.*
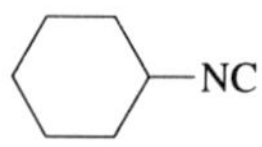	$CHCl_3$, reflux, 16 h	Not given	63ACS1459
R–NC (R = Me, Et, Pr^n, Pr^i)	petrol (40–60 °C) reflux, 1 week	5–80	65TL3003
Bu^t–NC	petrol (60–80 °C)	64	66ACS278
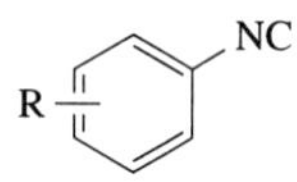 (R = H, 2-Me, 3-Me, 4-Me, 2-MeO, 4-MeO, 4-EtO, 4-Cl, 4-Me_2N)	$CHCl_3$, reflux, 2–3 h	42–87 4–26 (after recryst.)	67CB1367 72CCC3066
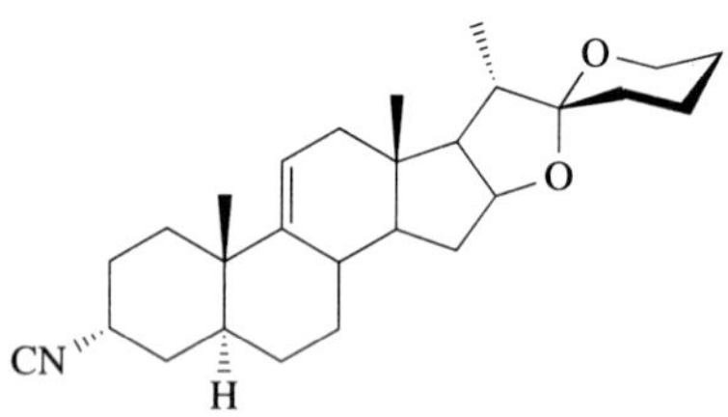	Heptane, reflux, 2 h	82	80JCS(P1)2657
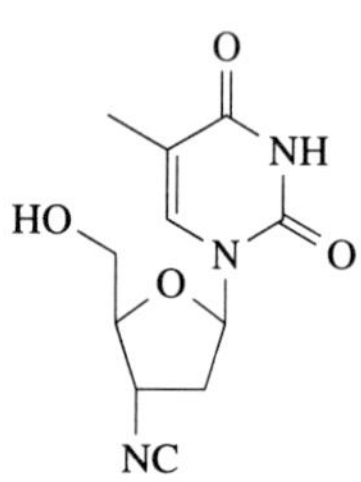	Pyridine, 80 °C, 4 h	48	90MI 527-01
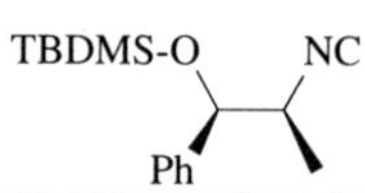	$CHCl_3$, reflux, 5 h	100	91JCS(P1)2495

Table 18 Isoselenocyanates from isocyanides with base catalysis.

$$R{-}NC \; + \; Se \xrightarrow[\text{THF, reflux}]{\text{Triethylamine}} R{-}NCSe$$

R	*Reaction time* (h)	*Yield* (%)
c-Hexyl	4	96
n-Butyl	1.5	83
t-Butyl	6	97

either by heating or irradiation, although this approach does not appear to be general. Heating diphenylmethyl or bis(4-tolyl)methyl selenocyanates in benzene at 120 °C in a sealed tube gave partial conversion to the isoselenocyanates ⟨63AC(R)1113⟩—purification by crystallisation from isopropanol allowed isolation of 24% of the required isomer. Similar thermal isomerisation of allyl selenocyanate gave only a mixture of the isomers on attempted purification by vacuum distillation. Benzyl selenocyanates have been successfully isomerised by irradiation in dry acetic acid with a high-pressure mercury lamp (Equation (57); Ar = Ph, 4-Cl-C_6H_4, 4-Br-2,3,5,6-Me_4-C_6, 3-Cl-2,4,5,6-

Me_4-C_6, Me_5Ph, 2,3,4-Me_3-1-naphthyl) ⟨79S705⟩—unfortunately, the method was not applicable to aliphatic or aryl selenocyanates.

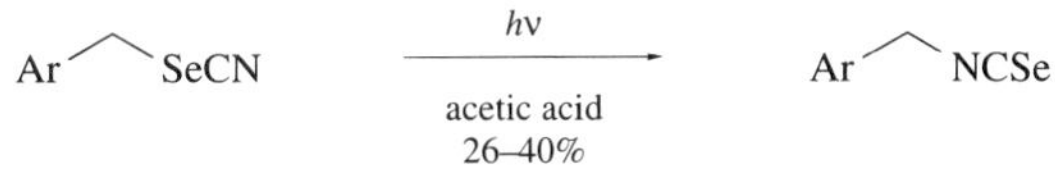

(57)

Direct reaction of alcohols with an organophosphorus selenocyanate ($Ph_3P(SeCN)_2$), corresponding to the thiocyanate analogue employed for isothiocyanation (see Section 5.27.2.1.9), gives exclusively selenocyanates with primary alcohols, while secondary alcohols give a mixture of selenocyanates (**31**) and isoselenocyanates (**32**) (Equation (58); $n = 1$, ratio of **31** to **32** = 66 : 13%; $n = 4$, **31** : **32** = 66 : 12%) ⟨79TL2251⟩. The reagent failed to react with tertiary alcohols.

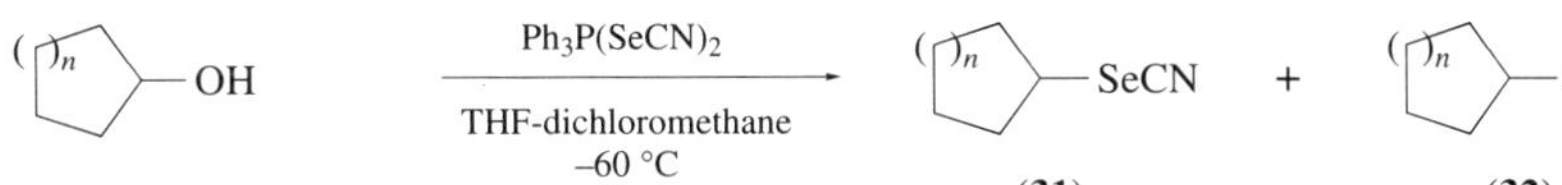

(58)

(**31**) (**32**)

5.27.2.2.3 *Isoselenocyanates from amines and their derivatives*

Despite the obvious comparison with the well-known formation of isothiocyanates from amines *via* their dithiocarbamates (see Section 5.27.2.1.1.(i)), little attention has been paid to the reaction of amines with carbon diselenide as an approach to the synthesis of isoselenocyanates. Henriksen first showed the general utility of this approach in 1973, demonstrating the ready formation of alkylammonium *N*-alkyldiselenocarbamates (Equation (59); R = Me, Et, Pr^i, Bu^i, isopentyl, cyclooctyl, Bn) from primary amines and carbon diselenide ⟨73IJS389⟩. Treatment of these diselenocarbamates with an electrophilic reagent, followed by a weak base, caused elimination to the isoselenocyanate—typical reagents were lead(II) nitrate–sodium bicarbonate, ethyl chloroformate–alumina, or tosyl chloride–sodium carbonate (Table 19, entries 1–3). Alternatively, additional carbon diselenide causes decomposition to the isoselenocyanates.

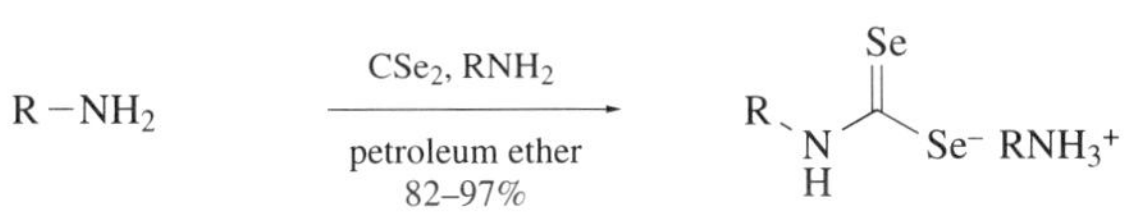

(59)

Table 19 Isoselenocyanates from diselenocarbamates with various electrophiles.

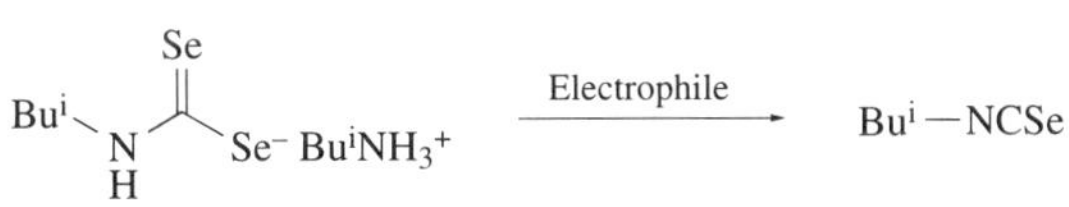

Entry	*Reagent*	*Yield* (%)
1	$PbNO_3$, $NaHCO_3$	70
2	$ClCO_2Et$, Al_2O_3	66
3	TsCl, Na_2CO_3	71
4	H_2O_2, AcOH	66

The pure diselenocarbamates are reasonably stable as solids in dry air, but in the presence of moisture or in solution, rapidly oxidise, giving elemental selenium and the isoselenocyanate. The isoselenocyanate may then go on to react with the liberated amine to yield the selenourea. This side reaction can be prevented by oxidation with hydrogen peroxide in aqueous acetic acid (Table 19, entry 4).

Henriksen and Ehrbar adapted this general approach to give a one-step synthesis of isoselenocyanates ⟨76S519⟩, avoiding the isolation of the intermediate diselenocarbamate and allowing extension of the procedure to *t*-alkylamines and arylamines where the low nucleophilicity of the amine function could otherwise lead to side reactions. Thus, primary amines were reacted with mercury(II) chloride to form a 1:1 adduct ($RNH_2 \cdot HgCl_2$), which then reacted with carbon diselenide in the presence of triethylamine to give the isoselenocyanates in good yield (Equation (60); R = Me, Et, Pr^i, Bu^t, *c*-hexyl, Bn, Ph, 4-MeO-C_6H_4, 2-naphthyl). The reaction was only limited by the presence of electron-withdrawing groups on the aromatic ring when yields were decreased considerably (R = 4-Cl-C_6H_4 (36%), 3-NO_2-C_6H_4 (10%), 4-NO_2-C_6H_4 (<1%)).

$$R{-}NH_2 \xrightarrow[\text{EtN}_3\text{, acetonitrile} \; 47\text{–}73\%]{HgCl_2\text{, }CSe_2} R{-}NCSe \qquad (60)$$

Carbonimidoyl dichlorides may be reacted with alkali metal selenocyanates to give moderate yields of the isoselenocyanate. First reported as an approach in 1886, crude phenyl isoselenocyanate in solution was used immediately for condensation with amines ⟨1886CB2350⟩ although containing unreacted imidoyl dichloride ⟨35JCS1762⟩. Okano and co-workers examined the reaction in more detail and demonstrated the need for two equivalents of the inorganic selenocyanate to remove the cyanogen chloride formed (Scheme 38; R = Ph, *c*-hexyl) ⟨77BCJ722⟩. A modification of Stolte's early method of preparation of phenyl isoselenocyanate has been employed in the synthesis of potential antimalarial agents ⟨81EJM317⟩.

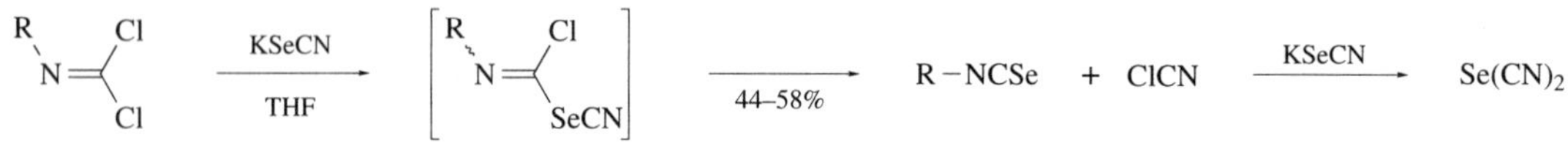

Scheme 38

Insertion reactions of diazo compounds with selenium species have been shown to yield isoselenocyanates ⟨85JOM(286)171⟩. Thus, diazomethane and ethyl diazoacetate react with potassium selenocyanate in the presence of benzeneselenenyl chloride to yield the isoselenocyanates (Equation (61); R = H, CO_2Et).

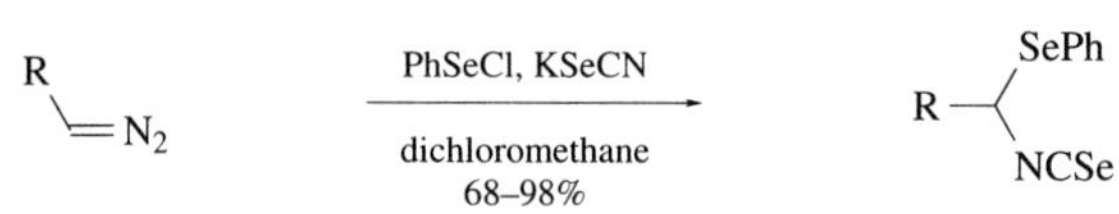

(61)

5.27.2.2.4 *Isoselenocyanates from isocyanates*

By comparison with the formation of isothiocyanates from isocyanates and phosphorus pentasulfide, Collard-Charon and Renson ⟨62BSB531⟩ proposed an analogous approach for the synthesis of isoselenocyanates. They found that heating phenyl isocyanate with phosphorus pentaselenide at 160 °C for 3.5–4 h in the absence of solvent gave an orange-brown oil that on reaction with hydrazine gave a low yield of the phenyl selenosemicarbazide (Scheme 39).

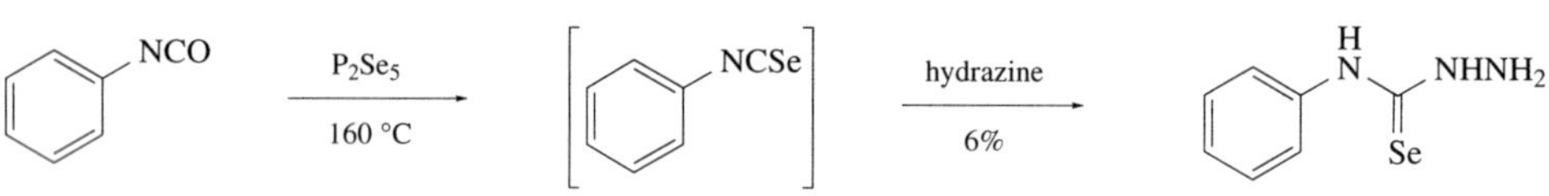

Scheme 39

5.27.2.2.5 *Acyl and imidoyl isoselenocyanates*

Isoselenocyanates of general formula (**33**) are only known where X = O or N—R.

A number of alkanoyl and aroyl isoselenocyanates have been prepared by reaction of the acyl

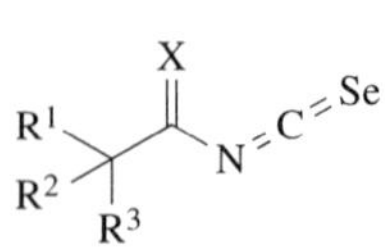

(33)

chloride with potassium selenocyanate, but in all cases so far reported, the product has not been isolated but further reacted in solution immediately. Douglass proposed that a high percentage of the product from this reaction (Equation (62); R = Ph or Me) was the isoselenocyanate rather than the selenocyanate, as the resultant product gave good yields of the selenourea on reaction with amines ⟨37JA740⟩. The identity of the unstable acyl isoselenocyanate was confirmed by IR spectroscopy ($\nu_{NCSe} = 1960\ cm^{-1}$ for R = Me) ⟨75S785⟩.

(62)

The limited range of acyl isoselenocyanates so far prepared is exemplified in Table 20. As stated above, the products were either used without isolation or only the yield of crude product quoted. Arylimidoyl isoselenocyanates have been successfully isolated from the reaction of the corresponding imidoyl chloride with sodium selenocyanate (Equation (63); R = H, Cl) ⟨70UKZ372⟩.

Table 20 Acyl isoselenocyanates from acyl chlorides.

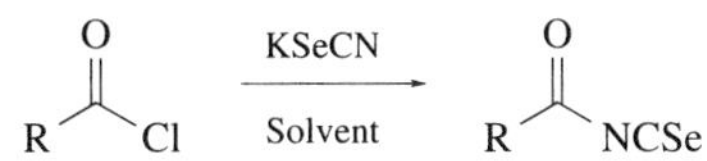

Reactant	*Conditions*	*Yield (%)*	*Ref.*
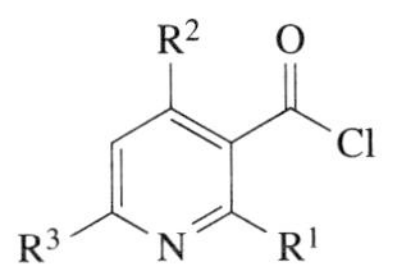 (R^1 = Cl, R^2, R^3 = H or R^1, R^3 = Me, R^2 = Cl)	Acetone, 0–5 °C, 5 min	Not isolated	83CCC3567
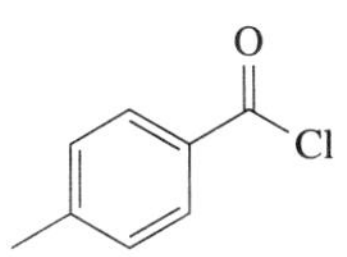	THF, 0 °C, 15 min	Not isolated	84S512
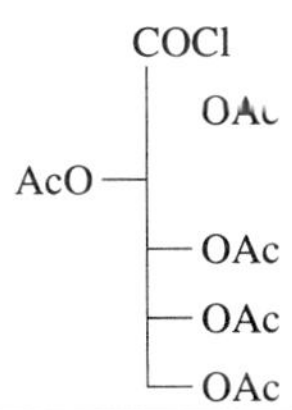	Acetone, RT, 2 h	62	85T4781

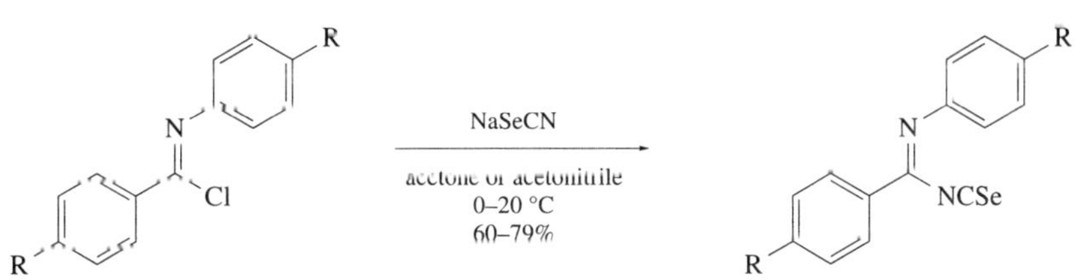

(63)

5.27.2.2.6 *Isotellurocyanates*

The successful synthesis of isotellurocyanates had not been reported up to 1994. Studies using methods employed for the preparation of isoselenocyanates from isocyanides have only given quantitative recovery of unreacted tellurium ⟨59M41, 65TL3003⟩.

5.27.3 PHOSPHORUS, ARSENIC, ANTIMONY OR BISMUTH FUNCTIONS

5.27.3.1 Functions Incorporating Phosphorus

A phosphathioketene, containing trivalent phosphorus, has been proposed as a transient intermediate in the reaction of a substituted phenyl bis(trimethylsilyl)phosphine with thiophosgene ⟨84AG(E)970⟩. This thioketene intermediate dimerises, in the same manner observed with ketenes, in an unsymmetrical [2 + 2]-cycloaddition (Scheme 40)—the structure of the product was confirmed by x-ray crystallography. The intermediacy of the phosphathioketene was also indicated by irradiation of the dimer in the presence of 2,4,6-tri-*t*-butylphenylphosphine (Scheme 41; Ar = 2,4,6-tri-*t*-butylphenyl), which yielded the 1,3-diphosphapropene (**34**), presumably by addition of the phosphane to the thioketene formed by retro-cycloaddition. The product (**34**) was again confirmed by x-ray crystallography.

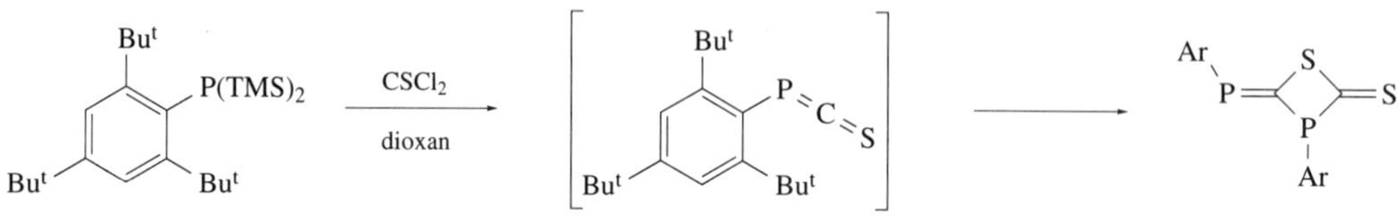

Scheme 40

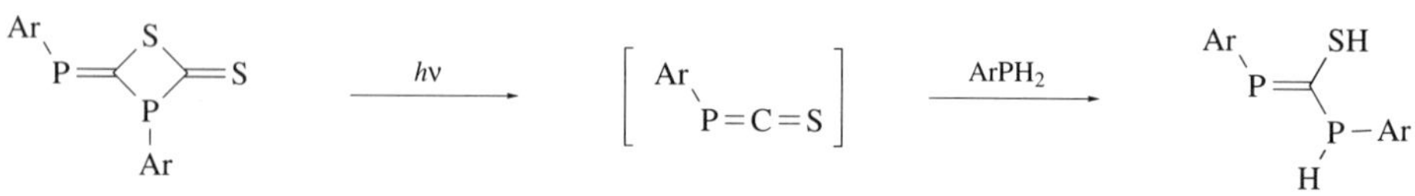

(34)

Scheme 41

A pentavalent thiocarbonyl phosphorane (Scheme 42) has been proposed as the product in the reaction of tetrakis[(phenylsodiophosphino)methyl]methane (**35**) and carbon disulfide at 60 °C ⟨69AG(E)203⟩. This differs from the dithioformate (Equation (64)) formed from alkali metal diphenylphosphide and thiophosgene at −50 °C. The former is unique in its ability to absorb nitrogen in the stoichiometric proportion of 1 : 2 to give the proposed spirocyclic complex (**36**).

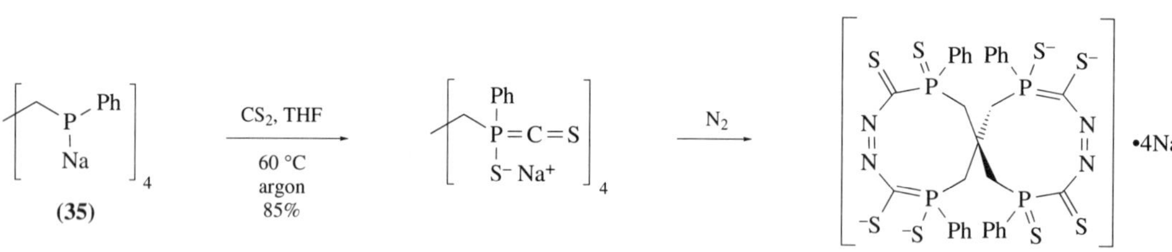

(35) **(36)**

Scheme 42

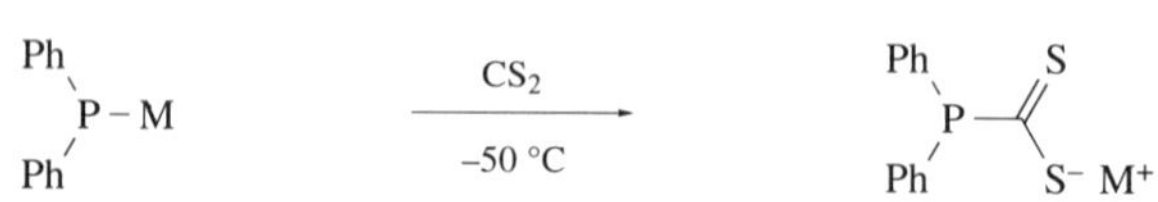 (64)

M = Li, Na, K

5.27.3.2 Functions Incorporating Arsenic, Antimony or Bismuth

Compounds covered in this section would be of the general formula (**37**), where the atom X may be optionally trivalent or pentavalent. A computer search of *Chemical Abstracts* revealed no examples of such structures.

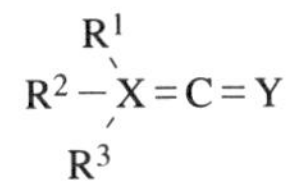

(37)

X = As, Sb or Bi
Y = S, Se, Te

5.27.4 METALLOID FUNCTIONS

Compounds in this section would be of the general structures (**38**) and (**39**). A computer search of *Chemical Abstracts* revealed no examples of such structures.

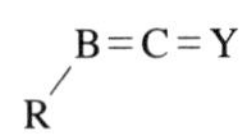

(38)

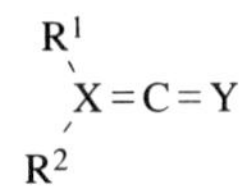

(39)

X = Si or Ge
Y = S, Se, Te

5.27.5 METAL FUNCTIONS

Compounds containing the thiocarbonyl, selenocarbonyl or tellurocarbonyl moiety doubly bonded to a metal (itself bonded to an organic function) were not found in a computer search of *Chemical Abstracts*.

5.28
Functions with at Least One Nitrogen and No Chalcogens

RAMAIAH MUTHYALA
3M Research Center, St. Paul, MN, USA

5.28.1 FUNCTIONS BASED ON NITROGEN: CARBODIIMIDES (RN=C=NR, RN=C=N(O)R etc.)

5.28.1.1 Introduction

Carbodiimides (CDI) are important condensing reagents for the preparation of functional groups, for example amides, anhydrides, amidines, carboxylic acids, sulfonic acids, esters, and are mild oxidizing agents as well. Substituted CDIs are useful intermediates for the preparation of a variety of heterocycles. Most industrial processes for the preparation of CDIs involve the extrusion of carbon dioxide from isocyanates. The synthesis, properties, and reactions of CDIs have been reviewed ⟨67CRV107, B-70MI 528-01, B-71MI 528-01, B-80MI 528-01, 81CRV589, 81AG855, 81T233, 84HOU(4)883⟩.

Carbodiimide (**1**) is isomeric with cyanamide (**2**) (Equation (1)). Many synthetic methods are common for symmetrical and unsymmetrical CDIs. The usual starting materials are ureas, thioureas, isocyanates, and isothiocyanates. Some aromatic and aliphatic CDIs are liquids and can be distilled. However, upon heating or even prolonged standing at room temperature, they decompose or polymerize. In general the stability of CDIs increases with the increase in the degree of branching on the alkyl substituents attached to the nitrogen atom ($RCH_2 < R_2CH < R_3C$).

$$\underset{(\mathbf{1})}{HN{=}\bullet{=}NH} \rightleftharpoons \underset{(\mathbf{2})}{H_2N{-}{\equiv}N} \qquad (1)$$

5.28.1.2 Unsymmetrical Carbodiimides ($R^1N{=}C{=}NR^2$)

Most commonly, unsymmetrical CDIs are prepared from unsymmetrical ureas and thioureas which are generally prepared from the corresponding isocyanates, or isothiocyanates and primary amines.

5.28.1.2.1 Via metal oxides

Desulfurization of thioureas (**3**) with yellow HgO gives stable CDIs (**4**) in good yields (Equation (2)) ⟨89ZOR2196⟩. Water soluble CDIs ⟨61JOC2525⟩, for example (**5**), are also prepared using HgO in 70–90% yield ⟨88LA787⟩.

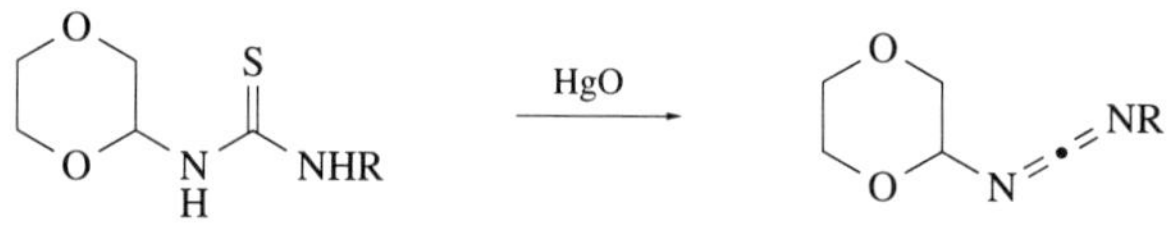

(**3**) (**4**) (2)

R = Pr^i, Ph, 2-pyridyl, CH_2CH_2OH

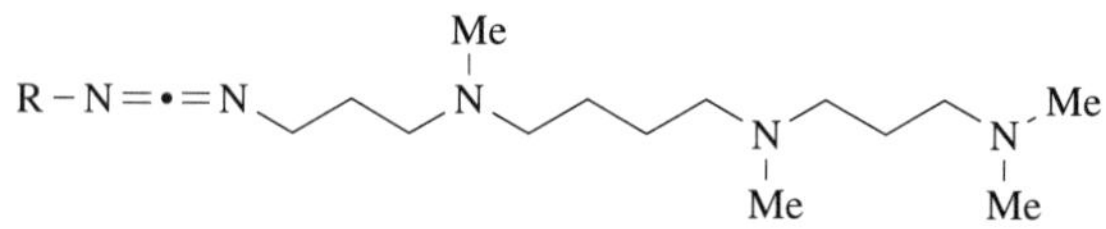

(**5**)

R = ethyl, cyclohexyl, *t*-butyl, 1-adamantyl, phenyl

Good yields of hydrolytically unstable, aryl- or alkylphosphinic or -phosphonic CDIs can be obtained from unsymmetrical thioureas and HgO in the presence of dehydrating agents such as molecular sieves, $CaCl_2$, or Na_2SO_4. This method has the disadvantage in that large quantities of expensive and toxic mercuric oxide and long reaction times are required; and products are generally contaminated with sulfur-containing impurities. Phosgene in the presence of triethylamine (TEA) is also employed to prepare *N*-phosphoryl or phosphonyl-*N′*-alkyl CDIs (**7**) from the corresponding thioureas (**6**) (Equation (3)) ⟨69TL3191⟩.

(**6**) (**7**) (3)

R^1	*R^2*	*Yield (%)*
Ph	Me	71
Ph	Et	89
Ph	Pr^n	85
Ph	Pr^i	91
Ph	C_6H_{11}	92
Ph	Ph	98
OEt	Pr^i	91
OEt	C_6H_{11}	87

5.28.1.2.2 *Via phosphorus reagents*

Dehydration of ureas or thioureas by organic sulfur and phosphorus reagents such as Ph_3P/CCl_4 ⟨71CB1335⟩, $PPh_3 \cdot Br_2$, and *p*-toluenesulfonyl chloride overcome some of the difficulties associated with heavy metal oxides. Phosphorus halide reagents, acyl halides, and sulfonyl halides often give better yields. Much better yields, especially for relatively unstable CDIs, can be obtained by slow addition of ureas to a mixture of $Ph_3P \cdot Br_2$ and TEA in cold dichloromethane or in benzene (Equation (4)) ⟨81S373⟩.

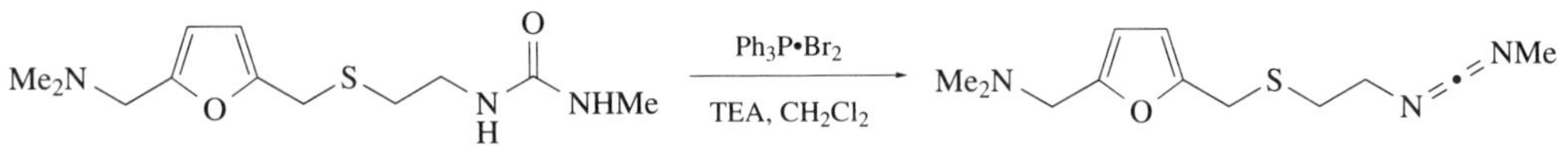

(8) (9) (4)

Some other phosphorus reagents which effect the dehydration of ureas to CDI are $POCl_3$, PCl_5, P_2O_5/pyridine etc. The action of a mixture of $PCl_5/POCl_3$ (93 : 1) on ureas of the structure (**10**) at 140–160 °C gives CDI (**11**) in 88% yield (Equation (5)) ⟨82ZOR668⟩.

(10) (11) (5)

N,N′-Bis(alkylidene)ureas (**12**) react with PCl_5 to give good yields of *N,N′*-bis(1-chloroalkyl)carbodiimides (**13**) (Equation (6)) ⟨79S738, 81ZOR1420⟩

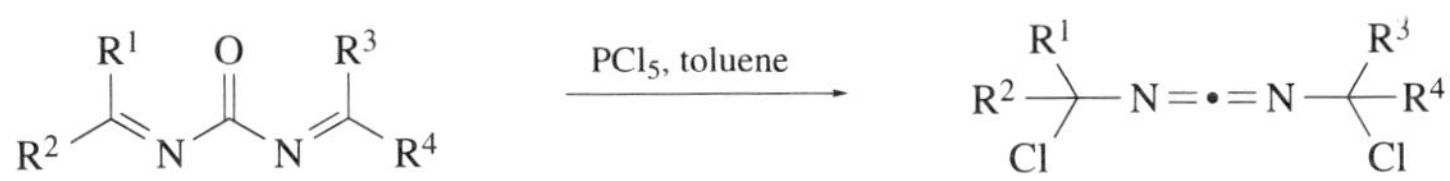

(12) (13) (6)

R^1	R^2	R^3	R^4	*Yield* (%)
Ph	CF_3	CF_3	Ph	85
Ph	CF_3	Bu^t	Bu^t	70
Ph	Bu^t	Bu^t	Ph	70
Ph	Bu^t	Bu^t	Bu^t	70

5.28.1.2.3 *Via acid chlorides and related reagents*

Treatment of alkoxycarbonyl thioureas (**14**) with phosgene in the presence of TEA in benzene at reflux temperature affords the unsymmetrical CDIs (**15**) (Equation (7)) ⟨65TL149, 86JOC1277⟩. Some acyl CDIs rearrange to the *N*-substituted cyanamide at 80 °C.

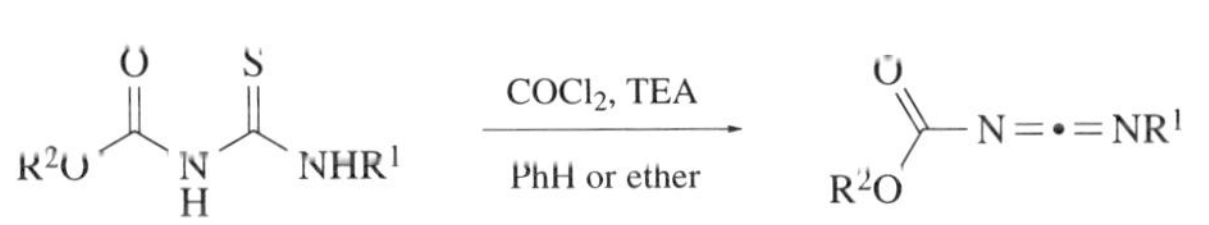

(14) (15) (7)

R^1	R^2
Et	Bn
Et	$4\text{-}MeOC_6H_4CH_2$
Bn	$4\text{-}MeOC_6H_4CH_2$
Et	$(4\text{-}MeOC_6H_4)_2CH$
Et	$c\text{-}C_6H_{11}$
Et	Pr^i

N-(Dialkylamino)CDIs are highly reactive compounds since they contain both nucleophilic and electrophilic centers ⟨72JOC3372⟩. The *N*-(dimethylamino)*N′*-phenylCDI, for example, cannot be isolated. When the phenyl group in (**16**; R = Ph) is replaced with a *t*-butyl substituent, the resulting *N*-*t*-butyl-*N′*-dimethylaminoCDI (**17**) showed enough stability to be characterized (Equation (8)) ⟨89JOC1048⟩.

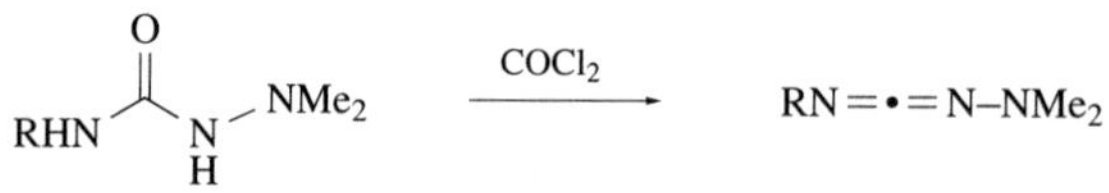

(16) (17)

R = Ph, Bu^t

(8)

N-Sulfonyl CDIs (**18**) when prepared by the reaction of chloramine-T and isonitriles such as cyclohexylisonitrile undergo ready polymerization. They can, however, be conveniently prepared from *N*-sulfonyl-*N*′-substituted thioureas (**18**) and PCl_5 or even better, with phosgene (Equation (9)) ⟨64AG781, 65TL149⟩.

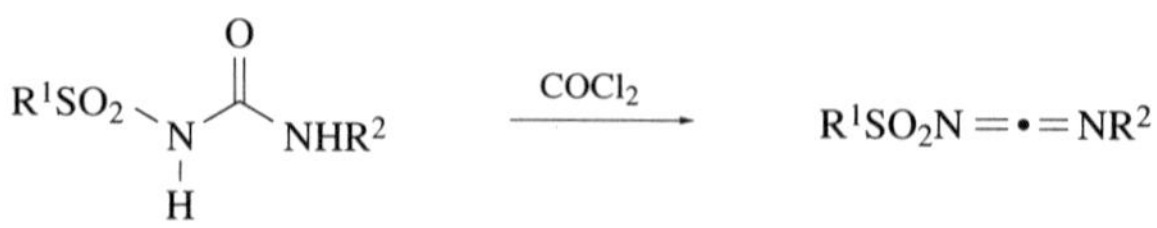

(18) (19)

(9)

R¹	*R²*	*Yield (%)*
Me	Et	68
Me	Pr^n	73
p-MeC_6H_4	Pr^n	47
Ph	Pr^n	40
p-MeC_6H_4	Pr^n	47
p-MeC_6H_4	Ph	12
Me	Ph	18

Cyanuric chloride has also been used as a reagent to prepare unsymmetrical imidoyl-, carbamoyl-, and thiocarbamoylCDIs (**21**) ⟨71TL2765, 77CB2996, 80CB79, 80CB1095, 86CB3737, 88CB271⟩ from the corresponding ureas (Equation (10)) (**20**).

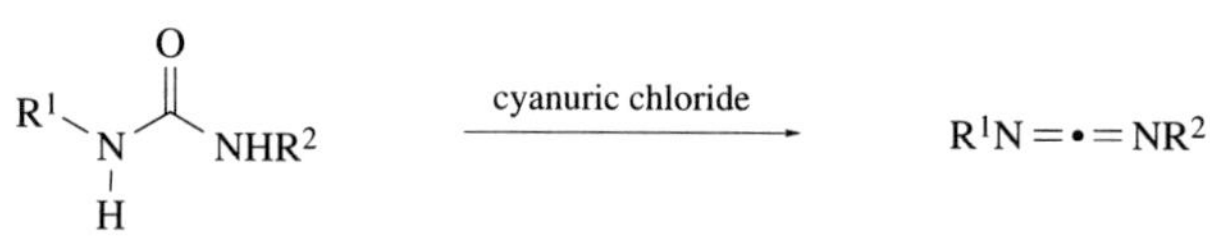

(20) (21)

(10)

R¹	*R²*	*Yield (%)*
Ph(C=N–C_6H_3-2,6-Me_2)	C_6H_{11}	75
p-ClC_6H_4(C=N–C_6H_3-2,6-Me_2)	C_6H_{11}	85
p-$NO_2C_6H_4$(C=N–C_6H_3-2,6-Me_2)	Me	81
(c-C_6H_{11})$_2$N(C=O)	Bu^t	82
(c-C_6H_{11})$_2$N(C=S)	Bu^t	72
(c-C_6H_{11})$_2$N(C=S)	2,5-$Me_2C_6H_3$	78

The carbodiimides of the type (**23**) have been prepared in good yields by the reaction of unsymmetrical thioureas such as (**22**) with 2-chloro-1-methylpyridinium iodide (Equation (11)) ⟨89HCA556⟩. These CDIs are susceptible to base attack on the isoxazole ring and the yield of CDIs is largely dependent on the reaction conditions.

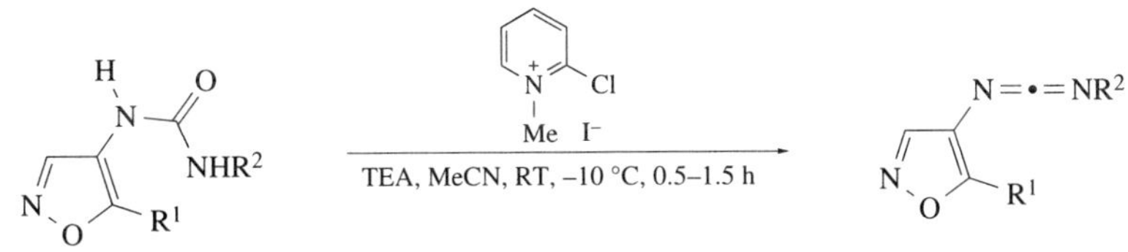

(22) (23) (11)

R^1	R^2	Yield (%)
H	Pr^i	89
H	Bu^t	74
Me	Me	91
Me	Pr	77
Me	Pr^i	89
Me	Bu^t	88
Me	2,6-$Me_2C_6H_3$	93
Me	4-$MeOC_6H_4$	88

Like chloropyridinium salts, chloropyrimidine derivatives are useful as active dehydrating and desulfhydrating reagents for the preparation of CDIs. They are particularly useful when the reagent is attached to polymers (polystyrene) due to the advantage of isolating the product by simple filtration ⟨81MAC3411⟩.

The dehydration of *N,N′*-disubstituted ureas can be effected in high yields with *p*-toluenesulfonyl chloride in pyridine and methylene chloride as solvent. This method has the advantage over desulfurization of thioureas with HgO, and is particularly useful for water soluble CDIs such as (**25**) when conducted under phase transfer catalytic (PTC) conditions using potassium carbonate as the base and a lipophilic quaternary ammonium salt, for example triethylbenzylammonium chloride, as catalyst (Equation (12)) ⟨87S520⟩. When the urea starting materials are poorly or not soluble in the solvents that are commonly used in the PTC technique, the reaction is slow and the yields are usually poor, or zero, in some cases.

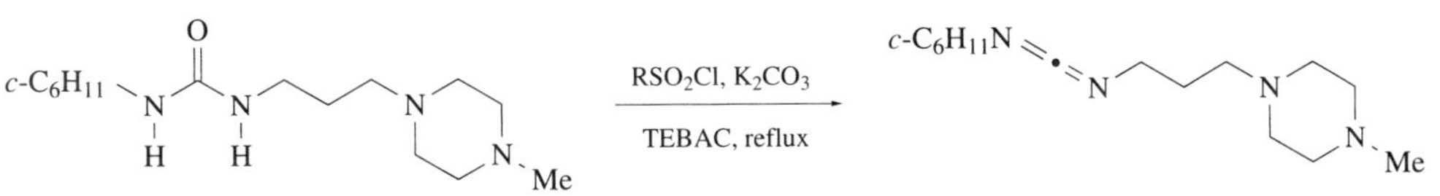

(24) (25) (12)

The conversion of ureas (**26**) (*p*- and *m*-isomers, R = alkyl) into the corresponding polymerizable CDIs (**27**) has been accomplished in low yields with the *p*-toluenesulfonyl chloride and pyridine system, while no CDIs have been obtained with phenyl derivatives (**27**; R = Ph) (Equation (13)) ⟨79BCJ533⟩.

(26) (27) (13)

R = Ph, *c*-C_6H_{11}, Pr^i

5.28.1.2.4 *Via isocyanates and isothiocyanates*

Phosphorus catalyzed formation of symmetrical carbodiimides from isocyanates and isothiocyanates involves the iminophosphorane intermediates (**28**) ⟨78AG761⟩. These intermediates can be prepared *in situ* or can be isolated as stable compounds either by the reaction of primary amines with triphenylphosphine dibromide or by the reaction of organic azides with trivalent phosphorus compounds (Equation (14)) ⟨93JOC5264⟩.

$$RNH_2 + PPh_3Br_2 \longrightarrow \underset{\textbf{(28)}}{R{-}N{=}PPh_3} \longleftarrow R{-}N_3 + PPh_3 \qquad (14)$$

The reactivities of trivalent phosphorus compounds (Ph_3P, $Bu^n{}_3P$, and $P(OEt)_3$) with 1-, or 2-adamantyl azides, and 3-homoadamantyl azides have been compared ⟨87NKK1280⟩. 2-Adamantyl azide reacts 10 times faster than 1-adamantyl azide, whereas the 3-homoadamantyl azide does not react at all. $Bu^n{}_3P$ reacts with 1-adamantyl azides 16 times faster than $P(OEt)_3$, whereas PPh_3 is only 8 times faster.

There are very few methods available for the synthesis of conjugated unsymmetrical CDIs ⟨70TL3453, 80CB1095, 94TL3817⟩. Iminophosphoranes (**29**) have been reacted with aryl or alkyl isocyanates in benzene at room temperature to give *N*-substituted *N′*-(1,2-diarylethenyl)CDIs (**30**) ⟨86CL135, 88CB271, 89CB307⟩. Although these CDIs are stable in air, they cannot be distilled (Equation (15)).

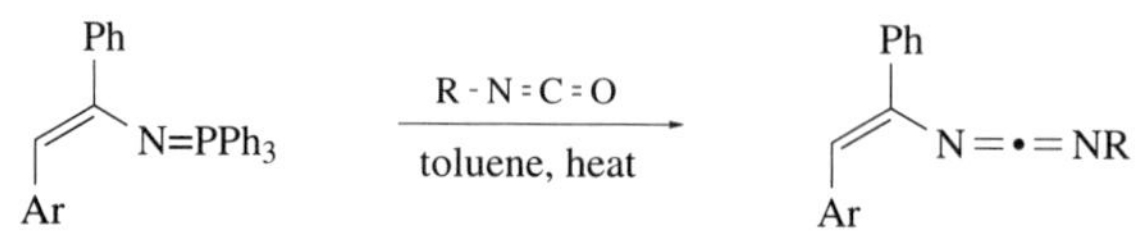

(29) (30) (15)

Ar	*R*	*Yield (%)*
Ph	Ph	80
Ph	*p*-MeC_6H_4	97
Ph	*p*-$MeOC_6H_4$	46
Ph	*p*-C_6H_4	64
Ph	*c*-C_6H_{11}	84
Ph	Me	51
p-$MeOC_6H_4$	Ph	90

CDIs of the type (**31**) are isolable at room temperature but undergo heterocyclic ring formation at high temperatures (refluxing toluene) ⟨86CB2114, 87S45, 88TL379, 88TL4863, 89CB307, 90CC7, 90JCS(P1)1727⟩.

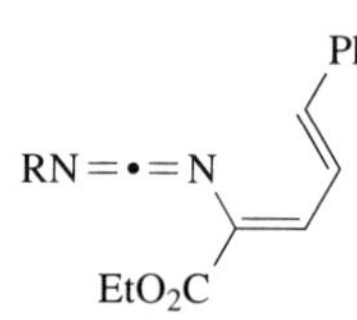

(31)

R = C_6H_5, 4-MeC_6H_4, 4$MeOC_6H_4$, 4-ClC_6H_4, C_2H_5, Bu^t

Conjugated iminophosphoranes of the type (**32**) react with isocyanates in methylene chloride at room temperature giving stable CDIs (**33**) (Equation (16)).

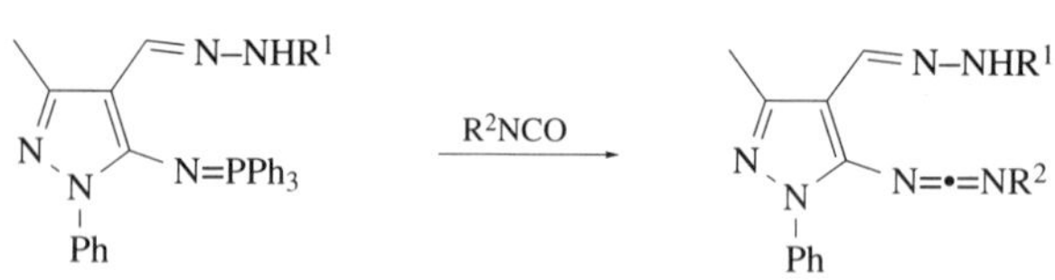

(32) (33) (16)

However, when the —CH═N—NH—R^1 group in structure (**32**) is replaced with —CH═N—R^1, —CH═O, ⟨88JOC4654⟩, or with —CH═CH—R^1 (except when $R^1 = Bu^t$) ⟨88TL379, 89CB307⟩ the derived heterocyclic CDIs undergo further cyclization even at room temperature and none of the corresponding CDIs have been isolated. Similarly, (**34**) and (**35**) undergo reactions with aryl isocyanates in refluxing toluene to give the corresponding CDIs (**36**) and (**37**) respectively (Scheme 1) ⟨87S45⟩. In only one case has the CDI (**36**; X = S, Ar = Ph) been isolated; in all cases these CDIs cyclize to give thieno- and furopyrimidines under the reaction conditions.

Iminophosphoranes (**38**) derived from *N*-substituted *o*-azidobenzamides react with aryl isocyanates giving highly reactive CDIs, and only CDI (**39**; $R^1 = R^2$ = Me, Ar = *p*-Me-Ph) is isolable in high yields (Equation (17)) ⟨89T4263⟩.

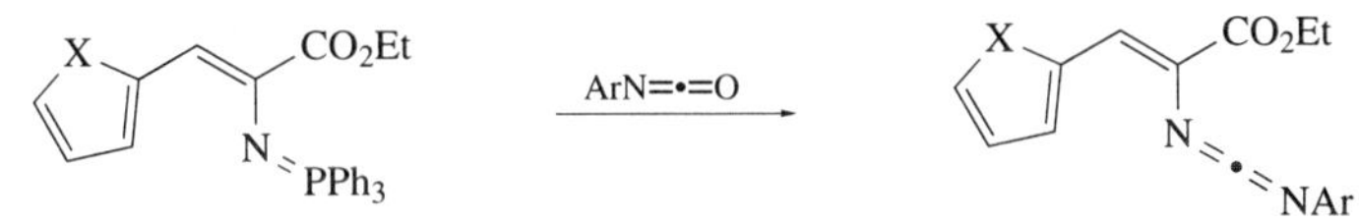

(34) **(36)**

X = O, S

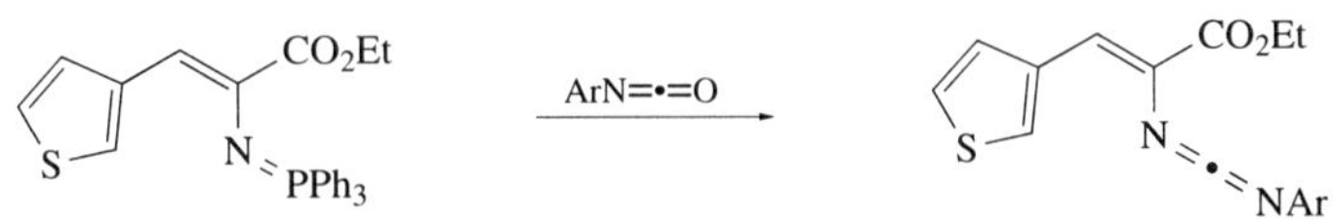

(35) **(37)**

Scheme 1

(38) **(39)** (17)

Iminophosphorane (**40**) with aromatic or aliphatic iso(thio)cyanates at room temperature gives CDIs (**41**) which can be isolated as viscous liquids, whereas iminophosphorane (**42**) with an isothiocyanate give CDIs (**43**) which are very reactive intermediates and cannot be isolated. When the pyrrole ring is not conjugated to the aromatic phosphinimine function, for example (**44**), the reaction with aryl isocyanates give very stable CDIs (**45**) (Scheme 2) ⟨89TL2847, 90T1063⟩.

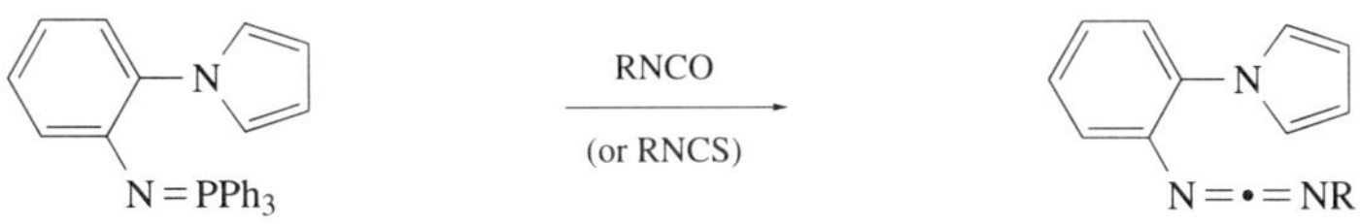

(40) **(41)**

R = Pr^i, 4-MeC_6H_4, Ph

(42) **(43)**

(44) **(45)**

Ar = 4-MeC_6H_4, 4-$MeOC_6H_4$

Scheme 2

Although several methods for the preparation of iminophosphoranes are available, one carbon homologation of iminophosphoranes has value. Reaction of benzotriazoloiminophosphorane (**46**) with Grignard reagents at room temperature or at 0 °C gives one carbon homologated iminophosphoranes, which in principle can be isolated but can react *in situ* with alkyl or aryl isocyanates to give good yields (65–85%) of unsymmetrical CDIs (**47**) (Equation (18)) ⟨90S565⟩.

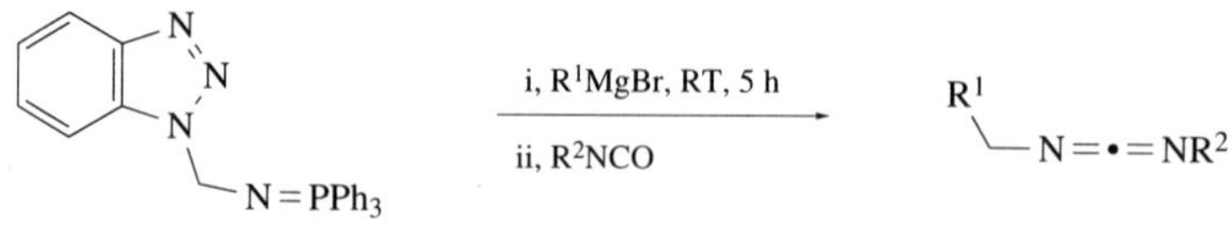

(46) (47) (18)

R^1	R^2	Yield (%)
Me	Bu^t	61
Me	Ph	62
Ph	Bu^t	65
Ph	1-naphthyl	80

The iminophosphorane (**48**) reacts with alkyl isocyanates to give the heteroarylcarbodiimide (**49**) when R = Bu^t, whereas when R = Me, ethyl, or *n*-propyl, the CDIs (**49**) react with the starting iminophosphoranes to give predominantly [2 + 2]-adducts followed by ring chain tautomerism resulting in betaine formation (Equation (19)) ⟨69CB1848, 79ZC59, 89JA355⟩.

(48) (49) (19)

The reaction of 2′-hydroxy or TMS derivatives of *β*-ribofuranose phosphorus ylides (**50**) with alkyl isocyanates give predominantly CDIs (**51**) along with trace amounts of the corresponding glycosyl isocyanates (Equation (20)) ⟨86M(117)1437⟩.

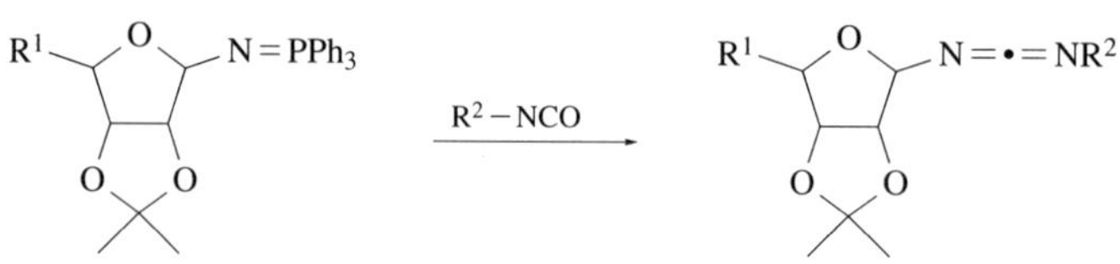

(50) (51) (20)

R^1 = CH_2OH or CH_2O-TMS

R^2	Yield (%)
Me	65
Pr^i	81
Bu^t	64
$H_2C{=}CH{-}CH_2$	44
$ClCH_2CH_2$	60

The *α*-chloroalkylcarbodiimides (**53**) are readily formed in the reaction of *α*-chloroalkyl isocyanates (**52**) with iminophosphoranes in benzene (Equation (21)) ⟨76ZOR2472, 89ZOR759⟩. These carbodiimides can be distilled under vacuum without decomposition. On the other hand, *α*,*α*-dichloroalkyl isocyanates (**54**) react with iminophosphoranes under even milder conditions (−20 °C) ⟨89ZOR1572⟩, but the product distribution depends on the *N*-substitution of iminophosphorane ⟨84ZOR2543⟩. For example, in the case of isopropyl- or phenyl-substituted iminophosphorane, the isomeric halogen substituted diazadiene (**56**) is formed whereas in the case of the *t*-butyl substituent, the CDI (**55**; R = Bu^t) is formed (Scheme 3).

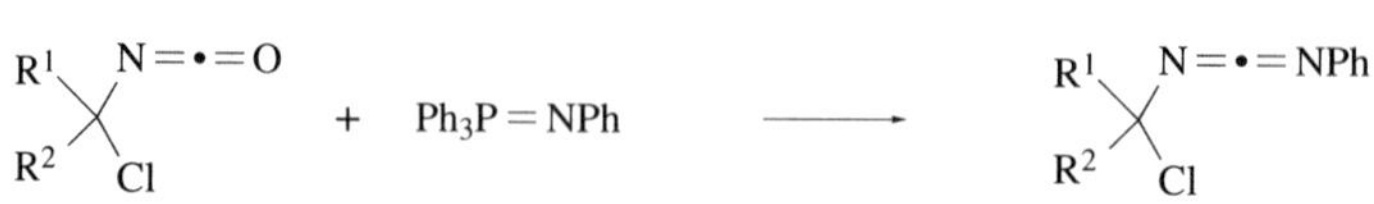

(52) (53) (21)

R^1	R^2	Yield (%)
H	CCl_3	45
CF_3	Ph	70
CF_3	*p*-MeOPh	60

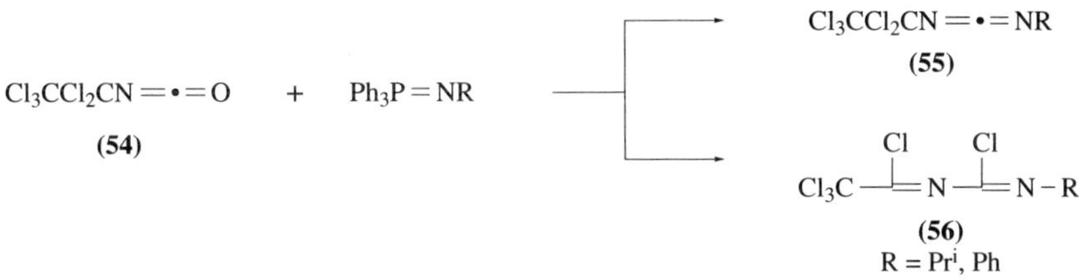

Scheme 3

This type of [1,3]-chlorotropic shift is observed for α-chloroalkyl isocyanates ⟨80S85⟩ in which the accumulation of chlorine atoms at the α-carbon tends to enhance the rearrangement. For CDI (**57**), no such equilibrium is observed between 25 °C and 125 °C. These transformations are not typical of the perfluoroalkyl CDIs. Thus the reaction of (**58**) with iminophosphoranes (Equation (22)) takes place under mild conditions (−10 °C) in methylene chloride ⟨84ZOR2543, 88ZOR986⟩ to produce (**59**).

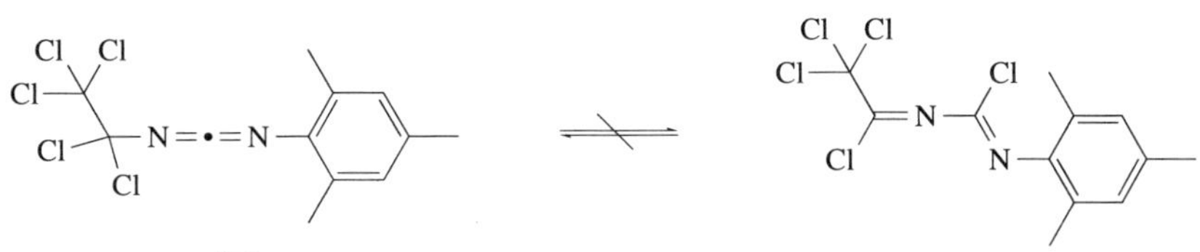

(57)

$$Cl_3CCF_2N{=}\bullet{=}O + Ph_3P{=}NR \longrightarrow Cl_3CCF_2N{=}\bullet{=}NR \qquad (22)$$

(58) **(59)**

R¹	*Yield (%)*
Pr^i	40
Ph	
2,4,6-$Me_3C_6H_2$	60

The 1-chloroalkyl isocyanates (**52**) are useful compounds ⟨66AG(E)704⟩ for the preparation of a variety of functionalized alkyl isocyanates which in turn are starting materials for CDIs (**60**) (Scheme 4) ⟨89ZOR759⟩. These CDIs are viscous liquids and do not distil under vacuum.

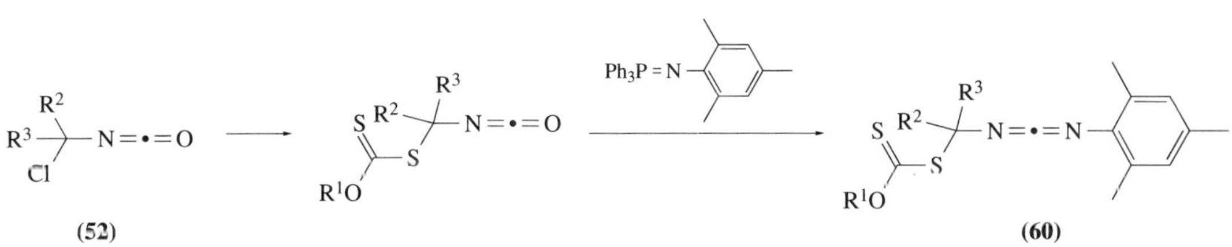

R¹	*R²*	*R³*	*Yield (%)*
Et	CF_3	Ph	38
Bu	CCl_3	H	43

Scheme 4

Like iminophosphoranes, phosphoramidate anions react with carbonyl compounds. Imidate anions are generally prepared with sodium hydride ⟨64JOC2816⟩ and can be reacted with isocyanates to give CDIs. This procedure works reasonably well with aromatic isocyanates but is not particularly good for aliphatic isocyanates. However, a more general procedure is the reaction of certain phosphoramidates (**61**) with aliphatic isocyanates in boiling xylene in the presence of solid potassium carbonate to give high yields of CDIs (**62**) (Equation (23)) ⟨88S397⟩. This procedure is also convenient for water soluble CDIs.

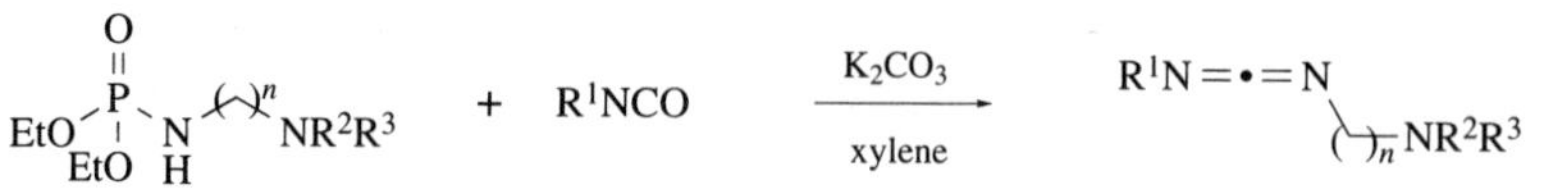

(61) (62)

(23)

R^1	R^2	R^3	n	Yield (%)
c-C_6H_{11}	Me	Me	2	90
c-C_6H_{11}	Me	Me	3	89
Bu^n	Me	Me	3	75
c-C_6H_{11}	$-CH_2CH_2OCH_2CH_2-$		2	82
c-C_6H_{11}	$-CH_2CH_2N(Me)CH_2CH_2-$		3	80
c-C_6H_{11}	$-(CH_2)_5-$		3	84
c-C_6H_{11}	$-(CH_2)_4-$		3	86

In this reaction, the intermediates such as (**63**) can be isolated and can be decomposed by heat or strong base to CDI. In solution, the thermal decomposition (160 °C) of intermediate (**63**) is considerably accelerated with strong bases such as sodium hydride ⟨87IZV1829⟩.

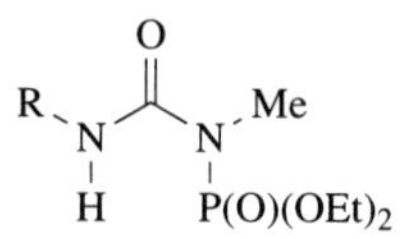

(**63**)

5.28.1.2.5 *Via imidoyl dichlorides*

N-Arylimidoyl dichlorides are formed by the reaction of aryl isocyanates and PCl_5 (see Section 6.20.1). It should be noted that these compounds are also known as carbonimidic dichlorides or iminocarbonyl dichlorides. These dichlorides are stable and can be isolated. However, with or without isolation these can be reacted with another mole of isocyanate (same or different) to give symmetrical or unsymmetrical CDIs. The reaction generally occurs between −20 °C and 0 °C. In many cases, the yields are good to excellent ⟨63JCS5558⟩. In general, isothiocyanates are better reagents than isocyanates ⟨87AG(E)359⟩ for the formation of imidoyl dichlorides.

Aliphatic primary amines react with *N*-arylimidoyl dichlorides (**64**) to give either N^1-alkyl-N^2-aryl CDIs (**65**) or guanidines, whereas with aromatic amines they form guanidines preferentially. Nevertheless, with aromatic amines, the CDI can be obtained by using less basic hydrochlorides of the amines ⟨69AG(E)20⟩ or bis(trimethylsilyl) amines ⟨70OCS23⟩ rather than amines themselves (Scheme 5). In the preparation of (**66**) using lithium hexamethyldisilazide, the elimination of TMS-Cl occurs at about 100 °C ⟨69TL1421⟩. The reaction of phenylisocyanide dichloride and disodium cyanamide in acetonitrile at room temperature for 30 h gives *N*-cyano-*N′*-phenylCDI (**67**) in low yield (Scheme 5) ⟨85ZC305⟩.

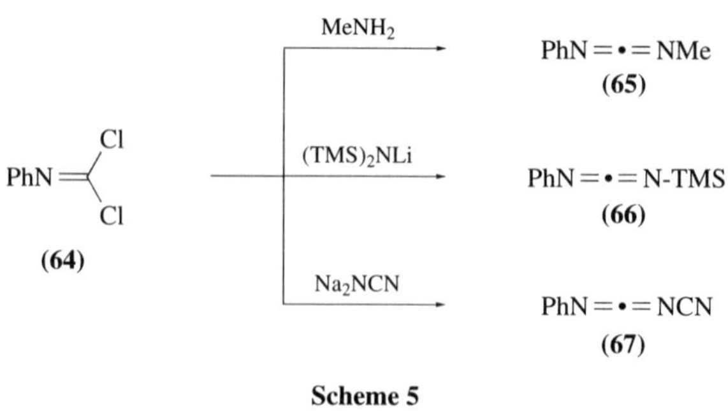

Scheme 5

The pentafluorophenylcarbonimidoyl dichloride (**68**) reacts with 2 to 3 molar excess of aliphatic amines to give greater than 70% yield of the CDIs (**69**) (Equation (24)) ⟨88JFC(40)217, 89ZOR1523⟩. The reactivities of (**68**) and its unfluorinated analogue with primary amines and aromatic amines are extremely similar ⟨89ZOR1689⟩.

(68) (69)

R = Bu^n, Bu^t, Ph (24)

Unlike carboimidoyl dichlorides, *N*-ethoxycarbonylimidoyl dichloride (**70**) reacts under very mild conditions to give CDIs (**71**) without formation of guanidines (Equation (25)) ⟨87ZOR2023⟩.

(70) (71) (25)

R	*Yield* (%)
Pr^i	54
Bu^i	50
Bu^t	60
c-C_6H_{11}	52

N-Dichloromethylenesulfonamides (**72**) which can be prepared by the chlorination of *N*-bis(alkylmercaptomethylene)benzosulfonamide, give CDIs (**73**) when treated with amine hydrochlorides (Equation (26)) ⟨65AG(E)430, 66CB1252⟩.

(72) (73) (26)

R	*Yield* (%)
c-C_6H_{11}	46
Bu^n	51
Pr^i	59
Pr^n	39
Bu^t	33
Bu^i	54

5.28.1.2.6 Via nitrilimine and isonitrile reagents

Nitrilimines are unstable intermediates and in general cannot be isolated. However, certain nitrilimines, for example (**75**), can be generated from the precursors (**74**) and a variety of electrophiles, and rearrange to CDIs (**76**) upon irradiation (Scheme 6) in good yields ⟨91JOC1801⟩.

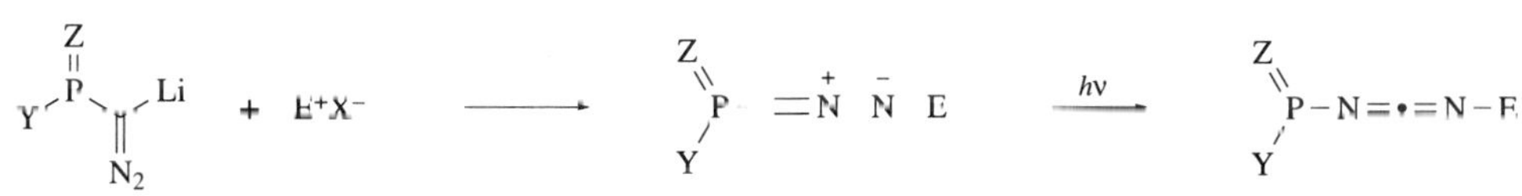

(74) (75) (76)

Y = $(Pr^i_2N)_2$, Me, Ph, Pr^i; Z = S, O; X E = $ClSiPh_3$, $ClSiPr^i_3$

$(Pr^i_2N)_2P-N{=}\bullet{=}N-SiPr^i_3$ 72%

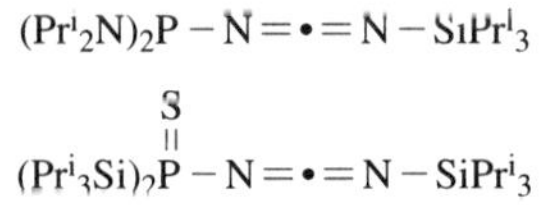

$(Pr^i_3Si)_2P(=S)-N{=}\bullet{=}N-SiPr^i_3$ 10%

Scheme 6

The metal carbonyl catalyzed reaction of an organic azide with isocyanide provides simple unsymmetrical CDIs ⟨70JOC3995⟩. The photochemical addition of nitrenes (generated from azides) to isonitriles leads to CDIs (Equation (27)) ⟨82TL1585⟩. The reaction is usually complete in 24–36 h.

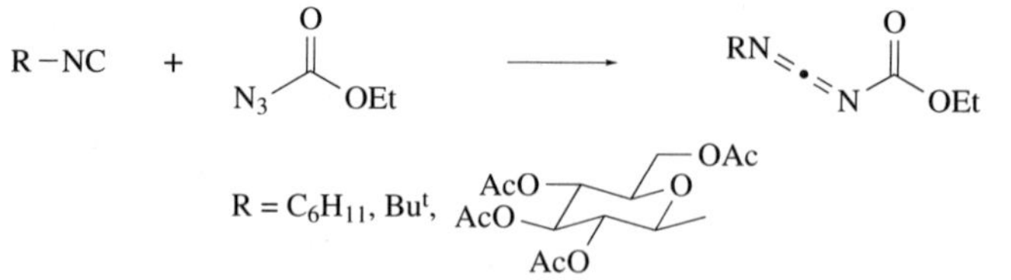

(27)

N,*N*′-Disubstituted diaminocarbene Pd(II) complexes which can readily be prepared from primary amines and Pd(II)Cl_2(R—NC)$_2$, are oxidized with Ag_2O to give symmetrical or unsymmetrical CDIs (Scheme 7). The isolation of carbene Pd(II) complex **(77)** is not necessary. In practice, the synthesis of CDI is accomplished by stirring a heterogeneous mixture of primary amine, isonitrile, Ag_2O and a catalytic amount of Pd(II)Cl_2 at room temperature or above, preferably in the presence of a dehydrating agent such as molecular sieves or sodium sulfate ⟨75JOC2981⟩.

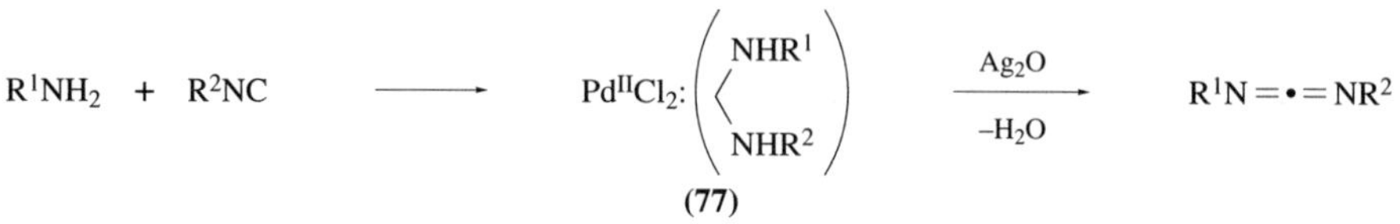

R¹	*R²*	*Yield (%)*
Bu^n	Bu^t	93
c-C_6H_{11}	Bu^t	94
Ph	Bu^t	80
H_2C=CH–CH_2	Bu^t	77
CH(Me)CO_2Et	Bu^t	80

Scheme 7

Alternatively, when aryl- or alkylamines are treated, for example with *t*-pentyl isocyanide in the presence of 10 mol% of $NiCl_2$ and 4A molecular sieves, while bubbling oxygen through the reaction mixture, the coupling reaction proceeds to give CDIs in good yields (Equation (28)) ⟨93JOC5118⟩. Air can also be used in place of oxygen. Although O_2 is a more effective oxidant for this procedure, the use of stoichiometric amounts of Ag_2O or HgO afford good yields of CDIs. The disadvantage with this method is that it needs 3–4 equiv. of isonitrile due to ready polymerization of isonitrile during the reaction. Although bidentate phosphines such as dppp and dppe inactivate the Ni(II), the Ph_3P does not appear to effect the yield.

$$R^1NH_2 + R^2NC \xrightarrow[\text{dehydrating agent}]{NiCl_2\ (10\ mol\%),\ O_2\ \text{or air}} R^1N{=}\bullet{=}NR^2 \quad (28)$$

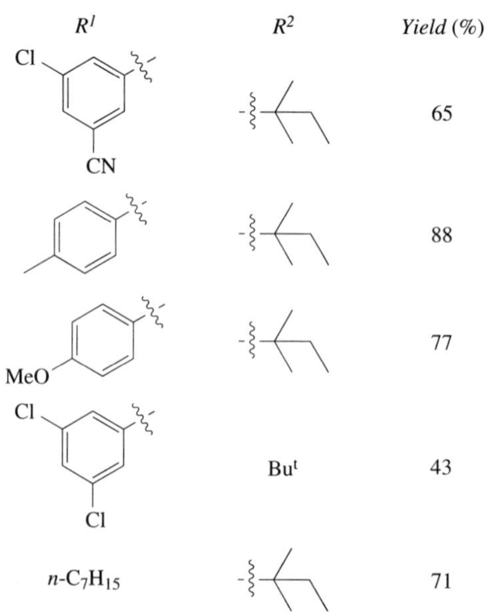

This method is applicable to the synthesis of aromatic and aliphatic, unsaturated, *N,N'*-disubstituted CDIs. In particular, aniline derivatives bearing electron donating groups provide CDIs in good yields.

5.28.1.2.7 Via heterocycles

Certain classes of heterocyclic compounds yield CDIs under thermal or photochemical conditions. Using this methodology, potentially unstable and highly reactive bifunctional CDIs are accessible due to easily removable by-products during the fragmentation. For example, treatment of 1,4-disubstituted 5-unsubstituted tetrazolium salts (**78**) with 1 equiv. of TEA in a suitable solvent, gives CDIs (Equation (29)). These have been isolated in solution and identified by spectroscopic data ⟨70TL3453⟩. Although acyl carbodiimides could not be prepared through tetrazolium salts, the oxadiazolium salt (**79**) with TEA gives unstable benzoyl CDI ⟨70TL3453⟩.

(29)

(**78**)

R¹	*R²*
$H_2C{=}CH$	Me
Et_2N	Me
EtO_2CCH_2	Me

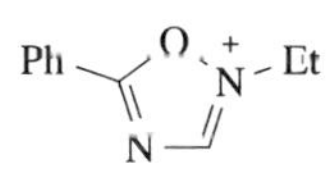

(**79**)

Thermolysis of other heterocycles ⟨65JOC3369, 77JOC3372, 85JCS(P1)1471, 89T259⟩, for example (**80**)–(**82**), also give CDIs but in much reduced yield and at much higher temperatures.

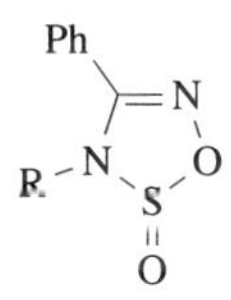

(**80**)

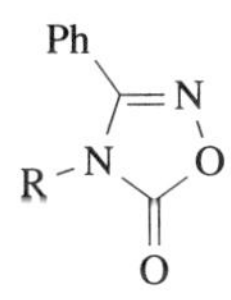

(**81**)

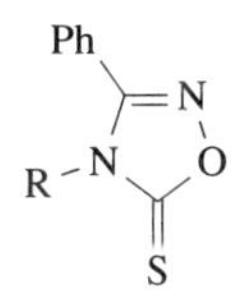

(**82**)

N-Aryl-1-aziridinecarboimidoyl chloride (**83**), as a neat liquid, at 40–60 °C or in solution in the presence of catalytic amounts of a Bronsted acid, rearranges to CDIs (**84**) (Equation (30)). The acid-catalyzed method appears to be superior to the thermal procedure ⟨71JOC2142⟩.

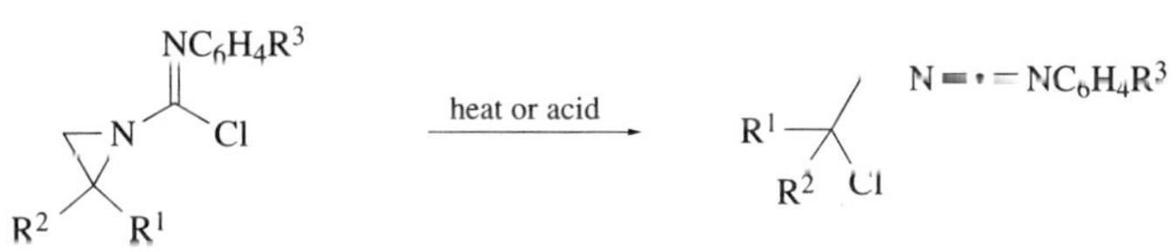

(30)

(**83**) (**84**)

R¹	*R²*	*R³*
H	H	H
H	H	*p*-ClPh
H	H	*m*-NO_2Ph
Me	Me	*m*-NO_2Ph

Iminothiadiazolines (**85**), or 1,2,4-oxathiazoline (**86**), upon reaction with a mixture of triphenylphosphine and tri-*n*-butylphosphine, give good yields of imidoyl CDIs (**87**) (Scheme 8) ⟨86CB3737⟩.

Similarly, tetrazolothiones (**88**) also give CDIs upon irradiation, whereas the corresponding tetrazolones (**89**) give diaziridinone (Equation (31)) ⟨75AG(E)428, 85CB526⟩.

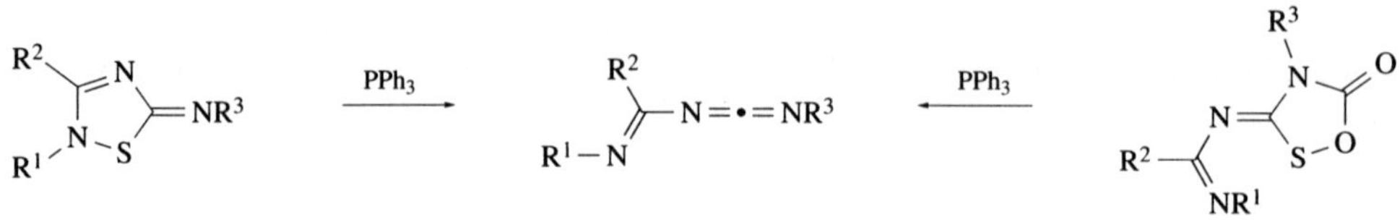

(85) **(87)** **(86)**

R¹	*R²*	*R³*
Me	Ph	Bu^t
Me	Ph	$2,6\text{-}Me_2C_6H_3$
Bn	Ph	Me
Bn	Ph	Et
Bn	Ph	Bu^t
Bn	Ph	Bu^i
Bn	Ph	Ph
Bn	Ph	$2,6\text{-}Me_2C_6H_3$

Scheme 8

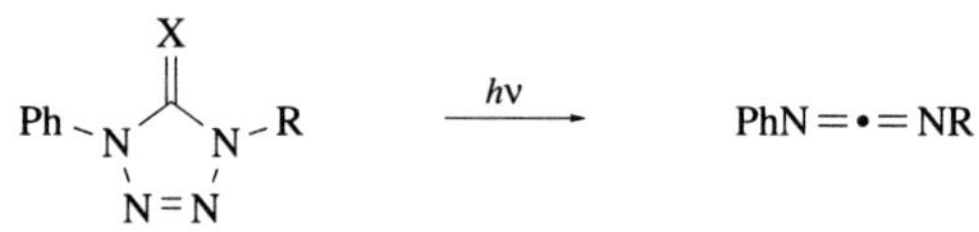

(88) X = S
(89) X = O

(31)

Under thermolytic conditions, the heterocycle **(90)** (R is bulky, e.g. cyclohexyl, *t*-butyl, and adamantyl) completely dissociates to CDIs, but partial isomerization takes place with simple alkyl groups ⟨88JOC1744⟩ with no CDI formation.

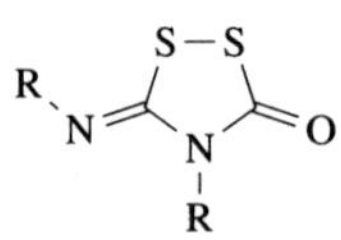

(90)

5.28.1.2.8 Miscellaneous

Alkoxy- or aryloxyphosphoryl chlorides react with $Ag_2(CN)_2$ very rapidly in DME or in THF at room temperature or below to give *N*-cyano *N'*-(alkoxy or aryloxy)phosphoryl CDIs **(91)** (Equation (32)). However, if 2 equiv. of alkoxy (or aryloxy)phosphoryl chlorides are employed, dicyanamides give bis(organophosphoryl)CDIs ⟨84ZC379, 89ZAAC(568)35, 89ZAAC(577)165, 89ZC145⟩.

(91)

(32)

M = Na, K, Ag, Ca; R = Me, Et, Pr^i, 2-MePh, 4-ClPh

Bis(trimethylsilyl)carbodiimide (see Section 5.28.1.3.1) is a useful precursor to other CDIs ⟨78ZN(B)950⟩. POF_3 reacts with TMS-N=C=N-TMS to give difluorophosphoryl trimethylsilyl CDI **(92)** (Equation (33)) ⟨67ZN(B)785⟩.

(92)

(33)

The Si—Cl bond is substantially more polar in chloro(chloromethyl)dimethylsilanes than in chlorotrialkylsilanes, and hence the reaction of bis(trimethylsilyl)CDI with **(93)** occurs readily giving mono- and bis(trimethylsilyl)CDIs (**94**) and (**95**) respectively by partial or complete *trans*-silation (Scheme 9) ⟨85ZOB1308⟩. Reaction of bis(trimethylsilyl)CDI with (**96**) gives CDI (**97**) in 68% yield (Equation (34)) ⟨70CB1664⟩.

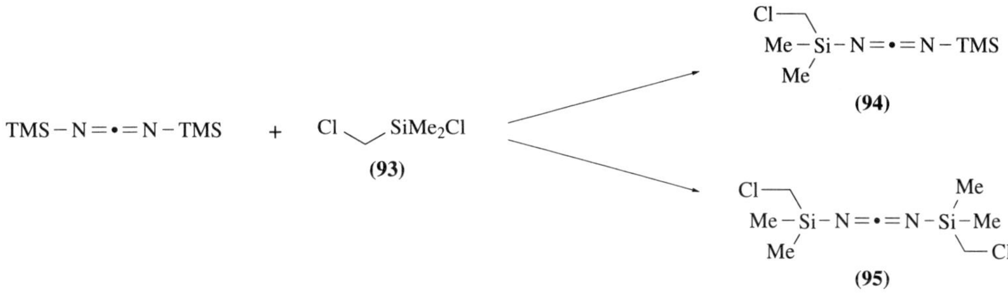

Scheme 9

 (34)

Unsymmetrical dichlorodiazadienes such as (**56**) are very reactive compounds towards suitable nucleophiles. In the presence of a base, for example TEA, they give a variety of CDIs in excellent yields (Scheme 10) ⟨87ZOR2385⟩. Aliphatic primary amines react similarly with (**56**) at 0°C, in the presence of TEA to give the corresponding imidoyl CDIs (**98**). While the aromatic imidoyl CDIs (**99**) are relatively stable, these aliphatic CDIs (**98**) are not very stable and undergo sigmatropic rearrangement to give 1,2-dihydrotriazines ⟨87ZOR2390⟩.

The 1-chloroalkylcarbodiimides (**100**) react with a variety of carboxylic acids at room temperature

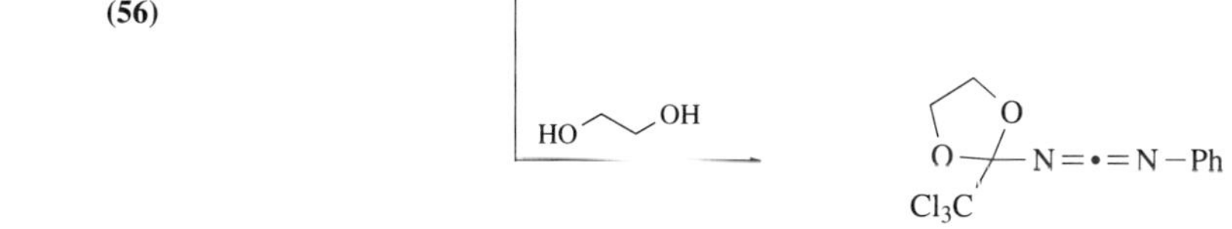

Scheme 10

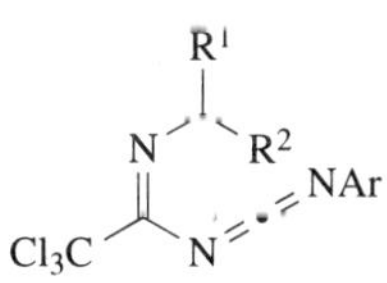

(98)

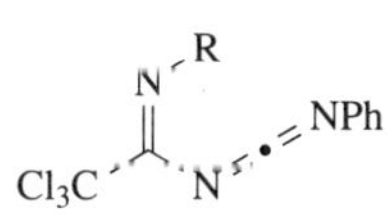

(99)

Ar	*R¹*	*R²*	*Yield (%)*
Ph	Me	Me	80
2,4,6-$Me_3C_6H_2$	Me	Me	96
Ph	$-(CH_2)_5-$		78
Ph	Ph	H	68

R	*Yield (%)*
Ph	81
Pr^i	40
3,4,5-$Me_3C_6H_2$	86

in benzene in the presence of TEA to form *N*-(1-acyloxyalkyl)-*N*[1]-aryl carbodiimides (**101**) (Equation (35)) ⟨86ZOR1099⟩.

(35)

(100) **(101)**

Ar	*R*	*Yield* (%)
Ph	4-BrC_6H_4	62
Ph	Me	55
2,4,6-$Me_3C_6H_2$	Me	71
2,4,6-$Me_3C_6H_2$	Et	75

Perfluoro-2-azapropene (**102**) reacts with primary aliphatic amines to give trifluoroalkyl-2-aza-propenylamines (**104**) which upon treatment with BF_3·etherate give the corresponding *N*′-trifluoromethyl-*N*-alkylCDIs (**105**). Alternatively, the reaction of (**102**) with α-aminoalkylphosphonates (**103**) occurs in the presence of KF at −10 to −15°C to give CDIs (**105**) directly (Scheme 11) ⟨87IZV1177⟩.

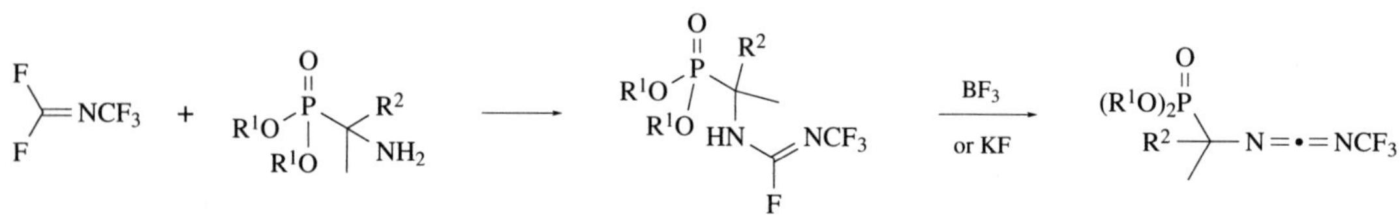

(102) **(103)** **(104)** **(105)**

R[1]	*R*[2]	*Yield* (%)
Me	Et	58
Et	Et	60
Et	Pr	40
Et	Bu	50
Bu[t]	Et	72

Scheme 11

By analogy to acyl azides, compound (**106**) undergoes Curtius rearrangement to give symmetrical and unsymmetrical CDIs (**107**) (Equation (36)) ⟨85IZV700⟩.

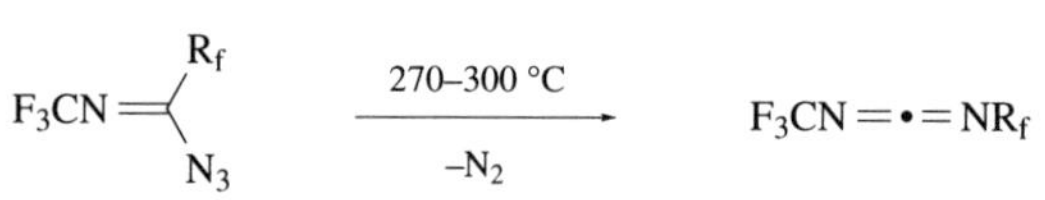

(36)

(106) **(107)**

5.28.1.3 Symmetrical Carbodiimides

5.28.1.3.1 Carbodiimides of the type RN═C═NR, R = SiR′₃

Bis(trimethylsilyl) carbodiimide (**108**) is an important member of CDIs and can be prepared by a variety of methods (Scheme 12) ⟨67CB3368, 78JOM(148)225, 78ZN(B)575, 78ZOB1943, 85ZOB621, 87JOM(327)303, 89AG(E)1250⟩. Many of these reactions take place under either forcing conditions or require unstable starting materials. However, (**108**) has been synthesized in 90% yield by the reaction of TMS-CN with cyanamide (Equation (37)) ⟨87JOC275⟩.

$$\text{TMS–CN} + H_2NCN \xrightarrow{\text{1 min}} \text{TMS–N}{=}\bullet{=}\text{N–TMS} \qquad (37)$$

(108)

Bis(di- and trisilanyl)CDIs and oligomeric CDIs have been synthesized by the reaction of silver cyanamide and the corresponding chlorodi- and trisilanes. Pentamethylchlorodisilane and 2-chloro-

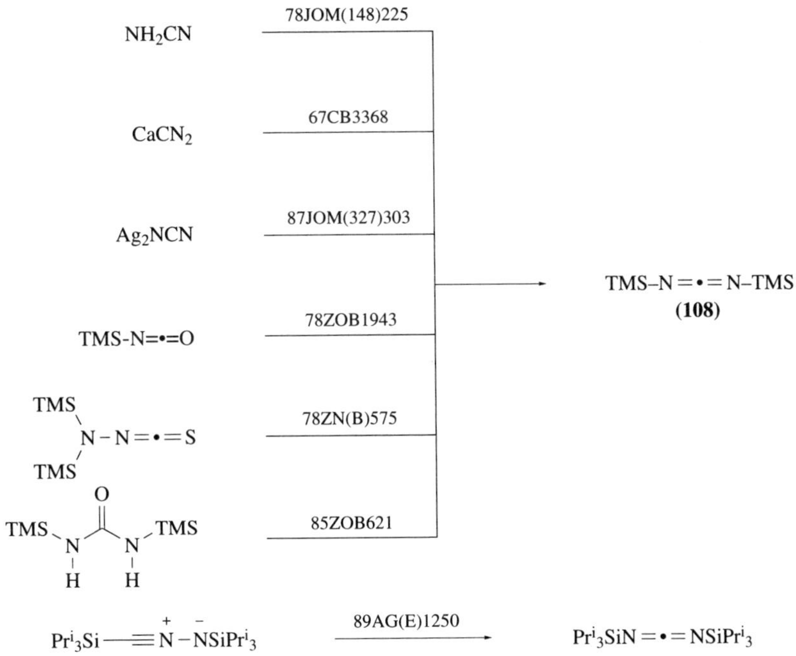

Scheme 12

heptamethyltrisilane interact with Ag_2NCN more rapidly than trimethylchlorosilane (Equation (38)) ⟨85IZV2824, 87JOM(327)303⟩.

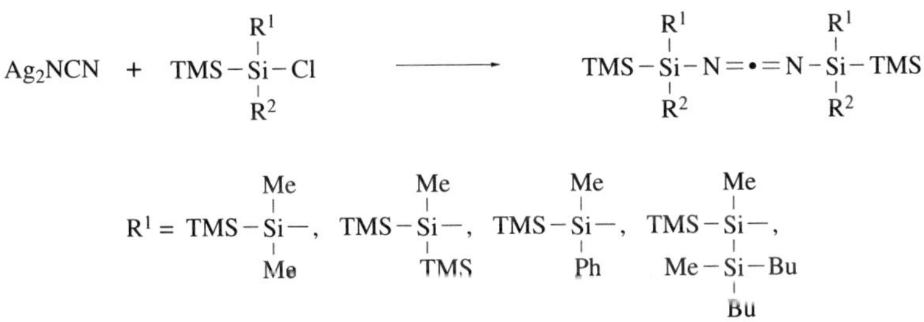

(38)

The oligomeric cyclic silyl carbodiimides (**109**) have been synthesized in moderate yields by the reaction of the respective dichlorosilanes and cyanamides (Equation (39)) ⟨93CB2569⟩.

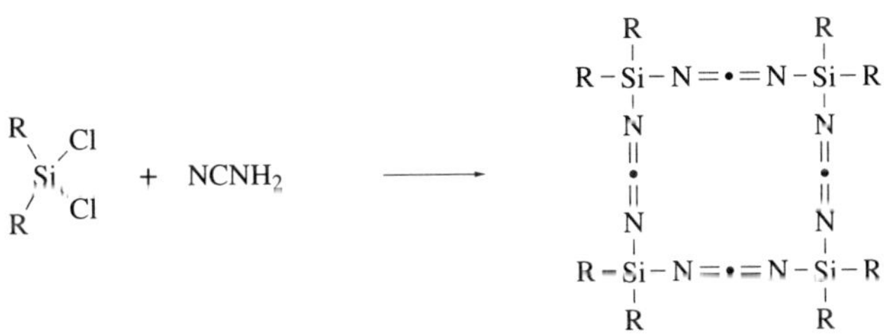

(39)

(109)

R = H, Me, $H_2C{=}CH$

5.28.1.3.2 Carbodiimides of the type RN=C=NR, R = alky or aryl

Dehydration of *N,N′*-disubstituted ureas gives disubstituted CDIs (see Section 5.28.1.1). The dehydrating agents are P_2O_5, $COCl_2$, metal carbonyls such as $Fe(CO)_5$, $Fe(CO)_9$, $Fe(CO)_4$, $W(CO)_6$, and $Mo(CO)_6$ etc. When ureas have no other reactive groups, yields are generally good with these reagents. For better results, other milder reagents, for example $Ph_3P/Br_2/TEA$, arenesulfonyl chlorides, or constant current electrolysis of PPh_3/CH_2Cl_2 ⟨68LA(718)24, 85CPB373, 85S509⟩ can also be used for the preparation of CDIs from ureas. Phosphorus pentoxide in pyridine is a particularly good reagent for simple CDIs ⟨67JOC2895⟩; $Ph_2P(S)Cl$ and $Ph_2P(O)Cl$ react with $Ag_2(CN)_2$ to give symmetrical phosphorus substituted CDIs (Equation (40)) ⟨68M2498⟩. The *N*,*N*′-bis(diphenylphosphine)CDI is stable only as its AgX complex ⟨89ZC145⟩.

X = O, S (40)

For the preparation of simple CDIs from disubstituted thioureas, metal oxides are often used as desulfurizing agents (see Section 5.28.1.1). The most frequently used metal oxide is yellow mercuric oxide but various other oxides or salts or complexes of other metals, for example PbO, As_2O_3, ZnO, $ZnCl_2$, and others have also been used. The reaction is generally carried out in a variety of solvents, for example ether, benzene, acetone, toluene, xylene, or CS_2. In many cases a suitable dehydrating agent such as $CaCl_2$, $MgSO_4$, or Na_2SO_4 is also used. Using HgO, water soluble CDIs of the type (**110**) have been prepared in good yields ⟨87MI 528-01⟩.

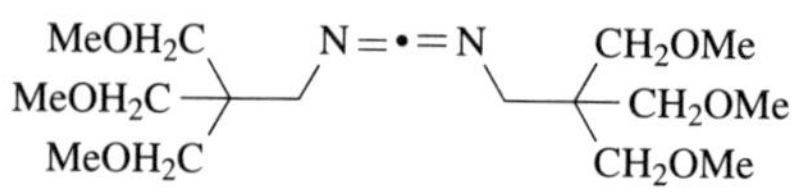

(110)

Other oxidizing agents such as *p*-methoxybenzenetellurinic acid anhydride ⟨86TL6099⟩ or tellurimides, for example $Ar_2Te{=}NSO_2C_6H_4CH_3$ ⟨88ZOR1581⟩ convert thioureas into CDIs in reasonable yields under mild conditions.

Cyanuric chloride can also be used as both dehydrating and desulfurizing agent for ureas and thioureas respectively (see Section 5.28.1.2.3). Similar to cyanuric chloride, 2,4-dichloro-, 2,4,6-trichloropyrimidine, and 2,4-dichloro-5-nitropyrimidine ⟨85SC171⟩ are also effective reagents.

Di-2-pyridyl sulfite (**111**) ⟨86TL1925⟩, and 2,2′-carbonyldipyridine (**112**) ⟨86JOC2613⟩ and di-2-pyridyl thiocarbonate (**113**) ⟨85TL1661⟩ have successfully been utilized for CDI formation from ureas and thioureas. Although these reagents work under neutral conditions and give high yields of CDIs, they have the common disadvantage that *N*,*N*′-primary alkyl disubstituted thioureas cannot be converted into CDIs.

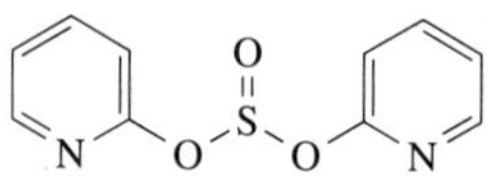

(111)

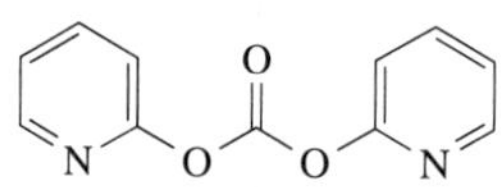

(112)

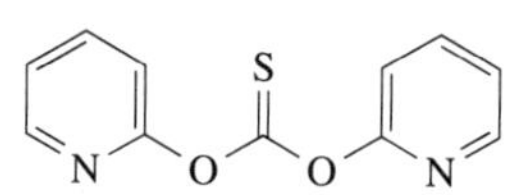

(113)

Hypochlorite oxidizes thioureas into CDIs. When the hypochlorite oxidation is conducted under phase transfer conditions, the reaction is complete in a few minutes in good yields ⟨85IJ219⟩.

Aromatic and aliphatic CDIs are prepared in good yields by treating *N*,*N*′-disubstituted thioureas with 2-chloro-1-methylpyridinium salt in the presence of TEA ⟨77CL575⟩ (see Section 5.28.1.2.3).

The Mitsunobu reaction (Ph_3P, azodicarboxylate) has also been used for the dethiolation of thioureas ⟨69TL2473, 70T5731⟩. Yields are generally in the range of 60–80%. Polymer supported alkyl azodicarboxylates have also been used in this reaction ⟨89JA3973⟩.

Thermal decomposition (170–200 °C, under vacuum) of *N*,*N*′-dilithio- or bis(bromomagnesio)thioureas which have bulky groups give dialkyl or diaryl CDIs in 33–61% yield ⟨76CL811⟩. The decomposition of *N*,*N*′-dilithiothioureas has been accelerated by the addition of carbon disulfide ⟨76CL811⟩ or sulfur dioxide ⟨77S889⟩; and under these conditions the CDIs have been formed below room temperature in 67–87% yield.

The preparation of CDIs from isocyanates or isothiocyanates has been reviewed ⟨B-71MI 528-01, B-80MI 528-01, 81AG855⟩. Generally these substrates give symmetrical CDIs (see Section 5.28.1.2.4). The condensation/decarboxylation of alkenyl isocyanates give good yields of *N*,*N*′-bis(1-alkenyl)CDIs (Equation (41)). The reaction is carried out in boiling toluene in the presence of catalytic amounts of 3-methyl-1-phenyl-3-phospholine oxide (**114**). In spite of the high degree of unsaturation, these CDIs are thermally stable and distillable under vacuum, without decomposition ⟨81ZOB2452⟩.

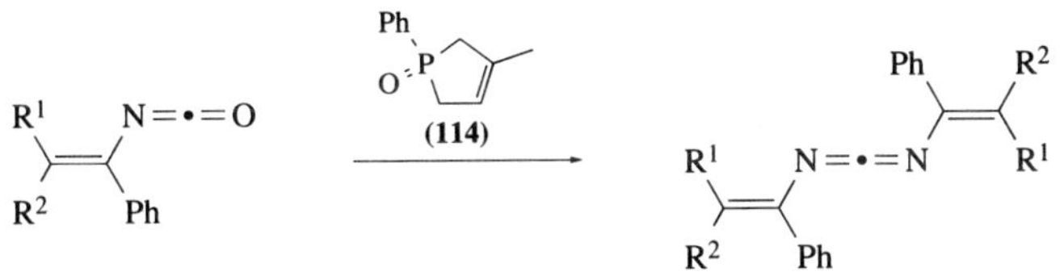

(41)

R^1	R^2	*Yield* (%)
Me	Me	80
Et	H	80
Pr	H	82

Other catalysts that have been used to obtain carbodiimides from isocyanates in high yields under very mild conditions are **(115)**–**(119)**.

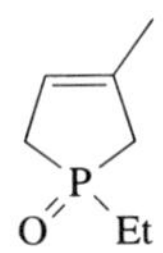

(115)

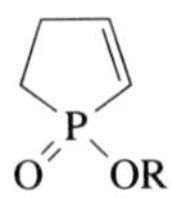

(116)

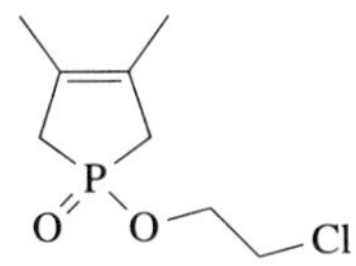

(117)

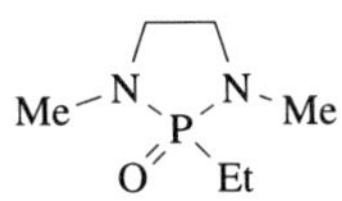

(118)

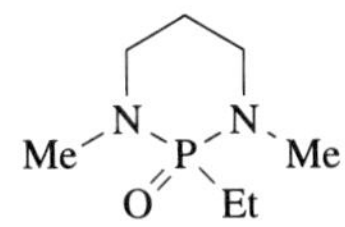

(119)

In the catalytic cycle of CDI formation from isocyanates, vanadium oxo and imido complexes are isoelectronic and isostructural with the phosphorus complexes, and therefore effect the same conversion (Equation (42)) ⟨93OM1023⟩.

$$Ph{-}N{=}\bullet{=}O \quad + \quad (MeC_6H_4N)V(OBu^t)_3 \longrightarrow PhN{=}\bullet{=}NPh \qquad (42)$$

Other catalysts used are simple phosphine oxides, alkoxides of titanium and zirconium, and triarylarsenes, and various metal carbonyls ⟨67TL1731⟩. In the case of aromatic isocyanates, the nature of the isocyanates has great influence on the formation of CDIs. The presence of an electron-releasing group on the aromatic ring tends to inhibit the reaction while the presence of an electron-withdrawing group tends to increase the reaction rate.

N,N′-Disubstituted ureas are converted into the corresponding CDIs at room temperature by treating with 1 equiv. of TEA or Py, and 0.5 equiv. of a reagent such as **(120)**–**(122)** in dichloromethane or benzene. The resulting crude materials are almost pure even though some CDIs prepared by this method are thermally unstable ⟨81BSF(2)361⟩.

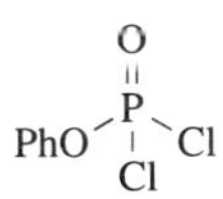

(120)

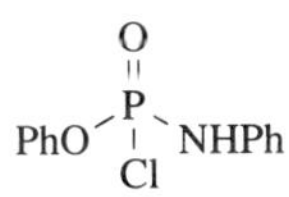

(121)

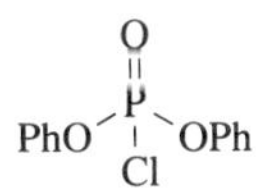

(122)

Iminophosphoranes react with CO_2 or CS_2 to give isocyanates or isothiocyanates respectively, which in turn, under appropriate conditions, can be treated with another mole of a different iminophosphorane to yield unsymmetrical CDIs (see Section 5.28.1.2.4). Sometimes, there is a tendency for an abnormal aza-Wittig reaction to occur to give symmetrical carbodiimides (Scheme 13) ⟨64AG(E)228, 82SC573, 87CAR(166)101, 90T1063⟩.

5.28.1.3.3 *Miscellaneous*

Thermolysis of trimethylsilyl *N*-trimethylsilyl-*N*-trimethylsiloxy carbamate (**123**) in liquid phase gives a very low yield of bis(trimethylsiloxy)CDI (**124**). The structure has been confirmed by mass spectrometry (Equation (43)) ⟨86ZOB1525⟩.

(43)

(123) **(124)**

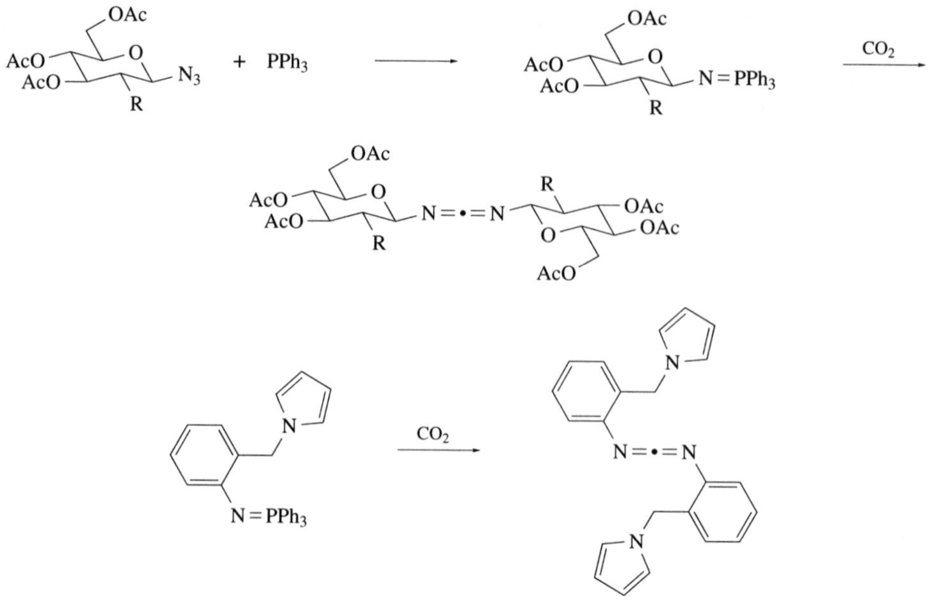

Scheme 13

At room temperature, in the presence of caesium fluoride, hexafluoro-2-4-diazapenta-1,4-diene (**125**) isomerizes rapidly to bis(trifluoromethyl)CDI (**126**) (Equation (44)) ⟨67CC59, 67JA5007⟩.

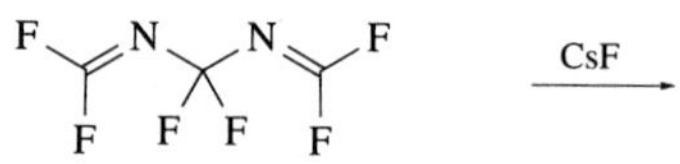

$F_3CN{=}\bullet{=}NCF_3$ (44)

(125) **(126)**

The alkenes (**127**) and (**128**), when treated with 4 equiv. of cyanamide and 1.1 equiv. of NBS in methylene chloride for three days, at room temperature, produces a small amount of CDIs (**129**) and (**130**) respectively (Scheme 14) ⟨85JA2931⟩.

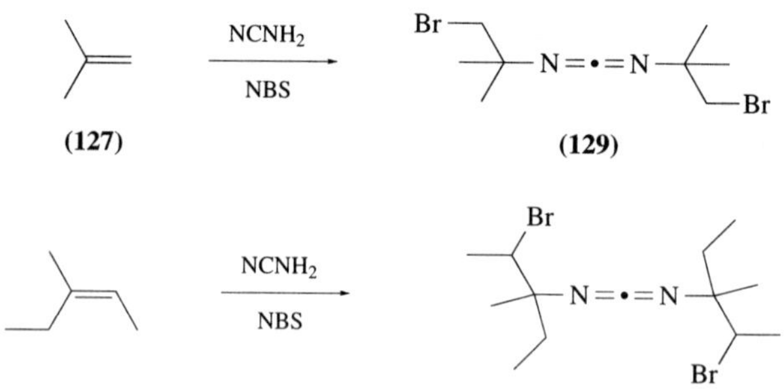

Scheme 14

Anionotropic transformations to the corresponding diazadienes are uncharacteristic of perfluoroalkyl carbodiimides (see Section 5.28.1.2.4). As a result, compound (**131**) (Scheme 15) does not rearrange but is extremely reactive towards nucleophiles. Even though (**131**) contains several reactive sites at which nucleophilic attack can occur, reaction with proton-containing nucleophilic reagents takes place at the α-carbon and leads to the functionality with 1,1-disubstituted CDIs (Scheme 15).

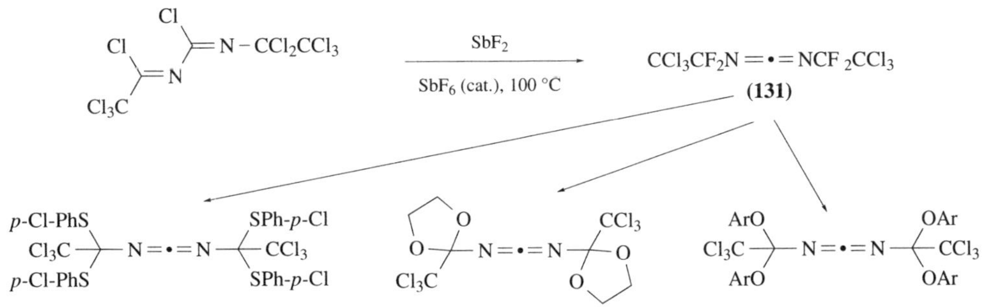

Scheme 15

5.28.1.4 Other Related Derivatives

Reactions of organometallic CDIs, for example M—N=C=N—M differ considerably with their organic counterparts R^1—N=C=N—R^2. In the hydrolysis of organic CDIs, the central carbon atom provides the electropositive center for nucleophilic attack with no fission of a C—N bond. On the other hand, in the metal carbodiimide derivatives, the metal atom provides the focus for nucleophilic attack with cleavage of the M—N bond.

5.28.1.4.1 Carbodiimides of the type MN=C=NM (M = Si, Ge, Sn, Pb)

Several methods for the preparation of fully substituted organometallic CDIs of the type R_3M—N=C=N—MR_3 (R = alkyl or aryl; M = Si, Ge, Sn, or Pb) are described ⟨88MI 528-01⟩. The chemistry of the most important member of this class, bis(trimethylsilyl)CDI (**108**), has already been discussed.

By analogy to alkyl iodides, the reaction of R_3GeCl with silver cyanamide in an ether medium at room temperature for 150 h gives bis(germanyl)CDI (**132**) in 89% yield (Equation (45)) ⟨87ZOB405⟩. Similarly, dialkyltin dichlorides give exclusively CDIs ⟨72JCS(D)869, 72ZC147⟩.

$$Et_3GeCl + Ag_2NCN \longrightarrow \underset{(\mathbf{132})}{Et_3GeN{=}\bullet{=}NGeEt_3} \qquad (45)$$

The preparation of organotin CDI, from dicyandiamide and the corresponding distannoxane generally leads to problems in isolation. However, methoxy tributylstannane is a good reagent for the preparation of stannane derivatives ⟨82ZOB921⟩ due to its good solubility, and the ready removal of methanol from the reaction.

Due to the thermal instability, hydridic species such as (**133**; X = Cl, $n = 2$, M = Si), are not suitable for the preparation of CDIs. However, metathetical reaction of Pb(II)cyanamide and alkylchlorohydridosilanes or -germanes lead to the satisfactory yields of the corresponding CDIs (Equation (46)). The comparable reaction of (**133**) with Ag(I)cyanamide gives significantly reduced yields of CDIs ⟨76JCS(D)366⟩.

$$\underset{(\mathbf{133})}{Me_nH_{3-n}MX} + Pb_2NCN \longrightarrow (Me_nH_{3-n}M)N{=}\bullet{=}N(MH_{3-n}Me_n) \qquad (46)$$

$$M = Si, Ge; n = 1\text{–}3 \text{ or } 0$$

Alternatively, germanium CDIs, for example $CH_3H_2GeN{=}C{=}NGeH_2CH_3$ can be synthesized in good yields from bis(trimethylsilyl)CDI and MeH_2GeF, although the preparation of latter reagent is inconvenient ⟨76JCS(D)366⟩.

Organometallation of thioureas can give many compounds depending on experimental conditions, metal, and the structure of the thioureas. While hexaalkyldiorganogermoxanes give no CDIs even at elevated temperatures, hexamethyldisilazane in the presence of $(NH_4)_2SO_4$ at 120–160°C gives bis(trimethylsilyl)CDI (Equation (47)) as a major product ⟨82ZOB885⟩.

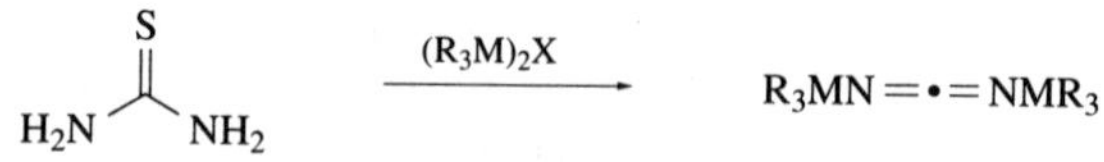

(47)

X = O, NH; M = Si, Ge; R = Me, Et

Hexaalkyldistannane with thiourea gives a mixture of CDI and trialkylstannylthiocyanate. The alkyl Ta, Ge, and Si compounds do not give CDIs with substituted allylthioureas; however, amidinothioureas (**134**), with trialkylstannanes and germanes produce a mixture of CDIs and isothiocyanatometal compounds (Equation (48)) ⟨82ZOB885⟩.

(48)

(**134**)

X = NH; R = Me, Et, Bu

M	*Yield* (%)
Sn	80
Ge	13

Dimeric and polymeric CDIs have been prepared from bis(trimethylsilyl)CDI and metal chlorides such as $MoCl_5$, WCl_6, $TaCl_5$ ⟨86ZAAC(538)36⟩, $SbCl_5$, and $SnCl_4$ ⟨86ZN(B)1223⟩. When $TaCl_5$ reacts with bis(trimethylsilyl)CDI in a 1 : 1 molar ratio, a dimeric carbodiimide, [Cl_4Ta—N═C═N—TMS] is formed, whereas with WCl_6, polymeric CDI [Cl_4W—N═C═N] ⟨86ZAAC(538)36⟩ is formed.

Bis(trialkyltin or germanium) oxides react readily with thiourea in the presence of HgO in benzene or ether to give CDIs. Good to excellent yields are obtained when the reaction is performed in the presence of a dehydrating agent such as $MgSO_4$. In the absence of HgO, isothiocyanato-bis(tributyl)metal sulfides are formed exclusively. In place of trialkyl metal oxides, and HgO, methoxytrialkyl metals, and Ph_3P/CCl_4/TEA can also be used ⟨88IZV739⟩ respectively.

5.28.1.4.2 *Carbodiimides of the type M—N═C═N—R (M = Si, Ge, Sn, Sb, Tl, Hg, Co, Cu, As, Ti, Cd; R = alkyl or aryl)*

N-Silylcyanamides exist predominantly in equilibrium with monosilylated CDIs ⟨77AG(E)311⟩. However, *N′*-acyl *N*-silyl CDIs (**135**) which can be prepared from the reaction of *N*-trimethylsilylacylamides and cyanogen chloride at 20–50 °C in methylene chloride, predominantly exist as CDIs with little or no cyanamide form (Scheme 16) ⟨77AG(E)311, 77TL1987, 78ZOB1313⟩.

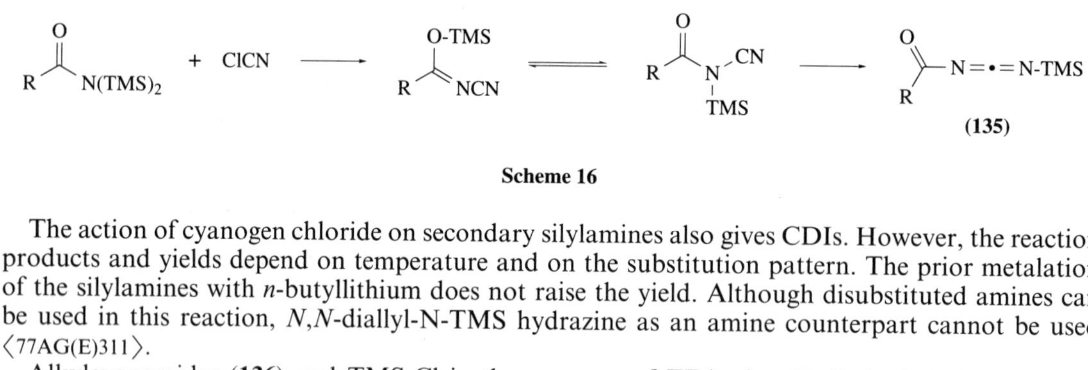

(**135**)

Scheme 16

The action of cyanogen chloride on secondary silylamines also gives CDIs. However, the reaction products and yields depend on temperature and on the substitution pattern. The prior metalation of the silylamines with *n*-butyllithium does not raise the yield. Although disubstituted amines can be used in this reaction, *N*,*N*-diallyl-N-TMS hydrazine as an amine counterpart cannot be used ⟨77AG(E)311⟩.

Alkylcyanamides (**136**) and TMS-Cl in the presence of TEA give *N*-alkyl-*N′*-silyl CDIs (**137**) (Equation (49)) ⟨77TL1987⟩. In place of TMS-Cl, TMS-NMe_2 can be used as silylating agent.

(49)

(**136**) (**137**)

R	*Ref.*
Me	77TL1987
Et	77TL1987
Bu^t	77TL1987
Ph_3C	77TL1987
Ph	77TL1987
1,3,5-Me_3Ph	78ZN950

Similarly functionalized CDIs, for example (**138**), have been prepared from the substituted functionalized cyanamide and TMS-Cl (Equation (50)) ⟨86JOC1277, 89TL2379⟩.

(138) (50)

In the reaction of lithioaminofluorosilane with trimethylsilyl isocyanate, lithium fluoride is eliminated and CDI is formed in good yield (Equation (51)) ⟨78ZN(B)575⟩.

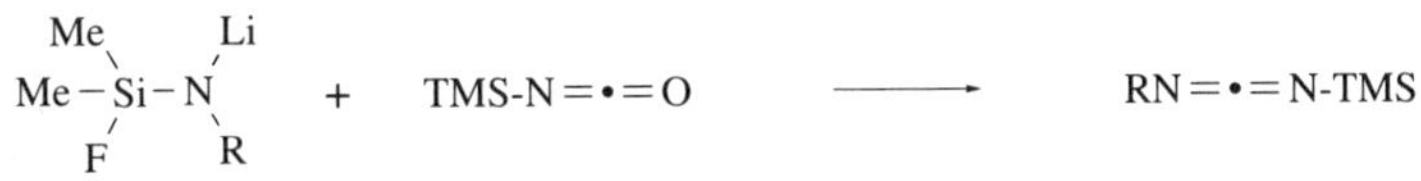

(51)

$R = 3,4,5\text{-}Me_3C_6H_2$

The phenylCDI and its halogenated derivative are thermodynamically stable under normal conditions and are sufficiently acidic to form stable salts with cobalt(II), nickel(II), copper(II), and zinc(II) ⟨71JCS(A)332⟩ etc. The synthesis of organometallic CDIs involves the reaction of metal chlorides with 0.5 or 1 mol equiv. of phenylcyanamide (Scheme 17) ⟨71ZAAC(386)197⟩.

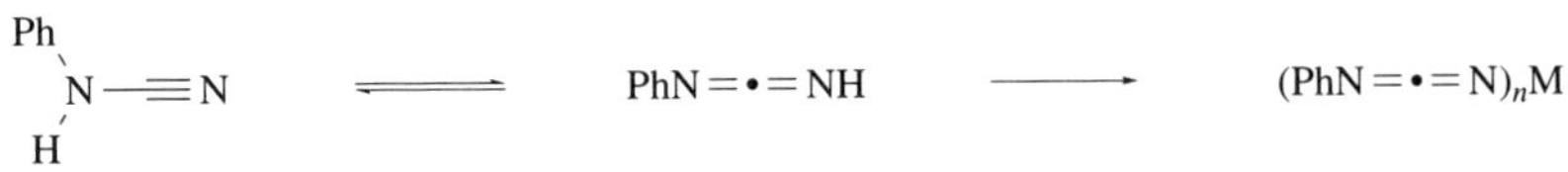

M	*n*
Si	1
Sb, Tl, Hg, Co Cu, As, Ti, Cd	2

Scheme 17

The monomeric bis(carbodiimidoyl)phthalocyanatogermanium (**139**) and its tetra *t*-butyl-substituted analogues have been prepared by the reaction of the corresponding phthalocyanatogermanium dihydro derivatives with cyanamide (Equation (52)) ⟨85SM347⟩.

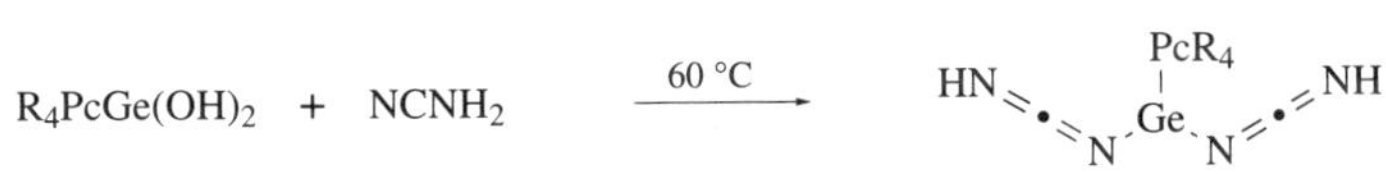

(139) (52)

Pc = phthalocyanine; R = H, Bu^t

The introduction of the electron-withdrawing benzoyl group into heterocumulene systems influences the structures of organometallic adducts and it frequently induces facile elimination of the first formed insertion product (**140**) according to Scheme 18 ⟨74JOM(69)353⟩ leading ultimately to monometal-substituted CDIs.

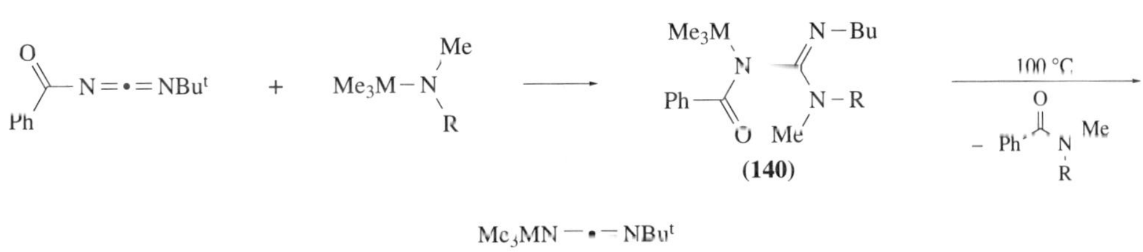

R = Me, TMS; M = Si, Ge

Scheme 18

The *N*,*N'*-bis(trimethyl or -butylsilyl) CDIs react at room temperature with hexafluoroacetone with the retention of the CDI group as shown in Equation (53) ⟨80ZOB1678⟩. Introduction of a second mole of hexafluoroacetone in (**141**) does not occur even upon prolonged heating at 80°C.

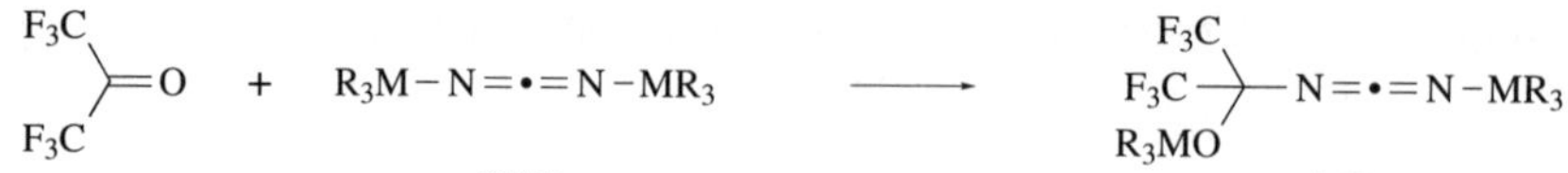

(53)

(141) **(142)**

M	*R*	*Yield* (%)
Si	Me	78
Sn	Bu	63

Organometallation reaction of Na or Ag diphenylcyanamidophosphinates, (**143**) with triorganometal chlorides, R_3MCl (M = Si, Sn, Pd) give exclusively *N*-phenylphosphonoyl-*N′*-triorganometalloyl CDIs (**144**) (Equation (54)) ⟨88ZAAC(564)85⟩.

(54)

(143) Z = Na, Ag; Y = O, S; M = Si, Sn, Pb **(144)**

R_3MCl	*Product*	*Yield* (%)
Me_3SnCl	$Ph_2P(=O)-N=\bullet=N-SnMe_3$	100
Ph_3SiCl	$Ph_2P(=S)-N=\bullet=N-SiPh_3$	100
Ph_3SnCl	$Ph_2P(=O)-N=\bullet=N-SnPh_3$	90
Ph_3SnCl	$Ph_2P(=S)-N=\bullet=N-SnPh_3$	82
Ph_3PbCl	$Ph_2P(=O)-N=\bullet=N-PbPh_3$	100
Ph_3PbCl	$Ph_2P(=S)-N=\bullet=N-PbPh_3$	100

5.28.1.4.3 Carbodiimide adducts with In, Ga, Ti, and Al

Bis(trimethylsilyl)CDI forms distillable liquid 1 : 1 adducts with $GaMe_3$ and $InMe_3$, whereas with Me_3Al it forms solid monomeric insertion products (Equation (55)) ⟨89JOM(359)1⟩.

$$TMS-N=\bullet=N-TMS \xrightarrow{MMe_3} Me_3M(TMS)N=\bullet=N-TMS \quad (55)$$

M	*Yield* (%)
In	90
Ga	90
Al	–

The reaction of $Me(Cp)Ti(NCO))_2$ or $Cp_2Ti(NCO)$ with $LiN(TMS)_2$ gives CDIs (**145**) (Scheme 19). With 2 equiv. of $LiN(TMS)_2$ both isocyanate groups are converted into CDIs (**146**) ⟨89ZN(B)94⟩. These are thermally and photolytically unstable red solids which have been characterized by mass spectrometry, ^1H NMR and elemental analysis.

$$\underset{\textbf{(146)}}{(Cp)_2Ti(N=\bullet=N\text{-}TMS)_2} \xleftarrow{(Cp)_2Ti(NCO)_2} LiN(TMS)_2 + R_2Ti(NCO)_2 \longrightarrow \underset{\textbf{(145)}}{R_2Ti(NCO)(N=\bullet=N\text{-}TMS)}$$

R	*Yield* (%)
MeCp	14
Cp	5

Scheme 19

5.28.1.4.4 Carbodiimides of the type M—N=C=N—M (M = P, S, Te, and B)

The reaction of bromodifluorophosphorus with silver cyanamides gives CDIs of the type $F_2PN{=}C{=}NPF_2$ ⟨72JCS(D)869, 72ZC147⟩ in 52% yield. The purified (**147**) has been rearranged to CDIs (**148**) with fluoride ion (Equation (56)) ⟨85IC4171⟩.

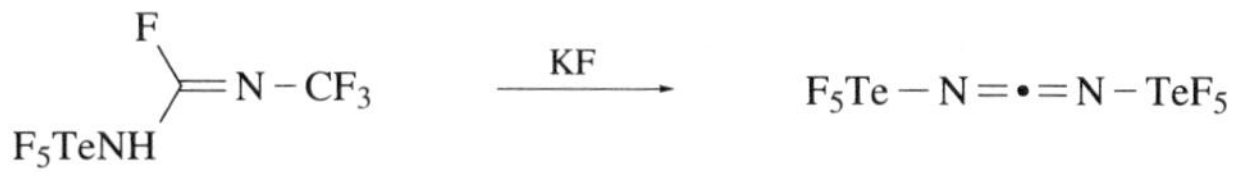

(**147**) (**148**) (56)

Compounds containing the pentafluorosulfanyl group possess similar properties to sulfur hexafluorides providing steric bulk, and high thermal and hydrolytic stability ⟨76JFC(7)65⟩. Thus bis(pentafluorosulfanyl)CDI (**150**) has been prepared from bis(pentafluorosulfanyl)urea (**149**) (Scheme 20) ⟨82IC1616, 89MI 528-01⟩.

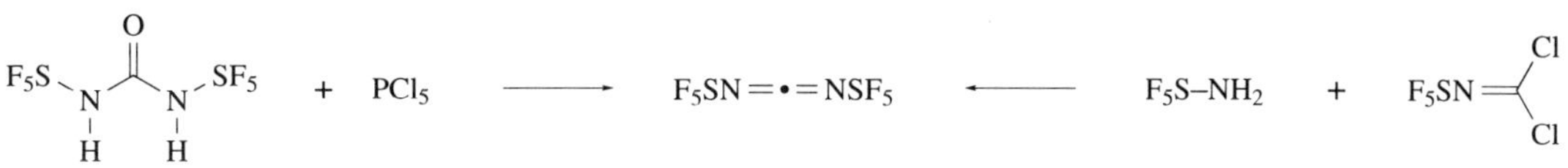

(**149**) (**150**)

Scheme 20

The diboron-substituted CDI (**151**) is ultimately obtained in 35% yield from *t*-butyliminotetramethylpiperidino borane (**152**) and thiourea via the triborated isothiourea derivative (Equation (57)) ⟨90CB455⟩.

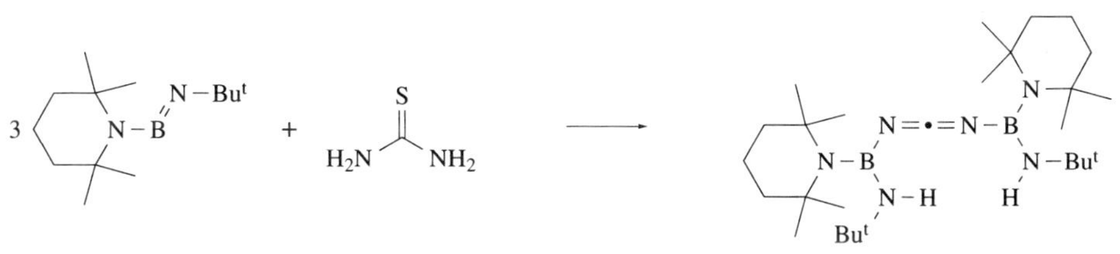

(**152**) (**151**) (57)

In the reaction of the acyclic haloboranes R_3BHal (R = Ph, Me, Me_2N) with TMS-N=C=N-TMS, polymeric CDIs $(R_2BCCN)_n$ are formed. With 9-borabicyclo[3.3.1]nonyl (9-BBN) the cyanamide forms monoborylated cyanamide ⟨86ZN(B)1367⟩. The borylated CDIs can be synthesized by the reaction of the 2-chloro-1,3,2-diazaborolidines (**153**), (**154**) with bis(trimethylsilyl)CDI (Equation (58)). The chloroboralane (**155**) leads to polymeric diborylated CDIs.

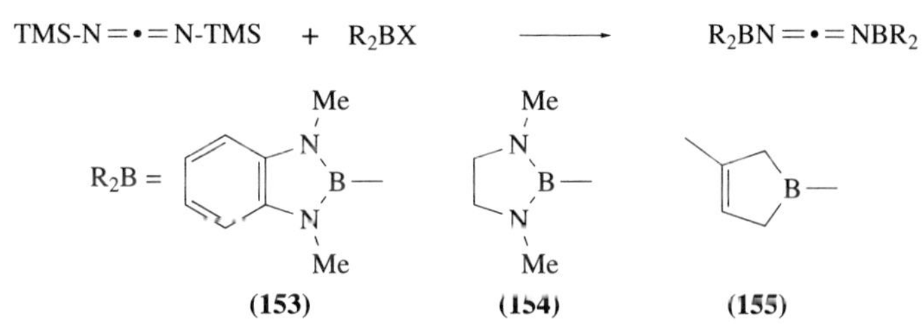

(**153**) (**154**) (**155**) (58)

5.28.2 METALLOID DERIVATIVES R[1]—M=C=N—R[2] (M = Si, Sn)

5.28.2.1 Carbodiimides with Silicon

Unsaturated higher period compounds are generally unstable. The reaction of (**156**) (generated photolytically from hexa-*t*-butylcyclotrisilane) with phenylisocyanide (**157**) gives silaketenimine (**158**) which dimerizes spontaneously (Equation (59)) ⟨90AG(E)90, 91POL1147⟩.

$[Bu^t_2S:]$ + ArN=C ⟶ $Bu^t_2Si{=}\bullet{=}NAr$ (59)

(156) **(157)** **(158)**

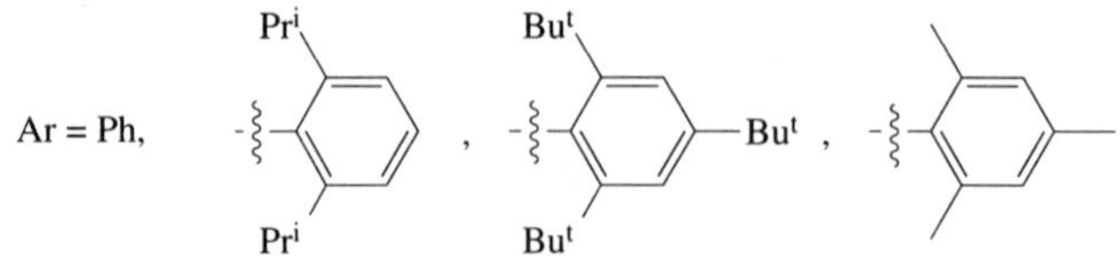

5.28.2.2 Carbodiimides with Tin

Similarly to silicon-containing CDIs (see above), a solution of equimolar amounts of diarylstannadiyl (**159**) and mesitylisocyanide, cooled to $-78\,^{\circ}C$ provide the solid (**160**) which can be isolated as pale yellow compound (Equation (60)) ⟨92AG(E)437⟩.

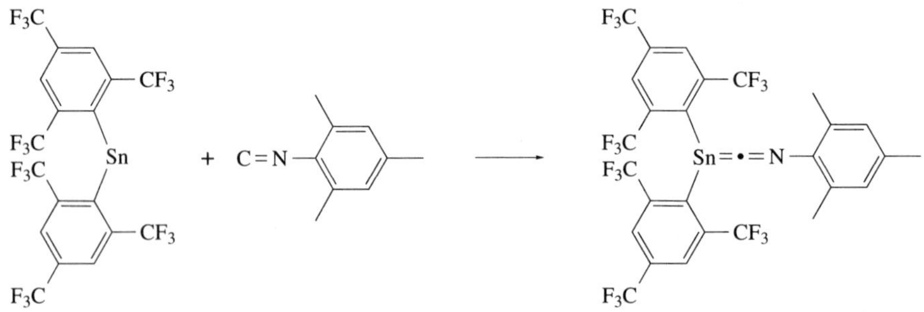

(60)

(159) **(160)**

5.28.3 PHOSPHORUS, ARSENIC, ANTIMONY, AND BISMUTH FUNCTIONS, $RN{=}C{=}PR_3$

By analogy with —Si=C, (iminomethylidene)phosphines (sometimes referred as carbimidophosphenes) possessing two-coordinate phosphorus atoms (—P=C), are generally unstable unless they have bulky groups on phosphorus, and to some extent, on nitrogen atoms. Unstable (iminomethylidene)phosphines have been observed by IR, mass spectrometry and by their chemical reactions (dimerization) ⟨87ZAAC(555)23⟩.

The reaction of halogenated aryl isocyanide dichlorides with phenyl bis(trimethylsilyl)phosphines yields 1,3-diphosphetanes via initially formed monophosphacarbodiimides (**161**) (Equation (61)) ⟨80TL2497⟩.

(61)

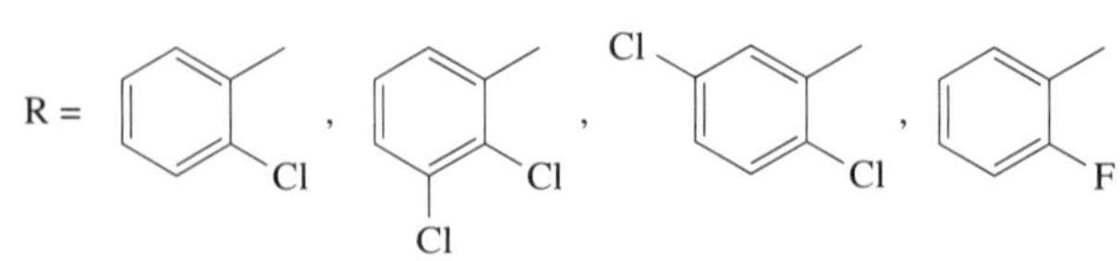

More conveniently, they are prepared from carbomoylphosphines (**162**) (Scheme 21) in the presence of solid KOH or NaOH at $25\,^{\circ}C$ ⟨82TL4933, 83PS(18)39, 83ZOB1093, 90BCJ2736⟩. This reaction is convenient in that it leads to the formation of hexamethyldisiloxane which can readily be removed in vacuum ⟨83JA7194⟩.

Sterically protected 1-phenyl and *t*-butyl-3-(2,4,6-tri-*t*-butylphenyl)-1-aza-3-phosphaallenes have been prepared (Equation (62)) ⟨80TL2497, 83PS(18)7, 84TL1809, 88POL2213⟩. Due to their bulkiness they are stable enough to be isolated.

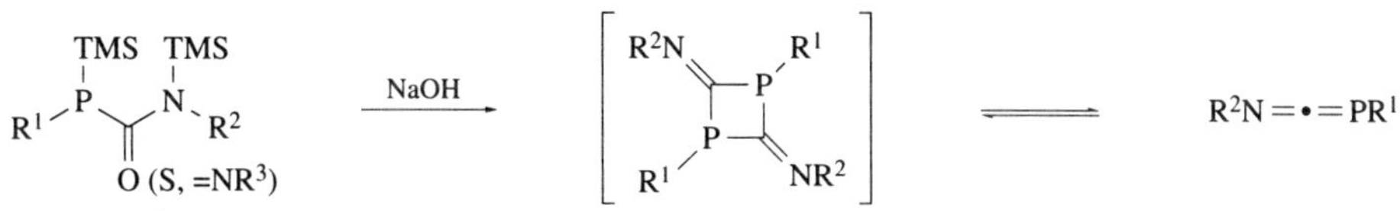

(162)

R^1	R^2	*Ref.*
Bu^t	Bu^t	83ZOB1093
		82TL4933
Ph	Bu^t	82TL4933
Bu^t	Ph	83JA7194
		83PS7
Ph	Ph	83PS39
		83JA7194
2,4,6-$Bu^tC_6H_2$	Ph	84TL1809
Et	Ph	83JA7194
$PhCH_2$	Ph	83JA7149
2,4,6-$Bu^tC_6H_2$	Bu^t	88POL2213
2,4,6-$Bu^tC_6H_2$	Pr^n	87ZAAC23
CF_3	Bu^t	91ZN(B)978
CF_3	Pr^i	91ZN(B)978

Scheme 21

(62)

The reaction of perfluoro-2-phosphapropene, $F_3CP{=}CF_2$ with *t*-butylamine or isopropylamine in 1 : 3 molar ratio leads to the iminomethylene phosphanes (Equation (63)). Unlikc other CDIs, fluorinated iminomethylenephosphanes do not dimerize to 1,4- or 1,2-phosphotanes ⟨91ZN(B)978⟩.

(63)

5.28.4 METAL DERIVATIVES R^1—M—C—N—R^2

α-Halogenoalkyl isocyanides are unstable. However, they can be synthesized as "protected metal" as shown in Scheme 22 ⟨87CB461, 88AG(E)1344, 89CB1907⟩. Thus radical alkylation of cyano complexes (**163**) with halogeno compounds (**164**) in the presence of the diazonium salt (**165**) gives a good yield of complex—halogenoalkylisocyanidemetal complex (**166**). In general, (**164**; X = H, Hal) is the solvent which provides the side chain of the isocyanide.

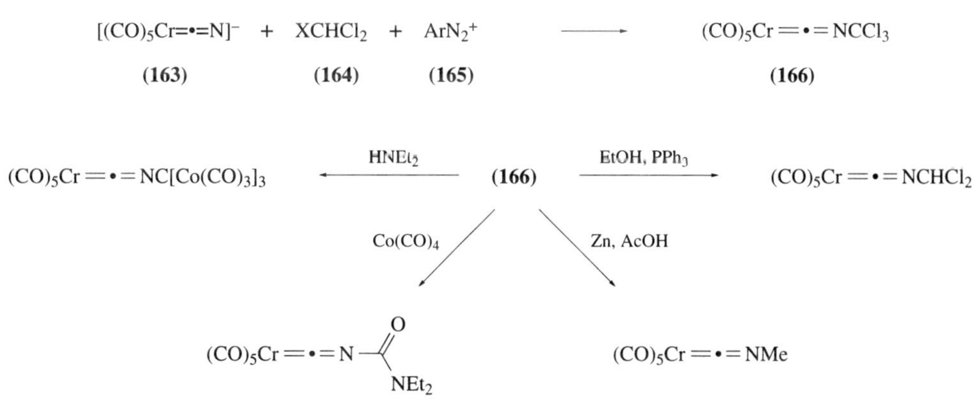

Scheme 22

Pentacarbonyl(trichloromethylisocyanide)chromium (**166**) affords a broad variety of carbene isocyanide complexes. Most reactions occur at the two electrophilic centers (=C=, M=C). However, at least in some cases it is possible to limit the reaction to the trichloromethyl group of (**166**), thereby obtaining functionalized isocyanide metal complexes (Scheme 22).

The *in situ* generated hydrogen isocyanate complexes of Cr, W, and Fe from (**167**) and (**168**) react with cyclohexylisocyanate to give metal-stabilized isocyanocarboxamides (**169**) and (**170**) respectively (Scheme 23) ⟨88JOM(347)101⟩. To clarify the influence of a single loan pair heteroatom attached directly to the carbene of transition-metal (chromium) compounds of the type (**171**) have been prepared as shown in Equation (64) ⟨75AG(E)760⟩.

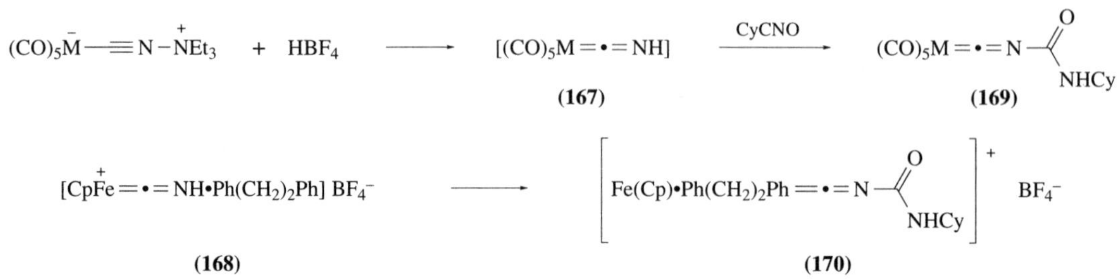

M	*m.p.* °C	*Yield* (%)
Cr	105	25
W	100	23

Scheme 23

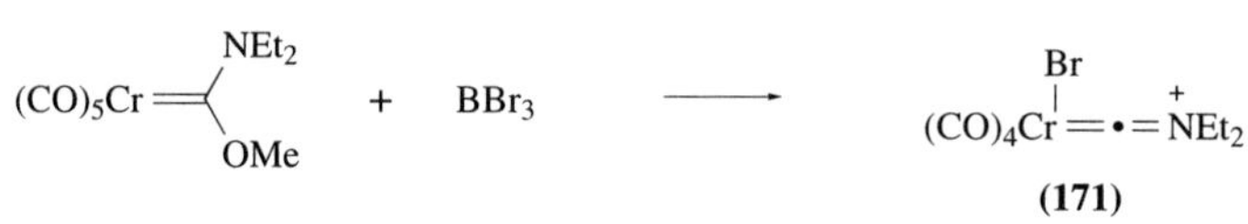

(64)

The successive reaction of $(CO)_6M$ with Na(—N=CR_2) and Et_2OBF_4 yields pentacarbonyl-1-alkoxy-2-aza-3,3-diarylallenylmetal (**172**) (Scheme 24). These complexes have been reacted with $AlBr_3$ to give complexes (**173**) ⟨74JOM(65)53, 85JOM(287), 86CB2080⟩.

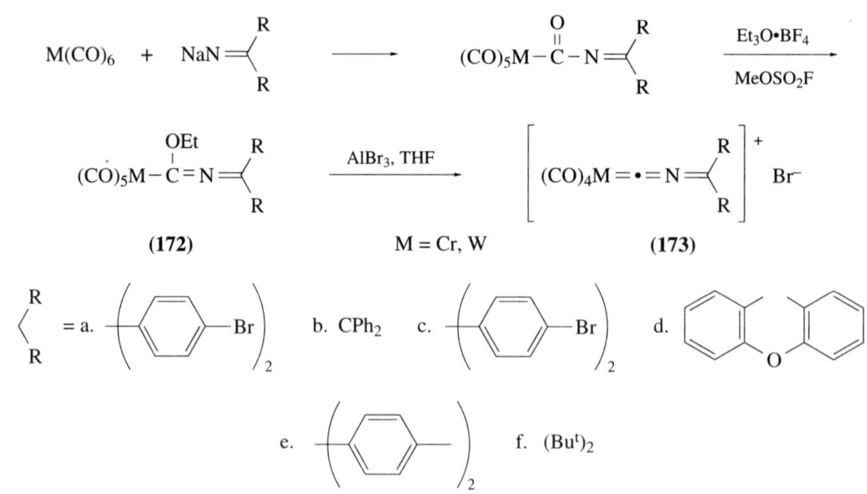

Scheme 24

The nitrogen terminus of $W(CO)_5CN$— displaces Me_2S in the alkylidene bimetallic complex (**174**) due to the presence of the electron-withdrawing —CN and nucleofugal Me_2S substituent at the bridging carbon atom, giving (**175**) (Equation (65)) ⟨91AG(E)847⟩.

(174) (175) (65)

Isonitriles have a structure, isoelectronic with carbon monoxide, and can coordinate to a variety of metals in both low and high oxidation states. The platinum and palladium complexes have been made by ligand exchange as shown in Equation (66) in 70–90% yield ⟨92CL1515⟩. Similarly, palladium complexes have been prepared from $PdCl_2$/KI/acetone with the isonitriles (**176**).

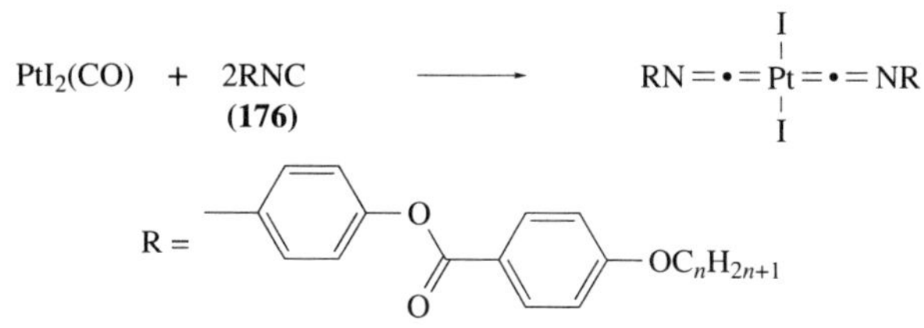

(66)

n = 6–8

The monomeric pentacarbonyl(cyanoisocyanide)chromium (**177**) can be prepared by bubbling ClCN through a cold solution of (**178**) at 0 °C for 5 min. It is purified by sublimation (Scheme 25) ⟨82CC184⟩. Alternatively, (**177**) can be prepared by the slow addition of isocyanogen (CNCN) to (**179**), which in turn is generated from the photolysis of hexacarbonylchromium in 2MeTHF, at −20 °C. The yield of (**177**) by this method is much lower ⟨88JOM(347)101, 90JOM(390)C1⟩.

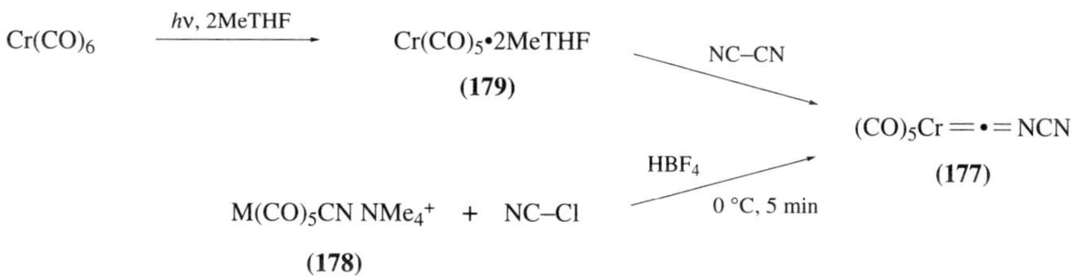

Scheme 25

The metallonitrile ylide (**181**) can be prepared by the reaction of the complex $(CO)_5M \cdot THF$ with isocyanomethylphosphonium chloride (**180**) in ether (Scheme 26). These salts, upon treatment with sodium amide give stable ylides (**182**) containing the CDI unit ⟨87JOM(331)193⟩.

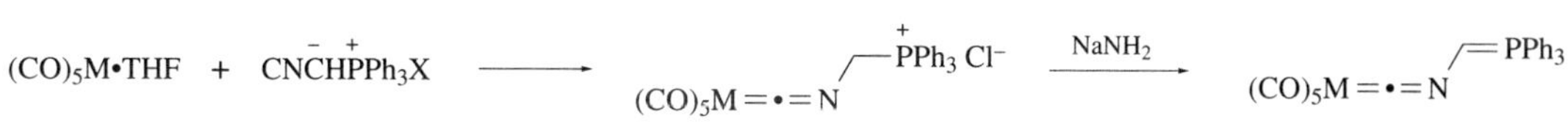

(180) (181) (182)

X = Cl, PF_6

M	*Yield (%)*	*m.p.* °C
Cr	55	95
W	61	108

Scheme 26

5.29
Functions with Heteroatoms Other Than Chalcogen or Nitrogen (Y=C=Y′)

KAMYAR AFARINKIA
Kings' College, London, UK

and

M. VICTORIA VINADER
Glaxo-Wellcome Medicines Research Center, Stevenage, UK

5.29.1 FUNCTIONS WITH AT LEAST ONE PHOSPHORUS, ARSENIC, ANTIMONY OR BISMUTH (AND NO CHALCOGEN OR NITROGEN)

5.29.1.1 Group 15 Functions

5.29.1.1.1 Dicoordinated phosphorus species

Due to the linear arrangement of phosphorus and carbon atoms, these carbodiphosphine species have been called phosphaallenes, with the substituents on phosphorus being *trans*. The chosen phosphorus substituents are invariably bulky so as to stabilise kinetically these highly reactive molecules which are otherwise too unstable to be isolated. Commonly used groups are 2,4,6-tris-*t*-butylphenyl and 2,4,6-tris(2′-methylbutyl)phenyl, and these phosphaallenes are air and moisture stable solids. Unlike the analogous PhP=C=NAr ⟨80TL2497, 83JA7194⟩, these species are monomeric.

Two synthetic approaches to these species have been established. One involves 1,2-elimination of H—X from a P—C—P ⟨84TL3687⟩ or a P═C—P skeleton ⟨84CC689⟩. Usually elimination of H—X from P—C—P is accomplished with organic amine bases though stronger basic conditions are required for P═C—P (Scheme 1).

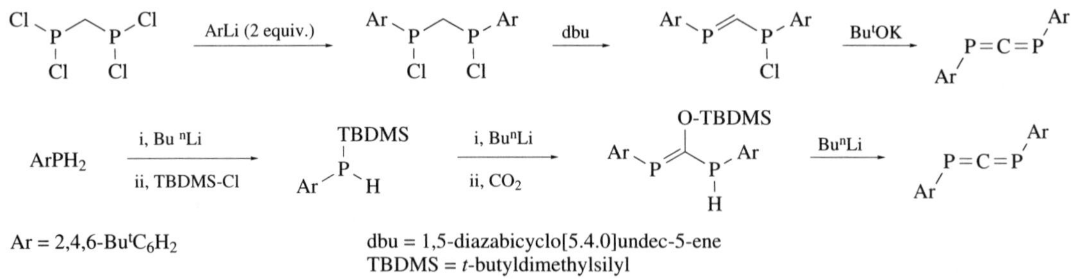

Scheme 1

Interestingly, in a closely related experiment, treatment of the analogous *t*-butyl substituted bisphosphine with mild base is reported to afford only 1,2-diphosphetane, though the authors did not attempt further treatment with strong base (Scheme 2) ⟨87JGU2350⟩.

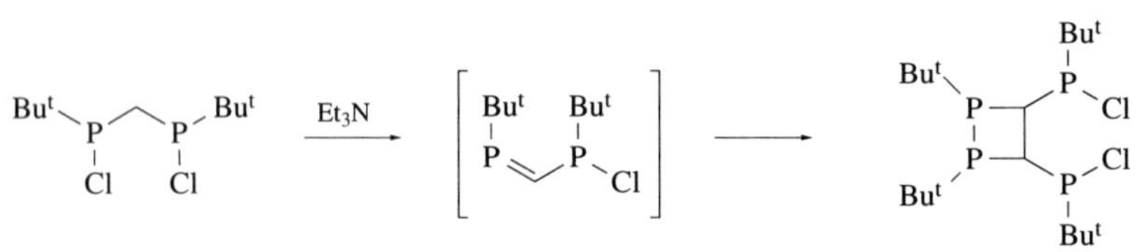

Scheme 2

A variation on the above assembly of a P═C—P skeleton, using addition to a phosphaketene, has been reported (Scheme 3) ⟨84AG(E)619⟩.

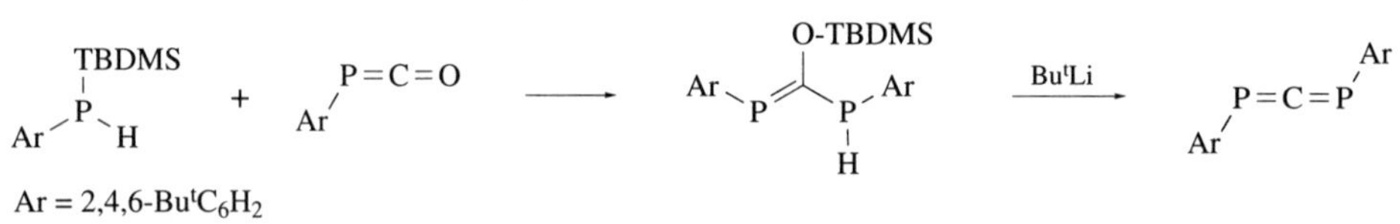

Scheme 3

An alternative synthesis which has been applied to the synthesis of unsymmetrically substituted 1,3-diphosphaallenes, involves an unusual carbene-type insertion to unsymmetrical diphosphenes. The reaction proceeds most probably via bis-radicals, in accordance with results from [2 + 2] cycloadditions of diphosphenes ⟨90CB733⟩. Halogen–metal exchange followed by skeletal rearrangement affords carbodiphosphine (Scheme 4) ⟨89CC1732, 89TL177, 89TL187, 89TL5433⟩.

A similar approach allows preparation of unsymmetrical 1,3-diphosphaallenes without resort to unsymmetrical diphosphenes (Scheme 5) ⟨89TL839⟩.

5.29.1.1.2 *Tetracoordinated phosphorus and arsenic species*

(i) Species with carbon attached to phosphorus

These carbodiphosphoranes have attracted considerable interest as versatile reagents in organic synthesis ⟨77AG(E)349⟩, as ligands for both main group (metalloids) and transition metals ⟨83AG(E)907⟩, and as materials with intriguing physical properties ⟨78ACR289⟩.

Results of structural studies have indicated that the P—C—P bond angle is quite flexible and can vary from 180° to at least 116° ⟨81CB3070, 83CB2275⟩. Thus formulae (**1a–c**) are used for a general

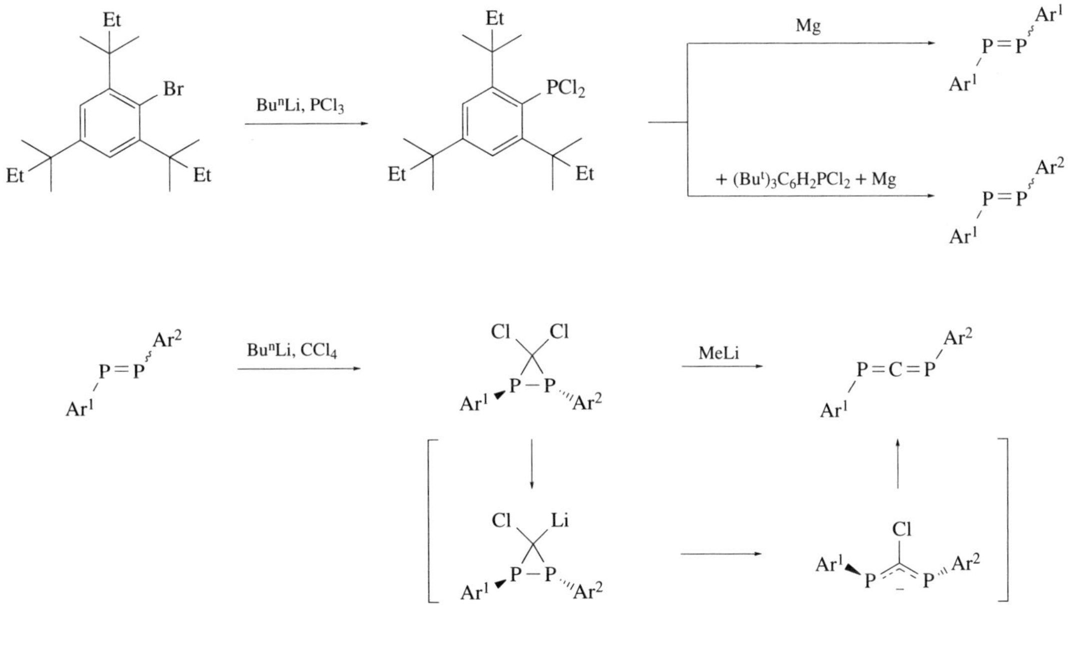

Scheme 4

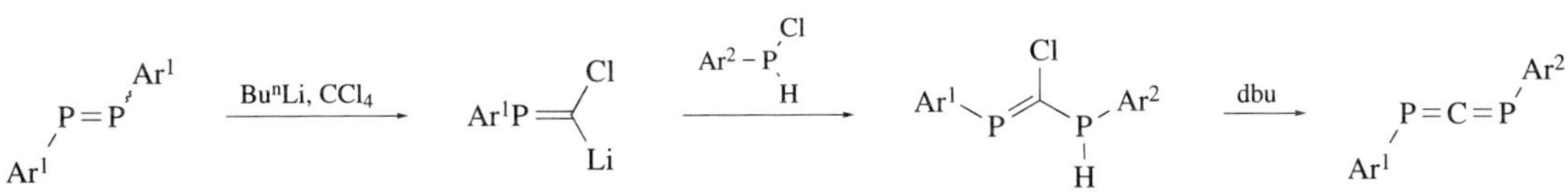

Scheme 5

description of structure and bonding in carbodiphosphoranes, with the main emphasis on the semiylidic formula (**1b**) as bisphosphoranes of naked (atomic) carbon.

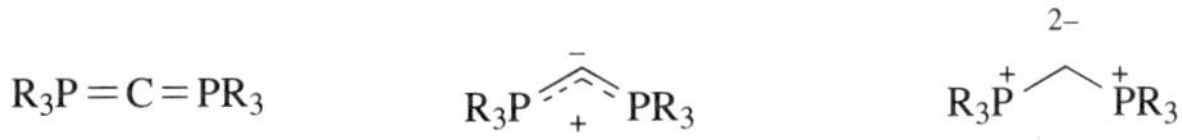

(**1a**) (**1b**) (**1c**)

This type of phosphorane compound is formally related to carbodiimides and allenes. The first example of the series, hexaphenylcarbodiphosphorane (**3**) was prepared from methylene-bis(triphenylphosphonium) bromide (**2**) which itself is produced from treatment of dibromomethane with triphenylphosphine (Scheme 6) ⟨61JA3539, 62JA4349, 64JOC2427⟩.

A similar procedure involves carbon tetrachloride and triphenylphosphine, using hexamethylphosphorus triamide as a dechlorinating agent (Scheme 7) ⟨76AG(E)702, 83CB2275⟩.

The synthesis of alkyl derivatives, for example (**4**), can also be carried out via the methods employed for the hexaphenyl analogue (**3**). Difluorotrimethylphosphorane (**5**) ⟨67SA(A)681, 74CB2169⟩ reacts smoothly with (**6**) under mild conditions to afford fluorotrimethylsilane and (**7**) (Equation (1)) ⟨65IC1458, 67CB1032, 68CB595⟩. Hexamethylcarbodiphosphorane (**4**) can be prepared by treatment of (**7**) with base. Reactions of chlorides (**8**) and (**9**) with organometallic base also affords (**4**) (Scheme 8) ⟨75JA6281, 77CB3501, 78TL2689⟩.

(**5**) (**6**) (**7**) (1)

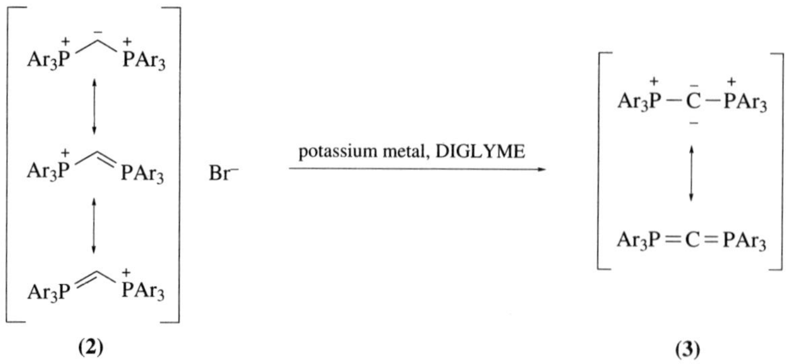

Scheme 6

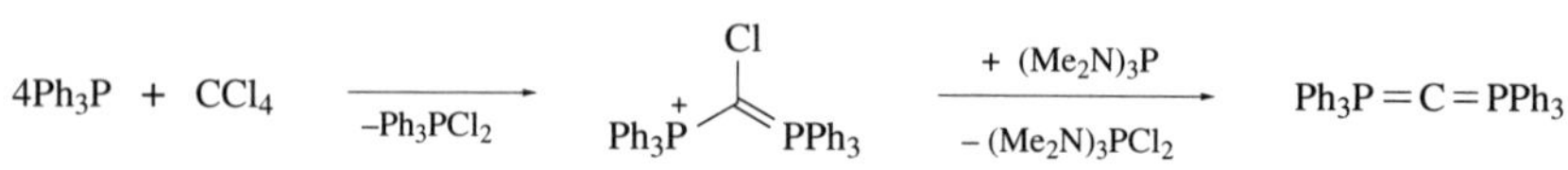

Scheme 7

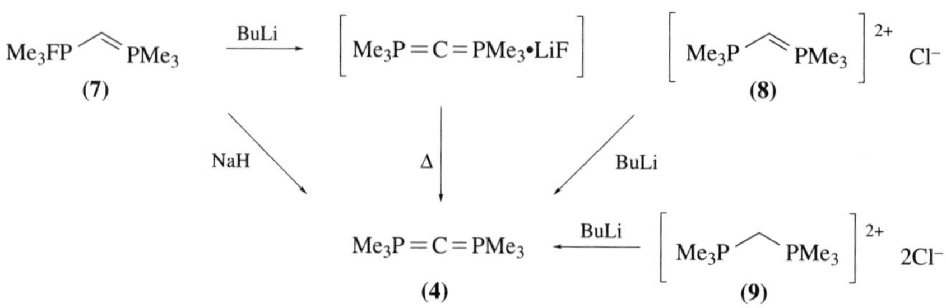

Scheme 8

Hexamethylcarbodiphosphorane (**4**) is one of the most air and moisture sensitive compounds in the series. Its general chemical behaviour is characterised by strong basicity, which makes it a powerful deprotonating agent, and strong nucleophile. The compounds do react with water and, for instance, compound (**3**) reacts readily with water to afford phosphine oxide (**10**) (Equation (2)). Compound (**3**) is also unusual in showing the phenomenon of triboluminiscence, that is it emits light when the crystals are touched ⟨73JA7510⟩.

(2)

Mixed alkyl–aryl carbodiphosphoranes have been obtained on deprotonation of conjugated phosphonium salts ⟨77AG(E)417, 77CB3501⟩ as discussed previously, or via an unusual molecular rearrangement (Scheme 9) ⟨79CB510⟩.

The synthetic procedures employed successfully for the preparation of carbodiphosphoranes are not applicable to the corresponding arsenic compounds. Double quaternization of dihalomethanes using tertiary arsines is an extremely slow process and is accompanied by a variety of side reactions. The use of haloarsines is also unattractive because of the inherent lability of the monoylides. Double quaternization of tertiary arsine (**11**) with a strong methylating agent to give the bis-quartenary salt (**12**), followed by treatment with strong base afforded carbodiarsane (**13**) ⟨85OM344, 87CB1281⟩. The mixedphosphorus–arsenic analogue (**14**) was prepared in a similar fashion (Scheme 10) ⟨87CB1281⟩.

Scheme 9

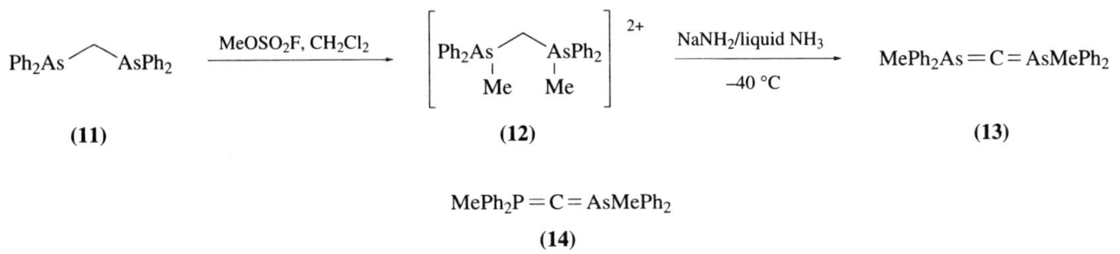

Scheme 10

(ii) Species with a heteroatom attached to phosphorus

A number of carbodiphosphoranes with ligands other than carbon attached to phosphorus have been isolated ⟨78CB2054, 83CB2275⟩. These species are prepared according to the same general synthetic routes discussed in Section 5.29.1.1.2.(i) and as shown in Scheme 11. *P*-Chloro substituted carbodiphosphoranes undergo displacement reaction at the phosphorus atom readily, though this aspect of their chemistry does not appear to have been exploited beyond nitrogen nucleophiles.

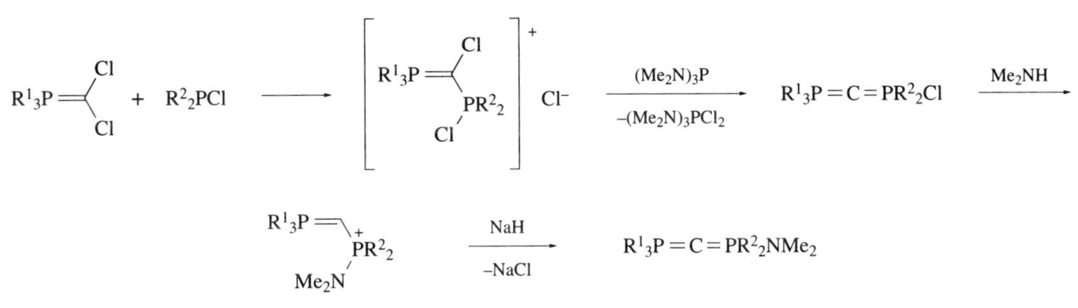

Scheme 11

5.29.1.2 Metalloid Functions

The attempted synthesis of P=C=Si species by application of a similar methodology to that successfully applied to the synthesis of P=C=P species failed ⟨87CB789⟩. The 1,2-elimination product is absent and instead an internal nucleophilic displacement affords a cyclic ylide. If phosphorus is substituted with aryl groups so that this reaction is not possible, a disilatane is isolated ⟨70CB97, 87CB789⟩ which is, at least nominally, a dimer of a P=C=Si species (Equation (3)). Of course this product could also result from a double displacement reaction, but there is not enough known about the chemical behaviour of P=C and Si=C bonds to decide the mechanism of the formation of this dimer.

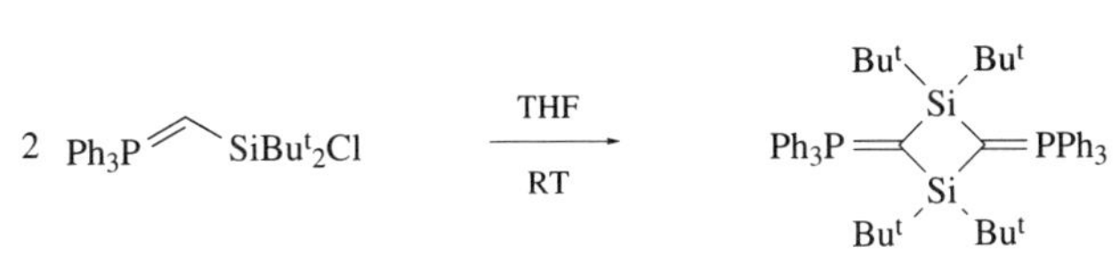

(3)

5.29.1.3 Metal Functions

Molecular species of this type are illusive. Rational synthesis of these species can derive from the two possible disconnections at either side of the central carbon atom. In fact both the addition of phosphorane ylide to an electrophilic metal, and addition of a phosphine to a metal carbide are documented but as of 1994 no monomeric species of type M=C=P is reported by either route. Closely related compounds do exist; for instance (**15**) ⟨77ZN(B)858⟩, a dimer related to the P=C=Si dimer (see Equation (3)), and (**16**) have been synthesized ⟨80IC2564⟩. The x-ray structure of (**16**) suggests some double bond character in both metal–carbon and phosphorus–carbon bonds. This subject has been reviewed and the reader should refer to this for further information (Scheme 12) ⟨83AG(E)907⟩.

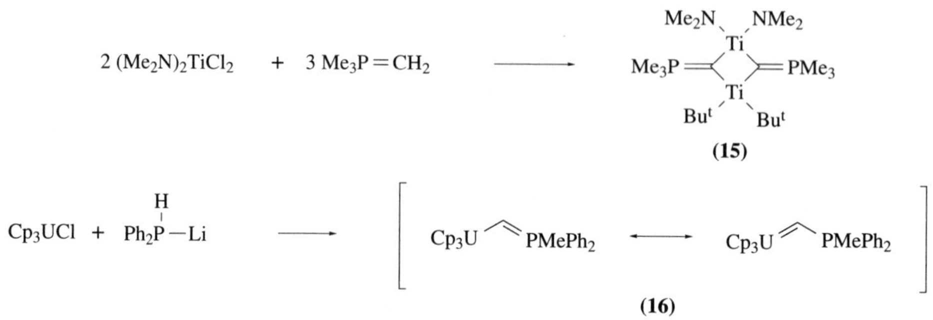

Scheme 12

A monomeric species containing the M=C=P unit is claimed to be produced according to the reaction sequence shown in Scheme 13 ⟨88JA4855, 88JA6926⟩. However, the x-ray crystal structure data is not published as of 1995.

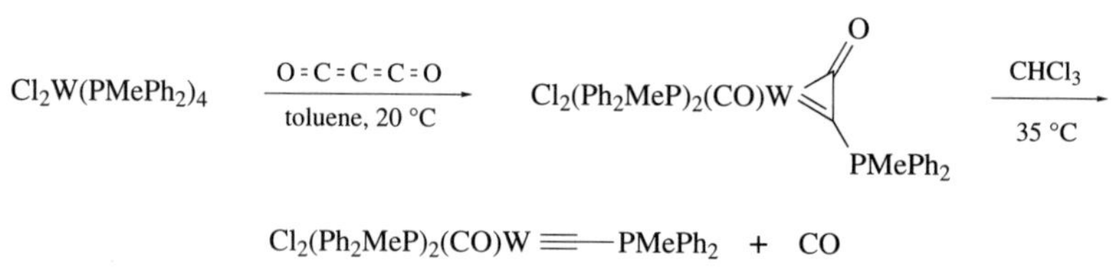

Scheme 13

5.29.2 FUNCTIONS WITH AT LEAST ONE METALLOID FUNCTION (AND NO CHALCOGEN OR GROUP 15 ELEMENT FUNCTION)

5.29.2.1 Derivatives With Two Metalloid Functions

Matrix isolation and high resolution FT–IR characterisation of products from Pulsed Laser Assisted reactions of vaporised boron atoms with methane has indicated the formation of a number of interesting organoboron molecules ⟨90JA3332⟩. Investigations in the early 1990s provided evidence of a linear HB=C=BH species with the double boron–carbon bond characterised from matrix IR spectrum, and supported by *ab initio* calculations ⟨92JA9239⟩.

Matrix isolation and IR characterisation of the silicon analogue, prepared from co-condensation of vaporised silicon and carbon on solid argon, is also reported ⟨83JCP(87)797, 91JCP(95)5612⟩. *Ab initio* theoretical calculations have supported the proposed bent structure for this molecule ⟨91JCP(95)5609⟩.

5.29.3 COMPOUNDS BEARING TWO METAL FUNCTIONS

An atomic carbon sandwiched between two metals has been the subject of intense investigation since the 1980s. These investigations have been from different standpoints. Metal carbides of this type are studied to unravel the chemical molecular processes that go on during metal surface catalysis ⟨82AG(E)117⟩, though no evidence has been presented that the analogy is accurate. On the other hand, these molecular species contain 'atomic' carbon ⟨93JA3824⟩, which provides the opportunity to extend the organic chemistry of carbon further.

Certainly, carbon is not the only nonmetal element which can be found sandwiched between two metals in this manner. The authors shall only consider carbon, but for other elements the reader is referred to the following sources: lead ⟨85AG(E)1062, 89CB445⟩, silicon ⟨88AG(E)583⟩, germanium (and lead references to phosphorus, arsenic, sulfur and tellurium compounds) ⟨81JOM(213)451, 84JOM(263)67, 85CB340⟩. Organometallic compounds also exist with more than one carbon atom in an unsupported linear bridge between two metals ⟨92JA5890, 93JA3824⟩.

Rational synthesis of this category of organometallic compounds (referred to as μ carbides) falls into two broad classes. One employs a reductive process and commonly uses halomethanes as a source of carbon. The other involves addition–elimination of a metal nucleophile to a metal carbon triple bond. Examples of each class will be given shortly.

A structure of type M=C=M′ naturally assumes an equal bonding order between carbon and each of the metals. Equal atomic distances between carbon and metals is documented when the two metals are the same. However, with different metals, this distance is usually shorter for one of them indicating a structure of type M≡C—M′. In this chapter we will consider both types but we shall exclude references to metallacycles, in which the three atoms in question are nonlinear, or clusters.

Theoretical calculations based on the known μ-nitrido and μ-oxo iron tetraphenylporphyrin complexes predicted that a carbon sandwich of two iron tetraphenylporphyrins, $(TPP)_2Fe$, could exist before any such species actually synthesized ⟨80CC509, 81JA3328⟩. Shortly after, the first synthesis appeared in the literature ⟨80PAC681, 81IC3119⟩, and the structure was later confirmed by x-ray crystallography ⟨82CC607⟩. This synthesis, which was later extended to the preparation of μ-carbido iron phthalocyanine dimer $(Pc)_2Fe$ ⟨88CC46⟩, employs a reducing agent, sodium dithionite, and carbon tetraiodide as the source of the carbon atom (Scheme 14). The reaction presumably proceeds via a diiodocarbene complex intermediate which is further reduced and reacts with another iron complex. Indeed if iron powder is used as the reducing agent and carbon tetrachloride as the source of carbon, only the dichlorocarbene complex is isolated ⟨77CC648, 78AG(E)781⟩.

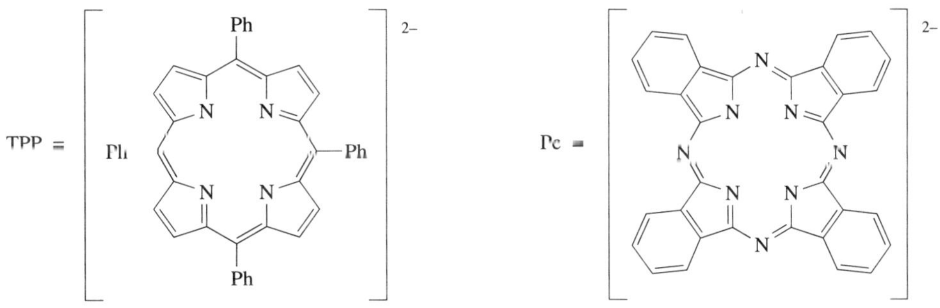

Scheme 14

Of course only symmetrical molecules can be prepared by this method. Soon afterwards, the chemistry of the dihalocarbene metal complexes was supplemented by the first isolation of a chloromethylidyne complex ⟨83CC157⟩. The process is similar to that mentioned above for the dihalocarbene metal complexes (Scheme 15). An oxidising agent, such as a diazonium salt, is used and dichloromethane is the source of carbon. After the formation of chlorocarbene metal complex, reductive elimination affords the chloromethylidyne metal complex. Later it was shown that nucleophilic conjugate addition to carbon followed by ejection of the chloride anion allows replacement of the chloride by a range of nucleophiles ⟨84CC75⟩. The method was soon extended to metal anion nucleophiles and thus unsymmetrical M=C=M′ molecules, **(18)** ⟨87JA4731⟩, **(19)** ⟨90AG(E)318⟩ and **(20)** ⟨91JA2324⟩ were prepared (Scheme 16).

It should be noted that, as mentioned before, the bonding order of the central carbon to the two metals is not the same. This is confirmed by x-ray crystal structure determination of all known examples of this species.

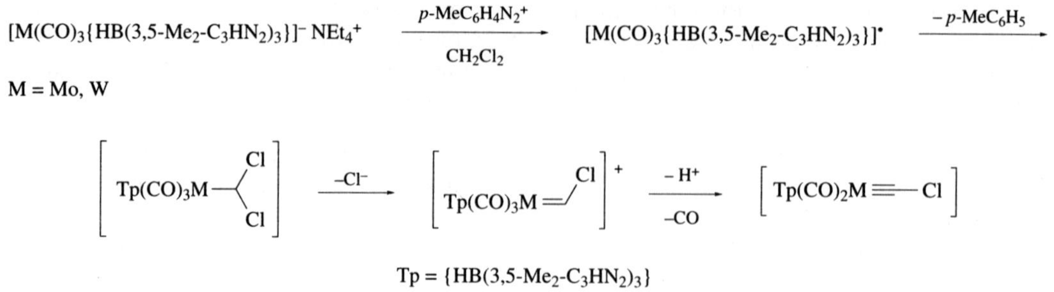

Scheme 15

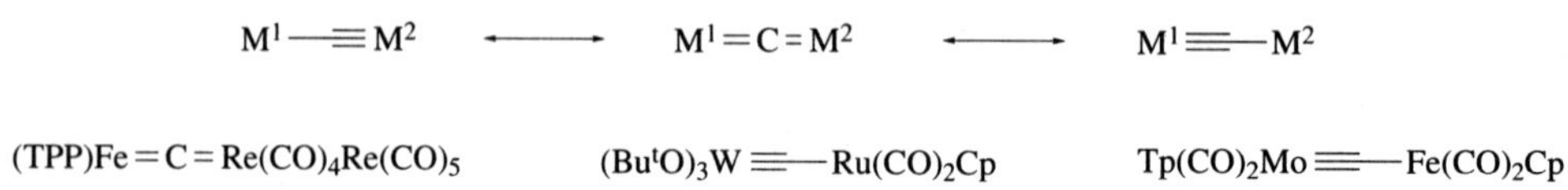

(18) (19) (20)

Scheme 16

5.30
Nitriles with a Heteroatom Attached to the Cyanocarbon

IAN K. BODDY
AgrEvo UK Ltd., Saffron Walden, UK

and

MARK J. FORD
AgrEvo UK Ltd., Cambridge, UK

5.30.1 SINGLY BONDED OXYGEN DERIVATIVES—RO—C≡N (CYANATES)

There are a number of reviews on singly bonded oxygen derivatives, most notably by Grigat ⟨67AG(E)206, 72AG(E)949⟩, Hedayatullah ⟨68BSF1472⟩, Martin ⟨67ZC123⟩, Prejzner ⟨68MI 530-01⟩ and Jensen ⟨B-77MI 530-01⟩. Whilst these articles have already covered the subject in some detail up until 1977, this review, for completeness, has included some material already discussed by these authors.

The chemistry of organic cyanates first gained acceptance with the synthesis by Stroh and Gerber of cyanates derived from sterically hindered phenols ⟨60AG1000⟩, and other pioneering work by Hedayatullah ⟨64MI 530-01⟩ and Kauer ⟨64JA4732⟩ soon followed. The latter had shown that it was possible to prepare previously inaccessible cyanates of normal alcohols. In subsequent years, a great deal of work on cyanates was carried out so that the chemistry of the cyanate group —OCN is well understood.

The most general methods for the synthesis of cyanates are via the thermolysis of thiatriazoles or the reactions of phenols or alcohols with cyanogen halides. There are, however, a number of other synthetic routes available; thus, cyanates may also be prepared from *O*-alkylthiocarbamates, *O*-alkyl-*N*-hydroxythiocarbamates, thiocyanates or from cyanic acid itself. These methods will be discussed in full.

5.30.1.1 Cyanate Formation via the Thermolysis of Thiatriazoles

Jensen *et al.* ⟨64ACS825⟩ have described the synthesis of 5-alkoxy-1,2,3,4-thiatriazoles and reported on their decomposition to nitrogen and sulfur. In a subsequent paper ⟨64ACS826⟩, Jensen and Holm describe how this instability allows for the isolation of ethyl cyanate in 84% yield. Using this method it was possible to prepare a range of alkyl cyanates ⟨65ACS438, 66ACS2107⟩; the reaction proceeds in good yield in all cases except for methyl cyanate (Equation (1)). This may be explained by the extreme lability of this particular compound. The *t*-butyl analogue could also be prepared using this route ⟨69ACS1567⟩. However, in this case the reaction took place in petroleum ether and required heating to 45 °C (Equation (2)). Similar reactions have been reported by Martin ⟨64TL2829⟩. Similarly, aryl cyanates have been prepared via the decomposition of the corresponding 5-aryloxy-1,2,3,4-thiatriazoles in benzene at room temperature (Equation (3)) ⟨64AG(E)311, 65MI 530-01, 67BSF422⟩. A number of other groups have been investigated, but the instability of benzyl cyanate, phenethyl cyanate and allyl cyanate has prevented the isolation of such compounds ⟨69ACS1567, 70ACS1512⟩. It has been shown, however, that 5-propargyloxythiatriazole will decompose quantitatively to the allenylisocyanate via a [3,3]-sigmatropic rearrangement of the transient propargyl cyanate in solution at room temperature ⟨92AG(E)866⟩. If the decomposition is allowed to proceed in the presence of hydrogen sulfide, the intermediate cyanate is trapped as the thiocarbamate (Scheme 1).

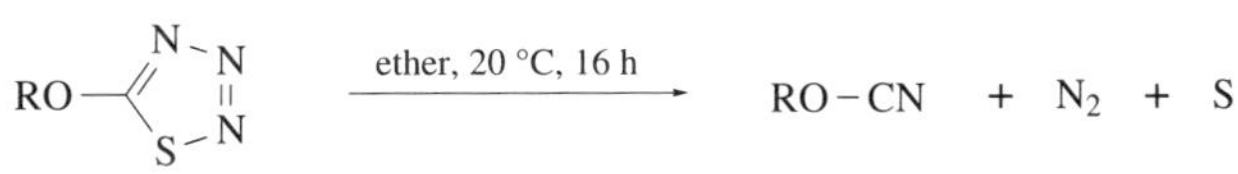

(1)

$R = Me, Et, Pr, Pr^i, Bu, Bu^i, Bu^s, C_5H_{11}, C_6H_{13}, C_7H_{15}$

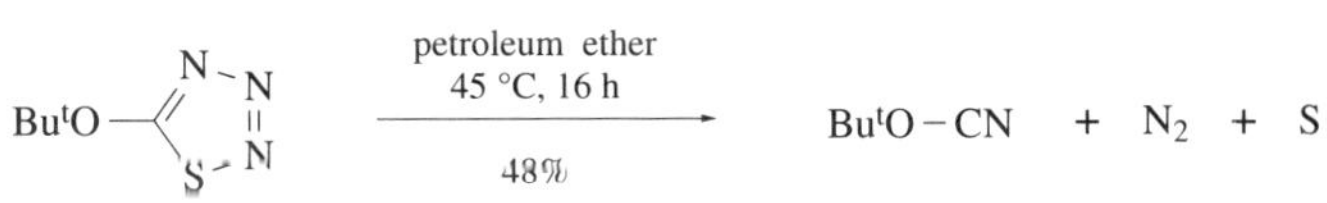

(2)

(3)

$Ar = C_6H_5, p\text{-}C_6H_5—C_6H_4$

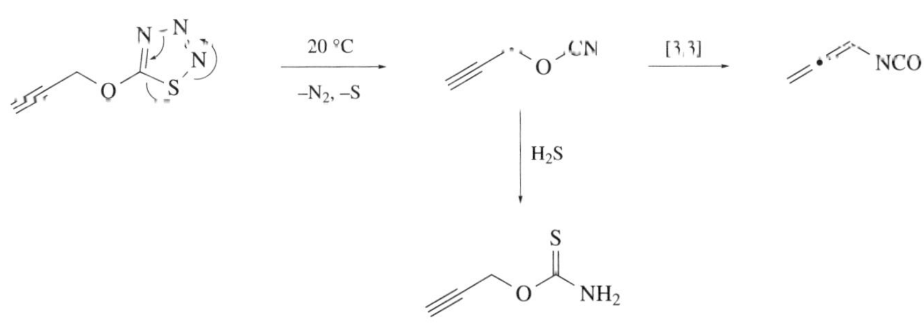

Scheme 1

5.30.1.2 Cyanate Formation via the Reaction of Phenols or Alcohols with Cyanogen Halides

Initially it was found ⟨60AG1000, 67BSF416⟩ that alkali metal phenolates would only give the corresponding cyanates when both *ortho* positions on the aromatic ring were blocked by bulky substituents, for example the *t*-butyl or phenyl group (Equation (4)), whilst stable alkyl cyanates (R—OCN) may be prepared if the alkyl group is suitably bulky to prevent trimerization and the carbonium ion R^+ is of sufficiently high energy to prevent solvolysis and/or rearrangement under the reaction conditions. Thus, Kauer *et al.* ⟨64JA4732⟩ have prepared cyanates from 1,4-dihydroxybicyclo[2.2.2]octane (**1**), 3-ethoxy-2,2,4,4-tetramethylcyclobutan-1-ol (**2**) or neopentyl alcohol (**3**) via the preformed metal salt and cyanogen chloride. When the reacting alcohol has no such sterically hindering groupings, the cyanate molecule formed reacts with excess phenolate to form the imidocarbonate ⟨B-77MI 530-01⟩. It is, however, possible to prepare aryl cyanates by addition of a phenolate to a solution of cyanogen halide ⟨70BSF2637⟩ or by gradual addition of triethylamine to a solution of the phenol and cyanogen halide in a suitable solvent (Equation (5)) ⟨64CB3012⟩. Whilst not generally applicable to the preparation of alkyl cyanates, a number of diols have been reacted as their alkali metal salts with cyanogen halides at low temperatures to give monocyanates along with some of the rearranged isocyanates (Equation (6)) ⟨79USP4250105⟩. A number of patents describing the synthesis of polyfluorocyanates exist in the literature. These compounds are useful intermediates in the preparation of thermosetting resins used in the electronics industry. Thus, phenyl-1,4-bis(hexafluoroisopropanol) reacts with cyanogen bromide in dichloromethane in the presence of triethylamine to give the biscyanate in moderate yield (Equation (7)) ⟨89JAP03145452⟩. Similarly, 2-methylhexafluoroisopropanol as either the sodium or potassium salt reacts with cyanogen chloride under anhydrous conditions (Equation (8)) ⟨71USP3761501⟩, whilst 2,2,3,3,4,4-hexafluoro-1,6-hexanediol reacts with cyanogen chloride in the presence of triethylamine in acetone at low temperatures to give the dicyanate (Equation (9)) ⟨68USP3733349⟩. Sodium hydride in *p*-xylene has also been used as a base to effect these transformations (Equation (10)) ⟨69JOC1278⟩.

(4)

R = Bu^t, Ph; M = K, Na; X = Cl, Br

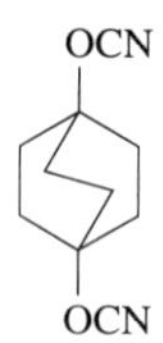

(**1**)

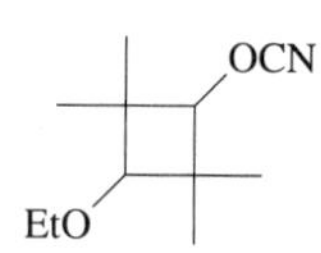

(**2**)

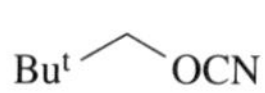

(**3**)

$$\text{ArOH} + \text{XCN} \xrightarrow[\text{acetone}]{\text{Et}_3\text{N}} \text{ArOCN} + \text{HX} \qquad (5)$$

X = Cl, Br

$$\text{NaO(Z)ONa} + \text{XCN} \xrightarrow[-30\,^\circ\text{C to } 0\,^\circ\text{C}]{\text{ether}} \text{NCO(Z)OH} + \text{OCN(Z)OH} \qquad (6)$$

Z = alkylene, oxaalkylene, thioalkylene; X = Cl, Br

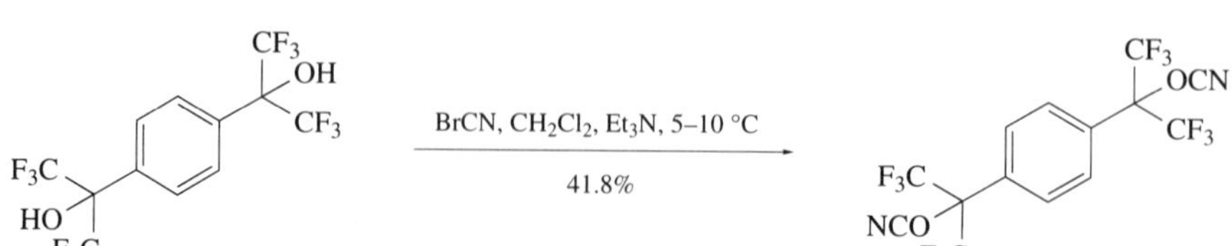

(7)

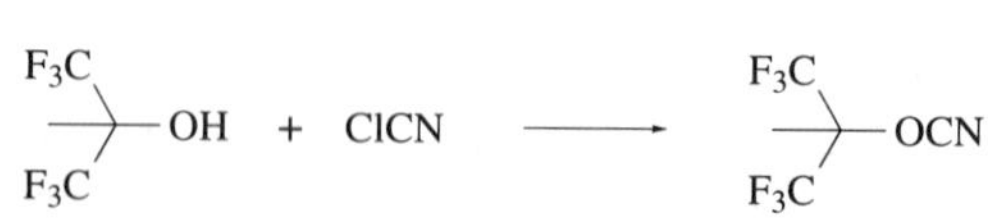

(8)

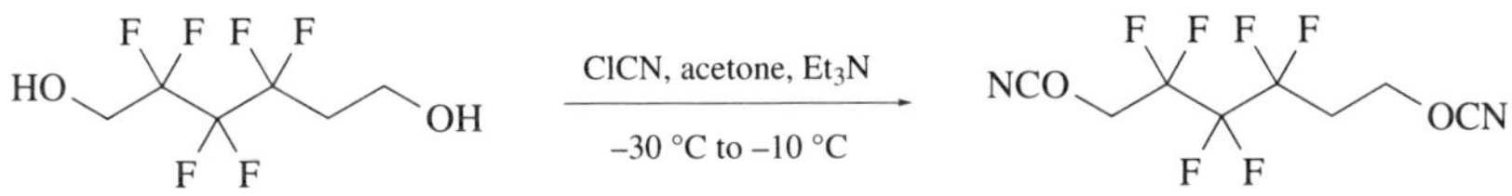

(9)

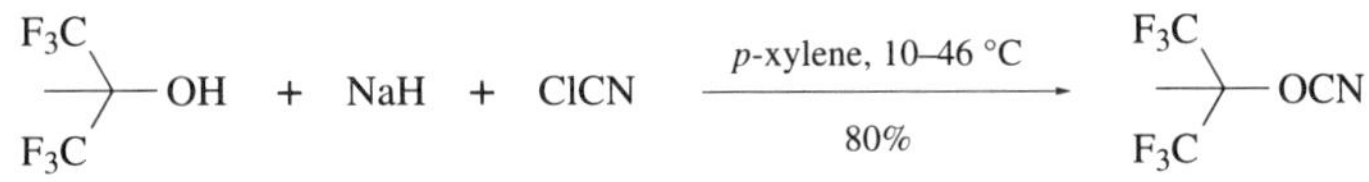

(10)

Thallium(I) alkoxides have also been used in the preparation of alkyl cyanates. Thus, 1,6-dicyanatohexane was obtained by the reaction of the bis-thallium salt of 1,6-hexanediol with cyanogen chloride (Equation (11)) ⟨79USP42163165⟩. Similarly, thallium(I) ethoxide reacted with cyanogen chloride in ether at 0 °C to give ethyl cyanate, which readily isomerized to the isocyanate on warming (Equation (12)) ⟨79USP42163165⟩.

(11)

$$\mathrm{TlOEt} + \mathrm{ClCN} \xrightarrow{\text{ether, 0 °C, 1.5 h}} \mathrm{EtOCN} \quad (12)$$

Cyanates can also be prepared by the reaction of a phenol or an alcohol with bromine and an alkali cyanide at low temperature, presumably via the *in situ* formation of cyanogen bromide (Equation (13)) ⟨79GEP142333⟩.

$$\mathrm{Br_2} + \mathrm{NaCN} \xrightarrow[\text{85 95\%}]{\text{i, H}_2\text{O, –5 °C to 5 °C; ii, ROH, CCl}_4\text{, Et}_3\text{N, –5 °C to 5 °C}} \mathrm{R{-}OCN} \quad (13)$$

R = alkyl, aryl

5.30.1.3 Cyanate Formation from *O*-Alkyl Thiocarbamates and Thiocyanates

Jensen and Holm have shown that alkyl cyanates can be prepared by the reaction of *O*-alkyl thiocarbamates with either mercury(II) or silver(I) oxide (Equation (14)) ⟨64ACS2417⟩. The reaction was carried out in ether with magnesium sulfate present to remove the water formed in the reaction. It is also reported ⟨64ACS2417⟩ that silver(I) oxide gives slightly higher yields (≈5%) than mercury(II) oxide, while lead(II) oxide, lead(IV) oxide, copper(I) oxide and copper(II) oxide react in a similar manner but at a slower rate. Similarly, *O*-ethyl selenocarbamate reacts rapidly with mercury(II) oxide, but the yield of ethyl cyanate is low (Equation (15)) ⟨64ACS2417⟩. Methyl cyanate has been prepared by passing *O*-methyl thiocarbamate through a U tube packed with yellow mercury(II) oxide at room temperature. The product was identified by both mass and microwave spectroscopy (Equation (16)) ⟨90JSP62⟩.

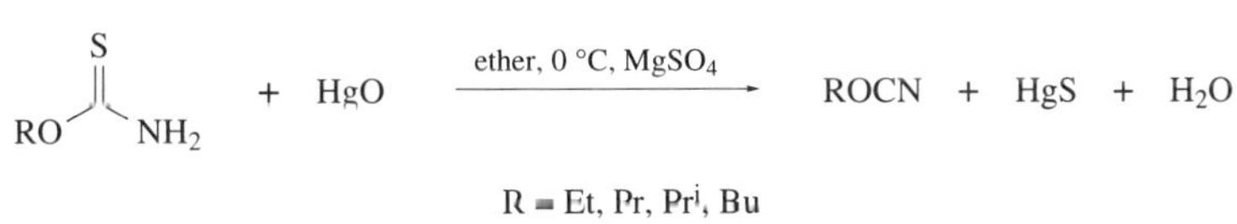

(14)

R = Et, Pr, Pr^i, Bu

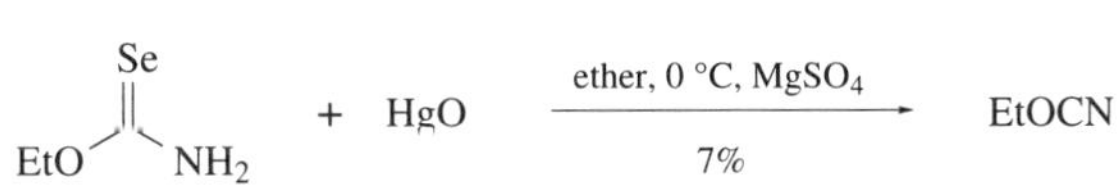

(15)

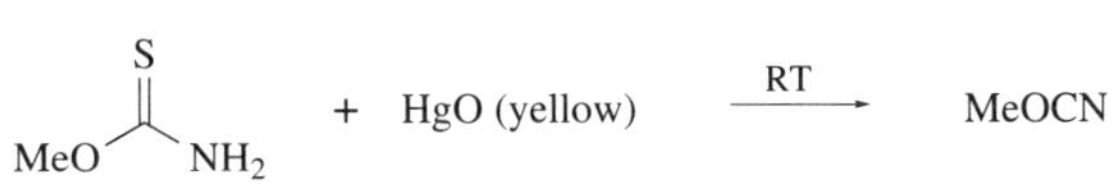

(16)

O-Aryl thiocarbamates are reported to react with silver(I) oxide but not mercury(II) oxide to generate aryl cyanates. The aryl cyanate thus formed was identified by infrared spectroscopy (Equation (17)) ⟨65MI 530-01, 67BSF422⟩.

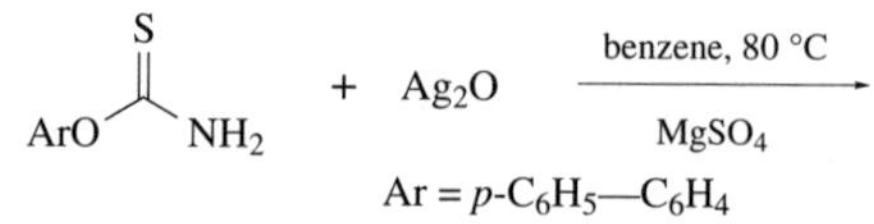

(17)

Imidothiocarbamates, which may be readily prepared from thiocyanates ⟨68AG(E)67⟩, also react with mercury(II) oxide in the presence of magnesium sulfate, to give alkyl, aralkyl, cycloalkyl and aryl cyanates (Scheme 2) ⟨68AG(E)67⟩.

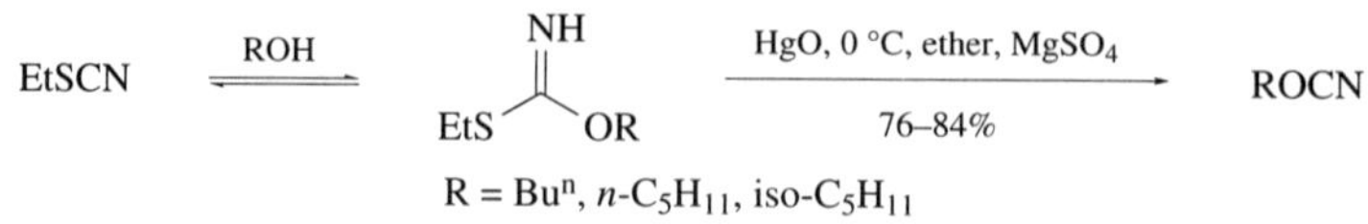

Scheme 2

5.30.1.4 Cyanate Formation from *O*-Alkyl *N*-Hydroxythiocarbamates

O-Alkyl *N*-hydroxythiocarbamates upon reaction with an acylating agent, preferably ethyl chloroformate, rapidly break down to give alkyl cyanates, sulfur, carbon dioxide and ethanol (Equation (18)). The yields of higher alkyl cyanates are generally good, and in some cases the yields are better than those obtained via the thiatriazole method (see Section 5.30.1.1) ⟨68ACS2019⟩.

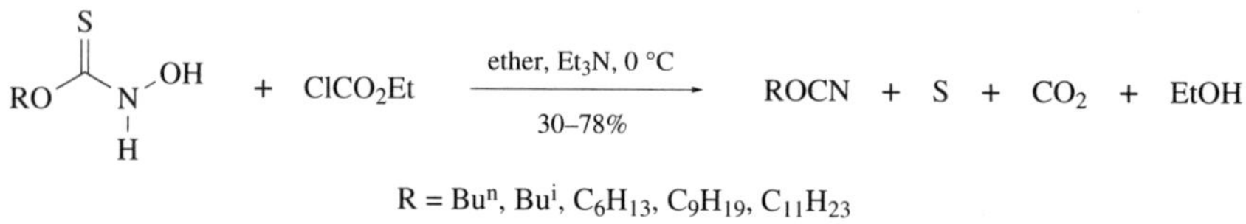

(18)

5.30.1.5 Cyanate Formation from Cyanic Acid or Metal Cyanates

It has been reported that diazoisobutane reacts with cyanic acid to give a low yield of butyl cyanate (4%) ⟨66ACS2123⟩. Isopropyl iodide and *s*-butyl iodide gave mixtures of the cyanates and isocyanates when treated with silver cyanate (Equation (19)) ⟨66ACS2123⟩, although as can be seen from the yields, this method is of no practical use for the preparation of alkyl cyanates.

$$\text{R—I} + \text{AgOCN} \xrightarrow{\text{solvent}} \text{ROCN} + \text{RNCO} \qquad (19)$$

R	*Solvent*	*Yield (%)*	
		ROCN	*RNCO*
Pr^i	$MeNO_2$	20	19
Bu^s	$MeNO_2$	8	8
Bu^s	ether or pentane	4	30

n-Pentyl cyanate has been prepared by the reaction of 1-chloropentane and sodium cyanate using a surfactant-catalysed biphasic solid–liquid system. The reaction is carried out using either cetyltrimethylammonium bromide (CTAB) or cetylpyridinium chloride as the catalyst (Equation (20)) ⟨89JCR(S)104⟩.

$$n\text{-}C_5H_{11}Cl + \text{NaOCN} \xrightarrow[31\%]{\text{H}_2\text{O, NaOAc, CTAB, 100 °C, 48 h}} n\text{-}C_5H_{11}OCN \qquad (20)$$

5.30.2 THIOCYANATES—RSCN

Organic thiocyanates (R = carbon) can be readily prepared by a wide range of methods, most commonly by the formation of the R—S bond or, to a lesser degree, the S—CN bond. The most recent review of their syntheses was by Guy ⟨B-77MI 530-02⟩. This section will deal mostly with the synthetic progress following this review. However, many of the syntheses and observations are still pertinent, and earlier work will be referenced where appropriate.

5.30.2.1 Formation of the R—SCN Bond

This section is divided according to the nature of the thiocyanate source.

5.30.2.1.1 Formation of the R—SCN bond by reaction with nucleophilic sulfur

The formation of the R—SCN bond by reaction with nucleophilic sulfur is often the method of choice for the preparation of thiocyanates. However, simple displacement of a leaving group by the thiocyanate anion (or its equivalent) suffers from the ambident nature of the nucleophile. Thus, a potential and often serious by-product is the isothiocyanate. While modern spectroscopic methods allow for the ready detection of this product, it is not always apparent from the literature whether its formation has been avoided, detected and removed by purification, or if it in fact remains an impurity.

Furthermore, even if the synthetic procedure initially results in exclusive formation of the thiocyanate, this may subsequently isomerize to the thermodynamically more stable isothiocyanate ⟨65JA4791, 74BCJ2745⟩.

Historically, the most common method for preparing alkyl thiocyanates is by the reaction of a metal thiocyanate with an alkyl halide. With a traditional reagent, sodium, potassium or ammonium thiocyanate, optimum yields were obtained in dipolar aprotic solvents, such as dimethylformamide, diethylformamide, dimethyl sulfoxide and sulfolane, as expected for an S_N2 reaction. The reactivity of the halide is, as expected, I > Br > Cl > > F ⟨78S577⟩. The degree of isothiocyanate formation increases with temperature, increasing substitution on carbon and increasing aromatic substitution ⟨61JCS2428, 61JCS2436, 61JCS2447⟩. However, this remains a very practical method for preparing primary and secondary thiocyanates; Equations (21) and (22) show example reactions ⟨92SC617, 92TL5047⟩.

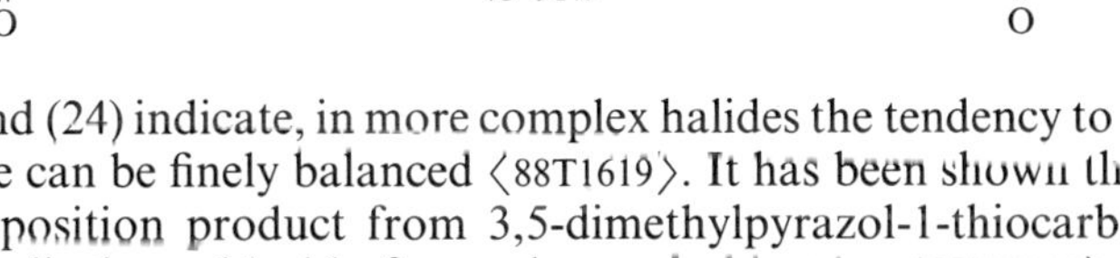

(21)

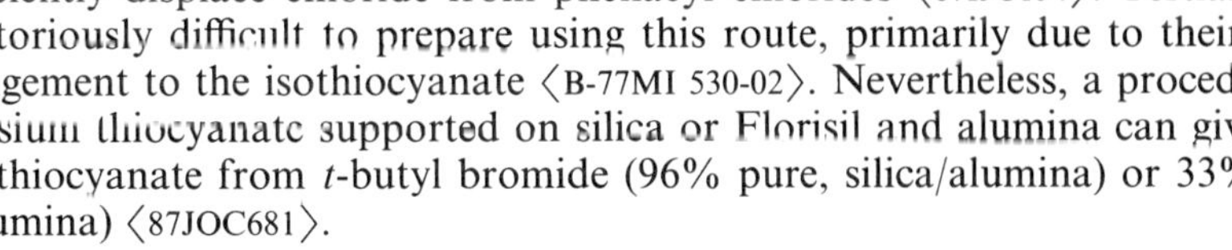

(22)

As Equations (23) and (24) indicate, in more complex halides the tendency to form the thiocyanate over the isothiocyanate can be finely balanced ⟨88T1619⟩. It has been shown that isothiocyanic acid generated as a decomposition product from 3,5-dimethylpyrazol-1-thiocarboximide in refluxing ethanol can efficiently displace chloride from phenacyl chlorides ⟨87IJC154⟩. Tertiary alkyl thiocyanates are notoriously difficult to prepare using this route, primarily due to their tendency to undergo rearrangement to the isothiocyanate ⟨B-77MI 530-02⟩. Nevertheless, a procedure based on the use of potassium thiocyanate supported on silica or Florisil and alumina can give up to 45% yield of *t*-butyl thiocyanate from *t*-butyl bromide (96% pure, silica/alumina) or 33% yield (99% pure, Florisil/alumina) ⟨87JOC681⟩.

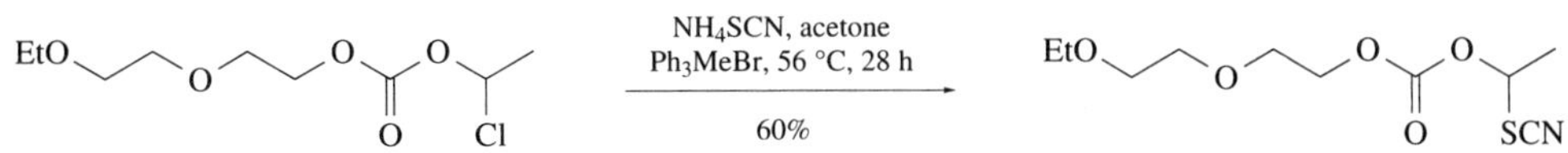

(23)

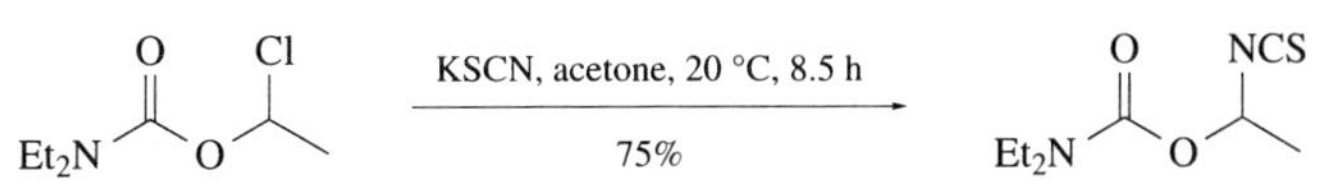

(24)

The reaction is also amenable to phase transfer catalysis ⟨72CPB427⟩, an interesting example being the use of the sulfoxide (**4**) as a neutral phase transfer catalyst ⟨81TL4409⟩. Many other different classes of phase transfer catalysts have been used for this reaction ⟨85JOC3717, 90ZN(B)1091⟩.

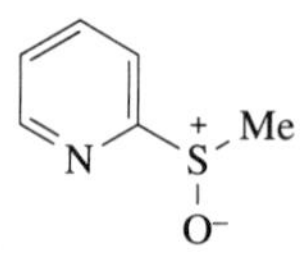

(4)

Trimethylsilyl isothiocyanate has also been used as a thiocyanating agent by nucleophilic displacement of halides. However, the product distribution is very dependent on the structure of the alkyl halide and the reaction conditions, and this, together with the use of hexamethylphosphoramide (HMPA) as the solvent of choice, has probably limited its use ⟨87BCJ2692⟩.

Other leaving groups, such as $ROSO_2R$ ⟨74LA901, 86CB1594⟩, $ROSO_3R$ ⟨88JA7538⟩ and $ROSO_2NR_2$ ⟨90TA881⟩, are also readily displaced by the thiocyanate anion. The reaction conditions, solvent and product distributions are similar to those found for the alkyl halides (see Equations (25)–(27)). The reaction may also be performed under aqueous conditions by use of the 2-trimethylammoniumethyl sulfonate group ⟨78JA1637⟩.

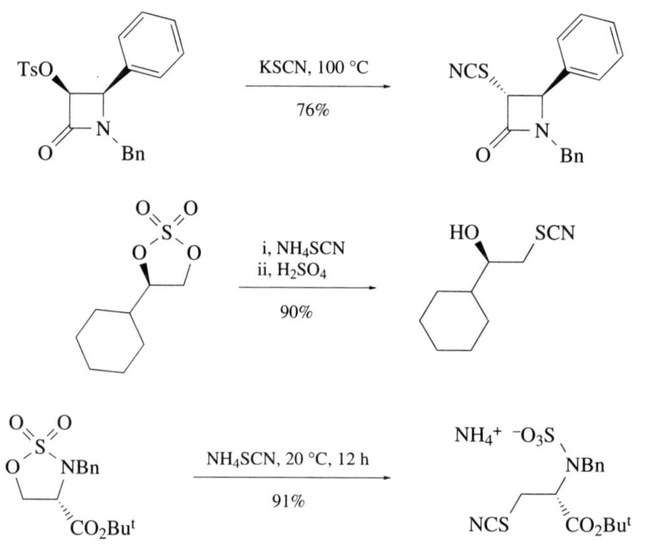

(25)

(26)

(27)

Thiocyanates may also be prepared by the displacement of the —CO_2 group from β-lactones ⟨52JA1323⟩, and by the displacement of activated nitronate esters ⟨72CCC3144⟩.

Epoxides are readily attacked by the thiocyanate anion. Under standard basic conditions the resulting β-alkoxy thiocyanate is rapidly converted into an episulfide ⟨B-77MI 530-02⟩. However, under acidic conditions the β-hydroxy thiocyanate may be isolated ⟨63MI 530-01⟩. The reaction may be catalysed by the use of Lewis acids (Equation (28)) ⟨91T5193⟩, or by palladium(0) (Equation (29)) ⟨90SC2313⟩. The palladium(0) reaction, however, gives far greater regioselectivity in the opening of 2,3-epoxyalcohols, with predominantly the 3-thiocyanato derivative being observed.

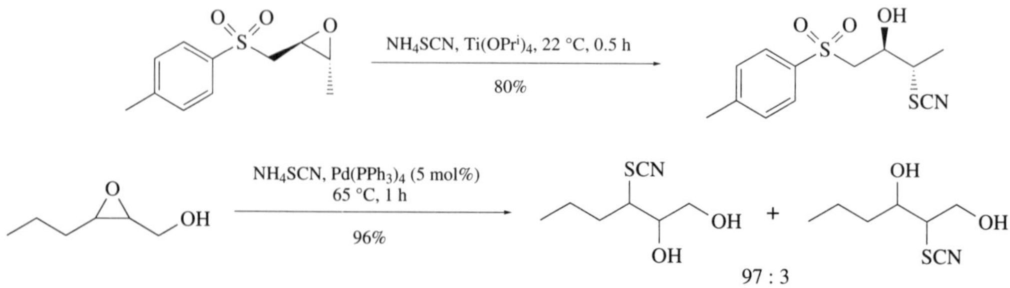

(28)

(29)

α,β-Epoxyketones are converted into α-thiocyanato α,β-enones by triphenyl(dithiocyanato)-phosphine (Equation (30)) ⟨79TL1129⟩. For simple epoxides, reaction with triphenylphosphine/thiocyanogen gives the dithiocyanato or the hydroxythiocyanato product, depending on the epoxide structure and the reaction conditions ⟨81JCS(P1)1577⟩. An alternative route to the β-hydroxy thiocyanates involves the reaction of trimethylsilyl isothiocyanate with cyclic *ortho*-esters at reflux. The product is obtained as a thiocyanato ester (Equation (31)) ⟨79GEP2817455⟩.

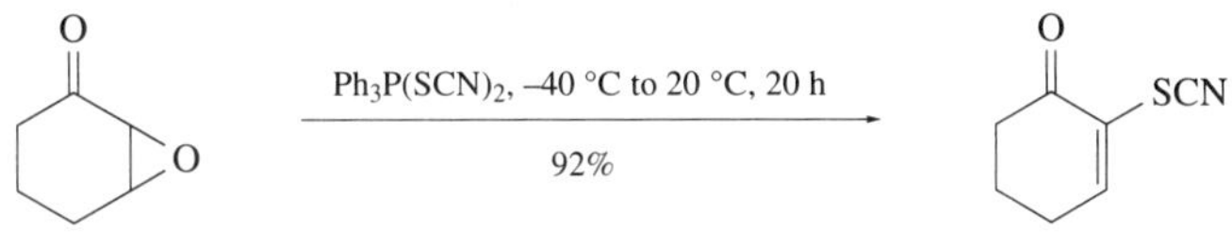

(30)

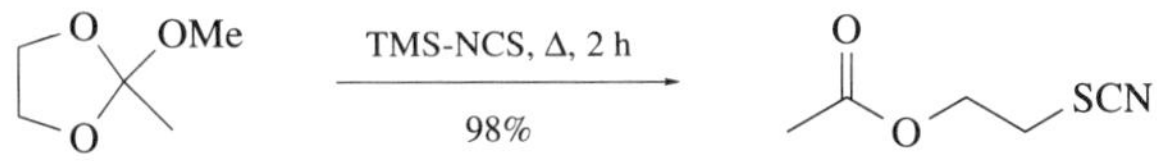

(31)

Aziridines and oxaziridines (C—O bond cleavage) also undergo ring opening with isothiocyanic acid ⟨B-77MI 530-02⟩.

Alcohols ⟨76HCA2100, 77TL4417, 83T4175⟩ and amines ⟨78CC133, 80JCS(P1)865, 82S472⟩ may be converted directly into thiocyanates by prior or *in situ* activation and displacement (e.g., Equations (32)–(34)). Simple benzylic thiocyanates have been prepared from the corresponding alcohols by heating with dilute acid and sodium thiocyanate ⟨48M316⟩. However, a more controlled reaction is the conversion of benzylic or allylic alcohols into thiocyanates by electrochemical oxidation with sodium thiocyanate and triphenyl phosphite ⟨90CPB1389⟩.

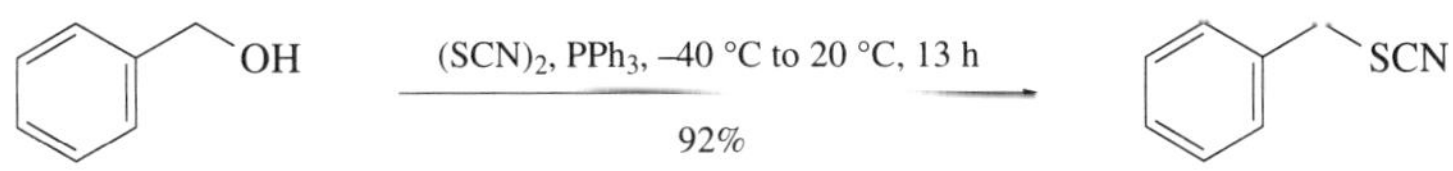

(32)

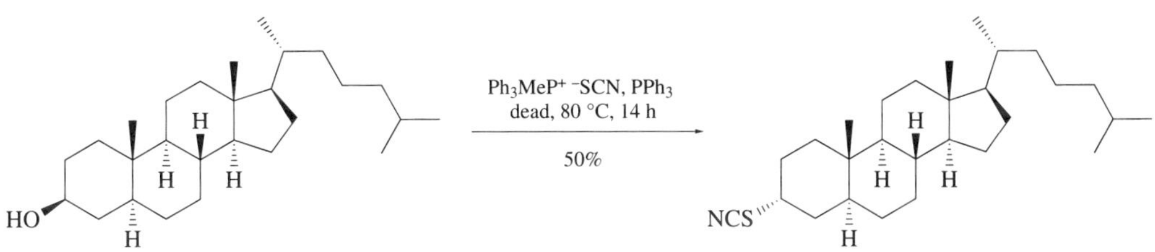

(33)

dead = diethyl azodicarboxylate

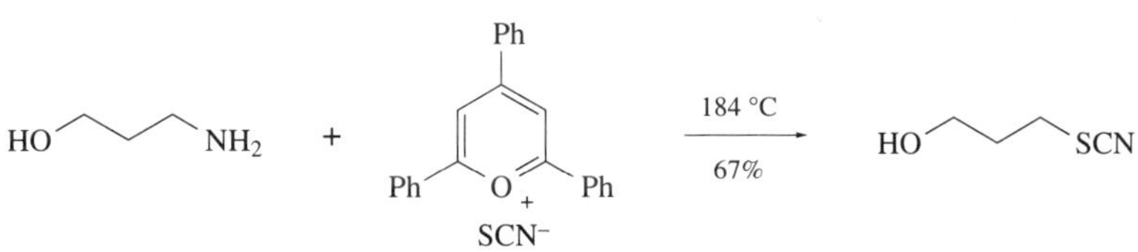

(34)

The product distribution resulting from the conjugate addition of isothiocyanic acid to α,β-unsaturated carbonyl compounds depends on the degree of substitution at the reacting centre ⟨B-77MI 530-02⟩. However, addition to propiolates and propiolamides affords the (*E*)-β-thiocyanato product exclusively ⟨65JOC2660, 92JOC5270⟩. The thiocyanate anion will also displace the halide from a β-halonitroalkene by a conjugate addition/elimination reaction ⟨89JOC5967⟩.

While the addition of isothiocyanic acid across double bonds gives at best mixtures of the thiocyanate and isothiocyanate ⟨B-77MI 530-02⟩, halogen-promoted thiocyanation can be controlled to give the β-halothiocyanates exclusively. For example, the reaction of potassium thiocyanate with bromine can give either the thiocyanate or isothiocyanate, depending on the reaction conditions ⟨81JCS(P1)58⟩, and while potassium thiocyanate and iodine give mixtures, the use of calcium fluoride-supported potassium thiocyanate and iodine gives the thiocyanate exclusively ⟨87CC1301⟩. In a similar manner, the reaction of alkenes with iodine and tri-*n*-butylstannyl isothiocyanate results in the thiocyanate only (Equation (35)) ⟨84S1017⟩. In this case only the sulfur is available to undergo nucleophilic attack on the iodonium intermediate. The addition of phenylselenenyl thiocyanate to alkenes affords β-thiocyanatoalkyl phenylselenides and β-isothiocyanatoalkyl phenylselenides, where mono- and disubstituted alkenes favoured formation of the thiocyanato species ⟨79CJC2145⟩.

Interestingly, addition to 1-phenylpropene yields exclusively the thiocyanate from the (*Z*) isomer and the isothiocyanate from the (*E*) isomer ⟨79CJC2180⟩. Butadienes can be converted into 1,4-dithiocyanato-2-butenes with potassium thiocyanate/iodine. Elimination with 1,5-diazabicyclo-[5.4.0]undec-5-ene (dbu) then affords 1-thiocyanatobutadienes ⟨85CC533, 87JCS(P1)2729⟩. Meerwein arylation can also be used to generate thiocyanates from alkenes ⟨B-77MI 530-02⟩.

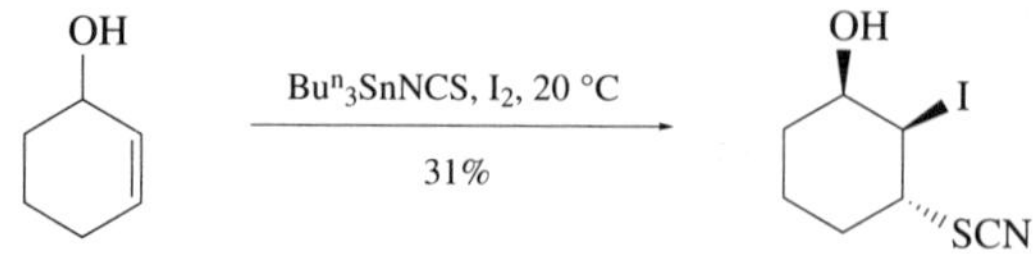

(35)

Isothiocyanic acid adds to terminal alkynes regioselectively in the presence of mercury(II) thiocyanate to give 2-thiocyanatoalkenes ⟨79CC1026⟩. As with alkenes, the addition of thiocyanate to alkynes can be promoted by an initial addition of an electrophile either by an intramolecular (Equation (36)) ⟨88JA612⟩ or an intermolecular reaction (Equation (37)) ⟨90TL7375⟩. Thiocyanatoalkynes can be prepared in a short sequence from a trimethylsilylalkyne or tri-*n*-butylstannylalkyne. Either compound is converted into a phenyliodonium alkyne, which reacts with sodium thiocyanate to give the desired product (Equation (38)) ⟨92SL535⟩.

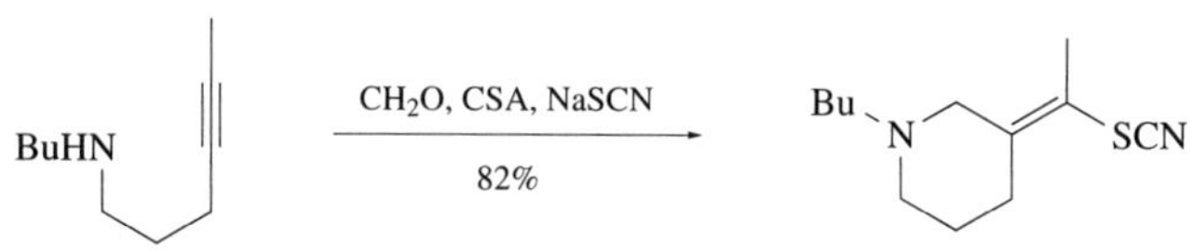

(36)

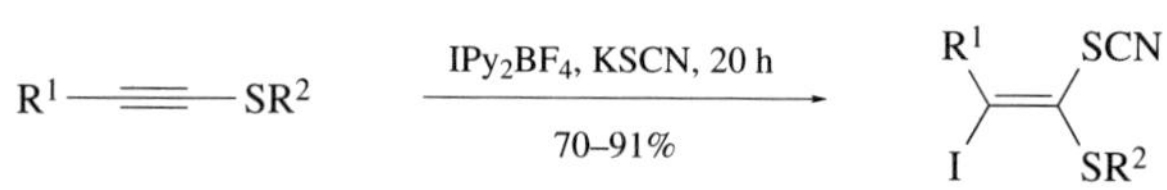

(37)

(38)

α-Diazo esters can be converted into α-phenylseleneno-α-thiocyanato esters by reaction with phenylselenenothiocyanate ⟨85JOM(286)171⟩.

Although many new procedures have been developed to introduce the thiocyanate group as a nucleophile into aliphatic systems, there has been substantially less work on aromatic systems. The standard procedure for preparing an aryl thiocyanate from a source of thiocyanate anions is by the displacement of nitrogen from a diazonium salt. The best yields and thiocyanate-to-isothiocyanate ratios are obtained under Sandmeyer-type reaction conditions using a source of copper(I) thiocyanate as the catalyst (Equation (39)) ⟨73CJC1741⟩. Other transition metal catalysts which have been used include iron(II) sulfate and cobalt(II) chloride ⟨53BSF771, 79BSF237⟩. While doubly activated diazonium salts avoid the need for transition metal salts (Equation (40)), these activating groups themselves may be displaced (Equation (41)) ⟨53JCS959⟩. A doubly activated *o*-nitrobenzenesulfonate may also be displaced directly by potassium thiocyanate ⟨58MI 530-01⟩. 2-Acetamidothiophene can be thiocyanated in the 5 position by treatment with sodium thiocyanate and copper(II) sulfate in a refluxing aqueous system ⟨51JA613⟩. An activated *N*-alkyl benzimidazole has been thiocyanated by treatment of the *N*-oxide with *p*-toluenesulfonyl chloride followed by reaction with sodium thiocyanate (Equation (42)) ⟨73JCS(P1)705⟩.

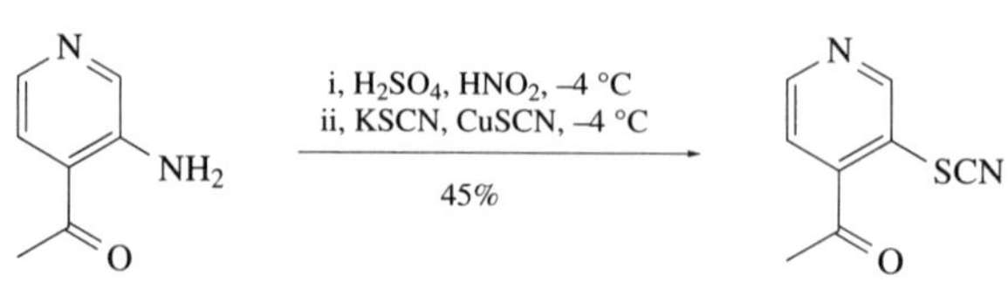

(39)

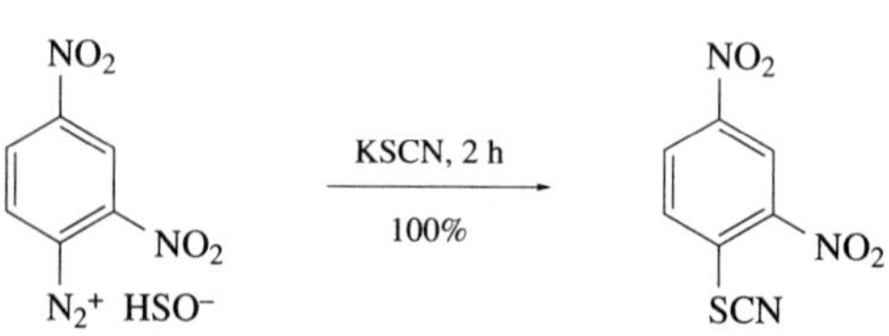

(40)

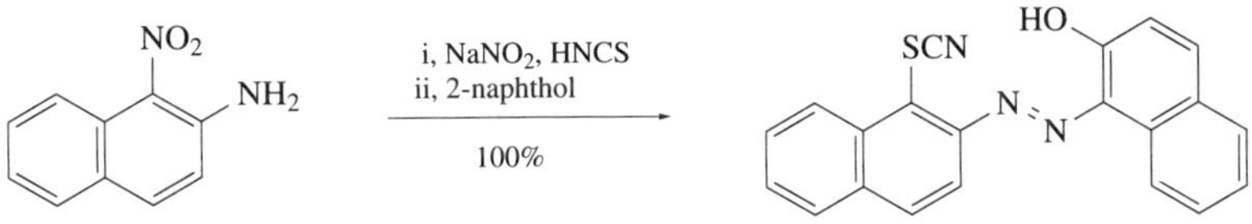

(41)

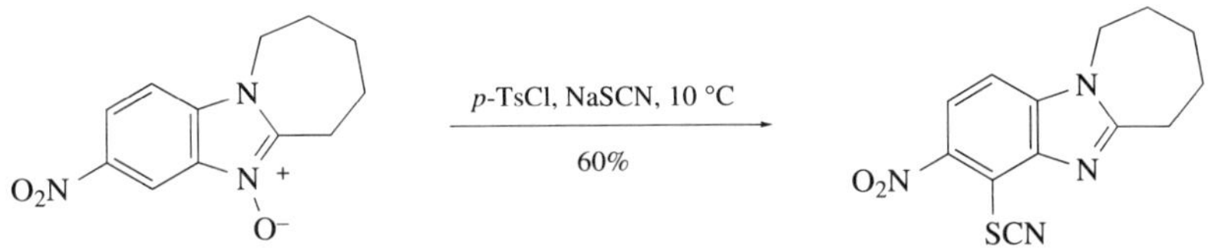

(42)

5.30.2.1.2 *Formation of the R—SCN bond by reactions with electrophilic sulfur*

The primary source of electrophilic 'SCN' is thiocyanogen or the thiocyanogen halides (XSCN, X = Cl or Br).

The reagents behave in a similar fashion to the halogens. Under heterolytic conditions they act as a source of NCS^+, which can, for example, add to electron-rich alkenes, electron-rich aromatics (where the yields can be dramatically increased by the use of Friedel–Crafts catalysts), activated methylene compounds and some transition metal organometallics. The two most commonly used reagents are thiocyanogen and thiocyanogen bromide. Thiocyanogen is made from lead(II) thiocyanate and bromine, and thiocyanogen bromide is prepared from ammonium thiocyanate and bromine. The order of reactivity of these compounds is $(SCN)_2$/Friedel Crafts catalyst > ClSCN > $(SCN)_2$ > BrSCN. Unfortunately, thiocyanogen bromide can give brominated by-products which are derived from the bromine with which the reagent is in equilibrium (Equation (43)) ⟨81JCS(P1)58⟩.

$$2BrSCN \rightleftharpoons (SCN)_2 + Br_2 \qquad (43)$$

This chemistry has been extensively reviewed ⟨B-77MI 530-02⟩. In many addition reactions of thiocyanogen it is difficult to prevent competing radical processes (see Section 5.30.2.1.3). The heterolytic reaction of thiocyanogen with alkenes suffers from a low thiocyanate-to-isothiocyanate ratio. However, it has been shown that the addition of a catalytic quantity of iron(III) thiocyanate results in a dramatic increase in rate and a shift in the product distribution, to give exclusive formation of the dithiocyanato product ⟨78MI 530-01⟩.

The reaction of thiocyanogen or thiocyanogen halides with aromatic compounds is extremely useful as a preparative technique ⟨B-77MI 530-02⟩. The aromatic system must be relatively electron-rich, so that, for example, anilines, phenols, pyrroles and polyaromatics ⟨65JOC184, 68JOC823⟩ react readily and thiophene reacts but gives vastly improved yields in the presence of a Friedel–Crafts catalyst ⟨54ACS1851⟩, whereas pyridine or pyrimidine requires two electron-donating groups attached to the ring ⟨73JOC4383⟩. A few examples of the reaction are shown in Equations (44)–(47).

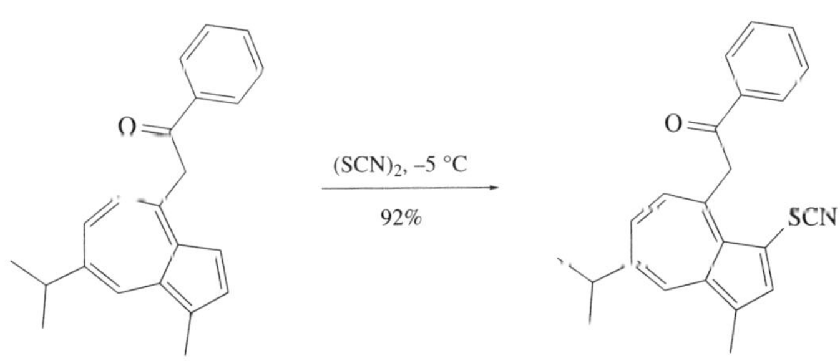

(44)

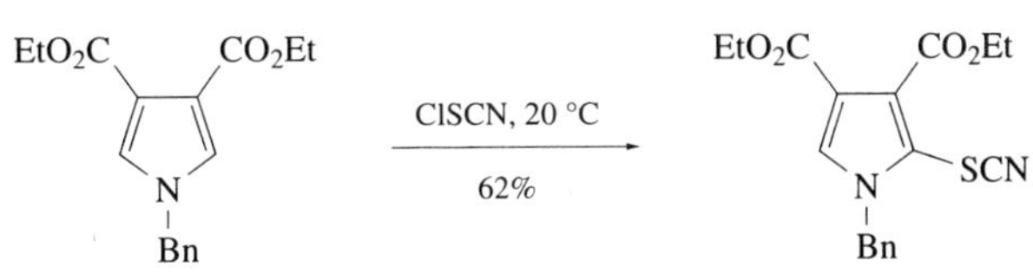

(45)

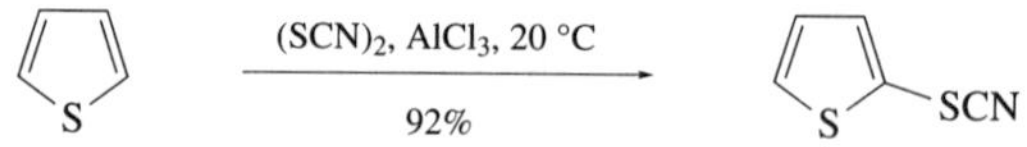

(46)

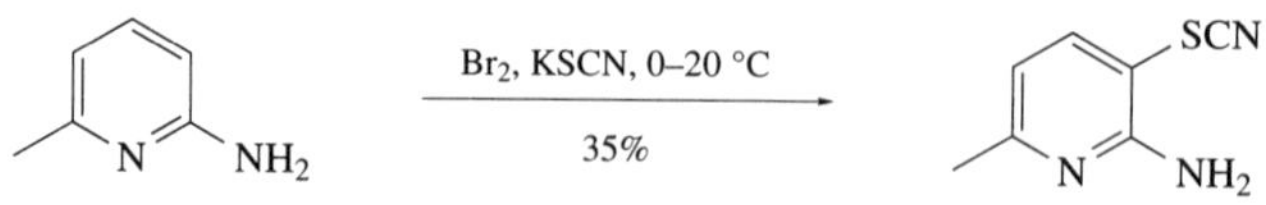

(47)

Interestingly, thiocyanation of a benzene ring by ipso substitution of a trimethylsilyl group has been achieved in the presence of a Friedel–Crafts catalyst (Equation (48)) ⟨84JOC5198⟩.

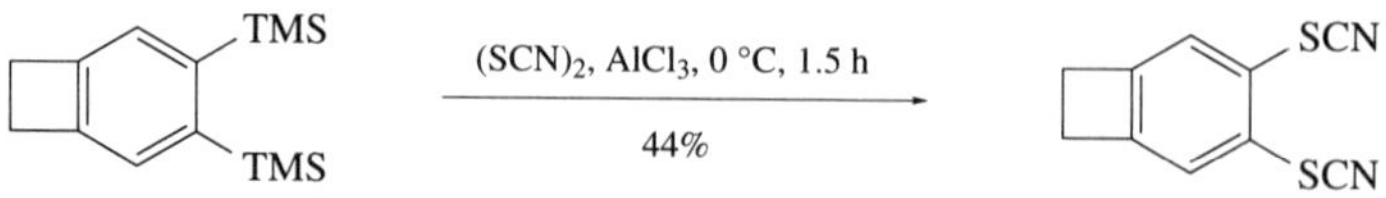

(48)

Enolizable ketones may be thiocyanated upon treatment with a methanolic suspension of copper(II) thiocyanate ⟨81JCR(S)234⟩. It has been shown that aryl, alkynyl or alkyl thiocyanates may be prepared from the corresponding organolithium species by treatment with zinc thiocyanate and NCS ⟨92CL509⟩. Silver acetylides will react with thiocyanogen to give thiocyanatoalkynes ⟨90MI 530-01⟩.

5.30.2.1.3 *Formation of the R—SCN bond by homolytic reactions*

The reaction of thiocyanogen or thiocyanogen chloride with alkenes or alkynes under homolytic conditions has been extensively studied ⟨B-77MI 530-02⟩. The addition of thiocyanogen to an alkene in the presence of a peroxide initiator or ultraviolet light results in the formation of α,β-dithiocyanates or allylic isothiocyanates, depending on the structure of the alkene ⟨78T541⟩. Alkynes afford mono addition products as a mixture of (E)- and (Z)-dithiocyanatoalkenes ⟨80T1839⟩. Thiocyanogen chloride reacts with alkenes under similar conditions to give predominantly the anti-Markovnikov products ⟨B-77MI 530-02⟩.

Thiocyanogen and the thiocyanogen halides will react with compounds containing benzylic or allylic hydrogen atoms to afford the α-thiocyanatoarylalkanes ⟨61JCS2428, 61JCS2436⟩. The reaction is effective for primary and secondary α-hydrogen atoms, but tertiary positions give mixtures of the thiocyanate and isothiocyanate products (*vide supra*).

Radicals generated by other means have also been trapped by a suitable thiocyanate source. For example, Barton esters can be converted into the corresponding thiocyanates upon irradiation in the presence of methanesulfonyl isothiocyanate (Equation (49)) ⟨92T2613⟩.

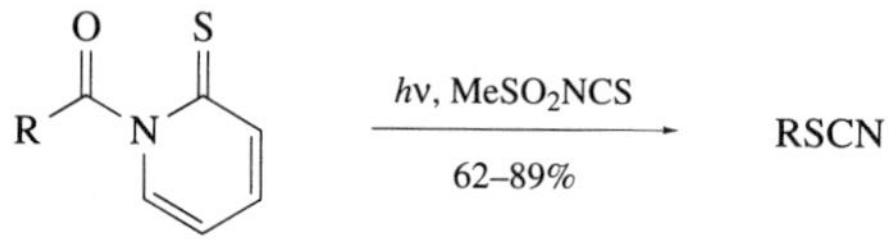

(49)

ω-Thiocyanatocarboxylic acids may be prepared from 1,1-hydroxyhydroperoxides by reaction with copper(I) thiocyanate and isothiocyanic acid (Equation (50)) ⟨59G2428⟩.

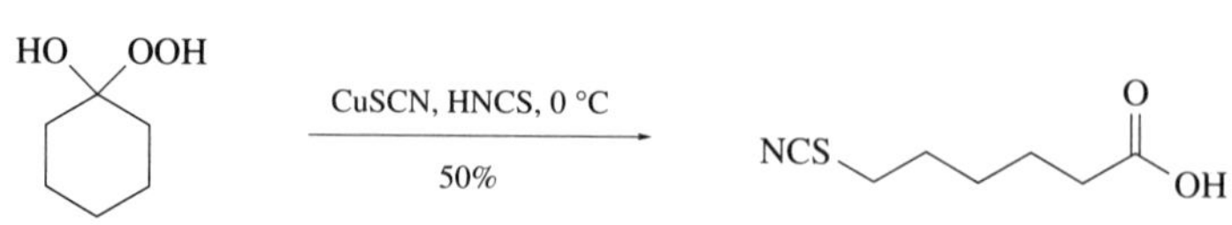

(50)

5.30.2.2 Formation of the RS—CN Bond

Formation of the RS—CN bond to produce thiocyanates has the obvious advantage that it avoids the competing formation of isothiocyanate. However, it does not preclude any subsequent isomerization reaction.

The thiocyanates are synthesized either from a source of cyanide ions and an electrophilic sulfur derivative or a cyanogen halide and a nucleophilic organosulfur compound. In the 1980s, a reductive procedure was developed which starts from a sulfinate or sulfonyl chloride (via the sulfinate).

5.30.2.2.1 *Nucleophilic cyanation of sulfur*

In the nucleophilic cyanation of sulfur a stable leaving group on sulfur is displaced by cyanide ion from a metal cyanide or trimethylsilyl cyanide. The groups that have been displaced include halide (from a sulfenyl halide, formed *in situ* ⟨70JCS(C)1530, 79S181⟩ or preformed under aqueous ⟨56JCS96⟩ and nonaqueous conditions ⟨67AP583, 87TL4189⟩), thiocyanate (from a sulfenyl thiocyanate ⟨64AG861⟩), sulfite (from a thiosulfate ⟨25JCS2887⟩), sulfide (from a disulfide, formed *in situ* ⟨48CB210⟩ or preformed ⟨92B7922⟩), hydantoin (from an *N*-arylthiohydantoin ⟨90H(30)367⟩) and isatin (from an *N*-arylthioisatin ⟨74CL881⟩). A few examples are shown in Equations (51)–(53).

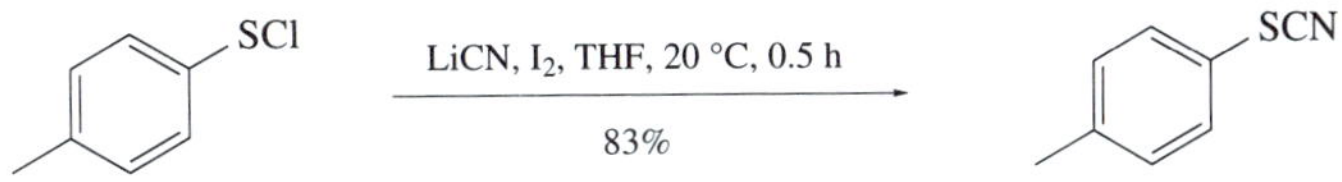

(51)

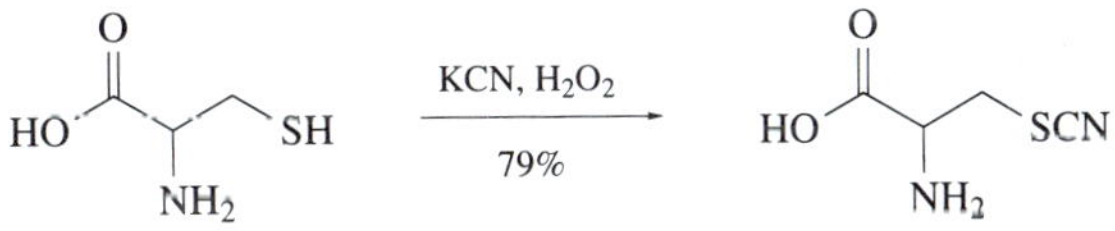

(52)

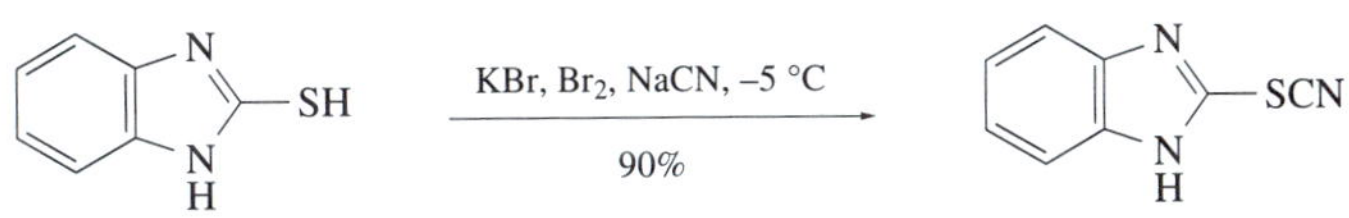

(53)

Interestingly, the use of potassium cyanide and iodine can convert a disulfide into two equivalents of thiocyanate ⟨83MI 530-01⟩. This has been done previously by treatment of a disulfide with a cyanide source followed by a cyanogen halide ⟨71JOC2727⟩.

5.30.2.2.2 *Electrophilic cyanation of sulfur*

In the electrophilic cyanation of sulfur a nucleophilic sulfur compound can displace a halide ion from a cyanogen halide ⟨53JA4655, 91TL2505⟩. An alternative to a cyanogen halide is 2-nitro-5-thiocyanatobenzoic acid, which is a mild and selective cyanating agent for thiols ⟨70JA6969⟩.

Cyanogen chloride and bromide are reactive enough to cleave a thioether, resulting in a thiocyanate and an alkyl halide ⟨78JOC1604⟩. In the case of aryl alkyl thioethers only one product is possible (Equation (54)).

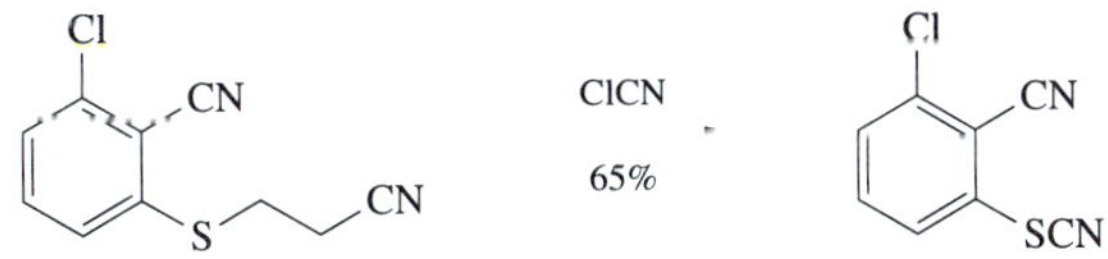

(54)

The introduction of sulfur and then the nitrile moiety may be achieved in a two-step process (Equations (55) and (56)) ⟨80JCS(P1)1029, 89T7411⟩. Displacement of the halide in thiophene (**5**) directly with the thiocyanate anion had been unsuccessful.

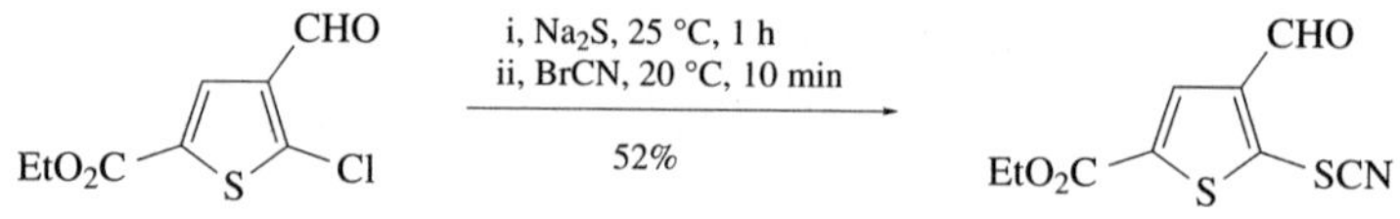

(5)

(55)

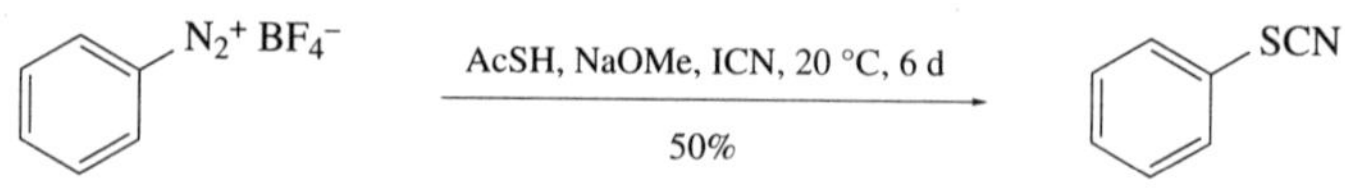

(56)

5.30.2.2.3 Reductive cyanation of sulfur

Arylsulfinate salts react with trimethylsilyl cyanide in HMPA at 0 °C to give the corresponding arylthiocyanates ⟨88CC1485⟩. The sulfinate salt may be formed *in situ* by the reaction of a sulfonyl chloride with sodium sulfite, and the resulting sulfinate converted into the thiocyanate by trimethylsilyl cyanide in refluxing acetonitrile ⟨88CC1485⟩. Arylsulfinic acids may be converted into arylthiocyanates by the use of diethyl phosphoryl cyanide ⟨82TL447⟩.

5.30.2.3 Formation of Acyl Thiocyanates

Acyl thiocyanates can be prepared by the reaction of thiocarboxylate salts with cyanogen chloride ⟨76T745⟩.

Alternatively, activation of a carboxylic acid by pyrocatechyl phosphoryl chloride and subsequent reaction with thiocyanogen affords the corresponding acyl thiocyanate ⟨83T4175⟩.

The reaction of thiourea with butyllithium and methyl iodide results in the formation of *N*,*N*-dimethylthiocarbamoyl thiocyanate ⟨80TL635⟩.

5.30.2.4 Formation of Sulfonyl Cyanides

The reaction of sulfinic acids with cyanogen chloride in water results in the corresponding sulfonyl cyanides ⟨88OSC(4)727⟩. Since sulfonyl chlorides can be readily converted into sulfinates by the action of sulfite/bicarbonate, these may also act as precursors for this reaction ⟨77OS88⟩.

5.30.2.5 Formation of RSCN by Fragmentation/Rearrangement Reactions

There have been several syntheses of thiocyanates which are derived from fragmentation or rearrangement reactions. For example, *N*-arylthioformamides can be converted into arylthiocyanates by phosgene and triethylamine ⟨73TL211⟩.

2-Imino-5,6-dihydro-4*H*-1,3-thiazine **(6)** reacts with diethyl phosphoryl chloride or sulfonyl chlorides to give the corresponding γ-thiocyanatophosphorylamides or sulfonamides (Equation (57)) ⟨89H(29)841⟩.

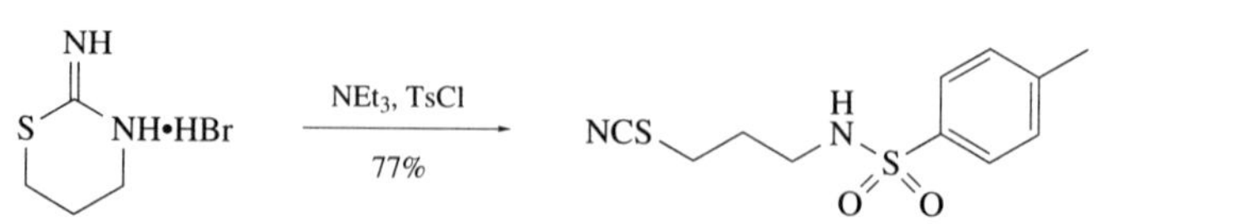

(6)

(57)

An appropriately substituted thiazole may undergo ring cleavage to afford thiocyanates ⟨68HCA1723, 71JCS(C)3262⟩.

5.30.3 SELENO- AND TELLUROCYANATES—RSeCN AND RTeCN

Organic selenocyanates and tellurocyanates may be prepared by a number of routes. This area of chemistry has previously been reviewed by Bulka ⟨B-77MI 530-03⟩ and, more recently, by Toshimitsu and Uemura ⟨B-87MI 530-01⟩.

5.30.3.1 Selenocyanates—RSeCN

Organic selenocyanates are predominantly made by the formation of the R—SeCN bond or, to a lesser extent, the RSe—CN bond.

5.30.3.1.1 Formation of the R—SeCN bond

(i) Formation of the R—SeCN bond by nucleophilic substitution reactions

Nucleophilic substitution reactions are the most common method for the preparation of selenocyanates: usually an alkyl halide is reacted with an alkali metal selenocyanate in a suitable solvent. Some examples of this reaction type are shown in Table 1. When the displacement is attempted on 1-bromocyclobutane-1-carboxylic acid in neutral solution (Equation (58)) the only product isolated is that arising from rearrangement ⟨70ACS3751⟩.

Table 1 Selenocyanates prepared from alkyl halides and potassium selenocyanate, KSeCN.

RX	*Solvent/temperature*	*Yield* (%)	*Ref.*
$ArCH_2X$ (X = Cl, Br)	EtOH/RT	75–88	80SC595
$ArCH_2Cl$	DMF/RT	66–84	79S705
$BrCH_2CH_2Br$	DMSO/RT[a]	77	80JOC2632
$Br(CH_2)_nOH$ ($n = 2, 3$)	Acetone/reflux	70–80	88S133
PhCH(Cl)COPh	Acetone/reflux	48[b]	84JOC2057
EtO_2CCH_2Br	EtOH/reflux	65	86CC1152

[a] Potassium selenocyanate prepared *in situ* from KCN and selenium metal in DMSO at 140 °C. [b] Product obtained is unstable.

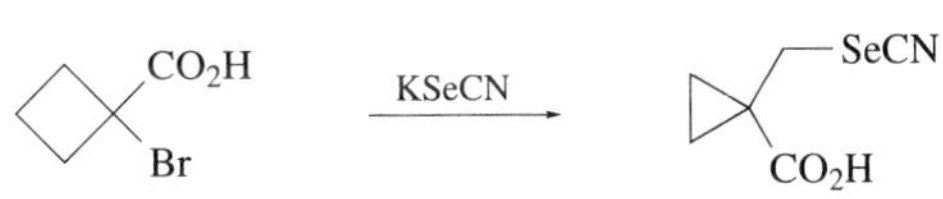

(58)

Displacement of other groups is also known, for example displacement of a tosyl group from the α-silyl tosylates (**7**) with KSeCN under phase transfer conditions (Equation (59)) gave the desired α-silylselenocyanates in good yields ⟨86JA1314⟩.

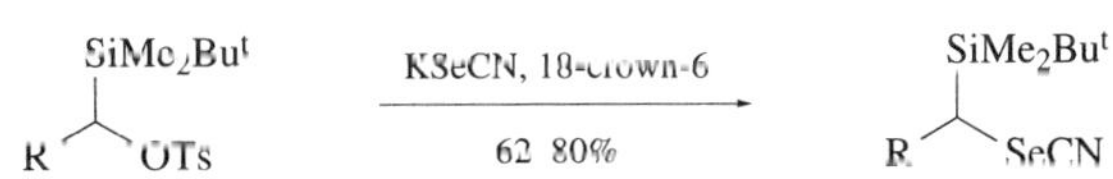

(**7**)

R = alkyl, aryl, benzyl

(59)

The routes described above are generally applicable only to alkyl selenocyanates or to electron-deficient aryl selenocyanates. As a result, when R is an aryl moiety the reaction is usually carried out on suitable diazonium species (e.g., see ⟨62CB1237, 84TL3387⟩). Uemura *et al.* have described the preparation of aryl selenocyanates via selenocyanatodethallation of arylthallium(III) species ⟨75BCJ1925, 80BCJ553⟩ using potassium selenocyanate and copper(II) sulfate in dioxane (Equation (60)).

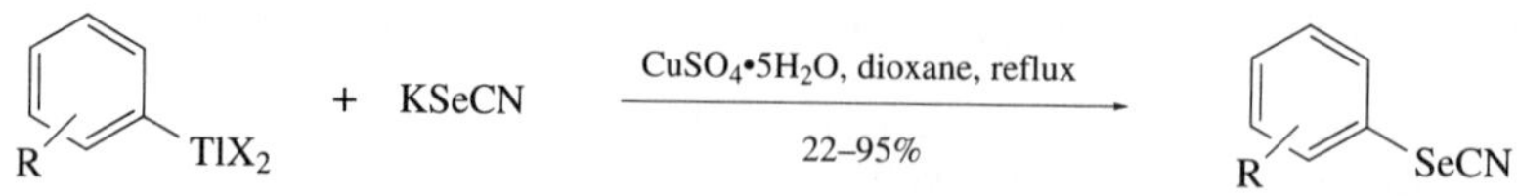

(60)

R = 2-Me, 2,3-Me, 2-OMe, 4-Cl, etc.

The direct conversion of alcohols to alkyl selenocyanates can be achieved using the novel reagent $Ph_3P(SeCN)_2$ (Equation (61)), prepared from equimolar amounts of triphenylphosphine and selenocyanogen. The reagent works best with primary alcohols, with secondary alcohols it gives mixtures of selenocyanates and isoselenocyanates, while tertiary alcohols fail to react ⟨79TL2251⟩.

$$R{-}OH \;+\; Ph_3P(SeCN)_2 \xrightarrow[51\text{–}86\%]{CH_2Cl_2,\ THF} R{-}SeCN \qquad (61)$$

R = alkyl, benzyl

The reaction of trimethylsilyl selenocyanate with oxiranes and oxetanes (Equation (62)) gives trimethylsiloxyalkyl selenocyanates in high yields ⟨90BCJ825⟩, without the release of selenium as observed in the addition to aldehydes ⟨89JOC2015⟩. The rate of reaction is solvent-dependent in the order $CH_3CN >> \text{hexane} > CH_2Cl_2$, while the addition of KSeCN on Amberlite resin, or with catalytic 18-crown-6, leads to some rate enhancement. Similarly, the addition of Lewis acids, for example $ZnCl_2$ or $Ti(OPr^i)_4$, also speeds up the reaction. The reaction is generally regiospecific.

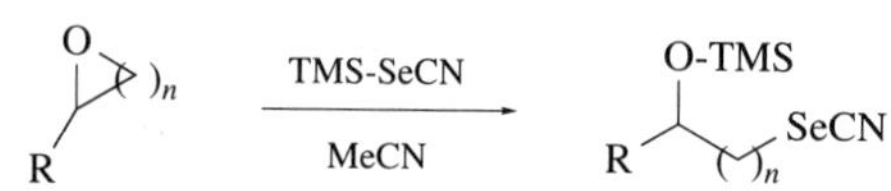

(62)

$n = 1, 2$

The reaction of β-lactones with KSeCN in aqueous medium gives rise to β-selenocyanatopropionic acids ⟨67ACS837⟩.

Mixed trialkylboranes prepared by the hydroboration of internal alkenes react with iron(III) selenocyanate in aqueous tetrahydrofuran to give the corresponding primary or secondary alkyl selenocyanates. This is an important reaction as it allows for the facile synthesis of secondary alkyl selenocyanates. The reaction works best using two equivalents of iron(III) selenocyanate. In boranes containing both secondary and primary alkyl groups the product observed is that of addition to the secondary groups (Scheme 3) ⟨76CL785, 76CL1115⟩.

$$R_3B \xrightarrow[H_2O]{Fe^{3+}\ (2\ \text{equiv.}),\ SeCN^-} R{-}SeCN$$

$$R^1{}_2R^2B \longrightarrow 2R^1SeCN \;+\; (HO)_2BR^2$$

R^1 = secondary, R^2 = primary

Scheme 3

The reaction of halogen-substituted selenides with potassium selenocyanate leads to a product in which both nucleophilic displacement of the halo groups and extrusion of the selenium has taken place ⟨73ACS726⟩. When an analogous reaction is carried out with 2-bromoethylphenyl selenide, only the aryl selenocyanates are isolated (Scheme 4) ⟨76ACS(B)941, 77ACS(B)1⟩.

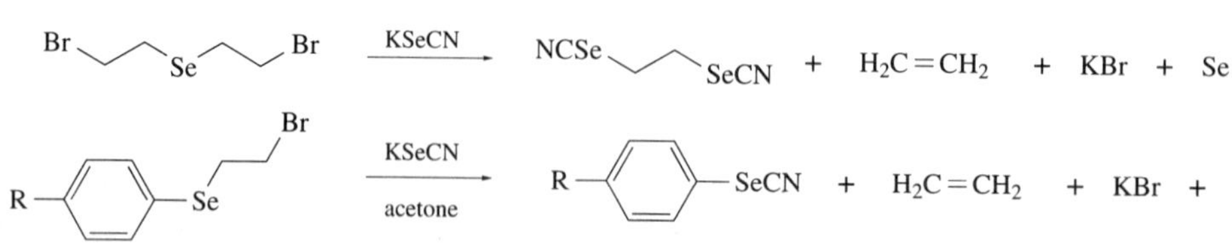

Scheme 4

(ii) Reaction with electrophilic selenium

Krafft and co-workers ⟨88JOC3632⟩ have developed a route to selenocyanates via the reaction of an organocopper species with the electrophilic selenium reagent selenocyanogen (Scheme 5). Copper was chosen to attenuate the nucleophilic reactivity after both alkyllithium and Grignard reagents had failed to react with selenocyanogen in the desired fashion. Okazaki ⟨89JA5949⟩ has also used this methodology to synthesize an α-trimethylsilyl selenocyanate as a key intermediate in his synthesis of 2,4,6-tri-*t*-butylselenobenzaldehyde, the first stable seleno aldehyde (Equation (63)).

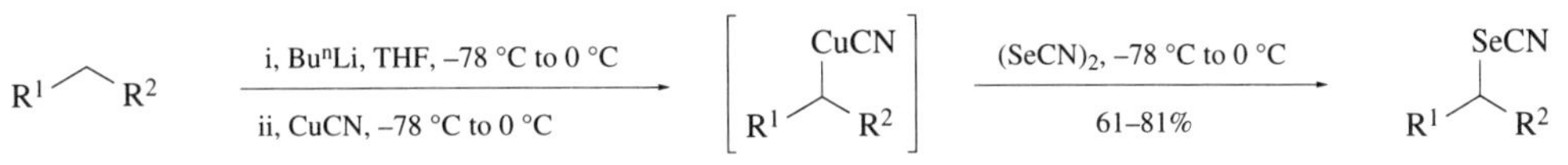

Scheme 5

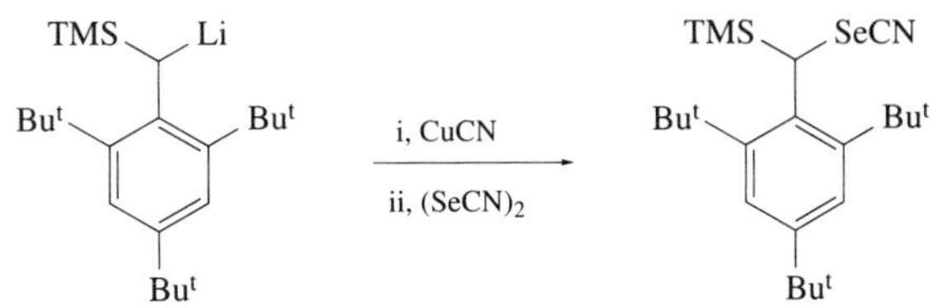

(63)

Guram has shown that allylic silanes react rapidly with $KSeCN \cdot 2CuBr_2$ in methanol (Equation (64)), to afford allylic selenocyanates in good yields ⟨93SL259⟩.

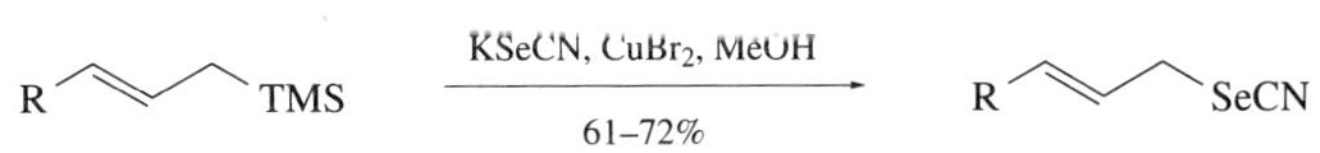

(64)

(iii) Additions to double bonds

Cambie *et al.* ⟨78S152⟩ have reported that a mixture of iodine and potassium selenocyanate reacts with cyclohexene in chloroform to give a mixture of the isomeric *vicinal* iodo selenocyanates (Equation (65)), while Uemura *et al.* ⟨75BCJ1925⟩ have shown that *in situ* oxyselenocyanation of terminal alkenes occurs readily to give the corresponding alkyl selenocyanates upon treatment of the alkene with thallium(III) acetate and potassium selenocyanate using an alcohol as the solvent (Equation (66)).

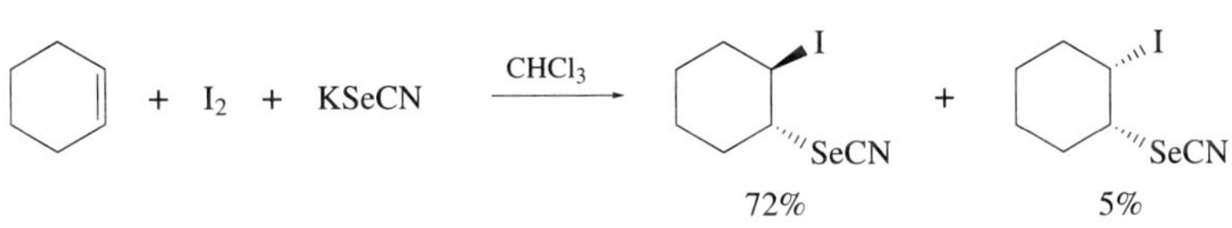

(65)

(66)

It has also been shown ⟨78JCS(P1)1273⟩ that alkenes undergo selenocyanation reactions with copper(II) halides and potassium selenocyanate in suitable solvents. Thus, if the reaction is carried out in an alcohol the products are *vicinal* alkoxy selenocyanates, whilst in acetonitrile only halo selenocyanates are isolated (Scheme 6).

(iv) Homolytic reactions

Barton *et al.* have described concise syntheses of both L-selenomethionine and L-selenocystine ⟨86T4983⟩. The key step in these syntheses was the introduction of the selenocyanate group via a photochemically induced radical decarboxylative selenocyanation reaction (Equation (67)). In this

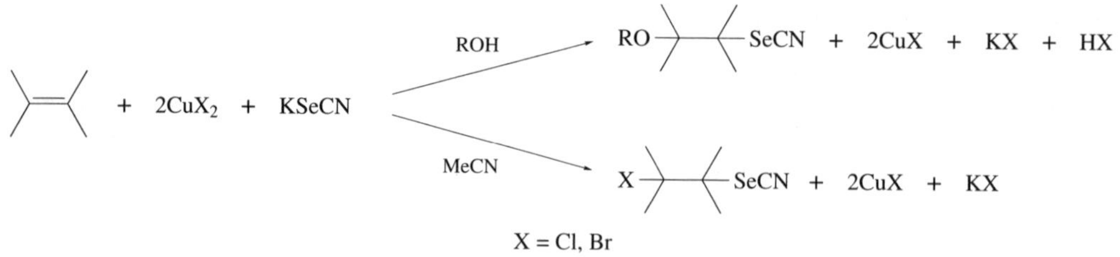

Scheme 6

reaction the selenium source is dicyanogen triselenide, a more convenient 'SeCN' source than selenocyanogen.

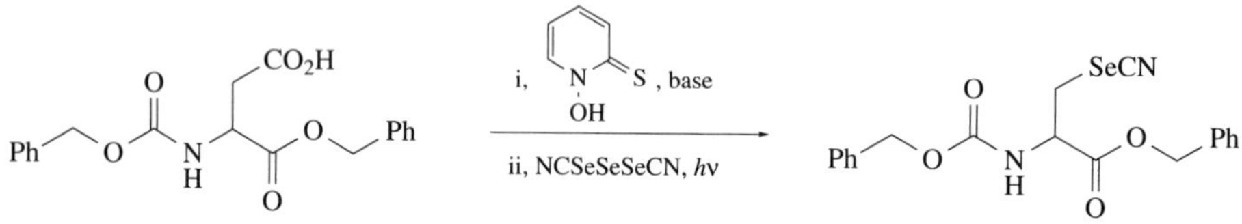 (67)

5.30.3.1.2 *Formation of the RSe—CN bond*

Whilst the routes previously described involve the introduction of the selenocyanate as an entire entity, it is possible to form these compounds via the reaction of suitable selenium compounds with cyanide to form the Se—CN bond. The most common reaction type involves the reaction of a selenenyl halide with a metal cyanide ⟨87TL4189⟩, or alternatively the reaction of a selenenyl halide with trimethylsilyl cyanide in tetrahydrofuran ⟨81CL1069⟩. However, it is possible to obtain selenocyanates from selenenyl thiocyanates ⟨B-77MI 530-03⟩, alkyl selenenylsulfinates ⟨47JA2236⟩ or arylselenenyl selenocyanates (Scheme 7) ⟨B-77MI 530-03⟩.

$$\mathrm{ArSeX} + \mathrm{MCN} \longrightarrow \mathrm{ArSeCN} + \mathrm{MX}$$
$$\mathrm{ArSeSCN} + \mathrm{CN^-} \longrightarrow \mathrm{ArSeCN} + \mathrm{SCN^-}$$
$$\mathrm{ArSeSO_2R} + \mathrm{CN^-} \longrightarrow \mathrm{ArSeCN} + \mathrm{RSO_2^-}$$
$$\mathrm{ArSeSeCN} \xrightarrow{\Delta} \mathrm{ArSeCN} + \mathrm{Se}$$

Scheme 7

5.30.3.2 Tellurocyanates—RTeCN

The notorious instability of tellurocyanates has meant that until the 1980s only a few stable derivatives were known, most notably the tellurocyanatoazulene **(8)** ⟨68MI 530-02⟩ and the *o*-formylphenyltellurenylcyanate **(9)** ⟨72BSF3559⟩.

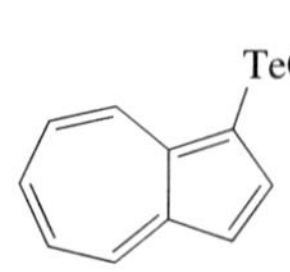

(8)

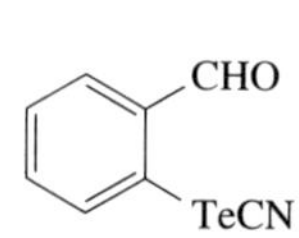

(9)

Since then a number of new routes to tellurocyanates have been developed, and these are discussed below.

5.30.3.2.1 *Formation of the R—TeCN bond*

The difficulties in preparing organic tellurocyanates are primarily due to the weakness of the carbon–tellurium bond ⟨77ACS(A)884⟩. Cava and co-workers ⟨77JA1470, 81JOC4194⟩ have shown that this problem can be overcome by using a solution of potassium tellurocyanate, made *in situ*, as the source of the tellurocyanate anion in reactions with benzyl halides. This method allows benzyl tellurocyanates to be readily synthesized in acceptable yields. Using this method, Songstad and co-workers ⟨81ACS(A)151, 89ACS518⟩ have been able to prepare a number of substituted benzyl tellurocyanates (Equation (68)).

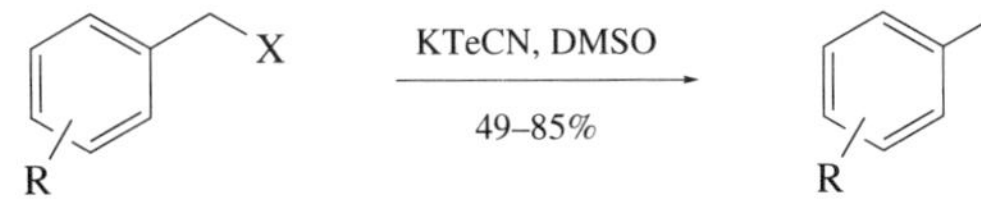

(68)

R = H, 2-CO_2Et, 2-NO_2, 3-NO_2; X = Cl, Br, I

5.30.3.2.2 *Formation of the RTe—CN bond*

It has been reported in the literature that phenyl tellurocyanates may be formed by the reaction of an appropriate diphenyl ditelluride ⟨25JCS2307⟩ with potassium cyanide in dimethyl sulfoxide ⟨80JOC1044⟩, albeit in low yield (Equation (69)).

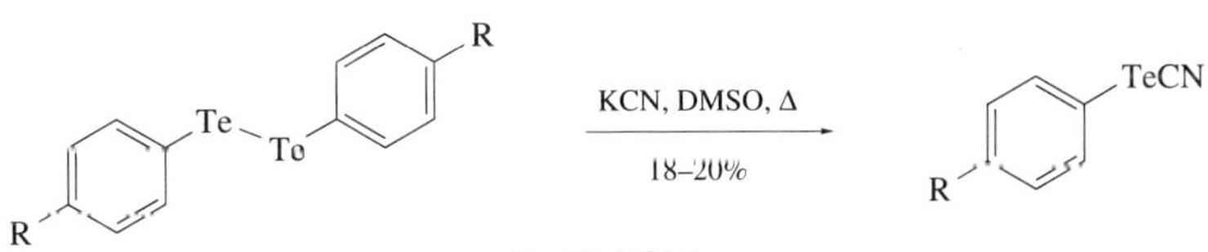

(69)

R = H, 4-OMe

Another route to phenyl tellurocyanates is also described in one of the reports ⟨80JOC1044⟩, via the reaction of a phenyltellurium halide with potassium cyanide in aqueous methanol. Similarly, 2-carboethoxybenzyl tellurocyanate has been prepared in 69% yield from the reaction of the corresponding benzyltellurenyl halide with potassium cyanide in dimethyl sulfoxide (Equations (70) and (71)) ⟨81JOC4194⟩, while tris(trimethylsilyl)methanetellurenyl iodide reacts cleanly with silver cyanide in refluxing benzene (Equation (72)) to give tris(trimethylsilyl)methane tellurocyanate in 72% yield ⟨91CB1131⟩.

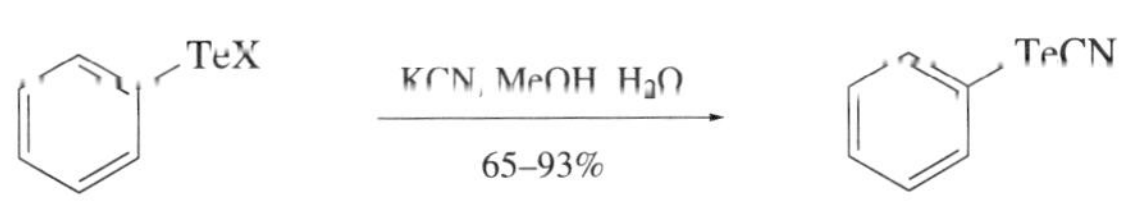

(70)

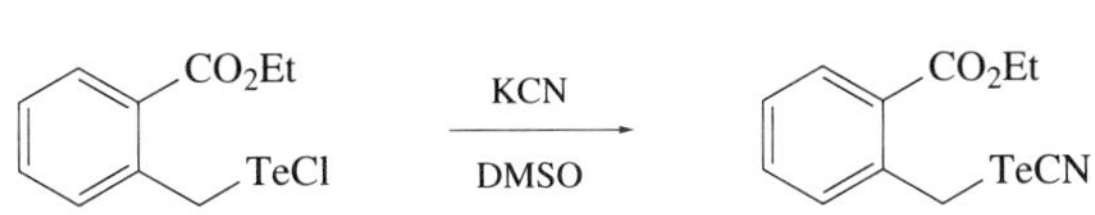

(71)

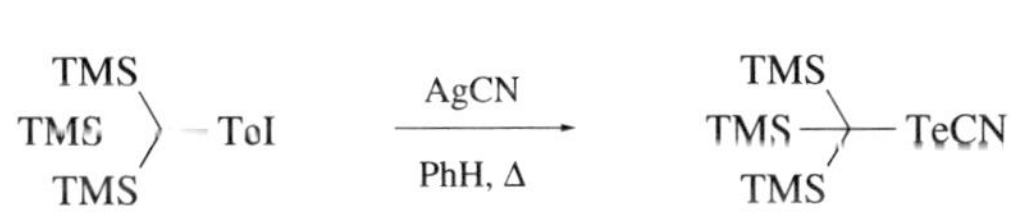

(72)

An alternative route to phenyl tellurocyanates via the reductive cleavage of the Te—Te bond in diphenyl ditelluride and subsequent trapping of the phenyl tellurate anion with cyanogen bromide (Equation (73)) has been reported by Ogura *et al.* ⟨82SC131⟩.

$$\text{PhTeTePh} \xrightarrow[\text{ii, BrCN}]{\text{i, NaBH}_4\text{, MeOH}} \text{PhTeCN} \qquad (73)$$

5.30.4 SINGLY BONDED NITROGEN DERIVATIVES—R_2N—C≡N (CYANAMIDES)

Cyanamides have been prepared by a wide range of synthetic methods, most commonly by the formation of an R—N bond or the N—CN bond, and to a lesser extent by the modification of an intact R—N—C—N system. The discussion of these routes will form the main part of this section. Synthetic routes leading to dicyanamides (RN(CN)$_2$) have been included alongside the appropriate cyanamide synthesis. Dicyanamide (HN(CN)$_2$) itself can be prepared from calcium cyanamide and cyanogen bromide ⟨66JHC387⟩. However, a convenient source of sodium dicyanamide, which avoids the use of cyanogen halides, is *N*-cyano-*S*-methylisothiourea and sodium hydroxide ⟨75S332⟩.

5.30.4.1 Formation of an R—NCN Bond

5.30.4.1.1 Formation of an R—NCN bond leading to alkyl cyanamides

The formation of an R—NCN bond is usually achieved by the reaction of cyanamide (or dicyanamide), or its salt, with the appropriate electrophile. Simple alkylation of cyanamide can give either the mono- or dialkylated product depending on the reaction conditions. However, monoalkylcyanamides are relatively unstable, having a tendency to dimerize or trimerize ⟨57NKZ567, 60JOC1043⟩.

Cyanamide can be converted into monomethylcyanamide upon reaction with sodium methoxide (one equivalent) and dimethyl sulfate (one equivalent) in methanol. The resulting monomethylcyanamide can be reacted directly with an aryl diazonium salt, to afford the 1-aryl-3-cyano-3-methyltriazine (Equation (74)) ⟨82CPB3125⟩. The reaction of cyanamide with alkyl halides under phase transfer catalysis affords dialkylcyanamides in 25–91% yield ⟨78S882⟩. Highly reactive electrophiles do not require basic catalysis, and monosubstituted cyanamides will react with a substituted aziridinone to give α-cyanamidoamides (Equation (75)) ⟨74TL2939⟩. A 2-cyanamidopyrazine has been methylated, albeit in low yield and low regioselectivity with diazomethane, to give both the side-chain and ring-methylated products (Equation (76)) ⟨83MI 530-02⟩.

(74)

(75)

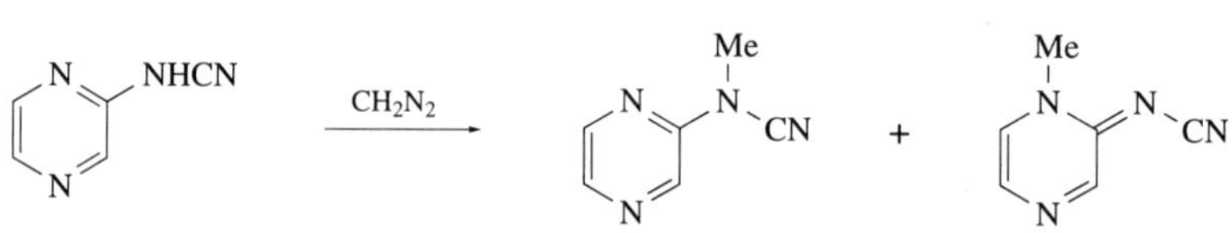

(76)

N-Alkyl-*N*-trimethylsilylcyanamides will react with 1-adamantyl chloride in the presence of titanium tetrachloride to give adamantylcyanamides ⟨81JOC5445⟩.

Cyanamide adds to an alkene in the presence of a source of Br$^+$ or PhSe$^+$ to give the corresponding *β*-bromoalkyl- or *β*-phenylselenenylalkylcyanamide. Thus, treatment of an alkene with cyanamide and NBS results in the formation of the monosubstituted cyanamide in 21–93% yield (Equation (77)) ⟨70TL1125, 84TL399, 85JA2931⟩. Cyanamide in the presence of phenylselenenyl chloride and silver tetrafluoroborate or *N*-phenylselenenylphthalimide and *p*-toluenesulfonic acid adds to alkenes to give monosubstituted cyanamides in 44–92% yield ⟨90JCS(P1)2417⟩. These can be further manipulated into allylcyanamides or *N*-cyanovinylimines, depending on the choice of oxidant and reaction conditions (Scheme 8).

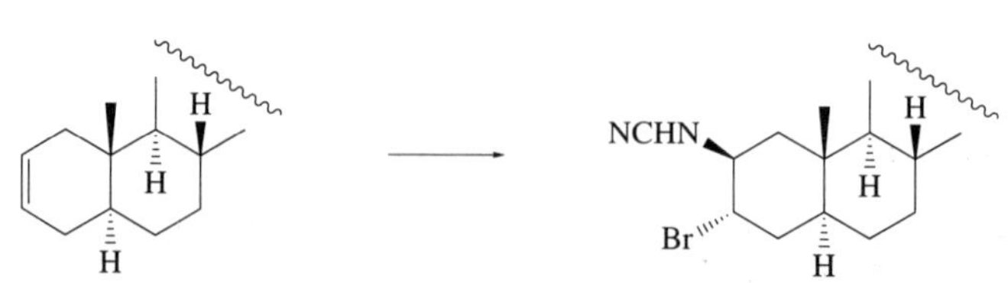

(77)

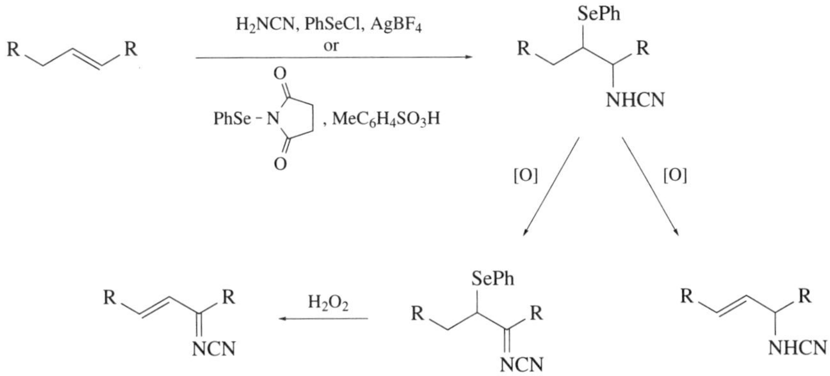

Scheme 8

Mono- and disubstituted alkylcyanamides can be prepared by the action of cyanogen azide (caution) (prepared from sodium azide and cyanogen bromide) on alkenes or alkanes. For example, it will react with 1,2-dihydropyridines and *N*-acyl-1,2-dihydropyridines to give *N*-cyanoaziridines (Equation (78)) ⟨78CJC1026, 79JHC409⟩. Cyanogen azide is also reported to react with simple alkanes to give monoalkylcyanamides ⟨65JA2296⟩. The reaction product from the cycloaddition of cyanogen azide to alkynes does not lose nitrogen, so that the 1-cyano-1,2,3-triazole is obtained. However, this shows a temperature-dependent equilibrium with the ring-open α-diazo-*N*-cyanoimine tautomer (Equation (79)) ⟨67JA4760⟩.

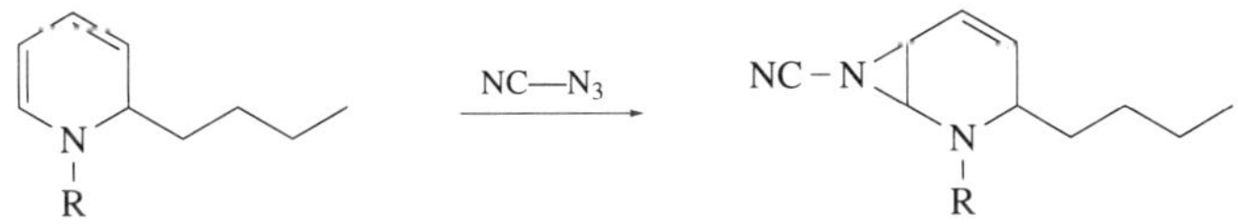

(78)

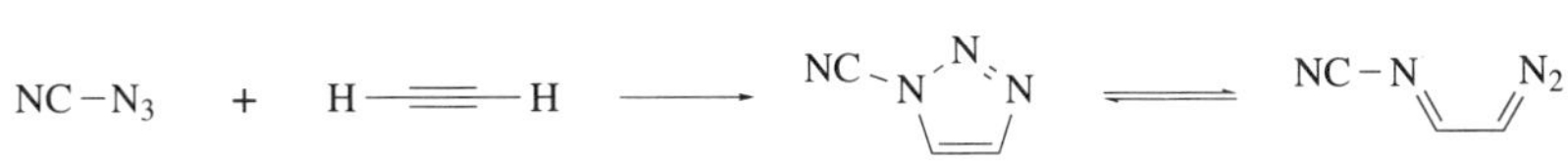

(79)

The reaction of the potassium salt of alkyl *N*-cyanocarbamates with alkylating agents affords reasonable yields of dialkyl *N*-cyanocarbamates ⟨87PS(29)219⟩. However, in the presence of excess alkylating agent, dialkylcyanamides are obtained.

Sodium nitrocyanamide will react with a straight-chain alkyl iodide in the presence of mcpba to give the corresponding alkyl nitrocyanamide as the major product ⟨89JCR(S)331⟩.

Azonitriles undergo Diels–Alder reactions with dienes to give *N*-cyanotetrahydropyridazines; 1,3-siloxydienes react further to give *N*-cyanodihydropyridazin-4-ones (Equation (80)) ⟨87CB1597⟩. The cyanamide **(10)** has been prepared by the action of calcium cyanamide on the sulfonic acid **(11)** (Equation (81)) ⟨48JA3212⟩.

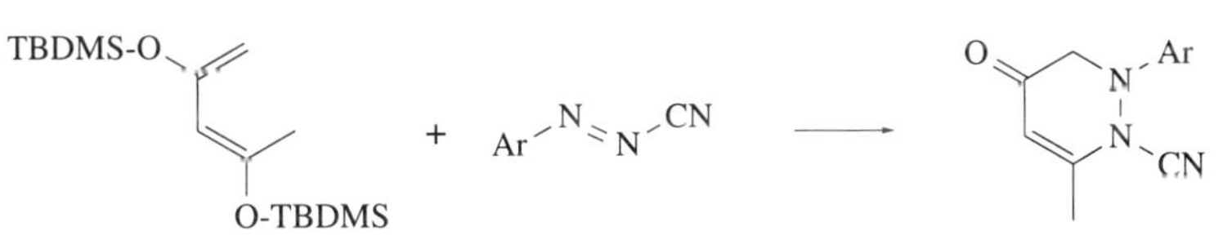

(80)

(11) **(10)**

(81)

Monosubstituted alkylcyanamides with a hydrogen on an sp^3-hybridized δ carbon can be cyclized to *N*-cyanopyrrolidines by irradiation ⟨87JCS(P1)937⟩.

N-Substituted-*N*-chlorocyanamides react with alkenes to give mainly the β-chloroalkylcyanamide addition products. However, the reaction between *N*-*t*-butyl-*N*-chlorocyanamide and cyclohexene afforded a β-chloroalkyldiimide ⟨69JOC1808⟩.

N-Cyano-*N*,*N*,*N*-trialkylammonium salts can be prepared by the action of a powerful alkylating agent such as trimethyloxonium tetrafluoroborate on a dialkylcyanamide ⟨74JOC1494⟩.

5.30.4.1.2 Formation of an R—NCN bond leading to aryl cyanamides

N-Arylcyanamides can be prepared from cyanamide and an aryl group bearing a substituent which readily undergoes nucleophilic displacement. For example, a trialkylammoniumtriazine will react with calcium cyanamide to give the corresponding cyanamidotriazine (Equation (82)) ⟨80CA(93)95215⟩. Similarly, treatment of a 2-methanesulfonylpyrimidine with cyanamide and potassium carbonate affords a 2-cyanamidopyrimidine (Equation (83)) ⟨87EUP209779⟩, while a 2-chloro-1,3,4-thiadiazole gives a 2-cyanamidothiadiazole ⟨87JMC951⟩.

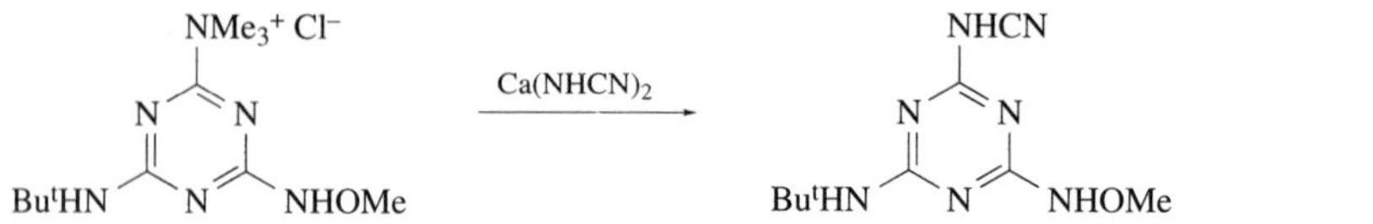

(82)

(83)

5.30.4.1.3 Formation of an R—NCN bond leading to acyl- or iminocyanamides

Simple unsubstituted acylcyanamides are not stable in the solid state but slowly decompose to a mixture of components, including *N*-acyl-*N*-cyanoguanidines ⟨90JOC3403⟩. They can be formed by the reaction of cyanamide with an acylating agent. For example, the treatment of the *N*-hydroxysuccinimide ester of *N*-benzyloxycarbonylleucine with sodium hydroxide and cyanamide affords an *N*-cyanoamide (Equation (84)) ⟨87SC1677⟩.

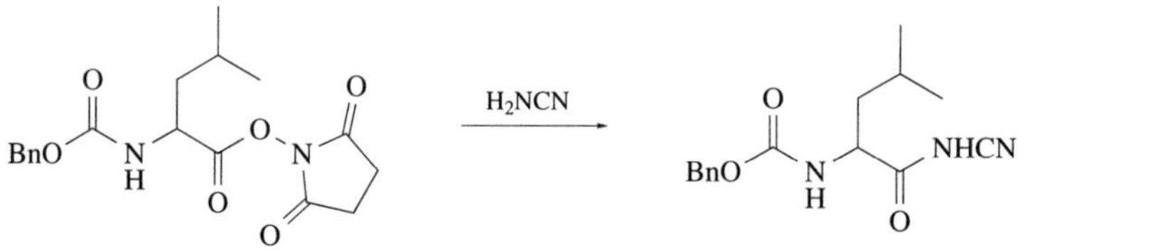

(84)

Isocyanates react with cyanamide to give the corresponding *N*-cyanoureas ⟨63OSC(4)213⟩. The conversion of an amide into an *N*-cyanourea via the isocyanate can be achieved with sodium hypochlorite and cyanamide ⟨81USP4304935⟩.

The reaction of 2-amino-1,3-benzoxazin-4-one (**12**) with sodium cyanamide affords 3-cyano-quinazoline-2,4-dione (**13**) (Equation (85)) ⟨83JCS(P1)813⟩.

(**12**) (**13**)

(85)

3-Alkyl-2-(*N*-cyanoimino)thiazolidines react with sodium alkoxides to give *N*-cyano-*N*′,*N*′-disubstituted ureas (Equation (86)) ⟨92CC1379⟩.

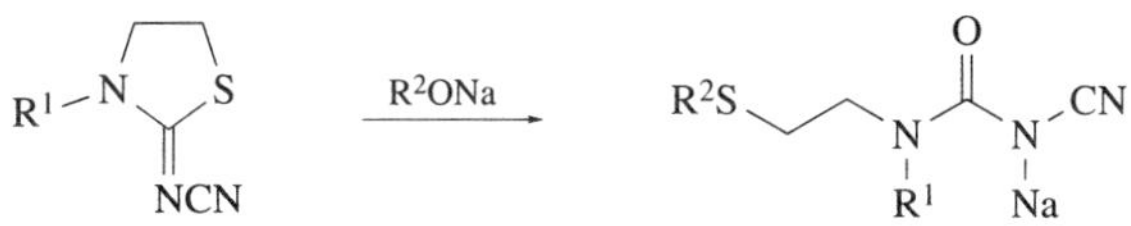

(86)

N-Acyl-*N*-(trimethylstannyl)cyanamides have been prepared by the action of acid chlorides or chloroformates with bis(trimethylstannyl)carbodiimide ⟨76SRI345⟩.

N-Substituted acyl cyanamides can be prepared by the alkylation of the unsubstituted acyl cyanamides. Thus, treatment of the potassium salt of benzyl *N*-cyanocarbamate with alkyl halides affords benzyl *N*-alkyl-*N*-cyanocarbamates in high yield ⟨73CL1225⟩. In a similar fashion, alkyl or aryl *N*-cyanothiolcarbamates afford the corresponding *N*-substituted *N*-cyanothiolcarbamates ⟨86PS(26)1⟩. However, it has been shown that the reaction of the potassium salt of alkyl *N*-cyanocarbamates with excess alkyl halide can also give dialkylcyanamides ⟨87PS(29)219⟩.

Sequential treatment of ureas with triphenylphosphine, carbon tetrachloride and triethylamine, followed by cyanamide and ethyl diisopropylamine, affords the corresponding cyanoguanidines in good yields ⟨93JMC2004⟩.

Amidines will react with cyanamide by displacement of ammonia to give *N*-cyanoamidines (Equation (87)) ⟨81EJM175⟩. Substituted *N*-cyanoguanidines have been prepared by a variety of methods. Bis(triphenylstannyl)carbodiimide reacts with 1,1,3-trisubstituted thioureas to give the corresponding *N*,*N*,*N*′-trisubstituted-*N*″-cyanoguanidines, and ethyl-*N*-arylthiocarbamates to give *N*-aryl-*N*′-triphenylstannyl-*N*′,*N*″-dicyanoguanidines (Scheme 9) ⟨81JCED105⟩. Cyanoguanidine itself will react with an amine and formaldehyde to give a 5-substituted 1-cyano-2-iminohexahydrotriazine (Equation (88)) ⟨60JOC147⟩.

(87)

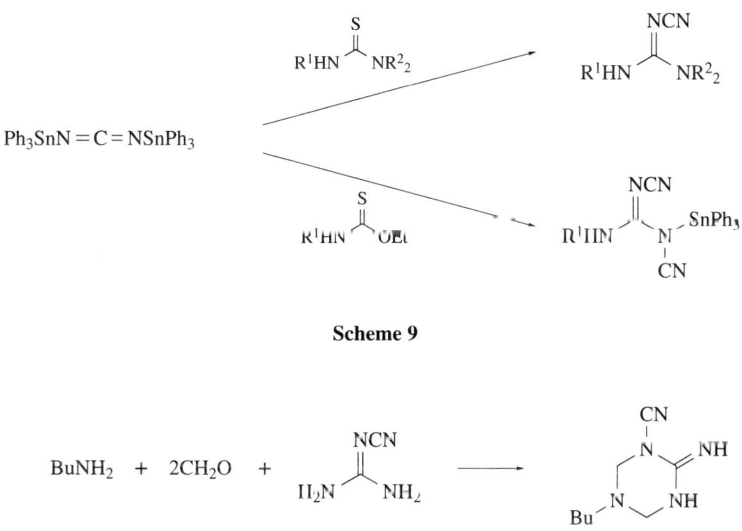

Scheme 9

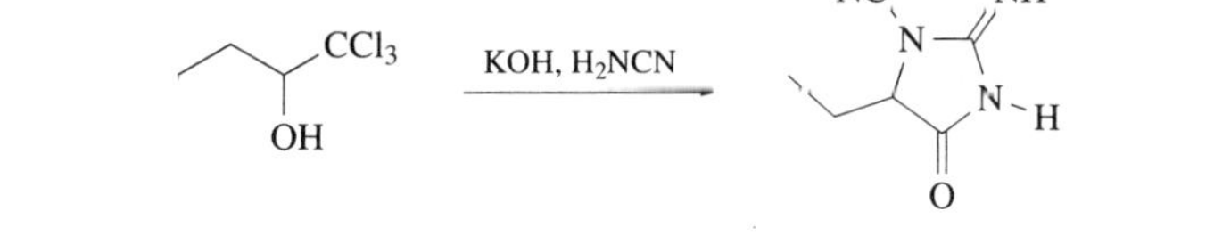

(88)

Cyanamide also reacts with trichloromethyl carbinols to give 5-substituted-1-cyano-2-iminoimidazol-4-ones (Equation (89)) ⟨80CJC2784⟩.

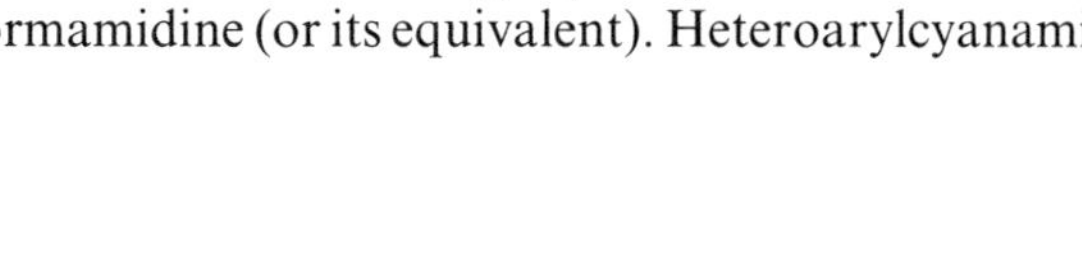

(89)

5.30.4.2 Dehydration (or Equivalent) Reactions Leading to Cyanamides

The following reactions fall into two basic types, dehydration of a urea (or its equivalent) or dehydration of a hydroxyformamidine (or its equivalent). Heteroarylcyanamides have been prepared from the corresponding hydroxyformamidines upon reaction with phosphorus oxychloride at room temperature or refluxing ethanolic sodium hydroxide (14–81% yield) ⟨83MI 530-02⟩. The dehydration has also been achieved with dimethylformamide dimethyl acetal, although in this case concomitant methylation of the resulting cyanamide occurred to give an *N*-aryl-*N*-methylcyanamide (Equation (90)) ⟨84MI 530-01⟩. Hydroxyformamidines have also been dehydrated to cyanamides with phosgene and triethylamine ⟨79TL3131⟩.

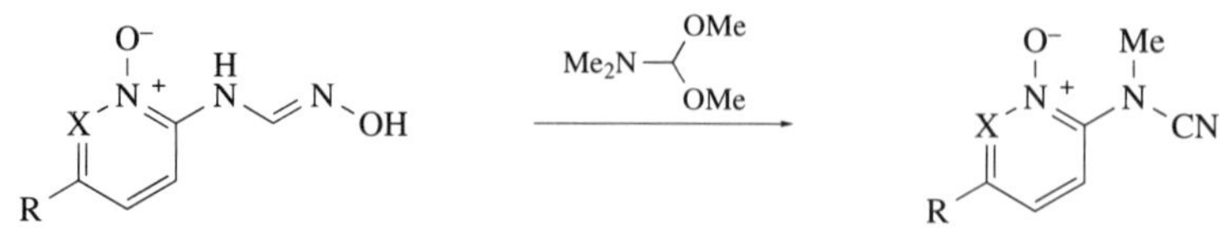
(90)

The treatment of a diarylamine with *N*-(chlorosulfenyl)iminocarbonyl chloride or *N*-phosphoryliminocarbonyl chloride in refluxing benzene affords a diarylcyanamide (Equation (91)) ⟨91IZV451⟩.

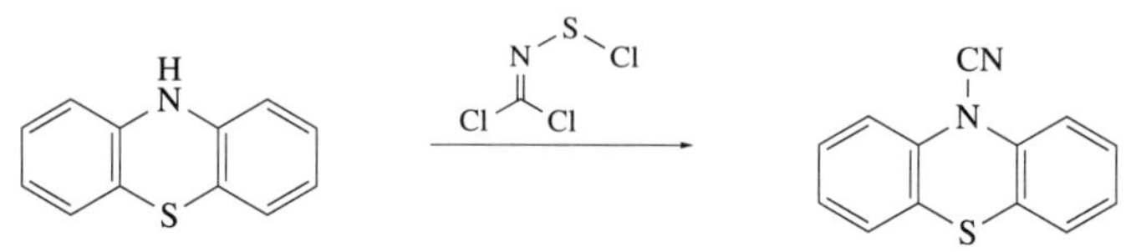
(91)

Monosubstituted thioureas can be converted in excellent yield into monosubstituted cyanamides by reaction with potassium superoxide at 60 °C in aprotic media. Thiourea itself is converted under these reaction conditions into cyanoguanidine (73% yield) ⟨84CC53⟩.

N,*N*-Diethylurea has been converted into diethylcyanamide by reaction with trichloromethyl chloroformate ⟨86SC1823⟩.

Bromodenitrogenation of N^2-benzoylarginine ethyl ester with *N*-bromosuccinimide affords N^2-benzoyl-N^5-cyanoornithine ethyl ester as the major reaction product (no yield given) (Equation (92)) ⟨70BCJ2539⟩.

(92)

The reaction of 4-amino-3,5-di(methylthio)-1-methyl-1,2,4-triazolium iodide and *p*-toluene isothiocyanate affords 5-(*N*-cyano-*N*-(4-methylphenyl)amino)-1-methyl-3-methylthio-1,2,4-triazole (Equation (93)) ⟨84TL5427⟩.

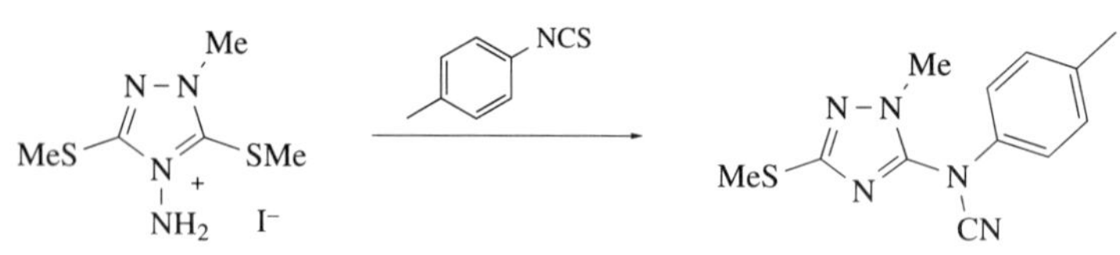
(93)

5.30.4.3 Rearrangement and Fragmentation Reactions Which Give Cyanamides

Several different types of fragmentation reactions have been discovered which result in the formation of cyanamides. One of the most synthetically accessible is the thermolysis of the ylide generated by the deprotonation of the [1,2,4]triazolo[1,5-*a*]pyrimidinium bromide (**14**) with triethylamine (Equation (94)) ⟨85JCS(P1)2333, 85TL1321, 87JCS(P1)2531⟩. Flash vacuum pyrolysis or heating in a sealed tube has converted [1,2,3,4]tetrazolo[1,5-*c*]pyrimidines into 1-cyanoimidazoles by the loss of nitrogen and ring contraction (Equation (95)) ⟨72HCA565, 80AG(E)566⟩.

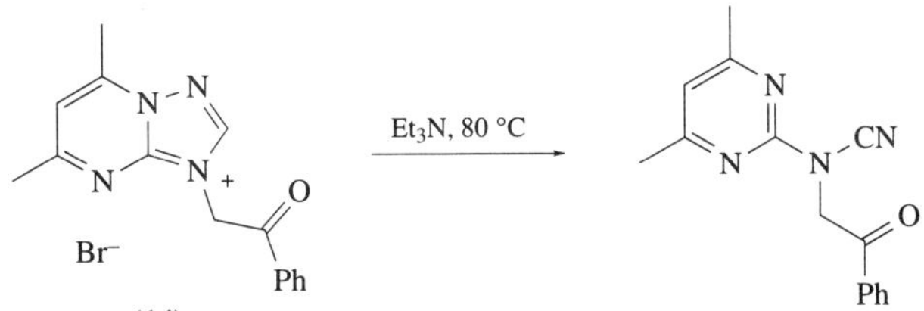

(14)

(94)

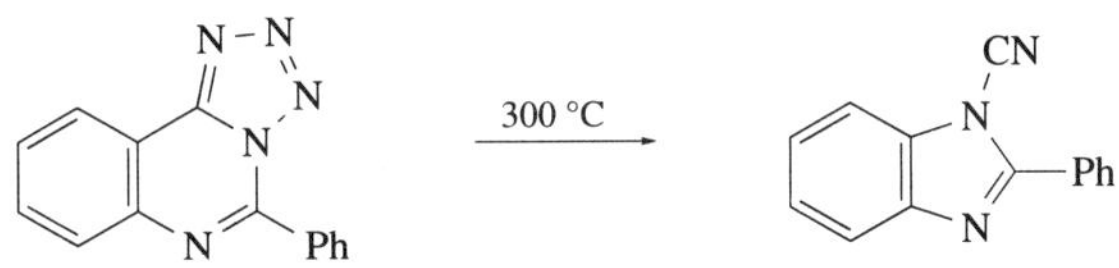

(95)

1-Methyl-5-tetrazoloyllithium (1-methyltetrazole and *n*-butyllithium, $-60\,^{\circ}$C to $-65\,^{\circ}$C), undergoes fragmentation above $-50\,^{\circ}$C to nitrogen and lithium methylcyanamide. This anion was readily alkylated with benzyl bromide (71% yield) ⟨71CJC2139⟩. 1-Phenyl-5-tetrazolyllithium rapidly decomposed to lithium phenylcyanamide even at $-65\,^{\circ}$C to $-70\,^{\circ}$C. Reaction with acetyl chloride afforded *N*-cyanoacetanilide (80% yield).

Flash vacuum pyrolysis of the phenylhydrazone (**15**) affords phenylcyanamide in 58% yield. The reaction is believed to proceed via a secondary isocyanamine (Equation (96)) ⟨81JOC1045⟩.

(15)

(96)

The treatment of quinoline with the *N*-phenylaminochloroimidate (**16**) at 170–180 °C results in the formation of the 2-(*N*-cyanophenylamino)quinoline in 40% yield (Equation (97)) ⟨68G511⟩.

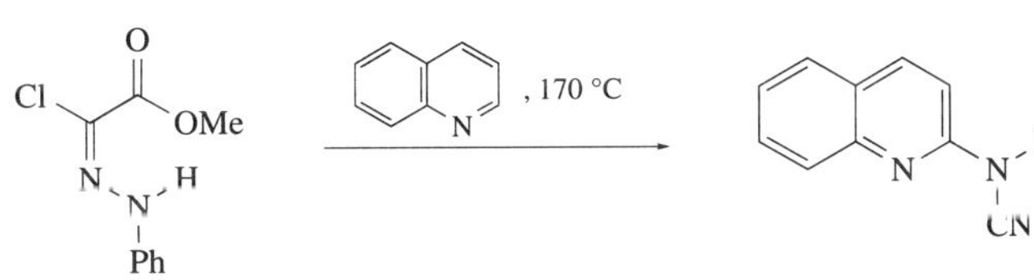

(16)

(97)

The rearrangement of the 1,3-dipolar cycloaddition product of acylnitrile oxides with 4,5-dihydro-2-methyloxazole affords *N*-acyloxyethyl-*N*-cyanoacetamides in good yield. Reaction with 4,5-dihydro-2-methylthiazole gives the corresponding *N*-acylthioethyl-*N*-cyanoacetamides (Equation (98)) ⟨92T7703⟩.

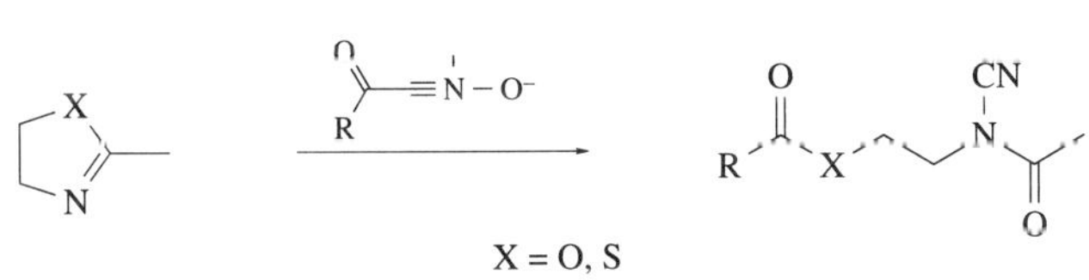

(98)

Functionalized diaziridinimines formed from sulfonylguanidines, *t*-butyl hypochlorite and potassium *t*-butoxide thermolize to *N*-*t*-butyl-*N*-cyano-*N*′-toluenesulfonylhydrazines (Equation (99)) ⟨79TL49, 81JOC4478⟩.

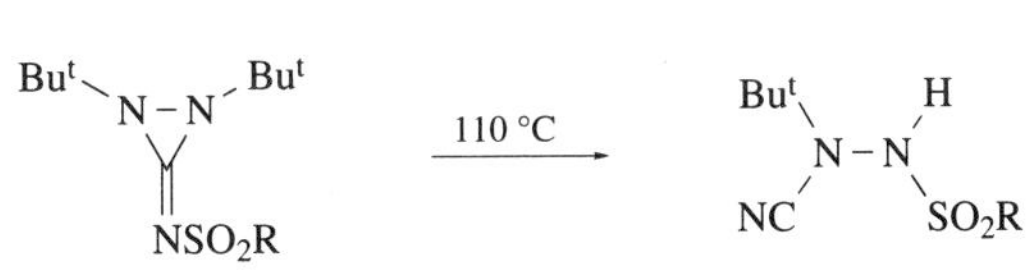

(99)

5.30.4.4 Formation of an RN—CN Bond by Reaction With Nucleophilic Cyanide

In contrast to the formation of cyanamides by reaction with electrophilic cyanide, few examples can be found where reaction with the cyanide anion results in the formation of a cyanamide.

However, aromatic nitroso compounds react readily with sodium cyanide in aprotic solvents to give *N*-aryl-*N*-hydroxycyanamides in excellent yields (Equation (100)) ⟨75TL3125⟩.

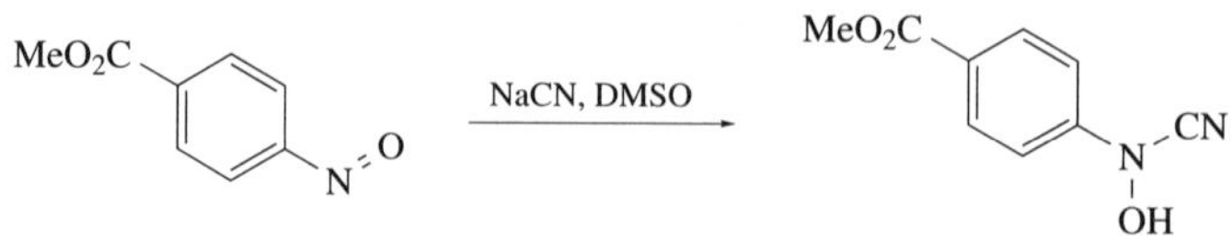

(100)

Aryl azides will react with potassium cyanide to give 1-aryl-3-cyanotriazenes as their potassium salts; subsequent alkylation of these salts with dimethyl sulfate under phase transfer conditions affords a mixture of the 1-methyl- and 3-methyl-1-aryl-3-cyanotriazenes in low yield (Equation (101)) ⟨82CPB3125⟩.

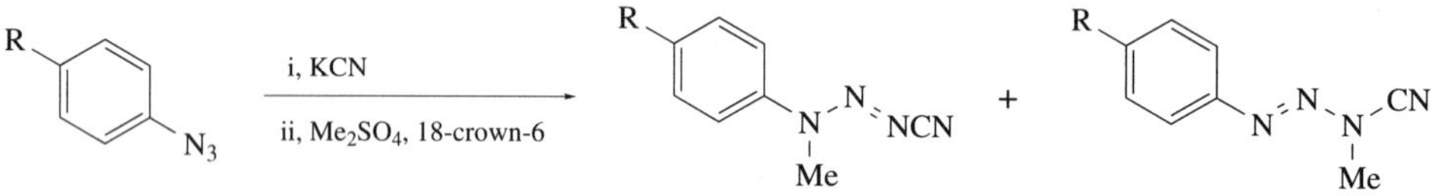

(101)

5.30.4.5 Reactions on the Carbon α to NCN Leading to Cyanamides

The *N*-cyanocarboimidate (**17**) will react with nucleophiles such as methyl cyanoacetate in a Michael–retro-Michael reaction to give the α-methoxycyanamide (**18**) as its sodium salt in excellent yield (Equation (102)) ⟨85HCA1155⟩.

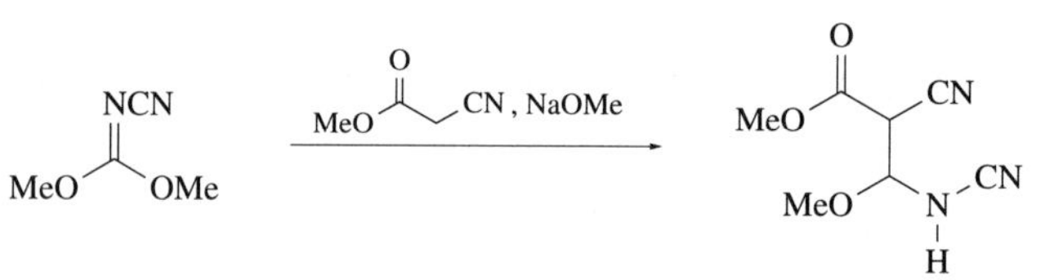

(102)

(17) **(18)**

The reaction of diaminobenzenes with the *N*-cyanocarbodithioimidate (**19**) affords 2-cyanamidobenzimidazoles in good yield. The reaction is favoured by electron-donating groups and no characterizable products were obtained with chloro or nitro substituents (Equation (103)) ⟨75JHC37⟩.

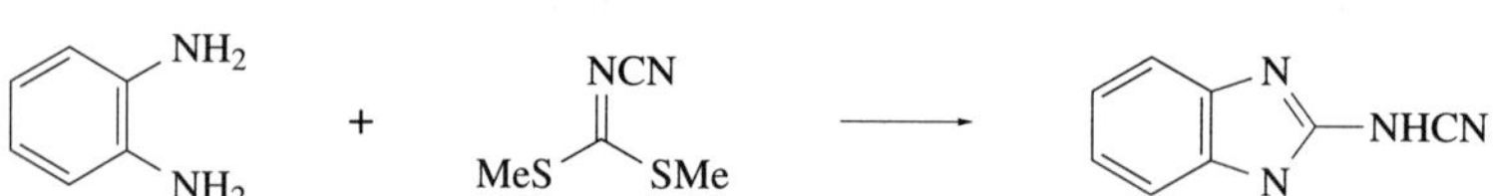

(103)

(19)

Sequential treatment of (**19**) with ammonia and sodium hydroxide affords sodium dicyanamide ⟨75S332⟩. Subsequent reaction with arylamines affords *N*-aryl-*N*′-cyanoguanidines.

Alkyldicyanamides react with aqueous sulfuric acid or hydrogen sulfide to give the corresponding *N*-alkyl-*N*-cyanoureas or thioureas, respectively ⟨78S775⟩.

5.30.4.6 Formation of an RN—CN Bond by Reaction with Electrophilic Cyanide

The formation of an RN—CN bond by reaction with electrophilic cyanide is one of the most widely used methods for preparing cyanamides. The general reaction is that of a cyanogen halide (or halide equivalent) with a suitably nucleophilic nitrogen. The section is subdivided by the nature of the nitrogen nucleophile.

5.30.4.6.1 Reaction of primary and secondary nitrogen with electrophilic cyanide

In this case a nucleophilic primary or secondary nitrogen can displace a halide or other stable leaving group from a cyanogen halide (or equivalent). In the case of a primary nitrogen there is the possibility of reaction occurring twice to give the dicyanamide. On the whole this can be readily controlled so that for any particular amine either the cyanamide or dicyanamide can be obtained at will. The most numerous examples in the literature are for reaction with cyanogen chloride and bromide under a range of different reaction conditions. For example, anilines have been converted into arylcyanamides ⟨46JA2102⟩ or aryldicyanamides ⟨61JOC4122⟩ by reaction with cyanogen bromide and aqueous sodium hydroxide. They have also been converted into arylcyanamides by reaction with cyanogen bromide ⟨47JA365⟩ or cyanogen chloride ⟨48JOC924⟩ and aqueous sodium bicarbonate, or by reaction with cyanogen bromide (formed *in situ* from sodium cyanide and bromine), phenol (catalyst) and calcium carbonate ⟨86JAP1022060⟩.

Additionally, primary alkyl amines have been converted into cyanamides by reaction with cyanogen chloride, or dicyanamides by reaction with cyanogen chloride and triethylamine in diethyl ether ⟨72RTC343⟩. They may also be converted into cyanamides with cyanogen bromide in ether ⟨62CJC1403⟩, with cyanogen bromide and triethylamine in tetrahydrofuran ⟨89TL1679⟩ or with sodium cyanide and methanesulfonyl chloride ⟨86MI 530-01⟩.

Amidines and guanidines will also react with cyanogen halides. For example, the amidine (**20**) is converted into the bis adduct (**21**) in 77% yield by reaction with potassium hydroxide and cyanogen chloride (Equation (104)) ⟨62JOC3890⟩. However, alkylideneaminoguanidines (prepared from aminoguanidine and an aldehyde or ketone) react with cyanogen bromide to give the N^4,N^4-dicyano adducts only (72% yield) (Equation (105)) ⟨86MI 530-02⟩.

(**20**) (**21**) (104)

(105)

A similarly wide range of reagents and conditions has been used to convert secondary aryl and alkyl amines into cyanamides. Thus, disubstituted cyanamides have been synthesized from secondary alkyl amines by reaction with the following: cyanogen bromide in chloroform ⟨86HCA246⟩ and diethyl ether ⟨82ZN(B)1481⟩; cyanogen bromide and magnesium carbonate ⟨82CC444, 84JCS(P1)147⟩; ethyl thiocyanate and cadmium(II) oxide in chloroform ⟨82ZOR1813⟩; and aqueous sodium cyanide and chlorine ⟨86JAP1280464⟩.

Aryl secondary amines have been converted into cyanamides by reaction with cyanogen bromide in chloroform ⟨88IJC583⟩. Phenylhydrazines are cyanated on the substituted nitrogen to give *N*-amino-*N*-phenylcyanamides with sodium hydroxide and cyanogen bromide in aqueous ethanol (Equation (106)) ⟨83JIIC1533⟩. The indole NH group can be cyanated by treatment with sodium hydride and phenyl cyanate in an aprotic solvent (Equation (107)) ⟨90TL3681⟩.

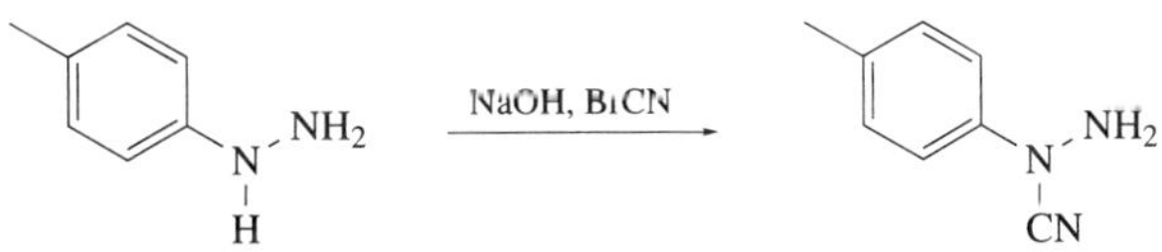

(106)

(107)

Amides and sulfonamides may also be *N*-cyanated under similar reaction conditions. Thus, the reaction of the sulfonamide (**22**) with potassium cyanide, sodium bicarbonate and bromine affords the cyanamide (**23**) (Equation (108)) ⟨80MIP-DE283404⟩. The treatment of the acylhydrazide (**24**) with sodium hydride and cyanogen bromide affords the cyanated adduct in 85% yield (Equation (109)) ⟨89USP4857550⟩. Sodium hydride and cyanogen bromide also convert hydantoins into *N*-cyanohydantoins (41–82% yield) ⟨85JHC1025⟩. Phthalimides and succinimides can be converted into the corresponding *N*-cyano derivatives by reaction with triethylamine and cyanogen bromide in acetone ⟨91SC2025⟩.

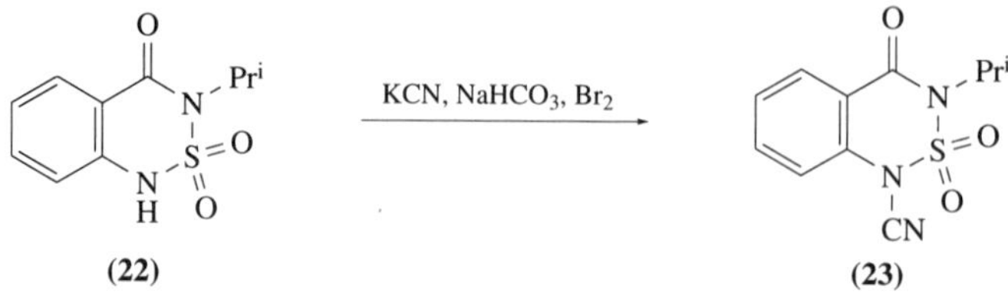

(108)

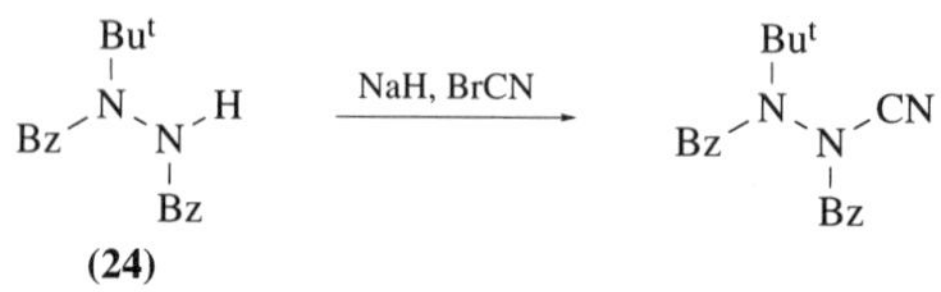

(109)

Reaction of the sulfimide and alkyne addition product (**25**) with cyanogen bromide affords the cyanamide with concomitant dealkylation of the sulfonium intermediate (Equation (110)) ⟨76JCS(P1)716⟩.

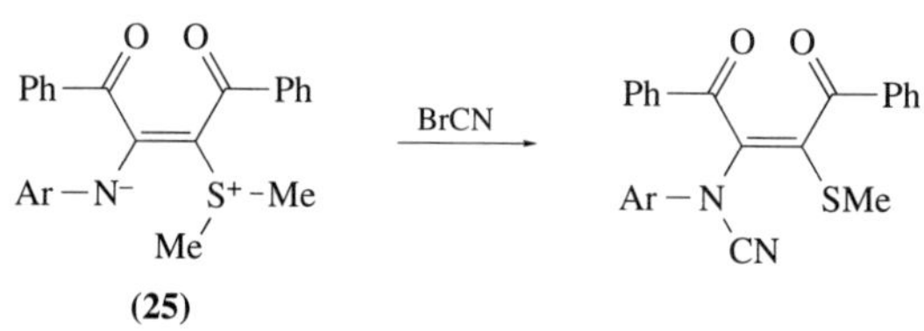

(110)

5.30.4.6.2 *Reaction of tertiary nitrogen with electrophilic cyanide*

The reaction of a tertiary amine with cyanogen halides (usually bromide) affords the cyanoammonium salt. However, under the normal reaction conditions this suffers nucleophilic attack by the liberated halide ion to afford the disubstituted cyanamide and an alkyl halide. This is known as the von Braun reaction. The nucleophilic dealkylation of the intermediate cyanoammonium ion usually shows a high degree of selectivity. The kinetics and mechanisms of the reaction of cyanogen bromide reactions with *N*-alkylanilines have been investigated ⟨76JCS(P2)524⟩. The cyanoammonium intermediate has been isolated ⟨70TL4731⟩.

One of the more common uses for the von-Braun reaction is the demethylation of a tertiary amine with the general formula R_2NCH_3 ⟨47OS(27)56, 55JA4079, 67JA3281, 85TL501, 87AP1088⟩. The bromodealkylation reaction has also been used to cleave ring systems, including aziridine ⟨88H(27)2609⟩ and pyrrolidine (Equations (111) and (112)) ⟨49JOC605⟩.

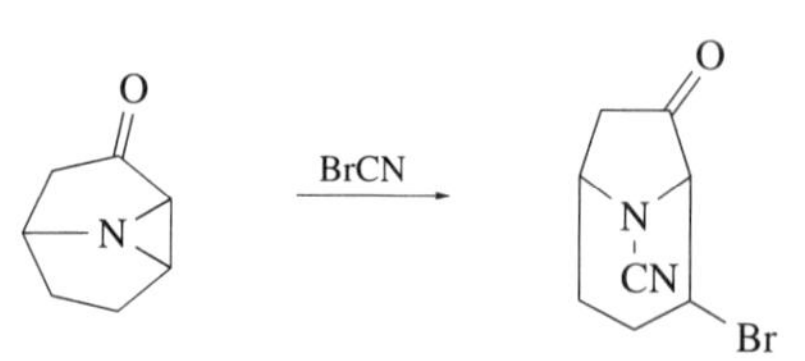

(111)

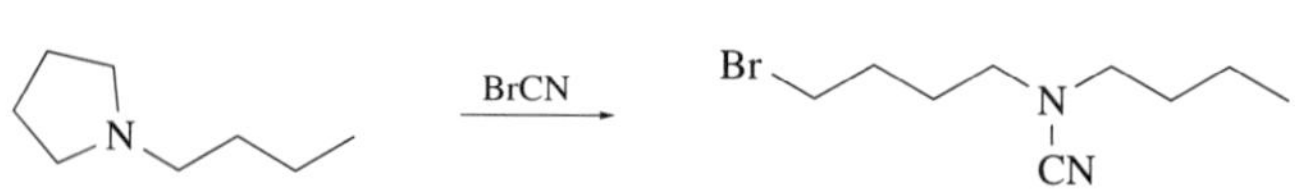

(112)

The cyanoammonium intermediate can be intercepted by alternative nucleophiles. For example, methanol ⟨79CI(L)319, 84AJC367⟩ and water ⟨72JPR382, 84AJC367, 87ZC64⟩ give the methyl ether and alcohol, respectively (Equations (113) and (114)).

(113)

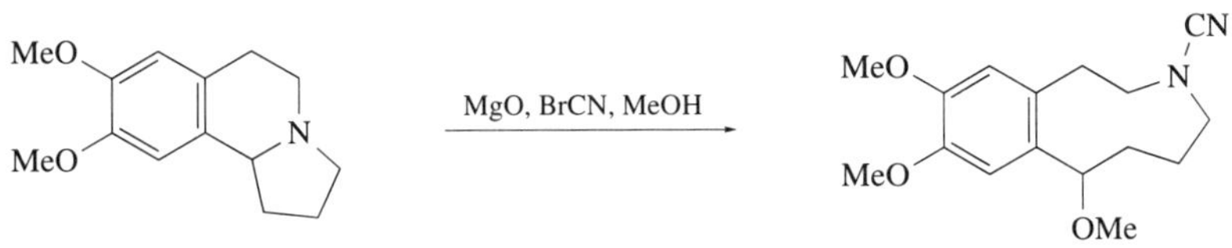

(114)

The cyanoammonium intermediate can also be trapped by internal nucleophiles such as alcohols (and silyl ethers ⟨90TL1287⟩) to generate small- ⟨81AP(314)577⟩ and medium-ring ⟨84AJC1043⟩ ethers (Equation (115)). Ether formation has been shown to occur with inversion of stereochemistry ⟨81MI 530-01⟩. The intermediate formed in the reaction may also undergo elimination reactions to liberate a disubstituted cyanamide and an alkene (Equation (116)) ⟨85JA3910, 90TL1287, 91AJC135⟩.

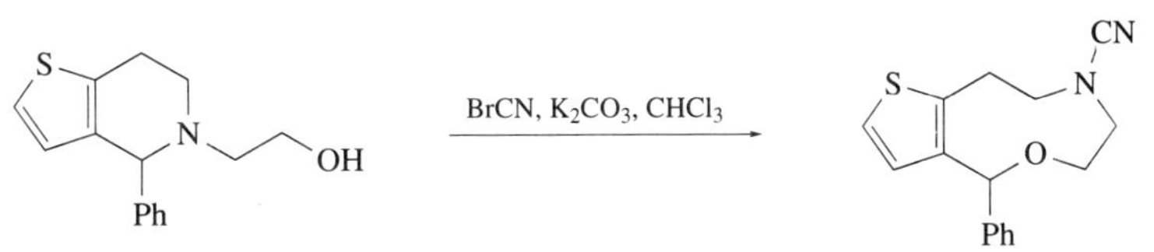

(115)

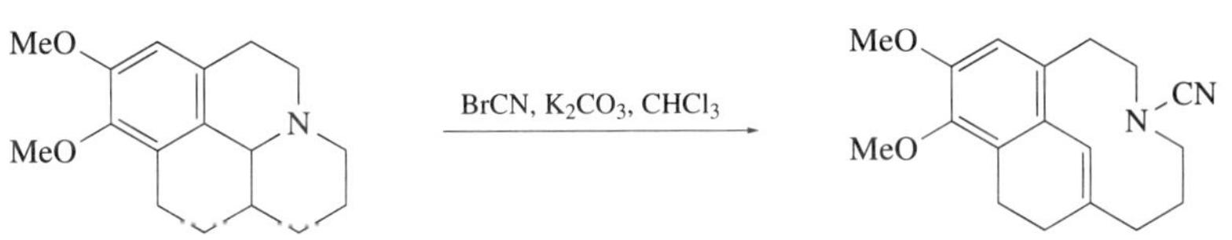

(116)

The reaction has also been used to convert bis(trimethylsilyl)methylamine into methyl(trimethylsilyl)cyanamide ⟨84URP570267⟩, and *N*-cyanomethylpiperidine into *N*-cyanopiperidine and bromoacetonitrile ⟨85MI 530-01⟩.

5.30.4.6.3 Reaction of a tertiary aromatic nitrogen with electrophilic cyanide

The reaction of a tertiary aromatic nitrogen with the cyanogen halide produces an aromatic quaternary ammonium ion. This can suffer one of two fates. First, addition of a nucleophile, for example methanol, to the aromatic system. For quinoline this process results in the 1,2-addition product (**26**) (Equation (117)) ⟨79H(12)164⟩. 1,2-Addition also occurs with a pyrido[2,3-*c*]pyridine to give the adduct (**27**) (Equation (118)) ⟨78YZ1361⟩. Alternatively, deprotonation of the initially formed salt may occur to give an ylide. For example, an N-1-substituted imidazole will react with cyanogen chloride at N-3 to give the ylide (**28**) (Equation (119)) ⟨85TL6273⟩.

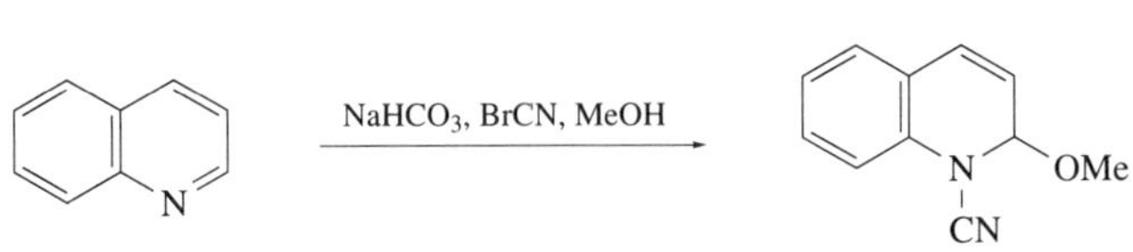

(117)

(26)

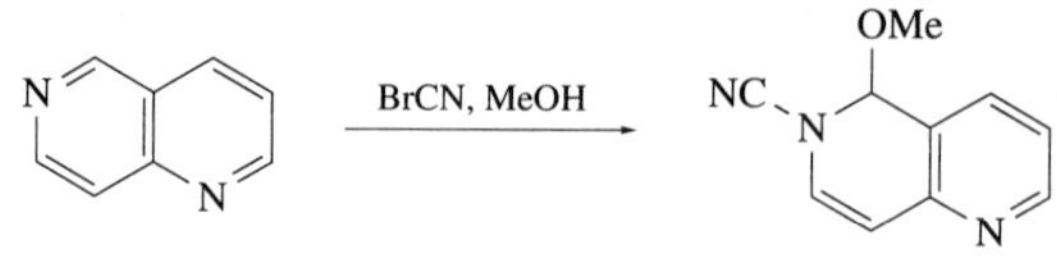

(27) (118)

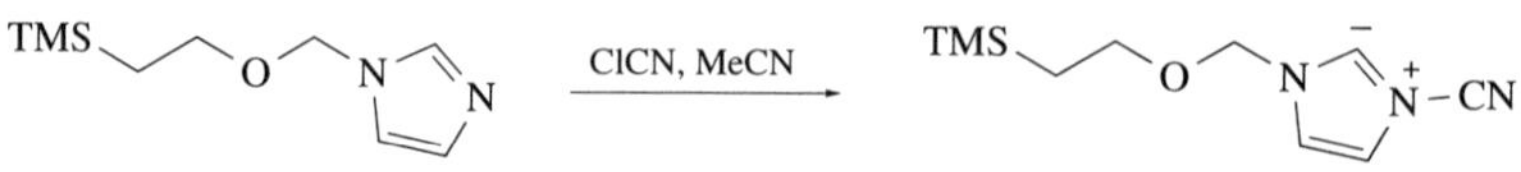

(28) (119)

5.30.5 SINGLY BONDED PHOSPHORUS, ANTIMONY, ARSENIC AND BISMUTH DERIVATIVES—$R_2P—C{\equiv}N$, etc.

5.30.5.1 Phosphorus(III) Derivatives

5.30.5.1.1 Formation of the C—PCN bond

The most straightforward approach to the synthesis of the C—PCN bond involves the simple mixing of phosphorus tricyanide with the appropriate trialkyl phosphite. Thus, phosphorus tricyanide reacts with trimethyl phosphite in carbon tetrachloride at 40°C to give methyl dicyanophosphite in 50% yield (Equation (120)) ⟨77AG747⟩.

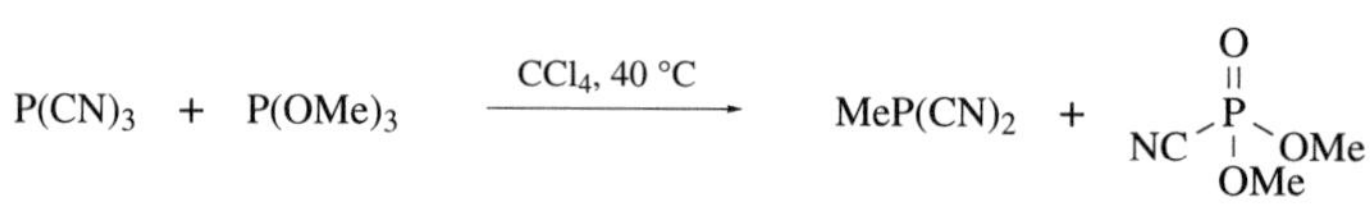

(120)

Schmidpeter *et al.* have reported the synthesis of a novel cyanophosphorus-substituted heterocycle. This involves the :P—CN group acting as a pseudochalcogen in a nucleophilic displacement of a halogen from a quaternized heterocycle (Equation (121)) ⟨80AG767⟩. The treatment of these compounds with sodium diethyl phosphite gave the monocyanophosphinidene derivative (**29**) in 20–40% yield ⟨80AG767⟩.

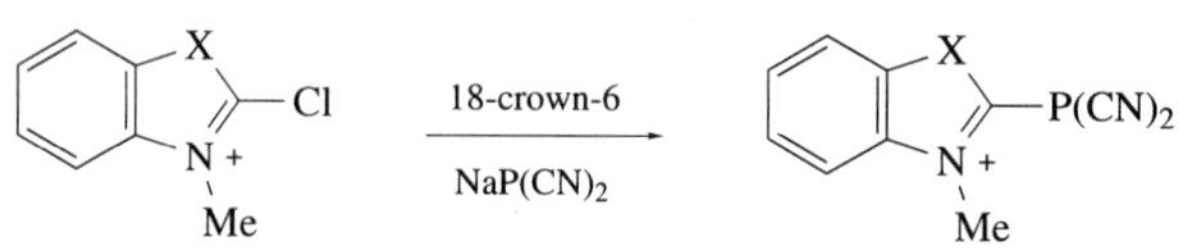

(121)

X = O, N

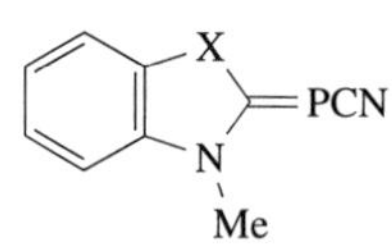

(29)

5.30.5.1.2 Formation of the P—CN bond

As might be expected, the formation of the P—CN bond is the most common method for the synthesis of this type of compound. The reaction usually involves the treatment of a suitably functionalized phosphorus compound with a source of cyanide.

The reaction of alkoxydimethylsilyl cyanides with phosphorus halides has been studied, and it was found that, depending on the nature of the phosphine, either one or both of the chlorines are substituted by cyano and alkoxy groups, respectively (Equation (122)) ⟨91IZV2847, 91MI 530-02⟩. Trimethylsilyl cyanide may also be used as the cyanide source in the reaction with phosphorus halides (Equation (123)) ⟨85CB676⟩. In some instances the reaction is carried out in the presence of a Lewis acid, for example $TiCl_4$ (Equation (124)) ⟨82IZV939⟩.

$$Me_2Si(OR)CN \;+\; MePCl_2 \longrightarrow MeP(Cl)CN \;+\; MeP(OR)CN \qquad (122)$$

R = Me, Et, Bu

(123)

R = Me, Ph

$$EtP(SPr^i)Cl \;+\; TMS\text{-}CN \xrightarrow[80\%]{TiCl_4} EtP(SPr^i)CN \qquad (124)$$

The use of metal cyanides, most noticeably silver cyanide, to introduce a nitrile group into a phosphorus compound is well known. A range of different alkyl groups are tolerated, and the reaction has been carried out in a range of solvents, for example CH_2Cl_2, $CHCl_3$, CCl_4, Et_2O and CH_3CN. It appears, however, that acetonitrile is the solvent of choice for reactions of this type (Equation (125)) ⟨60IS84, 68JA5236, 69JCS(A)2190, 71IC1536, 88JCS(D)2077⟩. Cyanobis(pentafluorophenyl)phosphine may be prepared by simply stirring the corresponding chloride with silver cyanide in dichloromethane for 1 h (Equation (126)) ⟨90JCS(D)2539⟩. Surprisingly then perhaps, iodobis(pentafluoroethyl)phosphine reacts only sluggishly with silver cyanide to give cyanobis(pentafluoroethyl)phosphine in 87% yield (Equation (127)) ⟨72AJC493⟩. In a similar type of reaction, iodotrifluoromethylphosphine reacts with mercury(II) cyanide to give only hydrogen cyanide and a black residue. Subsequent heating of this residue gave dicyanotrifluoromethylphosphine. This result suggests the intermediacy of a stable mercury–phosphorus complex (Scheme 10) ⟨75JCS(D)2368⟩.

$$R_nP(X)_{3-n} \;+\; AgCN \longrightarrow R_nP(CN)_{3-n} \qquad (125)$$

R = Me, Et, Bu^t, Ph, 4-FC_6H_4; X = Cl, Br; n = 0–2

$$(C_6F_5)_2PCl \;+\; AgCN \xrightarrow{CH_2Cl_2,\ 1\ h} (C_6F_5)_2PCN \qquad (126)$$

$$(C_2H_5)_2PI \;+\; AgCN \xrightarrow{8\ d} (C_2H_5)_2PCN \qquad (127)$$

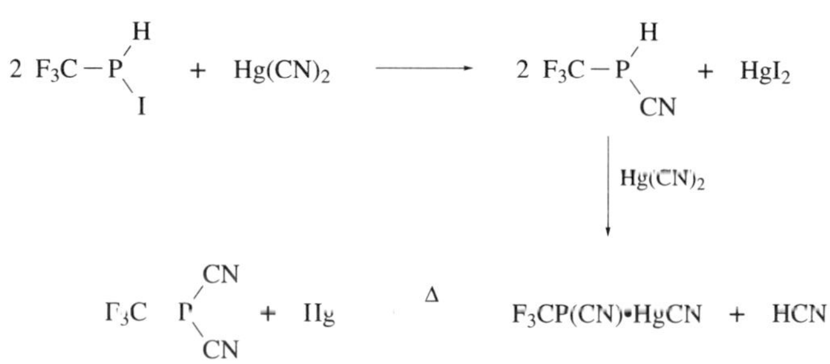

Scheme 10

Cyanophosphines may also be prepared by the reaction of halo-, dialkylaminomethyl- or dialkylaminophosphines with either a cyanogen halide or hydrogen cyanide.

When methyldifluorophosphine is treated with neat cyanogen bromide at low temperatures, both dicyanomethylphosphine and bromocyanomethylphosphine are isolated (Equation (128)) ⟨70IC1571⟩. The same paper also reports that the reaction of dibromomethylphosphine with dicyanomethylphosphine gives methylbromocyanophosphine (Equation (129)) ⟨70IC1571⟩.

$$MePF_2 + BrCN \longrightarrow MeP(CN)_2 + MeP(Br)CN \tag{128}$$

$$MePBr_2 + MeP(CN)_2 \longrightarrow 2MeP(Br)CN \tag{129}$$

Kostyanovskii and co-workers showed that dimethylaminomethylbis(isopropyl)phosphine would react with cyanogen chloride to give cyanobis(isopropyl)phosphine in reasonable yield (Equation (130)) ⟨77IZV249⟩. The same authors later reported that in this reaction the electrophilic attack is at phosphorus with subsequent P—C bond cleavage generating the dialkylcyanophosphine ⟨82IZV1611⟩.

(130)

Dimethyldimethylaminophosphine, prepared by either methylmagnesium halide or methyllithium addition to a dihalodimethylaminophosphine (Equation (131)), reacts with neat hydrogen cyanide to give cyanodimethylphosphine in 84% yield (Equation (132)) ⟨73IC241⟩.

$$Me_2NPX_2 + MeMgX \text{ or } MeLi \longrightarrow Me_2NPMe_2 \tag{131}$$

$$X = Br, Cl$$

$$Me_2NPMe_2 + HCN \xrightarrow[84\%]{} Me_2PCN \tag{132}$$

The *P*-cyanophosphaethene $H_2C{=}PCN$ has been prepared from cyanochlorotrimethylsilylmethylphosphine (**30**) by pyrolysis at 850 °C (Equation (133)). This novel phosphine (**30**) is itself prepared in three steps from dichlorotrimethylsilylmethylphosphine ⟨87JSP304⟩.

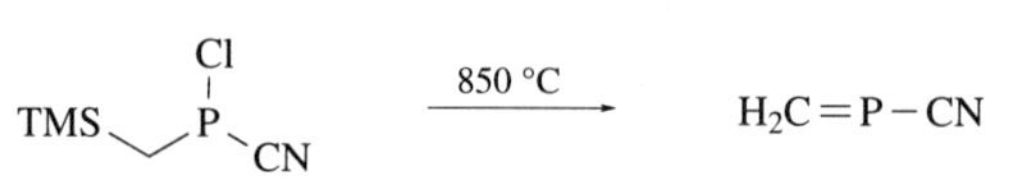
(133)

(**30**)

It is known that when certain isocyanides (i.e., *t*-butyl and benzyl) are reacted with either diethyl or diphenylphosphine in benzene containing 2,2′-azobisisobutyronitrile (AIBN) the products obtained are the corresponding dialkyl- or diarylcyanophosphines (Equation (134)). Strangely, when cyclohexyl or *n*-hexyl isocyanides are used under the same conditions, the only product obtained is the *N*-alkylformidoyldialkylphosphine (Equation (135)) ⟨70JOC4238⟩.

$$R^1NC + R^2{}_2PH \xrightarrow[70\ ^\circ C,\ 18\ h]{AIBN,\ benzene} R^2{}_2PCN + R^1H \tag{134}$$

$$R^1 = Bu^t, Bn;\ R^2 = Et, Ph$$

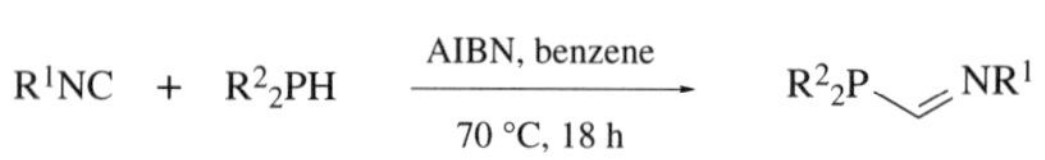
(135)

$$R^1 = C_6H_{11}, C_6H_{13}$$

5.30.5.2 Phosphorus(V) Derivatives

The synthesis of phosphorus derivatives is similar in many respects to that described earlier for phosphorus(III) analogues.

5.30.5.2.1 Formation of the C—PCN bond

The formation of the C—PCN bond is the least common method for the preparation of phosphorus(V) cyanides, and only a few examples are known.

When the phosphine (**31**) is reacted with a range of aldehydes and α-keto esters in the presence of a catalytic amount of boron trifluoride etherate the products obtained are those of oxidative addition involving the formation of a P—C bond between the phosphine and the keto or aldehydic group (Equation (136)) ⟨91IZV237⟩.

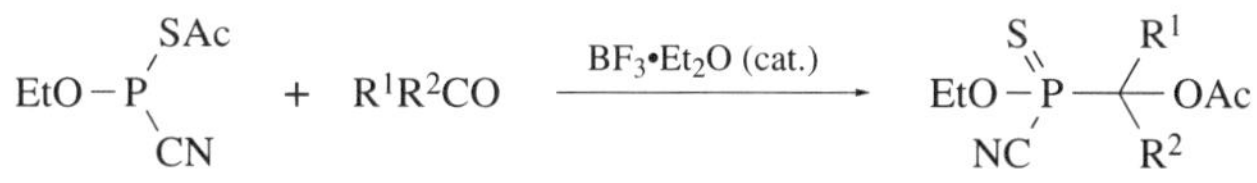

(136)

(31)

R^1 = Me, CCl_3, Ph; R^2 = CO_2Et, H

An alternative method involves the reaction of a nitrile oxide with triphenylphosphine. The oxide is formed *in situ* from the reaction of an *N*-silylated (diazomethyl)phosphine sulfide with nitrosyl chloride. It is proposed that the reaction proceeds via the formation of an α-nitroso compound which decomposes with loss of nitrogen to generate the nitrile oxide (Equation (137)) ⟨88AG289⟩.

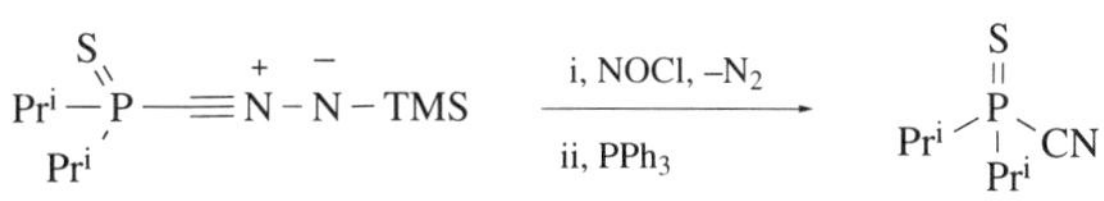

(137)

5.30.5.2.2 *Formation of the P—CN bond*

In general the formation of the P—CN bond involves the reaction of an aryl- or alkylphosphonyl halide with a source of cyanide anions in a similar manner to that described earlier (see Section 5.30.5.1.2).

Alkoxydimethylsilyl cyanides react with phosphorus(V) dichlorides to displace both chlorines with the cyano and alkoxy groups, respectively (Equation (138)) ⟨92IZV187⟩.

(138)

When the same reaction is carried out using silver cyanide as the cyanide source in refluxing acetonitrile, both of the chlorines are replaced by cyanide groups (Equation (139)) ⟨69AC144⟩. Das and co-workers have carried out similar nucleophilic displacements on a range of *O*-alkylphosphinothiochloridates. Using potassium cyanide in a biphasic system of dichloromethane/water with cetyltrimethylammonium bromide as a phase transfer catalyst, the reaction proceeds smoothly in quite acceptable yields (Equation (140)) ⟨84IJC(B)254⟩.

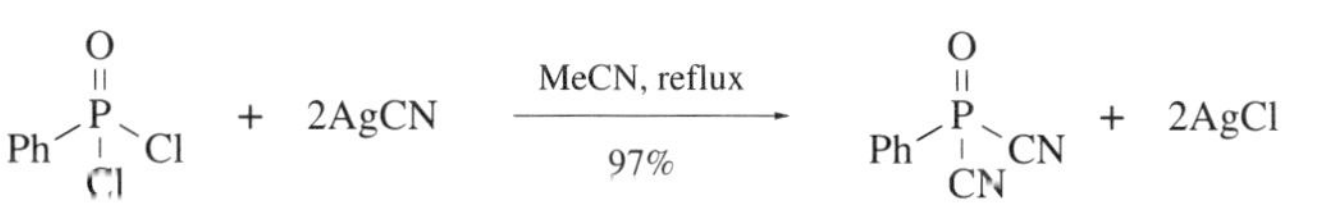

(139)

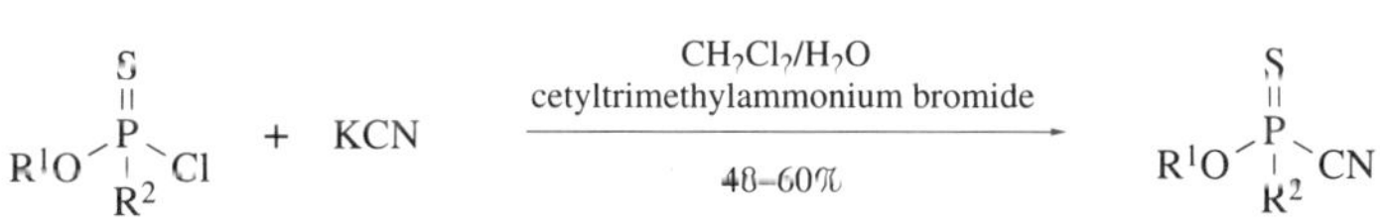

(140)

R^1 = Me, Et; R^2 = Ph

All the examples outlined so far in this section have described simple conversion of phosphorus(V) halides to their corresponding nitriles. However, when chlorodimethylphosphine is reacted with cyanogen chloride at 0°C for 1 h, the only product identified was dichlorocyanodimethylphosphorane (**32**) (Equation (141)) ⟨67MI 530-01⟩. Acid hydrolysis of (**32**) with acetic acid at low temperature gave cyanodimethylphosphine oxide in 50% yield (Equation (142)) ⟨67MI 530-01⟩. One other example involving the oxidative addition of cyanide to a phosphorus(III) compound is reported. In this case, the alkoxydialkyl phosphinite is reacted with hydrogen cyanide in a chlorinated hydrocarbon containing a base (triethylamine) (Equation (143)) ⟨92EGP300936A7⟩. The reaction also proceeds using alkylthiodialkyl phosphinite.

$$Me_2PCl + ClCN \xrightarrow{0\ °C,\ ether,\ 1\ h} \underset{(\mathbf{32})}{Me_2PCl_2CN} \quad (141)$$

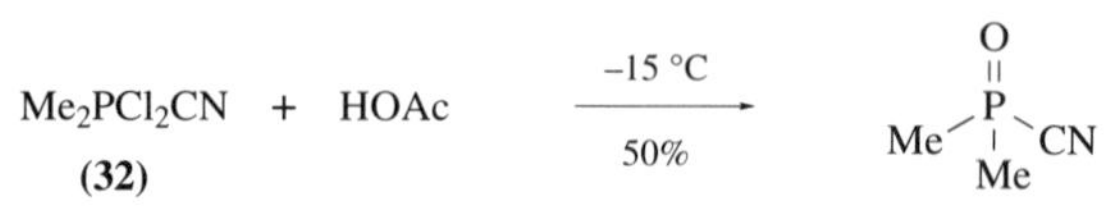
(142)

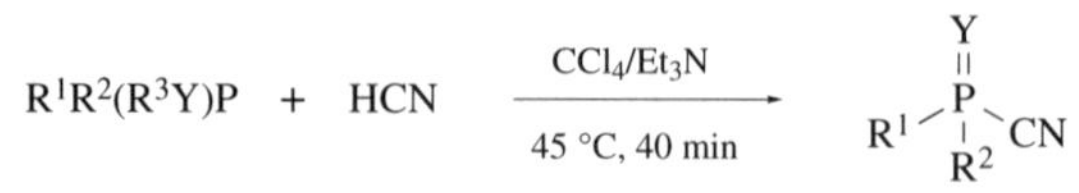
(143)

R^1, R^2 = alkyl, aryl; R^3 = alkyl; Y = O, S

5.30.5.3 Antimony Derivatives

5.30.5.3.1 Formation of the C—SbCN bond

Both potassium trichlorodicyanoantimonate(III) and chlorodicyanoantimony(III) react with sodium cyclopentadienide in THF. The former ionic antimony(III) complex reacts with two equivalents of the anion (Equation (144)), while the latter covalent cyanoantimony derivative reacts only once (Equation (145)) ⟨87SRI525⟩.

$$K_2[SbCl_3(CN)_2] + 2NaC_5H_5 \xrightarrow{THF} (C_5H_5)_2SbCN + 2KCl + NaCl + NaCN \quad (144)$$

$$SbCl(CN)_2 + NaC_5H_5 \xrightarrow{THF} (C_5H_5)_2Sb(CN)_2 + NaCl \quad (145)$$

5.30.5.3.2 Formation of the Sb—CN bond

Cyanogen halides have been found to add oxidatively to symmetrical triarylstibines to give excellent yields of the corresponding triorganostibine(V) derivatives (Equation (146)) ⟨78IJC(A)778⟩.

$$Ar_3M + XCN \xrightarrow[74\text{–}84\%]{MeCN,\ -5\ °C} Ar_3MXCN \quad (146)$$

Ar = Ph, 4-MeC_6H_4; X = I, Br; M = As, Sb

5.30.5.4 Arsenic Derivatives

Triarylarsines have also been shown to react with cyanogen halides as described for antimony compounds (see Section 5.30.5.3) (see Equation (146)) ⟨78IJC(A)778⟩. In contrast to antimony, arsenic(III) nitriles have been synthesized by simply heating the corresponding chloroarsines with potassium cyanide (Equation (147)) ⟨74MI 530-01, 74MI 530-02⟩.

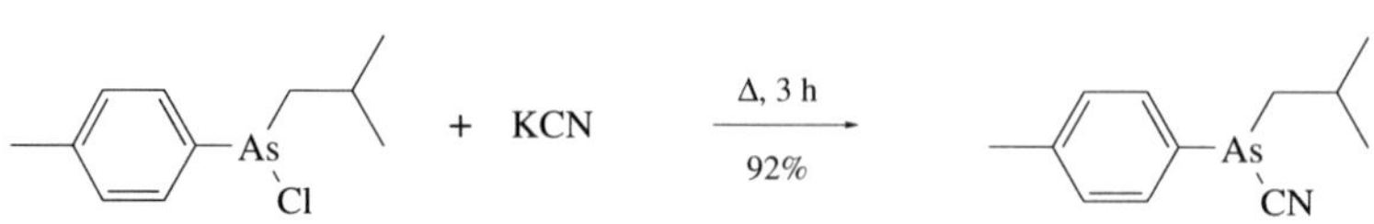
(147)

5.30.5.5 Bismuth Derivatives

In contrast to both antimony (see Section 5.30.5.3) and arsenic (see Section 5.30.5.4), triarylbismuth derivatives did not undergo oxidative additions of cyanogen iodide, instead one aryl–bismuth bond was cleaved in preference to oxidation. Thus, the products obtained were the diarylbismuth nitriles and an aryl iodide (Equation (148)) ⟨78IJC(A)778⟩.

$$Ar_3Bi + ICN \xrightarrow{MeCN} ArI + Ar_2BiCN \qquad (148)$$

$$Ar = Ph,\ 4\text{-}MeC_6H_4$$

Bismuth(V) cyanides may, however, be obtained by the reaction of the appropriate triarylbismuth(V) halide with potassium cyanide. The reaction is carried out in aqueous solution at low temperature (Equation (149)) ⟨73JOM(50)129⟩. Alternatively, anionic bismuth(V) complexes react with sodium cyanide in ethanol to give the expected bismuth(V) dicyanides (Equation (150)) ⟨79JOM(166)365⟩ in good yield.

$$Ph_3BiCl_2 + KCN \xrightarrow{H_2O,\ 5\ °C,\ 8\ h} Ph_3Bi(CN)_2 \qquad (149)$$

$$Et_4N[Ph_2BiCl_2] + 2NaCN \xrightarrow[70\%]{EtOH,\ RT} Et_4N[Ph_2Bi(CN)_2] \qquad (150)$$

5.30.6 SINGLY BONDED METALLOID DERIVATIVES—$R_3Si—C{\equiv}N$, $R_2B—C{\equiv}N$ AND $R_3Ge—C{\equiv}N$

The chemistry of group 3 and 4 pseudohalides has previously been reviewed by Lappert *et al.* ⟨66MI 530-01⟩. This current overview includes material from that source where applicable.

5.30.6.1 $R_3Si—CN$ Compounds

Compounds of the general formula $R_3Si—CN$ are amongst the most versatile of silicon reagents. Whilst simple compounds, for example TMS-CN, are now commercially available, more complex analogues must still be synthesized prior to use. This section looks at the synthetic routes available for the synthesis of both TMS—CN and more varied analogues.

Early workers in this area (e.g., see ⟨58JA4151⟩) named Si—CN-containing compounds as (iso)cyanosilanes due to the uncertainty of the structure. More recently ⟨87JA6532⟩, this facile equilibrium has been investigated more fully using *ab initio* calculations of the silyl cyanide/isocyanide energies.

5.30.6.1.1 Formation of the Si—CN bond

The earliest reports on the synthesis of trialkylsilyl cyanides ⟨49JCS2755, 50JCS3077, 52JA5247⟩ describe the reaction of halotrialkylsilanes with silver cyanide under reflux to give the corresponding cyanides (Equation (151)), although yields with the chlorotrimethylsilanes were considerably lower ⟨52JA5247⟩. Similarly, dialkyldihalosilanes gave the dialkylsilyl dicyanide derivatives under these conditions ⟨52JA5247⟩.

$$R_3SiX + AgCN \xrightarrow{reflux} R_3SiCN + AgX \qquad (151)$$

$$R = Me,\ Et,\ Pr^n;\ X = Cl,\ Br,\ I$$

A number of alternative methods for the synthesis of trialkylsilyl cyanides have been reported by Bither *et al.* ⟨58JA4151⟩. These involve: the reaction of potassium mercury(II) cyanide with a halotrialkylsilane in *N,N*-dimethylformamide (Scheme 11, reaction (a)); the treatment of hexaalkyldisilazane with hydrogen cyanide in ether (Scheme 11, reaction (b)); the aluminum chloride-catalysed reaction of a trialkylmethoxysilane and trimethylacetyl cyanide (Scheme 11, reaction (c)); and an exchange reaction between trimethylsilyl cyanide and a halotrialkylsilane (Scheme 11, reaction (d)). Becu ⟨87BSB115⟩ has described a simple modification of this latter *trans*-cyanogenation reaction which involves catalysis of the exchange reaction by potassium fluoride (Equation (152)).

This method works best (yields of 90–100%) for the preparation of silyl cyanides which have boiling points higher than those of the starting materials.

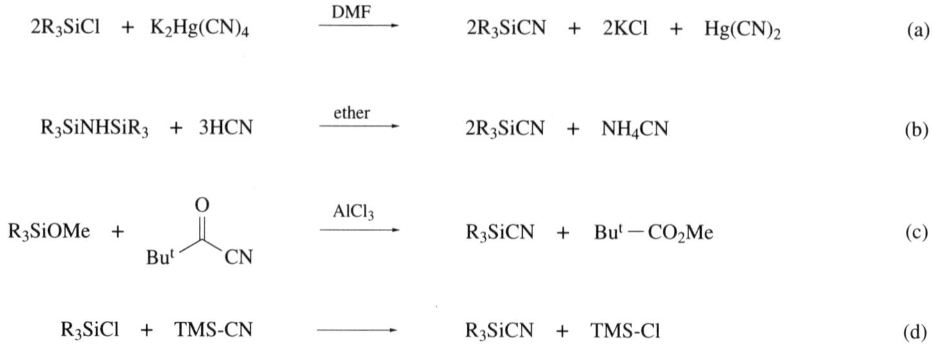

Scheme 11

$$\text{TMS-CN} + \text{RSiMe}_2\text{Cl} \xrightleftharpoons{\text{KF}} \text{TMS-Cl} + \text{RSiMe}_2\text{CN} \qquad (152)$$

Evans *et al.* ⟨74JOC914⟩ have developed a route to trimethylsilyl cyanide which is amenable to large-scale preparation. This involves the initial reaction of lithium hydride and hydrogen cyanide with subsequent addition of chlorotrimethylsilane and work up, giving the trimethylsilyl cyanide in 71–84% yield (Equation (153)). Mai and Patil ⟨86JOC3545⟩ have also used lithium cyanide as the source of cyanide in reactions with chloroalkylsilanes; in this case the lithium cyanide is first formed by the reaction of *n*-butyllithium with trimethylsilyl cyanide (Equation (154)).

$$\text{LiH} + \text{HCN} \xrightarrow[\text{71–84\%}]{\text{TMS-Cl}} \text{TMS-CN} \qquad (153)$$

$$\text{Bu}^n\text{Li} + \text{TMS-CN} \xrightarrow[\text{58–79\%}]{\text{R}_{4-n}\text{SiX}_n} \text{R}_{4-n}\text{Si(CN)}_n \qquad (154)$$

$n = 1$ or 2

The use of phase transfer catalysts to enhance the formation of silyl cyanides from halosilanes is known, with both 18-crown-6 ⟨75TL71, 78ZN(B)1465, 84S1020⟩ and Adogen ⟨79S522⟩ being reported (Scheme 12). The reaction of chlorotrimethylsilane with an alkali metal cyanide is also known to take place in *N*-pyrrolidone ⟨79S523⟩, in acetonitrile containing catalytic amounts of sodium iodide and pyridine ⟨82S212⟩, or in sulfolane with catalytic amounts of copper(I) cyanide or zinc cyanide (Scheme 13) ⟨81GEP3018821⟩.

$$\text{TMS-Cl} + \text{KCN} \xrightarrow[\text{45\%}]{\text{CH}_2\text{Cl}_2\text{, 18-crown-6, 24 h}} \text{TMS-CN}$$

$$\text{Bu}^t\text{Me}_2\text{SiCl} + \text{KCN} \xrightarrow[\text{75\%}]{\text{CH}_2\text{Cl}_2\text{, 18-crown-6, reflux, 72 h}} \text{Bu}^t\text{Me}_2\text{SiCN}$$

$$\text{TMS-Cl} + \text{NaCN} \xrightarrow[\text{60–70\%}]{N\text{-methylpyrrolidinone, Adogen (1 mol\%), 90–100 °C, 36 h}} \text{TMS-CN}$$

Scheme 12

A process for the preparation of silyl cyanides has been reported involving treatment of an appropriate alkylhalosilane with alkali metal cyanides supported on macroporous polymers (Equation (155)) ⟨88JAP63135393⟩.

$$\text{TMS-Cl} + \text{KCN} \xrightarrow[71\%]{N\text{-methylpyrrolidinone, reflux, 16 h}} \text{TMS-CN}$$

$$\text{TMS-Cl} + \text{KCN} \xrightarrow[\text{MeCN}]{\text{NaI (cat.), pyridine}} \text{TMS-CN (not isolated)}$$

$$\text{TMS-Cl} + \text{KCN or NaCN} \xrightarrow[90\%]{\text{CuCN or Zn(CN)}_2\text{ (cat.), }\Delta\text{, sulfolane}} \text{TMS-CN}$$

Scheme 13

$$\text{R}_{4-n}\text{SiCl}_n + \text{NaCN/Amberlite XAD-4} \xrightarrow[25\ ^\circ\text{C, 0.5 h}]{N\text{-methylpyrrolidine}} \text{R}_{4-n}\text{Si(CN)}_n \quad (155)$$

$n = 1$ or 2

The reaction of chlorotrimethylsilane with hydrogen cyanide in ether containing triethylamine gives trimethylsilyl cyanide in 70% yield (Equation (156)) ⟨78S154⟩. The reaction of primary, secondary and tertiary isocyanides with tris(trimethylsilyl)silane under radical conditions gives the corresponding silyl cyanides (Equation (157)) ⟨91JOC678⟩.

$$\text{TMS-Cl} + \text{HCN} \xrightarrow[70\%]{\text{Et}_3\text{N, Et}_2\text{O}} \text{TMS-CN} \quad (156)$$

(157)

The use of bis(trimethylsilyl) sulfate as an organosilicon synthon has been previously documented in the literature ⟨74URP449912⟩. The reaction of this reagent with potassium cyanide gives trimethylsilyl cyanide in 89% yield (Equation (158)) ⟨82S277⟩, while dialkyltrimethylsilyl phosphites react with alkyl thiocyanates at 110 °C to give trimethylsilyl cyanide in very good yield (Equation (159)) ⟨90ZOB476⟩.

$$\text{(TMS-O)}_2\text{SO}_2 + 2\text{KCN} \xrightarrow[89\%]{} 2\text{TMS-CN} + \text{K}_2\text{SO}_4 \quad (158)$$

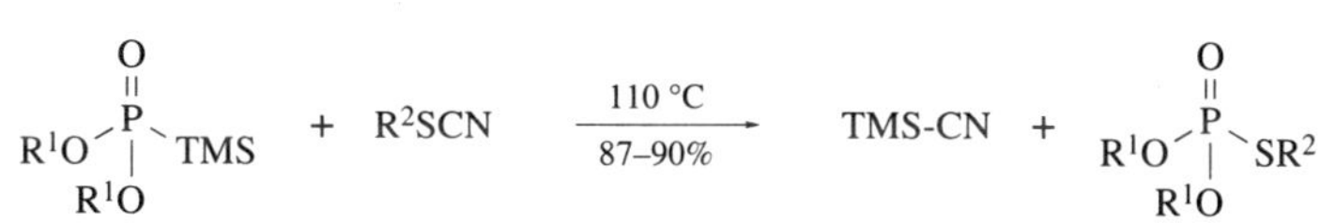

(159)

R^1 = Et, Pr^i, Bu^n; R^2 = Et, Me, Pr^i

Finally, chlorotrialkylsilanes will react with thallium(III) cyanide under reflux ⟨78JOC2280⟩ or with mercury(II) cyanide in refluxing benzene ⟨77IZV1839⟩ to give trialkylsilyl cyanides (Scheme 14).

$$\text{TMS-Cl} + \text{Tl(CN)}_3 \xrightarrow[84\%]{\text{reflux, 10 h}} \text{TMS-CN} + \text{TlCl}_3$$

$$\text{R}_3\text{SiCl} + \text{Hg(CN)}_2 \xrightarrow[62\text{–}89\%]{\text{benzene, reflux}} \text{R}_3\text{SiCN}$$

R = Me, Et

Scheme 14

5.30.6.2 R_2B—CN Compounds

Dialkylboron cyanides are generally prepared by the reaction of a dialkylboron halide and a source of cyanide ions. Thus, di-*n*-butylboron chloride reacts smoothly with trimethylsilyl cyanide to give the di-*n*-alkylboron cyanide in almost quantitative yield (Equation (160)) ⟨59JA5106⟩. Similarly, the reaction of di-*n*-butylboron chloride with silver cyanide in ether under reflux also gives di-*n*-butylboron cyanide (Equation (161)) ⟨59JA5106⟩. Other dialkylboron cyanides have been formed by the reaction of the appropriatethioborane with hydrogen cyanide (Equation (162)) ⟨59DOK(127)571⟩.

$$Bu^n{}_2BCl + TMS\text{-}CN \xrightarrow[98\%]{\text{pentane, } -196\,°C \text{ to } -5\,°C,\ 1\ h} Bu^n{}_2BCN + TMS\text{-}Cl \quad (160)$$

$$Bu^n{}_2BCl + AgCN \xrightarrow{\text{ether, reflux}} Bu^n{}_2BCN + AgCl \quad (161)$$

$$R^1{}_2BSR^2 + HCN \longrightarrow R^1{}_2BCN + R^2SH \quad (162)$$

$$R^1 = Bu^n,\ Bu^i,\ \text{isoamyl}$$

Zweifel *et al.* ⟨72JA6560⟩ have reported the intermediacy of a dialkylboron cyanide in their work directed at the stereospecific conversion of alkynes into *trans*-alkenes (Equation (163)). This product was not isolated, but rapidly underwent *cis* elimination to give the desired *trans*-alkene.

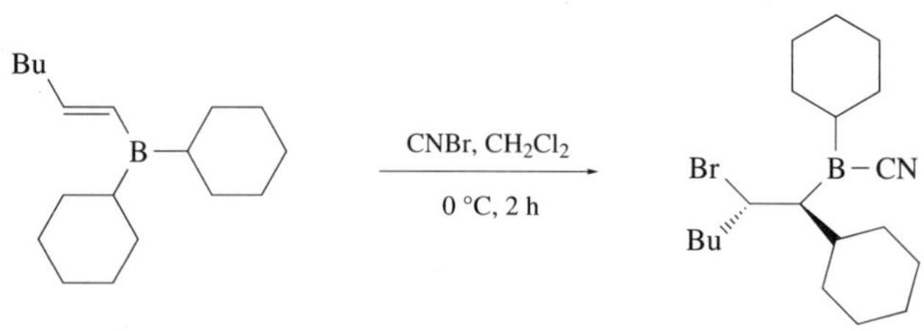

(163)

Dimethylboron cyanide may be prepared by the reaction of dimethylboron iodide with iodocyanogen in carbon disulfide (Equation (164)). The product is reported to exist in an oligomeric form ⟨75ZAAC(418)273⟩.

$$Me_2BI + ICN \xrightarrow[78\%]{CS_2} Me_2BCN + I_2 \quad (164)$$

5.30.6.3 R_3Ge—CN Compounds

Anderson ⟨51JA5439⟩ has reported that trialkylgermanium oxides react with hydrogen cyanide to form trialkylgermanium cyanides, which may be isolated by simple distillation (Equation (165)). Other reports suggest that this conversion can also be achieved by the reaction of organogermanium halides with silver or mercury(II) cyanide (Equation (166)) ⟨56JA1692, 51JA5440, 60JOC809, 74MI 530-03⟩, while reaction of triethylgermanium ethylsulfonate with potassium cyanide gave triethylgermanium cyanide in good yield (Equation (167)) ⟨64IC910⟩.

$$(R_3Ge)_2O + 2HCN \longrightarrow 2R_3GeCN + H_2O \quad (165)$$

$$R = Pr^n,\ Et$$

$$R_3GeX + AgCN \longrightarrow R_3GeCN + AgX \quad (166)$$

$$R = Et,\ Pr^i,\ Pr^n;\ X = Cl,\ Br,\ I$$

$$Et_3GeOSO_2Et + KCN \xrightarrow[\sim 86\%]{} Et_3GeCN + EtSO_3K \quad (167)$$

Triarylgermanium cyanides have been prepared in modest yields ⟨68JINC1399⟩ by the reaction of the corresponding triarylgermanium bromides with a hot solution of either silver or lead cyanide in benzene or carbon tetrachloride (Equation (168)). An alternative route to triphenylgermanium cyanide involves an exchange reaction between a bis(triphenylgermanium) mercury(II) complex and copper(II) cyanide at room temperature (Equation (169)) ⟨69ZOB2365⟩.

$$Ar_3GeBr + AgCN \xrightarrow[40\text{–}50\%]{\text{benzene or } CCl_4,\ \Delta} Ar_3GeCN + AgBr \quad (168)$$

$$Ar = Ph,\ 4\text{-}MeC_6H_4$$

$$(Ph_3Ge)_3Hg + 2Cu(CN)_2 \xrightarrow[\sim 80\text{–}90\%]{} 2Ph_3GeCN \quad (169)$$

5.30.7 SINGLY BONDED METAL DERIVATIVES

This section describes the synthesis of derivatives which contain the subunit R_n—M—CN, where M is a metal atom and R is a carbon atom. However, compounds where R is a CO group have been excluded to avoid the inclusion of purely coordination complexes. No examples of group 1 metals or actinide derivatives have been found.

5.30.7.1 Group 2 Derivatives

Of the group 2 metals, only alkylberyllium cyanides have been prepared. Thus, if dimethylberyllium and hydrogen cyanide are added simultaneously to ether a precipitate forms which may be removed by filtration. The methylberyllium cyanide remains in the filtrate, and is obtained as a solid on removal of the solvent. The methylberyllium cyanide coordinates ether, which may be removed under vacuum (Equation (170)) ⟨63JCS229⟩.

$$Me_2Be + HCN \xrightarrow{Et_2O} MeBeCN + CH_4 \quad (170)$$

5.30.7.2 Transition Metal Derivatives

Surprisingly, only suitable derivatives of nickel, palladium, platinum, copper, gold and mercury have been prepared. In general, the preparation of these compounds involves displacement of a suitable leaving group, such as a halogen or acetate, with a cyanide group derived from an alkali metal cyanide. Specific examples are discussed below.

5.30.7.2.1 Nickel derivatives

The chlorovinyl nickel complexes (**33**) react readily with potassium cyanide in acetone to give the expected nickel cyanide derivative. In order to avoid decomposition during the reactions, triphenylphosphine is added to the reaction mixture (Equation (171)) ⟨83JCS(D)1841, 84OM996⟩. The reactions proceed at room temperature, and are complete in minutes. In a similar manner, bis(triethylphosphine)pentachlorophenylnickel chloride reacts readily with silver cyanide to give a cyanonickel derivative. In this case the reaction is carried out at low temperature, and the yield is somewhat more modest (Equation (172)) ⟨92ICA173⟩. In some reactions of this type when an excess of cyanide reagent is used, not only is the halogen displaced (Scheme 15) but subsequent Ni—C bond cleavage occurs to eject the pentahalophenyl group and give a dicyanonickel compound (Scheme 15) ⟨87JOM(137)229⟩. Tetrahydrofuran has also been used as a solvent for such halogen displacement reactions on vinylnickel halides using potassium cyanide (Equation (173)), although in this case the yield of product was poor (27%) ⟨88CB1115⟩. Acetate groups may also be displaced from pentahalophenylnickel acetate complexes using sodium cyanide in refluxing acetone. Again the yields are good (Equation (174)) ⟨76IC2346⟩.

$$\underset{\textbf{(33)}}{ClNi(CR^1{=}CCl_2)(PR^2{}_3)_2} \xrightarrow[60\text{–}90\%]{KCN,\ PPh_3,\ acetone,\ RT,\ 10\ min} (Cl_2C{=}CR^1)Ni(CN)(PR^2{}_3)_2 \quad (171)$$

R^1 = H, Cl; R^2 = Ph, Me

$$NiCl(C_6Cl_5)(PEt_3)_2 \xrightarrow[low\ temperature]{AgCN,\ acetone} Ni(CN)C_6Cl_5(PEt_3)_2 \quad (172)$$

$$NiBr(C_6F_5)(PPh_3)_2 \xrightarrow[3\ h]{KCN,\ acetone} (C_6F_5)Ni(CN)(PPh_3)_2$$

$$NiBr(C_6F_5)(PPh_3)_2 \xrightarrow{KCN\ (excess),\ acetone} Ni(CN)_2(PPh_3)_2$$

Scheme 15

$$Ni(CR{=}CR_2)Cl(PMe_3)_2 \xrightarrow[RT,\ 120\ h]{KCN,\ THF} Ni(CR{=}CR_2)CN(PMe_3)_2 \quad (173)$$

R = Ph, Me

$$[C_6Cl_5Ni(PPh_2Me)_2OAc] \xrightarrow[reflux]{NaCN,\ acetone} [C_6Cl_5Ni(PPh_2Me)CN] \quad (174)$$

The complexation of a bis(pentafluorophenyl)bis(dioxane)nickel(II) compound with cyanide gave the bridged species (**34**), which on treatment with triphenylphosphine breaks down to give bis(pentafluorophenyl)triphenylphosphinenickel(II) cyanide (Equation (175)) ⟨78MI 530-02⟩.

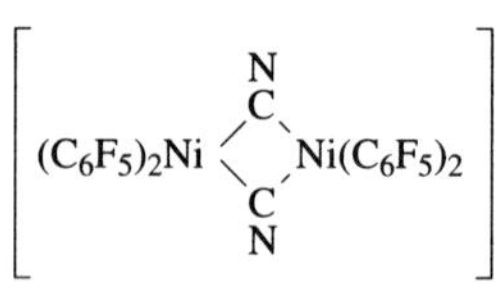

(34)

$$\xrightarrow{2PPh_3} (C_6F_5)_2Ni(CN)(PPh_3) \quad (175)$$

Oxidative addition reactions of aryl nitriles and tetrakis(triethylphosphine)nickel are known ⟨74JA2360⟩. The reactions are analogous to those reported for both palladium and platinum (*vide infra*). However, in the case of nickel the reactions are more facile, often occurring at room temperature. The product in these reactions is often contaminated with the dicyano compound (Equation (176)).

$$Ni(PEt_3)_4 \ + \ PhCN \xrightarrow{toluene,\ RT} C_6H_5NiCN(PEt_3)_2 \quad (176)$$

5.30.7.2.2 Palladium derivatives

In general, cyanopalladium compounds are formed either by the displacement of a suitable leaving group or by an oxidative insertion reaction with an aryl nitrile.

The most common leaving groups are the halogens. Thus chloro(cyanomethyl)bis(triphenylphosphine)palladium(II) reacts smoothly with potassium cyanide in methanol to give the desired cyano compound (Equation (177)) ⟨76BCJ464⟩. The reaction may also be carried out in acetone or benzene ⟨73SRI237⟩. In this case, the reactions work best when the metal cyanide is chosen to give the most insoluble metal chloride, which drives the equilibrium towards the cyano complex. Both bromide ⟨76SRI217⟩ and iodide ⟨84ICA215, 84JOM(276)431⟩ can be displaced by cyanide (Scheme 16).

$$ClPd(CH_2CN)(PPh_3)_2 \; + \; KCN \xrightarrow[60\ ^\circ C,\ 2\ h]{MeOH} (PPh_3)_2Pd(CN)(CH_2CN) \qquad (177)$$

$$PdBr(C_6Cl_5)(PPh_3)_2 \; + \; KCN \longrightarrow Pd(CN)C_6Cl_5(PPh_3)_2$$

$$MePdI(PEt_3)_2 \; + \; KCN \xrightarrow{ethanol} MePd(CN)(PEt_3)_2$$

Scheme 16

As in the case of nickel, tris(trialkylphosphine)- or tris(triarylphosphine)palladium(II) compounds undergo oxidative insertion reactions when reacted with aryl nitriles. The reactions may be carried out in toluene or benzene under reflux (Equation (178)) ⟨74JA2360, 76IC208⟩. Rieke and co-workers have reported similar insertion reactions of benzonitrile using a highly reactive palladium slurry. The metal slurry is prepared *in situ* from potassium metal (two equivalents), palladium chloride (one equivalent) and triethylphosphine (two equivalents) in GLYME. The resulting slurry is then reacted with benzonitrile in GLYME under reflux for 37 h. The resulting product was shown by NMR to have the *trans* orientation of ligands (Equation (179)) ⟨79JOC3069⟩.

$$Pd(PR_3)_3 \; + \; PhCN \xrightarrow[reflux]{toluene} (PR_3)_2Pd(Ph)(CN) \qquad (178)$$

$$2K \; + \; PdCl_2 \; + \; 2PEt_3 \xrightarrow[27\%]{GLYME,\ PhCN,\ reflux,\ 37\ h} (PEt_3)_2Pd(CN)(Ph) \qquad (179)$$

Acylcyanolpalladium complexes (**35**) have been postulated to be formed in the palladium-catalysed decarbonylation of acyl cyanides (Equation (180)) ⟨86JOC898⟩.

$$Pd(CN)(COAr)(PPh_3)_n$$

(35)

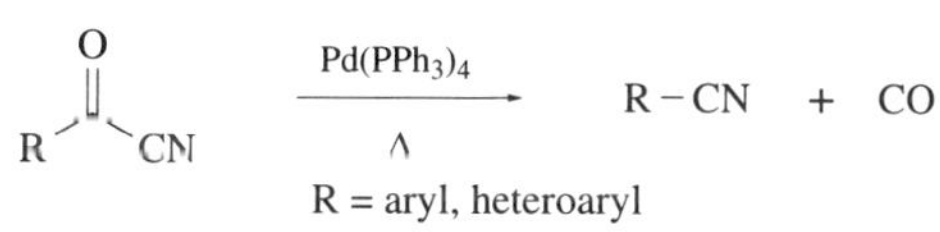

(180)

When *trans*-1,2-dicyano-1,2-bis(trifluoromethyl)ethylene reacts with stilbenebis(triphenylphosphine)palladium, the product obtained is that in which a cyano group has migrated from carbon to palladium (Equation (181)) ⟨70JCS(A)1847⟩.

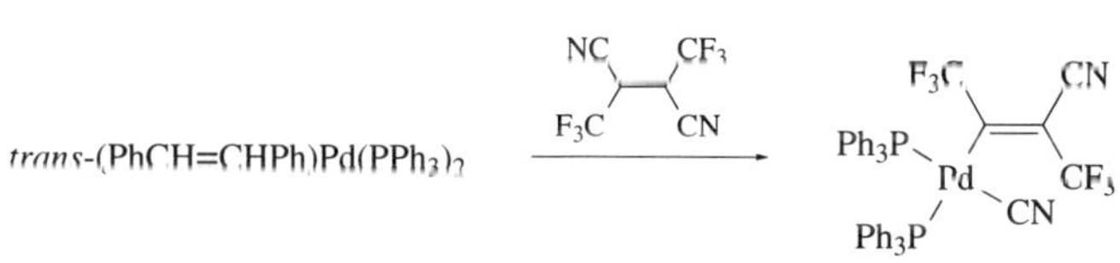

(181)

5.30.7.2.3 *Platinum derivatives*

Many of the reactions leading to cyanoplatinum compounds are analogous to those described for palladium. Thus, iodopentachlorophenylbis(triethylphosphine)platinum reacts with potassium cyanide in either acetone or chloroform to yield the expected cyanide derivative (Equation (182)) ⟨77JOM(128)291⟩. The displacement of trifluoroacetate from a platinum complex is also known ⟨82OM64⟩. In this instance the platinum complex is formed by the reaction of the bimetallic compound (**36**) with trifluoroacetic acid (Scheme 17). While displacement of halide and acyl groups from platinum(II) phosphine complexes is well known, the displacement of neutral ligands is not as common. However, potassium cyanide in the presence of a stoichiometric amount of 18-crown-6 will displace either cyclooctodiene (COD) (Equation (183)) or triethylphosphine (Equation (184)) from platinum complexes ⟨77JCS(D)1444⟩. The products are isolated as their potassium 18-crown-6 salts.

$$(C_6Cl_5)(PEt_3)_2PtI + KCN \xrightarrow{\text{acetone or } CHCl_3} (C_6Cl_5)(PEt_3)_2PtCN \quad (182)$$

$$Pt(PPh_3)_3 + HgAr_2 \xrightarrow{\text{benzene}} [(PPh_3)_2ArPt(HgAr)] \; \mathbf{(36)}$$

$$\mathbf{(36)} \xrightarrow{CF_3CO_2H,\ \text{benzene}} [Pt(O_2CCF_3)Ar(PPh_3)_2] + Hg + ArH$$

$$[Pt(O_2CCF_3)Ar(PPh_3)_2] \xrightarrow[\text{RT, 5 h}]{\text{KCN, acetone}} Pt(CN)Ar(PPh_3)_2$$

Ar = chlorinated phenyl

Scheme 17

$$PtR_2(COD) + KCN \xrightarrow[\text{RT, 18 h}]{\text{18-crown-6, benzene}} [K(18\text{-crown-6})]_2[Pt(CN)_2R_2] \quad (183)$$

R = 2-thienyl, benzofuranyl, methyl, 4-$MeOC_6H_4$

$$cis\text{-}PtMe_2(PEt_3)_2 + KCN \xrightarrow[\text{RT, 18 h}]{\text{18-crown-6, benzene}} [K(18\text{-crown-6})]_2[PtMe_2(CN)(PEt_3)] \quad (184)$$

Platinum, as with both nickel and palladium, will undergo oxidative insertion reactions with aryl nitriles ⟨74JA2360, 76IC208⟩. Thus, tetrakis(triphenylphosphine)platinum(0) undergoes an oxidative addition reaction with 1,1,1-tricyanoethane in benzene (Equation (185)) ⟨71JCS(A)1663⟩. Acylcyanoplatinum complexes have been isolated from the oxidative insertion of arylacyl cyanides and stilbenebis(triphenylphosphine)platinum(0) (Equation (186)) ⟨79JOM(175)303⟩.

$$Pt(PPh_3)_4 + MeC(CN)_3 \xrightarrow{\text{benzene, reflux}} Pt(CN)(PPh_3)_2(C(CN)_2Me) \quad (185)$$

$$Pt(trans\text{-}PhCH{=}CHPh)(PPh_3)_2 + PhCOCN \xrightarrow[32\%]{\text{RT, 12 h}}$$

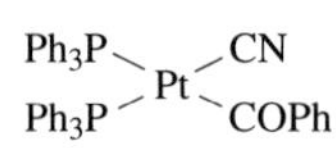

(186)

5.30.7.2.4 *Copper derivatives*

Cyanocuprates are not discussed in this section due to the uncertainty of their structure.

Copper(I) cyanide has been shown to react with phenylacetylene to give the addition product (**37**) (Equation (187)). The reaction is carried out in ethanol, but no exact experimental details are given ⟨86ICAL17⟩.

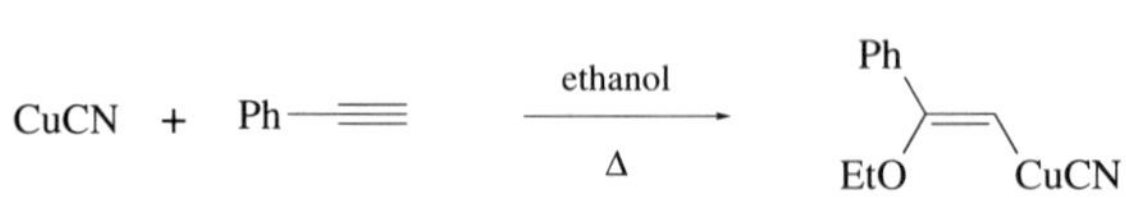

(37) (187)

5.30.7.2.5 *Gold derivatives*

As with those metals previously discussed, the main synthetic strategy to cyanogold compounds involves nucleophilic displacement of a suitable leaving group by an alkali metal cyanide. Displacement of chloride from the pentafluorophenylgold complex (**38**) proceeds as expected (Equation (188)).

However, if the same reaction is attempted on the acetate equivalent then only cationic species are obtained ⟨73MI 530-01⟩. Cyanodimethylgold has been prepared from the corresponding chloride using potassium cyanide in the presence of silver nitrate. The product is reportedly explosive in pure form, and is reacted *in situ* with 1,8-naphthyridine to form the naphthyrine–gold coordination complex (Scheme 18) ⟨73JA4855⟩. The formation of cyanodimethylgold from iododimethylgold reportedly proceeds smoothly, with the product reportedly existing as a cyclic tetramer in solution. The treatment of this cyclic bridged species with triphenylphosphine allows for the isolation of cyanodimethyl(triphenylphosphine)gold(III) (Scheme 19) ⟨72MI 530-01⟩. Neutral cyanobis(pentafluorophenyl)triphenylphosphinegold(III) may be prepared from the perchlorate-containing complex ⟨73MI 530-01⟩.

$$\underset{\textbf{(38)}}{(C_6F_5)_2AuCl(PPh_3)} + MCN \xrightarrow[M = K, Na]{} (C_6F_5)_2Au(CN)(PPh_3) \quad (188)$$

Scheme 18

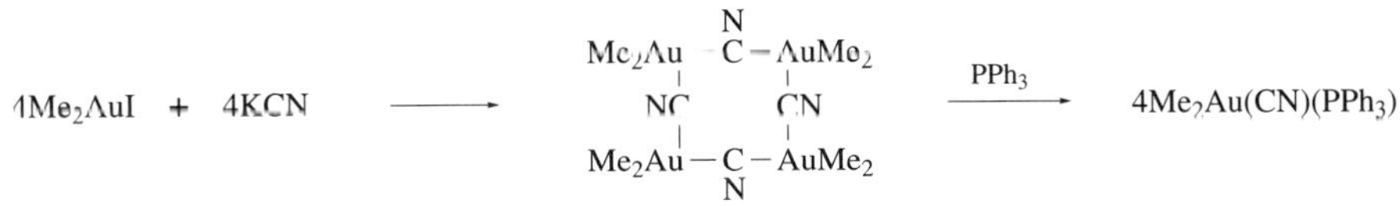

Scheme 19

Triphenylphosphoniomethanide, prepared from methyltriphenylphosphonium iodide and sodium amide ⟨70LA(739)211⟩, reacts with gold cyanide in THF to give cyano(triphenylphosphoniomethide)gold(I) in good yield (Equation (189)). This compound undergoes oxidative addition reactions with halogens to give *trans*-cyanodihalo(triphenylphosphoniomethide)gold(III) compounds (Equation (190)) ⟨84JCS(D)2839⟩.

$$AuCN + H_2C{=}PPh_3 \xrightarrow[75\%]{THF,\ RT,\ 22\ h} [Au(CN)(CH_2PPh_3)] \quad (189)$$

$$[Au(CN)(CH_2PPh_3)] + X_2 \xrightarrow{CHCl_3,\ RT} [AuX_2(CN)(CH_2PPh_3)] \quad (190)$$

X = Cl, 96%; Br, 92%; I, 92%

5.30.7.2.6 *Mercury derivatives*

Bis(cyanomercury)methane has been prepared by the displacement of both acetate groups from bis(acetoxymercury)methane in aqueous acetic acid using potassium cyanide (Equation (191)). Tetra(cyanomercury)methane is prepared in an analogous fashion ⟨83JOM(243)245⟩. Similarly, 4-(dimethylamino)phenylmercury(II) acetate reacts with potassium cyanide in refluxing aqueous ethanol to give the corresponding mercury(II) cyanide (Equation (192)) ⟨68IZV289⟩. Trifluoromethylmercury(II) iodide, formed in 37% yield from mercury and trifluoromethyl iodide, reacts with silver cyanide to give trifluoromethylmercury(II) cyanide (Equation (193)) ⟨76ZAAC(426)288⟩.

(191)

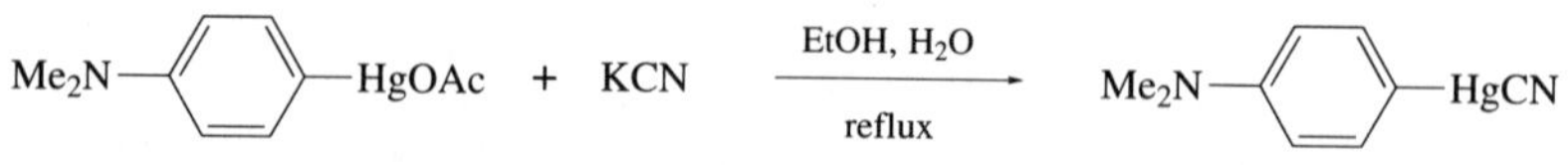
(192)

$$F_3CHgI + AgCN \longrightarrow F_3CHgCN \quad (193)$$

The reaction of mercury(II) cyanide with phenyltrifluorosilanes results in cleavage of the C—Si bond to generate either the phenylmercury(II) cyanide or the bis(phenyl)mercury(II) compound (Equation (194)). The reactions are carried out in benzene, and it has been found that substituents on the phenyl ring have a marked effect on the reaction. Thus, electron-donating groups facilitate the reaction while electron-withdrawing groups have an adverse effect ⟨88JOM(341)225⟩. The same paper reports that mercury salts also cleave C—Si bonds in chloromethyltrifluorosilanes to yield chloromethylmercury(II) cyanides. In a separate paper the same authors report that phenylsilanes other than the trifluoro derivative also undergo this C—Si bond cleavage in the presence of mercury(II) cyanide ⟨76DOK(230)853⟩.

$$Hg(CN)_2 + 4\text{-}XC_6H_4SiF_3 \xrightarrow{\text{benzene}} 4\text{-}XC_6H_4HgCN \text{ or } (4\text{-}XC_6H_4)_2Hg \quad (194)$$
$$X = H, Me, Cl, Br, NO_2$$

Bis(phenylacetylene)mercury and mercury(II) cyanide react in ethanol under reflux to give phenylacetylenemercury(II) cyanide (Equation (195)) ⟨82ICA127⟩.

$$(Ph{-}{\equiv}{-})_2Hg + Hg(CN)_2 \xrightarrow[\text{8 h}]{\text{ethanol, reflux}} 2Ph{-}{\equiv}{-}HgCN \quad (195)$$

One novel approach to the synthesis of mercury(II) cyanides involves the initial formation of a charge transfer complex between bisbenzylmercury and tetracyanoethylene. This highly coloured purple complex rapidly breaks down at room temperature to give benzylmercury(II) cyanide in quantitative yield (Scheme 20) ⟨79JOM(177)101⟩.

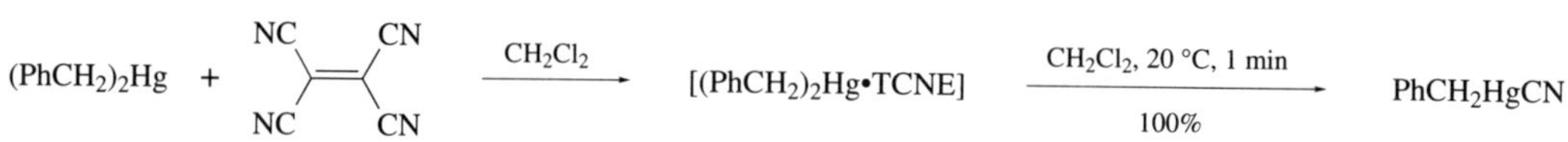

Scheme 20

5.30.7.3 Group 13 Derivatives

There are examples of cyanide derivatives of the group 13 metals aluminum, gallium, indium and thallium, and these will be discussed below.

5.30.7.3.1 Aluminum derivatives

The synthesis of organoaluminum cyanides usually involves treatment of a suitable alkylaluminum species with hydrogen cyanide or trimethylsilyl cyanide. Thus, dialkylaluminum cyanides, useful as hydrocyanation agents ⟨72JA4635⟩, may be prepared by the reaction of dialkylaluminum compounds containing a suitable leaving group (e.g., chlorine) with trimethylsilyl cyanide. The reaction takes place under an inert atmosphere over a wide range of temperatures. Yields are reported to be moderate to excellent (Equation (196)) ⟨84GEP3430019⟩.

$$R_2AlCl + TMS\text{-}CN \xrightarrow{10\text{–}200\ °C} R_2AlCN + TMS\text{-}Cl \quad (196)$$
$$R = C_{1-8} \text{ alkyl}$$

Trialkylaluminum species will react with hydrogen cyanide to generate dialkylaluminum cyanides in good yields (Equation (197)). The reaction will proceed in a variety of solvents and over a range of temperatures ⟨63JCS229, 64FRP1469481, 65USP3444161, 70JCS(A)2675, 72OS(52)90⟩.

$$R_3Al + HCN \longrightarrow R_2AlCN + RH \quad (197)$$

R	*Solvent*	*Temperature* (°C)	*Yield* (%)
Et	Benzene	RT	60–80
Et	Toluene	−196–RT	60
Et	THF	10–15	53
Et	THF/ether	10–15	75
Et	Benzene	10–15	92
Me	Benzene	10–15	74
Bu^i	Benzene	10–15	70

Diethylaluminum hydride will react similarly with hydrogen cyanide in benzene to give diethylaluminum cyanide in excellent yield (Equation (198)) ⟨64FRP1469481, 65USP3444161⟩.

$$Et_2AlH + HCN \xrightarrow[90\%]{\text{benzene}} Et_2AlCN \quad (198)$$

5.30.7.3.2 *Gallium derivatives*

Only dimethylgallium cyanide is reported in the literature. It may be prepared either by the reaction of trimethylgallium with hydrogen cyanide (Equation (199)) ⟨63JCS229⟩ or alternatively by the reaction of trimethylgallium with trimethylgermanyl cyanide (Equation (200)) ⟨75CB664⟩.

$$Me_3Ga + HCN \xrightarrow{\text{benzene}} Me_2GaCN \quad (199)$$

$$Me_3Ga + Me_3GaCN \xrightarrow[70\%]{\text{ether, 2 h}} Me_2GaCN + Me_4Ga \quad (200)$$

5.30.7.3.3 *Indium derivatives*

Only dimethylindium cyanide is reported in the literature. It has been prepared by the reaction of trimethylindium with hydrogen cyanide in ether (Equation (201)) ⟨63JCS229⟩.

$$Me_3In + HCN \xrightarrow{\text{ether}} Me_2InCN \quad (201)$$

5.30.7.3.4 *Thallium derivatives*

Reaction of bis(pentachlorophenyl)thallium(III) perchlorate with potassium cyanide gives the expected thallium(III) cyanide via the displacement of the perchlorate group (Equation (202)) ⟨77JOM(136)309⟩. Similar reactions are observed with diarylthallium iodides ⟨72JIC1143⟩.

$$(C_6Cl_5)_2TlClO_4 + KCN \xrightarrow[55\%]{\text{MeOH, 30–40 °C, 2 h}} (C_6Cl_5)_2TlCN + KClO_4 \quad (202)$$

Cyanomethylthallium(III) acetate can be formed by the reaction of thallium(III) acetate with a mixture of ammonium methylpentafluorosilicate, ammonium fluoride and sodium cyanide in water. The reaction involves both methyl transfer and displacement by cyanide (Equation (203)) ⟨69JAP47019769, 69MI 530-01⟩.

$$Tl(OAc)_3 + NH_4^+MeSiF_5^- + NH_4F + NaCN \xrightarrow{H_2O} MeTl(CN)(OAc) \quad (203)$$

Organothallium cyanides may also be prepared by the reaction of suitable thallium salts with other metal cyanides. Thus, diphenylthallium(III) chloride will react with mercury(II) cyanide to generate both thallium(III) and mercury(II) cyanides. The reaction will take place in methanol, acetone or a mixture of both (Equation (204)) ⟨76ICA-L64⟩. Thallium salts are also reported to react with trialkyltin cyanides to give alkylthallium cyanides (Scheme 21). The reactions are reported to occur with ⟨69JAP47016293, 70JOM(22)627⟩ and without added sodium cyanide ⟨68JOM(12)241⟩.

$$Ph_2TlCl + Hg(CN)_2 \xrightarrow{\text{MeOH or acetone, reflux}} PhTlCl(CN) + PhHgCN \quad (204)$$

$$Tl(OAc)_3 + R_mSnY_{4-m} + NaCN \xrightarrow[\text{30–40\%}]{\text{MeOH, } H_2O} RTl(CN)(OAc)$$

$$MeTl(OAc)_2 + Me_3SnCN \xrightarrow[\text{70\%}]{\text{MeOH}} MeTl(CN)(OAc)$$

Scheme 21

5.30.7.4 Group 14 Derivatives

The synthesis of trialkyltin cyanides has previously been reviewed by Lappert and Pyszora ⟨66MI 530-02⟩, while the corresponding lead compounds have been covered by Shapiro and Frey ⟨B-68MI 530-03⟩. This section will concentrate on work published after these two reviews.

5.30.7.4.1 Tin derivatives

Trialkyl and triaryltin cyanides are widely used for the conversion of acyl chlorides and acid anhydrides into acyl cyanides ⟨79S204⟩. Their synthesis commonly involves the reaction of a dialkyltin halide with a source of cyanide. The displacement of iodide using silver cyanide proceeds smoothly, and yields are generally good (Equation (205)) ⟨60JOC809, 72JIC1143, 88JOM(355)33⟩. The corresponding displacement of chloride, however, is facilitated by the addition of 18-crown-6 to the reaction mixture ⟨80TL2959⟩. The reactions of trialkyltin isopropoxides and dialkyltin diisopropoxides with acetyl cyanide gave compounds arising from the displacement of the isoproxy groups. In some cases it was possible to displace both of the alkoxy groups to give dicyano compounds (Scheme 22) ⟨77IJC(A)1099⟩.

$$R_3SnI + AgCN \xrightarrow[\text{R = alkyl, aryl}]{CHCl_3\text{, RT}} R_3SnCN + AgI \quad (205)$$

$$R_3SnOPr^i + MeCOCN \xrightarrow{\text{cyclohexane}} R_3SnCN + MeCO_2Pr^i$$

$$R_2Sn(OPr^i)_2 + MeCOCN \xrightarrow{\text{cyclohexane}} R_2Sn(OPr^i)(CN)$$

$$R_2Sn(OPr^i)_2 + 2MeCOCN \xrightarrow{\text{cyclohexane}} R_2Sn(CN)_2$$

Scheme 22

Trialkyltin and triaryltin cyanides may be conveniently prepared by the reaction of acyloxystannanes with trimethylsilyl cyanide (Equation (206)) ⟨86MI 530-03⟩. The yields for this reaction are excellent.

$$R^1{}_3SnO_2CR^2 + \text{TMS-CN} \xrightarrow{\text{88–97\%}} R^1{}_3SnCN + \text{TMS-}O_2CR^2 \quad (206)$$

R^1 = Me, Et, Ph, Bu; R^2 = Bu^t, Pr, Me, CH_2Cl, Et, Ph

In a similar fashion to dibenzylmercury (see Section 5.30.7.2.6), benzyltrimethylstannane reacts with tetracyanoethylene to give a charge transfer complex, which at 20 °C decomposes to give trimethyltin cyanide (see Scheme 20) ⟨79JOM(177)101⟩.

5.30.7.4.2 Lead derivatives

Organolead cyanides are derived from hexaphenyldilead by reaction with iodocyanogen in carbon tetrachloride (Equation (207)) ⟨86POL677⟩. Alternatively, the reaction of hexaphenyldilead with either hydrogen cyanide or an alkali metal cyanide in the presence of a peracid gives the desired triphenyllead cyanide (Equation (208)) ⟨68SAP6903273⟩.

$$(Ph_3Pb)_2 + ICN \xrightarrow{CCl_4,\ reflux,\ 30\ min} 2Ph_3PbCN \qquad (207)$$

$$(Ph_3Pb)_2 + HCN\ or\ MCN \xrightarrow{RCO_3H} 2Ph_3PbCN \qquad (208)$$

5.30.8 HETERONITRILIUM COMPOUNDS WITH THE GENERAL FORMULA RX—CN$^+$—Y

There are many species of the type RX—CN$^+$—Y which are known as transient intermediates. When the group Y bears a formal negative charge they are often synthetically useful in 1,3-dipolar cycloaddition reactions, an example of which would be the benzenesulfonylcarbonitrile oxide (**39**) ⟨81JOC5425⟩. This species is prepared by one or two routes (Scheme 23) and undergoes 1,3-dipolar addition reactions with alkenes to form the cycloadducts (**40**) in 44–96% yield. However, in the absence of an alkene, work-up affords the nitrile oxide dimer (**41**) in 41% yield. Such examples will not be discussed further in this section. As a consequence of this reactivity, the range of functional groups which fit the general structure RX—CN$^+$—Y that are isolable or indeed characterizable (other than by inference based on their subsequent reaction products) is relatively limited. This section is subdivided by the nature of the substituent Y, and further by the heteroatom attached to the nitrilium carbon atom.

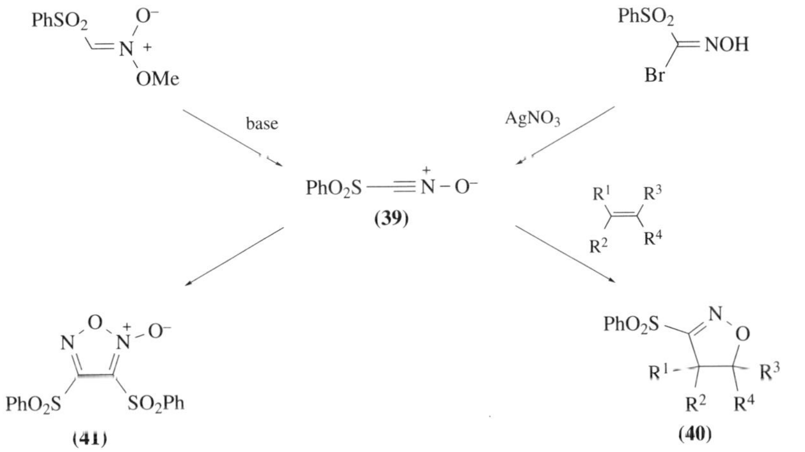

Scheme 23

5.30.8.1 *N*-Protonated Heteronitrilium Salts—RX—CN$^+$—H

No examples of *N*-protonated heteronitrilium salts have been found in the literature.

5.30.8.2 *N*-Carbon-linked Heteronitrilium Salts—RX—CN$^+$—CR$_3$

5.30.8.2.1 Alkylthionitrilium salts—RS—CN$^+$—CR$_3$ X$^-$

Only one example of alkylthionitrilium salts has been found in the literature. However, although the compound was isolated, no details were published as to its characterization. Thus, the treatment of one equivalent of methyl thiocyanate with one equivalent of benzoyl chloride and one equivalent of antimony(V) chloride results in the formation of the *N*-benzoyl-*C*-(methylthio)nitrilium hexachloroantimonate (**42**) (Equation (209)) ⟨81LA1198⟩. The ratio of reagents is important since excess thiocyanate results in the formation of the oxadiazonium salt (**43**).

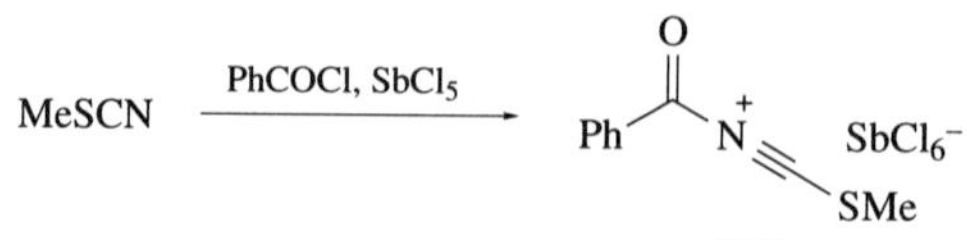

(**42**) (209)

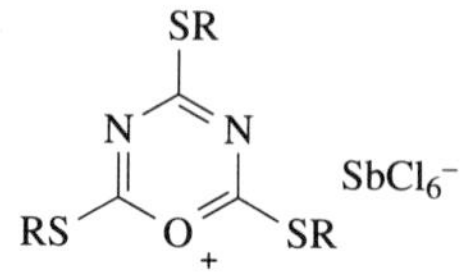

(**43**)

5.30.8.2.2 *Cyanamidium salts—R_2N—CN^+—CR_3 X^-*

There are only a small number of reports in the literature regarding cyanamidium salts. Cyanamidium salts can be prepared by the action of alkyl halides on a disubstituted carbodiimide (Scheme 24). Unfortunately, the cyanamidium salt reacts with the carbodiimide starting material to give the adduct (**44**), and thus isolation is not possible. However, it has been shown that cyanamidium species can be isolated in 60–80% yield as their hexachloroantimonates ⟨83CB1834⟩. Thus, reaction of alkyl or aryl isocyanides with *N*-chlorodialkylamines and antimony pentachloride at low temperatures affords the cyanamidium salts (**45**) (Equation (210)). The procedure requires the presence of zinc or mercury(II) chloride to coordinate to the isocyanide and thereby prevent polymerization under the influence of the antimony pentachloride.

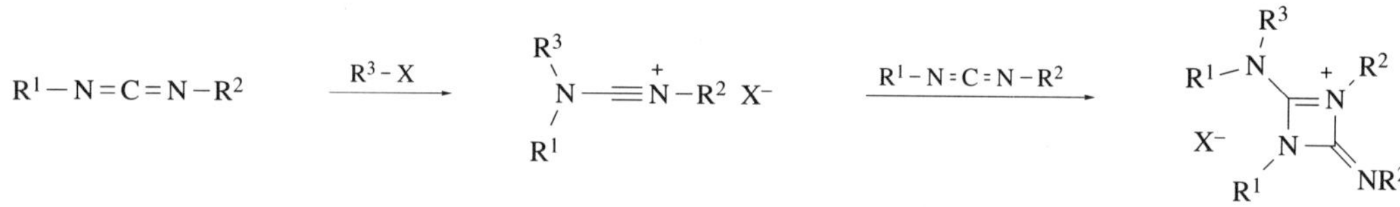

(**44**)

Scheme 24

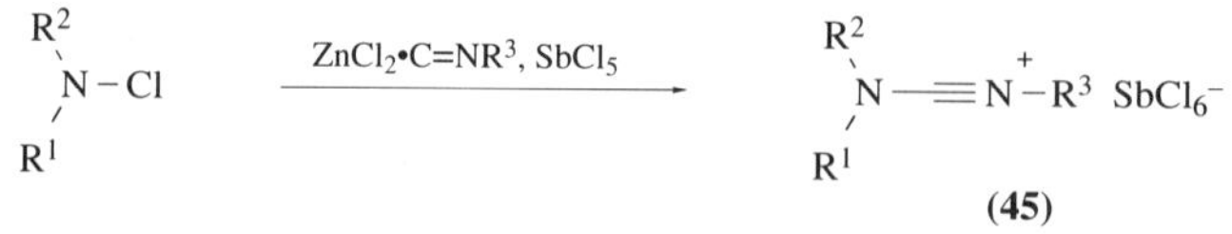

(**45**) (210)

Alternatively, cyanamidinium hexachloroantimonates can be prepared by the reaction of dialkyl cyanamides with *t*-alkyl chlorides in the presence of antimony pentachloride ⟨81CB3655⟩. However, it has been shown that these sensitive compounds can be obtained in a more pure and more stable state by the reaction of the *t*-alkyl halide with an isolated dialkyl cyanamide antimony(V) chloride complex (**46**) (Equation (211)) ⟨84CB1161⟩. Interestingly, the complex (**46**) was shown to have a structure that is consistent with the cyanamidium formulation (**46**) rather than the carbodiimidium structure (**47**) (see Section 5.30.8.4.3). In a similar manner the trichloroferrate complexes were shown to react with the *t*-alkyl chlorides to give the trialkyl cyanamidinium salts. The tetrafluoroborate salts can be prepared by the reaction of dialkyl cyanamides, *t*-butyl chloride and silver tetrafluoroborate.

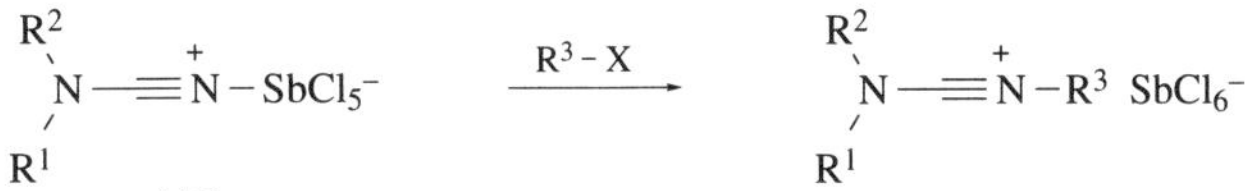

(46) (211)

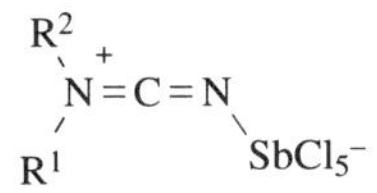

(47)

A stable acyl cyanamidinium salt has been prepared, in 95% yield, by the addition of the adduct resulting from the reaction of diisopropylcarbodiimide and acetyl chloride in dichloromethane to antimony pentachloride in dichloromethane, followed by precipitation with diethyl ether (Equation (212)) ⟨90CB1537⟩.

(212)

5.30.8.2.3 Nitriliumborates—R_3B^-—CN^+—CR_3

Nitriliumborates have been prepared by one of two possible routes, the alkylation of a cyanoborate or the reaction of a borane with an isonitrile. Thus, the reaction of benzylcyanoborane quinuclidine with an excess of triethyloxonium tetrafluoroborate affords the stable nitrilium salt **(48)** in 95% yield (Equation (213)) ⟨89CC900⟩. The salt **(48)** was not affected by moist air for short periods of time and could be stored for up to 2 weeks under nitrogen without noticeable decomposition. Other alkylcyanoborane tertiary amine complexes undergo the same reaction ⟨89MI 530-01, 91IC1046⟩. In a similar manner, potassium cyanotriphenylborate reacts with triphenylmethyl chloride and *t*-butyl iodide to give the salts **(49)** and **(50)** in yields of 78% and 1.3%, respectively (Equation (214)) ⟨70LA(737)80⟩. Alternatively, the salt **(49)** could be prepared from triphenylmethyl isonitrile and triphenylborane in 59% yield ⟨70LA(737)80⟩. Silyloxylalkyl isonitriles will also react with triphenylborane to give nitriliumborates ⟨89ZN(B)419⟩.

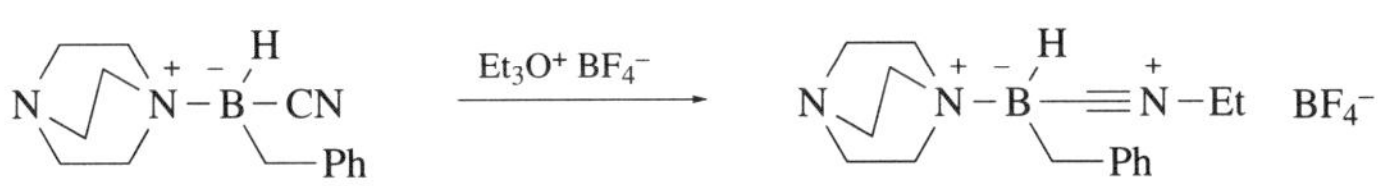

(213)

(48)

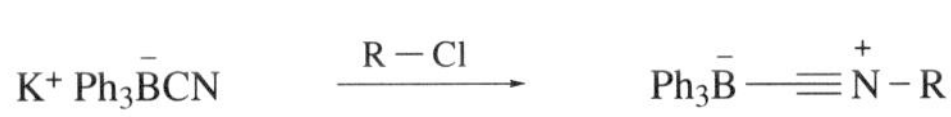

(214)

(49) R = CPh_3
(50) R = Bu^t

5.30.8.2.4 Nitrilium organometallates—R_nM^-—CN^+—CR_3

Isonitriles readily form complexes with transition metals, many of which would fit the general formula R_nM^-—CN^+—CR_3 ⟨B-70MI 530-01⟩. These do not fall within the scope of this work. Few examples outside of this structure are known; however, trimethylaluminum will react with methylisonitrile in pentane to give the (methylnitrilium)trimethylaluminate ⟨70M627⟩.

5.30.8.3 Heteronitrile Oxides—RX—CN$^+$—O$^-$

Since heteronitrile oxides would be expected to be a highly reactive 1,3-dipole it is not surprising to find that no examples of this functional group have been isolated. The only example of this functional group as an intermediate, benzenesulfonylcarbonitrile oxide (**39**) ⟨81JOC5425⟩, is discussed in the introduction to Section 5.30.8.

5.30.8.4 Other Derivatives—RX—CN$^+$—Y

5.30.8.4.1 Silylnitrilimines—R_3Si—CN$^+$—N$^-$—R—and phosphoranylnitrilimines—$R_2P(X)$—CN$^+$—N$^-$—R

Only a limited amount of literature exists for the silyl- and phosphoranylnitrilimine functional groups. Bis(triisopropylsilyl)nitrilimine (**51**), the first bis(silylated)nitrilimine to be reported, has been prepared by two synthetic methods. First, the silylnitrilimine (**51**) can be prepared in 90% yield, by the reaction of two equivalents of chlorotriisopropylsilane with bis(trimethylstannyl)diazomethane in the absence of solvent (Equation (215)) ⟨92JA6059⟩.

(51) (215)

Alternatively, the reaction of the triisopropylsilyldiazomethane with butyllithium affords the diazolithium salt. Subsequent reaction with chlorotriisopropylsilane affords the nitrilimine (**51**) in 80% yield (Equation (216)) ⟨89AG(E)1250, 91JOC1801⟩. Interestingly, this compound could be purified by distillation. The bulky nature of the silyl groups is critical to the formation of (**51**) since the lithium salt of trimethylsilyldiazomethane reacts with chlorotrimethylsilane to give bis(trimethylsilyl)diazomethane ⟨71JOM(29)C25⟩. The nitrilimine (**51**) readily undergoes 1,3-dipolar cycloaddition reactions, and thus the facile cleavage of both silicon–nitrogen and carbon–silicon bonds makes (**51**) a synthetic equivalent of the parent compound (**52**). The lithium salt of triisopropylsilyldiazomethane will also react with chlorobis(diisopropylamino)borane to give the *N*-borylnitrilimine (**53**) (Equation (216)) ⟨91OM3205⟩.

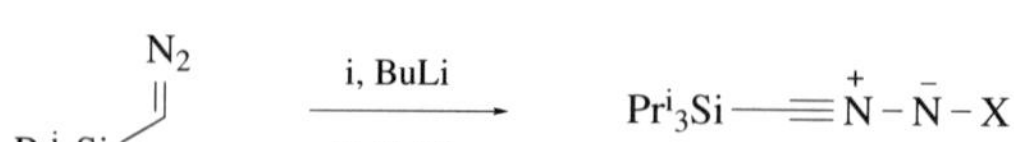

(216)

(**51**) X = Pr^i_3Si
(**53**) X = $(Pr^i_2N)_2B$

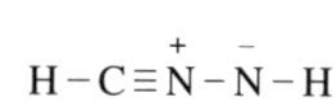

(52)

Similar methodology can be applied to phosphoranylnitrilimines. Thus, deprotonation of di-*t*-butylthioxophosphoranyldiazomethane with butyllithium and subsequent reaction with chlorobis(diisopropylamino)phosphine afforded the phosphoranylnitrilimine (**54**) (Equation (217)) ⟨90JA6277⟩. Unfortunately, this compound could only be spectroscopically characterized in solution, and within 72 h at room temperature it had completely transformed into the corresponding diazo derivative (**55**).

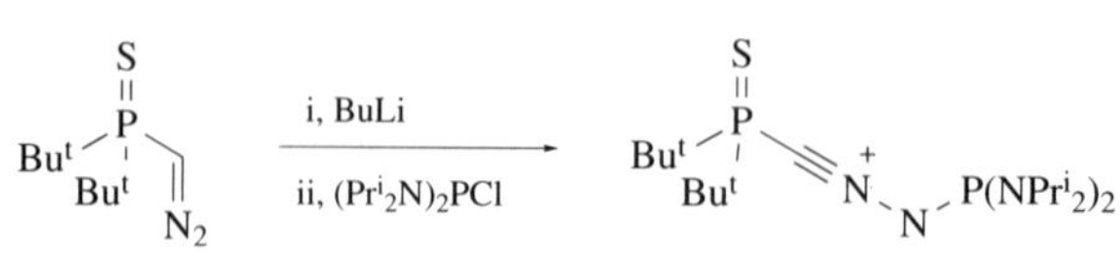

(54) (217)

5.30.8.4.2 Diazonitriliumborates—R_3B^-—CN$^+$—N_2R

Potassium triphenylcyanoborate reacts with benzenediazonium tetrafluoroborate to give the diazonium salt (**56**), which rearranges to the diazonitriliumborate (**57**) in 67% yield (Scheme 25) ⟨70LA(737)80⟩.

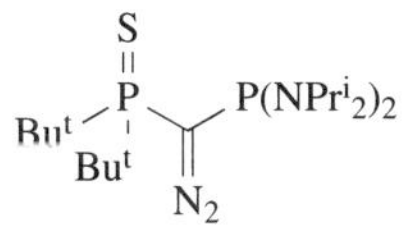

(55)

$$K^+ Ph_3BCN^- + Ph_2N_2^+ BF_4^- \longrightarrow \underset{\textbf{(56)}}{PhN_2^+{\bullet}Ph_3BCN^-} \longrightarrow \underset{\textbf{(57)}}{Ph_3\overset{-}{B}{-}{\equiv}\overset{+}{N}{-}N{=}N{-}Ph}$$

Scheme 25

5.30.8.4.3 Cyanamidiummetallates—R_2N—CN^+—MX_n^-

The pentachloroantimonates and trichloroferrates have been prepared as intermediates in the synthesis of trisubstituted cyanamidium salts. As part of this work the cyanamidinium trifluoroborates, trichloroaluminates, tetrachlorotitanates and dichlorozincates, and dicyanamidinium dichlorozincates and tetrachlorostannates were also prepared ⟨84CB1161⟩.

5.31
Triple-bonded Heteroatom Derivatives Other Than Nitriles with Another Heteroatom Attached to the *sp*-Carbon Atom

WALTER-GEORG VEECK and MANFRED REGITZ
Universität Kaiserslautern, Germany

5.31.1 TRIPLY-BONDED PHOSPHORUS ATTACHED TO THE *sp*-CARBON ATOM

The replacement of the CH-unit in terminal alkynes, R—C≡CH, by the isolobal P-fragment leads to the so-called "phosphaalkynes" R—C≡P (in accordance with IUPAC nomenclature regulations, the equivalent term "methylidynephosphines" is also used sometimes for this structure). Phosphaalkynes without further heteroatom substituents attached to the *sp*-carbon are also included here for comparison. See also Chapter 3.20.

Phosphalkynes constitute one example of the low-coordinated phosphorus compounds since they contain a monocoordinated (σ^1), trivalent (λ^3) phosphorus atom. Depending on the substituent R, these compounds can be classified into two groups: those with relatively small substituents such as H, Me, CH_2=CH—, Cl, F ((**1**), (**3a–c**), (**5**)) are so reactive that they undergo polymerization at room temperature whereas those with sterically hindering substituents (*t*-butyl is an outstanding example) are stable up to 293 K ⟨88AG(E)1484⟩. However, at higher temperatures, even these kinetically stabilized phosphaalkynes are susceptible to cyclooligomerization. In certain cases, this property has been utilized for the preparation of defined products, for example, the tetraphosphacubane system ⟨92AG(E)758⟩. Even so, both the short-lived and the kinetically stabilized phosphaalkynes have one feature in common: they both have to be handled and stored under an atmosphere of dry argon in order to avoid decomposition.

Phosphaalkynes are generally characterized by [^{13}C]- and [^{31}P]-NMR spectroscopy. The *sp*-hybridized carbon atom resonates at extremely low field ($\delta_C = 154$–201 ppm; e.g. Bu^t—C≡P, $\delta_C = 184.8$ ppm) and the signals have a doublet structure with widely varying magnitudes of the $^1J_{C,P}$ coupling constant ($^1J_{C,P} = 14$–56 Hz; e.g. Bu^t—C≡P, $^1J_{C,P} = 38.5$ Hz). The [^{31}P]-NMR signals occur in the range $\delta_P = +96$ ppm (for TMS—C≡P) to $\delta_P = -205$ ppm (for F—C≡P), e.g. Bu^t—C≡P, $\delta_P = -62.2$ ppm ⟨88AG(E)1484, 90CRV191, B-90MI 531-01⟩.

5.31.1.1 Elimination Reactions

5.31.1.1.1 Elimination of hydrogen halides

(i) Intramolecular

(a) FVP (flash vacuum pyrolysis) conditions. The parent compound of the phosphaalkynes, phosphaethyne (H—C≡P;**1**), was incidentally also the first member of the series ever to be generated. The original process involved passing a stream of phosphine gas through a rotating arc struck between graphite electrodes and produced (**1**), ethyne, and ethene (Equation (1)) ⟨61JA1769⟩. This procedure has now been replaced by a flash vacuum pyrolysis (FVP) method which also enables the preparation of phosphaalkynes other than (**1**). The new technique comprises brief exposure of a substituted methyldichlorophosphine (**2**) to temperatures of 800–1000 °C under reduced pressure. Phosphaethyne (**1**) as well as its methyl-, vinyl-, and chloro-substituted derivatives (**3a–c**) have been synthesized in this manner (Equation (2)). The reaction must be performed in the presence of a basic absorption agent, for example potassium hydroxide at −78 °C ⟨80CC333⟩, 1,3,5-tricyclohexylhexahydro-*s*-triazine at 20 °C ⟨86T15723⟩, or potassium hydroxide/potassium carbonate at −78 °C ⟨87JOM(338)329⟩ in order to prevent the retroreaction of the phosphaalkyne with the concomitantly formed hydrogen chloride. Alternatively, the hydrogen chloride can be frozen out of the reaction mixture immediately ⟨76CPL(42)460⟩. A summary of the required reaction conditions and characterization of the above-mentioned phosphaalkynes is given in Table 1.

$$PH_3 \xrightarrow{\text{arc struck between graphite electrodes}} \underset{(\mathbf{1})}{HC{\equiv}P} + HC{\equiv}CH + H_2C{=}CH_2 \qquad (1)$$

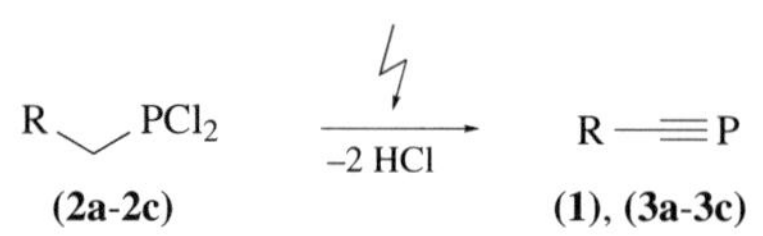

(2)

(2a-2c) **(1), (3a-3c)**

(2), (3) **a**; R = Me
b; R = CH=CH$_2$
c; R = Cl

(b) At room temperature. Fluorophosphaethyne (**5**) constitutes an exception since elimination of the hydrogen halide from trifluoromethylphosphine (**4**) proceeds at room temperature in the presence of potassium hydroxide ⟨78JA446, 79CC653⟩. The product (**5**) was characterized by photoelectron ⟨78JA446⟩, IR ⟨82CL981, 85JSP(111)415⟩, and microwave spectroscopy ⟨80JSP(82)185⟩.

The reaction of trifluoromethylphosphine (**4**) with three equivalents of diisopropylamine gives rise to diisopropylaminophosphaethyne (**7**) and hydrogen fluoride; the intermediate (**6**) possessing a C=P double bond is formed during the reaction sequence (Scheme 1) ⟨93PS(76)1⟩.

(c) Vacuum gas–solid reactions (VGSR). When a vaporized (dichloroalkyl)phosphine (**10**) is passed at reduced pressure through a glass tube filled with potassium carbonate heated at 350 °C, elimination of hydrogen chloride yields the transient *C*-chlorophosphaalkene (**11**) which immediately undergoes elimination of a further molecule of hydrogen chloride to produce the phosphaalkynes (**1**), (**3a**), (**3c**), or (**12a–c**) (Scheme 2) ⟨91AG(E)196⟩. The required (dichloroalkyl)phosphines (**10**) can be obtained from the corresponding diisopropyl dichloroalkanephosphonates (**9**) by reduction with dichloroaluminum hydride in diethyl ether at −80 °C. The substrates (**9**), in turn, are prepared by the reaction of diisopropyl trichloromethanephosphonate (**8**) with *n*-butyllithium and the respective alkyl halide. The phosphaalkynes prepared by this method are listed in Table 2.

Table 1 Reaction conditions for the generation of the short-lived phosphaalkynes (**1**), (**3a–c**) and their spectroscopic characterization.

R	*Temperature/pressure*	*Removal of HCl*	*Characterization*	*Ref.*
H (**1**)	(a) 1000°C/0.03 mbar	KOH/−78°C	NMR; trapping reactions with 1,3-dipoles (e.g. diazo-methyl compounds)	80CC333
	(b) 900°C/1.3×10^{-3} mbar	1,3,5-tricyclohexylhexahydro-*s*-triazine/20°C		86TL5723
	(c) 1100°C/5.3×10^{-5} mbar	KOH/K_2CO_3/−78°C		87JOM(338)329
Me (**3a**)	(a) 900°C/0.04 mbar	frozen out	NMR; trapping reactions	76MI 533-01
	(b) 750°C/1.3×10^{-3} mbar	1,3,5-tricyclohexylhexahydro-*s*-triazine/20°C		86TL5723
	(c) 500–700°C (temperature gradient in oven)/ 1.3×10^{-4} mbar	NEt_3		89PS(29)31
—CH=CH_2 (**3b**)	1000°C/0.04–0.07 mbar		microwave spectroscopy	79JCS(D)1405, 81JSP(90)507, 82JST(79)215
Cl (**3c**)	850°C/3×10^{-4} mbar		trapping with ethyl diazoacetate	89MI 531-01

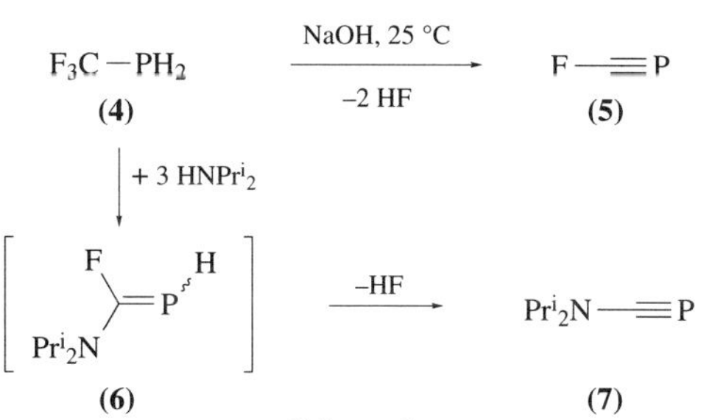

Scheme 1

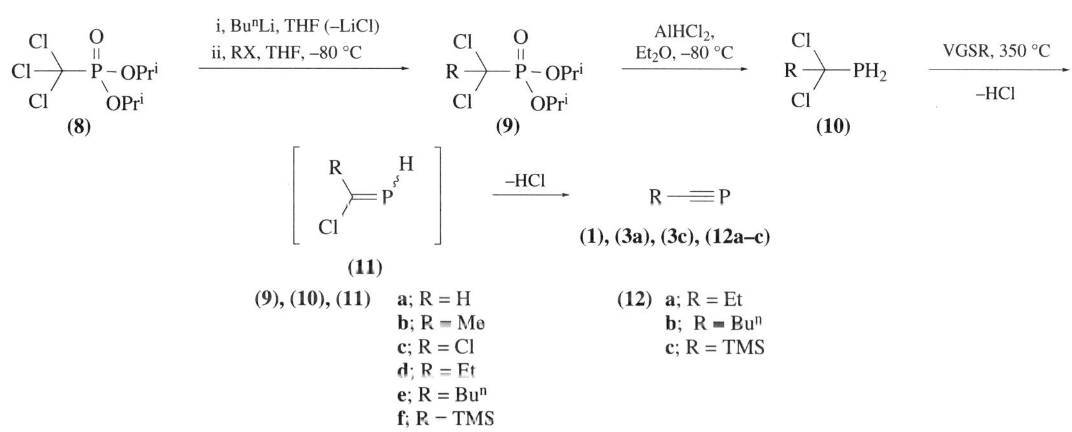

Scheme 2

(ii) Intermolecular

Intermolecular eliminations of hydrogen chloride from mixtures of 3-chloropropyne (**13**) (Equation (3)) or 1-cyanopropyne (**15**) (Equation (4)) with trichlorophosphine under flash vacuum pyrolysis conditions have been utilized for the syntheses of the phosphabutadiyne (**14**) ⟨81JSP(90)512,

Table 2 Phosphaalkynes synthesized by vacuum gas–solid reactions (VGSR).

Phosphaalkyne	*R*	*Yield (%)*
(1)	H	80
(3a)	Me	80
(3b)	Cl	60
(12a)	Et	80
(12b)	Bu^n	70
(12c)	Me_3Si	65

82JST(79)215〉 or the azaphosphahexatriyne (**16**) 〈85JSP(109)8〉. These compounds have, respectively, one or two triple bonds in conjugation with the C≡P triple bond.

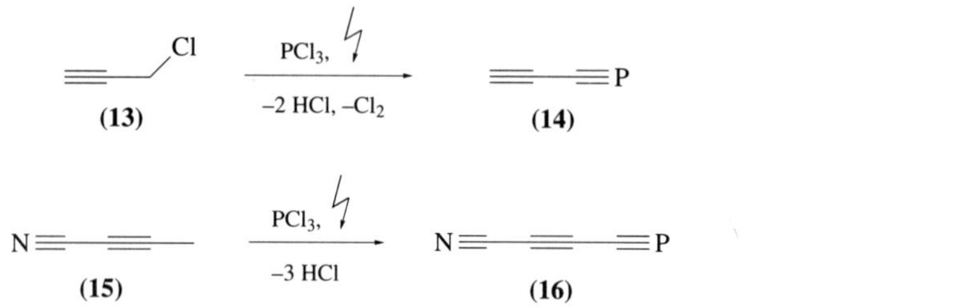

5.31.1.1.2 *Elimination of chlorotrimethylsilane*

Starting materials for eliminations of chlorotrimethylsilane under reduced pressure at high temperatures are either a *P*-chloro(trimethylsilyl)phosphaethene (**17**) 〈81AG(E)197, 81TL2159, 82JSP(92)158〉 or the tris(trimethylsilyl)methyldichlorophosphine (**19**) (Scheme 3) 〈89MI 531-01〉. An advantageous feature of this strategy is that no neutralizing reagents are required.

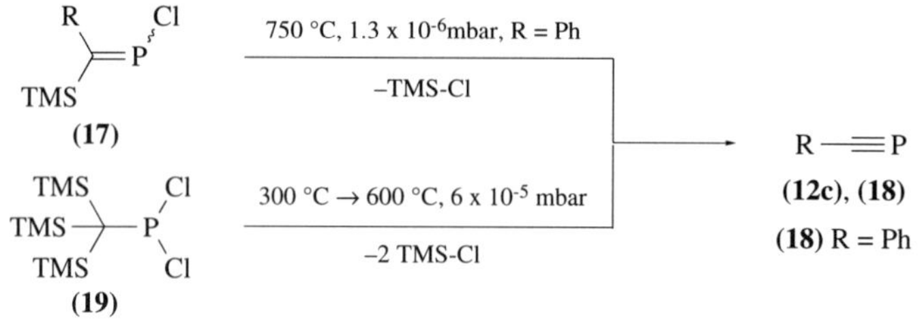

Scheme 3

With regard to their stabilities, however, it should be mentioned that the two phosphaalkynes (**12c**) and (**18**) prepared to date by this method can be considered as representing a transition stage between the short-lived and the kinetically stabilized phosphaalkynes. Their half-lives of $\tau = 50$ min at 20°C for (**12c**) and $\tau = 7$ min at 0°C for (**18**) 〈81AG(E)731〉 illustrate this characteristic.

5.31.1.1.3 *Elimination of hexamethyldisiloxane*

(i) In the presence of sodium hydroxide

The elimination of hexamethyldisiloxane from suitable precursors provides a general access to the kinetically stabilized phosphaalkynes 〈88AG(E)1484〉; *t*-butylphosphaethyne (**24a**) being the first compound to be synthesized by this approach 〈81ZN(B)16〉. The generalized method starts from tris(trimethylsilyl)phosphine (**20**) which is readily accessible from the reaction of white phosphorus with a sodium–potassium alloy and chlorotrimethylsilane 〈75CB2484〉. Acyl chlorides (**21**) bearing the required stabilizing substituent are allowed to react with the phosphine (**20**) to furnish the bis(trimethylsilyl)acylphosphines (**22**) which cannot be isolated since they undergo rapid [1,3]-

trimethylsilyl shifts to produce the corresponding phosphaalkenes (**23**). These compounds are formed predominantly as the (*Z*)-isomers; however, an influence of the configuration of the phosphaalkene on the subsequent formation of the phosphaalkyne has not yet been observed. The elimination of hexamethyldisiloxane from the phosphaalkenes (**23**) is catalyzed by potassium hydroxide at temperatures between 120°C and 200°C and yields the phosphaalkynes (**24a–o**) (Scheme 4). Performance of the elimination reaction in the absence of a solvent together with rapid removal of the generated phosphaalkyne from the reaction vessel have proved to be necessary for success ⟨86JOM(306)39, 87CB1645, 87PS(30)479⟩. In practice this is achieved by heating small amounts of the phosphaalkene (**23**) with the solid alkali base under reduced pressure. Under these conditions, the freshly formed phosphaalkyne (**24**) together with the eliminated hexamethyldisiloxane evaporate from the reaction mixture, pass through a condenser to reclaim the unreacted (**23**) and are frozen out in a cold trap. The two components are either separated by distillation or the mixture is used as such for subsequent reactions after prior determination of the hexamethyldisiloxane content by NMR spectroscopy. In some special cases, lithium bis(trimethylsilyl)phosphide·2 THF has been used as an alternative to tris(trimethylsilyl)phosphine (**20**) for the introduction of the phosphorus increment ⟨76ZAAC(422)104, 89MI 531-01⟩.

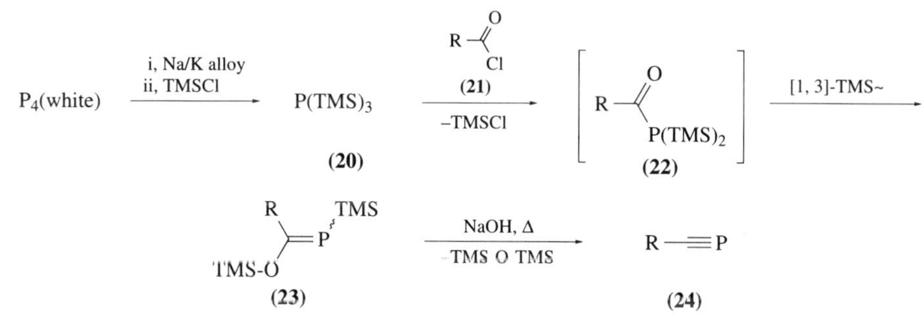

(**21**)–(**24**): Substituent variations see Table 3

Scheme 4

The kinetically stabilized phosphaalkynes (**24a–o**) prepared to date by this route are summarized in Table 3 together with the reaction conditions applied for the decisive elimination step and characteristic NMR spectroscopic data for the products.

(ii) In the presence of eliminating agents other than sodium hydroxide

In some particular cases, agents other than sodium hydroxide have been employed with success in the elimination step: for example, tetra(*n*-butyl)ammonium fluoride together with silica gel to prepare 1-adamantylphosphaethyne (1-Ad—C≡P, (**24f**)) ⟨87CC980⟩ or diiron nonacarbonyl to form Bu^t—C≡P (**24a**) ⟨87CC980⟩. The major advantage in favour of the use of the metal carbonyl seems to be the smooth reaction conditions.

(iii) In the absence of an auxiliary reagent

The syntheses of 1-tripticylphosphaethyne (**24g**) and 2,4,6-tris(*t*-butyl)phenylphosphaethyne (**24n**) do not require the presence of an elimination reagent and have been successfully performed in dimethyl ether as solvent at −85°C ⟨85TL5507, 86TL171⟩.

5.31.1.1.4 Elimination of lithium trimethylsilanolate

When, in place of the silylated phosphine (**20**) or lithium bis(trimethylsilyl)phosphide·2 THF, the lithium phosphide·DME complex is allowed to react with an acyl chloride (**21**), a dimeric complex (**25**) containing two molecules of solvent is formed ⟨77ZAAC(428)222, 85MI 531-01⟩. Subsequent silylation and lithiation lead, via the (trimethylsiloxy)methylidenephosphine (**26**), to the *P*-lithiated phosphaalkene (**27**) which undergoes elimination of lithium trimethylsilanolate to produce the desired phosphaalkyne (**24**) (Scheme 5).

Table 3 Kinetically stabilized phosphaalkynes (**24a–o**) prepared by elimination of hexamethyldisiloxane from phosphaalkenes (**23a–o**). NMR data (C_6D_6, J in Hz). DEG: diethylene glycol; DME: 1,2-dimethoxyethane

Phosphaalkyne		*Reaction conditions*		*Yield (%)*	$\delta(^{31}P)$	$\delta(^{13}C)$ ($^1J(C,P)$)	*Ref.*
(24a)	Bu^t—≡P	DEG	NaOH/20 °C	76	− 69.2	184.8 (38.5)	81ZN(B)16
(24b)	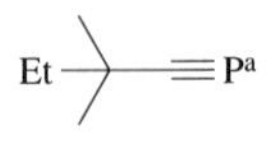	— —	NaOH/160 °C NaOH/150 – 200 °C	96 80	− 62.0	183.8 (37.6)	87PS479 87CB1645
(24c)	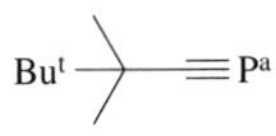	—	NaOH/150 – 200 °C	83	− 57.7	184.4 (37.2)	86MI 531-01
(24d)	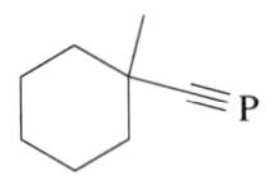	—	NaOH/120 – 160 °C	76	− 57.0	184.9 (37.2)	86JOM(306)39
(24e)	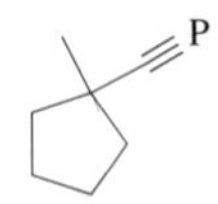	—	NaOH/120 – 160 °C	66	− 66.0	184.9 (37.9)	86JOM(306)39
(24f)	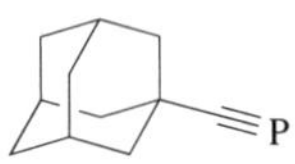	DME —	NaOH/20 °C $nBu_4N^+F^-$ on silica gel/20 °C	83	− 66.9	184.7 (39.0)	86S31
(24g)	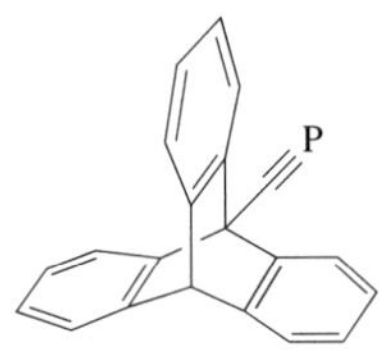	DME	− 85 °C	12 – 18	− 17.5	164.6[b] (46.5)	85TL5507
(24h)	Pr^i—≡P[c]	—	NaOH/120 – 160 °C	63	− 64.3	183.4 (41.3)	86JOM(306)39
(24i)		—	NaOH/120 – 160 °C	75	− 51.4	173.7 (45.5)	86JOM(306)39
(24j)	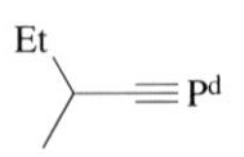	—	NaOH/160 – 200 °C	41	− 59.3	—	86MI 531-01
(24k)	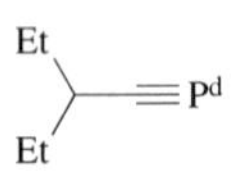	—	NaOH/160 – 200 °C	49	− 53.9	179.4 (41.3)	86MI 531-01
(24l)	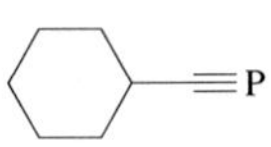	—	NaOH/160 – 200 °C	28	− 60.9	179.7 (41.3)	86MI 531-01
(24m)	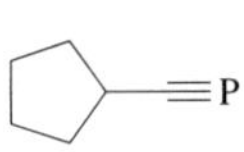	—	NaOH/160 – 200 °C	29	− 62.9	179.9 (41.6)	86MI 531-01
(24n)	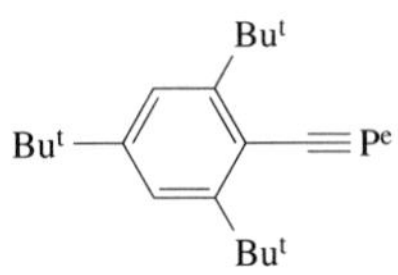	DME	− 85 °C	10 – 15[e]	+34.4[b]	168.7[b] (53.2)	86TL171
(24o)	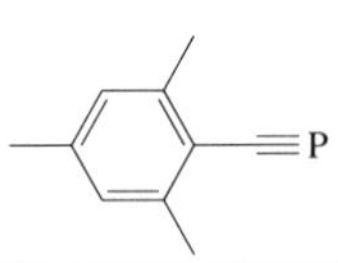	—	NaOH/140 °C	43	+2.5	163.7 (46.0)	86MI 531-01

[a]Stable at 20 °C, but complete removal of hexamethyldisiloxane has not yet been achieved. This is of no significance for further reactions.
[b]Measured in $CDCl_3$.
[c]Stored at −30 °C.
[d]Mixed with hexamethyldisiloxane; stored at −30 °C.
[e]The modest yield is partly the result of the competing formation of a 2,3-diphospha-1,3-butadiene.

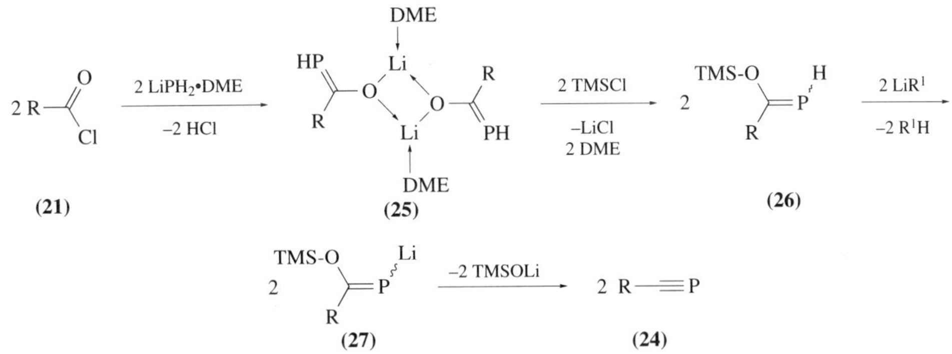

Scheme 5

In comparison to the elimination of hexamethyldisiloxane, cleavage of lithium trimethylsilanolate occurs under much milder conditions. At present, however, this methodology does not play a significant role in the chemistry of the phosphaalkynes.

5.31.1.2 Rearrangement Reactions

5.31.1.2.1 Amino-substituted phosphaalkynes

In an early Russian report ⟨74CA121069⟩, it was claimed that cyanophosphine rearranged in liquid ammonia in the presence of sodium amide to give aminophosphaethyne via a hydrogen shift. However, other investigators ⟨92CC1799⟩ were not able to repeat this work and merely observed a polymerization of the cyanophosphine.

One example of a genuine *N*-substituted phosphaalkyne is *N*-trimethylsilyl-*N*-isopropylaminophosphaethyne (**31**). The first step in the preparation of this compound is the formation of a tautomeric mixture of the phosphines (**29**) and (**30**) from the reaction of tris(trimethylsilyl)phosphine (**20**) with isopropyl isocyanate (**28**). When this mixture is immediately subjected to conditions appropriate for the elimination of hexamethyldisiloxane at 25°C, a substituted azaphosphaallene (**32**) is presumably formed initially (Scheme 6); however, this species cannot be detected since it undergoes a rapid [1,3]-silyl shift to furnish the target aminophosphaethyne.

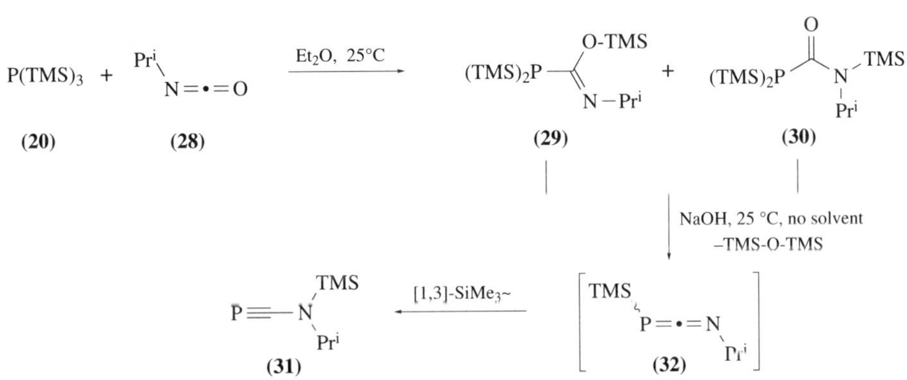

Scheme 6

5.31.1.2.2 Lewis base-induced rearrangements of 1-alkynylphosphines

Low temperature treatment of the 1-alkynylphosphines (**33**) with a catalytic amount (<5%) of a Lewis base (NEt_3, dbu (1,5-diazabicyclo[5.4.0]undec-5-ene)) results in the formation of the phosphaalkynes (**3a**), (**12a**), or (**35**) ⟨92CC415⟩. The reaction mechanism involves the *P*-unsubstituted phosphaallene intermediates (**34**) which undergo [1,3]-hydrogen shifts to provide the phosphaalkyne products (Scheme 7).

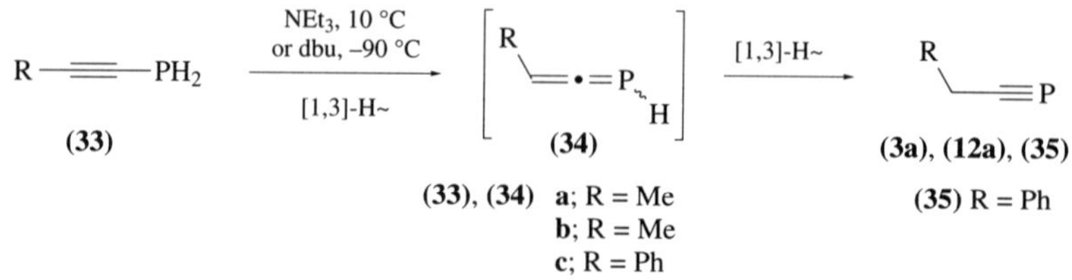

Scheme 7

5.31.1.2.3 *Isomerization under flash vacuum pyrolysis conditions*

Phosphapropyne (**3a**) can be prepared from 1-vinylphosphirane (**36**) as well as from divinylphosphine (**38**) under FVP conditions ⟨92CC1799⟩. The phosphaalkyne formation is presumed to be the result of an isomerization process involving a vinylphosphinidene intermediate (**37**) (Scheme 8).

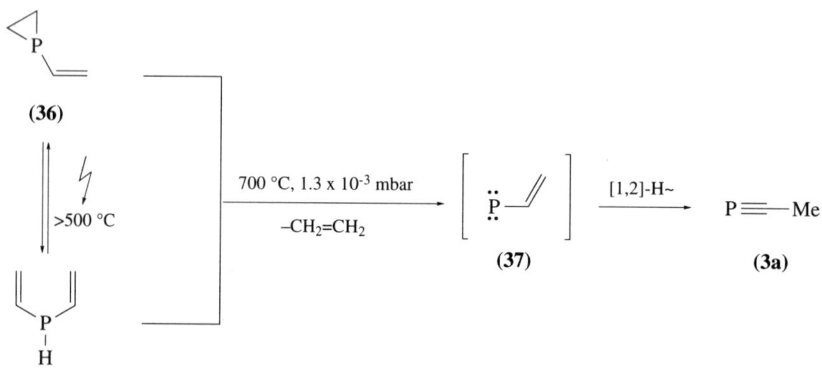

Scheme 8

By means of the same methodology, triallylphosphine (**39**) was transformed to phosphaethyne (**1**) ⟨93CC1295⟩ together with the phosphinines (**40**) and (**41**) (Equation (5)). The latter species are the main products when the pressure is less than 1.3×10^{-3} mbar.

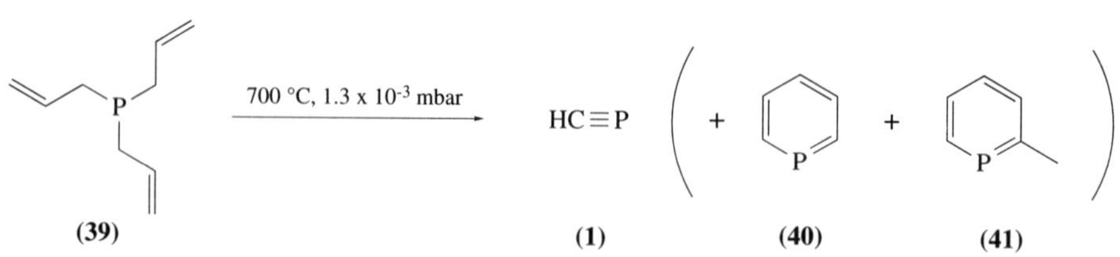

(5)

5.31.1.2.4 *Palladium-catalyzed rearrangements of phosphaalkenes*

Reaction of the readily available phosphaalkene (**42**) with the complex tetrakis(triphenylphosphine)palladium(0) produces the phosphaalkyne (**24n**) in almost quantitative yield ⟨92TL2981⟩. The reaction mechanism was interpreted as a multistage process involving the generation of the phosphaisocyanide (**45**) which rearranged into the phosphaalkyne (**24n**). No direct evidence is available for the participation of (**45**) although the η^2-Pd^0 complex (**43**) and the η^1-Pd^I complex (**44**) were discussed as intermediates in the formation of (**45**) (Scheme 9).

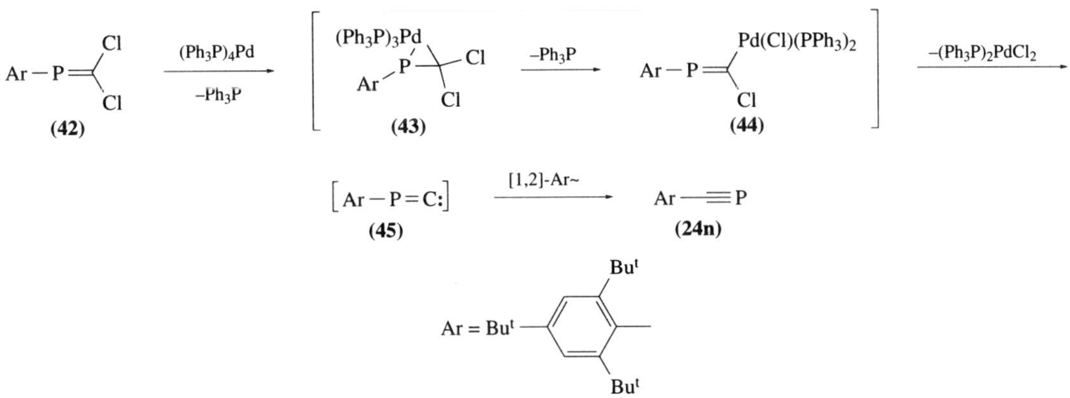

Scheme 9

5.31.1.3 From Other Phosphaalkynes

When *t*-butylphosphaethyne (**24a**) is subjected to flash vacuum pyrolysis, it is cleaved to produce isobutylene and phosphaethyne (**1**) (Equation (6)) ⟨87JOM(338)329⟩. In the light of the low yield achieved and the expenditure of effort required to prepare the starting phosphaalkyne (**24a**), this method has little or no practical value.

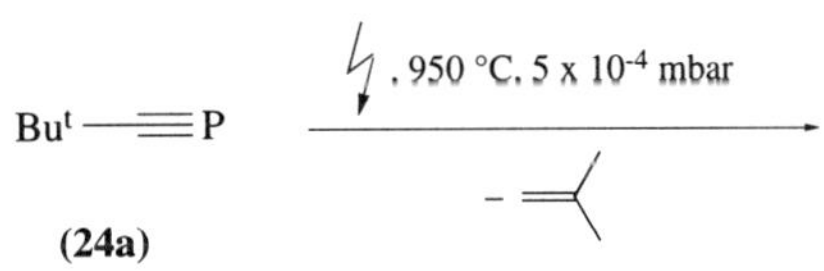

(6)

Flash vacuum pyrolysis of phosphaethyne (**1**) gives rise to further phosphaalkynes with the C≡P triple bond in conjugation with other triple bond systems (cf. Section 5.31.1.1.1(ii)). While flash vacuum pyrolysis of the pure compound (**1**) gives rise to diphosphabutadiyne (**47**) by dimerization of ·C≡P radicals ⟨90CRV191⟩, the azaphosphabutadiyne (**50**) is obtained when cyanogen azide (**48**) is fed into the pyrolysis oven concomitantly with (**1**) ⟨80CC333⟩. The cyanonitrene (**49**) is assumed to be an intermediate in the latter reaction (Scheme 10).

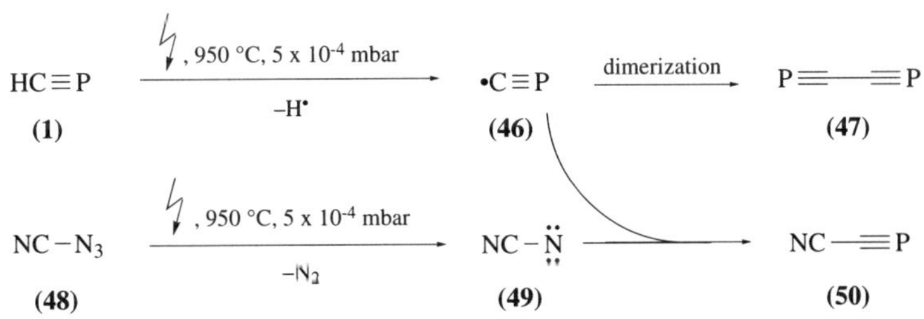

Scheme 10

5.31.1.4 From Phosphinodiazoalkanes

In this strategy, the phosphorus atom of the target phosphaalkyne is introduced as chlorobis-(*t*-butyl)phosphine (**51**) and the C—R unit is contributed by the corresponding, lithiated diazo compound (**52**). The reaction of these two components first gives rise to the phosphinodiazoalkanes (**53**) which, when subjected to flash vacuum pyrolysis, initially generate the λ^5-phosphaalkynes (**54**). The latter species then lose isobutylene and hydrogen to produce the target phosphaalkynes (**24a**) and (**12c**) ⟨88AG(E)1484⟩ (Scheme 11).

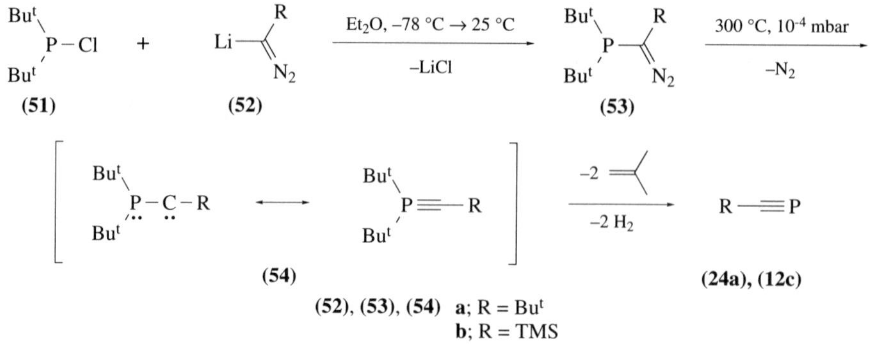

Scheme 11

5.31.2 TRIPLY-BONDED ARSENIC ATTACHED TO THE *sp*-CARBON ATOM

As a consequence of their lower stability in comparison to the phosphaalkynes, only three examples of the arsaalkynes are known to date. The arsenic analogue of (**24a**), *t*-butylarsaethyne (**56**), has not been isolated but has been postulated as a highly reactive intermediate in the formation of the tetraarsacubane (**57**) from the arsaalkene (**55**), lithium bis(trimethylsilyl)phosphide · 2 THF, and cobalt chloride (Scheme 12) ⟨93AG(E)103⟩.

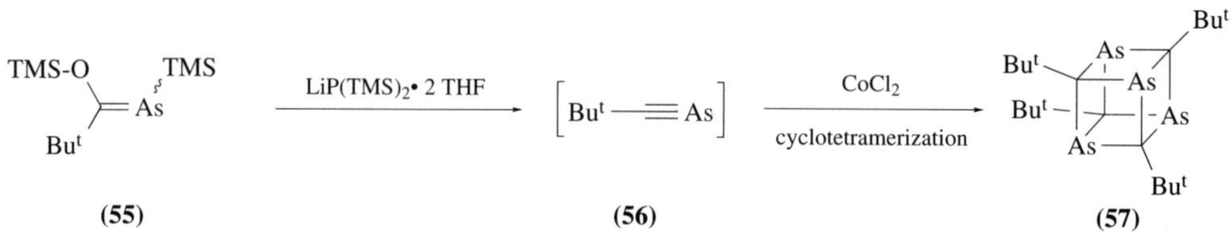

Scheme 12

Arsapropyne (**59**) has been prepared by the vacuum gas–solid reaction of 1-ethynylarsine (**58**) with sodium carbonate at 70°C (Equation (7)) ⟨94MI 531-01⟩. The reaction seems to be completely analogous to that described in Section 5.31.1.2.2.

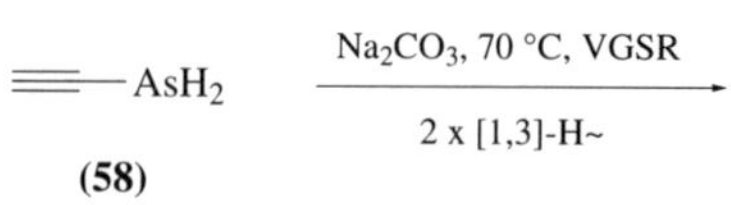

(7)

The first kinetically stabilized arsaalkyne, 2,4,6-tris(*t*-butyl)phenylarsaethyne (**63**), was prepared by elimination of hexamethyldisiloxane from a mixture of 2,4,6-tris(*t*-butyl)benzoyl chloride (**62**) and lithium bis(trimethylsilyl)arsine (**61**) at room temperature (Equation (9)). The starting material (**61**) was synthesized by the reaction of tris(trimethylsilyl)arsine (**60**) with methyllithium in THF (Equation (8)) ⟨86AG(E)264⟩.

$$\underset{(\mathbf{60})}{As(TMS)_3} \xrightarrow{\text{MeLi, THF}} \underset{(\mathbf{61})}{LiAs(TMS)_2 \cdot 2\,THF} \qquad (8)$$

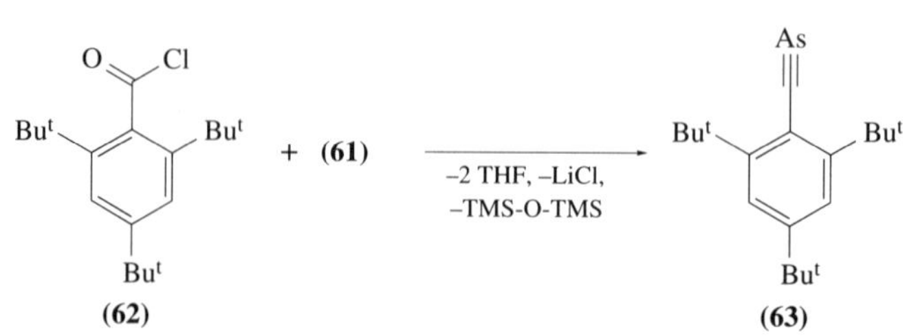

(9)

References

EXPLANATION OF THE REFERENCE SYSTEM

Throughout this work, references are designated by a number–lettering coding of which the first two numbers denote tens and units of the year of publication, the next one to three letters denote the journal, and the final numbers denote the page. This code appears in the text each time a reference is quoted; the advantages of this system are outlined in the Introduction. The system has been used previously in "Comprehensive Heterocyclic Chemistry," eds A. R. Katritzky and C. W. Rees, Pergamon, Oxford, 1984 and is based on that used in the following two monographs: (a) A. R. Katritzky and J. M. Lagowski, "Chemistry of the Heterocyclic N-Oxides," Academic Press, New York, 1971; (b) J. Elguero, C. Marzin, A. R. Katritzky and P. Linda, "The Tautomerism of Heterocycles," in "Advances in Heterocyclic Chemistry," Supplement 1, Academic Press, New York, 1976.

The following additional notes apply:

1. A list of journal codes in alphabetical order, together with the journals to which they refer, is given immediately following these notes. Journal names are abbreviated throughout using the CASSI (Chemical Abstracts Service Source Index) system.
2. Each volume contains all the references cited *in that volume*; no separate lists are given for individual chapters.
3. The list of references is arranged in order of (a) year, (b) journal in alphabetical order of journal code, (c) part letter or number if relevant, (d) volume number if relevant, (e) page number.
4. In the reference list the code is followed by (a) the complete literature citation in the conventional manner and (b) the number(s) of the page(s) on which the reference appears, whether in the text or in tables, schemes, etc.
5. For nontwentieth-century references the year is given in full in the code.
6. For journals which are published in separate parts, the part letter or number is given (when necessary) in parentheses immediately after the journal code letters.
7. Journal volume numbers are *not* included in the code numbers unless more than one volume was published in the year in question, in which case the volume number is included in parentheses immediately after the journal code letters.
8. Patents are assigned appropriate three-letter codes.
9. Frequently cited books are assigned codes.
10. Less common journals and books are given the code "MI" for miscellaneous with the whole code for books prefixed by the letter "B-".
11. Where journals have changed names, the same code is used throughout, e.g. CB refers to both *Chem. Ber.* and to *Ber. Dtsch. Chem. Ges.*

Journal Codes

AAC	Antimicrob. Agents Chemother.
ABC	Agric. Biol. Chem.
AC	Appl. Catal.
AC(P)	Ann. Chim. (Paris)
AC(R)	Ann. Chim. (Rome)
ACH	Acta Chim. Acad. Sci. Hung.

ACR	Acc. Chem. Res.
ACS	Acta Chem. Scand.
ACS(A)	Acta Chem. Scand., Ser. A
ACS(B)	Acta Chem. Scand., Ser. B
AF	Arzneim.-Forsch.
AFC	Adv. Fluorine Chem.
AG	Angew. Chem.
AG(E)	Angew. Chem., Int. Ed. Engl.
AHC	Adv. Heterocycl. Chem.
AHCS	Adv. Heterocycl. Chem. Supplement
AI	Anal. Instrum.
AJC	Aust. J. Chem.
AK	Ark. Kemi
AKZ	Arm. Khim. Zh.
AM	Adv. Mater. (Weinheim, Ger.)
AMLS	Adv. Mol. Spectrosc.
AMS	Adv. Mass. Spectrom.
ANC	Anal. Chem.
ANL	Acad. Naz. Lncei
ANY	Ann. N. Y. Acad. Sci.
AOC	Adv. Organomet. Chem.
AP	Arch. Pharm. (Weinheim, Ger.)
APO	Adv. Phys. Org. Chem.
AQ	An. Quim.
AR	Annu. Rep. Prog. Chem.
AR(A)	Annu. Rep. Prog. Chem., Sect. A
AR(B)	Annu. Rep. Prog. Chem., Sect. B
ARP	Annu. Rev. Phys. Chem.
ASI	Acta Chim. Sin. Engl. Ed.
ASIN	Acta Chim. Sin.
AX	Acta Crystallogr.
AX(A)	Acta Crystallogr., Part A
AX(B)	Acta Crystallogr., Part B
B	Biochemistry
BAP	Bull. Acad. Pol. Sci., Ser. Sci. Chim.
BAU	Bull. Acad. Sci. USSR, Div. Chim. Sci.
BBA	Biochim. Biophys. Acta
BBR	Biochim. Biophys. Res. Commun.
BCJ	Bull. Chem. Soc. Jpn.
BEP	Belg. Pat.
BJ	Biochem. J.
BJP	Br. J. Pharmacol.
BMC	Bioorg. Med. Chem. Lett.
BP	Biochem. Biopharmacol.
BPJ	Br. Polym. J.
BRP	Br. Pat.
BSB	Bull. Soc. Chim. Belg.
BSF	Bull. Soc. Chim. Fr.
BSF(2)	Bull. Soc. Chim. Fr., Part 2
C	Chimia
CA	Chem. Abstr.
CAN	Cancer
CAR	Carbohydr. Res.
CAT	Chim. Acta Turc.
CB	Chem. Ber.

CBR	Chem. Br.
CC	J. Chem. Soc., Chem. Commun.
CCA	Croat. Chem. Acta
CCC	Collect. Czech. Chem. Commun.
CCR	Coord. Chem. Rev.
CE	Chem. Express
CEN	Chem. Eng. News
CHE	Chem. Heterocycl. Compd. (Engl. Transl.)
CHEC	Comp. Heterocycl. Chem.
CI(L)	Chem. Ind. (London)
CI(M)	Chem. Ind. (Milan)
CJC	Can. J. Chem.
CJS	Can. J. Spectrosc.
CL	Chem. Lett.
CLY	Chem. Listy
CM	Chem. Mater.
CMC	Comp. Med. Chem.
COC	Comp. Org. Chem.
COMC-I	Comp. Organomet. Chem., 1st edn.
COS	Comp. Org. Synth.
CP	Can. Pat.
CPB	Chem. Pharm. Bull.
CPH	Chem. Phys.
CPL	Chem. Phys. Lett.
CR	C. R. Hebd. Seances Acad. Sci.
CR(A)	C. R. Hebd. Seances Acad. Sci., Ser. A
CR(B)	C. R. Hebd. Seances Acad. Sci., Ser. B
CR(C)	C. R. Hebd. Seances Acad. Sci., Ser. C
CRAC	Crit. Rev. Anal. Chem.
CRV	Chem. Rev.
CS	Chem. Scr.
CSC	Cryst. Struct. Commun.
CSR	Chem. Soc. Rev.
CT	Chem. Tech.
CZ	Chem.-Ztg.
CZP	Czech. Pat.
DIS	Diss. Abstr.
DIS(B)	Diss. Abstr. Int. B.
DOK	Dokl. Akad. Nauk SSSR
DP	Dyes Pigm.
E	Experientia
EC	Educ. Chem.
EF	Energy Fuels
EGP	Ger. (East) Pat.
EJM	Eur. J. Med. Chem.
EUP	Eur. Pat.
FCF	Forschr. Chem. Forsch.
FCR	Fluorine Chem. Rev.
FES	Farmaco Ed. Sci.
FOR	Forschr. Chem. Org. Naturst.
FRP	Fr. Pat.
G	Gazz. Chim. Ital.
GAK	Gummi Asbest Kunstst.
GEP	Ger. Pat.
GEP(O)	Ger. Pat. Offen.

GSM	Gen. Synth. Methods
H	Heterocycles
HAC	Heteroatom Chem.
HC	Chem. Heterocycl. Compd.
HCA	Helv. Chim. Acta
HOU	Methoden Org. Chem. (Houben-Weyl)
HP	Hydrocarbon Process
IC	Inorg. Chem.
ICA	Inorg. Chim. Acta
IEC	Ind. Eng. Chem. Res.
IJ	Isr. J. Chem.
IJC	Indian J. Chem.
IJC(A)	Indian J. Chem., Sect. A
IJC(B)	Indian J. Chem., Sect. B
IJM	Int. J. Mass Spectrom. Ion Phys.
IJQ	Int. J. Quantum Chem.
IJS	Int. J. Sulfur Chem.
IJS(A)	Int. J. Sulfur Chem., Part A
IJS(B)	Int. J. Sulfur Chem., Part B
IS	Inorg. Synth
IZV	Izv. Akad. Nauk SSSR Ser. Khim.
JA	J. Am. Chem. Soc.
JAN	J. Antibiot.
JAP	Jpn. Pat.
JAP(K)	Jpn. Kokai
JBC	J. Biol. Chem.
JC	J. Chromatogr.
JCC	J. Coord. Chem.
JCE	J. Chem. Ed.
JCED	J. Chem. Eng. Data
JCI	J. Chem. Inf. Comput. Sci.
JCP	J. Chem. Phys.
JCPB	J. Chim. Phys. Physico-Chim. Biol.
JCR(M)	J. Chem. Res. (M)
JCR(S)	J. Chem. Res. (S)
JCS	J. Chem. Soc.
JCS(A)	J. Chem. Soc. (A)
JCS(B)	J. Chem. Soc. (B)
JCS(C)	J. Chem. Soc. (C)
JCS(D)	J. Chem. Soc., Dalton Trans.
JCS(F1)	J. Chem. Soc., Faraday Trans. 1
JCS(F2)	J. Chem. Soc., Faraday Trans. 2
JCS(P1)	J. Chem. Soc., Perkin Trans. 1
JCS(P2)	J. Chem. Soc., Perkin Trans. 2
JCS(S2)	J. Chem. Soc. (Suppl. 2)
JEC	J. Electroanal. Chem. Interfacial Electrochem.
JEM	J. Energy Mater.
JES	J. Electron. Spectrosc.
JFA	J. Sci. Food. Agri.
JFC	J. Fluorine Chem.
JGU	J. Gen. Chem. USSR (Engl. Transl.)
JHC	J. Heterocycl. Chem.
JIC	J. Indian Chem. Soc.
JINC	J. Inorg. Nucl. Chem.
JLC	J. Liq. Chromatogr.

JMAS	J. Mat. Sci.
JMC	J. Med. Chem.
JMOC	J. Mol. Catal.
JMR	J. Magn. Reson.
JMS	J. Mol. Sci.
JOC	J. Org. Chem.
JOM	J. Organomet. Chem.
JOU	J. Org. Chem. USSR (Engl. Transl.)
JPC	J. Phys. Chem.
JPJ	J. Pharm. Soc. Jpn.
JPO	J. Phys. Org. Chem.
JPP	J. Pharm. Pharmacol.
JPR	J. Prakt. Chem.
JPS	J. Pharm. Sci.
JPS(A)	J. Polym. Sci., Polym. Chem., Part A
JPU	J. Phys. Chem. USSR (Engl. Transl.)
JSC	J. Serbochem. Soc.
JSP	J. Mol. Spectrosc.
JST	J. Mol. Struct.
K	Kristallografiya
KFZ	Khim. Farm. Zh.
KGS	Khim. Geterotsikl. Soedin.
KO	Kirk-Othmer Encyc.
KPS	Khim. Prir. Soedin.
L	Langmuir
LA	Liebigs Ann. Chem.
LC	Liq. Cryst.
LS	Life Sci.
M	Monatsh. Chem.
MAC	Macromol. Chem.
MC	Mendeleev Chem. J. (Engl. Transl.)
MCLC	Mol. Cryst. Liq. Cryst.
MI	Miscellaneous [book/journal]
MIP	Miscellaneous Pat.
MM	Macromolecules
MP	Mol. Phys.
MRC	Magn. Reson. Chem.
N	Naturwissenschaften
NAT	Nat.
NEP	Neth. Pat.
NJC	Nouv. J. Chim.
NKK	Nippon Kagaku Kaishi (J. Chem. Soc. Jpn.)
NKZ	Nippon Kagaku Zasshi
NZJ	N. Z. J. Sci. Technol.
OCS	Organomet. Synth.
OM	Organometallics
OMR	Org. Magn. Reson.
OMS	Org. Mass Spectrom.
OPP	Org. Prep. Proced. Int.
OR	Org. React.
OS	Org. Synth.
OSC	Org. Synth., Coll. Vol.
P	Phytochemistry
PA	Polym. Age
PAC	Pure Appl. Chem.

PAS	Pol. Acad. Sci.
PB	Polym. Bull.
PC	Personal Communication
PCS	Proc. Chem. Soc.
PHA	Pharmazi
PHC	Prog. Heterocycl. Chem.
PIA	Proc. Indian Acad. Sci.
PIA(A)	Proc. Indian Acad. Sci., Sect. A
PJC	Pol. J. Chem.
PJS	Pak. J. Sci. Ind. Res.
PMH	Phys. Methods Heterocycl. Chem.
PNA	Proc. Natl. Acad. Sci. USA
POL	Polyhedron
PP	Polym. Prepr.
PRS	Proceed. Roy. Soc.
PS	Phosphorus Sulfur
QR	Q. Rev., Chem. Soc.
QRS	Quart. Rep. Sulfur. Chem.
QSAR	Quant. Struct. Act. Relat. Pharmacol. Chem. Biol.
RC	Rubber Chem. Technol.
RCM	Rapid Commun. Mass Spectrom.
RCP	Rec. Chem. Prog.
RCR	Russ. Chem. Rev. (Engl. Transl.)
RHA	Rev. Heteroatom Chem.
RJ	Rubber J.
RP	Rev. Polarogr.
RRC	Rev. Roum. Chim.
RS	Ric. Sci.
RTC	Recl. Trav. Chim. Pays-Bas
RZC	Rocz. Chem.
S	Synthesis
SA	Spectrochim. Acta
SA(A)	Spectrochim. Acta, Part A
SAP	S. Afr. Pat.
SC	Synth. Commun.
SCI	Science
SL	Synlett
SM	Synth. Met.
SR	Sulfur Reports
SRI	Synth. React. Inorg. Metal-Org. Chem.
SS	Sch. Sci. Rev.
SST	Org. Compd. Sulphur, Selenium, Tellurium [R. Soc. Chem. series]
SUL	Sulfur Letters
SZP	Swiss Pat.
T	Tetrahedron
T(S)	Tetrahedron, Suppl.
TA	Tetrahedron Asymmetry
TAL	Talanta
TCA	Theor. Chim. Acta
TCC	Top. Curr. Chem.
TCM	Tetrahedron, Comp. Method
TFS	Trans. Faraday Soc.
TH	Thesis
TL	Tetrahedron Lett.
TS	Top. Stereochem.

UK	Usp. Khim.
UKZ	Ukr. Khim. Zh. (Russ. Ed.)
UP	Unpublished Results
URP	USSR Pat.
USP	US Pat.
WCH	Wiadom. Chem.
YGK	Yuki Gosei Kagaku Kyokaishi
YZ	Yakugaku Zasshi (J. Pharm. Soc. Jpn.)
ZAAC	Z. Anorg. Allg. Chem.
ZAK	Zh. Anal. Khim.
ZC	Z. Chem.
ZN	Z. Naturforsch.
ZN(A)	Z. Naturforsch., Teil A
ZN(B)	Z. Naturforsch., Teil B
ZOB	Zh. Obshch. Khim.
ZOR	Zh. Org. Khim.
ZPC	Hoppe–Seyler's Z. Physiol. Chem.
ZPK	Zh. Prikl. Khim.

VOLUME 5 REFERENCES

1832MI 501-01 F. Wöhler and J. Liebig; *J. Ann. Chem. Pharm.*, 1832, **3**, 267. 19

1853MI 504-01 C. Gerhardt; *Annales de Chimie et de Physique*, *Pt* 3, 1853, **37**, 285. 183

1856LA(97)329 W. Merck; *Justus Liebigs Ann. Chem.*, 1856, **97**, 329. 222

1857LA(104)111 A. Cahours; *Justus Liebigs Ann. Chem.*, 1857, **104**, 111. 19

1869CB120 A. W. Hofmann; *Ber. Dtsch. Chem. Ges.*, 1869, **2**, 120. 1026
1869CB305 L. Henry; *Ber. Dtsch. Chem. Ges.*, 1869, **2**, 305. 570
1869CB494 L. Henry; *Ber. Dtsch. Chem. Ges.*, 1869, **2**, 494. 570

1872CB799 A. Rathke; *Ber. Dtsch. Chem. Ges.*, 1872, **5**, 799. 1026
1872LA(161)347 H. Lossen; *Justus Liebigs Ann. Chem.*, 1872, **161**, 347. 999

1873CB210 W. Weith; *Ber. Dtsch. Chem. Ges.*, 1873, **6**, 210. 1036
1873CB1492 V. Meyer; *Ber. Dtsch. Chem. Ges.*, 1873, **6**, 1492. 838

1874CB425 V. Meyer; *Ber. Dtsch. Chem. Ges.*, 1874, **7**, 425. 838
1874CB670 V. Meyer and J. Locher; *Ber. Dtsch. Chem. Ges.*, 1874, **7**, 670. 838
1874CB790 E. Demole; *Ber. Dtsch. Chem. Ges.*, 1874, **7**, 790. 838
1874CB1228 E. Sell and G. Zierold; *Ber. Dtsch. Chem. Ges.*, 1874, **7**, 1228. 1004

1875CB114 J. Tscherniak; *Ber. Dtsch. Chem. Ges.*, 1875, **8**, 114. 838
1875LA(175)93 V. Meyer; *Justus Liebigs Ann. Chem.*, 1875, **175**, 93.

1877CB327 P. Hepp; *Ber. Dtsch. Chem. Ges.*, 1877, **10**, 327. 285
1877CB1095 A. W. Hofmann; *Ber. Dtsch. Chem. Ges.*, 1877, **10**, 1095. 609
1877CB1238 A. Bernthsen; *Ber. Dtsch. Chem. Ges.*, 1877, **10**, 1238. 609, 619
1877CB1889 A. Pinner and F. Klein; *Ber. Dtsch. Chem. Ges.*, 1877, **10**, 1889. 748
1877LA(184)79 O. Wallach; *Justus Liebigs Ann.*, 1877, **184**, 79. 652
1877LA(185)191 J. Kannonikoff and M. Saytzeff; *Justus Liebigs Ann. Chem.*, 1877, **185**, 191. 652
1877LA(186)1 H. Lossen; *Justus Liebigs Ann. Chem.*, 1877, **186**, 1. 999
1877MI 527-01 P. Miquel; *Ann. Chim. Phys.*, 1877, **11**, 289. 1047

1878CB4 A. Pinner and F. Klein; *Ber. Dtsch. Chem. Ges.*, 1878, **11**, 4. 744
1878CB338 A. W. Hofmann; *Ber. Dtsch. Chem. Ges.*, 1878, **11**, 338. 566
1878CB1825 A. Pinner and F. Klein; *Ber. Dtsch. Chem. Ges.*, 1878, **11**, 1825. 727

1879LA1 L. Schreiner; *Justus Liebigs Ann. Chem.*, 1879, 1. 24

1882CB985 A. W. Hofmann; *Ber. Dtsch. Chem. Ges.*, 1882, **15**, 985. 1035

1882LA(214)329 V. Meyer and E. J. Constam; *Justus Liebigs Ann. Chem.*, 1882, **214**, 329. 652

1884CB126 F. Tiemann; *Ber. Dtsch. Chem. Ges.*, 1884, **17**, 126. 833
1884CB183 A. Pinner; *Ber. Dtsch. Chem. Ges.*, 1884, **17**, 183. 835, 859
1884CB780 M. Freund; *Ber. Dtsch. Chem. Ges.*, 1884, **17**, 780. 32
1884CB1289 B. Radziszewski; *Ber. Dtsch. Chem. Ges.*, 1884, **17**, 1289. 709
1884CB1428 G. Luckenbach; *Ber. Dtsch. Chem. Ges.*, 1884, **17**, 1428. 756
1884CB1587 W. Lossen; *Ber. Dtsch. Chem. Ges.*, 1884, **17**, 1587. 835
1884CB1685 F. Tiemann and P. Krüger; *Ber. Dtsch. Chem. Ges.*, 1884, **17**, 1685. 825

1885CB355 B. R. Radziszewski; *Ber. Dtsch. Chem. Ges.*, 1885, **18**, 355. 275
1885CB727 F. Tiemann and P. Krüger; *Ber. Dtsch. Chem. Ges.*, 1885, **18**, 727. 825

1886CB1491 G. Müller; *Ber. Dtsch. Chem. Ges.*, 1886, **19**, 1491. 833
1886CB1668 F. Tiemann; *Ber. Dtsch. Chem. Ges.*, 1886, **19**, 1668. 830
1886CB2350 H. Stolte; *Ber. Dtsch. Chem. Ges.*, 1886, **19**, 2350. 1056

1887CB252 L. Claisen; *Ber. Dtsch. Chem. Ges.*, 1887, **20**, 252. 823
1887CB3422 A. Pictet; *Ber. Dtsch. Chem. Ges.*, 1887, **20**, 3422. 285
1887JPR449 A. Michael; *J. Prakt. Chem.*, 1887, **35**, 449. 593, 595

1888LA(249)244 L. Gattermann; *Justus Liebigs Ann. Chem.*, 1888, **249**, 244. 298

1889CB1936 F. Tiemann; *Ber. Dtsch. Chem. Ges.*, 1889, **22**, 1936. 833
1889CB2946 W. Zinkeisen; *Ber. Dtsch. Chem. Ges.*, 1889, **22**, 2946. 833
1889CB2958 F. Sembritzki; *Ber. Dtsch. Chem. Ges.*, 1889, **22**, 2958. 833
1889CB2973 G. Eichelbaum; *Ber. Dtsch. Chem. Ges.*, 1889, **22**, 2973. 833
1889LA(252)170 H. Lossen; *Justus Liebigs Ann. Chem.*, 1889, **252**, 170. 999
1889LA(252)214 W. Lossen; *Justus Liebigs Ann. Chem.*, 1889, **252**, 214. 835

1890CB2917 A. Pinner; *Ber. Dtsch. Chem. Ges.*, 1890, **23**, 2917. 755
1890CB3023 T. Curtius; *Ber. Dtsch. Chem. Ges.*, 1890, **23**, 3023. 989
1890CB3490 R. Scholl; *Ber. Dtsch. Chem. Ges.*, 1890, **23**, 3490. 838
1890JA499 W. J. Comstock and F. Cleeburg; *J. Am. Chem. Soc.*, 1890, **12**, 499. 723

1891CB803 D. Vorländer; *Ber. Dtsch. Chem. Ges.*, 1891, **24**, 803. 833
1891CB3021 S. M. Losanitsch; *Ber. Dtsch. Chem. Ges.*, 1891, **24**, 3021. 1023
1891CB3420 F. Tiemann; *Ber. Dtsch. Chem. Ges.*, 1891, **24**, 3420. 833
1891CB3426 F. Garny; *Ber. Dtsch. Chem. Ges.*, 1891, **24**, 3426. 833
1891CB3453 F. Tiemann; *Ber. Dtsch. Chem. Ges.*, 1891, **24**, 3453. 825
1891CB3658 O. Goldbeck; *Ber. Dtsch. Chem. Ges.*, 1891, **24**, 3658. 834
1891JCS396 E. A. Werner; *J. Chem. Soc.*, 1891, 396. 1035

1892CB705 A. Hantzsch; *Ber. Dtsch. Chem. Ges.*, 1892, **25**, 705. 822
1892CB1311 P. Douglas; *Ber. Dtsch. Chem. Ges.*, 1892, **25**, 1311. 619
1892CB2533 H. Biltz; *Ber. Dtsch. Chem. Ges.*, 1892, **25**, 2533. 654
1892CB3175 H. von Pechmann; *Ber. Dtsch. Chem. Ges.*, 1892, **25**, 3175. 857, 858, 859
1892CB3547 E. Bamberger; *Ber. Dtsch. Chem. Ges.*, 1892, **25**, 3547. 858
1892LA(270)267 J. U. Nef; *Justus Liebigs Ann. Chem.*, 1892, **270**, 267. 747

1893CB2385 E. Bamberger and P. de Gruyter; *Ber. Dtsch. Chem. Ges.*, 1893, **26**, 2385. 852
1893CB2978 E. Bamberger and F. Kuhlemann; *Ber. Dtsch. Chem. Ges.*, 1893, **26**, 2978. 857
1893LA(274)95 L. Calisen and O. Manasse; *Justus Liebigs Ann. Chem.*, 1893, **274**, 95. 821

1894CB320 H. von Pechmann; *Ber. Dtsch. Chem. Ges.*, 1894, **27**, 320. 842, 859
1894CB778 T. Curtius; *Ber. Dtsch. Chem. Ges.*, 1894, **27**, 778. 989
1894CB1733 K. Auwers and C. Beger; *Ber. Dtsch. Chem. Ges.*, 1894, **27**, 1733. 592
1894CB2121 H. von Pechmann and L. Seeberger; *Ber. Dtsch. Chem. Ges.*, 1894, **27**, 2121. 842
1894CB2193 A. Werner and H. Buss; *Ber. Dtsch. Chem. Ges.*, 1894, **27**, 2193. 821, 834
1894CB2816 R. Scholl; *Ber. Dtsch. Chem. Ges.*, 1894, **27**, 2816. 819
1894CB2846 A. Werner; *Ber. Dtsch. Chem. Ges.*, 1894, **27**, 2846. 821
1894LA(280)305 J. U. Neff; *Justus Liebigs Ann. Chem.*, 1894, **280**, 305. 819
1894MI 501-01 M. M. Meslans; *Ann. Chim. Phys.*, 1894, **1**, 411. 4

1895JCS565 A. E. Dixon and R. E. Doran; *J. Chem. Soc.*, 1895, **67**, 565. 1047
1895JCS1041 A. E. Dixon; *J. Chem. Soc.*, 1895, **67**, 1041. 1047
1895LA(281)169 W. Lossen; *Justus Liebigs Ann. Chem.*, 1895, **281**, 169. 826, 999

1896CB1161 A. Werner and A. Gemeseus; *Ber. Dtsch. Chem. Ges.*, 1896, **29**, 1161. 826
1896CB2166 H. von Pechmann; *Ber. Dtsch. Chem. Ges.*, 1896, **29**, 2166. 858
1896JA751 J. Stieglitz; *J. Am. Chem. Soc.*, 1896, **18**, 751. 698
1896JCS855 A. E. Dixon; *J. Chem. Soc.*, 1896, **69**, 855. 1047

1896JCS1593	A. E. Dixon; *J. Chem. Soc.*, 1896, **69**, 1593.	1047
1896JPR54	R. Engelhardt; *J. Prakt. Chem.*, 1896, **54**, 143.	848
1896JPR475	R. Walter; *J. Prakt. Chem.*, 1896, **53**, 475.	859
1896MI 526-01	J. Stieglitz; *Am. Chem. J.*, 1896, **18**, 751.	989
1897JA129	H. S. Wheeler and P. T. Waldon; *J. Am. Chem. Soc.*, 1897, **19**, 129.	813
1897LA(289)117	C. Kjellin and K. G. Kuylenstjevna; *Justus Liebigs Ann. Chem.*, 1897, **289**, 117.	831
1897MI 526-01	O. Folin; *Am. Chem. J.*, 1897, **19**, 323.	976
1898CB2854	O. Graul and A. Hantzsch; *Ber. Dtsch. Chem. Ges.*, 1898, **31**, 2854.	838
1898JPR223	O. Zwinger and R. Walter; *J. Prakt. Chem.*, 1898, **57**, 223.	851
1899CB1975	A. Werner and C. Bloch; *Ber. Dtsch. Chem. Ges.*, 1899, **32**, 1975.	834
1899CB3137	A. Hantzsch and H. Kissel; *Ber. Dtsch. Chem. Ges.*, 1899, **32**, 3137.	838
00CB751	F. Pichter and E. Schiess; *Ber. Dtsch. Chem. Ges.*, 1900, **33**, 751.	857
00CB2741	A. Wohl and H. Schiff; *Ber. Dtsch. Chem. Ges.*, 1900, **33**, 2741.	856
00LA(310)11	H. C. Briddle; *Justus Liebigs Ann. Chem.*, 1900, **310**, 11.	825
00LA(314)200	R. Behrend, F. C. Meyer and Y. Bucholz; *Justus Liebigs Ann. Chem.*, 1900, **314**, 200.	652
01CB168	W. Autenrieth; *Ber. Dtsch. Chem. Ges.*, 1901, **34**, 168.	184
01CB445	E. Fischer; *Ber. Dtsch. Chem. Ges.*, 1901, **34**, 445.	24
01CB1615	J. Stieglitz and J. H. Ransom; *Ber. Dtsch. Chem. Ges.*, 1901, **34**, 1615.	698
01CB2001	E. Bamberger and O. Schmidt; *Ber. Dtsch. Chem. Ges.*, 1901, **34**, 2001.	858
01JPR199	E. Bamberger and J. Müller; *J. Prakt. Chem.*, 1901, **64**, 199.	857
01JPR222	E. Bamberger and P. de Gruyter; *J. Prakt. Chem.*, 1901, **64**, 222.	858
02CB1084	E. Bamberger and J. Frei; *Ber. Dtsch. Chem. Ges.*, 1902, **35**, 1084.	855
02CB3101	O. Piloty and H. Steinbock; *Ber. Dtsch. Chem. Ges.*, 1902, **35**, 3101.	821
02CB3884	E. Bamberger and R. Seligman; *Ber. Dtsch. Chem. Ges.*, 1902, **35**, 3884.	838
02CR1312	G. Favrel; *C. R. Hebd. Seances Acad. Sci.*, 1902, **132**, 1312.	843
03CB585	F. Sachs and H. Loevy; *Ber. Dtsch. Chem. Ges.*, 1903, **36**, 585.	595
03CB1007	F. Weigert; *Ber. Dtsch. Chem. Ges.*, 1903, **36**, 1007.	232
03CB2484	H. Voswinckel; *Ber. Dtsch. Chem. Ges.*, 1903, **36**, 2484.	859
03CB3213	O. C. Billeter; *Ber. Dtsch. Chem. Ges.*, 1903, **36**, 3213.	994
03G226	L. Francesconi and G. De Plato; *Gazz. Chim. Ital.*, 1903, **33**, 226.	652
03JCS766	G. D. Lander and F. T. Jewson; *J. Chem. Soc.*, 1903, 766.	755
03LA(329)341	R. Behrend and P. Hesse; *Justus Liebigs Ann. Chem.*, 1903, **329**, 341.	597, 599
04CB874	F. Sachs and H. Loevy; *Ber. Dtsch. Chem. Ges.*, 1904, **37**, 874.	652
04CB3708	A. Reissert; *Ber. Dtsch. Chem. Ges.*, 1904, **37**, 3708.	566, 567
05CB2013	O. C. Billeter; *Ber. Dtsch. Chem. Ges.*, 1905, **38**, 2013.	994
05CB2986	W. Dieckamnn and L. Platz; *Ber. Dtsch. Chem. Ges.*, 1905, **38**, 2986.	843
05JCS747	A. Findlay and W. E. S. Turner; *J. Chem. Soc.*, 1905, 747.	24
05JPR366	M. Busch and W. Wolbring; *J. Prakt. Chem.*, 1905, **71**, 366.	858
05LA(338)360	B. Prager; *Justus Liebigs Ann. Chem.*, 1905, **338**, 360.	858
06CB1480	H. Wieland and H. Bauer; *Ber. Dtsch. Chem. Ges.*, 1906, **39**, 1480.	838
06M1221	R. Andreasch; *Monatsh. Chem.*, 1906, **27**, 1221.	1023
07CB1303	J. Houben and H. Pohl; *Ber. Dtsch. Chem. Ges.*, 1907, **40**, 1303.	546
07CR1125	M. Delepine; *C. R. Hebd. Seances Acad. Sci.*, 1907, **144**, 1125.	1023
08AC(P)84	A. Conduché; *Ann. Chim.* (*Paris*), 1908, **13**, 84.	837
08JCS621	S. Ruhemann; *J. Chem. Soc.*, 1908, **93**, 621.	593, 595
08JPR198	W. Steinkopf; *J. Prakt. Chem.*, 1908, **81**, 198.	822
08JPR489	R. Walter and A. Grossmann; *J. Prakt. Chem.*, 1908, **78**, 489.	851
09CB3484	A. Kaufmann and A. Luterbacher; *Ber. Dtsch. Chem. Ges.*, 1909, **42**, 3484.	184
09JCS1004	A. Findlay and E. M. Hickmans; *J. Chem. Soc.*, 1909, 1004.	24
09JPR192	Willgerodt and Merk; *J. Prakt. Chem.*, 1909, **80**, 192.	276
09LA(371)201	R. Anschutz; *Justus Liebigs Ann. Chem.*, 1909, **371**, 201.	1023
09M717	L. Kaluza; *Monatsh. Chem.*, 1909, **30**, 717.	1023
09MI 521-01	L. Cambi; *Atti R. Acad. dei Lincei*, 1909, **18**, 687 (*Chem. Zentralblatt*, *Roma*, 1909, **II**, 1552).	831
10G77	G. Ponzio; *Gazz. Chim. Ital.*, 1910, **40**, 77.	854
10JPR209	W. Steinkopf; *J. Prakt. Chem.*, 1910, **81**, 209.	838
10JPR216	W. Steinkopf; *J. Prakt. Chem.*, 1910, **81**, 216.	820
10JPR224	W. Steinkopf; *J. Prakt. Chem.*, 1910, **81**, 224.	822
11CB1693	C. Kelber and A. Schwarz; *Ber. Dtsch. Chem. Ges.*, 1911, **44**, 1693.	547

11CR(153)279 M. Delepine; *C. R. Hebd. Seances Acad. Sci.*, 1911, **153**, 279. 586
11G166 L. Cambi; *Gazz. Chim. Ital.*, 1911, **41**, 166. 831
11JCS1493 A. W. Titherley and E. C. Hughes; *J. Chem. Soc.*, 1911, 1493. 758
11JCS1507 A. W. Titherley and E. C. Hughes; *J. Chem. Soc.*, 1911, 1507. 758
11JPR453 W. Steinkopf and B. Jürgens; *J. Prakt. Chem.*, 1911, **83**, 453. 819, 821, 822
11JPR687 W. Steinkopf and B. Jürgens; *J. Prakt. Chem.*, 1911, **84**, 687. 819

12CB2188 J. von Braun and H. Deutsch; *Ber. Dtsch. Chem. Ges.*, 1912, **45**, 2188. 1025
12CB3732 C. Bülow and P. Neber; *Ber. Dtsch. Chem. Ges.*, 1912, **45**, 3732. 841
12M363 L. Kaluza; *Monatsh. Chem.*, 1912, **33**, 363. 1023

13CB1423 H. Staudinger and E. Anthes; *Ber. Dtsch. Chem. Ges.*, 1913, **46**, 1423. 19
13CB2032 C. Bülow and P. Neber; *Ber. Dtsch. Chem. Ges.*, 1913, **46**, 2032. 841
13CB2370 C. Bülow and P. Neber; *Ber. Dtsch. Chem. Ges.*, 1913, **46**, 2370. 841
13CB3616 J. Houben and E. Schmidt; *Ber. Dtsch. Chem. Ges.*, 1913, **46**, 3616. 698, 813
13G58 F. C. Palazzo and V. Egidi; *Gazz. Chim. Ital.*, 1913, **43**, 58. 822
13JCS2272 P. May, *J. Chem. Soc.*, 1913, **103**, 2272. 566

14CB2938 H. Ley and M. Ulrich; *Ber. Dtsch. Chem. Ges.*, 1914, **47**, 2938. 838, 855
14G72 G. Pelizarri and A. Gaiter; *Gazz. Chim. Ital.*, 1914, **44**, 72. 847

15CB470 A. Albert; *Ber. Dtsch. Chem. Ges.*, 1915, **48**, 470. 579
15JPR1 M. Busch, F. Ackerfeldt and R. Seufert; *J. Prakt. Chem.*, 1915, **92**, 1. 856, 860

17CB1035 H. Staudinger; *Ber. Dtsch. Chem. Ges.*, 1917, **50**, 1035. 286
17CB1478 C. Bülow and R. Huss; *Ber. Dtsch. Chem. Ges.*, 1917, **50**, 1478. 841

18JA415 D. E. Worrall; *J. Am. Chem. Soc.*, 1918, **40**, 415. 593, 595

20CR(171)38 J. Bougault and P. Robin; *C. R. Hebd. Seances Acad. Sci.*, 1920, **171**, 38. 763
20HCA824 H. Staudinger and J. Siegwart; *Helv. Chim. Acta*, 1920, **3**, 824. 476, 477
20JA599 R. Adams and L. H. Ulich; *J. Am. Chem. Soc.*, 1920, **42**, 599. 10
20JA1055 D. E. Worrall; *J. Am. Chem. Soc.*, 1920, **42**, 1055. 593, 595

21CR(172)452 J. Bougault and P. Robin; *C. R. Hebd. Seances Acad. Sci.*, 1921, **172**, 452. 763
21HCA861 H. Staudinger and E. Hauser; *Helv. Chim. Acta*, 1921, **4**, 861. 974
21JPR1 J. Tröger and A. Bernd; *J. Prakt. Chem.*, 1921, **102**, 1. 858

22JCS2169 E. L. Hirst and A. K. Macbeth; *J. Chem. Soc.*, 1922, 2169. 316
22MI 527-01 F. B. Dains, R. Q. Brewster and C. P. Olander; *Univ. Kansas Sci. Bull.*, 1922, **13**, 1 (*Chem. Abstr.*, 1923, **17**, 543). 1023

23CB633 E. Wedekind, D. Schenk and R. Stüsser; *Ber. Dtsch. Chem. Ges.*, 1923, **56**, 633. 488, 489
23CB1557 E. Wedekind and R. Stüsser; *Ber. Dtsch. Chem. Ges.*, 1923, **56**, 1557. 652
23JA1744 R. J. Kaufmann and R. Adams; *J. Am. Chem. Soc.*, 1923, **45**, 1744. 728
23JCS3359 F. L. Pyman; *J. Chem. Soc.*, 1923, 3359. 743, 744, 755
23LA(431)187 K. Kindler; *Justus Liebigs Ann. Chem.*, 1923, **431**, 187. 566, 579, 625

24CB981 A. Reissert and K. Bruggemann; *Ber. Dtsch. Chem. Ges.*, 1924, **57**, 981. 580, 592
24G389 G. Bruni and T. G. Levi; *Gazz. Chim. Ital.*, 1924, **54**, 389. 548
24HCA390 H. Staudinger, O. Muntwyler, L. Ruzicka and S. Seibt; *Helv. Chim. Acta*, 1924, 390. 145
24JA493 H. Gilman and C. R. Kinney; *J. Am. Chem. Soc.*, 1924, **46**, 493. 595, 729
24JA2834 D. E. Worrall; *J. Am. Chem. Soc.*, 1924, **46**, 2834. 593
24JCS1702 G. M. Dyson and H. J. George; *J. Chem. Soc.*, 1924, 1702. 1027
24LA(439)211 C. Bülow and E. King; *Justus Liebigs Ann. Chem.*, 1924, **439**, 211. 843

25CB1418 D. Holde and R. Gentner; *Ber. Dtsch. Chem. Ges.*, 1925, **58**, 1422. 184
25JA2974 D. E. Worrall; *J. Am. Chem. Soc.*, 1925, **47**, 2974. 595
25JCS279 H. Burgess and T. M. Lowry; *J. Chem. Soc.*, 1925, **127**, 279. 652
25JCS975 F. D. Chattaway and A. J. Walker; *J. Chem. Soc.*, 1925, **127**, 975. 840, 860
25JCS1304 J. E. Humphries, H. Humble and R. Evans; *J. Chem. Soc.*, 1925, **127**, 1304. 840
25JCS1687 F. D. Chattaway and A. J. Walker; *J. Chem. Soc.*, 1925, **127**, 1687. 841
25JCS2307 G. T. Morgan and H. D. K. Drew; *J. Chem. Soc.*, 1925, **127**, 2307. 1117
25JCS2887 H. B. Footner and S. Smiles; *J. Chem. Soc.*, 1925, **127**, 2887. 1111
25LA(444)113 H. Rheinboldt and O. Schmitz-Dumont; *Justus Liebigs Ann. Chem.*, 1925, **444**, 113. 823

26CB1162 M. Busch and H. Pfeiffer; *Ber. Dtsch. Chem. Ges.*, 1926, **59**, 1162. 856
26JA732 A. J. Hill and I. Rabinowitz; *J. Am. Chem. Soc.*, 1926, **48**, 732. 748
26JA3214 A. J. Hill and M. V. Cox; *J. Am. Chem. Soc.*, 1926, **48**, 3214.
26JCS2296 A. W. Chapman; *J. Chem. Soc.*, 1926, 2296. 728
26JCS3041 G. M. Dyson, H. J. George and R. F. Hunter; *J. Chem. Soc.*, 1926, 3041. 1027

27BSF1494 G. Favrel; *Bull. Soc. Chim. Fr.*, 1927, **41**, 1494. 843

27BSF1603	F. Favrel and J. Chrz; *Bull. Soc. Chim. Fr.*, 1927, **41**, 1603.	843
27JCS436	G. M. Dyson, H. J. George and R. F. Hunter; *J. Chem. Soc.*, 1927, 436.	1027
27JCS2318	C. Chew and F. L. Pyman; *J. Chem. Soc.*, 1927, 2318.	757
27LA(451)161	H. Rheinboldt; *Justus Liebigs Ann. Chem.*, 1927, **451**, 161.	821
27LA(451)453	J. v. Braun, F. Jostes and W. Munch; *Justus Liebigs Ann. Chem.*, 1927, **451**, 453.	751
27RTC600	G. Berger and S. C. J. Olivier; *Recl. Trav. Chim. Pays-Bas*, 1927, **46**, 600.	33, 108
28BSF683	L. Malaprade; *Bull. Soc. Chim. Fr.*, 1928, **43**, 683.	28
28CB32	H. Rheinboldt and O. Schmitz-Dumont; *Ber. Dtsch. Chem. Ges.*, 1928, **61**, 32.	823
28CR(186)382	L. Malaprade; *C. R. Hebd. Seances Acad. Sci.*, 1928, **186**, 382.	28
28JA1456	D. E. Worrall; *J. Am. Chem. Soc.*, 1928, **50**, 1456.	593, 595
28JA3106	A. S. Wheeler and C. L. Thomas; *J. Am. Chem. Soc.*, 1928, **50**, 3106.	549
29HCA554	P. Karrer and E. Weiss; *Helv. Chim. Acta*, 1929, **12**, 554.	592
29JA260	C. S. Marvel and E. R. Berkhimer; *J. Am. Chem. Soc.*, 1929, **51**, 260.	57, 64
29JA2262	W. H. Hartung and J. C. Munch; *J. Am. Chem. Soc.*, 1929, **51**, 2262.	823
30CB888	K. H. Slotta and H. Dressler; *Ber. Dtsch. Chem. Ges.*, 1930, **63**, 888.	1023
30CB2528	L. Birckenbach and M. Lindhard; *Ber. Dtsch. Chem. Ges.*, 1930, **63B**, 2528.	984
30CB2544	L. Birckenbach and M. Lindhard; *Ber. Dtsch. Chem. Ges.*, 1930, **63**, 2544.	984
30G700	I. de Paolini; *Gazz. Chim. Ital.*, 1930, **60**, 700.	821
30G703	I. de Paolini; *Gazz. Chim. Ital.*, 1930, **60**, 703.	822
30JA332	R. W. Bost and W. J. Mattox; *J. Am. Chem. Soc.*, 1930, **52**, 332.	549
30JA3647	T. W. Evans and W. M. Dehn; *J. Am. Chem. Soc.*, 1930, **52**, 3647.	74
30JCS94	F. D. Chattaway and L. H. Farinholt; *J. Chem. Soc.*, 1930, 94.	841
30LA(484)236	H. Wieland and B. Rosenfeld; *Justus Liebigs Ann. Chem.*, 1930, **484**, 236.	819
30MI 511-01	T. Hasselström; *Ann. Acad. Sci. Fennicae, Ser. A*, 1930, **30**, 3 (*Chem. Abstr.*, 1931, **25**, 3640).	652
30RZC652	L. Szperl; *Rocz. Chem.*, 1930, **10**, 652 (*Chem. Abstr.*, 1931, **25**, 938).	528, 529
31G561	G. Ponzio; *Gazz. Chim. Ital.*, 1931, **61**, 561.	819, 837
31JA1879	J. A. Mitchell and E. E. Reid; *J. Am. Chem. Soc.*, 1931, **53**, 1879.	275, 276
31M106	W. Oberhummer; *Monatsh. Chem.*, 1931, **57**, 106.	769
32CB65	G. Endres; *Ber. Dtsch. Chem. Ges.*, 1932, **65**, 65.	821, 822
32JA263	F. D. Chattaway and H. Irving; *J. Am. Chem. Soc.*, 1932, **54**, 263.	841
32JA3738	J. M. Zeavin and A. M. Fisher; *J. Am. Chem. Soc.*, 1932, **54**, 3738.	198
32M61	K. Morsch; *Monatsh. Chem.*, 1932, **60**, 61.	24
32M65	K. Morsch; *Monatsh. Chem.*, 1932, **60**, 65.	24
32M68	K. Morsch; *Monatsh. Chem.*, 1932, **60**, 68.	24
32OSC(1)14	H. T. Clarke and E. E. Dreger; *Org. Synth., Coll. Vol.*, 1932, **1**, 14.	109
32OSC(1)18	B. A. Ellis; *Org. Synth., Coll. Vol.*, 1932, **1**, 18.	46
32OSC(1)21	E. C. Kendall and B. F. McKenzie; *Org. Synth., Coll. Vol.*, 1932, **1**, 21.	85
32OSC(1)23	C. S. Marvel and E. E. Dreger; *Org. Synth., Coll. Vol.*, 1932, **1**, 238.	898
32OSC(1)23	W. C. Tobie and G. B. Ayres; *Org. Synth., Coll. Vol.*, 1932, **1**, 23.	89
32OSC(1)48	C. S. Marvel and V. Du Vigneaud; *Org. Synth., Coll. Vol.*, 1932, **1**, 48	89
32OSC(1)89	D. A. Ballard and W. M. Dehn; *Org. Synth., Coll. Vol.*, 1932, **1**, 89.	71
32OSC(1)115	H. T. Clarke and E. R. Taylor; *Org. Synth., Coll. Vol.*, 1932, **1**, 115.	60
32OSC(1)131	E. C. Kendall and B. F. McKenzie; *Org. Synth., Coll. Vol.*, 1932, **1**, 131.	56
32OSC(1)159	F. C. Whitmore and G. E. Woodward; *Org. Synth., Coll. Vol.*, 1932, **1**, 159.	115
32OSC(1)165	G. M. Dyson; *Org. Synth., Coll. Vol.*, 1932, **1**, 165.	1027
32OSC(1)166	C. Moureu and R. Chaux; *Org. Synth., Coll. Vol.*, 1932, **1**, 166.	59
32OSC(1)168	S. G. Powell, E. H. Huntress and E. B. Hershberg; *Org. Synth., Coll. Vol.*, 1932, **1**, 168.	59
32OSC(1)181	A. Lapworth and W. Baker; *Org. Synth., Coll. Vol.*, 1932, **1**, 181.	95
32OSC(1)248	C. S. Marvel and F. D. Hager; *Org. Synth., Coll. Vol.*, 1932, **1**, 248.	76
32OSC(1)252	C. S. Marvel and W. B. King; *Org. Synth., Coll. Vol.*, 1932, **1**, 252.	105
32OSC(1)254	J. K. H. Inglis; *Org. Synth., Coll. Vol.*, 1932, **1**, 254.	95
32OSC(1)276	W. C. Wilson; *Org. Synth., Coll. Vol.*, 1932, **1**, 276.	74
32OSC(1)289	C. S. Marvel and W. F. Tuley; *Org. Synth., Coll. Vol.*, 1932, **1**, 289.	34
32OSC(1)290	T. J. Otterbacher; *Org. Synth., Coll. Vol.*, 1932, **1**, 290.	30
32OSC(1)298	W. K. Anslow and H. King; *Org. Synth., Coll. Vol.*, 1932, **1**, 298.	85
32OSC(1)300	J. M. Orten and R. M. Hill; *Org. Synth., Coll. Vol.*, 1932, **1**, 300	89
32OSC(1)321	R. R. Read; *Org. Synth., Coll. Vol.*, 1932, **1**, 321.	64
32OSC(1)336	B. B. Corson, R. A. Dodge, S. A. Harris and J. S. Yeaw; *Org. Synth., Coll. Vol.*, 1932, **1**, 336.	63
32OSC(1)361	H. Gilman and R. H. Kirby; *Org. Synth., Coll. Vol.*, 1932, **1**, 361.	43
32OSC(1)379	G. D. Beal; *Org. Synth., Coll. Vol.*, 1932, **1**, 379.	30
32OSC(1)391	O. Kamm and J. B. Segur; *Org. Synth., Coll. Vol.*, 1932, **1**, 391.	106
32OSC(1)392	O. Kamm and A. O. Matthews; *Org. Synth., Coll. Vol.*, 1932, **1**, 392.	115
32OSC(1)398	F. K. Thayer; *Org. Synth., Coll. Vol.*, 1932, **1**, 398.	98
32OSC(1)406	G. R. Robertson; *Org. Synth., Coll. Vol.*, 1932, **1**, 406.	94
32OSC(1)408	P. J. Culhane and G. E. Woodward; *Org. Synth., Coll. Vol.*, 1932, **1**, 408.	109
32OSC(1)436	R. Adams and A. F. Thal; *Org. Synth., Coll. Vol.*, 1932, **1**, 436.	34
32OSC(1)447	F. B. Dains, R. Q. Brewster and C. P. Olander; *Org. Synth., Coll. Vol.*, 1932, **1**, 447.	1023

32OSC(1)451	A. Lapworth and W. Baker; *Org. Synth., Coll. Vol.*, 1932, **1**, 451.	35
32OSC(1)475	J. W. Howard and W. A. Fraser; *Org. Synth., Coll. Vol.*, 1932, **1**, 475.	73
32OSC(1)523	H. T. Clarke and T. F. Murray; *Org. Synth., Coll. Vol.*, 1932, **1**, 523.	30
32OSC(1)524	S. V. Puntambeker and E. A. Zoellner; *Org. Synth., Coll. Vol.*, 1932, **1**, 524.	43
32OSC(1)526	L. T. Sandborn and E. N. Bousquet; *Org. Synth., Coll. Vol.*, 1932, **1**, 526.	48, 49
32OSC(1)543	H. T. Clarke and W. W. Hartman; *Org. Synth., Coll. Vol.*, 1932, **1**, 543.	115
33HCA999	M. Battegay and E. Hégazi, *Helv. Chim. Acta*, 1933, **16**, 999.	592, 593
33HCA1323	L. Ruzicka, M. W. Goldberg, M. Hürbin and H. A. Boekenoogen; *Helv. Chim. Acta*, 1933, **16**, 1323.	566
33JA1262	H. Gilman and F. Breuer; *J. Am. Chem. Soc.*, 1933, **55**, 1262.	595
33JA2966	S. A. Karjala and S. M. McElvain; *J. Am. Chem. Soc.*, 1933, **55**, 2966.	551
33JA3672	J. Ross; *J. Am. Chem. Soc.*, 1933, **55**, 3672.	593, 595
33JA4079	L. P. Hammett and H. L. Pfluger; *J. Am. Chem. Soc.*, 1933, **55**, 4079.	31, 106
33JCS1389	F. D. Chattaway and D. R. Ashworth; *J. Chem. Soc.*, 1933, 1389.	842
33M285	W. Oberhummer; *Monatsh. Chem.*, 1933, **63**, 285.	769
34AG611	J. v. Braun; *Angew. Chem.*, 1934, **47**, 611.	654, 751
34CB1762	J. v. Braun and W. Rudolph; *Ber. Dtsch. Chem. Ges.*, 1934, **67**, 1762.	674, 767
34CRV275	R. C. Fuson and B. A. Bull; *Chem. Rev.*, 1934, **15**, 275.	38
34JA719	I. B. Douglass and F. B. Dains; *J. Am. Chem. Soc.*, 1934, **56**, 719.	1049
34JA1609	I. B. Douglass and L. E. Forman; *J. Am. Chem. Soc.*, 1934, **56**, 1609.	1049
34JCS822	J. M. Connolly and G. M. Dyson; *J. Chem. Soc.*, 1934, 822.	1028
34LA(512)97	W. Borsche and R. M. Manteuffel; *Justus Liebigs Ann. Chem.*, 1934, **512**, 97.	858
34MI 521-01	F. Feigl and V. Anger; *Mikrochimie*, 1934, **15**, 23.	864
35BJ(29)606	H. Raistrick and G. Smith; *Biochem. J.*, 1935, **29**, 606.	73
35CB163	A. Schönberg; *Ber. Dtsch. Chem. Ges.*, 1935, **68**, 163.	243
35CB751	E. Thielepape; *Ber. Dtsch. Chem. Ges.*, 1935, **68**, 751.	723
35JCS679	J. M. Connolly and G. M. Dyson; *J. Chem. Soc.*, 1935, 679.	1028
35JCS1467	P. W. Clutterbuck and F. Reuter; *J. Chem. Soc.*, 1935, 1467.	114
35JCS1762	C. Hasan and R. F. Hunter; *J. Chem. Soc.*, 1935, 1762.	1056
36CB2352	J. Houben and R. Zivadinovitsch; *Ber. Dtsch. Chem. Ges.*, 1936, **69**, 2352.	830
36G116	G. Ponzio; *Gazz. Chim. Ital.*, 1936, **66**, 116.	819
36JA1304	E. Jordan and C. R. Hauser; *J. Am. Chem. Soc.*, 1936, **58**, 1304.	65
36JA2608	H. Wenker; *J. Am. Chem. Soc.*, 1936, **58**, 2608.	976, 977
36JCS1788	G. King; *J. Chem. Soc.*, 1936, 1788.	49
36ZAAC(230)31	K. A. Jensen and E. Frederiksen; *Z. Anorg. Allg. Chem.*, 1936, **230**, 31.	1051
37CR(204)1342	F. Bloch; *C. R. Hebd. Seances Acad. Sci.*, 1937, **204**, 1342.	476, 477
37HCA691	H. Rivier and J. Zeltner; *Helv. Chem. Acta.*, 1937, **20**, 691.	566, 618, 735
37JA121	C. R. Hauser and W. B. Renflow, Jr.; *J. Am. Chem. Soc.*, 1937, **59**, 121.	998
37JA206	L. P. Kyrides; *J. Am. Chem. Soc.*, 1937, **59**, 206.	15
37JA401	J. R. Ruhoff and E. E. Reid; *J. Am. Chem. Soc.*, 1937, **59**, 401.	276
37JA740	I. B. Douglass; *J. Am. Chem. Soc.*, 1937, **59**, 740.	1057
37JA1486	D. E. Worrall; *J. Am. Chem. Soc.*, 1937, **59**, 1486.	595
37JPR251	H. Meerwein and H. Sönke; *J. Prakt. Chem.*, 1937, **147**, 251.	227
B-37MI 521-01	F. Feigl; "Spot Tests," Nordemann Publishing, New York, 1937, p. 22.	864
38BRP495510	S. Goldschmidt and K. Martin; *Br. Pat.* 495 510 (1938) (*Chem. Abstr.*, 1939, **33**, 2912).	1023
38JA235	C. D. Hurd and J. C. Sowden; *J. Am. Chem. Soc.*, 1938, **60**, 235.	66
38JA540	H. M. Singleton and W. R. Edwards, Jr.; *J. Am. Chem. Soc.*, 1938, **60**, 540.	272, 293, 304
38JA579	L. L. Fellinger and L. F. Audrieth; *J. Am. Chem. Soc.*, 1938, **60**, 579.	298
38JA1325	H. C. Brown; *J. Am. Chem. Soc.*, 1938, **60**, 1325.	15
39JA104	D. E. Worrall and E. Lavin; *J. Am. Chem. Soc.*, 1939, **61**, 104.	595
39JA105	D. E. Worrall, M. Lerner and J. Washnock, Jr.; *J. Am. Chem. Soc.*, 1939, **61**, 105.	595, 617
39JA632	J. C. Ambelang and T. B. Johnson; *J. Am. Chem. Soc.*, 1939, **61**, 632.	1048
40AG321	R. Crigee; *Angew. Chem.*, 1940, **53**, 321.	28
40CB843	O. Mumm and H. Richter; *Ber. Dtsch. Chem. Ges.*, 1940, **73B**, 843.	1040
40JA761	G. S. Forbes and H. H. Anderson; *J. Am. Chem. Soc.*, 1940, **62**, 761.	1015, 1016, 1017
40JA1595	A. J. Hill and W. M. Degnan; *J. Am. Chem. Soc.*, 1940, **62**, 1595.	994, 995
40JA3495	C. S. Marvel, J. Dec, H. G. Cooke, Jr. and J. C. Cowan; *J. Am. Chem. Soc.*, 1940, **62**, 3495.	55
40JCS831	D. T. Lewis; *J. Chem. Soc.*, 1940, 831.	528
40JOC606	A. Burger and S. Avakian; *J. Org. Chem.*, 1940, **5**, 606.	276
41AG51	J. Mittermaier; *Angew. Chem.*, 1941, **54**, 51.	62
41AG146	G. Stein; *Angew. Chem.*, 1941, **54**, 146.	28
41JA314	A. A. Morton, A. R. Olson and J. W. Blattenberger; *J. Am. Chem. Soc.*, 1941, **63**, 314.	595
41JA1151	A. O. Zoss and G. F. Hennion; *J. Am. Chem. Soc.*, 1941, **63**, 1151.	304
41JA1674	J. F. Lane and E. S. Wallis; *J. Am. Chem. Soc.*, 1941, **63**, 1674.	1000

41JA2431	M. S. Newman; *J. Am. Chem. Soc.*, 1941, **63**, 2431.	106
41JCS263	J. Kenyon and D. P. Young; *J. Chem. Soc.*, 1941, 263.	1000
B-41MI 502-01	A. Morel; "Traite de Chimie Organique," ed. V. Grignard, G. Dupont and R. Locquin, Masson et Cie, Paris, 1941, vol. 13, pp. 1–323.	83
41OSC(1)5	A. W. Dox; *Org. Synth., Coll. Vol.*, 1941, **1**, 5.	748
41OSC(1)82	C. N. Webb; *Org. Synth., Coll. Vol.*, 1941, **1**, 82.	652
41OSC(1)99	C. S. Marvel and W. A. Lazier; *Org. Synth., Coll. Vol.*, 1941, **1**, 99.	299
41OSC(1)111	J. R. Johnson and L. T. Sandborn; *Org. Synth., Coll. Vol.*, 1941, **1**, 111.	290
41OSC(1)153	W. A. Jacobs and M. Heidelberger; *Org. Synth., Coll. Vol.*, 1941, **1**, 153.	275
41OSC(1)457	W. A. Noyes and P. K. Porter; *Org. Synth., Coll. Vol.*, 1941, **1**, 457.	305
42CB75	R. Kuhn and D. Jerchel; *Ber. Dtsch. Chem. Ges.*, 1941, **74**, 75.	856
42CB1585	H. P. Kaufmann, J. Budwig and K. Mohnke; *Ber. Dtsch. Chem. Ges.*, 1942, **75**, 1585.	753
42CRV97	E. R. Littmann and J. R. M. Klotz; *Chem. Rev.*, 1942, **30**, 97.	24
42JA329	M. S. Kharasch and H. C. Brown; *J. Am. Chem. Soc.*, 1942, **64**, 329.	16
42JBC(144)203	W. Sakami and G. Toennies; *J. Biol. Chem.*, 1942, **144**, 203.	140
42JCS374	G. M. Dyson and T. Harrington; *J. Chem. Soc.*, 1942, 374.	1027
42JOC408	N. Levin and W. H. Hartung; *J. Org. Chem.*, 1942, **7**, 408.	823
42MI 502-01	R. Truchet; *La. Rev. Scientifique*, 1942, **80**, 265.	28
42MI 502-02	M. Triffeneau; *La. Rev. Scientifique*, 1942, **80**, 297.	62
42OR(1)210	J. R. Johnson; *Org. React.*, 1942, **1**, 210.	98
42OR(1)266	C. R. Hauser and B. E. Hudson, Jr.; *Org. React.*, 1942, **1**, 266.	46
42OR(1)38	W. E. Bachmann and W. S. Struve; *Org. React.*, 1942, **1**, 38.	276
43CRV209	H. L. Yale; *Chem. Rev.*, 1943, **33**, 209.	864
43HCA1722	H. Rivier and M. Langer; *Helv. Chim. Acta*, 1943, **26**, 1722.	476, 477
43JA262	R. R. Burtner and J. W. Cusic; *J. Am. Chem. Soc.*, 1943, **65**, 262.	43, 62, 114
43JA745	L. I. Smith and G. F. Rouault; *J. Am. Chem. Soc.*, 1943, **65**, 745.	76
43JA1783	A. W. Laubengayer and L. Reggel; *J. Am. Chem. Soc.*, 1943, **65**, 1783.	1016
43JCS101	H. J. Barber; *J. Chem. Soc.*, 1943, 101.	703, 767, 813
B-43MI 502-01	R. J. Williams; in "Advanced Enzymology and Related Subjects of Biochemistry," ed. C. H. Werkmann, Interscience, New York, 1943, vol. 3, p. 253.	62
B-43MI 502-02	H. T. Clarke; in "Organic Chemistry. An Advanced Treatise," ed. H. Gilman, Wiley, New York, 1943, vol. 2, pp. 1079–1165.	83
43OSC(2)10	T. W. Abbott, R. T. Arnold and R. B. Thompson; *Org. Synth., Coll. Vol.*, 1943, **2**, 10.	103
43OSC(2)11	R. M. Herbst and D. Shemin; *Org. Synth., Coll. Vol.*, 1943, **2**, 11.	652
43OSC(2)12	W. F. Bruce; *Org. Synth., Coll. Vol.*, 1943, **2**, 12.	98
43OSC(2)19	H. T. Clarke and L. D. Behr; *Org. Synth., Coll. Vol.*, 1943, **2**, 19.	86
43OSC(2)25	C. C. De Witt; *Org. Synth., Coll. Vol.*, 1943, **2**, 25.	85
43OSC(2)28	J. C. Eck; *Org. Synth., Coll. Vol.*, 1943, **2**, 28.	86
43OSC(2)29	H. T. Clarke and H. J. Bean; *Org. Synth., Coll. Vol.*, 1943, **2**, 29.	85
43OSC(2)44	J. S. Buck and W. S. Ide; *Org. Synth., Coll. Vol.*, 1943, **2**, 44.	275
43OSC(2)53	J. W. Hill and W. L. McEwen; *Org. Synth., Coll. Vol.*, 1943, **2**, 53.	52
43OSC(2)74	J. C. Eck and C. S. Marvel; *Org. Synth., Coll. Vol.*, 1943, **2**, 74.	57
43OSC(2)93	C. S. Marvel and V. du Vigneaud; *Org. Synth., Coll. Vol.*, 1943, **2**, 93.	33
43OSC(2)126	R. Riegel and F. Zwilgmeyer; *Org. Synth., Coll. Vol.*, 1943, **2**, 126.	62
43OSC(2)135	H. T. Clarke and E. R. Taylor; *Org. Synth., Coll. Vol.*, 1943, **2**, 135.	115
43OSC(2)140	R. L. Shriner, S. G. Ford and L. J. Roll; *Org. Synth., Coll. Vol.*, 1943, **2**, 140.	99
43OSC(2)177	H. S. Rhinesmith; *Org. Synth., Coll. Vol.*, 1943, **2**, 177.	60
43OSC(2)181	A. C. Cope, J. R. Clarke and R. Connor; *Org. Synth., Coll. Vol.*, 1943, **2**, 181.	59
43OSC(2)260	R. C. Fuson and B. H. Wojcik; *Org. Synth., Coll. Vol.*, 1943, **2**, 260.	76
43OSC(2)292	J. R. Ruhoff; *Org. Synth., Coll. Vol.*, 1943, **2**, 292.	35
43OSC(2)299	G. Schiemann and W. Winkelmuller; *Org. Synth., Coll. Vol.*, 1943, **2**, 299.	106
43OSC(2)302	N. A. Milas; *Org. Synth., Coll. Vol.*, 1943, **2**, 302.	102
43OSC(2)310	C. S. Marvel; *Org. Synth., Coll. Vol.*, 1943, **2**, 310.	85
43OSC(2)315	J. R. Ruhoff; *Org. Synth., Coll. Vol.*, 1943, **2**, 315.	49
43OSC(2)333	H. R. Snyder, J. S. Buck and W. S. Ide; *Org. Synth., Coll. Vol.*, 1943, **2**, 333.	51
43OSC(2)368	R. L. Shriner, S. G. Ford and L. J. Roll; *Org. Synth., Coll. Vol.*, 1943, **2**, 368.	99
43OSC(2)374	J. C. Eck and C. S. Marvel; *Org. Synth., Coll. Vol.*, 1943, **2**, 374.	88
43OSC(2)376	N. Weiner; *Org. Synth., Coll. Vol.*, 1943, **2**, 376.	35
43OSC(2)382	R. L. Shriner, S. G. Ford and L. J. Roll; *Org. Synth., Coll. Vol.*, 1943, **2**, 382.	89, 99
43OSC(2)384	G. Barger and T. E. Weichselbaum; *Org. Synth., Coll. Vol.*, 1943, **2**, 384.	90
43OSC(2)416	E. B. Vliet, C. S. Marvel and C. M. Hsueh; *Org. Synth., Coll. Vol.*, 1943, **2**, 416.	33
43OSC(2)422	J. J. Donleavy and M. A. Kise; *Org. Synth., Coll. Vol.*, 1943, **2**, 422.	93
43OSC(2)425	H. Gilman, N. B. St. John and F. Schulze; *Org. Synth., Coll. Vol.*, 1943, **2**, 425.	112
43OSC(2)428	M. S. Newman and H. L. Holmes; *Org. Synth., Coll. Vol.*, 1943, **2**, 428.	114
43OSC(2)457	E. H. Huntress, E. L. Shloss, Jr. and P. Ehrlich; *Org. Synth., Coll. Vol.*, 1943, **2**, 457.	109
43OSC(2)474	E. E. Reid and J. R. Ruhoff; *Org. Synth., Coll. Vol.*, 1943, **2**, 474.	33
43OSC(2)489	H. B. Gillespie and H. R. Snyder; *Org. Synth., Coll. Vol.*, 1943, **2**, 489.	87
43OSC(2)491	R. M. Herbst and D. Shemin; *Org. Synth., Coll. Vol.*, 1943, **2**, 491.	90
43OSC(2)515	T. W. Abbott; *Org. Synth., Coll. Vol.*, 1943, **2**, 515.	103
43OSC(2)519	R. M. Herbst and D. Shemin; *Org. Synth., Coll. Vol.*, 1943, **2**, 519.	66
43OSC(2)523	J. H. Gardner and C. A. Naylor, Jr.; *Org. Synth., Coll. Vol.*, 1943, **2**, 523.	74

43OSC(2)531 H. R. Snyder, L. A. Brooks and S. H. Shapiro; *Org. Synth., Coll. Vol.*, 1943, **2**, 531. 31
43OSC(2)538 R. L. Shriner and E. C. Kleiderer; *Org. Synth., Coll. Vol.*, 1943, **2**, 538. 114
43OSC(2)557 M. Nierenstein and D. A. Clibbens; *Org. Synth., Coll. Vol.*, 1943, **2**, 557. 110
43OSC(2)560 L. F. Fieser and E. L. Martin; *Org. Synth., Coll. Vol.*, 1943, **2**, 560. 199
43OSC(2)562 H. T. Clarke and L. D. Behr; *Org. Synth., Coll. Vol.*, 1943, **2**, 562. 305
43OSC(2)586 C. R. Noller; *Org. Synth., Coll. Vol.*, 1943, **2**, 586. 298
43OSC(2)588 H. T. Clarke and E. R. Taylor; *Org. Synth., Coll. Vol.*, 1943, **2**, 588. 108

44AG135 L. Birkofer; *Angew. Chem.*, 1944, **57**, 135. 84
44CRV351 R. L. Shriner and F. W. Neumann; *Chem. Rev.*, 1944, **35**, 351. 742, 751, 755
44MI 502-01 H. Neurath and J. P. Greenstein; *Annu. Rev. Biochem.*, 1944, **13**, 117. 83
44OR(2)94 T. A. Geissman; *Org. React.*, 1944, **2**, 94. 48
44OR(2)341 E. L. Jackson; *Org. React.*, 1944, **2**, 341. 28, 46
44USP2362648 J. G. Lichty and N. V. Seeger (Wingfoot Corp.); *US Pat.* 2 362 648 (1944) (*Chem. Abstr.*, 1945, **39**, 2764). 965

45BSF17 L. Velluz; *Bull. Soc. Chim. Fr.*, 1945, 17. 62
45BSF506 J. Vene; *Bull. Soc. Chim. Fr.*, 1945, 506. 28
45CRV235 G. R. Waitkins and C. W. Clark; *Chem. Rev.*, 1945, **36**, 235. 28
45JA1178 H. A. Bruson and T. W. Riener; *J. Am. Chem. Soc.*, 1945, **67**, 1178. 1043
45JA1783 L. Cavalieri, D. B. Pattison and M. Carmack; *J. Am. Chem. Soc.*, 1945, **67**, 1783. 272, 276
45JA2176 H. H. Anderson; *J. Am. Chem. Soc.*, 1945, **67**, 2176. 1017
45JCS601 W. R. Boon; *J. Chem. Soc.*, 1945, 601. 579
45JCS660 F. R. Atherton, H. T. Openshaw and A. R. Todd; *J. Chem. Soc.*, 1945, 660. 975
45JCS686 E. B. Knott; *J. Chem. Soc.*, 1945, 686. 652
45JCS751 F. G. Mann and J. W. G. Porter; *J. Chem. Soc.*, 1945, 751. 26, 94
45MI 502-01 J. Steinhardt; *Annu. Rev. Biochem.*, 1945, **14**, 145. 83
B-45MI 502-01 M. Montagne; in "Traite de Chemie Organique," ed. V. Grignard, C. Dupont and R. Lacquin, Masson et Cie, Paris, 1945, vol. 11, pp. 1–495. 62
B-45MI 502-02 P. Weill and R. Duchert; in "Traite de Chimie Organique," ed. V. Grignard, G. Dupont and R. Lacquin, Masson et Cie, Paris, 1945, vol. 11, pp. 531–1075. 105
45MI 502-02 S. W. Fox; *Adv. Protein Chem.*, 1945, **2**, 156. 84
B-45MI 502-03 H. J. Goudet; in "Traite de Chimie Organique," ed. V. Grignard, C. Dupont and R. Lacquin, Masson et Cie, Paris, 1945, vol. 11, pp. 1075–1147. 105
B-45MI 502-04 H. Gault; in "Traite de Chimie Organique," ed. V. Grignard, C. Dupont and R. Lacquin, Masson et Cie, Paris, 1945, vol. 11, pp. 1149–1233. 62
B-45MI 502-05 H. Gault and L. Piaux; in "Traite de Chimie Organique," ed. V. Grignard, C. Dupont and R. Lacquin, Masson et Cie, Paris, 1945, vol. 11, pp. 1235–1456. 62

46ANY(48)255 Symposium; *Ann. N. Y. Acad. Sci.*, 1946, **48**, 255. 92
46CRV501 R. J. Block; *Chem. Rev.*, 1946, **38**, 501. 83
46JA525 J. A. King and F. H. McMillan; *J. Am. Chem. Soc.*, 1946, **68**, 525. 272, 276
46JA588 D. B. Sharp and C. S. Hamilton; *J. Am. Chem. Soc.*, 1946, **68**, 588. 842
46JA1844 K. N. Campbell, J. F. Kerwin, R. A. LaForge and B. K. Campbell; *J. Am. Chem. Soc.*, 1946, **68**, 1844. 25, 108
46JA2029 M. Carmack and D. F. DeTar; *J. Am. Chem. Soc.*, 1946, **68**, 2029. 599
46JA2102 C. K. Banks, J. Controulis and W. F. Holcomb; *J. Am. Chem. Soc.*, 1946, **68**, 2102. 1125
46JA2506 R. L. McKee and R. W. Bost; *J. Am. Chem. Soc.*, 1946, **68**, 2506. 1027
46JCS25 A. Campbell and J. Kenyon; *J. Chem. Soc.*, 1946, 25. 214
46JCS147 P. Oxley and W. F. Short; *J. Chem. Soc.*, 1946, 147. 749, 757
46JCS454 J. Forrest, D. A. Liddell and S. H. Tucker; *J. Chem. Soc.*, 1946, 454. 293
46JCS599 A. G. Caldwell and E. R. H. Jones; *J. Chem. Soc.*, 1946, 599. 276
46MI 502-01 T. L. McMeekin and R. C. Warner; *Annu. Rev. Biochem.*, 1946, **15**, 119. 83
46MI 502-02 F. Smith; *Adv. Carbohydr. Chem.*, 1946, **2**, 79. 98
46OR(3)83 M. Carmack and M. A. Spielman; *Org. React.*, 1946, **3**, 83. 599
46OR(3)267 E. S. Wallis and J. F. Lane; *Org. React.*, 1946, **3**, 267. 998
46OR(3)307 H. Wolff; *Org. React.*, 1946, **3**, 307. 283, 1001
46OR(3)337 P. A. S. Smith; *Org. React.*, 1946, **3**, 337. 989

47AG65 R. Tochesche; *Angew. Chem.*, 1947, **59**, 65. 652
47BCF406 J. Guy and J. Paris; *Bull. Soc. Chim. Fr.*, 1947, 406. 755
47CRV585 K. L. Waters; *Chem. Rev.*, 1947, **41**, 585. 62
47JA200 A. P. Phillips and R. Baltzly; *J. Am. Chem. Soc.*, 1947, **69**, 200. 140
47JA365 G. B. Bachman, D. E. Welton, G. L. Jenkins and J. F. Christian; *J. Am. Chem. Soc.*, 1947, **69**, 365. 1125
47JA875 R. A. Turner; *J. Am. Chem. Soc.*, 1947, **69**, 875. 363, 364
47JA925 E. J. Cragoe, Jr., R. J. Andres, R. F. Coles, B. Elpern, J. F. Morgan and C. S. Hamilton; *J. Am. Chem. Soc.*, 1947, **69**, 925. 221
47JA1207 F. H. McMillan and J. A. King; *J. Am. Chem. Soc.*, 1947, **69**, 1207. 566, 608
47JA1211 E. Campaigne and P. V. Rutan; *J. Am. Chem. Soc.*, 1947, **69**, 1211. 599
47JA1241 G. S. Forbes and H. H. Anderson; *J. Am. Chem. Soc.*, 1947, **69**, 1241. 1016
47JA2007 R. B. Carlin and L. O. Smith, Jr.; *J. Am. Chem. Soc.*, 1947, **69**, 2007. 293
47JA2236 O. Foss; *J. Am. Chem. Soc.*, 1947, **69**, 2236. 1116

47JA2350	R. G. Jones; *J. Am. Chem. Soc.*, 1947, **69**, 2350.	76
47JA2568	S. Ställberg-Stenhagen; *J. Am. Chem. Soc.*, 1947, **69**, 2568.	13
47JA3048	G. S. Forbes and H. H. Anderson; *J. Am. Chem. Soc.*, 1947, **69**, 3048.	1016
47JCS160	F. H. S. Curd and C. G. Raison; *J. Chem. Soc.*, 1947, 160.	757
47JCS390	M. W. Partridge and W. F. Short; *J. Chem. Soc.*, 1947, 390.	750
47JCS738	G. Newbury and W. Webster; *J. Chem. Soc.*, 1947, 738.	744, 750, 754
47JCS1110	P. Oxley, M. W. Partridge and W. F. Short; *J. Chem. Soc.*, 1947, 1110.	750, 751
B-47MI 502-01	F. Salmon-Legagneur; in "Traite de Chimie Organique," ed. V. Grignard, C. Dupont and R. Lacquin, Masson et Cie, Paris, 1947, vol. 10, pp. 1–443.	24
47MI 502-01	M. S. Dunn and L. B. Rockland; *Adv. Protein Chem.*, 1947, **3**, 295.	84
47MI 502-02	E. Brand and J. T. Edsall; *Annu. Rev. Biochem.*, 1947, **16**, 223.	83
B-47MI 502-02	F. Salmon-Legagneur; in "Traite de Chimie Organique," ed. V. Grignard, C. Dupont and R. Lacquin, Masson et Cie, Paris, 1947, vol. 10, pp. 445–630.	24
47MI 502-03	P. Chauchard; *La. Rev. Scientifique*, 1947, **85**, 683.	92
B-47MI 502-03	R. Abderhalden; in "Zeitschrift Vitamin, Hormon, Ferment-Forschung," ed. E. Abderhalden, Urban-Schwarzenberg Gmbh, Wien, 1947–48, vol. 1, p. 163.	92
B-47MI 502-04	K. Bernhard; in "Zeitschrift Vitamin, Hormon, Ferment-Forschung," ed. E. Abderhalden, Urban-Schwarzenberg Gmbh, Wien, 1947–48, vol. 1, p. 199.	28
47OS(27)56	H. W. J. Cressman; *Org. Synth.*, 1947, **27**, 56.	1126
48AG169	G. Wittig; *Angew. Chem.*, 1948, **60**, 169.	48
48AG308	A. Schoberl; *Angew. Chem.*, 1948, **60**, 308.	84
48AR198	J. F. W. McOmie, *Annu. Rep. Prog. Chem.*, 1948, **45**, 198.	588
48BSF597	E. Gryszkiewicz-Trochimowski, K. Dymowski and E. Schmidt; *Bull. Soc. Chim. Fr.*, 1948, 597.	822
48CB210	A. Schoberl and R. Hamm; *Chem. Ber.*, 1948, **81**, 210.	1111
48G332	R. Fusco and R. Romani; *Gazz. Chim. Ital.*, 1948, **78**, 332.	858
48JA317	M. S. Newman and H. L. Gildenhorn; *J. Am. Chem. Soc.*, 1948, **70**, 317.	1002
48JA868	F. H. McMillan; *J. Am. Chem. Soc.*, 1948, **70**, 868.	621
48JA1043	G. S. Forbes and H. H. Anderson; *J. Am. Chem. Soc.*, 1948, **70**, 1043.	1016
48JA1153	C. Weizmann, M. Sulzbacher and E. Bergmann; *J. Am. Chem. Soc.*, 1948, **70**, 1153.	66
48JA1222	G. S. Forbes and H. H. Anderson; *J. Am. Chem. Soc.*, 1948, **70**, 1222.	1016
48JA1904	E. Lorz and R. Baltzly; *J. Am. Chem. Soc.*, 1948, **70**, 1904.	756
48JA2425	W. J. Horton and J. Van Den Berghe; *J. Am. Chem. Soc.*, 1948, **70**, 2425.	599
48JA2824	C. Schuerch, Jr. and E. H. Huntress; *J. Am. Chem. Soc.*, 1948, **70**, 2824.	34, 43
48JA2837	J. W. Corse, R. G. Jones, Q. F. Soper, C. W. Whitehead and O. K. Behrens; *J. Am. Chem. Soc.*, 1948, **70**, 2837.	584, 586
48JA2964	F. H. Carpenter; *J. Am. Chem. Soc.*, 1948, **70**, 2964.	184
48JA3091	N. Sperber, D. Papa and E. Schwenk; *J. Am. Chem. Soc.*, 1948, **70**, 3091.	36
48JA3212	L. F. Fieser, M. T. Leffler and co-workers; *J. Am. Chem. Soc.*, 1948, **70**, 3212.	1119
48JA3484	F. A. Hochstein and W. G. Brown; *J. Am. Chem. Soc.*, 1948, **70**, 3484.	47
48JA3781	S. M. McElvain and M. J. Curry; *J. Am. Chem. Soc.*, 1948, **77**, 3781.	139
48JA3945	A. Galat; *J. Am. Chem. Soc.*, 1948, **70**, 3945.	298
48JA4045	J. J. Ritter and P. P. Minieri; *J. Am. Chem. Soc.*, 1948, **70**, 4045.	281, 282
48JA4143	F. H. McMillan and J. A. King; *J. Am. Chem. Soc.*, 1948, **70**, 4143.	599
48JCS1314	P. Oxley and W. F. Short; *J. Chem. Soc.*, 1948, 1314.	689, 710
48JCS1618	P. Oxley, D. A. Peak and W. F. Short; *J. Chem. Soc.*, 1948, 1618.	710
48JCS1919	A. A. Goldberg and W. Kelly; *J. Chem. Soc.*, 1948, 1919.	622
48JOC924	S. R. Safir, S. Kushner, L. M. Brancone and Y. Subbarow; *J. Org. Chem.*, 1948, **13**, 924.	1125
48M316	E. Ziegler and H. Ludde; *Monatsch. Chem.*, 1948, **79**, 316.	1107
48MI 502-01	A. Neuberger; *Adv. Protein Chem.*, 1948, **4**, 297.	84
B-48MI 502-01	R. Truchet; in "Traite de Chimie Organique," ed. V. Grignard, C. Dupont and R. Lacquin, Masson et Cie, Paris, 1948, vol. 9, p. 1–92 and 145–687.	24
48MI 502-02	R. B. Carey; *Adv. Protein Chem.*, 1948, **4**, 385.	84
B-48MI 502-02	R. W. Ralston; "Fatty Acids and Their Derivatives," Wiley, New York, 1948.	28
48MI 502-03	K. O. Pedersen; *Annu. Rev. Biochem.*, 1948, **17**, 169.	83
48PCS243	T. P. Hilditch; *Proc. Chem. Soc.*, 1948, 243.	97
49BSF251	P. Desnuelle; *Bull. Soc. Chim. Fr.*, 1949, 251.	652
49BSF272	P. Chabrier and S. H. Renard; *Bull. Soc. Chim. Fr.*, 1949, 272.	565, 579
49BSF297	D. Billet; *Bull. Soc. Chim. Fr.*, 1949, 297.	652
49CRV385	J. E. Leffler; *Chem. Rev.*, 1949, **45**, 385.	52
49HCA647	R. Casanova and T. Reichstein, *Helv. Chem. Acta*, 1949, **32**, 647.	144
49JA1691	E. Shaw and J. McDowell; *J. Am. Chem. Soc.*, 1949, **71**, 1691.	330
49JA2215	E. T. Roe, J. T. Scanlan and D. Swern; *J. Am. Chem. Soc.*, 1949, **71**, 2215.	275, 276, 277, 278
49JA2240	J. N. LeConte and L. H. Chance; *J. Am. Chem. Soc.*, 1949, **71**, 2240.	1035
49JA2242	H. Adkins and Q. E. Thompson; *J. Am. Chem. Soc.*, 1949, **71**, 2242.	195
49JA3746	P. F. Wiley; *J. Am. Chem. Soc.*, 1949, **71**, 3746.	754
49JA3848	R. M. Roberts; *J. Am. Chem. Soc.*, 1949, **71**, 3848.	746
49JCS42	A. Lambert, J. D. Rose and B. C. L. Weedon; *J. Chem. Soc.*, 1949, 42.	820
49JCS1308	M. W. Partridge and H. A. Turner; *J. Chem. Soc.*, 1949, 1308.	751, 753
49JCS1555	N. Campbell and R. S. Gow; *J. Chem. Soc.*, 1949, 1555.	199
49JCS2755	C. Eaborn; *J. Chem. Soc.*, 1949, 2755.	1133

49JCS3029 F. Kurzer; *J. Chem. Soc.*, 1949, 3029. 344
49JCS3043 M. W. Partridge; *J. Chem. Soc.*, 1949, 3043. 757
49JCS3319 C. G. Raison; *J. Chem. Soc.*, 1949, 3319. 757
49JOC277 R. M. Roberts; *J. Org. Chem.*, 1949, **14**, 277. 746
49JOC605 R. C. Elderfield and H. A. Hageman; *J. Org. Chem.*, 1949, **14**, 605. 1126
49JOC962 G. Alliger, G. E. P. Smith Jr., E. L. Carr and H. P. Stevens; *J. Org. Chem.*, 1949, **14**, 962. 589, 595, 620
49JOC1099 W. S. Fones; *J. Org. Chem.*, 1949, **14**, 1099. 285
49MI 502-01 H. P. Lundgren and W. H. Ward; *Annu. Rev. Biochem.*, 1949, **18**, 115. 83
B-49MI 502-01 J. Lichtenberger and J. Paulus; in "Traite de Chimie Organique," ed. V. Grignard, C. Dupont and R. Lacquin, Masson et Cie, Paris, 1949, pp. 291–384. 105
49MI 519-01 F. E. Di Gangi and O. Gisvold; *J. Am. Pharm. Assoc.*, 1949, **38**, 154 (*Chem. Abstr.*, 1950, **44**, 2466e). 749
49OR(5)79 H. A. Bruson; *Org. React.*, 1949, **5**, 79. 46
49OR(5)229 E. Berliner; *Org. React.*, 1949, **5**, 229. 98
49OR(5)331 N. Rabjohn; *Org. React.*, 1949, **5**, 331. 48
49OR(5)413 M. S. Newman and B. G. Magerlein; *Org. React.*, 1949, **5**, 413. 46
49PCS2099 I. Heilbron; *Proc. Chem. Soc.*, 1949, 2099. 83
49TFS693 N. Sheppard; *Trans. Faraday Soc.*, 1949, **45**, 693. 506
49USP2475966 D. C. Hull; *US Pat.* 2 475 966 (1949) (*Chem. Abstr.*, 1949, **43**, 7954). 3

50ACS1080 N. A. Sörensen and K. Stavholt; *Acta Chem. Scand.*, 1950, **4**, 1080. 168
50BSF35 R. Jacquier; *Bull. Soc. Chim. Fr.*, 1950, 35. 652
50CB390 C. Schopf and R. Kuhl; *Chem. Ber.*, 1950, **83**, 390. 73
50CI(L)383 S. M. Partridge; *Chem. Ind. (London)*, 1950, 383. 84
50JA193 H. H. Anderson; *J. Am. Chem. Soc.*, 1950, **72**, 193. 1017
50JA1357 G. VanZyl and E. E. van Tamelen; *J. Am. Chem. Soc.*, 1950, **72**, 1357. 166
50JA2761 H. H. Anderson; *J. Am. Chem. Soc.*, 1950, **72**, 2761. 1017
50JA2998 H. E. Zaugg; *J. Am. Chem. Soc.*, 1950, **72**, 2998. 161
50JA3045 H. Gilman, B. Hofferth and H. W. Melvin; *J. Am. Chem. Soc.*, 1950, **72**, 3045. 595, 620, 1016
50JA3874 H. S. Angel and A. R. Day; *J. Am. Chem. Soc.*, 1950, **72**, 3874. 194
50JA4091 R. E. Lutz and D. F. Hinckley; *J. Am. Chem. Soc.*, 1950, **72**, 4091. 25, 34
50JA4270 W. A. Bonner; *J. Am. Chem. Soc.*, 1950, **72**, 4270. 243
50JA4284 C. R. Hauser and S. W. Kantor; *J. Am. Chem. Soc.*, 1950, **72**, 4284. 998
50JA4697 C. D. Hurd, M. E. Nilson and D. M. Wikholm; *J. Am. Chem. Soc.*, 1950, **72**, 4697. 820
50JA5012 E. D. Bergmann, D. Ginsburg and D. Lavie; *J. Am. Chem. Soc.*, 1950, **72**, 5012. 66
50JA5137 E. F. Degering and L. G. Boatright; *J. Am. Chem. Soc.*, 1950, **72**, 5137. 59, 89
50JA5163 M. S. Newman and P. F. Beal, III; *J. Am. Chem. Soc.*, 1950, **72**, 5163. 229
50JA5426 C. G. Swain, W. H. Stockmayer and J. T. Clarke; *J. Am. Chem. Soc.*, 1950, **72**, 5426. 207
50JCS325 R. E. Bowman; *J. Chem. Soc.*, 1950, 325. 76
50JCS747 M. Lamchen; *J. Chem. Soc.*, 1950, 747. 318
50JCS1582 E. Hoggarth and E. H. P. Young; *J. Chem. Soc.*, 1950, 1582. 1036
50JCS3077 C. Eaborn; *J. Chem. Soc.*, 1950, 3077. 1133
50JCS3213 A. C. Farthing; *J. Chem. Soc.*, 1950, 3213. 972
50JCS3389 W. Baker, W. D. Ollis and V. D. Poole; *J. Chem. Soc.*, 1950, 3389. 639
50JCS3425 J. Shorter; *J. Chem. Soc.*, 1950, 3425. 49
50JOC617 J. Cason and J. D. Wordie; *J. Org. Chem.*, 1950, **15**, 617. 108
50JOC800 R. H. Wiley and H. S. Morgan, Jr.; *J. Org. Chem.*, 1950, **15**, 800. 298
50JPP409 E. L. Smith; *J. Pharm. Pharmacol.*, 1950, **2**, 409. 92
50LA(566)210 H. Bestian; *Justus Liebigs Ann. Chem.*, 1950, **566**, 210. 1010
50LA(568)1 K. Ziegler and H. Wilms; *Justus Liebigs Ann. Chem.*, 1950, **568**, 1. 199
50LA(568)185 D. Jerchel and R. Kuhn; *Justus Liebigs Ann. Chem.*, 1950, **568**, 185. 855
50MI 502-01 R. K. Cannan and M. Levy; *Annu. Rev. Biochem.*, 1950, **19**, 125. 83
50MI 521-01 A. M. Rutenburg, R. Gofstein and A. M. Seligman; *Cancer Res.*, 1950, **10**, 113. 857

51AG7 T. Wieland; *Angew. Chem.*, 1951, **63**, 7. 84
51BSB54 M. F. Hebbelynck and R. H. Martin; *Bull. Soc. Chim. Belg.*, 1951, **60**, 54. 15
51C169 H. Jones; *Chimia*, 1951, **5**, 169. 24
51HCA2172 M Häusermann; *Helv. Chim. Acta*, 1951, **34**, 2172. 141
51JA25 R. W. Bost and O. L. Shealy; *J. Am. Chem. Soc.*, 1951, **73**, 25. 549
51JA436 M. Orchin and L. Reggel; *J. Am. Chem. Soc.*, 1951, **73**, 436. 25, 108
51JA613 C. D. Hurd and J. Moffat; *J. Am. Chem. Soc.*, 1951, **73**, 613. 1108
51JA4470 H. Gilman and H. L. Yale; *J. Am. Chem. Soc.*, 1951, **73**, 4470. 222
51JA4495 F. L. Greenwood and W. J. James; *J. Am. Chem. Soc.*, 1951, **73**, 4495. 1039
51JA4911 J. A. King and F. H. McMillan; *J. Am. Chem. Soc.*, 1951, **73**, 4911. 183
51JA4958 I. J. Pachter and M. C. Kloetzel; *J. Am. Chem. Soc.*, 1951, **73**, 4958. 205
51JA5439 H. H. Anderson; *J. Am. Chem. Soc.*, 1951, **73**, 5439. 1136
51JA5440 H. H. Anderson; *J. Am. Chem. Soc.*, 1951, **73**, 5440. 1136
51JA5553 J. R. Vaughan, Jr. and R. L. Osato; *J. Am. Chem. Soc.*, 1951, **73**, 5553. 193
51JCS485 P. T. Charlton, G. K. Maliphant, P. Oxley and D. A. Peak; *J. Chem. Soc.*, 1951, 485. 749, 750
51JCS1208 H. C. Brimelow, R. L. Jones and T. P. Metcalfe; *J. Chem. Soc.*, 1951, 1208. 114, 115
51JOC131 E. C. Kornfeld; *J. Org. Chem.*, 1951, **16**, 131. 566, 579, 584, 586, 612, 618
51LA(574)85 D. Jerchel and H. Fischer; *Justus Liebigs Ann. Chem.*, 1951, **574**, 85. 772, 854, 855, 859

B-51MI 502-01 A. W. Johnson, C. E. Dalgliesh and J. Walker; in "Chemistry of Carbon Compounds," ed. E. H. Rodd, Elsevier, Amsterdam, 1951, vol. 1A, pp. 537–647. 24

51MI 502-01 E. Katchalski; *Adv. Protein Chem.*, 1951, **6**, 123. 84

B-51MI 502-02 D. M. Greenberg; "Amino Acids and Proteins," Thomas, Springfield, Massachusetts, 1951. 83

51MI 502-02 K. Bailey and F. Sanger; *Annu. Rev. Biochem.*, 1951, **20**, 103. 83

51OR(6)1 W. S. Johnson and G. H. Daub; *Org. React.*, 1951, **6**, 1. 166

51RTC638 H. Backer and W. L. Wanmaker; *Recl. Trav. Chim. Pays-Bas*, 1951, **70**, 638. 749

52ACS327 A. Kjaer; *Acta Chem. Scand.*, 1952, **6**, 327. 584, 586, 587

52AP(285)375 H. Wojahn; *Arch. Pharm.* (*Weinheim, Ger.*), 1952, **285**, 375. 735

52BJ163 G. D. Hunter and G. Popjak; *Biochem. J.*, 1952, **50**, 163. 99

52BSF821 M. Jutisz; *Bull. Soc. Chim. Fr.*, 1952, 821. 83

52CB239 H. Albers, W. Künzel and W. Scheeber; *Chem. Ber.*, 1952, **85**, 239. 397

52CB344 H. Brintzinger and R. Titzmann; *Chem. Ber.*, 1952, **85**, 344. 821

52CB470 W. Ried and F. Müller; *Chem. Ber.*, 1952, **85**, 470. 855

52DOK(83)243 A. I. Titov and V. V. Smirnov; *Dokl. Akad. Nauk SSSR*, 1952, **83**, 243 (*Chem. Abstr.*, 1953, **47**, 4298). 838

52HCA1429 W. Jenny; *Helv. Chim. Acta*, 1952, **35**, 1429. 210

52JA394 L. Field; *J. Am. Chem. Soc.*, 1952, **74**, 394. 995

52JA710 J. A. Cuculo and L. A. Bigelow; *J. Am. Chem. Soc.*, 1952, **74**, 710. 807

52JA1010 J. D. Park, R. D. Englert and J. S. Meek; *J. Am. Chem. Soc.*, 1952, **74**, 1010. 273, 285

52JA1321 I. J. Pachter and M. C. Kloetzel; *J. Am. Chem. Soc.*, 1952, **74**, 1321. 285

52JA1323 T. L. Gresham, J. E. Jansen, F. W. Shaver, M. R. Frederick, F. T. Fiedorek, R. A. Bankert, J. T. Gregory and W. L. Beears; *J. Am. Chem. Soc.*, 1952, **74**, 1323. 1106

52JA1730 F. F. Blicke and H. Raffelson; *J. Am. Chem. Soc.*, 1952, **74**, 1730. 76

52JA2189 J. D. Park, H. J. Gerjovich, W. R. Lycan and J. R. Lacher; *J. Am. Chem. Soc.*, 1952, **74**, 2189. 318

52JA4083 S. J. Cristol and R. F. Helmreich; *J. Am. Chem. Soc.*, 1952, **74**, 4083. 166

52JA4496 G. Natta, P. Pino and R. Ercoli; *J. Am. Chem. Soc.*, 1952, **74**, 4496. 293

52JA4700 T. I. Bieber; *J. Am. Chem. Soc.*, 1952, **74**, 4700. 995

52JA4876 H. C. Stevens and O. Grummitt; *J. Am. Chem. Soc.*, 1952, **74**, 4876. 815

52JA5247 J. J. McBride, Jr. and H. C. Beachell; *J. Am. Chem. Soc.*, 1952, **74**, 5247. 1133

52JCS742 A. E. S. Fairfull, J. L. Lowe and D. A. Peak; *J. Chem. Soc.*, 1952, 742. 579

52JCS1363 S. F. Birch, V. E. Gripp, D. T. McAllan and W. S. Nathan; *J. Chem. Soc.*, 1952, 1363. 199

52JCS1607 C. W. Davies and G. G. Thomas; *J. Chem. Soc.*, 1952, 1607. 30

52JCS4521 F. Bergmann and A. Kalmus; *J. Chem. Soc.*, 1952, 4521. 98

52LA(576)61 H. Böhme and M. Clement; *Justus Liebigs Ann. Chem.*, 1952, **576**, 61. 244, 247, 254

52LA(577)153 S. Goldschmidt, W. Leicher and H. Haas; *Justus Liebigs Ann. Chem.*, 1952, **577**, 153. 208

52LA(577)234 F. Micheel and W. Flitsch; *Justus Liebigs Ann. Chem.*, 1952, **577**, 234. 753

52MI 502-01 F. Sanger; *Adv. Protein Chem.*, 1952, **7**, 1. 85

52MI 502-02 G. B. B. M. Sutherland; *Adv. Protein Chem.*, 1952, **7**, 291. 84

52MI 502-03 G. H. Beaven and E. R. Holiday; *Adv. Protein Chem.*, 1952, **7**, 319. 84

52MI 502-04 H. B. Bull; *Annu. Rev. Biochem.*, 1952, **21**, 179. 83

B-52MI 502-01 C. E. Dalgliesh, A. N. Johnson and C. Buchannan; in "Chemistry of Carbon Compounds," ed. E. H. Rodd, Elsevier, Amsterdam, 1952, vol. 1B, pp. 950–1011. 24

B-52MI 502-02 C. E. Dalgliesh, A. W. Johnson and L. J. Haynes; in "Chemistry of Carbon Compounds," ed. E. H. Rodd, Elsevier, Amsterdam, 1952, vol. 1B, pp. 779–849. 84

B-52MI 502-03 I. D. Morton and E. Hoggarth; in "Chemistry of Carbon Compounds," ed. E. H. Rodd, Elsevier, Amsterdam, 1952, vol. 1B, pp. 886–949. 62

B-52MI 502-04 C. E. Dalgliesh, A. W. Johnson, A. G. Long and C. J. Tyler; in "Chemistry of Carbon Compounds," ed. E. H. Rodd, Elsevier, Amsterdam, 1952, vol. 1B, pp. 1048–1154. 62

B-52MI 502-05 L. J. Haynes; in "Chemistry of Carbon Compounds," ed. E. H. Rodd, Elsevier, Amsterdam, 1952, vol. 1B, pp. 1155–1196. 62

B-52MI 502-06 L. K. N. Jones; in "Chemistry of Carbon Compounds," ed. E. H. Rodd, Elsevier, Amsterdam, 1952, vol. 1B, pp. 1197–1223. 62

B-52MI 502-07 L. K. N. Jones; in "Chemistry of Carbon Compounds," ed. E. H. Rodd, Elsevier, Amsterdam, 1952, vol. 1B, pp. 1224–1286. 62

B-52MI 502-08 C. E. Dalgliesh, A. W. Johnson and C. Buchannan; in "Chemistry of Carbon Compounds," ed. E. H. Rodd, Elsevier, Amsterdam, 1952, vol. 1B, pp. 850–885. 62

B-52MI 502-09 K. Bailey; in "Chemistry of Carbon Compounds," ed. E. H. Rodd, Elsevier, Amsterdam, 1952, vol. 1B, pp. 1327–1357. 85

52QR380 B. C. L. Weeden; *Q. Rev., Chem. Soc.*, 1952, **6**, 380. 625

52ZN(B)634 M. Goehring and W. Koch; *Z. Naturforsch., Teil B*, 1952, **7**, 634. 338

53AG442 R. Kuhn and H. Kainer; *Angew. Chem.*, 1953, **65**, 442. 857

53BSF771 L. Szabo; *Bull. Soc. Chim. Fr.*, 1953, 771. 1108

53CB400 J. Goerdeler and K. Wember; *Chem. Ber.*, 1953, **86**, 400. 762, 815

53CB697 G. Zemplen, L. Mester and A. Messmer; *Chem. Ber.*, 1953, **86**, 697. 641

53CB890 J. Goerdeler and D. Loevenich; *Chem. Ber.*, 1953, **86**, 890. 763

53CRV237 N. O. V. Sonntag; *Chem. Rev.*, 1953, **52**, 237. 278, 304

53JA285 C. D. Hurd and J. Patterson; *J. Am. Chem. Soc.*, 1953, **75**, 285. 214

53JA1761 G. L. Webster and J. S. Rodia; *J. Am. Chem. Soc.*, 1953, **75**, 1761. 751, 757

53JA1933	H. L. Yale, K. Losee, J. Martins, M. Holsing, F. M. Perry and J. Bernstein; *J. Am. Chem. Soc.*, 1953, **75**, 1933.	354
53JA3493	C. Djerassi and C. T. Lenk; *J. Am. Chem. Soc.*, 1953, **75**, 3493.	320
53JA3961	K. B. Wiberg; *J. Am. Chem. Soc.*, 1953, **75**, 3961.	709
53JA4655	R. E. Koeppe and J. L. Wood; *J. Am. Chem. Soc.*, 1953, **75**, 4655.	1111
53JA5309	F. L. Scott, D. A. O'Sullivan and J. Reilly; *J. Am. Chem. Soc.*, 1953, **75**, 5309.	857
53JA5905	D. E. Pearson, K. N. Carter and C. M. Greer; *J. Am. Chem. Soc.*, 1953, **75**, 5905.	303
53JCS107	K. H. Pausacker; *J. Chem. Soc.*, 1953, 107.	204
53JCS255	F. C. Cooper and M. W. Partridge; *J. Chem. Soc.*, 1953, 255.	756
53JCS959	A. Burawoy and C. Turner; *J. Chem. Soc.*, 1953, 959.	1108
53JCS1409	V. Gold and E. G. Jefferson; *J. Chem. Soc.*, 1953, 1409.	26, 37
53JCS2117	W. Gerrard and A. M. Thrush; *J. Chem. Soc.*, 1953, 2117.	184
53JCS3352	D. Brown and R. F. Hudson; *J. Chem. Soc.*, 1953, **883**, 3352.	130
53JOC64	E. D. Bergmann, H. Bendas and U. d'Avilla; *J. Org. Chem.*, 1953, **18**, 64.	833
53JOC432	H. L. Cohen and G. F. Wright; *J. Org. Chem.*, 1953, **18**, 432.	293
53JOC748	T. L. Cairns, A. W. Larchar and B. C. McKusick; *J. Org. Chem.*, 1953, **18**, 748.	579, 580
53LA(579)28	S. Hünig and O. Boes; *Justus Liebigs Ann. Chem.*, 1953, **579**, 28.	857, 858
53LA(581)23	W. Ried and R. Hoffschmidt; *Justus Liebigs Ann. Chem.*, 1953, **581**, 23.	858
53LA(581)44	W. Ried and H. Hillebrad; *Justus Liebigs Ann. Chem.*, 1953, **581**, 44.	857
53MI 502-01	C. L. Mehltretter; *Adv. Carbohydr. Chem.*, 1953, **8**, 231.	62
B-53MI 502-01	G. E. W. Wolstenholme and M. P. Cameron (eds.); "CIBA Foundation Symposia," Little, Brown and Co., Boston, 1953.	85
53MI 502-02	C. Fromagest and M. Jutisz; *Annu. Rev. Biochem.*, 1953, **22**, 629.	83
53MI 519-01	M. W. Partridge and H. A. Turner; *J. Pharm. Pharmacol.*, 1953, **5**, 111 (*Chem. Abstr.*, 1953, **47**, 11 156c).	748
53QR175	F. D. Gunstone; *Q. Rev. Chem. Soc.*, 1953, **7**, 175.	652
53QR255	G. Allen and E. F. Caldin; *Q. Rev. Chem. Soc.*, 1953, **7**, 255.	24
54ACS295	A. Kjaer and J. Conti; *Acta Chem. Scand.*, 1954, **8**, 295.	1029
54ACS1851	E. Söderbäck; *Acta Chem. Scand.*, 1954, **8**, 1851.	1109
54AG507	T. Wieland; *Angew. Chem.*, 1954, **66**, 507.	84
54BCJ624	T. Sato and M. Ohta; *Bull. Chem. Soc. Jpn.*, 1954, **27**, 624.	846
54CB825	H.-B. Konig, W. Siefken and H. A. Offe; *Chem. Ber.*, 1954, **87**, 825.	636
54CB1294	D. Klamann and E. Schaffer; *Chem. Ber.*, 1954, **87**, 1294.	300
54CI(L)785	J. N. Baxter, J. Cymerman-Craig, M. Moyle and R. A. White; *Chem. Ind.* (*London*), 1954, 785.	1035
54CI(L)1154	C. A. Bunton, T. A. Lewis and D. R. Llewellyn; *Chem. Ind.* (*London*), 1954, 1154.	652
54CLY1210	J. O. Jilek and V. Michajlyszyn; *Chem. Listy*, 1954, **48**, 1210.	749
54JA920	A. J. Hill and J. V. Johnston; *J. Am. Chem. Soc.*, 1954, **76**, 920.	757
54JA1372	G. M. Coppinger; *J. Am. Chem. Soc.*, 1954, **76**, 1372.	299
54JA1613	L. H. Sommer, D. L. Bailey, G. M. Goldberg, C. E. Buck, T. S. Bye, F. J. Evans and F. C. Whitmore; *J. Am. Chem. Soc.*, 1954, **76**, 1613.	421
54JA2411	R. M. Roberts and R. H. DeWolfe; *J. Am. Chem. Soc.*, 1954, **76**, 2411.	743
54JA2881	W. Sakami and G. Toennies; *J. Am. Chem. Soc.*, 1954, **76**, 2881.	140
54JA3039	H. R. Snyder and C. T. Elston; *J. Am. Chem. Soc.*, 1954, **76**, 3039.	25, 275
54JA3978	H. G. Mandel and A. J. Hill; *J. Am. Chem. Soc.*, 1954, **76**, 3978.	744, 745
54JA5803	A. W. Schrecker and P. B. Maury; *J. Am. Chem. Soc.*, 1954, **76**, 5803.	194
54JA6208	M. E. Speeter and W. C. Anthony; *J. Am. Chem. Soc.*, 1954, **76**, 6208.	15
54JCS1188	H. L. Bassett and C. R. Thomas; *J. Chem. Soc.*, 1954, 1188.	277
54JCS1860	G. Eglinton, E. R. H. Jones, B. L. Shaw and M. C. Whiting; *J. Chem. Soc.*, 1954, 1860.	189
54JCS2006	E. J. Bourne, M. Stacey, J. C. Tatlow and R. Worrall; *J. Chem. Soc.*, 1954, 2006.	193, 199
54JCS2968	H. J. Cottrell, D. L. Pain and R. Slack; *J. Chem. Soc.*, 1954, 2968.	858
54JCS3319	M. R. Atkinson and J. B. Polya; *J. Chem. Soc.*, 1954, 3319.	849
54JCS3722	R. P. Linstead and M. Whalley; *J. Chem. Soc.*, 1954, 3722.	199
54JCS4018	C. L. Arcus and B. S. Prydal; *J. Chem. Soc.*, 1954, 4018.	994
54JCS4251	K. Clarke; *J. Chem. Soc.*, 1954, 4251.	763, 833
54JOC1300	H. H. Anderson and J. A. Vasta; *J. Org. Chem.*, 1954, **19**, 1300.	1017
54JPJ821	T. Sato and M. Ohta; *J. Pharm. Soc. Jpn.*, 1954, **74**, 821 (*Chem. Abstr.*, 1955, **49**, 9537).	639, 846
54LA(585)230	E. Schmidt, F. Zaller, F. Moosmuller and E. Kammerl; *Justus Liebigs Ann. Chem.*, 1954, **585**, 230.	1025
54LA(587)93	L. Horner and W. Naumann; *Justus Liebigs Ann. Chem.*, 1954, **587**, 93.	366
54LA(590)123	W. Reid, H. Hillenbrand and G. Oertel; *Justus Liebigs Ann. Chem.*, 1954, **590**, 123.	1033
54MI 502-01	J. E. Courtois; *J. Chem. Soc. Jpn.*, 1954, **8**, 513.	28
54MI 502-02	G. O. Aspinall; *Adv. Carbohydr. Chem.*, 1954, **9**, 131.	62
54MI 502-03	W. L. Hughes and F. M. Sinex; *Annu. Rev. Biochem.*, 1954, **23**, 177.	83
54MI 502-04	J. P. Greenstein; *Adv. Protein Chem.*, 1954, **9**, 121.	84
B-54MI 502-01	W. J. Hickinbottom; in "Chemistry of Carbon Compounds," ed. E. H. Rodd, Elsevier, Amsterdam, 1954, vol. 3A, pp. 541–601.	105
54OR(8)28	D. A. Shirley; *Org. React.*, 1954, **8**, 28.	98
54OR(8)258	H. Gilman and J. W. Morton, Jr.; *Org. React.*, 1954, **8**, 258.	652
54USP2786865	J. W. Copenhaver; *US Pat.* 2 786 865 (1954) (*Chem. Abstr.*, 1957, **51**, 13 920).	829
54ZOB1033	A. V. Kirsanov; *Zh. Obshch. Khim.*, 1954, **24**, 1033 (*Chem. Abstr.*, 1955, **49**, 8787a).	988

55AG381 A. I. Virtanen; *Angew. Chem.*, 1955, **67**, 381. 83

55AG769 H. Stetter; *Angew. Chem.*, 1955, **67**, 769. 48

55CB130 G. Hesse and G. Krehbiel; *Chem. Ber.*, 1955, **88**, 130. 823, 824

55CB301 F. Weygand and R. Mitgau; *Chem. Ber.*, 1955, **88**, 301. 304

55CB1284 D. Jerchel and W. Elder; *Chem. Ber.*, 1955, **88**, 1284. 772

55CI(L)1087 H. G. Khorana; *Chem. Ind.* (*London*), 1955, 1087. 279

55CI(L)1183 F. R. Atherthon, A. L. Morrison, R. J. W. Cremlyn, G. W. Kenner, A. Todd and R. F. Webb; *Chem. Ind.* (*London*), 1955, 1183. 712

55CRV181 D. C. Schroeder; *Chem. Rev.*, 1955, **55**, 181. 1022

55CRV283 M. Ballester; *Chem. Rev.*, 1955, **55**, 283. 46

55CRV355 A. W. Nineham; *Chem. Rev.*, 1955, **55**, 355. 855, 857

55CRV659 R. E. Buckles, G. V. Mock and L. Locatell, Jr.; *Chem. Rev.*, 1955, **55**, 659. 98

55FOR349 J. Roche and R. Michel; *Fortschr. Chem. Org. Naturst.*, 1955, **12**, 349. 84

55HCA312 E. Hardegger and H. Ott; *Helv. Chim. Acta.*, 1955, **38**, 312. 262

55HCA1560 P. Schmidt and J. Druey; *Helv. Chim. Acta*, 1955, **38**, 1560. 769, 852

55HOU(9)745 A. Schöberl and A. Wagner; *Methoden Org. Chem.* (*Houben-Weyl*), 1955, **9**, 745. 232

55HOU(9)951 H. Rheinboldt; *Methoden Org. Chem.* (*Houben-Weyl*), 1955, **9**, 951. 1051

55JA469 C. F. Huebner, R. Lucas, H. B. MacPhillamy and H. A. Troxell; *J. Am. Chem. Soc.*, 1955, **77**, 469. 298

55JA1067 J. C. Sheehan and G. P. Hess; *J. Am. Chem. Soc.*, 1955, **77**, 1067. 279

55JA1831 M. G. Ettlinger and J. E. Hodgkins; *J. Am. Chem. Soc.*, 1955, **77**, 1831. 1024

55JA2519 K. B. Wiberg; *J. Am. Chem. Soc.*, 1955, **77**, 2519. 275

55JA3149 J. E. Noll; *J. Am. Chem. Soc.*, 1955, **77**, 3149. 1024

55JA4037 W. E. Parker, C. Ricciuti, C. L. Ogg and D. Swern; *J. Am. Chem. Soc.*, 1955, **77**, 4037. 205

55JA4062 R. I. Meltzer, A. D. Lewis and J. A. King; *J. Am. Chem. Soc.*, 1955, **77**, 4062. 566, 579, 591, 618, 621

55JA4079 B. D. Astill and V. Boekelheide; *J. Am. Chem. Soc.*, 1955, **77**, 4079. 1126

55JA4297 L. Mester and A. Major; *J. Am. Chem. Soc.*, 1955, **77**, 4297. 857

55JA4301 L. Mester; *J. Am. Chem. Soc.*, 1955, **77**, 4301. 857

55JA4571 S. M. McElvain and R. E. Starn; *J. Am. Chem. Soc.*, 1955, **77**, 4571. 706, 845

55JA5997 C. S. Marvel, P. de Radzitzky and J. J. Brader; *J. Am. Chem. Soc.*, 1955, **77**, 5997. 551

55JA6011 E. H. White; *J. Am. Chem. Soc.*, 1955, **77**, 6011. 36, 108

55JA6214 J. H. Brewster and C. J. Ciotti, Jr.; *J. Am. Chem. Soc.*, 1955, **77**, 6214. 123, 126, 302

55JA6370 D. F. Detar and L. A. Carpino; *J. Am. Chem. Soc.*, 1955, **77**, 6370. 206

55JA6559 C. Grundmann and A. Kreutzberger; *J. Am. Chem. Soc.*, 1955, **77**, 6559. 747

55JA6682 C. T. Peng and T. C. Daniels; *J. Am. Chem. Soc.*, 1955, **78**, 6682. 25, 108

55JOC1448 C. R. Hauser and D. S. Hoffenberg; *J. Org. Chem.*, 1955, **20**, 1448. 275

55LA(591)117 L. Horner and A. Gross; *Justus Liebigs Ann. Chem.*, 1955, **591**, 117. 90

55LA(591)190 R. Huisgen and H. J. Koch; *Justus Liebigs Ann. Chem.*, 1955, **591**, 190. 844

55LA(591)218 R. Huisgen and J. Koch; *Justus Liebigs Ann. Chem.*, 1955, **591**, 218. 677

55LA(591)226 R. Huisgen and J. Koch; *Justus Liebigs Ann. Chem.*, 1955, **591**, 226. 677

55LA(594)185 A. Dornow and W. Sassenberg; *Justus Liebigs Ann. Chem.*, 1955, **594**, 185. 822

55LA(594)233 E. Schmidt, E. Kammerl, D. Ross and F. Zaller; *Justus Liebigs Ann. Chem.*, 1955, **594**, 233. 1025

55LA(597)95 W. Theilacker and W. Schmidt; *Justus Liebigs Ann. Chem.*, 1955, **597**, 95. 199

B-55MI 502-01 D. Swern; in "Progress in Chemistry of Fats and Other Lipids," eds. R. T. Halman, W. O. Lindberg and T. Malkin, Pergamon, Oxford, 1955, vol. 3, p. 213. 28

55MI 502-01 L. W. Mapson; *Vitamins and Hormones*, 1955, **13**, 71. 98

B-55MI 502-02 J. Q. Griffith, C. F. Krewson and J. Naghski; "Amino Acids and Proteins, Chemistry, Pharmacology, Clinical Applications," Mack, Easton, Pennsylvania, 1955. 83

55MI 502-02 J. R. Bissier and J. F. Miquel; *Ann. Pharm. Francaises*, 1955, **13**, 143. 81

B-55MI 502-03 A. D. Jenkins and D. F. Elliott (eds.); "Peptide Chemistry, Symposium, London, 1955," The Chemical Society, Special Publications, London. 85

55MI 502-03 A. G. Ogston; *Annu. Rev. Biochem.*, 1955, **24**, 181. 83

55MI 515-01 Ferrosan; *Danish Pat.*, 79776 (1955) (*Chem. Abstr.*, 1955, **50**, 8730g). 639, 642

55MI 519-01 P. G. Bay and O. Gisvold; *J. Am. Pharm. Assoc.*, 1955, **44**, 585. 748

55OSC(3)28 C. F. H. Allen, C. J. Kibler, D. M. McLachlin and C. V. Wilson; *Org. Synth., Coll. Vol.*, 1955, **3**, 28. 184

55OSC(3)34 J. H. Ford; *Org. Synth., Coll. Vol.*, 1955, **3**, 34. 85

55OSC(3)46 C. E. Rehberg; *Org. Synth., Coll. Vol.*, 1955, **3**, 46. 122

55OSC(3)66 R. E. Steiger; *Org. Synth., Coll. Vol.*, 1955, **3**, 66. 85, 275

55OSC(3)88 R. E. Steiger; *Org. Synth., Coll. Vol.*, 1955, **3**, 88. 85, 275

55OSC(3)108 C. J. Kibler and A. Weissberger; *Org. Synth., Coll. Vol.*, 1955, **3**, 108. 652

55OSC(3)114 T. S. Oakwood and C. A. Weisgerber; *Org. Synth., Coll. Vol.*, 1955, **3**, 114. 64

55OSC(3)116 R. L. Frank and J. R. Blegen; *Org. Synth., Coll. Vol.*, 1955, **3**, 116. 251

55OSC(3)151 L. I. Smith and O. H. Emerson; *Org. Synth., Coll. Vol.*, 1955, **3**, 151. 305

55OSC(3)165 E. C. Horning, M. G. Horning and D. A. Dimmig; *Org. Synth., Coll. Vol.*, 1955, **3**, 165. 185

55OSC(3)169 J. Cason; *Org. Synth., Coll. Vol.*, 1955, **3**, 169. 37

55OSC(3)213 G. B. Heisig and F. H. Stodola; *Org. Synth., Coll. Vol.*, 1955, **3**, 213. 33

55OSC(3)221 C. M. McCloskey and G. H. Coleman; *Org. Synth., Coll. Vol.*, 1955, **3**, 221. 25

55OSC(3)267 D. S. Tarbell, J. W. Wilson and P. E. Fanta; *Org. Synth., Coll. Vol.*, 1955, **3**, 267. 106

55OSC(3)270 O. Grummitt, A. Buck and R. Egan; *Org. Synth., Coll. Vol.*, 1955, **3**, 270. 38

55OSC(3)272 C. C. Price and R. M. Roberts; *Org. Synth., Coll. Vol.*, 1955, **3**, 272. 106

55OSC(3)302 L. I. Smith, W. W. Prichard and L. J. Spillane; *Org. Synth., Coll. Vol.*, 1955, **3**, 302. 103

55OSC(3)326 J. L. Riebsomer and J. Irvine; *Org. Synth., Coll. Vol.*, 1955, **3**, 326. 62

55OSC(3)334 W. D. Langley; *Org. Synth., Coll. Vol.*, 1955, **3**, 334. 115

55OSC(3)420 G. Rieveschl, Jr. and F. E. Ray; *Org. Synth., Coll. Vol.*, 1955, **3**, 420. 114

55OSC(3)490 R. E. Kent and S. M. McElvain; *Org. Synth., Coll. Vol.*, 1955, **3**, 490. 275

55OSC(3)495 C. S. Marvel; *Org. Synth., Coll. Vol.*, 1955, **3**, 495. 33

55OSC(3)510 L. Friedman and E. Kosower; *Org. Synth., Coll. Vol.*, 1955, **3**, 510. 62

55OSC(3)516 J. Kleinberg and L. F. Audrieth; *Org. Synth., Coll. Vol.*, 1955, **3**, 516. 275

55OSC(3)538 J. G. Aston, J. D. Newkirk, D. M. Jenkins and J. Dorsky; *Org. Synth., Coll. Vol.*, 1955, **3**, 538. 66

55OSC(3)547 M. T. Leffler and A. E. Calkins; *Org. Synth., Coll. Vol.*, 1955, **3**, 547. 10

55OSC(3)553 D. M. Bowen; *Org. Synth., Coll. Vol.*, 1955, **3**, 553. 112

55OSC(3)555 R. P. Barnes; *Org. Synth., Coll. Vol.*, 1955, **3**, 555. 8, 112

55OSC(3)557 R. C. Fuson and N. Rabjohn; *Org. Synth., Coll. Vol.*, 1955, **3**, 557. 34

55OSC(3)560 R. H. Wiley and W. E. Waddey; *Org. Synth., Coll. Vol.*, 1955, **3**, 560. 25

55OSC(3)590 L. F. Fieser and J. E. Jones; *Org. Synth., Coll. Vol.*, 1955, **3**, 590. 290

55OSC(3)623 P. C. Guha and D. K. Sankaran; *Org. Synth., Coll. Vol.*, 1955, **3**, 623. 103

55OSC(3)637 K. Meyer and H. S. Bloch; *Org. Synth., Coll. Vol.*, 1955, **3**, 637. 106

55OSC(3)649 C. C. Price and E. Krebs; *Org. Synth., Coll. Vol.*, 1955, **3**, 649. 652

55OSC(3)650 W. W. Hartman and E. J. Rahrs; *Org. Synth., Coll. Vol.*, 1955, **3**, 650. 140

55OSC(3)661 P. E. Fanta and D. S. Tarbell; *Org. Synth., Coll. Vol.*, 1955, **3**, 661. 652

55OSC(3)705 C. S. Marvel; *Org. Synth., Coll. Vol.*, 1955, **3**, 705. 33, 89

55OSC(3)740 A. W. Singer and S. M. McElvain; *Org. Synth., Coll. Vol.*, 1955, **3**, 740. 115

55OSC(3)745 I. A. Pearl; *Org. Synth., Coll. Vol.*, 1955, **3**, 745. 114

55OSC(3)765 C. F. H. Allen and J. Van Allan; *Org. Synth., Coll. Vol.*, 1955, **3**, 765. 302

55OSC(3)791 C. F. Koelsch; *Org. Synth., Coll. Vol.*, 1955, **3**, 791. 114

55OSC(3)820 H. E. Zaugg and R. T. Rapala; *Org. Synth., Coll. Vol.*, 1955, **3**, 820. 115

55OSC(3)822 W. F. Tuley and C. S. Marvel; *Org. Synth., Coll. Vol.*, 1955, **3**, 822. 115

55RTC257 W. J. van der Burg; *Rec. Trav. Chim. Pays-Bas*, 1955, **74**, 257. 769, 852

55RTC1262 G. J. M. van der Kerk, C. W. Pluygers and G. de Vries; *Recl. Trav. Chim. Pays-Bas*, 1955, **74**, 1262. 1026

55TFS1105 G. Hargreaves and L. H. Sutcliffe; *Trans. Faraday Soc.*, 1955, **51**, 1105. 49

56AG313 H. Musso; *Angew. Chem.*, 1956, **68**, 313. 83

56AG349 E. Klenk; *Angew. Chem.*, 1956, **68**, 349. 62

56BRP758149 Bayer AG; *Br. Pat.*, 758 149 (1956) (*Chem. Abstr.*, 1956, **51**, 9712f). 639

56BSB291 R. Hart; *Bull. Soc. Chim. Belg.*, 1956, **65**, 291. 989

56CB133 F.-P. van de Kamp and F. Micheel; *Chem. Ber.*, 1956, **89**, 133. 1040

56CB563 D. Jerchel and H. Fischer; *Chem. Ber.*, 1956, **89**, 563. 857

56CB862 G. Olah, S. Kuhn and S. Beke; *Chem. Ber.*, 1956, **89**, 862. 4

56CB1081 H. Hauptmann and A. C. de Mello Périssé; *Chem. Ber.*, 1956, **89**, 1081. 856

56CB1110 R. Mecke and H. Spiesecke; *Chem. Ber.*, 1956, **89**, 1110. 506

56CI(L)548 I. Bell, E. R. H. Jones and M. C. Whiting; *Chem. Ind.* (*London*), 1956, 548. 102

56CRV27 P. H. Leake; *Chem. Rev.*, 1956, **56**, 27. 98

56HCA1253 K. Ronco, B. Prijs and H. Erlenmeyer; *Helv. Chim. Acta*, 1956, **39**, 1253. 365

56JA450 M. S. Newman and R. M. Wise; *J. Am. Chem. Soc.*, 1956, **78**, 450. 25, 35

56JA842 A. C. Haven, Jr.; *J. Am. Chem. Soc.*, 1956, **78**, 842. 1017, 1018

56JA1207 A. J. Havlik and N. Kharasch; *J. Am. Chem. Soc.*, 1956, **78**, 1207. 208

56JA1367 J. C. Sheehan, M. Goodman and G. P. Hess; *J. Am. Chem. Soc.*, 1956, **78**, 1367. 140, 279

56JA1461 H. C. Godt, Jr. and J. F. Quinn; *J. Am. Chem. Soc.*, 1956, **78**, 1461. 765

56JA1689 H. Gilman, C. G. Brannen and R. K. Ingham; *J. Am. Chem. Soc.*, 1956, **78**, 1689. 114, 115

56JA1692 H. H. Anderson; *J. Am. Chem. Soc.*, 1956, **78**, 1692. 1136

56JA2255 F. L. M. Pattison, J. B. Stothers and R. G. Woolford; *J. Am. Chem. Soc.*, 1956, **78**, 2255. 55, 57, 59, 62

56JA2489 M. L. Wolfrom and J. M. Bobbitt; *J. Am. Chem. Soc.*, 1956, **78**, 2489. 49

56JA3819 J. Collette, D. McGreer, R. Crawford, F. Chubb and R. B. Sandin; *J. Am. Chem. Soc.*, 1956, **78**, 3819. 205

56JA3825 R. Adams and W. Reifschneider; *J. Am. Chem. Soc.*, 1956, **78**, 3825. 767

56JA4135 W. G. Dauben and J. B. Rogan; *J. Am. Chem. Soc.*, 1956, **78**, 4135. 599

56JA4965 L. S. Luskin, G. E. Gantert and W. E. Craig; *J. Am. Chem. Soc.*, 1956, **78**, 4965. 1043

56JA5416 S. Sarel and M. S. Newman; *J. Am. Chem. Soc.*, 1956, **78**, 5416. 34

56JA6032 W. L. Reilly and H. C. Brown; *J. Am. Chem. Soc.*, 1956, **78**, 6032. 815

56JCS96 A. Burawoy, A. Chaudhuri and W. I. Hyslop; *J. Chem. Soc.*, 1956, 96. 1111

56JCS659 J. N. Baxter, J. Cymerman-Craig, M. Moyle and R. A. White; *J. Chem. Soc.*, 1956, 659. 1035

56JCS1231 G. W. Kenner, A. R. Todd and R. F. Webb; *J. Chem. Soc.*, 1956, 1231. 712

56JCS2253 D. D. Libman and R. Slack; *J. Chem. Soc.*, 1956, 2253. 769, 847, 848

56JCS3428 D. A. Barr and R. N. Haszeldine; *J. Chem. Soc.*, 1956, 3428. 988

56JGU2239 A. V. Kirsanov and G. I. Derkach; *J. Gen. Chem. USSR* (*Engl. Transl.*), 1956, **26**, 2239. 683

56JGU2935 A. V. Kirsanov and G. I. Derkach; *J. Gen. Chem. USSR* (*Engl. Transl.*), 1956, **26**, 2935. 683, 707

56JOC204 M. G. Ettlinger and J. E. Hodgkins; *J. Org. Chem.*, 1956, **21**, 204. 1024

56JOC439 J. C. Sheehan and J. J. Hlavka; *J. Org. Chem.*, 1956, **21**, 439. 279

56JOC520 G. Rosenkranz, O. Mancera, F. Sondheimer and C. Djerassi; *J. Org. Chem.*, 1956, **21**, 520. 288

56JOC772 P. L. de Benneville, J. S. Strong and V. T. Elkind; *J. Org. Chem.*, 1956, **21**, 772. 580

56JOC1182 L. Bauer; *J. Org. Chem.*, 1956, **21**, 1182. 90

56JOC1430 L. S. Luskin, A. J. McFaull and G. E. Gantert; *J. Org. Chem.*, 1956, **21**, 1430. 1044

56JOC1473	H. Shechter and H. C. Barker; *J. Org. Chem.*, 1956, **21**, 1473.	201
56MI 502-01	H. Fraenkel-Conrat; *Annu. Rev. Biochem.*, 1956, **25**, 291.	83
B-56MI 502-01	W. J. Hickinbottom and R. F. Garwood; in "Chemistry of Carbon Compounds," ed. E. H. Rodd, Elsevier, Amsterdam, 1956, vol. 3B, pp. 730–815.	105
B-56MI 502-02	W. J. Hickinbottom and R. F. Garwood; in "Chemistry of Carbon Compounds," ed. E. H. Rodd, Elsevier, Amsterdam, 1956, vol. 3B, pp. 816–869.	62
B-56MI 502-03	W. J. Hickinbottom, R. F. Garwood and M. F. Ansell; in "Chemistry of Carbon Compounds," ed. E. H. Rodd, Elsevier, Amsterdam, 1956, vol. 3B, pp. 870–954.	46
56MI 519-01	G. Tsatsas and R. Delaby; *Ann. Pharmac. Franc.*, 1956, **14**, 621 (*Chem. Abstr.*, 1957, **51**, 8670d).	751, 753
56MI 519-02	G. Tsatsas, R. Delaby, A. Quevauviller, R. Damiens and O. Blanpin; *Ann. Pharm. Franc.*, 1956, **14**, 607 (*Chem. Abstr.*, 1957, **51**, 8669e).	751, 753
B-56MI 526-01	C. H. Bamford, A. Elliot and W. E. Hanby; "Synthetic Polypeptides," Academic Press, New York, 1956.	972
56USP2733254	T. C. Allen and D. H. Chadwick (Monsanto Chemical Co.); *US Pat.*, 2 733 254 (1956) (*Chem. Abstr.*, 1956, **50**, 13 093b).	965
56USP2817662	R. A. Carboni; *US Pat.* 2 817 662 (1956) (*Chem. Abstr.*, 1958, **52**, 7360).	839
56ZOB2009	A. V. Kirsanov and G. I. Derkach; *Zh. Obshch. Khim.*, 1956, **26**, 2009 (*Chem. Abstr.*, 1957, **51**, 1821i).	683
56ZOB2631	A. V. Kirsanov and G. I. Derkach; *Zh. Obshch. Khim.*, 1956, **26**, 2631 (*Chem. Abstr.*, 1957, **51**, 1821b).	683, 707
57ACS582	P. K. Christensen; *Acta Chem. Scand.*, 1957, **11**, 582.	168
57AG135	J. Klosa; *Angew. Chem.*, 1957, **69**, 135.	126
57AG189	G. Baunitzer; *Angew. Chem.*, 1957, **69**, 189.	85
57AG362	T. Wieland and B. Heincke; *Angew. Chem.*, 1957, **69**, 362.	84
57AG600	K. Heyns and H. Paulsen; *Angew. Chem.*, 1957, **69**, 600.	28
57ANY(69)94	R. J. Koegel, R. A. McCallum, J. P. Greenstein, M. Winitz and S. M. Birnbaum; *Ann. N. Y. Acad. Sci.*, 1957, **69**, 94.	84
57BSF633	F. Hodosan; *Bull. Soc. Chim. Fr.*, 1957, 633.	735
57CB1590	F. Micheel and R. Habendorff; *Chem. Ber.*, 1957, **90**, 1590.	1040
57CB1638	J. Goerdeler, D. Krause-Loevenich and B. Wedekind; *Chem. Ber.*, 1957, **90**, 1638.	765, 766
57CB1837	H. Bredereck, R. Gompper and H. Seiz; *Chem. Ber.*, 1957, **90**, 1837.	753
57CB2291	W. Ziegenbein and W. Franke; *Chem. Ber.*, 1957, **90**, 2291.	289
57CCA319	V. Hahn, Ž. Stojanac, O. Ščedrov, N. Pravdić-Sladović, S. Tomasić and D. Emer; *Croat. Chem. Acta*, 1957, **29**, 319 (*Chem. Abstr.*, 1959, **53**, 16101g).	566, 618
57CI(L)736	R. F. Collins; *Chem. Ind.* (*London*), 1957, 736.	95
57CJC832	E. Lieber, C. N. Pillai and R. D. Hites; *Can. J. Chem.*, 1957, **35**, 832.	1046
57CRV47	R. G. Arnold, J. A. Nelson and J. J. Verbanc; *Chem. Rev.*, 1957, **47**.	962
57CRV191	W. J. Gensler; *Chem. Rev.*, 1957, **57**, 191.	28
57CRV583	A. S. Lindsey and H. Jeskey; *Chem. Rev.*, 1957, **57**, 583.	105
57CRV641	R. E. Buckles; *Chem. Rev.*, 1957, **57**, 641.	24
57CRV867	R. A. Benkeser, D. J. Foster, D. M. Sauve and J. F. Nobis; *Chem. Rev.*, 1957, **57**, 867.	27
57HCA852	M. Häring and T. Wagner-Jauregg; *Helv. Chim. Acta*, 1957, **40**, 852.	201
57JA213	K. G. Rutherford and M. S. Newman; *J. Am. Chem. Soc.*, 1957, **79**, 213.	1002
57JA419	R. T. Arnold and J. S. Showell; *J. Am. Chem. Soc.*, 1957, **79**, 419.	201
57JA637	H. B. Milne, J. E. Halver, D. S. Ho and M. S. Mason; *J. Am. Chem. Soc.*, 1957, **79**, 637.	87
57JA645	H. B. Milne and C. H. Peng; *J. Am. Chem. Soc.*, 1957, **79**, 645.	87
57JA725	C. R. Hauser and C. J. Eby; *J. Am. Chem. Soc.*, 1957, **79**, 725.	275
57JA1236	D. J. Beaver, D. P. Roman and P. J. Stoffel; *J. Am. Chem. Soc.*, 1957, **79**, 1236.	1027
57JA1764	M. G. Ettlinger and A. J. Lundeen; *J. Am. Chem. Soc.*, 1957, **79**, 1764.	636, 831
57JA2530	L. Tsai, T. Miwa and M. S. Newman; *J. Am. Chem. Soc.*, 1957, **79**, 2530.	275
57JA2544	T. J. Prosser and E. L. Eliel; *J. Am. Chem. Soc.*, 1957, **79**, 2544.	998
57JA3508	G. N. Walker; *J. Am. Chem. Soc.*, 1957, **79**, 3508.	184
57JA4373	A. G. Brook; *J. Am. Chem. Soc.*, 1957, **79**, 4373.	410, 417
57JA4759	F. W. Hoffman and H. D. Weiss; *J. Am. Chem. Soc.*, 1957, **79**, 4759.	122
57JA5739	W. D. Emmons; *J. Am. Chem. Soc.*, 1957, **79**, 5739.	272, 283
57JA5982	B. Phillips, F. C. Frostick Jr. and P. S. Starcher; *J. Am. Chem. Soc.*, 1957, **79**, 5982.	205
57JCS1091	C. L. Arcus and B. S. Prydal; *J. Chem. Soc.*, 1957, 1091.	994
57JCS3032	H. B. Henbest and A. Thomas; *J. Chem. Soc.*, 1957, 3032.	293
57JCS4397	A. Zilkha and Y. Liwschitz; *J. Chem. Soc.*, 1957, 4397.	199
57JGU1065	A. V. Kirsanov and G. I. Derkach; *J. Gen. Chem. USSR* (*Engl. Transl.*), 1957, **27**, 1065.	707
57JGU2301	N. N. Yarovenko, S. P. Motornyi, L. I. Kirenskaya and A. S. Vasilyeva; *J. Gen. Chem. USSR* (*Engl. Transl.*), 1957, **27**, 2301.	476
57JGU2481	A. N. Nesmeyanov, R. K. Freidlina, L. I. Zakharkin, E. I. Vasilyeva, V. N. Kost and T. T. Vasilyeva; *J. Gen. Chem. USSR* (*Engl. Transl.*), 1957, **27**, 2481.	38, 89
57JGU3284	A. V. Kirsanov and G. I. Derkach; *J. Gen. Chem. USSR* (*Engl. Transl.*), 1957, **27**, 3284.	683, 707
57JOC598	T. H. Coffield, J. Kozikowski and R. D. Closson; *J. Org. Chem.*, 1957, **22**, 598.	446
57JOC692	F. L. Scott, W. N. Morrish and J. Reilly; *J. Org. Chem.*, 1957, **22**, 692.	841
57JOC1680	G. B. Payne and C. W. Smith; *J. Org. Chem.*, 1957, **22**, 1680.	51
57LA(605)191	D. Jerchel and W. Woticky; *Justus Liebigs Ann. Chem.*, 1957, **605**, 191.	855
57MAC25	F. L. M. Schouteden; *Macromol. Chem.*, 1957, **24**, 25.	833

B-57MI 502-01	J. W. Green; in "The Carbohydrates—Chemistry, Biochemistry and Physiology," ed. W. Pigman, Academic Press, New York, 1957, pp. 299–366.	24
B-57MI 502-02	F. D. Gunstone; in "Progress in the Chemistry of Fats and Other Lipids," eds. R. T. Holman, W. O. Lundberg and T. Malkin, Pergamon Press, Oxford, 1957, vol. 4, p. 1.	28
57MI 502-01	J. C. Sowden; *Adv. Carbohydr. Chem.*, 1957, **12**, 35.	62
57MI 502-02	M. Goodman and G. W. Kenner; *Adv. Protein Chem.*, 1957, **12**, 465.	84
57MI 502-03	T. Wieland and G. Pfleiderer; *Adv. Enzymol. Rel. Subj. Biochem.*, 1957, **19**, 235.	83
57MI 502-04	L. J. Reed; *Adv. Enzymol. Rel. Subj. Biochem.*, 1957, **18**, 319.	81
57MI 502-05	D. Steinberg and E. Mihalyi; *Annu. Rev. Biochem.*, 1957, **26**, 373.	83
57MI 511-01	T. Bacchetti and A. Alemagna; *Rend. Ist. Lomb. Sci. Lett. Cl. Sci. Mat. Nat.*, **91**, 30 (*Chem. Abstr.*, 1958, **52**, 11 749).	476, 477
57MI 511-02	T. Bacchetti and A. Alemagna; *Rend. Ist. Lomb. Sci. Lett. Cl. Sci. Mat. Nat.*, **91**, 574 (*Chem. Abstr.*, 1958, **52**, 16 309).	476, 477
57NKZ567	R. Kitawake, M. Yamashita and K. Sugino; *Nippon Kagaku Zasshi*, 1957, **78**, 567.	1118
57OR(9)107	A. C. Cope, H. L. Holmes and H. O. House; *Org. React.*, 1957, **9**, 107.	46
57OR(9)388	J. C. Sheehan and E. J. Corey; *Org. React.*, 1957, **9**, 388.	652
57RTC969	L. Heslinga, G. J. Katerberg and J. F. Arens; *Recl. Trav. Chim. Pays-Bas*, 1957, **76**, 969.	14
57ZOB1080	A. V. Kirsanov and G. I. Derkach; *Zh. Obshch. Khim.*, 1957, **27**, 1080 (*Chem. Abstr.*, 1958, **52**, 3715d).	707
57ZOB3248	A. V. Kirsanov and G. I. Derkach; *Zh. Obshch. Khim.*, 1957, **27**, 3248 (*Chem. Abstr.*, 1958, **52**, 8997d).	683, 707

58AG1	B. Raecke; *Angew. Chem.*, 1958, **70**, 1.	
58AG173	R. Criegee; *Angew. Chem.*, 1958, **70**, 173.	28
58AG271	R. Appel and H. Gerber; *Angew. Chem.*, 1958, **70**, 271.	994
58ANY(70)870	M. Saunders and A. Wishnia; *Ann. N. Y. Acad. Sci.*, 1958, **70**, 870.	85
58BRP800807	Farbenfabriken Bayer A.-G., *Br. Pat.* 800807 (1958) (*Chem. Abstr.*, 1959, **53**, 5134).	1049
58BSF17	G. Dupont; *Bull. Soc. Chim. Fr.*, 1958, 17.	24
58CB437	L. Horner and H. Oediger; *Chem. Ber.*, 1958, **91**, 437.	921
58CB683	T. Wieland, H. Merz and A. Rennecke; *Chem. Ber.*, 1958, **91**, 683.	1038
58CB972	R. Huisgen and J. Witte; *Chem. Ber.*, 1958, **91**, 972.	611, 710
58CB1057	H. Krimm; *Chem. Ber.*, 1958, **91**, 1057.	292
58CB1129	R. Huisgen and J. Witte; *Chem. Ber.*, 1958, **91**, 1129.	710
58CB1200	R. Appel and H. Gerber; *Chem. Ber.*, 1958, **91**, 1200.	994, 1018
58CB1660	M. Lipp, F. Dallacker and G. Koenen; *Chem. Ber.*, 1958, **91**, 1660.	1047
58CB2553	F. Seel and J. Langer; *Chem. Ber.*, 1958, **91**, 2553.	4
58CR(246)1877	M. Olomucki; *C. R. Hebd. Seances Acad. Sci.*, 1958, **246**, 1877.	101
58CRV403	J. W. Ladbury and C. F. Cullis; *Chem. Rev.*, 1958, **58**, 403.	28
58CRV895	G. R. Ames; *Chem. Rev.*, 1958, **58**, 895.	46
58HOU(9)39	H. Soll; *Methoden Org. Chem.* (*Houben-Weyl*), 1958, **9**, 39.	742, 743, 748, 749
58JA217	E. Wenkert and B. G. Jackson; *J. Am. Chem. Soc.*, 1958, **80**, 217.	31
58JA1236	C. Djerassi and J. S. Mills; *J. Am. Chem. Soc.*, 1958, **80**, 1236.	116
58JA1895	R. L. Hinman and D. Fulton; *J. Am. Chem. Soc.*, 1958, **80**, 1895.	350
58JA2829	W. J. Middleton and V. A. Engelhardt; *J. Am. Chem. Soc.*, 1958, **80**, 2829.	845
58JA3079	E. E. van Tamelen and S. R. Bach; *J. Am. Chem. Soc.*, 1958, **80**, 3079.	166
58JA3332	D. L. Garmaise, R. Schwartz and A. F. McKay; *J. Am. Chem. Soc.*, 1958, **80**, 3332.	1024
58JA3711	C. D. Gutsche, M. W. Wendt and K. L. Seligman; *J. Am. Chem. Soc.*, 1958, **80**, 3711.	199
58JA4151	T. A. Bither, W. H. Knoth, R. V. Lindsey, Jr. and W. H. Sharkey; *J. Am. Chem. Soc.*, 1958, **80**, 4151.	1133
58JA4529	D. Wittenberg and H. Gilman; *J. Am. Chem. Soc.*, 1958, **80**, 4529.	417
58JA5793	R. A. Carboni and R. V. Lindsey, Jr.; *J. Am. Chem. Soc.*, 1958, **80**, 5793.	839
58JA6442	D. C. England, R. V. Lindsey Jr. and L. R. Melby; *J. Am. Chem. Soc.*, 1958, **80**, 6442.	373
58JA6682	L. M. Berkowitz and P. N. Rylander; *J. Am. Chem. Soc.*, 1958, **80**, 6682.	307
58JCS153	G. Shaw and R. N. Warrener; *J. Chem. Soc.*, 1958, 153.	1048
58JCS1141	D. T. Elmore and J. R. Ogle; *J. Chem. Soc.*, 1958, 1141.	1048
58JCS1313	I. Bell, E. R. H. Jones and M. C. Whiting; *J. Chem. Soc.*, 1958, 1313.	74, 102, 103
58JCS3648	W. Gerrard, M. F. Lappert and R. Shafferman; *J. Chem. Soc.*, 1958, 3648.	228
58JGU1930	A. V. Kirsanov and G. I. Derkach; *J. Gen. Chem. USSR* (*Engl. Transl.*), 1958, **28**, 1930.	683
58JGU2283	A. V. Kirsanov and G. I. Derkach; *J. Gen. Chem. USSR* (*Engl. Transl.*), 1958, **28**, 2283.	707
58JIC893	G. C. Singhal and M. M. Bokadia; *J. Indian Chem. Soc.*, 1958, **35**, 893.	821
58JOC318	H. J. Shine; *J. Org. Chem.*, 1958, **23**, 318.	215
58JOC1107	H. Feuer, H. Rubinstein and A. T. Nielsen; *J. Org. Chem.*, 1958, **23**, 1107.	1003
58JOC1517	L. W. Kissinger and H. E. Ungnade; *J. Org. Chem.*, 1958, **23**, 1517.	825
58JOC1931	R. Rätz and H. Schroeder; *J. Org. Chem.*, 1958, **23**, 1931.	852
58MAC246	F. Schouteden; *Macromol. Chem.*, 1958, **27**, 246.	833
B-58MI 502-01	G. E. W. Wolstenholme and M. P. Cameron (eds.); "CIBA Foundation Symposia," Little, Brown and Co., Boston, 1958.	84
58MI 502-01	J. D. Crum; *Adv. Carbohydr. Chem.*, 1958, **13**, 169.	62
58MI 502-02	F. Zilliken and M. W. Whitehouse; *Adv. Carbohydr. Chem.*, 1958, **13**, 237.	62
58MI 502-03	B. Boettcher and A. Meister; *Methods Enzymol.*, 1958, **113**, 458.	81
58MI 502-04	E. Katchalski and M. Sela; *Adv. Protein Chem.*, 1958, **13**, 243.	84
58MI 504-01	G. Young; *J. Am. Oil Chem. Soc.*, 1958, **35**, 416.	652

58MI 521-01 T. Bacchetti and A. Alemagna; *Atti Acad. naz. Linzei, Roma, Rend. Cl. Sci. fisiche, mat. e natur.*, 1958, **24**, 161 (*Chem. Abstr.*, 1958, **52**, 18 299). 831

58MI 530-01 R. V. Vizgert and Y. P. Berkman; *Zur. Russ. Fiz-Chim.*, 1958, **28**, 2119. 1108

58T1 R. B. Woodward, F. E. Bader, H. Bickel, A. J. Frey and R. W. Kierstead; *Tetrahedron*, 1958, **2**, 1. 164

58USP2851489 H. C. Godt, J. F. Quinn and H. Raffelson; *US Pat.* 2 851 489 (1958) (*Chem. Abstr.*, 1959, **53**, 3069c). 838

58ZOB1864 Y. V. Svetkin and G. Kretsu; *Zh. Obshch. Khim. Lett.*, 1958, **28**, 1864. 193

58ZOB1887 A. V. Kirsanov and G. I. Derkach; *Zh. Obshch. Khim.*, 1958, **28**, 1887 (*Chem. Abstr.*, 1959, **53**, 1208). 683

58ZOB2247 A. V. Kirsanov and G. I. Derkach; *Zh. Obshch. Khim.*, 1958, **28**, 2247 (*Chem. Abstr.*, 1959, **53**, 2134g). 707

58ZOB2305 S. S. Novikov, O. V. Lebedev, L. I. Khmelnitskii and Yu. P. Egorov; *Zh. Obshch. Khim.*, 1958, **28**, 2305 (*Chem. Abstr.*, 1959, **53**, 4112). 838

59CB528 F. Weygand and H. J. Bestmann; *Chem. Ber.*, 1959, **92**, 528. 271, 292

59CB837 H. Bredereck, R. Gompper, K. Klemm and H. Rempfer; *Chem. Ber.*, 1959, **92**, 837. 744, 745, 751

59CB910 H. Behringer, L. Hauser and K. Kohl; *Chem. Ber.*, 1959, **92**, 910. 581

59CB2563 J. Goerdeler and J. Vitt; *Chem. Ber.*, 1959, **92**, 2563. 766

59CB2813 B. Helferich and H. Boshagen; *Chem. Ber.*, 1959, **92**, 2813. 279

59CB2887 W. Groebel; *Chem. Ber.*, 1959, **92**, 2887. 337

59CB3183 K. Issleib and E. Priebe; *Chem. Ber.*, 1959, **92**, 3183. 402

59CCC2748 S. Hermanek; *Collect. Czech. Chem. Commun.*, 1959, **24**, 2748. 199

59CRV89 D. G. M. Diaper and A. Kuksis; *Chem. Rev.*, 1959, **59**, 89. 24

59CRV841 T. K. Brotherton and J. W. Lynn; *Chem. Rev.*, 1959, **59**, 841. 847

59DOK(127)337 I. L. Knunyants, E. G. Perevalova and N. Frosin; *Dokl. Akad. Nauk SSSR*, 1959, **127**, 337. 821

59DOK(127)571 B. M. Mikhailov and Y. N. Bubnov; *Dokl. Akad. Nauk SSSR*, 1959, **127**, 571. 1136

59DOK(128)1201 Y. A. Ol'Dekop, A. N. Sevchenko, I. P. Zyat'kov, G. S. Bylina and A. P. El'nitskiĭ; *Dokl. Akad. Nauk SSSR*, 1959, **128**, 1201 (*Chem. Abstr.*, 1960, **54**, 7630). 208

59G526 P. Papini and M. Ridi; *Gazz. Chim. Ital.*, 1959, **89**, 526. 583

59G2428 F. Minisci; *Gazz. Chim. Ital.*, 1959, **89**, 2428. 1110

59HCA1653 H. H. Bosshard, R. Mory, M. Schmid and H. Zollinger; *Helv. Chim. Acta*, 1959, **42**, 1653. 9

59IZV484 G. Y. Kondrateva; *Izv. Akad. Nauk SSSR, Ser. Khim.*, 1959, 484. 201

59JA1587 J. A. Young, W. S. Durrell and R. D. Dresdner; *J. Am. Chem. Soc.*, 1959, **81**, 1587. 989

59JA1751 W. Sakami and G. Toennies; *J. Am. Chem. Soc.*, 1959, **81**, 1751. 140

59JA2374 R. M. Keefer and L. J. Andrews; *J. Am. Chem. Soc.*, 1959, **81**, 2374. 204

59JA3089 J. C. Sheehan and K. R. Henery-Logan; *J. Am. Chem. Soc.*, 1959, **81**, 3089. 81

59JA3599 J. A. Attaway, R. H. Groth and L. A. Bigelow; *J. Am. Chem. Soc.*, 1959, **81**, 3599. 808

59JA5106 E. C. Evers, W. O. Freitag, W. A. Kriner and A. G. MacDiarmid; *J. Am. Chem. Soc.*, 1959, **81**, 5106. 1136

59JA5817 K. Sisido, Y. Kazama, H. Kodama and H. Nozaki; *J. Am. Chem. Soc.*, 1959, **81**, 5817. 29

59JCS227 H. B. Henbest and B. Nicholls; *J. Chem. Soc.*, 1959, 227. 168

59JCS1313 I. Bill, E. H. R. Jones and M. C. Whiting; *J. Chem. Soc.*, 1959, 1313. 168

59JGU244 G. I. Derkach; *J. Gen. Chem. USSR* (*Engl. Transl.*), 1959, **29**, 244. 652

59JGU596 A. V. Kirsanov and G. I. Derkach; *J. Gen. Chem. USSR* (*Engl. Transl.*), 1959, **29**, 596. 707

59JOC107 J. R. Slagle and H. J. Shine; *J. Org. Chem.*, 1959, **24**, 107. 208

59JOC388 C. D. Hurd and A. G. Prapas; *J. Org. Chem.*, 1959, **24**, 388. 187

59JOC1293 L. Bauer and S. V. Miarka; *J. Org. Chem.*, 1959, **24**, 1293. 1000

59JOC1309 A. K. Bose, F. Greer, J. S. Gots and C. C. Price; *J. Org. Chem.*, 1959, **24**, 1309. 748

59JOC1354 G. B. Payne; *J. Org. Chem.*, 1959, **24**, 1354. 205

59LA(621)1 E. Schmidt and L. Fehr; *Justus Liebigs Ann. Chem.*, 1959, **621**, 1. 1026

59LA(627)142 L. Horner and H. Oediger; *Justus Liebigs Ann. Chem.*, 1959, **627**, 142. 975

59M41 M. Lipp, F. Dallacker and I. Meier zu Köcker; *Monatsh. Chem.*, 1959, **90**, 41. 1057

59MI 502-01 P. Edman; *Annu. Rev. Biochem.*, 1959, **28**, 69. 83

59MI 502-02 R. Cecil and J. R. McPhee; *Adv. Protein Chem.*, 1959, **14**, 255. 85

59MI 503-01 M. Julia, S. Julia, B. Bérmont and G. Tchernoff; *Compt. Rend.*, 1959, **248**, 242. 145

59N35 H. Tuppy; *Naturwissenschaften*, 1959, **46**, 35. 652

59NAT675 L. C. Case; *Nat.*, 1959, **183**, 675. 1009

59OR(10)179 E. D. Bergman, D. Ginsburg and R. Pappo; *Org. React.*, 1959, **10**, 179. 46

59OS(39)37 D. W. Goheen and W. R. Vaughan; *Org. Synth.*, 1959, **39**, 37. 146

59T24 G. Gaudiano, A. Quilico and A. Ricca; *Tetrahedron*, 1959, **7**, 24. 821

59USP2873171 L. H. Jenkins and D. S. Sears; *US Pat.*, 2 873 171 (1959) (*Chem. Abstr.*, 1959, **53**, 18 864e). 1017

59USP2884435 J. J. Tazuma; *US Pat.*, 2 884 435 (1959) (*Chem. Abstr.*, 1959, **53**, 19 973d). 965

59USP2902514 P. L. De Beneville and Ch. S. Hollander; *US Pat.* 2 902 514 (1959) (*Chem. Abstr.*, 1960, **54**, 1323d). 833

59USP3053876 H. Malz, E. Kuehle and O. Bayer (Bayer A. G.); *US Pat.*, 3 053 876 (1959) (*Chem. Abstr.*, 1963, **58**, 4470). 738

59ZN(B)814 R. Riemschneider; *Z. Naturforsch., Teil B*, 1959, **14**, 814. 1043

59ZOB241 G. I. Derkach; *Zh. Obshch. Khim*, 1959, **29**, 241 (*Chem. Abstr.*, 1959, **53**, 21 750e). 683

59ZOB600 A. V. Kirsanov and G. I. Derkach; *Zh. Obshch. Khim.*, 1959, **29**, 600 (*Chem. Abstr.*, 1960, **54**, 382d). 707

59ZOB1333 I. Y. Postovskii and M. I. Ermakova; *Zh. Obshch. Khim.*, 1969, **29**, 1333 (*Chem. Abstr.*, 1969, **54**, 8796c). 641

59ZOB2256 A. V. Kirsanov and M. S. Marenets; *Zh. Obshch. Khim.*, 1959, **29**, 2256 (*Chem. Abstr.*, 1960, **54**, 10 855e). 988

60AC(P)845 M. Olomucki; *Ann. Chim.* (*Paris*), 1960, **5**, 845. 101
60AG268 W. Ried and W. von der Emden; *Angew. Chem.*, 1960, **72**, 268. 586
60AG836 H. Eilingsfeld, M. Seefelder and H. Weidinger; *Angew. Chem.*, 1960, **72**, 836. 609, 611
60AG1000 R. Stroh and H. Gerber; *Angew. Chem.*, 1960, **72**, 1000. 1100, 1101
60ANC1896 S. T. Ross and D. B. Denney; *Anal. Chem.*, 1960, **32**, 1896. 402
60ANY(88)533 Symposium; *Ann. N. Y. Acad. Sci.*, 1960, **88**, 533. 83
60ANY(88)676 G. W. Anderson; *Ann. N. Y. Acad. Sci.*, 1960, **88**, 676. 84
60AP(293)150 H. P. Kaufmann and K. Lüthje; *Arch. Pharm.* (*Weinheim, Ger.*), 1960, **293**, 150. 1048
60AQ93 J. E. Courtois; *An. Quim.*, 1960, **56B**, 93. 28
60BSF1350 R. Granger, P. Nau and J. Nau; *Bull. Soc. Chim. Fr.*, 1960, 1350. 142
60BSF1412 V. F. Koutchérov and G. M. Ségal; *Bull. Soc. Chim. Fr.*, 1960, 1412. 168
60C293 K. Biemann; *Chimia*, 1960, **14**, 293. 84
60CB397 J. Goerdeler and H. Haubrich; *Chem. Ber.*, 1960, **93**, 397. 763
60CB519 G. Drefahl and K. Ponsold; *Chem. Ber.*, 1960, **93**, 519. 984
60CB663 J. Goerdeler and H. Horstmann; *Chem. Ber.*, 1960, **93**, 663. 630, 631
60CB671 J. Goerdeler and H. Horstmann; *Chem. Ber.*, 1960, **93**, 671. 630
60CB901 W. Hieber, G. Braun and W. Beck; *Chem. Ber.*, 1960, **93**, 901. 449
60CB1379 J. Goubeau and H. Gräbner; *Chem. Ber.*, 1960, **93**, 1379. 1015
60CB1511 W. Walter and J. Curts; *Chem. Ber.*, 1960, **93**, 1511. 579, 580, 623, 624
60CB2804 L. Birkhofer, P. Richter and A. Ritter; *Chem. Ber.*, 1960, **93**, 2804. 374
60CI(L)1096 P. W. Sherwood; *Chem. Ind.* (*London*), 1960, 1096. 24
60CI(M)1243 S. Rossi, A. Riva and B. Piantanida; *Chem. Ind.* (*Milan*), 1960, **42**, 1243. 980
60CJC472 W. R. Cullen and L. G. Walker; *Can. J. Chem.*, 1960, **38**, 472. 222
60CSR45 R. N. Hurd and G. DeLaMater; *Chem. Soc. Rev.*, 1960, 45. 565, 579
60HCA113 F. Elsinger, J. Schreiber and A. Eschenmoser; *Helv. Chim. Acta*, 1960, **43**, 113. 78
60HCA1473 H. Hopff and A. K. Wick; *Helv. Chim. Acta.*, 1960, **43**, 1473. 298
60IS84 P. A. Staats and H. W. Morgan; *Inorg. Synth.*, 1960, **6**, 84. 1129
60IZV1783 L. I. Khmelnitskii, S. S. Novikov and O. V. Lebedev; *Izv. Akad. Nauk SSSR, Ser. Khim.*, 1960, 1783 (*Chem. Abstr.*, 1961, **55**, 19 832). 838
60JA543 W. R. Hasek, W. C. Smith and V. A. Engelhardt; *J. Am. Chem. Soc.*, 1960, **82**, 543. 5
60JA2498 M. S. Newman, A. Arkell and T. Fukunaga; *J. Am. Chem. Soc.*, 1960, **82**, 2498. 47, 275
60JA2656 E. C. Taylor and J. A. Zoltewicz; *J. Am. Chem. Soc.*, 1960, **82**, 2656. 580
60JA3076 P. A. S. Smith and D. W. Emerson; *J. Am. Chem. Soc.*, 1960, **82**, 3076. 1040
60JA4204 K. A. Keblys and A. H. Filbey; *J. Am. Chem. Soc.*, 1960, **82**, 4204. 448
60JA4700 H. C. Brown and D. Pilipovich; *J. Am. Chem. Soc.*, 1960, **82**, 4700. 847, 850
60JA5102 A. G. Brook, M. A. Quigley, G. J. D. Peddle, N. V. Schwartz and C. M. Warner; *J. Am. Chem. Soc.*, 1960, **82**, 5102. 431
60JCS997 M. R. Truter; *J. Chem. Soc.*, 1960, 997. 725
60JCS1593 C. A. Bunton and V. J. Shiner, Jr.; *J. Chem. Soc.*, 1960, 1593. 49
60JCS3070 C. A. Bunton and S. G. Perry; *J. Chem. Soc.*, 1960, 3070. 26, 37
60JOC147 E. H. Sheers; *J. Org. Chem.*, 1960, **25**, 147. 1121
60JOC352 C. King; *J. Org. Chem.*, 1960, **25**, 352. 652
60JOC560 J. M. Bobbitt and D. A. Scola; *J. Org. Chem.*, 1960, **25**, 560. 298
60JOC809 D. Seyferth and N. Kahlen; *J. Org. Chem.*, 1960, **25**, 809. 1136, 1144
60JOC1043 R. Kitawaki and K. Sugino; *J. Org. Chem.*, 1960, **25**, 1043. 1118
60JOC1064 H. Gilman and G. D. Lichtenwalter; *J. Org. Chem.*, 1960, **25**, 1064. 254
60JOC1265 Y. Inouye, K. Onodera, S. Kitaoka and S. Hirano; *J. Org. Chem.*, 1960, **25**, 1265. 183, 192
60JOC1837 L. Rand and M. J. Albinak; *J. Org. Chem.*, 1960, **25**, 1837. 998
60JOC2016 C. W. Tullock and D. D. Coffman; *J. Org. Chem.*, 1960, **25**, 2016. 808
60JOC2245 M. L. Weiner; *J. Org. Chem.*, 1960, **25**, 2245. 745, 754
60LA(633)35 W. Walter; *Justus Liebigs Ann. Chem.*, 1960, **633**, 35. 623, 624
B-60MI 502-01 K. S. Markley (ed.); "Fatty Acids: Their Chemistry, Properties, Production and Use," Wiley-Interscience, New York, 1960, vol. 1; 1961, vol. 2; 1964, vol. 3. 28
B-60MI 502-02 B. C. L. Weedon; in "Advanced Organic Chemistry, Methods and Results," eds. R. A. Raphael, E. C. Taylor and H. Wynberg, Interscience, New York, 1960, vol. 1, p. 1. 28
B-60MI 502-03 A. Gottschalk; "The Chemistry and Biology of Sialic Acids and Related Substances," Cambridge University Press, London, 1960. 62
60MI 502-01 R. H. Girdwood; *Adv. Clin. Chem.*, 1960, **3**, 235. 28
60MI 502-02 R. Schwyzer; *Annu. Rev. Biochem.*, 1960, **29**, 183. 83
60OR(11)1 L. G. Donaruma and W. Z. Heldt; *Org. React.*, 1960, **11**, 1. 215, 275, 282, 288, 292, 298
60OR(11)261 A. S. Kende; *Org. React.*, 1960, **11**, 261. 48
60QR292 L. J. Haynes and J. R. Plimmer; *Q. Rev., Chem. Soc.*, 1960, **14**, 292. 62
60RCP243 G. W. Thiessen; *Rec. Chem. Prog.*, 1960, **21**, 243. 28
60RTC401 H. Breederveld; *Recl. Trav. Chim. Pays-Bas*, 1960, **79**, 401. 288
60T39 M. T. Leplawy, D. S. Jones, G. W. Kenner and R. C. Sheppard; *Tetrahedron*, 1960, **11**, 39. 278
60T67 K. Heyns and L. Blazejewicz; *Tetrahedron*, 1960, **9**, 67. 48
60ZOB506 L. Zalukaev and E. V. Vanag; *Zh. Obshch. Khim.*, 1960, **30**, 506. 823
60ZOB1937 L. M. Kozlov, E. A. Markovich and B. G. Liorber; *Zh. Obshch. Khim.*, 1960, **30**, 1937 (*Chem. Abstr.*, 1961, **55**, 6366b). 820

60ZOB2409 V. A. Ginsburg, N. F. Privezentseva, V. A. Shpanskii, N. P. Rodionova, S. S. Dubov, A. M. Khokhlova, S. P. Makarov and A. Ya. Yakubovich; *Zh. Obshch. Khim.*, 1960, **30**, 2409 (*Chem. Abstr.*, 1961, **55**, 10 299h). 822

60ZOB3006 A. V. Kirsanov and V. A. Shokol; *Zh. Obshch. Khim.* (*Engl. Trans.*), 1960, **30**, 3006. 644

61ACS1087 K. A. Jensen and C. Pedersen; *Acta Chem. Scand.*, 1961, **15**, 1087. 636

61ACS1097 K. A. Jensen and C. Pedersen; *Acta Chem. Scand.*, 1961, **15**, 1097. 639, 640

61ACS1104 K. A. Jensen and C. Pedersen; *Acta Chem. Scand.*, 1961, **15**, 1104. 634

61ACS1109 K. A. Jensen, H. R. Baccaro, O. Buchardt, G. E. Olsen, C. Pedersen and J. Toft; *Acta Chem. Scand.*, 1961, **15**, 1109. 640

61ACS1124 K. A. Jensen and C. Pedersen; *Acta Chem. Scand.*, 1961, **15**, 1124. 641

61AG449 L. Jaenicke; *Angew. Chem.*, 1961, **73**, 449. 92

61AG621 A. Reiche, E. Schmitz and E. Gundermann; *Angew. Chem.*, 1961, **73**, 621. 276, 286

61AG656 R. Huisgen, W. Mack and E. Aneser; *Angew. Chem.*, 1961, **73**, 656. 832

61AG657 H. A. Staab and W. Benz; *Angew. Chem.*, 1961, **73**, 657. 973

61BCF2067 R. Delaby, P. Reynaud and F. Lilly; *Bull. Soc. Chim. Fr.*, 1961, 2067. 750

61BSB519 M. Renson and J. Bidaine; *Bull. Soc. Chim. Belg.*, 1961, **70**, 519. 476

61BSF2001 E. Bricas; *Bull. Soc. Chim. Fr.*, 1961, 2001. 84

61BSF2360 A. Kergomard; *Bull. Soc. Chim. Fr.*, 1961, 2360. 318

61CB480 H. Beyer and J. Körösi; *Chem. Ber.*, 1961, **94**, 480. 318

61CB1263 L. Birkofer, W. Konkol and A. Ritter; *Chem. Ber.*, 1961, **94**, 1263. 376

61CB1373 G. Wittig and M. Schlosser; *Chem. Ber.*, 1961, **94**, 1373. 897

61CB1762 G. Losse and E. Demuth; *Chem. Ber.*, 1961, **94**, 1762. 193

61CB1868 H. Hellmann and W. Schwiersch; *Chem. Ber.*, 1961, **94**, 1868. 854

61CB2856 J. Goerdeler and H. Porrmann; *Chem. Ber.*, 1961, **94**, 2856. 630, 633

61CB2996 G. Märkl; *Chem. Ber.*, 1991, **94**, 2996. 895, 896

61CI(M)1243 S. Rossi, A. Riva and B. Piantanida; *Chem. Ind.* (*Milan*), 1961, **42**, 1243. 652

61CJC761 J. C. Grivas and A. Taurins; *Can. J. Chem.*, 1961, **39**, 761. 749

61CRV45 R. N. Hurd and G. DeLaMater; *Chem. Rev.*, 1961, **61**, 45. 735

61CRV179 R. Roger and D. G. Neilson; *Chem. Rev.*, 1961, **61**, 179. 685, 687, 690, 726, 742, 769, 849

61GEP1100612 J. Hasselmann, *Ger. Pat.* 1 100 612 (1961) (*Chem. Abstr.*, 1961, **55**, 25 760). 192

61HCA1546 M. Walter, H. Besendorf and O. Schnider; *Helv. Chim. Acta*, 1961, **44**, 1546. 276

61HCA1991 R. Schwyzer and H. Kappeler; *Helv. Chim. Acta*, 1961, **44**, 1991. 32

61IZV477 L. I. Khmelnitskii, S. S. Novikov and O. V. Lebedev; *Izv. Akad. Nauk SSSR, Ser. Khim.*, 1961, 477 (*Chem. Abstr.*, 1961, **55**, 23 389). 838

61JA606 W. S. Johnson, V. J. Bauer, J. L. Margrave, M. A. Frisch, L. H. Dreger and W. N. Hubbard; *J. Am. Chem. Soc.*, 1961, **83**, 606. 164

61JA1191 E. H. White and D. W. Grisley Jr.; *J. Am. Chem. Soc.*, 1961, **83**, 1191. 361

61JA1492 M. E. Kuehe; *J. Am. Chem. Soc.*, 1961, **83**, 1492. 285

61JA1769 T. E. Gier; *J. Am. Chem. Soc.*, 1961, **83**, 1769. 1152

61JA1983 L. Field, P. B. Hughmark, S. H. Shumaker and W. S. Marshall; *J. Am. Chem. Soc.*, 1961, **83**, 1983. 276, 298

61JA2589 W. J. Middleton, E. G. Howard and W. H. Sharkey; *J. Am. Chem. Soc.*, 1961, **83**, 2589. 474

61JA2729 A. Iliceto, A. Fava, U. Mazzucato and O. Rossetto; *J. Am. Chem. Soc.*, 1961, **82**, 2729. 1038

61JA3539 F. Ramirez, N. B. Desai, B. Hansen and N. McKelvie; *J. Am. Chem. Soc.*, 1961, **83**, 3539. 911, 1093

61JA3912 R. C. Petry and J. P. Freeman; *J. Am. Chem. Soc.*, 1961, **83**, 3912. 314

61JA4023 R. F. Heck and D. S. Breslow; *J. Am. Chem. Soc.*, 1961, **83**, 4023. 456

61JA4285 S. G. Smith; *J. Am. Chem. Soc.*, 1961, **83**, 4285. 476, 477

61JA4469 H. E. Baumgarten, R. L. Zey and U. Krolls; *J. Am. Chem. Soc.*, 1961, **83**, 4469. 315

61JCE217 J. S. Roth; *J. Chem. Ed.*, 1961, **38**, 217. 85

61JCS808 A. K. Barbour, M. W. Buxton, P. L. Coe, R. Stephens and J. C. Tatlow; *J. Chem. Soc.*, 1961, 808. 34, 63, 100

61JCS1194 B. Milligan and J. M. Swan; *J. Chem. Soc.*, 1961, 1194. 601, 621

61JCS2428 R. G. R. Bacon and R. G. Guy, *J. Chem. Soc.*, 1961, 2428. 1105, 1110

61JCS2436 R. G. R. Bacon, R. G. Guy and R. S. Irwin; *J. Chem. Soc.*, 1961, 2436. 1105, 1110

61JCS2447 R. G. R. Bacon and R. S. Irwin; *J. Chem. Soc.*, 1961, 2447. 1105

61JGU473 K. A. Petrov, A. A. Neimysheva, M. G. Fomenko, L. M. Chernushevich and H. D. Kuntsevich; *J. Gen. Chem. USSR* (*Engl. Transl.*), 1961, **31**, 473. 708

61JGU3192 G. I. Derkach, A. M. Lepesa and V. A. Kirsanov; *J. Gen. Chem. USSR* (*Engl. Transl.*), 1961, **31**, 3192. 707

61JOC307 C. E. Frank, J. R. Leebrick, L. F. Moormeier, J. A. Scheben and O. Homberg; *J. Org. Chem.*, 1961, **26**, 307. 42

61JOC412 F. C. Schaefer and G. A. Peters; *J. Org. Chem.*, 1961, **26**, 412. 749

61JOC651 G. B. Payne and P. H. Williams; *J. Org. Chem.*, 1961, **26**, 651. 709

61JOC659 G. B. Payne, P. H. Derning and P. H. Williams; *J. Org. Chem.*, 1961, **26**, 659. 709

61JOC782 T. Mukaiyama and H. Nohira; *J. Org. Chem.*, 1961, **26**, 782. 214, 1001

61JOC1132 P. A. S. Smith and J. M. Sullivan; *J. Org. Chem.*, 1961, **26**, 1132. 620

61JOC1514 T. S. Gardner, E. Wenis and J. Lee; *J. Org. Chem.*, 1961, **26**, 1514. 10

61JOC1759 M. S. Newman and H. Boden; *J. Org. Chem.*, 1961, **26**, 1759. 110, 115

61JOC2525 J. C. Sheehan, P. A. Cruickshank and G. L. Boshart; *J. Org. Chem.*, 1961, **26**, 2525. 279, 1062

61JOC2563 G. R. Pettit, M. V. Kalnins, T. M. H. Liu, E. G. Thomas and K. Parent; *J. Org. Chem.*, 1961, **26**, 2563. 745

61JOC3002 B. V. Shetty; *J. Org. Chem.*, 1961, **26**, 3002. 282

61JOC3511 J. Weinstock; *J. Org. Chem.*, 1961, **26**, 3511. 991
61JOC3980 R. N. Hurd, G. De La Mater, G. C. McElheny, R. J. Turner and V. H. Wallingford; *J. Org. Chem.*, 1961, **26**, 3980. 591
61JOC4122 K. Sugino, K. Shirai and R. Kitawaki; *J. Org. Chem.*, 1961, **26**, 4122. 1125
61JOC4381 T. Mukaiyama, M. Tokizawa, H. Nohira and H. Takei; *J. Org. Chem.*, 1961, **26**, 4381. 1001
61LA(641)1 H. Meerwein, W. Florian, N. Schön and G. Stopp; *Justus Liebigs Ann. Chem.*, 1961, **641**, 1. 852
61LA(643)29 W. Walter, J. Curts and H. Pawelzik; *Justus Liebigs Ann. Chem.*, 1961, **643**, 29. 636
61LA(647)11 H. Balli and F. Kersting; *Justus Liebigs Ann. Chem.*, 1961, **647**, 11. 869
61LA(647)1 H. Balli and F. Kersting; *Justus Liebigs Ann. Chem.*, 1961, **647**, 1. 869
61LA(648)72 H. A. Staab and W. Benz; *Justus Liebigs Ann. Chem.*, 1961, **648**, 72. 973
61MAC62 M. E. B. Jones, D. A. Thornton and R. F. Webb; *Macromol. Chem.*, 1961, **49**, 62. 823
61MI 502-01 L. J. Reed; *Org. Sulfur Compounds*, 1961, **1**, 443. 81
B-61MI 502-01 T. P. Greenstein and M. Winnitz; "Chemistry of the Amino Acids," Wiley, New York, 1961. 83
B-61MI 502-02 R. G. Johl; in "Pharmaceutical Analysis," eds. T. Higuchi and E. Brochmann-Hanssen, Interscience, New York, 1961, pp. 265–311. 83
B-61MI 502-03 A. J. Zimmer and C. L. Huyck; in "Pharmaceutical Analysis," eds. T. Higuchi and E. Brochmann-Hanssen, Interscience, New York, 1961, pp. 11–29. 105
B-61MI 527-01 S. J. Assony; in "Organic Sulfur Compounds," ed. N. Kharasch, Pergamon, Oxford, 1961, vol. 1, ch. 28, p. 326. 1022
61T134 I. Tinoco, Jr., R. W. Woody and K. Yamaoka; *Tetrahedron*, 1961, **13**, 134. 85
61T143 J. T. Yang; *Tetrahedron*, 1961, **13**, 143. 85
61T176 J. Strem, Y. S. R. Krishna-Prasad and J. A. Schellman; *Tetrahedron*, 1961, **13**, 176. 84
61TL724 H. Hellmann and J. Bader; *Tetrahedron Lett.*, 1961, **2**, 724. 901
61TL807 G. Märkl; *Tetrahedron Lett.*, 1961, **2**, 807. 904, 905
61USP3013045 F. J. Sowa; *US Pat.*, 3 013 045 (1961) (*Chem. Abstr.*, 1962, **56**, 8632b). 979
61ZOB516 K. A. Petrov, A. A. Neimysheva, M. G. Fomenko, L. M. Chernushevich and A. D. Kuntsevich; *Zh. Obshch. Khim.*, 1961, **31**, 516 (*Chem. Abstr.*, 1961, **55**, 22111a). 708
61ZOB582 A. V. Kirsanov and V. A. Shokol; *Zh. Obshch. Khim.*, 1961, **31**, 582. 644
61ZOB665 V. P. Gol'Mov, Z. P. Malevannaya; *Zh. Obshch. Khim.*, 1961, **31**, 665 (*Chem. Abstr.*, 1961, **55**, 22 162). 199
61ZOB1607 A. V. Kirsanov and M. S. Marenets; *Zh. Obshch. Khim.*, 1961, **31**, 1607 (*Chem. Abstr.*, 1961, **55**, 25 752a). 988
61ZOB1621 M. O. Lozinsky and P. S. Pelkis; *Zh. Obshch. Khim.*, 1961, **31**, 1621. 843
61ZOB2275 G. I. Derkach, V. A. Shokol and A. V. Kirsanov; *Zh. Obshch. Khim.*, 1961, **31**, 2275. 1053
61ZOB3062 V. I. Shevschenko, A. S. Stepanek and A. V. Kirsanov; *Zh. Obshch. Khim.*, 1961, **31**, 3062. 988
61ZOB3424 G. I. Derkach, A. M. Lepesa and A. V. Kirsanov; *Zh. Obshch. Khim.*, 1961, **31**, 3424 (*Chem. Abstr.*, 1962, **57**, 3353d). 707, 861

62AC(R)79 T. Tarantelli and C. Pecile; *Ann. Chim.* (*Rome*), 1962, **52**, 79. 1053
62AG(E)351 H. A. Staab; *Angew. Chem., Int. Ed. Engl.*, 1962, **1**, 351. 278
62AG(E)361 H. A. Staab; *Angew. Chem., Int. Ed. Engl.*, 1962, **1**, 361. 973
62AG(E)595 H. Ulrich and A. A. Sayigh; *Angew. Chem., Int. Ed. Engl.*, 1962, **1**, 595. 982
62AG(E)617 H. Holtschmidt and G. Oertel; *Angew. Chem., Int. Ed. Engl.*, 1962, **1**, 617. 968, 970, 1009
62AG(E)626 F. Krohnke and N. Zecher; *Angew. Chem., Int. Ed. Engl.*, 1962, **1**, 626. 46
62AG(E)632 H. Holtschmidt; *Angew. Chem., Int. Ed. Engl.*, 1962, **1**, 632. 1005
62BCJ1104 T. Mukaiyama; *Bull. Chem. Soc. Jpn.*, 1962, **35**, 1104. 1005
62BEP612359 ICI; *Belg. Pat.*, 612 359 (1962) (*Chem. Abstr.*, 1962, **57**, 13 411h). 1017
62BRP907029 H. C. Fielding; *Br. Pat.*, 907 029 (1962) (*Chem. Abstr.*, 1963, **58**, 279d). 1017
62BSB260 M. Renson and C. Draguet; *Bull. Soc. Chim. Belg.*, 1962, **71**, 260. 238
62BSB531 C. Collard-Charon and M. Renson; *Bull. Soc. Chim. Belg.*, 1962, **71**, 531. 1051, 1056
62BSB563 C. Collard-Charon and M. Renson; *Bull. Soc. Chim. Belg.*, 1962, **71**, 563. 515, 516
62BSF186 E. Moczar and L. Mester; *Bull. Soc. Chim. Fr.*, 1962, 186. 855
62C385 J. Meinhoffer; *Chimia*, 1962, **16**, 385. 84
62CB126 I. Ugi, F. Beck and U. Fetzer; *Chem. Ber.*, 1962, **95**, 126. 654, 655, 657
62CB627 J. Goerdeler and H. Porrmann; *Chem. Ber.*, 1962, **95**, 627. 630
62CB926 S. Hünig, K. Hübner and E. Benzing; *Chem. Ber.*, 1962, **95**, 926. 597, 599, 614
62CB1237 G. Gosselck and E. Wolters; *Chem. Ber.*, 1962, **95**, 1237. 1113
62CB2073 H. A. Staab, G. Walther and W. Rohr; *Chem. Ber.*, 1961, **95**, 2073. 196
62CHE179 E. P. Oliveto; *Chem. Heterocycl. Compd.* (*Engl. Transl.*), 1962, **14**, 179. 105
62CHE347 J. C. Godfrey; *Chem. Heterocycl. Compd.* (*Engl. Transl.*), 1962, **14**, 347. 105
62CI(L)362 D. Elad; *Chem. Ind.* (*London*), 1962, 362. 286
62CI(L)1304 D. Swern, E. S. Rothman, L. S. Silbert and J. S. Showell; *Chem. Ind.* (*London*), 1962, 1304. 28
62CJC739 T. N. Srivastava, J. E. Griffiths and M. Onyszchuk; *Can. J. Chem.*, 1962, **40**, 739. 1016
62CJC1403 C. P. Podesva, E. J. Tarlton and A. F. McKay; *Can. J. Chem.*, 1962, **40**, 1403. 1125
62CRV99 H. J. Harwood; *Chem. Rev.*, 1962, **62**, 99. 264
62CRV155 F. Eloy and R. Lenaers; *Chem. Rev.*, 1962, **62**, 155. 322, 763, 765, 833, 864
62CRV283 C. W. Bird; *Chem. Rev.*, 1962, **62**, 283. 27
62HCA441 R. Lenaers, C. Moussebois and F. Eloy; *Helv. Chim. Acta*, 1962, **45**, 441. 219, 833
62HOU(5/3)1 E. Forsche; *Methoden Org. Chem.* (*Houben-Weyl*), 1962, **5/3**, 1. 821, 839
62HOU(5/3)503 R. Stroh; *Methoden Org. Chem.* (*Houben-Weyl*), 1962, **5/3**, 503. 810
62JA1316 W. S. Wadsworth, Jr. and W. D. Emmons; *J. Am. Chem. Soc.*, 1962, **84**, 1316. 975

62JA2337	W. J. Chambers, C. W. Tullock and D. D. Coffman; *J. Am. Chem. Soc.*, 1962, **84**, 2337.	808, 989
62JA2499	R. F. Heck and D. S. Breslow; *J. Am. Chem. Soc.*, 1962, **84**, 2499.	456
62JA4275	F. S. Fawcett, C. W. Tullock and D. D. Coffman; *J. Am. Chem. Soc.*, 1962, **84**, 4275.	1003
62JA4349	C. N. Matthews, J. S. Driscoll, J. E. Harris and R. J. Wineman; *J. Am. Chem. Soc.*, 1962, **84**, 4349.	1093
62JA4350	G. L. Closs and J. J. Coyle; *J. Am. Chem. Soc.*, 1962, **84**, 4350.	844, 866
62JA4987	L. A. Paquette; *J. Am. Chem. Soc.*, 1962, **84**, 4987.	271, 288
62JCS186	B. Gething, C. R. Patrick and J. C. Tatlow; *J. Chem. Soc.*, 1962, 186.	116
62JCS2824	P. A. Finan and G. A. Fothergill; *J. Chem. Soc.*, 1962, 2824.	295
62JCS4722	D. Hartley; *J. Chem. Soc.*, 1962, 4722.	64
62JGU167	G. I. Derkach, A. M. Lepesa and A. V. Kirsanov; *J. Gen. Chem. USSR* (*Engl. Transl.*), 1962, **32**, 167.	708
62JGU3330	A. Spassov and E. Golovinskii; *J. Gen. Chem. USSR* (*Engl. Transl.*), 1962, **32**, 3330.	770, 852
62JGU3689	G. I. Derkach, V. A. Sholov, L. I. Samarai and A. V. Kirsanov; *J. Gen. Chem. USSR* (*Engl. Transl.*), 1962, **32**, 3689.	860
62JOC304	R. E. McClure and A. Ross; *J. Org. Chem.*, 1962, **27**, 304.	831
62JOC413	B. T. Gillis and K. F. Schimmel; *J. Org. Chem.*, 1962, **27**, 413.	356
62JOC629	H. Rapoport and W. Nilsson; *J. Org. Chem.*, 1962, **27**, 629.	54
62JOC998	D. B. Denney and S. T. Ross; *J. Org. Chem.*, 1962, **27**, 998.	895, 896
62JOC1042	W. B. Wright, Jr.; *J. Org. Chem.*, 1962, **27**, 1042.	302
62JOC1255	F. C. Schaefer and A. P. Krapcho; *J. Org. Chem.*, 1962, **27**, 1255.	750
62JOC1383	S. M. Gadekar and J. L. Frederick; *J. Org. Chem.*, 1962, **27**, 1383.	368
62JOC1597	K. Nakagawa, R. Konaka and T. Nakata; *J. Org. Chem.*, 1962, **27**, 1597.	48, 113
62JOC2205	H. E. Johnson and D. G. Crosby; *J. Org. Chem.*, 1962, **27**, 2205.	273, 285
62JOC2433	M. A. McCall; *J. Org. Chem.*, 1962, **27**, 2433.	580, 615, 616
62JOC2585	H. Ulrich, E. Kober, H. Schroeder, R. Rätz and C. Grundmann; *J. Org. Chem.*, 1962, **27**, 2585.	656
62JOC3146	A. C. Duckworth; *J. Org. Chem.*, 1962, **27**, 3146.	199
62JOC3240	H. C. Brown and M. T. Cheng; *J. Org. Chem.*, 1962, **27**, 3240.	854
62JOC3671	H. G. Mautner, S.-H. Chu and C. M. Lee; *J. Org. Chem.*, 1962, **27**, 3671.	625
62JOC3701	J. C. Howard and G. Klein; *J. Org. Chem.*, 1962, **27**, 3701.	1028
62JOC3742	A. J. Speziale and L. R. Smith; *J. Org. Chem.*, 1962, **27**, 3742.	995
62JOC3890	J. T. Shaw; *J. Org. Chem.*, 1962, **27**, 3890.	1125
62JOC4361	A. J. Speziale and L. R. Smith; *J. Org. Chem.*, 1962, **27**, 4361.	996
62JOC4412	N. B. Mehta and J. Z. Strelitz; *J. Org. Chem.*, 1962, **27**, 4412.	595
62JOC4532	B. Altenkirk and S. S. Israelstam; *J. Org. Chem.*, 1962, **27**, 4532.	316
62JPR223	J. Runge and W. Treibs; *J. Prakt. Chem.*, 1962, **15**, 223.	362
62LA(657)104	H. A. Staab and G. Walther; *Justus Liebigs Ann. Chem.*, 1962, **657**, 104.	1029
62LA(660)60	W. Walter, J. Voss, J. Curts and H. Pawelzik; *Justus Liebigs Ann. Chem.*, 1962, **660**, 60.	620, 623, 624
62LA(660)74	W. Walter and K.-D. Bode; *Justus Liebigs Ann. Chem.*, 1962, **660**, 74.	591, 601, 623, 624
62M352	J. Pump and U. Wannagat; *Monatsh. Chem.*, 1962, **93**, 352.	375, 378
62MI 502-01	D. B. Wetlaufer; *Adv. Protein Chem.*, 1962, **17**, 303.	85
B-62MI 502-01	J. G. Wallace; "Hydrogen Peroxide in Organic Chemistry," E. I. du Pont de Nemours Co., New York, 1962.	46
B-62MI 502-02	G. T. Young (ed.); "Proceedings Fifth European Symposium on Peptides, Oxford, Sept. 1962," Pergamon Press, New York.	84
B-62MI 502-03	H. D. Springall; in "Chemistry of Carbon Compounds," ed. E. H. Rodd, Elsevier, Amsterdam, 1962, p. 185.	85
62MI 527-01	S. Yoneda, H. Kitano and K. Fukui; *Kogyo Kagaku Zasshi*, 1962, **65**, 1816 (*Chem. Abstr.*, 1963, **59**, 2679).	1039
62OR(12)157	N. F. Albertson; *Org. React.*, 1962, **12**, 157.	278
62PCS117	R. Filler, L. Gorelic and B. Taqui-Khan; *Proc. Chem. Soc.*, 1962, 117.	55, 89
62RCP3	J. Rudinger; *Rec. Chem. Prog.*, 1962, **23**, 3	84
62RTC430	W. W. Limburg and H. W. Post; *Recl. Trav. Chim. Pays-Bas*, 1962, **81**, 430.	379
62T3	R. Huisgen, M. Seidel, G. Wallbillich and H. Knupfer; *Tetrahedron*, 1962, **17**, 3.	201, 842
62T69	A. N. Nesmeyanov, R. K. Freidlina, V. N. Kost, T. T. Vasilyeva and B. V. Kopylova; *Tetrahedron*, 1962, **17**, 69.	55, 60, 89
62USP3017420	W. D. Schaeffer (Union Oil Co. and Cal); *US Pat.*, 3 017 420 (1962) (*Chem. Abstr.*, 1962, **57**, 664i).	964
62USP3023208	W. J. Chambers and D. D. Coffman; *US Pat.* 3 023 208 (1962) (*Chem. Abstr.*, 1962, **57**, 11 216a).	808
62USP3048629	K. V. Martin; *US Pat.* 3 048 629 (1962) (*Chem. Abstr.*, 1963, **58**, 454).	473, 479
62USP3061590	D. C. Remy; *US Pat.* 3 061 590 (1962) (*Chem. Abstr.*, 1963, **58**, 8057).	847
62USP3069395	W. J. Middleton; *US Pat.* 3 069 395 (1962) (*Chem. Abstr.*, 1963, **59**, 1493).	473, 474, 475, 479
62USP3070618	F. E. Drummond (Basic Research Corp.); *US Pat.*, 3 070 618 (1962) (*Chem. Abstr.*, 1963, **59**, 9886c).	974
62ZN(B)135	E. Tobler and D. J. Foster; *Z. Naturforsch., Teil B*, 1962, **17**, 135.	388
62ZOB171	G. I. Derkach, A. M. Lepesa and A. V. Kirsanov; *Zh. Obshch. Khim.*, 1962, **32**, 171 (*Chem. Abstr.*, 1962, **57**, 16 472e).	708
62ZOB1873	G. I. Derkach, E. S. Gubnitskaya, V. A. Shokol and A. V. Kirsanov; *Zh. Obshch. Khim.*, 1962, **32**, 1873 (*Chem. Abstr.*, 1963, **58**, 6857).	860
62ZOB2245	G. I. Derkach and A. V. Kirsanov; *Zh. Obshch. Khim.*, 1962, **32**, 2245.	861

62ZOB3759 G. I. Derkach, L. I. Samarai, A. S. Shtepanek and A. V. Kirsanov; *Zh. Obshch. Khim.*, 1962, **32**, 3759 (*Chem. Abstr.*, 1963, **58**, 13 985e). 988

63AC(P)315 L. Duhamel; *Ann. Chim.* (*Paris*), 1963, **8**, 315. 62
63AC(R)1113 T. Tarantelli and D. Leonesi; *Ann. Chim.* (*Rome*), 1963, **53**, 1113. 1054
63ACS95 J. Sandström; *Acta Chem. Scand.*, 1963, **17**, 95. 846
63ACS1459 C. T. Pedersen; *Acta Chem. Scand.*, 1963, **17**, 1459. 1053, 1054
63AG93 L. Birkofer, A. Ritter and W. Gießler; *Angew. Chem.*, 1963, **75**, 93. 379
63AG96 L. Birkofer, A. Ritter and W. Giessler; *Angew. Chem.*, 1963, **75**, 96. 718
63AG(E)211 H. Vorbrüggen; *Angew. Chem., Int. Ed. Engl.*, 1963, **75**, 211. 131
63AG(E)212 H. Brechbühler, H. Büchi, E. Hatz, J. Schreiber and A. Eschenmoser; *Angew. Chem., Int. Ed. Engl.*, 1963, **75**, 212. 131
63AG(E)213 C. Heathcock and A. Hassner; *Angew. Chem., Int. Ed. Engl.*, 1963, **4**, 213. 984
63AG(E)217 T. Kauffmann, S. Spaude and D. Wolf; *Angew. Chem., Int. Ed. Engl.*, 1963, **2**, 217. 848
63AG(E)309 C. Grundmann; *Angew. Chem., Int. Ed. Engl.*, 1963, **2**, 309. 770, 855
63AG(E)358 T. Wieland and H. Determann; *Angew. Chem., Int. Ed. Engl.*, 1963, **2**, 358. 84
63AG(E)397 F. Asinger and F. Gentz; *Angew. Chem., Int. Ed. Eng.*, 1963, **2**, 397. 579
63AG(E)552 A. Haas; *Angew. Chem., Int. Ed. Engl.*, 1963, **2**, 552. 995, 996, 1018
63AG(E)719 K. E. Moller; *Angew. Chem., Int. Ed. Engl.*, 1963, **2**, 719. 27
63AP(296)57 G. Zinner; *Arch. Pharm.* (*Weinheim, Ger.*), 1963, **296**, 57. 213
63AP(296)420 G. Zinner; *Arch. Pharm.* (*Weinheim, Ger.*), 1963, **296**, 420. 213
63BCJ95 T. Mukaiyama, H. Takei and Y. Koma; *Bull. Chem. Soc. Jpn.*, 1963, **36**, 95. 997
63BCJ1214 A. Takamizawa, K. Hirai and K. Matsui; *Bull. Chem. Soc. Jpn.*, 1963, **36**, 1214. 1048
63BCJ1264 K. Saotome and T. Yamazaki; *Bull. Chem. Soc. Jpn.*, 1963, **36**, 1264. 57, 59, 89
63BEP631961 H. Pelster, E. Muehlbauer and D. Delfs (Farbenfabriken Bayer A.-G.); *Belg. Pat.*, 631 961 (1963) (*Chem. Abstr.*, 1964, **61**, 6919f). 977
63BRP923581 H. C. Fielding and J. M. Pollock; *Br. Pat.*, 923 581 (1963) (*Chem. Abstr.*, 1964, **60**, 2758). 1016
63BSB304 C. Collard-Charon and M. Renson; *Bull. Soc. Chim. Belg.*, 1963, **72**, 304. 624, 625
63CB526 J. Goerdeler and H. W. Pohland; *Chem. Ber.*, 1963, **96**, 526. 597, 599
63CB1387 H. Gross and J. Gloede; *Chem. Ber.*, 1963, **96**, 1387. 976
63CB1473 L. Birkofer, A. Ritter and H. Dickopp; *Chem. Ber.*, 1963, **96**, 1473. 718
63CB1551 J. Goerdeler and H. Horn; *Chem. Ber.*, 1963, **96**, 1551. 581, 597, 599
63CB2138 C. Krüger, E. G. Rochow and U. Wannagat; *Chem. Ber.*, 1963, **96**, 2138. 721
63CB2671 H. Eilingsfeld, M. Seefelder and H. Weidinger; *Chem. Ber.*, 1963, **96**, 2671. 577, 608, 609, 610, 728
63CB3120 M. Regitz and B. Eistert; *Chem. Ber.*, 1963, **96**, 3120. 855
63CI(L)775 G. G. Price and M. C. Whiting; *Chem. Ind.* (*London*), 1963, 775. 106
63CI(L)1280 J. L. McClanahan and J. L. Harper; *Chem. Ind.* (*London*), 1963, 1280. 985
63CJC2351 A. G. Brook and G. J. D. Peddle; *Can. J. Chem.*, 1963, **41**, 2351. 410
63CRV625 C. W. Huffman and W. G. Skelly; *Chem. Rev.*, 1963, **63**, 625. 84
63DOK(151)856 N. A. Nesmeyanov, S. T. Zhuzhlikova and O. A. Reutov; *Dokl. Akad. Nauk SSSR*, 1963, **151**, 856 (*Chem. Abstr.*, 1963, **59**, 12 838a). 902
63JA616 H. L. Finkbeiner and M. Stiles; *J. Am. Chem. Soc.*, 1963, **85**, 616. 70, 95
63JA643 S. Wolfe, J. C. Godfrey, C. T. Holdrege and Y. G. Perron; *J. Am. Chem Soc.*, 1963, **85**, 643. 9
63JA749 J. F. Harris, Jr. and F. W. Stacey; *J. Am. Chem. Soc.*, 1963, **85**, 749. 474, 475
63JA827 H. O. Larson and E. K. W. Wat; *J. Am. Chem. Soc.*, 1963, **85**, 827. 332
63JA1918 R. B. King; *J. Am. Chem. Soc.*, 1963, **85**, 1918. 452
63JA1922 R. B. King; *J. Am. Chem. Soc.*, 1963, **85**, 1922. 943
63JA1997 D. H. Rammler and H. G. Khorana; *J. Am. Chem. Soc.*, 1963, **85**, 1997. 185
63JA2181 A. Hassner and J. Larkin; *J. Am. Chem. Soc.*, 1963, **85**, 2181. 332
63JA2676 J. F. King and T. Durst; *J. Am. Chem. Soc.*, 1963, **85**, 2676. 489
63JA2743 H. W. Heine; *J. Am. Chem. Soc.*, 1963, **85**, 2743. 1010
63JA3231 W. E. Truce and J. R. Norell; *J. Am. Chem. Soc.*, 1963, **85**, 3231. 496
63JA3288 L. A. Paquette; *J. Am. Chem. Soc.*, 1963, **85**, 3288. 271, 288
63JCS229 G. E. Coates and R. N. Mukherjee; *J. Chem. Soc.*, 1963, 229. 1137, 1142, 1143
63JCS491 E. G. Brain, F. P. Doyle, M. D. Mehta, D. Miller, J. H. C. Nayler and E. R. Stove; *J. Chem. Soc.*, 1963, 491. 9
63JCS1272 H. J. Emeléus and A. Haas; *J. Chem. Soc.*, 1963, 1272. 1018
63JCS1927 J. M. Theobald, M. W. Williams and G. T. Young; *J. Chem. Soc.*, 1963, 1927. 122
63JCS2918 C. A. Bunton, N. A. Fuller, S. G. Perry and V. J. Shiner; *J. Chem. Soc.*, 1963, 2918. 26
63JCS4096 J. A. McCleverty and G. Wilkinson; *J. Chem. Soc.*, 1963, 4096. 455
63JCS4918 J. E. Colchester, P. Tavs and H. J. Schulze-Steinen; *J. Chem. Soc.*, 1963, 4918. 338
63JCS5558 H. Ulrich and A. A. R. Sayigh; *J. Chem. Soc.*, 1963, 5558. 982, 1070
63JGU551 G. I. Derkach, E. S. Gubnitskaya, L. I. Samarai and V. A. Shokol; *J. Gen. Chem. USSR* (*Engl. Transl.*), 1963, **33**, 551. 708
63JGU1006 G. I. Derkach, G. K. Fedorova and E. S. Gubnitskaya; *J. Gen. Chem. USSR* (*Engl. Transl.*), 1963, **33**, 1006. 708
63JGU2880 G. F. Dregval and G. I. Derkach; *J. Gen. Chem. USSR* (*Engl. Transl.*), 1963, **33**, 2880. 861
63JGU2984 V. A. Shokol, G. I. Derkach and E. S. Gubnitskaya; *J. Gen. Chem. USSR* (*Engl. Transl.*), 1963, **33**, 2984. 861
63JOC194 E. Lieber, E. Oftedahl and C. N. R. Rao; *J. Org. Chem.*, 1963, **28**, 194. 1046
63JOC465 A. J. Speziale and K. W. Ratts; *J. Org. Chem.*, 1963, **28**, 465. 895, 896
63JOC586 P. R. Steyermark; *J. Org. Chem.*, 1963, **28**, 586. 994, 1016

63JOC1108 E. C. Taylor and W. A. Ehrhart; *J. Org. Chem.*, 1963, **28**, 1108. 743, 746, 753
63JOC1600 S. E. Ellzey Jr. and C. H. Mack; *J. Org. Chem.*, 1963, **28**, 1600. 592, 609
63JOC1673 D. T. Manning and H. A. Stansbury, Jr.; *J. Org. Chem.*, 1963, **28**, 1673. 823
63JOC1805 A. J. Speziale and L. R. Smith; *J. Org. Chem.*, 1963, **28**, 1805. 995
63JOC1825 F. W. Hoover, H. B. Stevenson and H. S. Rothrock; *J. Org. Chem.*, 1963, **28**, 1825. 986
63JOC1905 J. S. Nelson, L. A. Goldblatt and T. H. Applewhite; *J. Org. Chem.*, 1963, **28**, 1905. 189
63JOC2076 J. N. Tilley and A. A. R. Sayigh; *J. Org. Chem.*, 1963, **28**, 2076. 971
63JOC2082 F. W. Hoover and H. S. Rothrock; *J. Org. Chem.*, 1963, **28**, 2082. 985
63JOC2168 F. D. Greene and J. Kazan; *J. Org. Chem.*, 1963, **28**, 2168. 208
63JOC2184 W. L. Meyer and A. S. Levinson; *J. Org. Chem.*, 1963, **28**, 2184. 39
63JOC2260 M. S. Chang and A. J. Matuszko; *J. Org. Chem.*, 1963, **28**, 2260. 821
63JOC2294 R. Campbell and C. J. Peterson; *J. Org. Chem.*, 1963, **28**, 2294. 361
63JOC2915 R. J. De Feoand and P. D. Strickler; *J. Org. Chem.*, 1963, **28**, 2915. 277
63JOC3492 A. J. Speziale and L. R. Smith; *J. Org. Chem.*, 1963, **28**, 3492. 566, 577, 611
63JOC3512 T. L. Westman, R. Paredes and W. S. Brey, Jr.; *J. Org. Chem.*, 1963, **28**, 3512. 144
63JPR214 R. Mayer and S. Scheithauer; *J. Prakt. Chem.*, 1963, **21**, 214. 476, 477, 510
63LA(669)66 W. Walter and G. Maerten; *Justus Liebigs Ann. Chem.*, 1963, **669**, 66. 590, 609
63MI 502-01 A. T. James; *The Analyst*, 1963, **88**, 572. 97
B-63MI 502-01 D. E. Green; in "Progress in the Chemistry of Fats and Other Lipids," eds. R. T. Holman, W. O. Lundberg and T. Malkin, Pergamon Press, Oxford, 1963, vol. 6, p. 88. 28
B-63MI 502-02 R. G. Kadesch; in "Progress in the Chemistry of Fats and Other Lipids," eds. R. T. Holman, W. O. Lundberg and T. Malkin, Pergamon Press, Oxford, 1963, vol. 6, p. 292. 28
B-63MI 502-03 H. Neurath (ed.); "Composition, Structure and Function of Proteins," Academic Press, New York, 2nd Ed., 1963. 28
63MI 504-01 T. H. Applewhite, J. S. Nelson and L. A. Goldblatt; *J. Am. Oil Chem. Soc.*, 1963, **40**, 101. 189
63MI 514-01 V. Grinsteins and L. Serina; *Latvijas P.S.R. Zinatnu Akad. Vestis, Kim. Ser.*, 1963, 469 (*Chem. Abstr.*, 1964, **60**, 5391h). 615
B-63MI 526-01 P. A. S. Smith; in "Molecular Rearrangements," ed. P. de Mayo, Interscience, New York, 1963, p. 528. 989
63MI 530-01 Y. Ueda and E. Mosetlig; *Steroids*, 1963, **1**, 361. 1106
63OSC(4)5 R. H. Wiley and O. H. Borum; *Org. Synth., Coll. Vol.*, 1963, **4**, 5. 264
63OSC(4)19 J. R. Schaeffer and A. O. Snoddy; *Org. Synth., Coll. Vol.*, 1963, **4**, 19. 73
63OSC(4)39 C. Y. Meyers and L. E. Miller; *Org. Synth., Coll. Vol.*, 1963, **4**, 39. 86
63OSC(4)55 M. S. Dunn and B. W. Smart; *Org. Synth., Coll. Vol.*, 1963, **4**, 55. 90
63OSC(4)58 E. L. Eliel and J. P. Freeman; *Org. Synth., Coll. Vol.*, 1963, **4**, 58. 63, 66, 275
63OSC(4)93 F. S. Prout, R. J. Hartman, E. P. Y. Huang, C. J. Korpics and G. R. Tichelaar; *Org. Synth., Coll. Vol.*, 1963, **4**, 93. 34
63OSC(4)104 E. P. Oliveto and C. Gerold; *Org. Synth., Coll. Vol.* 1963, **4**, 104. 318
63OSC(4)106 T. O. Soine and M. R. Buchdahl; *Org. Synth., Coll. Vol.*, 1963, **4**, 106. 305
63OSC(4)110 J. J. Klingenberg; *Org. Synth., Coll. Vol.*, 1963, **4**, 110. 66
63OSC(4)169 E. L. Eliel, M. T. Fisk and T. Prosser; *Org. Synth., Coll. Vol.*, 1963, **4**, 169. 55
63OSC(4)213 F. Kurzer and J. R. Powell; *Org. Synth. Coll. Vol.*, 1963, **4**, 213. 1120
63OSC(4)242 R. L. Shriner and C. Furrow, Jr.; *Org. Synth., Coll. Vol.*, 1963, **4**, 242. 199
63OSC(4)339 A. C. Cope and E. Ciganek; *Org. Synth., Coll. Vol.*, 1963, **4**, 339. 278
63OSC(4)441 H. H. Farmer and N. Rabjohn; *Org. Synth., Coll. Vol.*, 1963, **4**, 441. 37
63OSC(4)444 K. L. Rinehart, Jr. and E. G. Perkins; *Org. Synth., Coll. Vol.*, 1963, **4**, 444. 98
63OSC(4)484 R. E. Dessy and M. S. Newman; *Org. Synth., Coll. Vol.*, 1963, **4**, 484. 115
63OSC(4)493 R. J. Harrisson and M. Moyle; *Org. Synth., Coll. Vol.*, 1963, **4**, 493. 74
63OSC(4)496 G. Paris, L. Berlinguet and R. Gaudry; *Org. Synth., Coll. Vol.*, 1963, **4**, 496. 34, 275, 305
63OSC(4)499 J. English, Jr. and J. E. Dayan; *Org. Synth., Coll. Vol.*, 1963, **4**, 499. 46
63OSC(4)513 J. L. Guthrie and N. Rabjohn; *Org. Synth., Coll. Vol.*, 1963, **4**, 513. 275
63OSC(4)532 J. H. Boyer and W. Schoen; *Org. Synth., Coll. Vol.*, 1963, **4**, 532. 106
63OSC(4)608 C. F. Allen and M. J. Kalm; *Org. Synth., Coll. Vol.*, 1963, **4**, 608. 98
63OSC(4)616 C. F. Allen and M. J. Kalm; *Org. Synth., Coll. Vol.*, 1963, **4**, 616. 99
63OSC(4)628 D. M. Burness; *Org. Synth., Coll. Vol.*, 1963, **4**, 628. 62
63OSC(4)633 G. A. Reynolds and J. A. VanAllan; *Org. Synth., Coll. Vol.*, 1963, **4**, 633. 62
63OSC(4)635 L. J. Durham, D. J. McLeod and J. Cason; *Org. Synth., Coll. Vol.*, 1963, **4**, 635. 32
63OSC(4)700 J. Cymerman-Craig, M. Moyle and R. A. White; *Org. Synth., Coll. Vol.*, 1963, **4**, 700. 1035
63OSC(4)760 W. Wenner; *Org. Synth., Coll. Vol.*, 1963, **4**, 760. 275
63OSC(4)766 C. S. Rondestvedt, Jr.; *Org. Synth., Coll. Vol.*, 1963, **4**, 766. 201
63OSC(4)790 E. C. Horning and A. F. Finelli; *Org. Synth., Coll. Vol.*, 1963, **4**, 790. 35, 199
63OSC(4)804 C. F. H. Allen and H. B. Johnson; *Org. Synth., Coll. Vol.*, 1963, **4**, 804. 34
63OSC(4)824 R. G. Jones and K. C. McLaughlin; *Org. Synth., Coll. Vol.*, 1963, **4**, 824. 115
63OSC(4)890 A. C. Cope and E. C. Herrick; *Org. Synth., Coll. Vol.*, 1963, **4**, 890. 199
63OSC(4)919 E. Campaigne and W. M. LeSuer; *Org. Synth., Coll. Vol.*, 1963, **4**, 919. 83
63OSC(4)924 P. Noble, Jr. and D. S. Tarbell; *Org. Synth., Coll. Vol.*, 1963, **4**, 924. 232
63OSC(4)928 E. K. Ellingboe; *Org. Synth., Coll. Vol.*, 1963, **4**, 928. 232
63OSC(4)972 I. A. Pearl; *Org. Synth., Coll. Vol.*, 1963, **4**, 972. 114
63PAC335 J. Rudinger; *Pure Appl. Chem.*, 1963, **7**, 335. 84
63SA1621 A. Finch and E. J. Pearn; *Spectrochim. Acta*, 1963, **19**, 1621. 1015
63TL77 D. Elad; *Tetrahedron Lett.*, 1963, 77. 298
63TL585 J. F. King and T. Durst; *Tetrahedron Lett.*, 1963, **4**, 585. 488, 489
63TL1437 J. Tsuji, M. Morikawa and J. Kiji; *Tetrahedron Lett.*, 1963, 1437. 142

63USP3076843 F. W. Stacey; *US Pat.* 3 076 843 (1963) (*Chem. Abstr.*, 1963, **59**, 2655d). 808

63USP3097236 D. M. Marquis; *US Pat.* 3 097 236 (1963) (*Chem. Abstr.*, 1963, **59**, 13 825). 473

63USP3113936 W. J. Middleton; *US Pat.* 3 113 936 (1963) (*Chem. Abstr.*, 1964, **60**, 4012). 479, 480

63ZOB557 G. I. Derkach, E. S. Gubnitskaya, L. I. Samarai and V. A. Shokol; *Zh. Obshch. Khim.*, 1963, **33**, 557 (*Chem. Abstr.*, 1963, **59**, 2684c). 708

63ZOB1017 G. I. Derkach, G. K. Fedorova and E. S. Gubnitskaya; *Zh. Obshch. Khim.*, 1963, **33**, 1017 (*Chem. Abstr.*, 1963, **59**, 8783b). 860

63ZOB1587 G. I. Derkach and L. I. Samarai; *Zh. Obshch. Khim.*, 1963, **33**, 1587 (*Chem. Abstr.*, 1963, **59**, 12 669d). 988

63ZOB3293 A. V. Fokin, Y. N. Studnev and A. A. Skladnev; *Zh. Obshch. Khim.* (*Engl. Trans.*), 1963, **33**, 3293. 645

63ZOB3917 R. G. Dubenko and P. S. Pelkis; *Zh. Obshch. Khim.*, 1963, **33**, 3917 (*Chem. Abstr.*, 1964, **60**, 9186). 843

64ACS825 K. A. Jensen, A. Holm and B. Thorkilsen; *Acta Chem. Scand.*, 1964, **18**, 825. 1101

64ACS826 K. A. Jensen and A. Holm; *Acta Chem. Scand.*, 1964, **18**, 826. 1101

64ACS1567 L. Eberson and K. Nyberg; *Acta Chem. Scand.*, 1964, **18**, 1567. 1102

64ACS2417 K. A. Jensen and A. Holm; *Acta Chem. Scand.*, 1964, **18**, 2417. 1103

64AG440 G. Zinner and H. Günther; *Angew. Chem.*, 1964, **76**, 440. 837

64AG781 K. Hartke; *Angew. Chem.*, 1964, **76**, 781. 1064

64AG861 R. Mayer and H. J. Frey; *Angew. Chem.*, 1964, **76**, 861. 1111

64AG(E)19 F. Asinger, W. Schäfer, K. Halcour, A. Saus and H. Triem; *Angew. Chem., Int. Ed. Engl.*, 1964, **3**, 19. 599

64AG(E)132 H. A. Staab and A. P. Datta; *Angew. Chem., Int. Ed. Engl.*, 1964, **3**, 132. 477

64AG(E)228 A. Messmer, I. Pintér and F. Szegö; *Angew. Chem., Int. Ed. Engl.*, 1964, **3**, 228. 1079

64AG(E)250 L. D. Bergelson and M. M. Shemyakin; *Angew. Chem., Int. Ed. Engl.*, 1964, **3**, 250. 100

64AG(E)287 M. Schlosser; *Angew. Chem., Int. Ed. Engl.*, 1964, **3**, 287. 436

64AG(E)311 D. Martin; *Angew. Chem., Int. Ed. Engl.*, 1964, **3**, 311. 1101

64AG(E)384 K. Dimroth and P. Hoffmann; *Angew. Chem., Int. Ed. Engl.*, 1964, **3**, 384. 876, 882

64AG(E)580 E. O. Fischer and A. Maasböl; *Angew. Chem., Int. Ed. Engl.*, 1964, **3**, 580. 932, 933, 934, 948

64AG(E)585 H. Ulrich and A. A. R. Sayigh; *Angew. Chem., Int. Ed. Engl.*, 1964, **3**, 585. 1003

64AG(E)705 R. Mayer and J. Wehl; *Angew. Chem., Int. Ed. Engl.*, 1964, **3**, 705. 599

64BSB419 M. Renson and R. Collienne; *Bull. Soc. Chim. Belg.*, 1964, **73**, 419. 515, 516

64BSF542 G. Schroeder and R. Lombard; *Bull. Soc. Chim. Fr.*, 1964, 542. 208

64CB483 R. Albrecht, G. Kresze and B. Mlakar; *Chem. Ber.*, 1964, **97**, 483. 652

64CB627 J. Pump and E. G. Rochow; *Chem. Ber.*, 1964, **97**, 627. 718

64CB849 R. Appel and H. Rittersbacher; *Chem. Ber.*, 1964, **97**, 849. 1018

64CB1361 W. Jentzsch; *Chem. Ber.*, 1964, **97**, 1361. 744

64CB1886 R. Mayer and H. Hartmann; *Chem. Ber.*, 1964, **97**, 1886. 559

64CB2209 J. Goerdeler and U. Keuser; *Chem. Ber.*, 1964, **97**, 2209. 593, 597

64CB2661 H. Reinheckel and D. Jahnke; *Chem. Ber.*, 1964, **97**, 2661. 595

64CB3012 E. Grigat and E. Pütter; *Chem. Ber.*, 1964, **97**, 3012. 1102

64CB3106 J. Goerdeler and U. Keuser; *Chem. Ber.*, 1964, **97**, 3106. 595, 597, 599, 614

64CB3234 W. Haaf; *Chem. Ber.*, 1964, **97**, 3234. 282

64CB3436 T. Kauffmann, S. Spaude and D. Wolf; *Chem. Ber.*, 1964, **97**, 3436. 769, 848

64CI(L)193 B. Loev; *Chem. Ind.* (*London*), 1964, 193. 29, 30, 92

64CI(L)544 S. M. Morehouse, A. R. Powell, J. P. Heffer, T. A. Stephenson and G. Wilkinson; *Chem. Ind.* (*London*), 1964, **13**, 544. 229

64CI(L)1867 J. Kollonitsch and A. Rosegay; *Chem. Ind.* (*London*), 1964, 1867. 199

64CJC2393 M. H. Benn; *Can. J. Chem.*, 1964, **42**, 2393. 832

64CPB403 K. Nakagawa, K. Igano and J. Sugita; *Chem. Pharm. Bull.*, 1964, **12**, 403. 48, 114

64CR4067 P. Reynaud, R. C. Moreau and T. Gousson; *C. R. Acad. Sci.*, 1964, **259**, 4067. 753

64FRP1469481 Shionogi and Co. Ltd; *Fr. Pat.* 1 469 481 (1964) (*Chem. Abstr.*, 1968, **68**, 96 052w). 1142, 1143

64GEP1154090 Farbenfabriken Bayer A.-G.; *Ger. Pat.*, 1 154 090 (1964) (*Chem. Abstr.*, 1964, **60**, 2758). 981

64GEP1170398 Farbenfabriken Bayer A.-G.; *Ger. Pat.*, 1 170 398 (1964) (*Chem. Abstr.*, 1964, **64**, 4267a). 966

64GEP1205087 K. F. Zenner, G. Oertel and H. Holtschmidt; *Ger. Pat.*, 1 205 087 (1964) (*Chem. Abstr.*, 1966, **64**, 19 413). 964

64IC910 H. H. Anderson; *Inorg. Chem.*, 1964, **3**, 910. 1136

64JA357 C. W. Tullock, D. D. Coffman and E. L. Muetterties; *J. Am. Chem. Soc.*, 1964, **86**, 357. 675, 975

64JA837 J. K. Kirsch and W. P. Jencks; *J. Am. Chem. Soc.*, 1964, **86**, 837. 31

64JA965 R. K. Hill and M. Rabinovitz; *J. Am. Chem. Soc.*, 1964, **86**, 965. 201

64JA1267 R. B. King and M. B. Bisnette; *J. Am. Chem. Soc.*, 1964, **86**, 1267. 448

64JA1839 G. W. Anderson, J. E. Zimmerman and F. M. Callahan; *J. Am. Chem. Soc.*, 1964, **86**, 1839. 277

64JA3994 R. J. Mawby, F. Basolo and R. G. Pearson; *J. Am. Chem. Soc.*, 1964, **86**, 3994. 447

64JA4732 J. C. Kauer and W. W. Henderson; *J. Am. Chem. Soc.*, 1964, **86**, 4732. 1100, 1102

64JCS755 R. K. Smalley and H. Suschitzky; *J. Chem. Soc.*, 1964, 755. 183, 194, 198

64JCS1189 J. T. Lumb and G. H. Whitham; *J. Chem. Soc.*, 1964, 1189. 709

64JCS1190 A. G. Long and A. Tulley; *J. Chem. Soc.*, 1190. 570

64JCS2108 R. H. Cragg, M. F. Lappert and B. P. Tilley; *J. Chem. Soc.*, 1964, 2108. 1038

64JCS2944 K. V. Martin; *J. Chem. Soc.*, 1964, 2944. 473, 479

64JCS5679 P. Hope and L. A. Wiles; *J. Chem. Soc.*, 1964, 5679. 334

64JGU1152 G. I. Derkach and L. I. Samarai; *J. Gen. Chem. USSR* (*Engl. Transl.*), 1964, **34**, 1152. 708

64JOC143 F. W. Hoover and H. S. Rothrock; *J. Org. Chem.*, 1964, **29**, 143. 985

64JOC246	H. W. Johnson, Jr. and P. H. Daughhetee, Jr.; *J. Org. Chem.*, 1964, **29**, 246.	1003
64JOC492	I. H. Skoog; *J. Org. Chem.*, 1964, **29**, 492.	228
64JOC1115	G. DeStevens, B. Smolensky and L. Dorfman; *J. Org. Chem.*, 1964, **29**, 1115.	631
64JOC1385	I. Kuwajima and T. Mukaiyama; *J. Org. Chem.*, 1964, **29**, 1385.	190
64JOC1855	D. Elad and J. Rokach; *J. Org. Chem.*, 1964, **29**, 1855.	276
64JOC2006	J. C. Sheehan and G. D. Daves, Jr.; *J. Org. Chem.*, 1964, **29**, 2006.	106
64JOC2091	W. E. Erner; *J. Org. Chem.*, 1964, **29**, 2091.	1009, 1037
64JOC2099	J. R. Moyer and N. C. Manley; *J. Org. Chem.*, 1964, **29**, 2099.	205
64JOC2261	P. A. S. Smith and R. O. Kan; *J. Org. Chem.*, 1964, **29**, 2261.	1047
64JOC2427	J. S. Driscoll, D. W. Grisley, Jr., J. V. Pustinger, J. E. Harris and C. N. Matthews; *J. Org. Chem.*, 1964, **29**, 2427.	1093
64JOC2592	J. E. Franz and C. Osuch; *J. Org. Chem.*, 1964, **29**, 2592.	996
64JOC2810	C. H. dePuy, F. W. Breitbeil and K. L. Eilers; *J. Org. Chem.*, 1964, **29**, 2810.	166
64JOC2813	C. H. DePuy, G. M. Dappen, K. L. Eilers and R. A. Klein; *J. Org. Chem.*, 1964, **29**, 2813.	142
64JOC2816	W. S. Wadsworth, Jr. and W. D. Emmons; *J. Org. Chem.*, 1964, **29**, 2816.	975, 1031, 1069
64JOC2854	L. A. Paquette; *J. Org. Chem.*, 1964, **29**, 2854.	496, 497
64JOC2932	D. H. Clemens, A. J. Bell and J. L. O'Brien; *J. Org. Chem.*, 1964, **29**, 2932.	599
64JOC2993	H. Haubenstock and C. VanderWerf; *J. Org. Chem.*, 1964, **29**, 2993.	203
64JOC3098	J. E. Hodgkins and W. P. Reeves; *J. Org. Chem.*, 1964, **29**, 3098.	1024
64JOC3344	A. A. R. Sayigh, J. N. Tilley and H. Ulrich; *J. Org. Chem.*, 1964, **29**, 3344.	980
64JOC3697	A. Larizza, G. Brancaccio and G. Lettieri; *J. Org. Chem.*, 1964, **29**, 3697.	746
64JOM(2)15	R. B. King and M. B. Bisnette; *J. Organomet. Chem.*, 1964, **2**, 15.	452
64LA(673)124	W. Ried and B. M. Beck; *Justus Liebigs Ann. Chem.*, 1964, **673**, 124.	482
64LA(673)132	W. Ried and W. Käppeler; *Justus Liebigs Ann. Chem.*, 1964, **673**, 132.	597, 599, 614
64LA(676)114	W. Ried and E. Schmidt; *Liebigs Ann. Chem.*, 1964, **676**, 114.	726
64LA(678)132	W. Ried and H. Kramm; *Justus Liebigs Ann. Chem.*, 1964, **678**, 132.	856
64MI 502-01	C. H. W. Hirs; *Annu. Rev. Biochem.*, 1964, **33**, 597.	84
B-64MI 502-01	R. Stewart (ed.); "Oxidation Mechanisms. Applications to Organic Chemistry," W. A. Benjamin, New York, 1964.	28
B-64MI 502-02	W. A. Waters; "Mechanism of Oxidation of Organic Compounds," Wiley, New York, 1964.	28
B-64MI 502-03	G. A. Olah and J. A. Olah; "Acid Synthesis in Friedel–Crafts and Related Reactions," ed. G. A. Olah, Interscience, New York, 1964, vol. 3, pp. 1257–1317.	105
64MI 506-01	T. Clarke, J. Devine and D. W. Dicker; *J. Am. Oil Chemists Soc.*, 1964, **41**, 78 (*Chem. Abstr.*, 1964, **60**, 6733).	652
64MI 530-01	M. Hedayatullah and L. Denivelle; *C. R. Acad. Sci. Paris*, 1964, **258**, 606.	1100
64OS(44)41	L. A. Paquette; *Org. Synth.*, 1964, **44**, 41.	271, 288
64RTC94	J. Strating; *Recl. Trav. Chim. Pays-Bas*, 1964, **83**, 94.	488, 489
64TL605	J. Tsuji, J. Kiji and S. Hosaka; *Tetrahedron Lett.*, 1964, 605.	158
64TL799	J. Goerdeler and D. Weber; *Tetrahedron Lett.*, 1964, **5**, 799.	1050
64TL855	H. M. Blatter and H. Lukaszewski; *Tetrahedron Lett.*, 1964, **5**, 855.	1050
64TL1087	H. M. Blatter and H. Lukaszewski; *Tetrahedron Lett.*, 1964, **5**, 1087.	1050
64TL1125	A. Hassner and C. C. Heathcock; *Tetrahedron Lett.*, 1964, 1125.	984
64TL2829	D. Martin; *Tetrahedron Lett.*, 1964, 2829.	1013, 1101
64TL2969	F. C. Chang and N. F. Wood; *Tetrahedron Lett.*, 1964, 2969.	31
64TL3363	S. G. Smith and J. P. Petrovich; *Tetrahedron Lett.*, 1964, **5**, 3363.	476, 477
64UKZ68	M. O. Lozinsky; *Ukr. Khim. Zh. (Russ. Ed.)*, 1964, **30**, 68.	843
64USP3136781	W. J. Middleton; *US Pat.* 3 136 781 (1964) (*Chem. Abstr.*, 1964, **61**, 5612).	476
64USP3155700	P. R. Steyermark (W. R. Grace & Co.); *US Pat.*, 3 155 700 (1964) (*Chem. Abstr.*, 1965, **62**, 444h).	994
64ZC457	R. Mayer and J. Orgis; *Z. Chem.*, 1964, **4**, 457.	584, 586, 609
64ZOB1161	G. I. Derkach and L. I. Samarai; *Zh. Obshch. Khim.*, 1964, **34**, 1161 (*Chem. Abstr.*, 1964, **61**, 1784f).	708, 988
64ZOB2812	A. V. Kirsanov, G. I. Derkach and N. I. Liptuga; *Zh. Obshch. Khim.*, 1964, **34**, 2812 (*Chem. Abstr.*, 1964, **61**, 14 514a).	1005, 1010
64ZOB3609	L. I. Bai, A. Y. Yakubovich and L. I. Muler; *Zh. Obshch. Khim.*, 1964, **34**, 3609 (*Chem. Abstr.*, 1965, **62**, 9168f).	1018
65ACS438	K. A. Jensen, M. Duc and A. Holm; *Acta Chem. Scand.*, 1965, **19**, 438.	1013, 1101
65ACS1768	N. Groving and A. Holm; *Acta Chem. Scand.*, 1965, **19**, 1768.	652
65ACS2395	J. Sandström; *Acta Chem. Scand.*, 1965, **19**, 2395.	846
65AG(E)370	A. Hassner; *Angew. Chem., Int. Ed. Engl.*, 1965, **4**, 370.	984
65AG(E)430	B. Anders and E. Kühle; *Angew. Chem., Int. Ed. Engl.*, 1965, **4**, 430.	1071
65AG(E)449	A. Hassner and C. Heathcock; *Angew. Chem., Int. Ed. Engl.*, 1965, **5**, 449.	984
65AG(E)846	U. Schmidt, P. Grafen and H. W. Goedde; *Angew. Chem., Int. Ed. Engl.*, 1965, **4**, 846.	81
65AG(E)1082	O. S. Mills and A. D. Redhouse; *Angew. Chem., Int. Ed. Engl.*, 1965, **4**, 1082.	932
65AG(E)1095	W. Walter, K.-D. Bode and P. Hell; *Angew. Chem., Int. Ed. Engl.*, 1965, **4**, 1095.	623
65AHC(5)123	F. Kurzer; *Adv. Heterocycl. Chem.*, 1965, **5**, 123.	635
65AP(298)51	R. Pohloudek-Fabini and F. Friedrich; *Arch. Pharm. (Weinheim, Ger.)*, 1965, **298**, 51.	1027
65BCJ320	Y. Hirabayashi, M. Mizuta and T. Mazume; *Bull. Chem. Soc. Jpn.*, 1965, **38**, 320.	232
65BCJ893	T. Suga, K. Kihara and T. Matsûura; *Bull. Chem. Soc. Jpn.*, 1965, **38**, 893.	47
65BSF1454	Y. H. Suen, A. Horeau and H. B. Kagan; *Bull. Soc. Chim. Fr.*, 1965, 1454.	290
65BSF3623	P. Reynaud, R. C. Moreau and J.-P. Samama; *Bull. Soc. Chim. Fr.*, 1965, 3623.	584, 586, 612
65BSF3628	P. Reynaud, R. C. Moreau and J.-P. Samama; *Bull. Soc. Chim. Fr.*, 1965, 3628.	586

65CB829	R. Mayer and S. Scheithauer; *Chem. Ber.*, 1965, **98**, 829.	476, 477, 484
65CB1097	G. Müller and R. Merten; *Chem. Ber.*, 1965, **98**, 1097.	979
65CB1308	H. Eilingsfeld; *Chem. Ber.*, 1965, **98**, 1308.	640, 641
65CB1531	J. Goerdeler and J. Gnad; *Chem. Ber.*, 1965, **98**, 1531.	597, 599
65CB1556	J. Goerdeler and K. Stadelbauer; *Chem. Ber.*, 1965, **98**, 1556.	630
65CB2128	J. C. Jochims; *Chem. Ber.*, 1965, **98**, 2128.	746
65CB2425	D. Martin, E. Beyer and H. Gross; *Chem. Ber.*, 1965, **98**, 2425.	1023
65CB2954	J. Goerdeler and H. Schenk; *Chem. Ber.*, 1965, **98**, 2954.	996
65CB3819	U. Schmidt, K. Kabitzke, I. Boie and C. Osterroht; *Chem. Ber.*, 1965, **98**, 3819.	612
65CC449	W. Wolf and L. Steinberg; *J. Chem. Soc., Chem. Commun.*, 1965, 449.	321
65CI(L)648	N. J. Doorenbos and M. T. Wu; *Chem. Ind.* (*London*), 1965, 648.	199
65CJC126	R. J. Crawford and R. Raap; *Can. J. Chem.*, 1965, **43**, 126.	845
65CJC1175	A. G. Brook, R. Kivisikk and G. E. LeGrow; *Can. J. Chem.*, 1965, **43**, 1175.	410
65CJC2640	G. R. Pettit and L. R. Garson; *Can. J. Chem.*, 1965, **43**, 2640.	745
65CRV369	H. Ulrich; *Chem. Rev.*, 1965, **65**, 369.	1018
65CRV435	B. A. Bohm; *Chem. Rev.*, 1965, **65**, 435.	62
65HOU(6/3)295	H. Meerwein; *Methoden Org. Chem.* (*Houben Weyl*), 1965, **6/3**, 295.	687
65HOU(10)67	R. Pütter; *Methoden Org. Chem.* (*Houben-Weyl*), 1965, **10**, 67.	360
65HOU(10/3)538	E. Enders; *Methoden Org. Chem.* (*Houben-Weyl*), 1965, **10/3**, 538.	843
65IC1458	N. E. Miller; *Inorg. Chem.*, 1965, **4**, 1458.	912, 913, 921, 1093
65IC1747	D. A. Nicholson and A. L. Allred; *Inorg. Chem.*, 1965, **4**, 1747.	432
65JA328	R. A. Mitsch; *J. Am. Chem. Soc.*, 1965, **87**, 328.	808
65JA390	N. E. Miller; *J. Am. Chem. Soc.*, 1965, **87**, 390.	913
65JA1141	H. E. Baumgarten and A. Staklis; *J. Am. Chem. Soc.*, 1965, **87**, 1141.	999
65JA2296	A. G. Anastassiou, H. E. Simmons and F. D. Marsh; *J. Am. Chem. Soc.*, 1965, **87**, 2296.	1119
65JA2492	J. C. Sheehan, J. Preston and P. A. Cruickshank; *J. Am. Chem. Soc.*, 1965, **87**, 2492.	279
65JA2776	H. E. Winberg and D. D. Coffman; *J. Am. Chem. Soc.*, 1965, **87**, 2776.	548
65JA3958	M. Miyano; *J. Am. Chem. Soc.*, 1965, **87**, 3958.	12
65JA4270	G. L. Closs and J. J. Coyle; *J. Am. Chem. Soc.*, 1965, **87**, 4270.	844
65JA4396	W. H. Graham; *J. Am. Chem. Soc.*, 1965, **87**, 4396.	763, 866
65JA4791	A. Fava, A. Iliceto and S. Bresadola; *J. Am. Chem. Soc.*, 1965, **87**, 4791.	1038, 1105
65JA5068	R. D. Partos and A. J. Speziale; *J. Am. Chem. Soc.*, 1965, **87**, 5068.	860
65JA5603	A. J. Speziale and K. W. Ratts; *J. Am. Chem. Soc.*, 1965, **87**, 5603.	901
65JCS1290	W. Roberts and M. C. Whiting; *J. Chem. Soc.*, 1965, 1290.	106, 298
65JCS2999	I. T. Barnish and M. S. Gibson; *J. Chem. Soc.*, 1965, 2999.	356
65JCS3312	L. A. Summers, P. F. H. Freeman and D. J. Shields; *J. Chem. Soc.*, 1965, 3312.	532
65JCS4256	M. F. Lappert, H. Pyszora and M. Rieber; *J. Chem. Soc.*, 1965, 4256.	1016
65JCS4594	P. F. E. Cook and A. J. Showler; *J. Chem. Soc.*, 1965, 4594.	123
65JCS5055	A. S. Mufti and R. C. Poller; *J. Chem. Soc.*, 1965, 5055.	1017
65JCS5230	N. Bacon, A. J. Boulton, R. T. C. Brownlee, A. R. Katritzky and R. D. Topsom; *J. Chem. Soc.*, 1965, 5230.	725
65JCS6655	P. D. G. Dean; *J. Chem. Soc.*, 1965, 6655.	78
65JGU1014	G. I. Derkach and E. S. Gubnitskaya; *J. Gen. Chem. USSR* (*Engl. Transl.*), 1965, **35**, 1014.	708, 709
65JGU1019	G. I. Derkach, E. S. Gubnitskaya and V. A. Shokol; *J. Gen. Chem. USSR* (*Engl. Transl.*), 1965, **35**, 1019.	683
65JGU1206	G. I. Derkach and V. P. Rudavskii; *J. Gen. Chem. USSR* (*Engl. Transl.*), 1965, **35**, 1206.	683
65JOC48	L. Rand and R. J. Dolinski; *J. Org. Chem.*, 1965, **30**, 48.	998
65JOC64	H. B. Milne and W. Kilday; *J. Org. Chem.*, 1965, **30**, 64.	366
65JOC184	R. K. Olsen and H. R. Snyder; *J. Org. Chem.*, 1965, **30**, 184.	1109
65JOC259	V. J. Grenda, R. E. Jones, G. Gal and M. Sletzinger; *J. Org. Chem.*, 1965, **30**, 259.	762, 816
65JOC318	J. F. Geldard and F. Lions; *J. Org. Chem.*, 1965, **30**, 318.	848
65JOC693	W. Stamm; *J. Org. Chem.*, 1965, **30**, 693.	1017
65JOC927	R. C. Parish and L. M. Stock; *J. Org. Chem.*, 1965, **30**, 927.	123, 126
65JOC931	F. H. Case; *J. Org. Chem.*, 1965, **30**, 931.	769, 847
65JOC1158	Y. Iwakura, K. Uno and S. Kang; *J. Org. Chem.*, 1965, **30**, 1158.	970
65JOC1162	S. G. Cohen and J. Nicholson; *J. Org. Chem.*, 1965, **30**, 1162.	366
65JOC1259	S. Herbstman; *J. Org. Chem.*, 1965, **30**, 1259.	222
65JOC1260	J. Smith, Jr., T. K. Brotherton and J. W. Lynn; *J. Org. Chem.*, 1965, **30**, 1260.	997
65JOC1282	E. J. Hedgley and H. G. Fletcher, Jr.; *J. Org. Chem.*, 1965, **30**, 1282.	476, 477
65JOC1311	W. Metlesics, R. F. Tavares and L. H. Sternbach; *J. Org. Chem.*, 1965, **30**, 1311.	354
65JOC1375	W. J. Middleton, E. G. Howard and W. H. Sharkey; *J. Org. Chem.*, 1965, **30**, 1375.	473, 474, 479, 480, 494
65JOC1390	W. J. Middleton; *J. Org. Chem.*, 1965, **30**, 1390.	492
65JOC1748	A. Hassner and C. Heathcock; *J. Org. Chem.*, 1965, **30**, 1748.	984
65JOC2386	G. A. Olah and J. A. Olah; *J. Org. Chem.*, 1965, **30**, 2386.	36, 108
65JOC2465	A. A. R. Sayigh, H. Ulrich and J. S. Potts; *J. Org. Chem.*, 1965, **30**, 2465.	1029
65JOC2566	A. G. Brook and J. B. Pierce; *J. Org. Chem.*, 1965, **30**, 2566.	412
65JOC2660	W. D. Crow and N. J. Leonard; *J. Org. Chem.*, 1965, **30**, 2660.	1107
65JOC2779	H. Ulrich and A. A. R. Sayigh; *J. Org. Chem.*, 1965, **30**, 2779.	982
65JOC2809	C. Grundmann and J. M. Dean; *J. Org. Chem.*, 1965, **30**, 2809.	819
65JOC2882	S. G. Smith and J. P. Petrovich; *J. Org. Chem.*, 1965, **30**, 2882.	476, 477
65JOC3361	D. Elad and J. Rokach; *J. Org. Chem.*, 1965, **30**, 3361.	276
65JOC3369	P. Rajagopalan and B. G. Advani; *J. Org. Chem.*, 1965, **30**, 3369.	1073

65JOC3718	C. L. Stevens, R. C. Freeman and K. Noll; *J. Org. Chem.*, 1965, **30**, 3718.	770, 849
65JOC3724	H. C. Brown and C. R. Wetzel; *J. Org. Chem.*, 1965, **30**, 3724.	711
65JOC3739	H. C. Brown and R. Pater; *J. Org. Chem.*, 1965, **30**, 3739.	641, 846
65JOC4306	A. J. Speziale, L. R. Smith and J. E. Fedder; *J. Org. Chem.*, 1965, **30**, 4306.	995
65JOM(4)202	N. A. Nesmeyanov, V. M. Novikov and O. A. Reutov; *J. Organomet. Chem.*, 1965, **4**, 202.	919
65JPS1373	B. B. Wylie, E. I. Isaacson and J. N. Delgado; *J. Pharm. Sci.*, 1965, **54**, 1373.	215
65LA(681)64	W. Walter and K.-D. Bode; *Justus Liebigs Ann. Chem.*, 1965, **681**, 64.	592, 623
65LA(684)103	G. Opitz and H. Schempp; *Justus Liebigs Ann. Chem.*, 1965, **684**, 103.	496, 497, 500
65LA(686)226	K. Rühlmann and B. Rupprich; *Justus Liebigs Ann. Chem.*, 1965, **686**, 226.	375
65LA(687)191	C. Grundmann, V. Mini, J. M. Dean and H.-D. Frommeld; *Justus Liebigs Ann. Chem.*, 1965, **687**, 191.	828, 837
65LA(688)14	W. P. Neumann, R. Sommer and H. Lind; *Justus Liebigs Ann. Chem.*, 1965, **688**, 14.	780
65M1278	F. Asinger, W. Schäfer and A. Saus; *Monatsh. Chem.*, 1965, **96**, 1278.	579, 602
65M1474	F. Asinger and F. Gentz; *Monatsh. Chem.*, 1965, **96**, 1474.	579
65MI 502-01	A. W. Langer, Jr.; *Trans. N. Y. Acad. Sci.*, 1965, **27**, 741.	43
B-65MI 527-01	P. A. S. Smith and C. Paulmier; in "Chemistry of Open-Chain Nitrogen Organic Compounds," Benjamin, New York, 1965, vol. 1, ch. 6, p. 233.	1000
65MI 530-01	M. Hedayatullah and L. Denivelle; *C. R. Acad. Sci. Paris*, 1965, **260**, 2839.	1101, 1103
65OR(14)1	J. W. Schulenberg and S. Archer; *Org. React.*, 1965, **14**, 1.	303
65TL149	R. Neidlein and E. Heukelbach; *Tetrahedron Lett.*, 1965, 149.	1063, 1064
65TL841	F. L. Scott and J. B. Aylward; *Tetrahedron Lett.*, 1965, **6**, 841.	356, 703, 845
65TL1857	E. Schmitz, R. Ohme and S. Schramm; *Tetrahedron Lett.*, 1965, **6**, 1857.	353
65TL2681	W. C. Agosta; *Tetrahedron Lett.*, 1965, 2681.	204
65TL3003	W. J. Franklin and R. L. Werner; *Tetrahedron Lett.*, 1965, **6**, 3003.	1054, 1057
65TL3801	A. F. Hegarty and F. L. Scott; *Tetrahedron Lett.*, 1965, 3801.	840
65TL4007	L. Birkofer and H. Dickopp; *Tetrahedron Lett.*, 1965, **6**, 4007.	378
65TL4039	B. Acott, A. L. J. Beckwith, A. Hassanali and J. W. Redmond; *Tetrahedron Lett.*, 1965, 4039.	999
65TL4645	P. Vita Finzi and M. Arbasino; *Tetrahedron Lett.*, 1965, 4645.	1004
65USP3218345	N. B. Rainer (Coastal Interchemical Co.); *US Pat.*, 3 218 345 (1965) (*Chem. Abstr.*, 1966, **64**, 3351f).	1003
65USP3444161	W. Nagata; *US Pat.* 3 444 161 (1965) (*Chem. Abstr.*, 1969, **71**, 39 296s).	1142, 1143
65ZC302	H. Rohnert; *Z. Chem.*, 1965, **5**, 302.	354
65ZOB248	J. V. Martynov and Y. L. Kruglyak; *Zh. Obshch. Khim.*, 1965, **35**, 248 (*Chem. Abstr.*, 1965, **62**, 14 490b).	822
65ZOB923	M. M. Aleksankin, L. I. Samarai and G. I. Derkach; *Zh. Obshch. Khim.*, 1965, **35**, 923 (*Chem. Abstr.*, 1965, **63**, 6813h).	988
65ZOB1009	G. I. Derkach and E. S. Gubnitskaya; *Zh. Obshch. Khim.*, 1965, **35**, 1009 (*Chem. Abstr.*, 1965, **63**, 9801a).	708, 709
65ZOB1014	G. I. Derkach, E. S. Gubnitskaya and V. A. Shokol; *Zh. Obshch. Khim.*, 1965, **35**, 1014 (*Chem. Abstr.*, 1965, **63**, 9981a).	683
65ZOB1202	G. I. Derkach and V. P. Rudavskii; *Zh. Obshch. Khim.*, 1965, **35**, 1202 (*Chem. Abstr.*, 1965, **63**, 11 398a).	683
65ZOR1415	M. O. Lozinskii and P. S. Pel'kis; *Zh. Org. Khim.*, 1965, **1**, 1415.	1049
66ACS278	K. A. Jensen, G. Felbert, C. T. Pedersen and U. Svanholm; *Acta Chem. Scand.*, 1966, **20**, 278.	1054
66ACS597	K. A. Jensen and P. H. Nielsen; *Acta Chem. Scand.*, 1966, **20**, 597.	584, 586, 587, 624, 625, 726, 732
66ACS689	U. Berg and J. Sandstrom; *Acta Chem. Scand.*, 1966, **20**, 689.	631, 632
66ACS2107	K. A. Jensen, A. Holm and C. Wentrup; *Acta Chem. Scand.*, 1966, **20**, 2107.	1101
66ACS2123	A. Holm and C. Wentrup; *Acta Chem. Scand.*, 1966, **20**, 2123.	1104
66AG908	D. Pawellek; *Angew. Chem.*, 1966, **78**, 908.	809
66AG(E)447	W. Walter and K.-D. Bode; *Angew. Chem., Int. Ed. Engl.*, 1966, **5**, 447.	565, 579
66AG(E)704	H. Ulrich and A. A. R. Sayigh; *Angew. Chem., Int. Ed. Engl.*, 1966, **5**, 704.	980, 981, 982, 1013, 1069
66AG(E)725	G. Ottmann and H. Hooks, Jr.; *Angew. Chem., Int. Ed. Engl.*, 1966, **5**, 725.	1037
66AG(E)870	K. Bott and H. Hellmann; *Angew. Chem., Int. Ed. Engl.*, 1966, **5**, 870.	26, 38
66AG(E)963	R. Gottfried; *Angew. Chem., Int. Ed. Engl.*, 1966, **5**, 963.	1029
66AG(E)1047	A. Senning; *Angew. Chem., Int. Ed. Engl.*, 1966, **5**, 1047.	995
66BAU1432	P. I. Alimov and L. A. Antokhina; *Bull. Acad. Sci. USSR, Div. Chem. Sci.*, 1966, 1432.	404
66BCJ852	E. Haruki, T. Fujii and E. Imoto; *Bull. Chem. Soc. Jpn.*, 1966, **39**, 852.	31
66BCJ2263	H. Sakurai, A. Hosomi, J. Nakajima and M. Kumada; *Bull. Chem. Soc. Jpn.*, 1966, **39**, 2263.	378
66BSF882	J. Armand: *Bull. Soc. Chim. Fr.*, 1966, 882.	637
66BSF1376	G. Gliksmans, B. Torck, M. Hellin and F. Coussemant; *Bull. Soc. Chim. Fr.*, 1966, 1376.	710
66BSF3828	L. Legrand and N. Lozac'h; *Bull. Soc. Chim. Fr.*, 1966, 3828.	586
66C175	N. Stojanac and V. Hahn; *Chimia*, 1966, **20**, 175.	630
66CB11	H. Reinheckel, D. Jahnke and G. Kretzschmar; *Chem. Ber.*, 1966, **99**, 11.	595
66CB110	E. Wünsch and F. Drees; *Chem. Ber.*, 1966, **99**, 110.	279
66CB239	R. Neidlein and W. Haussmann; *Chem. Ber.*, 1966, **22**, 239.	658
66CB850	I. Hagedorn and H.-D. Winkelmann; *Chem. Ber.*, 1966, **99**, 850.	852
66CB1252	R. Neidlein, W. Haussmann and E. Heukelbach; *Chem. Ber.*, 1966, **99**, 1252.	1071
66CB1258	H. Schenk; *Chem. Ber.*, 1966, **99**, 1258.	996
66CB1325	K. Dimroth and P. Hoffmann; *Chem. Ber.*, 1966, **99**, 1325.	882

66CB1393 R. Mayer, S. Scheithauer and D. Kunz; *Chem. Ber.*, 1966, **99**, 1393. 515, 516

66CB2127 H. Böhme and G. Berg; *Chem. Ber.*, 1966, **99**, 2127. 289

66CB2600 T. Kauffmann and L. Bán; *Chem. Ber.*, 1966, **99**, 2600. 769, 848

66CB3572 J. Goerdeler and K. Jonas; *Chem. Ber.*, 1966, **99**, 3572. 996

66CB3734 A. Spassov, E. Golovinsky and G. Demirov; *Chem. Ber.*, 1966, **99**, 3734. 852

66CC17 K. Nakagawa, H. Onoue and K. Minami; *J. Chem. Soc., Chem. Commun.*, 1966, 17. 295, 299

66CC121 M. J. Cook, E. J. Forbes and G. M. Khan; *J. Chem. Soc., Chem. Commun.*, 1966, 121. 275

66CC336 J. L. Cotter and G. J. Knight; *J. Chem. Soc., Chem. Commun.*, 1966, 336. 1013

66CC828 J. Tsuji and N. Iwamoto; *J. Chem. Soc., Chem. Commun.*, 1966, 828. 974

66CI(L)1529 J. Kopecký and J. Šmejkal; *Chem. Ind. (London)*, 1966, 1529. 280

66CJC297 W. E. Truce and A. R. Naik; *Can. J. Chem.*, 1966, **44**, 297. 670, 819

66CJC819 J. F. King and T. Durst; *Can. J. Chem.*, 1966, **44**, 819. 488, 489

66CJC2111 M. V. Kalnins; *Can. J. Chem.*, 1966, **44**, 2111. 336

66CPB1249 S. Mizukami and K. Nagata; *Chem. Pharm. Bull.*, 1966, **14**, 1249. 636

66CPB1263 K. Nagata and S. Mizukami; *Chem. Pharm. Bull.*, 1966, **14**, 1263. 830, 831

66CR(C)761 J. P. Battioni and W. Chodkiewicz; *C. R. Hebd. Seances Acad. Sci., Ser. C.*, 1966, **263**, 761. 59

66CR(C)788 P. Reynaud, R.-C. Moreau and P. Fodor; *C. R. Hebd. Seances Acad. Sci., Ser. C*, 1966, **263**, 788. 611

66FCF62 C. Grundmann; *Fortschr. Chem. Forsch.*, 1966, **7**, 62. 826, 828

66FRP1440591 J. L. Harper and A. H. Thomas (W. R. Grace & Co.); *Fr. Pat.*, 1 440 591 (1966) (*Chem. Abstr.*, 1967, **66**, 28 446m). 985

66G125 P. Papini and G. Auzzi; *Gazz. Chim. Ital.*, 1966, **96**, 125. 583

66GEP1215144 Farbenfabriken Bayer A.-G.; *Ger. Pat.* 1215144 (1966) (*Chem. Abstr.*, 1966, **65**, 3908). 1048

66GEP1228249 H. Krimm and H. Schnell (Farbenfabriken Bayer A.-G.); *Ger. Pat.*, (1966) (*Chem. Abstr.*, 1967, **66**, 28 541p). 966

66HCA1425 R. Buyle; *Helv. Chim. Acta*, 1966, **49**, 1425. 220

66HCA1716 F.-H. Marquardt; *Helv. Chim. Acta*, 1966, **49**, 1716. 1029

66IC889 J. P. Bibler and A. Wojcicki; *Inorg. Chem.*, 1966, **5**, 889. 452

66IC1177 P. Treichel, R. L. Shubkin, K. W. Barnett and D. Reichard; *Inorg. Chem.*, 1966, **5**, 1177. 453

66IC1795 D. H. Dybvig; *Inorg. Chem.*, 1966, **5**, 1795. 807

66JA3051 R. S. Neale, N. L. Marcus and R. G. Schepers; *J. Am. Chem. Soc.*, 1966, **88**, 3051. 318

66JA3163 R. G. Denkewalter, H. Schwam, R. G. Strachan, T. E. Beesley, D. F. Veber, E. F. Schoenewaldt, H. Barkemeyer, W. J. Paleveda, Jr., T. A. Jacob and R. Hirschmann; *J. Am. Chem. Soc.*, 1966, **88**, 3163. 972

66JA3318 M. S. Dewar and A. P. Marchand; *J. Am. Chem. Soc.*, 1966, **88**, 3318. 112, 114, 116

66JA3390 J. F. Klebe, H. Finkbeiner and D. M. White; *J. Am. Chem. Soc.*, 1966, **88**, 3390. 375, 378, 718

66JA3440 J. B. Lee; *J. Am. Chem. Soc.*, 1966, **88**, 3440. 13

66JA3452 J. Tsuji and K. Ohno; *J. Am. Chem. Soc.*, 1966, **88**, 3452. 458

66JA4996 R. D. Partos and K. W. Ratts; *J. Am. Chem. Soc.*, 1966, **88**, 4996. 860

66JCS(A)634 G. Booth and J. Chatt; *J. Chem. Soc. (A)*, 1966, 634. 463, 464

66JCS(A)1584 R. Jefferson, M. F. Lappert, B. Prokai and B. P. Tilley; *J. Chem. Soc. (A)*, 1966, 1584. 779

66JCS(C)840 D. J. Byron, G. W. Gray and R. C. Wilson; *J. Chem. Soc. (C)*, 1966, 840. 108

66JCS(C)950 R. S. McElhinney; *J. Chem. Soc. (C)*, 1966, 950. 1023

66JCS(C)1477 J. F. Elsworth and M. Lamchen; *J. Chem. Soc. (C)*, 1966, 1477. 325

66JCS(C)1514 R. E. Banks, R. N. Haszeldine and J. P. Lalu; *J. Chem. Soc. (C)*, 1966, 1514. 313

66JGU1101 G. I. Derkach and Yu. V. Piven; *J. Gen. Chem. USSR (Engl. Transl.)*, 1966, **36**, 1101. 861

66JGU1276 V. I. Shevchenko and V. P. Kukhar; *J. Gen. Chem. USSR (Engl. Transl.)*, 1966, **36**, 1276. 683

66JGU1340 B. L. Dyatkin, A. A. Gevorkyan and I. L. Knunyants; *J. Gen. Chem. USSR (Engl. Transl.)*, 1966, **66**, 1340. 671

66JGU1945 I. I. Lapkin, G. Y. Anvarova and T. N. Povarnitsyna; *J. Gen. Chem. USSR (Engl. Transl.)*, 1966, **36**, 1945. 438

66JHC155 J. K. Stille, F. W. Harris and M. A. Bedford; *J. Heterocycl. Chem.*, 1966, **3**, 155. 842

66JHC384 D. A. Tomalia; *J. Heterocycl. Chem.*, 1966, **3**, 384. 1033

66JHC387 A. Rosowsky and E. J. Modest; *J. Heterocycl. Chem.*, 1966, **3**, 387. 1118

66JIC481 K. M. Joshi and K. K. Shah; *J. Indian Chem. Soc.*, 1966, **43**, 481. 998

66JIC615 S. P. Inamdar and R. N. Usgaonkar; *J. Indian Chem. Soc.*, 1966, **43**, 615. 184

66JOC142 Y. Iwakura, K. Uno and S. Kang; *J. Org. Chem.*, 1966, **31**, 142. 970

66JOC157 C. Grundmann and H.-D. Frommeld; *J. Org. Chem.*, 1966, **31**, 157. 637, 819, 832, 837

66JOC596 E. W. Stern and M. L. Spector; *J. Org. Chem.*, 1966, **31**, 596. 974

66JOC722 H. M. Blatter and H. Lukaszewski; *J. Org. Chem.*, 1966, **31**, 722. 1052

66JOC1079 J. H. Hall and R. Kellogg; *J. Org. Chem.*, 1966, **31**, 1079. 1011

66JOC1202 W. H. H. Günther; *J. Org. Chem.*, 1966, **31**, 1202. 239

66JOC1689 K. W. Ratts and A. N. Yao; *J. Org. Chem.*, 1966, **31**, 1689. 495

66JOC2316 A. G. Pittman and D. L. Sharp; *J. Org. Chem.*, 1966, **31**, 2316. 4

66JOC2398 F. H. Case; *J. Org. Chem.*, 1966, **31**, 2398. 847

66JOC2658 H. Ulrich, B. Tucker and A. A. R. Sayigh; *J. Org. Chem.*, 1966, **31**, 2658. 981

66JOC3038 E. H. White, M. C. Chen and L. A. Dolak; *J. Org. Chem.*, 1966, **31**, 3038. 361

66JOC3439 G. A. Gornowicz and J. W. Ryan; *J. Org. Chem.*, 1966, **31**, 3439. 254, 536, 539

66JOC3689 A. L. Logothetis and G. N. Sausen; *J. Org. Chem.*, 1966, **31**, 3689. 698, 761, 812, 814, 817, 819

66JOC3708 H. E. Baumgarten, J. E. Dirks, J. M. Petersen and R. L. Zey; *J. Org. Chem.*, 1966, **31**, 3708. 698, 813

66JOC3829 D. P. Gush, N. S. Marans, F. Wessells, W. D. Addy and S. J. Olfky; *J. Org. Chem.*, 1966, **31**, 3829. 276

66JOC3877 J. J. Pappas and E. Gancher; *J. Org. Chem.*, 1966, **31**, 3877. 559

66JOC4232 R. S. Koshar, D. R. Husted and R. A. Meiklejohn; *J. Org. Chem.*, 1966, **31**, 4232. 815

66JOM(5)106 A. G. Brook and G. J. D. Peddle; *J. Organomet. Chem.*, 1966, **5**, 106. 431

66JOM(5)267 D. Seyferth, J. K. Heeren, G. Singh, S. O. Grim and W. B. Hughes; *J. Organomet. Chem.*, 1966, **5**, 267. 896

66JOM(5)341 R. B. King, M. B. Bisnette and A. Fronzaglia; *J. Organomet. Chem.*, 1966, **5**, 341. 449

66JOM(5)383 J. S. Thayer and D. P. Strommen; *J. Organomet. Chem.*, 1966, **5**, 383. 1016, 1017

66JOM(5)486 G. J. D. Peddle; *J. Organomet. Chem.*, 1966, **5**, 486. 431

66JOM(5)590 A. G. Davies and R. J. Puddephatt; *J. Organomet. Chem.*, 1966, **5**, 590. 716, 717

66JOM(6)265 F. Calderazzo, K. Novak and U. Schaerer; *J. Organomet. Chem.*, 1966, **6**, 265. 447

66JOU1809 V. L. Dubina and S. I. Burmistrov; *J. Org. Chem. USSR* (*Engl. Transl.*), 1966, **2**, 1809. 674

66LA(693)132 H. J. Bestmann and L. Mott; *Justus Liebigs Ann. Chem.*, 1966, **693**, 132. 17

66LA(695)217 W. Ried and E. Schmidt; *Justus Liebigs Ann. Chem.*, 1966, **695**, 217. 586

66LA(698)131 W. Walter and K.-D. Bode; *Justus Liebigs Ann. Chem.*, 1966, **698**, 131. 579, 581, 592, 601

66LA(699)40 K. Issleib and R. Lindner; *Justus Liebigs Ann. Chem.*, 1966, **699**, 40. 908

66M62 R. Oppenhauer; *Monatsh. Chem.*, 1966, **97**, 62. 129

66M115 E. Nachbaur and W. Gottardi; *Monatsh. Chem.*, 1966, **97**, 115. 1015

66M180 E. Zbiral; *Monatsh. Chem.*, 1966, **97**, 180. 898, 903

66M361 E. Nachbaur; *Monatsh. Chem.*, 1966, **97**, 361. 1015

66M1797 E. Zbiral and E. Werner; *Monatsh. Chem.*, 1966, **97**, 1797. 897, 898

66MI 502-01 F. Salmon-Legagneur; *Ind. Chem. Belg.*, 1966, **31**, 993. 24

66MI 507-01 H. Tani, T. Araki and H. Yasuda; *J. Polym. Sci.*, *Polym. Lett. Ed.*, 1966, **4**, 727. 383

66MI 512-01 B. Martel and N. Duffaut; *Compt. Rend.*, *Ser. C*, 1966, **263**, 74 (*Chem. Abstr.*, 1966, **65**, 13 746). 536

B-66MI 523-01 A. W. Johnson; "Ylid Chemistry," Academic Press, New York, 1966. 894

B-66MI 526-01 P. A. S. Smith; in "Open-Chain Nitrogen Compounds," W. A. Benjamin, Inc., New York, 1966, vol. 2, p. 220. 989

66MI 530-01 M. F. Lappert and H. Pyszora; *Adv. Inorg. Chem. Radiochem.*, 1966, **9**, 142. 1133

66MI 530-02 M. F. Lappert and H. Pyszora; *Adv. Inorg. Radiochem.*, 1966, **9**, 133. 1144

66NEP6603215 W. J. Middleton (E. I. du Pont de Nemours & Co.); *Neth. Pat.*, 6 603 215 (1966) (*Chem. Abstr.*, 1967, **66**, 115 706). 973

66NEP6606456 Farbenfabriken Bayer A.-G.; *Neth. Pat.*, 6 606 456 (1966) (*Chem. Abstr.*, 1967, **67**, 21 620a). 965

66RTC889 P. J. W. Schuijl, L. Brandsma and J. F. Arens; *Recl. Trav. Chim. Pays-Bas*, 1966, **85**, 889. 550, 563

66T975 K. D. Berlin and M. A. R. Khayat; *Tetrahedron*, 1966, **22**, 975. 709, 861

66TL3537 C. L. Coon, D. L. Ross and M. E. Hill; *Tetrahedron Lett.*, 1966, **30**, 3537. 763, 815

66TL5263 G. Opitz and D. Bücher; *Tetrahedron Lett.*, 1966, **7**, 5263. 495

66TL6199 U. Schöllkopf and P. Markusch; *Tetrahedron Lett.*, 1966, 6199. 868

66URP181085 V. L. Dubina and S. I. Burmistrov; *USSR Pat.* 181085 (1966) (*Chem. Abstr.*, 1966, **65**, 8831a). 638

66ZAAC(346)241 K. Issleib and O. Löw; *Z. Anorg. Allg. Chem.*, 1966, **346**, 241. 395, 397, 784

66ZC313 H. Brachwitz; *Z. Chem.*, 1966, **6**, 313. 830

66ZOB461 G. I. Derkach and N. I. Liptuga; *Zh. Obshch. Khim.*, 1966, **36**, 461 (*Chem. Abstr.*, 1966, **65**, 634g). 1010

66ZOB952 V. A. Chauzov, O. V. Litvinova and Y. I. Baukov; *Zh. Obshch. Khim.*, 1966, **36**, 952 (*Chem. Abstr.*, 1966, **65**, 8955). 1016

66ZOB1260 V. I. Shevchenko and V. P. Kukhar; *Zh. Obshch. Khim.*, 1966, **36**, 1260 (*Chem. Abstr.*, 1966, **65**, 16 815h). 683

66ZOB1326 B. L. Dyatkin, A. A. Gevorkyan and I. L. Knunyants; *Zh. Obshch. Khim.*, 1966, **36**, 1326 (*Chem. Abstr.*, 1966, **65**, 16 855). 671

66ZOB1639 G. I. Derkach and E. I. Slysarenko; *Zh. Obshch. Khim.*, 1966, 36, 1639 (*Chem. Abstr.*, 1967, **66**, 55 135r). 1008

66ZOB1707 L. V. Nifant'eva, K. D. Shvetsova-Shilovskaya and N. N. Mel'nikov; *Zh. Obshch. Khim.*, 1966, **36**, 1707 (*Chem. Abstr.*, 1967, **66**, 55 134q). 966

66ZOR1841 V. L. Dubina and S. I. Burmistrov; *Zh. Org. Khim.* 1966, **10**, 1841 (*Chem. Abstr.*, 1967, **66**, 55 171z). 652

67ACS837 L.-B. Agenäs and B. Persson; *Acta Chem. Scand.*, 1967, **21**, 837. 1114

67ACS1936 K. A. Jensen, O. Buchardt and C. Christophersen; *Acta Chem. Scand.*, 1967, **21**, 1936. 635, 636

67AG(E)107 G. Opitz; *Angew. Chem., Int. Ed. Engl.*, 1967, **6**, 107. 472, 488, 495, 497, 500

67AG(E)168 D. Martin and A. Weise; *Angew. Chem., Int. Ed. Engl.*, 1967, **6**, 168. 1004

67AG(E)169 W. Beck and W. P. Fehlhammer; *Angew. Chem., Int. Ed. Engl.*, 1967, **6**, 169. 983

67AG(E)174 J. C. Jochims and A. Seeliger; *Angew. Chem., Int. Ed. Engl.*, 1967, **6**, 174. 1023

67AG(E)206 E. Grigat and R. Pütter; *Angew. Chem., Int. Ed. Engl.*, 1967, **6**, 206. 1100

67AG(E)249 G. Gosselck, H. Schenk and H. Ahlbrecht; *Angew. Chem., Int. Ed. Engl.*, 1967, **6**, 249. 900

67AG(E)261 U. Schöllkopf and N. Rieber; *Angew. Chem., Int. Ed. Engl.*, 1967, **6**, 261. 871, 874

67AG(E)294 R. Mayer and K. Gewald; *Angew. Chem., Int. Ed. Engl.*, 1967, **6**, 294. 599

67AG(E)368 W. Walter and K.-J. Reubke; *Angew. Chem., Int. Ed. Engl.*, 1967, **6**, 368. 640

67AG(E)636 H. Ulrich, B. Tucker and A. A. R. Sayigh; *Angew. Chem., Int. Ed. Engl.*, 1967, **6**, 636. 977

67AG(E)733 M. Regitz; *Angew. Chem., Int. Ed. Engl.*, 1967, **6**, 733. 869

67AG(E)864 L. I. Samarai, V. A. Bondar and G. I. Derkach; *Angew. Chem., Int. Ed. Engl.*, 1967, **6**, 864. 967, 987

67AG(E)884 U. Schöllkopf and N. Rieber; *Angew. Chem., Int. Ed. Engl.*, 1967, **6**, 884. 873

67AG(E)907 F. Asinger and H. Offermanns; *Angew. Chem., Int. Ed. Engl.*, 1967, **6**, 907. 599

67AG(E)961 E. O. Fischer and V. Kiener; *Angew. Chem., Int. Ed. Engl.*, 1967, **6**, 961. 952

67AP241 H. Böhme and H.-H. Hotzel; *Arch. Pharm.* (*Weinheim, Ger.*), 1967, **300**, 241. 606

67AP583	K. Kotte, F. Friedrich and R. Pohloudek-Fabini; *Arch. Pharm.* (*Weinheim, Ger.*), 1967, **300**, 583.	1111
67BAU324	A. V. Fokin, A. A. Skladnev, Y. N. Studnev and L. L. Knunyants; *Bull. Acad. Sci. USSR, Div. Chim. Sci.*, 1967, 324.	474, 484
67BCJ2209	F. S. Okumura and T. Moritani; *Bull. Chem. Soc. Jpn.*, 1967, **40**, 2209.	582
67BCJ2705	A. Kaneda, M. Nagatsuka and R. Sudo; *Bull. Chem. Soc. Jpn.*, 1967, **40**, 2705.	731
67BSF416	M. Hedayatullah; *Bull. Soc. Chim. Fr.*, 1967, 416.	1101
67BSF422	M. Hedayatullah; *Bull. Soc. Chim. Fr.*, 1967, 422.	1101, 1103
67BSF985	J. Martel and C. Huynh; *Bull. Soc. Chim. Fr.*, 1967, 985.	145
67BSF2011	J. E. Dubois, M. Chastrette and E. Schunk; *Bull. Soc. Chim. Fr.*, 1967, 2011.	29, 34, 43
67CB71	R. Huisgen and V. Weberndörfer; *Chem. Ber.*, 1967, **100**, 71.	842
67CB142	E. Schmitz, D. Habisch and C. Gründemann; *Chem. Ber.*, 1967, **100**, 142.	365
67CB480	A. Haas and D. Y. Oh; *Chem. Ber.*, 1967, **100**, 480.	1018
67CB698	R. Neidlein and R. Bottler; *Chem. Ber.*, 1967, **100**, 698.	996
67CB1032	H. Schmidbaur and W. Tronich; *Chem. Ber.*, 1967, **100**, 1032.	913, 916, 920, 1093
67CB1082	O. Glemser, U. Biermann and M. Fild; *Chem. Ber.*, 1967, **100**, 1082.	1007, 1018
67CB1248	K. H. Büchel and A. Conte; *Chem. Ber.*, 1967, **100**, 1248.	337
67CB1367	E. Bulka, K.-D. Ahlers and E. Tuček; *Chem. Ber.*, 1967, **100**, 1367.	1054
67CB2445	E. O. Fischer and A. Maasböl; *Chem. Ber.*, 1967, **100**, 2445.	444, 445, 446, 936
67CB3368	J. Stenzel and W. Sundermeyer; *Chem. Ber.*, 1967, **100**, 3368.	1076, 1077
67CC59	R. A. Mitsch and P. H. Ogden; *J. Chem. Soc., Chem. Commun.*, 1967, 59.	1080
67CC836	M. F. Lappert and J. Lorberth; *J. Chem. Soc., Chem. Commun.*, 1967, 836.	874
67CC1024	J. A. Connor and E. O. Fischer; *J. Chem. Soc., Chem. Commun.*, 1967, 1024.	953
67CI(L)1406	S. Hyden and G. Wilbert; *Chem. Ind.* (*London*), 1967, 1406.	990, 991
67CJC1	I. D. Rae; *Can. J. Chem.*, 1967, **45**, 1.	725
67CJC301	C. D. Cook and G. S. Jauhal; *Can. J. Chem.*, 1967, **45**, 301.	464
67CJC1539	Y. Ahmad, M. I. Qureshi and M. I. Baig; *Can. J. Chem.*, 1967, **45**, 1539.	360, 361
67CJC2625	R. C. Slagel and A. E. Bloomquist; *Can. J. Chem.*, 1967, **45**, 2625.	997
67CPB61	K. Nagata and S. Mizukami; *Chem. Pharm. Bull.*, 1967, **15**, 61.	830
67CPB1183	A. Takamizawa, Y. Sato and H. Sato; *Chem. Pharm. Bull.*, 1967, **15**, 1183.	404
67CR(C)99	J. Sotiropoulos, A.-M. Lamazouère and P. Bédos; *C. R. Hebd. Seances Acad. Sci., Ser. C*, 1967, **265**, 99.	593
67CRV107	F. Kurzer and K. Douraghi-Zadeh; *Chem. Rev.*, 1967, **67**, 107.	849, 1061
67CRV623	A. K. Vijh and B. E. Conway; *Chem. Rev.*, 1967, **67**, 623.	28
67G1664	P. Bravo, G. Gaudiano and A. Umani-Ronchi; *Gazz. Chim. Ital.*, 1967, **97**, 1664.	832
67GEP1228249	H. Krimm and H. Schnell (Farbenfabriken Bayer A.-G.); *Ger. Pat.*, 1 228 249 (1967) (*Chem. Abstr.*, 1967, **66**, 28 541p).	652
67HOU(10/2)123	E. Müller; *Methoden Org. Chem.* (*Houben-Weyl*), 1967, **10/2**, 123.	350
67HOU(10/2)378	E. Enders; *Methoden Org. Chem.* (*Houben-Weyl*), 1967, **10/2**, 378.	842
67HOU(10/2)467	E. Enders; *Methoden Org. Chem.* (*Houben-Weyl*), 1967, **10/2**, 467.	844
67IZV1060	R. A. Bekker, B. L. Dyatkin and I. L. Knunyants; *Izv. Akad. Nauk SSSR, Ser. Khim.*, 1967, 1060 (*Chem. Abstr.*, 1968, **68**, 12 419k).	811
67IZV1820	R. M. Khomutor, E. S. Severin, N. V. Gnuchev and T. Ya. Derevyanko; *Izv. Akad. Nauk SSSR, Ser. Khim.*, 1967, 1820 (*Chem. Abstr.*, 1968, **68**, 39 053y).	826
67JA183	J. Kovacs, L. Kisfaludy and M. Q. Ceprini; *J. Am. Chem. Soc.*, 1967, **89**, 183.	277
67JA291	G. Zweifel and H. Arzoumanian; *J. Am. Chem. Soc.*, 1967, **89**, 291.	52
67JA431	A. G. Brook, J. M. Duff, P. F. Jones and N. R. Davis; *J. Am. Chem. Soc.*, 1967, **89**, 431.	410, 431
67JA434	E. J. Corey, D. Seebach and R. Freedman; *J. Am. Chem. Soc.*, 1967, **89**, 434.	410
67JA637	D. Y. Curtin and L. L. Miller; *J. Am. Chem. Soc.*, 1967, **89**, 637.	714
67JA2240	T. Saegusa, Y. Ito, S. Kobayashi and K. Hirota; *J. Am. Chem. Soc.*, 1967, **89**, 2240.	792
67JA3281	K. W. Bentley and D. G. Hardy; *J. Am. Chem. Soc.*, 1967, **89**, 3281.	1126
67JA4760	M. E. Hermes and F. D. Marsh; *J. Am. Chem. Soc.*, 1967, **89**, 4760.	1119
67JA4809	J. P. Collman, M. Kubota and J. W. Hosking; *J. Am. Chem. Soc.*, 1967, **89**, 4809.	983
67JA4811	D. Seyferth, P. Hilbert and R. S. Marmor; *J. Am. Chem. Soc.*, 1967, **89**, 4811.	868
67JA5007	P. H. Ogden and R. Mitsch; *J. Am. Chem. Soc.*, 1967, **89**, 5007.	1080
67JA5287	S. Wolfe and D. V. C. Awang; *J. Am. Chem. Soc.*, 1967, **89**, 5287.	320
67JA7014	S. Trofimenko; *J. Am. Chem. Soc.*, 1967, **89**, 7014.	779
67JA7017	C. J. Pedersen; *J. Am. Chem. Soc.*, 1967, **89**, 7017.	106
67JA7131	F. W. Wassmundt and S. J. Padegimas; *J. Am. Chem. Soc.*, 1967, **89**, 7131.	294
67JA7141	U. Klabunde and E. O. Fischer; *J. Am. Chem. Soc.*, 1967, **89**, 7141.	951, 953
67JA7151	J. E. Zimmerman and G. W. Anderson; *J. Am. Chem. Soc.*, 1967, **89**, 7151.	279
67JCED151	E. Marcus, J. K. Chan and J. L. Hughes; *J. Chem. Eng. Data*, 1967, **12**, 151.	217
67JCS(A)865	M. C. Baird and G. Wilkinson; *J. Chem. Soc.* (*A*), 1967, 865.	464
67JCS(B)1151	F. E. Hardy and P. Robson; *J. Chem. Soc.* (*B*), 1967, 1151.	709
67JCS(C)58	M. A. Rehman and G. Proctor; *J. Chem. Soc.* (*C*), 1967, 58.	347
67JCS(C)690	W. Hoyle; *J. Chem. Soc.* (*C*), 1967, 690.	772, 851
67JCS(C)1309	A. J. Bloodworth, A. G. Davies and S. C. Vasishtha; *J. Chem. Soc.* (*C*), **1967**, 1309.	716
67JCS(C)2128	L. S. Kaminsky and M. Lamchen; *J. Chem. Soc.* (*C*), 1967, 2128.	283
67JGU826	V. S. Abramov and M. G. Gubaidullin; *J. Gen. Chem. USSR*, 1968, **38**, 826.	404, 405
67JGU2606	A. I. Razumov and M. B. Gazizov; *J. Gen. Chem. USSR*, 1967, **37**, 2606.	404
67JHC54	G. Adembri and R. Nesi; *J. Heterocycl. Chem.*, 1967, **4**, 54.	532
67JHC178	D. A. Tomalia and J. N. Paige; *J. Heterocycl. Chem.*, 1967, **4**, 178.	1010
67JHC483	F. H. Case and L. Kennon; *J. Heterocycl. Chem.*, 1967, **4**, 483.	847

67JMC1123 B. R. Baker and E. H. Erickson; *J. Med. Chem.*, 1967, **10**, 1123. 748, 757

67JOC392 J. R. Schaeffer, C. T. Goodhue, H. A. Risley and R. E. Stevens; *J. Org. Chem.*, 1967, **32**, 392. 735

67JOC416 R. F. Merritt and F. A. Johnson; *J. Org. Chem.*, 1967, **32**, 416. 659

67JOC540 A. Hassner, M. E. Lorber and C. Heathcock; *J. Org. Chem.*, 1967, **32**, 540. 985

67JOC869 T. Inukai and T. Kojima; *J. Org. Chem.*, 1967, **32**, 869. 146

67JOC1279 W. S. Wadsworth, Jr. and W. D. Emmons; *J. Org. Chem.*, 1967, **32**, 1279. 975

67JOC1508 C. K. Johnson; *J. Org. Chem.*, 1967, **32**, 1508. 1010

67JOC1552 D. Y. Curtin and J. D. Druliner; *J. Org. Chem.*, 1967, **32**, 1552. 365

67JOC1800 L. D. Hatfield and W. D. Huntsman; *J. Org. Chem.*, 1967, **32**, 1800. 325, 333

67JOC1939 H. W. Johnson, Jr. and H. Krutzsch; *J. Org. Chem.*, 1967, **32**, 1939. 1004

67JOC2895 C. L. Stevens, G. H. Singhal and A. B. Ash; *J. Org. Chem.*, 1967, **32**, 2895. 1078

67JOC2925 W. H. Mueller and P. E. Butler; *J. Org. Chem.*, 1967, **32**, 2925. 248

67JOC3467 L. Field and J. D. Buckman; *J. Org. Chem.*, 1967, **32**, 3467. 247

67JOC3665 B. E. Grimwood and D. Swern; *J. Org. Chem.*, 1967, **32**, 3665. 985

67JOC3709 C. M. Sharts and D.-W. Fong; *J. Org. Chem.*, 1967, **32**, 3709. 528

67JOC3859 R. J. Koshar, D. R. Husted and C. D. Wright; *J. Org. Chem.*, 1967, **32**, 3859. 811, 815

67JOC3929 W. H. H. Günther; *J. Org. Chem.*, 1967, **32**, 3929. 239

67JOC4052 H. Ulrich, B. Tucker and A. A. R. Sayigh; *J. Org. Chem.*, 1967, **32**, 4052. 995

67JOM(10)511 H. Takahashi and J. Tsuji; *J. Organomet. Chem.*, 1967, **10**, 511. 356

67JPR86 E. Schmitz, S. Schramm and R. Ohme; *J. Prakt. Chem.*, 1967, **36**, 86. 353

67JPR294 R. Mayer, B. Hunger, R. Prousa and A.-K. Müller; *J. Prakt. Chem.*, 1967, **35**, 294. 589, 603, 611

67JPR302 G. Barnikow, K. Krüger and G. Hilgetag; *J. Prakt. Chem.*, 1967, **35**, 302. 595

67LA(707)112 K. Issleib and R. Lindner; *Justus Liebigs Ann. Chem.*, 1967, **707**, 112. 910

67LA(710)118 M. Regitz and A. Liedhegener; *Justus Liebigs Ann. Chem.*, 1967, **710**, 118. 595

B-67MI 502-01 C. W. Bird; "Transition Metal Intermediates in Organic Synthesis," Legos Press, London, 1967. 27

B-67MI 520-01 A. J. Kirby and S. G. Warren (eds.); "The Organic Chemistry of Phosphorus," Elsevier, New York, 1967, p. 59. 786

67MI 530-01 B. M. Gladshtein and V. M. Zimin; *Zh. Obshch. Khim. Lett.*, 167, **37**, 2055. 1131

67OR(15)1 J. P. Schaefer and J. J. Bloomfield; *Org. React.*, 1967, **15**, 1. 46

67PP1149 F. D. Trischler and J. Hollander; *Polym. Prepr.*, 1967, **8**, 1149. 967

67QR3 B. Sklarz; *Q. Rev., Chem. Soc.*, 1967, **21**, 3. 49

67RCR260 L. L. Rodina and I. K. Korobitsyna; *Russ. Chem. Rev.* (*Engl. Transl.*), 1967, **36**, 260. 271, 292

67RTC641 J. Strating, L. Thijs and B. Zwanenburg; *Recl. Trav. Chim. Pays-Bas*, 1967, **86**, 641. 489

67SA(A)681 A. J. Downs and R. Schmutzler; *Spectrochim. Acta.*, *Part A*, 1967, **23**, 681. 1093

67T2137 Y. Ito, M. Okano and R. Oda; *Tetrahedron*, 1967, **23**, 2137. 902

67T3327 Y. Ogata and Y. Sawaki; *Tetrahedron*, 1967, **23**, 3327. 205

67T3785 E. J. Smutny, W. Turner, E. D. Morgan and R. Robinson; *Tetrahedron*, 1967, **23**, 3785. 559

67TL327 G. Casnati and A. Ricca; *Tetrahedron Lett.*, 1967, 327. 821

67TL961 W. B. Hardy and R. P. Bennett; *Tetrahedron Lett.*, 1967, 961. 984

67TL1013 H. Behringer and D. Deichmann; *Tetrahedron Lett.*, 1967, **8**, 1013. 1049

67TL1731 H. Ulrich, B. Tucker and A. A. R. Sayigh; *Tetrahedron Lett.*, 1967, 1731. 1079

67TL1783 G. Costa and G. Mestroni; *Tetrahedron Lett.*, 1967, 1783. 457

67TL3453 B. Zwanenburg, L. Thys and J. Strating; *Tetrahedron Lett.*, 1967, 3453. 562

67ZC123 D. Martin; *Z. Chem.*, 1967, **7**, 123. 1100

67ZN(B)784 K. Issleib and R. Kümmel; *Z. Naturforsch. B.*, 1967, **22B**, 784. 396

67ZN(B)785 E. Niecke and J. Stenzel; *Z. Naturforsch.*, *Teil B*, 1967, **22**, 785. 1074

67ZOB1413 V. A. Ginsburg and K. N. Smirnov; *Zh. Obshch. Khim.*, 1967, **37**, 1413 (*Chem. Abstr.*, 1967, **67**, 108 115p). 807

67ZOR1207 Yu. V. Markova, N. G. Ostroumova and M. N. Shchukina; *Zh. Org. Khim.*, 1967, 1207 (*Chem. Abstr.*, 1967, **67**, 90 767). 827

68ACS2019 A. Holm; *Acta Chem. Scand.*, 1968, **22**, 2019. 1104

68AG352 K. Bittler, N. V. Kutepow, D. Neubauer and H. Reis; *Angew. Chem.*, 1968, **80**, 352. 142

68AG942 H. Holtschmidt, E. Degener, H. G. Schmelzer, H. Tarnow and W. Zecher; *Angew. Chem.*, 1968, **80**, 942. 654

68AG(E)67 W. D. Habicher and R. Mayer; *Angew. Chem., Int. Ed. Engl.*, 1968, **7**, 67. 1104

68AG(E)172 R. Graf; *Angew. Chem., Int. Ed. Engl.*, 1968, **7**, 172. 298

68AG(E)220 H. Schmidbaur and W. Tronich; *Angew. Chem., Int. Ed. Engl.*, 1968, **7**, 220. 912

68AG(E)278 H. Bestian; *Angew. Chem., Int. Ed. Engl.*, 1968, **7**, 278. 84

68AG(E)293 H. Schildknecht and G. Hatzmann; *Angew. Chem., Int. Ed. Engl.*, 1968, **7**, 293. 987

68AG(E)329 K. Bittler, N. V. Kutepow, D. Neubauer and H. Reis; *Angew. Chem., Int. Ed. Engl.*, 1968, **7**, 329. 27

68AG(E)350 F. Lingens; *Angew. Chem., Int. Ed. Engl.*, 1968, **7**, 350. 83

68AG(E)381 U. v. Gizycki and G. Oertel; *Angew. Chem., Int. Ed. Engl.*, 1968, **7**, 381. 974

68AG(E)390 C. G. Kreiter; *Angew. Chem., Int. Ed. Engl.*, 1968, **7**, 390. 955

68AG(E)391 H. Gross and B. Costisella; *Angew. Chem., Int. Ed. Engl.*, 1968, **7**, 391. 42

68AG(E)465 P. Rambacher and S. Mäke; *Angew. Chem., Int. Ed. Engl.*, 1968, **7**, 465. 183

68AG(E)856 H. Holtschmidt, E. Degener, H.-G. Schmelzer, H. Tarnow and W. Zecher; *Angew. Chem., Int. Ed. Engl.*, 1968, **7**, 856. 652

68AG(E)941 G. Greber and H. R. Kricheldorf; *Angew. Chem., Int. Ed. Engl.*, 1968, **7**, 941. 977

68AJC2756 A. L. J. Beckwith and R. Hickman; *Aust. J. Chem.*, 1968, **21**, 2756. 999
68AOC(7)95 A. G. Brook; *Adv. Organomet. Chem.*, 1968, **7**, 95. 409, 412, 421, 431
68BCJ1258 T. Sasaki and Y. Yoshioka; *Bull. Chem. Soc. Jpn.*, 1968, **41**, 1258. 990
68BCJ2521 R. Harada and H. Kondo; *Bull. Chem. Soc. Jpn.*, 1968, **41**, 2521. 351
68BSF1472 M. Hedayatullah; *Bull. Soc. Chim. Fr.*, 1968, 1472. 1100
68BSF3657 J. M. Kornprobst, A. Laurent and E. Laurent-Dieuzeide; *Bull. Soc. Chim. Fr.*, 1968, 3657. 282
68BSF4093 J. Satgé and M. Rivière-Baudet; *Bull. Soc. Chim. Fr.*, 1968, 4093. 389
68CB502 F. Effenberger, R. Gleiter, L. Heider and R. Niess; *Chem. Ber.*, 1968, **101**, 502. 301
68CB595 H. Schmidbaur and W. Tronich; *Chem. Ber.*, 1968, **101**, 595. 1093
68CB1428 H. Behringer, D. Bender, J. Falkenberg and R. Wiedenmann; *Chem. Ber.*, 1968, **101**, 1428. 605, 616, 634
68CB1746 J. C. Jochims; *Chem. Ber.*, 1968, **101**, 1746. 1023
68CB2028 R. Kraft, H. Paul and G. Hilgetag; *Chem. Ber.*, 1968, **101**, 2028. 773, 849
68CB2502 G. Schmid and H. Nöth; *Chem. Ber.*, 1968, **101**, 2502. 433
68CB2533 O. J. Scherer and P. Hornig; *Chem. Ber.*, 1968, **101**, 2533. 777, 780
68CB2585 L. Birkofer and H. Dickopp; *Chem. Ber.*, 1968, **101**, 2585. 722, 814
68CB3475 J. Goerdeler and D. Weber; *Chem. Ber.*, 1968, **101**, 3475. 1052
68CB3517 H. Viola, S. Scheithauer and R. Mayer; *Chem. Ber.*, 1968, **101**, 3517. 513, 592, 593, 620
68CB3545 H. Schmidbaur and W. Tronich; *Chem. Ber.*, 1968, **101**, 3545. 908, 911, 912, 916
68CB3734 M. Regitz, W. Anschütz and A. Liedhegener; *Chem. Ber.*, 1968, **101**, 3734. 869
68CB3947 H. Neunhoeffer and H. Hennig; *Chem. Ber.*, 1968, **101**, 3947. 852
68CC6 P. Fitton, M. P. Johnson and J. E. McKeon; *J. Chem. Soc., Chem. Commun.*, 1968, 6. 463
68CC206 C. Burridge and D. P. G. Hamon; *J. Chem. Soc., Chem. Commun.*, 1968, 206. 73
68CC499 A. R. McCarthy, W. D. Ollis and C. A. Ramsden; *J. Chem. Soc., Chem. Commun.*, 1968, 499. 641
68CC549 G. V. Boyd and A. J. H. Summers; *J. Chem. Soc., Chem. Commun.*, 1968, 549. 642
68CI(L)1568 H. R. Harrison, W. M. Haynes, P. Arthur and E. J. Eisenbraun; *Chem. Ind. (London)*, 1968, 1568. 122
68CJC1873 O. Exner, M. H. Benn and F. Willis; *Can. J. Chem.*, 1968, **46**, 1873. 828, 832
68CJC2549 S. Wolfe, J. C. Godfrey, C. T. Holdrege and Y. G. Perron; *Can. J. Chem.*, 1968, **46**, 2549. 9
68CR(C)1241 Y. Charbonell, R. Burgada and J. Barrans; *C. R. Hebd. Seances Acad. Sci.*, Ser. C, 1968, **266**, 1241. 861
68CR(C)1303 R. Rigny and S. Samne; *C. R. Hebd. Seances Acad. Sci., Ser. C*, 1968, **266**, 1303. 215
68DOK(181)115 V. F. Mironov, V. P. Kozyukov, V. D. Sheludyakov and G. D. Khatuntsev; *Dokl. Akad. Nauk SSSR*, 1968, **181**, 115 (*Chem. Abstr.*, 1969, **70**, 4189m). 1014
68FES642 E. Testa and L. Fontanella; *Farmaco Ed. Sci.*, 1968, **23**, 642. 566
68G511 R. Fusco, P. Dalla Croce and A. Salvi; *Gazz. Chim. Ital.*, 1968, **98**, 511. 1123
68HCA1723 A. O. Ilvespää; *Helv. Chim. Acta*, 1968, **51**, 1723. 1112
68HOU(10/4)1 H. Metzger; *Methoden Org. Chem. (Houben-Weyl)*, 1968, **10/4**, 1. 819
68HOU(10/4)184 H. Metzger; *Methoden Org. Chem. (Houben-Weyl)*, 1968, **10/4**, 184. 215
68IC27 J. P. Collman and C. T. Sears Jr.; *Inorg. Chem.*, 1968, **7**, 27. 460
68IJ587 H. Feuer, E. P. Rosenquist and F. Brown Jr.; *Isr. J. Chem.*, 1968, **6**, 587. 350
68IZV289 D. N. Kravtsov and B. A. Faingor; *Izv. Akad. Nauk SSSR, Ser. Khim.*, 1968, 289 (*Chem. Abstr.*, 1968, **69**, 66 717v). 1141
68IZV657 B. L. Dyatkin, K. N. Makarov and I. L. Knunyants; *Izv. Akad. Nauk SSSR, Ser. Khim.*, 1968, 657 (*Chem. Abstr.*, 1968, **69**, 18 567z). 811
68IZV1134 B. L. Dyatkin, K. N. Makarov and I. L. Knunyants; *Izv. Akad. Nauk SSSR, Ser. Khim.*, 1968, 1134 (*Chem. Abstr.*, 1968, **69**, 51 468r). 811
68JA1651 B. Belleau and G. Malek; *J. Am. Chem. Soc.*, 1968, **90**, 1651. 652
68JA2282 J. P. Collman, F. D. Vastine and W. R. Roper; *J. Am. Chem. Soc.*, 1968, **90**, 2282. 460
68JA2422 E. C. Taylor, G. W. McLay and A. McKillop; *J. Am. Chem. Soc.*, 1968, **90**, 2422. 183
68JA3295 R. P. Bennett and W. B. Hardy; *J. Am. Chem. Soc.*, 1968, **90**, 3295. 984
68JA4706 J. R. Beck, R. Kwok, R. N. Booher, A. C. Brown, L. E. Patterson, P. Pranc, B. Rockey and A. Pohland; *J. Am. Chem. Soc.*, 1968, **90**, 4706. 291
68JA5236 J. W. Rakshys, R. W. Taft and W. A. Sheppard; *J. Am. Chem. Soc.*, 1968, **90**, 5236. 1129
68JA5616 E. J. Corey, N. W. Gilman and B. E. Ganem; *J. Am. Chem. Soc.*, 1968, **90**, 5616. 51, 154
68JA6247 K. R. Ramaprasad, R. Varma and R. Nelson; *J. Am. Chem. Soc.*, 1968, **90**, 6247. 1016
68JAP6827842 S. Minakimi and K. Nagata, *Jpn. Pat.* 6827842 (1968) (*Chem. Abstr.*, 1969, **70**, 57472e). 636
68JCS(A)2535 J. R. Jennings, K. Wade and B. K. Wyatt; *J. Chem. Soc. (A)*, 1968, 2535. 383
68JCS(C)431 J. H. Davies, R. H. Davis and P. Kirby; *J. Chem. Soc. (C)*, 1968, 431. 828
68JCS(C)1375 F. Kurzer and D. R. Hanks; *J. Chem. Soc. (C)*, 1968, 1375. 849
68JCS(C)1612 A. M. Hamid and S. Trippett; *J. Chem. Soc (C)*, 1968, 1612. 900
68JCS(C)1777 R. Hull; *J. Chem. Soc. (C)*, 1968, 1777. 1033
68JCS(C)2201 D. Hodson, G. Holt and D. K. Wall; *J. Chem. Soc. (C)*, 1968, 2201. 872
68JGU1791 N. F. Orlov and B. L. Kaufman; *J. Gen. Chem. USSR*, 1968, **38**, 1791. 404
68JGU1815 V. A. Shokol, G. I. Derkach and V. F. Gamaleya; *J. Gen. Chem. USSR (Engl. Transl.)*, 1968, **38**, 1815. 707, 861
68JGU2501 V. V. Moskva, A. I. Maikova and A. I. Razumov; *J. Gen. Chem. USSR (Engl. Transl.)*, 1968, **38**, 2501. 707
68JHC223 F. H. Case; *J. Heterocycl. Chem.*, 1968, **5**, 223. 769
68JIC729 F. A. Hussein and A. A. Kadir; *J. Ind. Chem. Soc.*, 1968, **45**, 729. 770, 850
68JIC1109 A. N. Kost, P. B. Terentev and L. V. Moshentseva; *J. Indian Chem. Soc.*, 1968, **45**, 1109. 116
68JINC1399 T. N. Srivastava and S. K. Tandon; *J. Inorg. Nucl. Chem.*, 1968, **30**, 1399. 1136

68JMC534 F. J. McCarty, P. D. Rosenstock, J. P. Paolini, D. D. Micucci, L. Ashton, W. W. Bennetts and F. P. Palopoli; *J. Med. Chem.*, 1968, **11**, 534. 652

68JOC75 J. K. Kochi and T. W. Bethea, III; *J. Org. Chem.*, 1968, **33**, 75. 229

68JOC441 W. C. Firth, Jr.; *J. Org. Chem.*, 1968, **33**, 441. 995

68JOC753 A. F. Hegarty and F. L. Scott; *J. Org. Chem.*, 1968, **33**, 753. 356

68JOC762 B. T. Gillis and M. P. LaMontagne; *J. Org. Chem.*, 1968, **33**, 762. 363

68JOC823 L. L. Replogle, K. Katsumoto, T. C. Morrill and C. A. Minor; *J. Org. Chem.*, 1968, **33**, 823. 1109

68JOC1533 W. H. Mueller and P. E. Butler; *J. Org. Chem.*, 1968, **33**, 1533. 248

68JOC1679 L. Weintraub, S. R. Oles and N. Kalish; *J. Org. Chem.*, 1968, **33**, 1679. 852

68JOC1783 K. J. Ryan, E. M. Acton and L. Goodman; *J. Org. Chem.*, 1968, **33**, 1783. 476, 477

68JOC2143 F. S. Alvarez and A. N. Watt; *J. Org. Chem.*, 1968, **33**, 2143. 132

68JOC2330 G. N. Sausen and A. L. Logothetis; *J. Org. Chem.*, 1968, **33**, 2330. 811

68JOC2357 J. P. Chupp and E. R. Weiss; *J. Org. Chem.*, 1968, **33**, 2357. 597

68JOC2525 A. Nishihara and I. Koboto; *J. Org. Chem.*, 1968, **33**, 2525. 141

68JOC2565 E. L. Compere, Jr.; *J. Org. Chem.*, 1968, **33**, 2565. 67

68JOC2738 E. U. Elam, F. H. Rash, J. T. Dougherty, V. W. Goodlett and K. C. Brannock; *J. Org. Chem.*, 1968, **33**, 2738. 511

68JOC2819 W. C. Bailey, Jr., A. K. Bose, R. M. Ikeda, R. H. Newman, H. V. Secor and C. Varsel; *J. Org. Chem.*, 1968, **33**, 2819. 168

68JOC2887 H. Ulrich, B. Tucker and A. A. R. Sayigh; *J. Org. Chem.*, 1968, **33**, 2887. 977

68JOC3090 K. D. Berlin, R. T. Claunch and E. T. Gaudy; *J. Org. Chem.*, 1968, **33**, 3090. 403, 404

68JOC3517 G. B. Payne; *J. Org. Chem.*, 1968, **33**, 3517. 494

68JOC3727 K. J. Ryan, E. M. Acton and L. Goodman; *J. Org. Chem.*, 1968, **33**, 3727. 476, 477

68JOM(12)241 H. Kurosawa, M. Tanaka and R. Okawara; *J. Organomet. Chem.*, 1968, **12**, 241. 1144

68JOM(13)209 E. Moser and E. O. Fischer; *J. Organomet. Chem.*, 1968, **13**, 209. 932

68JOM(13)387 E. Moser and E. O. Fischer; *J. Organomet. Chem.*, 1968, **13**, 387. 953

68JOM(14)P9 E. O. Fischer, E. Louis, W. Bathelt, E. Moser and J. Müller; *J. Organomet. Chem.*, 1968, **14**, P9. 952

68JOM(15)147 E. Moser and E. O. Moser; *J. Organomet. Chem.*, 1968, **15**, 147. 952

68JOM(15)397 C. S. Kraihanzel and L. G. Herman; *J. Organomet. Chem.*, 1968, **15**, 397. 447

68JPS(B)389 H. Taru, H. Araki and H. Yasuda; *J. Polym. Sci., Polym. Chem., Part B*, 1968, **6**, 389. 718

68JSC(C)1602 D. Hodson and G. Holt, *J. Chem. Soc.* (*C*), 1968, 1602. 867

68LA(712)46 W. Walter and G. Maerten; *Justus Liebigs Ann. Chem.*, 1968, **712**, 46. 623, 624

68LA(712)58 W. Walter and G. Maerten; *Justus Liebigs Ann. Chem.*, 1968, **712**, 58. 590, 609, 623

68LA(716)204 U. Schöllkopf, F. Gerhart, M. Reetz, H. Frasnelli and H. Schumacher; *Justus Liebigs Ann. Chem.*, 1968, **716**, 204. 844, 866, 874

68LA(718)24 H. J. Bestmann, J. Lienert and L. Mott; *Justus Liebigs Ann. Chem.*, 1968, **718**, 24. 1078

68M1289 K. Hohenlohe-Oehringen and L. Call; *Monatsh. Chem.*, 1968, **99**, 1289. 342

68M2498 A. Weisz and K. Utvary; *Monatsh. Chem.*, 1968, **99**, 2498. 1078

B-68MI 502-01 S. R. Sandler and W. Karo; "Organic Functional Group Preparations," Academic Press, New York, 1968, vol. 12-I, p. 578. 24

B-68MI 504-01 "Beilstein Handbuch Der Organischen Chemie, 4th edn., Mainwork," Springer-Verlag, Berlin, 1968, vol. 2, p. 106. 229

68MI 509-01 S. Hoff, L. Brandsma and J. F. Arens; *Recueil*, 1968, **87**, 916. 423

B-68MI 517-01 H. Ulrich; "The Chemistry of Imidoyl Halides," Plenum Press, New York, 1968, p. 55. 654

B-68MI 517-02 H. Ulrich; "The Chemistry of Imidoyl Halides," Plenum Press, New York, 1968, p. 157. 664, 665, 669

B-68MI 517-03 H. Ulrich; "The Chemistry of Imidoyl Halides," Plenum Press, New York, 1968, p. 165. 675, 676, 678, 679, 682

B-68MI 517-04 H. Ulrich; "The Chemistry of Imidoyl Halides," Plenum Press, New York, 1968, p. 173. 699

B-68MI 521-01 H. Ulrich; "The Chemistry of Imidoyl Halides," Plenum Press, New York, 1968, p. 158. 800

68MI 526-01 J. C. Wright and A. Fry; *J. Chem. Eng. News*, 1968, (**1**), 28. 998

68MI 527-01 R. Pohloudek-Fabini and E. Schropl, *Pharm. Zentral.*, 1968, **107**, 277. 1047

68MI 530-01 J. Prejzner; *Wiadomosci. Chem.*, 1968, **22**, 187. 1100

68MI 530-02 V. A. Nefedov; *Zh. Obshch. Khim. Lett.*, 1968, **38**, 2191 (*Chem. Abstr.*, 1969, **70**, 28 556e). 1116

B-68MI 530-03 H. Shapiro and F. W. Frey; in "The Organic Compounds of Lead," Wiley, New York, 1968. 1144

68RCR363 M. O. Lozinskii and P. S. Pel'kis; *Russ. Chem. Rev. (Engl. Transl.)*, 1968, **37**, 363. 1022

68RTC38 P. J. W. Schuijl and L. Brandsma; *Recl. Trav. Chim. Pays-Bas*, 1968, **87**, 38. 589, 590, 603

68RTC929 P. J. W. Schuijl and L. Brandsma; *Recl. Trav. Chim. Pays-Bas*, 1968, **87**, 929. 552

68RTC1236 H. E. Wijers, C. H. D. Van Ginkel, P. J. W. Schuijl and L. Brandsma; *Recl. Trav. Chim. Pays-Bas*, 1968, **87**, 1236. 589, 614

68SAP6903273 H. O. Wirth, R. Maul, H. H. Friedrich and W. Hess; *S. Afr. Pat.* 6 903 273 (1968). 1145

68T2443 D. Taub, N. N. Girotra, R. D. Hoffsommer, C. H. Kuo, H. L. Slates, S. Weber and N. L. Wendler; *Tetrahedron*, 1968, **24**, 2443. 173

68T2583 O. Tsuge, T. Itoh and M. Tashiro; *Tetrahedron*, 1968, **24**, 2583. 996

68T2995 R. G. Kostyanovsky, V. V. Yakshin and S. L. Zimont; *Tetrahedron*, 1968, **24**, 2995. 395, 397, 402

68TL319 J. Sauer and K. K. Mayer; *Tetrahedron Lett.*, 1968, 319. 1013

68TL633 Y. Fujiwara, I. Moritani, M. Matsuda and S. Teranishi; *Tetrahedron Lett.*, 1968, 633. 143

68TL935 T. Saegusa, Y. Ito and S. Kobayashi; *Tetrahedron Lett.*, 1968, 935. 785, 792

68TL2463 F. L. Scott, F. A. Groeger and A. F. Hegarty; *Tetrahedron Lett.*, 1968, 2463. 846

68TL2651 T. Minami and T. Agawa; *Tetrahedron Lett.*, 1968, **9**, 2651. 1033

68TL2681 R. Misra, R. C. Pandey and S. Dev; *Tetrahedron Lett.*, 1968, 2681. 52
68TL2871 B. Zwanenburg, L. Thijs and J. Strating; *Tetrahedron Lett.*, 1968, 2871. 562
68TL3277 H. H. Wasserman, K. Stiller and M. B. Floyd; *Tetrahedron Lett.*, 1968, 3277. 759
68TL3341 V. D. Parker and B. E. Burgert; *Tetrahedron Lett.*, 1968, 3341. 986
68TL3569 T. Mukaiyama and S. Ono; *Tetrahedron Lett.*, 1968, 3569. 774, 850
68TL3787 T. Mukaiyama, S. Fukuyama and T. Kumamoto; *Tetrahedron Lett.*, 1968, **9**, 3787. 898
68TL4405 M. Sprecher and E. Nativ; *Tetrahedron Lett.*, 1968, 4405. 403
68TL4973 K. D. Kaufmann, B. Auräth, P. Träger and K. Rühlmann; *Tetrahedron Lett.*, 1968, 4973. 873
68TL5237 T. DoMinh, O. P. Strausz and H. E. Gunning; *Tetrahedron Lett.*, 1968, 5237. 874
68TL5731 G. H. Jones, E. K. Hamamura and J. G. Moffatt; *Tetrahedron Lett.*, 1968, **9**, 5731. 910
68TL5973 W. Walter and K.-J. Reubke; *Tetrahedron Lett.*, 1968, 5973. 641
68USP3361815 J. J. Hoekstra; *US Pat.* 3 361 815 (1968) (*Chem. Abstr.*, 1968, **68**, 41 777f). 815
68USP3366662 Olin Mathieson Chemical Corp.; *US Pat.*, 3 366 662 (1968) (*Chem. Abstr.*, 1968, **69**, 26 764). 976
68USP3374260 J. B. Buchanan; *US Pat.* 3 374 260 (1968) (*Chem. Abstr.*, 1968, **69**, 43 482a). 830
68USP3733349 B. L. Loudas and H. A. Vogel; 1968, *US Pat.* 3 733 349 (*Chem. Abstr.*, 1973, **79**, 92 790q). 1102
68YZ1039 H. Saikachi and S. Nakamura; *Yakugaku Zasshi* (*J. Pharm. Soc. Jpn.*), 1968, **88**, 1039 (*Chem. Abstr.*, 1969, **70**, 11 435). 898
68ZC142 G. Barnikow and T. Gabrio; *Z. Chem.*, 1968, **4**, 142. 476, 477, 486, 1050
68ZC219 J. Gloede and H. Groß; *Z. Chem.*, 1968, **8**, 219. 714
68ZC335 G. Barnikow and G. Strickmann; *Z. Chem.*, 1968, **8**, 335. 516
68ZC376 G. Wagner and S. Leistner; *Z. Chem.*, 1968, **8**, 376. 587
68ZOB326 V. I. Shevchenko, N. K. Kulibaba and A. V. Kirsanov; *Zh. Obshch. Khim.*, 1968, 326. 861
68ZOB850 V. I. Shevchenko, N. K. Kulibaba and A. V. Kirsanov; *Zh. Obshch. Khim.*, 1968, 850. 861
68ZOB1179 V. P. Kozyukov, V. D. Sheludyakov and V. D. Mironov; *Zh. Obshch. Khim.*, 1968, **38**, 1179. 1016
68ZOB1408 A. S. Filatov and M. A. Englin; *Zh. Obshch. Khim.*, 1968, **38**, 1408. 807
68ZOB1867 V. A. Shokol, V. F. Gamaleya and G. I. Derkach; *Zh. Obshch. Khim.*, 1968, **38**, 1867. 707
68ZOB2586 V. V. Moskva, A. I. Maikova and A. I. Razumov; *Zh. Obshch. Khim.*, 1968, **38**, 2586. 707
68ZOR658 R. G. Dubenko, V. M. Neplyuev and P. S. Pel'kis; *Zh. Org. Khim.*, 1968, **4**, 658. 1049

69ACR65 H. C. Brown; *Acc. Chem. Res.*, 1969, **2**, 65. 433
69ACS818 A. Pilotti, A. Reuterhäll, K. Torsell and C. G. Lindblad; *Acta Chem. Scand.*, 1969, **23**, 818. 652
69ACS1567 K. A. Jensen, A. Holm and J. Wolff-Jensen; *Acta Chem. Scand.*, 1969, **23**, 1567. 1101
69AG572 H. Hagemann, D. Arlt and I. Ugi; *Angew. Chem.*, 1969, **81**, 572. 809
69AG(E)20 E. Kühle; *Angew. Chem., Int. Ed. Engl.*, 1969, **8**, 20. 1004, 1032, 1070
69AG(E)203 J. Ellermann, F. Poersch, R. Kunstmann and R. Kramolowsky; *Angew. Chem., Int. Ed. Engl.*, 1969, **8**, 203. 1058
69AG(E)278 R. Tull and L. M. Weinstock; *Angew. Chem., Int. Ed. Engl.*, 1969, **8**, 278. 579, 580
69AG(E)421 G. I. Derkach; *Angew. Chem., Int. Ed. Engl.*, 1969, **8**, 421. 988, 1010, 1017
69AG(E)607 E. Grigat; *Angew. Chem., Int. Ed. Engl.*, 1969, **8**, 607. 1005
69AG(E)750 H. Eilingsfeld and M. Patsch; *Angew. Chem., Int. Ed. Engl.*, 1969, **8**, 750. 580
69AG(E)760 G. Ottmann and E. Kober; *Angew. Chem., Int. Ed. Engl.*, 1969, **8**, 760. 1033
69AG(E)761 C. G. Kreiter and E. O. Fischer; *Angew. Chem., Int. Ed. Engl.*, 1969, **8**, 761. 932
69AJC161 M. T. W. Hearn and A. D. Ward; *Austr. J. Chem.*, 1969, **22**, 161. 714, 723
69AP335 H. Böhme and W. Pasche; *Arch. Pharm. (Weinheim, Ger.)*, 1969, **302**, 335. 965
69AP617 H. Böhme and W. Pasche; *Arch. Pharm. (Weinheim, Ger.)*, 1969, **302**, 617. 965
69BCJ3310 Takeo Saegusa, Y. Ito, S. Kobayashi, K. Hirota and H. Yoshioka; *Bull. Chem. Soc. Jpn.*, 1969, **42**, 3310. 747
69BRP1142991 D. Wright (Imperial Chemical Industries Ltd.); *Br. Pat.*, 1 142 991 (1969) (*Chem. Abstr.*, 1969, **70**, 87 134p). 985
69BRP1152877 E. G. Gazzard, J. N. Greenshields and J. M. Phillipson (Imperial Chemical Industries Ltd.); *Br. Pat.*, 1 152 877 (1969) (*Chem. Abstr.*, 1969, **71**, 48 896q). 966
69BRP1153261 Upjohn Co.; *Br. Pat.*, 1 153 261 (1969) (*Chem. Abstr.*, 1969, **71**, 49 543j). 966
69BSF911 J. P. Battioni and W. Chodkiewicz; *Bull. Soc. Chim. Fr.*, 1969, 911. 59
69BSF2871 G. Le Guillanton; *Bull. Soc. Chim. Fr.*, 1969, 2871. 75
69CB269 R. R. Schmidt and D. Schwille; *Chem. Ber.*, 1969, **102**, 269. 634
69CB488 U. Schöllkopf and N. Rieber; *Chem. Ber.*, 1969, **102**, 488. 844, 871, 874
69CB1641 H. Stetter and D. Theisen; *Chem. Ber.*, 1969, **102**, 1641. 701, 827
69CB1848 R. Huisgen and J. Wulff; *Chem. Ber.*, 1969, **102**, 1848. 1068
69CB2117 W. Walter and K.-J. Reubke; *Chem. Ber.*, 1969, **102**, 2117. 640, 846
69CB2216 M. Regitz and W. Anschütz; *Chem. Ber.*, 1969, **102**, 2216. 869
69CB2640 W. Walter and K.-P. Ruess; *Chem. Ber.*, 1969, **102**, 2640. 746
69CB2718 A. Haas and H. Reinke; *Chem. Ber.*, 1969, **102**, 2718. 1018
69CB2889 T. Eicher, S. Hünig and H. Hansen; *Chem. Ber.*, 1969, **102**, 2889. 366
69CB3347 E. Lindner and U. Kunze; *Chem. Ber.*, 1969, **102**, 3347. 546
69CB3786 W. Walter and J. Krohn; *Chem. Ber.*, 1969, **102**, 3786. 606
69CC343 R. Mechoulam and Z. Ben-Zvi; *J. Chem. Soc., Chem. Commun.*, 1969, 343. 112
69CI(L)1702 R. G. Guy, R. Bonnett and D. Lanigan; *Chem. Ind. (London)*, 1969, 1702. 1044
69CPB550 T. Hino, K. Tsuneoka, M. Nakagawa and S. Akaboshi; *Chem. Pharm. Bull.*, 1969, **17**, 550. 567
69CPB653 K. Nagata; *Chem. Pharm. Bull.*, 1969, **17**, 653. 831
69CPB1564 H. Shirai, T. Yashiro and T. Sato; *Chem. Pharm. Bull.*, 1969, **17**, 1564. 25, 64
69CR(C)1325 J. F. Normant, J. P. Fowler and H. Deshayes; *C. R. Hebd. Seances Acad. Sci., Ser. C.*, 1969, **269**, 1325. 10

69CRV591 F. Freeman; *Chem. Rev.*, 1969, **69**, 591. 46

69CRV673 H. W. Gibson; *Chem. Rev.*, 1969, **69**, 673. 24

69CZ474 F. Becke and H. Hagen; *Chem. Ztg.*, 1969, **93**, 474 601

69CZP131547 J. Kalamar, J. Hrivnak, A. Kovac and J. Obertas, *Czech Pat.* 131 547 (1969) (*Chem. Abstr.*, 1970, **72**, 100 264). 1039

69DOK(181)115 V. F. Mironov, V. D. Sheludyakov, V. P. Kozyukov and G. D. Khatuntsev; *Dokl. Akad. Nauk SSSR*, 1969, **181**, 115. 977

69FRP1563380 V. P. Kozyukov, V. F. Mironov and V. D. Scheludyakov; *Fr. Pat.*, 1 563 380 (1969) (*Chem. Abstr.*, 1969, **71**, 81 516p). 968

69FRP1573102 J. B. Buchanan; *Fr. Pat.* 1 573 102 (1969) (*Chem. Abstr.*, 1970, **72**, 100 080). 829

69G69 R. Fusco and P. Dalla Croce; *Gazz. Chim. Ital.*, 1969, **99**, 69. 774, 839, 854, 855

69GEP1289526 H. Bestian, W. Blau and D. Guenther; *Ger. Pat.*, 1 289 526 (1969) (*Chem. Abstr.*, 1969, **70**, 87 312). 1006

69GEP1910303 E. Smith (Olin Mathieson Chemical Corp.); *Ger. Pat.*, 1 910 303 (1969) (*Chem. Abstr.*, 1969, **71**, 112 603w). 984

69HCA569 R. Blaser, P. Imfeld and O. Schindler; *Helv. Chim. Acta*, 1969, **52**, 569. 700

69ICA495 M. F. Farona, J. G. Grasselli, H. Grossman and W. M. Ritchey; *Inorg. Chim. Acta*, 1969, **3**, 495. 389

69IZV958 K. N. Makarov, B. L. Dyatkin and I. L. Knunyants; *Izv. Akad. Nauk SSSR, Ser. Khim.*, 1969, 958. 813

69JA1233 E. J. Corey and L. S. Hegedus; *J. Am. Chem. Soc.*, 1969, **91**, 1233. 273, 296

69JA3075 Y. F. Shealy and J. D. Clayton; *J. Am. Chem. Soc.*, 1969, **91**, 3075. 199

69JA3197 R. J. Angelici and L. Busetto; *J. Am. Chem. Soc.*, 1969, **91**, 3197. 975

69JA3992 A. R. Doumaux, Jr., J. E. McKeon and D. J. Trecker; *J. Am. Chem. Soc.*, 1969, **91**, 3992. 307

69JA4778 J. P. Freeman, R. C. Petry and T. E. Stevens; *J. Am. Chem. Soc.*, 1969, **91**, 4778. 314

69JA5647 H. C. Brown and J. T. Kurek; *J. Am. Chem. Soc.*, 1969, **91**, 5647. 282

69JA5674 J. A. Webber, E. M. Van Heyningen and R. T. Vasileff; *J. Am. Chem. Soc.*, 1969, **91**, 5674. 81

69JA5675 E. J. Corey, N. M. Weinshenker, T. K. Schaaf and W. Huber; *J. Am. Chem. Soc.*, 1969, **91**, 5675. 168

69JA7196 S. Otsuka, A. Nakamura and T. Yoshida; *J. Am. Chem Soc.*, 1969, **91**, 7196. 463, 800, 801

69JA7778 H. M. Walborsky and G. E. Niznik; *J. Am. Chem. Soc.*, 1969, **91**, 7778. 796

69JAP4427365 Mitsui Toatsu Chemicals; *Jpn. Pat.*, 44-27 365 (1969). 652

69JAP47016293 H. Kurosawa and K. Yasuda; *Jpn. Pat.* 47 016 293 (1969) (*Chem. Abstr.*, 1972, **77**, 48 619a). 1144

69JAP47019769 H. Kurosawa and R. Ogawara; *Jpn. Pat.* 47 019 769 (1969) (*Chem. Abstr.*, 1972, **77**, 101 868j). 1143

69JCED110 W. U. Malik, P. K. Srivastava and S. C. Mehra; *J. Chem. Eng. Data*, 1969, **14**, 110. 1038

69JCS(A)992 T. A. George and M. F. Lappert; *J. Chem. Soc.* (*A*), 1969, 992. 277

69JCS(A)2190 P. G. Kirk and T. D. Smith; *J. Chem. Soc.* (*A*), 1969, **14**, 2190. 1129

69JCS(C)874 T. Kametani, T. Terui, T. Ogino and K. Fukumoto; *J. Chem. Soc.* (*C*), 1969, 874. 276

69JCS(C)1117 G. C. Barrett and A. R. Khokhar; *J. Chem. Soc.* (*C*), 1969, 1117. 586

69JCS(C)2192 F. M. Dean, J. Goodchild and A. W. Hill; *J. Chem. Soc.* (*C*), 1969, 2192. 575

69JCS(C)2565 G. Chandra, T. A. George and M. F. Lappert; *J. Chem. Soc.* (*C*), 1969, 2565. 277

69JCS(C)2587 W. A. F. Gladstone, J. B. Aylward and R. O. C. Norman; *J. Chem. Soc.* (*C*), 1969, 2587. 356

69JGU597 V. S. Abramov and N. I. D'yakonova; *J. Gen. Chem. USSR* (*Engl. Transl.*), 1969, **39**, 597. 403

69JGU1445 A. S. Shtepanek, E. N. Tkachenko and A. V. Kirsanov; *J. Gen. Chem. USSR* (*Engl. Transl.*), 1969, **7**, 1443. 373

69JGU2541 M. B. Gazizov and A. I. Razumov; *J. Gen. Chem. USSR* (*Engl. Transl.*), 1969, **39**, 2541. 403

69JIC1096 P. H. Deshpande; *J. Indian Chem. Soc.*, 1969, **46**, 1096. 735

69JOC51 M. Behforouz and J. E. Kerwood; *J. Org. Chem.*, 1969, **34**, 51. 336

69JOC112 D. M. Piatak, H. B. Bhat and E. Caspi; *J. Org. Chem.*, 1969, **34**, 112. 49

69JOC116 D. M. Piatak, G. Herbst, J. Wicha and E. Caspi; *J. Org. Chem.*, 1969, **34**, 116. 49

69JOC616 W. R. Diveley, G. A. Buntin and A. D. Lohr; *J. Org. Chem.*, 1969, **34**, 616. 1043

69JOC730 T. Takeshima, M. Yokoyama, T. Imamoto, M. Akano and H. Asaba; *J. Org. Chem.*, 1969, **34**, 730. 547

69JOC1154 C. M. Orlando Jr. and M. H. Gianni; *J. Org. Chem.*, 1969, **34**, 1154. 314

69JOC1162 W. J. Schnabel and E. Kober; *J. Org. Chem.*, 1969, **34**, 1162. 966, 1013

69JOC1278 C. G. Krespan; *J. Org. Chem.*, 1969, **34**, 1278. 1011, 1102

69JOC1454 M. Debono and R. M. Molloy; *J. Org. Chem.*, 1969, **34**, 1454. 48

69JOC1631 K. D. Kopple, T. Saito and M. Ohnishi; *J. Org. Chem.*, 1969, **34**, 1631. 972

69JOC1677 C. G. Gebelein, S. Rosen and D. Swern; *J. Org. Chem.*, 1969, **34**, 1677. 985

69JOC1808 R. S. Neale and N. L. Marcus; *J. Org. Chem.*, 1969, **34**, 1808. 1120

69JOC2486 E. S. Rothman, G. G. Moore and S. Serota; *J. Org. Chem.*, 1969, **34**, 2486. 3

69JOC2543 J. A. Dale, D. L. Dull and H. S. Mosher; *J. Org. Chem.*, 1969, **34**, 2543. 12

69JOC2766 T. H. Chan and L. T. L. Wong; *J. Org. Chem.*, 1969, **34**, 2766. 291

69JOC2908 F. A. Daniher; *J. Org. Chem.*, 1969, **34**, 2908. 1000

69JOC3200 H. Ulrich, B. Tucker and A. A. R. Sayigh; *J. Org. Chem.*, 1969, **34**, 3200. 968

69JOC3204 J. S. Matthews and J. P. Cookson; *J. Org. Chem.*, 1969, **34**, 3204. 652

69JOC4032 U. Tonellato, O. Rossetto and A. Fava; *J. Org. Chem.*, 1969, **34**, 4032. 1038

69JOC4088 R. A. Falk and J. D. Readio; *J. Org. Chem.*, 1969, **34**, 4088. 1011

69JOC4173 W. V. Rochat and G. L. Gard; *J. Org. Chem.*, 1969, **34**, 4173. 246

69JOM(17)241 I. F. Lutsenko, Y. I. Baukov, A. S. Kostyuk, N. I. Savelyeva and V. K. Krysina; *J. Organomet. Chem.*, 1969, **17**, 241. 376

69JOM(19)P33 R. Calas, J.-P. Picard, J. Dunoguès and N. Duffaut; *J. Organomet. Chem.*, 1969, **19**, 33. 419

69JOM(16)29 E. O. Fischer, B. Heckl, K. H. Dötz, J. Müller and H. Werner; *J. Organomet. Chem.*, 1969, **16**, 29. 652

69JOU252 V. L. Dubina, S. I. Burmistrov, Y. N. Ishchuk and V. S. Emel'yanova; *J. Org. Chem. USSR* (*Engl. Transl.*), 1969, **5**, 252. 703

69JPS(A)945 P. M. Hergenrother; *J. Polym. Sci., Polm. Chem., Part A*, 1969, **7**, 945. 847

69JPS(B)17 H. Taru and H. Yasuda; *J. Polym. Sci., Polym. Chem., Part B*, 1969, **7**, 17. 718

69LA(721)121 E. Zbiral and H. Hengstberger; *Justus Liebigs Ann. Chem.*, 1969, **721**, 121. 898

69LA(723)129 A. Treibs; *Justus Liebigs Ann. Chem.*, 1969, **723**, 129. 549

69LA(725)234 W. Walter and R. F. Becker; *Justus Liebigs Ann. Chem.*, 1969, **725**, 234. 609

69LA(727)22 W. Walter and P.-M. Hell; *Justus Liebigs Ann. Chem.*, 1969, **727**, 22. 638, 735

69LA(727)50 W. Walter and P.-M. Hell; *Justus Liebigs Ann. Chem.*, 1969, **727**, 50. 623, 735

69LA(730)121 H. Böhme and H.-P. Steudel; *Justus Liebigs Ann. Chem.*, 1969, **730**, 121. 244

69M959 H. Wittmann, E. Ziegler, H. Sterk and G. Dworak; *Monatsh. Chem.*, 1969, **100**, 959. 340

69MAC144 C. E. Carraher, Jr. and T. Brandt; *Macromol. Chem.*, 1969, **123**, 144. 652

B-69MI 502-01 R. P. A. Sneeden; in "The Chemistry of Carboxylic Acids and Esters," ed. S. Patai, Wiley, New York, 1969, chap. 4, pp. 137–173. 27

B-69MI 502-02 V. F. Kucherov and L. A. Yanovskaya; in "The Chemistry of Carboxylic Acids and Esters," ed. S. Patai, Wiley, New York, 1969, chap. 5, pp. 175–209. 24

B-69MI 502-03 A. S. Perlin; in "Oxidation," ed. R. L. Augustine, Marcel Dekker, New York, 1969, vol. 1, p. 189. 28

B-69MI 511-01 M. J. Janssen; in "The Chemistry of Carboxylic Acids and Esters," ed. S. Patai, Wiley, New York, 1969, p. 705. 472

69MI 521-01 A. I. Busev, T. N. Zholondkovskaya, L. S. Krysina and L. I. Korobkina; *Sovrem. Metody Anal. Mater.*, 1969, 153 (*Chem. Abstr.*, 1970, **73**, 94 224h). 864

69MI 527-01 P. Kristian; *Chem. Zvesti*; 1969, **23**, 371. 1040

69MI 530-01 H. Hurosawa, T. Fukumoto and R. Okawara; *Inorg. Nucl. Chem. Lett.*, 1969, **5**, 473. 1143

69OR(17)213 L. I. Krimen and D. J. Cota; *Org. React.*, 1969, **17**, 213. 282

69RCR21 K. A. Petrov and L. N. Andreev; *Russ. Chem. Rev.* (*Engl. Transl.*), 1969, **38**, 21. 565, 579

69RTC597 P. J. W. Schuiji, H. J. T. Bos and L. Brandsma; *Recl. Trav. Chim. Pays-Bas*, 1969, **88**, 597. 554

69S1 J. Tsuji and K. Ohno; *Synthesis*, 1969, **1**, 157. 459

69S75 M. Schlosser and M. Zimmermann; *Synthesis*, 1969, 75. 894

69T5437 G. Schwenker and R. Kolb; *Tetrahedron*, 1969, **25**, 5437. 760

69T5745 H. Mohrle and R. Kilian; *Tetrahedron*, 1969, **25**, 745. 289

69TL117 K. D. Kampe; *Tetrahedron Lett.*, 1969, 117. 998

69TL699 Y. Kikugawa and S. Yamada; *Tetrahedron Lett.*, 1969, **9**, 699. 745

69TL1019 T. Sakakibara, S. Nishimura and Y. Odaira; *Tetrahedron Lett.*, 1969, 1019. 113

69TL1047 R. C. Blume; *Tetrahedron Lett.*, 1969, 1047. 161

69TL1421 K. Itoh, A. Nozawa and Y. Ishii; *Tetrahedron Lett.*, 1969, 1421. 1070

69TL1573 D. A. Evans; *Tetrahedron Lett.*, 1969, 1573. 277

69TL1591 S. H. Schroeter; *Tetrahedron Lett.*, 1969, 1591. 166

69TL2281 M. Fischer; *Tetrahedron Lett.*, 1969, 2281. 292

69TL2465 H. Tomisawa and H. Hongo; *Tetrahedron Lett.*, 1969, 2465. 199

69TL2473 O. Mitsunobu, K. Kato and F. Kakese; *Tetrahedron Lett.*, 1969, 2473. 1078

69TL3191 G. Tomaschewski, B. Breitfeld and D. Zanke; *Tetrahedron Lett.*, 1969, 3191. 1062

69TL4563 M. Sprecher and D. Levy; *Tetrahedron Lett.*, 1969, **10**, 4563. 191

69TL5151 A. I. Meyers and H. W. Adickes; *Tetrahedron Lett.*, 1969, 5151. 1014

69URP241428 B. F. Malichenko and A. V. Yaslovitskii; *USSR Pat.*, 241 428 (1969) (*Chem. Abstr.*, 1969, **71**, 80 913d). 967

69URP256779 A. E. Mil'grom and B. B. Paleev; *USSR Pat.* 256 779 (*Chem. Abstr.*, 1970, **72**, 132 703c). 754

69URP256781 A. E. Mil'grom and B. B. Paleev; *USSR Pat.* 256 781 (*Chem. Abstr.*, 1970, **72**, 132 752t). 754

69USP3465024 A. M. Brownstein and J. P. Sibilia; *US Pat.*, 3 465 024 (1969) (*Chem. Abstr.*, 1969, **71**, 112 602v). 1001

69USP3468924 E. I. du Pont de Nemours & Co.; *US Pat.*, 3 468 924 (1969) (*Chem. Abstr.*, 1969, **71**, 112 604x). 967

69USP3470228 D. H. Heinert (Dow Chemical Co.); *US Pat.*, 3 470 228 (1969) (*Chem. Abstr.*, 1969, **71**, 112 401d) 1015

69ZN(B)314 F. Weygand and C. Di Bello; *Z. Naturforsch., Teil B*, 1969, **24**, 314. 185, 190

69ZOB1282 M. I. Shevchuk, I. S. Zabrodskaya and A. V. Dombrovskii; *Zh. Obshch. Khim.*, 1969, **39**, 1282. 597

69ZOB1712 L. I. Samarai, O. I. Kolodyazhnyi and G. I. Derkach; *Zh. Obshch. Khim.*, 1969, **39**, 1712. 997

69ZOB2365 O. A. Kruglaya, B. I. Petrov and N. S. Vyazankin; *Zh. Obshch. Khim.*, 1969, **39**, 2365. 1137

69ZOB2598 V. F. Mironov, V. D. Sheludyakov and V. P. Kozyukov; *Zh. Obshch. Khim.*, 1969, **39**, 2598. 968

69ZOR119 L. I. Samarai, V. A. Bondar and G. I. Derkach; *Zh. Org. Khim.*, 1969, **5**, 119. 967

69ZOR263 V. L. Dubina, S. I. Burmistrov, Y. N. Ishchuk and V. S. Emel'yanova; *Zh. Org. Khim.*, 1969, **5**, 263. 703

69ZOR279 M. M. Kremlev and M. T. Plotnikova; *Zh. Org. Khim.*, 1969, **5**, 279. 345

69ZOR1285 I. A. Kirsanova and G. I. Derkach; *Zh. Org. Khim.*, 1969, **5**, 1285. 1052

69ZOR1404 G. S. Gol'din, V. G. Poddubnyi, A. A. Simonova, G. S. Shor and E. A. Rybakov; *Zh. Org. Khim.*, 1969, **5**, 1404. 850

69ZOR1411 G. S. Gol'din, V. G. Poddubnyi, A. A. Simonova, E. V. Orlova and G. S. Shor; *Zh. Org. Khim.*, 1969, **5**, 1411. 854

69ZOR1888 V. G. Sinyavskii and V. F. Kovaleva; *Zh. Org. Khim.*, 1969, **5**, 1888. 967

70ACS1367 D. Kjersgaard and A. Kjaer; *Acta Chem. Scand.*, 1970, **24**, 1367. 828, 831

70ACS1512 C. Christophersen and A. Holm; *Acta Chem. Scand.*, 1970, **24**, 1512. 1013, 1101

70ACS1852 C. Christophersen and A. Holm; *Acta Chem. Scand.*, 1970, **24**, 1852. 1013

70ACS3751 A. Fredga, L. B. Agenas and E. Jonsson; *Acta Chem. Scand.*, 1970, **24**, 3751. 1113

70AF790 H. Zimmer and L. J. Trauth; *Arzneim.-Forsch.*, 1970, **20**, 790. 370

70AG(E)54 G. Zumach and E. Kühle; *Angew. Chem., Int. Ed. Engl.*, 1970, **9**, 54. 1036

70AG(E)301 U. Schöllkopf and H. Frasnelli; *Angew. Chem., Int. Ed. Engl.*, 1970, **9**, 301. 873, 874

70AG(E)457 E. Öhler, K. Reininger and U. Schmidt; *Angew. Chem., Int. Ed. Engl.*, 1970, **9**, 457. 166

70AG(E)526 H. R. Kricheldorf; *Angew. Chem., Int. Ed. Engl.*, 1970, **9**, 526. 1026

70AP(303)905 C. H. Brieskorn and H. Sax; *Arch. Pharm.* (*Weinheim, Ger.*), 1970, **303**, 905. 1048

70BCF200 J.-A. Gautier, M. Miocque, C. Fauran and A.-Y. Le Cloarec; *Bull. Soc. Chim. Fr.*, 1970, 200. 756

70BCJ1890 J. Tsurugi, Y. Abe and S. Kawamura; *Bull. Chem. Soc. Jpn.*, 1970, **43**, 1890. 247

70BCJ2539 K. Anzai; *Bull. Chem. Soc. Jpn.*, 1970, **43**, 2539. 1122

70BCJ2870 T. Mukaiyama, T. Kumamato, S. Fukuyama and T. Taguchi; *Bull. Chem. Soc. Jpn.*, 1970, **43**, 2870 (*Chem. Abstr.*, 1970, **73**, 109 839). 898

70BSB301 F. Eloy and A. Deryckere; *Bull. Soc. Chim. Belg.*, 1970, **79**, 301. 990

70BSB383 J. L. Piette and M. Renson; *Bull. Soc. Chim. Belg.*, 1970, **79**, 383. 241

70BSF1363 Y. Charbonell, J. Barrans and R. Burgada; *Bull. Soc. Chim. Fr.*, 1970, 1363. 708, 861

70BSF2637 M. Hedayatullah, M. Brancourt and L. Denivelle; *Bull. Soc. Chim. Fr.*, 1970, 2637. 1102

70CB32 W. Ried, W. Merkel and R. Oxenius; *Chem. Ber.*, 1970, **103**, 32. 739

70CB37 W. Meyer zu Reckendorf and N. Wassiliadou-Micheli; *Chem. Ber.*, 1970, **103**, 37. 140

70CB82 L. Capuano, W. Ebner and J. Schrepfer; *Chem. Ber.*, 1970, **103**, 82. 831

70CB97 H. Schmidbaur and W. Malisch; *Chem. Ber.*, 1970, **103**, 97. 914, 1095

70CB553 S. P. von Halasz and O. Glemser; *Chem. Ber.* 1970, **103**, 553. 718

70CB583 W. Beck and M. Bauder; *Chem. Ber.*, 1970, **103**, 583. 386

70CB788 W. König and R. Geiger; *Chem. Ber.*, 1970, **103**, 788. 279

70CB1664 H. W. Roesky and L. F. Grimm; *Chem. Ber.*, 1970, **103**, 1664. 1075

70CB2024 W. König and R. Geiger; *Chem. Ber.*, 1970, **103**, 2024. 279

70CB2041 W. König and R. Geiger; *Chem. Ber.*, 1970, **103**, 2041. 285

70CB2388 D. Breitinger, K. Brodersen and J. Limmer; *Chem. Ber.*, 1970, **103**, 2388. 384, 387

70CB3007 H. Schmidbaur and W. Malisch; *Chem. Ber.*, 1970, **103**, 3007. 912

70CB3448 H. Schmidbaur and W. Malisch; *Chem. Ber.*, 1970, **103**, 3448. 912

70CC100 W. C. Kaska, C. Sutton and E. Serros; *J. Chem. Soc., Chem. Commun.*, 1970, 100. 386

70CC198 I. Harvie and R. D. W. Kemmitt; *J. Chem. Soc., Chem. Commun.*, 1970, 198. 466

70CC763 M. H. Chisholm and H. C. Clark; *J. Chem. Soc., Chem. Commun.*, 1970, 763. 939, 944

70CC1220 C. P. Casey; *J. Chem. Soc., Chem. Commun.*, 1970, 1220. 943

70CC1310 S. M. Hiscock, D. A. Swann and J. H. Turnbull; *J. Chem. Soc. D*, 1970, 1310. 238

70CI(L)238 J. V. Burakevich, A. M. Lore and G. P. Volpp; *Chem. Ind.* (*London*), 1970, 238. 832

70CJC983 J. D. Wilson and H. Weingarten; *Can. J. Chem.*, 1970, 983. 652

70CJC1991 M. Ahmed, J. M. Buchshriber and D. M. McKinnon; *Can. J. Chem.*, 1970, **48**, 1991. 634

70CJC3662 W. E. Elias; *Can. J. Chem.*, 1970, **48**, 3662. 727

70CRV151 D. G. Neilson, R. Roger, J. W. M. Heatlie and L. R. Newlands; *Chem. Rev.*, 1970, **70**, 151. 768, 769, 770, 847, 849, 850, 855

70CRV439 M. K. Hargreaves, J. G. Pritchard and H. R. Dave; *Chem. Rev.*, 1970, **70**, 439. 305

70DOK(190)110 V. F. Mironov, V. D. Sheludyakov and V. P. Kozyukov; *Dokl. Akad. Nauk SSSR*, 1970, **190**, 110. 968, 969

70DOK(194)595 B. M. Mikhailov, V. A. Dorokhov, V. S. Bogdanov, I. P. Yakovlev and A. D. Naumov; *Dokl. Akad. Nauk SSSR*, 1970, **194**, 595. 381, 382

70DOK(195)98 N. A. Nesmeyanov, A. V. Kalinin and O. A. Reutov; *Dokl. Akad. Nauk. SSSR*, 1970, **195**, 98. 919

70GEP2011937 M. L. Evans; *Ger. Pat.* 2 011 937 (1970) (*Chem. Abstr.*, 1971, **74**, 22 997k). 863

70GEP(O)1913273 A. Botta and H. Krimm (Farbenfabriken Bayer A.-G.); *Ger. Pat. Offen.*, 1 913 273 (1970) (*Chem. Abstr.*, 1970, **73**, 120 074f). 1012

70GEP(O)2024393 P. F. H. Freeman, J. M. Pedlar and I. T. Kay; *Ger. Pat. Offen.*, 2024393 (1970) (*Chem. Abstr.*, 1971, **74**, 53 830). 532

70HCA1091 F. L'Eplattenier and C. Pélichet; *Helv. Chim. Acta*, 1970, **53**, 1091. 455

70IC1571 R. Foester and K. Cohn; *Inorg. Chem.*, 1970, **9**, 1571. 1129

70IJC759 G. S. Johar, U. Agarwala and P. B. Rao; *Indian J. Chem.*, 1970, **8**, 759. 1026

70IZV1157 K. A. Andrianov, A. I. Nogaideli, L. M. Khananshvili, D. Sh. Akhobadze and R. Sh. Tkeshlashvili; *Izv. Akad. Nauk SSSR, Ser. Chim.*, 1970, 1157. 720

70JA397 E. J. Corey, T. K. Schaaf, W. Huber, U. Koelliker and N. M. Weinshenker; *J. Am. Chem. Soc.*, 1970, **92**, 397. 168

70JA741 W. S. Johnson, L. Werthemann, W. R. Bartlett, T. J. Brocksom, T. Li, D. J. Faulkner and M. R. Peterson; *J. Am. Chem. Soc.*, 1970, **92**, 741. 159

70JA1326 A. Hassner, R. P. Hoblitt, C. Heathcock, J. E. Kropp and M. Lorber; *J. Am. Chem. Soc.*, 1970, **92**, 1326. 984

70JA3515 L. Cassar, P. E. Eaton and J. Halpern; *J. Am. Chem. Soc.*, 1970, **92**, 3515. 460

70JA6080 M. P. Cooke Jr; *J. Am. Chem. Soc.*, 1970, **92**, 6080. 451

70JA6675 H. M. Walborsky, W. H. Morrison III and G. E. Niznik; *J. Am. Chem. Soc.*, 1970, **92**, 6675. 792, 796, 797

70JA6969 Y. Degani, H. Neumann and A. Patchornik; *J. Am. Chem. Soc.*, 1970, **92**, 6969. 1111

70JA7629 S. J. Brois, J. F. Pilot and H. W. Barnum; *J. Am. Chem. Soc.*, 1970, **92**, 7629. 247

70JAP26894 Hokko Chemical Industry Co., Ltd; *Jpn. Pat.*, 26 894 (1970) (*Chem. Abstr.*, 1970, **73**, 98 374e). 652

70JAP7019893 T. Kodera (Takeda Chemical Industries, Ltd.); *Jpn. Pat.*, 70 19 893 (1970) (*Chem. Abstr.*, 1970, **73**, 98 607h). 997

70JCED351 M. S. Heller; *J. Chem. Eng. Data*, 1970, **15**, 351. 348

70JCS(A)1847 J. Ashley-Smith, M. Green and D. C. Wood; *J. Chem. Soc. (A)*, 1970, 1847. 1139

70JCS(A)2675 C. A. Smith and M. G. H. Wallbridge; *J. Chem. Soc. (A)*, 1970, 2675. 1142

70JCS(C)875 J. Geevers, J. Th. Hackmann and W. P. Trompen; *J. Chem. Soc. (C)*, 1970, 875. 810

70JCS(C)878 C. Brown, M. Murray and R. Schmutzler; *J. Chem. Soc. (C)*, 1970, 878. 407

70JCS(C)968 J. B. Aylward and F. L. Scott; *J. Chem. Soc. (C)*, 1970, 968. 845

70JCS(C)1263 A. P. Stewart and S. Trippett; *J. Chem. Soc. (C)*, 1970, 1263. 403

70JCS(C)1429 A. Lawson and R. B. Tinkler; *J. Chem. Soc. (C)*, 1970, 1429. 767

70JCS(C)1530 E. Ager, B. Iddon and H. Suschitzky; *J. Chem. Soc. (C)*, 1970, 1530. 1111

70JCS(C)2106 P. D. Callaghan and M. S. Gibson; *J. Chem. Soc. (C)*, 1970, 2106. 641

70JHC177 E. H. Burk and D. D. Carlos; *J. Heterocycl. Chem.*, 1970, **7**, 177. 1013

70JHC863 R. A. Long, J. F. Gerster and L. B. Townsend; *J. Heterocycl. Chem.*, 1970, **7**, 863. 619

70JMC214 S. H. Chu and H. G. Mautner; *J. Med. Chem.*, 1970, **13**, 214. 563

70JMC948 H. C. Neumann, G. O. Potts, W. T. Ryan and F. W. Stonner; *J. Med. Chem.*, 1970, **13**, 948. 205

70JOC47 D. A. Tomalia, D. P. Sheetz and G. E. Ham; *J. Org. Chem.*, 1970, **35**, 47. 1010

70JOC806 R. V. Kendall and R. A. Olofson; *J. Org. Chem.*, 1970, **35**, 806. 626

70JOC825 R. E. Harmon, B. L. Jensen, S. K. Gupta and J. D. Nelson; *J. Org. Chem.*, 1970, **35**, 825. 295

70JOC1219 J. S. Bradshaw, R. D. Knudsen and E. L. Loveridge; *J. Org. Chem.*, 1970, **35**, 1219. 291

70JOC1275 R. D. Temple; *J. Org. Chem.*, 1970, **35**, 1275. 51, 75

70JOC1545 V. Grakauskas and K. Baum; *J. Org. Chem.*, 1970, **35**, 1545. 313, 314, 989

70JOC1619 J. R. Norell; *J. Org. Chem.*, 1970, **35**, 1619. 658

70JOC2002 B. R. O'Connor and F. N. Jones; *J. Org. Chem.*, 1970, **35**, 2002. 476, 487, 582, 613, 614

70JOC2056 T. R. Beebe and J. W. Wolfe; *J. Org. Chem.*, 1970, **35**, 2056. 203, 320

70JOC2102 J. V. Burakevich, A. M. Lore and G. P. Volpp; *J. Org. Chem.*, 1970, **35**, 2102. 580

70JOC2118 T. Saegusa, S. Kobayashi and Y. Ito; *J. Org. Chem.*, 1970, **35**, 2118. 1037

70JOC2438 T. Takeshima, M. Yokoyama, N. Fukada and M. Akano; *J. Org. Chem.*, 1970, **35**, 2438. 580

70JOC3093 D. L. Ross, C. L. Coon and M. E. Hill; *J. Org. Chem.*, 1970, **35**, 3093. 813, 814, 815

70JOC3470 M. S. Raasch; *J. Org. Chem.*, 1970, **35**, 3470. 511

70JOC3995 T. Saegusa, Y. Ito and T. Shimizu; *J. Org. Chem.*, 1970, **35**, 3995. 1071

70JOC4238 T. Saegusa, Y. Ito, N. Yasuda and T. Hotaka; *J. Org. Chem.*, 1970, **35**, 4238. 785, 1130

70JOM(22)269 G. Baum; *J. Organomet. Chem.*, 1970, **22**, 269. 228

70JOM(22)321 R. J. P. Corriu and J. P. Masse; *J. Organomet. Chem.*, 1970, **22**, 321. 410

70JOM(22)627 T. Fukumoto, H. Kurosawa and R. Okawara; *J. Organomet. Chem.*, 1970, **22**, 627. 1144

70JOM(24)595 J. J. Miller; *J. Organomet. Chem.*, 1970, **24**, 595. 383

70JOM(25)33 T. Hirabayashi, K. Itoh, S. Sakai and Y. Ishii; *J. Organomet. Chem.*, 1970, **25**, 33. 262

70JPB1696 S. Kubota, Y. Koida, T. Kosaka and O. Kirino; *J. Pharm. Bull.*, 1970, **18**, 1696. 847

70JPR601 H. G. O. Becker, G. Görmer and H.-J. Timpe; *J. Prakt. Chem.*, 1970, **312**, 601. 773

70JPR610 H. G. O. Becker, G. Görmar and H.-J. Timpe; *J. Prakt. Chem.*, 1970, **312**, 610. 856

70JPR(312)2 C. Bischoff and K. H. Platz; *J. Prakt. Chem.*, 1970, **312**, 2. 206

70JPS105 G. C. Wright, R. P. Halliday and C. S. Davis; *J. Pharm. Sci.*, 1970, **59**, 105. 849

70LA(733)195 W. Walter and R. F. Becker; *Justus Liebigs Ann. Chem.*, 1970, **733**, 195. 608

70LA(736)166 W. Walter, H. Weiss and K.-J. Reubke; *Justus Liebigs Ann. Chem.*, 1970, **736**, 166. 852, 853

70LA(737)80 E. Brehm, A. Haag and G. Hesse; *Justus Liebigs Ann. Chem.*, 1970, **737**, 80. 1147, 1148

70LA(737)179 A. Treibs and R. Friess; *Justus Liebigs Ann. Chem.*, 1970, **737**, 179. 561

70LA(739)201 W. Walter and M. Radke; *Justus Liebigs Ann. Chem.*, 1970, **739**, 201. 477, 584, 587, 620, 641

70LA(739)211 R. Köster, D. Simic and M. A. Grassberger; *Justus Liebigs Ann. Chem.*, 1970, **739**, 211. 1141

70M627 A. Meller and H. Batka; *Monatsh. Chem.*, 1970, **101**, 627. 1147

70M655 W. Gottardi; *Monatsh. Chem.*, 1970, **101**, 655. 1015

70MI 502-01 A. S. Nekrasov, A. N. Bashkirov and V. B. Abramovich; *Tr. Nauch.-Issled. Inst. Neftekkim. Proizvod.*, 1970, **2**, 75 (*Chem. Abstr.*, 1971, **74**, 87 283u). 46

B-70MI 502-01 C. A. Buehler and D. E. Pearson; "Survey of Organic Syntheses," Wiley-Interscience, New York, 1970, chap. 13, p. 744. 24

B-70MI 506-01 A. L. J. Beckwith; in "The Chemistry of Amides," ed. J. Zabicky, Interscience, New York, 1970, p. 73. 259

B-70MI 506-02 O. H. Wheeler and O. Rosado; in "The Chemistry of Amides," ed. J. Zabicky, Interscience, New York, 1970, p. 335. 305, 307

B-70MI 506-03 B. C. Challis and J. Challis; in "The Chemistry of Amides," ed. J. Zabicky, Interscience, New York, 1970, p. 731. 305

B-70MI 507-01 J. F. Bieron and F. J. Dinan; in "The Chemistry of Amides," ed. J. Zabicky, Interscience, New York, 1970, p. 245. 313, 360, 361

70MI 513-01 K. A. Jensen; *Quarterly Reports on Sulfur Chemistry*, 1970, **5**, 45. 549

70MI 514-01 B. Tornetta, G. Scapini, F. Guerrera and A. Bernardini; *Boll. Sedute. Accad. Gioenia Sci. Natur. Catania*, 1970, **10**, 353 (*Chem. Abstr.*, 1973, **78**, 620n). 615

B-70MI 514-01 W. Walter and J. Voss; in "The Chemistry of Amides," ed. J. Zabicky, Interscience, New York, 1970, p. 383. 565

B-70MI 515-01 W. Walter and J. Voss; in "The Chemistry of Amides," ed. J. Zabicky, Interscience, New York, 1970, Chapter 8, p. 383. 629, 638, 639, 640

B-70MI 517-01 M. B. Robin, F. A. Bovey and H. Basch; in "The Chemistry of Functional Groups: The Chemistry of the Amide Group," ed. J. Zabicky, Wiley, London, 1970, p. 1. 685

B-70MI 517-02 B. C. Challis and J. Challis; in "The Chemistry of Functional Groups: The Chemistry of Amides," ed. J. Zabicky, Wiley, London, 1970, p. 733. 723

B-70MI 517-03 R. B. Homer and C. D. Johnson; in "The Chemistry of Functional Groups: The Chemistry of Amides," ed. J. Zabicky, Interscience, London, 1970, p. 238. 723

B-70MI 518-01 W. Walter and J. Voss; in "The Chemistry of Amides," ed. J. Zabicky, Interscience, New York, 1970, p. 383. 725, 726, 733

70MI 519-01 S. Kubato, Y. Koida, T. Kosaka and O. Kirino; *J. Pharm. Bull.* (*Japan*), 1970, **18**, 1696. 769

B-70MI 520-01 S. Trippett (ed.); "Organophosphorus Chemistry" (Specialist Periodical Reports), Chemical Society, London, 1970, Vol. 1. 786

70MI 520-02 D. D. Davies and C. E. Gray; *Organomet. Chem. Rev. A.*, 1970, **6**, 283. 792, 794

70MI 521-01 G. Kamey, A. D. Nikolaeva and V. S. Perekhod'ko; *Izv. Vyssh. Ucheb. Zaved. Khim. Tekhnol.*, 1970, **13**, 227 (*Chem. Abstr.*, 1970, **73**, 34 750). 827

B-70MI 521-02 C. Grundmann and Z. Rappoport; in "The Chemistry of the Cyano Group," eds. S. Patai and J. Zabicky, Wiley-Interscience, London, 1970, p. 129. 833

B-70MI 521-03 W. Walter and K. J. Reubke; in "The Chemistry of Amides," ed. J. Zabicky, Interscience, New York, 1970, p. 509. 833

B-70MI 528-01 H. Ulrich and A. A. R. Sayigh; "Neuere Methoden der Preparativen Organischen Chemie," Weinheim/Bergstr., 1970, vol. 6, p. 211. 1061

B-70MI530-01 Z. Rappoport; "The Chemistry of the Cyano Group," Interscience, London, 1970. 652

70OCS23 K. Itoh, A. Nozowa and Y. Ishi; *Organomet. Synth.*, 1970, **1**, 23. 1070

70OPP275 E. A. Sedor, R. E. Freis and H. J. Richards; *Org. Prep. Proced. Int.*, 1970, **2**, 275. 359

70OS77 V. Lee and M. S. Newman; *Org. Synth.*, 1970, **50**, 77. 144

70RTC687 B. Zwanenburg, L. Thijs and J. Strating; *Recl. Trav. Chim. Pays-Bas*, 1970, **89**, 687. 490

70S20 D. Arlt; *Synthesis*, 1970, 20. 1004

70S99 H. H. Vogel; *Synthesis*, 1970, 99. 28, 286

70S539 H. R. Kricheldorf; *Synthesis*, 1970, 539. 1030

70S589 R. Merten and C. Weber; *Synthesis*, 1970, 589. 322, 971

70S590 H. J. Bestmann and R. Armsen; *Synthesis*, 1970, 590. 894

70S615 B. Blagoev and D. Ivanov; *Synthesis*, 1970, 615. 27

70S628 P. L. Southwick; *Synthesis*, 1970, 628. 27

70S649 H. R. Kricheldorf; *Synthesis*, 1970, 649. 978

70SST83 G. C. Barrett; *Org. Compd. Sulphur, Selenium, Tellurium* (*R. Soc. Chem. series*), 1970, **1**, 83. 1022

70SST181 S. McKenzie; *Org. Compd. Sulphur, Selenium and Tellurium* (*R. Soc. Chem. Series*), 1970, **1**, 181. 472

70SST201 S. McKenzie; *Org. Compd. Sulphur, Selenium, Tellurium* (*R. Soc. Chem. Series*), 1970, **1**, 201. 565

70T1539 A. Pelter, T. E. Levitt and P. Nelson; *Tetrahedron*, 1970, **26**, 1539.

70T5731 O. Mitsunobu, K. Kato and M. Tomari; *Tetrahedron*, 1970, **26**, 5731. 1078

70TL127 G. Lohaus; *Tetrahedron Lett.*, 1970, **11**, 127. 348

70TL311 E. J. Corey and R. Noyori; *Tetrahedron Lett.*, 1970, 311. 171

70TL1125 K. Ponsold and W. Ihn; *Tetrahedron Lett.*, 1970, 1125. 1118

70TL1791 K.-W. Yang, J. G. Cannon and J. G. Rose; *Tetrahedron Lett.*, 1970, 1791. 277, 299

70TL1901 T. Mukaiyama, R. Matsueda and M. Suzuki; *Tetrahedron Lett.*, 1970, 1901. 652

70TL2493 D. Seyferth and R. S. Marmor; *Tetrahedron Lett.*, 1970, 2493. 868

70TL3453 D. M. Zimmerman and R. A. Olofson; *Tetrahedron Lett.*, 1970, 3453. 1066, 1073

70TL3763 M. S. A. Vrijland and J. Th. Hackmann; *Tetrahedron Lett.*, 1970, 3763. 809

70TL4011 J. L. Marshall, K. C. Erickson and T. K. Folsom; *Tetrahedron Lett.*, 1970, 4011. 122

70TL4459 P. A. Bartlett and W. S. Johnson; *Tetrahedron Lett.*, 1970, 4459. 39

70TL4731 J. V. Paukstelis and M. Kim; *Tetrahedron Lett.*, 1970, 4731. 1126

70UKZ372 N. A. Kirsanova and G. I. Derkach; *Ukr. Khim. Zh.* (*Russ. Ed.*), 1970, **36**, 372 (*Chem. Abstr.*, 1970, **73**, 45 417q). 761, 1057

70USP3483242 A. M. Brownstein and J. P. Sibilia (Allied Chemical Corp.); *US Pat.*, 3 483 242 (1970) (*Chem. Abstr.*, 1970, **72**, 54 984d). 998

70USP3492331 A. A. R. Sayigh and H. Ulrich; *US Pat.*, 3 492 331 (1970) (*Chem. Abstr.*, 1970, **72**, 78 652v). 965

70USP3509199 Velsicol Chemical Corp; *US Pat.*, 3 509 199 (1970) (*Chem. Abstr.*, 1970, **73**, 3652g). 967

70USP3541120 W. C. Firth, Jr. (American Cyanamide Co.); *US Pat.*, 3 541 120 (1970) (*Chem. Abstr.*, 1971, **74**, 31 545p). 995

70USP3558684 Casella Farbwerke Mainkur A.-G.; *US Pat.*, 3 558 684 (1970) (*Chem. Abstr.*, 1970, **72**, 21 300z). 964

70ZOB707 Y. I. Baukov, G. S. Burlachenko, A. S. Kostyuk and I. F. Lutsenko; *Zh. Obshch. Khim.*, 1970, **40**, 707. 390

70ZOB1933 B. S. Drach, A. D. Sinitsa and A. V. Kirsanov; *Zh. Obshch. Khim.*, 1970, **40**, 1933. 965

70ZOB1949 V. I. Shevchenko, M. El Dik and A. M. Pinchuk; *Zh. Obshch. Khim.*, 1970, **40**, 1949. 373

70ZOB2464 V. I. Barabanov, Z. Y. Sazonova and V. S. Kaushaner; *Zh. Obshch. Khim.*, 1970, **40**, 2464. 370

70ZOR610 M. B. Kolesova, L. I. Maksimova and A. V. El'tsov; *Zh. Org. Khim.*, 1970, **6**, 610. 601

70ZOR805 E. P. Nesynov, M. M. Besprozvannaya and P. S. Pel'kis; *Zh. Org. Khim.*, 1970, **6**, 805. 727

71ACR193 E. M. Kosower; *Acc. Chem. Res.*, 1971, **4**, 193. 366

71AG(E)402 U. v. Gizycki; *Angew. Chem., Int. Ed. Engl.*, 1971, **10**, 402. 973

71AG(E)403 U. v. Gizycki; *Angew. Chem., Int. Ed. Engl.*, 1971, **10**, 403. 973

71AG(E)416 W. Gottardi; *Angew. Chem., Int. Ed. Engl.*, 1971, **10**, 416. 1015

71BAU723 L. A. Antokhina and P. I. Alimov; *Bull. Acad. Sci. USSR, Div. Chem. Sci.*, 1971, 723. 404, 405
71BAU959 L. I. Ragulin, P. P. Ropalo, G. A. Sokolskii and I. L. Knunyants; *Bull. Acad. Sci. USSR, Div. Chim. Sci.*, 1971, 959. 502
71BCF478 J.-A. Gautier, M. Miocque, C. Fauran and A.-Y. Le Cloarec; *Bull. Soc. Chim. Fr.*, 1971, 478. 754
71BCJ1362 O. Mitsunobu, K. Kato and M. Wada; *Bull. Chem. Soc. Jpn.*, 1971, **44**, 1362. 735
71BCJ1632 T. Imamoto, Y. Tsuno and Y. Yukawa; *Bull. Chem. Soc. Jpn.*, 1971, **44**, 1632. 318, 998
71BCJ1639 T. Imamoto, Y. Tsuno and Y. Yukawa; *Bull. Chem. Soc. Jpn.*, 1971, **44**, 1639. 998
71BCJ1644 T. Imamoto, Y. Tsuno and Y. Yukawa; *Bull. Chem. Soc. Jpn.*, 1971, **44**, 1644. 998
71BRP1235156 V. P. Kozyukov, V. D. Sheludyakov and V. F. Mironov; *Br. Pat.*, 1 235 156 (1971) (*Chem. Abstr.*, 1971, **75**, 77 492). 968
71BRP1252099 D. Arlt; *Br. Pat.*, 1 252 099 (1971) (*Chem. Abstr.*, 1972, **76**, 24 929). 1012
71CB639 H. Reimlinger, W. R. F. Lingier and J. J. M. Vandewalle; *Chem. Ber.*, 1971, **104**, 639. 773
71CB1199 A. Schmidpeter and W. Zeiss; *Chem. Ber.*, 1971, **104**, 1199. 708, 709, 861
71CB1335 R. Appel, R. Kleinstück and K.-D. Ziehn; *Chem. Ber.*, 1971, **104**, 1335. 1063
71CB2588 N. Bayreuther and A. Haas; *Chem. Ber.*, 1971, **104**, 2588. 1018
71CB2637 D. Hausigk; *Chem. Ber.*, 1971, **104**, 2637. 114
71CB2709 H. Bredereck, G. Simchen and B. Funke; *Chem. Ber.*, 1971, **104**, 2709. 586, 614, 616
71CB3131 H. R. Kricheldorf and G. Greber; *Chem. Ber.*, 1971, **104**, 3131. 1012
71CB3329 W. Ried and P. Weidemann; *Chem. Ber.*, 1971, **104**, 3329. 755
71CB3831 L. Birkofer and G. Schmidtberg; *Chem. Ber.*, 1971, **104**, 3831. 377
71CC10 S. W. Pelletier and D. L. Herald, Jr.; *J. Chem. Soc., Chem. Commun.*, 1971, 10. 51
71CC488 J. W. Patterson and J. E. McMurry; *J. Chem. Soc., Chem. Commun.*, 1971, 488. 164
71CC498 B. Singh and J. C. Collins; *J. Chem. Soc., Chem. Commun.*, 1971, 498. 756
71CC570 J. A. Connor and E. M. Jones; *J. Chem. Soc., Chem. Commun.*, 1971, 570. 949
71CC667 P. S. Manchand; *J. Chem. Soc., Chem. Commun.*, 1971, 667. 40, 107
71CC689 A. F. Hegarty, J. A. Kearny, M. P. Cashman and F. L. Scott; *J. Chem. Soc., Chem. Commun.*, 1971, 689. 703, 704
71CC733 N. W. Gilman; *J. Chem. Soc., Chem. Commun.*, 1971, 733. 295
71CC788 R. A. De Marco and J. M. Shreeve; *J. Chem. Soc., Chem. Commun.*, 1971, 788. 652
71CC1113 R. Mukherjee; *J. Chem. Soc., Chem. Commun.*, 1971, 1113. 358
71CC1514 T. H. Whitesides and R. A. Budnik; *J. Chem. Soc. Chem. Commun.*, 1971, 1514. 652
71CC1609 S. T. Reid and J. N. Tucker; *J. Chem. Soc., Chem. Commun.*, 1971, 1609. 332
71CI(L)545 G. G. Allan and A. N. Neogi; *Chem. Ind. (London)*, 1971, 545. 199
71CJC1011 A. F. Janzen and E. A. Kramer; *Can. J. Chem.*, 1971, **49**, 1011. 378
71CJC1384 S. Wolfe and D. V. C. Awang; *Can. J. Chem.*, 1971, **49**, 1384. 813
71CJC1598 T. R. Lynch, F. N. MacLachlan and Y. K. Siu; *Can. J. Chem.*, 1971, **49**, 1598. 354
71CJC1792 R. Raap; *Can. J. Chem.*, 1971, **49**, 1792. 730
71CJC2139 R. Raap; *Can. J. Chem.*, 1971, **49**, 2139. 1123
71CPB1606 S. Takemura, H. Niizato and Y. Keno; *Chem. Pharm. Bull.*, 1971, **19**, 1606. 701
71DOK(199)103 V. F. Mironov, V. P. Kozyukov and V. D. Sheludyakov; *Dokl. Akad. Nauk SSSR*, 1971, **199**, 103. 722, 863
71DOK(199)1328 B. M. Mikhailov, V. A. Dorokhov and V. I. Seredenko; *Dokl. Acad. Nauk. SSSR*, 1971, **199**, 1328. 778
71GEP2044574 J. A. Drobnick, P. R. Ktnecs and T. P. Chen; *Ger. Pat.* 2 044 574 (1971). 864
71HCA1939 F. Näf, P. Degen; *Helv. Chim. Acta*, 1971, **54**, 1939. 151
71IC407 J. M. Shreeve and L. M. Zaborowski; *Inorg. Chem.*, 1971, **10**, 407. 807
71IC911 R. A. De Marco and J. M. Shreeve; *Inorg. Chem.*, 1971, **10**, 911. 315
71IC1536 C. E. Jones and K. J. Coskran; *Inorg. Chem.*, 1971, **10**, 1536. 1129
71IC1635 R. F. Swindell, L. A. Zaborowski and J. M. Shreeve; *Inorg. Chem.*, 1971, **10**, 1635. 807
71IC1711 M. H. Chisholm and H. C. Clark; *Inorg. Chem.*, 1971, **10**, 1711. 940
71JA1368 M. Kubota and D. M. Blake; *J. Am. Chem. Soc.*, 1971, **93**, 1368. 460
71JA2672 F. A. Cotton and C. M. Lukehart; *J. Am. Chem. Soc.*, 1971, **93**, 2672. 943
71JA3554 C. P. Casey and R. L. Anderson; *J. Am. Chem. Soc.*, 1971, **93**, 3554. 943
71JA5821 L. Eberson and H. Welinder; *J. Am. Chem. Soc.*, 1971, **93**, 5821. 26
71JA6318 G. C. Windridge and E. C. Jorgensen; *J. Am. Chem. Soc.*, 1971, **93**, 6318. 279
71JCS(A)332 B. R. Hollebone and R. S. Nyholm; *J. Chem. Soc. (A)*, 1971, 332. 1083
71JCS(A)367 M. Green and D. J. Westlake; *J. Chem. Soc. (A)*, 1971, 367. 452
71JCS(A)794 M. L. H. Green, L. C. Mitchard and M. G. Swanick; *J. Chem. Soc. (A)*, 1971, 794. 937
71JCS(A)850 C. K. Brown, W. Mowat, G. Yagupsky and G. Wilkinson; *J. Chem. Soc. (A)*, 1971, 850. 461
71JCS(A)1663 J. L. Burmeister and L. M. Edwards; *J. Chem. Soc. (A)*, 1971, 1663. 1140
71JCS(A)3368 J. A. Connor and E. M. Jones; *J. Chem. Soc. (A)*, 3368. 949
71JCS(C)409 G. V. Boyd and A. J. H. Summers; *J. Chem. Soc. (C)*, 1971, 409. 854
71JCS(C)1114 S. Trippett and M. A. Walker; *J. Chem. Soc.(C)*, 1971, 1114. 920
71JCS(C)1704 J. C. Danilewicz, M. J. Sewell and J. C. Thurman; *J. Chem. Soc. (C)*, 1971, 1704. 767
71JCS(C)1870 I. Matsuda, K. Itoh and Y. Ishii; *J. Chem. Soc. (C)*, 1971, 1870. 390, 721
71JCS(C)1917 G. R. Fosker, K. D. Hardy, J. H. C. Nayler, P. Seggery and E. R. Stove; *J. Chem. Soc. (C)*, 1971, 1917. 219
71JCS(C)2944 I. J. Boulton, R. G. Harrison, B. Lythgoe and R. S. Manwaring; *J. Chem. Soc. (C)*, 1971, 2944. 168
71JCS(C)3262 D. E. L. Carrington, K. Clarke and R. M. Scrowston; *J. Chem. Soc. (C)*, 1971, 3262. 1112
71JCS(C)3540 D. H. R. Barton, F. Comer, D. G. T. Greig, P. G. Sammes, C. M. Cooper, G. Hewitt and W. G. E. Underwood; *J. Chem. Soc. (C)*, 1971, 3540. 278

71JCS(C)3550 M. Behforouz, H. Firouzabadi and A. A. Ardakani; *J. Chem. Soc.* (*C*), 1971, 3550. 336
71JGU2019 M. I. Shevchuk, E. M. Volynskaya and A. V. Dombrovskii; *J. Gen. Chem. USSR* (*Engl. Transl.*), 1971, **41**, 2019. 904
71JGU2513 S. Kh. Nurtdinov, N. M. Ismagilova, T. V. Zykova, R. A. Salakhutdinov, V. S. Tsivunin and G. Kh. Kamai; *J. Gen. Chem. USSR*, 1971, **41**, 2513. 403
71JMC535 P. E. Aldrich, E. C. Hermann, W. E. Meier, M. Paulshock, W. W. Prichard, J. A. Snyder and J. C. Watts; *J. Med. Chem.*, 1971, **14**, 535. 285
71JOC128 R. S. Marmor and D. Seyferth; *J. Org. Chem.*, 1971, **36**, 128. 870
71JOC284 J. E. Johnson, J. R. Springfield, J. S. Hwang, L. J. Hayes, W. C. Cunningham and D. C. McClaugherty; *J. Org. Chem.* 1971, **36**, 284. 723, 827
71JOC350 J. A. Krimmel, J. J. Schmidt-Collerus, J. A. Young, G. E. Bohner and D. N. Gray; *J. Org. Chem.*, 1971, **36**, 350. 361
71JOC1155 R. F. Smith, D. S. Johnson, C. L. Hyde, T. C. Rosenthal and A. C. Bates; *J. Org. Chem.*, 1971, **36**, 1155. 772, 850, 852, 855
71JOC1305 L. E. Barstow and V. J. Hruby; *J. Org. Chem.*, 1971, **36**, 1305. 300
71JOC1379 D. Seyferth, R. S. Marmor and P. Hilbert; *J. Org. Chem.*, 1971, **36**, 1379. 868
71JOC1613 J. D. Wilson, J. S. Wager and H. Weingarten; *J. Org. Chem.*, 1971, **36**, 1613. 745, 753
71JOC1742 M. von Strandtmann, S. Klutchko, D. Connor and J. Shavel Jr.; *J. Org. Chem.*, 1971, **36**, 1742. 596
71JOC2009 M. Yokoyama; *J. Org. Chem.*, 1971, **36**, 2009. 580
71JOC2142 D. A. Tomalia, T. J. Giacobbe and W. A. Sprenger; *J. Org. Chem.*, 1971, **30**, 2142. 1073
71JOC2155 Y. H. Chiang; *J. Org. Chem.*, 1971, **36**, 2155. 1001
71JOC2387 K. Sakai and J. P. Anselme; *J. Org. Chem.*, 1971, **36**, 2387. 991
71JOC2548 G. F. Vesley and V. I. Stenberg; *J. Org. Chem.*, 1971, **36**, 2548. 122
71JOC2721 S. Fukuoka, M. Ryang and S. Tsutsumi; *J. Org. Chem.*, 1971, **36**, 2721. 273, 286, 296
71JOC2727 Y. Degani and A. Patchornik; *J. Org. Chem.*, 1971, **36**, 2727. 1111
71JOC2735 L. Field, W. S. Hanley and I. McVeigh; *J. Org. Chem.*, 1971, **36**, 2735. 247
71JOC3048 J. E. McIsaac, Jr., R. E. Ball and E. J. Behrman; *J. Org. Chem.*, 1971, **36**, 3048. 275
71JOC3145 W. G. Phillips and K. W. Ratts; *J. Org. Chem.*, 1971, **36**, 3145. 478, 490
71JOC3238 H. L. Yale; *J. Org. Chem.*, 1971, **36**, 3238. 652
71JOC3542 M. Ohoka, S. Yanagida and S. Komori; *J. Org. Chem.*, 1971, **36**, 3542. 1003
71JOC3546 T. Fujisawa and N. Kobayashi; *J. Org. Chem.*, 1971, **36**, 3546. 248
71JOC3677 S. Kawamura, T. Horii and J. Tsurugi; *J. Org. Chem.*, 1971, **36**, 3677. 247
71JOC3835 W. B. Lutz; *J. Org. Chem.*, 1971, **36**, 3835. 332
71JOM(26)183 J. P. Picard, R. Calas, J. Dunoguès and N. Duffaut; *J. Organomet. Chem.*, 1971, **26**, 183. 419, 420
71JOM(28)237 E. O. Fischer, C. G. Kreiter, H. J. Kollmeier, J. Müller and R. D. Fischer; *J. Organomet. Chem.*, 1971, **28**, 237. 652
71JOM(28)359 E. O. Fischer, B. Heckl and H. Werner; *J. Organomet. Chem.*, 1971, **28**, 359. 953
71JOM(29)C25 D. Seyferth and T. C. Flood; *J. Organomet. Chem.*, 1971, **29**, C25. 1148
71JOM(31)C3 H. Leimeister and K. Dehnicke; *J. Organomet. Chem.*, 1971, **31**, C3-C4. 1016
71JOM(31)C68 L. Knauss and E. O. Fischer; *J. Organomet. Chem.*, 1971, **31**, C68. 953
71LA(743)154 W. Walter and E. Schaumann; *Justus Liebigs Ann. Chem.*, 1971, **743**, 154. 636
71LA(748)101 H. R. Kricheldorf; *Justus Liebigs Ann. Chem.*, 1971, **748**, 101. 1027
71LA(748)207 M. Regitz, A. Liedhegener, U. Eckstein, M. Martin and W. Anschütz; *Justus Liebigs Ann. Chem.*, 1971, **748**, 207. 871, 874
71LA(748)59 H. Böhme and W. Höver; *Justus Liebigs Ann. Chem.*, 1971, **748**, 59. 599
71LA(750)12 H. Neunhoeffer, F. Weischedel and V. Böhnisch; *Justus Liebigs Ann. Chem.*, 1971, **750**, 12. 856
71LA(752)115 W. Walter and K. Wohlers; *Justus Liebigs Ann. Chem.*, 1971, **752**, 115. 623, 624
71LA(752)142 K. D. Kampe; *Justus Liebigs Ann. Chem.*, 1971, **752**, 142. 998
71LA(753)169 W. Walter and C. O. Meese; *Justus Liebigs Ann. Chem.*, 1971, **753**, 169. 736
71LA(753)187 W. Walter and R. F. Becker; *Justus Liebigs Ann. Chem.*, 1971, **753**, 187. 592, 609
71LA(755)145 W. Walter and R. F. Becker; *Justus Liebigs Ann. Chem.*, 1971, **755**, 145. 608
71M121 A. Meller and W. Maringgele; *Monatsh. Chem.*, 1971, **102**, 121. 863, 864
71M264 W. Gottardi; *Monatsh. Chem.*, 1971, **102**, 264. 1015
71MI 503-01 E. Sung, W. Umbach and H. Baumann; *Fett, Seifen, Anstrichmittel*, 1971, **73**, 88. 138
B-71MI 507-01 W. Lwowski; in "The Chemistry of the Azido Group," ed. S. Patai, Interscience, New York, 1971, p. 503. 367
B-71MI 507-02 E. S. Levchenko and L. N. Markovskii; "Advances in the Chemistry of Organophosphorus and Organosulfur," 1971, p. 181. 349
71MI 515-01 Y. Ito, K. Umino, T. Sekguchi, T. Miyagishima and Y. Egawa; *J. Antibiotics-Tokyo*, 1971, **24**, 131. 635, 636
71MI 519-01 V. P. Rudavskii and D. M. Zagnibida; *Farm. Zh.* (*Kiev*), 1971, **26**, 29 (*Chem. Abstr.*, 1971, **75**, 109 914a). 774
B-71MI 520-01 I. Ugi (ed.); "Isonitrile Chemistry," Academic Press, New York, 1971, pp. 81 and 109. 795
71MI 520-02 J. Jappy and P. N. Preston; *Inorg. Nucl. Chem. Lett.*, 1971, **7**, 181. 794, 803
B-71MI 521-01 C. Grundmann and P. Grünanger; "The Nitrile Oxides," Springer Verlag, Berlin, 1971, p. 173. 819
B-71MI 526-01 D. V. Banthorpe; in "The Chemistry of the Azido Group," ed. S. Patai, Interscience, New York, 1971, p. 398. 989
B-71MI 526-02 D. V. Banthorpe; in "The Chemistry of the Azido Group," ed. S. Patai, Interscience, New York, 1971, p. 405. 989

B-71MI 526-03 E. Kühle, B. Anders and G. Zumach; in "Newer Methods of Preparative Organic Chemistry," ed. W. Foerst, Academic Press, New York, 1971, vol. VI, pp. 127–165. 989

B-71MI 528-01 S. R. Sandler; "Organic Functional Group Preparations," ed. W. Karo, Academic Press, New York, 1971, vol. 2, p. 233. 1061, 1078

71OS(51)31 G. E. Niznik, W. H. Morrison, III and H. M. Walborsky; *Org. Synth.*, 1971, **51**, 31. 796, 797

71PHA271 G. Wagner and L. Rothe; *Pharmazie*, 1971, **26**, 271. 567, 584, 618

71QR411 J. B. Aylward; *Q. Rev., Chem. Soc.*, 1971, **25**, 411. 999

71RCR835 G. I. Koldobskii, G. F. Tereshchenko, E. S. Gerasimova and L. I. Bagal; *Russ. Chem. Rev. (Engl. Transl.)*, 1971, **40**, 835. 270, 283

71RTC458 W. P. Trompen, J. Geevers and J. Th. Hackmann; *Recl. Trav. Chim. Pay-Bas*, 1971, **90**, 458. 662, 809, 810

71RTC463 W. P. Trompen, J. Geevers and J. Th. Hackmann; *Recl. Trav. Chim. Pay-Bas*, 1971, **90**, 463. 662

71RTC614 B. Zwanenburg, L. Thijs and J. Strating; *Recl. Trav. Chim. Pays-Bas*, 1971, **90**, 614. 562

71S111 W. Walter and E. Schaumann; *Synthesis*, 1971, 111. 635, 636, 637

71S131 W. Reeve; *Synthesis*, 1971, 131. 27, 62

71S266 G. L. Isele and A. Luttringhaus; *Synthesis*, 1971, 266. 285, 292

71S378 A. Battaglia, A. Dondoni and G. Mazzanti; *Synthesis*, 1971, 378. 1038

71S431 E. Vilsmaier, W. Sprügel and W. Böhm; *Synthesis*, 1971, 431. 898

71S435 H. R. Kricheldorf and E. Leppert; *Synthesis*, 1971, 435. 537

71T51 B. L. Dyatkin, K. N. Makarov and I. L. Knunyants; *Tetrahedron*, 1971, **27**, 51. 663, 685, 698, 699, 813

71T3795 T. Saegusa, I. Murase and Y. Ito; *Tetrahedron*, 1971, **27**, 3795. 760

71TL321 B. Singh; *Tetrahedron Lett.*, 1971, 321. 277

71TL591 Y. Leroux and R. Mantione; *Tetrahedron Lett.*, 1971, 591. 427

71TL1761 H. Kise, G. F. Whitfield and D. Swern; *Tetrahedron Lett.*, 1971, **12**, 1761. 350

71TL2765 J. Goerdeler, H. Lohmann, R. Losch and S. Raddatz; *Tetrahedron Lett.*, 1971, 2765. 1064

71TL3445 D. W. Emerson and H. Wynberg; *Tetrahedron Lett.*, 1971, 3445. 411

71TL3453 W. W. Lawrance, Jr.; *Tetrahedron Lett.*, 1971, 3453. 122

71TL3595 S. Yamada and Y. Takeuchi; *Tetrahedron Lett.*, 1971, 3595. 652

71USP3631071 M. Raasch, *US Pat.* 3 631 071 (1971) (*Chem. Abstr.*, 1972, **76**, 114 541). 736, 736

71USP3761501 C. Woolf, R. E. A. Dear and D. E. Young; *US Pat.* 3 761 501 (1971) (*Chem. Abstr.*, 1973, **79**, 136 560w). 1102

71ZAAC(386)197 V. H. Köhler and H. V. Döhler; *Z. Anorg. Allg. Chem.*, 1971, **386**, 197. 1083

71ZC60 H. Hartmann; *Z. Chem.*, 1971, **11**, 60. 626

71ZC110 L. Krasnec, Jr.; *Z. Chem.*, 1971, **11**, 110. 33

71ZC420 G. Barnikow and H. Ebeling; *Z. Chem.*, 1971, **11**, 420. 1051

71ZN(B)296 K. H. Linke, R. Taubert, K. Bister, W. Bornatsch and B. J. Liem; *Z. Naturforsch., Teil B*, 1971, **26**, 296. 848

71ZOB2340 V. D. Sheludyakov, E. S. Rodionov, G. D. Khatuntsev and V. F. Mironov; *Zh. Obshch. Khim.*, 1971, **41**, 2340. 1015

71ZOB2501 V. A. Dorokhov, L. I. Lavrinovich, I. P. Yakovelev and B. M. Mikhailov; *Zh. Obshch. Khim.*, 1971, **41**, 2501. 381

71ZOR1120 M. L. Petrov, B. S. Kupin and A. A. Petrov; *Zh. Org. Khim.*, 1971, **7**, 1120. 589, 590, 603, 611

71ZOR2352 V. M. Neplyuev, M. G. Lekar, R. G. Dubenko and P. S. Pel'kis; *Zh. Org. Khim.*, 1971, **7**, 2352. 595

72ACS1465 K. A. Jensen, L. Bøje and L. Henriksen; *Acta Chem. Scand.*, 1972, **26**, 1465. 233, 240, 243, 506, 508

72AG(E)128 H. R. Kricheldorf; *Angew. Chem., Int. Ed. Engl.*, 1972, **11**, 128. 978

72AG(E)583 I. L. Knunyants and G. A. Sokolski; *Angew. Chem., Int. Ed. Engl.*, 1972, **11**, 583. 502

72AG(E)949 E. Grigat; *Angew. Chem., Int. Ed. Engl.*, 1972, **11**, 949. 1100

72AG(E)1011 H. Hagemann and K. Ley; *Angew. Chem., Int. Ed. Engl.*, 1972, **11**, 1011. 1008

72AJC1731 G. I. Feutrill and R. N. Mirrington; *Aust. J. Chem.*, 1972, **25**, 1731. 106

72BCJ2937 N. Sonoda, G. Yamamoto and S. Tsutsumi; *Bull. Chem. Soc. Jpn.*, 1972, **45**, 2937. 1052

72BCJ3492 S. Kato and M. Mizuta; *Bull. Chem. Soc. Jpn.*, 1972, **45**, 3492. 546, 550

72BCJ3653 S. Kato, T. Mitani and M. Mizuta; *Bull. Chem. Soc. Jpn.*, 1972, **45**, 3653. 584, 620

72BRP1298049 Allied Chemical Corp.; *Br. Pat.*, 1972, 1 298 049 (*Chem. Abstr.*, 1973, **78**, 83 806b). 987

72BSF3559 J. L. Piette, R. Lysy and M. Renson; *Bull. Soc. Chim. Fr.*, 1972, 3559. 1116

72CAR(23)251 A. Stoffyn, P. Stoffyn and J. C. Orr; *Carbohydr. Res.*, 1972, **23**, 251. 285

72CB150 E. O. Fischer, M. Leupold, C. G. Kreiter and J. Müller; *Chem. Ber.*, 1972, **105**, 150. 950, 951

72CB1084 H. Schmidbaur, H. Stühler and W. Vornberger; *Chem. Ber.*, 1972, **105**, 1084. 913

72CB1258 R. Huisgen, H. Stangl, H. Sturm, R. Raab and K. Bunge; *Chem. Ber.*, 1972, **105**, 1258. 657

72CB2791 G. Lohaus; *Chem. Ber.*, 1972, **105**, 2791. 1005

72CB2800 R. Lattrell and G. Lohaus; *Chem. Ber.*, 1972, **105**, 2800. 1005

72CB2815 R. Huisgen and W. Mack; *Chem. Ber.*, 1972, **105**, 2815. 1046

72CB2975 G. Himbert and M. Regitz; *Chem. Ber.*, 1972, **105**, 2975. 827

72CB3173 H. Schmidbaur and W. Vornberger; *Chem. Ber.*, 1972, **105**, 3173. 913, 914

72CB3357 H. Scherer, A. Hartmann, M. Regitz, B. D. Tunggal and H. Günther; *Chem. Ber.*, 1972, **105**, 3357. 403, 404

72CB3958 H. R. Kricheldorf; *Chem. Ber.*, 1972, **105**, 3958. 992

72CC345 S. R. Ramadas and P. S. Srinivasan; *J. Chem. Soc., Chem. Commun.*, 1972, 345. 561

72CC354 J. E. Baldwin and J. A. Walker; *J. Chem. Soc., Chem. Commun.*, 1972, 354. 553

72CC472 W. E. Barnett and W. H. Sohn; *J. Chem. Soc., Chem. Commun.*, 1972, 472. 161

72CC672 W. Wong, S. J. Singer, W. D. Pitts, S. F. Watkins and W. H. Baddley; *J. Chem. Soc., Chem. Commun.*, 1972, 672. 465

72CC752 C. G. Screttas; *J. Chem. Soc., Chem. Commun.*, 1972, 752. 43, 110

72CC1222 S. A. Matlin and P. G. Sammes; *J. Chem. Soc., Chem. Commun.*, 1972, 1222. 325

72CCC3066 P. Kristián and G. Suchár; *Collect. Czech. Chem. Commun.*, 1972, **37**, 3066. 1054

72CCC3144 A. Jurášek, J. Kováč, A. Krutošiková and M. Hrdina; *Collect. Czech. Chem. Commun.*, 1972, **37**, 3144. 1106

72CJC2047 P. Soucy, T.-L. Ho and P. Deslongchamps; *Can. J. Chem.*, 1972, **50**, 2047. 95

72CJC3741 D. G. Lee, D. T. Hall and J. H. Cleland; *Can. J. Chem.*, 1972, **50**, 3741. 101

72CPB427 N. Sugimoto, T. Fujita, N. Shigematsu and A. Ayada; *Chem. Pharm. Bull. (Tokyo)*, 1972, **10**, 427. 1106

72CR(C)1215 G. L. Coustumer, Y. Mollier and M. H. Normant; *C. R. Acad. Sc. Paris, Ser. C*, 1972, 1215. 559

72CRV457 S. Ozaki; *Chem. Rev.*, 1972, **72**, 457. 962, 964, 965, 985, 986

72CRV545 D. J. Cardin, B. Cetinkaya and M. F. Lappert; *Chem. Rev.*, 1972, **72**, 545. 931, 940

72FRP2081556 J. A. Gautier, M. Miocque, A.-Y. Le Cloarec, C. Furan, G. Raynaud and B. Pourrias; *Fr. Pat.* 2 081 556 (1972) (*Chem. Abstr.*, 1972, **77**, 61 615r). 759

72GEP2146192 J. Perronet and P. Girault; *Ger. Pat.* 2 146 192 (1972) (*Chem. Abst.*, 1972, **77**, 34 128c). 771, 854

72HCA565 C. Wentrup; *Helv. Chim. Acta*, 1972, **55**, 565. 1122

72IC503 H. C. Clark and L. E. Manzer; *Inorg. Chem.*, 1972, **11**, 503. 940

72IJS89 T. Matsuda and Y. Takada; *Int. J. Sulfur Chem.*, 1972, **2**, 89. 602

72IJS139 V. T. Ramakrishnan and J. H. Boyer; *Int. J. Sulfur Chem., Part A*, 1972, **2**, 139. 1036

72IJS279 S. Kato, W. Akada and M. Mizuta; *Int. J. Sulfur Chem., Part A*, 1972, **2**, 279. 537

72IJS319 J. W. McFarland; *Int. J. Sulfur Chem., Part B*, 1972, **7**, 319. 1018

72IZV1150 K. A. Andrianov, A. I. Nogaideli, D. Sh. Akhobadze, L. M. Khananashvili and R. Sh. Tkeshelashvili; *Izv. Akad. Nauk SSSR, Ser. Khim.*, 1972, 1150. 720

72JA1532 M. H. Chisholm and H. C. Clark; *J. Am. Chem. Soc.*, 1972, **94**, 1532. 466, 940

72JA1788 J. P. Collman, S. R. Winter and D. R. Clark; *J. Am. Chem. Soc.*, 1972, **94**, 1788. 452

72JA2000 W. Adam, J. Baeza and J. C. Liu; *J. Am. Chem. Soc.*, 1972, **94**, 2000. 161

72JA4024 D. J. Sam and H. E. Simmons; *J. Am. Chem. Soc.*, 1972, **94**, 4024. 47, 52, 74, 114, 116

72JA4635 W. Nagata, M. Yoshioka and S. Hirai; *J. Am. Chem. Soc.*, 1972, **94**, 4635. 1142

72JA5285 A. Kamoriya and C. H. Yoder; *J. Am. Chem. Soc.*, 1972, **94**, 5285. 720

72JA6190 E. J. Corey and A. Venkateswarlu; *J. Am. Chem. Soc.*, 1972, **94**, 6190. 223, 374

72JA6203 T. Shioiri, K. Ninomiya and S. Yamada; *J. Am. Chem. Soc.*, 1972, **94**, 6203. 993

72JA6226 J. Schwartz and J. B. Cannon; *J. Am. Chem. Soc.*, 1972, **94**, 6226. 460

72JA6560 G. Zweifel, R. P. Fisher, J. T. Snow and C. C. Whitney; *J. Am. Chem. Soc.*, 1972, **94**, 6560. 1136

72JA7586 E. J. Corey and C. U. Kim; *J. Am. Chem. Soc.*, 1972, **94**, 7586. 339

72JA8776 D. S. Ross, T. Mill and M. B. Hill; *J. Am. Chem. Soc.*, 1972, **94**, 8776. 812

72JA8947 C. P. Casey, R. A. Boggs and R. L. Anderson; *J. Am. Chem. Soc.*, 1972, **94**, 8947. 955

72JCS(D)359 S. Cradock, E. A. V. Ebsworth and H. F. Jessep; *J. Chem. Soc., Dalton Trans.*, 1972, 359. 539, 541

72JCS(D)653 R. J. Hoare and O. S. Mills; *J. Chem. Soc., Dalton Trans.*, 1972, 653. 950

72JCS(D)869 D. W. H. Rankin; *J. Chem. Soc., Dalton Trans.*, 1972, 869. 1081, 1085

72JCS(P1)1692 R. O. C. Norman, R. Purchase, C. B. Thomas and J. B. Aylward; *J. Chem. Soc., Perkin Trans. 1*, 1972, 1692. 366

72JCS(P1)2326 M. D. Bachi and M. Rothfield; *J. Chem. Soc., Perkin Trans. 1*, 1972, 2326. 731

72JCS(P1)2728 D. W. Jones; *J. Chem. Soc., Perkin Trans. 1*, 1972, 2728. 364

72JCS(P1)2927 E. E. Glover, K. T. Rowbottom and D. C. Bishop; *J. Chem. Soc., Perkin Trans. 1*, 1972, 2927. 849

72JCS(P2)1043 M. Fukui, K. Itoh and Y. Ishii; *J. Chem. Soc., Perkin Trans. 2*, 1972, 1043. 719, 720

72JCS(P2)1908 A. Dondoni and O. Exner; *J. Chem. Soc., Perkin Trans. 2*, 1972, **13**, 1908. 657

72JCS(P2)1911 A. Battaglia, A. Dondoni and O. Exner; *J. Chem. Soc., Perkin Trans. 2*, 1972, 1911. 665

72JGU1256 B. N. Laskorin, V. V. Yakshin and L. I. Sokal'skaya; *J. Gen. Chem. USSR*, 1972, **42**, 1256. 404

72JGU1637 A. N. Pudovik and T. M. Sudakova; *J. Gen. Chem., USSR (Engl. Transl.)*, 1972, **42**, 1637, 1715. 786,1715

72JGU1857 V. A. Chauzov and Yu. I. Baukov; *J. Gen. Chem. USSR (Engl. Transl.)*, 1972, **42**, 1857. 716

72JGU1859 V. D. Sheludyakov, V. V. Shcherbinin, E. S. Rodionov and V. F. Mironov; *J. Gen. Chem. USSR (Engl. Transl.)*, 1972, **42**, 1859. 722

72JHC1267 E. Dyer, T. J. Nycz and M. B. Long; *J. Heterocycl. Chem.*, 1972, **9**, 1267. 990

72JIC1143 T. N. Srivastava, S. N. Bhattacharya and K. K. Bajpai; *J. Indian Chem. Soc.*, 1972, **49**, 1143. 1143, 1144

72JMC760 S.-H. Chu, G. R. Hillman and H. G. Mautner; *J. Med. Chem.*, 1972, **15**, 760. 516, 520, 522

72JOC724 G. J. Mikol and J. H. Boyer; *J. Org. Chem.*, 1972, **37**, 724. 989

72JOC1347 M. S. Raasch; *J. Org. Chem.*, 1972, **37**, 1347. 487, 511, 551, 563

72JOC1360 J. H. Boyer and V. T. Ramakrishnan; *J. Org. Chem.*, 1972, **37**, 1360. 1036, 1037

72JOC1526 W. G. Phillips and K. W. Ratts; *J. Org. Chem.*, 1972, **37**, 1526. 245, 478, 490, 491

72JOC1727 B. Bordás, P. Sohár, G. Matolcsy and P. Berencsi; *J. Org. Chem.*, 1972, **37**, 1727. 547

72JOC1732 M. W. Rathke and H. Yu; *J. Org. Chem.*, 1972, **37**, 1732. 466

72JOC1738 S. S. Washburne, W. R. Peterson, Jr. and D. A. Berman; *J. Org. Chem.*, 1972, **37**, 1738. 992

72JOC1907 P. L. Creger; *J. Org. Chem.*, 1972, **37**, 1907. 76, 166

72JOC1947 P. E. Eaton, G. F. Cooper, R. C. Johnson and R. H. Mueller; *J. Org. Chem.*, 1972, **37**, 1947. 73

72JOC2080 C. T. Goralski and T. E. Evans; *J. Org. Chem.*, 1972, **37**, 2080. 496, 497

72JOC3030 M. Ohoka, S. Yanagida and S. Komori; *J. Org. Chem.*, 1972, **37**, 3030. 967

72JOC3372	T. Tenforde, R. A. Fawwaz, N. K. Freeman and N. Castagnoli; *J. Org. Chem.*, 1972, **37**, 3372.	1063
72JOC3504	H. Kohn and R. A. Olofson; *J. Org. Chem.*, 1972, **37**, 3504.	855
72JOC3818	W. G. Phillips and K. W. Ratts; *J. Org. Chem.*, 1972, **37**, 3818.	490, 491
72JOC3820	M. S. Raasch; *J. Org. Chem.*, 1972, **37**, 3820.	253, 736
72JOC3958	E. C. Taylor and S. F. Martin; *J. Org. Chem.*, 1972, **37**, 3958.	769, 849
72JOC4136	R. W. Begland and D. R. Hartter; *J. Org. Chem.*, 1972, **37**, 4136.	818
72JOM(35)303	S. Freireich, D. Gertner and A. Zilkha; *J. Organomet. Chem.*, 1972, **35**, 303.	318
72JOM(36)227	D. N. Kravtsov, B. A. Kvasov, L. S. Golovchenko and E. I. Fedin; *J. Organomet. Chem.*, 1972, **36**, 227.	385
72JOM(36)C18	H. Brunner and E. Schmidt; *J. Organomet. Chem.*, 1972, **36**, C18.	453
72JOM(40)159	E. O. Fischer and S. Fontana; *J. Organomet. Chem.*, 1972, **40**, 159.	936
72JOM(42)33	T. Hirabayashi, H. Imaeda, K. Itoh and Y. Ishii; *J. Organomet. Chem.*, 1972, **42**, 33.	235
72JOM(43)157	J. Dunoguès, M. Bolourtchian, R. Calas, N. Duffaut and J.-P. Picard; *J. Organomet. Chem.*, 1972, **43**, 157.	416
72JOM(44)1	K. W. Barnett and D. W. Slocum; *J. Organomet. Chem.*, 1972, **44**, 1.	445
72JOM(46)C15	E. O. Fischer and W. Kalbus; *J. Organomet. Chem.*, 1972, **46**, C15.	936
72JPJ275	S. Kubota, O. Kirino and Y. Koida; *J. Pharm. Soc. Jpn.*, 1972, **92**, 275.	848
72JPR382	H. Ronsch; *J. Prakt. Chem.*, 1972, **314**, 382.	1127
72LA(755)101	M. Brugger, H. Wamhoff and F. Korte; *Justus Liebigs Ann. Chem.*, 1972, **755**, 101.	849
72LA(757)147	W. Ried, W. Ochs, H. Liebig and K. Wagner; *Justus Liebigs Ann. Chem.*, 1972, **757**, 147.	599
72LA(758)162	W. Walter and H. Weiss; *Justus Liebigs Ann. Chem.*, 1972, **758**, 162.	770, 771, 839, 850
72LA(760)102	H. Neunhoeffer and H.-W. Frühauf; *Justus Liebigs Ann. Chem.*, 1972, **760**, 102.	849
72LA(761)137	U. Schöllkopf, B. Bánhidai and H.-U. Scholz; *Justus Liebigs Ann. Chem.*, 1972, **761**, 137.	873, 874
B-72MI 501-01	S. Patai (ed.); "The Chemistry of Acyl Halides," Interscience, New York, 1972.	1, 16, 17
B-72MI 502-01	H. O. House; "Modern Synthetic Reactions," W. A. Benjamin, Menlo Park, California, 2nd Ed., 1972, chap. 10, pp. 629–733.	98
72MI 502-01	W. F. Brill and J. T. Baker; *High Polym.*, 1972, **27**, 477.	24
72MI 502-02	J. K. Stille, M. E. Freeburger, W. B. Alston and E. L. Mainen; *High Polym.*, 1972, **27**, 689.	24
B-72MI 503-02	M. F. Ansell; in "The Chemistry of Acyl Halides," ed. S. Patai, Interscience, London, 1972, p. 1.	130
B-72MI 503-03	A. Kivinen; in "The Chemistry of Functional Groups," ed. S. Patai, Interscience, London, 1972, p. 187.	130
B-72MI 507-01	H. R. Allcock; "Phosphorus–Nitrogen Compounds," Academic Press, New York, 1972.	368
B-72MI 511-01	K. T. Potts and C. Sapino; in "The Chemistry of Acyl Halides," ed. S. Patai, Wiley, Chichester, 1972, p. 349.	500
72MI 512-01	H. R. Kricheldorf and E. Leppert; *Makromol. Chem.*, 1972, **158**, 223.	538, 540
72MI 513-01	S. Oae, T. Yagihara and A. Nakanishi; *Kagaku*, 1972, **27**, 589.	546
B-72MI 517-01	S. R. Sandler and W. Karo; in "Organic Functional Group Preparations," ed. E. C. Taylor, Academic Press, New York, 1972, vol. 3, p. 268.	685, 687, 690
B-72MI 518-01	S. R. Sandler and W. Karo; in "Organic Functional Group Preparations," ed. E. C. Taylor, Academic Press, New York, 1972, vol. 9, p. 314.	726
B-72MI 520-01	G. M. Kosolapoff and L. Maier (eds.); "Organic Phosphorus Compounds," Wiley–Interscience, New York, 1972–1974.	800
72MI 521-01	J. Plenkiewicz and Z. Eckstein; *Przem. Chem.*, 1972, **51**, 785 (*Chem. Abstr.*, 1973, **78**, 83 988n).	828
72MI 523-01	G. M. Kosolapoff and L. Maier; *Organophosphorus Compounds*, 1972, **3**, 1.	900
72MI 526-01	Y. Iwakura, S. Shinsaku and Y. Masahiro; *Yuki Gosei Kagaku Kyokai Shi*, 1972, **30**, (10), 889 (*Chem. Abstr.*, 1973, **79**, 5289s).	990
72MI 530-01	G. C. Stocco, F. Stocco Gattuso and R. S. Tobias; *Atti. Acad. Sci., Lett. Arti Palermo, Parte 1*, 1972, **31**, 199 (*Chem. Abstr.*, 1973, **79** 105 364f).	1141
72OS5290	W. Nagata and M. Yoshioka; *Org. Synth.*, 1972, **52**, 90.	1142
72RTC37	A. M. Van Leusen, B. A. Reith, A. J. W. Iedema and J. Strating; *Recl. Trav. Chim. Pays Bas*, 1972, **91**, 37.	898, 901, 902
72RTC343	P. H. Benders and J. T. Hackmann; *Recl. Trav. Chim. Pays-Bas.*, 1972, **91**, 343.	1125
72RTC578	J. Meijer and L. Brandsma; *Recl. Trav. Chim. Pays-Bas*, 1972, **91**, 578.	589, 614, 618
72S34	C. Giordano, *Synthesis*, 1972, 34.	606
72S314	N. H. Nilsson, A. Senning, S. Karlsson and J. Sandström; *Synthesis*, 1972, 314.	592, 593
72S464	G. Fodor, J.-P. Fumeaux and V. Sankaran; *Synthesis*, 1972, 464.	90
72S551	H. R. Kricheldorf; *Synthesis*, 1972, 551.	991
72S622	J. Poupaert, A. Bruylants and P. Crooy; *Synthesis*, 1972, 622.	727
72S695	H. R. Kricheldorf; *Synthesis*, 1972, 695.	373
72SC79	A. H. Alberts, H. Wynberg and J. Strating; *Synth. Commun.*, 1972, **2**, 79.	51
72SC389	J. E. McMurray and G. B. Wong; *Synth. Commun.*, 1972, **2**, 389.	107
72SC423	W. Vaalburg, J. Strating, M. G. Woldring and H. Wynberg; *Synth. Commun.*, 1972, **2**, 423.	88
72SZP563109	K. Gaetzi and H. Fischer, Ciba-Geigy Corp.; *Swiss Pat.* 563 109 (1975) (*Chem. Abstr.*, 1975, **83**, 189 327).	744
72T199	H. Möhrle and P. Spillmann; *Tetrahedron*, 1972, **28**, 199.	566, 590, 606, 609, 610
72T1275	A. Hirshfeld, W. Taub and E. Glotter; *Tetrahedron*, 1972, **28**, 1275.	164
72T3203	S. Oae, T. Yagihara and T. Okabe; *Tetrahedron*, 1972, **28**, 3203.	550
72T4877	A. Kreutzberger; *Tetrahedron*, 1972, **28**, 4877.	747
72T5341	F. C. V. Larsson and S.-O. Lawesson; *Tetrahedron*, 1972, **28**, 5341.	547

72T5903	A. S. Shawali and H. M. Hassaneen; *Tetrahedron*, 1972, **28**, 5903.	358
72TL1513	J. Goerdeler and W. Teller; *Tetrahedron Lett.*, 1972, **13**, 1513.	1050
72TL3751	H. Ishihara and S. Kato; *Tetrahedron Lett.*, 1972, 3751.	255, 541
72TL3799	W. Welter and M. Regitz; *Tetrahedron Lett.*, 1972, 3799.	871
72TL3987	D. H. Miles and E. J. Parish; *Tetrahedron Lett.*, 1972, 3987.	78
72TL4191	D. St. C. Black and K. G. Watson; *Tetrahedron Lett.*, 1972, 4191.	605
72TL4747	B. Castro and J. R. Dormoy; *Tetrahedron Lett.*, 1972, **13**, 4747.	191
72TL4779	B. C. Pant; *Tetrahedron Lett.*, 1972, **13**, 4779.	210, 210
72TL4785	P. G. Nair and C. P. Joshua; *Tetrahedron Lett.*, 1972, 4785.	652
72TL5313	D. N. Harpp and T. G. Back; *Tetrahedron Lett.*, 1972, **13**, 5313.	342
72UKZ171	V. S. Garkusha-Bozhko, S. M. Baranov and O. P. Shvaika; *Ukr. Khim. Zh.* (*Russ. Ed.*), 1972, **38**, 171 (*Chem. Abstr.*, 1972, **77**, 34 426y).	848
72USP3637749	B. R. O'Connor; *US Pat.* 3 637 749 (1972) (*Chem. Abstr.*, 1972, **76**, 142 392).	476, 487
72USP3647856	A. D. Josey; *US Pat.* 3 647 856 (1972) (*Chem. Abstr.*, 1972, **76**, 141 023).	900
72USP3666784	A. F. Clifford, T. C. Rhyne and J. W. Thompson; *US Pat.*, 3 666 784 (1972) (*Chem. Abstr.*, 1972, **77**, 100 776r).	1003
72USP3674785	R. A. Mitsch; *US Pat.*, 3 674 785 (1972) (*Chem. Abstr.*, 1972, **77**, 151 998x).	808
72ZAAC(388)78	V. R. Engler and G. Gattow; *Z. Anorg. Allg. Chem.*, 1972, **388**, 78.	506, 507
72ZAAC(389)145	J. R. Engler and G. Gattow; *Z. Anorg. Allg. Chem.*, 1972, **389**, 145.	546
72ZAAC(394)26	G. Czieslik and O. Glemser; *Z. Anorg. Allg. Chem.*, 1972, **394**, 26.	383
72ZC133	N. S. Tao, S. Scheithauer and R. Mayer; *Z. Chem.*, 1972, **12**, 133.	586
72ZC147	H. Kohler and U. Lange; *Z. Chem.*, 1972, **12**, 147.	1081, 1085
72ZOB868	G. S. Gol'din, M. V. Maksakova, V. I. Kuz'micheva and A. N. Koltsova; *Zh. Obshch. Khim.*, 1972, **42**, 868.	863
72ZOB879	V. D. Sheludyakov, F. N. Vishnevskii, E. S. Rodionov and V. F. Mironov; *Zh. Obshch. Khim.*, 1972, **42**, 879.	1015
72ZOB1868	V. A. Chauzov and Y. I. Baukov; *Zh. Obshch. Khim.*, 1972, **42**, 1868.	716, 1006
72ZOB1870	V. D. Sheludyakov, V. V. Shcherbinin, E. S. Rodionov and V. F. Mironov; *Zh. Obshch. Khim.*, 1972, **42**, 1870.	722, 978
72ZOB2501	Y. I. Dergunov, V. D. Sheludyakov, I. A. Vostokov, Y. I. Mushkin, V. F. Mironov and V. P. Kozyukov, *Zh. Obshch. Khim.*, 1972, **42**, 2501.	969
72ZOR528	A. F. Grabenko, M. N. Danchenko and P. S. Pel'kis; *Zh. Org. Khim.*, 1972, **8**, 528.	592
72ZOR695	V. P. Kukhar', N. G. Pavlenko and A. V. Kirsanov; *Zh. Org. Khim.*, 1972, **8**, 695.	810
72ZOR1246	B. V. Passet, G. N. Kul'bitskii, N. A. Kalashnikova and T. I. Voropaeva; *Zh. Org. Khim.*, 1972, **8**, 1246.	742, 751
72ZOR1822	L. N. Markovskii, T. N. Dubinina, E. S. Levchenko, V. P. Kukhar and A. V. Kirsanov; *Zh. Org. Khim.*, 1972, **8**, 1822.	1032
73ACR202	M. H. Chisholm and H. C. Clark; *Acc. Chem. Res.*, 1973, **6**, 202.	939
73ACR422	G. P. Chiusoli; *Acc. Chem. Res.*, 1973, **6**, 422.	27
73ACS726	B. Lindgren; *Acta Chem. Scand.*, 1973, **27**, 726.	1114
73ACS3368	T. Greibrokk; *Acta Chem. Scand.*, 1973, **27**, 3368.	985
73AG(E)77	W. Beck, W. Danzer and R. Höfer; *Angew. Chem., Int. Ed. Engl.*, 1973, **12**, 77.	446, 448
73AG(E)402	H. F. Klein; *Angew. Chem., Int. Ed. Engl.*, 1973, **12**, 402.	461
73AG(E)842	W. I. Gorbatenko, W. A. Bondar and L. I. Samaraj; *Angew. Chem., Int. Ed. Engl.*, 1973, **12**, 842.	1032
73AG(E)923	D. Hoppe; *Angew. Chem., Int. Ed. Engl.*, 1973, **12**, 923.	1046
73AG(E)999	H. Hagemann; *Angew. Chem., Int. Ed. Engl.*, 1973, 999.	1004
73AG(E)1014	D. Enders and D. Seebach; *Angew. Chem., Int. Ed. Engl.*, 1973, **12**, 1014.	597
73AJC273	D. E. Giles and A. J. Parker; *Aust. J. Chem.*, 1973, **26**, 273.	1040
73BCJ244	S. Kato, W. Akada, M. Mizuta and Y. Ishii; *Bull. Chem. Soc. Jpn.*, 1973, **46**, 244.	254, 537, 539, 540
73BCJ1275	M. Ohoka, S. Yanagida, K. Sugahara, T. Fujita, M. Okahara and S. Komori; *Bull. Chem. Soc. Jpn.*, 1973, **46**, 1275.	1003
73BCJ1803	T. Osaki, J. Otera and Y. Kawasaki; *Bull. Chem. Soc. Jpn.*, 1973, **46**, 1803.	402
73BCJ2820	K. Kudo, M. Sato, M. Hidai and Y. Uchida; *Bull. Chem. Soc. Jpn.*, 1973, **46**, 2820.	463
73BCJ2921	K. Sugimoto and M. Ohta; *Bull. Chem. Soc. Jpn.*, 1973, **46**, 2921.	642
73BSF581	F. Clesse and H. Quiniou; *Bull. Soc. Chim. Fr.*, 1973, 581.	597, 599, 614
73BSF3349	G. Le Coustumer and Y. Mollier; *Bull. Soc. Chim. Fr.*, 1973, 3349.	584, 586, 614
73CB1062	U. Schubert and E. O. Fischer; *Chem. Ber.*, 1973, **106**, 1062.	897
73CB1277	K. Weiss and E. O. Fischer; *Chem. Ber.*, 1973, **106**, 1277.	953
73CB1752	C. Jäckh and W. Sundermeyer; *Chem. Ber.*, 1973, **106**, 1752.	1015
73CB3753	H. R. Kricheldorf and W. Regel; *Chem. Ber.*, 1973, **106**, 3753.	992
73CB3910	E. O. Fischer, G. Kreis, F. R. Kreissl, C. G. Kreiter and J. Muller; *Chem. Ber.*, 1973, **106**, 3910.	952
73CC158	J. A. Evans, G. F. Everitt, R. D. W. Kemmitt and D. R. Russell; *J. Chem. Soc., Chem. Commun.*, 1973, 158.	465
73CC500	P. A. Grieco and K. Hiroi; *J. Chem. Soc., Chem. Commun.*, 1973, 500.	68
73CC693	G. I. Nikishin, M. G. Vinogradov and T. M. Fedorova; *J. Chem. Soc., Chem. Commun.*, 1973, 693.	168
73CC932	M. F. Betkouski and W. Lwowski; *J. Chem. Soc., Chem. Commun.*, 1973, 932.	845
73CJC1741	A. Taurins and V. T. Khouw; *Can. J. Chem.*, 1973, **51**, 1741.	1108
73CJC4115	A. J. Elliott, M. S. Gibson, M. M. Kayser and G. A. Pawelchak; *Can. J. Chem.*, 1973, **51**, 4115.	358

73CL1143	M. Oya, R. Katakai, H. Nakai and Y. Iwakura; *Chem. Lett.*, 1973, 1143.	972
73CL1225	T. Taguchi, Y. Sato and T. Mukaiyama; *Chem. Lett.*, 1973, 1225.	1121
73CR(C)1039	H. P. Husson, C. Poupat and P. Potier; *C. R. Hebd. Seances Acad. Sci., Ser. C.*, 1973, **276**, 1039.	212
73CRV75	H. Babad and A. G. Zeiler; *Chem. Rev.*, 1973, **73**, 75.	965
73CZP148911	Z. Bukac and J. Sebenda; *Czech. Pat.*, 148 911 (1973) (*Chem. Abstr.*, 1973, **79**, 136 775v).	997
73DOK(211)113	A. N. Pudovik, R. A. Cherkasov, T. M. Sudokova and G. I. Evstafev; *Dokl. Akad. Nauk SSSR*, 1973, **211**, 113.	737
73GEP2242772	J. C. Petitpierre and C. Weis; *Ger Pat.* 2 242 772 (1973) (*Chem. Abstr.*, 1973, **78**, 147 390w).	816
73GEP(O)2145514	A. Botta; *Ger. Pat. Offen.*, 2 145 514 (1973) (*Chem. Abstr.*, 1973, **78**, 159 182).	1012
73GEP(O)2153795	A. Haas and E. Behrend (Bayer A.-G.); *Ger. Pat. Offen.*, 2 153 795 (1973) (*Chem. Abstr.*, 1973, **79**, 18 152s).	1005
73GEP(O)2225365	C. H. Hearsey; *Ger. Pat. Offen.*, 2 225 365 (1973) (*Chem. Abstr.*, 1973, **78**, 97 306a).	980
73HCA1826	R. A. Wohl; *Helv. Chim. Acta*, 1973, **56**, 1826.	346
73IC241	E. A. Dietz, Jr. and D. R. Martin; *Inorg. Chem.*, 1973, **12**, 241.	1130
73IJC735	A. F. M. Fahmy and H. A. Fadel; *Indian J. Chem.*, 1973, **11**, 735.	351
73IJS359	S. Kato, T. Mitani and M. Mizuta; *Int. J. Sulfur Chem.*, 1973, **8**, 359.	549
73IJS389	L. Henriksen; *Int. J. Sulfur Chem.*, 1973, **8**, 389.	1055
73IZV2159	A. V. Fokin, V. P. Stolyarov, V. S. Galakov and A. T. Uzun; *Izv. Acad. Nauk SSSR, Ser. Khim.*, 1973, 2159.	762
73JA279	E. M. Burgess and H. R. Penton Jr.; *J. Am. Chem. Soc.*, 1973, **95**, 279.	335, 336
73JA2017	J. A. Franz and J. C. Martin; *J. Am. Chem. Soc.*, 1973, **95**, 2017.	689
73JA3440	I. Shahak and Y. Sasson; *J. Am. Chem. Soc.*, 1973, **95**, 3440.	232, 1030
73JA3957	T. H. Koch, R. J. Sluski and R. H. Moseley; *J. Am. Chem. Soc.*, 1973, **92**, 3957.	1012
73JA4855	H. Schmidbaur and K. C. Dash; *J. Am. Chem. Soc.*, 1973, **95**, 4855.	1141
73JA5649	D. A. Buckingham, F. R. Keene and A. M. Sargeson; *J. Am. Chem. Soc.*, 1973, **95**, 5649.	385
73JA5803	H. Takahashi, K. Oshima, H. Yamamoto and H. Nozaki; *J. Am. Chem. Soc.*, 1973, **95**, 5803.	552
73JA5829	E. J. Corey and D. J. Beames; *J. Am. Chem. Soc.*, 1973, **95**, 5829.	235
73JA7510	J. I. Zink and W. C. Kaska; *J. Am. Chem. Soc.*, 1973, **95**, 7510.	1094
73JA8118	L. S. Trzupek, T. L. Newirth, E. G. Kelly, N. E. Sbarbati and G. M. Whitesides; *J. Am. Chem. Soc.*, 1973, **95**, 8118.	436
73JAP7300455	O. Shoichiro, T. Toru and K. Yoshiyasu (Mitsui Toatsu Chemicals Co., Ltd.); *Jpn. Pat.*, 7 300 455 (1973) (*Chem. Abstr.*, 1973, **78**, 97 205s).	985
73JAP7301653	K. Umeda; *Jpn. Pat.*, 7 301 653 (1973) (*Chem. Abstr.*, 1973, **78**, 158 906).	981
73JAP7320534	K. Umeda and T. Shigetomi; *Jpn. Pat.*, 7 320 534 (1973) (*Chem. Abstr.*, 1973, **79**, 13 649).	981
73JCS(D)975	J. R. Moss, M. Gree and F. G. A. Stone; *J. Chem. Soc., Dalton Trans.*, 1973, 975.	948, 949
73JCS(D)1267	M. Kubota, R. K. Rothrock and J. Geibel; *J. Chem. Soc., Dalton Trans.*, 1973, 1267.	
73JCS(P1)453	M. W. Partridge and A. Smith; *J. Chem. Soc., Perkin Trans. 1*, 1973, 453.	750, 751, 753
73JCS(P1)705	R. Fielden, O. Meth-Cohn and H. Suschitzky; *J. Chem. Soc., Perkin Trans. 1*, 1973, 705.	1108
73JCS(P1)1009	T. Takeshima, T. Miyauchi, N. Fukada, S. Koshizawa and M. Muraoka; *J. Chem. Soc., Perkin Trans. 1*, 1973, 1009.	547
73JCS(P1)1571	D. H. R. Barton, C. Chavis, M. K. Kaloustian, P. D. Magnus, G. A. Poulton and P. J. West; *J. Chem. Soc., Perkin Trans. 1*, 1973, 1571.	560
73JCS(P1)2540	P. Del Buttero and S. Maiorana; *J. Chem. Soc., Perkin Trans. 1*, 1973, 2540.	497
73JCS(P1)2614	J. S. Davies, C. H. Hassall and K. H. Hopkins; *J. Chem. Soc., Perkin Trans. 1*, 1973, 2614.	718
73JCS(P1)2796	D. Brewster, M. Myers, J. Ormerod, P. Otter, A. C. B. Smith, M. E. Spinner and S. Turner; *J. Chem. Soc., Perkin Trans. 1*, 1973, 2796.	142
73JCS(P2)534	E. Ciuffarin, L. Senatore and L. Sagramora; *J. Chem. Soc., Perkin Trans. 2*, 1973, 534.	766
73JCS(P2)1422	A. F. Hegarty, J. A. Kearney and F. L. Scott; *J. Chem. Soc., Perkin Trans. 2*, 1973, 1422.	358
73JFC(3)383	A. Haas, J. Helmbrecht, W. Klug, B. Koch, H. Reinke and J. Sommerhoff; *J. Fluorine Chem.*, 1973, **3**, 383.	246
73JGU10	V. A. Shokol, B. N. Kozhushko, V. V. Doroshenko and A. V. Kirsanov; *J. Gen. Chem., USSR* (*Engl. Transl.*), 1973, **43**, 10.	787
73JGU317	G. S. Gol'din, M. V. Maksakova and A. N. Kol'tsova; *J. Gen. Chem. USSR* (*Engl. Transl.*), 1973, **43**, 317.	850
73JGU2635	M. A. Vasyanina and V. K. Khairullin; *J. Gen. Chem. USSR* (*Engl. Transl.*), 1973, 2635.	534
73JMC1	M. Israel, E. C. Zoll, N. Muhammad and E. J. Modest; *J. Med. Chem.*, 1973, **16**, 1.	
73JOC414	S. S. Simons, Jr.; *J. Org. Chem.*, 1973, **38**, 414.	999
73JOC457	A. Berger, M. Smolarsky, N. Kurn and H. R. Bosshard; *J. Org. Chem.*, 1973, **38**, 457.	90
73JOC667	E. P. Papadopoulos; *J. Org. Chem.*, 1973, **38**, 667.	592
73JOC675	J. M. Vandensavel, G. Smets and G. L'Abbe; *J. Org. Chem.*, 1973, **38**, 675.	1014
73JOC1065	J. L. Zollinger, C. D. Wright, J. J. McBrady, D. H. Dybvig, F. A. Fleming, G. A. Kurhajec, R. A. Mitsch and E. W. Neuvar; *J. Org. Chem.*, 1973, **38**, 1065.	814
73JOC1075	C. D. Wright and J. L. Zollinger; *J. Org. Chem.*, 1973, **38**, 1075.	762, 811, 817
73JOC1241	J. R. Schaeffer and R. E. Stevens; *J. Org. Chem.*, 1973, **38**, 1241.	117
73JOC1316	F. J. Weigert; *J. Org. Chem.*, 1973, **38**, 1316.	984
73JOC1344	R. F. Smith, D. S. Johnson, R. A. Abgott and M. J. Madden; *J. Org. Chem.*, 1973, **38**, 1344.	772, 850, 855
73JOC1491	G. B. Trimitsis, A. Tuncay, R. D. Beyer and K. J. Ketterman; *J. Org. Chem.*, 1973, **38**, 1491.	43
73JOC1677	D. W. Slocum and F. E. Stonemark; *J. Org. Chem.*, 1973, **38**, 1677.	273, 285
73JOC1941	H. R. Nace and A. H. Smith; *J. Org. Chem.*, 1973, **38**, 1941.	146
73JOC2242	S. L. Razniak, E. M. Flagg and F. Siebenthall; *J. Org. Chem.*, 1973, **38**, 2242.	727

73JOC2982 J. H. MacMillan and S. S. Washburne; *J. Org. Chem.*, 1973, **38**, 2982. 992
73JOC3172 Y. Okamoto, K. L. Chellappa and R. Homsany; *J. Org. Chem.*, 1973, **38**, 3172. 210
73JOC3437 A. K. Bose, J. C. Kapur, J. L. Fahey and M. S. Manhas; *J. Org. Chem.*, 1973, **38**, 3437. 274, 286
73JOC3937 S. J. Valenty and P. S. Skell; *J. Org. Chem.*, 1973, **38**, 3937. 874
73JOC3951 M. W. Barker, S. C. Lauderdale and J. R. West; *J. Org. Chem.*, 1973, **38**, 3951. 730
73JOC4044 F. J. McEvoy and G. R. Allen; *J. Org. Chem.*, 1973, **38**, 4044. 25, 64
73JOC4071 P. E. Eaton, G. R. Carlson and J. T. Lee; *J. Org. Chem.*, 1973, **38**, 4071. 282
73JOC4073 M. S. Newman and W. M. Hung; *J. Org. Chem.*, 1973, **38**, 4073. 294
73JOC4086 E. J. Corey and R. H. K. Chen; *J. Org. Chem.*, 1973, **38**, 4086. 68, 97
73JOC4328 D. N. Harpp and T. G. Back; *J. Org. Chem.*, 1973, **38**, 4328. 342
73JOC4383 C. O. Okafor; *J. Org. Chem.*, 1973, **38**, 4383. 1109
73JOM(47)C17 H. C. Clark and L. E. Manzer; *J. Organomet. Chem.*, 1973, **47**, C17. 940
73JOM(49)149 H. Kunzek, M. Braun, E. Nesener and K. Rühlmann; *J. Organomet. Chem.*, 1973, **49**, 149. 397
73JOM(50)129 R. G. Goel and H. S. Prasad; *J. Organomet. Chem.*, 1973, **50**, 129. 1133
73JOM(51)C11 U. Schubert and E. O. Fischer; *J. Organomet. Chem.*, 1973, **51**, C11. 952
73JOM(56)C37 E. O. Fischer, G. Kreis and F. R. Kreissl; *J. Organomet. Chem.*, 1973, **56**, C37. 948
73JOM(61)415 P. M. Treichel and K. P. Wagner; *J. Organomet. Chem.*, 1973, **61**, 415. 802
73JOM(61)C15 W. Malisch; *J. Organomet. Chem.*, 1973, **61**, C15. 920
73JOM(63)221 D. P. Gaur, G. Srivastava and R. C. Mehrotra; *J. Organomet. Chem.*, 1973, **63**, 221. 389
73LA207 A. Treibs, L. Schulze, F.-H. Kreuzer and H.-G. Kolm; *Justus Liebigs Ann. Chem.*, 1973, 207. 638
73LA462 W. Walter and H.-W. Meyer; *Liebigs Ann. Chem.*, 1973, 462. 735, 736
73LA476 W. Walter and J. Krohn; *Justus Liebigs Ann. Chem.*, 1973, 476. 631
73LA578 U. Heep; *Justus Liebigs Ann. Chem.*, 1973, 578. 869
73LA636 W. Walter and M. Radke; *Justus Liebigs Ann. Chem.*, 1973, 636. 584, 587, 641
73LA772 H. R. Kricheldorf; *Justus Liebigs Ann. Chem.*, 1973, 772. 1012
73M421 W. Gottardi; *Monatsh. Chem.*, 1973, **104**, 421. 320, 321
73MI 503-02 E. N. Frankel and F. L. Thomas; *J. Am. Oil Chem. Soc.*, 1973, **50**, 39. 142
73MI 507-01 W. J. Hickinbottom; *Egypt. J. Chem.*, 1973, 189. 315
B-73MI 511-01 W. C. Cooper (ed.); "Tellurium," Nostrand-Reinhold, New York, 1973. 472
B-73MI 511-02 K. A. Jensen; in "Organic Selenium Compounds: Their Chemistry and Biology," eds. D. L. Klayman and W. H. H. Günther, Wiley, New York, 1973, p. 263. 472
73MI 512-01 H. R. Kricheldorf and E. Leppert; *Makromol. Chem.*, 1973, **167**, 47. 537
B-73MI 513-01 K. A. Jensen; in "Organic Selenium Compounds: their Chemistry and Biology," eds D. L. Klayman, W. H. H. Günther, Wiley, New York, 1973, p. 263. 563
73MI 530-01 A. Laguna, P. Royo and R. Uson; *Rev. Acad. Cienc. Exactus, Fis-Quim. Natur. Zaragoza*, 1973, **28**, 71 (*Chem. Abstr.*, 1974, **80**, 60 012s). 1141
73OSC(5)8 J. Staunton and E. J. Eisenbraun; *Org. Synth., Coll. Vol.*, 1973, **5**, 8. 66, 74
73OSC(5)51 D. H. Hey and M. J. Perkins; *Org. Synth., Coll. Vol.*, 1973, **5**, 51. 116
73OSC(5)73 C. L. Parris; *Org. Synth., Coll. Vol.*, 1973, **5**, 73.
73OSC(5)179 M. Regitz, J. Hocker and A. Liedhegener; *Org. Synth., Coll. Vol.*, 1973, **5**, 179. 872
73OSC(5)239 D. C. England and L. R. Melby; *Org. Synth., Coll. Vol.*, 1973, **5**, 239. 56
73OSC(5)251 A. W. Burgstahler and L. R. Worden; *Org. Synth., Coll. Vol.*, 1973, **5**, 251. 76
73OSC(5)367 A. S. Pagano and W. D. Emmons; *Org. Synth., Coll. Vol.*, 1973, **5**, 367. 205
73OSC(5)373 A. J. Zambito and E. E. Howe; *Org. Synth., Coll. Vol.*, 1973, **5**, 373. 90
73OSC(5)376 W. H. Hartung, J. H. R. Beaujon and G. Cocolas; *Org. Synth., Coll. Vol.*, 1973, **5**, 376. 90
73OSC(5)379 E. H. Larsen and S. O. Lawesson; *Org. Synth., Coll. Vol.*, 1973, **5**, 379. 76
73OSC(5)381 P. Block, Jr.; *Org. Synth., Coll. Vol.*, 1973, **5**, 381. 76
73OSC(5)384 M. Guha and D. Nasipuri; *Org. Synth., Coll. Vol.*, 1973, **5**, 384. 76
73OSC(5)387 A. J. Speziale and R. C. Freeman; *Org. Synth., Coll. Vol.*, 1973, **5**, 387. 76
73OSC(5)393 M. S. Raasch and J. E. Castle; *Org. Synth., Coll. Vol.*, 1973, **5**, 393. 59
73OSC(5)493 P. S. Bailey and R. E. Erickson; *Org. Synth., Coll. Vol.*, 1973, **5**, 493. 115
73OSC(5)509 M. J. Jorgenson and A. F. Thacher; *Org. Synth., Coll. Vol.*, 1973, **5**, 509. 103
73OSC(5)547 W. S. Wadsworth, Jr. and W. D. Emmons; *Org. Synth., Coll. Vol.*, 1973, **5**, 547. 103
73OSC(5)564 W. R. Dunnavant and C. R. Hauser; *Org. Synth., Coll. Vol.*, 1973, **5**, 564. 76
73OSC(5)580 F. H. Dean, J. H. Amin and F. L. M. Pattison; *Org. Synth. Coll. Vol.*, 1973, **5**, 580. 59
73OSC(5)586 A. T. Moore and H. N. Rydon; *Org. Synth., Coll. Vol.*, 1973, **5**, 586. 90
73OSC(5)612 P. A. S. Smith and R. O. Kan; *Org. Synth., Coll. Vol.*, 1973, **5**, 612. 37
73OSC(5)617 D. Todd and A. E. Martell; *Org. Synth., Coll. Vol.*, 1973, **5**, 617. 115
73OSC(5)635 E. J. Cragoe, Jr. and C. M. Robb; *Org. Synth., Coll. Vol.*, 1973, **5**, 635. 60
73OSC(5)650 R. Huisgen and K. Bast; *Org. Synth., Coll. Vol.*, 1973, **5**, 650. 360
73OSC(5)654 H. E. Johnson and D. G. Crosby; *Org. Synth., Coll. Vol.*, 1973, **5**, 654. 95
73OSC(5)660 J. G. Sharefkin and H. Saltzman; *Org. Synth., Coll. Vol.*, 1973, **5**, 660. 652
73OSC(5)663 W. R. Benson, E. T. McBee and L. Rand; *Org. Synth., Coll. Vol.*, 1973, **5**, 663. 320
73OSC(5)687 E. M. Bottorff and L. L. Moore; *Org. Synth., Coll. Vol.*, 1973, **5**, 687. 652
73OSC(5)706 P. E. Sokol; *Org. Synth., Coll. Vol.*, 1973, **5**, 706. 116
73OSC(5)762 J. Munch-Petersen; *Org. Synth., Coll. Vol.*, 1973, **5**, 762. 30
73OSC(5)810 L. Friedman; *Org. Synth., Coll. Vol.*, 1973, **5**, 810. 115
73OSC(5)813 B. Raecke and H. Schirp; *Org. Synth., Coll. Vol.*, 1973, **5**, 813. 109
73OSC(5)890 D. E. Pearson and D. Cowan; *Org. Synth., Coll. Vol.*, 1973, **5**, 890. 112
73OSC(5)904 L. S. Silbert, E. Siegel and D. Swern; *Org. Synth., Coll. Vol.*, 1973, **5**, 904. 205
73OSC(5)944 M. P. Cava, A. A. Deana, K. Muth and M. J. Mitchell; *Org. Synth., Coll. Vol.*, 1973, **5**, 944. 95, 306
73OSC(5)973 A. K. Bose; *Org. Synth., Coll. Vol.*, 1973, **5**, 973. 305

73OSC(5)1031 P. Block, Jr. and M. S. Newman; *Org. Synth., Coll. Vol.*, 1973, **5**, 1031. 93
73OSC(5)1043 J. C. Kauer and M. Brown; *Org. Synth., Coll. Vol.*, 1973, **5**, 1043. 100
73OSC(5)1046 F. Kurzer and A. Lawson; *Org. Synth., Coll. Vol.*, 1973, **5**, 1046. 83
73PAC145 R. B. Woodward; *Pure Appl. Chem.*, 1973, **33**, 145. 304
73PS287 C. Brown, R. F. Hudson and R. J. G. Searle; *Phosphorus*, 1973, **2**, 287. 405
73RCR939 B. I. Zapadinskii, B. I. Liogon'kii and A. A. Berlin; *Russ. Chem. Rev.* (*Engl. Transl.*), 1973, **42**, 939. 24
73RTC601 J. Meijer, P. Vermeer and L. Brandsma; *Recl. Trav. Chim. Pays-Bas*, 1973, **92**, 601. 550
73S149 J. W. Scheeren, P. H. J. Ooms and R. J. F. Nivard; *Synthesis*, 1973, 149. 512, 566, 568, 570, 608
73S297 H. Böhme and F. Ziegler; *Synthesis*, 1973, 297. 558, 587, 593
73S327 A. K. Mukerjee and R. C. Srivastava; *Synthesis*, 1973, 327. 28
73S347 T. L. Ho; *Synthesis*, 1973, 347. 28
73S413 J. W. van Reijendam and F. Baardman; *Synthesis*, 1973, 413. 991
73S416 J. F. W. McOmie and D. H. Perry; *Synthesis*, 1973, 416. 26, 37
73S487 G. A. Olah, M. Nojima and I. Kerekes; *Synthesis*, 1973, 487. 5
73S509 L. Cassar, G. P. Chiusoli and F. Guerrieri; *Synthesis*, 1973, 509. 27
73S545 V. P. Kukhar, N. P. Pisanenko and V. I. Shevtchenko; *Synthesis*, 1973, 545. 57
73SA(A)285 J.-L. Piette and M. Renson; *Spectrochim. Acta, Part A*, 1973, **29**, 285. 241
73SA(A)613 J. E. Devia and J. C. Carter; *Spectrochim. Acta, Part A*, 1973, **29**, 613. 317
73SC63 T.-L. Ho and C. M. Wong; *Synth. Commun.*, 1973, **3**, 63. 191
73SRI237 R. Uson, P. Royo and A. Laguna; *Synth. React. Inorg. Metal-Org. Chem.*, 1973, **3**, 237. 1138
73SST73 G. C. Barrett; *Org. Compd. Sulphur, Selenium, Tellurium* (*R. Soc. Chem. series*), 1973, **2**, 73. 1022
73SST200 F. Duus; *Org. Compd. Sulphur, Selenium and Tellurium* (*R. Soc. Chem. Series*), 1973, **2**, 200. 472
73SST228 F. Duus; *Org. Compd. Sulphur, Selenium, Tellurium* (*R. Soc. Chem. Series*), 1973, **1**, 228. 565
73T297 R. Fuks and M. Hartemink; *Tetrahedron*, 1973, **29**, 297. 963
73T1559 H. R. Ansari; *Tetrahedron*, 1973, **29**, 1559. 142
73T2077 L. Dalgaard, H. Kolind-Andersen and S.-O. Lawesson; *Tetrahedron*, 1973, **29**, 2077. 552, 555
73T2147 R. Fuks; *Tetrahedron*; 1973, **29**, 2147. 751
73TL211 C. Christophersen and P. Carlsen; *Tetrahedron Lett.*, 1973, 211. 1112
73TL1569 P. T. Lansbury and R. C. Stewart; *Tetrahedron Lett.*, 1973, 1569. 26
73TL1595 S. Yamada, Y. Kasai and T. Shioiri; *Tetrahedron Lett.*, 1973, 1595. 300
73TL1627 M. B. Shambhu and G. A. Digenis; *Tetrahedron Lett.*, 1973, 1627. 302
73TL1677 S. Polanc, B.Verček, B. Stanovnik and M. Tišler; *Tetrahedron Lett.*, 1973, 1677. 836
73TL1995 A. Nohara, T. Umetani and Y. Sanno; *Tetrahedron Lett.*, 1973, 1995. 74
73TL2585 Y. Leroux and C. Roman; *Tetrahedron Lett.*, 1973, 2585. 427
73TL2711 N. Singh and K. Krishan; *Tetrahedron Lett.*, 1973, **14**, 2711. 331
73TL2771 J. Leroy; *Tetrahedron Lett.*, 1973, **29**, 2771. 659
73TL2783 J. K. Rasmussen and A. Hassner; *Tetrahedron Lett.*, 1973, **14**, 2783. 348
73TL4091 F. D. Greene, R. L. Camp, V. P. Abegg and G. O. Pierson; *Tetrahedron Lett.*, 1973, **14**, 4091. 354, 359
73TL4839 J. H. P. Tyman and A. A. Durrani; *Tetrahedron Lett.*, 1973, 4839. 110
73TL5121 A. G. Anderson, Jr. and D. H. Kono; *Tetrahedron Lett.*, 1973, 5121. 17
73TL5207 A. M. van Leusen, J. Strating and D. van Leusen; *Tetrahedron Lett.*, 1973, 5207. 867
73UK1451 V. P. Kozyukov, V. D. Sheludyakov and V. F. Mironov; *Usp. Khim.*, 1973, **42**, 1451 (*Chem. Abstr.*, 1973, **79**, 126 547h). 1000
73USP3726903 R. J. Koshar, C. D. Wright, J. L. Zollinger, D. H. Dybvig and D. R. Husted; *US Pat.* 3 726 903 (1973) (*Chem. Abstr.*, 1973, **79**, 41 987b). 815, 817, 818
73USP3751312 C. D. Wright, J. L. Zollinger, R. A. Mitsch and D. H. Dybvig; *US Pat.* 3 751 312 (1973) (*Chem. Abstr.*, 1973, **79**, 104 746v). 808
73USP3772347 W. H. Atwell and G. N. Bokerman; *US Pat.* 3 772 347. Fr. Demande, 2 163 579 (1973) (*Chem. Abstr.*, 1973, **80**, 3625) 419
73USP3954441 B. Cross and C. P. Grasso; *US Pat.*, 3 954 441 (1973) (*Chem. Abstr.*, 1976, **85**, 159 725). 652
73ZOB603 A. S. Kostyuk, Y. I. Baukov and I. F. Lutsenko; *Zh. Obshch. Khim.*, 1973, **43**, 603. 389
73ZOB623 I. A. Vostokov, Y. I. Dergonov, V. P. Kozyukov, V. F. Mironov, V. D. Sheludyakov; *Zh. Obshch. Khim.*, 1973, **43**, 623. 968, 969
73ZOB1044 V. D. Sheludyakov, E. S. Rodionov and V. F. Mironov; *Zh. Obshch. Khim.*, 1973, **44**, 1044. 1015
73ZOB1451 A. N. Pudovik, É. S. Batyeva, A. V. Il'yasov, V. D. Nesterenko, A. S. Mukhtarov and N. P. Anoshina; *Zh. Obshch. Khim.*, 1973, **43**, 1451. 373
73ZOB1502 V. F. Mironov, V. D. Sheludyakov and E. S. Rodionov; *Zh. Obshch. Khim.*, 1973, **44**, 1502. 968, 969
73ZOB1753 A. S. Kostyuk, N. I. Savel'eve, Y. I. Baukov and I. F. Lutsenko; *Zh. Obshch. Khim.*, 1974, **44**, 1753. 376
73ZOB2089 V. F. Mironov, V. P. Kozyukov and V. P. Bulatov; *Zh. Obshch. Khim.*, 1973, **43**, 2089. 978
73ZOR347 V. M. Neplyuev, T. A. Sinenko, R. G. Dubenko and P. S. Pel'kis; *Zh. Org. Khim.*, 1973, **9**, 347. 595

74ACR77 A. G. Brook; *Acc. Chem. Res.*, 1974, **7**, 77. 413
74ACS(B)699 A. Brändström, B. Lamm and I. Palmertz; *Acta Chem. Scand., Ser. B*, 1974, **28** (6), 699. 991
74AG651 E. O. Fischer; *Angew. Chem.*, 1974, **86**, 651. 931
74AG(E)186 R. Acher; *Angew. Chem., Int. Ed. Engl.*, 1974, **13**, 186. 85
74AG(E)273 H. J. Bestmann and G. Schmid; *Ang. Chem., Int. Ed. Engl.*, 1974, **13**, 273. 481
74AG(E)376 L. Bauer and O. Exner; *Angew. Chem., Int. Ed. Engl.*, 1974, **13**, 376. 1000
74AG(E)419 L. Bauer and O. Exner; *Angew. Chem., Int. Ed. Engl.*, 1974, 419. 332
74AG(E)540 W. Malisch and H. Schmidbaur; *Angew. Chem., Int. Ed. Engl.*, 1974, **13**, 540. 914

74AG(E)599	W. A. Herrmann; *Angew. Chem., Int. Ed. Engl.*, 1974, **13**, 599.	943
74AJC2635	R. L. N. Harris; *Aust. J. Chem.*, 1974, **27**, 2635.	731
74AP(307)801	D. Matthies and E. D. Setiakusuma; *Arch. Pharm.* (*Weinheim, Ger.*), 1974, **307**, 801.	606
74BAU1593	B. M. Mikhailov, V. A. Dorokhov, V. I. Serendenko and I. P. Yakovlev; *Bull. Acad. Sci. USSR, Div Chim. Sci.* (*Eng. Transl.*), 1974, **23**, 1593.	778
74BCF994	J. Pornet and L. Miginiac; *Bull. Soc. Chim. Fr.*, 1974, 994.	754
74BCJ2745	N. Watanabe, M. Okano and S. Uemura; *Bull. Chem. Soc. Jpn.*, 1974, **47**, 2745.	1105
74BSF1965	P. Coutrot; *Bull. Soc. Chim. Fr.*, 1974, 1965.	692
74BSF(2)1949	G. Merault, P. Bourgeois and N. Duffaut; *Bull. Soc. Chim. Fr., Part* 2, 1974, 1949.	297
74CA121069	I. S. Mateev; *Chem. Abstr.*, 1974, **80**, 121069.	1157
74CB183	F. R. Kreissl and E. O. Fischer; *Chem. Ber.*, 1974, **107**, 183.	953
74CB502	J. Goerdeler, J. Haag, C. Lindner and R. Losch; *Chem. Ber.*, 1974, **107**, 502.	616, 1050
74CB508	H. W. Roesky and S. Tutkunkardes; *Chem. Ber.*, 1974, **107**, 508.	995
74CB915	E. O. Fischer and S. Riedmüller; *Chem. Ber.*, 1974, **107**, 915.	516, 520
74CB2169	R. Appel and A. Gilak; *Chem. Ber.*, 1975, **107**, 2169.	1093
74CB3326	E. O. Fischer and D. Plabst; *J. Organomet. Chem.*, 1974, **107**, 3326.	953
74CB3554	E. O. Fischer, K. Weiss and C. G. Kreiter; *Chem. Ber.*, 1974, **107**, 3554.	949
74CB3589	E. Schaumann, S. Sieveking and W. Walter; *Chem. Ber.*, 1974, **107**, 3589.	729
74CC466	E. T. Blues, D. Bryce-Smith, J. G. Irwin and I. W. Lawston; *J. Chem. Soc., Chem. Commun.*, 1974, 466.	874
74CC516	G. Fachinetti and C. Floriani; *J. Chem. Soc., Chem. Commun.*, 1974, 516.	444
74CC622	S. E. Diamond and H. Taube; *J. Chem. Soc., Chem. Commun.*, 1974, 622.	386
74CC841	E. R. Hamner, R. D. W. Kemmitt and M. A. R. Smith; *J. Chem. Soc., Chem. Commun.*, 1974, 841.	465
74CI(L)803	K. M. Doyle and F. Kurzer; *Chem. Ind.* (*London*), 1974, 803.	652
74CJC879	P. Wolkoff, S. Hammerum, P. D. Callaghan and M. S. Gibson; *Can. J. Chem.*, 1974, **52**, 879.	641
74CJC1762	R. J. Balahura; *Can. J. Chem.*, 1974, **52**, 1762.	385
74CJC4050	A. R. Stein and S.-H. Tan; *Can. J. Chem.*, 1974, **52**, 4050.	723
74CL187	T. Mukaiyama, T. Takeda and K. Atsumi; *Chem. Lett.*, 1974, 187.	236
74CL247	T. Fuchigami and K. Odo; *Chem. Lett.*, 1974, 247.	766
74CL881	M. Furukawa, T. Suda and S. Hayashi; *Chem. Lett.*, 1974, 881.	1111
74CL1035	T. Ito and Y. Takami; *Chem. Lett.*, 1974, 1035.	70
74CR(C)213	M. Paul and E. Frainnet; *C. R. Hebd. Seances Acad. Sci. Ser. C*, 1974, **279**, 213.	720
74IC438	Y. Yamamoto and H. Yamazaki; *Inorg. Chem.*, 1974, **13**, 438.	801
74IC1595	D. M. Blake, S. Shields and L. Wyman; *Inorg. Chem.*, 1974, **13**, 1595.	460
74IC2705	K. E. Peterman and J. M. Shreeve; *Inorg. Chem.*, 1974, **13**, 2705.	313
74IZV55	S. A. Novruzov, V. M. Ismailov, V. V. Moskva, S. T. Akhmedov and A. M. Razumov; *Izv. Akad. Nauk. SSSR, Ser. Khim.*, 1974, 55.	484
74JA925	G. A. Olah, M. Nojima and I. Kerekes; *J. Am. Chem. Soc.*, 1974, **96**, 925.	5
74JA1230	C. P. Casey and R. L. Anderson; *J. Am. Chem. Soc.*, 1974, **96**, 1230.	955
74JA1518	J. K. Stille, F. Huang and M. T. Regan; *J. Am. Chem. Soc.*, 1974, **96**, 1518.	458
74JA2250	C. L. Liotta and H. P. Harris; *J. Am. Chem. Soc.*, 1974, **96**, 2250.	4
74JA2360	G. W. Parshall; *J. Am. Chem. Soc.*, 1974, **96**, 2360.	1138, 1139, 1140
74JA2847	W. C. Kaska, D. K. Mitchell, R. F. Reichelderfer and W. D. Korte; *J. Am. Chem. Soc.*, 1974, **96**, 2847.	919
74JA3319	J. C. Martin and M. M. Chau; *J. Am. Chem. Soc.*, 1974, **96**, 3319.	204
74JA4283	C. H. Yoder, W. C. Copenhafer and B. DuBeshter; *J. Am. Chem. Soc.*, 1974, **96**, 4283.	379
74JA4983	K. S. Y. Lau, R. W. Fries and J. K. Stille; *J. Am. Chem. Soc.*, 1974, **96**, 4983.	463
74JA6469	H. M. Relles and R. W. Schluenz; *J. Am. Chem. Soc.*, 1974, **96**, 6469.	14
74JA7807	S. Danishefsky and T. Kitahara; *J. Am. Chem. Soc.*, 1974, **96**, 7807.	199
74JA8113	J. Bertin, H. B. Kagan, J.-L. Luche and R. Setton; *J. Am. Chem. Soc.*, 1974, **96**, 8113.	122
74JCS(D)421	C. J. Wilson, M. Green and R. J. Mawby; *J. Chem. Soc., Dalton Trans.*, 1974, 421.	465
74JCS(D)2302	O. M. Abu Salah and M. I. Bruce; *J. Chem. Soc., Dalton Trans.*, 1974, 2302.	452
74JCS(P1)627	A. R. McCarthy, W. D. Ollis and C. A. Ramsden; *J. Chem. Soc., Perkin Trans. 1*, 1974, 627.	641
74JCS(P1)732	D. H. R. Barton, R. H. Hesse, M. M. Pechet and H. T. Toh; *J. Chem. Soc., Perkin Trans. 1*, 1974, 732.	313, 314, 316
74JCS(P1)853	J. Almog, D. H. R. Barton, P. D. Magnus and R. K. Norris; *J. Chem. Soc., Perkin Trans. 1*, 1974, 853.	735
74JCS(P1)1499	G. S. King and E. S. Waight; *J. Chem. Soc., Perkin Trans. 1*, 1974, 1499.	
74JCS(P1)1541	F. T. Boyle and R. Hull; *J. Chem. Soc., Perkin Trans. 1*, 1974, 1541.	1035
74JCS(P1)1864	R. C. Cambie, R. C. Hayward, J. L. Roberts and P. S. Rutledge; *J. Chem. Soc. Perkin Trans. 1*, 1974, 1864.	168
74JCS(P2)546	S.-O. Chua, M. J. Cook and A. R. Katritzky; *J. Chem. Soc., Perkin Trans. 2*, 1974, 546.	758, 767
74JCS(P2)1483	P. Del Buttero, S. Maiorana, D. Pocar, G. D. Andretti, G. Bocelli and P. Sgarabotto; *J. Chem. Soc., Perkin Trans. 2*, 1974, 1483.	497, 499
74JCS(P2)1676	S. Bradamante, P. Del Buttero, D. Landini and S. Maiorana; *J. Chem. Soc., Perkin Trans. 2*, 1974, 1676.	499
74JGU1677	N. N. Magdesieva and R. A. Kyandzhetsian; *J. Gen. Chem. USSR* (*Engl. Transl.*), 1974, **44**, 1677.	903, 904
74JGU1685	B. N. Laskorin, V. V. Yakshin and L. I. Sokal'skaya; *J. Gen. Chem. USSR* (*Engl. Transl.*), 1974, **44**, 1685.	403, 404

74JGU2371 A. N. Pudovik, T. M. Sudakova and G. I. Evstaf'ev *J. Gen. Chem. USSR* (*Engl. Transl.*), 1974, **44**, 2371. 652

74JHC103 S. Polanc, B. Verček, B. Stanovnik and M. Tišler; *J. Heterocycl. Chem.*, 1974, 103. 836

74JHC397 W. Christ, D. Rakow and S. Strauss; *J. Heterocycl. Chem.*, 1974, **11**, 397. 567, 619, 625

74JMC549 W. L. Matier, W. T. Comer and A. W. Gomoll; *J. Med. Chem.*, 1974, **17**, 549. 767

74JOC511 S. Wawzonek and S. M. Heilmann; *J. Org. Chem.*, 1974, **39**, 511. 554

74JOC600 G. E. Niznik, W. H. I. Morrison III and H. M. Walborsky; *J. Org. Chem.*, 1974, **39**, 600. 68, 420, 792, 796, 797

74JOC604 N. Hirowatari and H. M. Walborsky; *J. Org. Chem.*, 1974, **39**, 604. 88

74JOC914 D. A. Evans, G. L. Carroll and L. K. Truesdale; *J. Org. Chem.*, 1974, **39**, 914. 1134

74JOC922 D. J. Scharf; *J. Org. Chem.*, 1974, **39**, 922. 372

74JOC957 M. Komatsu, Y. Ohshiro, K. Yasuda, S. Ichijima and T. Agawa; *J. Org. Chem.*, 1974, **39**, 957. 1032

74JOC1494 J. V. Paukstelis and M.-G. Kim; *J. Org. Chem.*, 1974, **39**, 1494. 1120

74JOC1597 W. H. Daly and H. J. Hölle; *J. Org. Chem.*, 1974, **39**, 1597. 979

74JOC1676 J. Martin, P. C. Watts and F. Johnson; *J. Org. Chem.*, 1974, **39**, 1676. 70

74JOC1968 J. E. Shaw and D. C. Kumerth; *J. Org. Chem.*, 1974, **39**, 1968. 132

74JOC1970 S. Sakai, T. Aizawa and T. Fujinami; *J. Org. Chem.*, 1974, **39**, 1970. 1024

74JOC2143 S. Polanc, B. Verček, B. Šek, B. Stanovnik and M. Tišler; *J. Org. Chem.*, 1974, 2143. 836

74JOC2472 W. J. Kauffman; *J. Org. Chem.*, 1974, **39**, 2472. 1013

74JOC2540 E. P. Papadopoulos; *J. Org. Chem.*, 1974, **39**, 2540. 592

74JOC2728 B. A. Reith, J. Strating and A. M. Van Leusen; *J. Org. Chem.*, 1974, **39**, 2728. 902

74JOC2740 M. P. Doyle, D. J. DeBruyn, S. J. Donnelly, D. A. Kooistra, A. A. Odubela, C. T. West and S. M. Zonnerbelt; *J. Org. Chem.*, 1974, **39**, 2740. 282

74JOC2897 H. Ulrich, R. Richter, P. J. Whitman and A. A. R. Sayigh; *J. Org. Chem.*, 1974, **39**, 2897. 971

74JOC3043 V. I. Cohen; *J. Org. Chem*, 1974, **39**, 3043. 631

74JOC3080 W. L. Matier and W. T. Comer; *J. Org. Chem.*, 1974, **39**, 3080. 767

74JOC3136 C. Bachand, H. Driguez, J. M. Paton, D. Touchard and J. Lessard; *J. Org. Chem.*, 1974, **39**, 3136. 316, 318

74JOC3144 B. J. Whitlock and H. W. Whitlock, Jr.; *J. Org. Chem.*, 1974, **39**, 3144. 70

74JOC3273 Y. Ito, K. Nakayama, K. Yonezawa and T. Saegusa; *J. Org. Chem.*, 1974, **39**, 3273. 146

74JOC3302 S. Yamada, Y. Yokoyama and T. Shioiri; *J. Org. Chem.*, 1974, **39**, 3302. 235

74JOC3318 A. Schoenberg, I. Bartoletti and R. F. Heck; *J. Org. Chem.*, 1974, **39**, 3318. 143

74JOC3327 A. Schoenberg and R. F. Heck; *J. Org. Chem.*, 1974, **39**, 3327. 273, 296, 301, 304

74JOC3924 I. Yamamoto, S. Yanagi, A. Mamba and H. Gotoh; *J. Org. Chem.*, 1974, **39**, 3924. 303

74JOM(65)51 B. C. Pant; *J. Organomet. Chem.*, 1974, **65**, 51. 550

74JOM(65)53 E. O. Fischer, G. Kreis, F. R. Kreissl, W. Kalbfus and E. Winkler; *J. Organomet. Chem.*, 1974, **65**, C53. 1088

74JOM(67)269 J. D. Wilkins; *J. Organomet. Chem.*, 1974, **67**, 269. 797

74JOM(69)353 I. Matsuda, K. Itoh and Y. Ishii; *J. Organomet. Chem.*, 1974, **69**, 353. 377, 388, 390, 1083

74JOM(70)273 C. T. Lam, C. V. Senoff and J. E. H. Ward; *J. Organomet. Chem.*, 1974, **70**, 273. 950, 951

74JOM(71)453 J. E. Parks and A. L. Balch; *J. Organomet. Chem.*, 1974, **71**, 453. 696

74JOM(71)C5 G. Fachinetti and C. Floriani; *J. Organomet. Chem.*, 1974, **71**, C5. 441

74JOM(72)C11 I. Ojima, S.-I. Inaba, and Y. Nagai; *J. Organomet. Chem.*, 1974, **72**, C11. 777

74JOM(73)319 J. A. Feiccabrino and E. J. Kupchik; *J. Organomet. Chem.*, 1974, **73**, 319. 390

74JOM(73)C28 C. P. Casey and R. L. Anderson; *J. Organomet. Chem.*, 1974, **73**, C28. 946

74JOM(76)355 G. La Monica, S. Cenini and M. Freni; *J. Organomet. Chem.*, 1974, **76**, 355. 386

74JOM(76)C1 P. Bourgeois; *J. Organomet. Chem.*, 1974, **76**, C1. 419, 793

74JOM(77)345 C. P. Casey and W. R. Brunsvold; *J. Organomet. Chem.*, 1974, **77**, 345. 958

74JOM(78)339 K. Itoh, H. Hayashi, M. Fukui and Y. Ishii; *J. Organomet. Chem.*, 1974, **78**, 339. 645

74JOM(78)405 S. Kato, M. Wakamatsu and M. Mizuta; *J. Organomet. Chem.*, 1974, **78**, 405. 549

74JOM(80)349 J. D. Wilkins; *J. Organomet. Chem.*, 1974, **80**, 349. 781

74JOM(80)C25 P. Bourgeois, J. Dunoguès, N. Duffaut and P. Lapouyade; *J. Organomet. Chem.*, 1974, **80**, C25. 416

74JOM(82)223 S. Kato, A. Hori, H. Shiotani, M. Mizuta, N. Hayashi and T. Takakuwa; *J. Organomet. Chem.*, 1974, **82**, 223. 550

74JPR534 G. Barnikow and G. Saeling; *J. Prakt. Chem.*, 1974, **316**, 534. 595

74JST123 H. S. Randhawa and C. N. R. Rao; *J. Mol. Struct.*, 1974, **21**, 123. 506, 507

74LA112 R. Köster and E. Rothgery; *Justus Liebigs Ann. Chem.*, 1974, 112. 228

74LA561 K. Clauss, H.-J. Friedrich and H. Jensen; *Justus Liebigs Ann. Chem.*, 1974, 561. 344

74LA776 W. Walter and H.-W. Meyer; *Liebigs Ann. Chem.*, 1974, 776. 736

74LA901 R. Lattrell and G. Lohaus; *Justus Liebigs Ann. Chem.*, 1974, 901. 1106

74LA1767 U. Schöllkopf, B. Bánhidai, R. Meyer, H. Beckhaus and H. Frasnelli; *Justus Liebigs Ann. Chem.*, 1974, **763**, 1767. 872

74M611 W. Gottardi; *Monatsh. Chem.*, 1974, **105**, 611. 321

74MI 507-01 R. Miller; *Chim. Ther.*, 1974, **9**, 301. 370

74MI 530-01 L. B. Ionov, Yu. F. Gatilov, I. P. Mukanov and L. G. Kokorina; *Zh. Obshch. Khim. Lett.*, 1974, **44**, 1737 (*Chem. Abstr.*, 1975, **82**, 4384u). 1132

74MI 530-02 L. B. Ionov, Yu. F. Gatilov, I. P. Mukanov and L. G. Kokorina; *Zh. Obshch. Khim. Lett.*, 1974, **44**, 2502 (*Chem. Abstr.*, 1975, **82**, 86 349y). 1132

74MI 530-03 T. N. Srivastava; *Labdev. J. Sci. Tech.*, 1974, **12**, 91. 1136

74RTC51 P. Vermeer, J. Meijer, H. J. T. Bos and L. Brandsma; *Recl. Trav. Chim. Pays-Bas*, 1974, **93**, 51. 510

74RTC258	F. C. V. Larsson, L. Brandsma and S.-O. Lawesson; *Recl. Trav. Chim. Pays-Bas*, 1974, **93**, 258.	556
74S36	H. Stetter and K. Elfert; *Synthesis*, 1974, 36.	29
74S229	A. J. Fatiadi; *Synthesis*, 1974, 229.	28
74S433	N. H. Nilsson and J. Sandström; *Synthesis*, 1974, 433.	593
74S443	Y. Fukuyama, Y. Kawashima, T. Miwa and T. Tokoroyama; *Synthesis*, 1974, 443.	68
74S501	L. Henriksen; *Synthesis*, 1974, 501.	626
74S549	Y. S. Klausner and M. Bodanszky; *Synthesis*, 1974, 549.	278
74S574	S. Paraskewas; *Synthesis*, 1974, 574.	275
74S583	K. M. Doyle and F. Kurzer; *Synthesis*, 1974, 583.	769
74S715	T.-L. Ho; *Synthesis*, 1974, 715.	39
74SC1	H. Christensen; *Synth. Commun.*, 1974, **4**, 1.	110
74T93	L. Dalgaard, L. Jensen and S.-O. Lawesson; *Tetrahedron*, 1974, **30**, 93.	547, 555
74T195	S. Conde, C. Corral and R. Madroñero; *Tetrahedron*, 1974, **15**, 195.	652
74T999	J. Perchais and J.-P. Fleury; *Tetrahedron*, 1974, **30**, 999.	826
74T2151	K. Nimomiya, T. Shiori and S. Yamada; *Tetrahedron*, 1974, **30**, 2151.	265, 278, 993
74T2899	J. B. Chattopadhyaya and A. V. Rama Rao; *Tetrahedron*, 1974, **30**, 2899.	276, 298
74T4147	E. Schaumann, S. Sieveking and W. Walter; *Tetrahedron*, 1974, **30**, 4147.	638
74TL91	D. S. L. Blackwell, P. de Mayo, R. Suau; *Tetrahedron Lett.*, 1974, 91.	634
74TL1187	A. Nohara; *Tetrahedron Lett.*, 1974, 1187.	363, 367, 774
74TL1275	H. J. Bestman, T. Denzel and H. Salbaum; *Tetrahedron Lett.*, 1974, 1275.	43
74TL1391	W. R. Roush, D. Feitler and J. Rebek; *Tetrahedron Lett.*, 1974, 1391.	869
74TL1511	A. W. Herriott and D. Picker; *Tetrahedron Lett.*, 1974, 1511.	47, 52, 115
74TL1785	L. Kisfaludy, I. Schön, T. Szirtes, O. Nyeki and M. Low; *Tetrahedron Lett.*, 1974, 1785.	277
74TL1965	S. Bittner, S. Grinberg and I. Kartoon; *Tetrahedron Lett.*, 1974, 1965.	1000
74TL2439	E. J. Corey, K. C. Nicolaou and D. J. Beames; *Tetrahedron Lett.*, 1974, 2439.	171
74TL2445	D. Döpp and H. Weiler; *Tetrahedron Lett.*, 1974, 2445.	1011
74TL2633	A. K. Bose, B. Lal, B. Dayal and M. S. Manhas; *Tetrahedron Lett.*, 1974, 2633.	294
74TL2939	Gy. Simig and K. Lempert; *Tetrahedron Lett.*, 1974, **34**, 2939.	1118
74TL3135	A. K. Bose, B. Dayal, J. C. Kapur, B. Lal and M. S. Manhas; *Tetrahedron Lett.*, 1974, 3135.	
74TL3319	P. K. Claus, P. Hofbauer and W. Rieder; *Tetrahedron Lett.*, 1974, 3319.	766
74TL3377	W. J. J. M. Sprangers, A. P. van Sweiten and R. Louw; *Tetrahedron Lett.*, 1974, **15**, 3377.	438
74TL3995	H. Yazawa, K. Tanaka and K. Kariyone; *Tetrahedron Lett.*, 1974, 3995.	277
74URP449912	M. G. Voronkov, E. A. Maletina and V. K. Roman; *USSR Pat.* 449 912 (1974) (*Chem. Abstr.*, 1975, **82**, 57 926).	1135
74ZC56	J. Bodeker and H. Zartner; *Z. Chem.*, 1974, **14**, 56.	643
74ZC265	M. Schulz and U. Missol; *Z. Chem.*, 1974, **14**, 265.	357
74ZC356	G. Barnikow and H. Ebeling; *Z. Chem.*, 1974, **14**, 356.	1052
74ZN(B)275	E. Lindner and H. Berke; *Z. Naturforsch, Teil B*, 1974, **29**, 275.	449
74ZOB553	V. P. Kozyukov and V. F. Mironov; *Zh. Obshch. Khim.*, 1974, **44**, 553.	1014
74ZOB1753	A. S. Kostyuk, N. I. Savel'eve, Y. I. Baukov and I. F. Lutsenko; *Zh. Obshch. Khim.*, 1974, **44**, 1753.	376, 389
74ZOR139	N. Y. Derkach, T. V. Lyapina and E. S. Levchenko; *Zh. Org. Khim.*, 1974, **10**, 139.	341
74ZOR667	M. M. Kremlev and V. G. Dolyuk; *Zh. Org. Khim.*, 1974, **10**, 667.	345
74ZOR1428	N. P. Aktaev, V. A. Pashinin and G. A. Sokol'skii; *Zh. Org. Khim.*, 1974, **10**, 1428.	386
74ZOR1992	I. M. Bazavova, R. G. Dubenko and P. S. Pel'kis; *Zh. Org. Khim.*, 1974, **10**, 1992.	593
74ZOR2200	B. I. Buzykin, L. P. Sysoeva and Y. P. Kitaev; *Zh. Org. Khim.*, 1974, **10**, 2220.	860
74ZOR2301	D. A. Kazlauskas and É. B. Ramoshkene; *Zh. Org. Khim.*, 1974, **10**, 2301.	361

75ACR10	J. F. King; *Acc. Chem. Res.*, 1975, **8**, 10.	472, 495
75ACR62	H. Schmidbaur; *Acc. Chem. Res.*, 1975, **8**, 62.	894
75ACR342	J. P. Collman; *Acc. Chem. Res.*, 1975, **8**, 342.	451
75ACS(B)538	J. Perregaard, I. Thomsen and S.-O. Lawesson; *Acta. Chem. Scand., Ser. B*, 1975, **29**, 538.	600, 621
75ACS(B)604	J. Perregaard and S.-O. Lawesson; *Acta. Chem. Scand., Ser. B*, 1975, **29**, 604.	600, 621
75AG710	G. Seybold; *Angew. Chem.*, 1975, **87**, 710.	550
75AG(E)53	M. Parmantier, J. Galloy, M. van Meerssche and H. G. Viehe; *Angew. Chem. Int. Ed. Engl.*, 1975, **14**, 53.	481
75AG(E)248	G. Seybold and C. Heibl; *Angew. Chem., Int. Ed. Engl.*, 1975, **14**, 248.	589
75AG(E)427	W. Walter and H.-W. Lüke; *Angew. Chem., Int. Ed. Engl.*, 1975, **14**, 427.	645
75AG(E)428	H. Quast and L. Bieber; *Angew. Chem., Int. Ed. Engl.*, 1975, **14**, 428.	1073
75AG(E)502	H. R. Kricheldorf; *Angew. Chem., Int. Ed. Engl.*, 1975, **14**, 502.	1031
75AG(E)703	G. Seybold; *Angew. Chem., Int. Ed. Engl.*, 1975, **14**, 703.	480
75AG(E)760	E. O. Fischer, G. Huttner, W. Kleine and A. Frank; *Angew. Chem., Int. Ed. Engl.*, 1975, **14**, 760.	1088
75AG(E)765	U. Schöllkopf and I. Hoppe; *Angew. Chem., Int. Ed. Engl.*, 1975, **14**, 765.	162
75AP(308/12)954	B. Unterhals and D. Thamer; *Arch. Pharm.* (*Weinheim, Ger.*), 1975, **308/12**, 954.	122
75BAU816	R. G. Kostyanovskii, Yu. I. El'natanov and L. M. Zagurskaya; *Bull. Acad. Sci. USSR, Div. Chem. Sci.*, 1975, 816.	397
75BCJ365	A. S. Shawali, N. F. Eweiss, H. M. Hassaneen and M. Sami; *Bull. Chem. Soc. Jpn.*, 1975, **48**, 365.	358
75BCJ1925	S. Uemura, A. Toshimitsu, M. Okano and K. Ichikawa; *Bull. Chem. Soc. Jpn.*, 1975, **48**, 1925.	1113, 1115

75BCJ2075 M. Hidai, T. Hikita, Y. Wada, Y. Fujikura and Y. Uchida; *Bull. Chem. Soc. Jpn.*, 1975, **48**, 2075. 143

75BCJ2091 T. Ito, K. Mori, T. Mizoroki and A. Ozaki; *Bull. Chem. Soc. Jpn.*, 1975, **48**, 2091. 143, 304

75BCJ2981 S. Sakai, T. Fujinami and T. Aizawa; *Bull. Chem. Soc. Jpn.*, 1975, **48**, 2981. 1024

75BCJ2993 S. Kawamura, T. Horii, T. Nakabayashi and M. Hamada; *Bull. Chem. Soc. Jpn.*, 1975, **48**, 2993. 247

75BCJ3205 N. Watanabe, S. Uemura and M. Okano; *Bull. Chem. Soc. Jpn.*, 1975, **48**, 3205. 1042

75BCJ3259 T. Minami, F. Takimoto and T. Agawa; *Bull. Chem. Soc. Jpn.*, 1975, **48**, 3259. 652

75BCJ3682 Y. Yamamoto, K. Kondo and I. Moritani; *Bull. Chem. Soc. Jpn.*, 1975, **48**, 3682. 795, 796

75BSB61 J. Poupaert and A. Bruylants; *Bull. Soc. Chim. Belg.*, 1975, **84**, 61. 558

75BSF829 F. Darré, A.-M. Lamazouère and J. Sotiropoulos; *Bull. Soc. Chim. Fr.*, 1975, 829. 593

75BSF1439 D. Paquer; *Bull. Soc. Chim. Fr.*, 1975, 1439. 232

75CB664 J. Müller, F. Schmock, A. Klopsch and K. Dehnicke; *Chem. Ber.*, 1975, **108**, 664. 1143

75CB2484 G. Becker and W. Hölderich; *Chem. Ber.*, 1975, **108**, 2484. 1154

75CB2547 H. Poisel and U. Schmidt; *Chem. Ber.*, 1975, **108**, 2547. 315

75CB2554 W. Malisch and P. Panster; *Chem. Ber.*, 1975, **108**, 2554. 913

75CB2834 H. Schmidbaur and W. Wolf; *Chem. Ber.*, 1975, **108**, 2834. 915

75CB2842 H. Schmidbaur and W. Wolf; *Chem. Ber.*, 1975, **108**, 2842. 915

75CB2851 H. Schmidbaur and W. Wolf; *Chem. Ber.*, 1975, **108**, 2851. 913

75CB3066 J. Goerdeler and K. Nandi; *Chem. Ber.*, 1975, **108**, 3066. 566, 592, 593, 610

75CB3071 J. Goerdeler, F. M. Panshiri and W. Vollrath; *Chem. Ber.*, 1975, **108**, 3071. 1052

75CC71 N. Tangari and V. Tortorella; *J. Chem. Soc., Chem. Commun.*, 1975, 71. 307

75CC267 F. Canziani, M. C. Malatesta and G. Longoni; *J. Chem. Soc., Chem. Commun.*, 1975, 267. 461

75CC320 L. Bernardi, R. de Castiglione and U. Scarponi; *J. Chem. Soc., Chem. Commun.*, 1975, 320. 300

75CC617 M. P. Cava and L. E. Saris; *J. Chem. Soc., Chem. Commun.*, 1975, 617. 626, 732

75CC622 P. Hodge and G. Richardson; *J. Chem. Soc., Chem. Commun.*, 1975, 622. 13

75CC719 K. Smith and K. Swaminathan; *J. Chem. Soc., Chem. Commun.*, 1975, 719. 433

75CC725 P. Y. Johnson, C. E. Hatch III and N. R. Schmuff; *J. Chem. Soc., Chem. Commun.*, 1975, 725. 354

75CC974 I. Creaser and A. M. Sargeson; *J. Chem. Soc., Chem. Commun.*, 1975, 974. 386

75CI(L)745 A. Kasahara and T. Saito; *Chem. Ind. (London)*, 1975, 745. 296

75CJC3211 P. Wolkoff, S. T. Nemeth and M. S. Gibson; *Can. J. Chem.*, 1975, **53**, 3211. 641

75CJC3693 S. Masamune, S. Kamata, J. Diakur, Y. Sugihara and G. S. Bates; *Can. J. Chem.*, 1975, **53**, 3693. 235

75CL197 H. Suzuki, K. Itoh, I. Matsuda and Y. Ishii; *Chem. Lett.*, 1975, 197. 387

75CL641 H. Suzuki, K. Itoh, I. Matsuda and Y. Ishii; *Chem. Lett.*, 1975, 641. 387

75CL1231 S. Kawamura, A. Sato, T. Nakabayashi and M. Hamada; *Chem. Lett.*, 1975, 1231. 247

75CR(C)1327 V. Ratovelomanana, C. Huynh, S. Julia and H. Normant; *C. R. Hebd. Seances Acad. Sci., Ser. C*, 1975, **280**, 1327. 554

75CRV491 M. V. George and K. S. Balachandran; *Chem. Rev.*, 1975, **75**, 491. 28

75GEP(O)2409357 A. G.; *Ger. Pat. Offen.*, 2 409 357 (1975) (*Chem. Abstr.*, 1976, **84**, 31 240). 481

75GEP(O)2410157 E. O. Fischer and S. Riedmüller; *Ger. Pat. Offen.*, 2410157 (1975) (*Chem. Abstr.*, 1976, **84**, 30690b). 520

75H(3)1081 O. Tsuge and A. Inaba; *Heterocycles*, 1975, **3**, 1081. 631

75HCA587 W. Oppolzer and W. Fröstl; *Helv. Chim. Acta.*, 1975, **58**, 587. 288

75HCA1316 C. Couret, F. Couret, J. Satgé and J. Escudie; *Helv. Chim. Acta*, 1975, **58**, 1316. 720

75IC42 A. W. Zanella and P. C. Ford; *Inorg. Chem.*, 1975, **14**, 42. 386

75JA228 C. A. Bertelo and J. Schwartz; *J. Am. Chem. Soc.*, 1975, **97**, 228. 441

75JA1180 R. E. White and P. Kovacic; *J. Am. Chem. Soc.*, 1975, **97**, 1180. 317

75JA1184 L. S. Hegedus and K. Siirala-Hansén; *J. Am. Chem. Soc.*, 1975, **97**, 1184. 464

75JA1608 T. Hayashi and H. Baba; *J. Am. Chem. Soc.*, 1975, **97**, 1608. 614

75JA2287 E. J. Corey, K. C. Nicolaou and T. Toru; *J. Am. Chem. Soc.*, 1975, **97**, 2287. 173

75JA3238 C. G. Venier, H. J. Barager and M. A. Ward; *J. Am. Chem. Soc.*, 1975, **97**, 3238. 867

75JA3515 S. Masamune, S. Kamata and W. Schilling; *J. Am. Chem. Soc.*, 1975, **97**, 3515. 234, 277

75JA4144 F.-C. Huang, L. F. H. Lee, R. S. D. Mittal, P. R. Ravikumar, J. A. Chan, C. J. Sih, E. Caspi and C. R. Eck; *J. Am. Chem. Soc.*, 1975, **97**, 4144. 138

75JA5628 P. W. Wojtkowski, J. E. Dolfini, O. Kocy and C. M. Cimarusti; *J. Am. Chem. Soc.*, 1975, **97**, 5628. 575

75JA6281 O. Gasser and H. Schmidbaur; *J. Am. Chem. Soc.*, 1975, **97**, 6281. 1093

75JAP89344 J. Saito, T. Tamura and N. Morishima; *Jpn. Pat.*, 89 344 (1975) (*Chem. Abstr.*, 1975, **83**, 205 987n). 971

75JCS(D)930 T. Inglis and M. Kilner; *J. Chem. Soc., Dalton Trans.*, 1975, 930. 781

75JCS(D)2368 R. C. Dobbie, P. D. Gosling and B. P. Straughan; *J. Chem. Soc., Dalton Trans.*, 1975, 2368. 1129

75JCS(P1)270 R. Okazaki, F. Ishii, K. Okawa, K. Ozawa and N. Inamoto; *J. Chem. Soc., Perkin Trans. 1*, 1975, 270. 634

75JCS(P1)922 R. Hull, P. J. van den Broek and M. L. Swain; *J. Chem. Soc., Perkin Trans. 1*, 1975, 922. 1035

75JCS(P1)1237 R. B. Boar, J. F. McGhie, M. Robinson, D. H. R. Barton, D. C. Horwell and R. V. Stick; *J. Chem. Soc., Perkin Trans. 1*, 1975, 1237. 289

75JCS(P1)1574 D. H. R. Barton and S. W. McCombie; *J. Chem. Soc., Perkin Trans. 1*, 1975, 1574. 516, 522, 526, 627

75JCS(P1)1964 T. L. Gilchrist, C. J. Moody and C. W. Rees; *J. Chem. Soc., Perkin Trans. 1*, 1975, 1964. 766

75JCS(P2)916 A. Tangerman and B. Zwanenburg; *J. Chem. Soc., Perkin Trans. 2*, 1975, 916. 562

75JGU935	V. E. Shishkin, Yu. M. Yukhno and B. I. No; *J. Gen. Chem. USSR* (*Engl. Transl*), 1975, **45**, 935.	722
75JGU2421	M. G. Voronkov, R. G. Mirskov, O. S. Ishchenko and S. P. Sitnikova; *J. Gen. Chem. USSR* (*Engl. Transl.*), 1975, 2421 (*Zh. Obshch. Khim.*, 1974, **44**, 2462).	540
75JGU2509	O. I. Kolodyazhnyi and V. P. Kukhar; *J. Gen. Chem. USSR* (*Engl. Transl.*), 1975, **45**, 2509.	903
75JGU2512	I. V. Konovalova, L. A. Burnaeva and A. N. Pudovik; *J. Gen. Chem. USSR* (*Engl. Transl.*), 1975, **45**, 2512.	786
75JHC37	L. S. Wittenbrook; *J. Heterocycl. Chem.*, 1975, **12**, 37.	1124
75JIC148	Y. E. Moharir; *J. Indian Chem. Soc.*, 1975, **52**, 148.	1026
75JMC581	Y. Lefebre and C. Revesz; *J. Med. Chem.*, 1975, **18**, 581.	652
75JOC66	J. P. Chupp and K. L. Leschinsky; *J. Org. Chem.*, 1975, **40**, 66.	632
75JOC129	W. J. Middleton; *J. Org. Chem.*, 1975, **40**, 129.	473
75JOC335	J. K. Stille and P. K. Wong; *J. Org. Chem.*, 1975, **40**, 335.	142
75JOC532	J. K. Stille and P. K. Wong; *J. Org. Chem.*, 1975, **40**, 532.	652
75JOC1186	A. I. Meyers and E. D. Mihelich; *J. Org. Chem.*, 1975, **40**, 1186.	166
75JOC2029	J. C. Grivas; *J. Org. Chem.*, 1975, **40**, 2029.	335
75JOC2129	D. G. Naae and J. Z. Gougoutas; *J. Org. Chem.*, 1975, **40**, 2129.	321
75JOC2667	J. M. Thompson and R. F. Heck; *J. Org. Chem.*, 1975, **40**, 2667.	464
75JOC2796	D. J. Burton and P. E. Greenlimb; *J. Org. Chem.*, 1975, **40**, 2796.	894
75JOC2981	Y. Ito, T. Hirao and T. Saegusa; *J. Org. Chem.*, 1975, **20**, 2981.	1072
75JOC2990	M. Vaultier, R. Danion-Bougot, D. Danion, J. Hamelin and R. Carrié; *J. Org. Chem.*, 1975, **40**, 2990.	494
75JOC3026	D. J. Burton and W. M. Koppes; *J. Org. Chem.*, 1975, **40**, 3026.	14
75JOC3227	R. C. Larock; *J. Org. Chem.*, 1975, **40**, 3227.	142
75JOC3450	J. B. Hendrickson and D. D. Sternbach; *J. Org. Chem.*, 1975, **40**, 3450.	363
75JOC3554	H. E. Baumgarten, H. L. Smith and A. Staklis; *J. Org. Chem.*, 1975, **40**, 3554.	999
75JOM(87)223	H. Brunner and M. Langer; *J. Organomet. Chem.*, 1975, **87**, 223.	448
75JOM(87)263	A. G. Brook and H. W. Kucera; *J. Organomet. Chem.*, 1975, **87**, 263.	410, 411
75JOM(88)237	S. Cenini, M. Pizzotti, F. Porta and G. La Monica; *J. Organomet. Chem.*, 1975, **88**, 237.	386
75JOM(92)139	T. Hirabayashi, K. Inoue, K. Yokota and Y. Ishii; *J. Organomet. Chem.*, 1975, **92**, 139.	578, 608, 618
75JOM(93)51	J.-P. Picard, A. Ekouya, J. Dunoguès, N. Duffaut and R. Calas; *J. Organomet. Chem.*, 1975, **93**, 51.	420
75JOM(93)325	E. J. Kupchik and J. A. Feiccabrino; *J. Organomet. Chem.*, 1975, **93**, 325.	390
75JOM(94)62	R. L. Bennett, M. I. Bruce and F. G. A. Stone; *J. Organomet. Chem.*, 1975, **94**, 62.	447
75JOM(97)39	E. J. Kupchik and H. E. Hanke; *J. Organomet. Chem.*; 1975, **97**, 39.	739
75JOM(99)379	A. G. Brook, J. W. Harris and A. R. Bassindale; *J. Organomet. Chem.*, 1975, **99**, 379.	421
75JOM(99)421	W. Malisch and P. Panster; *J. Organomet. Chem.*, 1975, **99**, 421.	912
75JOM(99)C5	I. Ojima, S. Inaba and Y. Nagai, *J. Organomet. Chem.*, 1975, **99**, C5.	777
75JOM(102)175	C. P. Casey and W. R. Brunsvold; *J. Organomet. Chem.*, 1975, **102**, 175.	955
75JOU1077	L. N. Kurkovskaya, N. N. Shapet'ko, N. B. Sokolova and I. Ya. Kvitko; *J. Org. Chem. USSR* (*Engl. Transl.*), 1975, 1077.	516, 521
75JOU1719	V. L. Dubina and V. I. Skripets; *J. Org. Chem. USSR* (*Engl. Transl.*), 1975, **8**, 1719.	712, 714
75JOU1926	B. S. Drach, V. A. Kovalev and T. Ya. Lavrenyuk; *J. Org. Chem. USSR* (*Engl. Transl.*), 1975, **11**, 1926.	685, 694
75LA266	L. Birkofer and W. Kaiser; *Justus Liebigs Ann. Chem.*, 1975, 266.	377
75LA275	W. Walter, T. Fleck, J. Voss and M. Gerwin; *Justus Liebigs Ann. Chem.*, 1975, 275.	618, 619
75LA305	W. Walter and O. H. Bauer; *Justus Liebigs Ann. Chem.*, 1975, 305.	624, 697
75LA1124	H. Hoberg and P. Bukowski; *Justus Liebigs Ann. Chem.*, 1975, **6**, 1124.	803
75LA1264	H. Behringer and R. Ramert; *Justus Liebigs Ann. Chim.*, 1975, 1264.	354
75LA1808	W. Walter, H.-W. Lüke and J. Voss; *Justus Liebig Ann. Chem.*, 1975, 1808.	645, 739
75LA1984	B. Zeeh and H. Kiefer; *Justus Liebigs Ann. Chem.*, 1975, 1984.	566
75LA2318	J. Curtze and K. Thomas; *Justus Liebigs Ann. Chem.*, 1975, 2318.	606
75M1019	W. Gottardi; *Monatsh. Chem.*, 1975, **106**, 1019.	320, 321
B-75MI 502-01	B. Cornils; in "Ullmanns Encykl. Tech. Chem., 4. Arefl.," eds. E. Bartholome, E. Biekert and H. Hellmann, Weinheim, Germany, 1975, vol. 10, pp. 135–143.	24
75MI 502-01	M. Bubner; *Zentrolinst. Kernforsch. Rossendorf Dresden* (*Ber.*), 1975, **294**, 86 (*Chem. Abstr.*, 1977, **86**, 120 120m).	24
75MI 505-01	R. L. Kirchmeier, G. H. Sprenger and J. M. Shreeve, *Inorg. Nucl. Chem. Lett.*, 1975, **11**, 699.	246
75MI 513-01	R. Haraoubia, J.-C. Gressier and G. Levesque; *Makromol. Chem.*, 1975, **176**, 2143.	561
B-75MI 517-01	G. Fodor and B. A. Phillips; in "The Chemistry of Functional Groups: The Chemistry of Amidines and Imidates," ed. S. Patai, Wiley, London, 1975, vol. 1, p. 85.	685
B-75MI 517-02	D. G. Neilson; in "The Chemistry of Functional Groups: The Chemistry of Amidines and Imidates," ed. S. Patai, Wiley, London, 1975, vol. 1, p. 385.	685, 687, 688, 690, 691, 692, 694
B-75MI 517-03	D. G. Neilson; in "The Chemistry of Functional Groups: The Chemistry of Amidines and Amidates," ed. S. Patai, Wiley, London, 1975, vol. 1, p. 462.	698
B-75MI 517-04	D. G. Neilson; in "The Chemistry of Functional Groups: The Chemistry of Amidines and Imidates," ed. S. Patai, Wiley, London, 1975, vol. 1, p. 385.	699
B-75MI 519-01	J.-A. Gautier, M. Miocque and C. C. Farnoux; in "The Chemistry of Amidines and Imidates," ed. S. Patai, Wiley, New York, 1975, vol. 1, p. 283.	742, 748, 751, 754, 755
B-75MI 519-02	D. G. Neilson; in "The Chemistry of Amidines and Imidates," ed. S. Patai, Wiley, New York, 1975, vol. 1, p. 385.	742, 769

B-75MI 519-03	K. M. Watson and D. G. Neilson; in "The Chemistry of Amidines and Imidates," ed. S. Patai, Wiley, New York, 1975, vol. 1, p. 491.	768, 769, 772, 773
B-75MI 521-01	D. G. Neilson; in "The Chemistry of the Functional Groups," ed. S. Patai, Wiley, London, 1975, p. 387.	826
B-75MI 521-02	K. M. Watson and D. G. Neilson; in "The Chemistry of Amidines and Imidates," ed. S. Patai, Wiley, New York, 1975, p. 491.	830, 847
B-75MI 521-03	J. Ševčik and F. Grambal; in "The Chemistry of Amidines and Imidates," ed. S. Patai, Wiley, London, 1975, p. 601.	864
B-75MI 525-01	R. P. A. Sneedon; "Organochromium Compounds," Academic Press, New York, 1975.	932
75MI 527-01	K. K. De, G. T. Shiau and R. E. Harmon; *J. Carbohydr. Nucleoside Nucleotide*, 1975, **2**, 171.	1040
75RTC83	J. Meijer, P. Vermeer and L. Brandsma; *Recl. Trav. Chim. Pays-Bas*, 1975, **94**, 83.	550
75S47	R. Zielke and H. Mägerlein; *Synthesis*, 1975, 47.	584, 589
75S291	T. Saegusa and Y. Ito; *Synthesis*, 1975, 291.	146, 692
75S332	R. W. Turner; *Synthesis*, 1975, 332.	1118, 1124
75S358	E. V. Brown; *Synthesis*, 1975, 358.	276, 599
75S456	M. Itoh, D. Hagiwara and J. Notani; *Synthesis*, 1975, 456.	125
75S510	T.-L. Ho; *Synthesis*, 1975, 510.	39
75S785	W. J. Stec, K. Lesiak and M. Sudol; *Synthesis*, 1975, 785.	1057
75S801	L. N. Markovski and V. E. Pashinnik; *Synthesis*, 1975, 801.	5
75SST65	G. C. Barrett; *Org. Compd. Sulphur, Selenium, Tellurium* (*R. Soc. Chem. series*), 1975, **3**, 65.	1022
75SST219	F. Duus; *Org. Compd. Sulphur, Selenium and Tellurium* (*R. Soc. Chem. Series*), 1975, **3**, 219.	472
75SST254	F. Duus; *Org. Compd. Sulphur, Selenium, Tellurium* (*R. Soc. Chem. Series*), 1975, **3**, 254.	565
75T171	M. Fetizon, M. Golfier and J.-M. Louis; *Tetrahedron*, 1975, **31**, 171.	171
75T597	A. M. van Leusen, B. A. Reith and D. van Leusen; *Tetrahedron*, 1975, **31**, 597.	867, 870
75T1209	J. Petermann and H. Plieninger; *Tetrahedron*, 1975, **31**, 1209.	577, 590, 608, 609
75T1245	T. Matsuura and Y. Ito; *Tetrahedron*, 1975, **31**, 1245.	584, 612
75TL51	J. R. Norton, K. E. Shenton and J. Schwartz; *Tetrahedron Lett.*, 1975, 51.	170
75TL71	J. W. Zubrick, B. I. Dunbar and H. D. Durst; *Tetrahedron Lett.*, 1975, 71.	1134
75TL277	J. B. Hendrickson and S. M. Schwartzman; *Tetrahedron Lett.*, 1975, 277.	123, 280, 291
75TL649	S. Tamagaki, S. Oae and K. Sakaki; *Tetrahedron Lett.*, 1975, 649.	349
75TL695	A. Krantz and B. Hoppe; *Tetrahedron Lett.*, 1975, 695.	1012, 1015
75TL981	K. A. Parker and E. G. Gibbons; *Tetrahedron Lett.*, 1975, 981.	272, 299, 304
75TL1219	B. Castro, J. R. Dormoy, G. Evin and C. Selve; *Tetrahedron Lett.*, 1975, 1219.	267
75TL1739	J. R. Grunwell and S. L. Dye; *Tetrahedron Lett.*, 1975, **16**, 1739.	201
75TL2689	Y. Yamamoto, K. Kondo and I. Moritani; *Tetrahedron Lett.*, 1975, **31**, 2689.	795, 796
75TL2701	K. Achiwa and S.-I. Yamada; *Tetrahedron Lett.*, 1975, **16**, 2701.	354
75TL3063	H. Tanida, R. Muneyuki and T. Tsushima; *Tetrahedron Lett.*, 1975, 3063.	588
75TL3125	J. C. Trisler, P. H. Deland and M. M. Goodgame; *Tetrahedron Lett.*, 1975, 3125.	1124
75TL3255	J. Wemple; *Tetrahedron Lett.*, 1975, 3255.	237
75TL3975	R. E. Ireland and A. K. Willard; *Tetrahedron Lett.*, 1975, **16**, 3975.	227
75USP3914252	R. D. Gordon; *US Pat.*, 3 914 252 (1975) (*Chem. Abstr.*, 1975, **85**, 5489).	321
75YZ980	Y. Tominaga, Y. Matsuda and G. Kobayashi; *Yakugaku Zasshi*, 1975, **95**, 980.	560
75ZAAC(415)115	H. Leimeister and K. Dehnicke; *Z. Anorg. Allg. Chem.*, 1975, **415**, 115.	316
75ZAAC(418)273	Von W. Siebert, R. Full, Th. Renk and A. Ospici; *Z. Anorg. Allg. Chem.*, 1975, **418**, 273.	1136
75ZC56	J. Bodeker and H. Zartner; *Z. Chem.*, 1975, **15**, 56.	1031
75ZC348	H. Viola and R. Mayer; *Z. Chem.*, 1975, **15**, 348.	476, 477
75ZOB477	V. F. Mironov, V. P. Kozyukov, A. S. Tkachev and E. K. Dobrovinskaya; *Zh. Obshch. Khim.*, 1975, **45**, 477.	1006
75ZOB949	V. E. Shishkin, Yu. M. Yukhno and B. I. No; *Zh. Obshch. Khim.*, 1975, **45**, 949.	722
75ZOB2007	V. F. Mironov, V. P. Kozyukov, A. D. Kirilin, V. D. Scheludyakov, Y. I. Dergunov and I. A. Vostokov; *Zh. Obshch. Khim.*, 1975, **45**, 2007.	978
75ZOB2025	I. A. Vostokov, Y. I. Dergonov, V. F. Mironov, D. O. Sheludyakov and V. P. Kozyukov; *Zh. Obshch. Khim.*, 1975, **45**, 2025.	968, 969
75ZOB2116	A. F. Tolochko, I. V. Megera, L. V. Zykova and M. I. Shevchuk; *Zh. Obshch. Khim.* 1975, **45**, 2116.	631
75ZOB2185	V. F. Mironov, M. V. Tsotadze, T. K. Gar and I. I. Gverdtsiteli; *Zh. Obshch. Khim.*, 1975, **45**, 2185.	969
75ZOB2684	Y. I. Dergunov and N. I. Mysin; *Zh. Obshch. Khim.*, 1975, **45**, 2684.	391
75ZOR173	B. I. Buzykin, L. P. Sysoeva and Y. P. Kitaev; *Zh. Org. Khim.*, 1975, **11**, 173.	356
75ZOR348	L. P. Sysoeva, B. I. Buzykin and Y. P. Kitaev; *Zh. Org. Khim.*, 1975, **11**, 348.	356
75ZOR1724	V. L. Dubina and V. I. Skripets; *Zh. Org. Khim.*, 1975, **8**, 1724.	712, 714
75ZOR1913	B. S. Drach, V. A. Kovalev and T. Ya. Lavrenyuk; *Zh. Org. Khim.*, 1975, **11**, 1913.	685, 694
75ZOR1981	N. S. Zefirov, N. K. Chapovskaya, L. Ya. D'yachova and S. S. Trach; *Zh. Org. Khim.*, 1975, **11**, 1981.	810
75ZOR2227	V. J. Gorbatenko and N. W. Mel'nichenko; *Zh. Org. Khim.*, 1975, **11**, 2227.	1008
76ACS(B)941	B. Lindgren; *Acta Chem. Scand., Ser. B*, 1976, **30**, 941.	1114
76AG(E)61	W.-H. Kunau; *Angew. Chem., Int. Ed. Engl.*, 1976, **15**, 61.	97
76AG(E)294	H. Poisel and U. Schmidt; *Angew. Chem., Int. Ed. Engl.*, 1976, **15**, 294.	315
76AG(E)313	D. Seebach and W. Lubosch; *Angew. Chem., Int. Ed. Engl.*, 1976, **15**, 313.	606

76AG(E)702 R. Appel, F. Knoll, H. Schöler and H.-D. Wihler; *Angew. Chem., Int. Ed. Engl.*, 1976, **15**, 702. 1093

76AG(E)774 T. L. Ho and G. A. Olah; *Angew. Chem., Int. Ed. Engl.*, 1976, **15**, 774. 225

76AJC357 L. H. Briggs, R. C. Cambie, I. C. Dean and P. S. Rutledge; *Aust. J. Chem.*, 1976, **29**, 357. 764

76AP(308)954 B. Unterhalt and D. Thamer; *Arch. Pharm.* (*Weinheim, Ger.*), 1975, **308**, 954. 94

76AR(B)199 E. W. Colvin; *Annu. Rep. Prog. Chem., Sect. B*, 1976, **72**, 199. 24

76BCF1119 H. Person and A. Foucaud; *Bull. Soc. Chim. Fr.*, 1976, 1119. 364

76BCJ464 K. Suzuki, J. Ooyama and M. Sakurai; *Bull. Chem. Soc. Jpn.*, 1976, **49**, 464. 1138

76BCJ546 T. Saruyama, T. Yamamoto and A. Yamamoto; *Bull. Chem. Soc. Jpn*, 1976, **49**, 546. 461

76BCJ3607 T. Fuchigami and K. Odo; *Bull. Chem. Soc. Jpn.*, 1976, **49**, 3607. 758, 762

76BEP831934 J. R. McCarthy, Jr.; *Belg. Pat.* 831 934 (1976) (*Chem. Abstr.*, 1976, **85**, 116 917). 751

76BSF120 D. Paquer and R. Pou; *Bull. Soc. Chim. Fr.*, 1976, 120. 634

76BSF301 P. Reynaud, J.-D. Brion and G. Ménard; *Bull. Soc. Chim. Fr.*, 1976, 301. 743

76BSF667 F. Raulin and G. Toupance; *Bull. Soc. Chim. Fr.*, 1976, 667. 728

76BSF1145 G. Levesque and J.-C. Gressier; *Bull. Soc. Chim. Fr.*, 1976, 1145. 547, 550

76C199 W. J. J. M. Sprangers, A. P. van Sweiten and R. Louw; *Chimia*, 1976, **30**, 199. 438

76CB41 H. Wolf, M. Kolleck, K. Claussen and W. Rascher; *Chem. Ber.*, 1976, **109**, 41. 49

76CB154 J. C. Jochims and A. Abu-Taha; *Chem. Ber.*, 1976, **109**, 154. 1046

76CB359 D. Fenske, E. Langer, M. Heymann and H. J. Becher; *Chem. Ber.*, 1976, **109**, 359. 400

76CB546 W. Disteldorf and M. Regitz; *Chem. Ber.*, 1976, **109**, 546. 871

76CB740 H. Gotthardt, M. C. Weisshuhn and B. Christl; *Chem. Ber.*, 1976, **109**, 740. 560

76CB906 E. Schaumann; *Chem. Ber.*, 1976, **109**, 906. 604

76CB956 C. O. Meese, W. Walter, H. Mrotzek and H. Mirzai; *Chem. Ber.*, 1976, **109**, 956. 624, 625, 732

76CB1120 K. Weiss and E. O. Fischer; *Chem. Ber.*, 1976, **109**, 1120. 949

76CB1128 E. O. Fischer and K. Weiss; *Chem. Ber.*, 1976, **109**, 1128. 949

76CB1309 D. Seebach, W. Lubosch and D. Enders; *Chem. Ber.*, 1976, **109**, 1309. 566, 584, 586, 590, 597, 608, 609

76CB1491 W. Lidy and W. Sundermeyer; *Chem. Ber.*, 1976, **109**, 1491. 994

76CB1779 U. Dabrowska and J. Dabrowski; *Chem. Ber.*, 1976, **109**, 1779. 534

76CB1868 K. Weiss and E. O. Fischer; *Chem. Ber.*, 1976, **109**, 1868. 953

76CB2039 G. Maas and M. Regitz; *Chem. Ber.*, 1976, **109**, 2039. 869, 871

76CB2542 W. Lidy and W. Sundermeyer; *Chem. Ber.*, 1976, **109**, 2542. 721, 722

76CB3243 G. Gattow and K. Hanewald; *Chem. Ber.*, 1976, **109**, 3243. 556

76CC240 J. A. Chapman, J. Crosby, C. A. Cummings and R. A. C. Rennie; *J. Chem. Soc., Chem. Commun.*, 1976, 240. 652

76CC522 G. Fachinetti, C. Floriani, F. Marchetti and S. Merlino; *J. Chem. Soc., Chem. Commun.*, 1976, 522. 442

76CC562 R. Okazaki, M. O-Oka, N. Inamoto; *J. Chem. Soc., Chem. Commun.*, 1976, 562. 634

76CC718 A. E. Baydar and G. V. Boyd; *J. Chem. Soc., Chem. Commun.*, 1976, 718. 367

76CC776 S. Takano, E. Yoshida, M. Hirama and K. Ogasawara; *J. Chem. Soc., Chem. Commun.*, 1976, 776. 616

76CC885 H. Blancou, P. Moreau and A. Commeyras; *J. Chem. Soc., Chem. Commun.*, 1976, 885. 58

76CC907 T. F. Murray, V. Varma and J. R. Norton; *J. Chem. Soc., Chem. Commun.*, 1976, 907. 170

76CC1008 S. J. Harris and D. R. M. Walton; *J. Chem. Soc., Chem. Commun.*, 1976, 1008. 511, 589

76CC1032 C. F. J. Barnard, J. A. Daniels and R. J. Mawby; *J. Chem. Soc., Chem. Commun.*, 1976, 1032. 455

76CCC1388 M. Dzurilla and P. Kristian; *Collect. Czech. Chem. Commun.*, 1976, **41**, 1388. 1049

76CCC3085 E. Jedlovská, J. Kovác, A. Piklerová and P. Zálupsky; *Collect. Czech. Chem. Commun.*, 1976, **41**, 3085. 665

76CJC647 P. E. Titus, P. A. Forte and J. M. Prokipcak; *Can. J. Chem.*, 1976, **54**, 647. 689

76CJC2652 J. F. King and D. R. K. Harding; *Can. J. Chem.*, 1976, **54**, 2652. 488, 489

76CL57 T. Mukaiyama, Y. Aikawa and S. Kobayashi; *Chem. Lett.*, 1976, 57. 291

76CL203 H. Ishihara and Y. Hirabayashi; *Chem. Lett.*, 1976, 203. 240, 252, 541

76CL267 T. Mukaiyama and K. Hojo; *Chem. Lett.*, 1976, 267. 125

76CL741 Y. Watanabe, S. Shoda and T. Mukaiyama; *Chem. Lett.*, 1976, 741. 234

76CL785 A. Arase and Y. Masuda; *Chem. Lett.*, 1976, 785. 1114

76CL811 S. Sakai, T. Fujinami, N. Otani and T. Aizawa; *Chem. Lett.*, 1976, 811. 1078

76CL993 I. Kuwajima, T. Abe and N. Minami; *Chem. Lett.*, 1976, 993. 412, 422

76CL1115 A. Arase and Y. Masuda; *Chem. Lett.*, 1976, 1115. 1114

76CPL(42)460 M. J. Hopkinson, H. W. Kroto, J. F. Nixon and N. P. C. Simmons; *Chem. Phys. Lett.*, 1976, **42**, 460. 1152

76CR(C)483 H. Normand and C. Laurenco; *C. R. Hebd. Seances Acad. Sci., Ser. C*, 1976, **283**, 483. 132, 139

76CRV389 J. K. Rasmussen and A. Hassner; *Chem. Rev.*, 1976, **76**, 389. 344

76CSR181 N. S. Isaacs; *Chem. Soc. Rev.*, 1976, **5**, 181. 652

76CZ3 H. Liebig and K. Uem; *Chem.-Ztg.*, 1976, **100**, 3. 54

76CZ340 V. K. Burger, K. Einhellig and M. Frohler; *Chem.-Ztg.*, 1976, **100**, 340. 789

76DOK(230)853 M. G. Voronkov, N. F. Chernov and T. A. Dekina; *Dokl. Acad. Nauk SSSR*, 1976, **230**, 853. 1142

76GEP(O)2539048 Shell; *Ger. Pat. Offen.*, 2 539 048 (1976) (*Chem. Abstr.*, 1977, **85**, 20 680). 145

76HCA350 V. I. Cohen; *Helv. Chim. Acta.*, 1976, **59**, 350. 631

76HCA1307 P. Dubs and R. Stüssi; *Helv. Chim. Acta*, 1976, **59**, 1307. 247

76HCA2100 H. Loibner and E. Zbiral; *Helv. Chim. Acta*, 1976, **59**, 2100. 1107

76IC208 R. A. Schunn; *Inorg. Chem.*, 1976, **15**, 208. 1139, 1140

76IC2346 M. Wada and K. Oguro; *Inorg. Chem.*, 1976, **15**, 2346. 1137
76ICA-L64 K. K. Bajpai and B. Bajpai; *Inorg. Chim. Acta*, 1976, **19**, L64. 1143
76IJC794 S. R. Ramadas and P. S. Srinivasan; *Indian J. Chem.*, 1976, **14B**, 794. 550
76IJS519 I. D. Rae and M. J. Wade; *Int. J. Sulfur Chem.*, 1976, **8**, 519. 624, 625, 626
76JA262 C. A. Bertelo and J. Schwartz; *J. Am. Chem. Soc.*, 1976, **98**, 262. 442
76JA437 C. P. Casey and C. A. Bunnell; *J. Am. Chem. Soc.*, 1976, **98**, 437. 448
76JA441 T. F. Block, R. F. Fenske and C. P. Casey; *J. Am. Chem. Soc.*, 1976, **98**, 441. 932
76JA1099 J. H. Boyer and J. Kooi; *J. Am. Chem. Soc.*, 1976, **98**, 1099. 658
76JA1166 C. P. Casey, C. A. Bunnell and J. C. Calabrese; *J. Am. Chem. Soc.*, 1976, **98**, 1166. 447
76JA1278 M. W. Grayston and D. M. Lamal; *J. Am. Chem. Soc.*, 1976, **98**, 1278. 866
76JA1583 G. Stork and S. Raucher; *J. Am. Chem. Soc.*, 1976, **98**, 1583. 159
76JA1743 J. P. Behr and J. M. Lehn; *J. Am. Chem. Soc.*, 1976, **98**, 1743. 114
76JA2352 L. E. Overman and L. A. Clizbe; *J. Am. Chem. Soc.*, 1976, **98**, 2352. 271, 288
76JA2365 C. M. Lukehart and J. V. Zeile; *J. Am. Chem. Soc.*, 1976, **98**, 2365. 450
76JA2868 R. E. Ireland, R. H. Mueller and A. K. Willard; *J. Am. Chem. Soc.*, 1976, **98**, 2868. 227
76JA3381 A. G. Brook and J. W. Harris; *J. Am. Chem. Soc.*, 1976, **98**, 3381. 924
76JA5832 K. S. Y. Lau, P. K. Wong and J. K. Stille; *J. Am. Chem. Soc.*, 1976, **98**, 5832. 143
76JA6529 J. X. McDermott, M. E. Wilson and G. M. Whitesides; *J. Am. Chem. Soc.*, 1976, **98**, 6529. 441
76JA6545 C. A. Burton and J. M. Shreeve; *J. Am. Chem. Soc.*, 1976, **98**, 6545. 248
76JA7084 S. Takano, M. Hirama, T. Araki and K. Ogasawara; *J. Am. Chem. Soc.*, 1976, **98**, 7084. 607
76JA7874 S. Masamune, Y. Hayase, W. K. Chan and R. L. Sobczak; *J. Am. Chem. Soc.*, 1976, **98**, 7874. 161
76JCS(D)203 G. Fachinetti, S. Del Nero and C. Floriani; *J. Chem. Soc., Dalton Trans.*, 1976, 203. 444
76JCS(D)327 S. Otsuka and K. Ataka; *J. Chem. Soc., Dalton Trans.*, 1976, 327. 801, 802
76JCS(D)366 J. E. Drake, R. T. Hemmings and E. Henderson; *J. Chem. Soc., Dalton Trans.*, 1976, 366. 1081
76JCS(D)1661 S. Cradock, E. A. V. Ebsworth, D. W. H. Rankin and W. J. Savage; *J. Chem. Soc., Dalton Trans.*, 1976, 1661. 538, 541
76JCS(D)1777 G. La Monica, S. Ceinini, F. Porta and M. Pizzotti; *J. Chem. Soc., Dalton Trans.*, 1976, 1777. 386
76JCS(P1)33 M. W. Gittos, M. R. Robinson, J. P. Verge, R. V. Davies, B. Iddon and H. Suschitzky; *J. Chem. Soc., Perkin Trans. 1*, 1976, 33. 593
76JCS(P1)38 D. H. R. Barton, J. W. Ducker, W. A. Lord and P. D. Magnus; *J. Chem. Soc., Perkin Trans. 1*, 1976, 38. 846
76JCS(P1)291 A. Albert; *J. Chem. Soc., Perkin Trans. 1*, 1976, 291. 690
76JCS(P1)564 M. Mikolajczyk, P. Kielbasiński and H. M. Schiebel; *J. Chem. Soc., Perkin Trans. 1*, 1976, 564. 243, 540
76JCS(P1)716 P. Barraclough, M. Edwards, T. L. Gilchrist and C. J. Harns; *J. Chem. Soc., Perkin Trans. 1*, 1976, 716. 1126
76JCS(P1)983 A. C. W. Curran and R. G. Shepherd; *J. Chem. Soc., Perkin Trans. 1*, 1976, 983. 595
76JCS(P1)1150 D. W. Jones; *J. Chem. Soc., Perkin Trans. 1*, 1976, 1150. 365
76JCS(P1)1708 S. Bittner and S. Grinberg; *J. Chem. Soc., Perkin Trans. 1*, 1976, 1708. 219
76JCS(P1)2079 T. G. Back, D. H. R. Barton, M. R. Britten-Kelly and F. S. Guziec, Jr.; *J. Chem. Soc., Perkin Trans. 1*, 1976, 2079. 1053
76JCS(P1)2112 D. H. R. Barton, R. V. Stick and R. Subramanian; *J. Chem. Soc. Perkin Trans. 1*, 1976, 2112. 516, 522, 526
76JCS(P1)2166 T. L. Gilchrist, C. J. Harris, D. G. Hawkins, C. J. Moody and C. W. Rees; *J. Chem. Soc., Perkin Trans. 1*, 1976, 2166. 764
76JCS(P1)2205 K. Takeuchi, M. Nojima and N. Tokura; *J. Chem. Soc., Perkin Trans. 1*, 1976, 2205. 282
76JCS(P1)2452 M. Evers, R. Weber, P. Thibaut, L. Christiaens, M. Renson, A. Croisy and P. Jacquignon; *J. Chem. Soc., Perkin Trans. 1*, 1976, 2452. 239
76JCS(P2)524 I. Bacaloglu, R. Bacaloglu, H. Glatt, C. Viragh and G. Ostrogovich; *J. Chem. Soc., Perkin Trans. 2*, 1976, 524. 1126
76JCS(P2)1895 W. J. J. M. Sprangers and R. Louw; *J. Chem. Soc., Perkin Trans. 2*, 1976, 1895. 438
76JFC(7)65 A. F. Clifford and A. Shanzer; *J. Fluorine Chem.*, 1976, **7**, 65. 1085
76JGU239 V. P. Kukhar and T. N. Kasheva; *J. Gen. Chem. USSR (Engl. Transl.)*, 1976, **46**, 239. 683
76JGU714 Yu. I. Dergunov, V. F. Gerega, V. G. Vodop'yanov and E. N. Boitsov; *J. Gen. Chem. USSR (Engl. Transl.)*, 1976, **46**, 714. 716
76JHC487 K. Babič, S. Molan, S. Polanc, B. Stanovnik, J. Stres-Bratoš, M. Tišler and B. Verček; *J. Heterocycl. Chem.*, 1976, **13**, 487. 764
76JIC1230 R. P. Verma and T. N. Misra; *J. Indian Chem. Soc.*, 1976, **53**, 1230. 384
76JOC148 A. E. Feiring; *J. Org. Chem.*, 1976, **41**, 148. 592, 593
76JOC252 J. E. Johnson, E. A. Nalley, Y. K. Kunz and J. R. Springfield; *J. Org. Chem.*, 1976, **41**, 252. 824
76JOC725 J. Auerbach, M. Zamore and S. M. Weinreb; *J. Org. Chem.*, 1976, **41**, 725. 273, 296
76JOC729 F. Malek-Yazdi and M. Yalpani; *J. Org. Chem.*, 1976, **41**, 729. 516, 519, 522
76JOC793 J. F. Knifton; *J. Org. Chem.*, 1976, **41**, 793. 142
76JOC962 E. P. Papadopoulos; *J. Org. Chem.*, 1976, **41**, 962. 592, 593, 620, 631
76JOC2071 R. L. Blankespoor and C. M. Snavely; *J. Org. Chem.*, 1976, **41**, 2071. 972
76JOC2102 A. Hassner, D. Tang and J. Keogh; *J. Org. Chem.*, 1976, **41**, 2102. 905
76JOC2639 L. Benati and P. C. Montevecchi; *J. Org. Chem.*, 1976, **41**, 2639. 561
76JOC3233 B. George and E. P. Papadopoulos; *J. Org. Chem.*, 1976, **41**, 3233. 631
76JOC3769 J. H. Hall and M. Gisler; *J. Org. Chem.*, 1976, **41**, 3769. 275, 298
76JOC4029 J. J. Fitt and H. W. Gschwend; *J. Org. Chem.*, 1976, **41**, 4029. 607
76JOM(107)63 A. Roubineau and J. C. Pommier; *J. Organomet. Chem.*, 1976, **107**, 63. 388

76JOM(107)229	E. M. Dexheimer and L. Spialter; *J. Organomet. Chem.*, 1976, **107**, 229.	416
76JOM(110)195	P. Krommes and J. Lorberth; *J. Organometallic Chem.*, 1976, **110**, 195.	871
76JOM(112)209	D. F. Christian, H. C. Clark and R. F. Stepaniak; *J. Organomet. Chem.*, 1976, **112**, 209.	802
76JOM(114)281	N. Petragnani, R. Rodrigues and J. V. Comasseto; *J. Organomet. Chem.*, 1976, **114**, 281.	903
76JOM(118)309	C. P. Casey and W. R. Brunsvold; *J. Organomet. Chem.*, 1975, **118**, 309.	955
76JOU60	V. L. Dubina, V. P. Pedan and I. K. Dembinskii; *J. Org. Chem. USSR* (*Engl. Transl.*), 1976, **12**, 60.	694
76JOU663	B. S. Drach, V. A. Kovalev and A. V. Kirsanov; *J. Org. Chem. USSR* (*Engl. Transl.*), 1976, **12**, 663.	657
76JOU1789	I. M. Dolgopol'skii, G. B. Fedorova, M. I. Bognitskii and L. G. Parshina; *J. Org. Chem. USSR* (*Engl. Transl.*), 1976, **12**, 1789.	694
76JOU2148	Yu. L. Yagupol'skii, B. K. Kerzhner and L. M. Yagupol'skii; *J. Org. Chem. USSR* (*Engl. Transl.*), 1976, **12**, 2148.	657
76JOU2252	B. S. Drach and V. A. Kovalev; *J. Org. Chem. USSR* (*Engl. Transl.*), 1976, **12**, 2252.	655, 657, 659
76JPR272	V. H. Gross, L. Brennecke and B. Costisella; *J. Prakt. Chem.*, 1976, **318**, 272.	788
76JPR343	K. Gewald and M. Hentschel; *J. Prakt. Chem.*, 1976, **318**, 343.	593
76JPR409	V. A. Tzschach and J. Heinicke; *J. Prakt. Chem.*, 1976, **318**, 409.	790
76KFZ41	N. N. Suvorov, L. K. Vinograd, G. N. Pershin, T. N. Zykova, T. A. Gus'kova and N. F. Lomanova; *Khim. Farm. Zh.*, 1976, **10**, 41 (*Chem. Abstr.*, 1976, **85**, 62 904m).	567, 581
76LA1584	W. Walter and O. H. Bauer; *Justus Liebigs Ann. Chem.*, 1976, 1584.	623, 624
B-76MI 501-01	J. A. Scheben and I. L. Mador; in "5th Conference on Catalysis in Organic Synthesis, 1975," ed. P. N. Rylander and H. Greenfield, Academic Press, New York, 1976, p. 181 (*Chem. Abstr.*, 1976, **85**, 176 785).	16
B-76MI 502-01	B. J. Coffin and C. N. C. Drey; in "Rodd's Chemistry of Carbon Compounds," ed. S. Coffey, Elsevier, Amsterdam, 2nd Ed., 1976, vol. 1, Part E., pp. 83–215 and 217–348.	62
76MI 502-01	D. Ivanov, B. Blagoev, L. Yavashev and A. Kabakov; *Doklady Bolgarsko Akad. Nauk.*, *Compt. Rend. Acad. Bulg.*, 1976, **29**, 1459 (*Chem. Abstr.*, 1977, **86**, 106 122g).	98
B-76MI 502-02	I. I. Lapkin and V. N. Kashinskii; in "Org. Khim.," ed. I. I. Lapkin, 1976, p. 1, Permsk. Gos. Univ. im., A. M. Gorkogo: Perm. USSR (*Chem. Abstr.*, 1978, **88**, 152 001f).	62
B-76MI 502-03	J. Grimshaw; in "Rodd's Chemistry of Carbon Compounds," ed. S. Coffey, Elsevier, Amsterdam, 2nd edn., 1976, vol. 3 Part D, pp. 141.	105
76MI 503-01	B. Boutevin and Y. Pietrasanata; *Eur. Polym. J.*, 1976, **231**, 231.	122
76MI 512-01	J. E. Drake, R. T. Hemmings and H. E. Henderson; *Inorg. Nucl. Chem. Lett.*, 1976, **12**, 563.	540
76MI 526-01	W. D. Fuller, M. S. Verlander and M. Goodman; *Biopolymers*, 1976, **15**, 1869.	972
76MI 527-01	G. I. Tesser and J. J. M. Lamberts; *Int. J. Peptide Protein Res.*, 1976, **8**, 559.	1028
76PS75	T. Matsuda and Y. Takada; *Phosphorus Sulfur*, 1976, **1**, 75.	601, 602
76RCR1035	E. Jansons; *Rus. Chem. Rev.* (*Engl. Transl.*), 1976, **45**, 1035.	545
76S32	S. R. Jones and J. M. Mellor; *Synthesis*, 1976, 32.	282
76S46	E. Vilsmaier and R. Bayer; *Synthesis*, 1976, 46.	689
76S113	R. Brehme; *Synthesis*, 1976, 113.	285
76S123	A. C. Ranade, R. S. Mali, S. R. Bhide and S. R. Mehta; *Synthesis*, 1976, 123.	110
76S133	A. J. Fatiadi; *Synthesis*, 1976, 133.	48
76S241	B. Unterhalt and D. Thamer; *Synthesis*, 1976, 241.	361
76S259	G. Simchen and W. Kober; *Synthesis*, 1976, 259.	718
76S308	M. Haake and H. Benack; *Synthesis*, 1976, 308.	339
76S329	H. R. Kricheldorf and E. Leppert; *Synthesis*, 1976, 329.	992
76S519	L. Henriksen and U. Ehrbar; *Synthesis*, 1976, 519.	1055
76S728	V. I. Cohen, P. B. Mokhtary and N. Hamraz; *Synthesis*, 1976, 728.	631
76S736	E. I. Sanchez and M. J. Fumarola; *Synthesis*, 1976, 736.	318
76S750	Y. Yamamoto and H. Yamazaki; *Synthesis*, 1976, 750.	356
76S811	M. A. Neirabeyeh, J.-C. Ziegler and B. Gross; *Synthesis*, 1976, 811.	141
76SC387	E. Campaigne, G. Skowronski, R. A. Forsch and J. C. Beckman; *Synth. Commun.*, 1976, **6**, 387.	597
76SC453	J. R. Grunwell and D. L. Foerst; *Synth. Commun.*, 1976, **6**, 453.	234
76SRI217	J. M. Coronas, G. Muller and J. Sales; *Synth. React. Inorg. Metal-Org. Chem.*, 1976, **6**, 217.	1138
76SRI345	E. J. Kupchik and D. K. Parikh; *Synth. React. Inorg. Metal-Org. Chem.*, 1976, **6**, 345.	389, 1121
76T745	C. Christophersen and P. Carlsen; *Tetrahedron*, 1976, **32**, 745.	1112
76T1031	K. M. Doyle and F. Kurzer; *Tetrahedron*, 1976, **32**, 1031.	769
76T1599	R. V. Stevens; *Tetrahedron*, 1976, **32**, 1599.	821
76T2507	D. A. Konen, P. E. Pfeffer and L. S. Silbert; *Tetrahedron*, 1976, **32**, 2507.	555
76TL143	H. A. Houwing, J. Wildeman and A. M. van Leusen; *Tetrahedron Lett.*, 1976, 143.	566, 618
76TL209	A. C. Brouwer and H. J. T. Bos; *Tetrahedron Lett.*, 1976, 209.	559, 605, 614
76TL451	S. Sarel, A. Felzenstein and J. Yovell; *Tetrahedron Lett.*, 1976, **17**, 451.	344
76TL609	I. Hoppe, D. Hoppe and U. Schollkopf; *Tetrahedron Lett.*, 1976, 609.	634
76TL967	R. G. Micetich, C. G. Chin and R. B. Morin; *Tetrahedron Lett.*, 1976, 967.	577
76TL1591	I. Kuwajima, T. Sato, N. Minami and T. Abe; *Tetrahedron Lett.*, 1976, **19**, 1591.	414
76TL1839	D. Pocar, R. Stradi and B. Gioia; *Tetrahedron Lett.*, 1976, 1839.	762
76TL1951	B. Ganem; *Tetrahedron Lett.*, 1976, **17**, 1951.	324
76TL2577	R. A. Gorski, G. J. Wolber and J. Wemple; *Tetrahedron Lett.*, 1976, 2577.	237
76TL2691	J. B. Bapat, R. J. Blade, A. J. Boulton, J. Epsztajn, A. R. Katritzky, J. Lewis, P. Molina-Buendía, P. Nie and C. A. Ramsden; *Tetrahedron Lett.*, 1976, 2691.	1002
76TL2783	K. Oka and S. Hara; *Tetrahedron Lett.*, 1976, **17**, 2783.	485, 583
76TL3089	L. E. Overman, G. F. Taylor and P. J. Jessup; *Tetrahedron Lett.*, 1976, 3089.	288

76TL3347 J. Becher and E. G. Frandsen; *Tetrahedron Lett.*, 1976, 3347. 593

76TL3489 G. Märkl and F. Lieb; *Tetrahedron Lett.*, 1976, 3489. 882

76TL4423 G. S. Bates, J. Diakur and S. Masamune; *Tetrahedron Lett.*, 1976, 4423. 234, 239

76URP534448 V. L. Dubina, V. I. Skripets and T. G. Shemonaeva; *USSR Pat* 534,448 (1976) (*Chem. Abstr.*, 1977, **86**, 121 004p). 674

76USP3954441 B. Cross and C. P. Grasso (American Cyanamid); *US Pat.*, 3 954 441 (1976) (*Chem. Abstr.*, 1976, **85**, 159 725). 965

76ZAAC(422)104 G. Fritz and W. Hölderich; *Z. Anorg. Allg. Chem.*, 1976, **422**, 104. 1155

76ZAAC(423)242 G. Becker; *Z. Anorg. Allg. Chem.*, 1976, **423**, 242. 878, 879

76ZAAC(426)288 K.-H. Flegler and A. Haas; *Z. Anorg. Allg. Chem.*, 1976, **426**, 288. 1141

76ZC270 J. Faust; *Z. Chem.*, 1976, **16**, 270. 532

76ZC318 T. Kissel, R. Matusch and K. Hartke; *Z. Chem.*, 1976, **16**, 318. 554

76ZC355 H. Viola and R. Mayer; *Z. Chem.*, 1976, **16**, 355. 550

76ZN(B)281 L. Roesch, W. Erb and H. Mueller; *Z. Naturforsch., Teil B*, 1976, **316**, 281. 793

76ZN(B)1546 W. Maringgele and A. Meller; *Z. Naturforsch., Teil B*, 1976, **31**, 1546. 380

76ZOB75 T. N. Sumarokova, I. G. Litvyak, E. M. Vorob'eva and L. A. Arykova; *Zh. Obshch. Khim.*, 1976, **46**, 75. 389

76ZOB243 V. P. Kukhar and T. N. Kasheva; *Zh. Obshch. Khim.*, 1976, **46**, 243 (*Chem. Abstr.*, 1976, **84**, 180 176v). 683

76ZOB594 S. A. Lebedev, L. L. Gervits, S. V. Ponomarev and I. F. Lutsenko; *Zh. Obshch. Khim.*, 1976, **46**, 594. 376, 389

76ZOB714 Yu. I. Dergunov, V. F. Gerega, V. G. Vodop'yanov and E. N. Boitsov; *Zh. Obshch. Khim.*, 1976, **46**, 714 (*Chem. Abstr.*, 1976, 85, 5818n). 716

76ZOB1531 V. P. Kozyukov, E. K. Dobrovinskaya and V. F. Mironov; *Zh. Obshch. Khim.*, 1976, **46**, 1531 (*Chem. Abstr.*, 1976, **85**, 177 525f). 1006

76ZOR64 V. L. Dubina, V. P. Pedan and I. K. Dembinskii; *Zh. Org. Khim.*, 1976, **12**, 64 (*Chem. Abstr.*, 1976, **84**, 135 254r). 694

76ZOR69 I. M. Bazavova, R. G. Dubenko and P. S. Pel'kis; *Zh. Org. Khim.*, 1976, **12**, 69 (*Chem. Abstr.*, 1976, **84**, 105 468w). 593

76ZOR673 B. S. Drach, V. A. Kovalev and A. V. Kirsanov; *Zh. Org. Khim.*, 1976, **12**, 673 (*Chem. Abstr.*, 1976, **85**; 21 300t). 657

76ZOR681 I. M. Bazavova, R. G. Dubenko and P. S. Pel'kis; *Zh. Org. Khim.*, 1976, **12**, 681 (*Chem. Abstr.*, 1976, **85**, 32 785f). 593

76ZOR1557 L. P. Sysoeva, B. I. Buzykin and Y. P. Kitaev; *Zh. Org. Khim.*, 1976, **12**, 1557. 652

76ZOR1826 I. M. Dolgopol'skii, G. B. Fedorova, M. I. Bognitskii and L. G. Parshina; *Zh. Org. Khim.*, 1976, **12**, 1826 (*Chem. Abstr.*, 1976, **85**, 159 364y). 694

76ZOR1963 V. I. Gorbatenko, L. F. Lurè and L. I. Samarai; *Zh. Org. Khim.*, 1976, **12**, 1963 (*Chem. Abstr.*, 1977, **86**, 16 269f). 1008

76ZOR2213 Y. L. Yagupol'skii, B. K. Kerzhner and L. M. Yagupol'skii; *Zh. Org. Khim.*, 1976, **12**, 2213 (*Chem. Abstr.*, 1977, **86**, 71 870v). 566, 567, 610, 657

76ZOR2319 B. S. Drach and V. A. Kovalev; *Zh. Org. Khim.*, 1976, **12**, 2319 (*Chem. Abstr.*, 1977, **86**, 89 771y). 655, 657, 659

76ZOR2472 V. I. Gorbatenko, V. N. Fetyukhin and L. I. Samarai; *Zh. Org. Khim.*, 1976, **12**, 2472. 1068

76ZOR2473 V. I. Gorbatenko, L. F. Lurè and L. I. Samarai; *Zh. Org. Khim.*, 1976, **12**, 2473 (*Chem. Abstr.*, 1977, **86**, 72 093f). 1008

77ACS(A)884 P. Klaeboe, C. J. Nielsen and J. Songstad; *Acta Chem. Scand., Ser. A*, 1977, **31**, 884. 1117

77ACS(B)1 B. Lindgren; *Acta Chem. Scand., Ser. B*, 1977, **31**, 1. 1114

77ACS(B)261 E. B. Pedersen; *Acta Chem. Scand., Ser. B*, 1977, **31**, 261. 745

77ACS(B)460 J. Perregaard, B. S. Pedersen and S.-O. Lawesson; *Acta Chem. Scand., Ser. B*, 1977, **31**, 460. 551

77AG743 C.-P. Klages and J. Voss; *Angew. Chem.*, 1977, **89**, 743. 523

77AG747 A. Schmidpeter and F. Zwaschka; *Angew. Chem.*, 1977, **89**, 747. 1128

77AG(E)180 R. Appel and W. Morbach; *Angew. Chem., Int. Ed. Engl.*, 1977, **16**, 180. 894

77AG(E)244 H.-J. Gais; *Angew. Chem., Int. Ed. Engl.*, 1977, **16**, 244. 239

77AG(E)264 D. Seebach, B. Seuring, H.-O. Kalinowski, W. Lubosch and B. Renger; *Angew. Chem., Int. Ed. Engl.*, 1977, **16**, 264. 177

77AG(E)269 E. Lindner, H.-D. Ebert, H. Lesiecki and G. Vordermaier; *Angew. Chem. Int. Ed. Engl.*, 1977, **16**, 269. 397

77AG(E)299 F. Calderazzo; *Angew. Chem., Int. Ed. Engl.*, 1977, **16**, 299. 441

77AG(E)311 I. Ruppert; *Angew. Chem., Int. Ed. Engl.*, 1977, **4**, 311. 1082

77AG(E)318 G. Heppke, J. Martens, K. Praefcke and H. Simon; *Angew. Chem., Int. Ed. Engl.*, 1977, **16**, 318. 238

77AG(E)349 H. J. Bestmann; *Angew. Chem., Int. Ed. Engl.*, 1977, **16**, 349. 1092

77AG(E)417 A. Wohlleben-Hammel and H. Schmidbaur; *Angew. Chem., Int. Ed. Engl.*, 1977, **16**, 417. 1094

77AG(E)463 G. Becker and G. Gutekunst; *Angew. Chem. Int. Ed. Engl.*, 1977, **16**, 463. 399

77AG(E)550 H. R. Kricheldorf, G. Schwarz and J. Kaschig; *Angew. Chem., Int. Ed. Engl.*, 1977, **16**, 550. 993

77AG(E)555 G. Trickes and H. Meier; *Angew. Chem., Int. Ed. Engl.*, 1977, **16**, 555. 1003, 1013

77AG(E)639 H. Schmidbaur and B. Zimmer-Gasser; *Angew. Chem., Int. Ed. Engl.*, 1977, **16**, 639. 913

77AG(E)725 C.-P. Klages and J. Voss; *Angew. Chem., Int. Ed. Engl.*, 1977, **16**, 725. 623

77AG(E)728 H. Gölz, B. Glatz, G. Haas, G. Helmchen and H. Muxfeldt; *Angew. Chem., Int. Ed. Engl.*, 1977, **16**, 728. 209

77AG(E)743 H. Hagemann; *Angew. Chem., Int. Ed. Engl.*, 1977, **16**, 743. 1007, 1008

77AG(E)786	S. Kato, T. Takagi, T. Katada and M. Mizuta; *Angew. Chem., Int. Ed. Engl.*, 1977, **16**, 786.	551
77AG(E)787	S. Kato, A. Hori, T. Takagi and M. Mizuta; *Angew. Chem., Int. Ed. Engl.*, 1977, **16**, 787.	609
77AP(310)614	R. Neidlein and W. Friederich; *Arch. Pharm.* (*Weinheim, Ger.*), 1977, **310**, 614.	596
77AP(310)1001	D. Matthies and K. Hain; *Arch. Pharm.* (*Weinheim, Ger.*), 1977, **310**, 1001.	606
77BAU206	I. L. Knunyants, L. T. Lantseva, E. P. Lur'e and Y. V. Zeifman; *Bull. Acad. Sci. USSR, Div. Chim. Sci.*, 1974, 206.	475
77BCJ220	S. Uemura, S. Tanaka and M. Okano; *Bull. Chem. Soc. Jpn.*, 1977, **50**, 220.	251
77BCJ425	S. Sakai, T. Fujinami and T. Aizawa; *Bull. Chem. Soc. Jpn.*, 1977, **50**, 425.	1035
77BCJ553	J. Tsuji and H. Yasuda; *Bull. Chem. Soc. Jpn.*, 1977, **50**, 553.	158
77BCJ722	S. Tanaka, S. Uemura and M. Okano; *Bull. Chem. Soc. Jpn.*, 1977, **50**, 722.	1032, 1056
77BCJ1639	A. Kasahara, T. Izumi and A. Suzuki; *Bull. Chem. Soc. Jpn.*, 1977, **50**, 1639.	142
77BCJ1863	K. Saigo, M. Usui, K. Kikuchi, E. Shimada and T. Mukaiyama; *Bull. Chem. Soc. Jpn.*, 1977, **50**, 1863.	125
77BCJ2021	T. Kawamura, K. Kikukawa, M. Takagi and T. Matsuda; *Bull. Chem. Soc. Jpn.*, 1977, **50**, 2021.	222
77BCJ2406	S. Ozaki, Y. Ike, H. Mizuno, K. Ishikawa and H. Mori; *Bull. Chem. Soc. Jpn.*, 1977, **50**, 2406.	972
77BCJ2785	S. Tanaka, S. Uemura and M. Okano; *Bull. Chem. Soc. Jpn.*, 1977, **50**, 2785.	1036
77BCJ3007	H. Ishihara, S. Sato and Y. Hirabayashi; *Bull. Chem. Soc. Jpn.*, 1977, **50**, 3007.	240, 244, 252
77BRP1466047	J. R. McCarthy, Jr.; *Br. Pat.* 1 466 047 (1977) (*Chem. Abstr.*, 1977, **87**, 117 663).	751
77BSB321	J. Perregaard, I. Thomsen and S.-O. Lawesson; *Bull. Soc. Chim. Belg.*, 1977, **86**, 321.	570, 575, 608, 610, 618
77BSB693	B. S. Pedersen and S.-O. Lawesson; *Bull. Soc. Chim. Belg.*, 1977, **86**, 693.	570, 575, 608, 610, 618
77CB285	J. Goerdeler, R. Büchler and S. Sólyom; *Chem. Ber.*, 1977, **110**, 285.	595, 597, 599
77CB677	H. Schmidbaur, J. Eberlein and W. Richter; *Chem. Ber.*, 1977, **110**, 677.	920, 922
77CB805	E. O. Fischer, A. Ruhs and F. R. Kreissl; *Chem. Ber.*, 1977, **110**, 805.	919
77CB841	R. Bürstinghaus and D. Seebach; *Chem. Ber.*, 1977, **110**, 841.	411
77CB1140	E. O. Fischer, E. W. Meineke and F. R. Kreissl; *Chem. Ber.*, 1977, **110**, 1140.	942
77CB1225	G. Seybold and C. Heibl; *Chem. Ber.*, 1977, **110**, 1225.	511
77CB1312	W. Richter, Y. Yamamoto and H. Schmidbaur; *Chem. Ber.*, 1977, **110**, 1312.	922
77CB2114	K. Burger, R. Ottlinger and J. Albanbauer; *Chem. Ber.*, 1977, **110**, 2114.	634
77CB2574	E. O. Fischer, T. Selmayr, F. R. Kreissl and U. Schubert; *Chem. Ber.*, 1977, **110**, 2574.	936
77CB2947	E. O. Fischer, T. Selmayr and F. R. Kreissl; *Chem. Ber.*, 1977, **110**, 2947.	949
77CB2996	J. Goerdeler and H. Lohmann; *Chem. Ber.*, 1977, **110**, 2996.	657, 1064
77CB3040	F. R. Kreissl, P. Stückler and E. W. Meinecke; *Chem. Ber.*, 1977, **110**, 3040.	919
77CB3501	H. Schmidbaur, O. Gasser and M. S. Hussain; *Chem. Ber.*, 1977, **110**, 3501.	911, 1093, 1094
77CB3733	H.-O. Kalinowski, W. Lubosch and D. Seebach; *Chem. Ber.*, 1977, **110**, 3733.	597
77CC73	E. Colomer, R. J. P. Corriu and J. C. Young; *J. Chem. Soc., Chem. Commun.*, 1977, 73.	421
77CC597	M. Benn and A. C. M. Meesters; *J. Chem. Soc., Chem. Commun.*, 1977, 597.	637
77CC648	D. Mansuy, M. Lange, J.-C. Chottard, P. Guerin, P. Morliere, D. Brault and M. Rougee; *J. Chem. Soc., Chem. Commun.*, 1977, 648.	1097
77CCC560	R. Straka and M. Zaoral; *Collect. Czech. Chem. Commun.*, 1977, **42**, 560.	263
77CL483	N. Ishikawa and S. Sasaki; *Chem. Lett.*, 1977, 483.	234
77CL573	T. Shibanuma, M. Shiono and T. Mukaiyama; *Chem. Lett.*, 1977, 573.	1025
77CL575	T. Shibanuma, M. Shiono and T. Mukaiyama; *Chem. Lett.*, 1977, 575.	1078
77CL1345	Y. Echigo, Y. Watanabe and T. Mukaiyama; *Chem. Lett.*, 1977, 1345.	979
77CPB1651	H. Saikachi and T. Kitagawa; *Chem. Pharm. Bull.*, 1977, **25**, 1651.	367
77CRV1	G. W. Kirby; *Chem. Rev.*, 1977, **6**, 1.	326
77CRV409	T. L Gilchrist and C. J. Moody; *Chem. Rev.*, 1977, **77**, 409.	767
77CZ154	G. Zinner and U. Krüger; *Chem. Ztg.*, 1977, **101**, 154.	698
77CZ508	G. Entenmann; *Chem. Ztg.*, 1977, **101**, 508.	595
77G499	G. Bottaccio, M. Marchi and G. P. Chiusoli; *Gazz. Chim. Ital.*, 1977, **107**, 499.	68, 95
77GEP751220	R. A. Sulzbach and F. Heller; *Ger. Pat.* 75-2557655, 751 220, 1977, (*Chem. Abstr.*, 1977, **87**, 151 997c).	709
77IC391	C. P. Casey and W. R. Brunsvold; *Inorg. Chem.*, 1977, **16**, 391.	958
77IC677	R. A. Bell, M. H. Chisholm, D. A. Couch and L. A. Rankel; *Inorg. Chem.*, 1977, **16**, 677.	940
77IC3059	C. P. Casey, W. R. Brunsvold and D. M. Scheck; *Inorg. Chem.*, 1977, **16**, 3059.	951, 958
77IJC(A)1099	B. L. Gorsi and R. C. Mehrotra; *Indian J. Chem., Sect. A*, 1977, **15**, 1099.	1144
77IS95	C. T. Lam, C. D. Malkiewich and C. V. Senoff; *Inorg. Synth.*, 1977, **17**, 95.	652
77IZV249	R. G. Kostyanovskii, Y. I. El'natanov and S. M. Shakhaliev; *Izv. Akad. Nauk SSSR, Ser. Khim.*, 1977, 249.	1130
77IZV393	V. Kh. Ivanova, B. I. Buzykin and Yu. P. Kitaev; *Izv. Akad. Nauk SSSR, Ser. Khim.*, 1977, 393.	681
77IZV1839	M. G. Voronkov, N. F. Chernov and A. A. Tatarinova; *Izv. Akad. Nauk SSSR, Ser. Khim.*, 1977, 1839.	1135
77JA182	S. E. Dinizo, R. W. Freerksen, W. E. Pabst and D. S. Watt; *J. Am. Chem. Soc.*, 1977, **99**, 182.	272, 283, 284
77JA646	Y. Fukuyama, C. L. Kirkemo and J. D. White; *J. Am. Chem. Soc.*, 1977, **99**, 646.	177
77JA968	M. E. Jung and M. A. Lyster; *J. Am. Chem. Soc.*, 1977, **99**, 968.	20, 225
77JA1470	H. K. Spencer, M. V. Lakshmikantham and M. P. Cava; *J. Am. Chem. Soc.*, 1977, **99**, 1470.	1117
77JA1649	B. M. Trost and P. E. Strege; *J. Am. Chem. Soc.*, 1977, **99**, 1649.	158
77JA2501	D. R. Fahey and J. E. Mahan; *J. Am. Chem. Soc.*, 1977, **99**, 2501.	461

77JA3184 G. Zweifel and S. J. Backlund; *J. Am. Chem. Soc.*, 1977, **99**, 3184. 417
77JA4181 I. Kuwajima, M. Arai and T. Sato; *J. Am. Chem. Soc.*, 1977, **99**, 4181. 414
77JA5453 D. A. Evans, C. L. Sims and G. C. Andrews; *J. Am. Chem. Soc.*, 1977, **99**, 5453. 553
77JA5827 T. Sato, M. Arai and I. Kuwajima; *J. Am. Chem. Soc.*, 1977, **99**, 5827. 413
77JA6995 W. Ando, A. Sekiguchi, A. J. Rothschild, R. A. Gallucci, M. Jones, Jr., T. J. Barton and J. A. Kilgour; *J. Am. Chem. Soc.*, 1977, **99**, 6995. 873
77JAP73023033 J. Suzuki (Nippon Soda Co.); *Jpn Pat.*, 73 023 033 (1977) (*Chem. Abstr.*, 1977, **87**, 102 058). 735
77JAP(K)77/102237 T. Morimoto, I. Hashimoto and H. Yamaoka; *Jpn. Kokai* 77/102 237 (1977) (*Chem. Abst.*, 1978, **88**, 22 373t). 134
77JC355 K. R. Leimer, R. H. Rice and C. W. Gehrke; *J. Chromatogr.*, 1977, **141**, 355. 226
77JCR(S)119 D. Rehn and I. Ugi; *J. Chem. Res.* (*S*), 1977, 119. 652
77JCR(S)144 J. Voss, W. Schmüser and K. Schlapkohl; *J. Chem. Res.* (*S*), 1977, 144. 561
77JCS(D)1444 M. E. Fakley and A. Pidcock; *J. Chem. Soc., Dalton Trans.*, 1977, 1444. 1140
77JCS(D)1946 G. Fachinetti, G. Fochi and C. Floriani; *J. Chem. Soc., Dalton Trans.*, 1977, 1946. 443
77JCS(D)2297 G. Fachinetti, C. Floriani and H. Stoeckli-Evans; *J. Chem. Soc., Dalton Trans.*, 1977, 2297. 441
77JCS(P1)1273 M. Muraoka, T. Yamamoto, S. Yamaguchi, F. Tonosaki, T. Takeshima and N. Fukada; *J. Chem. Soc., Perkin Trans. 1*, 1977, 1273. 547
77JCS(P1)1672 A. Pelter, T. E. Levitt, K. Smith and A. Jones; *J. Chem. Soc., Perkin Trans. 1*, 1977, 1672. 234
77JCS(P1)1723 D. H. R. Barton, P.-E. Hansen and K. Picker; *J. Chem. Soc., Perkin Trans. 1*, 1977, 1723. 516, 521, 522
77JCS(P1)1791 S. Crook and P. Sykes; *J. Chem. Soc., Perkin Trans. 1*, 1977, 1791. 693, 758, 759
77JCS(P1)1806 L. Lopez and J. Barrans; *J. Chem. Soc., Perkin Trans. 1*, 1977, 1806. 775
77JCS(P1)2298 D. H. Bremner and M. M. Campbell; *J. Chem. Soc., Perkin Trans. 1*, 1977, 2298. 867
77JCS(P1)2357 R. V. Davies, B. Iddon, M. W. Pickering, H. Suschitzky, P. T. Gallagher, M. W. Gittos and M. D. Robinson; *J. Chem. Soc., Perkin Trans. 1*, 1977, 2357. 593
77JCS(P2)1080 D. G. McCarthy and A. F. Hegarty; *J. Chem. Soc., Perkin Trans. 2*, 1977, 1080. 714, 723
77JCS(P2)1085 D. G. McCarthy and A. F. Hegarty; *J. Chem. Soc., Perkin Trans. 2*, 1977, 1085. 714
77JCS(P2)1136 A. F. Hegarty, M. T. McCormack, B. J. Hathaway and L. Hulett; *J. Chem. Soc., Perkin Trans. 2*, 1977, 1136. 703
77JFC(10)437 R. K. Mackie, S. Mhatre and J. M. Tedder; *J. Fluorine Chem.*, 1977, **10**, 437. 110
77JGU441 B. S. Drach and V. A. Kovalev; *J. Gen. Chem. USSR* (*Engl. Transl.*), 1977, **47**, 441. 787, 788
77JGU505 É. S. Batyeva, V. A. Al'fonsov, V. D. Nesternko and A. N. Pudovik; *J. Gen. Chem. USSR*, 1977, **47**, 505. 405
77JGU658 O. I. Kolodyazhnyi, L. I. Shevchuk and V. P. Kukhar; *J. Gen. Chem. USSR* (*Engl. Transl.*), 1977, **47**, 658. 902
77JGU916 A. N. Pudovik, V. I. Nikitina, M. G. Zimin and I. N. Leonteva; *J. Gen. Chem. USSR* (*Engl. Trans.*), 1977, **47**, 916. 737
77JHC749 J. P. Majoral, M. Revel, P. Kraemer, H. Germa and J. Navech; *J. Heterocycl. Chem.*, 1977, **14**, 749. 705, 706
77JHC765 J. H. Boyer and P. S. Ellis; *J. Heterocycl. Chem.*, 1977, **14**, 765. 656
77JHC1321 V. I. Cohen and S. Pourabass; *J. Heterocycl. Chem.*, 1977, **14**, 1321. 522
77JOC1508 W. C. McCarthy and L. E. Foss; *J. Org. Chem.*, 1977, **42**, 1508. 990, 1037
77JOC1523 E. C. Taylor and J. L. LaMattina; *J. Org. Chem.*, 1977, **42**, 1523. 746
77JOC1610 A. A. Watson; *J. Org. Chem.*, 1977, **42**, 1610. 826
77JOC1773 H. J. Reich and S. K. Shah; *J. Org. Chem.*, 1977, **42**, 1773. 412
77JOC1786 C. S. Rondestvedt, Jr.; *J. Org. Chem.*, 1977, **42**, 1786. 723
77JOC1862 R. F. Smith, L. L. Kinder, D. G. Walker, L. A. Buckley and J. M. Hammond; *J. Org. Chem.*, 1977, **42**, 1862. 772
77JOC2091 H. Feuer and L. F. Spinicelli; *J. Org. Chem.*, 1977, **42**, 2091. 772
77JOC2639 J. J. Fitt and H. W. Gschwend; *J. Org. Chem.*, 1977, **42**, 2639. 746
77JOC2645 V. I. Cohen; *J. Org. Chem.*, 1977, **42**, 2645. 516, 522, 626
77JOC3301 R. Y. Ning, R. I. Fryer and B. C. Sluboski; *J. Org. Chem.*, 1977, **42**, 3301. 327
77JOC3372 A. Dondoni, G. Barbaro and A. Battaglia; *J. Org. Chem.*, 1977, 3372. 1073
77JOC3383 T. Takeshima, M. Muraoka, N. Fukada, A. Takayama and T. Yamamoto; *J. Org. Chem.*, 1977, **42**, 3383. 555
77JOC3749 A. P. Krapcho, J. R. Larsen and J. M. Eldridge; *J. Org. Chem.*, 1977, **42**, 3749. 52
77JOC3903 B. A. Patel, C. B. Ziegler, N. A. Cortese, J. E. Plevyak, T. C. Zebovitz, M. Terpko and R. F. Heck; *J. Org. Chem.*, 1977, **42**, 3903. 157
77JOC3960 R. P. Hatch and S. M. Weinreb; *J. Org. Chem.*, 1977, **42**, 3960. 235
77JOM(124)1 N. Petragnani, J. V. Comasseto, R. Rodrigues and T. J. Brockson; *J. Organomet. Chem.*, 1977, **124**, 1. 903
77JOM(128)291 J. M. Coronas, C. Peruyero and J. Sales; *J. Organomet. Chem.*, 1977, **128**, 291. 1139
77JOM(129)189 T. Katada, M. Nishida, S. Kato and M. Mizuta; *J. Organomet. Chem.*, 1977, **129**, 189. 243, 252
77JOM(131)C1 A. Hassner and J. A. Soderquist; *J. Organomet. Chem.*, 1977, **131**, C1. 417, 425
77JOM(134)21 H. Jancke, G. Engelhardt, S. Wagner, W. Dirnens, G. Herzog, E. Thieme and K. Rühlmann; *J. Organomet. Chem.*, 1977, **134**, 21. 720
77JOM(135)183 W. H. De Roode, K. Vrieze, E. A. Von Gustorf, E. A. Koerner and A. Ritter; *J. Organomet. Chem.*, 1977, **135**, 183. 781
77JOM(136)309 P. Royo and R. Serrano; *J. Organomet. Chem.*, 1977, **136**, 309. 1143
77JOM(140)97 I. Ojima and S. Inaba; *J. Organomet. Chem.*, 1977, **140**, 97. 759, 777
77JOM(141)241 W. Maringgele and A. Meller; *J. Organomet. Chem.*, 1977, **141**, 241. 739

77JOM(142)165 K. K. Sharma, S. K. Mehrotra and R. C. Mehrotra; *J. Organomet. Chem.*, 1977, **142**, 165. 389

77JOU1471 B. S. Drach, V. A. Kovalev, A. D. Gordeev and G. B. Soifer; *J. Org. Chem. USSR* (*Engl. Transl.*), 1977, **13**, 1471. 655, 659

77JOU1474 B. S. Drach and V. A. Kovalev; *J. Org. Chem. USSR* (*Engl. Transl.*), 1977, **13**, 1474. 685, 694

77JPC915 D. Forrest, K. U. Ingold and D. H. R. Barton; *J. Phys. Chem.*, 1977, **81**, 915. 516, 526

77JPR169 S. R. Ramadas and P. S. Srinivasan; *J. Prakt. Chem.*, 1977, **319**, 169. 560

77JPR841 K. Unverferth, C. Rüger and K. Schwetlick; *J. Prakt. Chem.*, 1977, **319** (6), 841. 984

77JPS2163 H. Kurita, Y. Kusayama and Y. Iwakura; *J. Polym. Sci., Polym. Chem.*, 1977, **15**, 2163. 753

77JST25 H. S. Randhawa, C. O. Meese and W. Walter; *J. Mol. Struct.*, 1977, **36**, 25. 506

77KFZ38 A. S. Noravyan, A. P. Mkrtchyan, I. A. Dzhagatspanyan, N. E. Akopyan, R. A. Akopyan and S. A. Vartanyan; *Khim. Farm. Zh.*, 1977, **11**, 38 (*Chem. Abstr.*, 1978, **88**, 37 739r). 690

77LA1291 H. Böhme and W. Krack; *Justus Liebigs Ann. Chem.*, 1977, 1291. 566, 610, 726

77LA1751 E. Ziegler, G. Kollenz, G. Kriwetz and W. Ott; *Liebigs Ann. Chem.*, 1977, 1751. 556

77M1027 W. Maringgele and A. Meller; *Monatsh. Chem.*, 1977, **108**, 1027. 381

77M1279 A. Meller, W. Maringgele, G. Beer and H. Fetzer; *Monatsh. Chem.*, 1977, **108**, 1279. 381, 382

B-77MI 501-01 C. A. Buehler and D. E. Pearson; "Survey of Organic Synthesis," Wiley, New York, 1977, vol. 2. 1

77MI 502-01 A. Matsuda; *Tokoshi Nyusu, Kagaku Kogyo Shiryo*, 1977, **11**, 127 (*Chem. Abstr.*, 1978, **88**, 88 992a). 27

B-77MI 502-01 C. A. Buehler and D. E. Pearson; "Survey of Organic Syntheses," Wiley, New York, 1977, chap. 13, p. 655. 24

77MI 502-02 Y. Suwara and S. Miyazaki; *Tokashi Nyusu, Kagaku Kogyo Shiryo*, 1977, **12**, 55 (*Chem. Abstr.*, 1978, **88**, 73 734u). 28

77MI 507-01 H. Miyazaki; *Denki Kagaku Oyobi Kogyo Butsuri Kagaku*, 1977, **45**, 553. 317

77MI 519-01 D. Dzharov; *Farmatsiya* (*Sofia*), 1977, **27**, 1 (*Chem. Abstr.*, 1978, **89**, 108 048y). 745

77MI 519-02 W. H. De Roode and K. Vrieze; *J. Less Common Met.*, 1977, **54**, 209 (*Chem. Abstr.*, 1977, **87**, 201 683v). 781

B-77MI 521-01 M. Regitz; "Diazoalkane," Georg Thieme Verlag, Stuttgart, 1977. 844

B-77MI 526-01 R. Richter and H. Ulrich; in "The Chemistry of Cyanates and Their Thio Derivatives," ed. S. Patai, Interscience, New York, 1977, p. 619. 962, 1017, 1018

B-77MI 527-01 L. Drobnica, P. Kristián and J. Augustín; in "The Chemistry of Cyanates and Their Thio Derivatives", Part 2, ed. S. Patai, Wiley, Chichester, 1977, chapter 22, p. 1003. 1022, 1038

B-77MI 527-02 O. Tsuge; in "The Chemistry of Cyanates and Their Thio Derivatives," Part 1, ed. S. Patai, Wiley, Chichester, 1977, ch. 13, p. 445. 1022

B-77MI 527-03 E. Bulka; in "The Chemistry of Cyanates and Their Thio Derivatives," Part 2, ed. S. Patai, Wiley, Chichester, 1977, ch. 19, p. 908. 1051

B-77MI 530-01 K. A. Jensen and A. Holm; in "Chemistry of Cyanates and their Thio Derivatives," ed. S. Patai, Wiley, Chichester, 1977, Part 1, p. 569. 1100, 1102

B-77MI 530-02 R. G. Guy; in "Chemistry of Cyanates and their Thio Derivatives," ed. S. Patai, Wiley, New York, 1977, Part 1 p. 819. 1105, 1106, 1107, 1108, 1109, 1110

B-77MI 530-03 E. Bulka; in "The Chemistry of Cyanates and their Thio Derivatives," ed. S. Patai, Wiley, Chichester, 1977, Part 2, p. 887. 1113, 1116

77NKK109 K. Matsumoto; *Nippon Kagaku Kaishi* (*J. Chem. Soc. Jpn.*), 1977, **51**, R109. 84

77OPP57 T. Korpela, J. Lundell and P. Pasanen; *Org. Prep. Proced. Int.*, 1977, **9**, 57. 701

77OS88 M. S. A. Vrijland; *Org. Synth.*, 1977, 88. 1112

77PAC749 B. M. Mikhailov; *Pure Appl. Chem.*, 1977, **49**, 749. 778

77RRC1111 M. Vlassa and F. Hodosan; *Rev. Roum. Chim.*, 1977, **22**, 1111. 735

77RTC118 D. Ladurée and D. Paquer; *Recl. Trav. Chim. Pays-Bas*, 1977, **96**, 118. 550

77RTC139 G. E. Veenstra, N. M. Bronold, J. F. M. Smits, A. Tangerman and B. Zwanenburg; *Recl. Trav. Chim. Pays-Bas*, 1977, **96**, 139. 478, 491, 492, 562

77RTC254 D. Ladurée, D. Paquer and P. Rioult; *Recl. Trav. Chim. Pays-Bas*, 1977, **96**, 254. 558

77S31 R. Bayles, M. C. Johnson, R. F. Maisey and R. W. Turner; *Synthesis*, 1977, 31. 292

77S33 R. Bayles, M. C. Johnson, R. F. Maisey and R. W. Turner; *Synthesis*, 1977, 33. 292

77S250 R. Erbea, S. Panossian and C. Giordano; *Synthesis*, 1977, 250 606

77S328 F. Malek-Yazdi and M. Yalpani; *Synthesis*, 1977, 328. 589, 626

77S339 K. Dimroth and W. Tüncher; *Synthesis*, 1977, 339. 366

77S457 A. J. Kolar and R. K. Olsen; *Synthesis*, 1977, 457. 315

77S487 D. Binder, G. Habison and C. R. Noe; *Synthesis*, 1977, 487. 354

77S574 L. M. Yagupolskii, I. I. Maletina, N. V. Kondratenko and V. V. Orda; *Synthesis*, 1977, 574. 204

77S582 D. J. Le Count, D. J. Dewsbury and W. Grundy; *Synthesis*, 1977, 582. 1026

77S607 S. E. J. Glue and I. T. Kay; *Synthesis*, 1977, 607. 592

77S626 H. Yoshida, T. Ogata, S. Inokawa; *Synthesis*, 1977, 626. 728, 729

77S632 M. Ohsugi, Y. Inamoto, N. Takaishi, Y. Fujikura and K. Aigami; *Synthesis*, 1977, 632. 282

77S777 Y. Tohda, K. Sonogashira and N. Hagihara; *Synthesis*, 1977, 777. 304

77S888 R. S. Sukhai, R. de Jong and L. Brandsma; *Synthesis*, 1977, 888. 589, 590, 603, 611, 626,

77S889 T. Fujinami, N. Otani and S. Sakai; *Synthesis*, 1977, 889. 1078

77SC57 M. K. Sahni, I. K. Sharma, C. K. Narang and N. K. Mathur; *Synth. Commun.*, 1977, **7**, 57. 652

77SC251 K. Horiki; *Synth. Commun.*, 1977, **7**, 251. 235

77SST59 G. C. Barrett; *Org. Compd. Sulphur, Selenium, Tellurium* (*R. Soc. Chem. Series*), 1977, **4**, 59. 1022

77SST136 D. R. Hogg; *Org. Compd. Sulphur, Selenium and Tellurium* (*R. Soc. Chem. Series*), 1977, **4**, 136. 472

77SST141 W. Walter and J. Voss; *Org. Compd. Sulphur, Selenium and Tellurium* (*R. Soc. Chem. Series*), 1977, **4**, 141. 472

77SST159 W. Walter and J. Voss; *Org. Compd. Sulphur, Selenium, Tellurium* (*R. Soc. Chem. Series*), 1977, **4**, 159. 565

77T341 J. Becher and E. G. Frandsen; *Tetrahedron*, 1977, **33**, 341. 593

77T449 H. Bühl, B. Seitz and H. Meier; *Tetrahedron*, 1977, **33**, 449. 511

77T3137 P. Jakobsen and S. Treppendahl; *Tetrahedron*, 1977, **33**, 3137. 652

77TL155 C. E. McKenna, M. T. Higa, N. H. Cheung and M.-C. McKenna; *Tetrahedron Lett.*, 1977, 155. 404

77TL1037 M. Mori, K. Chiba and Y. Ban; *Tetrahedron Lett.*, 1977, 1037. 157

77TL1721 F. A. Davis and U. K. Nadir; *Tetrahedron Lett.*, 1977, **18**, 1721. 347

77TL1839 P. Beak, B. G. McKinnie and D. B. Reitz; *Tetrahedron Lett.*, 1977, 1839. 237

77TL1987 I. Ruppert; *Tetrahedron Lett.*, 1977, 1987. 1082

77TL2139 K. Friedrich and M. Zamkanei; *Tetrahedron Lett.*, 1977, 2139. 592, 630

77TL2145 D. Hoppe and M. Kloft; *Tetrahedron Lett.*, 1977, 2145. 592, 609

77TL2235 P. G. Gassman and R. J. Balchunis; *Tetrahedron Lett.*, 1977, **18**, 2235. 338

77TL2337 S. Yanagida, Y. Noji and M. Okahara; *Tetrahedron Lett.*, 1977, 2337. 6

77TL2893 S. Yanagida, Y. Noji and M. Okahara; *Tetrahedron Lett.*, 1977, 2893. 6

77TL3379 P. J. Beeby; *Tetrahedron Lett.*, 1977, 3379. 13

77TL4171 A. Basha, M. Lipton and S. M. Weinreb; *Tetrahedron Lett.*, 1977, 4171. 277

77TL4307 E. Schaumann and F.-F. Grabley; *Tetrahedron Lett.*, 1977, 4307. 589, 614

77TL4417 Y. Tamura, T. Kawasaki, M. Adachi, M. Tanio and Y. Kita; *Tetrahedron Lett.*, 1977, **18**, 4417. 1041, 1107

77TL4607 Y. Ueno, M. Bahry and M. Okawara; *Tetrahedron Lett.*, 1977, 4607. 560

77ZAAC(428)222 H. Schäfer, G. Fritz and W. Hölderich; *Z. Anorg. Allg. Chem.*, 1977, **428**, 222. 1155

77ZAAC(430)66 G. Becker; *Z. Anorg. Allg. Chem.*, 1977, **430**, 66. 396, 878, 879, 880

77ZAAC(430)77 G. Becker and H. P. Beck; *Z. Anorg. Allg. Chem.*, 1977, **430**, 77. 396

77ZAAC(430)91 G. Becker and H. P. Beck; *Z. Anorg. Allg. Chem.*, 1977, **430**, 91. 396

77ZAAC(433)94 W. Maringgele and A. Meller; *Z. Anorg. Allg. Chem.*, 1977, **433**, 94. 381, 660, 717

77ZAAC(436)173 W. Maringgele and A. Meller; *Z. Anorg. Allg. Chem.*, 1977, **436**, 173. 382

77ZC92 W. Thiel, H. Viola and R. Mayer; *Z. Chem.*, 1977, **17**, 92. 561

77ZC366 W. Thiel, H. Viola and R. Mayer; *Z. Chem.*, 1977, **17**, 366. 554

77ZN(B)98 F. P. Schmidtchen, P. Rauschenbach and H. Simon; *Z. Naturforsch., Teil B*, 1977, **32**, 98. 361

77ZN(B)858 H. Schmidbaur, W. Scharf and H. J. Füller; *Z. Naturforsch., Teil B*, 1977, **B32**, 858. 1096

77ZOB968 D. I. Makhon'kov, L. Y. D'yachkova and N. S. Zefirov; *Zh. Obshch. Khim.*, 1977, **13**, 968. 316

77ZOB2259 V. D. Sheludyakov, A. S. Tkachev, S. V. Sheludyakova, V. P. Kozyukov and V. F. Mironov; *Zh. Obshch. Khim.*, 1977, **47**, 2259. 1006

77ZOB2294 V. I. Shcherbakov, N. E. Stolyarova, O. S. D'yachkovskaya and T. I. Chulkova; *Zh. Obshch. Khim.*, 1977, **47**, 2294. 390

77ZOR100 N. Ya. Derkach and N. P. Tishchenko; *Zh. Org. Khim.*, 1977, **13**, 100. 238

77ZOR1144 S. A. Andreev, I. A. Savaev, B. A. Lebedev, I. V. Tselinskii and B. V. Gidaspov; *Zh. Org. Khim.*, 1977, **13**, 1144. 361

77ZOR1594 B. S. Drach, V. A. Kovalev, A. D. Gordeev and G. B. Soifer; *Zh. Org. Khim.*, 1977, **13**, 1594. 655, 659

77ZOR1597 B. S. Drach and V. A. Kovalev; *Zh. Org. Khim.*, 1977, **13**, 1597. 685, 694

77ZOR2055 N. Markovski, E. A. Stukalo and G. P. Kunitskaya; *Zh. Org. Khim.*, 1977, **13**, 2055. 349

78ACR289 J. I. Zink; *Acc. Chem. Res.*, 1978, **11**, 289. 1092

78ACS(B)780 A. Senning and K. Schank; *Acta Chem. Scand., Ser. B*, 1978, **32**, 780. 528

78AG761 H. Ulrich and A. A. R. Sayigh; *Angew. Chem.*, 1978, **16**, 761. 1065

78AG(E)27 R. C. Larock; *Angew. Chem., Int. Ed. Engl.*, 1978, **17**, 27. 142

78AG(E)176 Y. Izumi, I. Chibata and T. Itoh; *Angew. Chem., Int. Ed. Engl.*, 1978, **17**, 176. 84

78AG(E)267 H.-J. Gais and T. Lied; *Angew. Chem., Int. Ed. Engl.*, 1978, **17**, 267. 236, 239

78AG(E)455 E. Schaumann, J. Ehlers and U. Behrens; *Angew. Chem., Int. Ed. Engl.*, 1978, **17**, 455. 605

78AG(E)569 G. Höfle; W. Steglich and H. Vorbrüggen; *Angew. Chem., Int. Ed. Engl.*, 1978, **17**, 567. 129

78AG(E)689 H. Alper, J. K. Currie and R. Sachdeva; *Angew. Chem., Int. Ed. Engl.*, 1978, **17**, 689. 568, 613

78AG(E)781 D. Mansuy, M. Lange, J.-C. Chottard, J.-F. Bartoli, B. Chevrier and R. Weiss; *Angew. Chem., Int. Ed. Engl.*, 1978, **17**, 781. 1097

78AG(E)939 J. Neumeister, H. Keul, M. P. Saxena and K. Griesbaum; *Angew. Chem., Int. Ed. Engl.*, 1978, **17**, 939. 141

78BAU2304 B. M. Mikhailov, V. A. Dorokhov and L. I. Lavrinovich; *Bull. Acad. Sci. USSR, Div. Chim. Sci.* (*Eng. Transl.*), 1978, **27**, 2304. 778

78BCJ301 M. O-Oka, A. Kitamura, R. Okazaki and N. Inamoto; *Bull. Chem. Soc. Jpn.*, 1978, **51**, 301. 634

78BCJ512 A. S. Shawali, H. M. Hassaneen and S. Almousawi; *Bull. Chem. Soc. Jpn.*, 1978, **51**, 512. 358

78BCJ659 S. Tanaka, K. Masue and M. Okano; *Bull. Chem. Soc. Jpn.*, 1978, **51**, 659. 1036

78BCJ1267 A. Sekiya and N. Ishikawa; *Bull. Chem. Soc. Jpn.*, 1978, **51**, 1267. 4

78BCJ1818 S. Uemura, N. Watanabe, A. Toshimitsu and M. Okano; *Bull. Chem. Soc. Jpn.*, 1978, **51**, 1818. 1053

78BCJ1866 K. Inomata, H. Kinoshita, H. Fukuda, K. Tanabe and H. Kotake; *Bull. Chem. Soc. Jpn.*, 1978, **51**, 1866. 291

78BCJ2391 I. Kuwajima, N. Minami, T. Abe and T. Sato; *Bull. Chem. Soc. Jpn.*, 1978, **51**, 2391. 420

78BCJ2401 N. Ono, T. Yamada, T. Saito, K. Tanaka and A. Kaji; *Bull. Chem. Soc. Jpn.*, 1978, **51**, 2401. 132

78BCJ2909 M. Kubota and A. Yamamoto; *Bull. Chem. Soc. Jpn.*, 1978, **51**, 2909. 385

78BSB229 S. Scheibye, B. S. Pedersen and S.-O. Lawesson; *Bull. Soc. Chim. Belg.*, 1978, **87**, 229. 570, 571, 608, 610, 618

78BSB293 B. S. Pedersen, S. Scheibye, K. Clausen and S.-O. Lawesson; *Bull. Soc. Chim. Belg.*, 1978, **87**, 293. 484, 551, 570

78BSB299 S. Scheibye, B. S. Pedersen and S.-O. Lawesson; *Bull. Soc. Chim. Belg.*, 1978, **87**, 299. 618

78BSB525 H. Fritz, P. Hug, S.-O. Lawesson, E. Logemann, B. S. Pedersen, H. Sauter, S. Scheibye and T. Winkler; *Bull. Soc. Chim. Belg.*, 1978, **87**, 525. 570, 608, 610

78BSB717 M. Strebelle and R. Fuks; *Bull. Soc. Chim. Belg.*, 1978, **87**, 717. 751

78CB538 W. Maringgele and A. Meller; *Chem. Ber.*, 1978, **112**, 538. 380, 382

78CB2054 R. Appel and H.-D. Wihler; *Chem. Ber.*, 1978, **111**, 2054. 1095

78CB2206 H.-D. Scharf and J. Mattay; *Chem. Ber.*, 1978, **111**, 2206. 145

78CB2696 H. Schmidbaur and M. Heimann; *Chem. Ber.*, 1978, **111**, 2696. 914

78CB2702 H. Schmidbaur and G. Hasslberger; *Chem. Ber.*, 1978, **111**, 2702. 922

78CB3283 F. R. Kreissl, W. Uedelhoven and G. Kreis; *Chem. Ber.*, 1978, **111**, 3283. 919

78CB3750 A. Q. Hussein, A. Abu-Taha and J. C. Jochims; *Chem. Ber.*, 1978, **111**, 3750. 1046

78CC133 A. R. Katritzky, U. Gruntz, N. Mongelli and M. C. Rezende; *J. Chem. Soc., Chem. Commun.*, 1978, 133. 1107

78CC285 M. Sekine and T. Hata; *J. Chem. Soc., Chem. Commun.*, 1978, 285. 403

78CC351 F. D. King, S. Pike and D. R. M. Walton; *J. Chem. Soc., Chem. Commun.*, 1978, 351. 1000

78CC393 D. H. R. Barton, N. J. Cussans and S. V. Ley; *J. Chem. Soc., Chem. Commun.*, 1978, 393. 523

78CC478 I. Kuwajima, M. Kato and T. Sato; *J. Chem. Soc., Chem. Commun.*, 1978, 478. 414

78CC762 R. N. Perutz; *J. Chem. Soc., Chem. Commun.*, 1978, 762. 794

78CC838 J. A. Gladysz, V. K. Wong and B. S. Jick; *J. Chem. Soc., Chem. Commun.*, 1978, 838. 242, 252

78CC870 G. M. Blackburn and D. Ingleson; *J. Chem. Soc., Chem. Commun.*, 1978, 870. 404

78CC946 S. Neumann, D. Schomburg, G. Richtarsky and R. Schmutzler; *J. Chem. Soc., Chem. Commun.*, 1978, 946. 407

78CC961 T. Mah, H. M. Sirat and E. J. Thomas; *J. Chem. Soc., Chem. Commun.*, 1978, 961. 63

78CI(L)579 P. R. H. Speakman; *Chem. Ind. (London)*, 1978, 579. 208

78CI(L)806 Y. Tamura, T. Kawasaki, M. Tanio and Y. Kita; *Chem. Ind. (London)*, 1978, 806. 1048

78CJC1026 T. A. Ondrus, E. E. Knaus and C. S. Giam; *Can. J. Chem.*, 1978, **56**, 1026. 1119

78CJC1183 J. S. Grossert, M. M. Bharadwaj, R. F. Langler, T. S. Cameron and R. E. Cordes; *Can. J. Chem.*, 1978, **56**, 1183. 348

78CJC2286 A. G. Brook, C. Golino and E. Matern; *Can. J. Chem.*, 1978, **56**, 2286. 792, 796

78CL1007 H. Ishihara and Y. Hirabayashi; *Chem. Lett.*, 1978, 1007. 240

78CPB2246 K. Kaji, H. Matsubara, H. Nagashima, Y. Kikugawa and S.-I. Yamada; *Chem. Pharm. Bull.*, 1978, **26**, 2246. 746

78CR(C)69 M. Chihaoui and B. Baccar; *C. R. Hebd. Seances Acad. Sci., Ser. C.*, 1978, **287**, 69. 705

78CRV97 J. D. Coyle; *Chem. Rev.*, 1978, **78**, 97. 24

78CZ1 M. El-Chaharvi and U. Prange; *Chem.-Ztg.*, 1978, **102**, 1. 27

78DOK(239)330 R. A. Bekker, V. Ya. Popkova and I. L. Knunyants; *Dokl. Akad. Nauk SSSR*, 1978, **239**, 330. 59

78GEP2701309 H. Hoffmann, B. Homeyer and I. Hammann; *Ger. Pat.* 2 701 309 (1978) (*Chem. Abstr.*, 1979, **89**, 146 417). 774

78GEP2725519 T. Tsuji, M. Yoshioka, T. Kataoka, Y. Sendo, S. Hirai, T. Maeda and W. Nagata; *Ger. Pat.* 2 725 519 (1978) (*Chem. Abstr.*, 1978, **91**, 157 765e). 81

78GEP2758173 E. Sturm; *Ger. Pat.* 2 758 173 (1978) (*Chem. Abstr.*, 1979, **89**, 146 448m). 774

78GEP2822472 S. W. Bland and D. McKay; *Ger. Pat.* 2 822 472 (1978) (*Chem. Abstr.*, 1979, **96**, 121 233). 134

78GSM111 D. W. Knight; *Gen. Synth. Methods*, 1978, **1**, 111. 24

78H(9)1327 B. Verček, B. Stanovnik and M. Tišler; *Heterocycles*, 1978, **9**, 1327. 834

78H(11)139 M. Ogata and H. Matsumoto; *Heterocycles*, 1978, **11**, 139. 578, 608

78HCA2428 M. Neuenschwander, U. Lienhard, H.-P. Fahrni and B. Hurni; *Helv. Chim. Acta*, 1978, **61**, 2428. 280

78HCA2437 M. Neuenschwander, H.-P. Fahrni and U. Lienhard; *Helv. Chim. Acta*, 1978, **61**, 2437. 280

78IC2369 C. M. Lukehart and J. V. Zeile; *Inorg. Chem.*, 1978, **17**, 2369. 801

78IJ132 H. W. Roesky and K. Ambrosius; *Isr. J. Chem.*, 1978, **17**, 132. 389

78IJC(A)778 S. N. Bhattacharya and M. Singh; *Indian J. Chem., Sect. A*, 1978, **16**, 778. 1132

78IZV2094 V. M. Smagin, V. V. Gavrilenko, L. I. Zakharkin and V. K. Golubev; *Izv. Akad. Nauk SSSR, Ser. Khim.*, 1978, 2094. 44

78IZV2412 K. V. Nikonorov, É. A. Gurylev and L. A. Antokhina; *Izv. Akad. Nauk SSSR, Ser. Khim.*, 1978, 2412. 369

78IZV2802 A. V. Fokin, V. P. Stolyarov, A. M. Gukov, A. T. Uzan and V. A. Komarov; *Izv. Akad. Nauk SSSR, Ser. Khim.*, 1978, 2802 (*Chem. Abstr.*, 1979, **90**, 12 079g). 698, 762, 813

78JA446 H. W. Kroto, J. F. Nixon, N. P. C. Simmons and N. P. C. Westwood; *J. Am. Chem. Soc.*, 1978, **100**, 446. 1152

78JA640 J. W. Suggs; *J. Am. Chem. Soc.*, 1978, **100**, 640. 459

78JA1222 D. N. Harpp, K. Steliou and T. H. Chan; *J. Am. Chem. Soc.*, 1978, **100**, 1222. 338

78JA1637 J. F. King, S. M. Loosmore, J. D. Lock and M. Aslam; *J. Am. Chem. Soc.*, 1978, **100**, 1637. 1106

78JA1687 R. B. King, A. D. King Jr, M. Z. Iqbal and C. C. Frazier; *J. Am. Chem. Soc.*, 1978, **100**, 1678. 445, 453

78JA1923 Y. Tamaru, T. Harada and Z. Yoshida; *J. Am. Chem. Soc.*, 1978, **100**, 1923. 607

78JA4618 E. J. Corey, E. J. Trybulski, L. S. Melvin Jr., K. C. Nicolaou, J. A. Secrist, R. Lett, P. W. Sheldrake, J. R. Falck, D. J. Brunelle, M. F. Haslanger, S. Kim and S. Yoo; *J. Am. Chem. Soc.*, 1978, **100**, 4618. 173

78JA4620 E. J. Corey, S. Kim, S. Yoo, K. C. Nicolaou, L. S. Melvin, Jr., D. J. Brunelle, J. R. Falck, E. J. Trybulski, R. Lett, and P. W. Sheldrake; *J. Am. Chem. Soc.*, 1978, **100**, 4620. 173, 204

78JA4858 P. A. Bartlett and F. R. Green, III; *J. Am. Chem. Soc.*, 1978, **100**, 4858. 174

78JA5396 G. A. Olah and J. Welch; *J. Am. Chem. Soc.*, 1978, **100**, 5396. 6

78JA7112 J. M. Manriquez, P. J. Fagan, T. J. Marks, C. S. Day and V. W. Day; *J. Am. Chem. Soc.*, 1978, **100**, 7112. 469

78JA8214 I. Ernest, J. Gosteli, G. W. Greengrass, W. Holick, D. E. Jackman, H. R. Pfaendler and R. B. Woodward; *J. Am. Chem. Soc.*, 1978, **100**, 8214. 905

78JCR(M)5501 A. Étienne and B. Desmazieres; *J. Chem. Res.* (*M*), 1978, 5501. 499

78JCR(S)132 B. Decroix and P. Pastour; *J. Chem. Res.* (*S*), 1978, 132. 756

78JCR(S)198 M. Utjes-LeGall, F. Salmon-Legagneur and H. Le Moal; *J. Chem. Res.* (*S*), 1978, 198. 64

78JCR(S)484 A. Étienne and B. Desmazieres; *J. Chem. Res.* (*S*), 1978, 484. 499

78JCS(P1)309 L. Citerio, D. Pocar, R. Stradi and B. Giocă; *J. Chem. Soc., Perkin Trans. 1*, 1978, 309. 762, 816

78JCS(P1)325 G. E. Castillo and H. E. Bertorello; *J. Chem. Soc., Perkin Trans. 1*, 1978, 325. 589

78JCS(P1)558 N. Fukada, M. Kato, H. Seki, M. Kawana and T. Takeshima; *J. Chem. Soc., Perkin Trans. 1*, 1978, 558. 652

78JCS(P1)629 J. S. A. Brunskill, A. De and D. F. Ewing; *J. Chem. Soc., Perkin Trans. 1*, 1978, 629. 580, 615

78JCS(P1)1273 A. Toshimitsu, Y. Kozawa, S. Uemura and M. Okano; *J. Chem. Soc., Perkin Trans. 1*, 1978, 1273. 1115

78JCS(P1)1338 G. V. Boyd and R. L. Monteil; *J. Chem. Soc., Perkin Trans. 1*, 1978, 1338. 716

78JCS(P2)1167 R. F. Hudson and K. A. F. Record; *J. Chem. Soc., Perkin Trans. 2*, 1978, 1167. 531

78JFC(11)327 S. D. Morse, K. A. Laurence, G. H. Sprenger and J. M. Shreeve; *J. Fluorine Chem.*, 1978, **11**, 327. 246

78JGU414 M. A. Pudovik, L. K. Kibardina and A. N. Pudovik; *J. Gen. Chem. USSR* (*Engl. Transl.*), 1978, **48**, 414. 402

78JGU1046 V. F. Gerega, Yu. I. Mushkin, Yu. I. Baukov and Yu. I. Dergunov; *J. Gen. Chem. USSR* (*Engl. Transl.*), 1978, **48**, 1046. 717

78JGU1249 G. S. Zaitseva, L. L. Vasil'eva, N. G. Vinokurova, O. A. Safronara and Y. I. Bankow; *J. Gen. Chem. USSR* (*Engl. Transl.*), 1978, **48**, 1249. 162

78JGU1313 V. P. Kozyukov, S. V. Sheludyakova, V. S. Nikitin, N. M. Bizyukova and V. F. Mironov; *J. Gen. Chem. USSR* (*Engl. Transl.*), 1978, **48**, 1313. 721, 722

78JHC237 V. I. Cohen; *J. Heterocycl. Chem.*, 1978, **15**, 237. 516, 522, 704

78JHC439 S. W. Schneller and D. G. Bartholomew; *J. Heterocycl. Chem.*, 1978, **15**, 439. 350

78JHC697 U. Timm, H. Bühl and H. Meier; *J. Heterocycl. Chem.*, 1978, **15**, 697. 511

78JHC1113 V. I. Cohen; *J. Heterocycl. Chem.*, 1978, **15**, 1113. 642, 704

78JHC1169 G. M. Coppola and R. I. Mansukhani; *J. Heterocycl. Chem.*, 1978, **15**, 1169. 370

78JHC1415 V. I. Cohen; *J. Heterocycl. Chem.*, 1978, **15**, 1415. 522

78JHC1513 A. Nuvole and G. A. Pinna; *J. Heterocycl. Chem.*, 1978, **15**, 1513. 657

78JOC371 H. W. Pinnick and N. H. Lajis; *J. Org. Chem.*, 1978, **43**, 371. 141

78JOC381 A. Padwa, H. Ku and A. Mazza; *J. Org. Chem.*, 1978, **43**, 381. 657

78JOC388 K. J. Dignam, A. F. Hegarty and P. L. Quain; *J. Org. Chem.*, 1978, **43**, 388. 699

78JOC731 H. M. Walborsky and P. Ronman; *J. Org. Chem.*, 1978, **43**, 731. 796, 802

78JOC1283 P. A. Grieco, Y. Yokoyama and E. Williams; *J. Org. Chem.*, 1978, **43**, 1283. 239

78JOC1604 J. R. Beck and J. A. Yahner; *J. Org. Chem.*, 1978, **43**, 1604. 1111

78JOC1684 M. Mori, K. Chiba and Y. Ban; *J. Org. Chem.*, 1978, **43**, 1684. 301

78JOC2164 L. E. Overman, G. F. Taylor, C. B. Petty and P. J. Jessup; *J. Org. Chem.*, 1978, **43**, 2164. 991

78JOC2280 E. C. Taylor, J. G. Andrade, K. C. John and A. McKillop; *J. Org. Chem.*, 1978, **43**, 2280. 1135

78JOC2702 E. L. Compere, Jr. and A. Shockravi; *J. Org. Chem.*, 1978, **43**, 2702. 67

78JOC2735 A. P. Kozikowski and A. Ames; *J. Org. Chem.*, 1978, **43**, 2735. 239

78JOC2952 N. A. Cortese, C. B. Ziegler, Jr., B. J. Hrnjez and R. F. Heck; *J. Org. Chem.*, 1978, **43**, 2952. 157

78JOC3713 D. D. Keith, R. Yang, J. A. Tortora and M. Weigele; *J. Org. Chem.*, 1978, **43**, 3713. 315

78JOC3750 J. Lessard, R. Cote, P. Mackiewicz, R. Furstoss and B. Waegell; *J. Org. Chem.*, 1978, **43**, 3750. 316

78JOC3972 A. Wissner and C. V. Grudzinskas; *J. Org. Chem.*, 1978, **43**, 3972. 15

78JOC4393 D. B. Collum, S.-C. Chen and B. Ganem; *J. Org. Chem.*, 1978, **43**, 4393. 652

78JOM(145)C1 N. Minami, T. Abe and I. Kuwajima; *J. Organomet. Chem.*, 1978, **145**, C1. 414, 425

78JOM(148)225 A. Ekouya, J. Dunogues and N. Duffaut; *J. Organomet. Chem.*, 1978, **148**, 225. 1076, 1077

78JOM(159)137 D. Seyferth, D. P. Duncan, H. Schmidbaur and P. Holl, *J. Organomet. Chem.*, 1978, **159**, 137. 914

78JOM(159)417 K. Oguro, M. Wada and R. Okawara; *J. Organomet. Chem.*, 1978, **159**, 417. 940

78JOM(159)73 T. J. Collins and W. R. Roper; *J. Organomet. Chem.*, 1978, **159**, 73. 650

78JOM(160)47 K. Issleib, H. Schmidt and H. Meyer; *J. Organomet. Chem.*, 1978, **160**, 47. 784, 881

78JOM(161)313 C. L. Hauman and C. H. Yoder; *J. Organomet. Chem.*, 1978, **161**, 313. 720

78JOM(161)C23 N. Duffaut, J. Dunoguès, C. Biran and R. Calas; *J. Organomet. Chem.*, 1978, **161**, C23. 417

78JOU758 V. B. Piskov and V. P. Kasperovich; *J. Org. Chem. USSR* (*Engl. Transl.*), 1978, **14**, 758. 655

78JOU881 B. S. Drach and G. N. Mis'kevich; *J. Org. Chem. USSR* (*Engl. Transl.*), 1978, **14**, 881. 655

78JOU1457 A. D. Sinitsa and S. V. Bonadyk; *J. Org. Chem. USSR* (*Engl. Transl.*), 1978, **14**, 1457. 657

78JPR313 R. Mayer, H. Viola, J. Reichert and W. Krause; *J. Prakt. Chem.*, 1978, **320**, 313. 601

78KGS991 V. P. Shchipanov, E. O. Siderov, L. S. Podenko and G. D. Kadochnikova; *Khim. Geterotsikl. Soedin.*, 1978, 991. 860

78KGS1497	V. S. Velezheva, A. V. Yarosh, T. A. Kozik and N. N. Suvorov; *Khim. Geterotsikl. Soedin.*, 1978, **11**, 1497.	1025
78LA227	P. C. Thieme and E. Hädicke; *Justus Liebigs Ann. Chem.*, 1978, 227.	1044
78LA617	R. Albrecht; *Justus Liebigs Ann. Chem.*, 1978, 617.	94
78LA635	G. Steiner; *Justus Liebigs Ann. Chem.*, 1978, 635.	347
B-78MI 502-01	P. S. Bailey; "Ozonization in Organic Chemistry," Academic Press, New York, 1978, vol. 1.	28
78MI 502-01	S. Mercier; *Pet. Tech.*, 1978, **257**, 26.	27
B-78MI 502-02	L. Eberson and K. Nyberg; in "Encyclopedia of Electrochemistry of the Elements," ed. A. J. Bard, Marcel Dekker, New York, 1978, vol. 12, pp. 261–328.	24, 28
B-78MI 502-03	J. Q. Chambers; in "Encyclopedia of Electrochemistry of the Elements," ed. A. J. Bard, Marcel Dekker, New York, 1978, vol. 12, pp. 329–502.	81
78MI 503-01	J. F. Knifton; *J. Am. Oil Chem. Soc.*, 1978, **55**, 496.	142
B-78MI 522-01	M. Regitz; in "The Chemistry of Diazonium and Diazo Groups," ed. S. Patai, Wiley, London, 1978, ch. 17, p. 751.	865
B-78MI 522-02	M. Regitz; in "The Chemistry of Diazonium and Diazo Groups," ed. S. Patai, Wiley, London, 1978, ch. 17, p. 800.	869
B-78MI 522-03	M. Regitz; in "The Chemistry of Diazonium and Diazo Groups," ed. S. Patai, Wiley, London, 1978, ch. 17, p. 799.	869
78MI 530-01	R. J. Maxwell and L. S. Silbert; *J. Am. Oil Chem. Soc.*, 1978, **55**, 583.	1109
78MI 530-02	A. Arcas and P. Royo; *An. Univ. Murcia, Cienc.*, 1978, **31**, 61.	1138
78NKK265	T. Fujinami, N. Otani and S. Sakai; *Nippon Kagaku Kaishi* (*J. Chem. Soc. Jpn.*), 1978, 265.	1031
78OPP289	J. Bergman and L. Engman; *Org. Prep. Proced. Int.*, 1978, **10**, 289.	244
78PAC19	H. Schmidbaur; *Pure Appl. Chem.*, 1978, **50**, 19.	894
78PHA375	E. Hannig and G. Beyer; *Pharmazi*, 1978, **33**, 375.	652
78PHA568	G. Wagner and B. Voigt; *Pharmazi*, 1978, **33**, 568.	755, 758
78PS(4)317	J. P. Majoral, M. Revel and J. Navech; *Phosphorus Sulfur*, 1978, **4**, 317.	705
78S38	K. Findeisen and K. Wagner; *Synthesis*, 1978, 38.	658
78S50	L. W. Christensen and D. J. Heacock; *Synthesis*, 1978, 50.	251
78S152	P. D. Woodgate, Ho. H. Lee, P. S. Rutledge and R. C. Cambie; *Synthesis*, 1978, 152.	1115
78S154	B. Uznanski and W. J. Stec; *Synthesis*, 1978, 154.	1135
78S303	S. Linke; *Synthesis*, 1978, 303.	282, 300
78S439	M. Schoufs, J. Meijer, P. Vermeer, L. Brandsma; *Synthesis*, 1978, 439.	547, 552
78S489	G. R. Lenz; *Synthesis*, 1978, 489.	288, 297, 302
78S577	W. Verboom, J. Meijer and L. Brandsma; *Synthesis*, 1978, 577.	1105
78S668	V. I. Cohen; *Synthesis*, 1978, 668.	625
78S752	U. Schmidt and R. Schölm; *Synthesis*, 1978, 752.	262
78S775	P. H. Benders and P. A. E. van Erkelens; *Synthesis*, 1978, 775.	1124
78S803	S. Sharma; *Synthesis*, 1978, 803.	1027
78S849	J. Voss and H. Günther; *Synthesis*, 1978, 849.	554
78S882	A. Jonczyk, Z. Ochal and M. Makosza; *Synthesis*, 1978, 882.	1118
78S917	L. Cassar, S. Panossian and C. Giordano; *Synthesis*, 1978, 917.	580
78S928	F. M. F. Chen, K. Kuroda and N. L. Benoiton; *Synthesis*, 1978, 928.	185
78SC19	P. Savignac, M. Snoussi and P. Coutrot; *Synth. Commun.*, 1978, **8**, 19.	103
78SC497	R. Mohd. Ali and N. H. Khan; *Synth. Commun.*, 1978, **8**, 497.	
78SC541	R. A. Volkmann; *Synth. Commun.*, 1978, **8**, 541.	1041
78T541	R. G. Guy and J. J. Thompson; *Tetrahedron*, 1978, **34**, 541.	1110
78T1731	A. K. Mukerjee and A. K. Singh; *Tetrahedron*, 1978, **34**, 1731.	
78TL259	T. Sato, T. Abe and I. Kuwajima; *Tetrahedron Lett.*, 1978, 259.	413
78TL807	B. Zwanenburg, C. G. Venier, P. A. T. W. Porskamp and M. van der Leij; *Tetrahedron Lett.*, 1978, 807.	562
78TL811	M. van der Leij, P. A. T. W. Porskamp, B. H. M. Lammerink and B. Zwanenburg; *Tetrahedron Lett.*, 1978, 811.	562
78TL1269	V. Dourtoglou, J.-C. Ziegler and B. Gross; *Tetrahedron Lett.*, 1978, 1269.	279
78TL1429	H. Böhme and H.-J. Drechsler; *Tetrahedron Lett.*, 1978, 1429.	655, 660
78TL1951	G. Himbert and W. Schwickerath; *Tetrahedron Lett.*, 1978, 1951.	596, 618, 739
78TL2167	Y. Tamaru, T. Harada and Z. Yoshida; *Tetrahedron Lett.*, 1978, 2167.	607, 616
78TL2171	M. J. Sousa Lobo and H. J. Chaves das Neves; *Tetrahedron Lett.*, 1978, **19**, 2171.	209
78TL2175	L. Citerio, M. Garufi and R. Stradi; *Tetrahedron Lett.*, 1978, 2175.	762
78TL2689	R. Appel and G. Erbelding; *Tetrahedron Lett.*, 1978, 2689.	1093
78TL2715	P. Gosselin, S. Masson and A. Thuillier; *Tetrahedron Lett.*, 1978, 2715.	553, 559, 729
78TL3523	J. S. Nimitz and R. H. Wollenberg; *Tetrahedron Lett.*, 1978, 3523.	234
78TL3847	G. Levesque, A. Mahjoub and A. Thuillier; *Tetrahedron Lett.*, 1978, 3847.	546
78TL4065	B. Cazes and S. Julia; *Tetrahedron Lett.*, 1978, 4065.	553
78TL4319	T. Cohen and R. E. Gapinski; *Tetrahedron Lett.*, 1978, 4319.	235
78TL4461	H. J. Liu; W. H. Chang and S. P. Lee; *Tetrahedron Lett.*, 1978, 4461.	131
78TL4475	A. Hassner and V. Alexanian; *Tetrahedron Lett.*, 1978, 4475.	124
78TL4657	A. I. Meyers, T. A. Tait and D. L. Comins; *Tetrahedron Lett.*, 1978, 4657.	557
78TL4839	A. C. Brouwer, A. V. E. George, D. Seykens and H. J. T. Bos; *Tetrahedron Lett.*, 1978, 4839.	559
78TL5211	M. Node, K. Nishide, M. Sai and E. Fujita; *Tetrahedron Lett.*, 1978, 5211.	40, 106
78YGK1090	I. Okuda and H. Sugiyama; *Yuki Gosei Kagaku Kyokaishi*, 1978, **36**, 1090.	1025
78YZ1361	Y. Hamada, M. Sugiura and M. Hirota; *Yakugaku Zasshi*, 1978, **98**, 1361.	1127

78ZAAC(443)148 W. Maringgele and A. Meller; *Z. Anorg. Allg. Chem.*, 1978, **443**, 148. 383, 722

78ZAAC(445)107 W. Maringgele and A. Meller; *Z. Anorg. Allg. Chem.*, 1978, **445**, 107. 777, 778

78ZC90 R. Mayer, H. Viola and B. Hopf; *Z. Chem.*, 1978, **18**, 90. 554

78ZC452 J. Heinicke and A. Tzschach; *Z. Chem.*, 1978, **18**, 452. 791

78ZN(B)24 F. Gerstner and J. Weidlein; *Z. Naturforsch., Teil B.*, 1978, **33**, 24. 780

78ZN(B)75 N. A. Kassab, S. O. Abd Allah and S. A. Elbahaii; *Z. Naturforsch., Teil B: Anorg. Chem., Org. Chem.*, 1978, **33B**, 75. 532

78ZN(B)575 N. Wiberg and G. Hübler; *Z. Naturforsch., Teil B*, 1978, **33**, 575. 1076, 1077, 1083

78ZN(B)636 H. Böhme, K. H. Ahrens, H. J. Drechsler and G. Rumbaur; *Z. Naturforsch., Teil B*, 1978, **33**, 636. 685

78ZN(B)950 U. Klingebiel; *Z. Naturforsch., Teil B*, 1978, **33**, 950. 1074

78ZN(B)1457 E. Lindner and G. Vordermaier; *Z. Naturforsch. Teil B.*, 1978, **33**, 1457. 397

78ZN(B)1465 M. Weidenbruch and H. Pesel; *Z. Naturforsch., Teil B*, 1978, **33**, 1465. 1134

78ZOB78 V. E. Shishkin, Y. Zotov and B. I. No; *Zh. Obshch Khim.*, 1978, **48**, 78. 774

78ZOB695 M. A. Pudovik, L. K. Kibardina and A. N. Pudovik; *Zh. Obshch. Khim.*, 1978, **48**, 695. 369, 370

78ZOB843 A. N. Pudovik, R. A. Cherkasov, M. G. Zimin and N. G. Zabirov; *Zh. Obshch. Khim.*, 1978, **48**, 843. 643

78ZOB1146 V. F. Gerega, Y. I. Mushkin, Y. I. Baukov and Y. I. Dergunov; *Zh. Obshch. Khim.*, 1978, **48**, 1146. 389, 717

78ZOB1313 V. P. Kozyukov, S. V. Sheludyalova, V. S. Nikitin, N. M. Bizyukova and V. F. Mironov; *Z. Obsch. Khim.*, 1978, **48**, 1313. 1082

78ZOB1429 V. P. Kozyukov, S. V. Sheludyakova, V. S. Nikitin, N. M. Bizyukova and V. F. Mironov; *Zh. Obshch. Khim.*, 1978, **48**, 1429. 721, 722

78ZOB1594 V. I. Boev; *Zh. Obshch. Khim.*, 1978, **48**, 1594. 371

78ZOB1943 V. F. Mironov, V. P. Kozyukov, S. V. Sheludyakov and Vik. P. Kozyukov; *Z. Obshch. Khim.*, 1978, **48**, 1943. 1076, 1077

78ZOB2465 N. M. Mironovskaya, S. F. Dymova, I. L. Vladimirova and N. N. Mel'nikov; *Zh. Obshch. Khim.*, 1978, **48**, 2465. 774

78ZOB2468 M. A. Pudovik, L. K. Kibardina, M. D. Medvedeva, N. P. Anoshina and A. N. Pudovik; *Zh. Obshch. Khim.*, 1978, **48**, 2468. 369

78ZOR280 N. Y. Derkach, T. V. Lyapina and E. S. Levchenko; *Zh. Org. Khim.*, 1978, **14**, 280. 341

78ZOR820 V. B. Piskov and V. P. Kasperovich; *Zh. Org. Khim.*, 1978, **14**, 820. 655

78ZOR943 B. S. Drach and G. N. Mis'kevich; *Zh. Org. Khim.*, 1978, **14**, 943. 655

78ZOR1559 A. D. Sinitsa and S. V. Bonadyk; *Zh. Org. Khim.*, 1978, **14**, 1559. 657

78ZOR1846 E. S. Levchenko, L. V. Budnik and T. N. Dubinina; *Zh. Org. Khim.*, 1978, **14**, 1846. 317, 349

79ACS137 K. A. Jensen and E. Larsen; *Acta Chem. Scand.*, 1979, **33**, 137. 641

79ACS(B)544 E. H. Mørkved; *Acta Chem. Scand., Ser. B*, 1979, **33**, 544. 716

79AG(E)622 J. Mulzer, G. Brüntrup and A. Chucholowski; *Angew. Chem., Int. Ed. Engl.*, 1979, **18**, 622. 161

79AG(E)689 H. R. Kricheldorf; *Angew. Chem., Int. Ed. Engl.*, 1979, **18**, 689. 225

79AG(E)707 T. Mukaiyama; *Angew. Chem., Int. Ed. Engl.*, 1979, **18**, 707. 979

79AG(E)793 J. Mulzer, G. Brüntrup; *Angew. Chem., Int. Ed. Engl.*, 1979, **18**, 793. 170

79AG(E)941 A. Senning; *Angew. Chem., Int. Ed. Engl.*, 1979, **18**, 941. 247

79AHC(24)63 E. Schmitz; *Adv. Heterocycl. Chem.*, 1979, **24**, 63. 283

79AP822 K. Gottschalk and P. Pachaly; *Arch. Pharm. (Weinheim, Ger.)*, 1979, **312**, 822. 384

79BCJ533 H. Kamogawa, M. Nanasawa, S. Uehara and K. Osawa; *Bull. Chem. Soc. Jpn.*, 1979, **52**, 533. 1065

79BCJ1541 H. Yoshida, T. Ogata and S. Inokawa; *Bull. Chem. Soc. Jpn.*, 1979, **52**, 1541. 775

79BCJ1989 J. Inanaga, K. Hirata, H. Saeki, T. Katsuki and M. Yamaguchi; *Bull. Chem. Soc. Jpn.*, 1979, **52**, 1989. 173

79BCJ3377 A. Takaoka, H. Iwakiri and N. Ishikawa; *Bull. Chem. Soc. Jpn.*, 1979, **52**, 3377. 6

79BCJ3611 N. Watanabe, S. Uemura and M. Okano; *Bull. Chem. Soc. Jpn.*, 1979, **52**, 3611. 1042

79BSF237 C. Paulmier; *Bull. Soc. Chim. Fr.*, 1979, 237. 1108

79CB510 H. Schmidbaur and A. Wohlleben-Hammer; *Chem. Ber.*, 1979, **112**, 510. 908, 1094

79CB708 U. Schubert, M. Wiener and F. H. Köhler; *Chem. Ber.*, 1979, **112**, 708. 936

79CB763 G. Frey, H. Lesiecki, E. Lindner and G. Vordermaier; *Chem. Ber.*, 1979, **112**, 763. 403

79CB793 H. Lesiecki, E. Lindner and G. Vordermaier; *Chem. Ber.*, 1979, **112**, 793. 403

79CB834 H. Werner, R. Feser and W. Buchner; *Chem. Ber.*, 1979, **112**, 834. 458

79CB1000 E. Schaumann and J. Ehlers; *Chem. Ber.*, 1979, **112**, 1000. 604, 623

79CB1093 R. Appel, M. Wander and F. Knoll; *Chem. Ber.*, 1979, **112**, 1093. 910

79CB1456 E. Lindner and G. Vordermaier; *Chem. Ber.*, 1979, **112**, 1456. 403

79CB1650 H. Gotthardt, O. M. Huss and C. M. Weisshuhn; *Chem. Ber.*, 1979, **112**, 1650. 551

79CB2158 W. Lutz and W. Sundermeyer; *Chem. Ber.*, 1979, **112**, 2158. 721

79CC179 M. Yamada Y. Watabe, T. Sakakibara and R. Sudoh; *J. Chem. Soc., Chem. Commun.*, 1979, 179. 183, 199

79CC215 G. Khan, K. A. Pover and F. Scheinmann; *J. Chem. Soc., Chem. Commun.*, 1979, 215. 101

79CC645 A. G. M. Barrett, D. H. R. Barton and R. W. Read; *J. Chem. Soc., Chem. Commun.*, 1979, 645. 526

79CC653 H. Eshtiagh-Hosseini, H. W. Kroto and J. F. Nixon; *J. Chem. Soc., Chem. Commun.*, 1979, 653. 1152

79CC755 D. H. R. Barton, S. V. Ley and C. A. Meerholz; *J. Chem. Soc., Chem. Commun.*, 1979, 755. 523

79CC970 A. Nishinaga, T. Shimizu and T. Matsuura; *J. Chem. Soc., Chem. Commun.*, 1979, 970. 299

79CC1026 M. Giffard and J. Cousseau; *J. Chem. Soc., Chem. Commun.*, 1979, 1026. 1108

79CC1102 K. A. Lerstrup and L. Henriksen; *J. Chem. Soc., Chem. Commun.*, 1979, 1102. 627, 642, 726

79CC1180 A. Devos, J. Remion, A. M. Frisque-Hesbain, A. Colens and L. Ghosez; *J. Chem. Soc., Chem. Commun.*, 1979, 1180. 2

79CI(L)319 J. B. Bremner and K. N. Winzenberg; *Chem. Ind. (London)*, 1979, 319. 1127

79CJC2145 D. G. Garratt, M. D. Ryan and M. Ujjainwalla; *Can. J. Chem.*, 1979, **57**, 2145. 1107

79CJC2180 D. G. Garratt; *Can. J. Chem.*, 1979, **57**, 2180. 1108

79CL1019 A. Nakama, S. Takimoto, I. Inanaga, T. Katsuki, S. Ouchida, K. Inoue, M. Aiga, N. Okukado and M. Yamaguchi; *Chem. Lett.*, 1979, 1019. 173

79CL1021 I. Inanaga, T. Katsuki, S. Takimoto, S. Ouchida, K. Inoue, A. Nakama, N. Okukado and M. Yamaguchi; *Chem. Lett.*, 1979, 1021. 173

79COC(2)513 G. Tennant; *Comp. Org. Chem.*, 1979, **2**, 513. 1022

79COC(2)841 I. O. Sutherland; *Comp. Org. Chem.*, 1979, **2**, 841.

79COC(2)869 I. O. Sutherland; *Comp. Org. Chem.*, 1979, **2**, 869. 122

79COC(3)305 D. R. Hogg; *Comp. Org. Chem.*, 1979, **3**, 305. 1022

79COC(3)373 F. Duus; *Comp. Org. Chem.*, 1979, **3**, 373. 472, 510, 546

79COC(3)440 F. Duus; *Comp. Org. Chem.*, 1979, **3**, 440. 565

79COC(3)461 F. Duus; *Comp. Org. Chem.*, 1979, **3**, 461. 624, 1022

79COC(3)522 P. D. Magnus; *Comp. Org. Chem.*, 1979, **3**, 522. 1051

79CP1051008 R. G. Micetich and R. B. Morin (Connlab Holdings Ltd); *Can. Pat.* 1 051 008 (1979) (*Chem. Abstr.*, 1979, **91**, 74 450). 736

79CPB1137 H. Takahashi, K. Takeda, N. Nimura and H. Ogura; *Chem. Pharm. Bull.*, 1979, **27**, 1137. 1049

79CPB1352 M. Obayashi, S. Kuzuna and S. Noguchi; *Chem. Pharm. Bull.*, 1979, **27**, 1352. 601

79CPB1931 H. Yasuo, M. Suzuki and N. Yoneda; *Chem. Pharm. Bull.*, 1979, **27**, 1931. 93

79GEP142333 D. Martin and M. Bauer; *Ger. Pat.* 142 333 (1979) (*Chem. Abstr.*, 1981, **94**, 191 923b). 1103

79GEP2817455 W. Hartmann and H. G. Heine; *Ger. Pat.* 2 817 455 (1979) (*Chem. Abstr.*, 1980, **92**, 128 468a). 1107

79GSM67 D. W. Knight; *Gen. Synth. Methods*, 1979, **2**, 67. 24

79H(12)164 Y. Hamada and M. Sugiura; *Heterocycles*, 1979, **12**, 164. 1127

79H(12)485 H. Okuda, Y. Tominaga, Y. Matsuda and G. Kobayashi; *Heterocycles*, 1979, **12**, 485. 559

79H(13)235 T. Sakasai, T. Sakamoto and H. Yamanaka; *Heterocycles*, 1979, **13**, 235. 116

79HCA1401 J. M. J. Tronchet and O. R. Martin; *Helv. Chim. Acta*, 1979, **62**, 1401. 895, 896

79HCA2091 J. M. J. Tronchet and B. Gentile; *Helv. Chim. Acta*, 1979, **62**, 2091. 155

79HCA2825 G. Fráter; *Helv. Chim. Acta*, 1979, **62**, 2825. 148

79HCA2829 G. Fráter; *Helv. Chim. Acta*, 1979, **62**, 2829. 148

79IC937 R. J. Balahura and W. L. Purcell; *Inorg. Chem.*, 1979, **18**, 937. 386

79IC2928 R. A. Anderson; *Inorg. Chem.*, 1979, **18**, 2928. 797

79ICA345 R. Bardi, A. Del Pra, A. M. Piazzesi and L. Toniolo; *Inorg. Chim. Acta*, 1979, **35**, 345. 142

79ICAL99 T. N. Srivastava, R. C. Srivastava and M. Singh; *Inorg. Chim. Acta*, 1979, **33**, L99. 342

79IS164 E. O. Fischer, U. Schubert, W. Kleine and H. Fischer; *Inorg. Synth.*, 1979, **19**, 164. 652

79IZV805 A. N. Pudovik, R. A. Cherkasov, M. G. Zimin and N. G. Zabirov; *Izv. Akad. Nauk SSSR, Ser. Khim.*, 1979, **4**, 805. 643, 644

79IZV861 A. N. Pudovik, R. A. Cherkasov, M. G. Zimin and N. G. Zabirov; *Izv. Akad. Nauk SSSR, Ser. Khim.*, 1979, 861. 581

79JA83 A. G. Brook, J. W. Harris, J. Lennon and E. El Sheikh; *J. Am. Chem. Soc.*, 1979, **101**, 83. 924

79JA259 G. Schmid, T. Fukuyama, K. Akasaka and Y. Kishi; *J. Am. Chem. Soc.*, 1979, **101**, 259. 155

79JA489 J. W. Suggs; *J. Am. Chem. Soc.*, 1979, **101**, 489. 800

79JA1284 B. M. Trost, C. D. Shuey, F. deNinno, Jr. and S. S. McElvain; *J. Am. Chem. Soc.*, 1979, **101**, 1284. 164

79JA2225 H. J. Reich, J. J. Rusek and R. E. Olson; *J. Am. Chem. Soc.*, 1979, **101**, 2225. 410

79JA2249 F. Khouri and M. K. Kaloustian; *J. Am. Chem. Soc.*, 1979, **101**, 2249. 512

79JA3521 D. B. Carr and J. Schwartz; *J. Am. Chem. Soc.*, 1979, **101**, 3521. 467

79JA4245 H. Alper and M. Tanaka; *J. Am. Chem. Soc.*, 1979, **101**, 4245. 655

79JA4732 A. I. Meyers, D. M. Roland, D. L. Comins, R. Henning, M. P. Fleming, K. Shimizu; *J. Am. Chem. Soc.*, 1979, **101**, 4732. 557

79JA6750 A. G. Brook, S. C. Nyburg, W. F. Reynolds, Y. C. Poon, Y.-M. Chang, J.-S. Lee and J.-P. Picard; *J. Am. Chem. Soc.*, 1979, **101**, 6750. 924

79JA7104 A. I. Meyers, D. L. Comins, D. M. Roland, R. Henning and K. Shimizu; *J. Am. Chem. Soc.*, 1979, **101**, 7104. 557

79JA7131 E. J. Corey, P. B. Hopkins, S. Kim, S. Yoo, K. P. Nambiar and J. R. Falck; *J. Am. Chem. Soc.*, 1979, **101**, 7131. 173

79JAP7946762 T. Sasaki; *Jpn. Pat.* 7 946 762 (1979) (*Chem. Abstr.*, 1979, **91**, 174 915w). 652

79JCR(S)140 T. Takeshima, A. Yano, N. Fukada, Y. Hirose and M. Muraoka; *J. Chem. Res. (S)*, 1979, 140. 547

79JCR(S)148 A. W. Faull and R. Hull; *J. Chem. Res. (S)*, 1979, 148. 1035

79JCR(S)160 C.-P. Klages, W.-D. Malmberg and J. Voss; *J. Chem. Res. (S)*, 1979, 160. 510, 516, 523

79JCS(D)1246 G. K. Anderson and R. J. Cross; *J. Chem. Soc., Dalton Trans.*, 1979, 1246. 465

79JCS(D)1405 N. P. C. Westwood, H. W. Kroto, J. F. Nixon and N. P. C. Simmons; *J. Chem. Soc., Dalton Trans.*, 1979, 1405. 1153

79JCS(P1)446 A. R. Katritzky, J. Lewis and P. Nie; *J. Chem. Soc., Perkin Trans. 1*, 1979, 446. 1002

79JCS(P1)757 R. C. Cambie, H. H. Lee, P. S. Rutledge and P. D. Woodgate; *J. Chem. Soc., Perkin Trans. 1*, 1979, 757. 1044

79JCS(P1)771 M. Mori, S. Kudo and Y. Ban; *J. Chem. Soc., Perkin Trans. 1*, 1979, 771. 157

79JCS(P1)1255 A. J. Pearson; *J. Chem. Soc., Perkin Trans. 1*, 1979, 1255. 146

79JCS(P1)1762 S. Rajappa, M. D. Nair, B. G. Advani, R. Sreenivasan and J. A. Desai; *J. Chem. Soc., Perkin Trans., 1*, 1979, 1762. 760

79JCS(P1)1957 A. R. Katritzky and P. Molina-Buendia; *J. Chem. Soc., Perkin Trans. 1*, 1979, 1957. 1002

79JCS(P1)2001 S. Rajappa, T. G. Rajagopalan, R. Sreenivasan and S. Kanal; *J. Chem. Soc., Perkin Trans. 1*, 1979, 2001. 1035

79JCS(P1)2079 A. A. Durrani and J. H. P. Tyman; *J. Chem. Soc., Perkin Trans. 1*, 1979, 2079. 110

79JCS(P1)2478 S. A. Matlin, P. G. Sammes and R. M. Upton; *J. Chem. Soc., Perkin Trans. 1*, 1979, 2478. 720

79JCS(P2)634 B. C. Challis, J. A. Challis and I. R. McDermott; *J. Chem. Soc., Perkin Trans. 2*, 1979, 634. 714

79JGU88 O. I. Kolodyazhnyi; *J. Gen. Chem. USSR (Engl. Transl.)*, 1979, **49**, 88. 902, 903

79JGU940 E. P. Levedev, V. A. Baburina and M. D. Mizhiritskii; *J. Gen. Chem. USSR (Engl. Transl.)*, 1979, 940. 538, 539

79JGU1740 N. N. Magdesieva and V. A. Danilenko; *J. Gen. Chem. USSR (Engl. Transl.)*, 1979, **49**, 1740. 898, 903

79JGU2323 M. G. Zimin, M. M. Afanasev and A. N. Pudovik; *J. Gen. Chem. USSR (Engl. Trans.)*, 1979, **49**, 2323. 738

79JHC365 V. I. Cohen; *J. Heterocycl. Chem.*, 1979, **16**, 365. 522

79JHC409 T. A. Ondrus, E. E. Knaus and C. S. Giam; *J. Heterocycl. Chem.*, 1979, **16**, 409. 1119

79JHC1295 U. Timm and H. Meier; *J. Heterocycl. Chem.*, 1979, **16**, 1295. 511

79JIC545 H. Singh; *J. Indian Chem. Soc.*, 1979, **56**, 545. 597

79JOC420 J.-P. Picard, R. Calas, J. Dunoguès, N. Duffaut, J. Gerval and P. Lapouyade; *J. Org. Chem.*, 1979, **44**, 420. 419

79JOC569 D. F. Aycock and G. R. Jurch, Jr.; *J. Org. Chem.*, 1979, **44**, 569. 244, 546

79JOC633 C. D. Bedford and A. T. Nielsen; *J. Org. Chem.*, 1979, **44**, 633. 690

79JOC1016 J. Cuomo and R. A. Olofson; *J. Org. Chem.*, 1979, **44**, 1016. 4

79JOC1177 L. A. Carpino and J. R. Williams; *J. Org. Chem.*, 1979, **44**, 1177. 490

79JOC1247 G. A. Olah, S. C. Narang, B. G. B. Gupta and R. Malhotra; *J. Org. Chem.*, 1979, **44**, 1247. 226

79JOC1447 T. M. Balthazor, D. E. Godar and B. R. Stults; *J. Org. Chem.*, 1979, **44**, 1447. 321

79JOC1554 S. Torii, H. Tanaka and M. Ukida; *J. Org. Chem.*, 1979, **44**, 1554. 336

79JOC1695 B. Verček, I. Leban, B. Stanovnik and M. Tišler; *J. Org. Chem.*, 1979, **44**, 1695. 834

79JOC1999 A. Hassner, T. K. Morgan Jr. and A. R. McLaughlin; *J. Org. Chem.*, 1979, **44**, 1999. 384

79JOC2288 M. E. Landis and J. C. Mitchell; *J. Org. Chem.*, 1979, **44**, 2288. 366

79JOC2807 P. Gosselin, S. Masson and A. Thuillier; *J. Org. Chem.*, 1979, **44**, 2807. 561

79JOC2863 P. K. Claus, F. W. Vierhapper and R. L. Willer; *J. Org. Chem.*, 1979, **44**, 2863. 338

79JOC2945 P. A. Grieco, D. S. Clark and G. P. Withers; *J. Org. Chem.*, 1979, **44**, 2945. 652

79JOC3069 R. D. Rieke and A. V. Kavaliunas; *J. Org. Chem.*, 1979, **44**, 3069. 1139

79JOC3601 T. Cohen and R. B. Weisenfeld; *J. Org. Chem.*, 1979, **44**, 3601. 101

79JOC3726 R. C. Schnur; *J. Org. Chem.*, 1979, **44**, 3726. 757

79JOC3872 G. A. Olah, J. T. Welch, Y. D. Vankar, M. Nojima, I. Kerekes and J. A. Olah; *J. Org. Chem.*, 1979, **44**, 3872. 4

79JOC4160 Y. Lin, S. A. Lang, Jr., M. F. Lovell and N. A. Perkinson; *J. Org. Chem.*, 1979, **44**, 4160. 759

79JOC4727 N. Kornblum and S. Singaram; *J. Org. Chem.*, 1979, **44**, 4727. 275

79JOM(166)175 J. Heinicke and A. Tzschach; *J. Organomet. Chem.*, 1979, **166**, 175. 791, 892

79JOM(166)365 T. Allman, R. G. Goel and H. S. Prasad; *J. Organomet. Chem.*, 1979, **166**, 365. 1133

79JOM(166)373 B. Kohne, W. Lohner, K. Praefcke, H. J. Jakobsen and B. Villadsen; *J. Organomet. Chem.*, 1979, **166**, 373. 241

79JOM(168)97 M. Tanaka and H. Alper; *J. Organomet. Chem.*, 1979, **168**, 97. 801

79JOM(172)177 M. Dilgassa and M. D. Curtis; *J. Organomet. Chem.*, 1979, **172**, 177. 447

79JOM(172)251 R. D. Adams, D. F. Chodosh, N. M. Golembeski and E. C. Weissman; *J. Organomet. Chem.*, 1979, **172**, 251. 798, 799, 800

79JOM(175)273 A. R. Bassindale and T. B. Posner; *J. Organomet. Chem.*, 1979, **175**, 273. 720

79JOM(175)303 D. A. Clarke, M. M. Hunt and D. W. Raymond; *J. Organomet. Chem.*, 1979, **175**, 303. 1140

79JOM(175)33 F. Gerstner, W. Schwarz, H. D. Hausen and J. Weidlein; *J. Organomet. Chem.*, 1979, **175**, 33. 780

79JOM(175)C1 A. Hosomi, H. Hashimoto and H. Sakurai; *J. Organomet. Chem.*, 1979, **175**, C1. 419

79JOM(177)101 O. A. Reutov, V. I. Rozenberg, G. V. Gamlova and V. A. Nikanorov; *J. Organomet. Chem.*, 1979, **177**, 101. 1142, 1145

79JOM(182)465 H. Weichmann, B. Ochsler, I. Duchek and A. Tzschach; *J. Organomet. Chem.*, 1979, **182**, 465. 390

79JOU32 B. S. Drach, V. A. Kovalev, T. P. Popovich, A. D. Gordeev, G. B. Soifer, V. N. Kalinin and A. A. Kisilenko; *J. Org. Chem. USSR (Engl. Transl.)*, 1979, **15**, 32. 657, 659

79JPR928 K. Unverferth, R. Höntsch and K. Schwetlick; *J. Prakt. Chem.*, 1979, **321** (6), 928. 984

79JPR1034 S. M. M. Elshafie; *J. Prakt. Chem.*, 1979, **321**, 1034. 632

79LA74 H. Quast, R. Frank, B. Freudenreich, P. Schäfer and E. Schmitt; *Justus Liebigs Ann. Chem.*, 1979, 74. 657

79LA219 I. Hoppe and U. Schöllkopf; *Justus Liebigs Ann. Chem.*, 1979, 219. 162

79LA1547 K. Burger and M. Eggersdorfer; *Justus Liebigs Ann. Chem.*, 1979, 1547. 79

79LA1702 E. Schaumann and F.-F. Grabley; *Justus Liebigs Ann. Chem.*, 1979, 1702. 589, 604

79LA1715 E. Schaumann and F.-F. Grabley; *Justus Liebigs Ann. Chem.*, 1979, 1715. 589, 593

79LA1746 E. Schaumann and F.-F. Grabley; *Justus Liebigs Ann. Chem.*, 1979, 1746. 589, 614

79M473 W. Maringgele and A. Meller; *Monatsh. Chem.*, 1979, **110**, 473-81. 381, 382

79MI 501-01 E. L. Vulakh, V. M. Ivanova, L. V. Rad'ko, M. N. Chernobrovkina and D. S. Bark; *Khim. Prom-st (MOSCOW)*, 1979, 16 (*Chem. Abstr.*, 1979, **91**, 5008). 13

79MI 502-01 L. M. Bova; *Khim. Prom-st. Rubezhom*, 1979, 31 (*Chem. Abstr.*, 1979, **91**, 74 139c). 24

B-79MI 502-01	M. A. Ogliaruso and J. F. Wolfe; in "The Chemistry of Functional Groups," Supplement B. Part 1, ed. S. Patai, Wiley-Interscience, New York, 1979, chap. 7, pp. 270–490.	24
79MI 502-02	M. Ohta; *Yushi*, 1979, **32**, 50 (*Chem. Abstr.*, 1980, **93**, 25 800u).	24
79MI 502-03	R. V. Kucher, I. A. Opeida and L. G. Nechitailo; *Visn. Akad. Nauk Urk RSR*, 1979, 48 (*Chem. Abstr.*, 1980, **92**, 112 828g).	28
79MI 502-04	H. J. Schaefer; *Chem. Phys. Lipids*, 1979, **24**, 32.	28
79MI 502-05	N. Gerti and E. Avni; *Polim. Vehomorim Plast.*, 1979, **9**, 7 (*Chem. Abstr.*, 1980, **93**, 95 952u).	28
79MI 502-06	N. Ishikawa; *Kagaku* (*Kyoto*), 1979, **34**, 158 (*Chem. Abstr.*, 1979, **91**, 156 954d).	57
79MI 502-07	D. Jyst and A. V. S. Rao; *Curr. Sci.*, 1979, **48**, 439.	67
79MI 502-08	H. Parnes; *J. Labelled Cmpd. Radiopharm.*, 1979, **16**, 771.	82
79MI 502-09	A. D. Kagarlitskii and B. V. Suvorov; *Tr. Inst. Khim. Nauk, Adad. Nauk. Kaz SSR*, 1979, **49**, 3 (*Chem. Abstr.*, 1980, **92**, 147 230h).	105
B-79MI 511-01	J. Voss; in Supplement B, "The Chemistry of Acid Derivatives," ed. S. Patai, Wiley, New York, 1979, part 2, p. 1021.	480
B-79MI 511-02	S. Scheithauer and R. Mayer; in "Topics in Sulfur Chemistry," ed. A. Senning, Georg Thième Verlag, Stuttgart, 1979, vol. 4, p. 10.	480
B-79MI 513-01	J. Voss; in "Supplement B: the Chemistry of Acid Derivatives," ed. S. Patai, Wiley, New York, 1979, p. 1021.	546
B-79MI 513-02	S. Scheithauer and R. Mayer; in "Topics in Sulfur Chemistry," ed. A. Senning, Georg Thieme Publishers, Stuttgart, 1979, vol. 4.	546, 549
B-79MI 517-01	W. Kantlehner; in "Iminium Salts in Organic Chemistry," eds H. Böhme and H. G. Viehe, Wiley, New York, 1979, 181.	688
B-79MI 518-01	W. Kantlehner; in "Advances in Organic Chemistry," ed. H. Bohme and H. G. Viehe, Wiley, Chichester, 1979, vol. 9, p. 279.	726
79NAT854	K. C. Rice, A. Brossi, J. Tallman, S. M. Paul and P. Skolnick; *Nat.*, 1979, **278**, 854.	1029
79NKK1224	Y. Nakamura, N. Yoshioka, T. Ouchi and M. Imoto; *Nippon Kagaku Kaishi*, 1979, 1224 (*Chem. Abstr.*, 1980, **92**, 22 875).	541
79OPP93	C. H. Wang and H. H. Tso; *Org. Prep. Proced. Int.*, 1979, **11**, 93.	140
79OS(59)26	B. Haveaux, A. Dekoker, M. Rens, A. R. Sidani, J. Toye and L. Ghosez; *Org. Synth.*, 1979, **59**, 26.	2
79PS(7)69	L. Morin, J. Lebaud, D. Paquer, R. Chaussin and D. Barillier; *Phosphorus Sulfur*, 1979, **7**, 69.	552
79RCR473	V. F. Mironov, V. D. Sheludyakov and V. P. Kozyukov; *Russ. Chem. Rev.* (*Engl. Transl.*), 1979, **48**, 473.	968
79RTC55	R. S. Sukhai and L. Brandsma; *Recl. Trav. Chim. Pays-Bas*, 1979, **98**, 55.	626
79S97	M. Haake, H. Gebbing and H. Benack; *Synthesis*, 1979, 97.	339
79S100	J. M. Simpson, D. C. Kapp and T. M. Chapman; *Synthesis*, 1979, 100.	360
79S102	H. Kristinsson; *Synthesis*, 1979, 102.	668
79S181	D. N. Harpp, B. T. Friedlander and R. A. Smith; *Synthesis*, 1979, 181.	1111
79S204	K. Herrmann and G. Simchen; *Synthesis*, 1979, 204.	1144
79S228	G. Sosnovsky and J. A. Krogh; *Synthesis*, 1979, 228.	334
79S274	G. A. Olah, B. G. B. Gupta and S. C. Narang; *Synthesis*, 1979, 274.	270, 282, 300
79S299	P. Dalla Croce, P. Del Buttero, E. Licandro and S. Maiorana; *Synthesis*, 1979, 299.	905
79S432	H. Westmijze, H. Kleijn, J. Meijer and P. Vermeer; *Synthesis*, 1979, 432.	548, 550, 559, 560
79S522	S. Hünig and G. Wehner; *Synthesis*, 1979, 522.	1134
79S523	J. K. Rasmussen and S. M. Heilmann; *Synthesis*, 1979, 523.	1134
79S527	A. Koziara, S. Zawadzki and A. Zwierzak; *Synthesis*, 1979, 527.	273, 285
79S537	G. A. Olah and A. P. Fung; *Synthesis*, 1979, 537.	282
79S546	E. B. Pedersen; *Synthesis*, 1979, 546.	745
79S549	T. Gajda, A. Koziara, S. Zawadzki and A. Zwierzak; *Synthesis*, 1979, 549.	285
79S633	S. F. Martin; *Synthesis*, 1979, 633.	28
79S691	A. Zwierzak and E. Slusarska; *Synthesis*, 1979, 691.	707
79S705	H. Suzuki, M. Usuki and T. Hanafusa; *Synthesis*, 1979, 705.	1055, 1113
79S726	P. Kumar and A. K. Mukerjee; *Synthesis*, 1979, 726.	291
79S738	V. N. Fetyukhin, M. V. Vovk and L. I. Samarai; *Synthesis*, 1979, 738.	1063
79S887	Y. Tamura, T. Kawasaki, M. Adachi and Y. Kita; *Synthesis*, 1979, 887.	595
79S946	M. Golfier and R. Milcent; *Synthesis*, 1979, 946.	704
79SC17	N. C. Gonnella, M. V. Lakshmikantham and M. P. Cava; *Synth. Commun.*, 1979, **9**, 17.	561
79SC31	H.-J. Liu, W. H. Chan and S. P. Lee; *Synth. Commun.*, 1979, **9**, 31.	280
79SC363	M. M. Kidwai, N. H. Khan and M. Ali; *Synth. Commun.*, 1979, **9**, 363.	94
79SC477	R. Hull, *Synth. Commun.*, 1979, **9**, 477.	1035
79SC757	T.-Y. Luh and S. H. Fung; *Synth. Commun.*, 1979, **9**, 757.	285
79SST52	G. C. Barrett; *Org. Compd. Sulphur, Selenium, Tellurium* (*R. Soc. Chem. Series*), 1979, **5**, 52.	1022
79SST133	D. R. Hogg; *Org. Compd. Sulphur, Selenium, Tellurium* (*R. Soc. Chem. Series*), 1979, **5**, 133.	472
79SST139	W. Walter and J. Voss; *Org. Compd. Sulphur, Selenium, Tellurium* (*R. Soc. Chem. Series*), 1979, **5**, 139.	472
79SST156	W. Walter and J. Voss; *Org. Compd. Sulphur, Selenium, Tellurium* (*R. Soc. Chem. Series*), 1979, **5**, 156.	565
79T43	P. Milliet and X. Lusinchi; *Tetrahedron*, 1979, **35**, 43.	789
79T1087	M. Helbert, J. P. Renou and M. L. Martin; *Tetrahedron*, 1979, **35**, 1087.	577, 608, 609
79T1675	R. F. Abdulla and R. S. Brinkmeyer; *Tetrahedron*, 1979, **35**, 1675.	746
79T2077	C. Lion, J. E. Dubois, J. A. MacPhee and Y. Bonzougou; *Tetrahedron*, 1979, **35**, 2077.	32

79T2169 R. A. W. Johnstone and M. E. Rose; *Tetrahedron*, 1979, **35**, 2169. 273, 285, 300
79T2329 J. A. Gladysz, V. K. Wong and B. S. Jick; *Tetrahedron*, 1979, **35**, 2329. 242, 252
79TL49 G. L'abbé and A. Verbruggen; *Tetrahedron Lett.*, 1979, 49. 1123
79TL99 S. K. Chaudhary and O. Hernandez; *Tetrahedron Lett.*, 1979, **20**, 99. 374
79TL133 A. Cowell and J. K. Stille; *Tetrahedron Lett.*, 1979, 133. 143
79TL703 A. M. Sarpeshkar, G. J. Gossick and J. Wemple; *Tetrahedron Lett.*, 1979, 703. 232
79TL871 A. Sekiguchi, Y. Kabe and W. Ando; *Tetrahedron Lett.*, 1979, 871. 413, 430
79TL1039 Y. Ito, K. Kobayashi and T. Saegusa; *Tetrahedron Lett.*, 1979, 1039. 595, 620
79TL1129 Y. Tamura, T. Kawasaki, N. Gohda and Y. Kita; *Tetrahedron Lett.*, 1979, 1129. 1107
79TL1133 M. Mori and Y. Ban; *Tetrahedron Lett.*, 1979, 1133. 157
79TL1185 G. E. Keck and R. Webb; *Tetrahedron Lett.*, 1979, **20**, 1185. 326
79TL2251 Y. Tamura, M. Adachi, T. Kawasaki and Y. Kita; *Tetrahedron Lett.*, 1979, **20**, 2251. 1055, 1114
79TL2253 Y. Ito, K. Kyono and T. Matsuura; *Tetrahedron Lett.*, 1979, **20**, 2253. 356
79TL2271 M. Pohmakotr and D. Seebach; *Tetrahedron Lett.*, 1979, 2271. 552
79TL2517 Y. Nii, K. Okano, S. Kobayashi and M. Ohno; *Tetrahedron Lett.*, 1979, **20**, 2517. 753
79TL3131 R. A. Olofson and K. D. Lotts; *Tetrahedron Lett.*, 1979, 3131. 1122
79TL3809 Y.-F. Cheung; *Tetrahedron Lett.*, 1979, 3809. 18
79TL3851 Y. Yonezawa, C.-G. Shin, M. Kiyohara and J. Yoshimura; *Tetrahedron Lett.*, 1979, **20**, 3851. 371
79TL4311 Y. Kita, J. Haruta, J. Segawa and Y. Tamura; *Tetrahedron Lett.*, 1979, **20**, 4311. 226, 375, 378
79TL4627 G. E. Keck and J. B. Yates; *Tetrahedron Lett.*, 1979, **20**, 4627. 326
79TL4671 H. Nishiyama, H. Nagase and K. Ohno; *Tetrahedron Lett.*, 1979, 4671. 726
79TL4761 A. Bigotto, M. Forchiassin, A. Risaliti and C. Russo; *Tetrahedron Lett.*, 1979, **20**, 4761. 354
79TL4943 Y. Hamashima, T. Kubota, K. Ishikura and W. Nagata; *Tetrahedron Lett.*, 1979, **51**, 4943. 655
79UKZ624 A. F. Shivanyuk; *Ukr. Khim. Zh.* (*Russ. Ed.*), 1979, **45**, 624. 788
79USP771219 T.-S. Chou and I. G. Wright; *US Pat.*, 771 219 (1979) (*Chem. Abstr.*, 1979, **91**, 157 467). 316
79USP4216316 E. E. Flagg; *US Pat.* 4 216 316 (1979) (*Chem. Abstr.*, 1980, **93**, 187 001m). 226
79USP4250105 E. E. Flagg; *US Pat.* 4 250 105 (1979) (*Chem. Abstr.*, 1981, **95**, 6500w). 1102
79USP4302238 M. J. Konz; *US Pat.* 4302238 (1979) (*Chem. Abstr.*, 1982, **96**, 217824p). 533
79ZC59 J. Bodeker, P. Kockritz and K. Courault; *Z. Chem.*, 1979, **19**, 59. 1068
79ZN(B)31 H. H. Karsch; *Z. Naturforsch.*, *Teil B*, 1979, **34**, 31. 908
79ZN(B)768 M. Bahadir, S. Nitz, H. Parlar and F. Korte; *Z. Naturforsch.*, *Teil B*, 1979, **34**, 768. 1037
79ZN(B)1171 H. H. Karsch; *Z. Naturforsch.*, *Teil B*, 1979, **34**, 1171. 910
79ZOB784 V. P. Kozyukov, N. V. Mironova and V. F. Mironov; *Zh. Obshch. Khim.* 1979, **49**, 784 (*Chem. Abstr.*, 1979, **91**, 20 592b). 969
79ZOB2323 M. G. Zimin, M. M. Afanas'ev and A. N. Pudovik; *Zh. Obshch. Khim.* (*Engl. Trans.*), 1979, **49**, 2323. 644
79ZOR38 B. S. Drach, V. A. Kovalev, T. P. Popovich, A. D. Gordeev, G. B. Soifer, V. N. Kalinin and A. A. Kisilenko; *Zh. Org. Khim.*, 1979, **15**, 38 (*Chem. Abstr.*, 1979, **90**, 168 228n). 657, 659
79ZOR2396 N. N. Magdasieva and R. A. Kyandzhetsian; *Zh. Org. Khim.*, 1979, **15**, 2396 (*Engl. Transl.*, 1979, **15**, 2169). 900, 904

80ACS(B)313 E. H. Mørkved; *Acta Chem. Scand., Ser. B*, 1980, **34**, 313. 716
80ACS(B)597 A. A. El-Barbary, S. Scheibye, S. O. Lawesson and H. Fritz; *Acta Chem. Scand., Ser. B*, 1980, **34**, 597. 484
80AG169 K. Bott; *Angew. Chem.*, 1980, **92**, 169. 24
80AG767 A. Schmidpeter, W. Gebler, F. Zwaschka and W. S. Sheldrick; *Angew. Chem.*, 1980, **92**, 767. 1128
80AG(E)133 H. Wissmann and H. J. Kleiner; *Angew. Chem., Int. Ed. Engl.*, 1980, **19**, 133. 264
80AG(E)171 K. Bott; *Angew. Chem., Int. Ed. Engl.*, 1980, **19**, 171. 28
80AG(E)466 J. Mulzer and T. Kerkmann; *Angew. Chem., Int. Ed. Engl.*, 1980, **19**, 466. 729
80AG(E)556 R. Appel and A. Westerhaus; *Angew. Chem., Int. Ed. Engl.*, 1980, **19**, 556. 887
80AG(E)563 G. Adiwidjaja, H. Gunther and J. Voss; *Angew. Chem., Int. Ed. Engl.*, 1980, **19**, 563. 485
80AG(E)566 C. Wentrup, C. Thétaz, E. Tagliaferri, H. J. Lindner, B. Kitschke, H.-W. Winter and H. P. Reisenauer; *Angew. Chem., Int. Ed. Engl.*, 1980, **19**, 566. 1122
80AG(E)726 V. Rautenstrauch and F. Delay; *Angew. Chem., Int. Ed. Engl.*, 1980, **19**, 726. 356
80AG(E)1029 G. Quinkert, U. Schwartz, H. Stark, W.-D. Weber, H. Baier, F. Adam and G. Düner; *Angew. Chem., Int. Ed. Engl.*, 1980, **19**, 1029. 148
80AJC2605 M. I. Bruce, A. G. Swincer, B. J. Thomson and R. C. Wallis; *Aust. J. Chem.*, 1980, **33**, 2605. 944
80ANC967 M. B. Collela, S. Siggia and R. M. Barnes; *Anal. Chem.*, 1980, **52**, 967. 864
80AP(313)1029 K. Hartke and J. Quante; *Arch. Pharm.* (*Weinheim, Ger.*), 1980, **313**, 1029. 558
80BCJ553 S. Uemura, H. Miyoshi, M. Wakasugi, M. Okano, O. Itoh, T. Izumi and K. Ichikawa; *Bull. Chem. Soc. Jpn.*, 1980, **53**, 553. 1113
80BCJ1171 K. Hiraki, Y. Fuchita and T. Masumoto; *Bull. Chem. Soc. Jpn.*, 1980, **53**, 1171. 801
80BCJ1299 T. Yamamoto, Y. Ehara, M. Kubota and A. Yamamoto; *Bull. Chem. Soc. Jpn.*, 1980, **53**, 1299. 384, 387
80BCJ1625 M. Yamashita, M. Kojima, H. Yoshida, T. Ogata and S. Inokawa; *Bull. Chem. Soc. Jpn.*, 1980, **53**, 1625. 403, 404
80BCJ3225 H. Suginome and T. Uchida; *Bull. Chem. Soc. Jpn.*, 1980, **53**, 3225. 362
80CA(93)95215 V. V. Dovlatyan, V. A. Pivazyan and K. A. Eliazyan; *Chem. Abstr.*, 1980, **93**, 95 215. 1120
80CAR(79)205 S. H. Gordon and W. M. Doane; *Carbohydr. Res.*, 1980, **79**, 205. 317
80CB79 J. Goerdeler and R. Losch; *Chem. Ber.*, 1989, **113**, 79. 1064

80CB902 H. Schmidbaur, U. Deschler, B. Zimmer-Gasser, D. Neugebauer and U. Schubert; *Chem. Ber.*, 1980, **113**, 902. 908, 910, 911

80CB912 H. J. Bestmann, G. Schmid and D. Sandmeier; *Chem. Ber.*, 1980, **113**, 912. 343, 730

80CB919 R. Hoffmann and K. Hartke; *Chem. Ber.*, 1980, **113**, 919. 548, 559, 584, 586, 614

80CB1010 E. O. Fischer, F. J. Gammel and D. Neugebauer; *Chem. Ber.*, 1980, **113**, 1010. 226

80CB1095 J. Goerdeler and S. Raddatz; *Chem. Ber.*, 1980, **113**, 1095. 1064, 1066

80CB1394 K. Hartke and H.-M. Wolff; *Chem. Ber.*, 1980, **113**, 1394. 776

80CB1637 E. A. V. Ebsworth, D. W. H. Rankin, B. Zimmer-Gasser and H. Schmidbaur; *Chem. Ber.*, 1980, **113**, 1637. 913

80CB1898 K. Hartke, T. Kissel, J. Quante and R. Matusch; *Chem. Ber.*, 1980, **113**, 1898. 554

80CB2100 F. Effenberger and E. Bessey; *Chem. Ber.*, 1980, **113**, 2100. 236

80CB2519 U. Timm, U. Merkle and H. Meier; *Chem. Ber.*, 1980, **113**, 2519. 511

80CB2769 E. Lindner and G. Frey; *Chem. Ber.*, 1980, **113**, 2769. 403

80CB3010 E. Schaumann, J. Ehlers and F.-F. Grabley; *Chem. Ber.*, 1980, **113**, 3010. 604

80CB3024 E. Schaumann and F.-F. Grabley; *Chem. Ber.*, 1980, **113**, 3024. 589, 604, 605, 614

80CB3268 E. Lindner and G. Frey; *Chem. Ber.*, 1980, **113**, 3268. 397, 398

80CB3318 W. Malisch, M. Kuhn, W. Albert and H. Rössner; *Chem. Ber.*, 1980, **113**, 3318. 911

80CB3605 W. S. Sheldrick, A. Schönberg, E. Singer and P. Eckert; *Chem. Ber.*, 1980, **113**, 3605. 548

80CC130 A. F. Hegarty and A. Chandler; *J. Chem. Soc., Chem. Commun.*, 1980, 130. 748

80CC327 N. Matsumura, Y. Sakaguchi, T. Ohba and H. Inoue; *J. Chem. Soc., Chem. Commun.*, 1980, 327. 70, 73

80CC333 T. A. Cooper, H. W. Kroto, J. F. Nixon and O. Ohashi; *J. Chem. Soc., Chem. Commun.*, 1980, 333. 1152, 1153, 1159

80CC341 D. F. Marten; *J. Chem. Soc., Chem. Commun.*, 1980, 341. 944

80CC429 A. J. G. Baxter, R. J. Ponsford and R. Southgate; *J. Chem. Soc., Chem. Commun.*, 1980, 429. 235

80CC509 K. Tatsumi, R. Hoffmann and M. H. Whangbo; *J. Chem. Soc. Chem. Commun.*, 1980, 509. 1097

80CC1029 Y. Kiso, T. Miyazaki, M. Satomi, H. Hiraiwa and T. Akita; *J. Chem. Soc., Chem. Commun.*, 1980, 1029. 226

80CC1085 R. J. Ponsford and R. Southgate; *J. Chem. Soc., Chem. Commun.*, 1980, 1085. 234

80CJC1316 D. D. McIntyre, D. Mackay and L. L. Wong; *Can. J. Chem.*, 1980, **58**, 1316. 366

80CJC2784 W. Reeve and T. F. Steckel; *Can. J. Chem.*, 1980, **58**, 2784. 1121

80CL651 S. Hayashi, T. Nakai and N. Ishikawa; *Chem. Lett.*, 1980, 651. 38

80CPB2116 T. Sugawara, H. Masuya, T. Matsuo and T. Miki; *Chem. Pharm. Bull.*, 1980, **28**, 2116. 640, 642

80CZ365 Y. Inanici and H. Parlar; *Chem.-Ztg.*, 1980, **104**, 365. 1037

80GSM75 D. W. Knight; *Gen. Synth. Methods*, 1980, **3**, 75. 24

80HCA1383 G. Fráter; *Helv. Chim. Acta*, 1980, **63**, 1383. 148

80HOU(4)736 H. Muth and M. Sauerbier; *Methoden Org. Chem.* (*Houben-Weyl*), 1980, **4**, 736. 353

80HOU(6/1a)874 J. Thiem; *Methoden Org. Chem.* (*Houben-Weyl*), 1980, **6/1a**, 874. 141

80IC2564 R. E. Cramer, R. B. Maynard and J. W. Gilje; *Inorg. Chem.*, 1980, **19**, 2564. 1096

80IC3207 K. I. Gell and J. Schwartz; *Inorg. Chem.*, 1980, **19**, 3207. 917

80IJC(A)69 K. R. Nahar, A. K. Solanki and A. M. Bhandari; *Indian J. Chem., Sect. A*, 1980, **19**, 69. 226

80IZV2623 A. V. Fokin, Yu. N. Studnev, A. I. Rapkin and K. I. Pasevina; *Izv. Akad. Nauk SSSR, Ser. Khim.*, 1980, 2623 (*Chem. Abstr.*, 1981, **94**, 191 645n). 813

80JA1577 J. A. Soderquist and A. Hassner; *J. Am. Chem. Soc.*, 1980, **102**, 1577. 425, 431, 468

80JA1584 T. J. Barton and S. K. Hoekman; *J. Am. Chem. Soc.*, 1980, **102**, 1584. 926, 927

80JA2723 C. P. Casey and D. M. Scheck; *J. Am. Chem. Soc.*, 1980, **102**, 2723. 450

80JA3964 P. T. Lansbury, D. G. Hangauer, Jr. and J. P. Vacca; *J. Am. Chem. Soc.*, 1980, **102**, 3964. 164

80JA4193 A. Cowell and J. K. Stille; *J. Am. Chem. Soc.*, 1980, **102**, 4193. 143, 170

80JA4537 A. F. Hegarty and D. G. McCarthy; *J. Am. Chem. Soc.*, 1980, **102**, 4537. 226

80JA5093 S. B. Butts, S. H. Strauss, E. M. Holt, R. E. Stimson, N. W. Alcock and D. F. Shriver; *J. Am. Chem. Soc.*, 1980, **102**, 5093. 445, 447

80JA5393 P. J. Fagan, J. M. Manriquez, T. J. Marks, V. W. Day, S. H. Vollmer and C. S. Day; *J. Am. Chem. Soc.*, 1980, **102**, 5393. 469

80JA6744 L. W. Messerle, P. Jennische, R. R. Schrock and G. Stucky; *J. Am. Chem. Soc.*, 1980, **102**, 6744. 918

80JA6900 T. Cohen and J. R. Matz; *J. Am. Chem. Soc.*, 1980, **102**, 6900. 414

80JA7026 M. J. Miller, P. G. Mattingly, M. A. Morrison and J. F. Kerwin, Jr.; *J. Am. Chem. Soc.*, 1980, **102**, 7026. 273, 283

80JA7578 K. Steliou, A. Szczygielska-Nowosielska, A. Favre, M. A. Poupart and S. Hanessian; *J. Am. Chem. Soc.*, 1980, **102**, 7578. 277

80JA7579 M. K. Kaloustian and F. Khouri; *J. Am. Chem. Soc.*, 1980, **102**, 7579. 512

80JA7806 Y. Tamaru, T. Harada, S. Nishi, M. Mizutani, T. Hioki and Z. Yoshida; *J. Am. Chem. Soc.*, 1980, **102**, 7806. 607

80JA7811 P. K. Baker, G. K. Barker, M. Green and A. J. Welch; *J. Am. Chem. Soc.*, 1980, **102**, 7811. 952

80JA7929 B. M. Trost, M. Vaultier and M. L. Santiago; *J. Am. Chem. Soc.*, 1980, **102**, 7929. 731

80JCS(D)712 G. K. Anderson and R. J. Cross; *J. Chem. Soc., Dalton Trans.*, 1980, 712. 465

80JCS(D)1768 E. A. V. Ebsworth, D. W. H. Rankin, W. Steger and J. G. Wright; *J. Chem. Soc., Dalton Trans.*, 1980, 1768. 369, 713

80JCS(D)1845 J. A. S. Howell and A. J. Rowan; *J. Chem. Soc., Dalton Trans.*, 1980, 1845. 455

80JCS(P1)665 A. G. M. Barrett, D. H. R. Barton and R. Colle; *J. Chem. Soc., Perkin Trans. 1*, 1980, 665. 735

80JCS(P1)865 A. R. Katritzky and S. S. Thind; *J. Chem. Soc., Perkin Trans. 1*, 1980, 865. 1107

80JCS(P1)1029 K. Clarke, W. R. Fox and R. M. Scrowston; *J. Chem. Soc., Perkin Trans. 1*, 1980, 1029. 1111

80JCS(P1)1150 G. M. Blackburn and D. Ingleson; *J. Chem. Soc., Perkin Trans. 1*, 1980, 1150. 404

80JCS(P1)1212 D. H. R. Barton, D. J. Lester and S. V. Ley; *J. Chem. Soc., Perkin Trans. 1*, 1980, 1212. 366

80JCS(P1)1627 P.-E. Hansen; *J. Chem. Soc. Perkin Trans. 1*, 1980, 1627. 516, 521, 522

80JCS(P1)1650 N. J. Cussans, S. V. Ley and D. H. R. Barton; *J. Chem. Soc., Perkin Trans. 1*, 1980, 1650. 523

80JCS(P1)1658 A. A. Durrani and J. H. P. Tyman; *J. Chem. Soc., Perkin Trans. 1*, 1980, 1658. 110

80JCS(P1)2122 M. Akiyama, K. Shimizu, S. Aiba and F. Banba; *J. Chem. Soc., Perkin Trans. 1*, 1980, 2122. 332

80JCS(P1)2184 A. G. M. Barrett, R. W. Read and D. H. R. Barton; *J. Chem. Soc., Perkin Trans. 1*, 1980, 2184. 526, 685, 694

80JCS(P1)2191 A. G. M. Barrett, R. W. Read and D. H. R. Barton; *J. Chem. Soc., Perkin Trans. 1*, 1980, 2191. 526

80JCS(P1)2338 K. A. Pover and F. Scheinmann; *J. Chem. Soc., Perkin Trans. 1*, 1980, 2338. 101

80JCS(P1)2657 D. H. R. Barton, G. Bringmann, G. Lamotte, W. B. Motherwell, R. S. Hay Motherwell and A. E. A. Porter; *J. Chem. Soc., Perkin Trans. 1*, 1980, 2657. 1054

80JCS(P2)121 K. Brady and A. F. Hegarty; *J. Chem. Soc., Perkin Trans. 2*, 1980, 121. 716

80JCS(P2)1557 L. Carlsen, H. Egsgaard, E. Schaumann, H. Mrotzek and W.-R. Klein; *J. Chem. Soc., Perkin Trans. 2*, 1980, 1557. 539

80JFC(16)19 S. Rozen and Y. Menahem; *J. Fluorine Chem.*, 1980, **16**, 19. 202

80JHC621 P. Sarti-Fantoni, D. Donati, M. Fiorenza, E. Moschi and V. Dal Piaz; *J. Heterocycl. Chem.*, 1980, **17**, 621. 99

80JHC767 G. Duguay, J.-P. Guémas, J.-C. Meslin, J.-P. Pradère, F. Reliquet, A. Reliquet, C. Tea-Gokou, H. Quiniou, J. Bourrigaud, G. Nourrisson and C. Rabiller; *J. Heterocycl. Chem.*, 1980, **17**, 767. 760

80JHC877 A. S. Shawali, M. Sami, S. M. Sherif and C. Parkanyi; *J. Heterocycl. Chem.*, 1980, **17**, 877. 771

80JHC1457 S. Nagubandi and G. Fodor; *J. Heterocycl. Chem.*, 1980, **17**, 1457. 655

80JHC1745 A. S. Shawali, H. M. Hassaneen and S. M. Sherif; *J. Heterocycl. Chem.*, 1980, **17**, 1745. 704

80JOC80 M. R. Detty and G. P. Wood; *J. Org. Chem.*, 1980, **45**, 80. 728, 732

80JOC165 R. D. Bach, T. H. Taaffee and S. J. Rajan; *J. Org. Chem.*, 1980, **45**, 165. 282

80JOC347 C. D. Bedford and K. Baum; *J. Org. Chem.*, 1980, **45**, 347. 967

80JOC672 S. Rozen and O. Lerman; *J. Org. Chem.*, 1980, **45**, 672. 202

80JOC1044 S. J. Falcone and M. P. Cava; *J. Org. Chem.*, 1980, **45**, 1044. 1117

80JOC1717 D. H. Hua, N. J. Peacock and C. Y. Meyers; *J. Org. Chem.*, 1980, **45**, 1717. 867

80JOC2080 G. Saucy, N. Cohen, B. L. Banner and D. P. Trullinger; *J. Org. Chem.*, 1980, **45**, 2080. 159

80JOC2139 J. K. Stille and Y. Becker; *J. Org. Chem.*, 1980, **45**, 2139. 290

80JOC2485 E. C. Taylor and D. J. Dumas; *J. Org. Chem.*, 1980, **45**, 2485. 746

80JOC2601 E. Vedjes, M. J. Arnost, J. M. Dolphin and J. Eustache; *J. Org. Chem.*, 1980, **45**, 2601. 554

80JOC2632 M. V. Lakshmikantham and M. P. Cava; *J. Org. Chem.*, 1980, **45**, 2632. 1113

80JOC2766 G. E. Wilson, Jr. and A. Hess; *J. Org. Chem.*, 1980, **45**, 2766. 237

80JOC2785 A. I. Meyers and J. Slade; *J. Org. Chem.*, 1980, **45**, 2785. 80

80JOC3413 C. Gluchowski, L. Cooper, D. E. Bergbreiter and M. Newcomb; *J. Org. Chem.*, 1980, **45**, 3413. 274, 294

80JOC3750 Y. Lin, S. A. Lang, Jr. and S. R. Petty; *J. Org. Chem.*, 1980, **45**, 3750. 760

80JOC3916 K. C. Liu, B. R. Shelton and R. K. Howe; *J. Org. Chem.*, 1980, **45**, 3916. 664, 821

80JOC4162 M. Sekine, M. Satoh, H. Yamagata and T. Hata; *J. Org. Chem.*, 1980, **45**, 4162. 403, 404

80JOC4366 G. L'Abbé, J.-P. Dekerk, C. Martens and S. Toppet; *J. Org. Chem.*, 1980, **45**, 4366. 638

80JOC5052 T. Kubota, S. Miyashita, T. Kitazume and N. Ishikawa; *J. Org. Chem.*, 1980, **45**, 5052. 280

80JOM(186)39 J. Heinicke, B. Raap and A. Tzschach; *J. Organomet. Chem.*, 1980, **186**, 39. 791

80JOM(191)371 E. Glozbach and J. Lorberth; *J. Organomet. Chem.*, 1980, **191**, 371. 871, 873, 874

80JOM(197)111 S. A. Gardner and H. J. Gysling; *J. Organomet. Chem.*, 1980, **197**, 111. 241

80JOM(202)107 D. P. Arnold and M. A. Bennett; *J. Organomet. Chem.*, 1980, **202**, 107. 386

80JSP(82)185 H. W. Kroto, J. F. Nixon and N. P. C. Simmons; *J. Mol. Spectrosc.*, 1980, **82**, 185. 1152

80LA344 W. Kantlehner F. Wagner and H. Bredereck; *Liebigs Ann. Chem.*, 1980, 344. 747

80LA483 K. Hartke and R. Hoffmann; *Liebigs Ann. Chem.*, 1980, 483. 555

80LA791 J. C. Stowell and J. R. Christenson; *Liebigs Ann. Chem.*, 1980, 791. 665

80LA1055 E. Vilsmaier, M. Huber and J. Schütz; *Liebigs Ann. Chem.*, 1980, 1055. 702

80LA1064 E. Vilsmaier, R. Schneider and J. Schütz; *Liebigs Ann. Chem.*, 1980, 1064. 339, 689

80LA1075 E. Vilsmaier and K. Reichert; *Liebigs Ann. Chem.*, 1980, 1075. 339, 689

80LA1216 H. Hagen, R.-D. Kohler and H. Fleig; *Liebigs Ann. Chem.*, 1980, 1216. 690

80LA1376 G. Heubach; *Liebigs Ann. Chem.*, 1980, 1376. 771

80M(111)193 W. Bett and S. Cradock; *Monatsh. Chem.*, 1980, **111**, 193. 537

B-80MI 501-01 J. Falbe; "New Syntheses with Carbon Monoxide," Springer-Verlag, Berlin, 1980. 16

B-80MI 502-01 B. Kurtev; in "Organometalliques Fonct. Ambidents, Recl. Commun., Colloq. Fr.-Bulg.," eds. B. Blagoev, F. Gaudemar-Bardone and M. Mladenova, Acad. Bulg. Sci., Inst. Chim. Org., Sofia, 1980, p. 1 (*Chem. Abstr.*, 1981, **95**, 60 729r). 62, 84

80MI 502-01 E. P. Kovsman, B. G. Soldatov, G. N. Freidlin and G. V. Motsak; *Elektrosint. Monomerov*, 1980, 145 (*Chem. Abstr.*, 1980, **93**, 122 347j). 105

80MI 503-01 A. Kh. Mamishov and Kh. D. Khalilov; *Azerb. Khim. Zh.*, 1980, 58 (*Chem. Abstr*, 1981, **94**, 208 320). 139

B-80MI 507-01 F. A. Cotton and G. Wilkinson; "Advanced Inorganic Chemistry," Wiley, New York, 1980. 368

80MI 507-02 H. Miyazaki; *Denki Kagaku oyobi Kogyo Butsuri Kagaku*, 1980, **48**, 453. 317

80MI 509-01 E. S. Balenkova, M. I. Kengerli and S. N. Anfilogova; *Vestn. Mosk. Univ., Ser. 2; Khim*, 1980, **21**, 402 (*Chem. Abstr.*, 1981, **94**, 83656). 433

80MI 514-01 Y. Tamaru; *Kagaku no Ryoiki*, 1980, **34**, 191. (*Chem. Abstr.*, 1980, **93**, 25 350x). 607, 616

80MI 525-01 F. J. Brown; *Prog. Inorg. Chem.*, 1980, **27**, 1. 934

B-80MI 528-01 Y. Wolman; "The Chemistry of Ketenes, Allenes, and Related Compounds," ed. S. Patai, Wiley, New York, 1971, vol. 2, p. 721. 1061, 1078

80MIP-DE283404 F. Linhart, G. Stubenrauch and G. Hamprecht; *Miscellaneous Pat.* DE283 404 (1980) (*Chem. Abstr.*, 1980, **92**, 215 469v). 1126

80NJC47 R. Shabana, S. Scheibye, K. Clausen, S. O. Olesen and S.-O. Lawesson; *Nouv. J. Chim.*, 1980, **4**, 47. 570, 571, 610

80NJC127 C. M. Alpoim, A. G. M. Barrett, D. H. R. Barton and P. C. Hiberty; *Nouv. J. Chem.*, 1980, **4**, 127. 1004

80OS(59)49 M. F. Lipton and A. Basha; *Org. Synth.*, 1980, **59**, 49. 263

80OS(59)195 K. Kurita and Y. Iwakura; *Org. Synth.*, 1980, **59**, 195. 972

80PAC681 D. Mansuy; *Pure Appl. Chem.*, 1980, **52**, 681. 1097

80PAC1057 H. Schmidbaur; *Pure Appl. Chem.*, 1980, **52**, 1057. 894

80PJC745 W. Zielinski; *Pol. J. Chem.*, 1980, **54**, 745. 656

80PS(8)1 J. S. Pizey and K. Symeonides; *Phosphorus Sulfur*, 1980, **8**, 1. 485

80PS(8)27 M. Mulliez; *Phosphorus Sulfur*, 1980, **8**, 27. 370

80RCR357 O. A. Krugalya and N. S. Vyazankin; *Russ. Chem. Rev.*, 1980, **49**, 357. 874

80RCR843 Yu. A. Zhdanov, L. A. Uzlova and Z. I. Glebova; *Russ. Chem. Rev.* (*Engl. Transl.*), 1980, **49**, 843. 394, 404

80S85 V. I. Gorbatenko and L. I. Samarai; *Synthesis*, 1980, 85. 1069

80S114 J. E. Rowe; *Synthesis*, 1980, 114. 685, 728

80S243 S. Cacchi, D. Misiti and F. La Torre; *Synthesis*, 1980, 243. 275

80S385 S. Bernasconi, A. Comini, A. Corbella, P. Gariboldi and M. Sisti; *Synthesis*, 1980, 385. 295

80S403 S. Takano, A. Kurotaki, M. Takahashi and K. Ogasawara; *Synthesis*, 1980, 403. 164

80S547 J. Diago-Meseguer, A. L. Palomo-Coll, J. R. Fernández-Lizarbe and A. Zugaza-Bilbao; *Synthesis*, 1980, 547. 126, 267, 291

80S566 N. Fukada, K. Arai, T. Takeshima and M. Muraoka; *Synthesis*, 1980, 566. 547

80S661 G. A. Olah, S. C. Narang and A. Garcia-Luna; *Synthesis*, 1980, 661. 226

80S721 M. Mizuta, T. Katada, E. Itoh, S. Kato and K. Miyagawa; *Synthesis*, 1980, 721. 252, 254

80S736 K. Ogura, Y. Ito and G. Tsuchihashi; *Synthesis*, 1980, 736. 552

80S883 J. Bésán, L. Kulcsár and M. Kovács; *Synthesis*, 1980, 883. 745

80S887 Y. Tamura, Y. Yoshimoto, K. Sakai, J. Haruta and Y. Kita; *Synthesis*, 1980, 887. 294

80S907 E. Schaumann, U. Wriede and J. Ehlers; *Synthesis*, 1980, 907. 552

80S933 T. Agawa, M. Ishida and Y. Ohshiro; *Synthesis*, 1980, 933. 274, 286

80SA(A)769 J.-L. Piette, D. Debergh, M. Baiwir and G. Llabrès; *S. Petrochim. Acta, Part A*, 1980, **36**, 769. 241

80SC305 K. Nakagawa, S. Shiba, M. Horikawa, K. Sato, H. Nakamura, N. Harada and F. Harada; *Synth. Commun.*, 1980, **10**, 305. 251

80SC489 R. Hull and T. P. Seden; *Synth. Commun.*, 1980, **10**, 489. 1035

80SC595 T. Otsubo, F. Ogura, H. Yamaguchi, H. Higuchi and S. Misumi; *Synth. Commun.*, 1980, **10**, 595. 1113

80T1839 R. G. Guy, S. Cousins, D. M. Farmer, A. D. Henderson and C. L. Wilson; *Tetrahedron*, 1980, **36**, 1839. 1110

80T2409 E. Haslam; *Tetrahedron*, 1980, **36**, 2409. 122

80TH158 W. Uhl, Ph.D. Thesis, University of Marburg, 1980, p. 158. 879

80TL31 R. W. Ratcliffe, T. N. Salzmann and B. G. Christensen; *Tetrahedron Lett.*, 1980, **21**, 31. 374

80TL635 M. Yokoyama and T. Takeshima; *Tetrahedron Lett.*, 1980, **21**, 635. 1112

80TL841 Y. Nagao, K. Seno, K. Kawabata, T. Miyasaka, S. Takao and E. Fujita; *Tetrahedron Lett.*, 1980, **21**, 841. 277, 295

80TL849 J. Tsuji, M. Takahashi and T. Takahashi; *Tetrahedron Lett.*, 1980, **21**, 849. 160

80TL885 A. F. Hegarty and A. Chandler; *Tetrahedron Lett.*, 1980, **21**, 885. 748

80TL917 R. Gompper and B. Kohl; *Tetrahedron Lett.*, 1980, **21**, 917. 552

80TL1141 H. Oehme, E. Leissring and H. Meyer; *Tetrahedron Lett.*, 1980, **21**, 1141. 883

80TL1459 D. Marmet, P. Boullanger and G. Descotes; *Tetrahedron Lett.*, 1980, **21**, 1459. 747

80TL1467 H. Ogura, S. Nagai and K. Takeda; *Tetrahedron Lett.*, 1980, **21**, 1467. 226

80TL1653 K. Yamamoto, S. Suzuki and J. Tsuji; *Tetrahedron Lett.*, 1980, **21**, 1653. 419

80TL1771 H. Westmijze, K. Ruitenberg, J. Meijer and P. Vermeer; *Tetrahedron Lett.*, 1980, **21**, 1771. 554

80TL2013 L. Cama and B. G. Christensen; *Tetrahedron Lett.*, 1980, **21**, 2013. 233

80TL2165 M. Saquet and A. Thuillier; *Tetrahedron Lett.*, 1980, **21**, 2165. 552

80TL2223 A. O. Sy and J. W. Raksis; *Tetrahedron Lett.*, 1980, **21**, 2223. 998

80TL2421 P. Gosselin, S. Masson and A. Thuillier; *Tetrahedron Lett.*, 1980, **21**, 2421. 558, 559

80TL2443 A. Bladé-Font; *Tetrahedron Lett.*, 1980, **21**, 2443. 276

80TL2497 R. Appel and B. Laubach; *Tetrahedron Lett.*, 1980, **21**, 2497. 1086, 1092

80TL2857 S. Hashimoto, I. Furukawa and T. Kuroda; *Tetrahedron Lett.*, 1080, **21**, 2857. 134

80TL2877 K. Kikukawa, K. Kono, K. Nagira, F. Wada and T. Matsuda; *Tetrahedron Lett.*, 1980, **21**, 2877. 197, 198

80TL2959 M. Tanaka; *Tetrahedron Lett.*, 1980, **21**, 2959. 1144

80TL3085 J. Marchand-Brynaert, L. Ghosez and E. Cossement; *Tetrahedron Lett.*, 1980, **21**, 3085. 233

80TL3367 J. Rigaudy, E. Lytwyn, P. Wallach and N. K. Cuong; *Tetrahedron Lett.*, 1980, **21**, 3367. 380

80TL3927 A. Degl'Innocenti, D. R. M. Walton, G. Seconi, G. Pirazzini and A. Ricci; *Tetrahedron Lett.*, 1980, **21**, 3927. 410

80TL3987 J.-C. Clinet and G. Linstrumelle; *Tetrahedron Lett.*, 1980, **21**, 3987. 427

80TL4061 S. Raucher and P. Klein; *Tetrahedron Lett.*, 1980, **21**, 4061. 570, 610, 618

80TL4247 E. Schaumann, A. Röhr and G. Adiwidjaja; *Tetrahedron Lett.*, 1980, **21**, 4247. 344, 639
80TL4251 E. Schaumann and F.-F. Grabley; *Tetrahedron Lett.*, 1980, **21**, 4251. 626
80TL4461 T. Aoyama and T. Shioiri; *Tetrahedron Lett.*, 1980, **21**, 4461. 873
80TL4611 M. Asaoka, N. Yanagida and H. Takei; *Tetrahedron Lett.*, 1980, **21**, 4611. 179
80TL4619 N. Hashimoto, T. Aoyama and T. Shioiri; *Tetrahedron Lett.*, 1980, **21**, 4619. 873
80TL5071 A. J. G. Baxter, P. Davis, R. J. Ponsford and R. Southgate; *Tetrahedron Lett.*, 1980, **21**, 5071. 235
80USP781220 *US Pat.* 781 220 (1980) (*Chem. Abstr.*, 1980, **94**, 47 134). 316
80ZAAC(468)55 G. Becker, O. Mundt and M. Rössler; *Z. Anorg. Allg. Chem.*, 1980, **468**, 55. 400
80ZAAC(470)131 G. Becker and G. Gutekunst; *Z. Anorg. Allg. Chem.*, 1980, **470**, 131. 891
80ZAAC(470)144 G. Becker and G. Gutekunst; *Z. Anorg. Allg. Chem.*, 1980, **470**, 144. 891
80ZC97 G. Barnikow and D. Richter; *Z. Chem.*, 1980, **20**, 97. 674, 1053
80ZN(B)217 J. Bergman and L. Engman; *Z. Naturforsch., Teil B*, 1980, **35**, 217. 242
80ZN(B)1289 J. C. Baldwin, N. L. Keder, C. E. Strouse and W. C. Kaska; *Z. Naturforsch. Teil B*, 1980, **35**, 1289. 917
80ZOB620 V. P. Kozyukov and N. V. Mironova; *Zh. Obshch. Khim.*, 1980, **50**, 620 (*Chem. Abstr.*, 1981, **94**, 30 278e). 969
80ZOB886 V. I. Shcherbakov, N. E. Stolyarova, O. S. D'yachkovskaya and V. I. Kuzina; *Zh. Obshch. Khim.*, 1980, **50**, 886. 390
80ZOB960 V. P. Kozyukov, G. I. Orlov and V. F. Mironov; *Zh. Obshch. Khim.*, 1980, **50**, 960 (*Chem. Abstr.*, 1980, **93**, 71 171n). 983
80ZOB1678 V. N. Fetyukhin, M. V. Voek, Yu. I. Dergunov and L. I. Samarai; *Zh. Obshch. Khim.*, 1980, **51**, 1678. 1083
80ZOB2413 A. D. Sinitsa, V. S. Krishtal, V. I. Kal'chenko and L. N. Markovskii; *Zh. Obshch. Khim.*, 1980, **50**, 2413 (*Chem. Abstr.*, 1981, **94**, 121 646q). 863
80ZOR1330 Yu. L. Yagupol'skii and T. I. Savina; *Zh. Org. Khim.*, 1981, **17**, 1330. 314

81ACS(A)151 K. Maartmann-Moe, K. A. Sanderud and J. Songstad; *Acta Chem. Scand., Ser. A*, 1981, **35**, 151. 1117
81ACS(B)239 S. Scheibye, S.-O. Lawesson and C. Rømming; *Acta Chem. Scand., Ser. B*, 1981, **35**, 239. 570, 571, 613
81AG855 K. Wagner, K. Findeisen, W. Schäfer and W. Dietrich; *Angew. Chem.*, 1981, **93**, 855. 1061, 1078
81AG(E)131 E. Niecke, W. W. Schoeller and D. A. Wildbredt; *Angew. Chem., Int. Ed. Engl.*, 1981, **20**, 131. 887
81AG(E)197 R. Appel, G. Maier, H. P. Reisenauer and A. Westerhaus; *Angew. Chem., Int. Ed. Engl.*, 1981, **20**, 197. 1154
81AG(E)207 R. Schmierer, G. Grotemeier, G. Helmchen and A. Selim; *Angew. Chem., Int. Ed. Engl.*, 1981, **20**, 207. 148
81AG(E)703 D. Arlt, M. Jautelat and R. Lantzsch; *Angew. Chem., Int. Ed. Engl.*, 1981, **20**, 703. 28
81AG(E)731 R. Appel, F. Knoll and I. Ruppert; *Angew. Chem., Int. Ed. Engl.*, 1981, **20**, 731. 876, 1154
81AG(E)971 D. Seebach and D. Wasmuth; *Angew. Chem., Int. Ed. Engl.*, 1981, **20**, 971. 148
81AG(E)1055 H. G. Raubenheimer, E. O. Fischer, U. Schubert, C. Krüger and Y.-H. Tsay; *Angew. Chem., Int. Ed. Engl.*, 1981, **20**, 1055. 951
81AP(314)577 S. Prior and W. Wiegrebe; *Arch. Pharm.* (*Weinheim, Ger.*), 1981, **314**, 577. 1127
81BCJ790 T. Horii, S. Kawamura, T. Nakabayashi and Y. Abe; *Bull. Chem. Soc. Jpn.*, 1981, **54**, 790. 540
81BCJ1267 R. Nakao, K. Oka and T. Fukumoto; *Bull. Chem. Soc. Jap.*, 1981, **54**, 1267. 123
81BCJ1720 T. Yamamoto, T. Kohara and A. Yamamoto; *Bull. Chem. Soc. Jpn.*, 1981, **54**, 1720. 226
81BCJ2161 T. Yamamoto, T. Kohara and A. Yamamoto; *Bull. Chem. Soc. Jpn*, 1981, **54**, 2161. 462
81BCJ3863 S. Shiraishi, T. Shigemoto, M. Miyahara and S. Ogawa; *Bull. Chem. Soc. Jpn.*, 1981, **54**, 3863. 699
81BRP1589128 E. J. Hedgley; *Br. Pat.*, 1 589 128 (1981) (*Chem. Abstr.*, 1981, **95**, 150 227). 510
81BSF(2)361 C. Palomo and R. Mestres; *Bull. Soc. Chim. Fr., Part 2*, 1981, **9–10**, 361. 1079
81CB80 R. Neidlein and W. Lehr; *Chem. Ber.*, 1981, **114**, 80. 341
81CB173 F. Effenberger, K. Drauz and S. Förster; *Chem. Ber.*, 1981, **114**, 173. 964
81CB802 H. Quast and F. Kees; *Chem. Ber.*, 1981, **114**, 802. 364
81CB1428 H. Schmidbaur, T. Costa and B. Milewski-Mahrla; *Chem. Ber.*, 1981, **114**, 1428. 910, 911
81CB2272 E. Lindner and J. C. Wuhrmann; *Chem. Ber.*, 1981, **114**, 2272. 220, 221
81CB2491 H. Schmidbaur and U. Deschler; *Chem. Ber.*, 1981, **114**, 2491. 908
81CB3070 U. Schubert, C. Kappenstein, B. Milewski-Mahrla and H. Schmidbaur; *Chem. Ber.*, 1981, **114**, 3070. 1092
81CB3655 J. Lambrecht, L. Zsolnai, G. Huttner and J. C. Jochims; *Chem. Ber.*, 1981, **114**, 3655. 1146
81CB3684 T. Kauffmann, L. Bán and D. Kuhlmann; *Chem. Ber.*, 1981, **114**, 3684. 773
81CC191 A. G. Brook, F. Abdesaken, B. Gutekunst, G. Gutekunst and R. K. Kallury; *J. Chem. Soc., Chem. Commun.*, 1981, 191. 923, 924
81CC510 Y. Uchida and S. Kozuka; *J. Chem. Soc., Chem. Commun.*, 1981, 510. 334, 336
81CC546 A. Toshimitsu, H. Owada, T. Aoai, S. Uemura and M. Okano; *J. Chem. Soc., Chem. Commun.*, 1981, 546. 282
81CC670 J. Barluenga, C. Jiménez, C. Nájera and M. Yus; *J. Chem. Soc., Chem. Commun.*, 1981, 670. 284
81CC706 W. J. Evans, A. L. Wayda, W. E. Hunter and J. L. Atwood; *J. Chem. Soc., Chem. Commun.*, 1981, 706. 440
81CC822 B. F. Bonini, G. Mazzanti, S. Sarti, P. Zanirato and G. Maccagnani; *J. Chem. Soc., Chem. Commun.*, 1981, 822. 647

81CC1121	J. E. Baldwin and J. C. Bottaro; *J. Chem. Soc., Chem. Commun.*, 1981, 1121.	675
81CC1178	J. Barluenga, C. Jiménez, C. Nájera and M. Yus; *J. Chem. Soc., Chem. Commun.*, 1981, 1178.	284
81CHE315	G. P. Ellis; *Chem. Heterocycl. Compd.* (*Engl. Transl.*), 1981, **36**, 315.	226
81CI(L)164	A. V: R. Rao, M. N. Deshmukh and L. Sivadasan; *Chem. Ind.* (*London*), 1981, 164.	226
81CL285	Y. Watanabe and T. Mukaiyama; *Chem. Lett.*, 1981, 285.	226
81CL879	N. Miyaura and A. Suzuki; *Chem. Lett.*, 1981, 879.	157
81CL1069	S. Tomoda, Y. Takeuchi and Y. Nomura; *Chem. Lett.*, 1981, 1069.	1116
81CL1143	J. Yamawaki, T. Ando and T. Hanafusa; *Chem. Lett.*, 1981, 1143.	285
81CL1367	T. Mukaiyama, K. Kamekawa and Y. Watanabe; *Chem. Lett.*, 1981, 1367.	226
81CL1415	J. Junga, Y. Kawanami and M. Yamaguchi; *Chem. Lett.*, 1981, 1415.	173
81CPB1063	H. Takahata, Y. Ohnishi, H. Takehara, K. Tsuritani and T. Yamaki; *Chem. Pharm. Bull.*, 1981, **29**, 1063.	292
81CPB1475	N. Hashimoto, T. Aoyama and T. Shioiri; *Chem. Pharm. Bull.*, 1981, **29**, 1475.	132, 873
81CPB3249	T. Aoyama and T. Shioiri; *Chem. Pharm. Bull.*, 1981, **29**, 3249.	873
81CRV589	A. Williams and I. T. Ibrahim; *Chem. Rev.*, 1981, **81**, 589.	1061
81DOK(256)1401	I. F. Lutsenko, A. A. Prishchenko, A. A. Borisenko and Z. S. Novikova; *Dokl. Akad. Nauk SSSR*, 1981, **256**, 1401 (*Chem. Abstr.*, 1981, **95**, 81 109d).	884
81EJM175	J. B. Pream, C. W. Picard and T. G. White; *Eur. J. Med. Chem.*, 1981, **16**, 175.	1121
81EJM317	D. L. Klayman, J. P. Scovill, J. F. Bartosevich and C. J. Mason; *Eur. J. Med. Chem.*, 1981, **16**, 317.	1056
81EUP54516	J. Drabek; *Eur. Pat.* 54 516 A2 (1981) (*Chem. Abstr.*, 1981, **98**, 16 854).	390
81G285	A. Alemagna, P. Del Buttero, E. Licandro and S. Maiorana; *Gazz. Chim. Ital.*, 1981, **111**, 285.	905
81GEP3018821	K. Findeisen and K. H. Linker; *Ger. Pat.* 3 018 821 (1981) (*Chem. Abstr.*, 1982, **96**, 123 008y).	1134
81GSM87	D. W. Knight; *Gen. Synth. Methods*, 1981, **4**, 87.	24
81H(15)165	S. Nagabundi and G. Fodor; *Heterocycles*, 1981, **15**, 165.	710, 716
81H(15)293	M. Kočevar, B. Stanovnik and M. Tišler; *Heterocycles*, 1981, **15**, 293.	837
81H(15)975	N. Hashimoto, T. Aoyama and T. Shioiri; *Heterocycles*, 1981, **15**, 975.	873
81H(16)381	S. Takano; *Heterocycles*, 1981, **16**, 381.	166
81H(16)717	S. Tanaka and H. Terada; *Heterocycles*, 1981, **16**, 717.	678
81H(16)951	S. Takano; *Heterocycles*, 1981, **16**, 951.	166
81H(16)2173	B. Stanovnik, J. Žmitek and M. Tišler; *Heterocycles*, 1981, **16**, 2173.	836
81HCA1322	A. H. Schultess and H. J. Hansen; *Helv. Chim. Acta*, 1981, **64**, 1322.	696
81HCA2808	W. Oppolzer and H. J. Löher; *Helv. Chim. Acta*, 1981, **64**, 2808.	152
81IC1679	R. H. Neilson; *Inorg. Chem.*, 1981, **20**, 1679.	887
81IC1823	G. A. Brewer and E. Sinn; *Inorg. Chem.*, 1981, **20**, 1823.	325
81IC2466	R. E. Cramer, R. B. Maynard and J. W. Gilje; *Inorg. Chem.*, 1981, **20**, 2466.	920
81IC3119	D. Mansuy, J.-P. Lecomte, J.-C. Chottard and J.-F. Bartoli; *Inorg. Chem.*, 1981, **20**, 3119.	1097
81ICA165	J. D. Cotton, G. T. Crisp and V. A. Daly; *Inorg. Chim. Acta*, 1981, **47**, 165.	445
81IJC(B)411	G. Sinha and U. R. Ghatak; *Indian J. Chem. Sect. B*, 1981, **20**, 411.	144
81IZV2370	A. V. Fokin, Yu. N. Studnev, A. I. Rapkin, I. N. Krotovich and O. V. Verenikin; *Izv. Akad. Nauk SSSR, Ser. Khim.*, 1981, 2370 (*Chem. Abstr.*, 1982, **96**, 51 783x).	710
81JA861	J. Huang and J. Meinwald; *J. Am. Chem. Soc.*, 1981, **103**, 861.	235
81JA1467	A. J. Baskar, C. M. Lukehart and K. Srinivasan; *J. Am. Chem. Soc.*, 1981, **103**, 1467.	801
81JA1568	S. Masamune, M. Hirama, S. Mori, S. A. Ali and D. S: Garvey; *J. Am. Chem. Soc.*, 1981, **103**, 1568.	173
81JA3173	G. E. Keck and R. E. Webb, II; *J. Am. Chem. Soc.*, 1981, **103**, 3173.	326
81JA3328	K. Tatsumi and R. Hoffmann; *J. Am. Chem. Soc.*, 1981, **103**, 3328.	1097
81JA3559	I. I. Creaser, J. M. B. Harrowfield, F. R. Keene and A. M. Sargeson; *J. Am. Chem. Soc.*, 1981, **103**, 3559.	386
81JA4186	H. Yamamoto and K. Maruoka; *J. Am. Chem. Soc.*, 1981, **103**, 4186.	777
81JA5183	W. H. Kruizinga and R. M. Kellogg; *J. Am. Chem. Soc.*, 1981, **103**, 5183.	777
81JA5573	W. Ando, A. T. Sekiguchi and T. Sato; *J. Am. Chem. Soc.*, 1981, **103**, 5573.	925
81JA6164	R. A. Moss, W. Guo, D. Z. Denney, K. N. Houk and N. G. Rondon; *J. Am. Chem. Soc.*, 1981, **103**, 6164.	866
81JA6217	J. A. Miller and G. Zweifel; *J. Am. Chem. Soc.*, 1981, **103**, 6217.	426
81JA7055	E. R. Cullen, F. S. Guziec, Jr., C. J. Murphy, T. C. Wong and K. K. Andersen; *J. Am. Chem. Soc.*, 1981, **103**, 7055.	521
81JAP39050	Toray Industries, Inc.; *Jpn. Pat.* 56/39 050 (1981) (*Chem. Abstr.*, 1982, **96**, 68 344c).	86
81JAP81125320	Toshiba Silicon Co., Ltd.; *Jpn. Pat.* 81 125 320 (*Chem. Abstr.*, 1982, **96**, 69 203).	419
81JAP81125395	Toshiba Silicon Co., Ltd.; *Jpn. Pat.* 81 125 395 (*Chem. Abstr.*, 1982, **96**, 69 202).	419
81JCED105	E. J. Kupchik, H. E. Hanke, J. P. DiMarco and R. J. Chessari; *J. Chem. Eng. Data*, 1981, **26**, 105.	1121
81JCED418	R. S. Tewari and P. Parihar; *J. Chem. Eng. Data*, 1981, **26**, 418.	678, 679
81JCR(S)58	A. Bladé-Font, S. Aguilá, T. de Mas and J.-M. Torres; *J. Chem. Res.* (*S*), 1981, 58.	570, 618
81JCR(S)129	B. J. Wakefield and D. J. Wright; *J. Chem. Res.* (*S*), 1981, 129.	664
81JCR(S)234	S. Mahmoud Ali, D. Clarke, G. R. Cliff and G. A. Morrison; *J. Chem. Res.* (*S*), 1981, 234.	1110
81JCS(D)2088	K. W. Chiu, R. A. Jones, G. Wilkinson, A. M. R. Galas and M. B. Hursthouse; *J. Chem. Soc., Dalton Trans.*, 1981, 2088.	797, 801
81JCS(P1)40	R. C. Cambie, D. Chambers, P. S. Rutledge and P. D. Woodgate; *J. Chem. Soc., Perkin Trans. 1*, 1981, 40.	729

81JCS(P1)58 R. C. Cambie, D. S. Larsen, P. S. Rutledge and P. D. Woodgate; *J. Chem. Soc., Perkin Trans. 1*, 1981, 58. 1107, 1109
81JCS(P1)1577 Y. Tamura, T. Kawasaki, H. Yasuda, N. Gohda and Y. Kita; *J. Chem. Soc., Perkin Trans. 1*, 1981, 1577. 1107
81JCS(P1)2245 L. Bruché, L. Garanti and G. Zecchi; *J. Chem. Soc., Perkin Trans. 1*, 1981, 2245. 677
81JCS(P1)2435 W. R. McKay and G. R. Proctor; *J. Chem. Soc., Perkin Trans. 1*, 1981, 2435 696
81JCS(P1)3284 M. J. P. Harger; *J. Chem. Soc., Perkin Trans. 1*, 1981, 3284. 360
81JCS(P2)1043 G. Seconi, G. Pirazzini, A. Ricci, M. Fiorenza and C. Eaborn; *J. Chem. Soc., Perkin Trans. 2*, 1981, 1043. 793, 803
81JFC(18)553 A. F. Fokin, Yu. N. Studnev and A. I. Rapkin; *J. Fluorine Chem.*, 1981, **18**, 553. 699, 710
81JGU1015 G. A. Pensionerova, V. G. Rozinov and V. I. Glukhikh; *J. Gen. Chem. USSR* (*Engl. Transl.*), 1981, **51**, 1015. 656
81JGU1883 O. I. Kolodyazhnyi and V. P. Kukhar; *J. Gen. Chem. USSR* (*Engl. Transl.*), 1981, **51**, 1883. 879
81JGU2055 V. F. Mironov, V. P. Kozyukov and V. P. Kozyukov; *J. Gen. Chem. USSR* (*Engl. Transl.*), 1981, **51**, 2055. 722
81JGU2125 O. I. Kolodyazhnyi; *J. Gen. Chem. USSR* (*Engl. Transl.*), 1981, **51**, 2125. 898, 915
81JHC555 A. Ohta, S. Masano, M. Tsutsui, F. Yamamoto, S. Suzuki, H. Makita, H. Tamamura and Y. Akita; *J. Heterocycl. Chem.*, 1981, **18**, 555. 327
81JHC691 G. R. Lenz and C.-M. Woo; *J. Heterocycl. Chem.*, 1981, **18**, 691. 288
81JMC532 D. L. Rector, S. D. Folz, R. D. Conklin, L. H. Nowakowski and G. Kaugars; *J. Med. Chem.*, 1981, **24**, 532. 675, 678
81JMC1124 R. A. Kenley, R. A. Howd, C. W. Mosher and J. S. Winterle; *J. Med. Chem.*, 1981, **24**, 1124. 830
81JOC196 R. Kaya and N. R. Beller; *J. Org. Chem.*, 1981, **46**, 196. 546
81JOC1045 C. Wentrup and H.-W. Winter; *J. Org. Chem.*, 1981, **46**, 1045. 1123
81JOC1215 P. A. Grieco, J. Y. Jaw, D. A. Claremon and K. C. Nicolaou; *J. Org. Chem.*, 1981, **46**, 1215. 239, 281
81JOC1229 A. K. Bose, D. P. Sahu and M. S. Manhas; *J. Org. Chem.*, 1981, **46**, 1229. 273, 285
81JOC1277 A. Sekiya and D. D. DesMarteau; *J. Org. Chem.*, 1981, **46**, 1277. 807
81JOC1374 N. E. Heimer, L. Field and R. A. Neal; *J. Org. Chem.*, 1981, **46**, 1374. 247
81JOC1402 A. Padwa and S. Nahm; *J. Org. Chem.*, 1981, **46**, 1402. 677
81JOC1538 J. Bosch and J. Bonjoch; *J. Org. Chem.*, 1981, **46**, 1538. 146
81JOC1769 S. Hirokami, T. Takahashi, M. Nagata, Y. Hirai and T. Yamazaki; *J. Org. Chem.*, 1981, **46**, 1769. 693
81JOC2148 Y. Shvo and E. D. Kaufman; *J. Org. Chem.*, 1981, **46**, 2148. 213
81JOC2455 P. L. Barker, P. L. Gendler and H. Rapoport; *J. Org. Chem.*, 1981, **46**, 2455. 742, 758
81JOC3558 S. Raucher and P. Klein; *J. Org. Chem.*, 1981, **46**, 3558. 568
81JOC3911 J. C. Martin, P. G. Gott, R. H. Meen and P. W. Raynolds; *J. Org. Chem.*, 1981, **46**, 3911. 481
81JOC3936 P. H. J. Carlsen, T. Katsuki, V. S. Martin and K. B. Sharpless; *J. Org. Chem.*, 1981, **46**, 3936. 52
81JOC4188 J. C. Chang, M. El-Sheikh, A. Harmon, K. Avasthi and J. M. Cook; *J. Org. Chem.*, 1981, **46**, 4188. 322
81JOC4194 L. Engman and M. P. Cava; *J. Org. Chem.*, 1981, **46**, 4194. 242, 1117
81JOC4304 J. March and J. S. Engenito, Jr.; *J. Org. Chem.*, 1981, **46**, 4304. 294
81JOC4413 K. Kikukawa, K. Kono, K. Nagira, F. Wada and T. Matsuda; *J. Org. Chem.*, 1981, **46**, 4413. 198
81JOC4478 G. L'abbé, A. Verbruggen, T. Minami and S. Toppet; *J. Org. Chem.*, 1981, **46**, 4478. 1123
81JOC4629 O. Lerman, Y. Tor and S. Rozen; *J. Org. Chem.*, 1981, **46**, 4629. 202
81JOC4727 A. Toshimitsu, T. Aoai, H. Owada, S. Uemura and M. Okano; *J. Org. Chem.*, 1981, **46**, 4727. 282, 300
81JOC5039 A. B. Pepperman; *J. Org. Chem.*, 1981, **46**, 5039. 102
81JOC5163 M. Node, K. Nishide, M. Ochiai, K. Fuji and E. Fujita; *J. Org. Chem.*, 1981, **46**, 5163. 83
81JOC5351 W. J. Greenlee and E. D. Thorsett; *J. Org. Chem.*, 1981, **46**, 5351. 36
81JOC5425 P. A. Wade and M. K. Pillay; *J. Org. Chem.*, 1981, **46**, 5425. 1145, 1148
81JOC5445 T. Sasaki, A. Nakanishi and M. Ohno; *J. Org. Chem.*, 1981, **46**, 5445. 1118
81JOM(205)31 G. Cerveau, E. Colomer, R. J. P. Corriu and J. C. Young; *J. Organomet. Chem.*, 1981, **205**, 31. 421
81JOM(205)167 W. Lohner and K. Praefcke; *J. Organomet. Chem.*, 1981, **205**, 167. 241
81JOM(206)C11 A. Mantovani, G. Facchin, T. Boschi and B. Crociani; *J. Organomet. Chem.*, 1981, **206**, C11. 801, 802
81JOM(213)C31 T. Bodnar and A. R. Cutler; *J. Organomet. Chem.*, 1981, **213**, C31. 937
81JOM(213)451 W. Gäde and E. Weiss; *J. Organomet. Chem.*, 1981, **213**, 451. 1097
81JOU1527 Yu. L. Yagupol'skii, I. I. Gerus and V. M. Cherkasov; *J. Org. Chem. USSR* (*Engl. Transl.*), 1981, **17**, 1527. 657
81JOU1676 R. A. Kyandzhetsian, V. A. Danilenko and T. V. Magdasieva; *J. Org. Chem. USSR* (*Engl. Transl.*), 1981, **17**, 1676. 920
81JOU1986 V. V. Mezheritskii, L. G. Minyaeva, L. V. Mezheritskaya, A. L. Pilkus and G. N. Dorofeenko; *J. Org. Chem. USSR* (*Engl. Transl.*), 1981, **17**, 1986. 688
81JSP(90)507 K. Ohno, H. W. Kroto and J. F. Nixou; *J. Mol. Spectrosc.*, 1981, **90**, 507. 1153
81JSP(90)512 H. W. Kroto, J. F. Nixon and K. Ohno; *J. Mol. Spectrosc.*, 1981, **90**, 512. 1153
81LA65 M. A. Pérez, M. Rössert and G. Kresze; *Liebigs Ann. Chem.*, 1981, 65. 731
81LA1198 D. Herrmann, J. Jachmann and S. Neuhaus; *Liebigs Ann. Chem.*, 1981, 1198. 1146

81LA1798 S. Kato, K. Sugino, Y. Matsuzawa, T. Katada, I. Noda, M. Mizuta, M. Goto and M. Ishida; *Liebigs Ann. Chem.*, 1981, 1798. 584

81LA1844 G. Himbert and W. Schwickerath; *Liebigs Ann. Chem.*, 1981, 1844. 304

81M(112)959 J. Beger, R. Neumann and D. Scheller; *Monatsh. Chem.*, 1981, **112**, 959. 658

81MAC3411 K. Kondo, K. Hyodo, M. Murakami and K. Takemoto; *Macromol. Chem.*, 1981, **182**, 3411. 1065

81MI 502-01 T. Yasui; *Shokubai*, 1981, **23**, 23 (*Chem. Abstr.*, 1981, **95**, 23 607d). 27

81MI 502-02 J. Mulzer; *Nachr. Chem. Tech. Lab.*, 1981, **29**, 614 and 619 (*Chem. Abstr.*, 1981, **95**, 149 307z). 62

81MI 502-03 S. S. Oganesyan; *Prom-st. Arm.*, 1981, 25 (*Chem. Abstr.*, 1981, **95**, 168 397v). 28

81MI 502-04 V. V. Antonova, A. M. Bespalova, G. V. Krylova and V. K. Promonenkov; *Deposited Doc.*, 1981, SPSTL369 Khp-D81 (*Chem. Abstr.*, 1983, **98**, 71 834j). 105

81MI 502-05 C. H. Wang; *Kó Hsueh T'ung Pao*, 1981, **26**, 703 (*Chem. Abstr.*, 1981, **95**, 132 441r). 108

81MI 502-06 J. Hasegawa, S. Hamaguchi, M. Ogura and K. Watanabe; *J. Ferment. Technol.*, 1981, **59**, 257. 118

81MI 502-07 P. Brodelius, K. Nilsson and K. Mosbach; *Appl. Biochem. Biotechnol.*, 1981, **6**, 293. 118

81MI 514-01 K. A. Jensen; *Arch. Pharm. Chemi., Sci. Ed.*, 1981, **9**, 93 (*Chem. Abstr.*, 1982, **96**, 67 913a). 624

81MI 518-01 B. D. Podolesov and B. Bogdanov; *Glas. Hem. Dras. Beograd.*, 1981, **46**, 11 (*Chem. Abstr.*, 1981, **95**, 61 700). 735

B-81MI 521-01 A. Haas; "Gmelin Handbook of Inorganic Chemistry," Springer Verlag, New York, 1981, pt. 9, p. 125. 812

81MI 530-01 S. Prior and W. Wiegrebe; *Sci. Pharm.*, 1981, **49**, 503. 1127

81NKK1333 M. Yamashita, S. Ikegawa and R. Suemitsu; *Nippon Kagaku Kaishi* (*J. Chem. Soc. Jpn.*), 1981, 1333 (*Chem. Abstr.*, 1981, **95**, 219 850x). 43, 112

81OPP71 J. Ramachandran, S. R. Ramadas and C. N. Pillai; *Org. Prep. Proced. Int.*, 1981, **13**, 71. 560

81OPP81 J. Ramachandran, S. R. Ramadas and C. N. Pillai; *Org. Prep. Proced. Int.*, 1981, **13**, 81. 547

81PAC1109 D. A. Evans, J. M. Takacs, L. R. McGee, M. D. Ennis, D. J. Mathre and J. Bartroli; *Pure Appl. Chem.*, 1981, **53**, 1109. 148

81PJC655 J. Kaminski and Z. Eckstein; *Pol. J. Chem.*, 1981, **55**, 655. 671

81PS(10)153 S. Oae, M. Fukumura and N. Furukawa; *Phosphorus Sulfur*, 1981, **10**, 153. 349

81PS(10)213 J. Voss and H. Gunther; *Phosphorus Sulfur*, 1981, **10**, 213. 485

81PS(10)261 E. E. Nifantiev, A. I. Zavalishina and E. I. Smirnova; *Phosphorus Sulfur*, 1981, **10**, 261. 370

81RTC414 V. H. M. Elferink, R. G. Visser and H. J. T. Bos; *Recl. Trav. Chim. Pays-Bas*, 1981, **100**, 414. 605, 614

81S1 O. Mitsunobu; *Synthesis*, 1981, 1. 123, 131, 138

81S121 P. F. Alewood, I. C. Calder and R. L. Richardson; *Synthesis*, 1981, 121. 328

81S165 A. J. Mancuso and D. Swern; *Synthesis*, 1981, 165. 209

81S216 A. H. Schmidt, M. Russ and D. Grosse; *Synthesis*, 1981, 216. 17, 19

81S218 R. Mestres and C. Palomo; *Synthesis*, 1981, 218. 189

81S287 K. Ito, T. Iida, T. Fujita and S. Tsuji; *Synthesis*, 1981, 287. 281

81S288 J. A. Miller and G. Zweifel; *Synthesis*, 1981, **4**, 288. 417, 426

81S299 M. Tišler, B. Stanovnik, Z. Zrimšek and C. Stropnik; *Synthesis*, 1981, 299. 609, 626, 836

81S370 S. Kato, H. Watarai, T. Katada, M. Mizuta, K. Miyagawa and M. Ishida; *Synthesis*, 1981, 370. 254

81S373 C. Palomo and R. Mestres; *Synthesis*, 1981, 373. 1063

81S413 A. Akelah; *Synthesis*, 1981, 413. 122

81S451 Y. Kita, J. Haruta, T. Fujii, J. Segawa and Y. Tamura; *Synthesis*, 1981, 451. 378

81S487 I. G. C. Coutts, M. Edwards and D. J. Richards; *Synthesis*, 1981, 487. 81, 107

81S610 C. Yijima, F. Hino and K. Suda; *Synthesis*, 1981, 610. 346

81S616 J. Cabré-Castellvi, A. Palomo-Coll and A. L. Palomo-Coll; *Synthesis*, 1981, 616. 188, 195, 199

81S637 M. Kodomari, M. Fukuda and S. Yoshitomi; *Synthesis*, 1981, 637. 252

81S715 H. M. R. Hoffmann and K. Haase; *Synthesis*, 1981, 715. 19

81S729 R. Beugelmans and M. Bois-Choussy; *Synthesis*, 1981, 729. 302

81S746 S. Kato, K. Miyagawa, R. Hattori, M. Mizuta and M. Ishida; *Synthesis*, 1981, 746. 337

81S753 M. Petrzilka and J. I. Grayson; *Synthesis*, 1981, 753. 199

81S790 G. A. Olah, S. C. Navany and A. Garcia-Luna; *Synthesis*, 1981, 790. 226

81S801 M. F. Haslanger and J. Heikes; *Synthesis*, 1981, 801. 488

81S963 G. Levesque, J.-C. Gressier and M. Proust; *Synthesis*, 1981, 963. 561

81S977 D. P. Chakraborty, A. K. Mandal and S. K. Roy; *Synthesis*, 1981, 977. 298

81S991 M. Ueda, K. Seki and Y. Imai; *Synthesis*, 1981, 991. 278, 291, 303

81S993 C. Palomo; *Synthesis*, 1981, 993. 372

81S1005 T. Gajda and A. Zwierzak; *Synthesis*, 1981, 1005. 300

81SC763 M. Kolb and J. Barth; *Synth. Commun.*, 1981, **11**, 763. 225

81SC781 W. P. Reeves, A. Simmons, Jr., J. A. Rudis and T. C. Bothwell; *Synth. Commun.*, 1981, **11**, 781. 1047

81SR215 J. W. McFarland; *Sulfur Reports*, 1981, **1**, 215. 344

81SST66 G. C. Barrett; *Org. Compd. Sulphur, Selenium, Tellurium* (*R. Soc. Chem. series*), 1981, **6**, 66. 1022

81SST160 D. R. Hogg; *Org. Compd. Sulphur, Selenium, Tellurium* (*R. Soc. Chem. Series*), 1981, **6**, 160. 472

81SST165 J. K. Landquist; *Org. Compd. Sulphur, Selenium, Tellurium* (*R. Soc. Chem. Series*), 1981, **6**, 165. 472

81SST177 J. K. Landquist; *Org. Compd. Sulphur, Selenium, Tellurium* (*R. Soc. Chem. Series*), 1981, **6**, 177. 565

81T233 M. Mikolajczyk and P. Kielbasiński; *Tetrahedron*, 1981, **37**, 233. 1061

81T437 Y. G. Gololobov, I. N. Zhmurova and L. F. Kasukhin; *Tetrahedron*, 1981, **37**, 437. 373, 862

81T1283 G. R. Krow; *Tetrahedron*, 1981, **37**, 1283. 283

81T2181 J. E. Baldwin, A. L. J. Beckwith, A. P. Davis, G. Procter and K. A. Singleton; *Tetrahedron*, 1981, **37**, 2181. 341

81T2697 G. R. Krow; *Tetrahedron*, 1981, **37**, 2697. 141

81T3115 T. Morishita, N. Furukawa and S. Oae; *Tetrahedron*, 1980, **37**, 3115. 209

81T4059 H. H. Wasserman, R. J. Gambale and M. J. Pulwer; *Tetrahedron*, 1981, **37**, 4059. 174

81TL413 M. K. Kaloustian and F. Khouri; *Tetrahedron Lett.*, 1981, **22**, 413. 512

81TL1257 T. Kunieda, Y. Abe, T. Higuchi and M. Hirobe; *Tetrahedron Lett.*, 1981, **22**, 1257. 226

81TL1655 F. M. Menger and C. Lee; *Tetrahedron Lett.*, 1981, 1655. 47, 49, 52, 113

81TL1851 S. A. Benner; *Tetrahedron Lett.*, 1981, **22**, 1851. 581, 611, 643

81TL1881 D. Schinzer and C. H. Heathcock; *Tetrahedron Lett.*, 1981, **22**, 1881. 437

81TL2159 R. Appel and A. Westerhaus; *Tetrahedron Lett.*, 1981, **22**, 2159. 1154

81TL2731 P. G. Williard, L. A. Grab and J. P. Springer; *Tetrahedron Lett.*, 1981, **22**, 2731. 705

81TL2827 R. G. Visser, L. Brandsma and H. J. T. Bos; *Tetrahedron Lett.*, 1981, **22**, 2827. 427

81TL3409 Y. Tamaru, M. Kagotani and Z. Yoshida; *Tetrahedron Lett.*, 1981, **22**, 3409. 595, 614

81TL3413 Y. Tamaru, M. Kagotani, Y. Furukawa, Y. Amino and Z. Yoshida; *Tetrahedron Lett.*, 1981, **22**, 3413. 607

81TL3455 E. J. Corey, H. Cho, C. Rücker and D. H. Hua; *Tetrahedron Lett.*, 1981, **22**, 3455. 375

81TL3635 K. Clausen, M. Thorsen and S.-O. Lawesson; *Tetrahedron Lett.*, 1981, **37**, 3635. 573

81TL3925 A. I. Meyers and J. P. Hudspeth; *Tetrahedron Lett.*, 1981, **22**, 3925. 557

81TL4245 Y. Tamaru, M. Kagotani and Z. Yoshida; *Tetrahedron Lett.*, 1981, **22**, 4245. 607, 614

81TL4263 U. Redeker, N. Engel and W. Steglich; *Tetrahedron Lett.*, 1981, **22**, 4263. 290

81TL4405 J. M'Pati, P. Mangeney and Y. Langlois; *Tetrahedron Lett.*, 1981, **22**, 4405. 214

81TL4409 N. Furukawa, K. Kishimoto, S. Ogawa, T. Kawai, H. Fujihara and S. Oae; *Tetrahedron Lett.*, 1981, **22**, 4409. 1106

81T(S)213 S. V. Ley, C. A. Meerholz and D. H. R. Barton; *Tetrahedron Suppl.*, 1981, **37**, 213. 523

81USP4304935 J. V. Brammer-Petersen, C. L. Hakanson and P. T. Lindgren; *US Pat.* 4 304 935 (1981) (*Chem. Abstr.*, 1982, **96**, 85 278b). 1120

81YZ265 Y. Souma and H. Sano; *Yakugaku Zasshi*, (*J. Pharm. Soc. Jpn.*), 1981, **30**, 265 (*Chem. Abstr.*, 1981, **95**, 79 368u). 27

81ZAAC(473)7 G. Becker, M. Rössler and W. Uhl; *Z. Anorg. Allg. Chem.*, 1981, **473**, 7. 395, 879

81ZAAC(475)18 R. Appel, J. Peters and R. Schmitz; *Z. Anorg. Allg. Chem.*, 1981, **475**, 18. 888

81ZAAC(479)41 G. Becker, W. Uhl and H. J. Wessely; *Z. Anorg. Allg. Chem.*, 1981, **479**, 41. 882

81ZAAC(480)21 G. Becker; *Z. Anorg. Allg. Chem.*, 1981, **480**, 21. 879

81ZC271 H. Hartmann, E. Uhlemann, W. Hubner, E. Ludwig, L. Beyer, M. Walter and E. Hoyer; *Z. Chem.*, 1981, **21**, 271. 633

81ZN(B)16 G. Becker, G. Gresser and W. Uhl; *Z. Naturforsch., Teil B* 1981, **36**, 16. 1154

81ZN(B)1080 G. Becker, A. Münch and H. J. Wessely; *Z. Naturforsch., Teil B*, 1981, **36b**, 1080. 892

81ZOB239 V. P. Kozyukov and N. V. Mironova; *Zh. Obshch. Khim.*, 1981, **51**, 239 (*Chem. Abstr.*, 1981, **95**, 7390x). 1007

81ZOB482 A. A. Prishchenko, A. A. Borisenko, N. F. Nifant'ev, Z. S. Novikova and I. F. Lutsenko; *Zh. Obshch. Khim.*, 1981, **51**, 482 (*Chem. Abstr.*, 1981, **95**, 62 314a). 884

81ZOB1201 G. A. Pensionerova, V. G. Rozinov and V. I. Glukhikh; *Zh. Obshch. Khim.*, 1981, **51**, 1201 (*Chem. Abstr.*, 1981, **95**, 150 772d). 656

81ZOB1318 G. A. Razuvaev, V. I. Shcherbakov, N. E. Stolyarova, T. V. Trukhina, O. S. D'yachkovskaya and R. P. Zakharova; *Zh. Obshch. Khim.*, 1981, **51**, 1318. 390

81ZOB1814 V. P. Kozyukov, G. I. Orlov and V. F. Mironov; *Zh. Obshch. Khim.*, 1981, **51**, 1814 (*Chem. Abstr.*, 1981, **95**, 186 574e). 978

81ZOB1842 O. P. Shvaika and V. F. Lipnitskii; *Zh. Obshch. Khim.*, 1981, **51**, 1842 (*Chem. Abstr.*, 1982, **96**, 20 035a). 627

81ZOB2017 V. P. Kozyukov, G. I. Orlov and V. F. Mironov; *Zh. Obshch. Khim.*, 1981, **51**, 2017 (*Chem. Abstr.*, 1982, **96**, 6793q). 983

81ZOB2382 V. F. Mironov, V. P. Kozyukov and V. P. Kozyukov; *Zh. Obshch. Khim.*, 1981, **51**, 2382 (*Chem. Abstr.*, 1982, **96**, 20 169x). 722

81ZOR401 V. I. Naddaka and V. P. Gar'kin; *Zh. Org. Khim.*, 1981, **17**, 401. 349

81ZOR529 N. Y. Derkach and T. V. Lyapina; *Zh. Org. Khim.*, 1981, **17**, 529. 341

81ZOR533 I. V. Koval', A. I. Tarasenko, M. M. Kremlev and N. R. Molchanova; *Zh. Org. Khim.*, 1981, **17**, 533. 334

81ZOR1420 V. N. Fetyukhin, M. V. Vovk and L. I. Samarai; *Zh. Org. Khim.*, 1981, **17**, 1420. 1063

81ZOR1720 Yu. L. Yagupol'skii, I. I. Gerus and V. M. Cherkasov; *Zh. Org. Khim.*, 1981, **17**, 1720 (*Chem. Abstr.*, 1982, **96**, 104 147b). 657

81ZOR2064 V. Z. Laishev, M. L. Petrov and A. A. Petrov; *Zh. Org. Khim.*, 1981, **17**, 2064 (*Chem. Abstr.*, 1982, **96**, 122 329s). 627

81ZOR2153 E. Y. Posyagina, G. S. Posyagin, I. S. Berdinskii and R. A. Kusyakov; *Zh. Org. Khim.*, 1981, **17**, 2153. 354

81ZOR2225 V. V. Mezheritskii, L. G. Minyaeva, L. V. Mezheritskaya, A. L. Pilkus and G. N. Dorofeenko; *Zh. Org. Khim.*, 1981, **17**, 2225 (*Chem. Abstr.*, 1982, **96**, 68 865y). 688

81ZOR2452 V. N. Fetyukhin, V. V. Kobrzhitskii and L. I. Samarai; *Zh. Org. Khim.*, 1981, **17**, 2454. 226

82AG375 R. Höfelmeier and S. Blechert; *Angew. Chem.*, 1982, **94**, 375. 711

82AG(E)117 W. A. Herrmann; *Angew. Chem., Int. Ed. Engl.*, 1982, **21**, 117. 1097

82AG(E)150 S. Kato, E. Hattori, M. Mizuta and M. Ishida; *Angew. Chem., Int. Ed. Engl.*, 1982, **21**, 150. 246

82AOC(21)241 J. Satgé; *Adv. Organomet. Chem.*, 1982, **21**, 241. 927

82AQ(B)327 V. Cerdá, E. Casassas and F. Borull; *An. Quim.* (*B*), 1982, **78**, 327. 226

82BCJ643 T. Inamoto, T. Kusumoto and M. Yokoyama; *Bull. Chem. Soc. Jpn.*, 1982, **55**, 643. 286

82BCJ2186 M. Nakahata, M. Imaida, H. Ozaki, T. Harada and A. Tai; *Bull. Chem. Soc. Jpn.*, 1982, **55**, 2186. 63

82BSF(2)167 I. Ganboa and C. Palomo; *Bull. Soc. Chim. Fr., Part* 2, 1982, 167. 299

82BSF(2)369 J. A. Andrés and C. Palomo; *Bull. Soc. Chim. Fr. Part* 2, 1982, 369. 371

82CB804 G. Maier and W. A. Jung; *Chem. Ber.*, 1982, **115**, 804. 14

82CB1488 H. Blau, W. Malisch and P. Weickert; *Chem. Ber.*, 1982, **115**, 1488. 448

82CC132 Y. Fujiwara, I. Kawata, T. Kawauchi and H. Taniguchi; *J. Chem. Soc., Chem. Commun.*, 1982, 132. 197

82CC184 G. Christian, H. Stolzenberg and W. P. Fehlhammer; *J. Chem. Soc., Chem. Commun.*, 1982, 184. 1089

82CC215 N. A. Bailey, P. L. Chell, A. Mukhopadhyay, H. E. Tabbron and M. J. Winter; *J. Chem. Soc., Chem. Commun.*, 1982, 215. 943

82CC335 P. Metzner; *J. Chem. Soc., Chem. Commun.*, 1982, 335. 556

82CC444 M. D. Brown, D. W. Gillon, G. D. Meakins and G. H. Whitham; *J. Chem. Soc., Chem. Commun.*, 1982, 444. 1125

82CC457 H. Davy; *J. Chem. Soc., Chem. Commun.*, 1982, 457. 551

82CC607 V. L. Goedken, M. R. Deakin and L. A. Bottomley; *J. Chem. Soc., Chem. Commun.*, 1982, 607. 1097

82CC741 M. Ishikura, M. Mori, M. Terashima and Y. Ban; *J. Chem. Soc., Chem. Commun.*, 1982, 741. 301

82CC839 A. Waterfeld and R. Mews; *J. Chem. Soc., Chem. Commun.*, 1982, 839. 662, 663

82CC1178 K. Miki, H. Taniguchi, Y. Kai, N. Kasai, K. Nishiwaki and M. Wada; *J. Chem. Soc., Chem. Commun.*, 1982, 1178. 948

82CC1273 G. W. Kabalka, M. C. Delgado, U. Sastry and K. A. R. Sastry; *J. Chem. Soc., Chem. Commun.*, 1982, 1273. 45, 71, 97

82CC1303 N. Aktogu, H. Felkin and S. G. Davies; *J. Chem. Soc., Chem. Commun.*, 1982, 1303. 453

82CHE1302 Yu. N. Bulychev, J. A. Korbukh and M. N. Preobrazhenskaya; *Chem. Heterocycl. Compd.* (*Engl. Transl.*), 1982, 1302. 690

82CJC199 A. C. Storer, Y. Ozaki and P. R. Carey; *Can. J. Chem.*, 1982, **60**, 199. 558

82CL123 T. Fujinami, K. Wakuda, N. Takahashi and S. Sakai; *Chem. Lett.*, 1982, 123. 557

82CL529 S. Takano, M. Hirama and K. Ogasawara; *Chem. Lett.*, 1982, 529. 616

82CL885 W. Ando, T. Furuhata, H. Tsumaki and A. Sekiguchi; *Chem. Lett.*, 1982, 885. 234, 242

82CL981 K. Ohno, H. Matsuura, H. W. Kroto and H. Murata; *Chem. Lett.*, 1982, 981. 1152

82CL1255 T. Nakano, Y. Shimada, R. Sako, M. Kayama, M. Mikio, N. Hideyuki and Y. Nagai; *Chem. Lett.*, 1982, 1255. 57

82CL1607 M. Utaka, Y. Fujita and A. Takeda; *Chem. Lett.*, 1982, 1607. 357

82CL1891 T. Fujisawa, T. Mari, K. Fukumoto and T. Sato; *Chem. Lett.*, 1982, 1891. 131

82COMC-I G. Wilkinson, F. G. A. Stone and E. W. Abel; *Comp. Organomet. Chem.*, 1*st edn.*, 1982. 937

82CPB119 N. Hashimoto, T. Aoyama and T. Shioiri; *Chem. Pharm. Bull.*, 1982, **30**, 119. 873

82CPB899 Y.-C. Kuo, T. Aoyama and T. Shioiri; *Chem. Pharm. Bull.*, 1982, **30**, 899. 867

82CPB3125 M. Tanno, S. Sueyoshi and S. Kamiya; *Chem. Pharm. Bull.*, 1982, **30**, 3125. 1118, 1124

82CPB3380 S. Mori, I. Sakai, T. Aoyama and T. Shioiri; *Chem. Pharm. Bull.*, 1982, **30**, 3380. 873

82CPB4521 T. Higashino, S. Yoshida and E. Hayashi; *Chem. Pharm. Bull.*, 1982, **30**, 4521. 769

82CZ345 J. Graumann and W. Kliegel; *Chem.-Ztg.*, 1982, **106**, 345. 92

82EUP46194 B. Ruge; *Eur. Pat.* 4619 4A1 (1982) (*Chem. Abstr.*, 1982, **97**, 72 089m). 114

82GEP3116474 C. Ruechardt, J. Jaehme and U. Salz; *Ger. Pat.* 3 116 474 (1982) (*Chem. Abstr.*, 1983, **98**, 88 980b). 29

82GSM100 D. W. Knight; *Gen. Synth. Methods*, 1982, **5**, 100. 24

82H(19)57 A. S. Shawali, H. M. Hassaneen, A.-F. Shetta, A. Osman and F. Abdel-Galil; *Heterocycles*, 1982, **19**, 57. 676

82H(19)319 H. M. Hassaneen, A. O. Abdelhamid, A. S. Shawali and R. Pagni; *Heterocycles*, 1982, **19**, 319. 676

82H(19)1477 H. M. Hassaneen, A. H. Shetta, N. M. Elwan and A. S. Shawali; *Heterocycles*, 1982, **19**, 1477. 704

82HCA2299 R. Wälchli and M. Hesse; *Helv. Chim. Acta*, 1982, **65**, 2299. 272, 283

82HOU(E1)27 M. Regitz and B. Weber; *Methoden Org. Chem.* (*Houben-Weyl*), 1982, **E1**, 27. 876

82HOU(E1)616 H. J. Bestmann and R. Zimmermann; *Methoden Org. Chem.* (*Houben-Weyl*), 1982, **E1**, 616. 894

82IC1278 C. M. Lukehart and M. Raja; *Inorg. Chem.*, 1982, **21**, 1278. 801

82IC1616 J. S. Thrasher, J. L. Howell and A. F. Clifford; *Inorg. Chem.*, 1982, **21**, 1616. 1085

82ICA127 E. Delgado, M. Cano and A. Santos; *Inorg. Chim. Acta*, 1982, **62**, 127. 1142

82IJC(B)589 G. H. Sayed, M. S. Elhalim, M. Y. El-Kady and L. M. Elwahab; *Indian J. Chem., Sect. B*, 1982, **21**, 589. 201

82IJC(B)989 V. Sudarsanam, K. Nagarajan, V. P. Arya, A. P. Kaulgud, S. J. Shenoy and R. K. Shah; *Indian J. Chem., Sec. B*, 1982, **21**, 989. 690

82IZV937 A. V. Fokin, Yu. N. Studnev, V. P. Stolyarov and N. N. Baranov; *Izv. Akad. Nauk SSSR, Ser. Khim.*, 1982, 937 (*Chem. Abstr.*, 1982, **97**, 38 475z). 698

82IZV939 A. N. Pudovik, G. V. Romanov and V. N. Volkova; *Izv. Acad. Nauk SSSR, Ser. Khim.*, 1982, 939 (*Chem. Abstr.*, 1982, **97**, 55 913h). 1128

82IZV949 D. A. Bravo-Zhivotovskii, S. D. Pigarev, O. A. Vyazankina, I. D. Kalikhman and N. S. Vyazankin; *Izv. Akad. Nauk SSSR, Ser. Khim.*, 1982, **4**, 949 (*Chem. Abstr.*, 1982, **97**, 110122). 432

82IZV1611 R. G. Kostyanovskii, Y. I. El'natanov, S. M. Shikhaliev and S. M. Ignatov; *Izv. Akad. Nauk SSSR, Ser. Khim.*, 1982, 1611 (*Chem. Abstr.*, 1983, **98**, 16770k). 1130

82IZV2121 I. I. Kandror and I. O. Bragina; *Izv. Akad. Nauk SSSR, Ser. Khim.*, 1982, **9**, 2121 (*Chem. Abstr.*, 1983, **98**, 16383). 727

82JA166 H. Wynberg and E. G. J. Staring, *J. Am. Chem. Soc.*, 1982, **104**, 166. 162

82JA1096 S. Quinn, A. Shaver and V. W. Day; *J. Am. Chem. Soc.*, 1982, **104**, 1096. 459

82JA1119 H. J. Reich and M. J. Kelly; *J. Am. Chem. Soc.*, 1982, **104**, 1119. 423, 427

82JA1428 F. Sainte, B. Serckx-Poncin, A. M. Hesbain-Frisque and L. Ghosez; *J. Am. Chem. Soc.*, 1982, **104**, 1428. 718

82JA1735 H. U. Reissig and I. Böhm; *J. Am. Chem. Soc.*, 1982, **104**, 1735. 150

82JA3104 K. Steliou and M. Mrani; *J. Am. Chem. Soc.*, 1982, **104**, 3104. 512, 575

82JA3233 A. J. Biloski, R. D. Wood and B. Ganem; *J. Am. Chem. Soc.*, 1982, **104**, 3233. 285

82JA3879 N. M. Kostić and R. F. Fenske; *J. Am. Chem. Soc.*, 1982, **104**, 3879. 932

82JA4015 A. B. Smith, III, S. R. Schow, J. D. Bloom, A. S. Thompson and K. N. Winzenberg; *J. Am. Chem. Soc.*, 1984, **104**, 4015. 173

82JA4018 Y. Tamaru, Y. Amino, Y. Furukawa, M. Kagotani and Z. Yoshida; *J. Am. Chem. Soc.*, 1982, **104**, 4018. 607

82JA4686 G. Stork, I. Paterson and F. K. C. Lee; *J. Am. Chem. Soc.*, 1982, **104**, 4686. 173

82JA4708 D. R. Williams, B. A. Barner, K. Nishitani and J. G. Phillips; *J. Am. Chem. Soc.*, 1982, **104**, 4708. 173

82JA5521 S. Masamune, T. Kaiho and D. S. Garvey; *J. Am. Chem. Soc.*, 1982, **104**, 5521. 173

82JA5523 S. Masamune, L. D. L. Lu, W. P. Jackson, T. Kaiho and T. Toyoda; *J. Am. Chem. Soc.*, 1982, **104**, 5523. 173, 235

82JA5534 D. Seyferth and R. M. Weinstein; *J. Am. Chem. Soc.*, 1982, **104**, 5534. 418

82JA5667 A. G. Brook, S. C. Nyburg, F. Abdesaken, B. Gutekunst, G. Gutekunst, R. Krishna, M. R. Kallury, Y. C. Poon, Y.-M. Chang and W. Wong-Ng; *J. Am. Chem. Soc.*, 1982, **104**, 5667. 924

82JA6769 H. F. Gilbert and W. P. Jencks; *J. Am. Chem. Soc.*, 1982, **104**, 6769. 699

82JA6818 E. J. Corey, D. H. Hua, B.-C. Pan and S. P. Seitz; *J. Am. Chem. Soc.*, 1982, **106**, 6818. 179

82JCED97 W. H. Gilligan and M. E. Sitzmann; *J. Chem. Eng. Data*, 1982, **27**, 97. 366

82JCED103 H. S. El Kashef, K. U. Sadek, M. H. Elnagdi and H. H. Alnima; *J. Chem. Eng. Data*, 1982, **27**, 103. 676, 677

82JCS(D)793 M. Wada, K. Sameshima, K. Nishiwaki and Y. Kawasaki; *J. Chem. Soc., Dalton Trans.*, 1982, 793. 948

82JCS(D)1131 M. J. Barrow, E. A. V. Ebsworth, C. M. Huntley and D. W. H. Rankin; *J. Chem. Soc., Dalton Trans.*, 1982, 1131. 539

82JCS(P1)1763 B. M. R. Bandara, A. J. Birch and W. D. Raverty; *J. Chem. Soc. Perkin Trans. 1*, 1982, 1763. 96

82JCS(P1)2663 H. A. Elfahham, K. U. Sadek, G. E. H. Elgemeie and M. H. Elnagdi; *J. Chem. Soc., Perkin Trans. 1*, 1982, 2663. 771

82JCS(P1)2909 S. Terao, M. Shiraishi, K. Kato, S. Ohkawa, Y. Ashida and Y. Maki; *J. Chem. Soc., Perkin Trans. 1*, 1982, 2909. 278

82JCS(P2)473 E. R. Cullen, F. S. Guziec, Jr., C. J. Murphy, T. C. Wong and K. K. Andersen; *J. Chem. Soc. Perkin Trans. 2*, 1982, 473. 521, 523

82JGU980 O. P. Lobanov and B. S. Drach; *J. Gen. Chem. USSR* (*Engl. Transl.*), 1982, **52**, 980. 656

82JGU1222 V. P. Kozyukov, V. P. Kozyukov and V. F. Mironov; *J. Gen. Chem. USSR* (*Engl. Transl.*), 1982, **52**, 1222. 722

82JHC73 A. S. Shawali, A. O. Abdelhamid, H. M. Hassaneen and A. Shetta; *J. Heterocycl. Chem.*, 1982, **19**, 73. 676, 677

82JHC577 B. Stanovnik, A. Štimac, M. Tišler and B. Verček; *J. Heterocycl. Chem.*, 1982, **19**, 577. 836

82JHC1397 M. Kočevar, B. Stanovnik and M. Tišler; *J. Heterocycl. Chem.*, 1982, **19**, 1397. 824, 834

82JOC52 M. J. Marks and H. M. Walborsky; *J. Org. Chem.*, 1982, **47**, 52. 796

82JOC154 D. Landini and F. Rolla; *J. Org. Chem.*, 1982, **47**, 154. 29, 106

82JOC176 D. M. Floyd, A. W. Fritz and C. M. Cimarusti; *J. Org. Chem.*, 1982, **47**, 176. 342

82JOC179 C. M. Cimarusti, H. E. Applegate, H. W. Chang, D. M. Floyd, W. H. Koster, W. A. Slusarchyk and M. G. Young; *J. Org. Chem.*, 1982, **47**, 179. 346

82JOC747 S. R. Wilson, M. S. Hague and R. N. Misra; *J. Org. Chem.*, 1982, **47**, 747. 413

82JOC2251 K. Baum, C. D. Bedford and R. J. Hunadi; *J. Org. Chem.*, 1982, **47**, 2251. 967

82JOC2785 W. H. Rastetter, D. M. Spero and J. Adams D. N. Harpp and D. K. Ash; *J. Org. Chem.*, 1982, **47**, 2785. 338

82JOC3336 T. P. Mawhinney and M. A. Madson; *J. Org. Chem.*, 1982, **47**, 3336. 719

82JOC3358 R. Esmond, B. Fraser-Reid and B. B. Jarvis; *J. Org. Chem.*, 1982, **47**, 3358. 179

82JOC3450 T. J. Brocksom and M. G. Constantino; *J. Org. Chem.*, 1982, **47**, 3450. 199

82JOC3533 V. Mizrahi and T. A. Modro; *J. Org. Chem.*, 1982, **47**, 3533. 370

82JOC3658 J. J. Fitt, A. W. Gschwend, A. Hamdan, S. K. Boyer and H. M. Haider; *J. Org. Chem.*, 1982, **47**, 3658. 657

82JOC3787 G. S. Poindexter; *J. Org. Chem.*, 1982, **47**, 3787. 302

82JOC3876 P. W. Jeffs, G. Molina, N. A. Cortese, P. R. Hauck and J. Wolfram; *J. Org. Chem.*, 1982, **47**, 3876. 282

82JOC3881 P. W. Jeffs, N. A. Cortese and J. Wolfram; *J. Org. Chem.*, 1982, **47**, 3881. 282

82JOC3966 C. A. Bruynes and T. K. Jurriens; *J. Org. Chem.*, 1982, **47**, 3966. 224, 536
82JOC4384 E. Vedejs, M. J. Arnost, J. M. Eustache and G. A. Krafft; *J. Org. Chem.*, 1982, **47**, 4384. 554
82JOC4605 R. K. Olsen, W. J. Hennen and R. B. Wardle; *J. Org. Chem.*, 1982, **47**, 4605. 315
82JOC4812 M. Ravindranathan, N. Kalyanam and S. Sivaram; *J. Org. Chem.*, 1982, **47**, 4812. 275, 295
82JOC5042 W. G. Dauben and R. A. Bunce; *J. Org. Chem.*, 1982, **47**, 5042. 346
82JOC5051 J. J. Eisch, J. E. Galle, A. Piotrowski and M.-R. Tsai; *J. Org. Chem.*, 1982, **47**, 5051. 413
82JOC5160 D. M. Floyd, A. W. Fritz, J. Pluscec, E. R. Weaver and C. M. Cimarusti; *J. Org. Chem.*, 1982, **47**, 5160. 346
82JOC5283 K. Yamada, S. Tanaka, K. Naruchi and M. Yamamoto; *J. Org. Chem.*, 1982, **47**, 5283. 325
82JOM(213)C12 T. Kobayashi and M. Tanaka; *J. Organomet. Chem.*, 1982, **213**, C12. 226
82JOM(224)153 W. P. Fehlhammer, P. Hirschmann and A. Mayr; *J. Organomet. Chem.*, 1982, **224**, 153. 799
82JOM(225)331 C. Eaborn, R. W. Griffiths and A. Pidcock; *J. Organomet. Chem.*, 1982, **225**, 331. 419
82JOM(231)191 V. Sicker, A. Meller and W. Maringgele; *J. Organomet. Chem.*, 1982, **231**, 191. 795, 796
82JOM(235)243 K. Kikukawa and T. Matsuda; *J. Organomet. Chem.*, 1985, **235**, 243. 220
82JOM(236)C1 M. Luksza, K. Fiederling, U. Wanka and W. Malisch; *J. Organomet. Chem.*, 1982, **236**, C1. 911
82JOM(236)C37 A. Mantovani and B. Crocciani; *J. Organomet. Chem.*, 1982, **236**, C37. 801
82JPR(324)667 J. Gloede; *J. Prakt. Chem.*, 1982, **324**, 667. 710, 712, 714
82JSP(92)158 J. C. T. R. Burckett-St. Laurent, H. W. Kroto, J. F. Nixon and K. Ohno; *J. Mol. Spectrosc.*, 1982, **92**, 158. 1154
82JST(79)215 J. C. T. R. Burckett-St. Laurent, T. A. Cooper, H. W. Kroto, J. F. Nixon, O. Ohashi and K. Ohno; *J. Mol. Struct.*, 1982, **79**, 215. 1153
82KGS1682 Yu. N. Bulychev, I. A. Korbukh and M. N. Preobrdzhenskaya; *Khim. Geterotsikl. Soedin.*, 1982, 1682 (*Chem. Abstr.*, 1983, **98**, 120 6016b). 690
82LA1229 S. Kato, H. Shibahashi, T. Katada, T. Takagi, I. Noda, M. Mizuta and M. Goto; *Liebigs Ann. Chem.*, 1982, 1229. 510, 551, 583, 620
82LA1451 W. Krasuski, D. Nikolaus and M. Regitz; *Liebigs Ann. Chem.*, 1982, 1451. 548
82LA1582 W. Kantlehner, E. Haug and M. Farkas; *Liebigs Ann. Chem.*, 1982, 1582. 609
82LA1927 G. Blotny; *Liebigs Ann. Chem.*, 1982, 1927. 1026
82M(113)593 A. Wiatschka and K.-H. Ongania; *Monatsh. Chem.*, 1982, **113**, 593. 586, 609
82MI 501-01 Anon.; *Res. Discl.*, 1982, **219**, 239 (*Chem. Abstr.*, 1982, **97**, 144 508). 15
82MI 502-01 K Nilsson, P. Brodelius and K. Mosbach; *Appl. Biochem. Biotechnol.*, 1982, **7**, 47. 118
B-82MI 502-01 P. S. Bailey; "Ozonization in Organic Chemistry," Academic Press, New York, 1982, vol. 2. 28
82MI 502-02 M. Hronec and J. Ilavsky; *Ropa Uhlie*, 1982, **24**, 464 (*Chem. Abstr.*, 1982, **97**, 218 392d). 24, 118
82MI 504-01 S. Takimoto, K. Ogihara, Y. Kodera and H. Ohta; *Fukukoa Univ. Sci. Reports*, 1982, **12**, 95. 196
B-82MI 507-01 J. W. Munson; in "Chemistry and Biology of Hydroxamic Acids," eds H. Kehl and S. Karger, New York, 1982. 322
82MI 512-01 A. A. Turovskii, R. V. Kucher, N. A. Turovskii, A. A. Andrianov and V. V. Petrenko; *Dopov. Akad. Nauk. Ukr. RSR, Ser. B: Geol., Khim. Biol. Nauki*, 1982, 52 (*Chem. Abstr.*, 1982, **97**, 55 111). 530
82MI 517-01 S. Szakacz and F. Nagy; *Magy. Kem. Foly.*, 1982, **88**, 193 (*Chem. Abstr.*, 1982, **97**, 182 250q). 709
82MI 519-01 P. Navarrete, H. Rodriguez, H. Pavez and A. Marquez; *Bol. Soc. Chil. Quim.*, 1982, **27**, 227 (*Chem. Abstr.*, 1982, **97**, 23 411b). 759
82MI 520-01 M. R. Kilbourn and M. J. Welch; *Int. J. Appl. Radiat. Isot.*, 1982, **33**, 359. 796
82MI 521-01 B. Stanovnik; *Chem. Zvesti* (*Chemical Papers*), 1982, **36**, 693. 835
B-82MI 523-01 L. Weber; in "The Chemistry of the Metal Carbon Bond," eds F. R. Hartley and S. Patai, Wiley, Chichester, 1982, vol. 1, p. 91. 894
82NKK811 R. Dohno, Y. Yamauchi and T. Hakushi; *Nippon Kagaku Kaishi* (*J. Chem. Soc. Jpn.*), 1982, **56**, 811. 70
82OM64 G. K. Anderson, H. C. Clark and J. A. Davies; *Organometallics*, 1982, **1**, 64. 1139
82OM628 B. E. Boland-Lussier, M. R. Churchill, R. P. Hughes and A. L. Rheingold; *Organometallics*, 1982, **1**, 628. 950
82OM635 B. E. Boland-Lussier and R. P. Hughes; *Organometallics*, 1982, **1**, 635. 950
82OM830 J. A. Söderquist and G. J.-H. Hsu; *Organometallics*, 1982, **1**, 830. 416
82OM994 A. G. Brook, F. Abdesaken, G. Gutekunst and N. Plavac; *Organometallics*, 1982, **1**, 994. 418, 924
82OM1488 A. Marinetti and F. Mathey; *Organometallics*, 1982, **1**, 1488. 407
82OM1540 M. D. Fryzuk and P. A. MacNeil; *Organometallics*, 1982, **1**, 1540. 380
82OM1730 M. Brookhart, A. R. Pinkas and A. Lukacs; *Organometallics*, 1982, **1**, 1730. 448
82OPP249 L. Oshry and S. M. Rosenfeld; *Org. Prep. Proced. Int.*, 1982, **14**, 249 and 251. 62
82OPP396 A. D. Dunn, M. J. Mills and W. Henry; *Org. Prep. Proced. Int.*, 1982, **14**, 396. 275
82OS(61)147 S. Danishefsky, T. Kitahara and P. F. Schuda; *Org. Synth.*, 1982, **61**, 147. 199
82POL129 S.-C. Chang and D. D. DesMarteau; *Polyhedron*, 1982, **1**, 129. 313
82PS(12)325 J.-P. Guemas, M. Lees, A. Reliquet and J. Villieras; *Phosphorus Sulfur*, 1982, **12**, 325. 606
82RTC1 B. Zwanenburg; *Recl. Trav. Chim. Pays-Bas*, 1982, **101**, 1. 472, 488, 562
82RTC310 R. L. P. de Jong, J. Meijer, R. S. Sukhai and L. Brandsma; *Recl. Trav. Chim. Pays-Bas*, 1982, **101**, 310. 589
82S212 F. Duboudin, P. Cazeau, F. Moulines and O. Laporte; *Synthesis*, 1982, 212. 1134
82S237 H. M. R. Hoffmann, K. Haase and P. M. Geschwinder; *Synthesis*, 1982, 237. 19
82S277 M. G. Voronkov, V. K. Roman and E. A. Maletina; *Synthesis*, 1982, 277. 1135
82S288 R. Mestres and C. Palomo; *Synthesis*, 1982, 288. 300
82S297 J. P. Gillet, R. Sauvetre and J. F. Normant; *Synthesis*, 1982, 297. 58
82S457 S. Kato, S. Chiba, M. Goto, M. Mizuta and M. Ishida; *Synthesis*, 1982, 457. 547

82S472 P. Molina, M. Alajarin, A. Ferao, M. J. Lidon, P. M. Fresneda and M. J. Vilaplana; *Synthesis*, 1982, 472. 1107
82S506 K. S. Keshavamurthy, Y. D. Vankar and D. N. Dhar; *Synthesis*, 1982, 506. 125, 189, 196, 297, 303
82S521 N. Petragnani and M. Yonashiro; *Synthesis*, 1982, 521. 28, 46, 148
82S592 W. Klötzer, H. Baldinger, E. M. Karpitschka and J. Knoflach; *Synthesis*, 1982, 592. 360
82S596 P. Molina, M. Alajarin and A. Arques; *Synthesis*, 1982, 596 975, 1031
82S684 J. M. Aizpurua and C. Palomo; *Synthesis*, 1982, 684. 17
82S693 G. Singh, S. S. Bhattacharjee, H. Ila and H. Junjappa; *Synthesis*, 1982, 693. 555
82S824 N. A. Rodios and N. E. Alexandrou; *Synthesis*, 1982, 824. 356
82S833 S. Ohta, S. Shimabayashi, M. Aono and M. Okamoto; *Synthesis*, 1982, 833. 125, 126
82S933 M. Ueda, N. Kawaharasaki and Y. Imai; *Synthesis*, 1982, 933. 303
82S958 R. R. Schmidt and A. Wagner; *Synthesis*, 1982, 958. 199
82S1013 S. Kato, K. Miyagawa, S. Kawabata and M. Ishida; *Synthesis*, 1982, 1013. 245, 252
82SC131 F. Ogura, H. Yamaguchi, T. Otsubo and K. Chikamatsu; *Synth. Commun.*, 1982, **12**, 131. 1117
82SC177 C. G. Rao; *Synth. Commun.*, 1982, **12**, 177. 226
82SC453 A. Mohacsi; *Synth. Commun.*, 1982, **12**, 453. 132
82SC513 M. Tordeux and C. Wakselman; *Synth. Commun.*, 1982, **12**, 513. 4
82SC573 P. Molina, M. Alajarín, A. Arques and J. Sáez; *Synth Commun.*, 1982, **12**, 573. 1079
82SC709 X. Huang, C.-C. Chan and Q.-S. Zhou; *Synth. Commun.*, 1982, **12**, 709. 226
82SC787 J. J. Bodine and M. K. Kaloustian; *Synth. Commun.*, 1982, **12**, 787. 576
82SC829 A. S. Kende and D. A. Becker; *Synth. Commun.*, 1982, **12**, 829. 555
82SC865 A. Corsaro, A. Compagnini, M. Tarantello, S. Barbaro and G. Purrello; *Synth. Commun.*, 1982, **12**, 865. 734
82SC897 S. Juliá, G. Tagle and J. C. Vega; *Synth. Commun.*, 1982, **12**, 897. 242
82SC989 J. I. Levin, E. Turos and S. M. Weinreb; *Synth. Commun.*, 1982, **12**, 989. 329
82SUL69 A.-B. Ghattas, A. G. Ghattas, E.-E. A. M. El-Khrisky and S.-O. Lawesson; *Sulfur Letters*, 1982, **1**, 69. 560
82T1509 H. Malmberg and M. Nilsson; *Tetrahedron*, 1982, **38**, 1509. 151
82T3087 S. M. Weinreb and R. R. Staib; *Tetrahedron*, 1982, **38**, 3087. 326, 330
82TL447 S. Harusawa and T. Shioiri; *Tetrahedron Lett.*, 1982, **23**, 447. 1112
82TL881 S. Takano, M. Hirama and K. Ogasawara; *Tetrahedron Lett.*, 1982, **23**, 881. 607, 616
82TL1259 G. Boireau, D. Abenhaim, A. Deberly and B. Sabourault; *Tetrahedron Lett.*, 1982, 1259. 63
82TL1391 S. Prabhakar, A. M. Lobo and M. M. Marques; *Tetrahedron Lett.*, 1982, **23**, 1391. 212
82TL1585 E. Kozlowska-Gramsz and G. Descotes; *Tetrahedron Lett.*, 1982, 1585. 1071
82TL2023 G. Morel, E. Marchand, K. H. Nguyen Thi and A. Foucaud; *Tetrahedron Lett.*, 1982, **23**, 2023. 728
82TL2347 J. I. Levin and S. M. Weinreb; *Tetrahedron Lett.*, 1982, 2347. 30
82TL3101 K. J. H. Kruithof and G. W. Klumpp; *Tetrahedron Lett.*, 1982, **23**, 3101. 427
82TL3255 G. Boche, F. Bosold and M. Niessner; *Tetrahedron Lett.*, 1982, **23**, 3255. 302
82TL3357 A. McKillop, A. Henderson, P. S. Ray, C. Avendano and E. G. Molinero; *Tetrahedron Lett.*, 1982, **23**, 3357. 773
82TL4087 S. Masson and A. Thuillier; *Tetrahedron Lett.*, 1982, **23**, 4087. 557, 559
82TL4933 O. I. Kolodiazhnyi; *Tetrahedron Lett.*, 1982, 4933. 1086, 1087
82TL5513 D. Domínguez and M. P. Cava; *Tetrahedron Lett.*, 1982, **23**, 5513. 162
82USP4335058 N. Rizkalla (Halcon Corp.); *US Pat.* 4 335 058 (1982) (*Chem. Abstr.*, 1982, **97**, 71 957). 192
82USP4347379 J. T. Lai; *US Pat.* 4 347 379 (1982) (*Chem. Abstr.*, 1982, **97**, 215 564a). 67
82YGK123 T. Takeshima, M. Muraoka and N. Fukada; *Yuki Gosei Kagaku Kyokaishi*, 1982, 123. 547
82ZAAC(486)61 F. Weller, J. Pebler, K. Dehnicke, K. Hartke and H. M. Wolff; *Z. Anorg. Allg. Chem.*, 1982, **486**, 61. 776
82ZAAC(495)73 V. G. Becker, M. Rössler and G. Uhl; *Z. Anorg. Allg. Chem.*, 1982, **495**, 73. 395
82ZC223 P. Jeroschewski and P. Hansen; *Z. Chem.*, 1982, **22**, 223. 554
82ZN(B)950 H. Schmidbaur, U. Deschler and D. Seyferth; *Z. Naturforsch., Teil B*, 1982, **37**, 950. 908
82ZN(B)1481 S. Siddiqui, S. Begum and B. S. Siddiqui; *Z. Naturforsch., Teil B*, 1982, **37**, 1481. 1125
82ZOB885 A. S. Gordetsov, V. N. Latyaeva, Yu. A. Bogdarin, G. A. Boyarinov and Yu. I. Dergunov; *Zh. Obsch. Khim.*, 1982, **52**, 885. 1081, 1082
82ZOB921 N. I. Mysin, Yu. I. Dergunov and V. G. Vodop'yanov; *Zh. Obshch. Khim.*, 1982, **32**, 921. 1081
82ZOB1122 O. P. Lobanov and B. S. Drach; *Zh. Obshch. Khim.*, 1982, **52**, 1122 (*Chem. Abstr.*, 1982, **97**, 92 419x). 656
82ZOB1386 V. P. Kozyukov, V. P. Kozyukov and V. F. Mironov; *Zh. Obshch. Khim.*, 1982, **52**, 1386 (*Chem. Abstr.*, 1982, **97**, 163 077x). 722
82ZOB2654 N. V. Mironova and V. F. Mironov; *Zh. Obshch. Khim.*, 1982, **52**, 2654 (*Chem. Abstr.*, 1983, **98**, 179 452x). 969
82ZOR668 A. D. Sinitsa and N. A. Parkhomenko; *Zh. Org. Chem.*, 1982, **18**, 668. 1063
82ZOR1813 A. G. Rybin; *Zh. Org. Khim.*, 1982, **52**, 1813. 1125

83ACS(A)631 H. Saarinen, M. Orama, T. Raikas and J. Korvenranta; *Acta Chem. Scand., Ser. A*, 1983, **37**, 631. 864
83AG(E)55 E. Schaumann, U. Wriede and G. Rühter; *Angew. Chem., Int. Ed. Engl.*, 1983, **22**, 55. 604
83AG(E)63 J. Mulzer and M. Kappert; *Angew. Chem., Int. Ed. Engl.*, 1983, **95**, 63. 145, 155
83AG(E)785 R. Appel and W. Paulen; *Angew. Chem., Int. Ed. Engl.*, 1983, **22**, 785. 1019
83AG(E)877 H. Klusik and A. Berndt; *Angew. Chem.*, 1983, **95**, 895. *Angew. Chem., Int. Ed. Engl.*, 1983, **22**, 877. 927, 928
83AG(E)907 H. Schmidbaur; *Angew. Chem., Int. Ed. Engl.*, 1983, **22**, 907. 894, 1092, 1096

83AOC(22)59 M. I. Bruce and A. G. Swincer; *Adv. Organomet. Chem.*, 1983, **22**, 59. 941

83AP(316)21 J. Schneckenburger and H. Riedel; *Arch. Pharm.* (*Weinheim, Ger.*), 1983, **316**, 21. 674

83BAU1373 I. D. Kalikhman, B. A. Gostevskii, O. B. Bannikova, M. F. Larin, O. A. Vyazankina and N. S. Vyazankin; *Bull. Acad. Sci. USSR. Div. Chem. Sci.* 1983, 1373. 719, 720

83BCJ545 S. Ito, Y. Tanaka, A. Kakehi, T. Fukuyama, N. Osawa and N. Sayo; *Bull. Chem. Soc. Jpn.*, 1983, **56**, 545. 681

83BSF(2)277 A. Dibo, M. Stavaux and N. Lozac'h; *Bull. Soc. Chim. Fr., Part 2*, 1983, 277. 605, 613

83CB724 A. Waterfeld, W. Isenberg, R. Mews, W. Clegg and G. M. Sheldrick; *Chem. Ber.*, 1983, **116**, 724. 663

83CB938 R. Appel, V. Barth and F. Knoch; *Chem. Ber.*, 1983, **116**, 938. 880

83CB1183 F. Effenberger, G. Epple, J. K. Eberhard, U. Bühler and E. Sohn; *Chem. Ber.*, 1983, **116**, 1183. 209

83CB1386 H. Schmidbaur and U. Deschler; *Chem. Ber.*, 1983, **116**, 1386. 908

83CB1834 R. Abu-El-Halawa and J. C. Jochims; *Chem. Ber.*, 1983, **116**, 1834. 1146

83CB1955 H. Schmidbaur, C. Zybill, C. Krüger and H. J. Kraus; *Chem. Ber.*, 1983, **116**, 1955. 897

83CB2037 R. Appel and E. Hiester; *Chem. Ber.*, 1983, **116**, 2037. 275

83CB2275 R. Appel, U. Baumeister and F. Knoch; *Chem. Ber.*, 1983, **116**, 2275. 1092, 1093, 1095

83CB3413 W. Ladner; *Chem. Ber.*, 1983, **116**, 3413. 150

83CC74 R. Hull, J. A. H. MacBride, M. Wardleworth and P. M. Wright; *J. Chem. Soc., Chem. Commun.*, 1983, 74. 1035

83CC210 T. Livinghouse and R. Smith; *J. Chem. Soc., Chem. Commun.*, 1983, 210. 728

83CC239 K. J. H. Kruithof, R. F. Schmitz and G. W. Klumpp; *J. Chem. Soc., Chem. Commun.*, 1983, 239. 424, 468

83CC337 A. Alemagna, P. Del Buttero, E. Licandro, S. Maiorana and C. Guastini; *J. Chem. Soc., Chem. Commun.*, 1983, 337. 906

83CC388 S. H. Bertz, L. W. Jelinski and G. Dabbagh; *J. Chem. Soc., Chem. Commun.*, 1983, 388. 556

83CC392 D. Wenkert, K. M. Eliasson and R. Rudisill; *J. Chem. Soc., Chem. Commun.*, 1983, 392. 48, 49, 83, 103, 114

83CC457 T. Desmond, F. J. Lalor, G. Ferguson and M. Parvez; *J. Chem. Soc., Chem. Commun.*, 1983, 457. 1097

83CC880 P. A. Chaloner, G. D. Glick and R. A. Moss; *J. Chem. Soc., Chem. Commun.*, 1983, 880. 943

83CC1113 M. Koreeda and L. Brown; *J. Chem. Soc., Chem. Commun.*, 1983, 1113. 209

83CC1202 G. J. Baird, J. A. Bandy, S. G. Davies and K. Prout; *J. Chem. Soc., Chem. Commun.*, 1983, 1202 453

83CC1487 N. Matsumura, H. Asai and S. Yoneda; *J. Chem. Soc., Chem. Commun.*, 1983, 1487. 70, 73

83CCC596 E. A. Varella, E. D. Micromastoras, N. E. Alexandrou and O. Exner; *Collect. Czech. Chem. Commun.*, 1983, **48**, 596. 765

83CCC3421 V. Bobošík, A. Piklerová and A. Martvoň; *Collect. Czech. Chem. Commun.*, 1983, **48**, 3421. 1028

83CCC3567 P. Kristian, D. Koščik and J. Gonda; *Collect. Czech. Chem. Commun.*, 1983, **48**, 3567. 1057

83CI(L)391 G. Darnault, M. Saquet and A. Thuillier; *Chem. Ind.* (London), 1983, 391. 729

83CJC1105 Y. Y. C. Yeung Lam Ko and R. Carrie; *Can. J. Chem.*, 1983, **61**, 1105. 785, 792

83CL115 K. Sano, T. Yamamoto and A. Yamamoto; *Chem. Lett.*, 1983, 115. 465

83CL449 H. Suzuki, J. Tsuji, Y. Hiroi and N. Sato; *Chem. Lett.*, 1983, 449. 281

83CL683 T. Mukaiyama, J. Ichikawa and M. Asami; *Chem. Lett.*, 1983, 683. 282

83CL835 T. Fujisawa, T. Mori and T. Sato; *Chem. Lett.*, 1983, 835. 14

83CL871 A. Sekiguchi and W. Ando; *Chem. Lett.*, 1983, 871. 873

83CL885 Y. Miura, T. Kunishi and M. Kinoshita; *Chem. Lett.*, 1983, 885. 766

83CL1653 M. Yoshifuji, K. Toyota, K. Shibayama and N. Inamoto; *Chem. Lett.*, 1983, 1653. 879

83CL1901 T. Fujisawa, T. Itoh and T. Sato; *Chem. Lett.*, 1983, 1901. 553

83CPB2564 N. Kawai and T. Shioiri; *Chem. Pharm. Bull.*, 1983, **31**, 2564. 775

83CPB3139 N. Kawai, N. Kato, Y. Hamada and T. Shioiri; *Chem. Pharm. Bull.*, 1983, **31**, 3139. 775

83CRV321 A. J. L. Cooper, J. Z. Ginos and A. Meister; *Chem. Rev.*, 1983, **83**, 321. 62

83CZ121 H. Bollmacher and P. Sartori; *Chem.-Ztg.*, 1983, **107**, 121. 221

83CZ257 H. W. Hennig and P. Sartori; *Chem.-Ztg.*, 1983, **107**, 257. 221

83CZ363 H. W. Hennig and P. Sartori; *Chem.-Ztg.*, 1983, **107**, 363. 221

83EJM413 P.-A. Bonnet, C. Sablayrolles, J. P. Chapat, B. Soulie, M. Simeon de Buochberg, G. Dusart and M. Attisso; *Eur. J. Med. Chem.*, 1983, **18**, 413 (*Chem. Abstr.*, 1984, **100**, 139 063z). 772

83EUP66456 W. E. Bondinell, J. J. Lafferty and C. L. Zirkle; *Eur. Pat.* 6 645 6A1 (1982) (*Chem. Abstr.*, 1983, **98**, 198 025b). 92

83EUP127276A2 R. A. Head and A. Ibbotson; *Eur. Pat.* 127 276 A2 (1983) (*Chem. Abstr.*, 1985, **102**, 113 312w) 143

83FES101 A. Ranise, F. Bondavalli, P. Schenone, A. Bargagna, M. Scafuro, A. Marfella, L. Berrino and E. Marmo; *Farmaco Ed. Sci.*, 1983, **38**, 101. 597

83G421 F. Bonadies and M. L. Scarpati; *Gazz. Chim. Ital.*, 1983, **113**, 421. 79, 105

83GEP3134144 G. Hamprecht; *Ger. Pat.* 3 134 144 (1973) (*Chem. Abstr.*, 1983, **99**, 5226g). 758, 760

83GEP3144791 K. Findeisen (Bayer), *Ger. Pat.* 3 144 791 (1983) (*Chem. Abstr.*, 1983, **99**, 53 379). 189, 196

83GEP3148422 W. Wykypiel and W. Tronich; *Ger. Pat.* 3 148 422 (1983) (*Chem. Abstr.*, 1983, **99**, 158 032r). 24, 105

83GSM98 P. R. Jenkins; *Gen. Synth. Methods*, 1983, **6**, 98. 24

83H(20)1531 M. Shibuya, Y. Jinbo and S. Kubota; *Heterocycles*, 1983, **20**, 1531. 346

83H(20)1591 M. Tišler; *Heterocycles*, 1983, **20**, 1591. 835

83HCA1618 M. Schieß and D. Seebach; *Helv. Chim. Acta*, 1983, **66**, 1618. 657

83HCA2501 P. Mohr, N. Waespe-Šarčevič, C. Tamm, K. Gawronska and J. Gawronski; *Helv. Chim. Acta.*, 1983, **66**, 2501. 135, 138

83HOU(13)87 R. Köster; *Methoden Org. Chem.* (*Houben-Weyl*), 1983, **13**, 87. 380

83HOU(E4)738 K. Findeisen, K. Köning and R. Sundermann; *Methoden Org. Chem.* (*Houben-Weyl*), 1983, 738. 962

83HOU(E4)834 A. Hartmann; *Methoden Org. Chem.* (*Houben-Weyl*), 1983, **E4**, 834. 1022

83IC3954 D. Afzal and C. M. Lukehart; *Inorg. Chem.*, 1983, **22**, 3954. 801

83IJC(A)120 P. Raj, A. Ranjan and A. K. Saxena; *Indian J. Chem., Sect. A.*, 1983, **22**, 120. 318, 372

83IJC(A)307 G. S. Sanyal, A. B. Modak and A. K. Mudi; *Indian J. Chem., Sect. A*, 1983, **22**, 307. 864

83IJC(A)503 T. N. Schrivastava, R. C. Srivastava and V. K. Srivastava; *Indian J. Chem., Sect. A*, 1983, **22**, 503. 341, 342

83IJC(B)51 B. P. Pant and D. K. Jaiswal; *Ind. J. Chem., Sect. B*, 1983, **22**, 51. 826

83IZV1413 A. V. Kalinin, É. T. Apasov, S. V. Bugaeva, S. L. Ioffe and V. A. Tartakovskii; *Izv. Akad. Nauk SSSR, Ser. Khim.*, 1983, **6**, 1413. 380

83IZV1515 I. D. Kalikhman, B. A. Gostevskii, O. B. Bannikova, M. F. Larin, O. A. Vyazankina and N. S. Vyazankin; *Izv. Akad. Nauk SSSR, Ser. Khim.*, 1983, 1515 (*Chem. Abstr.*, 1983, **99**, 195 070x). 719, 720

83IZV1675 L. I. Rybin, D. V. Gendin, O. A. Vyazankina and N. S. Vyazankin; *Izv. Akad. Nauk SSSR, Ser. Khim.*, 1983, **7**, 1675 (*Chem. Abstr.*, 1984, **100**, 6715). 432

83JA258 M. Brookhart, J. R. Tucker and G. R. Husk; *J. Am. Chem. Soc.*, 1983, **105**, 258. 937

83JA426 H. Nakatsuji, J. Ushio, S. Han and T. Yonezawa; *J. Am. Chem. Soc.*, 1983, **105**, 426. 933

83JA2831 K. Maruoka, T. Miyazaki, M. Ando, Y. Matsumura, S. Sakane, K. Hattori and H. Yamamoto; *J. Am. Chem. Soc.*, 1983, **105**, 2831. 731

83JA3743 W. H. Koster, R. Zahler, H. W. Chang, C. M. Cimarusti, G. A. Jacobs and M. Perri; *J. Am. Chem. Soc.*, 1983, **105**, 3743. 370

83JA3999 P. S. Skell and D. D. May; *J. Am. Chem. Soc.*, 1983, **105**, 3999. 203, 205

83JA4099 M. F. Semmelhack and R. Tamura; *J. Am. Chem. Soc.*, 1983, **105**, 4099. 935

83JA5015 A. I. Meyers, K. A. Babiak, A. L. Campbell, D. L. Comins, M. P. Fleming, R. Henning, M. Heuschmann, J. D. Hudspeth, J. M. Kane *et al.*; *J. Am. Chem. Soc.*, 1983, **105**, 5015. 557

83JA7130 K. Steliou and M. A. Poupart; *J. Am. Chem. Soc.*, 1983, **105**, 7130. 174, 277

83JA7194 C. Wentrup, H. Briehl, G. Becker, G. Uhl, H.-J. Wessely, A. Maquestiau and R. Flammang; *J. Am. Chem. Soc.*, 1983, **105**, 7194. 1086, 1087, 1092

83JA7200 R. Scheffold and R. Orlinski; *J. Am. Chem. Soc.*, 1983, **105**, 7200. 152

83JA7345 M. Ihara, Y. Haga, M. Yonekura, T. Ohsawa, K. Fukumoto and T. Kametani; *J. Am. Chem. Soc.*, 1983, **105**, 7345. 285

83JA7755 S. Wolfe and C. F. Ingold; *J. Am. Chem. Soc.*, 1983, **105**, 7755. 347

83JAP8052 Sumitono Chemical Co; *Jpn. Pat.*, 8 052 (1983) (*Chem. Abstr.*, 1983, **98**, 160 403k). 972

83JAP165793 Bio Research Center Co. Ltd.; *Jpn. Pat.*, 58/165 793 (1983) (*Chem. Abstr.*, 1984, **100**, 50 061z). 117

83JAP165794 Bio Research Center Co. Ltd.; *Jpn. Pat.*, 58/165 794 (1983) (*Chem. Abstr.*, 1984, **100**, 50 063b). 117

83JAP53/35194 M. Hozumi, Y. Nomura and Y. Yoshioka; *Jpn. Pat.*, 53/35 194 (1983) (*Chem. Abstr.*, 1983, **99**, 195 215y). 993

83JAP5813530 Unitika Ltd., Indemitsu Kosan Co. Ltd.; *Jpn. Pat.* 58/13 530 (1983) (*Chem. Abstr.*, 1983, **99**, 87 651d). 68

83JCP(87)797 Z. H. Kafafi, R. H. Hauge, L. Fredin and J. L. Margrave; *J. Chem. Phys.*, 1983, **87**, 797. 1096

83JCS(D)359 R. J. Cross and A. J. McLennan; *J. Chem. Soc., Dalton Trans.*, 1983, 359. 465

83JCS(D)1841 M. Wada and K. Nishiwaki; *J. Chem. Soc., Dalton Trans.*, 1983, 1841. 1137

83JCS(D)2397 N. A. Bailey, P. L. Chell, C. P. Manuel, A. Mukhopadhyay, D. Rogers, H. E. Tabbron and M. J. Winter; *J. Chem. Soc., Dalton Trans.*, 1983, 2397. 944

83JCS(P1)49 T. L. Gilchrist, C. W. Rees and D. Vaughan; *J. Chem. Soc., Perkin Trans. 1*, 1983, 49. 226

83JCS(P1)813 D. J. Le Count; *J. Chem. Soc., Perkin Trans. 1*, 1983, **4**, 813. 1120

83JCS(P1)1067 B. L. Booth, K. O. Jibodu and M. F. J. R. P. Proenca; *J. Chem. Soc., Perkin Trans. 1*, 1983, 1067. 728

83JCS(P1)3019 J. Barluenga, J. Florez and M. Yus; *J. Chem. Soc., Perkin Trans. 1*, 1983, 3019. 71

83JCS(P2)1471 T. C. Wong, F. S. Guziec, Jr. and C. A. Moustakis; *J. Chem. Soc. Perkin Trans. 2*, 1983, 1471. 226

83JFC(22)91 T. Umemoto and O. Miyano; *J. Fluorine Chem.*, 1983, **22**, 91. 360

83JFC(23)293 M. G. Barlow, D. Bell, N. J. O'Reilly and A. E. Tipping; *J. Fluorine Chem.*, 1983, **23**, 293. 681

83JGU200 B. A. Gostevskii, O. A. Vyazankina, I. D. Kalikhman, O. B. Bannikova and N. S. Vyazankin; *J. Gen. Chem. USSR* (*Engl. Transl.*), 1983, **53**, 200. 719

83JGU1513 G. I. Orlov and V. F. Mironov; *J. Gen. Chem. USSR* (*Engl. Transl.*), 1983, **53**, 1513. 719

83JGU1851 B. A. Gostevskii, O. A. Vyazankina, I. D. Kalikhman, O. B. Bannikova and N. S. Vyazankin; *J. Gen. Chem. USSR* (*Engl. Transl.*), 1983, **53**, 1851. 719

83JGU1944 V. P. Kozyukov, A. E. Feoktistov and V. F. Mironov; *J. Gen. Chem. USSR* (*Engl. Transl.*), 1983, **53**, 1944. 722

83JHC69 R. F. Smith, K. J. Coffman and S. M. Geer; *J. Heterocycl. Chem.*, 1983, **20**, 69. 774

83JHC225 L. Baiocchi and M. Giannangeli; *J. Heterocycl. Chem.*, 1983, **20**, 225. 681, 771

83JHC305 M. Forchiassin, G. Pitacco, A. Risaliti, C. Russo and E. Valentin; *J. Heterocycl. Chem.*, 1983, **20**, 305. 354

83JHC719 A. O. Abdelhamid, H. M. Hassaneen and A. S. Shawali; *J. Heterocycl. Chem.*, 1983, **20**, 719. 677

83JHC1533 K. H. Pilgram and R. D. Skiles; *J. Heterocycl. Chem.*, 1983, **20**, 1533. 1125

83JHC1549 A. Bargagna, P. Schenone, G. Bignardi and M. Longobardi; *J. Heterocycl. Chem.*, 1983, **20**, 1549. 501

83JMC445	P. C. Srivastava and R. K. Robins; *J. Med. Chem.*, 1983, **26**, 445.	625
83JMC1187	A. Omodei-Salé, P. Consonni and G. Galliani; *J. Med. Chem.*, 1983, **26**, 1187.	773
83JOC3	R. Balicki, R. S. Hosmane and N. J. Leonard; *J. Org. Chem.*, 1983, **48**, 3.	690
83JOC24	B. H. Lee and M. J. Miller; *J. Org. Chem.*, 1983, **48**, 24.	328
83JOC366	A. P. Kozikowski and M. Adamzyk; *J. Org. Chem.*, 1983, **48**, 366.	666, 671
83JOC758	W. R. Roush and T. A. Blizzard; *J. Org. Chem.*, 1983, **48**, 758.	179
83JOC761	S. T. Purrington and W. A. Jones; *J. Org. Chem.*, 1983, **48**, 761.	313
83JOC771	S.-C. Chang and D. D. DesMarteau; *J. Org. Chem.*, 1983, **48**, 771.	313
83JOC888	L. A. Arias, S. Adkins, C. J. Nagel and R. D. Bach; *J. Org. Chem.* 1983, **48**, 888.	709
83JOC1144	D. Seyferth, R. M. Weinstein and W-L. Wang; *J. Org. Chem.*, 1983, **48**, 1144.	437
83JOC1544	D. Anderson, P. Zinke and S. L. Razniak; *J. Org. Chem.*, 1983, **48**, 1544.	727
83JOC2635	B. A. Jones, J. S. Bradshaw, P. R. Brown, J. J. Christensen and R. M. Izatt; *J. Org. Chem.*, 1983, **48**, 2635.	511
83JOC3358	G. K. S. Prakash, P. S. Iyer, M. Arvanaghi and G. A. Olah; *J. Org. Chem.*, 1983, **48**, 3358.	367, 992
83JOC3667	G. A. Olah, A. Husain, B. P. Singh and A. K. Mehrora; *J. Org. Chem.*, 1983, **48**, 3667.	39
83JOC4058	M. Mori, Y. Washioka, T. Urayama, K. Yoshiura, K. Chiba and Y. Ban; *J. Org. Chem.*, 1983, **48**, 4058.	296
83JOC4076	J. Kagan, S. K. Arora and A. Ustunol; *J. Org. Chem.*, 1983, **48**, 4076.	82
83JOC4111	J. R. Bowser, P. J. Williams and K. Kurz; *J. Org. Chem.*, 1983, **48**, 4111.	376
83JOM(243)245	D. K. Breitinger, K. Dietrich, W. Kress, R. Sendelbeck and K. Ishiwada; *J. Organomet. Chem.*, 1983, **243**, 245.	1141
83JOM(248)C1	G. J. Baird and S. G. Davies; *J. Organomet. Chem.*, 1983, **248**, C1.	453
83JOM(248)51	D. A. Bravo-Zhivotoskii, S. D. Pigarev, I. D. Kalikhman and O. A. Vyazankina; *J. Organomet. Chem.*, 1983, **248**, 51.	432
83JOM(251)175	J-P. Quintard, B. Elissondo and D. Mouko-Mpegna; *J. Organomet. Chem.*, 1983, **251**, 175.	468
83JOM(252)359	H. Hoberg and B. W. Oster; *J. Organomet. Chem.*, 1983, **252**, 359.	386
83JOM(255)295	R. Soundararajan; *J. Organomet. Chem.*, 1983, **255**, 295.	318, 320, 390
83MI 502-01	B. Wirz and W. Kueng; *J. Labelled Cmpd. Radiopharm.*, 1983, **20**, 635.	97
B-83MI 502-01	I. O. Sutherland (ed.); "General Organic Chemistry, Carboxylic Acids and Their Derivatives. Phosphorus Compounds," Khimiya, Moscow, 1983, vol. 4 (*Chem. Abstr.*, 1984, **100**, 138 213m).	96
B-83MI 507-01	H. W. Moore and D. M. Goldish; in "The Chemistry of Halides, Pseudo-halides, and Azides," ed. S. Patai and Z. Rappoport, Wiley, New York, 1983, part 1, p. 321.	367
83MI 511-01	S. Samejima; *Reports Res. Lab. Asahi Glass Corp., Ltd.*, 1983, **33**, 45 (*Chem. Abstr.*, 1984, **100**, 67 549).	494
83MI 530-01	S. Kawamura, T. Nakabayashi, Y. Abe and T. Horii; *Annu. Rep. Radiat. Cent. Osaka Prefect.*, 1983, **24**, 63.	1111
83MI 530-02	B. Vercek, B. Ogorevc, B. Stanounik and M. Tisler; *Vestn. Slov. Kem. Drus.*, 1983, **30**, 51.	1118, 1122
83OM1852	W. E. Buhro, A. Wong, J. H. Merrifield, G.-Y. Lin, A. C. Constable and J. A. Gladysz; *Organometallics*, 1983, **2**, 1852.	451, 937
83OPP315	N. Kamigata, S. Hashimoto and M. Kobayashi; *Org. Prep. Proced. Int.*, 1983, **15**, 315.	334
83OS(61)1	H. C. J. Ottenheijm and M. W. Tijhuis; *Org. Synth.*, 1983, **61**, 1.	14
83OS(61)8	M. E. Neubert and D. L. Fishel; *Org. Synth.*, 1983, **61**, 8.	15
83PJC647	R. Shabana and A. A. El-Kateb; *Pol. J. Chem.*, 1983, **57**, 647 (*Chem. Abstr.*, 1983, **101**, 130 086x).	610
83PS(15)143	A. Reliquet, F. Reliquet, J.-C. Meslin and H. Quiniou; *Phosphorus Sulfur*, 1983, **15**, 143.	760
83PS(18)7	G. Becker, W. Becker, G. Uhl, W. Uhl and H.-J. Wessely; *Phosphorus Sulfur*, 1983, **18**, 7.	1086
83PS(18)39	O. I. Kolodiazhnyi; *Phosphorus Sulfur*, 1983, **18**, 39.	1086
83RCR377	V. G. Granik; *Russ. Chem. Rev.* (*Engl. Transl.*), 1983, **52**, 377.	742
83RTC83	A. C. Brouwer, A. V. E. George and H. J. T. Bos; *Recl. Trav. Chim. Pays-Bas*, 1983, **102**, 83.	559
83RTC91	A. C. Brouwer and H. J. T. Bos; *Recl. Trav. Chim. Pays-Bas*, 1983, **102**, 91.	559
83RTC103	A. C. Brouwer and H. J. T. Bos; *Recl. Trav. Chim. Pays-Bas*, 1983, **102**, 103.	559
83RTC400	P. A. T. W. Porskamp, M. van der Leij, B. H. M. Lammerink and B. Zwanenburg; *Recl. Trav. Chim. Pays-Bas*, 1983, **102**, 400.	562
83S35	E. Haug, W. Kantlehner, P. Speh and H.-J. Bräuner; *Synthesis*, 1983, 35.	744
83S135	U. Widmer; *Synthesis*, 1983, 135.	126
83S207	L. Hadjiarapoglou, S. Spyroudis and A. Varvoglis; *Synthesis*, 1983, 207.	318, 321, 322
83S220	J. D. Brion, P. Raynaud and S. Kirkiacharian; *Synthesis*, 1983, 220.	827
83S225	H. Takahata, M. Nakano and T. Yamazaki; *Synthesis*, 1983, 225.	597, 599, 614
83S478	S. M. Fahmy and R. M. Mohareb; *Synthesis*, 1983, 478.	365
83S537	A. J. Biloski and B. Ganem; *Synthesis*, 1983, **7**, 537.	213
83S552	S. Kato, Y. Komatsu, K. Miyagawa and M. Ishida; *Synthesis*, 1983, 552.	245, 246
83S605	S. R. Ramadas, P. S. Srinivasan, J. Ramachandran and V. V. S. K. Sastry; *Synthesis*, 1983, 605.	81, 546, 549, 550
83S649	J. Barluenga, L. Alonso-Cires, P. J. Campos and G. Asensio; *Synthesis*, 1983, 649.	132
83S730	J. O. Amupitan; *Synthesis*, 1983, 730.	600, 621
83S733	B. Giese, H. Harnisch and S. Lachhein; *Synthesis*, 1983, 733.	35
83S814	S. Kato, K. Itoh, K. Miyagawa and M. Ishida; *Synthesis*, 1983, 814.	245, 246
83S1013	G. Trapani, A. Reho and A. Latrofa; *Synthesis*, 1983, 1013.	282
83SC289	J. M. Lago, A. Arrieta and C. Palomo; *Synth. Commun.*, 1983, **13**, 289.	367
83SC753	M. A. Casadei and B. Di Rienzo; *Synth. Commun.*, 1983, **13**, 753.	726

83SC1053	W. Ando and H. Tsumaki; *Synth. Commun.*, 1983, **13**, 1053.	323
83SC1083	J. T. Gupton, J. P. Idoux, R. Leonard and G. DeCrescenzo; *Synth. Commun.*, 1983, **13**, 1083.	755
83SUL191	K. Hartke and W. Hoederath; *Sulfur Letters*, 1983, **1**, 191.	558
83T949	H. J. Reich, M. J. Kelly, R. E. Olson and R. C. Holtan; *Tetrahedron*, 1983, **39**, 949.	423, 427
83T1075	D. W. Brown, M. M. Campbell and C. V. Walker; *Tetrahedron*, 1983, **39**, 1075.	573
83T1187	T. P. Mohandas, A. S. Mamman and P. M. Nair; *Tetrahedron*, 1983, **39**, 1187.	15
83T1475	V. Balasubramaniyan, V. G. Bhatia and S. B. Wagh; *Tetrahedron*, 1983, **39**, 1475.	131
83T2571	M. J. Miller, A. Biswas and M. A. Krook; *Tetrahedron*, 1983, **39**, 2571.	322
83T2863	J. Barluenga, J. López Prado, P. J. Campos and G. Asensio; *Tetrahedron*, 1983, **39**, 2863.	152
83T2869	D. L. Boger; *Tetrahedron*, 1983, **39**, 2869.	201
83T2995	G. H. L. Nefkens and B. Zwanenburg; *Tetrahedron*, 1983, **39**, 2995.	228
83T3073	K. J. H. Kruithof, R. F. Schmitz and G. W. Klumpp; *Tetrahedron*, 1983, **39**, 3073.	424
83T3419	T. P. Andersen, A.-B. A. G. Ghattas and S.-O. Lawesson; *Tetrahedron*, 1983, **39**, 3419.	573
83T3429	M. Thorsen, B. Yde, U. Pedersen, K. Clausen and S.-O. Lawesson; *Tetrahedron*, 1983, **39**, 3429.	558, 573, 587
83T4175	J. Burski, J. Kieszkowski, J. Michalski, M. Pakulski and A. Skowronska; *Tetrahedron*, 1983, **39**, 4175.	1107, 1112
83TL813	M. Catellani, G. P. Chiusoli and C. Peloso; *Tetrahedron Lett.*, 1983, **24**, 813.	192
83TL1477	G. Märkl and M. Horn; *Tetrahedron Lett.*, 1983, **24**, 1477.	924
83TL1679	M. A. Whitesell and E. P. Kyba; *Tetrahedron Lett.*, 1983, **24**, 1679.	789
83TL2231	W. R. Roush and B. M. Lesur; *Tetrahedron Lett.*, 1983, **24**, 2231.	152
83TL2639	R. Appel and W. Paulen; *Tetrahedron Lett.*, 1983, 2639.	1019
83TL3255	Y. Ishida, S. Sasatani, K. Maruoka and H. Yamamoto; *Tetrahedron Lett.*, 1983, **24**, 3255.	661, 728
83TL3617	P. Beslin, P. Metzner, Y. Vallée and J. Vialle; *Tetrahedron Lett.*, 1983, **24**, 3617.	552
83TL3815	G. Lajoie, F. Lépine, L. Maziak and B. Belleau; *Tetrahedron Lett.*, 1983, **24**, 3815.	574
83TL4001	C. Goasdoue, N. Goasdoue and M. Gaudemar; *Tetrahedron Lett.*, 1983, **24**, 4001.	607
83TL4195	F. Roulleau, D. Plusquellec and E. Brown; *Tetrahedron Lett.*, 1983, **24**, 4195.	183
83TL4203	P. Metzner and R. Rakotonirina; *Tetrahedron Lett.*, 1983, **24**, 4203.	556
83TL4245	W. Ando, H. Tanikawa and A. Sekiguchi; *Tetrahedron Lett.*, 1983, **24**, 4245.	873
83TL4355	A. F. Sviridov, M. S. Ermolenko, D. V. Yashunsky and N. K. Kochetkov; *Tetrahedron Lett.*, 1983, **24**, 4355.	239
83TL4927	D. N. Harpp and J. G. MacDonald; *Tetrahedron Lett.*, 1983, **24**, 4927.	605, 622
83TL5885	J. Navech, J. P. Majoral and R. Kraemer; *Tetrahedron Lett.*, 1983, **24**, 5885.	883
83TL5889	D. H. R. Barton and G. Kretzschmar; *Tetrahedron Lett.*, 1983, 5889.	727
83USP4382143	R. G. Shepherd; *US Pat.* 4 382 143 (1983) (*Chem. Abstr.*, 1983, **99**, 53 589z).	82, 92
83ZAAC(501)57	M. Hesse and U. Klingebiel; *Z. Anorg. Allg. Chem.*, 1983, **501**, 57.	887
83ZC99	K. Issleib, E. Leissring, M. Riemer and H. Oehme; *Z. Chem.*, 1983, **23**, 99.	883
83ZOB229	B. A. Gostevskii, O. A. Vyazankina, I. D. Kalikhman, O. B. Bannikova and N. S. Vyazankin; *Zh. Obshch. Khim.*, 1983, **53**, 229 (*Chem. Abstr.*, 1983, **98**, 126 216s).	719
83ZOB1093	O. I. Kolodiazhnyi; *Zh. Obsch. Khim.*, 1978, **53**, 1093.	1086, 1087
83ZOB1096	V. P. Kozyukov, E. V. Muzovskaya and V. F. Mironov; *Zh. Obshch. Khim.*, 1983, **53**, 1096 (*Chem. Abstr.*, 1983, **99**, 88 264s).	1006
83ZOB1434	V. P. Kozyukov and V. F. Mironov; *Zh. Obshch. Khim.*, 1983, **53**, 1434 (*Chem. Abstr.*, 1983, **99**, 88 088n).	1006
83ZOB1683	G. I. Orlov and V. F. Mironov; *Zh. Obshch. Khim.*, 1983, **53**, 1983 (*Chem. Abstr.*, 1983, **99**, 139 269g).	719
83ZOB2051	B. A. Gostevskii, O. A. Vyazankina, I. D. Kalikhman, O. B. Bannikova and N. S. Vyazankin; *Zh. Obshch. Khim.*, 1983, **53**, 2051 (*Chem. Abstr.*, 1984, **100**, 85 769p).	719, 720
83ZOB2155	V. P. Kozyukov, A. E. Feoktistov and V. F. Mironov; *Zh. Obshch. Khim.*, 1983, **53**, 2155 (*Chem. Abstr.*, 1984, **100**, 51 682q).	722, 997
83ZOR1764	A. P. Zelnitski, K. F. Serrano and O. S. Polyuyan; *Zh. Org. Khim.*, 1983, **19**, 1764.	208
83ZOR1838	D. A. Bravo-Zhivotovskii, S. D. Pigarev, I. D. Kalikhman, O. A. Vyazankina and N. S. Vyazankin; *Zh. Obshch. Khim.*, 1983, **53**, 1838. (*Chem. Abstr.*, 1984, **100**, 5721.	432
84ACR67	G. K. Anderson and R. J. Cross; *Acc. Chem. Res.*, 1984, **17**, 67.	464
84ACR103	G. Erker; *Acc. Chem. Res.*, 1984, **17**, 103.	443
84AG(E)142	H.-J. Gais and K. L. Lukas; *Angew. Chem., Int. Ed. Engl.*, 1984, **23**, 142.	135, 138
84AG(E)143	H.-J. Gais; *Angew. Chem., Int. Ed. Engl.*, 1984, **23**, 143.	154, 174
84AG(E)145	H.-J. Gais and T. Lied; *Angew. Chem., Int. Ed. Engl.*, 1984, **23**, 145.	174
84AG(E)369	R. Wehrmann, H. Klusik and A. Berndt; *Angew. Chem., Int. Ed. Engl.*, 1984, **23**, 369.	928
84AG(E)371	M. Hildenbrand, H. Pritzkow, U. Zenneck and W. Siebert; *Angew. Chem., Int. Ed. Engl.*, 1984, **23**, 371.	928
84AG(E)372	R. Wehrmann, C. Pues, H. Klusik and A. Berndt; *Angew. Chem., Int. Ed. Engl.*, 1984, **23**, 372.	928
84AG(E)381	H. J. Bestmann and T. Arenz; *Angew. Chem., Int. Ed. Engl.*, 1984, **23**, 381.	916
84AG(E)413	C. Giordano, G. Castaldi and F. Uggeri; *Angew. Chem., Int. Ed. Engl.*, 1984, **23**, 413.	28
84AG(E)587	K. H. Dötz; *Angew. Chem., Int. Ed. Engl.*, 1984, **23**, 587.	444, 931, 932
84AG(E)619	R. Appel, P. Fölling, B. Josten, M. Siray, V. Winkhaus and F. Knoch; *Angew. Chem., Int. Ed. Engl.*, 1984, **23**, 619.	1092
84AG(E)717	E. Vogel, H. Wieland, L. Schmalstieg and J. Lex; *Angew. Chem., Int. Ed. Engl.*, 1984, **23**, 717.	867

84AG(E)825 P. H. M. Budzelaar, P. von R. Schleyer and K. Krogh-Jespersen; *Angew. Chem., Int. Ed. Engl.*, 1984, **23**, 825. 927

84AG(E)826 R. Wehrmann, H. Klusik and A. Berndt; *Angew. Chem., Int. Ed. Engl.*, 1984, **23**, 826. 928

84AG(E)903 A. Schmidpeter and A. Willhalm; *Angew. Chem., Int. Ed. Engl.*, 1984, **23**, 903. 882

84AG(E)970 R. Appel, P. Fölling, L. Krieger, M. Siray and F. Knoch; *Angew. Chem., Int. Ed. Engl.*, 1984, **23**, 970. 1058

84AJC367 E. J. Browne; *Aust. J. Chem.*, 1984, **37**, 367. 1127

84AJC1043 J. B. Bremner, E. J. Browne, V. Chohan and B. F. Yates; *Aust. J. Chem.*, 1984, **37**, 1043. 1127

84AQ49 L. Trippetta and R. O. Torres; *An. Quim.*, 1984, **72**, 49. 320, 321

84BCJ539 S. Ito, Y. Tanaka and A. Kakehi; *Bull. Chem. Soc. Jpn.*, 1984, **57**, 539. 681

84BCJ1650 G. E. H. Elgemeie, H. A. Elfahham, S. A. S. Ghozlan and M. H. Elnagdi; *Bull. Chem. Soc. Jpn.*, 1984, **57**, 1650. 771

84BCJ1667 G. S. Singh, S. B. Singh and K. N. Mehrotra; *Bull. Chem. Soc. Jpn.*, 1984, **57**, 1667. 362

84BCJ2011 Y. Uchida and S. Kozuka; *Bull. Chem. Soc. Jpn.*, **57**, 2011. 666

84BCJ2167 T. Sato, K. Matsumoto, T. Abe and I. Kuwajima; *Bull. Chem. Soc. Jpn.*, 1984, **57**, 2167. 423

84BCJ2184 K. Tanaka, H. Masuda and K. Mitsuhashi; *Bull. Chem. Soc. Jpn.*, 1984, **57**, 2184. 664

84BSB405 G. L'Abbé, P. Vangheluwe, S. Toppet, G. S. D. King and L. Van Meervelt; *Bull. Soc. Chim. Belg.*, 1984, **93**, 405. 604

84BSF(2)204 K. A. Jørgensen, A.-B. Ghattas and S.-O. Lawesson; *Bull. Soc. Chim. Fr., Part 2*, 1984, 204. 572

84CB1161 J. C. Jochims, R. Abu-El-Halawa, L. Zsolnai and G. Huttner; *Chem. Ber.*, 1984, **117**, 1161. 1146, 1149

84CB1935 N. Neumann and P. Boldt; *Chem. Ber.*, 1984, **117**, 1935. 692

84CB2112 W. Maringgele, G. M. Sheldrick, A. Meller and M. Noltemeyer; *Chem. Ber.*, 1984, **117**, 2112. 381

84CB2205 E. Schaumann, U. Wriede and G. Adiwidjaja; *Chem. Ber.*, 1984, **117**, 2205. 581, 589, 604

84CC53 G. Crank and M. I. H. Makin; *J. Chem. Soc., Chem. Commun.*, 1984, 53. 1122

84CC75 T. Desmond, F. J. Lalor, G. Ferguson and M. Parvez; *J. Chem. Soc., Chem. Commun.*, 1984, 75. 1097

84CC157 M. Lees, M. Chehna, M. A. Riahi, G. Duguay and H. Quinlou; *J. Chem. Soc., Chem. Commun.*, 1984, 157. 760

84CC420 J. T. Gauntlett, B. F. Taylor and M. J. Winter; *J. Chem. Soc., Chem. Commun.*, 1984, 420. 446

84CC670 B. J. Banks, A. G. M. Barrett and M. A. Russell; *J. Chem. Soc., Chem. Commun.*, 1984, 670. 238

84CC689 M. Yoshifuji, K. Toyota and N. Inamoto; *J. Chem. Soc., Chem. Commun.*, 1984, 689. 1092

84CC747 P. A. Curtis and S. G. Davies; *J. Chem. Soc., Chem. Commun.*, 1984, 747. 943, 944

84CC768 Y. Apeloig and M. Karni; *J. Chem. Soc., Chem. Commun.*, 1984, 768. 924

84CC824 F. Ungváry; *J. Chem. Soc., Chem. Commun.*, 1984, 824. 456

84CC827 M. J. Begley, A. G. Cameron and D. W. Knight; *J. Chem. Soc., Chem. Commun.*, 1984, 827. 227

84CC956 S. G. Davies, I. M. Dordor and P. Warner; *J. Chem. Soc., Chem. Commun.*, 1984, 956. 453

84CC1010 R. Okazaki, T. Fujii and N. Inamoto; *J. Chem. Soc., Chem. Commun.*, 1984, 1010. 560

84CCC295 P. Kristian and J. Gonda; *Collect. Czech. Chem. Commun.*, 1984, **49**, 295. 1035

84CCC1505 L. Janda, Z. Voticky, J. Svetlik, J. Grimová and E. Maturová; *Coll. Czech. Chem. Commun.*, 1984, **49**, 1505 655

84CHEC *Comp. Heterocycl. Chem.*, ed. A. R. Katritzky and C. W. Rees, Pergamon, Oxford, 1984. 742, 850

84CJC574 D. A. Holden; *Can. J. Chem.*, 1984, **62**, 574. 366

84CL369 K. Ikeda, Y. Yoshinaga, K. Achiwa and M. Sekiya; *Chem. Lett.*, 1984, 369. 329

84CL2045 S. Kajigaeshi, T. Nakagawa and S. Fujisaki; *Chem. Lett.*, 1984, 2045. 318

84CPB1658 H. Takahata, T. Nakajima and T. Yamazaki; *Chem. Pharm. Bull.*, 1984, **32**, 1658. 597, 599

84CPB4437 K. Kaji, H. Nagashima, S. Nagao, K. Tabashi and H. Oda; *Chem. Pharm. Bull.*, 1984, **32**, 4437. 678

84CPB4694 H. Sawanishi, K. Tajima, M. Osada and T. Tsuchiya; *Chem. Pharm. Bull.*, 1984, **32**, 4694 693

84CPL521 G. Frenking and H. F. Schaefer, III; *Chem. Phys. Lett.*, 1984, **109**, 521. 928

84CRV17 B. A. Jones and J. S. Bradshaw; *Chem. Rev.*, 1984, **84**, 17. 510, 511, 512

84CS209 P. Jurič, M. Kočevar, B. Stanovnik, M. Tišler and B. Verček; *Chem. Scr.*, 1984, **23**, 209. 837

84EUP115085 N. Ishikawa and M. Takahashi; *Eur. Pat.* 115 085A1 (1984) (*Chem. Abstr.*, 1984, **102**, 24 117u). 59

84FES788 P. V. Plazzi, F. Bordi, F. Vitali, C. Silva, M. Chiavarini, G. Morini and M. Impicciatore; *Farmaco Ed. Sci.*, 1984, **39**, 788. 347

84G133 C. A. Coda and G. Tacconi; *Gazz. Chim. Ital.*, 1984, **114**, 133. 664

84GEP3430019 R. Fauss, K. Findeisen and D. Haebich; *Ger. Pat.*, 3 430 019 (1984) (*Chem. Abstr.*, 1986, **105**, 134 115m). 1142

84H(21)191 W. A. Slusarchyk, T. Dejneka, E. M. Gordon, E. R. Weaver and W. H. Koster; *Heterocycles*, 1984, **21**, 191. 346

84H(22)1545 B. Stanovnik, O. Bajt, B. Belčič, B. Koren, M. Perhavc, A. Štimac and M. Tišler; *Heterocycles*, 1984, **22**, 1545. 836

84H(22)2487 D. O. Spry and A. R. Bhala; *Heterocycles*, 1984, **22**, 2487. 697

84HCA534 T. Büchel, R. Prewo, J. H. Bieri and H. Heimgartner; *Helv. Chim. Acta*, 1984, **67**, 534. 622

84HOU(4)883 W. Rasshofer; *Methoden Org. Chem.* (*Houben-Weyl*), 1984, **4**, 883. 1061

84IC2217 P. Huppmann, G. Klöter, J. S. Thrasher, K. Seppelt and D. D. DesMarteau; *Inorg. Chem.*, 1984, **23**, 2217. 348, 767

84ICA85 J. A. Clark, M. Kilner and A. Pietrzykowski; *Inorg. Chim. Acta*, 1984, **82**, 85. 781

84ICA215 R. Uson, J. Forniés, P. Espinet and R. Navarro; *Inorg. Chem. Acta*, 1984, **82**, 215. 1138

84IJC(A)952 M. S. Raizada and S. N. Bhattacharya; *Indian J. Chem., Sect. A*, 1984, **23**, 952. 390

84IJC(B)254 S. K. Das and S. N. Balasubrahmanyam; *Indian J. Chem., Sect. B*, 1984, **23**, 254. 1131

84IJC(B)728 B. N. Misra, A. S. Singha, G. S. Chauhan and R. P. Sharma; *Ind. J. Chem. Sect. B*, 1984, **B23**, 728. 665

84IZV1567 A. A. Grinfeld, G. A. Artamkina and I. P. Beletokaya; *Izv. Akad. Nauk SSSR, Ser. Khim.*, 1984, 1567. 113

84IZV1897 D. A. Bravo-Zhivotovskii, S. D. Pigarev, O. A. Vyazankina, A. S. Medvedeva, L. P. Safronova and N. S. Vyazankin; *Izv. Akad. Nauk SSSR, Ser. Khim*, 1984, **8**, 1897 (*Chem. Abstr.*, 1985, **102**, 6710). 226

84JA434 W. D. Wulff and P.-C. Tang; *J. Am. Chem. Soc.*, 1984, **106**, 434. 934

84JA625 N. J. Curtis and A. M. Sargeson; *J. Am. Chem. Soc.*, 1984, **106**, 625. 386

84JA1133 M. Yoshioka, H. Nakai and M. Ohno; *J. Am. Chem. Soc.*, 1984, **106**, 1133. 345

84JA1146 A. I. Meyers, M. Harre and R. Garland; *J. Am. Chem. Soc.*, 1984, **106**, 1146. 79

84JA1496 M. F. Semmelhack and C. Bodurow; *J. Am. Chem. Soc.*, 1984, **106**, 1496. 170

84JA1543 C. F. Hammer and S. Chandrasegaran; *J. Am. Chem. Soc.*, 1984, **106**, 1543. 313, 314

84JA1576 T. E. Taylor and M. B. Hall; *J. Am. Chem. Soc.*, 1984, **106**, 1576. 932

84JA1813 J. Rebek, Jr., D. F. Tai and Y.-K. Shue; *J. Am. Chem. Soc.*, 1984, **106**, 1813. 166

84JA2440 S. Murai, I. Ryu, J. Iriguchi and N. Sonoda; *J. Am. Chem. Soc.*, 1984, **106**, 2440. 415

84JA2469 B. M. Trost and P. Quayle; *J. Am. Chem. Soc.*, 1984, **106**, 2469. 286

84JA2687 B. Cambou and A. M. Klibanov; *J. Am. Chem. Soc.*, 1984, **106**, 2687. 135

84JA2853 T. C. Flood and K. D. Campbell; *J. Am. Chem. Soc.*, 1984, **106**, 2853. 453

84JA2954 K. Narasaka, T. Sakakura, T. Uchimaru and D. Guédin-Vuony; *J. Am. Chem. Soc.*, 1984, **106**, 2954. 173

84JA3214 E. Carmona, L. Sánchez, J. M. Marin, M. L. Poveda, J. L. Atwood, R. D. Priester and R. D. Rogers; *J. Am. Chem. Soc.*, 1984, **106**, 3214. 445

84JA4266 B. A. O'Brien, J. S. Thrasher, C. W. Bauknight, Jr., M. L. Robin and D. D. DesMarteau; *J. Am. Chem. Soc.*, 1984, **106**, 4266. 809

84JA5892 J. Ushio, H. Nakatsuji and T. Yonezawa; *J. Am. Chem. Soc.*, 1984, **106**, 5892. 932

84JA6082 R. M. Moriarty, I. Prakash, O. Prakash and W. A. Freeman; *J. Am. Chem. Soc.*, 1984, **106**, 6082. 897

84JA6084 M. V. Lakshmikantham, P. Carroll, G. Furst, M. I. Levinson and M. P. Cava; *J. Am. Chem. Soc.*, 1984, **106**, 6084. 528

84JA6414 P. A. Grieco, D. L. Flynn and R. E. Zelle; *J. Am. Chem. Soc.*, 1984, **106**, 6414. 282

84JA6451 L. S. Liebeskind, J. P. Leeds, S. L. Baysdon and S. Iyer; *J. Am. Chem. Soc.*, 1984, **106**, 6451. 457

84JA6993 M. D. Fryzuk and P. A. MacNeil; *J. Am. Chem. Soc.*, 1984, **106**, 6993. 380

84JA7250 W. E. Ladner and G. M. Whitesides; *J. Am. Chem. Soc.*, 1984, **106**, 7250. 138

84JCR(S)100 A. Nangia and S. Chandrasekaran; *J. Chem. Res. (S)*, 1984, 100. 189, 196

84JCR(S)404 A. Corsaro, A. Compagnini and G. Perrini; *J. Chem. Res. (S)*, 1984, 404. 734

84JCS(D)399 G. S. Bristow, M. F. Lappert, T. R. Martin, J. L. Atwood and W. F. Hunter; *J. Chem. Soc., Dalton Trans.*, 1984, 399. 443

84JCS(D)2301 E. A. V. Ebsworth, C. M. Huntley and D. W. H. Rankin; *J. Chem. Soc., Dalton Trans.*, 1984, 2301. 535

84JCS(D)2859 G. A. Bowmaker and H. Schmidbaur; *J. Chem. Soc., Dalton Trans.*, 1984, **12**, 2859. 1141

84JCS(P1)85 M. Yokoyama, K. Tsuji, M. Hayashi and T. Imamoto; *J. Chem. Soc., Perkin Trans. 1*, 1984, 85. 558

84JCS(P1)147 T. N. Birkinshaw, D. W. Gillon, S. A. Harkin, G. D. Meakins and M. D. Tirel; *J. Chem. Soc., Perkin Trans. 1*, 1984, 147. 1125

84JCS(P1)785 K. Clausen, M. Thorsen, S.-O. Lawesson and A. F. Spatola; *J. Chem. Soc., Perkin Trans. 1*, 1984, 785. 558, 573, 587, 611

84JCS(P1)859 K. R. Lawson, A. Singleton and G. H. Whitham; *J. Chem. Soc., Perkin Trans. 1*, 1984, 859. 558, 559

84JCS(P1)865 K. R. Lawson, A. Singleton and G. H. Whitham; *J. Chem. Soc., Perkin Trans. 1*, 1984, 865. 552, 559

84JCS(P1)897 A. Corsaro, A. Compagnini, G. Perrini and G. Purrello; *J. Chem. Soc., Perkin Trans. 1*, 1984, 897. 734, 760

84JCS(P1)993 S. P. Langdon, R. J. Simmonds and M. F. G. Stevens; *J. Chem. Soc., Perkin Trans. 1*, 1984, 993. 770

84JCS(P1)1531 R. Ramage, G. J. Griffiths, F. E. Shutt and J. N. A. Sweeney; *J. Chem. Soc., Perkin Trans. 1*, 1984, 1531. 79

84JCS(P1)2535 L. Bruche, L. Garanti and G. Zecchi; *J. Chem. Soc., Perkin Trans. 1*, 1984, 2535. 676, 677

84JGU255 V. S. Brovarets, O. P. Lobanov, T. K. Vinogradova and B. S. Drach; *J. Gen. Chem. USSR (Engl. Transl.)*, 1984, **54**, 255. 655, 656

84JGU426 I. D. Kalikhman, E. N. Medvedeva, O. B. Bannikova, N. G. Fabina, M. F. Larin, V. A. Lopyrev and M. G. Voronkov; *J. Gen. Chem. USSR (Engl. Transl.)*, 1984, **54**, 426. 718

84JGU1925 D. M. Malenko, L. A. Repina and D. A. Sinitsa; *J. Gen. Chem. USSR (Engl. Transl.)*, 1984, **54**, 1925. 713

84JGU1969 I. V. Konovalova, L. A. Burnaeva, R. D. Gareev, T. M. Moshkina and A. N. Pudovik; *J. Gen. Chem. USSR (Engl. Transl.)*, 1984, **54**, 1969. 709

84JHC91 M. Robba and J.-C. Lancelot; *J. Heterocycl. Chem.*, 1984, **21**, 91. 706

84JHC253 S. O. Abd Allah, M. R. Elmoghayar and S. E. Abdou; *J. Heterocycl. Chem.*, 1984, **21**, 253. 532

84JHC725 S. Shiotani, H. Morita, M. Inoue, T. Ishida, Y. Iitaka and A. Itai; *J. Heterocycl. Chem.*, 1984, **21**, 725. 671

84JHC797 H. M. Hassaneen, A. A. Fahmi, H. Abdelhamid, A. A. Yassin and A. S. Shawali; *J. Heterocycl. Chem.*, 1984, **21**, 797. 676

84JHC1685 J. S. Duceppe and J. Gauthier; *J. Heterocycl. Chem.*, 1984, **21**, 1685. 1024

84JHC1689 F. Malbec, R. Milcent and G. Barbier; *J. Heterocycl. Chem.*, 1984, **21**, 1689. 705

84JMC1201 R. A. Kenley, C. D. Bedford, O. D. Dailey, Jr., R. A. Howd and A. Miller; *J. Med. Chem.*, 1984, **27**, 1201. 664

84JOC106 D. C. Berndt, N. Utrapiromsuk and D. E. Conran; *J. Org. Chem.*, 1984, **49**, 106. 322

84JOC174 G. W. Kabalka, M. C. Delgado, U. S. Kunda and S. A. Kunda; *J. Org. Chem.*, 1984, **49**, 174. 51

84JOC263 P. A. T. W. Porskamp, B. H. M. Lammerink and B. Zwanenburg; *J. Org. Chem.*, 1984, **49**, 263. 562, 645

84JOC546 J. E. Mills, R. M. Cosgrove, R. D. Shah, C. A. Maryanoff and V. Paragamian; *J. Org. Chem.*, 1984, **49**, 546. 657

84JOC714 A. Couture, R. Dubiez and A. Lablache-Combier; *J. Org. Chem.*, 1984, **49**, 714. 572, 618

84JOC806 O. Lerman, Y. Tor, D. Hebel and S. Rozen; *J. Org. Chem.*, 1984, **49**, 806. 202

84JOC919 W. J. Middleton; *J. Org. Chem.*, 1984, **49**, 919. 665, 672

84JOC1212 R. S. Hosmane, F. N. Burnett and M. S. Albert; *J. Org. Chem.*, 1984, **49**, 1212. 690

84JOC1467 B. A. O'Brien and D. D. DesMarteau; *J. Org. Chem.*, 1984, **49**, 1467. 662, 808

84JOC2015 P. C. Kuzma, L. E. Brown and T. M. Harris; *J. Org. Chem.*, 1984, **49**, 2015. 374

84JOC2057 J. Gramza, R. B. Mitchell and D. C. Dittmer; *J. Org. Chem.*, 1984, **49**, 2057. 1113

84JOC2139 A. Kume, M. Fujii, M. Sekine and T. Hata; *J. Org. Chem.*, 1984, **49**, 2139. 403

84JOC2688 O. Tsuge, S. Kanemasa and K. Matsuda; *J. Org. Chem.*, 1984, **49**, 2688. 1031

84JOC3083 J. E. Rowe and A. F. Hegarty; *J. Org. Chem.*, 1984, **49**, 3083. 675, 703

84JOC3216 D. P. Cox, J. Terpinski and W. Lawrynowicz; *J. Org. Chem.*, 1984, **49**, 3216. 4

84JOC3322 J. Garcia, J. González, R. Segura, F. Urpí and J. Vilarrasa; *J. Org. Chem.*, 1984, **49**, 3322. 361

84JOC3466 D. Keus and J. Warkentin; *J. Org. Chem.*, 1984, **49**, 3466. 707

84JOC3875 L. Brown and M. Koreeda; *J. Org. Chem.*, 1984, **49**, 3875. 209

84JOC4272 G. M. Loudon, A. S. Radhakrishna, M. R. Almond, J. K. Blodgett and R. H. Boutin; *J. Org. Chem.*, 1984, **49**, 4272. 999

84JOC4277 R. H. Boutin and G. M. Loudon; *J. Org. Chem.*, 1984, **49**, 4277. 999

84JOC4460 R. Yamaguchi and M. Kawanisi; *J. Org. Chem.*, 1984, **49**, 4460 460

84JOC4541 W. J. Middleton; *J. Org. Chem.*, 1984, **49**, 4541. 322, 719

84JOC4569 J. H. Rigby and N. Balasubramanian; *J. Org. Chem.*, 1984, **49**, 4569. 993

84JOC4784 A. R. Katritzky, G. R. Khan, D. E. Leahy and M. De Rosa; *J. Org. Chem.*, 1984, **49**, 4784. 316

84JOC4786 R. A. Earl and K. P. C. Vollhardt; *J. Org. Chem.*, 1984, **49**, 4786. 386

84JOC5198 W. C. Christopfel and L. L. Miller; *J. Org. Chem.*, 1984, **49**, 5198. 1110

84JOC5225 G. R. Underwood and P. E. Dietze; *J. Org. Chem.*, 1984, **49**, 5225. 315

84JOC5282 H. J. Reich and E. K. Eisenhart; *J. Org. Chem.*, 1984, **49**, 5282. 413

84JOM207 V. F. Mironov; *J. Organomet. Chem.*, 1984, **271**, 207. 962, 968, 969, 978, 1006, 1012, 1015

84JOM(262)49 N. Aktagu, H. Felkin, G. J. Baird, S. G. Davies and O. Watts; *J. Organomet. Chem.*, 1983, **262**, 49. 453

84JOM(263)67 D. Melzer and E. Weiss; *J. Organomet. Chem.*, 1984, **263**, 67. 1097

84JOM(266)313 H. Hoberg, D. Schalfer and B. W. Oster; *J. Organomet. Chem.*, 1984, **266**, 313. 101

84JOM(267)277 R. Märkel and H. Fischer; *J. Organomet. Chem.*, 1984, **267**, 277. 660

84JOM(267)309 T. Y. Orlova, V. N. Setkina and D. N. Kursanov; *J. Organomet. Chem.*, 1984, **267**, 309. 97

84JOM(276)431 A. Morvillo and A. Turco; *J. Organomet. Chem.*, 1984, **276**, 431. 1138

84JOU1963 N. P. Pel'kis and E. S. Levchenko; *J. Org. Chem. USSR* (*Engl. Transl.*), 1984, **20**, 1963. 718

84JPR(326)311 J. Liebscher and A. Rumler; *J. Prakt. Chem.*, 1984, **326**, 311. 660

84LA450 Y. Köksal, P. Raddatz and E. Winterfeldt; *Liebigs Ann. Chem.*, 1984, 450. 174

84LA462 Y. Köksal, P. Raddatz and E. Winterfeldt; *Liebigs Ann. Chem.*, 1984, 462. 174

84LA474 K. H. Marx, P. Raddatz and E. Winterfeldt; *Liebigs Ann. Chem.*, 1984, 474. 174

84MI 502-01 A. G. Ismailov, I. D. Sadykhov, M. A. Rustamov and A. A. Akhmedov; *Azerb. Khim. Zh.*, 1984, 84 (*Chem. Abstr.*, 1985, **103**, 53 720m). 63

B-84MI 502-01 P. Brodelius; in "Advances in Biochemical Engineering," eds. T. X. Ghose, A. Fiechter and N. Blakebrough, Springer-Verlag, New York, 1984, p. 75. 117

84MI 502-02 S. Inoue; *Yukagaku*, 1984, **33**, 676 (*Chem. Abstr.*, 1985, **102**, 44 262r). 117

84MI 502-03 T. Sugai and K. Mori; *Agric. Biol. Chem.*, 1984, **48**, 2501. 118

84MI 502-04 B. Cambou and A. M. Klibanov; *Appl. Biochem. Biotechnol.*, 1984, **9**, 255. 118

84MI 503-01 J. Mulzer; *Nachr. Chem. Tech. Lab.*, 1984, **32**, 226. 168

B-84MI 503-01 K. A. Lutomsky and A. I. Meyers; in "Asymmetric Synthesis," ed. J. D. Morrison, Academic Press, Orlando, 1984, vol. 3B, p. 213. 172

B-84MI 503-02 P. A. Bartlett and A. I. Meyers; in "Asymmetric Synthesis," ed. J. D. Morrison, Academic Press, Orlando, 1984, vol. 3B, p. 411 168, 172

B-84MI 506-01 I. T. Harrison, S. Harrison, L. S. Hegedus and L. G. Wade, Jr. (eds.); "Compendium of Organic Synthetic Methods, vols. 1–5," Interscience, New York, 1971–84. 259

84MI 507-02 R. Vadde, S. S. Jagannatha, S. Somu and L. Puri; *Transition Met. Chem.*, 1984, **9**, 103. 272

84MI 509-01 D. A. Bravo-Zhivotovskii, S. D. Pigarev, I. D. Kalikhman, O. A. Kruglaya and N. S. Vyazankin; *Kremniiorg. Soedin. Mater. Ikh Osn., Tr. Soveshch. Khim. Prakt. Primen. Kremniiorg. Soedin.*, 5th, *Meeting Date*, 1984. (*Chem. Abstr.*, 1985, **102**, 6717). 432

84MI 530-01 A. Gartner, B. Koren, B. Stanovnik and M. Tisler; *Vestn. Slov. Kem. Drus.*, 1984, **31**, 1. 1122

84MIP84/1376A1 H. Alper, B. Despeyroux, D. J. H. Smith and J. B. Woell; *PCT Int. Patent.* WO84/1376A1, 1984 (*Chem. Abstr.*, 1984, **101/13**, 110 369p). 142

84OM327 D. Seyferth and R. C. Hui; *Organometallics*, 1984, **3**, 327. 238

84OM996 J. Carvajal, G. Muller, J. Sales, X. Solans and C. Miravitlles; *Organometallics*, 1984, **3**, 996. 1137
84OM1764 G. J. Bird, S. G. Davies and T. R. Maberly; *Organometallics*, 1984, **3**, 1764. 937
84OMR724 U. Chiacchio, A. Corsaro and F. A. Bottino; *Org. Magn. Reson.*, 1984, **22**, 724. 734
84OS(62)14 M. Alderdice, F. W. Sum and L. Weiler; *Org. Synth.*, 1984, **62**, 14. 103
84OS(62)187 T. Shioiri and S.-I. Yamada; *Org. Synth.*, 1984, **62**, 187. 869
84OS(62)191 Y. Hamada and T. Shioiri; *Org. Synth.*, 1984, **62**, 191. 42
84OS(62)202 R. W. Lang and H.-J. Hansen; *Org. Synth.*, 1984, **62**, 202. 103
84PS(19)189 V. D. Romanenko, L. K. Polyachenko and L. N. Markovki; *Phosphorus Sulfur*, 1984, **19**, 189. 888
84PS(19)311 A. Marinetti and F. Mathey; *Phosphorus Sulfur*, 1984, **19**, 311. 407
84RTC342 B. G. Lenz and B. Zwanenburg; *Recl. Trav. Chim. Pays-Bas*, 1984, **103**, 342. 734, 736
84S51 A. Knoll and J. Liebscher; *Synthesis*, 1984, 51. 760
84S146 F. Pochat; *Synthesis*, 1984, 146. 675, 682
84S413 J. Cabre and A. L. Palomo; *Synthesis*, 1984, 413. 280, 295
84S509 M. J. Camarasa, P. Fernández-Resa, M. T. García-López, F. G. de las Heras, P. P. Méndez-Castrillón and A. San Felix; *Synthesis*, 1984, 509. 1039
84S512 A. Shafiee and G. Fanaii; *Synthesis*, 1984, 512. 1057
84S572 V. Dourtoglou and B. Gross; *Synthesis*, 1984, 572. 279
84S797 M. Yokoyama and T. Imamoto; *Synthesis*, 1984, 797. 546
84S827 M. Yokoyama, Y. Hasegawa, H. Hatanaka, Y. Kawazoe and T. Imamoto; *Synthesis*, 1984, 827. 551, 560, 574, 575
84S829 S. Prabhakar, A. M. Lobo, M. A. Santos and H. S. Rzepa; *Synthesis*, 1984, 829. 214, 635
84S876 P. Gosselin, C. Maignan and F. Rouessac; *Synthesis*, 1984, 876. 82
84S920 H. C. Brown, D. Basavaiah and S. M. Singh; *Synthesis*, 1984, 920. 229
84S946 Y. Lin, M. N. Jennings, D. R. Sliskovic, T. L. Fields and S. A. Lang, Jr; *Synthesis*, 1984, 946. 631
84S1017 P. D. Woodgate, S. J. Janssen, P. S. Rutledge, S. D. Woodgate and R. C. Cambie; *Synthesis*, 1984, 1017. 1107
84S1020 J. R. Hwu, J. G. Lazar and P. F. Corless; *Synthesis*, 1984, 1020. 1134
84S1037 D. N. Harpp, T. Aida, J. DeCesare, P. Tisnes and T. H. Chan; *Synthesis*, 1984, 1037. 337
84SC227 D. Wilkening and B. P. Mundy; *Synth. Commun.*, 1984, **14**, 227. 146
84SC1159 E. Mohacsi and E. M. Gordon; *Synth. Commun.*, 1984, **14**, 1159. 727
84SC1257 H. Takahata, T. Nakajima, K. Matoba and T. Yamazaki; *Synth. Commun.*, 1984, **14**, 1257. 597, 599
84SRI269 P. Raj, A. Ranjan, K. Singhal and R. Rastogi; *Synth. React. Inorg. Metal-Org. Chem.*, 1984, **14**, 269. 372
84SUL39 J. Kreutzmann; *Sulfur Letters*, 1984, **2**, 39. 587, 614
84SUL147 M. E. Jazouli, S. Masson and A. Thuillier; *Sulfur Letters*, 1984, **2**, 147. 551
84T145 M. Ohno, Y. Ito, M. Arita, T. Shibata, K. Adachi and H. Sawai; *Tetrahedron*, 1984, **40**, 145. 138
84T405 L. Beyer, J. Hartung and R. Widera; *Tetrahedron*, 1984, **40**, 405. 657
84T1075 G. Morel, E. Marchand, K. H. Nguyen Thi and A. Foucaud; *Tetrahedron*, 1984, **40**, 1075. 728
84T1573 S. Masson, V. Mothes and A. Thuillier; *Tetrahedron.*, 1984, **40**, 1573. 551, 729, 731
84T2165 A. Alemagna, L. Garanti, E. Licandro and G. Zecchi; *Tetrahedron*, 1984, **40**, 2165. 905
84T2211 J. Mulzer, P. de Lasalle, A. Chucholowski, U. Blaschek, G. Brüntrup, I. Jibril and G. Huttner; *Tetrahedron*, 1984, **40**, 2211. 166
84T2663 N. M. Yousif, U. Pedersen, B. Yde and S.-O. Lawesson; *Tetrahedron*, 1984, **40**, 2663. 551
84T2935 T. Kitahara and K. Mori; *Tetrahedron*, 1984, **40**, 2935. 173
84T3215 K. Taira and D. G. Gorenstein; *Tetrahedron*, 1984, **40**, 3215. 402
84T3273 R. Benn, R. Mynott, W. J. Richter and G. Schroth; *Tetrahedron*, 1984, **40**, 3273. 908
84TL223 H. Nishiyama, K. Sakuta and K. Itoh; *Tetrahedron Lett.*, 1984, **25**, 223. 215
84TL233 Y. D. Vankar and G. Kumaravel; *Tetrahedron Lett.*, 1984, **25**, 233. 321
84TL399 S. H. Jung and H. Kohn; *Tetrahedron Lett.*, 1984, **25**, 399. 1118
84TL1249 C. L. Liotta, M. L. McLaughlin and B. A. O'Brien; *Tetrahedron Lett.*, 1984, **25**, 1249. 395, 396
84TL1651 D. Seyferth, W.-L. Wang and R. C. Hui; *Tetrahedron Lett.*, 1984, **25**, 1651 437
84TL1797 M. Deng, Y. Tang and W. Xu; *Tetrahedron Lett.*, 1984, 1797. 101
84TL1809 M. Yoshifuji, K. Toyota, K. Shibayama and N. Inamoto; *Tetrahedron Lett.*, 1984, 1809. 1086, 1087
84TL2163 H. Tsutsui and O. Mitsunobu; *Tetrahedron Lett.*, 1984, **25**, 2163. 177
84TL2183 M. A. Blanchette, W. Choy, J. T. Davis, A. P. Essenfeld, S. Masamune, W. R. Roush and T. Sakai; *Tetrahedron Lett.*, 1984, **25**, 2183. 155
84TL2557 S. Kobayashi, K. Kamiyama, T. Iimori and M. Ohno; *Tetrahedron Lett.*, 1984, **25**, 2557. 135, 138
84TL2639 E. J. Corey and S. W. Wright; *Tetrahedron Lett.*, 1984, 2639. 636
84TL2709 S. G. Davies, I. M. Dordor, J. C. Walker and P. Warner; *Tetrahedron Lett.*, 1984, **25**, 2709. 453
84TL2793 A. Yoshida, Y. Tajima, N. Takeda and S. Oida; *Tetrahedron Lett.*, 1984, **25**, 2793. 234
84TL3091 Y. Ito, H. Imai, T. Matsuura and T. Saegusa; *Tetrahedron Lett.*, 1984, **25**, 3091. 796, 802
84TL3365 A. Arrieta, J. M. Aizpurua and C. Palomo; *Tetrahedron Lett.*, 1984, **25**, 3365. 367
84TL3387 C. A. Loeschorn and C. J. Kelly; *Tetrahedron Lett.*, 1984, **25**, 3387. 1113
84TL3651 M. J. O'Donnell, W. A. Bruder, B. W. Daugherty, D. Liu and K. Wojciechowski; *Tetrahedron Lett.*, 1984, **25**, 3651. 746
84TL3687 H. H. Karsch, F. H. Köhler and H. U. Reisacher; *Tetrahedron Lett.*, 1984, **25**, 3687. 884, 1092
84TL3701 P. Laszlo and E. Polla; *Tetrahedron Lett.*, 1984, **25**, 3701. 368
84TL4109 R. Appel and C. Casser; *Tetrahedron Lett.*, 1984, **25**, 4109. 878
84TL4683 I. Mori, K. Oshima and H. Nozaki; *Tetrahedron Lett.*, 1984, **25**, 4683. 1043
84TL4825 R. Ramage, C. P. Ashton, D. Hopton and M. J. Parrott; *Tetrahedron Lett.*, 1984, **25**, 4825. 280

84TL4989 G. A. Artamkina, A. A. Grinfeld and A. A. Beletskaya; *Tetrahedron Lett.*, 1984, 4989. 116

84TL5031 M. Braun and R. Devant; *Tetrahedron Lett.*, 1984, **25**, 5031. 138

84TL5139 Y. Ito, E. Nakajo and T. Saegusa; *Tetrahedron Lett.*, 1984, **25**, 5139. 793

84TL5251 D. Seyferth and R. C. Hui; *Tetrahedron Lett.*, 1984, **25**, 5251. 596, 719, 729

84TL5427 P. Molina, A. Lorenzo, R. M. Claramunt and J. Elguero; *Tetrahedron Lett.*, 1984, **25**, 5427. 1122

84TL5789 K. Kikukawa, K. Ikenaga, F. Wada and T. Matsuda; *Tetrahedron Lett.*, 1984, **25**, 5789. 220, 222

84TL5793 Y. Tamaru, T. Hioki and Z. Yoshida; *Tetrahedron Lett.*, 1984, **25**, 5793. 607

84TL6027 Y. Kita, S. Akai, M. Yoshigi, Y. Nakajima, H. Yasuda and Y. Tamura; *Tetrahedron Lett.*, 1984, **25**, 6027. 189, 199

84URP570267 V. P. Kozyokov; *USSR Pat.* 570 267 (1984). 1127

84USP4446068 R. A. Mitsch; *US Pat.* 4 446 068 (1984) (*Chem. Abstr.*, 1984, **101**, 90 750a). 816

84ZC99 K. Geisler, H. Below, A. Moller and E. Bulka; *Z. Chem.*, 1984, **24**, 99. 624, 625

84ZC379 H. Kohler and L. Jager; *Z. Chem.*, 1984, **24**, 379. 1074

84ZC438 G. Barnikow and A. A. Martin; *Z. Chem.*, 1984, **24**, 438. 632

84ZN(B)1456 H. Schmidbaur, P. Nußstein and G. Müller; *Z. Naturforsch., Teil B*, 1984, **39**, 1456. 911

84ZOB288 V. S. Brovarets, O. P. Lobanov, T. K. Vinogradova and B. S. Drach; *Zh. Obshch. Khim.*, 1984, **54**, 288 (*Chem. Abstr.*, 1984, **100**, 210 009u). 655, 656

84ZOB477 I. D. Kalikhman, E. N. Medvedeva, O. B. Bannikova, N. G. Fabina, M. F. Larin, V. A. Lopyrev and M. G. Voronkov; *Zh. Obshch. Khim.*, 1984, **54**, 477 (*Chem. Abstr.*, 1984, **101**, 55 158s). 718

84ZOB876 A. E. Feoktistov, R. I. Orlov, V. P. Kozyukov and V. F. Mironov; *Zh. Obshch. Khim.*, 1984, **54**, 876 (*Chem. Abstr.*, 1984, **101**, 72 811s). 1001

84ZOB1520 A. A. Prishchenko, A. V. Gromov, Y. N. Luzikov, A. A. Borisenko, E. I. Lazhko, K. Klaus and I. F. Lutsenko; *Zh. Obshch. Khim.*, 1984, **54**, 1520 (*Chem. Abstr.*, 1985, **102**, 6660m). 876, 877

84ZOB2154 D. M. Malenko, L. A. Repina and A. D. Sinitsa; *Zh. Obshch. Khim.*, 1984, **54**, 2154 (*Chem. Abstr.*, 1985, **102**, 95 727d). 713

84ZOB2203 I. V. Konovalova, L. A. Burnaeva, R. D. Gareev, T. M. Moshkina and A. N. Pudovik; *Zh. Obshch. Khim.*, 1984, **54**, 2203 (*Chem. Abstr.*, 1985, **102**, 166 844c). 709

84ZOR1784 G. S. Borovikova, I. F. Tsymbal, E. S. Levchenko and E. I. Borovik; *Zh. Org. Khim.*, 1984, **20**, 1784 (*Chem. Abstr.*, 1985, **102**, 62 331s). 775

84ZOR2153 N. P. Pel'kis and E. S. Levchenko; *Zh. Org. Khim.*, 1984, **20**, 2153, (*Chem. Abstr.*, 1985, **102**, 131 967t). 718

84ZOR2543 V. I. Gorbatenko, Yu. I. Matveev, M. N. Gertsyuk and L. I. Samarai; *Zh. Org. Khim.*, 1984, **20**, 2543. 1068, 1069

85AG1051 T. Severengiz, W.-W. du Mont, D. Lenoir and H. Voss; *Angew. Chem.*, 1985, **97**, 1051. 543

85AG(E)180 W. Dürckheimer, J. Blumbach, R. Lattrell and K. H. Scheunemann; *Angew. Chem., Int. Ed. Engl.*, 1985, **14**, 180. 226

85AG(E)788 R. Wehrmann, H. Meyer and A. Berndt; *Angew. Chem., Int. Ed. Engl.*, 1985, **24**, 788. 927, 928

85AG(E)961 H. Hoberg and E. Hernandez; *Angew. Chem., Int. Ed. Engl.*, 1985, **24**, 961. 386

85AG(E)979 G. Zinner and W. P. Fehlhammer; *Angew. Chem., Int. Ed. Engl.*, 1985, **24**, 979. 904, 906

85AG(E)1041 T. Severengiz, W.-W. du Mont, D. Lenoir and H. Voss; *Angew. Chem., Ent. Ed. Engl.*, 1985, **24**, 1041. 244

85AG(E)1062 W. A. Herrmann, H.-J. Kneuper and E. Herdtweck; *Angew. Chem., Int. Ed. Engl.*, 1985, **24**, 1062. 1097

85AG(E)1068 A. Ricci, M. Fiorenza, A. Degl'Innocenti, G. Seconi, P. Dembech, K. Witzgall and H. J. Bestmann; *Angew. Chem., Int. Ed. Engl.*, 1985, **97**, 1068. 413

85AJC447 G. Crank and H. R. Khan; *Aust. J. Chem.*, 1985, **38**, 447. 620

85AP(318)977 G. Zinner and J. Grünefeld; *Arch. Pharm.* (*Weinheim, Ger.*), 1985, **318**, 977. 212

85AX(C)1632 S. C. Nyburg, A. G. Brook, F. Abdesaken, G. Gutekunst and W. Wong-Ng; *Acta Crystallogr., Sect. C*, 1985, **41**, 1632. 923, 924

85BCJ352 K. Inami and T. Shiba; *Bull. Chem. Soc. Jpn.*, 1985, **58**, 352. 368

85BCJ769 S. Kajigaeshi, T. Nakagawa, S. Fujisaki and A. Nishida; *Bull. Chem. Soc. Jpn.*, 1985, **58**, 769. 318

85BCJ995 Y. Takikawa, K. Shimada, K. Sato, S. Sato and S. Takizawa; *Bull. Chem. Soc. Jpn.*, 1985, **58**, 995. 635

85BCJ1448 A. Ogawa, J. Miyake, N. Kambe, S. Murai and N. Sonoda; *Bull. Chem. Soc. Jpn.*, 1985, **58**, 1448 625, 626

85BCJ1903 K. Kawashiro, S. Morimoto and H. Yoshida; *Bull. Chem. Soc. Jpn.*, 1985, **58**, 1903. 226

85BCJ2721 Y. Inoue, M. Taguchi and H. Hashimoto; *Bull. Chem. Soc. Jpn.*, 1985, **58**, 2721. 726

85BCJ3395 M. Iwata and H. Kuzuhara; *Bull. Chem. Soc. Jpn.*, 1985, **58**, 3395. 344

85BSB29 A. Kharrat, C. Gardrat and B. Maillard; *Bull. Soc. Chim. Belg.*, 1985, **94**, 29. 206

85BSF881 S. Berrada, P. Metzner and R. Rakotonirina; *Bull. Soc. Chim. Fr.*, 1985, 881. 556

85BSF891 C.-B. Plaisance, J. C. Grossier, G. Levesque and A. Mahjoub; *Bull. Soc. Chim. Fr.*, 1985, 891. 546, 550

85BSF1211 M. Mulliez; *Bull. Soc. Chim. Fr.*, 1985, 1211. 370

85CA(103)104568c E. J. McGarry, R. J. Conway and M. J. Cooke; *Chem. Abstr.*, 1985, **103** (13), 104 568c. 126

85CAR375 J. M. Tronchet, E. Winter-Mihaly, J. Rupp, F. Barbalat-Rey and M. Geoffroy; *Carbohydrate Res.*, 1985, **136**, 375. 212, 213

85CB340 J. D. Korp, I. Bernal, R. Hörlein, R. Serrano and W. A. Herrmann; *Chem. Ber.*, 1985, **118**, 340. 1097

85CB526 H. Quast and U. Nahr; *Chem. Ber.*, 1985, **118**, 526. 1073

85CB676	L. Horner and H. Lindel; *Chem. Ber.*, 1985, **118**, 676.	1128
85CB851	J. Goerdeler and M. Yunis; *Chem. Ber.*, 1985, **118**, 851.	616
85CB1008	R. Helwig and M. Hanack; *Chem. Ber.*, 1985, **118**, 1008.	675
85CB1352	R. Appel, U. Kündgen and F. Knoch; *Chem. Ber.*, 1985, **118**, 1352.	889
85CB1517	H. Werner, B. Heiser, U. Schubert and K. Ackermann; *Chem. Ber.*, 1985, **118**, 1517.	800
85CB1632	R. Appel, H. Janssen, M. Siray and F. Knoch; *Chem. Ber.*, 1985, **118**, 1632.	675
85CB1668	S. Kato, M. Goto, R. Hattori, K. Nishiwaki, M. Mizuta and M. Ishida; *Chem. Ber.*, 1985, **118**, 1668.	337, 584
85CB2208	U. Rheude and W. Sundermeyer; *Chem. Ber.*, 1985, **118**, 2208.	502
85CB3089	R. Kupfer, M. Nagel, E.-U. Würthwein and R. Allmann; *Chem. Ber.*, 1985, **118**, 3089.	694
85CB4459	M. Geisel, A. Waterfeld and R. Mews; *Chem. Ber.*, 1985, **118**, 4459.	662, 809
85CC171	Y. N. Belokon', N. I. Chernoglazova, C. A. Kochetkov, N. S. Garbalinskaya and V. M. Belikov; *J. Chem. Soc., Chem. Commun.*, 1985, 171.	385
85CC209	S. G. Davies and J. C. Walker; *J. Chem. Soc., Chem. Commun.*, 1985, 209.	453
85CC473	S. Kim and Y. K. Ko; *J. Chem. Soc., Chem. Commun.*, 1985, 473.	126
85CC533	S. Huber, P. Stamouli and R. Neier; *J. Chem. Soc., Chem. Commun.*, 1985, 533.	1041, 1108
85CC1151	H. Arzoumanian, A. Baldy, R. Lai, J. Metzger, M. L. Nkeng Peh and M. Pierrot; *J. Chem. Soc., Chem. Commun.*, 1985, 1151.	918
85CC1386	S. D. A. Street, C. Yeates, P. Kocienski and S. F. Campbell; *J. Chem. Soc., Chem. Commun.*, 1985, 1386.	173
85CC1418	Y. Nagao, S. Yamada, T. Kumagai, M. Ochiai and E. Fujita; *J. Chem. Soc., Chem. Commun.*, 1985, 1418.	329, 330
85CC1478	M. Z. Deng, D. A. Lu and W. H. Xu; *J. Chem. Soc., Chem. Commun.*, 1985, 1478.	71
85CC1563	W. Kasel, P. G. Hultin and J. B. Jones; *J. Chem. Soc., Chem. Commun.*, 1985, 1563.	138
85CC1764	P. Pöchlauer, W. Petter, P. Peringer and E. P. Müller; *J. Chem. Soc., Chem. Commun.*, 1985, 1764.	373
85CI(L)824	H. Davy and P. Metzner; *Chem. Ind.* (*London*), 1985, 824.	551, 560
85CJC951	D. N. Harpp, J. G. MacDonald and C. Larsen; *Can. J. Chem.*, 1985, **63**, 951.	605, 622
85CJC3089	G. Sauvé, V. S. Rao, G. Lajoie and B. Belleau; *Can. J. Chem.*, 1985, **63**, 3089.	219, 574
85CL123	M. Miyake, M. Kirisawa and N. Tokutake; *Chem. Lett.*, 1985, 123.	280
85CL701	T. Miyasaka and S. Noguchi; *Chem. Lett.*, 1985, 701.	215
85CL1829	S. Kiyooka and A. Miyauchi; *Chem. Lett.*, 1985, 1829.	432
85CPB373	H. Ohmori, K. Sakai, N. Nagai, Y. Mizuki and M. Masui; *Chem. Pharm. Bull.*, 1985, **33**, 373.	1078
85CPB532	K. Shishido, K. Tanaka, K. Fukumoto and T. Kamelani; *Chem. Pharm. Bull.*, 1985, **33**, 532.	179
85CPB998	K. Iwasa, M. Sugiura and N. Takao; *Chem. Pharm. Bull.*, 1985, **33**, 998.	212
85CPB3479	T. Okawara, H. Kashihara and M. Furukawa; *Chem. Pharm. Bull.*, 1985, **33**, 3479.	630
85CPB5515	S. Yoshifuji, K.-I. Tanaka, T. Kawai and Y. Nitta; *Chem. Pharm. Bull.*, 1985, **33**, 5515.	307
85CRV419	G. Raabe and J. Michl; *Chem. Rev.*, 1985, **85**, 419.	923
85CZ389	G. Zinner and M. Perner; *Chem.-Ztg.*, 1985, **109**, 389.	212
85FES120	A. Ranise, F. Bondavalli, P. Schenone, L. Berrino, C. Matera, N. Persico, A. R. Romano and E. Marmo; *Farmaco Ed. Sci.*, 1985, **40**, 120.	597
85FRP2549470	Roussel-Uclaf; *Fr. Pat.* 2 549 470 (1985) (*Chem. Abstr.*, 1986, **104**, 69 032).	145
85G265	C. Carpanelli, G. Gaiani and F. Sancassan; *Gazz. Chim. Ital.*, 1985, **115**, 265.	348
85G663	F. Benedetti, M. Forchiassin, C. Russo and A. Risaliti; *Gazz. Chim. Ital.*, 1985, **115**, 663.	354
85GEP3501452	O. Kristiansen and J. Drabek; *Ger. Pat.* 3 501 452 (1985) (*Chem. Abstr.*, 1985, **103**, 196 255z).	775
85GSM96	P. R. Jenkins; *Gen. Synth. Methods*, 1985, **7**, 96.	24
85H(23)281	Y. Kurasawa, M. Ichikawa, I. Kamata, Y. Okamoto and A. Takada; *Heterocycles*, 1985, **23**, 281.	706
85H(23)549	H. Nemoto, E. Shirata, K. Fukumoto and T. Kametani; *Heterocycles*, 1985, **23**, 549.	227
85H(23)649	C. de Diego, E. Gomez and C. Avendano; *Heterocycles*, 1985, **23**, 649.	584, 587
85H(23)677	M. Komatsu, S. Yamamoto, Y. Ohshiro and T. Agawa; *Heterocycles*, 1985, **23**, 677.	364
85H(23)2213	H. Takahata, T. Suzuki and T. Yamazaki; *Heterocycles*, 1985, **23**, 2213.	607
85H(23)2819	B. Stanovnik, U. Urleb and M. Tišler; *Heterocycles*, 1985, **23**, 2819.	837
85H(23)3069	T. Shibata, Y. Sugimura, S. Sato and K. Kawazoe; *Heterocycles*, 1985, **23**, 3069.	1039
85HCA216	W. Oppolzer and P. Dudfield; *Helv. Chim. Acta*, 1985, **68**, 216.	63
85HCA484	R. Wälchli, S. Bienz and M. Hesse; *Helv. Chim. Acta*, 1985, **68**, 484.	272, 284
85HCA1155	H. Kristinsson, T. Winkler, G. Rihs and H. Fritz; *Helv. Chim. Acta*, 1985, **68**, 1155.	1124
85HCA1730	R. Huber, A. Knierzinger, J. P. Obrecht and A. Vasella; *Helv. Chim. Acta*, 1985, **68**, 1730.	212
85HCA1893	J. M. J. Tronchet, K. Mekhael, J. Graf-Poncet, R. Benhamza and M. Geoffroy; *Helv. Chim. Acta*, 1985, **68**, 1893.	214
85HOU(E5)659	H. Pielartzik, B. Irmisch-Pielartzik and T. Eicher; *Methoden Org. Chem.*, (*Houben-Weyl*), 1985, **E5**, 659.	122
85HOU(E5/1)812	H. Pielartzik, B. Irmisch-Pielartzik and T. Eicher; *Methoden Org. Chem.* (*Houben Weyl*), 1985, **E5/1**, 812.	692, 699, 700
85HOU(E5)832	W. Bauer and K. Kühlein; *Methoden Org. Chem.* (*Houben-Weyl*), 1985, **E5**, 832.	232, 233
85HOU(E5)891	R. Mayer and S. Scheithauer; *Methoden Org. Chem.* (*Houben-Weyl*), 1985, **E5**, 891.	546, 549
85HOU(E5)1135	D. Döpp and H. Döpp; *Methoden Org. Chem.* (*Houben-Weyl*), 1985, **E5**, 1135.	313, 322, 350, 384
85HOU(E5)1266	W. Bauer and K. Kühlein; *Methoden Org. Chem.* (*Houben-Weyl*), 1985, **E5**, 1266.	629
85HOU(E5)1279	W. Bauer and K. Kühlein; *Methoden Org. Chem.* (*Houben-Weyl*), 1985, **E5**, 1279.	635, 636
85HOU(E5)1291	W. Bauer and K. Kühlein; *Methoden Org. Chem.* (*Houben-Weyl*), 1985, **E5**, 1291.	639
85HOU(E5)1300	W. Bauer and K. Kühlein; *Methoden Org. Chem.* (*Houben-Weyl*), 1985, **E5**, 1300.	846

85HOU(E5)1304 D. Schumann; *Methoden Org. Chem.* (*Houben-Weyl*), 1985, **E5**, 1304. 850

85HOU(E5)1305 D. Schumann; *Methoden Org. Chem.* (*Houben-Weyl*), 1985, **E5**, 1305. 742

85HOU(E5/1)587 *Methoden Org. Chem.* (*Houben-Weyl*), 1985, **E5/1**, 587. 1

85HOU(E5/1)615 R. Mayer and S. Scheithauer; *Methoden Org. Chem.* (*Houben-Weyl*), 1985, **E5/1**, 615. 226

85HOU(E5/1)628 R. Sustmann and H.-G. Korth; *Methoden Org. Chem.* (*Houben-Weyl*), 1985, **E5/1**, 628. 654

85HOU(E5/1)631 R. Sustmann and H.-G. Korth; *Methoden Org. Chem.* (*Houben-Weyl*), 1985, **E5/1**, 631. 664

85HOU(E5/1)632 R. Sustmann and H.-G. Korth; *Methoden Org. Chem.* (*Houben-Weyl*), 1985, **E5/1**, 632. 654

85HOU(E5/2)1218 W. Bauer and K. Kuhlein; *Methoden Org. Chem.* (*Houben-Weyl*), 1985, **E5/2**, 1218. 699

85HOU(E5/2)1591 C. Grundmann; *Methoden Org. Chem.*, (*Houben Weyl*), 1985, **E5/2**, 1591. 699

85HOU(E11)1 D. Klamann; *Methoden Org. Chem.* (*Houben-Weyl*), 1985, **E11**, 1. 333

85HOU(E11)251 E. Schaumann; *Methoden Org. Chem.* (*Houben-Weyl*), 1985, **E11**, 251. 605

85HOU(E11/1)911 B. Zwanenburg and B. G. Lenz; *Methoden Org. Chem.* (*Houben-Weyl*), 1985, **E11/1**, 911. 472, 488

85HOU(E11/2)1326 B. G. Lenz and B. Zwanenburg; *Methoden Org. Chem.* (*Houben-Weyl*), 1985, **E11/2**, 1326. 472, 495

85IC1993 R. R. Ford, B. L. Li, R. H. Neilson and R. J. Thoma; *Inorg. Chem.*, 1985, **24**, 1993. 893

85IC4171 J. S. Thrasher and K. Seppelt; *Inorg. Chem.*, 1985, **24**, 4171. 1085

85IJ219 W. Broda and E. V. Dehmlow; *Israel J. Chem.*, 1985, **26**, 219. 1078

85IJC(B)977 M. Ramzan Ali, R. Singh and V. K. Verma; *Indian J. Chem., Sect. B*, 1985, **24**, 977. 631

85IZV173 B. E. Ivanov, S. S. Krokhina, T. V. Chichkanova and É. M. Kosacheva; *Izv. Akad. Nauk SSSR, Ser. Khim.*, 1985, 173. 369

85IZV700 A. F. Gontar, E. N. Glotov, A. S. Vinogradov, E. G. Bykhovskaya and I. L. Knunyants; *Izv. Akad. Nauk SSSR, Ser. Khim.*, 1985, **3**, 700. 1076

85IZV2824 G. A. Razuvaev, A. S. Gordestov, T. N. Brevnova, V. V. Semenov, A. N. Konev, E. A. Mishina and Yu. I. Dergunov; *Izv. Akad. Nauk SSSR, Ser. Khim.*, 1985, **12**, 2824. 1077

85JA503 W. D. Wulff and S. R. Gilbertson; *J. Am. Chem. Soc.*, 1985, **107**, 503. 956

85JA1695 R. E. Dolle and K. C. Nicolaou; *J. Am. Chem. Soc.*, 1985, **107**, 1695. 277

85JA1993 D. S. Marynick and C. M. Kirkpatrick; *J. Am. Chem. Soc.*, 1985, **107**, 1993. 932

85JA2773 P. H. M. Budzelaar, K. Krogh-Jespersen, T. Clark and P. von R. Schleyer; *J. Am. Chem. Soc.*, 1985, **107**, 2773. 928

85JA2931 S.-H. Jung and H. Kohn; *J. Am. Chem. Soc.*, 1985, **107**, 2931. 1080, 1118

85JA3285 R. E. Ireland, D. W. Norbeck, G. S. Mandel and N. S. Mandel, *J. Am. Chem. Soc.*, 1985, **107**, 3285. 239

85JA3910 J. B. Bremner, L. M. Engelhardt, A. H. White and K. N. Winzenberg; *J. Am. Chem. Soc.*, 1985, **107**, 3910. 1127

85JA4252 Y. N. Belokon', A. G. Bulychev, S. V. Vitt, *et al.*; *J. Am. Chem. Soc.*, 1985, **107**, 4252. 385

85JA4338 A. G. Brook, K. D. Safa, P. D. Lickiss and K. M. Baines; *J. Am. Chem. Soc.*, 1985, **107**, 4338. 925

85JA4342 R. D. Tung and D. H. Rich; *J. Am. Chem. Soc.*, 1985, **107**, 4342. 280

85JA4440 K. Tatsumi, A. Nakamura, P. Hofmann, P. Stauffer and R. Hoffmann; *J. Am. Chem. Soc.*, 1985, **107**, 4440. 441

85JA4474 K. R. Birdwhistell, T. L. Tonker and J. L. Templeton; *J. Am. Chem. Soc.*, 1985, **107**, 4474. 940

85JA4577 M. F. Semmelhack, Y. Appapillai and T. Sato; *J. Am. Chem. Soc.*, 1985, **107**, 4577. 301

85JA5310 H. Nishiyama, H. Arai, T. Ohki and K. Itoh; *J. Am. Chem. Soc.*, 1985, **107**, 5310. 325

85JA5490 T. Kaneko; *J. Am. Chem. Soc.*, 1985, **107**, 5490. 334, 335

85JA5732 V. B. Rao, C. F. George, S. Wolff and W. C. Agosta; *J. Am. Chem. Soc.*, 1985, **107**, 5732. 271

85JA6404 K. Fuji, M. Node, S. Terada, M. Murata, H. Nagasawa, T. Taga and K. Machida; *J. Am. Chem. Soc.*, 1985, **107**, 6404. 171

85JA7072 G. Kirchner, M. P. Scollar and A. M. Klibanov; *J. Am. Chem. Soc.*, 1985, **107**, 7072. 135

85JA7788 P. DeShong, G. A. Slough, V. Elango and G. L. Trainor; *J. Am. Chem. Soc.*, 1985, **107**, 7788. 446

85JAP6638 Seitetsu Chemical Industry Co., Ltd., *Jpn. Pat.*, 85/6638 (1985) (*Chem. Abstr.*, 1985, **102**, 203 609c). 63

85JAP114195 Agency of Industrial Sciences and Technology; *Jpn. Pat.*, 60/114 195 (1985) (*Chem. Abstr.*, 1986, **104**, 49 870p). 117

85JCR(S)132 E. De Carvalho, F. Norberto, E. Rosa, J. Iley and P. Patel; *J. Chem. Res.* (*S*), 1985, 132. 361, 362, 711

85JCR(S)272 H. Davy and P. Metzner; *J. Chem. Res.* (*S*), 1985, 272. 551

85JCS(P1)153 C. J. Easton; *J. Chem. Soc., Perkin Trans. 1*, 1985, 153. 334

85JCS(P1)409 D. I. Gasking and G. H. Whitham; *J. Chem. Soc., Perkin Trans. 1*, 1985, 409. 431

85JCS(P1)461 R. Ramage, D. Hopton, M. J. Parrott, R. S. Richardson, G. W. Kenner and G. A. Moore; *J. Chem. Soc., Perkin Trans. 1*, 1985, 461. 264

85JCS(P1)641 S. Apparao, S. S. Bhattacharjee, H. Ila and H. Junjappa; *J. Chem. Soc., Perkin Trans. 1*, 1985, 641. 555

85JCS(P1)831 R. Brettle, S. M. Shibib and K. J. Wheeler; *J. Chem. Soc., Perkin Trans. 1*, 1985, 831. 288

85JCS(P1)883 J. E. T. Corrie, G. W. Kirby and J. W. M. Mackinnon; *J. Chem. Soc., Perkin Trans. 1*, 1985, 883. 326

85JCS(P1)907 N. Harrit, S. Rosenkilde, B. D. Larsen and A. Holm; *J. Chem. Soc., Perkin Trans. 1*, 1985, 907. 626

85JCS(P1)1471 P. G. Houghton, D. F. Pipe and C. W. Rees; *J. Chem. Soc., Perkin Trans. 1*, 1985, 1471. 1073

85JCS(P1)1875 C. T. Gokou, J.-P. Pradère, H. Quiniou and L. Toupet; *J. Chem. Soc., Perkin Trans. 1*, 1985, 1875. 633

85JCS(P1)1927 M. J. Pearson and J. W. Tyler; *J. Chem. Soc., Perkin Trans. 1*, 1985, 1927. 223

85JCS(P1)2277 P. Moutevelis-Minakakis and I. Photaki; *J. Chem. Soc., Perkin Trans. 1*, 1985, 2277. 367

85JCS(P1)2333 M. Hori, K. Tanaka, T. Kataoka, H. Shimizu, E. Imai, K. Kimura and Y. Hashimoto; *J. Chem. Soc., Perkin Trans. 1*, 1985, 2333. 1122

85JCS(P1)2361 Y. Nagao, T. Kumagai, S. Yamada, E. Fujita, Y. Inoue, Y. Nagase, S. Aoyagi and T. Abe; *J. Chem. Soc., Perkin Trans. 1*, 1985, 2361. 223

85JFC(27)157 H. W. Hennig and P. Sartori; *J. Fluorine Chem.*, 1985, **27**, 157. 221

85JFC(30)269 K. J. L. Paciorek, J. H. Nakahara and R. H. Kratzer; *J. Fluorine Chem.*, 1985, **30**, 269. 786

85JFC(30)289 K. J. L. Paciorek, J. H. Nakahara and R. H. Kratzer; *J. Fluorine Chem.*, 1985, **30**, 289. 786

85JGU402 P. P. Onys'ko, E. I. Kiseleva, T. V. Kim and A. D. Sinitsa; *J. Gen. Chem. USSR* (*Engl. Transl.*), 1985, **55**, 402. 788

85JGU1526 M. G. Zimin, E. V. Fomakhin, A. R. Sadykov, V. N. Smirnov and A. N. Pudovik; *J. Gen. Chem. USSR* (*Engl. Transl.*), 1985, **55**, 1526. 713

85JGU1956 V. A. Al'fonsov, D. A. Pudovik, É. S. Batyeva and A. N. Pudovik; *J. Gen. Chem. USSR*, 1985, **55**, 1956. 406

85JHC1025 J. Zinczuk, O. O. Orazi, R. A. Corral and H. Roncaglia; *J. Heterocycl. Chem.*, 1985, **22**, 1025. 1126

85JHC1065 J. Kant, F. D. Popp and B. C. Uff; *J. Heterocycl. Chem.*, 1985, **22**, 1065. 558

85JHC1121 G. D. Madding, D. W. Smith, R. I. Sheldon and B. Lee; *J. Heterocycl. Chem.*, 1985, **22**, 1121. 655

85JHC1353 A. Monge, V. Martinez-Merino and E. Fernandez-Alvarez; *J. Heterocycl. Chem.*, 1985, **22**, 1353. 334

85JHC1429 L. C. Vishwakarma and J. W. Sowell, Sr.; *J. Heterocycl. Chem.*, 1985, **22**, 1429. 106

85JHC1435 P. Scheiner, S. Arwin, M. Eliacin and J. Tu; *J. Heterocycl. Chem.*, 1985, **22**, 1435. 706

85JHC1635 A. Tsolomitis and C. Sandris; *J. Heterocycl. Chem.*, 1985, **22**, 1635. 334

85JMC1661 K. H. Baggaley, P. D. English, L. J. A. Jennings, B. Morgan, B. Nunn and A. W. R. Tyrrell; *J. Med. Chem.*, 1985, **28**, 1661. 334

85JMC1772 R. Okachi, H. Niino, K. Kitaura, K. Mineura, Y. Nakamizo, Y. Murayama, T. Ono and A. Nakamizo; *J. Med. Chem.*, 1985, **28**, 1772. 334

85JOC130 A. Ricci, A. Degl'Innocenti, S. Chimichi, M. Fiorenza, G. Rossini and H. J. Bestmann; *J. Org. Chem.*, 1985, **50**, 130. 419

85JOC169 A. G. M. Barrett, M. J. Betts and A. Fenwick; *J. Org. Chem.*, 1985, **50**, 169. 344, 1005

85JOC273 K. Takeda, K. Tsuboyama, K. Yamaguchi and H. Ogura; *J. Org. Chem.*, 1985, **50**, 273. 235

85JOC384 A. Ogawa, J. Miyake, Y. Karasaki, S. Murai and N. Sonoda; *J. Org. Chem.*, 1985, **50**, 384. 625

85JOC560 S. Kim, J. I. Lee and Y. C. Kim; *J. Org. Chem.*, 1985, **50**, 560. 126

85JOC707 R. L. Funk and J. D. Munger, Jr.; *J. Org. Chem.*, 1985, **50**, 707. 227

85JOC715 R. Katakai and Y. Iizuka; *J. Org. Chem.*, 1985, **50**, 715. 972

85JOC760 M. Ueda and H. Oikawa; *J. Org. Chem.*, 1985, **50**, 760. 126, 280, 300

85JOC771 G. Morel, E. Marchand and A. Foucaud; *J. Org. Chem.*, 1985, **50**, 771. 728

85JOC811 G. P. Mullen, N. P. Luthra, R. B. Dunlap and J. D. Odom; *J. Org. Chem.*, 1985, **50**, 811. 238

85JOC993 J. E. Johnson, A. Ghafouripour, Y. K. Haug, A. W. Cordes, W. T. Pennington and O. Exner; *J. Org. Chem.*, 1985, **50**, 993. 665, 672

85JOC1112 J. S. McMurray and D. F. Dyckes; *J. Org. Chem.*, 1985, **50**, 1112. 185

85JOC1405 C. L. Perrin, C. P. Lollo and C.-S. Hahn; *J. Org. Chem.*, 1985, **50**, 1405. 723

85JOC1681 M. Machida, K. Oda, E. Yoshida and Y. Kanaoka; *J. Org. Chem.*, 1985, **50**, 1681. 631, 632

85JOC1712 S. Bittner, Y. Assaf, P. Krief, M. Pomerantz, B. T. Ziemnicka and C. G. Smith; *J. Org. Chem.*, 1985, **50**, 1712. 374

85JOC1741 W. J. Hennen, B. C. Hinshaw, T. A. Riley, S. G. Wood and R. K. Robins; *J. Org. Chem.*, 1985, **50**, 1741. 516, 522

85JOC1751 S. Kim, H. Chang and W. J. Kim; *J. Org. Chem.*, 1985, **50**, 1751. 138

85JOC1818 H. Iida, Y. Watanabe and C. Kibayashi; *J. Org. Chem.*, 1985, **50**, 1818. 330

85JOC1992 C. H. Wang and J. R. Matos; *J. Org. Chem.*, 1985, **50**, 1992. 117

85JOC2198 R. E. Ireland and D. W. Norbeck; *J. Org. Chem.*, 1985, **50**, 2198. 412, 413, 422

85JOC2220 J. W. Tilley, J. W. Clader, M. Wirkus and J. F. Blount; *J. Org. Chem.*, 1985, **50**, 2220. 284

85JOC2224 M. Abou-Gharbia, D. M. Ketcha, D. E. Zacharias and D. Swern; *J. Org. Chem.*, 1985, **50**, 2224. 350

85JOC2719 Y. C. Hwang and F. W. Fowler; *J. Org. Chem.*, 1985, **50**, 2719. 8

85JOC2930 B. G. Lenz, H. Regeling, H. L. M. van Rozendaal and B. Zwanenburg; *J. Org. Chem.*, 1985, **50**, 2930. 562

85JOC3462 C. C. Wei, S. De Bernardo, J. P. Tengi, J. Borgese and M. Weigele; *J. Org. Chem.*, 1985, **50**, 3462. 346

85JOC3698 W. E. Barnette, R. C. Wheland, W. J. Middleton and S. Rozen; *J. Org. Chem.*, 1985, **50**, 3698. 203

85JOC3717 G. Soula; *J. Org. Chem.*, 1985, **50**, 3717. 1106

85JOC4164 N. E. Heimer, L. Field and J. A. Waites; *J. Org. Chem.*, 1985, **50**, 4164. 244

85JOC4415 N. J. Turro, Y. Cha, I. R. Gould, A. Padwa, J. R. Gasdaska and M. Tomas; *J. Org. Chem.*, 1985, **50**, 4415. 728

85JOC4877 R. E. Tirpak, R. S. Olsen and M. W. Rathke; *J. Org. Chem.*, 1985, **50**, 4877. 70

85JOC4991 M. Ramaiah; *J. Org. Chem.*, 1985, **50**, 4991. 131

85JOC4993 H. T. Nagasawa, W. E. Smith, C.-H. Kwon and D. J. W. Goon; *J. Org. Chem.*, 1985, **50**, 4993. 344

85JOC5393 R. L. Danheiser, D. M. Fink, K. Okano, Y. M. Tsai and S. W. Szczepanski; *J. Org. Chem.*, 1985, **50**, 5393. 413, 425

85JOC5604 T. Sheradsky and R. Moshenberg; *J. Org. Chem.*, 1985, **50**, 5604. 366

85JOM(280)355 W. Fromberg and G. Erker; *J. Organomet. Chem.*, 1985, **280**, 355. 797

85JOM(281)135 R. F. Cunico and L. Bedell; *J. Organomet. Chem.*, 1985, **281**, 135. 765

85JOM(282)421 N. A. Bumagin, Y. V. Gulevich and I. P. Beletskaya; *J. Organomet. Chem.*, 1985, **282**, 421. 467

85JOM(285)213 S. G. Davies, I. M. Dordor-Hedgecock, P. Warner, R. H. Jones and K. Prout; *J. Organomet. Chem.*, 1985, **285**, 213. 453

85JOM(285)293 M. Foa, F. Francalanchi, E. Bencini and A. Gardano; *J. Organomet. Chem.*, 1985, **285**, 293. 143

85JOM(286)171 T. G. Back and R. G. Kerr; *J. Organomet. Chem.*, 1985, **286**, 171. 1056, 1108

85JOM(287)87 F. Seitz, H. Fischer and J. Riede; *J. Organomet. Chem.*, 1985, **287**, 87. 108, 887

85JOM(288)237 H. Hoberg, K. Suemmermann and A. Milchereit; *J. Organomet. Chem.*, 1985, **288**, 237. 386

85JOM(294)C29 D. C. Cupertino, M. M. Harding and D. J. Cole-Hamilton; *J. Organomet. Chem.*, 1985, **294**, C29. 220

85JOM(294)139 J. L. Moreau and R. Couffignal; *J. Organomet. Chem.*, 1985, **294**, 139. 223

85JOM(295)C21 H. Hoberg, E. Hernandez and K. Suemmermann; *J. Organomet. Chem.*, 1985, **295**, C21. 386

85JOU1166 E. S. Levchenko, T. N. Dubina, L. V. Budnik and E. A. Romanenko; *J. Org. Chem. USSR* (*Engl. Transl.*), 1985, **21**, 1166. 703

85JPR(327)455 A. Knoll and J. Liebscher; *J. Prakt. Chem.*, 1985, **327**, 455. 633

85JPR(327)893 K. Kirschke, A. Möller and E. Schmitz; *J. Prakt. Chem.*, 1985, **327**, 893. 357

85JSP(109)8 M. C. Durrant, H. W. Kroto, D. McNaughton and J. F. Nixon; *J. Mol. Spectrosc.*, 1985, **109**, 8. 1154

85JSP(111)415 K. Ohno, H. Matsuura, D. McNaughton and H. W. Kroto; *J. Mol. Spectrosc.*, 1980, **111**, 415. 1152

85LA1669 G. Himbert; *Liebigs Ann. Chem.*, 1985, 1669. 739

85LA1910 H. Wamhoff, M. Ertas and S. M. S. Atta; *Liebigs Ann. Chem.*, 1985, 1910. 690

85LA1935 S. Scholz, H. Marschall-Weyerstahl and P. Weyerstahl; *Liebigs Ann. Chem.*, 1985, 1935. 184

85LA2073 S. Blechert; *Liebigs Ann. Chem.*, 1985, 2073. 214

85LA2305 L. Capuano and K. Djokar; *Liebigs Ann. Chem.*, 1985, 2305. 634

85LA2371 G. Himbert; *Liebigs Ann. Chem.*, 1985, 2371. 596, 617

85MAC1803 W. Kawanobe, K. Yamaguchi, S. Nakahama and N. Yamazaki; *Macromol. Chem.*, 1985, **186**, 1803. 368

85MI 501-01 H. Parnes and S. De Keczer; *J. Labelled Compd. Radiopharm.*, 1985, **22**, 287 (*Chem. Abstr.*, 1985, **103**, 122 962). 13

85MI 502-01 F. Ozawa; *Kagaku To Kogyu* (*Tokyo*), 1985, **38**, 926 (*Chem. Abstr.*, 1987, **106**, 32 070w). 27

85MI 502-02 J. Simbera; *Folia Fac. Sci. Nat. Univ. Pukyninal Brun*, 1985, **26**, 49 (*Chem. Abstr.*, 1986, **105**, 114 936b). 62

85MI 523-01 G. Becker, W. Becker, R. Knebl, H. Schmidt, U. Weber and M. Westerhausen; *Nova Acta Leopoldina* (*Neue Folge*), 1985, **59**, 55. 879, 880

85MI 523-02 A. Schmidpeter; *Nova Acta Leopoldina* (*Neue Folge*), 1985, **59**, 69. 881

85MI 530-01 K. H. Rohm and R. L. Vanetten; *J. Labelled Compd. Rad.*, 1985, **22**, 909. 1127

85MI 531-01 G. Becker, W. Becker, R. Knebl, H. Schidt, U. Weber and M. Westerhausen; *Nova Acta Leopold.*, 1985, **59**, 39. 1155

85OM275 H. G. Raubenheimer, G. J. Kruger, A. van Lombard, L. Linford and J. C. Viljoen; *Organometallics*, 1985, **4**, 275. 951

85OM344 H. Schmidbaur and M. S. Hussain; *Organometallics*, 1985, **4**, 344. 921, 1094

85OM736 C. P. Casey and A. J. Shusterman; *Organometallics*, 1985, **4**, 736. 942

85OM943 M. Brookhart, W. B. Studabaker and G. R. Husk; *Organometallics*, 1985, **4**, 943. 937

85OM944 J. R. Lisko and W. M. Jones; *Organometallics*, 1985, **4**, 944. 937

85OM1247 T. C. Forschner and A. R. Cutler; *Organometallics*, 1985, **4**, 1247. 453

85OM1310 G. Erker, P. Czisch, R. Mynott, Y. H. Tsay and C. Krüger; *Organometallics*, 1985, **4**, 1310. 917

85OM1749 L. S. Hegedus, A. Kramer and Y. Chen; *Organometallics*, 1985, **4**, 1749. 696

85OM2102 M. J. McGeary, T. L. Tonker and J. L. Templeton; *Organometallics*, 1985, **4**, 2102. 952, 953

85OPP385 O. Attanasi, M. Grossi and F. Serra-Zanetti; *Org. Prep. Proced. Int.*, 1985, **17**, 385. 367

85OS(63)175 H. P. Fahrni, U. Lienhard and M. Neuenschwander; *Org. Synth.*, 1985, **63**, 175. 280

85OS(63)183 B. Neises and W. Steglich; *Org. Synth.*, 1985, **63**, 183. 124

85OS(63)188 G. A. Olah and A. P. Fung; *Org. Synth.*, 1985, **63**, 188. 282

85OS(64)1 R. S. Iyer and P. Helquist; *Org. Synth.*, 1985, **64**, 1. 100

85OS(64)63 P. Ballesteros and B. W. Roberts; *Org. Synth.*, 1985, **64**, 63. 104

85OS(64)108 M. J. Toschner, T. Rosen and C. H. Heathcock; *Org. Synth.*, 1985, **64**, 108. 103

85POL251 P. Raj, A. K. Saxena, K. Singhal and A. Ranjan; *Polyhedron*, 1985, **4**, 251. 372

85POL1151 R. P. Narain and M. Z. Siddiqui; *Polyhedron*, 1985, **4**, 1151. 785

85PS(22)135 M. Ishida, H. Nakanishi and S. Kato; *Phosphorus Sulfur*, 1985, **22**, 135. 560

85PS(23)253 A. Thuillier; *Phosphorus Sulfur*, 1985, **23**, 253. 550

85PS(25)63 W. Walter and C. R. Saha; *Phosphorus Sulfur*, 1985, **25**, 63. 630, 733

85PS(25)279 F. Reliquet, A. Reliquet, F. Sharrard, J.-C. Meslin and H. Quiniou; *Phosphorus Sulfur*, 1985, **25**, 279. 633

85PS(25)297 R. Shabana, H. J. Meyer and S.-O. Lawesson; *Phosphorus Sulfur*, 1985, **25**, 297. 581, 611, 643, 644

85RTC74 R. Peerboom, S. Ingemann and N. M. M. Nibbering; *Recl. Trav. Chim. Pays-Bas*, 1985, **104**, 74. 437

85S40 J. Barry, G. Bram, G. Decodts, A. Loupy, C. Orange, A. Petit and J. Sansoulet; *Synthesis*, 1985, 40. 132

85S77 F. Dutron-Woitrin, R. Merényi and H. G. Viehe; *Synthesis*, 1985, 77. 601, 612

85S79 F. Dutron-Woitrin, R. Merényi and H. G. Viehe; *Synthesis*, 1985, 79. 601

85S86 A. Lorente, J. L. Garcia Navio, L. Fuentes and J. L. Soto; *Synthesis*, 1985, 86. 606

85S92 T. Takajo, S. Kambe and W. Ando; *Synthesis*, 1985, 92. 606

85S172 E. Schaumann and S. Fittkau; *Synthesis*, 1985, 172. 729

85S212 S. Tomoda, Y. Takeuchi and Y. Nomura; *Synthesis*, 1985, 212. 335
85S253 C. Narayana and M. Periasamy; *Synthesis*, 1985, 253. 27
85S276 Y. Ohshiro, N. Ando, M. Komatsu and T. Agawa; *Synthesis*, 1985, 276. 989
85S331 D. Sicker, D. Reifegerste, S. Hauptmann, H. Wilde and G. Mann; *Synthesis*, 1985, 331. 324
85S403 T. Oesterle and G. Simchen; *Synthesis*, 1985, 403. 718
85S493 S. Randriamahefa, P. Deschamps, R. Gallo and H. Grangette; *Synthesis*, 1985, 493. 38
85S509 F. Moimas, C. Angeli, G. Comisso, P. Zanon, E. Decorte and V. Sunjić; *Synthesis*, 1985, 509. 1078
85S519 S. Kato, H. Kabuto, M. Kimura, H. Ishihara and T. Murai; *Synthesis*, 1985, 519. 248, 249
85S520 S. Kato, H. Kabuto, H. Ishihara and T. Murai; *Synthesis*, 1985, 520. 248, 249
85S665 S. Rozen and M. Brand; *Synthesis*, 1985, 665. 203
85S674 S. Kajigaeshi, T. Nakagawa, N. Nagasaki and S. Fujisaki; *Synthesis*, 1985, 674. 48, 49
85S681 G. H. Cleland and F. S. Bennett; *Synthesis*, 1985, 681. 183
85S773 I. Arenal, M. Bernabe, E. Fernandez-Alvarez and S. Penacles; *Synthesis*, 1985, 773. 87
85S878 T. Toru, M. Nishigaki, T. Seko, T. Kanefusa and E. Maekawa; *Synthesis*, 1985, 878. 248
85S929 L. Nakonieczna, M. Milewska, T. Kolasa and A. Chimiak; *Synthesis*, 1985, 929. 322
85S960 K. Hartke, O. Kunze and W. Hoederath; *Synthesis*, 1985, 960. 513, 559
85SC171 K. Kondo, C. Komamura, M. Murakami and K. Takemoto; *Synth. Commun.* 1985, **15**, 171. 1078
85SC643 P. Molina, A. Arques, I. Cartagena and M. V. Valcarcel; *Synth. Commun.*, 1985, **15**, 643. 711
85SC759 H.-J. Liu and I.-S. Han; *Synth. Commun.*, 1985, **15**, 759. 47
85SC1159 D. A. Brown, R. A. Geraty, J. D. Glennon and N. N. Choileain; *Synth. Commun.*, 1985, **15**, 1159. 322
85SC1233 C. Belaud, C. Roussakis, Y. Letourneux, N. El Alami and J. Villieras; *Synth. Commun.*, 1985, **15**, 1233. 296
85SM347 M. Hanack and K. Fischer; *Synth. Met.*, 1985, **10**, 347. 1083
85SRI321 C. H. Yoder, S. L. Tesno, S. M. Heaney and C. Bohan; *Synth. React. Inorg. Metal-Org. Chem.*, 1985, **15**, 321. 224
85SUL11 E. Fischer and G. Boldt; *Sulfur Letters*, 1985, **3**, 11. 340
85SUL177 A. Senning, M. A. Chevallier and B. Jensen; *Sulfur Letters*, 1985, **3**, 177. 478, 494
85T575 G. V. Shustov, N. B. Tavakalyan and R. G. Kostyanovsky; *Tetrahedron*, 1985, **41**, 575. 317
85T693 G. Motte-Coppe, F. Dutron-Woitrin, T. G. C. Bird, H. G. Viehe, J. P. Declercq, G. Germain and M. Van Meerssche; *Tetrahedron*, 1985, **41**, 693. 601
85T811 S. G. Khbeis, G. Maas and M. Regitz; *Tetrahedron*, 1985, **41**, 811. 871
85T825 M. Böhshar, G. Maas, H. Heydt and M. Regitz; *Tetrahedron*, 1985, **41**, 825. 871
85T1289 P. Metzner and R. Rakotonirina; *Tetrahedron*, 1985, **41**, 1289. 616
85T2441 M. Chmielewski, Z. Kaluza, C. Belzecki, P. Salanski, J. Jurczak and H. Adamowicz; *Tetrahedron*, 1985, **41**, 2441. 344
85T2503 A. Alemagna, P. Del Buttero, E. Licandro, A. Papagni and M. Ballabio; *Tetrahedron*, 1985, **41**, 2503. 676, 677, 906
85T2625 W. Theis and M. Regitz; *Tetrahedron*, 1985, **41**, 2625. 870
85T2691 P. Beslin and Y. Vallée; *Tetrahedron*, 1985, **41**, 2691. 552
85T3321 A. Alemagna, P. Del Buttero, E. Licandro, S. Maiorana and A. Papagni; *Tetrahedron*, 1985, **41**, 3321. 906
85T3331 C. Lambert, B. Caillaux and H. G. Viehe; *Tetrahedron*, 1985, **41**, 3331. 577, 617
85T3625 F. Matsuda, S. Itoh, N. Hattori, M. Yanagiya and T. Matsumoto; *Tetrahedron*, 1985, **41**, 3625. 9
85T4047 B. Maillard, A. Kharrat, F. Rakotomanana, E. Montaudon and C. Gardrat; *Tetrahedron*, 1985, **41**, 4047. 206
85T4195 M. Feldhues and H. J. Schäfer; *Tetrahedron*, 1985, **41**, 4195. 207
85T4213 M. Feldhues and H. J. Schäfer; *Tetrahedron*, 1985, **41**, 4213. 207
85T4781 Z. J. Witczak; *Tetrahedron*, 1985, **41**, 4781. 1053, 1057
85T4821 A. P. Kozikowski and A. Ames; *Tetrahedron*, 1985, **41**, 4821. 239
85T4835 K. C. Nicolaou, N. A. Petasis and D. A. Claremon; *Tetrahedron*, 1985, **41**, 4835. 239, 336, 337
85T4929 M. Cinquini, A. Manfredi, H. Molinari and A. Restelli; *Tetrahedron*, 1985, **41**, 4929. 607
85T5061 M. P. Cava and M. I. Levinson; *Tetrahedron*, 1985, **41**, 5061. 512, 570
85T5089 A. I. Meyers and R. D. Walkup; *Tetrahedron*, 1985, **41**, 5089. 557
85T5133 A. Palomo-Coll, A. L. Palomo-Coll and C. P. Nicolau; *Tetrahedron*, 1985, **41**, 5133. 193
85T5145 H. C. Hansen, A. Senning and R. G. Hazell; *Tetrahedron*, 1985, **41**, 5145. 336
85T5331 B. Calas, J. Mery, J. Parello and A. Cave; *Tetrahedron*, 1985, **41**, 5331. 185
85T5393 J. D. Coyle; *Tetrahedron*, 1985, **41**, 5393. 609
85T5595 O. E. Jensen, S.-O. Lawesson, R. Bardi, A. M. Piazzesi and C. Toniolo; *Tetrahedron*, 1985, **41**, 5595. 573
85T6063 G. H. L. Nefkens and B. Zwanenburg; *Tetrahedron*, 1985, **41**, 6063. 229
85TH86 C. Casser; Ph.D. Thesis, University of Bonn, 1985, p. 86. 878
85TL215 J.-P. Behr, C. J. Burrows, R. Heng and J.-M. Lehn; *Tetrahedron Lett.*, 1985, **26**, 215. 360
85TL475 J. M. Aizpurua and C. Palomo; *Tetrahedron Lett.*, 1985, **26**, 475. 223, 374
85TL501 A. J. Birch, L. F. Kelly and A. J. Liepa; *Tetrahedron Lett.*, 1985, **26**, 501. 1126
85TL673 Y. Tajima, A. Yoshida, N. Takeda and S. Oida; *Tetrahedron Lett.*, 1985, **26**, 673. 236
85TL977 M. Jung and M. J. Miller; *Tetrahedron Lett.*, 1985, **26**, 977. 329
85TL1109 S. Cacchi, E. Morera and G. Ortar; *Tetrahedron Lett.*, 1985, **26**, 1109. 296, 331
85TL1175 J. M. Renga and P.-C. Wang; *Tetrahedron Lett.*, 1985, **26**, 1175. 375
85TL1321 M. Hori, K. Takana, T. Kataoka, H. Shimizu, E. Imai, K. Kimura and Y. Hashimoto; *Tetrahedron Lett.*, 1985, **26**, 1321. 1122

85TL1341 S. Kim, H. Chang and Y. K. Ko; *Tetrahedron Lett.*, 1985, **26**, 1341. 280

85TL1503 D. Voisin and B. Gastambide; *Tetrahedron Lett.*, 1985, **26**, 1503. 191, 196

85TL1509 O. Sock, M. Troupel and J. Perichon; *Tetrahedron Lett.*, 1985, 1509. 44

85TL1623 B. Schaub and M. Schlosser; *Tetrahedron Lett.*, 1985, **26**, 1623. 896

85TL1631 J. Mulzer, M. Kappert, G. Huttner and J. Jibril; *Tetrahedron Lett.*, 1985, **26**, 1631. 146

85TL1661 S. Kim and K. Y. Yi; *Tetrahedron Lett.*, 1985, 1661. 1030, 1078

85TL1803 S. Knapp, K. E. Rodriques, A. T. Levorse and R. M. Ornaf; *Tetrahedron Lett.*, 1985, **26**, 1803. 285, 377

85TL1845 G. Maier, K. A. Reuter and L. Franz; *Tetrahedron Lett.*, 1985, **26**, 1845. 874

85TL1857 G. Langrand, M. Secchi, G. Buono, J. Baratti and C. Triantaphylides; *Tetrahedron Lett.*, 1985, **26**, 1857. 226

85TL2125 S. G. Davies, I. M. Dordor-Hedgecock and P. Warner; *Tetrahedron Lett.*, 1985, **26**, 2125. 453

85TL2217 M. Shibasaki, Y. Ishida and N. Okabe; *Tetrahedron Lett.*, 1985, **26**, 2217. 375

85TL2337 A. Sekiguchi, W. Ando and K. Honda; *Tetrahedron Lett.*, 1985, **26**, 2337. 925

85TL2509 R. L. Danheiser and D. M. Fink; *Tetrahedron Lett.*, 1985, **26**, 2509. 410, 431

85TL2513 R. L. Danheiser and D. M. Fink; *Tetrahedron Lett.*, 1985, **26**, 2513. 424, 429

85TL2609 S. Antebi and H. Alper; *Tetrahedron Lett.*, 1985, **26**, 2609. 237

85TL2675 T. Mandai, M. Yamaguchi, Y. Nakayama, J. Otera and M. Kawada; *Tetrahedron Lett.*, 1985, **26**, 2675. 411

85TL3167 C. J. Moody, C. J. Pearson and G. Lawton; *Tetrahedron Lett.*, 1985, **26**, 3167. 354

85TL3171 C. J. Moody, C. J. Pearson and G. Lawton; *Tetrahedron Lett.*, 1985, **26**, 3171. 351

85TL3207 Y. Tamaru, T. Kobayashi, S. Kawamura, H. Ochiai, M. Hojo and Z. Yoshida; *Tetrahedron Lett.*, 1985, **26**, 3207. 170

85TL3551 R. Appel, C. Casser and M. Immenkeppel; *Tetrahedron Lett.*, 1985, **26**, 3551. 877, 878

85TL3891 S. R. Woulfe, H. Iwagami and M. J. Miller; *Tetrahedron Lett.*, 1985, **26**, 3891. 334

85TL4349 U. Hildebrand, K. Taraz and H. Budzikiewicz; *Tetrahedron Lett.*, 1985, **26**, 4349. 246

85TL4439 C. Lambert and H. G. Viehe; *Tetrahedron Lett.*, 1985, **26**, 4439. 132

85TL4855 C.-N. Hsiao, S. P. Ashburn and M. J. Miller; *Tetrahedron Lett.*, 1985, **26**, 4855. 329

85TL5269 E. Schaumann and S. Scheiblich; *Tetrahedron Lett.*, 1985, **26**, 5269. 589

85TL5281 R. Höfelmeier and S. Blechert; *Tetrahedron Lett.*, 1985, **26**, 5281. 711

85TL5385 G. Rajendra and M. J. Miller; *Tetrahedron Lett.*, 1985, **26**, 5385. 285, 327, 328

85TL5505 H. Fritz and W. Sundermeyer; *Tetrahedron Lett.*, 1985, **26**, 5505. 492

85TL5507 G. Markl and H. Sejpka; *Tetrahedron Lett.*, 1985, **26**, 5507. 1155, 1157

85TL5515 H. Wetter and K. Oertle; *Tetrahedron Lett.*, 1985, **26**, 5515. 224, 375

85TL5739 D. Ranganathan, S. Bamezai, H. Cun-heng and J. Clardy; *Tetrahedron Lett.*, 1985, **26**, 5739. 355

85TL6051 G. Helmchen and G. Wegner; *Tetrahedron Lett.*, 1985, 6051. 32

85TL6075 J.-B. Verlhac, E. Chanson, B. Jousseaume and J.-P. Quintard; *Tetrahedron Lett.*, 1985, **26**, 6075. 467

85TL6273 J. R. McCarthy, D. P. Matthews and J. P. Whitten; *Tetrahedron Lett.*, 1985, **26**, 6273. 1127

85ZC219 G. Barnikow, J. Lehner and A. A. Martin; *Z. Chem.*, 1985, **25**, 219. 588, 632

85ZC305 H. Kohler and L. Jager; *Z. Chem.*, 1985, **28**, 305. 1070

85ZC324 J. Bernat, L. Kniezu, P. Kristian and A. Dondoni; *Z. Chem.*, 1985, **25**, 324. 612

85ZC364 D. Briel and G. Wagner; *Z. Chem.*, 1985, **25**, 364. 622

85ZN(B)124 H. W. Roesky, H. Hofmann, M. Noltemeyer and G. M. Sheldrick; *Z. Naturforsch., Teil B*, 1985, **40**, 124. 606

85ZN(B)1077 F. Zurmühlen, W. Rösch and M. Regitz; *Z. Naturforsch., Teil B*, 1985, **40**, 1077. 839, 880

85ZN(B)1177 G. Erker, P. Czisch and R. Mynott; *Z. Naturforsch., Teil B*, 1985, **40**, 1177. 917

85ZN(B)1563 U. Hildebrand, K. Taraz and H. Budzikiewicz; *Z. Naturforsch., Teil B*, 1985, **40B**, 1563. 246

85ZOB621 N. N. Vlasova, A. E. Pestunovich and M. G. Voronkov; *Zh. Obshch. Khim.*, 1985, **55**, 621. 1076, 1077

85ZOB776 S. K. Tupchienko, T. N. Dudchenko and A. D. Sinitsa; *Zh. Obshch. Khim.*, 1985, **55**, 776 (*Chem. Abstr.*, 1985, **103**, 196 160). 776

85ZOB1071 A. S. Gordetsov, V. V. Pereshein, V. N. Latyaeva, V. P. Mar'in, O. S. Kukevich and E. S. Zakharova; *Zh. Obshch. Khim.*, 1985, **55**, 1071. 388

85ZOB1080 S. I. Shaposhnikov, G. N. Koidan, A. P. Marchenko and A. M. Pinchuk; *Zh. Obshch. Khim.*, 1985, **55**, 1080. 379

85ZOB1308 V. D. Sheludyakov, A. B. Lebedeva, A. Kh. Shukyurov and A. D. Kirilin; *Zh. Obshch. Khim.*, 1985, **56**, 1308. 1075

85ZOB1526 M. Zimin, E. V. Fomakhin, A. R. Sadykov, V. N. Smirnov and A. N. Pudovik; *Zh. Obshch. Khim.* (*Engl. Trans.*), 1985, **55**, 1526. 644

85ZOB1718 M. G. Zimin, E. V. Fomakhin, A. R. Sadykov, V. N. Smirnov and A. N. Pudovik; *Zh. Obshch. Khim.*, 1985, **55**, 1718 (*Chem. Abstr.*, 1986, **105**, 42 579n). 713

85ZOB2100 N. G. Luk'yanenko, V. V. Limich, S. V. Shcherbakov and T. I. Kirichenko; *Zh. Obshch. Khim.*, 1985, **55**, 2100. 1024

85ZOB2318 O. P. Shvaika and V. F. Lipnitskii; *Zh. Obshch. Khim.* (*Engl. Trans.*), 1985, **55**, 2318. 642

85ZOB2483 M. N. Dimukhametov and I. A. Nuretdinov; *Zh. Obshch. Khim.*, 1985, **55**, 2483. 371

85ZOR196 A. Y. Tikhonov, N. V. Belova, L. B. Volodarskii and Y. V. Gatilov; *Zh. Org. Khim.*, 1985, **21**, 196. 212

85ZOR559 V. A. Al'fonsov, D. A. Pudovik, E. S. Batyeva and A. N. Pudovik; *Zh. Org. Khim.*, 1985, **55**, 559 (*Chem. Abstr.*, 1985, **103**, 105 067). 737

85ZOR769 M. M. Kremlev, L. I. Moklyschuck, Y. A. Fialkov, N. V. Ignatev and L. M. Yagupolskii; *Zh. Org. Khim.*, 1985, **21**, 769. 58

85ZOR1281 E. S. Levchenko, T. N. Dubinina, L. V. Budnik and E. A. Romanenko; *Zh. Org. Khim.*, 1985, **21**, 1281 (*Chem. Abstr.*, 1986, **104**, 148 441f). 703

85ZOR2407 Y. A. Fialkov, I. M. Zalesskaya, S. V. Shelyazhenko and L. M. Yagupolskii; *Zh. Org. Khim.*, 1985, **21**, 2407. 112

86ACR49 M. J. Miller; *Acc. Chem. Res.*, 1986, **19**, 49. 322

86ACS(B)466 E.-L. Lindstedt and M. Nilsson; *Acta Chem. Scand. B*, 1986, **40**, 466. 151

86ACS(B)534 R. Carlson, T. Lundstedt and R. Shabana; *Acta. Chem. Scand., Ser. B*, 1986, **40**, 534. 599

86ACS(B)694 R. Carlson, T. Lundstedt and R. Shabana; *Acta. Chem. Scand., Ser. B*, 1986, **40**, 694. 599

86AG213 R. R. Schmidt; *Angew. Chem.*, 1986, **98**, 213. 692

86AG(E)212 R. R. Schmidt; *Angew. Chem. Int. Ed. Engl.*, 1986, **25**, 212. 692

86AG(E)264 G. Märkl and H. Sejpka; *Angew. Chem., Int. Ed. Engl.*, 1986, **25**, 264. 1160

86AG(E)454 H. H. Karsch, H.-U. Reisacher and G. Müller; *Angew. Chem., Int. Ed. Engl.*, 1986, **25**, 454. 887

86AG(E)559 H. J. Bestmann and T. Arenz; *Angew. Chem., Int. Ed. Engl.*, 1986, **25**, 559. 916

86AG(E)994 H. J. Bestmann and T. Moenius; *Angew. Chem., Int. Ed. Engl.*, 1986, **25**, 994. 223

86AG(E)1028 K. Ditrich, T. Bube, R. Stürmer and R. W. Hoffmann; *Angew. Chem., Int. Ed. Engl.*, 1986, **98**, 1028. 173

86AG(E)1093 G. Maier, I. Bauer, D. Born and H.-O. Kalinowski; *Angew. Chem., Int. Ed. Engl.*, 1986, **25**, 1093. 874

86AOC(25)1 A. G. Brook and K. M. Baines; *Adv. Organomet. Chem.*, 1986, **25**, 1. 418, 923

86AOC(25)121 M. A. Gallop and W. R. Roper; *Adv. Organomet. Chem.*, 1986, **25**, 121. 942

86AP(319)203 H.-J. Bergmann, R. Mayrhofer and H.-H. Otto; *Arch. Pharm.* (*Weinheim, Ger.*), 1986, **319**, 203. 374

86AP(319)216 H.-J. Bergmann and H.-H. Otto; *Arch. Pharm.* (*Weinheim, Ger.*), 1986, **319**, 216. 384

86AP(319)325 H. Möhrle and B. Grimm; *Arch. Pharm.* (*Weinheim, Ger.*), 1986, **319**, 325. 215

86AP(319)682 H. Egg, I. Ganzera, H. Leibetseder, A. Patzak and U. Sperl; *Arch. Pharm.* (*Weinheim, Ger.*), 1986, **319**, 682. 344

86AP(319)1037 H. Grigat and G. Zinner; *Arch. Pharm.* (*Weinheim, Ger.*), 1986, **319**, 1037. 211, 322

86BCJ1403 M. Ishida, K. Kaga, H. Sato, M. Yokoi and S. Kato; *Bull. Chem. Soc. Jpn.*, 1986, **59**, 1403. 560

86BCJ1575 P. P. Reddy, C. K. Reddy and P. S. N. Reddy; *Bull. Chem. Soc. Jpn.*, 1986, **59**, 1575. 363

86BCJ2017 I. Shibuya and K. Yonemoto; *Bull. Chem. Soc. Jpn.*, 1986, **59**, 2017. 335

86BCJ2203 K. Harada, S. Suzuki, H. Narita and H. Watanabe; *Bull. Chem. Soc. Jpn.*, 1986, **59**, 2203. 385

86BCJ2529 A. Ishii, R. Okazaki and N. Inamoto; *Bull. Chem. Soc. Jpn.*, 1986, **59**, 2529. 516, 523, 524

86BCJ3696 Y. Yonezawa, A. Kisuno and C. G. Shin; *Bull. Chem. Soc. Jpn.*, 1986, **59**, 3696. 371

86BCJ3871 T. Tokumitsu; *Bull. Chem. Soc. Jpn.*, 1986, **59**, 3871. 597, 599

86BSF101 M. Mulliez and R. Wolf; *Bull. Soc. Chim. Fr.*, 1986, 101. 370

86BSF817 S. Berrada and P. Metzner; *Bull. Soc. Chim. Fr.*, 1986, 817. 556

86CAR(150)111 H. W. Pauls and B. Fraser-Reid; *Carbohydr. Res.*, 1986, **150**, 111. 226

86CB744 D. Geffken; *Chem. Ber.*, 1986, **119**, 744. 211

86CB1127 U. Jäger, M. Schwab and W. Sundermeyer; *Chem. Ber.*, 1986, **119**, 1127. 340

86CB1594 U. Burkard and F. Effenberger; *Chem. Ber.*, 1986, **119**, 1594. 1106

86CB1857 L. Weber, K. Reizig and M. Frebel; *Chem. Ber.*, 1986, **119**, 1857. 400

86CB2080 H. Fischer, F. Seitz and J. Riede; *Chem. Ber.*, 1986, **119**, 2080. 1088

86CB2114 H. Wamhoff, F.-J. Fassbender, G. Hendrikx, H. Puff and P. Woller; *Chem. Ber.*, 1986, **119**, 2114. 1066

86CB2191 R. Devant and M. Braun; *Chem. Ber.*, 1986, **119**, 2191. 66

86CB2678 A. Onistschenko, B. Buchholz and H. Stamm; *Chem. Ber.*, 1986, **119**, 2678. 345

86CB3405 U. Jäger and W. Sundermeyer; *Chem. Ber.*, 1986, **119**, 3405. 340

86CB3737 J. Goerdeler and W. Eggers; *Chem. Ber.*, 1986, **119**, 3737. 1064, 1073

86CB3801 R. Aumann and H. Heinen; *Chem. Ber.*, 1986, **119**, 3801. 953

86CB3812 K. Diehl and G. Himbert; *Chem. Ber.*, 1986, **119**, 3812. 597, 630

86CC71 R. Okazaki, A. Ishii and N. Inamoto; *J. Chem. Soc., Chem. Commun.*, 1986, 71. 524

86CC344 S. E. de Laszlo, S. V. Ley and R. A. Porter; *J. Chem. Soc., Chem. Commun.*, 1986, 344. 278

86CC495 S. G. Davies and J. C. Walker; *J. Chem. Soc., Chem. Commun.*, 1986, 495. 453

86CC514 A. Parlier and H. Rudler; *J. Chem. Soc., Chem. Commun.*, 1986, 514. 944

86CC607 S. G. Davies, I. M. Dordor-Hedgecock, K. H. Sutton *et al.*; *J. Chem. Soc., Chem. Commun.*, 1986, 607. 453

86CC609 S. G. Davies and J. C. Walker; *J. Chem. Soc., Chem. Commun.*, 1986, 609. 453

86CC719 S. Kim and S. S. Kim; *J. Chem. Soc., Chem. Commun.*, 1986, 719. 280

86CC791 J. H. Clark, A. J. Hyde and D. K. Smith; *J. Chem. Soc., Chem. Commun.*, 1986, 791. 4

86CC841 M. Mori, Y. Uozumi and Y. Ban; *J. Chem. Soc., Chem. Commun.*, 1986, 841. 301

86CC1152 G. W. Kirby and A. N. Trethewey; *J. Chem. Soc., Chem. Commun.*, 1986, 1152. 336, 1113

86CC1196 A. F. Hegarty and I. Cunningham; *J. Chem. Soc., Chem. Commun.*, 1986, 1196. 675

86CC1754 P. Laurent, G. Tanguy and H. des Abbayes; *J. Chem. Soc., Chem. Commun.*, 1986, 1754. 452

86CC1756 M. Ladlow, P. M. Cairns and P. Magnus; *J. Chem. Soc., Chem. Commun.*, 1986, 1756. 574

86CCC112 J. Gonda, P. Kristian and L. Mikler; *Collect. Czech. Chem. Commun.*, 1986, **51**, 112. 1035

86CCC2167 L. Fišera, N. D. Kozhina, P. Oravec, H.-J. Timpe, L. Štibrányi and L. A. Badovskaya; *Collect. Czech. Chem. Commun.*, 1986, **51**, 2167. 350

86CJC308 T. G. Back and R. G. Kerr; *Can. J. Chem.*, 1986, **64**, 308. 335, 338

86CL81 K. Navasaki and Y. Ukaji; *Chem. Lett.*, 1986, 81. 148

86CL135 T. Saito, M. Nakane, M. Endo, H. Yamashita, Y. Oyamada and S. Motoki; *Chem. Lett.*, 1986, 135. 1066

86CL177 T. Nakajima, M. Tanabe, K. Ohno, M. Segi and S. Suga; *Chem. Lett.*, 1986, 177. 429

86CL871 T. Miyasaka, H. Ishizu, A. Sawada, A. Fujimoto and Noguchi; *Chem. Lett.*, 1986, 871. 134

86CL915 T. Mukaiyama, H. Suzucki and T. Yamada; *Chem. Lett.*, 1986, 915. 237, 277

86CL935 S. Kato, Y. Komatsu, T. Murai, H. Hamaguchi, M. Tasumi and R. Okazaki; *Chem. Lett.*, 1986, 935. 246

86CL967 T. Imamoto and H. Koto; *Chem. Lett.*, 1986, 967. 205

86CL977 K. Sasaki, Y. Aso, T. Otsubo and F. Ogura; *Chem. Lett.*, 1986, 977. 242

86CL1013 N. Yamasaki, M. Murakami and T. Mukaiyama; *Chem. Lett.*, 1986, 1013. 236

86CL1251 M. A. Palominos, R. Rodriguez and J. C. Vega; *Chem. Lett.*, 1986, 1251. 513

86CL1817 T. Mukaiyama, M. Tamura and S. Kobayashi; *Chem. Lett.*, 1986, 1817. 236

86CL2025 A. Sekiguchi and W. Ando; *Chem. Lett.*, 1986, 2025. 925

86CPB4545 T. Kurihara, M. Hanakawa, S. Harusawa and R. Yoneda; *Chem. Pharm. Bull.*, 1986, **34**, 4545. 347

86CPB4984 Y. Tagawa, N. Honjo, Y. Goto and T. Kato; *Chem. Pharm. Bull.*, 1986, **34**, 4984. 215

86CRV697 H. C. J. Ottenheijm and J. D. M. Hersheid; *Chem. Rev.*, 1986, **86**, 697. 84

86CS347 F. Johannsen, A. Jørgensen and A. B. Pedersen; *Chem. Scr.*, 1986, **26**, 347. 690

86CZ377 D. Geffken; *Chem.-Ztg.*, 1986, **110**, 377. 211

86EUP268820 M. Kondo, M. Minai,, S. Kai, T. Higashi and Y. Ueda (Sumitomo Chemical Co. Ltd.); *Eur. Pat.* 268 820 (1986) (*Chem. Abstr.*, 1988, **109**, 170 222). 193

86FES539 G. Romussi, B. Parodi, G. Bignardi, G. Menozzi and P. Schenone; *Farmaco Ed. Sci.*, 1986, **41**, 539. 344

86FES676 T. Zawaszi and W. Malinka; *Farmaco Ed. Sci.*, 1986, **41**, 676. 347

86GEP3511050 G. Luft and G. Ritter (Hoechst AG.); *Ger. Pat.* 3 511 050 (1986) (*Chem. Abstr.*, 1987, **106**, 17 909). 192

86GEP3522304 J. Perichon, O. Sock and M. Troupel; *Ger. Pat.*, 3 522 304 (1986) (*Chem. Abstr.*, 1986, **104**, 158 203u). 44

86GSM131 D. W. Knight; *Gen. Synth. Methods*, 1986, **8**, 131. 24

86H(24)1247 H. Takahata, T. Suzuki and T. Yamazaki; *Heterocycles*, 1986, **24**, 1247. 607

86H(24)1899 U. Urleb, B. Stanovnik, V. Stibilj and M. Tišler; *Heterocycles*, 1986, **24**, 1899. 845

86H(24)2813 J. H. Boyer and T. Manimaran; *Heterocycles*, 1986, **24**, 2813. 963

86HCA44 M. E. Scheller and B. Frei; *Helv. Chim. Acta*, 1986, **69**, 44. 223, 412, 429

86HCA174 C. Jenny and H. Heimgartner; *Helv. Chim. Acta*, 1986, **69**, 174. 605

86HCA236 P. Huber, H. Leuenberger and W. Keller-Schierlein; *Helv. Chim. Acta*, 1986, **69**, 236. 214

86HCA246 R. Nordmann and P. Gull; *Helv. Chim. Acta*, 1986, **69**, 246. 1125

86HCA333 H. Kristinsson, T. Winkler and M. Mollenkopf; *Helv. Chim. Acta*, 1986, **69**, 333. 355

86HCA1224 H. Moser, A. Fliri, A. Steiger, G. Costello, J. Schreiber and A. Eschenmoser; *Helv. Chim. Acta*, 1986, **69**, 1224. 749

86HCA1378 M. E. Scheller, G. Iwasaki and B. Frei; *Helv. Chim. Acta*, 1986, **69**, 1378. 410, 424

86IC9 D. Hedden and D. M. Roundhill; *Inorg. Chem.*, 1986, **25**, 9. 385

86ICA67 C. M. Lukehart and M. Raja; *Inorg. Chem. Acta*, 1986, **122**, 67. 801

86ICA107 C. C. Chang and N. K. Sung-Yu; *Inorg. Chem. Acta*, 1986, **119**, 107. 469

86ICAL15 J. H. Davis, Jr. and C. M. Lukehart; *Inorg. Chem. Acta*, 1986, **119**, L15. 801

86ICAL17 T. A. Emokpae and A. C. Ukwueze; *Inorg. Chem. Acta*, 1986, **112**, L17. 1140

86IJC57 Y. D. Kulkarni, S. Srivastava and M. Athar; *Indian J. Chem.*, 1986, **25A**, 57. 734

86IJC433 A. K. Bag, S. R. Gupta and D. N. Dhar; *Indian J. Chem., Sect. B*, 1986, **25**, 433. 107

86IJC(A)57 Y. D. Kulkarni, S. Srivastava and M. Athar; *Indian J. Chem., Sect. A*, 1986, **25**, 57. 341

86IJC(B)373 W. S. El-Hamouly and A. F. Aboulezz; *Indian J. Chem., Sect B*, 1986, **25B**, 373. 215

86IZV2032 E. I. Klabunovskii, E. S. Levitina, L. N. Kaigorodova, L. F. Godunova, D. D. Gogoladze, E. I. Karpeiskaya and S. A. Novikova; *Izv. Akad. Nauk SSSR, Ser. Khim.*, 1986, 2032. 215

86IZV2535 E. B. Ivanov, S. S. Krokhina, T. V. Chichkanova and A. B. Ageeva; *Izv. Akad. Nauk SSSR, Ser. Khim.*, 1986, 2535. 221

86JA182 L. Maziak, G. Lajoie and B. Belleau; *J. Am. Chem. Soc.*, 1986, **108**, 182. 574

86JA284 B. M. Trost, J. Lynch, R. Renaut and D. H. Steinmann; *J. Am. Chem. Soc.*, 1986, **108**, 284. 146

86JA520 W. D. Wulff, S. R. Gilbertson and J. P. Springer; *J. Am. Chem. Soc.*, 1986, **108**, 520. 934

86JA1039 R. V. Stevens, N. Beaulieu, W. H. Chan, A. R. Daniewski, T. Takeda, A. Waldner, P. G. Williard and U. Zutter; *J. Am. Chem. Soc.*, 1986, **108**, 1039. 278

86JA1314 G. A. Krafft and P. T. Meinke; *J. Am. Chem. Soc.*, 1986, **108**, 1314. 1113

86JA1462 G. D. Vaughn, C. E. Strouse and J. A. Gladysz; *J. Am. Chem. Soc.*, 1986, **108**, 1462. 449

86JA1722 W. J. Evans, L. A. Hughes, D. K. Drummond, H. Zhang and J. L. Atwood; *J. Am. Chem. Soc.*, 1986, **108**, 1722. 439

86JA2173 F. Volatron and O. Eisenstein; *J. Am. Chem. Soc.*, 1986, **108**, 2173. 932

86JA2180 E. A. Carter and W. A. Goddard, III; *J. Am. Chem. Soc.*, 1986, **108**, 2180. 932

86JA2451 R. Zibuck, N. J. Liverton and A. B. Smith, III; *J. Am. Chem. Soc.*, 1986, **108**, 2451. 179

86JA3318 J. D. Meinhart, B. D. Santarsiero and R. H. Grubbs; *J. Am. Chem. Soc.*, 1986, **108**, 3318. 441

86JA4746 E. A. Carter and W. A. Goddard, III; *J. Am. Chem. Soc.*, 1986, **108**, 4746. 932

86JA5655 C. R. Johnson and T. D. Penning; *J. Am. Chem. Soc.*, 1986, **108**, 5655. 138

86JA6054 D. J. Hart, C.-S. Lee, W. H. Pirkle, M. H. Hyon and A. Tsipouras; *J. Am. Chem. Soc.*, 1986, **108**, 6054. 274, 294

86JA6394 C. Gennari, L. Colombo and G. Bertolini; *J. Am. Chem. Soc.*, 1986, **108**, 6394. 88

86JA8279 S. Masamune, T. Sato, B. M. Kim and T. A. Wollmann; *J. Am. Chem. Soc.*, 1986, **108**, 8279. 237

86JAN1257 A. Makita, Y. Yamada and H. Okada; *J. Antibiot.*, 1986, **39**, 1257. 199

86JAP47675 Agency of Industrial Sciences and Technology; *Jpn. Pat.*, 60/47 675 (1986) (*Chem. Abstr.*, 1985, **103**, 140 323c). 117

86JAP1022060 H. Kawachi, S. Monno, M. Wajima and A. Takahashi; *Jpn. Pat.* 61 022 060 (1986) (*Chem. Abstr.*, 1986, **105**, 79 507p). 1125

86JAP1280464 M. Sasaki, S. Niizeki and M. Yoshida; *Jpn. Pat.* 61 280 464 (1986) (*Chem. Abstr.*, 1987, **107**, 39 236j). 1125

86JCE1013 J. Pouliquen, M. Heintz, O. Sock and M. Troupel; *J. Chem. Ed.*, 1986, **63**, 1013. 44

86JCR(S)210 L. Bruché and G. Zecchi; *J. Chem. Res.* (*S*), 1986, 210. 681

86JCS(D)813 H. Adams, N. A. Bailey, T. N. Briggs, J. A. McCleverty, H. M. Colquhoun and D. J. Williams; *J. Chem. Soc., Dalton Trans*, 1986, 813. 387, 388

86JCS(D)1359 J. Barker, N. Cameron, M. Kilner, M. M. Mahoud and S. C. Wallwork; *J. Chem. Soc., Dalton Trans.*, 1986, 1359. 781

86JCS(P1)39 D. H. R. Barton, D. Crich and G. Kretzschmar; *J. Chem. Soc., Perkin Trans. 1*, 1986, 39. 727

86JCS(P1)381 G. Barbaro, A. Battaglia, P. Giorgianni, G. Maccagnani, D. Macciantelli, B. F. Bonini, G. Mazzanti and P. Zani; *J. Chem. Soc., Perkin Trans. 1*, 1986, 381. 647, 648

86JCS(P1)521 A. L. Gutman and V. Ribon; *J. Chem. Soc., Perkin Trans. 1*, 1986, 521. 45

86JCS(P1)795 O. Menkisoglou-Spyroudi and A. Varvoglis; *J. Chem. Soc., Perkin Trans. 1*, 1986, 795. 205

86JCS(P1)927 G. Doleschall, P. Seres, L. Párkányi, G. Tóth, A. Almásy and E. Bihátsi-karsai; *J. Chem. Soc., Perkin Trans. 1*, 1986, 927. 199

86JCS(P1)1655 A. S. Dutta, M. B. Giles and J. C. Williams; *J. Chem. Soc., Perkin Trans. 1*, 1986, 1655. 351

86JCS(P1)1837 K. Terao, A. Toshimitsu and S. Uemura; *J. Chem. Soc., Perkin Trans. 1*, 1986, 1837. 285

86JCS(P1)1873 F. Kurzer and K. M. Doyle; *J. Chem. Soc., Perkin Trans. 1*, 1986, 1873. 640

86JCS(P1)2177 L. Bruché, L. Garanti and G. Zecchi; *J. Chem. Soc., Perkin Trans. 1*, 1986, 2177. 862

86JCS(P1)2207 B. Hanlon and D. I. John; *J. Chem. Soc., Perkin Trans. 1*, 1986, 2207. 360

86JCS(P2)995 A. F. Hegarty and M. Mullane; *J. Chem. Soc., Perkin Trans. 2*, 1986, 995. 672, 821, 825

86JCS(P2)1211 M. R. Banks and R. F. Hudson; *J. Chem. Soc., Perkin Trans. 2*, 1986, 1211. 347

86JFC(30)439 A. Haas and K. Peterman; *J. Fluorine Chem.*, 1986, **30**, 439. 784

86JFC(31)37 T. Umemoto, Y. Kuriu, H. Shuyama, O. Miyano and S. Nakayama; *J. Fluorine Chem.*, 1986, **31**, 37. 205

86JGU636 Yu. A. Veits, E. G. Neganova and V. L. Foss; *J. Gen. Chem. USSR*, 1986, **56**, 636. 400

86JGU1096 I. V. Konovalova, L. A. Burnaeva, E. A. Khusnitidinova and A. N. Pudovik; *J. Gen. Chem. USSR* (*Engl. Transl.*), 1986, **56**, 1096. 708

86JGU1113 A. D. Sinitsa, D. M. Malenki, L. A. Repina, R. A. Loktionsova and A. K. Shurubura; *J. Gen. Chem. USSR* (*Engl. Transl.*), 1986, **56**, 1113. 713, 788

86JGU1467 D. M. Malenko and A. D. Sinitsa; *J. Gen. Chem. USSR* (*Engl. Transl.*), 1986, **56**, 1467. 788

86JGU2141 O. I. Kolodyazhnyi; *J. Gen. Chem. USSR*, 1986, **56**, 2141. 397

86JGU2372 A. D. Sinitsa, P. P. Onys'ko, T. V. Kim, E. I. Kiseleva and V. V. Pirozhenko; *J. Gen. Chem. USSR* (*Engl. Transl.*), 1986, **56**, 2372. 659, 788, 789

86JGU2483 O. B. Smolii, V. S. Brovarets and B. S. Drach; *J. Gen. Chem. USSR* (*Engl. Transl.*), 1986, **56**, 2483. 655

86JHC183 B. Rigo, C. Lespagnol and M. Pauly; *J. Heterocycl. Chem.*, 1986, **23**, 183. 718

86JHC455 G. Menozzi, A. Bargagna, L. Mosti and P. Schenone; *J. Heterocycl. Chem.*, 1986, **23**, 455. 501

86JHC1253 W. S. Saari and J. E. Schwering; *J. Heterocycl. Chem.*, 1986, **23**, 1253. 342

86JHC1411 W. Pendergast and W. R. Hall; *J. Heterocycl. Chem.*, 1986, **23**, 1411. 690

86JHC1657 E. Kawashima and K. Tabei; *J. Heterocycl. Chem.*, 1986, **23**, 1657. 220

86JHC1757 W.-Y. Ren, K. V. B. Rao and R. S. Klein; *J. Heterocycl. Chem.*, 1986, **23**, 1757. 690

86JHC1861 J. E. Johnson, J. A. Maia, K. Tan, A. Ghafouripour, P. de Meester and S. S. C. Chu; *J. Heterocycl. Chem.*, 1986, **23**, 1861. 765, 824

86JMC199 S. Agrawal, N. K. Singh, R. C. Aggarwal, A. Sodhi and P. Tandon; *J. Med. Chem.*, 1986, **29**, 199. 385

86JMC359 J. P. Yevich, J. S. New, D. W. Smith, W. G. Lobeck, J. D. Catt, J. L. Minielli, M. S. Eison, D. P. Taylor, L. A. Riblet and D. L. Temple, Jr.; *J. Med. Chem.*, 1986, **29**, 359. 334

86JMC1689 C. D. Bedford, M. Miura, J. C. Bottaro, R. A. Howd and H. W. Nolen, III; *J. Med. Chem.*, 1986, **29**, 1689. 668, 830

86JMC2280 M. R. Stillings, A. P. Welbourn and D. S. Walter; *J. Med. Chem.*, 1986, **29**, 2280. 639

86JOC167 J. M. Lin, K. Koch and F. W. Fowler; *J. Org. Chem.*, 1986, **51**, 167. 214

86JOC546 T. Shono, O. Ishige, H. Uyama and S. Kashimura; *J. Org. Chem.*, 1986, **51**, 546. 132

86JOC558 H. Fritz, P. Sutter and C. D. Weis; *J. Org. Chem.*, 1986, **51**, 558. 375

86JOC567 E. Dalcanale and F. Mantanari; *J. Org. Chem.*, 1986, **51**, 567. 102

86JOC898 S. Murahashi, T. Naota and N. Nakajima; *J. Org. Chem.*, 1986, **51**, 898. 1139

86JOC1065 M. P. Groziak, J.-W. Chern and L. B. Townsend; *J. Org. Chem.*, 1986, **51**, 1065. 275

86JOC1277 M. P. Groziak and L. B. Townsend; *J. Org. Chem.*, 1986, **51**, 1277. 1063, 1083

86JOC1374 J. H. Rigby and F. J. Burkhardt; *J. Org. Chem.*, 1986, **51**, 1374. 993

86JOC1724 A. Toshimitsu, K. Terao and S. Uemura; *J. Org. Chem.*, 1986, **51**, 1724. 285

86JOC1929 D. A. Burnett, D. J. Hart and J. Liu; *J. Org. Chem.*, 1986, **51**, 1929. 335

86JOC2391 Y. Nagao, Y. Hagiwara, T. Kumagai, M. Ochiai, T. Inoue, K. Hashimoto and E. Fujita; *J. Org. Chem.*, 1986, **51**, 2391. 329

86JOC2613 S. Kim and K. Y. Yi; *J. Org. Chem.*, 1986, **51**, 2613. 1030, 1078

86JOC2669 I. M. Lazbin and G. F. Koser; *J. Org. Chem.*, 1986, **51**, 2669. 999

86JOC3133 S. R. Woulfe and M. J. Miller; *J. Org. Chem.*, 1986, **51**, 3133. 334, 336

86JOC3376 W. A. Nugent and F. W. Hobbs, Jr.; *J. Org. Chem.*, 1986, **51**, 3376. 151

86JOC3390 S. Abdel-Baky and R. W. Giese; *J. Org. Chem.*, 1986, **51**, 3390. 196

86JOC3545 K. Mai and G. Patil; *J. Org. Chem.*, 1986, **51**, 3545. 226, 1134

86JOC3744 W. K. Fife and Z.-D. Zhang; *J. Org. Chem.*, 1986, **51**, 3744. 184

86JOC4077 D. F. Taber, R. E. Ruckle, Jr. and M. J. Hennessy; *J. Org. Chem.*, 1986, **51**, 4077. 869

86JOC4150 Y. Kita, S. Akai, N. Ajimura, M. Yoshigi, T. Tsugoshi, H. Yasuda and Y. Tamura; *J. Org. Chem.*, 1986, **51**, 4150. 189, 195, 282

86JOC4354 P. Li and H. Alper; *J. Org. Chem.*, 1986, **51**, 4354. 38

86JOC4840 A. G. M. Barrett, R. A. E. Carr, S. V. Attwood, G. Richardson and N. D. A. Walshe; *J. Org. Chem.*, 1986, **51**, 4840. 177

86JOC4941 J. M. Aizpurua, F. P. Cossío and C. Palomo; *J. Org. Chem.*, 1986, **51**, 4941. 17

86JOC4959 J. C. Day, N. Govindaraj, D. S. McBain, P. S. Skell and J. M. Tanko; *J. Org. Chem.*, 1986, **51**, 4959. 318

86JOC5123 R. S. Glass, W. W. McConnell and S. W. Andruski; *J. Org. Chem.*, 1986, **51**, 5123. 991

86JOC5198 R. M. Coates and S. J. Firsan; *J. Org. Chem.*, 1986, **51**, 5198. 327, 636

86JOC5332 L. Liu, R. S. Tanke and M. J. Miller; *J. Org. Chem.*, 1986, **51**, 5332. 336

86JOM(306)39 W. Rösch, U. Vogelbacher, T. Allspach and M. Regitz; *J. Organomet. Chem.*, 1986, **306**, 39. 878, 879, 1155

86JOM(309)333 Y. Tsuji, T. Ohsumi, T. Kondo and Y. Watanabe; *J. Organomet. Chem.*, 1986, **309**, 333. 286

86JOM(311)269 L. Weber and D. Bungardt; *J. Organomet. Chem.*, 1986, **311**, 269. 400

86JOM(315)245 E. Hernandez and H. Hoberg; *J. Organomet. Chem.*, 1986, **315**, 245. 386

86JPR(328)812 W. Thiel, J. Fabian and B. Friebe; *J. Prakt. Chem.*, 1986, **328**, 812. 559

86JST(139)255 Y. Xie, J. E. Boggs and L. S. Khaikin; *J. Mol. Struct.* (*Theochem*), 1986, **139**, 255. 395

86KGS1405 T. Yagodzinski, T. Dzembovska, E. Yagodzinskaya and Z. Yablonski; *Khim. Geterotsikl. Soedin.*, 1986, 1405. 595

86LA334 F. Effenberger and U. Burkard; *Liebigs Ann. Chem.*, 1986, 334. 92

86LA1997 W. Ried and K. Schöpke; *Liebigs Ann. Chem.*, 1986, 1997. 658

86M(117)221 M. Merslavič, B. Stanovnik and M. Tišler; *Monatsh. Chem.*, 1986, **117**, 221. 836

86M(117)491 D. Knittel; *Monatsh. Chem.*, 1986, **117**, 491. 215

86M(117)1437 H. Knotz and E. Zbiral; *Monatsh. Chem.*, 1986, **117**, 1437. 1068

86MI 502-01 Y. Tang, M. Deng and W. Xu; *Huaxue Xuebao*, 1986, **44**, 276 (*Chem. Abstr.*, 1986, **106**, 84 695y). 82

86MI 504-01 R. Franz and H. G. Neumann; *Carcinogenesis*, 1986, **7**, 183. 215

B-86MI 504-01 S. R. Sandler, W. Karo; "Organic Functional Group Preparations," Academic Press, New York, 1986. 182

B-86MI 507-01 S. R. Sandler; in "Organic Functional Group Preparations," ed. H. H. Wasserman, Academic Press, New York, 1986, vol. 2, p. 499. 360

B-86MI 511-01 C. Paulmier; "Selenium Reagents and Intermediates in Organic Synthesis" (Tetrahedron Organic Chemistry Series Volume 4), Pergamon, Oxford, 1986, chap. 3. 472

86MI 513-01 S. Perlier, G. Levesque and P. Depraétère; *Makromol. Chem.*, 1986, **187**, 2369. 558

B-86MI 514-01 C. Paulmier; in "Selenium Reagents and Intermediates in Organic Synthesis," Pergamon, Oxford, 1986, p. 77. 624

86MI 517-01 B. A. O'Brien and D. D. DesMarteau; *Rev. Chim. Miner.*, 1986, **23**, 621 (*Chem. Abstr.*, 1987, **107**, 217 058n). 662

86MI 517-02 T. D. Truskanova, N. V. Vasil'ev, A. F. Kolomiets and G. A. Sokol'skii; *Zh. Vses. Khim. O.-va*, 1986, **31**, 594 (*Chem. Abstr.*, 1987, **107**, 39 704k). 666, 671, 692

B-86MI 518-01 C. Paulmier; "Selenium Reagents and Intermediates in Organic Synthesis," Pergamon, Oxford, 1986. 732

86MI 521-01 B. A. O'Brien and D. D. DesMarteau; *Rev. Chim. Min.*, 1986, **23**, 621 (*Chem. Abstr.*, 1987, **107**, 217 058n). 808, 861

86MI 521-02 H. Omichi, A. Katakai, T. Sugo and J. Okamoto; *Sep. Sci. Technol.*, 1986, **21**, 563. 864

B-86MI 522-01 M. Regitz and G. Maas; "Diazo Compounds: Properties and Synthesis," Academic Press, London, 1986. 865, 868, 874

B-86MI 522-02 M. Regitz and G. Maas; "Diazo Compounds: Properties and Synthesis," Academic Press, London, 1986, ch. 6, p. 221. 870

B-86MI 522-03 M. Regitz and G. Maas; "Diazo Compounds: Properties and Synthesis," Academic Press, London, 1986, ch. 8, p. 233. 870

B-86MI 522-04 M. Regitz and G. Maas; "Diazo Compounds: Properties and Synthesis," Academic Press, London, 1986, ch. 9, p. 257. 868

86MI 527-01 H. Chang and S. Kim; *Bull. Korean Chem. Soc.*, 1986, **7**, 407 (*Chem. Abstr.*, 1987, **107**, 58940). 1030

B-86MI 527-01 P. A. S. Smith and C. Paulmier; "Selenium Reagents and Intermediates in Organic Synthesis," Pergamon, Oxford, 1986, p. 70. 1051

86MI 530-01 R. Carrau, R. Freire, R. Hernandez and E. Suarez; *Synthesis—Stuttgart*, 1986, 1055. 1125

86MI 530-02 Y. Kobayashi, L. Minung, T. Kumamoto and K. Shirai; *Proc. Fac. Sci. Tokai Univ.*, 1986, **21**, 59. 1125

86MI 530-03 S. Kozuka, S. Tamura and W. Tagaki; *Mem. Fac. Eng., Osaka City Univ.*, 1986, **27**, 143 (*Chem. Abstr.*, 1988, **109**, 211 152h). 1144

86MI 531-01 T. Allspach; Dissertation, Universität Kaiserslautern, 1986. 1148

86OM178 Y. Stenstrom and W. M. Jones; *Organometallics*, 1986, **5**, 2155. 937

86OM336 D. Hedden, D. M. Roundhill, W. C. Fultz and A. L. Rheingold; *Organometallics*, 1986, **5**, 336. 385

86OM596 S. Antebi and H. Alper; *Organometallics*, 1986, **5**, 596. 238

86OM1369 B. Crociani, M. Sala, A. Polo and G. Bombieri; *Organometallics*, 1986, **5**, 1369. 801

86OM1976 G. Consiglio, F. Morandini, G. F. Ciani and A. Sironi; *Organometallics*, 1986, **5**, 1976. 940

86OPP269 W. A. Henderson and L. W. Chang; *Org. Prep. Proced. Int.*, 1986, **18**, 269. 316

86OPP369 S. Kato, T. Murai and M. Ishida; *Org. Prep. Proced. Int.*, 1986, **18**, 369. 472, 510

86POL677 P. Raj, K. Singhal and R. Rastogi; *Polyhedron*, 1986, **5**, 677. 1145

86POL1277 F. Borrul, J. Guasch, E. Casassas and V. Cerdá; *Polyhedron*, 1986, **5**, 1277. 864

86PS(26)1 T. Schafer, L. A. Suba, P. G. Ruminski and J. J. D'Amico; *Phosphorous Sulfur*, 1986, **29**, 1. 1121

86PS(26)169 J. H. Hillhouse, I. A. Blair and L. Field; *Phosphorus Sulfur*, 1986, **26**, 169. 610, 623, 624

86PS(27)289 K. Karaghiosoff, C. Cleve and A. Schmidpeter; *Phosphorus Sulfur*, 1986, **27**, 289. 635

86PS(27)327 C. T. Gokou, M. Chehna, J.-P. Pradere, G. Duguay and L. Toupet; *Phosphorus Sulfur*, 1986, **27**, 327. 633

86S31 T. Allspach, M. Regitz, G. Becker and W. Becker; *Synthesis*, 1986, 31. 839, 878, 879

86S128 H. Ishihara, S. Muto and S. Kato; *Synthesis*, 1986, 128. 240

86S142 Z. Machon and J. Cieplik; *Synthesis*, 1986, 142. 528, 529

86S437 D. N. Dhar and K. S. K. Murthy; *Synthesis*, 1986, 437. 344

86S482 J. E. Mills; *Synthesis*, 1986, 482. 591, 609

86S768 K. Oda, M. Machida and Y. Kanaoka; *Synthesis*, 1986, 768. 632

86S772 L. Bruché, L. Garanti and G. Zecchi; *Synthesis*, 1986, 772. 371

86S807 B. Stanovnik, V. Stibilj and M. Tišler; *Synthesis*, 1986, 803. 834

86S817 A. Vass and G. Szalontai; *Synthesis*, 1986, 817. 1039

86S843 P. Molina, M. Alajarin and A. Ferao; *Synthesis*, 1986, 843. 220, 862

86S968 A. Sawluk and J. Voss; *Synthesis*, 1986, 968. 554

86S970 H. J. Kim and Y. H. Kim; *Synthesis*, 1986, 970. 361

86S1021 W. J. Spillane and P. O. Burke; *Synthesis*, 1986, 1021. 342

86S1063 H. Takahata, K. Yamabe and T. Yamazaki; *Synthesis*, 1986, 1063. 607

86SC499 T. Amano, K. Yoshikawa, T. Sano, Y. Ohuchi, M. Shiono, M. Ishigaro and Y. Fujita; *Synth. Commun.*, 1986, **16**, 499. 30, 35

86SC763 C. J. Peake and J. H. Strickland; *Synth. Commun.*, 1986, **16**, 763. 664

86SC1423 M. Saroja and T. N. B. Kaimal; *Synth. Commun.*, 1986, **16**, 1423. 18

86SC1823 K. Mai and G. Patil; *Synth. Commun.*, 1986, **16**, 1823. 1122

86SR1 S. Sharma; *Sulfur Reports*, 1986, **5**, 1. 1027

86SRI113 G. S. Sanyal and A. B. Modak; *Synth. React. Inorg. Metal-Org. Chem.*, 1986, **16**, 113 (*Chem. Abstr.*, 1986, **104**, 6057). 864

86T739 A. Senning, H. C. Hansen, M. F. Abdel-Megeed, W. Mazurkiewicz and B. Jensen; *Tetrahedron*, 1986, **42**, 739. 483, 590

86T885 O. Piccolo, L. Filippini, E. Valoti, A. Citterio and L. Tinucci; *Tetrahedron*, 1986, **42**, 885. 301

86T1285 M. Feldhues and H. J. Schäfer; *Tetrahedron*, 1986, **42**, 1285. 208

86T1815 M. Böhshar, H. Heydt and M. Regitz; *Tetrahedron*, 1986, **42**, 1815. 871

86T2025 P. Metzner, T. N. Phan and J. Vialle; *Tetrahedron*, 1986, **42**, 2025. 553

86T2275 A. Robert, S. Jaquelin and J. L. Guinamant; *Tetrahedron*, 1986, **42**, 2275. 25, 56

86T2863 M. Isobe, Y. Ichikawa, Y. Funabashi, S. Mio and T. Goto; *Tetrahedron*, 1986, **42**, 2863. 152

86T2971 I. Minami, K. Takahashi, I. Shimizu, T. Kimura and J. Tsuji; *Tetrahedron*, 1986, **42**, 2971. 154

86T3029 R. K. Dieter; *Tetrahedron*, 1986, **42**, 3029. 547, 555

86T4577 L. Testaferri, M. Tiecco, M. Tingoli and D. Chianelli; *Tetrahedron*, 1986, **42**, 4577. 238

86T4593 T. Murai, S. Oida, S. Min and S. Kato; *Tetrahedron Lett.*, 1986, **27**, 4593. 245

86T4983 D. H. R. Barton, D. Bridon, Y. Hervé, P. Potier, J. Thierry and S. Z. Zard; *Tetrahedron*, 1986, **42**, 4990. 1115

86T5961 J. Mulzer and C. Brand; *Tetrahedron*, 1986, **42**, 5961. 138

86T5969 U. Hildebrand, J. Hübner and H. Budzikiewicz; *Tetrahedron*, 1986, **42**, 5969. 244, 246

86TL167 G. Opitz, K. Rieth and G. Rimmler; *Tetrahedron Lett.*, 1986, **27**, 167. 489, 490

86TL171 G. Märkl and H. Sejpka; *Tetrahedron Lett.*, 1986, **27**, 171. 1155, 1157

86TL501 J. Einhorn and J.-L. Luche; *Tetrahedron Lett.*, 1986, **27**, 501. 272, 301

86TL623 S. L. Brown, S. G. Davies, D. F. Foster, J. I. Seeman and P. Warner; *Tetrahedron Lett.*, 1986, **27**, 623. 453

86TL683 D. L. Boger and M. Patel; *Tetrahedron Lett.*, 1986, **27**, 683. 330

86TL919 T. Kometani, T. Fitz and D. S. Watt; *Tetrahedron Lett.*, 1986, **27**, 919. 306

86TL1411 H. Hiemstra, W. J. Klaver and W. N. Speckamp; *Tetrahedron Lett.*, 1986, **27**, 1411. 374

86TL1505 K. Kpegba, P. Metzner and R. Rakotonirina; *Tehrahedron Lett.*, 1986, **27**, 1505. 556

86TL1695 P. Andreoli, G. Cainelli, M. Contento, D. Giacomini, G. Martelli and M. Panunzio; *Tetrahedron Lett.*, 1986, **27**, 1695. 287

86TL1925 S. Kim and K. Y. Yi; *Tetrahedron Lett.*, 1986, 1925. 1078

86TL1995 H. J. Bestmann and T. Arenz; *Tetrahedron Lett.*, 1986, **27**, 1995. 896

86TL2005 T. Aoyama and T. Shioiri; *Tetrahedron Lett.*, 1986, **24**, 2005. 413, 873

86TL2527 P. C. B. Page and S. Rosenthal; *Tetrahedron Lett.*, 1986, **27**, 2527. 424

86TL2743 K. Hartke, W. Richter, W. Massa and G. Baum; *Tetrahedron Lett.*, 1986, **27**, 2743. 685

86TL2761 P. J. Garratt and A. Tsotinis; *Tetrahedron Lett.*, 1986, **27**, 2716. 199

86TL2913 M. Taddei and A. Mann; *Tetrahedron Lett.*, 1986, **27**, 2913. 199

86TL3095 H. Iwagami, S. R. Woulfe and M. J. Miller; *Tetrahedron Lett.*, 1986, **27**, 3095. 336

86TL3119 D. A. Evans and E. B. Sjogren; *Tetrahedron Lett.*, 1986, **27**, 3119. 329

86TL3239 M. Hasebe and T. Tsuchiya; *Tetrahedron Lett.*, 1986, **27**, 3239. 215

86TL3445 T. B. Rauchfuss and G. A. Zank; *Tetrahedron Lett.*, 1986, **27**, 3445. 513

86TL3787 S. G. Davies, I. M. Dordor-Hedgecock, K. H. Sutton and J. C. Walker; *Tetrahedron Lett.*, 1986, **27**, 3787. 453

86TL3907 M. Ubukata and K. Isono; *Tetrahedron Lett.*, 1986, **27**, 3907. 342

86TL3931 S. Cacchi, P. G. Ciattini, E. Morera and G. Ortar; *Tetrahedron Lett.*, 1986, **27**, 3931. 301

86TL4437 M. Kolb and B. Neises; *Tetrahedron Lett.*, 1986, **27**, 4437. 183

86TL4537 A. Abiko, J. C. Roberts, T. Takemasa and S. Masamune; *Tetrahedron Lett.*, 1986, 4537. 74

86TL4593 T. Murai, S. Oida, S. Min and S. Kato; *Tetrahedron Lett.*, 1986, **27**, 4593. 530, 540

86TL4609 H. Niwa, Y. Miyachi, Y. Uosaki, A. Kuroda, H. Ishiwata and K. Yamada; *Tetrahedron Lett.*, 1986, **27**, 4609. 199

86TL4623 F. Urpí and J. Vilarrasa; *Tetrahedron Lett.*, 1986, **27**, 4623. 234

86TL4701 D. J. Goldsmith and J. J. Soria; *Tetrahedron Lett.*, 1986, **27**, 4701. 702

86TL4749 Y. H. Kim, K. Kim and S. B. Shim; *Tetrahedron Lett.*, 1986, **27**, 4749. 368

86TL4933 W. K. Fife and Z.-D. Zhang; *Tetrahedron Lett.*, 1986, **27**, 4933. 184, 198

86TL4937 W. K. Fife and Z.-D. Zhang; *Tetrahedron Lett.*, 1986, **27**, 4937. 184

86TL4957 D. A. Evans, E. B. Sjogren, J. Bartroli and R. L. Dow; *Tetrahedron Lett.*, 1986, **27**, 4957. 329

86TL4961 D. A. Evans and E. B. Sjogren; *Tetrahedron Lett.*, 1986, **27**, 4961. 329

86TL4995 A. A. Galan, T. V. Lee and C. B. Chapleo; *Tetrahedron Lett.*, 1986, **27**, 4995. 223

86TL5147 L. Crombie, R. C. F. Jones and D. Haigh; *Tetrahedron Lett.*, 1986, **27**, 5147. 374

86TL5197 J. A. Marshall and R. C. Andrews; *Tetrahedron Lett.*, 1986, **27**, 5197. 207

86TL5285 R. C. F. Jones and A. D. Bates; *Tetrahedron Lett.*, 1986, **27**, 5285. 374, 375

86TL5405 T. Itoh, Y. Yonekawa, T. Sato and T. Fujisawa; *Tetrahedron Lett.*, 1986, **27**, 5405. 557

86TL5421 P. C. B. Page and S. Rosenthal; *Tetrahedron Lett.*, 1986, **27**, 5421. 423

86TL5683 D. A. Evans and R. P. Polniaszek; *Tetrahedron Lett.*, 1986, **27**, 5683. 229

86TL5723 B. Pellerin, J. M. Denis, J. Perrocheau and R. Carrie; *Tetrahedron Lett.*, 1986, **27**, 5723. 1153

86TL6099 N. X. Hu, Y. Aso, T. Otsubo and F. Ogura; *Tetrahedron Lett.*, 1986, 6099. 1078

86TL6339 F. Henin, J. Muzart and J.-P. Pete; *Tetrahedron Lett.*, 1986, **27**, 6339. 296

86ZAAC(533)55 W. Höneise, W. Schwarz, G. Heckmann and A. Schmidt; *Z. Anorg. Allg. Chem.*, 1986, **533**, 55. 776

86ZAAC(538)36 G. Rajca and J. Weidlein; *Z. Anorg. Allg. Chem.*, 1986, **538**, 36. 1082

86ZC179 S. Kato, T. Kakigano and M. Ishida; *Z. Chem.*, 1986, **26**, 179. 244

86ZC407 P. Mading; *Z. Chem.*, 1986, **26**, 407. 784, 785, 791

86ZN(B)859 H.-W. Hennig, P. Neu and P. Sartori; *Z. Naturforsch., Teil B.*, 1986, **41B**, 859. 220, 403

86ZN(B)1096 L. Weber and D. Bungardt, K. Reizig and R. Boese; *Z. Naturforsch., Teil B*, 1986, **41B**, 1096. 400

86ZN(B)1223 H.-D. Hausen, W. Schwarz, G. Rajca and J. Weidlein; *Z. Naturforsch., Teil B*, 1986, **41**, 1223. 1082

86ZN(B)1367 W. Einholz and W. Haubold; *Z. Naturforsch., Teil B*, 1986, **41**, 1367. 1085

86ZOB613 V. F. Pozdnev; *Zh. Obshch. Khim.* 1986, **56**, 613. 189

86ZOB1196 M. G. Voronkov and V. Y. Vvedenskii; *Zh. Obshch. Khim.*, 1986, **56**, 1196. 376

86ZOB1245 I. V. Konovalova, L. A. Burnaeva, E. A. Khusnutdinova and A. N. Pudovik; *Zh. Obshch. Khim.*, 1986, **56**, 1245 (*Chem. Abstr.*, 1987, **106**, 156 573g). 708

86ZOB1262 A. D. Sinitsa, D. M. Malenko, L. A. Repina, R. A. Loktionova and A. K. Shurubura; *Zh. Obshch. Khim.*, 1986, **56**, 1262 (*Chem. Abstr.*, 1987, **106**, 50 311e). 713

86ZOB1467 D. M. Malenko and A. D. Sinitsa; *Zh. Obshch. Khim.* (*Engl. Trans.*), 1986, **56**, 1467. 643

86ZOB1525 V. D. Sheludyakov, A. B. Lebedeva, A. V. Kisin, I. S. Nikishina, A. V. Lebedev and A. D. Kirilin; *Zh. Obshch. Khim.*, 1986, **56**, 1525. 1079

86ZOB1547 M. D. Mizhiritskii and V. O. Reikhsfel'd; *Zh. Obshch. Khim.*, 1986, **56**, 1547. (*Chem. Abstr.*, 1987, **106**, 137 656v). 583

86ZOB2359 M. G. Zimin, R. M. Kamalov, M. M. Afanas'ev and A. N. Pudovik; *Zh. Obshch. Khim.* (*Engl. Trans.*), 1986, **56**, 2359. 644

86ZOB2660 M. G. Zimin, N. G. Zabirov, R. M. Kamalov and A. N. Pudovik; *Zh. Obshch. Khim.*, 1986, **56**, 2660 (*Chem. Abstr.*, 1987, **107**, 236 844g). 581

86ZOB2681 A. D. Sinitsa, P. P. Onys'ko, T. V. Kim, E. I. Kiseleva and V. V. Pirozhenko; *Zh. Obshch. Khim.*, 1986, **56**, 2681 (*Chem. Abstr.*, 1988, **108**, 6088e). 659

86ZOB2802 O. B. Smolii, V. S. Brovarets and B. S. Drach; *Zh. Obshch. Khim.*, 1986, **56**, 2802 (*Chem. Abstr.*, 1987, **107**, 236 847k). 655

86ZOR100 G. S. Borovikova, E. S. Levchenko and E. I. Kaminskaya; *Zh. Org. Khim.*, 1986, **22**, 100. 337

86ZOR1099 M. V. Vovk, V. I. Dorokhov, V. V. Momot and L. I. Samarai; *Zh. Org. Khim.*, 1986, **22**, 1099 1076

86ZOR1189 G. G. Barashenkov and N. Y. Derkach; *Zh. Org. Khim.*, 1986, **22**, 1189. 341

86ZOR1654 D. M. Malenko and A. D. Sinitsa; *Zh. Org. Khim.*, 1986, **56**, 1654 (*Chem. Abstr.*, 1987, **106**, 176 507). 737

87ABC1345 T. Inagaki and H. Ueda; *Agric. Biol. Chem.*, 1987, **51**, 1345. 138

87ACS(B)29 A. Kjaer and T. Skrydstrup; *Acta Chem. Scand., Ser. B*, 1987, **41**, 29. 832

87ACS(B)157 T. Lundstedt, R. Carlson and R. Shabana; *Acta. Chem. Scand., Ser. B*, 1987, **41**, 157. 599

87ACS(B)164 R. Carlson and T. Lundstedt; *Acta. Chem. Scand., Ser. B*, 1987, **41**, 164. 599

87AG(E)359 V. T. Hassel and H. P. Müller; *Angew. Chem., Int. Edn. Engl.*, 1987, **26**, 359. 1070

87AG(E)362 G. Quinkert, N. Heim, J. Glenneberg, U. M. Billhardt, V. Autze, J. W. Bats and G. Dürner; *Angew. Chem., Int. Ed. Engl.*, 1987, **26**, 362. 174

87AG(E)546 H. Meyer, G. Baum, W. Massa, S. Berger and A. Berndt; *Angew. Chem., Int. Ed. Engl.*, 1987, **26**, 546. 928

87AG(E)776 A. Schäfer, M. Weidenbruch, W. Saak and S. Pohl; *Angew. Chem., Int. Ed. Engl.*, 1987, **29**, 776. 889

87AG(E)779 R. Appel, J. Hünerbein and N. Siabalis; *Angew. Chem., Int. Ed. Engl.*, 1987, **26**, 779. 397, 880

87AG(E)798 H. Meyer, G. Baum, W. Massa and A. Berndt; *Angew. Chem., Int. Ed. Engl.*, 1987, **26**, 798. 928

87AG(E)894 H. Eckert and B. Forster; *Angew. Chem., Int. Ed. Engl.*, 1987, **26**, 894. 972

87AG(E)1253 R. Lomölder and H. J. Schäfer; *Angew. Chem., Int. Ed. Engl.*, 1987, **26**, 1253. 207
87AJC701 S. A. Glover and A. L. J. Beckwith; *Aust. J. Chem.*, 1987, **40**, 701. 317, 698, 722
87AJC1919 M. J. Milewska and A. Chimiak; *Aust. J. Chem.*, 1987, **40**, 1919. 214
87AOC(27)51 H. P. Kim and R. J. Angelici; *Adv. Organomet. Chem.*, 1987, **27**, 51. 942
87AP(320)850 S. A. S. Ghozlan and E. A. A. Hafez; *Arch. Pharm. (Weinheim, Ger.)*, 1987, **320**, 850. 677
87AP(320)1088 J. A. Jensen and E. B. Pedersen; *Arch. Pharm. (Weinheim, Ger.)*, 1987, **320**, 1088. 1126
87BAU204 A. N. Ivanov, T. A. Epishina, T. V. Goreva, V. B. Sokolov and I. V. Martynov; *Bull. Acad. Sci., USSR. Div. Chem. Sci.*, 1987, 204. 669
87BAU1000 I. V. Martynov, A. N. Ivanov, T. A. Epishina and V. B. Sokolov; *Bull. Acad. Sci., USSR. Div. Chem. Sci.*, 1987, 1000. 669, 672
87BAU1314 V. B. Sokolov, A. N. Ivanov, T. A. Epishina and I. V. Martynov; *Bull. Acad. Sci. USSR. Div. Chem. Sci.*, 1987, 1314. 683
87BCJ1037 A. Ishii, R. Okazaki and N. Inamoto; *Bull. Chem. Soc. Jpn.*, 1987, **60**, 1037. 524
87BCJ2289 K. Nishiyama and M. Oba; *Bull. Chem. Soc. Jpn.*, 1987, **60**, 2289. 1042
87BCJ2463 O. Tsuge, Sh. Kanemasa, H. Suga and N. Nakagawa; *Bull. Chem. Soc. Jpn.*, 1987, **60**, 2463. 821
87BCJ2692 K. Nishiyama and M. Oba; *Bull. Chem. Soc. Jpn.*, 1987, **60**, 2692. 1106
87BCJ3307 T. Umemoto and Y. Gotoh; *Bull. Chem. Soc. Jpn.*, 1987, **60**, 3307. 205
87BCJ3462 M. Kosugi, H. Naka, H. Sano and T. Migita; *Bull. Chem. Soc. Jpn*, 1987, **60**, 3462. 468
87BCJ4325 H. Takeshita, A. Mori, T. Kusaba and H. Watanabe; *Bull. Chem. Soc. Jpn.*, 1987, **60**, 4325. 184
87BSB51 B. Ascherl, G. Kresze, J. L. Vaerman, B. Vandenbulcke-Coyette and H. G. Viehe; *Bull. Soc. Chim. Belg.*, 1987, **96**, 51. 670, 671
87BSB115 Chr. Becu and M. J. O. Anteunis; *Bull. Soc. Chim. Belg.*, 1987, **96**, 115. 1133
87BSF997 P. Babin, B. Clin and J. Dunogues; *Bull. Soc. Chim. Fr.*, 1987, 997. 373
87CAR(166)101 J. Kovacs, I. Pinter, A. Messmer, G. Toth and H. Duddeck; *Carbohydr. Res.*, 1987, **166**, 101. 1079
87CB39 H. Schmidbaur, R. Pichl and G. Müller; *Chem. Ber.*, 1987, **120**, 39. 226
87CB61 H. Gotthardt and W. Pflaumbaum; *Chem. Ber.*, 1987, **120**, 61. 561
87CB411 H. Gotthardt and W. Pflaumbaum; *Chem. Ber.*, 1987, **120**, 411. 550, 560
87CB429 A. Haas and W. Wanzke; *Chem. Ber.*, 1987, **120**, 429. 561
87CB461 W. P. Fehlhammer, F. Degel and G. Beck; *Chem. Ber.*, 1987, **120**, 461. 1087
87CB537 R. Aumann and H. Heinen; *Chem. Ber.*, 1987, **120**, 537. 946, 957
87CB657 W. Ried and U. Reiher; *Chem. Ber.*, 1987, **120**, 657. 718
87CB751 H. Gotthardt, T. Löhr and D. J. Brauer; *Chem. Ber.*, 1987, **120**, 751. 605
87CB789 H. Schmidbaur, R. Pichl and G. Müller; *Chem. Ber.*, 1987, **120**, 789. 1095
87CB1281 H. Schmidbaur and P. Nußstein; *Chem. Ber.*, 1987, **120**, 1281. 921, 1094
87CB1297 R. Aumann and H. Heinen; *Chem. Ber.*, 1987, **120**, 1297. 954
87CB1499 R. Henn, W. Sundermeyer and H. Pritzkow; *Chem. Ber.*, 1987, **120**, 1499. 476
87CB1597 W. Ried and U. Reiher; *Chem. Ber.*, 1987, **120**, 1597. 1119
87CB1645 W. Rösch, U. Hees and M. Regitz; *Chem. Ber.*, 1987, **120**, 1645. 1155
87CC83 R. Yamaguchi, K. Honda and M. Kawanisi; *J. Chem. Soc., Chem. Commun.*, 1987, 83. 146
87CC114 S. Ahmad and J. Iqbal; *J. Chem. Soc., Chem. Commun.*, 1987, 114. 264, 275
87CC311 H. Kogen and T. Nishi; *J. Chem. Soc., Chem. Commun.*, 1987, 311. 157
87CC559 H. Fischer, A. Tiriliomis, U. Gerbing, B. Huber and G. Müller; *J. Chem. Soc., Chem. Commun.*, 1987, 559. 605, 614, 627
87CC820 T. Severengiz and W.-W. du Mont; *J. Chem. Soc., Chem. Commun.*, 1987, 820. 255, 542
87CC980 A. R. Barron, A. H. Cowley and S. W. Hall; *J. Chem. Soc., Chem. Commun.*, 1987, 980. 878, 1155
87CC1233 T. G. Back and K. Brunner; *J. Chem. Soc., Chem. Commun.*, 1987, 1233. 335
87CC1301 T. Ando, J. H. Clark, D. G. Cork, M. Fujita and T. Kimura; *J. Chem. Soc., Chem. Commun.*, 1987, 1301. 1107
87CC1471 S.-I. Murahashi, T. Oda, T. Sugahara and Y. Masui; *J. Chem. Soc., Chem. Commun.*, 1987, 1471. 324
87CC1491 J. Barluenga, M. A. Rodríguez, P. J. Campos and G. Asensio; *J. Chem. Soc., Chem. Commun.*, 1987, 1491. 1044
87CC1523 M. Hirama, T. Iwakuma and S. Ito; *J. Chem. Soc., Chem. Commun.*, 1987, 1523. 152
87CC1627 H. Takahata, T. Takamatsu, M. Mozumi, Y. S. Chen, T. Yamazaki and K. Aoe; *J. Chem. Soc., Chem. Commun.*, 1987, 1627. 284
87CC1753 A. R. Barron, S. W. Hall and A. H. Cowley; *J. Chem. Soc. Chem. Commun.*, 1987, 1753. 396
87CCC2087 J. Mollin, K. Polaskova, Z. Odlerova and V. Bekarek; *Collect. Czech. Chem. Commun.*, 1987, **52**, 2087. 630
87CHE907 S. Tumkevicius and R. Matuliauskiene; *Chem. Heterocycl. Compd. (Engl. Transl.)*, 1987, 907. 690
87CJC2390 G. R. Pettit and S. B. Singh; *Can. J. Chem.*, 1987, **65**, 2390. 223
87CL293 T. Mukaiyama, H. Suzucki and T. Yamada; *Chem. Lett.*, 1987, 293. 237, 277
87CL475 T. Kakigano, T. Kanda, M. Ishida and S. Kato; *Chem. Lett.*, 1987, 475. 242, 250, 506, 509
87CL495 S. Kiyooka, F. Shiota and T. Shibuya; *Chem. Lett.*, 1987, 495. 432
87CL743 T. Mukaiyama, M. Tamura and S. Kobayashi; *Chem. Lett.*, 1987, 743. 236
87CPB364 K.-I. Tanaka, S. Yoshifuji and Y. Nitta; *Chem. Pharm. Bull.*, 1987, **35**, 364. 307
87CPB1262 T. Kitigawa, H. Kuroda and H. Sasaki; *Chem. Pharm. Bull.*, 1987, **35**, 1262. 189, 196
87CPB1741 K. Sakai; *Chem. Pharm. Bull.*, 1987, **35**, 1741. 138
87CPB1903 N. Ohi, B. Aoki, T. Shinozaki, K. Moro, T. Kuroki, T. Noto, T. Nehashi, M. Matsumoto, H. Okazaki and I. Matsunaga; *Chem. Pharm. Bull.*, 1987, **35**, 1903. 223
87CPB2698 T. Shioiri, M. Murata and Y. Hamada; *Chem. Pharm. Bull.*, 1987, **35**, 2698. 293

87CPB3139 H. Takahata, E.-C. Wang, T. Nakajima and T. Yamazaki; *Chem. Pharm. Bull.*, 1987, **35**, 3139. 597, 599

87CPB3453 T. Yamazaki, T. Itooka, M. Chintani, T. Momose, O. Muraoka and K. Matoba; *Chem. Pharm. Bull.*, 1987, **35**, 3453. 199

87CRV411 M. Brookhart and W. B. Studabaker; *Chem. Rev.*, 1987, **87**, 411. 942

87CZ61 H. W. Hennig, P. Neu and P. Sartori; *Chem.-Ztg.*, 1987, **111**, 61. 221

87DOK114 V. L. Foss, Yu. A. Veits, E. G. Neganova, A. A. Borisenko and I. F. Lutsenko; *Dokl. Akad. Nauk. SSSR*, 1987, **297**, 114 (*Chem. Abstr.*, 1987, **109**, 170 545t). 396, 399

87EUP209779 J. Kluth, K. H. Mueller and T. Pfister; *Eur. Pat.* 209 779 (1987) (*Chem. Abstr.*, 1987, **106**, 119 902t). 1120

87EUP243970 D. R. Baker and K. H. Brownell; *Eur. Pat.* 243 970 (1987) (*Chem. Abstr.*, 1988, **108**, 94 402). 738

87GSM130 D. W. Knight; *Gen. Synth. Methods*, 1987, **9**, 130. 24

87H(25)79 T. Toda, M. Yoshida, T. Katayama and M. Minabe; *Heterocycles*, 1987, **25**, 79. 226

87H(25)183 R. Okazaki, M. Unno and N. Inamoto; *Heterocycles*, 1987, **25**, 183. 635

87H(25)221 N. V. Shah and L. D. Cama; *Heterocycles*, 1987, **25**, 221. 334

87H(25)569 A. Tsolomitis and C. Sandris; *Heterocycles*, 1987, **25**, 569. 334

87H(26)163 A. Q. Hussein; *Heterocycles*, 1987, **26**, 163. 220

87H(26)359 S. Kano, T. Yokomatsu, H. Iwasawa and S. Shibuya; *Heterocycles*, 1987, **26**, 359. 570, 613

87H(26)437 P. A. van Elburg and D. N. Reinhoudt; *Heterocycles*, 1987, **26**, 437. 325

87H(26)1051 A. Kamal and P. B. Sattur; *Heterocycles*, 1987, **26**, 1051. 344

87H(26)2081 T. Otsuka and T. Tezuka; *Heterocycles*, 1987, **26**, 2081. 357

87H(26)359 S. Kano, T. Yokomatsu, H. Iwasawa and S. Shibuya; *Heterocycles*, 1987, **26**, 359. 284

87H(26)777 J. S. Shandu and B. Sain; *Heterocycles*, 1987, **26**, 777. 226

87HCA1001 P. Wipf, C. Jenny and H. Heimgartner; *Helv. Chim. Acta*, 1987, **70**, 1001. 575

87HCA1320 A. Griesbeck and D. Seebach; *Helv. Chim. Acta*, 1987, **70**, 1320. 161

87HCA1666 W. Oppolzer and J.-P. Barras; *Helv. Chim. Acta*, 1987, **70**, 1666. 342

87IJC154 S. P. Singh, S. Sehgal, L. Singh and S. N. Dhawan; *Indian J. Chem.*, 1987, **26**, 154. 1105

87IJC(A)146 K. Singhal, R. Rastogi and P. Raj; *Indian J. Chem., Sect. A*, 1987, **26**, 146. 372

87IJC(A)260 T. N. Srivasatva and J. D. Singh; *Indian J. Chem., Sect. A*, 1987, **26**, 260. 210

87IJC(A)1046 K. Singhal; *Indian J. Chem., Sect. A*, 1987, **26**, 1046. 391

87IJC(B)407 C. Someswara Rao, M. Rambabu and P. S. Srinivasan; *Indian J. Chem. Sect. B*, 1987, **26**, 407. 189

87IJC(B)620 B. E. V. S. Rao, S. S. Kuman and K. S. R. K. M. Rao; *Indian J. Chem. Sect. B*, 1987, **26**, 620. 215

87IZV165 V. A. Dorokhov, A. R. Amamchyan, M. N. Bochkareva, I. A. Teslya and Z. A. Starikova; *Izv. Akad. Nauk. SSSR, Ser. Khim.*, 1987, **1**, 165 (*Chem. Abstr.*, 1987, **107**, 115 616a). 778

87IZV226 A. N. Ivanov, T. A. Epishina, T. V. Goreva, V. B. Sokolov and I. V. Martynov; *Izv. Akad. Nauk SSSR, Ser. Khim.*, 1987, 226 (*Chem. Abstr.*, 1988, **108**, 6104g). 669

87IZV1086 I. V. Martynov, A. N. Ivanov, T. A. Epishina and V. B. Sokolov; *Izv. Akad. Nauk SSSR, Ser. Khim.*, 1987, 1086 (*Chem. Abstr.*, 1988, **109**, 6597r). 669, 672

87IZV1177 A. Yu. Aksinenko, A. N. Pushin, V. B. Sokolov, A. F. Gontar and I. V. Martynov; *Izv. Akad. Nauk SSSR, Ser. Khim.*, 1987, **5**, 1177. 1076

87IZV1422 V. B. Sokolov, A. N. Ivanov, T. A. Epishina and I. V. Martynov; *Izv. Akad. Nauk SSSR, Ser. Khim.*, 1987, 1422 (*Chem. Abstr.*, 1988, **109**, 23 034x). 683

87IZV1829 V. V. Tkachev, V. D. Sen and L. O. Atovmyan; *Izv. Akad. Nauk SSSR, Ser. Khim.*, 1987, **8**, 1829. 1070

87IZV2857 A. G. Korepin, S. A. Eksanov and L. T. Eremenko; *Izv. Akad. Nauk SSSR, Ser. Khim.*, 1987, 2857. 314

87JA913 P. R. Jones and T. F. Bates; *J. Am. Chem. Soc.*, 1987, **109**, 913. 926

87JA1564 G. Stork and S. D. Rychnovsky; *J. Am. Chem. Soc.*, 1987, **109**, 1564. 173

87JA1565 G. Stork and S. D. Rychnovsky; *J. Am. Chem. Soc.*, 1987, **109**, 1565. 173

87JA1584 J. B. Sheridan, G. L. Geoffroy and A. L. Rheingold; *J. Am. Chem. Soc.*, 1987, **109**, 1584. 451

87JA1798 I. Ojima, K. Nakahashi, S. M. Brandstadter and N. Hatanaka; *J. Am. Chem. Soc.*, 1987, **109**, 1798. 274, 286

87JA2049 B. K. Campion, J. Falk and T. D. Tilley; *J. Am. Chem. Soc.*, 1987, **109**, 2049. 422

87JA2426 J. Rebek, Jr., B. Askew, M. Killoran, D. Nemeth and F.-T. Lin; *J. Am. Chem. Soc.*, 1987, **109**, 2426. 151

87JA2553 E. Kaufmann, P. von Ragué Schleyer, S. Gronert, A. Streitwieser, Jr. and M. Halpern; *J. Am. Chem. Soc.*, 1987, **109**, 2553. 436

87JA2759 S. Iyer and L. S. Liebeskind; *J. Am. Chem. Soc.*, 1987, **109**, 2759. 457

87JA3402 K. Nomura, K. Okazaki, K. Hori and E. Yoshii; *J. Am. Chem. Soc.*, 1987, **109**, 3402. 301

87JA3953 J. Kress and J. A. Osborn; *J. Am. Chem. Soc.*, 1987, **109**, 3953. 932

87JA4731 S. L. Latesky and J. P. Selegue; *J. Am. Chem. Soc.*, 1987, **109**, 4731. 1097

87JA4837 E. J. O'Connor, M. Kobayashi, H. G. Floss and J. A. Gladysz; *J. Am. Chem. Soc.*, 1987, **109**, 4837. 451

87JA5145 B. J. Rappoli, M. R. Churchill, T. S. Janik, W. M. Rees and J. D. Atwood; *J. Am. Chem. Soc.*, 1987, **109**, 5145. 461

87JA5549 H. J. Reich and C. P. Jasperse; *J. Am. Chem. Soc.*, 1987, **109**, 5549. 334

87JA5844 C. C. Tso, A. R. Cutler and R. K. Kullnig; *J. Am. Chem. Soc.*, 1987, **109**, 5844. 457

87JA6532 W. R. Hertler, D. A. Dixon, E. W. Matthews, F. Davidson and F. G. Kitson; *J. Am. Chem. Soc.*, 1987, **109**, 6532. 1133

87JA7188 A. Sakamoto, Y. Yamamoto and J. Oda; *J. Am. Chem. Soc.*, 1987, **109**, 7188. 172

87JA7578 J. M. O'Connor, L. Pu and A. Rheingold; *J. Am. Chem. Soc.*, 1987, **109**, 7578. 944

87JA7888 Y. Ito, T. Matsuura and M. Murakami; *J. Am. Chem. Soc.*, 1987, **109**, 7888. 793, 796

87JAP6261987 J. Odera, M. Kawada and T. Bandai; *Jpn. Pat.* 6261987 (1987) (*Chem. Abstr.*, 1987, **107**, 78069). 411

87JAP6289642 T. Fuchigami, H. Urata and A. Murata (Sagami Chemical Research Center); *Jpn. Pat.* 6 289 642 (1987) (*Chem. Abstr.*, 1987, **107**, 153 937). 194

87JCS(D)2687 J. Barker, M. Kilner and R. O. Gould; *J. Chem. Soc., Dalton Trans.*, 1987, 2687. 781

87JCS(P1)845 A. Arrieta, B. Lecea and C. Palomo; *J. Chem. Soc., Perkin Trans. 1*, 1987, 845. 329

87JCS(P1)937 R. Carrau, R. Hernandez, E. Saurez and C. Betancor; *J. Chem. Soc., Perkin Trans. 1*, 1987, 937. 1119

87JCS(P1)1153 J. C. Brindley, J. M. Caldwell, G. D. Meakins, S. J. Plackett and S. J. Price; *J. Chem. Soc., Perkin Trans. 1*, 1987, 1153. 655

87JCS(P1)1389 M. Casey, C. J. Moody and C. W. Rees; *J. Chem. Soc., Perkin Trans. 1*, 1987, 1389. 655, 657

87JCS(P1)1635 A. R. Evans, R. Martin, G. A. Taylor and C. H. M. Yap; *J. Chem. Soc., Perkin Trans. 1*, 1987, 1635. 325

87JCS(P1)2163 J. H. Boyer, T. Manimaran and V. T. Ramakrishnan; *J. Chem. Soc., Perkin Trans. 1*, 1987, 2163. 963

87JCS(P1)2511 T. L. Gilchrist, R. C. Wasson, F. D. King and G. Wootton; *J. Chem. Soc., Perkin Trans. 1*, 1987, 2511. 362, 367

87JCS(P1)2531 M. Hori, T. Kataoka, H. Shimizu, E. Imai, K. Tanaka, K. Kimura, Y. Hashimoto, S. Inagaki, N. Goto and M. Kido; *J. Chem. Soc., Perkin Trans. 1*, 1987, 2531. 1122

87JCS(P1)2729 E. H. M. Abd Elall, M. I. Al Ashmawy and J. M. Mellor; *J. Chem. Soc., Perkin Trans. 1*, 1987, 2729. 1108

87JCS(P2)1489 B. C. Challis and J. N. Iley; *J. Chem. Soc., Perkin Trans. 2*, 1987, 1489. 370

87JFC(35)523 T. Nguyen and C. Wakselman; *J. Fluorine Chem.*, 1987, **35**, 523. 481, 487, 582

87JGU415 A. D. Sinitsa, T. V. Kolodka and A. K. Shurnbura; *J. Gen. Chem. USSR* (*Engl. Transl.*), 1987, **57**, 416. 655, 788, 790

87JGU1073 L. I. Belousova, I. D. Kalikhman, O. A. Vyazankina, E. I. Brodskaya and N. S. Vyazankin; *J. Gen. Chem. USSR* (*Engl. Transl.*), 1987, **57**, 1073. 718

87JGU1104 P. P. Onys'ko, T. V. Kim, E. I. Kiseleva and A. D. Sinitsa; *J. Gen. Chem. USSR* (*Engl. Transl.*), 1987, **57**, 1104. 657

87JGU1254 A. A. Prishchenko, M. V. Livantsov and A. F. Lutsenko; *J. Gen. Chem. USSR*, 1987, **57**, 1254–1255. 401

87JGU1349 E. A. Suvalova, T. I. Chudakova, P. P. Onys'ko and A. D. Sinitsa; *J. Gen. Chem. USSR* (*Engl. Transl.*), 1987, **57**, 1349. 660, 788

87JGU1478 V. B. Sokolov, T. A. Epishina, A. N. Ivanov, A. V. Kharitonov, V. K. Brel and I. V. Martynov; *J. Gen. Chem. USSR* (*Engl. Transl.*), 1987, **57**, 1478. 669

87JGU1479 V. B. Sokolov, A. N. Ivanov, T. A. Epishina and I. V. Martynov; *J. Gen. Chem. USSR* (*Engl. Transl.*), 1987, **57**, 1479. 669

87JGU1710 Yu. M. Dmitrichenko, V. G. Rozinov, I. V. Donskikh, G. V. Dolgushin, V. P. Feshin and A. V. Kalabina; *J. Gen. Chem. USSR* (*Engl. Transl.*), 1987, **57**, 1710. 656

87JGU2350 Z. S. Novikova, E. A. Monin, A. A. Borisenko, K. S. Zavadsku and I. F. Lutsenko; *J. Gen. Chem. USSR* (*Engl. Trans 1.*), 1987, **57**, 2350. 1092

87JHC325 C. K. Kim, F. Debellis and C. A. Maggiulli; *J. Heterocycl. Chem.*, 1987, **24**, 325. 219, 694

87JHC577 H. M. Hassaneen, A. M. Farag, A. S. Shawali and M. S. Algharib; *J. Heterocycl. Chem.*, 1987, **24**, 577. 703

87JHC597 Y. Kang, R. Soyka and W. Pfleiderer; *J. Heterocycl. Chem.*, 1987, **24**, 597. 215

87JHC633 G. Menozzi, A. Bargagna, L. Mosti and P. Schenone; *J. Heterocycl. Chem.*, 1987, **24**, 633. 501

87JHC773 R. Fusco, A. Marchesini and F. Sannicolò; *J. Heterocycl. Chem.*, 1987, **24**, 773. 354

87JHC789 F. Z. Basha and J. F. De Bernardis; *J. Heterocycl. Chem.*, 1987, **24**, 789. 687

87JHC1391 K. Tanaka, O. Honda, K. Minoguchi and K. Mitsuhashi; *J. Heterocycl. Chem.*, 1987, **24**, 1391. 681

87JMC24 J.-C. Aloup, J. Bouchandon, D. Farge, C. James, J. Deregnaucourt and M. Hardy-Houis; *J. Med. Chem.*, 1987, **30**, 24. 558

87JMC498 D. E. Butler, J. D. Leonard, B. W. Caprathe, Y. J. L'Italien, M. R. Pavia, F. M. Hershenson, P. H. Poschel and J. G. Marriott; *J. Med. Chem.*, 1987, **30**, 498. 306

87JMC951 C. B. Chapleo, P. L. Myers, A. C. B. Smith, I. F. Tulloch and D. S. Walter; *J. Med. Chem.*, 1987, **30**, 951. 1120

87JOC169 N. Z. Huang, M. V. Lakshmikantham and M. P. Cava; *J. Org. Chem.*, 1987, **52**, 169. 336

87JOC275 K. Mai and G. Patil; *J. Org. Chem.*, 1987, **52**, 275. 1076

87JOC476 I. M. Lazbin and G. F. Koser; *J. Org. Chem.*, 1987, **52**, 476. 321

87JOC681 T. Ando, J. H. Clark, D. G. Cork, M. Fujita and T. Kimura; *J. Org. Chem.*, 1987, **52**, 681. 1105

87JOC689 F. X. Webster, J. Rivas-Enterrios and R. M. Silverstein; *J. Org. Chem.*, 1987, **52**, 689. 53

87JOC1156 E. K. Moltzen, M. P. Kramer, A. Senning and K. J. Klabunde; *J. Org. Chem.*, 1987, **52**, 1156. 550

87JOC2018 A. Toshimitsu, K. Terao and S. Uemura; *J. Org. Chem.*, 1987, **52**, 2018. 285

87JOC2033 T. Högberg, P. Ström, M. Ebner and S. Rämsby; *J. Org. Chem.*, 1987, **52**, 2033. 277

87JOC2327 K. Yamada, K. Kishikawa and M. Yamamoto; *J. Org. Chem.*, 1987, **52**, 2327. 325

87JOC2442 J. D. Friedrich; *J. Org. Chem.*, 1987, **52**, 2442. 630

87JOC2523 O. Tsuge, S. Kanemasa, T. Yamada and K. Matsuda; *J. Org. Chem.*, 1987, **52**, 2523. 729

87JOC3263 Y. C. Xu and W. D. Wulff; *J. Org. Chem.*, 1987, **52**, 3263. 955, 956

87JOC3468 M. C. Wani, D. H. Rector and C. E. Cook; *J. Org. Chem.*, 1987, **52**, 3468. 131

87JOC3477 P. E. Sonnet; *J. Org. Chem.*, 1987, **52**, 3477. 138

87JOC3777 K. Luthman, M. Orbe, T. Waglund and A. Claesson; *J. Org. Chem.*, 1987, **52**, 3777. 132

87JOC3784 J. Mulzer, L. Autenrieth-Ansorge, H. Kirstein, T. Matsuoka and W. Münch; *J. Org. Chem.*, 1987, **52**, 3784. 131

87JOC3937 F. Vedejs, S. Ahmad, S. D. Larsen and S. Westwood; *J. Org. Chem.*, 1987, **52**, 3937. 177

87JOC4319 S. I. Murahashi, T. Naota, K. Ito, Y. Maeda and H. Taki; *J. Org. Chem.*, 1987, **52**, 4319. 171

87JOC4379 D. D. Davey; *J. Org. Chem.*, 1987, **52**, 4379. 275

87JOC4471 G. Rajendra and M. J. Miller; *J. Org. Chem.*, 1987, **52**, 4471. 285, 322

87JOC4531 H.-O. Kim, R. K. Olsen and O.-S. Choi; *J. Org. Chem.*, 1987, **52**, 4531. 237, 277

87JOC4570 M. E. Jung, D. D. Grove and S. I. Khan; *J. Org. Chem.*, 1987, **52**, 4570. 826, 829, 830

87JOC4665 D. J. Hart, W.-P. Hong and L.-Y. Hsu; *J. Org. Chem.*, 1987, **52**, 4665. 277

87JOC4875 T. N. Wheeler and J. Ray; *J. Org. Chem.*, 1987, **52**, 4875. 989

87JOC4933 A. C. Huggett, J. L. Cone, S. S. Thorgeirsson and P. P. Roller; *J. Org. Chem.*, 1987, **52**, 4933. 214

87JOC5378 R. D. Clark and Jahangir; *J. Org. Chem.*, 1987, **52**, 5378. 301

87JOC5745 R. B. Ruggeri and C. H. Heathcock; *J. Org. Chem.*, 1987, **52**, 5745. 277

87JOM(137)229 F. Caballero and P. Royo; *J. Organomet. Chem.*, 1977, **137**, 229. 1137

87JOM(325)185 D. Barratt, S. J. Davies, G. P. Elliott, J. A. K. Howard, D. B. Lewis and F. G. A. Stone; *J. Organomet. Chem.*, 1987, **325**, 185. 927

87JOM(327)303 G. A. Razuvaev, A. S. Gordestov, A. P. Kozina, T. N. Brevnova, V. V. Seminov, S. E. Skobeleva, N. A. Boxer and Yu. I. Dergunov; *J. Organomet. Chem.*, 1987, **327**, 303. 1076, 1077

87JOM(327)339 A. Parlier, H. Rudler, J. C. Daran, C. Alvarez and F. D. Reyes; *J. Organomet. Chem.*, 1987, **327**, 339. 798

87JOM(327)429 E. Hernandez and H. Hoberg; *J. Organomet. Chem.*, 1987, **327**, 429. 386

87JOM(328)403 E. Hernandez and H. Hoberg; *J. Organomet. Chem.*, 1987, **328**, 403. 386

87JOM(331)161 R. T. Boere, R. T. Oakley and R. W. Reed; *J. Organomet. Chem.*, 1987, **331**, 161. 777, 778

87JOM(331)193 W. P. Fehlhammer, G. Zinner and M. Bakola-Christanopoulou; *J. Organomet. Chem.*, 1987, **331**, 193. 1089

87JOM(332)69 M. S. Samples and C. H. Yoder; *J. Organomet. Chem.*, 1987, **332**, 69. 379

87JOM(332)105 H. Fischer, U. Gerbing and A. Tiriliomis; *J. Organomet. Chem.*, 1987, **332**, 105. 627

87JOM(332)371 M. Backmann, I. Hawkins and M. P. Sloan; *J. Organomet. Chem.*, 1987, **332**, 371. 800

87JOM(333)245 A. Parlier, H. Rudler, J. C. Daran and C. Alvarez; *J. Organomet. Chem.*, 1987, **333**, 245. 798, 952

87JOM(334)C43 H. Takahashi, K. Ohe, S. Uemura and N. Sugita, *J. Organomet. Chem.*, 1987, **334**, C43. 242

87JOM(334)C45 H. Takahashi, K. Ohe, S. Uemura and N. Sugita; *J. Organomet. Chem.*, 1987, **334**, C45. 241

87JOM(334)255 E. I. Lindstedt and M. Nilsson; *J. Organomet. Chem.*, 1987, **334**, 255. 652

87JOM(338)329 E. P. O. Fuchs, M. Hermesdorf, W. Schnurr, W. Rösch, H. Heydt and M. Regitz; *J. Organomet. Chem.*, 1987, **338**, 329. 1152, 1159

87JOU75 E. S. Levchenko, T. N. Dubinina, S. V. Sereda, M. Yu. Antipin, Yu. T. Struchkov and I. E. Boldeshul; *J. Org. Chem. USSR* (*Engl. Transl.*), 1987, **23**, 75. 703

87JPR1039 H. Ripperger; *J. Prakt. Chem.*, 1987, **329**, 1039. 574

87JSP304 M. C. Durrant, H. W. Kroto and D. R. M. Walton; *J. Mol. Spectrosc.*, 1987, **121**, 304. 1130

87JST381 R. Fausto and J. J. C. Teixeira-Dias; *J. Mol. Struct.* 1987, **150**, 381. 507

87KGS845 V. A. Bakulev, E. F. Dankova, V. S. Mokrushin, E. O. Sidorov and A. T. Lebedev; *Khim. Geterotsikl. Soedin.*, 1987, 845 (*Chem. Abstr.*, 1988, **108**, 150 373m). 567, 571

87KGS1131 S. Tumkevicius and R. Matulianskiene; *Khim Geterotsikl. Soedin.*, 1987, 1131 (*Chem. Abstr.*, 1988, **108**, 167 254w). 690

87LA189 H. Gotthardt and T. Löhr; *Liebigs Ann. Chem.*, 1987, 189. 605, 623

87LA483 W. Gruszecki, M. Gruszecka and H. Bradaczek, *Liebigs Ann. Chem.*, 1987, 483. 716

87LA687 Th. Oesterle and G. Simchen; *Liebigs Ann. Chem.*, 1987, 687. 718

87LA1101 P. B. Sulay and I. C. Ivanov; *Liebigs Ann. Chem.*, 1987, 1101. 599

87M601 B. Stanovnik, U. Urleb and M. Tišler; *Monatsh. Chem.*, 1987, **118**, 601. 836

B-87MI 503-01 O. W. Webster and G. D. Y. Sogah; in "Recent Advances in Mechanistic and Synthetic Aspects of Polymerization," eds. M. Fantaville and A. Guyrt, Reidel, Dordrecht, 1987, p. 3. 152

87MI 504-01 D. Yamashiro; *Int. J. Peptide and Protein Res.*, 1987, **30**, 9. 185

87MI 504-02 J. M. Tronchet, N. Bizzozero, F. Barbalat-Rey and M. Geoffroy; *J. Carbohydrate Chem.*, 1987, **6**, 155. 211

B-87MI 507-01 G. Kirsch and L. Christiaens; "Compounds with Se–N and Te–N bonds," Wiley, New York, 1987, p. 421. 333

87MI 507-02 P. S. R. Anjaneyulu and J. V. Staros; *Int. J. Pept. Protein Res.*, 1987, **30**, 117. 332

87MI 507-03 K. Sreekumar and V. N. R. Pillai; *Polymer*, 1987, **28**, 1599. 317

B-87MI 514-01 F. S. Guziec, Jr; in "The Chemistry of Organic Selenium and Tellurium Compounds," ed. S. Patai, Wiley, New York, 1987, vol. 2, p. 234. 624

B-87MI 515-01 F. S. Guziec Jr.; in "The Chemistry of Organic Selenium and Tellurium Compounds", ed. S. Patai, Wiley, New York, 1987, vol. 2, p. 238. 631

87MI 517-01 B. Wegmann and R. R. Schmidt; *J. Carbohydr. Chem.*, 1987, **6**, 357. 692

87MI 517-02 L. I. Kas'yan, M. F. Galafeeva, G. B. Levashova and V. Trachevskii; *Vopr. Khim. Khim. Tekhnol.*, 1987, **83**, 70 (*Chem. Abstr.*, 1989, **110**, 7962p). 709

87MI 519-01 C. Soula, A. Marsura and C. Luu-Duc; *J. Pharm. Belg.*, 1987, **42**, 293 (*Chem. Abstr.*, 1988, **109**, 109 969t). 749

87MI 519-02 M. A. Glushkova, M. A. Chumaevskii, L. V. Khmelevskaya, M. M. Ershova and Y. A. Buslaev; *Koord. Khim.*, 1987, **13**, 349 (*Chem. Abstr.*, 1987, **108**, 55 482). 781

B-87MI 521-01 B. Stanovnik; in "Organic Synthesis, Modern Trends," ed. O. Chizhov, Blackwell Scientific, Oxford, 1987, p. 315. 835

87MI 521-02 B. Stanovnik; *Ö. Chem. Z.*, 1987, **88**, 216. 835

87MI 521-03 H. Omichi, A. Katakai, T. Sugo, J. Okamoto, S. Katoh, K. Sakane, K. Sugasake and T. Itagaki; *Sep. Sci. Technol.*, 1987, **22**, 1313. 864

B-87MI 527-01 F. S. Guziec, Jr.; in "The Chemistry of Organic Selenium and Tellurium Compounds," ed. S. Patai, Wiley, Chichester, 1987, vol. 2, p. 245. 1051

B-87MI 527-02 A. Toshimitsu and S. Uemura; in "The Chemistry of Organic Selenium and Tellurium Compounds," ed. S. Patai, Wiley, Chichester, 1987, vol. 2, p. 541. 1051

87MI 528-01 W. Daoquan, L. Liangzhu and Z. Pang; *Scienta Sinica* (*B*), 1987, **30**, 449. 1078

B-87MI 530-01 A. Toshimitsu and S. Uemura; in "The Chemistry of Organic Selenium and Tellurium Compounds," ed. S. Patai, 1987, Wiley, New York, Vol. 2, p. 541. 1113

87MIP8707604 H. F. Deluca, A. Kutner and H. K. Schnoes; *World Pat.* 8 707 604 (1987) (*Chem. Abstr.*, 1989, **110**, 115 144). 185

87NKK1280 S. Eguchi, H. Takeuchi and N. Watanabe; *Nippon Kagaku Kaishi*, 1987, **7**, 1280. 1066

87NKK1304 Y. Matusbara, T. Ban, T. Nakamura, M. Yoshihara and T. Maeshima; *Nippon Kagaku Kaishi* (*J. Chem. Soc. Jpn.*), 1987, 1304 (*Chem. Abstr.*, 1988, **108**, 131 138x). 694

87NKK1414 T. Itoh, T. Sato and T. Fujisawa; *Nippon Kagaku Kaishi*, 1987, 1414. 557

87NKK1430 Y. Hirabayashi, H. Ishihara and E. Echigo; *Nippon Kagaku Kaishi* (*J. Chem. Soc. Jpn.*), 1987, 1430. 233, 506, 509

87NKK1442 T. Toru, Y. Yamada, T. Oshima, S. Imao and E. Maekawa; *Nippon Kagaku Kaishi* (*J. Chem. Soc. Jpn.*), 1987, 1442. 248, 249

87NKK1457 S. Kato, Y. Ono, T. Miyazaki, Y. Itoh, T. Aoyama, M. Ishida and T. Murai; *Nippon Kagaku Kaishi* (*J. Chem. Soc. Jpn.*), 1987, 1457. 252

87NKK1475 Y. Hirabayashi, H. Ishihara, Y. Ogi, T. Miyake, H. Oida, N. Kondo and H. Goji; *Nippon Kagaku Kaishi* (*J. Chem. Soc. Jpn.*), 1987, 1475. 253

87NKK1479 Y. Hirabayashi, H. Ishihara, Y. Ogi, T. Miyake, H. Oida, N. Kondo and H. Goji; *Nippon Kagaku Kaishi* (*J. Chem. Soc. Jpn.*), 1987, 1479. 249

87OM974 K. Yamamoto, A. Hayashi, S. Suzuki and J. Tsuji; *Organometallics*, 1987, **6**, 974. 419

87OM988 C. Knors, G. F.-H. Kuo, J. W. Lauher, C. Eigenbrot and P. Helquist; *Organometallics*, 1987, **6**, 988. 951

87OM1246 A. G. Brook, W. J. Chatterton, J. F. Sawyer, D. W. Hughes and K. Vorspohl; *Organometallics*, 1987, **6**, 1246. 924

87OM1424 K. H. Dötz, W. Sturm and H. G. Alt; *Organometallics*, 1987, **6**, 1424. 944

87OM1839 M. F. Semmelhack and G. R. Lee; *Organometallics*, 1987, **6**, 1839. 444, 937

87OM2128 A. G. Brook, K. Vorspohl, R. R. Ford, M. Hesse and W. J. Chatterton; *Organometallics*, 1987, **6**, 2128. 924

87OM2249 A. G. M. Barrett and N. E. Carpenter; *Organometallics*, 1987, **6**, 2249. 937

87OM2270 W. Ando, K. Hagiwara and A. Sekiguchi; *Organometallics*, 1987, **6**, 2270. 1020

87OM2337 A. Sekiguchi, T. Sato and W. Ando; *Organometallics*, 1987, **6**, 2337. 925

87OM2617 G. R. Gillette, G. H. Noren and R. West; *Organometallics*, 1987, **6**, 2617. 1020

87OPP52 Y. Akcamur and G. Kollenz; *Org. Prep. Proceed. Int.*, 1987, **19**, 52. 217

87OS(65)47 M. Olomucki and T. Y. Le Gall; *Org. Synth.*, 1987, **65**, 47. 103

87OS(65)236 B. Giese, J. Dupuis and M. Nix; *Org. Synth.*, 1987, **65**, 236. 152

87OS(66)29 D. E. Vogel and G. H. Buchi; *Org. Synth.*, 1987, **66**, 29. 105

87OS(66)37 D. Spitzner and A. Engler; *Org. Synth.*, 1987, **66**, 37. 62

87OS(66)87 I. Kuwajima and H. Urabe; *Org. Synth.*, 1987, **66**, 87. 12, 96

87OS(66)121 T. Fujisawa and T. Sato; *Org. Synth.*, 1987, **66**, 121. 14

87OS(66)151 B. Koppenhoefer and V. Schurig; *Org. Synth.*, 1987, **66**, 151. 60

87OS(66)173 O. Hormi; *Org. Synth.*, 1987, **66**, 173. 55, 99

87OS(66)180 D. Bankston; *Org. Synth.*, 1987, **66**, 180. 54

87OS(66)220 J. Villieras and M. Rambaud; *Org. Synth.*, 1987, **66**, 220. 103

87PS(29)219 J. J. D'Amico, L. Suba, T. Schafer and P. G. Ruminski; *Phosphorous Sulfur*, 1987, **29**, 219. 1119, 1121

87PS(30)479 M. Regitz, W. Rösch, T. Allspach, U. Annen, K. Blatter, J. Fink, M. Hermesdorf, H. Heydt, U. Vogelbacher and O. Wagner; *Phosphorus Sulfur*, 1987, **30**, 479. 1155

87PS(30)463 R. H. Neilson, B. A. Boyd, D. A. Dubois, R. Hani, G. M. Scheide, J. T. Shore and U. G. Wettermark; *Phosphorus Sulfur*, 1987, **30**, 463. 887

87PS(31)33 U. Kunze and V. Kruppa; *Phosphorus Sulfur*, 1987, **31**, 33. 566, 595, 610

87PS(32)51 N. M. Yousif and M. A. Salama; *Phosphorus Sulfur*, 1987, **32**, 51. 590, 619

87RCR1125 T. D. Petrova and V. E. Platonov; *Russ. Chem. Rev.*, 1987, **56**, 1125. 654

87S45 P. Molina, P. M. Fresneda and F. Hurtado; *Synthesis*, 1987, 35. 1066

87S85 A. J. Fatiadi; *Synthesis*, 1987, 85. 48

87S162 O. P. Goel and U. Krolls; *Synthesis*, 1987, 162. 569, 570

87S183 T. Okawara, Y. Kanazawa, T. Yamasaki and M. Furukawa; *Synthesis*, 1987, 183. 353

87S228 R. M. Mohareb, A. Habashi, N. S. Ibrahim and S. M. Sherif; *Synthesis*, 1987, 228. 631

87S262 M. Barreau and G. Ponsinet; *Synthesis*, 1987, 262. 688

87S306 P. Legrel, M. Baudy-Floc'h and A. Robert; *Synthesis*, 1987, 306. 652

87S363 A. Couture, P. Grandclaudon and E. Huguerre; *Synthesis*, 1987, 363. 516, 522

87S371 H. Ishihara, N. Matsunami and Y. Yamada; *Synthesis*, 1987, 371. 250

87S418 L. Nakonieczna and A. Chimiak; *Synthesis*, 1987, 418. 323

87S421 A. L. Campbell and G. R. Lenz; *Synthesis*, 1987, 421. 288, 290, 297, 302

87S517 L. Meier and H. D. Scharf; *Synthesis*, 1987, 517. 206

87S520 Z. M. Jászay, I. Petneházy, L. Töke and B. Szajáni; *Synthesis*, 1987, 520. 1065

87S525 G. L'abbé; *Synthesis*, 1987, 525. 962, 1022

87S547 D. Pooranchand, H. Ila and H. Junjappa; *Synthesis*, 1987, 547. 731

87S628 R. Madronero and S. Vega; *Synthesis*, 1987, **7**, 628. 655

87S689	W. Rösch and M. Regitz; *Synthesis*, 1987, 689.	839
87S801	V. Jäger and B. Häfele; *Synthesis*, 1987, 801.	223
87S831	R. V. Hoffman and G. A. Buntain, *Synthesis*, 1987, 831.	219
87S835	P. Beslin, A. Dlubala and G. Levesque; *Synthesis*, 1987, 835.	550
87S850	J. Oren, M. Demuth and B. Fuchs; *Synthesis*, 1987, 850.	199
87S907	H. Boyer and T. Manimaran; *Synthesis*, 1987, 907.	963
87S931	E. Anders, A. Stankowiak and R. Riemer; *Synthesis*, 1987, 931.	226
87S1122	D. N. Harpp, M. Gingras, T. Aida and T. H. Chan; *Synthesis*, 1987, 1122.	242, 337
87SC409	R. M. Moriarty, I. Prakash and R. Penmasta; *Synth. Commun.*, 1987, **17**, 409.	366
87SC657	A. M. McKillop and S. P. Brown; *Synth. Commun.*, 1987, **17**, 657.	718
87SC1677	C.-H. Kwon and H. T. Nagasawa; *Synth. Commun.*, 1987, **17**, 1677.	1120
87SC1741	R. A. Rebelo, M. C. Rezende, F. Nome and C. Zucco; *Synth. Commun.*, 1987, **17**, 1741.	652
87SRI525	M. K. Rastogi; *Syn. React. Inorg. Metal-Org. Chem.*, 1987, **17**, 525.	1132
87SUL105	B. Y. Riad, S. E. Abdou, F. A. Attaby and S. A. Mansour; *Sulfur Letters*, 1987, **6**, 105.	615
87SUL153	E. A. Laifa, M. F. Abdel-Mageed, A. B. A. G. Ghattas, B. Jensen and A. Senning; *Sulfur Letters*, 1987, **5**, 153.	478, 494, 582
87T215	M. S. Baird and H. H. Hussain; *Tetrahedron*, 1987, **43**, 215.	873
87T2045	H. W. Hoppe, M. Kaiser, D. Müller and P. Welzel; *Tetrahedron*, 1987, **43**, 2045.	223
87T2557	P. Ballesteros, R. M. Claramunt and J. Elguero; *Tetrahedron*, 1987, **43**, 2557.	327
87T3395	M. Fujii, K. Ozaki, M. Sekine and T. Hata; *Tetrahedron*, 1987, **43**, 3395–3407.	403
87T4185	A. Maquestiau and J. J. Vanden Eynde; *Tetrahedron*, 1987, **43**, 4185.	773
87T4249	O. A. Attanasi, M. Grossi, F. Serra-Zanetti and E. Foresti; *Tetrahedron*, 1987, **43**, 4249.	354
87T4785	E. Rossi, R. Stradi and A. Benedusi; *Tetrahedron*, 1987, **43**, 4785.	762
87T5095	A. I. Meyers, D. A. Dickman and M. Boes; *Tetrahedron*, 1987, **43**, 5095.	747
87T5341	M. Akhtar and D. Gani; *Tetrahedron*, 1987, **43**, 5341.	207
87TL59	C. Belaud and J. Villieras; *Tetrahedron Lett.*, 1987, **28**, 59.	296, 297
87TL285	L. N. Jungheim, S. K. Sigmund and J. W. Fisher; *Tetrahedron Lett.*, 1987, **28**, 285.	363
87TL325	Y. Tamaru, M. Hojo and Z. Yoshida; *Tetrahedron Lett.*, 1987, **28**, 325.	170
87TL341	J. Garcia and J. Vilarrasa; *Tetrahedron Lett.*, 1987, **28**, 341.	353, 354
87TL409	S. Berrada and P. Metzner; *Tetrahedron Lett.*, 1987, **28**, 409.	556
87TL655	K. Nishitani and K. Yamakawa; *Tetrahedron Lett.*, 1987, **28**, 655.	166
87TL913	A. Padwa, D. N. Kline and J. Perumattam; *Tetrahedron Lett.*, 1987, **28**, 913.	425
87TL1281	N. X. Hu, Y. Aso, T. Otsubo and F. Ogura; *Tetrahedron Lett.*, 1987, **28**, 1281.	210
87TL1817	M. J. Milewska and A. Chimiak; *Tetrahedron Lett.*, 1987, **28**, 1817.	214
87TL1945	F. P. Cossio, I. Ganboa, J. M. Garcia, B. Lecea and C. Palomo; *Tetrahedron Lett.*, 1987, **28**, 1945.	294
87TL2087	G. Stork and N. A. Saccomano; *Tetrahedron Lett.*, 1987, **28**, 2087.	152
87TL2135	J. C. Carretero, S. De Lombaert and L. Ghosez; *Tetrahedron*, 1987, **28**, 2135.	166
87TL2171	D. W. Brown, M. M. Campbell, M. S. Chambers and C. V. Walker; *Tetrahedron Lett.*, 1987, **28**, 2171.	573
87TL2265	H. Muratake and M. Natsume; *Tetrahedron Lett.*, 1987, **28**, 2265.	572
87TL2595	P. Duhamel, B. Goument and J. C. Plaquevent; *Tetrahedron Lett.*, 1987, **28**, 2595.	283
87TL3019	S. V. Ley and P. R. Woodward; *Tetrahedron Lett.*, 1987, **28**, 3019.	277
87TL3261	J. Kang, J. H. Lee, K. S. Kim, J. U. Jeong and C. Pyun; *Tetrahedron Lett.*, 1987, **28**, 3261.	418, 428
87TL3377	M. Wada, H. Aiura and K. Akiba; *Tetrahedron Lett.*, 1987, **28**, 3377.	287
87TL3815	Y. Zhang, J. Jiang and Y. Chen; *Tetrahedron Lett.*, 1987, **28**, 3815.	272, 293
87TL4093	A. Ricci, A. Degl'Innocenti, G. Borselli and G. Reginato; *Tetrahedron Lett.*, 1987, **28**, 4093.	416, 423
87TL4169	M. Vaultier, F. Truchet, B. Carboni, R. W. Hoffmann and I. Denne; *Tetrahedron Lett.*, 1987, **28**, 4169.	199
87TL4189	S. Harusawa, R. Yoneda, Y. Omori and T. Kurihara; *Tetrahedron Lett.*, 1987, **28**, 4189.	1111, 1116
87TL4319	J. E. Baldwin, S. C. Ng and A. J. Pratt; *Tetrahedron Lett.*, 1987, **28**, 4319.	346
87TL4441	J. Bergman and B. Pelcman; *Tetrahedron Lett.*, 1987, **28**, 4441.	199
87TL4569	H. Tone, T. Nishi, Y. Oikawa, M. Hikota and O. Yonemitsu; *Tetrahedron Lett.*, 1987, **28**, 4569.	173, 199
87TL4711	S. Fukuoka, S. Takimoto, T. Katsuki and M. Yamaguchi; *Tetrahedron Lett.*, 1987, **28**, 4711.	125
87TL5091	V. Pozsgay and H. J. Jennings; *Tetrahedron Lett.*, 1987, **28**, 5091.	368
87TL5099	S. Rajeswari, R. J. Jones and M. P. Cava; *Tetrahedron Lett.*, 1987, **28**, 5099.	264
87TL5369	G. Cainelli, D. Giacomini, M. Panunzio, G. Martelli and G. Spunta; *Tetrahedron Lett.*, 1987, **28**, 5369.	287
87TL5431	G. F. Koser, P. B. Kokil and M. Shah; *Tetrahedron Lett.*, 1987, **28**, 5431.	339
87TL5801	W. P. Dailey; *Tetrahedron Lett.*, 1987, **28**, 5801.	866
87TL5823	M. Denarie, D. Grenouillat, T. Malfroot, J.-P. Senet, G. Sennyey and P. Wolf; *Tetrahedron Lett.*, 1987, **28**, 5823.	282
87TL5937	A. Bathgate and J. R. Malpass; *Tetrahedron Lett.*, 1987, **28**, 5937.	330
87TL6257	G. Rajendra and M. J. Miller; *Tetrahedron Lett.*, 1987, **28**, 6257.	285
87UKZ397	V. Ya. Semenii, G. F. Solodushchenko, V. P. Kukhar and V. A. Bondar; *Ukr. Khim. Zh* (*Russ. Ed.*), 1987, **53**, 397 (*Chem. Abstr.*, 1988, **109**, 110 503t).	683, 707
87ZAAC(553)7	R. Appel and M. Immenkeppel; *Z. Anorg. Allg. Chem.*, 1987, **553**, 7.	876, 877, 878
87ZAAC(555)23	R. Appel and C. Behnke; *Z. Anorg. Allg. Chem.*, 1987, **555**, 23.	1086
87ZC8	J. Liebscher, A. Knoll and B. Abegaz; *Z. Chem.*, 1987, **27**, 8.	565, 614
87ZC27	M. Schulz, R. Kluge, S. Willscher and D. Leistner; *Z. Chem.*, 1987, **27**, 27.	365
87ZC64	H. Rousch; *Z. Chem.*, 1987, **27**, 64.	1127
87ZC90	A. A. Martin and G. Barnikow; *Z. Chem.*, 1987, **27**, 90.	583

87ZOB405 G. A. Razuvaev, A. S. Gordetsov, V. N. Latyaeva, N. F. Cherepennkova, T. N. Brevnova, S. E. Skobeleva, T. N. Konkina and A. P. Kozina; *Zh. Obshch. Khim.*, 1987, **57**, 405. 1081

87ZOB475 A. D. Sinitsa, T. V. Kolodka and A. K. Shurubura; *Zh. Obshch. Khim.*, 1985, **57**, 475 (*Chem. Abstr.*, 1988, **108**, 94 642j). 655

87ZOB685 V. A. Kozlov, T. Yu. Dol'nikova, A. F. Grapov and N. N. Mel'nikov; *Zh. Obshch. Khim.* (*Engl. Trans.*), 1987, **57**, 685. 644

87ZOB686 A. D. Voitenko, D. P. Fel'dman, M. Gavars, E. Liepins, G. Konstante, M. V. Shimanskaya and M. Lidaks; *Zh. Obshch. Khim. Lett.*, 1987, **57**, 686. 224

87ZOB956 M. G. Voronkov and V. Y. Vvedenskii; *Zh. Obshch. Khim.*, 1987, **57**, 956. 376

87ZOB1199 L. I. Belousova, I. D. Kalikhman, O. A. Vyazankina, E. I. Brodskaya and N. S. Vyazankin; *Zh. Obshch. Khim.*, 1987, **57**, 1199, (*Chem. Abstr.*, 1988, **108**, 56 246a). 391, 718

87ZOB1233 P. P. Onys'ko, T. V. Kim, E. I. Kiseleva and A. D. Sinitsa; *Zh. Obshch. Khim.*, 1987, **57**, 1233 (*Chem. Abstr.*, 1988, **108**, 186 856k). 657

87ZOB1514 E. A. Suvalova, T. I. Chudakova, P. P. Onys'ko and A. D. Sinitsa; *Zh. Obshch. Khim.*, 1987, **57**, 1514 (*Chem. Abstr.*, 1988, **108**, 204 712s). 660

87ZOB1643 V. I. Shcherbakov, N. E. Stolyarova and T. N. Konkina; *Zh. Obshch. Khim.*, 1987, **57**, 1643. 391

87ZOB1658 V. B. Sokolov, T. A. Epishina, A. N. Ivanov, A. V. Kharitonov, V. K. Brel and I. V. Martynov; *Zh. Obshch. Khim.*, 1987, **57**, 1658 (*Chem. Abstr.*), 1988, **108**, 167 594g). 669

87ZOB1659 V. B. Sokolov, A. N. Ivanov, T. A. Epishina and I. V. Martynov; *Zh. Obshch. Khim.*, 1987, **57**, 1659 (*Chem. Abstr.*, 1988, **109**, 6618y). 669

87ZOB1912 Yu. M. Dmitrichenko, V. G. Rozinov, V. I. Donskikh, G. V. Dolgushin, V. P. Feshin and A. V. Kalabina; *Zh. Obshch. Khim.*, 1987, **57**, 1912 (*Chem. Abstr.*, 1988, **109**, 73 526e). 656

87ZOR86 E. S. Levchenko, T. N. Dubinina, S. V. Sereda, M. Yu. Antipin, Yu. I. Struchkov and I. E. Boldeshul; *Zh. Org. Khim.*, 1987, **23**, 86 (*Chem. Abstr.*, 1988, **108**, 5669h). 703

87ZOR2023 M. V. Vovk; *Zh. Org. Khim.*, 1987, **23**, 2023. 1071

87ZOR2385 V. I. Gorbatenco, Yu. I. Matveev and L. I. Samarai; *Zh. Org. Khim.*, 1987, **23**, 2385. 1075

87ZOR2390 Yu. I. Mateev, V. I. Gorbatenko, L. I. Samarai, S. V. Sereda, M. Yu. Antipin and Yu. T. Struchkov; *Zh. Org. Khim.*, 1987, **23**, 2390. 1075

88ACS(B)605 S. Jacobsen; *Acta Chem. Scand., Ser. B*, 1988, **42**, 605. 692

88AG289 G. Sicard, A. Baceiredo, G. Crocco and G. Bertrand; *Angew. Chem.*, 1988, **100**, 289. 1131

88AG(E)268 T. Kunz and H.-U. Reissig; *Angew. Chem., Int. Ed. Engl.*, 1988, **27**, 268. 164

88AG(E)272 U. J. Vogelbacher, M. Ledermann, T. Schach, G. Michels, U. Hees and M. Regitz; *Angew. Chem., Int. Ed. Engl.*, 1988, **27**, 272. 883

88AG(E)583 C. Zybill, D. L. Wilkinson and G. Müller; *Angew. Chem., Int. Ed. Engl.*, 1988, **27**, 583. 1097

88AG(E)838 A. H. Cowley, S. W. Hall, C. M. Nunn and J. M. Power; *Angew. Chem., Int. Ed. Engl.*, 1988, **27**, 838. 890

88AG(E)960 H. Lukasch, G. Schmidt-Lukasch, U. Lippold and A. Berndt; *Angew. Chem., Int. Ed. Engl.*, 1988, **27**, 960. 928, 930

88AG(E)961 R. Hunold, J. Allwohn, G. Baum, W. Massa and A. Berndt; *Angew. Chem., Int. Ed. Engl.*, 1988, **27**, 961 (*Angew. Chem.*, 1988, **100**, 961). 928

88AG(E)1331 S. Helm and H. Nöth; *Angew. Chem.*, 1988, **100**, 1378. *Angew. Chem., Int. Ed. Engl.*, 1988, **27**, 1331. 927

88AG(E)1344 G. Beck and W. P. Fehlhammer; *Angew. Chem., Int. Ed. Engl.*, 1988, **27**, 1344. 1087

88AG(E)1456 R. Aumann; *Angew. Chem., Int. Edn. Engl.*, 1988, **27**, 1456. 931

88AG(E)1484 M. Regitz and P. Binger; *Angew. Chem.*, 1988, **27**, 1484. 883, 1151, 1152, 1154, 1159

88AP(321)25 H. J. Duchstein, G. Ruch-Zaske, G. Holzmann, E. Wollenberg and H. Weber; *Arch. Pharm.* (*Weinheim, Ger.*), 1988, **321**, 25. 357

88AP(321)41 J. Dusemund and T. Hardt; *Arch. Pharm.* (*Weinheim, Ger.*), 1988, **321**, 41. 652

88AP(321)235 D. Geffken; *Arch. Pharm.* (*Weinheim, Ger.*), 1988, **321**, 235. 327

88AP(321)311 A. Burchardt and D. Geffken; *Arch. Pharm.* (*Weinheim, Ger.*), 1988, **321**, 311. 323

88AP(321)551 H. Weber and E. Wollenberg; *Arch. Pharm.* (*Weinheim, Ger.*), 1988, **321**, 551. 357

88AP(321)903 K. Hartke and A. Afrashteh; *Arch. Pharm.* (*Weinheim, Ger.*), 1988, **321**, 903. 555

88AQ120 A. Monge, J. A. Palop, A. Ocho de Retana, P. Parrado and E. Fernandez-Alvarez; *An. Quim.*, 1988, **84**, 120. 199

88BAP139 Z. Matacz, H. Piotrowska and M. Poplowska; *Bull. Acad. Pol. Sci., Ser. Sci. Chim.*, 1988, **36**, 139. 692

88BAU378 I. D. Kalikhman, O. B. Bannikova, L. I. Volkova, B. A. Gostevskii, T. T. Lopyrev and M. G. Voronkov; *Bull. Akad. Sci. USSR. Div. Chem. Sci*, 1988, 378. 718

88BAU1506 V. B. Sokolov, T. A. Epishina and I. V. Martynov; *Bull. Acad. Sci. USSR. Div. Chem. Sci.*, 1988, 1506. 669

88BBA269 I. D. Wlassics, C. Stille and V. E. Anderson; *Biochim. Biophys. Acta*, 1988, **952**, 269. 551

88BCJ2205 T. Toru, Y. Yamada, E. Maekawa and Y. Ueno; *Bull. Chem. Soc. Jpn.*, 1988, **61**, 2205. 249

88BCJ3649 M. Murakami, M. Inouye, M. Suginome and Y. Ito; *Bull. Chem. Soc. Jpn.*, 1988, **61**, 3649. 802

88BCJ3653 M. Ueki, T. Saito J. Sasaya, S. Ikeda and H. Oyamada; *Bull. Chem. Soc. Jpn.*, 1988, **61**, 3653. 221

88BCJ3731 Y. Nakano, H. Yokota, M. Igarashi and S. Sato; *Bull. Chem. Soc. Jpn.*, 1988, **61**, 3731. 692

88BCJ3931 S. Kato, E. Yasui, K. Terashima, H. Ishihara and T. Murai; *Bull. Chem. Soc. Jpn.*, 1988, **61**, 3931. 249, 563, 640

88BCJ4467 T. Inazu, H. Hosokawa and M. Amemiya; *Bull. Chem. Soc. Jpn.*, 1988, **61**, 4467. 221

88C230 T. Wirth and C. Rüchardt; *Chimia*, 1988, **42**, 230. 657

88CB75 M. Schwab and W. Sundermeyer; *Chem. Ber.*, 1988, **121**, 75. 358

88CB271 L. Capuano, B. Dahm, V. Port, R. Schnur and V. Schramm; *Chem. Ber.*, 1988, **121**, 271. 1064, 1066

88CB313 H. Gotthardt, W. Pflaumbaum and P. Gutowski; *Chem. Ber.*, 1988, **121**, 313. 561

88CB507 D. Seebach, G. Adam, T. Gees, M. Schiess and W. Weigand; *Chem. Ber.*, 1988, **121**, 507. 658

88CB689 E. Schaumann, M. Möller and G. Adiwidjaja; *Chem. Ber.*, 1988, **121**, 689. 344

88CB701 E. Schaumann and M. Möller; *Chem. Ber.*, 1988, 701. 639

88CB723 E. M. Bakó, K. Horváth, E. Pálosi and D. Korbonits; *Chem. Ber.*, 1988, **121**, 723. 219, 833

88CB833 S. Cosack and G. Himbert; *Chem. Ber.*, 1988, **121**, 833. 365

88CB1115 H.-F. Klein and L. Reitzel; *Chem. Ber.*, 1988, **121**, 1115. 1137

88CB1363 H. Richter, S. Arenz, G. Michels, J. Schneider, O. Wagner and M. Regitz; *Chem. Ber.*, 1988, **121**, 1363. 924

88CB1435 U. Hartwig, H. Pritzkow and W. Sundermeyer; *Chem. Ber.*, 1988, **121**, 1435. 502

88CB1891 A. Belforte, D. B. Dell'Amico and F. Calderazzo; *Chem. Ber.*, 1988, **121**, 1891. 975, 976

88CB2095 H. Fischer, U. Gerbing, A. Tiriliomis, G. Müller, B. Huber, J. Riede, J. Hofmann and P. Burger; *Chem. Ber.*, 1988, **121**, 2095. 605, 614, 627

88CC46 G. Rossi, V. L. Goedken and C. Ercolani; *J. Chem. Soc.*, *Chem. Commun.*, 1988, 46. 1097

88CC171 A. M. Arif, A. R. Barron, A. H. Cowley and S. W. Hall; *J. Chem. Soc.*, *Chem. Commun.*, 1988, 171. 884

88CC282 M. Ashwell and R. F. W. Jackson; *J. Chem. Soc.*, *Chem. Commun.*, 1988, 282. 238

88CC365 B. F. Bonini, G. Mazzanti, P. Zani and G. Maccagnani; *J. Chem. Soc.*, *Chem. Commun.*, 1988, 365. 648

88CC390 P. Metzner and T. N. Pham; *J. Chem. Soc.*, *Chem. Commun.*, 1988, 390. 248, 562

88CC419 T. Miyazawa, T. Otomatsu, Y. Fukui, T. Yamada and S. Kuwata; *J. Chem. Soc.*, *Chem. Commun.*, 1988, 419. 279

88CC561 H. Ogawa, T. Joh and S. Takahashi; *J. Chem. Soc.*, *Chem. Commun.*, 1988, 561. 801

88CC710 W. Chamchaang and A. R. Pinhas; *J. Chem. Soc.*, *Chem. Commun.*, 1988, 710. 346

88CC753 A. H. Cowley, S. W. Hall, C. M. Nunn and J. M. Power; *J. Chem. Soc.*, *Chem. Commun.*, 1988, 753. 889

88CC1249 E. Carvalho, J. Iley and E. Rosa; *J. Chem. Soc.*, *Chem. Commun.*, 1988, 1249. 711

88CC1485 S. Kagabu, M. Maehara, K. Sawahara and K. Saito; *J. Chem. Soc.*, *Chem. Commun.*, 1988, 1485. 1112

88CC1531 T. Suzuki, E. Sato, Y. Matsuda, H. Tada, S. Koizumi, K. Unno and T. Kametani; *J. Chem. Soc.*, *Chem. Commun.*, 1988, 1531. 159

88CE215 I. Furukawa, N. Abe and S. Hashimoto; *Chem. Express*, 1988, **3**, 215 (*Chem. Abstr.*, 1989, **110**, 74938). 652

88CHE1297 S. P. Tumkevicius; *Chem. Heterocycl. Compd.* (*Engl. Transl.*), 1988, 1297. 690

88CJC1537 A. Martel, J.-P. Daris, C. Bachand, J. Corbeil and M. Menard; *Can. J. Chem.*, 1988, **66**, 1537. 236

88CL377 H. Matsuzaka, Y. Hiroe, M. Iwasaki, Y. Ishii, Y. Koyasu and M. Hidai; *Chem. Lett.*, 1988, 377. 463

88CL1021 T. Yura, N. Iwasawa and T. Mukaiyama; *Chem. Lett.*, 1988, 1021. 557

88CL1025 T. Yura, N. Iwasawa, K. Narasaka and T. Mukaiyama; *Chem. Lett.*, 1988, 1025. 557

88CL1141 T. Koga, S. Makinouchi and N. Okukada; *Chem. Lett.*, 1988, 1141. 432

88CL1145 K. Okuma, Y. Komiya and H. Ohta; *Chem. Lett.*, 1988, 1145. 603

88CL1385 T. Nagase, K. Akama and S. Oraki; *Chem. Lett.*, 1988, 1385. 995

88CL1733 M. Yoshifuji, T. Niitsu and N. Inamoto; *Chem. Lett.*, 1988, 1733. 877

88CPB178 K. Okamoto, M. Watanabe, H. Morimoto and I. Imada; *Chem. Pharm. Bull.*, 1988, **36**, 178. 206

88CPB877 T. Ohsaki, T. Kuriki, T. Ueda, J. Sakakibara and M. Asano; *Chem. Pharm. Bull.*, 1988, **36**, 877. 340

88CPB3125 K. I. Tanaka, S. Yoshifuji and Y. Nitta; *Chem. Pharm. Bull.*, 1988, **36**, 3125. 652

88CPB3354 N. Katagiri, H. Watanabe and C. Kaneko; *Chem. Pharm. Bull.*, 1988, **36**, 3354. 652

88CZ106 P. Aslanidis and F. Kober; *Chem.-Ztg.*, 1988, **112**, 106. 222

88CZ239 E. Pfleging and K. Hartke; *Chem.-Ztg.*, 1988, **112**, 239. 639

88CZ349 K. Burger and E. Hoess; *Chem.-Ztg.*, 1988, **112**, 349. 631

88EJM257 M. Fatome, H. Sentenac-Roumanou, C. Lion, J. Satge and G. Rima; *Eur. J. Med. Chem.*, 1988, **23**, 257. 388, 389

88GSM75 D. W. Knight; *Gen. Synth. Methods*, 1988, **10**, 75. 24

88H(27)155 M. Takahashi, H. Miyahara and N. Yoshida; *Heterocycles*, 1988, **27**, 155. 657

88H(27)695 H. M. Hassaneen, H. A. H. Mousa, N. M. Abed and A. S. Shawali; *Heterocycles*, 1988, **27**, 695. 703

88H(27)867 A. Saba and A. Selva; *Heterocycles*, 1988, **27**, 867. 343

88H(27)1881 T. Okawara, T. Shono, T. Yamasaki and M. Furukawa; *Heterocycles*, 1988, **27**, 1881. 351

88H(27)1953 H. Takahata and T. Yamazaki; *Heterocycles*, 1988, **27**, 1953. 597

88H(27)2077 T. Fujii, M. Ohba and M. Sakari; *Heterocycles*, 1988, **27**, 2077.

88H(27)2609 S. Furuya and T. Okamoto; *Heterocycles*, 1988, **27**, 2609. 1126

88H(27)1935 R. Molina, A. Lorenzo, J. Vilaplana, E. Aller and J. Planes; *Heterocycles*, 1988, **27**, 1935. 680

88IC3100 E. O. John and J. M. Shreeve; *Inorg. Chem.*, 1988, **27**, 3100. 692

88IJC583 S. S. Ahmad, F. Hafecz, S. I. Haider and T. Mahmood; *Indian J. Chem.*, 1988, **27**, 583. 1125

88IJC(B)472 S. A. El-Sharabasy, S. M. Hussain, S. M. A. Gawad and H. A. Daboun; *Ind. J. Chem.*, *Sect. B*, 1988, **27**, 472. 615

88IJC(B)1159 R. P. Sharma, N. Sharma and B. N. Misra; *Indian J. Chem.*, *Sect. B*, 1988, **27**, 1159. 214

88IZV460	I. D. Kalikhman, O. B. Bannikova, L. I. Volkova, B. A. Gostevskii, T. I. Yushmanova, V. A. Lopyrev and M. G. Voronkov; *Izv. Akad. Nauk SSSR Ser. Khim.*, 1988, 460 (*Chem. Abstr.*, 1989, **110**, 75 608c).	718
88IZV739	G. A. Razuvaev, O. S. D'yachkovskaya, I. P. Malysheva and S. E. Skobeleva; *Izv. Akad. Nauk SSSR, Ser. Khim.*, 1988, **4**, 739.	1082
88IZV859	G. A. Razuvaev, O. S. D'yachkovskaya, I. P. Malysheva and S. E. Skobeleva; *Izv. Akad. Nauk SSSR, Ser. Khim.*, 1988, 859.	390, 391
88IZV1691	V. B. Sokolov, T. A. Epishina and I. V. Martynov; *Izvest. Akad. Nauk SSSR, Ser. Khim.*, 1988, 1691 (*Chem. Abstr.*, 1989, **110**, 231 728z).	652
88IZV1947	I. S. Biltueva, D. A. Bravo-Zhivotovskii, I. D. Kalikhman, N. S. Vyazankin and M. G. Voronkov; *Izv. Akad. Nauk SSSR, Ser. Khim.*, 1988, 1947 (*Chem. Abstr.*, 1989, **110**, 231758).	924
88JA307	D. C. Yang, V. Dragisich, W. D. Wulff and J. C. Huffman; *J. Am. Chem. Soc.*, 1988, **110**, 307.	696
88JA313	F. H. Elsner, H. G. Woo and T. D. Tilley; *J. Am. Chem. Soc.*, 1988, **110**, 313.	422
88JA612	L. E. Overman and M. J. Sharp; *J. Am. Chem. Soc.*, 1988, **110**, 612.	1108
88JA957	C. S. Shiner, A. H. Berks and A. M. Fisher; *J. Am. Chem. Soc.*, 1988, **110**, 957.	437
88JA1631	D. J. Hart and F. L. Seely; *J. Am. Chem. Soc.*, 1988, **111**, 1631.	152
88JA1638	M. Kahn, S. Wilke, B. Chen and K. Fujita; *J. Am. Chem. Soc.*, 1988, **110**, 1638.	366
88JA2122	L. S. Hegedus, G. de Weck and S. D'Andrea; *J. Am. Chem. Soc.*, 1988, **110**, 2122.	953
88JA2237	L. D. Arnold, R. G. May and J. C. Vederas; *J. Am. Chem. Soc.*, 1988, **110**, 2237.	85
88JA2341	T. P. Burkholder and P. L. Fuchs; *J. Am. Chem. Soc.*, 1988, **110**, 2341.	330
88JA2663	G. Sicard, A. Baceiredo and G. Bertrand; *J. Am. Chem. Soc.*, 1988, **110**, 2663.	871, 872, 873
88JA3670	S. Hashimoto, S. Sakata, M. Sonegawa and S. Ikegami; *J. Am. Chem. Soc.*, 1988, **110**, 3670.	138
88JA3672	E. J. Corey, P. DaSilva Jardine and J. C. Rohloff; *J. Am. Chem. Soc.*, 1988, **110**, 3672.	138
88JA4640	J. Mulzer, U. Steffen, L. Zorn, C. Schneider, E. Weinhold, W. Münch, R. Rudert, P. Luger and H. Hartl; *J. Am. Chem. Soc.*, 1988, **110**, 4640.	151, 164
88JA4855	A. K. List, G. L. Hillhouse and A. L. Rheingold; *J. Am. Chem. Soc.*, 1988, **110**, 4855.	1096
88JA6110	G. L. Crocco and J. A. Gladysz; *J. Am. Chem. Soc.*, 1988, **110**, 6110.	451
88JA6172	S. G. Gilbreath, C. M. Harris and T. M. Harris; *J. Am. Chem. Soc.*, 1988, **110**, 6172.	330
88JA6240	C. A. Arrington, J. T. Petty, S. E. Payne and W. C. K. Haskins; *J. Am. Chem. Soc.*, 1988, **110**, 6240.	1019
88JA6463	A. Igau, H. Grutzmacher, A. Baceiredo and G. Bertrand; *J. Am. Chem. Soc.*, 1988, **110**, 6463.	871, 872, 873
88JA6926	A. K. List, G. L. Hillhouse and A. L. Rheingold; *J. Am. Chem. Soc.*, 1988, **110**, 6926.	1096
88JA7228	M. A. Pearsall and R. West; *J. Am. Chem. Soc.*, 1988, **110**, 7228.	1019
88JA7538	Y. Gao and K. B. Sharpless; *J. Am. Chem. Soc.*, 1988, **110**, 7538.	1106
88JA8413	A. Hafner, L. S. Hegedus, G. de Weck, B. Hawkins and K. H. Dötz; *J. Am. Chem. Soc.*, 1988, **110**, 8413.	932
88JAP63135393	K. Sugata; *Jpn. Pat.* 63 135 393 (1988) (*Chem. Abstr.*, 1988, **109**, 170 637z).	1134
88JCR(S)376	Shafiullah, Shamsuzzaman and R. K. Pathak; *J. Chem. Res.* (*S*), 1988, 376.	533
88JCS(D)27	J. L. Davidson and W. F. Wilson; *J. Chem. Soc., Dalton Trans.*, 1988, 27.	798
88JCS(D)53	A. C. Sullivan, G. Wilkinson, M. Motevalli and M. B. Hursthouse; *J. Chem. Soc., Dalton Trans.*, 1988, 53.	444
88JCS(D)1591	V. Riera, J. Ruiz, Y. Jeannin and M. Philoche-Levisalles; *J. Chem. Soc., Dalton Trans.*, 1988, 1591.	799
88JCS(D)2077	R. Ali and K. B. Dillon; *J. Chem. Soc., Dalton Trans.*, 1988, 2077.	1129
88JCS(P1)169	J. N. Vishwakarma, A. Thomas, S. Apparao, H. Ila and H. Junjappa; *J. Chem. Soc., Perkin Trans. 1*, 1988, 169.	731
88JCS(P1)305	Y. N. Belokon', V. I. Bakhmutov, N. I. Chernoglazova, K. A. Kochetkov, S. V. Vitt, N. S. Garbalinskaya and V. M. Belikov; *J. Chem. Soc., Perkin Trans* 1, 1988, 305.	652
88JCS(P1)529	C. Kashima, S. Hibi, K. Harada and Y. Omote; *J. Chem. Soc., Perkin Trans. 1*, 1988, 529.	197
88JCS(P1)1499	B. F. Bonini, G. Mazzanti, P. Zani, G. Maccagnani and E. Foresti; *J. Chem. Soc., Perkin Trans. 1*, 1988, 1499.	648
88JCS(P1)1913	G. W. Kirby and A. N. Trethewey; *J. Chem. Soc., Perkin Trans. 1*, 1988, 1913.	336
88JCS(P1)2137	J. H. Boyer, T. Manimaran and L. T. Wolford; *J. Chem. Soc., Perkin Trans. 1*, 1988, 2137.	963
88JCS(P1)2185	R. Brettle and A. J. Mosedale; *J. Chem. Soc., Perkin Trans. 1*, 1988, 2185.	288
88JCS(P1)2203	M. Sakamoto, S. Watanabe, T. Fujita, M. Tohnishi, H. Aoyama and Y. Omote; *J. Chem. Soc., Perkin Trans. 1*, 1988, 2203.	630
88JCS(P1)3103	D. M. B. Hickey, P. D. Leeson, R. Novelli *et al.*; *J. Chem. Soc., Perkin Trans. 1*, 1988, 3103.	205
88JCS(P2)821	M. Muhlstadt, L. Weber and P. Birner; *J. Chem. Soc., Perkin Trans. 2*, 1988, 821.	641
88JCS(P2)2029	E. Breuer, R. Karaman, A. Goldblum and H. Leader; *J. Chem. Soc., Perkin Trans. 2*, 1988, 2029–2034.	403, 404
88JFC(40)217	I. V. Kolesnikova, T. D. Petrova, V. E. Platonov, V. A. Mikhailov, A. A. Popov and V. A. Savelova; *J. Fluorine Chem.*, 1988, **40**, 217.	1070
88JGU30	P. P. Onys'ko, T. V. Kim, E. I. Kiseleva, V. P. Prokopenko and A. D. Sinitsa; *J. Gen. Chem. USSR* (*Engl. Transl.*), 1988, **58**, 30.	789
88JGU201	E. G. Neganova, R. V. Brusov, Yu. A. Veits, V. L. Foss and I. F. Lutsenko; *J. Gen. Chem. USSR*, 1988, **58**, 201.	395, 396, 399, 400
88JGU1548	V. A. Al'fonsov, D. A. Pudovik, R. Z. Musin, V. N. Nazmutdinova, Yu. Ya. Efremov, É. S. Batyeva and A. N. Pudovik; *J. Gen. Chem. USSR*, 1988, **58**, 1548.	406
88JGU1965	O. G. Sinyashin, I. Yu Gorshunov, A. N. Chernov, A. V. Il'yasov, E. S. Betyeva and A. N. Pudovik; *J. Gen. Chem. USSR*, 1988, **58**, 1965.	398

88JGU2192 O. B. Smolii, V. S. Brovarets, V. V. Pirozhenko and B. S. Drach; *J. Gen. Chem. USSR* (*Engl. Transl.*), 1988, **58**, 2192. 655

88JGU2372 V. L. Foss, Yu. A. Veits, E. G. Neganova, A. A. Borisenko and I. F. Lutsenko; *J. Gen. Chem. USSR*, 1988, **58**, 2372. 395, 396, 399, 400

88JHC23 A. Monge, V. Martinez-Merino and E. Fernández-Alvarez; *J. Heterocycl. Chem.*, 1988, **25**, 23. 334

88JHC49 B. Rigo, C. Lespagnol and M. Pauly; *J. Heterocycl. Chem.*, 1988, **25**, 49. 374

88JHC59 B. Rigo, C. Lespagnol and M. Pauly; *J. Heterocycl. Chem.*, 1988, **25**, 59. 223, 374

88JHC1873 V. J. Lee, W. V. Curran, T. F. Fields and K. Learn; *J. Heterocycl. Chem.*, 1988, **25**, 1873. 639, 640

88JMC362 S. G. Mislankar, D. L. Gildersleeve, D. Wieland, C. C. Massin, K. G. Mulholland and S. A. Toorongian; *J. Med. Chem.*, 1988, **31**, 362. 203

88JMC499 W. P. Jackson, P. J. Islip, G. Kneen, A. Pugh and P. J. Wates; *J. Med. Chem.*, 1988, **31**, 499. 214

88JMC1031 S. F. Campbell, J. D. Hardstone and M. J. Palmer; *J. Med. Chem.*, 1988, **31**, 1031. 690

88JMC1960 J. B. Summers, B. P. Gunn, J. G. Martin, M. B. Martin, H. Mazdiyasni, A. O. Stewart, P. R. Young, J. B. Bouska and A. M. Goetze; *J. Med. Chem.*, 1988, **31**, 1960. 214

88JMC2066 H. Bundgaard and J. D. Larsen; *J. Med. Chem.*, 1988, **31**, 2066. 703

88JOC22 A. Hassner, R. Fibiger and A. S. Amarasekara; *J. Org. Chem.*, 1988, **53**, 22. 697

88JOC382 R. S. Hosmane, B. B. Lim and F. N. Burnett; *J. Org. Chem.*, 1988, **53**, 382. 690

88JOC845 L. V. Dunkerton, N. K. Adair, J. M. Euske, K. T. Brady and P. D. Robinson; *J. Org. Chem.*, 1988, **53**, 845. 540

88JOC1203 M. R. Detty and L. W. McGarry; *J. Org. Chem.*, 1988, **53**, 1203. 934

88JOC1569 R. J. Linderman and Y. Suhr; *J. Org. Chem.*, 1988, **53**, 1569. 412, 413, 425

88JOC1744 J. Y. Becker and S. Yatziv; *J. Org. Chem.*, 1988, **53**, 1744. 1074

88JOC1922 C. H. Heathcock, B. L. Finkelstein, E. T. Jarvi, P. A. Radel and C. R. Hadley; *J. Org. Chem.*, 1988, **53**, 1922. 159

88JOC2371 T. H. Black, J. A. Hall and R. G. Sheu; *J. Org. Chem.*, 1988, **53**, 2371. 170

88JOC2440 C. Brückner and H.-U. Reissig; *J. Org. Chem.*, 1988, **53**, 2440. 166

88JOC3612 A. Capperucci, A. Degl'Innocenti, C. Faggi, A. Ricci, P. Dembech and G. Seconi; *J. Org. Chem.*, 1988, **53**, 3612. 417

88JOC3632 P. T. Meinke, G. A. Krafft and A. Guram; *J. Org. Chem.*, 1988, **53**, 3632. 1115

88JOC3852 L. L. Melhado and J. L. Brodsky; *J. Org. Chem.*, 1988, **53**, 3852. 690

88JOC4006 S. Knapp and A. T. Levorse; *J. Org. Chem.*, 1988, **53**, 4006. 378, 379

88JOC4098 J. Mulzer, T. Schulze, A. Strecker and W. Denzer; *J. Org. Chem.*, 1988, **53**, 4098. 164, 168

88JOC4158 M. Murakami, H. Ito and Y. Ito; *J. Org. Chem.*, 1988, **53**, 4158. 802

88JOC4421 P. Prasit and J. Rokach; *J. Org. Chem.*, 1988, **53**, 4421. 172

88JOC4443 C. W. Bauknight, Jr. and D. D. DesMarteau; *J. Org. Chem.*, 1988, **53**, 4443. 809

88JOC4654 P. Molina, A. Arques, M. V. Vinander, J. Becher and K. Brondum; *J. Org. Chem.*, 1988, **53**, 4654. 1066

88JOC4736 M. A. Henderson and C. H. Heathcock; *J. Org. Chem.*, 1988, **53**, 4736. 159

88JOC4773 S. Knapp, A. T. Levorse and J. A. Potenza; *J. Org. Chem.*, 1988, **53**, 4773. 379

88JOC4923 M. H. Block and D. E. Cane; *J. Org. Chem.*, 1988, **53**, 4923. 235

88JOC5158 A. Vasudevan and G. F. Koser; *J. Org. Chem.*, 1988, **53**, 5158. 999

88JOC5309 R. S. Hosmane, B. B. Lim, M. F. Summers, U. Siriwardane, N. S. Hosmane and S. S. C. Chu; *J. Org. Chem.*, 1988, **53**, 5309. 690

88JOC5415 M. Fujita and T. Hiyama; *J. Org. Chem.*, 1988, **53**, 5415. 213

88JOC5507 P. S. Manchand, K. C. Luk, P. S. Belica, S. C. Choudhry, C. C. Wei and M. Soukup; *J. Org. Chem.*, 1988, **53**, 5507. 345, 346

88JOC5854 A. R. Katritzky, M. Drewniak and P. Lue; *J. Org. Chem.*, 1988, **53**, 5854. 606

88JOM(334)139 R. Tacke and R. Rohr-Aehle; *J. Organomet. Chem.*, 1988, **334**, 139. 224

88JOM(339)213 H. Hoberg, E. Hernandez and D. Guhl; *J. Organomet. Chem.*, 1988, **339**, 213. 386

88JOM(341)225 M. G. Voronkov, N. F. Chernov and E. M. Perlova; *J. Organomet. Chem.*, 1988, **341**, 225. 1142

88JOM(341)95 U. Wannagat and R. Schrader; *J. Organomet. Chem.*, 1988, **341**, 95. 379

88JOM(341)C9 A. G. Brook and A. MacMillan; *J. Organomet. Chem.*, 1988, **341**, C9. 924

88JOM(342)C5 J. T. Gauntlett, B. E. Mann, M. J. Winter and S. Woodward; *J. Organomet. Chem.*, 1988, **342**, C5. 446

88JOM(344)29 B. Gyori, I. Lazar and J. Emri; *J. Organomet. Chem.*, 1988, **344**, 29. 795

88JOM(344)71 S. Sato, I. Matsuda and Y. Izumi; *J. Organomet. Chem.*, 1988, **344**, 71. 419

88JOM(344)C35 H. Hoberg, K. Sümmermann, E. Hernandez, C. Ruppin and D. Guhl; *J. Organomet. Chem.*, 1988, **344**, C35. 386

88JOM(345)C1 M. Wedler, H. W. Roesky and F. Edelmann; *J. Organomet. Chem.*, 1988, **345**, C1. 782

88JOM(347)101 E. Bar, W. P. Fehlhammer, W. Weigand and W. Beck; *J. Organomet. Chem.*, 1988, **347**, 101. 1088, 1089

88JOM(353)355 D. S. Bohle, G. R. Clark, C. E. F. Rickard and W. R. Roper; *J. Organomet. Chem.*, 1988, **353**, 355. 400

88JOM(355)337 H. G. Raubenheimer, G. J. Kruger, C. F. Marais, J. T. Z. Hattingh, L. Linford and P. H. van Rooyen; *J. Organomet. Chem.*, 1988, **355**, 337. 651

88JOM(355)33 S. M. Dhaher, C. Eaborn and D. J. Smith; *J. Organomet. Chem.*, 1988, **355**, 33. 1144

88JOU1049 I. G. Ostroumov, A. E. Tsil'ko, I. A. Maretina and A. A. Petrov; *J. Org. Chem. USSR* (*Engl. Transl.*), 1988, **24**, 1049. 691

88JOU2149 V. P. Semenov, M. A. Mishin and K. A. Ogloblin; *J. Org. Chem. USSR* (*Engl. Transl.*), 1988, **24**, 2149. 702

88JOU2154 V. P. Semenov and K. A. Ogloblin; *J. Org. Chem. USSR* (*Engl. Transl.*), 1988, **24**, 2154. 702

88JPR241 G. Weber, J. Hartung and L. Beyer; *J. Prakt. Chem.*, 1988, **330**, 241. 657, 1053
88JPR473 A. Koziara, *J. Prakt. Chem.*, 1988, **330**, 473. 1031
88JPR825 S. Nielek and T. Lesiak; *J. Prakt. Chem.*, 1988, **330**, 825. 991
88KGS1559 S. P. Tumkyavichus; *Khim. Geterotsikl. Soedin.*, 1988, 1559 (*Chem. Abstr.*, 1988, **111**, 39 302k). 652
88LA465 C. Brückner and H.-U. Reissig; *Liebigs Ann. Chem.*, 1988, 465. 553
88LA471 C. Brückner, B. Suchland and H.-U. Reissig; *Liebigs Ann. Chem.*, 1988, 471. 729
88LA787 G. V. Keidrowski and F. Z. Dorwald; *Liebigs Ann. Chem*; 1988, 787. 1062
88LA833 G. Schlüter and W. Meise; *Leibigs Ann. Chem.*, 1988, 833. 655
88LA891 J. Mulzer, K.-D. Graske and B. Kirste; *Liebigs Ann. Chem.*, 1988, 891. 159
88LA1107 W. Ried and J. Laoutidis; *Liebigs Ann. Chem.*, 1988, 1107. 690
B-88MI 502-01 K. Ingvorsen, B. Yde, S. E. Godtfredsen and R. T. Tsuchiya; in "Cyanide Compounds in Biology," CIBA Foundation Symposia, Wiley, Chichester, 1988, vol. 140. 118
B-88MI 507-01 A. Pelter, K. Smith and H. C. Brown; in "Borane Reagents," eds. A. R. Katritzky, O. Meth-Cohn and C. W. Rees, Academic Press, New York, 1988, p. 71. 381
B-88MI 507-02 A. H. Haines; "Methods for the Oxidation of Organic Compounds," Academic Press, New York, 1988, p. 222. 325
88MI 520-01 B. Gyori, I. Lazar and J. Emri; *Magyar Kemiai Folyoirat.*, 1988, **94**, 100. 652
88MI 528-01 A. S. Gordetzov, A. P. Koziora, S. E. Skobeleva and A. N. Egocochkin; *MetalloOrg. Khim.*, 1988, **1**, 107. 1081
88NJC547 R. Regragui and P. H. Dixneuf; *Nouv. J. Chim.*, 1988, **12**, 547. 400
88OM366 M. Akita, A. Kondoh, T. Kawahara, T. Takagi, and Y. Morooka; *Organometallics*, 1988, **7**, 366. 223
88OM558 W. Ando, K. Hagiwara, A. Sekiguchi, A. Sakakibara and H. Yoshida; *Organometallics*, 1988, **7**, 558. 1020
88OM702 R. Imwinkerlried and L. S. Hegedus; *Organometallics*, 1988, **7**, 702. 953
88OM2245 A. G. Brook, Y. K. Kong, A. K. Saxena and J. F. Sawyer; *Organometallics*, 1988, **7**, 2245. 924
88OM2553 A. G. M. Barrett, J. Mortier, M. Sabat and M. A. Sturgess; *Organometallics*, 1988, **7**, 2553. 955
88OPP521 A. M. Farag and M. S. Algharib; *Org. Prep. Proced. Int.*, 1988, **20**, 521. 676, 703
88OR(35)1 R. E. Gawley; *Org. React.*, 1988, **35**, 1. 270, 282, 288
88OS(66)14 R. L. Danheiser, D. M. Fink, K. Okano, Y. M. Tsai and S. W. Szczepanski; *Org. Synth.*, 1988, **66**, 14. 412
88OS(67)76 K. Furuta, K. Iwanaga and H. Yamamoto; *Org. Synth.*, 1988, **67**, 76. 30
88OSC(4)727 M. S. A. Vrijland; *Org. Synth. Coll. Vol.*, 1988, **4**, 727. 1112
88OSC(6)190 D. N. Harpp, L. Q. Bao, C. Coyle, J. G. Gleason and S. Horovitch; *Org. Synth., Coll. Vol.*, 1988, **6**, 190. 8
88OSC(6)549 E. C. Taylor, A. McKillop and G. H. Hawks; *Org. Synth., Coll. Vol.*, 1988, **6**, 549. 8
88POL2213 M. Yoshifuji, T. Niitsu, K. Toyota, N. Inamoto, K. Hirotsu, Y. Odagaki, T. Higuchi and S. Nagase; *Polyhedron*, **7**, 2213. 1086, 1087
88PS(35)1 A. Schwoebel, R. Lux and G. Kresze; *Phosphorus Sulfur*, 1988, **35**, 1. 341
88PS(40)263 S. Zawadzki; *Phosphorus Sulfur*, 1988, **40**, 263. 652
88RCR234 T. D. Petrova and V. E. Platonov; *Russ. Chem. Rev.*, 1988, **57**, 234. 712
88RTC381 P. A. van Elburg and D. N. Reinhoudt; *Recl. Trav. Chim. Pays-Bas*, 1988, **107**, 381. 325
88S133 S. I. Kang and C. P. Spears; *Synthesis*, 1988, **2**, 133. 1113
88S242 R. S. Hosmane and B. B. Lim; *Synthesis*, 1988, 242. 690
88S303 R. K. Dieter and A. G. Lugade; *Synthesis*, 1988, 303. 561
88S349 W. Sundermeyer; *Synthesis*, 1988, 349. 492, 494
88S397 Z. M. Jaszay, I. Petnehazy, L. Toke and B. Szajáni; *Synthesis*, 1988, 397. 1069
88S475 M. J. Milewska and T. Połoński; *Synthesis*, 1988, 475. 7, 15
88S525 T. Eicher, K. Tiefensee and R. Pick; *Synthesis*, 1988, 525. 8
88S621 P. C. B. Page, S. Rosenthal and R. V. Williams; *Synthesis*, 1988, 621. 304
88S743 S. C. Suri and R. D. Chapman; *Synthesis*, 1988, 743. 361, 362
88S807 W. Hartmann; *Synthesis*, 1988, 807. 353
88S990 W. Mormann and G. Leukel; *Synthesis*, 1988, 990. 224, 1007
88S992 L. Grehn, M. L. S. Almeida and U. Ragnarsson; *Synthesis*, 1988, 992. 977
88SC323 A. Oliva and L. Arancibia; *Synth. Commun.*, 1988, **18**, 323. 513
88SC545 P. S. Reddy, P. Yadagiri, S. Lumin, D.-S. Shin and J. R. Falck; *Synth. Commun.*, 1988, **18**, 545. 368
88SC1331 E. Wenkert and E. C. Angell; *Synth. Commun.*, 1988, **18**, 1331. 325
88SC2123 C. S. Pande and N. Jain; *Synth. Commun.*, 1988, **18**, 2123. 205
88SC2179 P. S. N. Reddy and P. P. Reddy; *Synth. Commun.*, 1988, **18**, 2179. 219
88SR1 W. Thiel and R. Mayer; *Sulfur Reports*, 1988, **8**, 1. 565
88SR155 S. Kato and M. Ishida; *Sulfur Reports*, 1988, **8**, 155. 550, 560
88T139 M. Bakasse, G. Duguay, H. Quiniou and L. Toupet; *Tetrahedron*, 1988, **44**, 139. 633
88T613 D. Tanner and P. Somfai; *Tetrahedron*, 1988, **44**, 613. 342
88T1619 A. Riondel, P. Caubere, J. P. Senet and S. Lecolier; *Tetrahedron*, 1988, **44**, 1619. 1105
88T1667 A. Thomas, J. N. Vishwakarma, S. Apparao, H. Ila and H. Junjappa; *Tetrahedron*, 1988, **44**, 1667. 731
88T1827 E. Schaumann; *Tetrahedron*, 1988, **44**, 1827. 550, 589
88T3501 D. H. R. Barton, N. Ozbalik and B. Vacher; *Tetrahedron*, 1988, **44**, 3501. 728
88T3575 S. Berrada, S. Desert and P. Metzner; *Tetrahedron*, 1988, **44**, 3575. 556
88T4113 J. S. Nowick and R. L. Danheiser; *Tetrahedron*, 1988, **44**, 4113. 423, 430
88T4157 E. W. Colvin, D. McGarry and M. J. Nugent; *Tetrahedron*, 1988, **44**, 4157. 377

88T5507 Y. N. Belokon, A. S. Sagyan, S. M. Djamgaryan, V. I. Bakhmutov and V. M. Belikov; *Tetrahedron*, 1988, **44**, 5507. 385

88T5615 A. G. M. Barrett and M. A. Sturgess; *Tetrahedron*, 1989, **44**, 5615. 652

88T6079 M. Z. Deng, P. Caubere, J. P. Senet and S. Lecolier; *Tetrahedron*, 1988, 6079. 994

88T6841 P. Metz and C. Mues; *Tetrahedron*, 1988, **44**, 6841. 685

88T7271 M. F. Aly and R. Grigg; *Tetrahedron*, 1988, **44**, 7271. 603, 621

88TH17 B. Biedenbach; Ph.D. Thesis, University of Kaiserslautern, 1988, p. 17. 890

88TL333 M. Yoshifuji, T. Niitsu, K. Toyota, N. Inamoto, H. H. Karsch and H. U. Reisacher; *Tetrahedron Lett.*, 1988, **29**, 333. 884

88TL379 P. Molina, P. M. Fresneda and P. Alarcón; *Tetrahedron Lett.*, 1988, 379. 1066

88TL739 A. Benderly and S. Stavchansky; *Tetrahedron Lett.*, 1988, **29**, 739. 350

88TL1595 S. A. Ahmed, E. Bardshiri and T. J. Simpson; *Tetrahedron Lett.*, 1988, **29**, 1595. 199

88TL1653 J. T. Drummond and G. Johnson; *Tetrahedron Lett.*, 1988, **29**, 1653. 342

88TL1657 J. A. Marshall and W. Y. Gung; *Tetrahedron Lett.*, 1988, **29**, 1657. 467

88TL1755 A. R. Katritzky and M. Drewniak; *Tetrahedron Lett.*, 1988, **29**, 1755. 606

88TL2033 R. T. Taylor and T. A. Stevenson; *Tetrahedron Lett.*, 1988, **29**, 2033. 205

88TL2661 P. Jouin, J. Poncet, M.-N. Dufour, I. Maugras, A. Pantaloni and B. Castro; *Tetrahedron Lett.*, 1988, **29**, 2661. 333

88TL3475 G. Weber, J. Hartung and L. Beyer; *Tetrahedron Lett.*, 1988, **29**, 3475. 655

88TL4085 S. L. Schreiber and Z. Wang; *Tetrahedron Lett.*, 1988, **29**, 4085. 625

88TL4411 A. Alexakis, R. Sedrani, P. Mangeney and J. F. Normant; *Tetrahedron Lett.*, 1988, **29**, 4411. 152

88TL4419 S. Jegham and B. C. Das; *Tetrahedron Lett.*, 1988, **29**, 4419. 157

88TL4435 H. Urata, A. Fujita and T. Fuchikami; *Tetrahedron Lett.*, 1988, **29**, 4435. 194

88TL4863 J. Barluenga, M. Ferrero and F. Palacios; *Tetrahedron Lett.*, 1988, 4863. 1066

88TL4909 M. Rowley and Y. Kishi; *Tetrahedron Lett.*, 1988, **29**, 4909. 223

88TL5369 J. L. Luche and C. Allavena; *Tetrahedron Lett.*, 1988, **29**, 5369. 152

88TL5583 G. Zhi-Wei, T. K. Ngooi, A. Scilimati, G. Fülling and C. J. Sih; *Tetrahedron Lett.*, 1988, **29**, 5583. 179

88TL5719 P. F. O'Bannon and W. P. Dailey; *Tetrahedron Lett.*, 1988, **29**, 5719. 326

88TL6173 G. W. Kirby and M. Nazeer; *Tetrahedron Lett.*, 1988, **29**, 6173. 330

88TL6509 G. Bashiardes and S. G. Davies; *Tetrahedron Lett.*, 1988, **29**, 6509. 329, 330

88TL6823 T. H. Webb, L. M. Thomasco, S. T. Schlachter, J. J. Gaudino and C. S. Wilcox; *Tetrahedron Lett.*, 1988, **29**, 6823. 155

88UK405 T. D. Petrova and V. E. Platonov; *Usp. Khim.*, 1988, **57**, 405. 712

88USP4762945 F. A. Golec, Jr. (USV Pharmaceutical Corporation); *US Pat.* 4 762 945 (1988) (*Chem. Abstr.*, 1988, **109**, 190 037). 206

88ZAAC(564)85 L. Jäger, H. Köhler, A. I. Brusilovec and V. V. Skopenko; *Z. Anorg. Allg. Chem.*, 1988, **564**, 85. 1084

88ZAAC(566)151 R. Minkwitz and J. Sawatzki; *Z. Anorg. Allg. Chem.*, 1988, **566**, 151. 246, 254

88ZC24 C. Csongár, L. Grubert and G. Tomaschewski; *Z. Chem.*, 1988, **28**, 24. 358, 707

88ZC141 J. Bernat, L. Kniezo, D. Petercakova, J. Imrich and K. Kutschy; *Z. Chem.*, 1988, **28**, 141. 1043

88ZC182 F. Pragst, J. Liebscher, B. Roloff and E. Boche; *Z. Chem.*, 1988, **28**, 182. 631

88ZC233 W. Thiel, R. Mayer and H. Viola; *Z. Chem.*, 1988, **28**, 233. 601

88ZC269 H. Dunken, L. Kolditz, R. Mayer, H. H. Emons, S. R. Geyer, W. Herzog, H. A. Lehmann, S. Rapoport, G. Rienäcker, H. Sackmann, *et al.*; *Z. Chem.*, 1988, **28**, 269. 560

88ZN(B)801 H. Meyer, G. Schmidt-Lukasch, G. Baum, W. Massa and A. Berndt; *Z. Naturforsch., Teil B*, 1988, **43**, 801. 928

88ZN(B)1495 A. Blaschette, F. Safari, K. Linoh and D. Koch; *Z. Naturforsch., Teil B*, 1988, **43**, 1495. 348

88ZN(B)1628 S. Maier, W. Hiller, J. Strähle, C. Ergezinger and K. Dehnicke; *Z. Naturforsch. Teil B*, 1988, **43**, 1628. 781

88ZN(B)1679 E. Hey, C. Ergezinger and K. Dehnicke; *Z. Naturforsch. Teil B*, 1988, **43**, 1679. 781

88ZOB62 É. G. Perevalova, Y. T. Struchkov, D. N. Kravtsov, L. G. Kuz'mina, E. I. Smysloma, K. I. Grandberg, O. N. Kalinina, V. P. Dyadchenko and T. I. Voevodskaya; *Zh. Obshch. Khim.*, 1988, **58**, 62. 385

88ZOB491 O. I. Kolodyazhnyi and D. V. Golokhov; *Zh. Obshch. Khim.*, 1988, **58**, 491 (*Chem. Abstr.*, 1989, **110**, 23 999m). 895

88ZOB616 B. D. Grishchuk, P. M. Gorbovoi and N. I. Ganushchak; *Zh. Obshch. Khim.*, 1988, **58**, 616. 1044

88ZOB706 I. S. Zal'tsman, A. P. Marchenko and A. M. Pinchuk; *Zh. Obshch. Khim.*, 1988, **58**, 706. 369

88ZOB829 G. I. Kokorev, F. D. Yambushev and L. A. Al'metkina; *Zh. Obshch. Khim.*, 1988, **58**, 829. 369, 373

88ZOB1056 V. P. Kozyukov, A. E. Feoktistov and V. F. Mironov; *Zh. Obshch. Khim.*, 1988, **58**, 1056. 323

88ZOB1093 E. P. Kramarova, A. G. Shipov, N. A. Orlova, O. B. Artamkina, I. Y. Belavin and Y. I. Baukov; *Zh. Obshch. Khim.*, 1988, **58**, 1093. 374

88ZOB1299 V. P. Kozyukov, A. E. Feoktistov and V. F. Mironov; *Zh. Obshch. Khim.*, 1988, **58**, 1299. 323

88ZOB1934 V. A. Dodonov, A. V. Gushkin and M. B. Ezhova; *Zh. Obshch. Khim. Lett.*, 1984, **58**, 1934. 222

88ZOB2424 V. N. Solov'ev, A. N. Chekhlov, N. G. Zabirov, R. A. Cherkasov, I. V. Martynov and A. N. Pudovik; *Zh. Obshch. Khim.*, 1988, **58**, 2424. 370

88ZOB2465 O. B. Smolii, V. S. Brovarets, V. V. Pirozhenko and B. S. Drach; *Zh. Obshch. Khim.*, 1988, **58**, 2465 (*Chem. Abstr.*, 1989, **111**, 57 663h). 655

88ZOB2628 A. A. Krolevets, A. V. Adamov, A. G. Popov and I. V. Martynov; *Zh. Obshch. Khim.*, 1988, **58**, 2628. 369

88ZOR593 A. A. Mironova, I. V. Soloshonok, I. I. Maletina, V. V. Orda and L. M. Yagupol'skii; *Zh. Org. Khim.*, 1988, **24**, 593. 205

88ZOR986 Yu. I. Matveev, V. I. Gorbatenko, L. I. Samarai, E. A. Romanenko and A. V. Turov; *Zh. Org. Khim.*, 1988, **24**, 986. 1069

88ZOR1165 I. G. Ostroumov, A. E. Tsil'ko, I. A. Maretina and A. A. Petrov; *Zh. Org. Khim.*, 1988, **24**, 1165 (*Chem. Abstr.*, 1989, **110**, 114 808q). 691

88ZOR1581 V. I. Naddaka, K. V. Avanesyan, M. L. Cherkinskaya and V. I. Minkin; *Zh. Org. Khim*, 1988, **24**, 1581. 1078

88ZOR1771 B. A. Trofimov, N. A. Nedolya, N. P. Papsheva, V. V. Gerasimova and M. G. Voronkov; *Zh. Org. Khim.*, 1988, **24**, 1771. 1025

88ZOR2384 V. P. Semenov, M. A. Mishin and K. A. Ogloblin; *Zh. Org. Khim.*, 1988, **24**, 2384 (*Chem. Abstr.*, 1989, **111**, 23 144z). 702

88ZOR2389 V. P. Semenov and K. A. Ogloblin; *Zh. Org. Khim.*, 1988, **214**, 2389 (*Chem. Abstr.*, 1989, **111**, 23 145a). 702

89ACS518 K. E. Grung, C. Rømming and J. Songstad; *Acta Chem. Scand.*, 1989, **43**, 518. 1117

89AG(E)221 U. Hartwig, H. Pritzkow, K. Roll and W. Sundermeyer; *Angew. Chem., Int. Ed. Engl.*, 1989, **28**, 221. 652

89AG(E)337 M. Famulok, F. Bosold and G. Boche; *Angew. Chem., Int. Ed. Engl.*, 1989, **28**, 337. 212

89AG(E)397 G. Erker; *Angew. Chem., Int. Ed. Engl.*, 1989, **28**, 397. 931

89AG(E)445 A. Recknagel, D. Stalke, H. W. Roesky and F. T. Edelmann; *Angew. Chem., Int. Ed. Engl.*, 1989, **28**, 445. 893

89AG(E)468 M. Famulok and G. Boche; *Angew. Chem., Int. Ed. Engl.*, 1989, **28**, 468. 212

89AG(E)740 H. Grützmacher and H. Pritzkow; *Angew. Chem., Int. Ed. Engl.*, 1989, **28**, 740. 886

89AG(E)1173 B. M. Trost; *Angew. Chem., Int. Ed. Engl.*, 1989, **28**, 1173. 158

89AG(E)1250 F. Castan, A. Baceiredo and G. Bertrand; *Angew. Chem., Int. Ed. Engl.*, 1989, **28**, 1250. 1076, 1077, 1148

89AG(E)1706 M. T. Reetz and D. Röhrig; *Angew. Chem., Int. Ed. Engl.*, 1989, **28**, 1706. 152

89AJC2275 I. R. Bryant and L. K. Dyall; *Aust. J. Chem.*, 1989, **42**, 2275. 214

89AOC(29)163 H. Le Bozec, D. Touchard and P. H. Dixneuf; *Adv. Organomet. Chem.*, 1989, **29**, 163. 941

89AP(322)343 C. E. Müller and H. J. Roth; *Arch. Pharm.* (*Weinheim, Ger.*), 1989, **322**, 343. 337

89BAU672 V. B. Sokolov, T. A. Epishina and I. V. Martynov; *Bull. Acad. Sci. USSR. Div. Chem. Sci.*, 1989, 672. 669

89BAU2201 V. B. Sokolov, O. V. Korenchenko, T. A. Epishina and I. V. Martynov; *Bull. Acad. Sci. USSR. Div. Chem. Sci.*, 1989, 2201. 672

89BAU2612 V. B. Sokolov, A. N. Chekhlov, T. A. Epishina and I. V. Martynov; *Bull. Acad. Sci. USSR. Div. Chem. Sci.*, 1989, 2612. 669, 672

89BAU2628 Yu. E. Lyashenko, V. B. Sokolov and I. V. Martynov; *Bull. Acad. Sci. USSR. Div. Chem. Sci.*, 1989, 2628. 669, 672

89BCJ539 Y. Saegusa, T. Watanabe and S. Nakamura; *Bull. Chem. Soc. Jpn.*, 1989, **62**, 539. 225

89BCJ2111 K. Yamamoto and M. Takemae; *Bull. Chem. Soc. Jpn.*, 1989, **62**, 2111. 225, 375

89BCJ2968 S.-I. Murahashi, Y. Imada and Y. Hirai; *Bull. Chem. Soc. Jpn.*, 1989, **62**, 2968. 324

89BSB289 J. Garin, E. Melendez, F. Merchan, P. Merino and T. Tejero; *Bull. Soc. Chim. Belg.*, 1989, **98**, 289. 1026

89BSB395 M. Mbuyi, L. Christiaens and M. Renson; *Bull. Soc. Chim. Belg.*, 1989, **98**, 395. 334

89CAR(188)35 J. F. Mota, J. M. G. Fernandez, C. O. Mellet, M. A. P. Adrian and R. B. Caballero; *Carbohydr. Res.*, 1989, **188**, 35. 1028

89CB307 P. Molina, A. Arques, P. M. Fresneda, M. V. Vinader, M. C. Foces-Foces and F. H. Cano; *Chem. Ber.*, 1989, **122**, 307. 1066

89CB365 R. Aumann and P. Hinterding; *Chem. Ber.*, 1989, **122**, 365. 957

89CB389 W. Kirmse and G. Hellwig; *Chem. Ber.*, 1989, **122**, 389. 360

89CB445 W. A. Herrmann, H.-J. Kneuper and E. Herdtweck; *Chem. Ber.*, 1989, **122**, 445. 1097

89CB565 S. Arenz, M. Böhshar, H. Heydt and M. Regitz; *Chem. Ber.*, 1989, **122**, 565. 871

89CB657 K. Hartke, W. Richter, W. Massa and G. Baum; *Chem. Ber.*, 1989, **122**, 657. 685

89CB669 K. Hartke, W. Richter and T. Kämpchen; *Chem. Ber.*, 1989, **122**, 669. 685

89CB1107 D. Korbonits, I. Kanzel-Szvoboda, C. Gönczi, K. Simon and P. Kolonits; *Chem. Ber.*, 1989, **122**, 1107. 220

89CB1147 A. Berkessel and M. Voges; *Chem. Ber.*, 1989, **122**, 1147. 867

89CB1757 H. Fritz and W. Sundermeyer; *Chem. Ber.*, 1989, **122**, 1757. 488, 492

89CB1907 W. P. Fehlhammer and G. Beck; *Chem. Ber.*, 1989, **122**, 1907. 1087

89CC31 M. Larchevêgue, G. Tamagnan and Y. Petit; *J. Chem. Soc., Chem. Commun.*, 1989, 31. 152

89CC414 P. Beslin and S. Perrio; *J. Chem. Soc., Chem. Commun.*, 1989, 414. 557

89CC717 I. Savage and E. J. Thomas; *J. Chem. Soc., Chem. Commun.*, 1989, 717. 234

89CC900 W. J. Mills, L. J. Todd and J. C. Huffman; *J. Chem. Soc., Chem. Commun.*, 1989, 900. 1147

89CC1160 B.-W. A. Yeung, J. L. M. Contelles and B. Fraser-Reid; *J. Chem. Soc., Chem. Commun.*, 1989, 1160. 157

89CC1265 M. Yamamoto, T. Uruma, S. Iwasa, S. Kohmoto and K. Yamada; *J. Chem. Soc., Chem. Commun.*, 1989, 1265. 513

89CC1640 F. Dasgupta and P. J. Garegg; *J. Chem. Soc., Chem. Commun.*, 1989, 1640. 354

89CC1732 M. Yoshifuji, S. Sasaki and N. Inamoto; *J. Chem. Soc., Chem. Commun.*, 1989, 1732. 1092

89CCC2738 N. Sharma and B. N. Misra; *Collect. Czech. Chem. Commun.*, 1989, **54**, 2738. 328, 332, 710, 714

89CCC3245 B. Tinant, J. Dupont-Fenfau, J.-P. Declercq, J. Podlaka and O. Exner; *Collect. Czech. Chem. Commun.*, 1989, **54**, 3245. 655

89CI(L)389 M. P. Kaushik and R. Vaidyanathaswamy; *Chem. Ind.* (*London*), 1989, 389. 371

89CL1001 T. Mukaiyama, H. Uchiro and S. Kobayashi; *Chem. Lett.*, 1989, 1001. 236

89CL1261 Y. Ito, M. Inouye and M. Murakami; *Chem. Lett.*, 1989, 1261. 728

89CL1603 M. Murakami, H. Ito, W. A. Baker, A. B. Baba and Y. Ito; *Chem. Lett.*, 1989, 1603. 802, 803

89CL1775 H. Yamada, S. Ohsawa, T. Sugai and H. Ohta; *Chem. Lett.*, 1989, 1775. 179

89CL1825 Y. Nishiyama, A. Katsuura, Y. Okamoto and S. Hamanaka; *Chem. Lett.*, 1989, 1825. 253

89CL2181 S. Kagabu and T. Inoue; *Chem. Lett.*, 1989, 2181. 344

89CPB1179 K. Ikeda, K. Achiwa and M. Sekiya; *Chem. Pharm. Bull.*, 1989, **37**, 1179. 329

89CPB2334 K. Takeda, K. Tsuboyama, H. Takanagi, R. Shirokami, M. Takeura and H. Ogura; *Chem. Pharm. Bull.*, 1989, **37**, 2334. 1025

89CPB2952 J. Sakaki, S. Kobayashi, M. Sato and C. Kaneko; *Chem. Pharm. Bull.*, 1989, **37**, 2952. 652

89CRV1447 D. J. Hart and D.-C. Ha; *Chem. Rev.*, 1989, 1447. 377

89CRV1927 P. Braunstein and D. Nobel; *Chem. Rev.*, 1989, **89**, 1927. 962

89GEP3811344 G. Luft and P. Trabold (Hoechst AG); *Ger. Pat.* 3 811 344 (1989) (*Chem. Abstr.*, 1990, **112**, 235 610). 192

89GSM89 D. W. Knight; *Gen. Synth. Methods*, 1989, **11**, 89. 24

89H(28)565 J. P. Kutney, F. J. Lopez, S. P. Huang and H. Kurobe; *Heterocycles*, 1989, **28**, 565. 199

89H(29)435 Y. Kosugi, M. Kosugi and F. Hamaguchi; *Heterocycles*, 1989, **29**, 435. 631, 632

89H(29)841 M. Ichinari and Y. Hayase; *Heterocycles*, 1989, **29**, 841. 1112

89H(29)2225 M. J. Brown; *Heterocycles*, 1989, **29**, 2225. 652

89HCA264 M. E. Scheller, W. B. Schweizer and B. Frei, *Helv. Chim. Acta*, 1989, **72**, 264. 430

89HCA556 A. Pascual; *Helv. Chim. Acta*, 1989, **72**, 556. 1064

89HCA1522 V. I. Ognyanov and M. Hesse; *Helv. Chim. Acta*, 1989, **72**, 1522. 307

89HOU(5)633 H. Pielartzik and B. Irmisch-Pielartzik and T. Eicher; *Methoden Org. Chem.* (*Houben-Weyl*), 1985, **5**, 633. 182

89IC815 T. G. Appleton, J. R. Hall and P. D. Prenzler; *Inorg. Chem.*, 1989, **28**, 815. 385

89IJC(A)1528 A. Sandhu, S. Singh, K. K. Bhasin and R. D. Verma; *Indian J. Chem., Sect A*, 1989, **28**, 1528. 210

89IJC(B)186 M. T. Thorat, R. B. Mane, J. G. Deshmukh, M. M. Salunkhe and P. P. Wadgaonkar; *Indian J. Chem., Sect. B.*, 1989, **28**, 186. 367

89IJC(B)673 P. K. Bhattacharya, S. Bandyopadhyay and S. C. Pakrashi; *Indian J. Chem., Sect. B*, 1989, **28**, 673. 601, 621

89IZV418 A. N. Pudovik, V. N. Nazmutdinova and L. P. Chirkova; *Izv. Akad. Nauk SSSR, Ser. Khim.*, 1989, 418. 221

89IZV745 V. B. Sokolov, T. A. Epishina and I. V. Martynov; *Izv. Akad. Nauk SSSR, Ser. Khim.*, 1989, 745 (*Chem. Abstr.*, 1989, **111**, 97 351m). 669

89IZV1319 U. M. Dzhemilev, N. Z. Baibulatova, R. V. Kunakova, T. K. Tkachenko and G. A. Tolstikov; *Izv. Akad. Nauk SSSR, Ser. Khim.*, 1989, 1319 (*Chem. Abstr.*, 1990, **112**, 98 003g). 603

89IZV2353 V. A. Dorokhov, A. R. Amamchyan and V. S. Bogdanov; *Izv. Akad. Nauk SSSR, Ser. Khim.* 1989, **10**, 2353 (*Chem. Abstr.*, 1990, **112**, 216 800q). 778

89IZV2390 V. B. Sokolov, O. V. Korenchenko, T. A. Epishina and I. V. Martynov; *Izv. Akad. Nauk SSSR Ser. Khim.*, 1989, 2390 (*Chem. Abstr.*, 1990, **112**, 234 756j). 672

89IZV2849 V. B. Sokolov, A. N. Chekhlov, T. A. Epishina and I. V. Martynov; *Izv. Akad. Nauk SSSR, Ser. Khim.*, 1989, 2849 (*Chem. Abstr.*, 1990, **113**, 5694r). 669

89IZV2865 Yu. F. Lyashenko, V. B. Sokolov and I. V. Martynov; *Izv. Akad. Nauk SSSR, Ser. Khim.*, 1989, 2865 (*Chem. Abstr.*, 1990, **113**, 40 822c). 669, 672

89JA149 J. Arnold, T. D. Tilley, A. L. Rheingold, S. J. Geib and A. M. Arif; *J. Am. Chem. Soc.*, 1989, **111**, 149. 444

89JA355 P. Molina, M. Alajarín, C. L. Leonardo, R. M. Claramunt, M. C. Foces-Foces, F. H. Cano, J. Catalán, J. L. de Paz and J. Elguero; *J. Am. Chem. Soc.*, 1989, **111**, 355. 1068

89JA1285 K. I. Goldberg and R. G. Bergman; *J. Am. Chem. Soc.*, 1989, **111**, 1285. 450, 935

89JA1886 M. D. Bachi and D. Denenmark; *J. Am. Chem. Soc.*, 1989, **111**, 1886. 739

89JA2984 W. R. Roush, M. R. Michaelides, D. F. Tai, B. M. Lesur, W. K. M. Chong and D. J. Harris; *J. Am. Chem. Soc.*, 1989, **111**, 2984. 152

89JA3258 J. Chang, M. D. Seidler and R. G. Bergman; *J. Am. Chem. Soc.*, 1989, **111**, 3258. 386

89JA3973 L. D. Arnold, H. I. Assil and J. C. Vederas; *J. Am. Chem. Soc.*, 1989, **111**, 3973. 1078

89JA4129 J. M. O'Connor, L. Pu and A. Rheingold; *J. Am. Chem. Soc.*, 1989, 4129. 941

89JA4864 T. Ibuka, M. Tanaka, S. Nishii and Y. Yamamoto; *J. Am. Chem. Soc.*, 1989, **111**, 4864. 158

89JA5949 R. Okazaki, N. Kumon and N. Inamoto; *J. Am. Chem. Soc.*, 1989, **111**, 5949. 1115

89JA8018 E. Negishi, Y. Zhang, I. Shimoyama and G. Wu; *J. Am. Chem. Soc.*, 1989, **111**, 8018. 464

89JA8333 J. Cioslowski; *J. Am. Chem. Soc.*, 1989, **111**, 8333. 1020

89JA9194 L. S. Hegedus and B. R. Lundmark; *J. Am. Chem. Soc.*, 1989, **111**, 9194. 356

89JAP3145452 H. Kawachi, M. Saeki, T. Maruta and T. Nishinomiya; *Jpn. Pat.* 03 145 452 (1989) (*Chem. Abstr.*, 1991, **115**, 256 883n). 1102

89JCP1031 T. P. Hamilton and H. F. Schaefer, III; *J. Chem. Phys.*, 1989, **90**, 1031. 1020

89JCR(S)56 A. Sygula; *J. Chem. Res.* (*S*), 1989, 56. 685

89JCR(S)104 B. Juršíc; *J. Chem. Res.* (*S*), 1989, 104. 1104

89JCR(S)155 F. S. Guziec, Jr. and L. M. Wasmund; *J. Chem. Res.* (*S*), 1989, 155. 575

89JCR(S)260 E. Carvalho, J. Iley, F. Norberto and E. Rosa; *J. Chem. Res.* (*S*), 1989, **8**, 260. 361

89JCR(S)262 P. Molina, M. A. Lorenzo and E. Aller; *J. Chem. Res.* (*S*), 1989, 262. 680

89JCR(S)331 T. Manimaran, L. T. Wolford and J. H. Boyer; *J. Chem. Res.* (*S*), 1989, 331. 964, 1119

89JCR(S)337 A. O. Fitton, J. Hill, D. E. Jane and R. Millar; *J. Chem. Res.* (*S*), 1989, 337. 692

89JCS(D)345 R. D. W. Kemmitt, P. McKenna, D. R. Russell and L. J. S. Prouse; *J. Chem. Soc., Dalton Trans.*, 1989, 345. 802

89JCS(D)947 F. P. Fanizzi, F. P. Intini and G. Natile; *J. Chem. Soc., Dalton Trans*, 1989, 947. 696

89JCS(F2)1945 R. Fausto, L. A. E. Batista de Carvalho, J. J. C. Teixeira-Dias and M. N. Ramos; *J. Chem. Soc., Faraday Trans. 2*, 1989, **85**, 1945. 506, 507

89JCS(P1)765 R. Karaman, A. Goldblum, E. Breuer and H. Leader; *J. Chem. Soc., Perkin Trans. 1*, 1989, 765. 403, 404

89JCS(P1)1319 P. P. McCleery and B. Tuck; *J. Chem. Soc., Perkin Trans. 1*, 1989, 1319. 401

89JCS(P1)1702 S. Kajigaeshi, K. Murakawa, K. Asano, S. Fujisaki and T. Kakinami; *J. Chem. Soc. Perkin 1*, 1989, 1702. 318

89JCS(P1)1815 P. Barraclough, A. G. Caldwell, C. J. Harris, W. P. Jackson and B. J. R. Whittle; *J. Chem. Soc., Perkin Trans. 1*, 1989, 1815. 342

89JCS(P1)1862 Y. Kita, O. Tamura, N. Shibata and T. Miki; *J. Chem. Soc., Perkin Trans. 1*, 1989, 1862. 377

89JCS(P1)1927 T. Toru, T. Seko, E. Maekawa and Y. Ueno; *J. Chem. Soc., Perkin Trans. 1*, 1989, 1927. 249

89JCS(P1)2083 B. F. Bonini, G. Mazzanti, P. Zani and G. Maccagnani; *J. Chem. Soc., Perkin Trans. 1*, 1989, 2083. 648, 650

89JCS(P1)2207 M. R. Harnden and R. L. Jarvest; *J. Chem. Soc., Perkin Trans. 1*, 1989, 2207. 764

89JFC(45)401 W. Tyrra and D. Naumann; *J. Fluorine Chem.*, 1989, **45**, 401. 205, 321

89JGU1123 P. P. Onys'ko, T. V. Kim, E. I. Kiseleva and A. D. Sinitsa; *J. Gen. Chem. USSR (Engl. Transl.)*, 1989, **59**, 1123. 789

89JGU1129 P. P. Onys'ko, T. V. Kim, E. I. Kiseleva and A. D. Sinitsa; *J. Gen. Chem. USSR (Engl. Transl.)*, 1989, **59**, 1129. 788

89JGU1496 P. P. Onys'ko, T. V. Kim, E. I. Kiseleva, M. I. Povolotskii and A. D. Sinitsa; *J. Gen. Chem. USSR (Engl. Trans.)*, 1989, **59**, 1496. 787

89JGU1548 V. A. Al'fonsov, D. A. Pudovik, R. Z. Musin, V. N. Nazmutdinova, Yu. Ya. Efremov, É. S. Batyeva and A. N. Pudovik; *J. Org. Chem. USSR (Engl. Transl.)*, 1989, 1548 (*Zh. Obshch. Khim.*, 1988, **58**, 1734). 535

89JGU1817 Yu. A. Veits, E. G. Neganova, A. A. Borisenko, V. L. Foss and I. F. Lutsenko; *J. Gen. Chem. USSR*, 1989, **59**, 1817. 395, 396, 399

89JGU1977 I. V. Shevchenko, M. V. Furmanova, V. P. Kukhar and O. I. Kolodyazhnyi; *J. Gen. Chem. USSR*, 1989, **59**, 1977. 401

89JGU2055 L. I. Livantsova, O. P. Novikova, G. S. Zaitseva and Yu. I. Baukov; *J. Gen. Chem. USSR (Engl. Transl.)*, 1989, **59**, 2055. 723

89JGU2228 A. D. Sinitsa, N. K. Maidanovich, P. P. Onys'ko and A. K. Shurobura; *J. Gen. Chem. USSR (Engl. Transl.)*, 1989, **59**, 2228. 713, 788

89JGU2388 V. A. Al'fonsov, I. S. Nizamov, A. G. Trusenev, É. S. Batyeva and A. N. Pudovik; *J. Gen. Chem. USSR*, 1989, **59**, 2388. 405

89JHC1319 M. Carmack; *J. Heterocycl. Chem.*, 1989, **26**, 1319. 599

89JHC1341 A. Shafiee, B. Yazdikarimy and S. Sadrai; *J. Heterocycl. Chem.*, 1989, **26**, 1341. 639

89JHC1735 B. A. Dreikorn and P. Unger; *J. Heterocycl. Chem.*, 1989, **26**, 1735. 1029

89JMC1024 M. Abou-Gharbia, J. A. Moyer, U. Patel, M. Webb, G. Schiehser, T. Andree and J. T. Haskins; *J. Med. Chem.*, 1989, **32**, 1024. 334

89JMC1886 P. K. Chakravarty, W. J. Greenlee, W. H. Parsons, A. A. Patchett, P. Combs, A. Roth, R. D. Busch and T. N. Mellin; *J. Med. Chem.*, 1989, **32**, 1886. 370

89JMC2493 J. Wrobel, J. Millen, J. Sredy, A. Dietrich, J. M. Kelly, B. J. Gorham and K. Sestanj; *J. Med. Chem.*, 1989, **32**, 2493. 567, 630

89JOC19 A. Ricci, A. Degl'Innocenti, A. Capperucci and G. Reginato; *J. Org. Chem.*, 1989, **54**, 19. 648, 650

89JOC145 C. Mapelli, G. Turocy, F. L. Switzer and C. H. Stammer; *J. Org. Chem.*, 1989, **54**, 145. 367

89JOC224 J. H. Rigby and N. Balasubramanian; *J. Org. Chem.*, 1989, **54**, 224. 990

89JOC617 P. Jouin, J. Poncet, M.-N. Dufour, A. Pantaloni and B. Castro; *J. Org. Chem.*, 1989, **54**, 617. 333

89JOC1048 J. H. Cooley, E. J. Evain, R. D. Willett and J. T. Blanchette; *J. Org. Chem.*, 1989, **54**, 1048. 1063

89JOC1144 J. B. Hendrickson and M. S. Hussoin; *J. Org. Chem.*, 1989, **54**, 1144. 196

89JOC1212 J. C. Stowell, B. M. Ham, M. A. Esslinger and A. J. Duplantier; *J. Org. Chem.*, 1989, **54**, 1212. 586, 609

89JOC1234 M. D. Bachi and E. Bosch; *J. Org. Chem.*, 1989, **54**, 1234. 513

89JOC1584 T. R. Wagler, Y. Fang and C. J. Burrows; *J. Org. Chem.*, 1989, **54**, 1584. 385

89JOC1904 T. G. Back and K. Brunner; *J. Org. Chem.*, 1989, **54**, 1904. 335

89JOC2015 K. Sukata; *J. Org. Chem.*, 1989, **54**, 2015. 1114

89JOC2798 J. S. Nowick and R. L. Danheiser; *J. Org. Chem.*, 1989, **54**, 2798. 423, 426

89JOC2964 L. Engman and A. Hallberg; *J. Org. Chem.*, 1989, **54**, 2964. 334

89JOC2966 A. Capperucci, A. Degl'Innocenti, C. Faggi, G. Reginato, A. Ricci, P. Dembech and G. Seconi; *J. Org. Chem.*, 1989, **54**, 2966. 467

89JOC3096 P. E. O'Bannon and W. P. Dailey; *J. Org. Chem.*, 1989, **54**, 3096. 326

89JOC3209 P. Romea, F. Urpí and J. Vilarrasa; *J. Org. Chem.*, 1989, **54**, 3209. 360

89JOC3249 W. E. Bauta, W. D. Wulff, S. F. Pavkovic and E. J. Zaluzec; *J. Org. Chem.*, 1989, **54**, 3249. 934

89JOC3394 M. Kawase, T. Kitamura and Y. Kikugawa; *J. Org. Chem.*, 1989, **54**, 3394. 317

89JOC3913 M. R. Angelastro, N. P. Peet and P. Bey; *J. Org. Chem.*, 1989, **54**, 3913. 327

89JOC4019 J. H. Rigby, D. D. Holsworth and K. James; *J. Org. Chem.*, 1989, **54**, 4019. 993

89JOC4088 Y. Watanabe, H. Iida and C. Kibayashi; *J. Org. Chem.*, 1989, **54**, 4088. 322, 330

89JOC4263 A. L. Gutman and T. Bravdo; *J. Org. Chem.*, 1989, **54**, 4263. 652

89JOC4335 A. J. Gutierrez, K. J. Shea and J. J. Svoboda; *J. Org. Chem.*, 1989, **54**, 4335. 688, 694

89JOC4391 N. X. Hu, Y. Aso, T. Otsubo and F. Ogura; *J. Org. Chem.*, 1989, **54**, 4391. 210

89JOC4481 A. Yamashita, A. Toy, N. B. Ghazal and C. R. Muchmore; *J. Org. Chem.*, 1989, **54**, 4481. 951

89JOC4511 I. Ojima, K. Kato, K. Nakahashi, T. Fuchikami and M. Fujita; *J. Org. Chem.*, 1989, **54**, 4511. 224

89JOC4649 G. M. Wieber, L. S. Hegedus, B. Åkermark and E. T. Michalson; *J. Org. Chem.*, 1989, **54**, 4649. 375

89JOC4723 A. G. M. Barrett, J. A. Flygare and C. D. Spilling; *J. Org. Chem.*, 1989, **54**, 4723. 238

89JOC4746 M. V. Lakshmikantham, W. Chen and M. P. Cava; *J. Org. Chem.*, 1989, **54**, 4746. 528, 529

89JOC5746 M. A. Weidner-Wells, A. DeCamp and P. H. Mazzocchi; *J. Org. Chem.*, 1989, **54**, 5746. 325

89JOC5852 J. H. Rigby and M. Qabar; *J. Org. Chem.*, 1989, **54**, 5852. 325

89JOC5945 H. Suginome and Y. Kurokawa; *J. Org. Chem.*, 1989, **54**, 5945. 332

89JOC5967 Z. Rappoport and A. Topol; *J. Org. Chem.*, 1989, **54**, 5967. 1107

89JOM(359)1 R. Lechler, H.-D. Hausen and J. Weidlein; *J. Organomet. Chem.*, 1989, **359**, 1. 1084

89JOM(359)215 J. Zakrzewski; *J. Organomet. Chem.*, 1989, **359**, 215. 387

89JOM(361)63 S. Uemura, H. Takahashi, K. Ohe and N. Sugita; *J. Organomet. Chem.*, 1989, **361**, 63. 242

89JOM(364)C17 J. Heinicke; *J. Organomet. Chem.*, 1989, **364**, C17. 796

89JOM(366)281 A. Asthana and R. C. Srivastava; *J. Organomet. Chem.*, 1989, **366**, 281. 402

89JOM(368)163 I. S. Biltueva, D. A. Bravo-Zhivotovskii, I. D. Kalikhman, V. Y. Vitkovskii, S. G. Shevchenko, N. S. Vyazankin and M. G. Voronkov; *J. Organomet. Chem.*, 1989, **368**, 163. 924

89JOM(375)245 H. Hoberg and D. Guhl; *J. Organomet. Chem.*, 1989, **375**, 245. 386

89JOM(377)157 E. Carmona, J. M. Marin, P. Palma and M. L. Poveda; *J. Organomet. Chem.*, 1989, **377**, 157. 800

89JOM(377)281 M. Herberhold, P. Leitner, C. Dörnhöfer and J. Ott-Lastic; *J. Organomet. Chem.*, 1989, **377**, 281. 241

89JOM(378)279 H. Hoberg and D. Guhl; *J. Organomet. Chem.*, 1989, **378**, 279. 386

89JOU1365 Yu. I. Matveev and V. I. Gorbatenko; *J. Org. Chem. USSR* (*Engl. Transl.*), 1989, **25**, 1365. 658

89JOU2329 V. A. Chervonyi, V. S. Zyabrev, A. V. Kharchenko and B. S. Drach; *J. Org. Chem. USSR* (*Engl. Transl.*), 1989, **25**, 2329. 655

89JPR149 M. A. R. Khayat and F. S. Al-Isa; *J. Prakt. Chem.*, 1989, **331**, 149. 774

89JPR243 W. Thiel and R. Mayer; *J. Prakt. Chem.*, 1989, **331**, 243. 554

89JPS(A)4341 W. Mormann and G. Leukel; *J. Polym. Sci., Polym. Chem., Part A*, 1989, **27** (13), 4341. 1006, 1007

89KFZ1317 V. B. Sokolov, Yu. Ya. Ivanov, T. A. Epishina, R. S. Agabekyan and I. V. Martynov; *Khim. Farm. Zh.*, 1989, **23**, 1317 (*Chem. Abstr.*, 1990, **112**, 191 876d). 669

89LA321 K. Hartke and O. Kunze; *Liebigs Ann. Chem.*, 1989, 321. 557, 559

89LA387 K. Barlos, J. Hondrelis, G. Lonergan, J. Matsoukas and C. Sanida; *Liebigs Ann. Chem.*, 1989, 387. 223

89LA665 J. Danklmaier and H. Hönig; *Liebigs Ann. Chem.*, 1989, 665. 694

89LA739 P. Binger, A. Brinkmann, P. Roefke and B. Schäfer; *Liebigs Ann. Chem.*, 1989, 739. 343

89LA1233 D. Seebach, G. Adam, R. Zibuck, W. Simon, M. Rouilly, W. L. Meyer, J. F. Hinton, T. A. Privett, G. F. Templeton, D. K. Heiny, U. Gisi and H. Binder; *Liebigs Ann. Chem.*, 1989, 1233. 179

B-89MI 502-01 R. M. Williams; "Synthesis of Optically Active α-Amino Acids," Pergamon Press, Oxford, 1989. 84

B-89MI 506-01 R. C. Larock; "Comprehensive Organic Transformations," VCH, New York, 1989. 259

B-89MI 506-02 H. G. Davies, R. H. Green, D. R. Kelly and S. M. Roberts; "Biotransformations in Preparative Organic Chemistry," Academic Press, London, 1989. 262

89MI 506-03 N. J. Turner; *Nat. Prod. Rep.*, 1989, **6**, 625. 262

B-89MI 506-04 H.-D. Jakubke; in "The Peptides, Vol. 9," eds S. Udenfriend and J. Meienhofer, Academic Press, London, 1989. 262

B-89MI 506-05 W. Kullman; "Enzymatic Peptide Synthesis," CRC Press, Florida, 1989. 262

B-89MI 507-01 S. R. Sandler and W. Karo; in "Organic Functional Group Preparation," ed. H. H. Wasserman, Academic Press, New York, 1989, vol. 3, p. 482. 322

89MI 512-01 N. T. Abdel Ghani, I. M. Abbas and A. L. El-Ansary; *Egypt. J. Chem.*, 1989, **32**, 301. 532

89MI 519-01 L. Vio, M. G. Mamolo, G. Ronsisvalle, S. Oliveri and A. M. Grego; *Farmaco*, 1989, **44**, 819 (*Chem. Abstr.*, 1990, **112**, 155 075h). 769

89MI 520-01 A. V. Seleznev; *Organomet. Chem. USSR*, 1989, **2**, 238. 794

89MI 523-01 R. Appel and F. Knoll; *Adv. Inorg. Chem.*, 1989, **33**, 259. 876

89MI 526-01 V. A. Zlobin and A. K. Tarasov; *Izv. Vyssh. Uchebn. Zaved., Khim. Khim. Tekhnol.*, 1989, **32** (12), 35 (*Chem. Abstr.*, 1990, **113**, 131 686c). 989

89MI 528-01 A. S. Gordetzov, V. P. Kozyukov, I. A. Vostokov, Yu. I. Mushkin, R. P. Zakharova and Yu. I. Dergunov; *MetalloOrg. Khim.*, 1989, **2**, 1129. 1085

89MI 530-01 A. Sood and B. F. Spielvogel; *Main Group Met. Chem.*, 1989, **12**, 143. 1147

89MI 531-01 M. Regitz and M. Hermesdorf, Unpublished Results, Kaiserslautern, 1989. 1154, 1155

89NKK822 I. Furukawa, N. Abe and S. Hashimoto; *Nippon Kagaku Kaishi* (*J. Chem. Soc. Jpn.*), 1989, 822. 1025

89NKK1881 S. Norio, H. Kiyotaka, I. Shota, A. Takasi and T. Takashi; *Nippon Kagaku Kaishi* (*J. Chem. Soc. Jpn.*), 1989, **11**, 1881 (*Chem. Abstr.*, 1990, **112**, 216 202w). 1012

89OM655 G. W. Kabalka and D. E. Bierer; *Organometallics*, 1989, **8**, 655. 223

89OM693 K. M. Baines, A. G. Brook, R. R. Ford, P. D. Lickiss, A. K. Saxena, W. J. Chatterton, J. F. Sawyer and B. A. Behnam; *Organometallics*, 1989, **8**, 693. 418, 924, 925

89OM709 K. M. Baines, A. G. Brook, P. D. Lickiss and J. F. Sawyer; *Organometallics*, 1989, **8**, 709. 925

89OM850 A. G. Brook, A. K. Saxena and J. F. Sawyer; *Organometallics*, 1989, **8**, 850. 924

89OM1275 D. W. Macomber, P. Madhukar and R. D. Rogers; *Organometallics*, 1989, **8**, 1275. 955

89OM1523 P. L. Motz, J. P. Williams, J. J. Alexander and D. M. Ho; *Organometallics*, 1989, **8**, 1523. 799

89OM2075 P. M. Zizelman and J. M. Stryker; *Organometallics*, 1989, **8**, 2075. 948

89OM2253 C. Alvarez, A. Parlier, H. Rudler, R. Yefsah, J. C. Daran and C. Knobler; *Organometallics*, 1989, **8**, 2253. 798
89OM2282 J. R. Gandler and C. Bernasconi; *Organometallics*, 1989, **8**, 2282. 955
89OM2589 P. L. Motz, J. J. Alexander and D. M. Ho; *Organometallics*, 1989, **8**, 2589. 799
89OPP228 S. Muthusamy and V. T. Ramakrishnan; *Org. Prep. Proceed. Int.*, 1989, **21**, 228. 1039
89OPP517 I. Karatas and C. Tuzun; *Org. Prep. Proced. Int.*, 1989, **21**, 517. 664, 821
89OS(67)69 O. P. Goel, U. Krolls, M. Stier and S. Kesten; *Org. Synth.*, 1989, **67**, 69. 327
89OS(68)198 R. T. Taylor, M. W. Pelter and L. A. Paquette; *Org. Synth.*, 1989, **68**, 198. 30
89OS(68)41 J.-P. Depres and A. E. Greene; *Org. Synth.*, 1989, **68**, 41. 49
89OS(68)83 J. R. Gage and D. A. Evans; *Org. Synth.*, 1989, **68**, 83. 80
89PAC1723 G. L. Geoffroy, J. B. Sheridan, S. L. Bassner and C. Kelley; *Pure Appl. Chem.*, 1989, **61**, 1723. 449
89PJC265 R. Guryn; *Pol. J. Chem.*, 1989, **63**, 265. 688
89POL581 P. Raj, A. K. Aggarwal and N. Misra; *Polyhedron*, 1989, **8**, 581. 318, 369
89PS(29)31 M. Hermesdorf, H. Heydt and M. Regitz; *Phosphorus Sulfur*, 1989, **29**, 31. 1153
89PS(43)1 B. Zwanenburg; *Phosphorus Sulfur*, 1989, **43**, 1. 562
89PS(43)129 J. Voss; *Phosphorus Sulfur*, 1989, **43**, 129. 550
89RHA152 P. Metzner; *Rev. Heteroatom Chem.*, 1989, 152. 550, 556
89RTC51 C. Winkel, E. G. Buitenhuis and J. Lugtenburg; *Recl. Trav. Chim. Pays-Bas*, 1989, **108**, 51. 15
89S97 T. Schrader and W. Steglich; *Synthesis*, 1989, 97. 790
89S202 W. Schroth and J. Andersch; *Synthesis*, 1989, 202. 600, 621
89S256 U. Schmidt, M. Respondek, A. Lieberknecht, J. Werner and P. Fischer; *Synthesis*, 1989, 256. 692
89S272 W. Hartmann; *Synthesis*, 1989, 272. 10
89S300 B. Olejniczak and A. Zwierzak; *Synthesis*, 1989, 300. 1031
89S334 Y. Kita, S. Akai, M. Yamamoto, M. Taniguchi and Y. Tamura; *Synthesis*, 1989, 334. 215
89S506 T. Janecki and R. Bodalski; *Synthesis*, 1989, 506. 12
89S566 G. A. Olah, A-h. Wu and O. Farooq; *Synthesis*, 1989, 566. 8
89S598 R. J. Alabaster, I. F. Cottrell, D. Hands, G. R. Humphrey, D. J. Kennedy and S. H. B. Wright; *Synthesis*, 1989, 598. 8
89S618 R. Chen and R. Bao; *Synthesis*, 1989, 618. 370
89S647 A. Ricci and A. Degl'Innocenti; *Synthesis*, 1989, **9**, 647. 409
89S832 U. Schmidt, M. Kroner and H. Griesser; *Synthesis*, 1989, 832. 15
89S843 P. Molina, A. Lorenzo and E. Aller; *Synthesis*, 1989, 843. 680
89S929 H. Ishihara, S. Muto, T. Endo, M. Komada and S. Kato; *Synthesis*, 1989, 929. 250
89SC275 P. Cuadrado, A. M. González, B. González and F. J. Pulido; *Synth. Commun.*, 1989, **19**, 275. 417
89SC561 S. Ohiro; *Synth. Commun.*, 1989, **19**, 561. 872
89SC1181 J. R. Salim, F. Nome and M. C. Rezende; *Synth. Commun.*, 1989, **19**, 1181. 652
89SC1533 P. Wepplo; *Synth. Commun.*, 1989, **19**, 1533. 233
89SC3113 D. Mondeshka, V. Parashikov and C. Angelov; *Synth. Commun.*, 1989, **19**, 3113. 344
89SC3331 S. G. Burton and P. T. Kaye; *Synth. Commun.*, 1989, **19**, 3331. 197
89SL30 T. V. Lee, A. J. Leigh and C. B. Chapleo; *Synlett*, 1989, 30. 199
89SL32 A. Miller, T. M. Paterson and G. Procter; *Synlett*, 1989, 32. 326, 330
89SUL117 P. A. Harris, A. Jackson and J. A. Joule; *Sulfur Letters*, 1989, **10**, 117. 603
89T259 I. R. Dunkin, C. J. Shields and H. Quast; *Tetrahedron*, 1989, **45**, 259. 1073
89T363 E. Piers, J. M. Chong and H. E. Morton; *Tetrahedron*, 1989, **45**, 363. 160
89T2041 K. Kpegba, P. Metzner and R. Rakotonirina; *Tetrahedron*, 1989, **45**, 2041. 556
89T3299 A. McKillop and D. Kemp; *Tetrahedron*, 1989, **45**, 3299. 204
89T3535 A. Hassner and B. Fischer; *Tetrahedron*, 1989, **45**, 3535. 355
89T4263 P. Molina, M. Alajarín, A. Vidal, M. C. Foces-Foces and F. H. Cano; *Tetrahedron*, 1989, **45**, 4263. 1066
89T4445 P. Beslin and M.-C. Houtteville; *Tetrahedron*, 1989, **45**, 4445. 557
89T4599 N. M. Yousif; *Tetrahedron*, 1989, **45**, 4599. 581, 643
89T5979 S. A. Ali, J. H. Khan, M. I. M. Wazeer and H. P. Perzanowski; *Tetrahedron*, 1989, **45**, 5979. 201
89T6019 L. N. Markovski and V. D. Romanenko; *Tetrahedron*, 1989, **45**, 6019. 876
89T6771 I. T. Barnish, C. W. G. Fishwick, D. R. Hill and C. Szantay Jr.; *Tetrahedron*, 1989, **45**, 6771. 566, 613
89T7411 G. Petrillo, M. Novi, G. Garbarino and M. Filiberti; *Tetrahedron*, 1989, **45**, 7411. 1111
89T7427 L. Bruché and G. Zecchi; *Tetrahedron*, 1989, **45**, 7427. 677
89TL177 M. Gouygou, C. Tachon, R. El Ouatib, O. Ramarijaona, G. Etemad-Moghadam and M. Koenig; *Tetrahedron Lett.*, 1989, **30**, 177. 1092
89TL187 M. Yoshifuji, S. Sasaki, T. Niitsu and N. Inamoto; *Tetrahedron Lett.*, 1989, **30**, 187. 1092
89TL459 E. Niecke and D. Barion; *Tetrahedron Lett.*, 1989, **30**, 459. 885
89TL515 S. R. Angle and D. O. Arnaiz; *Tetrahedron Lett.*, 1989, **30**, 515. 227
89TL839 M. Yoshifuji, S. Sasaki and N. Inamoto; *Tetrahedron Lett.*, 1989, **30**, 839. 1092
89TL863 N. Torra, F. Urpí and J. Vilarrasa; *Tetrahedron*, 1989, **45**, 863. 360, 361
89TL1073 K. Hartke, H.-D. Gerber and U. Roesrath; *Tetrahedron Lett.*, 1989, 1073. 513, 549, 560
89TL1633 W. Flitsch, J. Lauterwein and W. Micke; *Tetrahedron Lett.*, 1989, **30**, 1633. 934
89TL1679 E. Bruni, G. Cardillo, M. Orena, S. Sandri and C. Tomasini; *Tetrahedron Lett.*, 1989, **30**, 1679. 1125
89TL1773 D. J. Hlasta and J. J. Court; *Tetrahedron Lett.*, 1989, **30**, 1773. 331

89TL1829 T. Kanda, S. Nakaiida, T. Murai and S. Kato; *Tetrahedron Lett.*, 1989, **30**, 1829. 242, 506, 509

89TL1927 R. Knorr, A. Trzeciak, W. Bannworth and D. Gillessen; *Tetrahedron Lett.*, 1989, **30**, 1927. 279, 281

89TL1997 M. M. Campbell, B. C. Ross and G. Semple; *Tetrahedron Lett.*, 1989, **30**, 1997. 574

89TL2173 G. Stork and K. Zhao; *Tetrahedron Lett.*, 1989, **30**, 2173. 896

89TL2317 M. Kahn and S. Bertenshaw; *Tetrahedron Lett.*, 1989, **30**, 2317. 366

89TL2379 F. Jung, A. Olivier, D. Boucherot and F. Loftus; *Tetrahedron Lett.*, 1989, 2379. 1083

89TL2743 F. G. Fang, G. B. Feigelson and S. J. Danishefsky; *Tetrahedron Lett.*, 1989, **30**, 2743. 631

89TL2847 P. Molina, M. Alajarín and A. Vidal; *Tetrahedron Lett.*, 1989, 2847. 622, 1067

89TL3069 D. Person and M. Le Corre; *Tetrahedron Lett.*, 1989, **30**, 3069. 694

89TL3073 B. Rigo, D. Fasseur, P. Cauliez and D. Couturier; *Tetrahedron Lett.*, 1989, **30**, 3073. 227

89TL3189 P. A. Harris, A. Jackson and J. A. Joule; *Tetrahedron Lett.*, 1989, **30**, 3189. 603

89TL3217 R. C. F. Jones and J. M. Patience; *Tetrahedron Lett.*, 1989, **30**, 3217. 342

89TL3219 C. L. Branch, S. C. Finch and M. J. Pearson; *Tetrahedron Lett.*, 1989, **30**, 3219. 345

89TL3295 R. F. Menezes, C. A. Zezza, J. Sheu and M. B. Smith; *Tetrahedron Lett.*, 1989, **30**, 3295. 199

89TL3799 M. Yanagida, K. Hashimoto, M. Ishida, H. Shinozaki and H. Shirahama; *Tetrahedron Lett.*, 1989, **30**, 3799. 157

89TL4061 W. D. Wulff, B. A. Anderson and L. D. Isaacs; *Tetrahedron Lett.*, 1989, **30**, 4061. 956

89TL4443 A. I. D. Alanine and C. W. G. Fishwick; *Tetrahedron Lett.*, 1989, **30**, 4443. 685, 726

89TL4449 I. T. Barnish, C. W. G. Fishwick, D. R. Hill and C. Szantay Jr.; *Tetrahedron Lett.*, 1989, **30**, 4449. 613

89TL4845 S. Hatakeyama, K. Osanai, H. Numata and S. Takano; *Tetrahedron Lett.*, 1989, **30**, 4845. 155

89TL5261 H. J. Bestmann, H. C. Rippel and R. Dostalek; *Tetrahedron Lett.*, 1989, **30**, 5261. 896

89TL5433 M. Yoshifuji, T. Niitsu, D. Shiomi and N. Inamoto; *Tetrahedron Lett.*, 1989, **30**, 5433. 1092

89TL5821 A. S. Kende, K. Kawamura and M. J. Orwat; *Tetrahedron Lett.*, 1989, **30**, 5821. 223

89TL5841 J.-C. Cuevas, P. Patil and V. Snieckus; *Tetrahedron Lett.*, 1989, **30**, 5841. 289

89TL6009 W. Oppolzer, R. Moretti and S. Thomi; *Tetrahedron Lett.*, 1989, **30**, 6009. 342

89TL6095 S. Stavber, Z. Planinsek and M. Zupan; *Tetrahedron Lett.*, 1989, **30**, 6095. 6

89TL6529 T. Nakata, T. Suenaga, K. Nakashima and T. Oishi; *Tetrahedron Lett.*, 1989, **30**, 6529. 157

89TL7061 A. Brouillard-Poichet, A. Defoin and J. Streith; *Tetrahedron Lett.*, 1989, **30**, 7061. 330

89UKZ1071 M. V. Vovk, E. A. Romanenko, V. V. Pirozhenko, A. A. Esipenko, V. I. Bolko, V. N. Fetuykhin and L. I. Samarai; *Ukr. Khim. Zh.* (*Russ. Ed.*), 1989, **55**, 1071 (*Chem. Abstr.*, 1990, **113**, 5855u). 669

89USP4857550 V. Kameswaran, D. P. Wright, Jr. and R. A. Herman; *US Pat.* 4 857 550 (1989) (*Chem. Abstr.*, 1990, **112**, 55 266L). 1126

89ZAAC(568)35 H. Köhler, M. Kretschmann, L. Jäger, W. Gebel, R. Uebel, P. Heine and R. Fischer; *Z. Anorg. Allg. Chem.*, 1989, **568**, 35. 1074

89ZAAC(574)130 M. Domagala, F. Huber and H. Preut; *Z. Anorg. Allg. Chem.*, 1989, **574**, 130. 222

89ZAAC(575)10 E. Hartmann, K. Dehnicke and D. Fenske; *Z. Anorg. Allg. Chem.*, 1989, **575**, 10. 781

89ZAAC(576)203 R. Köhler, J. Sieler, R. Richter, E. Hoyer, L. Beyer and K. Maartmann-Moe; *Z. Anorg. Allg. Chem.*, 1989, **576**, 203. 657

89ZAAC(577)165 H. Köhler, M. Kretschmann, R. Skirl, D. Glanz and L. Jäger; *Z. Anorg. Allg. Chem.*, 1989, **577**, 165. 1074

89ZC71 J. Heinicke; *Z. Chem.*, 1989, **29**, 71. 790

89ZC145 H. Kohler, L. Jager and W. Gebel; *Z. Chem.*, 1989, **29**, 145. 1074, 1078

89ZC396 O. I. Kolodiaznyi; *Z. Chem.*, 1989, **29**, 396. 894

89ZN(B)94 H. Plenio and H. W. Roesky, *Z. Naturforsch.*, *Teil B*, 1989, **44**, 94. 1084

89ZN(B)419 W. P. Fehlhammer, H. Hoffmeister, H. Stolzenberg and B. Boyadjiev; *Z. Naturforsch.*, *Teil B*, 1989, **44**, 419. 1147

89ZN(B)1050 H. Kageyama, H. Tsutsumi, T. Murai and S. Kato; *Z. Naturforsch.*, *Teil B*, 1989, **44**, 1050. 243

89ZN(B)1414 W. P. Fehlhammer and G. Beck; *Z. Naturforsch.*, *Teil B*, 1989, **44**, 1414. 658

89ZOB611 V. I. Naddaka, A. P. Kinyazev and K. V. Avanesyan; *Zh. Obshch. Khim.*, 1989, **59**, 611. 340

89ZOB1917 P. P. Onys'ko, E. A. Suvalova, T. I. Chudakova, T. V. Kim, E. I. Kiseleva and A. D. Sinitsa; *Zh. Obshch. Khim.* (*Engl. Trans.*), 1989, **59**, 1917. 644

89ZOB2293 L. I. Livantsova, O. P. Novikova, G. S. Zaitseva and Yu. I. Baukov; *Zh. Obshch. Khim.*, 1989, **59**, 2293, (*Chem. Abstr.*, 1990, **112**, 118 929u). 723

89ZOB2392 A. R. Burilov, I. L. Nikolaeva, M. A. Pudovik and A. N. Pudovik; *Zh. Obshch. Khim.*, 1989, **59**, 2392. 369

89ZOB2492 A. D. Sinitsa, N. K. Maidanovich, P. P. Onys'ko and A. K. Shurubara; *Zh. Obshch. Khim.*, 1989, **59**, 2492 (*Chem. Abstr.*, 1990, **112**, 217 078x). 713

89ZOR759 M. V. Vovk, M. M. Bretski and V. I. Dorokhov; *Zh. Org. Khim.*, 1989, **25**, 759. 1068, 1069

89ZOR885 D. N. Rakhimov, N. P. Abdullaev and Kh. M. Shakhidoyatov, *Zh. Org. Khim.*, 1989, **25** (4), 885 (*Chem. Abstr.*, 1990, **112**, 76 557p). 977

89ZOR1323 Y. A. Sharanin, A. M. Shestopalov, V. N. Nesterov, S. N. Melenchuk, V. K. Promonenkov, V. E. Shklover, Y. T. Struchkov and V. P. Litvinov; *Zh. Org. Khim.*, 1989, **25**, 1323 (*Chem. Abstr.*, 1990, **112**, 76 896g). 615

89ZOR1523 I. V. Kolesnikova, T. D. Petrova, V. E. Platonov, T. G. Ryabicheva, V. A. Mikhailov, A. A. Popov and V. A. Savelova; *Zh. Org. Khim.*, 1989, **25**, 1523. 1070

89ZOR1572 Yu. I. Matveev and V. I. Gorbatenko; *Zh. Org. Chim.*, 1989, **25**, 1572. 658, 1068

89ZOR1689 I. V. Kolesnikova, T. D. Petriva, V. E. Platonova, T. G. Ryabicheva, V. A. Mikhailov, A. A. Popov and V. A. Savelova; *Zh. Org. Khim.*, 1989, **25**, 1689. 1070

89ZOR2196 I. I. Krasavtsev, Yu. V. Korotkii and M. O. Lozinskii; *Zh. Org. Khim.*, 1989, **25**, 2196. 1062

89ZOR2597 V. A. Chervonyi, V. S. Zyabrev, A. V. Kharchenko and B. S. Drach; *Zh. Org. Khim.*, 1989, **25**, 2597 (*Chem. Abstr.*, 1990, **113**, 23 863z). 655, 1053

90ACH423 A. M. A. El-Khamry, M. M. Habashy, S. El-Nagdy and F. A. El-Bassiouny; *Acta Chim. Acad. Sci. Hung.*, 1990, **3**, 423. 652
90AF1030 V. Kalcheva, Z. Mincheva and P. Andreeva; *Arzneim.-Forsch*, 1990, **40**, 1030. 376
90AG(E)60 H. Pritzkow, K. Rall, S. Reimann-Andersen and W. Sundermeyer; *Angew. Chem., Int. Ed. Engl.*, 1990, **29**, 60. 502
90AG(E)90 M. Weidenbruch, B. Brand-Roth, S. Pohl and W. Saak; *Angew. Chem., Int. Ed. Engl.*, 1990, **29**, 90. 1085
90AG(E)199 S. Channareddy, G. Linti and H. Nöth; *Angew. Chem., Int. Ed. Engl.*, 1990, **29**, 199. 927
90AG(E)282 K. Smith and G. J. Pritchard; *Angew. Chem., Int. Ed. Engl.*, 1990, **29**, 282. 438
90AG(E)318 W. Beck, W. Knauer and C. Robl; *Angew. Chem., Int. Ed. Engl.*, 1990, **29**, 318. 1097
90AG(E)398 C. Wieczorek, J. Allwohn, G. Schmidt-Lukasch, R. Hunold, W. Massa and A. Berndt; *Angew. Chem., Int. Ed. Engl.*, 1990, **29**, 398. 928
90AG(E)399 M. Pilz, J. Allwohn, W. Massa and A. Berndt; *Angew. Chem., Int. Ed. Engl.*, 1990, **29**, 399. 929
90AG(E)1050 W. Günther and H. Kunz; *Angew. Chem., Int. Ed. Engl.*, 1990, **29**, 1050. 1042
90AG(E)1147 G. Besenyei, S. Németh and L. I. Simándi; *Angew. Chem., Int. Ed. Engl.*, 1990, **29**, 1147. 987
90AOC(30)1 S. G. Davies, J. P. McNally and A. J. Smallridge; *Adv. Organomet. Chem.*, 1990, **30**, 1. 941
90AP(323)181 A. Burchardt and D. Geffken; *Arch. Pharm. (Weinheim, Ger.)*, 1990, **323**, 181. 323
90BAU294 O. A. Luk'yanov, Yu. B. Salamonov, A. G. Bass, Yu. A. Strelenko and N. F. Karpenko; *Bull. Acad. Sci. USSR. Div. Chem. Sci.*, 1990, 294. 680
90BAU415 T. A. Epishina, V. B. Sokolov and I. V. Martynov; *Bull. Acad. Sci. USSR. Div. Chem. Sci.*, 1990, 415. 669
90BAU1007 V. B. Sokolov, A. N. Ivanov, T. V. Goreva, T. A. Epishina and I. V. Martynov; *Bull. Acad. Sci. USSR. Div. Chem. Sci.*, 1990, 1007. 669, 672
90BAU1538 D. N. Belen'kii and M. M. Krayushkin; *Bull. Acad. Sci. USSR. Div. Chem. Sci.*, 1990, 1538. 666
90BAU1725 R. A. Bekker and V. Y. Popkova; *Bull. Acad. Sci. USSR, Div. Chem. Sci.*, 1990, 1725. 475
90BAU2424 A. S. Ionkin, S. N. Ignateva and B. A. Arbuzov; *Bull. Acad. Sci. USSR, Div. Chem. Sci.*, 1990, 2424. 395
90BCJ638 S. Kato, T. Fukushima, H. Ishihara and T. Murai; *Bull. Chem. Soc. Jpn.*, 1990, **63**, 638. 510, 561, 563
90BCJ825 K. Sukata; *Bull. Chem. Soc. Jpn.*, 1990, **63**, 825. 1114
90BCJ835 H. Ishihara, M. Yoshimi, N. Hara, H. Ando and S. Kato; *Bull. Chem. Soc. Jpn.*, 1990, **63**, 835. 516, 519, 522, 523, 626
90BCJ2728 T. Oshikawa and M. Yamashita; *Bull. Chem. Soc. Jpn.*, 1990, **63**, 2728. 403, 404
90BCJ2736 T. Niitsu, N. Inamoto, K. Toyota and M. Yoshifuji; *Bull. Chem. Soc. Jpn.*, 1990, **63**, 2736. 1086
90BSB281 G. L'Abbé; *Bull. Soc. Chim. Belg.*, 1990, **99**, 281. 622
90BSF446 A. Guigné and P. Metzner; *Bull. Soc. Chim. Fr.*, 1990, **127**, 446. 556
90CB63 A. Brandl, G. Geisberger and H. Nöth; *Chem. Ber.*, 1990, **123**, 63. 382
90CB455 G. Geisberger, K. Neukirchinger and H. Nöth; *Chem. Ber.*, 1990, **123**, 455. 381, 382, 1085
90CB611 R. Aumann and P. Hinterding; *Chem. Ber.*, 1990, **123**, 611. 957
90CB733 L. Weber, M. Frebel and R. Boese; *Chem. Ber.*, 1990, **123**, 733. 1092
90CB1537 J. C. Jochims and M. O. Glocker; *Chem. Ber.*, 1990, **123**, 1537. 1147
90CB1563 G. Opitz, K. Rieth and T. Ehlis; *Chem. Ber.*, 1990, **123**, 1563. 489
90CB1617 P. Binger, B. Biedenbach, A. T. Herrmann, F. Langhauser, P. Betz, R. Goddard and C. Krüger; *Chem. Ber.*, 1990, **123**, 1617. 890
90CB1989 G. Opitz, T. Ehlis and K. Rieth; *Chem. Ber.*, 1990, **123**, 1989. 489
90CB2047 R. Aumann and P. Hinterding; *Chem. Ber.*, 1990, **123**, 2047. 957
90CB2053 R. Aumann and J. Schröder; *Chem. Ber.*, 1990, **123**, 2053. 950, 951
90CC7 P. Molina, M. Alajarín and A. Vidal; *J. Chem. Soc., Chem. Commun.*, 1990, 7. 1066
90CC641 D. Devanne and P. H. Dixneuf; *J. Chem. Soc., Chem. Commun.*, 1990, 641. 940, 941
90CC648 P. Honicke-Schmidt and M. P. Schneider; *J. Chem. Soc., Chem. Commun.*, 1990, 648. 118
90CC720 J. E. Baldwin, R. M. Adlington, D. W. Gollins and C. J. Schofield; *J. Chem. Soc., Chem. Commun.*, 1990, 720. 329
90CC1213 T. Kimura, M. Fujita and T. Ando; *J. Chem. Soc., Chem. Commun.*, 1990, 1213. 1039
90CHE1170 G. K. Nalbandyan, A. P. Mkrtchyan, R. A. Akopyan and S. A. Vartanyan; *Chem. Heterocycl. Compd. (Engl. Transl.)*, 1990, 1170. 690
90CJC502 A. P. Masters and T. S. Sorensen; *Can. J. Chem.*, 1990, **68**, 502. 447
90CL559 H. Suzuki, H. Shimizu, T. Inamatsu, H. Tani and R. Tamura; *Chem. Lett.*, 1990, 559. 665
90CL905 K. Saigo, S. Shimada and M. Hasegawa; *Chem. Lett.*, 1990, 905. 344
90CL1019 T. Mukaiyama, H. Uchiro, I. Shiina and S. Kobayashi; *Chem. Lett.*, 1990, 1019. 236
90CL1403 K. Shimada, S. Hikage, Y. Takeishi and Y. Takikawa; *Chem. Lett.*, 1990, 1403. 625
90CL1549 Y. Kamiya, K. Kawato and H. Ohta; *Chem. Lett.*, 1990, 1549. 164
90CL1651 T. Ogawa, K. Miyazaki and H. Suzuki; *Chem. Lett.*, 1990, 1651. 630
90CPB1389 H. Maeda, T. Kawaguchi, M. Masui and H. Ohmori; *Chem. Pharm. Bull.*, 1990, **38**, 1389. 1107
90CPB1836 Y. Kita, M. Kirihara, J. Sekihachi, R. Okunaka, M. Sasho, S. Mohri, T. Honda, S. Akai, Y. Tamura and K. Shimooka; *Chem. Pharm. Bull.*, 1990, **38**, 1836. 223
90CPB2352 A. Gueiffier, J. C. Milhavet, Y. Blache, O. Chavignon, J. C. Teulade, M. Madesclaire, H. Viols, G. Dauphin and J. P. Chapat; *Chem. Pharm. Bull.*, 1990, **38**, 2352. 360
90CPB3184 N. Katagiri, H. Sato and C. Kaneko; *Chem. Pharm. Bull.*, 1990, **38**, 3184. 367
90CRV191 M. Regitz; *Chem. Rev.*, 1990, **90**, 191. 1152, 1159
90CRV283 J. Barrau, J. Escudié and J. Satgé; *Chem. Rev.*, 1990, **90**, 283. 927
90CSR147 P. C. B. Page, S. S. Klair and S. Rosenthal; *Chem. Soc. Rev.*, 1990, **19**, 147. 409, 411, 412, 417
90CZ71 P. Aslanidis and F. Kober; *Chem.-Ztg.*, 1990, **114**, 71. 222
90CZ295 K. Dehnicke; *Chem.-Ztg.*, 1990, **114**, 295. 777

90GSM91 D. W. Knight; *Gen. Synth. Methods*, 1990, **12**, 91. 24

90H(30)367 K. Takeda and K. Horiki; *Heterocycles*, 1990, **30**, 367. 1111

90H(31)247 H. M. Hassaneen, A. S. Shawali and N. M. Elwan; *Heterocycles*, 1990, **31**, 247. 675

90IC20 R. L. Angel, D. P. Fairlie and W. G. Jackson; *Inorg. Chem.*, 1990, **29**, 20. 385

90IC140 D. P. Fairlie and W. G. Jackson; *Inorg. Chem.*, 1990, **29**, 140. 780

90IC309 A. A. Khatib and H. H. Sisler; *Inorg. Chem.*, 1990, **29**, 309. 354

90IC3218 M. Mittakanti, M. R. M. D. Charandabi and K. W. Morse; *Inorg. Chem.*, 1990, **29**, 3218. 688

90IC4255 G. Sarwar, R. L. Kirchmeier and J. M. Shreeve; *Inorg. Chem.*, 1990, **29**, 4255. 679, 703

90ICA203 D. P. Fairlie and W. G. Jackson; *Inorg. Chim. Acta*, 1990, **175**, 203. 385

90IZV352 O. A. Luk'yanov, Yu. B. Salamonov, A. G. Bass, Yu. A. Strelenko and N. F. Karpenko; *Izv. Akad. Nauk SSSR, Ser., Khim*, 1990, 352 (*Chem. Abstr.*, 1990, **113**, 131 641j). 680

90IZV477 T. A. Epishina, V. B. Sokolov and I. V. Martynov; *Izv. Akad. Nauk SSSR, Ser. Khim.*, 1990, 477 (*Chem. Abstr.*, 1990, **113**, 114 602e). 669

90IZV1122 V. B. Sokolov, A. N. Ivanov, T. V. Goreva, T. A. Epishina and I. V. Martynov; *Izv. Akad. Nauk SSSR, Ser. Khim.*, 1990, 1122 (*Chem. Abstr.*, 1990, **113**, 115 438t). 669, 672

90IZV1139 G. M. Khusnullina, N. A. Buina and I. A. Nuretdinov; *Isv. Akad. Nauk SSSR, Ser. Khim.*, 1990, **5**, 1139. 371

90IZV1692 D. B. Brokhovetskii, L. I. Belen'kii and M. M. Krayushkin; *Izv. Akad. Nauk SSSR, Ser. Khim.*, 1990, 1692 (*Chem. Abstr.*, 1991, **115**, 279 557y). 666

90IZV1898 R. A. Bekker and V. Y. Popkova; *Izv. Akad. Nauk SSSR, Ser. Khim.*, 1990, 1898 (*Chem. Abstr.*, 1991, **114**, 42 041s). 583

90JA455 T. Hiiro, Y. Morita, T. Inoue, N. Kambe, A. Ogawa, I. Ryu and N. Sonoda; *J. Am. Chem. Soc.*, 1990, **112**, 455. 437

90JA462 P. Magnus and B. Mugrage; *J. Am. Chem. Soc.*, 1990, **112**, 462. 344

90JA464 J. F. Hoover and J. M. Stryker; *J. Am. Chem. Soc.*, 1990, **112**, 464. 944

90JA2437 M. Murakami, T. Kawano and Y. Ito; *J. Am. Chem. Soc.*, 1990, **112**, 2347. 803, 804

90JA2849 A. L. Margolin, D. L.: Delinck and M. R. Whalon; *J. Am. Chem. Soc.*, 1990, **112**, 2849. 135

90JA3029 K. C. Nicolaou, S. A. DeFrees, C.-K. Hwang, N. Stylianides, P. J. Carroll and J. P. Synder; *J. Am. Chem. Soc.*, 1990, **112**, 3029. 513

90JA3040 K. C. Nicolaou, C.-K. Hwang, B. E. Marron, S. A. DeFrees, E. A. Couladouros, Y. Abe, P. J. Carroll and J. P. Synder; *J. Am. Chem. Soc.*, 1990, **112**, 3040. 513

90JA3220 N. Lugan, C. Kelley, M. R. Terry, G. L. Geoffroy and A. L. Rheingold; *J. Am. Chem. Soc.*, 1990, **112**, 3220. 946

90JA3332 G. H. Jeong, R. Boucher and K. J. Klabunde; *J. Am. Chem. Soc.*, 1990, **112**, 3332. 1096

90JA4083 C. P. Casey, P. C. Vosejpka and J. A. Gavney; *J. Am. Chem. Soc.*, 1990, **112**, 4083. 450

90JA4183 J. Cioslowski, T. Hamilton, G. Scuseria, B. Andes Hess, Jr., J. Hu, L. J. Schaad and M. Dupuis; *J. Am. Chem. Soc.*, 1990, **112**, 4183. 1020

90JA5677 M. Ochiai, Y. Takaoka, Y. Masaki, Y. Nagao and M. Shiro; *J. Am. Chem. Soc.*, 1990, **112**, 5677. 204

90JA6232 J. M. O'Connor, L. Pu and A. L. Rheingold; *J. Am. Chem. Soc.*, 1990, **112**, 6232. 944

90JA6255 J. Montgomery, G. M. Wieber and L. S. Hegedus; *J. Am. Chem. Soc.*, 1990, **112**, 6255. 953

90JA6263 K. C. Nicolaou, D. G. McGarry, P. K. Somers, B. H. Kim, W. W. Ogilvie, G. Yiannikouros, C. V. C. Prasad, C. A. Veale and R. R. Hark; *J. Am. Chem. Soc.*, 1990, **112**, 6263. 513

90JA6277 M. Granier, A. Baceiredo, Y. Dartiguenave, M. Dartiguenave, M. J. Menu and G. Bertrand; *J. Am. Chem. Soc.*, 1990, **112**, 6277. 871, 872, 873, 1148

90JA6439 D. B. Damon and D. J. Hoover; *J. Am. Chem. Soc.*, 1990, **112**, 6439. 688

90JA7061 I. Ryu, Y. Hayama, A. Hirai, N. Sonoda, A. Orita, K. Ohe and S. Murai; *J. Am. Chem. Soc.*, 1990, **112**, 7061. 438

90JA7434 K. Burgess and L. D. Jennings; *J. Am. Chem. Soc.*, 1990, **112**, 7434. 135

90JA7791 K. Maruoka, H. Banno and H. Yamamoto; *J. Am. Chem. Soc.*, 1990, **112**, 7791. 421

90JA7806 D. F. Harvey and M. F. Brown; *J. Am. Chem. Soc.*, 1990, **112**, 7806. 935, 936

90JA8369 R. W. Marshman and P. A. Shapley; *J. Am. Chem. Soc.*, 1990, **112**, 8369. 456, 801

90JA9586 P. Tecilla, S.-K. Chang and A. D. Hamilton; *J. Am. Chem. Soc.*, 1990, **112**, 9586. 11

90JA9601 T. P. Burkholder and P. L. Fuchs; *J. Am. Chem. Soc.*, 1990, **112**, 9601. 330

90JA9646 K. Ohe, T. Ishihara, N. Chatani and S. Murai; *J. Am. Chem. Soc.*, 1990, **112**, 9646. 346

90JA9651 L. A. Carpino, D. Sadat-Aalaee, H. G. Chao and R. H. DeSelms; *J. Am. Chem. Soc.*, 1990, **112**, 9651. 5

90JAP280254 M. Zenbayashi and K. Shuto (Toshiba Silicone Co. Ltd.); *Jpn. Pat.* 280 254 (1990) (*Chem. Abstr.*, 1993, **117**, 171 682). 225

90JCR(S)56 H. Suzuki, T. Inamasu, T. Ogawa and H. Tani; *J. Chem. Res.* (*S*), 1990, 56. 242, 253

90JCR(S)84 J.-X. Wang, C. H. Wang, Y.-L. Hu and W. F. Cui; *J. Chem. Res.* (*S*), 1990, 84. 183, 195

90JCR(S)106 I. Jabre, M. Saquet and A. Thuillier; *J. Chem. Res.* (*S*), 1990, 106. 589, 612

90JCR(S)230 J.-X. Wang, W. F. Cui, Y. L. Hu, and S.-S. Zang; *J. Chem. Res.* (*S*), 1990, 230. 253

90JCS(D)781 W. Henderson, R. D. W. Kemmitt, L. J. S. Prouse and D. R. Russell; *J. Chem. Soc., Dalton Trans.*, 1990, 781. 802

90JCS(D)2593 R. Ali and K. B. Dillon; *J. Chem. Soc., Dalton Trans.*, 1990, 2593. 652

90JCS(D)3193 J. Dupont and M. Pfeffer; *J. Chem. Soc., Dalton Trans.*, 1990, 3193. 801

90JCS(P1)1267 G. E. H. Elgemeie, H. A. Regaila and N. Shehata; *J. Chem. Soc., Perkin Trans. 1*, 1990, 1267. 615

90JCS(P1)1623 D. Alker, W. D. Ollis and H. Shahriari-Zavareh; *J. Chem. Soc., Perkin Trans. 1*, 1990, 1623. 326

90JCS(P1)1705 M. J. Alves, B. L. Booth and M. F. J. R. P. Proença; *J. Chem. Soc., Perkin Trans. 1*, 1990, 1705. 690

90JCS(P1)1727 P. Molina, A. Tárraga and M. J. Lidon; *J. Chem. Soc., Perkin Trans. 1*, 1990, 1727. 1066
90JCS(P1)1869 J. V. Greenhill, J. Hanaee and P. J. Steel; *J. Chem. Soc., Perkin Trans. 1*, 1990, 1869. 597
90JCS(P1)2193 J. Barluenga, M. Ferrero and F. Palacios; *J. Chem. Soc., Perkin. Trans. 1*, 1990, 2193. 371
90JCS(P1)2417 C. G. Francisco, R. Hernández, E. I. León, J. A. Salazar and E. Suárez; *J. Chem. Soc., Perkin Trans. 1*, 1990, 2417. 1118
90JCS(P1)2559 C.-S. Chen, Y.-C. Liu and M. Marsella; *J. Chem. Soc., Perkin Trans. 1*, 1990, 2559. 135
90JCS(P1)2721 D. J. Coveney, V. F. Patel, G. Pattenden and D. M. Thompson; *J. Chem. Soc., Perkin Trans., 1*, 1990, 2721. 457
90JCS(P1)3103 B. C. Challis, J. R. Milligan and R. C. Mitchell; *J. Chem. Soc., Perkin Trans. 1*, 1990, 3103. 360
90JCS(P1)3263 E. Breuer, A. Schlossman, M. Safadi, D. Gibson, M. Chorev and H. Leader; *J. Chem. Soc., Perkin Trans. 1*, 1990, 3263. 371, 403
90JFC(46)423 H. Sawada, M. Nakayama, M. Yoshida, T. Yoshida, K. Tatsuro and N. Kamigata; *J. Fluorine Chem.*, 1990, **46**, 423. 208
90JFC(48)367 Qui Chi Mir and D. D. DesMarteau; *J. Fluorine Chem.*, 1990, **48**, 367. 663
90JFC(48)395 N. R. Patel, R. L. Kirchmeyer and J. M. Shreeve; *J. Fluorine Chem.*, 1990, **48**, 395. 679
90JFC(48)407 J. B. Nielsen and J. S. Thrasher; *J. Fluorine Chem.*, 1990, **48**, 407. 675
90JGU849 S. E. Pipko, Yu. V. Balitskii, T. V. Kolodka, A. D. Sinitsa and M. I. Povlotskii; *J. Gen. Chem. USSR* (*Engl. Transl.*), 1990, **60**, 849. 713
90JGU864 B. N. Kozhushko, A. V. Lomakina, and V. A. Shokol; *J. Gen. Chem. USSR* (*Engl. Transl.*), 1990, **60**, 864. 787
90JGU1057 D. M. Malenko, L. I. Nesterova, S. N. Luk'yanenko and A. D. Sinitsa; *J. Gen. Chem. USSR* (*Engl. Transl.*), 1990, **60**, 1057. 713
90JGU1059 D. M. Malenko, L. I. Nesterova, S. N. Lukyanenko and A. D. Sinitsa; *J. Gen. Chem. USSR* (*Engl. Transl.*), 1990, **60**, 1059. 713, 788
90JGU1165 P. P. Onys'ko, T. V. Kim, E. I. Kiseleva, A. D. Sinitsa and M. Y. Kornilov; *J. Gen. Chem. USSR* (*Engl. Transl.*), 1990, **60**, 1165. 652
90JGU1389 P. P. Onysko, E. A. Suvalova, T. I. Chudakova and A. D. Sinitsa; *J. Gen. Chem. USSR* (*Engl. Transl.*), 1990, **60**, 1389. 738
90JGU1718 Yu. Lyashenko, V. B. Sokolov and I. V. Martynov; *J. Gen. Chem. USSR* (*Engl. Transl.*), 1990, 60, 1718. 672
90JGU1810 Yu. A. Veits, E. G. Neganova, A. A. Borisenko and V. L. Foss; *J. Gen. Chem. USSR*, 1990, **60**, 1810. 398, 400
90JGU1817 E. G. Neganova, Yu. A. Veits, V. K. Byistro, N. M. Semenko, A. A. Borisenko and V. L. Foss; *J. Gen. Chem. USSR*, 1990, **60**, 1817. 400
90JGU1822 Yu. A. Veits, E. G. Neganova, A. A. Borisenko and V. L. Foss; *J. Gen. Chem. USSR*, 1990, **60**, 1822. 400
90JGU1940 T. Kh. Gazizov and L. K. Sal'keeva; *J. Gen. Chem. USSR*, 1990, **60**, 1940. 406
90JHC407 U. Urleb, B. Stanovnik and M. Tišler; *J. Heterocycl. Chem.*, 1990, **27**, 407. 1028
90JHC433 U. Urleb, R. Neidlein and W. Kramer; *J. Heterocycl. Chem.*, 1990, **27**, 433. 1028
90JHC567 N. Okajima and Y. Okada; *J. Heterocycl. Chem.*, 1990, **27**, 567. 589, 641
90JHC1059 G. L'abbé, P. Brems and E. Albrecht; *J. Heterocycl. Chem.*, 1990, **27**, 1059. 1040
90JHC1741 A. B. Theocharis, N. E. Alexandrou and A. Terzis; *J. Heterocycl. Chem.*, 1990, **27**, 1741. 366
90JIC172 H. S. Bose and K. K. Dutta; *J. Indian Chem. Soc.*, 1990, **67**, 172. 367
90JIC260 T. T. Raj and S. Y. Ambekar; *J. Indian Chem. Soc.*, 1990, **67**, 260. 690
90JMC2192 K. Yoshino, K. Goto, K. Yoshiizumi, T. Morita and G. Tsukamoto; *J. Med. Chem.*, 1990, **33**, 2192. 370
90JMC2621 V. G. Matassa, F. J. Brown, P. R. Bernstein, *et al.*; *J. Med. Chem.*, 1990, **33**, 2621. 342
90JOC560 M. R. Pavia, W. H. Moos and F. M. Hershenson; *J. Org. Chem.*, 1990, **55**, 560. 306
90JOC831 F. G. Fang, M. E. Maier, S. J. Danishefsky and G. Schulte; *J. Org. Chem.*, 1990, **55**, 831. 572
90JOC1679 H. Harada, T. Yamaguchi, A. Iyobe, A. Tsubaki, T. Kamijo, K. Iizuka, K. Ogura and Y. Kiso; *J. Org. Chem.*, 1990, **55**, 1679. 199
90JOC1744 S.-I. Murahashi, T. Oda, T. Sugahara and Y. Masui; *J. Org. Chem.*, 1990, **55**, 1744. 324
90JOC2196 S. D. Sharma and S. B. Pandhi; *J. Org. Chem.*, 1990, **55**, 2196. 355
90JOC2311 A. Thenappan and D. J. Burton; *J. Org. Chem.*, 1990, **55**, 2311. 894
90JOC2576 J. N. Haseltine and S. J. Danishefsky; *J. Org. Chem.*, 1990, **55**, 2576. 336
90JOC2621 S. Rozen and D. Hebel; *J. Org. Chem.*, 1990, **55**, 2621. 202
90JOC2943 W. Chamchaang and A. R. Pinhas; *J. Org. Chem.*, 1990, **55**, 2943. 346
90JOC3374 J. S. Moore and S. I. Stupp; *J. Org. Chem.*, 1990, **55**, 3374. 314
90JOC3403 C.-H. Kwon and H. T. Nagasawa; *J. Org. Chem.*, 1990, **55**, 3403. 1120
90JOC3442 M. D. Bachi and D. Denenmark; *J. Org. Chem.*, 1990, **55**, 3442. 593
90JOC3682 D. H. Hua, S. W. Miao, S. N. Bharathi, T. Katsuhira and A. A. Bravo; *J. Org. Chem.*, 1990, **55**, 3682. 720
90JOC3744 G. Barbaro, A. Battaglia, P. Giorgianni, B. F. Bonini, G. Maccagnani and P. Zani; *J. Org. Chem.*, 1990, **55**, 3744. 648
90JOC3749 H. Kurzmeier and F. P. Schmidtchen; *J. Org. Chem.*, 1990, **55**, 3749. 1029
90JOC3922 E. Keinan and M. Sahai; *J. Org. Chem.*, 1990, **55**, 3922. 20
90JOC3967 G. H. Posner and C. M. Kinter; *J. Org. Chem.*, 1990, **55**, 3967. 159
90JOC4282 O. Farooq, Q. Wang, A. Wu and G. A. Olah; *J. Org. Chem.*, 1990, **55**, 4282. 992
90JOC4763 B. Gong and D. G. Lynn; *J. Org. Chem.*, 1990, **55**, 4763. 199
90JOC6037 S. T. Perri, S. C. Slater, S. G. Toske and J. D. White; *J. Org. Chem.*, 1990, **55**, 6037. 351
90JOC6091 K. E. DeBruin and E. E. Boros; *J. Org. Chem.*, 1990, **55**, 6091. 370, 590, 614, 643
90JOM(338)21 M. Wedler, F. Knoesel, M. Noltemeyer, F. T. Edelmann and U. Behrens; *J. Organomet. Chem.*, 1990, **338**, 21. 782

90JOM(381)191	A. Parlier, R. Yefsah, M. Rudler, H. Rudler, J. C. Daran and J. Vaissermann; *J. Organomet. Chem.*, 1990, **381**, 191.	798
90JOM(382)C6	H. Hoberg and M. Nohlen; *J. Organomet. Chem.*, 1990, **382**, C6.	386
90JOM(383)C1	L. L. Padolik, J. Gallucci and A. Wojcicki; *J. Organomet. Chem.*, 1990, **383**, C1.	451
90JOM(386)313	S. Kato, H. Ishihara, K. Ibe, H. Kageyama and T. Murai; *J. Organomet. Chem.*, 1990, **386**, 313.	250
90JOM(388)47	R. Tacke; *J. Organomet. Chem.*, 1990, **388**, 47.	411, 416
90JOM(390)361	C.-H. Jun; *J. Organomet. Chem.*, 1990, **390**, 361.	459
90JOM(390)C1	G. C. Schoemaker, D. J. Stufkens, S. J. Goede, T. V. D. Does and F. B. Bickelhaupt; *J. Organomet. Chem.*, 1990, **390**, C1.	1089
90JOM(390)C57	J. Montoya, A. Santos, A. M. Echavarren and J. Ros; *J. Organomet. Chem.*, 1990, **390**, C57.	455
90JOM(391)13	J. P. Picard, J. M. Aizpurua, A. Elyusufi and P. Kowalski; *J. Organomet. Chem.*, 1990, **391**, 13.	420, 794
90JOM(397)7	A. Castel, P. Riviere, J. Satge, Y. H. Ko and D. Desor; *J. Organomet. Chem.*, 1990, **397**, 7.	432
90JOU287	V. N. Fetyukhin, A. A. Esipenko and L. I. Samarai; *J. Org. Chem. USSR* (*Engl. Transl.*), 1990, **26**, 287.	669
90JOU1408	A. F. Shivanyuk, E. V. Dashkovskaya, A. B. Rozhenko, M. O. Lozinskii, N. P. Onishchuk, A. I. Bukarev, O. A. D'Yachenko and L. O. Atovmyan; *J. Org. Chem. USSR* (*Engl. Transl.*), 1990, **26**, 1408.	685
90JPR731	S. Witek, T. Czekanski and A. J. Kaczmarek; *J. Prakt. Chem.*, 1990, **332**, 731.	671
90JSP62	T. Sakaizumi, H. Mure, O. Ohashi and I. Yamaguchi; *J. Mol. Spectrosc.*, 1990, **140**, 62.	1103
90JST67	R. Fausto, L. A. E. Batista de Carvalho and J. J. C. Teixeira-Dias; *J. Mol. Struct.*, 1990, **207**, 67.	506, 508
90JST171	S. Liedle, H. Oberhammer, H. Fritz and W. Sundermeyer; *J. Mol. Struct.*, 1990, **216**, 171.	488, 492
90KGS1403	G. K. Nalbandyan, A. P. Mkrtchyan, R. A. Akopyan and S. A. Vartanyan; *Khim. Geterotsikl. Soedin*, 1990, 1403 (*Chem. Abstr.*, 1991, **114**, 185425f).	690
90KGS1430	T. V. Manaeva and V. P. Perevalov; *Khim. Geterotsikl. Seodin.*, 1990, 1430.	652
90MI 501-01	K. Gamoh, H. Sawamoto and J. Iida; *Anal. Chim. Acta*, 1990, **228**, 307.	11
90MI 503-01	J. Mulzer and K.-D. Graske; Ph.D. Thesis, Freie Universität, Berlin, 1990.	138
90MI 505-01	J. Wang, W. Cui, Y. Hu and K. Zhao; *Chin. Chem. Lett.*, 1990, **1**, 193.	252
B-90MI 507-01	J. G. Tillett; in "The Chemistry of Sulphinic Acids, Esters and Their Derivatives," ed. S. Patai, Wiley, New York, 1990, p. 603.	340
90MI 517-01	R. I. I. Webb and T. Kaneko; *Bioconjugate Chem.*, 1990, **1**, 96.	700, 701
90MI 519-01	S. Zawadzki and A. Wijata; *Phosphorus, Sulfur, Silicon, Relat. Elem.*, 1990, **53**, 307 (*Chem. Abstr.*, 1991, **114**, 42902).	774
90MI 520-01	C. H. Jun; *Bull. Korea Chem. Soc.*, 1990, **11**, 187.	800
B-90MI 523-01	A. Schmidpeter; in "Multiple Bonds and Low Coordination in Phosphorus Chemistry," ed. M. Regitz and O. J. Scherer, Thieme, Stuttgart, 1990, p. 149.	882
B-90MI 523-02	R. Appel; in "Multiple Bonds and Low Coordination in Phosphorus Chemistry," ed. M. Regitz and O. J. Scherer, Thieme, Stuttgart, 1990, p. 157.	876
90MI 524-01	O. Halova; *Sb. Vys. Sk. Chem.-Technol. Praze, P: Fyz. Mater. Merici Tech.*, 1990, **P11**, 39. (*Chem. Abstr.*, 1992, **117**, 171515k).	900
90MI 527-01	A. Matsuda, M. Satoh, T. Ueda, H. Machida and T. Sasaki; *Nucleosides Nucleotides*, 1990, **9**, 587.	1054
90MI 530-01	A. P. Yuvchenko, K. L. Moiseichuk, E. A. Dikusar, N. A. Zhukovskaya and Yu. A. Ol'dekop; *Zh. Obshch. Khim. Lett.*, 1990, **60**, 1587.	1110
B-90MI 531-01	M. Regitz and O. J. Scherer; "Multiple Bonds and Low Coordination in Phosphorus Chemistry," Thiéme Verlag, Stuttgart, 1990.	1152
90NJC433	K. H. Dötz; *Nouv. J. Chim.*, 1990, **14**, 433.	931
90OM560	C. H. Sun, T. J. Chow and L. K. Liu; *Organometallics*, 1990, **9**, 560.	384
90OM583	E. Carmona, P. Palma, M. Paneque and M. L. Poveda; *Organometallics*, 1990, **9**, 583.	800
90OM1245	D. Roberto and H. Alper; *Organometallics*, 1990, **9**, 1245.	356
90OM1557	J. C. Otter, C. L. Adamson, C. H. Yoder and A. L. Rheingold; *Organometallics*, 1990, **9**, 1557.	375, 376
90OM2621	H. Adams, N. A. Bailey, M. Grayson, C. Ridgway, A. J. Smith, P. Taylor, M. J. Winter and C. E. Housecroft; *Organometallics*, 1990, **9**, 2621.	943
90OM2836	J. Garner, A. Irving and J. R. Moss; *Organometallics*, 1990, **9**, 2836.	943
90OPP755	T. Jagodzinski; *Org. Prep. Proceed. Int.*, 1990, **22**, 755.	631
90OS(68)1	T. Shioiri, T. Aoyama and S. Mori; *Org. Synth.*, 1990, **68**, 1.	873
90OS(68)25	J. A. Soderquist; *Org. Synth.*, 1990, **68**, 25.	416
90OS(69)10	P. Kalaritis and R. W. Regenye; *Org. Synth.*, 1990, **69**, 10.	55
90OS(69)173	M.-A. Poupart, G. Lassalle and L. A. Paquette; *Org. Synth.*, 1990, **69**, 173.	76
90PAC1441	T. Nagasawa and H. Yamada; *Pure Appl. Chem.*, 1990, **62**, 1441.	118
90PS(49)337	P. Binger, B. Biedenbach, A. T. Herrmann, R. Milczarck and R. Schneider; *Phosphorus Sulfur Silicon*, 1990, **49**, 337.	890
90PS(50)139	K. E. DeBruin and E. E. Boros; *Phosphorus Sulfur*, 1990, **50**, 139.	370
90PS(53)145	F. F. Abdel-Latif; *Phosphorus Sulfur*, 1990, **53**, 145.	616
90S525	G. Simchen and E. Purkner; *Synthesis*, 1990, 525 (*Chem. Abstr.*, 1990, **113**, 191764m).	718
90S565	A. R. Katritzky, J. Jiang and L. Urogdi; *Synthesis*, 1990, 565.	1031, 1067
90S646	G. A. Olah, R. Krishnamurti and G. K. S. Prakash; *Synthesis*, 1990, 646.	1045
90S661	W. Richter and I. Ugi; *Synthesis*, 1990, 661.	370
90S697	R. Rieger and E. Breitmaier; *Synthesis*, 1990, 697.	199
90S704	M. I. Younes, S. A. M. Metwally and A. H. Atta; *Synthesis*, 1990, 704.	690

90S1001	T. N. Mitchell and K. Kwetkat; *Synthesis*, 1990, 1001.	467
90S1008	M. H. Jakobsen, O. Buchardt, A. Holm and M. Meldal; *Synthesis*, 1990, 1008.	323
90S1095	R. S. Hosmane, V. S. Bhadti and B. B. Lim; *Synthesis*, 1990, 1095.	351
90S1129	J. C. Jochims, S. Hehl and S. Herzberger; *Synthesis*, 1990, 1129.	215
90S1143	R. G. Wallace, J. M. Barker and M. L. Wood; *Synthesis*, 1990, 1143.	8
90SC23	P. S. N. Reddy and V. G. Reddy; *Synth. Commun.*, 1990, **20**, 23.	706
90SC509	L. Kniežo and J. Bernát; *Synth. Commun.*, 1990, **20**, 509.	1041, 1049
90SC887	M. Kawase, T. Kitamura, M. Shimada and Y. Kikugawa; *Synth. Commun.*, 1990, **20**, 887.	328
90SC1105	J. Einhorn, C. Einhorn and J.-L. Luche; *Synth. Commun.*, 1990, **20**, 1105.	327
90SC1733	S. Subramanian, U. R. Desai and G. K. Trivedi; *Synth. Commun.*, 1990, **20**, 1733.	324
90SC1811	M. Al-Talib, H. Tashtoush and N. Odeh; *Synth. Commun.*, 1990, **20**, 1811.	350
90SC2083	I. A. Atanassova, J. S. Petrov, V. H. Ognjanova and N. M. Mollov; *Synth. Commun.*, 1990, **20**, 2083.	345
90SC2235	M. W. Bredenkamp, C. W. Holzapfel and W. J. van Zyl; *Synth. Commun.*, 1990, **20**, 2235.	573
90SC2313	B. M. Choudary, S. S. Rani and M. L. Kantam; *Synth. Commun.*, 1990, **20**, 2313.	1106
90SC2901	D. J. Park, A. G. Stern and R. L. Willer; *Synth. Commun.*, 1990, **20**, 2901.	215
90SC3085	D. Brillon; *Synth. Commun.*, 1990, **20**, 3085.	568, 569
90SC3245	M. Kabaki, S. Inoue, Y. Nagata and Y. Sato; *Synth. Commun.*, 1990, **20**, 3245.	431
90SL690	V. Gotor and E. Menéndez; *Synlett*, 1990, 690.	217
90SRI437	H. I. Ucan and R. Mirzaoglu; *Synth. React. Inorg. Metal-Org. Chem.*, 1990, **20**, 437.	664
90SRI1413	H. A. Hodali and A. Q. Hussein; *Synth. React. Inorg. Met.-Org. Chem.*, 1990, **20**, 1413 (*Chem. Abstr.*, 1991, **114**, 122 559g).	776
90T46	B. Christenson, G. Halnemer and C. Ullenius; *Tetrahedron*, 1990, 46.	151
90T1057	V. Alcázar, I. Tapia and J. R. Morán; *Tetrahedron*, 1990, **46**, 1057.	554
90T1063	P. Molina, M. Alajarin and A. Vidal; *Tetrahedron*, 1990, **46**, 1063.	1067, 1079
90T1235	K. H. Dötz, R. Noack, K. Harms and G. Müller; *Tetrahedron*, 1990, **46**, 1235.	932
90T2573	P. C. B. Page and S. Rosenthal; *Tetrahedron*, 1990, **46**, 2573.	410, 423, 424
90T4733	J. E. Baldwin, R. M. Adlington, D. W. Gollins and C. J. Schofield; *Tetrahedron*, 1990, **46**, 4733.	376
90T5423	H. Junjappa, H. Ila and C. V. Asokan; *Tetrahedron*, 1990, **46**, 5423.	555
90T6753	N. De Kimpe and C. Stevens; *Tetrahedron*, 1990, **46**, 6753.	697
90T7247	S. A. Glover, C. A. Rowbottom, A. P. Scott and J. L. Schoonraad; *Tetrahedron*, 1990, **46**, 7247.	317
90T7413	J. R. Hwu and S.-C. Tsay; *Tetrahedron*, 1990, **46**, 7413.	637, 829
90T7429	M. Ashwell, R. F. W. Jackson and J. M. Kirk; *Tetrahedron*, 1990, **46**, 7429.	238
90TA141	G. Cevasco, E. Narisano and S. Thea; *Tetrahedron Asymmetry*, 1990, **1**, 141.	336
90TA881	J. E. Baldwin, A. C. Spivey and C. J. Schofield; *Tetrahedron Asymmetry*, 1990, **1**, 881.	1106
90TH29	A. T. Herrmann; Ph.D. Thesis, University of Kaiserslautern, 1990, p. 29.	890
90TL75	G. Jeanneret-Gris and P. Pousaz; *Tetrahedron Lett.*, 1990, **31**, 75.	12
90TL123	A. B. N. Luheshi and R. K. Smalley; *Tetrahedron Lett.*, 1990, **31**, 123.	688
90TL243	Y. Kikugawa, K. Mitsui, T. Sakamoto, M. Kawase and H. Tamiya; *Tetrahedron Lett.*, 1990, **31**, 243.	332
90TL709	K. Tatsuta, T. Ishiyama, S. Tajima, Y. Koguchi and H. Gunji; *Tetrahedron Lett.*, 1992, **31**, 709.	199
90TL873	T. Hosay, L. Christiaens, G. Anthoine and P. Moniotte; *Tetrahedron Lett.*, 1990, **31**, 873.	337
90TL1245	W. Adam and X. Wang; *Tetrahedron Lett.*, 1990, **31**, 1245.	227
90TL1287	H. Takayama, M. Horigome, N. Aimi and S.-I. Sakai; *Tetrahedron Lett.*, 1990, **31**, 1287.	1127
90TL1327	S. G. Manjunatha, K. V. Reddy and S. Rajappa; *Tetrahedron Lett.*, 1990, **31**, 1327.	611
90TL1417	M. C. Viaud and P. Rollin; *Tetrahedron Lett.*, 1990, **31**, 1417.	665
90TL1559	S. Jew, H. G. Park, H. Park, M. Park and Y. Cho; *Tetrahedron Lett.*, 1990, **31**, 1559.	999
90TL1853	K. Kpegba and P. Metzner; *Tetrahedron Lett.*, 1990, **31**, 1853.	557
90TL1969	R. S. Garigipati; *Tetrahedron Lett.*, 1990, **31**, 1969.	749
90TL2139	F. Backenstrass, J. Streith and T. Tschamber; *Tetrahedron Lett.*, 1990, **31**, 2139.	330
90TL2259	W. Adam, E. Kades and X. Wang; *Tetrahedron Lett.*, 1990, **31**, 2259.	227
90TL2529	D. F. Harvey and M. F. Brown; *Tetrahedron Lett.*, 1990, **31**, 2529.	935
90TL3497	J. Barluenga, M. Ferrero and F. Palacios; *Tetrahedron Lett.*, 1990, **31**, 3497.	371
90TL3587	S. Kato, H. Kageyama, T. Kanda, T. Murai and T. Kawamura; *Tetrahedron Lett.*, 1990, **31**, 3587.	255, 506, 509, 543
90TL3591	T. Sato, J. Otera and H. Nozaki; *Tetrahedron Lett.*, 1990, **31**, 3591.	251
90TL3603	K. Nakamura, K. Ishihara, A. Ohno, M. Uemura, H. Nishimura and Y. Hayashi; *Tetrahedron Lett.*, 1990, 3603.	652
90TL3681	G. I. Dmitrienko, K. E. Nielsen, C. Steingart, N. S. Ming, J. M. Willson and G. Weeratunga; *Tetrahedron Lett.*, 1990, **31**, 3681.	1125
90TL3723	L. S. Liebeskind, R. Chidambaram, S. Nimkar and D. Liotta; *Tetrahedron Lett.*, 1990, **31**, 3723.	457
90TL3895	Y. Yamazaki and K. Hosono; *Tetrahedron Lett.*, 1990, 3895.	652
90TL4117	W. Oppolzer, M. Wills, C. Starkemann and G. Bernardinelli; *Tetrahedron Lett.*, 1990, **31**, 4117.	347
90TL4821	V. V. Zhdankin, C. M. Crittell, P. J. Stang and N. S. Zefirov; *Tetrahedron Lett.*, 1990, **31**, 4821.	205, 321
90TL4875	J. D. Fourneron, M. Chiche and G. Piéroni; *Tetrahedron Lett.*, 1990, 4875.	135
90TL5201	P. D. Davis and R. A. Bit; *Tetrahedron Lett.*, 1990, **31**, 5201.	306
90TL5297	M. J. Taschner and P. T. Cyr; *Tetrahedron Lett.*, 1990, **31**, 5297.	329

90TL5327 M. Komatsu, T. Yagii and Y. Ohshiro; *Tetrahedron Lett.*, 1990, **31**, 5327. 359

90TL6339 C. Maliverney and H. G. Viehe; *Tetrahedron Lett.*, 1990, **31**, 6339. 601

90TL6547 A. Sera, K. Itoh and H. Yamaguchi; *Tetrahedron Lett.*, 1990, **31**, 6547. 325

90TL7223 M. A. Cohen, J. N. Sawden and N. J. Turner; *Tetrahedron Lett.*, 1990, 7223. 118

90TL7229 G. Palmisano, E. Brenna, B. Danieli, G. Lesma, B. Vodopivec and G. Fiori; *Tetrahedron Lett.*, 1990, **31**, 7229. 1045

90TL7237 I. E. Markó and A. Mekhalfia; *Tetrahedron Lett.*, 1990, **31**, 7237. 18, 276, 293, 299

90TL7341 P. E. O'Bannon and W. P. Dailey; *Tetrahedron*, 1990, **46**, 7341. 868

90TL7375 J. Barluenga, P. J. Campos, F. López, I. Llorente and M. A. Rodríguez; *Tetrahedron Lett.*, 1990, **31**, 7375. 1108

90TL7513 I. Paterson and A. N. Hulme; *Tetrahedron Lett.*, 1990, **31**, 7513. 161

90TL7633 T. Tezuka, T. Otsuka and H. Kasuga; *Tetrahedron Lett.*, 1990, **31**, 7633. 357

90UKZ1313 M. V. Vovk and L. I. Samarai; *Ukr. Khim. Zh.* (*Russ. Ed.*), 1990, **56** (12), 1313 (*Chem. Abstr.*, 1990, **113**, 152 058q). 982

90ZAAC(582)37 M. Domagala, F. Huber and H. Preut; *Z. Anorg. Allg. Chem.*, 1990, **582**, 37. 222

90ZAAC(588)59 L. Jäger and H. Köhler; *Z. Anorg. Allg. Chem.*, 1990, **588**, 59. 690, 723

90ZC67 S. Kato, H. Masumoto, S. Ikeda, M. Itoh, T. Murai and H. Ishihara; *Z. Chem.*, 1990, **30**, 67. 561, 563, 588

90ZN(B)1091 E. V. Dehmlow and G. O. Torossian; *Z. Naturforsch., Teil B*, 1990, **45**, 1091. 1106

90ZN(B)1273 D. Fenske, G. Baum, A. Zinn and K. Dehnicke; *Z. Naturforsch. Teil B.*, 1990, **45**, 1273. 781

90ZOB464 N. G. Zabirov, F. M. Shamsevaleev and R. A. Cherkasov; *Zh. Obshch. Khim.* (*Engl. Trans.*), 1990, **60**, 464. 644

90ZOB476 R. M. Kamalov, G. S. Stepanov, M. A. Pudovik and A. N. Pudovik; *Zh. Obshch. Khim.*, 1990, **60**, 476 (*Chem. Abstr.*, 1990, **113**, 78 520r). 1135

90ZOB963 S. E. Pipko, Yu. V. Balitskii, T. V. Kolodka, A. D. Sinitsa and M. I. Povlotskii; *Zh. Obshch. Khim.*, 1990, **60**, 963 (*Chem. Abstr.*, 1990, **113**, 132 340x). 713

90ZOB1183 S. A. Mamedov, F. Y. Aliev and O. V. Karaeva; *Zh. Obshch. Khim.*, 1990, **60**, 1183. 369

90ZOB1186 D. M. Malenko, L. I. Nesterova, S. N. Luk'yanenko and A. D. Sinitsa; *Zh. Obshch. Khim.*, 1990, **60**, 1186 (*Chem. Abstr.*, 1990, **113**, 172 208s). 713

90ZOB1187 D. M. Malenko, L. I. Nesterova, S. N. Luk'yanenko and A. D. Sinitsa; *Zh. Obshch. Khim.*, 1990, **60**, 1187 (*Chem. Abstr.*, 1990, **113**, 172 209t). 369, 713

90ZOB1389 P. P. Onys'ko, E. A. Suvalova, T. I. Chudakova and A. D. Sinitsa; *Zh. Obshch. Khim.* (*Engl. Trans.*), 1990, **60**, 1389. 644

90ZOB1923 Yu. Lyashenko, V. B. Sokolov and I. V. Martynov; *Zh. Obshch. Khim.*, 1990, **60**, 1923 (*Chem. Abstr.*, 1991, **114**, 42 940j). 672

90ZOB1928 I. I. Kandror, B. D. Lavrukhin, I. O. Bragina, M. A. Galkina and Y. G. Gololbov; *Zh. Obshch. Khim. Lett.*, 1990, **60**, 1928. 223

90ZOB1997 N. G. Zabirov, F. M. Shamsevaleev *et al.*; *Zh. Obshch. Khim.*, 1990, **60**, 1997. 384

90ZOR214 N. M. Przheval'skii and V. N. Drozd; *Zh. Org. Khim.*, 1990, **26**, 214. 736

90ZOR339 V. N. Fetyukhin, A. A. Esipenko and L. I. Samarai; *Zh. Org. Khim.*, 1990, **26**, 339 (*Chem. Abstr.*, 1990, **113**, 78 272m). 669, 672

90ZOR1629 O. A. D'yachenco, E. V. Dashkovskaya, A. B. Rozhenko, M. O. Lozinskii, N. P. Onishchuk, A. I. Bukarev and L. O. Atovmyan; *Zh. Org. Khim.*, 1990, **26**, 1629 (*Chem. Abstr.*, 1991, **114**, 81 711e). 685

90ZOR2059 N. A. Nedolya and V. V. Gerasimova; *Zh. Org. Khim.*, 1990, **26**, 2059. 1026

91ACH35 S. El-Nagdy, A. Soliman and F. A. A. El-Bassiouny; *Acta Chim. Acad. Sci. Hung.*, 1991, **128**, 35. 652

91AG(E)196 J. C. Guillemin, T. Janati, P. Juenot, P. Savignac and J. M. Denis; *Angew. Chem. Int. Ed. Engl.*, 1991, **30**, 196. 1152

91AG(E)310 P. Binger, J. Haas, A. T. Herrmann, F. Langhauser and C. Krüger; *Angew. Chem., Int. Ed. Engl.*, 1991, **30**, 310. 885

91AG(E)847 V. G. Albano, S. Bordoni, D. Braga, L. Busetto, A. Palazzi and V. Zanotti, *Angew. Chem., Int. Ed. Engl.*, 1091, **30**, 847. 1088

91AG(E)1154 J. M. Sotiropoulos, A. Baceiredo, K. H. Locquenghien, F. Dahan and G. Bertrand; *Angew. Chem., Int. Ed. Engl.*, 1991, **30**, 1154. 872

91AG(E)1658 F. Stein, M. Duetsch, R. Lackmann, M. Noltemeyer and A. de Meijere; *Angew. Chem., Int. Ed. Engl.*, 1991, **30**, 1658. 934

91AJC135 J. B. Bremner and W. Jaturonrusmee; *Aust. J. Chem.*, 1991, **44**, 135. 1127

91AOC(32)227 A. Mayr and H. Hoffmeister; *Adv. Organomet. Chem.*, 1991, **32**, 227. 942

91BAU382 V. A. Reznikov, L. A. Vishnivetskaya and L. B. Volodarskii; *Bull. Acad. Sci. USSR. Div. Chem. Sci.*, 1991, 382. 669

91BAU438 A. S. Kulikov, N. N. Makhova and L. I. Khmel'nitskil; *Bull. Acad. Sci. USSR. Div. Chem. Sci.*, 1991, 438. 667, 671

91BAU661 A. N. Pudovik, V. N. Nazmutdinova, L. P. Chirkova and R. Z. Musin; *Bull. Acad. Sci. USSR, Div. Chem. Sci.*, 1991, 661. 406

91BAU1705 M. D. Bargamova, S. I. Pletnev and I. L. Knunyants; *Bull. Acad. Sci. USSR. Div. Chem. Sci.*, 1991, 1705. 22

91BCJ106 S. Kagabu, K. Saito, H. Watanabe, K. Takahashi and K. Wada; *Bull. Chem. Soc. Jpn.*, 1991, **64**, 106. 344

91BCJ766 H. Okamoto and S. Kato; *Bull. Chem. Soc. Jpn.*, 1991, **64**, 766. 691

91BCJ2402 K. Okuma, Y. Komiya and H. Ohta; *Bull. Chem. Soc. Jpn.*, 1991, **64**, 2402. 603, 622

91BCJ2720 K. Takatoh and M. Sakamoto; *Bull. Chem. Soc. Jpn.*, 1991, **64**, 2720. 246

91BCJ3732 K. Yonemoto, I. Shibuya, M. Yasumoto, Y. Taguchi and T. Tsuchiya; *Bull. Chem. Soc. Jpn.*, 1991, **64**, 3732. 320, 335

91BMC733 C. Subramanyam and M. R. Bell; *Bioorg. Med. Chem. Lett.*, 1991, **1**, 733. 342

91BSF574 S. A. Toure, A. Voglozin, E. Degny, R. Danion-Bougot, D. Danion, J. P. Pradére, L. Toupet, Y. T. N'Guessan; *Bull. Soc. Chim. Fr.*, 1991, **128**, 574. 633

91BSF760 P. Hudhomme and G. Duguay; *Bull. Soc. Chim. Fr.*, 1991, **128**, 760. 580

91CAR283 D. Mostowicz, O. Zegrocka and M. Chmielewski; *Carbohydr. Res.*, 1991, **212**, 283. 344

91CB493 J. Heinicke, I. Kovacs and L. Nyulaszi; *Chem. Ber.*, 1991, **124**, 493. 880

91CB957 A. Wienand and H.-U. Reissig; *Chem. Ber.*, 1991, **124**, 957. 957

91CB989 L. Weber, U. Nolte, H. G. Stammler and B. Neumann; *Chem. Ber.*, 1991, **124**, 989. 893

91CB1121 H. Werner, W. Knaup and M. Schulz; *Chem. Ber.*, 1991, **124**, 1121. 944

91CB1131 W. Fimml and F. Sladky; *Chem. Ber.*, 1991, **124**, 1131. 1117

91CB1311 C. Maletzo and W. Sundermeyer; *Chem. Ber.*, 1991, **124**, 1311. 478, 492

91CB1635 K. H. Dötz and R. Weber; *Chem. Ber.*, 1991, **124**, 1635. 953

91CB1997 M. Arnswald and W. P. Neumann; *Chem. Ber.*, 1991, **124**, 1997. 1006

91CB2431 J. Breimair, T. Weidmann, B. Wagner and W. Beck; *Chem. Ber.*, 1991, **124**, 2431. 955

91CB2677 S. J. Goede and F. Bickelhaupt; *Chem. Ber.*, 1991, **124**, 2677. 878, 889, 890

91CB2843 E. Fuchs, B. Breit, H. Heydt, W. Schoeller, T. Busch, C. Krüger, P. Betz and M. Regitz; *Chem. Ber.*, 1991, **124**, 2843. 395

91CC116 M.-E. Eleftheriou, J. Novosad, D. J. Williams and J. D. Woollins; *J. Chem. Soc., Chem. Commun.*, 1991, 116. 575

91CC302 U. Fleischer, H. Grützmacher and U. Krüger; *J. Chem. Soc., Chem. Commun.*, 1991, 302. 887

91CC881 H. Le Bozec, C. Cosset and P. H. Dixneuf; *J. Chem. Soc., Chem. Commun.*, 1991, 881. 941

91CC964 M. Khalid, J.-L. Ripoll and Y. Vallée; *J. Chem. Soc., Chem. Commun.*, 1991, 964. 563

91CC1127 L. M. Ferreira, A. M. Lobo, S. Prabhakar, M. J. Marcelo-Curto, H. S. Rzepa and M. Yi; *J. Chem. Soc., Chem. Commun.*, 1991, 1127. 212

91CC1237 Y. Shishido and C. Kibayashi; *J. Chem. Soc., Chem. Commun.*, 1991, 1237. 330

91CCC1317 J. Svoboda and J. Paleček; *Collect. Czech. Chem. Commun.*, 1991, **56**, 1317. 324

91CHE299 G. A. Shvekhgeimer, K. I. Kobrakov, O. S. Kartseva and L. V. Balabanova; *Chem. Heterocycl. Comp.* (*Engl. Transl.*), 1991, 299. 687

91CHE424 S. E. Tolchinskii, M. V. Kormer and I. A. Maretina; *Chem. Heterocycl. Compd.* (*Engl. Transl.*), 1991, 424. 691

91CI(L)353 N. Borthakur, A. Goswami and R. C. Rastogi; *Chem. Ind.* (*London*), 1991, 353. 370

91CL61 M. Hatanaka and I. Ueda; *Chem. Lett.*, 1991, 61. 208

91CL105 H. Suzuki, C. Nakaya, Y. Matano and T. Ogawa; *Chem. Lett.*, 1991, 105. 344

91CL989 T. Mukaiyama, M. Furuya, A. Ohtsubo and S. Kobayashi; *Chem. Lett.*, 1991, 989. 236

91CL1413 K. Narasaka, H. Kusama and Y. Hayashi; *Chem. Lett*, 1991, 1413. 427

91CL1443 T. Ogawa, T. Kiji, K. Hayami and H. Suzuki; *Chem. Lett.*, 1991, **8**, 1443. 289, 302

91CL1479 K. Takai, Y. Kataoka, K. Yoshizumi, Y. Oguchi and K. Utimoto; *Chem. Lett.*, 1991, **8**, 1479. 344

91CL1901 T. Mukaiyama, I. Shiina and S. Kobayashi; *Chem. Lett.*, 1991, 1901. 236

91COS419 E. Schaumann; *Comp. Org. Synth.*, 1991, **6**, 419. 565

91COS(5)1065 W. D. Wulff; *Comp. Org. Synth.*, 1991, **5**, 1065. 955, 959

91COS(6)33 R. R. Schmidt; *Comp. Org. Synth.*, 1991, **6**, 33. 692

91COS(6)301 R. Sustmann; in *Comp. Org. Synth.*, 1991, 6, 301. 1

91COS(6)323 J. Mulzer; *Comp. Org. Synth.*, 1991, **6**, 323. 122

91COS(6)419 E. Schaumann; *Comp. Org. Synth.*, 1991, **6**, 419. 472

91COS(6)435 J. Voss; *Comp. Org. Synth.*, 1991, **6**, 435. 233, 510, 512, 513

91COS(6)461 A. Ogawa and N. Sonoda; *Comp. Org. Synth.*, 1991, **6**, 461. 510

91COS(6)476 A. Ogawa and N. Sonoda; *Comp. Org. Synth.*, 1991, **6**, 476. 624

91COS(6)485 W. Kantlehner; *Comp. Org. Synth.*, 1991, **6**, 485. 685, 687, 688, 689, 690, 691, 692, 693, 694

91COS(6)523 W. Kantlehner; *Comp. Org. Synth.*, 1991, **6**, 523. 654

91COS(6)536 W. Kantlehner; *Comp. Org. Synth.*, 1991, **6**, 536. 726

91CRV197 M. I. Bruce; *Chem. Rev.*, 1991, **91**, 197. 941

91CSR503 H. Rudler, M. Audouin, E. Chelain, B. Denise, R. Goumont, A. Massoud, A. Parlier, A. Pacreau, M. Rudler, R. Yefsah, C. Alvarez and F. Delgado-Reyes; *Chem. Soc. Rev.*, 1991, **20**, 503. 952

91CZ135 F. Dallacker and A. Peisker; *Chem.-Ztg*, 1991, **115**, 135. 334

91CZ215 J. Gante and H. Kahlenberg; *Chem.-Ztg.*, 1991, **115**, 215. 354

91FES1033 A. Ranise, F. Bondavalli, O. Bruno, P. Schenone, M. L. Cenicola, D. Donnoli, E. Lampa and F. Rossi; *Farmaco Ed. Sci.*, 1991, **46**, 1033. 597

91G101 G. Bellachioma, G. Cardaci, A. Macchioni and G. Reichenbach; *Gazz. Chim. Ital.*, 1991, **121**, 101. 799

91G401 F. Benedetti, S. Bozzini, S. Fatutta, M. Forchiassin, P. Nitti, G. Pitacco and C. Russo; *Gazz. Chim. Ital.*, 1991, **121**, 401. 355

91H(32)221 T. Kawasaki, A. Kodama, T. Nishida, K. Shimizu and M. Somei; *Heterocycles*, 1991, **32**, 221. 324

91H(32)481 T. G. Back, E. K. Y. Lai and J. W. Morzycki; *Heterocycles*, 1991, **32**, 481. 316

91H(32)1687 H. R. Bravo and H. M. Niemeyer; *Heterocycles*, 1991, **32**, 1687. 325

91HCA1477 S. Bienz and A. Chapeaurouge; *Helv. Chim. Acta*, 1991, **74**, 1477. 416

91IC535 Qui Chi Mir and D. D. DesMarteau; *Inorg. Chem.*, 1991, **30**, 535. 663

91IC1046 W. J. Mills, C. H. Sutton, M. W. Baize and L. J. Todd; *Inorg. Chem.*, 1991, **30**, 1046. 1147

91IC1174 R. B. King, N. K. Bhattacharyya and E. M. Holt; *Inorg. Chem.*, 1991, **30**, 1174. 407

91IC1564 D. P. Fairlie, P. M. Angus, M. D. Fenn and W. G. Jackson; *Inorg. Chem.*, 1991, **30**, 1564. 385, 386

91IC1862 H.-Y. Huang, W.-J. Chen, C.-C. Yang and A. Yeh; *Inorg. Chem.*, 1991, **30**, 1862. 386

91IC4806 P. M. Angus and W. G. Jackson; *Inorg. Chem.*, 1991, **30**, 4806. 387

91IJC(B)717 M. B. Hogale; *Indian J. Chem., Sect. B*, 1991, **30**, 717. 355

91IZV237 A. N. Pudovik, V. N. Nazmutdinova and L. P. Chirkova; *Izv. Akad. Nauk SSSR, Ser. Khim.*, 1991, 237 (*Chem. Abstr.*, 1991, **114**, 185 645c). 1131

91IZV444 V. A. Reznikov, L. A. Vishnivetskaya and L. B. Volodarskii; *Izv. Akad. Nauk SSSR, Ser. Khim*, 1991, 444 (*Chem. Abstr.*, 1991, **115**, 8668n). 669

91IZV451 R. M. Kamalov; *Izv. Akad. Nauk SSSR, Ser. Khim.*, 1991, 451. 1122

91IZV506 A. S. Kulikov, N. N. Makhova and L. I. Khmel'nitskii; *Izv. Akad. Nauk SSSR, Ser. Khim.*, 1991, 506 (*Chem. Abstr.*, 1991, **115**, 8059g). 667, 671

91IZV1108 A. V. Kalinin, É. T. Apasov, S. L. Ioffe and V. A. Tartakovskii; *Izv. Akad. Nauk SSSR, Ser. Khim.*, 1991, **5**, 1108. 362

91IZV1605 O. N. Nuretdinova and F. F. Guseva; *Izv. Akad. Nauk SSSR, Ser. Khim.*, 1991, 1605 (*Chem. Abstr.*, 1991, **115**, 232 368x). 576

91IZV1918 M. D. Bargamova, S. I. Pletnev and I. L. Knunyants; *Izv. Akad. Nauk SSSR, Ser. Khim.*, 1991, 1918 (*Chem. Abstr.*, 1992, **116**, 41 028x). 679

91IZV2103 V. A. Al'fonsov, A. G. Trusenev, É. S. Batyeve, A. N. Pudovik and M. A. Pudovik; *Isv. Akad. Nauk SSSR, Ser. Khim.*, 1991, 2103. 369

91IZV2847 A. A. Krolevets, V. V. Antipova and I. V. Martynov; *Izv. Akad. Nauk SSSR, Ser. Khim.*, 1991, 2847 (*Chem. Abstr.*, 1992, **116**, 129 066u). 1128

91JA910 J. Mulzer, H. M. Kirstein, J. Buschmann, C. Lehmann and P. Luger; *J. Am. Chem. Soc.*, 1991, **113**, 910. 138

91JA927 M. A. Brookhart, Y. Liu, E. W. Goldman, D. A. Timmers and G. D. Williams; *J. Am. Chem. Soc.*, 1991, **113**, 927. 937

91JA2228 U. Kirchgässner, H. Piana and U. Schubert; *J. Am. Chem. Soc.*, 1991, **113**, 2228. 954

91JA2324 M. Etienne, P. S. White and J. L. Templeton; *J. Am. Chem. Soc.*, 1991, **113**, 2324. 1097

91JA5085 J. M. Fevig, R. W. Marquis Jr and L. E. Overman; *J. Am. Chem. Soc.*, 1991, **113**, 5085. 331

91JA5089 S. B. King and B. Ganem; *J. Am. Chem. Soc.*, 1991, **113**, 5089. 330

91JA5742 F. H. Van der Steen, G. P. M. Van Mier, A. L. Spek, J. Kroon and G. Van Koten; *J. Am. Chem. Soc.*, 1991, **113**, 5742. 377

91JA5856 M.-P. Arthur, A. Baceiredo and G. Bertrand; *J. Am. Chem. Soc.*, 1991, **113**, 5856. 872

91JA6318 Y. Takeuchi, N. Itoh, H. Note, T. Koizumi and K. Yamaguchi; *J. Am. Chem. Soc.*, 1991, **113**, 6318. 360

91JA6673 C. P. Casey, H. Sakaba and T. L. Underiner; *J. Am. Chem. Soc.*, 1991, **113**, 6673. 949

91JA7277 D. B. Dess and J. C. Martin; *J. Am. Chem. Soc.*, 1991, **113**, 7277. 204

91JA8972 M. E. Kopach, J. Gonzalez and W. D. Harman; *J. Am. Chem. Soc.*, 1991, **113**, 8972. 201

91JA9796 H. Kuniyasu, A. Ogawa, S.-I. Miyazaki, I. Ryu, N. Kambe and N. Sonoda; *J. Am. Chem. Soc.*, 1991, **113**, 9796. 238, 241

91JAP(K)3193760 T. Yamamoto and K. Kikuchi (Sankyo Kagaku K. K.); *Jpn. Kokai JP* 03 193 760 (1991) (*Chem. Abstr.*, 1992, **116**, 58 775). 1024

91JCP(95)5609 C. M. L. Rittby; *J. Chem. Phys.*, 1991, **95**, 5609. 1096

91JCP(95)5612 J. D. Presilla-Marquez and W. R. M. Graham; *J. Chem. Phys.*, 1991, **95**, 5612. 1096

91JCS(D)1531 K. Onitsuka, H. Ogawa, T. Joh, S. Takahashi, Y. Yamamoto and H. Yamazaki; *J. Chem. Soc., Dalton Trans.*, 1991, 1531. 801

91JCS(P1)223 D. Armesto, W. M. Horspool, F. Langa and A. Ramos; *J. Chem. Soc., Perkin Trans. 1*, 1991, 223. 215

91JCS(P1)273 M. C. Cano, F. Gómez-Contreras and M. Lora-Tamayo; *J. Chem. Soc., Perkin Trans. 1*, 1991, 273. 366

91JCS(P1)607 G. L'Abbé, E. Vanderstede, W. Dehaen, P. Delbeke and S. Toppet; *J. Chem. Soc., Perkin Trans. 1*, 1991, 607. 622

91JCS(P1)897 M. Ashwell, W. Clegg and R. F. W. Jackson; *J. Chem. Soc., Perkin Trans. 1*, 1991, 897. 423

91JCS(P1)1001 A. A. Freer, M. A. Islam, G. W. Kirby and M. P. Mahajan; *J. Chem. Soc., Perkin Trans. 1*, 1991, 1001. 330

91JCS(P1)1173 S. Iwasa, M. Yamamoto, S. Kohmoto and K. Yamada; *J. Chem. Soc., Perkin Trans. 1*, 1991, 1173. 513

91JCS(P1)2435 S. Hashiguchi, A. Kawada and H. Natsugari; *J. Chem. Soc., Perkin Trans. 1*, 1991, 2435. 351

91JCS(P1)2495 L. A. Silks, III, J. Peng, J. D. Odom and R. B. Dunlap; *J. Chem. Soc., Perkin Trans. 1*, 1991, 2495. 1054

91JCS(P1)3103 C. T. Hewkin and R. F. W. Jackson; *J. Chem. Soc., Perkin Trans. 1*, 1991, 3103. 423

91JCS(P1)3383 M. Wills, R. J. Butlin, I. D. Linney and R. W. Gibson; *J. Chem. Soc., Perkin Trans. 1*, 1991, 3383. 340

91JCS(P1)3389 N. Lage, S. Masson and A. Thuillier; *J. Chem. Soc., Perkin Trans. 1*, 1991, 3389. 731

91JCS(P2)249 C. Egan, M. Clery, A. F. Hegarty and A. J. Welch; *J. Chem. Soc., Perkin Trans. 2*, 1991, 249. 838

91JFC(52)389 R. E. Banks, V. Murtagh and E. Tsiliopoulos; *J. Fluorine Chem.*, 1991, **52**, 389. 22

91JFC(53)1 E. Bollens, F. Szonyi and A. Cambon; *J. Fluorine Chem.*, 1991, **53**, 1. 1022

91JFC(55)47 I. M. Cook and D. D. DesMarteau; *J. Fluorine Chem.*, 1991, **55**, 47. 659, 660

91JGU108 Yu. A. Veits, M. V. Filippov, E. G. Neganova, A. A. Borisenko and V. L. Foss; *J. Gen. Chem. USSR (Engl. Transl.)*, 1991, **61**, 108. 399

91JGU114 Yu. A. Veits, E. G. Neganova, M. V. Filippov, A. A. Borisenko and V. L. Foss; *J. Gen. Chem. USSR* (*Engl. Transl.*), 1991, **61**, 114. 398
91JGU1152 V. B. Sokolov, T. Epishina, A. N. Ivanov and I. V. Martynov; *J. Gen. Chem. USSR* (*Engl. Transl.*), 1991, **61**, 1152. 672
91JGU1268 G. S. Zaitseva, O. P. Nivikova, L. I. Livantsova, V. S. Petrosyan and Yu. I. Baukov; *J. Gen. Chem. USSR* (*Engl. Transl.*), 1981, **61**, 1268. 702
91JGU2372 L. A. Antokhina, H. V. Chernova, G. M. Doroshkina, R. R. Shagidullin and F. F. Ganiev; *J. Gen. Chem. USSR* (*Engl. Transl.*), 1991, **61**, 2372. 679
91JHC593 A. Bojilova and N. A. Rodios; *J. Heterocycl. Chem.*, 1991, **28**, 593. 355
91JHC1003 M. S. South; *J. Heterocycl. Chem.*, 1991, **28**, 1003. 990
91JHC1091 J. Rouvinen, P. J. Mälkönen, P. Vainiotalo, K. Hanninen, M. Ahlgrén and J. Vepsäläinen; *J. Heterocycl. Chem.*, 1991, **28**, 1091. 1028
91JHC1511 R. Milcent, G. Barbier, B. Yver and F. Mazouz; *J. Heterocycl. Chem.*, 1991, **28**, 1511. 22
91JMC387 I. O. Edafiogho, K. R. Scott, J. A. Moore, V. A. Farrar and J. M. Nicholson; *J. Med. Chem.*, 1991, **34**, 387. 332, 333
91JOC353 J. T. Welch, J. S. Plummer and T. S. Chou; *J. Org. Chem.*, 1991, **56**, 353. 227
91JOC528 M. Shiozaki; *J. Org. Chem.*, 1991, **56**, 528. 208
91JOC678 M. Ballestri, C. Chatgilialoglu, K. B. Clark, D. Griller, B. Giese and B. Kopping; *J. Org. Chem.*, 1991, **56**, 678. 1135
91JOC1176 R. L. Danheiser and J. S. Nowick; *J. Org. Chem.*, 1991, **56**, 1176. 161
91JOC1307 J. Yoshida, S. Matsunaga, Y. Ishichi, T. Maekawa and S. Isoe; *J. Org. Chem.*, 1991, **56**, 1307. 417, 421, 423
91JOC1801 F. Castan, A. Baceiredo, D. Bigg and G. Bertrand; *J. Org. Chem.*, 1991, **56**, 1801. 871, 872, 873, 1071, 1148
91JOC1933 M. J. Brown and L. E. Overman; *J. Org. Chem.*, 1991, **56**, 1933. 287
91JOC2143 A. R. Katritzky, Z. Yang and J. N. Lam; *J. Org. Chem.*, 1991, **56**, 2143. 596
91JOC2258 P. E. O'Bannon and W. P. Dailey; *J. Org. Chem.*, 1991, **56**, 2258. 868
91JOC2849 S. Iwasa, M. Yamamoto, S. Kohmoto and K. Yamada; *J. Org. Chem.*, 1991, **56**, 2849. 513
91JOC2911 W. L. Whipple and H. J. Reich; *J. Org. Chem.*, 1991, **56**, 2911. 331
91JOC3043 A. A. Malik, D. Tzeng, P. Cheng and K. Baum; *J. Org. Chem.*, 1991, **56**, 3043. 967
91JOC3118 J. Collin, J.-L. Namy, F. Dallemer and H. Kagan; *J. Org. Chem.*, 1991, **56**, 3118. 439
91JOC3210 J. Sisko and S. M. Weinreb; *J. Org. Chem.*, 1991, **56**, 3210. 345
91JOC3744 D. S. Karanewsky, M. F. Malley and J. Z. Gougoutas; *J. Org. Chem.*, 1991, **56**, 3744. 360
91JOC3750 S. V. Frye, M. C. Johnson and N. L. Valvano; *J. Org. Chem.*, 1991, **56**, 3750. 330
91JOC3776 Y. Nishiyama, A. Katsuura, A. Negoro, S. Hamanaka, N. Miyoshi, Y. Yamana, A. Ogawa and N. Sonoda; *J. Org. Chem.*, 1991, **56**, 3776. 240, 243, 253
91JOC4241 E. B. Nadler, E. S. Zipory and Z. Rappoport; *J. Org. Chem.*, 1991, **56**, 4241. 406
91JOC4791 E. Breuer, M. Mahajna, L. D. Quin and G. S. Quin; *J. Org. Chem.*, 1991, **56**, 4791. 403
91JOC5256 M. I. Cook and D. D. DesMarteau; *J. Org. Chem.*, 1991, **56**, 5256. 315, 662
91JOC5480 G. D. Maynard and L. A. Paquette; *J. Org. Chem.*, 1991, **56**, 5480. 425
91JOC5482 S. K. Davidsen, P. D. May and J. B. Summers; *J. Org. Chem.*, 1991, **56**, 5482. 329
91JOC6697 D. H. R. Barton, R. H. Hesse, A. C. O'Sullivan and M. M. Pechet; *J. Org. Chem.*, 1991, **56**, 6697. 735
91JOC7038 P. Romea, M. Aragonès, J. Garcia and J. Vilarrasa; *J. Org. Chem.*, 1991, **56**, 7038. 361
91JOM(403)401 H. Hoberg and D. Baerhausen; *J. Organomet. Chem.*, 1991, **403**, 401. 386
91JOM(412)225 H. Hoberg and M. Nohlen; *J. Organomet. Chem.*, 1991, **412**, 225. 386
91JOM(420)387 R. C. Cambie, S. J. Janssen, P. S. Rutledge and P. D. Woodgate; *J. Organomet. Chem.*, 1991, **420**, 387. 384
91JPS(A)1097 G. Zhou and J. Smid; *J. Polym. Sci., Polym. Chem., Part A*, 1991, **29** (8), 1097. 1007
91KFZ33 V. B. Skolov, Yu. Ya. Ivanov, T. A. Epishina and I. V. Martynov; *Khim. Farm. Zh.*, 1991, **25**, 33 (*Chem. Abstr.*, 1991, **115**, 48 828w). 672
91KGS369 G. A. Shvekhgeimer, K. I. Kobrakov, O. S. Kartseva and L. V. Balabanova; *Khim. Geterotsikl. Soedin*, 1991, 369; (*Chem. Abstr.*, 1991, **115**, 159 049n). 687
91KGS532 S. E. Tolchinskii, M. V. Kormer and I. A. Maretina; *Khim. Geterotsikl. Soedin*, 1991, 532 (*Chem. Abstr.*, 1991, **115**, 207 935x). 691
91KGS1205 D. A. Tikhomirov, O. S. Slyadevskaya and A. V. Eremeev; *Khim. Geterotsikl. Soedin*, 1991, 1205 (*Chem. Abstr.*, 1992, **116**, 194 210a). 661
91LA47 F. Effenberger, B. Gutterer, T. Ziegler, E. Eckhardt and R. Aichholz; *Liebigs Ann. Chem.*, 1991, 47. 135
91LA903 K. Hartke, H. D. Gerber and U. Roesrath; *Liebigs Ann. Chem.*, 1991, 903. 560, 572, 597, 617
B-91MI 502-01 M. A. Ogliaruso and J. F. Wolfe; "Synthesis of Carboxylic Acids, Esters and Their Derivatives, Updates From the Chemistry of Functional Groups," eds. S. Patai and Z. Rappoport, Wiley-Interscience, New York, 1991. 24
B-91MI 504-01 T. W. Greene and P. G. M. Wuts; "Protective Groups in Organic Synthesis," 2nd edn, Wiley, New York, 1991. 223
91MI 504-02 O. E. Kompan, A. A. Maksimenko, Z. Galdecki and I. D. Sadekov; *Metalloorg. Khim.*; 1991, **4**, 571. 211
B-91MI 506-01 J. Jones; "The Chemical Synthesis of Peptides," Clarendon, Oxford, 1991. 259, 262
B-91MI 506-02 T. W. Greene and P. G. M. Wuts; "Protective Groups in Organic Synthesis," 2nd edn, Wiley, New York, 1991. 260
B-91MI 507-01 T. W. Greene and P. G. M. Wuts; "Protective Groups in Organic Synthesis," Wiley, New York, 1991. 22

B-91MI 507-02	J. Hoyle; in "The Chemistry of Sulphonic Acids, Esters and Their Derivatives," ed. S. Patai and Z. Rappoport, Wiley, New York, 1991, p. 351.	342
B-91MI 507-03	D. A. Armitage; in "The Silicon–Heteroatom Bond," ed. S. Patai and Z. Rappoport, Wiley, New York, 1991, p. 365.	374
91MI 507-04	S. S. A. Rizvi; *Bull. Chem. Soc. Ethiop.*, 1991, **5**, 7.	22
91MI 507-05	Y. Nagao and S. Katagiri; *Sci. Rep. Hirosaki, Univ.*, 1991, **38**, 20.	316
91MI 507-06	V. A. Tartakovskii, S. L. Ioffe, A. V. Kalinin, E. T. Apasov and Y. N. Strelenko; *Mendeleev Commun.*, 1991, **3**, 95.	362
91MI 515-01	Y. Matsbara, K. Litano; *Chemistry Express*, 1991, **6**, 411.	639
91MI 516-01	K.-T. Kang, J.-S. U, I.-N. Yoon and C. H. Park; *J. Korean Chem. Soc.*, 1991, **35**, 292 (*Chem. Abstr.*, 1991, **115**, 159047k).	648
91MI 517-01	A. M. Churakov, E. L. Goncharova, A. Yu. Tyurin, S. L. Ioffe, Yu. A. Strelenko and V. A. Tartakovskii; *Mendeleev Commun.*, 1991, 141 (*Chem. Abstr.*, 1992, **117**, 47999g).	680
91MI 517-02	W. M. F. Fabian; *J. Comput. Chem.*, 1991, **12**, 17 (*Chem. Abstr.*, 1991, **115**, 48429s).	685
B-91MI 517-03	D. G. Neilson; in "The Chemistry of Functional Groups: The Chemistry of Amidines and Imidates," ed. S. Patai, Wiley, London, 1991, vol. 2, p. 427.	685, 687
B-91MI 517-04	D. G. Neilson in "The Chemistry of Functional Groups: The Chemistry of Amidines and Imidates," ed. S. Patai, Wiley, London, 1991, vol. 2, p. 425.	698
B-91MI 517-05	D. G. Neilson; in "The Chemistry of Functional Groups: The Chemistry of Amidines and Imidates," ed. S. Patai, Wiley, London, 1991, vol. 2, p. 427.	699
B-91MI 518-01	D. G. Neilson; in "The Chemistry of Amidines and Imidates," ed. S. Patai and Z. Rappoport, Wiley, New York, 1991, vol. 2, p. 425.	726
B-91MI 519-01	G. V. Boyd; in "The Chemistry of Amidines and Imidates," ed. S. Patai and Z. Rappoport, Wiley, New York, 1991, vol. 2, p. 339.	742, 755
B-91MI 521-01	L. C. Perrin; in "The Chemistry of Amidines and Imidates," eds. S. Patai and Z. Rappoport, Wiley, New York, 1991, p. 147.	847
B-91MI 521-02	J. Oszczapowicz; in "The Chemistry of Amidines and Imidates," ed. S. Patai and Z. Rappoport, Wiley, New York, 1991, vol. 2, p. 682.	864
91MI 530-02	A. A. Krolevets and I. V. Martynov; *Zh. Obshch. Khim. Lett.*, 1991, **61**, 259 (*Chem. Abstr.*, 1991, **115**, 71703k).	1128
91OM11	B. Alcaide, G. Dominguez, J. Plumet and M. A. Sierra; *Organometallics*, 1991, **10**, 11.	696
91OM737	D. W. Macomber, M. Hung, P. Madhukar, M. Liang and R. D. Rogers; *Organometallics*, 1991, **10**, 737.	958
91OM807	L. Lattuada, E. Licandro, S. Mariorana, H. Molinari and A. Papagni; *Organometallics*, 1991, **10**, 807.	946
91OM1201	G. Erker, M. Mena, C. Kruger and R. Noe; *Organometallics*, 1991, **10**, 1201.	797
91OM2121	D. W. Macomber, P. Madhukar and R. D. Rogers; *Organometallics*, 1991, **10**, 2121.	958
91OM2665	R. Fandos, A. Meetsma and J. H. Teuben; *Organometallics*, 1991, **10**, 2665.	797
91OM2752	A. G. Brook, S. S. Hu, W. J. Chatterton and A. J. Lough; *Organometallics*, 1991, **10**, 2752.	924
91OM2758	A. G. Brook, S. S. Hu, A. K. Saxena and A. J. Lough; *Organometallics*, 1991, **10**, 2758.	924
91OM2768	H. Le Bozec, K. Ouzzine and P. H. Dixneuf; *Organometallics*, 1991, **10**, 2768.	940
91OM3205	M.-P. Arthur, H. P. Goodwin, A. Baceiredo, K. B. Dillon and G. Bertrand; *Organometallics*, 1991, **10**, 3205.	872, 1148
91OM3951	B. C. Söderberg and M. J. Turbeville; *Organometallics*, 1991, **10**, 3951.	949
91OPP435	Q. D. You, H. Y. Zhou, Q.-Z. Wang and X.-H. Lei; *Org. Prep. Proced. Int.*, 1991, **23**, 435.	600, 621
91OS(70)1	S. V. Pansare, G. Huyer, L. D. Arnold and J. C. Vederas; *Org. Synth.*, 1991, **70**, 1.	85
91OS(70)79	D. L. Boger, J. S. Panek and M. Patel; *Org. Synth.*, 1991, **70**, 79.	93
91OS(70)101	S. Knapp and F. S. Gibson; *Org. Synth.*, 1991, **70**, 101.	11
91POL1147	M. Weidenbruch, B. Brand-Roth, S. Pohl and W. Saak; *Polyhedron*, 1991, **10**, 1147.	1085
91PS(59)1	P. Metzner; *Phosphorus Sulfur*, 1991, **59**, 1.	552
91PS(59)125	G. Himbert, K.-P. Pfeifer and C. E. Finkele; *Phosphorus Sulfur*, 1991, **59**, 125.	645
91PS(59)141	N. Lage, S. Masson and A. Thuillier; *Phosphorus Sulfur*, 1991, **59**, 141.	731
91PS(60)85	Yu. E. Lyashenko, V. B. Sokolov and I. V. Martynov; *Phosphorus Sulfur Silicon Relat. Elem.*, 1991, **60**, 85.	669, 672
91PS(60)119	M. K. A. Ibrahim, A. H. El Ghandour and K. Abbou-Hadeed; *Phosphorus Sulfur Silicon Relat. Elem.*, 1991, **60**, 119.	677
91PS(60)159	A. H. A.-E. Rahman, A. M. Khalil and M. A. Hanna; *Phosphorus Sulfur*, 1991, **60**, 159.	22
91PS(60)239	F. Breuer, M. Safadi, M. Chorev and D. Gibson; *Phosphorus Sulfur*, 1991, **60**, 239.	22
91PS(62)15	L. Maier and P. J. Diel; *Phosphorus Sulfur Silicon*, 1991, **62**, 15.	404
91PS(63)95	A. Kockritz, G. Rohr and M. Schnell; *Phosphorus Sulfur*, 1991, **63**, 95.	790
91PS(56)283	I. Jabre, M. Saquet and A. Thuillier; *Phosphorus Sulfur*, 1991, **56**, 283.	589, 612
91PS(63)373	M. F. Ismail, S. A. Emara and O. E. A. Mustafa; *Phosphorus Sulfur*, 1991, **63**, 373.	355
91PS(55)133	A. S. Ionkin, L. F. Chertanova and B. A. Arbuzov; *Phosphorus Sulfur Silicon*, 1991, **55**, 133.	395, 396
91RCR1128	N. G. Zabirov, F. M. Shamsevaleev and R. A. Cherkasov; *Russ. Chem. Rev.* (*Engl. Transl.*), 1991, **60**, 1128.	642, 644
91S18	W. Murray, M. Wachter, D. Barton and Y. Forero-Kelly; *Synthesis*, 1991, 18.	11
91S23	Ph. Coutrot, C. Charbonnier and C. Grison; *Synthesis*, 1991, 23.	233
91S350	C. Astorga, F. Rebolledo and V. Gotor; *Synthesis*, 1991, 350.	353
91S382	H. J. Cristau, E. Manginot and E. Torreilles; *Synthesis*, 1991, 382.	373
91S441	K. A. Burdett; *Synthesis*, 1991, 441.	8
91S750	T. Sakamoto, H. Mori, M. Takizawa and Y. Kikugawa; *Synthesis*, 1991, 750.	665, 666, 824
91S793	N. Petragnani and J. V. Comasseto; *Synthesis* 1991, 793.	472

91S897	N. Petragnani and J. V. Comasseto; *Synthesis* 1991, 897.	472
91S920	M. K. Mao and J. E. Franz; *Synthesis*, 1991, 920.	235
91S951	Y. Seo, K. R. Mun, K. Kim; *Synthesis*, 1991, 951.	637
91S954	R. J. P. Corriu, G. F. Lanneau and M. Perrot-Petta; *Synthesis*, 1991, 954.	759, 777
91S1001	D. Albanese and M. Penso; *Synthesis*, 1991, 1001.	1030
91S1099	E. Fuchs, B. Breit, U. Bergsträsser, J. Hoffmann, H. Heydt and M. Regitz; *Synthesis*, 1991, 1099.	395
91S1163	T. Aoyama, T. Nakano, K. Marumo, Y. Uno and T. Shioiri; *Synthesis*, 1991, 1163.	874
91S1173	T. Eicher, K. Massonne and M. Herrmann; *Synthesis*, 1991, 1173.	11
91S1177	G. A. Olah and A-h. Wu; *Synthesis*, 1991, 1177.	10
91SC85	C. Lambert, M. Hilbert, L. Christiaens and N. Dereu; *Synth. Commun.*, 1991, **21**, 85.	
91SC625	J. J. Ares; *Synth. Commun.*, 1991, **21**, 625.	595
91SC1889	U. R. Kalkote, A. R. Choudhary, A. A. Natu, R. J. Lahoti and N. R. Ayyangar; *Synth. Commun.*, 1991, **21**, 1889.	384
91SC2025	R. W. Stephens and L. A. Domeier; *Synth. Commun.*, 1991, **21**, 2025.	1126
91SC2097	H.-J. Liu and W. Luo; *Synth. Commun.*, 1991, **21**, 2097.	234
91SL243	S. Busato, G. Cainelli, M. Panunzio, E. Bandini, G. Martelli and G. Spunta; *Synlett*, 1991, 243.	377
91SL381	D. B. Grotjahn and K. H. Dotz; *Synlett.*, 1991, 381.	931
91SL654	A. P. Kozikowski, A. H. Fauq, W. Tuckmantel, S. O. Casalotti and K. E. Krueger; *Synlett.*, 1991, 654.	1052
91SL781	S. C. Case-Green, S. G. Davies and C. J. R. Hedgecock; *Synlett*, 1991, 781.	161
91SL793	K. Maruoka and H. Yamamoto; *Synlett*, 1991, **11**, 793.	421
91SL919	M. Ohno, H. Tanaka, M. Komatsu and Y. Ohshiro; *Synlett*, 1991, **12**, 919.	10, 367
91T447	L. I. Belen'kii, D. B. Brokhovetskii and M. M. Krayushkin; *Tetrahedron*, 1991, **47**, 447.	363
91T5193	C. Nájera and J. M. Sansano; *Tetrahedron*, 1991, **47**, 5193.	1106
91T5481	C. Alexandre, C. Bertho, B. Tabti and F. Rouessac; *Tetrahedron*, 1991, **47**, 5481.	199
91T6127	D. H. R. Barton, C. Chen and G. M. Wall; *Tetrahedron*, 1991, **47**, 6127.	247
91T6275	P. Beslin and S. Perrio; *Tetrahedron*, 1991, **47**, 6275.	557
91T6673	F. Nief and F. Mathey; *Tetrahedron*, 1991, **47**, 6673.	397
91T7447	A. A. Gakh, S. V. Romaniko, B. I. Ugrak and A. A. Fainzilberg; *Tetrahedron*, 1991, **47**, 7447.	314, 384
91T7503	F. H. van der Steen and G. van Koten; *Tetrahedron*, 1991, **47**, 7503.	22
91T8767	M. Cinquini, F. Cozzi, P. G. Cozzi and E. Consolandi; *Tetrahedron*, 1991, **47**, 8767.	233
91T9053	C. Lambert and L. Christiaens; *Tetrahedron*, 1991, **47**, 9053.	22
91T9599	E. C. Taylor and D. M. Sobieray; *Tetrahedron*, 1991, **47**, 9599.	355
91T9929	R. W. Bates, R. Fernández-Moro and S. V. Ley; *Tetrahedron*, 1991, **47**, 9929.	164
91T10053	S. J. Allcock, T. L. Gilchrist, S. J. Shuttleworth and F. D. King; *Tetrahedron*, 1991, **47**, 10053.	351, 363
91TA647	K. Maruoka, H. Banno and H. Yamamoto; *Tetrahedron Asymmetry*, 1991, **2**, 647.	421, 431
91TA1209	A. Defoin, J. Pires, I. Tissot, T. Tschamber, D. Bur, M. Zehnder and J. Streith; *Tetrahedron Asymmetry*, 1991, **2**, 1209.	330
91TL133	T. Sheradsky, J. Milvitskaya and I. E. Pollak; *Tetrahedron Lett.*, 1991, **32**, 133.	351, 366
91TL283	P. Somfai, H. M. He and D. Tanner; *Tetrahedron Lett.*, 1991, **32**, 283.	342, 345
91TL341	N. Klempier, A. de Raadt, K. Faber and H. Griengl; *Tetrahedron Lett.*, 1991, 341.	118
91TL405	I. T. Barnish, C. W. G. Fishwick and D. R. Hill; *Tetrahedron Lett.*, 1991, **32**, 405.	613
91TL1035	M. Newcomb and J. L. Esker; *Tetrahedron Lett.*, 1991, **32**, 1035.	655, 712
91TL1825	S. S. Elmorsy, M. A. Nour, E. M. Kandeel and A. Pelter; *Tetrahedron Lett.*, 1991, **32**, 1825.	282
91TL2457	D. Goldsmith and J. J. Soria; *Tetrahedron Lett.*, 1991, **32**, 2457.	702
91TL2505	T. T. Nikiforov and B. A. Connolly; *Tetrahedron Lett.*, 1991, **32**, 2505.	1111
91TL2577	C. Ma and M. J. Miller; *Tetrahedron Lett.*, 1991, **32**, 2577.	329
91TL2703	D. H. R. Barton, J. Cs. Jaszberenyi and C. Tachdjian; *Tetrahedron Lett.*, 1991, **32**, 2703.	516, 526
91TL2765	A. Lecoq, M. Marraud and A. Aubry; *Tetrahedron Lett.*, 1991, **32**, 2765.	22
91TL3449	J. A. Moore and J.-H. Kim; *Tetrahedron Lett.*, 1991, **32**, 3449.	359
91TL3503	S. Fujiwara, T. Shin-Ike, N. Sonoda, M. Aoki, K. Okada, N. Miyoshi and N. Kambe; *Tetrahedron Lett.*, 1991, **32**, 3503.	1036
91TL3515	Y.-M. Tsai and C.-D. Cherng; *Tetrahedron Lett.*, 1991, **32**, 3515.	411
91TL4193	J. Baraniak and W. J. Stec; *Tetrahedron Lett.*, 1991, **32**, 4193.	694
91TL4279	H. J. Bestmann and H. Lehnen; *Tetrahedron Lett.*, 1991, **32**, 4279.	691
91TL4325	S. Aoyagi, Y. Shishido and C. Kibayashi; *Tetrahedron Lett.*, 1991, **32**, 4325.	330
91TL4679	K. Hartke and N. Rettberg; *Tetrahedron Lett.*, 1991, **32**, 4679.	548, 559
91TL4929	M. Nakada, S. Nakamura, S. Kobayashi and M. Ohno; *Tetrahedron Lett.*, 1991, **32**, 4929.	418
91TL5223	D. F. Harvey and M. F. Brown; *Tetrahedron Lett.*, 1991, **32**, 5223.	935
91TL6507	S. H. Rosenberg, S. A. Boyd and R. A. Mantei; *Tetrahedron. Lett.*, 1991, **32**, 6507.	236
91TL7165	D. E. Ward and C. K. Rhee; *Tetrahedron Lett.*, 1991, **32**, 7165.	12
91TL7411	M. Lemarié, T.-N. Pham and P. Metzner; *Tetrahedron Lett.*, 1991, **32**, 7411.	527
91TL7431	I. Matsuda, J. Sakakibara and H. Nagashima; *Tetrahedron Lett.*, 1991, **32**, 7431.	346
91TL7683	T. Mandai, K. Ryoden, M. Kawada and J. Tsuji; *Tetrahedron Lett.*, 1991, **32**, 7683.	346
91TL7759	J. D. King and P. Quayle; *Tetrahedron Lett.*, 1991, **32**, 7759.	934
91UKZ89	B. D. Grishchuk, P. M. Gorbovoi, M. V. Vouk, A. A. Dyachuk, I. E. Martymyak and N. I. Ganushchak; *Ukr. Khim. Zh.* (*Russ. Ed.*), 1991, **57**, 89.	1044
91UKZ202	M. V. Vovk, A. A. Esipenko, V. N. Fetyukhin, V. I. Boiki and L. I. Samarai; *Ukr. Khim. Zh* (*Russ. Ed.*), 1991, **57**, 202 (*Chem. Abstr.*, 1991, **115**, 71 055a).	669

91UKZ431 M. V. Vouk, M. K. Bratenko and L. I. Samarai; *Ukr. Khim. Zh.* (*Russ. Ed.*), 1991, **57**, 431. 1035

91UKZ530 E. A. Rudenko, S. O. Seferov, V. A. Chervonyi, A. V. Kharchenko and B. S. Drach; *Ukr. Khim. Zh* (*Russ. Ed.*), 1991, **57**, 530 (*Chem. Abstr.*, 1991, **115**, 207901h). 656

91USP4997967 R. Haessig (Ciba-Geigy Corp.); *US Pat.* 4997967, (1991) (*Chem. Abstr.*, 1991, **115**, 71160). 1035

91ZAAC(596)47 A. Zinn, K. Dehnicke, D. Fenske and G. Baum; *Z. Anorg. Allg. Chem.*, 1991, **596**, 47. 780

91ZAAC(603)89 P. Mamos and N. Klouras; *Z. Anorg. Allg. Chem.*, 1991, **603**, 89. 374, 376, 718

91ZAAC(605)7 G. Becker, W. Becker, M. Schmidt, W. Schwarz and M. Westerhausen; *Z. Anorg. Allg. Chem.*, 1991, **605**, 7. 396, 879, 891

91ZN(B)978 J. Grobe, D. L. Van and T. Grobpietsch; *Z. Naturforsch. B.*, 1991, 978. 786, 1087

91ZN(B)1328 J. K. Buijink, M. Noltemeyer and F. T. Edelmann; *Z. Naturforsch. Teil B*, 1991, **46**, 1328. 781

91ZOB450 G. I. Kokorev, I. A. Litvinov and V. A. Naumov; *Zh. Obshch. Khim.*, 1991, **61**, 450. 370

91ZOB1268 V. B. Sokolov, T. Epishina, A. N. Ivanov and I. V. Martynov; *Zh. Obshch. Khim.*, 1991, **61**, 1268 (*Chem. Abstr.*, 1992, **116**, 6643x). 672

91ZOB1389 G. S. Zaitseva, O. P. Novikova, L. I. Livantsova, V. S. Petrosyan and Yu. I. Baukov; *Zh. Obshch. Khim.*, 1991, **61**, 1389 (*Chem. Abstr.*, 1992, **116**, 21105s). 702

91ZOB1581 E. E. Nifant'ev, V. V. Negrebetskii and M. K. Grachev; *Zh. Obshch. Khim.*, 1991, **61**, 1581 (*Chem. Abstr.*, 1992, **116**, 129074v). 775

91ZOB2257 L. A. Antokhina, A. V. Chernova, G. M. Doroshkina, R. R. Shagidullin and F. F. Ganiev; *Zh. Obshch. Khim.*, 1991, **61**, 2257 (*Chem. Abstr.*, 1992, **116**, 194461h). 679

91ZOR340 V. R. Kartashov, P. S. Afanas'ev, E. V. Skorobogatova, V. A. Chertkov, N. M. Sergeev and N. S. Zefirov; *Zh. Org. Khim.*, 1991, **27**, 340. 1044

92ACH357 V. Bojinov and A. Bogdanova; *Acta Chim. Acad. Sci. Hung.*, 1992, **129**, 357. 348

92AG(E)90 K. Banert, H. Hückstädt and K. Vrobel; *Angew. Chem., Int. Ed. Engl.*, 1992, **31**, 90. 1041

92AG(E)316 M. Driess and H. Pritzkow; *Angew. Chem., Int. Ed. Engl.*, 1992, **31**, 316. 892

92AG(E)437 H. Grützmacher, S. Freitag, R. Herbst-Irmer and G. S. Sheldrick; *Angew. Chem., Int. Ed. Engl.*, 1992, **31**, 437. 1086

92AG(E)758 T. Wettling, B. Geissler, S. Barth, P. Binger, R. Schneider and M. Regitz; *Angew. Chem., Int. Ed. Engl.*, 1992, **31**, 758. 1152

92AG(E)851 K. Onitsuka, T. Joh and S. Takahashi; *Angew. Chem. Int. Ed. Engl.*, 1992, **31**, 851. 801

92AG(E)866 K. Banert and S. Groth; *Angew. Chem., Int. Ed. Engl.*, 1992, **31**, 866. 1101

92AG(E)1055 B. Breit, U. Bergsträsser, G. Maas and M. Regitz; *Angew. Chem., Int. Ed. Engl.*, 1992, **31**, 1055. 885

92AG(E)1238 P. Willershausen, C. Kybart, N. Stamatis, W. Massa, M. Bühl, P. v. R. Schleyer and A. Berndt; *Angew. Chem., Int. Ed. Engl.*, 1992, **31**, 1238. 928, 929

92AG(E)1384 P. Willershausen, G. Schmidt-Lukasch, C. Kybart, J. Allwohn, W. Massa, M. L. McKee, P. v. R. Schleyer and A. Berndt; *Angew. Chem., Int. Ed. Engl.*, 1992, **31**, 1384. 929

92AG(E)1603 G. Gabbert, W. Weinmann, H. Pritzkow and W. Siebert; *Angew. Chem., Int. Ed. Engl.*, 1992, **31**, 1603. 927

92AP325 K. Kostka and M. Porada; *Arch. Pharm.* (*Weinheim, Ger.*), 1992, **325**, 325. 370

92B7922 G. M. Doherty, R. Motherway, S. G. Mayhew and J. P. G. Malthouse; *Biochemistry*, 1992, **31**, 7922. 1111

92BAU236 V. B. Sokolov, A. N. Chekhlov, T. A. Epishina and I. V. Martynov; *Bull. Akad. Sci. USSR, Div. Chem. Sci.*, 1992, 236. 672

92BCJ674 H. Okamoto, T. Kobutani and S. Kato, *Bull. Chem. Soc. Jpn.*, 1992, 674. 691

92BCJ1636 M. Ueda and H. Mori; *Bull. Chem. Soc. Jpn.*, 1992, **65**, 1636. 303

92BSB257 J. K. Mulengi and N. Fatmi; *Bull. Soc. Chim. Belg.*, 1992, **101**, 257. 324

92BSB445 B. Hajjem, M. L. Ben Koud and B. Baccar; *Bull. Soc. Chem. Belg.*, 1992, **101**, 445. 690

92BSF360 A. Daich and B. Decroix; *Bull. Soc. Chim. Fr.*, 1992, **129**, 360. 367

92BSF367 J. M. Sotiropoulos, A. Baceiredo and G. Bertrand; *Bull. Soc. Chim. Fr.*, 1992, **129**, 367. 872

92CB73 S. W. Helm, G. Linti, H. Nöth, S. Channareddy and P. W. Hofmann; *Chem. Ber.*, 1992, **125**, 73. 927

92CB417 S. Kato, H. Kageyama, Y. Kawahara, T. Murai and H. Ishihara; *Chem. Ber.*, 1992, **125**, 417. 541, 542

92CB607 R. Tacke, H. Hengelsberg, E. Klingner and H. Henke; *Chem. Ber.*, 1992, **125**, 607. 416

92CB1141 R. Aumann, *Chem. Ber.*, 1992, **125**, 1141. 955

92CB1571 C. M. Stegmair, W. Ullrich, W. Schütt, P. Kiprof and F. R. Kreissl; *Chem. Ber.*, 1992, **125**, 1571. 955

92CB1583 A. Mayr, C. M. Bastos, N. Daubenspeck and G. A. McDermott; *Chem. Ber.*, 1992, **125**, 1583. 446

92CB1889 J. Gradel and W. Sundermeyer; *Chem. Ber.*, 1992, **125**, 1889. 476, 492

92CB2051 M. Duetsch, F. Stein, R. Lackmann, E. Pohl, R. Herbst-Irmer and A. de Meijere; *Chem. Ber.*, 1992, **125**, 2051. 958

92CB2137 K. H. Dötz, A. Rau and K. Harms; *Chem. Ber.*, 1992, **125**, 2137. 953

92CB2171 M. Wedler, F. Knösal, U. Pieper, D. Stalke, F. T. Edelmann and H.-D. Amberger; *Chem. Ber.*, 1992, **125**, 2171. 782

92CB2765 R. Aumann and P. Hinterding; *Chem. Ber.*, 1992, **125**, 2765. 957

92CC415 J. C. Guillemin, T. Janati and J. M. Denis; *J. Chem. Soc., Chem. Commun.*, 1992, 415. 1158

92CC1186 U. Schmidt and B. Riedl; *J. Chem. Soc., Chem. Commun.*, 1992, 1186. 363

92CC1220 D. Pilette, H. Le Bozec, A. Romero and P. H. Dixneuf; *J. Chem. Soc., Chem. Commun.*, 1992, 1220. 941

92CC1379 C. Iwata, M. Fujimoto, M. Watanabe, T. Kawakami, Y. Nakamoto, M. Sakae, M. Katsurada, T. Imanishi and T. Tanaka; *J. Chem. Soc., Chem. Commun.*, 1992, 1379. 1120

92CC1522 J. Nakayama and I. Akiyama; *J. Chem. Soc., Chem. Commun.*, 1992, 1522. 548

92CC1571 T. Mosandl, C. O. Kappe, R. Flammang and C. Wentrup; *J. Chem. Soc., Chem. Commun.*, 1992, 1571. 698

92CC1623 E. Licandro, S. Maiorana, A. Papagni and A. Zanotti-Gerosa; *J. Chem. Soc., Chem. Commun.*, 1992, 1623. 943

92CC1631 J. Sundermeyer, K. Weber and O. Nürnberg; *J. Chem. Soc., Chem. Commun.*, 1992, 1631. 918, 919

92CC1799 S. Haber, P. Le Floch, F. Mathey; *J. Chem. Soc., Chem. Commun.*, 1992, 1799. 1157, 1158

92CCC540 S. Lakhanpal, G. C. Rao and B. N. Misra; *Collect. Czech. Chem. Commun.*, 1992, **57**, 540. 710, 714

92CCC1505 V. Jakus and E. Bojsova; *Collect. Czech. Chem. Commun.*, 1992, **57**, 1505. 984

92CHE579 D. A. Tikhomirov, O. S. Slyadenskaya and A. V. Eremeev; *Chem. Heterocycl. Compd. (Engl. Transl.)*, 1992, 579. 661

92CL131 K. Okuma, K. Ikari and H. Ohta; *Chem. Lett.*, 1992, 131. 603, 622, 627

92CL509 K. Takagi, H. Takachi and N. Hayama; *Chem. Lett.*, 1992, 509. 1110

92CL1437 H. Fujihara, R. Saito, M. Yabe and N. Furukawa; *Chem. Lett.*, 1992, 1437. 210

92CL1515 T. Kaharu and S. Takahashi; *Chem. Lett.*, 1992, 1515. 1089

92CL1843 K. Shimada, N. Jin, M. Fujimura, Y. Nagano, E. Kudoh and Y. Takikawa; *Chem. Lett.*, 1992, 1843. 516, 520

92CPB602 T. Naito, Y. Habu, O. Miyata, I. Ninomiya and H. Ohishi; *Chem. Pharm. Bull.*, 1992, **40**, 602. 726

92CPB1177 Y. Ishihara, H. Yukimasa, M. Miyamoto and G. Goto; *Chem. Pharm. Bull.*, 1992, **40**, 1177. 332, 333

92CRV113 J. Muzart; *Chem. Rev.*, 1992, **92**, 113. 28

92EGP300936A7 E. Herrmann, S. Steinbach and L. Riesel; *Ger. (East) Pat.* 30 093 6A7 (1992) (*Chem. Abstr.*, **118**, 7196g). 1131

92GSM79 D. W. Knight; *Gen. Synth. Methods*, 1992, **13**, 79. 24

92H(33)121 H. Takayama, N. Seki, M. Kitajima, N. Aimi, H. Seki and S. Sakai; *Heterocycles*, 1992, **33**, 121. 324

92H(33)657 B. K. Jordan, B. Stanovnik and M. Tišler; *Heterocycles*, 1992, **33**, 657. 22

92H(33)893 T. Ohsumi and H. Neunhoeffer; *Heterocycles*, 1992, **33**, 893. 362

92H(34)369 K. Afarinkia; *Heterocycles*, 1992, **34**, 369. 394, 396

92H(34)739 P. Pecquet, F. Huet, M. Legraverend and E. Bisagni; *Heterocycles*, 1992, **34**, 739. 199

92H(34)1573 A. Lorente, M. L. García, M. Fernández and J. L. Soto; *Heterocycles*, 1992, **34**, 1573. 606

92HCA62 T. Koch, V. I. Ognyanov and M. Hesse; *Helv. Chim. Acta*, 1992, **75**, 62. 307

92HCA109 A. Defoin, A. Brouillard-Poichet and J. Streith; *Helv. Chim. Acta*, 1992, **75**, 109. 330, 332

92IC2318 M. H. Chou, C. Creutz and N. Sutin; *Inorg. Chem.*, 1992, **31**, 2318. 385

92IC4917 Y. Chen, N. R. Patel, R. L. Kirchmeier and J. M. Shreeve; *Inorg. Chem.*, 1992, **31**, 4917. 659

92ICA173 M. Anton, J. M. Coronas, G. Muller, J. Sales and M. Seco; *Inorg. Chim. Acta*, 1992, **36**, 173. 1137

92ICA527 G. Sarwar, N. R. Patel, Y. Y. Zheng, E. O. John, R. L. Kirchmeier and J. M. Shreeve; *Inorg. Chim. Acta*, 1992, **198–200**, 527. 679, 703

92IJC(B)906 V. Gotor, C. Astorga and F. Rebolledo; *Indian J. Chem., Sect. B.*, 1992, **31**, 906. 353

92IZV187 A. A. Krolevets, V. V. Antipova, P. V. Petrovskii and I. V. Martynov; *Izv. Akad. Nauk SSSR, Ser. Khim.*, 1992, **187** (*Chem. Abstr.*, 1993, **116**, 255 714s). 1131

92IZV220 M. A. Pudovik, L. K. Kibardina and A. N. Pudovik; *Izv. Akad. Nauk SSSR, Ser. Khim.*, 1992, 220. 369

92IZV315 V. B. Sokolov, A. N. Chekhlov, T. A. Epishina and I. V. Martynov; *Izv. Akad. Nauk SSSR, Ser. Khim.*, 1992, 315 (*Chem. Abstr.*, 1993, **118**, 101 614p). 672

92JA539 T. R. Cundari and M. S. Gordon; *J. Am. Chem. Soc.*, 1992, **114**, 539. 932

92JA1094 J. Li and M. Jones; *J. Am. Chem. Soc.*, 1992, **114**, 1094. 873

92JA2985 S. Aoki, T. Fujimura and E. Nakamura; *J. Am. Chem. Soc.*, 1992, **114**, 2985. 953, 956

92JA4910 W. Ziche, N. Auner and P. Kiprof; *J. Am. Chem. Soc.*, 1992, **114**, 4910. 925

92JA5890 J. A. Ramsden, W. Weng, A. M. Arif and J. A. Gladysz; *J. Am. Chem. Soc.*, 1992, **114**, 5890. 1097

92JA6059 R. Réau, G. Veneziani and G. Bertrand; *J. Am. Chem. Soc.*, 1992, **114**, 6059. 871, 872, 1148

92JA6266 M. J. Burk and J. E. Feaster; *J. Am. Chem. Soc.*, 1992, **114**, 6266. 354

92JA6735 C. Kelley, N. Lugan, M. R. Terry, G. L. Geoffroy, B. S. Haggerty and A. L. Rheingold; *J. Am. Chem. Soc.*, 1992, **114**, 6735. 946

92JA7287 D. P. Martin and D. G. Drueckhammer; *J. Am. Chem. Soc.*, 1992, **114**, 7287. 235

92JA7288 O. Boutry, E. Gutiérrez, A. Monge, M. C. Nicasio, P. J. Pérez and E. Carmona; *J. Am. Chem. Soc.*, 1992, **114**, 7288. 948

92JA8088 E. Chelain, R. Goumont, L. Hamon, A. Parlier, M. Rudler, H. Rudler, J.-C. Daran and J. Vaissermann; *J. Am. Chem. Soc.*, 1992, **114**, 8088. 953

92JA8313 C. Chen, D. Crich and A. Papadatos; *J. Am. Chem. Soc.*, 1992, **114**, 8313. 241

92JA8735 B. L. Balzer, M. Cazanoue and M. G. Finn; *J. Am. Chem. Soc.*, 1992, **114**, 8735. 936

92JA9239 P. Hassanzadeh and L. Andrews; *J. Am. Chem. Soc.*, 1992, **114**, 9239. 1096

92JA9710 N. Chatani, S. Ikeda, K. Ohe and S. Murai; *J. Am. Chem. Soc.*, 1992, **114**, 9710. 416

92JA9715 E. Nakamura, K. Tanaka and S. Aoki; *J. Am. Chem. Soc.*, 1992, **114**, 9715. 958

92JA10078 J. A. Soderquist and E. I. Miranda; *J. Am. Chem. Soc.*, 1992, **114**, 10078. 422

92JA10665 S. L. B. Wang, J. Su, W. D. Wulff and K. Hoogsteen; *J. Am. Chem. Soc.*, 1992, **114**, 10 665. 949

92JAP(K)4224595 M. Yagihara, H. Sasaki and S. Kato; *Jpn. Kokai Tokkyo Koho, Jpn. Pat.* 04224595 (*Chem. Abstr.*, 1993, **118**, 21 947). 506, 509

92JCS(D)19 S. U. Koschmieder, G. Wilkinson, B. Hussain-Bates and M. B. Hursthouse; *J. Chem. Soc., Dalton Trans.*, 1992, 19. 798

92JCS(D)53	G. López, J. Ruiz, G. García, C. Vicente, J. M. Martí, J. A. Hermoso, A. Vegas and M. Martínez-Ripoll; *J. Chem. Soc., Dalton Trans.*, 1992, 53.	386
92JCS(D)3199	T. Clark, S. D. Robinson and D. A. Tocher; *J. Chem. Soc., Dalton Trans.*, 1992, 3199.	781
92JCS(P1)1101	M. Toselli and P. Zanirato; *J. Chem. Soc., Perkin Trans. 1*, 1992, 1101.	992
92JCS(P1)1671	B. de Costa, C. George and C. Dominguez; *J. Chem. Soc., Perkin Trans. 1*, 1992, 1671.	1029
92JCS(P1)2127	M. Yokoyama, M. Irie, K. Sujino, T. Kagemoto, H. Togo and M. Funabashi; *J. Chem. Soc., Perkin Trans. 1*, 1992, 2127.	700, 701
92JCS(P1)2789	A. H. M. Al Shaar, R. K. Chambers, D. W. Gilmour, D. J. Lythgoe, I. McClenaghan and C. A. Ramsden; *J. Chem. Soc., Perkin Trans. 1*, 1992, 2789.	690
92JCS(P2)281	J. Iley, E. Carvalho, F. Norberto and E. Rosa; *J. Chem. Soc., Perkin Trans. 2*, 1992, 281.	362, 657, 711
92JFC(56)85	A. M. Jouani, F. Szonyi and A. Cambon; *J. Fluorine Chem.*, 1992, **56** (1), 85.	991
92JFC(56)335	E. D. John, R. L. Kirchmeier and J. M. Shreeve; *J. Fluorine Chem.*, 1992, **56**, 335.	684
92JHC1081	R. Milcent and G. Barbier; *J. Heterocycl. Chem.*, 1992, **29**, 1081.	355
92JOC1009	P. Dussault and A. Sahli; *J. Org. Chem.*, 1992, **57**, 1009.	206
92JOC1605	M. J. Kim and Y. K. Choi; *J. Org. Chem.*, 1992, **57**, 1605.	135
92JOC1948	L. J. Wilson and D. C. Liotta; *J. Org. Chem.*, 1992, **57**, 1948.	334
92JOC2419	M. Sakamoto, T. Ishida, T. Fujita and S. Watanabe; *J. Org. Chem.*, 1992, **57**, 2419.	570, 613
92JOC3245	T. Sakamoto, K. Okamoto and Y. Kikugawa; *J. Org. Chem.*, 1992, **57**, 3245.	665, 672, 824, 825
92JOC4648	J. E. Johnson, S. L. Todd, S. M. Dutson, A. Ghafouripour, R. M. Alderman and M. R. Hotema; *J. Org. Chem.*, 1992, **57**, 4648.	835
92JOC4877	J. Yoshida, M. Itoh, S. Matsunaga and S. Isoe; *J. Org. Chem.*, 1992, **57**, 4877.	343
92JOC5270	T. G. Kutateladze, J. L. Kice and N. S. Zefirov; *J. Org. Chem.*, 1992, **57**, 5270.	1107
92JOC5328	C. K. F. Hermann, J. A. Campbell, T. D. Greenwood, J. A. Lewis and J. F. Wolfe; *J. Org. Chem.*, 1992, **57**, 5328.	342
92JOC6216	H. Tomioka, K. Komatsu and M. Shimizu; *J. Org. Chem.*, 1992, **57**, 6216.	404
92JOC6339	R. Annunziata, M. Cinquini, F. Cozzi and A. L. Borgia; *J. Org. Chem.*, 1992, **57**, 6339.	237
92JOC6371	L. A. Carpino and E.-S. M. E. Mansour; *J. Org. Chem.*, 1992, **57**, 6371.	264
92JOC6649	J. N. Kim and E. K. Ryu; *J. Org. Chem.*, 1992, **57**, 6649.	664
92JOC6662	W. V. Murray and S. K. Hadden; *J. Org. Chem.*, 1992, **57**, 6662.	
92JOC6803	M. D. Bachi, E. Bosch, D. Denenmark and D. Girsh; *J. Org. Chem.*, 1992, **57**, 6803.	513
92JOC7001	F. E. McDonald and S. J. Danishefsky; *J. Org. Chem.*, 1992, **57**, 7001.	342
92JOC7010	Y. M. Tsai, H. C. Nieh and C. D. Cherng; *J. Org. Chem.*, 1992, **57**, 7010.	410, 411
92JOC7140	R. J. Bergeron and O. Phanstiel IV; *J. Org. Chem.*, 1992, **57**, 7140.	328
92JOC7359	M. Komatsu, Y. Kajihara, M. Kobayashi, S. Itoh and Y. Ohshiro; *J. Org. Chem.*, 1992, **57**, 7359.	359
92JOC7364	J. S. Nowick, N. A. Powell, T. M. Nguyen and G. Noronha; *J. Org. Chem.*, 1992, **57**, 7364.	971
92JOM(423)C5	M. Riviere-Baudet, A. Morére, J. F. Britten and M. Onyszchuk; *J. Organomet. Chem.*, 1992, **423**, C5.	389
92JOM(426)343	P. M. Fritz, J. Breimair, B. Wagner and W. Beck; *J. Organomet. Chem.*, 1992, **426**, 343.	936
92JOM(426)49	C. Silvestru, M. Curtui, I. Haiduc, M. J. Begley and D. B. Sowerby; *J. Organomet. Chem.*, 1992, **426**, 49.	222
92JOM(426)C28	P. Blenkiron, M. L. Lavender and M. J. Morris; *J. Organomet. Chem.*, 1992, **426**, C28.	400
92JOM(431)247	P. J. Stang and Y. Huang; *J. Organomet. Chem.*, 1992, **431**, 247.	944
92JOM(433)49	A. Castel, P. Rivière, J. Satgé and D. Desor; *J. Organomet. Chem.*, 1992, **433**, 49.	432
92JOM(437)C13	J. B. Verlhac, H. Kwon and M. Pereyre; *J. Organomet. Chem.*, 1992, **437**, C13.	415
92JOM(441)63	E. Lindner, M. Pabel, R. Fawzi and M. Steimann; *J. Organomet. Chem.*, 1992, **441**, 63.	943
92JOU1399	R. A. Karakhanov, V. I. Kelarev, A. S. Kokosova, V. A. Malyshev and V. I. Zav'yalov; *J. Org. Chem. USSR* (*Engl. Transl.*), 1992, **28**, 1399.	691
92JPC10195	T. R. Burkholder and L. Andrews; *J. Phys. Chem.*, 1992, **96**, 10 195.	1020
92JPR187	J. Nowakowski; *J. Prakt. Chem.*, 1992, **334**, 187.	972
92LA595	A. Kohrt and K. Hartke; *Liebigs Ann. Chem.*, 1992, 595.	558
92LA921	A. Brutsche and K. Hartke; *Liebigs Ann. Chem.*, 1992, 921.	584, 586, 612
92LA997	S. Shatzmiller and S. Bercovici; *Liebigs Ann. Chem.*, 1992, 997.	701
92LA1081	R. Köster and R. Kucznierz; *Liebigs Ann. Chem.*, 1992, 1081.	578, 610
B-92MI 503-01	M. A. Ogliaruso and J. F. Wolfe; in "The Chemistry of Functional Groups," ed. S. Patai, Supplement B, The Chemistry of Acid Derivatives, part 1, Wiley, New York, 1979, Supplement 1992, p. 411.	122
92MI 504-01	V. F. Pozdnev; *Int. J. Peptide and Protein Res.*, 1987, **40**, 407.	189
92MI 504-02	M. Dodd, S. L. Grundy, K. J. Reimer and W. R. Cullen; *Appl. Organomet. Chem.*, 1992, **6**, 207	222
92MI 507-01	A. Morére, M. Riviére-Baudet, J. F. Britten and M. Onyszchuk; *Main Group Met. Chem.*, 1992, **15**, 141.	388
92MI 507-02	V. A. Petrosyan, B. V. Lyalin and S. D. Shamshinov; *Electrokhimiya*, 1992, **28**, 523 (*Chem. Abstr.*, 1992, **117**(18), 180 506).	317
92MI 509-01	M. R. Ibrahim, M. Buehl, R. Knab and P. v. R. Schleyer; *J. Comput. Chem.*, 1992, **13**, 423.	433
92MI 512-01	S. Kato, H. Sasaki and M. Yagihara; *Phosphorus, Sulfur, Silicon Relat. Elem.*, 1992, **67**, 27.	506, 509
B-92MI 512-01	S. Kato and T. Murai; in "The Chemistry of Acid Derivatives," ed. S. Patai, John Wiley & Sons, 1992, suppl. B, vol. 2, p. 803.	509, 510, 512
92MI 519-01	K. B. Faitondjieva, M. G. Dimitrova and T. P. Pantev; *Dokl. Bulg. Acad. Nauk.* (German), 1992, **45**, 37 (*Chem. Abstr.*, 1993, **119**, 27 795a).	770
92MI 519-02	K. B. Faitondjieva, M. G. Dimitrova and T. P. Pantev; *Dokl. Bulg. Acad. Nauk.* (German), 1992, **45**, 41 (*Chem. Abstr.*, 1993, **119**, 28 052t).	770

92MI 526-01 C. V. Avadhani, P. P. Wadgaonkar and S. P. Vernekar; *J. App. Polym. Sci.*, 1992, **45** (8), 1335. 990

92NKK1344 T. Suyama and Y. Fujiyama; *Nippon Kagaku Kaishi*, 1992, **11**, 1344 (*Chem. Abstr.*, 1993, **118**, 124 637n). 774

92OM622 D. J. Crowther, Z. Zhang, G. J. Palenik and W. M. Jones; *Organometallics*, 1992, **11**, 622. 937

92OM809 D. Pilette, K. Ouzzine, H. Le Bozec, P. H. Dixneuf, C. E. F. Rickard and W. Roper; *Organometallics*, 1992, **11**, 809. 941

92OM1242 D. L. Ramage, G. L. Geoffroy, A. L. Rheingold and B. S. Haggerty; *Organometallics*, 1992, **11**, 1242. 988

92OM1442 R. L. Trace, J. Sanchez, J. Yang, J. Yin and W. M. Jones; *Organometallics*, 1992, **11**, 1442. 938

92OM1759 B. L. Balzer, M. Cazanoue, M. Sabat and M. G. Finn; *Organometallics*, 1992, **11**, 1759. 936

92OM2024 D. R. Schaad and C. R. Landis; *Organometallics*, 1992, **11**, 2024. 384

92OM2198 H. G. Woo, W. P. Freeman and T. D. Tilley; *Organometallics*, 1992, **11**, 2198. 422, 431

92OM2317 S. Zhang, R. T. Conlin, P. F. McGarry and J. C. Scaiano; *Organometallics*, 1992, **11**, 2317. 924

92OM2494 W. Ziche, N. Auner and J. Behm; *Organometallics*, 1992, **11**, 2494. 926

92OM3088 A. G. Brook, A. Baumegger and A. J. Lough; *Organometallics*, 1992, **11**, 3088. 923, 924

92OM3844 S. J. Goede, H. P. van Schaik, F. Bickelhaupt, H. Kooijman and A. L. Spek; *Organometallics*, 1992, **11**, 3844. 890

92OPP147 J. Sheu, M. B. Smith, T. R. Oeschger and J. Satchell; *Org. Prep. Proced. Int.*, 1992, **24**, 147. 346

92OPP346 T. Yamamoto, S. Sugiyama, K. Akimoto and K. Hayashi; *Org. Prep. Proceed. Int.*, 1992, **24**, 346. 1023

92OPP352 G. Broggini, L. Bruche and G. Zecchi; *Org. Prep. Proced. Int.*, 1992, **24**, 352. 677

92OPP553 P. F. Cirillo and J. S. Panek; *Org. Prep. Proceed. Int.*, 1992, **24**, 553. 409

92OR(42)335 D. L. Hughes; *Org. React.*, 1992, **42**, 335. 123, 131

92OS(71)48 H. Gerlach, D. Kappes, R. K. Boeckman, Jr. and G. N. Maw; *Org. Synth.*, 1992, **71**, 48. 66

92OS(71)200 C. Muhlemann, P. Hartmann and J.-P. Obrecht; *Org. Synth.*, 1992, **71**, 200. 92

92PS(66)87 T. S. Hafez; *Phosphorus Sulfur*, 1992, **66**, 87. 731

92PS(67)27 S. Kato, H. Sasaki and M. Yagihara; *Phosphorus Sulfur*, 1992, **67**, 27. 242

92PS(69)163 A. Ben Akacha, S. Barkallah and B. Baccar; *Phosphorus Sulfur*, 1992, **69**, 163. 22

92PS(70)75 M. Riviere-Baudet, A. Morére, M. Onyszchuk and J. Satgé; *Phosphorus Sulfur*, 1992, **70**, 75. 389

92PS(70)109 A. Bentolila, J. Katzhendler and E. Breuer; *Phosphorus Sulfur*, 1992, **70**, 109. 403

92PS(72)201 S. Bauermeister and T. A. Modro; *Phosphorus Sulfur Silicon Rel. Elem.*, 1992, **72**, 201. 714

92RCT211 J. Legters, L. Thijs and B. Zwanenburg; *Recl. Trav. Chim. Pays-Bas*, 1992, **111**, 211. 229

92S358 D. Häbich; *Synthesis*, 1992, 358. 975

92S439 S. S. Bari, I. R. Trehan, A. K. Sharma and M. S. Manhas; *Synthesis*, 1992, 439. 274, 286

92S482 U. Schmidt, F. Stäbler and A. Lieberknecht; *Synthesis*, 1992, 482. 351

92S582 C. Thom and P. Kocienski; *Synthesis*, 1992, 582. 343

92S637 H. Neunhoeffer, U. Karafiat and M. Vey; *Synthesis*, 1992, 637. 640

92S705 C. I. Fannes, L. Meerpoel, S. Toppet and G. Hoornaer; *Synthesis*, 1992, 705. 661

92S778 H. Takechi, M. Machida and Y. Kanaoka; *Synthesis*, 1992, 778. 572

92S1145 K. Burger, M. Gold, H. Neuhauser, M. Rudolph and E. Höss; *Synthesis*, 1992, 1145. 22

92S1185 P. Metzner; *Synthesis*, 1992, 1185. 550

92S1199 K. Hartke and A. Brutsche; *Synthesis*, 1992, 1199. 558, 586, 727

92S1219 P.-Y. Lin, W.-S. Ku and M.-J. Shiao; *Synthesis*, 1992, 1219. 582

92SC1 J. M. Kane and M. A. Staeger; *Synth. Commun.*, 1992, **22**, 1. 351

92SC411 G. A. Roth and E. L. McClymont; *Synth. Commun.*, 1992, **22** (3), 411. 993

92SC453 V. Andrianov and A. Eremeev; *Synth. Commun.*, 1992, **22**, 453. 668

92SC617 M. C. Berthe, Y. Fort and P. Caubère; *Synth. Commun.*, 1992, **22**, 617. 1105

92SC871 S. W. Djuric; *Synth. Commun.*, 1992, **22**, 871. 327, 328

92SC1199 H. Egg, J. Hauzinger, K. Kraus and E. Spanyar; *Synth. Commun.*, 1992, **22**, 1199. 359

92SC1397 D. Brillon; *Synth. Commun.*, 1992, **22**, 1397. 582, 611

92SC1619 M. J. Dabdoub and L. H. Viana; *Synth. Commun.*, 1992, **22**, 1619. 241

92SC1985 A. B. H. Amor, H. Mrabet and B. Baccar; *Synth. Commun.*, 1992, **22**, 1985. 700, 701

92SC2729 T. H. Black, J. T. Olson and D. C. Abt; *Synth. Commun.*, 1992, **22**, 2729. 344

92SC2741 P. Krishnan and V. G. Gurjar; *Synth. Commun.*; 1992, **22**, 2741. 1045

92SC3067 T. F. Braish and D. E. Fox; *Synth. Commun.*, 1992, **22**, 3067. 322

92SL41 M. B. Berry and D. Craig; *Synlett*, 1992, 41. 223

92SL329 A. Degl'Innocenti, E. Stucchi, A. Capperucci, A. Mordini, G. Reginato and A. Ricci; *Synlett*, 1992, **4**, 329. 427, 428

92SL332 A. Degl'Innocenti, E. Stucchi, A. Capperucci, A. Mordini, G. Reginato and A. Ricci; *Synlett*, 1992, **4**, 332. 427

92SL535 D. R. Fischer, B. L. Williamson and P. J. Stang; *Synlett*, 1992, 535. 1108

92SL537 J. A. Marshall and G. S. Welmaker; *Synlett*, 1992, 537. 467

92SL631 D. P. Curran and M. Palovich; *Synlett*, 1992, 631. 432

92SR297 D. Brillon; *Sulfur Reports*, 1992, **12**, 297. 565

92SRI851 H. Sevendir and R. Mirzaoglu; *Synth. React. Inorg. Metal-Org. Chem.*, 1992, **22**, 851. 664

92SUL199 G. Derbesy, D. N. Harpp, B. Rather and G. Carroll; *Sulfur Letters*, 1992, **14**, 199. 247

92T531 F. Fülöp, M. Simeonov and K. Pihlaja; *Tetrahedron*, 1992, **48**, 531. 363

92T651 T. Ohsumi and H. Neunhoeffer; *Tetrahedron*, 1992, **48**, 651. 362

92T1465 R. T. Lewis and W. B. Motherwell; *Tetrahedron*, 1992, **48**, 1465. 868

92T1595 E. Differding and P. M. Bersier; *Tetrahedron*, 1992, **48**, 1595. 314

92T2613 D. H. R. Barton, J. Cs. Jaszberenyi and E. A. Theodorakis; *Tetrahedron*, 1992, **48**, 2613. 1110

92T4135 P. Beslin and S. Perrio; *Tetrahedron*, 1992, **48**, 4135. 557

92T4863 S. O. de Silva, J. N. Reed, R. J. Billedeau, X. Wang, D. J. Norris and V. Snieckus; *Tetrahedron*, 1992, **48**, 4863. 199

92T7109 D. H. R. Barton and C. Tachdjian; *Tetrahedron*, 1992, **48**, 7109. 516, 523

92T7345 N. De Kimpe and B. De Corte; *Tetrahedron*, 1992, **48**, 7345. 697

92T7629 S. E. Thomas, G. J. Tustin and A. Ibbotson; *Tetrahedron*, 1992, **48**, 7629. 425

92T7703 D. J. Miller, R. M. Scrowston, P. D. Kennewell and R. Westwood; *Tetrahedron*, 1992, **48**, 7703. 1123

92T8271 A. E.-H. Ismail, A. Hamed, I. Zeid and J. C. Jochims; *Tetrahedron*, 1992, **48**, 8271. 661

92T8601 J. Jurayj and M. Cushman; *Tetrahedron*, 1992, **48**, 8601. 573, 587

92T8615 W. H. W. Lunn and J. K. Shadle; *Tetrahedron*, 1992, **48**, 8615. 340

92TL153 S. W. Wright, M. M. Abelman, L. L. Bostrom and R. L. Corbett; *Tetrahedron Lett.*, 1992, **33**, 153. 334, 336

92TL273 J. R. Malpass and C. Smith; *Tetrahedron Lett.*, 1992, **33**, 273. 330

92TL471 A. Schmidpeter and G. Jochem; *Tetrahedron Lett.*, 1992, **33**, 471. 907, 908

92TL593 R. R. Srivastava and G. W. Kabalka; *Tetrahedron Lett.*, 1992, **33**, 593. 184, 195, 198

92TL1355 K. Suda, J. Watanabe and T. Takanami; *Tetrahedron Lett.*, 1992, **33**, 1355. 410

92TL1455 T. Satoh, K. Oguro, J. Shishikura, N. Kanetaka, R. Okada and K. Yamakawa; *Tetrahedron Lett.*, 1992, **33**, 1455. 272, 283

92TL1507 A. Degl'Innocenti, A. Capperucci, G. Reginato, A. Mordini and A. Ricci; *Tetrahedron Lett.*, 1992, **33**, 1507. 427

92TL2049 C. C. Fannes and G. J. Hoornaer; *Tetrahedron Lett.*, 1992, **33**, 2049. 661

92TL2067 E. Breuer, H. Zaher and Z. Tashma; *Tetrahedron Lett.*, 1992, 2067. 371, 403

92TL2981 V. D. Romanenko, M. Sanchez, T. V. Sarina, M.-R. Mazières and R. Wolf; *Tetrahedron Lett.*, 1992, **33**, 2981. 1158

92TL3165 J. G. Lee and K. H. Kwak; *Tetrahedron Lett.*, 1992, **33**, 3165. 368

92TL3261 E. Vedejs and H. Sano; *Tetrahedron Lett.*, 1992, **33**, 3261. 331

92TL3863 P. V. Jacquemin, L. E. Christiaens, M. J. Renson, M. J. Evers and N. Dereu; *Tetrahedron Lett.*, 1992, **33**, 3863. 334

92TL4137 M. S. Smyth, H. Ford, Jr. and T. R. Burke, Jr.; *Tetrahedron Lett.*, 1992, **33**, 4137. 404

92TL5047 D. Guillerm and G. Guillerm; *Tetrahedron Lett.*, 1992, **33**, 5047. 1105

92TL5055 J. M. Altenburger, C. Mioskowski, H. d'Orchymont, D. Schirlin, C. Schalk and C. Tarnus; *Tetrahedron Lett.*, 1992, **33**, 5055. 328

92TL5117 T. W. Hart, D. Guillochon, G. Perrier, B. W. Sharp and B. Vacher; *Tetrahedron Lett.*, 1992, 5117. 727

92TL5717 M. Scartozzi, R. Grondin and Y. Leblanc; *Tetrahedron Lett.*, 1992, **33**, 5717. 351

92TL6131 M. Lemarié, Y. Vallée and M. Worrell; *Tetrahedron Lett.*, 1992, 6131. 563

92TL6383 G. Mazzanti, R. Ruinaard, L. A. V. Vliet, P. Zani, B. F. Bonini and B. Zwanenburg; *Tetrahedron Lett.*, 1992, 6383. 562

92TL6751 R. F. Cunico and C. Zhang; *Tetrahedron Lett.*, 1992, **33**, 6751. 426

92TL6915 C. L. Anderson, J. A. Soderquist and G. W. Kabalka; *Tetrahedron Lett.*, 1992, **33**, 6915. 897

92TL7865 M. Segi, T. Takahashi, H. Ichinose, G. M. Li and T. Nakajima; *Tetrahedron Lett.*, 1992, **33**, 7865. 516, 520

92TL7957 K. V. Reddy and S. Rajappa; *Tetrahedron Lett.*, 1992, **33**, 7957. 616

92ZAAC(607)101 V. G. Becker, M. Schmidt and M. Westerhausen; *Z. Anorg. Allg. Chem.*, 1992, **607**, 101. 396, 400

92ZAAC(608)33 G. Becker, M. Schmidt, W. Schwarz and M. Westerhausen; *Z. Anorg. Allg. Chem.*, 1992, **608**, 33. 396

92ZAAC(612)72 G. Becker, W. Schwarz, N. Seidler and M. Westerhausen; *Z. Anorg. Allg. Chem.*, 1992, **612**, 72. 880

92ZN(B)453 M. Westerhausen and W. Schwarz; *Z. Naturforsch. Teil B*, 1992, **47**, 453. 777

92ZN(B)558 S. Kato, K. Ibi, H. Kageyama, H. Ishihara and T. Murai; *Z. Naturforsch., Teil B*, 1992, **47**, 558. 249

92ZN(B)983 P. Willershausen, A. Hoefner, J. Allwohn, M. Pilz, W. Massa and A. Berndt; *Z. Naturforsch., Teil B*, 1992, **47**, 983 928, 930

92ZN(B)1165 W. Weigand and G. Bosl; *Z. Naturforsch., Teil B*, 1992, **47**, 1165. 388

92ZOB71 N. G. Zabirov, F. M. Shamsevaleev and R. A. Cherkasov; *Zh. Obshch. Khim.*, 1992, **62**, 71 (*Chem. Abstr.*, 1992, **117**, 130 882j). 581, 619

92ZOR568 Z. I. Tyukhteneva and L. A. Badovskaya; *Zh. Org. Khim.*, 1992, **28**, 568. 350

92ZOR1750 R. A. Karakhanov, V. I. Kelarev, A. S. Kokosova, V. A. Malyshev and V. I. Zav'yalov; *Zh. Org. Khim.*, 1992, **28**, 1750 (*Chem. Abstr.*, 1993, **119**, 8779g).

92ZPK1430 K. Kh. Dzhumakaev, N. S. Bekova and I. I. Mai; *Zh. Prikl. Khim.* (*Leningrad*), 1992, **65**, 1430 (*Chem. Abstr.*, 1993, **118**, 83 245n). 691

93ACS1004 H. R. El-Seedi, H. M. Jensen, N. Kure, I. Thomsen and K. B. G. Torsell; *Acta. Chem. Scand.*, 1993, **47**, 1004. 700

93ACS1141 S. M. Neset, T. Benneche and K. Undheim; *Acta Chem. Scand.*, 1993, **47**, 1141. 324

93AG(E)103 P. B. Hitchcock, J. A. Johnson and J. F. Nixon; *Angew. Chem., Int. Ed. Engl.*, 1993, **32**, 103. 1160

93AG(E)731 J. Sundermeyer, K. Weber and H. Pritzkow; *Angew. Chem., Int. Ed. Engl.*, 1993, **32**, 731. 918

93AG(E)1089 G. Jochem, H. Nöth and A. Schmidpeter; *Angew. Chem., Int. Ed. Engl.*, 1993, **32**, 1089. 907, 908

93AG(E)1191 H. Schick, R. Ludwig, K.-H. Schwarz, K. Kleiner and A. Kunath; *Angew. Chem., Int. Ed. Engl.*, 1993, **32**, 1191. 161

93AJC1213 S. A. Glover, G. P. Hammond, D. G. Harman, J. G. Mills and C. A. Rowbottom; *Aust. J. Chem.*, 1993, **46**, 1213. 687, 698

93BCJ327 H. Kimura, H. Konno and N. Takahashi; *Bull. Chem. Soc. Jpn.*, 1993, **66**, 327.

93BCJ421 E. Leporati; *Bull. Chem. Soc. Jpn.*, 1993, **66**, 421. 322

93BCJ623 K. S. Choi, I. Akiyama, M. Hoshino and J. Nakayama; *Bull. Chem. Soc. Jpn.*, 1993, **66**, 623. 554, 563, 605

93BCJ819 N. Iwasawa, S. Hayakawa, M. Funahashi, K. Isobe and K. Narasaka; *Bull. Chem. Soc. Jpn.*, 1993, **66**, 819. 430

93BCJ2426 S. Fujisaki, S. Hamura, H. Eguchi and A. Nishida; *Bull. Chem. Soc. Jpn.*, 1993, **66**, 2426. 318

93BMC619 B. Zacharie, R. Martel, G. Sauve and B. Belleau; *Bioorg. Med. Chem. Lett.*, 1993, **3**, 619. 588

93CB565 M. Yalpani, R. Köster and R. Boese; *Chem. Ber.*, 1993, **126**, 565. 381

93CB657 J. Schubert, S. Mock and U. Schubert; *Chem. Ber.*, 1993, **126**, 657. 954

93CB1119 S. Channareddy, B. Glaser, E. P. Mayer, H. Nöth and S. W. Helm; *Chem. Ber.*, 1993, **126**, 1119. 381, 534

93CB1849 S. Köbbing, J. Mattay and G. Raabe; *Chem. Ber.*, 1993, **126**, 1849. 951

93CB1945 B. Breit and M. Regitz; *Chem. Ber.*, 1993, **126**, 1945. 885

93CB2569 A. Kienzle, A. Obermeyer, R. Riedel, F. Aldinger and A. Simon; *Chem. Ber.*, 1993, **126**, 2569. 1077

93CB2573 K. Öfele, F. Dyckhoff and H. G. Krist; *Chem. Ber.*, 1993, **126**, 2573. 934

93CC267 S. S. Al-Juaid, P. B. Hitchcock, R. M. Matos and J. F. Nixon; *J. Chem. Soc., Chem. Commun.*, 1993, 267. 886

93CC277 Y. Kawahara, S. Kato, T. Kanda, T. Murai and H. Ishihara; *J. Chem. Soc., Chem. Commun.*, 1993, 277. 542

93CC673 U. Heim, H. Pritzkow, H. Schönberg and H. Grützmacher; *J. Chem. Soc., Chem. Commun.*, 1993, 673. 888, 893

93CC1068 J. Barluenga, J. M. Montserrat and J. Flórez; *J. Chem. Soc., Chem. Commun.*, 1993, 1068. 946, 958

93CC1295 P. Lefloch and F. Mathey; *J. Chem. Soc. Chem. Commun.*, 1993, 1295. 1158

93CC1354 G. Alcaraz, R. Reed, A. Baceiredo and G. Bertrand; *J. Chem. Soc., Chem. Commun.*, 1993, 1354. 888

93CCC2150 R. P. Sharma, K. Sharma, S. Lakhanpal, N. Sharma and B. N. Misra; *Collect. Czech. Chem. Commun.*, 1993, **58**, 2150. 710, 714

93CL521 T. Chiba and M. Okimoto; *Chem. Lett.*, 1993, 521. 698

93CL1469 T. Murai, A. Hayashi, T. Kanda and S. Kato; *Chem. Lett.*, 1993, 1469. 563

93H(35)315 N. P. Peet and E. W. Huber; *Heterocycles*, 1993, **35**, 315. 22

93H(36)1263 P. Molina, A. Tárraga, I. Díaz, A. Espinosa and C. Gaspar; *Heterocycles*, 1993, **36**, 1263. 350

93H(36)1979 Y. El-Ahmad, J.-D. Brion and P. Reynaud; *Heterocycles*, 1993, **36**, 1979. 615

93HAC11 S. Bauermeister, T. A. Modro and A. Zwierzak; *Heteroatom Chem.*, 1993, **4**, 11. 371, 373, 707

93HCA957 W. Oppolzer and E. Merifield; *Helv. Chim. Acta*, 1993, **76**, 957. 342

93IC287 E. O. John, H. G. Mack, H. Oberhammer, R. L. Kirchmeier and J. M. Shreeve; *Inorg. Chem.*, 1993, **32**, 287. 675

93IC5285 P. M. Angus, W. G. Jackson and A. M. Sargeson; *Inorg. Chem.*, 1993, **32**, 5285. 385

93JA1341 C. F. Bernasconi and M. W. Stronach; *J. Am. Chem. Soc.*, 1993, **115**, 1341. 953

93JA3000 S. Kato, T. Komuro, T. Kanda, H. Ishihara and T. Murai; *J. Am. Chem. Soc.*, 1993, **115**, 3000. 563, 626

93JA3806 C. S. Yi, G. L. Geoffroy, C. A. White and A. L. Rheingold; *J. Am. Chem. Soc.*, 1993, **115**, 3806. 946, 957

93JA3824 W. Weng, J. A. Ramsden, A. M. Arif and J. A. Gladysz; *J. Am. Chem. Soc.*, 1993, **115**, 3824. 1097

93JA4389 I. Lee and N. J. Cooper; *J. Am. Chem. Soc.*, 1993, **115**, 4389. 444

93JA4397 L. A. Carpino; *J. Am. Chem. Soc.*, 1993, **115**, 4397. 279

93JA4602 B. A. Anderson, W. D. Wulff and A. Rahm; *J. Am. Chem. Soc.*, 1993, **115**, 4602. 952, 958

93JA5123 R. Cini, F. P. Fanizzi, F. P. Intini, L. Maresca and G. Natile; *J. Am. Chem. Soc.*, 1993, **115**, 5123. 386

93JA5302 T. Sunazuka, T. Nagamitsu, K. Matsuzaki, H. Tanaka and S. Ômura and A. B. Smith III; *J. Am. Chem. Soc.*, 1993, **115**, 5302. 235

93JA5815 M. S. Congreve, A. B. Holmes, A. B. Hughes and M. G. Looney; *J. Am. Chem. Soc.*, 1993, **115**, 5815. 179

93JA5823 T. Murai, T. Mizutani, T. Kanda and S. Kato; *J. Am. Chem. Soc.*, 1993, **115**, 5823. 516, 523, 563

93JA7584 X. Creary, A. F. Sky, G. Phillips and D. E. Alonso; *J. Am. Chem. Soc.*, 1993, **115**, 7584. 765, 766

93JA8898 V. Dupont, A. Lecoq, J.-P. Mangeot, A. Aubry, G. Boussard and M. Marraud; *J. Am. Chem. Soc.*, 1993, **115**, 8898. 327, 351

93JA9015 E. Nakamura, K. Tanaka, T. Fujimura, S. Aoki and P. G. Williard; *J. Am. Chem. Soc.*, 1993, **115**, 9015. 958

93JA9986 L. L. Padolik, J. C. Gallucci and A. Wojcicki; *J. Am. Chem. Soc.*, 1993, **115**, 9986. 943

93JA11434 P. L. Wong and K. D. Moeller; *J. Am. Chem. Soc.*, 1993, **115**, 11434. 126

93JA12526 C. F. Bernasconi and W. Sun; *J. Am. Chem. Soc.*, 1993, **115**, 12 526. 955

93JCR(S)244 D. Kumar, O. Prakash and S. P. Singh; *J. Chem. Res. (S)*, 1993, **6**, 244. 356

93JCS(D)2827 T. Clark and S. D. Robinson; *J. Chem. Soc., Dalton Trans.*, 1993, 2827. 781

93JCS(P1)451 P. D. Bailey, S. P. Hollinshead, N. R. McLay, J. H. Everett, C. D. Reynolds, S. D. Wood and F. Giordano; *J. Chem. Soc., Perkin Trans. 1*, 1993, 2451. 263

93JCS(P1)459 G. Bashiardes, G. J. Bodwell and S. G. Davies; *J. Chem. Soc., Perkin Trans. 1*, 1993, 459. 328, 329

93JCS(P1)505 V. A. Petrov and D. D. DesMarteau; *J. Chem. Soc., Perkin Trans. 1*, 1993, 505. 662, 663, 674, 824, 825

93JCS(P1)1397 G. W. Kirby and M. Nazeer; *J. Chem. Soc., Perkin Trans. 1*, 1993, 1397. 330

93JCS(P1)1581 R. J. Butlin, I. D. Linney, D. J. Critcher, M. F. Mahon, K. C. Molloy and M. Wills; *J. Chem. Soc. Perkin Trans. 1*, 1993, 1581. 340, 347

93JCS(P1)1719 G. L'Abbé, M. Verbeke, W. Dehaen and S. Toppet; *J. Chem. Soc., Perkin Trans. 1*, 1993, 1719. 622

93JCS(P1)1835 A. M. Thompson, M. Boyd and W. A. Denny; *J. Chem. Soc., Perkin Trans. 1*, 1993, 1835. 613, 615

93JCS(P2)655 J. Koppel, I. Koppel, I. Leito, V. Pihl, A. Wallin, L. Grehn and U. Ragnarsson; *J. Chem. Soc., Perkin Trans. 2*, 1993, 655. 689

93JFC(65)101 A. O. A. Eltoum, N. J. O'Reilly and A. E. Tipping; *J. Fluorine Chem.*, 1993, **65**, 101. 681, 703

93JFC(65)157 A. O. A. Eltoum, N. J. O'Reilly and A. E. Tipping; *J. Fluorine Chem.*, 1993, **65**, 157. 681

93JHC61 A. U. Siddiqui, A. H. Siddiqui and T. S. Ramaiah; *J. Heterocycl. Chem.*, 1993, **30**, 61. 355

93JHC559 L. Garanti, G. Zecchi and L. Bruche; *J. Heterocycl. Chem.*, 1993, **30**, 559. 677

93JHC887 D. G. Hehemann and W. Winnik; *J. Heterocycl. Chem.*, 1993, **30**, 887. 593

93JHC1079 P. Borgna, M. Pregnolato, A. G. Invernizzi and G. Mellerio; *J. Heterocycl. Chem.*, 1993, **30**, 1079. 588

93JMC1157 F. Mazouz, S. Gueddari, C. Burstein, D. Mansuy and R. Milcent; *J. Med. Chem.*, 1993, **36**, 1157. 640

93JMC1604 T. J. Brown, R. F. Chapman, J. S. Mason, M. N. Palfreyman, N. Vicker and R. J. A. Walsh; *J. Med. Chem.*, 1993, **36**, 1604. 587, 593

93JMC2004 P. W. Manley, U. Quast, H. Andres and K. Bray; *J. Med. Chem.*, 1993, **36**, 2004. 1121

93JMC2459 A. M. Thompson, G. W. Rewcastle, M. Tercel, E. M. Dobrusin, D. W. Fry, A. J. Kraker and W. A. Denny; *J. Med. Chem.*, 1993, **36**, 2459. 567

93JOC32 K. Tamura, H. Mizukami, K. Maeda, H. Watanabe and K. Kneyama; *J. Org. Chem.*, 1993, **58**, 32. 659

93JOC3042 N. W. Boaz and K. M. Fox; *J. Org. Chem.*, 1993, **58**, 3042. 560

93JOC3886 B. Alcaide, L. Casarrubios, G. Dominguez and M. A. Sierra; *J. Org. Chem.*, 1993, **58**, 3886. 696

93JOC4742 M.-J. Shiao, L.-L. Lai, W.-S. Ku, P.-Y. Lin and J. R. Hwu; *J. Org. Chem.*, 1993, **58**, 4742. 582, 611

93JOC4908 E. A. Castro, F. Ibañez, J. G. Santos and C. Ureta; *J. Org. Chem.*, 1993, **58**, 4908. 586

93JOC5118 T. Kiyoi, N. Seko, K. Yoshino and Y. Ito; *J. Org. Chem.*, 1993, **58**, 5118. 1072

93JOC5264 P. Molina, A. Arques and A. Alías; *J. Org. Chem.*, 1993, **58**, 5264. 1065

93JOC6076 G. Kalaus, I. Greiner, M. Kajtár-Peredy, J. Brlik, L. Szabó and C. Szántay; *J. Org. Chem.*, 1993, **58**, 6076. 567

93JOC6083 G. E. Keck and D. R. Romer; *J. Org. Chem.*, 1993, **58**, 6083. 322, 330

93JOC6952 F. E. McDonald, C. B. Connolly, M. M. Gleason, T. B. Towne and K. D. Treiber; *J. Org. Chem.*, 1993, **58**, 6952. 944

93JOC7001 H. Lee and K. Kim; *J. Org. Chem.*, 1993, **58**, 7001. 736

93JOC7022 M. Arnswald and W. P. Neumann; *J. Org. Chem.*, 1993, **58**, 7022. 344, 593

93JOM(445)143 J. R. Tinder, R. R. Kohl, R. S. Bly and M. M. Hossain; *J. Organomet. Chem.*, 1993, **445**, 243. 942

93JOM(446)25 A. Ansorge, D. J. Brauer, H. Bürger, B. Krumm and G. Pawelke; *J. Organomet. Chem.*, 1993, **446**, 25. 382

93JOM(454)133 H. Fischer, G. Roth, D. Reindl and C. Troll; *J. Organomet. Chem.*, 1993, **454**, 133. 954

93JOM(459)169 K. H. Dötz, F. Kroll and K. Harms; *J. Organomet. Chem.*, 1993, **459**, 169. 953

93JOM(460)C25 F. Fritsch, T. Kerscher, K. Polborn and W. Beck; *J. Organomet. Chem.*, 1993, **460**, C25. 944

93JPR294 T. Lesiak and L. Maciejewski; *J. Prakt. Chem.*, 1993, **335**, 294. 990

93M721 K. Bogdanowicz-Szwed, M. Krasodomska, M. Lipowska, B. Rys and A. Skonecka; *Monatsh. Chem.*, 1993, **124**, 721. 597, 607

93MI 505-01 A. Ogawa and N. Sonoda; *J. Syn. Org. Chem. Jpn.*, 1993, **51**, 815. 241

93MI 507-01 K. Kim and Y. H. Kim; *Arch. Pharmacol. Res.*, 1993, **16**, 94. 368

93MI 517-01 R. Singh, V. K. Pandey, R. C. Srimal and P. R. Dua; *Acta Pharm.*, 1993, **43**, 195. 685

93MI 517-02 F. Gatta, M. R. Delgiudice, A. Boroni, P. A. Borea, S. Dionisotti and E. Ongini; *Eur. J. Med. Chem.*, 1993, **28**, 569. 22

B-93MI 523-01 A. W. Johnson; "Ylides and Imines of Phosphorus," Wiley, Chichester, 1993. 894

93OM597 A. Rahm, W. D. Wulff and A. Rheingold; *Organometallics*, 1993, **12**, 597. 956

93OM1023 K. R. Birdwhistel, T. Boucher, M. Ensminger, S. Harris, M. Johnson and S. Toprek; *Organometallics*, 1993, **12**, 1023. 1079

93OM3510 R. J. Hinkle, P. J. Stang and A. M. Arif; *Organometallics*, 1993, **12**, 3510. 944

93OM4056 W. Petz and F. Weller; *Organometallics*, 1993, **12**, 4056. 900

93PJC711 J. Mlochowski, F. F. Abdel Latif and E. Kubicz; *Pol. J. Chem.*, 1993, **67**, 711. 366

93POL563 M. B. Hursthouse, M. A. Mazid, T. Clark and S. D. Robinson; *Polyhedron*, 1993, **12**, 563. 781

93PS(72)153 Yu. F. Lyashenko and V. B. Skolov; *Phosphorus Sulfur Silicon Rel. Elem.*, 1993, **78**, 153. 22

93PS(76)1 J. Grobe, Duc Le Van, M. Hegemann, B. Leith and J. Nientiedt; *Phosphorus Sulfur*, 1993, **76**, 1. 1152

93PS(76)21 H. Grützmacher, U. Heim, H. Schönberg and H. Pritzkow; *Phosphorus Sulfur Silicon*, 1993, **76**, 21. 893

93PS(76)41 G. Etemad-Moghadam, R. El Quatib, D. Ballivet-Tkatchenko and M. Koenig; *Phosphorus Sulfur Silicon*, 1993, **76**, 41. 886

93PS(77)252 S. J. Goede, H. P. van Schaik and F. Bickelhaupt; *Phosphorus Sulfur Silicon*, 1993, **77**, 252. 890

93PS(78)97 S. Pascual, M.-C. Escudier, A.-M. Lamazouere, J. Sotiropoulos, L. Dupont, O. Dideberg and G. Germain; *Phosphorus Sulfur*, 1993, **78**, 97. 595

93PS(78)161 P. Frøyen; *Phosphorus Sulfur*, 1993, **78**, 161. 367, 373

93PS(80)31 H. Makomo, S. Masson and M. Saquet; *Phosphorus Sulfur*, 1993, **80**, 31. 589

93S117 V. P. Kukhar, Y. N. Belokon, V. A. Soloshonok, N. Y. Svistunova, A. B. Rozhenko and N. A. Kuz'mina; *Synthesis*, 1993, 117. 385
93S287 C. Astorga, F. Rebolledo and V. Gotor; *Synthesis*, 1993, 287. 359
93S441 A. Pommier and J.-M. Pons; *Synthesis*, 1993, 441. 161
93S463 N. M. Przheval'skii, I. V. Magedov and V. N. Drozd; *Synthesis*, 1993, 463. 639
93S483 S. Gronowitz, A.-B. Hörnfeldt and M. Temciuc; *Synthesis*, 1993, 483. 595
93S870 L.-L. Lai and D. H. Reid; *Synthesis*, 1993, 870. 625
93SC335 J. W. Gilman and Y. A. Otonari; *Synth. Commun.*, 1993, **23** (3), 335. 994
93SC419 D. Villemin, B. Labiad and A. Loupy; *Synth. Commun.*, 1993, **23**, 419. 199
93SC829 M. Urbansky and P. Drašar; *Synth. Commun.*, 1993, **23**, 829. 567
93SC1061 Y. Kikugawa, L. H. Fu and T. Sakamoto; *Synth. Commun.*, 1993, **23**, 1061. 666
93SC1403 X. Jia, Y. Zhang and X. Zhou; *Synth. Commun.*, 1993, **23**, 1403. 253
93SC2139 T. Mizuno, M. Matsumoto, I. Nishiguchi and T. Hirashima; *Synth. Commun.*, 1993, **23**, 2139. 596
93SC2259 D. Villemin and B. Martin; *Synth. Commun.*, 1993, **23**, 2259. 615
93SC2423 L. Duhamel, G. Plé, P. Angibaud and J. R. Desmurs; *Synth. Commun.*, 1993, **23**, 2423. 315, 316, 318
93SL148 D. Villemin and F. Thibault-Starzyk; *Synlett*, 1993, 148. 554
93SL259 A. S. Guram; *Synlett*, 1993, 259. 1115
93SL486 F. Stein, M. Duetsch, M. Noltemeyer and A. de Meijere; *Synlett*, 1993, 486. 946
93SL801 S. Alleman and P. Vogel; *Synlett.*, 1993, 801. 584
93SL850 L. A. Smith, W.-B. Wang, C. Burnell-Curty and E. J. Roskamp; *Synlett*, 1993, 850. 329
93SRI1011 K. Singhal and P. Raj; *Synth. React. Inorg. Metal.-Org. Chem.*, 1993, **23**, 1011. 372
93T1053 B. Schäfer; *Tetrahedron*, 1993, **49**, 1053. 894
93T3131 P. Beslin and S. Perrio; *Tetrahedron*, 1993, **49**, 3131. 557
93T4007 S. Puertas, R. Brieva, F. Rebolledo and V. Gotor; *Tetrahedron*, 1993, **49**, 4007. 263
93T7215 J. M. Villalgordo and H. Heimgartner; *Tetrahedron*, 1993, **49**, 7215. 611
93T9137 G. C. Mulqueen, G. Pattenden and D. A. Whiting; *Tetrahedron*, 1993, **49**, 9137. 327
93T10099 S. Geresh, Y. Giron, Y. Gilboa and R. Glaser; *Tetrahedron*, 1993, **49**, 10 099. 118
93T10489 B. Zacharie, G. Sauvé and C. Penney; *Tetrahedron*, 1993, **49**, 10 489. 568, 571
93T11029 P. R. Birkett, H. King, C. B. Chapleo, D. F. Ewing and G. McKenzie; *Tetrahedron*, 1993, **49**, 11 029. 690
93T11065 P. Lloyd-Williams, F. Albericio and E. Giralt; *Tetrahedron*, 1993, **49**, 11 065. 259
93TA457 J. Mulzer, N. Salimi and H. Hartl; *Tetrahedron Asymmetry*, 1993, **4**, 457. 138
93TA1573 M. Miyashita, Y. Toshimitsu, T. Shiratani and H. Irie; *Tetrahedron Asymmetry*, 1993, **4**, 1573. 329
93TL115 J. Nakayama, K. S. Choi, I. Akiyama and M. Hoshino; *Tetrahedron Lett.*, 1993, **34**, 115. 605
93TL227 C. A. Merlic, D. Xu, M. C. Nguyen and V. Truong; *Tetrahedron Lett.*, 1993, **34**, 227. 946
93TL457 D. Batty and Y. Langlois; *Tetrahedron Lett.*, 1993, **34**, 457. 351
93TL1247 M. Schuster, B. Munoz, W. Yuan and C.-H. Wong; *Tetrahedron Lett.*, 1993, **36**, 1247. 263
93TL1925 X. Zhang and M. Schlosser; *Tetrahedron Lett.*, 1993, **34**, 1925. 894, 896
93TL2605 Y. Vallée, M. Khalid and J.-L. Ripoll; *Tetrahedron Lett.*, 1993, **34**, 2605. 559
93TL2677 G. Pattenden, M. Tarkard and P. C. Cherry; *Tetrahedron Lett.*, 1993, **36**, 2677. 304
93TL2839 G. Besenyei and L. I. Simándi; *Tetrahedron Lett.*, 1993, 2839. 987
93TL3703 T. L. Curphey and H. H. Joyner; *Tetrahedron Lett.*, 1993, **34**, 3703. 547
93TL5397 J. H. Rigby, G. Ahmed and M. D. Ferguson; *Tetrahedron Lett.*, 1993, **34**, 5397. 606
93TL5627 N. Martín, A. Martínez-Grau, C. Seoane and J. L. Marco; *Tetrahedron Lett.*, 1993, **34**, 5627. 343
93TL5983 R. W. Friesen, A. Giroux and K. L. Cook; *Tetrahedron Lett.*, 1993, **34**, 5983. 692
93TL6395 G. Rousselet, P. Capdevielle and M. Maumy; *Tetrahedron Lett.*, 1993, **34**, 6395. 750
93TL6669 C. M. Adams, C. M. Sharts and S. A. Shackelford; *Tetrahedron Lett.*, 1993, **34**, 6669. 361, 362
93TL6717 C. C. da Silva, V. Almagro and A. J. Marsaioli; *Tetrahedron Lett.*, 1993, **34**, 6717. 1044
93TL6741 F. Cerreta, C. Leriverend and P. Metzner; *Tetrahedron Lett.*, 1993, **34**, 6741. 587, 611
93TL6821 M. A. Rachita and G. A. Slough; *Tetrahedron Lett.*, 1993, **34**, 6821. 342
93TL6877 J. L. Esker and M. Newcomb; *Tetrahedron Lett.*, 1993, **34**, 6877. 334
93TL7371 M. Bergdahl, R. Hett, T. L. Friebe, A. R. Gangloff, J. Iqbal, Y. Wu and P. Helquist; *Tetrahedron Lett.*, 1993, **34**, 7371. 329
93TL7933 Y. Ueda, H. Watanabe, J. Uemura and K. Uneyama; *Tetrahedron Lett.*, 1993, **34**, 7933. 733

94AG(E)303 H.-G. Schmalz; *Angew. Chem., Int. Ed. Engl.*, 1994, **33**, 303. 932
94AG(E)456 H. Grützmacher, W. Deck, H. Pritzkow and M. Sander; *Angew. Chem., Int. Ed. Engl.*, 1994, **33**, 456. 919
94BMC1865 M. C. Desai, S. L. Lefkowitz, D. K. Bryce and S. McLean; *Bioorg. Med. Chem. Lett.*, 1994, **4**, 1865. 377
94CB725 R. Aumann; *Chem. Ber.*, 1994, **127**, 725. 958
94CB2475 R. Aumann, B. Jasper, M. Läge and B. Krebs; *Chem. Ber.*, 1994, **127**, 2475. 958
94CC805 R. A. Aitken, M. J. Drysdale and B. M. Ryan; *J. Chem. Soc., Chem. Commun.*, 1994, 805. 900
94CC1437 A. Rolfs and J. Liebscher; *J. Chem. Soc., Chem. Commun.*, 1994, 1437. 614
94CC2457 R. Streubel, M. Hobbold, J. Jeske and P. G. Jones; *J. Chem. Soc., Chem. Commun.*, 1994, 2457. 955
94CL77 J. Nakayama, A. Mizumura, I. Akiyama, T. Nishio and I. Iida; *Chem. Lett.*, 1994, 77. 627
94CL777 M. Iyoda, L. Zhao and H. Matsuyama; *Chem. Lett.*, 1994, 777. 953
94CL1537 T. Nakamura, H. Matsuyama and M. Iyoda; *Chem. Lett.*, 1994, 1537. 936
94H(38)159 T. G. Back and J. H.-L. Chau; *Heterocycles*, 1994, **38**, 159. 315

94HCA554	W. Oppolzer, J. Deerberg and O. Tamura; *Helv. Chim. Acta*, 1994, **77**, 554.	342
94ICA215	W. D. Wulff, B. A. Anderson, A. J. Toole and Y. Xu; *Inorg. Chim. Acta*, 1994, **220**, 215.	956
94ICA233	E. Licandro, S. Maiorana, A. Papagni, A. Z. Gerosa, F. Cariati, S. Bruni, M. Moret and A. C. Villa; *Inorg. Chim. Acta*, 1994, **220**, 233.	946, 959
94JA562	S. B. King and B. Ganem; *J. Am. Chem. Soc.*, 1994, **116**, 562.	330
94JA791	E. Gutiérrez, A. Monge, M. C. Nicasio, M. L. Poveda and E. Carmona; *J. Am. Chem. Soc.*, 1994, **116**, 791.	948
94JA916	A. Bagno, C. Comuzzi and G. Scorrano; *J. Am. Chem. Soc.*, 1994, **116**, 916.	322
94JA5473	S. Tsunoi, I. Ryu and N. Sonoda; *J. Am. Chem. Soc.*, 1994, **116**, 5473.	172
94JA5511	K. R. Buszek, N. Sato and Y. Jeong; *J. Am. Chem. Soc.*, 1994, **116**, 5511.	173
94JA7455	M. F. Semmelhack, W. R. Epa, A. W.-H. Cheung, Y. Gu, C. Kim, N. Zhang and W. Lew; *J. Am. Chem. Soc.*, 1994, **116**, 7455.	142
94JA9009	J. Aube, S. Gosh and M. Tanol; *J. Am. Chem. Soc.*, 1994, **116**, 9009.	129
94JA9363	F. E. McDonald and C. C. Schultz; *J. Am. Chem. Soc.*, 1994, **116**, 9363.	957
94JA10921	M. F. Gross and M. G. Finn; *J. Am. Chem. Soc.*, 1994, **116**, 10921.	936
94JA11584	Z.-W. Li and H. Taube; *J. Am. Chem. Soc.*, 1994, **116**, 11 584.	948
94JCS(D)1671	R. W. Grime and M. Whiteley; *J. Chem. Soc., Dalton Trans.*, 1994, 1671.	944
94JCS(P1)767	A. R. Ramesha and S. Chandresekaran; *J. Chem. Soc., Perkin Trans. 1*, 1994, 767.	252
94JCS(P1)989	J. Bloxham and C. P. Dell; *J. Chem. Soc., Perkin Trans. 1*, 1994, 989.	615
94JCS(P1)1083	H. Kageyama, K. Kido, S. Kato and T. Murai; *J. Chem. Soc., Perkin Trans. 1*, 1994, 1083.	244, 250
94JMC170	G. H. Posner, C.-G. Cho, J. V. Green, Y. Zhang and P. Talalay; *J. Med. Chem.*, 1994, **37**, 170.	1044
94JOC348	D. C. Smith, S. W. Lee and P. L. Fuchs; *J. Org. Chem.*, 1994, **59**, 348.	576
94JOC806	C. R. Williams, J. F. Britten and D. N. Harpp; *J. Org. Chem.*, 1994, **59**, 806.	336
94JOC929	T. Sakamoto and Y. Kikugawa; *J. Org. Chem.*, 1994, **59**, 929.	328
94JOC1129	K. S. Feldman and K. Schildknegt; *J. Org. Chem.*, 1994, **59**, 1129.	572
94JOC1160	J. Mulzer and J.-T. Mohr; *J. Org. Chem.*, 1994, **59**, 1160.	129
94JOC1231	D. R. Sidler, T. C. Lovelace, J. M. McNamara and P. J. Reider; *J. Org. Chem.*, 1994, **59**, 1231.	329
94JOC1257	T. Høeg-Jensen, C. E. Olsen and A. Holm; *J. Org. Chem.*, 1994, **59**, 1257.	589
94JOC1358	M. Naruse, S. Aoyagi and C. Kibayashi; *J. Org. Chem.*, 1994, **59**, 1358.	330
94JOC4602	A. R. Ritter and M. J. Miller; *J. Org. Chem.*, 1994, **59**, 4602.	330
94JOC4749	H. Tabuchi, T. Hamamoto, S. Miki, T. Tejima and A. Ichihara; *J. Org. Chem.*, 1994, **59**, 4749.	129
94JOC4862	P. R. Guzzo and M. J. Miller; *J. Org. Chem.*, 1994, **59**, 4862.	329
94JOC4972	P.-J. Colson and L. S. Hegedus; *J. Org. Chem.*, 1994, **59**, 4972.	162
94JOC5122	Y. Shi and W. D. Wulff; *J. Org. Chem.*, 1994, **59**, 5122.	958
94JOC5189	N. De Kimpe and M. Boeykens; *J. Org. Chem.*, 1994, **59**, 5189.	348
94JOC5292	P. A. Jacobi, L. M. Armacost, H. L. Brielmann, R. O. Cann, J. I. Kravitz and M. J. Martinelli; *J. Org. Chem.*, 1994, **59**, 5292.	329
94JOC5532	T. Kan, S. Hosekama, S. Nara, M. Oikawa, S. Ito, F. Matsuda and H. Shirakama; *J. Org. Chem.*, 1994, **59**, 5532.	129
94JOC6882	T. S. Powers, Y. Shi, K. J. Wilson, W. D. Wulff and A. L. Rheingold; *J. Org. Chem.*, 1994, **59**, 6882.	953, 956
94JOM(472)C15	U. Schubert and S. Seebald; *J. Organomet. Chem.*, 1994, **472**, C15	954
94JOM(476)C25	F. Rose-Munch, C. Susanne, F. Balssa and E. Rose; *J. Organomet. Chem.*, 1994, **476**, C25.	955
94JOM(479)C12	J. G. Toerien, M. Desmet, G. J. Kruger and H. G. Raubenheimer; *J. Organomet. Chem.*, 1994, **479**, C12.	951
B-94MI 506-01	B. R. Brown (ed.); "The Organic Chemistry of Aliphatic Nitrogen Compounds," Claredon Press, Oxford, 1994.	259
94MI 506-02	J. J. Kilama; personal communication at "Mona Symposium" 1994.	262
B-94MI 506-03	M. Bodansky and A. Bodansky; "The Practice of Peptide Synthesis," Springer-Verlag, Berlin, Heidelberg, 1994.	262
94MI 506-04	M. North; *Contemp. Org. Synth.*, 1994, **1**, 475.	259
94MI 531-01	J. C. Guillemin, L. Lasalle, P. Dre'an, G. Wlodarczak and J. Demaison; Unpublished Results, Rennes, 1994.	1160
94MIP3494	M. Kahn, *WO Pat.* 03 494 (1994) (*Chem. Abstr.*, **122**, 56580).	353
94OM971	J. Yang, J. Yin, K. A. Abboud and W. M. Jones; *Organometallics*, 1994, **13**, 971.	938
94OM3502	R. Aumann, B. Jasper, M. Läge and B. Krebs; *Organometallics*, 1994, **13**, 3502.	958
94OM3510	R. Aumann, B. Jaspar, M. Läge and B. Krebs; *Organometallics*, 1994, **13**, 3510.	946, 957
94OM4189	J. Christoffers and K. H. Dötz; *Organometallics*, 1994, **13**, 4189.	935
94OM4676	A. Rabier, N. Lugan, R. Mathieu and G. L. Geoffroy; *Organometallics*, 1994, **13**, 4676.	955
94OPP111	T. Hemmi, T. Sakamoto and Y. Kikugawa; *Org. Prep. Proced. Int.*, 1994, **26**, 111.	333
94PS(89)57	P. Freyen; *Phosphorus Sulfur*, 1994, **89**, 57.	367
94S282	T. Kanda, K. Mizoguchi, T. Koike, T. Murai and S. Kato; *Synthesis*, 1994, 282.	250
94S699	F. Freeman, B.-G. Huang and R. I.-S. Lin; *Synthesis*, 1994, 699.	247
94S761	C. Leriverend and P. Metzner; *Synthesis*, 1994, 761.	248
94S805	G. Cainelli, M. Panunzio, D. Giacomini, B. Di Simone and R. Camerini; *Synthesis*, 1994, 805.	377
94S909	Y.-J. Chun, J.-H. Park, G.-M. Oh, S.-I. Hong and Y.-J. Kim; *Synthesis*, 1994, 909.	378
94S1057	K.-J. Lee, S. H. Kim, S. Kim, H. Park, Y. R. Cho, B. Y. Chung and E. E. Schweizer; *Synthesis*, 1994, 1057.	862
94SC1761	X.-R. Zhao, M.-D. Ruan, W.-Q. Fan and X.-J. Zhou; *Synth. Commun.*, 1994, **24**, 1761.	625

94SL197	J. Y. Lee, Y. J. Chung and B. H. Kim; *Synlett*, 1994, 197.	343
94T2003	K. S. Chan and C. C. Mak; *Tetrahedron*, 1994, **50**, 2003.	934
94T7193	L. Z. Viteva, T. S. Gospodova and Y. N. Stefanovsky; *Tetrahedron*, 1994, **50**, 7193.	607
94T11091	P. R. Guzzo and M. J. Miller; *Tetrahedron*, 1994, **50**, 11 091.	315
94TA149	T. Polonski; *Tetrahedron Asymmetry*, 1994, **5**, 149.	572
94TL561	T. Albrecht, J. Sauer and H. Nöth; *Tetrahedron Lett.*, 1994, **35**, 561.	934
94TL1135	B. H. Lee and F. Han; *Tetrahedron Lett.*, 1994, **35**, 1135.	572
94TL1247	H. Kigoshi, M. Ojika, K. Suenaga, T. Mutou, J. Hirano, A. Sakakura, T. Ogawa, M. Nisiwaki and K. Yamada; *Tetrahedron Lett.*, 1994, **35**, 1247.	329
94TL1669	G. R. Cook, L. G. Beholz and J. R. Stille; *Tetrahedron Lett.*, 1994, **35**, 1669.	572
94TL1977	W. Schroth, M. Felicetti, E. Hintzsche, R. Spitzner and M. Pink; *Tetrahedron Lett.*, 1994, **35**, 1977.	616
94TL2057	I. S. Blagbrough and E. Moya; *Tetrahedron Lett.*, 1994, **35**, 2057.	572
94TL3651	M. J. Sharp and C. H. Heathcock; *Tetrahedron Lett.*, 1994, **35**, 3651.	572
94TL3801	P. Quayle, S. Rahman, E. L. M. Ward and J. Herbert; *Tetrahedron Lett.*, 1994, **35**, 3801.	944
94TL3817	P. Molina, E. Aller, A. L. Lopez-Lazaro, M. Alajarín and A. Lorenzo; *Tetrahedron Lett.*, 1994, 3817.	1066
94TL4865	A. P. Davis and J. J. Walsh; *Tetrahedron Lett.*, 1994, **37**, 4865.	264
94TL4947	V. H. Rawal and H. M. Zhong; *Tetrahedron Lett.*, 1994, **35**, 4947.	367
94TL5157	N. M. Koshti, H. K. Jacobs, P. A. Martin, P. H. Smith and A. S. Gopalan; *Tetrahedron Lett.*, 1994, **35**, 5157.	322
94TL6021	M. A. Staszak and C. W. Doecke; *Tetrahedron Lett.*, 1994, **35**, 6021.	328
94TL6377	C. Baldoli, P. Del Buttero, E. Licandro, S. Maiorana, A. Papagni and A. Zanotti-Gerosa; *Tetrahedron Lett.*, 1994, **35**, 6377.	956
94TL6717	C. M. Marson, L. Randall and M. J. Winter; *Tetrahedron Lett.*, 1994, **35**, 6717.	944
94TL8883	P. Quayle, E. L. M. Ward and P. Taylor; *Tetrahedron Lett.*, 1994, **35**, 8883.	944
94UP0	T. Murai, Y. Ogino and S. Kato; Unpublished Results.	563
95AG(E)350	F. E. McDonald and M. M. Gleason; *Angew. Chem., Int. Ed. Engl.*, 1995, **34**, 350.	944
95AOC(37)169	J. A. Andersen and J. R. Moss; *Adv. Organomet. Chem.*, 1995, **37**, 169.	943
95CB49	H. Werner, A. Stark, P. Steinert, C. Grünwald and J. Wolf; *Chem. Ber.*, 1995, **128**, 49.	944
95CB157	J. Christoffers and K. H. Dötz; *Chem. Ber.*, 1995, **128**, 157.	944
95CB289	F. R. Kreissl, J. Ostermeier and C. Ogric; *Chem. Ber.*, 1995, **128**, 289.	952
95CB373	H. Fischer, J. Schleu and G. Roth; *Chem. Ber.*, 1995, **128**, 373.	942
95CB645	K. H. Dötz, C. Christoffers, J. Christoffers, D. Böttcher, M. Nieger and S. Kotila; *Chem. Ber.*, 1995, **128**, 645.	934
95JOM(486)279	C. Baldoli, P. Del Buttero, E. Licandro, S. Maiorana, A. Pagni and A. Zanotti-Gerosa; *J. Organomet. Chem.*, 1995, **486**, 279.	953, 958
95JOM(488)C9	O. Lavastre and P. Dixneuf; *J. Organomet. Chem.*, 1995, **488**, C9.	940, 941
95JOM(494)43	B. Denise, A. Parlier, H. Rudler and J. Vaissermann; *J. Organomet. Chem.*, 1995, **494**, 43.	953
95JOM(494)C12	H. Adams, N. A. Bailey, C. M. Marson, L. Randall and M. J. Winter; *J. Organomet. Chem.*, 1995, **494**, C12.	955
95OM547	S. R. Amin and A. Sarkar; *Organometallics*, 1995, **14**, 547.	955
95OM570	D. Dvorak; *Organometallics*, 1995, **14**, 570.	953
95OM1095	N. Ruiz, D. Peron and P. H. Dixneuf; *Organometallics*, 1995, **14**, 1095.	944
95OM1938	C. Cosset, I. Del Rio and H. Le Bozec; *Organometallics*, 1995, **14**, 1938.	941
95OM2102	J. M. O'Connor, R. Merwin, A. L. Rheingold and M. L. Adams; *Organometallics*, 1995, **14**, 2102.	944
95OM2447	R. Aumann, B. Jasper and R. Fröhlich; *Organometallics*, 1995, **14**, 2447.	953
95OM2522	I. Merino and L. S. Hegedus; *Organometallics*, 1995, **14**, 2522.	953
95OM3163	H. Stadmüller and P. Knochel; *Organometallics*, 1995, **14**, 3163.	937
95TL197	S. S. Nikam, B. E. Kornberg, D. R. Johnson and A. M. Doherty; *Tetrahedron Lett.*, 1995, **36**, 197.	327
95TL3699	M. Iyoda, L. Zhao and H. Matsuyama; *Tetrahedron Lett.*, 1995, **36**, 3699.	946, 958
95TL3937	J. Barluenga, P. L. Bernad, Jr. and J. M. Concellón; *Tetrahedron Lett.*, 1995, **36**, 3937.	958
95UP 525-01	P. Quayle, S. Rahman and J. Herbert; *Tetrahedron Lett.*, 1995, **36**, (in press).	944

Subject Index

Every effort has been made to index as comprehensively as possible, and to standardize the terms used in the index in line with the IUPAC Recommendations. In view of the diverse nature of the terminology employed by the different authors, the reader is advised to search for related entries under the appropriate headings.

The index entries are presented in letter-by-letter alphabetical sequence. Compounds are normally indexed under the parent compound name, with the substituent component separated by a comma of inversion. An entry with a prefix/locant is filed after the same entry without any attachments, and in alphanumerical sequence. For example, 'dienes', '1,3-dienes', and '1-bromo-1,4-dienes' will be filed as:

Dienes
1,3-Dienes
1,4-Dienes, 1-bromo-

The Index is arranged in set-out style, with a maximum of three levels of heading. Location references refer to page number; major coverage of a subject is indicated by bold, elided page numbers: for example,

Halides, alkynyl, synthesis of, **1234–55**
properties of, 345

See cross-references direct the user to the preferred term; for example,

Mercaptans *See* Thiols

See also cross-references provide the user with guideposts to terms of related interest, from the broader term to the narrower term, and appear at the end of the main heading to which they refer, e.g.

Thiones
See also
Thioketone